W9-AGM-900

Dictionary of Organophosphorus Compounds

Dictionary of Organophosphorus Compounds

Edited and compiled by
R.S. Edmundson
Formerly of Bradford University

LONDON NEW YORK
CHAPMAN AND HALL

First published in 1988 by Chapman and Hall Ltd
11 New Fetter Lane, London EC4P 4EE
29 West 35th Street, New York NY 10001

Phototypeset in the United States of America by
Mack Printing Company, Easton, Pennsylvania 18042
Printed in Great Britain at the University Press, Cambridge

ISBN 0 412 25790 4

British Library
Cataloguing in Publication Data

Edmundson, R.S.

Dictionary of organophosphorus compounds.
1. Organophosphorus compounds – Dictionaries
I. Title

547'.07'00321 QD412.P1

ISBN 0-412-25790-4

Library of Congress
Cataloguing in Publication Data

Edmundson, R. S.
Dictionary of organophosphorus compounds.

Includes indexes.
1. Organophosphorus compounds – Dictionaries.
I. Title.
QD412.P1E36 1987 547'.07'0321 87-23936
ISBN 0-412-25790-4

Contents

Foreword by J.I.G. Cadogan *page* vii

Introduction ix

Entries in alphanumeric order 1

Name Index 881

Molecular Formula Index 989

CAS Registry Number Index 1119

Type of Compound Index 1205

HEILBRON

The Chemical Properties Database

The contents of this and Chapman & Hall's other chemical dictionaries together form HEILBRON, the chemical properties database.

HEILBRON contains full property data and key references for the most important chemicals (and biochemicals) such as amino acids, carbohydrates, lipids, pharmaceuticals, antibiotics, solvents, reagents, organometallics and simple organic compounds.

The unique structure and content of HEILBRON, together with Dialog's sophisticated software, means that you have alternative and complementary ways of finding data that is vital to your research by using its powerful and versatile range of search facilities:

- full searching on seven kinds of numerical data— melting point · boiling point · freezing point · dissociation constant · density · optical rotation · molecular weight.
- comprehensive full text searching eg. source, uses, toxicity and hazard information.
- searching on name fragments as well as complete names.
- keyword searching.
- searching by references eg. author, journal name, patent number.
- ready cross-file searching by automatic matching of CAS registry numbers.

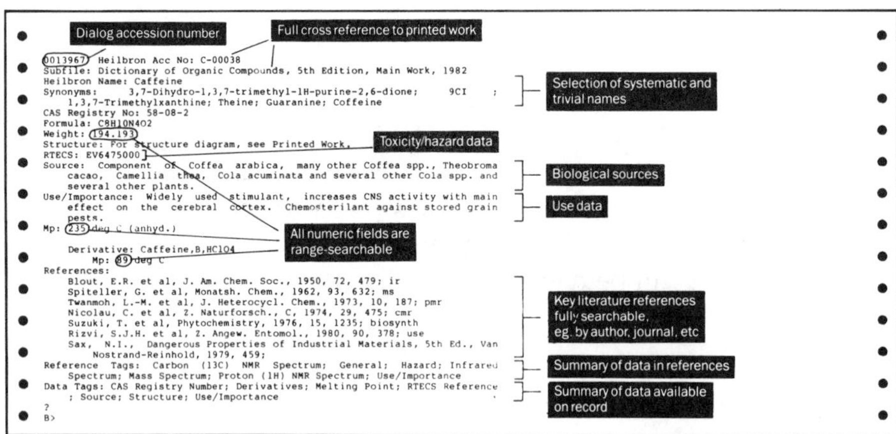

```
0013967 Heilbron Acc No: C-00038
Subfile: Dictionary of Organic Compounds, 5th Edition, Main Work, 1982
Heilbron Name: Caffeine
Synonyms:     3,7-Dihydro-1,3,7-trimethyl-1H-purine-2,6-dione;     9CI    ;
    1,3,7-Trimethylxanthine; Theine; Guaranine; Coffeine
CAS Registry No: 58-08-2
Formula: C8H10N4O2
Weight: 194.193
Structure: For structure diagram, see Printed Work.
RTECS: EV6475000
Source:  Component  of  Coffea  arabica,  many other Coffea spp., Theobroma
    cacao,  Camellia  thea,  Cola acuminata and several other Cola spp. and
    several other plants.
Use/Importance:  Widely  used  stimulant,  increases CNS activity with main
    effect  on  the  cerebral  cortex.  Chemosterilant against stored grain
    pests.
Mp: 235 deg C (anhyd.)

    Derivative: Caffeine,B,HClO4
        Mp: 89 deg C
References:
    Blout, E.R. et al, J. Am. Chem. Soc., 1950, 72, 479; ir
    Spiteller, G. et al, Monatsh. Chem., 1962, 93, 632; ms
    Twanmoh, L.-M. et al, J. Heterocycl. Chem., 1973, 10, 187; pmr
    Nicolau, C. et al, Z. Naturforsch., C, 1974, 29, 475; cmr
    Suzuki, T. et al, Phytochemistry, 1976, 15, 1235; biosynth
    Rizvi, S.J.H. et al, Z. Angew. Entomol., 1980, 90, 378; use
    Sax,  N.I.,  Dangerous Properties of Industrial Materials, 5th Ed., Van
        Nostrand-Reinhold, 1979, 459;
Reference  Tags:  Carbon  (13C)  NMR  Spectrum;  General;  Hazard; Infrared
    Spectrum; Mass Spectrum; Proton (1H) NMR Spectrum; Use/Importance
Data Tags: CAS Registry Number; Derivatives; Melting Point; RTECS Reference
    ; Source; Structure; Use/Importance
?
B>
```


HEILBRON contains information on over 175,000 substances. It currently encompasses the *Dictionary of Organic Compounds* and its Supplements; the *Dictionary of Organometallic Compounds* and its Supplements; *Carbohydrates* (a Chapman and Hall Sourcebook); *Dictionary of Organophosphorus Compounds; Dictionary of Antibiotics and Related Substances; Amino Acids and Peptides* (a Chapman and Hall Sourcebook). More titles will be added to the database in the coming years. HEILBRON is completely updated and considerably expanded twice a year, and is available online via Dialog Information Services. For more information contact the HEILBRON manager at the address given below.

Compiled by

CHAPMAN AND HALL
11 New Fetter Lane, London EC4P 4EE

Foreword

Organophosphorus compounds pervade not only the science of organic synthesis but also everyday living. They seem to appear everywhere. The extraordinary value of organophosphorus reagents in synthesis stems, of course, from the ease with which it is possible to progress from the lowest to the highest co-ordination number [P(III) → P(IV) → P(V) → P(VI)], and occasionally back again [P(V) → P(IV)]. The reactivities of each of these types of reagent can be varied by suitable changes in substituents. All of this makes for a very large number of usable reagents. But when we add to this the increasingly widespread applications of organophosphorus compounds, we have a long inventory of substances which the non-specialist chemist, biochemist and general industrialist need to know about. And what a list it is:

- crop protection agents and growth regulators
- flame retardants
- polymer additives
- compounds with pharmacological and chemotherapeutic effects
- radiodiagnostic agents
- metal extraction agents
- corrosion inhibitors
- emulsifiers
- antistatic agents
- complexing agents
- phase transfer and regular catalysts
- ligand modifiers of catalysts

This present *Dictionary of Organophosphorous Compounds* is therefore timely and welcome. Produced in the well-tried style and format of the *Dictionary of Organic Compounds*, this volume has very wide coverage, from which it is clear that Dr Edmundson has borne in mind the needs of the specialist and generalist. I am sure that both will be comfortable with it.

J.I.G. Cadogan

Caution

Treat all organic compounds as if they have dangerous properties.

The information contained in this volume has been compiled from sources believed to be reliable. However, no warranty, guarantee, or representation is made by the Publisher as to the correctness or sufficiency of any information herein, and the Publisher assumes no responsibility in connection therewith.

The specific information in this publication on the hazardous and toxic properties of certain compounds is included to alert the reader to possible dangers associated with the use of those compounds. The absence of such information should not however be taken as an indication of safety in use or misuse.

Introduction

1. Using the Dictionary

The Dictionary is arranged alphabetically by entry name. Every entry is numbered to assist ready location. Many compounds are included as derivatives of main entry compounds: the extensive and reliable indexing of the Dictionary means that these can readily be located through the Name, Molecular Formula or Type of Compound Indexes.

If a compound cannot be located immediately in the main body of the entries it is important to use the indexes.

Indexes

There are four printed indexes: a Name Index which lists every compound name or synonym in alphabetical order: a Molecular Formula Index which lists all molecular formulae, including those of derivatives, in Hill convention order; a CAS Registry Number Index listing all CAS numbers included in the Dictionary in serial order, and a Type of Compound Index in which compounds are listed by structural type.

All indexes refer to the entry number. In the Name Index an entry number which follows immediately upon an index term means that the term itself is used as the entry name but an entry which is preceded by the work '*see*' means that the term is a synonym to an entry name. In all indexes an entry number which is preceded by the word '*in*' refers the reader to a specified stereoisomer or derivative which is to be found embedded within the entry.

2. Literature Coverage

In compiling this Dictionary the primary literature has been carefully surveyed to 1985 and important information to the end of 1987 has also been incorporated.

A small proportion (less than 10%) of the entries in this dictionary first appeared in the *Dictionary of Organic Compounds,* 5th Edition, (1982); all of these have been carefully updated and expanded with the addition of a large proportion of new information.

3. Further Information on Presentation of Data

For further information about the presentation of data and the organisation of entries in this and related dictionaries, see the introduction to the *Dictionary of Organic Compounds,* Fifth Edition, 1982. In all important respects this dictionary is uniform with the DOC.

4. Toxicity

Many organophosphorus compounds are highly toxic and **some are among the most toxic substances known.** In compiling this dictionary every attempt has been made to report details of toxicity where these are documented in the primary literature, but the *absence of such information cannot be taken as indication of safety in use or misuse,* and the Publishers cannot be held responsible for any deficiencies in the toxicity/hazard information provided in the Dictionary.

In view of the fact that many organic compounds of high potential toxicity have not been evaluated, intending workers with organophosphorus compounds are strongly advised to acquaint themselves with the literature on toxic hazards associated with their use, so that they can be aware of the likely toxicological properties of newly synthesised compounds.

5. Nomenclature of Organophosphorus Compounds

Rules for the naming of organophosphorus compounds were agreed upon by the American Chemical Society and the Chemical Society of London in 1956 [1]. Most such compounds are now named in accordance with later additions and modifications by the Chemical Abstracts Service (CAS) [2] and the International Union of Pure and Applied Chemistry (IUPAC) [3].

In this dictionary CA terminology is generally preferred, although IUPAC names are often included particularly when they are substantially different from the CA name. The Dictionary also contains names taken from the literature which may be variants of the CA and IUPAC nomenclatures, and also trivial names.

The wide range of structural types in organophosphorus chemistry has led to the increasing use of the co-ordination number σ and the valency term λ in the description of compounds. For λ^3 compounds (tervalent phosphorus), σ can be 1, 2, or 3, as in the systems $RC{\equiv}P$, $R_2C{=}PR$, and R_3P, respectively. In λ^5 compounds (quinquevalent phosphorus), σ may be 3 as in $RP({=}CR_2)_2$, 4 as in R_4P^+ and $R_3P{=}CR_2$, 5 as in R_5P, and even 6 as in R_6P^-. Valency states are also commonly differentiated by the use of the symbols (III) and (V) suitably placed in the name, e.g., phosph(III)orinan.

Acyclic compounds

Phosphines and phosphoranes

Phosphines are the simplest organophosphorus compounds formed by the replacement of hydrogen atoms in phosphine PH_3. The primary, RPH_2, secondary, R_2PH, and tertiary, R_3P compounds are named using a substitutive nomenclature. When the phosphorus atom is inserted into a long carbon chain molecule to which more intrinsic 'importance' is attached, the compound may be named by the use of the term *phospha*. Tetraco-ordinate derivatives of phosphines, e.g. the chalcogenides, are named additively. The same operations can be applied in principle to diphosphine, H_2PPH_2 and diphosphene, $HP{=}PH$. Molecules consisting of longer chains of phosphorus atoms are termed phosphanes.

Organophosphoranes are based on PH_5 and are also named by the substitution method, e.g. Ph_5P and $Ph_3P{=}CH_2$ are named pentaphenylphosphorane and methylenetriphenylphosphorane, respectively. Compounds based on H_4PPH_4 are diphosphoranes.

Table 1 gives the names of the common phosphorus radicals used in substitutive nomenclature.

Table 1 *Common phosphorus radicals used in substitutive nomenclature*

Radical	Term	
	CA usage	IUPAC usage
H_2P—	phosphino	phosphino
HP<	phosphinidene	phosphinediyl
P⋖	phosphinidyne	phosphinetriyl
$H_2(O)P$—	phosphinyl	phosphinoyl
$H_2(S)P$—	phosphinothioyl	thiophosphinoyl
H(O)P<	phosphinylidene	phosphonoyl
(O)P⋖	phosphinylidyne	phosphoroyl
H_4P—	phosphoranyl	phosphoranyl
H_3P<	phosphoranylidene	phosphoranediyl
H_2P⋖	phosphoranylidyne	phosphoranetriyl

Examples of the use of these terms and of the types of nomenclature applied to phosphines and phosphoranes are given in Table 2.

Table 2 *Examples of the use of CA and IUPAC nomenclature applied to phosphines and phosphoranes*

Structure	Name
$Ph_3P(O)$	triphenylphosphine oxide (CA; substitutive and additive) oxotriphenylphosphorane (replacement) oxotriphenylphosphorus (co-ordination)
$Me_2PC_6H_4PMe_2$ (*o*-)	1,2-phenylenebis(dimethylphosphine) (CA) 1,2-bis(dimethylphosphino)benzene
H_2PCOOH	phosphinecarboxylic acid (CA)
$H_2PCH_2CH_2NH_2$	2-phosphinoethylamine (note preference)
H_2PCH_2COOH	phosphinoacetic acid (CA)* (carboxymethyl)phosphine
$HP(CH_2COOH)_2$	2,2′-phosphinidenediacetic acid (CA)* 2,2′-phosphinediyldiacetic acid (IUPAC)
$P(CH_2COOH)_3$	2,2′,2″-phosphinidynetrisacetic acid (CA)* 2,2′,2″-phosphinetriyltrisacetic acid (IUPAC)
$H_2P(O)CH_2COOH$	phosphinylacetic acid (CA)* phosphinoylacetic acid (IUPAC)
$(S)P(CH_2COOH)_3$	phosphinothioylidynetrisacetic acid (CA)* thiophosphoroyltriacetic acid (IUPAC)
$H_3CCH_2PPhCH_2CH_2PPh(O)CH_2COOH$	3,6-diphenyl-3,6-diphosphaoctanoic acid, 3-oxide
$Ph_4PC_6H_4COOH$ (*p*-)	4-(tetraphenylphosphoranyl)benzoic acid (4-carboxyphenyl)tetraphenylphosphorus

* Indexed in CA under acetic acid.

Mononuclear oxyacids of tervalent phosphorus

Derivatives of the oxyacids possessing one atom of tervalent phosphorus are based on the $\lambda^3\sigma^3$ structures H_2POH (phosphinous acid), $HP(OH)_2$ (phosphonous acid), and $P(OH)_3$ (phosphorous acid). Organoacids are named by prefixing the acid name with the hydrocarbon name, e.g. methanephosphonous acid or, preferably, with the hydrocarbon

radical term, i.e. methylphosphonous acid. Exceptionally, phosphenous acid HOPO is structurally based on $\sigma^2\lambda^3$ phosphorus.

Esters of tervalent oxyacids are named by replacement of the ending *ous* by *ite,* and the ending *nite* suggests (with phosphenous derivatives being exceptional) the ester of a tervalent phosphorus acid with at least one phosphorus–carbon bond. Table 3 lists some of the possibilities for the replacement of phosphorus-bound OH groups by other functions. Such replacing groups are entered into the final name in the decreasing order of priority: hydrazide, halide (alphabetical order), azide, amide, cyanide, isocyanide, cyanate, thiocyanate (etc.), isocyanate, isothiocyanate (etc.), nitride, imide (the same order of priority is applied to derivatives of quinquevalent oxyacids). The use of these terms and their modifications in the light of their position in the names is illustrated in Table 4.

Table 3 *Replacement nomenclature for substitution of P–O by other functions*

Group replaced	Group replacing	Term
=O;—OH	=S,—SH;=S,—OH; =O,—SH	(di)thio
=O;—OH	=Se,—SeH;=Se,—OH; =O,—SeH	(di)seleno
—OH	—OOH	peroxo
—OH	$—NH_2$	amid(o)
—OH	—Cl	chlorid(o)
—OH	—CN	cyanid(o)
=O	=NH	imid(o)
—OH	$—NHNH_2$	hydrazid(o)

Table 4 *Examples of substitutive, replacement and co-ordination nomenclature*

Structure	Name (nomenclature)
Ph_2POEt	ethoxydiphenylphosphine (a) ethyl diphenylphosphinite (b; CA) ethoxodiphenylphosphorus (c)
$EtPBr_2$	dibromo(ethyl)phosphine (a) ethylphosphonous dibromide (b; CA) dibromo(ethyl)phosphorus (c)
PhOPClF	chlorofluoro(phenoxy)phosphine (a) phenyl phosphorochloridofluoridite (b; CA) chlorofluoro(phenoxo)phosphorus (c)
$Cl_2PC_6H_4PCl_2$ (*p*-)	P,P,P′,P′-tetrachloro-1,4-phenylenebis(phosphine) (a) 1,4-phenylenebis(phosphonous dichloride) (b; CA) μ-1,4-phenylenebis(dichlorophosphorus) (c)

Note: (a) substitutive nomenclature (b) replacement nomenclature (c) co-ordination nomenclature.

Organo-phosphinous and -phosphonous acids, and phosphorous acid derivatives with at least one free phosphorus-bound OH group exist almost exclusively in the tautomeric $\sigma^4\lambda^5$ phosphoryl forms $R_2P(O)H$ (secondary phosphine oxide), RHP(O)OH (known both as hypophosphorous acid when R = H, and as a phosphinic acid), and $HP(O)(OH)_2$ (phosphonic acid).

Mononuclear oxyacids of quinquevalent phosphorus

Derivatives of those oxyacids possessing one atom of quinquevalent phosphorus are based on the $\sigma^4\lambda^5$ structures $H_2P(O)OH$ (phosphinic acid), $HP(O)(OH)_2$ (phosphonic acid), and $(O)P(OH)_3$ (phosphoric acid). There is, in addition, the $\sigma^3\lambda^5$ phosphenic (metaphosphoric) acid $HOP(O)_2$.

Derivatives of the quinquevalent oxyacids are named in accordance with Table 3 and the order of priority given earlier. Esters receive the ending *ate*. The terms thiolate and thionate for the groupings P(O)SR and P(S)OR, and also thionothiolate for P(S)SR, have been superseded by the *S*—R and *O*—R terminology. The very common moiety $(EtO)_2P(O)$— which could be referred to as a diethyl ester, is widely also referred to as diethylphosphoryl, or diethoxyphosphoryl, but in this dictionary the preferred term is diethoxyphosphinyl.

Table 5 lists useful and recommended terms when the phosphorus-containing group is not part of the 'principal' structure. Differentiation between the ionized and the fully protonated forms of an acidic compound is achieved by the introduction of the word hydrogen for the latter. Thus, $MeOP(O)(O^-)_2$ is methyl phosphate; $MeOP(O)(OH)_2$ is methyl dihydrogen phosphate (IUPAC) or monomethyl phosphate (CA).

Table 5 *Substitutive nomenclature applied to P-containing groups*

Radical	Term
$—P(O((OH)_2$	phosphono phospho (commonly used in biochemical terminology when the group is connected to a hetero atom and not to carbon) [4]
$—P(O)(OH)O^-$	phosphonato
>P(O)OH	phosphinico
$>P(O)O^-$	phosphinato

Examples of the use of these terms and of the various nomenclatures as applied to derivatives of quinquevalent oxyacids are given in Table 6.

Table 6 *Nomenclature of quinquevalent oxyacids*

Structure	Name
$(MeO)_2P(O)Cl$	dimethyl phosphorochloridate (CA name)
	phosphorochloridic acid, diMe ester (CA entry)
$MeP(O)Cl(NMe_2)$	trimethylphosphonamidic chloride (CA name)
	phosphonamidic chloride, trimethyl (CA entry)
$PhP(O)(OH)Cl$	phenylchlorophosphonic acid
	phenylphosphonochloridic acid (CA name)
	phosphonochloridic acid, phenyl (CA entry)
	hydrogenchlorodioxophenyl-phosphate(V)
$PhPH(S)OMe$	*O*-methyl phenylthio-phosphinate
	O-methyl phenylphos-phinothioate (CA name)
	hydridomethoxophenylthio-phosphorus
$Me_2NP(O)(OH)_2$	dimethylamidophosphoric acid
	dimethylphosphoramidic acid (CA name)
	phosphoramidic acid, dimethyl (CA entry)
	dihydrogen(methylamido)-trioxophosphate(V)
$(ClCH_2CH_2O)_3P(O)$	tris(2-chloroethyl) phosphate (IUPAC)
	ethanol, 2-chloro, phosphate 3 : 1 (CA entry)*
$(Me_3SiO)_3P(O)$	tris(trimethylsilyl) phosphate
	silanol, trimethyl, phosphate 3 : 1 (CA entry)*
$(Me_2NNH)_3P(S)$	thiophosphoric tris-(2,2-dimethylhydrazide)
	2,2,2′,2′,2″,2″-hexamethyl-phosphorothioic trihydrazide
	tris(2,2-dimethylhydrazido)-thiophosphorus
$(HO)_2P(O)C_6H_{10}COOH$-4	4-phosphonocyclohexane-carboxylic acid
$(HO)_2P(O)C_6H_{10}COOEt$-4	4-(ethoxycarbonyl)cyclohexyl phosphonic acid

* See [4] for classification of alcohols.

Derivatives of dinuclear oxyacids

Compounds with P–P bonds. From the nomenclature point of view, the acid $(HO)_2PP(OH)_2$ is unusual in being called hypo*di*phosphorous acid, possibly to distinguish its esters from those of the acid $H_2P(O)OH$ also known as hypophosphorous acid. By contrast, the acids (HO)PHPH(OH), $(HO)_2P(O)P(O)(OH)_2$ and (OH)PH(O)PH(O)OH are named hypophosphonous acid, hypophosphoric acid, and hypophosphonic acid (CA; IUPAC). Replacement of one or more oxygen atoms by sulfur affords thio acids and derivatives, potentially isomeric, but which are all grouped under the term thiohypo compounds.

Compounds with P–X–P bonds. Compounds with the P—O—P structure are formally anhydrides of the appropriate mononuclear acids and in the past have been referred to as *pyro* acids and derivatives. This prefix has now been superseded by *di*. However, in spite of its present classification as tetraethyl diphosphate, the ester $(EtO)_2P(O)—O—P(O)(OEt)_2$ is still widely known as tetraethyl pyrophosphate.

Certain attempts to simplify nomenclature have led to confusion. Thus the recommended name for each of the following pairs of isomeric arrangements $(EtO)_2P(O)SP(O)(OEt)_2$, $(EtO)_2P(S)OP(O)(OEt)_2$, $(EtO)_2P(S)OP(S)(OEt)_2$ and $(EtO)_2(S)SP(O)(OEt)_2$ is tetraethyl thiodiphosphate. However, members of each pair are easily differentiated by the use of the terms *sym* and *unsym*–a means employed in the Dictionary.

The replacement of one or more phosphorus-bound OH groups in a dinuclear oxyacid can be indicated by use of the prefixes *P*- or *P′*-. Thus $(HO)_2P(O)NHP(O)(OH)_2$ is imidodiphosphoric acid, and $(H_2N)_2P(O)NHP(O)(OH)_2$ is *P,P*-diamidoimidodiphosphoric acid.

Compounds with phosphorus–nitrogen double bonds. The compound $Ph_3P{=}NH$ is triphenylphosphine imide (CA; IUPAC) but this and similar structures with $\sigma^4\lambda^5$ phosphorus have been widely termed phosphazenes, phosphinimines and phosphoranimines. Strictly, the term phosphazene applies to the structure HN=PH with $\sigma^2\lambda^3$ phosphorus and is alternatively named iminophosphine. For compounds having the structure $RN{=}PR_3$ some further indication of valency might seem appropriate. In the Dictionary the term phosphazene is retained as a non-preferred nomenclature for compounds of the structure $R_3P{=}NR$. The reduced structure H_2NPH_2 is called phosphazane (aminophosphine).

Hexaco-ordinate phosphorus compounds. These are named in terms of the charge carried by the central atom or its valency. Thus, $Na^+ (Ph_6P)^-$ is sodium hexaphenylphosphate(1-) or hexaphenylphosphate(V).

Cyclic Compounds

According to the extended Hantzsch–Widman system (CA; IUPAC) the names of monocyclic ring systems possessing up to 10 ring atoms including only one phosphorus atom are those listed in Table 7. The presence of more than one phosphorus atom may be indicated by the use of the appropriate numerical prefix. The name of a partially saturated ring is to be derived from that of the fully unsaturated ring, although in some cases, a subtractive nomenclature, using the prefix *de-* may be applicable to the fully saturated system. IUPAC does allow the use of the terms 2-(or 3-)phospholene for the appropriate dihydrophosphole (CA).

In a recent revision of the IUPAC rules on nomenclature of ring systems [5], most of the names are retained but the ending *-e,* formerly used when nitrogen is also present, becomes optional for the nitrogen-free rings. The most important change is that covering the six-membered ring which now becomes phosphinine for the unsaturated ring, and phosphinane for the saturated ring.

Some trivial names have been retained; they include phosphindole, isophosphindole, phosphinoline and isophosphinoline, phosphanthrene, phosphanthridine, phenoxaphosphine, phenothiaphosphine, and phenophosphazine. These are preferred names in the Dictionary.

For ring systems possessing more than 10 ring atoms, the replacement term *phospha-* is accepted.

References

1. Handbook for Chemical Society Authors, 2nd edn, The Chemical Society, London, 1961.
2. Chemical Abstracts Service, Index Guide, 1985, Appendix IV.
3. International Union of Pure and Applied Chemistry, Organic Chemistry Division, Nomenclature of Organic Compounds, Sections A–F and H, 382, Pergamon, Oxford, 1979.
4. Nomenclature of phosphorus-containing compounds of biochemical importance, IUPAC Bulletin No. 66, 1977.
5. Rules of the extended Hantzsch–Widman system of nomenclature for heteromonocycles, 1982, *Pure Appl. Chem.,* 1983, **55,** 409.

6. Sources of Further Information

The following books and reviews provide more information about general aspects of the organic chemistry of phosphorus and also about particular groups of compounds. Reviews concerned with specific organophosphorus compounds are to be found under those particular entries. The list is not intended to be exhaustive but represents extensive coverage of synthetic and structural organophosphorus chemistry from about 1960.

General

General texts

Structure and Mechanism in Organophosphorus Chemistry, Hudson, R.F., Academic Press, London, 1965.

The Organic Chemistry of Phosphorus, Kirby, A.J. and Warren, S.G., Elsevier, Amsterdam, 1967.

Table 7 *Nomenclature of P-containing ring systems*

No. of ring atoms	Nitrogen absent		Nitrogen present	
	Unsaturated[a]	Saturated[b]	Unsaturated[a]	Saturated[b]
3	phosphirene	phosphirane	phosphirine	phosphiridene
4	phosphete	phosphetane	phosphete	phosphetidine
5	phosphole	phospholane	phosphole	phospholidine
6	phosphorin	phosphorinane	phosphorine	perhydrophosphorine
7	phosphepin	phosphepane	phosphepine	perhydrophosphepine
8	phosphocin	phosphocane	phosphocine	perhydrophosphocine
9	phosphonin	phosphonane	phosphonine	perhydrophosphonine
10	phosphecin	phosphecane	phosphecine	perhydrophosphecine

[a] Used when maximum number of double bonds is present.
[b] Fully saturated.

A Comprehensive Treatise on Inorganic and Theoretical Chemistry, Mellor, J.W., Vol. VIII, Supplement III–Phosphorus, Longmans, London, 1971.

Organophosphorus Chemistry, Walker, B.J., Penguin, London, 1972.

The Chemistry of Phosphorus, Emsley, J. and Hall, D., Harper and Row, 1976.

Phosphorus Chemistry in Everyday Living, Toy, A.D.F., The American Chemical Society, Washington, DC, 1976.

Comprehensive Organic Chemistry, Vol. 2, Sutherland, I.O. (ed.) Edmundson, R.S., pp. 1189, 1257; Smith, D.J.H., pp. 1121, 1127, 1233, 1301, Pergamon, Oxford, 1979.

Introduction to Phosphorus Chemistry, Goldwhite, H., Cambridge University Press, 1981.

Phosphorus: An Outline of the Chemistry, Biochemistry, and Technology, 3rd edn, Corbridge, D.E.C., Elsevier, Amsterdam, 1985.

Comprehensive texts

Houben-Weyl Methoden der Organischen Chemie, 4th edn, Muller, E. (ed.) Vol. 12, bands I and II, Sasse, K., bands E1 and E2, Regitz, M. (ed.) Thieme-Verlag, Stuttgart, 1963 ('*Houben-Weyl*').

Organophosphorus Compounds, 2nd edn, Kosolapoff, G.M. and Maier, L., (eds) Vols 1–7, Wiley, New York, 1972–76 ('*Kosolapoff and Maier*').

Specialist Periodical Reports; Organophosphorus Chemistry, Trippett, S., Hutchinson, D.W., Miller, J. and Walker, B.J. (eds) RSC, London. Essentially an annual survey of organophosphorus chemistry with regular chapters on phosphines and phosphonium salts, quinquecovalent compounds, phosphine oxides and related compounds, tervalent phosphorus acids, quinquevalent phosphorus acids, phosphates and phosphonates of biochemical interest, nucleotides, ylides, and physical methods, with occasional chapters on other topics. ('*Organophosphorus Chem*').

The following books and reviews cover aspects of the chemistry of more than one type of organophosphorus compound.

Some Aspects of the Chemistry and Toxic Action of Organic Compounds containing Phosphorus and Fluorine, Saunders, B.C., Cambridge University Press, 1957.

Toxic Phosphorus Esters, O'Brien, R.D., Academic Press, New York, 1960.

Organophosphorus Poisons: Anticholinesterases and Related Compounds, Heath, D.F., Pergamon, Oxford, 1961.

Cadogan, J.I.G., *Quart. Rev.,* 1962, **16,** 208 (oxidation of tervalent phosphorus compounds).

Organophosphorus Monomers and Polymers, Gefter, Ye L., Pergamon, Oxford, 1962.

Berlin, K.D., *et al., Top. Phosphorus Chem.,* 1964, **1,** 17 (nucleophilic displacements in esters and halides by organometallic reagents).

Schmutzler, R., *Adv. Fluorine Chem.,* 1965, **5,** 31 (phosphorus fluorides).

Horner, L., *Fortschr. Chem. Forsch.,* 1966–67, **7,** 1 (preparative phosphorus chemistry).

Chernyshev, E.A., *et al., Organomet. Chem. Rev.,* 1968, **A3,** 469 (organosilicon compounds containing phosphorus).

Promonenkov, V.K., *et al., Russ. Chem. Rev. (Engl. Transl.),* 1968, **37,** 670 (compounds with unsaturated groups at phosphorus).

Pudovik, A.N., *et al., Russ. Chem. Rev. (Engl. Transl.),* 1968, **37,** 317 (reactions of phosphorus chlorides with unsaturated compounds).

Drodz, G.J., *Russ. Chem. Rev. (Engl. Transl.),* 1970, **39,** 1 (phosphorus fluorides).

Ivanov, B.E., *et al., Russ. Chem. Rev. (Engl. Transl.),* 1970, **39,** 358 (reactivity of tervalent phosphorus compounds).

Schmidbaur, H., *Organometallic Chem.,* 1970, **9,** 260 (isoelectronic organo-phosphorus, -silicon, and -aluminium series).

Maier, L., *Fortschr. Chem. Forsch.,* 1971, **19,** 1 (compounds derived from elemental phosphorus).

Redmore, D., *Chem. Rev.,* 1971, **71,** 315 (heterocyclic systems bearing phosphorus substituents).

Shaw, M.A., *et al., Top. Phosphorus Chem.,* 1972, **7,** 1 (addition reactions of tervalent phosphorus compounds with electrophilic olefins and acetylenes).

The Identification of Functional Groups in Organophosphorus Compounds, Thomas, L.C., Academic Press, London, 1974.

Razumov, A.I., *et al., Khim. Geterotsikl. Soedin,* 1976, 867 (phosphorylated indoles).

Slotin, L.A., *Synthesis,* 1977, 737 (methods of phosphorylation).

Brestkin, A.B., *et al., Russ. Chem. Rev.(Engl. Transl.),* 1978, **47,** 859 (organophosphorus compounds as anticholinesterases).

Organophosphorus Reagents in Organic Synthesis, Cadogan, J.I.G. (ed.) Academic Press, New York, 1979 (uses).

Pudovik, A.N., *et al., Synthesis,* 1979, 81 (addition reactions of tervalent phosphorus acids with unsaturated systems).

Phosphorus Chemistry Directed Towards Biology, Stec, W.J. (ed.) Pergamon, Oxford, 1980.

Phosphorus Chemistry; Proceedings of the International Conference, Durham, NC, 1981, Quin, L.D. and Verkade, J.G. (eds) American Chemical Society Sym. Ser., 1981, No. 171.

Konieczny, M., *et al., Chem. Rev.,* 1981, **81,** 49 (organophosphorus compounds with a peroxide bond).

Konieczny, M., *et al., Synthesis,* 1981, 682 (spin-labelled compounds).

Shulyndin, S.V., *et al., Russ. Chem. Rev., (Engl. Transl.),* 1981, **50,** 865 (copolymerisation of unsaturated organophosphorus monomers).

Petrov, K.A., *et al., Russ. Chem. Rev. (Engl. Transl.),* 1982, **51,** 234 (alkoxyalkyl compounds of phosphorus).

Razvodovskaya, L.V., *et al., Russ. Chem. Rev. (Engl. Transl.),* 1982, **51,** 135 (phosphorylated and thiophosphorylated aminoheterocycles).

Proceedings of the International Conference on Phosphorus Chemistry, Nice, 1983, Riess, J.G., Mathey, F., Robert, D., and Wolf, R., (eds) *Phosphorus Sulfur,* 1983, **18,** 1.

Kirpichnikov, P.A., *et al., Russ. Chem. Rev. (Engl. Transl.),* 1983, **52,** 1051 (organophosphorus stabilizers of polymers).

Sadykov, A.S., *et al. Russ. Chem. Rev. (Engl. Transl.),* 1983, **52,** 918 (phosphorylated alkaloids and other nitrogen heterocycles).

Dankowski, M., *Chem. Ztg.,* 1984, **108,** 303 (photochemistry of organophosphorus compounds).

Khaskin, B.A., *Russ. Chem. Rev. (Engl. Transl.),* 1984, **53,** 768 (phosphorus-containing polysulfides).

Khokhlov, P.S., *Russ. Chem. Rev. (Engl. Transl.),* 1984, **53,** 463 (exchange reactions in tri- and tetraco-ordinate phosphorus compounds).

Kutyrev, G.A., *et al., Russ. Chem. Rev. (Engl. Transl.),* 1984, **53,** 971 (addition reactions of organosilicon compounds containing phosphorus).

Cherkasov, R.A., *et al., Tetrahedron,* 1985, **41,** 2567 (sulfur-containing organophosphorus reagents in organic synthesis).

Timokhin, B.V., *Russ. Chem. Rev. (Engl. Transl.),* 1985, **54,** 1201 (halogenophilic reactions of tetra- and pentaco-ordinate phosphorus compounds).

Gareev, R.D., *et al., Zh. Obshch. Khim.,* 1986, **56,** 241 (Engl. Transl. p. 211) (reactions of trico-ordinate phosphorus compounds with 1-nitro-1-alkenes).

Ugi, I., *et al., Chem. Scr.,* 1986, **26,** 205 (new phosphorylating agents).

Zimin, M.G., *et al., Zh. Obshch. Khim.,* 1986, **56,** 977 (Engl. Transl., p. 859) (addition reactions of P–H compounds to multiple bonds).

Structure

McEwen, W.E., *Top. Phosphorus Chem.,* 1965, **2,** 1 (stereochemistry of reactions of organophosphorus compounds).

Michalski, J., *Bull. Soc. Chim. Fr.* 1967, 1109 (stereochemistry of optically active phosphorus compounds).

Schlosser, M., *Top. Stereochem.,* 1970, **5,** 1 (stereochemistry of Wittig reaction).

Tsvetkov, E.N., *et al., Russ. Chem. Rev. (Engl. Transl.),* 1971, **40,** 97 (conjugation in compounds of tervalent phosphorus).

Zon, G. *et al., Fortschr. Chem. Forsch.,* 1971, **19,** 61 (phosphorus stereochemistry).

Gallagher, M.J., *et al., Top. Stereochem.,* 1972, **3,** 1 (stereochemistry of phosphorus compounds).

Dynamic Stereochemistry of Pentaco-ordinated Phosphorus and Related Elements, Luckenbach, R.G., Thieme, Stuttgart, 1973.

The Structural Chemistry of Phosphorus, Corbridge, D.E.C., Elsevier, Amsterdam, 1974.

Organophosphorus Stereochemistry: I (tervalent and tetracoordinate compounds), and II (pentacoordinate compounds). Benchmark Papers in Organic Chemistry 3 and 4, McEwen, W.E. and Berlin, K.D. (eds) Dowden, Hutchinson, and Ross, Stroudesburg, Pennsylvania, 1975.

Mikolajczyk, M., *et al., Russ. Chem. Rev. (Engl. Transl.),* 1975, **44,** 670 (stereochemistry of optically active thio derivatives of phosphorus acids).

Trippett, S., *Phosphorus Sulfur,* 1976, **1,** 89 (apicophilicity and ring strain in phosphoranes).

Verkade, J.G., *Phosphorus Sulfur,* 1976, **2,** 251 (steric and electronic effects in cyclic phosphorus esters).

d-Orbital Involvement in the Organochemistry of Silicon, Phosphorus and Sulfur, Kwart, H. and King, K.G., Springer-Verlag, Berlin, 1977.

Gilyarov, V.A., *Russ. Chem. Rev. (Engl. Transl.),* 1978, **47,** 870. (imide–amide and imide–imide rearrangements of imido compounds).

Ishmaeva, E.A., *Russ. Chem. Rev. (Engl. Transl.),* 1978, **47,** 896 (polarity and structure of unsaturated derivatives of pentaco-ordinate phosphorus).

Hall, C.R., *et al., Phosphorus Sulfur,* 1979, **7,** 171 (resolution of chiral compounds of tetraco-ordinate phosphorus).

Pentacoordinate Phosphorus I Structure and Spectroscopy, Holmes, R.R., American Chemical Society Monograph No. 175, ACS, Washington, DC, 1980.

Lutsenko, I.F., *et al., Pure Appl. Chem.,* 1980, **52,** 917 (phosphorotropic tautomerism).

Mastryukova, T.A., *et al., Pure Appl. Chem.,* 1980, **52,** 945 (phosphorus–carbon dyad tautomerism).

Solodovnikov, S.P., *et al., Russ. Chem. Rev. (Engl. Transl.),* 1980, **49,** 1 (stereochemistry of reactions of phosphorus-containing radicals).

Wolf, R., *Pure Appl. Chem.,* 1980, **52,** 1141 (ring-chain tautomerism).

Frey, P.A., *Tetrahedron,* 1982, **38,** 1541 (stereochemistry of enzymic reactions of phosphates).

Istomin, B.I., *et al., Russ. Chem. Rev. (Engl. Transl.),* 1982, **51,** 223 (influence of structure on reactivity of tetraco-ordinate compounds).

Mastryukova, T.A., *et al., Russ. Chem. Rev. (Engl. Transl.),* 1983, **52,** 1012 (phosphorus–carbon prototropy).

Pudovik, A.N., *et al., Russ. Chem. Rev. (Engl. Transl.),* 1983, **52,** 1036 (migrations of phosphoryl groups between nucleophilic sites).

Raevskii, O.A., *et al., Russ. Chem. Rev. (Engl. Transl.),* 1983, **52,** 1156 (conformational analysis of phosphoryl compounds with P–C bonds).

Arshinova, R.P., *Russ. Chem. Rev. (Engl. Transl.),* 1984, **53,** 351 (electronic interactions in mainly cyclic systems).

See also references on pseudorotation of pentacoordinate phosphorus.

Spectroscopy

Mavel, G., *Prog. in NMR Spectroscopy,* 1966, **1,** 251 (data for nuclei other than P in phosphorus compounds).

Crutchfield, M.M., *et al., Top. Phosphorus Chem.,* 1967, **5,** 1 (measurement and interpretation of high resolution ^{31}P nmr spectra).

Mark, V., *et al., Top. Phosphorus Chem.,* 1967, **5,** 227 (list of ^{31}P nmr data).

Van Wazer, J.R. *et al., Top Phosphorus Chem.,* 1967, **5,** 169 (interpretation of ^{31}P chemical shifts and coupling constants).

Corbridge, D.E.C., *Top. Phosphorus Chem.,* 1969, **6,** 235 (infrared spectroscopy).

Finer, E.G., *et al., Prog. in NMR Spectroscopy,* 1971, **6,** 61 (phosphorus–phosphorus coupling constants).

Dimroth, K., *Top. Curr. Chem.,* 1973, **38,** 1 (spectroscopy of delocalized phosphorus–carbon double bonds).

Mass Spectrometry of Inorganic and Organometallic Compounds, Litzow, M.R. and Spalding, T.R., Elsevier, Amsterdam, 1973, 439.

Mavel, G., *Annu. Rep. NMR Spectroscopy,* 1973, **5B,** 1 (^{1}H and ^{31}P chemical shift data from the period 1965–69).

Interpretation of the Infrared Spectra of Organophosphorus Compounds, Thomas, L.C., Heyden, London, 1974.

Bock, H., *Pure Appl. Chem.,* 1975, **44,** 343 (photoelectron spectra and bonding).

Fluck, E., *et al., Pure Appl. Chem.,* 1975, **44,** 373 (photoelectron spectroscopy and application in phosphorus chemistry).

Goubeau, J., *Pure Appl. Chem.,* 1975, **44,** 393 (vibrational spectroscopy and force constants for phosphorus compounds).

Brazier, J.F., *et al., Top. Phosphorus Chem.,* 1976, **8,** 99 (nmr data for compounds with P–H bonds).

Granoth, J., *Top. Phosphorus Chem.,* 1976, **8,** 41 (mass spectroscopy of organophosphorus compounds).

Schipper, P., *et al., Top. Phosphorus Chem.,* 1977, **9,** 407 (esr of phosphorus compounds).

Zverev, V.V., *et al., Russ. Chem. Rev. (Engl. Transl.),* 1977, **46,** 791 (photoelectron spectroscopy).

Kabachnik, M.I., *et al., Russ. Chem. Rev. (Engl. Transl.),* 1978, **47,** 821 (nmr spectroscopy of optical isomers).

Albrand, J.P., *et al., Pure Appl. Chem.,* 1980, **52,** 1047 (nmr spectroscopy of organophosphorus compounds in liquid crystals).

Fluck, E., *Top Phosphorus Chem.,* 1980, **10,** 193 (spectroscopy of compounds with co-ordination number 2).

Appel, R., *et al., Angew. Chem. Int. Ed. Engl.,* 1981, **20,** 731 (spectroscopy of phosphaalkenes and phosphaalkynes).

Chapman, J.R., *Organophosphorus Chem.,* 1983, **14,** 278 (mass spectrometry).

Phosphorus 31-P NMR. Principles and Applications, Gorenstein, D.G. (ed.) Academic Press, London, 1984.

Phosphines

Houben-Weyl, 1963, band I, 17.

Maier, L., *Prog. Inorg. Chem.,* 1963, **5,** 27 (synthesis and properties of primary, secondary, and tertiary phosphines).

Berlin, K.D., *et al., Top. Phosphorus Chem.,* 1964, **1,** 17 (nucleophilic displacements in halides and esters by organometallic reagents).

Horner, L., *Pure Appl. Chem.,* 1964, **9,** 225 (synthesis and properties of optically active tertiary phosphines).

Petrov, K.A., *et al., Russ. Chem. Rev. (Engl. Transl.),* 1968, **37,** 532 (hydroxyalkyl phosphines and their oxides).

Burg, A.B., *Acc. Chem. Res.,* 1969, **2,** 353 (fluoroalkylphosphines).

Abel, E.W., *et al., Organometallic Chem. Rev.,* 1970, **5A,** 143 (phosphines, arsines, stibines and bismuthines containing Si, Ge, Sn or Pb).

Maier, L. in *Kosolapoff and Maier,* 1972, **1,** 1.

Horner, L., *Pure Appl. Chem.,* 1980, **52,** 843 (synthesis, properties and uses of optically active phosphines in asymmetric hydrogenations).

Elsner, G. in *Houben-Weyl,* 1982, band E1, 106.

See also references on carbon-phosphorus heterocycles and on compounds with low co-ordination numbers.

Phosphine chalcogenides

Berlin, K.D., *et al., Chem. Rev.,* 1960, **60,** 243 (secondary and tertiary phosphine oxides).

Houben-Weyl, 1963, band I, 127.

Harvey, R.G., *et al., Top. Phosphorus Chem.,* 1964, **1,** 57.

Maier, L., *Top. Phosphorus Chem.,* 1965, **2,** 43 (primary and secondary phosphine sulfides, their thio acids and anhydrides, and tertiary phosphine sulfides).

Hays, H.R. and Peterson, D.J. in *Kosolapoff and Maier,* 1972, **3,** 341 (phosphine oxides).

Maier, L. in *Kosolapoff and Maier,* 1972, **4,** 1 (phosphine sulfides, selenides, and tellurides).

Bhattacharya, A.K.. *et al., Chem. Rev.,* 1980, **81,** 415 (the Michaelis–Arbuzov reaction).

Pudovik, A.N., *et al., Pure Appl. Chem.,* 1980, **52,** 989 (addition reactions of partially esterified phosphorus acids; rearrangements of α-hydroxyalkyl phosphorus esters and their α-mercapto and α-amino analogues).

Heydt, H. and Regitz, M., in *Houben-Weyl,* 1982, band E2, 1 (tertiary phosphine chalcogenides).

Hennig, H.-W. *et al., Chem. Ztg.,* 1985, **109,** 239 (carboxyphosphine derivatives).

See also references on phosphines.

Compounds with phosphorus–phosphorus bonds

Banks, R.E., *et al., Adv. Inorg. Chem. Radiochem.,* 1961, **3,** 338.

Huheey, J.E., *J. Chem. Educ.,* 1963, **40,** 153.

Maier, L., *Prog. Inorg. Chem.,* 1963, **5,** 27.

Wiberg, E., *et al., Angew. Chem.,* 1963, **75,** 814 (polyphosphines).

Houben-Weyl, 1963, band I, 182 (biphosphines and cyclopolyphosphines).
Issleib, K., *Pure Appl. Chem.,* 1964, **9,** 205.
Cowley, A.H., *Chem. Rev.,* 1965, **65,** 617 (chemistry of the P–P bond).
Cowley, A.H., *et al., Top Phosphorus Chem.,* 1967, **4,** 1 (structure and reactions of cyclopolyphosphines).
Maier, L., *Top. Curr. Chem.,* 1967, **8,** 1 (cyclopolyphosphines).
Fluck, E. in *Preparative Inorganic Reactions,* Jolly, W.L. (ed.) 1968, **5,** 103 Interscience, New York.
Maier, L. in *Kosolapoff and Maier,* 1972, **1,** 289.
Lutsenko, J.F. *et al., Russ. Chem. Rev. (Engl. Transl.),* 1978, **47,** 880.
Baudler, M. in *Kosolapoff and Maier,* 1973, **5,** 1 (hypodiphosphorous and hypodiphosphoric acids and their derivatives).
Baudler, M., *Pure Appl. Chem.,* 1980, **52,** 755 (three-membered ring compounds).
Lutsenko, J.F., *et al., Pure Appl. Chem.,* 1980, **52,** 917 (rearrangements of diphosphine oxides and anhydrides of phosphorus acids: phosphorotropic tautomerism).
Baudler, M., *Angew. Chem. Int. Ed. Engl.,* 1982, **21,** 492 (chain and ring phosphorus compounds: analogies between phosphorus and carbon).
Heydt, H. and Regitz, M. in *Houben-Weyl,* 1982, band E1, 183.

Phosphonium salts

Houben-Weyl, 1963, band I, 79.
Hoffmann, H., *et al., Angew. Chem. Int. Ed. Engl.,* 1964, **3,** 737.
Maercker, A., *Organic Reactions,* 1965, **14,** 270.
Beck, P. in *Kosolapoff and Maier,* 1972, **2,** 189.
Jodden, K. in *Houben-Weyl,* 1982, band E1, 491.
Hudson, H.R., *Top Phosphorus Chem.,* 1983, **11,** 339 (quasi-phosphonium intermediates and compounds).

See also references concerned with phosphine-methylenes and their uses.

Phosphinous and phosphonous acids and their derivatives

Frank, A.W., *Chem. Rev.,* 1961, **61,** 389 (phosphonous acids).
Houben-Weyl, 1963, band I, 193 (phosphinous acids).
Houben-Weyl, 1963, band I, 294 (phosphonous acids).
Petrov, K.A., *et al., Russ. Chem. Rev. (Engl. Transl.),* 1966, **35,** 622 (reactions of phosphites and phosphonites with hydroxy compounds).
Nixon, J.F., *Adv. Inorg. Chem. Radiochem.,* 1970, **13,** 364 (phosphonous phosphinous fluorides).
Fild, M. and Schmutzler, R., in *Kosolapoff and Maier,* 1972, **4,** 75 (phosphinous and phosphonous halides).
Frank, A.W., in *Kosolapoff and Maier,* 1972, **4,** 255 (phosphonous acids and their thio and seleno analogues).
Hamilton, L.A. and Landis, P.S., in *Kosolapoff and Maier,* 1972, **4,** 463 (phosphinous acids).
Krasil'nikova, E.A. *et al., Russ. Chem. Rev. (Engl. Transl.),* 1977, **46,** 861 (structure and reactivity of esters of thio acids of trico-ordinate phosphorus).
Mikolajczyk, M., *Pure Appl. Chem.,* 1980, **52,** 959 (optically active tervalent phosphorus acid esters).
Kleiner, H.-J. in *Houben-Weyl,* 1982, band E1, 240 (phosphinous acids).
Neumaier, B. in *Houben-Weyl,* 1982, band E1, 271 (phosphonous acids).

Phosphorous acid derivatives

Houben-Weyl, 1964, band II, 5.
Petrov, K.A., *et al., Russ. Chem. Rev. (Engl. Transl.),* 1966, **35,** 622 (phosphite and phosphonites, reactions with hydroxy compounds).
Nifant'ev, E.E., *et al., Russ. Chem. Rev. (Engl. Transl.),* 1970, **39,** 1050 (acyl phosphites).
Nixon, J.F. *Adv. Inorg. Chem. Radiochem.,* 1970, **13,** 354 (phosphorous fluorides).
Gerrard, W. and Hudson, H.R., in *Kosolapoff and Maier,* 1973, **5,** 21.
Konovalova, I.V. *et al., Russ. Chem. Rev. (Engl. Transl.),* 1972, **41,** 411 (reactions of tervalent derivatives with carbonyl compounds).
Nifant'ev, E.E., *Russ. Chem. Rev. (Engl. Transl.).* 1978, **47,** 835 (secondary phosphites and related compounds).
Block, H.-D. in *Houben-Weyl,* 1982, band I, 313.

Phosphoric acid derivatives

Cramer, F., *Angew. Chem.,* 1960, **72,** 236 (esters, amides, and anhydrides in phosphorus chemistry).
Lichtenthaler, F.W., *Chem. Rev.,* 1961, **61,** 607 (enol phosphates).
Recent Developments in the Chemistry of Phosphate Esters of Biological Interest, Khorana, H.G., Wiley, New York, 1961.
Brown, D.M., *Adv. Org. Chem.,* 1963, **3,** 75 (phosphorylation).
Die Entwicklung Neuer Insectiziden Phosphorsaure-Ester, Schrader, G., Verlag Chemie, Weinheim, 1963.
Cox, J.R. *et al., Chem. Rev.,* 1964, **64,** 317 (reactions of phosphoryl compounds).
Hilgetag, G. *et al., Angew. Chem. Int. Ed., Engl.,* 1965, **4,** 914 (alkylating properties of thiophosphate esters).
Clark, V.M. *et al., Prog. Org. Chem.,* 1968, **7,** 75 (phosphoryl transfer).
Westheimer, F.H., *Acc. Chem. Res.,* 1968, **1,** 70 (pseudorotation in the hydrolysis of phosphate esters).
Blackburn, G.M. *et al., Top. Phosphorus Chem.,* 1969, **6,** 187 (oxidative phosphorylation).
Organophosphorus Pesticides: Organic and Biological Chemistry, Eto, M., CRC Press, 1974.
Cherbuliez, E., in *Kosolapoff and Maier,* 1975, **6,** 211.

Ailman, D.E. and Magee, R.J. in *Kosolapoff and Maier,* 1975, **7,** 487 (thio-, seleno- and telluro-phosphoric acid derivatives).

Ramirez, F., *et al., Phosphorus Sulfur,* 1976, **1,** 231 (phosphoric diesters).

Bel'skii, V.E., *Russ. Chem. Rev. (Engl. Transl.),* 1977, **46,** 828 (hydrolysis of phosphate esters).

Westheimer, F.H., *Pure Appl. Chem.,* 1977, **49,** 1059 (hydrolysis of phosphate esters).

Michalski, J. *et al., Russ. Chem. Rev. (Engl. Transl.),* 1978, **47,** 814 (monothiopyrophosphates).

Ramirez, F., *et al., Pure Appl. Chem.,* 1980, **52,** 1021 (phosphoryl transfer).

Frey, P.A., *Tetrahedron,* 1982, **38,** 1541 (Tetrahedron Report 127) (enzymic reactions of phosphate esters).

Gallenkamp, B., *et al.,* in *Houben-Weyl,* 1982, band E2, 487.

Phosphinic and phosphonic acids

Doak, G.O. *et al., Chem. Rev.,* 1961, **61,** 31 (dialkyl phosphonates).

Houben-Weyl, 1963, band I, 220 (phosphinic acids).

Houben-Weyl, 1963, band I, 338 (phosphonic acids).

Arbuzov, B.A., *Pure Appl. Chem.,* 1964, **9,** 307 (Arbuzov and Perkow reactions).

Harvey, R.G. *et al., Top. Phosphorus Chem.,* 1964, **1,** 57 (Arbuzov reaction).

Curry, J.D. *et al., Top. Phosphorus Chem.,* 1972, **7,** 37 (oligophosphonates).

Fild, M. in *Kosolapoff and Maier,* 1972, **4,** 155 (phosphinic and phosphonic halides and their thio derivatives).

Baudler, M. in *Kosolapoff and Maier,* 1973, **5,** 1 (esters of phosphinic acid).

Razumov, A.I. *et al., Russ. Chem. Rev. (Engl. Transl.),* 1973, **42,** 538 (phosphonoacetaldehydes).

Petrov, K.A. *et al., Russ. Chem. Rev. (Engl. Transl.),* 1974, **43,** 984 (aminoalkyl -phosphonic and -phosphinic acid derivatives).

Prajer, K. *et al., Z. Chem.,* 1975, **15,** 209 (1-aminophosphonic acids).

Crofts, P.C. in *Kosolapoff and Maier,* 1976, **6,** 1 (phosphinic acid derivatives).

Worms, K.H. and Schmidt-Dunker, M. in *Kosolapoff and Maier,* 1976, **7,** 1 (phosphonic acids).

Engel, R., *Chem. Rev.,* 1977, **77,** 349 (phosphonate analogues of natural phosphates).

Grob, H., *Z. Chem.,* 1977, **17,** 281.

Pudovik, A.N. *et al., Synthesis,* 1979, 81 (additions of esters of tervalent phosphorus acids to unsaturated compounds).

Maier, L., *Top. Phosphorus Chem.,* 1980, **10,** 129 (thiophosphonic anhydrides).

Yudevich, V.I. *et al., Russ. Chem. Rev. (Engl. Transl.),* 1980, **49,** 46 (phosphinic esters).

Zhdanov, Yu. A. *et al., Russ. Chem. Rev. (Engl. Transl.),* 1980, **49,** 843 (1-ketophosphonic acid esters).

Bhattacharya, A.K. *et al., Chem. Rev.,* 1981, **81,** 415 (Arbuzov reaction).

Felcht, U.-H. in *Houben-Weyl,* 1982, band E2, 123 (phosphinic acids).

Gallenkamp, B., *et al.,* in *Houben-Weyl,* 1982, band E2, 300 (phosphinic acids).

Hildebrand, R.L. *et al., Top. Phosphorus Chem.,* 1983, **11,** 297 (biology of naturally-occurring phosphonic acid derivatives).

Maier, L., *Phosphorus Sulfur,* 1983, **14,** 295 (aminoalkylphosphinic acids).

Petrov. A.A., *et al., Russ. Chem. Rev. (Engl. Transl.),* 1983, **52,** 1030 (Arbuzov reactions with halogenoacetylenes).

Wolfsberger, W., *Chem. Ztg.,* 1985, **109,** 317 (additions of dialkyl phosphonates and related compounds to unsaturated systems).

Phosphorus–nitrogen compounds

Derivatives of phosphorus acids

Burgada, R., *Bull. Soc. Chim. Fr.* 1963, 2335 (tervalent phosphorus–nitrogen compounds).

Burgada, R., *Ann. Chim. (Paris),* 1966, 15 (tervalent phosphorus–nitrogen compounds).

Derkatsch, G.I., *Angew. Chem. Int. Ed. Engl.,* 1969, **8,** 42 (phosphorus isocyanates).

Derkatsch, G.I., *et al., Z. Chem.,* 1969, **9,** 369 (phosphorus isocyanates).

Fluck, E. and Haubold, W., in *Kosolapoff and Maier,* 1973, **6,** 579 (tetraco-ordinate quinquevalent phosphorus acid derivatives).

Vilceanu, R. *et al., Pure Appl. Chem.,* 1975, **44,** 285 (compounds of tri- and tetraco-ordinate phosphorus).

Cremlyn, R.J.W., *et al., Top. Phosphorus Chem.,* 1976, **8,** 1 (phosphorus acid azides and hydrazides).

Mathis, F., *Phosphorus Sulfur,* 1976, **1,** 109 (properties of the P–N bond).

Gilyarov, V.A., *Russ. Chem. Rev. (Engl. Transl.),* 1982, **51,** 909 (azides).

Kukhar', V.P., *et al., Pure Appl. Chem.,* 1980, **52,** 89 (nitrogen-containing phosphaspirans).

Matevosyan, G.L. *et al., Zh. Obshch. Khim.,* 1982, **52,** 1441 (Engl. Transl. p. 1275) (phosphorylated benzimidazoles).

Kamalev, R.M. *et al., Russ. Chem. Rev. (Engl. Transl.),* 1985, **54,** 1210 (isothiocyanates).

Ovrutskii, V.M. *et al., Russ. Chem. Rev. (Engl. Transl.),* 1986, **55,** 343 (hydrazides).

Phosphazenes (linear and cyclic)

Shaw, R.A. *et al., Chem. Rev.,* 1962, **62,** 247 (phosphonitrilic derivatives).

Houben-Weyl, 1963, band I, 175.

Fluck, E., *Top. Phosphorus Chem.,* 1967, **4,** 291.

Singh, G. *et al., Organometallic Chem. Rev.,* 1967, **2,** 279 (tertiary phosphine imides and phosphazenes).

Kabatchnik, M.I., *Phosphorus,* 1971, **1,** 117 (phosphorimides).

Bestmann, H.J. and Zimmermann, R., in *Kosolapoff and Maier,* 1972, **3,** 1.
Allcock, H.R., *Chem. Rev.,* 1972, **72,** 315 (phosphazenes).
Phosphorus–Nitrogen Chemistry, Allcock, H.R., Academic Press, New York, 1972.
Bermann, M., *Top. Phosphorus Chem.,* 1972, **7,** 311 (physical data for phosphazotrihalides).
Bermann, M., *Adv. Inorg. Chem. Radiochem.,* 1972, **14,** 1 (phosphazotrihalides).
Keat, R., in *Kosolapoff and Maier,* 1973, **6,** 833 (phosphonitrilic compounds and related systems).
Utvary, K., in *Method. Chim.,* **7B,** 447, Zimmer, H. and Niedenzu, K. (eds) Academic Press, New York, 1978 (phosphazenes).
Kukhar', V.P. *et al., Pure Appl. Chem.,* 1980, **52,** 891 (synthesis and properties).
Shaw, R.A., *Pure Appl. Chem.,* 1980, **52,** 1063.
Kireev, V.V. *et al., Russ. Chem. Rev. (Engl. Transl.),* 1981, **50,** 1186 (oligomeric alkoxyphosphazenes).
Kireev, V.V. *et al., Russ. Chem. Rev. (Engl. Transl.),* 1982, **51,** 149 (polyphosphazenes).
Vinogradova, S.V., *et al., Russ. Chem. Rev. (Engl. Transl.),* 1984, **53,** 49 (polyfluoroalkylphosphazenes).

Methylenephosphoranes

Bestmann, H.J., *Pure Appl. Chem.,* 1964, **9,** 285.
Trippett, S., *Pure Appl. Chem.,* 1964, **9,** 255 (Wittig reaction).
Schöllkopf, U., in *Newer Methods of Preparative Organic Chemistry,* **3,** 111, Foerst, W. (ed.) Academic Press, New York, 1964. (Wittig reaction).
Wittig, G., *Pure Appl. Chem.,* 1964, **9,** 245.
Ylid Chemistry, Johnson, A.W., Academic Press, New York, 1966.
Bestmann, H.J. and Zimmermann, R., in *Kosolapoff and Maier,* 1972, **3,** 1.
Bestmann, H.J., *Angew. Chem. Int. Ed. Engl.,* 1977, **16,** 349 (phosphaallenes and phosphacumulenes).
Bestmann, H.J. and Zimmermann, R., in *Carbon–Carbon Bond Formation,* Augustine, R.L. (ed.) **1,** 353, M. Dekker, New York, 1977.
Schlosser, M., in *Method. Chim.,* Zimmer, H. and Niedenzu, K. (eds) **7B,** 506, Academic Press, New York, 1978.
Bestmann, H.J., *Pure Appl. Chem.,* 1980, **52,** 771 (old and new ylide chemistry).
Tyuleneva, V.V. *et al., Russ. Chem. Rev. (Engl. Transl.),* 1981, **50,** 280 (fluorine-containing halogenomethylenephosphoranes).
Bestmann, H.J. and Zimmermann, R., in *Houben-Weyl,* 1982, band E1, 616.
Tyuleneva, V.V. *et al., Russ. Chem. Rev. (Engl. Transl.),* 1982, **51,** 1 (fluorine-containing ylides).
Kolodyazhnyi, O.I. *et al., Russ. Chem. Rev. (Engl. Transl.),* 1983, **52,** 1096 (heterosubstituted ylides).

See also references on the uses of organophosphorus compounds.

Compounds with penta- or hexaco-ordinate phosphorus

Ramirez, F., *Pure Appl. Chem.,* 1964, **9,** 337 (reactions of carbonyl compounds with phosphite esters).
Ramirez, F., *Bull. Soc. Chim. Fr.,* 1966, 2443; 1970, 3491 (oxyphosphoranes).
Schmutzler, R., in *Halogen Chemistry,* Gutman, V. (ed.) **2,** 31, Academic Press, New York, 1967 (fluorophosphoranes).
Gallagher, M.J. *et al., Top. Phosphorus Chem.,* 1968, **3,** 1.
Ramirez, F., *Acc. Chem. Res.,* 1968, **1,** 168 (oxyphosphoranes).
Westheimer, F.H., *Acc. Chem. Res.,* 1968, **1,** 70.
Mislow, K., *Acc. Chem. Res.,* 1970, **3,** 321 (pseudorotation in nucleophilic displacements).
Gillespie, P. *et al., Angew. Chem. Int. Ed. Engl.,* 1971, **10,** 687 (pseudorotation).
Ramirez, F. *et al., Phosphorus,* 1971, **1,** 1 (pseudorotation in cage polycyclic oxyphosphoranes).
Ugi, I. *et al., Acc. Chem. Res.,* 1971, **4,** 288 (pseudorotation).
Hellwinkel, D., in *Kosolapoff and Maier,* 1972, **3,** 185.
Holmes, R.R., *Acc. Chem. Res.,* 1972, **5,** 296 (spectroscopy of pentaco-ordinate molecules).
Dynamic Stereochemistry of Pentaco-ordinated Phosphorus and Related Elements, Luckenbach, R., G. Thieme, Stuttgart, 1973.
Gillespie, P., *Angew. Chem. Int. Ed. Engl.,* 1973, **12,** 91.
Arbuzov, B.A., *Russ. Chem. Rev. (Engl. Transl.),* 1974, **43,** 414 (cyclic oxyphosphoranes).
Ramirez, F., *Synthesis,* 1974, 90 (uses).
Burgada, R., *Phosphorus Sulfur,* 1976, **2,** 237 (structure and reactivity of spirophosphoranes).
Ramirez, F. *et al., Phosphorus Sulfur,* 1976, **1,** 231 (use in synthesis of phosphoric diesters).
Trippett, S., *Phosphorus Sulfur,* 1976, **1,** 89 (apicophilicity and ring strain in phosphoranes).
d-Orbitals in the Chemistry of Silicon, Phosphorus and Sulfur, Kwart, H. and King, K.G., Springer, Berlin, 1977.
Sheldrick, W.S., *Top. Curr. Chem.,* 1978, **73,** 1 (stereochemistry).
Pentaco-ordinated Phosphorus I Structure and Spectroscopy, ACS Monograph No 175, *II Reaction Mechanism,* ACS Monograph No 176, Holmes, R.R., American Chemical Society, Washington, DC, 1980.
Kukhar', V.P., *et al., Pure Appl. Chem.,* 1980, **52,** 891 (nitrogen-containing spirans).
Wolf, R., *Pure Appl. Chem.,* 1980, **52,** 1141 (ring-chain tautomerism).
Luckenbach, R., in *Houben-Weyl,* 1982, band E2, 833 (compounds with co-ordination number 5); 895 (compounds with co-ordination number 6).
Hellwinkel, D., *Top Curr. Chem.,* 1983, **109,** 1 (penta- and hexaco-ordinate compounds of main group V elements).
Polozhaeva, N.A., *et al., Russ. Chem. Rev. (Engl. Transl.),*

1985, **54,** 1126 (1,3,2-diheterophosphacyclohexanes containing pentaco-ordinate phosphorus).

See also references on phosph(V)orines under Cyclic Compounds.

Phosphorus-containing heterocyclic compounds

Compounds with phosphorus bonded to carbon only

Märkl, G., *Angew. Chem. Int. Ed. Engl.*, 1965, **4,** 1023.
Berlin, K.D., *et al., Top. Phosphorus Chem.*, 1969, **6,** 1.
Dimroth, K., *Top. Curr. Chem.*, 1971, **38,** 1 (phosphorines).
Bokanov, A.J., *et al., Russ. Chem. Rev. (Engl. Transl.)*, 1977, **46,** 861 (dihydrophenophosphazines).
Shvetsov-Shilovskii, N.I. *et al., Russ. Chem. Rev. (Engl. Transl.)*, 1977, **46,** 514 (phosphorines).
Venkataremu, S.D. *et al., Chem. Rev.*, 1977, **77,** 121 (polycyclic systems).
Kassner, J.E. and Zimmer, H. in *Method. Chim.*, **7B,** 537, Zimmer, H. and Niedenzu, K. (eds) Academic Press, 1978.
Edmundson, R.S. in *Rodd's Chemistry of Carbon Compounds,* 2nd edn, **IVK,** Coffey, S. (ed.) 1 (six-membered rings containing unusual heteroatoms); 315 (compounds with seven-membered and larger rings containing unusual heteroatoms), Elsevier, Amsterdam, 1979.
Fluck, E., *Top. Phosphorus Chem.*, 1980, **10,** 193 (compounds of two-co-ordinate phosphorus; phosphorines).
Mathey, F., *Top. Phosphorus Chem.*, 1980, **10,** 1 (phospholes).
Quin, L.D., *et al., Pure Appl. Chem.*, 1980, **52,** 1013 (large ring compounds).
Appel, R., *et al., Angew. Chem. Int. Ed. Engl.*, 1981, **20,** 731.
The Heterocyclic Chemistry of Phosphorus, Quin, L.D., Wiley, New York, 1981.
Dimroth, K., *Acc. Chem. Res.*, 1982, **15,** 58 (phosphorines).
Dimroth, K., in *Houben-Weyl,* 1982, band E1, 783 (phospho(V)orines).
Markl, G., in *Houben-Weyl,* 1982, band E1, 72 (phosphorines).
Freedman, L.D., *et al., Chem. Rev.*, 1987, **87,** 289 (dihydrophenophosphazines).

Compounds with phosphorus bonded to atoms other than carbon

Ramirez, F., *Pure Appl. Chem.*, 1964, **9,** 337 (oxyphosphoranes).
Quin, L.D., in *1,4-Cycloaddition Reactions,* Hamer, J. (ed.) Academic Press, New York, 1967 (tervalent phosphorus compounds as dienophiles).
The Heterocyclic Compounds of Phosphorus, Arsenic, Antimony, and Bismuth, Mann, F.G., 2nd edn, Interscience, 1970.
Grapov, A.F. *et al., Russ. Chem. Rev. (Engl. Transl.)*, 1970, **39,** 39 (cyclodiphosphazanes).
Hudson, R.F. *et al., Acc. Chem. Res.*, 1972, **5,** 204 (reactivity of four and five-membered ring systems).
Arbuzov, B.A. *et al., Russ. Chem. Rev. (Engl. Transl.)*, 1974, **43,** 414 (cyclic oxyphosphoranes).
Burgada, R., *Phosphorus Sulfur,* 1976, **2,** 237 (structure and reactivity of spirophosphoranes).
Penczek, S., *Pure Appl. Chem.*, 1976, **48,** 363 (polymerization of cyclic esters of phosphoric acid).
Verkade, J.G., *Phosphorus Sulfur,* 1976, **2,** 251 (steric and electronic effects in cyclic phosphorus esters).
Majoral, J.-P., *Synthesis,* 1978, 557 (ring hydrazides).
Edmundson, R.S. in *Rodd's Chemistry of Carbon Compounds,* 2nd edn, **IVK,** Coffey, S. (ed.) 33 (six-membered ring compounds with one or more unusual heteroatoms); 349 (seven-membered or larger rings with unusual heteroatoms), Elsevier, Amsterdam, 1979.
Maryanoff, B.E., *et al., Top. Stereochem.*, 1979, **11,** 187 (stereochemistry of phosphorus-containing cyclohexanes).
Hall, C.R., *et al., Tetrahedron,* 1980, **36,** 2059 (mechanistic implications of stereochemistry at phosphorus on reactivity of five- and six-membered cyclic esters of phosphorus acids).
Verkade, J.G., *Pure Appl. Chem.*, 1980, **52,** 1131 (ligation of tervalent phosphorus to H, Se, and metals).
Wolf, R. *Pure Appl. Chem.*, 1980, **52,** 1141 (ring-chain tautomerism; spirophosphoranes).
Grapov, A.F., *et al., Russ. Chem. Rev. (Engl. Transl.)*, 1981, **50,** 324 (diazadiphosphetidines).
Cherkasov, R.A., *et al., Russ. Chem. Rev. (Engl. Transl.)*, 1982, **51,** 746 (reactivity of 1,3,2-diheterophospholans and -phosphorinans with tervalent phosphorus).
Keat, R., *Top. Curr. Chem.*, 1982, **102,** 89 (compounds with tervalent phosphorus to nitrogen bonds).
Nifant'ev, E.E. *et al., Russ. Chem. Rev. (Engl. Transl.)*, 1982, **51,** 921 (1,3,2-diheterophosph(III)orinans).
Arbuzov, B.A., *et al., Bull. Acad. Sci. USSR, Ser. Chim.*, 1983, 2254 (synthesis and properties of compounds with ring elements from Groups V and VI).
Navech, J., *et al., Phosphorus Sulfur,* 1983, **15,** 51 (phosphorus-containing adamantanes).
Pudovik, M.A., *et al., Russ. Chem. Rev. (Engl. Transl.)*, 1983, **52,** 361 (reactivity of 1,3,2-diheterophosph(III)olans).
Polozhaeva, N.A., *et al., Russ. Chem. Rev. (Engl. Transl.)*, 1985, **54,** 1126 (1,3,2-diheterophospha(V)cyclanes).
Schmidpeter, A., *et al., Nachr. Chem. Tech. Lab.*, 1985, **33,** 793 (azaphospholes).
Gevaza, Yu. I. *et al., Khim. Geterotsikl. Soedin.*, 1986, 29 (electrophilic heterocyclization of unsaturated organosilicon and -phosphorus compounds).
Ugi, I., *et al., Chem. Scr.*, 1986, **26,** 205 (heterophospholes as phosphorylating agents).

Reactive intermediates and compounds with unusual co-ordination numbers

Dimroth, K., *Top. Curr. Chem.*, 1973, **38,** 1 (compounds with delocalized phosphorus–carbon double bonds).

Niecke, E., *et al., Nachr. Chem. Techn. Lab.*, 1975, **23,** 395 (new phosphorus–nitrogen ylides).

Regitz, M., *Angew. Chem. Int. Ed. Engl.*, 1975, **14,** 222 (phosphorylcarbene).

Schmidt, U., *Angew. Chem. Int. Ed. Engl.*, 1975, **14,** 523 (phosphinidines).

Shvestov-Shilovskii, N.I. *et al., Russ. Chem. Rev. (Engl. Transl.),* 1977, **46,** 514 (compounds of two-co-ordinate phosphorus).

Fluck, E., *Top. Phosphorus Chem.*, 1980, **10,** 193 (compounds of two-co-ordinate phosphorus).

Appel, R. *et al., Angew. Chem. Int. Ed. Engl.*, 1981, **20,** 731 (phosphaalkenes and phosphaalkynes).

Regitz, M. *et al., Top. Curr. Chem.*, 1981, **97,** 71 (short lived phosphorus(V) compounds having co-ordination number 3).

Westheimer, F.H., *Chem. Rev.*, 1981, **81,** 313 (monomeric metaphosphates).

Regitz, M., in *Houben-Weyl,* 1982, band E1, 23 (compounds with co-ordination number 1).

Regitz, M., in *Houben-Weyl,* 1982, band E1, 583 (pentacovalent compounds with co-ordination number 3).

Weber, B. and Regitz, M., in *Houben-Weyl,* 1982, band E1, 27 (compounds with co-ordination number 2).

Cowley, A.H. *et al., Chem. Rev.*, 1985, **85,** 367 (stable two-co-ordinate phosphorus cations).

Niecke, E. *et al., Nova Acta Leopold,* 1985, **59,** 83 (alkene, carbene and nitrene analogues in phosphorus chemistry).

Germa, H. *et al., Phosphorus Sulfur,* 1986, **26,** 327 (compounds of trico-ordinate pentavalent phosphorus).

Markovskii, L.N. *et al., Zh. Obshch. Khim.*, 1986, **56,** 253 (Engl. Transl. p. 221) (compounds with phosphorus–carbon double bonds).

Becker, G. *et al., Nova Acta Leopold,* 1985, **59,** 55 (phosphines with coordination numbers 1 and 2).

Complexes

Booth, G., in *Kosolapoff and Maier,* 1972, **1,** 433 (phosphine complexes).

Levason, W., *Adv. Inorg. Chem. Radiochem.*, 1972, **42,** 173 (complexes of bidentate phosphine ligands).

Verkade, J.G. and Coskran, H.J., in *Kosolapoff and Maier,* 1972, **2,** 1 (phosphite, phosphonite, and aminophosphine complexes).

Transition Metal Complexes of Phosphorus, Arsenic and Antimony, McAuliffe, C.A. (ed.) MacMillan, 1973.

Wasson, J.R., *et al., Fortschr. Chem. Forsch.*, 1973, **35,** 65 (thio and seleno phosphate complexes of transition metals).

Kabachnik, M.I., *et al., Russ. Chem. Rev. (Engl. Transl.),* 1974, **43,** 733, 1554 (organophosphorus complexones).

Issleib, K., *Pure Appl. Chem.*, 1975, **44,** 237 (co-ordination chemistry of tervalent phosphorus).

Kabachnik, M.I., *et al., Pure Appl. Chem.*, 1975, **44,** 269 (organophosphorus complexones).

Stelzer, O., *Top. Phosphorus Chem.*, 1977, **9,** 1 (transition metal complexes).

Tolman, C.A., *Chem. Rev.*, 1977, **77,** 313 (steric effects on catalyst ligands).

Mason, R., *et al., Angew. Chem. Int. Ed. Engl.*, 1978, **17,** 183 (phosphine ligands).

Schmidbaur, H., *Pure Appl. Chem.*, 1978, **50,** 19.

Phosphine, Arsine and Stibine Complexes of the Transition Elements, McAuliffe, C.A. and Levenson, W., Elsevier, Amsterdam, 1979.

Marko, L., *Pure Appl. Chem.*, 1979, **51,** 2211 (phosphine complexes).

Schmidbaur, H., *Pure Appl. Chem.*, 1980, **52,** 1057 (complexes from new organophosphorus ligands).

Venanzi, M.L., *Pure Appl. Chem.*, 1980, **52,** 1117 (phosphine complexes).

Bashkirov, Sh. Sh. *et al., Russ. Chem. Rev. (Engl. Transl.),* 1981, **50,** 749 (Mossbauer spectroscopy of Sn halide complexes with organophosphorus ligands).

Garrou, P.E., *Chem. Rev.*, 1981, **81,** 229 (^{31}P nmr spectroscopy of transition metal complexes).

de Bolster, M.W.G., *Top. Phosphorus Chem.*, 1983, **11,** 69 (phosphoryl co-ordination chemistry, 1975–81).

Mathey, F., *et al., Structure and Bonding,* 1983, **55,** 154 (phosphole complexes).

Brill, T.B. *et al., Chem. Rev.*, 1984, **84,** 577 (metal phosphite complexes).

Walther, B., *Coord. Chem. Rev.*, 1984, **60,** 67 (co-ordination chemistry of compounds of the general type $R_2P(X)H$).

Uses of organophosphorus compounds

Bergelson, L.D. *et al., Pure Appl. Chem.*, 1964, **9,** 271 (carbonyl olefination).

Bestmann, H.J. *et al., Angew Chem Int. Ed. Engl.*, 1965, **4,** 583, 645, 830 (Wittig synthesis).

Maercker, A., *Organic Reactions,* 1965, **14,** 270 (Wittig reaction).

Dombrovskii, A.V. *et al., Russ. Chem. Rev. (Engl. Transl.),* 1966, **35,** 733 (olefin synthesis with phosphoryl activated reagents).

Bergelson, L.D. and Shemyakin, M.M., in *Newer Methods of Preparative Organic Chemistry,* **5,** Foerst, W. (ed.) Academic Press, 1969 (fatty acid synthesis).

Bestmann, H.J., in *Newer Methods of Preparative Organic Chemistry,* **5,** Foerst, W. (ed.) Academic Press, 1968 (alkylidenephosphoranes).

Cadogan, J.I.G., *Synthesis,* 1969, 11 (nitrogen heterocycles from aromatic nitro compounds and phosphites).

Bestmann, H.J., *et al., Fortschr. Chem. Forsch.*, 1971, **20,** 1 (phosphinealkenes and their uses).

Bestmann, H.J., *et al., Chem. Ztg.*, 1972, **96,** 649 (synthesis of cyclic compounds).

Boutagy, J., *et al., Chem. Rev.*, 1974, **74,** 87 (alkene synthesis with phosphonate carbanions).

Cadogan, J.I.G., *et al., Chem. Soc. Rev.,* 1974, **3,** 87 (tervalent phosphorus compounds in organic synthesis).
Ramirez, F., *Synthesis,* 1974, 90 (oxyphosphoranes in synthesis).
Zbiral, E., *Synthesis,* 1974, 775 (use of phosphine-methylenes, phosphine oxides and phosphine imides in organic synthesis).
Vollhardt, K.P.C., *Synthesis,* 1975, 765 (bisWittig reagents in the synthesis of nonbenzenoid aromatic systems).
Grob, H., *et al., Z. Chem.,* 1977, **17,** 281 (synthesis and applications of phosphoric and phosphonic acid derivatives).
Wadsworth, W.S., *Organic Reactions,* 1977, **25,** 73 (olefin synthesis using phosphoryl-activated carbanions).
Organophosphorus Reagents in Organic Synthesis, Cadogan, J.I.G. (ed.) Academic Press, London, 1979.
Becker, K.B., *Tetrahedron,* 1980, **36,** 1717 (Wittig synthesis of cycloalkenes).
Bestmann, H.J. and Zimmermann, R., in *Method. Chim.,* **4,** 67, 137 Falbe, J. (ed.) G. Thieme, Stuttgart, 1980 (synthesis of dienes and polyenes).
Bestmann, H.J., *et al., Top. Curr. Chem.,* 1983, **109,** 83 (Wittig reagents in the synthesis of natural products).
Pommer, H. *et al., Top. Curr. Chem.,* 1983, **109,** 163 (industrial applications of the Wittig reaction).
Cherkasov, R.A. *et al., Tetrahedron,* 1985, **41,** 2567 (sulfur-containing organophosphorus reagents in organic synthesis).

See also references on methylenephosphoranes.

7. Abbreviations

[α]	specific rotation
abs. config.	absolute configuration
Ac	Acetyl
AcOH	Acetic acid
Ac_2O	Acetic anhydride
alk.	alkaline
amorph.	amorphous
anal.	analytical applications, analysis or detection
anhyd.	anhydrous
aq.	Aqueous
asym.	asymmetrical, unsymmetrical
B	base
bibl.	bibliography
biosynth.	biosynthesis
Bp	boiling point
BAN	British Approved Name
B.P.	British Patent
c.	concentration
cmr	^{13}C nuclear magnetic resonance
ca.	(*circa*) about
cd	circular dichromism
chromatog.	chromatography
col.	colour, coloration
conc.	concentrated
config.	configuration
conformn	conformation
constit.	constituent
compd.	compound
cryst. struct.	x-ray crystal structure determination
d.	density
dec.	decomposes, decomposition
degradn.	degradation
deg.	degree
deriv(s).	derivative(s)
descr.	described
dil.	dilute, dilution
dimorph.	dimorphic
diss.	dissolves, dissolved
dist.	distil, distillation
equilib.	equilibrium
esp.	especially
esr.	electron spin resonance
Et	Ethyl
EtOAc	Ethyl acetate
evapn	evaporation
exp.	exposure, experimental
fl. p.	flash point
fluor.	fluoresces, fluorescence
formn	formation
Fp	freezing point
glc	gas liquid chromatography
Glc	β-D-glucopyranosyl
haz.	hazard
hydrol.	hydrolyses, hydrolysed, hydrolysis
ir.	infra-red spectrum
isol.	isolated, isolation
isom.	isomerises
LD	lethal dose: LD_{50}, a dose which is lethal to 50% of the animals tested
M	molecular weight (formula weight)
manuf.	manufacturer, manufactured
max.	maximum
Me	Methyl
metab.	metabolite
misc.	miscible
mixt.	mixture
mod.	moderately
Mp	melting point
ms	mass spectrum
n	index of refraction e.g. (n_D^{20} for 20° and sodium light)
nmr	nuclear magnetic resonance spectrum
obt.	obtained
ord	optical rotatory dispersion
pet. ether	Petroleum ether (light petroleum)
Ph	Phenyl (C_6H_5)
pharmacol	pharmacology
pmr	proton (^1H) nuclear magnetic resonance
Pnmr	phosphorus (^{31}P) nuclear magnetic resonance

polarog.	polarography	subl.	sublimation, sublimes
ppd.	precipitated	synth.	synthesis
props.	properties	tautom.	tautomerism
purifn.	purification	tlc	thin layer chromatography
Py	Pyridine	tox.	toxicity
ref.	reference	TLV	Threshold Limit Value
resoln.	resolution	unsatd.	unsaturated
rev.	review	USAN	United States Approved Name
r.t.	room temperature	U.S.P.	United States Patent
sepn.	separation	uv	ultraviolet spectrum
sl.	slightly	v.	very
sol.	soluble	var.	variety
soln.	solution	vis.	visible
solv.	solvent	vol.	volume

A

(5,6-Acenaphthenylenedimethylene)-bis[triphenylphosphonium](2+), 8CI — A-00001

5,6-Bis(triphenylphosphoniomethyl)acenaphene(2+)

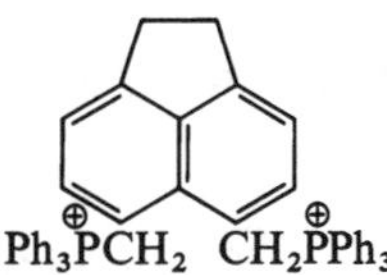

$C_{50}H_{42}P_2^{\oplus\oplus}$ M 704.829 (ion)

Dibromide: [23055-15-4].
$C_{50}H_{42}Br_2P_2$ M 864.637
Solid. Mp 270-272°.

Diperiodate:
$C_{50}H_{42}I_2O_8P_2$ M 1086.634
Solid. Mp 150-152° dec. LiOEt → 1,2-dihydrocyclopent[*f*,*g*]acenaphthylene.

Bestmann, H.J. *et al*, *Chem. Ber.*, 1969, **102**, 2259 (*synth, use*)

1-Acenaphthenylidenetriphenylphosphorane — A-00002

1-Triphenylphosphoranylideneacenaphthene

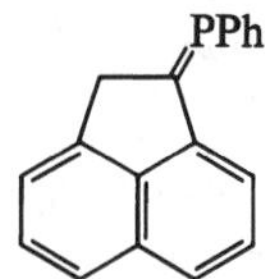

$C_{30}H_{23}P$ M 414.485
Intermed. for Wittig reactions. With H_2O yields acenaphthene.

Bestmann, H.J. *et al*, *Z. Naturforsch.*, *B*, 1962, **17**, 787.

Acephate, BSI, ISO — A-00003

O,S-*Dimethyl acetylphosphoramidothioate. Ortho 12420. Orthene*

[30560-19-1]

MeO(MeS)P(=O)NHAc

$C_4H_{10}NO_3PS$ M 183.162

▷TB4760000.

(±)-*form*

Contact and systemic insecticide of mod. persistence. Solid. V. sol. H_2O, spar. sol. aromatic solvs. Mp 93°. Dec. on dist. Rel. non-toxic (weak anticholinesterase).

USP, 3 716 600, (*1973*); 3 845 172, (*1974*) (*synth*)
Magee, P.S., *Residue Rev.*, 1974, **53**, 3 (*analogues*)
Umetzu, N. *et al*, *J. Agric. Food Chem.*, 1977, **25**, 946 (*tlc, glc, tox*)
Schneider, P. *et al*, *J. Prakt. Chem.*, 1982, **324**, 1063 (*props*)
Alawi, M.A., *Fresenius' Z. Anal. Chem.*, 1983, **315**, 358 (*hplc*)
Stan, H-J. *et al*, *J. Chromatogr.*, 1983, **279**, 173 (*glc*)
Pesticide Manual, 6th ed., 1; 7th ed., 1.
Agrochemicals Handbook, Royal Society of Chemistry, London, 1983, A001.

Acetophos — A-00004

Ethyl [(diethoxyphosphinothioyl)thio]acetate, 9CI. Acethion. Ethoxyphos

[919-54-0]

$(EtO)_2P(S)SCH_2COOEt$

$C_8H_{17}O_4PS_2$ M 272.314
Insecticide. Liq. d_4^{20} 1.18. $Bp_{1.5}$ 137°, $Bp_{0.08}$ 104.5-105.5°. n_D^{20} 1.5011.

▷AI6825000.

B.P., 803 441, (*1958*); *CA*, **54**, 4502 (*synth*)
Lui, H., *CA*, 1966, **64**, 14896 (*synth*)

1-Acetoxy-1,1-ethanediphosphonic acid — A-00005

[1-(Acetyloxy)ethylidene]bisphosphonic acid

H_3C, AcO–C[P(=O)(OH)_2]_2

$C_4H_{10}O_8P_2$ M 248.066

Tetra-Me ester: Tetramethyl [1-(acetoxy)ethylidene]-bisphosphonate.
$C_8H_{18}O_8P_2$ M 304.173
Liq. $Bp_{0.02}$ 126-128°.

Chopard, P.A., *Helv. Chim. Acta*, 1967, **50**, 1021 (*synth, ir*)

(2-Acetoxyethyl)phosphonic acid — A-00006

[2-(Acetyloxy)ethyl]phosphonic acid, 9CI

[32541-80-3]

$AcOCH_2CH_2P(O)(OH)_2$

$C_4H_9O_5P$ M 168.086
Syrup.

Monoanilinium salt: [75502-81-7]. Solid. Mp 110-116°.

Mono-dicyclohexylammonium salt: [75502-82-8]. Solid. Mp 160-162°.

Di-Me ester: [39118-50-8]. *Dimethyl (2-acetyloxyethyl)phosphonate.*
$C_6H_{13}O_5P$ M 196.139
Liq. d_4^{20} 1.19-1.22. Bp_5 104-105°, $Bp_{5\text{-}6}$ 182-194°. n_D^{20} 1.4298.

Di-Et ester: [15300-97-7]. *Diethyl (2-acetyloxyethyl)-phosphonate.*
$C_8H_{17}O_5P$ M 224.193
Liq. d_4^{20} 1.13. $Bp_{3.5\text{-}4.4}$ 127-128°. n_D^{20} 1.4318.

Bis(trimethylsilyl) ester: [72563-41-8]. *Bis(trimethylsilyl) (2-acetyloxyethyl)phosphonate.*
$C_{10}H_{25}O_5PSi_2$ M 312.449
Liq. $Bp_{0.2}$ 80°.

Pudovik, A.N., *Zh. Obshch. Khim.*, 1952, **22**, 473 (*Engl. transl.* p. 537) (*esters, synth*)
McConnell, R.L. *et al*, *J. Am. Chem. Soc.*, 1957, **79**, 1961 (*ester, synth*)
Brel', A.K. *et al*, *Zh. Obshch. Khim.*, 1979, **49**, 714 (*Engl. transl.* p. 619) (*esters, synth, ir, pmr*)
Gloede, J. *et al*, *J. Prakt. Chem.*, 1980, **322**, 327 (*synth, derivs, P nmr*)

(Acetoxymethyl)phosphonic acid **A-00007**

[(*Acetyloxy*)*methyl*]*phosphonic acid, 9CI*

$AcOCH_2P(O)(OH)_2$

$C_3H_7O_5P$ M 154.059

Di-Me ester: [24630-57-7]. *Dimethyl* [(*acetyloxy*)-*methyl*]*phosphonate.*
$C_5H_{11}O_5P$ M 182.113
Liq. Bp_{13} 108°, $Bp_{2.5}$ 92-95°.

Di-Et ester: [7016-78-6]. *Diethyl* [(*acetyloxy*)*methyl*]-*phosphonate.*
$C_7H_{15}O_5P$ M 210.166
Liq. d_4^{20} 1.14. Bp_4 106.5-107°. n_D^{20} 1.4300.

Bis(*triethylsilyl*) *ester: Bis*(*triethylsilyl*) [(*acetyloxy*)-*methyl*]*phosphonate.*
$C_{15}H_{35}O_5PSi_2$ M 382.583
Liq. d_4^{20} 1.01. $Bp_{5.5}$ 155°. n_D^{20} 1.4470.

Pudovik, A.N. *et al*, *Zh. Obshch. Khim.*, 1963, **33**, 3201 (*Engl. transl.* p. 3128) (*synth*)
Orlov, N.F. *et al*, *Zh. Obshch. Khim.*, 1966, **36**, 699 (*Engl. transl.* p. 713) (*silyl ester*)
Bel'skii, V.E. *et al*, *Izv. Akad. Nauk SSSR, Ser. Khim.*, 1975, 1047 (*Engl. transl.* p. 958) (*pmr, P nmr*)
Bel'skii, V.E. *et al*, *Izv. Akad. Nauk SSSR, Ser. Khim.*, 1975, 1624 (*Engl. transl. p. 1511*) (*ir, nmr*)
Griffiths, W.R. *et al*, *Phosphorus Sulfur*, 1978, **5**, 101 (*ms*)
Holy, A. *et al*, *Collect. Czech. Chem. Commun.*, 1982, **47**, 3447 (*synth, ms*)

(1-Acetoxypropyl)phosphonic acid **A-00008**

[*1*-(*Acetyloxy*)*propyl*]*phosphonic acid, 9CI*
[53621-86-6]

$H_3CCH_2CH(OAc)P(O)(OH)_2$

$C_5H_{11}O_5P$ M 182.113

Di-Et ester: Diethyl [*1*-(*acetyloxy*)*propyl*]*phosphonate.*
$C_9H_{19}O_5P$ M 238.220
Liq. d_4^{20} 1.09. Bp_3 104-105°. n_D^{20} 1.4302.

Pudovik, A.N. *et al*, *Zh. Obshch. Khim.*, 1963, **33**, 3201 (*Engl. transl.* p. 3128) (*synth*)

(2-Acetoxypropyl)phosphonic acid **A-00009**

[*2*-(*Acetyloxy*)*propyl*]*phosphonic acid, 9CI*

$H_3CCH(OAc)CH_2P(O)(OH)_2$

$C_5H_{11}O_5P$ M 182.113

(±)-*form*

Di-Et ester: Diethyl [*2*-(*acetyloxy*)*propyl*]*phosphonate.*
$C_9H_{19}O_5P$ M 238.220
Liq. $Bp_{0.5}$ 89-93°. n_D^{25} 1.4301.

Preis, S. *et al*, *J. Am. Chem. Soc.*, 1955, **77**, 6225 (*synth*)

3-Acetoxypropylphosphonic acid **A-00010**

[(*3-Acetyloxy*)*propyl*]*phosphonic acid, 9CI*
[73186-67-1]

$AcOCH_2CH_2CH_2P(O)(OH)_2$

$C_5H_{11}O_5P$ M 182.113

Di-Me ester: [39118-51-9]. *Dimethyl* (*3-acetyloxypropyl*)*phosphonate. 3-Dimethoxyphosphinyl-1-propanol acetate.*
$C_7H_{15}O_5P$ M 210.166
Liq. d_4^{20} 1.18. $Bp_{1.5\text{-}2.0}$ 92-94°. n_D^{20} 1.4412.

Di-Et ester: [1186-23-8]. *Diethyl* (*3-acetyloxypropyl*)-*phosphonate. 3-Diethoxyphosphinyl-1-propanol acetate.*
$C_9H_{19}O_5P$ M 238.220
Liq. d_4^{20} 1.10. $Bp_{1.3\text{-}2.0}$ 110-112°. n_D^{20} 1.4349.

Brel', A.K. *et al*, *Zh. Obshch. Khim.*, 1979, **49**, 714 (*Engl. transl.* p. 619)

(1-Acetoxy-2,2,2-trichloroethyl)phosphonic acid **A-00011**

[*1*-(*Acetyloxy*)*-2,2,2-trichloroethyl*]*phosphonic acid, 9CI*

$Cl_3CCH(OAc)P(O)(OH)_2$

$C_4H_6Cl_3O_5P$ M 271.421

Di-Me ester: [5952-41-0]. *Dimethyl 1-acetoxy-2,2,2-trichloroethylphosphonate. Chloracetophos.*
$C_6H_{10}Cl_3O_5P$ M 299.475
Has been used for the treatment of fungal diseases in man. A weak insecticide. Liq. d_4^{20} 1.47. $Bp_{1.5}$ 112-114°. n_D^{20} 1.4700.
▷AH3535000.

Di-Et ester: [5952-42-1]. *Diethyl 1-acetoxy-2,2,2-trichloroethylphosphonate.*
$C_8H_{14}Cl_3O_5P$ M 327.528
Liq. d_4^{20} 1.36. $Bp_{0.15}$ 127-127.5°. n_D^{20} 1.4660.
▷AH3531000.

Diisopropyl ester: Diisopropyl 1-acetoxy-2,2,2-trichloroethylphosphonate.
$C_{10}H_{18}Cl_3O_5P$ M 355.582
Liq. d_4^{20} 1.28. $Bp_{0.03}$ 95-98°. n_D^{20} 1.4608.

Nikonorov, K.V. *et al*, *Izv. Akad. Nauk SSSR, Ser. Khim.*, 1962, 1882 (*Engl. transl.* p. 1795) (*synth*)
Antokhina, L.A. *et al*, *Izv. Akad. Nauk SSSR, Ser. Khim.*, 1966, 2135 (*Engl. transl.* p. 2067) (*synth*)
Gazizov, T.Kh. *et al*, *Dokl. Akad. Nauk SSSR, Ser. Sci. Khim.*, 1966, **166**, 615 (*Engl. transl.* p. 124) (*synth, pmr*)

[(Acetylamino)methyl]phosphonic acid, 9CI **A-00012**

(*Acetamidomethyl*)*phosphonic acid*
[57637-97-5]

$AcNHCH_2P(O)(OH)_2$

$C_3H_8NO_4P$ M 153.074

Di-Me ester: [20495-30-1]. *Dimethyl* (*acetamidomethyl*)*phosphonate.*
$C_5H_{12}NO_4P$ M 181.128
Oil.

Di-Et ester: [20495-31-2]. *Diethyl* (*acetamidomethyl*)-*phosphonate.*
$C_7H_{16}NO_4P$ M 209.181
Solid. Mp 39-40°. $Bp_{0.01}$ 136-137°. n_D^{25} 1.4533.

Diisopropyl ester: Diisopropyl (*acetamidomethyl*)-*phosphonate.*
$C_9H_{20}NO_4P$ M 237.235
Liq. $Bp_{0.006}$ 132-134°. n_D^{20} 1.4575.

Bis(*trimethylsilyl*)*ester:* [55108-80-0]. *Bis*(*trimethylsilyl*) (*acetamidomethyl*)*phosphonate.*
$C_9H_{24}NO_4PSi_2$ M 297.438
No phys. props. reported.

Ivanov, B.E. *et al*, *Izv. Akad. Nauk SSSR, Ser. Khim.*, 1971, 2493 (*Engl. transl.* p. 2364) (*synth*)
Harvey, D.J. *et al*, *Org. Mass. Spectrom.*, 1974, **9**, 955 (*silyl ester, ms*)

Griffiths, W.R. *et al*, *Phosphorus*, 1975, **5**, 273 (*ms*)
Scharf, D.J., *J. Org. Chem.*, 1976, **41**, 28 (*synth, ir, pmr, P nmr*)
Costisella, B. *et al*, *J. Prakt. Chem.*, 1979, **321**, 361 (*synth, pmr*)

Acetyl diethyl phosphite **A-00013**
Acetic acid anhydride with diethyl phosphite, 9CI, 8CI
[3266-66-8]

$MeCOOP(OEt)_2$

$C_6H_{13}O_4P$ M 180.140
Liq. d_4^{20} 1.05-1.07. Bp_{10} 69°. n_D^{20} 1.4195. Relatively thermally stable.

Petrov, K.A. *et al*, *Zh. Obshch. Khim.*, 1961, **31**, 2373 (*Engl. transl.* p. 2211) (*synth*)
Bilevich, K.A. *et al*, *Zh. Obshch. Khim.*, 1965, **35**, 365 (*Engl. transl.* p. 364) (*synth, props*)
Pudovik, A.N. *et al*, *Zh. Obshch. Khim.*, 1975, **45**, 2621 (*Engl. transl.* p. 2852) (*props*)
Gazizov, T.Kh. *et al*, *Zh. Obshch. Khim.*, 1982, **52**, 706 (*Engl. transl.* p. 614) (*synth*)

***P*-Acetyl-*N,N*-diethylphosphonamidic acid, 9CI** **A-00014**

$H_3CC(O)P(=O)(NEt_2)OH$

$C_6H_{14}NO_3P$ M 179.155
Me ester: [14114-67-1]. *Methyl* P-*acetyl*-N,N-*diethylphosphonamidate.*
$C_7H_{16}NO_3P$ M 193.182
Liq. d_4^{20} 1.07. $Bp_{0.1}$ 79-80°. n_D^{20} 1.4555.
Et ester: [41392-00-1]. *Ethyl* P-*acetyl*-N,N-*diethylphosphonamidate.*
$C_8H_{18}NO_3P$ M 207.209
Liq. $Bp_{0.1}$ 35-40°.

Alimov, P.I. *et al*, *Izv. Akad. Nauk SSSR, Ser. Khim.*, 1966, 1486 (*Engl. transl.* p. 1432) (*synth*)
Brown, C. *et al*, *Phosphorus*, 1973, **2**, 287 (*synth, pmr, ir*)

Acetyldimethylphosphine, 9CI, 8CI **A-00015**
1-(Dimethylphosphino)ethanone, 9CI
[18983-86-3]

Me_2PCOCH_3

C_4H_9OP M 104.088
Liq., highly sensitive to air. Bp_{20} 38-39°.
Oxide: [18938-26-6]. *1-(Dimethylphosphinyl)ethanone.*
$C_4H_9O_2P$ M 120.088
Liq. $Bp_{0.01}$ 67-69°.

Kostyanovskii, R.G. *et al*, *Izv. Akad. Nauk SSSR, Ser. Khim.*, 1975, **24**, 901 (*Engl. transl.* p. 816) (*synth, ir, pmr, nmr*)
Pudovik, A.N. *et al*, *Zh. Obshch. Khim.*, 1975, **45**, 1895 (*Engl. transl.* p. 1857) (*sulfide*)
Grikina, O.E. *et al*, *J. Mol. Struct.*, 1977, **37**, 251 (*struct*)
Khaikin, L.S. *et al*, *J. Mol. Struct.*, 1977, **37**, 237 (*struct*)
Razumova, E.R. *et al*, *Izv. Akad. Nauk SSSR, Ser. Khim.*, 1978, **27**, 97 (*Engl. transl.* p. 85) (*ir, raman*)
Frey, G. *et al*, *Chem. Ber.*, 1979, **112**, 763 (*oxide, ir, pmr*)

Acetyldiphenylphosphine, 9CI, 8CI **A-00016**
1-(Diphenylphosphino)ethanone, 10CI
[18629-57-7]

Ph_2PCOCH_3

$C_{14}H_{13}OP$ M 228.230
Intermed. in reaction between Ph_2PCl and AcOH.
Oxide: [27384-09-4]. *1-(Diphenylphosphinyl)ethanone.*
$C_{14}H_{13}O_2P$ M 244.229
Cryst. solid. Mp 23-28°. Sensitive to H_2O and to heat.

Lindner, E. *et al*, *Angew. Chem.*, 1977, **89**, 276 (*oxide, ir, nmr, ms*)
Miller, J.A. *et al*, *J. Chem. Soc., Perkin Trans. 1*, 1977, 1898 (*oxide, cmr, nmr*)
Frey, G. *et al*, *Chem. Ber.*, 1979, **112**, 763 (*oxide, ms, pmr*)
Lesiecki, H. *et al*, *Chem. Ber.*, 1979, **112**, 793 (*oxide*)
Lindner, E. *et al*, *Chem. Ber.*, 1981, **114**, 2272 (*synth, ir, nmr*)

Acetylethylphosphinic acid, 8CI **A-00017**

$H_3CCOPEt(O)OH$

$C_4H_9O_3P$ M 136.087
Et ester: [18788-65-3]. *Ethyl acetylethylphosphinate.*
$C_6H_{13}O_3P$ M 164.141
d_4^{20} 1.11. $Bp_{0.1}$ 117-118°. n_D^{20} 1.4440.
Propyl ester: Propyl acetylethylphosphinate.
$C_7H_{15}O_3P$ M 178.167
Liq. d_4^{20} 1.09. $Bp_{0.05}$ 118-119°. n_D^{20} 1.4530.

Nurtdinov, S.Kh. *et al*, *Zh. Obshch. Khim.*, 1971, **41**, 2486 (*Engl. transl.* p. 2513)

Acetylmethylphosphinic acid, 8CI **A-00018**

$H_3CCOPMe(O)OH$

$C_3H_7O_3P$ M 122.060
Me ester: [33945-42-5]. *Methyl acetylmethylphosphinate.*
$C_4H_9O_3P$ M 136.087
Liq. Bp_1 53-54°. n_D^{20} 1.4439.

Laskorin, B.N. *et al*, *Zh. Obshch. Khim.*, 1972, **42**, 1261 (*Engl. transl.* p. 1256) (*synth, ir, uv, pmr, P nmr*)

[[2-[(Acetyloxy)methyl]phenyl]methyl]-triphenylphosphonium(1+), 10CI **A-00019**
[(2-Acetoxymethyl)benzyl]triphenylphosphonium(1+)

$C_6H_4(CH_2PPh_3^{\oplus})(CH_2OAc)$

$C_{28}H_{26}O_2P^{\oplus}$ M 425.486 (ion)
Bromide: [67219-46-9].
$C_{28}H_{26}BrO_2P$ M 505.390
Cryst. Mp 216-219°. On treatment with RONa yields isochromene *via* the ylide.

Begasse, B. *et al*, *Tetrahedron Lett.*, 1979, 2149 (*synth, use*)

[(Acetyloxy)phenylmethyl]phosphonic acid, 9CI **A-00020**
(α-Acetoxybenzyl)phosphonic acid

PhCH(OAc)P(O)(OH)$_2$

$C_9H_{11}O_5P$ M 230.157

(±)-*form*

Di-Me ester: [16965-84-7]. *Dimethyl [(acetyloxy)-phenylmethyl]phosphonate.*
$C_{11}H_{15}O_5P$ M 258.210
Liq. d_4^{20} 1.23. $Bp_{0.05}$ 122-124°. n_D^{20} 1.5002. Pyrolysis at 270° → MeCOOMe.

Di-Et ester: [16153-59-6]. *Diethyl [(acetyloxy)-phenylmethyl]phosphonate.*
$C_{13}H_{19}O_5P$ M 286.264
Liq. d_4^{20} 1.16. Bp_2 159-160°, $Bp_{0.015}$ 116°. n_D^{20} 1.4896. Pyrolysis at 270° → MeCOOEt.

Pudovik, A.N. *et al, Zh. Obshch. Khim.*, 1963, **33**, 3201 (*Engl. transl.* p. 3128) (*synth*)
Okamoto, Y. *et al, Bull. Chem. Soc. Jpn.*, 1973, **46**, 643 (*props*)
Nesterov, L.V. *et al, Zh. Obshch. Khim.*, 1976, **46**, 1974 (*Engl. transl.* p. 1904) (*synth, pmr*)
Lebedev, E.P. *et al, Zh. Obshch. Khim.*, 1979, **49**, 1731 (*Engl. transl.* p. 1517) (*synth, pmr*)

Acetylphenylphosphinic acid, 8CI A-00021

[67864-83-9]

$H_3CCOPPh(O)OH$

$C_8H_9O_3P$ M 184.131

Me ester: [7078-91-3]. *Methyl acetylphenylphosphinate.*
$C_9H_{11}O_3P$ M 198.158
Acylating agent in heterocyclic synth. Oil. $Bp_{0.15}$ 89.5-91°.

Et ester: [7016-64-0]. *Ethyl acetylphenylphosphinate.*
Oil. d_4^{20} 1.15. $Bp_{0.5}$ 107°. n_D^{20} 1.5161.

Takamizawa, A. *et al, Chem. Pharm. Bull.*, 1967, **15**, 1183 (*ir, use*)
Felcht, U. *et al, Chem. Ber.*, 1975, **108**, 2040 (*use*)
Gareev, R.D. *et al, Zh. Obshch. Khim.*, 1979, **49**, 503 (*Engl. transl.* p. 442) (*ir, pmr*)

O-(2-Acetylphenyl) phosphorodichloridothioate A-00022

O-(*2-Acetylphenyl) dichlorothiophosphate.* O-(*2-Acetylphenyl) thiophosphoryl dichloride*

S
‖
OPCl$_2$
COCH$_3$

$C_8H_7Cl_2O_2PS$ M 269.082
Liq. $Bp_{0.1}$ 125-127°. n_D^{20} 1.5810.

Yagnyukova, Z.I. *et al, Zh. Obshch. Khim.*, 1971, **41**, 84 (*Engl. transl.* p. 80)

O-(3-Acetylphenyl) phosphorodichloridothioate A-00023

O-(*3-Acetylphenyl) dichlorothiophosphate.* O-(*3-Acetylphenyl) thiophosphoryl dichloride*

$C_8H_7Cl_2O_2PS$ M 269.082
Liq. $Bp_{0.1}$ 143-145°. n_D^{20} 1.5830.

Yagnyukova, Z.I. *et al, Zh. Obshch. Khim.*, 1971, **41**, 84 (*Engl. transl.* p. 80)

O-(4-Acetylphenyl) phosphorodichloridothioate A-00024

O-(*4-Acetylphenyl) dichlorothiophosphate.* O-(*4-Acetylphenyl) thiophosphoryl dichloride*

$C_8H_7Cl_2O_2PS$ M 269.082
Liq. $Bp_{0.25}$ 135-137°. n_D^{20} 1.5860.

Yagnyukova, Z.I. *et al, Zh. Obshch. Khim.*, 1971, **41**, 84 (*Engl. transl.* p. 80)

Acetyl phosphate A-00025

[590-54-5]

AcOP(O)(OH)$_2$

$C_2H_5O_5P$ M 140.032

Unstable in soln. Isol. as Ag or Li salt. pK_{a1} 1.2, pK_{a2} 4.8.

Di-NH_4 salt: [55660-58-7]. Plates. Mp 128-130°.

Monoanilinium salt: [74785-77-6]. Solid. Mp 93-94°.

Dianilinium salt: [55660-59-8]. Solid. Mp 104-105°.

Mono-Me ester: Acetic acid monoanhydride with methyl dihydrogen phosphate, 9CI. Methyl hydrogen acetyl phosphate.
$C_3H_7O_5P$ M 154.059
As Na salt, acts as neutral acetylating agent.

Di-Me ester: [27744-98-5]. *Acetic acid anhydride with dimethyl hydrogen phosphate, 9CI. Acetyl dimethyl phosphate.*
$C_4H_9O_5P$ M 168.086
Acetylating agent. Liq. $Bp_{0.05}$ 51-52°.

Di-Et ester: [4526-20-9]. *Acetic acid anhydride with diethyl hydrogen phosphate, 9CI. Acetyl diethyl phosphate.*
$C_6H_{13}O_5P$ M 196.139
Liq. $Bp_{0.2}$ 70°. n_D^{25} 1.4115.

Dibenzyl ester: Acetic acid anhydride with dibenzyl hydrogen phosphate, 9CI. Acetyl dibenzyl phosphate.
$C_{16}H_{17}O_5P$ M 320.281
Oil.

Bentley, R., *J. Am. Chem. Soc.*, 1948, **70**, 2183 (*synth*)
Cramer, F. *et al, Chem. Ber.*, 1958, **91**, 704 (*ester, synth, ir*)
Michalski, J. *et al, Chem. Ber.*, 1962, **95**, 1629 (*ester*)
Oestreich, C.H. *et al, Biochemistry*, 1967, **6**, 1515 (*ester, props, ir*)
Fatiadi, A., *Carbohydr. Res.*, 1968, **6**, 237 (*use*)
Phillips, D.R. *et al, J. Am. Chem. Soc.*, 1968, **90**, 6803 (*props*)
Schmidt, P.G. *et al, J. Biol. Chem*, 1969, **244**, 1860 (*pmr, props*)
Klinman, J.P. *et al, Biochemistry*, 1971, **10**, 2126 (*props*)
Kluger, R. *et al, Biochemistry*, 1972, **12**, 1544 (*ester, pmr, ir, props*)
Kluger, R. *et al, J. Am. Chem. Soc.*, 1975, **97**, 4298 (*complexes*)
Whitesides, G.M. *et al, J. Org. Chem.*, 1975, **40**, 2516 (*synth*)
Methado, L.L. *et al, J. Am. Chem. Soc.*, 1978, **100**, 1850 (*props*)
Lewis, J.M. *et al, J. Org. Chem.*, 1979, **44**, 864 (*synth*)
Yamaguchi, K. *et al, J. Am. Chem. Soc.*, 1980, **102**, 4534 (*synth, pmr*)
Neels, J. *et al, Z. Anorg. Allg. Chem.*, 1982, **495**, 65 (*synth, props*)
Crans, D.C. *et al, J. Org. Chem.*, 1983, **48**, 3130 (*synth*)
Vogel, H.J. *et al, Can. J. Biochem. Cell Biol.*, 1983, **61**, 363 (*P nmr*)

Acetylphosphonic acid, 9CI A-00026

(*1-Oxoethyl)phosphonic acid*

[6881-54-5]

(HO)$_2$P(O)COCH$_3$

$C_2H_2O_4P$ M 121.009

Small plates. Mp ca. 110° dec.

Bis(cyclohexylammonium) salt: Cryst. (MeOH-/Me_2CO). Mp 159-162° dec.

Dicyclohexylammonium salt: Mp 197-208°.

Di-Me ester: [17674-28-1]. *Dimethyl acetylphosphonate. Dimethyl (1-oxoethyl)phosphonate.*
$C_4H_9O_4P$ M 149.063
Liq. Bp_6 72°.

Di-Et ester: [919-19-7]. *Diethyl acetylphosphonate. Diethyl (1-oxoethyl)phosphonate.*
$C_6H_{13}O_4P$ M 180.140
Liq. d_4^{20} 1.10. Bp_8 98-100°. n_D^{25} 1.4280.

Diisopropyl ester: Diisopropyl acetylphosphonate.
$C_8H_{14}O_4P$ M 205.170
Liq. Bp_4 73-74°.

Di-Ph ester: [22950-58-9]. *Diphenyl acetylphosphonate. Diphenyl (1-oxoethyl)phosphonate.*
$C_{14}H_{13}O_4P$ M 276.228
Liq. d_4^{20} 1.22. $Bp_{1.5}$ 165-167°, $Bp_{0.01}$ 128-130°. n_D^{25} 1.5400.

Bis(trimethylsilyl) ester: Bis(trimethylsilyl) acetylphosphonate. Bis(trimethylsilyl) (1-oxoethyl)phosphonate.
$C_8H_{18}O_4PSi_2$ M 265.373
Liq. $Bp_{0.3}$ 53-54°.

Amide: see P-Acetylphosphonic diamide, A-00027

Takamizawa, A. *et al*, *Chem. Pharm. Bull.*, 1967, **15**, 1183 (*esters, synth, ir, use*)
Ogata, J. *et al*, *J. Org. Chem.*, 1970, **35**, 596 (*esters, synth, ir*)
Nikonorov, K.V. *et al*, *Zh. Obshch. Khim.*, 1972, **42**, 2661 (*Engl. transl.* p. 2650) (*ester, synth*)
Laskorin, B.N. *et al*, *Zh. Obshch. Khim.*, 1976, **46**, 2545 (*Engl. transl.* p. 2434) (*ester, ms*)
Zygmunt, J. *et al*, *Synthesis*, 1978, 609 (*synth, silyl ester, ir, pmr, deriv*)

P-Acetylphosphonic diamide, 9CI A-00027

$H_3CCOP(O)(NH_2)_2$

$C_2H_7N_2O_2P$ M 122.063

N,N,N′,N′-*Tetra-Et:* [14114-74-0]. P-*Acetyl-*N,N,N′,N′-*tetraethylphosphonic diamide.*
$C_{10}H_{23}N_2O_2P$ M 234.278
Liq. d_4^{20} 1.03. Bp_1 98-99°. n_D^{20} 1.4790.

Alimov, P.I. *et al*, *Izv. Akad. Nauk SSSR, Ser. Khim.*, 1966, 1486 (*Engl. transl.* p. 1432)
Al'fonsov, V.A. *et al*, *Zh. Obshch. Khim.*, 1982, **52**, 19 (*Engl. transl.* p. 16)

Acetylphosphonothioic acid, 9CI A-00028

Acetylthiophosphonic acid

$H_3CCOP(S)(OH)_2 \rightleftharpoons H_3CCOP(O)(SH)(OH)$

$C_2H_5O_3PS$ M 140.093

O,O-*Di-Et ester:* O,O-*Diethyl acetylphosphonothioate.*
$C_6H_{13}O_3PS$ M 196.201
Liq. d_4^{20} 1.09. Bp_2 92-93°. n_D^{20} 1.4500.

O,O-*Dibutyl ester:* O,O-*Dibutyl acetylphosphonothioate.*
$C_{10}H_{21}O_3PS$ M 252.308
Liq. d_4^{20} 1.04. Bp_2 94-96°. n_D^{20} 1.4630.

Pudovik, A.N. *et al*, *Zh. Obshch. Khim.*, 1964, **34**, 3946 (*Engl. transl.* p. 4007) (*diethyl ester*)
Lutsenko, I.F. *et al*, *Phosphorus*, 1974, **4**, 57 (*dibutyl ester*)

Acetylphosphoramidic acid, 9CI A-00029

$AcNHP(O)(OH)_2$

$C_2H_6NO_4P$ M 139.047

Di-Me ester: [85046-80-6]. *Dimethyl acetylphosphoramidate.* N-*(Dimethoxyphosphinyl)acetamide.*
$C_4H_{10}NO_4P$ M 167.101
Liq. d_4^{20} 1.29. $Bp_{0.035}$ 128-130°. n_D^{20} 1.4410.

Di-Et ester: [1559-55-3]. *Diethyl acetylphosphoramidate.* N-*(Diethoxyphosphinyl)acetamide.*
$C_6H_{14}NO_4P$ M 195.155
Liq. or needles (Et_2O/pet. ether). d_4^{20} 1.18. Mp 52-53° (49-49.5°). Bp_1 122-123°. n_D^{20} 1.4420.

Kabachnik, M.I. *et al*, *Izv. Akad. Nauk SSSR, Ser. Khim.*, 1961, 1022 (*Engl. transl.* p. 945) (*diethyl ester, ir*)
Matrosov, E.I. *et al*, *Izv. Akad. Nauk SSSR, Ser. Khim.*, 1967, 1465 (*Engl. transl.* p. 1417) (*uv, ir*)
Alimov, P.I. *et al*, *Izv. Akad. Nauk SSSR, Ser. Khim.*, 1972, 147 (*Engl. transl.* p. 132) (*diethyl ester*)
Nikonorov, K.V. *et al*, *Zh. Obshch. Khim.*, 1982, **52**, 2645 (*Engl. transl.* p. 2341) (*esters, synth, ir, P nmr*)

9-Acridinephosphonic acid A-00030

9-Acridinylphosphonic acid, 9CI. 9-Phosphonoacridine

[19656-41-8]

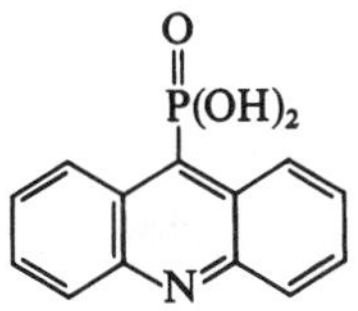

$C_{13}H_{10}NO_3P$ M 259.201

Lustrous green or yellow cryst. Mp 247-249° dec., >300°. The green form, Mp 247-9°, is metastable.

Di-Et ester: [19656-40-7]. *Diethyl 9-acridinylphosphonate.*
$C_{17}H_{18}NO_3P$ M 315.308
Yellow feathery cryst. Mp 95-96°, 165-167°.

9,10-Dihydro: see (9,10-Dihydro-9-acridinyl)-phosphonic acid, D-00438

Kosolapoff, G.M., *J. Am. Chem. Soc.*, 1947, **69**, 1002 (*synth*)
Redmore, D., *J. Org. Chem.*, 1969, **34**, 1420 (*synth, ir, ester, uv, pmr*)
Sheinkmann, A.K. *et al*, *Dokl. Akad. Nauk SSSR, Ser. Sci. Khim.*, 1971, **196**, 1377 (*synth, ir*)

Acridophosphine, 8CI A-00031

Dibenzo[b,e]*phosphorin. Dibenzo*[b,e]*phosphinine. 9-Phosphaanthracene*

[398-14-1]

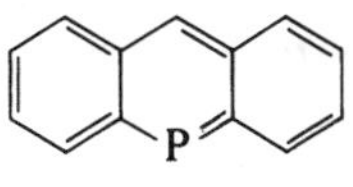

$C_{13}H_9P$ M 196.188

Obt. only in soln. Pale yellow. Immediately dec. by O_2.

de Koe, P. *et al*, *Angew. Chem., Int. Ed. Engl.*, 1967, **6**, 567 (*formn, uv*)

N-(1-Adamantyl)-P,P-bis(aziridinyl)-phosphinic amide A-00032

P,P-Bis(1-aziridinyl)-N-tricyclo[3.3.1.1^{3,7}]dec-1-yl-phosphinic amide, 9CI

[53743-43-4]

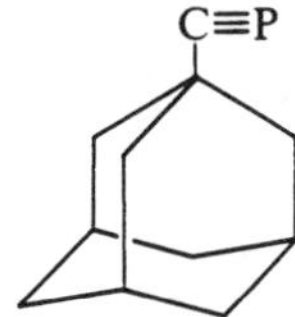

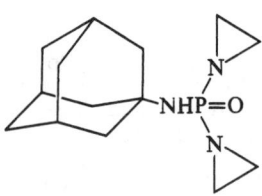

$C_{14}H_{24}N_3OP$ M 281.337

Exhibits high antileukemic activity.

▷SZ5790000.

Cates, L.A. *et al*, *J. Med. Chem.*, 1977, **21**, 143 (*synth, pharmacol*)

1-Adamantylmethylidenephosphine A-00033

$C_{11}H_{15}P$ M 178.213

Cryst. Mp 69-70°.

Allspach, T. *et al*, *Synthesis*, 1986, 31 (*synth, ir, raman, ms, pmr, cmr, P nmr*)

1-Adamantylphosphonic acid, 8CI A-00034

(Tricyclo[3.3.1.1^{3,7}]dec-1-yl)phosphonic acid, 9CI

[23906-88-9]

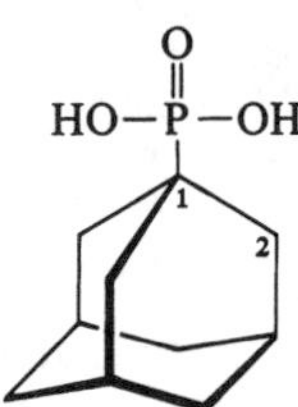

$C_{10}H_{17}O_3P$ M 216.216

Solid. Mp 308-310° dec. (297-304°).

Monoanilinium salt: [37516-10-2]. Solid. Mp 297-300°.

Di-Me ester: [37516-07-7]. *Dimethyl 1-adamantylphosphonate. Dimethyl (tricyclo[3.3.1.1^{3,7}]dec-1-yl)-phosphonate.*
 $C_{12}H_{21}O_3P$ M 244.270
 Solid. Mp 59-60°.

Di-Et ester: [37516-08-8]. *Diethyl 1-adamantylphosphonate. Diethyl (tricyclo[3.3.1.1^{3,7}]dec-1-yl)-phosphonate.*
 $C_{14}H_{25}O_3P$ M 272.323
 Solid. Mp 12-17°.

Diisopropyl ester: [37516-09-9]. *Diisopropyl 1-adamantylphosphonate. Bis(1-methylethyl) (tricyclo[3.3.1.1^{3,7}]dec-1-yl)phosphonate.*
 $C_{16}H_{29}O_3P$ M 300.377
 Solid. Mp 38-40°.

Dichloride: [23906-87-8].
 $C_{10}H_{15}Cl_2OP$ M 253.108
 Cryst. (Me_2CO). Mp 102-103°. Bp_2 135-136°. Subl. at 70-80°/1 mm.

Stetter, H. *et al*, *Chem. Ber.*, 1969, **102**, 3364 (*synth, deriv*)

Shepeleva, E.S. *et al*, *Dokl. Akad. Nauk SSSR, Ser. Sci. Khim.*, 1972, **203**, 608 (*Engl. transl.* p. 275) (*synth, derivs, ms*)

2-Adamantylphosphonic acid, 8CI A-00035

(Tricyclo[3.3.1.1^{3,7}]dec-2-yl)phosphonic acid, 9CI

[37516-02-2]

$C_{10}H_{17}O_3P$ M 216.216

Monoanilinum salt: [37515-81-4]. Solid. Mp 236-237°.

Di-Me ester: [37516-03-3]. *Dimethyl 2-adamantylphosphonate. Dimethyl (tricyclo[3.3.1.1^{3,7}]dec-2-yl)-phosphonate.*
 $C_{12}H_{21}O_3P$ M 244.270
 Solid. Mp 49.5-50°.

Dichloride: [37516-01-1].
 $C_{10}H_{15}Cl_2OP$ M 253.108
 Solid. Mp 75-76°.

Shepeleeva, E.S. *et al*, *Dokl. Akad. Nauk SSSR, Ser. Sci. Khim.*, 1972, **203**, 608 (*Engl. transl.* p. 275) (*synth, derivs, ms*)

1-Adamantylphosphoramidic acid A-00036

Tricyclo[3.3.1.1.^{3,7}]dec-1-ylphosphoramidic acid, 9CI

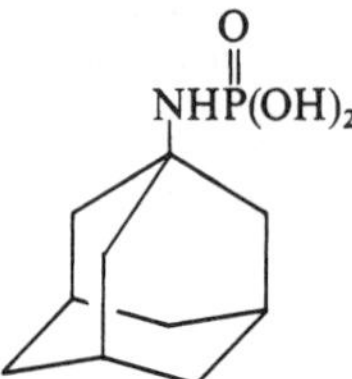

$C_{10}H_{18}NO_3P$ M 231.231

Di-Me ester: Dimethyl 1-adamantylphosphoramidate.
 $C_{12}H_{22}NO_3P$ M 259.284
 Solid. Mp 69-71°.

Di-Et ester: [49802-18-8]. *Diethyl 1-adamantylphosphoramidate.*
 $C_{14}H_{26}NO_3P$ M 287.338
 Cryst. (pentane). Mp 97.5-102°.

Di-Ph ester: [49802-24-6]. *Diphenyl 1-adamantylphosphoramidate.*
 $C_{22}H_{26}NO_3P$ M 383.426
 Cryst. (EtOH aq.). Mp 123.5-125°.

Dibenzyl ester: Dibenzyl 1-adamantylphosphoramidate.
 $C_{24}H_{30}NO_3P$ M 411.480
 Solid. Mp 96-97.5°.

Dichloride:
 $C_{10}H_{16}Cl_2NOP$ M 268.122
 Cryst. (Et_2O or C_6H_6/pet. ether). Mp 129-131.5°.

Cates, L.A. *et al*, *J. Pharm. Sci.*, 1973, **62**, 1719; 1974, **63**, 1480 (*esters, synth, ir, pmr, dichloride*)

Warner, V.D. *et al*, *J. Med. Chem.*, 1973, **16**, 1185 (*diethyl ester, synth*)

Yurchenko, R.I. *et al*, *Zh. Obshch. Khim.*, 1983, **53**, 2445; 1984, **54**, 1201 (*Engl. transl.* pp. 2206, 1075) (*diethyl ester, dichloride, synth, ir, pmr*)

2-Adamantylphosphoramidic acid A-00037

$C_{10}H_{18}NO_3P$ M 231.231

Dichloride: [93032-39-4].
 $C_{10}H_{16}Cl_2NOP$ M 268.122
 Cryst. (pet. ether). Mp 126-128°.

Yurchenko, R. *et al*, *Zh. Obshch. Khim.*, 1984, **54**, 1201 (*Engl. transl.* p. 1075) (*synth, ir, pmr, P nmr*)

Adenosine diphosphate **A-00038**

Adenosine 5′-(trihydrogen diphosphate), 9CI. *Adenosine 5′-pyrophosphate. ADP*

[58-64-0]

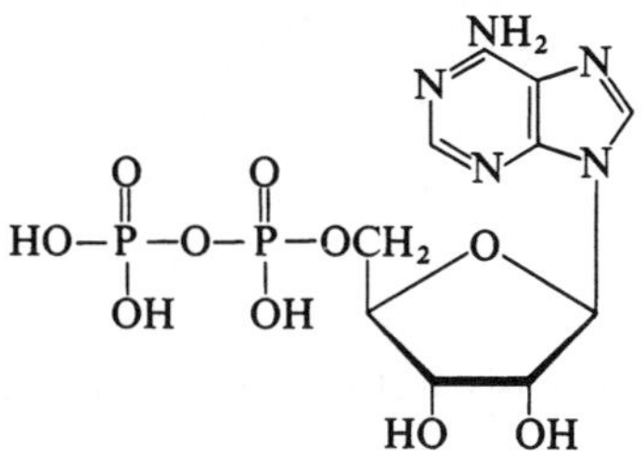

$C_{10}H_{15}N_5O_{10}P_2$ M 427.204

Formed from ATP in the muscle by the enzyme adenosinetriphosphatase. λ_{max} 259 nm (ϵ 15 400) (H_2O).

▷AU7467000.

Acridine salt: Mp 215° dec.

6N-Benzoyl, tri-Na salt: λ_{max} 281 nm (ϵ 19 400) (H_2O).

Baddiley, J. *et al*, *J. Chem. Soc.*, 1947, 648.
Biochem. Prep., 1949, **1**, 1.
Chambers, R.W. *et al*, *J. Am. Chem. Soc.*, 1960, **82**, 970 (*synth*)
Wieland, T. *et al*, *Chem. Ber.*, 1968, **107**, 3031 (*synth*)
Sarma, R.H. *et al*, *J. Chem. Soc., Chem. Commun.*, 1973, 140 (*pmr, nmr*)
Hampton, A. *et al*, *Biochemistry*, 1975, **14**, 5438.

Adenosine 2′,5′-diphosphate **A-00039**

2′-Adenylic acid 5′-(dihydrogen phosphate), 9CI. *Adenosine 2′,5′-bis(phosphate)*, 8CI

[3805-37-6]

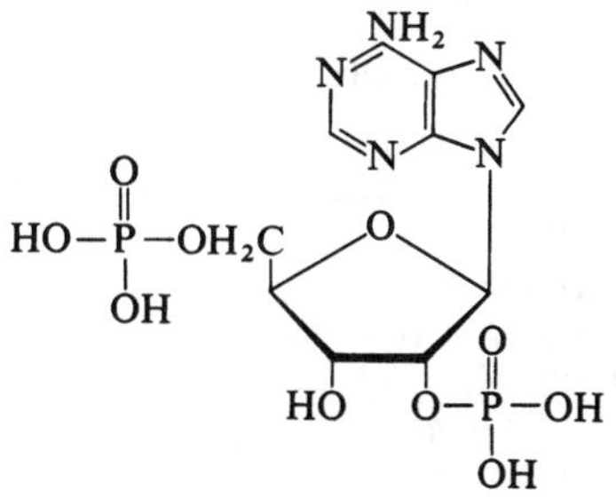

$C_{10}H_{15}N_5O_{10}P_2$ M 427.204

Component of coenzyme II. Amorph. powder.

Di-Ca salt: Octahydrate. λ_{max} 258 nm (ϵ 14 250) (H_2O).

Wang, T.P. *et al*, *J. Biol. Chem.*, 1954, **206**, 299.
Baddiley, J. *et al*, *J. Chem. Soc.*, 1958, 1000 (*synth*)
Japan. Pat., 536, (*1967*); *CA*, **66**, 64391y (*synth*)
Takaku, H. *et al*, *Chem. Pharm. Bull.*, 1973, **21**, 1844 (*synth*)
Roeder, S.B.W., *Physiol. Chem. Phys.*, 1975, **7**, 115 (*cmr*)

Adenosine 3′,5′-diphosphate **A-00040**

3′-Adenylic acid 5′-(dihydrogen phosphate), 9CI. *Adenosine 3′,5′-bis(phosphate)*, 8CI

[1053-73-2]

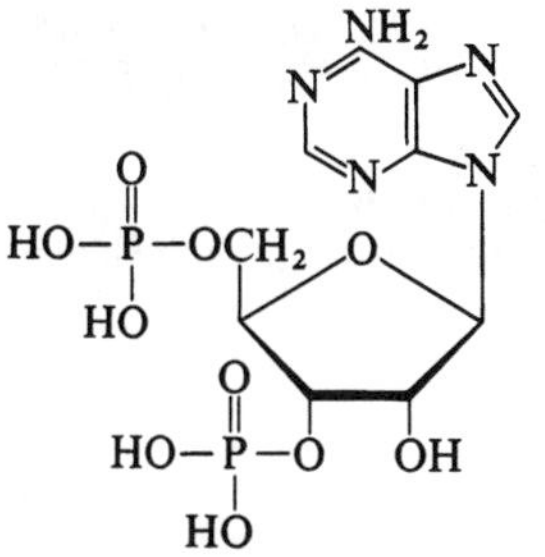

$C_{10}H_{15}N_5O_{10}P_2$ M 427.204

Component of coenzyme *A*. Amorph. powder.

Ca salt: Pentahydrate. λ_{max} 258 nm (ϵ 14 050) (H_2O).

Wang, T.P. *et al*, *J. Biol. Chem.*, 1954, **206**, 299.
Baddiley, J. *et al*, *J. Chem. Soc.*, 1958, 1000 (*synth*)
Takaku, H. *et al*, *Chem. Pharm. Bull.*, 1973, **21**, 1844 (*synth*)
Lee, C. *et al*, *FEBS Lett.*, 1974, **43**, 271; *CA*, **82**, 12479u (*pmr*)
Roeder, S.B.W., *Physiol. Chem. Phys.*, 1975, **7**, 115 (*cmr*)

Adenosine diphosphate ribose **A-00041**

Adenosine 5′-(trihydrogen diphosphate) 5′→5-ester with D-ribose, 9CI. *ADPR*

[20762-30-5]

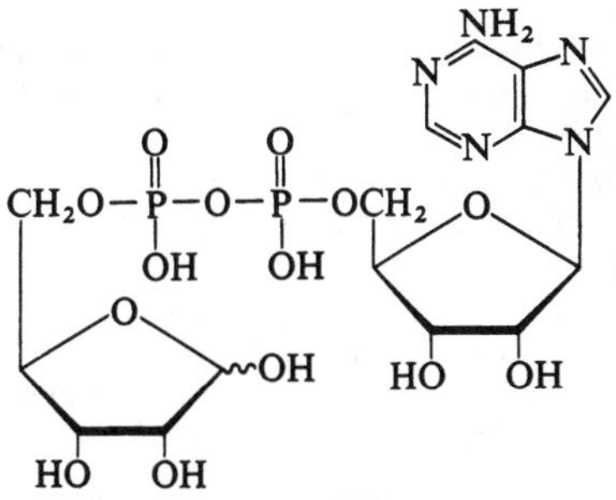

$C_{15}H_{23}N_5O_{14}P_2$ M 559.319

Prod. of enzymic or chemical hydrol. of diphosphopyridine nucleotide. Inhibits coenzyme action of diphosphopyridine nucleotide.

Rosenberg, S. *et al*, *J. Biol. Chem.*, 1954, **211**, 763 (*isol*)
Blumenstein, M. *et al*, *Biochemistry*, 1972, **11**, 1643 (*cmr, nmr*)
Sarma, R.H. *et al*, *J. Am. Chem. Soc.*, 1973, **95**, 7470 (*conformn, pmr*)
Abdallah, M.A. *et al*, *Eur. J. Biochem.*, 1975, **50**, 475 (*cryst struct*)

Adenosine 5′-diphosphoglucose **A-00042**

Adenosine 5′-(trihydrogen diphosphate) mono-α-D-glucopyranosyl ester, 9CI. *ADPG*

[2140-58-1]

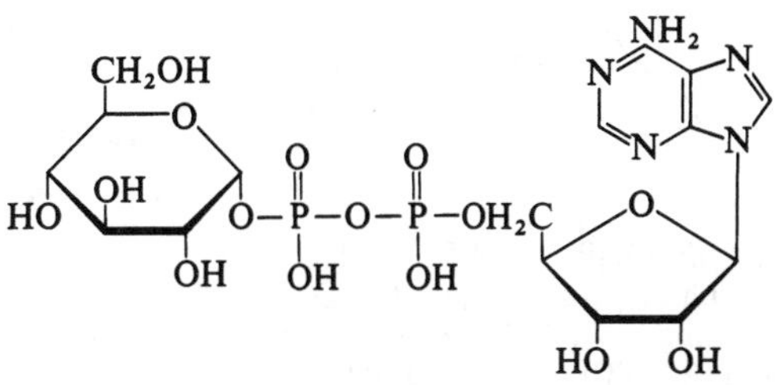

$C_{16}H_{25}N_5O_{15}P_2$ M 589.346

Present in ripening cereal grains. Glucose donor in glycogen synth., sucrose synth. and glucoside formn. λ_{max} 257 nm (pH 2).

Di-K salt: λ_{max} 259 nm (ϵ 15 400) (H_2O).

Roseman, S. *et al*, *J. Am. Chem. Soc.*, 1961, **83**, 659.
Murata, T. *et al*, *Arch. Biochem. Biophys.*, 1964, **106**, 371 (*isol, synth*)
Krauss, G. *et al*, *J. Chromatogr.*, 1973, **76**, 248 (*chromatog*)
Sarma, R.H. *et al*, *FEBS Lett.*, 1973, **36**, 157 (*nmr*)
Lee, C.H. *et al*, *Biochemistry*, 1976, **15**, 697 (*conformn, pmr*)

Adenosine triphosphate A-00043

Adenosine 5′-(tetrahydrogen triphosphate), 9CI, 8CI. *ATP*
[56-65-5]

$C_{10}H_{16}N_5O_{13}P_3$ M 507.183
Isol. from muscle extracts. Mammalian skeletal muscle at rest contains 0.3-0.4 g of ATP/100 g. Important metabolic coenzyme. Mp 143-145° dec. $[\alpha]_D^{22}$ −26.7° (c, 3.1 in H_2O). λ_{max} 259 nm (ϵ 15 400) (H_2O).
▷AU7416000.

Di-Na salt: Trinosin. Adetphos. Inhibits enzymatic browning of apples, potatoes etc. Mp 188-190° dec. (hydrate).
Triacridine salt: Mp 208-209° dec.
6N-Benzoyl, tetra-Na salt: λ_{max} 281 nm (ϵ 19 400) (H_2O).

Baddiley, J. *et al*, *J. Chem. Soc.*, 1949, 582 (*synth*)
Biochem. Prep., 1949, **1**, 5 (*isol*)
Michelson, A.M., *The Chemistry of Nucleosides and Nucleotides*, 1963, Academic Press, N.Y. and London, 153 (*rev*)
Feldman, I. *et al*, *J. Am. Chem. Soc.*, 1968, **90**, 7329 (*pmr*)
Dorman, D.E. *et al*, *Proc. Natl. Acad. Sci. U.S.A.*, 1970, **65**, 19 (*cmr*)
Kennard, O. *et al*, *Nature (London)*, 1970, **225**, 333 (*cryst struct*)
Hampton, A. *et al*, *Biochemistry*, 1975, **14**, 5438.

Adenosine 5′-uridine 5′-phosphate A-00044

Uridylyl-(5′,5′)-adenosine

$C_{19}H_{24}N_7O_{12}P$ M 573.412
Amorph. hygroscopic powder. Dec. at 180-200° without melting. λ_{max} 259.5 (ϵ 21 300) (0.01N H_2SO_4), 260.5 nm (20 300) (0.01N NaOH).

Elmore, D.T. *et al*, *J. Chem. Soc.*, 1952, 3681 (*synth*)
Michelson, A.M., *CA*, 1962, **56**, 14622b.

2′-Adenylic acid, 9CI, 8CI A-00045

Adenosine 2′-(dihydrogen phosphate), 9CI. *Adenylic acid a*
[130-49-4]

$C_{10}H_{14}N_5O_7P$ M 347.224
Isol. from yeast ribonucleic acid. Mp 187° dec., 205-215° dec.

Acridine salt: Mp 215° dec.
Dibrucine salt: Mp 165-175°.

Brown, D.M. *et al*, *J. Chem. Soc.*, 1952, 44 (*synth*)
Brown, D.M. *et al*, *Nature (London)*, 1953, **172**, 1184 (*struct*)
Kotowycz, G. *et al*, *Biochemistry*, 1973, **12**, 517 (*cmr*)
Son, T. *et al*, *Biochim. Biophys. Acta*, 1974, **335**, 1 (*conformn, pmr*)

3′-Adenylic acid, 9CI, 8CI A-00046

Adenosine 3′-(dihydrogen phosphate), 9CI. *Yeast adenylic acid. Adenylic acid b*

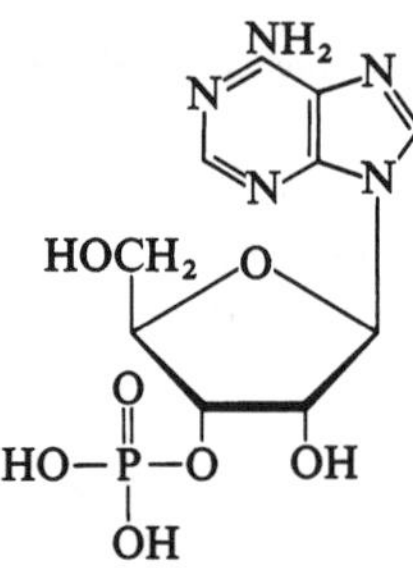

$C_{10}H_{14}N_5O_7P$ M 347.224
Isol. from yeast ribonucleic acid. Mp 197° dec. (dihydrate).

Acridine salt: Mp 175° dec.
Diacridine salt: Mp 184° dec.
Dibrucine salt: Mp 177° (225° dec.).

Levene, P.A. *et al*, *J. Biol. Chem.*, 1932, **98**, 9.
Brown, D.M. *et al*, *J. Chem. Soc.*, 1952, 44 (*synth*)
Sundaralingham, M., *Acta Crystallogr.*, 1966, **21**, 495 (*cryst struct*)
Ts'O, P.O.P. *et al*, *Biochemistry*, 1969, **8**, 997 (*conformn, pmr*)
Kotowycz, G. *et al*, *Biochemistry*, 1973, **12**, 517 (*cmr*)
Takaku, H. *et al*, *Chem. Pharm. Bull.*, 1973, **21**, 1844 (*synth*)

Adenylosuccinic acid **A-00047**

N-*[9-(5-O-Phosphono-β-*D*-ribofuranosyl)-9*H*-purin-6-yl]-*L*-aspartic acid, 9CI.* N*-(9-β-*D*-Ribofuranosyl-9*H*-purin-6-yl)-*L*-aspartic acid 5′-(dihydrogen phosphate), 8CI*

[19046-78-7]

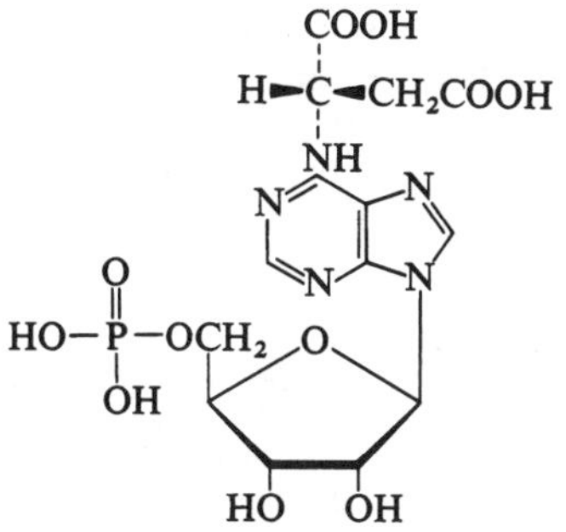

$C_{14}H_{18}N_5O_{11}P$ M 463.297

Found with adenosinesuccinic acid in the mycelium of *Penicillium chrysogenum* and in *Fusarium nivale*. Intermed. in the formn. of adenylic acid. $[\alpha]_D^{25}$ −3.4° (c, 1.04 in H_2O at pH 7). The diastereoisomer from D-aspartic acid has $[\alpha]_D^{25}$ −77.3° (c, 1.15 in H_2O at pH 7).

Lieberman, I., *J. Biol. Chem.*, 1956, **223**, 327.
Ballio, A. *et al*, *Arch. Biochem. Biophys.*, 1963, **101**, 311.
Mansurova, S.E. *et al*, *Biokhimiya*, 1966, **31**, 1057.
Ballio, A. *et al*, *Ann. Ist. Super. Sanita*, 1967, **3**, 149; *CA*, **68**, 59824d (*synth*)
Van der Weyden, M.B. *et al*, *J. Biol. Chem.*, 1974, **249**, 7282.

Agrocin 84, 9CI **A-00048**

[59111-78-3]

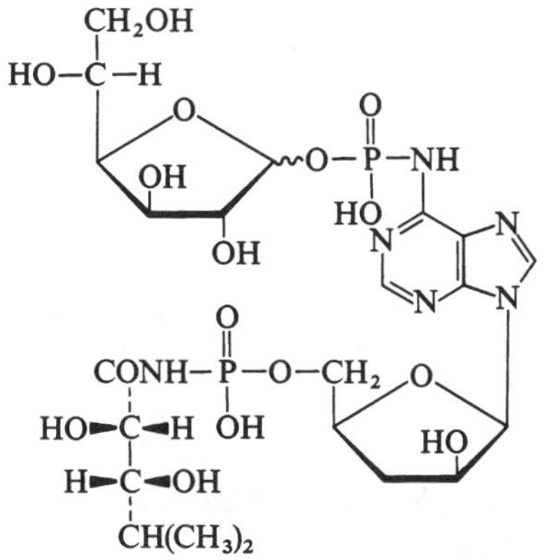

$C_{22}H_{36}N_6O_{16}P_2$ M 702.505

Nucleotide antibiotic. Isol. from *Agrobacterium radiobacter*. Bacteriocin active against *Agrobacterium tumefaciens*.

Heip, J. *et al*, *Arch. Int. Physiol. Biochem.*, 1975, **83**, 974 (*isol*)
Roberts, W.P. *et al*, *Nature (London)*, 1977, **265**, 379 (*struct*)
Das, P.K. *et al*, *J. Antibiot.*, 1978, **31**, 490 (*props*)
Thompson, R.J. *et al*, *Antimicrob. Agents Chemother.*, 1979, **16**, 293 (*isol, props*)

Alafosfalin, BAN **A-00049**

[1-[(2-Amino-1-oxopropyl)amino]ethyl]phosphonic acid, 9CI. N*-Alanyl-(2-aminoethyl)phosphonic acid. [2-(Alanylamino)ethyl]phosphonic acid. Alafosfin. Ro 03-7008*

[54788-43-1]

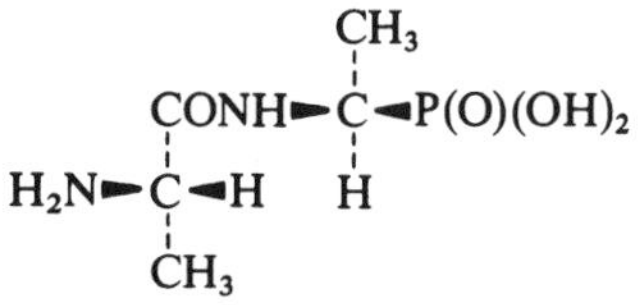

$(S)_c,(R)_p$-*form*

$C_5H_{13}N_2O_4P$ M 196.142

The names Alafosfalin and Alafosfin strictly refer only to the (*R*,*S*)-diastereoisomer.

(R)_C(S)_P-form [54772-83-7]

D,D-form

Mp 293-294°. $[\alpha]_D$ +46.5° (c, 1.0 in H_2O).

N-*Benzyloxycarbonyl:* [98820-76-9].
$C_{13}H_{19}N_2O_6P$ M 330.277
Mp 223-226° (as benzylamine salt). $[\alpha]_D^{20}$ +36.5° (c, 1.0 in AcOH) (benzylamine salt).

(R)_C(R)_P-form [98857-06-8]

D,L(P)-form

Mp 295-297°. $[\alpha]_D^{20}$ −81.1° (c, 1.0 in H_2O).

N-*Benzyloxycarbonyl:* [98820-75-8]. Mp 219-221° (as benzylamine salt). $[\alpha]_D^{20}$ −21.1° (c, 1.0 in AcOH) (benzylamine salt).

(S)_C(S)_P-form [66023-94-7]

L,D(P)-form

Cryst. (H_2O). Mp 296-298°. $[\alpha]_D^{20}$ +84.1° (c, 1.0 in H_2O).

N-*Benzyloxycarbonyl:* [60668-66-8]. Mp 221-223° (as benzylamine salt). $[\alpha]_D^{20}$ +21.0° (c, 1.0 in AcOH) (benzylamine salt).

(S)_C,(R)_P-form [60668-24-8]

L,L(P)-form

Antibacterial phosphonodipeptide inhibiting growth of various pathogenic organisms: useful in chemotherapy of bacterial gastrointestinal infections. Cryst. (EtOH aq.). Mp 294-295°. $[\alpha]_D^{20}$ −45.6° (−49°) (c, 1.0 in H_2O).

N-*Benzyloxycarbonyl:* [60668-26-0]. Mp 229-231° (as benzylamine salt). $[\alpha]_D^{20}$ −34.2° (c, 1.0 in AcOH) (as benzylamine salt).

(S)_C,(RS)_P-form

L,(±) (P)-form

Cryst. + $1H_2O$ (EtOH aq.). Mp 260-265° dec. $[\alpha]_D^{25}$ +15.0° (c, 0.5 in H_2O).

(R)_C,(±)_P-form

D, (±)-(P)-form

Cryst. + $0.5H_2O$ (EtOH aq.). Mp 267-269° dec. $[\alpha]_D^{25}$ −17.5° (c, 1.0 in H_2O).

Allen, J.G. *et al*, *Nature (London)*, 1978, **272**, 56 (*props*)
Allen, J.G. *et al*, *Antimicrob. Agents Chemother.*, 1978, **15**, 684; 1979, **16**, 306 (*props, metab*)
Atherton, F.R. *et al*, *Antimicrob. Agents Chemother.*, 1978, **15**, 677, 696 (*props*)
Okada, Y. *et al*, *Chem. Pharm. Bull.*, 1980, **28**, 1320 (*synth, pmr, tlc*)
Kametani, T. *et al*, *Heterocycles*, 1981, **16**, 1205 (*ester, synth, pmr*)
Kafarski, P. *et al*, *Can. J. Chem.*, 1982, **60**, 3081 (*synth*)
Lejczak, B. *et al*, *Synthesis*, 1982, 412 (*synth, ir, pmr*)
Atherton, F.R. *et al*, *J. Med. Chem.*, 1986, **29**, 29 (*synth, props, ir, pmr*)

Aldophosphamide A-00050

3-Oxopropyl N,N-*bis(2-chloroethyl)-phosphorodiamidate, 10CI*

[35144-64-0]

$$O{=}P(NH_2)(N(CH_2CH_2Cl)_2)OCH_2CH_2CHO$$

$C_7H_{15}Cl_2N_2O_3P$ M 277.087

Exists in equilib. with *trans*-4-Hydroxycyclophosphamide, H-00115 in aq. soln. Metab. of Cyclophosphamide, C-00283 . Syrup. Unstable at −10°. Dec. rapidly at r.t.

O-*Methyl oxime:* Cryst. ($CHCl_3$/pet. ether). Mp 72-75°. Mixt. of (*Z*) and (*E*)-forms.

Semicarbazone: Solid. Mp 127-128°.

Przybyski, M. *et al, Cancer Treat. Rep.*, 1976, **60**, 509 (*deriv, ms*)
Fenselau, C. *et al, Cancer Res.*, 1977, **37**, 2538.
Myles, A. *et al, Tetrahedron Lett.*, 1977, 2475 (*synth, ir, props, deriv*)
Zon, G. *et al, J. Pharm. Sci.*, 1982, **71**, 443 (*deriv, ms, pmr*)
Borch, R.F. *et al, J. Med. Chem.*, 1984, **27**, 490 (*synth*)
Zon, G. *et al, J. Med. Chem.*, 1984, **27**, 466 (*deriv*)

Amidithion A-00051

S-*[2-[(2-Methoxyethyl)amino]-2-oxoethyl]* O,O-*dimethyl phosphorothioate*

[919-76-6]

$(MeO)_2P(S)SCH_2CONHCH_2CH_2OMe$

$C_7H_{16}NO_4PS_2$ M 273.301

Insecticide, now superseded. Solid.

▷TE1575000.

Rohrbaugh, W.J. *et al, J. Agric. Food Chem.*, 1977, **25**, 588 (*cryst struct*)
Stan, H.-J. *et al, Fresenius' Z. Anal. Chem.*, 1977, **287**, 271 (*glc, ms*)
Stan, H.-J. *et al, Biomed. Mass Spectrom.*, 1982, **9**, 483 (*ms*)

Amifostine, INN A-00052

2-[(3-Aminopropyl)amino]ethanethiol dihydrogen phosphate (ester), 9CI. S-*[2-[(3-Aminopropyl)amino]ethyl]-dihydrogen phosphorothioate, 8CI. Ethiofos. Gammaphos. NSC 296961. WR 2721*

[20537-88-6]

$H_2NCH_2CH_2CH_2NHCH_2CH_2SP(O)(OH)_2$

$C_5H_{15}N_2O_3PS$ M 214.219

Radioprotective agent. Monohydrate. Mp 160-161°.

▷TE6491000.

Piper, J.R. *et al, J. Org. Chem.*, 1968, **33**, 636.
Piper, J.R. *et al, J. Med. Chem.*, 1969, **12**, 236; 1975, **18**, 803 (*synth, pharmacol*)
Czerwinski, A.W. *et al, U.S. Nat. Tech. Inform. Serv. AD Rep.*, 1972, No. 758432 (*tox*)
Phillips, T.L., *Cancer Clin. Trials*, 1980, **3**, 165 (*rev*)
Tong, Z.S. *et al, CA*, 1982, **96**, 6137 (*synth*)
Laduranty, J. *et al, Bull. Soc. Chim. Belg.*, 1984, **93**, 903 (*synth, bibl*)
Kim, Y.S. *et al, CA*, 1985, **102**, 149347 (*synth*)

(Aminobenzylidene)bisphosphonic acid, 9CI A-00053

Benzylamine-α,α-diphosphonic acid

[4712-06-5]

$[(HO)_2P(O)]_2C(NH_2)Ph$

$C_7H_{11}NO_6P_2$ M 267.115

Cryst. + H_2O. Mp 225° dec., 240-242° dec. pK_{a1} 2.60, pK_{a2} 5.62, pK_{a3} 8.81, pK_{a4} 11.20.

Tetra-Et ester: [81928-62-3]. *Tetraethyl (aminophenylmethyl)bisphosphonate.*
$C_{15}H_{27}NO_6P_2$ M 379.329
Solid. Mp 58.5-59.5°. $Bp_{0.05}$ 170-172°.

Tetrabutyl ester: [81928-64-5]. *Tetrabutyl (aminophenylmethyl)bisphosphonate.*
$C_{25}H_{43}NO_6P_2$ M 515.565
Liquid or solid. Mp 10-15°. $Bp_{0.05}$ 180-185°.

Kreutzkamp, N. *et al, Justus Liebigs Ann. Chem.*, 1959, **623**, 103 (*esters, synth, ir*)
Plöger, W. *et al, Z. Anorg. Allg. Chem.*, 1972, **389**, 119 (*synth*)
B.P., 2 079 285, (*1980*); *CA*, **96**, 218031 (*esters, synth, ir*)
Grogorovich, I.I. *et al, Zh. Obshch. Khim.*, 1984, **54**, 1007 (*Engl. transl.* p. 898) (*synth, ir, props*)

(1-Aminobutyl)phosphonic acid, 9CI A-00054

[13138-36-8]

$H_3CCH_2CH_2CH(NH_2)P(O)(OH)_2$

$C_4H_{12}NO_3P$ M 153.117

Free acid exists in dipolar forms.

(±)-***form*** [20263-10-9]

Solid. Mp 262-264°, 298-299°. pK_{a1} 1.95, pK_{a2} 5.83, pK_{a3} 10.32.

Di-Et ester: [83245-93-6]. *Diethyl (1-aminobutyl)-phosphonate.*
$C_8H_{20}NO_3P$ M 209.225
Characterised spectroscopically.

N-*Benzoyl: (1-Benzoylaminobutyl)phosphonic acid.*
$C_{11}H_{16}NO_4P$ M 257.225
Solid. Mp 186-187°.

N-*Benzyl: [1-(Benzylamino)butyl]phosphonic acid.*
$C_{11}H_{18}NO_3P$ M 243.242
Solid. Mp 214-216°.

Berry, J.P. *et al, J. Org. Chem.*, 1972, **37**, 4396 (*synth, props, derivs*)
Harvey, D.J., *Org. Mass Spectrom.*, 1974, **9**, 111 (*silyl derivs, ms*)
Gancarz, R. *et al, Synthesis*, 1977, 625 (*synth*)
Kudzin, Z.H. *et al, Synthesis*, 1978, 469 (*synth*)
Rachan, J. *et al, Justus Liebigs Ann. Chem.*, 1981, 709 (*synth, ester, pmr, ir*)

(2-Aminobutyl)phosphonic acid, 9CI A-00055

[67264-34-0]

$H_3CCH_2CH(NH_2)CH_2P(O)(OH)_2$

$C_4H_{12}NO_3P$ M 153.117

Free acid exists in dipolar form.

(±)-***form***

Mp >260°.

Di-Et ester: Diethyl (2-aminobutyl)phosphonate.
$C_8H_{20}NO_3P$ M 209.225
No phys. props. reported.

Brigot, D. *et al, Tetrahedron*, 1979, **35**, 1345 (*synth, pmr*)
Varlet, J.M. *et al, Tetrahedron*, 1981, **37**, 3713 (*synth, ester, pmr, ir*)

(3-Aminobutyl)phosphonic acid, 9CI A-00056

[70519-90-3]

$H_3CCH(NH_2)CH_2CH_2P(O)(OH)_2$

$C_4H_{12}NO_3P$ M 153.117

(±)-form

Solid. Mp >260°.

Et ester: Diethyl (3-aminobutyl)phosphonate.
$C_8H_{20}NO_3P$ M 209.225
Characterised spectroscopically.

Savignac, P. *et al, Synth. Commun.*, 1979, **9**, 287 (*synth, ester, pmr*)

(4-Aminobutyl)phosphonic acid, 9CI — A-00057

[35622-27-6]

$H_2N(CH_2)_4P(O)(OH)_2$

$C_4H_{12}NO_3P$ M 153.117

Cryst. (H_2O). Sl. sol. EtOH. Mp 133-134°. pK_{a1} 2.55, pK_{a2} 7.55, pK_{a3} 10.9.

Di-Et ester: [53253-54-6]. *Diethyl (4-aminobutyl)-phosphonate.*
$C_8H_{20}NO_3P$ M 209.225
Liq. $Bp_{0.002}$ 77°.

Chavane, V., *C.R. Hebd. Seances Acad. Sci.*, 1947, **224**, 406.

[(Aminocarbonyl)amino]methylphosphonic acid, 9CI — A-00058

Ureidomethylphosphonic acid

$H_2NCONHCH_2P(O)(OH)_2$

$C_2H_7N_2O_4P$ M 154.062

Di-Et ester: Diethyl [(aminocarbonyl)amino]-methylphosphonate.
$C_6H_{15}N_2O_4P$ M 210.169
Solid. Mp 127-128°.

Shokol, V.A. *et al, Zh. Obshch. Khim.*, 1970, **40**, 1458 (*Engl. transl.* p. 1445) (*synth, ir, derivs*)

P-Aminocarbonyl-*N*,*N*-diethylphosphonamidic acid, 9CI — A-00059

P-*Carbamoyl*-N,N-*diethylphosphonamidic acid*

$H_2NCO-P(=O)(NEt_2)(OH)$

$C_5H_{13}N_2O_3P$ M 180.143

Esters are plant growth regulators.

(±)-form

Et ester: Ethyl P-*aminocarbonyl*-N,N-*diethylphosphonamidate.*
$C_7H_{17}N_2O_3P$ M 208.197
Solid. Mp 98-101°.

U.S.P., 3 712 936, (*1973*); *CA*, **78**, 120236 (*derivs, synth, use*)

(Aminocarbonylmethyl)phosphonic acid, 9CI — A-00060

(Carbamoylmethyl)phosphonic acid. (2-Amino-2-oxoethyl)phosphonic acid

$H_2NCOCH_2P(O)(OH)_2$

$C_2H_6NO_4P$ M 139.047

Monoanilinium salt: Cryst. (MeOH/Me_2CO). Mp 131-133°.

Di-Et ester: [5464-68-6]. *Diethyl (2-amino-2-oxoethyl)-phosphonate. Diethyl (carbamoylmethyl)phosphonate.*
$C_6H_{14}NO_4P$ M 195.155
Cryst. (EtOAc/pet. ether). Mp 80-82°.

Diisopropyl ester: Bis(1-methylethyl) (2-amino-2-oxoethyl)phosphonate. Diisopropyl (carbamoylmethyl)phosphonate.
$C_8H_{18}NO_4P$ M 223.208
Solid. Mp 98-99°.

Speziale, A.J. *et al, J. Org. Chem.*, 1958, **23**, 1883 (*diethyl ester*)
Bodnarchuk, N.V. *et al, Zh. Obshch. Khim.*, 1969, **39**, 1707 (*Engl. transl.* p. 1673) (*diisopropyl ester*)
Morita, T. *et al, Bull. Chem. Soc. Jpn.*, 1978, **51**, 2169 (*bistrimethylsilyl ester, pmr*)

[(Aminocarbonyl)methyl]-triphenylphosphonium(1+), 9CI — A-00061

(Carbamoylmethyl)triphenylphosphonium(1+), 8CI

$Ph_3P^{\oplus}CH_2CONH_2$

$C_{20}H_{19}NOP^{\oplus}$ M 320.350 (ion)

pK_a 11.

Chloride: [25361-54-0].
$C_{20}H_{19}ClNOP$ M 355.803
Source of ylide. Solid. Mp 227-229°. Alkali → ylide.
▷TA1860000.

Ylide: [(Aminocarbonyl)methylene]-triphenylphosphorane. (Carbamoylmethylene)-triphenylphosphorane.
$C_{20}H_{18}NOP$ M 319.342
Wittig reagent. Solid. Mp 177-178°. Dec. in contact with hydroxylic solvents.

Trippett, S. *et al, J. Chem. Soc.*, 1959, 3874 (*synth, props*)

Aminocarbonylphosphoramidic acid, 9CI — A-00062

Carbamoylphosphoramidic acid. Phosphonourea. Ureidophosphonic acid

$H_2NCONHP(O)(OH)_2$

$CH_5N_2O_4P$ M 140.035

Di-Me ester: [3135-77-1]. *Dimethyl aminocarbonylphosphoramidate.*
$C_3H_9N_2O_4P$ M 168.089
Needles (H_2O or MeCN). Sol. H_2O. Mp 143-145°, Mp 186-187°.

Di-Et ester: [3135-78-2]. *Diethyl aminocarbonylphosphoramidate.*
$C_5H_{13}N_2O_4P$ M 196.142
Solid or needles (H_2O). Sol. H_2O. Mp 136°, Mp 208-209°.

Diisopropyl ester: Diisopropyl aminocarbonylphosphoramidate.
$C_7H_{17}N_2O_4P$ M 224.196
Solid. Mp 151-155°.

Di-Ph ester: [1817-85-2]. *Diphenyl aminocarbonylphosphoramidate.*
$C_{13}H_{13}N_2O_4P$ M 292.230
Cryst. (EtOH). Sol. aq. alkali. Mp 199-200° (191-194°).

Dichloride:
$CH_3Cl_2N_2O_2P$ M 176.927
No phys. data reported.

Wiesboeck, R.A., *J. Org. Chem.*, 1965, **30**, 3161 (*esters, synth*)
Markalous, F. *et al, Collect. Czech. Chem. Commun.*, 1972, **37**, 725 (*dichloride, esters, synth, ir*)
Schaffrath, W. *et al, Z. Chem.*, 1978, **18**, 180 (*diphenyl ester, synth, P nmr*)

2-Amino-3-(2-chloroethyl)tetrahydro-4*H*-1,3,2-oxazaphosphorine 2-oxide **A-00063**

*3-(2-Chloroethyl)tetrahydro-4*H*-1,3,2-oxazaphosphorin-2-amine 2-oxide*, 9CI. *2-Amino-3-(2-chloroethyl)-1,3,2-oxazaphosphorinane 2-oxide*

[53459-55-5]

(*R*)-*form*

$C_5H_{12}ClN_2O_2P$ M 198.589

(*R*)-*form* [83802-21-5]

Solid. Mp 103-104°. $[\alpha]^{25}_{D}$ +16.4° (c, 3.0 in MeOH).

(*S*)-*form* [83802-22-6]

Solid. Mp 103-104°. $[\alpha]^{25}_{D}$ −18.0° (c, 3.0 in MeOH).

(±)-*form*

Prisms (Me_2CO/Et_2O). Mp 99-100°.

N-*Ac:*

$C_7H_{14}ClN_2O_3P$ M 240.626

Cryst. (Me_2CO/Et_2O). Mp 118-119°.

Norpoth, K. *et al*, *Arzneim.-Forsch.*, 1975, **25**, 1331; 1976, **26**, 1376.
Takamizawa, A. *et al*, *Chem. Pharm. Bull.*, 1977, **25**, 2900 (*synth, deriv*)
Su, C.N. *et al*, *J. Am. Chem. Soc.*, 1982, **104**, 7343 (*cd*)
Misiura, K. *et al*, *J. Med. Chem.*, 1983, **26**, 674 (*synth, ms, tlc, P nmr*)

(10-Aminodecyl)phosphonic acid, 9CI **A-00064**

$H_2N(CH_2)_{10}P(O)(OH)_2$

$C_{10}H_{24}NO_3P$ M 237.278

Cryst. (EtOH). Insol. H_2O. Mp 35-36°. pK_{a2} 8.0, pK_{a3} 11.25.

Chavane, V., *C.R. Hebd. Seances Acad. Sci.*, 1947, **224**, 406 (*synth*)
Rumpf, P. *et al*, *C.R. Hebd. Seances Acad. Sci.*, 1947, **224**, 919 (*props*)

2-Amino-2-deoxygalactose 1-(dihydrogen phosphate), 8CI **A-00065**

Galactosamine 1-phosphate. Chondrosamine 1-phosphate

α-pyranose-*form*

$C_6H_{14}NO_8P$ M 259.152

α-D-*Pyranose-form* [35946-79-3]

2-Amino-2-deoxy-α-D-galactopyranosyl phosphate

$[\alpha]^{25}_{D}$ +142.6° (c, 2.0 in H_2O).

N-*Ac:*

$C_8H_{16}NO_9P$ M 301.189

$[\alpha]_D$ +178° (H_2O).

N-*Ac, Di-K salt:* $[\alpha]^{25}_{D}$ +112.4° (c, 2.9 in H_2O).

Biochem. Prep., 1971, **13**, 3 (*synth*)
Bauer, C. *et al*, *Hoppe - Seylers Z. Physiol. Chem.*, 1972, **353**, 1053; *CA*, **77**, 85209n (*pharmacol*)
MacDonald, D.L., *The Carbohydrates*, Academic Press, 1972, 2nd Ed., **1A**, 253 (*rev*)

2-Amino-2-deoxyglucitol, 9CI, 8CI **A-00066**

Glucosaminol. Glucosaminitol

$C_6H_{10}NO_5$ M 176.149

D-form

B,HCl: Mp 159-161°. $[\alpha]^{19}_{365}$ −6.8° (c, 0.76 in H_2O).

N-*2,4-Dinitrophenyl:* Yellow needles. Mp 163-164°.

N-*Ac:*

$C_8H_{12}NO_6$ M 218.186

Mp 153°. $[\alpha]^{18}_{D}$ −11° (H_2O).

N-*Ac, 3,4:5,6-di-O-isopropylidene:*

$C_{14}H_{20}NO_6$ M 298.315

Mp 155-156°. $[\alpha]^{20}_{D}$ −19.9° (c, 1.72 in H_2O).

3-Phosphate:

$C_6H_9NO_8P$ M 254.113

Mp 193-195°. $[\alpha]^{22}_{D}$ −20.5° (c, 0.25 in H_2O).

Karrer, P. *et al*, *Helv. Chim. Acta*, 1937, **20**, 626 (*synth*)
Lambert, R. *et al*, *Chem. Ber.*, 1963, **96**, 2350.
Karkkainen, J. *et al*, *Carbohydr. Res*, 1969, **10**, 113 (*glc, ms*)
Paulsen, H. *et al*, *Chem. Ber.*, 1969, **102**, 459 (*synth*)
Kuzuhara, H. *et al*, *Carbohydr. Res.*, 1975, **45**, 245.

2-Amino-2-deoxyglucose 1-(dihydrogen phosphate) **A-00067**

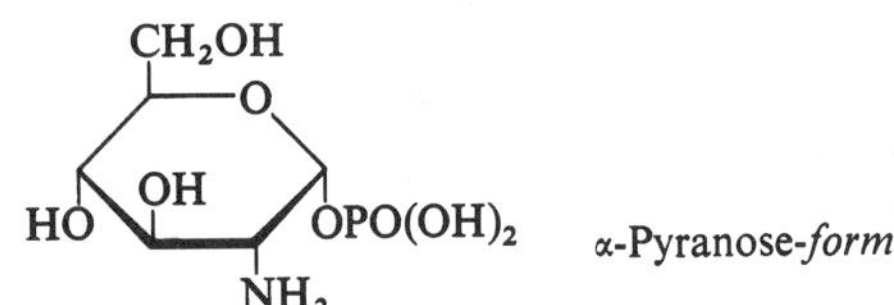

α-Pyranose-*form*

$C_6H_{14}NO_8P$ M 259.152

D-form

Mp 178-179° dec. $[\alpha]_D$ −20° (calc) (H_2O).

α-D-form

N-*Ac: 2-Acetamido-2-deoxy-α-D-glucose 1-(dihydrogen phosphate).*

$C_8H_{16}NO_9P$ M 301.189

$[\alpha]_D$ +79° (H_2O) (as mono-K salt).

N-*Ac, di-K salt:* Monohydrate. $[\alpha]^{25}_{D}$ +76.1° (c, 3.4 in H_2O).

3,4,6-Tri-Ac, diphenyl ester; B,HCl: Mp 137-138°. $[\alpha]^{23}_{D}$ +110° (c, 2.44 in MeOH).

β-D-form

N-*Ac: 2-Acetamido-2-deoxy-β-D-glucose 1-(dihydrogen phosphate).*

$C_8H_{16}NO_9P$ M 301.189

Mp 170-171° dec. (as mono-Na salt). $[\alpha]_D$ −1.7° (H_2O).

N-*Ac, di-Na salt:* $[\alpha]^{25}_{D}$ −1.6° (c, 2.9 in H_2O).

Maley, F. *et al*, *J. Am. Chem. Soc.*, 1956, **78**, 5303.
Buluja, G. *et al*, *J. Chem. Soc.*, 1960, 4678.
O'Brien, P.J., *Biochim. Biophys. Acta*, 1964, **86**, 628.

2-Amino-2-deoxyglucose 3-(dihydrogen phosphate) **A-00068**

$C_6H_{14}NO_8P$ M 259.152

D-form
Mp 180°. $[\alpha]_D^{20}$ +70° (c, 0.03 in H_2O).
α-D-Pyranose-form
*Benzyl glycoside, 4,6-*O*-benzylidene: Benzyl 2-amino-4,6-*O*-benzylidene-2-deoxy-α-D-glucopyranoside 3-(dihydrogen phosphate).*
$C_{20}H_{24}NO_8P$ M 437.385
Mp 212-213°. $[\alpha]_D^{20}$ +40° (c, 0.25 in DMF).
Benzyl glycoside, N*-benzyloxycarbonyl, 4,6-*O*-benzylidene, di-Ph ester:* Mp 98°. $[\alpha]_D^{20}$ +37.5° ($CHCl_3$).
Benzyl glycoside, N*-benzyloxycarbonyl, 4,6-*O*-benzylidene, di-Ph ester:* Mp 125°. $[\alpha]_D$ −39° ($CHCl_3$).

Lambert, R. *et al, Chem. Ber.*, 1963, **96**, 2350.
Westphal, O. *et al, Angew. Chem., Int. Ed. Engl.*, 1963, **2**, 327.

2-Amino-2-deoxyglucose 6-(dihydrogen phosphate) A-00069

Glucosamine 6-(dihydrogen phosphate)
$C_6H_{14}NO_8P$ M 259.152
D-form [3616-42-0]
$[\alpha]_D^{20}$ +60° (c, 2.0 in 1N H_2SO_4).
Ba salt: $[\alpha]_D^{20}$ +49.6° (c, 4.0 in H_2O at pH 2).
1,3,4-Tri-Ac:
$C_{12}H_{20}NO_{11}P$ M 385.264
Mp 166-167°. $[\alpha]_D^{20}$ +48.9° (c, 2.7 in HCl).
1,3,4-Tri-Ac, di-Ph ester:
$C_{24}H_{28}NO_{11}P$ M 537.459
Mp 190-191°. $[\alpha]_D^{20}$ +49.6° (c, 4 in MeOH).
N*-Ac:*
$C_8H_{16}NO_8P$ M 285.190
Mp 147-148° dec. (as Mono-NH_4 salt).
*1,2*N*,3,4-Tetra-Ac:*
$C_{14}H_{22}NO_{12}P$ M 427.301
Mp 166-168°. $[\alpha]_D^{20}$ +25° (c, 1 in H_2O).
*1,2*N*,3,4-Tetra-Ac, di-Ph ester:*
$C_{26}H_{30}NO_{12}P$ M 579.496
Mp 144-145°. $[\alpha]_D$ +25° (c, 0.7 in $CHCl_3$).
N*-Anisylidene, 1,3,4-tri-Ac, diphenyl ester:* Mp 134-135°. $[\alpha]_D^{20}$ +96.8° (c, 1 in $CHCl_3$).

Anderson, J.M. *et al, J. Chem. Soc.*, 1956, 814.
Maley, F. *et al, J. Am. Chem. Soc.*, 1956, **78**, 1393.

2-Amino-1,3-di-*tert*-butylazaphosphiridine A-00070

1,3-Bis(1,1-dimethylethyl)-2-azaphosphiridinamine, 10CI

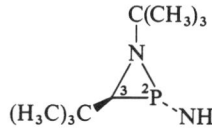

$C_9H_{21}N_2P$ M 188.252
(2RS,3SR)-form
(±)-trans-*form*
N,N*-Diisopropyl:* [78342-55-9]. Liq. $Bp_{0.01}$ 53-55°.

Niecke, E. *et al, Angew. Chem., Int. Ed. Engl.*, 1981, **20**, 675 (*synth, cmr, P nmr, pmr*)

3-Amino-2,3-di-*tert*-butyl-1,2,4,3-thiadiazaphosphetidine A-00071

2,4-Bis(1,1-dimethylethyl)-1,2,4,3-thiadiazaphosphetidin-3-amine, 10CI

$C_8H_{20}N_3PS$ M 221.300
N,N*-Di-Me, 1,1-dioxide:* [76037-04-2].
$C_{10}H_{24}N_3O_2PS$ M 281.352
Waxy solid. $Bp_{0.5}$ 86-87°.

Cowley, A.H. *et al, Inorg. Chem.*, 1981, **20**, 712 (*synth, pmr, P nmr, cmr*)

2-Amino-2,3-dihydro-5-methyl-3-phenyl-3*H*-1,3,4,2-thiadiazaphosphole 2-sulfide A-00072

*2,3-Dihydro-5-methyl-3-phenyl-1,3,4,2-thiadiazaphosphole-2(3*H*)-amine 2-sulfide,* 9CI

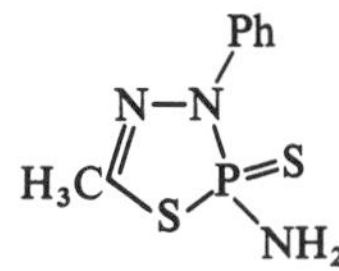

$C_8H_{10}N_3PS_2$ M 243.281
N,N*-Di-Me:* [35169-12-1]. *2,3-Dihydro-*N,N*,5-trimethyl-3-phenyl-1,3,4,2-thiadiazaphosphole-2(3*H*)-amine-2-sulfide.*
$C_{10}H_{14}N_3PS_2$ M 271.334
Solid. Mp 67-69°. $Bp_{0.12}$ 145-147°.
N,N*-Di-Et:* [35169-13-2]. N,N*-Diethyl-2,3-dihydro-5-methyl-3-phenyl-1,3,4,2-thiadiazaphosphole-2(3*H*)-amine 2-sulfide.*
$C_{12}H_{18}N_3PS_2$ M 299.388
Mp 64-66°. $Bp_{0.2}$ 150-152°.
N*-Ph:* [35169-16-5]. *2,3-Dihydro-5-methyl-*N*,3-diphenyl-1,3,4,2-thiadiazaphosphole-2(3*H*)amine 2-sulfide.*
$C_{14}H_{14}N_3PS_2$ M 319.378
Mp 144°. $Bp_{0.25}$ 139-142°.

Italinskaya, T.L. *et al, Zh. Obshch. Khim.*, 1971, **41**, 1980 (*Engl. transl.* p. 1998) (*synth*)

1-Amino-1,1-ethanediphosphonic acid A-00073

(1-Aminoethylidene)bisphosphonic acid, 9CI. *(1-Aminoethylidene)diphosphonic acid,* 8CI. *Ethylamine-1,1-diphosphonic acid*
[15049-85-1]

$H_3CC(NH_2)[P(O)(OH)_2]_2$

$C_2H_9NO_6P_2$ M 205.044
Exists in zwitterionic form. Cryst. (Me_2CO/dioxane, or EtOH/dioxane aq.). Mp 283-285° (254-255°).
Tetra-Et ester: [59081-07-1]. *Tetraethyl (1-aminoethylidene)bisphosphonate.*
$C_{10}H_{25}NO_6P_2$ M 317.258
Liq. $Bp_{0.005}$ 98-99°. n_D^{20} 1.4541.

Orlovskii, V.V. *et al, Zh. Obshch. Khim.*, 1976, **46**, 297 (*Engl. transl.* p. 294) (*synth, ester*)
Fukuda, M. *et al, Chem. Lett.*, 1977, 529 (*pmr*)
Koch, U. *et al, Z. Chem.*, 1980, **20**, 66 (*complexes*)
Serebrennikova, G.A. *et al, Zh. Obshch. Khim.*, 1985, **55**, 440; *J. Gen. Chem. USSR (Engl. Transl.)* 390 (*synth, ir, P nmr*)

(2-Aminoethyl)butylphenylphosphine A-00074
2-(Butylphenylphosphino)ethanamine, 10CI

$H_3C(CH_2)_3PPhCH_2CH_2NH_2$

$C_{12}H_{20}NP$ M 209.270
Ligand for Co and Pd.
(+)-***form*** [67747-68-6]
Liq.
(−)-***form*** [67747-69-7]
Liq.
(±)-***form*** [67747-67-5]
Liq. Bp_{80} 115°. Resolved as Pd(II) PF_6^- complex.

Issleib, K. *et al, Chem. Ber.*, 1976, **100**, 2685 (*synth*)
Kashiwabara, K. *et al, Chem. Lett.*, 1978, 673 (*resoln, cd, config*)
Kinoshita, I. *et al, ACS Symp. Ser.*, 1980, **119**, 207; *Bull. Chem. Soc. Jpn.*, 1980, **53**, 3715 (*synth, cmr, cd, resoln*)
Kashiwabara, K. *et al, Bull. Chem. Soc. Jpn.*, 1981, **54**, 725 (*complexes*)

2-Aminoethyl dihydrogen phosphate A-00075
Ethanolamine O-phosphate. O-Phosphoethanolamine

$H_2NCH_2CH_2OP(O)(OH)_2$

$C_2H_8NO_4P$ M 141.063
Component of eye lens and toxins of gram-negative bacteria. Important in phosphoglyceride synth. in mammals. Cryst. (EtOH aq.). Mp 242°.
Mono-Et ester: Ethyl hydrogen 2-aminoethyl phosphate.
$C_4H_{12}NO_4P$ M 169.117
Cryst. (EtOH/Me_2CO). Mp 238°.
Diisopropyl ester: [14646-04-9]. *Diisopropyl 2-aminoethyl phosphate.*
$C_8H_{20}NO_4P$ M 225.224
Liq. $Bp_{0.06}$ 75°.
Mono-Ph ester: Phenyl hydrogen 2-aminoethyl phosphate.
$C_8H_{12}NO_4P$ M 217.161
Solid. V. sol. H_2O, spar. sol. EtOH, insol. C_6H_6, $CHCl_3$, Et_2O. Mp 243-244° dec.

Plapinger, R.E. *et al, J. Am. Chem. Soc.*, 1953, **75**, 5757.
Tsizin, Y.S. *et al, Zh. Obshch. Khim.*, 1963, **33**, 2873 (*Engl. transl.* p. 2800) (*synth, ir*)
Greenhalgh, R. *et al, Can. J. Chem.*, 1967, **45**, 495 (*diisopropyl ester, synth, ir, pmr, P nmr*)
Glonek, T. *et al, J. Neurochem.*, 1982, **39**, 1210 (*P nmr*)
Greiner, J.V. *et al, Exp. Eye Res.*, 1982, **34**, 545 (*P nmr*)
Trigalo, F. *et al, J. Chem. Soc., Perkin Trans. 1*, 1982, 1733 (*monoethyl ester*)
Cates, L.A. *et al, J. Med. Chem.*, 1984, **27**, 654 (*synth*)

(1-Aminoethyl)methylphosphinic acid, 9CI A-00076
[67398-15-6]

HO–P(=O)(Me)–CH(NH₂)CH₃

$C_3H_{10}NO_2P$ M 123.091
(±)-***form*** [74891-95-5]
Suckering agent for tobacco. Cryst. solid. Mp 258-260°.

Oleksyszyn, J. *et al, Synthesis*, 1978, 479 (*synth, ir, pmr*)
Wasielewski, C. *et al, Pol. J. Chem.*, 1978, **52**, 1315.

(1-Aminoethyl)phosphinic acid, 9CI A-00077
[74333-44-1]

$H_3CCH(NH_2)PH(O)OH$

$C_2H_8NO_2P$ M 109.064
(±)-***form*** [21309-23-9]
Solid. Mp 223-224°.
N-*(Diphenylmethyl):* [*1-(Diphenylmethylamino)ethyl*]-*phosphinic acid.*
$C_{15}H_{18}NO_2P$ M 275.286
Solid. Mp 220-221°.

Ger. Pat., 2 722 162, (*1977*); *CA*, **88**, 105559 (*synth*)

(1-Aminoethyl)phosphonic acid, 9CI A-00078
[6323-97-3]

$H_2NCH(CH_3)P(O)(OH)_2$

$C_2H_8NO_3P$ M 125.064
Exists in dipolar form.
(***R***)-***form*** [60687-36-7]
Component of alafosfalin.
(±)-***form*** [16606-65-8]
Cryst. Mp 272-274°, Mp 283-285° dec.
Di-Et ester: [74292-99-2]. *Diethyl (1-aminoethyl)-phosphonate.*
$C_6H_{16}NO_3P$ M 181.171
Liq. Bp_3 70-73°, $Bp_{0.05}$ 100°.
Di-Ph ester: Diphenyl (1-aminoethyl)phosphonate.
$C_{14}H_{16}NO_3P$ M 277.259
Solid. Mp 36-39°.
N-*Benzoyl:* [*1-(Benzoylamino)ethyl*]*phosphonic acid.* [*1-(Benzamido)ethyl*]*phosphonic acid.*
$C_9H_{12}NO_4P$ M 229.172
Cryst. (EtOAc). Mp 183° dec.
N-*Benzyl:* [*(1-Benzylamino)ethyl*]*phosphonic acid.*
$C_9H_4NO_3P$ M 205.109
Solid. Mp 236-238°.

Chambers, J.R. *et al, J. Org. Chem.*, 1964, **29**, 832 (*synth*)
Tyka, R., *Tetrahedron Lett.*, 1965, 3071; 1970, 677 (*synth, deriv*)
Harvey, D.J. *et al, J. Chromatogr.*, 1973, **79**, 65 (*silyl, derivs, ms*)
Stec, W.J. *et al, J. Org. Chem.*, 1976, **41**, 3757 (*synth, P nmr, pmr*)
Oleksyszn, J. *et al, Synthesis*, 1979, 985 (*diphenyl ester, synth, ir, pmr*)
Wozniak, M. *et al, Talanta*, 1979, **26**, 1135 (*complexes*)
Kudzin, Z.H. *et al, Synthesis*, 1980, 1028 (*synth, diethyl ester, pmr, P nmr, ms, tlc*)
Rachon, J. *et al, Justus Liebigs Ann. Chem.*, 1981, 709 (*synth, pmr, ir, diethyl ester, synth, deriv*)

2-Aminoethylphosphonic acid, 9CI A-00079
Ciliatine
[2041-14-7]

$H_2NCH_2CH_2P(O)(OH)_2$

$C_2H_8NO_3P$ M 125.064
Constit. as glyceryl esters of rumen protozoan, *Tetrahymena pyroformis*, sea anemones, bovine brain, mycobacteria, and the fungus *P. prolatum*. Rhombic cryst. or needles (EtOH aq.). Mp 295-297° dec. pK_{a1} 2.45, pK_{a2} 7.0 (6.25), pK_{a3} 10.8 (11.05). The metastable rhombic form is difficult to prepare. First known natural compd. contg. the C—P bond.

Di-Et ester: [41468-36-4].
$C_6H_{16}NO_3P$ M 181.171
Liq. $Bp_{0.025}$ 54-56°.
N-*Benzoyl: 2-(Benzoylamino)ethylphosphonic acid.*
$C_9H_{12}NO_4P$ M 229.172
Solid. Mp 191-192°.
N-*Me:* [14596-55-5]. *2-Methylaminoethylphosphonic acid.*
$C_3H_{10}NO_3P$ M 139.091
Isol. from the sea anemone *Anthopleura xanthogrammica.* Needles (MeOH aq.). Mp 291° dec.
N-*Di-Me: 2-Dimethylaminoethylphosphonic acid.*
$C_4H_{12}NO_3P$ M 153.117
Isol. from *A. xanthogrammica.* Pale-yellow cryst. (MeOH aq.). Mp 249.5° dec.

Kosolapoff, G.M., *J. Am. Chem. Soc.*, 1947, **69**, 2112 (*synth*)
Top. Phosphorus Chem., (*Grayson, M. et al*, Eds.), 1966, Interscience, **4**, 23: **11**, 297 (*rev*)
Barycki, J. *et al*, *Tetrahedron Lett.*, 1970, 3147 (*synth*)
Isbell, A.F. *et al*, *J. Org. Chem.*, 1972, **37**, 4399 (*synth, bibl*)
Harvey, D.J. *et al*, *J. Chromatogr.*, 1973, **79**, 65 (*ms*)
Lagrange, C.G., *Can. J. Chem.*, 1978, **56**, 663 (*ir, raman*)
Brigot, D. *et al*, *Tetrahedron*, 1979, **35**, 1345 (*synth, pmr, derivs*)
Fabre, G. *et al*, *Can. J. Chem.*, 1981, **59**, 2864 (*synth, diethyl ester, ir, pmr*)

(2-Aminoethyl)phosphoramidic acid, 9CI, 8CI — A-00080

$H_2NCH_2CH_2NHP(O)(OH)_2$

$C_2H_9N_2O_3P$ M 140.078
Di-Et ester: [52551-78-7]. *Diethyl (2-aminoethyl)-phosphoramidate.* N-*(Diethoxyphosphinyl)-ethylenediamine.*
$C_6H_{17}N_2O_3P$ M 196.186
Liq. $Bp_{0.15}$ 108-109°.
Mono-Ph ester: [26404-96-6]. *Phenyl hydrogen (2-aminoethyl)phosphoramidate.*
$C_8H_{13}N_2O_3P$ M 216.176
Cryst. (dioxan aq. or Me_2CO aq.). Mp 200.5-201° (after subl.). Sublimes. Forms a hemihydrate, Mp 195-196° and a monohydrate Mp 219-221°.

Edmundson, R.S., *J. Chem. Soc. (C)*, 1969, 2730 (*phenyl ester, synth, ir*)
Cates, L.A. *et al*, *J. Med. Chem.*, 1984, **27**, 654 (*diethyl ester*)

2-Amino-2,2,4,4,6,6-hexahydro-2-methyl-4,4,6,6-tetraphenyl-1,3,5-triaza-2,4,6-triphosphorine, 9CI — A-00081

2-Amino-2,2,4,4,6,6-hexahydro-2-methyl-4,4,6,6-tetraphenylcyclotriphosphazene
[50457-21-1]

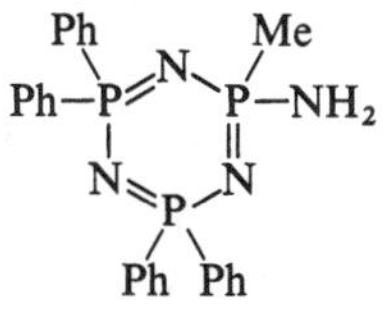

$C_{25}H_{25}N_4P_3$ M 474.421
Needles (MeCN). Mp 180-182°.

Schmidpeter, A. *et al*, *Chem. Ber.*, 1976, **109**, 2340 (*synth, pmr, P nmr*)

2-Amino-2,2,4,4,6,6-hexahydro-2,4,4,6,6-pentaphenoxy-1,3,5-triaza-2,4,6-triphosphorine, 9CI — A-00082

2-Amino-2,2,4,4,6,6-hexahydro-2,4,4,6,6-pentaphenoxycyclotriphosphazene

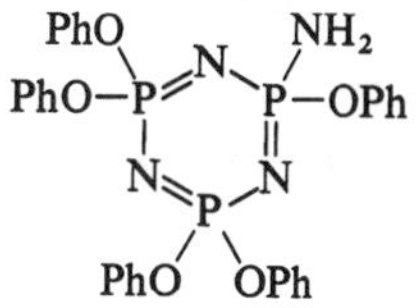

$C_{30}H_{27}N_4O_5P$ M 554.541
Solid. Mp 60°.

Volodin, A.A. *et al*, *Zh. Obshch. Khim.*, 1973, **43**, 2206; 1975, **45**, 37 (*Engl. transl.* pp. 2198, 32) (*synth*)

(1-Amino-2-hydroxyethyl)phosphonic acid — A-00083

Phosphonoserine

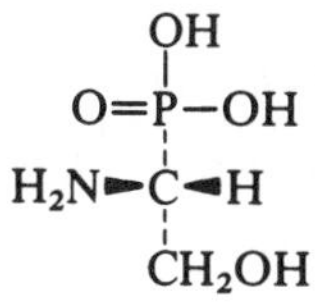

(*R*)-*form*

$C_2H_8NO_4P$ M 141.063
(*R*)-*form*
Hygroscopic solid, melting to an oil. Mp ca. 70°. $[\alpha]_D^{20}$ −30° (c, 1 in 1M NaOH aq.). Dec. 198-200°.
(*S*)-*form*
Hygroscopic solid melting to an oil. Mp ca. 70°. $[\alpha]_D$ +35° (c, 1 in 1M NaOH aq.). Dec. 186-188°.
(±)-*form*
Cryst. + $0.5H_2O$. Mp ca. 80°. Dec. 165-168°.
O-*Ac, Di-Ph ester:*
$C_{16}H_8NO_5P$ M 325.217
Solid (as hydrobromide). Mp 137-139° (hydrobromide).

Lejczak, B. *et al*, *Synthesis*, 1984, 577 (*synth, derivs, ir, pmr*)

(1-Amino-2-mercaptoethyl)phosphonic acid — A-00084

Phosphonocysteine

$HSCH_2CH(NH_2)P(O)(OH)_2$

$C_2H_8NO_3PS$ M 157.124
(±)-*form*
Solid. Mp 228-234°, Mp 251.5-252.5°.

Kudzin, Z.H. *et al*, *Synthesis*, 1983, 812 (*synth, ir, pmr, P nmr, deriv, ms*)

(Aminomethyl)diphenylphosphine oxide — A-00085

1-(Diphenylphosphinyl)methanamine, 9CI. *1-(Diphenylphosphinyl)methylamine*
[5276-94-8]

$Ph_2P(O)CH_2NH_2$

$C_{13}H_{14}NOP$ M 231.233
Source of diazomethyldiphenylphosphine oxide. Cryst. (EtOH). Mp 102-103°.
B,HCl: [55422-17-8].
$C_{13}H_{15}ClNOP$ M 267.694

Cryst. (EtOH aq.). Mp 169-169.5°.

Popoff, I.C. *et al*, *J. Org. Chem.*, 1963, **28**, 2898 (*synth*)
Kreutzkamp, N. *et al*, *Angew. Chem., Int. Ed. Engl.*, 1965, **4**, 1078 (*use*)
Mukhacheva, O.A. *et al*, *Zh. Obshch. Khim.*, 1975, **45**, 526 (*Engl. transl.* p. 520) (*synth*)

(Aminomethylene)bisphosphonic acid, 9CI **A-00086**

Aminomethanediphosphonic acid. Methylamine-1,1-diphosphonic acid

[29712-28-5]

$H_2NCH[P(O)(OH)_2]_2$

$CH_7NO_6P_2$ M 191.017

Free acid probably exists in dipolar form. Solid. Mp 258-259°. pK_{a1} 1.7, pK_{a2} 5.59, pK_{a3} 8.06, pK_{a4} 10.42.

Tetra-Me ester: Tetramethyl (aminomethylene)-bisphosphonate.
$C_5H_{15}NO_6P_2$ M 247.124
Oil.

Plöger, W. *et al*, *Z. Anorg. Allg. Chem.*, 1972, **389**, 119 (*synth*)
Rusina, M.N. *et al*, *Zh. Obshch. Khim.*, 1977, **47**, 1721 (*Engl. transl.* p. 1574) (*synth, ir, complexes*)
Worms, K.-H. *et al*, *Justus Liebigs Ann. Chem.*, 1982, 275 (*synth, derivs*)

(1-Amino-1-methylethyl)phosphonic acid, 9CI **A-00087**

(α-Aminoisopropyl)phosphonic acid, 8CI

[5035-79-0]

$(H_3C)_2C(NH_2)P(O)(OH)_2$

$C_3H_{10}NO_3P$ M 139.091

Free acid exists in dipolar form. Cryst. pK_{a1} 2.09, pK_{a2} 6.05, pK_{a3} 10.43 (H_2O).

Monohydrate: Mp 256-257°.

Di-Me ester: [53753-41-6]. *Dimethyl (1-amino-1-methylethyl)phosphonate.*
$C_5H_{14}NO_3P$ M 167.144
Solid. Mp 40°. $Bp_{2.5}$ 73-74°.

Mono-Et ester: Ethyl hydrogen (1-amino-1-methylethyl)phosphonate.
$C_5H_{14}NO_3P$ M 167.144
Solid. Mp 274-275°.

Di-Et ester: [16814-09-8]. *Diethyl (1-amino-1-methylethyl)phosphonate.*
$C_7H_{18}NO_3P$ M 195.198
Liq. $Bp_{0.2}$ 70-72°.

Di-Et ester; B,HCl: Solid. Mp 137-138°.

N-*Benzoyl: (1-Benzoylamino-1-methylethyl)phosphonic acid.*
$C_{10}H_{14}NO_4P$ M 243.199
Solid. Mp 207-208°.

N-*Benzyl: (1-Benzylamino-1-methylethyl)phosphonic acid. (1-Phenylmethylamino-1-methylethyl)-phosphonic acid.*
$C_{10}H_{16}NO_3P$ M 229.215
Cryst. (EtOH aq.). Mp 223°.

Berry, J.P. *et al*, *J. Org. Chem.*, 1972, **37**, 4396 (*deriv*)
Petrov, K.A. *et al*, *Zh. Obshch. Khim.*, 1976, **46**, 1246 (*Engl. transl.* p. 1226) (*ester, synth, props*)
Stec, W.J. *et al*, *J. Org. Chem.*, 1976, **41**, 3757 (*synth, pmr, P nmr*)
Szczepaniak, W. *et al*, *Phosphorus Sulfur*, 1979, **7**, 337 (*synth, deriv*)
Rachon, J. *et al*, *Justus Liebigs Ann. Chem.*, 1981, 709 (*synth, ir, pmr, ester*)
Russell, G.A. *et al*, *J. Org. Chem.*, 1982, **47**, 1480 (*ester, synth, ms, pmr, cmr, P nmr*)

(2-Amino-1-methylethyl)phosphonic acid, 9CI **A-00088**

[28660-34-6]

$H_2NCH_2CH(CH_3)P(O)(OH)_2$

$C_3H_{10}NO_3P$ M 139.091

Free acid probably exists in dipolar form.

(±)-*form*

Solid. Mp 282-286° dec.

Di-Et ester: [42591-68-4]. *Diethyl (2-amino-1-methylethyl)phosphonate.*
$C_7H_{18}NO_3P$ M 195.198
Liq. d_4^{20} 1.07. Bp_1 65-66°. n_D^{20} 1.4470.

Barycki, J. *et al*, *Tetrahedron Lett.*, 1970, 3147 (*synth*)
Lazareva, M.V. *et al*, *Izv. Akad. Nauk SSSR, Ser. Khim.*, 1973, 1382 (*Engl. transl.* p. 1342) (*ester*)

(Aminomethyl)methylphosphinic acid, 9CI **A-00089**

[15901-11-8]

$MeP(O)(OH)CH_2NH_2$

$C_2H_8NO_2P$ M 109.064

Probably exists as zwitterion. Solid. Mp 296-298°.

Popoff, I.C. *et al*, *J. Org. Chem.*, 1963, **28**, 2898 (*synth, ir*)
Drozdzowski, P.M. *et al*, *Bull. Acad. Pol. Sci., Ser. Sci. Chim.*, 1977, **25**, 209 (*ir, raman*)
Glawiak, T. *et al*, *Acta Crystallogr., Sect. B*, 1977, **33**, 1522 (*cryst struct*)
Latajka, Z. *et al*, *J. Mol. Struct.*, 1981, **70**, 49 (*struct*)

***O*-(4-Amino-3-methylphenyl) *O,O*-dimethyl phosphorothioate, 10CI** **A-00090**

Aminofenitrothion

[13306-69-9]

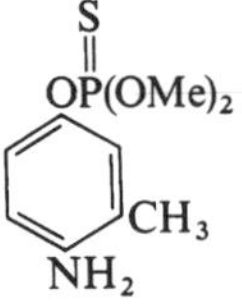

$C_9H_{14}NO_3PS$ M 247.248

Degradn. product of Fenitrothion, F-00005 .

Greenhalgh, R. *et al*, *J. Agric. Food Chem.*, 1980, **28**, 102.
Adhya, T.K. *et al*, *Pestic. Biochem. Physiol.*, 1981, **16**, 14.
Volpé, G. *et al*, *Chromatographia*, 1981, **14**, 333 (*hplc*)

(Aminomethyl)phosphonic acid, 9CI **A-00091**

[1066-51-9]

$H_2NCH_2P(O)(OH)_2$

CH_6NO_3P M 111.037

Free acid exists in dipolar form. Metab. of Glyphosine, G-00098 and Glyphosate, G-00097 . Cryst. Mp 308-310°. pK_{a1} 0.44 (1.85), pK_{a2} 5.39, pK_{a3} 10.05 (H_2O, 25°).

Di-Et ester: [50915-72-1]. *Diethyl (aminomethyl)-phosphonate.*
$C_5H_{14}NO_3P$ M 167.144
Yellowish liq. Dec. on attempted distn.

N-*Benzyl, di-Et ester: Diethyl [(benzylamino)methyl]-phosphonate.*
$C_{12}H_{20}NO_3P$ M 257.269
Liq. $Bp_{0.02}$ 124-125°. n_D^{20} 1.5010.
N-*Benzoyl:* (N-*Benzoylaminomethyl)phosphonic acid.*
$C_8H_{10}NO_4P$ M 215.145
Solid. Mp 182°.

Chambers, J.R. *et al, J. Org. Chem.*, 1964, **29**, 832 (*synth*)
Garrigan-Lagrange, C. *et al, J. Chim. Phys.*, 1970, **67**, 1646 (*ir*)
Destrade, C. *et al, J. Chim. Phys.*, 1970, **67**, 2013 (*ir*)
Harvey, D.J. *et al, J. Chromatogr.*, 1973, **79**, 65 (*silyl derivs, ms*)
Bel'skii, V.E. *et al, Izv. Akad. Nauk SSSR, Ser. Khim.*, 1975, **24**, 1047 (*Engl. transl.* p. 958) (*derivs, synth, pmr, nmr*)
Darriet, M. *et al, Acta Crystallogr., Sect. B*, 1975, **31**, 469 (*cryst struct*)
Yamauchi, K. *et al, Bull. Chem. Soc. Jpn.*, 1975, **48**, 3285 (*diethyl ester, ir, pmr*)
Rueppel, M.L. *et al, Org. Magn. Reson.*, 1976, **8**, 19 (*pmr, P nmr*)
Kahovec, J. *et al, Org. Prep. Proced. Int.*, 1978, **10**, 285 (*synth*)
Wozniak, W. *et al, Talanta*, 1979, **26**, 1135 (*complexes*)
Oleksyszyn, J. *et al, Synthesis*, 1980, 906 (*synth*)
Latajka, Z. *et al, J. Mol. Struct.*, 1981, **70**, 49 (*struct*)
Kurguzova, A.M. *et al, Izv. Akad. Nauk SSSR, Ser. Khim.*, 1982, **31**, 1265 (*Engl. transl.* p. 1126) (*derivs, props*)

(1-Amino-2-methylpropyl)phosphinic acid, 9CI A-00092

[67896-52-0]

$(H_3C)_2CHCH(NH_2)PH(O)OH$

$C_4H_{12}NO_2P$ M 137.118

(+)-*form*
Cryst. Mp 209°. $[\alpha]_D^{25}$ +3.5° (c, 1.5 in H_2O).
N-*Benzyloxycarbonyl:* Solid. Mp 164-167°. $[\alpha]_D^{25}$ +19°.

(−)-*form*
Cryst. Mp 209°. $[\alpha]_D^{25}$ −3.6° (c, 1.5 in H_2O).

(±)-*form*
Solid. Mp 198-198.5° dec.
B,HBr: Solid. Mp 134-136°.
N-*(Diphenylmethyl):*
$C_{17}H_{22}NO_2P$ M 303.340
Solid. Mp 188-192°.
N-*Benzyloxycarbonyl:*
$C_{12}H_{18}NO_4P$ M 271.252
Solid. Mp 108-111°.

Ger. Pat., 2 722 162, (*1977*); *CA*, **88**, 105559 (*synth, redn, derivs*)

(1-Amino-1-methylpropyl)phosphonic acid, 9CI A-00093

[79014-65-6]

$$H_3CCH_2C(CH_3)(NH_2)P(O)(OH)_2$$

$C_4H_{12}NO_3P$ M 153.117

(+)-*form* [62879-51-0]
Mp 250-251° (monohydrate). $[\alpha]_D^{20}$ +4.8° (c, 2-3 in 5M HCl aq.).
Mono-Et ester: Ethyl hydrogen (1-amino-1-methylpropyl)phosphonate.
$C_6H_{16}NO_3P$ M 181.171
Cryst. (MeOH/Et_2O). Mp 209°. $[\alpha]_D^{20}$ +2.1° (c, 2.4 in 5MHCl aq.).
Di-Et ester: [63126-94-3]. *Diethyl (1-amino-1-methylpropyl)phosphonate.*
$C_8H_{20}NO_3P$ M 209.225
Liq. $[\alpha]_D^{20}$ +2.2° (5M HCl aq.).

(±)-*form*
Solid +1 H_2O. Mp 259-260° (250-251°).
Di-Et ester: [64228-49-5]. Liq. d_4^{20} 1.04. Bp_1 65°. n_D^{20} 1.4419.

Belov, Yu.P. *et al, Izv. Akad. Nauk SSSR, Ser. Khim.*, 1977, 1596 (*Engl. transl.* p. 1467) (*ester, resoln*)
Rachon, J. *et al, Justus Liebigs Ann. Chem.*, 1981, 709 (*synth, ir, pmr*)

(1-Amino-2-methylpropyl)phosphonic acid, 9CI A-00094

(1-Aminoisobutyl)phosphonic acid
[18108-24-2]

$$O{=}P(OH)_2-C(H)(NH_2)-CH(CH_3)_2$$

(*R*)-*form*

$C_4H_{12}NO_3P$ M 153.117
Free acid exists in dipolar form.

(*R*)-*form* [66254-56-6]
L-*form*
Solid. Mp 272-273°. $[\alpha]_D^{20}$ +10° (c, 2.0 in 1M NaOH aq.).

(*S*)-*form* [66254-55-5]
D-*form*
Solid. Mp 277-278°. $[\alpha]_D^{20}$ −10° (c, 2.0 in 1M NaOH aq.).

(±)-*form* [37100-67-7]
Cryst. (EtOH aq.). Mp 262-264°, 274-278°. pK_{a1} 2.04 (0.62), pK_{a2} 6.00, pK_{a3} 10.45.
Di-Et ester: Diethyl (1-amino-2-methylpropyl)-phosphonate.
$C_8H_{20}NO_3P$ M 209.225
Liq. $Bp_{0.5}$ 86-88°. n_D^{20} 1.4400.
Di-Et ester; B,HCl: Cryst. (EtOH/Et_2O). Mp 116°.
Di-Ph ester: [73270-42-5]. *Diphenyl (1-amino-2-methylpropyl)phosphonate.*
$C_{16}H_{20}NO_3P$ M 305.313
Solid. Mp 60-63°.
Di-Ph ester; B,HBr: Mp 172-175°.
N-*Benzoyl: (1-Benzoylamino-2-methylpropyl)-phosphonic acid.*
$C_{10}H_{16}NO_4P$ M 245.214
Solid. Mp 153-154°.
N-*Benzoyl, Di-Et ester: Diethyl (1-*N*-benzoylamino-2-methylpropyl)phosphonate.*
$C_{15}H_{24}NO_4P$ M 313.333
Cryst. (C_6H_6/hexane). Mp 95.5-96.5°.

Berry, J.P. *et al, J. Org. Chem.*, 1972, **37**, 4396.
Asano, S. *et al, Agric. Biol. Chem.*, 1973, **37**, 1193 (*derivs, ir, pmr*)
Harvey, D.J. *et al, Org. Mass Spectrom.*, 1973, **9**, 111 (*derivs, ms*)
Stec, W.J. *et al, J. Org. Chem.*, 1976, **41**, 3757 (*synth, pmr, P nmr*)
Glowiak, T. *et al, Tetrahedron Lett.*, 1977, 3965 (*resoln, config*)
Huber, J.W. *et al, Synthesis*, 1977, 883 (*synth, ir, pmr*)
Oleksyszyn, J. *et al, Synthesis*, 1979, 985 (*ester, synth, ir, pmr*)
Wozniak, M. *et al, Talanta*, 1979, **26**, 1135 (*complexes*)
Kudzin, Z.H. *et al, Synthesis*, 1980, 1028 (*synth, ester, pmr, nmr, tlc*)
Kafarski, P. *et al, Can. J. Chem.*, 1983, **61**, 2425 (*resoln, config*)

(1-Amino-3-methylthiopropyl)phosphonic acid A-00095

Phosphonomethionine

$MeSCH_2CH_2CH(NH_2)P(O)(OH)_2$

$C_4H_{12}NO_3PS$ M 185.177

(±)-***form***
Solid. Mp 274-275°.
B,MeI:
$C_5H_{15}INO_3PS$ M 327.117
Solid. Mp 287-289°.
S-*Oxide:*
$C_4H_{12}NO_4PS$ M 201.177
Cryst. (MeOH/Me_2CO aq.). Mp 188-190°.
S,S-*Dioxide:*
$C_4H_{12}NO_5PS$ M 217.176
Cryst. (MeOH aq.). Mp 258-260°.

Tam, C.T. *et al*, *Synthesis*, 1982, 188 (*synth, derivs, pmr*)

(1-Aminopentyl)phosphonic acid, 9CI **A-00096**
[13138-37-9]

$H_3C(CH_2)_3CH(NH_2)P(O)(OH)_2$

$C_5H_{14}NO_3P$ M 167.144

(±)-***form*** [33439-83-7]
Solid. Mp 263-264°, 285-287°. pK_{a1} 1.83 (0.58), pK_{a2} 5.82, pK_{a3} 10.35 (H_2O, 25°).
Di-Et ester: Diethyl (1-aminopentyl)phosphonate.
$C_9H_{22}NO_3P$ M 223.251
Characterised spectroscopically.
N-*Benzoyl: (1-Benzoylaminopentyl)phosphonic acid.*
$C_{12}H_{18}NO_4P$ M 271.252
Solid. Mp 106-107°.
N-*Benzyl: [1-(Benzylamino)pentyl]phosphonic acid.*
$C_{12}H_{20}NO_3P$ M 257.269
Solid. Mp 205-207°.

Tyka, R., *Tetrahedron Lett.*, 1970, 677 (*synth, deriv*)
Berry, J.P. *et al*, *J. Org. Chem.*, 1972, **37**, 4397 (*synth, props, derivs*)
Harvey, D.J., *Org. Mass Spectrom.*, 1974, **9**, 111 (*silyl derivs, ms*)
Wozniak, M. *et al*, *Talanta*, 1979, **26**, 1135 (*complexes*)
Kudzin, Z.H. *et al*, *Synthesis*, 1980, 1028 (*synth, diethyl ester, pmr, P nmr, ms*)
Rachoń, J. *et al*, *Justus Liebigs Ann. Chem.*, 1981, 709 (*synth, ir, pmr*)

(5-Aminopentyl)phosphonic acid, 9CI **A-00097**

$H_2N(CH_2)_5P(O)(OH)_2$

$C_5H_{14}NO_3P$ M 167.144
Sol. EtOH. pK_{a1} 2.6, pK_{a2} 7.6, pK_{a3} 11.0.
Di-Et ester: [53253-55-7]. *Diethyl (5-aminopentyl)-phosphonate.*
$C_9H_{22}NO_3P$ M 223.251
Liq. $Bp_{0.003}$ 90°.

Rumf, P. *et al*, *C.R. Hebd. Seances Acad. Sci.*, 1947, **224**, 919.

4-Aminophenyl diethyl phosphate, 9CI **A-00098**
1-Amino-4-[(diethoxyphosphinyl)oxy]benzene. 4-[(Diethoxyphosphinyl)oxy]aniline
[14984-58-8]

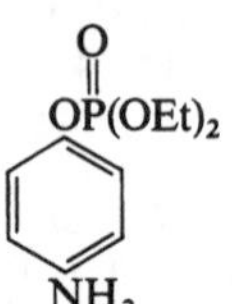

$C_{10}H_{16}NO_4P$ M 245.214
Metab. of Parathion, P-00002 Paraoxon, P-00001 and *O*-(4-Aminophenyl) *O,O*-diethyl phosphorothioate, A-00099 . Solid. Mp 52-53°.

Hitchcock, M. *et al*, *Biochem. Pharmacol.*, 1967, **16**, 1801.
Lenz, D.E. *et al*, *Biochim. Biophys. Acta*, 1973, **321**, 189.
Lichtenstein, P.E. *et al*, *J. Agric. Food Chem.*, 1973, **21**, 416.

***O*-(4-Aminophenyl) *O,O*-diethyl phosphorothioate, 9CI** **A-00099**
1-Amino-4-[(diethoxyphosphinothioyl)oxy]benzene. 4-[(Diethoxyphosphinothioyl)oxy]aniline
[3735-01-1]

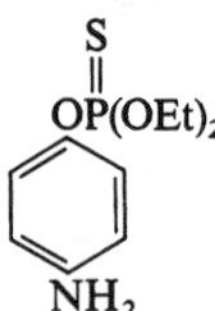

$C_{10}H_{16}NO_3PS$ M 261.275
Metab. of Parathion, P-00002 in plants and mammals, incl. humans.

Lichtenstein, E.P. *et al*, *J. Econ. Entomol.*, 1964, **57**, 618.
Katan, J. *et al*, *J. Agric. Food Chem.*, 1977, **25**, 1404 (*metab*)
Adhya, T.K. *et al*, *Pestic. Biochem. Physiol.*, 1981, **16**, 14.
Yoneyama, K. *et al*, *Pestic. Biochem. Physiol.*, 1981, **15**, 213 (*metab*)

(1-Amino-1-phenylethyl)phosphonic acid, 9CI **A-00100**
[79014-69-0]

$H_3CCPh(NH_2)P(O)(OH)_2$

$C_8H_{12}NO_3P$ M 201.161
Acid exists in dipolar form.

(±)-***form***
Cryst. (AcOH aq.). Mp 241°.
Di-Et ester: Diethyl (1-amino-1-phenylethyl)-phosphonate.
$C_{12}H_{20}NO_3P$ M 257.269
Liq.-solid. Cryst. (Me_2CO at low temp). Mp ~−10°. $Bp_{0.1}$ 132°.

Rachoń, T. *et al*, *Justus Liebigs Ann. Chem.*, 1981, 709 (*synth, ir, pmr*)

(1-Amino-2-phenylethyl)phosphonic acid, 9CI **A-00101**
[6324-00-1]

$PhCH_2CH(NH_2)P(O)(OH)_2$

$C_8H_{12}NO_3P$ M 201.161
Free acid exists in dipolar form.

(±)-*form*
Solid. Mp 268-270°.
Di-Et ester, N-*benzoyl: Diethyl (1-benzoylamino-2-phenylethyl)phosphonate.*
$C_{19}H_{24}NO_4P$ M 361.377
Prisms (C_6H_6/hexane). Mp 156-157.5°.

Asano, S. *et al*, *Agric. Biol. Chem.*, 1973, **37**, 1193 (*derivs, ir, pmr*)
Harvey, D.J., *Org. Mass Spectrom.*, 1974, **9**, 111 (*bistrimethylsilyl ester, ms*)
Stec, W.J. *et al*, *J. Org. Chem.*, 1976, **41**, 3757 (*synth, P nmr, pmr*)
Gancarz, R. *et al*, *Synthesis*, 1977, 625 (*synth*)
Kudzin, K.H. *et al*, *Synthesis*, 1978, 469 (*synth*)

(Aminophenylmethyl)phenylphosphinic acid, 9CI A-00102

(*α-Aminobenzyl*)*phenylphosphinic acid*
[25891-89-8]

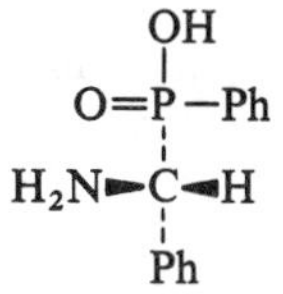

(*R*)$_C$-*form*

$C_{13}H_{14}NO_2P$ M 247.233
(*R*)$_C$-*form* [74841-98-8]
$[\alpha]^{20}_D$ −28.8° (c, 0.9 in HCl aq.).
Et ester: [74841-97-7]. Mp 85-89°. $[\alpha]^{20}_D$ +14.0° (c, 5.2 in EtOH).
(±)-*form* [74806-16-9]
Cryst. + $1H_2O$ (EtOH aq.). Mp 246-247° (234-235°).
Et ester: Ethyl (aminophenylmethyl)phenylphosphinate.
$C_{15}H_{18}NO_2P$ M 275.286
Cryst. (EtOH). Mp 183-184°.
N-*Benzyl: [(Phenylmethylamino)phenylmethyl]-phenylphosphinic acid. (α-Benzylaminobenzyl)-phenylphosphinic acid.*
$C_{20}H_{20}NO_2P$ M 337.357
Solid. Mp 252-253°.

Tyka, R., *Tetrahedron Lett.*, 1970, 677 (*synth*)
Belov, Yu.P. *et al*, *Izv. Akad. Nauk SSSR, Ser. Khim.*, 1980, 1125 (*Engl. transl.* p. 932) (*synth, resoln, cryst struct*)

(2-Aminophenyl)methylphosphinic acid, 9CI A-00103

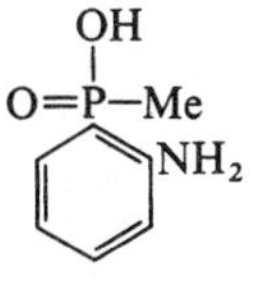

$C_7H_{10}NO_2P$ M 171.135
(−)-*Menthyl ester:* (−)-*Menthyl (2-aminophenyl)-methylphosphinate.* $[\alpha]_D$ −40.8° (c, 2.5 in MeOH) (80% o.p.).

Horner, L. *et al*, *Phosphorus Sulfur*, 1978, **4**, 155 (*props, pmr*)

(4-Aminophenyl)methylphosphinic acid, 9CI A-00104

$C_7H_{10}NO_2P$ M 171.135
Cryst. (EtOH aq.). Mp 167-168°.

Mastalerz, P., *Rocz. Chem.*, 1963, **37**, 187; *CA*, **59**, 6435.

(Aminophenylmethyl)phosphonic acid, 9CI A-00105

(*α-Aminobenzyl*)*phosphonic acid*
[18108-22-0]

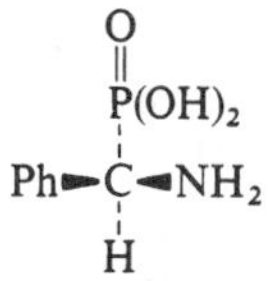

(*R*)-*form*

$C_7H_{10}NO_3P$ M 187.135
Free acid exists in dipolar form.
(*R*)-*form* [37714-05-9]
L-*form*
Mp 280-282°. $[\alpha]^{20}_D$ +18° (1 in 1M NaOH aq.).
(*S*)-*form* [37714-06-0]
D-*form*
Mp 278-279°. $[\alpha]^{20}_D$ −18° (1 in 1M NaOH aq.).
(±)-*form* [59284-70-7]
Cryst. (MeOH aq. or EtOH aq.). Mp 299° (268-220°, 280-282°). pK_{a1} 1.80, pK_{a2} 5.60, pK_{a3} 9.50 (H_2O).
B,HCl: Cryst. (MeOH aq.). Mp 287-289°.
Mono-Et ester: Ethyl hydrogen (α-aminobenzyl)-phosphonate.
$C_9H_{14}NO_3P$ M 215.188
Solid. Mp 247°.
Di-Et ester: [42077-97-4]. *Diethyl (aminophenylmethyl)phosphonate.*
$C_{11}H_{18}NO_3P$ M 243.242
Cryst. (C_6H_6/EtOH or C_6H_6). Mp 134-135°, Mp 147-148°. $Bp_{0.001}$ 98-100°. n_D^{20} 1.5160.
Di-Et ester; B,HCl: [42077-94-1]. Mp 158° dec.
Di-Ph ester: Diphenyl (aminophenylmethyl)-phosphonate.
$C_{19}H_{18}NO_3P$ M 339.330
Solid. Mp 63-65°.
Di-Ph ester; B,HBr: Mp 194-195°.
N-*Benzoyl: [(Benzoylamino)phenylmethyl]phosphonic acid. (Benzamidophenylmethyl)phosphonic acid. [α-(Benzamido)benzyl]phosphonic acid.*
$C_{14}H_{14}NO_4P$ M 291.243
Mp 103.5-129°.
N-Benzyl: *see [Phenyl[(phenylmethyl)amino]methyl]-phosphonic acid, P-00188*

Kreutzkamp, N. *et al*, *Justus Liebigs Ann. Chem.*, 1959, **623**, 103 (*ester*)
Rogozhin, S.V. *et al*, *Izv. Akad. Nauk SSSR, Ser. Khim.*, 1973, 955 (*diethyl ester, resoln*)
Harvey, D.J. *et al*, *Org. Mass Spectrom.*, 1974, **9**, 111 (*derivs, ms*)
Glowiak, T. *et al*, *Tetrahedron Lett.*, 1977, 3965 (*config*)
Luksco, T. *et al*, *Synthesis*, 1977, 239 (*synth, ester*)
Hoffmann, M., *Pol. J. Chem.*, 1978, **52**, 851 (*resoln, derivs*)
Oleksyszyn, J. *et al*, *Synthesis*, 1978, 479 (*synth*)
Kudzin, Z.H. *et al*, *Synthesis*, 1980, 1028 (*synth, ester, ms, pmr, nmr*)
Rachón, J. *et al*, *Justus Liebigs Ann. Chem.*, 1981, 709 (*synth, ir, pmr*)

[(2-Aminophenyl)methyl]phosphonic acid, 9CI A-00106

o-*Aminobenzylphosphonic acid. 2-(Phosphonomethyl)-aniline*

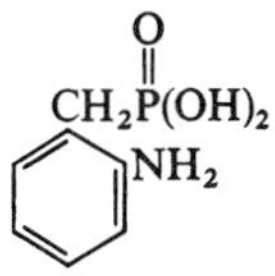

$C_7H_{10}NO_3P$ M 187.135

Di-Me ester: [54006-10-9]. *Dimethyl* [(*2-aminophenyl*)-*methyl*]*phosphonate.*
$C_9H_{14}NO_3P$ M 215.188
Solid. Mp 78-80°.

Dibutyl ester: Dibutyl [(*2-aminophenyl*)*methyl*]-*phosphonate.*
$C_{15}H_{26}NO_3P$ M 299.349
Liq. $Bp_{0.05}$ 148-151°.

Issleib, K. *et al*, *Synth. React. Inorg. Metal-Org. Chem.*, 1974, **4**, 191.
Collins, D.J. *et al*, *Tetrahedron Lett.*, 1982, **23**, 1117.

[(4-Aminophenyl)methyl]phosphonic acid, 8CI — A-00107

p-*Aminobenzylphosphonic acid.* 4-(*Phosphonomethyl*)-*aniline*

$C_7H_{10}NO_3P$ M 187.135
Cryst. Mp 324°.

Di-Me ester: Dimethyl [(*4-aminophenyl*)*methyl*]-*phosphonate. Dimethyl* p-*aminobenzylphosphonate.*
$C_9H_{14}NO_3P$ M 215.188
Cryst. (C_6H_6/pet. ether). Mp 103°.

Di-Et ester: [20074-79-7]. *Diethyl* [(*4-aminophenyl*)-*methyl*]*phosphonte. Diethyl* p-*aminobenzylphosphonate.*
$C_{11}H_{18}NO_3P$ M 243.242
Employed in affinity-chromatogr. sepn. of alkaline phosphatase from human liver. Solid. Mp 91-94°.

▷SZ6480000.

Kreutzkamp, N. *et al*, *Arch. Pharm.* (*Weinheim, Ger.*), 1961, **294**, 49 (*synth*)
Ernst, L., *Org. Magn. Reson.*, 1977, **9**, 35 (*ester, cmr*)
Seargeant, L.E. *et al*, *J. Chromatogr.*, 1979, **173**, 101 (*ester, use*)
Rewcastle, G.W. *et al*, *J. Med. Chem.*, 1982, **25**, 1231 (*synth*)

(4-Aminophenyl)phosphinic acid, 9CI — A-00108

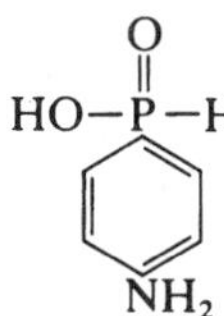

$C_6H_8NO_2P$ M 157.108
Tautomeric with (4-aminophenyl)phosphonous acid, but free acid probably exists in the phosphinic acid form. Shows antibacterial activity, less active than sulfanilamide against *E. coli*. Solid. Mp 169°. pK_a 3.7.

Hydrate: Mp 210°.

Klotz, I.M. *et al*, *J. Am. Chem. Soc.*, 1947, **69**, 473.

(2-Aminophenyl)phosphonic acid, 9CI — A-00109

o-*Phosphonoaniline*
[7472-16-4]

$C_6H_8NO_3P$ M 173.108
Cryst. (EtOH or EtOH aq.). Mp 199-200°. pK_{a1} <1, pK_{a2} 4.0, pK_{a3} 7.29 (H_2O), 8.34 (50% EtOH aq.). Br_2 aq. → 2,4,6-Tribromoaniline.

▷SZ6561200.

Mono-Me ester: [61107-67-3]. *Methyl hydrogen* (*2-aminophenyl*)*phosphonate.*
$C_7H_{10}NO_3P$ M 187.135
Cryst. (2-propanol). Mp 153-154°.

Di-Me ester: Dimethyl (*2-aminophenyl*)*phosphonate.*
$C_8H_{12}NO_3P$ M 201.161
$Bp_{0.25}$ 117-118°.

Mono-Et ester: [61107-65-1]. *Ethyl hydrogen* (*2-aminophenyl*)*phosphonate.*
$C_8H_{12}NO_3P$ M 201.161
Cryst. (EtOAc). Mp 78-79°.

Di-Et ester: [31238-50-3]. *Diethyl* (*2-aminophenyl*)-*phosphonate.*
$C_{10}H_{16}NO_3P$ M 229.215
Liq. $Bp_{0.01}$ 124-128°.

Freedman, L.D. *et al*, *J. Org. Chem.*, 1964, **29**, 2450 (*synth*)
Obrycki, R. *et al*, *J. Org. Chem.*, 1968, **33**, 632 (*ester, ir*)
Zhadanov, B.V. *et al*, *Dokl. Akad. Nauk SSSR, Ser. Sci. Khim.* (*Phys. Chem.*), 1973, **208**, 124 (*Engl. transl.* p. 9) (*props, uv, ir*)
Naylor, R.A. *et al*, *J. Chem. Soc., Perkin Trans. 2*, 1976, 1908 (*ester*)
Dolzhnikova, E.N. *et al*, *Zh. Obshch. Khim.*, 1978, **48**, 525 (*Engl. transl.* p. 474) (*synth, esters*)

(3-Aminophenyl)phosphonic acid, 9CI — A-00110

m-*Phosphonoaniline*
[5427-30-5]

$C_6H_8NO_3P$ M 173.108
Greyish microcryst. powder. Mp 290° dec. pK_a 5.66 (H_2O), 6.43 (50% EtOH aq.).

▷SZ6561100.

Kosolapoff, G.M., *J. Am. Chem. Soc.*, 1948, **70**, 3465 (*synth*)
Williams, A. *et al*, *J. Chem. Soc., Perkin Trans. 2*, 1973, 25 (*synth*)

(4-Aminophenyl)phosphonic acid, 9CI — A-00111

Phosphanilic acid. p-*Phosphonoaniline*
[5337-17-7]

$C_6H_8NO_3P$ M 173.108
Mp 245°, 254-256°. pK_a 6.03 (H_2O).

▷SZ6561300.

Mono-Me ester: Methyl hydrogen (*4-aminophenyl*)-*phosphonate.*
$C_7H_{10}NO_3P$ M 187.135
Cryst. (MeOH). Mp 196.5-197°. pK_a 3.8.

Di-Me ester: Dimethyl (*4-aminophenyl*)*phosphonate.*
$C_8H_{12}NO_3P$ M 201.161
Solid. Mp 107-108°.

Mono-Et ester: Ethyl hydrogen (*4-aminophenyl*)-*phosphonate.*
$C_8H_{12}NO_3P$ M 201.161
Cryst. (MeOH). Mp 208-210°. pK_a 3.9.

Di-Et ester: [42822-57-1]. *Diethyl* (*4-aminophenyl*)-*phosphonate.*
$C_{10}H_{16}NO_3P$ M 229.215
Solid. Mp 120-122°.

Mono-Ph ester: Phenyl hydrogen (*4-aminophenyl*)-*phosphonate.*
$C_{12}H_{12}NO_3P$ M 249.205
Cryst. (MeOH). Mp 245.5-246°. pK_a 4.0.

Di-Ph ester: [85599-24-2]. *Diphenyl* (*4-aminophenyl*)-*phosphonate.*
$C_{18}H_{16}NO_3P$ M 325.303
Solid. Mp 130-132°.

Bis(*trimethylsilyl*) *ester:* [99136-07-9]. *Bis*(*trimethylsilyl*)(*4-aminophenyl*)*phosphonate.*
$C_{12}H_{24}NO_3PSi_2$ M 317.471

Liq. $Bp_{0.0001}$ 150°.

Amide: see P-(4-Aminophenyl)phosphonic diamide, A-00112

N,N-*Di-Me: see (4-Dimethylaminophenyl)phosphonic acid, D-00706*

Doak, G.O. *et al, J. Am. Chem. Soc.*, 1952, **74**, 753 (*synth*)
Burger, A. *et al, J. Am. Chem. Soc.*, 1957, **79**, 3575 (*monoesters*)
Obrycki, R. *et al, J. Org. Chem.*, 1968, **33**, 632 (*ester, ir*)
Williams, A. *et al, J. Chem. Soc., Perkin Trans. 2*, 1973, 25 (*synth*)
Zhadanov, B.V. *et al, Dokl. Akad. Nauk SSSR, Ser. Sci. Khim. (Phys. Chem.)*, 1973, **208**, 124 (*Engl. transl.* p. 9) (*props, ir, uv*)
Naylor, R.A. *et al, J. Chem. Soc., Perkin Trans. 2*, 1976, 1908 (*ester*)
Osuka, A. *et al, Synthesis*, 1983, 69 (*ester, ir, pmr*)
Issleib, K. *et al, Z. Anorg. Allg. Chem.*, 1985, **529**, 151 (*silyl ester, synth, P nmr*)

P-(4-Aminophenyl)phosphonic diamide, 9CI — A-00112

Phosphanilic diamide

$C_6H_{10}N_3OP$ M 171.138

N,N′-*Di-Me:* [27453-00-5]. P-*(4-Aminophenyl)*-N,N′-*dimethylphosphonic diamide. Phosphanilic dimethylamide.* Shows antibacterial activity against *Salmonella schottmuelleri*. Mp 152-154°.

N,N′-*Di-Ph:* P-*(4-Aminophenyl)*-N,N′-*diphenylphosphonic diamide. Phosphanilic dianilide.*
$C_{18}H_{18}N_3OP$ M 323.333
Solid. Mp 210-213°.

Doak, G.O. *et al, J. Am. Chem. Soc.*, 1954, **76**, 1621 (*synth*)
Fr. Pat., 1 573 919 (*1967*); *CA*, **73**, 35513

(2-Aminophenyl)phosphoramidic acid, 9CI, 8CI — A-00113

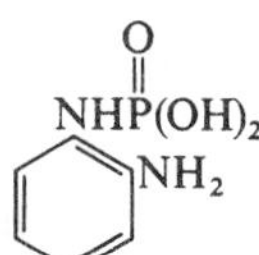

$C_6H_9N_2O_3P$ M 188.122

Mono-Ph ester: [13795-84-1]. *Phenyl hydrogen (2-aminophenyl)phosphoramidate.*
$C_{12}H_{13}N_2O_3P$ M 264.220
Cryst. (EtOH). Mp 160°. Forms a sesquihydrate, Mp 175-176° (EtOH), and a hemihydrate, Mp 155-156°.

Edmundson, R.S., *J. Chem. Soc. (C)*, 1969, 2730; 1971, 3614 (*synth, ir*)

(3-Aminophenyl)phosphoramidic acid — A-00114

$C_6H_9N_2O_3P$ M 188.122

Di-Ph ester: Diphenyl (3-aminophenyl)-phosphoramidate.
$C_{18}H_{17}N_2O_3P$ M 340.318
Cryst. (EtOH). Mp 166-168°.

Edmundson, R.S., *J. Chem. Soc. (C)*, 1971, 3614 (*synth, ir*)

(4-Aminophenyl)phosphoramidic acid — A-00115

$C_6H_9N_2O_3P$ M 188.122

Di-Me ester: Dimethyl (4-aminophenyl)-phosphoramidate.
$C_8H_{13}N_2O_3P$ M 216.176
Cryst. (MeOH/EtOAc). Mp 125°.

Rewcastle, G.W. *et al, J. Med. Chem.*, 1984, **27**, 1053 (*ester*)

2-Amino-3-phosphonobutanoic acid, 10CI — A-00116

[72217-80-2]

$(HO)_2P(O)CH(CH_3)CH(NH_2)COOH$

$C_4H_{10}NO_5P$ M 183.100

No phys. props. reported.

Varlet, J.M. *et al, Can. J. Chem.*, 1979, **57**, 3216 (*synth*)

2-Amino-4-phosphonobutanoic acid, 9CI — A-00117

[6323-99-5]

(*R*)-*form*

$C_4H_{10}NO_5P$ M 183.100

Inhibits glutamine synthetase and glutamate dehydrogenase.

(*R*)-*form* [78739-01-2]
D-form
Cryst. (EtOH aq.). Mp 229-231°. $[\alpha]^{25}_{546}$ −8.8° (H_2O).

(*S*)-*form* [23052-81-5]
L-form
$[\alpha]^{25}_{546}$ +5.7° (H_2O).

(±)-*form* [20263-07-4]
N-*Benzoyl: 2-Benzoylamino-4-phosphonobutanoic acid.*
$C_{11}H_{14}NO_6P$ M 287.208
Cryst. (dioxan aq.). Mp 197°.

Chalmers, J.R. *et al, J. Org. Chem.*, 1964, **29**, 832 (*synth, deriv*)
Varlet, J.M. *et al, Tetrahedron*, 1981, **37**, 1877 (*synth, deriv, ir, pmr*)
Wasielewski, C. *et al, Synthesis*, 1981, 540 (*synth, ir, pmr*)
Croucher, M.J. *et al, Science*, 1982, **216**, 899 (*pharmacol*)

3-Amino-3-phosphonobutanoic acid, 9CI — A-00118

[73870-70-9]

$C_4H_{10}NO_5P$ M 183.100

(±)-*form*
Solid. Mp 236-237°.

Me ester: Methyl 3-amino-3-phosphonobutanoate.
$C_5H_{12}NO_5P$ M 197.127
Mp 231-232° dec.

Amide: 3-Amino-3-phosphonobutanamide.
$C_4H_{11}N_2O_4P$ M 182.116

Solid. Mp 235-237° dec.

Gruszecka, E. *et al*, *Pol. J. Chem.*, 1979, **53**, 2327 (*synth, ir, pmr*)
Oleksyszyn, J. *et al*, *Monatsh. Chem.*, 1982, **113**, 59 (*synth, derivs, pmr*)

4-Amino-4-phosphonobutanoic acid, 9CI A-00119

[18865-31-1]

$(HO)_2P(O)CH(NH_2)CH_2CH_2COOH$

$C_4H_{10}NO_5P$ M 183.100

Free acid probably exists in dipolar form.

(±)-form

Solid; cryst (2-propanol aq.). Mp 167-169°, Mp 185-187°.

Me ester: Methyl 4-amino-4-phosphonobutanoate.
$C_5H_{12}NO_5P$ M 197.127
Solid. Mp 174-175°.

Khomutov, R.M. *et al*, *Izv. Akad. Nauk SSSR, Ser. Khim.*, 1979, **28**, 2118 (*Engl. transl.* p. 1949) (*synth*)
Oleksyszyn, J. *et al*, *Monatsh. Chem.*, 1982, **113**, 59 (*synth, deriv, pmr*)

2-Amino-3-phosphonooxybutanoic acid A-00120

Threonine phosphate. Threonine dihydrogen phosphate. O-Phosphothreonine

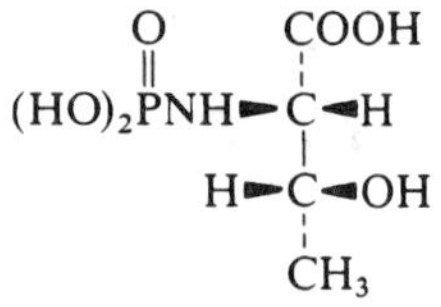

```
   O       COOH
   ‖        |
(HO)2PNH ▸ C ◂ H
            |
       H ▸ C ◂ OH
            |
           CH3
```

$C_4H_{10}NO_6P$ M 199.100

(2S,3R)-form [1114-81-4]

L-form

Formed in many cellular processes. Mp 194° dec. $[\alpha]_D^{24}$ −7.4° (c, 2.58 in H_2O).

(2RS,3SR)-form [27530-80-9]

Solid. Mp 150-152° dec.

P,P-*Di-Ph ester*, C-*Et ester: Ethyl 2-amino-3-[(diphenoxyphosphinyl)oxy]butanoate.*
$C_{18}H_{22}NO_6P$ M 379.349
Solid (as hydrobromide). Mp 88-89° (hydrobromide).

P,P-*Di-Ph ester*, C-*Et ester*, N-*benzyloxycarbonyl: Ethyl 2-(benzyloxycarbonyl)amino-3-[(diphenoxyphosphinoyl)oxy]butanoate.*
$C_{26}H_{28}NO_8P$ M 513.483
Cryst. (Et_2O/pet. ether). Mp 56-57°.

De Verdier, C.-H., *Acta Chem. Scand.*, 1953, **7**, 196 (*isol, synth*)
Riley, G. *et al*, *J. Chem. Soc.*, 1957, 1373.
Dubra, M.S. *et al*, *J. Chromatogr.*, 1982, **250**, 124 (*uv, tlc*)
Bolton, P.H. *et al*, *J. Magn. Reson.*, 1983, **52**, 326 (*pmr*)
Tuy, P.D. *et al*, *Nature (London)*, 1983, **305**, 435

2-Amino-3-phosphonopropanoic acid A-00121

3-Phosphonoalanine, 9CI

[5652-28-8]

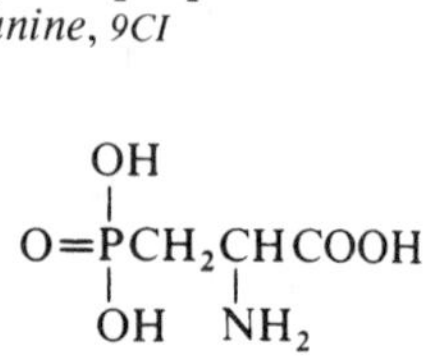

```
   OH
   |
O=PCH2CHCOOH
   |    |
   OH   NH2
```

$C_3H_8NO_5P$ M 169.074

Constit. of *Zoanthus sociatus* and the cilate, *Tetrahymena pyriformis*.

(±)-form [20263-06-3]

Solid (EtOH aq.). Mp 226° dec.

N-*Benzoyl:* Cryst. (H_2O or dioxan). Mp 197°.

Chambers, J.R. *et al*, *J. Org. Chem.*, 1964, **29**, 832 (*synth*)
Kittredge, J.S. *et al*, *Biochemistry*, 1964, **3**, 991 (*isol*)
Quin, L.D., *Top. Phosphorus Chem.*, 1967, **4**, 23 (*rev*)
Soroka, M. *et al*, *Rocz. Chem.*, 1976, **50**, 661 (*synth, ir, pmr*)
Varlet, J.-M. *et al*, *Can. J. Chem.*, 1979, **57**, 3216 (*synth, pmr*)
Varlet, J.-M. *et al*, *Tetrahedron*, 1981, **37**, 1377 (*synth, pmr*)
Sawka-Dubrowalska, W. *et al*, *Acta Crystallogr., Sect. C*, 1985, **41**, 453 (*cryst struct*)

3-Amino-3-phosphonopropanoic acid A-00122

[5652-40-4]

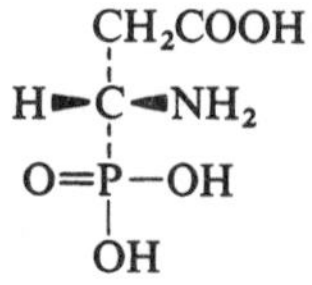

```
     CH2COOH
      |
  H ▸ C ◂ NH2
      |
   O=P—OH         (R)-form
      |
      OH
```

$C_3H_8NO_5P$ M 169.074

(R)-form [85249-44-1]

Solid. Mp 234-237°. $[\alpha]_D^{25}$ −32.6° (c, 1 in H_2O).

Me ester: [85207-39-2]. *Methyl 3-amino-3-phosphonopropanoate.*
$C_4H_{10}NO_5P$ M 183.100
Solid. Mp 156-160°. $[\alpha]_D^{25}$ −26.6° (c, 1 in H_2O).

Amide: (3-Amino-3-phosphono)propanoamide. Solid. Mp 268-273°. $[\alpha]_D^{25}$ −33.0° (c, 1 in H_2O).

(±)-form

Solid. Mp 224° dec.

Me ester: Cryst. (MeOH aq.). Mp 236-237° dec.

Amide:
$C_3H_9N_2O_4P$ M 168.089
Cryst. (EtOH aq.). Mp 247-252° dec.

Tri-Me ester: Methyl [3-amino-3-(dimethoxyphosphinyl)]propanoate.
$C_6H_{14}NO_5P$ M 211.154
Oil.

P,P-*Di-Et*, C-*Me triester*, N,N-*di-Me: Methyl [3-dimethylamino-3-(diethoxyphosphinyl)]propanoate.*
$C_{10}H_{22}NO_5P$ M 267.261
Liq. $Bp_{0.01}$ 98-100°. n_D^{20} 1.4465.

Soroka, M. *et al*, *Rocz. Chem.*, 1976, **50**, 661 (*synth, derivs, ir, pmr*)
Siatecki, Z. *et al*, *Org. Magn. Reson.*, 1981, **17**, 172 (*pmr, P nmr*)
Burgada, R. *et al*, *Phosphorus Sulfur*, 1982, **13**, 85 (*deriv, pmr, P nmr*)
Campbell, M.M. *et al*, *Tetrahedron*, 1982, **38**, 2513 (*synth, ir, ester, pmr*)
Vasella, A. *et al*, *Helv. Chim. Acta*, 1982, **65**, 1953 (*derivs, ir, cmr, pmr*)

2-Amino-5-phospho-3-pentenoic acid A-00123

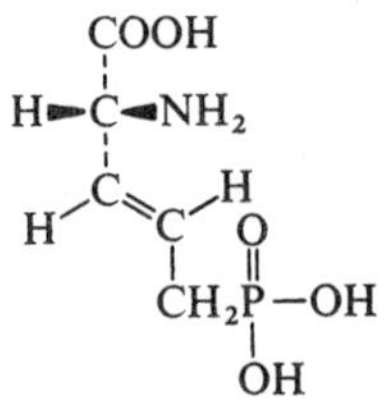

```
      COOH
       |
   H ▸ C ◂ NH2
       |
       C    H
     /  \\ /
    H    C    O
         |    ‖
        CH2P—OH
              |
              OH
```

$C_5H_{10}NO_5P$ M 195.111

(R)-form [60978-99-6]

D-form

Prod. by *Streptomyces plumbeus*. Threonine antagonist. Plates (H_2O). Sol. H_2O, prac. insol. org. solvs. Mp 183-185° dec. $[\alpha]_D^{25}$ +51.4° (c, 1.07 in H_2O). Occurs as a tripeptide with Alanine and Asparagine called N-1409.

Park, B.K. *et al*, *Agric. Biol. Chem.*, 1976, **40**, 1905.

(1-Aminopropylidene)bisphosphonic acid, 9CI **A-00124**

Propylamine-1,1-diphosphonic acid

[15049-86-2]

$H_3CCH_2C(NH_2)[P(O)(OH)_2]_2$

$C_3H_{11}NO_6P_2$ M 219.071

Solid. Mp 249-250° dec.

Plöger, W. *et al*, *Z. Anorg. Allg. Chem.*, 1972, **389**, 119 (*synth*)
Fukuda, M. *et al*, *Chem. Lett.*, 1977, 529 (*nmr*)

(1-Aminopropyl)phosphonic acid, 9CI **A-00125**

[14047-23-5]

$H_3CCH_2CH(NH_2)P(O)(OH)_2$

$C_3H_{10}NO_3P$ M 139.091

Free acid exists in dipolar form.

(±)-***form*** [16606-64-7]

Cryst. (MeOH aq.). Mp 285-286° (267-269°, 276-277°). pK_{a1} 1.95, pK_{a2} 5.75 (5.67), pK_{a3} 10.28 (25°, H_2O).

Di-Et ester: Diethyl (1-aminopropyl)phosphonate.

$C_7H_{18}NO_3P$ M 195.198

Liq. $Bp_{0.05}$ 100°.

N-*Benzoyl: [1-(*N*-Benzoylamino)propyl]phosphonic acid.*

$C_{10}H_{14}NO_4P$ M 243.199

Solid. Mp 183-185°.

Di-Et ester, N-*benzoyl: Diethyl [1-(*N*-benzoylamino)-propyl]phosphonate.*

$C_{14}H_{22}NO_4P$ M 299.306

Prisms (C_6H_6/hexane). Mp 114-115°.

N-*Benzyl: [1-(Benzylamino)propyl]phosphonic acid.*

$C_{10}H_{16}NO_3P$ M 229.215

Solid. Mp 222-224°.

Berry, J.P. *et al*, *J. Org. Chem.*, 1972, **37**, 4396 (*synth*)
Asano, S. *et al*, *Agric. Biol. Chem.*, 1973, **37**, 1193 (*synth, ir, derivs*)
Harvey, D.J. *et al*, *J. Chromatogr.*, 1973, **79**, 65 (*derivs, glc, ms*)
Harvey, D.J. *et al*, *Org. Mass Spectrom.*, 1974, **9**, 111 (*derivs, ms*)
Stec, W.J. *et al*, *J. Org. Chem.*, 1976, **41**, 3757 (*synth, pmr, P nmr*)
Wozniak, M. *et al*, *Talanta*, 1979, **26**, 1135 (*complexes*)
Kudzin, Z.H. *et al*, *Synthesis*, 1980, 1028 (*synth, ester, pmr, nmr, ms*)
Rachon, J. *et al*, *Justus Liebigs Ann. Chem.*, 1981, 709 (*synth, deriv, ir, pmr*)

(2-Aminopropyl)phosphonic acid, 9CI **A-00126**

[28660-33-5]

$H_3CCH(NH_2)CH_2P(O)(OH)_2$

$C_3H_{10}NO_3P$ M 139.091

(+)-***form*** [85653-24-3]

$[\alpha]^{25}_{D}$ +4.9° (1M NaOH aq.).

(±)-***form***

Mp 278-284°.

Di-Et ester: Diethyl (2-aminopropyl)phosphonate.

$C_7H_{18}NO_3P$ M 195.198

Oil.

N-*Me: (2-Methylaminopropyl)phosphonic acid.*

$C_4H_{12}NO_3P$ M 153.117

Solid. Mp 235°.

Barycki, J. *et al*, *Tetrahedron Lett.*, 1970, 3147 (*synth*)
Varlet, J.M. *et al*, *Synth. Commun.*, 1978, **8**, 335 (*synth, ester, deriv, pmr*)
Varlet, J.M. *et al*, *Tetrahedron*, 1981, **37**, 3713 (*synth, ir, pmr, ester*)
Sauveur, F. *et al*, *Phosphorus Sulfur*, 1983, **14**, 341 (*resoln*)

(3-Aminopropyl)phosphonic acid, 9CI, 8CI **A-00127**

[13138-33-5]

$H_2NCH_2CH_2CH_2P(O)(OH)_2$

$C_3H_{10}NO_3P$ M 139.091

Free acid probably exists in dipolar form. Prisms. Mp >260°.

Di-Et ester: [4402-24-8]. *Diethyl (3-aminopropyl)-phosphonate.*

$C_7H_{18}NO_3P$ M 195.198

Liq. $Bp_{0.01-00.05}$ 68-75°.

Diisopropyl ester: Diisopropyl (3-aminopropyl)-phosphonate.

$C_9H_{22}NO_3P$ M 223.251

Liq. $Bp_{0.018}$ 74°.

N-*Ac, Di-Et ester:*

$C_9H_{20}NO_4P$ M 237.235

Liq. $Bp_{0.05}$ 125°.

Harvey, D.J. *et al*, *J. Chromatogr.*, 1973, **79**, 65 (*derivs, glc, ms*)
Collins, D.J. *et al*, *Aust. J. Chem.*, 1974, **27**, 1759 (*diethyl ester, ir, pmr, deriv*)
Cassaigne, A. *et al*, *C.R. Hebd. Seances Acad. Sci., Ser. D*, 1976, **282**, 1637 (*metab*)
Brigot, D. *et al*, *Tetrahedron*, 1979, **35**, 1345 (*synth, pmr*)
Glowiak, T. *et al*, *Acta Crystallogr., Sect. B*, 1980, **36**, 961 (*cryst struct*)
Fabre, G. *et al*, *Can. J. Chem.*, 1981, **59**, 864 (*synth, ir, pmr*)
Appleton, T.G. *et al*, *Aust. J. Chem.*, 1984, **37**, 1833 (*pmr, cmr, P nmr, struct*)

(2-Amino-4-pyrimidinyl)phosphonic acid, 9CI **A-00128**

2-Amino-4-phosphonopyrimidine

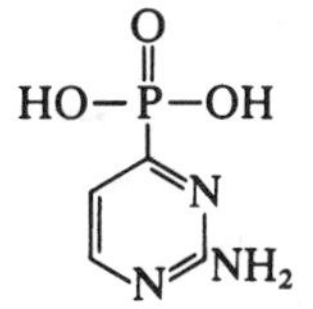

$C_4H_6N_3O_3P$ M 175.083

Solid. Subl. at 250°.

Diisopropyl ester: Diisopropyl (2-amino-4-pyrimidinyl)-phosphonate.

$C_{10}H_{18}N_3O_3P$ M 259.244

Cryst. (2-propanol aq.). Mp 163-165°.

Kosolapoff, G.M. *et al*, *J. Org. Chem.*, 1961, **26**, 1895 (*synth*)

N-(5-Amino-1-β-D-ribofuranosylimidazole-4-carbonyl)-L-aspartic acid 5′-phosphate A-00129

N-[[*5-Amino-1-(5-O-phosphono-β-D-ribofuranosyl)-1H-imidazol-4-yl]carbonyl]aspartic acid*, 9CI. N-[(*5-Amino-1-β-D-ribofuranosylimidazol-4-yl)carbonyl]-5′-(dihydrogen phosphate)-L-aspartic acid*, 8CI

[3031-95-6]

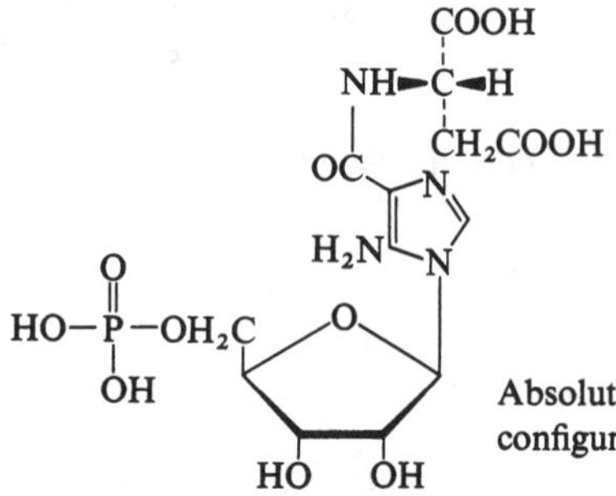

Absolute configuration

$C_{13}H_{19}N_4O_{12}P$ M 454.286

Nucleotide accumulated in mutants of *Escherichia coli* and *Salmonella typhimurium*. Important purine precursor. λ_{max} 269 (ε 11 850), 244 nm (9 580) (H_2O). Undergoes the enzyme-catalysed interconversion SAICAR ⇌ AICAR + Fumarate. In the presence of the Bratton-Marshall reagents and on cooling produces a purple chromophore with λ_{max} 550 nm.

Di-Ba salt: $[\alpha]^{21}_{D}$ −26.1° (c, 1.5 in 0.1N HCl).

Bratton, A.C. *et al*, *J. Biol. Chem.*, 1939, **128**, 537.
Lukens, L.N. *et al*, *J. Biol. Chem.*, 1959, **234**, 1806.
Burrows, I.E. *et al*, *J. Chem. Soc. (C)*, 1968, 40 (*synth*)
Brox, L.W., *Can. J. Biochem.*, 1973, **51**, 1072.

5-Amino-1-ribofuranosylimidazole-4-carboxamide 5′-(dihydrogen phosphate), 8CI A-00130

5-Amino-1-(5-O-phosphonoribofuranosyl)-1H-imidazole-4-carboxamide, 9CI. *AICAR*

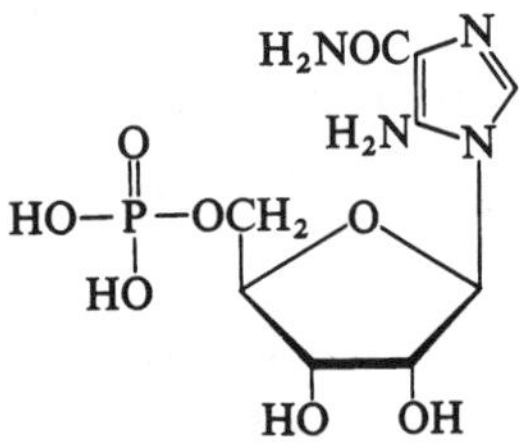

$C_9H_{15}N_4O_8P$ M 338.213

β-D-form [3031-94-5]

Nucleotide involved in the biosynthesis *de novo* of other purine nucleotides. AICAR and its nontoxic salts are potent flavour enhancers. λ_{max} 268 (ε 12 100), 245 nm (9 600) (pH 1). In the presence of Bratton-Marshall reagents, AICAR is converted to a purple dye with λ_{max} 540 nm (ε 26 400).

Ba salt: λ_{max} 269 nm (ε 12 600) (pH 7).

4-(N-Me carboxamide): λ_{max} 268 (ε 12 500), 244 nm (8 200) (pH 1); Bratton-Marshall λ_{max} 538 nm (ε 24 300).

4-(N-Di-Et carboxamide): λ_{max} 267, 244 nm (pH 1); Bratton-Marshall λ_{max} 540 nm.

Flaks, J.G. *et al*, *J. Biol. Chem.*, 1957, **228**, 201.
Lukens, L.N. *et al*, *J. Biol. Chem.*, 1959, **234**, 1791 (*biosynth*)
Shaw, E., *J. Am. Chem. Soc.*, 1961, **83**, 4770.
Burrows, I.E. *et al*, *J. Chem. Soc. (C)*, 1967, 1088 (*synth*)
U.S.P., 3 355 301, (*1967*); *CA*, **68**, 28636q

Amiprophos *M* A-00131

O-*Ethyl* O-*(4-methyl-2-nitrophenyl) (1-methylethyl)-phosphoramidothioate*, 9CI. O-*Ethyl* O-*(4-methyl-2-nitrophenyl) isopropylphosphoramidothioate*

[33857-23-7]

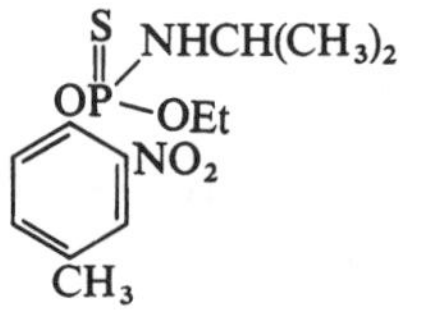

$C_{12}H_{19}N_2O_4PS$ M 318.327

Herbicide.

Pan, Z. *et al*, *CA*, 1982, **97**, 227914 (*cryst struct*)

Amiprophos-Methyl A-00132

O-*Methyl* O-*(4-methyl-2-nitrophenyl) (1-methylethyl)-phosphoramidothioate*. O-*Methyl* O-*(4-methyl-2-nitrophenyl) isopropylphosphoramidothioate*

[36001-88-4]

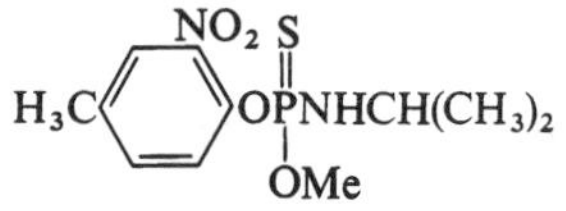

$C_{11}H_{17}N_2O_4PS$ M 304.300

▷TB5100000.

(±)-*form*

Herbicide. Has been resolved by glc.

Oi, N. *et al*, *Agric. Biol. Chem.*, 1979, **43**, 2403 (*glc*)
Ger. Pat., 2 841 881, (*1980*); *CA*, **93**, 2267 (*synth, use*)

Amiton A-00133

S-*[2-(Diethylamino)ethyl]* O,O-*diethyl phosphorothioate*, 9CI, 8CI. O,O-*Diethyl* S-*2-diethylaminoethyl thiophosphate. Tetram*

[78-53-5]

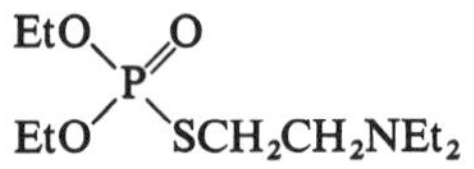

$C_{10}H_{24}NO_3PS$ M 269.338

Insecticide, no longer in widespread use. Cholinesterase inhibitor. $Bp_{0.2}$ 97° (110°), $Bp_{0.01}$ 76°. n_D^{21} 1.4743 (n_D^{25} 1.4666). More stable than thiophosphoryl isomer. At 165° → Et_3N. With H_2O at 150° → $Et_2NCH_2CH_2OH$ and $Et_2NCH_2CH_2NEt_2$.

▷TF0525000.

Hydrogen oxalate: [3734-97-2]. Cryst. (Me_2CO). Mp 98-99°.

▷TF1400000.

p-*Toluenesulfonate:* Cryst. (Me_2CO/Et_2O). Mp 105-106°.

B,PhCH₂Br: Solid. Mp 108° dec.

Fukuto, T.R. *et al*, *J. Am. Chem. Soc.*, 1957, **79**, 6083 (*synth, props*)
Calderbank, A. *et al*, *J. Chem. Soc.*, 1960, 637 (*synth*)
Cadogan, J.I.G., *J. Chem. Soc.*, 1962, 18 (*props*)
Sidky, M.M. *et al*, *Egypt. J. Chem.*, 1973, 43; *CA*, **81**, 49536.
Merck Index, 9th Ed., No. 8938.

Anilofos, BSI **A-00134**

S-[2-[4-Chlorophenyl(1-methylethyl)amino]-2-oxoethyl] O,O*-dimethyl phosphorodithioate,* 9CI. *S-4-Chloro-*N*-isopropylcarbaniloylmethyl* O,O*-dimethyl phosphorodithioate*

[64249-01-0]

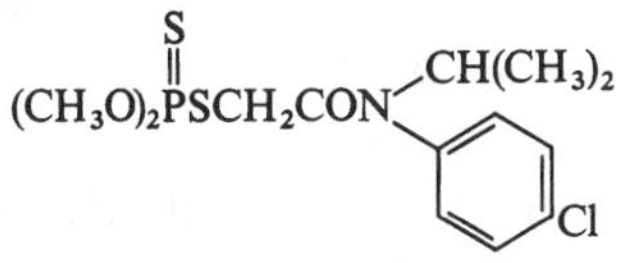

$(CH_3O)_2P(S)SCH_2CON(CH(CH_3)_2)C_6H_4Cl$

$C_{13}H_{19}ClNO_3PS_2$ M 367.845

Herbicide. Cryst. V. sol. org. solvs. Mp 50.5-52.5°.

Pesticide Manual, 7th Ed., 18.

Antibiotic BMG 59-R2 **A-00135**

BMG 59-R2

[90119-88-3]

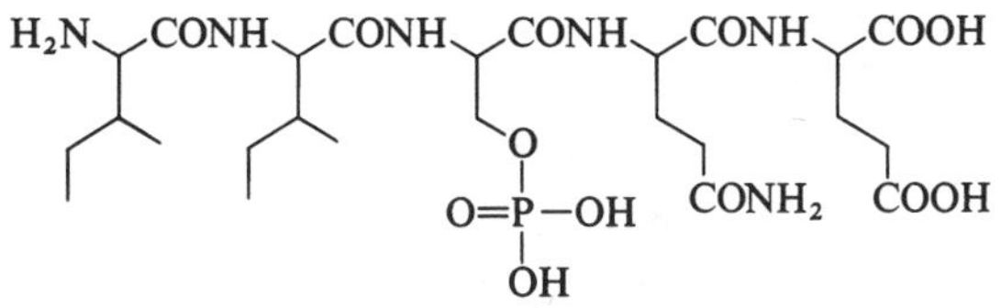

$C_{25}H_{45}N_6O_{13}P$ M 668.637

Peptide antibiotic. Prod. by *Bacillus megaterium.* Inhibits alkaline phosphatase. Tumour inhibitor. Powder.

Japan. Pat., 83 164 561, (*1983*); *CA*, **101**, 5548 (*isol, struct, props*)

Antibiotic CI 920 **A-00136**

*CI 920. PD 110161. Antibiotic PD 110161. CL 1565*A. *Antibiotic CL 1565*A

[87810-56-8]

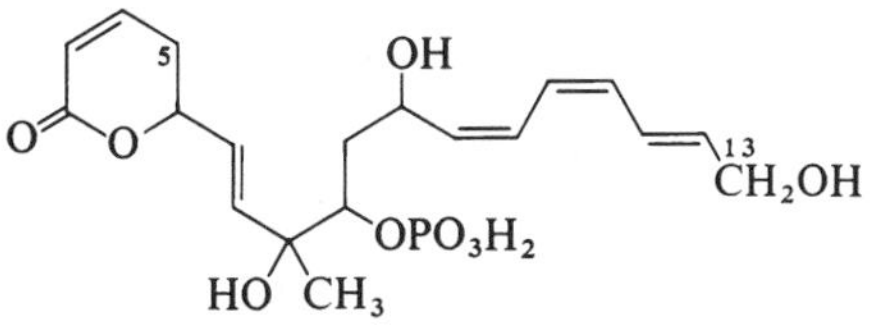

$C_{19}H_{27}O_9P$ M 430.391

Triene antibiotic. Prod. by *Streptomyces pulveraceus* sub. *fostreus.* Shows antitumour props. Also active against yeasts. Hygroscopic solid + $1\frac{1}{2}H_2O$ (as partial Na salt). $[\alpha]_D$ +33° (c, 1 in 0.1M phosphate buffer, pH 7).

13-Deoxy: [87860-37-5]. ***Antibiotic PD 113270***. *PD 113270. CL 1565*B. *Antibiotic CL 1565*B.

$C_{19}H_{27}O_8P$ M 414.391

Prod. by *S. pulveraceus* sub. *fostreus.* Antitumour agent. Solid. $[\alpha]_D$ +31° (phosphate buffer).

5-Hydroxy: [87860-38-6]. ***Antibiotic PD 113271***. *PD 113271. CL 1565*T. *Antibiotic CL 1565*T.

$C_{19}H_{27}O_{10}P$ M 446.390

Prod. by *S. pulveraceus* sub. *fostreus.* Antitumour agent. Pale-yellow solid. There is also a further component of the CL 1565 complex, Antibiotic CL 1565*C.*

Tunac, J.B. *et al, J. Antibiot.*, 1983, **36**, 1595 (*isol, props*)
Stampwala, S.S. *et al, J. Antibiot.*, 1983, **36**, 1601 (*isol*)
Hokanson, G.C. *et al, J. Org. Chem.*, 1985, **50**, 462 (*struct*)
U.S.P., 4 578 383, (*1986*); *CA*, **105**, 113609

Antibiotic FR 32863 **A-00137**

[3-(Formylhydroxyamino)-1-propenyl]phosphonic acid, 9CI. *FR 32863. FR 900136. Antibiotic FR 900136*

[66508-88-1]

$(HO)_2P(O)CH{=}CHCH_2N(OH)CHO$

$C_4H_8NO_5P$ M 181.085

Phosphonic acid antibiotic. Isol. from *Streptomyces lavendulae.* Active against gram-negative bacteria.

K salt: [66508-52-9]. Cryst. Mp 176-180°.

Hashimoto, M. *et al, Tetrahedron Lett.*, 1980, **21**, 99 (*synth*)
Okuhara, M. *et al, J. Antibiot.*, 1980, **33**, 24 (*isol*)
Kuroda, Y. *et al, J. Antibiot.*, 1980, **33**, 29 (*struct*)

Antibiotic FR 33289 **A-00138**

[3-(Acetylhydroxyamino)-2-hydroxypropyl]phosphonic acid, 10CI. *FR 33289*

[66508-89-2]

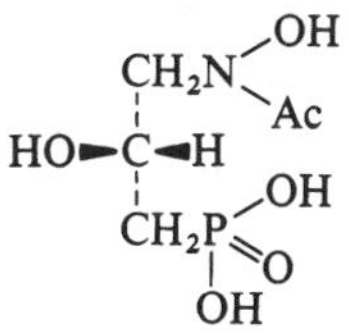

(*R*)-*form*

$C_5H_{12}NO_6P$ M 213.127

(R)-form [73240-15-0]

Produced in cultures of *Streptomyces rubellomurinus indigoferris.* Antibiotic.

Na salt: [73240-14-9]. Powder. Mp 193-194°.

Di-Me ester: [73240-16-1]. *Dimethyl [3-(acetylhydroxyimino)-2-hydroxypropyl]phosphonate.*

$C_7H_{16}NO_6P$ M 241.180

Oil. $[\alpha]^{25}_D$ −5.1° (c, 0.75 in MeOH).

Okuhara, M. *et al, J. Antibiot.*, 1980, **33**, 24 (*isol, ir, tlc, pmr*)
Kuroda, Y. *et al, J. Antibiot.*, 1980, **33**, 29 (*struct, ir, pmr, cmr*)
Hemmi, K. *et al, Chem. Pharm. Bull.*, 1981, **29**, 646 (*synth, struct, ir, cmr, pmr*)

Antibiotic I5*B* **A-00139**

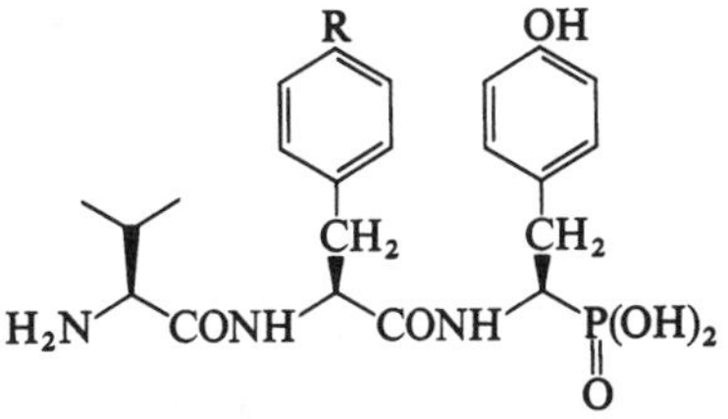

I5B1, R = H
I5B2, R = OH

Oligopeptide antibiotic complex. Inhibits angiotensin I converting enzyme and shows hypotensive activity.

Antibiotic I5B1 [84890-90-4]

I5B1. K 4. Antibiotic K 4

$C_{23}H_{32}N_3O_6P$ M 477.496

Isol. from *Actinomadura spiculospora* nov. sp. K-4 and *A.* sp. 937ZE-1. Mp >300°. $[\alpha]_D$ −110° (c, 0.075 in 0.01*M* NaOH).

Antibiotic I5B2 [93768-49-1]

I5B2

$C_{23}H_{32}N_3O_7P$ M 493.495

From *A.* sp. 937ZE-1. Inhibits angiotensin I converting enzyme. Mp >300°. $[\alpha]_D$ −55° (c, 0.0186 in 0.01*M* NaOH).

Eur. Pat., 61 172, (*1982*); *CA*, **98**, 107793 (*synth*)
Kido, Y. *et al*, *J. Antibiot.*, 1984, **37**, 965 (*isol, uv, ir, props*)

Antibiotic K 26 A-00140

K 26

[84890-91-5]

AcNH–CH(CH(CH₃)CH₂CH₃)–CONH–CH(CH₂C₆H₄OH)–CONH–CH(CH₂C₆H₄OH)–P(O)(OH)₂

$C_{25}H_{34}N_3O_8P$ M 535.533

Oligopeptide antibiotic. From *Actinomycetes* strain K-26. Shows hypotensive activity. Mp >300° (browning). $[\alpha]_D^{20}$ −4.8° (c, 0.1 in H_2O).

Kido, Y. *et al*, *J. Antibiot.*, 1984, **37**, 965 (*isol, uv, ir, props*)
Yamoto, M. *et al*, *J. Antibiot.*, 1986, **39**, 44.

Antibiotic SF 1293 A-00141

γ-(*Hydroxymethylphosphinyl*)-L-α-*aminobutyryl*-L-*alanyl*-L-*alanine. Phosphinothricylalanylalanine. Bialaphos. SF 1293*

[35597-43-4]

$$Me{-}P(O)(OH)CH_2CH_2CH(NH_2)CONHCH(CH_3)CONHCH(CH_3)COOH \quad (L, L)$$

$C_{11}H_{22}N_3O_6P$ M 323.285

Peptide antibiotic. From *Streptomyces viridochromogenes* and *S. hygroscopicus*. Antifungal antibiotic and herbicide. Amphoteric powder. Mp 159-161°. $[\alpha]_D^{25}$ −34° (c, 1 in H_2O).

▷AY6826000.

Bayer, E. *et al*, *Helv. Chim. Acta*, 1972, **55**, 224 (*isol, struct*)
Ogawa, Y. *et al*, *Meiji Seika Kenkyu Nempo*, 1973, 39, 42, 54; *CA*, **81**, 89705, 37806, 120994 (*isol, struct, synth*)
Seto, H. *et al*, *J. Antibiot.*, 1982, **35**, 1719 (*biosynth*)
Seto, H. *et al*, *Biochem. Biophys. Res. Commun.*, 1983, **111**, 1008 (*biosynth*)
Imai, S. *et al*, *J. Antibiot.*, 1985, **38**, 678 (*biosynth*)

Aphidan A-00142

S-[(*Ethylsulfinyl*)*methyl*] O,O-*bis*(*1-methylethyl*) *phosphorodithioate*, 9CI. S-*Ethylsulphinylmethyl* O,O-*diisopropyl phosphorodithioate. IPSP*

[5827-05-4]

$$[(H_3C)_2CHO]_2P(S)SCH_2S(O)Et$$

$C_9H_{21}O_3PS_3$ M 304.417

Agricultural systemic insecticide. Liq. Sl. sol. H_2O, misc. most org. solvs. except hexane. d_4^{20} 1.166. $n_D^{16.5}$ 1.5278.

▷TE4200000.

U.S.P., 3 408 426, (*1963*); *CA*, **65**, 14362f (*synth*)
Pesticide Manual, 6th Ed., 252; 7th Ed., 251.

β-Aspartyl phosphate A-00143

L-*Aspartic acid 4-monoanhydride with phosphoric acid*, 9CI

[22138-53-0]

H_2N–C(COOH)(H)–$CH_2COOPO_3H_2$

Absolute configuration

$C_4H_8NO_7P$ M 213.083

Biological intermediate in the synthesis of aspartic β-semialdehyde and homoserine. Formed enzymatically from aspartic acid and ATP by aspartokinase. Oil.

Katchalsky, A. *et al*, *J. Am. Chem. Soc.*, 1954, **76**, 6042 (*synth*)
Block, S. *et al*, *J. Biol. Chem.*, 1955, **213**, 27 (*synth*)

Aspon A-00144

Tetrapropyl thiodiphosphate, 9CI. O,O,O′,O′-*Tetrapropyl dithiopyrophosphate*. O,O-*Dipropyl phosphorothioate anhydride*

[3244-90-4]

$$(H_3CCH_2CH_2O)_2P(S)OP(S)(OCH_2CH_2CH_3)_2$$

$C_{12}H_{28}O_5P_2S_2$ M 378.418

Turf insecticide. Amber liq. Spar. sol. H_2O, pet. ether. d_4^{25} 1.12. $Bp_{0.1}$ 104°. n_D^{21} 1.4710, n_D^{25} 1.4712.

▷XN4550000.

Toy, A.D.F., *J. Am. Chem. Soc.*, 1951, **73**, 4670 (*synth*)
Babad, H. *et al*, *Anal. Chim. Acta*, 1968, **41**, 259 (*pmr*)
Turpin, R. *et al*, *Bull. Soc. Chim. Fr., Part I*, 1977, 999 (*synth*)
Pesticide Manual, 6th Ed., 507; 7th Ed., 521.

6-Azauridine 5′-phosphate A-00145

2-[5-O-(*Phosphonooxy*)-β-D-*ribofuranosyl*]-*1,2,4-triazine-3,5*(2H,4H)-*dione*, 9CI. 2-β-D-*Ribofuranosyl-5′-*(*dihydrogen phosphate*)-as-*triazine-3,5*(2H,4H)-*dione*, 8CI. *6-Azauridylic acid. AzaUMP*

[2018-19-1]

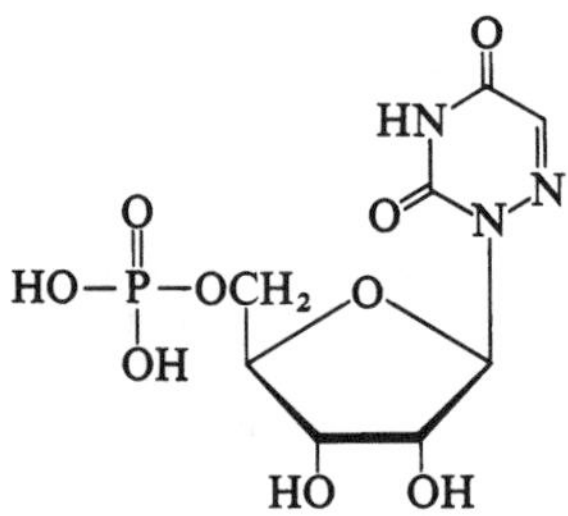

$C_8H_{12}N_3O_9P$ M 325.171

Formed by the action of *Sphaerotheca fuliginea* on Azauracil. Biologically active intermediate, essential for antitumour activity of Azauridine. Cryst. + ½H_2O (EtOH/Et_2O). Mp 139-141° (147° dec.). λ_{max} 261 nm (ε 6 100) (pH 2).

Dicyclohexylammonium salt: Mp 189-190°.

Smrt, J. *et al*, *Collect. Czech. Chem. Commun.*, 1960, **25**, 130 (*synth*)
Van't Land, B.G. *et al*, *Neth. J. Plant Pathol.*, 1972, **78**, 242; *CA*, **79**, 112267e (*synth*)
Saenger, W. *et al*, *Nature (London)*, 1973, **242**, 610 (*cryst struct*)
Wood, D.J. *et al*, *Can. J. Chem.*, 1973, **51**, 2571 (*conformn, pmr*)

Azemethephos, BSI **A-00146**

S-[(*6-Chloro-2-oxooxazolo*[*4,5*-b]*pyridin-3*(*2*H)-*yl*]-*methyl* O,O-*dimethyl phosphorothioate.* S-*6-Chloro-2,3-dihydro-2-oxooxazolo*[*4,5*-b]*pyridin-3-ylmethyl* O,O-*dimethyl phosphorothioate*

[35575-96-3]

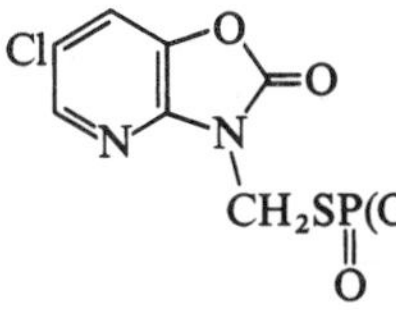

$C_9H_{10}ClN_2O_5PS$ M 324.675

Broad spectrum contact and stomach insecticide and acaricide. Cryst. Sl. sol. H_2O. Mp 89°. Unstable to strong acids and alkalis.

Egli, H., *J. Agric. Food Chem.*, 1982, **30**, 861 (*props*)
Pesticide Manual, 7th Ed., 24.
The Agrochemicals Handbook, Royal Society of Chemistry, 1983, A024.

1-Azetidinephosphonic acid, 8CI **A-00147**

1-Phosphonoazetidine. 1-Azetidinylphosphonic acid, 9CI

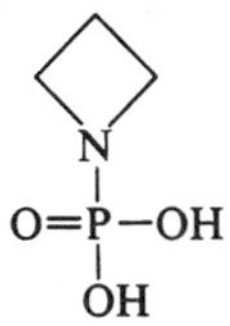

$C_3H_8NO_3P$ M 137.075

Di-Me ester: Dimethyl 1-azetidinylphosphonate.
$C_5H_{12}NO_3P$ M 165.128
Liq. Bp_{11} 108-109°, Bp_2 66°.

Grechkin, N.P. *et al*, *Dokl. Akad. Nauk SSSR, Ser. Sci. Khim.*, 1965, **162**, 1063 (*synth*)
Buchanan, G.W. *et al*, *Can. J. Chem.*, 1979, **59**, 21 (*synth*)
Gray, G.A. *et al*, *J. Org. Chem.*, 1979, **44**, 1768 (*N and P nmr*)
Duangthai, S. *et al*, *Org. Magn. Reson.*, 1982, **20**, 33 (*nmr, struct*)

2-Azido-1,3,2-benzodioxaphosphole, 9CI, 8CI **A-00148**

o-*Phenylenedioxyphosphinyl azide*

[79343-17-2]

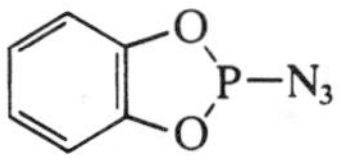

$C_6H_4N_3O_2P$ M 181.090

Oil. Dec. at 15° → resin.

2-Oxide: [85741-82-8]. o-*Phenylene phosphoryl azide.*
$C_6H_4N_3O_3P$ M 197.090
Deliquescent needles. $Bp_{0.06}$ 72-76°.

Gusar', N.I. *et al*, *Zh. Obshch. Khim.*, 1981, **51**, 1477 (*Engl. transl.* p. 1254) (*synth, ir, P nmr*)
Budilova, I.Yu. *et al*, *Zh. Obshch. Khim.*, 1983, **53**, 285 (*Engl. transl.* p. 247) (*oxide, ir, P nmr*)

Azidotris(dimethylamino)phosphorus(1+) **A-00149**

Azidotris(N-*methylmethanaminato*)*phosphorus*(*1*+), 9CI

$(Me_2N)_3PN_3^{\oplus}$

$C_6H_{18}N_6P^{\oplus}$ M 205.222 (ion)

Reagent for coupling of peptides, converts carboxylic acids into acyl azides.

Hexafluorophosphate: [50281-51-1].
$C_6H_{18}F_6N_6P_2$ M 350.186
Cryst. (Me_2CO/Et_2O). Mp >250°.

Castro, B. *et al*, *Bull. Soc. Chim. Fr. Part 2*, 1973, 3359 (*synth, ir, pmr, use*)
Castro, B. *et al*, *Tetrahedron Lett.*, 1973, 3243 (*synth*)

Azinphos-ethyl, BSI **A-00150**

O,O-*Diethyl* S-[(*4-oxo-1,2,3-benzotriazin-3*(*4*H)-*yl*)-*methyl*] *phosphorodithioate,* 9CI. S-(*3,4-Dihydro-4-oxobenzo*[d][*1,2,3*]*triazin-3-ylmethyl*) O,O-*diethyl phosphorodithioate. Gusathion* A

[2642-71-9]

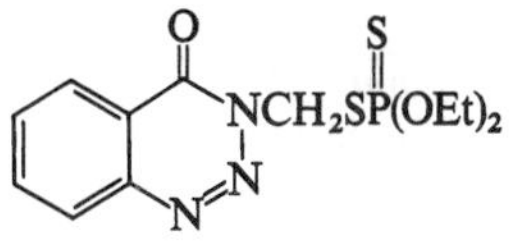

$C_{12}H_{16}N_3O_3PS_2$ M 345.370

Non-systemic agricultural insecticide and acaricide. Needles. d_4^{20} 1.284. Mp 53°. $Bp_{0.001}$ 111°. n_D^{53} 1.5928. Readily hydrolysed by alkali.

▷TD8400000.

U.S.P., 2 758 115, (*1955*); *CA*, **51**, 2888h
Stan, H.-J. *et al*, *Fresenius' Z. Anal. Chem.*, 1977, **287**, 271; *Biomed. Mass. Spectrom.*, 1982, **9**, 483 (*glc, ms*)
Busch, K.L. *et al*, *Appl. Spectrosc.*, 1978, **32**, 388 (*ms*)
Daldrup, T. *et al*, *Fresenius' Z. Anal. Chem.*, 1981, **308**, 413 (*tlc, glc, hplc*)
Stan, H.-J. *et al*, *J. Chromatogr.*, 1983, **279**, 173 (*glc*)
Pesticide Manual, 6th Ed., 23; 7th Ed., 25.
The Agrochemicals Handbook, Royal Society of Chemistry, 1983, A025.

Azinphos-Methyl, BSI **A-00151**

O,O-*Dimethyl* S-[(*4-oxo-1,2,3-benzotriazin-3*(*4*H)-*yl*)-*methyl*] *phosphorodithioate,* 9CI. S-(*3,4-Dihydro-4-oxobenzo*[d][*1,2,3*]*triazin-3-ylmethyl*) O,O-*dimethyl phosphorodithioate*

[86-50-0]

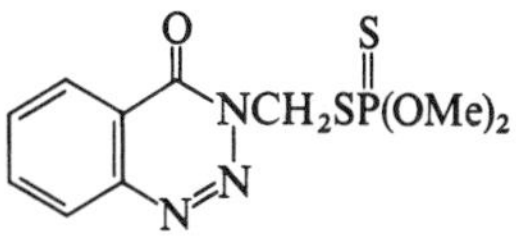

$C_{10}H_{12}N_3O_3PS_2$ M 317.317

Non-systemic insecticide and acaricide with long persistence. Cryst. V. spar. sol. H_2O, v. sol $CHCl_3$, toluene. Mp 73-74°. Rapidly hydrolysed by aq. alkali or acid. Unstable at 200°.

▷TE1925000.

Szalontai, G., *J. Chromatogr.*, 1976, **124**, 9 (*hplc*)
Stan, H.-J. *et al*, *Fresenius's Z. Anal. Chem.*, 1977, **287**, 271 (*glc, ms*)
Daldrup, T. *et al*, *Fresenius' Z. Anal. Chem.*, 1981, **308**, 413 (*tlc, hplc, glc*)
Stan, H.-J. *et al*, *Biomed. Mass. Spectrom.*, 1982, **9**, 483 (*ms*)
Ripley, B.D. *et al*, *J. Assoc. Off. Anal. Chem.*, 1983, **66**, 1084 (*glc*)
Pesticide Manual, 7th ed., 26.
Agrochemicals Handbook, Royal Society of Chemistry, London, 1983, A026.

1-Aziridinyl-*P*-methylphosphonamidic acid, 9CI **A-00152**

1-Aziridinylmethylphosphinic acid

$C_3H_8NO_2P$ M 121.075

*Et ester: Ethyl 1-aziridinyl-*P*-methylphosphonamidate. Ethyl 1-aziridinylmethylphosphinate.*
$C_5H_{12}NO_2P$ M 149.129
Liq. d_4^{20} 1.11. Bp_6 84-85°. n_D^{20} 1.4514.

Nifant'ev, É.E. *et al*, *Zh. Obshch. Khim.*, 1967, **37**, 1854 (*Engl. transl.* p. 1766) (*synth*)

1-Aziridinylphosphonic acid, 9CI **A-00153**

1-Aziridinephosphonic acid

$C_2H_6NO_3P$ M 123.048

Di-Me ester: [469-47-6]. *Dimethyl 1-aziridinylphosphonate.*
$C_4H_{10}NO_3P$ M 151.102
Liq. d_4^{20} 1.215. Bp_{16} 126-128°, Bp_4 85-86°. n_D^{20} 1.4407.

Di-Et ester: [470-27-9]. *Diethyl 1-aziridinylphosphonate.*
$C_6H_{14}NO_3P$ M 179.155
No phys. props. reported.

N,N,N′,N-*Tetramethyldiamide:* [1195-67-1]. P-*1-Aziridinyl-*N,N,N′,N′*-tetramethylphosphonic diamide. 1-Aziridinylphosphonic bis(dimethylamide).*
$C_6H_{16}N_3OP$ M 177.186
Insect sterilant. Liq. d_4^{20} 1.07. Bp_6 96-97°, $Bp_{0.15}$ 42-46°. n_D^{20} 1.4680.

▷TA1460000.

Sonnet, P.E. *et al*, *J. Org. Chem.*, 1966, **31**, 2962 (*amide, synth, pmr*)
Nifant'ev, E.É. *et al*, *Zh. Obshch. Khim.*, 1967, **37**, 1854 (*Engl. transl.* p. 1766) (*ester, amide, synth*)
Shagidullin, R.R. *et al*, *Zh. Obshch. Khim.*, 1968, **38**, 150 (*Engl. transl.* p. 148) (*ester, ir*)
Gray, G.A. *et al*, *J. Org. Chem.*, 1979, **44**, 1768 (*dimethyl ester, cmr, nmr*)
Davidowitz, B. *et al*, *S. Afr. J. Chem.*, 1982, **35**, 63 (*dimethyl ester, synth, pmr*)

1-Aziridinylphosphonous acid **A-00154**

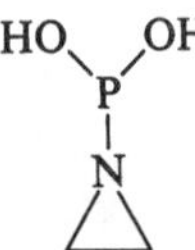

$C_2H_6NO_2P$ M 107.049

Di-Et ester: [5110-51-0]. *Diethyl 1-aziridinylphosphonite.*
$C_6H_{14}NO_2P$ M 163.156
Liq. d_4^{20} 1.01. Bp_{10} 57.5-58.5°. n_D^{20} 1.4458.

Diisopropyl ester: Diisopropyl 1-aziridinylphosphonite.
$C_8H_{18}NO_2P$ M 191.209
Liq. Bp_{12} 70-71°. n_D^{20} 1.4290.

Mono-Me, Mono-Ph ester: Methyl phenyl 1-aziridinylphosphonite.
$C_9H_{12}NO_2P$ M 197.173
Liq. d_4^{20} 1.14. $Bp_{0.08}$ 73°. n_D^{20} 1.5320.

Mono-Et, Mono-Ph ester: [23206-00-0]. *Ethyl phenyl 1-aziridinylphosphonite.*
$C_{10}H_{14}NO_2P$ M 211.200
Liq. d_4^{20} 1.10. $Bp_{0.08}$ 75.5°. n_D^{20} 1.5248.

Di-Ph ester: [14496-37-8]. *Diphenyl 1-aziridinylphosphonite.*
$C_{14}H_{14}NO_2P$ M 259.244
Liq. d^{20} 1.17. $Bp_{0.06}$ 122°. n_D^{20} 1.5780.

Diamide: see P-1-Aziridinylphosphonous diamide, A-00155

Grechkin, N.P., *CA*, 1958, **52**, 241 (*diethyl ester*)
Grechkin, N.P. *et al*, *Izv. Akad. Nauk SSSR, Ser. Khim.*, 1965, 1105 (*Engl. transl.* p. 1072) (*esters*)
Buina, N.A. *et al*, *Izv. Akad. Nauk SSSR, Ser. Khim.*, 1967, 217 (*Engl. transl.* p. 216) (*diphenyl ester, synth, ir*)
Grechkin, N.P. *et al*, *Izv. Akad. Nauk SSSR, Ser. Khim.*, 1969, 881 (*Engl. transl.* p. 800) (*mixed esters*)

***P*-1-Aziridinylphosphonous diamide** **A-00155**

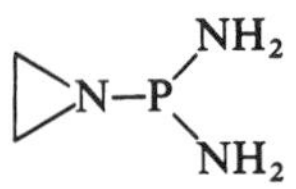

$C_2H_8N_3P$ M 105.079

N,N,N′,N′-*Tetra-Me:* [1195-18-2]. P-*1-Aziridinyl-*N,N,N′,N′*-tetramethylphosphonous diamide. 1-Aziridinylphosphonous bis(dimethylamide).*
$C_6H_{16}N_3P$ M 161.186
Liq. Bp_{30} 68-69°, $Bp_{1.5}$ 56-57°. n_D^{20} 1.4764.

N,N,N′,N′-*Tetra-Et:* P-*1-Aziridinyl-*N,N,N′,N′*-tetraethylphosphonous amide. 1-Aziridinylphosphonous bis(diethylamide).*
$C_{10}H_{24}N_3P$ M 217.293
Liq. Bp_7 85-86°. n_D^{20} 1.4823.

Nuretdinov, N.A. *et al*, *Izv. Akad. Nauk SSSR, Ser. Khim.*, 1964, 1883 (*Engl. transl.* p. 1784) (*derivs, synth*)
Shagidullin, R.R. *et al*, *Zh. Obshch. Khim.*, 1968, **38**, 150 (*Engl. transl.* p. 148) (*tetraethyl, ir*)
Nifant'ev, E.É. *et al*, *Zh. Obshch. Khim.*, 1969, **39**, 360 (*Engl. transl.* p. 337) (*tetramethyl, synth*)

(1-Azulenylmethyl)-triphenylphosphonium(1+) **A-00156**

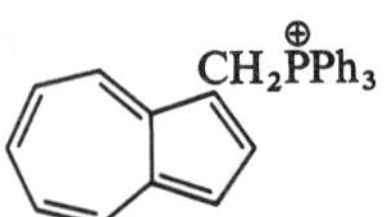

$C_{29}H_{24}P^{\oplus}$ M 403.482 (ion)

Iodide: [40450-19-9].
$C_{29}H_{24}IP$ M 530.387
Solid. Mp 172-176°. With PhLi, yields the ylide.

Ylide: [40450-24-6]. *(1-Azulenylmethylene)-triphenylphosphorane.*
$C_{29}H_{23}P$ M 402.474
Wittig reagent prepd. *in situ*. Red.

Currie, J.O. *et al*, *Justus Liebigs Ann. Chem.*, 1973, 166 (*synth, ylide, use*)

B

Benfosformin, INN **B-00001**

[Imino[[imino[(phenylmethyl)amino]methyl]amino]-methyl]phosphoramidic acid, 9CI. [(Benzylamidino)-amidino]phosphoramidic acid

$PhCH_2NHC(=NH)NHC(=NH)NHP(=O)(OH)_2$

$C_9H_{14}N_5O_3P$ M 271.215

Di-Na salt: [35282-33-8]. *JAV 852.* Antihyperglycaemic agent. Monohydrate. Mp 98.5-99.2°.

Ger. Pat., 2 130 303, (*1971*); *CA*, **76**, 59745u (*synth, pharmacol*)
Loiseau, G. *et al*, *Arzneim.-Forsch.*, 1973, **23**, 1576 (*pharmacol*)

Benfotiamine, INN **B-00002**

S-*[2-[[(4-Amino-2-methyl-5-pyrimidinyl)methyl]-formylamino]-1-[2-(phosphonoxy)ethyl]-1-propenyl]-benzenecarbothioate, 9CI. 8088 C.B.. Neurostop. Biotamin. Vitanervil*

[22457-89-2]

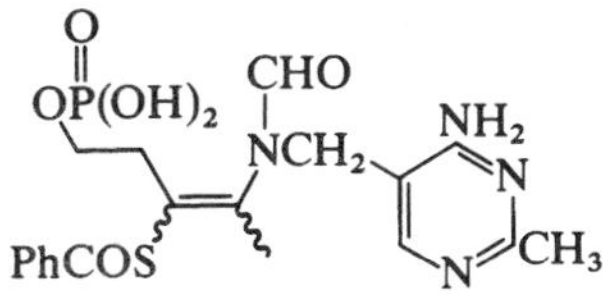

$C_{19}H_{23}N_4O_6PS$ M 466.448

Vitamin B_1 replacement. Dihydrate. Mp 165° dec.

▷DH6910000.

Ger. Pat. 1 130 811, (*1962*); *CA*, **57**, 13764h (*synth, pharmacol*)
Martindale, The Extra Pharmacopoeia, 28th Ed., 1982, Pharmaceutical Press, London,7831.

Benoxafos, INN **B-00003**

S-*[(5,7-Dichlorobenzoxazol-2-yl)methyl]* O,O-*diethyl-phosphorodithioate, 8CI. HOE 2910. Batestan*

[16759-59-4]

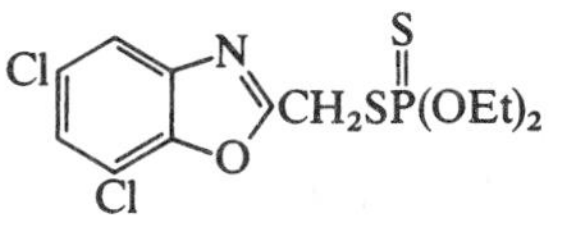

$C_{12}H_{14}Cl_2NO_3PS_2$ M 386.247

Pesticide. Brown oil. Low toxicity claimed.

Netherlands Pat., 6 607 822, (*1966*); *CA*, **68**, 21922w (*synth*)
Ger. Pat., 1 267 466, (*1968*); *CA*, **69**, 26297b (*tox, use*)

Bensulide, BSI, ISO **B-00004**

O,O-*Bis(1-methylethyl)* S-*[2-[(phenylsulfonyl)amino]-ethyl] phosphorodithioate, 9CI.* O,O-*Diisopropyl* S-*2-phenylsulfonylaminoethyl phosphorodithioate, 8CI.* S-*2-Benzenesulfonamidoethyl* O,O-*diisopropyl dithiophosphate. Betasan. Prefar*

[741-58-2]

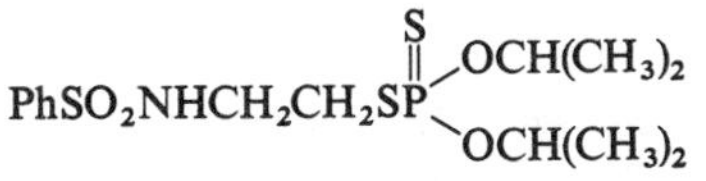

$PhSO_2NHCH_2CH_2SP(=S)(OCH(CH_3)_2)_2$

$C_{14}H_{24}NO_4PS_3$ M 397.502

Agricultural and turf herbicide. Visc. amber liq. or cryst. d^{22} 1.25. Mp 34.4°. n_D^{30} 1.5428.

▷TE0250000.

U.S.P., 3 205 253, (*1963*); *CA*, **64**, 2015c (*synth*)
Babad, H. *et al*, *Anal. Chim. Acta*, 1968, **41**, 259 (*pmr*)
Ross, R.T. *et al*, *Anal. Chim. Acta*, 1970, **52**, 139 (*P nmr*)
Ripley, B.D. *et al*, *J. Assoc. Off. Anal. Chem.*, 1983, **66**, 1084 (*glc*)
Pesticide Manual, 6th Ed., 33; 7th Ed., 38.
The Agrochemicals Handbook, Royal Society of Chemistry, 1983, A033.

1*H*-1,3-Benzazaphosphole, 9CI **B-00005**

[32881-50-8]

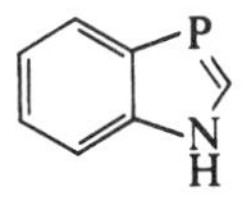

C_7H_6NP M 135.105

Mp 102-103°.

1-Me: [84759-25-1].

C_8H_8NP M 149.132

Liq. $Bp_{0.06}$ 77-80°.

Becker, G. *et al*, *Z. Anorg. Allg. Chem.*, 1980, **462**, 130 (*pmr, cmr, P nmr*)
Issleib, K. *et al*, *Z. Anorg. Allg. Chem.*, 1981, **481**, 22 (*synth, pmr, P nmr, derivs*)
Heinicke, J. *et al*, *Tetrahedron Lett.*, 1982, **23**, 3643 (*deriv, pmr, cmr, P nmr, uv*)

1,2-Benzenediphosphonic acid **B-00006**

1,2-Phenylenebisphosphonic acid, 9CI. 1,2-Diphosphonobenzene. 1,2-Bis(dihydroxyphosphinyl)benzene

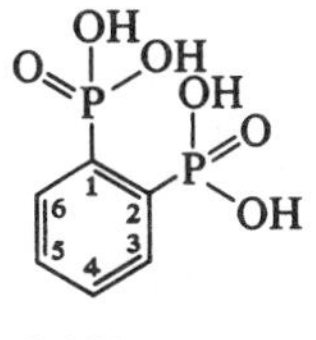

$C_6H_8O_6P_2$ M 238.073

Solid. Mp 203-204°.

Tetra-Me ester: Tetramethyl 1,2-phenylenebisphosphonate. 1,2-Bis(dimethoxyphosphinyl)benzene.

$C_{10}H_{16}O_6P_2$ M 294.180

Solid. Mp 80-81°.

Obrycki, R. *et al*, *Tetrahedron Lett.*, 1966, 5049 (*ester*)
Tavs, P., *Chem. Ber.*, 1970, **103**, 2428 (*synth*)
Kyba, E.P. *et al*, *Organometallics*, 1983, **2**, 1877 (*ester*)

1,3-Benzenediphosphonic acid **B-00007**

1,3-Phenylenebisphosphonic acid, 9CI. 1,3-Diphosphonobenzene. 1,3-Bis(dihydroxyphosphinyl)benzene

$C_6H_8O_6P_2$ M 238.073

Tetra-Me ester: [78271-46-2]. *Tetramethyl 1,3-phenylenebisphosphonate. 1,3-Bis(dimethoxyphosphinyl)-benzene.*

$C_{10}H_{16}O_6P_2$ M 294.180

High boiling liq.

Tetra-Et ester: [25944-79-0]. *Tetraethyl 1,3-phenylenebisphosphonate. 1,3-Bis(diethoxyphosphinyl)benzene.*

$C_{14}H_{24}O_6P_2$ M 350.288

Liq. $Bp_{0.04}$ 191-194°.

Obrycki, R. *et al*, *Tetrahedron Lett.*, 1966, 5049 (*synth*)
Tavs, P., *Chem. Ber.*, 1970, **103**, 2428 (*synth*)
Bunnett, J.F. *et al*, *J. Org. Chem.*, 1978, **43**, 1873, 1877 (*synth*)

1,4-Benzenediphosphonic acid B-00008

1,4-Phenylenebisphosphonic acid, 9CI. *1,4-Diphosphonobenzene. 1,4-Bis(dihydroxyphosphinyl)benzene*

[880-68-2]

$C_6H_8O_6P_2$ M 238.073

Cryst. (EtOH aq.). Mp 203-204°.

Tetra-Me ester: Tetramethyl 1,4-phenylenebisphosphonate. 1,4-Bis(dimethoxyphosphinyl)benzene.
$C_{10}H_{16}O_6P_2$ M 294.180
Mp 100-101°.

Tetra-Et ester: [21267-14-1]. *Tetraethyl 1,4-phenylenebisphosphonate. 1,4-Bis(diethoxyphosphinyl)benzene.*
$C_{14}H_{24}O_6P_2$ M 350.288
Needles (pet. ether). Mp 71-72°.

Obrycki, R. *et al*, *Tetrahedron Lett.*, 1966, 5049 (*ester*)
Tavs, P., *Chem. Ber.*, 1970, **103**, 2428 (*ester*)
Kovalova, T.V. *et al*, *Zh. Obshch. Khim.*, 1977, **47**, 318 (*Engl. transl.* p. 294) (*synth*)
Bunnett, J.F. *et al*, *J. Org. Chem.*, 1978, **43**, 1867 (*ester, synth, pmr, ir*)

1,2-Benzenediphosphonous acid B-00009

1,2-Phenylenebis(phosphonous acid), 9CI. o-*Phenylenediphosphonous acid. 1,2-Benzenediphosphinic acid*

$C_6H_8O_4P_2$ M 206.074

Free acid probably exists in the tautomeric bisphosphoryl form.

Tetra-Et ester: [65882-82-8]. *Tetraethyl 1,2-phenylenebisphosphonite.*
$C_{14}H_{24}O_4P_2$ M 318.289
Ligand for Cr, Co, Mo, Ni, and Ag. Distillable liq.

Bis(dichloride): [82495-67-8].
$C_6H_4Cl_4P_2$ M 279.857
Liq. $Bp_{0.001}$ 97-99°.

Bis[bis(dimethylamide)]: [82495-65-6].
N,N,N′,N′,N″,N″,N‴,N‴*-Octamethyl-*P,P′*-1,2-phenylenebis[phosphonous diamide].*
$C_{14}H_{28}N_4P_2$ M 314.350
Liq. $Bp_{0.001}$ 96-98°.

Meiners, J.H. *et al*, *J. Coord. Chem.*, 1977, **7**, 131 (*esters, complexes*)
Drewelies, K. *et al*, *Angew. Chem., Int. Ed. Engl.*, 1982, **21**, 638 (*tetrachloride, dimethylamide, synth, ir, pmr, nmr, cmr*)

1,4-Benzenediphosphonous acid B-00010

1,4-Phenylenebis(phosphonous acid), 9CI. p-*Phenylenediphosphonous acid. 1,4-Benzenediphosphinic acid*

$C_6H_8O_4P_2$ M 206.074

Free acid probably exists in the bisphosphoryl form.

Tetra-Me ester: [10498-59-6]. *Tetramethyl 1,4-phenylenebisphosphonite.*
$C_{10}H_{16}O_4P_2$ M 262.182
Liq., readily hydrol. and oxidised. $Bp_{0.1-0.2}$ 94-97°. n_D^{20} 1.5479.

Tetra-Et ester: [21267-11-8]. *Tetraethyl 1,4-phenylenebisphosphonite.*
$C_{14}H_{24}O_4P_2$ M 318.289
Liq., readily hydrol. and oxidised. d_4^{25} 1.12. Bp_2 146-148°. n_D^{25} 1.5181.

Bis(dichloride): [10498-56-8].
$C_6H_4Cl_4P_2$ M 279.857
Solid, readily hydrol. and oxidised. Mp 58-60°. $Bp_{0.15}$ 103-106°.

Bis(diiodide):
$C_6H_4I_4P_2$ M 645.663
Solid. Mp 138-142°.

Baldwin, R.A. *et al*, *J. Org. Chem.*, 1967, **32**, 2172 (*tetramethyl ester, synth, ir*)
Baranov, Yu.I. *et al*, *Dokl. Akad. Nauk SSSR, Ser. Sci. Khim.*, 1968, **182**, 337 (*Engl. transl.* p. 799) (*synth, chloride, esters*)
Kovaleva, T.V. *et al*, *Zh. Obshch. Khim.*, 1977, **47**, 318 (*Engl. transl.* p. 294) (*tetraiodide*)
Kovaleva, T.V. *et al*, *Zh. Obshch. Khim.*, 1979, **49**, 1228 (*Engl. transl.* p. 1077) (*tetrachloride*)
Cabelli, D.E. *et al*, *J. Am. Chem. Soc.*, 1981, **103**, 3286 (*derivs, pe*)

1,3-Benzenediphosphoramidic acid B-00011

1,3-Phenylenebis(phosphoramidic acid), 9CI

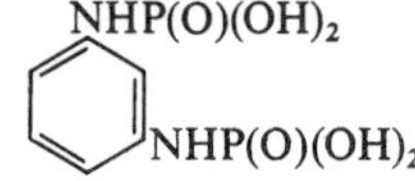

$C_6H_{10}N_2O_6P_2$ M 268.102

Tetra-Ph ester: [34714-88-0]. *Tetraphenyl 1,3-phenylenebisphosphoramidate.*
$C_{30}H_{26}N_2O_6P_2$ M 572.493
Cryst. (EtOH). Mp 183-184°.

Edmundson, R.S., *J. Chem. Soc. (C)*, 1971, 3614 (*synth, ir*)

1,4-Benzenediphosphoramidic acid B-00012

1,4-Phenylenebis(phosphoramidic acid), 9CI

$C_6H_{10}N_2O_6P_2$ M 268.102

Tetra-Et ester: [71402-66-9]. *Tetraethyl 1,4-phenylenebisphosphoramidate.*
$C_{14}H_{26}N_2O_6P_2$ M 380.317
Solid. Mp 145°.

Tetra-Ph ester: [34670-66-1]. *Tetraphenyl 1,4-phenylenebisphosphoramidate.*
$C_{30}H_{26}N_2O_6P_2$ M 572.493
Cryst. (EtOH aq.). Mp 210-211°.

Edmundson, R.S., *J. Chem. Soc. (C)*, 1971, 3614 (*phenyl ester, synth, ir*)
Ogata, N. *et al*, *J. Polym. Sci., Polym. Chem. Ed.*, 1979, **17**, 2401 (*ethyl ester*)

N-Benzenesulfonyl-*P,P*-diphenylphosphinimidic acid B-00013

P,P*-Diphenyl-*N*-phenylsulfonylphosphinimidic acid*

$Ph_2P(OH){=}NSO_2Ph$

$C_{18}H_{16}NO_3PS$ M 357.363

Acid exists only as tautomeric amide

Me ester: [17436-30-5]. *Methyl* P,P*-diphenyl-*N*-phenylsulfonylphosphinimidate.* P*-Methoxy-*P,P*-diphenyl-*N*-phenylsulfonylphosphazene.* P*-Methoxy-*P,P*-diphenyl-*N*-phenylsulfonylphosphine imide.*
$C_{19}H_{18}NO_3PS$ M 371.390

Solid. Mp 109-110°.

Et ester: Ethyl P,P-*diphenyl*-N-*phenylsulfonylphosphinimidate.* P-*Ethoxy*-P,P-*diphenyl*-N-*phenylsulfonylphosphazene.* P-*Ethoxy*-P,P-*diphenyl*-N-*phenylsulfonylphosphine imide.*

$C_{20}H_{20}NO_3PS$ M 385.417

Cryst. (EtOH). Insol. Et_2O, H_2O, CCl_4. Mp 117-118°. Stable in hot H_2O. Hydrol. in EtOH/$OH^{\ominus}$ aq.

Ph ester: Phenyl P,P-*diphenyl*-N-*phenylsulfonylphosphinimidate.* P-*Phenoxy*-P,P-*diphenyl*-N-*phenylsulfonylphosphazene.* P-*Phenoxy*-P,P-*diphenyl*-N-*phenylsulfonylphosphine imide.*

$C_{24}H_{20}NO_3PS$ M 433.461

Cryst. (EtOH). Insol. H_2O, C_6H_6. Mp 113-114°. Stable in H_2O and aq. alkali. Hydrol. in EtOH aq. $OH^{\ominus}$.

Chloride:

$C_{18}H_{15}ClNO_2PS$ M 375.809

Cryst. (C_6H_6 or EtOAc). Mp 109-111°.

Shevchenko, V.I. *et al*, *Zh. Obshch. Khim.*, 1960, **30**, 1566 (*Engl. transl.* p. 1573) (*chloride*)
Shevchenko, V.I. *et al*, *Zh. Obshch. Khim.*, 1960, **30**, 1958 (*Engl. transl.* p. 1937) (*esters*)
Lutskii, A.E. *et al*, *Zh. Obshch. Khim.*, 1967, **37**, 2034 (*Engl. transl.* p. 1930) (*synth, ester, uv*)
Haubold, W. *et al*, *Z. Anorg. Allg. Chem.*, 1970, **372**, 273 (*chloride, P nmr, props*)

10*H*-5,10[1′,2′]Benzenoacridophosphine, 9CI B-00014

*Phosphatriptycene. 9,10-Dihydro-9,10-*o*-benzeno-9-phosphaanthracene*

[55364-14-2]

$C_{19}H_{13}P$ M 272.285

Cryst. (EtOH). Mp 242-243°.

B,MeI: [55364-15-3]. *5-Methyl-10*H*-5,10[1′,2′]-benzenoacridophosphonium iodide.*

$C_{20}H_{16}IP$ M 414.225

Solid. Mp 185° dec.

*B,PhCH₂Br: 5-Benzyl-10*H*-5,10[1′,2′]-benzenoacridophosphonium bromide.*

$C_{26}H_{20}BrP$ M 443.322

Solid. Mp 316° dec.

5-Oxide: [55364-18-6].

$C_{19}H_{13}OP$ M 288.285

Cryst. (EtOH aq.). Mp 281-282°.

Jongsma, C. *et al*, *Tetrahedron*, 1974, **30**, 3465 (*synth, ir, uv, cmr, ms*)
Freijee, F.J.M. *et al*, *Acta Crystallogr., Sect. B*, 1980, **36**, 1247 (*cryst struct*)

5,10[1′,2′]-Benzenophenophosphazine, 9CI B-00015

*5,10-*o*-Benzenophosphazine, 8CI. Azaphosphatriptycene*

[24942-19-6]

$C_{18}H_{12}NP$ M 273.273

Cryst. (EtOH). Mp 255-255.5°.

B,MeI: 10-Methyl-5,10[1′,2′]-benzophenophosphazinium iodide.

$C_{19}H_{15}INP$ M 415.212

Solid. Mp 279-283°.

Oxide: [24942-20-9].

$C_{18}H_{12}NOP$ M 289.273

Cryst. (EtOH). Mp 290-295°.

Hellwinkel, D., *Chem. Ber.*, 1978, **111**, 1798 (*synth, pmr, ms, P nmr*)

5-Benzimidazolephosphonic acid B-00016

(Benzimidazol-5-yl)phosphonic acid, 9CI. 5-Phosphonobenzimidazole

$C_7H_7N_2O_3P$ M 198.118

Cryst. (H_2O). Mp >250°.

Bost, R.W. *et al*, *J. Org. Chem.*, 1953, **18**, 358 (*synth*)

Benzodepa, USAN B-00017

Phenylmethyl [bis(1-aziridinyl)phosphinyl]carbamate, 9CI. Benzyl [bis(1-aziridinyl)phosphinyl]carbamate. Dualar

[1980-45-6]

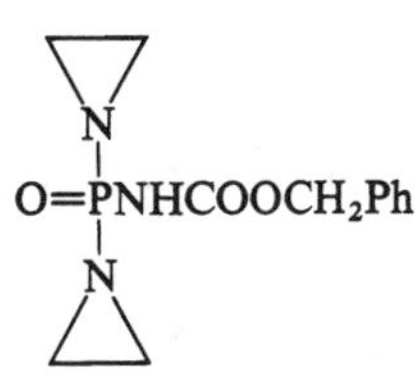

$C_{12}H_{16}N_3O_3P$ M 281.250

Antineoplastic, insect sterilant. Cryst. (C_6H_6/cyclohexane). Mp 134-135°.

Papanastassiou, Z.B. *et al*, *J. Med. Chem.*, 1962, **5**, 1000 (*synth, pharmacol*)
Borkovec, A.B. *et al*, *J. Med. Chem.*, 1966, **9**, 522 (*use*)
Gouck, H.K. *et al*, *J. Econ. Entomol.*, 1964, **57**, 663 (*use, props*)
Bardos, T.J. *et al*, *Ann. N.Y. Acad. Sci.*, 1969, **163**, 1006 (*pharmacol*)
Merck Index, 11th Ed., No. 1088.

1,3,2-Benzodioxaphosphole, 8CI B-00018

o-*Phenylene phosphonite*

$C_6H_5O_2P$ M 140.078

2-Oxide: [934-35-0]. o-*Phenylene phosphite.* o-*Phenylene phosphonate.*

$C_6H_5O_3P$ M 156.077

Mp 54-55°. $Bp_{0.03}$ 87-88°. pK_a 2.44 (50% EtOH aq.). Tautomeric; almost completely in phosphonate form.

Ovchinnikov, V.V. *et al*, *Zh. Obshch. Khim.*, 1978, **48**, 2424 (*Engl. transl.* p. 2199) (*oxide, P nmr*)
Rüger, C. *et al*, *J. Prakt. Chem.*, 1982, **324**, 706 (*oxide, synth, P nmr*)
Konovalova, I.V. *et al*, *Zh. Obshch. Khim.*, 1983, **53**, 1945 (*Engl. transl.* p. 1754)

(1,3-Benzodioxol-5-yl)phosphonic acid, 9CI **B-00019**

5-Phosphono-1,3-benzodioxole. 3,4-Methylenedioxybenzenephosphonic acid

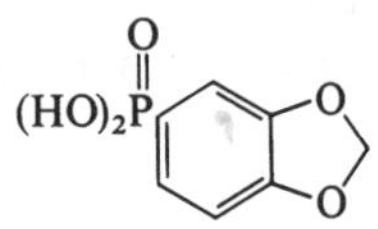

$C_7H_7O_5P$ M 202.103

Solid. Mp 160-161°. Isol. as di-*p*-toluidine salt.

Bost, R.W. *et al*, *J. Org. Chem.*, 1953, **18**, 362 (*synth*)

1,3-Benzodithiol-2-ylphosphonic acid, 9CI **B-00020**

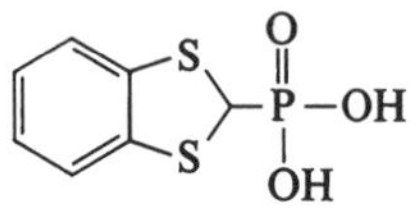

$C_7H_6O_3PS_2$ M 233.216

Di-Et ester: [62217-21-4].

$C_{11}H_{15}O_3PS_2$ M 290.331

Cryst. (toluene). Mp 108-111°.

Akiba, K. *et al*, *Synthesis*, 1977, 861 (*use*)
Gross, H. *et al*, *Synthesis*, 1977, 622 (*synth, pmr, P nmr, cmr*)
Akiba, K. *et al*, *Bull. Chem. Soc. Jpn.*, 1978, **51**, 2674 (*use*)

1,3-Benzodithiol-2-yltriphenylphosphonium(1+), 9CI **B-00021**

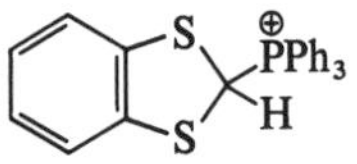

$C_{25}H_{20}PS_2^{\oplus}$ M 415.527 (ion)

Chloride: [66221-32-7].

$C_{25}H_{20}ClPS_2$ M 450.980

Ylide source, obt. with butyllithium. No phys. props. reported.

Tetrafluoroborate: [62217-34-9].

$C_{25}H_{20}BF_4PS_2$ M 502.330

Cryst. Mp 211.5-212.5° dec.

Ylide: [62217-36-1]. *(1,3-Benzodithiol-2-ylidene)-triphenylphosphorane. 2-Triphenylphosphoranylidene-1,3-benzodithiole.*

$C_{25}H_{19}PS_2$ M 414.519

Ylide used in synth. of benzodithiafulvenes. Red.

Ishikawa, K. *et al*, *Tetrahedron Lett.*, 1976, 3695 (*synth, props*)
Akiba, K. *et al*, *Bull. Chem. Soc. Jpn.*, 1978, **51**, 2674 (*synth, ir, pmr, use*)
Costisella, B. *et al*, *J. Prakt. Chem.*, 1978, **320**, 128 (*synth, nmr*)
Gonnella, N.C. *et al*, *J. Org. Chem.*, 1978, **43**, 369 (*synth, use*)
Nakayama, J. *et al*, *Bull. Chem. Soc. Jpn.*, 1981, **54**, 2845 (*synth, use*)

***7H*-Benzo[*c*]phenothiaphosphine, 9CI** **B-00022**

[64509-14-4]

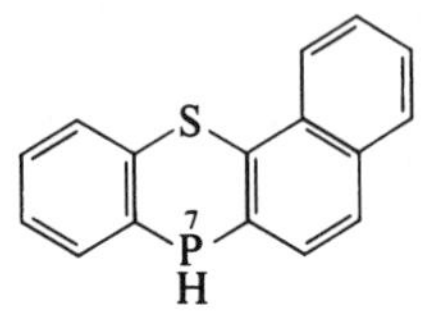

$C_{16}H_{11}PS$ M 266.297

7-Butyl, 7-oxide: [64278-33-7].

$C_{20}H_{19}OPS$ M 338.403

Needles (C_6H_6). Mp 108°.

Acharekar, A.R. *et al*, *Indian J. Chem., Sect. B*, 1977, **15**, 408 (*synth, ir, pmr*)

12*H*-Benzo[*c*]phenothiaphosphine, 9CI **B-00023**

[56087-62-8]

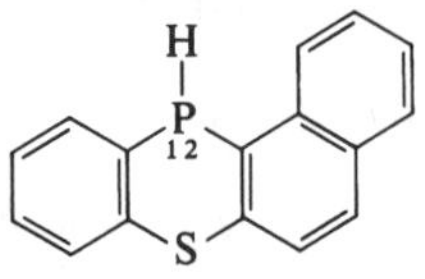

$C_{16}H_{11}PS$ M 266.297

12-Ethoxy, 7,7,12-trioxide: [55329-13-0].

$C_{18}H_{15}O_4PS$ M 358.348

Needles ($CHCl_3$/pet. ether). Mp 162°.

Dhaneshwar, N.N. *et al*, *Acta Crystallogr., Sect. B*, 1975, **31**, 750 (*cryst struct*)
Acherekar, A.R. *et al*, *Indian J. Chem., Sect. B*, 1977, **15**, 408 (*synth, ir, pmr*)

5,10-[1′,2′]Benzophosphenanthrene, 9CI **B-00024**

1,6-Diphosphatriptycene

[31634-70-5]

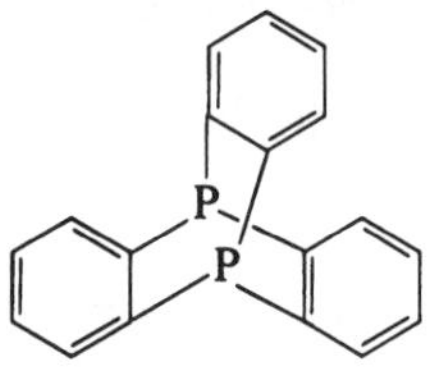

$C_{18}H_{12}P_2$ M 290.240

Cryst. (tetrachloroethene). Mp 323-325°.

5,10-Dioxide: [31634-72-7].

$C_{18}H_{12}O_2P_2$ M 322.239

Cryst. (propanol). Mp 488-490°.

5,10-Disulfide: [56783-21-2].

$C_{18}H_{12}P_2S_2$ M 354.360

Cryst. (AcOH). Mp 396-399°.

Schomburg, D. *et al*, *Acta Crystallogr., Sect. B*, 1975, **31**, 2427 (*cryst struct*)
Weinberg, K.G., *J. Org. Chem.*, 1975, **40**, 3586 (*synth, props, derivs, ir*)
Sørensen, S. *et al*, *Org. Magn. Reson.*, 1977, **9**, 101 (*cmr, P nmr*)

1,3-Benzothiaphosphole, 9CI **B-00025**

[59205-19-5]

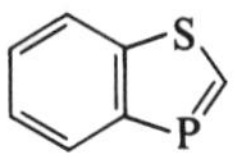

C_7H_5PS M 152.150

Yellow oil. $Bp_{0.1}$ 154-156°.

Issleib, K. *et al*, *Tetrahedron Lett.*, 1980, **21**, 3483 (*synth, pmr, cmr, P nmr*)

1,3,2-Benzoxathiaphosphole, 9CI **B-00026**

[15120-59-9]

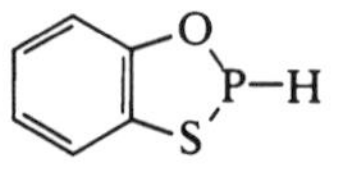

C_6H_5OPS M 156.139

2-Me: [13968-97-3].
C_7H_7OPS M 170.165
Liq. Bp_2 88°.
2-Ph: [51656-53-2].
$C_{12}H_9OPS$ M 232.236
Cryst. (hexane at 0°). Mp 59-61°. $Bp_{0.1}$ 65° subl.
2-Chloro: [23358-83-0].
C_6H_4ClOPS M 190.584
Oil. n_D^{21} 1.6631.
2-Bromo: [65669-35-4].
C_6H_4BrOPS M 235.035
Yellow liq. $Bp_{0.05}$ 84-87°.

Wieber, M. *et al*, *Chem. Ber.*, 1967, **100**, 974 (*methyl, derivs*)
Andrae, A. *et al*, *Z. Anorg. Allg. Chem.*, 1977, **434**, 127 (*bromo*)
Sau, A.C. *et al*, *J. Organomet. Chem.*, 1978, **156**, 153 (*phenyl, pmr, P nmr*)
Anchisi, C. *et al*, *J. Heterocycl. Chem.*, 1979, **16**, 1439 (*chloro, phenyl, synth, ir, pmr*)

6-Benzoxazolephosphonic acid B-00027

(1,3-Benzoxazol-6-yl)phosphonic acid, 9CI. 6-Phosphonobenzoxazole

$C_7H_6NO_4P$ M 199.102
Solid. Mp 204-205° dec.

Bost, R.W. *et al*, *J. Org. Chem.*, 1953, **18**, 358 (*synth*)

Benzoyl diethyl phosphite B-00028

Benzoic acid, anhydride with diethyl phosphite, 9CI
[40648-79-1]

$PhCOOP(OEt)_2$

$C_{11}H_{15}O_4P$ M 242.211
Liq. d_4^{20} 1.12. $Bp_{0.5}$ 113-115°. n_D^{20} 1.4962. Relatively thermally stable.

Petrov, K.A. *et al*, *Zh. Obshch. Khim.*, 1961, **31**, 2373 (*Engl. transl.* p. 2211) (*synth*)
Gazizov, T.Kh. *et al*, *Zh. Obshch. Khim.*, 1973, **43**, 2626 (*Engl. transl.* p. 2606) (*props*)
Pudovik, A.N. *et al*, *Zh. Obshch. Khim.*, 1975, **45**, 2621 (*Engl. transl.* p. 2582) (*props*)
Okamoto, Y. *et al*, *Bull. Chem. Soc. Jpn.*, 1984, **57**, 2693 (*props*)

O-Benzoyl *O,O*-diethylphosphorothioate B-00029

Benzoic acid, anhydride with O,O*-diethylphosphorothioate, 9CI, 8CI. Benzoic-*O,O*-diethyl phosphorothioic anhydride.* O*-Benzoyl* O,O*-diethyl thiophosphate*
[32119-41-8]

$PhCOOP(S)(OEt)_2$

$C_{11}H_{15}O_4PS$ M 274.271
Liq. d_4^{20} 1.19. $Bp_{0.009}$ 60-62°. n_D^{20} 1.5182.

Mastryukova, T.A. *et al*, *Zh. Obshch. Khim.*, 1971, **41**, 239 (*Engl. transl.* p. 236) (*synth, ir*)

Benzoyldiphenylphosphine, 9CI B-00030

[36838-04-7]

Ph_2PCOPh

$C_{19}H_{15}OP$ M 290.301
Cryst. (EtOH aq.). Mp 79°.
Oxide:
$C_{19}H_{15}O_2P$ M 306.300
Solid. Mp 145° dec.

Martens, J. *et al*, *Phosphorus*, 1976, **6**, 247 (*synth, ir, nmr*)
Lindner, E. *et al*, *Chem. Ber.*, 1979, **112**, 1456 (*oxide, synth, ir, ms*)
Dankowski, M. *et al*, *Phosphorus Sulfur*, 1980, **8**, 105 (*synth, ms, nmr*)

N-Benzoyl-*P*-methylphosphonamidic acid, 9CI B-00031

Me(HO)P(O)NHCOPh

$C_8H_{10}NO_3P$ M 199.146
NH_4 salt: Cryst. (EtOH aq.). Mp 260-262°.
Me ester: [21228-96-6]. *Methyl* N*-benzoyl-*P*-methylphosphonamidate.*
$C_9H_{12}NO_3P$ M 213.172
Prisms (MeOH). Mp 170-172°.
Chloride: [21229-87-8]. N*-Benzoyl-*P*-methylphosphonamidic chloride.*
$C_8H_9ClNO_2P$ M 217.591
Solid. Mp 112-114°.

Shokol, V.A. *et al*, *Zh. Obshch. Khim.*, 1968, **38**, 1867 (*Engl. transl.* p. 1815) (*chloride, ester, synth*)
Shokol, V.A. *et al*, *Zh. Obshch. Khim.*, 1969, **39**, 1485 (*Engl. transl.* p. 1455) (*chloride, ir*)

Benzoylphenylphosphinic acid, 9CI B-00032

PhCOPPh(O)OH

$C_{13}H_{11}O_3P$ M 246.202
Me ester: [18106-76-8]. *Methyl benzoylphenylphosphinate.*
$C_{14}H_{13}O_3P$ M 260.229
Used in synth. of heterocyclic compds. Cryst. (Et_2O/pet. ether), v. pale-yellow prisms (THF/pentane). Mp 78.5°. $Bp_{0.25}$ 147-147.5°.
Et ester: [18788-27-7]. *Ethyl benzoylphenylphosphinate.*
$C_{15}H_{15}O_3P$ M 274.255
Liq. d_4^{20} 1.19. $Bp_{0.08}$ 148.5-150°. n_D^{20} 1.5760.

Razumov, A.I. *et al*, *Zh. Obshch. Khim.*, 1967, **37**, 2738 (*Engl. transl.* p. 2606) (*synth*)
Takamizawa, A. *et al*, *Chem. Pharm. Bull.*, 1967, **15**, 1183 (*synth, ir, use*)
Musierowicz, S. *et al*, *Phosphorus Sulfur*, 1977, **3**, 345 (*synth, ir, pmr*)

N-Benzoyl-*P*-phenylphosphonimidic dichloride, 9CI B-00033

N*-Benzoyl-*P,P*-dichloro-*P*-phenylphosphazene.* N*-Benzoyl-*P,P*-dichloro-*P*-phenylphosphine imide*
[25239-79-6]

$PhPCl_2{=}NCOPh$

$C_6H_6Cl_2NP$ M 194.000
Thick liq.

Derkach, G.I. *et al*, *Zh. Obshch. Khim.*, 1963, **33**, 553 (*Engl. transl.* p. 547) (*synth*)

Benzoylphosphonic acid, 9CI **B-00034**

[6881-61-4]

$PhCOP(O)(OH)_2$

$C_7H_7O_4P$ M 186.104

Esters can act as benzoylating agents, e.g. $NH_3(l) \rightarrow PhCONH_2$. They also react with Wittig-Horner reagents to give 1-phosphonostyrenes.

Monoanilinium salt: [67472-26-8]. Solid. Mp 172° dec.

Di-Me ester: [18106-71-3]. *Dimethyl benzoylphosphonate.*
$C_9H_{11}O_4P$ M 214.157
Liq. $Bp_{0.25}$ 146°, $Bp_{1.1}$ 130-133°.

Di-Me ester, 2,4-dinitrophenylhydrazone: Mp 198-199°.

Di-Et ester: [3277-27-8]. *Diethyl benzoylphosphonate.*
$C_{11}H_{15}O_4P$ M 242.211
Liq. Bp_3 141-147°. Reaction with $CH_2N_2 \rightarrow$ diethyl phenacylphosphonate.

Di-Et ester, 2,4-dinitrophenylhydrazone: Mp 171-172°.

Diisopropyl ester: [22950-57-8]. *Diisopropyl benzoylphosphonate.*
$C_{13}H_{19}O_4P$ M 270.264
Liq. $Bp_{0.3}$ 117-120°.

Bis(trimethylsilyl) ester: [33876-85-6]. *Bis(trimethylsilyl) benzoylphosphonate.*
$C_{13}H_{23}O_3PSi_2$ M 314.468
Liq. Bp_3 140-145°.

Diamide: see P-Benzoylphosphonic diamide, B-00035

Berlin, K.D. *et al*, *J. Am. Chem. Soc.*, 1964, **86**, 3862 (*ester, synth, ir, pmr, deriv*)
Scherer, H. *et al*, *Chem. Ber.*, 1972, **105**, 3357 (*ester, synth, ir, use*)
Clark, P.E. *et al*, *Phosphorus*, 1973, **2**, 265 (*ester, pmr, conformn*)
Soroka, M. *et al*, *Zh. Obshch. Khim.*, 1974, **44**, 463 (*Engl. transl.* p. 446) (*props*)
Laskorin, B.N. *et al*, *Zh. Obshch. Khim.*, 1976, **46**, 2545 (*Engl. transl.* p. 2434) (*ester, ms*)
Blackburn, G.M. *et al*, *J. Chem. Soc., Perkin Trans. 1*, 1980, 1150 (*synth*)
Sekine, M. *et al*, *J. Org. Chem.*, 1980, **45**, 4162 (*ester, synth, props*)
Yamashita, M. *et al*, *Bull. Chem. Soc. Jpn.*, 1980, **53**, 1625 (*ester, use*)

***P*-Benzoylphosphonic diamide,** 9CI **B-00035**

$PhCOP(O)(NH_2)_2$

$C_7H_9N_2O_2P$ M 184.134

N,N,N′,N′-*Tetra-Et:* [14114-73-9]. P-*Benzoyl-*N,N,N′,N′-*tetraethylphosphonic diamide. Benzoylphosphonic bis(diethylamide).*
$C_{15}H_{25}N_2O_2P$ M 296.348
Liq. d_4^{20} 1.09. Bp_1 152-153°. n_D^{20} 1.5390.

Alimov, P.I. *et al*, *Izv. Akad. Nauk SSSR, Ser. Khim.*, 1966, 1486 (*Engl. transl.* p. 1432)

Benzoylphosphoramidic acid, 9CI **B-00036**

[36097-63-9]

$PhCONHP(O)(OH)_2$

$C_7H_8NO_4P$ M 201.118

Solid. Mp 136-138°, Mp 157-158°. pK_{a1} 1.99, pK_{a2} 6.42, pK_{a3} ca. 14.

Mono-cyclohexylammonium salt: Cryst. (MeOH aq.). Mp 150-151°.

Di-Me ester: [24856-23-3]. *Dimethyl benzoylphosphoramidate.*
$C_9H_{12}NO_4P$ M 229.172
Cryst. (Me_2CO/pet. ether, EtOH aq. or C_6H_6/ligroin). Mp 117-118°.

Di-Et ester: [16102-45-7]. *Diethyl benzoylphosphoramidate.*
$C_{11}H_{16}NO_4P$ M 257.225
Solid. Mp 78°.

Diisopropyl ester: [3808-08-0]. *Diisopropyl benzoylphosphoramidate.*
$C_{13}H_{20}NO_4P$ M 285.279
Solid or prisms (H_2O). Mp 107-109°.

Di-Ph ester: Diphenyl benzoylphosphoramidate.
$C_{19}H_{16}NO_4P$ M 353.313
Needles (Et_2O/ligroin), prisms (EtOH). Mp 147-149°. pK_a 7.7.

Dibenzyl ester: Dibenzyl benzoylphosphoramidate.
$C_{21}H_{20}NO_4P$ M 381.367
Cryst. (C_6H_6/heptane). Mp 123-124°. pK_a 8.9.

Dichloride: see Benzoylphosphoramidic dichloride, B-00037

Kirsanov, A.V. *et al*, *Zh. Obshch. Khim.*, 1958, **28**, 35, 1227, 2247 (*Engl. transl.* pp. 35, 1283, 2283) (*diphenyl ester*)
Derkach, G.I. *et al*, *Zh. Obshch. Khim.*, 1962, **32**, 1201; 1964, **34**, 604 (*Engl. transl.* pp. 1176, 605) (*esters*)
Zioudron, C., *Tetrahedron*, 1962, **18**, 197 (*synth, ir, props, esters*)
Almasi, L. *et al*, *Chem. Ber.*, 1967, **100**, 2625 (*diethyl ester*)
Glidewell, C. *et al*, *J. Organomet. Chem.*, 1976, **108**, 335 (*esters, ir*)
Mizrahi, V. *et al*, *Cryst. Struct. Commun.*, 1982, **11**, 627 (*dimethyl ester, cryst struct*)
Mizrahi, V. *et al*, *J. Org. Chem.*, 1982, **47**, 3533 (*esters, ms*)
Mizrahi, V. *et al*, *S. Afr. J. Chem.*, 1983, **36**, 111 (*dimethyl ester, cmr*)

Benzoylphosphoramidic dichloride, 8CI **B-00037**

[4737-14-8]

$PhCONHP(O)Cl_2$

$C_7H_6Cl_2NO_2P$ M 238.010

Solid. Mp 96-114° approx. Mp depends on rate of heating. At 150° $\rightarrow POCl_3$ and PhCN.

Titherley, A.W. *et al*, *J. Chem. Soc.*, 1909, **95**, 1143 (*synth*)
Kirsanov, A.V. *et al*, *Zh. Obshch. Khim.*, 1956, **26**, 905 (*Engl. transl.* p. 1029) (*synth, props*)
Shokol, V.A. *et al*, *Zh. Obshch. Khim.*, 1969, **39**, 1485 (*Engl. transl.* p. 1455) (*ir*)

Benzoylphosphoramidothioic acid, 9CI **B-00038**

$PhCONHP(S)(OH)_2 \rightleftharpoons PhCONHP(O)(OH)(SH)$

$C_7H_8NO_3PS$ M 217.179

O,O-*Di-Et ester:* [13604-50-7]. O,O-*Diethyl benzoylphosphoramidothioate.*
$C_{11}H_{16}NO_3PS$ M 273.286
Cryst. (pet. ether). Mp 113°. pK_a 8.30 (H_2O), pK_a 8.87 (80% EtOH aq.), pK_a 11.23 (EtOH).

O,O-*Diisopropyl ester:* [13604-51-8]. O,O-*Diisopropyl benzoylphosphoramidothioate.* O,O-*Bis(1-methylethyl) benzoylphosphoramidothioate.*
$C_{13}H_{20}NO_3PS$ M 301.340
Solid. Mp 99°.

O,O-*Di-Ph ester:* [13604-54-1]. O,O-*Diphenyl benzoylphosphoramidothioate.*
$C_{19}H_{16}NO_3PS$ M 369.374
Solid. Mp 124°.

Almasi, L. *et al*, *Chem. Ber.*, 1966, **99**, 3293 (*esters, synth, props*)
Zsakó, I. *et al*, *Monatsh. Chem.*, 1969, **100**, 587 (*diethyl ester, props*)

Benzoylphosphorimidic acid, 9CI, 8CI — B-00039

PhCON=P(OH)$_3$

$C_7H_8NO_4P$ M 201.118
Free acid is tautomeric with Benzoylphosphoramidic acid, B-00036 .
Tri-Me ester: [59658-83-2]. *Trimethyl benzoylphosphorimidate.* N-*Benzoyl*-P,P,P-*trimethoxyphosphazene.* N-*Benzoyl*-P,P,P-*trimethoxyphosphine imide.*
$C_{10}H_{14}NO_4P$ M 243.199
Liq. d_4^{20} 1.22. $Bp_{0.2}$ 135-137°. n_D^{20} 1.5102.
Tri-Et ester: [59658-82-1]. *Triethyl benzoylphosphorimidate.* N-*Benzoyl*-P,P,P-*triethoxyphosphazene.* N-*Benzoyl*-P,P,P-*triethoxyphosphine imide.*
$C_{13}H_{20}NO_4P$ M 285.279
Liq. d_4^{20} 1.13. $Bp_{0.5}$ 146-147°. n_D^{20} 1.5019.
Triisopropyl ester: Triisopropyl benzoylphosphorimidate. Tris(1-methylethyl) benzoylphosphorimidate. N-*Benzoyl*-P,P,P-*triisopropoxyphosphazene.*
$C_{16}H_{26}NO_4P$ M 327.359
Liq. d_4^{20} 1.06. $Bp_{0.3}$ 143-144°. n_D^{20} 1.4863.
Tri-Ph ester: [59658-86-5]. *Triphenyl benzoylphosphorimidate.* N-*Benzoyl*-P,P,P-*triphenoxyphosphazene.* N-*Benzoyl*-P,P,P-*triphenoxyphosphine imide.*
$C_{25}H_{20}NO_4P$ M 429.411
No phys. props. reported.
Trichloride: see Benzoylphosphorimidic trichloride, B-00040

Lutskii, A.E. *et al*, *Zh. Obshch. Khim.*, 1967, **37**, 2034 (*Engl. transl.* p. 1930) (*synth, uv*)

Benzoylphosphorimidic trichloride, 9CI — B-00040

N-*Benzoyl*-P,P,P-*trichlorophosphazene.* N-*Benzoyl*-P,P,P-*trichlorophosphine imide*
[17437-62-6]

PhCON=PCl$_3$

$C_7H_5Cl_3NOP$ M 256.455
Prisms. Sol. CCl_4, C_6H_6 dioxane, spar. sol. pet. ether. Mp 60-61°. When heated → PhCN.

Kirsanov, A.V. *et al*, *Zh. Obshch. Khim.*, 1956, **26**, 907 (*Engl. transl.* p. 1033) (*synth, props*)
Tarasevich, A.S. *et al*, *Teor. Eksp. Khim.*, 1971, **7**, 828 (*Engl. transl.* p. 676) (*struct, P nmr*)
Glidewell, C., *Inorg. Chim. Acta*, 1976, **18**, 51 (*ms*)

3-Benzyl-2-chlorotetrahydro-2*H*-1,3,2-oxazaphosphorine — B-00041

*2-Chlorotetrahydro-3-phenylmethyl-2*H*-1,3,2-oxazaphosphorine, 9CI. 3-Benzyl-2-chloro-1,3,2-oxazaphosphorinane*
[59758-17-7]

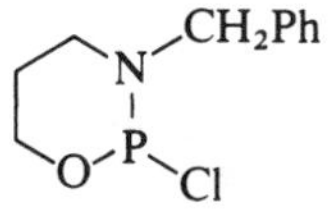

$C_{10}H_{13}ClNOP$ M 229.646
Liq. d_4^{20} 1.21. $Bp_{0.001}$ 130-135° (bath). n_D^{20} 1.5700.
2-Sulfide: [50742-51-3].
$C_{10}H_{13}ClNOPS$ M 261.706
Solid. Mp 110-111°.

Durrieu, J. *et al*, *Org. Magn. Reson.*, 1973, **5**, 407 (*sulfide, ir, pmr*)
Nifant'ev, É.E. *et al*, *Zh. Obshch. Khim.*, 1976, **46**, 477 (*Engl. transl.* p. 475) (*synth, P nmr*)

5-Benzyl-5,5-dihydro-5-phenylacridophosphine — B-00042

5,5-Dihydro-5-phenyl-5-(phenylmethyl)-acridophosphine, 9CI. 9-Benzyl-9-phenyl-9-phosphaanthracene. 5,5-Dihydro-5-phenyl-5-(phenylmethyl)dibenzo[b,e]*phosphinine. 5-Benzyl-5-phenyl-(5-P^V)-dibenzo*[b,e]*phosphorin*
[59590-74-8]

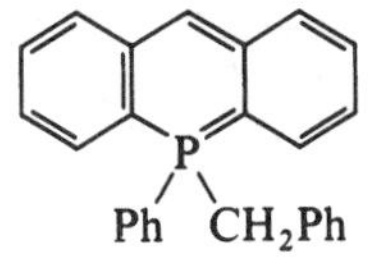

$C_{26}H_{21}P$ M 364.426
Mp 161-164°.

Jongsma, C. *et al*, *Tetrahedron Lett.*, 1976, 481 (*synth, pmr, uv, ms*)

5-Benzyl-5,6-dihydro-6-phenylphosphanthridine — B-00043

5,6-Dihydro-6-phenyl-5-(phenylmethyl)-phosphanthridine, 9CI

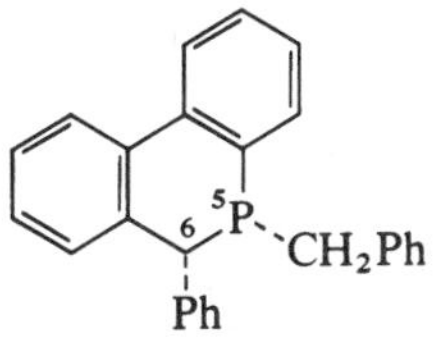

(5*RS*,6*RS*)-*form*

$C_{26}H_{21}P$ M 364.426
(*5RS,6RS*)-*form* [75543-51-0]
(±)-cis-*form*
Cryst. (MeOH). Mp 120°.
5-Sulfide: [75543-49-6].
$C_{26}H_{21}PS$ M 396.486
Cryst. (MeOH). Mp 144°.
(*5RS,6SR*)-*form* [75543-52-1]
(±)-trans-*form*
Oil.
5-Sulfide: [75543-48-5]. Cryst. (MeOH). Mp 182°.

Nief, F. *et al*, *Tetrahedron Lett.*, 1980, **21**, 1441 (*synth, ir, pmr, P nmr*)

O-Benzyl *O,O*-diisopropyl phosphorothioate, 8CI — B-00044

O,O-*Bis(1-methylethyl)* O-*phenylmethyl phosphorothioate, 9CI*
[42255-08-3]

$[(H_3C)_2CHO]_2P(S)OCH_2Ph$

$C_{13}H_{21}O_3PS$ M 288.341
Metab. and photoisomer of *S*-Benzyl *O,O*-diisopropyl phosphorothioate, B-00045 .

Yamamoto, H. *et al*, *Agric. Biol. Chem.*, 1973, **37**, 1553 (*glc*)

Murai, T. *et al, Agric. Biol. Chem.*, 1977, **41**, 803 (*glc, ir, ms*)

S-Benzyl *O,O*-diisopropyl phosphorothioate, 8CI — B-00045

O,O-*Bis(1-methylethyl)* S-*phenylmethyl phosphorothioate, 9CI. Kitazin* P. *Kitazin* L

[26087-47-8]

$PhCH_2SP(O)[OCH(CH_3)_2]_2$

$C_{13}H_{21}O_3PS$ M 288.341

Systemic rice fungicide. Extractant for Au, Ag, Hg, Pt and Pd. Yellow oil. $Bp_{0.04}$ 126°. n_D 1.5106.

▷TE6550000.

U.S.P., 4 064 200, (*1977*); *CA*, **88**, 89353 (*synth*)
Tomizawa, C. *et al, J. Environ. Sci., Health, Part B*, 1976, **11**, 231 (*metab*)
Torii, S. *et al, J. Org. Chem.*, 1979, **44**, 2938 (*synth, ir, pmr*)
Shigetomi, Y. *et al, Anal. Chim. Acta*, 1983, **152**, 301 (*use*)
Pesticide Manual, 6th Ed., 39; 7th Ed., 44.

Benzyldiphenylphosphine, 8CI — B-00046

Diphenyl(phenylmethyl)phosphine, 9CI

[1650-91-1]

Ph_2PCH_2Ph

$C_{19}H_{17}P$ M 276.317

Ligand for metals of Groups IB, IIB, VB, VIB, and VIII. Cryst. (O_2-free EtOH). Mp 74°, 141-142° dec. $Bp_{0.15}$ 146-148°.

B,MeI: Benzylmethyldiphenylphosphonium iodide. Methyldiphenyl(phenylmethyl)phosphonium iodide.
$C_{20}H_{20}IP$ M 418.256
Cryst. (EtOH). Mp 243-245°.

Oxide: [2959-74-2].
$C_9H_{13}OP$ M 168.175
Forms complexes with Co, Ni, Sn, and Nd. Cryst. (C_6H_6 or EtOH). Mp 198-199° (190°). pK_a 23.1 (diglyme), 27.1 (DMSO).

Sulfide: [15367-75-6].
$C_9H_{13}PS$ M 184.235
Cryst. (THF or EtOH). Mp 161-163° (156-157°). pK_a 26.5 (diglyme), 26.8 (DMSO).

Trippett, S., *J. Chem. Soc.*, 1961, 2813 (*synth, derivs*)
Browning, M.C. *et al, J. Chem. Soc.*, 1962, 693 (*synth, complexes*)
Mesyats, S.P. *et al, Izv. Akad. Nauk SSSR, Ser. Khim.*, 1974, 2489 (*Engl. transl.* p. 2399) (*derivs, props*)
Cauquis, G. *et al, Org. Mass Spectrom.*, 1975, **10**, 770 (*sulfide, ms*)
Gloyna, D. *et al, J. Prakt. Chem.*, 1975, **317**, 840 (*oxide, use*)
Inorg. Synth., 1976, **16**, 159 (*synth*)
Postle, S.R. *et al, Phosphorus Sulfur*, 1977, **3**, 269 (*derivs, pmr*)
Verstuyft, A.W. *et al, Inorg. Chem.*, 1977, **16**, 2776 (*pmr, cmr, P nmr, complexes, derivs*)
Holderegger, R. *et al, Helv. Chim. Acta*, 1979, **62**, 2154 (*synth, complexes*)
Kellner, K. *et al, Phosphorus Sulfur*, 1980, **8**, 269 (*synth, derivs*)
Regitz, M. *et al, Chem. Ber.*, 1980, **113**, 3303 (*oxide, use*)
Mitchie, J.K. *et al, J. Chem. Soc., Perkin Trans. 1*, 1981, 1744 (*oxide, props*)

Benzylethylphenylphosphine, 8CI — B-00047

Ethylphenyl(phenylmethyl)phosphine, 9CI

[25945-92-0]

Ph / :►P◄Et / CH_2Ph (*R*)-*form*

$C_{15}H_{17}P$ M 228.273

(*R*)-*form*

B,MeI: Benzylethylmethylphenylphosphonium iodide.
$C_{16}H_{20}IP$ M 370.212
Solid. Mp 167-168°. $[\alpha]_D^{25}$ −24.5° (c, 2.62 in MeOH). Has (*R*)-config.

(*S*)-*form*

B,MeI: Cryst. (H_2O). Mp 166-168°. $[\alpha]_D^{25}$ +25.0° ±1.0 (c, 2.82 in MeOH). Has (*S*)-config.

(±)-*form*

Oily liq. d_4^{20} 1.04. Bp_{13} 165-168°, Bp_1 115-120°. n_D^{20} 1.5960.

B,MeI: Cryst. (H_2O). Mp 170-171°.

Oxide: [53742-11-3].
$C_{15}H_{17}OP$ M 244.272
Cryst. (C_6H_6). Mp 115-116.5°. pK_a 29.3 (DMSO), 24.0 (diglyme).

Sulfide: [21989-72-0].
$C_{15}H_{17}PS$ M 260.333
Cryst. (2-propanol). Mp 75-76°. $Bp_{0.007}$ 157-159°.

McEwen, W.E. *et al, J. Am. Chem. Soc.*, 1964, **86**, 2378 (*synth, salts*)
Rizpolozhenskii, N.I. *et al, Izv. Akad. Nauk SSSR, Ser. Khim.*, 1969, 370; *CA*, **70**, 115234 (*sulfide*)
Harmon, R.E. *et al, Org. Prep. Proced.*, 1970, **2**, 19 (*synth*)
Mesyats, S.P. *et al, Izv. Akad. Nauk SSSR, Ser. Khim.*, 1974, 2497 (*Engl. transl.* p. 2406) (*oxide, props*)
Schnell, A. *et al, J. Chem. Soc., Perkin Trans. 1*, 1977, 1883 (*salts, resoln*)
Horner, L. *et al, Phosphorus Sulfur*, 1980, **8**, 225 (*salt*)

1-Benzylhexahydro-2-hydroxy-1,2-azaphosphorine 2-oxide — B-00048

Hexahydro-2-hydroxy-1-(phenylmethyl)-1,2-azaphosphorine 2-oxide, 9CI. 1-Benzylhexahydro-2-hydroxy-1,2-azaphosphinine 2-oxide

[63075-71-8]

N–P(=O)OH ring; N–CH_2Ph

$C_{11}H_{16}NO_2P$ M 225.227

V. hygroscopic oil. $Bp_{0.5}$ 200°.

Et ester: [63075-69-4]. *1-Benzyl-2-ethoxyhexahydro-1,2-azaphosphorine 2-oxide.*
$C_{13}H_{18}NO_2P$ M 251.264
Oil.

Hewitt, D.G. *et al, Aust. J. Chem.*, 1977, **30**, 579 (*synth, ms, pmr*)

1-(Benzylideneamino)ethylphosphonic acid — B-00049

[1-[(Phenylmethylene)amino]ethyl]phosphonic acid, 9CI

$PhCH{=}NCH(CH_3)P(O)(OH)_2$

$C_9H_{12}NO_3P$ M 213.172

(±)-*form*

Di-Et ester: Diethyl 1-benzylideneaminophosphonate.
$C_{13}H_{20}NO_3P$ M 269.280
A Wittig-Horner reagent useful for the synth. of ketones, and for their further conversion into αβ-cyclohexenones. Pale-yellow oil.

Dehnel, A. *et al, Synthesis*, 1977, 474 (*synth, ir, pmr, use*)
Dehnel, A. *et al, Bull. Soc. Chim. Fr. Part II*, 1978, 95 (*use*)
Martin, S.F. *et al, J. Am. Chem. Soc.*, 1980, **102**, 5866 (*synth, use*)

Benzylidenetriphenylphosphorane B-00050

Triphenyl(phenylmethylene)phosphorane, 9CI
[16721-45-2]

$Ph_3P{=}CHPh$

$C_{25}H_{21}P$ M 352.415
Reactive Wittig reagent. Isolable as salt-free, orange cryst.

Bergelson, L.D. *et al, Tetrahedron*, 1967, **23**, 2709 (*props, use*)
Albright, T.A. *et al, J. Am. Chem. Soc.*, 1976, **98**, 6249 (*cmr, P nmr*)
Ostoja Starzewski, K.A. *et al, J. Am. Chem. Soc.*, 1976, **98**, 8486 (*synth, pe*)
Ostoja Starzewski, K.A. *et al, Phosphorus*, 1976, **6**, 177 (*cmr*)
Butterfield, P.J. *et al, J. Chem. Soc., Perkin Trans. 1*, 1978, 1237 (*uv*)
Schlosser, M. *et al, Chimia*, 1982, **36**, 396 (*synth*)

Benzylmethylphenylphosphine B-00051

Methylphenyl(phenylmethyl)phosphine, 9CI
[23275-37-8]

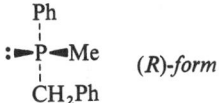

Ph
:►P◄Me (*R*)-*form*
CH₂Ph

$C_{14}H_{15}P$ M 214.246

(*R*)-*form* [25140-53-8]
Oil. $Bp_{0.05}$ 106-108°. $[\alpha]_D$ +62.0° (c, 6.1 in MeOH) (100% o.p.).
Sulfide: [19488-90-5]. Mp 97-100°. $[\alpha]_D$ −33.8° (c, 2.16 in MeOH) (66% o.p.). Sulfide has (*S*)-config.

(*S*)-*form* [34868-25-2]
$Bp_{0.02}$ 100°. $[\alpha]_D$ −26.3° (c, 5.98 in MeOH) (42.5% o.p.).
Oxide: [721-74-4].
$C_{14}H_{15}OP$ M 230.246
Cryst. (C_6H_6/pet. ether). Mp 145° (135-137°). $[\alpha]_D^{25}$ +38.3° (c, 2.6 in MeOH) (75% o.p.). Oxide has (*R*)-config.
Sulfide:
$C_{14}H_{15}PS$ M 246.306
$[\alpha]_D$ −32.90°. Has (*S*)-config.

(±)-*form* [64395-80-8]
Liq. Bp_5 120°.
B,MeI: Benzyldimethylphenylphosphonium iodide.
$C_{15}H_{18}IP$ M 356.185
Solid. Mp 158-160°.
Oxide: [51153-50-5]. Cryst. (C_6H_6/hexane). Mp 148-149° (136-141°). Bp_{15} 235°.

Balzer, W.-D., *Tetrahedron Lett.*, 1968, 1189 (*oxide, sulfide, cd*)
Naumann, K. *et al, J. Am. Chem. Soc.*, 1969, **91**, 2788, 7012.
Horner, L. *et al, Phosphorus*, 1971, **1**, 73 (*synth*)
Luckenbach, R., *Justus Liebigs Ann. Chem.*, 1974, 1618 (*synth, props*)
Wetzel, R.B. *et al, J. Org. Chem.*, 1974, **39**, 1531 (*oxide, pmr*)
Luckenbach, R., *Chem. Ber.*, 1975, **108**, 3522, 3533 (*props, sulfide, oxide*)
Horner, L. *et al, Phosphorus Sulfur*, 1980, **8**, 209 (*synth, sulfide*)
Payne, N.C. *et al, Can. J. Chem.*, 1980, **58**, 15.
Kogure, T. *et al, J. Organomet. Chem.*, 1982, **234**, 249 (*use*)
Moriyama, M. *et al, J. Am. Chem. Soc.*, 1983, **105**, 4727 (*synth, oxide, props*)

Benzylmethylphenylpropylphosphonium(1+) B-00052

Methylphenyl(phenylmethyl)propylphosphonium(1+)

CH₂Ph
Me►P⊕◄Ph (*R*)-*form*
CH₂CH₂CH₃

$C_{17}H_{22}P^{\oplus}$ M 257.335 (ion)
Converted by a strong base into the ylide.

(*R*)-*form*
Bromide: [839-28-1]. Solid. Mp 201°. $[\alpha]_D^{20}$ +36.7° (c, 1.58 in MeOH).

(*S*)-*form*
Bromide: [5137-89-3].
$C_{17}H_{22}BrP$ M 337.239
Solid. Mp 201°. $[\alpha]_D$ −36.8° (c, 1.38 in MeOH).
Ylide: [26255-46-9]. *Methylphenyl(phenylmethylene)-propylphosphorane. Benzylidenemethylphenylpropylphosphorane.*
$C_{17}H_{21}P$ M 256.327
Reactive Wittig reagent. Has the (*R*)-config.

(±)-*form*
Bromide: Solid. Mp 185°.

Horner, L. *et al, Tetrahedron Lett.*, 1961, 161 (*synth*)
Peerdeman, A.F. *et al, Tetrahedron Lett.*, 1965, 811 (*abs config*)
Bestmann, H.J. *et al, Tetrahedron*, 1968, **24**, 3299 (*ylide, props*)
Bestmann, H.J. *et al, Justus Liebigs Ann. Chem.*, 1974, 1684 (*ylide*)
Akasaki, T. *et al, Phosphorus Sulfur*, 1978, **4**, 211 (*synth*)
Bestmann, H.J. *et al, Chem. Ber.*, 1982, **115**, 3875 (*synth, resoln*)

Benzylmethylphosphinic acid, 8CI B-00053

Methyl(phenylmethyl)phosphinic acid, 9CI

OR
PhCH₂►P◄Me (*S*)ₚ-form of esters
O

$C_8H_{11}O_2P$ M 170.147

(*S*)-*form*
Me ester: [85185-99-5]. *Methyl benzylmethylphosphinate.*
$C_9H_{13}O_2P$ M 184.174
Solid. Mp 50-51°. $[\alpha]_D^{25}$ −8.5° (c, 1.96 in C_6H_6) (99% o.p.).

(±)-*form*
Solid. Mp 121-122°.
Et ester: Ethyl benzylmethylphosphinate.
$C_{10}H_{15}O_2P$ M 198.201
Liq. $Bp_{0.4}$ 120-125°.

Maier, L., *Chem. Ber.*, 1961, **94**, 3051 (*synth*)
Steininger, E., *Chem. Ber.*, 1963, **96**, 3184 (*ethyl ester*)

Moriyama, M. *et al*, *Tetrahedron Lett.*, 1982, **23**, 4547 (*methyl ester*)

Benzylmethylpropylphosphine **B-00054**

Methyl(phenylmethyl)propylphosphine, 9CI

CH_2Ph
Me–P–:
$CH_2CH_2CH_3$
(S)-form

$C_{11}H_{17}P$ M 180.229

(S)-form

Solid. $[\alpha]_D^{25}$ +21.5° (c, 0.97 in MeOH) (55% o.p.), +39.0° (extrapolated).

B,PhI: see Benzylmethylphenylpropylphosphonium(1+), B-00052

Oxide:

$C_{11}H_{17}OP$ M 196.228

Solid. Mp 54-59°. $[\alpha]_D^{25}$ −3.2° (c, 4.4 in MeOH) (55% o.p.). Has the (*R*)-config.

Korpium, O. *et al*, *J. Am. Chem. Soc.*, 1967, **89**, 4784 (*synth, oxide, salt*)
Horner, L. *et al*, *Tetrahedron Lett.*, 1968, 3157 (*salt*)
Moriyama, M. *et al*, *J. Am. Chem. Soc.*, 1983, **105**, 4727 (*synth, abs config, oxide*)

6-Benzyl-2-oxa-6-phosphatricyclo[3.3.1.1^{3,7}]decane **B-00055**

6-(Phenylmethyl)-2-oxa-6-phosphatricyclo[3.3.1.1^{3,7}]decane, 9CI. 6-Benzyl-6-phospha-2-oxaadamantane

[37759-12-9]

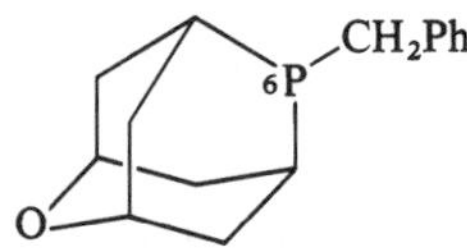

$C_{15}H_{19}OP$ M 246.288

Oil.

B,PhCH$_2$Cl: 6,6-Dibenzyl-6-phosphonia-2-oxaadamantane chloride.

$C_{22}H_{26}ClOP$ M 372.874

Cryst. (MeOH/EtOAc). Mp 248-250°.

6-Oxide: [37835-11-3].

$C_{15}H_{19}O_2P$ M 262.288

Cryst. (Me_2CO/hexane).

Kashman, Y. *et al*, *Tetrahedron*, 1972, **28**, 4091 (*synth, derivs, ir, pmr*)

Benzylphenylphosphinic acid **B-00056**

Phenyl(phenylmethyl)phosphinic acid, 9CI

[7282-98-6]

$PhCH_2P(O)(OH)Ph$

$C_{13}H_{13}O_2P$ M 232.218

Cryst. (EtOH). Mp 182.5-183.5°. pK_a 23.9 (1,2-dimethoxyethane), 27.8 (DMSO).

Me ester: [2129-79-5]. *Methyl benzylphenylphosphinate.*

$C_{14}H_{15}O_2P$ M 246.245

Cryst. (cyclohexane). Mp 94-95°.

Et ester: Ethyl benzylphenylphosphinate.

$C_{15}H_{17}O_2P$ M 260.272

Cryst. (pet. ether). Mp 62.5-63.5°. $Bp_{0.2}$ 180-182°.

Isopropyl ester: Isopropyl benzylphenylphosphinate.

$C_{16}H_{19}O_2P$ M 274.299

Needles (Et_2O). Mp 98°.

Chloride: [32395-29-2].

$C_{13}H_{12}ClOP$ M 250.664

No phys. props. reported.

Anilide: P-*Benzyl*-N,P-*diphenylphosphinic amide.*

$C_{19}H_{22}NOP$ M 311.363

Cryst. (EtOH). Mp 220-222°.

Morpholide: N-*[Phenyl(phenylmethyl)phosphinyl]-morpholine.*

$C_{17}H_{20}NO_2P$ M 301.324

Cryst. ($CHCl_3/Et_2O$). Mp 149-150°.

Horner, L. *et al*, *Chem. Ber.*, 1962, **95**, 581 (*ethyl ester*)
Blicke, F.F. *et al*, *J. Org. Chem.*, 1964, **29**, 204 (*synth, methyl ester*)
Gallagher, M.J., *Aust. J. Chem.*, 1968, **21**, 1197 (*pmr*)
Regitz, M. *et al*, *Chem. Ber.*, 1971, **104**, 2177 (*methyl ester, amides*)
Garst, M.E., *Synth. Commun.*, 1979, **9**, 261 (*synth*)

Benzyl phosphinate **B-00057**

Phenylmethyl phosphinate. Benzyl hyophosphite

[18108-16-2]

$PhCH_2OP(O)H_2 \rightleftharpoons PhCH_2OP(OH)H$

$C_7H_9O_2P$ M 156.121

Liq. d_4^{20} 1.15. $Bp_{0.05}$ 160-170°. n_D^{20} 1.5415.

Ivanov, B.E. *et al*, *Izv. Akad. Nauk SSSR, Ser. Khim.*, 1967, 1498 (*Engl. transl.* p. 1447) (*synth*)
Gallagher, M.J. *et al*, *J. Chem. Soc., Chem. Commun.*, 1978, 54 (*P nmr*)

Benzylphosphine, 8CI **B-00058**

(Phenylmethyl)phosphine, 9CI

[14990-01-3]

$PhCH_2PH_2$

C_7H_9P M 124.122

Liq. Bp 180°, Bp_{26} 83°. Forms cryst. salts with HCl and HI.

Horner, L. *et al*, *Chem. Ber.*, 1958, **91**, 1583 (*synth*)
Maier, L., *Helv. Chim. Acta*, 1966, **49**, 1718 (*nmr*)
Kostyanovskii, R.G. *et al*, *Izv. Akad. Nauk SSSR, Ser. Khim.*, 1975, 901 (*Engl. transl.* p. 816) (*synth, pmr*)
Kostyanovskii, R.G. *et al*, *Org. Mass Spectrom.*, 1976, **11**, 237 (*ms*)

Benzylphosphinic acid **B-00059**

Phenylmethylphosphinic acid, 9CI

[20394-86-9]

$PhCH_2PH(O)OH$

$C_7H_9O_2P$ M 156.121

Tautomeric with Benzylphosphonous acid, B-00068 . Syrup.

Weil, T. *et al*, *Helv. Chim. Acta*, 1953, **36**, 1314.

Benzylphosphonic acid, 8CI **B-00060**

(Phenylmethyl)phosphonic acid, 9CI

[6881-57-8]

$PhCH_2P(O)(OH)_2$

$C_7H_9O_3P$ M 172.120

Cryst. (H_2O). Mp 173-175°. pK_{a1} 2.38 (H_2O, 25°), 3.60 (50% EtOH pK_{a2} *aq.), pK_{a3} *pK_{a2} 5.98 (H_2O, pK_{a4} *25°), 6.52 (50% EtOH aq.).

Mono-Me ester: [63581-66-8]. *Methyl hydrogen benzylphosphonate.*

$C_8H_{11}O_3P$ M 186.147

Cryst. (heptane). Mp 96-97°.

Di-Me ester: [773-47-7]. *Dimethyl benzylphosphonate.*
$C_9H_{13}O_3P$ M 200.174
Liq. Bp_{25} 168-170°, $Bp_{0.5}$ 110-112°.

Mono-Et ester: [18933-98-7]. *Ethyl hydrogen benzylphosphonate.*
$C_9H_{13}O_3P$ M 200.174
Cryst. (heptane). Mp 63-65°.

Di-Et ester: see Diethyl benzylphosphonate, D-00272

Dibutyl ester: [3762-27-4]. *Dibutyl benzylphosphonate.*
$C_{15}H_{25}O_3P$ M 284.334
Liq. Bp_3 140-143°. n_D^{25} 1.4820.
▷SZ6582000.

Di-Ph ester: [10419-87-1]. *Diphenyl benzylphosphonate.*
$C_{19}H_{17}O_3P$ M 324.315
Needles (ligroin). Mp 61-62°.

Bis(trimethylsilyl) ester: [18406-56-9]. *Bis(trimethylsilyl) benzylphosphonate.*
$C_{13}H_{25}O_3PSi_2$ M 316.483
Liq. $Bp_{0.1}$ 95-97°.

Dichloride: see Benzylphosphonic dichloride, B-00062

Diamide: see P-Benzylphosphonic diamide, B-00061

Diisocyanate:
$C_9H_7N_2O_3P$ M 222.140
Liq. $Bp_{1\text{-}2}$ 145°.

Kosolapoff, G.M., *J. Am. Chem. Soc.*, 1945, **67**, 2259 (*synth, esters*)
Haven, A.C., *J. Am. Chem. Soc.*, 1956, **78**, 842 (*diisocyanate*)
Rabinowitz, R., *J. Am. Chem. Soc.*, 1960, **82**, 4564 (*synth, esters*)
Griffiths, W.R. *et al*, *Phosphorus*, 1975, **5**, 273 (*ms*)
Rosini, G. *et al*, *Synthesis*, 1975, 44 (*ester, synth, ir, pmr*)
Honig, M.L. *et al*, *J. Org. Chem.*, 1977, **42**, 379 (*ester, synth, pmr, P nmr*)
Griffiths, W.R., *Phosphorus Sulfur*, 1978, **5**, 101 (*esters, ms*)
Gross, H. *et al*, *J. Prakt. Chem.*, 1978, **320**, 344 (*synth, silyl ester, pmr*)
Blackburn, G.M. *et al*, *J. Chem. Soc., Perkin Trans. 1*, 1980, 1150 (*synth, pmr*)
Morita, T. *et al*, *Bull. Chem. Soc. Jpn.*, 1981, **54**, 267 (*synth, silyl ester, pmr, ir*)

P-Benzylphosphonic diamide, 8CI — B-00061

P-*(Phenylmethyl)phosphonic diamide, 9CI*

$PhCH_2P(O)(NH_2)_2$

$C_7H_{11}N_2OP$ M 170.150

N,N,N′,N′-*Tetra-Me:* [14655-73-3]. N,N,N′,N-*Tetramethyl*-P-*(phenylmethyl)phosphonic diamide. Benzylphosphonic bis(dimethylamide).*
$C_{11}H_{19}N_2OP$ M 226.258
Hygroscopic cryst. (Et_2O/hexane), cryst. (pet. ether). Mp 82.5-83.5°. Bp_2 141-142°. pK_a 25.1 (dimethoxyethane), pK_a 30.0 (DMSO).

N,N,N′,N′-*Tetra-Et:* [52175-06-1]. N,N,N′,N′-*Tetraethyl*-P-*(phenylmethyl)phosphonic diamide. Benzylphosphonic bis(diethylamide).*
$C_{15}H_{27}N_2OP$ M 282.365
pK_a 25.7 (dimethoxyethane), pK_a 30.9 (DMSO).

N,N′-*Di-Ph: Benzylphosphonic dianilide.* N,N′-*Diphenyl*-P-*(phenylmethyl)phosphonic diamide.*
$C_{19}H_{19}N_2OP$ M 322.346
Solid. Mp 185-186°.

Kinnear, A.M. *et al*, *J. Chem. Soc.*, 1952, 3437 (*dianilide*)
Normant, H. *et al*, *C.R. Hebd. Seances Acad. Sci., Ser. C*, 1967, **264**, 707 (*bisdimethylamide*)
Mesyats, S.P. *et al*, *Izv. Akad. Nauk SSSR, Ser. Khim.*, 1974, **23**, 2497 (*Engl. transl.* p. 2406) (*props*)
Collignon, N. *et al*, *Bull. Soc. Chim. Fr.*, 1977, 120 (*bisdimethylamide*)

Benzylphosphonic dichloride, 8CI — B-00062

Phenylmethylphosphonic dichloride, 9CI
[1499-19-0]

$PhCH_2P(O)Cl_2$

$C_7H_7Cl_2OP$ M 209.011
Solid. Mp 57.5°. Bp_2 130°.

Kinnear, A.M. *et al*, *J. Chem. Soc.*, 1952, 3437 (*synth*)
Petrov, K.A. *et al*, *Zh. Obshch. Khim.*, 1961, **31**, 3027 (*Engl. transl.* p. 2823) (*synth, props*)
Griffiths, W.R. *et al*, *Phosphorus Sulfur*, 1978, **4**, 341 (*ms*)
Glukhikh, V.I. *et al*, *Dokl. Akad. Nauk SSSR, Ser. Sci. Khim.*, 1979, **248**, 142 (*Phys. Chem., Engl. transl.* p. 744) (*cmr*)
Morita, T. *et al*, *Chem. Lett.*, 1980, 435 (*synth*)

Benzylphosphonochloridic acid — B-00063

(Phenylmethyl)phosphonochloridic acid, 9CI

$PhCH_2PCl(O)OH$

$C_7H_8ClO_2P$ M 190.566

Et ester: [41760-95-6]. *Ethyl benzylphosphonochloridate.*
$C_9H_{12}ClO_2P$ M 218.619
Liq. d_4^{20} 1.22. $Bp_{0.4}$ 90-100° (bath). n_D^{20} 1.5250.

Cyclohexyl ester: Cyclohexyl benzylphosphonochloridate.
$C_{13}H_{18}ClO_2P$ M 272.711
Liq. n_D^{25} 1.5309.

Hafner, L.S. *et al*, *J. Med. Chem.*, 1970, **13**, 1025 (*synth*)
Knunyants, I.L. *et al*, *Zh. Obshch. Khim.*, 1972, **42**, 2421 (*Engl. transl.* p. 2415) (*synth*)

Benzylphosphonochloridothioic acid, 8CI — B-00064

Phenylmethylphosphonochloridothioic acid, 9CI

$PhCH_2P(=S)(OH)Cl \rightleftharpoons PhCH_2P(=O)(SH)Cl$

C_7H_8ClOPS M 206.626

1-Naphthyl ester: [33801-18-2]. *1-Naphthyl benzylphosphonochloridothioate.*
$C_{17}H_{14}ClOPS$ M 332.784
Solid. Mp 108-113°.

2-Naphthyl ester: [33801-17-1]. *2-Naphthyl benzylphosphonochloridothioate.*
$C_{17}H_{14}ClOPS$ M 332.784
Oil. n_D^{20} 1.6495.

Ger. Pat., 1 956 187, (*1971*); *CA*, **75**, 77028

Benzylphosphonodithioic acid, 8CI — B-00065

Phenylmethylphosphonodithioic acid, 9CI

$PhCH_2P(=O)(SH)SH \rightleftharpoons PhCH_2P(=S)(OH)SH$

$C_7H_9OPS_2$ M 204.241

S,S-*Di-Et ester:* [22082-39-9]. S,S-*Diethyl benzylphosphonodithioate.*
$C_{11}H_{17}OPS_2$ M 260.348

Liq. d_4^{20} 1.17. $Bp_{0.065}$ 117-118°. n_D^{20} 1.5963.
S,S-*Dipropyl ester:* S,S-*Dipropyl benzylphosphonodithioate.*
$C_{13}H_{21}OPS_2$ M 288.402
Liq. d_4^{20} 1.12. $Bp_{0.04}$ 132-133°. n_D^{20} 1.5720.

Minich, D. *et al, Izv. Akad. Nauk SSSR, Ser. Khim.*, 1968, 1792 (*Engl. transl.* p. 1694)

Benzylphosphonofluoridic acid **B-00066**

(*Phenylmethyl*)*phosphonofluoridic acid*, 9CI
[55137-99-0]

$PhCH_2PF(O)OH$

$C_7H_8FO_2P$ M 174.111
Cyclohexyl ester: [28364-26-3]. *Cyclohexyl benzylphosphonofluoridate.*
$C_{13}H_{18}FO_2P$ M 256.256
Liq. $Bp_{0.03}$ 96°. n_D^{25} 1.5041.

Hafner, L.S. *et al, J. Med. Chem.*, 1970, **13**, 1025 (*synth*)

Benzylphosphonothioic acid, 8CI **B-00067**

Phenylmethylphosphonothioic acid, 9CI

$PhCH_2P(S)(OH)_2 \rightleftharpoons PhCH_2P(O)(OH)(SH)$

$C_7H_9O_2PS$ M 188.181
Mono-O-Et ester: [61707-02-6]. O-*Ethyl hydrogen benzylphosphonothioate. Monoethyl benzylthiophosphonate.*
$C_9H_{13}O_2PS$ M 216.234
Liq. d_4^{20} 1.19. n_D^{20} 1.5486.
O,O-*Di-Et ester:* [73178-33-3]. O,O-*Diethyl benzylphosphonothioate.*
$C_{11}H_{17}O_2PS$ M 244.288
Liq. d_4^{20} 1.10. Bp_3 125.5-127.5°. n_D^{20} 1.5305.
O,S-*Di-Et ester:* O,S-*Diethyl benzylphosphonothioate.*
$C_{11}H_{17}O_2PS$ M 244.288
Liq. d_4^{20} 1.13. Bp_2 134-136°. n_D^{20} 1.5350.
Dichloride: [6588-19-8].
$C_7H_7Cl_2PS$ M 225.072
Liq. d_4^{20} 1.33-1.36. Bp_1 130-133°. n_D^{20} 1.6010.
Dibromide:
$C_7H_7Br_2PS$ M 313.974
Liq. d_4^{20} 1.92. Bp_{10} 178-180°. n_D^{20} 1.6505.

Kabachnik, M.I. *et al, Izv. Akad. Nauk SSSR, Ser. Khim.*, 1956, 193 (*Engl. transl.* p. 185) (*esters, synth*)
Kuzamyshev, V.M. *et al, Izv. Akad. Nauk SSSR, Ser. Khim.*, 1976, 1885 (*Engl. transl.* p. 1775) (*monoethyl ester, synth, ir*)
Khokhlov, P.S. *et al, Zh. Obshch. Khim.*, 1978, **48**, 564 (*Engl. transl.* p. 514) (*dichloride, synth*)
Khokhlov, P.S. *et al, Zh. Obshch. Khim.*, 1984, **54**, 972 (*Engl. transl.* p. 867) (*dibromide, synth, ir, P nmr*)

Benzylphosphonous acid, 8CI **B-00068**

(*Phenylmethyl*)*phosphonous acid*, 9CI

$PhCH_2P(OH)_2$

$C_7H_9O_2P$ M 156.121
The free acid exists as the phosphoryl tautomer, Benzylphosphinic acid, B-00059 .
Di-Et ester: [36103-43-2]. *Diethyl benzylphosphonite.*
$C_{11}H_{17}O_2P$ M 212.228
Liq. d_4^{20} 1.02. Bp_2 92-94°. n_D^{20} 1.5032.
Diisopropyl ester: [14561-18-3]. *Diisopropyl benzylphosphonite.*
$C_{13}H_{21}O_2P$ M 240.281
Liq. Bp_{15} 128-131°.
Difluoride: [16141-66-5]. *Benzyldifluorophosphine.*
$C_7H_7F_2P$ M 160.103
Liq. Bp_{10} 43-45°. n_D^{20} 1.4974.
Dichloride: [4545-85-1]. *Benzyldichlorophosphine.*
$C_7H_7Cl_2P$ M 193.012
Liq. d_4^{20} 1.28. Bp_{12} 111-113°. n_D^{20} 1.5840.
Dibromide: [698-89-5]. *Benzyldibromophosphine.*
$C_7H_7Br_2P$ M 281.914
Liq. d_4^{20} 1.70. Bp_{10} 142-143°, Bp_1 94-96°. n_D^{20} 1.6391.

Sander, M., *Chem. Ber.*, 1960, **93**, 1220 (*diethyl ester*)
Petrov, K.A. *et al, Zh. Obshch. Khim.*, 1961, **31**, 3027 (*Engl. transl.* p. 2823) (*dichloride*)
Batkowski, T. *et al, Rocz. Chem.*, 1967, **41**, 471 (*diisopropyl ester, synth, props*)
Drozd, G.I. *et al, Zh. Obshch. Khim.*, 1967, **37**, 958, 1343 (*Engl. transl.* pp. 906, 1269) (*difluoride, synth, F and P nmr*)
Baranov, Yu.I. *et al, Zh. Obshch. Khim.*, 1969, **39**, 836 (*Engl. transl.* p. 799) (*dibromide*)

Benzylphosphoramidic acid, 8CI **B-00069**

(*Phenylmethyl*)*phosphoramidic acid*, 9CI

$PhCH_2NHP(O)(OH)_2$

$C_7H_{10}NO_3P$ M 187.135
pK_{a1} 2.89, pK_{a2} 8.84 (H_2O, 20°).
Di-Me ester: [74124-44-0]. *Dimethyl benzylphosphoramidate.*
$C_9H_{14}NO_3P$ M 215.188
Oil. $Bp_{0.04}$ 114°.
Di-Et ester: [53640-96-3]. *Diethyl benzylphosphoramidate.*
$C_{11}H_{18}NO_3P$ M 243.242
Oil. $Bp_{0.05}$ 120°. n_D^{20} 1.4940.
Diisopropyl ester: [59658-74-1]. *Diisopropyl benzylphosphoramidate. Bis(1-methylethyl) (phenylmethyl)phosphoramidate.*
$C_{13}H_{22}NO_3P$ M 271.295
Cryst. ($CHCl_3$ or pet. ether). Mp 57-58°.
*Di-*tert*-butyl ester:* [56884-00-5]. *Di-*tert*-butyl benzylphosphoramidate. Bis(1,1-dimethylethyl) (phenylmethyl)phosphoramidate.*
$C_{15}H_{26}NO_3P$ M 299.349
Needles (EtOH aq.). Mp 97-99°.
Di-Ph ester: [33985-75-0]. *Diphenyl benzylphosphoramidate.*
$C_{19}H_{18}NO_3P$ M 339.330
Needles (hexane). Mp 103-104° (92-94°).
Dibenzyl ester: [56883-97-7]. *Dibenzyl benzylphosphoramidate.*
$C_{21}H_{22}NO_3P$ M 367.383
Solid. Mp 84-85°.

Atherton, F.R. *et al, J. Chem. Soc.*, 1945, 382 (*dibenzyl ester*)
Benkovic, S.J. *et al, J. Am. Chem. Soc.*, 1971, **93**, 4009 (*synth, diphenyl ester, ir, pmr*)
Cremlyn, R.J.W., *Aust. J. Chem.*, 1973, **26**, 1591 (*diphenyl ester*)
Zwierzak, A., *Synthesis*, 1975, 507; 1982, 920; 1984, 332 (*esters, synth, ir, pmr, P nmr*)
Glidewell, C., *J. Organomet. Chem.*, 1976, **108**, 335 (*diisopropyl ester*)
Kunieda, T. *et al, Tetrahedron*, 1983, **39**, 3253 (*diphenyl ester, synth, pmr*)
Willert, A. *et al, Helv. Chim. Acta*, 1983, **66**, 2467 (*esters, synth, ir, pmr, cmr, P nmr*)
Davidowitz, B. *et al, Org. Mass. Spectrom.*, 1984, **19**, 128 (*diphenyl ester, ms*)

Benzyl phosphorodichloridite, 8CI **B-00070**

Phenylmethyl phosphorodichloridite, 9CI. Benzyl dichlorophosphite

[76101-29-6]

$PhCH_2OPCl_2$

$C_7H_7Cl_2OP$ M 209.011

Useful phosphylating agent in oligonucleotide synth. d_0^0 1.32, d^{23} 1.28. Bp_{11} 113-4°, $Bp_{0.2}$ 65°. $n_D^{19.4}$ 1.5584.

Razumov, A.I., *J. Gen. Chem. USSR*, 1944, **14**, 464; *CA*, **39**, 4568.
Ogilvie, K.K., *Can. J. Chem.*, 1980, **58**, 2686 (*synth, P nmr, use*)

***S*-Benzyl phosphorothioate, 8CI** **B-00071**

S-*(Phenylmethyl) phosphorothioate, 9CI.* S-*Benzyl thiophosphate.* S-*Benzyl dihydrogen thiophosphate*

$PhCH_2SP(O)(OH)_2$

$C_7H_9O_3PS$ M 204.180

Monodicyclocyclohexylammonium salt: Solid. Mp 151-152°.

Zwierzak, A. *et al*, *Z. Naturforsch., B*, 1971, **26**, 386 (*synth, ir, pmr*)
Meade, T.J. *et al*, *J. Org. Chem.*, 1985, **50**, 936 (*synth, P nmr*)

[(Benzylthio)methyl]phosphonic acid, 8CI **B-00072**

[[(Phenylmethyl)thio]methyl]phosphonic acid, 9CI

$PhCH_2SCH_2P(O)(OH)_2$

$C_8H_{11}O_3PS$ M 218.207

Di-Et ester: [41760-64-9]. *Diethyl [[(phenylmethyl)-thio]methyl]phosphonate, 9CI. Diethyl [(benzylthio)-methyl]phosphonate.*
$C_{12}H_{19}O_3PS$ M 274.314
Wittig-Horner reagent for the prep. of benzyl vinyl sulfides. n_D^{24} 1.5356.

Bel'ski, V.E. *et al*, *Zh. Obshch. Khim.*, 1972, **42**, 2427 (*Engl. transl.* p. 2421) (*P nmr*)
Mikolajczyk, M. *et al*, *J. Org. Chem.*, 1979, **44**, 2967 (*synth, pmr*)

Benzyltriphenylphosphonium(1+), 8CI **B-00073**

Triphenyl(phenylmethyl)phosphonium(1+), 9CI

[15853-35-7]

$PhCH_2P^{\oplus}Ph_3$

$C_{25}H_{22}P^{\oplus}$ M 353.423 (ion)

Converted by base to benzyltriphenylphosphorane, a Wittig reagent.

Chloride: [1100-88-5].
$C_{25}H_{22}ClP$ M 388.876
Cryst. ($CHCl_3$/pet. ether). Mp 343-345° (338°).

Bromide: [1449-46-3].
$C_{25}H_{22}BrP$ M 433.327
Cryst. (EtOH). Mp 295° (285°).

Iodide: [1243-97-6].
$C_{25}H_{22}IP$ M 480.327
Hexagonal prisms (EtOH).
▷TA1846500.

Triodide: [72974-99-3].
$C_{25}H_{22}I_3P$ M 734.136
Mp 164-166°.

Pentaiodide: [72975-00-9].
$C_{25}H_{22}I_5P$ M 987.945
Mp 92-96°.

Heptaiodide: [72974-99-3].
$C_{25}H_{22}I_7P$ M 1241.754
Mp 105-107°.

Kröehnke, F., *Chem. Ber.*, 1950, **83**, 291 (*bromide*)
Witschard, G. *et al*, *Spectrochim. Acta*, 1963, **19**, 1905 (*ir*)
Skapski, A.C. *et al*, *J. Cryst. Mol. Struct.*, 1974, **4**, 77 (*cryst struct*)
Albright, T.A. *et al*, *J. Am. Chem. Soc.*, 1975, **97**, 2946 (*nmr*)
Ostoja Starzewski, K.A. *et al*, *Phosphorus*, 1976, **6**, 177 (*cmr*)
Verstuyft, A.W. *et al*, *Inorg. Chem.*, 1977, **16**, 2776 (*pmr, cmr, nmr*)
Kostina, V.G. *et al*, *Zh. Obschch. Khim.*, 1979, **49**, 2452 (*Engl. transl.* p. 2165) (*polyiodides*)
Archer, S.J. *et al*, *Phosphorus Sulfur*, 1981, **11**, 101 (*iodide, cryst struct*)
Havens, S. *et al*, *J. Polym. Sci., Polym. Chem. Ed.*, 1981, **19**, 1349 (*chloride*)
Maccarone, E. *et al*, *Gazz. Chim. Ital.*, 1982, **112**, 25; *CA*, **97**, 23908.
Makovestskii, Yu.P. *et al*, *Zh. obshch. Khim.*, 1982, **52**, 2235 (*Engl. transl.* p. 1989) (*polyiodides, uv*)

2,2′-Bi(1,3,2-benzodithiaphosphole) **B-00074**

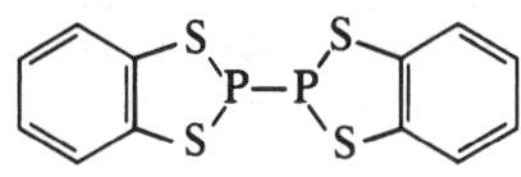

$C_{12}H_8P_2S_4$ M 342.383

Cryst. (pentane at −20°).

Baudler, M. *et al*, *Z. Naturforsch., B*, 1973, **28**, 363 (*synth, ir, raman*)

Bicyclo[2.2.1]hept-5-ene-2,3-diylbis(methylene)bis[diphenylphosphine], 10CI **B-00075**

2,3-Bis(diphenylphosphinomethyl)-5-norbornene

CH_2PPh_2
CH_2PPh_2 (2*R**,3*R**)*-form*

$C_{33}H_{32}P_2$ M 490.563

Ligand for asymmetric hydrogenation.

(2R*,3R*)-form [79426-29-2]
(+)-endo,exo-*form*
Solid. Mp 78-79°. $[\alpha]_D^{25}$ +12.0° (c, 1 in C_6H_6), $[\alpha]_D$ −25.7° (c, 1 in $CHCl_3$).
Dioxide: [79426-27-0]. Cryst. Mp 204-206°. $[\alpha]_D^{25}$ +77.8° (c, 1 in C_6H_6), $[\alpha]_D$ +33.4° (c, 1 in EtOH), $[\alpha]_D$ +28.2° (c, 1 in $CHCl_3$).

(2S*,3R*)-form [79426-28-1]
(−)-endo,exo-*form*
Solid. Mp 78-79°. $[\alpha]_D^{25}$ −12.1° (c, 1 in C_6H_6), $[\alpha]_D$ +25.2° (c, 1 in $CHCl_3$).
Dioxide: [79426-26-9]. Cryst. Mp 205-206°. $[\alpha]_D^{25}$ −76.4° (c, 1 in C_6H_6), $[\alpha]_D$ −33.4° (c, 1 in EtOH), $[\alpha]_D$ −27.0° (c, 1 in $CHCl_3$).

(2RS,3RS)-form [79426-24-7]
(±)-endo,exo-*form*
Solid.
Dioxide: [79426-25-8]. Cryst. (EtOH aq.). Mp 208-209°.

Döbler, C. *et al*, *J. Prakt. Chem.*, 1981, **323**, 667 (*synth, resoln*)

(1,8-Biphenylenediyldimethylidyne)-bis[triphenylphosphorane], 9CI B-00076

1,8-Bis(triphenylphosphoranylidenemethyl)biphenylene

[36230-19-0]

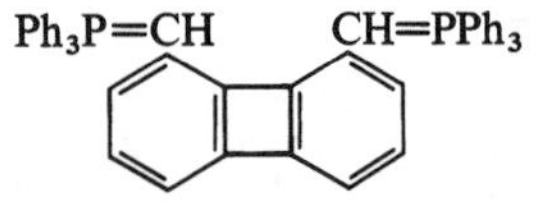

$Ph_3P{=}CH$ $CH{=}PPh_3$

$C_{50}H_{38}P_2$ M 700.798

Used in Wittig reactions. Blood-red ylide produced *in situ*.

Wilcox, C.F. *et al*, *J. Am. Chem. Soc.*, 1975, **97**, 1914 (*synth, use*)

4-Biphenylphosphonic acid, 8CI B-00077

[1,1′-Biphenyl]-4-ylphosphonic acid, 9CI

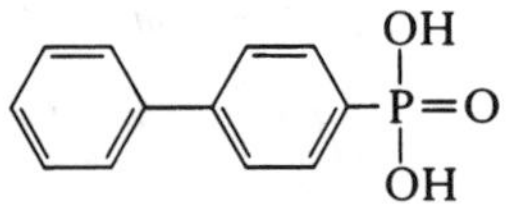

$C_{12}H_{11}O_3P$ M 234.191

Cryst. (EtOH/HCl aq.). Mp 218-220°, 235-240°. pK_{a1} 3.78 (50% EtOH aq.), pK_{a2} 8.13 (H_2O).

Di-Et ester: Diethyl 4-biphenylylphosphonate.

$C_{16}H_{19}O_3P$ M 290.298

Liq. $Bp_{0.1\text{-}0.4}$ 150° (bath).

Jaffé, H.H. *et al*, *J. Am. Chem. Soc.*, 1954, **76**, 1548 (*synth*)
Grabiak, R.C. *et al*, *Phosphorus Sulfur*, 1980, **9**, 197 (*synth, derivs, P nmr*)

[1,1′-Biphenyl]-4-yldiphenylphosphine, 9CI B-00078

4-(Diphenylphosphino)biphenyl

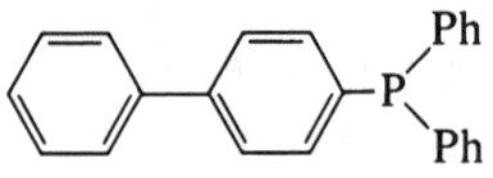

$C_{24}H_{19}P$ M 338.388

Needles. Mp 84°. $Bp_{0.05\text{-}00.06}$ 202-206°. Dec. at 384°.

Oxide: 4-(Diphenylphosphinyl)biphenyl.

$C_{24}H_{19}OP$ M 354.387

Solid. Mp 143-144°.

Aguiar, A.M. *et al*, *J. Org. Chem.*, 1963, **28**, 2091 (*oxide*)
Schindlbauer, H. *et al*, *Z. Anal. Chem.*, 1966, **221**, 394 (*ir*)
Mitterhofer, F. *et al*, *Monatsh. Chem.*, 1967, **98**, 206 (*synth*)

[(1,1′-Biphenyl)-2-ylmethyl]phosphonic acid, 9CI B-00079

2-(Phosphonomethyl)biphenyl

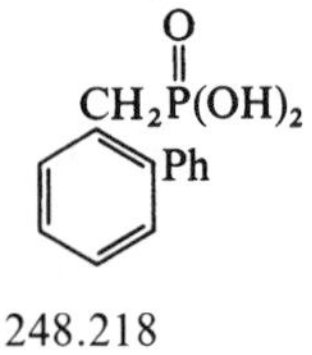

$C_{13}H_{13}O_3P$ M 248.218

Solid. Mp 167-169°.

Di-Et ester: Diethyl [(1,1′-biphenyl)-2-ylmethyl]-phosphonate.

$C_{17}H_{21}O_3P$ M 304.325

Wittig-Horner reagent. Liq. $Bp_{0.3}$ 148-152°. n_D^{25} 1.5488.

Dichloride:

$C_{13}H_{11}Cl_2OP$ M 285.109

Liq. $Bp_{0.2}$ 150°.

Dianilide: P-[(*1,1′-Biphenyl*)-*2-ylmethyl*]-N,N′-*diphenylphosphonic diamide.*

$C_{25}H_{23}N_2OP$ M 398.443

Cryst. (EtOH). Mp 194-195°.

Lynch, E.R., *J. Chem. Soc.*, 1962, 3729.

[(1,1′-Biphenyl)-4-ylmethyl]phosphonic acid, 9CI B-00080

4-(Phosphonomethyl)biphenyl

$C_{13}H_{13}O_3P$ M 248.218

Solid. Mp ca. 250° dec.

Di-Et ester: [30818-70-3]. *Diethyl[(1,1′-biphenyl)-4-ylmethyl]phosphonate.*

$C_{17}H_{21}O_3P$ M 304.325

Characterised spectroscopically.

Kosolapoff, G.M., *J. Am. Chem. Soc.*, 1945, **67**, 2259 (*synth*)
Ernst, L., *Org. Magn. Reson.*, 1977, **9**, 35 (*ester, cmr*)

[1,1′-Biphenyl]-4-yl-1-naphthalenylphenyl-phosphine, 9CI B-00081

[25076-73-7]

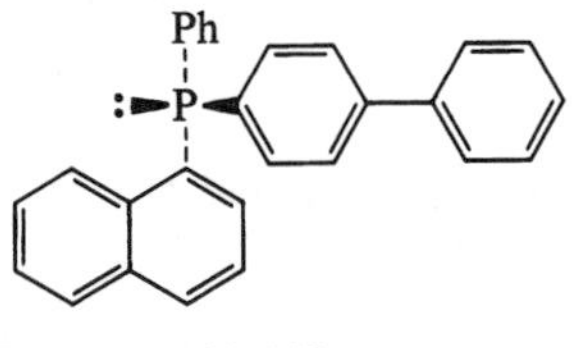

(*R*)-*form*

$C_{28}H_{21}P$ M 388.448

(R)-form [17430-39-6]

Solid. Mp 193-196°. $[\alpha]^{20}_{436}$+9.7° (c, 2.45 in CH_2Cl_2).

B,MeBr: [1,1′-Biphenyl]-4-ylmethyl-1-naphthalenylphenylphosphonium bromide.

$C_{29}H_{24}BrP$ M 483.386

Mp 162-164°. $[\alpha]_{436}$+0.41° (c, 22.48 in MeOH). Has (*S*)-config.

Oxide:

$C_{28}H_{21}OP$ M 404.447

Mp 237-240°. $[\alpha]_{436}$+0.71° (c, 14 in $CHCl_3$). Has (*S*)-config.

(S)-form [17622-57-0]

Solid. Mp 191-193°. $[\alpha]^{32}_{D}$ −27.0° (c, 2.50 in $CHCl_3$), $[\alpha]^{20}_{436}$−8.5° (c, 2.12 in CH_2Cl_2).

B,MeBr: Mp 162°. $[\alpha]_{436}$−0.23° (c, 22 in MeOH). (*R*)-config.

Oxide: Mp 240°. $[\alpha]_{440}$−1.8° (c, 5 in $CHCl_3$). Has (*R*)-config.

(±)-*form* [54458-92-3]

Cryst. (CH_2Cl_2/MeOH). Mp 192.5-193.5°.

B,MeI:

$C_{29}H_{24}IP$ M 530.387

Cryst. (EtOH aq.). Mp 158-159°.

Oxide: Cryst. (CH_2Cl_2/MeOH). Mp 247-248°.

Wittig, G. *et al*, *Justus Liebigs Ann. Chem.*, 1971, **751**, 17 (*synth, resoln*)
Luckenbach, R., *Justus Liebigs Ann. Chem.*, 1974, 1618 (*abs config*)
Tani, K. *et al*, *J. Am. Chem. Soc.*, 1977, **99**, 7876 (*resoln*)

(1,1′-Biphenyl)-3-yl phosphorodichloridate, 9CI **B-00082**
(1,1′-Biphenyl)-3-yl dichlorophosphate. m-Xenyl dichlorophosphate

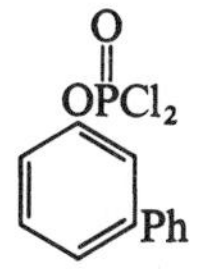

$C_{12}H_9Cl_2O_2P$ M 287.082
Oil. d_4^{25} 1.36. Bp_{9-11} 218-221°.

U.S.P., 2 117 290, (1938)

***O*-[(1,1′-Biphenyl)-2-yl] phosphorodichlorodothioate,** 9CI **B-00083**
O-2-Biphenylyl dichlorothiophosphate
[31139-62-5]

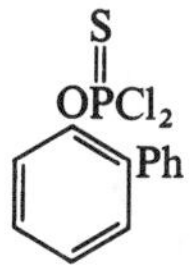

$C_{12}H_9Cl_2OPS$ M 303.142
Liq. $Bp_{0.12}$ 137°. n_D^{20} 1.6225.

Cremlyn, R.J.W. *et al*, *Aust. J. Chem.*, 1974, **27**, 1065 (*synth, ir*)

1,1′-Biphospholane, 9CI, 8CI **B-00084**

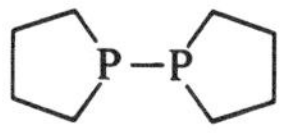

$C_8H_{16}P_2$ M 174.162
Liq. $Bp_{0.05}$ 50°.
Disulfide: [5958-55-4].
$C_8H_{16}P_2S_2$ M 238.282
Cryst. (toluene/MeOH). Mp 185°.

Schmutzler, R., *Inorg. Chem.*, 1964, **3**, 421 (*synth, deriv, ir, P nmr*)
Lee, J.D. *et al*, *Acta Crystallogr., Sect. B*, 1969, **25**, 2127 (*disulfide, cryst struct*)
Aime, S. *et al*, *J. Chem. Soc., Dalton Trans.*, 1976, 2144 (*disulfide, cmr, P nmr*)
Colquhoun, I.J. *et al*, *Org. Magn. Reson.*, 1979, **12**, 473 (*diselenide, nmr*)

1,1′-Biphosphorinane, 8CI **B-00085**
1,1′-Biphosphinane

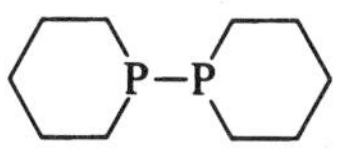

$C_{10}H_{20}P_2$ M 202.216
Disulfide: [5958-56-5].
$C_{10}H_{20}P_2S_2$ M 266.336
Cryst. ($CHCl_3$). Mp 225° (softens from 185°).

Schmutzler, R., *Inorg. Chem.*, 1964, **3**, 421 (*disulfide, synth, ir, nmr*)
Allen, D.W. *et al*, *J. Chem. Soc. (B)*, 1970, 1527 (*P nmr*)
Lee, J.D. *et al*, *Acta Crystallogr., Sect. B*, 1970, **26**, 507 (*disulfide, cryst struct*)

Bis(2-aminoethyl)phenylphosphine **B-00086**
2,2′-(Phenylphosphinidene)bis-1-ethanamine, 9CI

$PhP(CH_2CH_2NH_2)_2$

$C_{10}H_{17}N_2P$ M 196.231
Stable only as dihydrochloride; free base undergoes cyclisation. pK_{a1} 8.24, pK_{a2} 8.98 (50% EtOH aq./0.1M LiCl, 25°).
B,2HCl: Cryst. (EtOH). Mp 255° dec.

Issleib, K. *et al*, *Chem. Ber.*, 1967, **100**, 2685 (*synth, nmr*)

Bis(aminomethyl)phosphinic acid, 9CI **B-00087**
[68358-09-8]

$(H_2NCH_2)_2P(O)OH$

$C_2H_9N_2O_2P$ M 124.079
Powder. Mp 187-190° dec.
B,HCl: [68358-05-4]. Cryst. (EtOH aq.). Mp 307° dec.
B,HBr: [68358-06-5]. Solid. Mp 260-267° dec.
N,N-*dibenzyl: Bis(*N*-benzylaminomethyl)phosphinic acid.*
$C_{16}H_{21}N_2O_2P$ M 304.328
Cryst. (H_2O) as hydrochloride. Mp 257-258°.
N,N-*di*-tert-*butyl: Bis(*N-tert-*butylaminomethyl)-phosphinic acid.*
$C_{10}H_{25}N_2O_2P$ M 236.293
Solid as hydrochloride. Mp 260-262°.
N,N,O-*Tris(trimethylsilyl): Trimethylsilyl bis[*N-*(trimethylsilyl)aminomethyl]phosphinate.*
$C_{11}H_{31}N_2O_2PSi_3$ M 338.608
Liq. $Bp_{0.05}$ 98-105°.

Maier, L., *J. Organomet. Chem.*, 1979, **178**, 157 (*synth, derivs, ir, pmr, P nmr*)

Bis(3-aminophenyl)phosphinic acid, 9CI **B-00088**
[25806-71-7]

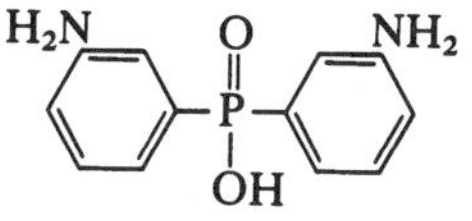

$C_{12}H_{13}N_2O_2P$ M 248.221
Solid. Mp 287-289° dec.
▷SZ4210100.

Doak, G.O. *et al*, *J. Am. Chem. Soc.*, 1952, **74**, 753.

Bis(4-aminophenyl)phosphinic acid, 9CI **B-00089**
$C_{12}H_{13}N_2O_2P$ M 248.221
Cryst. (AcOH). Mp 242-243°.

Robins, R.K. *et al*, *J. Org. Chem.*, 1951, **16**, 324.
Doak, G.O. *et al*, *J. Am. Chem. Soc.*, 1952, **74**, 753.

Bis(3-aminopropyl)phenylphosphine **B-00090**
3,3′-(Phenylphosphinidene)bis-1-propanamine, 9CI
[6775-01-5]

$PhP(CH_2CH_2CH_2NH_2)_2$

$C_{12}H_{21}N_2P$ M 224.285
Oil. d_4^{20} 1.03. Bp_1 148-149°. n_D^{20} 1.5740.
P-*Oxide:*
$C_{12}H_{21}N_2OP$ M 240.284
Liq. d^{20} 1.12. $Bp_{0.45}$ 290-302° (Bp_1 179-180°). n_D^{25} 1.5690.
P-*Sulfide:* [6779-49-3]. *3,3′-(Phenylphosphinothioylidene)bis-1-propanamine.*
$C_{12}H_{21}N_2PS$ M 256.345

Low-melting solid. Mp 41-42°. Bp_1 199-199.5°.
N-*Tetra-Me: see Bis[3-(dimethylamino)propyl]-phenylphosphine, B-00195*

Arbuzov, B.A. *et al*, *Izv. Akad. Nauk SSSR, Ser. Khim.*, 1962, 290 (*Engl. transl.* p. 267) (*synth, derivs*)
Pellon, J. *et al*, *J. Polym. Sci., Part A-1*, 1963, 863 (*synth*)
Oehme, H. *et al*, *J. Prakt. Chem.*, 1973, **315**, 526 (*synth*)
Uriarte, R. *et al*, Inorg. Chem., 1980, **19**, 79 (*synth, pmr, nmr, ir*)

Bis(1-aziridinyl)phosphinic acid, 9CI, 8CI B-00091

[13913-34-3]

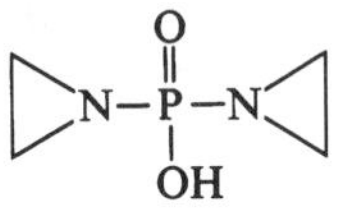

$C_4H_9N_2O_2P$ M 148.101
The esters are insect sterilants.
Me ester: [466-15-9]. *Methyl bis(1-aziridinyl)-phosphinate.*
$C_5H_{11}N_2O_2P$ M 162.128
No phys. props. reported.
▷SZ4348000.
Et ester: [469-35-2]. *Ethyl bis(1-aziridinyl)phosphinate.*
$C_6H_{13}N_2O_2P$ M 176.155
Oil. $Bp_{1.3\text{-}1.4}$ 95-97°.
▷SZ4345000.
Isopropyl ester: [35996-72-6]. *Isopropyl bis(1-aziridinyl)phosphinate.*
$C_7H_{15}N_2O_2P$ M 190.181
No phys. props. reported.
▷SZ4360000.
Butyl ester: Butyl bis(1-aziridinyl)phosphinate.
$C_8H_{17}N_2O_2P$ M 204.208
Liq. $Bp_{0.8}$ 106-107°.
Ph ester: [10537-55-0]. *Phenyl bis(1-aziridinyl)-phosphinate.*
$C_{10}H_{13}N_2O_2P$ M 224.199
Cryst. (Et_2O/pet. ether). Mp 58-59°.
Benzyl ester: [84681-48-1]. *Benzyl bis(1-aziridinyl)-phosphinate.*
$C_{11}H_{15}N_2O_2P$ M 238.225
No phys. props. reported.

Bestian, H., *Justus Liebigs Ann. Chem.*, 1950, **566**, 210 (*esters*)
Kropacheva, A.A. *et al*, *Zh. Obshch. Khim.*, 1959, **29**, 556 (*Engl. transl.* p. 553) (*phenyl ester*)
Chabrier, P. *et al*, *Ann. Pharm. Fr.*, 1978, **36**, 409; *CA*, **90**, 204035 (*synth*)
Tomilets, V.A. *et al*, *Khim.-Farm. Zh.*, 1981, **15**, 34; *CA*, **96**, 62577 (*synth*)
Misiura, K. *et al*, *J. Med. Chem.*, 1983, **26**, 674 (*benzyl ester*)

P,P-Bis(1-aziridinyl)phosphinic amide, 9CI B-00092

[6784-51-6]

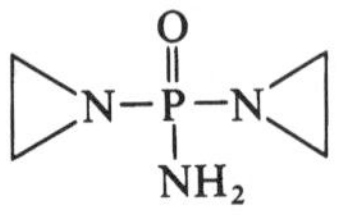

$C_4H_{10}N_3OP$ M 147.116
Simple *N*-alkyl derivatives are insect sterilants.
▷SZ5785000.
N-*Me:* [2275-61-8]. P,P-*Bis(1-aziridinyl)*-N-*methylphosphinic amide.*
$C_5H_{12}N_3OP$ M 161.143
Solid. Mp 105.5-106.5°.
▷SY9500000.
N,N-*Di-Me:* [1195-69-3]. P,P-*Bis(1-aziridinyl)*-N,N-*dimethylphosphinic amide.*
$C_6H_{14}N_3OP$ M 175.170
Liq. d_4 1.13. $Bp_{0.3}$ 60°. n_D 1.4810.
▷SZ5825000.
N-*Et:* [302-48-7]. P,P-*Bis(1-aziridinyl)*-N-*ethylphosphinic amide.*
$C_6H_{14}N_3OP$ M 175.170
Solid. Mp 57-61°. Bp_5 144°.
▷SY9320000.
N,N-*Di-Et:* [1907-75-1]. P,P-*Bis(1-aziridinyl)*-N,N-*diethylphosphinic amide.*
$C_8H_{18}N_3OP$ M 203.223
Insect sterilant eg. against boll weevil. Liq. Bp_1 98-100°.
N-*Ph:* [6784-53-8]. P,P-*Bis(1-aziridinyl)*-N-*phenylphosphinic amide. Phenidet.*
$C_{10}H_{14}N_3OP$ M 223.214
Cryst. (C_6H_6). Mp 143-144°. Reduces egg fertility for female quails.
N-*Benzoyl:* [4110-66-1]. N-*Benzoyl*-P,P-*bis(1-aziridinyl)phosphinic amide. Benzotef.*
$C_{11}H_{14}N_3O_2P$ M 251.224
Solid. Mp 93-94°.
▷CV2350000.
N-*2-Naphthyl:* P,P-*Bis(1-aziridinyl)*-N-*2-naphthylphosphinic amide.*
$C_{14}H_{16}N_3OP$ M 273.274
Cryst. (C_6H_6). Mp 148°.
N-*Adamantyl: see N-(1-Adamantyl)-P,P-bis(aziridinyl)phosphinic amide, A-00032*

Bestian, H., *Justus Liebigs Ann. Chem.*, 1950, **566**, 210 (*synth*)
Kropacheva, A.A. *et al*, *Zh. Obshch. Khim.*, 1959, **29**, 556 (*Engl. transl.* p. 553) (*synth*)
Mizuma, T. *et al*, *J. Pharm. Soc. Jpn.*, 1961, **81**, 48; *CA*, **55**, 13403 (*synth*)
Shagidullin, R.R. *et al*, *Zh. Obshch. Khim.*, 1968, **38**, 150 (*Engl. transl.* p. 148) (*ir*)
Protsenko, L.D. *et al*, *Zh. Obshch. Khim.*, 1971, **41**, 1933 (*Engl. transl.* p. 1949) (*uv*)
Bollinger, J.C. *et al*, *J. Mol. Struct.*, 1980, **69**, 273 (*deriv, struct*)

Bis(1-aziridinyl)phosphinothioic acid, 9CI B-00093

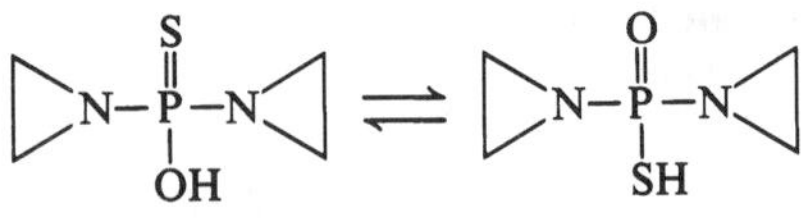

$C_4H_9N_2OPS$ M 164.162
Esters possess insect sterilant props.
O-*Me ester:* [13163-99-0]. O-*Methyl bis(1-aziridinyl)-phosphinothioate.*
$C_5H_{11}N_2OPS$ M 178.188
No phys. props. reported.
O-*Et ester:* [3677-97-2]. O-*Ethyl bis(1-aziridinyl)-phosphinothioate.*
$C_6H_{13}N_2OPS$ M 192.215
No phys. props. reported.
S-*Et ester:* [32429-65-5]. S-*Ethyl bis(1-aziridinyl)-phosphinothioate.*
$C_6H_{13}N_2OPS$ M 192.215

No phys. props. reported.

O-Propyl ester: [2590-44-5]. *O-Propyl bis(1-aziridinyl)-phosphinothioate.*

$C_7H_{15}N_2OPS$ M 206.242

No phys. props. reported.

Amide: see P,P-Bis(1-aziridinyl)phosphinothioic amide, B-00094

Borkovec, A.B. *et al, J. Econ. Entomol.*, 1972, **65**, 1543.

P,P-Bis(1-aziridinyl)phosphinothioic amide, 9CI, 8CI B-00094

Dimatif

[14465-96-4]

$C_4H_{10}N_3PS$ M 163.177

Derivs. possess insect sterilant props.

▷Highly mutagenic. SZ5990100.

N-Me: [13687-09-7]. *P,P-Bis(1-aziridinyl)-N-methyl-phosphinothioic amide. Bisazir.*

$C_5H_{12}N_3PS$ M 177.204

Solid. Mp 70-71.5°.

N-Et: [32364-85-5]. *P,P-Bis(1-aziridinyl)-N-ethylphosphinothioic amide.*

$C_6H_{14}N_3PS$ M 191.230

No phys. props. reported.

N-Ph: [25033-34-5]. *P,P-Bis(1-aziridinyl)-N-phenyl-phosphinothioic amide.*

$C_{10}H_{14}N_3PS$ M 239.274

No phys. props. reported.

Cheymol, J. *et al, Biol. Med. (Paris)*, 1967, **56**, 519; *CA*, **68**, 114173
Chang, S.C. *et al, J. Econ. Entomol.*, 1970, **63**, 1744 (*pharmacol*)
Protsenko, L.D. *et al, Zh. Obshch. Khim.*, 1971, **41**, 1933 (*Engl. transl.* p. 1949) (*deriv, uv*)
Bauman, M.C., *J. Chromatogr. Sci.*, 1975, **13**, 307 (*glc*)

Bis(1-aziridinyl)phosphinous acid B-00095

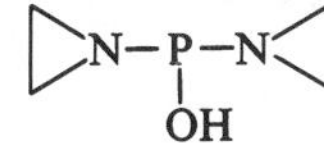

$C_4H_9N_2OP$ M 132.102

Me ester: [26120-40-0]. *Methyl bis(1-aziridinyl)-phosphinite.*

$C_5H_{11}N_2OP$ M 146.128

Liq. Bp_{13} 62°. n_D^{20} 1.4955.

Et ester: [3004-47-5]. *Ethyl bis(1-aziridinyl)-phosphinite.*

$C_6H_{13}N_2OP$ M 160.155

Liq. $Bp_{0.2}$ 41-42°. n_D^{20} 1.4844.

Isopropyl ester: Isopropyl bis(1-aziridinyl)phosphinite. 1-Methylethyl bis(1-aziridinyl)phosphinite.

$C_7H_{15}N_2OP$ M 174.182

Liq. d_4^{20} 1.01. Bp_{14} 86-86.5°. n_D^{20} 1.4749.

Ph ester: [14496-36-7]. *Phenyl bis(1-aziridinyl)-phosphinite.*

$C_{10}H_{13}N_2OP$ M 208.199

Liq. d_4^{20} 1.15. $Bp_{0.06}$ 88°. n_D^{20} 1.5600.

Amide: see P,P-Bis(1-aziridinyl)phosphinous amide, B-00096

Grechkin, N.P. *et al, Izv. Akad. Nauk SSSR, Ser. Khim.*, 1965, 1105 (*Engl. transl.* p. 1072) (*synth*)
Buina, N.A. *et al, Izv. Akad. Nauk SSSR, Ser. Khim.*, 1967, 217 (*Engl. transl.* p. 216) (*synth, ir*)
Shagidullin, R.R. *et al, Zh. Obshch. Khim.*, 1968, **38**, 150 (*Engl. transl.* p. 148) (*ir*)
Grechkin, N.P. *et al, Izv. Akad. Nauk SSSR, Ser. Khim.*, 1969, 881 (*Engl. transl.* p. 800) (*synth*)
Kostyanovskii, R.G. *et al, Izv. Akad. Nauk SSSR, Ser. Khim.*, 1969, 2588 (*Engl. transl.* p. 2429) (*synth, pmr, ms*)

P,P-Bis(1-aziridinyl)phosphinous amide B-00096

$C_4H_{10}N_3P$ M 131.117

N,N-Di-Me: [1194-52-1]. *P,P-Bis(1-aziridinyl)-N,N-dimethylphosphinous amide. Bis(1-aziridinyl)-phosphinous dimethylamide.*

$C_6H_{14}N_3P$ M 159.170

Liq. Bp_{70} 105-106°, Bp_5 54-55°. n_D^{20} 1.4914.

N,N-Di-Et: P,P-Bis(1-aziridinyl)-N,N-diethylphosphinous amide. Bis(1-aziridinyl)phosphinous diethylamide.

$C_8H_{18}N_3P$ M 187.224

Liq. Bp_{28} 113-114°. n_D^{20} 1.4946.

Nuretdinov, N.A. *et al, Izv. Akad. Nauk SSSR, Ser. Khim.*, 1964, 1883 (*Engl. transl.* p. 1784) (*derivs, synth*)
Nifant'ev, E.É., *et al, Zh. Obshch. Khim.*, 1969, **39**, 360 (*Engl. transl.* p. 337) (*dimethyl, synth*)

Bis(1*H*-benzimidazol-1-yl)phosphinic acid B-00097

[73373-57-6]

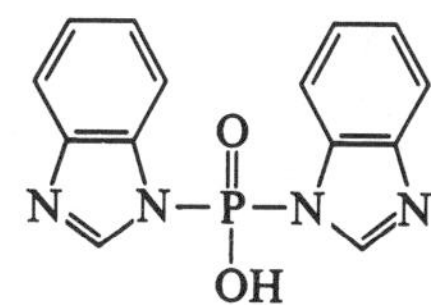

$C_{14}H_{11}N_4O_2P$ M 298.240

Both the acid and its halogenoaryl esters are root growth promotors.

Ph ester: [67059-56-7]. *Phenyl bis(1H-benzimidazol-1-yl)phosphinate.*

$C_{20}H_{15}N_4O_2P$ M 374.338

Solid. Mp 167-168°.

Anilide: P,P-Bis(1H-benzimidazol-1-yl)-N-phenylphosphinic amide. Bis(1H-benzimidazol-1-yl)phosphinic anilide.

$C_{20}H_{16}N_5OP$ M 373.353

Solid. Mp 165-167°.

Matevosyan, G.L. *et al, Zh. Obshch. Khim.*, 1978, **48**, 2433 (*Engl. transl.* p. 2208) (*synth, ir, uv*)
Matevosyan, G.L. *et al, Zh. Obshch. Khim.*, 1979, **49**, 1252 (*aryl esters, props*)

O,O-Bis(1-benzotriazolyl) 2-chlorophenyl phosphate **B-00098**

*1,1′-[(2-Chlorophenoxy)phosphinylidene]bis(oxy)-bis(1*H*-benzotriazole)*

[80817-46-5]

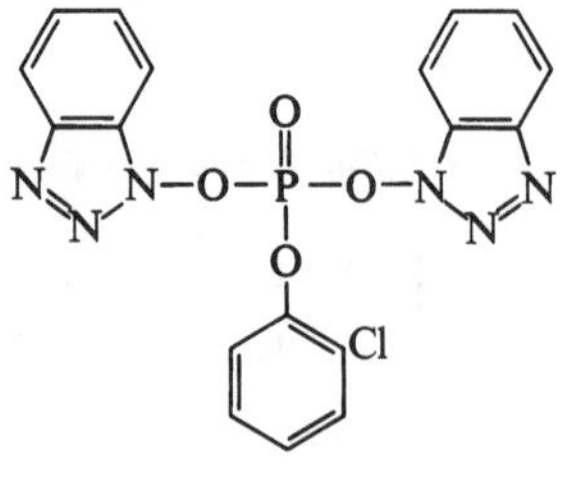

$C_{18}H_{12}ClN_6O_4P$ M 442.757

Phosphorylating agent for carbohydrates and nucleosides. Used in synth. of RNA fragments. Prepd. and employed *in situ*. May be stored in soln. in MeCN at −20°.

Oltvoort, J.J. *et al, Recl. Trav. Chim. Pays-Bas*, 1983, **102**, 523 (*use*)
Van Broeckel, C.A.A. *et al, Recl. Trav. Chim. Pays-Bas*, 1983, **102**, 526 (*use*)
Wreesman, C.T.J. *et al, Nucleic Acids Res.*, 1983, **11**, 8389 (*use*)
Marugg, J.E. *et al, Tetrahedron*, 1984, **40**, 73 (*use*)
Marugg, J.E. *et al, Nucleic Acids Res.*, 1984, **12**, 8639 (*synth, use*)

O,O-Bis(1-benzotriazolyl) morpholinyl phosphate **B-00099**

*1,1′-(4-Morpholinylphosphinylidene)bis[oxy]bis[1-*H*-benzotriazole]*, *10CI*

[80973-45-1]

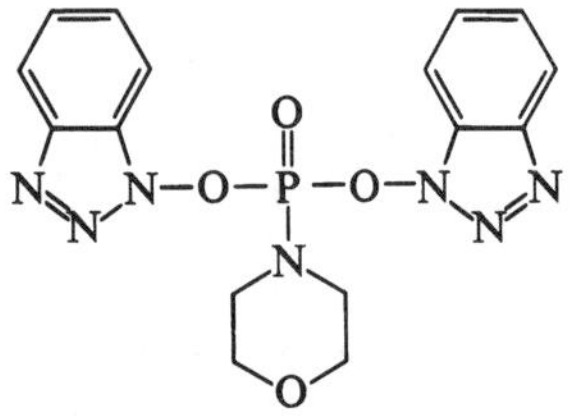

$C_{16}H_{16}N_7O_4P$ M 401.321

Phosphorylating agent for carbohydrates and nucleosides. Generally used *i*n situ. Solid. Mp 128° dec. May be stored as soln. in THF.

Van der Marel, G.A. *et al, Nucleic Acids Res.*, 1982, **10**, 2337 (*synth, use*)
Schattenkerk, C. *et al, Nucleic Acids Res.*, 1983, **11**, 7545 (*synth, pmr, P nmr, use*)
Visser, G.M. *et al, Recl. Trav. Chim. Pays-Bas*, 1984, **103**, 165 (*use*)

O,O-Bis(1-benzotriazolyl) 2,2,2-tribromoethyl phosphate **B-00100**

*1,1′-[(2,2,2-Tribromoethoxy)phosphinylidene]bis[oxy]-bis(1*H*-benzotriazole)]*, *10CI*

[80980-28-5]

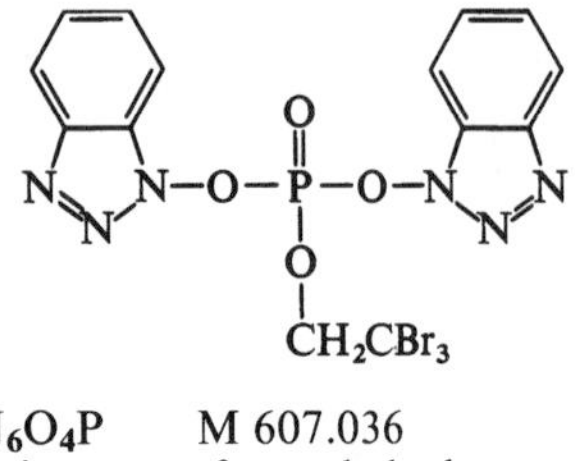

$C_{14}H_{20}Br_3N_6O_4P$ M 607.036

Phosphorylating agent for carbohydrates and nucleosides. Generally prepd. and used *in situ*.

Van der Marel, G.A. *et al, Nucleic Acids Res.*, 1982, **10**, 2337 (*synth, use*)
Van Broeckel, C.A.A. *et al, Recl. Trav. Chim. Pays-Bas*, 1983, **102**, 526 (*synth, use*)

Bis(benzylthiomethyl)phosphinic acid **B-00101**

Bis(phenylmethylthiomethyl)phosphinic acid, *9CI*

[21199-96-2]

$(PhCH_2SCH_2)_2P(O)OH$

$C_{16}H_{19}O_2PS_2$ M 338.419

Cryst. (C_6H_6/pet. ether). Mp 87-89°.

S,S,S,S-*Tetraoxide: Bis(phenylmethylsulfonylmethyl)-phosphinic acid. Bis(benzylsulfonylmethyl)phosphinic acid.*

$C_{16}H_{19}O_6PS_2$ M 402.416

Cryst. (EtOH). Mp 256-257°.

Ivasyuk, N.V. *et al, Izv. Akad. Nauk SSSR, Ser. Khim.*, 1968, 2388 (*Engl. transl.* p. 2262)

Bis[(1,1′-biphenyl)-2-yl] phosphorochloridate, 9CI **B-00102**

Bis[(1,1′-biphenyl)-2-yl] chlorophosphate

[36357-23-0]

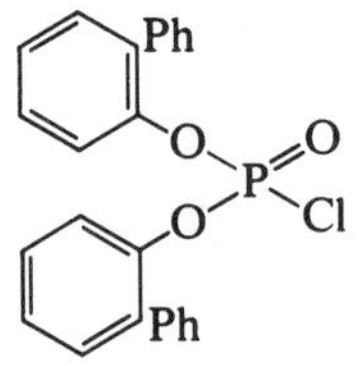

$C_{24}H_{18}ClO_3P$ M 420.831

Liq. Bp_2 246-250°. n_D^{20} 1.6265. At 600° undergoes internal phosphorylation.

Chernyshev, E.A. *et al, Zh. Obshch. Khim.*, 1971, **41**, 2189 (*Engl. transl.* p. 2214) (*synth, props*)

1,2-Bis[(bis-2-methylphenyl)phosphino]-ethane **B-00103**

1,2-Ethanediylbis[bis(2-methylphenyl)phosphine], *9CI*. *1,2-Ethanediylbis(di-*o*-tolylphosphine). 1,2-Bis(di-*o*-tolylphosphino)ethane*

[50396-26-4]

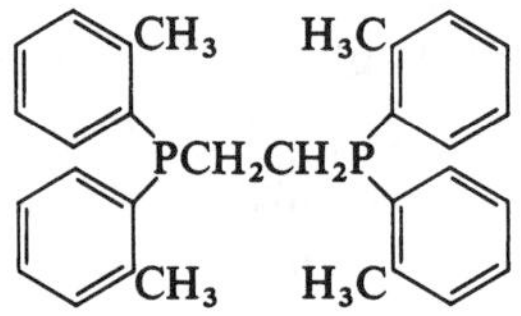

$C_{30}H_{32}P_2$ M 454.530

Solid. Mp 93-94°.

Thornhill, D.J. *et al, J. Chem. Soc., Dalton Trans.*, 1973, 2086 (*synth, complexes*)

1,2-Bis[(bis-4-methylphenyl)phosphino]-ethane **B-00104**

1,2-Ethanediylbis[bis(4-methylphenyl)phosphine], *9CI*. *1,2-Ethanediylbis[di-*p*-tolylphosphine]. 1,2-Bis(di-*p*-tolylphosphino)ethane*

[70320-30-8]

$C_{30}H_{32}P_2$ M 454.530

Forms Co and Mo complexes. Cryst. (toluene/hexane). Mp 135-136°, 147-148°.

Thornhill, D.J. *et al*, *J. Chem. Soc., Dalton Trans.*, 1973, 2086 (*synth, complexes*)
Archer, L.J. *et al*, *Inorg. Chem.*, 1979, **18**, 2079 (*synth, pmr, complexes*)
Casey, C.P. *et al*, *J. Am. Chem. Soc.*, 1983, **105**, 7574 (*synth, pmr, complexes*)

[Bis[bis(trimethylsilyl)methylene]]-chlorophosphorane, 11CI **B-00105**

[83438-72-6]

$(Me_3Si)_2C{=}PCl{=}C(SiMe_3)_2$

$C_{14}H_{36}ClPSi_4$ M 383.207
Cryst. (CH_2Cl_2). Mp 58°. $Bp_{0.001}$ 106°.

Appel, R. *et al*, *Z. Chem.*, 1984, **24**, 384 (*synth, cmr, P nmr*)
Appel, R. *et al*, *Chem. Ber.*, 1986, **119**, 535 (*cryst stuct*)

Bis[bis(trimethylsilyl)methylene]-9*H*-fluoren-9-ylphosphorane, 11CI **B-00106**

[96688-95-8]

(9-fluorenyl)–P(=C(SiMe₃)₂)₂

$C_{27}H_{45}PSi_4$ M 512.968
Cryst. (pentane). Mp 175°. BuLi in THF gives $[Li(THF)_4]^+$ complex.
$[Li(THF)_4]^+$ complex: [96705-27-0].
$C_{27}H_{44}LiPSi_4$ M 518.901
Unusual phosphorane anion with triple P=C bonding.
Cryst. (Et_2O). Mp 130° dec.

Appel, R. *et al*, *Angew. Chem., Int. Ed. Engl.*, 1985, **24**, 589 (*synth, pmr, cmr, P nmr, cryst struct*)

Bis[bis(trimethylsilyl)methylene]-phenylphosphorane **B-00107**

$(Me_3Si)_2C{=}PPh{=}C(SiMe_3)_2$

$C_{20}H_{41}PSi_2$ M 368.689
Liq. $Bp_{0.001}$ 136°.

Appel, R. *et al*, *Angew. Chem., Int. Ed. Engl.*, 1982, **21**, 80 (*synth*)

Bis(bromomethyl)phosphinic acid, 9CI **B-00108**

[26904-52-9]

$(BrCH_2)_2P(O)OH$

$C_2H_5Br_2O_2P$ M 251.842
Cyclohexylammonium salt: Cryst (C_6H_6 or EtOH). Mp 151-152°.
Bromide:
$C_2H_4Br_3OP$ M 314.739
Liq. $Bp_{0.05}$ 127-132°. n_D^{19} 1.6040.
Anhydride: [27445-88-1].
$C_4H_8Br_4O_3P_2$ M 485.669
Solid. Mp 51-52°. $Bp_{0.1}$ 220-226°.

Edmundson, R.S. *et al*, *J. Chem. Soc. (C)*, 1966, 1096 (*synth, ir*)

Bis[(4-bromophenyl)methyl] phosphate, 9CI **B-00109**

Bis-p-bromobenzyl phosphate. Bis-p-bromobenzyl phosphoric acid. Bis-p-bromobenzyl hydrogen phosphate

$BrC_6H_4CH_2OP(O)(OH)OCH_2C_6H_4Br$

$C_{14}H_{13}Br_2O_4P$ M 436.036
Needles (EtOH). Mp 155-156°. Forms spar. sol. Ag salt.
Anhydride:
$C_{28}H_{24}Br_4O_7P_2$ M 854.057
Cryst. Mp 105-106°.
Amide: Bis[(4-bromophenyl)methyl] phosphoramidate. Di-p-bromobenzyl phosphoric amide.
$C_{14}H_{14}Br_2NO_3P$ M 435.051
Cryst. ($CHCl_3$). Mp 130-131°.
Anilide: Bis[(4-bromophenyl)methyl] phenylphosphoramidate. Di-p-bromobenzyl phenylphosphoramidate.
$C_{20}H_{18}Br_2NO_3P$ M 511.149
Cryst. (EtOH aq.). Mp 119-120°.

Baddiley, J. *et al*, *J. Chem. Soc.*, 1949, 815 (*synth*)
Miyano, M. *et al*, *J. Am. Chem. Soc.*, 1955, **77**, 3522, 3524 (*synth, derivs*)
Zervas, L. *et al*, *Chem. Ber.*, 1956, **89**, 925 (*synth, deriv*)

Bis[(4-bromophenyl)methyl] phosphonate, 8CI **B-00110**

Di-p-bromobenzyl phosphonate. Bis[(4-bromophenyl)-methyl] phosphite. Di-p-bromobenzyl phosphite

$BrC_6H_4CH_2OP(O)(H)OCH_2C_6H_4Br$

$C_{14}H_{13}Br_2O_3P$ M 420.037
Tautomeric. Cryst. Mp 93-94°.

Miyano, M. *et al*, *J. Am. Chem. Soc.*, 1955, **77**, 3522 (*synth, use*)

Bis(4-bromophenyl) phosphate, 9CI **B-00111**

Bis(4-bromophenyl) hydrogen phosphate. Bis(4-bromophenyl) phosphoric acid

$BrC_6H_4{-}O{-}P(O)(OH){-}O{-}C_6H_4Br$

$C_{12}H_9Br_2O_4P$ M 407.982
Needles (H_2O or $CHCl_3$). Mp 199-201°. pK_a 0.95 (H_2O), pK_a 3.08 (95% EtOH aq.), pK_a 3.98 (EtOH).

Zetsche, F. *et al*, *Helv. Chim. Acta*, 1926, **9**, 420 (*synth*)

Bis(2-bromophenyl)phosphinic acid, 9CI **B-00112**

$(2\text{-}BrC_6H_4)_2P(O)OH$

$C_{12}H_9Br_2O_2P$ M 375.984
Solid. Mp 268.5-269.5°.
Et ester: Ethyl bis(2-bromophenyl)phosphinate.
$C_{14}H_{13}Br_2O_2P$ M 404.037

Solid. Mp 115-118°. $Bp_{0.1}$ 186-194°.

Doak, G.O. *et al*, *J. Am. Chem. Soc.*, 1952, **74**, 753 (*synth*)
Freedman, L.D. *et al*, *J. Am. Chem. Soc.*, 1953, **75**, 1379 (*ester*)

Bis(3-bromophenyl)phosphinic acid, 9CI **B-00113**

$C_{12}H_9Br_2O_2P$ M 375.984
Solid. Mp 186.5-189°.

Chloride:
$C_{12}H_8Br_2ClOP$ M 394.429
d_4^{20} 1.59. Bp_1 189-190°. n_D^{20} 1.6553.

Hydrazide:
$C_{12}H_{11}Br_2N_2OP$ M 390.013
Cryst. (C_6H_6 or C_6H_6/EtOH). Mp 178°.

Doak, G.O. *et al*, *J. Am. Chem. Soc.*, 1953, **75**, 683 (*synth*)
Shandruk, M.I. *et al*, *Zh. Obshch. Khim.*, 1973, **43**, 2194 (*Engl. transl.* p. 2186) (*chloride, hydrazide*)

Bis(4-bromophenyl)phosphinic acid, 9CI **B-00114**

$C_{12}H_9Br_2O_2P$ M 375.984
Solid. Mp 170.5-172.5°.

Ph ester: [65924-29-0]. *Phenyl bis(4-bromophenyl)-phosphinate.*
$C_{18}H_{13}Br_2O_2P$ M 452.081
Plates (EtOH). Mp 185-186°.

Chloride: [51103-90-3].
$C_{12}H_8Br_2ClOP$ M 394.429
Solid. Mp 73-74°. Bp_1 223-226°.

Hydrazide:
$C_{12}H_{11}Br_2N_2OP$ M 390.013
Cryst. (C_6H_6 or C_6H_6/EtOH). Mp 193°.

Doak, G.O. *et al*, *J. Am. Chem. Soc.*, 1953, **75**, 683 (*synth*)
Shandruk, M.I. *et al*, *Zh. Obshch. Khim.*, 1973, **43**, 2194 (*Engl. transl.* p. 2186) (*chloride, hydrazide*)
Karrasova, F.M. *et al*, *Zh. Obshch. Khim.*, 1978, **48**, 1046 (*Engl. transl.* p. 953) (*ester*)

Bis(2-*tert*-butylphenyl) phosphate **B-00115**

*Bis[2-(1,1-dimethylethyl)phenyl] phosphate, 9CI. Bis(2-*tert*-butylphenyl) hydrogen phosphate. Bis(2-*tert*-butylphenyl) phosphoric acid*

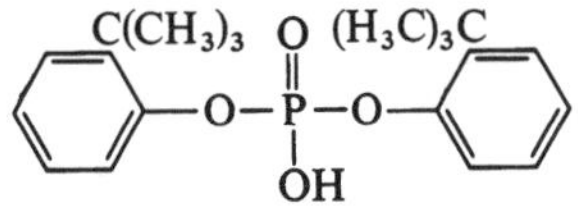

C(CH₃)₃ O (H₃C)₃C — C_6H_4–O–P(=O)(OH)–O–C_6H_4

$C_{20}H_{27}O_4P$ M 362.405

Chloride: [52553-42-7]. *Bis(2-*tert*-butylphenyl) phosphorochloridate. Bis(2-*tert*-butylphenyl) chlorophosphate.*
$C_{20}H_{26}ClO_3P$ M 380.850
Selective phosphorylating agent for nucleoside 5′-OH. No phys. props. reported. Prolonged heating at 160° → loss of *tert*-butyl groups.

Hes, J. *et al*, *J. Org. Chem.*, 1974, **39**, 3767 (*synth, use*)

Bis(4-*tert*-butylphenyl) phosphonate **B-00116**

*Bis[4-(1,1-dimethylethyl)phenyl] phosphonate, 9CI. Bis(4-*tert*-butylphenyl) phosphite. Bis[4-(1,1-dimethylethyl)phenyl] phosphite*
[73726-76-8]

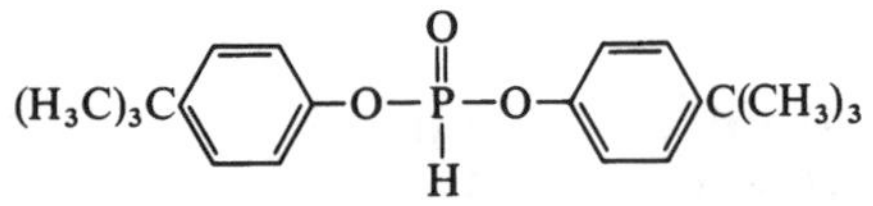

$(H_3C)_3C$–C_6H_4–O–P(=O)(H)–O–C_6H_4–$C(CH_3)_3$

$C_{20}H_{27}O_3P$ M 346.405
Tautomeric. Liq. which slowly cryst. Mp 30-33°. $Bp_{0.005}$ 160°.

Walsh, E.N., *J. Am. Chem. Soc.*, 1959, **81**, 3023.

Bis(2-carboxyphenyl)phosphinic acid **B-00117**

2,2′-Phosphinicobis[benzoic acid], 9CI
[66568-25-0]

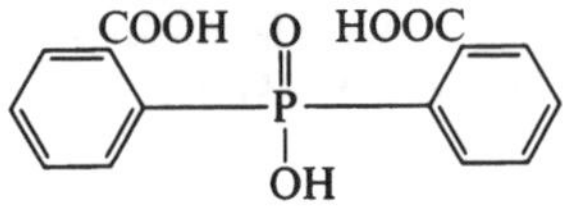

COOH O HOOC — C_6H_4–P(=O)(OH)–C_6H_4

$C_{14}H_{11}O_6P$ M 306.211
Solid. Mp 268°. Acted on by trifluoroacetic acid or by $SOCl_2$ to give a bicyclic phosphorane.

Tri-Et ester: Ethyl bis(2-ethoxycarbonylphenyl)-phosphinate.
$C_{20}H_{17}O_6P$ M 384.324
Cryst. (Et_2O/heptane). Mp 93°.

Petrov, K.A. *et al*, *Zh. Obshch. Khim.*, 1978, **48**, 91 (*Engl. transl.* p. 74) (*synth, ir*)
Segall, Y. *et al*, *J. Am. Chem. Soc.*, 1978, **100**, 5130; 1979, **101**, 3687 (*synth, props*)

Bis(4-carboxyphenyl)phosphinic acid **B-00118**

4,4′-Phosphinicobis[benzoic acid], 9CI

$C_{14}H_{11}O_6P$ M 306.211
Solid. Insol. Et_2O, C_6H_6, H_2O. Mp >330°.

Di-Me ester: Bis(4-methoxycarbonylphenyl)phosphinic acid.
$C_{16}H_{15}O_6P$ M 334.265
Cryst. (MeOH). Mp 191-192°.

Tri-Et ester: Ethyl bis(4-ethoxycarbonylphenyl)-phosphinate.
$C_{20}H_{23}O_6P$ M 390.372
Cryst. (EtOH aq.). Mp 162-164°.

Trichloride: Bis(4-chlorocarbonylphenyl)phosphinic chloride.
$C_{14}H_8Cl_3O_3P$ M 361.548
Solid. Mp 118-120° dec.

Freedman, L.D. *et al*, *J. Org. Chem.*, 1959, **24**, 638 (*synth, uv*)
Petrov, K.A. *et al*, *Zh. Obshch. Khim.*, 1960, **30**, 3000 (*Engl. transl.* p. 2972) (*synth, derivs*)

Bis(2-chloroethyl) phosphate, 9CI **B-00119**

Bis(2-chloroethyl) hydrogen phosphate. Bis(2-chloroethyl) phosphoric acid

$(ClCH_2CH_2O)_2P(O)OH$

$C_4H_9Cl_2O_4P$ M 222.992

Fluoride: Bis(2-chloroethyl) phosphorofluoridate. Bis(2-chloroethyl) phosphoryl fluoride.
$C_4H_8Cl_2FO_3P$ M 224.984
Liq. Bp_{23} 159-160°.

Chloride: [6087-94-1]. *Bis(2-chloroethyl) phosphorochloridate. Bis(2-chloroethyl) phosphoryl chloride.*
$C_4H_8Cl_3O_3P$ M 241.438

Liq. d_4^{20} 1.46. Bp_2 116-117°. n_D^{20} 1.4710.
▷Pungent
Amide: Bis(2-chloroethyl) phosphoramidate.
$C_4H_{10}Cl_2NO_3P$ M 222.008
Solid. Insol. $CHCl_3$. Mp 77°.

Cook, H.G. *et al, J. Chem. Soc.*, 1949, 635 (*chloride, fluoride, synth, tox*)
Goldwhite, H. *et al, J. Chem. Soc.*, 1955, 2040 (*fluoride*)
Walsh, E.N., *J. Am. Chem. Soc.*, 1959, **81**, 3028 (*amide*)
Pudovik, A.N. *et al, Zh. Obshch. Khim.*, 1966, **36**, 1454 (*Engl. transl.* p. 1461) (*chloride*)

Bis(2-chloroethyl) phosphonate, 9CI **B-00120**
Bis(2-chloroethyl)phosphite
[1070-42-4]

$$(ClCH_2CH_2O)_2P(O)H \rightleftharpoons (ClCH_2CH_2O)_2POH$$

$C_4H_9Cl_2O_3P$ M 206.993
Tautomeric. Liq. d_4^{20} 1.40. Bp_4 118-119°, $Bp_{0.01}$ 82-84°. n_D^{25} 1.4708. Density also reported as $d^{25}{}_{25}$ 1.10 (less probable value).
▷KK2800000.

Walsh, E.N., *J. Am. Chem. Soc.*, 1959, **81**, 3023 (*synth*)
Petrov, K.A. *et al, Zh. Obshch. Khim.*, 1962, **32**, 1277, 3723 (*Engl. transl.* pp. 1250, 3650) (*synth*)
Gloede, J. *et al, J. Prakt. Chem.*, 1976, **318**, 1043 (*ms*)
Markowska, A. *et al, Bull. Acad. Pol. Sci., Ser. Sci. Chim.*, 1979, **27**, 115 (*synth, P nmr*)

Bis(2-chloroethyl)phosphoramidic acid **B-00121**

$$(ClCH_2CH_2)_2NP(O)(OH)_2$$

$C_4H_{10}Cl_2NO_3P$ M 222.008
Mono-Et ester: Ethyl hydrogen bis(2-chloroethyl)-phosphoramidate.
$C_6H_{14}Cl_2NO_3P$ M 250.061
Needles (Me_2CO) (as cyclohexylammonium salt). Mp 146-147°.
*Mono-*tert*-butyl ester:* tert*-Butyl hydrogen bis(2-chloroethyl)phosphoramidate.*
$C_8H_{18}Cl_2NO_3P$ M 278.115
Needles (Me_2CO) (as cyclohexylammonium salt). Mp 131-133°.
Mono-Ph ester: Phenyl hydrogen bis(2-chloroethyl)-phosphoramidate.
$C_{10}H_{14}Cl_2NO_2P$ M 282.106
Needles (as cyclohexylammonium salt). Mp 184-185°.
Di-Ph ester: [1950-04-5]. *Diphenyl bis(2-chloroethyl)-phosphoramidate.*
$C_{16}H_{18}Cl_2NO_2P$ M 358.203
Exhibits significant antitumour activity. Cryst. (Me_2CO). Mp 104-109°. Readily hydrolysed.
Mono-Ph ester, monoanilide: Phenyl N,N*-bis(2-chloroethyl)-*N'*-phenylphosphorodiamidate.*
$C_{16}H_{19}Cl_2N_2O_2P$ M 373.218
Cryst. (C_6H_6). Mp 87-88.5°.
Monoamide: see N,N-Bis(2-chloroethyl)-phosphorodiamidic acid, B-00126
Dianilide: N,N*-Bis(2-chloroethyl)-*N',N''*-diphenyl-phosphoric triamide.*
$C_{16}H_{20}Cl_2N_3OP$ M 372.233
Cryst. (C_6H_6). Mp 185-186°.

Friedman, O.M. *et al, J. Am. Chem. Soc.*, 1954, **76**, 655 (*monoesters*)
Rapp, L.B. *et al, Zh. Obshch. Khim.*, 1963, **33**, 2277 (*Engl. transl.* p. 2219) (*derivs*)
Kuz'menko, I.I. *et al, Khim.-Farm. Zh.*, 1971, **5**, 7 (*Engl. transl.* p. 252) (*diphenyl ester, props*)
Zon, G. *et al, J. Am. Chem. Soc.*, 1977, **99**, 5785 (*diethyl ester*)

Bis(2-chloroethyl)phosphoramidic dichloride **B-00122**
[127-88-8]

$$(ClCH_2CH_2)_2NP(O)Cl_2$$

$C_4H_8Cl_3NOP$ M 223.446
Cryst. (Et_2O). Mp 54-56°. $Bp_{0.6}$ 123-125°. Stable to ice-cold H_2O.
▷May dec. vigorously when heated

Friedman, O.M. *et al, J. Am. Chem. Soc.*, 1954, **76**, 655 (*synth*)
Dorn, H. *et al, Pharmazie, 1967*, **22**, 558 (*synth*)
Ludeman, S.M. *et al, J. Med. Chem.*, 1975, **18**, 1251 (*synth, props*)

Bis(2-chloroethyl)phosphoramidothioic dichloride, 9CI, 8CI **B-00123**
[34492-32-5]

$$(ClCH_2CH_2)_2NP(S)Cl_2$$

$C_4H_8Cl_4NPS$ M 274.960
Cryst. (Me_2CO/pet. ether). Mp 33-34°. $Bp_{0.006}$ 91-92°.
▷TB5980000.

Dorn, H. *et al, J. Prakt. Chem.*, 1971, **313**, 218 (*synth*)

***O,O*-Bis(2-chloroethyl) phosphorochloridothioate, 9CI, 8CI** **B-00124**
O,O*-Bis(2-chloroethyl) thiophosphoryl chloride.* O,O*-Bis(2-chloroethyl) chlorothiophosphate*
[34819-51-7]

$$(ClCH_2CH_2O)_2P(S)Cl$$

$C_4H_8Cl_3O_2PS$ M 257.499
Liq. d_4^{20} 1.51. Bp_{17} 130°. n_D^{20} 1.5641.

Galashina, M.L. *et al, Zh. Obshch. Khim.*, 1953, **23**, 433 (*Engl. transl.* p. 441) (*synth*)

***N,N'*-Bis(2-chloroethyl)phosphorodiamidic acid** **B-00125**
Isophosphoramide mustard
[31645-39-3]

$$(ClCH_2CH_2NH)_2P(O)OH$$

$C_4H_{11}Cl_2N_2O_2P$ M 221.023
Active metab. of Ifosfamide, I-00001 and as active as Cyclophosphamide, C-00283 against leukomas and carcinomas. Cryst. Mp 106-107°, Mp 137-138°.
3-Butenyl ester: 3-Butenyl N,N'*-bis(2-chloroethyl)-phosphorodiamidate.*
$C_8H_{17}Cl_2N_2O_2P$ M 275.114
Oil.

Takamizawa, A. *et al, Chem. Pharm. Bull.*, 1977, **25**, 1877 (*ester, pmr*)
Misiura, K. *et al, J. Med. Chem.*, 1983, **26**, 674 (*synth, ester, ms, P nmr*)
Struck, R.F. *et al, Br. J. Cancer.*, 1983, **47**, 15 (*metab, ir, pmr*)

***N,N*-Bis(2-chloroethyl)phosphorodiamidic acid, 9CI, 8CI** **B-00126**
Phosphoramide mustard

[10159-53-2]

$(ClCH_2CH_2)_2NP(O)(OH)NH_2$

$C_4H_{11}Cl_2N_2O_2P$ M 221.023

Active metab. of Cyclophosphamide, C-00283 4-Hydroxycyclophosphamide, H-00115 and 4-Oxocyclophosphamide, O-00058 .

Mono(cyclohexylammonium) salt: [1566-15-0]. Cryst. ($EtOH/Et_2O$). Mp 109°, Mp 125-126°.

▷TD2530000.

Me ester: [57154-91-3]. *Methyl* N,N-*bis(2-chloroethyl)-phosphorodiamidate.*
$C_5H_{13}Cl_2N_2O_2P$ M 235.050
Plates (CCl_4). Mp 82-83° (75-78°).

3-Oxopropyl ester: see Aldophosphamide, A-00050

3-Hydroxypropyl ester: [52336-54-6]. *Alcophosphamide.*
$C_7H_{17}Cl_2N_2O_4P$ M 295.102
Cryst. Mp 40-41°.

3-Butenyl ester: 3-Butenyl N,N-*bis(2-chloroethyl)-phosphorodiamidate.*
$C_8H_{17}Cl_2N_2O_2P$ M 275.114
Oil or cryst. Mp 20°.

Ph ester: [18374-36-2]. *Phenyl* N,N-*bis(2-chloroethyl)-phosphorodiamidate.*
$C_{10}H_{15}Cl_2N_2O_2P$ M 297.120
Liq. or needles (C_6H_6/pet. ether). Mp 57-58°. $Bp_{0.6}$ 123-125°.

Friedman, O.M. *et al, J. Am. Chem. Soc.*, 1954, **76**, 655 (*phenyl ester*)
Eiden, F. *et al, Arch. Pharm.* (*Weinheim, Ger.*), 1973, **306**, 126 (*phenyl ester*)
Schulten, H.R. *et al, Biomed. Mass Spectrom.*, 1974, **1**, 223 (*ms*)
Griggs, L.J. *et al, J. Med. Chem.*, 1975, **18**, 1102 (*methyl ester, ms, pmr*)
Struck, R.S. *et al, Biomed. Mass Spectrom.*, 1975, **2**, 46 (*ms*)
Takamizawa, A. *et al, J. Med. Chem.*, 1975, **18**, 376 (*esters*)
Montgomery, J.A. *et al, Cancer Treat. Rep.*, 1976, **60**, 381 (*bibl*)
Engle, T.W. *et al, J. Med. Chem.*, 1979, **22**, 897 (*synth, P nmr*)
Chiu, F.T. *et al, J. Pharm. Sci.*, 1982, **71**, 542 (*methyl ester, ir, pmr*)
Zon, G. *et al, J. Med. Chem.*, 1984, **27**, 466 (*ester, cmr, pmr*)

Bis(chloromethyl)phosphinic acid, 9CI **B-00127**

[13274-83-4]

$(ClCH_2)_2P(O)OH$

$C_2H_5Cl_2O_2P$ M 162.940

Solid. Mp 80-81°.

Me ester: [14212-91-0]. *Methyl bis(chloromethyl)-phosphinate.*
$C_3H_7Cl_2O_2P$ M 176.967
Liq. d_4^{20} 1.46. $Bp_{0.05}$ 74-78°. n_D^{20} 1.4878.

Et ester: [13274-84-5]. *Ethyl bis(chloromethyl)-phosphinate.*
$C_4H_9Cl_2O_2P$ M 190.994
Liq. d_4^{20} 1.37. $Bp_{1.0}$ 104-106°. n_D^{20} 1.4809.

Isopropyl ester: Isopropyl bis(chloromethyl)-phosphinate.
$C_5H_{11}Cl_2O_2P$ M 205.020
Liq. d_4^{20} 1.30. $Bp_{0.07}$ 92-93°. n_D^{20} 1.4759.

Ph ester: [14212-98-7]. *Phenyl bis(chloromethyl)-phosphinate.* Liq. d_4^{20} 1.39. $Bp_{0.2}$ 117-120°. n_D^{20} 1.5485.

Trimethylsilyl ester: [31675-59-9]. *Trimethylsilyl bis(chloromethyl)phosphinate.*
$C_5H_{13}Cl_2O_2PSi$ M 235.122
Liq. with sl. camphoraceous odour. Mp 44-47° (sealed tube). Bp_{14} 132-136°. n_D^{20} 1.4645.

Chloride: [13482-64-9].
$C_2H_4Cl_3OP$ M 181.386
Liq. d^{20} 1.61. Bp_{11} 131-133°, Bp_1 98-101°. n_D^{20} 1.5203.

Anhydride: [27445-88-1].
$C_4H_8Cl_4O_3P_2$ M 307.865
Solid. Mp 74°. Bp_2 205-210°.

*Amide: see P,P-*Bis(chloromethyl)phosphinic amide, B-00128

Ivanov, B.E. *et al, Zh. Obshch. Khim.*, 1967, **37**, 1856 (*Engl. transl.* p. 1768) (*chloride, esters, ir*)
Maier, L. *et al, Helv. Chim. Acta*, 1969, **52**, 827 (*chloride, esters, P nmr, pmr*)
Goldwhite, H. *et al, Spectrochim. Acta, Part A*, 1970, **26**, 1403 (*chloride, ester, ir*)
Orlov, N.F. *et al, Zh. Obshch. Khim.*, 1970, **40**, 2335 (*Engl. transl.* p. 2321) (*silyl ester*)
Maier, L., *Helv. Chim. Acta*, 1971, **54**, 1651 (*anhydride, chloride, P nmr*)
Romanenko, V.D. *et al, Zh. Obshch. Khim.*, 1978, **48**, 2517 (*Engl. transl.* p. 2286) (*chloride, silyl ester*)
Maier, L., *J. Organomet. Chem.*, 1979, **178**, 157 (*synth, chloride, P nmr, pmr*)

P,P-Bis(chloromethyl)phosphinic amide, 9CI **B-00128**

$(ClCH_2)_2P(O)NH_2$

$C_2H_6Cl_2NOP$ M 161.955

N-*Me:* [17166-94-8]. P,P-*Bis(chloromethyl)*-N-*methyl-phosphinic amide.*
$C_3H_8Cl_2NOP$ M 175.982
Cryst. (C_6H_6). Mp 67-68°.

N,N-*Di-Me:* [17166-90-4]. P,P-*Bis(chloromethyl)*-N,N-*dimethylphosphinic amide.*
$C_4H_{10}Cl_2NOP$ M 190.009
Cryst. (C_6H_6). Mp 80-81°.

N-*Ph:* [17166-98-2]. P,P-*Bis(chloromethyl)*-N-*phenyl-phosphinic amide.*
$C_8H_{10}Cl_2NOP$ M 238.053
Cryst. (dioxan). Mp 142-143°.

N-*Benzyl:* N-*Benzyl*-P,P-*bis(chloromethyl)phosphinic amide.*
$C_9H_{12}Cl_2NOP$ M 252.080
Cryst. (C_6H_6). Mp 97-99°.

Ivanov, B.E. *et al, Zh. Obshch. Khim.*, 1967, **37**, 1856 (*Engl. transl.* p. 1768) (*synth, ir*)
Maier, L., *Helv. Chim. Acta*, 1971, **54**, 1651 (*deriv, P nmr*)

Bis(chloromethyl)phosphinothioic acid **B-00129**

$(ClCH_2)_2P(S)OH \rightleftharpoons (ClCH_2)_2P(O)SH$

$C_2H_5Cl_2OPS$ M 179.001

O-*Et ester:* O-*Ethyl bis(chloromethyl)phosphinothioate.*
$C_4H_9Cl_2OPS$ M 207.054
Liq. d_4^{20} 1.34. $Bp_{0.03}$ 73°. n_D^{25} 1.5342.

S-*Et ester:* S-*Ethyl bis(chloromethyl)phosphinothioate.*
$C_4H_9Cl_2OPS$ M 207.054
Liq. d_4^{20} 1.39. $Bp_{0.01}$ 82°. n_D^{20} 1.5483.

O-*Ph ester:* O-*Phenyl bis(chloromethyl)-phosphinothioate.*
C_8H_9ClOPS M 219.645
Cryst. (Et_2O/pet. ether). Mp 56-57°.

Chloride:
$C_2H_4Cl_3PS$ M 197.446

Liq. d_4^{20} 1.56. Bp_{10} 104-106°. n_D^{25} 1.5890.
Diethylamide: N,N-*Diethyl*-P,P-*bis(chloromethyl)-phosphinothioic amide.*
$C_6H_{14}Cl_2NPS$ M 234.123
Liq. d_4^{20} 1.27. $Bp_{0.001}$ 95°. n_D^{20} 1.5508.
Anilide: [20459-69-2]. N-*Phenyl*-P,P-*bis(chloromethyl)phosphinothioic amide.*
$C_8H_{10}Cl_2NPS$ M 254.113
Cryst. (Et_2O). Mp 102-103°.

Panteleeva, A.R. *et al*, *Izv. Akad. Nauk SSSR, Ser. Khim.*, 1968, 1644 (*Engl. transl.* p. 1557) (*synth*)
Bel'skii, V.E. *et al*, *Dokl. Akad. Nauk SSSR, Ser. Sci. Khim.*, 1971, **197**, 85 (*Engl. transl.* p. 171) (*synth*)

Bis[(4-chlorophenyl)methyl] phosphonate, 8CI — B-00130

Di-p-*chlorobenzyl phosphonate. Bis[(4-chlorophenyl)-methyl] phosphite. Di*-p-*chlorobenzyl phosphite*, 8CI
[65463-55-0]

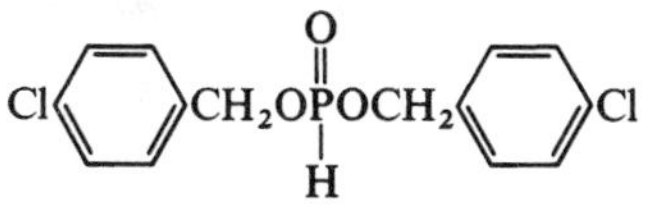

$C_{14}H_{13}Cl_2O_3P$ M 331.135
Tautomeric. Solid. Mp 75°.

Miyano, M. *et al*, *J. Am. Chem. Soc.*, 1955, **77**, 3522 (*synth*)

Bis(2-chlorophenyl) phosphate, 9CI — B-00131

Bis(2-chlorophenyl) hydrogen phosphate. Bis(2-chlorophenyl) phosphoric acid
[36400-49-4]

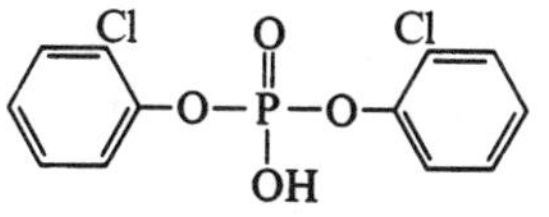

$C_{12}H_9Cl_2O_4P$ M 319.080
Needles. Mp 105-106°.
NH_4 salt: Needles (H_2O).
Chloride: [17776-78-2]. *Bis(2-chlorophenyl) phosphorochloridate. Bis(2-chlorophenyl) phosphoryl chloride. Bis(2-chlorophenyl) chlorophosphate.*
$C_{12}H_8Cl_3O_3P$ M 337.526
Phosphorylating agent. Oil. Bp_1 156°. n_D^{25} 1.5670.

Zetsche, F. *et al*, *Helv. Chim. Acta*, 1926, **9**, 420 (*synth*)
Walsh, E.N., *J. Am. Chem. Soc.*, 1959, **81**, 3023 (*chloride*)
Flockerzi, D. *et al*, *Helv. Chim. Acta*, 1983, **66**, 2069 (*chloride, use*)
Hayakawa, Y. *et al*, *Tetrahedron Lett.*, 1983, **24**, 1165 (*chloride, use*)

Bis(3-chlorophenyl) phosphate, 9CI, 8CI — B-00132

Bis(3-chlorophenyl) hydrogen phosphate. Bis(3-chlorophenyl) phosphoric acid
$C_{12}H_9Cl_2O_4P$ M 319.080
Chloride: [51103-92-5]. *Bis(3-chlorophenyl) phosphorochloridate. Bis(3-chlorophenyl) phosphoryl chloride. Bis(3-chlorophenyl) chlorophosphate.*
$C_{12}H_8Cl_3NO_3P$ M 351.533
Liq. d_4^{20} 1.51. Bp_2 128-133°. n_D^{20} 1.5382.

Shandruk, M.I. *et al*, *Zh. Obshch. Khim.*, 1973, **43**, 2194 (*Engl. transl.* p. 2186)

Bis(4-chlorophenyl) phosphate, 9CI — B-00133

Bis(4-chlorophenyl) hydrogen phosphate. Bis(4-chlorophenyl) phosphoric acid
[4795-31-7]
$C_{12}H_9Cl_2O_4P$ M 319.080
Needles or plates. Mp 133-135°.
Chloride: [15074-53-0]. *Bis(4-chlorophenyl) phosphorochloridate. Bis(4-chlorophenyl) phosphoryl chloride. Bis(4-chlorophenyl) chlorophosphate.*
$C_{12}H_8Cl_3O_3P$ M 337.526
Phosphorylating agent for nucleosides. Cryst. (ligroin). Mp 56-57°. Bp_8 219°, $Bp_{0.1}$ 164-176°.

Zetsche, F. *et al*, *Helv. Chim. Acta*, 1926, **9**, 420 (*synth*)
Rosenmund, K.W. *et al*, *Arch. Pharm.* (*Weinheim, Ger.*), 1943, **281**, 317 (*synth, chloride*)
Reimschüssel, W. *et al*, *Int. J. Chem. Kinet.*, 1980, **12**, 979 (*chloride, synth, props*)
Daskalov, H.P. *et al*, *Bull. Chem. Soc. Jpn.*, 1981, **54**, 3076 (*chloride, use*)

Bis(2-chlorophenyl)phosphine, 9CI — B-00134

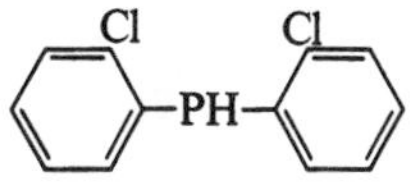

$C_{12}H_9Cl_2P$ M 255.083
Mp 67-68°. Bp 288-289°.
Oxide: Tautomeric with Bis(2-chlorophenyl)phosphinous acid, B-00140 .

Kosolapoff, G.M., *Organophosphorus Compds.*, 1950, Wiley, 1st Ed..

Bis(3-chlorophenyl)phosphine, 9CI — B-00135

$C_{12}H_9Cl_2P$ M 255.083
Oxide: [71360-03-7].
$C_{12}H_9Cl_2OP$ M 271.082
$Bp_{0.02}$ 160-180°. Tautomeric with Bis(3-chlorophenyl)-phosphinous acid, B-00141 .

Hengartner, U. *et al*, *J. Org. Chem.*, 1979, **44**, 3741 (*oxide*)

Bis(4-chlorophenyl)phosphine, 9CI — B-00136

$C_{12}H_9Cl_2P$ M 255.083
Solid. Mp 39-40°. Bp 314-315°.
Oxide: [15948-60-4].
$C_{12}H_9Cl_2OP$ M 271.082
pK_a 18.8 (DMSO, 25°). Tautomeric with bis(4-chlorophenyl)phosphinous acid.

Kosolapoff, G.M., *Organophosphorus Compds*, 1950, Wiley, New York, 1st Ed..
Tsvetkov, E.N. *et al*, *Izv. Akad. Nauk SSSR, Ser. Khim.*, 1978, 1981 (*Engl. transl.* p. 1743) (*oxide*)

Bis(2-chlorophenyl)phosphinic acid, 9CI — B-00137

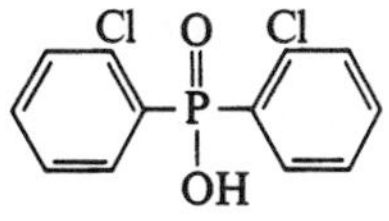

$C_{12}H_9Cl_2O_2P$ M 287.082
Solid. Mp 233-236°.
Chloride:
$C_{12}H_8Cl_3O_2P$ M 321.527
Liq. Bp >340°.

Doak, G.O. *et al*, *J. Am. Chem. Soc.*, 1951, **73**, 5658 (*synth*)

Jaffé, H.H. *et al*, *J. Am. Chem. Soc.*, 1952, **74**, 1069 (*uv*)

Bis(3-chlorophenyl)phosphinic acid, 9CI B-00138

$C_{12}H_9Cl_2O_2P$ M 287.082
Solid. Mp 164-165°.

Et ester: Ethyl bis(3-chlorophenyl)phosphinate.
$C_{14}H_{13}Cl_2O_2P$ M 315.135
$Bp_{0.7}$ 183-187°. n_D^{25} 1.5794.
Chloride: [57906-77-1].
$C_{12}H_8Cl_3O_2P$ M 321.527
Liq. $Bp_{0.3}$ 192-195°. n_D^{20} 1.6228.
Azide: [57906-70-4].
$C_{12}H_8Cl_2N_3O_2P$ M 328.094
Liq. $Bp_{0.01}$ 153-155°. n_D^{20} 1.6204.

Doak, G.O. *et al*, *J. Am. Chem. Soc.*, 1951, **73**, 5658 (*synth*)
Jaffé, H.H. *et al*, *J. Am. Chem. Soc.*, 1952, **74**, 1069 (*uv*)
Denham, J.M. *et al*, *J. Org. Chem.*, 1958, **23**, 1298 (*ester, chloride*)
Weissbach, F. *et al*, *J. Prakt. Chem.*, 1975, **317**, 394 (*chloride, azide, ir*)

Bis(4-chlorophenyl)phosphinic acid, 9CI B-00139

[13119-01-2]
$C_{12}H_9Cl_2O_2P$ M 287.082
Solid. Mp 133-135°, 145-146°. pK_a 1.68 (7% EtOH aq.), 3.48 (80% EtOH aq.), 3.18 (23°, 60% THF aq.).
▷SZ4368300.

Me ester: [21713-62-2]. *Methyl bis(4-chlorophenyl)-phosphinate.*
$C_{13}H_{11}Cl_2O_2P$ M 301.108
Solid. Mp 92.5-93.5°. $Bp_{0.15}$ 152-158° (oven).
Butyl ester: [41044-91-1]. *Butyl bis(4-chlorophenyl)-phosphinate.*
$C_{16}H_{17}Cl_2O_2P$ M 343.189
Insol. pet. ether. d_4^{20} 1.22. n_D^{20} 1.5698.
Chloride: [4129-39-9].
$C_{12}H_8Cl_3OP$ M 305.527
Solid. $Bp_{0.3}$ 190-191°. n_D^{20} 1.6339.
Azide: [4129-19-5].
$C_{12}H_8Cl_2N_3OP$ M 312.094
Liq. $Bp_{0.007}$ 165-166°. n_D^{20} 1.6230.
Amide: [83470-33-1].
$C_{12}H_{10}Cl_2NOP$ M 286.097
Cryst. (toluene). Mp 169-171.5°.
Hydrazide:
$C_{12}H_{11}Cl_2N_2OP$ M 301.111
Solid. Mp 203°.

Mastryukova, T.A. *et al*, *Zh. Obshch. Khim.*, 1959, **29**, 2178; *CA*, **54**, 10463 (*synth, props*)
Baldwin, R.A. *et al*, *J. Org. Chem.*, 1967, **32**, 2176 (*props*)
Petrov, K.A. *et al*, *Zh. Obshch. Khim.*, 1973, **43**, 37 (*Engl. transl.* p. 34) (*synth*)
Shandruk, M.I. *et al*, *Zh. Obshch. Khim.*, 1973, **43**, 2194 (*Engl. transl.* p. 2186) (*chloride*)
Weissbach, F. *et al*, *J. Prakt. Chem.*, 1975, **317**, 394 (*chloride, azide*)
Harger, M.J.P. *et al*, *Tetrahedron*, 1982, **38**, 1511 (*synth, ir, pmr, derivs, ms*)

Bis(2-chlorophenyl)phosphinous acid, 9CI B-00140

Cl
POH
Cl

$C_{12}H_9Cl_2OP$ M 271.082
Free acid exists as the phosphoryl tautomer.

Et ester: Ethyl bis(2-chlorophenyl)phosphinate.
$C_{14}H_{13}Cl_2OP$ M 299.136
Mp 26-29°. Bp_{15} 132-137°.
Chloride: [32186-89-3]. *Chlorobis(2-chlorophenyl)-phosphine.*
$C_{12}H_8Cl_3P$ M 289.528
Solid. Mp 38°, Mp 92-94°. Bp_7 208-210°, $Bp_{0.3}$ 148-153°.

Weinberg, K.G. *et al*, *J. Org. Chem.*, 1975, **40**, 3586.
Shvets, A.A. *et al*, *Zh. Obshch. Khim.*, 1978, **48**, 232 (*Engl. transl.* p. 208)

Bis(3-chlorophenyl)phosphinous acid B-00141

$C_{12}H_9Cl_2OP$ M 271.082
The free acid exists as the phosphoryl tautomer. See under Bis(3-chlorophenyl)phosphine, B-00135 .

Et ester: [13685-47-7]. *Ethyl bis(3-chlorophenyl)-phosphinite.*
$C_{14}H_{13}Cl_2OP$ M 299.136
Liq. d_4^{20} 1.24. Bp_3 152-153°. n_D^{20} 1.6030.
Chloride: [13685-27-3]. *Chlorobis(3-chlorophenyl)-phosphine.*
$C_{12}H_8Cl_3P$ M 289.528
Liq. d_4^{20} 1.37. Bp_2 148-150°. n_D^{20} 1.6460.
Diethylamide: P,P-*Bis(3-chlorophenyl)*-N,N-*diethyl-phosphinous amide.*
$C_{16}H_{18}Cl_2N_2P$ M 340.211
Liq. d_4^{20} 1.20. Bp_1 154-156°. n_D^{20} 1.6057.

Yudina, K.S. *et al*, *Izv. Akad. Nauk SSSR, Ser. Khim.*, 1966, 1954 (*Engl. transl.* p. 1889) (*derivs, synth*)
Muylle, E. *et al*, *Spectrochim. Acta, Part A*, 1976, **32**, 599 (*chloride, P nmr*)
Hengartner, U. *et al*, *J. Org. Chem.*, 1979, **44**, 3741 (*synth*)

Bis(4-chlorophenyl)phosphinous acid, 9CI B-00142

[28343-42-2]
$C_{12}H_9Cl_2OP$ M 271.082
The free acid exists as the phosphoryl tautomer. See under Bis(4-chlorophenyl)phosphine, B-00136 .

Me ester: [17106-25-1]. *Methyl bis(4-chlorophenyl)-phosphinite.*
$C_{13}H_{11}Cl_2OP$ M 285.109
Liq. d_4^{20} 1.278. Bp_8 176°. n_D^{20} 1.6178.
Et ester: [13685-46-6]. *Ethyl bis(4-chlorophenyl)-phosphinite.*
$C_{14}H_{13}Cl_2OP$ M 299.136
Liq. d_4^{20} 1.24. Bp_2 161-162°. n_D^{20} 1.6038.
Butyl ester: [17106-28-4]. *Butyl bis(4-chlorophenyl)-phosphinite.*
$C_{16}H_{17}Cl_2OP$ M 327.189
Liq. d_4^{20} 1.19. Bp_6 197°. n_D^{20} 1.5872.
Chloride: [13685-26-2]. *Chlorobis(4-chlorophenyl)-phosphine.*
$C_{12}H_{18}Cl_3P$ M 299.607

Solid. d_4^{20} 1.37. Mp 52-53°. Bp_2 134-136°. n_D^{20} 1.6455.
Diethylamide: P,P-*Bis(4-chlorophenyl)*-N,N-*diethylphosphinous amide.*
$C_{16}H_{18}Cl_2N_2P$ M 340.211
Liq. d_4^{20} 1.20. Bp_2 186-187°. n_D^{20} 1.6058.

Yudina, K.S. *et al, Izv. Akad. Nauk SSSR, Ser. Khim.*, 1966, 1954 (*Engl. transl.* p. 1889) (*derivs, synth*)
Kharrasova, F.M. *et al, Zh. Obshch. Khim.*, 1967, **37**, 687 (*Engl. transl.* p. 643) (*methyl ester*)
Ojima, I. *et al, Bull. Chem. Soc. Jpn.*, 1969, **42**, 2975 (*synth*)
Muylle, E. *et al, Spectrochim. Acta, Part A*, 1975, **31**, 1039 (*chloride, nmr*)
Muylle, E. *et al, Spectrochim. Acta, Part A*, 1976, **32**, 599 (*P nmr*)

Bis(2-chlorophenyl) phosphonate, 9CI **B-00143**
Di-o-*chlorophenyl phosphonate. Di*-o-*chlorophenyl phosphite*
[65475-24-3]

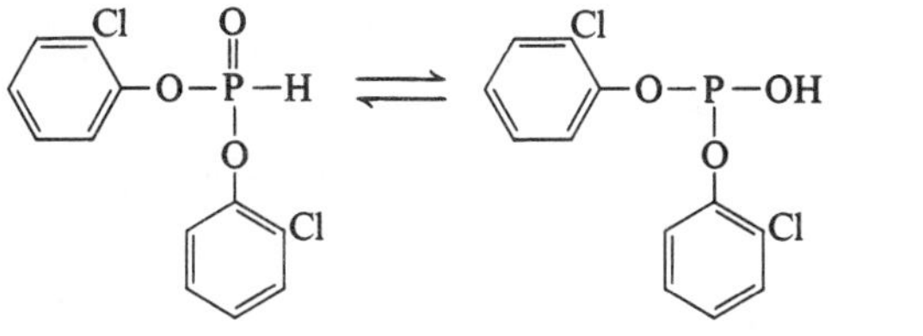

$C_{12}H_9Cl_2O_3P$ M 303.081
Tautomeric; almost completely in phosphonate form. Liq. $Bp_{0.005}$ 125°. n_D^{25} 1.5750.

Walsh, E.N., *J. Am. Chem. Soc.*, 1959, **81**, 3023 (*synth*)

Bis(4-chlorophenyl) phosphonate, 9CI **B-00144**
Di-p-*chlorophenyl phosphonate. Di*-p-*chlorophenyl phosphite*
[15516-41-3]
$C_{12}H_9Cl_2O_3P$ M 303.081
Tautomeric; almost completely in phosphonate form. Solid. Mp 44°. $Bp_{0.005}$ 125°.

Walsh, E.N., *J. Am. Chem. Soc.*, 1959, **81**, 3023 (*synth*)

Bis(2-chlorophenyl) phosphorochloridite, 9CI **B-00145**
Di-o-*chlorophenyl chlorophosphite*

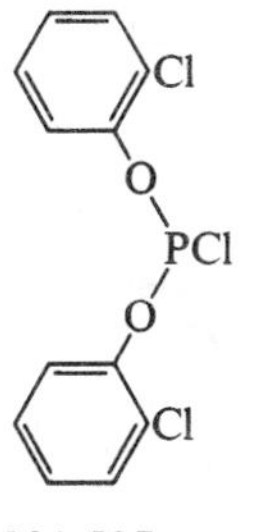

$C_{12}H_8Cl_3O_2P$ M 321.527
Liq. Bp_{10} 205-210°.

Cebrian, G.R., *CA*, 1957, **51**, 12020 (*synth*)

Bis(3-chlorophenyl) phosphorochloridite, 9CI **B-00146**
Di-m-*chlorophenyl chlorophosphite*
[51103-93-6]
$C_{12}H_8Cl_3O_2P$ M 321.527
Pungent liq. d_4^{20} 1.42. Bp_{10} 213-215°. n_D^{20} 1.5960.

Shandruk, M.I. *et al, Zh. Obshch. Khim.*, 1973, **43**, 2194 (*Engl. transl.* p. 2186) (*synth*)

Bis(4-chlorophenyl) phosphorochloridite, 9CI **B-00147**
Di-p-*chlorophenyl chlorophosphite*
[15520-16-8]
$C_{12}H_8Cl_3O_2P$ M 321.527
Pungent liq. Bp_{12} 205-215°.

Cebrian, G.R., *CA*, 1957, **51**, 12020 (*synth*)

***O,O*-Bis(3-chlorophenyl) phosphorochloridothioate, 9CI** **B-00148**
O,O-*Di*-m-*chlorophenyl chlorothiophosphate.* O,O-*Di*-m-*chlorophenyl thiophosphoryl chloride*
[51103-94-7]

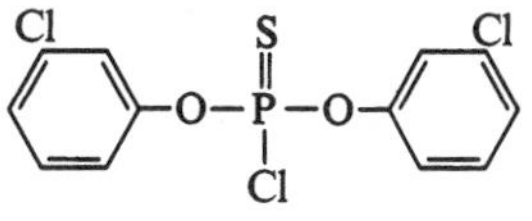

$C_{12}H_8Cl_3O_2PS$ M 353.587
Low-melting solid. Mp 28-29°. Bp_8 218-221°.

Shandruk, M.I. *et al, Zh. Obshch. Khim.*, 1973, **43**, 2194 (*Engl. transl.* p. 2186) (*synth*)

***O,O*-Bis(4-chlorophenyl) phosphorochloridothioate, 9CI** **B-00149**
O,O-*Di*-p-*chlorophenyl chlorothiophosphate.* O,O-*Di*-p-*chlorophenyl thiophosphoryl chloride*
[55526-70-0]
$C_{12}H_8Cl_3O_2PS$ M 353.587
Cryst. (EtOH). Mp 43-44°. Bp_{11} 243-245°.

Strecker, W. *et al, Ber.*, 1916, **49**, 63 (*synth*)
Reimschüssel, W. *et al, Int. J. Chem. Kinet.*, 1980, **12**, 979 (*synth, props*)

***O,O*-Bis(2-chlorophenyl) phosphorodithioate, 9CI** **B-00150**
O,O-*Bis(2-chlorophenyl) hydrogen dithiophosphate.* O,O-*Bis(2-chlorophenyl) dithiophosphoric acid*

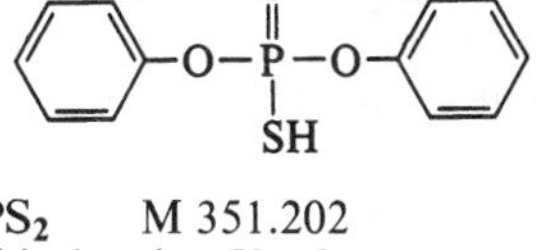

$C_{12}H_9Cl_2O_2PS_2$ M 351.202
Unstable acid isolated as K salt.
K salt: Cryst. (Me_2CO/C_6H_6). Dec. at 180°.

Zemlyanskii, N.I. *et al, Zh. Obshch. Khim.*, 1962, **32**, 1962 (*Engl. transl.* p. 1942) (*synth, props*)

***O,O*-Bis(4-chlorophenyl) phosphorodithioate, 9CI** **B-00151**
O,O-*Bis(4-chlorophenyl) hydrogen dithiophosphate.* O,O-*Bis(4-chlorophenyl) dithiophosphoric acid*
[14366-45-1]
$C_{12}H_9Cl_2O_2PS_2$ M 351.202
Unstable acid isolated as Na or K salt. Dec. → H_2S, H_3PO_4 and 4-chlorophenol.
Na salt: Solid. Dec. at 250°.

Na salt, trihydrate: Cryst. (Me_2CO/C_6H_6). Mp 97-99°.
K salt: Cryst. (Me_2CO/C_6H_6). Mp 210-212°.
Anilinium salt: Cryst. (H_2O). Mp 163-165°.

Zemlyanskii, N.I. *et al*, *Zh. Obshch. Khim.*, 1962, **32**, 1962; 1966, **36**, 2193 (*Engl. transl.* pp. 1942, 2187) (*synth, props*)

Bis(2-cyanoethyl)phenylphosphine **B-00152**

3,3′-(Phenylphosphinidene)bispropanenitrile, *10CI*
[15909-92-9]

$PhP(CH_2CH_2CN)_2$

$C_{12}H_{13}N_2P$ M 216.222
Cryst. (EtOH). Mp 72-73°. $Bp_{0.2}$ 195-205°. pK_a 3.20 (H_2O). n_D^{20} 1.5672.
B,MeI: Bis(2-cyanoethyl)methylphenylphosphonium iodide. Cryst. (EtOH). Mp 115°.
Oxide:
$C_{12}H_{13}N_2OP$ M 232.221
Cryst. (2-propanol). Mp 108-109°.
Sulfide:
$C_{12}H_{13}N_2PS$ M 248.282
Solid. Mp 74-75°.

Mann, F.G. *et al*, *J. Chem. Soc.*, 1952, 4453 (*synth, deriv*)
Rauhut, M.M. *et al*, *J. Am. Chem. Soc.*, 1959, **81**, 1103 (*oxide*)
Pellon, J. *et al*, *J. Polym. Sci., Part A*, 1963, 863 (*sulfide*)
Schindlbauer, H., *Monatsh. Chem.*, 1963, **94**, 99 (*uv*)

Bis(2-cyanoethyl)phosphine **B-00153**

3,3′-Phosphinidenebis(propanenitrile), *9CI*
[4023-49-8]

$HP(CH_2CH_2CN)_2$

$C_6H_9N_2P$ M 140.124
Obt. from PH_3 and $H_2C{=}CHCN$. Constit. of heat-stabilized polymers. $Bp_{0.3}$ 157-159°. pK_a 0.41 (H_2O). n_D^{25} 1.5070.
Oxide:
$C_6H_9N_2OP$ M 156.124
Solid. Mp 98-99°. Tautomeric with bis(2-cyanoethyl)-phosphinous acid.

Rauhut, M.M. *et al*, *J. Am. Chem. Soc.*, 1958, **80**, 6690 (*oxide*)
Rauhut, M.M. *et al*, *J. Am. Chem. Soc.*, 1959, **81**, 1103 (*synth*)

Bis(cyclenphosphorane) **B-00154**

Hexahydro-8b,8b′-bi-8bH-2a,4a,6a,8a-tetraaza-8b-phosphapentaleno[1,6-cd]pentalene, *9CI*

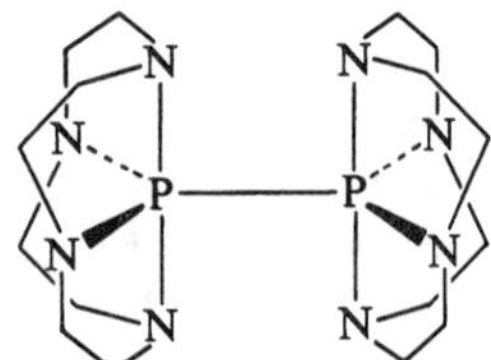

$C_{16}H_{32}N_8P_2$ M 398.430
Solid. Mp 320-330° (sealed tube). $Bp_{0.1}$ 180-190°.

Richman, J.E. *et al*, *Inorg. Chem.*, 1981, **20**, 3378 (*synth, ir, uv, pmr, P nmr, cryst struct*)

Bis(dibenzyloxyphosphinothioyl) disulfide **B-00155**

Tetrakis(phenylmethyl) thioperoxydiphosphate, *9CI*. *Tetrabenzyl thioperoxydiphosphate*
[7575-26-0]

$(PhCH_2O)_2P(S){-}S{-}S{-}P(S)(OCH_2Ph)_2$

$C_{28}H_{28}O_4P_2S_4$ M 618.714
Liq. n_D^{25} 1.6109.

Hu, P.-F. *et al*, *CA*, 1959, **53**, 3120.

Bis(2,5-dibromophenyl)phosphinic acid, *9CI* **B-00156**

$C_{12}H_7Br_4O_2P$ M 533.776
Cryst. (EtOAc). Mp 277-279°.

Denham, J.M. *et al*, *J. Org. Chem.*, 1958, **23**, 1298.

1,4-Bis(dibutylphosphino)butane **B-00157**

1,4-Butanediylbis[dibutylphosphine], *10CI*, *9CI*. *Tetramethylenebis[dibutylphosphine]*, *8CI*

$(H_3CCH_2CH_2CH_2)_2P(CH_2)_4P(CH_2CH_2CH_2CH_3)_2$

$C_{20}H_{44}P_2$ M 346.515
Dioxide: [4151-26-2]. *1,4-Bis(dibutylphosphinyl)butane.*
$C_{20}H_{44}O_2P_2$ M 378.514
Cryst. (C_6H_6). Mp 116-118°. Bp_2 270-271°.

Kosolapoff, G.M. *et al*, *J. Chem. Soc.*, 1959, 3950 (*oxide, synth*)
Kabachnik, M.I. *et al*, *Teor. Eksp. Khim.*, 1972, **8**, 361; *CA*, **77**, 157148.
Zvezdkina, L.I. *et al*, *CA*, 1982, **96**, 52409 (*synth*)

1,2-Bis(dibutylphosphino)ethane **B-00158**

1,2-Ethanediylbis[dibutylphosphine], *10CI*, *9CI*. *Ethylenebis[dibutylphosphine]*, *8CI*
[4141-59-7]

$(H_3CCH_2CH_2CH_2)_2PCH_2CH_2P(CH_2CH_2CH_2CH_3)_2$

$C_{18}H_{40}P_2$ M 318.462
Dioxide: [4141-63-3]. *1,2-Bis(dibutylphosphinyl)ethane.*
$C_{18}H_{40}O_2P_2$ M 350.460
Mp 174-175° (168-169°). $Bp_{0.8}$ 222-224°. Dec. on heating. Changes on subl.

Kosolapoff, G.M. *et al*, *J. Chem. Soc.*, 1961, 2423 (*oxide*)
Petrov, K.A. *et al*, *Zh. Obshch. Khim.*, 1965, **35**, 1602 (*Engl. transl.* p. 1606) (*oxide*)
Zvezdkina, L.I. *et al*, *CA*, 1982, **96**, 52409 (*synth*)

1,2-Bis(dibutylphosphino)ethylene **B-00159**

1,2-Ethenediylbis[dibutylphosphine], *9CI*

$(H_3CCH_2CH_2CH_2)_2PCH{=}CHP(CH_2CH_2CH_2CH_3)_2$

$C_{18}H_{38}P_2$ M 316.446

(E)-form
Liq. $Bp_{0.05}$ 123-125°.
Monoxide:
$C_{18}H_{38}OP_2$ M 332.445
Cryst. (hexane). Mp 107-109°.
Monoxide, monomethiodide:
Dibutylmethyl[(dibutylphosphinyl)ethenyl]-phosphonium iodide.
$C_{19}H_{41}IOP_2$ M 474.384
Solid. Mp 82-83°.

Dioxide: 1,2-Bis(dibutylphosphinyl)ethylene.
$C_{18}H_{38}O_2P_2$ M 348.445
Solid. Mp 266-267°.
Dimethiodide: 1,2-Ethenediylbis[(dibutylmethyl)-phosphonium] diiodide.
$C_{20}H_{44}I_2P_2$ M 600.324
Solid. Mp 130°.

Weiner, M.A. *et al*, *J. Org. Chem.*, 1969, **34**, 1130 (*synth, uv, ir, pmr, derivs*)

Bis(dibutylphosphino)methane B-00160

Methylenebis[dibutylphosphine], 9CI

$(H_3CCH_2CH_2CH_2)_2PCH_2P(CH_2CH_2CH_2CH_3)_2$

$C_{17}H_{38}P_2$ M 304.435
Dioxide: [21245-07-8]. *Bis(dibutylphosphinyl)methane.*
$C_{17}H_{38}O_2P_2$ M 336.434
Needles (hexane). Mp 174-176°. Forms complexes containing Fe, Co or Ni.

Kosolapoff, G.M. *et al*, *J. Chem. Soc.*, 1961, 2423.

1,5-Bis(dibutylphosphino)pentane B-00161

1,5-Pentanediylbis[dibutylphosphine], 10CI, 9CI. Pentamethylenebis[dibutylphosphine], 8CI

$(H_3CCH_2CH_2CH_2)_2P(CH_2)_5P(CH_2CH_2CH_2CH_3)_2$

$C_{21}H_{46}P_2$ M 360.542
Dioxide: 1,5-Bis(dibutylphosphinyl)pentane.
$C_{21}H_{46}O_2P_2$ M 392.541
Mp 106-107°. $Bp_{0.75}$ 259°. Not associated in C_6H_6.

Kosolapoff, G.M. *et al*, *J. Chem. Soc.*, 1959, 3950.

Bis(2,4-dichlorophenyl) ethyl phosphate, 9CI, 8CI B-00162

Phosphiden
[36519-00-3]

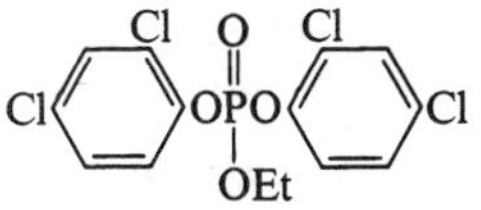

$C_{14}H_{11}Cl_4O_4P$ M 416.024
Fungicide. Yellow liq. (tech. grade). d_4^{25} 1.40. $Bp_{0.2}$ 175°. n_D^{20} 1.5628.

Pesticide Manual, 7th Ed., 53.

Bis(2,3-dichlorophenyl)phosphinic acid, 9CI B-00163

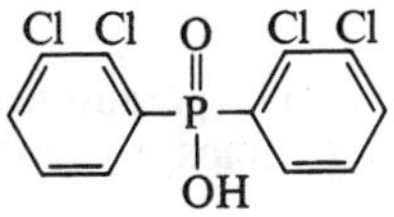

$C_{12}H_7Cl_4O_2P$ M 355.972
Cryst. (EtOH aq.). Mp 278-280°.

Denham, J.M. *et al*, *J. Org. Chem.*, 1958, **23**, 1298.

Bis(2,5-dichlorophenyl)phosphinic acid, 9CI B-00164

$C_{12}H_7Cl_4O_2P$ M 355.972
Cryst. (EtOH aq.). Mp 232-233° (221-222°).

Freedman, L.D. *et al*, *J. Am. Chem. Soc.*, 1953, **75**, 1379.
Denham, J.M. *et al*, *J. Org. Chem.*, 1958, **23**, 1298.

Bis(3,5-dichlorophenyl)phosphinic acid, 9CI B-00165

$C_{12}H_7Cl_4O_2P$ M 355.972
Cryst. (EtOH aq.). Mp 243-244.5°.

Denham, J.M. *et al*, *J. Org. Chem.*, 1958, **23**, 1298.

1,4-Bis(dicyclohexylphosphino)butane B-00166

1,4-Butanediylbis[dicyclohexylphosphine], 10CI, 9CI. Tetramethylenebis[dicyclohexylphosphine], 8CI
[65038-36-0]

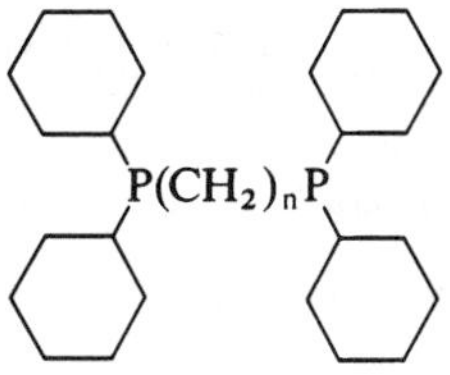

$P(CH_2)_nP$

n = 4

$C_{28}H_{52}P_2$ M 450.666
Used in catalysts for hydroformylation of 1-alkenes.
Cryst. (dioxan). Mp 98-100°.
Disulfide: 1,4-Bis(dicyclohexylphosphinothioyl)butane.
$C_{28}H_{52}P_2S_2$ M 514.786
Cryst. (Me_2CO). Mp 214-215°.

Issleib, K. *et al*, *Chem. Ber.*, 1959, **92**, 3175 (*synth*)
Issleib, K. *et al*, *Chem. Ber.*, 1961, **94**, 2244 (*disulfide*)
Hayashi, T. *et al*, *Bull. Chem. Soc. Jpn.*, 1981, **54**, 3438 (*use*)

1,2-Bis(dicyclohexylphosphino)ethane B-00167

1,2-Ethanediylbis[dicyclohexylphosphine], 9CI
[23743-26-2]
As 1,4-Bis(dicyclohexylphosphino)butane, B-00166 with

n = 2

$C_{26}H_{48}P_2$ M 422.613
Ligand for Cr, Ni and W. Cryst. (THF/Et_2O or THF/EtOH). Mp 82-85°, 96-97°. Forms red CS_2 adduct.
Dioxide: [16527-12-1]. *1,2-Bis(dicyclophosphinyl)-ethane.*
$C_{26}H_{48}O_2P_2$ M 454.612
Solid. Mp 192°. Forms complexes with Gp. VI metals.
Disulfide: 1,2-Bis(dicyclophosphinothioyl)ethane.
$C_{26}H_{48}P_2S_2$ M 486.733
Solid. Mp 195-197°.

Issleib, K. *et al*, *Chem. Ber.*, 1963, **96**, 1544, 2186 (*synth*)
Connor, J.A. *et al*, *J. Organomet. Chem.*, 1975, **94**, 55 (*dioxide, synth, ms, ir*)
Burt, R.J. *et al*, *J. Organomet. Chem.*, 1979, **182**, 203 (*synth, nmr*)

4,5-Bis[(dicyclohexylphosphino)methyl]-2,2-dimethyl-1,3-dioxolane **B-00168**

[(2,2-Dimethyl-1,3-dioxolane-4,5-diyl)bis[methylene]]-bis[dicyclohexylphosphine], 9CI. CyDIOP

(4R,5R)-form

$C_{31}H_{56}O_2P_2$ M 522.730

(4R,5R)-form [82239-68-7]

(−)-trans-*form*

Forms Rh complex, useful in enantioselective hydrogenations of ketones. Solid. Mp 90-92°. $[\alpha]_D^{20}$ −24.1° (c, 0.97 in C_6H_6).

Tani, K. *et al, Chem. Lett.*, 1982, 265 (*synth, pmr, complex, use*)
Tani, K. *et al, ACS Symp. Ser.*, 1982, **185**, 187, 283 (*use*)

1,3-Bis(dicyclohexylphosphino)propane **B-00169**

1,3-Propanediylbis[dicyclohexylphosphine], 10CI, 9CI. Trimethylenebis[dicyclohexylphosphine], 8CI

As 1,4-Bis(dicyclohexylphosphino)butane, B-00166 with

n = 3

$C_{27}H_{50}P_2$ M 436.640

Oil.

Disulfide: 1,3-Bis(dicyclohexylphosphinothioyl)-propane.

$C_{27}H_{50}P_2S_2$ M 500.760

Cryst. (Me_2CO). Mp 147-148°.

Issleib, K. *et al, Chem. Ber.*, 1959, **92**, 3175.

Bis(diethoxyphosphinothioyl) disulfide **B-00170**

Tetraethyl thioperoxydiphosphate, 9CI

[2901-90-8]

$(EtO)_2P(S)-S-S-P(S)(OEt)_2$

$C_8H_{20}O_4P_2S_4$ M 370.431

Low-melting solid. Mp 28-29° (23-24°). $Bp_{0.2}$ 135°. n_D^{25} 1.5600.

Zemlyanski, N.I. *et al, Zh. Obshch. Khim.*, 1961, **31**, 880 (*Engl. transl.* p. 811) (*synth*)
Vasil'ev, A.F. *et al, Zh. Obshch. Khim.*, 1964, **34**, 2322 (*Engl. transl.* p. 2333) (*ir*)
Lippman, A.E. *et al, J. Org. Chem.*, 1966, **31**, 471 (*P nmr*)
Oae, S. *et al, Tetrahedron*, 1972, **28**, 2981 (*synth*)
Harned, W.H. *et al, J. Agric. Food Chem.*, 1976, **24**, 689 (*ms, pmr, P nmr*)

Bis(diethoxyphosphinothioyl) trisulfide, 9CI **B-00171**

[6926-73-4]

$(EtO)_2P(S)-S-S-S-P(S)(OEt)_2$

$C_8H_{20}O_4P_2S_5$ M 402.491

Needles (EtOH). Mp 78° (73.5-75°).

Klement, R. *et al, Chem. Ber.*, 1964, **97**, 1716 (*synth, ir, raman*)
Almasi, L. *et al, Chem. Ber.*, 1966, **99**, 3288 (*synth*)
Tolmacheva, N.A. *et al, Zh. Obshch. Khim.*, 1978, **48**, 1078; 1982, **52**, 847 (*Engl. transl.* pp. 982, 736) (*synth*)
Khaskin, B.A. *et al, Zh. Obshch. Khim.*, 1980, **50**, 2233, 2700; 1983, **53**, 1219 (*Engl. transl.* pp. 1802, 2182, 1086) (*props*)

Bis(diethoxyphosphinyl) disulfide **B-00172**

Tetraethyl thioperoxydiphosphate, 9CI

[4403-51-4]

$(EtO)_2P(O)-S-S-P(O)(OEt)_2$

$C_8H_{20}O_6P_2S_2$ M 338.310

Liq. d_4^{20} 1.26. $Bp_{0.03}$ 110°. n_D^{20} 1.4872.

▷LD_{50} (rat) 40 mg/kg

Ettel, V. *et al, Chem. Listy*, 1956, **50**, 1261; *CA*, **50**, 16025.
Borecka, B. *et al, Bull. Acad. Pol. Sci., Ser. Sci. Chim.*, 1974, **22**, 201 (*synth, P nmr*)
Mel'nikov, N.N. *et al, Zh. Obshch. Khim.*, 1975, **45**, 1005 (*Engl. transl.* p. 992) (*props*)
Schrader, G. *et al, Angew. Chem.*, 1958, **70**, 690 (*synth, tox*)

2,3-Bis(diethoxyphosphinyl)propanoic acid **B-00173**

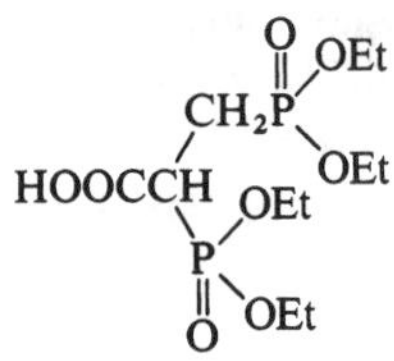

$C_{11}H_{24}O_8P_2$ M 346.253

(±)-form

Me ester: Methyl 2,3-bis(diethoxyphosphinyl)-propanoate.

$C_{12}H_{26}O_8P_2$ M 360.280

Liq. Bp_2 166°.

Et ester: Ethyl 2,3-bis(diethoxyphosphinyl)propanoate. Pentaethyl 2,3-diphosphonopropanoate.

$C_{13}H_{28}O_8P_2$ M 374.307

Liq. d_4^{20} 1.17. $Bp_{0.01}$ 150°. n_D^{20} 1.4495.

Nitrile: [7563-70-4]. *Tetraethyl (1-cyano-1,2-ethanediyl)bisphosphonate.*

$C_{11}H_{23}NO_6P_2$ M 327.253

Liq. d_4^{20} 1.17. Bp_2 173-174°, $Bp_{0.01}$ 144°. n_D^{20} 1.4465-1.4508.

Pudovik, A.N. *et al, Zh. Obshch. Khim.*, 1965, **35**, 354 (*Engl. transl.* p. 354) (*ester*)
Pudovik, A.N. *et al, Zh. Obshch. Khim.*, 1966, **36**, 1232 (*Engl. transl.* p. 1247) (*nitrile*)
Novikova, Z.S. *et al, Zh. Obshch. Khim.*, 1974, **44**, 276 (*Engl. transl.* p. 261) (*ester, nitrile*)
Pfeiffer, F.R. *et al, J. Med. Chem.*, 1974, **17**, 112 (*ester*)
Okamoto, Y. *et al, Bull. Chem. Soc. Jpn.*, 1981, **54**, 303 (*ester, synth, pmr, P nmr*)

N″-[Bis(diethylamino)phosphinothioyl]-N,N,N′,N′-tetraethylphosphorodiamidoimidic acid, 9CI **B-00174**

$(Et_2N)_2P(OH)=NP(S)(NEt_2)_2$

$C_{16}H_{41}N_5OP_2S$ M 413.540

Free acid tatuomeric.

Et ester: [59998-80-0]. *Ethyl* N″-*[bis(diethylamino)-phosphinothioyl]*-N,N,N′,N′-*tetraethylphosphorodiamidoimidate.* P-*Ethoxy*-P,P-*bis(diethylamino)*-N-*[bis(diethylamino)phosphinothioyl]phosphazene.*

$C_{18}H_{45}N_5OP_2S$ M 441.594

Liq. d_4^{20} 1.05. $Bp_{0.001}$ 106°. n_D^{20} 1.4980.
Butyl ester: [59998-81-1]. *Butyl* N″-[*bis(diethylamino)-phosphinothioyl*]-N,N,N′,N′-*tetraethylphosphorodiamidoimidate.* P-*Butoxy*-P,P-*bis(diethylamino)*-N-[*bis(diethylamino)phosphinothioyl*]*phosphazene.*
$C_{20}H_{49}N_5OP_2S$ M 469.648
Liq. d_4^{20} 1.00. $Bp_{0.001}$ 127°. n_D^{20} 1.4950.

Zaslavskaya, N.N. *et al, Izv. Akad. Nauk SSSR, Ser. Khim.*, 1976, 931 (*Engl. transl.* p. 911) (*esters, synth, ir*)

[Bis(diethylamino)phosphinyl]acetic acid B-00175

$(Et_2N)_2P(O)CH_2COOH$

$C_{10}H_{23}N_2O_3P$ M 250.277
pK_a 4.40 (H_2O), pK_a 5.72 (50% EtOH aq., 25°).
Et ester: Ethyl [bis(diethylamino)phosphinyl]acetate. (Ethoxycarbonylmethyl)phosphonic bis(diethylamide).
$C_{12}H_{27}N_2O_3P$ M 278.331
Liq. d_4^{20} 1.04, 1.07. Bp_4 152-153°, $Bp_{0.45-0.7}$ 142-147°. n_D^{20} 1.4668.
Ph ester:
$C_{16}H_{27}N_2O_3P$ M 326.375
Liq. d_4^{20} 1.06. $Bp_{0.25-0.4}$ 160-162°. n_D^{20} 1.4972.
N,N-*Diethylamide:* [(N,N-*Diethylcarbamoyl)methyl*]-*phosphonic bis(diethylamide).*
$C_{14}H_{32}N_3O_2P$ M 305.399
Liq. d_4^{20} 1.05. $Bp_{0.4}$ 139-142°. n_D^{20} 1.4775.

Mel'nikov, N.N. *et al, Zh. Obshch. Khim.*, 1961, **31**, 3949 (*Engl. Transl.* p. 3684) (*derivs, synth*)
Malevannaya, R.A. *et al, Zh. Obshch. Khim.*, 1972, **42**, 765 (*Engl. transl.* p. 757) (*ethyl ester, synth, pmr*)
Tsvetkov, E.N. *et al, Zh. Obshch. Khim.*, 1974, **44**, 1225 (*Engl. transl.* p. 1203) (*synth, pmr, props*)

[Bis(diethylamino)phosphinyl]-phosphorimidic acid, 9CI B-00176

$(Et_2N)_2P(O)N{=}P(OH)_3$

$C_8H_{23}N_3O_4P_2$ M 287.235
Free acid tautomeric with [bis(diethylamino)-phosphinyl]phosphoramidic acid.
Tri-Me ester: [59740-56-6]. *Trimethyl [bis(diethylamino)phosphinyl]phosphorimidate.* N-[*Bis(diethylamino)phosphinyl*]-P,P,P-*trimethoxy-phosphine imide.* N-[*Bis(diethylamino)phosphinyl*]-P,P,P-*trimethoxyphosphazene.*
$C_{11}H_{29}N_3O_4P_2$ M 329.315
Liq. d_4^{20} 1.10. n_D^{20} 1.4580.
Tri-Et ester: [59740-57-7]. *Triethyl [bis(diethylamino)-phosphinyl]phosphorimidate.* P,P,P-*Triethoxy*-N-[*bis(diethylamino)phosphinyl*]*phosphine imide.* P,P,P-*Triethoxy*-N-[*bis(diethylamino)phosphinyl*]-*phosphazene.*
$C_{14}H_{35}N_3O_4P_2$ M 371.396
Liq. d_4^{20} 1.05. $Bp_{0.002}$ 80°. n_D^{20} 1.4540.

Zaslavskaya, N.N. *et al, Izv. Akad. Nauk SSSR, Ser. Khim.*, 1976, 675 (*Engl. transl.* p. 662) (*synth, ir, pmr, P nmr*)

N″-[Bis(diethylamino)phosphinyl]-*N,N,N′,N′*-tetraethylphosphorodiamidoimidic acid B-00177

$(Et_2N)_2P(OH){=}NP(O)(NEt_2)_2$

$C_{16}H_{41}N_5O_2P_2$ M 397.480
Free acid tautomeric.
Et ester: [59998-70-8]. *Ethyl* N″-[*bis(diethylamino)-phosphinyl*]-N,N,N′,N′-*tetraethylphosphorodiamidoimidate.* P-*Ethoxy*-P,P-*bis(diethylamino)*-N-[*bis(diethylamino)phosphinyl*]*phosphazene.*
$C_{18}H_{45}N_5O_2P_2$ M 425.533
Liq. d_4^{20} 1.01. $Bp_{0.002}$ 112°. n_D^{20} 1.4720.
Butyl ester: [59998-72-0]. *Butyl* N″-[*bis(diethylamino)-phosphinyl*]-N,N,N′,N′-*tetraethylphosphorodiamidoimidate.* P-*Butoxy*-P,P-*bis(diethylamino)*-N-[*bis(diethylamino)phosphinyl*]*phosphazene.*
$C_{20}H_{49}N_5O_2P_2$ M 453.587
Liq. d_4^{20} 1.01. $Bp_{0.01}$ 145°. n_D^{20} 1.4710.

Zaslavskaya, N.N. *et al, Izv. Akad. Nauk SSSR, Ser. Khim.*, 1976, 931 (*Engl. transl.* p. 911) (*synth, ir, P nmr*)

Bis(diethylphosphino)acetylene B-00178

1,2-Ethynediylbis[diethylphosphine], 9CI

$Et_2PC{\equiv}CPEt_2$

$C_{10}H_{20}P_2$ M 202.216
Liq. $Bp_{0.07}$ 54°. n_D^{20} 1.5332.

Chatt, J. *et al, J. Chem. Soc.*, 1960, 1378.

2,2′-Bis(diethylphosphino)biphenyl B-00179

[1,1′-Biphenyl]-2,2′-diylbis(diethylphosphine), 9CI

PEt_2 Et_2P

$C_{20}H_{28}P_2$ M 330.389
Ligand for Ni and Pd. Low-melting solid. Mp 28-30°. $Bp_{0.25}$ 152°.
B,2MeI: [1,1′-Biphenyl]-2,2′-diylbis(diethylmethylphosphonium)diiodide.
$C_{22}H_{34}I_2P_2$ M 614.267
Cryst. (EtOH). Mp 255-256°.
Dioxide: 2,2-Bis(diethylphosphinyl)biphenyl.
$C_{20}H_{28}O_2P_2$ M 362.388
Mp 193-195° (sealed evac. tube).

Allen, D.W. *et al, J. Chem. Soc. (C)*, 1967, 1869 (*synth, dioxide*)
Allen, D.W. *et al, J. Chem. Soc. (A)*, 1969, 1097 (*pmr, complexes*)

1,2-Bis(diethylphosphino)ethane B-00180

1,2-Ethanediylbis[diethylphosphine], 9CI
[6411-21-8]

$Et_2PCH_2CH_2PEt_2$

$C_{10}H_{24}P_2$ M 206.247
Air-sensitive liq. Bp 220-230°, Bp 250-255°, Bp_{10} 124-126°. n_D^{20} 1.510. Forms complexes with Au, Ir, Fe, Mo and W compds.
B,2HBr: Mp 85° dec.
B,2HI: Mp 181-183°.
Dioxide: [13337-10-5]. *1,2-Bis(diethylphosphinyl)-ethane.*
$C_{10}H_{24}O_2P_2$ M 238.246
Solid. Mp 125-126°. $Bp_{2.5}$ 228-229°.
Disulfide: 1,2-Bis(diethylphosphinothioyl)ethane.
$C_{10}H_{24}P_2S_2$ M 270.367
Cryst. (Me_2CO or EtOH). Mp 84-85°.

Chatt, J. *et al*, *J. Chem. Soc.*, 1960, 1378 (*synth, complexes*)
Meriwether, L.S. *et al*, *J. Am. Chem. Soc.*, 1961, **83**, 3192 (*nmr*)
Issleib, K. *et al*, *Chem. Ber.*, 1963, **96**, 1544, 2186.
Tsivunin, V.S. *et al*, *Zh. Obshch. Khim.*, 1966, **36**, 1430 (*Engl. transl. p. 1436*) (*dioxide, disulfide*)
Burt, R.J. *et al*, *J. Organomet. Chem.*, 1979, **182**, 203 (*synth, nmr*)

1,2-Bis(diethylphosphino)ethylene B-00181

1,2-Ethenediylbis[diethylphosphine], 9CI

$Et_2PCH{=}CHPEt_2$

$C_{10}H_{22}P_2$ M 204.231

Liq. $Bp_{0.03}$ 67°. Prob. (*Z*–).

(*E*)-form

Dioxide:

$C_{10}H_{22}O_2P_2$ M 236.230

Solid. Mp 224-226°.

B.P., 859 391, (*1961*); *CA*, **55**, 23345 (*synth*)
Kabachnik, M.I. *et al*, *Dokl. Akad. Nauk SSSR, Ser. Sci. Khim.*, 1976, **230**, 1347 (*Engl. transl.* p. 647) (*dioxide*)

Bis[(dihydroxyphosphinyl)methyl]-phosphinic acid, 9CI B-00182

[(Hydroxyphosphinylidene)bis(methylene)]-bis(phosphonic acid), 9CI

$[(HO)_2P(O)CH_2]_2P(O)OH$

$C_2H_9O_8P_3$ M 254.010

Cryst. + $2H_2O$.

Triscyclohexylamine salt: Cryst. + $2H_2O$. Mp 205° dec.

Penta-Et ester: [18033-91-5]. *Tetraethyl [(ethoxyphosphinylidene)bis(methylene)]-bisphosphonate. Ethyl bis[(diethoxyphosphinyl)-methyl]phosphinate.*

$C_{12}H_{29}O_8P_3$ M 394.278

Wittig-Horner reagent for synth. of diethyl ethenylphosphonates. $Bp_{0.001}$ 153-157°. n_D^{20} 1.4610.

Maier, L., *Helv. Chim. Acta*, 1969, **52**, 827 (*synth, deriv, ir, P nmr, pmr*)
Gilmore, W.F. *et al*, *J. Org. Chem.*, 1973, **38**, 1423 (*ester, use*)

Bis(diisopropoxyphosphinyl) disulfide B-00183

Tetrakis(1-methylethyl) thioperoxydiphosphate, 9CI

[56341-78-7]

$[(H_3C)_2CHO]_2P(O)SSP(O)[OCH(CH_3)_2]_2$

$C_{12}H_{28}O_6P_2S_2$ M 394.417

Liq. d_4^{20} 1.15. $Bp_{0.01}$ 102°. n_D^{20} 1.4761. Rel. non-toxic.

Ettel, V. *et al*, *Chem. Listy*, 1956, **50**, 1261; *CA*, **50**, 16025.
Schrader, G. *et al*, *Angew. Chem.*, 1958, **70**, 690 (*synth, tox*)
Khaskin, B.A. *et al*, *Zh. Obshch. Khim.*, 1982, **52**, 597 (*Engl. transl.* p. 525) (*props*)

Bis(diisopropyoxyphosphinothioyl) disulfide B-00184

Tetrakis(1-methylethyl) thioperoxydiphosphate, 9CI. *Tetraisopropyl thioperoxydiphosphate*

[3031-21-8]

$[(H_3C)_2CHO]_2P(S){-}S{-}S{-}P(S)[OCH(CH_3)_2]_2$

$C_{12}H_{28}O_4P_2S_4$ M 426.538

Antioxidant; heat stabilizer for synth. rubbers. Cu complex used in estimation of tocopherols. Cryst. (EtOH). Mp 91-92°.

Zemlyanskii, N.I. *et al*, *Zh. Obshch. Khim.*, 1961, **31**, 880 (*Engl. transl.* p. 811) (*synth*)
Vasil'ev, A.F. *et al*, *Zh. Obshch. Khim.*, 1964, **34**, 2322 (*Engl. transl.* p. 2333) (*ir*)
Yordanov, N.D. *et al*, *Inorg. Nucl. Chem. Lett.*, 1976, **12**, 527 (*props, use*)

Bis(dimethoxyphosphinothioyl) disulfide B-00185

Tetramethyl thioperoxydiphosphate, 9CI

[5930-71-2]

$(MeO)_2P(S){-}S{-}S{-}P(S)(OMe)_2$

$C_4H_{12}O_4P_2S_4$ M 314.324

Solid. Mp 49-51°. Forms Pt complex.

▷Causes inflammation to eyes, corneal opacity, and swelling of eyelids (in rabbits)

Miller, B., *Tetrahedron*, 1964, **20**, 2069 (*synth*)
Vasil'ev, A.F. *et al*, *Zh. Obshch. Khim.*, 1964, **34**, 2322 (*Engl. transl.* p. 2333) (*ir*)
Lippman, A.E., *J. Org. Chem.*, 1966, **31**, 471 (*nmr*)
Wenzel, K.D. *et al*, *Z. Chem.*, 1971, **11**, 461 (*props, haz*)
Maekawa, K. *et al*, *CA*, 1974, **81**, 49205 (*props*)

Bis(dimethoxyphosphinyl) disulfide B-00186

Tetramethyl thioperoxydiphosphate, 9CI

[7439-49-8]

$(MeO_2P(O){-}S{-}S{-}P(O)(OMe)_2$

$C_4H_{12}O_6P_2S_2$ M 282.203

Liq. d_4^{20} 1.41. n_D^{20} 1.4968.

Ettel, V. *et al*, *Chem. Listy*, 1956, **50**, 1261; *CA*, **50**, 16025.

2,4-Bis(dimethylamino)-3-methyl-1,3,2,4-thiazadiphosphetidine B-00187

N,N,N′,N′,*3-Pentamethyl-1,3,2,4-thiazadiphosphetidin-2,4-diamine*, 9CI

Me₂N–P(–S–)(–N(Me)–)P–NMe₂ (four-membered P–S–P–N ring)

$C_5H_{15}N_3P_2S$ M 211.201

2,4-Disulfide: [55042-00-7].

$C_5H_{15}N_3P_2S_3$ M 275.321

Solid. Mp 130-150°. Consists of a single stereoisomer.

Bulloch, G. *et al*, *J. Chem. Soc., Dalton Trans.*, 1974, 2329 (*synth, pmr, P nmr*)

Bis(4-dimethylaminophenyl)phosphinic acid, 9CI B-00188

$(Me_2NC_6H_4)_2P(O)OH$

$C_{16}H_{21}N_2O_2P$ M 304.328

Plates (EtOH aq.). Mp 209-210°, 249-250°.

Me ester: [18593-23-2]. *Methyl bis(4-dimethylaminophenyl)phosphinate.*
$C_{17}H_{23}N_2O_2P$ M 318.355
Characterised spectroscopically.

Et ester: [61153-56-8]. *Ethyl bis(4-dimethylaminophenyl)phosphinate.*
$C_{18}H_{25}N_2O_2P$ M 332.381
Solid. Mp 128-129°.

Robins, R.K. *et al, J. Org. Chem.*, 1951, **16**, 324 (*synth*)
Haake, P. *et al, J. Org. Chem.*, 1969, **34**, 788 (*ester, ms*)
Cheng, C.Y. *et al, J. Chem. Soc., Perkin Trans. 1*, 1976, 1739 (*synth, pmr, ester*)

O,O-Bis(3-dimethylaminophenyl) phosphorothioate B-00189

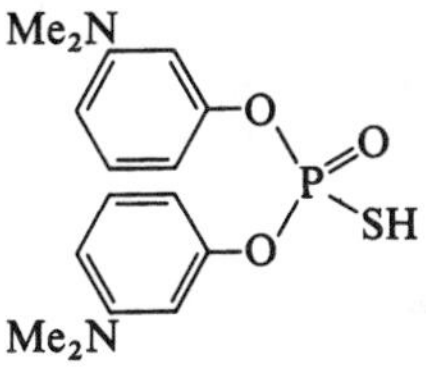

Me_2N ... O, P(=O)SH, O ... Me_2N

$C_{16}H_{21}N_2O_3PS$ M 352.387
Cryst. (EtOH or MeOH). Mp 170-172°.

Chloride:
$C_{16}H_{20}ClN_2O_2PS$ M 370.833
Pale-yellow oil. n_D^{25} 1.6135.

Anilide: O,O-*Bis(3-dimethylaminophenyl) phenylphosphoramidothioate.*
$C_{22}H_{16}N_3O_2PS$ M 417.421
Needles (EtOH aq.). Mp 97-98°.

Anilide; B, 2MeI: Cryst. + $1H_2O$ ($MeOH/Me_2CO$). Mp 130-131° dec.

U.S.P., 2 759 961, (*1956*); *CA*, **51**, 482

O,O-Bis(4-dimethylaminophenyl) phosphorothioate, 9CI B-00190

O,O-*Bis(4-dimethylaminophenyl) thiophosphoric acid*

$C_{16}H_{21}N_2O_3PS$ M 352.387

O-*Isopropyl ester:* O,O-*Bis(4-dimethylaminophenyl)* O-*isopropyl phosphorothioate.*
$C_{19}H_{27}N_2O_3PS$ M 394.468
Cryst. (MeOH). Mp 74-76°.

O-*Isopropyl ester; B,2MeI:* Cryst. + $3H_2O$. Mp 177-177.5° dec.

Fluoride: O,O-*Bis(4-dimethylaminophenyl) phosphorofluoridothioate.*
$C_{16}H_{20}FN_2O_2PS$ M 354.378
Needles (2-propanol). Mp 110.5-112°.

Fluoride; B, 2MeI: Plates. Mp 159-162° dec.

Chloride: O,O-*Bis(4-dimethylaminophenyl) phosphorochloridothioate.*
$C_{16}H_{20}ClN_2O_2PS$ M 370.833
Needles (C_6H_6/2-propanol). Mp 165.5-166.5°.

Chloride; B,2MeI: Cream plates + $1H_2O$. Mp 165-167°.

Bromide: O,O-*Bis(4-dimethylaminophenyl) phosphorobromidothioate.*
$C_{16}H_{20}BrN_2O_2PS$ M 415.284
Cryst. (toluene/pet. ether). Mp 158-164°.

Bromide; B, 2MeI: Yellow cryst. + $2H_2O$. Mp 137-141° dec.

U.S.P., 2 759 961, (*1956*); *CA*, **51**, 482

Bis(dimethylaminophosphinothioyl)-pentamethylthiophosphoric triamide B-00191

Nonamethyl thioimidodiphosphoramide, 9CI.
Bis[bis(dimethylamino)phosphinothioyl]methylamine
[34244-09-2]

$(Me_2N)_2P(S){-}NMe{-}P(S)(NMe_2)_2$

$C_9H_{27}N_5P_2S_2$ M 331.413
Note 9CI name identical with that of N″-Bis(dimethylamino)phosphinyl-*N,N,N′,N′,N″*-pentamethylphosphorotriamidothioate, N-00032 .
Solid. Mp 85-86°.

Haegele, G. *et al, J. Chem. Soc., Dalton Trans.*, 1974, 1985 (*pmr, cmr, P nmr*)
Bulloch, G. *et al, J. Chem. Soc., Dalton Trans.*, 1974, 2329 (*synth*)

[Bis(dimethylamino)phosphinothioyl]-phosphorimidic acid, 9CI B-00192

$(Me_2N)_2P(S)N{=}P(OH)_3$

$C_4H_{15}N_3O_3P_2S$ M 247.188
Free acid tautomeric with [bis(dimethylamino)-phosphinothioyl]phosphoramidic acid.

Tri-Et ester: [59998-75-3]. *Triethyl [bis(dimethylamino)phosphinothioyl]-phosphorimidate.* P,P,P-*Triethoxy*-N-*[bis(dimethylamino)phosphinothioyl]phosphine imide.* P,P,P-*Triethoxy*-N-*[bis(dimethylamino)-phosphinothioyl]phosphazene.*
$C_{10}H_{27}N_3O_3P_2S$ M 331.349
Liq. d_4^{20} 1.07. $Bp_{0.001}$ 92°. n_D^{20} 1.4843. No isom. during distn.

Tributyl ester: [59998-77-5]. *Tributyl [bis(dimethylamino)phosphinothioyl]-phosphorimidate.* P,P,P-*Tributoxy*-N-*[bis(dimethylamino)phosphinothioyl]phosphazene.* P,P,P-*Tributoxy*-N-*[bis(dimethylamino)-phosphinothioyl]phosphine imide.*
$C_{16}H_{39}N_3O_3P_2S$ M 415.510
Liq. d_4^{20} 1.02. $Bp_{0.001}$ 120°. n_D^{20} 1.4785 No isom. during distn.

Zaslavskaya, N.N. *et al, Izv. Akad. Nauk SSSR, Ser. Khim.*, 1976, 931 (*Engl. transl.* p. 911) (*synth, ir*)

[Bis(dimethylamino)phosphinyl]acetic acid B-00193

$(Me_2N)_2P(O)CH_2COOH$

$C_6H_{15}N_2O_3P$ M 194.170
pK_a 4.15 (H_2O), 5.37 (50% EtOH aq., 25°).

Et ester: Ethyl [bis(dimethylamino)phosphinyl]acetate.
$C_8H_{19}N_2O_3P$ M 222.223
Liq. d_4^{20} 1.10. Bp_4 140-141°. n_D^{20} 1.4660.

Malevannaya, R.A. *et al, Zh. Obshch. Khim.*, 1972, **42**, 765 (*Engl. transl.* p. 757) (*ester, pmr*)
Tsvetkov, E.N. *et al, Zh. Obshch. Khim.*, 1974, **44**, 1225 (*Engl. transl.* p. 1203) (*props*)

[Bis(dimethylamino)phosphinyl]-phosphorimidic acid, 9CI B-00194

$(Me_2N)_2P(O)N{=}P(OH)_3$

$C_4H_{15}N_3O_4P_2$ M 231.128
Free acid tautomeric with [bis(dimethylamino)-phosphinyl]phosphoramidic acid.

Tri-Et ester: [59833-34-0]. *Triethyl [bis(dimethylamino)phosphinyl]phosphorimidate.* P,P,P-*Triethoxy*-N-*[bis(dimethylamino)phosphinyl]-phosphazene.* P,P,P-*Triethoxy*-N-*[bis(dimethylamino)phosphinyl]phosphine imide.*
$C_{10}H_{27}N_3O_4P_2$ M 315.289
Liq. d_4^{20} 1.11. $Bp_{0.001}$ 73°. n_D^{20} 1.4527. Undergoes rearrangement at 130°.

Tri-Ph ester: [75944-12-6]. *Triphenyl [bis(dimethylamino)phosphinyl]phosphorimidate.* N-*[Bis(dimethylamino)phosphinyl]*-P,P,P-*triphenoxyphosphazene.* N-*[Bis(dimethylamino)phosphinyl]*-P,P,P-*triphenoxyphosphine imide.*
$C_{22}H_{27}N_3O_4P_2$ M 459.421
Liq. d_4^{20} 1.22. $Bp_{0.004}$ 168-170°. n_D^{20} 1.5567.

Zaslavskaya, N.N. *et al, Izv. Akad. Nauk SSSR, Ser. Khim.*, 1976, 675 (*Engl. transl.* p. 662) (*triethyl ester, ir, pmr, P nmr, props*)
Khodak, A.A. *et al, Izv. Akad. Nauk SSSR, Ser. Khim.*, 1980, 2379; *CA*, **94**, 10578 (*triphenyl ester, pmr, P nmr*)

Bis[3-(dimethylamino)propyl]phenylphosphine **B-00195**

*3,3'-(Phenylphosphinidene)bis[*N,N*-dimethyl-1-propanamine]*, 9CI
[32357-32-7]

$PhP(CH_2CH_2CH_2NMe_2)_2$

$C_{16}H_{29}N_2P$ M 280.392
Used in sulfide contraction reactions. Liq. $Bp_{0.005}$ 102-105°. n_D^{24} 1.5265.

Roth, M. *et al, Helv. Chim. Acta*, 1971, **54**, 710 (*synth, uv, ir, pmr, ms, use*)
Org. Synth., 1976, **55**, 127 (*synth*)
Ireland, R.E. *et al, J. Org. Chem.*, 1980, **45**, 1868 (*use*)

Bis(dimethylamino)trifluorophosphorane, 8CI **B-00196**

Trifluoro-N,N,N',N'-*tetramethylphosphoranediamine*, 9CI
[1735-83-7]

$(Me_2N)_2PF_3$

$C_4H_{12}F_3PN_2$ M 176.121
Liq. Mp −22°. Bp_8 49.5°. n_D^{23} 1.3837.

Brown, D.H. *et al, J. Chem. Soc. (A)*, 1966, 171 (*synth, ir, pmr, nmr*)
Blazer, T.A. *et al, Z. Naturforsch., B*, 1969, **24**, 1081 (*ms*)
Cowley, A.H. *et al, J. Am. Chem. Soc.*, 1973, **95**, 6506 (*pe*)
Inorg. Synth., 1978, **18**, 186 (*synth*)

2,2'-Bis(5,5-dimethyl-1,3,2-dioxaphosphorinane) **B-00197**

5,5,5',5'-Tetramethyl-2,2'-bis-1,3,2-dioxaphosphorinane. 2,2'-Bis(5,5-dimethyl-1,3-dioxa-2-phosphacyclohexane)
[54975-97-2]

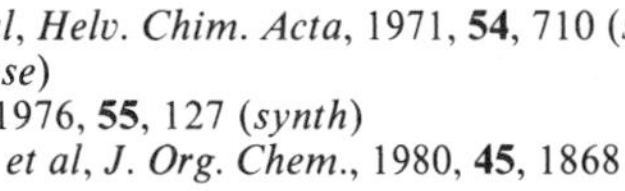

$C_{10}H_{20}O_4P_2$ M 266.213
Liq. $Bp_{0.01}$ 98-100°.

2,2'-Dioxide: [16368-06-2]. *2,2'-Bis(5,5-dimethyl-1,3,2-dioxaphosphorinane) 2,2'-dioxide. 2,2-Dimethyl-1,3-propanediol cyclic* P,P,P',P'-*hypophosphate. Hypophosphoric acid cyclic bis(2,2-dimethyltrimethylene) ester. 5,5,5',5'-Tetramethyl-2,2'-bis-1,3,2-dioxaphosphorinane 2,2'-dioxide.*
$C_{10}H_{20}O_6P_2$ M 298.212
Needles ($CHCl_3$/EtOAc). Mp 253-255°.

2-Oxide, 2'-Sulfide: [16368-07-3]. *Thiohypophosphoric acid cyclic* P,P,P',P'-*bis(2,2-dimethyltrimethylene) ester.*
$C_{10}H_{20}O_5P_2S$ M 314.273
Cryst ($CHCl_3$/EtOAc). Mp 243-245°.

2,2'-Disulfide: [16368-08-4]. *Thiohypophosphoric acid cyclic* P,P,P',P'-*bis(2,2-dimethyltrimethylene) ester.*
$C_{10}H_{20}O_4P_2S_2$ M 330.333
Cryst. ($CHCl_3$/EtOAc). Mp 234-235°. $Bp_{0.0001}$ 170-180°.

Stec, W.J. *et al, Can. J. Chem.*, 1967, **45**, 2513 (*synth, ir*)
Harris, R.K. *et al, J. Chem. Soc., Chem. Commun.*, 1970, 1391 (*nmr*)
Stec, W.J. *et al, J. Phys. Chem.*, 1971, **75**, 3975 (*pe*)
Nifant'ev, É.E. *et al, Zh. Obshch. Khim.*, 1975, **45**, 295 (*Engl. transl.* p. 282) (*disulfide, synth, P nmr*)
Stec, W.J. *et al, Org. Mass Spectrom.*, 1975, **10**, 485 (*ms*)
Karolak-Wojciechowska, J. *et al, Acta Crystallogr., Sect. B*, 1980, **36**, 1683 (*disulfide, cryst struct*)
Nifant'ev, É.E. *et al, Dokl. Akad. Nauk SSSR, Ser. Sci. Khim.*, 1982, **263**, 900 (*Engl. transl.* p. 123) (*synth, P nmr*)

Bis(2,5-dimethylphenyl)phosphinic acid, 9CI **B-00198**

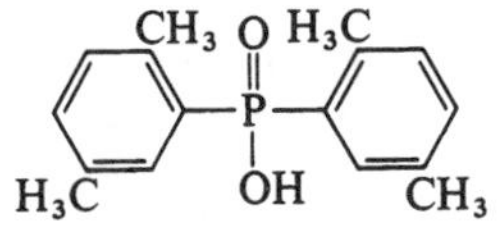

$C_{16}H_{19}O_2P$ M 274.299
Hygroscopic cryst. (EtOH aq.). Mp 184-186°.

Issleib, K. *et al, Chem. Ber.*, 1961, **94**, 392.
Horner, L. *et al, Chem. Ber.*, 1961, **94**, 2122.

1,4-Bis(dimethylphosphino)butane **B-00199**

1,4-Butanediylbis[dimethylphosphine], 10CI, 9CI. *Tetramethylenebis[dimethylphosphine]*, 8CI. *2,6-Dimethyl-2,7-diphosphaoctane*

$Me_2P(CH_2)_4PMe_2$

$C_8H_{20}P_2$ M 178.194

Dioxide: 1,4-Bis(dimethylphosphinyl)butane.
$C_8H_{20}O_2P_2$ M 210.192
Cryst. (C_6H_6). Mp 204-205°. $Bp_{0.35}$ 219-222°.

Disulfide: [70111-23-8]. *1,4-Bis(dimethylphosphinothioyl)butane.*
$C_8H_{20}P_2S_2$ M 242.314
Cryst. (MeOH). Mp 194-196°.

Kosolapoff, G.M. *et al, J. Chem. Soc.*, 1959, 3950 (*oxide, synth*)
Tsvetkov, E.N. *et al, Izv. Akad. Nauk SSSR, Ser. Khim.*, 1979, 426 (*Engl. transl.* p. 394)

1,2-Bis(dimethylphosphino)ethane **B-00200**

1,2-Ethanediylbis[dimethylphosphine], 9CI. *DMPE. 2,5-Dimethyl-2,5-diphosphahexane*
[23936-60-9]

$Me_2PCH_2CH_2PMe_2$

$C_6H_{16}P_2$ M 150.140
Ligand for metals of Groups VB, VIB, and VIII. Air-sensitive liq., not spontaneously flammable. Mp −1° to 0°. Bp 188.1°, Bp_{26} 81-82°.

Dioxide: [18724-97-5]. *1,2-Bis(dimethylphosphinyl)-ethane.*
$C_6H_{16}O_2P_2$ M 182.139

Solid. Mp 232-233°. Subl. *in vacuo*.
Disulfide: 1,2-Bis(dimethylphosphinothioyl)ethane.
$C_6H_{12}P_2S_2$ M 210.228
Cryst. (CH_2Cl_2 or DMF). Mp 257-258°, 273-275°.

Burg, A.B. *et al*, *J. Am. Chem. Soc.*, 1961, **83**, 2226 (*synth*)
Chatt, J. *et al*, *J. Chem. Soc.*, 1961, 896 (*synth*)
Kosolapoff, G.M. *et al*, *J. Chem. Soc.*, 1961, 2423 (*dioxide*)
Akhtar, M. *et al*, *Inorg. Chem.*, 1972, **11**, 2917 (*cmr, nmr, complexes*)
Connor, J.A. *et al*, *J. Chem. Soc., Dalton Trans.*, 1973, 347 (*dioxide, complexes*)
Burt, R.J. *et al*, *J. Organomet. Chem.*, 1979, **182**, 203 (*synth, props*)
Tsvetkov, E.N. *et al*, *Izv. Akad. Nauk SSSR, Ser. Khim.*, 1979, **28**, 426 (*Engl. transl.* p. 394) (*disulfide*)

1,6-Bis(dimethylphosphino)hexane B-00201

1,6-Hexanediylbis[dimethylphosphine], 10CI, 9CI. *Hexamethylenebis[dimethylphosphine]*, 8CI. *2,9-Dimethyl-2,9-diphosphadecane*

$Me_2P(CH_2)_6PMe_2$

$C_{10}H_{24}P_2$ M 206.247
Dioxide: 1,6-Bis(dimethylphosphinyl)hexane.
$C_{10}H_{24}O_2P_2$ M 238.246
Cryst. (C_6H_6). Mp 179-180°. $Bp_{0.35}$ 217°.

Kosolapoff, G.M. *et al*, *J. Chem. Soc.*, 1959, 3950.

Bis(dimethylphosphino)methane B-00202

Methylenebis[dimethylphosphine], 10CI, 9CI. *2,4-Dimethyl-2,4-diphosphosphapentane*
[64065-08-3]

$Me_2PCH_2PMe_2$

$C_5H_{14}P_2$ M 136.113
Ligand for Co, Au, Fe and Ag. Liq. Mp −59° to −56°. Bp_{12} 42°. Forms Li deriv.
Dioxide: [17535-01-2]. *Bis(dimethylphosphinyl)methane.*
$C_5H_{14}O_2P_2$ M 168.112
Hygroscopic needles (pet. ether). Mp 132-134°.

Richard, J.J. *et al*, *J. Am. Chem. Soc.*, 1961, **83**, 1722 (*oxide, synth*)
Karsch, H.H. *et al*, *Z. Naturforsch., B*, 1977, **32**, 762 (*synth, pmr, cmr, nmr, ms*)
Karsch, H.H., *ACS Symp. Ser.*, 1981, **171**, 141 (*rev*)
Rankin, D.W.H. *et al*, *J. Mol. Struct.*, 1981, **77**, 121 (*ed, struct*)
Karsch, H.H. *et al*, *Chem. Ber.*, 1982, **115**, 823 (*synth, pmr, cmr, nmr*)

[Bis(dimethylphosphino)methylene]trimethylphosphorane, 10CI B-00203

[82159-11-3]

$Me_3P{=}C(PMe_2)_2$

$C_8H_{21}P_3$ M 210.175
Neutral bidentate ligand. Cryst. Forms Co complexes.

Karsch, H.H., *Chem. Ber.*, 1982, **115**, 1956 (*synth, props, pmr, cmr, nmr*)

Bis[(dimethylphosphino)methyl]methylphosphine B-00204

2,4,6-Trimethyl-2,4,6-triphosphaheptane
[81626-09-7]

$Me_2PCH_2PMeCH_2PMe_2$

$C_7H_{19}P_3$ M 196.148
Liq. Mp −8° to −3°. $Bp_{0.1}$ 72-75°, $Bp_{0.04}$ 42°.
B,3MeI: Dimethylbis(trimethylphosphonimethyl)phosphonium triiodide.
$C_{10}H_{28}I_3P_3$ M 621.966
Cryst. Mp 250-255°.
P^1,P^3-Dioxide: Bis[(dimethylphosphinyl)methyl]methylphosphine.
$C_7H_{19}O_2P_3$ M 228.147
Solid. Mp >230° dec.

Hietkamp, S. *et al*, *Angew. Chem., Int. Ed. Engl.*, 1982, **21**, 376 (*synth, pmr, nmr*)
Karsch, H.H., *Z. Naturforsch., B*, 1982, **37**, 284 (*synth, derivs, pmr, cmr, nmr*)

1,5-Bis(dimethylphosphino)pentane B-00205

1,5-Pentanediylbis[dimethylphosphine], 10CI, 9CI. *Pentamethylenebis[dimethylphosphine]*, 8CI. *2,8-Dimethyl-2,8-diphosphanonane*

$Me_2P(CH_2)_5PMe_2$

$C_9H_{22}P_2$ M 192.220
Dioxide: 1,5-Bis(dimethylphosphinyl)pentane.
$C_9H_{22}O_2P_2$ M 224.219
Hygroscopic solid. Spar. sol. hydrocarbons. Mp 167-168°. Forms adduct (1:2) with EtOH.

Kosolapoff, G.M. *et al*, *J. Chem. Soc.*, 1959, 3950.

1,3-Bis(dimethylphosphino)propane B-00206

1,3-Propanediylbis[dimethylphosphine], 10CI, 9CI. *Trimethylenebis[dimethylphosphine]*, 8CI. *2,6-Dimethyl-2,6-diphosphaheptane*
[39564-18-6]

$Me_2PCH_2CH_2CH_2PMe_2$

$C_7H_{18}P_2$ M 164.167
Ligand for Co, Mo, Ni and W. Foul-smelling liq. Air-sensitive. Bp_4 62-63°.
Dioxide: 1,3-Bis(dimethylphosphinyl)propane.
$C_7H_{18}O_2P_2$ M 196.166
Mp 211-212°. Subl. *in vacuo*.
Disulfide: [50518-33-7]. *1,3-Bis(dimethylphosphinothioyl)propane.*
$C_7H_{18}P_2S_2$ M 228.287
Cryst. (CH_2Cl_2). Mp 226-227.5°.

Kosolapoff, G.M. *et al*, *J. Chem. Soc.*, 1961, 2423 (*oxide, synth*)
Cloyd, J.C. *et al*, *Inorg. Chim. Acta*, 1972, **6**, 480 (*synth, complexes*)
Inorg. Synth., 1973, **14**, 14 (*synth*)
Levason, W. *et al*, *J. Organomet. Chem.*, 1976, **105**, 195 (*ms*)

Bis(2,2-dimethylpropoxy)triphenylphosphorane B-00207

$Ph_3P[OCH_2C(CH_3)_3]_2$

$C_{28}H_{37}O_2P$ M 436.573
Reagent for cyclodehydration of diols, aminoalcohols and mercaptoalcohols. Cryst.

Kelly, J.W. *et al*, *J. Org. Chem.*, 1986, **51**, 5490 (*synth, pmr, P nmr, props, use*)

Bis(2,2-dimethylpropyl) phosphonate, 9CI **B-00208**

Dineopentyl phosphonate. Bis(2,2-dimethylpropyl) phosphite. Dineopentyl phosphite

[22289-00-5]

$[(H_3C)_3CCH_2O]_2P(O)H \rightleftharpoons [(H_3C)_3CCH_2O]_2POH$

$C_{10}H_{23}O_3P$ M 222.264

Tautomeric; almost completely in phosphonate form. Liq. Bp_{10} 109°. n_D^{13} 1.4200.

Gerrard, W. *et al*, *J. Chem. Soc.*, 1950, 2088 (*synth*)
Bellamy, L.J. *et al*, *J. Chem. Soc.*, 1952, 475 (*ir*)
Hudson, H.R., *J. Chem. Soc. (B)*, 1968, 664 (*P nmr*)

Bis(2,2-dimethylpropyl) phosphorochloridite, 9CI **B-00209**

Dineopentyl chlorophosphite

[53236-29-6]

$[(H_3C)_3CCH_2O]_2PCl$

$C_{10}H_{22}ClO_2P$ M 240.709

Pungent liq. Bp_{18} 105-107°. n_D^{20} 1.4370.

Bluj, S. *et al*, *Rocz. Chem.*, 1974, **48**, 329; *CA*, **81**, 63085 (*synth, ir, pmr*)

Bis(2,4-dinitrophenyl) phosphate, 9CI **B-00210**

Bis(2,4-dinitrophenyl) hydrogen phosphate. Bis(2,4-dinitrophenyl) phosphoric acid

[18962-97-5]

$C_{12}H_7N_4O_{12}P$ M 430.181

Pyridinium salt: Cryst. ($MeOH/Et_2O$ or Me_2CO/Et_2O). Mp 159-160°.

Azerad, R. *et al*, *Bull. Soc. Chim. Fr.*, 1963, 2078 (*synth*)
Bunton, C.A. *et al*, *J. Org. Chem.*, 1969, **34**, 767 (*synth, props*)
Kirby, A.J. *et al*, *J. Chem. Soc. (B)*, 1970, 510 (*props*)

Bis[(diphenoxyphosphino)thioyl] disulfide **B-00211**

Tetraphenyl thioperoxydiphosphate, 9CI

[36383-22-9]

$(PhO)_2P(S)-S-S-P(S)(OPh)_2$

$C_{24}H_{20}O_4P_2S_4$ M 562.607

Needles (heptane). Mp 74.5-75°.

Miller, B., *Tetrahedron*, 1964, **20**, 2069 (*synth*)
Trdlička, V. *et al*, *Collect. Czech. Chem. Commun.*, 1972, **37**, 896 (*synth*)
Khaskin, B.A. *et al*, *Zh. Obshch. Khim.*, 1973, **43**, 1916, 2083; 1974, **44**, 95 (*Engl. transl.* pp. 1901, 2065, 93) (*props*)

Bis[2-(diphenylarsino)ethyl]phenylphosphine, 9CI **B-00212**

Diarphos. 1,1,4,7,7-Pentaphenyl-1,7-diarsa-4-phosphaheptane

[23582-05-0]

$Ph_2AsCH_2CH_2PPhCH_2CH_2AsPh_2$

$C_{34}H_{33}As_2P$ M 622.452

Ligand for Mo, Pt, Os and Rh. Cryst. ($C_6H_6/MeOH$). Mp 160-162°.

▷Toxic

As,As,P-*Trioxide:*
$C_{34}H_{33}As_2O_3P$ M 670.450
Solid. Insol. C_6H_6, Et_2O. Mp 205-207° dec.

P-*Sulfide:*
$C_{34}H_{33}As_2PS$ M 654.512
Cryst. (CH_2Cl_2/hexane). Insol. hexane, Et_2O, MeOH. Mp 137-139° dec.

King, R.B. *et al*, *J. Am. Chem. Soc.*, 1971, **93**, 4158 (*synth, ir, pmr, nmr*)
King, R.B. *et al*, *Inorg. Chim. Acta*, 1972, **6**, 391 (*complexes*)
King, R.B. *et al*, *Phosphorus*, 1974, **3**, 209 (*derivs, pmr, nmr*)

[Bis(diphenylarsino)methylene]triphenylphosphorane, 11CI **B-00213**

[95531-55-8]

$Ph_3P{=}C(AsPh_2)_2$

$C_{43}H_{35}As_2P$ M 732.566

Yellow cryst. (toluene/pentane). Mp 103° dec., Mp 207°.

Schmidbaur, H. *et al*, *Z. Naturforsch., B*, 1984, **39**, 1456 (*synth, pmr, cmr, P nmr, cryst struct*)
Weber, L. *et al*, *Organometallics*, 1985, **4**, 841 (*synth, ir, P nmr, complexes*)

Bis[2-(diphenylarsino)phenyl]phenylphosphine, 9CI **B-00214**

$C_{42}H_{33}As_2P$ M 718.540

Ligand for Co and Ni. Cryst. (methoxybenzene). Mp 238-240°.

Howell, T.E.W. *et al*, *J. Chem. Soc.*, 1961, 3167 (*synth*)

2,2′-Bis(diphenylphosphinamido)-6,6′-dimethylbiphenyl **B-00215**

N,N′-*(6,6′-Dimethyl[1,1′-biphenyl]-2,2′-diyl)bis*[P,P-*diphenylphosphinous amide*]

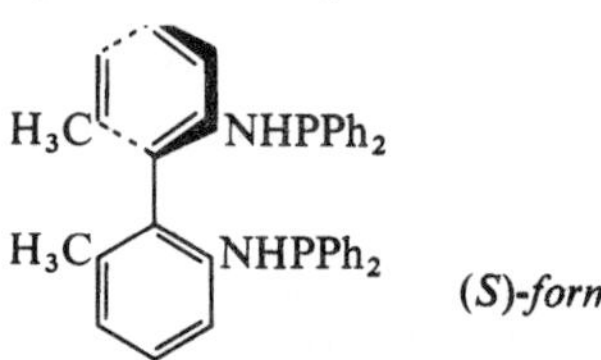

$C_{38}H_{34}N_2P_2$ M 580.648

(S)-form

Rh complexes useful for asymmetric hydrogenation. V. air-stable cryst. (EtOH aq.). Mp 98-100°. $[\alpha]_D^{25}$ −140° (c, 1.1 in C_6H_6).

Uehara, A. *et al*, *Chem. Lett.*, 1983, 441 (*synth, use*)

Bis(diphenylphosphino)acetylene **B-00216**

1,2-Ethynediylbis[diphenylphosphine], 9CI. 1,2-Bis(diphenylphosphino)ethyne

[5112-95-8]

$Ph_2PC{\equiv}CPPh_2$

$C_{26}H_{20}P_2$ M 394.392
Air-stable cryst. (EtOH). Mp 86°.
Dioxide: [22428-64-4]. *1,2-Bis(diphenylphosphinyl)-ethyne.*
$C_{26}H_{20}O_2P_2$ M 426.390
Cryst. (Me_2CO/pet. ether). Mp 164°.
Disulfide: [22428-65-5]. *1,2-Bis(diphenylphosphinothioyl)ethyne.*
$C_{26}H_{20}P_2S_2$ M 458.512
Needles ($CHCl_3$/EtOH). Mp 186°.

Charrier, C. *et al*, *Bull. Soc. Chim. Fr.*, 1966, 1002 (*synth, ir*)
Hartmann, H. *et al*, *Justus Liebigs Ann. Chem.*, 1968, **714**, 1 (*synth*)
Bart, J.C.J., *Acta Crystallogr., Sect. B*, 1969, **25**, 489 (*cryst struct*)
Schwartz, W.E. *et al*, *Spectrochim. Acta, Part A*, 1974, **30**, 1561 (*pe*)
Paasonen, R. *et al*, *Org. Magn. Reson.*, 1978, **11**, 42 (*cmr, nmr*)
Orama, O. *et al*, *Cryst. Struct. Commun.*, 1979, **8**, 409, 905 (*dioxide, disulfide, cryst struct*)

2,2′-Bis[(diphenylphosphino)amino]-1,1′-binaphthyl B-00217

N,N′-[*1,1′-Binaphthalene*]*-2,2′-diylbis*[P,P′*-diphenylphosphinous amide*], *10CI*

$NHPPh_2$ $NHPPh_2$ (*R*)-*form*

$C_{44}H_{34}N_2P_2$ M 652.714
Rh complexes used as catalysts for asymmetric hydrogenations.
(*R*)-*form* [74974-14-4]
Solid. Mp 150-151°. $[\alpha]_D^{23}$ +27.2° (c, 1.14 in C_6H_6). This can also be descr. as the (*S*)-form (by central chirality).
(*S*)-*form* [74974-15-5]
Solid. Mp 149-150°. $[\alpha]_D^{23}$ −25.8° (c, 0.776 in C_6H_6).
(±)-*form*
Solid. Mp 163-165°.

Miyano, S. *et al*, *Chem. Lett.*, 1980, 729 (*synth, ms, use*)

2,3-Bis[(diphenylphosphino)amino]butane B-00218

N,N′-(*1,2-Dimethyl-1,2-ethanediyl*)*bis*[P,P*-diphenylphosphinous amide*], *10CI*

CH_3
$H{-}C{-}NHPPh_2$ (2*S*,3*S*)-*form*
$Ph_2PNH{-}C{-}H$ Absolute configuration
CH_3

$C_{28}H_{30}N_2P_2$ M 456.506
(*2S,3S*)-*form*
Rh complexes are useful catalysts for asymmetric hydrogenations. Mp 83-85°. $[\alpha]_D^{15}$ +2.98° (c, 0.5 in C_6H_6).

Onuma, K. *et al*, *Bull. Chem. Soc. Jpn.*, 1980, **53**, 2012 (*synth, use*)

1,2-Bis[(*N*-diphenylphosphino)amino]-cyclohexane B-00219

N,N′*-1,2-Cyclohexanediylbis*[P,P*-diphenylphosphinous amide*], *10CI*

$NHPPh_2$ $NHPPh_2$ (1*R*,2*R*)-*form*

$C_{30}H_{32}N_2P_2$ M 482.544
Ligand for Rh in asymmetric hydrogenation catalysts.
(*1R,2R*)-*form* [70708-37-1]
Solid or cryst. (Me_2CO). Mp 130-132° (119-120°). $[\alpha]_D^{15}$ −4.51° (c, 1.0 in C_6H_6).
N,N′*-Di-Me:* [68683-75-0].
$C_{32}H_{36}N_2P_2$ M 510.597
Mp 137-139°. $[\alpha]_D^{22}$ −7.3° (c, 1.1 in $CHCl_3$).
(*1S,2S*)-*form* [72090-83-6]
Cryst. Mp 130-132°. $[\alpha]_D^{15}$ +4.48° (c, 1.0 in C_6H_6).

Fiorini, M. *et al*, *J. Mol. Catal.*, 1979, **5**, 303 (*use*)
Kashiwabara, K. *et al*, *Bull. Chem. Soc. Jpn.*, 1980, **53**, 2275 (*synth, deriv, pmr*)
Onuma, K. *et al*, *Bull. Chem. Soc. Jpn.*, 1980, **53**, 2012, 2016 (*synth, complexes, use*)

5,6-Bis(diphenylphosphino)bicyclo[2.2.1]-hept-2-ene B-00220

Bicyclo[*2.2.1*]*hept-5-ene-2,3-diylbis*[*diphenylphosphine*], *10CI*. *5,6-Bis(diphenylphosphino)norbornene. Norphos*

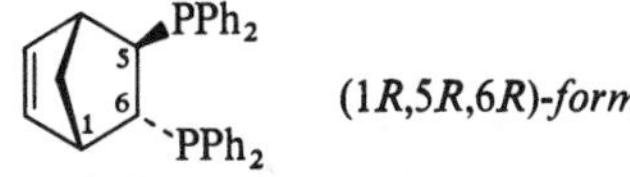

$C_{31}H_{28}P_2$ M 462.510
(*1R,5R,6R*)-*form* [71042-54-1]
(+)-endo,exo-*form*
Cryst. (Me_2CO). Mp 129-130°. $[\alpha]_D^{20}$ +45° (c, 1 in $CHCl_3$).
P,P′*-Dioxide:* [71075-23-5]. *5,6-Bis(diphenylphosphinyl)bicyclo*[*2.2.1*]*hept-2-ene.*
$C_{31}H_{28}O_2P_2$ M 494.509
Solid. $[\alpha]_D^{20}$ +58° (c, 1 in $CHCl_3$).
(*1S,5S,6S*)-*form* [71042-55-2]
(−)-endo,exo-*form*
Cryst. (Me_2CO). Mp 129-130°. $[\alpha]_D^{20}$ −43.5° (c, 1 in $CHCl_3$).
P,P′*-Dioxide:* [71075-24-6]. $[\alpha]_D^{20}$ −62° (c, 1 in $CHCl_3$).
5,6-Dihydro: 2,3-Bis(diphenylphosphino)bicyclo[*2.2.1*]*-heptane.*
$C_{31}H_{30}P_2$ M 464.526
Solid. Mp 95-97°. $[\alpha]_D^{25}$ −42.6° (c, 0.713 in $CHCl_3$).
(*1RS,5RS,6RS*)-*form* [76740-45-9]
(±)-endo,exo-*form*
No phys. props. reported.
P,P′*-Dioxide:* [71075-22-4]. No phys. props. reported.

Brunner, H. *et al*, *Angew. Chem., Int. Ed. Engl.*, 1979, **18**, 620 (*synth, dioxide, resoln*)
Brunner, H. *et al*, *Chem. Ber.*, 1981, **114**, 1137 (*abs config, complex, use*)
Kyba, E.P. *et al*, *Inorg. Chem.*, 1981, **20**, 3616 (*synth, complexes*)
Brunner, H. *et al*, *Inorg. Chim. Acta*, 1982, **61**, 129 (*pmr, dioxide*)

2,2′-Bis(diphenylphosphino)-1,1′-binaphthyl **B-00221**

[1,1′-Binaphthalene]-2,2′-diylbis[diphenylphosphine], 9CI. Binap

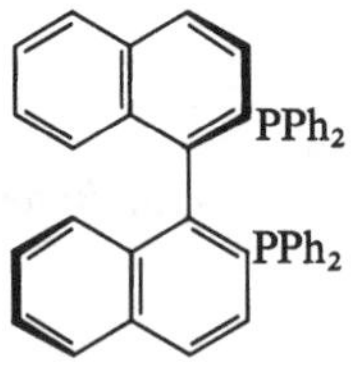

(*R*)-form

$C_{44}H_{32}P_2$ M 622.684

Ligand for Rh and Pd, important in asymmetric hydrogenations.

(R)-form [76189-55-4]

Mp 240-241°. $[\alpha]_D^{25}$ +229° (c, 0.32 in C_6H_6).

Dioxide:

$C_{44}H_{32}O_2P_2$ M 654.683

Solid. Mp 262-263°. $[\alpha]_D^{24}$ +399° (c, 0.5 in C_6H_6).

(S)-form [76189-56-5]

Mp 241-242°. $[\alpha]_D^{25}$ −229° (c, 0.31 in C_6H_6).

Dioxide: solid. Mp 241-242°. $[\alpha]_D^{24}$ −228° (c, 0.68 in C_6H_6).

(±)-form [76144-87-1]

No phys. props. reported.

Dioxide: Cryst. (toluene/hexane). Mp 304-306°.

Miyashita, A. *et al*, *J. Am. Chem. Soc.*, 1980, **102**, 7932 (*synth*)
Tani, K. *et al*, *ACS Symp. Ser.*, 1982, **185**, 187 (*use*)
Miyashita, A. *et al*, *ACS Symp. Ser.*, 1982, **185**, 274 (*synth, use*)
Takaya, H. *et al*, *J. Org. Chem.*, 1986, **51**, 629 (*dioxide, synth, ir, ms, pmr, uv, cd, cryst struct*)

3,3′-Bis(diphenylphosphino)biphenyl **B-00222**

[1,1′-Biphenyl]-3,3′-diylbis(diphenylphosphine), 9CI

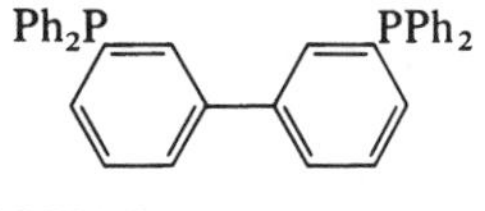

$C_{36}H_{28}P_2$ M 522.565

Cryst. Mp 125-127°.

Chiswell, B. *et al*, *J. Chem. Soc. (A)*, 1966, 901.

1,4-Bis(diphenylphosphino)-1,3-butadiyne **B-00223**

1,3-Butadiyne-1,4-diylbis[diphenylphosphine]

$Ph_2PC{\equiv}CC{\equiv}CPPh_2$

$C_{28}H_{20}P_2$ M 418.414

Cryst. (Et_2O). Mp 105°.

Charrier, C. *et al*, *Bull. Soc. Chim. Fr.*, 1966, 1002 (*synth*)

1,3-Bis(diphenylphosphino)butane **B-00224**

(1-Methyl-1,3-propanediyl)bis[diphenylphosphine], 10CI. Chairphos

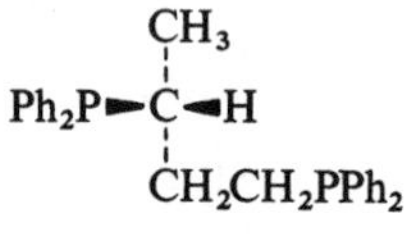

(*R*)-form

$C_{28}H_{28}P_2$ M 426.477

Ineffective as chiral ligand for Rh catalysts in asymm. hydrogenations.

(R)-form [74044-77-2]

Oil. $[\alpha]_D$ +18.7° (c, 1 in C_6H_6).

(S)-form [77876-46-1]

Viscous oil.

B,2MePF$_6$: (1-Methyl-1,3-propanediyl)-bis[methyldiphenylphosphonium] hexafluorophosphate. Solid. Mp 105-107°. $[\alpha]_D^{25}$ −8.7° (c, 1.2 in Me_2CO).

Kagan, H.B. *et al*, *Bull. Soc. Chim. Belg.*, 1979, **88**, 923 (*pmr, complex*)
MacNeil, P.A. *et al*, *J. Am. Chem. Soc.*, 1981, **103**, 2273 (*derivs*)

1,4-Bis(diphenylphosphino)butane **B-00225**

1,4-Butanediylbis[diphenylphosphine], 9CI

[7688-25-7]

$Ph_2P(CH_2)_4PPh_2$

$C_{28}H_{28}P_2$ M 426.477

Co-catalyst for the manuf. of 1,4-hexadiene from ethylene and butadiene. Ligand for Au, V, and Group VIII metals. Air-stable cryst. (EtOH). Mp 135-136°.

Dioxide: [4151-27-3]. *1,4-Bis(diphenylphosphinyl)-butane.*

$C_{28}H_{28}O_2P_2$ M 458.476

Cryst. (C_6H_6 or EtOH aq.). Mp 264-266°. $Bp_{1.6}$ 290-291°.

Disulfide: [61894-26-6]. *1,4-Bis(diphenylphosphinothioyl)butane.*

$C_{28}H_{28}P_2S_2$ M 490.597

Solid. Mp 223-224°.

Diselenide: 1,4-Bis(diphenylphosphinoselenoyl)butane.

$C_{28}H_{28}P_2Se_2$ M 584.397

Solid. Mp 190°.

Gough, S.T.D. *et al*, *J. Chem. Soc.*, 1961, 4263 (*synth*)
Horner, L. *et al*, *Chem. Ber.*, 1962, **95**, 581 (*dioxide, props*)
Sacconi, L. *et al*, *Inorg. Chem.*, 1968, **7**, 291 (*synth, complexes*)
Goda, K. *et al*, *Bull. Chem. Soc. Jpn.*, 1975, **48**, 2484 (*dioxide*)
Sandhu, S.S. *et al*, *Trans. Met. Chem. (Weinheim, Ger.)*, 1976, **1**, 155 (*disulfide, diselenide*)
Sanger, A.R., *J. Chem. Soc., Dalton Trans.*, 1977, 1971 (*ir, complexes*)
Dean, P.A.W., *Can. J. Chem.*, 1979, **57**, 754 (*diselenide, nmr*)
Lindsay, C.H. *et al*, *Inorg. Chem.*, 1980, **19**, 3503 (*pmr, nmr, complexes*)
Hayashi, T. *et al*, *Bull. Chem. Soc. Jpn.*, 1981, **54**, 3438 (*use*)
Gurusamy, N. *et al*, *J. Am. Chem. Soc.*, 1982, **104**, 3107 (*synth, deriv*)

2,3-Bis(diphenylphosphino)butane **B-00226**

(1,2-Dimethyl-1,2-ethanediyl)bis[diphenylphosphine], 10CI. Chiraphos

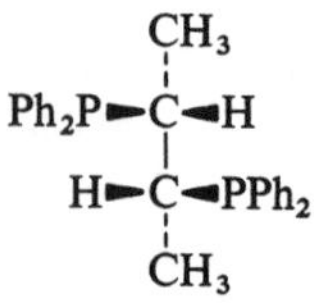

(2*R*,3*R*)-form

$C_{28}H_{28}P_2$ M 426.477

Bidentate ligand. Pd and Rh complexes used in asymm. cyclizations.

(2R,3R)-form

Cryst. (EtOH). Mp 106-107°. $[\alpha]_D^{27}$ +197° (c, 1.5 in $CHCl_3$).

(2S,3S)-form

Plates (EtOH). Mp 108-109° (sealed tube) (101-103°). $[\alpha]^{27}_{D}$ −211° (c, 1.5 in $CHCl_3$), −266°.

Fryzuk, M.D. *et al, J. Am. Chem. Soc.*, 1977, **99**, 6262 (*synth*)
Koettner, J. *et al, Chem. Ber.*, 1980, **113**, 2323 (*synth, pmr*)
Payne, N.C. *et al, J. Organomet. Chem.*, 1981, **221**, 203 (*pmr, nmr, complexes*)
Trost, B.M. *et al, Tetrahedron Lett.*, 1981, **22**, 4929 (*use*)
Maciel, G.E. *et al, Adv. Chem. Ser.*, 1982, **196**, 389 (*use*)
Yamamoto, K. *et al, Tetrahedron Lett.*, 1982, **23**, 3089 (*use*)

1,4-Bis(diphenylphosphino)-2-butyne B-00227

2-Butynediyl-1,4-bis[diphenylphosphine]
[27318-83-8]

$Ph_2PCH_2C{\equiv}CCH_2PPh_2$

$C_{28}H_{24}P_2$ M 422.445
Cryst. (C_6H_6/EtOH). Mp 79-80°.
Dioxide: 1,4-Bis(diphenylphosphinyl)-2-butyne.
$C_{28}H_{24}O_2P_2$ M 454.444
Cryst. (CH_2Cl_2/hexane). Mp 214-216°.
Disulfide: 1,4-Bis(diphenylphosphinothioyl)-2-butyne.
$C_{28}H_{24}P_2S_2$ M 486.565
Cryst. (CH_2Cl_2/hexane). Mp 139-141°.

King, R.B. *et al, Inorg. Chim. Acta*, 1970, **4**, 123 (*synth, ms, ir, complexes, derivs*)
Arthurs, M. *et al, Tetrahedron Lett.*, 1978, 1153 (*pmr, cmr*)

3,4-Bis(diphenylphosphino)-3-cyclobutene-1,2-dione, 9CI B-00228

[41006-27-3]

$C_{28}H_{20}O_2P_2$ M 450.412
Fine powder.

Becher, H.J. *et al, Chem. Ber.*, 1973, **106**, 177 (*synth, ir*)

1,1-Bis(diphenylphosphino)cyclopropane B-00229

Cyclopropylidenebis[diphenylphosphine], 11CI
[96921-88-9]

$C_{27}H_{24}P_2$ M 410.434
Ligand for Au. Air-stable needles (EtOH/hexane). Mp 109°.
B,PhCl: [1-(Diphenylphosphino)cyclopropyl]-triphenylphosphonium chloride.
$C_{33}H_{29}ClP_2$ M 522.993
Hygroscopic solid. Mp 125° dec.
Dioxide: [96921-89-0]. *1,1-Bis(diphenylphosphinyl)-cyclopropane.*
$C_{27}H_{24}O_2P_2$ M 442.433
Hygroscopic cryst. Mp 206°. Sublimes.
Disulfide: [96921-90-3].
$C_{27}H_{24}P_2S_2$ M 474.554
Cryst. (EtOH/THF). Mp 201°.

Schmidbaur, H. *et al, Helv. Chim. Acta*, 1984, **67**, 2175 (*synth, pmr, P nmr, ms*)
Schmidbaur, H. *et al, Chem. Ber.*, 1985, **118**, 3105 (*derivs, props, pmr, cmr, P nmr*)
Schmidbaur, H. *et al, Organometallics*, 1986, **5**, 566 (*cryst struct, complexes*)

1,2-Bis(diphenylphosphino)cyclopropane B-00230

1,2-Cyclopropanediylbis[diphenylphosphine]

$C_{27}H_{24}P_2$ M 410.434

(1RS,2RS)-form

(±)-trans-*form*
Ligand for Ag, Rh, and Ir. Solid. Mp 156-157°.
Dioxide: 1,2-Bis(diphenylphosphinyl)cyclopropane.
$C_{27}H_{24}O_2P_2$ M 442.433
Solid. Mp 268-270°.

Braun, J.M. *et al, J. Organomet. Chem.*, 1986, **314**, 241.

2,2′-Bis(diphenylphosphino)diethylamine B-00231

2-(Diphenylphosphino)-N-[2-(diphenylphosphino)-ethyl]ethanamine, 10CI. *Bis[(2-Diphenylphosphino)-ethyl]amine. 1,1,7,7-Tetraphenyl-4-aza-1,7-diphosphaheptane*
[66534-96-1]

$(Ph_2PCH_2CH_2)_2NH$

$C_{28}H_{29}NP_2$ M 441.491
Used for prep. of water-sol. diphosphines and metal complexes.
B,HCl: [66534-97-2]. Needles (MeOH). Mp 174.5-175.5°.
Dioxide:
$C_{28}H_{29}NO_2P_2$ M 473.490
Dense yellow oil.
N-*Ac:*
$C_{30}H_{31}NOP_2$ M 483.529
Glass.
N-*Et: see N,N-Bis[(2-diphenylphosphino)ethyl]-ethylamine, B-00238*

Sacconi, L. *et al, J. Chem. Soc. (A)*, 1968, 2997 (*oxide, complexes*)
Wilson, M.E. *et al, J. Am. Chem. Soc.*, 1978, **100**, 306, 2269 (*synth, use, pmr*)
Nuzzo, R.G. *et al, J. Am. Chem. Soc.*, 1979, **101**, 3683 (*use*)
Nuzzo, R.G. *et al, J. Org. Chem.*, 1981, **46**, 2861 (*complexes*)

Bis(diphenylphosphino)diphenylgermane B-00232

$Ph_2Ge(PPh_2)_2$

$C_{36}H_{30}GeP_2$ M 597.171
Colourless needles (methylcyclohexane). Oxidises and hydrolyses readily.

Brooks, E.H. *et al, J. Chem. Soc.*, 1965, 4283.

N,N-Bis(diphenylphosphino)-*P,P*-diphenylphosphinous amide, 11CI B-00233

Tris(diphenylphosphino)amine
[97322-65-1]

$(Ph_2P)_3N$

$C_{36}H_{30}NP_3$ M 569.561

Solid. Mp 180-225° dec.

Ellermann, J. *et al*, *J. Organomet. Chem.*, 1985, **281**, C29 (*synth, pmr, ms, ir, raman*)

1,2-Bis(diphenylphosphino)ethane B-00234

1,2-Ethanediylbis[diphenylphosphine], 9CI. *Diphos*

[1663-45-2]

$Ph_2PCH_2CH_2PPh_2$

$C_{26}H_{24}P_2$ M 398.423

Ligand for metals of Groups IB, IIB, VIB, VIIB and VIII. Air-stable cryst. (C_6H_6). Mp 143-144°.

▷Exhibits cytotoxic activity

B,2MeI: 1,2-Ethanediylbis[methyldiphenylphosphonium] diiodide.
$C_{28}H_{30}I_2P_2$ M 682.302
Cryst. (MeOH). Mp 305-307°.

Monooxide: Diphenyl[(2-diphenylphosphinyl)ethyl]-phosphine.
$C_{26}H_{24}OP_2$ M 414.423
Cryst. Mp 190.5-193°.

Dioxide: [4141-50-8]. *1,2-Bis(diphenylphosphinyl)-ethane. 1,2-Ethanediylbis[diphenylphosphine oxide].*
$C_{26}H_{24}O_2P_2$ M 430.422
Needles (toluene or DMF), plates (C_6H_6). Mp 252-254°, 276-278°.

Disulfide: [7615-15-1]. *1,2-Bis(diphenylphosphinothioyl)ethane.*
$C_{26}H_{24}P_2S_2$ M 462.543
Cryst. (Me_2CO). Insol. Et_2O. Mp 196-198°, 219-221°.

Diselenide: 1,2-Bis(diphenylphosphinoselenoyl)ethane.
$C_{26}H_{24}P_2Se_2$ M 556.343
Solid. Mp 192°.

Issleib, K. *et al*, *Chem. Ber.*, 1959, **92**, 3175 (*synth, derivs*)
Aguiar, A.M. *et al*, *J. Org. Chem.*, 1964, **29**, 1660; *J. Am. Chem. Soc.*, 1964, **86**, 5354 (*synth, pmr, monooxide*)
Kosolapoff, G.M. *et al*, *J. Chem. Soc. (C)*, 1967, 1789 (*dioxide*)
Colton, R. *et al*, *Aust. J. Chem.*, 1968, **21**, 2215 (*ms*)
Maier, L., *Helv. Chim. Acta*, 1968, **51**, 405 (*dioxide, ir, nmr*)
Yakutina, C.A. *et al*, *Zh. Obshch. Khim.*, 1972, **42**, 1733 (*Engl. transl.* p. 1722) (*uv, raman*)
Grim, S.O. *et al*, *Inorg. Chem.*, 1974, **13**, 1095 (*pmr, nmr, complexes*)
Swartz, W.E. *et al*, *Spectrochim. Acta, Part A*, 1974, **30**, 1561 (*pe*)
Dubois, D.L. *et al*, *J. Chem. Soc., Dalton Trans.*, 1975, 1011 (*synth, pmr*)
King, R.B. *et al*, *J. Chem. Soc., Perkin Trans. 2*, 1975, 938 (*cmr*)
Sandhu, S.S. *et al*, *Trans. Metal. Chem. (Weinheim, Ger.)*, 1976, **1**, 155 (*disulfide, diselenide, complexes*)
Dean, P.A.W., *Can. J. Chem.*, 1979, **57**, 754 (*diselenide, nmr*)
Bemi, L. *et al*, *J. Am. Chem. Soc.*, 1982, **104**, 438 (*nmr*)
Gurusamy, N. *et al*, *J. Am. Chem. Soc.*, 1982, **104**, 3107 (*synth, derivs*)

1,2-Bis(diphenylphosphino)-1,2-ethanedione B-00235

(1,2-Dioxo-1,2-ethanediyl)bis[diphenylphosphine], 10CI. *Oxaloylbis[diphenylphosphine]*

[41006-23-9]

$Ph_2PCOCOPPh_2$

$C_{26}H_{20}O_2P_2$ M 426.390

Solid. Unstable to oxidn. Mp 77°.

Becher, H.J. *et al*, *Chem. Ber.*, 1973, **106**, 177 (*synth, ir*)
Becher, H.J. *et al*, *Monatsh. Chem.*, 1980, **111**, 749 (*ir, struct*)
Lindner, E. *et al*, *Z. Naturforsch., B*, 1983, **38**, 790 (*props*)

1,1-Bis(diphenylphosphino)ethylene B-00236

Ethenylidenebis(diphenylphosphine), 9CI. *Vinylidenebis(diphenylphosphine)*

[84494-89-3]

$H_2C{=}C(PPh_2)_2$

$C_{26}H_{22}P_2$ M 396.407

Forms 1:1 and 1:2 Au complexes. Cryst. (EtOH). Mp 112-115°.

B,2MeI: 1,1-Bis(methyldiphenylphosphonio)ethene diiodide.
$C_{28}H_{28}I_2P_2$ M 680.286
Cryst. ($CHCl_3$). Mp 246° dec.

Dioxide: [29207-36-1]. *1,1-Bis(diphenylphosphinyl)-ethylene.*
$C_{26}H_{22}O_2P_2$ M 428.406
Cryst. (C_6H_6). Mp 205° (197-199°).

Disulfide: [97106-56-4]. *1,1-Bis(diphenylphosphinothioyl)ethylene.*
$C_{26}H_{22}P_2S_2$ M 460.527
Solid. Mp 184°.

Diselenide: [84494-91-7]. *1,1-Bis(diphenylphosphinoselenoyl)ethylene.*
$C_{26}H_{22}P_2Se_2$ M 554.327
Solid. Mp 188°.

Colquhoun, I.J. *et al*, *J. Chem. Soc., Dalton Trans.*, 1982, 1915 (*synth, selenide, cmr, P nmr*)
Schmidbaur, H. *et al*, *Chem. Ber.*, 1984, **117**, 2322; 1985, **118**, 3105 (*synth, cryst struct, derivs, props*)
Schmidbaur, H. *et al*, *Z. Chem.*, 1984, **24**, 378 (*synth, derivs*)
Schmidbaur, H. *et al*, *Organometallics*, 1985, **4**, 1208 (*complexes*)

1,2-Bis(diphenylphosphino)ethylene B-00237

1,2-Ethenediylbis[diphenylphosphine], 9CI. *1,2-Vinylenebis[diphenylphosphine]*

[17427-91-7]

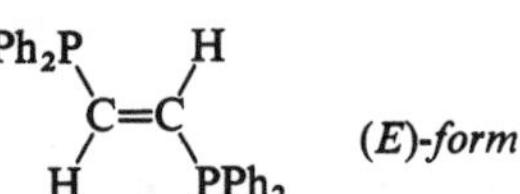

$C_{26}H_{22}P_2$ M 396.407

Ligand forming complexes contg. metals of groups IB, VIB, VIIB, and VIII.

***(E)*-form** [983-81-3]

Air-stable cryst. Mp 125-126°.

P,P'-*Dioxide:* [40612-18-8].
$C_{26}H_{22}O_2P_2$ M 428.406
Mp 310-311°. pK_a (conj. acid) 1.7, 1.7.

P,P'-*Disulfide:* [986-07-2].
$C_{26}H_{22}P_2S_2$ M 460.527
Cryst. (EtOH aq.). Mp 196-197°.

***(Z)*-form** [983-90-2]

Air-stable cryst. Mp 125-126°.

P,P'-*Dioxide:* [40468-55-1]. Mp 244-245°. pK_a (conj. acid in nitromethane) 7.8.

P,P'-*Disulfide:* [986-06-1]. Cryst. (EtOH aq.). Mp 196-197°.

Aguiar, A.M. *et al*, *J. Am. Chem. Soc.*, 1964, **86**, 2299 (*synth, ir, pmr, dioxide*)
Aguiar, A.M. *et al*, *J. Am. Chem. Soc.*, 1964, **86**, 5354 (*disulfides*)
Brophy, J.J. *et al*, *Aust. J. Chem.*, 1967, **20**, 503 (*methiodides*)
King, R.B. *et al*, *J. Am. Chem. Soc.*, 1971, **93**, 4158 (*synth, pmr, nmr*)

Cauquis, G. *et al*, *Org. Mass Spectrom.*, 1975, **10**, 1021 (*disulfides, ms*)
Lindsay, C.H. *et al*, *Inorg. Chem.*, 1980, **19**, 3503 (*complexes*)
Duncan, M. *et al*, *Org. Magn. Reson.*, 1981, **15**, 37 (*pmr, cmr, nmr*)

N,N-Bis[(2-diphenylphosphino)ethyl]-ethylamine B-00238

2-(Diphenylphosphino)-N-*[2-(diphenylphosphino)-ethyl]*-N-*ethylethanamine*

[7628-15-1]

$(Ph_2PCH_2CH_2)_2NEt$

$C_{30}H_{33}NP_2$ M 469.545

Ligand for Co, Ru, Pd, Pt and Rh. Solid. Mp 43-45°.

Dioxide:
$C_{30}H_{33}NO_2P_2$ M 501.544
Cryst. (EtOH aq.). Mp 190-191°.

Kabachnik, M.I. *et al*, *Izv. Akad. Nauk SSSR, Ser. Khim.*, 1962, 1584 (*Engl. transl.* p. 1499) (*dioxide*)
Dobson, G.R. *et al*, *J. Chem. Soc., Chem. Commun.*, 1966, 281 (*synth*)

Bis[2-(diphenylphosphino)ethyl]-phenylphosphine, 9CI B-00239

Triphos. 1,1,4,7,7-Pentaphenyl-1,4,7-triphosphaheptane

[23582-02-7]

$Ph_2PCH_2CH_2PPhCH_2CH_2PPh_2$

$C_{34}H_{33}P_3$ M 534.556

Ligand for Cu, Mo, Pt, and W. Rh complexes used in hydroformylation of butadiene. Air-sensitive cryst. (MeOH or EtOH). Mp 131-132° (121-126°).

B,3MeI: Bis[2-(methyldiphenylphosphonio)ethyl]-methylphenylphosphonium triiodide.
$C_{37}H_{42}I_3P_3$ M 960.374
Cryst. + $2H_2O$ ($MeOH/Et_2O$). Mp 162-163°.

Trioxide: [53889-34-2].
$C_{34}H_{33}O_3P_3$ M 582.554
Ligand for Cr, Mn, W, Co, Ni, and Pt. Solid. Insol. C_6H_6, Et_2O, pet. ether. Mp 294-298° dec.

Trisulfide:
$C_{34}H_{33}P_3S_3$ M 630.736
Ligand for Cd. Cryst. (CH_2Cl_2/hexane). Insol. Et_2O, MeOH, pet. ether. Mp 227-229°.

Triselenide:
$C_{34}H_{33}P_3Se_3$ M 771.436
Solid. Mp 208-210°.

Hewertson, W. *et al*, *J. Chem. Soc.*, 1962, 1490 (*synth*)
King, R.B. *et al*, *J. Am. Chem. Soc.*, 1971, **93**, 4158 (*synth, ir, ms, pmr*)
Cloyd, J.C. *et al*, *Inorg. Chim. Acta*, 1972, **6**, 607 (*synth, complexes*)
King, R.B. *et al*, *Phosphorus*, 1974, **3**, 209 (*derivs*)
DuBois, D.L. *et al*, *J. Chem. Soc., Dalton Trans.*, 1975, 1011 (*synth, nmr, pmr*)
King, R.B. *et al*, *Inorg. Chem.*, 1975, **14**, 1550 (*nmr*)
King, R.B. *et al*, *J. Chem. Soc., Perkin Trans. 2*, 1975, 938 (*cmr*)
Dean, P.A. *et al*, *Can. J. Chem.*, 1980, **58**, 180 (*nmr, derivs*)
Lindsay, C.H. *et al*, *Inorg. Chem.*, 1980, **19**, 3503 (*nmr, complexes*)

4,7-Bis[2-(diphenylphosphino)ethyl]-1,1,10,10-tetraphenyl-1,4,7,10-tetraphosphadecane, 10CI B-00240

Ethylenebis[bis[2-(diphenylphosphino)ethyl]-phosphine], 8CI. Bis[[bis(2-diphenylphosphino)ethyl]-phosphino]ethane. 1,2-Ethanediylbis[[2-(diphenylphosphino)ethyl]phosphine]

[23582-00-5]

$(Ph_2PCH_2CH_2)_2PCH_2CH_2P(CH_2CH_2PPh_2)_2$

$C_{58}H_{60}P_6$ M 942.955

First known hexadentate tertiary phosphine ligand. Cryst. (C_6H_6/EtOH). Mp 138-140°. Forms V complexes.

King, R.B. *et al*, *J. Am. Chem. Soc.*, 1971, **93**, 4158 (*synth, pmr, ir, nmr*)

1,1′-Bis(diphenylphosphino)ferrocene, 9CI B-00241

1,1′-Ferrocenediylbis(diphenylphosphine), 8CI

[12150-46-8]

Ph_2P–⟨C₅H₄⟩–Fe–⟨C₅H₄⟩–PPh_2

$C_{34}H_{28}FeP_2$ M 554.390

Rh complexes used in hydroformylation reactions. Yellow-orange needles (C_6H_6/pet. ether). Mp 186-188°.

Bis(methiodide): 1,1′-Bis(methyldiphenylphosphonio)-ferrocene diiodide. Ferrocene-1,1′-bis(methyldiphenylphosphonium) diiodide.
$C_{36}H_{32}FeI_2P_2$ M 836.252
Orange cryst. (EtOH). Mp 244-245°.

P,P′-*Dioxide: 1,1′-Bis(diphenylphosphinyl)ferrocene.*
$C_{34}H_{28}FeO_2P_2$ M 586.389
Ligand for Cr, Co, Cu, Mo, Ni, Ta, W and V. Dark-orange cryst. (EtOH). Mp 252-254.5°.

P,P′-*Disulfide: 1,1′-Bis(diphenylphosphinothioyl)-ferrocene.*
$C_{34}H_{28}FeP_2S_2$ M 618.510
Golden-yellow cryst. (butanol). Mp 244-246°.

Sollott, G.P. *et al*, *CA*, 1965, **63**, 18147 (*synth*)
Marr, G. *et al*, *J. Chem. Soc. (C)*, 1969, 1070 (*synth*)
Bishop, J.J. *et al*, *J. Organomet. Chem.*, 1971, **27**, 241 (*synth, pmr, ir*)
Hughes, O.R. *et al*, *J. Am. Chem. Soc.*, 1981, **103**, 6636 (*nmr*)
Unruh, J.D. *et al*, *J. Mol. Catal.*, 1982, **14**, 19 (*use, synth, pmr*)

1,7-Bis(diphenylphosphino)heptane B-00242

1,7-Heptanediylbis[diphenylphosphine], 9CI

[64730-63-8]

$Ph_2P(CH_2)_7PPh_2$

$C_{31}H_{34}P_2$ M 468.557

Ligand for Rh and Ir. Liq. $Bp_{0.1}$ 140°.

Dioxide: 1,7-Bis(diphenylphosphinyl)heptane.
$C_{31}H_{34}O_2P_2$ M 500.556
Cryst. (toluene). Mp 128-129°.

Kosolapoff, G.M. *et al*, *J. Chem. Soc. (C)*, 1967, 1789 (*oxide*)
Clark, P.W., *J. Organomet. Chem.*, 1977, **137**, 235 (*synth, complexes*)
Hill, W.E. *et al*, *J. Chem. Soc., Perkin Trans. 2*, 1982, 327 (*cmr, nmr*)

1,6-Bis(diphenylphosphino)hexane B-00243

1,6-Hexanediylbis[diphenylphosphine], 9CI. Hexamethylenebis[diphenylphosphine]

[19845-69-3]

$Ph_2P(CH_2)_6PPh_2$

$C_{30}H_{32}P_2$ M 454.530

Ligand for Au, Ir, Pt, Pd, Rh and V. Solid. Mp 124°.

Dioxide: [35387-44-1]. *1,6-Bis(diphenylphosphinyl)-hexane.*

$C_{30}H_{32}O_2P_2$ M 486.529

Metal extractant. Solid. Mp 196-198°. $Bp_{0.2}$ 292-294°.

Disulfide: [61894-27-7]. *1,6-Bis(diphenylphosphinothioyl)hexane.*

$C_{30}H_{32}P_2S_2$ M 518.650

Solid. Mp 225°.

Diselenide: 1,6-Bis(diphenylphosphinoselenoyl)hexane.

$C_{30}H_{32}P_2Se_2$ M 612.450

Solid. Mp 207°.

Sandhu, S.S. *et al, Transition Met. Chem. (Weinheim, Ger.)*, 1976, **1**, 155 (*derivs*)
Levason, W. *et al, J. Organomet. Chem.*, 1977, **125**, 209 (*ms*)
Clark, P.W., *J. Organomet. Chem.*, 1977, **137**, 235 (*synth, complexes*)
Hill, W.E. *et al, J. Chem. Soc., Perkin Trans. 2*, 1982, 327 (*cmr, nmr*)

Bis(diphenylphosphino)methane B-00244

Methylenebis[diphenylphosphine], 9CI

[2071-20-7]

$Ph_2PCH_2PPh_2$

$C_{25}H_{22}P_2$ M 384.396

Ligand for metals of Groups IB, VIB and VIII. Air-stable cryst. (EtOH or C_6H_6/Me_2CO). Mp 122°. Cleaved by PhN_3.

Dioxide: [2071-21-8]. *Bis(diphenylphosphinyl)methane.*

$C_{25}H_{22}O_2P_2$ M 416.395

Ni complex used in hydrosilylation of alkenes. Solid. Mp 178-179°. pK_a 6.23 ($MeNO_2$).

Disulfide: [14633-92-2]. *Bis(diphenylphosphinothioyl)-methane.*

$C_{25}H_{22}P_2S_2$ M 448.516

Cryst. (EtOH). Mp 147-148°, 178-180°. Forms Cd, Hg, Pt and Pd complexes.

Monoselenide: Diphenyl[(diphenylphosphinoselenoyl)-methyl]phosphine.

$C_{25}H_{22}P_2Se$ M 463.356

Cryst. (CH_2Cl_2/hexane). Mp 110-111°.

Diselenide: Bis(diphenylphosphinoselenoyl)methane.

$C_{25}H_{22}P_2Se_2$ M 542.316

Cryst. (butanol). Mp 182-184°.

Issleib, K. *et al, Chem. Ber.*, 1959, **92**, 3175 (*synth*)
Richard, J.J. *et al, J. Am. Chem. Soc.*, 1961, **83**, 1722 (*dioxide*)
Maier, L., *Helv. Chim. Acta*, 1965, **48**, 1034 (*synth, ir, nmr*)
Colton, R. *et al, Aust. J. Chem.*, 1968, **21**, 2215 (*ms*)
Marsmann, H. *et al, Z. Naturforsch., B*, 1972, **27**, 137 (*pmr, cmr, nmr*)
Wheatland, D.A. *et al, Inorg. Chem.*, 1972, **11**, 2340 (*disulfide, complexes*)
Cauguis, G. *et al, Org. Mass Spectrom.*, 1975, **10**, 1021 (*disulfide, ms*)
Sanger, A.R., *J. Chem. Soc., Dalton Trans.*, 1977, 1971 (*ir, complexes*)
Appel, R. *et al, Chem. Ber.*, 1979, **112**, 648 (*synth*)
Grim, S.O. *et al, Inorg. Chem.*, 1980, **19**, 1982 (*synth, derivs, pmr, cmr, nmr, complexes*)
Garusamy, N. *et al, J. Am. Chem. Soc.*, 1982, **104**, 3107 (*synth, use*)

Bis(diphenylphosphino)methanone B-00245

Carbonylbis[diphenylphosphine], 9CI. *Bis(diphenylphosphinyl) ketone*

[51342-28-0]

$Ph_2PCOPPh_2$

$C_{25}H_{20}OP_2$ M 398.380

Solid, fairly stable at r.t. in absence of O_2. Vigorously oxidized by O_2. Dec. on heating.

Becher, H.J. *et al, Angew. Chem., Int. Ed. Engl.*, 1973, **12**, 842 (*synth, props*)

2,11-Bis[(diphenylphosphino)methyl]-benzo[*c*]phenanthrene B-00246

[Benzo[c]phenanthrene-2,11-diylbis[methylene]]-bis[diphenylphosphine], 10CI

[61892-31-7]

CH_2PPh_2

CH_2PPh_2

$C_{44}H_{34}P_2$ M 624.700

Ligand for asymmetric hydrogenations. Cryst. (Me_2CO). Mp 173-176°. Forms complexes containing Gp. 1B or Gp. VIII metals.

Dioxide:

$C_{44}H_{34}O_2P_2$ M 656.699

Cryst. ($CHCl_3$/toluene). Mp 291-292°.

DeStefano, N.J. *et al, Helv. Chim. Acta*, 1976, **59**, 2674 (*synth, pmr, nmr, ms, dioxide*)
Reed, F.J.S. *et al, Helv. Chim. Acta*, 1977, **60**, 2804 (*complexes*)
Bachechi, F. *et al, Helv. Chim. Acta*, 1977, **60**, 2815 (*complexes*)
Venanzi, L.M., *Pure Appl. Chem.*, 1980, **52**, 1117 (*struct, complexes*)

5,6-Bis(diphenylphosphino)-2-methylbicyclo[2.2.1]hept-2-ene B-00247

(5-Methylbicyclo[2.2.1]hept-5-ene-2,3-diyl)-bis[diphenylphosphine], 10CI. *Methyl norphos*

H_3C 1 6 5 PPh_2 PPh_2 (1*R**,5*R**,6*R**)-*form*

$C_{32}H_{30}P_2$ M 476.537

Ligand for asymmetric hydrogenation. Resolved forms obtained by $HSiCl_3$ redn. of resolved dioxides.

(1R*,5R*,6R*)-form

(+)-endo,exo-*form*

$[\alpha]^{20}_D$ +31° (c, 1 in $CHCl_3$).

P,P′-*Dioxide:* $[\alpha]^{20}_D$ −64° (c, 1 in $CHCl_3$).

(1S*,5S*,6S*)-form [83462-66-2]

(−)-endo,exo-*form*

Cryst. Mp 129-130°. $[\alpha]^{20}_D$ −32° (c, 1 in $CHCl_3$).

P,P′-Dioxide: $[\alpha]^{20}_D$ +66° (c, 1 in $CHCl_3$).

Brunner, H. *et al, Inorg. Chim. Acta*, 1982, **61**, 129 (*synth, dioxide, resoln, pmr*)

2,2′-Bis[(diphenylphosphino)methyl]-1,1′-binaphthyl B-00248

[[1,1′-Binaphthalene]-2,2′-diylbis[methylene]]-bis[diphenylphosphine], 9CI. *Naphos[1,1]*

CH_2PPh_2 CH_2PPh_2 *(S)-form*

$C_{46}H_{36}P_2$ M 650.738

Ligand for asymmetric hydrogenation. Forms Ni and Rh complexes.

(S)-form [64091-23-2]

Glassy. $[\alpha]_D^{25}$ −69.6° (c, 0.5 in C_6H_6).

Dioxide: [64091-24-3].

$C_{46}H_{36}O_2P_2$ M 682.737

Solid. Mp 235-238°. $[\alpha]_D^{23}$ −152.1° (c, 0.4 in C_6H_6).

(±)-form [64075-33-8]

Needles, almost air-stable. Solns. are air-sensitive. Mp 160-161.5°.

Akimoto, H. *et al*, *Tetrahedron*, 1971, **27**, 5999 (*abs config*)
Tameo, K. *et al*, *Tetrahedron Lett.*, 1977, 1389 (*synth, ms, pmr*)

1,2-Bis(diphenylphosphinomethyl)-cyclobutane B-00249

[1,2-Cyclobutanediylbis[methylene]]-bis[diphenylphosphine], 9CI

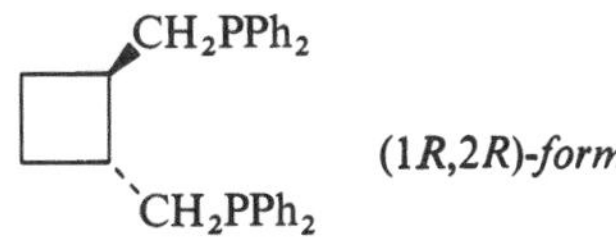

$C_{30}H_{30}P_2$ M 452.515

Ligand for asymmetric hydrogenations. Forms Cu, Pt and Rh complexes.

(1R,2R)-form [55629-68-0]

(−)-trans-*form*

Cryst. solid (EtOH). Mp 107°. $[\alpha]_D^{20}$ −18.6° (c, 1 in C_6H_6).

(1RS,2RS)-form [72151-42-9]

(±)-trans-*form*

Cryst. solid (EtOH). Mp 99-100°.

(1RS,2SR)-form [79768-77-7]

cis-*form*

Cryst. solid (EtOH). Mp 99.5-101°. Less active than *trans*-form in Rh-catalysed hydroformylations.

Glaser, R. *et al*, *Tetrahedron*, 1978, **34**, 3617 (*use*)
Aviron-Violet, P. *et al*, *J. Mol. Catal.*, 1979, **5**, 41 (*use*)
Hayashi, T. *et al*, *Bull. Chem. Soc. Jpn.*, 1981, **54**, 3438 (*use*)
Hughes, R.O. *et al*, *J. Am. Chem. Soc.*, 1981, **103**, 6636 (*complexes*)
Hughes, R.O. *et al*, *J. Mol. Catal.*, 1981, **12**, 71 (*use*)

1,2-Bis(diphenylphosphinomethyl)-cyclohexane B-00250

[1,2-Cyclohexanediylbis[methylene]]-bis[diphenylphosphine], 9CI

[73892-37-2]

CH_2PPh_2 CH_2PPh_2 *(1S,2S)-form*

$C_{32}H_{34}P_2$ M 480.568

Ligand for asymmetric hydrogenations.

(1R,2R)-form [70774-28-6]

(−)-trans-*form*

Oil. $[\alpha]_D^{25}$ −52.8° (c, 0.4 in C_6H_6).

(1S,2S)-form [70223-77-7]

(+)-trans-*form*

Solid. Mp 55-56°. $[\alpha]_D^{25}$ +52.7° (c, 1 in C_6H_6).

(1RS,2SR)-form [72151-45-2]

cis-*form*

Low-melting solid. $Bp_{0.05}$ 215-220° (bath).

Glaser, R. *et al*, *Tetrahedron*, 1978, **34**, 2405, 3617 (*synth, use*)
Aviron-Violet, P. *et al*, *J. Mol. Catal.*, 1979, **5**, 41 (*use*)
Glaser, R. *et al*, *Isr. J. Chem.*, 1980, **20**, 102 (*use*)
Hayashi, T. *et al*, *Bull. Chem. Soc. Jpn.*, 1981, **54**, 3438 (*use*)
Hughes, O.R. *et al*, *J. Mol. Catal.*, 1981, **12**, 71 (*use*)

4,5-Bis[(diphenylphosphino)methyl]-2,2-dimethyl-1,3-dioxolane B-00251

[(2,2-Dimethyl-1,3-dioxolane-4,5-diyl)bis[methylene]]-bis[diphenylphosphine], 9CI. *2,3-O-Isopropylidene-2,3-dihydroxy-1,4-bis(diphenylphosphino)butane. Diop*

[53531-20-7]

(R,R)-form

$C_{31}H_{32}O_2P_2$ M 498.540

Ligand for Al, Cu, and Group VIII metals.

(4R,5R)-form [32305-98-9]

(−)-trans-*form*

Used in asymmetric catalytic reductions, hydroformylations, cyclizations, and hydrocarboalkoxylations. Needles (MeOH). Mp 89-90°. $[\alpha]_D$ −12.4° (c, 0.71 in C_6H_6). Unstable in soln.

Kagan, H.B. *et al*, *J. Am. Chem. Soc.*, 1972, **94**, 6429 (*synth*)
Trost, B.M. *et al*, *J. Am. Chem. Soc.*, 1977, **99**, 1649 (*use*)
Fritschel, S.J. *et al*, *J. Org. Chem.*, 1979, **44**, 3152 (*nmr*)
Murrer, B.A. *et al*, *Synthesis*, 1979, 350 (*synth*)
Cavinato, G. *et al*, *J. Organomet. Chem.*, 1982, **229**, 93 (*use*)
Knowles, W.S. *et al*, *Adv. Chem. Ser.*, 1982, **196**, 325.
Kogure, T. *et al*, *J. Organomet. Chem.*, 1982, **234**, 249 (*use*)
Tani, K. *et al*, *ACS Symp. Ser.*, 1982, **185**, 187 (*rev*)

[Bis(diphenylphosphino)methylene]-triphenylphosphorane B-00252

$Ph_3P{=}C(PPh_2)_2$

$C_{43}H_{35}P_3$ M 644.671

Solid. Mp 239-243° dec.

Issleib, K. *et al*, *Justus Liebigs Ann. Chem.*, 1966, **699**, 40 (*synth*)
Schmidbaur, H. *et al*, *Chem. Ber.*, 1983, **116**, 1393 (*nmr, cryst struct*)

Bis[(diphenylphosphino)methyl]-phenylphosphine, 9CI B-00253

1,1,3,5,5-Pentaphenyl-1,3,5-triphosphapentane

[2071-24-1]

$PhP(CH_2PPh_2)_2$

$C_{32}H_{29}P_3$ M 506.502

Thermally stable cryst. (EtOH). Mp 107-110°, 122-123°.

Trioxide: [21851-89-8]. *Bis[(diphenylphosphinyl)-methyl]phenylphosphine oxide.*

$C_{32}H_{29}O_3P_3$ M 554.501

Reagent for extraction of Hf and Zr from acid soln. Mp 196-207°.
Trisulfide: Bis[(diphenylphosphinothioyl)methyl]-phenylphosphine sulfide.
$C_{32}H_{29}P_3S_3$ M 602.682
Cryst. (C_6H_6). Mp 173-174.5°.

Maier, L., *Helv. Chim. Acta*, 1965, **48**, 1034 (*synth, trisulfide*)
Maier, L., *Helv. Chim. Acta*, 1969, **52**, 845 (*pmr, nmr, synth*)
Appel, R. *et al*, *Chem. Ber.*, 1979, **112**, 648 (*synth, pmr, cmr, nmr*)
Sinyavskaya, E.L. *et al*, *Zh. Neorg. Khim.*, 1981, **26**, 1274 (*Engl. transl.* p. 686) (*complexes*)

1,4-Bis(diphenylphosphino)naphthalene **B-00254**
1,4-Naphthalenediylbis[diphenylphosphine], 10CI, 9CI

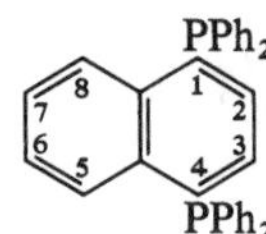

$C_{34}H_{26}P_2$ M 496.527
Solid. Mp 214-215°.
Dioxide: 1,4-Bis(diphenylphosphinyl)naphthalene.
$C_{34}H_{26}O_2P_2$ M 528.526
Solid. Mp 282-284°.

Schindlbauer, H. *et al*, *Monatsh. Chem.*, 1965, **96**, 285 (*uv*)
Zorn, H. *et al*, *Chem. Ber.*, 1965, **98**, 2431 (*synth, ir, dioxide*)

1,5-Bis(diphenylphosphino)naphthalene **B-00255**
1,5-Naphthalenediylbis[diphenylphosphine], 10CI, 9CI
$C_{34}H_{26}P_2$ M 496.527
Cryst. (C_6H_6/ligroin). Mp 261°.
Dioxide: 1,5-Bis(diphenylphosphinyl)naphthalene.
$C_{34}H_{26}O_2P_2$ M 528.526
Solid. Mp >330°.

Schindlbauer, H. *et al*, *Monatsh. Chem.*, 1965, **96**, 285 (*uv, dioxide*)
Zorn, H. *et al*, *Chem. Ber.*, 1965, **98**, 2431 (*synth, ir*)

2,6-Bis(diphenylphosphino)naphthalene **B-00256**
2,6-Naphthalenediylbis[diphenylphosphine], 10CI, 9CI
$C_{34}H_{26}P_2$ M 496.527
Cryst. (C_6H_6/pet. ether). Mp 221°.
Dioxide: 2,6-Bis(diphenylphosphinyl)naphthalene.
$C_{34}H_{26}O_2P_2$ M 528.526
Solid. Mp 276-277°.

Schindlbauer, H. *et al*, *Monatsh. Chem.*, 1965, **96**, 285 (*uv, oxide*)
Zorn, H. *et al*, *Chem. Ber.*, 1965, **98**, 2431 (*synth, ir*)

2,7-Bis(diphenylphosphino)naphthalene **B-00257**
2,7-Naphthalenediylbis[diphenylphosphine], 10CI, 9CI
$C_{34}H_{26}P_2$ M 496.527
Cryst. Mp 141°.
Dioxide: 2,7-Bis(diphenylphosphinyl)naphthalene.
$C_{34}H_{26}O_2P_2$ M 528.526
Solid. Mp 121-122°.

Schindlbauer, H. *et al*, *Monatsh. Chem.*, 1965, **96**, 285 (*uv, oxide*)
Zorn, H. *et al*, *Chem. Ber.*, 1965, **98**, 2431 (*synth, ir*)

1,8-Bis(diphenylphosphino)octane **B-00258**
1,8-Octanediylbis[diphenylphosphine], 9CI
[41625-30-3]

$Ph_2P(CH_2)_8PPh_2$

$C_{32}H_{36}P_2$ M 482.584
Ligand for Au, Pd, Pt and Ni. Solid. Mp 105°.
Dioxide: 1,8-Bis(diphenylphosphinyl)octane.
$C_{32}H_{36}O_2P_2$ M 514.583
Solid. Mp 170.5-172°.

Kosolapoff, G.M. *et al*, *J. Chem. Soc. (C)*, 1967, 1789 (*oxide*)
Levason, W. *et al*, *J. Organomet. Chem.*, 1977, **125**, 209 (*ms*)
Clark, P.W., *J. Organomet. Chem.*, 1977, **137**, 235 (*synth, complexes*)
Hill, W.E. *et al*, *J. Chem. Soc., Perkin Trans. 2*, 1982, 327 (*cmr, nmr*)

1,5-Bis(diphenylphosphino)pentane **B-00259**
1,5-Pentanediylbis[diphenylphosphine], 10CI, 9CI. *Pentamethylenebis(diphenylphosphine)*, 8CI
[27721-02-4]

$Ph_2P(CH_2)_5PPh_2$

$C_{29}H_{30}P_2$ M 440.504
Ligand for metals of Groups VB, VIIIB and Hg. Clear-yellow oil. $Bp_{0.1}$ 230°.
Dioxide: [51145-33-6]. *1,5-Bis(diphenylphosphinyl)-pentane.*
$C_{29}H_{30}O_2P_2$ M 472.502
Solid. Mp 124-126°. $Bp_{0.27}$ 326-327°.
Diselenide: 1,5-Bis(diphenylphosphinoselenoyl)pentane.
$C_{29}H_{30}P_2Se_2$ M 598.424
Solid.

Kosolapoff, G.M. *et al*, *J. Chem. Soc.*, 1959, 3950 (*oxide*)
Horner, L. *et al*, *Chem. Ber.*, 1962, **95**, 581 (*oxide*)
Sacconi, L. *et al*, *Inorg. Chem.*, 1968, **7**, 291 (*synth, complexes*)
Clark, P.W., *J. Organomet. Chem.*, 1977, **137**, 235 (*synth, complexes*)
Al-Saleem, N.A. *et al*, *J. Chem. Soc., Dalton Trans.*, 1979, 1972 (*synth, complexes*)
Dean, P.A.W., *Can. J. Chem.*, 1979, **57**, 754 (*selenide, nmr*)

2,4-Bis(diphenylphosphino)pentane **B-00260**
(1,3-Dimethyl-1,3-propanediyl)bis[diphenylphosphine], 9CI. *Skewphos*

CH_3
Ph_2P–C–H
CH_2
H–C–PPh_2 ***(R,R)-form***
CH_3

$C_{29}H_{30}P_2$ M 440.504
(R,R)-form
Oil or solid. Mp 78°. $[\alpha]^{25}_{D}$ +81° (c, 1.3 in $CHCl_3$), $[\alpha]^{20}_{D}$ +124° (c, 4.0 in $CHCl_3$). Not v. air-sensitive.

(S,S)-form
Oil or solid. Mp 81°. $[\alpha]_D^{25}$ −81° (c, 1.3 in $CHCl_3$), $[\alpha]_D^{20}$ −124° (c, 3.0 in $CHCl_3$). Not v. air-sensitive.

MacNeil, P.A. *et al, J. Am. Chem. Soc.*, 1981, **103**, 2273 (*synth, complexes, use*)
Bakos, J. *et al, J. Organomet. Chem.*, 1985, **279**, 23 (*synth, ms, pmr, cmr, P nmr, use*)

9,10-Bis(diphenylphosphino)phenanthrene B-00261

9,10-Phenanthrenediylbis[diphenylphosphine], 10CI
[39705-76-5]

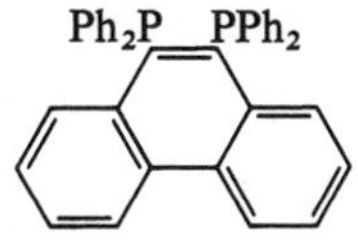

$C_{38}H_{28}P_2$ M 546.587
Ligand for Ni. Pale-yellow cryst. Mp 140°.

Chow, K.K. *et al, J. Chem. Soc., Dalton Trans.*, 1973, 147 (*synth, complexes*)

Bis[2-(diphenylphosphino)phenyl]-phenylphosphine, 9CI B-00262

[53103-03-0]

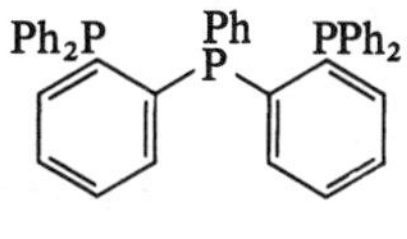

$C_{42}H_{33}P_3$ M 630.644
Complexing agent. Cryst. + 1DMF (DMF). Mp 226-228°.

Hartley, J.G. *et al, J. Chem. Soc.*, 1963, 3920 (*synth, complexes*)
Mynatt, R.J. *et al, J. Coord. Chem.*, 1973, **3**, 145 (*complexes*)

Bis[4-(diphenylphosphino)phenyl]-phenylphosphine, 9CI B-00263

$C_{42}H_{33}P_3$ M 630.644
Solid (EtOAc/EtOH). Mp 154-158°.
Trioxide:
$C_{42}H_{33}O_3P_3$ M 678.642
Mp >320°.
Trisulfide:
$C_{42}H_{33}P_3S_3$ M 726.824
Solid ($CHCl_3$/EtOH). Mp 295-298°.

Baldwin, R.A. *et al, J. Org. Chem.*, 1967, **32**, 1572.

1,2-Bis(diphenylphosphino)propane B-00264

(1-Methyl-1,2-ethanediyl)bis[diphenylphosphine], 10CI. *Prophos*
[15383-58-1]

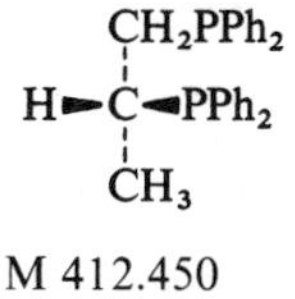

CH_2PPh_2
H►C◄PPh_2 ***(R)-form***
CH_3

$C_{27}H_{26}P_2$ M 412.450

(R)-form [67884-32-6]
Rh complexes used for asymmetric hydrogenations. Prisms. Mp 68.5° (sealed tube). $[\alpha]_D^{20}$ +186.0° (c, 1 in Me_2CO).

(S)-form [67884-33-7]
Ni complex used in asymmetric cross-coupling of Grignard reagents.

Fryzuk, M.D. *et al, J. Am. Chem. Soc.*, 1978, **100**, 5491 (*synth, use*)
Hayashi, T. *et al, Tetrahedron Lett.*, 1981, **22**, 137 (*use*)
Maciel, G.E. *et al, Adv. Chem. Ser.*, 1982, **196**, 389 (*use*)
Morandini, F. *et al, Inorg. Chim. Acta*, 1982, **57**, 15 (*pmr, complexes*)

1,3-Bis(diphenylphosphino)propane B-00265

1,3-Propanediylbis[diphenylphosphine], 9CI
[6737-42-4]

$Ph_2P(CH_2)_3PPh_2$

$C_{27}H_{26}P_2$ M 412.450
Catalyst for manuf. of 1,4-hexadiene from ethylene and butadiene. Ligand for metals of Groups VB, VIB, VIIB and VIII. Air-stable cryst. Mp 60-63°.
Dioxide: [16524-41-7]. *1,3-Bis(diphenylphosphinyl)-propane.*
$C_{27}H_{26}O_2P_2$ M 444.449
Forms complexes with rare earth and transuranic elements. Cryst. (toluene/cyclohexane). Mp 142-144°. $Bp_{0.1}$ 295-300°.
Disulfide: [61894-25-5]. *1,3-Bis(diphenylphosphinothioyl)propane.*
$C_{27}H_{26}P_2S_2$ M 476.570
Forms complexes containing Cd, Hg, Pt, Pd. Solid. Mp 110-112°, 126-127°.
Diselenide:
$C_{27}H_{26}P_2Se_2$ M 570.370
Forms Pd complexes. Solid. Mp 112°.

Horner, L. *et al, Chem. Ber.*, 1962, **95**, 581 (*dioxide*)
Kosolapoff, G.M., *J. Chem. Soc. (C)*, 1967, 1789 (*dioxide*)
Maier, L., *Helv. Chim. Acta*, 1968, **51**, 405 (*dioxide, nmr*)
Grim, S.O. *et al, J. Organomet. Chem.*, 1975, **94**, 327 (*synth*)
Langford, G.R. *et al, Inorg. Chem.*, 1975, **14**, 2937 (*nmr, complexes*)
Sandhu, S.S. *et al, Transition Met. Chem. (Weinheim, Ger.)*, 1976, **1**, 155 (*disulfide, diselenide*)
Sanger, A.R., *J. Chem. Soc., Dalton Trans.*, 1977, 1971 (*ir, complexes*)
Lindsay, C.H. *et al, Inorg. Chem.*, 1980, **19**, 3503 (*pmr, nmr, complexes*)
Gurusamy, N. *et al, J. Am. Chem. Soc.*, 1982, **104**, 3107 (*synth, deriv*)

Bis[3-(diphenylphosphino)propyl]-phenylphosphine B-00266

1,1,5,9,9-Pentaphenyl-1,5,9-triphosphanonane
[34989-06-5]

$PhP(CH_2CH_2CH_2PPh_2)_2$

$C_{36}H_{37}P_3$ M 562.610
Ligand for Co, Ni and Rh. Pale-yellow viscous oil.
5-Oxide:
$C_{36}H_{37}OP_3$ M 578.609
Pale-yellow viscous oil.

Nappier, T.E. *et al, J. Am. Chem. Soc.*, 1973, **95**, 4194 (*synth, complexes*)
Kuhlmann, E.J. *et al, Inorg. Chim. Acta*, 1979, **34**, 197 (*synth, pmr, ir, complexes*)
Uriarte, R. *et al, Inorg. Chem.*, 1980, **19**, 79 (*synth*)
Niewahner, J. *et al, Adv. Chem. Ser.*, 1982, **196**, 257 (*use*)

3,4-Bis(diphenylphosphino)pyrrolidine B-00267

3,4-Pyrrolidinediylbis[diphenylphosphine]

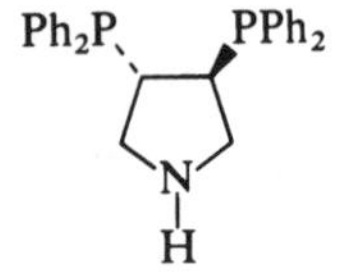

$C_{28}H_{27}NP_2$ M 439.476

(3R,4R)-form

Solid. Mp 120-121°. $[\alpha]_D^{20}$ +154° (c, 3.8 in EtOH).

B,HCl: Solid. Mp 207-208°. $[\alpha]_D^{20}$ +161° (c, 2.6 in EtOH).

N-*Benzoyl:*

$C_{35}H_{31}NOP_2$ M 543.584

Solid. Mp 180-183°. $[\alpha]_D^{20}$ +153° (c, 2.84 in toluene).

N-*Benzyl:*

$C_{35}H_{33}NP_2$ M 529.600

Solid. Mp 117°. $[\alpha]_D^{20}$ +76.5° (c, 1.7 in toluene).

Nagel, U. *et al, Chem. Ber.*, 1986, **119**, 3326 (*synth, derivs, P nmr*)

Bis[2-(diphenylphosphino)-3,4,5,6-tetrafluorophenyl]phenylphosphine, 9CI B-00268

[27384-35-6]

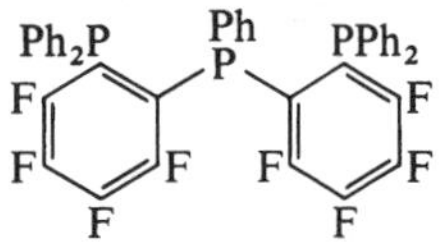

$C_{42}H_{25}F_8P_3$ M 774.568

Ligand for Ni. Yellow needles. Mp 173.5-174.5°.

Eller, P.G. *et al, J. Organomet. Chem.*, 1970, **22**, 631 (*synth*)
Eller, P.G. *et al, Inorg. Chem.*, 1972, **11**, 2518 (*complexes*)

Bis(*O,O*-diphenyl phosphoroselenoyl) diselenide B-00269

O,O,O′,O′-*Tetraphenyl selenoperoxydiphosphate, 8CI*

[34764-11-9]

$(PhO)_2P(Se){-}Se{-}Se{-}P(Se)(OPh)_2$

$C_{24}H_{20}O_4P_2Se_4$ M 750.207

Cryst. (EtOH). Mp 125-126°.

Gorak, R.D. *et al, Zh. Obshch. Khim.*, 1971, **41**, 1994 (*Engl. transl.* p. 2012) (*synth*)

[Bis(diphenylstibino)methylene]-triphenylphosphorane, 11CI B-00270

Triphenylphosphonium bis(diphenylstibino)methylide

[87482-32-4]

$Ph_3P{=}C(SbPh_2)_2$

$C_{43}H_{35}PSb_2$ M 826.223

V. air- and moisture-sensitive light-yellow solid. Spar. sol. most solvs. Mp 165° dec.

Schmidbaur, H. *et al, Organometallics*, 1984, **3**, 38 (*synth, ir, P nmr, cryst struct*)

***O,O*-Bis(2-ethoxyethyl) phosphorochloridothioate, 9CI, 8CI** B-00271

O,O-*Bis(2-ethoxyethyl) thiophosphoryl chloride.* O,O-*Bis(2-ethoxyethyl) chlorothiophosphate*

[13891-63-9]

$(EtOCH_2CH_2O)_2P(S)Cl$

$C_8H_{18}ClO_4PS$ M 276.715

Yellowish oil. Bp_5 160-162°.

Srivastava, K.C., *Aust. J. Chem.*, 1966, **19**, 2397 (*synth, props*)

Bis(ethoxymethyl)phosphinic acid, 9CI B-00272

$(EtOCH_2)_2P(O)OH$

$C_6H_{15}O_4P$ M 182.156

Liq. $Bp_{0.0001}$ 220°.

Et ester: [23081-65-4]. *Ethyl bis(ethoxymethyl)-phosphinate.*

$C_8H_{19}O_4P$ M 210.209

Liq. d_4^{20} 1.05. $Bp_{0.04}$ 78-80°. n_D^{20} 1.4353.

Mukhametzyanova, É.Kh. *et al, Izv. Akad. Nauk SSSR, Ser. Khim.*, 1969, 710 (*Engl. transl.* p. 644)

Bis(2-ethylhexyl) phosphonate, 9CI B-00273

Bis(2-ethylhexyl) phosphite

[3658-48-8]

$[H_3C(CH_2)_3CH(CH_2CH_3)CH_2O]_2P(O)H \rightleftharpoons [H_3C(CH_2)_3CH(CH_2CH_3)CH_2O]_2POH$

$C_{16}H_{35}O_3P$ M 306.424

Tautomeric. Used in extraction of U, Fe, and Mo from acid solns. Liq. d_{25}^{25} 0.93. Bp_1 148-151°. n_D^{25} 1.4420.

▷SZ6840000.

Walsh, E.N., *J. Am. Chem. Soc.*, 1959, **81**, 3023 (*synth*)
B.P., 1 298 156, (*1970*); *CA*, **78**, 83841 (*manuf*)

***O,O*-Bis(2-ethylhexyl) phosphorodithioate, 9CI, 8CI** B-00274

O,O-*Bis(2-ethylhexyl) hydrogen phosphorodithioate.* O,O-*Bis(2-ethylhexyl) phosphorodithioic acid.* O,O-*Bis(2-ethylhexyl) dithiophosphate*

[5810-88-8]

$[H_3C(CH_2)_3CH(CH_2CH_3)CH_2O]_2P(S)SH$

$C_{16}H_{35}O_2PS_2$ M 354.545

Extractant for Mo, Zn, Pd. Separates Ni from Co. Zn salt is important lubricant oil additive. pK_a 1.25 (20°, H_2O).

▷TD5075000.

Vorsina, I.A. *et al, Zh. Prikl. Spectrosc.*, 1974, **21**, 110 (*Engl. transl.* p. 926) (*ir, raman*)
Samoilov, Yu.M. *et al, Zh. Neorg. Khim.*, 1981, **26**, 3039 (*Engl. transl.* p. 1623) (*props, use*)
Rozen, A.M. *et al, Zh. Obshch. Khim.*, 1982, **52**, 1235 (*Engl. transl.* p. 1086) (*props, use*)
Kletenik, Yu.P. *et al, Zh. Prikl. Khim.*, 1983, **56**, 1510 (*Engl. transl.* p. 1414) (*props, use*)
Rumpf, T. *et al, Fresenius' Z. Anal. Chem.*, 1983, **315**, 350 (*props*)
Udalova, T.A. *et al, Zh. Neorg. Khim.*, 1983, **28**, 702 (*Engl. transl.* p. 396) (*props, use*)
Udalova, T.A. *et al, Zh. Anal. Khim.*, 1984, **39**, 659 (*Engl. transl.* p. 530) (*props, use*)

2,4-Bis(4-fluorophenyl)-1,3,2,4-dioxaphosphetane 2,4-dioxide, 9CI **B-00275**
[791-43-5]

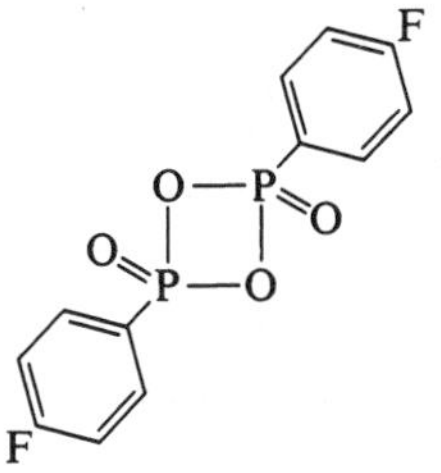

$C_{12}H_8F_2O_4P_2$ M 316.137
Cyclic dianhydride of (4-Fluorophenyl)phosphonic acid, F-00038 . Solid. Mp 109-111°.

Cherbuliez, E. *et al*, *Helv. Chim. Acta*, 1962, **45**, 2665 (*synth*)

Bis(3-fluorophenyl)phosphine, 9CI **B-00276**
[23587-34-0]

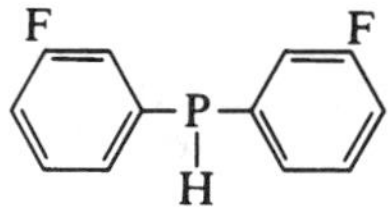

$C_{12}H_9F_2P$ M 222.174
Liq. Bp_{13} 144°. n_D^{20} 1.5355.

Schindlbauer, H. *et al*, *Chem. Ber.*, 1969, **102**, 2914 (*synth, nmr*)

Bis(4-fluorophenyl)phosphine **B-00277**
[25186-17-8]
$C_{12}H_9F_2P$ M 222.174
Liq. Bp_{12} 140-142°. n_D^{20} 1.5318.

Prikoszovich, W. *et al*, *Chem. Ber.*, 1969, **102**, 2922 (*synth, nmr*)

Bis(2-fluorophenyl)phosphinic acid, 9CI **B-00278**

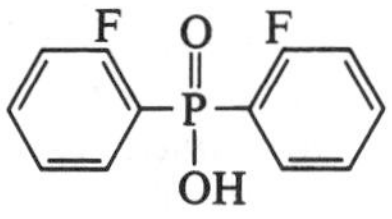

$C_{12}H_9F_2O_2P$ M 254.172
Cryst. (EtOH aq.). Mp 220-222°.

Freedman, L.D. *et al*, *J. Am. Chem. Soc.*, 1953, **75**, 1379.

Bis(3-fluorophenyl)phosphinic acid, 9CI **B-00279**
[1495-59-6]
$C_{12}H_9F_2O_2P$ M 254.172
Cryst. (EtOH). Mp 171°.
Chloride: [23588-23-0].
$C_{12}H_8ClF_2OP$ M 272.618
Liq. $Bp_{0.02}$ 130-131°. n_D^{20} 1.5768.

Schindlbauer, H. *et al*, *Chem. Ber.*, 1969, **102**, 2914 (*synth, nmr*)

Bis(4-fluorophenyl)phosphinic acid, 9CI **B-00280**
$C_{12}H_9F_2O_2P$ M 254.172
Cryst. (EtOH). Mp 175-177°.
Chloride: [25324-36-1].
$C_{12}H_8ClF_2OP$ M 272.618
$Bp_{0.08}$ 145-147°. n_D^{20} 1.5726.

Prikoszovich, W. *et al*, *Chem. Ber.*, 1969, **102**, 2922 (*synth, nmr*)

Bis(hydroxymethyl)phenylphosphine **B-00281**
(Phenylphosphinidene)bismethanol, 9CI
[3127-08-0]

$PhP(CH_2OH)_2$

$C_8H_{11}O_2P$ M 170.147
Used in prep. of heterocyclic phosphorus compds. Liq. Distils with sl. dec. d_4^{25} 1.19. $Bp_{0.1}$ 100°. n_D^{25} 1.6064.
Di-Ac: [70602-72-1]. *Bis(acetoxymethyl)phenylphosphine. Phenylphosphinidenebis[methyl acetate].*
$C_{12}H_{15}O_4P$ M 254.222
Liq. $Bp_{0.1}$ 124-125°. n_D^{20} 1.5380.
Oxide: [39118-56-4]. *(Phenylphosphinylidene)bismethanol.*
$C_8H_{11}O_3P$ M 186.147
Solid. Mp 100-102°.
Oxide, di-Ac: (Phenylphosphinyldene)bis[methyl acetate], 9CI.
$C_{12}H_{15}O_5P$ M 270.221
Solid. Mp 113°.
Sulfide: [3296-60-4]. *(Phenylphosphinothioylidene)bismethanol.*
$C_{18}H_{11}O_2PS$ M 322.317
Solid. Mp 84-86°.

Hellmann, H. *et al*, *Justus Liebigs Ann. Chem.*, 1962, **659**, 49 (*synth, derivs*)
Arbuzov, B.A. *et al*, *Izv. Akad. Nauk SSSR, Ser. Khim.*, 1979, **28**, 866 (*Engl. transl.* p. 810) (*synth, derivs, ir, nmr*)
Rampal, J.B. *et al*, *J. Am. Chem. Soc.*, 1981, **103**, 7602 (*use*)
Arbuzov, B.A. *et al*, *Izv. Akad. Nauk SSSR, Ser. Khim.*, 1982, **31**, 440 (*use*)

Bis(hydroxymethyl)phosphinic acid, 9CI **B-00282**
[2074-67-1]

$(HOCH_2)_2P(O)OH$

$C_2H_7O_4P$ M 126.049
Syrupy liq. pK_a 2.08.
Na salt: Decahydrate.
Cyclohexylammonium salt: Solid. Mp 180°.
Me ester: [67309-03-9]. *Methyl bis(hydroxymethyl)phosphinate.*
$C_3H_9O_4P$ M 140.075
Liq. d_4^{20} 1.36. n_D^{20} 1.4808.
Et ester: [67309-04-0]. *Ethyl bis(hydroxymethyl)phosphinate.*
$C_4H_{11}O_4P$ M 154.102
Liq. d_4^{20} 1.29. n_D^{20} 1.4765.
Di-Et ether: see Bis(ethoxymethyl)phosphinic acid, B-00272
Di-Ph ether: see Bis(phenoxymethyl)phosphinic acid, B-00384
Bis(trimethylsilyl)ether, trimethylsilyl ester: Trimethylsilyl bis(trimethylsilyloxymethyl)phosphinate.
$C_{11}H_{30}O_4PSi_3$ M 341.586
Liq. d_4^{20} 0.97. $Bp_{2.5}$ 114-115°. n_D^{20} 1.4290.

Hoffman, A., *J. Am. Chem. Soc.*, 1930, **52**, 2995 (*synth*)
Ivanov, B.E. *et al*, *Zh. Obshch. Khim.*, 1967, **37**, 1856 (*Engl. transl.* p. 1768) (*synth, use*)
Maier, L., *Z. Anorg. Allg. Chem.*, 1972, **394**, 117 (*synth, P nmr*)
Romanenko, V.D. *et al*, *Zh. Obshch. Khim.*, 1978, **48**, 1182 (*Engl. transl.* p. 1082) (*esters, ir, pmr*)
Sorokin, M.S. *et al*, *Zh. Obshch. Khim.*, 1978, **48**, 2053 (*Engl. transl.* p. 1868) (*deriv*)

Bis(2-hydroxyphenyl)phosphinic acid, 9CI **B-00283**

[99706-43-1]

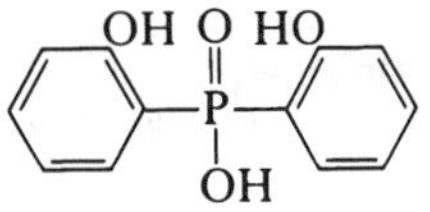

$C_{12}H_{11}O_4P$ M 250.190

Granular cryst. (EtOH aq.). Mp 171-172°.

Et ester: [99706-39-5]. *Ethyl bis(2-hydroxyphenyl)-phosphinate.*

$C_{14}H_{15}O_4P$ M 278.244

Cryst. (CH_2Cl_2/hexane). Mp 129-130°.

Dhawan, B. *et al*, *J. Org. Chem.*, 1986, **51**, 179 (*synth, pmr, cmr, P nmr*)

Bis(3-hydroxyphenyl)phosphinic acid, 9CI **B-00284**

[84251-29-6]

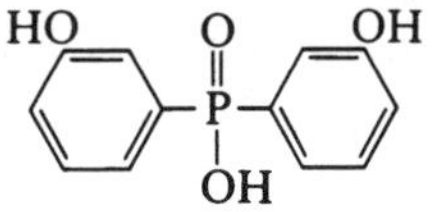

$C_{12}H_{11}O_4P$ M 250.190

Needles. Mp 239-240° (226-229°).

Di-Me ether: see Bis(3-methoxyphenyl)phosphinic acid, B-00300

Doak, G.O. *et al*, *J. Am. Chem. Soc.*, 1952, **74**, 753.
Cornforth, J. *et al*, *J. Chem. Soc., Perkin Trans. 1*, 1982, 2299.

Bis(4-hydroxyphenyl)phosphinic acid, 9CI **B-00285**

$C_{12}H_{11}O_4P$ M 250.190

Solid. Mp 213-215°.

Di-Me ether: see Bis(4-methoxyphenyl)phosphinic acid, B-00301

Di-Ph ether: Bis(4-phenoxyphenyl)phosphinic acid, 9CI.

$C_{24}H_{19}O_4P$ M 402.385

Solid. Mp 203-206°.

Doak, G.O. *et al*, *J. Am. Chem. Soc.*, 1952, **74**, 753 (*synth*)
Freedman, L.D. *et al*, *J. Med. Chem.*, 1965, **8**, 891 (*deriv*)

Bis(3-hydroxypropyl)phenylphosphine **B-00286**

3,3′-(Phenylphosphinidene)bis[1-propanol], 9CI

[2468-05-5]

$PhP(CH_2CH_2CH_2OH)_2$

$C_{12}H_{19}O_2P$ M 226.255

Yellowish oil. d_4^{20} 1.10. Bp_4 246-250°, Bp_1 175°. n_D^{20} 1.5740.

B,MeI: Bis(3-hydroxypropyl)methylphenylphosphonium iodide.

$C_{13}H_{22}IO_2P$ M 368.194

Solid. Mp 137°.

Di-Ac:

$C_{16}H_{23}O_4P$ M 310.329

d_4^{20} 1.01. $Bp_{1.5}$ 178-179°. n_D^{20} 1.5262.

Oxide: 3,3′-(Phenylphosphinylidene)bis[1-propanol], 9CI.

$C_{12}H_{19}O_3P$ M 242.254

Solid. Mp 77-78°.

Oxide, di-Ac: 3,3′-(Phenylphosphinylidene)bis[1-propyl acetate], 9CI.

$C_{16}H_{23}O_5P$ M 326.328

Solid. Mp 71-71.5°. Bp_2 201-203°.

Sulfide: 3,3′-(Phenylphosphinothioylidene)bis[1-propanol], 9CI.

$C_{12}H_{19}O_2PS$ M 258.315

Cryst. (Et_2O). Mp 102-103.5°.

Sulfide, di-Ac: 3,3′-(Phenylphosphinothioylidene)bis[1-propyl acetate], 9CI.

$C_{16}H_{23}O_4PS$ M 342.389

Liq. Bp_1 227°.

Arbuzov, B.A. *et al*, *Izv. Akad. Nauk SSSR, Ser. Khim.*, 1962, 290 (*Engl. transl.* p. 263) (*derivs*)
Issleib, K. *et al*, *Chem. Ber.*, 1965, **98**, 2091 (*synth*)

Bis(1-hydroxy-2,2,2-trichloroethyl)-phosphinic acid **B-00287**

$[Cl_3CCH(OH)]_2P(O)OH$

$C_4H_5Cl_6O_4P$ M 360.773

Cryst. (MeOH). Mp 205-213°.

Me ester: Methyl bis(1-hydroxy-2,2,2-trichloroethyl)-phosphinate.

$C_5H_7Cl_6O_4P$ M 374.800

Cryst. (EtOH/Me_2CO). Mp 180-181°.

Et ester: Ethyl bis(1-hydroxy-2,2,2-trichloroethyl)-phosphinate.

$C_6H_9Cl_6O_4P$ M 388.826

Cryst. (EtOH/Me_2CO). Mp 177-178°.

Ettel, V. *et al*, *Collect. Czech. Chem. Commun.*, 1961, **26**, 2087 (*synth*)
Ivanov, B.E. *et al*, *Izv. Akad. Nauk SSSR, Ser. Khim.*, 1968, 1633 (*Engl. transl.* p. 1544) (*esters*)

Bis(iodomethyl)phosphinic acid, 9CI **B-00288**

[17052-16-3]

$(ICH_2)_2P(O)OH$

$C_2H_5I_2O_2P$ M 345.843

Cryst. (H_2O). Mp 139-140°.

Et ester: [17052-17-4]. *Ethyl bis(iodomethyl)-phosphinate.*

$C_4H_9I_2O_2P$ M 373.897

Cryst. (pet. ether). Mp 99.5-100°.

Butyl ester: Butyl bis(iodomethyl)phosphinate.

$C_6H_{13}I_2O_2P$ M 401.950

Cryst. (pet. ether). Mp 78-79°.

Ph ester: [17052-19-6]. *Phenyl bis(iodomethyl)-phosphinate.*

$C_8H_9I_2O_2P$ M 421.941

Cryst. (pet. ether). Mp 86-87°.

Diethylamide: N,N-*Diethyl*-P,P-*bis(iodomethyl)-phosphinic amide.*

$C_4H_{10}I_2NOP$ M 372.912

Cryst. (pet. ether). Mp 88-89°.

Mukhametzyanova, E.Kh. *et al*, *Izv. Akad. Nauk SSSR, Ser. Khim.*, 1969, 1597 (*Engl. transl. p. 1535*) (*synth, derivs*)
Bel'skii, V.E. *et al*, *Izv. Akad. Nauk SSSR, Ser. Khim.*, 1970, 1542 (*Engl. transl.* p. 1454) (*props*)

Bis(2-iodophenyl)phosphinic acid, 9CI **B-00289**

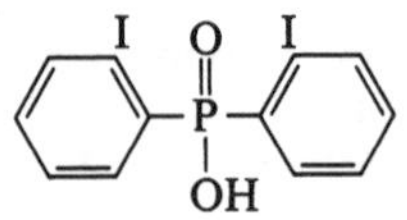

$C_{12}H_9I_2O_2P$ M 469.985

Solid. Mp 206-207.5°.

Freedman, L.D. *et al*, *J. Am. Chem. Soc.*, 1953, **75**, 1379.

Bis(3-iodophenyl)phosphinic acid, 9CI **B-00290**

$C_{12}H_9I_2O_2P$ M 469.985
Cryst. (EtOH aq.). Mp 212.5-213.5°.

Denham, J.M. *et al*, *J. Org. Chem.*, 1958, **23**, 1298.

Bis(4-iodophenyl)phosphinic acid, 9CI **B-00291**

$C_{12}H_9I_2O_2P$ M 469.985
Cryst. (EtOH aq.). Mp 209-210°.

Freedman, L.D. *et al*, *J. Med. Chem.*, 1965, **8**, 891.

Bis(mercaptomethyl)phosphinic acid, 9CI **B-00292**

[20428-23-3]

(HSCH₂)₂P(O)OH

$C_2H_7O_2PS_2$ M 158.170
Thick syrup.
S,S-*Dibenzoyl:* [21199-97-3]. *Bis(benzoylthiomethyl)-phosphinic acid.*
$C_{16}H_{15}O_4PS_2$ M 366.386
Cryst. (C_6H_6). Mp 140-141°.

Ivasyuk, N.V. *et al*, *Izv. Akad. Nauk SSSR, Ser. Khim.*, 1968, 1625, 2388 (*Engl. transl.* pp. 1532, 2262)

2,4-Bis(4-methoxyphenyl)-1,3,2,4-dithiadiphosphetane 2,4-disulfide, 9CI **B-00293**

p-*Methoxyphenylthionophosphine sulfide dimer.* p-*Methoxyphenyldithioxophosphorane dimer. Lawesson's reagent*

[19172-47-5]

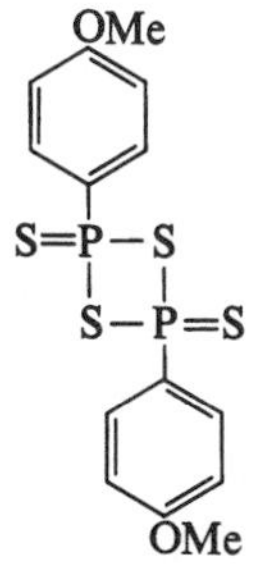

$C_{14}H_{14}O_2P_2S_4$ M 404.451
Highly effective reagent for thiation of ketones, amides and esters. Intermed. in the synth. of *S*- and *P*-containing heterocyclic compds. Peptide-coupling reagent inducing little or no racemisation. Cryst. (1,2-dichlorobenzene). Mp 229°. V. sensitive to moisture. May exist partly as the dithioxo monomer in soln.

Lecher, H.Z. *et al*, *J. Am. Chem. Soc.*, 1956, **78**, 5018 (*synth*)
Pedersen, B.S. *et al*, *Bull. Soc. Chim. Belg.*, 1978, **87**, 223 (*synth*)
Fritz, H. *et al*, *Bull. Chem. Soc. Belg.*, 1978, **87**, 525 (*use*)
Keck, H. *et al*, *Phosphorus Sulfur*, 1978, **4**, 173 (*ms*)
El-Barbary, A.A. *et al*, *Tetrahedron*, 1982, **38**, 405 (*use*)
Shridhar, D.R. *et al*, *Synthesis*, 1982, 1061 (*use*)
Brown, D.W. *et al*, *Tetrahedron*, 1983, **39**, 1075 (*use*)
Cherkasov, R.A. *et al*, *Tetrahedron*, 1985, **41**, 2567 (*rev, bibl*)
Thorson, M. *et al*, *Tetrahedron*, 1985, **41**, 5633 (*use*)
Anderen, T.P. *et al*, *Justus Liebigs Ann. Chem.*, 1986, 269 (*use*)

Bis(2-methoxyphenyl) phosphate, 9CI **B-00294**

Bis(2-methoxyphenyl) phosphoric acid. Bis(2-methoxyphenyl) hydrogen phosphate

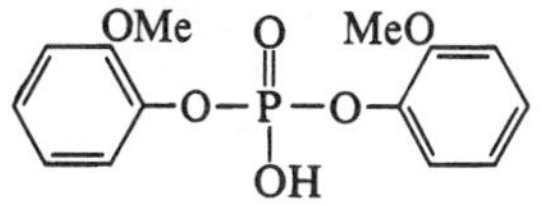

$C_{14}H_{15}O_6P$ M 310.243
Prisms. Mp 95-96°.
Cyclohexylammonium salt: Needles (EtOH). Mp 161-162°.
Chloride: [20464-81-7]. *Bis(2-methoxyphenyl) phosphorochloridate. Bis(2-methoxyphenyl)phosphoryl chloride. Bis(2-methoxyphenyl) chlorophosphate.*
$C_{14}H_{14}ClO_5P$ M 328.688
Solid. d_4^{22} 1.33. Mp 65-67°. Bp_{13} 239-241°. n_D^{22} 1.5592.

Kucherov, V.F. *et al*, *Zh. Obshch. Khim.*, 1949, **19**, 126; *CA*, **43**, 6178 (*derivs*)
Corby, N.S. *et al*, *J. Chem. Soc.*, 1952, 1234 (*synth, derivs*)
Kuz'menko, I.I. *et al*, *Zh. Obshch. Khim.*, 1968, **38**, 158 (*Engl. transl.* p. 156) (*chloride*)

Bis(4-methoxyphenyl) phosphate, 9CI **B-00295**

Bis(4-methoxyphenyl) hydrogen phosphate. Bis(4-methoxyphenyl) phosphoric acid

$C_{14}H_{15}O_6P$ M 310.243
Needles (CCl_4 or $CHCl_3$/pet. ether). Mp 92-94°. pK_a 2.00 (H_2O), pK_a 2.47 (75% v/v EtOH aq.).
Chloride: [20464-82-8]. *Bis(4-methoxyphenyl) phosphorochloridate. Bis(4-methoxyphenyl) phosphoryl chloride. Bis(4-methoxyphenyl) chlorophosphate.*
$C_{14}H_{14}ClO_5P$ M 328.688
Oil. d_D^{25} 1.55. Bp_5 232-234°, $Bp_{0.001}$ 108-110°. n_D^{25} 1.5540.
Amide: Bis(4-methoxyphenyl) phosphoramidate. Bis(4-methoxyphenyl)phosphoryl amide.
$C_{14}H_{16}NO_5P$ M 309.258
Needles ($CHCl_3$). Mp 179-181°.

Corby, N.S. *et al*, *J. Chem. Soc.*, 1952, 1234 (*synth, amide*)
Kuz'menko, I.I. *et al*, *Zh. Obshch. Khim.*, 1968, **38**, 158 (*Engl. transl.* p. 156) (*chloride*)
Kirby, A.J. *et al*, *J. Chem. Soc. (B)*, 1970, 510 (*synth*)
Reimschüssel, W. *et al*, *Int. J. Chem. Kinet.*, 1980, **12**, 979 (*chloride, synth, props*)

Bis(2-methoxyphenyl)phosphine, 9CI **B-00296**

Di-o-*anisylphosphine*

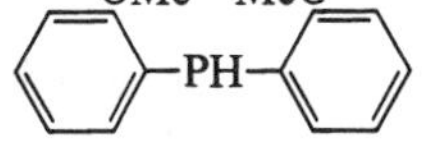

$C_{14}H_{15}O_2P$ M 246.245
Oxide: [71360-04-8].
$C_{14}H_{15}O_3P$ M 262.244
Cryst. (C_6H_6). Mp 132-136°. Tautomeric with bis(2-methoxyphenyl)phosphinous acid.

Ger. Pat., 1 223 838, (*1966*); *CA*, **65**, 18619 (*synth*)
Hengartner, U. *et al*, *J. Org. Chem.*, 1979, **44**, 3741 (*oxide*)
Brown, J.M. *et al*, *Tetrahedron Lett.*, 1980, **21**, 581 (*synth, use*)

Bis(3-methoxyphenyl)phosphine, 9CI **B-00297**

Di-m-*anisylphosphine*

$C_{14}H_{15}O_2P$ M 246.245

Oxide: [71360-01-5].

$C_{14}H_{15}O_3P$ M 262.244

Tautomeric with bis(3-methoxyphenyl)phosphinous acid.

Hengartner, U. *et al*, *J. Org. Chem.*, 1979, **44**, 3741 (*oxide, synth, deriv*)

Bis(4-methoxyphenyl)phosphine, 9CI **B-00298**

*Di-*p*-anisylphosphine*

[15754-51-5]

$C_{14}H_{15}O_2P$ M 246.245

Solid at r.t. $Bp_{0.15}$ 145-148°. Li deriv. exhibits yellow chemiluminescence.

Oxide: [15754-51-5].

$C_{14}H_{15}O_3P$ M 262.244

Cryst. (Me_2CO/Et_2O). Mp 115-116°. pK_a 21.8 (DMSO, 25°). Tautomeric with bis(4-methoxyphenyl)phosphinous acid.

Strecker, R.A. *et al*, *J. Am. Chem. Soc.*, 1973, **95**, 210 (*synth*)
Kapoor, P.N. *et al*, *Helv. Chim. Acta*, 1977, **60**, 2824 (*oxide, synth, ir*)
Tsvetkov, E.N. *et al*, *Izv. Akad. Nauk SSSR, Ser. Khim.*, 1978, 1981 (*Engl. transl.* p. 1743) (*oxide*)

Bis(2-methoxyphenyl)phosphinic acid, 9CI **B-00299**

Di-o-anisylphosphinic acid

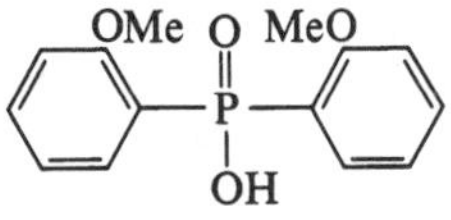

$C_{14}H_{15}O_2P$ M 246.245

Cryst. (6M HCl). Mp 234-238°.

Freedman, L.D. *et al*, *J. Am. Chem. Soc.*, 1953, **75**, 1379.

Bis(3-methoxyphenyl)phosphinic acid, 9CI **B-00300**

*Di-*m*-anisylphosphinic acid*

[84251-28-5]

$C_{14}H_{15}O_2P$ M 246.245

Plates. Mp 185-187°.

Me ester: Methyl bis(3-methoxyphenyl)phosphinate.

$C_{15}H_{17}O_2P$ M 260.272

Liq. $Bp_{0.15}$ 240° (oven).

Cornforth, J. *et al*, *J. Chem. Soc., Perkin Trans. 1*, 1982, 2299 (*synth, ms, pmr, ir*)

Bis(4-methoxyphenyl)phosphinic acid, 9CI **B-00301**

*Di-*p*-anisylphosphinic acid*

[20434-05-3]

$C_{14}H_{15}O_4P$ M 278.244

Needles (EtOH aq.). Mp 180-181°.

Me ester: [83470-31-9]. *Methyl bis(4-methoxyphenyl)-phosphinate.*

$C_{15}H_{17}O_4P$ M 292.271

Solid. Mp 106.5-108°. $Bp_{0.15}$ 192° (oven).

Chloride: [20434-06-4].

$C_{14}H_{14}ClO_3P$ M 296.690

Solid. Mp 59-63°. $Bp_{0.05}$ 216°.

Azide: [83470-25-1].

$C_{14}H_{14}N_3O_3P$ M 303.257

Liq. $Bp_{0.007}$ 130-136° (oven).

Amide: [83470-32-0].

$C_{14}H_{16}NO_3P$ M 277.259

Cryst. (toluene). Mp 138-139°.

Kabachnik, M.I. *et al*, *Zh. Obshch. Khim.*, 1963, **33**, 382 (*Engl. transl.* p. 375) (*synth*)
Tomaschewski, G. *et al*, *Arch. Pharm. (Weinheim, Ger.)*, 1968, **301**, 520 (*synth, chloride, isocyanate*)
Harger, M.J.P. *et al*, *Tetrahedron*, 1982, **38**, 1571 (*synth, ester, chloride, azide, amide*)

Bis(3-methylbutyl) phosphate, 9CI **B-00302**

Diisopentyl phosphate, 8CI. *Bis(3-methylbutyl) hydrogen phosphate. Bis(3-methylbutyl) phosphoric acid. Diisoamyl phosphate*

$[(H_3C)_2CHCH_2CH_2O]_2P(O)OH$

$C_{10}H_{23}O_4P$ M 238.263

Oil. d_4^{20} 1.03. n_D^{20} 1.4327.

Fluoride: Bis(3-methylbutyl) phosphorofluoridate. Diisopentyl phosphoryl fluoridate.

$C_{10}H_{22}FO_3P$ M 240.254

Liq. Bp_{23} 135-138°.

Chloride: [53378-32-8]. *Bis(3-methylbutyl) phosphorochloridate. Diisopentyl phosphoryl chloride.*

$C_{10}H_{22}ClO_3P$ M 256.709

Liq. Bp_1 122-124°, $Bp_{0.02}$ 74°.

Amide: Bis(3-methylbutyl) phosphoramidate. Diisopentyl phosphoramide.

$C_{10}H_{24}NO_3P$ M 237.278

Liq. d_4^{20} 1.01. $Bp_{0.2}$ 142-143°. n_D^{20} 1.4392.

Anilide: Bis(3-methylbutyl) phenylphosphoramidate. Diisopentyl phenylphosphoramide. Diisoamyl phosphoranilide.

$C_{16}H_{28}NO_3P$ M 313.376

Oil. Bp_2 191°. n_D^{20} 1.4863.

Cook, H.G. *et al*, *J. Chem. Soc.*, 1945, 873 (*chloride*)
Cook, H.G. *et al*, *J. Chem. Soc.*, 1949, 635 (*fluoride*)
Ramaswami, D. *et al*, *J. Am. Chem. Soc.*, 1953, **75**, 1763 (*chloride, anilide*)
Petrov, K.A. *et al*, *Zh. Obshch. Khim.*, 1960, **30**, 1233 (*Engl. transl.* p. 1256) (*amide*)
Grosse-Ruyken, H. *et al*, *J. Prakt. Chem.*, 1962, **18**, 287 (*synth, chloride*)

Bis(3-methylbutyl) phosphonate, 9CI **B-00303**

Diisopentyl phosphonate. Bis(3-methylbutyl) phosphite. Diisopentyl phosphite

[19201-23-1]

$[(H_3C)_2CHCH_2CH_2O]_2P(O)H \rightleftharpoons [(H_3C)_2CHCH_2CH_2O]_2POH$

$C_{10}H_{23}O_3P$ M 222.264

Tautomeric; almost completely in the phosphonate form. Liq. d^{20} 0.95-0.97. Bp_6 131-132°. n_D^{20} 1.4268-1.4331.

Cook, H.G. *et al*, *J. Chem. Soc.*, 1945, 873 (*synth*)
Petrov, K.A. *et al*, *Zh. Obshch. Khim.*, 1962, **32**, 1277, 3723 (*Engl. transl.* pp. 1250, 3650) (*synth*)
B.P., 1 298 156, (*1970*); *CA*, **78**, 83841 (*manuf*)

***O,O*-Bis(3-methylbutyl) phosphorodithioate,** 9CI **B-00304**

O,O-*Diisopentyl phosphorodithioate,* 8CI. O,O-*Diisopentyl hydrogen dithiophosphate.* O,O-*Diisopentyl dithiophosphoric acid*

$[(H_3C)_2CHCH_2CH_2O]_2P(S)SH$

$C_{10}H_{23}O_2PS_2$ M 270.384

Oil. d^{20} 1.04. $Bp_{0.3}$ 103°. n_D^{20} 1.4900.

Almasi, L. *et al*, *CA*, 1965, **62**, 2729.

O,O-Bis(3-methylbutyl) phosphorothioate, 9CI **B-00305**

O,O-Diisopentyl phosphorothioate. O,O-Diisopentyl hydrogen phosphorothioate. O,O-Diisopentyl hydrogen thiophosphate

$[(H_3C)_2CHCH_2CH_2O]_2P(O)SH \rightleftharpoons [(H_3C)_2CHCH_2CH_2O]_2P(S)OH$

$C_5H_{23}O_3PS$ M 194.269

NH_4 salt: Solid. Mp 167°.
K salt: Cryst. ($CHCl_3$).

Foss, O., *Acta Chem. Scand.*, 1947, **1**, 8 (*synth*)
Pesin, V.G. *et al*, *Zh. Obshch. Khim.*, 1961, **31**, 2508 (*Engl. transl.* p. 2337) (*synth*)

2,4-Bis(1-methylethylidene)-1,3,6,9-tetraoxa-5-phospha(5-P^V)spiro[4.4]nonane, 9CI **B-00306**

2,4-Diisopropylidene-1,3,6,9-tetraoxa-5-phospha(5-P^V)spiro[4.4]nonane

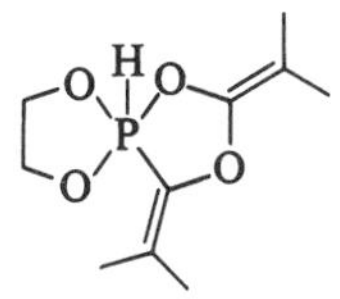

$C_{10}H_{17}O_4P$ M 232.216

5-Ph: [37521-15-6].
$C_{16}H_{21}O_4P$ M 308.313
Semisolid.

5-Dimethylamino: [37521-13-4]. *N,N-Dimethyl-2,4-bis(1-methylethylidene)-1,3,6,9-tetraoxa-5-phospha(5-P^V)spiro[4.4]nonan-5-amine. 5-Dimethylamino-2,4-diisopropylidene-1,3,6,9-tetraoxa-5-phospha(5-P^V)spiro[4.4]nonane.*
$C_{12}H_{22}NO_4P$ M 275.284
Liq. $Bp_{0.1}$ 68-70°.

Bentrude, W.G. *et al*, *J. Am. Chem. Soc.*, 1972, **94**, 3058 (*synth, ir, pmr, P nmr*)

Bis(1-methylheptyl) phosphonate **B-00307**

Di(2-octyl) phosphonate. Bis(1-methylheptyl) phosphite. Di(2-octyl) phosphite

$[H_3C(CH_2)_5CH(CH_3)O]_2P(O)H \rightleftharpoons [H_3C(CH_2)_5CH(CH_3)O]_2POH$

$C_8H_{35}O_3P$ M 210.336

(+)-*form*
Liq. $[\alpha]_D^{20}$ +7.98°.

(±)-*form*
Liq. $Bp_{0.6}$ 148-150°. n_D^{20} 1.4370.

Carlson, E.J. *et al*, *J. Chem. Soc. (B)*, 1969, 2364 (*synth, props*)

Bis(1-methylheptyl) phosphorochloridite, 9CI **B-00308**

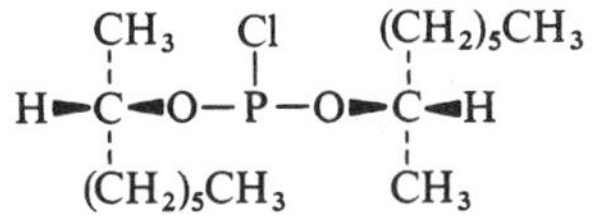

$C_{16}H_{34}ClO_2P$ M 324.870

(*S,S*)-*form*
Bp_2 135-142°. $[\alpha]_D^{18}$ +0.7° (neat). n_D^{20} 1.4430.

Gerrard, W., *J. Chem. Soc.*, 1944, 85.

Bis(2-methylphenyl)phenylphosphine, 9CI **B-00309**

Phenyldi-o-tolylphosphine

[18803-09-3]

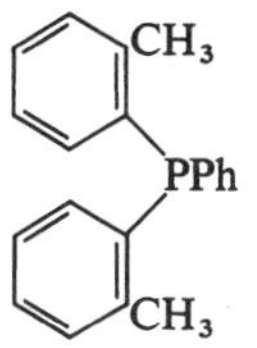

$C_{20}H_{19}P$ M 290.344

Ligand for Cr, Mo, Ni, Pt, Pd and W. Cryst. (EtOH). Mp 73°, 83°.

B,MeBr: Methylbis(2-methylphenyl)-phenylphosphonium bromide.
$C_{21}H_{22}BrP$ M 385.283
Solid. Mp 192-194°.

Oxide: [18803-11-7].
$C_{20}H_{19}OP$ M 306.343
Cryst. (C_6H_6 or EtOH aq.). Mp 135-136°, 193-194°.

Bennett, M.A. *et al*, *J. Am. Chem. Soc.*, 1969, **91**, 6266 (*synth, complexes*)
Grim, S.O. *et al*, *J. Org. Chem.*, 1977, **42**, 1236 (*derivs, nmr*)
Grim, S.O. *et al*, *Phosphorus Sulfur*, 1977, **3**, 191 (*nmr*)
Petrov, K.A. *et al*, *Zh. Obshch. Khim.*, 1978, **48**, 91 (*Engl. transl.* p. 74) (*oxide*)
Segall, Y. *et al*, *J. Am. Chem. Soc.*, 1978, **100**, 5130 (*synth, ir, pmr, ms, oxide*)
Kaplan, L.J. *et al*, *J. Org. Chem.*, 1979, **44**, 2226 (*oxide, synth*)

Bis(3-methylphenyl)phenylphosphine, 9CI **B-00310**

Phenyldi-m-tolylphosphine

[18803-08-2]

$C_{20}H_{19}P$ M 290.344

Additive to polyethylene electrical insulators. Cryst. (MeOH). Mp 53-53.5°. $Bp_{0.5}$ 210-220°.

Oxide: [18803-10-6].
$C_{20}H_{19}OP$ M 306.343
Additive to polyethylene insulators. Cryst. (pet. ether). Mp 108.5-109°.

Sulfide: [20676-76-0].
$C_{20}H_{19}PS$ M 322.404
Additive to polyethylene insulators.

Hart, F.A. *et al*, *J. Chem. Soc.*, 1955, 4107 (*synth, oxide*)

Bis(4-methylphenyl)phenylphosphine, 9CI **B-00311**

Phenyldi-p-tolylphosphine

[19934-95-3]

$C_{20}H_{19}P$ M 290.344

Additive to polyethylene electrical insulators. Cryst. (EtOH). Mp 56-57°. Bp_2 140-145°.

Oxide: [18957-70-5].
$C_{20}H_{19}OP$ M 306.343
Additive to polyethylene electrical insulators. Cryst. (heptane). Mp 81-83°. Forms hemihydrate Mp 91°.

Sulfide: [16822-31-4].
$C_{20}H_{19}PS$ M 322.404
Additive to polyethylene electrical insulators.

Morgan, P.W. *et al*, *J. Am. Chem. Soc.*, 1952, **74**, 4526 (*synth, oxide*)
Horner, L. *et al*, *Chem. Ber.*, 1968, **101**, 2899 (*synth*)
Timokhin, B.V. *et al*, *Zh. Obshch. Khim.*, 1977, **47**, 1267 (*Engl. transl.* p 1167) (*synth, nmr*)

Bis(2-methylphenyl) phosphate, 9CI, 8CI **B-00312**
Di-o-tolyl hydrogen phosphate. Di-o-tolyl phosphoric acid. Di-o-tolyl phosphate
[35787-74-7]

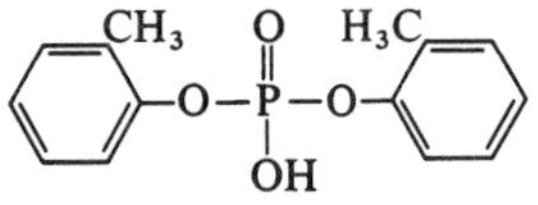

$C_{14}H_{15}O_4P$ M 278.244
▷TC0600000.
Chloride: [6630-13-3]. *Bis(2-methylphenyl) phosphorochloridate. Di-o-tolyl chlorophosphate.*
$C_{14}H_{14}ClO_3P$ M 296.690
Liq. d^{25} 1.25. Bp_{18} 234-236°, $Bp_{0.5}$ 158°. n_D^{25} 1.5373.

Bojars, N. *et al*, *CA*, 1968, **69**, 50951 (*chloride*)
Just, G. *et al*, *J. Prakt. Chem.*, 1971, **313**, 69 (*chloride*)

Bis(3-methylphenyl) phosphate, 9CI, 8CI **B-00313**
Di-m-tolyl hydrogen phosphate. Di-m-tolyl phosphoric acid. Di-m-tolyl phosphate
[36400-46-1]
$C_{14}H_{15}O_4P$ M 278.244
Viscous oil.
Chloride: [6630-14-4]. *Bis(3-methylphenyl) phosphorochloridate. Di-m-tolyl chlorophosphate.*
$C_{14}H_{14}ClO_3P$ M 296.690
Liq. Bp_1 176-180°.
Anhydride:
$C_{28}H_{28}O_7P_2$ M 538.473
Liq. $Bp_{0.0001}$ 180-190°.

Orloff, H.D. *et al*, *J. Am. Chem. Soc.*, 1958, **80**, 727 (*chloride*)
Boyden, J.W. *et al*, *Analyst (London)*, 1970, **95**, 935 (*derivs, synth, glc*)
Beever, W.H. *et al*, *J. Polym. Sci., Polym. Symp.*, 1978, **65**, 41 (*synth, pmr, P nmr*)
Reddy, C.D. *et al*, *Synthesis*, 1980, 1004 (*synth, pmr, P nmr, chloride*)

Bis(4-methylphenyl) phosphate, 9CI, 8CI **B-00314**
Bis(4-methylphenyl) hydrogen phosphate. Di-p-tolyl hydrogen phosphate. Di-p-tolyl phosphoric acid. Di-p-tolyl phosphate
[843-24-3]
$C_{14}H_{15}O_4P$ M 278.244
Cryst. Exists as dimer in cryst. phase.

Katzin, L.I. *et al*, *Spectrochim. Acta, Part A*, 1978, **34**, 51 (*ir*)
Gebert, E. *et al*, *J. Inorg. Nucl. Chem.*, 1981, **43**, 1451 (*cryst struct*)

Bis(2-methylphenyl)phosphine, 9CI **B-00315**
Di-o-tolylphosphine, 8CI
[29949-64-2]

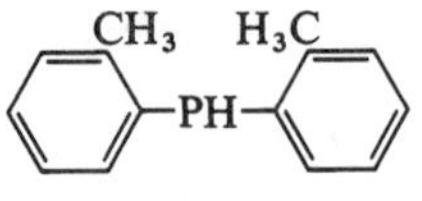

$C_{14}H_{15}P$ M 214.246
Odourous liq. Bp 306-307°.
▷Probably toxic
Oxide: [30309-80-9].
$C_{14}H_{15}OP$ M 230.246
Cryst. (Et_2O/heptane or toluene). Mp 94-95°. Tautomeric with Bis(2-methylphenyl)phosphinous acid, B-00323 .

Grim, S.O. *et al*, *J. Chem. Eng. Data*, 1970, **15**, 497 (*nmr*)
Suggs, J.L. *et al*, *J. Org. Chem.*, 1971, **36**, 2566 (*oxide, pmr*)
Grim, S.O. *et al*, *J. Org. Chem.*, 1977, **42**, 1236 (*Li salt, nmr*)
Petrov, K.A. *et al*, *Zh. Obshch. Khim.*, 1978, **48**, 91 (*Engl. transl.* p. 74) (*oxide*)

Bis(3-methylphenyl)phosphine, 9CI **B-00316**
Di-m-tolylphosphine, 8CI
[10177-78-3]
$C_{14}H_{15}P$ M 214.246
Odourous liq. $Bp_{0.01}$ 155°.
▷Toxic
Oxide: Tautomeric with Bis(3-methylphenyl)phosphinous acid, B-00324 .

Grim, S.O. *et al*, *J. Org. Chem.*, 1977, **42**, 1236 (*nmr*)

Bis(4-methylphenyl)phosphine, 9CI **B-00317**
Di-p-tolylphosphine, 8CI
[1017-60-3]
$C_{14}H_{15}P$ M 214.246
Ligand for Cr, Fe, Os, Rh, and Ru. Foul-smelling liq. Bp_2 121-122°. pK_a 24.4 (DMSO). Li deriv. exhibits strong yellow chemiluminescence.
Oxide: [2409-61-2].
$C_{14}H_{15}OP$ M 230.246
Solid. Mp 93-96°. pK_a 22.0 (DMSO, 25°). Tautomeric with Bis(4-methylphenyl)phosphinous acid, B-00325 .

Horner, L. *et al*, *Chem. Ber.*, 1968, **101**, 2899 (*synth*)
Strecker, R.A. *et al*, *J. Am. Chem. Soc.*, 1973, **95**, 210 (*Li salt, oxide*)
Grim, S.O. *et al*, *J. Org. Chem.*, 1977, **42**, 1236 (*nmr*)
Tsvetkov, E.N. *et al*, *Izv. Akad. Nauk SSSR, Ser. Khim.*, 1978, 1981 (*Engl. transl.* p. 1743) (*oxide*)
Terekhova, M.I. *et al*, *Zh. Obshch. Khim.*, 1982, **52**, 516 (*Engl. transl.* p. 452) (*props*)

Bis(2-methylphenyl)phosphinic acid, 9CI **B-00318**
Di-o-tolylphosphinic acid, 8CI
[18593-19-6]

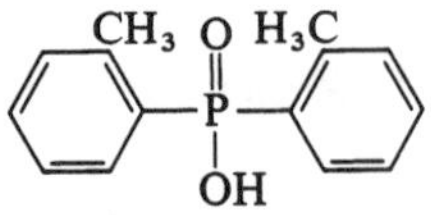

$C_{14}H_{15}O_2P$ M 246.245
Cryst. (EtOH or EtOH aq.). Mp 178°.
Me ester: [69928-10-5]. *Methyl bis(2-methylphenyl)-phosphinate. Methyl di-o-tolylphosphinate.*
$C_{15}H_{17}O_2P$ M 260.272

Oil.
Chloride: [59472-84-3].
$C_{14}H_{14}ClOP$ M 264.691
Solid. Mp 65-66°. Bp_5 207°, $Bp_{0.08}$ 150-160° (oven).
Amide: [59452-79-8].
$C_{14}H_{14}NOP$ M 243.244
Solid. Mp 156.5-157.5°.
Anhydride: [73720-00-0].
$C_{28}H_{28}O_3P$ M 443.501
Cryst. (EtOAc). Mp 194-196°.
Anilide: [59452-81-2]. P,P-*Bis(2-methylphenyl)*-N-*phenylphosphinic amide.* N-*Phenyl di*-o-*tolylphosphinic amide. Di*-o-*tolylphosphinic anilide.*
$C_{20}H_{20}NOP$ M 321.358
Solid. Mp 186-187°.

Haake, P. *et al*, *J. Org. Chem.*, 1969, **34**, 788 (*ms*)
Freedman, L.D. *et al*, *J. Org. Chem.*, 1971, **36**, 2566 (*synth, uv*)
Petrov, K.A. *et al*, *Zh. Obshch. Khim.*, 1978, **48**, 91 (*Engl. transl.* p. 74) (*synth, chloride*)
Kaplan, L.J. *et al*, *J. Org. Chem.*, 1979, **44**, 2226 (*synth, ir, pmr, derivs*)
Harger, M.J.P., *J. Chem. Soc., Perkin Trans. 2*, 1980, 154 (*synth, pmr, ir, derivs*)

Bis(3-methylphenyl)phosphinic acid, 9CI B-00319

Di-m-*tolylphosphinic acid*, 8CI
[57906-75-9]
$C_{14}H_{15}O_2P$ M 246.245
Cryst. (EtOH aq. or Me_2CO aq.). Mp 173.5-175°.
Chloride: [51103-89-0].
$C_{14}H_{14}ClOP$ M 264.691
Liq. d_4^{20} 1.20. Bp_1 178-180°. n_D^{20} 1.5978.
Azide: [57906-69-1].
$C_{14}H_{14}N_3OP$ M 271.258
Liq. $Bp_{0.01}$ 154-155°. n_D^{20} 1.5953.
Hydrazide:
$C_{14}H_{17}N_2OP$ M 260.275
Cryst. (C_6H_6 or C_6H_6/EtOH). Mp 122°.

Denham, J.M. *et al*, *J. Org. Chem.*, 1958, **23**, 1298 (*synth*)
Freedman, L.D. *et al*, *J. Org. Chem.*, 1959, **24**, 638 (*uv*)
Shandruk, M.I. *et al*, *Zh. Obshch. Khim.*, 1973, **43**, 2194 (*Engl. transl.* p. 2186) (*derivs*)
Weissbach, F. *et al*, *J. Prakt. Chem.*, 1975, **317**, 394 (*derivs*)

Bis(4-methylphenyl)phosphinic acid, 9CI B-00320

Di-p-*tolylphosphinic acid*
[1084-11-3]
$C_{14}H_{15}O_2P$ M 246.245
Can be employed in the sepn. of Co from Ni. Cryst. (Et_2O or EtOH aq.). Mp 135-136°. pK_a 2.47 (7% EtOH aq.), 3.66 (50% EtOH aq.), 4.45 (80% EtOH aq.), 4.32 (60% THF aq.), 9.36 (DMSO).
Me ester: [83470-30-8]. *Methyl bis(4-methylphenyl)-phosphinate. Methyl di*-p-*tolylphosphinate.*
$C_{15}H_{17}O_2P$ M 260.272
Liq. Mp 68-69°. $Bp_{0.5}$ 174° (oven).
Azide: [4129-18-4].
$C_{14}H_{14}N_3OP$ M 271.258
Liq. $Bp_{0.05}$ 155-160°. n_D^{20} 1.5760.
Chloride: [4129-40-2].
$C_{14}H_{14}ClOP$ M 264.691
Liq. d_4^{20} 1.19. Bp_3 221-223°, $Bp_{0.3}$ 194-196°. n_D^{20} 1.6022.
Amide: [57055-05-7].
$C_{14}H_{16}NOP$ M 245.260
Cryst. (C_6H_6). Mp 182-183°.
Anilide: [20846-04-2]. P,P-*Bis(4-methylphenyl)*-N-*phenylphosphinic amide.* N-*Phenyl di*-p-*tolylphosphinic amide. Di*-p-*tolylphosphinic anilide.*
$C_{20}H_{20}NOP$ M 321.358
Cryst. (EtOAc or EtOH/Et_2O). Mp 216-216.5°.
Hydrazide:
$C_{14}H_{17}N_2OP$ M 260.275
Cryst. (C_6H_6 or C_6H_6/EtOH). Mp 165°.

Freedman, L.D. *et al*, *J. Org. Chem.*, 1959, **24**, 638 (*synth, uv*)
Petrov, K.A. *et al*, *Zh. Obshch. Khim.*, 1960, **30**, 3000 (*Engl. transl.* p. 2972) (*synth*)
Shandruk, M.I. *et al*, *Zh. Obshch. Khim.*, 1973, **43**, 2194 (*Engl. transl.* p. 2186) (*chloride*)
Érré, É.A. *et al*, *Zh. Obshch. Khim.*, 1975, **45**, 1480 (*Engl. transl.* p. 1447) (*amide*)
Weissbach, F. *et al*, *J. Prakt. Chem.*, 1975, **317**, 394 (*chloride, azide*)
Harger, M.J.P., *J. Chem. Soc., Perkin Trans. 2*, 1980, 154 (*synth, pmr, amides*)
Harger, M.J.P. *et al*, *Tetrahedron*, 1982, **38**, 1511 (*derivs*)

1,3-Bis(methylphenylphosphino)propane B-00321

1,3-Propanediylbis[methylphenylphosphine], 10CI, 9CI.
Trimethylenebis[methylphenylphosphine], 8CI

$$MePhPCH_2CH_2CH_2PPhMe$$

$C_{17}H_{22}P_2$ M 288.308
Liq. $Bp_{4\text{-}5}$ 199-201°.
Disulfide: 1,3-Bis(methylphenylphosphinothioyl)-propane.
$C_{17}H_{22}P_2S_2$ M 352.428
Cryst. (MeOH). Mp 126-129°.

Issleib, K. *et al*, *Chem. Ber.*, 1963, **96**, 2186.

Bis(4-methylphenyl)phosphinothioic acid, 9CI B-00322

Di-p-*tolylphosphinothioic acid*
[60711-68-4]

H_3C–C_6H_4–P(=S)(OH)–C_6H_4–CH_3 ⇌ H_3C–C_6H_4–P(=O)(SH)–C_6H_4–CH_3

$C_{14}H_{15}OPS$ M 262.306
pK_a 2.14 (7% EtOH aq.), 7.90 (100% EtOH), 10.10 ($MeNO_2$).
Fluoride: [18975-70-7].
$C_{14}H_{14}FPS$ M 264.297
Liq. $Bp_{0.1}$ 170-173°.
Chloride: [16534-61-5].
$C_{14}H_{14}ClPS$ M 280.751
No phys. props. reported.

Maier, L., *Phosphorus*, 1974, **4**, 41 (*chloride, nmr*)
Kozachenko, A.G. *et al*, *Izv. Akad. Nauk SSSR, Ser. Khim.*, 1976, 1646 (*Engl. transl.* p. 1561) (*acid, props*)

Bis(2-methylphenyl)phosphinous acid, 9CI B-00323

Di-o-*tolylphosphinous acid*

(2-$CH_3C_6H_4$)$_2$P–OH

$C_{14}H_{15}OP$ M 230.246

The free acid exists as the phosphoryl tautomer. See Bis(2-methylphenyl)phosphine, B-00315 .

Et ester: [13685-43-3]. *Ethyl bis(2-methylphenyl)-phosphinite.*
$C_{16}H_{19}OP$ M 258.299
Liq.

Butyl ester: Butyl bis(2-methylphenyl)phosphinite.
$C_{18}H_{23}OP$ M 286.353
Liq. Bp_{2-3} 154°.

Chloride: [36042-94-1]. *Chlorodi-*o*-tolylphosphine.*
$C_{14}H_{14}ClP$ M 248.691
Solid. Mp 36°. Bp_7 179-183°, $Bp_{0.03}$ 120-122°.

Weinberg, K.G., *J. Org. Chem.*, 1975, **40**, 3586 (*chloride*)
McEwen, W.E. *et al*, *J. Am. Chem. Soc.*, 1978, **100**, 7304 (*chloride*)
Shvets, A.A. *et al*, *Zh. Obshch. Khim.*, 1978, **48**, 232 (*Engl. transl.* p. 208) (*chloride, butyl ester*)
Harger, M.J.P., *J. Chem. Soc., Perkin Trans. 2*, 1980, 154 (*synth*)
Clarke, P.W. *et al*, *J. Organomet. Chem.*, 1981, **217**, 51 (*chloride, synth, pmr, cmr, P nmr*)

Bis(3-methylphenyl)phosphinous acid B-00324

*Di-*m*-tolylphosphinous acid*
$C_{14}H_{15}OP$ M 230.246

Chloride: [13685-23-9]. *Chlorobis(3-methylphenyl)-phosphine.*
$C_{14}H_{14}ClP$ M 248.691
Oil. $Bp_{0.9}$ 135-138°.

Weinberg, K., *J. Org. Chem.*, 1975, **40**, 3586.

Bis(4-methylphenyl)phosphinous acid B-00325

*Di-*p*-tolylphosphinous acid*
[14655-66-4]
$C_{14}H_{15}OP$ M 230.246
The free acid exists in the tautomeric phosphoryl form. See Bis(4-methylphenyl)phosphine, B-00317 .

Me ester: [13917-52-7]. *Methyl bis(4-methylphenyl)-phosphinite.*
$C_{15}H_{17}OP$ M 244.272
Liq. d_4^{20} 1.06. Bp_3 155-156.5°. n_D^{20} 1.5912.

Et ester: [13648-75-4]. *Ethyl bis(4-methylphenyl)-phosphinite.*
$C_{16}H_{19}OP$ M 258.299
Liq. d_4^{20} 1.04. Bp_5 168.5-169.5°. n_D^{20} 1.5799.

Chloride: [1019-71-2]. *Chlorobis(4-methylphenyl)-phosphine.*
$C_{14}H_{14}ClP$ M 248.691
Viscous liq. d_4^{20} 1.14. Bp_7 175.5-177°, $Bp_{0.03}$ 125-127°.

Kharrasova, F.M. *et al*, *Zh. Obshch. Khim.*, 1966, **36**, 1987 (*Engl. transl.* p. 1979) (*chloride, esters, synth, ir*)
Weinberg, K.G., *J. Org. Chem.*, 1975, **40**, 3586 (*chloride*)
McEwen, W.E. *et al*, *J. Am. Chem. Soc.*, 1978, **100**, 7304 (*chloride*)

Bis(2-methylphenyl) phosphonate, 9CI B-00326

*Di-*o*-tolyl phosphonate. Bis(2-methylphenyl) phosphite. Di-*o*-tolyl phosphite*
[15516-38-8]

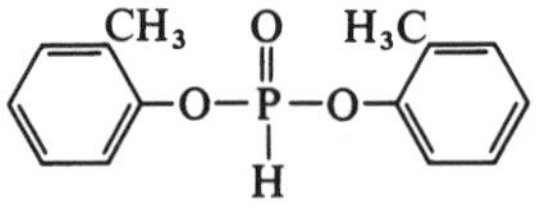

$C_{14}H_{15}O_3P$ M 262.244
Tautomeric. Liq. d_{25}^{25} 1.18. $Bp_{0.005}$ 100°. n_D^{25} 1.5495.

Houalla, D. *et al*, *C.R. Hebd. Seances Acad. Sci.*, 1958, **247**, 482 (*synth, ir*)
Walsh, E.N., *J. Am. Chem. Soc.*, 1959, **81**, 3023 (*synth*)

Bis(3-methylphenyl) phosphonate, 9CI B-00327

*Di-*m*-tolyl phosphonate. Bis(3-methylphenyl) phosphite. Di-*m*-tolyl phosphite*
[66076-07-1]
$C_{14}H_{15}O_3P$ M 262.244
Tautomeric. d_{25}^{25} 1.18. $Bp_{0.005}$ 90-100°. n_D^{25} 1.5468.

Houalla, D. *et al*, *C.R. Hebd. Seances Acad. Sci.*, 1958, **247**, 482 (*synth, ir*)
Walsh, E.N., *J. Am. Chem. Soc.*, 1959, **81**, 3023 (*synth*)

Bis(4-methylphenyl) phosphonate, 9CI B-00328

*Di-*p*-tolyl phosphonate. Bis(4-methylphenyl) phosphite. Di-*p*-tolyl phosphite*
[13869-19-7]
$C_{14}H_{15}O_3P$ M 262.244
Tautomeric. Solid. d_{25}^{25} 1.16. Mp 30°. $Bp_{0.005}$ 110°. n_D^{20} 1.547.

Houalla, D. *et al*, *C.R. Hebd. Seances Acad. Sci.*, 1958, **247**, 482 (*synth, ir*)
Walsh, E.N., *J. Am. Chem. Soc.*, 1959, **81**, 3023 (*synth*)

Bis(4-methylphenyl) phosphorochloridate, 9CI B-00329

*Bis(4-methylphenyl) phosphoryl chloride. Di-*p*-tolyl phosphoryl chloride. Di-*p*-tolyl chlorophosphate*
[6630-15-5]

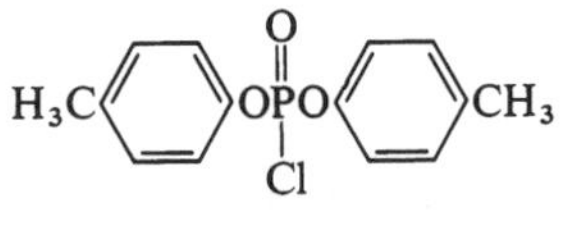

$C_{14}H_{14}ClO_3P$ M 296.690
Liq. $Bp_{1.25}$ 166-170°.

Orloff, H.D. *et al*, *J. Am. Chem. Soc.*, 1958, **80**, 727 (*synth*)
Just, G. *et al*, *J. Prakt. Chem.*, 1971, **313**, 69 (*synth*)
Reimschüssel, W. *et al*, *Int. J. Chem. Kinet.*, 1980, **12**, 979 (*synth, props*)

Bis(2-methylphenyl) phosphorochloridite, 9CI **B-00330**

*Di-*o*-tolyl phosphorochloridite, 8CI. Di-*o*-tolyl chlorophosphite*

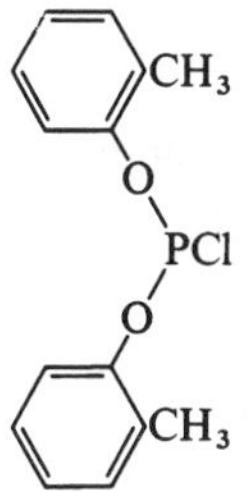

$C_{14}H_{14}ClO_2P$ M 280.690

Pungent oil. Bp_{11} 195-196°.

Strecker, W. *et al, Ber.*, 1916, **49**, 63 (*synth*)

Bis(3-methylphenyl) phosphorochloridite, 9CI **B-00331**

*Di-*m*-tolyl phosphorochloridite, 8CI. Di-*m*-tolyl chlorophosphite*

$C_{14}H_{14}ClO_2P$ M 280.690

Pungent liq. Bp_{11} 198°.

Broecker, W., *J. Prakt. Chem.*, 1928, **118**, 287 (*synth*)

Bis(4-methylphenyl) phosphorochloridite, 9CI **B-00332**

*Di-*p*-tolyl phosphorochloridite, 8CI. Di-*p*-tolyl chlorophosphite*

[20003-41-2]

$C_{14}H_{14}ClO_2P$ M 280.690

Pungent liq. d^{20} 1.20. Bp_{11} 206-208°, Bp_1 161-164°. n_D^{20} 1.5684.

Strecker, W. *et al, Ber.*, 1916, **49**, 63 (*synth, props*)
Petrov, K.A. *et al, Zh. Obshch. Khim.*, 1961, **31**, 2373 (*Engl. transl.* p. 2211) (*synth*)

***O,O*-Bis(2-methylphenyl) phosphorochloridothioate, 9CI** **B-00333**

O,O-*Di-*o*-tolyl phosphorochloridothioate, 8CI.* O,O-*Di-*o*-tolyl chlorothiophosphate.* O,O-*Di-*o*-tolyl thiophosphoryl chloride*

[81532-95-8]

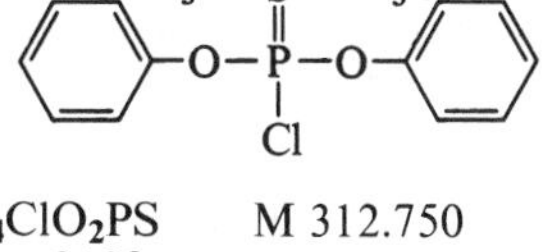

$C_{14}H_{14}ClO_2PS$ M 312.750

Oil. Bp_{11} 212°.

Strecker, W. *et al, Ber.*, 1916, **49**, 63 (*synth*)

***O,O*-Bis(3-methylphenyl) phosphorochloridothioate, 9CI** **B-00334**

O,O-*Di-*m*-tolyl phosphorochloridothioate, 8CI.* O,O-*Di-*m*-tolyl thiophosphoryl chloride.* O,O-*Di-*m*-tolyl chlorothiophosphate*

[81532-96-9]

$C_{14}H_{14}ClO_2PS$ M 312.750

Needles. Mp 33-34°. Bp_{11} 218°.

Broeker, W., *J. Prakt. Chem.*, 1928, **118**, 287; *CA*, **22**, 1964 (*synth*)

***O,O*-Bis(4-methylphenyl) phosphorochloridothioate, 9CI** **B-00335**

O,O-*Di-*p*-tolyl phosphorochloridothioate, 8CI.* O,O-*Di-*p*-tolyl chlorothiophosphate.* O,O-*Di-*p*-tolyl thiophosphoryl chloride*

[55526-69-7]

$C_{14}H_{14}ClO_2PS$ M 312.750

Cryst. (EtOH). Mp 54-55°.

Strecker, W. *et al, Ber.*, 1916, **49**, 63 (*synth*)
Autenrieth, W. *et al, Ber.*, 1925, **58**, 840 (*synth*)
Reimschüssel, W. *et al, Int. J. Chem. Kinet.*, 1980, **12**, 979 (*synth, props*)

***O,O*-Bis(4-methylphenyl) phosphorodiselenoate, 9CI, 8CI** **B-00336**

O,O-*Di-*p*-tolyl phosphorodiselenoic acid.* O,O-*Di-*p*-tolyl hydrogen phosphorodiselenoate.* O,O-*Bis(4-methylphenyl) phosphorodiselenoic acid*

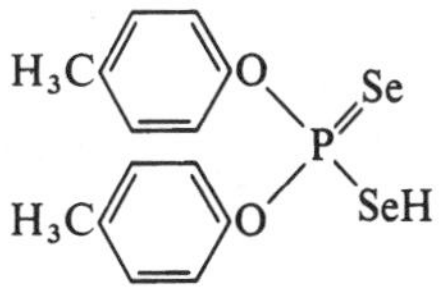

$C_{14}H_{15}O_2PSe_2$ M 404.165

Isol. as K salt.

K salt: [37443-15-5]. Cryst. (EtOH/pet. ether). Mp 105-106°.

Gorak, R.D. *et al, Zh. Obshch. Khim.*, 1972, **42**, 56 (*Engl. transl.* p. 52) (*synth*)
Zemlyanskii, N.I. *et al, Zh. Obshch. Khim.*, 1976, **46**, 1475 (*Engl. transl.* p. 1447) (*ir*)

***O,O*-Bis(2-methylphenyl) phosphorodithioate, 9CI** **B-00337**

O,O-*Di-*o*-tolyl phosphorodithioate, 8CI.* O,O-*Di-*o*-tolyl hydrogen dithiophosphate.* O,O-*Di-*o*-tolyl dithiophosphoric acid*

[34381-74-3]

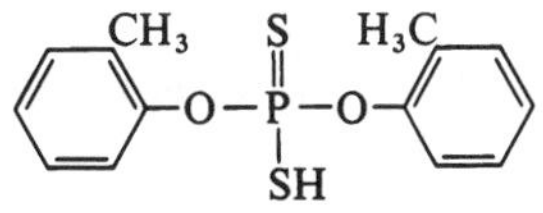

$C_{14}H_{15}O_2PS_2$ M 310.365

Oil. n_D^{25} 1.6028.

Triethylammonium salt: Solid. Mp 99-99.5°.

Komkov, I.P. *et al, CA*, 1968, **68**, 39241 (*synth*)
Mazitova, F.N. *et al, Zh. Obshch. Khim.*, 1980, **50**, 815 (*Engl. transl.* p. 652) (*synth, P nmr*)

***O,O*-Bis(3-methylphenyl) phosphorodithioate, 9CI** **B-00338**

O,O-*Di-*m*-tolyl phosphorodithioate, 8CI.* O,O-*Di-*m*-tolyl dithiophosphoric acid.* O,O-*Di-*m*-tolyl hydrogen dithiophosphate*

[7595-89-3]

$C_{14}H_{15}O_2PS_2$ M 310.365

Solid. Mp 45-47°. Best stored as K salt.

▷TE3740000.

K salt: Solid. Mp 205-207°.

Zemlyanskii, N.I. *et al, Zh. Obshch. Khim.*, 1966, **36**, 2193 (*Engl. transl.* p. 2187) (*synth*)

***O,O*-Bis(4-methylphenyl) phosphorodithioate, 9CI** **B-00339**

O,O-*Di-*p*-tolyl phosphorodithioate, 8CI.* O,O-*Di-*p*-tolyl hydrogen phosphorodithioate.* O,O-*Di-*p*-tolyl dithiophosphate. Di-*p*-tolyl dithiophosphoric acid*

[14366-43-9]

$C_{14}H_{15}O_2PS_2$ M 310.365

Solid. Mp 52-54°.

K salt: Cryst. (Me_2CO/C_6H_6). Mp 217-218°.

Et_3N salt: Solid. Mp 80-82°.

Zemlyanskii, N.I. *et al*, *Zh. Obshch. Khim.*, 1966, **36**, 2193 (*Engl. transl.* p. 2187) (*synth*)
Komkov, I.P. *et al*, *CA*, 1968, **68**, 39241 (*synth*)
Mazitova, F.N. *et al*, *Zh. Obshch. Khim.*, 1980, **50**, 815 (*Engl. transl.* p. 652) (*synth, P nmr*)
McCleverty, J.A. *et al*, *J. Chem. Soc., Dalton Trans.*, 1983, 627 (*salts, complexes, P nmr*)

Bis(1-methylpropyl)phenylphosphine, 9CI **B-00340**

*Di-*sec*-butylphenylphosphine*

[7650-80-8]

$PhP[CH(CH_3)CH_2CH_3]_2$

$C_{14}H_{23}P$ M 222.309

Constit. of catalysts for the hydrodimerization of 1,3-butadiene. Liq. Bp_2 107-110°. Presumably a mixt. of (±)- and *meso*-isomers.

Grim, S.O. *et al*, *J. Org. Chem.*, 1967, **32**, 781 (*synth, nmr*)
Grim, S.O. *et al*, *Phosphorus*, 1974, **4**, 189 (*synth*)

Bis(1-methylpropyl) phosphate, 9CI **B-00341**

*Di-*sec*-butyl phosphate. Di-*sec*-butyl hydrogen phosphate. Di-*sec*-butyl phosphoric acid*

[70069-20-4]

$[H_3CCH_2CH(CH_3)O]_2P(O)OH$

$C_8H_{19}O_4P$ M 210.209

*Fluoride: Bis(1-methylpropyl) phosphorofluoridate. Di-*sec*-butyl phosphorofluoridate. Di-*sec*-butyl fluorophosphate. Di-*sec*-butyl phosphoryl fluoride.*

$C_8H_{18}FO_3P$ M 212.201

Liq. $Bp_{0.8}$ 62-64°.

Chloride: [54757-38-9]. *Bis(1-methylpropyl) phosphorochloridate. Di-*sec*-butyl phosphorochloridate. Di-*sec*-butyl chlorophosphate. Di-*sec*-butyl phosphoryl chloride.*

$C_8H_{18}ClO_3P$ M 228.655

Liq. $Bp_{0.8}$ 84°. n_D^{20} 1.4280.

Anilide: see Phenylphosphoramidic acid, P-00257

Cook, H.G. *et al*, *J. Chem. Soc.*, 1949, 635 (*chloride, fluoride, synth, tox*)
De Roos, A.M. *et al*, *Recl. Trav. Chim. Pays-Bas*, 1958, **77**, 946 (*chloride*)

Bis(2-methylpropyl) phosphate, 9CI **B-00342**

Diisobutyl phosphate, 8CI. Bis(2-methylpropyl) hydrogen phosphate. Diisobutyl phosphoric acid

[6303-30-6]

$[(H_3C)_2CHCH_2O]_2P(O)OH$

$C_8H_{19}O_4P$ M 210.209

Oil. d^{20} 1.05. n_D^{20} 1.4248.

Chloride: [17158-87-1]. *Bis(2-methylpropyl) phosphorochloridate. Diisobutyl phosphoryl chloride.*

$C_8H_{18}ClO_3P$ M 228.655

Bp_{12} 111°, $Bp_{0.1}$ 57°. n_D^{25} 1.4262.

Anilide: see Phenylphosphoramidic acid, P-00257

Cook, H.G. *et al*, *J. Chem. Soc.*, 1945, 873 (*chloride, anilide*)
De Roos, A.M. *et al*, *Recl. Trav. Chim. Pays-Bas*, 1958, **77**, 946 (*chloride*)
Grosse-Ruyken, H. *et al*, *J. Prakt. Chem.*, 1962, **18**, 287 (*synth, chloride*)
Sosnovsky, G. *et al*, *J. Org. Chem.*, 1969, **34**, 968 (*chloride*)

Bis(1-methylpropyl)phosphine, 9CI **B-00343**

*Di-*sec*-butylphosphine, 8CI*

[2240-45-1]

$HP[CH(CH_3)CH_2CH_3]_2$

$C_8H_{19}P$ M 146.212

Liq. Mp −70°. Bp_{13} 85-86°, Bp_6 39-40°. Li deriv. exhibits weak light-blue chemiluminescence.

U.S.P., 2 893 202, (*1959*); *CA*, **53**, 22955 (*synth*)
Strecker, R.A. *et al*, *J. Am. Chem. Soc.*, 1973, **95**, 210 (*deriv*)
Grim, S.O. *et al*, *Phosphorus*, 1974, **4**, 189 (*nmr*)

Bis(2-methylpropyl)phosphine, 9CI **B-00344**

Diisobutylphosphine, 8CI

[4006-38-6]

$HP[CH_2CH(CH_3)_2]_2$

$C_8H_{19}P$ M 146.212

Liq. Bp 166-173°, Bp_8 47-48°. pK_a 4.11 (H_2O). n_D^{25} 1.4487.

Oxide:

$C_8H_{19}OP$ M 162.211

Cryst. Mp 31-33°. Bp_2 74-75°. Tautomeric with diisobutylphosphinous acid.

Sulfide:

$C_8H_{19}PS$ M 178.272

Cryst. (pet. ether). Mp 62.5-63.5°. Tautomeric with Bis(2-methylpropyl)phosphinothious acid, B-00352 .

Selenide: [5853-65-6].

$C_8H_{19}PSe$ M 225.172

Cryst. (pet. ether). Mp 77-78°. Tautomeric with diisobutylphosphinoselenous acid.

Rauhut, M.M. *et al*, *J. Org. Chem.*, 1961, **26**, 5138 (*synth*)
Kabachnik, M.I. *et al*, *Izv. Akad. Nauk SSSR, Ser. Khim.*, 1963, 1227 (*Engl. transl.* p. 1120) (*oxide*)
Maier, L., *Helv. Chim. Acta*, 1964, **47**, 2137 (*synth, nmr*)
Maier, L., *Helv. Chim. Acta*, 1966, **49**, 1000, 1249 (*selenide, sulfide, ir, nmr*)

Bis(1-methylpropyl)phosphinic acid, 9CI **B-00345**

*Di-*sec*-butylphosphinic acid*

[35210-27-6]

$[H_3CCH_2CH(CH_3)]_2P(O)OH$

$C_8H_{19}O_2P$ M 178.211

Prepns. presumably consist. of mixts. of the two possible diastereoisomers. Liq. $Bp_{0.03}$ 120°. pK_a 5.75 (75% EtOH aq., 25°). n_D^{20} 1.4610.

Chloride:

$C_8H_{18}ClOP$ M 196.656

$Bp_{0.1}$ 58°. n_D^{20} 1.4738.

Christen, P.J. *et al*, *Recl. Trav. Chim. Pays-Bas*, 1959, **78**, 543 (*chloride, synth*)
Blasse, R.G. *et al*, *Recl. Trav. Chim. Pays-Bas*, 1965, **84**, 267 (*chloride, ir*)
Cook, A.G. *et al*, *J. Org. Chem.*, 1972, **37**, 3342 (*synth, props*)
Katzin, L.I. *et al*, *Spectrochim. Acta, Part A*, 1978, **34**, 57 (*ir*)

Bis(2-methylpropyl)phosphinic acid, 9CI **B-00346**
Diisobutylphosphinic acid, 8CI
[15924-57-9]

$[(H_3C)_2CHCH_2]_2P(O)OH$

$C_8H_{19}O_2P$ M 178.211
Solid. Mp 46-47°. $Bp_{0.4}$ 140-141°. pK_a 3.70 (7% EtOH aq.), 5.63 (80% EtOH aq.).
2-Methylpropyl ester: 2-Methylpropyl bis(2-methylpropyl)phosphinate. Isobutyl diisobutylphosphinate.
$C_{12}H_{27}O_2P$ M 234.318
Liq. Bp_3 99-102°. n_D^{21} 1.4311.
Fluoride: [18358-25-3].
$C_8H_{18}FOP$ M 180.202
Characterised spectroscopically.
Chloride: [1118-29-2].
$C_8H_{18}ClOP$ M 196.656
Liq. d_4^{20} 1.02. Bp_2 110-110.5°. n_D^{20} 1.4595.
Amide: P,P-*Bis(2-methylpropyl)phosphinic amide. Diisobutylphosphinic amide.*
$C_8H_{20}NOP$ M 177.226
Solid. Mp 76-77° (sealed tube). Loses NH_3 on heating.

Razumov, A.I. *et al, Izv. Akad. Nauk SSSR, Ser. Khim.*, 1952, 894 (*Engl. transl.* p. 797); *CA*, **47**, 10466 (*ester*)
Razumov, A.I. *et al, Zh. Obshch. Khim.*, 1957, **27**, 754; *CA*, **51**, 16332 (*chloride*)
Christen, P.J. *et al, Recl. Trav. Chim. Pays-Bas*, 1959, **78**, 543 (*chloride*)
Petrov, K.A. *et al, Zh. Obshch. Khim.*, 1960, **30**, 2995 (*Engl. transl.* p. 2967) (*synth*)
Mukhacheva, O.A. *et al, Zh. Obshch. Khim.*, 1962, **32**, 2696 (*Engl. transl.* p. 2654) (*amide*)
Blasse, R.G., *Recl. Trav. Chim. Pays-Bas*, 1965, **84**, 267 (*chloride, ir*)
Cook, A.G. *et al, J. Org. Chem.*, 1972, **37**, 3342 (*synth, props*)
Fokin, A.V. *et al, Izv. Akad. Nauk SSSR, Ser. Khim.*, 1976, 2435 (*Engl. transl.* p. 2271) (*fluoride, nmr*)

Bis(1-methylpropyl)phosphinodithioic acid, 9CI **B-00347**
*Di-*sec-*butylphosphinodithioic acid*

$[H_3CCH_2CH(CH_3)]_2P(S)SH$

$C_8H_{19}PS_2$ M 210.332
Liq. $Bp_{0.15}$ 90°. Presumably a mixt. of stereoisomers.

Christen, P.J. *et al, Recl. Trav. Chim. Pays-Bas*, 1959, **78**, 161 (*synth, ir*)

Bis(2-methylpropyl)phosphinodithioic acid, 9CI **B-00348**
Diisobutylphosphinodithioic acid

$[(H_3C)_2CHCH_2]_2P(S)SH$

$C_8H_{19}PS_2$ M 210.332
Na salt acts as a flotation agent for Cu ores. Needles (heptane). Mp 39-40°.
Butyl ester: [7697-76-9]. *Butyl diisobutylphosphinodithioate.*
$C_{12}H_{27}PS_2$ M 266.439
Liq. $Bp_{0.05}$ 97-103°.

Peters, G., *J. Org. Chem.*, 1962, **27**, 2198 (*synth*)
Grayson, M. *et al, J. Org. Chem.*, 1967, **32**, 236 (*synth*)

Bis(1-methylpropyl)phosphinoselenoic acid, 9CI **B-00349**
*Di-*sec-*butylphosphinoselenoic acid*

$[H_3CCH_2CH(CH_3)]_2P(Se)OH \rightleftharpoons [H_3CCH_2CH(CH_3)]_2P(O)SeH$

$C_8H_{19}OPSe$ M 241.171
(±)-*form*
$Bp_{0.05}$ 82-83°. pK_a 2.98 (7% EtOH aq.), 5.56 (80% EtOH aq.). n_D^{20} 1.5430.

Markowska, A., *Bull. Acad. Pol. Sci., Ser. Sci. Chim.*, 1965, **13**, 149; *CA*, **63**, 9791.

Bis(1-methylpropyl)phosphinothioic acid, 9CI **B-00350**
*Di-*sec-*butylphosphinoithioic acid*

$[H_3CCH_2CH(CH_3)]_2P(S)OH \rightleftharpoons [H_3CCH_2CH(CH_3)]_2P(O)SH$

$C_8H_{19}OPS$ M 194.271
0.4% thiol form present in 7% EtOH aq., 0.1% in 80% EtOH aq. Solid. Mp 55.5-57°. $Bp_{0.8}$ 102-102.5°. pK_a 3.10 (7% EtOH aq.), 5.71 (80% EtOH aq.). Presumably a mixt. of *meso-* and (±)-diastereoisomers.

Kabachnik, M.I. *et al, Tetrahedron*, 1960, **9**, 10 (*synth, props*)

Bis(2-methylpropyl)phosphinothioic acid, 9CI **B-00351**
Diisobutylphosphinothioic acid

$[(H_3C)_2CHCH_2]_2P(S)OH \rightleftharpoons [(H_3C)_2CHCH_2]_2P(O)SH$

$C_8H_{19}OPS$ M 194.271
Solid. Mp 69.5-70.5°. pK_a 3.17 (7% EtOH aq.) (0.5% thiol form), 5.46 (80% EtOH aq.)(0% thiol form).
O-*Butyl ester:* O-*Butyl bis(2-methylpropyl)-phosphinothioate.* O-*Butyl diisobutylphosphinothioate.*
$C_{12}H_{27}OPS$ M 250.378
d_4^{20} 0.94. Bp_1 88-89°. n_D^{20} 1.4790.
Chloride: [4104-65-8].
$C_8H_{18}ClPS$ M 212.717
$Bp_{0.15}$ 69-72°. n_D^{20} 1.5108.
Bromide: [4652-20-4].
$C_8H_{13}BrPS$ M 252.128
$Bp_{0.05}$ 81-84°. n_D^{20} 1.5531.

Mastryokova, T.A. *et al, Zh. Obshch. Khim.*, 1959, **29**, 1450; *CA*, **54**, 9729 (*synth*)
Kabachnik, M.I. *et al, Dokl. Akad. Nauk SSSR*, 1960, **135**, 323; *CA*, **55**, 14288 (*ester*)
Kabachnik, M.I. *et al, Tetrahedron*, 1960, **9**, 10 (*props*)
Maier, L., *Helv. Chim. Acta*, 1964, **47**, 27 (*chloride, P nmr*)
Fokin, A.V. *et al, Izv. Akad. Nauk SSSR, Ser. Khim.*, 1976, 2435 (*Engl. transl.* p. 2271) (*halides, P nmr*)

Bis(2-methylpropyl)phosphinothious acid, 9CI **B-00352**
Diisobutylphosphinothious acid, 8CI

$[(H_3C)_2CHCH_2]_2PSH$

$C_8H_{19}PS$ M 178.272
Thought to exist in the tautomeric thiophosphoryl form; see under Bis(2-methylpropyl)phosphine, B-00344 .
Butyl ester: [7697-80-5]. *Butyl bis(2-methylpropyl)-phosphinothioite. Butyl diisobutylphosphinothioite.*
$C_{12}H_{27}PS$ M 234.379

Liq. $Bp_{0.17}$ 156-166°.
Ph ester: [7697-83-8]. *Phenyl bis(2-methylpropyl)-phosphinothioite. Phenyl diisobutylphosphinothioite.*
$C_{14}H_{23}PS$ M 254.369
Liq. $Bp_{0.12}$ 94-98°.

Grayson, M. *et al, J. Org. Chem.*, 1967, **32**, 236 (*esters, synth*)

Bis(1-methylpropyl) phosphonate, 9CI **B-00353**

*Di-*sec-*butyl phosphonate. Bis(1-methylpropyl) phosphite. Di-*sec-*butyl phosphite*

[2283-25-2]

$[H_3CCH_2CH(CH_3)O]_2P(O)H \rightleftharpoons [H_3CCH_2CH(CH_3)O]_2POH$

$C_8H_{19}O_3P$ M 194.210
Tautomeric.

(+)-***form***
Liq. $[\alpha]_D^{14.6}$ +26.0°. n_D^{20} 1.4182.
(−)-***form***
Liq. $Bp_{0.01}$ 41-43°. $[\alpha]_D^{27.3}$ −11.3°. n_D^{20} 1.4186.
(±)-***form***
$Bp_{0.7}$ 82.4°. n_D^{20} 1.4186. Prob. a mixt. of diastereoisomers.

Carlson, E.J. *et al, J. Chem. Soc.*, 1965, 2364 (*synth*)
Goodwin, D.G. *et al, J. Chem. Soc. (B)*, 1968, 1333 (*synth*)
Sekine, M. *et al, J. Org. Chem.*, 1980, **45**, 4162.

Bis(2-methylpropyl) phosphonate, 9CI **B-00354**

Diisobutyl phosphonate, 8CI. Bis(2-methylpropyl) phosphite. Diisobutyl phosphite

[1189-24-8]

$[(H_3C)_2CHCH_2O]_2P(O)H \rightleftharpoons [(H_3C)_2CHCH_2O]_2POH$

$C_8H_{19}O_3P$ M 194.210
Tautomeric; almost completely in phosphonate form. Extractant for rare earth metals from HNO_3 solns. Liq. Bp_7 108-109°. n_D^{20} 1.4195.

Coulson, E.J. *et al, J. Chem. Soc.*, 1965, 2364 (*synth*)

Bis(2-methylpropyl)phosphoramidic acid, 9CI **B-00355**

Diisobutylphosphoramidic acid

$[(H_3C)_2CHCH_2]_2NP(O)(OH)_2$

$C_8H_{20}NO_3P$ M 209.225
Diisopropyl ester: Diisopropyl diisobutylphosphoramidate. Bis(1-methylethyl) bis(2-methylpropyl)-phosphoramidate.
$C_{14}H_{32}NO_3P$ M 293.385
Liq. Bp_1 83°.
Diisobutyl ester: Bis(2-methylpropyl) bis(2-methylpropyl)phosphoramidate. Diisobutyl diisobutylphosphoramidate.
$C_{16}H_{36}NO_3P$ M 321.439
Liq. Bp_3 127°. n_D^{20} 1.4381.
Di-Ph ester: Diphenyl bis(2-methylpropyl)-phosphoramidate.
$C_{20}H_{28}NO_3P$ M 361.420
Cryst. (cyclohexane/EtOH). Mp 139-141°.

Ger. Pat., 1 033 200, (*1958*); *CA*, **54**, 14124 (*esters*)
Audrieth, L.F. *et al, J. Prakt. Chem.*, 1959, **8**, 117 (*diphenyl ester*)

***O,O*-Bis(2-methylpropyl) phosphorochloridoselenoate, 9CI** **B-00356**

O,O-*Diisobutyl phosphorochloridoselenoate.* O,O-*Diisobutyl chloroselenophosphate.* O,O-*Diisobutyl selenophosphoryl chloride*

[55205-95-3]

$[(H_3C)_2CHCH_2O]_2P(Se)Cl$

$C_8H_{18}ClO_2PSe$ M 291.616
Liq. d_4^{20} 1.26. $Bp_{0.04}$ 76°. n_D^{20} 1.4812. Sensitive to moisture.

Nuretdinov, I.A. *et al, Zh. Obshch. Khim.*, 1975, **45**, 533 (*Engl. transl.* p. 526) (*synth, ir, pmr, P nmr*)
Shagidullin, R.R. *et al, Izv. Akad. Nauk SSSR, Ser. Khim.*, 1976, 184 (*Engl. transl.* p. 174) (*uv*)

***O,O*-Bis(2-methylpropyl) phosphorochloridothioate, 9CI** **B-00357**

O,O-*Diisobutyl phosphorochloridothioate, 8CI.* O,O-*Diisobutyl thiophosphoryl chloride.* O,O-*Diisobutyl dichlorothiophosphate*

[51090-79-0]

$[(H_3C)_2CHCH_2O]_2P(S)Cl$

$C_8H_{18}ClO_2PS$ M 244.716
Liq. $Bp_{0.5}$ 76-82°. n_D^{25} 1.4624.

Fletcher, J.H. *et al, J. Am. Chem. Soc.*, 1950, **72**, 2461 (*synth*)

***N,N'*-Bis(2-methylpropyl)-phosphorodiamidodiselenoic acid** **B-00358**

N,N'-*Diisobutylphosphorodiamidodiselenoic acid*

$[(H_3C)_2CHCH_2NH]_2P(Se)SeH$

$C_8H_{21}N_2PSe_2$ M 334.161
Isobutylammonium salt: [19483-58-0]. Solid. Mp 164-166° dec.

Melton, R.G. *et al, Can. J. Chem.*, 1968, **46**, 1425 (*synth, ir, props*)

***O,O*-Bis(1-methylpropyl) phosphorodithioate, 9CI** **B-00359**

O,O-*Di-*sec-*butyl phosphorodithioate, 8CI.* O,O-*Di-*sec-*butyl hydrogen dithiophosphate.* O,O-*Di-*sec-*butyl dithiophosphoric acid*

[107-55-1]

$[H_3CCH_2CH(CH_3)O]_2P(S)SH$

$C_8H_{19}O_2PS_2$ M 242.331
K salt: [3287-85-2]. Selective flotation agent for Cu and Pb ores. Solid. Mp 134°. Presumably a mixt. of stereoisomers.

Bolotova, G.L. *et al, CA*, 1965, **63**, 6897.

***O,O*-Bis(2-methylpropyl) phosphorodithioate, 9CI** **B-00360**

O,O-*Diisobutyl phosphorodithioate, 8CI.* O,O-*Diisobutyl hydrogen dithiophosphate.* O,O-*Diisobutyl dithiophosphoric acid*

[2253-52-3]

$[(H_3C)_2CHCH_2O]_2P(S)SH$

$C_8H_{19}O_2PS_2$ M 242.331

Oil. d_4^{20} 1.06. $Bp_{1.5}$ 77.5-78°. pK_a 1.79 (7% EtOH aq.), pK_a 2.65 (80% EtOH aq.). n_D^{20} 1.4921.
K salt: [3549-52-8]. Solid. Mp 189-190°.

Kabachnik, M.I. *et al*, *Tetrahedron*, 1960, **9**, 10 (*synth*)
Almasi, L. *et al*, *CA*, 1965, **62**, 2729 (*synth*)
Bolotova, G.L. *et al*, *CA*, 1965, **63**, 6897 (*synth*)
Lefferts, J.L. *et al*, *Inorg. Chem.*, 1980, **19**, 1662 (*synth, complexes*)

S,S-Bis(2-methylpropyl) phosphorodithioate, 9CI **B-00361**

S,S-*Diisobutyl dithiophosphate.* S,S-*Diisobutyl hydrogen dithiophosphate.* S,S-*Diisobutyl dithiophosphoric acid*

$[(H_3C)_2CHCH_2S]_2P(O)OH$

$C_8H_{19}O_2PS_2$ M 242.331
Cyclohexylammonium salt: [72284-36-7]. Solid. Mp 185-187°.

Yamaguchi, K. *et al*, *Chem. Lett.*, 1979, 1057.

Bis(2-methylpropyl) phosphorofluoridate, 9CI **B-00362**

Diisobutyl phosphorofluoridate, 8CI. Diisobutyl phosphoryl fluoride. Diisobutyl fluorophosphate. Diisobutoxyphosphinyl fluoride
[563-22-4]

$[(H_3C)_2CHCH_2O]_2P(O)F$

$C_8H_{18}FO_3P$ M 212.201
Liq.

Sheluchenko, V.V. *et al*, *Dokl. Akad. Nauk SSSR, Ser. Sci. Khim.*, 1967, **177**, 376 (*Engl. transl.* p. 1050) (*F and P nmr*)
Landau, M.A. *et al*, *Zh. Strukt. Khim.*, 1970, **11**, 513 (*Engl. transl.* p. 467) (*struct*)
Ryzhikov, B.D. *et al*, *Zh. Strukt. Khim.*, 1975, **16**, 754 (*Engl. transl.* p. 700) (*F nmr*)

O,O-Bis(2-methylpropyl) phosphorothioate, 9CI **B-00363**

O,O-*Diisobutyl phosphorothioate.* O,O-*Diisobutyl hydrogen phosphorothioate.* O,O-*Diisobutyl phosphorothioic acid.* O,O-*Diisobutyl thiophosphoric acid*
[50614-16-8]

$[(H_3C)_2CHCH_2O]_2P(O)SH \rightleftharpoons [(H_3C)_2CHCH_2O]_2P(S)OH$

$C_8H_{19}O_3PS$ M 226.270
52% thiol form (7% EtOH aq.); 27% thiol form (80% EtOH aq.). Oil. d_4^{20} 1.04. $Bp_{0.05}$ 82.5-84°. n_D^{20} 1.4570.
NH_4 salt: Solid. Mp 172-176°.
*Chloride: see O,O-*Bis(2-methylpropyl) phosphorochloridothioate, B-00357

Kabachnik, M.I. *et al*, *Tetrahedron*, 1960, **9**, 10 (*synth, struct*)
Pesin, V.G. *et al*, *Zh. Obshch. Khim.*, 1961, **31**, 2508 (*Engl. transl.* p. 2337) (*synth*)

[Bis(methylthio)methyl]phosphonic acid, 9CI **B-00364**

$(MeS)_2CHP(O)(OH)_2$

$C_3H_9O_3PS_2$ M 188.196

Di-Me ester: [61779-87-1]. *Dimethyl [bis(methylthio)-methyl]phosphonate.*
$C_5H_{13}O_3PS_2$ M 216.250
Wittig-Horner reagent for the synth. of 1,1-bis(methylthio)-1-alkenes. Liq. $Bp_{0.05}$ 72°. n_D^{20} 1.5212.
Di-Et ester: [62999-70-6]. *Diethyl [bis(methylthio)-methyl]phosphonate.*
$C_7H_{17}O_3PS_2$ M 244.303
Wittig-Horner reagent for the synth. of 1,1-bis(methylthio)-1-alkenes. Liq. $Bp_{0.01}$ 78°. n_D^{20} 1.5030, 1.5200.

Mlotkowska, B. *et al*, *J. Prakt. Chem.*, 1977, **319**, 17 (*esters, synth, pmr, P nmr, cmr*)
Mikolajczyk, M. *et al*, *Tetrahedron*, 1978, **34**, 3081 (*use*)
Mikolajczyk, M. *et al*, *J. Org. Chem.*, 1979, **44**, 2967 (*synth, pmr, P nmr, use*)
Mikolajczyk, M. *et al*, *Synthesis*, 1980, 127 (*synth, use, pmr*)

Bis(4-methyl-2-thioxo-1,3,2-dioxaphosphorinan-2-yl) oxide **B-00365**

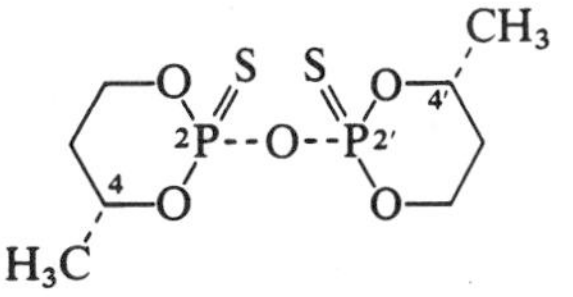

(2*RS*,2'*RS*,4*RS*,4'RS)-*form*

$C_8H_{16}O_5P_2S_2$ M 318.279
6 diastereoisomers characterised.

(2RS,2'RS,4RS,4'RS)-form
(±)-trans-trans-*isomer* b
Obt. only as a mixt. with *trans-trans*-isomer *a* (see below). Racemate.

(2RS,2'SR,4RS,2'SR)-form
trans-trans-*isomer* a
Mp 91-92.5°. Mp. refers to pure diastereoisomer obt. by crystallisation. Achiral.

(2RS,2'SR,4RS,4'RS)-form
cis-trans-*isomer* b
Obt. only as a mixt. with *cis-trans*-isomer *a* (see below). Racemate.

(2RS,2'RS,4RS,4'SR)-form
cis-trans-*isomer* a
Mp 80-84°. Obt. only as a mixt. with *cis-trans*-isomer *b*. Racemate.

(2RS,2'RS,4SR,4'SR)-form
cis,cis-*isomer* b
Obt. only as a mixt. with *cis-cis* isomer *a* (see below). Racemate.

(2RS,2'SR,4SR,4'RS)-form
cis-cis-*isomer* a
Mp 159-162°. Mp refers to pure diastereoisomer obt. by crystallisation. Achiral.

Mikolajczyk, M. *et al*, *J. Chem. Soc., Perkin Trans. 2*, 1983, 501 (*synth, ir, pmr, P nmr, cryst struct*)

Bis[(2-nitrophenyl)methyl] phosphate, 9CI **B-00366**

Di-o-nitrobenzyl phosphate, 8CI. Di-o-nitrobenzyl hydrogen phosphate

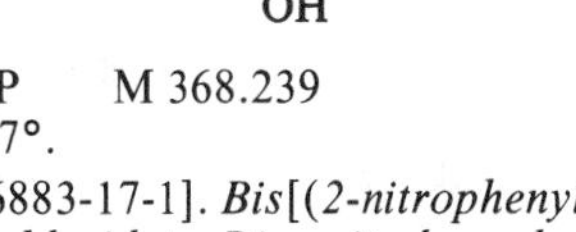

$C_{14}H_{13}N_2O_8P$ M 368.239
Solid. Mp 137°.
Chloride: [56883-17-1]. *Bis[(2-nitrophenyl)methyl] phosphorochloridate. Di-o-nitrobenzyl phosphorochloridate. Di-o-nitrobenzyl phosphoryl chloride.*
$C_{14}H_{12}ClN_2O_7P$ M 386.685

Phosphorylating agent. Solid. Mp 83-84°. Light-sensitive.

Rubinstein, M. *et al*, *Tetrahedron*, 1975, **31**, 2107 (*synth, pmr, chloride*)
Rapi, G. *et al*, *Gass. Chim. Ital.*, 1978, **108**, 135 (*chloride, use*)

Bis[(4-nitrophenyl)methyl] phosphate, 9CI **B-00367**

*Di-*p*-nitrobenzyl phosphate. Di-*p*-nitrobenzyl phosphoric acid. Di-*p*-nitrobenzyl hydrogen phosphate*

$C_{14}H_{13}N_2O_8P$ M 368.239

Yellow cryst. (EtOH). Mp 175°.

Chloride: [57188-46-2]. *Bis[(4-nitrophenyl)methyl] phosphorochloridate. Di-*p*-nitrobenzyl phosphoryl chloride.*

$C_{14}H_{12}ClN_3O_7P$ M 400.691

Phosphorylating agent. No phys. props. reported.

*Anhydride: Tetrakis[(4-nitrophenyl)methyl] diphosphate. Tetrakis-*p*-nitrobenzyl pyrophosphate.*

$C_{28}H_{24}N_4O_{15}P_2$ M 718.463

Phosphorylating agent. Prisms (EtOH). Mp 105-107°.

Baddiley, J. *et al*, *J. Chem. Soc.*, 1949, 815 (*synth*)
Zervas, L. *et al*, *Chem. Ber.*, 1956, **89**, 925 (*synth, derivs*)
Rubenstein, M. *et al*, *Tetrahedron*, 1975, **31**, 2107 (*synth, deriv*)
Rapi, G. *et al*, *Gazz. Chim. Ital.*, 1978, **108**, 135 (*chloride, use*)

Bis[(4-nitrophenyl)methyl]phosphonate, 8CI **B-00368**

*Di-*p*-nitrobenzyl phosphonate, 8CI. Bis[(4-nitrophenyl)-methyl] phosphite. Di-*p*-nitrobenzyl phosphite*

[65463-54-9]

O_2N–C₆H₄–$CH_2OP(O)(H)OCH_2$–C₆H₄–NO_2

$C_{14}H_{13}N_2O_{11}P$ M 416.237

Tautomeric. Cryst. ($CHCl_3$/cyclohexane). Mp 75°.

Fölsch, G., *Acta Chem. Scand.*, 1956, **10**, 686 (*synth, props*)

Bis(2-nitrophenyl) phosphate, 9CI **B-00369**

Bis(2-nitrophenyl) hydrogen phosphate. Bis(2-nitrophenyl) phosphoric acid

[78993-09-6]

$C_{12}H_9N_2O_8P$ M 340.185

Cryst. ($CHCl_3$). Mp 166-167°.

Kirby, A.J. *et al*, *J. Chem. Soc. (B)*, 1970, 510 (*synth, props*)

Bis(3-nitrophenyl) phosphate, 9CI, 8CI **B-00370**

Bis(3-nitrophenyl) hydrogen phosphate. Bis(3-nitrophenyl) phosphoric acid

[28022-29-9]

$C_{12}H_9N_2O_8P$ M 340.185

Cryst. ($CHCl_3$/pentane). Mp 109-110°.

Kirby, A.J. *et al*, *J. Chem. Soc. (B)*, 1970, 510 (*synth, props*)

Bis(4-nitrophenyl) phosphate, 9CI **B-00371**

Bis(4-nitrophenyl) hydrogen phosphate. Bis(4-nitrophenyl) phosphoric acid

[645-15-8]

$C_{12}H_9N_2O_8P$ M 340.185

Reagent for condensations, e.g. synth. of nucleoside acetonides. Yellow cryst. (EtOAc or H_2O). Mp 176-178°. pK_a 2.79 (H_2O).

Cyclohexylammonium salt: Solid. Mp 188-190°.

Chloride: see Bis(4-nitrophenyl) phosphorochloridate, B-00376

Anhydride: see Tetrakis(4-nitrophenyl) diphosphate, T-00183

Corby, N.S. *et al*, *J. Chem. Soc.*, 1952, 1234 (*synth*)
Moffatt, J.G. *et al*, *J. Am. Chem. Soc.*, 1957, **79**, 3741 (*synth, derivs, use*)
Stansley, P.G., *J. Org. Chem.*, 1958, **23**, 148 (*synth*)
Cabre-Castellvi, J. *et al*, *Afinidad*, 1982, **39**, 508; *CA*, **98**, 160816 (*synth, ir*)

Bis(4-nitrophenyl)phosphine, 9CI **B-00372**

O_2N–C₆H₄–PH–C₆H₄–NO_2

$C_{12}H_9N_2O_4P$ M 276.188

Yellow cryst. Mp 55-56°. Bp_{15} 210-220°.

Kosolapoff, G.M., *Organophosphorus Compds.*, 1950, John Wiley & Son, 1st Ed..

Bis(3-nitrophenyl)phosphinic acid, 9CI **B-00373**

[18593-20-9]

$C_{12}H_9N_2O_6P$ M 308.187

Solid. Mp 264-266.5°. pK_a 2.37 (80% EtOH aq.), 5.55 (DMSO).

▷SZ4400100.

NH_4 salt: Yellow prisms. Mp 260°.

Doak, G.O. *et al*, *J. Am. Chem. Soc.*, 1951, **73**, 5658 (*synth*)
Jaffé, H.H. *et al*, *J. Am. Chem. Soc.*, 1952, **74**, 1069 (*uv*)
Haake, P. *et al*, *J. Org. Chem.*, 1969, **34**, 788 (*ester, ms*)
Poskunova, O.G. *et al*, *Zh. Obshch. Khim.*, 1978, **48**, 1316 (*Engl. transl.* p. 1201) (*synth, props*)

Bis(4-nitrophenyl)phosphinic acid, 9CI **B-00374**

$C_{12}H_9N_2O_6P$ M 308.187

Solid. Mp 275-277°.

Doak, G.O. *et al*, *J. Am. Chem. Soc.*, 1951, **73**, 5658 (*synth*)
Jaffé, H.H. *et al*, *J. Am. Chem. Soc.*, 1952, **74**, 1069 (*uv*)

Bis(4-nitrophenyl) phosphonate, 9CI **B-00375**

Bis(4-nitrophenyl) phosphite. Di(4-nitrophenyl) phosphonate. Di-p-*nitrophenyl phosphite*

[65463-64-1]

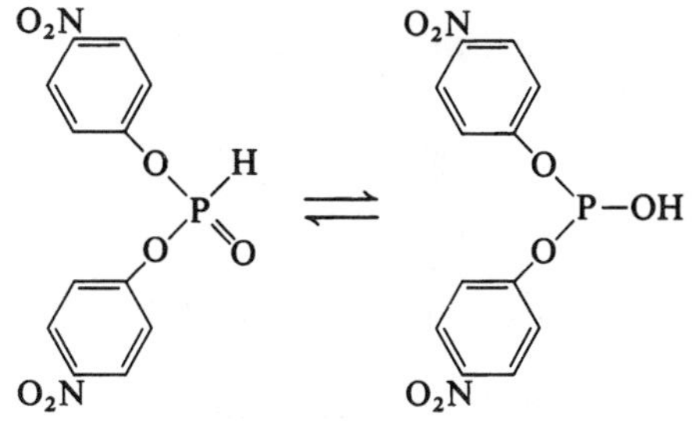

$C_{12}H_9N_2O_7P$ M 324.186

Tautomeric; almost completely in the phosphonate form. Solid. Mp 98-100°.

Ger. Pat., 1 078 136, (*1960*); *CA*, **56**, 3413

Bis(4-nitrophenyl) phosphorochloridate, 9CI, 8CI **B-00376**

Bis(4-nitrophenyl) phosphoryl chloride. Bis(4-nitrophenyl) chlorophosphate

[6546-97-0]

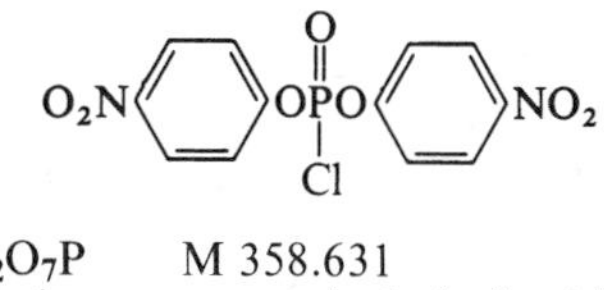

$C_{12}H_8ClN_2O_7P$ M 358.631

Phosphorylating agent, particularly for 5-$HOCH_2$ in otherwise protected carbohydrate. Greenish-yellow prisms ($CHCl_3$/pet. ether or $CHCl_3/C_6H_6$). Mp 94-95°.

Ukita, C. *et al*, *J. Am. Chem. Soc.*, 1962, **84**, 1879 (*synth, use*)
Murayama, A. *et al*, *J. Org. Chem.*, 1971, **36**, 3029 (*synth, use*)
Reimschüssel, W. *et al*, *Int. J. Chem. Kinet.*, 1980, **12**, 979 (*synth, props*)

***O,O*-Bis(4-nitrophenyl) phosphorochloridothioate, 9CI** **B-00377**

O,O-*Di*-p-*nitrophenyl phosphorochloridothioate, 8CI.* O,O-*Di*-p-*nitrophenyl chlorothiophosphate.* O,O-*Di*-p-*nitrophenyl thiophosphoryl chloride*

[52678-80-5]

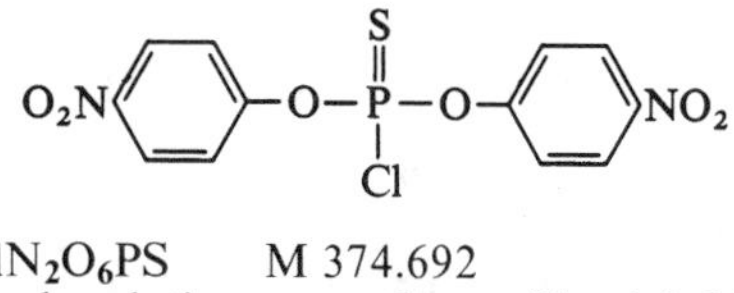

$C_{12}H_8ClN_2O_6PS$ M 374.692

Thiophosphorylating agent. Cryst. (ligroin). Mp 81-82°.

Morr, M. *et al*, *Chem. Ber.*, 1978, **111**, 2152 (*use*)
Reimschüssel, W. *et al*, *Int. J. Chem. Kinet.*, 1980, **12**, 979 (*synth, props*)

Bis(2-oxo-3-oxazolidinyl)phosphinic acid, 9CI **B-00378**

[79673-93-1]

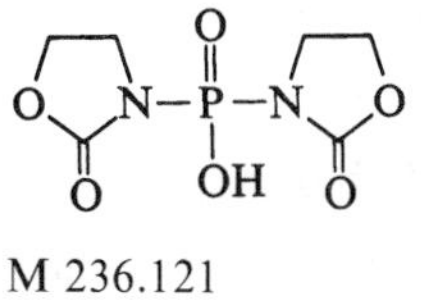

$C_6H_9N_2O_6P$ M 236.121

Cryst. (MeCN). Mp 153-154°. pK_a 3.34 (H_2O).

Chloride: [68641-49-6].

$C_6H_8ClN_2OP$ M 190.569

Converts carboxylic acids to anhydrides or (with amines) to amides, and imines to β-lactams. Employed in synth. of oligodeoxyribonucleotides. Cryst. (MeCN). Mp 195-198°.

Azide:

$C_6H_8N_5OP$ M 197.136

Converts carboxylic acids to anhydrides or (with amines added) to amides, and imines to β-lactams. Cryst. (EtOH or 1,2-dimethoxyethane). Mp 97-99°.

Anhydride: [85430-61-1].

$C_{12}H_{16}N_4O_{11}P_2$ M 454.226

Solid. Mp 212-213°.

Anhydride with diphenyl hydrogen phosphate: Bis(2-oxo-3-oxazolidinyl) diphenyl phosphoric anhydride.

$C_{18}H_{18}N_2O_9P$ M 437.322

Oil.

Diago-Meseguer, J. *et al*, *Synthesis*, 1980, 547 (*chloride, azide, use*)
Cabré-Castellví, J. *et al*, *Synthesis*, 1981, 616 (*chloride, use*)
Cabré-Castellví, J. *et al*, *Afinidad*, 1982, **39**, 508; *CA*, **98**, 160816 (*acid, anhydrides, salts*)
Shridhar, D.R. *et al*, *Synthesis*, 1982, 63 (*chloride, use*)
Katti, S.B. *et al*, *Tetrahedron Lett.*, 1985, **26**, 2547 (*use*)

Bis(pentachlorophenyl) phosphate **B-00379**

Bis(pentachlorophenyl) hydrogen phosphate. Bis(pentachlorophenyl) phosphoric acid

$(C_6Cl_5O)_2P(O)OH$

$C_{12}HCl_{10}O_4P$ M 594.641

Solid. Mp 320-325°.

Chloride: Bis(pentachlorophenyl) phosphorochloridate. Bis(pentachlorophenyl)phosphoryl chloride. Bis(pentachlorophenyl) chlorophosphate.

$C_{12}Cl_{11}O_3P$ M 613.087

Liq. $Bp_{0.5}$ 200-210°.

Bojars, N., *CA*, 1968, **68**, 50951 (*chloride*)
Cremlyn, R.J.W. *et al*, *J. Chem. Soc., Perkin Trans. 1*, 1972, 583 (*synth*)

Bis(pentafluorophenyl)phenylphosphine, 9CI **B-00380**

[5074-71-5]

$PhP(C_6F_5)_2$

$C_{18}H_5F_{10}P$ M 442.195

Calibration standard for ion-abundance scales in glc/ms. Ligand for Rh, Pt, and Pd. Cryst. Mp 69-70°. Protonated in SbF_5/HSO_3F.

Oxide: [5594-90-1].

$C_{18}H_5F_{10}OP$ M 458.195

Solid.

Sulfide: [19100-59-5].

$C_{18}H_5F_{10}PS$ M 474.255

Solid. Mp 110°.

Fild, M. *et al*, *Naturwissenschaften*, 1965, **52**, 590 (*synth, ir*)
Fild, M. *et al*, *Z. Naturforsch., B*, 1967, **22**, 248, 253 (*oxide, ir, nmr*)
Fild, M., *Z. Anorg. Allg. Chem.*, 1968, **358**, 57 (*sulfide, ir, nmr*)
Kemmitt, R.D.W. *et al*, *J. Chem. Soc. (A)*, 1968, 2149 (*synth*)
Hogben, M.G. *et al*, *J. Am. Chem. Soc.*, 1969, **91**, 283 (*nmr*)

Dua, S.S. *et al*, *J. Organomet. Chem.*, 1970, **24**, 703 (*uv, oxide*)
Eichelberger, J.W. *et al*, *Anal. Chem.*, 1975, **47**, 995 (*ms, use*)

Bis(pentafluorophenyl) phosphate, 9CI, 8CI **B-00381**

Bis(pentafluorophenyl) hydrogen phosphate. Bis(pentafluorophenyl) phosphoric acid

$(C_6F_5O)_2P(O)OH$

$C_{12}HF_{10}O_4P$ M 430.095

Cryst. (hexane). Mp 137-138°. pK_a 2.36 (50% EtOH aq.), pK_a 2.67 (80% EtOH aq.), pK_a 3.27 (EtOH), pK_a 4.02 ($MeNO_2$).

Chloride: [17788-08-8]. *Bis(pentafluorophenyl) phosphorochloridate. Bis(pentafluorophenyl) phosphoryl chloride. Bis(pentafluorophenyl) chlorophosphate.*
$C_{12}ClF_{10}O_3P$ M 448.541
Liq. Bp_{20} 181-183°.

U.S.P., 3 341 630, (*1967*); *CA*, **68**, 77961 (*derivs*)
Matveeva, A.G. *et al*, *Izv. Akad. Nauk SSSR, Ser. Khim.*, 1982, 1491 (*Engl. transl.* p. 1329) (*synth, props*)

Bis(pentafluorophenyl)phosphinic acid, 9CI **B-00382**

[60354-95-2]

$C_{12}HF_{10}O_2P$ M 398.096

Fluoride: [22474-69-7].
$C_{12}F_{11}OP$ M 400.088
Solid. Mp 82°. $Bp_{0.01}$ 100-105°.

Fild, M. *et al*, *J. Chem. Soc. (A)*, 1969, 840 (*synth, nmr*)
Jones, T.R.B. *et al*, *Org. Mass Spectrom.*, 1977, **12**, 317 (*ms*)

Bis(pentafluorophenyl)phosphinothioic acid **B-00383**

$(C_6F_5)_2P(S)OH \rightleftharpoons (C_6F_5)_2P(O)SH$

$C_{12}HF_{10}OPS$ M 414.157

O-*Me ester:* [19100-62-0]. O-*Methyl bis(pentafluorophenyl)phosphinothioate.*
$C_{13}H_3F_{10}OPS$ M 428.184
Solid. Mp 90°. $Bp_{0.1}$ 106-108°.

O-*Et ester:* [19100-63-1]. O-*Ethyl bis(pentafluorophenyl)phosphinothioate.*
$C_{14}H_5F_{10}OPS$ M 442.211
Solid. Mp 45°.

O-*Ph ester:* [19100-64-2]. O-*Phenyl bis(pentafluorophenyl)phosphinothioate.*
$C_{18}H_5F_{10}OPS$ M 490.255
Liq. $Bp_{0.2}$ 145-146°.

Fluoride: [53327-22-3].
$C_{12}F_{11}PS$ M 416.148
Liq. $Bp_{0.05}$ 79°.

Chloride: [53327-28-9].
$C_{12}ClF_{10}PS$ M 432.603
Liq. $Bp_{0.1}$ 90°.

Bromide:
$C_{12}BrF_{10}PS$ M 477.054
Liq. $Bp_{0.05}$ 105°.

Fild, M., *Z. Anorg. Allg. Chem.*, 1968, **358**, 257 (*esters, ir, nmr*)
Fild, M. *et al*, *Z. Naturforsch., B*, 1974, **29**, 206 (*halides, nmr*)
Jones, T.R.B. *et al*, *Org. Mass Spectrom.*, 1977, **12**, 317 (*halides, ms*)

Bis(phenoxymethyl)phosphinic acid, 9CI **B-00384**

[21993-10-2]

$(PhOCH_2)_2P(O)OH$

$C_{14}H_{15}O_4P$ M 278.244

Cryst. (H_2O). Mp 131-132°.

Et ester: [21993-11-3]. *Ethyl bis(phenoxymethyl)phosphinate.*
$C_{16}H_{19}O_4P$ M 306.297
Liq. d_4^{20} 1.19. $Bp_{0.0001}$ 170°. n_D^{20} 1.5548.

Zyablikova, T.A. *et al*, *Izv. Akad. Nauk SSSR, Ser. Khim.*, 1969, 373 (*Engl. transl.* p. 324)
Mukhamzyanova, É.Kh. *et al*, *Izv. Akad. Nauk SSSR, Ser. Khim.*, 1969, 951 (*Engl. transl.* p. 870) (*synth, ester*)
Bel'skii, V.E. *et al*, *Zh. Obshch. Khim.*, 1974, **44**, 2657 (*Engl. transl.* p. 2612) (*nmr*)

Bis(4-phenyl-1,3-butadienyl)phosphinic acid, 9CI **B-00385**

$(PhCH{=}CHCH{=}CH)_2P(O)OH$

$C_{20}H_{19}O_2P$ M 322.343

Plates. Mp 201-202°.

Ph ester: Phenyl bis(4-phenyl-1,3-butadienyl)phosphinate.
$C_{26}H_{23}O_2P$ M 398.440
Microcryst. powder. Mp 144-144.5°.

Chloride:
$C_{20}H_{18}ClOP$ M 340.788
Red oil which solidifies. Mp 114-116°.

Anilide: N-*Phenyl*-P,P-*bis(4-phenyl-1,3-butadienyl)phosphinic amide.*
$C_{26}H_{24}NOP$ M 397.455
Prisms (EtOH). Mp 196-197°.

Fedorova, G.K. *et al*, *Zh. Obshch. Khim.*, 1966, **36**, 1262 (*Engl. transl.* p. 1278) (*synth, derivs*)

Bis(2-phenylethenyl)phosphinic acid, 9CI **B-00386**

Distyrylphosphinic acid. Bis(2-phenylvinyl)phosphinic acid

[4895-75-4]

$(PhCH{=}CH)_2P(O)OH$

$C_{16}H_{15}O_2P$ M 270.267

Solid. Mp 157-158°.

Butylammonium salt: [65181-55-7]. Cryst. (dioxane). Mp 194-196°.

Chloride: [5003-99-6].
$C_{16}H_{14}ClOP$ M 288.713
Cryst. (Et_2O/pet. ether). Mp 78-80°. $Bp_{0.09}$ 229-231°.

Fedorova, G.K. *et al*, *Zh. Obshch. Khim.*, 1967, **37**, 2686 (*Engl. transl.* p. 2557) (*synth*)
Galeev, N.S. *et al*, *Zh. Obshch. Khim.*, 1972, **42**, 1496 (*Engl. transl.* p. 1487) (*chloride*)
Fedorova, G.K. *et al*, *Zh. Obshch. Khim.*, 1977, **47**, 2205 (*Engl. transl.* p. 2013) (*synth*)

Bis(2-phenylethyl)phosphine **B-00387**

$HP(CH_2CH_2Ph)_2$

$C_{16}H_{19}P$ M 242.300
Obt. from styrene and PH_3. Liq. $Bp_{0.5}$ 158°. pK_a 3.46 (H_2O). n_D^{25} 1.5815.

Rauhut, M.M. *et al*, *J. Org. Chem.*, 1961, **26**, 5138 (*synth*)

Bis(phenylethynyl)phosphinic acid, 9CI **B-00388**

[4895-86-7]

(PhC≡C)$_2$P(O)OH

$C_{16}H_{11}O_2P$ M 266.235
Prisms (C_6H_6). Mp 146-147°.
Et ester: [25411-71-6]. *Ethyl bis(phenylethynyl)-phosphinate.*
$C_{18}H_{15}O_2P$ M 294.289
Greenish, viscous liq. $Bp_{0.06}$ 190-193°.
Ph ester: Phenyl bis(phenylethynyl)phosphinate.
$C_{22}H_{15}O_2P$ M 342.333
Solid. Mp 83-84°.
Anilide: N-*Phenyl*-P,P-*bis(phenylethynyl)phosphinic amide.* P,P-*Bis(phenylethynyl)phosphinic anilide.*
$C_{22}H_{16}NOP$ M 341.348
Prisms (EtOH). Mp 179-179.5°.

Fedorova, G.K. *et al*, *Zh. Obshch. Khim.*, 1965, **35**, 1984 (*Engl. transl.* p. 1975) (*synth, derivs*)
Kirsanov, A.V. *et al*, *Zh. Obshch. Khim.*, 1969, **39**, 2596 (*Engl. transl.* p. 2535) (*ethyl ester, props*)
Fedorova, G.K. *et al*, *Zh. Obshch. Khim.*, 1974, **44**, 85 (*Engl. transl.* p. 83) (*synth*)
Fedorova, G.H. *et al*, *Zh. Obshch. Khim.*, 1977, **47**, 951 (*Engl. transl.* p. 866) (*props*)

[Bis(phenylmethoxy)phosphinyl]acetic acid, 9CI **B-00389**

Bis[(benzyloxy)phosphinyl]acetic acid. [Di(benzyloxy)-phosphinyl]acetic acid
[53243-58-6]

(PhCH$_2$O)$_2$P(O)CH$_2$COOH

$C_{16}H_{17}O_5P$ M 320.281
Wittig-Horner reagent for carboxyvinylations. Oil or solid. Insol. H_2O. Mp 58-59°.

Koppel, G.A. *et al*, *Tetrahedron Lett.*, 1974, 711 (*synth, pmr, use*)
Alewood, P.F. *et al*, *Synthesis*, 1984, 403 (*synth, ir, pmr, P nmr*)

1,2-Bis(phenylphosphino)ethane **B-00390**

1,2-Ethanediylbis[phenylphosphine], *9CI*. *Ethylenebis-[phenylphosphine]*, *8CI*
[18899-64-4]

PhPHCH$_2$CH$_2$PHPh

$C_{14}H_{16}P_2$ M 246.228
Air-sensitive liq. $Bp_{0.8}$ 161-162°.
(*RS,RS*)-*form*
(±)-*form*
Dioxide: [80358-25-4].
$C_{14}H_{16}O_2P_2$ M 278.227
Solid. Mp 105-106° or 156°. Config. of the two dioxides undetermined.
Disulfide: [80358-19-6].
$C_{14}H_{16}P_2S_2$ M 310.348
Cryst. (C_6H_6/Et_2O). Mp 105-106°.
(*RS,SR*)-*form*
meso-*form*
Dioxide: [73210-72-7]. Solid. Mp 156° or 105-106°.
Disulfide: [80358-18-5]. Solid. Mp 182-183°.

Issleib, K. *et al*, *Chem. Ber.*, 1963, **96**, 279; 1968, **101**, 2197 (*synth*)
Issleib, K. *et al*, *Z. Anorg. Allg. Chem.*, 1974, **406**, 178 (*deriv, use*)
King, R.B. *et al*, *J. Am. Chem. Soc.*, 1975, **97**, 46 (*props*)
Issleib, K. *et al*, *Z. Chem.*, 1979, **19**, 417 (*dioxide*)
Grossmann, G. *et al*, *Phosphorus Sulfur*, 1981, **11**, 259 (*derivs, pmr, nmr, ir*)
Andriamizake, J.D. *et al*, *Phosphorus Sulfur*, 1982, **12**, 265 (*deriv, use*)

1,4-Bis(phosphonomethyl)biphenyl, 8CI **B-00391**

[[1,1'-Biphenyl]-4,4'-diylbis(methylene)]bis(phosphonic acid), *9CI*

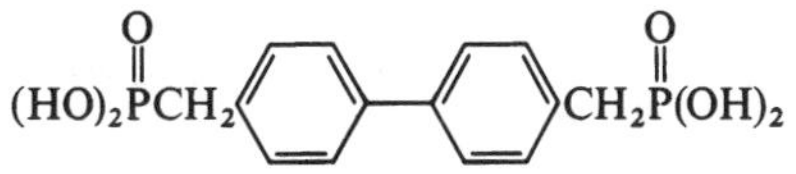

(HO)$_2$P(O)CH$_2$–C$_6$H$_4$–C$_6$H$_4$–CH$_2$P(O)(OH)$_2$

$C_{14}H_{16}O_6P_2$ M 342.224
Solid. Mp >500°.
Tetra-Et ester: Tetraethyl [[1,1'-biphenyl]-4,4'-diylbis(methylene)]bisphosphonate.
$C_{22}H_{32}O_6P_2$ M 454.439
Solid. Mp 109-110°.

Abramov, V.S. *et al*, *Zh. Obshch. Khim.*, 1967, **37**, 2243 (*Engl. transl.* p. 2129)

1,2-Bis[1-(2,2,6,6-tetramethylpiperidino)]-diphosphene **B-00392**

1,1'-(1,2-Diphosphenediyl)bis(2,2,6,6-tetramethylpiperidine), *11CI*

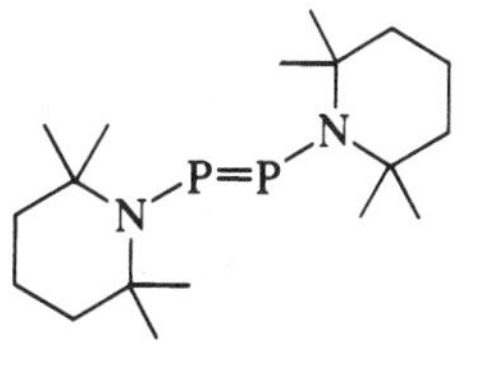

$C_{18}H_{36}N_2P_2$ M 342.443
(*E*)-*form* [89982-50-3]
Ruby-red liq. Sol. org. solvents. Stable at 0° in inert atmos. Monomeric.

Romanenko, V.D. *et al*, *Zh. Obshch. Khim.*, 1984, **54**, 465 (*Engl. transl.* p. 415) (*synth, pmr, P nmr*)

1,4-Bis(2,4,6-tri-*tert*-butylphenyl)-1,4-bis-(trimethylsilyloxy)-2,3-diphosphabutadiene **B-00393**

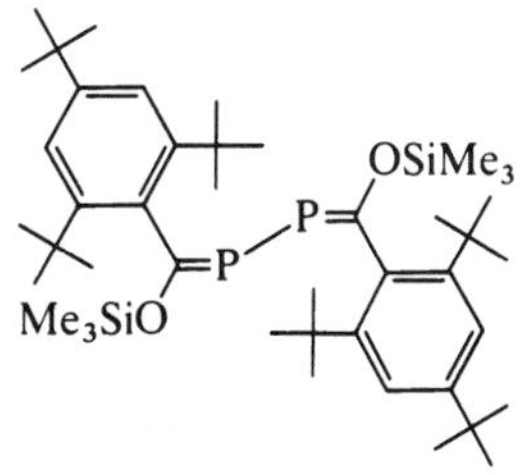

$C_{44}H_{76}O_2P_2Si_2$ M 755.202
Golden needles (MeCN). Mp 134-136.5°.

Märkl, G. *et al*, *Tetrahedron Lett.*, 1986, **27**, 171 (*synth, ms, pmr, cmr, P nmr*)

P,P'-Bis(2,4,6-tri-*tert*-butylphenyl)-1,3-diphosphaallene **B-00394**

Methanetetraylbis[2,4,6-tris(1,1-dimethylethyl)phenyl]-phosphine, 9CI

[91425-19-3]

$C_{37}H_{58}P_2$ M 564.813

Cryst. (hexane or toluene). Mp 206-207° (202-204°). Stable to air and moisture.

Appel, R. *et al, Angew. Chem., Int. Ed. Engl.*, 1984, **23**, 619 (*synth, ms, pmr, cmr, P nmr*)

Karsch, H.H. *et al, Angew. Chem., Int. Ed. Engl.*, 1984, **23**, 618 (*cryst struct*)

Karsch, H.H. *et al, Tetrahedron Lett.*, 1984, **25**, 3687 (*synth, cmr, pmr, P nmr*)

Yoshifuji, M. *et al, J. Chem. Soc., Chem. Commun.*, 1984, 689 (*synth, uv, cmr, pmr, P nmr*)

Bis(2,4,6-tri-*tert*-butylphenyl)diphosphene **B-00395**

Bis[2,4,6-tris(1,1-dimethylethyl)phenyl]diphosphene, 9CI

$C_{36}H_{58}P_2$ M 552.802

Forms complexes contg. Co or Mo.

(E)-form [83466-54-0]

Orange-red, thermally stable solid, which can be handled in air (pentane). Mp 175-176°, 208-210°.

Oxide:

$C_{36}H_{58}OP_2$ M 568.801

Yellow cryst. (pentane). Mp 174-176°. Reactive to water.

Sulfide:

$C_{36}H_{58}P_2S$ M 584.862

Cryst. (pentane). Mp 151.5-152°.

Yoshifuji, M. *et al, J. Am. Chem. Soc.*, 1981, **103**, 4587 (*synth, struct*)

Bertrand, G. *et al, Tetrahedron Lett.*, 1982, 3567.

Cowley, A.H. *et al, J. Am. Chem. Soc.*, 1982, **104**, 5820 (*nmr, deriv*)

Eseudie, J. *et al, J. Organomet. Chem.*, 1982, **228**, C76 (*use*)

Cetinkaya, B. *et al, J. Electron. Spectrosc. Relat. Phenom.*, 1983, **32**, 133 (*pe, struct*)

Yoshifuji, M. *et al, Angew. Chem., Int. Ed. Engl.*, 1983, **22**, 418.

Yoshifuji, M. *et al, Chem. Lett.*, 1983, 585.

Yoshifuji, M. *et al, J. Chem. Soc., Chem. Commun.*, 1983, 419, 862 (*derivs, ir, uv, ms, pmr, P nmr, cryst struct*)

Yoshifuji, M. *et al, J. Am. Chem. Soc.*, 1983, **105**, 2495 (*synth, cmr*)

Yoshifuji, M. *et al, Phosphorus Sulfur*, 1983, **16**, 157 (*esca, synth*)

Cowley, A.H. *et al, Inorg. Chem.*, 1984, **23**, 2552 (*bibliog*)

Hamaguchi, H. *et al, J. Am. Chem. Soc.*, 1984, **106**, 508 (*raman*)

Caminade, A.M. *et al, Phosphorus Sulfur*, 1986, **26**, 91 (*props*)

2,3-Bis(2,4,6-tri-*tert*-butylphenyl)-1,2,3-selenadiphosphirane **B-00396**

2,3-Bis[2,4,6-tris(1,1-dimethylethyl)phenyl]-selenadiphosphirane, 11CI

[90599-65-8]

$C_{36}H_{58}P_2Se$ M 631.762

Pale-yellow cryst. Mp 133-135° dec. Stable at r.t. Sensitive to light.

Yoshifuji, M. *et al, Chem. Lett.*, 1984, 603 (*synth, uv, pmr, P nmr, props*)

2,3-Bis(2,4,6-tri-*tert*-butylphenyl)-thiadiphosphirane **B-00397**

[98060-75-4]

$C_{36}H_{58}P_2S$ M 584.862

(2RS,3RS)-form [85421-75-6]

(±)-trans-*form*

Solid. Mp 131.5-132°.

Yoshifuji, M. *et al, Angew. Chem., Int. Ed. Engl.*, 1983, **22**, 418 (*synth, uv, ms, pmr, P nmr, cryst struct*)

Yoshifuji, M. *et al, J. Chem. Soc., Chem. Commun.*, 1983, 862 (*synth, P nmr*)

1,4-Bis(2,4,6-tri-*tert*-butylphenyl)-3-trimethylsilyloxy-1,2,4-triphospha-1,3-butadiene **B-00398**

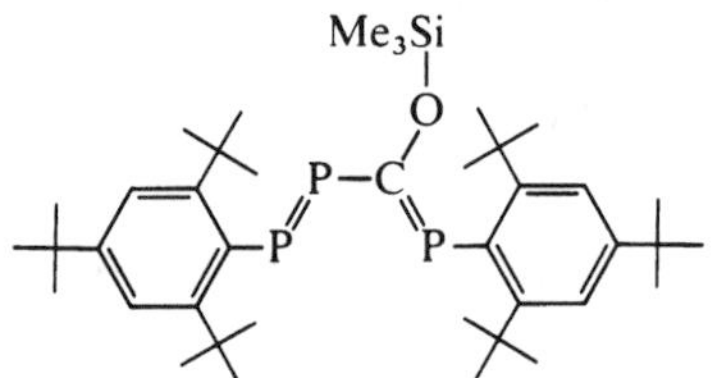

$C_{40}H_{67}OP_3Si$ M 684.975

First known compd. with conjugated P=P and P=C bonds. Red cryst. (dioxan/MeCN). Mp 168°.

Appel, R. *et al, Angew. Chem., Int. Ed. Engl.*, 1986, **25**, 932 (*synth, ms, pmr, P nmr, cryst struct*)

Bis(2,2,2-trichloro-1,1-dimethylethyl) phosphorochloridate **B-00399**

Bis(2-trichloromethyl-2-propyl) chlorophosphate

[17677-92-8]

$[Cl_3CC(CH_3)_2O]_2P(O)Cl$

$C_8H_{12}Cl_7O_3P$ M 435.326

Selective phosphorylating agent for nucleoside 5-OH groups. Solid. Mp 82°. $Bp_{0.0005}$ 140° (bath).

Kellner, H.A. *et al, Z. Naturforsch., B*, 1979, **34**, 1159 (*synth, pmr, P nmr*)

Kellner, H.A. *et al*, *Angew. Chem., Int. Ed. Engl.*, 1981, **20**, 577 (*synth, use*)
Schneiderwind-Stöcklein, R.G.K. *et al*, *Z. Naturforsch., B*, 1984, **39**, 968 (*synth, ir, P nmr*)

Bis(2,2,2-trichloroethyl) phosphate, 9CI — B-00400

Bis(2,2,2-trichloroethyl) hydrogen phosphate. Bis(2,2,2-trichloroethyl) phosphoric acid

$(Cl_3CCH_2O)_2P(O)OH$

$C_4H_5Cl_6O_4P$ M 360.773
Cryst. (hexane). Mp 81-84°.

Cyclohexylammonium salt: Solid. Mp 172-173°.
Chloride: [17672-53-6]. *Bis(2,2,2-trichloroethyl) phosphorochloridate. Bis(2,2,2-trichloroethyl) phosphoryl chloride.* Phosphorylating agent for nucleosides. Solid. Mp 48-49°.
Bromide: Bis(2,2,2-trichloroethyl) phosphorobromidate. Bis(2,2,2-trichloroethyl) phosphoryl bromide.
$C_4H_4BrCl_6O_3P$ M 423.670
Solid. Mp 51-53°.
Anilide: Bis(2,2,2-trichloroethyl) phenylphosphoramidate.
$C_{10}H_{10}Cl_6NO_3P$ M 435.886
Cryst. (pet. ether). Mp 135-136°.

Franke, A. *et al*, *Chem. Ber.*, 1968, **101**, 2998 (*chloride, use*)
Owen, G.R. *et al*, *J. Org. Chem.*, 1976, **41**, 3010 (*synth, chloride, use*)
Markowska, A. *et al*, *Bull. Acad. Pol. Sci., Ser. Sci. Chem.*, 1979, **27**, 115 (*synth*)
Markovskii, L.N. *et al*, *Zh. Obshch. Khim.*, 1980, **50**, 807 (*Engl. transl.* p. 644) (*derivs, synth, P nmr*)

Bis(2,2,2-trichloroethyl) phosphorochloridite, 9CI — B-00401

Bis(2,2,2-trichloroethyl) chlorophosphite
[41662-41-3]

$(Cl_3CCH_2O)_2PCl$

$C_4H_4Cl_7O_2P$ M 363.219
Useful reagent for phosphylation of nucleosides and oligonucleotides. Pungent liq. d_4^{25} 1.75. $Bp_{0.15}$ 98°. n_D^{22} 1.5167. Stable under dry conditions.

Gerrard, W. *et al*, *J. Chem. Soc.*, 1954, 1148 (*synth*)
Imai, J. *et al*, *J. Org. Chem.*, 1981, **46**, 4015 (*synth, use*)

Bis(trichloromethyl)phosphinic acid, 9CI — B-00402

[21089-19-0]

$(Cl_3C)_2P(O)OH$

$C_2HCl_6O_2P$ M 300.720
Cryst. + $1.5H_2O$ (C_6H_6), cryst. (CCl_4). Mp 141-142°.

Chloride: [23041-25-0].
C_2Cl_7OP M 319.166
Solid with camphoraceous odour. Mp 48.5-49°. Subl. at 70°/5 mm.
Amide: [23691-99-8]. P,P-*Bis(trichloromethyl)-phosphinic amide.*
$C_2H_2Cl_6NOP$ M 299.736
Cryst. (C_6H_6). Mp 194-196°.
Methylamide: [28387-30-6]. N-*Methyl*-P,P-*bis(trichloromethyl)phosphinic amide.*
$C_3H_4Cl_6NOP$ M 313.762
Needles (CCl_4). Mp 182-183°.
Cyclohexylamide: N-*Cyclohexyl*-P,P-*bis(trichloromethyl)phosphinic amide.*
$C_8H_{12}Cl_6NOP$ M 381.881
Needles (CCl_4). Mp 177-178°.

Frank, A.W., *Can. J. Chem.*, 1968, **46**, 3573 (*synth, ir*)
Kozlov, É.S. *et al*, *Zh. Obshch. Khim.*, 1969, **39**, 933 (*Engl. transl.* p. 902) (*synth, chloride*)
Kozlov, É.S. *et al*, *Zh. Obshch. Khim.*, 1969, **39**, 1648 (*Engl. transl.* p. 1616) (*amide*)
Kozlov, É.S. *et al*, *Zh. Obshch. Khim.*, 1970, **40**, 991 (*Engl. transl.* p. 976) (*amides*)
Kozlov, É.S. *et al*, *Zh. Obshch. Khim.*, 1975, **45**, 471 (*Engl. transl.* p. 458) (*amide*)

PP-Bis(trichloromethyl)phosphinimidic chloride — B-00403

P-*Chloro*-P,P-*bis(trichloromethyl)phosphazene.* P-*Chloro*-P,P-*bis(trichloromethyl)phosphine imide*
[23575-76-0]

$(Cl_3C)_2PCl{=}NH$

C_2HCl_7NP M 318.181
Solid. Mp 41.2°. $Bp_{0.03}$ 68-69°.

N-*Chloro:* [38047-30-2]. N-*Chloro*-P,P-*bis(trichloromethyl)phosphinimidic chloride.*
C_2Cl_8NP M 352.626
Greenish liq. d_4^{20} 1.91. $Bp_{0.03}$ 80-82°. n_D^{20} 1.6063.
N-*Me:* [28225-41-4]. N-*Methyl*-P,P-*bis(trichloromethyl)phosphinimidic chloride.*
$C_3H_3Cl_7NP$ M 332.208
No phys. props. reported.
N-*Ph:* [80166-20-7]. N-*Phenyl*-P,P-*bis(trichloromethyl)phosphinimidic chloride.*
$C_8H_5Cl_7NP$ M 394.279
No phys. props. reported.

Kozlov, E.S. *et al*, *Zh. Obshch. Khim.*, 1969, **39**, 1648 (*Engl. transl.* p. 1616) (*synth*)
Kozlov, E.S. *et al*, *Zh. Obshch. Khim.*, 1972, **42**, 106 (*Engl. transl.* p. 101) (*chloro, synth, ir*)
Tsymbal, I.F. *et al*, *Teor. Eksp. Khim.*, 1981, **17**, 692 (*Engl. transl.* p. 540) (*derivs, ir*)

[Bis(trichloromethyl)phosphinyl]-phosphorimidic trichloride, 9CI — B-00404

P,P,P-*Trichloro*-N-*[bis(trichloromethyl)phosphinyl]-phosphine imide.* P,P,P-*Trichloro*-N-*[bis(trichloromethyl)phosphinyl]phosphazene.* P,P,P-*Trichloro*-N-*[bis(trichloromethyl)phosphinyl]-iminophosphorane*
[38047-33-5]

$(Cl_3C)_2P(O)N{=}PCl_3$

$C_2Cl_9NOP_2$ M 435.053

Kyuntsel', I.A. *et al*, *Zh. Obshch. Khim.*, 1975, **45**, 1989 (*Engl. transl.* p. 1954) (*nqr*)

Bis(2,3,6-trichlorophenyl)phosphinic acid, 9CI — B-00405

Cl Cl O Cl Cl
P
Cl OH Cl

$C_{12}H_5Cl_6O_2P$ M 424.862
Cryst. (EtOH). Mp 287-288.5°.

Denham, J.M. *et al*, *J. Org. Chem.*, 1958, **23**, 1298.

Bis(2,4,5-trichlorophenyl)phosphinic acid, 9CI B-00406

$C_{12}H_5Cl_6O_2P$ M 424.862
Cryst. (EtOAc or EtOH aq.). Mp 244.5-246°.

Denham, J.M. *et al*, *J. Org. Chem.*, 1958, **23**, 1298.

Bis(triethylsilyl) phosphonate, 9CI B-00407

Bis(triethylsilyl) phosphite
[3663-51-2]

$$(Et_3SiO)_2P(O)H \rightleftharpoons (Et_3SiO)_2POH$$

$C_{12}H_{31}O_3PSi_2$ M 310.520
Tautomeric. Liq. Bp_5 160-162°, $Bp_{1.5}$ 122-124°. n_D^{20} 1.4412.

Orlov, N.F. *et al*, *Zh. Obshch. Khim.*, 1966, **36**, 920 (*Engl. transl.* p. 935) (*synth*)
Voronkov, M.G. *et al*, *Zh. Obshch. Khim.*, 1971, **41**, 1987 (*Engl. transl.* p. 2005) (*synth*)
D'yakov, V.M. *et al*, *Zh. Obshch. Khim.*, 1972, **42**, 1291 (*Engl. transl.* p. 1286) (*props*)

Bis(2,2,2-trifluoroethoxy)-triphenylphosphorane, 9CI B-00408

[67696-25-7]

$$Ph_3P(OCH_2CF_3)_2$$

$C_{22}H_{19}F_6O_2P$ M 460.355
By ligand exchange readily gives alkoxy or acyloxyphosphoranes which are good alkylating or acylating agents. Alcohols and carboxylic acids are converted into trifluoroethyl ethers or esters, respectively; mixts. of amines with alcohols or acids give (secy.) amines or amides; mixts. of carbanions with alcohols or carboxylic acids give alkanes or ketones. V. hygroscopic cryst. (pentane). Mp 127-129°. $Bp_{0.2}$ 110° subl.

Kubota, T. *et al*, *J. Org. Chem.*, 1980, **45**, 5052 (*synth, pmr, nmr, props, use*)
Denney, D.B. *et al*, *J. Am. Chem. Soc.*, 1981, **103**, 1785 (*synth, pmr, cmr, nmr*)

Bis(2,2,2-trifluoroethyl) phosphonate, 10CI B-00409

Bis(2,2,2-trifluoroethyl) phosphite

$$(F_3CCH_2O)_2P(O)H \rightleftharpoons (F_3CCH_2O)_2POH$$

$C_4H_5F_6O_3P$ M 246.046
Reagent for synth. of mono- and diesters of phosphorous acid and for phosphorylation of nucleosides. Liq. d_4^{25} 1.67. Bp_2 43-44°.
Cyclohexylammonium salt: Cryst. (EtOH/Me_2CO). Mp 160-160.5°.
S-*(4-Chlorobenzyl)thiouronium salt:* Cryst. (MeCN). Mp 149-150°.

Gibbs, D.E. *et al*, *Synthesis*, 1984, 410 (*synth, use, ir*)
Takaku, H. *et al*, *Chem. Lett.*, 1984, 1267 (*synth, ir use*)

1,2-Bis(trifluoromethyl)diphosphine, 9CI B-00410

[462-57-7]

$$\begin{array}{c} CF_3 \\ | \\ H\blacktriangleright P \blacktriangleleft : \\ | \\ :\blacktriangleright P \blacktriangleleft H \\ | \\ CF_3 \end{array}$$ (*1RS,2RS*)-*form*

$C_2H_2F_6P_2$ M 201.976
(1RS,2RS)-form [57028-24-7]
(±)-*form*
Characterised spectroscopically.
(1RS,2SR)-form [57028-20-3]
meso-*form*
Characterised spectroscopically.

Albrand, J.P. *et al*, *Inorg. Chem.*, 1975, **14**, 570 (*synth*)
Dobbie, R.C. *et al*, *J. Chem. Soc., Dalton Trans.*, 1975, 2368 (*synth*)
Albrand, J.P. *et al*, *Tetrahedron Lett.*, 1976, 949 (*pmr, nmr*)
Albrand, J.P. *et al*, *ACS Symp. Ser.*, 1981, **171**, 577 (*uv, nmr, derivs*)

Bis(trifluoromethyl)phosphinic acid, 9CI B-00411

[422-94-6]

$$(F_3C)_2P(O)OH$$

$C_2HF_6O_2P$ M 201.993
Liq. Bp 182°, Bp_{238} 137-138°. pK_a <1. Treatment with 15% NaOH aq. gives CHF_3.
Me ester: [25439-11-6]. *Methyl bis(trifluoromethyl)-phosphinate.*
$C_3H_3F_6O_2P$ M 216.020
Liq. Bp_{745} 40°, Bp_{25} 19°.
Fluoride: [34005-83-9].
C_2F_7OP M 203.984
Gas. Mp −110°. Bp 5°.
Chloride: [646-71-9].
C_2ClF_6OP M 220.439
Bp 52.5° (calc.).
Bromide: [754-37-0].
C_2BrF_6OP M 264.890
Mp −35.5°. Bp 78° (calc.).
Anhydride: [34043-22-6].
$C_4F_{12}O_3P_2$ M 385.971
Mp −42°. V.p. 8.2 mm at 0°.
Dimethylamide: N,N-*Dimethyl*-P,P-*bis(trifluoromethyl)phosphinic amide.* Mp −21.1°. Bp 154.6° (calc.).

Emeleus, H.J. *et al*, *J. Chem. Soc.*, 1955, 563 (*synth*)
Griffiths, J.E. *et al*, *J. Am. Chem. Soc.*, 1962, **84**, 3442 (*chloride, bromide, ir*)
Burg, A.B. *et al*, *J. Am. Chem. Soc.*, 1965, **87**, 238 (*amide, ir*)
Cavell, R.G. *et al*, *J. Am. Chem. Soc.*, 1971, **93**, 1130 (*amide, ms, ir*)
Dobbie, R.C., *J. Chem. Soc. (A)*, 1971, 2894 (*anhydride, ir, nmr*)
Cavell, R.G. *et al*, *Inorg. Chem.*, 1972, **11**, 2573, 2578 (*bromide, ir, F nmr, pmr, silyl ester*)
Dagnac, P. *et al*, *J. Chem. Soc., Dalton Trans.*, 1979, 155 (*chloride, cmr, nqr*)
Mahler, W., *Inorg. Chem.*, 1979, **18**, 352 (*fluoride*)
Mahmood, T. *et al*, *Inorg. Chem.*, 1986, **25**, 3128 (*synth, pmr, F and P nmr*)

Bis(trifluoromethyl)phosphinodithioic acid, 9CI B-00412

[18799-75-2]

$(F_3C)_2P(S)SH$

$C_2HF_6PS_2$ M 234.114

Liq. or cryst. Mp 14-14.5°. Bp_{262} 71°, Bp_{40} 27°. Monomeric.

Me ester: [18799-79-6]. *Methyl bis(trifluoromethyl)-phosphinodithioate.*

$C_3H_3F_6PS_2$ M 248.141

Liq. Mp −17.5°. $Bp_{95.5}$ 70°, $Bp_{11.5}$ 26°.

Anhydrosulfide:

$C_4F_{12}P_2S_3$ M 434.152

Liq.

Gosling, K. *et al, J. Am. Chem. Soc.*, 1968, **90**, 2011 (*synth, derivs, ir, props*)

Dobbie, R.C. *et al, J. Am. Chem. Soc.*, 1968, **90**, 2015 (*synth, pmr, nmr, ms*)

Doty, L.F. *et al, Inorg. Chem.*, 1974, **13**, 2722 (*derivs*)

Cavell, R.G. *et al, Inorg. Chem.*, 1979, **18**, 2901 (*synth, props, pmr, nmr*)

Bis(trifluoromethyl)phosphinoselenous acid, 9CI, 8CI B-00413

$(F_3C)_2PSeH \rightleftharpoons (F_3C)_2PH{=}Se$

C_2HF_6PSe M 248.954

Me ester: [73076-90-1]. *Methyl bis(trifluoromethyl)-phosphinoselenoite.*

$C_3H_3F_6PSe$ M 262.981

Liq. Bp 103°.

Trifluoromethyl ester: [671-65-8]. *Trifluoromethyl bis(trifluoromethyl)phosphinoselenoite.*

C_3F_9PSe M 316.952

Liq. Mp −98°. Bp 62° (est.).

Anhydroselenide: [51101-58-7].

$C_4F_{12}P_2Se$ M 416.932

Liq. with unpleasant penetrating odour. Mp −35°.

Emeléus, H.J. *et al, J. Chem. Soc.*, 1962, 2529 (*ester, synth, ir*)

Dobbie, R.C. *et al, J. Fluorine Chem.*, 1973, **3**, 367 (*anhydroselenide, synth, ir, props*)

Dobbie, R.C. *et al, J. Mol. Struct.*, 1974, **23**, 141 (*anhydroselenide, ir, raman*)

Denhert, P. *et al, Z. Naturforsch., B*, 1979, **34**, 1646 (*ester, synth, P nmr, pmr*)

Bis(trifluoromethyl)phosphinotellurous acid B-00414

$(F_3C)_2PTeH$

C_2HF_6PTe M 297.594

Me ester: [73076-91-2]. *Methyl bis(trifluoromethyl)-phosphinotelluroite.*

$C_3H_3F_6PTe$ M 311.621

Ligand for Cr, Mo, and W. Orange-yellow liq. Bp 118°.

Dehnert, P. *et al, Z. Naturforsch., B*, 1979, **34**, 1642; 1981, **36**, 48 (*ester, synth, props, pmr, P nmr*)

Bis(trifluoromethyl)phosphinothioic acid, 9CI B-00415

[35814-49-4]

$(F_3C)_2P(S)OH \rightleftharpoons (F_3C)_2P(O)SH$

C_2HF_6OPS M 218.053

Thermally unstable liq. Bp 116° (calc.). pK_a ~2.50 (H_2O). Stable in aq. soln., but alkali causes dec. to CHF_3.

O-*Me ester:* [71040-58-9]. O-*Methyl bis(trifluoromethyl)phosphinothioate.*

$C_3H_3F_6OPS$ M 232.080

Liq. Bp 89° (calc.).

S-*Me ester:* [71009-87-5]. S-*Methyl bis(trifluoromethyl)phosphinothioate.*

$C_3H_3F_6OPS$ M 232.080

Bp 157° (calc.).

O-*Et ester:* [71009-90-0]. O-*Ethyl bis(trifluoromethyl)-phosphinothioate.*

$C_4H_5F_6OPS$ M 246.107

No phys. props. reported.

S-*Et ester:* [71009-88-6]. S-*Ethyl bis(trifluoromethyl)-phosphinothioate.*

$C_4H_5F_6OPS$ M 246.107

No phys. props. reported.

O-*Isopropyl ester:* [71009-91-1]. O-*Isopropyl bis(trifluoromethyl)phosphinothioate.*

$C_5H_7F_6OPS$ M 260.134

Bp 132° (calc.).

S-*Isopropyl ester:* [71009-89-7]. S-*Isopropyl bis(trifluoromethyl)phosphinothioate.*

$C_5H_7F_6OPS$ M 260.134

Bp 155° (calc.).

O-*Trimethylsilyl ester:* [38562-84-4]. O-*Trimethylsilyl bis(trifluoromethyl)phosphinothioate.*

$C_5H_9F_6OPSSi$ M 290.235

Liq.

Fluoride:

C_2F_7PS M 220.045

Liq. or gas. Bp 18° (calc.).

Chloride: [18799-82-1].

C_2ClF_6PS M 236.499

Liq. Mp −21.5°. Bp 61° (calc.).

Bromide: [18799-77-4].

C_2BrF_6PS M 280.950

Liq. Bp 80° (calc.).

Iodide: [18799-76-3].

C_2F_6IPS M 327.951

Solid. Mp 30° dec. Slowly liberates iodine on standing.

Anhydride: [35814-50-7].

$C_4F_{12}P_2S_2$ M 402.092

Bp 122-126° (calc.). Stable at temps. up to 160°.

Dobbie, R.C. *et al, J. Am. Chem. Soc.*, 1968, **90**, 2015 (*halides, ir, ms, F nmr*)

Gosling, K. *et al, J. Am. Chem. Soc.*, 1968, **90**, 2011 (*halides, ir*)

Cavell, R.G. *et al, Inorg. Chem.*, 1972, **11**, 2573 (*silyl ester, F nmr*)

Pinkerton, A.A. *et al, J. Am. Chem. Soc.*, 1972, **94**, 1870 (*synth, anhydride, ir ms, props*)

Cavell, R.G. *et al, Inorg. Chem.*, 1979, **18**, 2901 (*esters, ir, pmr, F nmr, props*)

P,P-Bis(trifluoromethyl)phosphinothioic amide, 9CI B-00416

[18904-52-4]

$(F_3C)_2P(S)NH_2$

$C_2H_2F_6NPS$ M 217.069

Liq. Mp −28.5°. Bp 131° (extrap.).

Gosling, K. *et al, J. Am. Chem. Soc.*, 1968, **90**, 2011.

Dobbie, R.C. *et al, J. Am. Chem. Soc.*, 1968, **90**, 2015 (*ir, ms, F nmr*)

Bis(trifluoromethyl)phosphinothious acid, 9CI B-00417

[1486-19-7]

$(F_3C)_2PSH$

C_2HF_6PS M 202.054
Liq. Bp 55° (est.).

Me ester: [1486-18-6]. *Methyl bis(trifluoromethyl)-phosphinothioite.*
$C_3H_3F_6PS$ M 216.081
Liq. with foul odour. Mp –58°. Bp 89°.

Trifluoromethyl ester: [671-64-7]. *Trifluoromethyl bis(trifluoromethyl)phosphinothioite.*
C_3F_9PS M 270.052
Mp –107°. Bp 50° (est.).

tert-*Butyl ester:* [1733-46-6]. tert-*Butyl bis(trifluoromethyl)phosphinothioite.*
$C_6H_9F_6PS$ M 258.161
Liq. Mp –18°. Bp 144° (est.).

Anhydrosulfide: [1486-20-0].
$C_4F_{12}P_2S$ M 370.032
Liq. Mp –33°. Bp 112° (est.).

Cavell, R.G. *et al, J. Chem. Soc.*, 1964, 5825 (*synth, derivs, ir, pmr, F nmr*)
Burg, A.B. *et al, J. Am. Chem. Soc.*, 1965, **87**, 2113 (*ester, synth, ir*)
Cavell, R.G. *et al, Inorg. Chem.*, 1968, **7**, 690 (*anhydrosulfide, ms*)
Dobbie, R.C. *et al, Spectrochim. Acta, Part A*, 1971, **27**, 255 (*ir*)
Burg, A.B., *Inorg. Nucl. Chem. Lett.*, 1977, **13**, 199 (*ester, F and P nmr, cmr*)
Cavell, R.G. *et al, Inorg. Chem.*, 1979, **18**, 2901 (*esters, synth, ir, pmr*)
Dehnert, P. *et al, Z. Naturforsch., B*, 1979, **34**, 1646 (*methyl ester, pmr, P nmr*)

Bis(trifluoromethyl)phosphinous acid B-00418

[359-65-9]

$(F_3C)_2POH$

C_2HF_6OP M 185.993
Exists in the hydroxyl form. Liq. d_4^{20} 1.57. Mp –20.8° (est.).

Me ester: [684-25-3]. *Methyl bis(trifluoromethyl)-phosphinite. Methoxybis(trifluoromethyl)phosphine.*
$C_3H_3F_6OP$ M 200.020
Mp –78.5°. Bp 55°. Inert to BF_3.

tert-*Butyl ester:* tert-*Butyl bis(trifluoromethyl)-phosphinite.* tert-*Butoxybis(trifluoromethyl)-phosphine.*
$C_6H_9F_6OP$ M 242.101
Liq. Mp –26°. Bp 110°.

Fluoride: [1426-40-0]. *Fluorobis(trifluoromethyl)-phosphine.*
C_2F_7P M 187.985
Gas. Bp –11° to –8°.

Chloride: see Bis(trifluoromethyl)phosphinous chloride, B-00420

Bromide: [758-45-2]. *Bromobis(trifluoromethyl)-phosphine.*
C_2BrF_6P M 248.890
Liq. Bp 42° (est.).

Iodide: [359-64-8]. *Iodobis(trifluoromethyl)phosphine.*
C_2F_6IP M 295.891
Liq. d_4^{20} 2.04. Bp 72-73°. n_D^{25} 1.403.

Cyanide: [431-97-0]. *Cyanobis(trifluoromethyl)-phosphine.*
C_3F_6NP M 195.004
Liq. Bp 48°. n_D^{20} 1.3248.
▷Spont. flammable

Isocyanate:
C_3F_6NOP M 211.003
Liq. Bp 54° (est.).

Isothiocyanate:
C_3F_6NPS M 227.064
Liq. Bp 84° (est.).

Anhydride: [2728-67-8].
$C_4F_{12}OP_2$ M 353.972
Liq. d_4^{20} 1.61.

*Amide: see P,P-*Bis(trifluoromethyl)phosphinous amide, B-00419

Bennett, F.W. *et al, J. Chem. Soc.*, 1953, 1565 (*derivs, synth, ir*)
Issleib, K. *et al, Chem. Ber.*, 1959, **92**, 2681 (*iodide, synth*)
Kulakova, V.N. *et al, Zh. Obshch. Khim.*, 1959, **29**, 3957 (*Engl. transl.* p. 3916) (*fluoride*)
Griffiths, J.E. *et al, J. Am. Chem. Soc.*, 1962, **84**, 3442 (*props, esters*)
Packer, K.J., *J. Chem. Soc.*, 1963, 960 (*isocyanate, isothiocyanate, F and P nmr, synth, ir*)
Cavell, R.G. *et al, Inorg. Chem.*, 1968, **7**, 101, 690 (*derivs, ms*)
Dobbie, R.C. *et al, Spectrochim. Acta, Part A.*, 1971, **27**, 255 (*ir*)
Elbel, S. *et al, Z. Naturforsch., B*, 1976, **31**, 1472 (*cyanide, pe*)
Burg, A.B., *Inorg. Nucl. Chem. Lett.*, 1977, **13**, 199 (*derivs, cmr, F and P nmr*)
Virlichie, J.L. *et al, Rev. Chim. Minerol*, 1977, **14**, 355 (*synth*)

P,P-Bis(trifluoromethyl)phosphinous amide B-00419

[431-95-8]

$(F_3C)_2PNH_2$

$C_2H_2F_6NP$ M 185.009
Liq. Mp –87.6°. Bp 67.1°.
▷Spont. flammable

N-*Me:* [431-98-1]. N-*Methyl-*P,P-*bis(trifluoromethyl)-phosphinous amide.*
$C_3H_4F_6NP$ M 199.035
Liq. Mp –46.5°. Bp 73°.

N,N-*Di-Me:* [432-01-9]. N,N-*Dimethyl-*P,P-*bis(trifluoromethyl)phosphinous amide.*
$C_4H_6F_6NP$ M 213.062
Liq. Mp –81.5°. Bp 83°.

N-*Ph:* [348-71-0]. N-*Phenyl-*P,P-*bis(trifluoromethyl)-phosphinous amide. Bis(trifluoromethyl)phosphinous anilide.*
$C_8H_6F_6NP$ M 261.106
Liq. Mp 1.7°. Bp 182°.

Harris, G.S., *J. Chem. Soc.*, 1958, 512 (*synth, derivs, ir, props*)
Greenwood, N.N. *et al, J. Chem. Soc.*, 1968, 226, 230 (*derivs, ir, conformn*)
Cowley, A.H. *et al, J. Am. Chem. Soc.*, 1970, **92**, 1085, 5206 (*deriv, pmr, conformn*)

Bis(trifluoromethyl)phosphinous chloride B-00420

Chlorobis(trifluoromethyl)phosphine
[650-52-2]

$(F_3C)_2PCl$

C_2ClF_6P M 204.439
Liq. Bp 21-21.5°.
▷Spont. flammable

Bennett, F.W. *et al, J. Chem. Soc.*, 1953, 1565 (*synth, props, ir*)
Cavell, R.G. *et al, Inorg. Chem.*, 1968, **7**, 101 (*ms*)
Prons, V.N. *et al, Zh. Obshch. Khim.*, 1970, **40**, 589 (*Engl. transl.* p. 559)

Cowley, A.H. *et al, J. Am. Chem. Soc.*, 1975, **97**, 3653 (*pe*)
Demuth, R., *Z. Anorg. Allg. Chem.*, 1975, **418**, 149 (*ir, raman*)
Burg, A.B., *Inorg. Nucl. Chem. Lett.*, 1977, **13**, 199 (*cmr, F and P nmr*)
Dagnac, P. *et al, J. Chem. Soc., Dalton Trans.*, 1979, 155 (*cmr, nqr*)

4′,5′-Bis(trifluoromethyl)spiro[1,3,2-benzodioxaphosphole-2,2′-[1,3,2]-dioxaphosphole], 9CI — B-00421

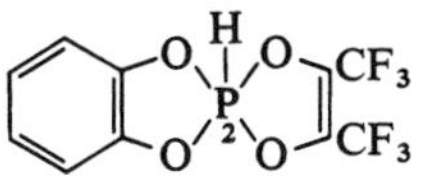

$C_{10}H_5F_6O_4P$ M 334.111

2-Methoxy: [53799-43-2].
$C_{11}H_7F_6O_5P$ M 364.137
Oil which cryst. Mp 34-35°.

2-Phenoxy: [53799-42-1].
$C_{16}H_9F_6O_5P$ M 426.208
Cryst. (hexane). Mp 85-86°.

2-Fluoro: [67348-17-8].
$C_{10}H_4F_7O_4P$ M 352.102
Solid. Mp 83°. Sublimes.

Ramirez, F. *et al, J. Am. Chem. Soc.*, 1974, **96**, 7269 (*derivs, synth, pmr, F and P nmr*)
Narayanan, P. *et al, J. Am. Chem. Soc.*, 1977, **99**, 3336 (*phenoxy, cryst struct*)
Eikmeier, H.B. *et al, Chem. Ber.*, 1978, **111**, 2077 (*fluoro, F and P nmr*)

4′,5′-Bis(trifluoromethyl)spiro[1,3,2-benzodioxaphosphole-2,2′-[1,3,2]-dithiaphosphole], 10CI — B-00422

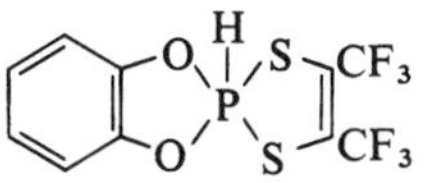

$C_{10}H_5F_6O_2PS_2$ M 366.232

2-Methoxy: [37893-34-8].
$C_{11}H_7F_6O_3PS_2$ M 396.259
Liq. $Bp_{0.05}$ 100-102°.

Burros, B.C. *et al, J. Am. Chem. Soc.*, 1978, **100**, 7300 (*synth, F and P nmr*)

4,5-Bis(trifluoromethyl)spiro[1,3,2-dithiaphosphole-2,1′-[2,6,7]trioxa[1]-phosphabicyclo[2.2.1]heptane, 9CI — B-00423

[60049-50-5]

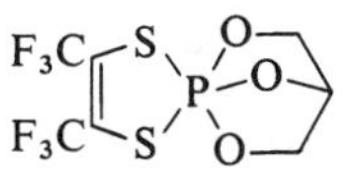

$C_7H_5F_6O_3PS_2$ M 346.199
Dec. on attempted isol.

Campbell, B.S. *et al, J. Am. Chem. Soc.*, 1976, **98**, 2924 (*synth, ms, pmr, P and F nmr*)

4,5-Bis(trifluoromethyl)spiro[1,3,2-dithiaphosphole-2,1′-[2,8,9]trioxa[1]-phosphatricyclo[3.3.1.1^{5,7}]decane], 9CI — B-00424

[60049-48-1]

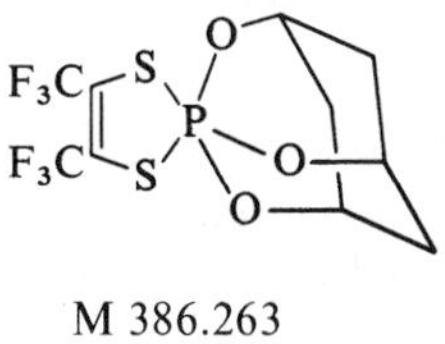

$C_{10}H_9F_6O_3PS_2$ M 386.263
Cryst. (MeOH). Mp 125-129° (sealed tube).

Campbell, B.S. *et al, J. Am. Chem. Soc.*, 1976, **98**, 2924 (*synth, pmr, F and P nmr*)

3,4-Bis(trifluoromethyl)-1,2,5-thiadiphosphole, 9CI — B-00425

[78172-51-7]

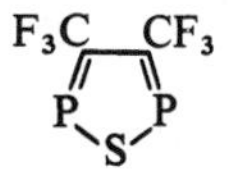

$C_4F_6P_2S$ M 256.042
Air-sensitive oil. Behaves as a diene in Diels-Alder reactions.

Kobayashi, Y. *et al, J. Am. Chem. Soc.*, 1981, **103**, 2465 (*synth, ir, F nmr, uv, props*)

Bis(2,4,6-trimethylphenyl)phosphine, 9CI — B-00426

Dimesitylphosphine, 8CI

[1732-66-7]

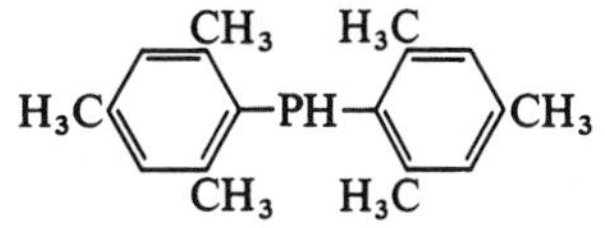

$C_{18}H_{23}P$ M 270.353
Cryst., insensitive to air (EtOH). Mp 74°. $Bp_{0.1}$ 140-160°.

Oxide: [23897-16-7].
$C_{18}H_{23}OP$ M 286.353
Cryst. (Et_2O). Mp 130-132°. Tautomeric with bis(2,4,6-trimethylphenyl)phosphinous acid.

Fritzsche, H. *et al, Chem. Ber.*, 1965, **98**, 1681 (*synth*)
Stepanov, B.I. *et al, Zh. Obshch. Khim.*, 1969, **39**, 1544 (*Engl. transl.* p. 1514) (*oxide*)
Lindner, E. *et al, Chem. Ber.*, 1980, **113**, 2769, 3268 (*oxide*)

Bis(2,4,5-trimethylphenyl)phosphinic acid, 9CI — B-00427

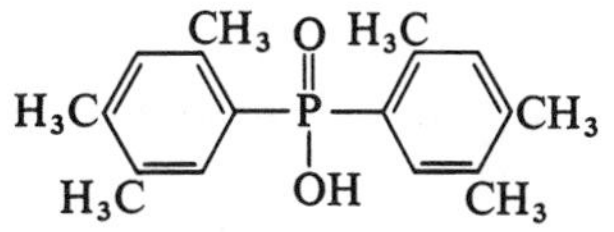

$C_{18}H_{23}O_2P$ M 302.352
Cryst. (EtOH). Mp 205-206°.

Stepanov, B.I. *et al, Zh. Obshch. Khim.*, 1970, **40**, 2217 (*Engl. transl.* p. 2204) (*synth, pmr, uv*)

Bis(2,4,6-trimethylphenyl)phosphinic acid, 9CI — B-00428

Dimesitylphosphinic acid, 8CI

$C_{18}H_{23}O_2P$ M 302.352
Cryst. (EtOH). Mp 210°.

Chloride: [31639-12-0].
$C_{18}H_{22}ClOP$ M 320.798
Cryst. (cyclohexane). Mp 140.5-141°.

Anhydride: [73731-22-3].
$C_{36}H_{44}O_3P_2$ M 586.689
Cryst. (C_6H_6/pet. ether). Mp 235-238°.

Amide: P,P-*Bis(2,4,6-trimethylphenyl)phosphinic amide.*
$C_{18}H_{24}NOP$ M 301.367
Solid. Mp 225.5-226.5°.

Anilide: N-*Phenyl*-P,P-*bis(2,4,6-trimethylphenyl)-phosphinic amide.*
$C_{24}H_{28}NOP$ M 377.465
Solid. Mp 167-169°.

Fritzsche, H. *et al*, *Chem. Ber.*, 1965, **98**, 1681 (*synth*)
Stepanov, B.I. *et al*, *Zh. Obshch. Khim.*, 1970, **40**, 2217 (*Engl. transl.* p. 2214) (*synth, pmr, chloride, ir, uv*)
Harger, M.J.P., *J. Chem. Soc., Perkin Trans. 2*, 1980, 154 (*synth, derivs, ir, pmr*)

Bis(trimethylphosphoranylidene)methane B-00429

Methanetetraylbis(trimethylphosphorane), *9CI*
[57437-91-9]

$$Me_3P{=}C{=}PMe_3$$

$C_7H_{18}P_2$ M 164.167
Air-sensitive liq. Dec. by water to give Me_3PO. Mp ca. 0°. $Bp_{0.1}$ 41°. Forms complexes containing Al, Ga, Ni, Pt, Pd, Au, Zn, or Cd. Very strongly basic powerful deprotonating agent.

Gasser, O. *et al*, *J. Am. Chem. Soc.*, 1975, **97**, 6281 (*synth, pmr, cmr, nmr, ms*)
Ebsworth, E.A.V. *et al*, *Chem. Ber.*, 1977, **110**, 3508 (*struct*)
Schmidbaur, H. *et al*, *Chem. Ber.*, 1977, **110**, 3501 (*synth, pe, pmr, cmr, P nmr, ir*)
Schmidbaur, H. *et al*, *Chem. Ber.*, 1977, **110**, 3517 (*complexes*)

2-[Bis(trimethylsilyl)amino]-2,2-dihydro-3,3-bis(trifluoromethyl)-2-[(trimethylsilyl)imino]oxaphosphirane, 10CI B-00430

[68222-38-8]

O, SiMe3, F3C, N, SiMe3, P, CF3, NSiMe3

$C_{12}H_{27}F_6N_2OPSi_3$ M 444.579
One of v. few oxaphosphiranes. Liq. $Bp_{0.2}$ 72°.

Röschenthaler, G.-V. *et al*, *Chem. Ber.*, 1978, **111**, 3105 (*synth, ms, ir, pmr, F, P and Si nmr, cmr*)

2-[Bis(trimethylsilyl)methylene]-1,1-di-*tert*-butyldiphosphine B-00431

2-[Bis(trimethylsilyl)methylene]-1,1-bis(1,1-dimethylethyl)diphosphine, *11CI*

$$[(H_3C)_3C]_2P{-}P{=}C(SiMe_3)_2$$

$C_{15}H_{36}P_2Si_2$ M 334.568
Orange solid. Mp 49-51°. $Bp_{0.05}$ 90-93°.

Kolodnyazhnyi, O.I. *et al*, *Zh. Obshch. Khim.*, 1985, **55**, 1862 (*Engl. transl.* p. 1655) (*synth, P nmr*)

[Bis(trimethylsilyl)methyl]-triphenylphosphonium(1+), 9CI B-00432

$$Ph_3P^{\oplus}CH(SiMe_3)_2$$

$C_{25}H_{34}PSi_2^{\oplus}$ M 421.688 (ion)
Bromide:
$C_{25}H_{34}BrPSi_2$ M 501.592
Treatment with butyllithium gives the ylide. Solid. Mp 141°.
Ylide: [36050-78-9]. *[Bis(trimethylsilyl)methylene]-triphenylphosphorane.*
$C_{25}H_{33}PSi_2$ M 420.680
Rel. stable ylide for use in Wittig reactions. Cryst. (Et_2O). Mp 139-140°.

Schmidbaur, H. *et al*, *Chem. Ber.*, 1972, **105**, 1084.

Bis(trimethylsilyloxy)phosphinecarboxylic acid oxide, 9CI B-00433

[Bis(trimethylsilyloxy)]phosphinylformic acid. Bis(trimethylsilyl) phosphonoformate

$$(Me_3SiO)_2P(O)COOH$$

$C_7H_{19}O_5PSi_2$ M 270.369
Trimethylsilyl group lost on methanolysis of esters.
Me ester: [67605-36-1]. *Methyl bis(trimethylsilyloxy)-phosphinecarboxylate oxide.*
$C_8H_{21}O_5PSi_2$ M 284.396
Liq. $Bp_{0.1}$ 60-62°.
Et ester: [66191-00-2]. *Ethyl bis(trimethylsilyloxy)-phosphinecarboxylate oxide.*
$C_9H_{23}O_5PSi_2$ M 298.422
Liq. $Bp_{0.45}$ 95-96°.

Morita, T. *et al*, *Bull. Chem. Soc. Jpn.*, 1978, **51**, 2169 (*synth, ir, pmr*)
Sekine, M. *et al*, *J. Chem. Soc., Chem. Commun.*, 1978, 285 (*synth*)
McKenna, C.E. *et al*, *J. Chem. Soc., Chem. Commun.*, 1979, 739 (*synth, use*)

[Bis(trimethylsilyloxy)phosphinyl]acetic acid B-00434

$$(Me_3SiO)_2P(O)CH_2COOH$$

$C_8H_{21}O_5PSi_2$ M 284.396
Me ester: Methyl [bis(trimethylsilyloxy)phosphinyl]-acetate.
$C_9H_{23}O_5PSi_2$ M 298.422
Liq. Bp_8 123-124°.
Et ester: Ethyl [bis(trimethylsilyloxy)phosphinyl]-acetate.
$C_{10}H_{25}O_5PSi_2$ M 312.449
Liq. $Bp_{0.1}$ 90-93°.
tert-*Butyl ester:* tert-*Butyl [bis(trimethylsilyloxy)-phosphinyl]acetate.*
$C_{12}H_{29}O_5PSi_2$ M 340.503
Liq. $Bp_{0.01}$ 82-84°.
Ph ester: Phenyl [bis(trimethylsilyloxy)phosphinyl]-acetate.
$C_{14}H_{25}O_5PSi_2$ M 360.493
Liq. $Bp_{0.5}$ 133-135°.
Benzyl ester: Benzyl [bis(trimethylsilyloxy)phosphinyl]-acetate.
$C_{15}H_{27}O_5PSi_2$ M 374.520
Liq. $Bp_{0.002}$ 123-125°.

Morita, T. *et al*, *Bull. Chem. Soc. Jpn.*, 1978, **51**, 2169; 1981, **54**, 267 (*synth, ir, pmr*)
Sekine, M. *et al*, *J. Org. Chem.*, 1981, **46**, 2097 (*ethyl ester*)

Bis(trimethylsilyl) phosphonate, 9CI B-00435

Bis(trimethylsilyl) phosphite
[3663-52-3]

$$(Me_3SiO)_2P(O)H \rightleftharpoons (Me_3SiO)_2POH$$

$C_6H_{19}O_3PSi_2$ M 226.359
Tautomeric. Liq. d_4^{20} 0.97-1.00. Bp_6 100-102°, Bp_1 74-76°, Bp_{10} 81°. n_D^{20} 1.4110.

Orlov, N.F. *et al*, *Zh. Obshch. Khim.*, 1966, **36**, 920 (*Engl. transl.* p. 935) (*synth*)
Voronkov, M.G. *et al*, *Zh. Obshch. Khim.*, 1971, **41**, 1987 (*Engl. transl.* p. 2005) (*synth, ir*)
Lebedev, E.P. *et al*, *Zh. Obshch. Khim.*, 1977, **47**, 765 (*Engl. transl.* p. 698) (*synth, ir, P nmr, props*)
Sekine, M. *et al*, *J. Org. Chem.*, 1981, **46**, 2097 (*synth, pmr, props*)

Bis(trimethylsilyl) phosphonite, 9CI — B-00436

[30148-50-6]

$(Me_3SiO)_2PH$

$C_6H_{19}O_2PSi$ M 182.274
Liq. Bp 164° dec., Bp_{10} 52°. n_D^{20} 1.4116.
▷Ignites spontaneously in air

Voronkov, M.G. *et al*, *Zh. Obshch. Khim.*, 1970, **40**, 2135 (*Engl. transl.* p. 2121) (*synth, nmr, props*)
Voronkov, M.G. *et al*, *Zh. Obshch. Khim.*, 1971, **41**, 1987 (*Engl. transl.* p. 2005) (*synth, props, ir, pmr*)

O,O-Bis(trimethylsilyl) phosphonothioate, 9CI — B-00437

O,O-*Bis(trimethylsilyl) thiophosphite*
[34734-82-2]

$(Me_3SiO)_2P(S)H \rightleftharpoons (Me_3SiO)_2PSH$

$C_6H_{19}O_2PSSi_2$ M 242.420
Probably exists largely in the thiophosphoryl form. Bp_4 72°.

Voronkov, M.G. *et al*, *Zh. Obshch. Khim.*, 1971, **41**, 1987 (*Engl. transl.* p. 2005) (*ir, pmr*)
Issleib, K. *et al*, *Z. Anorg. Allg. Chem.*, 1985, **530**, 16 (*synth, P nmr*)

Bis(trimethylsilyl)phosphoramidous acid, 9CI — B-00438

$(Me_3Si)_2NP(OH)_2$

$C_6H_{20}NO_2PSi$ M 197.289
Di-Et ester: [66628-81-7]. *Diethyl bis(trimethylsilyl)-phosphoramidite.*
$C_{10}H_{28}NO_2PSi$ M 253.396
Liq. Bp_4 78-80°.
Difluoride: [50732-22-4].
$C_6H_{18}F_2NPSi$ M 201.271
Bp_{12} 48-50°, Bp_4 66-7°.
Dichloride: [54036-90-7].
$C_6H_{18}Cl_2NPSi$ M 234.180
Liq. Mp −10° to −8°. $Bp_{0.01}$ 45-47°. Unstable at r.t. giving a yellow solid.

Scherer, O.J. *et al*, *J. Organomet. Chem.*, 1974, **82**, C3 (*dichloride, synth, ms, pmr, P nmr*)
Neilson, R.H. *et al*, *Inorg. Chem.*, 1977, **16**, 1455 (*difluoride, dichloride, synth, ms, pmr, F nmr*)
Gara, W.B. *et al*, *J. Chem. Soc., Perkin Trans. 2*, 1978, 150 (*ester, synth, esr*)

O,O-Bis(trimethylsilyl) phosphorothioate, 9CI, 8CI — B-00439

O,O-*Bis(trimethylsilyl) phosphorothioic acid.* O,O-*Bis-(trimethylsilyl) hydrogen phosphorothioate.* O,O-*Bis-(trimethylsilyl) thiophosphate*

$(Me_3SiO)_2P(O)SH \rightleftharpoons (Me_3SiO)_2P(S)OH$

$C_6H_{19}O_3PSSi$ M 230.334
Liq. d_4^{20} 0.98. Bp_3 93-95°. n_D^{20} 1.4200.

Volodina, L.N., *Zh. Obshch. Khim.*, 1967, **37**, 1842 (*Engl. transl.* p. 1755)

Bis(trimethylsilyl) selenophosphonate, 11CI — B-00440

O,O-*Bis(trimethylsilyl) phosphoroselenoite*
[104189-33-5]

$(Me_3SiO)_2P(Se)H \rightleftharpoons (Me_3SiO)_2P{-}SeH$

$C_6H_{19}O_2PSe$ M 233.149
Probably exists almost completely as the selenophosphoryl tautomer. Liq. $Bp_{0.06}$ 56°.

Issleib, K. *et al*, *Z. Anorg. Allg. Chem.*, 1985, **530**, 16 (*synth, P nmr*)

Bis(trimethylsilyl) tellurophosphonate, 11CI — B-00441

O,O-*Bis(trimethylsilyl) phosphorotelluroite*
[104189-34-6]

$(Me_3SiO)_2P(Te)H \rightleftharpoons (Me_3SiO)_2P{-}TeH$

$C_6H_{19}O_2PTe$ M 281.789
Probably exists almost completely as the tellurophosphoryl tautomer. Liq. $Bp_{0.03}$ 58°.

Issleib, K. *et al*, *Z. Anorg. Allg. Chem.*, 1985, **530**, 16 (*synth, P nmr*)

Bis(triphenylphosphine)iminium(1+) — B-00442

μ-Nitridobis(triphenylphosphorus)(1+)

$Ph_3P{=}N^{\oplus}{=}PPh_3$

$C_{36}H_{30}NP_2^{\oplus}$ M 538.587 (ion)
Fluoride:
$C_{36}H_{30}FNP_2$ M 557.586
Cryst. (Me_2CO/Et_2O). Mp 177-178°.
Chloride: [2156-68-5]. N-*Chlorotriphenylphosphoranyl-P,P,P-triphenylphosphine imide, 9CI.*
$C_{36}H_{30}ClNP_2$ M 574.040
Stable nonhygroscopic cryst. (H_2O or $PhNO_2$). Mp 272°.
Bromide:
$C_{36}H_{30}BrNP_2$ M 618.491
Cryst. (Me_2CO/Et_2O). Mp 253-255°.
Iodide:
$C_{36}H_{30}INP_2$ M 665.492
Cryst. (Me_2CO/Et_2O). Mp 252-254°.

Flück, E. *et al*, *Chem. Ber.*, 1963, **96**, 3085.
Inorg. Synth., 1974, **15**, 84 (*synth, props, use*)
Martinsen, A. *et al*, *Acta Chem. Scand., Ser. A*, 1977, **31**, 645 (*uv*)
Winter, S.R. *et al*, *J. Organomet. Chem.*, 1977, **133**, 339 (*complexes*)

1,8-Bis(triphenylphosphoniomethyl)-anthracene(2+) — B-00443

[1,8-Anthracenediylbis[methylene]]-bis[triphenylphosphonium](2+), 9CI

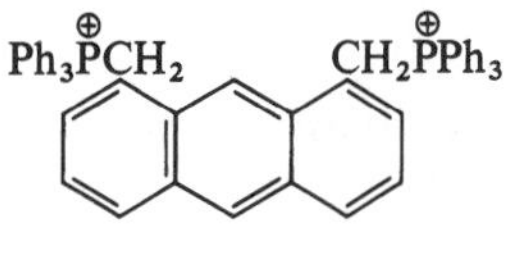

$Ph_3\overset{\oplus}{P}CH_2$ $CH_2\overset{\oplus}{P}Ph_3$

$C_{52}H_{42}P_2^{\oplus\oplus}$ M 728.851 (ion)

Dibromide: [34824-23-2].
$C_{52}H_{42}Br_2P_2$ M 888.659
Solid. Mp >360°. Bis-ylide formed using PhLi.
Bis-ylide: *[1,8-Anthracenediylbis[methylene]]-bis[triphenylphosphorane].*
$C_{52}H_{40}P_2$ M 726.836
Used in synth. of dianthr[14]annulene.

Akiyama, S. *et al*, *Bull. Chem. Soc. Jpn.*, 1971, **44**, 3158 (*synth, use*)

2,2′-Bis[(triphenylphosphonio)methyl]-1,1-binaphthyl(2+) B-00444

[[1,1′-Binaphthalene]-2,2′-diylbis[methylene]]-bis[triphenylphosphonium](2+), 9CI

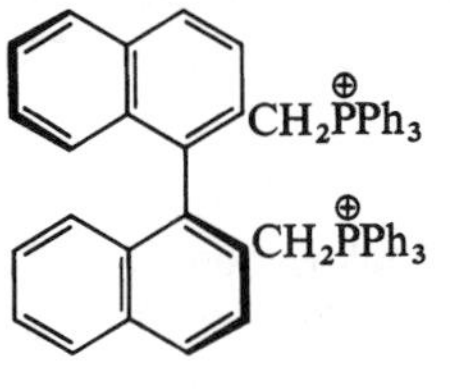

(S)-form

$C_{58}H_{46}P_2^{\oplus\oplus}$ M 804.949 (ion)

(S)-form

Dibromide: [37803-03-5].
$C_{58}H_{46}Br_2P_2$ M 964.757
Used in synth. of (+)-pentahelicene. Cryst. Mp 274-277° dec. $[\alpha]_D^{23}$ +112° (c, 5.48 in $CHCl_3$).
Diperiodate: [37812-59-2].
$C_{58}H_{46}I_2O_8P_2$ M 1186.753
Solid. Mp 174° dec.

(±)-form

Dibromide: Cryst. (DMF/Et_2O or $CH_2Cl_2/EtOAc$). Mp 289-290°.
Diperiodate: Mp 174°. With LiOEt, gives yellow bis-ylide.

Bestmann, H.J. *et al*, *Chem. Ber.*, 1969, **102**, 2259; 1974, **107**, 2923 (*synth, use*)

2,2′-Bis[(triphenylphosphonio)methyl]-biphenyl(2+) B-00445

[[1,1′-Biphenyl]-2,2′-diylbis[methylene]]-bis[triphenylphosphonium](2+), 9CI. (2,2′-Biphenylylenedimethylene)-bis[triphenylphosphonium](2+)

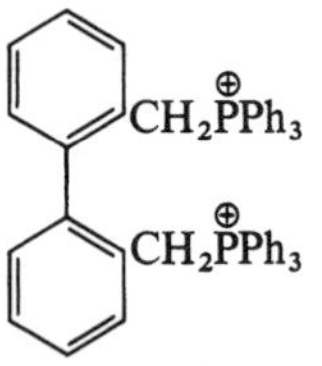

$C_{50}H_{42}P_2^{\oplus\oplus}$ M 704.829 (ion)
Dibromide: [4283-98-1].
$C_{50}H_{42}Br_2P_2$ M 864.637
Used in synth. of benzannelated annulenes. Cryst. (MeCN or DMF). Mp >300°. With LiOMe, yields the bis-ylide.
Diperiodate:
$C_{50}H_{42}I_2O_8P_2$ M 1086.634
Solid. Mp 158° dec. Gives phenanthrene when treated with $NaNH_2$.
Bis-ylide: [30012-71-6]. *[[1,1′-Biphenyl]-2,2′-diylbis[methylene]]bis[triphenylphosphorane].*
$C_{50}H_{40}P_2$ M 702.814
Used in Wittig reactions.

Bergmann, E.D. *et al*, *J. Chem. Soc. (C)*, 1967, 328 (*dibromide*)
Bestmann, H.J. *et al*, *Chem. Ber.*, 1969, **102**, 2259 (*periodate*)
Grohmann, K. *et al*, *J. Org. Chem.*, 1973, **38**, 808 (*synth, use*)
Auclair, F. *et al*, *Tetrahedron*, 1975, **31**, 2499 (*use*)

3,3′-Bis[(triphenylphosphonio)methyl]-biphenyl(2+) B-00446

[[1,1′-Biphenyl]-3,3′-diylbis[methylene]]-bis[triphenylphosphonium](2+), 9CI. (3,3′-Biphenylylenedimethylene)-bis[triphenylphosphonium](2+)

$C_{50}H_{42}P_2^{\oplus\oplus}$ M 704.829 (ion)
Dibromide: [61358-48-3].
$C_{50}H_{42}Br_2P_2$ M 864.637
Used in synth. of benzannelated annulenes. Cryst. (MeOH/EtOH). Mp 360° dec.
Bis-ylide: *[[1,1′-Biphenyl]-3,3′-diylbis[methylene]]-bis[triphenylphosphorane].*
$C_{50}H_{40}P_2$ M 702.814
Used in Wittig reactions.

Staab, H.A. *et al*, *Chem. Ber.*, 1976, **109**, 3875 (*synth, use*)
Thulin, B. *et al*, *Tetrahedron Lett.*, 1977, 929 (*use*)

4,4′-Bis[(triphenylphosphonio)methyl]-biphenyl(2+) B-00447

[[1,1′-Biphenyl]-4,4′-diylbis[methylene]]-bis[triphenylphosphonium](2+), 9CI. (4,4′-Biphenylylenedimethylene)-bis[triphenylphosphonium](2+)

$C_{50}H_{42}P_2^{\oplus\oplus}$ M 704.829 (ion)
Dichloride: [54050-02-1].
$C_{50}H_{42}Cl_2P_2$ M 775.735
Powder. NaOMe in DMSO yields bis-ylide.
Bis-ylide: *[[1,1′-Biphenyl]-4,4′-diylbis[methylene]]-bis[triphenylphosphorane].*
$C_{50}H_{40}P_2$ M 702.814
Used in Wittig reactions.

Märky, M., *Helv. Chim. Acta*, 1981, **64**, 957 (*synth, uv, pmr*)

2,5-Bis[(triphenylphosphonio)methyl]-furan(2+) B-00448

[2,5-Furandiylbis[methylene]]-bis[triphenylphosphonium](2+), 9CI

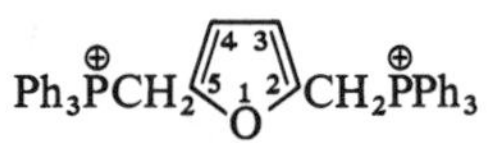

$C_{42}H_{36}OP_2^{\oplus\oplus}$ M 618.693 (ion)
Dichloride: [50738-79-9].
$C_{42}H_{36}Cl_2OP_2$ M 689.599
Source of ylide. Off-white prisms ($MeOH/Et_2O$). Dec. at 230° without melting.
Bis-ylide: *[2,5-Furandiylbis[methylene]]-bis[triphenylphosphorane].*
$C_{42}H_{34}OP_2$ M 616.678
Used in synth. of heteroannulenes.

Cresp, T.M. *et al*, *J. Chem. Soc., Perkin Trans. 1*, 1973, 1786, 2961 (*synth, use*)

3,4-Bis[(triphenylphosphonio)methyl]-furan(2+) B-00449

[3,4-Furandiylbis[methylene]]-bis[triphenylphosphonium](2+), 9CI

$C_{42}H_{36}OP_2^{\oplus\oplus}$ M 618.693 (ion)
Dichloride: [16401-29-9].
$C_{42}H_{36}Cl_2OP_2$ M 689.599
Source of ylide. Plates ($MeOH/Et_2O$). Dec. at 250° without melting.

Dibromide: [58776-20-8].
$C_{42}H_{36}Br_2OP_2$ M 778.501
No phys. props. reported.
Bis-ylide: [*3,4-Furandiylbis[methylene]]-bis[triphenylphosphorane]*.
$C_{42}H_{34}OP_2$ M 616.678
Used in Wittig reactions leading to furocyclooctenes and heteroannulenes.

Bindra, A.P. *et al*, *Tetrahedron Lett.*, 1968, 4335 (*use*)
Elix, J.A. *et al*, *J. Am. Chem. Soc.*, 1970, **92**, 973 (*chloride, synth, use*)
Kaplan, F.A. *et al*, *J. Am. Chem. Soc.*, 1977, **99**, 513 (*chloride, use*)

1,4-Bis(triphenylphosphoniomethyl)-naphthalene(2+) B-00450

1,4-[Naphthalenediylbis[methylene]]-bis(triphenylphosphonium)(2+)

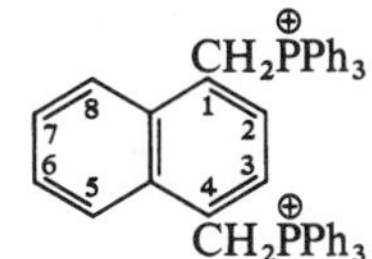

$C_{48}H_{40}P_2^{\oplus\oplus}$ M 678.792 (ion)
Dibromide: [72862-30-7].
$C_{48}H_{40}Br_2P_2$ M 838.600
Used in synth. of naphthalenophanes.

Tanner, D. *et al*, *Acta Chem. Scand., Ser. B*, 1979, **33**, 443 (*synth, use*)

1,5-Bis(triphenylphosphoniomethyl)-naphthalene(2+) B-00451

[1,5-Naphthalenediylbis[methylene]]-bis[triphenylphosphonium](2+), 9CI
$C_{48}H_{40}P_2^{\oplus\oplus}$ M 678.792 (ion)
Dibromide: [52045-12-2].
$C_{48}H_{40}Br_2P_2$ M 838.600
Used in synth. of naphthalenophanes. Cryst. Mp ca. 300° dec. With NaOEt or $NaCH_2SOMe$, yields the bis-ylide.
Bis-ylide: [*1,5-Naphthalenediylbis[methylene]]-bis[triphenylphosphorane]*.
$C_{48}H_{38}P_2$ M 676.776

Flemming, R.H. *et al*, *J. Am. Chem. Soc.*, 1974, **96**, 7738 (*use*)
Zander, M. *et al*, *Chem. Ber.*, 1974, **107**, 727 (*synth, use*)
Tanner, D. *et al*, *Acta Chem. Scand., Ser. B*, 1979, **33**, 443 (*use*)

1,8-Bis(triphenylphosphoniomethyl)-naphthalene(2+) B-00452

[1,8-Naphthalenediylbis[methylene]]-bis[triphenylphosphonium](2+), 9CI
$C_{48}H_{40}P_2^{\oplus\oplus}$ M 678.792 (ion)
Dibromide: [10038-36-5].
$C_{48}H_{40}Br_2P_2$ M 838.600
Cryst. (MeCN). Mp 322° (311-313°). With NaH in DMSO, yields acenaphthylene. LiOEt → bis-ylide.
Bis-ylide: [*1,8-Naphthalenediylbis[methylene]]-bis[triphenylphosphorane]*.
$C_{48}H_{38}P_2$ M 676.776

Bestmann, H.J. *et al*, *Chem. Ber.*, 1966, **99**, 2848 (*synth*)
Auclair, F. *et al*, *Tetrahedron*, 1975, **31**, 2499 (*synth, use*)
Mitchell, R.H. *et al*, *Can. J. Chem.*, 1977, **55**, 210 (*use*)
Kemp, W. *et al*, *J. Chem. Soc., Perkin Trans. 1*, 1980, 2812 (*use*)

2,3-Bis(triphenylphosphoniomethyl)-naphthalene(2+) B-00453

[2,3-Naphthalenediylbis[methylene]]-bis[triphenylphosphonium](2+), 9CI
$C_{48}H_{40}P_2^{\oplus\oplus}$ M 678.792 (ion)
LiOEt → bis-ylide.
Dibromide: [39013-98-4].
$C_{48}H_{40}Br_2P_2$ M 838.600
Cryst. (EtOH). Mp 316-320°.
Bis-ylide: [*(2,3-Naphthalenediylbis[methylene]]-bis[triphenylphosphorane]*.
$C_{48}H_{38}P_2$ M 676.776
Used in Wittig reactions.

Cava, M.P. *et al*, *J. Am. Chem. Soc.*, 1972, **94**, 6441 (*synth, use*)
Kemp, W. *et al*, *J. Chem. Res. (S)*, 1981, 28 (*use*)

2,6-Bis(triphenylphosphoniomethyl)-naphthalene(2+) B-00454

[2,6-Naphthalenediylbis[methylene]]-bis[triphenylphosphonium](2+), 9CI
$C_{48}H_{40}P_2^{\oplus\oplus}$ M 678.792 (ion)
Dibromide: [25075-81-4].
$C_{48}H_{40}Br_2P_2$ M 838.600
Used in synth. of naphthalenophanes. Solid.

Tanner, D. *et al*, *Acta Chem. Scand., Ser. B*, 1979, **33**, 443 (*use*)
Norinder, U. *et al*, *Acta Chem. Scand., Ser. B*, 1981, **35**, 403.

2,7-Bis(triphenylphosphoniomethyl)-naphthalene(2+) B-00455

[2,7-Naphthalenediylbis[methylene]]-bis[triphenylphosphonium](2+), 9CI
$C_{48}H_{40}P_2^{\oplus\oplus}$ M 678.792 (ion)
Dibromide: [41784-92-3].
$C_{48}H_{40}Br_2P_2$ M 838.600
Used in synth. of helicenes. Solid. Mp >350°. Yields bis-ylide when treated with NaOMe.
Bis-ylide: [*2,7-Naphthalenediylbis[methylene]]-bis[triphenylphosphorane]*.
$C_{48}H_{38}P_2$ M 676.776

Dopper, J.H. *et al*, *J. Am. Chem. Soc.*, 1973, **95**, 3692 (*synth, use*)
Laarhoven, W.H. *et al*, *Tetrahedron*, 1974, **30**, 1101 (*use*)

2,5-Bis[(triphenylphosphonio)methyl]-thiophene(2+) B-00456

[2,5-Thiophenediylbis[methylene]]-bis[triphenylphosphonium](2+), 9CI

$Ph_3\overset{\oplus}{P}CH_2$–(2,5-thiophenediyl)–$CH_2\overset{\oplus}{P}Ph_3$

$C_{42}H_{36}P_2S^{\oplus\oplus}$ M 634.754 (ion)
Dichloride: [25264-54-4].
$C_{42}H_{36}Cl_2P_2S$ M 705.660
Source of ylide. Prisms ($MeOH/Et_2O$). Dec. at 220° without melting. With LiOEt, yields 2,5-bis-ylide.
Bis-ylide: [*2,5-Thiophenediylbis[methylene]]-bis[triphenylphosphorane]*.
$C_{42}H_{34}P_2S$ M 632.738
Used in Wittig reactions for synth. of heteroannulenes.

Cresp, T.M. *et al*, *J. Chem. Soc., Perkin Trans. 1*, 1973, 2961 (*synth, use*)

Strand, A. *et al*, *Acta Chem. Scand., Ser. B*, 1977, **31**, 52 (*use*)
Kossmehl, G. *et al*, *Makromol. Chem.*, 1982, **183**, 2747, 2771 (*use*)

3,4-Bis[(triphenylphosphonio)methyl]-thiophene(1+) B-00457

[3,4-Thiophenediylbis[methylene]]-bis[triphenylphosphonium](1+), 9CI

$C_{42}H_{36}P_2S^{\oplus}$ M 634.754 (ion)

Dibromide: [58776-19-5].
$C_{42}H_{36}Br_2P_2S$ M 794.562
Source of ylide. Solid. Mp 295° dec. Yields the 3,4-bis-ylide when treated with BuLi in THF.

Bis-ylide: [3,4-Thiophenediylbis[methylene]]-bis[triphenylphosphorane].
$C_{42}H_{34}P_2S$ M 632.738
Used in Wittig reactions leading to heteroannulenes.

Wightman, R.H. *et al*, *Tetrahedron Lett.*, 1975, 4179 (*use*)
Jones, R.H. *et al*, *Tetrahedron Lett.*, 1975, 4183 (*synth, use*)

Bis(triphenylphosphoranylidene)hydrazine, 9CI B-00458

Azinobis(triphenylphosphorane), 8CI

[752-23-8]

$Ph_3P{=}NN{=}PPh_3$

$C_{36}H_{30}N_2P_2$ M 552.594
Solid. Mp 183-189° (184°).

B,2HCl: [55370-43-9]. N,N′-*Bis(triphenylphosphonio)-hydrazine dichloride.*
$C_{36}H_{32}Cl_2N_2P_2$ M 625.516
Solid. Mp 265°.

Appel, R. *et al*, *Z. Anorg. Allg. Chem.*, 1968, **363**, 176, 183 (*synth, props*)
Appel, R. *et al*, *Chem. Ber.*, 1975, **108**, 623 (*use*)

Bis(triphenylphosphoranylidene)methane B-00459

Methanetetraylbis[triphenylphosphorane], 9CI

[7533-52-0]

$Ph_3P{=}C{=}PPh_3$

$C_{37}H_{30}P_2$ M 536.592
Tribolumnescent diamond-shaped cryst. (diglyme, slow cryst.) or nontriboluminescent needles (diglyme, fast cryst.) (dimorph.). Mp 216-218° (both forms). Forms 1:1 adducts with S and Se. Triboluminescence is lost on standing.

Ramirez, F. *et al*, *J. Am. Chem. Soc.*, 1961, **83**, 3539 (*synth, ir, uv*)
Vincent, A.T. *et al*, *J. Chem. Soc., Dalton Trans.*, 1972, 617 (*cryst struct*)
Hardy, G.E. *et al*, *J. Am. Chem. Soc.*, 1977, **99**, 3552 (*synth*)
Hardy, G.E. *et al*, *J. Am. Chem. Soc.*, 1978, **100**, 8001 (*raman*)
Hardy, G.E. *et al*, *J. Am. Chem. Soc.*, 1981, **103**, 1074 (*cryst struct*)
Verma, S. *et al*, *Indian J. Chem., Sect. B*, 1981, **20**, 1096 (*synth*)
Schmidbaur, H. *et al*, *Angew. Chem., Int. Ed. Engl.*, 1982, **21**, 310 (*props*)
Houben-Weyl, Method. Org. Chem., Band E1, Ed. M. Regitz, 1982, Thieme Verlag, Stuttgart, 752 (*rev*)

1,2-Bis(triphosphoranylidene)-benzocyclobutene B-00460

Bicyclo[4.2.0]octa-1,3,5-trien-7,8-diylidenebis[triphenylphosphorane], 9CI. 1,2-Bis(triphosphoranylidene)-bicyclo[4.2.0]octa-1,3,5-triene

[1820-39-9]

PPh₃ / PPh₃

$C_{44}H_{34}P_2$ M 624.700
Used in Wittig reactions to synthesise polycyclic hydrocarbons. Deep-red compd., stable at −40° for several hrs.; dec. rapidly at r.t.

Blomquist, A.T. *et al*, *J. Am. Chem. Soc.*, 1964, **86**, 5041 (*synth, use*)
Garrett, P.J. *et al*, *J. Chem. Soc., Perkin Trans. 1*, 1973, 2253 (*use*)
Sondheimer, F. *et al*, *J. Org. Chem.*, 1981, **46**, 4594 (*use*)

Bomyl B-00461

Dimethyl 3-[(dimethoxyphosphinyl)oxy]-2-pentene-dioate, 9CI. 1,3-Bis(methoxycarbonyl)-1-propen-2-yl dimethyl phosphate. 3-Hydroxyglutaconic acid dimethyl ester dimethyl phosphate

[122-10-1]

$MeOOCCH{=}C(OP(O)(OMe)_2)CH_2COOMe$

$C_9H_{15}O_8P$ M 282.186
Agricultural and public health insecticide and acaricide. Yellow oil. Mod. sol. H_2O. $Bp_{1.7}$ 155-164° (mixture of isomers).

▷LZ9450000.

***(E)*-form** [15272-77-2]
$Bp_{0.05}$ 130-131°. n_D^{25} 1.4515.

***(Z)*-form** [15272-78-3]
Bp_1 148-150°.

U.S.P., 2 891 887, (*1957*); *CA*, **53**, 17415b (*synth*)
Newallis, P.E. *et al*, *J. Agric. Food Chem.*, 1967, **15**, 940 (*synth, pharmacol*)
Getz, M.E. *et al*, *J. Assoc. Off. Anal. Chem.*, 1968, **51**, 1101 (*tlc*)
Ross, R.T. *et al*, *Anal. Chim. Acta*, 1970, **52**, 139 (*P nmr*)
Pesticide Manual, 6th Ed., 201.

Bromfenvinfos B-00462

2-Bromo-1-(2,4-dichlorophenyl)ethenyl diethyl phosphate, 9CI. 2-Bromo-1-(2,4-dichlorophenyl)vinyl diethyl phosphate

[33399-00-7]

$(RO)_2P(O)O$, H, C=C, Br, Cl, Cl *(E)-form*

R = Et

$C_{12}H_{14}BrCl_2O_4P$ M 404.024
Insecticide. Liq. $Bp_{0.006}$ 123-126°.

▷TB8450000.

***(E)*-form** [58580-14-6]
Liq. Obt. only as mixt. with *(Z)*-form.

(Z)-form [58580-13-5]
Liq. $Bp_{0.008}$ 126-128°. n_D^{20} 1.5428. Contained 12% (*E*)-form.

Sledzinski, B. *et al*, *Organika*, 1977, 11, 19; 1979, 1; *CA*, **90**, 102961; **94**, 83719 (*synth, ir, pmr*)
Malinowski, H. *et al*, *Organika*, 1977, 96; *CA*, **89**, 124494 (*use*)
Zalucki, A., *Organika*, 1977, 131; *CA*, **89**, 71905 (*tox*)
Gwiazda, M., *Organika*, 1977, 140; *CA*, **95**, 168632 (*metab*)

Bromfenvinfos-methyl B-00463

2-Bromo-1-(2,4-dichlorophenyl)ethenyl dimethyl phosphate, 9CI. *2-Bromo-1-(2,4-dichlorophenyl)vinyl dimethyl phosphate*

[13104-21-7]

As Bromfenvinfos, B-00462 with

R = Me

$C_{10}H_{10}BrCl_2O_4P$ M 375.970
Insecticide. Cryst. (CCl_4/hexane). Mp 65.5-67°.
▷TB8460000.

(E)-form [68107-00-6]
Obt. only as mixt. with (*Z*)-form.

(Z)-form [68107-01-7]
Cryst. (cyclohexane or CCl_4/hexane). Mp 67-67.5°. Mixed with <7% (*E*)-form.

Śledziński, B. *et al*, *Organika*, 1977, 11, 19; *CA*, **90**, 22160, 102961 (*synth, pmr*)
Malinowski, H. *et al*, *Organika*, 1977, 96; *CA*, **89**, 124494 (*use*)
Galdecki, Z. *et al*, *Phosphorus Sulfur*, 1979, **5**, 299 (*cryst struct*)
Śledziński, B. *et al*, *Organika*, 1979, 1; *CA*, **94**, 83719 (*synth, ir, pmr*)

2-Bromo-1,3,2-benzodioxaphosphole, 9CI B-00464

o-*Phenylene phosphorobromidite*. o-*Phenylene bromophosphite*

[3583-02-6]

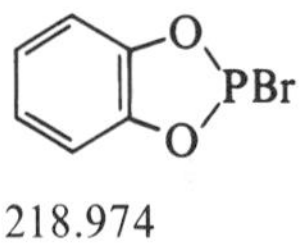

$C_6H_4BrO_2P$ M 218.974
Cleaves silyl ethers. Air-sensitive liq. d_4^{20} 1.719. Bp_{13} 105-106°, $Bp_{0.003}$ 48-52°. n_D^{20} 1.6150.

Oxide: [3492-46-4]. o-*Phenylene phosphorobromidate*.
$C_6H_4BrO_3P$ M 234.974
Solid. Mp 35-39°. Bp_{15} 132-135°. Sensitive to moisture.

Gross, H. *et al*, *J. Prakt. Chem.*, 1965, **29**, 315 (*synth, use, oxide*)
Fluck, E. *et al*, *Z. Naturforsch., B*, 1966, **21**, 1125 (*P nmr*)
Gloede, J. *et al*, *Z. Anorg. Allg. Chem.*, 1980, **471**, 147 (*synth, oxide, P nmr*)
Deng, R.M.K. *et al*, *J. Chem. Soc., Dalton Trans.*, 1984, 1917 (*complexes*)

2-Bromo-5-bromomethyl-5-methyl-1,3,2-dioxaphosphorinane B-00465

2-Bromo-5-bromomethyl-5-methyl-1,3-dioxa-2-phosphacyclohexane. *2-Bromo-5-bromomethyl-5-methyl-1,3,2-dioxaphosphinane*. *(2-Methyl-2-bromomethyl)-trimethylene bromophosphite*

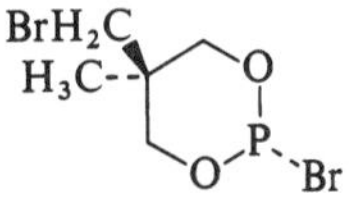

$C_5H_9Br_2O_2P$ M 291.907

trans-form

2-Oxide: [16979-13-8]. *(2-Methyl-2-bromomethyl)-trimethylene phosphoryl bromide*.
$C_5H_9Br_2O_3P$ M 307.906
Cryst. (hexane). Mp 81-82°. Also referred to as the *cis*-form.

Beinecke, T.A., *Acta Crystallogr., Sect. B*, 1969, **25**, 413 (*cryst struct*)
Corriu, R.J.P. *et al*, *Tetrahedron*, 1979, **35**, 2889; 1981, **37**, 3681 (*synth, props*)
Van Nuffel, P. *et al*, *J. Mol. Struct.*, 1984, **125**, 1 (*struct*)

(4-Bromobutyl)phosphonic acid, 9CI B-00466

$BrCH_2(CH_2)_3P(O)(OH)_2$

$C_4H_{10}BrO_3P$ M 216.999
Plates (Et_2O). Mp 126.5-127.5°.

Di-Et ester: [63075-66-1]. *Diethyl (4-bromobutyl)-phosphonate*.
$C_8H_{18}BrO_3P$ M 273.106
Liq. $Bp_{0.1}$ 120° (bath).

Helferich, B. *et al*, *Justus Liebigs Ann. Chem.*, 1962, **655**, 59 (*synth, derivs*)
Eberhard, A. *et al*, *J. Am. Chem. Soc.*, 1965, **87**, 253 (*synth, ir*)
Hewitt, D.G. *et al*, *Aust. J. Chem.*, 1977, **30**, 579 (*ester, synth, pmr*)

Bromo(diethoxyphosphinyl)acetic acid, 9CI B-00467

$(EtO)_2P(O)CHBrCOOH$

$C_6H_{12}BrO_5P$ M 275.036
Esters are Wittig-Horner reagents for synth. of α-bromo-α,β-unsatd. carboxylic acids and esters.

(±)-form

Me ester: [23755-72-8]. *Methyl bromo(diethoxyphosphinyl)acetate*.
$C_7H_{14}BrO_5P$ M 289.062
Liq. d_4^{20} 1.40. Bp_3 142-147°. n_D^{20} 1.4610.

Et ester: [23755-73-9]. *Ethyl bromo(diethoxyphosphinyl)acetate*.
$C_8H_{16}BrO_5P$ M 303.089
Liq. d_4^{20} 1.37. Bp_1 128-131°. n_D^{20} 1.4632.

Grinev, G.V. *et al*, *Zh. Obshch. Khim.*, 1969, **39**, 1253 (*Engl. transl.* p. 1223) (*synth*)
Semmelhack, M.F. *et al*, *J. Am. Chem. Soc.*, 1981, **103**, 3945 (*use*)

(Bromodifluoromethyl)-triphenylphosphonium(1+), 9CI B-00468

$Ph_3P^{\oplus}CF_2Br$

$C_{19}H_{15}BrF_2P^{\oplus}$ M 392.202 (ion)

Halomethyl transfer agent. In the presence of Cd, Zn, or Hg converts ketones or aldehydes into 1,1-difluoro-1-alkenes. In presence of CsF, is a source of difluorocarbene. Hydrol. yields Ph_3PO and CHF_2Br.

Bromide: [58201-66-4].
$C_{19}H_{15}Br_2F_2P$ M 472.106
Tan-coloured powder. Mp 206.5° dec.

Burton, D.J. *et al, J. Fluorine Chem.*, 1981, **18**, 573 (*props*)
Flynn, R.M. *et al, J. Fluorine Chem.*, 1981, **18**, 525 (*props*)
Burton, D.J. *et al, J. Fluorine Chem.*, 1982, **20**, 89 (*synth, props*)
Sohn, D. *et al, Chem. Ber.*, 1982, **115**, 3334 (*props*)

1-Bromo-2,5-dihydro-3,4-dimethyl-1*H*-phosphole, 9CI B-00469

1-Bromo-3,4-dimethyl-3-phospholene, 8CI
[28273-34-9]

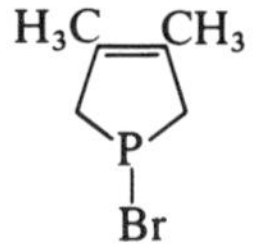

$C_6H_{10}BrP$ M 193.023
Liq. Bp_{26} 103°.
1-Oxide:
$C_6H_{10}BrOP$ M 209.022
Solid. Mp 85-87°. $Bp_{0.55}$ 102-104°.
1-Sulfide: [42202-33-5].
$C_6H_{10}BrPS$ M 225.083
Liq. d_4^{20} 1.49. $Bp_{0.02}$ 70-72°. n_D^{20} 1.6024.

Arbuzov, B.A. *et al, Dokl. Akad. Nauk SSSR, Ser. Sci. Khim.*, 1964, **159**, 582 (*Engl. transl.* p. 1205) (*oxide, synth, pmr*)
Myers, D.K. *et al, J. Org. Chem.*, 1971, **36**, 1285 (*synth, ir, pmr, P nmr*)
Arbuzov, B.A. *et al, Izv. Akad. Nauk SSSR, Ser. Khim.*, 1973, 1176 (*Engl. transl.* p. 1144) (*sulfide*)
Hammond, P.J. *et al, J. Chem. Soc., Perkin Trans. 2*, 1982, 205 (*synth, pmr, P nmr*)
Buchanan, G.W. *et al, Org. Magn. Reson.*, 1983, **21**, 436 (*derivs, cmr*)

2-Bromo-4,5-dimethyl-1,3,2-dioxaphospholane, 8CI B-00470

2,3-Butylene bromophosphite. 2,3-Butylene phosphorobromidite. 2-Bromo-4,5-dimethyl-1,3-dioxa-2-phosphacyclopentane
[21992-80-3]

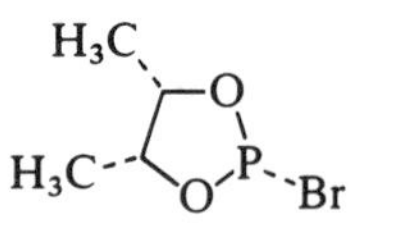

(2α,4α,5α)-*form*

$C_4H_8BrO_2P$ M 198.984

(2α,4α,5α)-form
meso-cis-*form*
2-Sulfide: [66242-31-7]. O,O-*2,3-Butylene bromothiophosphate.* O,O-*2,3-Butylene phosphorobromidothioate.* O,O-*2,3-Butylene thiophosphoryl bromide.*
$C_4H_8BrO_2PS$ M 231.044
Cryst. (C_6H_6/pet. ether). Mp 56-58°.

(2α,4β,5β)-form [66289-12-1]
meso-trans-*form*
Fuming liq. d_4^{20} 1.57. $Bp_{0.08}$ 45°. n_D^{20} 1.5095. 90% Diastereoisomeric purity. Easily hydrolysed.
▷Reacts violently with H_2O
2-Sulfide: [66289-11-0]. Liq. $Bp_{1.5}$ 60-65°.

Voznesenskaya, A.Kh. *et al, Zh. Obshch. Khim.*, 1969, **39**, 387 (*Engl. transl.* p. 365) (*synth, ir*)
Mikolajczyk, M. *et al, J. Chem. Soc., Perkin Trans. 1*, 1977, 2213 (*sulfide, props, P nmr, cmr*)

2-Bromo-5,5-dimethyl-1,3,2-dioxaphosphorinane, 9CI B-00471

2-Bromo-5,5-dimethyl-1,3,2-dioxaphosphinane. 2-Bromo-5,5-dimethyl-1,3-dioxa-2-phosphacyclohexane. Neopentylene phosphorobromidite

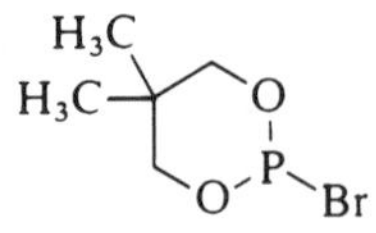

$C_5H_{10}BrO_2P$ M 213.011
Fuming liq. Bp_{18} 72°.
2-Oxide: [16368-12-0]. *Neopentylene phosphorobromidate.*
$C_5H_{10}BrO_3P$ M 229.010
Cryst. (EtOAc). Mp 98-100°.
2-Sulfide: [16368-13-1]. *Neopentylene phosphorobromidothioate.*
$C_5H_{10}BrO_2PS$ M 245.071
Pale-yellow cryst. (EtOAc). Mp 114-116°.

Stec, W. *et al, Can. J. Chem.*, 1967, **45**, 2513 (*derivs, synth, ir*)
White, D.W. *et al, J. Am. Chem. Soc.*, 1970, **92**, 7125 (*synth, pmr*)

(2-Bromoethyl)phosphonic acid, 9CI B-00472

[999-82-6]

$BrCH_2CH_2P(O)(OH)_2$

$C_2H_6BrO_3P$ M 188.945
Stimulates flow of latex from *Hevea brasiliensis*. Mp 86-87°. pK_{a1} 2.55, pK_{a2} 5.8 (H_2O, 25°C).
Di-Et ester: [5324-30-1]. *Diethyl (2-bromoethyl)-phosphonate.*
$C_6H_{14}BrO_3P$ M 245.053
Bp_2 86-87°, $Bp_{0.8}$ 101°. n_D^{25} 1.4555.
Bis(2-bromoethyl)ester: Bis(2-bromoethyl) (2-bromoethyl)phosphonate.
$C_6H_{12}Br_3O_3P$ M 402.845
Solid. Mp 48-49°.
Dichloride: [28482-01-1].
$C_2H_4BrCl_2OP$ M 225.837
Liq. d_4^{20} 1.82. Bp_{18} 119-120°. n_D^{20} 1.5210.
Dibromide: [25196-01-4].
$C_2H_4Br_3OP$ M 314.739
Liq. Bp_4 94°.
Dianilide: P-(*2-Bromoethyl*)-N,N′-*diphenylphosphonic diamide. (2-Bromoethyl)phosphonic dianilide.*
$C_{14}H_{16}BrN_2OP$ M 339.171
Solid. Mp 169-170°.

Ford-Moore, A.H. *et al, J. Chem. Soc.*, 1947, 1465 (*ester, synth*)
Kabachnik, M.I. *et al, Izv. Akad. Nauk SSSR, Ser. Khim.*, 1947, 389 (*synth, derivs*)
Kosolopoff, G.M. *et al, J. Am. Chem. Soc.*, 1948, **70**, 1971 (*ester, synth*)
Jones, G. *et al, Naturwissenschaften*, 1969, **56**, 637 (*dibromide, synth, pmr*)
Bel'skii, V.E. *et al, Izv. Akad. Nauk SSSR, Ser. Khim.*, 1975, 1047 (*Engl. transl.* p. 958) (*ester, pmr, nmr*)
Blackburn, G.M. *et al, J. Chem. Soc., Perkin Trans. 1*, 1980, 1150 (*synth*)

2-Bromo-4-methyl-1,3,2-dioxaphospholane, 9CI **B-00473**

Propylene bromophosphite. Propylene phosphorobromidite

[21992-79-0]

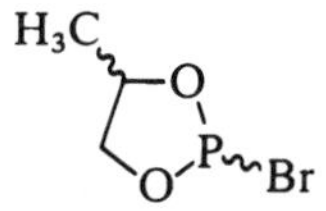

$C_3H_6BrO_2P$ M 184.957

Fuming liq. d_4^{20} 1.68. $Bp_{0.08}$ 48°. n_D^{20} 1.5269. Mixt. of *cis*- and *trans*-forms.

Voznesenskaya, A.Kh. *et al, Zh. Obshch. Khim.*, 1969, **39**, 387 (*Engl. transl.* p. 365) (*synth*)

2-Bromo-4-methyl-1,3,2-dioxaphosphorinane, 9CI **B-00474**

2-Bromo-4-methyl-1,3,2-dioxaphosphinane. 1,3-Butylene bromophosphite. 2-Bromo-4-methyl-1,3-dioxa-2-phosphacyclohexane

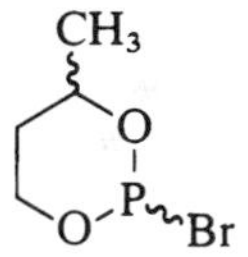

$C_4H_8BrO_2P$ M 198.984

2-Oxide: 1,3-Butylene phosphoryl bromide. 1,3-Butylene bromophosphate.
$C_4H_8BrO_3P$ M 214.983
No phys. props. described. *Cis*- and *trans*-forms characterized spectroscopically.

2-Sulfide: OO-*1,3-Butylene thiophosphoryl bromide.* O,O-*1,3-Butylene bromothiophosphate.*
$C_4H_8BrO_2PS$ M 231.044
No phys. props. described. *Cis*- and *trans*-forms characterised spectroscopically.

Stec, W. *et al, Tetrahedron*, 1973, **29**, 539 (*oxide, synth, ir, pmr, P nmr*)
Mikolajczyk, M. *et al, Tetrahedron Lett.*, 1975, 1607 (*sulfide, props*)
Lopusinski, A. *et al, Justus Liebigs Ann. Chem.*, 1977, 924 (*oxide, P nmr, props*)

(Bromomethyl)phosphonic acid, 9CI **B-00475**

[7582-40-3]

$BrCH_2P(O)(OH)_2$

CH_4BrO_3P M 174.919

Very hygroscopic solid. pK_{a1} 1.44, pK_{a2} 4.75 (H_2O, 25°).

Mono-anilinium salt: Cryst. (EtOH). Mp 187° dec.

Di-Et ester: [66197-72-6]. *Diethyl (bromomethyl)-phosphonate.*
$C_5H_{12}BrO_3P$ M 231.026
Liq. d_{20}^{20} 1.45. Bp_1 99°. n_D^{20} 1.4585.

Dichloride:
CH_2BrCl_2OP M 211.810
Liq. Bp_7 118-120°. n_D^{20} 1.512.

Dibromide: [6259-97-8].
CH_2Br_3OP M 300.712
Liq. d_4^{20} 2.68. $Bp_{0.3}$ 103°. n_D^{22} 1.6108.

Yakubovich, A.Ya. *et al, Zh. Obshch. Khim.*, 1952, **22**, 1534 (*Engl. transl.* p. 1575) (*synth, derivs*)
Cade, J.A., *J. Chem. Soc.*, 1959, 2266 (*dichloride*)
Edmundson, R.S. *et al, J. Chem. Soc. (C)*, 1966, 1096 (*dibromide*)
Griffiths, W.R. *et al, Phosphorus*, 1975, **5**, 273 (*ms*)
Griffiths, W.R. *et al, Phosphorus Sulfur*, 1978, **4**, 341; 1978, **5**, 101 (*dibromide, ester, ms*)

(Bromomethyl)phosphonous dibromide, 9CI **B-00476**

Dibromo(bromomethyl)phosphine

$BrCH_2PBr_2$

CH_2Br_3P M 284.713

Red liq. d_{20}^{20} 2.64. $Bp_{3.25}$ 63-65°. n_D^{21} 1.6607. Fumes in air.

Yakubovich, A.Y. *et al, Zh. Obshch. Khim.*, 1952, **22**, 1534; *CA*, **47**, 9254 (*synth*)
Edmundson, R.S. *et al, J. Chem. Soc. (C)*, 1966, 1096 (*synth*)
Prishchenko, A.A. *et al, Zh. Obshch. Khim.*, 1985, **55**, 340; *J. Gen. Chem. USSR (Engl. Transl.)*, 299 (*cmr*)

(Bromomethyl)triphenylphosphonium(1+), 9CI, 8CI **B-00477**

$Ph_3P^{\oplus}CH_2Br$

$C_{19}H_{17}BrP^{\oplus}$ M 356.221 (ion)

Treatment of salts with Li piperidide, butyllithium, or potassium *tert*-butoxide, yields the ylide.

Bromide: [1034-49-7].
$C_{19}H_{17}Br_2P$ M 436.125
Cryst. (H_2O). Mp 242-243.5°.

Tetrafluoroborate: [80249-44-1].
$C_{19}H_{17}BBrF_4P$ M 443.025
Solid. Mp 198°.

Tetraphenylborate:
$C_{43}H_{37}BBrP$ M 675.453
Solid. Mp 214-216°.

Ylide: [39598-55-5]. *(Bromomethylene)-triphenylphosphorane.*
$C_{19}H_{16}BrP$ M 355.213
Reactive Wittig reagent useful for conversion of aldehydes into alkenes and alkynes. Prepd. *in situ*.

Köbrich, G., *Angew. Chem.*, 1962, **74**, 33 (*ylide, synth, use*)
Seyferth, D. *et al, J. Organomet. Chem.*, 1966, **5**, 267 (*salts, ylide*)
Gallagher, M.J., *Aust. J. Chem.*, 1968, **21**, 1197 (*pmr*)
Matsumoto, M. *et al, Tetrahedron Lett.*, 1980, **21**, 4021 (*ylide, use*)
Regitz, M. *et al, Justus Liebigs Ann. Chem.*, 1981, 1865 (*synth, ir, pmr*)

(3-Bromo-2-oxopropyl)-triphenylphosphonium(1+), 9CI **B-00478**

(3-Bromoacetonyl)triphenylphosphonium, 8CI

$Ph_3P^{\oplus}CH_2COCH_2Br$

$C_{21}H_{19}BrOP^{\oplus}$ M 398.258 (ion)

Bromide: [19753-64-1].
$C_{21}H_{19}Br_2OP$ M 478.162
Solid. Mp 222-224° dec. pK_a 4.53 (25°, 67% EtOH aq.). Dil. $OH^{\ominus}$ yields the ylide.

Ylide: see 1-Bromo-3-(triphenylphosphoranylidene)-2-propanone, B-00506

Issleib, K. *et al, Justus Liebigs Ann. Chem.*, 1968, **713**, 19 (*synth, props*)
Zhdanov, Yu.A. *et al, Zh. Obshch. Khim.*, 1972, **42**, 759 (*Engl. transl.* p. 751) (*synth, ylide, ir, use*)

2-(4-Bromophenyl)-8,8-dihydro-4,6,8,8-tetraphenyl[1,2]oxaphospholo[2,3-*b*][1,2,5]dioxaphosphorin, 9CI **B-00479**

[34306-15-5]

$C_{35}H_{26}BrO_3P$ M 605.466

Yellow cryst. (CH_2Cl_2). Mp 170-172°. Sensitive to moisture.

Swank, D.D. *et al, J. Am. Chem. Soc.*, 1971, **93**, 5236 (*synth, ir, P nmr, cryst struct*)

(2-Bromophenyl)diphenylphosphine, 9CI **B-00480**

o-(*Diphenylphosphino*)*bromobenzene*

[62336-24-7]

$C_{18}H_{14}BrP$ M 341.186

Ligand for Pd. Cryst. (EtOH or MeOH). Mp 113-114°.

B,PhCH₂Cl: Benzyl-2-bromophenyldiphenylphosphonium chloride.

$C_{25}H_{21}BrClP$ M 467.772

Cryst. ($CHCl_3/CCl_4$). Mp 268-269°.

Luckenbach, R. *et al, Z. Naturforsch., B*, 1977, **32**, 1038 (*synth*)
McEwen, W.E. *et al, J. Am. Chem. Soc.*, 1978, **100**, 7304 (*synth, derivs, pmr*)
Talay, R. *et al, Z. Naturforsch., B*, 1981, **36**, 451 (*synth*)

(4-Bromophenyl)diphenylphosphine, 9CI **B-00481**

p-(*Diphenylphosphino*)*bromobenzene*

[734-59-8]

$C_{18}H_{14}BrP$ M 341.186

Cryst. ($CHCl_3/CCl_4$). Mp 79-80°. $Bp_{0.05}$ 188-195°. pK_a 2.09.

B,PhCH₂Cl: Benzyl-4-bromophenyldiphenylphosphonium bromide.

$C_{25}H_{21}BrClP$ M 467.772

Cryst. ($CHCl_3/CCl_4$).

Oxide:

$C_{18}H_{14}BrOP$ M 357.186

Cryst. (EtOH/pet. ether or MeOH aq.). Mp 133-134°, 154-156° (dimorph.).

Sulfide: [1474-09-5].

$C_{18}H_{14}BrPS$ M 373.246

Cryst. (MeOH or EtOH). Mp 135°, 142-144°.

Goetz, H. *et al, Justus Liebigs Ann. Chem.*, 1963, **665**, 1 (*synth, ir, uv, derivs*)
Kuhn, H.J. *et al, Naturwissenschaften*, 1966, **53**, 359 (*cryst struct*)
Schiemenz, G.P., *Chem. Ber.*, 1966, **99**, 504 (*synth, derivs*)
Dreissig, W. *et al, Acta Crystallogr., Sect. B*, 1971, **27**, 1140 (*oxide, cryst struct*)
Dreissig, W. *et al, Acta Crystallogr., Sect. B*, 1972, **28**, 3473 (*sulfide, cryst struct*)

Bychkov, N.N. *et al, Zh. Obshch. Khim.*, 1978, **48**, 2625 (*Engl. transl.* p. 2783) (*synth*)
McEwen, W.E. *et al, J. Am. Chem. Soc.*, 1978, **100**, 7304 (*synth, pmr, deriv*)

(Bromophenylmethyl)phosphonic acid, 9CI **B-00482**

(*α-Bromobenzyl*)*phosphonic acid*

[40962-35-4]

PhCHBrP(O)(OH)₂

$C_7H_8BrO_3P$ M 251.016

(±)-*form*

Solid. Mp 139-142°.

Di-Et ester: Diethyl (bromophenylmethyl)phosphonate. Diethyl (α-bromobenzyl)phosphonate.

$C_{11}H_{15}BrO_3P$ M 306.115

Liq. d_4^{20} 1.35. Bp_1 164-166°. n_D^{20} 1.5310.

Grinev, G.V. *et al, Zh. Obshch. Khim.*, 1969, **39**, 1253 (*Engl. transl.* p. 1223) (*ester*)
Chervenyuk, G.I. *et al, Zh. Obshch. Khim.*, 1972, **42**, 2183 (*Engl. transl.*, p. 2180) (*synth*)

[(2-Bromophenyl)methyl]phosphonic acid, 9CI **B-00483**

o-*Bromobenzylphosphonic acid*

[80395-13-7]

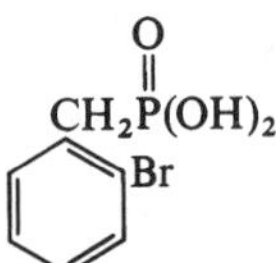

$C_7H_8BrO_3P$ M 251.016

Di-Et ester: [63909-55-7]. *Diethyl [(2-bromophenyl)methyl]phosphonate. Diethyl* o-*bromobenzylphosphonate.*

$C_{11}H_{16}BrO_3P$ M 307.123

Used in Wittig-Horner reactions. Liq. Bp_{34} 211-213°, $Bp_{0.02}$ 112-122°. n_D^{22} 1.5238.

Ernst, L., *Org. Magn. Reson.*, 1977, **9**, 35 (*synth, cmr, P nmr*)
Gronowitz, S. *et al, Heterocycles*, 1981, **15**, 947 (*synth, pmr, use*)

[(4-Bromophenyl)methyl]phosphonic acid, 9CI **B-00484**

p-*Bromobenzylphosphonic acid*

[40962-34-3]

$C_7H_8BrO_3P$ M 251.016

Di-Et ester: [38186-51-5]. *Diethyl [(4-bromophenyl)methyl]phosphonate.*

$C_{11}H_{16}BrO_3P$ M 307.123

Used in Wittig-Horner reactions. Liq. $Bp_{0.6}$ 183°.

Doty, J.C. *et al, J. Organomet. Chem.*, 1972, **38**, 229 (*synth, use*)

4-(4-Bromophenyl)-2′,2′,3′,4′,4′-pentamethyl-3,3,6,6-tetrakis(trifluoromethyl)spiro[2,7-dioxa-1-phosphabicyclo[3.2.0]hept-3-ene-1,1′-phosphetane] B-00485

[60797-48-0]

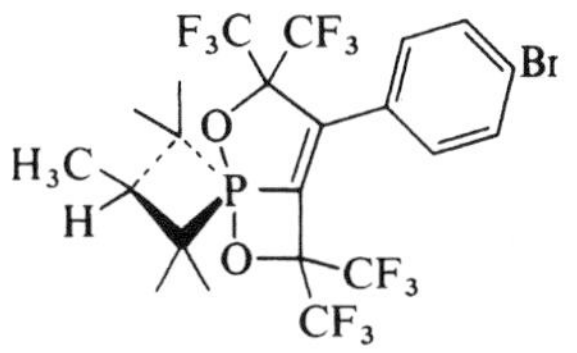

$C_{22}H_{20}BrF_{12}O_2P$ M 655.257

Solid. Mp 176-177°.

Aly Hassan, A.E. *et al*, *J. Chem. Soc., Chem. Commun.*, 1976, 449 (*synth, P nmr, cryst struct*)

(2-Bromophenyl)phosphine, 9CI B-00486

[53772-58-0]

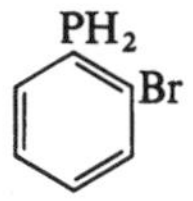

C_6H_6BrP M 188.991

Liq.

Maier, L., *Phosphorus*, 1974, **4**, 41 (*pmr, nmr*)

(4-Bromophenyl)phosphine, 9CI B-00487

[53772-55-7]

C_6H_6BrP M 188.991

Low-melting solid. Mp 40°, Mp 50°. Bp 195-196°, Bp_1 53-56°.

Maier, L., *Phosphorus*, 1974, **4**, 41 (*synth, pmr, nmr*)

(3-Bromophenyl)phosphinic acid, 9CI B-00488

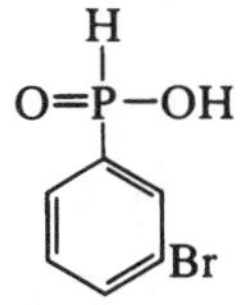

$C_6H_6BrO_2P$ M 220.990

Tautomeric with 3-bromophenylphosphonous acid, but free acid probably exists in phosphinic acid form. Cryst. (CCl_4). Mp 97-98°. pK_a 1.39 (H_2O).

Me ester: Methyl 3-bromophenylphosphinate.

$C_7H_8BrO_2P$ M 235.017

Liq. $Bp_{0.15}$ 123-124°.

Quin, L.D. *et al*, *J. Am. Chem. Soc.*, 1961, **83**, 4124 (*synth*)
Quin, L.D. *et al*, *J. Org. Chem.*, 1962, **27**, 1012 (*ester, ir*)

(4-Bromophenyl)phosphinic acid, 9CI B-00489

[39238-95-4]

$C_6H_6BrO_2P$ M 220.990

Tautomeric with (4-Bromophenyl)phosphonous acid, B-00493 but free acid exists in phosphinic acid form. Cryst. (EtOH aq.). Mp 140-140.5°. pK_a 1.25 (H_2O), ca. 3.0 (EtOH).

Et ester: [38766-19-7]. *Ethyl (4-bromophenyl)-phosphinate.*

$C_8H_{10}BrO_2P$ M 249.044

Characterised spectroscopically.

Weil, T. *et al*, *Helv. Chim. Acta*, 1953, **36**, 1314 (*synth, uv*)
Vinogradov, L.I. *et al*, *Zh. Obshch. Khim.*, 1972, **42**, 1724 (*Engl. transl.* p. 1712) (*ester, pmr, P nmr*)
Barabanov, V.P. *et al*, *Zh. Obshch. Khim.*, 1973, **43**, 1147, 1517 (*Engl. transl.* p. 1138, 1391) (*acid, props*)

(2-Bromophenyl)phosphonic acid, 9CI B-00490

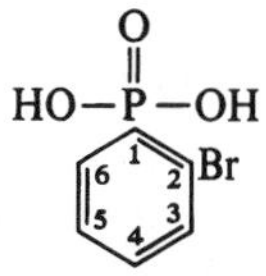

$C_6H_6BrO_3P$ M 236.989

Solid. Mp 199-201°. pK_{a1} 1.64 (H_2O), 2.91 (50% EtOH aq.), pK_{a2} 7.00 (H_2O), 8.22 (50% EtOH aq.).

Di-Me ester: Dimethyl (2-bromophenyl)phosphonate.

$C_8H_{10}BrO_3P$ M 265.043

Liq. $Bp_{0.2}$ 113-115°.

Di-Et ester: [77526-90-1]. *Diethyl (2-bromophenyl)-phosphonate.*

$C_{10}H_{14}BrO_3P$ M 293.097

Liq. $Bp_{0.1}$ 145°.

Dichloride: [67395-78-2].

$C_6H_4BrCl_2OP$ M 273.881

No phys. props. reported.

Doak, G.O. *et al*, *J. Am. Chem. Soc.*, 1952, **74**, 753 (*synth*)
Obrycki, R. *et al*, *Tetrahedron Lett.*, 1966, 5049 (*ester*)
Hirao, T. *et al*, *Synthesis*, 1981, 56 (*ester, ir*)

(3-Bromophenyl)phosphonic acid, 9CI B-00491

[6959-02-0]

$C_6H_6BrO_3P$ M 236.989

Plates. Mp 152-153°. pK_{a1} 1.84 (H_2O), 3.15 (50%, EtOH aq.), pK_{a2} 5.19 (H_2O), 6.08 (50%, EtOH aq.).

Di-Me ester: Dimethyl (3-bromophenyl)phosphonate.

$C_8H_{10}BrO_3P$ M 265.043

Liq. $Bp_{0.2}$ 104-105°.

Di-Et ester: [35125-65-6]. *Diethyl (3-bromophenyl)-phosphonate.*

$C_{10}H_{14}BrO_3P$ M 293.097

Liq. $Bp_{0.2}$ 110-114°. n_D^{25} 1.5200.

Dichloride: [65442-15-1].

$C_6H_4BrCl_2OP$ M 273.881

Liq. $Bp_{0.24}$ 94-100°.

Doak, G.O. *et al*, *J. Am. Chem. Soc.*, 1953, **75**, 683 (*synth*)
Obrycki, R. *et al*, *Tetrahedron Lett.*, 1966, 5049 (*ester*)
Bunnett, J.F. *et al*, *J. Org. Chem.*, 1978, **43**, 1873 (*synth, ester*)
Denham, J.M. *et al*, *J. Org. Chem.*, 1958, **23**, 1298 (*derivs*)

(4-Bromophenyl)phosphonic acid, 9CI B-00492

[16839-13-7]

$C_6H_6BrO_3P$ M 236.989

Solid. Mp 185-190°. pK_{a1} 3.18 (50%, EtOH aq.), 1.9 (H_2O), pK_{a2} 6.13 (50%, EtOH aq.), 5.1 (H_2O).

▷SZ6870300.

Di-Me ester: Dimethyl (4-bromophenyl)phosphonate.

$C_8H_{10}BrO_3P$ M 265.043

Liq. $Bp_{0.1}$ 95-96°.

Di-Et ester: [20677-12-7]. *Diethyl (4-bromophenyl)-phosphonate.*

$C_{10}H_{14}BrO_3P$ M 293.097

Liq. d_4^{20} 1.39. Bp_{21} 191.5-192.5°, $Bp_{0.4}$ 117-120°. n_D^{20} 1.5258.

Bis(trimethylsilyl) ester: [99136-08-0]. *Bis(trimethylsilyl)(4-bromophenyl)phosphonate.*
$C_{12}H_{22}BrO_3PSi_2$ M 381.353
Liq. $Bp_{0.001}$ 110-125°.

Dichloride: [4648-58-2].
$C_6H_4BrCl_2OP$ M 273.881
Solid. Mp 46-48°. Bp 290-291°, Bp_{10} 157-159°.

Mastalerz, M. *et al, Rocz. Chem.*, 1965, **39**, 951; *CA*, **64**, 3591 (*dichloride*)
Obrycki, R. *et al, Tetrahedron Lett.*, 1966, 5049 (*dimethyl ester*)
Kharrasova, F.M. *et al, Zh. Obshch. Khim.*, 1968, **38**, 1262 (*Engl. transl.* p. 1215) (*dimethyl ester*)
Williams, A. *et al, J. Chem. Soc., Perkin Trans. 2*, 1973, 25 (*synth*)
Allen, D.W. *et al, J. Chem. Soc., Perkin Trans. 2*, 1977, 789 (*ester, P nmr, pmr*)
Remizov, A.B., *Zh. Prikl. Spektrosk.*, 1978, **29**, 288 (*Engl. transl.* p. 958) (*dichloride, ir*)
Issleib, K. *et al, Z. Anorg. Allg. Chem.*, 1985, **529**, 151 (*silyl ester, synth, P nmr*)

(4-Bromophenyl)phosphonous acid, 9CI B-00493

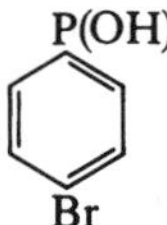

$C_6H_6BrO_2P$ M 220.990

The free acid exists as the phosphoryl tautomer (4-Bromophenyl)phosphinic acid, B-00489 .

Di-Me ester: Dimethyl (4-bromophenyl)phosphonite.
$C_8H_{10}BrO_2P$ M 249.044
Liq. d_4^{20} 1.435. Bp_{16} 134.5-135.5°. n_D^{20} 1.5660.

Mono-Et ester: see (4-Bromophenyl)phosphinic acid, B-00489

Di-Et ester: Diethyl (4-bromophenyl)phosphonite.
$C_{10}H_{14}BrO_2P$ M 277.097
Liq. d_4^{20} 1.32. Bp_{11} 149-150°. n_D^{20} 1.5432.

Di-Ph ester: [4762-37-2]. *Diphenyl (4-bromophenyl)phosphonite.*
$C_{18}H_{14}BrO_2P$ M 373.185
Liq. d_4^{20} 1.38. Bp_5 218-220°. n_D^{20} 1.6283.

Dichloride: [4762-31-6]. *4-Bromophenyldichlorophosphine.*
$C_6H_4BrCl_2P$ M 257.881
Liq. Bp_{14} 140°.

Dibromide: [70430-42-1]. *Dibromo-4-bromophenylphosphine.*
$C_6H_4Br_3P$ M 346.783
Liq.-solid. Mp 15-16.5°. $Bp_{0.01}$ 99-101°.

Diiodide: [26943-92-0]. *4-Bromophenyldiiodophosphine.*
$C_6H_4BrI_2P$ M 440.784
Liq. $Bp_{0.05}$ 109-110°.

Kharrasova, F.M. *et al, Zh. Obshch. Khim.*, 1965, **35**, 1993 (*Engl. transl.* p. 1983) (*esters*)
Kharrasova, F.M. *et al, Zh. Obshch. Khim.*, 1967, **37**, 902 (*Engl. transl.* p. 852) (*dichloride*)
Feshchenko, N.G. *et al, Zh. Obshch. Khim.*, 1969, **39**, 2184 (*Engl. transl.* p. 2133) (*diiodide*)
Hinke, A. *et al, Phosphorus Sulfur*, 1983, **15**, 93 (*dibromide, nmr*)

(2-Bromophenyl)phosphoramidic acid, 9CI B-00494

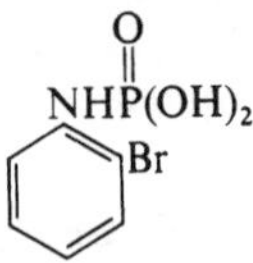

$C_6H_7BrNO_3P$ M 252.004

Di-Ph ester: Diphenyl (2-bromophenyl)phosphoramidate.
$C_{18}H_{15}BrNO_3P$ M 404.199
Cryst. (EtOH). Mp 122-123°.

Zhmurova, I.N. *et al, Zh. Obshch. Khim.*, 1961, **31**, 3741 (*Engl. transl.* p. 3495)

(3-Bromophenyl)phosphoramidic acid, 9CI B-00495

$C_6H_7BrNO_3P$ M 252.004

Di-Ph ester: Diphenyl (3-bromophenyl)phosphoramidate.
$C_{18}H_{15}BrNO_3P$ M 404.199
Cryst. (EtOH). Mp 117-119°.

Dichloride:
$C_6H_5BrCl_2NOP$ M 288.895
Solid. Mp 87°.

Michaelis, A., *Justus Liebigs Ann. Chem.*, 1903, **326**, 129 (*dichloride*)
Zhmurova, I.N. *et al, Zh. Obshch. Khim.*, 1961, **31**, 3741 (*Engl. transl.* p. 3495) (*diphenyl ester*)

(4-Bromophenyl)phosphoramidic acid, 9CI B-00496

[38874-30-5]

$C_6H_7BrNO_3P$ M 252.004

Solid. Mp 158°, Mp 272-274°.

Bis(cyclohexylammonium) salt: [35314-68-2]. Solid. Mp 215-225°.

Di-Et ester: [52912-57-9]. *Diethyl (4-bromophenyl)phosphoramidate.*
$C_{10}H_{15}BrNO_3P$ M 308.111
Cryst. (pet. ether). Mp 79-80°.

Di-Ph ester: [62569-09-9]. *Diphenyl (4-bromophenyl)phosphoramidate.*
$C_{18}H_{15}BrNO_3P$ M 404.199
Cryst. (EtOH). Mp 110-112°.

Dichloride:
$C_6H_5BrCl_2NOP$ M 288.895
Cryst. (C_6H_6). Mp 98°.

Goldwhite, H. *et al, J. Chem. Soc.*, 1957, 2409 (*synth*)
Zhmurova, I.N. *et al, Zh. Obshch. Khim.*, 1961, **31**, 3741 (*Engl. transl.* p. 3495) (*diphenyl ester*)
Williams, A. *et al, J. Chem. Soc. (B)*, 1971, 1973 (*synth*)
Zwierzak, A. *et al, Tetrahedron*, 1973, **29**, 3899 (*diethyl ester, dichloride, synth, ir, pmr, P nmr*)
Mollin, J. *et al, Collect. Czech. Chem. Commun.*, 1976, **41**, 3245 (*diphenyl ester, props*)

O-(4-Bromophenyl) phosphorodichloridothioate, 9CI, 8CI **B-00497**

O-(*4-Bromophenyl*) *thiophosphoryl dichloride.* O-(*4-Bromophenyl*) *dichlorothiophosphate*

$C_6H_4BrCl_2OPS$ M 305.941

Liq. Bp_4 120-122°. n_D^{20} 1.6080.

Protsenko, L.D. *et al*, *Zh. Obshch. Khim.*, 1964, **34**, 2233 (*Engl. transl.* p. 2244) (*synth*)

Bromophos, BSI **B-00498**

O-(*4-Bromo-2,5-dichlorophenyl*) O,O-*dimethyl phosphorothioate, 9CI. Nexion*

[2104-96-3]

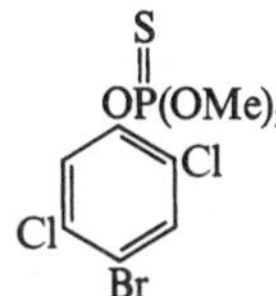

$C_8H_8BrCl_2O_2PS$ M 349.994

Non-systemic agricultural and public health insecticide. Yellowish cryst. V. spar. sol. H_2O, sol. CCl_4, Et_2O, toluene. Mp 53-54°. $Bp_{0.01}$ 140-142°.

▷*N*-methylating agent for DNA-bases. TE7175000.

B.P., 956 343, (*1961*); *CA*, **60**, 13187a (*synth*)
Eichler, D., *Residue Rev.*, 1972, **41**, 65 (*rev*)
Stan, H.-J. *et al*, *Fresenius' Z. Anal. Chem.*, 1977, **287**, 271 (*glc, ms*)
Busch, K.L. *et al*, *Appl. Spectrosc.*, 1978, **32**, 388 (*ms*)
Daldrup, T. *et al*, *Fresenius' Z. Anal. Chem.*, 1981, **308**, 413 (*tlc, glc, hplc*)
Stan, H.-J. *et al*, *Biomed. Mass. Spectrom.*, 1982, **9**, 483 (*ms*)
Pesticide Manual, 6th Ed., 52; 7th Ed., 63.
The Agrochemicals Handbook, Royal Society of Chemistry, 1983, A043.

Bromophos-Et, BSI **B-00499**

O-(*4-Bromo-2,5-dichlorophenyl*) O,O-*diethyl phosphorothioate, 9CI*

[4824-78-6]

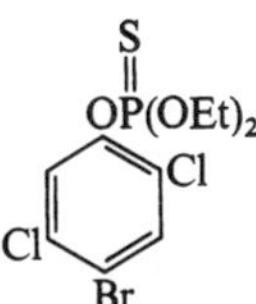

$C_{10}H_{12}BrCl_2O_3PS$ M 394.047

Acaricide and insecticide. Pale-yellow liq. d^{20} 1.52-1.55 (tech.). $Bp_{0.004}$ 122-123°. Unstable at >pH 9. n_D^{20} 1.5600.

▷TE7000000.

Stan, H.-J. *et al*, *Fresenius' Z. Anal. Chem.*, 1977, **287**, 271; *Biomed. Mass Spectrom.*, 1982, **9**, 483 (*glc, ms*)
Stan, H.-J. *et al*, *J. Chromatogr.*, 1983, **279**, 173 (*glc*)
Pesticide Manual, 7th Ed., 64.
The Agrochemicals Handbook, Royal Society of Chemistry, 1983, A044.

(1-Bromopropyl)phosphonic acid, 9CI **B-00500**

$H_3CCH_2CHBrP(O)(OH)_2$

$C_3H_8BrO_3P$ M 202.972

(±)-*form*

Di-Et ester: Diethyl (*1-bromopropyl*)*phosphonate.*
$C_7H_{16}BrO_3P$ M 259.079
Liq. $Bp_{0.02}$ 58.5-59.5°.

Baban, J.A. *et al*, *J. Chem. Soc., Perkin Trans. 2*, 1981, 161 (*synth, pmr*)

(3-Bromopropyl)phosphonic acid, 9CI **B-00501**

[1190-09-6]

$BrCH_2CH_2CH_2P(O)(OH)_2$

$C_3H_8BrO_3P$ M 202.972

Plates (H_2O or C_6H_6). Mp 113-113.5° (107-108°).

Di-Et ester: [1186-10-3]. *Diethyl* (*3-bromopropyl*)-*phosphonate.*
$C_7H_{16}BrO_3P$ M 259.079
Liq. Bp_2 94-95°.

Dichloride:
$C_3H_6BrCl_2OP$ M 239.864
Liq. Bp_{23} 153-156°.
▷Dec. vigorously at 180° releasing HBr

Kosolapoff, G.M., *J. Am. Chem. Soc.*, 1944, **66**, 1511 (*synth*)
Kosolapoff, G.M. *et al*, *J. Chem. Soc.*, 1957, 3739 (*dichloride*)
Helferich, B. *et al*, *Justus Liebigs Ann. Chem.*, 1962, **655**, 59 (*synth*)
Okamoto, Y. *et al*, *Bull. Chem. Soc. Jpn.*, 1975, **48**, 484 (*ester*)

(3-Bromopropyl)triphenylphosphonium(1+), 9CI **B-00502**

$Ph_3P^{\oplus}CH_2CH_2CH_2Br$

$C_{21}H_{21}BrP^{\oplus}$ M 384.275 (ion)

▷Salts are antiacetylcholinesterases in vertebrates and schistosomes

Bromide: [3607-17-8].
$C_{21}H_{21}Br_2P$ M 464.179
Acted on by NaOEt to give Cyclopropyltriphenylphosphonium(1+), C-00287 . Cryst. (EtOH). Mp 223°.

Derocque, J.-L. *et al*, *Justus Liebigs Ann. Chem.*, 1973, 419 (*synth, use*)
Utimoto, K. *et al*, *Tetrahedron*, 1973, **29**, 1169 (*synth*)
Komendantov, M.I. *et al*, *Zh. Org. Khim.*, 1979, **15**, 2076 (*Engl. transl.* p. 1876) (*props*)
Schmidbaur, H. *et al*, *Chem. Ber.*, 1982, **115**, 722 (*use*)
Willcockson, W.S. *et al*, *Comp. Biochem. Physiol. C*, 1982, **72**, 101 (*pharmacol*)

Bromotetraphenoxyphosphorane, 9CI, 8CI **B-00503**

[62395-00-9]

$(PhO)_4PBr$

$C_{24}H_{20}BrO_4P$ M 483.297

Main product when Br_2 added slowly to $(PhO)_3P$ in 1:1 ratio. Solid becoming liq. on standing.

Gloede, J. *et al*, *J. Prakt. Chem.*, 1979, **321**, 1029 (*synth, nmr*)
Tseng, C.K., *J. Org. Chem.*, 1979, **44**, 2793 (*synth, nmr*)
Gloede, J. *et al*, *Z. Anorg. Allg. Chem.*, 1980, **471**, 147 (*synth, nmr*)

Bromo(triphenylphosphoranylidene)acetic acid, 9CI B-00504

$Ph_3P{=}CBrCOOH$

$C_{20}H_{16}BrO_2P$ M 399.223

Me ester: [13504-77-3]. *Methyl bromo(triphenylphosphoranylidene)acetate.*
$C_{21}H_{18}BrO_2P$ M 413.250
Wittig reagent. Cryst. (EtOH/pet. ether). Mp 167-168°.

Et ester: [803-14-5]. *Ethyl bromo(triphenylphosphoranylidene)acetate.*
$C_{22}H_{20}BrO_2P$ M 427.277
Wittig reagent. Yellow cryst. (Me_2CO/hexane). Mp 157-158°. pK_a 6.7 (conj. phosphonium halide).

Märkl, G., *Chem. Ber.*, 1961, **94**, 2996 (*synth, uv*)
Denney, D.B. *et al*, *J. Org. Chem.*, 1962, **27**, 998 (*synth*)
Speziale, A.J. *et al*, *J. Am. Chem. Soc.*, 1963, **85**, 2790 (*ir*)
Martin, D. *et al*, *Chem. Ber.*, 1967, **100**, 187 (*synth*)
Roulet, C. *et al*, *Bull. Soc. Chim. Fr.*, 1974, 531 (*use*)
Tronchet, J.M.J. *et al*, *Helv. Chim. Acta*, 1975, **58**, 1735; 1979, **62**, 1303 (*use*)

1-Bromo-1-(triphenylphosphoranylidene)-2-butanone B-00505

$Ph_3P{=}CBrCOCH_2CH_3$

$C_{22}H_{20}BrOP$ M 411.277

Carbonyl-stabilized ylide, used in Wittig reactions. Solid. Mp 156-158°.

Märkl, G., *Chem. Ber.*, 1962, **95**, 3003 (*synth, ir, use*)

1-Bromo-3-(triphenylphosphoranylidene)-2-propanone, 9CI B-00506

[19753-68-5]

$Ph_3P{=}CHCOCH_2Br$

$C_{21}H_{18}BrOP$ M 397.250

Cryst. (EtOH aq. or EtOAc). Mp 170-172°, 185° dec. Half-life in 45 hr. in 10% EtOH at pH 10.

Issleib, K. *et al*, *Justus Liebigs Ann. Chem.*, 1968, **713**, 12 (*props*)
Zhdanov, Yu.A. *et al*, *Zh. Obshch. Khim.*, 1972, **42**, 759 (*Engl. transl.* p. 751) (*synth, ir, use*)
Le Corre, M., *Bull. Soc. Chim. Fr. Part II*, 1974, 1951 (*synth, ir, pmr, use*)
Altenbach, H.-J., *Angew. Chem., Int. Ed. Engl.*, 1979, **18**, 940 (*use*)
Magdesieva, N.N. *et al*, *Zh. Org. Khim.*, 1981, **17**, 340 (*Engl. transl.* p. 751) (*props*)

1,3-Butadiene-1,4-diphosphonic acid B-00507

1,3-Butadiene-1,4-diylbisphosphonic acid, 9CI. *1,4-Diphosphono-1,3-butadiene*

$(HO)_2P(O)CH{=}CHCH{=}CHP(O)(OH)_2$

$C_4H_8O_6P_2$ M 214.051

Tetra-Et ester: [16650-20-7]. *Tetraethyl 1,3-butadiene-1,4-diylbisphosphonate.*
$C_{12}H_{24}O_6P_2$ M 326.266
Liq. d_4^{20} 1.15. $Bp_{0.045}$ 162°. n_D^{20} 1.4664.

Pudovik, A.N. *et al*, *Zh. Obshch. Khim.*, 1963, **33**, 2509 (*Engl. trans.* p. 2446) (*synth*)
Shagidullin, R.R., *Adv. Mol. Relaxation Processes*, 1973, **5**, 157 (*uv, raman*)

***P*-(1,3-Butadienyl)-*N*,*N*-diethylphosphonamidic acid, 9CI** B-00508

$H_2C{=}CHCH{=}CHP(O)(OH)NEt_2$

$C_8H_{16}NO_2P$ M 189.194

Et ester: [32271-99-1]. *Ethyl* P-*(1,3-butadienyl)*-N,N-*diethylphosphonamidate.*
$C_{10}H_{20}NO_2P$ M 217.247
Liq. d_4^{20} 1.02. $Bp_{3.5}$ 109-111°. n_D^{20} 1.4852.

Mashlyakovskii, L.N. *et al*, *Zh. Obshch. Khim.*, 1971, **41**, 330 (*Engl. transl.* p. 325) (*synth, ir*)

(1,2-Butadienyl)phosphonic acid, 9CI B-00509

1-Phosphono-1,2-butadiene

[3095-05-4]

$H_3CCH{=}C{=}CHP(O)(OH)_2$

$C_4H_7O_3P$ M 134.071

Di-Et ester: [3095-05-4]. *Diethyl (1,2-butadienyl)-phosphonate.*
$C_8H_{15}O_3P$ M 190.178
Liq. d_4^{20} 1.05. Bp_{12} 123-125°. n_D^{20} 1.4637.

Pudovik, A.N. *et al*, *Zh. Obshch. Khim.*, 1965, **35**, 1210 (*Engl. transl.* p. 1214) (*synth*)
Shekhade, A.M. *et al*, *Zh. Obshch. Khim.*, 1978, **48**, 55 (*Engl. transl.* p. 45) (*props*)
Angelov, Kh. *et al*, *Zh. Obshch. Khim.*, 1980, **50**, 2448; 1982, **52**, 538 (*Engl. transl.* pp. 472, 1976) (*props*)
Altenbach, H.J. *et al*, *Tetrahedron Lett.*, 1981, **22**, 5175 (*use*)

(1,3-Butadienyl)phosphonic acid B-00510

1-Phosphono-1,3-butadiene

$H_2C{=}CHCH{=}CHP(O)(OH)_2$

$C_4H_7O_3P$ M 134.071

Di-Me ester: [4037-11-0]. *Dimethyl (1,3-butadienyl)-phosphonate.*
$C_6H_{11}O_3P$ M 162.125
Bp_1 74-75°. n_D^{20} 1.4840.

Di-Et ester: [7158-35-2]. *Diethyl (1,3-butadienyl)-phosphonate.*
$C_8H_{15}O_3P$ M 190.178
Liq. d_4^{20} 1.05. Bp_{13} 122-123°. n_D^{20} 1.4749.

Difluoride: [18026-39-6].
$C_4H_6F_2OP$ M 139.061
Liq. d^{20} 1.21. Bp_8 39-40°. n_D^{20} 1.4312.

Dichloride: [4707-93-1].
$C_4H_5Cl_2OP$ M 170.963
Liq. d_4^{20} 1.31. Bp_1 65-66°. n_D^{20} 1.5413.

***(E)*-form**

Di-Et ester: [25145-58-8]. d_4^{20} 1.04. $Bp_{0.03}$ 47-49°. n_D^{20} 1.4670.

Di-Ph ester: Diphenyl (1,3-butadienyl)phosphonate.
$C_{16}H_{15}O_3P$ M 286.266
Solid. Mp 62-63°.

Pudovik, A.N. *et al*, *Zh. Obshch. Khim.*, 1961, **31**, 1693 (*Engl. transl.* p. 1580) (*ester*)
Slovoktova, N.A. *et al*, *Izv. Akad. Nauk SSSR, Ser. Khim.*, 1961, 71 (*Engl. transl.* p. 62) (*esters, ir*)
Mashlyakovskii, L.N. *et al*, *Zh. Obshch. Khim.*, 1965, **35**, 1577 (*Engl. transl.* p. 1582) (*dichloride, ester*)
Darling, S.D. *et al*, *J. Org. Chem.*, 1975, **40**, 2851 (*ester, props*)
Mashlyakovskii, L.N. *et al*, *Vysokomol. Soedin., Ser.A.*, 1976, **18**, 308 (*Engl. transl.* p. 354) (*P nmr, polymers*)

Mashlyakovskii, L.N. *et al*, *Zh. Obshch. Khim.*, 1979, **49**, 54 (*Engl. transl.* p. 44) (*dichloride, esters, pmr, cmr*)
Zyablikova, T.A. *et al*, *Zh. Obshch. Khim.*, 1982, **52**, 287 (*Engl. transl.* p. 249) (*ester, synth, pmr*)

(1,3-Butadien-1-yl)phosphonothioic acid, 9CI — B-00511

$H_2C{=}CHCH{=}CHP(S)(OH)_2 \rightleftharpoons H_2C{=}CHCH{=}CHP(O)(OH)(SH)$

$C_4H_7O_2PS$ M 150.132

O,O-Di-Me ester: O,O-*Dimethyl (1,3-butadien-1-yl)-phosphonothioate.*
$C_6H_{11}O_2PS$ M 178.185
Liq. d_4^{20} 1.11. Bp_2 80°. n_D^{20} 1.5300.

O,O-Di-Et ester: [16726-81-1]. O,O-*Diethyl (1,3-butadien-1-yl)phosphonothioate.*
$C_8H_{15}O_2PS$ M 206.239
Liq. d_4^{20} 1.05. Bp_2 95°. n_D^{20} 1.5026.

Bis(diethylamide): P-*(1,3-Butadien-1-yl)*-N,N,N′,N-*tetraethylphosphonothioic diamide.*
$C_{12}H_{25}N_2PS$ M 260.377
Liq. d_4^{20} 1.04. $Bp_{0.5}$ 142°. n_D^{20} 1.5461.

Pudovik, A.N. *et al*, *Zh. Obshch. Khim.*, 1963, **33**, 2509 (*Engl. transl.* p. 2446) (*ester, amide, synth*)
Pudovik, A.N. *et al*, *Zh. Obshch. Khim.*, 1965, **35**, 358 (*Engl. transl.* p. 358) (*esters, synth*)

(1,3-Butadienyl)triphenylphosphonium(1+), 9CI — B-00512

$Ph_3P^{\oplus}CH{=}CHCH{=}CH_2$

$C_{22}H_{20}P^{\oplus}$ M 315.374 (ion)
Reagent for conversion of carbonyl compds. into 1,3-cyclohexadienes.

Chloride: [76373-02-9].
$C_{22}H_{20}ClP$ M 350.827
No phys. props. reported.

Bromide: [21310-07-6].
$C_{22}H_{20}BrP$ M 395.278
Yellow cryst. ($CHCl_3/Et_2O$ or EtOH).

Ford, J.A. *et al*, *J. Org. Chem.*, 1961, **26**, 1433 (*synth*)
Fuchs, P.L., *Tetrahedron Lett.*, 1974, 4055 (*synth, use*)
Schiess, P. *et al*, *Helv. Chim. Acta*, 1981, **64**, 787, 801 (*use*)

Butamifos, BSI — B-00513

O-*Ethyl* O-*(5-methyl-2-nitrophenyl) 1-methylpropylphosphoramidothioate*, 9CI. O-*Ethyl* O-*(6-nitro-*m*-tolyl)* sec-*butylphosphoramidothioate. Cremart*
[36335-67-8]

```
                           NO2
   CH3  S
   |    ||
H3CCH2CHNHPO-C6H3
          |
          OEt        CH3
```

$C_{13}H_{21}N_2O_4PS$ M 332.354
▷TB4920000.

(±)-form
Preemergent herbicide and fungicide. Brownish liq. Prac. insol. H_2O, sol. most org. solvs. d_{25}^{25} 1.188. $Bp_{0.1}$ 160°. n_D^{25} 1.5340.

B.P., 1 359 727, (*1970*); *CA*, **76**, 153375v (*synth*)
Horiba, M., *J. Chromatogr.*, 1984, **287**, 189 (*glc*)
Pesticide Manual, 7th Ed., 73.

1,4-Butanediphosphonic acid — B-00514

1,4-Butanediylbis(phosphonic acid), 9CI. *Tetramethylenediphosphonic acid*, 8CI. *1,4-Diphosphonobutane*
[4671-77-6]

$(HO)_2P(O)(CH_2)_4P(O)(OH)_2$

$C_4H_{12}O_6P_2$ M 218.083
Corrosion inhibitor. Cryst. (H_2O). Mp 217-220°.

Tetra-Et ester: [7203-67-0]. *Tetraethyl 1,4-butanediylbisphosphonate.*
$C_{12}H_{28}O_6P_2$ M 330.297
Liq. $Bp_{0.1}$ 171°. n_D^{25} 1.4455.

Bisdichloride: [1660-97-5].
$C_4H_8Cl_4O_2P_2$ M 291.866
Solid, mod. air-stable. Mp 110-112°.

Moedritzer, K. *et al*, *J. Inorg. Nucl. Chem.*, 1961, **22**, 297 (*ester, ir, P nmr*)
Maier, L., *Helv. Chim. Acta*, 1965, **48**, 133 (*chloride, synth, P nmr*)
Kosolapoff, G.M. *et al*, *J. Chem. Soc. (C)*, 1966, 757 (*synth, chloride*)
Van Haverbeke, L. *et al*, *Bull. Soc. Chim. Belg.*, 1972, **81**, 547; *CA*, **78**, 49963 (*ir, raman*)

1,4-Butanediphosphonous acid — B-00515

1,4-Butanediylbis(phosphonous acid), 9CI. P,P-*Tetramethylenediphosphonous acid*

$(HO)_2P(CH_2)_4P(OH)_2 \rightleftharpoons HO(O)PH(CH_2)_4PH(O)OH$

$C_4H_{12}O_4P_2$ M 186.084

Tetra-Me ester: Tetramethyl 1,4-butanediylbisphosphonite. Tetramethyl tetramethylenediphosphonite.
$C_8H_{22}O_4P_2$ M 244.207
Liq. Bp_1 87-9°. n_D^{22} 1.4693.

Tetra-Et ester: Tetraethyl 1,4-butanediylbisphosphonite. Tetraethyl tetramethylenediphosphonite.
$C_{12}H_{28}O_4P_2$ M 298.298
Liq. Bp_2 135-137°. n_D^{20} 1.4604.

Tetraisopropyl ester: Tetraisopropyl 1,4-butanediylbisphosphonite. Tetraisopropyl tetramethylenediphosphonite.
$C_{16}H_{36}O_4P_2$ M 354.406
Liq. Bp_1 132-135°. n_D^{21} 1.4515.

Bis(dichloride): [28240-71-3].
$C_4H_8Cl_4P_2$ M 259.867
Liq. $Bp_{2.5}$ 119-123°.

Bis(dibromide):
$C_4H_8Br_4P_2$ M 437.671
Liq. $Bp_{0.1}$ 123°.

Bis[bis(diethylamide)]: [86926-30-9]. P,P′-*1,4-Butanediylbis[bis* N,N-*diethylphosphonous diamide].*
$C_{20}H_{48}N_4P_2$ M 406.574
Liq. $Bp_{0.1}$ 125°.

Sander, M., *Chem. Ber.*, 1962, **95**, 473 (*tetrachloride, esters, synth*)
Sommer, K., *Z. Anorg. Allg. Chem.*, 1970, **376**, 37 (*tetrachloride, synth, P nmr*)
Diemert, K. *et al*, *Chem. Ber.*, 1982, **115**, 1947 (*tetrabromide, synth, P nmr*)
Diemert, K. *et al*, *Phosphorus Sulfur*, 1983, **15**, 155 (*diethylamide, synth, nmr*)

1,4-Butanediylbis[phenylphosphinodithioic acid], 9CI B-00516

[5689-46-3]

Ph►P(=S)◄SH, (CH₂)₄, HS►P(=S)◄Ph *(RS,RS)-form*

$C_{16}H_{20}P_2S_2$ M 338.402

Cryst. (C_6H_6). Mp 136.5-137.5°.

Di-Na salt: Solid. Mp 286° dec.

Di-Me ester: Dimethyl 1,4-butanediylbis[phenylphosphinodithioate].

$C_{22}H_{32}P_2S_2$ M 422.562

Cryst. (propanol). Mp 105-129°. Probably a mixt. of racemic and *meso*-forms.

(RS,RS)-form

(±)-form

Di-Et ester: Diethyl 1,4-butanediylbis[phenylphosphinodithioate].

$C_{24}H_{36}P_2S_2$ M 450.616

Mp 72°.

(RS,SR)-form

meso-*form*

Di-Et ester: Cryst. (pet. ether). Mp 115-119°.

Issleib, K. *et al*, *Chem. Ber.*, 1961, **94**, 107 (*synth*)
Issleib, K. *et al*, *Z. Anorg. Allg. Chem.*, 1966, **347**, 268 (*synth, derivs*)
Diemert, K.L. *et al*, *Chem. Ber.*, 1978, **111**, 629 (*synth*)

1,4-Butanediylbis[triphenylphosphonium](2+), 9CI B-00517

Tetramethylenebis[triphenylphosphonium](2+), 8CI. 1,4-Bis(triphenylphosphonio)butane(2+)

$Ph_3P^{\oplus}(CH_2)_4PPh_3^{\oplus}$

$C_{40}H_{38}P_2^{\oplus\oplus}$ M 580.688 (ion)

Used in Wittig synth. of annulenones, fused heterocyclics and 1,3-cyclohexadienes.

Dibromide: [15546-42-6].

$C_{40}H_{38}Br_2P_2$ M 740.496

Source of ylide obt. with $NaNH_2/NH_3$ or PhLi. Cryst. (Me_2CO). Mp 292-294°.

Perbromide:

$C_{40}H_{38}Br_6P_2$ M 1060.112

Dark-yellow cryst. Mp 216°.

Bisylide: [62486-05-9]. *1,4-Butanediylidenebis[triphenylphosphorane].*

$C_{40}H_{36}P_2$ M 578.672

Wittig reagent.

Mondon, A., *Justus Liebigs Ann. Chem.*, 1957, **603**, 115 (*synth, use*)
Wittig, G. *et al*, *Justus Liebigs Ann. Chem.*, 1958, **619**, 10 (*synth, use*)
Cresp, T.M. *et al*, *J. Chem. Soc., Perkin Trans. 1*, 1974, 2145 (*use*)
Wood, G.W. *et al*, *J. Org. Chem.*, 1975, **40**, 636 (*ms*)
Nicolaides, D.N., *Synthesis*, 1977, 127 (*use*)
Booth, B.L. *et al*, *J. Organomet. Chem.*, 1981, **220**, 229 (*ylide, complexes*)

1,2,3,4-Butanetetraphosphonic acid B-00518

1,2,3,4-Butanetetrayltetraphosphonic acid, 9CI. 1,2,3,4-Tetraphosphonobutane

[25404-73-3]

$CH_2P(O)(OH)_2$–$CHP(O)(OH)_2$–$CHP(O)(OH)_2$–$CH_2P(O)(OH)_2$

$C_4H_{14}O_{12}P_4$ M 378.042

Octa-Et ester: [25091-08-1]. *Octaethyl 1,2,3,4-butanetetrayltetraphosphonate.*

$C_{20}H_{46}O_{12}P_4$ M 602.471

Viscous liq.

Cilley, W.A. *et al*, *J. Am. Chem. Soc.*, 1970, **92**, 1685 (*synth*)
Nicholson, D.A. *et al*, *J. Org. Chem.*, 1970, **35**, 3149 (*synth, P nmr*)

2-Butene-1,4-diphosphonic acid B-00519

2-Butene-1,4-diylbisphosphonic acid, 9CI. 1,4-Diphosphono-2-butene

$(HO)_2P(O)CH_2CH{=}CHCH_2P(O)(OH)_2$

$C_4H_{10}O_6P_2$ M 216.067

Tetra-Me ester: [3858-16-0]. *Tetramethyl (2-butene-1,4-diyl)bisphosphonate.*

$C_8H_{18}O_6P_2$ M 272.174

Mp 56-58°. Bp_1 162-163°. Mixt. of *E*- and *Z*-forms.

Tetra-Et ester: [1112-96-5]. *Tetraethyl (2-butene-1,4-diyl)bisphosphonate.*

$C_{12}H_{26}O_6P_2$ M 328.281

Liq. d_4^{20} 1.13. Bp_2 181-182°. n_D^{20} 1.4585-1.4595. Mixt. of *E*- and *Z*-forms.

Pudovik, A.N. *et al*, *Zh. Obshch. Khim.*, 1961, **31**, 1693 (*Engl. transl.* p. 1580) (*synth, ir*)
Kruglov, S.V. *et al*, *Zh. Obshch. Khim.*, 1975, **45**, 1027 (*Engl. transl.* p. 1014) (*synth, pmr*)

(2-Buten-1-yl)diphenylphosphine B-00520

1-Diphenylphosphino-2-butene

[63322-25-8]

$H_3CCH{=}CHCH_2PPh_2$

$C_{16}H_{17}P$ M 240.284

(E)-form

Oxide: [17668-60-9].

$C_{16}H_{17}OP$ M 256.283

Cryst. (hexane). Mp 118-119°.

(Z)-form

Oxide: [58322-08-0]. Cryst. (cyclohexane). Mp 111-112.5°.

El-Deek, M. *et al*, *J. Org. Chem.*, 1976, **41**, 1403 (*synth, ir, pmr, nmr*)
Lythgoe, B. *et al*, *J. Chem. Soc., Perkin Trans. 1*, 1976, 2386 (*synth, ir, pmr*)
Khachatryan, R.A. *et al*, *CA*, 1981, **95**, 169286 (*synth*)
Darling, S.D. *et al*, *J. Org. Chem.*, 1982, **47**, 1413 (*synth, ir, pmr*)

3-Butenyldiphenylphosphine, 9CI B-00521

4-Diphenylphosphino-1-butene

[4124-07-6]

$Ph_2PCH_2CH_2CH{=}CH_2$

$C_{16}H_{17}P$ M 240.284
Ligand for Group VIII metals. Liq. $Bp_{0.1}$ 115°.
Oxide: [16958-43-3].
$C_{16}H_{17}OP$ M 256.283
Cryst. (EtOAc/hexane). Mp 103-105°.

Schweizer, E.E. *et al, J. Org. Chem.*, 1968, **33**, 3082 (*oxide*)
Clark, P.W. *et al, Can. J. Chem.*, 1974, **52**, 1714 (*synth, pmr, P nmr*)
Curtis, J.L.S. *et al, J. Chem. Soc., Dalton Trans.*, 1974, 1898 (*pmr, complexes*)
Clark, P.W. *et al, J. Organomet. Chem.*, 1977, **139**, 385 (*complexes*)

2-Butenylidenetriphenylphosphorane, 9CI B-00522

[41892-64-2]

$Ph_3P{=}CHCH{=}CHCH_3$

$C_{22}H_{21}P$ M 316.382
Used in Wittig reactions and in annelations reacns.leading to 1,3-cyclohexadienes. Deep-purple cryst. Obt. as a salt-free mixt. of (*Z*)- and (*E*)-forms in the ratio 3:1.
(E)-form [56374-57-3]
Not separately characterised.
(Z)-form [50375-97-8]
Not separately characterised.

Dauben, W.G. *et al, J. Am. Chem. Soc.*, 1973, **95**, 5088 (*use*)
Dauben, W.G. *et al, Tetrahedron Lett.*, 1973, 3711, 4425 (*use*)
Elliott, M. *et al, J. Chem. Soc., Perkin Trans. 1*, 1974, 2470 (*use*)
Bogdanović, B. *et al, Justus Liebigs Ann. Chem.*, 1975, 692 (*synth*)
Ostoja Starzewski, K.A. *et al, J. Am. Chem. Soc.*, 1976, **98**, 8486 (*pe*)
Ostoja Starzewski, K.A. *et al, Phosphorus*, 1976, **6**, 177 (*cmr*)

(2-Butenyl)methylphosphinic acid, 9CI B-00523

$H_3CCH{=}CHCH_2P(O)(OH)Me$

$C_5H_{11}O_2P$ M 134.114
Et ester: [20420-16-0]. *Ethyl (2-butenyl)-methylphosphinate.*
$C_7H_{15}O_2P$ M 162.168
Liq. $Bp_{0.1}$ 65-67°. n_D^{20} 1.4522.
Isopropyl ester: Isopropyl (2-butenyl)-methylphosphinate.
$C_8H_{17}O_2P$ M 176.195
Liq. Bp_{13} 116-118°, $Bp_{0.1}$ 64-66°. n_D^{20} 1.4419, 1.4503.

Liorber, B.G. *et al, Zh. Obshch. Khim.*, 1968, **38**, 878 (*Engl. transl.* p. 843)
Razumov, A.I. *et al, Zh. Obshch. Khim.*, 1976, **46**, 1237 (*Engl. transl.* p. 1218)

1-Butenylphosphonic acid, 9CI B-00524

1-Phosphono-1-butene

$H_3CCH_2CH{=}CHP(O)(OH)_2$

$C_4H_9O_3P$ M 136.087
Di-Et ester: [41760-80-9]. *Diethyl 1-butenylphosphonate.*
$C_8H_{17}O_3P$ M 192.194
Liq. mixt. cont. 85% *E*-form, 15% *Z*-form. $Bp_{0.05}$ 86°. n_D^{21} 1.4282.
Diisopropyl ester: Diisopropyl 1-butenylphosphonate. Bis(1-methylethyl) 1-butenylphosphonate.
$C_{10}H_{41}O_3P$ M 240.406
Liq. mixt. of *E*- and *Z*-forms (85:15). $Bp_{0.05}$ 78°. n_D^{21} 1.4312.
(E)-form
Di-Et ester: [18689-31-1]. Liq. Bp_{20} 130-140°.

Baboulene, M. *et al, Phosphorus Sulfur*, 1979, **7**, 101 (*synth, ir, pmr*)
Koizumi, T. *et al, Synthesis*, 1982, 917 (*ir, pmr*)

2-Butenylphosphonic acid B-00525

1-Phosphono-2-butene
[682-34-8]

$H_3CCH{=}CHCH_2P(O)(OH)_2$

$C_4H_9O_3P$ M 136.087
(E)-form
Di-Et ester: [26327-86-6]. *Diethyl 2-butenylphosphonate.*
$C_6H_{13}O_3P$ M 164.141
Liq. d_4^{20} 1.02. Bp_1 85-86°. n_D^{20} 1.4392.
(Z)-form
Di-Et ester: [26327-87-7]. d_4^{20} 1.04. Bp_1 69°. n_D^{20} 1.4325.

Pudovik, A.N. *et al, Zh. Obshch. Khim.*, 1961, **31**, 1693; 1963, **33**, 3096 (*Engl. transl.* pp. 1580, 3022) (*synth*)
Zakharov, V.I. *et al, Zh. Obshch. Khim.*, 1974, **44**, 98 (*Engl. transl.* p. 96) (*synth, pmr, P nmr*)
Canavet, C. *et al, Chem. Ber.*, 1980, **113**, 1115 (*use*)

3-Buten-1-ylphosphonic acid B-00526

4-Phosphono-1-butene

$H_2C{=}CHCH_2CH_2P(O)(OH)_2$

$C_4H_9O_3P$ M 136.087
Di-Et ester: Diethyl 3-buten-1-ylphosphonate.
$C_8H_{19}O_3P$ M 194.210
Liq. $Bp_{0.1}$ 51-53°.

Savignac, P. *et al, Synth. Commun.*, 1979, **9**, 487 (*synth, pmr*)

2-Butenylphosphonothioic acid, 9CI B-00527

$H_3CCH{=}CHCH_2P(S)(OH)_2 \rightleftharpoons H_3CCH{=}CHCH_2P(O)(OH)(SH)$

$C_4H_9O_2PS$ M 152.148
O,O-*Di-Et ester:* [20420-12-6]. O,O-*Diethyl 2-butenylphosphonothioate.*
$C_8H_{17}O_2PS$ M 208.255
Liq. d_4^{20} 1.02. Bp_9 103°. n_D^{20} 1.4772.

Pudovik, A.N. *et al, Zh. Obshch. Khim.*, 1963, **33**, 2560, 2924 (*Engl. transl.* pp. 2493, 2850)

(2-Butenyl)triphenylphosphonium(1+), 9CI B-00528

[48195-46-6]

$Ph_3P^{\oplus}CH_2CH{=}CHCH_3$

$C_{22}H_{22}P^{\oplus}$ M 317.390 (ion)
(E)-form
Bromide: [39741-81-6].
$C_{22}H_{22}BrP$ M 397.294

Solid. Mp 244-245°.
Ylide: see 2-Butenylidenetriphenylphosphorane, B-00522

(*Z*)-form

Bromide: [39616-20-1]. No phys. props. reported.
Ylide: see 2-Butenylidenetriphenylphosphorane, B-00522

Jaenicke, L. *et al*, *Justus Liebigs Ann. Chem.*, 1973, 1252 (*bromide, use*)
Wood, G.W., *J. Org. Chem.*, 1975, **40**, 636 (*ms*)
Dilbeck, G.A. *et al*, *J. Org. Chem.*, 1975, **40**, 1150 (*bromide, ir, pmr, P nmr*)
Ostoja Starzewski, K.A. *et al*, *Phosphorus*, 1976, **6**, 177 (*cmr*)
Hirai, M.F. *et al*, *J. Organomet. Chem.*, 1978, **160**, 25 (*complexes*)

3-Buten-1-ynylphosphonic acid, 9CI — B-00529

$H_2C{=}CHC{\equiv}CP(O)(OH)_2$

$C_4H_3O_3P$ M 130.040

Di-Me ester: Dimethyl 3-buten-1-ynylphosphonate.
$C_6H_9O_3P$ M 160.109
Liq. d_4^{20} 1.14. $Bp_{3.5}$ 110-112°. n_D^{20} 1.4740.
Di-Et ester: Diethyl 3-buten-1-ynylphosphonate.
$C_8H_8O_3P$ M 183.123
Liq. d_4^{20} 1.06. $Bp_{2.5}$ 107-108°. n_D^{20} 1.4696.
Dichloride: [4981-31-1].
$C_4H_3Cl_2OP$ M 168.947
Liq. d_4^{20} 1.38. $Bp_{2.5}$ 61-65°. n_D^{20} 1.5150.

Ionin, B.E. *et al*, *Zh. Obshch. Khim.*, 1963, **33**, 2863 (*Engl. transl.* p. 2791) (*synth, ir, pmr*)
Mashlyukovskii, L.N. *et al*, *Zh. Obshch. Khim.*, 1965, **35**, 1577 (*Engl. transl.* p. 1582) (*synth, ir, pmr*)

Butonate, USAN, BSI — B-00530

[2,2,2-Trichloro-1-(dimethoxyphosphinyl)ethyl] butanoate, 9CI. Dimethyl (2,2,2-trichloro-1-hydroxyethyl)-phosphonate butyrate, 8CI
[126-22-7]

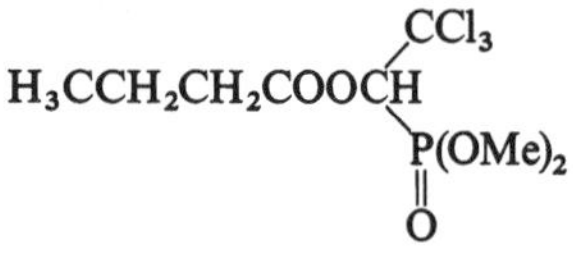

$C_8H_{14}Cl_3O_5P$ M 327.528
Vet. anthelmintic, insecticide, now superseded. $Bp_{0.5}$ 129°.
▷Nucleoside methylating agent. ET0175000.

East Ger. Pat., 40 097, (*1965*); *CA*, **64**, 756b (*synth*)
Knowles, C.O., *J. Agric. Food Chem.*, 1966, **14**, 566 (*pharmacol*)
Fechner, G. *et al*, *Fresenius' Z. Anal. Chem.*, 1969, **244**, 393 (*anal*)
Betteridge, D. *et al*, *Anal. Chem.*, 1972, **44**, 2005 (*pe*)
Krijgsman, W. *et al*, *J. Chromatogr.*, 1976, **117**, 201 (*glc*)

tert-Butoxycarbonyl diethyl phosphate — B-00531

Anhydride of 1,1-dimethylethyl carbonate and diethyl phosphate, 9CI. tert-*Butylcarbonate-diethyl phosphate anhydride. Carbonic acid anhydride with diethyl phosphate* tert-*butyl ester*
[14618-58-7]

$(EtO)_2P(O)OCOOC(CH_3)_3$

$C_9H_{19}O_6P$ M 254.219
Acylating agent for amines.

Popova, G.A. *et al*, *Zh. Org. Khim.*, 1981, **17**, 2011 (*Engl. transl.* p. 1795) (*synth, use*)
Org. Synth., 1970, **50**, 9 (*synth, ir, pmr*)

N-*tert*-Butoxycarbonyl-4-diphenylphosphino-2-diphenylphosphinomethylpyrrolidine — B-00532

1,1-Dimethylethyl 4-(diphenylphosphino)-2-[(diphenylphosphino)methyl]-1-pyrrolidinecarboxylate. BPPM

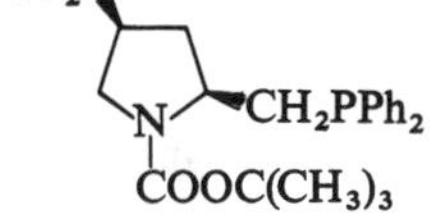

$C_{34}H_{37}NO_2P_2$ M 553.619

(*2S,4S*)-form [61478-28-2]

Effective ligand for highly asymmetric hydrogenation using a Rh catalyst. Also used in asymmetric synth. of peptides. Mp 104-105°. $[\alpha]_D^{20}$ −36° (c, 0.6 in C_6H_6).

Achiwa, K., *J. Am. Chem. Soc.*, 1976, **98**, 8265 (*synth*)
Achiwa, E.V. *et al*, *Chem. Lett.*, 1978, 297 (*use*)
Ojima, I. *et al*, *Chem. Lett.*, 1978, 567 (*use*)
Becker, Y. *et al*, *J. Org. Chem.*, 1980, **45**, 2145 (*use*)
Koenig, K.E. *et al*, *J. Org. Chem.*, 1980, **45**, 2362 (*use*)
Ojima, I. *et al*, *J. Org. Chem.*, 1980, **45**, 4728 (*synth*)
Knowles, W.S. *et al*, *Adv. Chem. Ser.*, 1982, **196**, 325.
Maciel, G.E. *et al*, *Adv. Chem. Ser.*, 1982, **196**, 389.
Ojima, I., *ACS Symp. Ser.*, 1982, **185**, 109.

2-*tert*-Butoxy-4,5-dimethyl-1,3,2-dioxaphospholane — B-00533

4,5-Dimethyl-2-(1,1-dimethylethoxy)-1,3,2-dioxaphospholane, 9CI. tert-*Butyl 2,3-butylene phosphite. 2-*tert*-Butoxy-4,5-dimethyl-1,3-dioxa-2-phosphacyclopentane*
[33835-25-5]

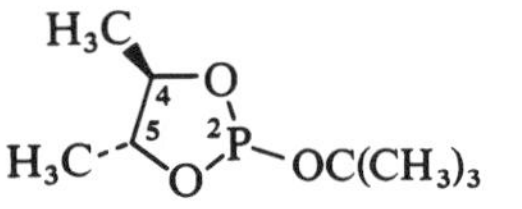

(4*RS*,5*RS*)-*form*

$C_8H_{17}O_3P$ M 192.194

(*4RS,5RS*)-form

Liq. Bp_{10} 38°.

(*4RS,5SR*)-form

meso-*form*
Liq. Bp_{10} 42°. Mixt. of epimers at P.

Cox, R.H. *et al*, *J. Org. Chem.*, 1972, **37**, 1557 (*synth, pmr*)

2-*tert*-Butoxy-5,5-dimethyl-1,3,2-dioxaphosphorinane, 8CI — B-00534

*5,5-Dimethyl-2-(1,1-dimethylethoxy)-1,3,2-dioxaphosphorinane, 9CI. 5,5-Dimethyl-2-(1,1-dimethylethoxy)-1,3,2-dioxaphosphinane. 2-*tert*-Butoxy-5,5-dimethyl-1,3-dioxa-2-phosphacyclohexane.* tert-*Butyl neopentylene phosphite*
[27275-47-4]

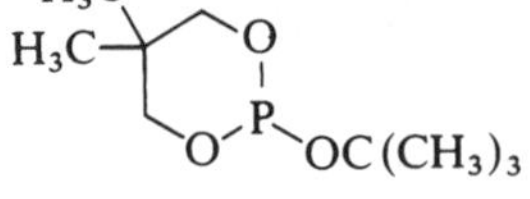

$C_9H_{19}O_3P$ M 206.221

Antioxidant. Liq. Bp_{15} 83-84°.

2-Oxide: [15762-00-2]. tert-*Butyl neopentylene phosphate.*

$C_9H_{19}O_4P$ M 222.220

Cryst. (pet. ether). Mp 78-79°.

Bartle, K.D. *et al*, *Tetrahedron*, 1967, **23**, 1701 (*oxide, synth, pmr*)
White, D.W. *et al*, *J. Magn. Reson.*, 1971, **4**, 123 (*pmr*)
Dale, A.J., *Acta Chem. Scand., Ser. B*, 1976, **30**, 255 (*pmr*)
Rueger, R. *et al*, *J. Prakt. Chem.*, 1984, **326**, 622 (*synth*)

2-*tert*-Butoxy-1,3,2-dioxaphospholane, 8CI B-00535

2-(1,1-Dimethylethoxy)-1,3,2-dioxaphospholane, 9CI. tert-*Butyl ethylene phosphite*

[28950-17-6]

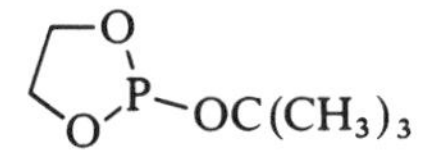

$C_6H_{13}O_3P$ M 164.141

Liq. with characteristic phosphite odour. d_4^{25} 1.059. Bp_{25} 74°. n_D^{25} 1.4368.

2-Oxide: [76819-63-1]. tert-*Butyl ethylene phosphate.*

$C_6H_{13}O_4P$ M 180.140

Solid. Mp 28°. Dec. at 85° evolving isobutene.

Lucas, H.J. *et al*, *J. Am. Chem. Soc.*, 1950, **72**, 5491 (*synth*)
Besserre, D. *et al*, *Org. Magn. Reson.*, 1980, **13**, 235, 313 (*cmr, conformn*)
Yasuda, H. *et al*, *Macromolecules*, 1981, **14**, 458 (*oxide, synth, props*)

P-(Butoxymethyl)-*N*,*N*-diethylphosphonamidic acid, 9CI B-00536

$H_3C(CH_2)_3OCH_2P(O)(OH)NEt_2$

$C_9H_{22}NO_3P$ M 223.251

Et ester: [59375-32-5]. *Ethyl* P-*(butoxymethyl)*-N,N-*diethylphosphonamidate.*

$C_{11}H_{26}NO_3P$ M 251.305

Liq. d_4^{20} 0.98. Bp_{15} 147-151°. n_D^{20} 1.4420.

Butyl ester: Butyl P-*(butoxymethyl)*-N,N-*diethylphosphonamidate.*

$C_{13}H_{30}NO_3P$ M 279.359

Liq. d_4^{20} 0.97. Bp_5 145°. n_D^{20} 1.4490.

Krutskii, L.N. *et al*, *Zh. Obshch. Khim.*, 1976, **46**, 507 (*Engl. transl.* p. 505)

2-*tert*-Butoxy-4-methyl-1,3,2-dioxaphosphorinane B-00537

4-Methyl-2-(1,1-dimethylethoxy)-1,3,2-dioxaphosphorinane, 9CI. tert-*Butyl 1,3-butylene phosphite.* 2-tert-*Butoxy-4-methyl-1,3-dioxa-2-phosphacyclohexane*

[32579-62-7]

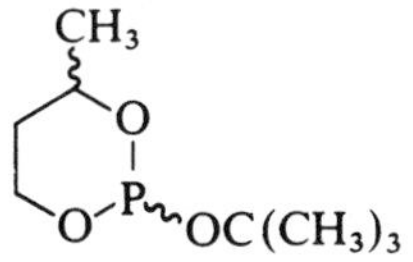

$C_8H_{17}O_3P$ M 192.194

2-Oxide: [76819-59-5]. tert-*Butyl 1,3-butylene phosphate.*

$C_8H_{17}O_4P$ M 208.194

Cryst. (toluene/hexane). Mp 39°. Stereoisomeric mixt.

Nifant'ev, É.E. *et al*, *Dokl. Akad. Nauk SSSR, Ser. Sci. Khim.*, 1973, **208**, 651 (*Engl. transl. Phys. Chem.* p. 100) (*cmr*)
Yasuda, H. *et al*, *Macromolecules*, 1981, **14**, 458 (*oxide, props*)

2-*tert*-Butylamino-5,5-dimethyl-1,3,2-dioxaphosphorinane, 8CI B-00538

*5,5-Dimethyl-*N*-(1,1-dimethylethyl)-1,3,2-dioxaphosphorinan-2-amine, 9CI. Neopentylene* tert-*butylphosphoramidite*

[56597-20-7]

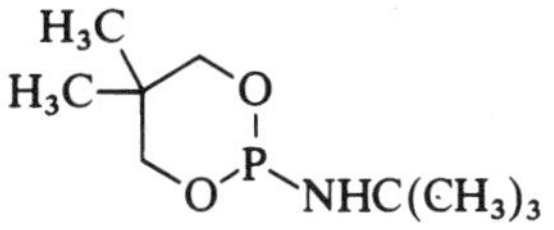

$C_9H_{20}NO_2P$ M 205.236

2-Oxide: [944-23-0]. *Neopentylene* tert-*butylphosphoramidate.*

$C_9H_{20}NO_3P$ M 221.236

Cryst. (C_6H_6 or CH_2Cl_2/cyclohexane). Mp 166-166.5° (159-161°).

2-Sulfide: [72542-64-4]. *Neopentylene* tert-*butylphosphoramidothioate.*

$C_9H_{20}NO_2PS$ M 237.296

Cryst. (MeOH). Mp 128-130°.

Edmundson, R.S., *Tetrahedron*, 1964, **20**, 2781 (*oxide, ir*)
Bartle, K.D. *et al*, *Tetrahedron*, 1967, **23**, 1701 (*pmr*)
Modro, T.A. *et al*, *J. Org. Chem.*, 1978, **43**, 5000 (*oxide, props*)
Dutton, F.E. *et al*, *J. Agric. Food Chem.*, 1981, **29**, 1114 (*sulfide, pmr*)
Edmundson, R.S., *Phosphorus Sulfur*, 1981, **9**, 307 (*oxide, ms*)

2-(*tert*-Butylamino)-1,3,2-dioxaphospholane B-00539

N*-(1,1-Dimethylethyl)-1,3,2-dioxaphospholan-2-amine, 9CI. Ethylene* tert-*butylphosphoramidite*

[51439-10-2]

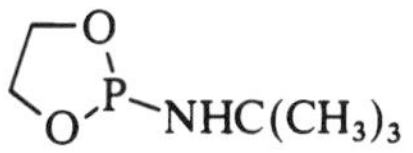

$C_6H_{13}NO_2P$ M 162.148

2-Oxide: Ethylene tert-*butylphosphoramidate.*

$C_6H_{13}NO_3P$ M 178.147

Cryst. (CH_2Cl_2/hexane). Mp 113-115°.

El-Borgi, A. *et al*, *C.R. Hebd. Seances Acad. Sci., Ser. C*, 1977, **284**, 983 (*derivs, ir*)
Modro, T.A. *et al*, *J. Org. Chem.*, 1978, **43**, 5000 (*oxide, synth, props*)
Al-Rawi, J.M.A. *et al*, *Org. Magn. Reson.*, 1984, **22**, 336 (*cmr*)
Al-Rawi, J.M.A. *et al*, *Magn. Reson. Chem.*, 1985, **23**, 285 (*sulfide, cmr*)

2-[(Butylamino)-1-hexenyl]phosphonic acid, 9CI B-00540

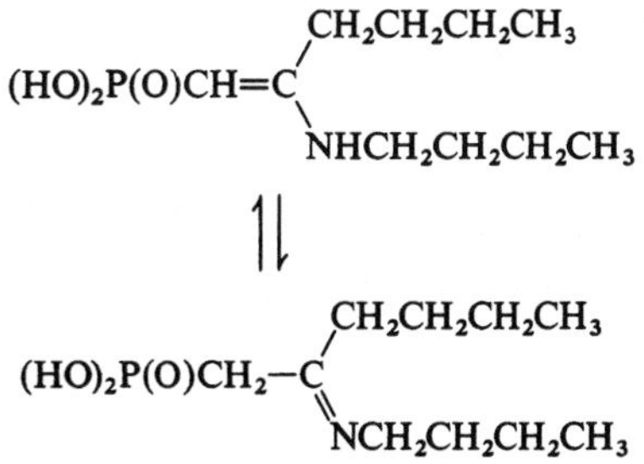

$C_{10}H_{22}NO_3P$ M 235.262

Esters exist in tautomeric forms.

Di-Et ester: Diethyl [2-(butylamino)-1-hexenyl]-phosphonate.

$C_{14}H_{30}NO_3P$ M 291.370

Forms an anion which has been used to convert aldehydes and ketones into butyl ethenyl ketones.

Diisopropyl ester: [78554-61-7]. *Diisopropyl [2-(butylamino)-1-hexenyl)]phosphonate.*

$C_{16}H_{34}NO_3P$ M 319.423

$Bp_{0.08}$ 111-112°. n_D^{20} 1.4695.

Chatta, M.S. *et al, Tetrahedron Lett.*, 1971, 1419 (*diethyl ester, use*)

Alikin, A.Yu. *et al, Zh. Obshch. Khim.*, 1982, **52**, 316 (*Engl. transl.* p. 274) (*diisopropyl ester, synth*)

2-*tert*-Butylamino-4-methyl-1,3,2-dioxaphosphorinane B-00541

*4-Methyl-*N*-(1,1-dimethylethyl)-1,3,2-dioxaphosphorinan-2-amine,* 9CI. *1,3-Butylene* tert-*butylphosphoramidite*

[70869-67-9]

CH₃ / O / P / NHC(CH₃)₃ (2RS,4RS)-form

$C_8H_{18}NO_2P$ M 191.209

Liq. $Bp_{0.6}$ 43-44°. n_D^{20} 1.4665. Mixt. of stereoisomers.

(2RS,4RS)-form [57733-32-1]

(±)-trans-*form*

2-Oxide: [93633-04-6]. *1,3-Butylene* tert-*butylphosphoramidate.*

$C_8H_{18}NO_3P$ M 207.209

Cryst. (C_6H_6). Mp 163-164°.

2-Selenide: [57733-33-2]. O,O-*1,3-Butylene* tert-*butylphosphoramidoselenoate.*

$C_8H_{18}NO_2PSe$ M 270.169

Solid. Mp 81-82°.

(2RS,4SR)-form [57733-31-0]

(±)-cis-*form*

2-Oxide: [93633-03-5]. Cryst. (CCl_4). Mp 166°.

2-Selenide: Solid. Mp 120-120.5°.

Bartczak, T.J. *et al, Tetrahedron Lett.*, 1975, 3243 (*synth, selenides, pmr, P nmr*)

Bartczak, T.J. *et al, Cryst. Struct. Commun.*, 1976, **5**, 21 (*selenide, cryst struct*)

Edmundson, R.S. *et al, J. Chem. Soc., Perkin Trans. 1*, 1984, 1943 (*oxides, selenides, ir, P nmr, cryst struct*)

[1-(Butylamino)-1-methylethyl]phosphinic acid, 8CI B-00542

Butafosfan, BAN. *Coforta*

[17316-67-5]

$H_3C(CH_2)_3NHC(CH_3)_2PH(O)(OH)$

$C_7H_{18}NO_2P$ M 179.198

Clinical phosphorus source. Also antioxidant for fats. Cryst. (MeOH/Me_2CO). Mp 219°.

Kreutzkamp, N. *et al, Arch. Pharm. (Weinheim, Ger.)*, 1967, **300**, 868; *CA*, **68**, 49689

1-*tert*-Butyl-1,2,4-azadiphosphetidine B-00543

1-(1,1-Dimethylethyl)-1,2,4-azadiphosphetidine, 9CI. *1-*tert-*Butylaza-2,4-diphosphacyclobutane*

HP—NC(CH₃)₃ / PH

$C_5H_{13}NP_2$ M 149.112

2,4-Dichloro, 2,4-dioxide: [60609-91-8].

$C_5H_{11}Cl_2NO_2P_2$ M 250.001

Oil which solidifies. $Bp_{0.7}$ 110°. Mixt. of stereoisomers.

2,4-Diisopropoxy: [85685-74-1].

$C_{11}H_{25}NO_2P_2$ M 265.272

Liq. $Bp_{0.006}$ 73-74°. n_D^{20} 1.4770.

2,4-Bis(dimethylamino), 2,4-dioxide: [60609-88-3].

$C_9H_{23}N_3O_2P_2$ M 267.247

Solid. Mp 137-139°. $Bp_{0.01}$ 160°. Consists of pure *trans*-isomer.

Bulloch, G. *et al, J. Chem. Soc., Dalton Trans.*, 1976, 1113 (*dichloride, diamide, synth, pmr, P nmr*)

Novikova, Z.S. *et al, Zh. Obshch. Khim.*, 1983, **53**, 474 (*Engl. transl.* p. 417) (*derivs, synth, P nmr*)

2-*tert*-Butyl-1,3-benzoxaphosphole B-00544

2-(1,1-Dimethylethyl)-1,3-benzoxaphosphole, 10CI

[77013-92-4]

O / C(CH₃)₃ / P

$C_{11}H_{13}OP$ M 192.197

Liq. $Bp_{0.5}$ 57-60°.

Heinecke, J. *et al, Z. Chem.*, 1980, **20**, 342 (*synth, cmr, pmr*)

Heinecke, J. *et al, Tetrahedron Lett.*, 1983, **24**, 5481 (*props*)

tert-Butylbis(trimethylsilyl)phosphine B-00545

(1,1-Dimethylethyl)bis(trimethylsilyl)phosphine, 10CI

[42491-33-8]

$(Me_3Si)_2PC(CH_3)_3$

$C_{10}H_{27}PSi_2$ M 234.468

Used for synth. of phosphorus-containing heterocyclic compds. Ligand for Cr, Co, Mn, Mo, Rh and W. Liq. $Bp_{0.2}$ 52-56°. Reacts slowly with DMF.

Baudler, M. *et al, Z. Naturforsch., B*, 1976, **31**, 1305 (*synth*)

Becker, G. *et al, Z. Anorg. Allg. Chem.*, 1978, **443**, 42 (*synth, pmr, nmr*)

Andriamizaka, J.D. *et al, Phosphorus Sulfur*, 1982, **12**, 265 (*use*)

Butyl-*tert*-butylphosphinic acid B-00546

Butyl(1,1-dimethylethyl)phosphinic acid, 9CI

$H_3C(CH_2)_3P(O)(OH)C(CH_3)_3$

$C_8H_{19}O_2P$ M 178.211

Low-melting solid. $Bp_{0.02}$ 138°.

Brown, A.D. *et al, J. Chem. Soc. (C)*, 1968, 839 (*synth, pmr, ir*)

5-*tert*-Butyl-2-chloro-1,3,2-dioxaphosphorinane, 8CI **B-00547**

2-Chloro-5-(1,1-dimethylethyl)-1,3,2-dioxaphosphorinane, 9CI

[21135-08-0]

$C_7H_{14}ClO_2P$ M 196.613

Fuming solid. Mp 59-60°. $Bp_{0.75}$ 60°.

Bentrude, W.G. *et al*, *J. Am. Chem. Soc.*, 1970, **92**, 7136 (*synth, ir, pmr*)

Mazhar-ul-Haque, *et al*, *J. Chem. Soc.* (*A*), 1970, 1786 (*synth*)

(4-*tert*-Butyl-2-chlorophenyl) 2,2,2-tribromoethyl phosphorochloridate **B-00548**

2-Chloro-4-(1,1-dimethylethyl)phenyl 2,2,2-tribromoethyl phosphorochloridate, 9CI

[69919-18-2]

$C_{12}H_{14}Br_3Cl_2O_3P$ M 547.833

Phosphorylating agent for free 3-OH groups in otherwise protected nucleosides. Cryst. Mp 85-87°.

Arentzen, R. *et al*, *Synthesis*, 1979, 137 (*synth, use*)

5-*tert*-Butyl-2-chlorotetrahydro-3-phenyl-2*H*-1,3,2-oxazaphosphorine **B-00549**

*2-Chloro-5-(1,1-dimethylethyl)tetrahydro-3-phenyl-2*H*-1,3,2-oxazaphosphorine, 9CI. 5-*tert*-Butyl-2-chlorotetrahydro-3-phenyl-1,3,2-oxazaphosphorinane*

[83096-42-8]

$C_{13}H_{19}ClNOP$ M 271.726

Oil. $Bp_{1.3}$ 155-158°. Possibly mixt. of stereoisomers.

Bajwa, G.S. *et al*, *J. Am. Chem. Soc.*, 1982, **104**, 6385 (*synth, pmr, P nmr*)

4-*tert*-Butylcyclohexylphosphonous acid **B-00550**

[4-(1,1-Dimethylethyl)cyclohexyl]phosphonous acid, 9CI

cis-form

$C_{10}H_{21}O_2P$ M 204.248

Free acid is prototropic.

Cis-form

Di-Me ester: [58359-96-9]. *Dimethyl 4-*tert*-butylcyclohexylphosphonite.*

$C_{12}H_{25}O_2P$ M 232.302

No phys. props. reporetd.

Dichloride: (*4-*tert*-Butylcyclohexyl)dichlorophosphine.*

$C_{10}H_{19}Cl_2P$ M 241.140

Liq. $Bp_{0.07}$ 65-65.5° (containing 20% trans).

trans-form

Di-Me ester: [58359-97-0]. Liq. $Bp_{0.3}$ 85-87° (contg. 21% cis).

Dichloride: Liq. $Bp_{0.02}$ 65-66° (contg. 12% cis).

Gordon, M.D. *et al*, *J. Org. Chem.*, 1976, **41**, 1690 (*derivs, cmr*)

Gordon, M.D. *et al*, *J. Am. Chem. Soc.*, 1976, **98**, 15 (*derivs, synth, ir, P nmr, cmr*)

Gordon, M.D. *et al*, *J. Magn. Reson.*, 1976, **22**, 149 (*deriv, P nmr*)

***OO-tert*-Butyl *O,O*-diethyl phosphoroperoxoate, 9CI** **B-00551**

[10160-45-9]

$(EtO)_2P(O)OOC(CH_3)_3$

$C_8H_{19}O_5P$ M 226.209

Liq. $Bp_{0.1}$ 75-77°. n_D^{20} 1.4175.

Sosnovsky, G. *et al*, *J. Org. Chem.*, 1969, **34**, 968; 1972, **37**, 2267 (*synth, props*)

Maslennikov, V.P. *et al*, *Kinet. Katal.*, 1971, **12**, 575; 1974, **15**, 38 (*CA*, **75**, 87912; **80**, 119971) (*props*)

Sosnovsky, G. *et al*, *Synthesis*, 1972, 202 (*synth*)

Sosnovsky, G. *et al*, *Z. Naturforsch, B*, 1975, **30**, 724, 732 (*synth, props*)

7-Butyl-2,3-dihydro-1,6-diphenyl-1*H*-1-benzophosphonin-2,3,4,5-tetracarboxylic acid **B-00552**

$C_{32}H_{29}O_8P$ M 572.550

Tetra-Me ester, oxide: [59502-08-8].

$C_{36}H_{37}O_9P$ M 644.657

Cryst. Mp 188°.

Hughes, A.N. *et al*, *J. Heterocycl. Chem.*, 1976, **13**, 65 (*synth, ir, pmr, ms*)

1-*tert*-Butyl-1,4-dihydro-4-phenyl-1,4-phospharsenin **B-00553**

1-(1,1-Dimethylethyl)-1,4-dihydro-4-phenyl-1,4-phospharsenin, 9CI

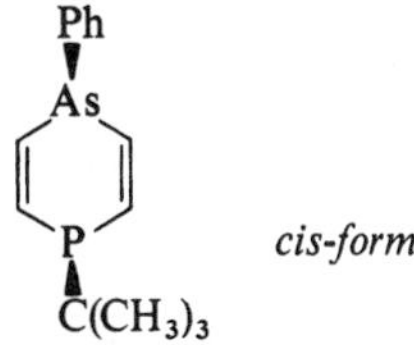

cis-form

$C_{14}H_{18}AsP$ M 292.192

Mp 47-52°. $Bp_{0.0001}$ 165°. Presumably a mixt. of *cis-* and *trans*-forms.

cis-form [57767-20-1]

Not obt. pure.

1-Oxide: [57767-22-3].

$C_{14}H_{18}AsOP$ M 308.191

Not obt. pure.

***trans*-form** [57767-19-8]

Solid. Mp 50-51°.

1-Oxide: [57767-21-2]. Cryst. (MeCN). Mp 128-130°.

Märkl, G. *et al*, *Tetrahedron Lett.*, 1975, 3171 (*synth, pmr, ms*)
Märkl, G. *et al*, *Synthesis*, 1977, 842 (*synth*)

OO-tert-Butyl *O,O*-diisopropyl phosphoroperoxoate, 9CI — B-00554

[10160-46-0]

$[(H_3C)_2CHO]_2P(O)OOC(CH_3)_3$

$C_{10}H_{23}O_5P$ M 254.262

Liq. $Bp_{0.3}$ 82-85°. n_D^{25} 1.4150.

Sosnovsky, G. *et al*, *J. Org. Chem.*, 1969, **34**, 968; 1972, **37**, 2267 (*synth, props*)
Sosnovsky, G. *et al*, *Synthesis*, 1972, 202 (*synth*)
Sosnovsky, G. *et al*, *Phosphorus*, 1973, **3**, 87 (*props*)

5-*tert*-Butyl-2-dimethylamino-1,3,2-dioxaphosphorinane, 8CI — B-00555

N,N-*Dimethyl-5-(1,1-dimethylethyl)-1,3,2-dioxaphosphorin-2-amine*, 9CI. 2-tert-*Butyltrimethylene dimethylphosphoroamidite*

$(H_3C)_3C$ ... O ... P ... NMe_2 *cis-form*

$C_9H_{20}NO_2P$ M 205.236

$Bp_{2.5}$ 80-92°. *Trans*-isomer the major component.

***cis*-form** [39536-73-7]

2-Oxide: [39536-50-0]. 2-tert-*Butyltrimethylene dimethylphosphoramidate*.
$C_9H_{20}NO_3P$ M 221.236
Cryst. (Et_2O/pet. ether). Mp 117.5-118°.

***trans*-form** [39536-56-6]

2-Oxide: [39536-49-7]. Cryst. (Et_2O). Mp 114.5-115°.

Bentrude, W.G. *et al*, *J. Am. Chem. Soc.*, 1972, **94**, 8222; 1973, **95**, 4666 (*synth, pmr, cmr, P nmr*)

SS-tert-Butyl *O,O*-dimethyl phosphoro(dithioperoxo)thioate — B-00556

O,O-*Dimethyl* SS-*(1,1-dimethylethyl) phosphoro(dithioperoxo)thioate*, 9CI. S-tert-*Butylthio* O,O-*dimethyl phosphorodithioate*

$(MeO)_2P(S){-}S{-}S{-}C(CH_3)_3$

$C_6H_{15}O_2PS_3$ M 246.337

Liq. n_D^{25} 1.4671.

U.S.P., 3 109 770, (*1963*); *CA*, **60**, 2841

Butyldiphenylphosphine, 9CI, 8CI — B-00557

[6372-41-4]

$Ph_2P(CH_2)_3CH_3$

$C_{16}H_{19}P$ M 242.300

Liq. Bp_{13} 172-174°, $Bp_{0.1}$ 88-90°. n_D^{20} 1.5926.

B,EtI: Butylethyldiphenylphosphonium iodide.
$C_{18}H_{24}IP$ M 398.266
Mp 153°.

Oxide: [4233-13-0].
$C_{16}H_{19}PO$ M 258.299
Cryst. (pet. ether or EtOAc/pet. ether)). Mp 93-94°.

Sulfide: [15367-52-9].
$C_{16}H_{19}PS$ M 274.360
Mp 45°.

Selenide:
$C_{16}H_{19}PSe$ M 321.260
Mp 63°.

Kuchen, W. *et al*, *Chem. Ber.*, 1959, **92**, 227 (*synth, oxide*)
Zingaro, R.A. *et al*, *J. Chem. Eng. Data*, 1963, **8**, 226 (*synth, derivs*)
Inorg. Synth., 1976, **16**, 158 (*synth*)
Gray, G.A. *et al*, *J. Am. Chem. Soc.*, 1976, **98**, 2109 (*pmr, cmr, nmr, oxide*)
Goff, S.D. *et al*, *Org. Mass Spectrom.*, 1977, **12**, 33 (*oxide, ms*)
Postle, S.R., *Phosphorus Sulfur*, 1977, **3**, 269 (*sulfide, cmr*)
Dmitriev, V.I. *et al*, *Zh. Obshch. Khim.*, 1978, **48**, 52 (*Engl. transl.* p. 42) (*synth, nmr*)

tert-Butyldiphenylphosphine, 8CI — B-00558

(1,1-Dimethylethyl)diphenylphosphine, 9CI

[6002-34-2]

$Ph_2PC(CH_3)_3$

$C_{16}H_{19}P$ M 242.300

Liq. Bp_2 144-146°, $Bp_{0.2}$ 118-122°.

B,MeI: tert-*Butylmethyldiphenylphosphonium iodide.*
$C_{17}H_{22}IP$ M 384.239
Cryst. (MeOH/Et_2O). Mp 185-187°.

Oxide: [56598-35-7]. Cryst. Mp 131-132°.

Sulfide: [66295-77-0].
$C_{16}H_{19}PS$ M 274.360
No phys. props. reported.

Hoffmann, H. *et al*, *Chem. Ber.*, 1966, **99**, 1134 (*synth*)
Grim, S.O. *et al*, *J. Org. Chem.*, 1967, **32**, 781 (*synth, pmr, nmr*)
Albright, T.A. *et al*, *J. Org. Chem.*, 1975, **40**, 3437 (*oxide, cmr, nmr*)
Goff, S.D. *et al*, *Org. Mass Spectrom.*, 1977, **12**, 33 (*oxide, ms*)
Postle, S.R., *Phosphorus Sulfur*, 1977, **3**, 269 (*oxide, sulfide, pmr*)
Petrov, K.A. *et al*, *Zh. Obshch. Khim.*, 1980, **50**, 1518 (*Engl. transl.* p. 1227) (*oxide, synth, pmr*)
Vincent, E. *et al*, *Spectrochim. Acta, Part A*, 1980, **36**, 699 (*pmr, cmr, nmr*)

P-tert-Butyl-*N,N'*-diphenylphosphonamidimidic chloride, 9CI, 8CI — B-00559

P-*(1,1-Dimethylethyl)*-N,N'-*diphenylphosphonamidimidic chloride*, 9CI. *Anilino*-tert-*butylchlorophosphine phenylimide*. P-tert-*Butyl*-P-*chloro*-N-*phenyl*-P-*phenylaminophosphine imide*

[51771-83-6]

$(H_3C)_3C(Cl)P(=NPh)NHPh$

$C_{16}H_{20}ClN_2P$ M 306.774

Solid. Mp 103-105°.

Scherer, O.J., *Chem. Ber.*, 1974, **107**, 552 (*synth, ir, ms, pmr*)

tert-Butylethylphosphinic acid — B-00560

Ethyl(1,1-dimethylethyl)phosphinic acid, 9CI

[16543-42-3]

$(H_3C)_3CP(O)(OH)Et$

$C_6H_{15}O_2P$ M 150.157
Cryst. (pet. ether). Mp 101-103°.
Chloride: [25781-14-0].
$C_6H_{14}ClOP$ M 168.603
Liq. Bp_{10} 118°.

Brown, A.D. *et al*, *J. Chem. Soc.* (*C*), 1968, 839 (*synth, pmr, ir*)
Crofts, P.C. *et al*, *J. Chem. Soc.* (*C*), 1970, 332 (*chloride, pmr, ir*)

tert-Butyl ethyl phosphonate, 8CI B-00561

Ethyl (1,1-dimethylethyl) phosphonate, 9CI. tert-*Butyl ethyl phosphite. Ethyl (1,1-dimethylethyl) phosphite*
[14540-45-5]

$(H_3C)_3COPH(O)OEt \rightleftharpoons (H_3C)_3COP(OH)OEt$

$C_6H_{15}O_3P$ M 166.156
Tautomeric. Thermally unstable liq. n_D^{22} 1.4197.

Zwierzak, A. *et al*, *Tetrahedron*, 1973, **29**, 1089 (*synth, analogues, ir, pmr*)

tert-Butylisopropylphosphinic acid B-00562

(1,1-Dimethylethyl)(1-methylethyl)phosphinic acid, 9CI
[25788-99-2]

$(H_3C)_3C(H_3C)_2CH-P(=O)OH$

$C_7H_{17}O_2P$ M 164.184
Cryst. (pet. ether). Mp 84°.
Chloride: [25781-15-1].
$C_7H_{16}ClOP$ M 182.630
Liq. Bp_{10} 120°.

Crofts, P.C. *et al*, *J. Chem. Soc.*, 1958, 2995 (*synth*)
Crofts, P.C. *et al*, *J. Chem. Soc.* (*C*), 1970, 332 (*synth, chloride, ir, pmr*)

5-*tert*-Butyl-2-methoxy-1,3,2-dioxaphosphorinane, 9CI B-00563

5-tert-*Butyl-2-methoxy-1,3-dioxa-2-phosphacyclohexane*

$(H_3C)_3C$ ... P–OMe *cis-form*

$C_8H_{17}O_3P$ M 192.194
Liq. $Bp_{0.075}$ 32°.
cis-form [23201-70-9]
2-Oxide: [26344-07-0].
$C_8H_{17}O_4P$ M 208.194
Cryst. (hexane). Mp 76-77°.
trans-form [23201-71-0]
2-Oxide: [26344-06-9]. Cryst. Mp 90-91°.

Bentrude, W.G. *et al*, *J. Am. Chem. Soc.*, 1970, **92**, 7136 (*synth, derivs, ir, pmr*)
Mazhar-ul-Haque, *et al*, *J. Chem. Soc.* (*A*), 1970, 1786 (*synth*)
Warrent, R.W. *et al*, *J. Org. Chem.*, 1978, **43**, 4266 (*oxide, cryst struct*)
Cullis, P.M., *J. Chem. Soc., Chem. Commun.*, 1984, 1510 (*oxide, pmr, P nmr*)
Van Nuffel, P. *et al*, *J. Mol. Struct.*, 1984, **125**, 1 (*struct*)

5-*tert*-Butyl-2-methyl-1,3,2-dioxaphosphorinane, 9CI B-00564

5-tert-*Butyl-2-methyl-1,3-dioxa-2-phosphacyclohexane*

$(H_3C)_3C$ ... P–Me *cis-form*

$C_8H_{17}O_2P$ M 176.195
cis-form [32511-23-2]
2-Oxide: [26344-10-5].
$C_8H_{17}O_3P$ M 192.194
Needles (hexane). Mp 127-128°.
trans-form [32511-62-9]
2-Oxide: [26344-11-6]. Cryst. (hexane). Mp 69-71°.

Bentrude, W.G. *et al*, *J. Am. Chem. Soc.*, 1970, **92**, 7136 (*synth, pmr, ir*)
Mazhar-ul-Haque, *et al*, *J. Chem. Soc.* (*A*), 1970, 1786 (*cryst struct, pmr*)
Bentrude, W.G. *et al*, *J. Am. Chem. Soc.*, 1972, **94**, 3264; 1973, **95**, 4666; 1975, **97**, 573 (*pmr, conformn*)
Finocchiaro, P. *et al*, *J. Am. Chem. Soc.*, 1976, **98**, 3537 (*pmr, conformn*)
Van Nuffel, P. *et al*, *J. Mol. Struct.*, 1984, **125**, 1 (*struct*)

tert-Butylmethylphenylphosphine, 8CI B-00565

Methyl(1,1-dimethylethyl)phenylphosphine, 9CI
[7621-16-1]

Ph / :–P–Me / $C(CH_3)_3$ (*R*)-*form*

$C_{11}H_{17}P$ M 180.229
(*R*)-*form* [25140-24-3]
$Bp_{0.05}$ 40°. $[\alpha]_D^{22}$ +28.4° (c, 1.1-2.0 in MeOH) (60% o.p.).
*B,PhCH₂Br: Benzyl-*tert-*butylmethylphenylphosphonium bromide.*
$C_{18}H_{24}BrP$ M 351.265
$[\alpha]_D$ −75° (c, 5.06 in MeOH) (50% o.p.), $[\alpha]_D$ −116° (c, 0.70 in MeOH) (61% o.p.). Has (*S*)-config.
Oxide: [38802-08-3].
$C_{11}H_{17}OP$ M 196.228
$Bp_{0.05}$ 115°. $[\alpha]_D$ −9.05° (c, 7.18 in MeOH) (43% o.p.). Has (*S*)-config.
Sulfide: [58159-96-9].
$C_{11}H_{17}PS$ M 212.289
Solid. Mp 100-105°. $[\alpha]_D$ −37° (c, 3.40 in MeOH) (60% o.p.). Has (*S*)-config.
(*S*)-*form* [38736-54-8]
$Bp_{0.05}$ 35-40°. $[\alpha]_D$ −13.4° (c, 8.18 in MeOH), $[\alpha]_D^{25}$ −43.1° (c, 4.56 in C_6H_6).
B,PhCH₂Br: Solid. Mp 245-250°. $[\alpha]_D$ +45° (MeOH) (30% o.p.). Has (*R*)-config.
Oxide: [21448-79-3]. V. hygroscopic solid. Mp 78°, Mp 98-99°. $Bp_{0.05}$ 98-100°. $[\alpha]_D^{25}$ +21.5° (c, 2.22 in MeOH), $[\alpha]_D$ +21.7° (c, 1.83 in C_6H_6). Has (*R*)-config.
Sulfide: $[\alpha]_D$ +17.3° (c, 5.80 in MeOH) (37% o.p.). Has (*R*)-config.
(±)-*form* [64839-73-2]
Liq. Bp_{15} 112-115°.
B,MeI: tert-*Butyldimethylphenylphosphonium iodide.*
$C_{12}H_{20}IP$ M 322.168

Cryst. ($MeOH/Et_2O$). Mp 160-161°.
Oxide: Solid. Mp 76-77°.

Hoffmann, H. *et al*, *Chem. Ber.*, 1966, **99**, 1134 (*synth*)
Luckenbach, R., *Phosphorus*, 1972, **1**, 293; 1973, **3**, 77 (*synth, oxide, salt*)
Luckenbach, R., *Chem. Ber.*, 1975, **108**, 803, 3533 (*oxide, sulfide*)
Kyba, E.P., *J. Am. Chem. Soc.*, 1976, **98**, 4805 (*synth, oxide*)
Luckenbach, R. *et al*, *Justus Liebigs Ann. Chem.*, 1976, 2305 (*synth, oxide*)

Butylmethylphosphinic acid, 9CI — B-00566

$H_3C(CH_2)_3P(O)(OH)Me$

$C_5H_{13}O_2P$ M 136.130
Mp 36-37°. $Bp_{0.0001}$ 142°. n_D^{40} 1.4439.
Butyl ester: Butyl butylmethylphosphinate.
$C_9H_{21}O_2P$ M 192.237
Liq. Bp_4 111-112°. n_D^{20} 1.4409.
2-Methylpropyl ester: Isobutyl butylmethylphosphinate.
$C_9H_{21}O_2P$ M 192.237
Liq. Bp_1 82°.
Fluoride: [18358-21-9]. Characterised spectroscopically.
Chloride: [13213-41-7].
$C_5H_{12}ClOP$ M 154.576
Liq. d_4^{20} 1.09. Bp_{20} 116° (Bp_{13} 151-157°). Bp_{13} 151-157°. n_D^{20} 1.4630.

Petrov, K.A. *et al*, *Zh. Obshch. Khim.*, 1960, **30**, 2995 (*Engl. transl.* p. 2967) (*ester*)
Maier, L., *Chem. Ber.*, 1961, **94**, 3051, 3056 (*synth, chloride*)
Neimysheva, A.A. *et al*, *Zh. Obshch. Khim.*, 1966, **36**, 1090 (*Engl. transl.* p. 1105) (*chloride*)
Knunyants, I.L. *et al*, *Dokl. Akad. Nauk SSSR, Ser. Sci. Khim.*, 1971, **201**, 862 (*Engl. transl.* p. 992) (*halides, nmr*)
Finke, M. *et al*, *Justus Liebigs Ann. Chem.*, 1974, 741 (*ester*)

tert-Butylmethylphosphinic acid — B-00567

(1,1-Dimethylethyl)methylphosphinic acid, 9CI
[18351-80-9]

$(H_3C)_3CP(O)(OH)Me$

$C_5H_{13}O_2P$ M 136.130
Cryst. (pet. ether). Mp 108-109°.
Me ester: Methyl tert-*butylmethylphosphinate.*
$C_6H_{15}O_2P$ M 150.157
No phys. props. reported.
Ph ester: [63027-78-1]. *Phenyl* tert-*butylmethylphosphinate.*
$C_{11}H_{17}O_2P$ M 212.228
Liq. Bp_8 120° (oven).
Chloride: [25788-98-1].
$C_5H_{12}ClOP$ M 154.576
Solid. Mp 37°. Bp_{10} 99°.
Amide: [32306-56-2]. P-*(1,1-Dimethylethyl)*-P-*methylphosphinic amide.* P-tert-*Butyl*-P-*methylphosphinic amide.*
$C_5H_{14}NOP$ M 135.145
Cryst. (toluene). Mp 99-100°, 114-116°. Subl. at 130°/0.05.
Azide:
$C_5H_{12}N_3OP$ M 161.143
Liq. $Bp_{0.6}$ 90-95° (oven).
Diethylamide: P-tert-*Butyl*-N,N-*diethyl*-P-*methylphosphinic amide.*
$C_9H_{22}NOP$ M 191.253
Liq. $Bp_{0.08}$ 80°.

Brown, A.D. *et al*, *J. Chem. Soc. (C)*, 1968, 839 (*synth, pmr, ir*)
Crofts, P.C. *et al*, *J. Chem. Soc. (C)*, 1970, 332 (*chloride, ir, pmr*)
Knunyants, I.L. *et al*, *Dokl. Akad. Nauk SSSR, Ser. Sci. Khim.*, 1971, **201**, 862 (*Engl. transl.* p. 992) (*halides, nmr*)
Scherer, O.J. *et al*, *Chem. Ber.*, 1971, **104**, 1490 (*amide, ir, pmr*)
Harger, M.J.P. *et al*, *Tetrahedron*, 1982, **38**, 3073 (*amide, azide, chloride, ir, pmr*)
Kolodyazhnyi, O.I., *Zh. Obshch. Khim.*, 1982, **52**, 1314 (*Engl. transl.* p. 1156) (*diethylamide*)

tert-Butylmethylphosphinodithioic acid, 8CI — B-00568

(1,1-Dimethylethyl)methylphosphinodithioic acid
[22069-92-7]

$(H_3C)_3CPMe(S)SH$

$C_5H_{13}PS_2$ M 168.251
Solid by subl. Mp 220.5°.
Na salt: [27509-01-9]. Solid. Mp 334-338° dec.
Dimethylammonium salt: [29049-29-4]. Needles (Me_2CO). Mp 136.5°.

Hägele, G. *et al*, *Chem. Ber.*, 1970, **103**, 2885 (*synth, deriv, ir, pmr, P nmr*)
Kuchen, W. *et al*, *Chem. Ber.*, 1970, **103**, 2276 (*deriv, pmr, P nmr, ir*)

tert-Butylmethylphosphinodithioic acid anhydrosulfide — B-00569

(1,1-Dimethylethyl)methylphosphinodithioic acid anhydrosulfide, 11CI

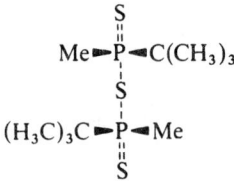

S
‖
Me–P–C(CH_3)$_3$
|
S
|
(H_3C)$_3$C–P–Me
‖
S

$C_{10}H_{24}P_2S_3$ M 302.427
(RS,RS)-form [95837-84-6]
(±)-form
Cryst. (MeOH). Mp 144°.

Haegele, G. *et al*, *Z. Naturforsch., B*, 1984, **39**, 1574 (*synth, P nmr*)
Haegele, G. *et al*, *J. Chem. Soc., Dalton Trans.*, 1984, 2803 (*P nmr, struct*)
Wunderlich, H. *et al*, *Z. Naturforsch., B*, 1984, **39**, 1581 (*cryst struct*)

tert-Butylmethylphosphinothioic acid — B-00570

(1,1-Dimethylethyl)methylphosphinothioic acid
[22069-94-9]

$[(H_3C)_3C]MeP(S)OH \rightleftharpoons [(H_3C)_3C]MeP(O)SH$

$C_5H_{13}OPS$ M 152.191
(±)-*form*
Cryst. by subl. Mp 116.5°.
Na salt: Solid. Mp 334-336°.
Quinine salt: [27490-04-6]. Cryst. (Me_2CO). Mp 205.5-206.5°.
O-*Me ester:* O-*Methyl* tert-*butylmethylphosphinothioate.*
$C_6H_{15}OPS$ M 166.218
Solid with camphor-like odour. Mp 90°. Bp_1 40-50° subl.

O-tert-*Butyl ester:* O-tert-*Butyl* tert-*butylmethylphosphinothioate.*
$C_9H_{21}OPS$ M 208.298
Liq. $Bp_{0.25}$ 77-78°. Pyrolysis yields the parent acid.
Chloride: [27509-18-8].
$C_5H_{12}ClPS$ M 170.637
Solid. Mp 188°. Sublimes.
Bromide: [22069-91-6].
$C_5H_{12}BrPS$ M 215.088
Solid. Mp 198-198.5°. Sublimes.
Anhydride: see tert-Butylmethylphosphinothioic anhydride, B-00571

Kuchen, W. *et al*, *Chem. Ber.*, 1970, **103**, 2274 (*synth, derivs, ir, pmr, P nmr*)
Hägele, G. *et al*, *Chem. Ber.*, 1970, **103**, 2885 (*derivs*)
Kuchen, W. *et al*, *Chem. Ber.*, 1970, **103**, 2114 (*bromide, synth, pmr, ir, P nmr*)

***tert*-Butylmethylphosphinothioic anhydride** B-00571

(1,1-Dimethylethyl)methylphosphinothioic anhydride

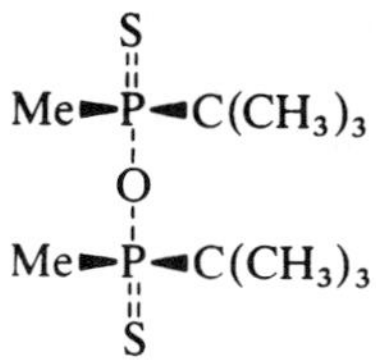

$C_{10}H_{24}OP_2S_2$ M 286.367
(*RS,SR*)-*form* [95888-14-5]
meso-*form*
Needles (ligroin). Mp 110.5-111°.

Haegele, G. *et al*, *J. Chem. Soc., Dalton Trans.*, 1984, 2803 (*struct, P nmr*)
Haegele, G. *et al*, *Phosphorus Sulfur*, 1985, **22**, 241 (*synth, pmr, P nmr, cryst struct*)

***N*-*tert*-Butyl-*P*-methylphosphonamidic acid, 8CI** B-00572

P-*Methyl*-N-*(1,1-dimethylethyl)phosphonamidic acid, 9CI*

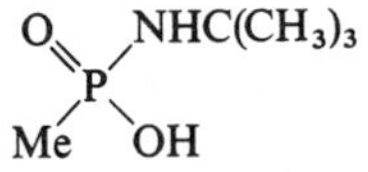

$C_5H_{14}NO_2P$ M 151.145
Me ester: [85656-09-3]. *Methyl* N-tert-*butyl*-P-*methylphosphonamidate.*
$C_6H_{16}NO_2P$ M 165.172
Cryst. (pet. ether). Mp 58-62°. $Bp_{0.8}$ 100-110° (oven).
Et ester: Ethyl N-tert-*butyl*-P-*methylphosphonamidate.*
$C_7H_{18}NO_2P$ M 179.198
Liq. Bp_{25} 134-140°.
Chloride: [88652-78-2].
$C_5H_{13}ClNOP$ M 169.591
Cryst. (C_6H_6/pet. ether). Mp 121-123°.

Keay, L., *J. Org. Chem.*, 1963, **28**, 329 (*ethyl ester, synth*)
Harger, M.J.P. *et al*, *Tetrahedron*, 1982, **38**, 3073 (*methyl ester, synth, ir, pmr*)
Harger, M.J.P., *J. Chem. Soc., Perkin Trans. 1*, 1983, 2127 (*chloride, synth, ir, ms, pmr, props*)

***O*-Butyl methylphosphonochloridothioate, 9CI** B-00573

[18005-38-4]

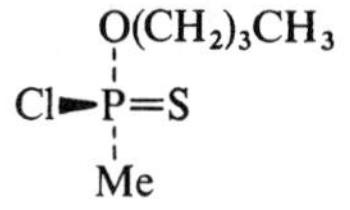

(*S*)-*form*

$C_5H_{12}ClOPS$ M 186.636
▷TA3750000.
(*S*)-*form* [38315-83-2]
Oil. $[\alpha]_D$ +59.20° (neat).
(±)-*form*
Liq. $Bp_{0.2}$ 40°. n_D^{20} 1.4870.

Hoffmann, F.W. *et al*, *J. Am. Chem. Soc.*, 1958, **80**, 3945 (*synth*)
Mikolajczyk, M. *et al*, *Tetrahedron*, 1972, **28**, 3855 (*synth*)
Pudovik, A.N. *et al*, *Zh. Obhsch. Khim.*, 1972, **42**, 317 (*Engl. transl.* p. 308) (*ir*)
Eliseenkov, V.N. *et al*, *Zh. Obshch. Khim.*, 1973, **43**, 2150 (*Engl. transl.* p. 2141) (*synth*)

***O*-Butyl *O*-methyl phosphorodithioate, 9CI, 8CI** B-00574

O-*Butyl* O-*methyl phosphorothioic acid.* O-*Butyl* O-*methyl hydrogen thiophosphate*

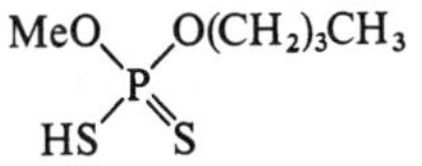

$C_5H_{13}O_2PS_2$ M 200.250
K salt: Solid. Mp 159-161°.

Kotovich, B.P. *et al*, *Zh. Obshch. Khim.*, 1968, **38**, 1282 (*Engl. transl.* p. 1235) (*synth*)

***O*-Butyl *O*-1-naphthalenyl phosphorothioate, 9CI** B-00575

O-*Butyl* O-*1-naphthyl hydrogen thiophosphate*

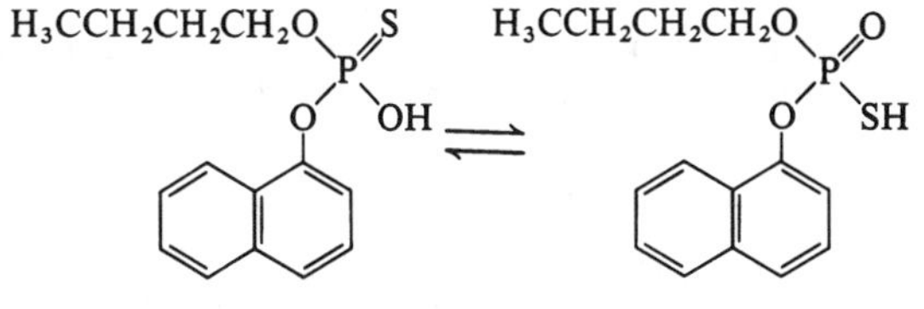

$C_{14}H_{17}O_3PS$ M 296.320
Tautomeric.
(+)-*form* [68144-09-2]
Solid. $[\alpha]_D$ +22.20°, $[\alpha]_{365}$+94.20° (c, 0.36 in $CHCl_3$).
Tetramethylammonium salt: [68198-84-5]. Solid. $[\alpha]_D$ +17.60°, $[\alpha]_{365}$+72.70° (c, 1.0 in $CHCl_3$).

Akintonwa, D.A.A., *Tetrahedron*, 1978, **34**, 959 (*synth, props, derivs*)

***tert*-Butyloxaphosphine** B-00576

(1,1-Dimethylethyl)oxaphosphine, 10CI
[68108-86-1]

$(H_3C)_3CP{=}O$

C_4H_9OP M 104.088
Reactive species which may be trapped with *o*-quinones.

Quast, H. *et al*, *Chem. Ber.*, 1982, **115**, 901.

5-*tert*-Butyl-2-phenyl-1,3,2-dioxaphosphorinane, 8CI B-00577

*5-(1,1-Dimethylethyl)-2-phenyl-1,3,2-dioxaphosphorinane, 9CI. 2-*tert*-Butyltrimethylene phenylphosphonite. 5-*tert*-Butyl-2-phenyl-1,3-dioxa-2-phosphacyclohexane*

$(H_3C)_3C$ ... O–P–Ph *cis-form*

$C_{13}H_{19}O_2P$ M 238.266

Low melting solid. $Bp_{0.05}$ 82-84°. Mixt. of stereoisomers.

cis-form [54655-28-6]

2-Oxide: [42295-63-6]. *2-*tert*-Butyltrimethylene phenylphosphonate.*

$C_{13}H_{19}O_3P$ M 254.265

Solid. Mp 106-107°.

trans-form [54655-29-7]

2-Oxide: [37555-31-0]. Solid. Mp 89-89.5°.

Bentrude, W.G. *et al*, *J. Am. Chem. Soc.*, 1972, **94**, 3264; 1973, **95**, 4666; 1975, **97**, 573; 1977, **99**, 4383 (*synth, pmr, deriv*)

Finocchiaro, P. *et al*, *J. Am. Chem. Soc.*, 1976, **98**, 3537 (*pmr, conformn*)

Butylphenylphosphinic acid, 9CI B-00578

[18629-24-8]

$H_3C(CH_2)_3P(O)(OH)Ph$

$C_{10}H_{15}O_2P$ M 198.201

Oil. $Bp_{0.0001}$ 240-250°. Forms cryst. hydrate.

Dicyclohexylammonium salt: [18629-29-3]. Cryst. ($CHCl_3/Et_2O$). Mp 145.5-147°.

Evdakov, V.P. *et al*, *Zh. Obshch. Khim.*, 1967, **37**, 2508 (*Engl. transl.* p. 2385) (*synth*)

Emmick, T.L. *et al*, *J. Am. Chem. Soc.*, 1968, **90**, 3459 (*synth*)

Giordano, F. *et al*, *Acta Crystallogr., Sect. B*, 1969, **25**, 1057 (*cryst struct*)

***tert*-Butylphenylphosphinic acid, 8CI** B-00579

(1,1-Dimethylethyl)phenylphosphinic acid, 9CI

[4923-86-8]

$(H_3C)_3C$–P(OR)(=O)–Ph (*R*)-form of esters

$C_{10}H_{15}O_2P$ M 198.201

(*R*)-form

Me ester: [69460-42-0]. *Methyl* tert-*butylphenylphosphinate.*

$C_{11}H_{17}O_2P$ M 212.228

$[\alpha]_D$ +58.08° (c, 0.6 in C_6H_6).

(–)-*Menthyl ester:* [63246-15-1]. *Menthyl* tert-*butylphenylphosphinate.*

$C_{20}H_{33}O_2P$ M 336.453

$[\alpha]^{22}_D$ –59.7° (c, 0.53 in MeOH). Could not be fully purified.

Chloride: [75213-02-4].

$C_{10}H_{14}ClOP$ M 216.647

$[\alpha]^{20}_D$ +40.6° (C_6H_6) (81% o.p.).

Isocyanate: [79157-92-9].

$C_{11}H_{14}NO_2P$ M 223.211

$Bp_{0.1}$ 135°. $[\alpha]^{20}_D$ +4.8°.

*Amide: see P-tert-*Butyl-*P*-phenylphosphinic amide, B-00580

(*S*)-form

(–)-*Menthyl ester:* [63283-74-9]. Solid. Mp 135-136°. $[\alpha]^{22}_D$ –103.4° (c, 1.0 in MeOH).

Chloride: [75213-01-3]. $[\alpha]^{25}_D$ –30.6° (C_6H_6) (61% o.p.).

Isocyanate: [79157-96-3]. Liq. $Bp_{0.1}$ 81-83°. $[\alpha]^{20}_D$ –9.33°. n_D^{20} 1.5360.

Isothiocyanate: [79158-00-2].

$C_{11}H_{14}NOPS$ M 239.271

$[\alpha]^{20}_D$ –30.3° (C_6H_6).

Thiocyanate: [82945-12-8].

$C_{11}H_{14}NOPS$ M 239.271

Solid. Mp 42-44°.

*Amide: see P-tert-*Butyl-*P*-phenylphosphinic amide, B-00580

(±)-form

Me ester: Liq. Bp_{125} 140° (oven).

Fluoride:

$C_{10}H_{14}FOP$ M 200.192

Solid. Mp 40-42°. $Bp_{0.03}$ 73°.

Chloride: [63246-16-2].

$C_{10}H_{14}ClOP$ M 216.647

Solid. Mp 28-29°, Mp 57-60°. $Bp_{0.1}$ 103-104°, $Bp_{0.15}$ 70°. n_D^{23} 1.5397.

Ph ester: Phenyl tert-*butylphenylphosphinate.*

$C_{16}H_{19}O_2P$ M 274.299

Liq. Bp_8 120° (oven).

*Amide: see P-tert-*Butyl-*P*-phenylphosphinic amide, B-00580

Azide:

$C_{10}H_{14}N_3OP$ M 223.214

Solid at r.t. $Bp_{0.1}$ 114-119° (oven).

Isothiocyanate: Oil. $Bp_{0.1}$ 94-97°. n_D^{21} 1.5750.

Imidazolide: N-(tert-*Butylphenylphosphinyl)imidazole.*

$C_{13}H_{18}N_2OP$ M 249.272

Cryst. (C_6H_6/hexane). Mp 134-136°.

Brown, A.D. *et al*, *J. Chem. Soc. (C)*, 1968, 839 (*synth, ir, pmr*)

Brooks, R.J. *et al*, *J. Org. Chem.*, 1975, **40**, 2059 (*chloride, fluoride, ir, pmr*)

Lopusinski, A. *et al*, *Justus Liebigs Ann. Chem.*, 1977, 924 (*isothiocyanate, pmr, ir, nmr*)

Luckenbach, R. *et al*, *Z. Naturforsch., B*, 1977, **32**, 584 (*chloride*)

Omelańczuk, J. *et al*, *J. Am. Chem. Soc.*, 1979, **101**, 7292 (*methyl ester*)

Haegela, G. *et al*, *Z. Naturforsch., B*, 1980, **35**, 1182 (*menthyl esters, cmr, nmr, pmr*)

Harger, M.J.P., *J. Chem. Soc., Perkin Trans. 2*, 1980, 1505 (*esters, pmr, ms*)

Krawiecka, B. *et al*, *J. Am. Chem. Soc.*, 1980, **102**, 6582 (*chloride*)

Petrov, K.A. *et al*, *Zh. Obshch. Khim.*, 1980, **50**, 1518 (*Engl. transl.* p. 1227) (*chloride*)

Lopusinski, A. *et al*, *Tetrahedron*, 1981, **37**, 2011 (*chloride, isocyanate, isothiocyanate*)

Sheldrick, W.S. *et al*, *J. Mol. Struct.*, 1981, **74**, 331 (*menthyl ester, abs config*)

Dabkowski, W. *et al*, *Chem. Ber.*, 1982, **115**, 1636 (*imidazolide, use*)

Lopusiński, A. *et al*, *Tetrahedron*, 1982, **38**, 679 (*thiocyanate, isothiocyanate*)

Harger, M.J.P. *et al*, *Tetrahedron*, 1982, **38**, 3073 (*azide, pmr*)

Krawiecka, B. *et al*, *J. Org. Chem.*, 1986, **51**, 4201 (*derivs, synth, pmr, P nmr*)

P-tert-Butyl-_P_-phenylphosphinic amide, 8CI B-00580

P-(*1,1-Dimethylethyl*)-P-*phenylphosphinic amide*, *9CI*

[51028-18-3]

$(H_3C)_3C$–P(=O)(Ph)–NH_2 (*R*)-*form*

$C_{10}H_{16}NOP$ M 197.216

(±)-***form*** [67213-43-8]

Cryst. (pet. ether). Mp 85-86°. (*R*)-form also known.

N-*Ph:* [51028-11-6]. P-tert-*Butyl*-N,P-*diphenylphosphinic amide.* P-tert-*Butyl*-P-*phenylphosphinic anilide.*

$C_{16}H_{20}NOP$ M 273.314

Cryst. (MeOH). Mp 263-264°.

Harger, M.J.P., *J. Chem. Soc., Perkin Trans. 1*, 1975, 514; 1977, 605 (*synth, deriv, ir, pmr*)

Harger, M.J.P., *J. Chem. Soc., Perkin Trans. 2*, 1977, 1882; 1978, 326 (*pmr*)

Harger, M.J.P., *Tetrahedron*, 1982, **38**, 3073.

tert-Butylphenylphosphinodithioic acid B-00581

(*1,1-Dimethylethyl*)*phenylphosphinodithioic acid*, *9CI*

$(H_3C)_3CPPh(S)SH$

$C_{10}H_{15}PS_2$ M 230.322

Cryst. (pet. ether). Mp 72-73°.

Hoffmann, H. *et al*, *Chem. Ber.*, 1966, **99**, 1134.

tert-Butylphenylphosphinoselenoic acid, 9CI B-00582

(*1,1-Dimethylethyl*)*phenylphosphinoselenoic acid*, *9CI*

[46141-44-0]

HSe–P(=O)($C(CH_3)_3$)(Ph) (*R*)-*form*

$C_{10}H_{15}OPSe$ M 261.162

(***R***)-***form*** [51584-27-1]

$[\alpha]_D$ +25.65° (MeOH).

Se-*Me ester:* Se-*Methyl* tert-*butylphenylphosphinoselenoate.*

$C_{11}H_{17}OPSe$ M 275.188

$[\alpha]_D$ +120° (C_6H_6).

(***S***)-***form*** [51584-28-2]

$[\alpha]_D$ −19.8° (c, 1.70 in MeOH), −30.05° (MeOH).

Se-*Me ester:* $[\alpha]_D$ −148.9° (c, 1.55 in C_6H_6).

Krawiecka, B. *et al*, *Phosphorus*, 1973, **3**, 177 (*synth, resoln, ester*)

Omelańczuk, J. *et al*, *J. Am. Chem. Soc.*, 1979, **101**, 7292 (*ester*)

tert-Butylphenylphosphinothioic acid, 8CI B-00583

(*1,1-Dimethylethyl*)*phenylphosphinothioic acid*, *9CI*

[6002-45-5]

HO–P(=S)(Ph)($C(CH_3)_3$) (*R*)-*form*

$C_{10}H_{15}OPS$ M 214.262

The enantiomers are used as chiral nmr shift regents.

(***R***)-***form*** [54100-47-9]

Solid. Mp 103-106° (softens at 96°). $[\alpha]_D$ +28.1° (c, 2.4 in MeOH).

S-*Me ester:* [51584-29-3]. S-*Methyl* tert-*butylphenylphosphinothioate.*

$C_{11}H_{17}OPS$ M 228.288

Solid. Mp 69-73°. $Bp_{0.1}$ 90° (oven). $[\alpha]_D$ +153.1° (c, 2.1 in C_6H_6).

Chloride:

$C_{10}H_{14}ClPS$ M 232.707

$[\alpha]_D^{20}$ +3.81°. Has (*S*)-chirality.

(***S***)-***form*** [55705-77-6]

Solid. Mp 96-98°. $[\alpha]_D$ −24.9° (c, 2.2 in MeOH).

O-*Me ester:* [55705-78-7]. O-*Methyl* tert-*butylphenylphosphinothioate.*

$C_{11}H_{17}OPS$ M 228.288

$[\alpha]_D$ +34.9° (c, 1.68 in C_6H_6).

S-*Me ester:* [51584-30-6]. $[\alpha]_D$ −106.90° (C_6H_6), −148° (91% o.p.).

Chloride: $[\alpha]_D^{20}$ −4.8°. Has (*R*)-chirality.

(±)-***form*** [67314-76-5]

Cryst. Mp 124-125°. $Bp_{0.2}$ 105° (oven).

O-*Me ester:* [76420-37-6]. Cryst. (pet. ether). Mp 75-75.5° (36-38°). $Bp_{0.03}$ 84°.

Chloride: [62839-84-3].

$C_{10}H_{14}ClPS$ M 232.707

Solid. Mp 77°.

Michalski, J. *et al*, *J. Organomet. Chem.*, 1975, **97**, C31 (*synth*)

Harger, M.J.P., *J. Chem. Soc., Perkin Trans. 2*, 1978, 326 (*resoln, pmr*)

Harger, M.J.P., *Tetrahedron Lett.*, 1978, 2927 (*use*)

Mikolajczyk, M. *et al*, *J. Am. Chem. Soc.*, 1978, **100**, 7003 (*pmr, cmr*)

Harger, M.J.P., *J. Chem. Soc., Perkin Trans. 2*, 1980, 1505 (*esters*)

Krawiecka, B. *et al*, *J. Am. Chem. Soc.*, 1980, **102**, 6582 (*esters*)

Kauslik, M. *et al*, *Indian J. Chem., Sect. B*, 1981, **20**, 932 (*chloride, ir, pmr*)

Omelańczuk, J. *et al*, *Angew. Chem.*, 1981, **93**, 875 (*config*)

Krawiecka, B. *et al*, *J. Org. Chem.*, 1986, **51**, 4201 (*synth, pmr, P nmr*)

tert-Butylphenylphosphinothioselenoic acid B-00584

(*1,1-Dimethylethyl*)*phenylphosphinothioselenoic acid*

[66499-14-7]

$(H_3C)_3CPPh(Se)SH \rightleftharpoons (H_3C)_3CPPh(S)SeH$

$C_{10}H_{15}PSSe$ M 277.222

Free acid dec. rapidly at r.t.

Triethylammonium salt: Cryst. (C_6H_6/pet. ether). Mp 100-101.5°.

Mastryukova, T.A. *et al*, *Zh. Obshch. Khim.*, 1978, **48**, 1447 (*Engl. transl.* p. 1329) (*synth, props, esters, pmr*)

Butyl phenyl phosphonate, 9CI **B-00585**

Butyl phenyl phosphite

[14609-91-7]

$H_3C(CH_2)_3OPH(O)OPh \rightleftharpoons H_3C(CH_2)_3OP(OH)OPh$

$C_{10}H_{15}O_3P$ M 214.200

Tautomeric. Liq. d_4^{20} 1.08-1.10. $Bp_{0.3}$ 124-126°. n_D^{20} 1.4950.

Wolf, R. *et al*, *Bull. Soc. Chim. Fr.*, 1960, 124 (*synth, ir*)
Houalla, D. *et al*, *Bull. Soc. Chim. Fr.*, 1960, 129 (*ir*)
Mandel'baum, Ya.A. *et al*, *Zh. Obshch. Khim.*, 1972, **42**, 502 (*Engl. transl.* p. 500) (*synth*)

4-*tert*-Butyl-1-phenylphosphorinane, 8CI **B-00586**

4-(1,1-Dimethylethyl)-1-phenylphosphorinane, 9CI. 4-(1,1-Dimethylethyl)-1-phenylphosphinane

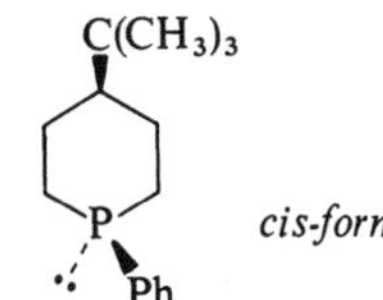

cis-form

$C_{15}H_{23}P$ M 234.320

cis-form [61332-73-8]

$Bp_{0.1}$ 110-120° (oven).

Oxide: [61332-82-9].
$C_{15}H_{23}OP$ M 250.320
Solid. Mp 160-161° (impure).

*B,MeBr: 4-*tert*-Butyl-1-methyl-1-phenylphosphorinanium bromide.*
$C_{16}H_{26}BrP$ M 329.259
Cryst. (EtOH/EtOAc). Mp 234.5-237°.

*B,PhCH₂Br: 1-Benzyl-4-*tert*-butyl-1-phenylphosphorinanium bromide. 4-(1,1-Dimethylethyl)-1-phenyl-1-phenylmethylethylphosphorinanium bromide.*
$C_{22}H_{30}BrP$ M 405.357
Cryst. (EtOH/EtOAc). Mp 224.5-226°.

trans-form [61332-72-7]

Liq. $Bp_{0.1}$ 110-120° (oven).

Oxide: [61332-81-8]. V. hygroscopic solid. Mp 88.5-89°.
B,MeBr: Cryst. (EtOH/EtOAc). Mp 178-179°.
B,PhCH₂Br: Cryst. (EtOH/EtOAc). Mp 268-270°.

Marsi, K., *J. Org. Chem.*, 1977, **42**, 1306 (*cmr, P nmr*)
MacDonell, G. *et al*, *J. Am. Chem. Soc.*, 1978, **100**, 4535 (*cryst struct*)

Butyl phosphinate, 9CI **B-00587**

Butyl hypophosphite

[18108-09-3]

$H_3C(CH_2)_3OP(O)H_2 \rightleftharpoons H_3C(CH_2)_3OP(OH)H$

$C_4H_{11}O_2P$ M 122.103

Liq. d_4^{20} 1.00. Bp_{10} 120-125°, $Bp_{1.5}$ 52-60°. n_D^{20} 1.4288.

Ivanov, B.E. *et al*, *Izv. Akad. Nauk SSSR, Ser. Khim.*, 1967, 1498 (*Engl. transl.* p. 1447); *CA*, **68**, 78359 (*synth*)
Karlstedt, N.B. *et al*, *Zh. Obshch. Khim.*, 1976, **46**, 2018 (*Engl. transl.* p. 1942) (*synth, P nmr*)

Butylphosphine, 9CI **B-00588**

[1732-74-7]

$H_3C(CH_2)_3PH_2$

$C_4H_{11}P$ M 90.105

Liq. with foul odour. Bp 54°, Bp 60°, Bp 76°. pK_a −0.03 (H_2O). n_D^{20} 1.4252. Forms Co and Ta complexes.

Na salt: Yellowish solid.
Di-Na salt: Brownish solid.

Schindlbauer, H. *et al*, *Monatsh. Chem.*, 1961, **92**, 868 (*synth, ir*)
Pass, F. *et al*, *Montash. Chem.*, 1962, **93**, 230 (*synth*)
Fritzsche, H. *et al*, *Chem. Ber.*, 1965, **98**, 1681 (*synth*)

***tert*-Butylphosphine, 8CI** **B-00589**

(1,1-Dimethylethyl)phosphine, 9CI

[2501-94-2]

$(H_3C)_3CPH_2$

$C_4H_{11}P$ M 90.105

Ligand for Co and Mo. Readily oxidizable, odourous liq. Bp 53-55°. n_D^{20} 1.4252.

Schindlbauer, H. *et al*, *Monatsh. Chem.*, 1961, **92**, 868 (*ir*)
Lappert, M.F. *et al*, *J. Chem. Soc., Dalton Trans.*, 1975, 1207 (*pe*)
Becker, G. *et al*, *Z. Anorg. Allg. Chem.*, 1978, **439**, 121; **443**, 42 (*nmr*)
Li, Y.S. *et al*, *J. Mol. Spectrosc.*, 1978, **70**, 34 (*microwave*)
Mosbo, J.A. *et al*, *Phosphorus Sulfur*, 1981, **11**, 11 (*struct*)

Butylphosphinic acid, 9CI **B-00590**

$H_3C(CH_2)_3PH(O)OH$

$C_4H_{11}O_2P$ M 122.103

Tautomeric with Butylphosphonous acid, B-00615 but free acid exists in the phosphinic acid form. Oil.

Me ester: [21661-51-8]. *Methyl butylphosphinate.*
$C_5H_{13}O_2P$ M 136.130
Liq. d^{20} 1.01. Bp_{10} 85-86°. n_D^{20} 1.4355.

Et ester: [21661-52-9]. *Ethyl butylphosphinate.*
$C_6H_{15}O_2P$ M 150.157
Liq. d_4^{20} 0.98. $Bp_{1.5}$ 49-50°. n_D^{20} 1.4350.

Isopropyl ester: [21661-54-1]. *Isopropyl butylphosphinate.*
$C_7H_{17}O_2P$ M 164.184
Liq. d_4^{20} 0.96. Bp_7 91-92°. n_D^{20} 1.4300.

Ph ester: [21655-99-2]. *Phenyl butylphosphinate.*
$C_{10}H_{15}O_2P$ M 198.201
Liq. d^{20} 1.28. Bp_6 179-180°, $Bp_{0.001}$ 110-120°. n_D^{20} 1.5495.

Sander, M., *Chem. Ber.*, 1960, **93**, 1220 (*synth*)
Abramov, V.S. *et al*, *CA*, 1968, **69**, 67469 (*esters*)

***tert*-Butylphosphinic acid, 8CI** **B-00591**

(1,1-Dimethylethyl)phosphinic acid, 9CI

[26920-54-7]

OR
H–P–C(CH₃)₃ *(R)ₚ-form* of esters
O

$C_4H_{11}O_2P$ M 122.103

Tautomeric with *tert*-Butylphosphonous acid, B-00616 .

(R)-form

Me ester: [79157-90-7]. *Methyl* tert-*butylphosphinate.*
Liq. Bp_{35} 87°. $[\alpha]_D^{20}$ +10.9° (neat).

(±)-form

Cryst. (pet. ether). Mp 86°.

Me ester:
$C_5H_{13}O_2P$ M 136.130

Liq. Bp_{40} 87-89°. n_D^{20} 1.4326.
Et ester: Ethyl tert-*butylphosphinate.*
$C_6H_{15}O_2P$ M 150.157
Liq. Bp_{10} 70°. n_D^{20} 1.4360.
Ph ester: Phenyl tert-*butylphosphinate.*
$C_{10}H_{15}O_2P$ M 198.201
Liq. Bp_1 95-98°. n_D^{20} 1.5069.

Crofts, P.C. *et al, J. Chem. Soc. (C)*, 1970, 332 (*synth, ester, ir, pmr*)
Krawiecka, B. *et al, Bull. Acad. Pol. Sci., Ser. Sci. Chim.*, 1971, **19**, 377 (*ester*)
Foss, V.L. *et al, Zh. Obshch. Khim.*, 1979, **49**, 559 (*Engl. transl.* p. 489) (*esters*)
Kopusiński, A. *et al, Tetrahedron*, 1981, **37**, 2011 (*ester, config*)
Quast, H. *et al, Chem. Ber.*, 1982, **115**, 901 (*ester*)

O-Butyl phosphinothioate, 9CI — B-00592

[61351-26-6]

$H_3CCH_2CH_2CH_2OP(S)H_2 \rightleftharpoons H_3CCH_2CH_2OP(SH)H$

$C_4H_{11}OPS$ M 138.164
Liq. $Bp_{1.5}$ 59-64°.

Karlstedt, N.B. *et al, Zh. Obshch. Khim.*, 1976, **46**, 2018 (*Engl. transl.* p. 1942) (*synth, nmr*)

Butylphosphinothioic acid — B-00593

$H_3C(CH_2)_3PH(S)OH \rightleftharpoons H_3C(CH_2)_3PH(O)SH \rightleftharpoons H_3C(CH_2)_3P(SH)OH$

$C_4H_{11}OPS$ M 138.164
Exists as tautomeric equilibrium between thione and thiol forms, in further equilibrium with butylphosphonothious acid.
O-*Propyl ester:* O-*Propyl butylphosphinothioate.*
$C_7H_{17}OPS$ M 180.244
Liq. Bp_2 118-120°. n_D^{20} 1.4780.

Abramov, V.S. *et al, Zh. Obshch. Khim.*, 1969, **39**, 1543 (*Engl. transl.* p. 1512) (*synth*)

Butylphosphonic acid — B-00594

[3321-64-0]

$H_3CCH_2CH_2CH_2P(O)(OH)_2$

$C_4H_{11}O_3P$ M 138.103
Needles (hexane), cryst. (C_6H_6). Mp 104-106°. Bp_{20} 160-162°.
Di-Me ester: [24475-23-8]. *Dimethyl butylphosphonate.*
$C_6H_{15}O_3P$ M 166.156
No phys. props. reported.
Mono-Et ester: [5284-11-7]. *Monoethyl butylphosphonate.*
$C_6H_{15}O_3P$ M 166.156
Liq. Bp_1 147-149°.
Di-Et ester: [2404-75-3]. *Diethyl butylphosphonate.*
$C_8H_{19}O_3P$ M 194.210
Liq. Bp_1 74°, $Bp_{0.1}$ 53-5°. n_D^{25} 1.4213.
Di-Ph ester: [3049-19-2]. *Diphenyl butylphosphonate.*
$C_{16}H_{19}O_3P$ M 290.298
Liq. d_4^{20} 1.14. $Bp_{0.3}$ 143.5-145°. n_D^{20} 1.5398.
Dibutyl ester: see Dibutyl butylphosphonate, D-00076
Difluoride: [690-97-1].
$C_4H_9F_2OP$ M 142.085
Liq. d_4^{20} 1.13. $Bp_{17.5}$ 51.5-52°. n_D^{20} 1.3702.
Dichloride: see Butylphosphonic dichloride, B-00598

Kosolapoff, G.M., *J. Am. Chem. Soc.*, 1945, **67**, 1180 (*synth*)
Myers, T.C. *et al, J. Am. Chem. Soc.*, 1954, 4122 (*synth, diethyl ester*)
Burger, A. *et al, J. Am. Chem. Soc.*, 1957, **79**, 3575 (*monoethyl ester*)
Petrov, K.A. *et al, Zh. Obshch. Khim.*, 1959, **29**, 3407 (*Engl. transl.* p. 3369) (*synth, derivs*)
Nixon, J.F. *et al, Spectrochim. Acta*, 1964, **20**, 1835 (*difluoride, P nmr*)
Bel'skii, V.E. *et al, Zh. Obshch. Khim.*, 1972, **42**, 2427 (*Engl. transl.* p. 2421) (*diethyl ester, P nmr*)
Dietze, U., *J. Prakt. Chem.*, 1974, **316**, 293 (*ir*)
Ernst, L., *Org. Magn. Reson.*, 1977, **9**, 35 (*diethyl ester, cmr*)
Kharrasova, F.M. *et al, Zh. Obshch. Khim.*, 1978, **48**, 1041 (*Engl. transl.* p. 948) (*diphenyl ester, P nmr, synth*)

tert-Butylphosphonic acid, 8CI — B-00595

(1,1-Dimethylethyl)phosphonic acid, 9CI
[4923-84-6]

$(H_3C)_3CP(O)(OH)_2$

$C_4H_{11}O_3P$ M 138.103
Solid or needles (pet. ether/AcOH or H_2O). Mp 159-160°, 191.5-192°. pK_{a1} 2.79, pK_{a2} 8.88 (H_2O, 25°).
Mono-Et ester: Monoethyl (1,1-dimethylethyl)-phosphonate. Ethyl hydrogen tert-*butylphosphonate.*
$C_6H_{15}O_3P$ M 166.156
Liq. $Bp_{0.03}$ 72-76°. n_D^{25} 1.4253.
Di-Et ester: [19935-93-4]. *Diethyl (1,1-dimethylethyl)-phosphonate. Diethyl* tert-*butylphosphonate.*
$C_8H_{19}O_3P$ M 194.210
Liq. $Bp_{1.5}$ 54°. n_D^{25} 1.4160.
Di-Ph ester: [59361-27-2]. *Diphenyl (1,1-dimethylethyl)phosphonate. Diphenyl* tert-*butylphosphonate.*
$C_{16}H_{19}O_3P$ M 290.298
Cryst. (CH_2Cl_2/pet. ether). Mp 107-108°.
*Difluoride: see tert-*Butylphosphonic difluoride, B-00600
Dichloride: see *tert*-Butylphosphonic dichloride, B-00599
Diamide: see *P-tert*-Butylphosphonic diamide, B-00597

Crofts, P.C. *et al, J. Am. Chem. Soc.*, 1953, **75**, 3379 (*synth, props*)
Cadogan, J.I.G. *et al, J. Chem. Soc. (B)*, 1971, 1988 (*monoethyl ester*)
Seyferth, D. *et al, J. Organomet. Chem.*, 1973, **59**, 237 (*diethyl ester, pmr*)
Griffiths, W.R. *et al, Phosphorus*, 1975, **5**, 273 (*ms*)
Murav'ev, I.V. *et al, Zh. Obshch. Khim.*, 1976, **46**, 1262 (*Engl. transl.* p. 1241) (*synth*)
Nesterov, L.V. *et al, Izv. Akad. Nauk SSSR, Ser. Khim.*, 1976, 475 (*Engl. transl.* p. 462) (*diphenyl ester, synth, pmr*)
Griffiths, W.R. *et al, Phosphorus Sulfur*, 1978, **5**, 101 (*esters, ms*)
Wozniak, M. *et al, J. Chem. Soc., Dalton Trans.*, 1981, 2423 (*synth, complexes*)

P-Butylphosphonic diamide, 9CI — B-00596

$H_3C(CH_2)_3P(O)(NH_2)_2$

$C_4H_{13}N_2OP$ M 136.133
Solid. Sl. sol. DMF, insol. C_6H_6, Et_2O, pet. ether. Mp 90-110°. Dec. in EtOH or H_2O at r.t.

N,N,N′,N′-*Tetra-Me:* [5277-10-1]. P-*Butyl*-N,N,N′,N′-*tetramethylphosphonic diamide. Butylphosphonic bis(dimethylamide).*
$C_8H_{21}N_2OP$ M 192.240
Liq. d_4^{21} 0.98. Bp_{33} 157°, $Bp_{0.4}$ 80°. n_D^{22} 1.4564.
N,N,N′,N′-*Tetra-Et:* [26348-78-7]. P-*Butyl*-N,N,N′,N′-*tetraethylphosphonic diamide. Butylphosphonic bis(diethylamide).*
$C_{12}H_{29}N_2OP$ M 248.348
Liq. d_4^{30} 0.93. $Bp_{3.5}$ 137°. n_D^{30} 1.4585.
N,N′-*Dibenzyl: Butylphosphonic* N,N′-*dibenzyldiamide.*
$C_{18}H_{25}N_2OP$ M 316.382
Solid. Insol. H_2O, Et_2O, pet. ether, sl. sol. EtOH, $CHCl_3$. Mp 91-93°.

Kosolapoff, G.M. *et al, J. Org. Chem.*, 1956, **21**, 413 (*synth*)
Helferich, B. *et al, Justus Liebigs Ann. Chem.*, 1963, **670**, 48 (*synth*)
Normant, H. *et al, Bull. Soc. Chim. Fr.*, 1965, 3441 (*synth*)
Normant, H. *et al, C.R. Hebd. Seances Acad. Sci., Ser. C*, 1967, **264**, 707 (*synth*)

P-tert-Butylphosphonic diamide, 9CI B-00597

P-(*1,1-Dimethylethyl*)*phosphonic diamide*, *9CI*

$(H_3C)_3CP(O)(NH_2)_2$

$C_4H_{13}N_2OP$ M 136.133
N,N′-*Di-Me:* N,N′-*Dimethyl*-P-(*1,1-dimethylethyl*)-*phosphonic diamide.* tert-*Butylphosphonic* N,N′-*dimethyldiamide.*
$C_6H_{17}N_2OP$ M 164.187
Cryst. (MeCN). Mp 130-131°.
N,N′-*Di*-tert-*butyl:* [55702-27-7]. N,N′,P-*Tris*(*1,1-dimethylethyl*)*phosphonic diamide.* tert-*Butylphosphonic* N,N′-*di*-tert-*butyldiamide.*
$C_{12}H_{29}N_2OP$ M 248.348
Cryst. (ligroin). Mp 181-182°.
N,N′-*Di-Ph:* P-(*1,1-Dimethylethyl*)-N,N′-*diphenylphosphonic diamide.* tert-*Butylphosphonic dianilide.*
$C_{16}H_{21}N_2OP$ M 288.328
Cryst. (DMF). Mp 256-257°, 267-268°.

Kinnear, A.M. *et al, J. Chem. Soc.*, 1952, 3437 (*dianilide*)
Harger, M.J.P., *Tetrahedron Lett.*, 1981, **22**, 4741 (*deriv, ms, ir, pmr*)
Quast, H. *et al, Justus Liebigs Ann. Chem.*, 1981, 943 (*derivs, synth, ir, pmr, P nmr, ms*)

Butylphosphonic dichloride, 9CI B-00598

1-(*Dichlorophosphinyl*)*butane*
[2302-80-9]

$H_3C(CH_2)_3P(O)Cl_2$

$C_4H_9Cl_2OP$ M 174.994
Liq. Bp_{23} 105-107°, Bp_7 81.5-82°.

Petrov, K.A. *et al, Zh. Obshch. Khim.*, 1959, **29**, 3407 (*Engl. transl.* p. 3369) (*synth*)
Geiseler, G. *et al, Ber. Bunsenges. Phys. Chem.*, 1967, **71**, 478 (*ir, raman*)
Tsvetkov, E.N. *et al, Izv. Akad. Nauk SSSR, Ser. Khim.*, 1967, 2375 (*Engl. transl.* p. 2267) (*synth, nqr*)
Grishina, O.N. *et al, Zh. Prikl. Khim.*, 1969, **42**, 2289 (*Engl. transl.* p. 2149) (*synth, props*)

tert-Butylphosphonic dichloride, 8CI B-00599

(*1,1-Dimethylethyl*)*phosphonic dichloride*, *9CI*
[4707-95-3]

$(H_3C)_3CP(O)Cl_2$

$C_4H_9Cl_2OP$ M 174.994
Solid with camphoraceous odour. Mp 110°, 123°. Bp_{25} 110° subl.

Kinnear, A.M. *et al, J. Chem. Soc.*, 1952, 3437 (*synth*)
Metzger, S.H. *et al, J. Org. Chem.*, 1964, **29**, 627 (*synth*)
Schmutzler, R. *et al, Z. Naturforsch., B*, 1965, **20**, 832 (*ir, pmr, P nmr*)
Bushweller, C.H. *et al, J. Am. Chem. Soc.*, 1973, **95**, 5949 (*pmr, P nmr, struct*)
Griffiths, W.R. *et al, Phosphorus Sulfur*, 1978, **4**, 341 (*ms*)

tert-Butylphosphonic difluoride, 8CI B-00600

(*1,1-Dimethylethyl*)*phosphonic difluoride*, *9CI*
[754-24-5]

$(H_3C)_3CP(O)F_2$

$C_4H_9F_2OP$ M 142.085
Solid. Mp 40°. Bp_{100} 97°.

Schmutzler, R. *et al, Z. Naturforsch., B*, 1965, **20**, 832 (*ir, pmr, F and P nmr*)
Fild, M. *et al, J. Chem. Soc. (A)*, 1970, 2359 (*synth*)
Holmes, R.R. *et al, Spectrochim. Acta, Part A*, 1971, **27**, 1525 (*synth, ir, raman*)
Patsanovskii, I.I. *et al, Zh. Obshch. Khim.*, 1981, **51**, 985 (*Engl. transl.* p. 822) (*synth*)

Butylphosphonisocyanatidic acid, 9CI B-00601

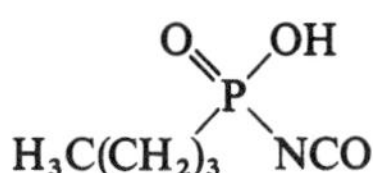

$C_5H_{10}NO_3P$ M 163.113
Et ester: [21959-85-3]. *Ethyl butylphosphonisocyanatidate.*
$C_7H_{14}NO_3P$ M 191.166
Liq. d_4^{20} 1.06. $Bp_{0.5}$ 122-124°. n_D^{20} 1.4346.

Gubnitskaya, E.S. *et al, Zh. Obshch. Khim.*, 1968, **38**, 1530 (*Engl. transl.* p. 1479)

tert-Butylphosphon(isothiocyanatidic) acid B-00602

(*1,1-Dimethylethyl*)*phosphon*(*isothiocyanatidic*) *acid*

NCS
O=P◄OR (*R*)-*form* of esters
C(CH₃)₃

$C_5H_{10}NO_2PS$ M 179.173

(*R*)-*form*

Me ester: [79157-98-5]. *Methyl* tert-*butylphosphon*(*isocyanatidate*).
$C_6H_{12}NO_2PS$ M 193.200
$Bp_{0.1}$ 58°. $[\alpha]_D^{20}$ +107.6° (C_6H_6).

Lopusinski, A. *et al, Tetrahedron*, 1981, **37**, 2011 (*synth, ir, P nmr*)
Lopusinski, A. *et al, Tetrahedron*, 1982, **38**, 679.

Butylphosphonochloridic acid, 9CI B-00603

$H_3C(CH_2)_3PCl(O)OH$

$C_4H_{10}ClO_2P$ M 156.549
Cyclohexyl ester: Cyclohexyl butylphosphonochloridate.
$C_{10}H_{20}ClO_2P$ M 238.694
Liq. $Bp_{0.1}$ 90°. n_D^{25} 1.4680.

Hafner, L.S. *et al, J. Med. Chem.*, 1970, **13**, 1025.

Butylphosphonochloridothioic acid, 9CI — B-00604

$S{=}P(OH)(Cl)(CH_2)_3CH_3 \rightleftharpoons HS{-}P({=}O)(Cl)(CH_2)_3CH_3$

$C_4H_{10}ClOPS$ M 172.609
O-Propyl ester: O-Propyl butylphosphonochloridothioate.
$C_7H_{16}ClOPS$ M 214.690
Liq. $Bp_{0.7}$ 70°. n_D^{25} 1.4853.

Chupp, J.P. *et al, J. Org. Chem.*, 1962, **27**, 3832.

Butylphosphonochloridous acid, 9CI — B-00605

$H_3C(CH_2)_3PCl(OH)$

$C_4H_{10}ClOP$ M 140.549
Et ester: [23654-71-9]. *Ethyl butylphosphonochloridite.*
$C_6H_{14}ClOP$ M 168.603
Liq. Bp_{13} 66°. n_D^{20} 1.4470.
Butyl ester: [41839-48-9]. *Butyl butylphosphonochoridite.*
$C_8H_{18}ClOP$ M 196.656
Liq. Bp_2 58-59°. n_D^{20} 1.4604.

Foss, V.L. *et al, Zh. Obshch. Khim.*, 1973, **43**, 1000 (*Engl. transl.* p. 994) (*butyl ester*)
Butkova, O.L. *et al, Izv. Akad. Nauk SSSR, Ser. Khim.*, 1982, 2390 (*Engl. transl.* p. 2106) (*ethyl ester*)

Butylphosphonodithioic acid, 9CI — B-00606

$O{=}P(SH)(SH)(CH_2)_3CH_3 \rightleftharpoons HO{-}P({=}S)(SH)(CH_2)_3CH_3$

$C_4H_{11}OPS_2$ M 170.224
O-Me ester: [13685-74-0]. *O-Methyl butylphosphonodithioate.*
$C_5H_{13}OPS_2$ M 184.251
Liq. d_4^{20} 1.11. $Bp_{0.02}$ 61.5°. n_D^{20} 1.5360.
O-Et ester: [5074-77-1]. *O-Ethyl butylphosphonodithioate.*
$C_6H_{15}OPS_2$ M 198.278
Liq. d_4^{20} 1.09. $Bp_{0.18}$ 66-67°. n_D^{20} 1.5281.
O-Isopropyl ester: [13685-75-1]. *O-Isopropyl butylphosphonodithioate.*
$C_7H_{17}OPS_2$ M 212.304
Liq. d_4^{20} 1.05. $Bp_{0.2}$ 74.5-75°. n_D^{20} 1.5100.
S,S-Dibutyl ester: [2797-54-8]. *S,S-Dibutyl butylphosphonodithioate.*
$C_{12}H_{27}OPS_2$ M 282.438
Defoliant. Liq. $Bp_{0.02}$ 125°. n_D^{26} 1.5112.
S,S-Di-Ph ester: [29703-23-9]. *S,S-Diphenyl butylphosphonodithioate.*
$C_{16}H_{19}OPS_2$ M 322.419
Fungicide. Solid. Mp 50-3°.

Chupp, J.P *et al, J. Org. Chem.*, 1962, **27**, 3832 (*synth*)
Grishina, O.N. *et al, Izv. Akad. Nauk SSSR, Ser. Khim.*, 1965, 2140 (*Engl. transl.* p. 2109) (*synth*)
U.S.P., 3 193 372, (*1965*); *CA*, **63**, 8976 (*dibutyl ester, synth, use*)
Grishina, O.N. *et al, Izv. Akad. Nauk SSSR, Ser. Khim.*, 1966, 1617 (*Engl. transl.* p. 1558) (*O-esters*)
Ger. Pat., 1 902 928, (*1970*); *CA*, **73**, 119693 (*diphenyl ester, synth, use*)

Butylphosphonodithious acid, 9CI — B-00607

$H_3CCH_2CH_2CH_2P(SH)_2$

$C_4H_{11}PS_2$ M 154.225
Di-Et ester: [38476-63-0]. *Diethyl butylphosphonodithioite.*
$C_8H_{19}PS_2$ M 210.332
Liq. d_4^{20} 1.00. Bp_{13} 128-130°. n_D^{20} 1.5368.
Dibutyl ester: [7697-74-7]. *Dibutyl butylphosphonodithioite.*
$C_{12}H_{27}PS_2$ M 266.439
Liq. $Bp_{0.25}$ 96-110°.
Di-Ph ester: [7697-73-6]. *Diphenyl butylphosphonodithioite.*
$C_{16}H_{19}PS_2$ M 306.420
Liq. $Bp_{0.3}$ 175-180°.

Sander, M., *Chem. Ber.*, 1960, **93**, 1220 (*dipropyl ester, synth*)
Grayson, M. *et al, J. Org. Chem.*, 1967, **32**, 236 (*dibutyl, diphenyl esters, synth, ir*)
Razumov, A.I. *et al, Zh. Obshch. Khim.*, 1972, **42**, 1250 (*Engl. transl.* p. 1245) (*diethyl ester, synth, nmr*)

Butylphosphonofluoridic acid, 9CI — B-00608

$H_3C(CH_2)_3PF(O)OH$

$C_4H_{10}FO_2P$ M 140.094
Cyclohexyl ester: [28364-22-9]. *Cyclohexyl butylphosphonofluoridate.*
$C_{10}H_{20}FO_2P$ M 222.239
Liq. $Bp_{0.15}$ 77°. n_D^{25} 1.4391.

Hafner, L.S. *et al, J. Med. Chem.*, 1970, **13**, 1025 (*synth*)
Landau, M.A. *et al, Zh. Strukt. Khim.*, 1970, **11**, 513 (*Engl. transl.* p. 467) (*struct*)

Butylphosphonoselenoic acid, 9CI — B-00609

$H_3C(CH_2)_3P(Se)(OH)_2 \rightleftharpoons H_3C(CH_2)_3P(O)(SeH)(OH)$

$C_4H_{11}O_2PSe$ M 201.063
O,O-Di-Et ester: O,O-Diethyl butylphosphonoselenoate.
$C_8H_{19}O_2PSe$ M 257.171
Liq. d_4^{20} 1.11. $Bp_{0.03}$ 136°. n_D^{20} 1.5182.
Dichloride: [55249-21-3].
$C_4H_9Cl_2PSe$ M 237.955
No phys. props. reported.

Nuretdinov, I.A. *et al, Zh. Obshch. Khim.*, 1974, **44**, 2588 (*Engl. transl.* p. 2548) (*ester, synth*)
Nuretdinov, I.A. *et al, Izv. Akad. Nauk SSSR, Ser. Khim.*, 1975, 327 (*Engl. transl.* p. 263) (*dichloride, nqr*)
Shagidullin, R.R. *et al, Izv. Akad. Nauk SSSR, Ser. Khim.*, 1976, 184 (*Engl. transl.* p. 174) (*dichloride, uv*)

tert-Butylphosphonoselenoic acid — B-00610

(1,1-Dimethylethyl)phosphonoselenoic acid

$(H_3C)_3C{-}P({=}Se)(OH)_2 \rightleftharpoons (H_3C)_3C{-}P({=}O)(SeH)(OH)$

$C_4H_{11}O_2PSe$ M 201.063
O-Me ester:
$C_5H_{13}O_2PSe$ M 215.090

$Bp_{0.1}$ 50-52°.

Mastryukova, T.A. *et al*, *Zh. Obshch. Khim.*, 1978, **48**, 1447 (*Engl. transl.* p. 1329) (*esters, pmr*)
Crofts, P.C. *et al*, *J. Chem. Soc. (B)*, 1968, 1416 (*synth*)
Crofts, P.C. *et al*, *J. Chem. Soc. (C)*, 1970, 332 (*ir*)
Kuchen, W. *et al*, *Chem. Ber.*, 1970, **103**, 2114 (*synth, ir, pmr, P nmr*)
Holmes, R.R. *et al*, *Spectrochim. Acta, Part A*, 1971, **27**, 1537 (*ir, raman*)

Butylphosphonothioic acid, 9CI — B-00611

$H_3C(CH_2)_3P(S)(OH)_2 \rightleftharpoons H_3C(CH_2)_3P(O)(OH)(SH)$

$C_4H_{11}O_2PS$ M 154.163

Mono-O-Et ester:
$C_6H_{15}O_2PS$ M 182.217
Solid (as Na salt). Mp 169-171° (Na salt).
O,O-*Di-Et ester:* O,O-*Diethyl butylphosphonodithioate.*
$C_8H_{19}O_2PS$ M 210.271
Liq. d_4^{20} 1.00. $Bp_{2.5}$ 74.5-75.5°. n_D^{20} 1.4600.
O,S-*Di-Et ester:* O,S-*Diethyl butylphosphonothioate.*
$C_8H_{19}O_2PS$ M 210.271
Liq. d_4^{20} 1.03. Bp_1 82.5-84°. n_D^{20} 1.4730.
O,O-*Dibutyl ester:* [17643-87-7]. O,O-*Dibutyl butylphosphonothioate.*
$C_{12}H_{27}O_2PS$ M 266.378
Liq. Bp_2 108-109°. n_D^{20} 1.4654.
Dichloride: [6588-22-3].
$C_4H_9Cl_2PS$ M 191.055
Liq. d^{20} 1.25. Bp_{12-14} 84-87°. n_D^{20} 1.5311.

Pudovik, A.N. *et al*, *Zh. Obshch. Khim.*, 1954, **24**, 307 (*Engl. transl.* p. 311) (*synth*)
Kabachnik, M.I. *et al*, *Zh. Obshch. Khim.*, 1956, **26**, 2228 (*Engl. transl.* p. 2491) (*esters, synth, ir*)
Kabachnik, M.I. *et al*, *Izv. Akad. Nauk SSSR, Ser. Khim.*, 1956, 193 (*Engl. transl.* p. 185) (*ester, synth*)
Grishina, O.N. *et al*, *Neftekhimiya*, 1968, **8**, 111; *CA*, **69**, 2980 (*dichloride, synth*)

tert-Butylphosphonothioic acid, 8CI — B-00612

(*1,1-Dimethylethyl*)*phosphonothioic acid*, 9CI
[57605-13-7]

$(H_3C)_3CP(S)(OH)_2 \rightleftharpoons (H_3C)_3P(O)(OH)(SH)$

$C_4H_{11}O_2PS$ M 154.163
Solid. Mp 159-160°.
*Mono-O-Me ester: see O-Methyl tert-*butylphosphonothioate, M-00107
O,S-Di-Me ester: see *O*-Methyl *tert*-butylphosphonothioate, M-00107
Dichloride: see *tert*-Butylphosphonothioic dichloride, B-00613
Dibromide: [27509-08-6].
$C_4H_9Br_2PS$ M 279.957
Cryst. (ligroin). Mp 215°. Bp_{2-3} 75° subl.

Kuchen, W. *et al*, *Chem. Ber.*, 1970, **103**, 2114 (*synth, ir, pmr, P nmr*)
Murav'ev, I.V. *et al*, *Zh. Obshch. Khim.*, 1976, **46**, 1262 (*Engl. transl.* p. 1241) (*synth*)

tert-Butylphosphonothioic dichloride, 8CI — B-00613

(*1,1-Dimethylethyl*)*phosphonothioic dichloride*, 9CI.
tert-*Butylthiophosphonic dichloride*
[21187-18-8]

$(H_3C)_3CP(S)Cl_2$

$C_4H_9Cl_2PS$ M 191.055
Cryst. (ligroin at −75°). Mp 177.5-179°. Bp_{2-3} 75° subl.

Butylphosphonotrithioic acid, 9CI — B-00614

$H_3C(CH_2)_3P(S)(SH)_2$

$C_4H_{11}PS_3$ M 186.285
Cyclic bis(anhydrosulfide): [1135-57-5]. *2,4-Dibutyl-1,3,2,4-dithiaphosphetane 2,4-disulfide.*
$C_8H_{18}P_2S_4$ M 304.418
Solid. Mp 49-52°, Mp 105-110°. $Bp_{0.13}$ 170-173°.
Dibutyl ester: [7697-78-1]. *Dibutyl butylphosphonotrithioate.*
$C_{12}H_{27}PS_3$ M 298.499
Liq. $Bp_{0.17}$ 138-144°.

Newallis, P.E. *et al*, *J. Org. Chem.*, 1962, **27**, 3829 (*anhydrosulfide*)
Grayson, M. *et al*, *J. Org. Chem.*, 1967, **32**, 236 (*ester, synth, ir*)
Grishina, O.N. *et al*, *CA*, 1969, **71**, 39085 (*anhydrosulfide*)

Butylphosphonous acid — B-00615

$H_3CCH_2CH_2CH_2P(OH)_2$

$C_4H_{11}O_2P$ M 122.103
Free acid exists as phosphoryl tautomer Butylphosphinic acid, B-00590 .
Mono-Me ester: see Butylphosphinic acid, B-00590
Di-Me ester: [17383-44-7]. *Dimethyl butylphosphonite.*
$C_6H_{15}O_2P$ M 150.157
Liq. Bp_{15} 89-92°.
Mono-Et ester: see Butylphosphinic acid, B-00590
Di-Et ester: [51503-25-4]. *Diethyl butylphosphonite.*
$C_8H_{19}O_2P$ M 178.211
Liq. Bp_{20} 78°. n_D^{20} 1.4310.
Monoisopropyl ester: see Butylphosphinic acid, B-00590
Diisopropyl ester: [76297-11-5]. *Diisopropyl butylphosphonite. Bis(1-methylethyl) butylphosphonite.*
$C_{10}H_{23}O_2P$ M 206.264
Liq. Bp_{18} 85-87°.
Dibutyl ester: [3058-20-6]. *Dibutyl butylphosphonite.*
$C_{12}H_{27}O_2P$ M 234.318
Liq. Bp_{10} 126-128°, $Bp_{0.5}$ 75°. n_D^{20} 1.4400.
Mono-Ph ester: see Butylphosphinic acid, B-00590
Di-Ph ester: [17383-46-9]. *Diphenyl butylphosphonite.*
$C_{16}H_{19}O_2P$ M 274.299
Liq. d_4^{20} 1.08. $Bp_{0.6}$ 126.5°. n_D^{20} 1.5532.
Difluoride: Butyldifluorophosphine.
$C_4H_9F_2P$ M 126.086
Liq. d_{20} 1.02. Bp 70-71°.
Dichloride: see Butylphosphonous dichloride, B-00618
Dibromide: Dibromobutylphosphine.
$C_4H_9Br_2P$ M 247.897

Liq. Bp_{10} 80-95°.
Diiodide: [89262-61-3]. *Butyldiiodophosphine.*
$C_4H_9I_2P$ M 341.898
Liq. $Bp_{0.08}$ 67°.
Bis(dibutylamide): [32596-70-6]. *Pentabutylphosphonous diamide.*
$C_{20}H_{45}N_2P$ M 344.563
Liq. $Bp_{0.1}$ 111°. n_D^{20} 1.4687.

Kulakova, V.N. *et al, Zh. Obshch. Khim.*, 1959, **29**, 3957 (*Engl. transl.* p. 3916) (*difluoride*)
Sander, M., *Chem. Ber.*, 1960, **93**, 1220 (*esters*)
Mastalerz, P., *Rocz. Chem.*, 1964, **38**, 61 (*diisopropyl ester*)
Voigt, D. *et al, Bull. Soc. Chim. Fr.*, 1964, 3087 (*dibutyl ester, synth*)
Zentil, M. *et al, Bull. Soc. Chim. Fr.*, 1971, 376 (*dibutylamide, synth, P nmr*)
Efimova, V.D. *et al, Zh. Obshch. Khim.*, 1974, **44**, 55 (*Engl. transl.* p. 51) (*esters, synth*)
Quin, L.D. *et al, Org. Magn. Reson.*, 1974, **6**, 503 (*dimethyl ester, cmr*)
Kharrasova, F.M. *et al, Zh. Obshch. Khim.*, 1978, **48**, 1041 (*Engl. transl.* p. 948) (*diphenyl ester, synth, P nmr*)

***tert*-Butylphosphonous acid, 8CI** **B-00616**

(1,1-Dimethylethyl)phosphonous acid, 9CI

$(H_3C)_3CP(OH)_2$

$C_4H_{11}O_2P$ M 122.103
The free acid exists in the tautomeric phosphoryl form *tert*-Butylphosphinic acid, B-00591 .
Mono-Me ester: see tert-Butylphosphinic acid, B-00591
Di-Me ester: [32045-17-3]. *Dimethyl* tert-*butylphosphonite.*
$C_6H_{15}O_2P$ M 150.157
Liq. Bp_{230} 100-105°.
Mono-Et ester: see tert-Butylphosphinic acid, B-00591
Di-Et ester: [25781-13-9]. *Diethyl* tert-*butylphosphonite.*
$C_8H_{19}O_2P$ M 178.211
Liq. Bp_{10} 85°. n_D^{20} 1.4162.
Diisopropyl ester: [76257-42-6]. *Diisopropyl* tert-*butylphosphonite. Bis(1-methylethyl) (1,1-dimethylethyl)-phosphonite.*
$C_{10}H_{23}O_2P$ M 206.264
Liq. Bp_1 57-58°.
Mono-Ph ester: see tert-Butylphosphinic acid, B-00591
Di-Ph ester: [66441-92-7]. *Diphenyl* tert-*butylphosphonite.*
$C_{16}H_{19}O_2P$ M 274.299
Solid. Mp 53-55°. Bp_{12-13} 95-97°, Bp_3 134-139°.
Difluoride: see tert-Butylphosphonous difluoride, B-00620
Dichloride: see *tert*-Butylphosphonous dichloride, B-00619
Dibromide: [80518-21-4]. *Dibromo*-tert-*butylphosphine.*
$C_4H_9Br_2P$ M 247.897
Solid. Mp 39.5-40°. Bp_{57} 105°.
Diiodide: [66517-44-0]. tert-*Butyldiiodophosphine.*
$C_4H_9I_2P$ M 341.898
Red-brown oil. $Bp_{0.08}$ 65-70°, $Bp_{0.04}$ 44-47°.
Diamide: see tert-Butylphosphonous diamide, B-00617

Crofts, P.C. *et al, J. Chem. Soc. (C)*, 1970, 332, 2342 (*synth, ethyl esters*)
Stewart, A.P. *et al, J. Chem. Soc., Chem. Commun.*, 1970, 1279 (*dimethyl ester, synth*)
Bartsch, R. *et al, Chem. Ber.*, 1978, **111**, 1420 (*diiodide, synth, pmr, P nmr*)
Feshchenko, N.G. *et al, Zh. Obshch. Khim.*, 1978, **48**, 365 (*Engl. transl.* p. 329) (*diphenyl ester, synth*)
Foss, V.L. *et al, Zh. Obshch. Khim.*, 1979, **49**, 559 (*Engl. transl.* p. 489) (*esters, synth, P nmr*)
Romanenko, V.D. *et al, Synthesis*, 1980, 823 (*diiodide, synth, P nmr*)
Hinke, A. *et al, Phosphorus Sulfur*, 1983, **15**, 93 (*dibromide, synth, nmr*)

***tert*-Butylphosphonous diamide, 8CI** **B-00617**

(1,1-Dimethylethyl)phosphonous diamide, 9CI
[19911-11-6]

$(H_3C)_3CP(NH_2)_2$

$C_4H_{13}N_2P$ M 120.134
Mp 77-79°. Bp_{12} 75-85° subl.
N,N′-*Di-Me:* [24090-56-0]. N,N′-*Dimethyl*-P-*(1,1-dimethylethyl)phosphonous diamide.* tert-*Butylphosphonous bis(methylamide).*
$C_6H_{17}N_2P$ M 148.187
Liq. Mp −26°. Bp_{12} 70-72°.
N,N′-*Bis(trimethylsilyl):* [19911-12-7]. P-tert-*Butyl*-N,N′-*bis(trimethylsilyl)phosphonous diamide.*
$C_{10}H_{29}N_2PSi_2$ M 264.497
Liq. Mp −1°. $Bp_{0.1}$ 51-55°.

Scherer, O.J. *et al, Z. Anorg. Allg. Chem.*, 1969, **370**, 171 (*synth, derivs, pmr*)

Butylphosphonous dichloride, 9CI **B-00618**

Butyldichlorophosphine
[6460-27-1]

$H_3C(CH_2)_3PCl_2$

$C_4H_9Cl_2P$ M 158.995
Pungent liq. d_4^{20} 1.16. Bp 156-159°, Bp_9 39-39.5°. n_D^{20} 1.4870. Easily hydrol.

Fox, R.B., *J. Am. Chem. Soc.*, 1950, **72**, 4147 (*synth*)
Sander, M., *Chem. Ber.*, 1960, **93**, 1220 (*synth*)
Bliznyuk, N.K. *et al, Zh. Obshch. Khim.*, 1967, **37**, 890 (*Engl. transl.* p. 840) (*synth*)
Quin, L.D. *et al, Org. Magn. Reson.*, 1974, **6**, 503 (*cmr, nmr*)

***tert*-Butylphosphonous dichloride** **B-00619**

(1,1-Dimethylethyl)phosphonous dichloride, 9CI. tert-*Butyldichlorophosphine*
[25979-07-1]

$(H_3C)_3CPCl_2$

$C_4H_9Cl_2P$ M 158.995
Air-sensitive solid. d_4^{20} 1.08. Mp 49-50°. Bp 140°, Bp_{15} 48-50°. n_D^{20} 1.4579.

Voskuil, W. *et al, Recl. Trav. Chim. Pays-Bas*, 1963, **82**, 302 (*synth*)
Crofts, P.C. *et al, J. Chem. Soc. (C)*, 1970, 332, 2342 (*synth, ir, pmr, P nmr*)
Holmes, R.R. *et al, Spectrochim. Acta, Part A*, 1971, **27**, 1537 (*ir, raman*)
Inorg. Synth., 1973, **14**, 4 (*synth, pmr, P nmr*)
Labarre, M.-C. *et al, J. Mol. Struct.*, 1975, **26**, 17 (*synth, P nmr*)
Lappert, M.F. *et al, J. Chem. Soc., Dalton Trans.*, 1975, 1207 (*pe*)
Naumov, V.A. *et al, Dokl. Akad. Nauk SSSR, Ser. Sci. Khim.*, 1980, **253**, 167 (*Engl. transl.* p. 543) (*struct*)

tert-Butylphosphonous difluoride B-00620

(1,1-Dimethylethyl)phosphonous difluoride. tert-*Butyldifluorophosphine*

[29149-32-4]

$(H_3C)_3CPF_2$

$C_4H_9F_2P$ M 126.086

Liq. Bp 41-44°, Bp 54°, Bp °.

▷Spontaneously flammable in air

Fild, M. *et al*, *J. Chem. Soc. (A)*, 1970, 2359 (*synth, pmr, F and P nmr*)

Holmes, R.R. *et al*, *Spectrochim. Acta, Part A*, 1971, **27**, 1525 (*ir, raman*)

Labarre, M.-C. *et al*, *J. Mol. Struct.*, 1975, **26**, 17 (*synth, P nmr*)

Inorg. Synth., 1978, **18**, 173 (*synth, props*)

Butylphosphoramidic acid, 9CI B-00621

$H_3C(CH_2)_3NHP(O)(OH)_2$

$C_4H_{12}NO_3P$ M 153.117

pK_{a1} 2.92, pK_{a2} 9.88 (H_2O, 20°).

Di-Me ester: [20465-01-4]. *Dimethyl butylphosphoramidate.*

$C_6H_{16}NO_3P$ M 181.171

Liq. d_4^{20} 1.08. Bp_9 143-154°. n_D^{20} 1.4326.

Di-Et ester: [20465-03-6]. *Diethyl butylphosphoramidate.*

$C_8H_{20}NO_3P$ M 209.225

Liq. d_4^{20} 1.02. Bp_{20} 161-162°, $Bp_{0.035}$ 101-103°. n_D^{20} 1.4300.

Dibutyl ester: [5756-07-0]. *Dibutyl butylphosphoramidate.*

$C_{12}H_{28}NO_3P$ M 265.332

Liq. $Bp_{0.03}$ 128-130°. n_D^{20} 1.4381.

Di-Ph ester: [5756-05-8]. *Diphenyl butylphosphoramidate.*

$C_{16}H_{20}NO_3P$ M 305.313

Cryst. (EtOH aq.). Mp 59-60°.

Difluoride:

$C_4H_{10}F_2NOP$ M 157.100

Liq. Bp_8 64°. n_D^{20} 1.3848.

Dichloride:

$C_4H_{10}Cl_2NOP$ M 190.009

Liq. Bp_2 115-117.5°.

Olah, G.A. *et al*, *Justus Liebigs Ann. Chem.*, 1959, **625**, 88 (*difluoride*)

Mizuma, T. *et al*, *J. Pharm. Soc. Jpn.*, 1961, **81**, 48; *CA*, **55**, 13403 (*dichloride*)

Stock, J.A. *et al*, *J. Chem. Soc. (C)*, 1966, 637 (*dibutyl ester, synth, props*)

Nikonorov, K.V. *et al*, *Izv. Akad. Nauk SSSR, Ser. Khim.*, 1968, 587 (*Engl. transl.* p. 569) (*esters, synth*)

Benkovic, S.J. *et al*, *J. Am. Chem. Soc.*, 1971, **93**, 4009 (*synth, diphenyl ester, ir, pmr*)

Laskarin, B.N. *et al*, *Izv. Akad. Nauk SSSR, Ser. Khim.*, 1978, 1201 (*Engl. transl.* p. 1045) (*dibutyl ester, ir*)

Zwierzak, A., *Synthesis*, 1982, 920; 1984, 223 (*diethyl ester, synth, ir, pmr, P nmr*)

tert-Butylphosphoramidic acid, 8CI B-00622

(1,1-Dimethylethyl)phosphoramidic acid, 9CI

$(H_3C)_3CNHP(O)(OH)_2$

$C_4H_{12}NO_3P$ M 153.117

Di-Me ester: [68036-32-8]. *Dimethyl* tert-*butylphosphoramidate.*

$C_6H_{16}NO_3P$ M 181.171

Cryst. (hexane). Mp 64-65°.

Di-Et ester: [22685-20-7]. *Diethyl* tert-*butylphosphoramidate.*

$C_8H_{20}NO_3P$ M 209.225

Liq. Bp_1 95°. n_D^{25} 1.4290.

Dibutyl ester: [5756-10-5]. *Dibutyl* tert-*butylphosphoramidate.*

$C_{12}H_{28}NO_3P$ M 265.332

Liq. $Bp_{7.5}$ 152-154°. n_D^{25} 1.4356.

Di-Ph ester: [3335-12-4]. *Diphenyl* tert-*butylphosphoramidate.*

$C_{16}H_{20}NO_3P$ M 305.313

Cryst. (pet. ether or EtOH). Mp 114-115°.

Difluoride:

$C_4H_{10}F_2NOP$ M 157.100

Liq. Bp_8 50-51°. n_D^{20} 1.3789.

Dichloride: [3456-71-1].

$C_4H_{10}Cl_2NOP$ M 190.009

Solid. Mp 110-111°. Subl. at 114-115°.

Olah, G.A. *et al*, *Justus Liebigs Ann. Chem.*, 1959, **625**, 88 (*difluoride*)

Wadsworth, M.S. *et al*, *J. Org. Chem.*, 1964, **29**, 2816 (*diethyl ester, synth, props*)

Zhmurova, I.N. *et al*, *Zh. Obshch. Khim.*, 1965, **35**, 1018 (*Engl. transl.* p. 1023) (*diphenyl ester, dichloride*)

Garrison, A.W. *et al*, *Spectrochim. Acta, Part A*, 1969, **25**, 77 (*diethyl ester, ir, nmr*)

Stock, J.A. *et al*, *J. Chem. Soc. (C)*, 1966, 637 (*esters, synth*)

Dagleish, W.H. *et al*, *J. Chem. Soc., Dalton Trans.*, 1977, 1505 (*dichloride, nqr*)

Modro, T.A. *et al*, *J. Org. Chem.*, 1978, **43**, 5000 (*esters, synth, props*)

Al-Rawi, J.M.P. *et al*, *Org. Magn. Reson.*, 1983, **21**, 75 (*esters, cmr*)

Davidowitz, B. *et al*, *Org. Mass. Spectrom.*, 1984, **19**, 128 (*esters, ms*)

tert-Butylphosphoramidothioic acid, 8CI B-00623

(1,1-Dimethylethyl)phosphoramidothioic acid, 9CI

$(H_3C)_3CNHP(S)(OH)_2 \rightleftharpoons (H_3C)_3CNHP(O)(OH)(SH)$

$C_4H_{12}NO_2PS$ M 169.178

O,O-*Di-Et ester:* [57673-94-6]. O,O-*Diethyl* tert-*butylphosphoramidothioate.*

$C_8H_{20}NO_2PS$ M 225.285

No phys. props. reported.

Difluoride:

$C_4H_{10}F_2NPS$ M 173.160

Liq. Bp_8 39-40°. n_D^{20} 1.4378.

Olah, G. *et al*, *Justus Liebigs Ann. Chem.*, 1959, **625**, 88 (*difluoride*)

El-Borgi, A. *et al*, *C.R. Hebd. Seances Acad. Sci., Ser. C*, 1977, **284**, 983 (*ester, ir*)

Al-Rawi, J.M.A. *et al*, *Org. Magn. Reson.*, 1983, **21**, 75 (*ester, cmr*)

tert-Butylphosphoramidous acid, 8CI B-00624

(1,1-Dimethylethyl)phosphoramidous acid, 9CI

$(H_3C)_3CNHP(OH)_2$

$C_4H_{12}NO_2P$ M 137.118

Di-Et ester: [59466-94-3]. *Diethyl* tert-*butylphosphoramidite.*

$C_8H_{20}NO_2P$ M 193.225

Liq. Bp_9 68-70°. n_D^{20} 1.4321.

Difluoride:

$C_4H_{10}F_2NP$ M 141.100

Liq. V.p. 38 mm at 25°.
Dichloride: [24335-36-2].
$C_4H_{10}Cl_2NP$ M 174.009
Liq. Mp −38°. Bp_6 53-58°.

Scherer, O.J. *et al, Angew. Chem., Int. Ed. Engl.*, 1969, **8**, 752 (*dichloride, synth, nmr*)
Harman, J.S. *et al, J. Chem. Soc.* (*A*), 1970, 1935 (*difluoride, synth, ir, pmr*)
Jefferson, R. *et al, J. Chem. Soc., Dalton Trans.*, 1973, 1414 (*dichloride, P nmr, pmr*)
Tupchienko, S.K. *et al, Zh. Obshch. Khim.*, 1981, **51**, 1015 (*Engl. transl.* p. 847) (*ester, synth, ir, pmr*)

tert-Butylphosphorimidic acid **B-00625**

(1,1-Dimethylethyl)phosphorimidic acid, 9CI

$(H_3C)_3CN{=}P(OH)_3$

$C_4H_{12}NO_3P$ M 153.117
Free acid tautomeric with *tert*-Butylphosphoramidic acid, B-00622 .
Tri-Me ester: [67353-94-0]. *Trimethyl* tert-*butylphosphorimidate.* N-tert-*Butyl*-P,P,P-*trimethoxyphosphazene.* N-tert-*Butyl*-P,P,P-*trimethoxyiminophosphorane.*
$C_7H_{18}NO_3P$ M 195.198
Liq. Bp_{12} 34-35°.
Trichloride: see Dicyclohexylphosphinous acid, D-00214
Tribromide: [39500-93-1]. P,P,P-*Tribromo*-N-tert-*butylphosphazene.* P,P,P-*Tribromo*-N-tert-*butylphosphine imide.* P,P,P-*Tribromo*-N-tert-*butyliminophosphorane.*
$C_4H_9Br_3NP$ M 341.808
No phys. props. reported.

Markovskii, L.H. *et al, Zh. Org. Khim.*, 1972, **8**, 2057 (*Engl. transl.* p. 2104) (*tribromide, ir*)
Scherer, O.J. *et al, Z. Naturforsch., B*, 1978, **33**, 652 (*ester, synth, pmr, P nmr*)

tert-Butylphosphorimidic trichloride, 8CI **B-00626**

(1,1-Dimethylethyl)phosphorimidic trichloride, 9CI. N-tert-*Butyl*-P,P,P-*trichlorophosphine imide.* N-tert-*Butyl*-P,P,P-*trichlorophosphazene.* N-tert-*Butyl*-P,P,P-*trichloroiminophosphorane*
[18854-80-3]

$(H_3C)_3CN{=}PCl_3$

$C_4H_9Cl_3NP$ M 208.455
Normally monomeric but can dimerise. Liq. d_4^{20} 1.22. Bp 153-154°.

Zhmurova, I.N. *et al, Zh. Obshch. Khim.*, 1964, **34**, 1441 (*Engl. transl.* p. 1446) (*synth*)
Glidewell, C., *Inorg. Chim. Acta*, 1976, **18**, 51 (*ms*)
Kovenya, V.A. *et al, Zh. Obshch. Khim.*, 1976, **46**, 2679 (*Engl. transl.* p. 2557) (*dimer*)
Kozlov, É.S. *et al, Zh. Obshch. Khim.*, 1978, **48**, 1263 (*Engl. transl.* p. 1155) (*nqr*)

Butyl phosphorodibromidite, 9CI, 8CI **B-00627**

Butyl dibromophosphite
[53764-96-8]

$H_3C(CH_2)_3OPBr_2$

$C_4H_9Br_2OP$ M 263.896
Fuming liq. Bp_6 70-72°, $Bp_{0.01}$ 38°. n_D^{20} 1.5441 (1.5240).

Gerrard, W. *et al, J. Chem. Soc.*, 1955, 277 (*synth, props*)
Novikova, Z.S. *et al, Zh. Obshch. Khim.*, 1974, **44**, 1857 (*Engl. transl.* p. 1805) (*synth, P nmr*)

Butyl phosphorodichloridate, 9CI, 8CI **B-00628**

Butyl phosphoryl dichloride. Butyl dichlorophosphate
[1498-52-8]

$H_3C(CH_2)_3OP(O)Cl_2$

$C_4H_9Cl_2O_2P$ M 190.994
Fuming liq. d_4^{25} 1.26. Bp_{13} 85°. n_D^{20} 1.4420.
▷Hydrolyses in moist air

Gerrard, W., *J. Chem. Soc.*, 1940, 1464 (*synth*)
Grunze, H., *Chem. Ber.*, 1959, **92**, 850 (*synth*)
Rafikov, S.R. *et al, Zh. Obshch. Khim.*, 1965, **25**, 591 (*Engl. transl.* p. 591) (*synth*)

Butyl phosphorodichloridite, 9CI, 8CI **B-00629**

Butyl dichlorophosphite
[10496-13-6]

$H_3CCH_2CH_2CH_2OPCl_2$

$C_4H_9Cl_2OP$ M 174.994
Pungent liq. d_4^{20} 1.17. Bp_{10} 47°. Easily hydrolysed.

Gerrard, W. *et al, J. Chem. Soc.*, 1940, 1464; 1953, 1920 (*synth*)
Vasil'ev, V.V. *et al, Zh. Obshch. Khim.*, 1981, **51**, 2134 (*Engl. transl.* p. 1836) (*O nmr*)

O-Butyl phosphorodichloridothioate, 9CI, 8CI **B-00630**

O-*Butyl thiophosphoryl dichloride.* O-*Butyl dichlorothiophosphate*
[17643-83-3]

$H_3C(CH_2)_3OP(S)Cl_2$

$C_4H_9Cl_2OPS$ M 207.054
Liq. Bp_{19} 91-93°, Bp_{10} 81-82°. n_D^{20} 1.4951.

Mastin, T.W. *et al, J. Am. Chem. Soc.*, 1945, **67**, 1662 (*synth*)
Kas'yanova, E.F. *et al, Zh. Obshch. Khim.*, 1969, **39**, 365 (*Engl. transl.* p. 342)

S-Butyl phosphorodichloridothioate, 9CI, 8CI **B-00631**

S-*Butyl dichlorothiophosphate*
[26155-88-4]

$H_3C(CH_2)_3SP(O)Cl_2$

$C_4H_9Cl_2OPS$ M 207.054
Liq. d_4^{20} 1.24. Bp_{15} 113-114°. n_D^{20} 1.5037.

Petrov, K.A. *et al, Zh. Obshch. Khim.*, 1961, **31**, 1361, 1366 (*Engl. transl.* pp. 1260, 1265) (*synth*)
Kobayashi, K. *et al, CA*, 1970, **72**, 100196 (*synth*)

Butyl phosphorodichloridothioite, 9CI **B-00632**

Butyl dichlorothiophosphite
[36696-26-1]

$H_3C(CH_2)_3SPCl_2$

$C_4H_9Cl_2PS$ M 191.055
Pungent, odorous liq. d_4^{20} 1.23. Bp_9 81-82°. n_D^{20} 1.5392. Slowly disproportionates to PCl_3 and dibutyl phosphorochloridodithioite.

Stepashkina, L.V. *et al*, *Izv. Akad. Nauk SSSR, Ser. Khim.*, 1972, 380 (*Engl. transl.* p. 330) (*synth*)

Butyl phosphorodifluoridite, 9CI, 8CI **B-00633**

Butyl difluorophosphite
[693-00-5]

$H_3C(CH_2)_3OPF_2$

$C_4H_9F_2OP$ M 142.085
Ligand for Co, Pt and Pd. d^{20} 1.03. Bp 74-75°. n_D^{20} 1.3631.

Schmutzler, R., *Chem. Ber.*, 1963, **96**, 2435 (*synth*)
Reddy, G.S. *et al*, *Z. Naturforsch.*, *B*, 1965, **20**, 104 (*pmr, F and P nmr*)
Ivanova, Zh.M., *Zh. Obshch. Khim.*, 1964, **34**, 858 (*Engl. transl.* p. 852) (*synth*)
Binder, H. *et al*, *Z. Naturforsch.*, *B*, 1972, **27**, 753 (*synth, P nmr*)

S-Butyl phosphorothioate, 9CI, 8CI **B-00634**

S-*Butyl dihydrogen thiophosphate.* S-*Butyl thiophosphoric acid*

$H_3C(CH_2)_3SP(O)(OH)_2$

$C_4H_{11}O_3PS$ M 170.163
Monodicyclohexylammonium salt: Solid. Mp 157°.
Dichloride: see S-Butyl phosphorodichloridothioate, B-00631

Zwierzak, A. *et al*, *Z. Naturforsch.*, *B*, 1971, **26**, 386 (*synth, ir, pmr*)

5-*tert*-Butyltetrahydro-2-dimethylamino-2*H*-1,3,2-oxazaphosphorine **B-00635**

N,N-*Dimethyl-5-(1,1-dimethylethyl)tetrahydro-2*H*-1,3,2-oxazaphosphorin-2-amine*, 9CI. 5-tert-*Butyl-2-dimethylamino-1,3,2-oxazaphosphorinane*

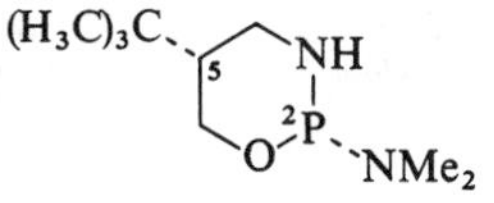

(*2RS,5RS*)-*form*

$C_9H_{21}N_2OP$ M 204.251

(2RS,5RS)-form
(±)-cis-*form*
2-Oxide: [77815-23-7].
$C_9H_{21}N_2O_2P$ M 220.251
Cryst. Mp 102-104°.
2-Sulfide: [83096-33-7].
$C_9H_{21}N_2OPS$ M 236.311
Solid. Mp 112-113°. Has (2*SR*)-config.

(2RS,5SR)-form
(±)-trans-*form*
2-Oxide: [77815-22-6]. Cryst. (EtOAc). Mp 121-123°.
2-Sulfide: [82757-16-2]. Solid. Mp 105-106°. Has (2*RS*)-config.

Bajwa, G.S. *et al*, *J. Am. Chem. Soc.*, 1982, **104**, 6385 (*synth, ir, pmr, P nmr*)
Newton, M.G. *et al*, *Tetrahedron Lett.*, 1982, **23**, 1527 (*sulfide, cryst struct*)

5-*tert*-Butyltetrahydro-2-dimethylamino-3-phenyl-2*H*-1,3,2-oxazaphosphorine **B-00636**

*5-(1,1-Dimethylethyl)tetrahydro-*N,N*-dimethyl-3-phenyl-2*H*-1,3,2-oxazaphosphorin-2-amine*, 9CI. 5-tert-*Butyl-2-dimethylamino-3-phenyl-1,3,2-oxazaphosphorinane*

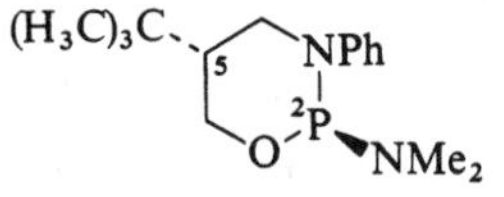

(*2RS,5RS*)-*form*

$C_{15}H_{25}N_2OP$ M 280.349
Liq. Bp_3 165-166°. Mixt. of stereoisomers.

(2RS,5RS)-form [83096-41-7]
(±)-trans-*form*
2-Oxide: [70219-43-1]. Cryst. (Et_2O). Mp 125-126°.
2-Sulfide: [83096-34-8]. Cryst. Mp 94-95°. Has (2*SR*)-config.

(2RS,5SR)-form [83096-40-6]
(±)-cis-*form*
2-Oxide: [70219-44-2].
$C_{15}H_{25}N_2O_2P$ M 296.348
Cryst. (C_6H_6). Mp 165-166°.
2-Sulfide: [83096-35-9].
$C_{15}H_{25}N_2OPS$ M 312.409
Cryst. (Et_2O or C_6H_6). Mp 145-145.5°. Has (2*SR*)-config.

Bajwa, G.S. *et al*, *J. Am. Chem. Soc.*, 1982, **104**, 6385 (*synth, pmr, P nmr, conformn*)
Bentrude, W.G. *et al*, *J. Am. Chem. Soc.*, 1984, **106**, 106 (*pmr*)

5-*tert*-Butyl-2,2,4,5-tetrahydro-2,2,4-triisopropoxy-1,2,4-oxadiphosphole **B-00637**

5-(1,1-Dimethylethyl)-2,2,4,5-tetrahydro-2,2,4-tris(1-methylethoxy)-1,2,4-oxadiphosphole 4-oxide, 10CI
[70346-98-4]

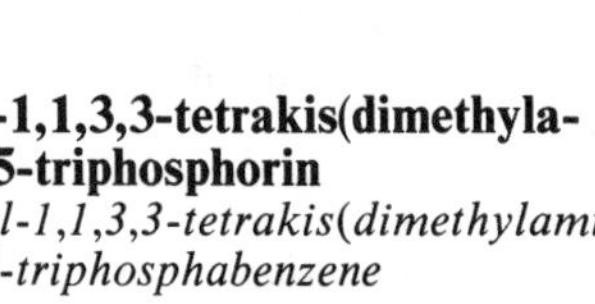

$C_{15}H_{32}O_5P_2$ M 354.362
A cyclic ylide. Liq. $Bp_{0.01}$ 110°. n_D^{20} 1.4530.

Novikova, Z.S. *et al*, *Zh. Obshch. Khim.*, 1979, **49**, 470 (*Engl. transl.* p. 412) (*synth, ir, P nmr*)

4-*tert*-Butyl-1,1,3,3-tetrakis(dimethylamino)-1,3,5-triphosphorin **B-00638**

4-tert-*Butyl-1,1,3,3-tetrakis(dimethylamino)-1λ⁵,3λ⁵,5λ³-triphosphabenzene*

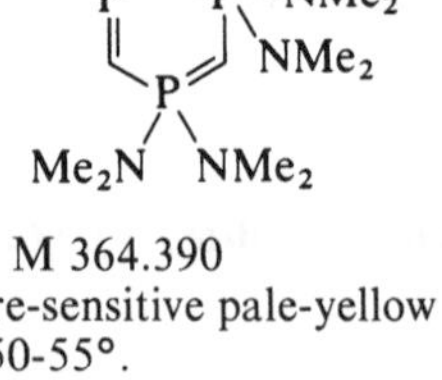

$C_{15}H_{35}N_4P_3$ M 364.390
Air- and moisture-sensitive pale-yellow cryst. Sol. C_6H_6, pentane. Mp 50-55°.

Fluck, E. *et al*, *Angew. Chem., Int. Ed. Engl.*, 1986, **25**, 1002 (*synth, cmr, P nmr, cryst struct*)

N-tert-Butyl-*P*-(2,4,6-tri-*tert*-butylphenyl)metaphosphonimidate B-00639

*2-Methyl-*N*-[[2,4,6-tris(1,1-dimethylethyl)phenyl]-phosphinidene]-2-propanamine*, 11CI. N-*(1,1-Dimethylethyl)*-P-*[2,4,6-tris(1,1-dimethylethyl)phenyl]-phosphazene*

P=NC(CH₃)₃ ; (H₃C)₃C, C(CH₃)₃ ; C(CH₃)₃

$C_{22}H_{38}NP$ M 347.523

(E)-form [95792-75-9]

Oxide: N-tert-*Butyl*-P-*(2,4,6-tri*-tert-*butylphenyl)-metaphosphonamidate. 2-Methyl-*N*-[[2,4,6-tris(1,1-dimethylethyl)phenyl]phosphinylidene]-2-propanamine*, 9CI.
$C_{22}H_{38}NOP$ M 363.522
Viscous oil.

Sulfide: N-tert-*Butyl*-P-*(2,4,6-tri*-tert-*butylphenyl)-metaphosphonamidothioate. 2-Methyl-*N*-[[2,4,6-tris(1,1-dimethylethyl)phenyl]phosphinothioylidene]-2-propanamine*, 9CI.
$C_{22}H_{38}NPS$ M 379.583
Pale-yellow cryst. (pentane). Mp 54-56°. Stable at 20°. Sensitive to moisture.

Selenide: N-tert-*Butyl*-P-*(2,4,6-tri*-tert-*butylphenyl)-metaphosphonamidoselenoate. 2-Methyl-*N*-[[2,4,6-tris(1,1,-dimethylethyl)phenyl]-phosphinoselenoylidene]-2-propanamine*, 11CI.
$C_{22}H_{38}NPSe$ M 426.483
Pale-yellow cryst. (pentane). Mp 68-70°. Stable at 20°. Sensitive to moisture.

Markovski, L.N. *et al*, *J. Chem. Soc., Chem. Commun.*, 1984, 1692 (*derivs, ir, pmr, P nmr*)

4-*tert*-Butyl-2,6,7-trioxa-1-phosphabicyclo[2.2.2]octane, 8CI B-00640

4-(1,1-Dimethylethyl)-2,6,7-trioxa-1-phosphabicyclo[2.2.2]octane, 9CI. tert-*Butyl bicyclic phosphite*

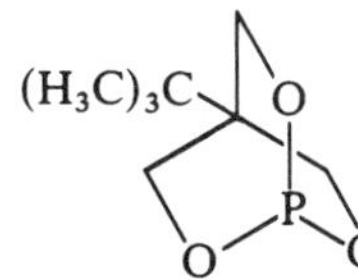

$C_8H_{15}O_3P$ M 190.178
Cryst. (pet. ether or $CHCl_3$/pet. ether). Mp 149-151° (95-97°). $Bp_{0.5}$ 75-80° subl.
▷Convalsant. Highly toxic

1-Oxide: [61481-19-4]. tert-*Butyl bicyclic phosphate.*
$C_8H_{15}O_4P$ M 206.178
Cryst. (C_6H_6 or H_2O). Mp 321-324° (306-309°).
▷Convulsant poison. TY3700000.

1-Sulfide: [70636-86-1]. tert-*Butyl bicyclic thiophosphate.*
$C_8H_{15}O_3PS$ M 222.238
Cryst. ($CHCl_3$). Mp 232-234°.
▷Convulsant poison

Cooper, G.H. *et al*, *Eur. J. Med. Chem.*, 1978, **13**, 207 (*synth, tox, oxide*)
Milbrath, D.S. *et al*, *Toxicol Appl. Pharmacol.*, 1979, **47**, 287 (*synth, sulfide, pmr*)
Edmundson, R.S. *et al*, *Phosphorus Sulfur*, 1980, **8**, 315 (*synth, props, oxide, ms, ir, pmr*)
Ozoe, Y. *et al*, *Agric. Biol. Chem.*, 1982, **46**, 411 (*oxide, pmr, ir*)

Butyltriphenylphosphonium(1+), 9CI B-00641

$Ph_3P^{\oplus}(CH_2)_3CH_3$

$C_{22}H_{24}P^{\oplus}$ M 319.405 (ion)
With butyllithium, salts yield the ylide.

Chloride: [13371-17-0].
$C_{22}H_{24}ClP$ M 354.858
No phys. props. reported.

Bromide: [1779-51-7].
$C_{22}H_{24}BrP$ M 399.309
Cryst. (butanol). Mp 249° (223°).
▷TA1855200.

Iodide: [22949-84-4].
$C_{22}H_{24}IP$ M 446.310
Solid. Mp 213-215°.
▷TA1856000.

Triiodide: [76835-82-0].
$C_{22}H_{24}I_3P$ M 700.119
Solid. Mp 139-141°.

Pentaiodide: [76835-83-1].
$C_{22}H_{24}I_5P$ M 953.928
Solid. Mp 74-76°.

Heptaiodide: [76835-84-2].
$C_{22}H_{24}I_7P$ M 1207.737
Solid. Mp 81-83°.

Ylide: [3728-50-5]. *Butylidenetriphenylphosphorane.*
$C_{22}H_{23}P$ M 318.397
Reactive Wittig reagent, prepd. *in situ.*

Mechoulam, R. *et al*, *J. Am. Chem. Soc.*, 1958, **80**, 4386 (*ylide*)
Keough, P. *et al*, *J. Am. Chem. Soc.*, 1960, **82**, 3919 (*props*)
Albright, T.A. *et al*, *J. Am. Chem. Soc.*, 1975, **97**, 2942, 2946 (*cmr, nmr*)
Makovetskii, Yu.P. *et al*, *Zh. Obshch. Khim.*, 1980, **50**, 2436 (*Engl. transl.* p. 1967); 1982, **52**, 2235 (*Engl. transl.*, p. 1989) (*iodides, uv*)
Schlosser, M. *et al*, *Chimia*, 1982, **36**, 396 (*ylide*)

2-*tert*-Butyl-1,3,6,2-trithiaphosphocane, 8CI B-00642

2-(1,1-Dimethylethyl)-1,3,6,2-trithiaphosphocane, 9CI. *2-*tert*-Butyl-1,3,6,2-trithiaphosphacyclooctane*

[77075-77-5]

S ; S ; 2P—C(CH₃)₃ ; S

$C_8H_{17}PS_3$ M 240.376
Unstable; readily dimerises.

2-Oxide: [77075-78-6].
$C_8H_{17}OPS_3$ M 256.375
Solid. Mp 132-134°.

2-Sulfide: [71156-57-5].
$C_8H_{17}PS_4$ M 272.436
Solid. Mp 144-145°.

Martin, J. *et al*, *Acta Crystallogr., Sect. B*, 1979, **35**, 1623 (*sulfide, cryst struct*)
Martin, J. *et al*, *Nouv. J. Chim.*, 1980, **4**, 515 (*synth, props, cmr, P nmr*)
Martin, J. *et al*, *Org. Magn. Reson.*, 1981, **15**, 87 (*pmr, P nmr*)

1-Butynylphosphonic acid, 9CI B-00643

1-Phosphono-1-butyne

[4851-54-1]

$H_3CCH_2C{\equiv}CP(O)(OH)_2$

$C_4H_7O_3P$ M 134.071

Di-Et ester: [4851-54-1]. *Diethyl 1-butynylphosphonate.*
$C_8H_{15}O_3P$ M 190.178
Liq. Bp_2 110-113°, $Bp_{0.4}$ 100°. n_D^{25} 1.4520, n_D^{23} 1.4437.

Sturtz, G., *Bull. Soc. Chim. Fr.*, 1966, 1707 (*synth*)
Kirilov, M. *et al*, *Monatsh. Chem.*, 1980, **111**, 1351 (*synth, ir*)

2-Butynylphosphonic acid, 9CI B-00644

1-Phosphono-2-butyne
[3095-06-5]

$$H_3CC{\equiv}CCH_2P(O)(OH)_2$$

$C_4H_7O_3P$ M 134.071
Derivatives may contain those of (1,2-Butadienyl)-phosphonic acid, B-00509 as impurities.

Di-Et ester: [3095-06-5]. *Diethyl 2-butynylphosphonate.*
$C_8H_{15}O_3P$ M 190.178
Liq. d_4^{20} 1.05. Bp_3 98°. n_D^{20} 1.4540.

Ionin, B.E. *et al*, *Zh. Obshch. Khim.*, 1963, **33**, 2863 (*Engl. transl.* p. 2791) (*synth, ir, pmr*)
Zakharov, V.I. *et al*, *Zh. Obshch. Khim.*, 1974, **44**, 98 (*Engl. transl.* p. 96) (*P nmr*)

C

Calyculin *A* **C-00001**

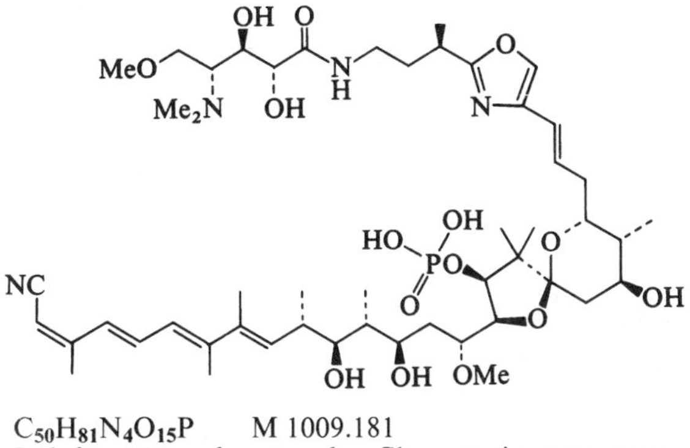

$C_{50}H_{81}N_4O_{15}P$ M 1009.181

Isol. from *Discodermia calyx*. Shows antitumour props. Needles (Me_2CO/Et_2O/hexane). Mp 247-249°. $[\alpha]_D^{15}$ +59.8° (c, 0.12 in EtOH).

Kato, Y. *et al*, *J. Am. Chem. Soc.*, 1986, **108**, 2780 (*isol, cryst struct, props*)

Carbamoyl dihydrogen phosphate **C-00002**

Carbamoyl phosphoric monoanhydride, 9CI. *Carbamoyl phosphate. Carbamyl phosphate*

[590-55-6]

$H_2NCOOP(O)(OH)_2$

CH_4NO_5P M 141.020

Intermed. in the fixation of NH_3 and urea and in biosynth. of citrullinic acid, ureidosuccinic acid and the pyrimidines. Unstable in soln., isol. as di-Li or di-NH_4 salt.

▷CNS toxin *via* biodegradn. to cyanate. Esters are anticanker agents

Di-Li salt: [1866-68-8]. Cryst. (EtOH aq.). As usually prepared is ca. 86% pure. Stable when in frozen soln.

Grisolía, S. *et al*, *Biochim. Biophys. Acta*, 1955, **17**, 150, 277 (*biochem*)
Jones, M.E. *et al*, *J. Am. Chem. Soc.*, 1955, **77**, 819 (*synth*)
Spector, L., *Methods Enzymol.*, 1957, **3**, 653 (*synth*)
Cramer, F. *et al*, *Chem. Ber.*, 1959, **92**, 2761 (*synth*)
Biochem. Prep., 1960, **7**, 23 (*synth, anal*)
Jencks, W.P. *et al*, *Arch. Biochem. Biophys.*, 1960, **88**, 193 (*ir*)
Halman, M. *et al*, *J. Chem. Soc.*, 1962, 1944 (*props*)
Allen, C.M. *et al*, *Biochemistry*, 1964, **3**, 1238 (*props*)
Oestreich, C.H. *et al*, *Biochemistry*, 1967, **6**, 1515 (*ir*)
Seel, F. *et al*, *Z. Naturforsch., B*, 1978, **33**, 374 (*synth*)

Carbamoylphosphonic acid, 8CI **C-00003**

Aminocarbonylphosphonic acid, 9CI. *Phosphonoformamide. Dihydroxyphosphinylformamide*

[6874-57-3]

$(HO)_2P(O)CONH_2$

CH_4NO_4P M 125.021

Metab. of fosamine. Monoalkyl esters (as their ammonium salts) are effective plant growth regulators.

Di-Me ester: [33534-83-7]. *Dimethyl aminocarbonylphosphonate. (Dimethoxyphosphinyl)formamide. Methafos.*
$C_3H_8NO_4P$ M 153.074
Very effective against tomato canker.

Mono-Et ester: see Monoethyl (aminocarbonyl)-phosphonate, M-00469

Di-Et ester: [31142-29-7]. *Diethyl aminocarbonylphosphonate. (Diethoxyphosphinyl)formamide. Etkafos.*
$C_5H_{12}NO_4P$ M 181.128
Useful in the treatment of tomato canker. Needles (C_6H_6 or EtOAc). Mp 138-139°.

Diisopropyl ester: Diisopropyl aminocarbonylphosphonate. (Diisopropoxyphosphinyl)formamide.
$C_7H_{16}NO_4P$ M 209.181
Prisms (EtOAc). Mp 95-97°.

Di-Ph ester: [41839-63-8]. *Diphenyl aminocarbonylphosphonate. (Diphenoxyphosphinyl)formamide.*
$C_{13}H_{12}NO_4P$ M 277.216
Prisms (C_6H_6/pet. ether). Mp 113.5-114.5°.

Bis(diethylamide): P-*Aminocarbonyl*-N,N,N′,N′-*tetraethylphosphonic diamide.*
$C_9H_{22}N_3O_2P$ M 235.265
Mp 119-120°.

Nylen, P., *Ber.*, 1924, **57**, 1023 (*diethyl ester*)
Grisely, D.W., *J. Org. Chem.*, 1961, **26**, 2544 (*esters, synth, ir, nmr*)
Bel'skii, V.E. *et al*, *Zh. Obshch. Khim.*, 1972, **42**, 2427 (*Engl. transl.* p. 2421) (*diethyl ester, P pmr*)
Gorbatenko, V.I. *et al*, *Zh. Obshch. Khim.*, 1973, **43**, 1043 (*Engl. transl.* p. 1035) (*diphenyl ester*)

(2-Carbethoxyphenyl)phosphoramidic acid **C-00004**

N-*Phosphono-2-aminobenzoic acid ethyl ester. 2-(Ethoxycarbonylphenyl)phosphoramidic acid. Ethyl* N-*phosphono-2-aminobenzoate*

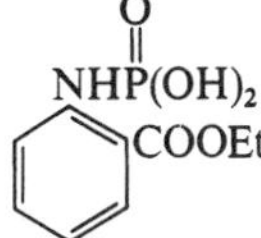

$C_9H_{12}NO_5P$ M 245.171

Di-Et ester: [31120-32-8]. *Ethyl 2-[(diethoxyphosphinyl)amino]benzoate. Triethyl* N-*phosphono-2-aminobenzoate. Triethyl* N-*phosphonoanthranilate.*
$C_{13}H_{20}NO_5P$ M 301.278
Liq. d_4^{20} 1.17. $Bp_{0.05}$ 129°. n_D^{20} 1.5068.

Di-Ph ester: Ethyl 2-[(diphenoxyphosphinyl)amino]-benzoate.
$C_{21}H_{20}NO_5P$ M 397.366

Solid. Mp 148.5-149°.

Kucherov, V.F., *Zh. Obshch. Khim.*, 1949, **19**, 126; *CA*, **43**, 6178 (*diphenyl ester*)
Cadogan, J.I.G. *et al*, *J. Chem. Soc.* (*C*), 1970, 2441 (*diethyl ester, synth, ir, pmr*)
Gusar, N.I. *et al*, *Zh. Obshch. Khim.*, 1979, **49**, 21 (*Engl. transl.* p. 16) (*diethyl ester, synth, ir, pmr*)

(3-Carbethoxyphenyl)phosphoramidic acid C-00005

N-*Phosphono-3-aminobenzoic acid ethyl ester. 3-(Ethoxycarbonylphenyl)phosphoramidic acid. Ethyl* N-*phosphono-3-aminobenzoate*

$C_9H_{12}NO_5P$ M 245.171

*Di-Et ester: Triethyl 3-(*N*-phosphono)benzoate. Ethyl 3-[(diethoxyphosphinyl)amino]benzoate.*
$C_{13}H_{20}NO_5P$ M 301.278
Liq. Bp 232-234°, Bp_{35} 130-135°.

Michaelis, A., *Justus Liebigs Ann. Chem.*, 1903, **326**, 129.

(4-Carbethoxyphenyl)phosphoramidic acid C-00006

N-*Phosphono-4-aminobenzoic acid ethyl ester. (4-Ethoxycarbonylphenyl)phosphoramidic acid. Ethyl* N-*phosphono-4-aminobenzoate*

$C_9H_{12}NO_5P$ M 245.171

Di-Ph ester: [75905-80-5]. *Diphenyl [4-(ethoxycarbonyl)phenyl]phosphoramidate.* N-*Diphenoxyphosphinyl-4-aminobenzoic acid ethyl ester. Ethyl 4-[(diphenoxyphosphinyl)amino]benzoate.*
$C_{21}H_{20}NO_5P$ M 397.366
Needles (EtOH/dil. HCl). Mp 151-152°.

Dichloride:
$C_9H_{10}Cl_2NO_3P$ M 282.063
Cryst. ($CHCl_3$). Mp 100-102°.

Kropacheva, A.A. *et al*, *Zh. Obshch. Khim.*, 1959, **29**, 556 (*Engl. transl.* p. 553) (*dichloride*)
Cates, L.A. *et al*, *J. Pharm. Sci.*, 1964, **53**, 691 (*synth*)
Omara, M.M. *et al*, *Rev. Roum. Chim.*, 1980, **25**, 253 (*synth, ir*)

Carbonylbis[phosphorimidic trichloride], 9CI C-00007

[59675-46-6]

$O{=}C(N{=}PCl_3)_2$

$CCl_6N_2OP_2$ M 330.689
When heated → $POCl_3$.

Glidewell, C., *Inorg. Chim. Acta*, 1976, **18**, 51 (*ms*)
Riesel, L. *et al*, *Z. Anorg. Allg. Chem.*, 1977, **435**, 268 (*synth, props*)

Carbophenothion, BSI, ISO C-00008

S-*[[(4-Chlorophenyl)thio]methyl]* O,O-*diethyl phosphorodithioate, 9CI.* S-*4-Chlorophenylthiomethyl* O,O-*diethyl phosphorodithioate. Trithion. Carbofenothion*

[786-19-6]

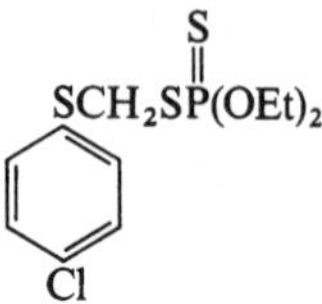

$C_{11}H_{16}ClO_2PS_3$ M 342.853

Non-systemic acaricide and insecticide. Liq. with mercaptan-like odour. $Bp_{0.01}$ 82°. n_D^{25} 1.597.

▷TD5250000.

Keith, L.H. *et al*, *J. Assoc. Off. Anal. Chem.*, 1968, **51**, 1063 (*pmr*)
Gore, R.C. *et al*, *J. Assoc. Off. Anal. Chem.*, 1971, **54**, 1040 (*ir, uv*)
Szalontai, G., *J. Chromatogr.*, 1976, **124**, 9 (*hplc*)
Stan, H.-J. *et al*, *Fresenius' Z. Anal. Chem.*, 1977, **287**, 271 (*glc, ms*)
Busch, K.L. *et al*, *Appl. Spectrosc.*, 1978, **32**, 388 (*ms*)
Stan, H.-J. *et al*, *Biomed. Mass Spectrom.*, 1982, **9**, 483 (*ms*)
Ripley, B.D. *et al*, *J. Assoc. Off. Anal. Chem.*, 1983, **66**, 1084 (*glc*)
Agrochemicals Handbook, Royal Society of Chemistry, London, 1983, A061.
Pesticide Manual, 7th Ed., 94.

(4-Carboxybutyl)-triphenylphosphonium(1+), 9CI C-00009

4-(Triphenylphosphonio)butanoic acid

[41264-12-4]

$Ph_3P^{\oplus}(CH_2)_4COOH$

$C_{23}H_{24}O_2P^{\oplus}$ M 363.415 (ion)

Chloride: [60633-19-4].
$C_{23}H_{24}ClO_2P$ M 398.868
Source of ylide.

Bromide: [17814-85-6].
$C_{23}H_{24}BrO_2P$ M 443.319
Source of ylide.

Iodide: [61168-05-6].
$C_{23}H_{24}IO_2P$ M 490.320
Source of ylide.

Et ester, iodide:
$C_{25}H_{28}IO_2P$ M 518.373
Pale-yellow oil.

Ylide: see 5-(Triphenylphosphoranylidene)pentanoic acid, T-00655

House, H.O. *et al*, *J. Org. Chem.*, 1963, **28**, 90 (*ester, iodide, props*)
Inoue, K. *et al*, *Bull. Chem. Soc. Jpn.*, 1978, **51**, 2361 (*bromide, ylide, use*)
Grieco, P.A. *et al*, *J. Org. Chem.*, 1979, **44**, 2189 (*bromide, ylide, use*)
Rokach, J. *et al*, *Tetrahedron Lett.*, 1981, **22**, 979 (*bromide, ylide, use*)

(1-Carboxyethyl)-triphenylphosphonium(1+), 8CI C-00010

2-(Triphenylphosphonio)propanoic acid

$Ph_3P^{\oplus}CH(CH_3)COOH$

$C_{21}H_{20}O_2P^{\oplus}$ M 335.362 (ion)

Et ester, chloride: [18480-27-8].
$C_{23}H_{24}ClO_2P$ M 398.868
Source of Ethyl 2-triphenylphosphoranylidenepropanoate, E-00199 . No phys. props. reported.

Et ester, bromide: [30018-16-7]. *(1-Ethoxycarbonyl)-triphenylphosphonium bromide.*
$C_{23}H_{24}BrO_2P$ M 443.319
Source of Ethyl 2-triphenylphosphoranylidenepropanoate, E-00199 . Solid. Mp 145°.

Ylide: see 2-(Triphenylphosphoranylidene)propanoic acid, T-00660

Plieninger, H. *et al*, *Justus Liebigs Ann. Chem.*, 1976, 1475 (*synth, use*)

(2-Carboxyethyl)-triphenylphosphonium(1+), 9CI C-00011

3-(Triphenylphosphonio)propanoic acid

$Ph_3P^{\oplus}CH_2CH_2COOH$

$C_{21}H_{20}O_2P^{\oplus}$ M 335.362 (ion)

Source of an ylide extensively used in prostaglandin synth. Salts + NaH, $NaNH_2$, $NaCH_2SOMe$ or butyllithium, yield the ylide.

Chloride: [36626-29-6].
$C_{21}H_{20}ClO_2P$ M 370.815
Cryst. ($CHCl_3/Et_2O$ or $EtOH/Et_2O$). Mp 196-197° (187°).

Bromide: [51114-94-4].
$C_{21}H_{20}BrO_2P$ M 415.266
Cryst. (EtOAc/2-propanol). Mp 196-197°.

Tribromide: [55985-85-8].
$C_{21}H_{20}Br_3O_2P$ M 575.074
Selectively brominates at position α to a carbonyl group in the presence of C=C. Cryst. (MeCN). Mp 139-141°.

Betaine:
$C_{21}H_{19}O_2P$ M 334.354
Obt. by treatment of the salts with $NaHCO_3$ aq. Solid. Mp 167-170°.

Ylide: see 3-(Triphenylphosphoranylidene)propanoic acid, T-00661

Denney, D.B. *et al*, *J. Org. Chem.*, 1962, **27**, 3404 (*chloride, betaine, ir, nmr*)
Findlay, J.A. *et al*, *Can. J. Chem.*, 1973, **51**, 3299 (*bromide, ir, pmr, use*)
Armstrong, V.W. *et al*, *Tetrahedron Lett.*, 1975, 376 (*tribromide, synth, use*)
Maercker, A. *et al*, *Chem. Ber.*, 1976, **109**, 2064 (*chloride, use*)
Bogoslovskii, N.A. *et al*, *Zh. Obshch. Khim.*, 1978, **48**, 897 (*Engl. transl.* p. 818) (*bromide, ir, pmr, use*)
Narayanan, K.S. *et al*, *J. Org. Chem.*, 1980, **45**, 2240 (*chloride, ir, pmr, cmr, P nmr, props*)
Hann, M.M. *et al*, *J. Chem. Soc., Perkin Trans. 1*, 1982, 307 (*chloride, use*)

(7-Carboxyheptyl)-triphenylphosphonium(1+), 9CI C-00012

7-(Triphenylphosphonio)octanoic acid

$Ph_3P^{\oplus}(CH_2)_7COOH$

$C_{26}H_{30}O_2P^{\oplus}$ M 405.496 (ion)

Used in synth. of pheromones and other fatty acids.

Bromide: [52956-93-1].
$C_{26}H_{30}BrO_2P$ M 485.400
Source of ylide. Microcryst. powder. Mp 119-120°.

Ylide: (7-Carboxyheptylidene)triphenylphosphorane.
$C_{26}H_{29}O_2P$ M 404.488
Used in Wittig reactions.

Kovaleva, A.S. *et al*, *Zh. Org. Khim.*, 1974, **10**, 696 (*Engl. transl.* p. 700) (*synth, use*)
Dawson, M.L. *et al*, *J. Org. Chem.*, 1977, **42**, 2783 (*synth, ir, pmr, use*)

(6-Carboxyhexyl)-triphenylphosphonium(1+) C-00013

6-(Triphenylphosphonio)keptanoic acid

$Ph_3P^{\oplus}(CH_2)_6COOH$

$C_{25}H_{27}O_2P^{\oplus}$ M 390.461 (ion)

Widely used in Wittig reactions leading to fatty acids and prostaglandins.

Chloride:
$C_{25}H_{27}ClO_2P$ M 425.914
Solid. Mp 161-161.5°.

Kovaleva, A.S. *et al*, *Zh. Org. Khim.*, 1974, **10**, 696 (*Engl. transl.* p. 700) (*synth, use*)
Bundy, G.L. *et al*, *J. Med. Chem.*, 1983, **26**, 790 (*use*)

(Carboxymethyl)triphenylphosphonium(1+) C-00014

(Triphenylphosphonio)acetic acid

[28486-27-8]

$Ph_3P^{\oplus}CH_2COOH$

$C_{20}H_{18}O_2P^{\oplus}$ M 321.335 (ion)

Chloride: [7343-26-2].
$C_{20}H_{18}ClO_2P$ M 356.788
Corrosion inhibitor for steel.

▷TA1870000.

Bromide: [1530-44-5].
$C_{20}H_{18}BrO_2P$ M 401.239
Used in flame-retardant prepns.

Tetrafluoroborate: [72918-87-7].
$C_{20}H_{18}BF_4O_2P$ M 408.138
Cryst. (EtOH aq.). Mp 156°.

Me ester: see (2-Methoxy-2-oxoethyl)-triphenylphosphonium(1+), M-00055

Et ester: see (2-Ethoxy-2-oxoethyl)-triphenylphosphonium, E-00047

Cavicchio, G. *et al*, *Gazz. Chim. Ital.*, 1979, **109**, 315; *CA*, **92**, 110573 (*synth*)

(5-Carboxypentyl)-triphenylphosphonium(1+) C-00015

6-(Triphenylphosphonio)hexanoic acid

$Ph_3P^{\oplus}(CH_2)_5COOH$

$C_{24}H_{26}O_2P^{\oplus}$ M 377.442 (ion)

Bromide: [50889-29-7].
$C_{24}H_{26}BrO_2P$ M 457.346
Cryst. ($MeOH/Et_2O$). Mp 202-203° (188-191°).
NaH in DMSO yields 5-carboxypentyldiphenylphosphine oxide.

Ylide: [53036-80-9]. *6-(Triphenylphosphoranylidene)-hexanoic acid.*
$C_{24}H_{25}O_2P$ M 376.434
Stated to be obtainable from the phosphonium salts, and to be widely used in prostaglandin synth.

Kovaleva, A.S. *et al*, *Zh. Org. Khim.*, 1974, **10**, 696 (*Engl. transl.* p. 700) (*pmr, ir, ylide*)
Narayanan, K.S. *et al*, *J. Org. Chem.*, 1980, **45**, 2240 (*synth, ir, pmr, cmr, nmr, props*)
Castellanos, L. *et al*, *Tetrahedron*, 1981, **37**, 1691 (*synth, use*)
Knight, D.W. *et al*, *Tetrahedron Lett.*, 1981, **22**, 5101 (*use*)

(2-Carboxy-2-propenyl)-triphenylphosphonium(1+), 9CI C-00016

2-(Triphenylphosphoniomethyl)-2-propenoic acid(1+)

$H_2C{=}C(COOH)CH_2PPh_3^{\oplus}$

$C_{22}H_{20}O_2P^{\oplus}$ M 347.373 (ion)

Bromide: [72707-67-6].
$C_{22}H_{20}BrO_2P$ M 427.277

Source of ylide, obt. by treatment with NaH. Solid. Mp 214-217°.
Ylide: (2-Carboxy-2-propenylidene)-triphenylphosphorane.
$C_{22}H_{19}O_2P$ M 346.365
Wittig reagent for prep. of (*Z*)-4-substd.-1,3-butadienes.
Me ester, bromide: [72707-68-7].
$C_{23}H_{22}BrO_2P$ M 441.303
Solid. Mp 166-170°.
Et ester, bromide: [38104-00-6].
$C_{24}H_{24}BrO_2P$ M 455.330
No phys. props. reported.

Garbers, C.F. *et al*, *Tetrahedron Lett.*, 1972, 1421 (*synth, use*)
Düttmann, H. *et al*, *Chem. Ber.*, 1979, **112**, 3480 (*synth, use*)

(3-Carboxypropyl)-triphenylphosphonium(1+) C-00017

$Ph_3P^{\oplus}(CH_2)_3COOH$

$C_{22}H_{22}O_2P^{\oplus}$ M 349.388 (ion)
Chloride: [60633-18-3].
$C_{22}H_{22}ClO_2P$ M 384.841
Solid. Mp 235-237°. With NaH/DMSO gives 4-(diphenylphosphinyl)butanoic acid.
Bromide: [17857-14-6].
$C_{22}H_{22}BrO_2P$ M 429.292
Widely used in prostaglandin synth. through the ylide. Solid. Mp 245-248°.
Ylide: [52134-55-1]. *4-(Triphenylphosphoranylidene)-butanoic acid.*
$C_{22}H_{21}O_2P$ M 348.380
Wittig reagent.

Seidel, W. *et al*, *Chem. Ber.*, 1977, **110**, 3544 (*bromide, pmr, ms, ir, use*)
Nicolaou, K.C. *et al*, *J. Chem. Soc., Chem. Commun.*, 1978, 1067 (*use*)
Nidy, E.G. *et al*, *J. Org. Chem.*, 1980, **45**, 1121 (*use*)
Narayanan, K.S. *et al*, *J. Org. Chem.*, 1980, **45**, 2240 (*chloride, pmr, cmr, nmr*)

Chlorfenvinphos, BSI C-00018

2-Chloro-1-(2,4-dichlorophenyl)ethenyl diethyl phosphate, 9CI. *2-Chloro-1-(2,4-dichlorophenyl)vinyl diethyl phosphate. Enolofos*
[470-90-6]

$(EtO)_2P(O)OC(=CHCl)$–2,4-dichlorophenyl (structure: CHCl, Cl, Cl)

$C_{12}H_{14}Cl_3O_4P$ M 359.573
The coml. prod. of which the phys. props. are given contains 10% (*E*)-form and 90% (*Z*)-. The (*Z*)-form is the more active isomer. Insecticide, esp. to combat soil pests and ectoparasites of farm animals. Amber liq. Mp −23° to −19°. $Bp_{0.5}$ 167-170° (tech.), $Bp_{0.008}$ 124-126°. n_D^{20} 1.5281 (tech. product).
▷Highly toxic. TB8750000.

Beynon, K.I., *Residue Rev.*, 1973, **47**, 55 (*metab, rev*)
Kita, K. *et al*, *Organika*, 1977, 177, 193; *CA*, **89**, 54388, 71802 (*tox*)
Śledziński, B. *et al*, *Organika*, 1977, 1, 11, 19; *CA*, **90**, 22159, 22160, 102961 (*synth*)
Stan, H.-J. *et al*, *Fresenius' Z. Anal. Chem.*, 1977, **287**, 271 (*ms*)
Zalucki, A., *Organika*, 1977, 131; *CA*, **89**, 71905 (*tox*)
Śledziński, B. *et al*, *Organika*, 1978, 1; 1979, 1; *CA*, **91**, 123292; **94**, 83719 (*synth, ir, pmr*)
Stan, H.-J. *et al*, *J. Chromatogr.*, 1983, **279**, 173 (*glc*)
Agrochemicals Handbook, Royal Society of Chemistry, London, 1983, A078.
Pesticide Manual, 6th Ed., 100; 7th Ed., 107.
Sax, N.I., *Dangerous Properties of Industrial Materials*, 6th Ed., Van Nostrand-Reinhold, 1984, 483.

Chlormephos, BSI, ISO C-00019

S-Chloromethyl O,O-diethyl phosphorodithioate, 9CI. *Chloromethyl O,O-diethyl dithiophosphate. Dotan*
[24934-91-6]

$(EtO)_2P(S)SCH_2Cl$

$C_5H_{12}ClO_2PS_2$ M 234.695
Soil insecticide. Oil. V. spar. sol. H_2O, misc. most org. solvs. d 1.260. $Bp_{0.1}$ 81-85°. n_D 1.5244.
▷TD5170000.

Ger. Pat., 1 015 794, (*1955*); *CA*, **53**, 19877a (*synth*)
Lynch, V.P., *Anal. Methods Pestic. Plant Growth Regul.*, 1978, **10**, 49 (*rev, uv, P nmr*)
Dougherty, R.C. *et al*, *Biomed. Mass Spectrom.*, 1980, **7**, 401 (*ms*)
Corkins, H.G. *et al*, *Phosphorus Sulfur*, 1981, **10**, 133 (*synth, ir, pmr, cmr*)
Pesticide Manual, 6th Ed., 103; 7th Ed., 110.
Agrochemicals Handbook, Royal Society of Chemistry, London, 1983, A081.

(Chloroacetyl)phosphonic acid C-00020

Fosfonochlorin
[89699-33-2]

$ClCH_2COP(O)(OH)_2 \rightleftharpoons ClCH{=}C(OH)P(O)(OH)_2$

$C_2H_4ClO_4P$ M 158.478
Prod. by *Talaromyces flavus* and *Fusarium* spp. Active against *Proteus* sp. and in the presence of glucose-6-phosphate gram-positive and -negative bacteria. Oil.
Di-Et ester: [25196-02-5]. *Diethyl(chloroacetyl)-phosphonate.*
$C_6H_{12}ClO_4P$ M 214.585
Liq. $Bp_{0.3}$ 75-80°.
Diisopropyl ester: [40612-55-3]. *Diisopropyl(chloroacetyl)phosphonate.*
$C_8H_{16}ClO_4P$ M 242.639
Liq. $Bp_{0.2}$ 80-100° (oven).

Japan. Pat., 83 209 986, (*1983*); *CA*, **100**, 173158 (*isol*)
Brittelli, D.R. *et al*, *J. Org. Chem.*, 1985, **50**, 1845 (*esters, ir, pmr*)

2-Chloro-1,3,2-benzodioxaphosphole, 9CI C-00021

o-Phenylene phosphorochloridite. o-Phenylene chlorophosphite
[1641-40-3]

(structure: benzene ring fused with O–P(Cl)–O ring)

$C_6H_4ClO_2P$ M 174.523
Needles. d_0^{20} 1.466. Mp 30°. Bp_{20} 80°. n_D^{20} 1.5711.

Anschütz, L. *et al*, *Chem. Ber.*, 1943, **76**, 218 (*synth*)
Crofts, P.C. *et al*, *J. Chem. Soc.*, 1958, 4250 (*synth*)
Zemlyanskii, N.I. *et al*, *Zh. Obshch. Khim.*, 1969, **39**, 602 (*Engl. transl.* p. 570) (*sulfide*)

Belski, V.E. *et al*, *Dokl. Akad. Nauk SSSR, Ser. Sci. Khim.*, 1974, **215**, 355 (*nqr*)
Ernst, L., *J. Magn. Reson.*, 1974, **16**, 190 (*pmr*)
Fieser, M. *et al*, *Reagents for Organic Synthesis*, Wiley, 1967-84, **5**, 516.

2-Chloro-4*H*-1,3,2-benzodioxaphosphorinan-4-one, 9CI C-00022

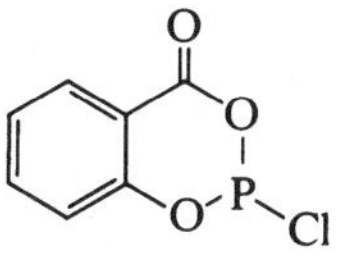

$C_7H_4ClO_3P$ M 202.534

Converts carboxylic acids into their chlorides. Solid. Mp 36-37°. Bp_{11} 127-128°.

2-Oxide: [5381-98-6].
$C_7H_4ClO_4P$ M 218.533
Prisms (CCl_4). Mp 90-93° (85-88°). $Bp_{0.1}$ 114-120°.

Young, R.W., *J. Am. Chem. Soc.*, 1952, **74**, 1672 (*synth*)
Montgomery, H.A.C., *et al*, *J. Chem. Soc.*, 1956, 4603 (*oxide*)
Cade, J.A. *et al*, *J. Chem. Soc.*, 1960, 1249 (*synth, props*)
Cremlyn, R. *et al*, *Phosphorus Sulfur*, 1981, **10**, 333 (*oxide, synth, ir, pmr, ms*)

2-Chloro-1,3,2-benzodithiaphosphole, 9CI C-00023

[52199-87-8]

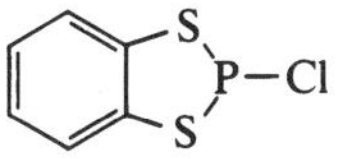

$C_6H_4ClPS_2$ M 206.644

Cryst. Mp 43°. $Bp_{0.0016}$ 85-87°.

Baudler, M. *et al*, *Z. Naturforsch, B*, 1973, **28**, 363 (*synth, P nmr*)
Denny, D.B. *et al*, *Phosphorus Sulfur*, 1982, **13**, 243 (*P nmr*)

P-(4-Chlorobutyl)-*N*-phenylphosphonamidic acid, 9CI C-00024

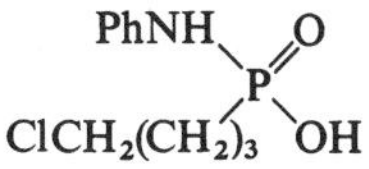

$C_{10}H_{15}ClNO_2P$ M 247.661

Intermed. for synth. of phostams. Cryst. + $1H_2O$ (EtOH aq.). Sol. MeOH, EtOH, $CHCl_3$, THF; mod. sol. Me_2CO, H_2O, Et_2O; insol. pet. ether. Mp 135-140°.

Anhydride:
$C_{20}H_{28}Cl_2N_2O_3P_2$ M 477.306
Needles ($CHCl_3$). V. sol. MeOH, Me_2CO; spar. sol. $CHCl_3$, Et_2O; insol. pet. ether. Mp 134-138°.

Helferich, B. *et al*, *Justus Liebigs Ann. Chem.*, 1962, **655**, 59 (*synth, props*)

(4-Chlorobutyl)phosphonic acid, 9CI C-00025

$ClCH_2(CH_2)_3P(O)(OH)_2$

$C_4H_{10}ClO_3P$ M 172.548

Plates. Mp 110-112°.

Monoanilinum salt: Solid. Mp 110-118°.

Di-Et ester: Diethyl (4-chlorobutyl)phosphonate.
$C_8H_{18}ClO_3P$ M 228.655
Oil. Bp_{12} 150-152°, $Bp_{0.05}$ 106°.

Dibutyl ester: [39968-54-2]. *Dibutyl (4-chlorobutyl)-phosphonate.*
$C_{12}H_{26}ClO_3P$ M 284.762
Oil. $Bp_{0.05}$ 120°.

Dichloride: [40657-94-1].
$C_4H_8Cl_3OP$ M 209.439
Liq. $Bp_{0.05}$ 90-92°.

Monoamide:
$C_4H_{11}ClNO_2P$ M 171.563
Cryst. (C_6H_6/EtOH) as mono-NH_4 salt. Sl. sol. H_2O, EtOH. Mp 120-140°.

Diamide: see P-(4-Chlorobutyl)phosphonic diamide, C-00026

Helferich, B. *et al*, *Justus Liebigs Ann. Chem.*, 1962, **655**, 59; **670**, 48 (*synth*)

P-(4-Chlorobutyl)phosphonic diamide, 9CI C-00026

$ClCH_2(CH_2)_3P(O)(NH_2)_2$

$C_4H_{12}ClN_2OP$ M 170.578

Cryst. (DMF). Insol. $CHCl_3$, C_6H_6, Et_2O, Me_2CO, pet. ether. Mp 95-100°.

N,N'-*Di-Ph: (4-Chlorobutyl)phosphonic dianilide. P-(4-Chlorobutyl)-*N,N'-*diphenylphosphonic diamide.*
$C_{16}H_{20}ClN_2OP$ M 322.774
Needles. Mp 120-121°.

N,N'-*Dibenzyl: (4-Chlorobutyl)phosphonic* N,N'-*dibenzylamide.*
$C_{18}H_{24}ClN_2OP$ M 350.827
Needles. V. sol. EtOH, $CHCl_3$, mod. sol. Et_2O, C_6H_6. Mp 64-68°.

Helferich, B. *et al*, *Justus Liebigs Ann. Chem.*, 1962, **655**, 59 (*dianilide*)
Helferich, B. *et al*, *Justus Liebigs Ann. Chem.*, 1963, **670**, 48 (*synth*)

2-(Chlorocarbonyl)phenyl phosphorodichloridate, 9CI, 8CI C-00027

2-(Chloroformyl)phenyl phosphoryl dichloride. 2-(Chloroformyl)phenyl dichlorophosphate

[6099-41-8]

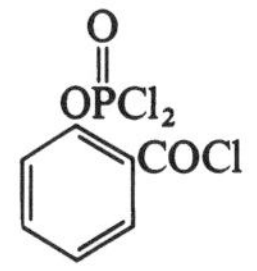

$C_7H_4Cl_3O_3P$ M 273.440

Of historical interest in connection with early theories of organic structure. Converts RCOOH→RCOCl. Fuming liq. Bp_3 162-164°. n_D^{20} 1.5555.

Couper, A.S., *C.R. Hebd. Seances Acad. Sci.*, 1858, **46**, 1157; *Justus Liebigs Ann. Chem.*, 1859, **109**, 369; *Edinburgh New Philosophical Journal*, 1858, **8**, 213.
Pinkus, A.G. *et al*, *J. Org. Chem.*, 1961, **26**, 682 (*synth, ir*)
Pinkus, A.G. *et al*, *J. Org. Chem.*, 1966, **31**, 575 (*ir, nmr, struct*)

3-(Chlorocarbonyl)phenyl phosphorodichloridate, 9CI, 8CI C-00028

3-(Chloroformyl)phenyl phosphorodichloridate. 3-(Chloroformyl)phenyl dichlorophosphate. 3-(Chloroformyl)phenyl phosphoryl dichloride

$C_7H_4Cl_3O_3P$ M 273.440

Reactive liq. $Bp_{6.5}$ 173°. n_D^{20} 1.5577.

Pinkus, A.G. *et al*, *J. Org. Chem.*, 1966, **31**, 575 (*synth, P nmr*)

4-(Chlorocarbonyl)phenyl phosphorodichloridate, 9CI, 8CI **C-00029**

4-(Chloroformyl)phenyl phosphorodichloridate. 4-(Chloroformyl)phenyl dichlorophosphate. 4-(Chloroformyl)phenyl phosphoryl dichloride

$C_7H_4Cl_3O_3P$ M 273.440

Reactive liq. Bp_7 170-175°. n_D^{20} 1.5573.

Pinkus, A.G. *et al*, *J. Org. Chem.*, 1966, **31**, 575 (*synth, P nmr*)

2-Chloro-3-(2-chloroethyl)tetrahydro-2*H*-1,3,2-oxazaphosphorine, 9CI **C-00030**

*2-Chloro-3-(2-chloroethyl)-1,3,2-oxazaphosphorinane. 2-Chloro-3-(2-chloroethyl)tetrahydro-2*H*-1,3,2-oxazaphosphinine*

$C_5H_{10}Cl_2NOP$ M 202.020

(±)-*form*

Ligand for Pt(II). Oil. $Bp_{0.07}$ 80°.

2-Oxide: [81485-04-3].
$C_5H_{10}Cl_2NO_2P$ M 218.019
Thick oil.

Okruszek, A. *et al*, *Phosphorus Sulfur*, 1979, **7**, 235 (*synth, pmr, P nmr, complex*)
Misiura, K. *et al*, *J. Med. Chem.*, 1983, **26**, 674 (*oxide, tlc, ms, P nmr*)

2-Chloro-5-chloromethyl-5-methyl-1,3,2-dioxaphosphorinane, 9CI **C-00031**

2-Chloro-5-chloromethyl-5-methyl-1,3,2-dioxaphosphinane. 2-Chloro-5-chloromethyl-5-methyl-1,3-dioxa-2-phosphacyclohexane

cis-form

$C_5H_9Cl_2O_2P$ M 203.005

cis-form

2-Oxide:
$C_5H_9Cl_2O_3P$ M 219.004
Not sepd. from *trans*-form; also known as *trans*-form.

2-Sulfide:
$C_5H_9Cl_2O_2PS$ M 235.065
Cryst. ($CHCl_3$/hexane). Mp 64-65°. Also known as *trans*-form.

trans-form

2-Oxide: Cryst. (CCl_4 or Et_2O/pet. ether). Mp 69.5-71.5°. $Bp_{0.5}$ 155°. Also known as *cis*-form. Isom. when heated at 150°.

2-Sulfide: Not sepd. from *cis*-form; also known as *cis*-form.

Edmundson, R.S. *et al*, *J. Chem. Soc. (C)*, 1968, 3033 (*oxide, synth, pmr, ir*)
Wadsworth, W.S., *J. Chem. Soc., Perkin Trans. 2*, 1972, 1686 (*oxide, synth, pmr*)
Wadsworth, W.S., *J. Org. Chem.*, 1973, **38**, 256 (*oxide, props*)
Wadsworth, W.S. *et al*, *J. Org. Chem.*, 1974, **39**, 984 (*sulfide, synth, pmr*)
Corriu, R.J.P. *et al*, *Tetrahedron*, 1979, **35**, 2889; 1981, **37**, 3681 (*derivs, props*)
Corriu, R.J.P. *et al*, *J. Am. Chem. Soc.*, 1984, **106**, 1060 (*oxide, props*)

6-Chlorodibenzo[*d,f*][1,3,2]dioxaphosphepin, 9CI **C-00032**

O,O′-*(2,2′-Biphenylyl) phosphorochloridite*

[16611-68-0]

$C_{12}H_8ClO_2P$ M 250.621

Cryst. (after long standing). d_4^{20} 1.36. Mp 62-63°. Bp_{15} 192°. n_D^{20} 1.6347. Easily hydrolysed.

6-Oxide: [52258-06-7]. O,O-*(2,2′-Biphenylyl) phosphorochloridate.*
$C_{12}H_8ClO_3P$ M 266.620
Cryst. (pet. ether). Mp 114-115°.

6-Sulfide: [61335-18-0]. O,O-*(2,2′-Biphenylyl) phosphorochloridothioate.*
$C_{12}H_8ClO_2PS$ M 282.681
Cryst. (pet. ether). Mp 108-109°.

6-Selenide: [77671-18-2]. O,O-*(2,2′-Biphenylyl) phosphorochloridoselenoate.*
$C_{12}H_8ClO_2PSe$ M 329.581
Cryst. (MeOH). Mp 105°.

Anschütz, L. *et al*, *Chem. Ber.*, 1956, **89**, 1119 (*synth, sulfide*)
Verizhnikov, L.V. *et al*, *Zh. Obshch. Khim.*, 1967, **37**, 1355 (*Engl. transl.* p. 1281) (*synth*)
Keck, H. *et al*, *Org. Mass Spectrom.*, 1980, **15**, 591 (*ms, derivs*)
Kuchen, W. *et al*, *Phosphorus Sulfur*, 1980, **8**, 139 (*derivs, P nmr*)

6-Chloro-6*H*-dibenzo[*c,e*][1,2]oxaphosphorin, 9CI, 8CI **C-00033**

[22749-43-5]

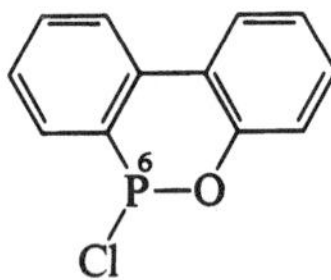

$C_{12}H_8ClOP$ M 234.621

Mp 45-48°, Mp 87°. Bp_{20} 195°, $Bp_{0.5}$ 130-132°.

6-Oxide: [32186-92-8].
$C_{12}H_8ClO_2P$ M 250.621
Solid. Mp 97-98°. Bp_2 172-174°.

6-Sulfide: [30343-12-5].
$C_{12}H_8ClOPS$ M 266.681
Cryst. (hexane). Mp 140-141°. Bp_7 203-208°.

Chernyshev, E.A. *et al*, *Zh. Obshch. Khim.*, 1970, **40**, 1423 (*Engl. transl.* p. 1409) (*sulfide*)
Chernyshev, E.A. *et al*, *Zh. Obshch. Khim.*, 1971, **41**, 2189 (*Engl. transl.* p. 2214) (*oxide, synth, ir*)
Chernyshev, E.A. *et al*, *Zh. Obshch. Khim.*, 1972, **42**, 93 (*Engl. transl.* p. 88) (*synth, sulfide*)
Bhatia, M.S. *et al*, *Indian J. Chem., Sect. B*, 1976, **14**, 811 (*sulfide, ir, uv, ms*)
Bhatia, M.S. *et al*, *Org. Mass Spectrom.*, 1977, **12**, 1 (*ms*)

5-Chloro-5*H*-dibenzophosphole, 8CI **C-00034**

*1-Chloro-1*H*-benzo*[b]*phosphindole, 10CI*

[33300-85-5]

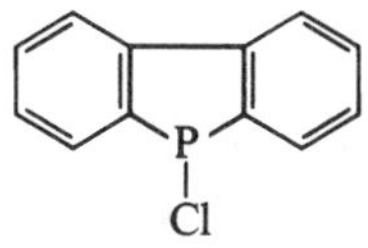

$C_{12}H_8ClP$ M 218.622

Solid. Mp 53-56°, 62-63°. Bp_2 145-146°.

5-Oxide: [33771-51-6].

$C_{12}H_8ClOP$ M 234.621

Solid. Mp 148-149°. Bp_6 222-224°.

5-Sulfide: [33771-55-0].

$C_{12}H_8ClPS$ M 250.682

Solid. Mp 192-193°.

Doak, G.O. *et al, J. Org. Chem.*, 1964, **29**, 2382 (*synth, derivs*)
Chernyshev, E.A. *et al, Zh. Obshch. Khim.*, 1971, **41**, 800 (*Engl. transl.* p. 806) (*synth, derivs*)
Bochkarev, V.N. *et al, Zh. Obshch. Khim.*, 1974, **44**, 1273 (*Engl. transl.* p. 1251) (*ms*)
Hellwinkel, D. *et al, Chem. Ber.*, 1978, **111**, 13 (*synth*)

2-Chloro-1-(2,4-dichlorophenyl)vinyl dimethyl phosphate **C-00035**

2-Chloro-1-(2,4-dichlorophenyl)ethenyl dimethyl phosphate , 9CI

[2274-67-1]

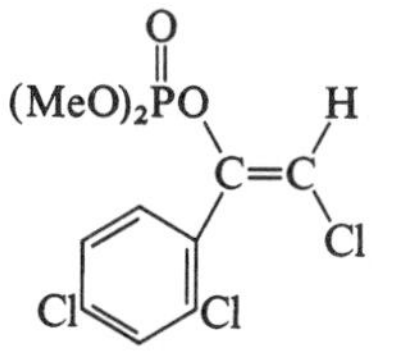

(*E*)-*form*

$C_{10}H_{10}Cl_3O_4P$ M 331.519

Insecticide. Cryst. (CCl_4/hexane). Sl. sol. H_2O. Mp 65-67°. $Bp_{0.05}$ 126°.

▷TB8805000.

(*E*)-*form* [71363-52-5]

Obt. only as mixt. with (*Z*)-form.

(*Z*)-*form* [67628-93-7]

Obt. only as mixt. with (*E*)-form.

Crawford, M.J. *et al, Xenobiotica*, 1976, **6**, 745; *CA*, **86**, 115834 (*metab*)
Malinowski, H. *et al, Organika*, 1977, 96; 1978, 152; *CA*, **89**, 124494; **91**, 103647 (*use, tox*)
Szalontai, G. *et al, Org. Magn. Reson.*, 1977, **10**, 63 (*cmr*)
Śledziński, B. *et al, Organika*, 1978, 1; 1979, 1; *CA*, **91**, 123292; **94**, 83719 (*synth, ir, pmr*)

Chloro(diethoxyphosphinyl)acetic acid **C-00036**

Diethyl carboxychloromethylphosphonate. 2-Chloro-2-diethylphosphonoacetic acid

[30094-33-8]

$(HO)_2P(O)CHClCOOH$

$C_6H_{12}ClO_5P$ M 230.585

The dianion reacts with carbonyl compds. to give on hydrol. α-chloro-α,β-unsaturated acids. Oil which slowly cryst. Mp 66°.

Et ester: [7071-12-7]. *Ethyl chloro(diethoxyphosphinyl)acetate. Diethyl [chloro(ethoxycarbonyl)methyl]phosphonate.*

$C_8H_{12}ClO_5P$ M 254.607

The carbanion reacts with aldehydes or ketones to give ethyl esters of α-chloro-α,β-unsaturated acids. Liq. d_4^{20} 1.21. $Bp_{0.2}$ 100°. n_D^{20} 1.4455.

Shevchenko, V.I. *et al, Zh. Obshch. Khim.*, 1962, **32**, 2994 (*Engl. transl.* p. 2945) (*ester*)
Grell, W. *et al, Justus Liebigs Ann. Chem.*, 1966, **693**, 134 (*ester, synth, ir, pmr, use*)
Eberlein, W. *et al, Chem. Ber.*, 1972, **105**, 3686 (*use*)
Savignac, P. *et al, Synth. Commun.*, 1978, **8**, 19 (*synth, use, pmr*)
Villieras, J. *et al, Synthesis*, 1978, 31 (*ester, synth, pmr, use*)

5-Chloro-5,10-dihydroacridophosphine, 9CI **C-00037**

5-Chloro-5,10-dihydrodibenzo[b,e]*phosphorin. 5-Chloro-5,10-dihydrodibenzo*[b,e]*phosphinine*

[15309-64-5]

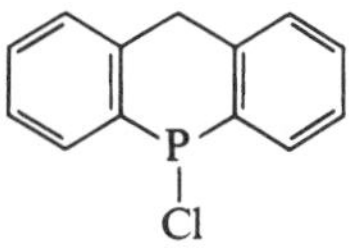

$C_{13}H_{10}ClP$ M 232.649

Cryst. (toluene or by subl.). Mp 102-109°.

Doak, G.O. *et al, J. Am. Chem. Soc.*, 1964, **29**, 2382 (*synth*)
de Koe, P. *et al, Angew. Chem., Int. Ed. Engl.*, 1967, **6**, 567 (*synth*)

3-Chloro-1,5-dihydro-2,4,3-benzodioxaphosphepin, 9CI **C-00038**

[69813-54-3]

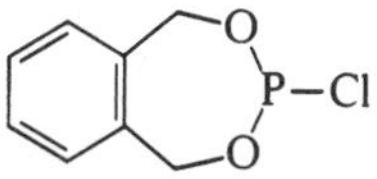

$C_8H_8ClO_2P$ M 202.577

3-Oxide: [49785-01-5].

$C_8H_8ClO_3P$ M 218.576

Solid. Mp 135-139°.

Sato, T. *et al, J. Chem. Soc., Chem. Commun.*, 1973, 494 (*oxide, synth, pmr*)
Guimanaes, A.C. *et al, Org. Magn. Reson.*, 1978, **11**, 411 (*sulfide, cmr, pmr, P nmr*)

2-Chloro-2,3-dihydro-3,5-dimethyl-1,3,4,2-oxadiazaphosphole, 9CI **C-00039**

2-Chloro-3,5-dimethyl-Δ^4-1,3,4,2-oxadiazaphospholine

[25130-52-3]

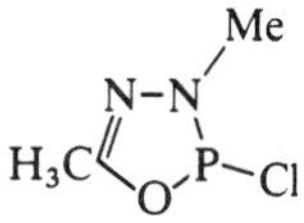

$C_3H_6ClN_2OP$ M 152.520

Liq. Bp_7 50-50.5°. Easily hydrolysed.

Vilkov, L.V. *et al, Dokl. Akad. Nauk SSSR, Ser. Sci. Khim.*, 1969, **187**, 1293 (*Engl. transl.* p. 659) (*synth, ir, raman, pmr, P nmr*)
Khaikin, L.S. *et al, J. Mol. Struct.*, 1982, **82**, 115 (*ed*)

2-Chloro-2,3-dihydro-4,5-dimethyl-1,2-oxaphosphole, 9CI C-00040

2-Chloro-4,5-dimethyl-1,2-oxaphosphol-4-ene, 8CI

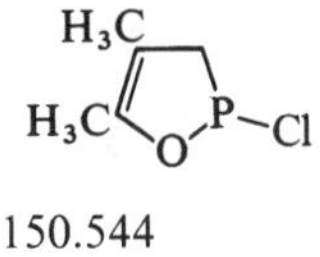

C_5H_8ClOP M 150.544

2-Oxide: [20342-03-4].

$C_5H_8ClO_2P$ M 166.544

Liq. d_4^{20} 1.28. $Bp_{0.4}$ 76°. n_D^{20} 1.4845.

Novitskii, N.I. *et al*, *CA*, 1968, **69**, 43981.

1-Chloro-2,5-dihydro-3,4-dimethyl-1*H*-phosphole, 9CI C-00041

1-Chloro-3,4-dimethyl-3-phospholene, 8CI

[40965-68-2]

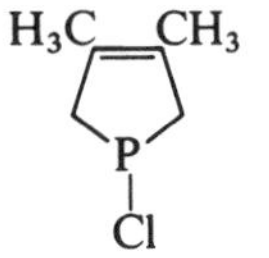

$C_6H_{10}ClP$ M 148.572

Liq. d_4^{20} 1.11. Bp_{11} 60-64°. n_D^{20} 1.5339.

1-Oxide: [873-16-5].

$C_6H_{10}ClOP$ M 164.571

Solid. Mp 83-6°. Bp_{10} 132°, $Bp_{1.0}$ 112°.

1-Sulfide: [42534-57-6].

$C_6H_{10}ClPS$ M 180.632

Liq. d_4^{20} 1.21. $Bp_{0.02}$ 54-55°. n_D^{20} 1.5705.

Arbuzov, B.A. *et al*, *Dokl. Akad. Nauk SSSR, Ser. Sci. Khim.*, 1964, **159**, 582 (*Engl. transl.* p. 1205) (*oxide, synth*)
Vizel', A.O. *et al*, *Zh. Obshch. Khim.*, 1973, **43**, 2137 (*Engl. transl.* p. 2128) (*synth*)
Felcht, U.-H. *et al*, *Phosphorus Sulfur*, 1982, **13**, 291 (*oxide*)
Buchanan, G.W. *et al*, *Org. Magn. Reson.*, 1983, **21**, 436 (*derivs, cmr*)

2-Chloro-2,3-dihydro-3,5-diphenyl-1,3,4,2-oxadiazaphosphole, 9CI C-00042

2-Chloro-3,5-diphenyl-Δ⁴-1,3,4,2-oxadiazaphospholine, 8CI

[19503-07-2]

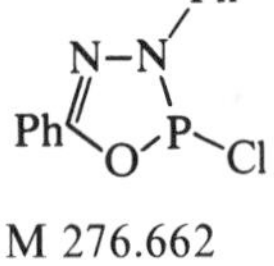

$C_{13}H_{10}ClN_2OP$ M 276.662

Liq. $Bp_{0.3}$ 204-210°.

2-Oxide: [33711-91-0].

$C_{13}H_{10}ClN_2O_2P$ M 292.661

Solid. Mp 79-82° (sealed tube under N_2).

2-Sulfide: [18720-17-7].

$C_{13}H_{10}ClN_2OPS$ M 308.722

Liq. $Bp_{0.12}$ 167-170°.

Italinskaya, T.L. *et al*, *Zh. Obshch. Khim.*, 1968, **38**, 2265 (*Engl. transl.* p. 2192) (*synth, sulfide*)
Shvetsov-Shilovskii, N.I. *et al*, *Zh. Obshch. Khim.*, 1971, **41**, 1200 (*Engl. transl.* p. 1210) (*oxide*)

2-Chloro-2,3-dihydro-3-methyl-1,3,2-benzothiazaphosphole, 9CI C-00043

[62128-57-8]

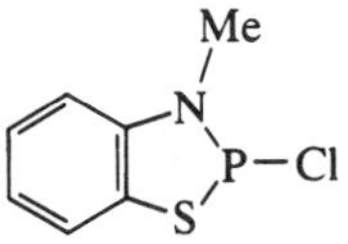

C_7H_7ClNPS M 203.626

Solid. Mp 48-49°. $Bp_{0.4}$ 115°.

2-Sulfide: [62128-60-3].

$C_7H_7ClNPS_2$ M 235.686

Solid. Mp 58-59°.

Pudovik, M.A. *et al*, *Izv. Akad. Nauk SSSR, Ser. Khim.*, 1976, 2837 (*Engl. transl.* p. 2648) (*synth, P nmr*)

2-Chloro-2,3-dihydro-5-methyl-3-phenyl-1,3,4,2-oxadiazaphosphole, 9CI C-00044

2-Chloro-5-methyl-3-phenyl-Δ⁴-1,3,4,2-oxadiazaphospholine, 8CI

[19525-44-1]

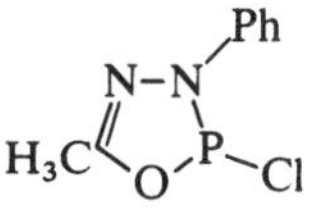

$C_8H_8ClN_2OP$ M 214.591

Liq. $Bp_{0.2}$ 97°.

2-Oxide: [33711-90-9].

$C_8H_8ClN_2O_2P$ M 230.590

Solid. Mp 58-60° (sealed tube under N_2). $Bp_{0.12}$ 83.5-84°.

2-Sulfide: [18655-72-6].

$C_8H_8ClN_2OPS$ M 246.651

Solid. Mp 45-47°. $Bp_{0.2}$ 100-102.5°.

Italinskaya, T.L. *et al*, *Zh. Obshch. Khim.*, 1968, **38**, 2265; 1971, **41**, 1980 (*Engl. transl.* pp. 2192, 1998) (*synth, sulfide*)
Shvetsov-Shilovskii, N.I. *et al*, *Zh. Obshch. Khim.*, 1971, **41**, 1200 (*Engl. transl.* p. 1210) (*oxide*)

1-Chloro-2,3-dihydro-4-methyl-1*H*-phosphole, 9CI C-00045

1-Chloro-3-methyl-2-phospholene, 8CI

[28273-37-2]

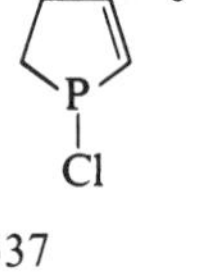

C_5H_7ClP M 133.537

Pungent liq. d_4^{20} 1.14. Bp_{18} 67-70°. n_D^{20} 1.5575.

1-Oxide: [823-14-3].

C_5H_7ClOP M 149.536

Liq. d_4^{20} 1.26. Bp_2 116-117°, $Bp_{0.01}$ 88-90°. n_D^{20} 1.5236.

Myers, D.K. *et al*, *J. Org. Chem.*, 1971, **36**, 1285 (*synth, oxide, ir, pmr, P nmr*)
Quin, L.D. *et al*, *J. Org. Chem.*, 1971, **36**, 1297 (*ir, P nmr, pmr*)
Vizel', A.O. *et al*, *Zh. Obshch. Khim.*, 1973, **43**, 2137 (*Engl. transl.* p. 2128) (*synth*)

1-Chloro-2,5-dihydro-2-methyl-1*H*-phosphole, 9CI C-00046

1-Chloro-2-methyl-3-phospholene, 8CI

[51090-80-3]

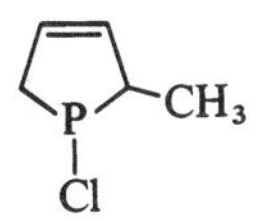

C_5H_7ClP M 133.537

Liq. d_4^{20} 1.13. Bp_{10} 48-49°. n_D^{20} 1.5275.

1-Oxide: [18943-90-3].

C_5H_7ClOP M 149.536

d_4^{20} 1.25. Bp_1 79-80°. n_D^{20} 1.5075, 1.5159.

1-Sulfide: [4552-75-4].

C_5H_7ClPS M 165.597

d_4^{20} 1.23. Bp_1 82-83°. n_D^{20} 1.5771.

Bliznyuk, N.K. *et al, Zh. Obshch. Khim.*, 1967, **37**, 1811 (*Engl. transl.* p. 1726) (*oxide*)
Razumova, N.A. *et al, Zh. Obshch. Khim.*, 1970, **40**, 2563 (*Engl. transl.* p. 2554) (*oxide, pmr*)
Zubtsova, L.I. *et al, Zh. Obshch. Khim.*, 1971, **41**, 2428 (*Engl. transl.* p. 2453) (*sulfide*)
Vizel', O.A. *et al, Zh. Obshch. Khim.*, 1973, **43**, 2137 (*Engl. transl.* p. 2128) (*synth*)

1-Chloro-2,5-dihydro-3-methyl-1*H*-phosphole, 9CI C-00047

1-Chloro-3-methyl-3-phospholene, 8CI

[28273-35-0]

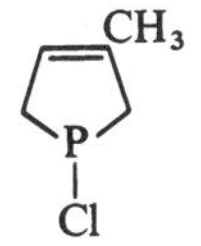

C_5H_7ClP M 133.537

Liq. d_4^{20} 1.14. Bp_{17} 61-63°.

1-Oxide: [18874-22-1].

C_5H_7ClOP M 149.536

Liq. $Bp_{1.8}$ 97-100°.

1-Sulfide: [4414-16-8].

C_5H_7ClPS M 165.597

Low-melting solid. Mp 40.4-41°. Bp_2 118-120°.

Arbusov, B.A. *et al, Izv. Akad. Nauk SSSR, Ser. Khim.*, 1971, 2489 (*Engl. transl.* p. 2360) (*sulfide*)
Myers, K. *et al, J. Org. Chem.*, 1971, **36**, 1285 (*oxide, ir, P nmr, pmr*)
Shagidullin, R.R. *et al, Izv. Akad. Nauk SSSR, Ser. Khim.*, 1974, 1611 (*Engl. transl.* p. 1533) (*oxide, sulfide, ir*)
Fazliev, D.F. *et al, Izv. Akad. Nauk SSSR, Ser. Khim.*, 1975, 1058 (*Engl. transl.* p. 967) (*ir, raman*)

4-Chloro-1,4-dihydro-2*H*-naphth[2,1-c][1,2]oxaphosphorin C-00048

[68257-69-2]

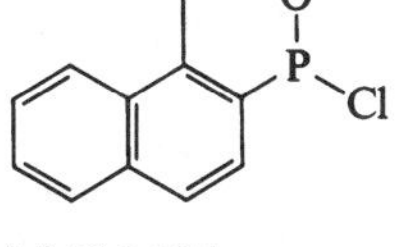

$C_{12}H_{10}ClOP$ M 236.637

Cryst. ($CHCl_3$). Mp 94°.

Bhatia, M.S. *et al, Indian J. Chem., Sect. B*, 1978, **16**, 638 (*synth, pmr*)

10-Chloro-5,10-dihydrophenophosphazine, 9CI, 8CI C-00049

[79735-27-6]

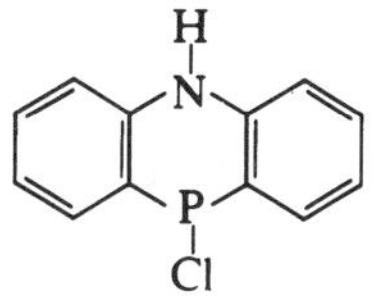

$C_{12}H_9ClNP$ M 233.637

10-Oxide: [17534-72-4].

$C_{12}H_9ClNOP$ M 249.636

Solid. Mp 290°.

10-Sulfide: [40074-54-2].

$C_{12}H_9ClNPS$ M 265.697

Yellowish solid. Mp 238.5-240° (229.5-237.5°).

Häring, M., *Helv. Chim. Acta*, 1960, **43**, 1826 (*derivs*)
Demidova, N.I. *et al, Zh. Obshch. Khim.*, 1982, **52**, 1099 (*Engl. transl.* p. 958) (*sulfide*)

2-Chloro-2,3-dihydro-3-phenyl-4*H*-1,3,2-benzoxazaphosphorin-4-one, 8CI C-00050

[15494-45-8]

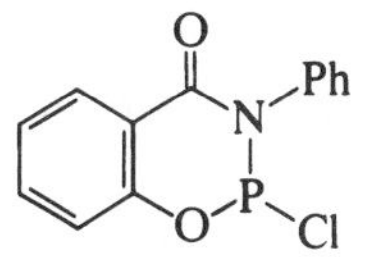

$C_{18}H_9ClNO_2P$ M 337.701

Cryst. (C_6H_6/pet. ether). Mp 97°.

2-Oxide:

$C_{13}H_9ClNO_3P$ M 293.646

Cryst. (C_6H_6). Mp 163-165°.

Sabirova, R.A. *et al, Zh. Obshch. Khim.*, 1967, **37**, 732 (*Engl. transl.* p. 686)

1-Chloro-2,3-dihydro-1*H*-phosphole, 9CI C-00051

1-Chloro-2-phospholene, 8CI

[28273-36-1]

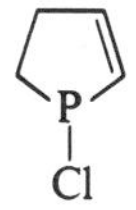

C_4H_6ClP M 120.518

Pungent liq. Bp_{32} 67-68°.

1-Oxide: [1003-18-5].

C_4H_6ClOP M 136.518

Bp_2 107-109.5°, $Bp_{0.1}$ 72-75°.

1-Sulfide: [36305-33-6].

C_4H_6ClPS M 152.578

$Bp_{0.1}$ 78°.

Arbuzov, B.A. *et al, Izv. Akad. Nauk SSSR, Ser. Khim.*, 1969, 460 (*Engl. transl.* p. 408) (*oxide*)
Fazliev, D.F. *et al, Izv. Akad. Nauk SSSR, Ser. Khim.*, 1975, 1058 (*Engl. transl.* p. 967) (*oxide, ir, raman*)
Moedritzer, K., *Synth. React. Inorg. Metal-Org. Chem.*, 1975, **5**, 45 (*oxide, ms, ir, pmr, P nmr*)
Moedritzer, K., *Z. Naturforsch., B*, 1976, **31**, 709 (*sulfide, ms, ir, cmr, pmr, nmr*)
Felcht, U-H. *et al, Phosphorus Sulfur*, 1982, **13**, 291 (*synth, oxide, nmr*)

1-Chloro-2,5-dihydro-1*H*-phosphole, 9CI **C-00052**
1-Chloro-3-phospholene, 8CI
[28278-56-0]

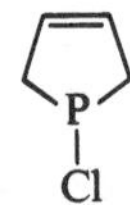

C_4H_6ClP M 120.518
1-Oxide: [822-47-9].
C_4H_6ClOP M 136.518
Solid. Mp 53°. $Bp_{1.0}$ 82-84°, $Bp_{0.1}$ 63-65°.
1-Sulfide: [18359-32-5].
C_4H_6ClPS M 152.578
Liq. d_4^{20} 1.32. Bp_{10} 138-143°, $Bp_{0.1}$ 65°. n_D^{20} 1.5946.

Arbuzov, B.A. *et al*, *Izv. Akad. Nauk SSSR, Ser. Khim.*, 1969, 460 (*Engl. transl.* p. 408) (*oxide*)
Fazliev, D.F. *et al*, *Izv. Akad. Nauk SSSR, Ser. Khim.*, 1975, 1058 (*Engl. transl.* p. 967) (*oxide, ir, raman*)
Moedritzer, K., *Synth. React. Inorg. Metal-Org. Chem.*, 1975, **5**, 45 (*oxide, ms, ir, nmr, pmr*)
Moedritzer, K., *Z. Naturforsch., B*, 1976, **31**, 709 (*sulfide, ir, ms, cmr, nmr, pmr*)
Felcht, U-H. *et al*, *Phosphorus Sulfur*, 1982, **13**, 291 (*oxide, ir, nmr, pmr*)

2-Chloro-3,6-dihydro-2*H*-1,2-thiaphosphorin 2-sulfide, 9CI **C-00053**
[76442-61-0]

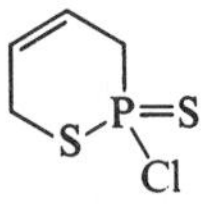

$C_4H_6ClPS_2$ M 184.638
Cryst. (heptane). Mp 63-66°.

U.S.P., 4 231 970, (*1980*); *CA*, **94**, 84303 (*synth, pmr, cmr, P nmr*)

2-Chloro-2,3-dihydro-3,3,5-trimethyl-1,2-oxaphosphole, 9CI **C-00054**
2-Chloro-3,3,5-trimethyl-1,2-oxaphosphol-4-ene, 8CI

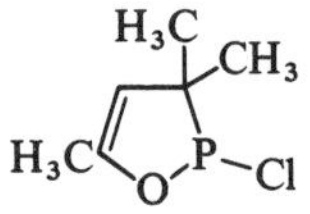

$C_6H_{10}ClOP$ M 164.571
2-Oxide:
$C_6H_{10}ClO_2P$ M 180.571
Liq. d_4^{20} 1.21. $Bp_{0.3}$ 87-89°, $Bp_{0.05}$ 53-55°. n_D^{20} 1.4758 (1.4800).

Nurtdinov, S.Kh. *et al*, *Zh. Obshch. Khim.*, 1970, **40**, 2189, 2377 (*Engl. transl.* pp. 2176, 2365) (*synth*)
Mukhametov, F.S. *et al*, *Izv. Akad. Nauk SSSR, Ser. Khim.*, 1972, 1827 (*Engl. transl.* p. 1765) (*synth, nmr*)
Arbuzov, B.A. *et al*, *Zh. Obshch. Khim.*, 1975, **45**, 512 (*Engl. transl.* p. 507) (*synth*)

2-Chloro-1,3-dimethyl-1,3,2-diazaphospholidine, 9CI, 8CI **C-00055**
[6069-36-9]

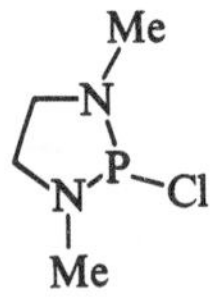

$C_4H_{10}ClN_2P$ M 152.563
Pungent liq. $Bp_{0.2}$ 70°.
2-Oxide: [6069-37-0].
$C_4H_{10}ClN_2OP$ M 168.563
No phys. props. reported.
2-Sulfide: [31755-40-5].
$C_4H_{10}ClN_2PS$ M 184.623
No phys. props. reported.

Ramirez, F. *et al*, *J. Am. Chem. Soc.*, 1967, **89**, 6276 (*synth, pmr, P nmr*)
Bousquet, A. *et al*, *C.R. Hebd. Seances Acad. Sci., Ser. C*, 1971, **272**, 246 (*derivs, pmr, P nmr*)
Bulloch, G. *et al*, *J. Chem. Soc., Dalton Trans.*, 1978, 764 (*pmr, cmr*)
Nuretdinov, I.A. *et al*, *Izv. Akad. Nauk SSSR, Ser. Khim.*, 1978, 950 (*Engl. transl.* p. 824) (*nqr*)
Gouesnard, J.P. *et al*, *J. Mol. Struct.*, 1980, **67**, 297 (*N and P nmr*)
Gonbeau, D. *et al*, *Inorg. Chem.*, 1981, **20**, 1966 (*pe, struct*)

2-Chloro-4,5-dimethyl-1,3,2-dioxaphospholane, 9CI **C-00056**
2-Chloro-4,5-dimethyl-1,3-dioxa-2-phosphacyclopentane. 2,3-Butylene chlorophosphite. 2,3-Butylene phosphorochloridite
[16352-28-6]

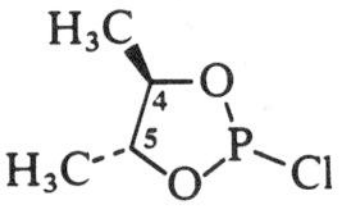

$C_4H_8ClO_2P$ M 154.533
(*4R,5R*)-*form* [89104-49-4]
Fuming liq. d_4^{25} 1.20. Bp_{10} 49°. $[\alpha]_D^{25}$ +97.12°. n_D^{25} 1.4604.
2-Oxide: [89104-48-3]. *2,3-Butylene chlorophosphate. 2,3-Butylene phosphoryl chloride. 2,3-Butylene phosphorochloridate.* Chiral derivatising agent. Liq. d 1.5. $Bp_{0.5}$ 105-110°.
(*4RS,5RS*)-*form* [15479-16-0]
(±)-*form*
Fuming liq. Bp_{22} 64.5°, Bp_{17} 87°.
2-Sulfide: O,O-*2,3-Butylene chlorothiophosphate.* O,O-*2,3-Butylene thiophosphoryl chloride.*
$C_4H_8ClO_2PS$ M 186.593
Liq. $Bp_{0.05}$ 55°.
(*4RS,5SR*)-*form*
meso-*form. 4,5*-cis-*form*
Fuming liq. d_4^{25} 1.22. Bp_{17} 90°, Bp_{25} 76°. n_D^{20} 1.4696. Epimers at P characterised spectroscopically.
2-Sulfide: Liq. $Bp_{0.08}$ 65.5-66°.

Garner, H.K. *et al*, *J. Am. Chem. Soc.*, 1950, **72**, 5497 (*synth*)
Lucas, H.J. *et al*, *J. Am. Chem. Soc.*, 1950, **72**, 5491 (*synth*)
Gagnaire, D. *et al*, *Bull. Soc. Chim. Fr.*, 1966, 3719 (*synth, nmr*)
Anderson, B.A. *et al*, *Dokl. Akad. Nauk SSSR, Ser. Sci. Khim.*, 1972, **204**, 1349 (*Engl. transl.* p. 523) (*sulfides, synth, P nmr*)
Bergesen, K. *et al*, *Acta Chem. Scand.*, 1972, **26**, 2153 (*synth, pmr*)
Osokin, D.Ya. *et al*, *Izv. Akad. Nauk SSSR, Ser. Khim.*, 1972, **21**, 1513 (*Engl. transl.* p. 1460) (*nqr*)

Shagidullin, R.R. *et al*, *Zh. Obshch. Khim.*, 1975, **45**, 1257 (*Engl. transl.* p. 1235) (*ir, raman*)
Shagidullin, R.R. *et al*, *Zh. Obshch. Khim.*, 1976, **46**, 1021 (*Engl. transl.* p. 1017) (*oxide, sulfide, ir, raman*)
Anderson, R.C. *et al*, *J. Org. Chem.*, 1984, **49**, 1304 (*oxide, synth, P nmr, use*)

2-Chloro-4,5-dimethyl-1,3,2-dioxaphosphole, 9CI **C-00057**

Dimethylvinylene chlorophosphite

[84383-01-7]

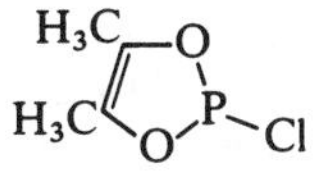

$C_4H_6ClO_2P$ M 152.517

Fuming liq. Bp_{12} 42-43°.

2-Oxide: [21949-38-2]. *Dimethylvinylene chlorophosphate.*
$C_4H_6ClO_3P$ M 168.516
Phosphorylating agent. Fuming liq., moisture-sensitive. Bp_5 82-83°, $Bp_{0.1}$ 48°.

Ramirez, F.R. *et al*, *Synthesis*, 1976, 819 (*oxide, synth, cmr, pmr, P nmr*)
Karlstedt, N.B. *et al*, *Zh. Obshch. Khim.*, 1982, **52**, 1974; 1984, **54**, 221 (*Engl. transl.* pp. 1754, 196) (*synth, props*)
Ramirez, F.R. *et al*, *J. Org. Chem.*, 1983, **48**, 847 (*oxide*)
Ramirez, F.R. *et al*, *Phosphorus Sulfur*, 1983, **17**, 67 (*oxide, use*)
Goetz, J. *et al*, *Heterocycles*, 1984, **21**, 265 (*synth, oxide, ir, cmr, pmr, P nmr*)

2-Chloro-5,5-dimethyl-1,3,2-dioxaphosphorinane, 9CI **C-00058**

2-Chloro-5,5-dimethyl-1,3,2-dioxaphosphinane. 2-Chloro-5,5-dimethyl-1,3-dioxa-2-phosphacyclohexane. Neopentylene phosphorochloridite

[2428-06-0]

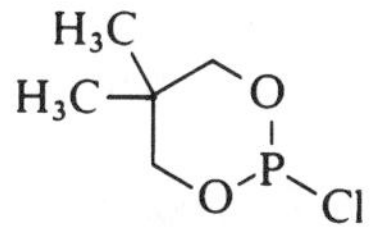

$C_5H_{10}ClO_2P$ M 168.560

Fuming mobile liq. Bp_{13} 66°. n_D^{22} 1.4745.

2-Oxide: see 2-Chloro-5,5-dimethyl-1,3,2-dioxaphosphorinane 2-oxide, C-00059

2-Sulfide: see 2-Chloro-5,5-dimethyl-1,3,2-dioxaphosphorinane 2-sulfide, C-00060

2-Selenide: [70532-50-2]. *Neopentylene phosphorochloridoselenoate.*
$C_5H_{10}ClO_2PSe$ M 247.520
Needles (C_6H_6/pet. ether). Mp 72-74°.

Edmundson, R.S., *Chem. Ind. (London)*, 1965, 1220 (*synth*)
White, D.W. *et al*, *J. Am. Chem. Soc.*, 1970, **92**, 7125 (*pmr*)
Osokin, D.Ya. *et al*, *Izv. Akad. Nauk SSSR, Ser. Khim.*, 1972, 1513 (*Engl. transl.* p. 1460) (*nqr*)
Nifant'ev, É.E. *et al*, *Dokl. Akad. Nauk SSSR, Ser. Sci. Khim.*, 1973, **208**, 651 (*Engl. transl. Phys. Chem.*, p. 100) (*cmr*)
Efremov, Yu.Ya. *et al*, *Khim. Geterotsikl. Soedin.*, 1974, 1620 (*Engl. transl.* p. 1424) (*ms*)
Fazliev, D.F. *et al*, *Zh. Obshch. Khim.*, 1976, **46**, 1832 (*Engl. transl.* p. 1776) (*ir*)
Michalska, M. *et al*, *Tetrahedron*, 1978, **34**, 2821 (*selenide, synth, P nmr*)
Shagidullin, R.R. *et al*, *Izv. Akad. Nauk SSSR, Ser. Khim.*, 1981, 1535 (*Engl. transl.* p. 1234) (*selenide, ir, raman*)

2-Chloro-5,5-dimethyl-1,3,2-dioxaphosphorinane 2-oxide, 9CI **C-00059**

Neopentylene phosphorochloridate. 2-Chloro-5,5-dimethyl-1,3,2-dioxaphosphinane 2-oxide. 2-Chloro-5,5-dimethyl-1,3-dioxa-2-oxo-2-phosphacyclohexane

[4090-55-5]

$C_5H_{10}ClO_3P$ M 184.559

Cryst. (1,2-dichloroethane or C_6H_6/cyclohexane). Mp 104.5-106°.

Edmundson, R.S., *Tetrahedron*, 1965, **21**, 2379 (*synth*)
Stec, W.J. *et al*, *Can. J. Chem.*, 1967, **45**, 2513 (*synth*)
Hall, L.D. *et al*, *Can. J. Chem.*, 1972, **50**, 2092 (*pmr, P nmr*)
Majoral, J.-P. *et al*, *Spectrochim. Acta, Part A*, 1972, **28**, 2247 (*ir*)
Silver, L. *et al*, *Acta Crystallogr., Sect. B*, 1972, **28**, 574 (*cryst struct*)
Dale, A.J., *Acta Chem. Scand., Ser. B*, 1976, **30**, 255 (*pmr*)
Francis, G.W. *et al*, *Acta Chem. Scand., Ser. B*, 1976, **30**, 31 (*ms*)
Van Nuffel, P. *et al*, *J. Mol. Struct.*, 1984, **125**, 1 (*struct*)

2-Chloro-5,5-dimethyl-1,3,2-dioxaphosphorinane 2-sulfide, 9CI **C-00060**

Neopentylene phosphorochloridothioate. 2-Chloro-5,5-dimethyl-1,3,2-dioxaphosphinane 2-sulfide. 2-Chloro-5,5-dimethyl-1,3-dioxa-2-thioxo-2-phosphacyclohexane

[873-98-3]

$C_5H_{10}ClO_2PS$ M 200.620

Cryst. (pet. ether). Mp 90-91.5°.

Edmundson, R.S., *Tetrahedron*, 1965, **21**, 2379 (*synth*)
Bartle, K.D. *et al*, *Tetrahedron*, 1967, **23**, 1701 (*pmr*)
Stec, W.J. *et al*, *Can. J. Chem.*, 1967, **45**, 2513 (*synth*)
Majoral, J.-P. *et al*, *Spectrochim. Acta, Part A*, 1972, **28**, 2247 (*ir*)
Omelanczuk, J. *et al*, *Tetrahedron*, 1975, **31**, 2809 (*synth, P nmr*)
Tabony, J., *Spectrochim. Acta, Part A*, 1979, **35**, 217 (*P nmr*)
Edmundson, R.S., *Phosphorus Sulfur*, 1981, **9**, 307 (*ms*)

2-Chloro-3,4-dimethyl-5-phenyl-1,3,2-oxazaphospholidine, 9CI, 8CI **C-00061**

2-Chloro-3,4-dimethyl-5-phenyl-1,3,2-oxazaphospholane

[64023-57-0]

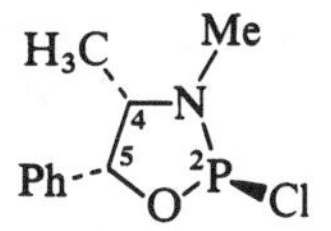

(2R,4S,5R)-form

$C_{10}H_{13}ClNOP$ M 229.646

(2R,4S,5R)-form [61739-41-1]

$Bp_{0.05}$ 120-125°.

2-Oxide: [54750-13-9].
$C_{10}H_{13}ClNO_2P$ M 245.645
Cryst. (diisopropyl ether or pet. ether). Mp 88-89°. $[\alpha]_D$ −64° ($CHCl_3$). Has 2*S*-config.

2-Sulfide: [57573-32-7].
$C_{10}H_{13}ClNOPS$ M 261.706
Reagent for synth. of chiral monothiophosphate esters. Cryst. ($CHCl_3$/pet. ether or cyclohexane). Mp 58°. $[\alpha]_D$ −23° ($CHCl_3$), $[\alpha]_D$ −47° (C_6H_6). Has 2*R*-config.

(2S,4S,5R)-form

2-Oxide: [54750-12-8]. Cryst. (diisopropyl ether). Mp 111-113°. $[\alpha]_D$ −26° ($CHCl_3$). Has 2*R*-config.

2-Sulfide: [57651-34-0]. Reagent for synth. of chiral monothiophosphate esters. Cryst. (diisopropyl ether or cyclohexane). Mp 125-128°. $[\alpha]_D$ −121° ($CHCl_3$), $[\alpha]_D$ −123° (C_6H_6). Has 2*S*-config.

Larizza, A. *et al*, *J. Med. Chem.*, 1966, **9**, 966 (*oxide*)
Devillers, J. *et al*, *Bull. Soc. Chim. Fr.*, 1970, 4341 (*oxide, sulfide, synth, pmr*)
Devillers, J. *et al*, *Org. Magn. Reson.*, 1974, **6**, 211 (*derivs, conformn*)
Hall, C.R. *et al*, *Tetrahedron Lett.*, 1976, 3645 (*synth*)
Prange, T. *et al*, *Bull. Soc. Chim. Fr., Pt. 1*, 1977, 185 (*cryst struct*)
Lesiàk, K. *et al*, *Z. Naturforsch., B*, 1978, **33**, 782 (*sulfide, pmr, P nmr*)
Hall, C.R. *et al*, *J. Chem. Soc., Perkin Trans. 1*, 1979, 1104 (*sulfide, use*)
Bartczak, T. *et al*, *Acta Crystallogr., Sect. C*, 1983, **39**, 219, 222 (*sulfide, cryst struct*)
Beer, P.D. *et al*, *Phosphorus Sulfur*, 1983, **17**, 283 (*synth, pmr*)

2-Chloro-1,3,2-dioxaphospholan-4,5-dicarboxylic acid, 9CI — C-00062

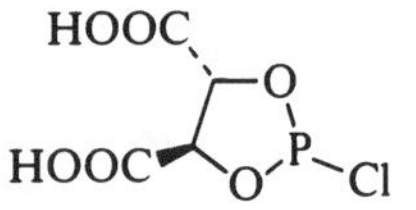

$C_4H_4ClO_6P$ M 214.499

(4RS,5RS)-form

Di-Me ester: [75045-09-9]. *Dimethyl 2-chloro-1,3,2-dioxaphospholan-4,5-dicarboxylate. 2-Chloro-4,5-bis(methoxycarbonyl)-1,3,2-dioxaphospholane.*
$C_6H_8ClO_6P$ M 242.552
Fuming liq. $Bp_{0.04}$ 89-91°. n_D^{20} 1.4825.

Samitov, Yu.Yu. *et al*, *Izv. Akad. Nauk SSSR, Ser. Khim.*, 1975, 1518 (*Engl. transl.* p. 1407) (*synth, pmr, P nmr*)

2-Chloro-1,3,2-dioxaphospholane, 9CI — C-00063

Ethylene phosphorochloridite, 8CI. Ethylene chlorophosphite
[822-39-9]

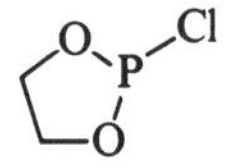

$C_2H_4ClO_2P$ M 126.479

Peptide coupling reagent, used for identification of asparaginyl and glutaminyl residues. Mobile fuming liq. d_0^{20} 1.42. Bp_{47} 66-68°, Bp_{10} 44-45°. n_D^{20} 1.4894. Rapidly hydrolysed.

2-Oxide: see 2-Chloro-1,3,2-dioxaphospholane 2-oxide, C-00064

2-Sulfide: see 2-Chloro-1,3,2-dioxaphospholane 2-sulfide, C-00065

Lucas, H.J. *et al*, *J. Am. Chem. Soc.*, 1950, **72**, 5491 (*synth*)
Cason, J. *et al*, *J. Org. Chem.*, 1959, **24**, 247 (*synth*)
Efremov, Yu.Ya. *et al*, *Khim. Geterosikl. Soedin.*, 1972, 1329 (*Engl. transl.* p. 1202) (*ms*)
Mathis, R. *et al*, *Spectrochim. Acta, Part A*, 1974, **30**, 357 (*ir*)
Shagidullin, R.R. *et al*, *Zh. Obshch. Khim.*, 1975, **45**, 1257 (*Engl. transl.* p. 1235) (*ir, raman*)
Nuretdinov, I.A. *et al*, *Izv. Akad. Nauk SSSR, Ser. Khim.*, 1978, 950 (*Engl. transl.* p. 824) (*nqr*)
Besserre, D. *et al*, *Org. Magn. Reson.*, 1980, **13**, 235, 313 (*cmr, nmr*)
Vasil'ev, V.V. *et al*, *Zh. Obshch. Khim.*, 1981, **51**, 2134 (*Engl. transl.* p. 1836) (*O nmr*)
Fieser, M. *et al*, *Reagents for Organic Synthesis*, Wiley, 1967-84, **1**, 372.

2-Chloro-1,3,2-dioxaphospholane 2-oxide, 9CI — C-00064

Ethylene phosphorochloridate, 8CI. 2-Chloro-1,3-dioxa-2-oxo-2-phosphacyclopentane. Ethylene chlorophosphate
[6609-64-9]

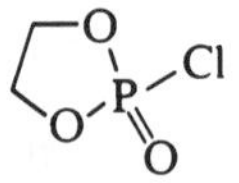

$C_2H_4ClO_3P$ M 142.479

Phosphorylating agent. Mobile fuming liq. $Bp_{0.4}$ 79°. n_D^{25} 1.4459. May polymerise during distn.

Cox, J.R. *et al*, *J. Am. Chem. Soc.*, 1958, **80**, 5441 (*synth, ir*)
Edmundson, R.S., *Chem. Ind. (London)*, 1962, 1828 (*synth*)
Naumov, V.A. *et al*, *Zh. Strukt. Khim.*, 1973, **14**, 787 (*ed*)
Vogt, W. *et al*, *Makromol. Chem.*, 1973, **163**, 89 (*props*)
Shagidullin, R.R. *et al*, *Zh. Obshch. Khim.*, 1976, **46**, 1021 (*Engl. transl.* p. 1017) (*ir, raman*)
Arshinova, R.P. *et al*, *Izv. Akad. Nauk SSSR, Ser. Khim.*, 1978, 609 (*Engl. transl.* p. 524) (*P nmr, conformn*)
Okahata, Y. *et al*, *J. Am. Chem. Soc.*, 1984, **106**, 4696 (*use*)

2-Chloro-1,3,2-dioxaphospholane 2-sulfide, 9CI — C-00065

O,O-Ethylene chlorothiophosphate. O,O-Ethylene phosphorochloridothioate. 2-Chloro-1,3-dioxa-2-thioxo-2-phosphacyclopentane
[32847-69-1]

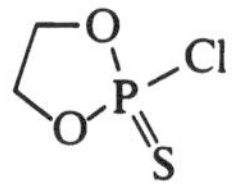

$C_2H_4ClO_2PS$ M 158.539

Solid. Mp 30.5°. $Bp_{3.5}$ 88°.

Yamasaki, T. *et al*, *CA*, 1956, **50**, 314 (*synth*)
Shagidullin, R.R. *et al*, *Zh. Obshch. Khim.*, 1976, **46**, 1021 (*Engl. transl.* p. 1017) (*ir, raman*)
Arshinova, R.P. *et al*, *Izv. Akad. Nauk SSSR, Ser. Khim.*, 1978, 609 (*Engl. transl.* p. 524) (*conformn, P nmr*)

2-Chloro-1,3,2-dioxaphosphorinane, 9CI — C-00066

2-Chloro-1,3,2-dioxaphosphinane. 2-Chloro-1,3-dioxa-2-phosphacyclohexane. Trimethylene chlorophosphite. Trimethylene phosphorochloridite
[6362-89-6]

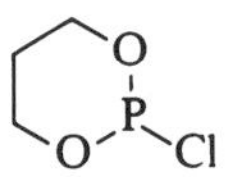

$C_3H_6ClO_2P$ M 140.506

Fuming liq. d_4^{15} 1.35. Bp_{15} 66.5-67.5°. n_D^{25} 1.4884.

▷Reacts violently with H_2O

2-Oxide: see 2-Chloro-1,3,2-dioxaphosphorinane 2-oxide, C-00067

2-Sulfide: see 2-Chloro-1,3,2-dioxaphosphorinane 2-sulfide, C-00068

2-Selenide: [77585-91-2]. *O,O-Trimethylene phosphorochloridoselenoate.*
$C_3H_6ClO_2PSe$ M 219.466
Solid. Mp 58-60°.

Lucas, H.J. *et al*, *J. Am. Chem. Soc.*, 1950, **72**, 5491 (*synth*)
Arbuzov, B.A. *et al*, *Khim. Geterotsikl. Soedin.*, 1971, **7**, 1324 (*Engl. transl.* p. 1237) (*struct*)
Bergesen, K. *et al*, *Acta Chem. Scand.*, 1971, **25**, 2257 (*pmr*)
Naumov, V.A. *et al*, *Zh. Strukt. Khim.*, 1972, **13**, 768 (*Engl. transl.* p. 722) (*ed*)

Nifant'ev, É.E. *et al*, *Dokl. Akad. Nauk SSSR, Ser. Sci. Khim.*, 1973, **208**, 651 (*Engl. transl. Phys. Chem.* p. 100) (*cmr*)
Efremov, Yu.Ya. *et al*, *Khim. Geterotsikl Soedin.*, 1974, 1620 (*Engl. transl.* p. 1424) (*ms*)
Fazliev, D.F. *et al*, *Zh. Obshch. Khim.*, 1976, **46**, 1832 (*Engl. transl.* p. 1776) (*ir*)
Nuretdinov, I.A. *et al*, *Izv. Akad. Nauk SSSR, Ser. Khim.*, 1978, 950 (*Engl. transl.* p. 824) (*nqr*)
Shagidullin, R.R. *et al*, *Izv. Akad. Nauk SSSR, Ser. Khim.*, 1981, 1535 (*Engl. transl.* p. 1234) (*selenide, synth, ir, raman*)
Hacklin, H. *et al*, *Phosphorus Sulfur*, 1985, **25**, 79 (*synth, ms, pmr, P nmr*)

2-Chloro-1,3,2-dioxaphosphorinane 2-oxide, 9CI **C-00067**

2-Chloro-1,3,2-dioxaphosphinane 2-oxide. Trimethylene phosphoryl chloride. Trimethylene phosphorochloridate. 2-Chloro-1,3-dioxa-2-oxo-2-phosphacyclohexane

[872-99-1]

$C_3H_6ClO_3P$ M 156.505

Phosphorylating agent. Fuming liq. Mp 39-42°. $Bp_{0.3}$ 106-109°. Easily hydrolysed. Liable to polymerise on distillation.

Edmundson, R.S., *Chem. Ind. (London)*, 1962, 1828 (*synth*)
Shakirov, I.Kh. *et al*, *Zh. Obshch. Khim.*, 1978, **48**, 508 (*Engl transl.* p. 458) (*ir, raman*)
Ashani, Y. *et al*, *Biochem. J*, 1979, **177**, 781 (*props*)
Arbuzov, B.A. *et al*, *Dokl. Akad. Nauk SSSR, Ser. Sci. Khim.*, 1971, **199**, 1062 (*Engl. transl.* p. 662) (*struct*)
Hunston, R.N. *et al*, *J. Med. Chem.*, 1984, **27**, 440 (*synth, pmr*)

2-Chloro-1,3,2-dioxaphosphorinane 2-sulfide, 9CI **C-00068**

2-Chloro-1,3,2-dioxaphosphinane 2-sulfide. O,O-Trimethylene thiophosphoryl chloride. O,O-Trimethylene phosphorochloridothioate. 2-Chloro-1,3-dioxa-2-thioxo-2-phosphacyclohexane

$C_3H_6ClO_2PS$ M 172.566

Phosphorylating agent. Cryst. (hexane). Mp 39.5-40°.

Zemlyanskii, H.N. *et al*, *Zh. Obshch. Khim.*, 1967, **37**, 1141 (*Engl. transl.* p. 1082) (*synth*)
Predvoditelev, D.A. *et al*, *Zh. Obshch. Khim.*, 1974, **44**, 748; 1976, **46**, 40 (*Engl. transl.* pp. 720, 39) (*synth, pmr, P nmr*)
Arshinova, R.P., *Dokl. Akad. Nauk SSSR, Ser. Sci. Khim.*, 1978, **238**, 858 (*Engl. transl.* p. 47) (*struct*)
Shakirov, I.Kh. *et al*, *Zh. Obshch. Khim.*, 1978, **48**, 508 (*Engl. transl.* p. 458) (*ir, raman*)

2-Chloro-4,5-diphenyl-1,3,2-dioxaphospholane, 8CI **C-00069**

[50597-18-7]

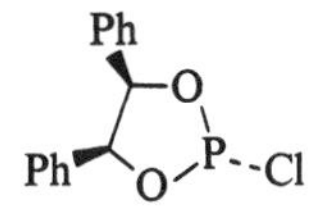

Ph, O, P, Cl, Ph, O (2α,4β,5β)-*form*

$C_{14}H_{12}ClO_2P$ M 278.674

(2α,4β,5β)-form

trans-*form*

2-Oxide: [79356-79-9].
$C_{14}H_{12}ClO_3P$ M 294.674
Phosphorylating agent. Cryst. (C_6H_6/Et_2O). Mp 160-162°.

Ukita, T. *et al*, *Chem. Pharm. Bull.*, 1961, **9**, 363 (*synth, use*)
Cullis, P.M. *et al*, *J. Chem. Soc., Perkin Trans. 1*, 1981, 2317 (*synth, use, P nmr*)

2-Chloro-4,5-diphenyl-1,3,2-dioxaphosphole, 9CI **C-00070**

[62128-53-4]

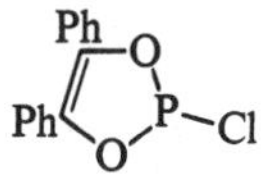

$C_{14}H_{10}ClO_2P$ M 276.659

$Bp_{0.01}$ 130°. n_D^{20} 1.6881.

Mukhametov, F.S. *et al*, *Izv. Akad. Nauk SSSR, Ser. Khim.*, 1976, 2841 (*Engl. transl.* p. 2652) (*synth, ir, nmr*)
Mukhametov, F.S. *et al*, *Zh. Obshch. Khim.*, 1981, **51**, 2674 (*Engl. transl.* p. 2306) (*synth, nmr*)

2-Chloro-1,3,2-dithiaphospholane, 9CI, 8CI **C-00071**

Ethylene phosphorochloridodithioite. Ethylene chlorodithiophosphite

[4669-51-6]

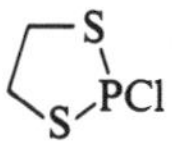

$C_2H_4ClPS_2$ M 158.600

Pungent liq. Easily hydrol. and oxidised. Bp_{10} 100°, $Bp_{0.5}$ 80°. n_D^{20} 1.6870.

2-Oxide: [35437-33-3]. S,S-*Ethylene phosphorochloridodithioate.*
$C_2H_4ClOPS_2$ M 174.600
No phys. props. reported.

2-Sulfide: [34303-19-0]. *Ethylene phosphorochloridotrithioate.*
$C_2H_4ClPS_3$ M 190.660
Cryst. (Et_2O/pet. ether). Mp 45°.

Kovalev, L.S. *et al*, *Zh. Obshch. Khim.*, 1968, **38**, 2277 (*Engl. transl.* p. 2203) (*synth*)
Lee, J.D., *et al*, *Acta Crystallogr., Sect. B*, 1971, **27**, 1055 (*sulfide, cryst struct*)
Ishmaeva, É.A. *et al*, *Zh. Obshch. Khim.*, 1972, **42**, 2642 (*Engl. transl.* p. 2633) (*sulfide, stereochem*)
Peake, S.C. *et al*, *J. Chem. Soc., Perkin Trans. 2*, 1972, 380 (*synth, derivs, P nmr, pmr*)
Schultz, G.Y. *et al*, *Tetrahedron*, 1974, **30**, 2365 (*ed*)
Davidson, G. *et al*, *Spectrochim. Acta, Part A*, 1983, **39**, 419 (*ir, raman*)
Gonbeau, D. *et al*, *J. Mol. Struct.*, 1983, **98**, 109 (*pe, struct, cmr*)

2-[(2-Chloroethyl)amino]tetrahydro-2*H*-1,3,2-oxazaphosphorine 2-oxide C-00072

N-(2-Chloroethyl)tetrahydro-4H-1,3,2-oxazaphosphorin-2-amine 2-oxide, 9CI. *2-[(2-Chloroethyl)amino]-1,3,2-oxazaphosphorinane 2-oxide*

[36761-83-8]

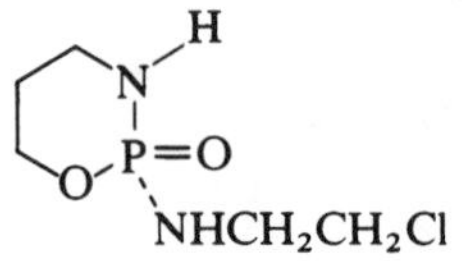

(*R*)-*form*

$C_5H_{12}ClN_2O_2P$ M 198.589

Metab. of Cyclophosphamide, C-00283 . Defleecer for sheep.

(***R***)-***form*** [72578-69-9]

Cryst. (Me_2CO/CCl_4). Mp 109-112°. $[\alpha]_D^{25}$ −15.1° (c, 3 in MeOH).

3-[(S)-1-Phenylethyl]:

$C_{13}H_{20}ClN_2O_2P$ M 302.740

Cryst. (Me_2CO). Mp 114-116°. $[\alpha]_D^{25}$ −72.5° (c, 3.2 in MeOH). Has the 2*S*-config.

(***S***)-***form*** [72578-70-2]

Cryst. Mp 107-110°. $[\alpha]_D^{25}$ +15.2° (c, 3.1 in MeOH).

Takamizawa, A. *et al*, *J. Med. Chem.*, 1974, **17**, 1237 (*synth, ir, pmr*)
Cox, P.J. *et al*, *Cancer. Treat. Rep.*, 1976, **60**, 321.
Pankiewicz, K. *et al*, *J. Am. Chem. Soc.*, 1979, **101**, 7712 (*synth, resoln, ms, P nmr*)
Su, C.N. *et al*, *J. Am. Chem. Soc.*, 1982, **104**, 7343 (*spectra*)

(1-Chloroethyl)phosphonic acid, 9CI C-00073

1-Chloro-1-phosphonoethane

$H_3CCHClP(O)(OH)_2$

$C_2H_6ClO_3P$ M 144.494

(±)-***form***

Di-Et ester: [10419-78-0]. *Diethyl (1-chloroethyl)-phosphonate.*

$C_6H_{14}ClO_3P$ M 200.602

Liq. $Bp_{0.2}$ 62°. n_D^{20} 1.4352. Dehydrochlorinated by Py at r.t.

Bel'skii, V.E. *et al*, *Zh. Obshch. Khim.*, 1972, **42**, 2427 (*Engl. transl.* p. 2421) (*P nmr*)
Coutrot, P. *et al*, *Synthesis*, 1977, 615 (*synth, pmr*)

(2-Chloroethyl)phosphonic acid, 9CI C-00074

Ethrel. Ethephon. Florel

[16672-87-0]

$ClCH_2CH_2P(O)(OH)_2$

$C_2H_6ClO_3P$ M 144.494

Plant hormone; used to induce ripening and abscission of fruit, liberating C_2H_4 on hydrol. Cotton defoliant. Needles (C_6H_6). Stable at pH <3. Sol. H_2O, EtOH, spar. sol. C_6H_6. Mp 74-75°. V. hygroscopic.

▷SZ7100000.

Dichloride: see (2-Chloroethyl)phosphonic dichloride, C-00076

Di-Me ester: [26119-41-5]. *Dimethyl (2-chloroethyl)-phosphonate.*

$C_6H_{10}ClO_3P$ M 196.570

Liq. d_4^{20} 1.27. Bp_1 65-67°. n_D^{20} 1.4490.

Di-Et ester: [10419-79-1]. *Diethyl (2-chloroethyl)-phosphonate.*

$C_6H_{14}ClO_3P$ M 200.602

Liq. d_4^{20} 1.16. $Bp_{2.5}$ 92-93°. n_D^{20} 1.4390.

▷SZ7110000.

Bis(2-chloroethyl) ester: [6294-34-4]. *Bis(2-chloroethyl) (2-chloroethyl)phosphonate.*

$C_6H_{12}Cl_3O_3P$ M 269.492

Solid. d_4^{26} 1.39. Mp 37°. Bp_5 170-172°. n_D^{26} 1.4828.

Bis(trimethylsilyl) ester: [67344-36-9]. *Bis(trimethylsilyl) (2-chloroethyl)phosphonate.*

$C_8H_{22}ClO_3PSi_2$ M 288.858

Liq. $Bp_{0.01}$ 61-63°.

Di-Ph ester: [53986-90-6]. *Diphenyl (2-chloroethyl)-phosphonate.*

$C_{14}H_{14}ClO_3P$ M 296.690

Liq. $Bp_{1-1.5}$ 176-178.5°, $Bp_{0.01}$ 146°. n_D^{20} 1.5574.

Diamide: see P-2-Chloroethylphosphonic diamide, C-00075

Kabachnik, M.I. *et al*, *Izv. Akad. Nauk SSSR, Ser. Khim.*, 1946, 403; *CA*, **42**, 7242 (*synth*)
Kabachnik, M.I. *et al*, *Izv. Akad. Nauk SSSR, Ser. Khim.*, 1947, 97; *CA*, **42**, 4132 (*esters, synth*)
Blizniyuk, N.K. *et al*, *Zh. Obshch. Khim.*, 1967, **37**, 1119 (*Engl. transl.* p. 1061) (*ester*)
Gloede, J. *et al*, *J. Prakt. Chem.*, 1976, **318**, 1043 (*esters, synth, pmr*)
Gross, H. *et al*, *J. Prakt. Chem.*, 1978, **320**, 344 (*synth, silyl ester, pmr*)
Pesticide Manual, 6th Ed., 241; 7th Ed., 241.

***P*-2-Chloroethylphosphonic diamide, 9CI** C-00075

$ClCH_2CH_2P(O)(NH_2)_2$

$C_2H_8ClN_2OP$ M 142.525

N,N,N′,N′-*Tetra-Me:* [14518-01-5]. P-*(2-Chloroethyl)-*N,N,N′,N′*-tetramethylphosphonic diamide. (2-Chloroethyl)phosphonic bis(dimethylamide).*

$C_6H_{16}ClN_2OP$ M 198.632

Liq. d_4^{20} 1.13. Bp_2 102-103°. n_D^{25} 1.4771.

N,N,N′,N′-*Tetra-Et:* [14605-34-6]. P-*(2-Chloroethyl)-*N,N,N′,N′*-tetraethylphosphonic diamide. (2-Chloroethyl)phosphonic bis(diethylamide).*

$C_{10}H_{24}ClN_2OP$ M 254.739

Liq. d_4^{20} 1.07. Bp_3 128-130°. n_D^{20} 1.4763.

Kabachnik, M.I. *et al*, *Izv. Akad. Nauk SSSR, Ser. Khim.*, 1966, 1365 (*Engl. transl.* p. 1312) (*derivs, synth*)

(2-Chloroethyl)phosphonic dichloride, 9CI C-00076

[690-12-0]

$ClCH_2CH_2P(O)Cl_2$

$C_2H_4Cl_3OP$ M 181.386

Synthetic intermed. Pungent liq. Mp <80°. Bp 213-217°, Bp_6 77°.

Kinnear, A.M. *et al*, *J. Chem. Soc.*, 1952, 3437 (*synth*)
Maynard, J.A. *et al*, *Aust. J. Chem.*, 1963, **16**, 596 (*synth*)
Maier, L., *Helv. Chim. Acta*, 1969, **52**, 1337 (*synth, P nmr, pmr*)
Steger, E. *et al*, *Spectrochim. Acta, Part A*, 1967, **23**, 2189 (*ir, raman*)
Maier, L, *Phosphorus*, 1971, **1**, 105 (*synth, pmr, P nmr*)

(2-Chloroethyl)phosphonochloridic acid, 9CI C-00077

[53711-17-4]

$ClCH_2CH_2PCl(O)OH$

$C_2H_5Cl_2O_2P$ M 162.940

Et ester: [38139-02-5]. *Ethyl (2-chloroethyl)-phosphonochloridate.*

$C_4H_9Cl_2O_2P$ M 190.994

Liq. Bp_2 80-81°.

Ph ester: Phenyl (2-chloroethyl)phosphonochloridate.

$C_8H_9Cl_2O_2P$ M 239.038

Liq. d_4^{20} 1.35. Bp_1 130-132°. n_D^{20} 1.5329.

Gefter, E.L., *Zh. Obshch. Khim.*, 1961, **31**, 3316 (*Engl. transl.* p. 3093) (*synth*)
Knunyants, I.L. *et al*, *Zh. Obshch. Khim.*, 1972, **42**, 2421 (*Engl. transl.* p. 2415) (*synth*)

(2-Chloroethyl)phosphonothioic acid, 9CI C-00078

$ClCH_2CH_2P(S)(OH)_2$

$C_2H_6ClO_2PS$ M 160.555

Difluoride:

$C_2H_4F_2ClPS$ M 164.537

Liq. d_4^{20} 1.43. Bp 127-130°. n_D^{20} 1.4565.

Dichloride: [20428-20-0].

$C_2H_4Cl_3PS$ M 197.446

Liq. with sharp unpleasant odour. d_4^{20} 1.51. Bp_{10} 92°. n_D^{20} 1.5670.

Diamide:

$C_2H_8ClN_2PS$ M 158.585

Solid. Mp 61-62°.

Ivanova, Zh.M. *et al*, *Zh. Obshch. Khim.*, 1968, **38**, 1334 (*Engl. transl.* p. 1284) (*difluoride*)
Bothner-By, A.A. *et al*, *J. Phys. Chem.*, 1969, **73**, 1830 (*dichloride, pmr, nmr*)
Maier, L., *Helv. Chim. Acta*, 1969, **52**, 1337 (*dichloride, pmr, P nmr*)
Levin, Ya.A. *et al*, *Zh. Obshch. Khim.*, 1973, **43**, 281 (*Engl. transl.* p. 280) (*dichloride*)
Fedorova, G.K. *et al*, *Zh. Obshch. Khim.*, 1978, **48**, 2015 (*Engl. transl.* p. 1833) (*diamide*)

1-Chloroethyl phosphorodichloridate, 9CI C-00079

1-Chloroethyl dichlorophosphate. 1-Chloroethyl phosphoryl dichloride

[41998-90-7]

$H_3CCHClOP(O)Cl_2$

$C_2H_4Cl_3O_2P$ M 197.385

(±)-*form*

Pungent liq. d_4^{20} 1.48. Bp_{12} 53°. n_D^{20} 1.4570. Easily hydrol.

Maier, L., *Helv. Chim. Acta*, 1973, **56**, 1257 (*pmr, P nmr*)
Moskva, V.V. *et al*, *Zh. Obshch. Khim.*, 1973, **43**, 677 (*Engl. transl.* p. 672) (*synth*)

O-(2-Chloroethyl) phosphorodichloridothioate, 9CI, 8CI C-00080

O-(*2-Chloroethyl) thiophosphoryl dichloride.* O-(*2-Chloroethyl) dichlorothiophosphate*

[52041-10-8]

$ClCH_2CH_2OP(S)Cl_2$

$C_2H_4Cl_3OPS$ M 213.446

Oil. d_4^{20} 1.47. Bp_{14} 104-108°. n_D^{20} 1.5362.

Galashina, M.L. *et al*, *Zh. Obshch. Khim.*, 1953, **23**, 433 (*Engl. transl.* p. 441) (*synth*)
Gay, D.C. *et al*, *J. Chem. Soc. (B)*, 1970, 1123 (*synth*)

S-(2-Chloroethyl) phosphorodichloridothioate, 9CI, 8CI C-00081

S-(*2-Chloroethyl) dichlorothiophosphate*

[22077-83-4]

$ClCH_2CH_2SP(O)Cl_2$

$C_2H_4Cl_3OPS$ M 213.446

Liq. d_4^{20} 1.55. $Bp_{0.02}$ 91-92°. n_D^{20} 1.5405.

Martynov, I.V. *et al*, *Zh. Obshch. Khim.*, 1969, **39**, 996 (*Engl. transl.* p. 966)

1-(2-Chloroethyl)tetrahydro-1*H*,5*H*-[1.3.2]diazaphospholo[2,1-*b*][1.3.2]-oxazaphosphorine 9-oxide C-00082

[64724-10-3]

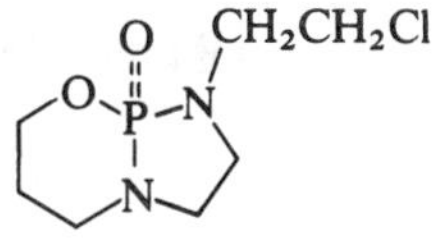

(*R*)-*form*

$C_7H_{14}ClN_2O_2P$ M 224.627

Dec. product of cyclophosphamide.

(*R*)-*form*

Solid. Mp 75-79°. $[\alpha]_D^{25}$ −10.2° (c, 2.5 in MeOH).

(*S*)-*form*

Solid. Mp 75-79°. $[\alpha]_D^{25}$ +10.8° (c, 2.0 in MeOH).

(±)-*form* [72598-28-8]

Waxy solid. Mp 75-79°, Mp 142-143°. Dec. slowly at 10°.

Zon, G. *et al*, *J. Am. Chem. Soc.*, 1977, **99**, 5785 (*synth, pmr, cmr*)
Pankiewicz, K. *et al*, *J. Am. Chem. Soc.*, 1979, **101**, 7712 (*synth, tlc, ms, P nmr*)

(2-Chloroethyl)triphenylphosphonium(1+), 9CI, 8CI C-00083

$Ph_3P^{\oplus}CH_2CH_2Cl$

$C_{20}H_{19}ClP^{\oplus}$ M 325.797 (ion)

Intermed. in prepn. of ethenyltriphenylphosphonium salts.

Bromide: [31238-20-7].

$C_{20}H_{19}BrClP$ M 405.701

Prisms (2-propanol/Et_2O). Mp 161-164° dec.

Swan, J.M. *et al*, *Aust. J. Chem.*, 1971, **24**, 777 (*synth, props*)

2-[(Chloroformyl)oxy]-ethyltriphenylphosphonium(1+) C-00084

$Ph_3P^{\oplus}CH_2CH_2OOCCl$

$C_{21}H_{19}ClO_2P^{\oplus}$ M 369.807 (ion)

Chloride: [61083-59-8].

$C_{21}H_{19}Cl_2O_2P$ M 405.260

Exceptionally acid-stable amino acid protecting group which also increases the amino acid's solubility in water. Cryst. ($CHCl_3/Et_2O$). Mp 136°.

Kunz, H., *Chem. Ber.*, 1976, **109**, 2670 (*synth, ir, use*)
Kunz, H., *Justus Liebigs Ann. Chem.*, 1976, 1674 (*use*)
Kunz, H., *Angew. Chem., Int. Ed. Engl.*, 1978, **17**, 67 (*use*)
Kunz, H., *Justus Liebigs Ann. Chem.*, 1982, 2068 (*props*)

2-Chlorohexahydro-4*H*-1,3,2-benzodioxaphosphorin C-00085

2-Chloro-5,6-tetramethylene-1,3,2 dioxaphosphorinane. 2-Chloro-1,3-dioxa-2-phosphadecalin

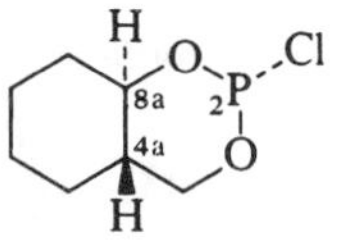

$C_7H_{12}ClO_2P$ M 194.597

(***2RS,4aSR,8aSR*)-form** [93381-80-7]

(2α,4aβ,8aα)-*form*

Oil. $Bp_{2.2}$ 102°.

2-Oxide: [74410-72-3].

$C_7H_{12}ClO_3P$ M 210.597

Cryst. (Et_2O/hexane at −78°). Mp 75-76°.

Gorenstein, D.G. *et al, J. Am. Chem. Soc.*, 1980, **102**, 5077 (*synth, ir, pmr*)
Corriu, R.J.P. *et al, J. Am. Chem. Soc.*, 1984, **106**, 1060 (*oxide, synth, P nmr*)
Taira, K. *et al, J. Am. Chem. Soc.*, 1984, **106**, 7831 (*synth, P nmr*)

2-Chloro-2,2,4,4,6,6-hexahydro-2-methyl-4,4,6,6-tetraphenyl-1,3,5-triaza-2,4,6-triphosphorine, 9CI C-00086

2-Chloro-2,2,4,4,6,6-hexahydro-2-methyl-4,4,6,6-tetraphenylcyclotriphosphazene

[50457-20-0]

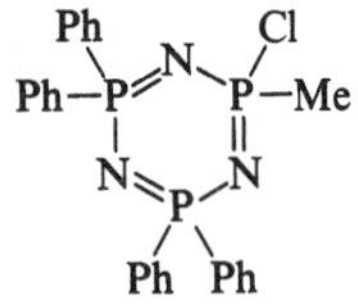

$C_{25}H_{23}ClN_3P_3$ M 493.851

Cryst. (CH_2Cl_2/Et_2O). Mp 220-222°.

Schmidpeter, A. *et al, Chem. Ber.*, 1975, **108**, 1454 (*synth, pmr, P nmr*)

2-Chloro-2,2,4,4,6,6-hexahydro-2,4,4,6,6-pentaphenoxy-1,3,5-triaza-2,4,6-triphosphorine, 9CI C-00087

2-Chloro-2,2,4,4,6,6-hexahydro-2,4,4,6,6-pentaphenoxycyclotriphosphazene

[5032-39-3]

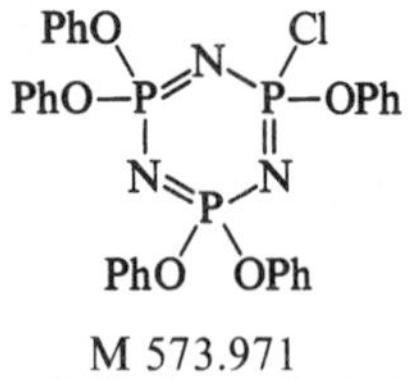

$C_{30}H_{25}ClN_3O_5P$ M 573.971

Cryst. (heptane or C_6H_6/pet. ether). Mp 69°.

Telkova, I.B. *et al, Zh. Obshch. Khim.*, 1973, **43**, 1257 (*Engl. transl.* p. 1247) (*synth, uv, P nmr*)
Allcock, H.R. *et al, Inorg. Chem.*, 1980, **19**, 1026 (*synth, P nmr*)
Sulkowski, W. *et al, Zh. Obshch. Khim.*, 1981, **51**, 1221 (*Engl. transl.* p. 1033) (*synth, ms, P nmr*)

2-Chloro-2,2,4,4,6,6-hexahydro-2,4,4,6,6-pentaphenyl-1,3,5-triaza-2,4,6-triphosphorine, 9CI C-00088

2-Chloro-2,2,4,4,6,6-hexahydro-2,4,4,6,6-pentaphenylcyclotriphosphazene

[3606-84-6]

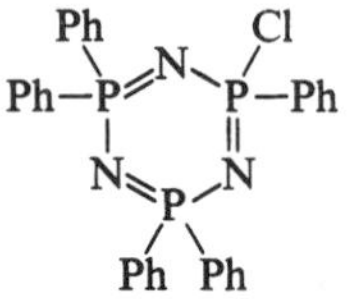

$C_{30}H_{25}ClN_3P_3$ M 555.922

Needles (MeCN). Sol. C_6H_6, MeCN, CH_2Cl_2, insol. Et_2O, pet. ether. Mp 150-151°.

Schmulbach, C.D. *et al, Inorg. Chem.*, 1966, **5**, 1621 (*synth, props*)
Shaw, R.A. *et al, Phosphorus Sulfur*, 1981, **10**, 121 (*P nmr*)

2-Chlorohexahydropyrano[3,2-*d*]-1,3,2-dioxaphosphorin, 9CI C-00089

3-Chloro-2,4,7-trioxa-3-phosphabicyclo[4.4.0]decane

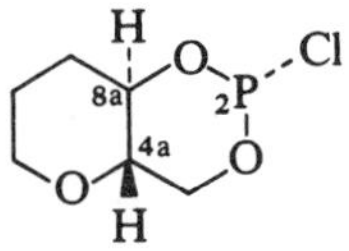

(2*RS*,4a*SR*,8a*RS*)-*form*

$C_6H_{10}ClO_3P$ M 196.570

(***2RS,4aSR,8aRS*)-form**

(2α,4aβ,8aα)-*form*

2-Sulfide: [61768-32-9].

$C_6H_{10}ClO_3PS$ M 228.630

Cryst. Mp 61-63°.

(***2RS,4aRS,8aSR*)-form**

(2α,4aα,8aβ)-*form*

2-Sulfide: [61826-18-4]. Liq.

Bouchu, D. *et al, Phosphorus Sulfur*, 1982, **13**, 25 (*synth, ir*)
Bouchu, D. *et al, Phosphorus Sulfur*, 1983, **15**, 33; 1983, **17**, 173 (*pmr, ms*)

2-Chloro-4-isopropyl-5,5-dimethyl-1,3,2-dioxaphosphorinane, 9CI C-00090

2-Chloro-5,5-dimethyl-4-(1-methylethyl)-1,3,2-dioxaphosphorinane. 2-Chloro-4-isopropyl-5,5-dimethyl-1,3-dioxa-2-phosphacyclohexane

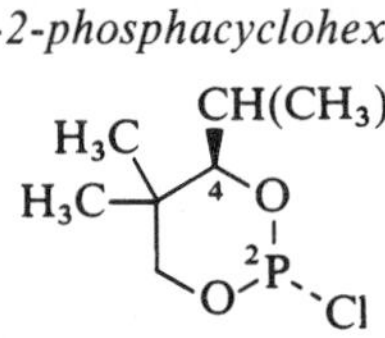

(2*RS*,4*RS*)-*form*

$C_8H_{16}ClO_2P$ M 210.640

(***2RS,4RS*)-form** [95115-80-3]

(±)-trans-*form*

Bp_1 71-74°. n_D^{23} 1.4772.

2-Oxide: [95115-81-4]. Cryst. (C_6H_6). Mp 120-121.5°.

2-Sulfide: Cryst. (CCl_4). Mp 75-76° (67-70°).

2-Selenide: Cryst. (pet. ether). Mp 80°.

(***2RS,4SR*)-form**

(±)-cis-*form*

2-Oxide: [95115-86-9].

$C_8H_{16}ClO_3P$ M 226.639

Cryst. (CCl_4/pet. ether). Mp 55-56°.

2-Sulfide:

$C_8H_{16}ClO_2PS$ M 242.700

Characterised spectroscopically.

2-Selenide:

$C_8H_{16}ClO_2PSe$ M 289.600

Characterised spectroscopically.

Majoral, J.-P. *et al, Bull. Soc. Chim. Fr.*, 1971, 95 (*derivs, ir, pmr, P nmr*)
Majoral, J.-P. *et al, Spectrochim. Acta, Part A*, 1972, **28**, 2247 (*sulfide, ir*)
Edmundson, R.S. *et al, J. Chem. Soc., Perkin Trans. 2*, 1985, 69 (*synth, deriv, ir, pmr, P nmr, cryst struct*)

P-Chloromethyl-*N*-(dichlorophosphinyl)-phosphonochloridimidic acid, 9CI C-00091

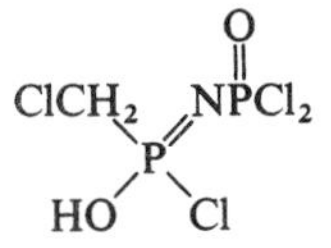

$CH_3Cl_4NO_2P_2$ M 264.800

(±)-*form*

Ph ester: Phenyl P-*chloromethyl*-N-*(dichlorophosphinyl)phosphonochloridimidate.*

$C_7H_7Cl_4NO_2P_2$ M 340.897

Solid. Mp 52-54°. $Bp_{0.02}$ 166-167°.

Gordeev, A.D. *et al, Zh. Obshch. Khim.*, 1973, **43**, 9 (*Engl. transl.* p. 7) (*nqr, struct*)
Shokol, V.A. *et al, Zh. Obshch. Khim.*, 1973, **43**, 267, 747 (*Engl. transl.* pp. 266, 745) (*ir, pmr, P nmr, props*)

2-Chloro-4-methyl-1,3,2-dioxaphospholane, 9CI C-00092

Propylene chlorophosphite. Propylene phosphorochloridite. 2-Chloro-4-methyl-1,3-dioxa-2-phosphacyclopentane

[6362-86-3]

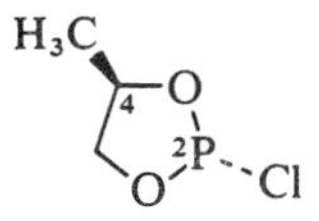

(*2RS,4RS*)-*form*

$C_3H_6ClO_2P$ M 140.506

Fuming liq. d_4^{25} 1.29. Bp_{50} 75-76°, Bp_{12} 43-44°. n_D^{20} 1.4707. Reacts violently with H_2O. Mixt. of stereoisomers.

2-Oxide: Propylene phosphorochloridate. Propylene chlorophosphate. Propylene phosphoryl chloride.

$C_3H_6ClO_3P$ M 156.505

Phosphorylating agent. Fuming liq. $Bp_{0.4}$ 74°. n_D^{25} 1.4389.

2-Sulfide: Propylene phosphorochloridothioate. Propylene thiophosphoryl chloride. Propylene chlorothiophosphate.

$C_3H_6ClO_2PS$ M 172.566

Oil. n_D^{20} 1.4770.

(*2RS,4RS*)-*form*

(±)-trans-*form*

Fuming liq. $Bp_{0.1}$ 32°.

2-Oxide:

$C_3H_6ClO_3P$ M 156.505

Fuming liq. $Bp_{0.05}$ 88°. n_D^{21} 1.4391. Readily hydrolysed.

Lucas, H.J. *et al, J. Am. Chem. Soc.*, 1950, **72**, 5491 (*synth*)
Edmundson, R.S., *Chem. Ind. (London)*, 1962, 1828 (*oxide*)
Edmundson, R.S. *et al, J. Chem. Soc. (C)*, 1966, 1997 (*sulfide*)
Bergesen, K. *et al, Acta Chem. Scand.*, 1972, **26**, 2153 (*synth*)
Nguyen Hoang Phuong, *et al, Bull. Soc. Chim. Fr., Part II*, 1975, 2326 (*oxide*)
Shagidullin, R.R. *et al, Zh. Obshch. Khim.*, 1975, **45**, 1257 (*Engl. transl.* p. 1235) (*ir, raman*)
Arshinova, R.P. *et al, Izv. Akad. Nauk SSSR, Ser. Khim.*, 1978, 609 (*Engl. transl.* p. 524) (*derivs, conformn*)
Nuretdinov, I.A. *et al, Izv. Akad. Nauk SSSR, Ser. Khim.*, 1978, 950 (*Engl. transl.* p. 824) (*nqr*)
Nuretdinov, I.A. *et al, Zh. Fiz. Khim.*, 1979, **53**, 126 (*Engl. trans.* p. 66) (*nqr*)

2-Chloro-4-methyl-1,3,2-dioxaphosphorinane, 9CI C-00093

2-Chloro-4-methyl-1,3,2-dioxaphosphinane. 1,3-Butylene chlorophosphite. 2-Chloro-4-methyl-1,3-dioxa-2-phosphacyclohexane

[6362-87-4]

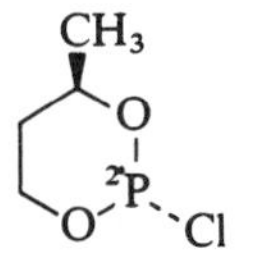

(*2RS,4RS*)-*form*

$C_4H_8ClO_2P$ M 154.533

2-Oxide: see 2-Chloro-4-methyl-1,3,2-dioxaphosphorinane 2-oxide, C-00094
2-Sulfide: see 2-Chloro-4-methyl-1,3,2-dioxaphosphorinane 2-sulfide, C-00095

(*2RS,4RS*)-*form*

(±)-trans-*form*

Fuming liq. Bp_{19-22} 78-84° (diastereomerically pure). n_D^{25} 1.4792.

2-Selenide: O,O-*Butylene phosphorochloridoselenoate.*

$C_4H_8ClO_2PSe$ M 233.493

Cryst. (EtOAc/hexane). Mp 50-51°. $Bp_{0.01}$ 90-92°. n_D^{19} 1.5605.

Zwierzak, A., *Can. J. Chem.*, 1967, **45**, 2501 (*synth*)
Bodkin, C.L. *et al, J. Chem. Soc. (B)*, 1971, 1137 (*pmr*)
Osokin, D.Ya. *et al, Izv. Akad. Nauk SSSR, Ser. Khim.*, 1972, 1513 (*Engl. transl.* p. 1460) (*nqr*)
Efremov, Yu.Ya. *et al, Khim. Geterosikl. Soedin.*, 1974, 1620 (*Engl. transl.* p. 1424) (*ms*)
Okruszek, A. *et al, Z. Naturforsch., B*, 1975, **30**, 430 (*selenide, synth, pmr, cmr, P nmr, ms*)
Fazliev, D.F. *et al, Zh. Obshch. Khim.*, 1976, **46**, 1832 (*Engl. transl.* p. 1776) (*ir, raman*)
Arshinova, R.P. *et al, Zh. Obshch. Khim.*, 1981, **51**, 1757 (*Engl. transl.* p. 1503) (*pe*)
Vasil'ev, V.V. *et al, Zh. Obshch. Khim.*, 1981, **51**, 2134 (*Engl. transl.* p. 1836) (*O nmr*)
Edmundson, R.S. *et al, J. Chem. Soc., Perkin Trans. 1*, 1984, 1943 (*synth, ir, P nmr*)
Samitov, Yu.Yu. *et al, Zh. Obshch. Khim.*, 1984, **54**, 805 (*Engl. transl.* p. 714) (*cmr*)

2-Chloro-4-methyl-1,3,2-dioxaphosphorinane 2-oxide, 9CI C-00094

1,3-Butylene chlorophosphate. 1,3-Butylene phosphoryl chloride. 2-Chloro-4-methyl-1,3,2-dioxaphosphinane 2-oxide

[31951-90-3]

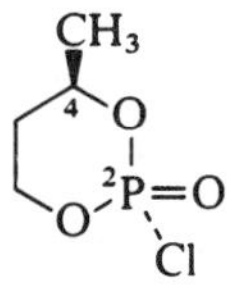

(*2RS,4RS*)-*form*

$C_4H_8ClO_3P$ M 170.532

(*2RS,4RS*)-*form*

(±)-trans-*form*

Fuming liq. d_4^{20} 1.38. $Bp_{0.1}$ 101-102°. n_D^{19} 1.4649. May dec. on distn.

(2RS,4SR)-form
(±)-cis-*form*
Fuming liq. n_D^{20} 1.4591 (94% diastereomeric purity).

Nifant'ev, É.E. *et al*, *Dokl. Akad. Nauk SSSR, Ser. Sci. Khim.*, 1972, **203**, 841 (*Engl. transl.* p. 304) (*synth, P nmr*)
Bodkin, C.L. *et al*, *J. Chem. Soc., Perkin Trans. 2*, 1973, 676 (*synth*)
Stec, W. *et al*, *Tetrahedron*, 1973, **29**, 539 (*synth, ir, pmr, P nmr*)
Nifant'ev, É.E. *et al*, *Zh. Obshch. Khim.*, 1981, **51**, 2428 (*Engl. transl.* p. 2092) (*synth, ir, raman*)

2-Chloro-4-methyl-1,3,2-dioxaphosphorinane 2-sulfide, 9CI C-00095
O,O-*1,3-Butylene chlorothiophosphate. 2-Chloro-4-methyl-1,3,2-dioxaphosphinane 2-sulfide.* O,O-*1,3-Butylene thiophosphoryl chloride*
[17377-20-7]

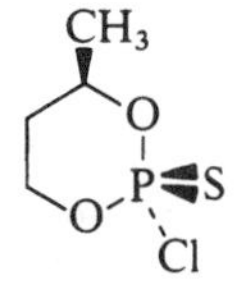

$C_4H_8ClO_2PS$ M 186.593
Liq. d_4^{20} 1.36. Bp_2 110-112°. n_D^{20} 1.5255. Mixt. of diastereoisomers.

(2RS,4RS)-form [86569-32-6]
(±)-trans-*form*
Liq. $Bp_{0.05}$ 78-80°. n_D^{20} 1.5220.

Zemlyanskii, H.N. *et al*, *Zh. Obshch. Khim.*, 1967, **37**, 1141 (*Engl. transl.* p. 1082) (*synth*)
Stec, W.J. *et al*, *Phosphorus*, 1973, **2**, 235 (*synth*)
Mikolajczyk, M. *et al*, *Phosphorus*, 1974, **5**, 67 (*synth, pmr, P nmr*)
Okruszek, A. *et al*, *Z. Naturforsch., B*, 1975, **30**, 430 (*synth, cmr, P nmr*)

5-Chloromethyl-2-fluoro-5-methyl-1,3,2-dioxaphosphorinane, 9CI C-00096
5-Chloromethyl-2-fluoro-5-methyl-1,3-dioxa-2-phosphacyclohexane. 5-Chloromethyl-2-fluoro-5-methyl-1,3,2-dioxaphosphinane

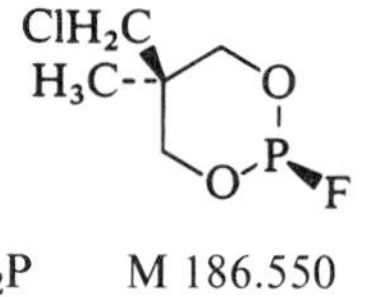

cis-form

$C_5H_9ClFO_2P$ M 186.550

cis-form
2-Oxide:
$C_5H_9ClFO_3P$ M 202.549
Cryst. (CCl_4). Mp 85-86°. Also referred to as *trans*-form.
2-Sulfide:
$C_5H_9ClFO_2PS$ M 218.610
Not sepd. from *trans*-form; also referred to as *trans*-form.

trans-form
2-Oxide: Not sepd. from *cis*-form; also referred to as *cis*-form.
2-Sulfide: Cryst. (hexane). Mp 62-63°. Also referred to as *cis*-form.

Corriu, R.J.P. *et al*, *Tetrahedron*, 1979, **35**, 2889; 1981, **37**, 3681 (*derivs, synth, P nmr, props*)
Corriu, R.J.P. *et al*, *J. Am. Chem. Soc.*, 1984, **106**, 1060 (*props*)

5-Chloromethyl-2-hydroxy-5-methyl-1,3,2-dioxaphosphorinane 2-oxide, 9CI C-00097
5-Chloromethyl-2-hydroxy-5-methyl-1,3,2-dioxaphosphinane. 5-Chloromethyl-2-hydroxy-5-methyl-1,3-dioxa-2-phosphacyclohexane

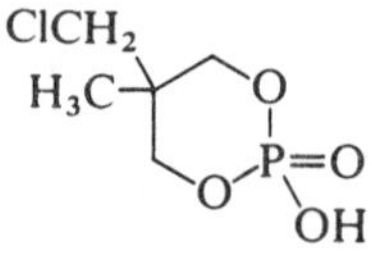

$C_5H_{10}ClO_4P$ M 200.558
Cryst. (MeCN). Mp 144-146°.
Me ester: see 5-Chloromethyl-2-methoxy-5-methyl-1,3,2-dioxaphosphorinane, C-00098
Ph ester: see 5-Chloromethyl-5-methyl-2-phenoxy-1,3,2-dioxaphosphorinane, C-00099
Fluoride: see 5-Chloromethyl-2-fluoro-5-methyl-1,3,2-dioxaphosphorinane, C-00096
Chloride: see 2-Chloro-5-chloromethyl-5-methyl-1,3,2-dioxaphosphorinane, C-00031

Wadsworth, W.S., *J. Chem. Soc., Perkin Trans. 2*, 1972, 1686 (*synth, pmr*)
Wadsworth, W.S. *et al*, *J. Org. Chem.*, 1973, **38**, 256 (*synth*)

5-Chloromethyl-2-methoxy-5-methyl-1,3,2-dioxaphosphorinane, 9CI C-00098
5-Chloromethyl-2-methoxy-5-methyl-1,3,2-dioxaphosphinane. 5-Chloromethyl-2-methoxy-5-methyl-1,3-dioxa-2-phosphacyclohexane

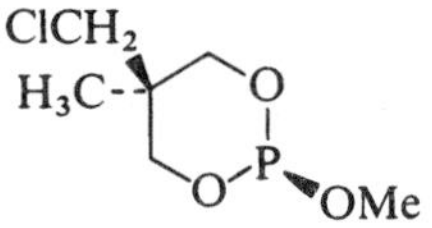

cis-form

$C_6H_{12}ClO_3P$ M 198.586

cis-form
2-Oxide:
$C_6H_{12}ClO_4P$ M 214.585
Phys. props. not reported. Also known as *trans*-form.

trans-form
2-Oxide: Cryst. (toluene). Mp 112-113°. Also known as *cis*-form.

Wadsworth, W.S. *et al*, *J. Org. Chem.*, 1973, **38**, 256 (*oxide, pmr*)
Gehrke, S.H. *et al*, *J. Org. Chem.*, 1980, **45**, 3921 (*oxide*)
Wadsworth, W.S. *et al*, *J. Am. Chem. Soc.*, 1983, **105**, 1631 (*oxide, synth, pmr*)
Corriu, R.J.P. *et al*, *J. Am. Chem. Soc.*, 1984, **106**, 1060 (*derivs, synth, pmr, P nmr*)

5-Chloromethyl-5-methyl-2-phenoxy-1,3,2-dioxaphosphorinane, 9CI C-00099
5-Chloromethyl-5-methyl-2-phenoxy-1,3,2-dioxaphosphinane. 5-Chloromethyl-5-methyl-2-phenoxy-1,3-dioxa-2-phosphacyclohexane

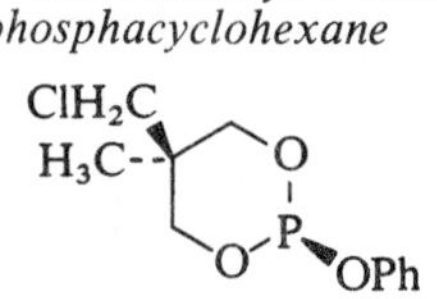

cis-form

$C_{11}H_{14}ClO_3P$ M 260.657
2-Oxide:
$C_{11}H_{14}ClO_4P$ M 276.656
No phys. props. reported. Both stereoisomers known; each is referred to as both *cis*- and *trans*-form.
2-Sulfide: Cryst. (CCl_4/hexane). Mp 110-111°. The cryst. isomer has the config. with =S *cis* to CH_2Cl. It is referred to as both *cis*- and *trans*-form. The other isomer is also known but no Mp is recorded.

Wagner, P. *et al*, *Cryst. Struct. Commun.*, 1973, **2**, 507 (*oxide, cryst struct*)
Wadsworth, W.S. *et al*, *J. Org. Chem.*, 1974, **39**, 984 (*sulfide, synth, pmr*)
Corriu, R.J.P. *et al*, *Tetrahedron*, 1979, **35**, 2889 (*derivs, synth, P nmr*)
Corriu, R.J.P. *et al*, *J. Am. Chem. Soc.*, 1984, **106**, 1060 (*derivs, synth*)

(Chloromethyl)methylphosphinic acid, 9CI — C-00100

[40207-47-4]

$ClCH_2PMe(O)OH$

$C_2H_6ClO_2P$ M 128.495

Me ester: [73013-46-4]. *Methyl (chloromethyl)-methylphosphinate.*
$C_3H_8ClO_2P$ M 142.522
Liq. Bp_{11} 87-88°.
Chloride: [26350-26-5].
$C_2H_5Cl_2OP$ M 146.941
Liq. $Bp_{0.8}$ 70°.
Anhydride:
$C_4H_{10}Cl_2O_3P_2$ M 238.975
Solid. Mp 63°. Bp_1 180°.

Moedritzer, K., *J. Am. Chem. Soc.*, 1961, **83**, 4381 (*synth, anhydride, ir, P nmr*)
Maier, L., *Z. Anorg. Allg. Chem.*, 1972, **394**, 117 (*chloride, P nmr, pmr*)
Finke, M. *et al*, *Justus Liebigs Ann. Chem.*, 1974, 741 (*chloride*)
Raevskii, O.A. *et al*, *Izv. Akad. Nauk SSSR, Ser. Khim.*, 1979, 2251 (*Engl. transl.* p. 2673) (*ester, ir, raman, struct*)

P-Chloromethyl-*N*-(1-methylpropyl)-phosphonamidothioic acid — C-00101

N-sec-*Butyl*-P-*chloromethylphosphoramidothioic acid*

$C_5H_{13}ClNOPS$ M 201.651

O-(*2-Chloro-4-methylphenyl*) *ester:* [42585-08-0]. P-*Chloromethyl*-O-(*2-chloro-4-methylphenyl*)-N-(*1-methylpropyl*) *phosphonamidothioate. Isophos-3.*
$C_{12}H_{18}Cl_2NOPS$ M 326.220
Herbicide. Liq. d_4^{25} 1.23. $Bp_{0.1}$ 147-148°. Presumably a mixt. of diastereoisomers. n_D^{25} 1.5564.

▷Toxic. Causes dystrophic changes in myocyardial cells

Grapov, A.F. *et al*, *Zh. Obshch. Khim.*, 1975, **45**, 280 (*Engl. transl.* p. 266) (*synth*)
Kozlova, T.F. *et al*, *Zh. Obshch. Khim.*, 1975, **45**, 744 (*Engl. transl.* p. 734) (*synth, ir*)
Kruglev, Yu.V. *et al*, *CA*, 1980, **93**, 108794 (*metab*)
Shitskova, A.P. *et al*, *CA*, 1983, **99**, 170987 (*tox*)

2-Chloro-3-methyl-1,3,2-oxazaphospholidine, 9CI — C-00102

2-Chloro-3-methyl-1,3,2-oxazaphospholane
[22082-71-9]

C_3H_7ClNOP M 139.521
Fuming liq. d_4^{20} 1.25. Bp_2 57-58°. n_D^{20} 1.5068. Easily hydrolysed.

Martnynov, I.V. *et al*, *Zh. Obshch. Khim.*, 1968, **38**, 2343 (*Engl. transl.* p. 2272) (*synth*)
Shagidullin, R.R. *et al*, *Khim. Geterotsikl. Soedin.*, 1971, **7**, 1612 (*Engl. transl.* p. 1498) (*ir, raman*)
Nuretdinov, I.A. *et al*, *Zh. Fiz. Khim.*, 1979, **53**, 126 (*Engl. transl.* p. 66) (*nqr*)
Pudovik, M.A. *et al*, *Izv. Akad. Nauk SSSR, Ser. Khim.*, 1983, 903, 1859 (*Engl. transl.* pp. 821, 1683) (*synth, P nmr*)

(Chloromethyl)phenylphosphinic acid, 9CI — C-00103

$ClCH_2PPh(O)OH$

$C_7H_8ClO_2P$ M 190.566
Cryst. (H_2O). Spar. sol. H_2O, C_6H_6, Et_2O, pet. ether. Mp 96°.
Me ester: Methyl (chloromethyl)phenylphosphinate.
$C_8H_{10}ClO_2P$ M 204.593
Liq. d_4^{20} 1.27. $Bp_{2.5}$ 125-126°. n_D^{20} 1.5409.
Et ester: Ethyl (chloromethyl)phenylphosphinate.
$C_9H_{12}ClO_2P$ M 218.619
Liq. d_4^{20} 1.23. $Bp_{2.5}$ 132-134°. n_D^{20} 1.5275.
Chloride: [40561-09-9].
$C_7H_7Cl_2OP$ M 209.011
Solid. Mp 48-50°. $Bp_{0.5}$ 125°.
Anhydride:
$C_{14}H_{14}Cl_2O_3P_2$ M 363.116
Liq. Bp_1 210°.

Kabachnik, M.I. *et al*, *Izv. Akad. Nauk SSSR, Ser. Khim.*, 1953, 862 (*Engl. transl.* p. 763) (*synth, esters, chloride*)
Moedritzer, K., *J. Am. Chem. Soc.*, 1961, **83**, 4381 (*synth, chloride, P nmr, ir*)
Aliev, R.Z. *et al*, *Zh. Obshch. Khim.*, 1976, **46**, 266 (*Engl. transl.* p. 263) (*chloride*)

(Chloromethyl)phosphine — C-00104

[7237-08-3]

$ClCH_2PH_2$

CH_4ClP M 82.469
Gas. Bp −68°.

Fontal, B. *et al*, *J. Org. Chem.*, 1966, **31**, 2424 (*synth, props, ir, pmr*)
Goldwhite, H. *et al*, *J. Phys. Chem.*, 1968, **72**, 2666 (*pmr*)

P-Chloromethylphosphonamidic acid, 9CI — C-00105

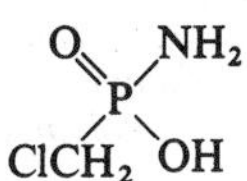

CH_5ClNO_2P M 129.483
Esters are employed as fireproofing agents.

Et ester: [19280-61-6]. *Ethyl* P-*chloromethylphosphonamidate.*
$C_3H_9ClNO_2P$ M 157.536
Cryst. (CCl_4). Mp 71-73°.
Ph ester: [19280-60-5]. *Phenyl* P-*chloromethylphosphonamidate.*
$C_7H_9ClNO_2P$ M 205.580
Cryst. (H_2O). Mp 96-97°.

Shokol, V.A. *et al*, *Zh. Obshch. Khim.*, 1968, **38**, 871 (*Engl. transl.* p. 836)

(Chloromethyl)phosphonic acid, 9CI C-00106

[2565-58-4]

$ClCH_2P(O)(OH)_2$

CH_4ClO_3P M 130.468
Needles ($C_6H_6/MeNO_2$). Mp 90-91°. pK_{a1} 1.7, pK_{a2} 5.22 (H_2O, 25°).
Mono-anilinium salt: Cryst. (EtOH). Mp 187-188° dec.
Bis(cyclohexyl)ammonium salt: Cryst. (EtOH). Mp 199-200°.
Mono-Me ester: [1929-41-5]. *Methyl hydrogen chloromethylphosphonate.*
$C_2H_6ClO_3P$ M 144.494
Liq. d_4^{20} 1.48. $Bp_{0.006}$ 130-135°. n_D^{20} 1.4595.
Di-Me ester: [6346-15-2]. *Dimethyl chloromethylphosphonate.*
$C_3H_8ClO_3P$ M 158.521
Liq. d_4^{20} 1.33. Bp_4 79.5-80°. n_D^{20} 1.4450.
Mono-Et ester: Monoethyl chloromethylphosphonate. Ethyl hydrogen chloromethylphosphonate.
$C_3H_8ClO_3P$ M 158.521
d_{20}^{20} 1.19. $Bp_{2.5}$ 86-87°. n_D^{20} 1.4360.
Di-Et ester: [3167-63-3]. *Diethyl chloromethylphosphonate.*
$C_5H_{12}ClO_3P$ M 186.575
Liq. d_4^{20} 1.20. $Bp_{2.5}$ 78-79°. n_D^{20} 1.4212.
Di-2-propenyl ester: Di-2-propenyl chloromethylphosphonate. Diallyl chloromethylphosphonate.
$C_7H_{12}ClO_3P$ M 210.597
Acts as a synergist for insecticides. Liq. Bp_{18} 100°. n_D^{20} 1.4682.
Diisopropyl ester: [6954-83-2]. *Diisopropyl chloromethylphosphonate.*
$C_7H_{16}ClO_3P$ M 214.628
Liq. d_4^{20} 1.10. Bp_2 72-73°. n_D^{20} 1.4333.
Di-Ph ester: [10419-85-9]. *Diphenyl chloromethylphosphonate.*
$C_{13}H_{12}ClO_3P$ M 282.663
Cryst. (Et_2O/pet. ether). Mp 42-42.5°. Bp_2 170-171°.
Difluoride: see (Chloromethyl)phosphonic difluoride, C-00109
Dichloride: see (Chloromethyl)phosphonic dichloride, C-00108
Diamide: see P-Chloromethylphosphonic diamide, C-00107

Yakubovich, A.Ya. *et al*, *Zh. Obshch. Khim.*, 1952, **22**, 1534 (*Engl. transl.* p. 1575) (*synth*)
Petrov, K.A. *et al*, *Zh. Obshch. Khim.*, 1960, **30**, 1602 (*Engl. transl.* p. 1604); 1965, **35**, 732 (*Engl. transl.* p. 731) (*esters*)
Tsvetkov, E.N. *et al*, *Zh. Obshch. Khim.*, 1969, **39**, 1520 (*Engl. transl.* p. 1490) (*derivs*)
Bel'skii, V.E. *et al*, *Zh. Obshch. Khim.*, 1972, **42**, 2427 (*Engl. transl.* p. 2421) (*ester, P nmr*)
Mango, L.A., *J. Polym. Sci., Polym. Chem. Ed.*, 1977, **15**, 513 (*ester, synth, ir, pmr*)
Coutrot, P. *et al*, *Synthesis*, 1978, 34 (*ester, use*)
Griffiths, W.R. *et al*, *Phosphorus Sulfur*, 1978, **5**, 101 (*esters, ms*)
Savignac, P. *et al*, *Synthesis*, 1978, 682 (*ester, use*)
Hall, C.R. *et al*, *J. Chem. Soc., Perkin Trans. 1*, 1984, 669 (*esters*)

P-Chloromethylphosphonic diamide, 9CI C-00107

[6326-70-1]

$ClCH_2P(O)(NH_2)_2$

CH_6ClN_2OP M 128.498
Solid. Mp 120-122°.
N,N,N′,N′-*Tetra-Me:* [7393-12-6]. P-*Chloromethyl-*N,N,N′,N′-*tetramethylphosphonic diamide. (Chloromethyl)phosphonic bis(dimethylamide).*
$C_5H_{14}ClN_2OP$ M 184.605
Cryst. (Et_2O/pet. ether). Mp 48.5-49.5°. Bp_4 117-118°.
N,N,N′,N′-*Tetra-Et:* P-*Chloromethyl-*N,N,N′,N′-*tetraethylphosphonic diamide. (Chloromethyl)-phosphonic bis(diethylamide).*
$C_9H_{22}ClN_2OP$ M 240.712
Liq. d_4^{20} 1.08. Bp_3 125-126°. n_D^{20} 1.4779.
N,N′-*Di-Ph: Chloromethylphosphonic dianilide.* P-*Chloromethyl-*N,N′-*diphenylphosphonic diamide.*
$C_{13}H_{14}ClN_2OP$ M 280.693
Solid. Mp 97-125°.

Kinnear, A.M. *et al*, *J. Chem. Soc.*, 1952, 3437 (*dianilide*)
Tsvetkov, E.N. *et al*, *Zh. Obshch. Khim.*, 1969, **39**, 1520 (*Engl. transl.* p. 1490) (*derivs*)
Cates, L.A., *J. Med. Chem.*, 1970, **13**, 301 (*synth*)

(Chloromethyl)phosphonic dichloride, 9CI C-00108

[1983-26-2]

$ClCH_2P(O)Cl_2$

CH_2Cl_3OP M 167.359
Pungent liq. d_4^{20} 1.64. Bp_{17} 92-94°, $Bp_{0.5}$ 50°. n_D^{20} 1.4930.

Kinnear, A.M. *et al*, *J. Chem. Soc.*, 1952, 3437 (*synth*)
Steger, E. *et al*, *Z. Phys. Chem.*, 1965, **229**, 110 (*synth, ir, raman*)
Tsvetkov, E.N. *et al*, *Izv. Akad. Nauk SSSR, Ser. Khim.*, 1967, 2375 (*Engl. transl.* p. 2267) (*synth, nqr*)
Maier, L., *Z. Anorg. Allg. Chem.*, 1972, **394**, 117 (*synth, P nmr*)
Raevskii, O.A. *et al*, *Izv. Akad. Nauk SSSR, Ser. Khim.*, 1977, 797 (*Engl. transl.* p. 726) (*pmr*)
Griffiths, W.R. *et al*, *Phosphorus Sulfur*, 1978, **4**, 341 (*ms*)
Prishchenko, A.A. *et al*, *Zh. Obshch. Khim.*, 1985, **55**, 340 (*Engl. transl.* p. 299) (*cmr*)

(Chloromethyl)phosphonic difluoride, 9CI C-00109

[1111-98-4]

$ClCH_2P(O)F_2$

CH_2ClF_2OP M 134.450
Liq. d_4^{20} 1.58. Bp_{146} 74°. n_D^{20} 1.3758.

Steger, E. *et al*, *Spectrochim. Acta, Part A*, 1967, **23**, 2189 (*ir, raman*)
Zakharov, V.I. *et al*, *Dokl. Akad. Nauk SSSR, Ser. Sci. Khim.*, 1973, **209**, 1343 (*Engl. transl.* p. 329) (*pmr, nmr*)
Komarov, V.Ya., *et al*, *Zh. Obshch. Khim.*, 1980, **50**, 1262 (*Engl. transl.*, p. 1020) (*nmr*)
Patsanovskii, I.I. *et al*, *Zh. Obshch. Khim.*, 1981, **51**, 985 (*Engl. transl.* p. 822) (*synth, P nmr*)
Van der Veken, B.J. *et al*, *J. Chem. Phys.*, 1985, **83**, 1517 (*ir, raman, microwave*)

(Chloromethyl)phosphonic diisocyanate, 9CI **C-00110**

[21783-55-1]

$(OCN)_2P(O)CH_2Cl$

$C_3H_2ClN_2O_3P$ M 180.487

Pale-yellow liq. d^{20} 1.56. Bp_{15} 128-129°, $Bp_{0.9}$ 80-82°. n_D^{20} 1.4955.

Haven, A.C., *J. Am. Chem. Soc.*, 1956, **78**, 842 (*synth*)
Derkach, G.I. *et al*, *Zh. Obshch. Khim.*, 1968, **38**, 1784 (*Engl. transl.* p. 1739) (*synth*)

(Chloromethyl)phosphonochloridic acid, 9CI **C-00111**

$ClCH_2PCl(O)OH$

$CH_3Cl_2O_2P$ M 148.913

Ph ester: [39200-99-2]. *Phenyl (chloromethyl)-phosphonochloridate.*
$C_7H_7Cl_2O_2P$ M 225.011
Liq. d_4^{20} 1.40. Bp_1 123-124°. n_D^{20} 1.5354.

Gefter, E.L., *Zh. Obshch. Khim.*, 1961, **31**, 3316 (*Engl. transl.* p. 3093) (*synth*)

(Chloromethyl)phosphonofluoridic acid, 9CI **C-00112**

$ClCH_2PF(O)OH$

CH_3ClFO_2P M 132.459

Anilinium salt: Solid. Mp 91-92°.

Reddy, G.S. *et al*, *Z. Naturforsch., B*, 1970, **25**, 1199 (*synth, pmr, F and P nmr*)

(Chloromethyl)phosphonothioic acid **C-00113**

$ClCH_2P(S)(OH)_2 \rightleftharpoons ClCH_2P(O)(OH)(SH)$

CH_4ClO_2PS M 146.528

Difluoride: [1426-00-2].
CH_2ClF_2PS M 150.510
Liq. Mp −84°. Bp 93.5-94°.

Dichloride: [1983-27-3].
CH_2Cl_3PS M 183.420
Liq. d_4^{20} 1.60. Bp_{20} 80-82°, Bp_{17} 96-98°. n_D^{20} 1.5770.

Steger, E. *et al*, *Z. Phys. Chem. (Leipzig)*, 1965, **229**, 110 (*dichloride, synth, ir, raman*)
Steger, E. *et al*, *Spectrochim. Acta, Part A*, 1967, **23**, 2189 (*difluoride, synth, ir, raman*)
Reddy, G.S. *et al*, *Z. Naturforsch, B*, 1970, **25**, 1199 (*difluoride, pmr, F and P nmr*)
Khaskin, L.S. *et al*, *Dokl. Akad. Nauk SSSR, Ser. Sci. Khim.*, 1972, **203**, 1090 (*Engl. transl.* p. 349) (*dichloride, ir, ed*)
Org. Synth., Coll. Vol., **5**, 218 (*dichloride, synth, P nmr*)
Khairullin, V.K. *et al*, *Zh. Obshch. Khim.*, 1978, **48**, 1993 (*Engl. transl.* p. 1813) (*dichloride, synth*)
Hall, C.R. *et al*, *J. Chem. Soc., Perkin Trans. 1*, 1984, 669.
Prishchenko, A.A. *et al*, *Zh. Obshch. Khim.*, 1985, **55**, 340 (*Engl. transl.* p. 299) (*dichloride, cmr*)

***P*-(Chloromethyl)phosphonothioic diamide,** 9CI **C-00114**

$ClCH_2P(S)(NH_2)_2$

CH_6ClN_2PS M 144.559

N,N'-Di-Ph: P-*Chloromethyl-*N,N'*-diphenylphosphonothioic diamide. (Chloromethyl)phosphonothioic dianilide.*
$C_{13}H_{14}ClN_2PS$ M 296.754
Solid. Mp 118-118.5°.

Kabachnik, M.I. *et al*, *Dokl. Akad. Nauk SSSR, Ser. Sci. Khim.*, 1956, **110**, 217; *CA*, **51**, 4982.

(Chloromethyl)phosphonous acid **C-00115**

$ClCH_2P(OH)_2$

CH_4ClO_2P M 114.468

Di-Et ester: Diethyl (chloromethyl)phosphonite.
$C_5H_{12}ClO_2P$ M 170.575
Liq. Bp_{20} 73-77°. Becomes hot on exp. to air. Polymerises when heated.

Dibutyl ester: Dibutyl (chloromethyl)phosphonite.
$C_9H_{20}ClO_2P$ M 226.683
Liq. Bp_5 85-90°. Easily oxid.

Difluoride: (Chloromethyl)difluorophosphine.
CH_2ClF_2P M 118.450
Forms Mo and Ni complexes. Liq. Bp 33.5-34.5°.
▷Spontaneously flammable

Dichloride: see (Chloromethyl)phosphonous dichloride, C-00116

Dibromide: [86012-34-2]. *Dibromo(chloromethyl)-phosphine.*
CH_2Br_2ClP M 240.262
Liq. Bp_{53} 98°.

Schmutzler, R., *Chem. Ind. (London)*, 1962, 1868 (*difluoride*)
Nixon, J.F. *et al*, *Spectrochim. Acta*, 1964, **20**, 1835 (*P nmr*)
Maier, L. *et al*, *Helv. Chim. Acta*, 1970, **53**, 1944 (*esters, synth, props, pmr*)
Hinke, A. *et al*, *Phosphorus Sulfur*, 1983, **15**, 93 (*dibromide, synth, P nmr*)

(Chloromethyl)phosphonous dichloride, 9CI **C-00116**

[2155-78-4]

$ClCH_2PCl_2$

CH_2Cl_3P M 151.360

Liq. with sharp, unpleasant odour. d_{20}^{20} 1.52. Bp 129°, Bp_{50} 63.5-64.5°. n_D^{20} 1.5247. Easily oxid. and hydrol.

Yakubovich, A.Y. *et al*, *Zh. Obshch. Khim.*, 1952, **22**, 1534; *CA*, **47**, 9254 (*synth*)
Uhing, E.H. *et al*, *J. Am. Chem. Soc.*, 1961, **83**, 2299 (*synth*)
Wieber, M. *et al*, *Chem. Ber.*, 1973, **106**, 2733 (*synth, pmr*)
Zverev, V.V. *et al*, *Izv. Akad. Nauk SSSR, Ser. Khim.*, 1975, 1051 (*Engl. transl.* p. 961) (*pe*)
Fishman, A.I. *et al*, *Spectrochim. Acta, Part A*, 1976, **32**, 651 (*ir, raman*)
Frenking, G. *et al*, *Phosphorus Sulfur*, 1980, **8**, 337, 343 (*pe, struct*)
Prishchenko, A.A. *et al*, *Zh. Obshch. Khim.*, 1985, **35**, 340; *J. Gen. Chem. USSR (Engl. Transl.)*, 299 (*cmr*)

(Chloromethyl)triphenylphosphonium(1+) **C-00117**

$Ph_3P^{\oplus}CH_2Cl$

$C_{19}H_{17}ClP^{\oplus}$ M 311.770 (ion)

Source of ylide used in Wittig reactions for chloromethylenation of aldehydes and ketones.

Chloride: [5293-84-5].
$C_{19}H_{17}Cl_2P$ M 347.223
Cryst. (CH_2Cl_2/Et_2O). Mp 253°.
▷TA1882000.

Iodide: [68089-86-1].
$C_{19}H_{17}ClIP$ M 438.675
Cryst. (EtOH). Mp 185-187°, 206°.
Tetrafluoroborate:
$C_{19}H_{17}BClF_4P$ M 398.574
Yellow cryst. (EtOH aq.). Mp 201-203°.
Tetraphenylborate:
$C_{43}H_{37}BClP$ M 631.002
Solid. Mp 190-193° dec.
Ylide: [29949-92-6].
$C_{19}H_{16}ClP$ M 310.762
Prepd. *in situ.*

Appel, R. *et al, Chem. Ber.*, 1976, **109**, 58 (*pmr, nmr, cmr*)
Appel, R. *et al, Synthesis*, 1977, 699 (*chloride, synth, pmr, nmr*)
Miyano, S. *et al, Bull. Chem. Soc. Jpn.*, 1979, **52**, 1197 (*synth, ylide, use*)
Appel, R. *et al, Chem. Ber.*, 1981, **114**, 858 (*chloride, use*)
Kukhar', V.P. *et al, Zh. Org. Khim.*, 1981, **17**, 180 (*Engl. transl.* p. 161) (*synth*)
Schlosser, M. *et al, Chimia*, 1982, **36**, 396 (*ylide, synth*)

2-Chloronaphtho[1,2-*d*]-1,3,2-dioxaphosphole, 9CI **C-00118**

1,2-Naphthylene chlorophosphite. 1,2-Naphthylene phosphorochloridite
[79788-73-1]

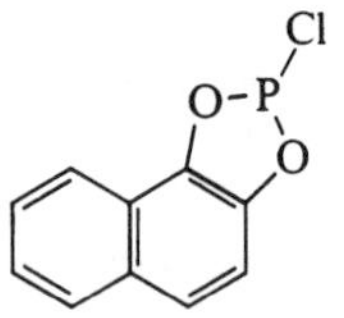

$C_{10}H_6ClO_2P$ M 224.583
Solid. Mp 72-73°. Bp_2 145-148°.

Nifant'ev, É.E. *et al, Zh. Obshch. Khim.*, 1981, **51**, 1528 (*Engl. transl.* p. 1295) (*synth, P nmr*)

2-Chloronaphtho[2,3-*d*]-1,3,2-dioxaphosphole, 9CI, 8CI **C-00119**

2,3-Naphthylene phosphorochloridite. 2,3-Naphthylene chlorophosphite

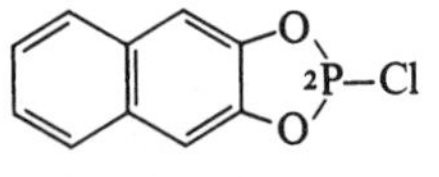

$C_{10}H_6ClO_2P$ M 224.583
Solid. Mp 153-154°.
2-Oxide: [62290-26-0]. *2,3-Naphthylene phosphorochloridate. 2,3-Naphthylene chlorophosphate.*
$C_{10}H_6ClO_3P$ M 240.582
Cryst. (THF). Mp >250°.
2-Sulfide: [22922-25-4]. *2,3-Naphthylene phosphorochloridothioate. 2,3-Naphthylene chlorothiophosphate.*
$C_{10}H_6ClO_2PS$ M 256.643
Cryst. (THF). Mp >250°.

Bhatia, M.S. *et al, Ann. Chim. (Paris)*, 1976, **1**, 239 (*derivs, synth, ir*)
Voropai, L.M. *et al, Zh. Obshch. Khim.*, 1985, **85**, 65 (*Engl. transl.* p. 55) (*synth, P nmr*)

2-Chloronaphtho[1,8-*de*]-1,3,2-dioxaphosphorin, 10CI, 8CI **C-00120**

1,8-Naphthylene phosphorochloridite. 1,8-Naphthylene chlorophosphite
[72310-28-2]

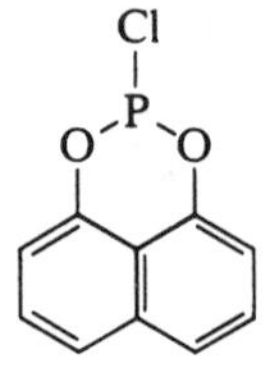

$C_{10}H_6ClO_2P$ M 224.583
Solid. Mp 135-137°. Bp_2 192-195°.

Nifant'ev, É.E. *et al, Zh. Obshch. Khim.*, 1981, **51**, 1528 (*Engl. transl.* p. 1295) (*synth, P nmr*)

(2-Chloro-3-oxobutyl)methylphosphinic acid, 9CI **C-00121**

$H_3CCOCHClCH_2PMe(O)OH$

$C_5H_{10}ClO_3P$ M 184.559
(±)-***form***
Me ester: Methyl (2-chloro-3-oxobutyl)-methylphosphinate.
$C_6H_{12}ClO_3P$ M 198.586
Liq. d_4^{20} 1.24. $Bp_{0.01}$ 97°. n_D^{20} 1.4762.

Pudovik, A.N. *et al, Zh. Obshch. Khim.*, 1967, **37**, 865 (*Engl. transl.* p. 814) (*synth, pmr*)

[(1-Chloro-2-oxo-2-phenyl)ethylidene]-triphenylphosphorane **C-00122**

2-Chloro-2-(triphenylphosphoranylidene)acetophenone

$Ph_3P{=}CClCOPh$

$C_{26}H_{20}ClOP$ M 414.870
Stabilized ylide which undergoes Wittig reactions. Monoclinic cryst. (Me_2CO/hexane). Mp 157-159°. pK_a 4.3 (conj. phosphonium salt). Isomorphous with bromo compd.

Denney, D.B. *et al, J. Org. Chem.*, 1962, **27**, 998 (*synth*)
Speziale, A.J. *et al, J. Org. Chem.*, 1963, **28**, 465 (*synth, ir, uv*)
Speziale, A.J. *et al, J. Am. Chem. Soc.*, 1965, **87**, 5603 (*cryst struct*)
Stephens, F.S., *J. Chem. Soc.*, 1965, 5658 (*cryst struct*)
Zbiral, E., *Monatsh. Chem.*, 1966, **97**, 180 (*synth*)

(3-Chloro-2-oxopropyl)-triphenylphosphonium(1+) **C-00123**

$Ph_3P^{\oplus}CH_2COCH_2Cl$

$C_{21}H_{19}ClOP^{\oplus}$ M 353.807 (ion)
pK_a 4.5 (80% EtOH aq.).
Chloride: [13605-65-7].
$C_{21}H_{19}Cl_2OP$ M 389.260
Cryst. (EtOH). Mp 213°, 177-178° (monohydrate). Aq. Na_2CO_3 yields the ylide.
Ylide: see 1-Chloro-3-(triphenylphosphoranylidene)-2-propanone, C-00191

Hudson, R.F. *et al, J. Org. Chem.*, 1963, **28**, 2446 (*synth, props*)
Issleib, K. *et al, Justus Liebigs Ann. Chem.*, 1968, **713**, 12 (*props*)

Nesmeyanov, N.A. *et al*, *J. Organomet. Chem.*, 1977, **129**, 41 (*synth, tautom, pmr, nmr*)
Henichart, J.P., *J. Mol. Struct.*, 1983, **99**, 283 (*cryst struct*)

1-Chloro-2,2,3,4,4-pentamethylphosphetane, 9CI, 8CI — C-00124

[42336-88-9]

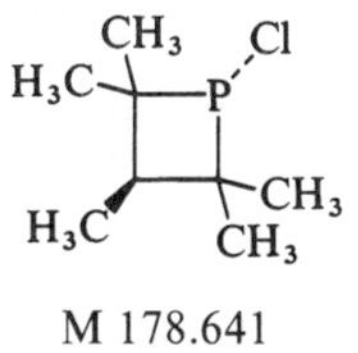

trans-form

$C_8H_{16}ClP$ M 178.641

Bp_{20} 87°. Mixture of stereoisomers.

cis-form [25145-23-7]

Oxide: [25145-33-9]. *1-Chloro-2,2,3,4,4-pentamethylphosphetane 1-oxide. 1,1,2,3,3-Pentamethyltrimethylenephosphinic chloride.*
$C_8H_{16}ClOP$ M 194.641
Cryst. Mp 55-59°.

Sulfide: 1-Chloro-2,2,3,4,4-pentamethylphosphetane 1-sulfide. 1,1,2,3,3-Pentamethyltrimethylenephosphinothioic chloride.
$C_8H_{16}ClPS$ M 210.701
Cryst. (pet. ether). Mp 121-122° (sealed tube).

trans-form [25145-24-8]

Oxide: [26674-18-0]. Solid. Mp 75-76°.
Sulfide: [35623-66-6]. Cryst. (pet. ether). Mp 118-120°.

McBride, J.J. *et al*, *J. Org. Chem.*, 1962, **27**, 1833 (*synth, props*)
Mazhar-ul-Haque, *J. Chem. Soc.* (*B*), 1970, 934 (*oxide, cryst struct*)
Corfield, J.R. *et al*, *J. Chem. Soc., Perkin Trans. 1*, 1972, 713 (*oxides, sulfides, ms, P nmr, pmr*)
Gray, G.A. *et al*, *J. Org. Chem.*, 1972, **37**, 3470 (*cmr, P nmr, sulfides*)
Ardray, R. *et al*, *J. Chem. Soc., Dalton Trans.*, 1973, 2641 (*sulfide, ms*)
Emsley, J. *et al*, *J. Chem. Soc., Dalton Trans.*, 1973, 2701 (*oxide, ir, raman*)
Wiseman, J. *et al*, *J. Am. Chem. Soc.*, 1974, **96**, 4262 (*oxides*)
Kemp, G. *et al*, *J. Chem. Soc., Perkin Trans. 1*, 1979, 879 (*oxide, props*)
Harger, M.J.P., *J. Chem. Soc., Perkin Trans. 1*, 1980, 705 (*oxide, props, pmr*)

4-Chlorophenyl 2-cyanoethyl phosphorochloridate, 9CI — C-00125

4-Chlorophenyl 2-cyanoethyl phosphoryl chloride

[69507-62-6]

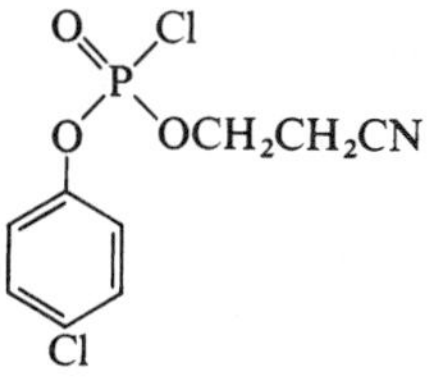

$C_9H_8Cl_2NO_3P$ M 280.047

Phosphorylating agent. Liq. Stable for 1 month at −15°.

De Bernardini, S. *et al*, *Helv. Chim. Acta*, 1981, **64**, 2142 (*synth, pmr, use*)
Seliger, H. *et al*, *J. Chromatogr.*, 1982, **253**, 65 (*use*)

2-Chloro-5-phenyl-1,3,4,2-dioxazaphosphole, 10CI — C-00126

Benzhydroxamatophosphorus(III) chloride

[66181-07-5]

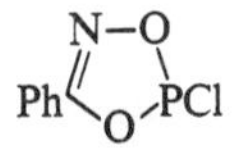

$C_7H_5ClNO_2P$ M 201.549

V. hygroscopic liq. $Bp_{0.05}$ 45° (bath 75°). Easily hydrol.

▷Explodes at bath temp. 90°

2-Oxide: [66181-06-4].
$C_7H_5ClNO_3P$ M 217.548
Liq. $Bp_{0.001}$ 84-86°.

2,2-Dichloro-2,2-dihydro: [66181-02-0]. *2,2,2-Trichloro-2,2-dihydro-5-phenyl-1,3,4,2-dioxazaphosphole.*
$C_7H_5Cl_3NO_2P$ M 272.455
V. hygroscopic liq. $Bp_{0.0001}$ 90°.

Fluck, E. *et al*, *Z. Anorg. Allg. Chem.*, 1977, **437**, 53 (*synth, derivs, P nmr*)
Von Hinrichs, E. *et al*, *J. Chem. Res.* (*S*), 1978, 338 (*synth, derivs, P nmr*)

(2-Chlorophenyl)diphenylphosphine, 9CI — C-00127

[35035-62-2]

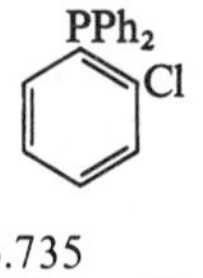

$C_{18}H_{14}ClP$ M 296.735

Ligand for Ni and Pd. Cryst. ($EtOH/CH_2Cl_2$). Mp 106-108°.

B,PhCH₂Cl: Benzyl(2-chlorophenyl)-diphenylphosphonium chloride.
$C_{25}H_{21}Cl_2P$ M 423.321
Cryst. ($CHCl_3/CCl_4$). Mp 288-290°.

Oxide: [61102-87-2].
$C_{18}H_{14}ClOP$ M 312.735
Cryst. (EtOH/hexane). Mp 95-97°, 123-124°.

Hart, F.A., *J. Chem. Soc.*, 1960, 3324 (*synth, complexes*)
Shvets, A.A. *et al*, *Zh. Obshch. Khim.*, 1976, **46**, 1701 (*Engl. transl.* p. 1654) (*oxide, ir*)
McEwen, W.E. *et al*, *J. Am. Chem. Soc.*, 1978, **100**, 7304 (*synth, deriv*)
Allen, D.W. *et al*, *J. Chem. Soc., Perkin Trans. 1*, 1979, 1499 (*oxide*)

(3-Chlorophenyl)diphenylphosphine, 9CI — C-00128

[23415-66-9]

$C_{18}H_{14}ClP$ M 296.735

Cryst. (2-propanol). Mp 30-31°. Bp_4 195-198°.

Oxide: [23415-68-1].
$C_{18}H_{14}ClOP$ M 312.735
Cryst. (C_6H_6/hexane). Mp 108-110°.

Sulfide: [23415-69-2].
$C_{18}H_{14}ClPS$ M 328.795
Cryst. (2-propanol). Mp 61°, 75-76.5°.

Selenide:
$C_{18}H_{14}ClPSe$ M 375.695
Cryst. Mp 81°.

Lobanov, D.I. *et al*, *Zh. Obshch. Khim.*, 1969, **39**, 841 (*Engl. transl.* p. 805) (*synth*)
De Ketalaere, R.F. *et al*, *J. Mol. Struct.*, 1974, **23**, 233 (*derivs, ir, raman*)

De Ketelaere, R.F. *et al*, *Phosphorus*, 1974, **5**, 43 (*derivs, ms*)
De Ketelaere, R.F. *et al*, *J. Mol. Struct.*, 1975, **27**, 363 (*derivs, nmr*)
Claeys, E.G. *et al*, *J. Mol. Struct.*, 1977, **40**, 89 (*struct*)

(4-Chlorophenyl)diphenylphosphine, 9CI C-00129

[734-60-1]

$C_{18}H_{14}ClP$ M 296.735
Cryst. (EtOH). Mp 44-45°. $Bp_{0.02}$ 155-157°. pK_a 2.18.
Oxide: [34303-18-9].
$C_{18}H_{14}ClOP$ M 312.735
Cryst. (MeOH aq. or 2-propanol aq.). Mp 143-144°.
Sulfide: [1474-08-4].
$C_{18}H_{14}ClPS$ M 328.795
Cryst. (MeOH or EtOH). Mp 129-131°.
Selenide:
$C_{18}H_{14}ClPSe$ M 375.695
Mp 149°.

Goetz, H. *et al*, *Justus Liebigs Ann. Chem.*, 1963, **665**, 1 (*synth, ir, uv, derivs*)
Tsvetkov, E.N. *et al*, *Zh. Obshch. Khim.*, 1968, **38**, 2285 (*Engl. transl.* p. 2211) (*synth, oxide*)
Dreissig, W. *et al*, *Acta Crystallogr., Sect. B*, 1971, **27**, 1146 (*oxide, cryst struct*)
Dreissig, W. *et al*, *Acta Crystallogr., Sect. B*, 1972, **28**, 3478 (*sulfide, cryst struct*)
De Ketelaere, R.F. *et al*, *J. Mol. Struct.*, 1974, **23**, 233 (*derivs, ir, raman*)
De Ketelaere, R.F. *et al*, *Phosphorus*, 1974, **5**, 43 (*derivs, ms*)
De Ketelaere, R.F. *et al*, *J. Mol. Struct.*, 1975, **27**, 363 (*nmr, derivs*)
McEwen, W.E. *et al*, *J. Am. Chem. Soc.*, 1978, **100**, 7304 (*pmr*)

2-Chlorophenyl (4-methoxyphenyl)-phosphoramidochloridate C-00130

2-Chlorophenyl phosphono-(4-anisido) chloridate
[82556-11-4]

O Cl / P / MeO–C6H4–NH, O–C6H4Cl

$C_{13}H_{12}Cl_2NO_3P$ M 332.122
Phosphorylating agent used in oligodeoxyribonucleotide synth. Solid. Mp 96°.

Ohtsuka, E. *et al*, *Chem. Pharm. Bull.*, 1981, **29**, 3440; 1983, **31**, 1910 (*synth, use, pmr*)

(4-Chlorophenyl)methylphosphinic acid, 9CI C-00131

[13114-10-8]

OH / O=P–Me / C6H4 / Cl

$C_7H_8ClO_2P$ M 190.566
Wax. Mp 94°. pK_a 2.39.
Me ester: [13114-08-4]. *Methyl (4-chlorophenyl)-methylphosphinate.*
$C_8H_{10}ClO_2P$ M 204.593
Liq. d_4^{20} 1.27. Bp_5 144°, $Bp_{0.15}$ 103-104°. n_D^{20} 1.5363.
Fluoride: [38169-15-2].
C_7H_7ClFOP M 192.557
No phys. props. reported.
Chloride: [13114-09-5].
$C_7H_7Cl_2OP$ M 209.011
Liq. d_4^{20} 1.39. Bp_5 155-156°. n_D^{20} 1.5727.

Neimysheva, A.A. *et al*, *Zh. Obshch. Khim.*, 1966, **36**, 500 (*Engl. transl.* p. 520) (*synth, chloride, ester*)
Knunyants, I.L. *et al*, *Dokl. Akad. Nauk SSSR, Ser. Sci. Khim.*, 1971, **201**, 862 (*Engl. transl.* p. 992) (*halides, nmr*)
Bunnett, J.F. *et al*, *J. Org. Chem.*, 1973, **38**, 2703.
Curci, P. *et al*, *J. Chem. Soc., Perkin Trans. 2*, 1973, 531 (*ester, pmr*)

[(2-Chlorophenyl)methyl]phosphonic acid, 9CI C-00132

o-*Chlorobenzylphosphonic acid*
[13249-95-1]

O / ‖ / $CH_2P(OH)_2$ / C6H4 / Cl

$C_7H_8ClO_3P$ M 206.565
Cryst. (H_2O). Mp 183°.
Mono-Me ester:
$C_8H_{10}ClO_3P$ M 220.592
Solid, as tetramethylammonium salt. Mp 196°.
Di-Me ester: Dimethyl [(2-chlorophenyl)methyl]-phosphonate.
$C_9H_{12}ClO_3P$ M 234.619
Wittig-Horner reagent. Liq. $Bp_{0.04}$ 103-105°. n_D^{20} 1.527.
Di-Et ester: [29074-98-4]. *Diethyl [(2-chlorophenyl)-methyl]phosphonate.*
$C_{11}H_{16}ClO_3P$ M 262.672
Wittig-Horner reagent. Liq. $Bp_{0.4}$ 140-146°.

Kreutzkamp, N. *et al*, *Arch. Pharm. (Weinheim, Ger.)*, 1962, **295**, 276 (*synth*)
Chabrier, P. *et al*, *C.R. Hebd. Seances Acad. Sci.*, 1964, **259**, 2244 (*esters*)
Ernst, L., *Org. Magn. Reson.*, 1977, **9**, 35 (*ester, cmr*)
Franke, A. *et al*, *Synthesis*, 1979, 712 (*ester, synth, pmr*)

[(3-Chlorophenyl)methyl]phosphonic acid, 9CI C-00133

m-*Chlorobenzylphosphonic acid*
[80395-11-5]
$C_7H_8ClO_3P$ M 206.565
Di-Et ester: [78055-64-8]. *Diethyl [(3-chlorophenyl)-methyl]phosphonate. Diethyl* m-*chlorobenzylphosphonate.*
$C_{11}H_{16}ClO_3P$ M 262.672
Wittig-Horner reagent. Liq. Bp_3 160-163°. n_D^{20} 1.5067.

Fresneda, P.M. *et al*, *Synthesis*, 1981, 222.

[(4-Chlorophenyl)methyl]phosphonic acid, 9CI C-00134

p-*Chlorobenzylphosphonic acid*
[39225-05-3]
$C_7H_8ClO_3P$ M 206.565
Cryst. (H_2O). Mp 168-171°.

Di-Et ester: [39225-17-7]. *Diethyl [(4-chlorophenyl)-methyl]phosphonate.*
$C_{11}H_{16}ClO_3P$ M 262.672
Wittig-Horner reagent. Liq. Bp_3 180°, $Bp_{0.01}$ 137-138°. n_D^{20} 1.4940.

Williams, A. *et al, J. Chem. Soc., Perkin Trans. 2,* 1973, 25 (*synth*)
Fresneda, P.M. *et al, Synthesis,* 1981, 222 (*ester*)

[(2-Chlorophenyl)methyl]-triphenylphosphonium(1+), 9CI C-00135

o-Chlorobenzyltriphenylphosphonium(1+)

$C_{25}H_{21}ClP^{\oplus}$ M 387.868 (ion)
Chloride: [18583-53-6].
$C_{25}H_{21}Cl_2P$ M 423.321
With EtONa gives the ylide. Needles (chlorobenzene). Mp 248-249°.
Bromide: [62640-67-9].
$C_{25}H_{21}BrClP$ M 467.772
No phys. props. reported.
Ylide: [59625-56-8]. *(2-Chlorophenylmethylene)-triphenylphosphorane. (o-Chlorobenzylidene)-triphenylphosphorane.*
$C_{25}H_{20}ClP$ M 386.860
Wittig reagent prepd. *in situ.*

Fisichella, S. *et al, Gazz. Chim. Ital.,* 1974, **104**, 1237; *CA*, **82**, 124599 (*synth, use*)
Knorr, H. *et al, Justus Liebigs Ann. Chem.,* 1977, 545 (*ylide, use*)
Kinnamon, H.E. *et al, J. Med. Chem.,* 1979, **22**, 452 (*pharmacol*)

[(3-Chlorophenyl)methyl]-triphenylphosphonium(1+), 9CI C-00136

(m-Chlorobenzyl)triphenylphosphonium(1+)

$C_{25}H_{21}ClP^{\oplus}$ M 387.868 (ion)
Chloride: [32597-92-5].
$C_{25}H_{21}Cl_2P$ M 423.321
With NaOEt gives the ylide. Cryst. (*o*-dichlorobenzene or EtOH/pet. ether). Mp 326-328° (310-311°).
Bromide: [28540-72-9].
$C_{25}H_{21}BrClP$ M 467.772
No phys. props. reported.
Ylide: [67535-53-9]. *(3-Chlorophenylmethylene)-triphenylphosphorane. (m-Chlorobenzylidene)-triphenylphosphorane.*
$C_{25}H_{20}ClP$ M 386.860
Wittig reagent. Prepd. *in situ.*

Baker, B.R. *et al, J. Med. Chem.,* 1971, **14**, 315 (*synth*)
Fisichella, S. *et al, Gazz. Chim. Ital.,* 1974, **104**, 1237; *CA*, **82**, 124599 (*synth, use*)

[(4-Chlorophenyl)methyl]-triphenylphosphonium(1+), 9CI C-00137

(p-Chlorobenzyl)triphenylphosphonium(1+)
[47522-13-4]

$C_{25}H_{21}ClP^{\oplus}$ M 387.868 (ion)
Chloride: [1530-39-8].
$C_{25}H_{21}Cl_2P$ M 423.321
Source of ylide with MeONa or butyllithium. Mp 270°, 284-287°, 307-308°.
Bromide: [51044-12-3].
$C_{25}H_{21}BrClP$ M 467.772
Source of ylide with butyllithium. Cryst. ($CHCl_3$). Mp 278-280°.
Ylide: [38897-99-3]. *(4-Chlorophenylmethylene)-triphenylphosphorane. (p-Chlorobenzylidene)-triphenylphosphorane.*
$C_{25}H_{20}ClP$ M 386.860
Wittig reagent prepd. *in situ.*

Leznoff, C.C. *et al, Can. J. Chem.,* 1972, **50**, 528 (*chloride, use*)
Kendurkar, P.S. *et al, Z. Naturforsch., B,* 1973, **28**, 475 (*bromide, pmr, use*)
LaMontagne, M.P., *J. Med. Chem.,* 1973, **16**, 68 (*chloride, use*)
Yamato, M. *et al, Chem. Pharm. Bull.,* 1977, **25**, 706 (*bromide, use*)
Lumbroso, H. *et al, J. Organomet. Chem.,* 1978, **161**, 347 (*ylide, struct*)
Kroppová, V. *et al, Collect. Czech. Chem. Commun.,* 1981, **46**, 515 (*chloride, use*)

2-Chloro-3-phenyl-1,3,2-oxazaphospholidine, 9CI C-00138

2-Chloro-3-phenyl-1,3,2-oxazaphospholane

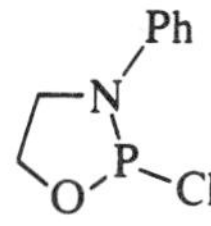

C_8H_9ClNOP M 201.592
Solid. Mp 59-60°. $Bp_{0.01}$ 99-101°. Easily hydrolysed.
2-Oxide:
$C_8H_9ClNO_2P$ M 217.591
Cryst. (C_6H_6). Mp 96-97°.
2-Sulfide: [32847-70-4].
$C_8H_9ClNOPS$ M 233.652
Cryst. (CH_2Cl_2). Mp 112-115°.

Ivanova, Zh.M. *et al, Zh. Obshch. Khim.,* 1970, **40**, 1942 (*Engl. transl.* p. 1927) (*sulfide*)
Shagidullin, R.R. *et al, Khim. Geterotsikl. Soedin.,* 1971, **7**, 1612 (*Engl. transl.* p. 1498) (*ir, oxide*)
Naumov, V.A. *et al, Zh. Strukt. Khim.,* 1975, **16**, 3 (*Engl. transl.* p. 1) (*ed*)
Kibardina, L.K. *et al, Izv. Akad. Nauk SSSR, Ser. Khim.,* 1977, 1932 (*Engl. transl.* p. 1796) (*synth, P nmr*)
Pudovik, M.A. *et al, Izv. Akad. Nauk SSSR, Ser. Khim.,* 1983, 903, 1859 (*Engl. transl.* pp. 821, 1685) (*synth, P nmr*)

(2-Chlorophenyl)phosphine, 9CI C-00139

[53772-57-9]

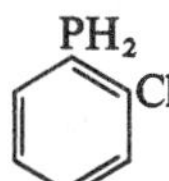

C_6H_6ClP M 144.540
Liq. Bp_3 54-55°.

Maier, L., *Phosphorus,* 1974, **4**, 41 (*pmr, nmr*)

(3-Chlorophenyl)phosphine, 9CI C-00140

[23415-73-8]
C_6H_6ClP M 144.540
Liq. d_4^{20} 1.18. Bp_{12} 77-78°. n_D^{20} 1.5953.

Lobanov, D.I. *et al, Zh. Obshch. Khim.,* 1969, **39**, 841 (*Engl. transl.* p. 805) (*synth*)
Kabachnik, M.I. *et al, Aust. J. Chem.,* 1975, **28**, 755 (*ir*)

(4-Chlorophenyl)phosphine, 9CI C-00141

[4538-32-3]

C_6H_6ClP M 144.540
Low-melting solid. Mp 17°. Bp_{35} 105-106°, Bp_{11} 68-70°.

Tsvetkov, E.N. *et al, Zh. Obshch. Khim.*, 1968, **38**, 2285 (*Engl. transl.* p. 2211) (*synth*)
Maier, L., *Phosphorus*, 1974, **4**, 41 (*synth, pmr, nmr*)
Kabachnik, M.I. *et al, Aust. J. Chem.*, 1975, **28**, 755 (*ir*)

(2-Chlorophenyl)phosphinic acid, 9CI — C-00142

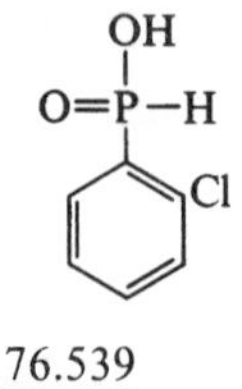

$C_6H_6ClO_2P$ M 176.539
Tautomeric with 2-chlorophenylphosphonous acid, but free acid probably exists in phosphinic acid form. Cryst. ($CHCl_3$ or H_2O). Mp 128.5-129°.

Quin, L.D. *et al, J. Am. Chem. Soc.*, 1961, **83**, 4124 (*synth*)
Bokanov, A.I. *et al, Zh. Obshch. Khim.*, 1965, **35**, 350 (*Engl. transl.* p. 350) (*synth*)

(3-Chlorophenyl)phosphinic acid, 9CI — C-00143

$C_6H_6ClO_2P$ M 176.539
Tautomeric with 3-chlorophenylphosphonous acid, but free acid probably exists in phosphinic acid form. Cryst. (CCl_4). Mp 90.5-91.5°. pK_a 1.35 (H_2O).
Me ester: Methyl (3-chlorophenyl)phosphinate.
$C_7H_8ClO_2P$ M 190.566
Liq. $Bp_{0.25}$ 88°.

Quin, L.D. *et al, J. Am. Chem. Soc.*, 1961, **83**, 4124 (*synth*)
Quin, L.D. *et al, J. Org. Chem.*, 1962, **27**, 1012 (*ester, ir*)
Lobanov, D.I. *et al, Zh. Obshch. Khim.*, 1969, **39**, 841 (*synth*)

(4-Chlorophenyl)phosphinic acid, 9CI — C-00144

[22336-21-6]
$C_6H_6ClO_2P$ M 176.539
Tautomeric with (4-Chlorophenyl)phosphonous acid, C-00153 but free acid exists in the phosphinic acid form. Cryst. (H_2O). Mp 131-132°. pK_a 1.57 (H_2O), ca. 3.0 (EtOH).
Et ester: [38766-20-8]. *Ethyl (4-chlorophenyl)-phosphinate.*
$C_8H_{10}ClO_2P$ M 204.593
Characterised spectroscopically.

Pudovik, A.N. *et al, Zh. Obshch. Khim.*, 1969, **39**, 1715 (*Engl. transl.* p. 1681) (*ir*)
Vinogradov, L.I. *et al, Zh. Obshch. Khim.*, 1972, **42**, 1724 (*Engl. transl.* p. 1712) (*ester, pmr, P nmr*)
Barabanov, V.P. *et al, Zh. Obshch. Khim.*, 1973, **43**, 1147, 1517 (*Engl. transl.* p. 1138, 1391) (*synth, props*)
Harger, M.J.P. *et al, Tetrahedron*, 1982, **38**, 1511 (*synth, ir*)

(2-Chlorophenyl)phosphonic acid, 9CI — C-00145

[5431-19-6]

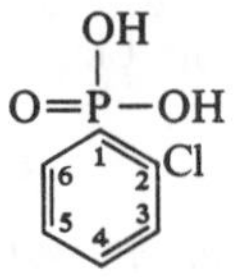

$C_6H_6ClO_3P$ M 192.538
Solid. Mp 182-184°. pK_{a1} 1.63 (H_2O), 2.94 (50% EtOH aq.), pK_{a2} 6.98 (H_2O), 8.21 (50% EtOH aq.).
▷SZ7175100.
Di-Me ester: Dimethyl (2-chlorophenyl)phosphonate.
$C_8H_{10}ClO_3P$ M 220.592
Liq. $Bp_{0.3}$ 106-107°.
Di-Et ester: [28036-18-2]. *Diethyl (2-chlorophenyl)-phosphonate.*
$C_{10}H_{14}ClO_3P$ M 248.646
Liq. $Bp_{0.3}$ 113-115°.
Dichloride: [4672-45-1].
$C_6H_4Cl_3OP$ M 229.430
No phys. props. reported.

Doak, G.O. *et al, J. Am. Chem. Soc.*, 1951, **73**, 5658 (*synth*)
Obrycki, R. *et al, Tetrahedron Lett.*, 1966, 5049 (*ester*)
Tars, P., *Chem. Ber.*, 1970, **103**, 2428 (*ester*)
Bard, R.R. *et al, J. Org. Chem.*, 1979, **44**, 4918 (*ester, ms*)

(3-Chlorophenyl)phosphonic acid, 9CI — C-00146

[5431-34-5]
$C_6H_6ClO_3P$ M 192.538
Needles (HCl aq.). Mp 136-137°. pK_{a1} 3.23 (50% EtOH aq.), pK_{a2} 6.11 (50% EtOH aq.).
▷SZ7175000.
Di-Me ester: Dimethyl (3-chlorophenyl)phosphonate.
$C_8H_{10}ClO_3P$ M 220.592
Liq. $Bp_{0.5}$ 109-110°.
Di-Et ester: [23415-71-6]. *Diethyl (3-chlorophenyl)-phosphonate.*
$C_{10}H_{14}ClO_3P$ M 248.646
Liq. $Bp_{0.3}$ 113-115°. n_D^{20} 1.5086.
Dichloride: [23415-70-5].
$C_6H_4Cl_3OP$ M 229.430
Liq. d^{20} 1.51. Bp_{27} 127°, $Bp_{0.4}$ 86-89°. n_D^{20} 1.5742.

Kosolapoff, G.M., *J. Am. Chem. Soc.*, 1948, **70**, 3465 (*synth*)
Denham, J.M. *et al, J. Org. Chem.*, 1958, **23**, 1298 (*ester, dichloride*)
Obrycki, R. *et al, Tetrahedron Lett.*, 1966, 5049 (*ester*)
Lobonov, D.I. *et al, Zh. Obshch. Khim.*, 1969, **39**, 841 (*Engl. transl.* p. 805) (*dichloride*)
Tavs, P., *Chem. Ber.*, 1970, **103**, 2428 (*ester*)
Williams, A. *et al, J. Chem. Soc., Perkin Trans. 2*, 1973, 25 (*synth*)
Szafraniec, L.L., *Org. Magn. Reson.*, 1974, **6**, 565 (*difluoride, F and P nmr*)
Daasch, L.W. *et al, Phosphorus*, 1975, **5**, 189 (*difluoride, ms*)
Allen, D.W. *et al, J. Chem. Soc., Perkin Trans. 2*, 1977, 789 (*diethyl ester, pmr, P nmr*)

(4-Chlorophenyl)phosphonic acid, 9CI — C-00147

[5431-35-6]
$C_6H_6ClO_3P$ M 192.538
Cryst. (Me_2CO/1,2-dichloroethane or EtOAc). Mp 193-194° (184-186°). pK_{a1} 1.96 (H_2O), 3.23 (50% EtOH aq.), pK_{a2} 5.25 (H_2O), 6.11 (50% EtOH aq.).
▷SZ7350000.

Di-Me ester: [13114-07-3]. *Dimethyl (4-chlorophenyl)-phosphonate.*
$C_8H_{10}ClO_3P$ M 220.592
Liq. d_4^{20} 1.30. Bp_{14} 154-155°, $Bp_{0.1}$ 95-96°. n_D^{20} 1.5231.

Di-Et ester: [2373-43-5]. *Diethyl (4-chlorophenyl)-phosphonate.*
$C_{10}H_{14}ClO_3P$ M 248.646
Liq. d_4^{20} 1.20. Bp_3 141-142°, $Bp_{0.15}$ 105-108°. n_D^{20} 1.5081.

Bis(trimethylsilyl) ester: [104412-65-9]. *Bis(trimethylsilyl)(4-chlorophenyl)phosphonate.*
$C_{12}H_{22}ClO_3PSi_2$ M 336.902
Liq. $Bp_{0.001}$ 95°.

Difluoride: see (4-Chlorophenyl)phosphonic difluoride, C-00149

Dichloride: see (4-Chlorophenyl)phosphonic dichloride, C-00148

*Bis(*N,N*-dimethylamide):* P*-4-Chlorophenyl-*N,N,N′,N′*-tetramethylphosphonic diamide.*
$C_{10}H_{16}ClN_2OP$ M 246.676
Liq. d_4^{20} 1.19. Bp_4 152°. n_D^{20} 1.5457.

Obrycki, R. *et al, Tetrahedron Lett.*, 1966, 5049 (*dimethyl ester*)
Kharrasova, F.M. *et al, Zh. Obshch. Khim.*, 1968, **38**, 1262 (*Engl. transl.* p. 1215) (*esters, synth*)
Tavs, P., *Chem. Ber.*, 1970, **103**, 2428 (*diethyl ester*)
Kharrasova, F.M. *et al, Zh. Obshch. Khim.*, 1973, **43**, 2642 (*Engl. transl.* p. 2621) (*esters, P nmr*)
Williams, A. *et al, J. Chem. Soc., Perkin Trans. 2*, 1973, 25 (*synth*)
Grabiak, R.C. *et al, Phosphorus Sulfur*, 1980, **9**, 197 (*synth, derivs, P nmr*)
Hirao, T. *et al, Synthesis*, 1981, 56; *Bull. Chem. Soc. Jpn.*, 1982, **55**, 909 (*diethyl ester*)
Issleib, K. *et al, Z. Anorg. Allg. Chem.*, 1985, **529**, 151 (*silyl ester, synth, P nmr*)

(4-Chlorophenyl)phosphonic dichloride, 9CI C-00148

[22585-81-5]

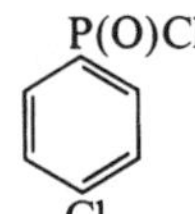

$C_6H_4Cl_3OP$ M 229.430
d_4^{20} 1.51. Mp 29.5-30.5°. Bp 284-285°, Bp_{11} 143-144°. n_D^{20} 1.5775.

Kharrasova, F.M. *et al, Zh. Obshch. Khim.*, 1968, **38**, 1262 (*synth*)
Tsvetkov, E.N. *et al, Zh. Obshch. Khim.*, 1968, **38**, 2285 (*Engl. transl.* p. 2211) (*synth*)
Timokhin, B.V. *et al, Zh. Obshch. Khim.*, 1971, **41**, 2658 (*Engl. transl.* p. 2691) (*raman*)
Kharrasova, F.M. *et al, Zh. Obshch. Khim.*, 1973, **43**, 2642 (*Engl. transl.* p. 2621) (*P nmr*)
Zakirova, D.U. *et al, Zh. Obshch. Khim.*, 1977, **47**, 1662 (*Engl. transl.* p. 1522) (*nqr*)
Dorokhova, V.V. *et al, Zh. Obshch. Khim.*, 1979, **49**, 83 (*Engl. transl.* p. 68) (*uv, raman*)

(4-Chlorophenyl)phosphonic difluoride, 9CI C-00149

[1535-37-1]

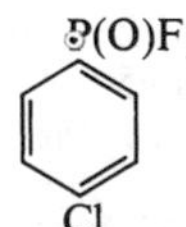

$C_6H_4ClF_2OP$ M 196.521
Liq. d_{20}^{20} 1.34. Bp 211-213°.

Yagupolskii, L.M. *et al, Zh. Obshch. Khim.*, 1959, **29**, 3766; *CA*, **54**, 19553 (*synth*)
Szafraniec, L.L., *Org. Magn. Reson.*, 1974, **6**, 565 (*F and P nmr*)
Daasch, L.W. *et al, Phosphorus*, 1975, **5**, 189 (*ms*)
Dorokhova, V.V. *et al, Zh. Obshch. Khim.*, 1979, **49**, 83 (*Engl. transl.* p. 68) (*uv, raman*)

(4-Chlorophenyl)phosphonothioic acid, 9CI C-00150

$C_6H_6ClO_2PS$ M 208.599

O,O*-Di-Et ester:* [22585-79-1]. O,O*-Diethyl (4-chlorophenyl)phosphonothioate.*
$C_{10}H_{14}ClO_2PS$ M 264.706
Liq. d_4^{20} 1.21. Bp_4 134°. n_D^{20} 1.5485.

Dichloride: [3064-55-9].
$C_6H_4Cl_3PS$ M 245.490
Liq. d_4^{20} 1.52. Bp_5 135°. n_D^{20} 1.6348.

Bis(dimethylamide): P*-4-Chlorophenyl-*N,N,N′,N′*-tetramethylphosphonothioic diamide.*
$C_{10}H_{16}ClN_2PS$ M 262.737
Cryst. (pet. ether). Mp 75-77°.

Tsvetkov, E.N. *et al, Zh. Obshch. Khim.*, 1968, **38**, 2285 (*Engl. transl.* p. 2211) (*synth*)
Dorokhova, V.V. *et al, Zh. Obshch. Khim.*, 1979, **49**, 83 (*Engl. transl.* p. 68) (*dichloride, difluoride, uv, raman*)

(2-Chlorophenyl)phosphonous acid C-00151

$C_6H_6ClO_2P$ M 176.539
The free acid exists as the phosphoryl tautomer (2-Chlorophenyl)phosphinic acid, C-00142 .

Dichloride: [1004-78-0]. *Dichloro(2-chlorophenyl)-phosphine.*
$C_6H_4Cl_3P$ M 213.430
Liq. Bp_9 113-117°, $Bp_{0.7}$ 70-71°. n_D^{23} 1.6118.

Bis(diethylamide): P*-2-Chlorophenyl-*N,N,N′,N′*-tetraethylphosphonous diamide.*
$C_{14}H_{24}ClN_2P$ M 286.784
Liq. Bp_{10} 164-167°.

Schindlbauer, H., *Monatsh. Chem.*, 1965, **96**, 1936 (*derivs, synth, ir*)
Weinberg, K.G., *J. Org. Chem.*, 1975, **40**, 3586 (*dichloride*)

(3-Chlorophenyl)phosphonous acid, 9CI C-00152

$C_6H_6ClO_2P$ M 176.539
The free acid exists as the phosphoryl tautomer (3-Chlorophenyl)phosphinic acid, C-00143 .

Di-Et ester: [23603-71-6]. *Diethyl (3-chlorophenyl)-phosphinite.*
$C_{10}H_{14}ClO_2P$ M 232.646
d_4^{20} 1.13. Bp_3 81°. n_D^{20} 1.5232.

Dichloride: [1718-21-4]. *Dichloro(3-chlorophenyl)-phosphine.*
$C_6H_4Cl_3P$ M 213.430

Liq. d_4^{20} 1.45. Bp_{35} 142-143°, Bp_9 113-116°. n_D^{20} 1.6082.
*Bis(dimethylamide): P-3-Chlorophenyl-*N,N,N′,N′-*tetramethylphosphonous diamide.*
$C_{10}H_{16}ClN_2P$ M 230.677
Oil which solidifies on standing. d_4^{20} 1.11. Mp 42-44°. Bp_3 97°. n_D^{20} 1.5583.
*Bis(diethylamide): P-3-Chlorophenyl-*N,N,N′,N′-*tetraethylphosphonous diamide.*
$C_{14}H_{24}ClN_2P$ M 286.784
Liq. Bp_{11} 168-172°.

Schindlbauer, H., *Monatsh. Chem.*, 1965, **96**, 1936 (*dichloride, diethylamide, synth, ir*)
Lobanov, D.I. *et al*, *Zh. Obshch. Khim.*, 1969, **39**, 841 (*Engl. transl.* p. 805) (*synth, ester, dimethylamide*)
Muylle, E. *et al*, *Spectrochim. Acta, Part A*, 1976, **32**, 599 (*dichloride, P nmr*)

(4-Chlorophenyl)phosphonous acid, 9CI C-00153

$C_6H_6ClO_2P$ M 176.539
The free acid exists as the phosphoryl tautomer (4-Chlorophenyl)phosphinic acid, C-00144 .
Di-Me ester: [19909-82-1]. *Dimethyl (4-chlorophenyl)-phosphonite.*
$C_8H_{10}ClO_2P$ M 204.593
Liq. d_4^{20} 1.21. Bp_8 107-109°. n_D^{20} 1.5430.
Mono-Et ester: see (4-Chlorophenyl)phosphinic acid, C-00144
Di-Et ester: [20355-97-9]. *Diethyl (4-chlorophenyl)-phosphonite.*
$C_{10}H_{14}ClO_2P$ M 232.646
Liq. d_0^{20} 1.13. Bp_{11} 129-130.5°, Bp_4 100-101°. n_D^{20} 1.5252.
Diisopropyl ester: Diisopropyl (4-chlorophenyl)-phosphonite.
$C_{12}H_{18}ClO_2P$ M 260.700
Liq. d_0^{20} 1.09. Bp_{11} 138.5-139°. n_D^{20} 1.5108.
Difluoride: [17056-34-7]. *(4-Chlorophenyl)-difluorophosphine.*
$C_6H_4ClF_2P$ M 180.521
Liq. d^{20} 1.32. Bp_5 33-35°. n_D^{20} 1.4957.
Dichloride: see (4-Chlorophenyl)phosphonous dichloride, C-00154
Dibromide: [70430-43-2]. *Dibromo(4-chlorophenyl)-phosphine.*
$C_6H_4Br_2ClP$ M 302.332
Solid-liq. Mp 16.5-18°. Bp_6 144-146°.
Diiodide: [24901-25-5]. *(4-Chlorophenyl)-diiodophosphine.*
$C_6H_4ClI_2P$ M 396.333
Liq. $Bp_{0.06}$ 107-108°.
Bis(dimethylamide): [22585-87-1]. *P-4-Chlorophenyl-*N,N,N′,N′-*tetramethylphosphonous diamide.*
$C_{10}H_{16}ClN_2P$ M 230.677
Liq. d_4^{20} 1.10. Bp_5 116°. n_D^{20} 1.5602.

Kamai, G. *et al*, *Zh. Obshch. Khim.*, 1961, **31**, 3550 (*Engl. transl.* p. 3311) (*esters*)
Neimysheva, A.A. *et al*, *Zh. Obshch. Khim.*, 1966, **36**, 500 (*Engl. transl.* p. 520) (*ester*)
Drozd, G.I. *et al*, *Zh. Obshch. Khim.*, 1967, **37**, 958, 1343 (*Engl. transl.* pp. 906, 1269) (*difluoride, synth, F and P nmr*)
Tsvetkov, E.N. *et al*, *Zh. Obshch. Khim.*, 1968, **38**, 2285 (*Engl. transl.* p. 2211) (*ester, dimethylamide, synth*)
Feshchenko, N.G. *et al*, *Zh. Obshch. Khim.*, 1969, **39**, 2184 (*Engl. transl.* p. 2133) (*diiodide*)
Parr, W.J.E., *J. Chem. Soc., Faraday Trans. 2*, 1978, **74**, 933 (*difluoride, pmr, cmr*)
Hinke, A. *et al*, *Phosphorus Sulfur*, 1983, **15**, 93 (*dibromide, synth, nmr*)

(4-Chlorophenyl)phosphonous dichloride, 9CI C-00154

Dichloro-4-chlorophenylphosphine
[1005-33-0]

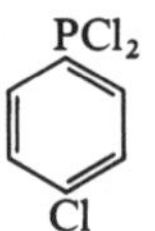

$C_6H_4Cl_3P$ M 213.430
Liq. Bp_{15} 121-122°, $Bp_{0.3}$ 67-69°.

Schindlbauer, H., *Monatsh. Chem.*, 1965, **96**, 1936 (*synth, ir*)
Kharrasova, F.M. *et al*, *Zh. Obshch. Khim.*, 1967, **37**, 902 (*Engl. transl.* p. 852) (*synth*)
Weinberg, K.G., *J. Org. Chem.*, 1975, **40**, 3586 (*synth*)
Muylle, E. *et al*, *Spectrochim. Acta, Part A*, 1976, **32**, 599 (*P nmr*)
Zakirov, D.U. *et al*, *Zh. Obshch. Khim.*, 1977, **47**, 1661 (*Engl. transl.* p. 1522) (*nqr*)
Parr, W.J.E., *J. Chem. Soc., Faraday Trans. 2*, 1978, **74**, 933 (*pmr, cmr, struct*)

(2-Chlorophenyl)phosphoramidic acid, 9CI C-00155

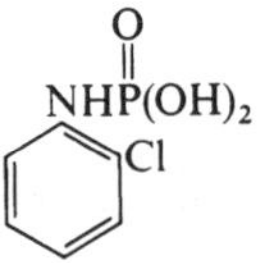

$C_6H_7ClNO_3P$ M 207.553
Di-Ph ester: Diphenyl (2-chlorophenyl)-phosphoramidate.
$C_{18}H_{15}ClNO_3P$ M 359.748
Cryst. (EtOH). Mp 120-122°.

Zhmurova, I.N. *et al*, *Zh. Obshch. Khim.*, 1961, **31**, 3741 (*Engl. transl.* p. 3495)

(3-Chlorophenyl)phosphoramidic acid, 9CI C-00156

[38874-31-6]
$C_6H_7ClNO_3P$ M 207.553
Bis(cyclohexylammonium) salt: [35314-67-1]. Solid. Mp indefinite.
Di-Ph ester: [62569-08-8]. *Diphenyl (3-chlorophenyl)-phosphoramidate.*
$C_{12}H_{15}ClNO_3P$ M 287.682
Cryst. (EtOH). Mp 100-102°.

Zhmurova, I.N. *et al*, *Zh. Obshch. Khim.*, 1961, **31**, 3741 (*Engl. transl.* p. 3495) (*diphenyl ester, synth*)
Williams, A. *et al*, *J. Chem. Soc. (B)*, 1971, 1973 (*synth*)
Mollin, J. *et al*, *Collect. Czech. Chem. Commun.*, 1976, **41**, 3245 (*diphenyl ester, props*)

(4-Chlorophenyl)phosphoramidic acid, 9CI C-00157

[1892-18-8]
$C_6H_7ClNO_3P$ M 207.553
Solid. Mp 268-270° dec.
Di-Me ester: [83470-29-5]. *Dimethyl (4-chlorophenyl)-phosphoramidate.*
$C_8H_{11}ClNO_3P$ M 235.607
Cryst. (toluene). Mp 95-96°.
Di-Et ester: [49802-15-5]. *Diethyl (4-chlorophenyl)-phosphoramidate.*
$C_{10}H_{15}ClNO_3P$ M 263.660
Cryst. (pentane). Mp 76-78°.

Di-Ph ester: [49802-22-4]. *Diphenyl (4-chlorophenyl)-phosphoramidate.*
$C_{18}H_{15}ClNO_3P$ M 359.748
Cryst. (EtOH or cyclohexane). Mp 117-118°.
Dichloride:
$C_6H_5Cl_3NOP$ M 244.444
Cryst. (C_6H_6). Mp 104-106°.

Goldwhite, H. *et al, J. Chem. Soc.*, 1957, 2409 (*synth*)
Zhmurova, I.N. *et al, Zh. Obshch. Khim.*, 1961, **31**, 3741 (*Engl. transl.* p. 3495) (*diphenyl ester*)
Williams, A. *et al, J. Chem. Soc. (B)*, 1971, 1973 (*synth*)
Warner, V.D. *et al, J. Med. Chem.*, 1973, **16**, 1185 (*esters*)
Harger, M.J.P. *et al, Tetrahedron*, 1982, **38**, 1511 (*dichloride, dimethyl ester, synth, ir, pmr*)
Zwierzak, A. *et al, Synthesis*, 1984, 223 (*diethyl ester, synth, P nmr*)

(2-Chlorophenyl)phosphoramidothioic acid, 9CI C-00158

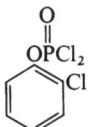

$C_6H_7ClNO_2PS$ M 223.614
O,O-*Di-Et ester:* [15950-60-4]. O,O-*Diethyl (2-chlorophenyl)phosphoramidothioate.*
$C_{10}H_{15}ClNO_2PS$ M 279.721
Liq. n_D^{29} 1.4940.

Japan. Pat., 62 4 458, (*1962*); *CA*, **67**, 73347

(3-Chlorophenyl)phosphoramidothioic acid, 9CI C-00159

$C_6H_7ClNO_2PS$ M 223.614
O,O-*Di-Et ester:* [15950-61-5]. O,O-*Diethyl (3-chlorophenyl)phosphoramidothioate.*
$C_{10}H_{15}ClNO_2PS$ M 279.721
Liq. $Bp_{0.3}$ 157-160°.

Ger. Pat., 1 154 669, (*1960*); *Chem. Zentralbl.*, 1964, **24**, 2167

(4-Chlorophenyl)phosphoramidothioic acid, 9CI C-00160

$C_6H_7ClNO_2PS$ M 223.614
O,O-*Di-Me ester:* O,O-*Dimethyl (4-chlorophenyl)-phosphoramidothioate.*
$C_8H_{11}ClNO_2PS$ M 251.667
Cryst. (EtOH aq.). Mp 78-80°.
O,O-*Di-Et ester:* O,O-*Diethyl (4-chlorophenyl)-phosphoramidothioate.*
$C_{10}H_{15}ClNO_2PS$ M 279.721
Liq. $Bp_{0.3}$ 153-156°.

Ger. Pat., 1 154 669, (*1960*); *Chem. Zentralbl.*, 1964, **24**, 2167

2-Chlorophenyl phosphorodichloridate, 9CI C-00161

2-Chlorophenyl dichlorophosphate. 2-Chlorophenyl phosphoryl dichloride
[15074-54-1]

$C_6H_4Cl_3O_2P$ M 245.429
Phosphorylating agent, often used in conjunction with 1,2,4-triazole or 1-hydroxybenzotriazole. Fuming liq. Bp_{12} 135-137°.

Anschütz, R. *et al, Justus Liebigs Ann. Chem.*, 1918, **415**, 64 (*synth*)
Oltvoort, J.J. *et al, Recl. Trav. Chim. Pays-Bas*, 1982, **101**, 87 (*use*)
Van Boeckel, C.A.A. *et al, Recl. Trav. Chim. Pays-Bas*, 1983, **102**, 526 (*use*)
Wreesman, C.T.J. *et al, Nucleic Acids Res.*, 1983, **11**, 8389 (*use*)

4-Chlorophenyl phosphorodichloridate, 9CI C-00162

4-Chlorophenyl phosphoryl dichloride. 4-Chlorophenyl dichlorophosphate
[772-79-2]
$C_6H_4Cl_3O_2P$ M 245.429
Phosphorylating agent, and precursor to other phosphorylating reagents, for use in oligonucleotide synth. Oily liq. Bp 265°, Bp_{12} 141°.

Anschütz, R. *et al, Justus Liebigs Ann. Chem.*, 1918, **415**, 51 (*synth*)
Orloff, H.D. *et al, J. Am. Chem. Soc.*, 1958, **80**, 727 (*synth*)
Narang, S.A. *et al, Methods Enzymol.*, 1980, **65**, 610 (*synth, use*)
De Bernardini, S. *et al, Helv. Chim. Acta*, 1981, **64**, 2142 (*use*)
Rosenthal, A. *et al, J. Prakt. Chem.*, 1982, **324**, 793 (*use*)
Wing, L.S., *J. Org. Chem.*, 1982, **47**, 3623 (*use*)

O-(4-Chlorophenyl) phosphorodichloridothioate, 9CI, 8CI C-00163

O-*(4-Chlorophenyl) dichlorothiophosphate.* O-*(4-Chlorophenyl) thiophosphoryl dichloride*
[19081-37-9]

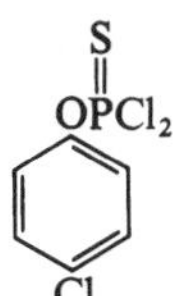

$C_6H_4Cl_3OPS$ M 261.490
Liq. Bp_5 98-100°. n_D^{20} 1.5850.

Protsenko, L.D. *et al, Zh. Obshch. Khim.*, 1964, **34**, 2233 (*Engl. transl.* p. 2244) (*synth*)

S-(4-Chlorophenyl) phosphorodichloridothioate, 9CI, 8CI C-00164

S-*(4-Chlorophenyl) dichlorothiophosphate*
[21186-91-4]

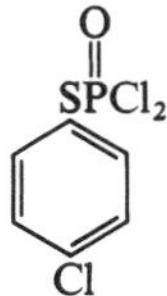

$C_6H_4Cl_3OPS$ M 261.490
Solid. Mp 65°. Bp_1 117-118°.

Shitov, L.N. *et al, Zh. Obshch. Khim.*, 1968, **38**, 2340 (*Engl. transl.* p. 2268) (*synth*)

N-[(4-Chlorophenyl)sulfonyl]-P-ethylphosphonimidothioic acid, 9CI C-00165

N-(4-Chlorobenzenesulfonyl)-P-ethylphosphonimidothioic acid

$C_8H_{11}ClNO_3PS_2$ M 299.727

Free acid tautomeric.

(±)-form

O-Et-S-Me ester: O-Ethyl S-methyl N-[(4-chlorophenyl)sulfonyl]-P-ethylphosphonamidothioate. N-[(4-Chlorophenyl)sulfonyl]-P-ethoxy-P-ethyl-P-methylthiophosphine imide.

$C_{11}H_{17}ClNO_3PS_2$ M 341.807

Syrup. d_4^{20} 1.34. n_D^{20} 1.5672.

Almasi, L. *et al*, *Tetrahedron*, 1977, **33**, 1327 (*synth*)
Barabas, A. *et al*, *Org. Magn. Reson.*, 1977, **10**, 35 (*pmr, P nmr*)

N-[(4-Chlorophenyl)sulfonyl]-P-phenylphosphonimidothioic acid, 9CI C-00166

$C_{12}H_{11}ClNO_3PS$ M 315.711

(±)-form

O-Et-S-Me ester: O-Ethyl S-methyl N-[(4-chlorophenyl)sulfonyl]-P-phenylphosphonimidothioate.

$C_{15}H_{17}ClNO_3PS_2$ M 389.851

Solid. Mp 56-57°.

Almasi, L. *et al*, *Chem. Ber.*, 1973, **106**, 1384 (*synth, props*)

1-Chlorophospholane, 8CI C-00167

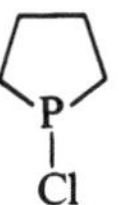

C_4H_8ClP M 122.534

Liq. Bp 165° (est.), Bp_{20} 65°.

Oxide: Tetramethylenephosphinic chloride.

C_4H_8ClOP M 138.533

Liq. Bp_{15} 119° (Bp_{12} 137-141°). n_D^{20} 1.5001.

Sulfide: Tetramethylenephosphinothioic chloride.

C_4H_8ClPS M 154.594

Liq. $Bp_{0.1}$ 106-110°.

Burg, A.B. *et al*, *J. Am. Chem. Soc.*, 1960, **82**, 2148 (*synth*)
Hunger, K. *et al*, *Tetrahedron*, 1964, **20**, 1593 (*oxide*)
Sommer, K., *Z. Anorg. Allg. Chem.*, 1970, **379**, 56 (*synth, nmr*)
Fell, B. *et al*, *Synthesis*, 1974, 119 (*use*)
Wolfsberger, W., *J. Organomet. Chem.*, 1986, **317**, 167 (*cmr, P nmr*)

1-Chlorophosphorinane, 9CI C-00168

1-Chlorophosphacyclohexane. 1-Chlorophosphinane

[30292-78-5]

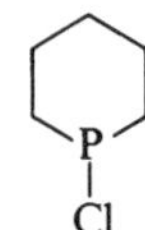

$C_5H_{10}ClP$ M 136.561

Liq. Bp_{20} 75-77°.

Oxide:

$C_5H_{10}ClOP$ M 152.560

Liq. Bp_{15} 128-130°.

Sulfide:

$C_5H_{10}ClPS$ M 168.621

Liq. $Bp_{0.1}$ 108-123°.

Sommer, K., *Z. Anorg. Allg. Chem.*, 1970, **379**, 56 (*synth, P nmr, derivs*)

(3-Chloropropyl)diphenylphosphine C-00169

[57137-55-0]

$Ph_2PCH_2CH_2CH_2Cl$

$C_{15}H_{16}ClP$ M 262.718

Ligand for Fe, Mn, Ni, W and Re; used in synth. of multidentate ligands. Viscous oil.

B,MeBPh₄: (3-Chloropropyl)-methyldiphenylphosphonium tetraphenylborate.

$C_{40}H_{39}BClP$ M 596.985

Mp 156-158°.

Grim, S.O. *et al*, *J. Organomet. Chem.*, 1975, **94**, 327 (*synth, pmr, nmr, derivs*)
Kuklmann, E.J. *et al*, *Inorg. Chim. Acta*, 1979, **34**, 197 (*pmr, ir*)
Arpac, E. *et al*, *Z. Naturforsch.*, *B*, 1980, **35**, 146 (*synth*)
Uriarte, R. *et al*, *Inorg. Chem.*, 1980, **19**, 79 (*synth*)

(1-Chloropropyl)phosphonic acid, 9CI C-00170

$H_3CCH_2CHClP(O)(OH)_2$

$C_3H_8ClO_3P$ M 158.521

(±)-form

Di-Et ester: [56436-86-3]. *Diethyl (1-chloropropyl)-phosphonate.*

$C_7H_{16}ClO_3P$ M 214.628

Liq. $Bp_{0.3}$ 70°. n_D^{20} 1.4390.

Okamoto, Y. *et al*, *Bull. Chem. Soc. Jpn.*, 1975, **48**, 484 (*synth*)
Coutrot, P. *et al*, *Synthesis*, 1977, 615 (*synth, pmr*)

(2-Chloropropyl)phosphonic acid, 9CI C-00171

[53589-30-3]

$H_3CCHClCH_2P(O)(OH)_2$

$C_3H_8ClO_3P$ M 158.521

(±)-form

Di-Et ester: Diethyl (2-chloropropyl)phosphonate.

$C_7H_{16}ClO_3P$ M 214.628

Liq. Bp_1 77-79°.

Dichloride:

$C_3H_6Cl_3OP$ M 195.413

d_4^{20} 1.46. $Bp_{0.1}$ 65-68°. n_D^{20} 1.4960.

Boyce, C.B.C. *et al*, *J. Chem. Soc., Perkin Trans. 1*, 1974, 1644, 1650 (*ester, dichloride, pmr*)
Nuretdinova, O.N. *et al*, *Izv. Akad. Nauk SSSR, Ser. Khim.*, 1974, **23**, 1364 (*Engl. transl.* p. 1283) (*dichloride*)
Okamoto, Y. *et al*, *Bull. Chem. Soc. Jpn.*, 1975, **48**, 484 (*ester*)

(3-Chloropropyl)phosphonic acid, 9CI C-00172

[13317-09-4]

$ClCH_2CH_2CH_2P(O)(OH)_2$

$C_3H_8ClO_3P$ M 158.521

Plates ($CHCl_3$); hygroscopic. Sl. sol. H_2O, EtOH, Me_2CO, insol. pet. ether. Mp 105-106°.

Monoanilinum salt: Cryst. (CCl_4). Mp 151-152°.

Di-Me ester: Dimethyl (3-chloropropyl)phosphonate.
$C_5H_{12}ClO_3P$ M 186.575
Liq. d_4^{20} 1.24. $Bp_{0.06}$ 74-75°. n_D^{20} 1.4520.

Di-Et ester: [23269-98-9]. *Diethyl (3-chloropropyl)-phosphonate.*
$C_7H_{16}ClO_3P$ M 214.628
Liq. d_4^{20} 1.14. $Bp_{0.04}$ 84-85°. n_D^{20} 1.4460.

Di-2-propenyl ester: Di-2-propenyl (3-chloropropyl)-phosphonate. Diallyl (3-chloropropyl)phosphonate.
$C_9H_{16}ClO_3P$ M 238.650
Liq. d_4^{20} 1.15. $Bp_{0.02}$ 102-103°. n_D^{20} 1.4695.

Di-Ph ester: [23270-01-1]. *Diphenyl (3-chloropropyl)-phosphonate.*
$C_{15}H_{16}ClO_3P$ M 310.716
Liq. d_4^{20} 1.24. $Bp_{0.04}$ 172-173°. n_D^{20} 1.5515.

Dichloride: [21510-86-1].
$C_3H_6Cl_3OP$ M 195.413
Liq. d_4^{20} 1.45. $Bp_{0.04}$ 63-64°. n_D^{20} 1.4982.

Diamide: see P-(3-Chloropropyl)phosphonic diamide, C-00173

Helferich, B. *et al*, *Justus Liebigs Ann. Chem.*, 1962, **655**, 59 (*synth, dichloride*)
Vinokurova, G.M. *et al*, *Izv. Akad. Nauk SSSR, Ser. Khim.*, 1969, 884 (*Engl. transl.* p. 803) (*derivs, synth, ir*)

P-(3-Chloropropyl)phosphonic diamide, 9CI C-00173

$ClH_2CCH_2CH_2P(O)(NH_2)_2$

$C_3H_{10}ClN_2OP$ M 156.552

N,N′-*Di-2-propenyl:* P-*(3-Chloropropyl)*-N,N′-*di-2-propenylphosphonic diamide.* N,N′-*Diallyl*-P-*(3-chloropropyl)phosphonic diamide.*
$C_9H_{16}ClN_2OP$ M 234.665
Liq. d_4^{20} 1.13. $Bp_{0.04}$ 134-135°. n_D^{20} 1.5233.

N,N,N′,N′-*Tetra-Et:* [23040-94-0]. P-*(3-Chloropropyl)*-N,N,N′,N′-*tetraethylphosphonic diamide.*
$C_{11}H_{26}ClN_2OP$ M 268.766
Liq. d_4^{20} 1.06. $Bp_{0.04}$ 146-147°. n_D^{20} 1.4810.

N,N′-*Di-Ph:* [23040-98-4]. P-*(3-Chloropropyl)*-N,N′-*diphenylphosphonic diamide. (3-Chloropropyl)-phosphonic dianilide.*
$C_{15}H_{18}ClN_2OP$ M 308.747
Solid. Mp 116-118°, 161-162°.

Helferich, B. *et al*, *Justus Liebigs Ann. Chem.*, 1962, **655**, 59 (*dianilide*)
Aleksandrova, I.A. *et al*, *Izv. Akad. Nauk SSSR, Ser. Khim.*, 1969, 1163 (*Engl. transl.* p. 1064) (*derivs*)

2-Chloro-2,2′-spirobi[1,3,2-benzodioxaphosphole], 9CI C-00174

[6857-81-4]

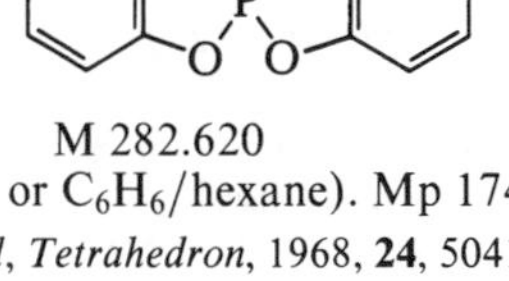

$C_{12}H_8ClO_4P$ M 282.620

Cryst. (hexane or C_6H_6/hexane). Mp 174-175°.

Ramirez, F. *et al*, *Tetrahedron*, 1968, **24**, 5041 (*synth, struct, P nmr*)
Binder, H. *et al*, *Z. Anorg. Allg. Chem.*, 1971, **384**, 193 (*synth*)
Brown, R.K. *et al*, *Inorg. Chem.*, 1977, **16**, 2294 (*cryst struct*)
Dillon, K.B. *et al*, *J. Chem. Soc., Dalton Trans.*, 1981, 2292 (*complexes*)
Gloede, J. *et al*, *Z. Anorg. Allg. Chem.*, 1981, **480**, 142; 1983, **500**, 59 (*synth, P nmr*)
Lattman, M. *et al*, *Inorg. Chim. Acta*, 1983, **76**, L139 (*complexes*)
Gloede, J., *Phosphorus Sulfur*, 1984, **20**, 15 (*props*)

2-(Chlorosulfonyl)phenyl phosphorodichloridate C-00175

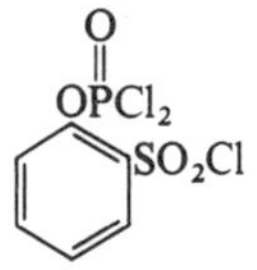

$C_6H_4Cl_3O_4PS$ M 309.488

Oil.

Anschütz, R., *Justus Liebigs Ann. Chem.*, 1918, **415**, 64 (*synth, struct*)

4-(Chlorosulfonyl)phenyl phosphorodichloridate, 9CI C-00176

$C_6H_4Cl_3O_4PS$ M 309.488

Solid. Mp 87-88°. $Bp_{13.5}$ 203°.

Anschütz, R., *Justus Liebigs Ann. Chem.*, 1908, **358**, 92 (*struct, synth*)

2-Chloro-1,2,3,4-tetrahydro-6-methyl-3-(4-methylphenyl)-1,3,2-benzodiazaphosphorine 2-oxide C-00177

[73086-51-8]

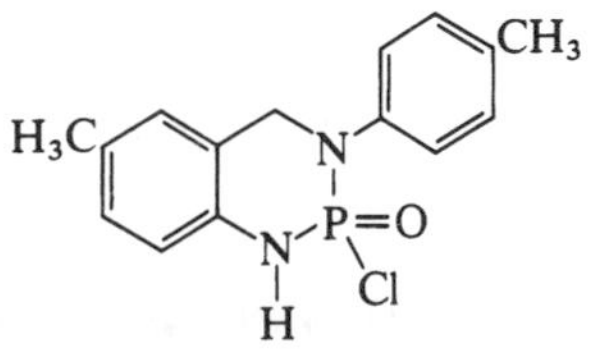

$C_{15}H_{16}ClN_2OP$ M 306.731

Cryst. Mp 158-163° dec.

Cameron, T.S. *et al*, *J. Chem. Soc., Perkin Trans. 1*, 1979, 2896 (*synth, pmr, cryst struct*)

2-Chlorotetrahydro-3-methyl-2*H*-1,3,2-oxazaphosphorine, 9CI C-00178

2-Chloro-3-methyl-1,3,2-oxazaphosphorinane. 2-Chloro-3-methyl-1-oxa-3-aza-2-phosphacyclohexane

[67105-50-4]

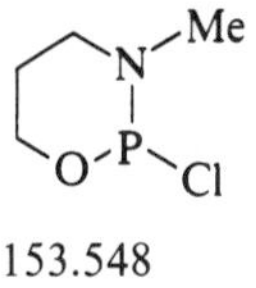

C_4H_9ClNOP M 153.548

Fuming liq. d_4^{20} 1.23. Bp_{13} 81-83°. n_D^{20} 1.5110.

Nuretdinov, I.A. *et al*, *Izv. Akad. Nauk SSSR, Ser. Khim.*, 1978, 950 (*Engl. transl.* p. 824) (*synth, nqr*)
Nuretdinov, I.A. *et al*, *Zh. Fiz. Khim.*, 1979, **53**, 126 (*Engl. transl.* p. 66) (*nqr*)

4-Chloro-1,2,3,4-tetrahydronaphth[2,1-*c*][1,2]azaphosphorine C-00179

$C_{12}H_{11}ClNP$ M 235.652

Sulfide: [76065-67-3].
$C_{12}H_{11}ClNPS$ M 267.712
Cryst. ($CHCl_3$). Mp 147°.

Grewal, G.S. *et al*, *Indian J. Chem., Sect. B*, 1980, **19**, 404 (*synth, ir, ms*)

2-Chlorotetrahydro-2*H*-1,3,2-oxazaphosphorine C-00180

2-Chloro-1,3,2-oxazaphosphorinane. 2-Chloro-1,3,2-oxazaphosphinane

C_3H_7ClNOP M 139.521

2-Oxide: [5638-58-4].
$C_3H_7ClNO_2P$ M 155.521
Phosphorylating agent. Solid. Mp 80-83°. V. sensitive to moisture.

2-Sulfide: [19733-70-1].
$C_3H_7ClNOPS$ M 171.581
Solid. Mp 84-86°.

Iwamoto, R.H. *et al*, *J. Org. Chem.*, 1961, **26**, 4743 (*oxide, ir*)
Durrieu, J. *et al*, *Org. Magn. Reson.*, 1973, **5**, 407 (*sulfide, synth, pmr*)
Zon, G. *et al*, *J. Am. Chem. Soc.*, 1977, **99**, 5785 (*oxide*)
Hunston, R.N. *et al*, *Tetrahedron*, 1980, **36**, 2337 (*oxide, synth, ms, pmr, P nmr*)
Mikolajczyk, M. *et al*, *Tetrahedron*, 1982, **38**, 2183 (*sulfide, props*)

2-Chlorotetrahydro-3-(1-phenylethyl)-2*H*-1,3,2-oxazaphosphorine 2-oxide, 10CI C-00181

2-Chloro-3-(α-methylbenzyl)-1,3,2-oxazaphosphorinane 2-oxide

(R_c,R_p)*-form*

$C_{11}H_{15}ClNO_2P$ M 259.672

(R_C,R_P)-form [73834-61-4]
Solid. Mp 71-73°. $[\alpha]_D^{20}$ −51.7° (c, 6.8 in EtOH).

(R_C,S_P)-form [73834-62-5]
Solid. Mp 69-71°. $[\alpha]_D^{20}$ −61.3° (c, 3.3 in EtOH).

(S_C,R_P)-form [72578-62-2]
Cryst. (Et_2O/hexane). Mp 69-71°. $[\alpha]_D^{20}$ −61.4° (c, 3.1 in EtOH).

(S_C,S_P)-form [72578-63-3]
Needles (toluene/hexane). Mp 71-73°. $[\alpha]_D^{25}$ +51.5° (c, 8.2 in EtOH).

Pankiewicz, K. *et al*, *J. Am. Chem. Soc.*, 1979, **101**, 7712 (*synth, P nmr*)
Sato, T. *et al*, *J. Org. Chem.*, 1983, **48**, 98 (*synth, resoln, pmr, ms*)

6-Chloro-5,6,7,12-tetrahydro-2,5,7,10-tetramethyldibenzo[*d,g*][1,3,2]-diazaphosphocine C-00182

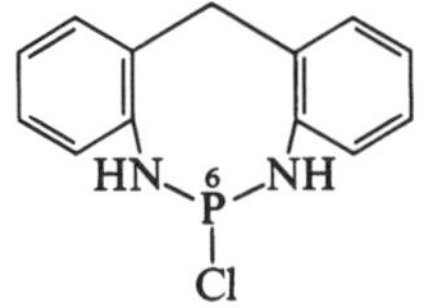

$C_{13}H_{12}ClN_2P$ M 262.678

6-Oxide: [19732-22-0].
$C_{13}H_{12}ClN_2OP$ M 278.677
Solid. Mp 170-171°.

6-Sulfide: [38151-54-1].
$C_{13}H_{12}ClN_2PS$ M 294.738
Cryst. (EtOAc or CCl_4). Mp 209-210°.

Cheng, C.Y. *et al*, *J. Chem. Soc., Chem. Commun.*, 1968, 616 (*synth, pmr*)
Cameron, T.S. *et al*, *J. Chem. Soc., Perkin Trans. 2*, 1972, 591 (*oxide, cryst struct*)
Cameron, T.S. *et al*, *Acta Crystallogr., Sect. B*, 1975, **31**, 2331 (*sulfide, cryst struct*)
Dagleish, W.H. *et al*, *J. Chem. Soc., Dalton Trans.*, 1977, 1505 (*nqr*)

2-Chlorotetrahydro-3,5,5-trimethyl-2*H*-1,3,2-oxazaphosphorine, 9CI C-00183

2-Chloro-3,5,5-trimethyl-1,3,2-oxazaphosphorinane. 2-Chloro-3,5,5-trimethyl-1-oxa-3-aza-2-phosphacyclohexane

$C_6H_{13}ClNOP$ M 181.602

2-Oxide: [85289-31-2].
$C_6H_{13}ClNO_2P$ M 197.601
Cryst. (cyclohexane). Mp 69.5-70°.

2-Sulfide: [85289-33-4].
$C_6H_{13}ClNOPS$ M 213.662
Cryst. (cyclohexane/pet. ether). Mp 69°.

Edmundson, R.S., *Org. Mass Spectrom.*, 1982, **17**, 558 (*synth, ir, ms*)

2-Chloro-4,4,5,5-tetrakis(trifluoromethyl)-1,3,2-dioxaphospholane C-00184

Tetrakis(trifluoromethyl)ethylene chlorophosphite. Tetrakis(trifluoromethyl)ethylene phosphorochloridite

[70311-64-7]

$C_6ClF_{12}O_2P$ M 398.472
Fuming liq. Bp 118°.

2-Oxide: Tetrakis(trifluoromethyl)ethylene chlorophosphate. Tetrakis(trifluoromethyl)ethylene phosphorochloridate.
$C_6ClF_{12}O_3P$ M 414.472
Liq. Bp 125°.

Roeschenthaler, G.V. *et al*, *Z. Anorg. Allg. Chem.*, 1979, **450**, 79 (*synth, ir, ms, F and P nmr*)
Bohlen, R. *et al*, *Z. Anorg. Allg. Chem.*, 1984, **513**, 199 (*props*)

2-Chloro-4,4,5,5-tetramethyl-1,3,2-dioxaphospholane, 9CI C-00185

2-Chloro-4,4,5,5-tetramethyl-1,3-dioxa-2-phosphacyclopentane. Tetramethylethylene chlorophosphite

[14812-59-0]

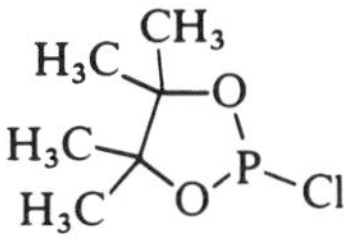

$C_6H_{12}ClO_2P$ M 182.586

Fuming liq. Bp_9 78-81°. n_D^{22} 1.4723.

2-Oxide: [60146-70-5]. *Tetramethylethylene chlorophosphate. Tetramethylethylene phosphoryl chloride.*
$C_9H_{12}ClO_3P$ M 234.619
No phys. props. reported.

2-Sulfide: [59523-65-8]. *Tetramethylethylene chlorothiophosphate. Tetramethylethylene thiophosphoryl chloride.*
$C_6H_{12}ClO_2PS$ M 214.646
Cryst. (pet. ether at 0°). Mp 62-64°.

Fontal, B. *et al*, *Tetrahedron*, 1966, **22**, 3275 (*synth, pmr*)
Shagidullin, R.R. *et al*, *Zh. Obshch. Khim.*, 1975, **45**, 1257; 1976, **46**, 1021 (*Engl. transl.* pp. 1235, 1021) (*ir, raman*)
Mikolajczyk, M. *et al*, *J. Chem. Soc., Perkin Trans. 1*, 1976, 371 (*sulfide*)
Plyamovotyi, A.Kh. *et al*, *Dokl. Akad. Nauk SSSR, Ser. Sci. Khim.*, 1977, **235**, 124 (*Engl. transl.* p. 388) (*struct*)
Arshinova, R.P. *et al*, *Izv. Akad. Nauk SSSR, Ser. Khim.*, 1978, 609 (*Engl. transl.*, p. 524) (*derivs, ir, P nmr, conformn*)
Besserre, D. *et al*, *Org. Magn. Reson.*, 1980, **13**, 235, 313 (*cmr, conformn*)
Kubayashi, S. *et al*, *Polym. Bull. (Berlin)*, 1980, **3**, 585 (*synth*)

Chlorotetraphenoxyphosphorane, 9CI, 8CI C-00186

[32394-39-1]

$(PhO)_4PCl$ or $(PhO)_4P^{\oplus}(PhO)_4PCl_2^{\ominus}$

$C_{24}H_{20}ClO_4P$ M 438.846

Has an ionic struct., possibly as shown. Converts alcohols into alkyl chlorides. Plates (chlorobenzene/ethylene dichloride). Mp 76-78°. Hydrolysed to triphenyl phosphate.

Rydon, H.N. *et al*, *J. Chem. Soc.*, 1956, 3043 (*synth, struct*)
Tseng, C.K., *J. Org. Chem.*, 1979, **44**, 2793 (*struct*)
Dennis, L.W. *et al*, *J. Am. Chem. Soc.*, 1982, **104**, 230 (*synth, nmr, struct*)

7-Chloro-6-thia-7-phosphabicyclo[3.2.1]octane, 10CI C-00187

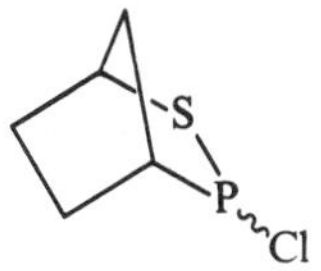

$C_6H_{10}ClPS$ M 180.632

7-Sulfide: [76442-57-4].
$C_6H_{10}ClPS_2$ M 212.692
Cryst. (heptane). Mp 75-77°. $Bp_{0.05}$ 135-140°. Probably a mixt. of stereoisomers.

U.S.P., 4 231 970, (*1980*); *CA*, **94**, 84303 (*synth, cmr, P nmr*)

3-Chloro-1,3-thiaphosphetane, 9CI, 8CI C-00188

Thiodimethylenephosphinous chloride

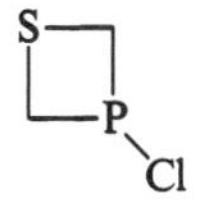

C_2H_4ClPS M 126.540

3-Oxide: [22585-76-8].
C_2H_4ClOPS M 142.540
Polycryst. solid. Mp 76°. Rapidly hydrolysed in air.

3-Sulfide: [35172-21-5].
C_2H_4ClPS M 126.540
Liq. d_4^{20} 1.53. Bp_{10} 114-5°. n_D^{20} 1.6635.

Gilyazov, M.M. *et al*, *Izv. Akad. Nauk SSSR, Ser. Khim.*, 1970, 1177 (*Engl. transl.* p. 1117) (*oxide, synth*)
Naumov, V.A. *et al*, *Dokl. Akad. Nauk SSSR, Ser. Sci. Khim.*, 1971, **200**, 882 (*Engl. transl.* p. 859) (*oxide, struct*)
Arshinova, R.P. *et al*, *Dokl. Akad. Nauk SSSR, Ser. Sci. Khim.*, 1972, **204**, 1118 (*Engl. transl.* p. 504) (*sulfide, synth, pmr, P nmr, oxide*)
Arshinova, R.P. *et al*, *Izv. Akad. Nauk SSSR, Ser. Khim.*, 1973, 2240 (*Engl. transl.* p. 2185) (*derivs, conformn*)
Shagidullin, R.R. *et al*, *Zh. Obshch. Khim.*, 1975, **45**, 536 (*Engl. transl.* p. 530) (*derivs, ir, raman*)

2-Chloro-4,4,6-trimethyl-1,3,2-dioxaphosphorinane, 9CI C-00189

2-Chloro-4,4,6-trimethyl-1,3-dioxa-2-phosphacyclohexane. 1,3,3-Trimethyltrimethylene chlorophosphite. 2-Methyl-2,4-pentanediyl chlorophosphite

[80812-97-1]

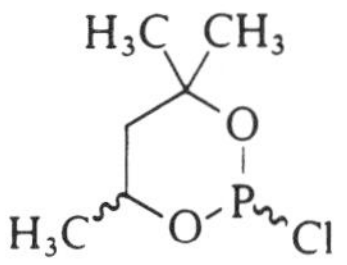

$C_6H_{12}ClO_2P$ M 182.586

Fuming liq. Bp_2 53-55°. n_D^{25} 1.473. Mixt. of *cis* and *trans* forms.

2-Sulfide: 1,3,3-Trimethyltrimethylene chlorothiophosphate. 2-Methyl-2,4-pentanediyl chlorothiophosphate.
$C_6H_{12}ClO_2PS$ M 214.646
Straw-coloured oil. n_D^{20} 1.5129.

Edmundson, R.S. *et al*, *J. Chem. Soc. (C)*, 1966, 2001 (*sulfide*)
Majoral, J.P. *et al*, *Spectrochim. Acta, Part A*, 1972, **28**, 2247 (*oxide, sulfide, ir*)
Kinas, R. *et al*, *Phosphorus Sulfur*, 1978, **4**, 295 (*synth, P nmr*)

Chloro(triphenylphosphoranylidene)acetic acid, 9CI C-00190

$Ph_3P{=}CClCOOH$

$C_{20}H_{16}ClO_2P$ M 354.772

Me ester: [31459-98-0]. *Methyl chloro(triphenylphosphoranylidene)acetate.*
$C_{21}H_{18}ClO_2P$ M 368.799
Stabilized Wittig reagent. Cryst. (EtOAc/pet. ether). Mp 151-152°, 172-173°.

Et ester: Ethyl chloro(triphenylphosphoranylidene)-acetate.
$C_{22}H_{20}ClO_2P$ M 382.826
Stabilised Wittig reagent. Cryst. (Me_2CO/hexane). Mp 147.5-148.5°, 176°.

Denney, D.B. *et al*, *J. Org. Chem.*, 1962, **27**, 998 (*synth*)
Märkl, G., *Chem. Ber.*, 1962, **95**, 3003 (*synth, ir*)
Speziale, A.J. *et al*, *J. Am. Chem. Soc.*, 1963, **85**, 2790; *J. Org. Chem.*, 1963, **28**, 465 (*ir, use, uv*)
Bestmann, H.J. *et al*, *Synthesis*, 1970, 590 (*synth*)
Elliott, M. *et al*, *J. Chem. Soc., Perkin Trans. 1*, 1974, 2470 (*use*)

1-Chloro-3-(triphenylphosphoranylidene)-2-propanone, 9CI **C-00191**

[13605-66-8]

$Ph_3P{=}CHCOCH_2Cl$

$C_{21}H_{18}ClOP$ M 352.799

Solid. Mp 178-180°.

Hudson, R.F. *et al*, *J. Org. Chem.*, 1963, **28**, 2446 (*synth, ir, pmr, props*)
Henichart, J.P., *J. Mol. Struct.*, 1983, **99**, 288 (*synth, cmr, pmr, nmr, ir*)

Chlorphonium, BSI **C-00192**

Tributyl[(2,4-dichlorophenyl)methyl]phosphonium, 9CI. *Tributyl(2,4-dichlorobenzyl)phosphonium. Phosphon*

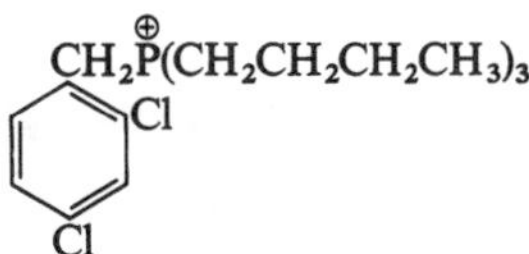

$C_{19}H_{32}Cl_2P^{\oplus}$ M 362.342 (ion)

Plant growth regulator.

Chloride: [115-78-6].

$C_{19}H_{32}Cl_3P$ M 397.795

Cryst. Sol. H_2O, alcs. Mp 114-120°.

▷TA2975000.

U.S.P., 3 268 323, (*1959*); *CA*, **65**, 12800b (*use, props*)
Nicholas, M.L. *et al*, *J. Assoc. Off. Anal. Chem.*, 1976, **59**, 1071 (*raman*)
Pesticide Manual, 6th Ed., 116; 7th Ed., 122.

Chlorphoxim, BSI **C-00193**

7-(2-Chlorophenyl)-4-ethoxy-3,5-dioxa-6-aza-4-phosphaoct-6-ene-8-nitrile 4-sulfide, 9CI. *2-Chloro-α-[(diethoxyphosphinothioyloxy)imino]-benzeneacetonitrile*, 9CI. o*-Chlorophenylglyoxylonitrile oxime* O,O*-diethyl phosphorothioate*, 8CI. O,O*-Diethyl 2-chloro-α-cyanobenzylideneaminooxyphosphonothioate*

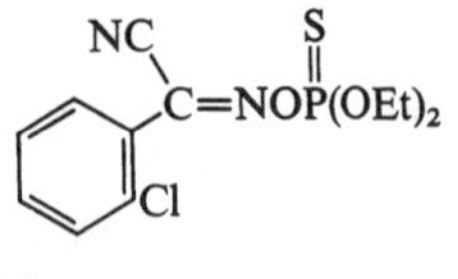

$C_{12}H_{14}ClN_2O_3PS$ M 332.741

Insecticide. Solid. Spar. sol. H_2O, v. sol. cyclohexane, toluene. Mp 66.5°.

Pesticide Manual, 7th Ed., 123.

Chlorpyrifos, BSI, ISO **C-00194**

O,O*-Diethyl* O*-(3,5,6-trichloro-2-pyridinyl) phosphorothioate*, 9CI. *Dursban*

[2921-88-2]

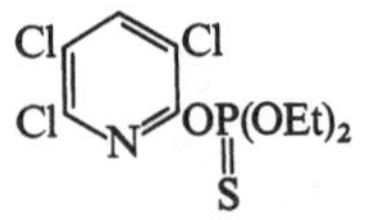

$C_9H_{11}Cl_3NO_3PS$ M 350.584

Broad range contact insecticide. Cryst. with mild mercaptan odour. V. spar. sol. H_2O, sol. most org. solvs. Mp 42-43.5°. Rel. stable to sunlight.

▷Cholinesterase inhibitor. *N*-Ethylating agent for RNA and DNA. TF6300000.

Babad, H. *et al*, *Anal. Chim. Acta*, 1968, **41**, 259 (*pmr*)
Ross, R.T. *et al*, *Anal. Chim. Acta*, 1970, **52**, 139 (*P nmr*)
Baughman, R.G. *et al*, *J. Agric. Food Chem.*, 1978, **26**, 576 (*cryst struct*)
Busch, K.L. *et al*, *Appl. Spectrosc.*, 1978, **32**, 388 (*ms*)
Lores, E.M. *et al*, *J. Agric. Food Chem.*, 1978, **26**, 118 (*metab*)
Stan, H.-J. *et al*, *Biomed. Mass. Spectrom.*, 1982, **9**, 483 (*ms*)
Stan, H.-J. *et al*, *J. Chromatogr.*, 1983, **279**, 173 (*glc*)
Nolan, R.J. *et al*, *J. Toxicol. Appl. Pharmacol.*, 1984, **73**, 8 (*pharmacol*)
Sultatos, L.G. *et al*, *J. Toxicol. Appl. Pharmacol.*, 1984, **73**, 60 (*tox*)
The Agrochemicals Handbook, Royal Society of Chemistry, 1983, A088.
Pesticide Manual, 7th Ed., 125.

Chlorpyrifos Methyl, BSI, ISO **C-00195**

O*-(3,5,6-Trichloro-2-pyridinyl)* O,O*-dimethyl phosphorothioate*, 9CI

[5598-13-0]

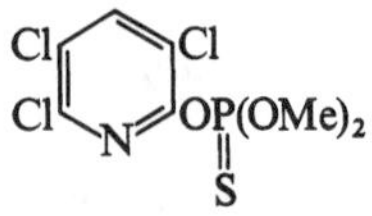

$C_7H_7Cl_3NO_3PS$ M 322.530

Broad range contact insecticide with little persistence. Solid. Spar. sol. H_2O, v. sol. org. solvs. Mp 44.5-45.5°. Stable under neutral conditions. Dec. at pH<6 and >8.

▷TG0700000.

Rigterink, R.H. *et al*, *J. Agric. Food Chem.*, 1966, **14**, 304 (*synth*)
Beckmann, D.E. *et al*, *J. Agric. Food Chem.*, 1979, **27**, 712 (*cryst struct*)
Ripley, B.D. *et al*, *J. Assoc. Off. Anal. Chem.*, 1983, **66**, 1084 (*glc*)
Pesticide Manual, 7th Ed., 126.
The Agrochemicals Handbook, Royal Society of Chemistry, 1983, A445.

Chlorthion **C-00196**

O*-(3-Chloro-4-nitrophenyl)* O,O*-dimethyl phosphorothioate*, 9CI, 8CI. *3-Chloro-4-nitrophenyl dimethyl thiophosphate*

[500-28-7]

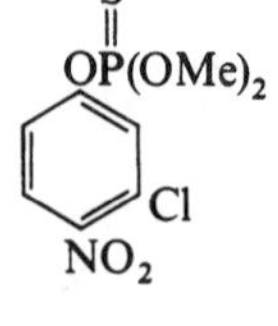

$C_8H_9ClNO_5PS$ M 297.650

Insecticide, esp. used for combating boll weevils, no longer in widespread use. Yellow cryst. (coml. prod. yellow oil). Misc. C_6H_6, insol. H_2O. d_4^{20} 1.437. Mp 21°. $Bp_{0.1}$ 125°. Rapidly hydrolysed in alkaline soln.

▷Toxic. TE8050000.

Schrader, G., *Angew. Chem.*, 1954, **66**, 265.
Bowman, M.C. *et al*, *J. Assoc. Off. Anal. Chem.*, 1970, **53**, 499 (*glc*)
Stan, H.-J. *et al*, *Fresenius' Z. Anal. Chem.*, 1977, **287**, 271; *Biomed. Mass Spectrom.*, 1982, **9**, 483 (*glc, ms*)
Stan, H.-J. *et al*, *J. Chromatogr.*, 1983, **279**, 173 (*glc*)
The Agrochemicals Handbook, Royal Society of Chemistry, 1983, A092.
Sax, N.I., *Dangerous Properties of Industrial Materials*, 6th Ed., Van Nostrand-Reinhold, 1984, 501.

Chlorthiophos, BSI — C-00197

O-[*2,5-Dichloro-4-(methylthio)phenyl*] O,O-*diethyl phosphorothiate. Celathion*

$C_{11}H_{15}Cl_2O_3PS_2$ M 361.237

Chlorthiophos is the name given to a technical mixture of the main component, together with isomeric compds. Non-systemic contact and stomach insecticide. Yellow-brown liq. d_4^{20} 1.35. $Bp_{0.013}$ 153-158°.

Agrochemicals Handbook, Royal Society of Chemistry, London, 1983, A093.
Pesticide Manual, 7th ed., 130.

Choline *O*-phosphate — C-00198

2-[(Dihydroxyphosphinyl)oxy]-N,N,N-*trimethylethanaminium hydroxide, inner salt, 9CI. Choline hydroxide hydrogen phosphate inner salt, 8CI*

$Me_3N^{\oplus}CH_2CH_2OP(O)(OH)O^{\ominus}$

$C_5H_{14}NO_4P$ M 183.144

Me ester: [2375-06-6]. *2-[(Hydroxymethoxyphosphinyl)oxy]*-N,N,N-*trimethylethanaminium hydroxide, inner salt.*
$C_6H_{16}NO_4P$ M 197.170
Solid. Mp 110°.

Ethyl ester: [22080-28-0]. *2-[(Ethoxyhydroxyphosphinyl)oxy]*-N,N,N-*trimethylethanaminium hydroxide, inner salt.*
$C_7H_{18}NO_4P$ M 211.197
Solid. Mp 160°.

Isopropyl ester: [21991-66-2]. *2-[(Hydroxyisopropoxyphosphinyl)oxy]*-N,N,N-*trimethylethanaminium hydroxide, inner salt.*
$C_8H_{20}NO_4P$ M 225.224
Solid. Mp 288°.

Ph ester: 2-[(Hydroxyphenoxyphosphinyl)oxy]-N,N,N-*trimethylethanaminium hydroxide, inner salt.*
$C_{11}H_{18}NO_4P$ M 259.241
Solid. Mp 208°.

Chabrier, P. *et al*, *C.R. Hebd. Seances Acad. Sci., Ser. C*, 1968, **267**, 732 (*synth*)
Weller, T. *et al*, *Chem. Phys. Lipids.*, 1975, **15**, 5 (*struct*)

2-Choro-1,3,2-dithiaphosphorinane, 9CI, 8CI — C-00199

2-Chloro-1,3,2-dithiaphosphinane. Trimethylene phosphorochloridodithioite

[28896-84-6]

$C_3H_6ClPS_2$ M 172.627

Pungent solid. Mp 45°. Bp_9 120-121°.

2-Oxide: [37442-78-7].
$C_3H_6ClOPS_2$ M 188.627
No phys. props. reported.

2-Sulfide: [57115-76-1].
$C_3H_6ClPS_3$ M 204.687
No phys. props. reported.

Nifant'ev, É.E. *et al*, *Zh. Obshch. Khim.*, 1974, **44**, 1694 (*Engl. transl.* p. 1664) (*synth, nmr*)
Grand, A. *et al*, *Acta Crystallogr., Sect. B*, 1976, **22**, 1244 (*sulfide, cryst struct*)
Martin, J. *et al*, *J. Phys. Chem.*, 1976, **80**, 2417 (*cmr, pmr, P nmr*)
Martin, J. *et al*, *Org. Magn. Reson.*, 1977, **9**, 637 (*sulfide, cmr, P nmr, pmr*)
Borisenko, A.A. *et al*, *Zh. Obshch. Khim.*, 1978, **48**, 1251 (*Engl. transl.* p. 1144) (*derivs, cmr, P nmr, pmr*)
Arshinova, R.P. *et al*, *Zh. Obshch. Khim.*, 1980, **50**, 829 (*Engl. transl.* p. 665) (*stereochem*)

Coenzyme *A* — C-00200

Adenosine 5'-(trihydrogen diphosphate) 3'-(dihydrogen phosphate) 5'-[3-hydroxy-4-[[3-[(2-mercaptoethyl)-amino]-3-oxopropyl]amino]-2,2-dimethyl-4-oxobutyl] ester, 9CI. CoA

[85-61-0]

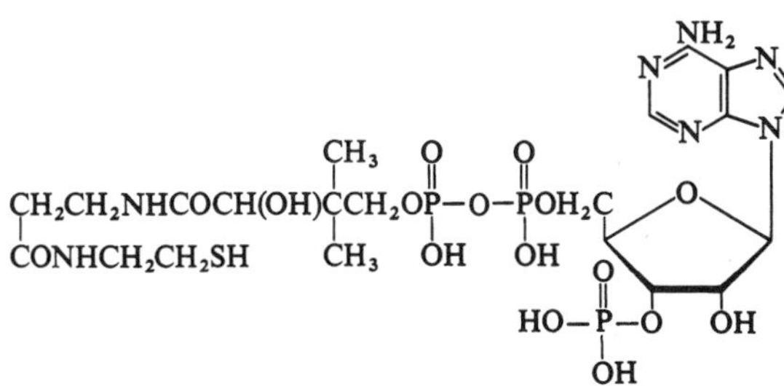

$C_{21}H_{36}N_7O_{16}P_3S$ M 767.534

Constit. of many microorganisms. A cofactor in enzymatic acetyl transfer reactions. Powder. Sol. H_2O. λ_{max} 259.5 nm (ϵ 16 800) (H_2O). Inactivated by phosphatase. Oxid. by air to inactive disulphide.

Lipmann, F. *et al*, *J. Biol. Chem.*, 1950, **186**, 235 (*isol*)
Lipmann, F., *Bacteriol. Rev.*, 1953, **17**, 1 (*rev*)
Baddiley, J., *Adv. Enzymol.*, 1955, **16**, 1 (*rev*)
Moffatt, J.G. *et al*, *J. Am. Chem. Soc.*, 1961, **83**, 663 (*synth*)
Lee, C. *et al*, *J. Am. Chem. Soc.*, 1975, **97**, 1225 (*conformn*)
Wilson, A. *et al*, *J. Am. Chem. Soc.*, 1975, **97**, 2907 (*pmr*)

Coenzyme F_{420} **C-00201**

F_{420}. Factor F_{420}

[64885-97-8]

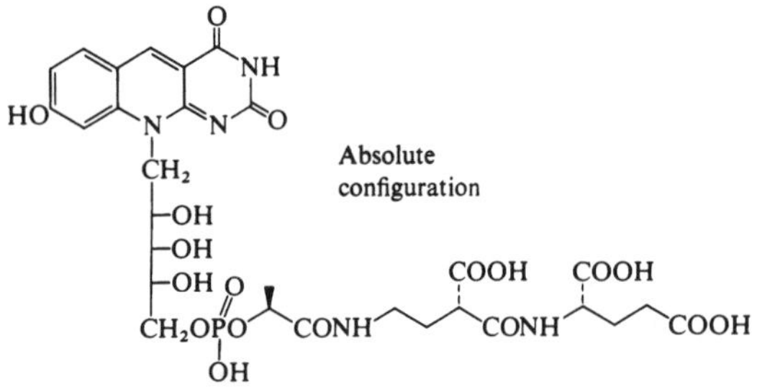

Absolute configuration

$C_{29}H_{36}N_5O_{18}P$ M 773.600

Isol. from *Methanobacterium* spp. Bacterial cofactor of all methanogenic bacteria. Contains the unusual 5-deazaflavin residue.

Eirich, L.D. *et al, Biochemistry*, 1978, **17**, 4583 (*struct, spectra*)
Ashton, W.T. *et al, J. Am. Chem. Soc.*, 1979, **101**, 4419 (*struct, stereochem*)

Coenzyme I, 9CI, 8CI **C-00202**

Adenosidine 5′-(trihydrogen diphosphate) (5′→5′) ester with 3-(aminocarbonyl)-1-β-D-ribofuranosylpyridinium hydroxide inner salt, 9CI. NAD. Nictinamide adenine dinucleotide. Nadide, BAN, USAN. DPN. Diphosphopyridine nucleotide. Cozymase. Codehydrase I. Codehydrogenase I. Enzopride. NRPPRA. Co I

[53-84-9]

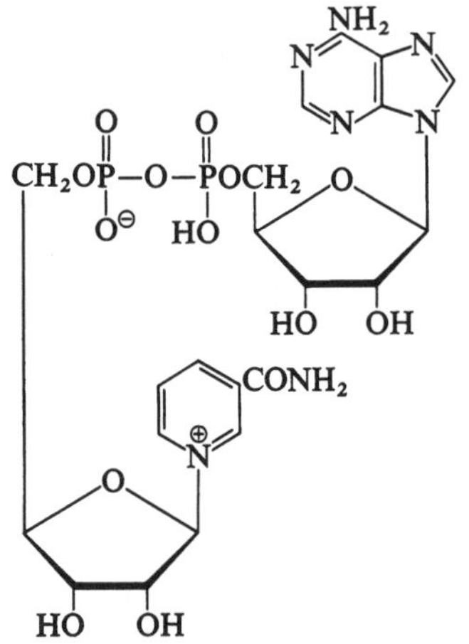

$C_{21}H_{27}N_7O_{14}P_2$ M 663.430

Isol. from bakers yeast. Antagonist to alcohol and narcotic analgesics. Biological hydrogen acceptor. Hygroscopic powder. Sol. H_2O. λ_{max} 260 (ϵ 17.6 × 10^6) (H_2O), 340 nm (6.2 × 10^6) (reduced form). Forms a complex with alkaline cyanide which may be used for its estimation. Stable for weeks in cold neutral soln. Less stable in acid soln., rapidly dec. by alkali.

▷UU3450000.

Biochem. Prep., 1949, **1**, 28 (*isol*)
Hayes, L.J. *et al, J. Chem. Soc.*, 1950, 303 (*synth*)
Biochem. Prep., 1953, **3**, 20 (*isol, synth*)
Lemieux, R.U. *et al, Can. J. Chem.*, 1963, **41**, 889 (*sterochem*)
Biochem. Prep., 1966, **11**, 84 (*purifn*)
Sarma, R.H. *et al, Biochemistry*, 1970, **9**, 557 (*conformn, pmr*)
Blumenstein, M. *et al, Biochemistry*, 1973, **12**, 3585 (*cmr*)

Coenzyme II, 9CI, 8CI **C-00203**

Adenosine 5′-(trihydrogen diphosphate) 2′-(dihydrogen phosphate) (5′→5′) ester with 3-(aminocarbonyl)-1-β-D-ribofuranosylpyridinium hydroxide inner salt, 9CI. Nicotinamide adenine dinucleotide phosphate. NADP. Codehydrase II. Codehydrogenase II. Triphosphopyridine nucleotide. TPN

[53-59-8]

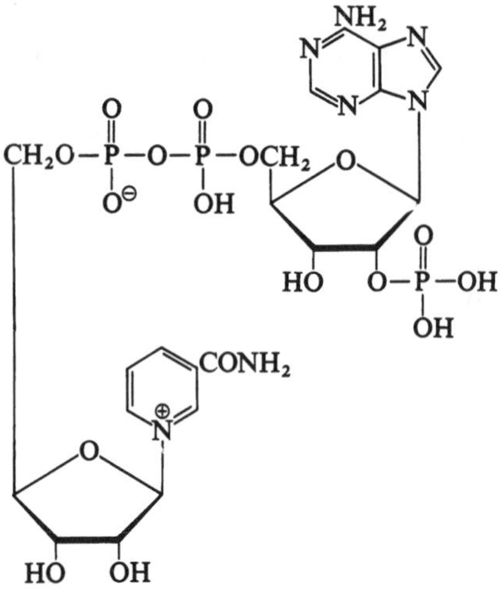

$C_{21}H_{28}N_7O_{17}P_3$ M 743.410

Wide occurence in living matter, particularly in the liver and in red blood corpuscles, mainly in the reduced form. A component of vitamin B_2 complex. Hydrogen carrier in biochemical redox systems. In the hexose monophosphoric acid system it is reduced to Dihydrocoenzyme II and reoxid. in the presence of flavoproteins. Greyish-white powder. Sol. H_2O. pK_{a1} 3.9, pK_{a2} 6.1. λ_{max} 260 (ϵ 18 × 10^6) (H_2O), 340 nm (6.2 × 10^6) (reduced form).

Todd, A.R., *J. Chem. Soc.*, 1941, 427.
Lepage, G.A., *J. Biol. Chem.*, 1949, **180**, 775 (*isol*)
Biochem. Prep., 1953, **3**, 24 (*isol, synth*)
Sund, R., *The Pyridine Nucleotide Coenzymes*, in *Biological Oxidations*, (*Singer, T., Ed.*), 1968, Interscience, N.Y., 603 (*rev*)
Blumenstein, M. *et al, Biochemistry*, 1973, **12**, 3585 (*cmr*)
Sarma, R.H. *et al, Can. J. Chem.*, 1973, **51**, 1843 (*nmr*)

Conen **C-00204**

O-*Butyl* S-*ethyl*-S-*(phenylmethyl) phosphorodithioate*, *9CI.* S-*Benzyl* O-*butyl* S-*ethyl phosphorodithioate, 8CI.* S-*Benzyl* O-*butyl* S-*ethyl dithiophosphate*

[27949-52-6]

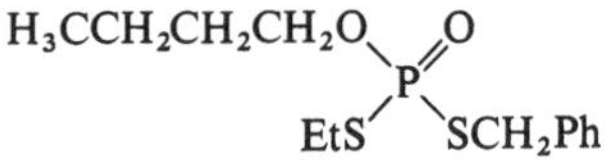

$H_3CCH_2CH_2CH_2O$, O, P, EtS, SCH_2Ph

$C_{13}H_{21}O_2PS_2$ M 304.401

▷TD5100000.

(±)-*form*

Fungicide, used against rice blast. Oil. n_D^{20} 1.5392.

Fr. Pat., 1 560 374, (*1969*); *CA*, **72**, 90043 (*synth*)
Kawamura, Y. *et al, CA*, 1971, **91**, 191459 (*glc*)
Mikami, A., *CA*, 1975, **82**, 81279 (*anal*)

Coumaphos, BSI **C-00205**

O-(3-Chloro-4-methyl-2-oxo-2H-1-benzoypran-7-yl) O,O-diethyl phosphorothioate, 9CI. *3-Chloro-7-hydroxy-4-methylcoumarin O-ester with O,O-diethyl phosphorothioate*, 8CI. *Coumafos. Asuntol. Co-ral. Muscatox. Resitox. Negasunt*

[56-72-4]

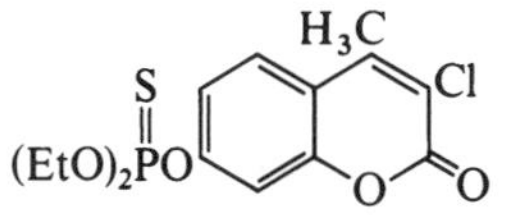

$C_{14}H_{16}ClO_5PS$ M 362.764

Insecticide, acaricide, e.g. for houseflies, cattle ticks. Mp 95°. Apparently no longer in widespread use.

▷Mod. toxic by skin absorption. GN6300000.

B.P., 713 142, (*1954*); *CA*, **50**, 411g (*synth*)
Keith, L.H., *J. Assoc. Off. Anal. Chem.*, 1968, **51**, 1063 (*pmr*)
Bowman, M.C. *et al*, *J. Assoc. Off. Anal. Chem.*, 1970, **53**, 499 (*glc*)
Ross, R.T. *et al*, *Anal. Chim. Acta*, 1970, **52**, 139 (*P nmr*)
Krijgsman, W. *et al*, *J. Chromatogr.*, 1976, **117**, 201 (*glc*)
Stan, H.-J. *et al*, *Fresenius' Z. Anal. Chem.*, 1977, **287**, 271; *Biomed. Mass Spectrom.*, 1982, **9**, 483 (*glc, ms*)
Ripley, B.D. *et al*, *J. Assoc. Off. Anal. Chem.*, 1983, **66**, 1084 (*glc*)
Sax, N.I., *Dangerous Properties of Industrial Materials*, 6th Ed., Van Nostrand-Reinhold, 1984, 519.

Creatinolfosfate, INN **C-00206**

N-Methyl-N-[2-(phosphonooxy)ethyl]guanidine, 9CI. *1-(2-Hydroxyethyl)-1-methylguanidine dihydrogen phosphate*, 8CI

[6903-79-3]

$HN{=}C(NH_2)NMeCH_2CH_2OP(O)(OH)_2$

$C_4H_{12}N_3O_4P$ M 197.130

Antianginal agent.

▷MF3600500.

Ferarri, G. *et al*, *Farmaco Ed. Sci.*, 1965, **20**, 879 (*synth*)
Marchetti, G. *et al*, *Arch. Int. Pharmacodyn. Ther.*, 1971, **191**, 337 (*pharmacol*)

Crotoxyphos, BSI, ISO **C-00207**

1-Phenylethyl 3-[(dimethoxyphosphinyl)oxy]-2-butenoate, 9CI. *3-Hydroxycrotonic acid α-methylbenzyl ester dimethylphosphate*, 8CI. *Dimethyl 1-methyl-2-(1-phenylethoxycarbonyl)vinyl phosphate. α-Methylbenzyl 3-(dimethoxyphosphinyloxy)isocrotonate. Ciodrin*

[326-12-5]

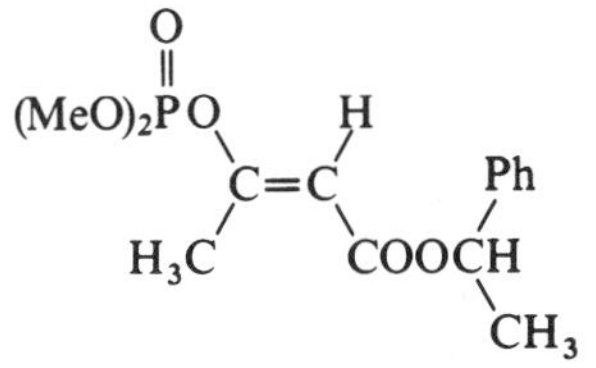

$C_{14}H_{19}O_6P$ M 314.274

Coml. samples contain ca. 80% (*E*)-form.

(±)-(*E*)-**form** [7700-17-6]

Insecticide for external use on livestock. Cholinesterase inhibitor. Straw-coloured liq. (tech.). d^{25} 1.19. $Bp_{0.03}$ 135°. n_D^{25} 1.4988.

▷GQ5075000.

U.S.P., 2 982 686, (*1961*); *CA*, **55**, 22240 (*synth*)
Keith, L.H. *et al*, *Anal. Chim. Acta*, 1968, **51**, 1063 (*pmr*)
Ross, R.T. *et al*, *Anal. Chim. Acta*, 1970, **52**, 139 (*P nmr*)
Beynon, K.I. *et al*, *Residue Rev.*, 1973, **47**, 55.
Szalontai, G., *J. Chromatogr.*, 1976, **124**, 9 (*hplc*)
Pesticide Manual, 6th Ed., 130; 7th Ed., 138.
Agrochemicals Handbook, Royal Society of Chemistry, London, 1983, A100.

Crufomate, BSI **C-00208**

2-Chloro-4-(1,1-dimethylethyl)phenyl methyl methylphosphoramidate, 9CI. *(4-tert-Butyl-2-chlorophenyl) methyl methylphosphoramidate*, 8CI. *Ruelene. Dowco 132. Montrel*

[299-86-5]

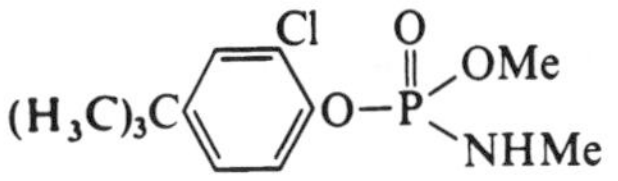

$C_{12}H_{19}ClNO_3P$ M 291.714

Insecticide, veterinary anthelmintic. Cryst. (pet. ether or CCl_4) (coml. prod. may be an oil). Sol. Me_2CO, CCl_4, C_6H_6, insol. H_2O. Mp 60-60.5°.

▷TB3850000.

U.S.P., 2 929, 762, (*1960*); *CA*, **54**, 18439 (*synth*)
Keith, L.H. *et al*, *J. Assoc. Off. Anal. Chem.*, 1968, **51**, 1063 (*pmr*)
Getz, M.E. *et al*, *J. Assoc. Off. Anal. Chem.*, 1968, **51**, 1101 (*tlc*)
Ross, R.T. *et al*, *Anal. Chim. Acta*, 1970, **52**, 139 (*P nmr*)
Bakke, J.E. *et al*, *Biomed. Mass Spectrom.*, 1976, **3**, 299 (*metab, ms*)
Feil, V.J. *et al*, *Biomed. Mass Spectrom.*, 1976, **3**, 316 (*ms, metab*)
Szalontai, G., *J. Chromatogr.*, 1976, **124**, 9 (*hplc*)
Baughman, R.G. *et al*, *J. Agric. Food Chem.*, 1978, **26**, 398 (*cryst struct*)
Ripley, B.D. *et al*, *J. Assoc. Off. Anal. Chem.*, 1983, **66**, 1084 (*glc*)
Pesticide Manual, 6th Ed., 131.

2-Cyano-5,5-dimethyl-1,3,2-dioxaphosphorinane **C-00209**

5,5-Dimethyl-1,3,2-dioxaphosphorinane-2-carbonitrile, 9CI

[57436-73-4]

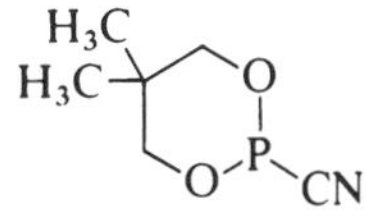

$C_6H_{10}NO_2P$ M 159.124

Liq. Bp_8 76-77°. n_D^{18} 1.4670.

2-Oxide: [55379-59-4].
$C_6H_{10}NO_3P$ M 175.124
Solid. Mp 79-81°.

2-Selenide: [57436-72-3].
$C_6H_{10}NO_2PSe$ M 238.084
Solid. Mp 114-115°. Sublimes.

Stec, W.J. *et al*, *J. Chem. Soc., Chem. Commun.*, 1974, 923; 1975, 467 (*synth, derivs, ir, P nmr, props*)
Uznanski, B. *et al*, *Synthesis*, 1975, 735 (*synth, ir, P nmr*)

(1-Cyanoethyl)phosphonic acid **C-00210**

$H_3CCH(CN)P(O)(OH)_2$

$C_3H_6NO_3P$ M 135.059

Esters are Wittig-Horner reagents for synth. of 2-cyano-2-alkenes.

(±)-form

Di-Et ester: Diethyl (1-cyanoethyl)phosphonate.

$C_7H_{14}NO_3P$ M 191.166

Liq. $Bp_{0.1}$ 83-86°.

Di-Ph ester: Diphenyl (1-cyanoethyl)phosphonate.

$C_{15}H_{14}NO_3P$ M 287.254

Liq. $Bp_{0.1}$ 174-175°. n_D^{20} 1.5452.

D'Incan, E. *et al, Synthesis*, 1975, 516 (*synth, use*)
D'Incan, E., *Tetrahedron*, 1977, **33**, 951 (*use*)
Comins, D.L. *et al, Synthesis*, 1978, 309 (*synth, use*)
Garaev, R.D. *et al, Zh. Obshch. Khim.*, 1982, **52**, 708 (*Engl. transl.* p. 616) (*synth, ir, pmr, P nmr*)

(2-Cyanoethyl)phosphonic acid, 9CI C-00211

[34549-99-0]

$NCCH_2CH_2P(O)(OH)_2$

$C_3H_6NO_3P$ M 135.059

Di-Me ester: [20580-36-3]. *Dimethyl (2-cyanoethyl)-phosphonate.*

$C_5H_{10}NO_3P$ M 163.113

Liq. d_4^{20} 1.196. Bp_{11} 158°. n_D^{20} 1.4432.

Di-Et ester: [10123-62-3]. *Diethyl (2-cyanoethyl)-phosphonate.*

$C_7H_{14}NO_3P$ M 191.166

Liq. d_4^{20} 1.09. Bp_{10} 159-160°, Bp_1 130°. n_D^{20} 1.4388.

▷KI6650000.

Diisopropyl ester: [52726-83-7]. *Diisopropyl (2-cyanoethyl)phosphonate.*

$C_9H_{18}NO_3P$ M 219.220

Liq. d_4^{20} 1.05. Bp_{13} 160°. n_D^{20} 1.4345.

Pudovik, A.N. *et al, Dokl. Akad. Nauk SSSR, Ser. Sci. Khim.*, 1950, **73**, 327; *CA*, **45**, 2853 (*esters, synth*)
Ginsburg, V.A. *et al, Zh. Obshch. Khim.*, 1960, **30**, 3987 (*Engl. transl.* p. 3944) (*ester, synth*)
Harvey, R.G., *Tetrahedron*, 1966, **22**, 2561 (*ester, synth*)
Bel'skii, V.E. *et al, Zh. Obshch. Khim.*, 1972, **42**, 2427 (*Engl. transl.* p. 2421) (*P nmr*)

(2-Cyanoethyl)triphenylphosphonium(1+) C-00212

[47252-66-4]

$Ph_3P^{\oplus}CH_2CH_2CN$

$C_{21}H_{19}NP^{\oplus}$ M 316.362 (ion)

Bromide: [57519-80-9].

$C_{21}H_{19}BrNP$ M 396.266

Cryst. ($EtOH/Et_2O$). Mp 128° dec. Treatment with $NaOEt/EtOH$ yields $EtOCH_2CH_2CN$.

Tetrafluoroborate: [31662-55-2].

$C_{21}H_{19}BF_4NP$ M 403.165

Cryst. (H_2O). Mp 132-133°.

Ylide: [75860-43-4]. *(2-Cyanoethylidene)-triphenylphosphorane. 3-Triphenylphosphoranylidenepropanenitrile.*

$C_{21}H_{18}NP$ M 315.354

Brophy, J.J. *et al, Aust. J. Chem.*, 1969, **22**, 1405 (*props*)
Schiemenz, G.P. *et al, Chem. Ber.*, 1971, **104**, 1219 (*synth*)
Marshall, D.P. *et al, J. Chem. Soc., Perkin Trans. 2*, 1977, 1898 (*synth*)
Hashimoto, S. *et al, Tetrahedron Lett.*, 1980, **21**, 2857 (*ylide*)

Cyanofenphos, BSI C-00213

O-4-Cyanophenyl O-ethyl phenylphosphonothioate, 9CI.
Surecide

[13067-93-1]

(Structure: 4-NC-C6H4-O-P(=S)(Ph)(OEt))

$C_{15}H_{14}NO_2PS$ M 303.315

▷TB1750000.

(±)-form [62421-62-9]

Agricultural insecticide. Cryst. Mp 83°. Rel. poor ChE inhibitor, but sp. inhibitor of brain neurotoxic esterase and monoamine oxidase.

▷Causes irreversible paralytic ataxia

B.P., 929 738, (*1960*); *CA*, **59**, 14025b (*synth*)
Steurbaut, W. *et al, Bull. Soc. Chim. Belg.*, 1975, **84**, 791 (*synth, ir, pmr*)
Steurbaut, W. *et al, Bull. Soc. Chim. Belg.*, 1977, **86**, 65 (*ms*)
Francis, B.M. *et al, J. Environ. Sci. Health, Part B*, 1982, **17**, 611 (*tox*)
Pesticide Manual, 6th Ed., 135; 7th Ed., 141.

Cyanomethylenebis(triphenylphosphonium)(2+) C-00214

$Ph_3P^{\oplus}CH(CN)P^{\oplus}Ph_3$

$C_{38}H_{31}NP_2^{\oplus\oplus}$ M 563.617 (ion)

Reagent for conversion of aldehydes into $\alpha\beta$-unsatd. nitriles.

Dibromide:

$C_{38}H_{31}Br_2NP_2$ M 723.425

Source of ylide. Intermed. for prep. of cyanomethylenetriphenylphosphorane. Solid. Mp 268-269°.

Ylide: [(Cyanomethylene)triphenylphosphoranyl]-triphenylphosphonium.

$C_{38}H_{30}NP^{\oplus}$ M 531.635 (ion)

Wittig reagent.

Wilt, J.W. *et al, J. Org. Chem.*, 1971, **36**, 2026 (*synth, use*)

(Cyanomethyl)phosphonic acid, 9CI C-00215

[35236-60-3]

$NCCH_2P(O)(OH)_2$

$C_2H_4NO_3P$ M 121.032

Mono-anilinium salt: Cryst. ($MeOH/Me_2CO$). Mp 172-174° dec.

Di-Et ester: [2537-48-6]. *Diethyl (cyanomethyl)-phosphonate.*

$C_6H_{12}NO_3P$ M 177.139

Wittig-Horner reagent for synth. of acrylonitriles (1-cyanoethenes). Liq. $Bp_{0.02}$ 88-94°.

Bis(trimethylsilyl) ester: Bis(trimethylsilyl) (cyanomethyl)phosphonate.

$C_8H_{20}NO_3PSi_2$ M 265.396

Liq. Bp_7 129-130°.

Kirilov, M. *et al, Chem. Ber.*, 1971, **104**, 3073 (*props*)
Mikolajczyk, M. *et al, Synthesis*, 1976, 396 (*use*)
Bottin-Strzalko, T. *et al, J. Org. Chem.*, 1978, **43**, 4346 (*cmr, pmr, P nmr*)
Morita, T. *et al, Bull. Chem. Soc. Jpn.*, 1978, **51**, 2169 (*silyl ester, ir, pmr*)
Wharton, C.J. *et al, J. Chem. Soc., Perkin Trans. 1*, 1981, 433 (*synth, ir, pmr, props*)

Ferreira, A.B.B. *et al*, *J. Chem. Soc.*, *Perkin Trans. 2*, 1982, 25 (*use*)

P-Cyanomethyl-*N*,*N*,*N*′,*N*′-tetramethylphosphonic diamide C-00216

[57648-42-7]

$(Me_2N)_2P(O)CH_2CN$

$C_6H_{14}N_3OP$ M 175.170

May be alkylated under phase-transfer conditions, and products reduced and then hydrol. to (2-aminoalkyl)-phosphonic acids. Cryst. (dimethoxyethane). Mp 93°.

Blanchard, J. *et al*, *Synthesis*, 1975, 655 (*synth*, *props*)
Blanchard, J. *et al*, *Tetrahedron*, 1976, **32**, 455 (*use*)

(Cyanomethyl)triphenylphosphonium(1+), 9CI C-00217

$Ph_3P^{\oplus}CH_2CN$

$C_{20}H_{17}NP^{\oplus}$ M 302.335 (ion)

Chloride: [4336-70-3].
$C_{20}H_{17}ClNP$ M 337.788
Cryst. ($MeNO_2$ or EtOH). Mp 252°, 278-279°. pK_a 7.6. Alkali yields the ylide Triphenylphosphoranylideneacetonitrile, T-00641 .

▷TA2075000.

Iodide: [77785-52-5].
$C_{20}H_{17}INP$ M 429.239
Cryst. (H_2O). Mp 224-226°.

Trippett, S. *et al*, *J. Chem. Soc.*, 1959, 3874 (*synth*, *props*)
Wagenknecht, J. *et al*, *J. Org. Chem.*, 1966, **31**, 3885.
Albright, T.A. *et al*, *J. Am. Chem. Soc.*, 1975, **97**, 2942 (*nmr*)
Fluck, E. *et al*, *Pure Appl. Chem.*, 1975, **44**, 373 (*pe*)
Kukhar', V.P. *et al*, *Zh. Obshch. Khim.*, 1979, **49**, 1025 (*Engl. transl.* p. 889) (*synth*, *ir*, *ylide*)
Tronchet, J.M.J. *et al*, *Helv. Chim. Acta*, 1979, **62**, 1401 (*props*)
Kukhar', V.P. *et al*, *Zh. Org. Khim.*, 1981, **17**, 180 (*Engl. transl.* p. 161) (*iodide*)

3-Cyano-2-oxo-3-(triphenylphosphoranylidene)propanoic acid C-00218

Cyano(triphenylphosphoranylidene)pyruvic acid

$Ph_3P{=}C(CN)COCOOH$

$C_{22}H_{16}NO_3P$ M 373.347

Me ester: Methyl cyano(triphenylphosphoranylidene)-pyruvate.
$C_{23}H_{18}NO_3P$ M 387.374
Stabilised Wittig reagent. Pale-yellow cryst. Mp 211-212°.

Et ester: [23853-25-0]. *Ethyl cyano(triphenylphosphoranylidene)pyruvate.*
$C_{24}H_{20}NO_3P$ M 401.401
Stabilised Wittig reagent. Cryst. (MeCN). Mp 215-216°.

Ciganek, E., *J. Org. Chem.*, 1970, **35**, 1725 (*esters*, *uv*, *ir*, *pmr*)

(Cyanophenylmethyl)phosphonic acid, 9CI C-00219

(α-Cyanobenzyl)phosphonic acid, 8CI

$PhCH(CN)P(O)(OH)_2$

$C_8H_8NO_3P$ M 197.130

(±)-*form*

Di-Et ester: [43055-48-7]. *Diethyl (cyanophenylmethyl)-phosphonate. Diethyl (α-cyanobenzyl)phosphonate.*
$C_{12}H_{16}NO_3P$ M 253.237
Wittig-Horner reagent. Liq. Bp_1 106-108°, $Bp_{0.03}$ 120-124°. n_D^{20} 1.4930.

Dipropyl ester: [51863-54-8]. *Dipropyl (cyanophenylmethyl)phosphonate. Dipropyl (α-cyanobenzyl)phosphonate.*
$C_{14}H_{20}NO_3P$ M 281.291
Liq. Bp_1 124-126°. n_D^{20} 1.4892.

Devlin, C.J. *et al*, *J. Chem. Soc.*, *Perkin Trans. 1*, 1973, 1428 (*synth*, *ir*, *pmr*, *ms*)
Kirilov, M. *et al*, *Bull. Soc. Chim. Fr. Part II*, 1973, 3053 (*synth*, *ir*)
Comins, D.L. *et al*, *Synthesis*, 1978, 309 (*synth*, *use*)

[(2-Cyanophenyl)methyl]phosphonic acid, 9CI C-00220

o-*Cyanobenzylphosphonic acid, 8CI. 2-(Phosphonomethyl)benzonitrile*

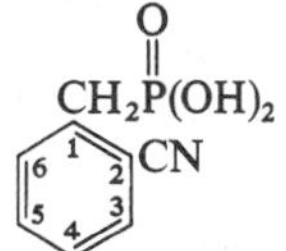

$C_8H_8NO_3P$ M 197.130

Di-Me ester: Dimethyl [(2-cyanophenyl)methyl]-phosphonate.
$C_{10}H_{12}NO_3P$ M 225.183
Wittig-Horner reagent. Solid. Mp 73-74°.

Arient, J., *Collect. Czech. Chem. Commun.*, 1981, **46**, 101 (*synth*, *use*)

[(4-Cyanophenyl)methyl]phosphonic acid, 9CI C-00221

p-*Cyanobenzylphosphonic acid, 8CI. 4-(Phosphonomethyl)benzonitrile*

$C_8H_8NO_3P$ M 197.130

Di-Et ester: [1552-41-6]. *Diethyl [(4-cyanophenyl)-methyl]phosphonate. Diethyl 4-cyanobenzylphosphonate.*
$C_{12}H_{16}NO_3P$ M 253.237
Solid-liquid. Mp 31°. $Bp_{0.6}$ 162°.

Dibutyl ester: Dibutyl [(4-cyanophenyl)methyl]-phosphonate. Dibutyl 4-cyanobenzylphosphonate.
$C_{16}H_{24}NO_3P$ M 309.344
Liq. Mp 3°. $Bp_{0.6}$ 185°. n_D^{22} 1.4975.

Kagan, F. *et al*, *J. Am. Chem. Soc.*, 1959, **81**, 3026.
Kreutzkamp, N. *et al*, *Arch. Pharm. (Weinheim, Ger.)*, 1961, **294**, 49
Franke, A. *et al*, *Synthesis*, 1979, 712.

(4-Cyanophenyl)phosphinic acid C-00222

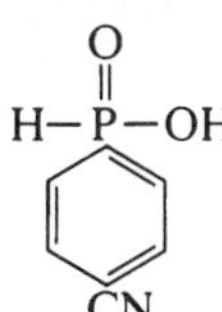

$C_7H_6NO_2P$ M 167.104

Tautomeric with (4-Cyanophenyl)phosphonous acid, but free acid prob. exists in phosphinic acid form. Cryst. (CCl_4/EtOH). Mp 166-167°. pK_a 1.19 (H_2O).

Quin, L.D. *et al*, *J. Am. Chem. Soc.*, 1961, **83**, 4124.

(2-Cyanophenyl)phosphonic acid, 9CI **C-00223**

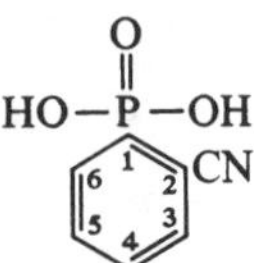

$C_7H_6NO_3P$ M 183.103

Di-Et ester: [34595-07-8]. *Diethyl (2-cyanophenyl)-phosphonate.*

$C_{11}H_{14}NO_3P$ M 239.210

Liq. Bp_7 163°.

Garner, G.V. *et al*, *J. Chem. Soc. (C)*, 1971, 3693 (*synth, ir*)

(3-Cyanophenyl)phosphonic acid **C-00224**

$C_7H_6NO_3P$ M 183.103

Di-Et ester: [85915-09-9]. *Diethyl (3-cyanophenyl)-phosphonate.*

$C_{11}H_{14}NO_3P$ M 239.210

No phys. props. reported.

Ewen, G.D. *et al*, *J. Chem. Res. (S)*, 1983, 14.

(4-Cyanophenyl)phosphonic acid, 9CI **C-00225**

$C_7H_6NO_3P$ M 183.103

Di-Me ester: [76659-90-0]. *Dimethyl (4-cyanophenyl)-phosphonate.*

$C_9H_{10}NO_3P$ M 211.157

No phys. props. reported.

Di-Et ester: [28255-72-3]. *Diethyl (4-cyanophenyl)-phosphonate.*

$C_{11}H_{14}NO_3P$ M 239.210

Low-melting solid. Mp 30-32°. $Bp_{0.01}$ 132°.

Tavs, P., *Chem. Ber.*, 1970, **103**, 2428 (*synth*)
Okawa, H. *et al*, *Biochem. Pharmacol.*, 1980, **29**, 2721 (*tox*)
Hirao, T. *et al*, *Bull. Chem. Soc. Jpn.*, 1982, **55**, 909 (*synth*)

Cyanophos, BSI **C-00226**

O-(*4-Cyanophenyl*) O,O-*dimethyl phosphorothioate*, *9CI*. O,O-*Dimethyl* O-*4-cyanophenyl thiophosphate*. *Ciafos*. *Cyanox*

[2636-26-2]

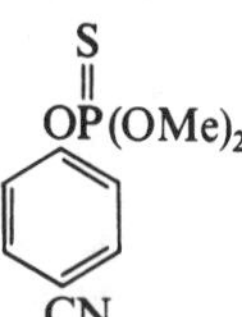

$C_9H_{10}NO_3PS$ M 243.217

Insecticide. Yellow to orange liq. Spar. sol. H_2O, pet. ether, v. sol. MeOH, xylene. Mp 14-15°. $Bp_{0.09}$ 119-120° sl. dec. Dec. in alkali; otherwise stable in storage. $n_D^{32.5}$ 1.5404. Converted by sunlight to the phosphate.

▷TF7600000.

Schrader, G. *et al*, *Angew. Chem.*, 1961, **73**, 331 (*synth, use*)
Nishizawa, Y. *et al*, *Agric. Biol. Chem.*, 1961, **25**, 597 (*synth, tox, use*)
Pesticide Manual, 6th Ed., 138; 7th Ed., 142.

Cyanophosphoramidic dichloride, 9CI **C-00227**

[59857-18-0]

$Cl_2P(O)NHCN$

$CHCl_2N_2OP$ M 158.911

Koehler, H. *et al*, *Z. Anorg. Allg. Chem.*, 1976, **423**, 21 (*synth, ir*)

Cyanophosphorimidic acid, 9CI, 8CI **C-00228**

$NCN{=}P(OH)_3$

$CH_3N_2O_3P$ M 122.020

Free acid tautomeric with cyanophosphoramidic acid.

Tri-Me ester: [17167-30-5]. *Trimethyl cyanophosphorimidate.* N-*Cyano*-P,P,P-*trimethoxyphosphine imide.* N-*Cyano*-P,P,P-*trimethoxyphosphazene.*

$C_4H_9N_2O_3P$ M 164.100

Liq. $Bp_{0.16}$ 69°.

Tri-Ph ester: [17233-33-9]. *Triphenyl cyanophosphorimidate.* N-*Cyano*-P,P,P-*triphenoxyphosphine imide.* N-*Cyano*-P,P,P-*triphenoxyphosphazene.*

$C_{19}H_{15}N_2O_3P$ M 350.313

Cryst. (butyl acetate). Mp 92.5-93.5°.

Mitsch, R.A., *J. Am. Chem. Soc.*, 1967, **89**, 6297 (*synth*)
Marsh, F.D., *J. Org. Chem.*, 1972, **37**, 2966 (*synth*)
U.S.P., 3 776 950, (*1973*); *CA*, **81**, 63769 (*synth*)

(1-Cyanopropyl)phosphonic acid, 9CI **C-00229**

$H_3CCH_2CH(CN)P(O)(OH)_2$

$C_4H_8NO_3P$ M 149.086

(±)-*form*

Di-Et ester: [34491-76-4]. *Diethyl (1-cyanopropyl)-phosphonate.*

$C_8H_{16}NO_3P$ M 205.193

Wittig-Horner reagent for synth. of 3-cyano-3-alkenes. Liq. d_4^{20} 1.10. Bp_2 127-128°. n_D^{20} 1.4340.

Kirilov, M. *et al*, *Chem. Ber.*, 1971, **104**, 3073 (*synth*)
D'Incan, E. *et al*, *Synthesis*, 1975, 516 (*synth, use*)

(3-Cyanopropyl)phosphonic acid, 9CI **C-00230**

$NCCH_2CH_2CH_2P(O)(OH)_2$

$C_4H_8NO_3P$ M 149.086

Di-Et ester: [53253-67-1]. *Diethyl (3-cyanopropyl)-phosphonate.*

$C_8H_{16}NO_3P$ M 205.193

Liq. d_4^{17} 1.09. Bp_8 163-164°.

Nylén, P., *Ber.*, 1926, **59**, 1119.

Cyano(triphenylphosphoranylidene)acetic acid, 9CI **C-00231**

$Ph_3P{=}C(CN)COOH$

$C_{21}H_{16}NO_2P$ M 345.337

Me ester: [13504-71-7]. *Methyl cyano(triphenylphosphoranylidene)acetate.*

$C_{22}H_{18}NO_2P$ M 359.363

Stabilised ylide. Cryst. (EtOH). Mp 217-218° (211-212°).

tert-*Butyl ester:* tert-*Butyl cyano(triphenylphosphoranylidene)acetate.*

$C_{25}H_{24}NO_2P$ M 401.444

Stabilised ylide. Solid. Mp 168-169°.

Horner, L. *et al, Chem. Ber.*, 1958, **91**, 437 (*synth, uv*)
Martin, D. *et al, Chem. Ber.*, 1967, **100**, 187 (*synth*)
Bestmann, H.J. *et al, Justus Liebigs Ann. Chem.*, 1974, 1688 (*synth*)

Cyanthoate, BSI C-00232

S-[2-[(1-Cyano-1-methylethyl)amino]-2-oxoethyl] O,O-diethyl phosphorothioate, 9CI. *O,O-Diethyl phosphorothioate, S-ester with N-(1-cyano-1-methylethyl-2-mercaptoacetamide)*, 8CI. *S-[N-(1-Cyano-1-methylethyl)carbamoylmethyl] O,O-diethylphosphorothioate*

[3734-95-0]

$(EtO)_2P(O)SCH_2CONHC(CH_3)_2CN$

$C_{10}H_{19}N_2O_4PS$ M 294.305

Insecticide and acaricide; no longer in use. Oil. Sol. H_2O. d_4^{19} 1.19. n_D^{20} 1.4845.

▷Toxic. TE8750000.

Krijgsman, W. *et al, J. Chromatogr.*, 1976, **117**, 201 (*glc*)
Pesticide Manual, 5th ed., 138.

Cyclic AMP C-00233

Adenosine cyclic 3′,5′-(hydrogen phosphate), 9CI, 8CI. *Acrasin. cAMP*

[60-92-4]

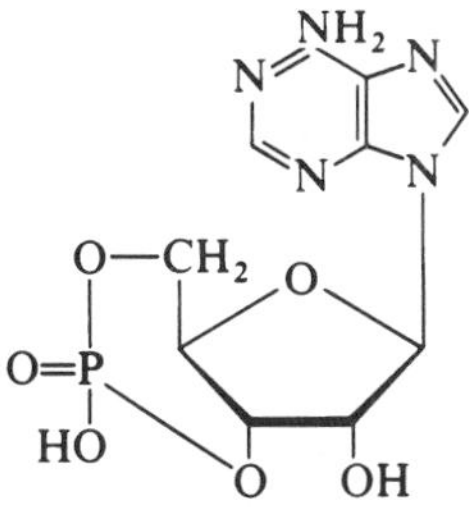

$C_{10}H_{12}N_5O_6P$ M 329.208

Found in several higher plants, bacteria and most animal cells. Excreted in human urine. Formed by action of adenate cyclase on ATP *in vivo*. Intracellular regulator of several cellular processes. Involved in hormone-mediated biological systems as a "second messenger". Cryst. + $1H_2O$. Mp 219-220°. $[\alpha]_D$ −43° (H_2O). λ_{max} 258 (ϵ 14 650) (pH 7) and 256 nm (14 500) (pH 2).

▷AU7357600.

Na salt: [37839-81-9]. Dihydrate. λ_{max} 256 nm (ϵ 14 500) (pH 2).

2-Propyl: λ_{max} 261 (ϵ 14 600) (pH 7), 256 (14 100) (pH 1), 261 nm (14 100) (pH 11).

2-Hexyl: λ_{max} 261 (ϵ 14 900) (pH 7), 256 (14 200) (pH 1), 261 nm (14 700) (pH 11) (H_2O).

N,O[2]-*Dibutanoyl:* [362-74-3]. ***Bucladesine, INN***. *Dibutyrylcyclic AMP.*

$C_{18}H_{24}N_5O_8P$ M 469.390

Vasodilator.

▷ES5055000.

Sutherland, E.W. *et al, J. Biol. Chem.*, 1958, **232**, 1077 (*isol*)
Lipkin, D. *et al, J. Am. Chem. Soc.*, 1959, **81**, 6075, 6198.
Posternak, T.L. *et al, Biochem. Biophys. Acta*, 1962, **65**, 558 (*Bucladesine*)
Borden, R.K., *J. Org. Chem.*, 1966, **31**, 3247 (*synth*)
Jost, J.P. *et al, Ann. Rev. Biochem.*, 1971, **40**, 741 (*biochem*)
Robison, G.A. *et al, Cyclic AMP*, Academic Press, New York, 1971.
Lapper, R.D. *et al, J. Am. Chem. Soc.*, 1972, **94**, 6243 (*pmr, nmr, Bucladesine*)
Kainosho, M. *et al, J. Am. Chem. Soc.*, 1975, **97**, 6839 (*pmr*)
Meyer, R.B. *et al, Biochemistry*, 1975, **14**, 3315.
Furusawa, K. *et al, J. Chem. Soc., Perkin Trans. 1*, 1976, 1711 (*synth*)
Kainosho, M., *Org. Magn. Reson.*, 1979, **12**, 548 (*cmr*)

Cyclic GMP C-00234

Guanosine cyclic 3′,5′-(hydrogen phosphate), 9CI, 8CI. *cGMP*

[7665-99-8]

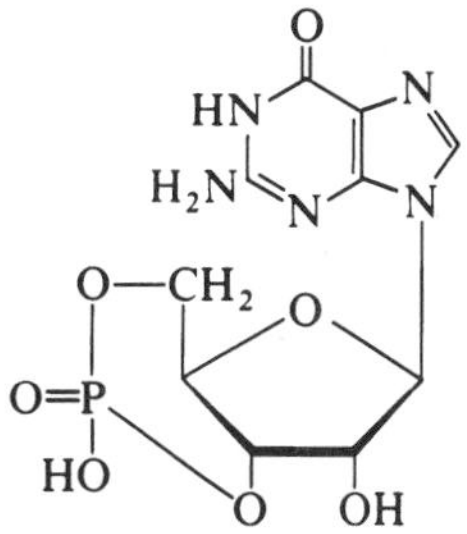

$C_{10}H_{12}N_5O_7P$ M 345.208

Found in animal and bacterial cells. Excreted in human urine. Formed *in vivo* by action of guanylate cyclase on GTP. Has also been isol. from *Evodiae fructus*. Intracellular regulator of cellular processes. Involved in hormone mediated biological systems as a "second messenger.".

Ca salt: Decahydrate. λ_{max} 254 (ϵ 12 950) (pH 7), 256.5 (11 350) (pH 1), 262 nm (12 400) (pH 12).

2′-Ac: [56879-79-9].

$C_{12}H_{14}N_5O_8P$ M 387.245

λ_{max} 252 (ϵ 16 000), 274 sh (10 600) (pH 7), 256 (14 200) (pH 1), 257 nm (14 400) (pH 11).

Borden, R.K., *J. Org. Chem.*, 1966, **31**, 3247 (*synth*)
Adv. Cyclic. Nucleotide Res., (*Greengard, P. et al*, Eds.), Raven Press, N.Y., 1973, Vol. 3, 155 (*rev*)
Anon, *Nature* (*London*), 1973, **246**, 186 (*rev*)
Chwang, A.K. *et al, Acta Crystallogr., Sect. B*, 1974, **30**, 1233 (*cryst struct*)
Kainosho, M. *et al, J. Am. Chem. Soc.*, 1975, **97**, 6839 (*pmr*)
Miller, J.P. *et al, Biochemistry*, 1976, **15**, 217 (*synth*)
Kainosho, M., *Org. Magn. Reson.*, 1979, **12**, 548 (*cmr*)
Cyong, J.C. *et al, Chem. Pharm. Bull.*, 1982, **30**, 2463 (*isol*)

Cyclobutyltriphenylphosphonium(1+), 9CI, 8CI C-00235

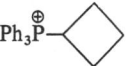

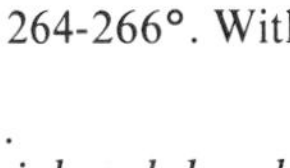

$C_{22}H_{22}P^{\oplus}$ M 317.390 (ion)

Bromide: [3666-89-5].

$C_{22}H_{22}BrP$ M 397.294

Solid (H_2O). Mp 264-266°. With MeLi or butyllithium → ylide.

Ylide: [53213-06-2].

Cyclobutylidenetriphenylphosphorane.

$C_{22}H_{21}P$ M 316.382

Wittig reagent. Solid. Mp 93°.

Bestmann, H.J. *et al, Chem. Ber.*, 1969, **102**, 1802 (*synth, ylide, use*)
Bestmann, H.J. *et al, Tetrahedron Lett.*, 1974, 1275 (*use*)
Albright, T.A. *et al, J. Am. Chem. Soc.*, 1975, **97**, 2942 (*cmr, nmr*)
Albright, T.A. *et al, J. Am. Chem. Soc.*, 1976, **98**, 6249 (*ylide, cmr, nmr*)
Schmidbaur, H. *et al, Chem. Ber.*, 1983, **116**, 2173 (*ylide, synth, pmr, nmr, cryst struct*)

(1,3,6-Cycloheptatrien-1-ylmethylene)-triphenylphosphorane C-00236

[80325-32-2]

$C_{26}H_{23}P$ M 366.441

Stabilised ylide. Deep-red. Mp 188°.

Ott, N. *et al*, *Angew. Chem., Int. Ed. Engl.*, 1982, **21**, 68 (*synth, pmr*)

(2,4,6-Cycloheptatrien-1-ylmethyl)-triphenylphosphonium(1+) C-00237

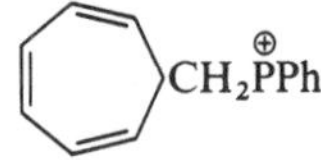

$C_{26}H_{24}P^{\oplus}$ M 367.449 (ion)

Tetrafluoroborate: [76640-63-4].

$C_{26}H_{24}BF_4P$ M 454.253

Cryst. (butanol). Mp 205°. When heated in DMSO or DMF, an isomeric salt, Mp 210-2°, is formed.

Cavicchio, G. *et al*, *Tetrahedron Lett.*, 1980, **21**, 2333 (*synth, pmr, ir, props*)
Ott, N. *et al*, *Angew. Chem., Int. Ed. Engl.*, 1982, **21**, 68 (*synth*)

Cycloheptyltriphenylphosphonium(1+), 9CI C-00238

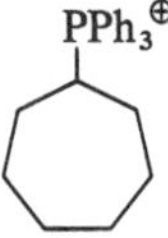

$C_{25}H_{28}P^{\oplus}$ M 359.470 (ion)

Bromide: [22836-06-2].

$C_{25}H_{28}BrP$ M 439.374

Source of ylide. Cryst. Mp 202-204°. $NaNH_2$ or LiOMe → ylide.

Ylide: [53213-07-3]. *Cycloheptylidenetriphenylphosphorane.*

$C_{25}H_{27}P$ M 358.462

Used in Wittig reactions.

Albright, T.A. *et al*, *J. Am. Chem. Soc.*, 1975, **97**, 2942 (*cmr, nmr*)
Saleh, G. *et al*, *Chem. Ber.*, 1979, **112**, 355 (*synth*)

Cyclohexanecarbonylphosphonic acid, 9CI C-00239

$C_7H_{13}O_4P$ M 192.151

Di-Me ester: [1490-12-6]. *Dimethyl (cyclohexylcarbonyl)phosphonate.*

$C_9H_{17}O_4P$ M 220.205

Liq. $Bp_{0.05}$ 80-82°. n_D^{25} 1.4680.

Di-Et ester: [1490-14-8]. *Diethyl (cyclohexylcarbonyl)-phosphonate.*

$C_{11}H_{21}O_4P$ M 248.258

Liq. $Bp_{0.2}$ 89-90°.

Diisopropyl ester: Diisopropyl (cyclohexylcarbonyl)-phosphonate.

$C_{13}H_{25}O_4P$ M 276.312

Liq. $Bp_{0.4}$ 100-102°.

Berlin, K.D. *et al*, *J. Org. Chem.*, 1965, **30**, 1265 (*ir, pmr*)
Marmor, R.S. *et al*, *J. Org. Chem.*, 1971, **36**, 128 (*ir, pmr*)
Yamashita, M. *et al*, *Bull. Chem. Soc. Jpn.*, 1980, **53**, 1625 (*synth, ir, pmr, use*)

1,2-Cyclohexanediphosphonic acid C-00240

1,2-Cyclohexanediylbisphosphonic acid, 9CI. 1,2-Diphosphonocyclohexane

[14655-81-3]

$C_6H_{14}O_6P_2$ M 244.121

Cryst. (AcOH). Mp 217-220°.

Tetracyclohexylammonium salt: Solid. Mp 271-279°.

Tetra-Et ester: Tetraethyl 1,2-cyclohexanediylbisphosphonate.

$C_{14}H_{30}O_6P_2$ M 356.335

Liq. $Bp_{0.3}$ 164-167°.

Eckert, R. *et al*, *Chem. Ber.*, 1967, **100**, 639 (*synth*)
Tavs, P., *Chem. Ber.*, 1967, **100**, 1571 (*synth, ester, ir*)
Tavs, P., *Tetrahedron*, 1967, **23**, 4677 (*synth*)
Ohms, G. *et al*, *Z. Anorg. Allg. Chem.*, 1982, **486**, 22 (*synth*)

1-Cyclohexenephosphonic acid C-00241

1-Cyclohexen-1-ylphosphonic acid, 9CI. 1-Phosphonocyclohexene

[10562-88-6]

$C_6H_{11}O_3P$ M 162.125

Di-Et ester: [31651-16-8]. *Diethyl 1-cyclohexen-1-ylphosphonate.*

$C_{10}H_{19}O_3P$ M 218.232

Liq. d_4^{20} 1.08. $Bp_{1.5}$ 98-98.5°. n_D^{20} 1.4680.

Dichloride: [20095-28-7].

$C_6H_9Cl_2OP$ M 199.016

Liq. d_4^{20} 1.35. $Bp_{0.5}$ 81°. n_D^{20} 1.5245.

Fay, P. *et al*, *J. Am. Chem. Soc.*, 1952, **74**, 4933 (*synth*)
Kenyon, G.L. *et al*, *J. Am. Chem. Soc.*, 1966, **88**, 3557 (*ir, pmr*)
Suminov, S.I. *et al*, *Zh. Obshch. Khim.*, 1972, **42**, 239 (*Engl. transl.* p. 233) (*synth, derivs, pmr*)
Hirao, T. *et al*, *Bull. Chem. Soc. Jpn.*, 1982, **55**, 909 (*ester, synth*)

3-Cyclohexenephosphonic acid C-00242

2-Cyclohexen-1-ylphosphonic acid, 9CI. 3-Phosphonocyclohexene

$C_6H_{11}O_3P$ M 162.125

(±)-*form*

Cryst. (C_6H_6). Sol. H_2O, Et_2O. Mp 104-106°.

Di-Me ester: Dimethyl 2-cyclohexen-1-ylphosphonate.

$C_8H_{15}O_3P$ M 190.178

Liq. d_4^{20} 1.14. $Bp_{1.5}$ 92.5-93.5°. n_D^{20} 1.4728.

Di-Et ester: Diethyl 2-cyclohexen-1-ylphosphonate.

$C_{10}H_{19}O_3P$ M 218.232

Liq. d_4^{20} 1.07. Bp_1 86.5-87°. n_D^{20} 1.4645.
Diisopropyl ester: Diisopropyl 2-cyclohexen-1-ylphosphonate.
$C_{12}H_{23}O_3P$ M 246.286
Liq. d_4^{20} 1.02. Bp_2 107.5-108°. n_D^{20} 1.4568.
Dichloride:
$C_6H_9Cl_2OP$ M 199.016
Liq. d_4^{20} 1.35. $Bp_{1.5}$ 94-95°. n_D^{20} 1.5210.

Fay, P. *et al, J. Am. Chem. Soc.*, 1952, **74**, 4933 (*synth*)
Arbuzov, B.A. *et al, Izv. Akad. Nauk SSSR, Ser. Khim.*, 1961, 1288 (*Engl. transl.* p. 1197) (*esters*)
Suminov, S.I. *et al, Zh. Obshch. Khim.*, 1972, **42**, 239 (*Engl. transl.* p. 233) (*derivs, synth, ir, pmr*)

4-Cyclohexenephosphonic acid C-00243

3-Cyclohexen-1-ylphosphonic acid, 9CI. 4-Phosphonocyclohexene

$C_6H_{11}O_3P$ M 162.125
Di-Et ester: Diethyl 3-cyclohexen-1-ylphosphonate.
$C_{10}H_{19}O_3P$ M 218.232
Liq. $Bp_{1.5}$ 114-115°.

Daniewski, W.M. *et al, J. Org. Chem.*, 1966, **31**, 3236 (*synth*)

1-Cyclohexyl-1,2-bis(diphenylphosphino)ethane C-00244

(1-Cyclohexyl-1,2-ethanediyl)bis[diphenylphosphine], 10CI. Cycphos

(*R*)-*form*

$C_{32}H_{34}P_2$ M 480.568
Rh complex effective for asymmetric hydrogenations.
(*R*)-*form* [75421-31-7]
Needles (EtOH). $[\alpha]_D^{25}$ +103.3° (c, 1.0 in THF).

Riley, D.P. *et al, J. Org. Chem.*, 1980, **45**, 5787.

Cyclohexyldiphenylphosphine, 9CI C-00245

Diphenylphosphinocyclohexane
[6372-42-5]

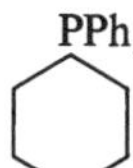

$C_{18}H_{21}P$ M 268.338
Cryst. (EtOH). Mp 60-61°.
Oxide: [13689-20-8]. *Diphenylphosphinylcyclohexane.*
$C_{18}H_{21}OP$ M 284.337
Cryst. (pet. ether or C_6H_6). Mp 164-166°, 187-188°.
Sulfide: [6591-42-0]. *Diphenylphosphinothioylcyclohexane.*
$C_{18}H_{21}PS$ M 300.398
Cryst. (MeCN). Mp 180-185°.

Issleib, K. *et al, Chem. Ber.*, 1961, **94**, 392 (*synth, oxide*)
Grim, S.O. *et al, Nature (London)*, 1965, **208**, 995 (*P nmr*)
Inorg. Synth., 1976, **16**, 159 (*synth*)
Davidson, A.H. *et al, J. Chem. Soc., Perkin Trans. 1*, 1976, 639 (*oxide, ir, pmr*)
Troy, D. *et al, Bull. Soc. Chim. Fr. Part I*, 1979, 241 (*uv*)
Bertz, S.H. *et al, J. Am. Chem. Soc.*, 1981, **103**, 5932 (*oxide, synth*)

N-Cyclohexyl-*P*-ethylphosphonamidic acid, 9CI C-00246

$C_8H_{18}NO_2P$ M 191.209
Cyclohexyl ester: Cyclohexyl N-*cyclohexyl*-P-*ethylphosphonamidate.*
$C_{14}H_{29}NO_2P$ M 274.362
Solid. Mp 117°.
Benzyl ester: Benzyl N-*cyclohexyl*-P-*ethylphosphonamidate.*
$C_{15}H_{24}NO_2P$ M 281.334
Solid. Mp 194-195°.

Ger. Pat.., 1 084 716, (*1960*); *CA*, **55**, 21052

Cyclohexylmethylphosphinic acid, 9CI C-00247

$C_7H_{15}O_2P$ M 162.168
Me ester: Methyl cyclohexylmethylphosphinate.
$C_8H_{17}O_2P$ M 176.195
Liq. $Bp_{1.2}$ 90°.
(−)-*Menthyl ester: Menthyl cyclohexylmethylphosphinate.*
$C_{17}H_{33}O_2P$ M 300.420
Mp 80-81°, 110-111°. $[\alpha]_D^{23-6}$ −54°, −59° (c, 1-3 in C_6H_6). Two diastereoisomers known of unknown chirality at phosphorus.
Fluoride: [824-71-5].
$C_7H_{14}FOP$ M 164.159
Liq. $Bp_{1-1.5}$ 77°.
Chloride: Liq. becoming solid at r.t. d^{20} 1.16. Bp_3 101-102°. n_D^{20} 1.4988.

Dawson, T.P. *et al, J. Org. Chem.*, 1957, **22**, 1671 (*chloride, fluoride*)
Korpiun, O. *et al, J. Am. Chem. Soc.*, 1968, **90**, 4842 (*chloride, esters*)
Knunyants, I.L. *et al, Dokl. Akad. Nauk SSSR, Ser. Sci. Khim.*, 1971, **201**, 862 (*Engl. transl.* p. 992) (*halides, nmr*)

Cyclohexylmethylphosphinothioic acid C-00248

$C_7H_{15}OPS$ M 178.229
Chloride:
$C_7H_{14}ClPS$ M 196.674
Liq. d_4^{20} 1.17. Bp_2 89-90°. n_D^{20} 1.5570.
Anilide: P-*Cyclohexyl*-P-*methyl*-N-*phenylphosphinothioic amide.*
$C_{13}H_{20}NPS$ M 253.341
Solid. Mp 135.5-136°.

Godovikov, N.N. *et al, Dokl. Akad. Nauk SSSR, Ser. Sci. Khim.*, 1956, **110**, 217 (*Engl. transl.* p. 549) (*synth*)

Cyclohexylphenylphosphine, 9CI C-00249

$C_{12}H_{17}P$ M 192.240

Oxide: [75354-35-7].
$C_{12}H_{17}OP$ M 208.239
No phys. props. reported. Tautomeric with cyclohexylphenylphosphinous acid.

Sulfide: [49749-59-9].
$C_{12}H_{17}PS$ M 224.300
Solid. Mp 107°. Tautomeric with cyclohexylphenylphosphinothious acid.

Issleib, K. *et al, J. Prakt. Chem.*, 1973, **315**, 471 (*sulfide*)
Lindner, E. *et al, Chem. Ber.*, 1980, **113**, 3268 (*oxide*)

Cyclohexylphenylphosphinic acid, 9CI C-00250

[2310-66-9]

(*R*)-*form* of esters

$C_{12}H_{17}O_2P$ M 224.239

Known in opt. active form as menthyl esters.

(±)-***form***

Cryst. (EtOH aq. or Et_2O). Mp 121-122° (107-109°). pK_a 5.02 (75% EtOH aq.), 5.60 (95% EtOH aq.).

Me ester: [85357-61-5]. *Methyl cyclohexylphenylphosphinate.* No phys. props. reported.

Et ester: [13689-17-3]. *Ethyl cyclohexylphenylphosphinate.*
$C_{14}H_{21}O_2P$ M 252.292
Liq. d_4^{20} 1.10. Bp_{10} 182-188°. n_D^{20} 1.5288.

Chloride:
$C_{12}H_{16}ClOP$ M 242.685
d^{20} 1.21. Bp_6 191-193°. n_D^{20} 1.5560.

Zinov'ev, Yu.M. *et al, Zh. Obshch. Khim.*, 1956, **26**, 3030 (*Engl. transl.* p. 3375) (*chloride*)
Pudovik, A.N. *et al, Zh. Obshch. Khim.*, 1960, **30**, 2348 (*Engl. transl.* p. 2328) (*synth, ester*)
Peppard, D.F. *et al, J. Inorg. Nucl. Chem.*, 1965, **27**, 697 (*synth, props*)
Farnham, W.B. *et al, J. Chem. Soc., Chem. Commun.*, 1971, 146 (*esters*)
Legin, G.Ya., *Zh. Obshch. Khim.*, 1973, **43**, 2202 (*Engl. transl.* p. 2194) (*esters*)
Yamashita, M. *et al, Bull. Chem. Soc. Jpn.*, 1983, 219 (*ester, pmr, ir*)

***N*-Cyclohexyl-*P*-phenylphosphonamidic acid**, 9CI C-00251

$C_{12}H_{18}NO_2P$ M 239.253

Me ester: [58245-39-9]. *Methyl* N-*cyclohexyl*-P-*phenylphosphonamidate.*
$C_{13}H_{20}NO_2P$ M 253.280
Cryst. (Me_2CO). Mp 152-155°.

Benzyl ester: Benzyl N-*cyclohexyl*-P-*phenylphosphonamidate.*
$C_{19}H_{24}NO_2P$ M 329.378
Needles (C_6H_6/pet. ether). Mp 119-120°.

Anand, N. *et al, J. Chem. Soc.*, 1951, 1867 (*synth*)
Ruveda, M.A. *et al, Phosphorus*, 1975, **5**, 217 (*synth, ir, pmr*)

Cyclohexylphosphine, 9CI C-00252

[822-68-4]

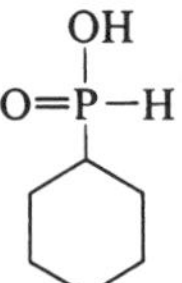

$C_6H_{13}P$ M 116.142

Ligand for metals of Groups VIB, VIIB and VIII. Liq. Bp 146°, Bp_{160} 97°, Bp_{15} 43°. n_D^{20} 1.4860. Li derivs. chemiluminesce.

▷Pyrophoric

Horner, L. *et al, Chem. Ber.*, 1959, **92**, 2088 (*synth*)
Rauhut, M.M. *et al, J. Org. Chem.*, 1961, **26**, 5138 (*synth*)
Pass, F. *et al, Monatsh. Chem.*, 1962, **93**, 230 (*synth, ir*)
Issleib, K. *et al, Chem. Ber.*, 1964, **97**, 721; 1965, **98**, 2091 (*synth, deriv*)
Gordon, M.D. *et al, J. Magn. Reson.*, 1976, **22**, 149 (*nmr*)
Gordon, M.D. *et al, J. Org. Chem.*, 1976, **41**, 1690 (*cmr*)
Jardine, I. *et al, J. Am. Chem. Soc.*, 1976, **98**, 5086 (*ms*)
Andreev, N.A. *et al, Zh. Obshch. Khim.*, 1981, **51**, 1743 (*Engl. transl.* p. 1491) (*synth, ir, nmr*)

Cyclohexylphosphinic acid, 9CI C-00253

[2310-71-6]

$C_6H_{13}O_2P$ M 148.141

Tautomeric with Cyclohexylphosphonous acid, C-00267 Free acid exists in phosphinic acid form. Viscous liq. d_4^{20} 1.16. pK_a 3.99 (75% EtOH aq.), 4.55 (95% EtOH aq.). n_D^{20} 1.4938.

Anilinium salt: Mp 106-110°.

Me ester: [16196-03-5]. *Methyl cyclohexylphosphinate.*
$C_7H_{15}O_2P$ M 162.168
Liq. d_4^{20} 1.08. n_D^{20} 1.4670.

Et ester: [14610-01-6]. *Ethyl cyclohexylphosphinate.*
$C_8H_{17}O_2P$ M 176.195
Liq. d_4^{20} 1.03. Bp_3 105-106°. n_D^{20} 1.4690.

Isopropyl ester: [16954-60-2]. *Isopropyl cyclohexylphosphinate.*
$C_9H_{19}O_2P$ M 190.222
No phys. props. reported.

Trimethylsilyl ester: [77339-71-0]. *Trimethylsilyl cyclohexylphosphinate.*
$C_9H_{21}O_2PSi$ M 220.323
Bp_3 95°. n_D^{20} 1.4446.

Diethylamide: P-*Cyclohexyl*-N,N-*diethylcyclohexylphosphinic amide.*
$C_{10}H_{22}NOP$ M 203.264
Liq. n_D^{20} 1.4909.

Peppard, D.F. *et al, J. Inorg. Nucl. Chem.*, 1965, **27**, 697 (*synth, props*)
Nifantev, É.E. *et al, Zh. Obshch. Khim.*, 1967, **37**, 1366 (*Engl. transl.* p. 1293) (*ester*)
Sanchez, M. *et al, Spectrochim. Acta, Part A*, 1967, **23**, 2617 (*esters, ir*)
Andreev, N.A. *et al, Zh. Obshch. Khim.*, 1979, **49**, 2230 (*Engl. transl.* p. 1959) (*synth, amide, ir, P nmr*)

Andreev, N.A. *et al*, *Zh. Obshch. Khim.*, 1980, **50**, 803 (*Engl. transl.* p. 641) (*synth, ester, ir, pmr*)
Nifantev, É.E. *et al*, *Zh. Obshch. Khim.*, 1980, **50**, 2676 (*Engl. transl.* p. 2159) (*silyl ester*)

P-Cyclohexylphosphonamidic fluoride, 9CI C-00254

[29070-41-5]

$C_6H_{13}FNOP$ M 165.147
Solid. Mp 112°.

Roesky, H.W. *et al*, *Z. Anorg. Allg. Chem.*, 1970, **375**, 140 (*synth, ir, ms, pmr, F nmr*)

P-Cyclohexylphosphonamidothioic chloride, 9CI C-00255

$C_6H_{13}ClNPS$ M 197.662

N,N-*Di-Me:* [17833-37-3]. P-*Cyclohexyl*-N,N-*dimethylphosphonamidothioic chloride.*
$C_8H_{17}ClNPS$ M 225.716
Solid. Mp 73°.

N,N-*Di-Et:* [17833-38-4]. P-*Cyclohexyl*-N,N-*diethylphosphonamidothioic chloride.*
$C_{10}H_{21}ClNPS$ M 253.769
Liq. d_4^{20} 1.12. $Bp_{0.9}$ 138°. n_D^{20} 1.5350.

Tolkmith, H. *et al*, *J. Med. Chem.*, 1967, **10**, 1074 (*synth*)
Andreev, N.A. *et al*, *Zh. Obshch. Khim.*, 1979, **49**, 718 (*Engl. transl.* p. 623) (*synth, ir, P nmr*)

Cyclohexylphosphonic acid, 9CI C-00256

Phosphonocyclohexane. Cyclohexanephosphonic acid
[1005-23-8]

$C_6H_{13}O_3P$ M 164.141
Cryst. Mp 166-167°.

Di-Me ester: [1641-61-8]. *Dimethyl cyclohexylphosphonate. Dimethoxyphosphinylcyclohexane.*
$C_8H_{17}O_3P$ M 192.194
Liq. d_4^{18} 1.09. Bp_2 96°. n_D^{18} 1.4720.

Mono-Et ester: [4546-15-0]. *Monoethyl cyclohexylphosphonate. Ethyl hydrogen cyclohexylphosphonate.*
$C_8H_{17}O_3P$ M 192.194
d_4^{20} 1.14. n_D^{20} 1.4777.

Di-Et ester: [7413-09-4]. *Diethyl cyclohexylphosphonate. Diethoxyphosphinylcyclohexane.*
$C_{10}H_{21}O_3P$ M 220.248
Liq. d_4^{20} 1.03. Bp_{13} 140-144°. n_D^{18} 1.4615.

Monoisopropyl ester: Monoisopropyl cyclohexylphosphonate. Isopropyl hydrogen cyclohexylphosphonate.
$C_9H_{19}O_3P$ M 206.221
Solid. Mp 50°.

Diisopropyl ester: Diisopropyl cyclohexylphosphonate. Diisopropoxyphosphinylcyclohexane.
$C_{12}H_{25}O_3P$ M 248.301
Liq. d_4^{18} 1.01. Bp_{11} 141-145°. n_D^{18} 1.4620.

Mono-Ph ester: [2596-41-0]. *Monophenyl cyclohexylphosphonate. Phenyl hydrogen cyclohexylphosphonate.*
$C_{12}H_{17}O_3P$ M 240.238
Solid + H_2O. Mp 80-105°.

Di-Ph ester: [13689-16-2]. *Diphenyl cyclohexylphosphonate. Diphenoxyphosphinylcyclohexane.*
$C_{18}H_{21}O_3P$ M 316.336
Solid. Mp 62°.

Difluoride:
$C_6H_{11}F_2OP$ M 168.123
Liq. d_4^{20} 1.19. Bp_{27} 83-84°. n_D^{20} 1.4177.

Dichloride: see Cyclohexylphosphonic dichloride, C-00258

Diamide: see P-Cyclohexylphosphonic diamide, C-00257

Graf, R., *Chem. Ber.*, 1952, **85**, 9 (*synth, esters*)
Pfeiffer, G. *et al*, *Bull. Soc. Chim. Fr.*, 1966, 1652 (*synth, derivs, esters, ir*)
Müller, E. *et al*, *Chem. Ber.*, 1967, **100**, 521 (*esters, synth*)
Yoshida, Z. *et al*, *Chem. Lett.*, 1975, 279 (*esters, synth, ir, pmr*)
Andreev, N.A. *et al*, *Zh. Obshch. Khim.*, 1979, **49**, 2230 (*Engl. transl.* p. 1959) (*synth, ir, P nmr*)
Wozniak, M. *et al*, *J. Chem. Soc., Dalton Trans.*, 1981, 2423 (*complexes*)

P-Cyclohexylphosphonic diamide, 9CI C-00257

[53782-53-9]

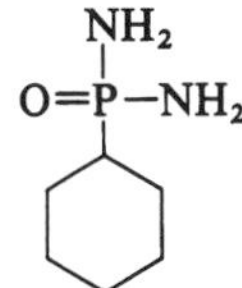

$C_6H_{15}N_2OP$ M 162.171
Solid. Mp 184-185°.

N,N,N′,N′-*Tetra-Me:* [32400-38-7]. P-*Cyclohexyl*-N,N,N′,N′-*tetramethylphosphonic diamide. Cyclohexylphosphonic bis(dimethylamide).*
$C_{10}H_{23}N_2OP$ M 218.278
Hygroscopic solid. $Bp_{0.45}$ 118°. BuLi → anion.

N,N′-*Di-Ph:* [37624-73-0]. P-*Cyclohexyl*-N,N′-*diphenylphosphonic diamide. Cyclohexylphosphonic dianilide.*
$C_{18}H_{23}N_2OP$ M 314.366
Cryst. (EtOH). Mp 235-236°.

Graf, R., *Chem. Ber.*, 1952, **85**, 9 (*synth*)
Kinnear, A.M. *et al*, *J. Chem. Soc.*, 1952, 3437 (*dianilide*)
Jones, J.B. *et al*, *Can. J. Chem.*, 1971, **49**, 1300 (*bisdimethylamide, synth, pmr, ir, ms, props*)

Cyclohexylphosphonic dichloride, 9CI C-00258

[1005-22-7]

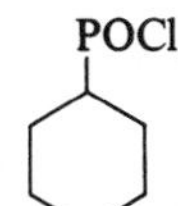

$C_6H_{11}Cl_2OP$ M 201.032
Solid. Mp 41°. Bp 260°, $Bp_{3\text{-}4}$ 100-102°.

Kinnear, A.M. *et al*, *J. Chem. Soc.*, 1952, 3437 (*synth*)

Isbell, A.F. *et al, J. Am. Chem. Soc.*, 1956, **78**, 6042 (*synth*)
Christol, C. *et al, J. Chim. Phys.*, 1965, **62**, 246 (*ir, raman*)
Nifant'ev, É.E. *et al, Zh. Obshch. Khim.*, 1979, **49**, 1905 (*Engl. transl.* p. 1678) (*synth, props*)
Morita, T. *et al, Chem. Lett.*, 1980, 435 (*synth*)

Cyclohexylphosphonochloridic acid, 9CI — C-00259

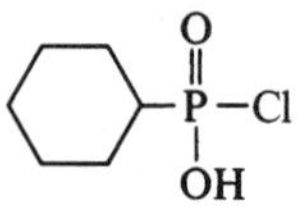

$C_6H_{12}ClO_2P$ M 182.586

Me ester: [16139-82-5]. *Methyl cyclohexylphosphonochloridate.*
$C_7H_{14}ClO_2P$ M 196.613
Liq. d_4^{20} 1.52. Bp_5 105-110°, $Bp_{0.03}$ 69°. n_D^{20} 1.4693.

Et ester: [15873-69-5]. *Ethyl cyclohexylphosphonochloridate.*
$C_8H_{16}ClO_2P$ M 210.640
Liq. Bp_3 160°.

Zinov'ev, J.M. *et al, Zh. Obshch. Khim.*, 1958, **28**, 1551.
Klamann, D. *et al, Monatsh. Chem.*, 1967, **98**, 911.

Cyclohexylphosphonochloridothioic acid, 9CI — C-00260

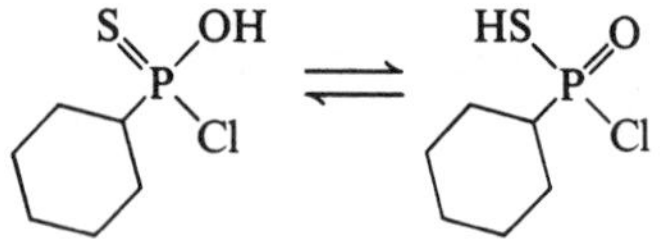

$C_6H_{12}ClOPS$ M 198.647

O-(*2-Methylpropyl*) *ester:* [16243-48-4]. O-(*2-Methylpropyl*) *cyclohexylphosphonochloridothioate.* O-*Isobutyl cyclohexylphosphonochloridothioate.*
$C_{10}H_{20}ClOPS$ M 254.754
Liq. $Bp_{0.1}$ 66°.

O-*Octyl ester:* O-*Octyl cyclohexylphosphonochloridothioate.*
$C_{14}H_{28}ClOPS$ M 310.861
Liq. $Bp_{0.1}$ 63-65°.

Klamann, D. *et al, Monatsh. Chem.*, 1967, **98**, 911.

Cyclohexylphosphonodithioic acid, 9CI — C-00261

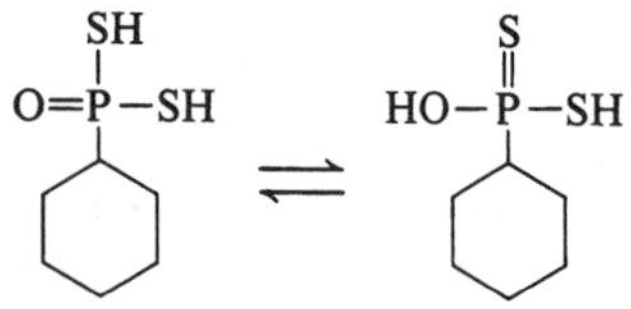

$C_6H_{13}OPS_2$ M 196.262

*Mono-*O-*Et ester:* [1007-92-7]. O-*Ethyl hydrogen cyclohexylphosphonodithioate.*
$C_8H_{17}OPS_2$ M 224.315
Isol. as amine salts.

O,S-*Di-Et ester:* [13685-87-5]. O,S-*Diethyl cyclohexylphosphonodithioate.*
$C_{10}H_{21}OPS_2$ M 252.369
Liq. d_4^{20} 1.09. $Bp_{0.01}$ 99-100°. n_D^{20} 1.5404.

S,S-*Di-Et ester:* [16139-87-0]. S,S-*Diethyl cyclohexylphosphonodithioate.*
$C_{10}H_{21}OPS_2$ M 252.369
Liq. $Bp_{1.7}$ 149-150°.

S-*Et*, O-*Ph ester:* [15374-50-2]. S-*Ethyl* O-*phenyl cyclohexylphosphonodithioate.*
$C_{14}H_{21}OPS_2$ M 300.413
Liq. d_4^{20} 1.15. $Bp_{0.01}$ 213-215°. n_D^{20} 1.5825.

O-*Cyclohexyl*, S-*Et ester:* O-*Cyclohexyl* S-*ethyl cyclohexylphosphonodithioate.*
$C_{14}H_{27}OPS_2$ M 306.460
Liq. d_4^{20} 1.09. $Bp_{0.02}$ 153-154°. n_D^{20} 1.5470.

Grishina, O.N. *et al, Zh. Obshch. Khim.*, 1967, **37**, 470 (*Engl. transl.* p. 438) (*diethyl ester, synth*)
Klamann, D. *et al, Monatsh. Chem.*, 1967, **98**, 911 (*dialkyl esters*)
Grishina, O.N. *et al, Zh. Obshch. Khim.*, 1970, **40**, 579 (*Engl. transl.* p. 548) (*monoethyl ester, synth*)
Yoshifuji, M. *et al, J. Chem. Soc., Perkin Trans. 1*, 1973, 2065 (*diethyl ester, synth, P nmr, pmr*)

Cyclohexylphosphonodithious acid, 9CI — C-00262

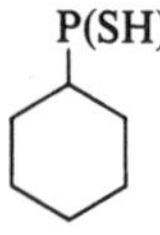

$C_6H_{13}PS_2$ M 180.262

Di-Ph ester: [7697-72-5]. *Diphenyl cyclohexylphosphonodithioite.*
$C_{18}H_{21}PS_2$ M 332.458
Liq. $Bp_{0.15}$ 183-189°.

Grayson, M. *et al, J. Org. Chem.*, 1967, **32**, 236 (*deriv, synth, ir*)

Cyclohexylphosphonoselenoic acid — C-00263

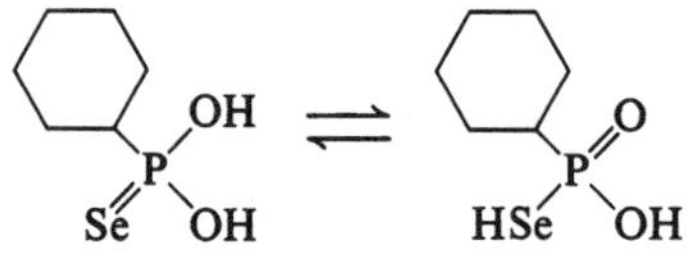

$C_6H_{13}O_2PSe$ M 227.101

Difluoride:
$C_6H_{11}F_2PSe$ M 231.083
Liq. $Bp_{0.01}$ 30°.

Dichloride: [53213-05-1].
$C_6H_{11}Cl_2PSe$ M 263.993
Liq. $Bp_{0.05}$ 72°.

Roesky, H.W. *et al, Z. Naturforsch., B*, 1973, **28**, 697 (*synth, ir, pmr, F nmr*)

Cyclohexylphosphonothioic acid, 9CI — C-00264

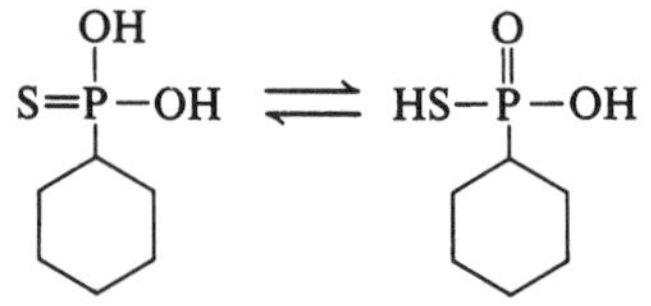

$C_6H_{13}O_2PS$ M 180.201
Solid. Mp 40°.

Dianilinium salt: Solid. Mp 150-150.5°.

O-*Mono-Et ester:* [58078-24-3]. O-*Ethyl hydrogen cyclohexylphosphonothioate.*
$C_8H_{17}O_2PS$ M 208.255
Liq. d_4^{20} 1.13. $Bp_{0.05}$ 120-123°. n_D^{20} 1.5130.

O,O-*Di-Et ester:* [73178-34-4]. O,O-*Diethyl cyclohexylphosphonothioate.*
$C_{10}H_{21}O_2PS$ M 236.308
Liq. d_4^{20} 1.06. Bp_2 116.5-117°. n_D^{20} 1.4888.

O,O-*Dibutyl ester:* O,O-*Dibutyl cyclohexylphosphonothioate.*
$C_{14}H_{29}O_2PS$ M 292.416

Liq. d_4^{20} 1.00. Bp_2 141-142.5°. n_D^{20} 1.4837.
O,O-*Bis(trimethylsilyl) ester:* [77339-64-1].
$C_{12}H_{29}O_2PSSi_2$ M 324.565
Liq. Bp_2 100-102°. n_D^{20} 1.4684.
Dichloride: see Cyclohexylphosphonothioic dichloride, C-00266

Pudovik, A.N. *et al*, *Zh. Obshch. Khim.*, 1960, **30**, 2348 (*Engl. transl.* p. 2328) (*esters*)
Berkhamov, M.Kh. *et al*, *Zh. Obshch. Khim.*, 1977, **47**, 989 (*Engl. transl.* p. 909) (*ester*)
Nifant'ev, E.E. *et al*, *Zh. Obshch. Khim.*, 1980, **50**, 2676 (*Engl. transl.* p. 2159) (*synth, derivs*)

P-Cyclohexylphosphonothioic diamide, 9CI C-00265

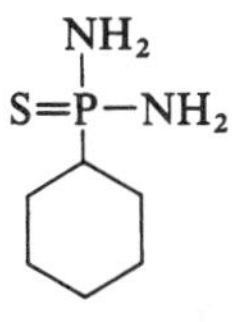

$C_6H_{15}N_2PS$ M 178.232
N,N,N',N'-*Tetra-Et:* [63135-71-7]. P-*Cyclohexyl-N,N,N',N'-tetraethylphosphonothioic diamide. Cyclohexylphosphonothioic bis(diethylamide).*
$C_{14}H_{31}N_2PS$ M 290.446
Liq. d_4^{20} 1.01. $Bp_{0.0001}$ 95-100°. n_D^{20} 1.5152.
N,N'-*Di-Ph:* P-*Cyclohexyl*-N,N'-*diphenylphosphonothioic diamide. Cyclohexylphosphonothioic dianilide.*
$C_{18}H_{23}N_2PS$ M 330.427
Solid. Mp 122-123°.

Kabachnik, M.I. *et al*, *Dokl. Akad. Nauk SSSR, Ser. Sci. Khim.*, 1956, **110**, 217; *CA*, **51**, 4982 (*diphenyl, synth*)
Nifant'ev, É.E. *et al*, *Zh. Obshch. Khim.*, 1977, **47**, 299 (*Engl. transl.* p. 276) (*tetraethyl, synth, P nmr*)

Cyclohexylphosphonothioic dichloride, 9CI C-00266

Cyclohexylthiophosphonic dichloride
[1498-63-1]

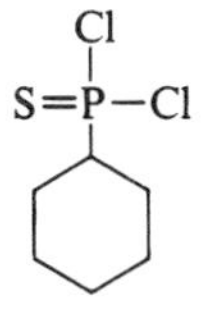

$C_6H_{11}Cl_2PS$ M 217.093
Liq. with pungent odour. d_4^{20} 1.30. Bp_{17} 139°, Bp_2 83-85°. n_D^{20} 1.5620.

Christol, C. *et al*, *J. Chim. Phys.*, 1965, **62**, 246 (*ir, raman*)
Al'zoba, T.G. *et al*, *Zh. Obshch. Khim.*, 1977, **47**, 1979 (*Engl. transl.* p. 1810) (*synth*)
Kaushik, M.P. *et al*, *J. Org. Chem.*, 1980, **45**, 2270 (*synth*)

Cyclohexylphosphonous acid, 9CI C-00267

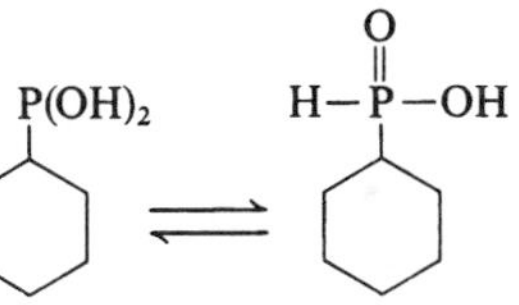

$C_6H_{13}O_2P$ M 148.141
The free acid exists as the phosphoryl tautomer Cyclohexylphosphinic acid, C-00253 .
Mono-Me ester: see Cyclohexylphosphinic acid, C-00253
Di-Me ester: [16195-98-5]. *Dimethyl cyclohexylphosphonite.*
$C_8H_{17}O_2P$ M 176.195
Liq. Bp_8 87-90°. n_D^{20} 1.4640.
Mono-Et ester: see Cyclohexylphosphinic acid, C-00253
Di-Et ester: [16195-99-6]. *Diethyl cyclohexylphosphonite.*
$C_{10}H_{21}O_2P$ M 204.248
Liq. n_D^{20} 1.4624.
Monoisopropyl ester: see Cyclohexylphosphinic acid, C-00253
Di-Ph ester: Diphenyl cyclohexylphosphonite.
$C_{18}H_{21}O_2P$ M 300.336
Liq. d_0^{20} 1.14. Bp_{10} 233-235°. n_D^{20} 1.5705.
Mono(trimethylsilyl) ester: see Cyclohexylphosphinic acid, C-00253
Dichloride: see Cyclohexylphosphonous dichloride, C-00268
Dibromide: [86012-32-0]. *Dibromocyclohexylphosphine.*
$C_6H_{11}Br_2P$ M 273.935
Liq. Bp_5 108-109°.
Diiodide: [24547-24-8]. *Cyclohexyldiiodophosphine.*
$C_6H_{11}I_2P$ M 367.936
Orange-red liq. Bp_{14} 171-172°, $Bp_{0.05}$ 101°.
Mono(diethylamide): see Cyclohexylphosphinic acid, C-00253
Bis(diethylamide): [3348-40-1]. P-*Cyclohexyl-N,N,N',N'-tetraethylphosphonous diamide.* Liq. Bp_3 125-126°, $Bp_{0.1}$ 87-89°. n_D^{20} 1.4900.

Kamai, G. *et al*, *CA*, 1957, **51**, 11273 (*ester, synth*)
Issleib, K. *et al*, *Z. Anorg. Allg. Chem.*, 1960, **303**, 155 (*halides*)
Feshchenko, N.G. *et al*, *Zh. Obshch. Khim.*, 1969, **39**, 2139 (*Engl. transl.* p. 2090) (*diiodide*)
Gordon, M.D. *et al*, *J. Org. Chem.*, 1976, **41**, 1690 (*dimethyl ester, cmr*)
Gordon, M.D. *et al*, *J. Am. Chem. Soc.*, 1976, **98**, 15 (*dimethyl ester, synth, pmr, P nmr*)
Andreev, N.A. *et al*, *Zh. Obshch. Khim.*, 1979, **49**, 2230 (*Engl. transl.* p. 1959) (*bisdiethylamide, synth, ir, P nmr*)
Andreev, N.A. *et al*, *Zh. Obshch. Khim.*, 1980, **50**, 803 (*Engl. transl.* p. 641) (*monoethyl ester, synth, nmr*)
Hinke, A. *et al*, *Phosphorus Sulfur*, 1983, **15**, 93 (*dibromide, synth, nmr*)

Cyclohexylphosphonous dichloride C-00268

Dichlorocyclohexylphosphine
[2844-89-5]

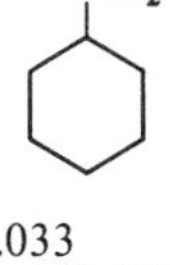

$C_6H_{11}Cl_2P$ M 185.033
Liq. d_4^{20} 1.21-1.22. Bp_{20} 105-107°, $Bp_{0.4}$ 50.5-51.5°. n_D^{20} 1.5242 (1.5040).

Issleib, K. *et al*, *Z. Anorg. Allg. Chem.*, 1960, **303**, 155 (*synth*)
Henderson, W.A. *et al*, *J. Org. Chem.*, 1961, **26**, 4770 (*synth*)
Nifant'ev, É.E. *et al*, *Zh. Obshch. Khim.*, 1967, **37**, 1366 (*Engl. transl.* p. 1293) (*synth*)
Babkina, É.I. *et al*, *Zh. Obshch. Khim.*, 1968, **38**, 1772 (*Engl. transl.* p. 1727) (*synth*)
Babkina, É.I. *et al*, *Zh. Obshch. Khim.*, 1969, **39**, 1651 (*Engl. transl.* p. 1620) (*ir*)
Gordon, M.D. *et al*, *J. Am. Chem. Soc.*, 1976, **98**, 15 (*P nmr*)
Gordon, M.D. *et al*, *J. Org. Chem.*, 1976, **41**, 1690 (*cmr*)

Cyclohexylphosphoramidic acid, 9CI, 8CI C-00269

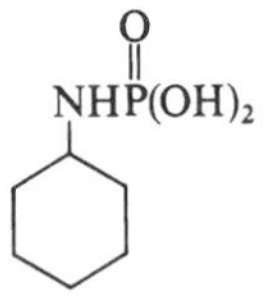

$C_6H_{14}NO_3P$ M 179.155

Di-Me ester: [32405-87-1]. *Dimethyl cyclohexylphosphoramidate.*
$C_8H_{18}NO_3P$ M 207.209
Solid. Mp 65.5-66°.

Di-Et ester: [32405-88-2]. *Diethyl cyclohexylphosphoramidate.*
$C_{10}H_{22}NO_3P$ M 235.262
Cryst. (pet. ether). Mp 71°, Mp 79-80°.

Diisopropyl ester: [54480-52-3]. *Diisopropyl cyclohexylphosphoramidate.*
$C_{12}H_{26}NO_3P$ M 263.316
Solid. Mp 53.5-55°.

*Di-*tert*-butyl ester: Di-*tert*-butyl cyclohexylphosphoramidate.*
$C_{14}H_{30}NO_3P$ M 291.370
Solid. Mp 81-83°.

Di-Ph ester: [6372-21-0]. *Diphenyl cyclohexylphosphoramidate.*
$C_{18}H_{22}NO_3P$ M 331.350
Cryst. (EtOH aq.). Mp 103-104°.

Dibenzyl ester: [7494-75-9]. *Dibenzyl cyclohexylphosphoramidate.*
$C_{20}H_{26}NO_3P$ M 359.404
Cryst. (pet. ether or cyclohexane). Mp 78-80°.

Dichloride: [13089-54-8].
$C_6H_{12}Cl_2NOP$ M 216.047
Cryst. (C_6H_6). Mp 85-86°.

Wagner-Jauregg, T. *et al*, *J. Am. Chem. Soc.*, 1951, **73**, 5202 (*esters, synth*)
Baumgarten, H.E. *et al*, *J. Am. Chem. Soc.*, 1959, **81**, 2132 (*dimethyl ester*)
Atkinson, R.E. *et al*, *J. Chem. Soc. (C)*, 1967, 1356 (*esters*)
Dorn, H. *et al*, *J. Prakt. Chem.*, 1971, **313**, 218 (*dichloride*)
Mathis, R. *et al*, *C.R. Hebd. Seances Acad. Sci., Ser. C*, 1975, **281**, 437; 1977, **284**, 767 (*esters, ir*)
Zwierzak, A. *et al*, *Synthesis*, 1975, 507; 1984, 223, 332 (*ester, synth, ir, pmr, P nmr*)
Al-Rawi, J.M.A. *et al*, *Org. Magn. Reson.*, 1983, **21**, 75 (*diphenyl ester, cmr*)

Cyclohexylphosphoramidochloridothioic acid, 9CI, 8CI C-00270

[19714-34-2]

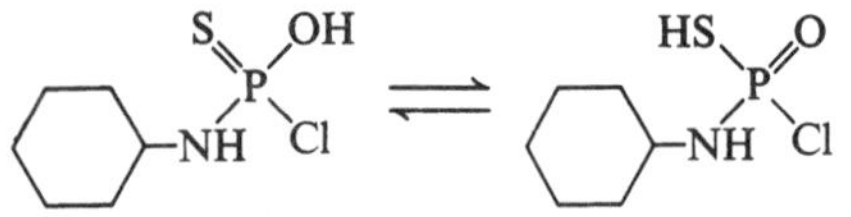

$C_6H_{13}ClNOPS$ M 213.662

(+)-*form*

O-Me ester: O-Methyl cyclohexylphosphoramidochloridothioate.
$C_7H_{15}ClNOPS$ M 227.688
Solid. Mp 5-15°. $[\alpha]_D$ +9.2° (pet. ether).

(±)-*form*

O-*Me ester:* Cryst. (pet. ether). Mp 31-31.5°.

Cooks, R.G. *et al*, *J. Chem. Soc. (B)*, 1968, 1327 (*ms*)
Gerrard, A.F. *et al*, *J. Chem. Soc. (B)*, 1968, 539; 1969, 369 (*synth, props*)

Cyclohexylphosphoramidothioic acid, 9CI, 8CI C-00271

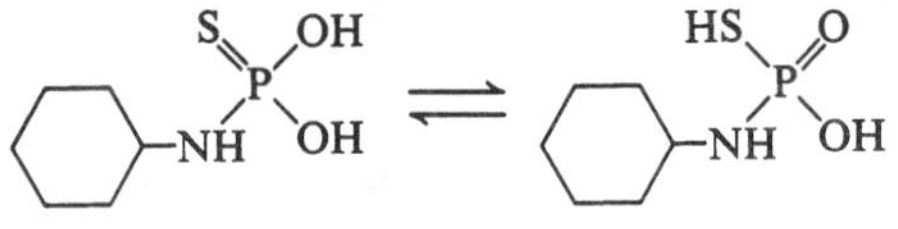

$C_6H_{14}NO_2PS$ M 195.216

O,O-*Di-Me ester:* [941-39-9]. O,O-*Dimethyl cyclohexylphosphoramidothioate.*
$C_8H_{18}NO_2PS$ M 223.269
No phys. props. reported.

O,O-*Di-Et ester:* [55739-97-4]. O,O-*Diethyl cyclohexylphosphoramidothioate.*
$C_{10}H_{22}NO_2PS$ M 251.323
Liq. $Bp_{0.75}$ 121°. n_D^{25} 1.4920.

O,S-*Di-Et ester:* O,S-*Diethyl cyclohexylphosphoramidothioate.*
$C_{10}H_{22}NO_2PS$ M 251.323
Solid. Mp 61°. $Bp_{0.25}$ 130°.

O-*Methyl* O-*(4-nitrophenyl) ester: see O-Methyl O-4-nitrophenyl cyclohexylphosphoramidothioate, M-00179*

Difluoride:
$C_6H_{12}F_2NPS$ M 199.198
Liq. Bp_8 73-74°.

Olah, G. *et al*, *Justus Liebigs Ann. Chem.*, 1959, **625**, 88 (*difluoride*)
Burn, A.J. *et al*, *J. Chem. Soc.*, 1961, 5532 (*esters*)
Cooks, R.G. *et al*, *J. Chem. Soc. (B)*, 1968, 1327 (*dimethyl ester, ms*)
El-Borgi, A. *et al*, *C.R. Hebd. Seances Acad. Sci., Ser. C*, 1977, **284**, 983 (*ir*)
Al-Rawi, J.M.A. *et al*, *Org. Magn. Reson.*, 1983, **21**, 75 (*diethyl ester, cmr*)

Cyclohexyl phosphorodichloridite, 9CI, 8CI C-00272

Cyclohexyl dichlorophosphite
[54921-67-4]

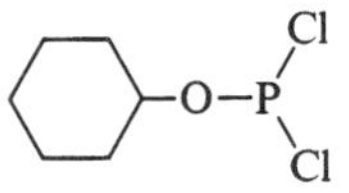

$C_6H_{11}Cl_2OP$ M 201.032
Pungent liq. d_4^{20} 1.21. $Bp_{0.3}$ 54-55°. n_D^{20} 1.5012.

Orudzheva, I.M. *et al*, *CA*, 1968, **69**, 86515 (*synth*)

Cyclohexyltriphenylphosphonium(1+), 9CI C-00273

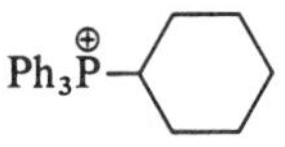

$C_{24}H_{26}P^{\oplus}$ M 345.443 (ion)
Treatment of salts with MeLi, butyllithium or $NaNH_2$, yields the ylide.

Bromide: [7333-51-9].
$C_{24}H_{26}BrP$ M 425.347
Cryst. ($CHCl_3/Et_2O$ or $CHCl_3$/EtOAc). Mp 218-219.5°, 265-267°, 273-276°.

4-Methylbenzenesulfonate: [55874-27-2].
$C_{31}H_{32}O_3PS$ M 515.626
Cryst. (H_2O). Mp 219-221°.

Ylide: [16666-81-2].
Cyclohexylidenetriphenylphosphorane.
$C_{24}H_{25}P$ M 344.435
Prepd. *in situ*. A reactive Wittig reagent.

Bestmann, H.J.R. *et al*, *Chem. Ber.*, 1963, **96**, 1899 (*bromide, ylide, props*)
Grim, S.O. *et al*, *J. Org. Chem.*, 1968, **33**, 2993 (*bromide, ylide, uv, nmr*)
Chasin, D.G., *Chem. Phys. Lipids*, 1971, **6**, 8 (*bromide*)
Albright, T.A. *et al*, *J. Am. Chem. Soc.*, 1975, **97**, 2942 (*iodide, cmr, P nmr*)
McLennan, D.G. *et al*, *J. Chem. Soc., Perkin Trans. 2*, 1977, 293 (*bromide, tosylate*)
Lythgoe, B. *et al*, *J. Chem. Soc., Perkin Trans. 1*, 1978, 590 (*ylide, use*)
Wheaton, G.A. *et al*, *J. Org. Chem.*, 1983, **48**, 917 (*ylide, use*)

2,4-Cyclopentadien-1-ylidenetriphenylphosphorane, 9CI, 8CI — C-00274

5-Triphenylphosphoranylidene-1,3-cyclopentadiene
[29473-30-1]

$C_{23}H_{19}P$ M 326.377
Cryst. struct. indicates 80% ylide character. Yellow cryst. ($EtOH/CHCl_3$ or toluene). Mp 235-236° (228-231°). Unreactive in Wittig reactions. Readily sol. in 5% aq. HCl and recovd. on basification. Forms complexes with Cr, Mo, W, Pt or Pd.

Ramirez, F. *et al*, *J. Am. Chem. Soc.*, 1957, **79**, 67 (*synth, ir, uv*)
Lord, E. *et al*, *J. Chem. Soc. (B)*, 1970, 1401 (*synth, pmr, nmr*)
Ammon, H.L. *et al*, *J. Am. Chem. Soc.*, 1973, **95**, 6158 (*cryst struct*)
Gray, G.A., *J. Am. Chem. Soc.*, 1973, **95**, 7736 (*nmr*)
Yoshida, Z. *et al*, *J. Org. Chem.*, 1973, **38**, 3537 (*props, struct*)
Albright, T.A. *et al*, *J. Am. Chem. Soc.*, 1976, **98**, 624 (*cmr, nmr*)
Ostoja, Starzewski, K.A. *et al*, *J. Am. Chem. Soc.*, 1976, **98**, 8486 (*pe*)
Dimroth, K. *et al*, *Phosphorus Sulfur*, 1981, **10**, 295, 305 (*cmr, pmr*)
Tresoldi, G. *et al*, *J. Organomet. Chem.*, 1982, **231**, 265 (*complexes*)
Caminade, A.M. *et al*, *Phosphorus Sulfur*, 1983, **14**, 381 (*props, nmr*)

Cyclopentadienylphosphonous acid, 9CI — C-00275

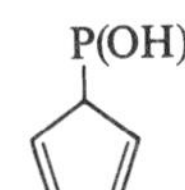

$C_5H_7O_2P$ M 130.083
Dibutyl ester: [58141-57-4]. *Dibutyl 2,4-cyclopentadienylphosphonite.*
$C_{13}H_{25}O_2P$ M 244.313
Liq. Bp_1 80-81° (slow dist.). n_D^{20} 1.4822. Forms a dimer on standing.
Dichloride: [89777-65-7].
$C_5H_5Cl_2P$ M 166.974
Forms a W complex.
Bromide: [90285-16-8].
$C_5H_5Br_2P$ M 255.876
Obt. as a W complex.

Kabachnik, M.I. *et al*, *Zh. Obshch. Khim.*, 1960, **30**, 3227 (*Engl. transl.* p. 3197) (*ester, synth, props*)
Deschamps, B. *et al*, *Phosphorus Sulfur*, 1983, **17**, 317 (*dichloride, dibromide, complexes*)
Schoeller, W., *Z. Naturforsch., B*, 1983, **38**, 1635 (*derivs, struct*)

Cyclopentanephosphonic acid — C-00276

Cyclopentylphosphonic acid, 9CI
[6869-04-1]

OH
O=P—OH

$C_5H_{11}O_3P$ M 150.114
Cryst. (C_6H_6/heptane). Mp 123-123.5°. pK_{a1} 4.37, pK_{a2} 7.47 (50% EtOH aq., 25°).
Anilinium salt: Cryst. (MeCN). Mp 181-183°.
Di-Me ester: [26580-25-6]. *Dimethyl cyclopentylphosphonate.*
$C_7H_{15}O_3P$ M 178.167
Liq. $Bp_{0.35}$ 57°. n_D^{25} 1.4518.
Dichloride:
$C_5H_9Cl_2OP$ M 187.005
Liq. d_4^{20} 1.32. Bp_6 98°. n_D^{20} 1.4973.

Isbell, A.F. *et al*, *J. Am. Chem. Soc.*, 1956, **78**, 6042 (*dichloride*)
Buckler, S.A. *et al*, *Tetrahedron*, 1962, **18**, 1211 (*synth*)
Marmor, R. *et al*, *J. Org. Chem.*, 1971, **36**, 128 (*synth*)
Yoneda, S. *et al*, *J. Org. Chem.*, 1978, **43**, 1980 (*ester, ir, pmr*)

1-Cyclopentenephosphonic acid — C-00277

1-Cyclopenten-1-ylphosphonic acid, 9CI. 1-Phosphonocyclopentene

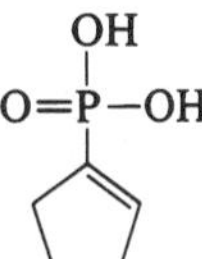

$C_5H_9O_3P$ M 148.098
Di-Me ester: Dimethyl 1-cyclopenten-1-ylphosphonate.
$C_7H_{13}O_3P$ M 176.152
Liq. $Bp_{1.7}$ 85-87°. n_D^{25} 1.4669.
Di-Et ester: Diethyl 1-cyclopenten-1-ylphosphonate.
$C_9H_{17}O_3P$ M 204.205
Liq. Bp_{12} 129°. n_D^{20} 1.4609.
Diisopropyl ester: [38868-17-6]. *Diisopropyl 1-cyclopenten-1-ylphosphonate.*
$C_{11}H_{21}O_3P$ M 232.259
Bp_1 104-105°, $Bp_{0.01}$ 85-87°.

Arbuzov, B.A. *et al*, *Dokl. Akad. Nauk SSSR, Ser. Sci. Khim.*, 1958, **121**, 641; *CA*, **53**, 1180.
Borowitz, I.J. *et al*, *J. Org. Chem.*, 1971, **36**, 3282; 1973, **38**, 1713 (*ester, synth, ir, pmr*)
Marmor, R.S. *et al*, *J. Org. Chem.*, 1971, **36**, 128 (*ester, synth, ir, pmr*)

3-Cyclopentenephosphonic acid — C-00278

2-Cyclopenten-1-ylphosphonic acid, 9CI. 3-Phosphonocyclopentene
$C_5H_9O_3P$ M 148.098

(±)-*form*

Di-Et ester: Diethyl 2-cyclopenten-1-ylphosphonate.
$C_9H_{17}O_3P$ M 204.205
Liq. d_4^{20} 1.07. Bp_3 100-100.5°. n_D^{20} 1.4579.
Diisopropyl ester: Diisopropyl 2-cyclopenten-1-ylphosphonate.
$C_{11}H_{21}O_3P$ M 232.259

Liq. d_0^{20} 1.02. Bp_2 96-96.5°. n_D^{20} 1.4483.

Arbuzov, B.A. *et al, Izv. Akad. Nauk SSSR, Ser. Khim.*, 1960, 1399 (*Engl. transl.* p. 1301) (*synth, props*)

Cyclopentylmethylphosphinic acid, 9CI — C-00279

[69387-01-5]

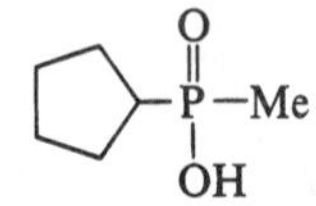

$C_6H_{13}O_2P$ M 148.141

Liq. d_4^{20} 1.14. Bp_2 171-172.5°. n_D^{20} 1.4900.

Butyl ester: Butyl cyclopentylmethylphosphinate.
$C_{10}H_{21}O_2P$ M 204.248
Liq. d_4^{20} 1.00. Bp_1 90-91°. n_D^{20} 1.4628.

Chloride:
$C_6H_{12}ClOP$ M 166.587
Liq. d^{20} 1.18. Bp_1 80-80.5°. n_D^{20} 1.4989.

Kabachnik, M.I. *et al, Zh. Obshch. Khim.*, 1960, **30**, 3227 (*Engl. transl.* p. 3127)

Cyclopentylphosphonothioic acid, 9CI — C-00280

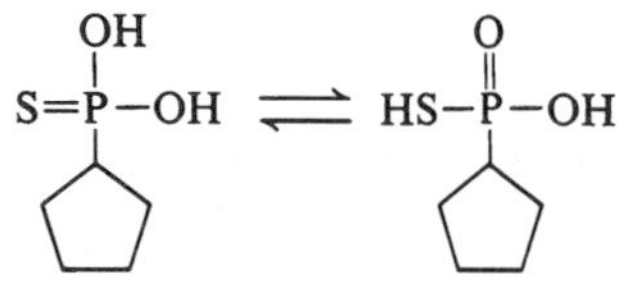

$C_5H_{11}O_2PS$ M 166.174

O,O-Dibutyl ester: O,O-*Dibutyl cyclopentylphosphonothioate.*
$C_{13}H_{27}O_2PS$ M 278.389
Liq. d_4^{20} 1.00. Bp_1 108-108.5°. n_D^{20} 1.4793.

Dichloride: [1498-62-0].
$C_5H_9Cl_2PS$ M 203.066
Liq. Bp_{13} 120°. n_D^{20} 1.5627.

Kabachnik, M.I. *et al, Zh. Obshch. Khim.*, 1960, **30**, 3227 (*Engl. transl.* p. 3197) (*ester, synth*)
Christol, C. *et al, J. Chim. Phys.*, 1965, **62**, 246 (*dichloride, ir, raman*)

Cyclopentylphosphonous acid, 9CI — C-00281

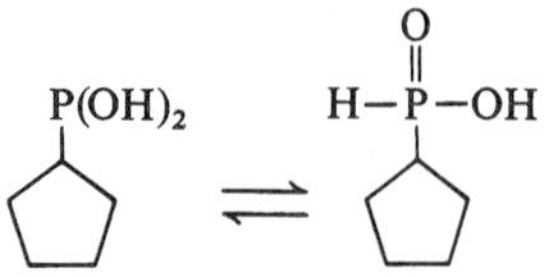

$C_5H_{11}O_2P$ M 134.114

Dibutyl ester: Dibutyl cyclopentylphosphonite.
$C_{13}H_{27}O_2P$ M 246.329
Liq. $Bp_{1.5}$ 83-83.5°. n_D^{20} 1.4620.

Kabachnik, M.I. *et al, Zh. Obshch. Khim.*, 1960, **30**, 3227 (*Engl. transl.* p. 3197)

Cyclopentyltriphenylphosphonium(1+), 9CI — C-00282

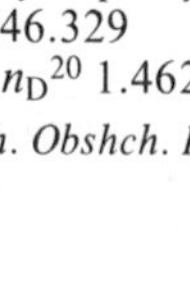

$C_{23}H_{24}P^{\oplus}$ M 331.416 (ion)

$NaNH_2 \rightarrow$ ylide.

Chloride: [72385-59-0].
$C_{23}H_{24}ClP$ M 366.869
No phys. props. reported.

Bromide: [7333-52-0].
$C_{23}H_{24}BrP$ M 411.320
Cryst. (H_2O). Mp 263°.

Ylide: [21482-00-8].
Cyclopentylidenetriphenylphosphorane.
$C_{23}H_{23}P$ M 330.408
Used in Wittig reactions.

Grim, S.O. *et al, J. Chem. Soc., Chem. Commun.*, 1967, 1191 (*bromide, nmr*)
Bestmann, H.J. *et al, Chem. Ber.*, 1969, **102**, 1802 (*synth, use*)
Albright, T.A. *et al, J. Am. Chem. Soc.*, 1975, **97**, 2942 (*bromide, cmr, nmr*)
Wheaton, G.A. *et al, J. Org. Chem.*, 1983, **48**, 917 (*use*)

Cyclophosphamide, BAN — C-00283

N,N-*Bis(2-chloroethyl)tetrahydro-2*H*-1,3,2-oxazaphosphorin-2-amine 2-oxide*, 9CI. *2-[Di(2-chloroethyl)-amino]-1-oxa-3-aza-2-phosphacyclohexane 2-oxide. Endoxane. Cytoxan. Procytox. Sendoxan*

[50-18-0]

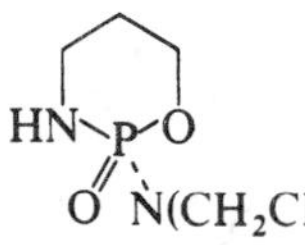

(*R*)-*form*

$C_7H_{15}Cl_2N_2O_2P$ M 261.087

▷Highly toxic. Exp. carcinogen and teratogen. RP5950000.

(*R*)-*form* [60030-72-0]
Selectively metab. in humans. Shows different biological props. from the (*S*)-form but the differences may vary with species. Cryst. (pet. ether/MeOH or EtOAc/diisopropyl ether). Mp 65-66.5°. $[\alpha]_D^{25}$ +2.48° (c, 10 in MeOH).
Monohydrate: Mp 68°.

(*S*)-*form* [60007-96-7]
Cryst. (EtOAc/diisopropyl ether). Mp 67-68°. $[\alpha]_D^{25}$ −2.46° (c, 10 in MeOH).
Monohydrate: Mp 68°.

(±)-*form* [60007-95-6]
Antitumour agent, immunosuppressant, sheep defleecing agent. Prisms (dry Et_2O). Sol. H_2O. Mp 51.5-52.5°.
Monohydrate: Mp 41-45°.

Garcia-Blanco, S. *et al, Acta Crystallogr., Sect. B*, 1972, **28**, 2647 (*cryst struct*)
Clardy, J.C. *et al, Phosphorus*, 1974, **4**, 151 (*cryst struct*)
Schulten, H.R. *et al, Biomed. Mass Spectrom.*, 1974, **1**, 223 (*ms*)
Struck, R.F. *et al, J. Am. Chem. Soc.*, 1974, **96**, 313 (*props, cmr*)
Voelker, G. *et al, Arzneim.-Forsch.*, 1974, **24**, 1172 (*tlc*)
Hill, D.L., *A Review of Cyclophosphamide*, 1975, Thomas, Springfield, Ill. (*rev*)
Przybylski, M. *et al, Cancer Treat. Rep.*, 1976, **60**, 509 (*ms*)
Adamiak, D.A. *et al, Z. Naturforsch., C*, 1977, **32**, 672 (*cryst struct*)
Cox, P.J. *et al, Biomed. Mass Spectrom.*, 1977, **4**, 371 (*ms*)
Karle, I.L. *et al, J. Am. Chem. Soc.*, 1977, **99**, 4803 (*abs config*)
Kawashima, T. *et al, J. Org. Chem.*, 1978, **43**, 1111 (*resoln, ir, pmr, P nmr*)
Bates, D.J. *et al, Biochem. Pharmacol.*, 1981, **30**, 3055 (*metab*)
Chiu, F.T. *et al, J. Pharm. Sci.*, 1982, **71**, 542 (*deriv, pmr, P nmr*)
Przybyski, M., *Arzneim. Forsch.*, 1982, **32**, 995 (*glc, ms, rev*)
Stec, W.J., *Organophosphorus Chemistry*, 1982, **13**, 145 (*rev*)
Su, C.N. *et al, J. Am. Chem. Soc.*, 1982, **104**, 7343 (*spectra*)

Zon, G., *Prog. Med. Chem.*, 1982, **19**, 205 (*analogues, rev*)
Sato, T. *et al*, *J. Org. Chem.*, 1983, **48**, 98 (*synth*)
Zon, G. *et al*, *J. Pharm. Sci.*, 1983, **72**, 687 (*cmr*)
Buess, M.L. *et al*, *Org. Magn. Reson.*, 1984, **22**, 67 (*nqr*)
Mruzek, M.N. *et al*, *Biomed. Mass Spectrom.*, 1984, **11**, 360 (*ms*)
Setzer, W.N. *et al*, *J. Am. Chem. Soc.*, 1985, **107**, 2083 (*pmr, P nmr, conformn*)
Sax, N.I., *Dangerous Properties of Industrial Materials*, 6th Ed., Van Nostrand-Reinhold, 1984, 640.

Cyclopropylidenetriphenylphosphorane C-00284

[14633-95-5]

Ph_3P=◁

$C_{21}H_{19}P$ M 302.355

Reactive Wittig reagent. Orange needles. Mp 103° dec. Reacts with water to give cyclopropyldiphenylphosphine oxide.

Schweizer, E. *et al*, *J. Org. Chem.*, 1968, **33**, 336 (*props*)
Albright, T.A. *et al*, *J. Am. Chem. Soc.*, 1976, **98**, 6249 (*cmr, nmr*)
Schmidbaur, H. *et al*, *Chem. Ber.*, 1982, **115**, 722 (*synth, pmr, cmr, cryst struct*)
Fitjer, L., *Chem. Ber.*, 1982, **115**, 1035, 1047 (*use*)

[α-Cyclopropyl(methoxycarbonyl)methyl]-triphenylphosphonium(1+) C-00285

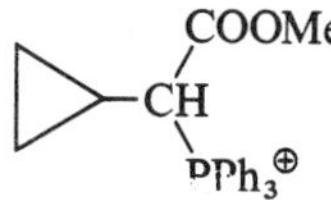

$C_{24}H_{24}O_2P^{\oplus}$ M 375.426 (ion)

Bromide:
$C_{24}H_{24}BrO_2P$ M 455.330
Source of ylide. Cryst. (CH_2Cl_2/EtOAc). Mp 183° (175-177°).

Iodide:
$C_{24}H_{24}IO_2P$ M 502.331
Source of ylide. Cryst. (CH_2Cl_2/EtOAc). Mp 178.5-179°.

Ylide: [14902-14-8]. *Methyl α-(triphenylphosphoranylidene)cyclopropaneacetate.*
$C_{24}H_{23}O_2P$ M 374.418
Reactive Wittig reagent, prepd. *in situ*.

Schweizer, E.E. *et al*, *J. Org. Chem.*, 1968, **33**, 3082 (*synth*)
Maercker, A., *Justus Liebigs Ann. Chem.*, 1970, **732**, 151 (*synth, pmr, ylide, props*)
Maercker, A. *et al*, *Justus Liebigs Ann. Chem.*, 1972, **759**, 132 (*ylide, props*)

(Cyclopropylmethyl)-triphenylphosphonium(1+), 9CI C-00286

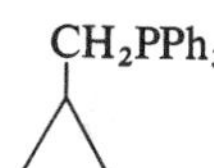

$C_{22}H_{22}P^{\oplus}$ M 317.390 (ion)

Intermed. for prepn. of cyclopropyl and cyclopropylidene compds. Reacts with butyllithium to give the ylide.

Bromide: [14799-82-7].
$C_{22}H_{22}BrP$ M 397.294
Source of ylide. Solid. Mp 175-177°.

Ylide: [14902-12-6]. *(Cyclopropylmethylene)-triphenylphosphorane.*
$C_{22}H_{21}P$ M 316.382
Reactive intermed. Used in Wittig reactions.

Maercker, A. *et al*, *Angew. Chem., Int. Ed. Engl.*, 1967, **6**, 557 (*synth, use*)
Maercker, A. *et al*, *Justus Liebigs Ann. Chem.*, 1972, **759**, 132 (*use*)

Cyclopropyltriphenylphosphonium(1+), 9CI C-00287

$Ph_3\overset{\oplus}{P}$–◁

$C_{21}H_{20}P^{\oplus}$ M 303.363 (ion)

When treated with alkali, salts yield cyclopropyldiphenylphosphine oxide.

Bromide: [14114-05-7].
$C_{21}H_{20}BrP$ M 383.267
Hygroscopic cryst. (EtOH/EtOAc). Mp 188-190° (176-177°).

Ylide: see Cyclopropylidenetriphenylphosphorane, C-00284

Schweizer, E. *et al*, *J. Org. Chem.*, 1968, **33**, 336 (*synth, pmr, use, props*)
Bestmann, H.J., *Chem. Ber.*, 1972, **105**, 2098 (*synth, props*)
Utimoto, K. *et al*, *Tetrahedron*, 1973, **29**, 1169 (*synth, use*)
Albright, T.A. *et al*, *J. Am. Chem. Soc.*, 1975, **97**, 2942; 1976, **98**, 6249 (*cmr, P nmr*)
Marino, J.P. *et al*, *J. Org. Chem.*, 1981, **46**, 1828 (*use*)
Fieser, M. *et al*, *Reagents for Organic Synthesis*, Wiley, 1967-84, **2**, 95 (*use*)

Cytidine 2′-(dihydrogen phosphate), 9CI, 8CI C-00288

2′-Cytidylic acid, 9CI, 8CI. *Cytidylic acid* a. *Cytosylic acid* a

[85-94-9]

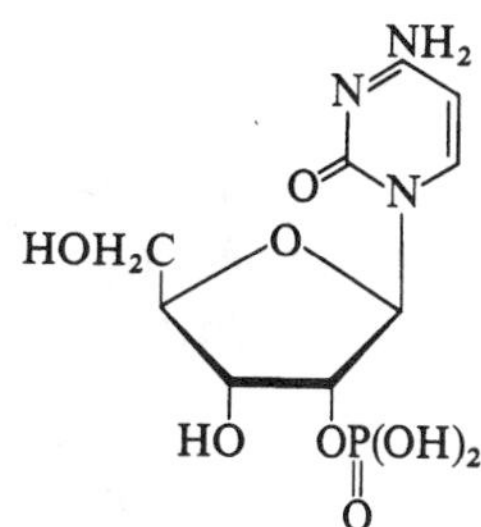

$C_9H_{14}N_3O_8P$ M 323.199

Hydrol. product of yeast ribonucleic acid. Ribonuclease inhibitor. Mp 238-240° dec. $[\alpha]_D^{20}$ +20.7° (H_2O). pK_{a1} 0.8, pK_{a2} 4.4, pK_{a3} 6.2. λ_{max} 278 (ε 12 700) (pH 2), 272 nm (8 600) (pH 12).

Na salt: $[\alpha]_D^{19}$ −4° (H_2O).
Di-Na salt: $[\alpha]_D^{24.5}$ −8° (H_2O).
Ba salt: $[\alpha]_D^{26}$ +5.5° (H_2O).

Harris, R.J.C. *et al*, *J. Chem. Soc.*, 1953, 489 (*isol*)
Baron, F. *et al*, *J. Chem. Soc.*, 1955, 2855 (*struct*)
Rammler, D.H. *et al*, *J. Am. Chem. Soc.*, 1962, **84**, 3112 (*synth*)
Saffhill, R., *J. Org. Chem.*, 1970, **35**, 2881 (*synth*)
Davies, D.B. *et al*, *Biochemistry*, 1975, **14**, 543 (*pmr*)

Cytidine 3′-(dihydrogen phosphate), 9CI, 8CI C-00289

3′-Cytidylic acid, 9CI, 8CI. *Cytidylic acid* b. *Cytosylic acid* b

[84-52-6]

$C_9H_{14}N_3O_8P$ M 323.199

Hydrolytic product of yeast ribonucleic acid. Ribonuclease inhibitor. Sol. H_2O. Mp 232-234° dec. $[\alpha]_D^{20}$ +49.4° (c, 1.0 in H_2O). pK_{a1} 0.8, pK_{a2} 4.3, pK_{a3} 6.0. λ_{max} 279 (ε 13 000) (pH 2), 272 nm (8 900) (pH 12).

Na salt: $[\alpha]_D^{19}$ +47° (H_2O).
Di-Na salt: $[\alpha]_D^{24.5}$ +46° (H_2O).

Harris, R.J.C. *et al*, *J. Chem. Soc.*, 1953, 489 (*isol*)
Baron, F. *et al*, *J. Chem. Soc.*, 1955, 2855 (*struct*)
Lohrmann, R. *et al*, *J. Am. Chem. Soc.*, 1964, **86**, 4188 (*synth*)
Bugg, C.E. *et al*, *J. Mol. Biol.*, 1967, **25**, 67; *CA*, **67**, 15833f (*cryst struct*)
Ts'O, P.O.P. *et al*, *Biochemistry*, 1969, **8**, 997 (*pmr*)
Saffhill, R., *J. Org. Chem.*, 1970, **35**, 2881 (*synth*)

Cytidine 5′-(dihydrogen phosphate), 9CI, 8CI **C-00290**
5′-Cytidylic acid, 9CI, 8CI
[63-37-6]
$C_9H_{14}N_3O_8P$ M 323.199
Mp 233° dec. (monohydrate).
▷HA3980000.
Ba salt: [13435-44-4]. $[\alpha]_D^{16}$ +11.4° (c, 0.33 in H_2O).
Dibrucine salt: Mp 215° dec.

Michelson, A.M. *et al*, *J. Chem. Soc.*, 1949, 2476 (*synth*)
Dorman, D.E. *et al*, *Proc. Natl.*, *Acad. Sci. U.S.A.*, 1970, **65**, 19; *CA*, **72**, 107074r (*cmr*)
Ohtsuka, E. *et al*, *Nucleic Acids Res.*, 1974, **1**, 323.
Takaku, O. *et al*, *Agric. Biol. Chem.*, 1975, **39**, 2373 (*pmr*)

Cytidine diphosphate choline **C-00291**
Cytidine 5′-(trihydrogen diphosphate) hydroxide mono[2-(trimethylammonio)ethyl ester. Choline hydroxide 5′ ester with cytidine 5′-(trihydrogen pyrophosphate), inner salt. Citicoline. Cereb. Colite. Cyscholin. CDP-Choline
[987-78-0]

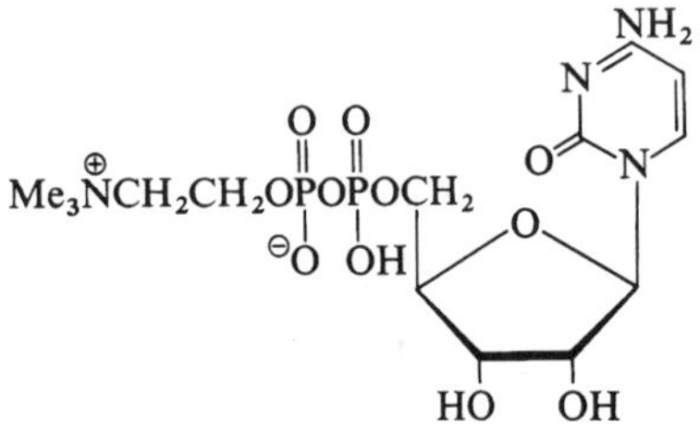

$C_{14}H_{26}N_4O_{11}P_2$ M 488.327
Antineoplastic agent. Involved in the biosynth. of lecithin and sphingomyelin and in the formation of plasmologen in the liver and brain. Hygroscopic powder + $3H_2O$. λ_{max} 271 (ε 9 100), 280 nm (13 700) (0.1M HCl).
▷GA4027000.

Kennedy, E.P. *et al*, *J. Biol. Chem*, 1956, **222**, 193.
Makishima, K. *et al*, *Arzneim.-Forsch.*, 1971, **21**, 1343 (*pharmacol*)
Viswamitra, M.A. *et al*, *Nature* (*London*), 1975, **258**, 497 (*cryst struct*)
Furusawa, K. *et al*, *J. Chem. Soc.*, *Perkin Trans. 1*, 1976, 1711 (*synth, uv*)

Cytidine diphosphate ethanolamine **C-00292**
Cytidine 5′-(trihydrogen diphosphate) mono(2-aminoethyl) ester, 9CI. Cytidine 5′-(trihydrogen pyrophosphate) mono(2-aminoethyl) ester, 8CI. CDP-ethanolamine
[3036-18-8]

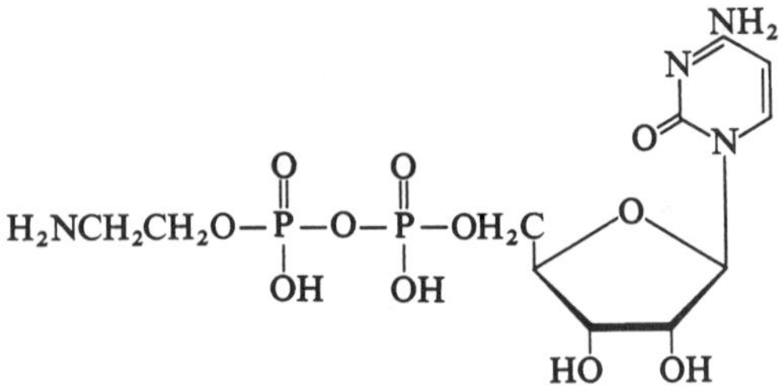

$C_{11}H_{20}N_4O_{11}P_2$ M 446.247
Biogenetic precursor of cephalins and lecithins in yeast. Amorph.

Kennedy, E.P. *et al*, *J. Am. Chem. Soc.*, 1955, **77**, 250.
Kennedy, E.P. *et al*, *J. Biol. Chem.*, 1956, **22**, 193.
Chojnacki, T. *et al*, *Nature* (*London*), 1966, **210**, 947 (*synth*)
Ger. Pat., 2 059 429, (*1972*); *CA*, **77**, 102141d (*synth*)

Cytidine diphosphate glycerol **C-00293**
Cytidine 5′-(trihydrogen diphosphate) mono(2,3-dihydroxypropyl) ester(S), *9CI. CDP-glycerol*
[6665-99-2]

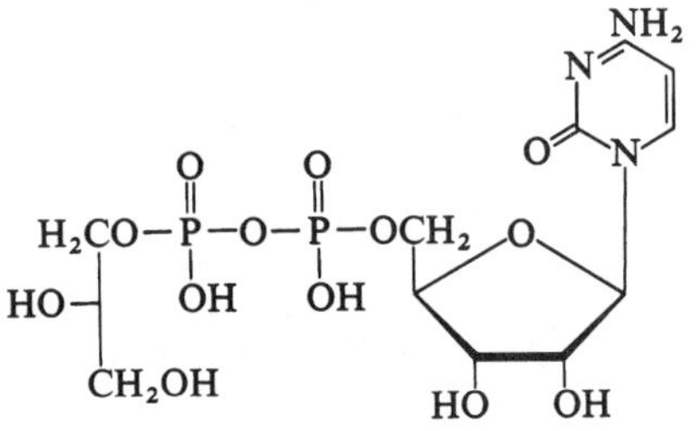

$C_{12}H_{21}N_3O_{13}P_2$ M 477.258
The biosynthetic precursor of the tri(glycerol phosphate) which links teichoic acid to the peptidoglycan in *Staphylococcus aureus*, *Bacillus* and *Micrococcus* spp.

Baddiley, J. *et al*, *J. Chem. Soc.*, 1958, 3107.
Roseman, S. *et al*, *J. Am. Chem. Soc.*, 1961, **83**, 659 (*synth*)
Baddiley, J. *et al*, *Biochem. J.*, 1968, **110**, 565.
Hancock, I. *et al*, *J. Bacteriol.*, 1976, **125**, 880.

Cytidine 2′,3′-phosphate **C-00294**
Cyclic 2′,3′-(hydrogen phosphate)cytidine, 9CI, 8CI
[633-90-9]

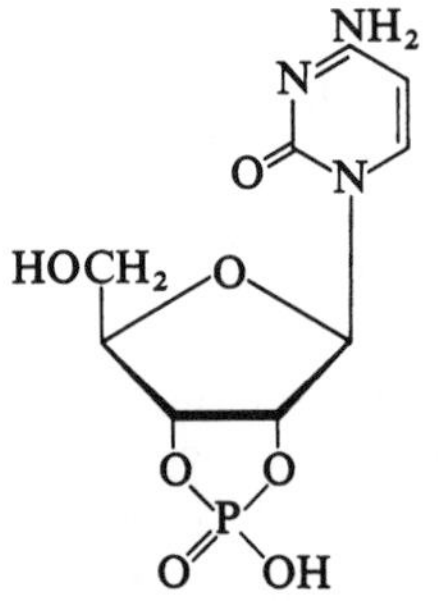

$C_9H_{12}N_3O_7P$ M 305.183
Intermed. formed during digestion of ribonucleic acid by ribonuclease. Synthesised from either Cytidylic acid *a* or *b*.

K salt: [20486-23-1]. Hygroscopic cryst. + $1H_2O$. λ_{max} 268 nm (ϵ 8 650) (pH 7.5).
Ba salt: Amorph. powder. λ_{max} 268 (ϵ 8 400), 232 nm (ϵ 8 500) (H_2O).

Brown, D.M. *et al*, *J. Chem. Soc.*, 1952, 2708.
Biochem. Prep., 1968, **12**, 107 (*synth*)
Coulter, C.L., *J. Am. Chem. Soc.*, 1973, **95**, 570 (*cryst struct*)
Lapper, R.D. *et al*, *J. Am. Chem. Soc.*, 1973, **95**, 2880 (*pmr*)
Davies, D.B., *Stud. Biophys.*, 1976, **55**, 29; *CA*, **85**, 59065h (*cmr*)

D

7,8,10,11,13,14,16,17,19,24-Decahydro-5*H*-dibenzo[*o,r*][1,4,7,10,13,17]-pentaoxaphosphacycloeicosin D-00001

2,3:19,20-Dibenzo-5,8,11,14,17-pentaoxa-1-phospha-2,19-cycloeicosadiene

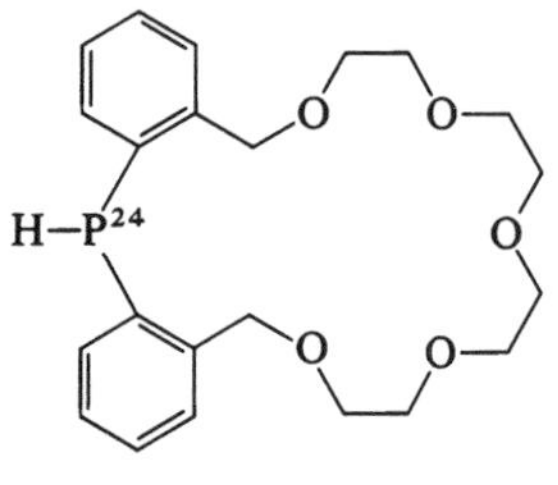

$C_{22}H_{29}O_5P$ M 404.442

Derivs. form complexes with group I metals.

24-Methoxy, 24-oxide: [69928-19-4].

$C_{23}H_{31}O_7P$ M 450.467

Oil.

24-Ph, 24-oxide: [69928-15-0].

$C_{28}H_{33}O_6P$ M 496.539

Plates (Et_2O). Mp 75-76°.

Kaplan, L.J. *et al*, *J. Org. Chem.*, 1979, **44**, 2226 (*synth, pmr, ir*)

Decahydro-1,5-dihydroxy-1*H*-1,5-benzodiphosphepin 1,5-dioxide, 10CI D-00002

1,5-Dihydroxy-1,5-diphosphabicyclo[5.4.0]undecane 1,5-dioxide

[63708-67-6]

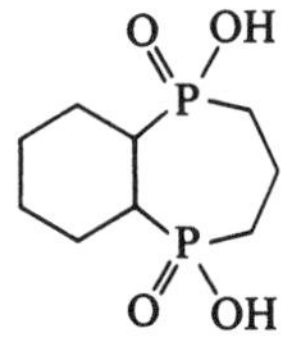

$C_9H_{18}O_4P_2$ M 252.186

Solid. Sol. H_2O. Mp >310° dec. pK_{a1} 2.42, pK_{a2} 3.62 (H_2O, 25°).

Polikarpov, Yu.M. *et al*, *Izv. Akad. Nauk SSSR, Ser. Khim.*, 1977, 1188 (*Engl. transl.* p. 1094)

Decahydro-1,4-dihydroxy-1,4-benzodiphosphorin 1,4-dioxide, 10CI D-00003

1,4-Dioxo-1,4-dihydroxy-1,4-diphosphabicyclo[4.4.0]-decane

[63708-56-5]

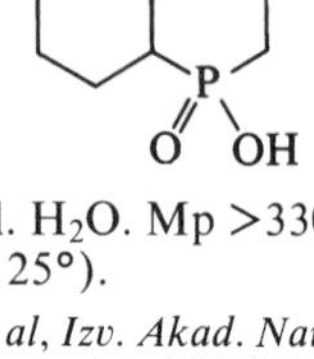

Cryst. (EtOH). Sol. H_2O. Mp >330° dec. pK_{a1} 2.58, pK_{a2} 2.86 (H_2O, 25°).

Polikarpov, Yu.M. *et al*, *Izv. Akad. Nauk SSSR, Ser. Khim.*, 1977, 1188 (*Engl. transl.* p. 1094) (*synth, props*)

5,6,7,8,9,14,15,16,17,18-Decahydro-9,14-dimethyl-5,18-diphenyldibenzo[*b,i*][1,4,8,11]-diphosphadiarsacyclotetradecin, 10CI D-00004

2,17-Dimethyl-6,13-diphenyl-2,17-diarsa-6,13-diphosphatricyclo[16.4.0.0^{7,12}]docosa-7(12),8,10,18(1),19,21-hexaene

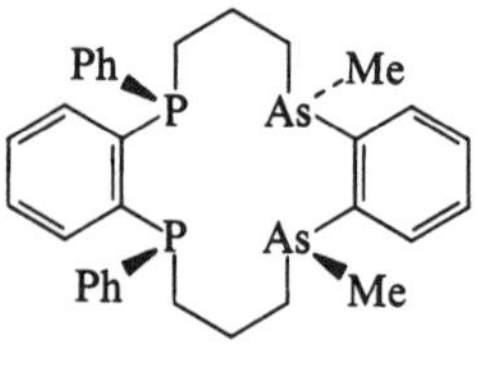

$C_{32}H_{36}As_2P_2$ M 632.427

6 possible stereoisomers, only one characterised.

(*5RS,9RS,14RS,18SR*)-form [76010-93-0]

Cryst. (toluene/hexane). Mp 138.5-139.5°. Probable config.

Kyba, E.P. *et al*, *J. Org. Chem.*, 1981, **46**, 860 (*synth, pmr, cmr, P nmr, ms*)

1,3,4,5,5,6,7,8,9,9*a*-Decahydro-5-methylene-2*H*-phosphinolizine, 10CI D-00005

[75780-26-6]

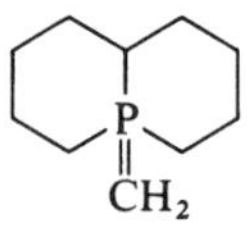

$C_{10}H_{19}P$ M 170.234

Schmidbaur, H. *et al*, *Z. Naturforsch., B*, 1980, **35**, 990.

5,6,7,8,9,14,15,16,17,18-Decahydro-5,9,14,18-tetraphenyldibenzo[*b,i*][1,4,8,11]-tetraphosphacyclotetradecin, 10CI D-00006

2,6,13,17-Tetraphenyl-2,6,13,17-tetraphosphatricyclo[16.4.0.0^{7,12}]docosa-7(12),8,10,1(18),19,21-hexaene

[65113-30-6]

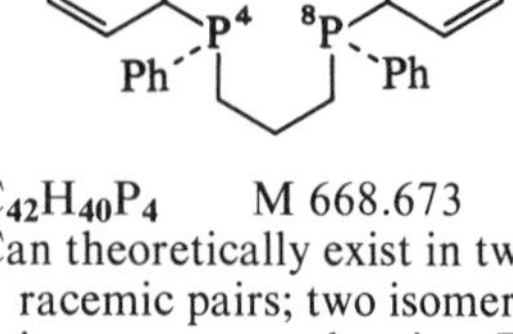

(1α,4α,8α,11α)-*form*

$C_{42}H_{40}P_4$ M 668.673

Can theoretically exist in two meso forms and three racemic pairs; two isomers are known which interconvert on heating. Tetradentate ligand for Ni.

(1α,4α,8α,11α)-form

cis,cis,cis-*form*

Tetragonal prisms. Mp 160-164°.

(1α,4α,8β,11β)-form
cis,trans,cis-*form*
Triclinic prisms. Mp 227-229°.

Kyba, E.P. *et al*, *J. Am. Chem. Soc.*, 1977, **99**, 8053; 1981, **103**, 3868 (*synth, pmr, cryst struct*)
Davis, R.E. *et al*, *J. Am. Chem. Soc.*, 1978, **100**, 3642 (*synth, cryst struct, P nmr*)

Decahydro-1,3,5-trihydroxy-1*H*-1,3,5-benzotriphosphepin 1,3,5-trioxide, 10CI D-00007

[63708-58-7]

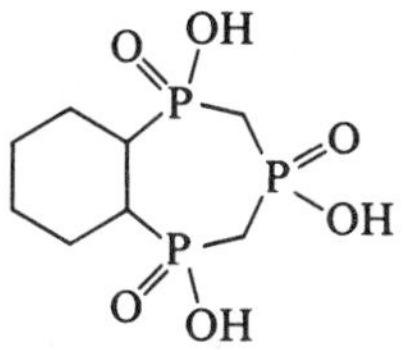

Solid. Sol. H_2O. Mp >312° dec. pK_{a1} 2.18, pK_{a2} 2.43, pK_{a3} 4.12 (H_2O).

Polikarpov, Yu.M. *et al*, *Izv. Akad. Nauk SSSR, Ser. Khim.*, 1977, 1188 (*Engl. transl.* p. 1094) (*synth, props*)

1,10-Decanediphosphonic acid D-00008

1,10-Decanediylbisphosphonic acid, 9CI. Decamethylenediphosphonic acid, 8CI. 1,10-Diphosphonodecane
[5943-21-5]

$(HO)_2P(O)(CH_2)_{10}P(O)(OH)_2$

$C_{10}H_{24}O_6P_2$ M 302.244
Cryst. (EtOH aq.). Mp 208-210°.
Tetra-Et ester: [5943-62-4]. *Tetraethyl 1,10-decanediylbisphosphonate. 1,10-Bis(diethoxyphosphinyl)decane.*
$C_{18}H_{40}O_6P_2$ M 414.458
Liq. d_4^{25} 1.04. $Bp_{0.8}$ 225°. n_D^{25} 1.4453.
Bis(dichloride):
$C_{10}H_{22}Cl_4O_2P_2$ M 378.042
Solid, mod. stable in air. Mp 70-72°. $Bp_{0.25}$ 195-196°.

Moedritzer, K. *et al*, *J. Inorg. Nucl. Chem.*, 1961, **22**, 297 (*synth, ir, P nmr*)
Kosolapoff, G.M. *et al*, *J. Chem. Soc.* (*C*), 1966, 757 (*synth, chloride, ir*)

Decylmethylphosphinic acid, 9CI D-00009

$H_3C(CH_2)_9P(O)(OH)Me$

$C_{11}H_{25}O_2P$ M 220.291
Low-melting solid. Mp 28°. Bp_4 173-175°. n_D^{30} 1.4412.
2-Methylpropyl ester: 2-Methylpropyl decylmethylphosphinate. Isobutyl decylmethylphosphinate.
$C_{15}H_{33}O_2P$ M 276.398
Liq. $Bp_{0.5}$ 138°.

Petrov, K.A. *et al*, *Zh. Obshch. Khim.*, 1960, **30**, 2995 (*Engl. transl.* p. 2967) (*synth*)
Finke, M. *et al*, *Justus Liebigs Ann. Chem.*, 1974, 741 (*ester*)

Decylphenylphosphinic acid, 9CI D-00010

$H_3C(CH_2)_9P(O)(OH)Ph$

$C_{16}H_{27}O_2P$ M 282.362
Cryst. (Et_2O). Mp 55.5-56°.
Et ester: Ethyl decylphenylphosphinate.
$C_{18}H_{31}O_2P$ M 310.415
Liq. $Bp_{5.5}$ 199-200°. n_D^{20} 1.4920.

Pudovik, A.N. *et al*, *Zh. Obshch. Khim.*, 1960, **30**, 2348 (*Engl. transl.* p. 2328)

Decylphosphine, 9CI D-00011

[23561-01-5]

$H_3C(CH_2)_9PH_2$

$C_{10}H_{23}P$ M 174.265
Liq. Bp 203°, Bp_{100} 145°. n_D^{20} 1.4591.

Schindlbauer, H. *et al*, *Monatsh. Chem.*, 1961, **92**, 868 (*synth, ir*)
Pass, F. *et al*, *Monatsh. Chem.*, 1962, **93**, 230 (*synth*)
Hays, H.R., *J. Org. Chem.*, 1966, **31**, 3817 (*pmr, nmr*)

Decylphosphonic acid D-00012

[6874-60-8]

$H_3C(CH_2)_9P(O)(OH)_2$

$C_{10}H_{23}O_3P$ M 222.264
Solid. Mp 102-102.5°.
▷SZ7525000.
Di-Me ester: Dimethyl decylphosphonate. 1-(Dimethoxyphosphinyl)decane.
$C_{12}H_{27}O_3P$ M 250.317
Nematocide. Liq. Bp_{16} 182°. n_D^{20} 1.4388.
Di-Et ester: [16165-68-7]. *Diethyl decylphosphonate. 1-(Diethoxyphosphinyl)decane.*
$C_{14}H_{31}O_3P$ M 278.371
Liq. Bp_{17} 186-193°.
Dibutyl ester: Dibutyl decylphosphonate. 1-(Dibutoxyphosphinyl)phosphonate.
$C_{18}H_{39}O_3P$ M 334.478
Liq. Bp_1 161°. n_D^{25} 1.4402.
Di-Ph ester: Diphenyl decylphosphonate. 1-(Diphenoxyphosphinyl)decane.
$C_{22}H_{31}O_3P$ M 374.459
Liq. $Bp_{0.1}$ 178°. n_D^{25} 1.5202.
Difluoride:
$C_{10}H_{21}F_2OP$ M 226.246
Liq. d_4^{20} 1.00. Bp 235-236°, Bp_6 113-115°.
Dichloride: [14576-66-0].
$C_{10}H_{21}Cl_2OP$ M 259.155
Liq. d_4^{20} 1.08. Bp_8 155-157°. n_D^{20} 1.4680.

Kosolapoff, G.M., *J. Am. Chem. Soc.*, 1945, **67**, 1180 (*synth, esters*)
Pudovik, A.N. *et al*, *Zh. Obshch. Khim.*. 1959, **29**, 3342 (*Engl. transl.* p. 3305) (*esters*)
Laughlin, R.C., *J. Org. Chem.*, 1962, **27**, 3644 (*ester*)
Feshchenko, N.G. *et al*, *Zh. Obshch. Khim.*, 1967, **37**, 473 (*Engl. transl.* p. 441) (*dichloride, difluoride*)
Geiseler, G. *et al*, *Ber. Bunsenges. Phys. Chem.*, 1967, **71**, 478 (*dichloride, ir, raman*)
Dietze, U., *J. Prakt. Chem.*, 1974, **316**, 293 (*ir*)

Decylphosphoramidic acid, 9CI D-00013

$H_3C(CH_2)_9NHP(O)(OH)_2$

$C_{10}H_{24}NO_3P$ M 237.278
Dibutyl ester: Dibutyl decylphosphoramidate.
$C_{18}H_{40}NO_3P$ M 349.493
Liq. Bp_{1-2} 105-110°. n_D^{20} 1.4353.
Dichloride:
$C_{10}H_{22}Cl_2NOP$ M 274.170

Liq. $Bp_{1.5}$ 166-169°.

Mizuma, T. *et al*, *J. Pharm. Soc. Jpn.*, 1961, **81**, 48; *CA*, **55**, 13403 (*dichloride*)
Bebikh, G. *et al*, *Zh. Obshch. Khim.*, 1977, **47**, 2196 (*Engl. transl.* p. 2005) (*ester, synth, ir*)

Decyltriphenylphosphonium(1+), 9CI, 8CI D-00014

$Ph_3P^{\oplus}(CH_2)_9CH_3$

$C_{26}H_{36}P^{\oplus}$ M 379.544 (ion)
Bromide: [32339-43-8].
$C_{26}H_{36}BrP$ M 459.448
Source of ylide, obt. with NaH or $MeSOCH_2^{\ominus}$. Solid. Mp 93.5-94°.
Ylide: [79827-34-2]. *Decylidenetriphenylphosphorane.*
$C_{26}H_{35}P$ M 378.536
Used to synthesise fatty acid esters.

Chasin, D.G. *et al*, *Chem. Phys. Lipids*, 1971, **6**, 8 (*synth, use*)
Bestmann, H.J. *et al*, *Angew. Chem., Int. Ed. Engl.*, 1981, **20**, 1038 (*props*)

Defosfamide, INN D-00015

*2-Chloroethyl-*N,N*-bis(2-chloroethyl)-*N′*-(3-hydroxypropyl)phosphorodiamidate*
[3733-81-1]

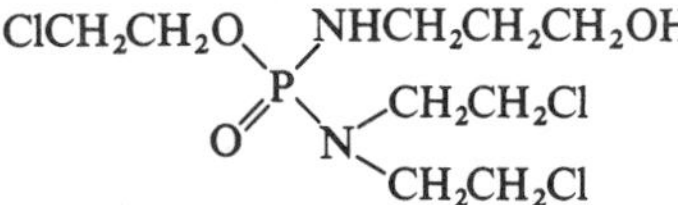

$C_9H_{20}Cl_3N_2O_3P$ M 341.601
Cytostatic agent. Oily liq. Insol. in H_2O.
▷TD2534700.

Ger. Pat., 1 061 780, (*1959*); *CA*, **55**, 11302i (*synth, pharmacol*)

Demanyl phosphate D-00016

Phosphoric acid mono[2-(dimethylamino)ethyl] ester, 9CI. *Mono-2-dimethylaminoethanol dihydrogen phosphate. Phosphoryldimethylaminoethanol. Phosphoryldimethylcholamine. Phosphoryldimethylethanolamine. PDMEA. Panclar*
[6909-62-2]

$Me_2NCH_2CH_2OP(O)(OH)_2$

$C_4H_{12}NO_4P$ M 169.117
Isol. from a choline-requiring mutant of *Neurospora crassa*. Psychotonic agent. Cryst. + $1H_2O$ (EtOH). Mp 78-81° (anhyd. 175-176°).

Cherbuliez, E. *et al*, *Helv. Chim. Acta*, 1956, **41**, 1168 (*synth*)
Wolf, B. *et al*, *J. Biol. Chem*, 1959, **234**, 1068 (*synth, isol*)
Ansell, G.B. *et al*, *Biochem. J.*, 1966, **98**, 303 (*pharmacol*)
Taguchi, Y. *et al*, *J. Org. Chem.*, 1975, **40**, 2310 (*synth*)
Miyazaki, H. *et al*, *Chem. Pharm. Bull.*, 1976, **24**, 763 (*metab*)

Demephion-*O* D-00017

O,O-*Dimethyl* O-*[2-(methylthio)ethyl] phosphorothioate*
[682-80-4]

$(MeO)_2P(S)OCH_2CH_2SMe$

$C_5H_{13}O_3PS_2$ M 216.250
See also Demephion-*S*, D-00018 . Systemic insecticide and acaricide. Component of coml. Demephion (Tinox). Now superseded. Liq. V. spar. sol. H_2O. $Bp_{0.05}$ 51-53°. At 100-150°, isom. to thiol form, and also dec. to Me_2S and *O,O,S*-Trimethyl phosphorothioate, T-00546 Methylates thiourea at high temp.

Grübner, P. *et al*, *J. Prakt. Chem.*, 1970, **312**, 603 (*synth, props*)
Pesticide Manual, 6th Ed., 153.

Demephion-*S* D-00018

O,O-*Dimethyl* S-*[(2-methylthio)ethyl] phosphorothioate,* 9CI
[2587-90-8]

$(MeO)_2P(O)SCH_2CH_2SMe$

$C_5H_{13}O_3PS_2$ M 216.250
See also Demephion-*O*, D-00017 . Systemic insecticide and acaricide, component of coml. Demephion (Tinox). Now superseded. Liq. Mod. sol. H_2O. d_4^{25} 1.218. $Bp_{0.1}$ 65° dec. n_D^{25} 1.508. Forms Ni and Co complexes.
▷TF9450000.

Grübner, P. *et al*, *J. Prakt. Chem.*, 1970, **312**, 603 (*synth*)
Stan, H.-J. *et al*, *Fresenius' Z. Anal. Chem.*, 1977, **287**, 271 (*glc, ms*)
Bloodworth, B.C., *Inorg. Chim. Acta*, 1979, **34**, L197 (*complexes*)
Stan, H.-J. *et al*, *Biomed. Mass Spectrom.*, 1982, **9**, 483 (*ms*)
Stan, H.-J. *et al*, *J. Chromatogr.*, 1983, **279**, 173 (*glc*)
Pesticide Manual, 6th Ed., 153.

Demeton-*O* D-00019

O,O-*Diethyl* O-*[2-(ethylthio)ethyl] phosphorothioate,* 9CI, 8CI
[298-03-3]

$(Et_2O)_2P(S)OCH_2CH_2SEt$

$C_8H_{19}O_3PS_2$ M 258.330
Powerful systemic agricultural insecticide. Liq. d_4^{21} 1.119. $Bp_{0.15}$ 92-93°. n_D^{22} 1.4865. On heating at 130° is converted to *O,O*-Diethyl *S*-[2-(ethylthio)ethyl] phosphorothioate, D-00294 A mixt. of the isomers is used commercially as Demeton or Systox.
▷Highly toxic by skin absorption and orally, exp. teratogen and carcinogen, TLV 0.1. TF3125000.

Fukuto, T.R. *et al*, *J. Am. Chem. Soc.*, 1954, **76**, 5103 (*props*)
Hoffman, F.W. *et al*, *J. Am. Chem. Soc.*, 1958, **80**, 1150 (*synth*)
Gore, R.C. *et al*, *J. Assoc. Off. Anal. Chem.*, 1971, **54**, 1040 (*ir, uv*)
Stan, H.-J. *et al*, *Biomed. Mass Spectrom.*, 1982, **9**, 483 (*ms*)
Ripley, B.D. *et al*, *J. Assoc. Off. Anal. Chem.*, 1983, **66**, 1084 (*glc*)
Pesticide Manual, 6th Ed., 154; 7th Ed., 162.
Sax, N.I., *Dangerous Properties of Industrial Materials*, 6th Ed., Van Nostrand-Reinhold, 1984, 538.

Demeton-*S*-methyl, BSI D-00020

O,O-*Dimethyl 5-[2-(ethylthio)ethyl] phosphorothioate,* 9CI
[919-86-8]

$(MeO)_2P(O)SCH_2CH_2SEt$

$C_6H_{15}O_3PS_2$ M 230.276
Systemic and contact insecticide. Pale-yellow oil. Sl. sol. H_2O. d_4^{20} 1.21. $Bp_{0.15}$ 89°. n_D^{20} 1.5110.
▷TG1750000.
S-*Oxide: see Oxydemeton-methyl, O-00098*
S,S-*Dioxide: see Phosmet, P-00343*

Fukuto, T.R., *J. Econ. Entomol.*, 1955, **48**, 348; 1956, **49**, 147; 1957, **50**, 399 (*metab, pharmacol*)
U.S.P., 3 082 240, (*1963*); *CA*, **59**, 5077 (*synth*)

Stan, H.-J. *et al, Biomed. Mass. Spectrom.*, 1982, **9**, 483 (*ms*)
Stan, H.-J. *et al, J. Chromatogr.*, 1983, **279**, 173 (*glc*)
Pesticide Manual, 7th Ed., 163.
The Agrochemicals Handbook, Royal Society of Chemistry, 1983, A124.

Demeton-*S*-methyl sulfone, BSI, ISO D-00021

S-[*2-(Ethylsulfonyl)ethyl*] O,O-*dimethyl phosphorothioate, 9CI*

[17040-19-6]

$(MeO)_2P(O)SCH_2CH_2SO_2Et$

$C_6H_{15}O_5PS_2$ M 262.275

Systemic insecticide. Cryst. Mp 60°. $Bp_{0.03}$ 120°.

▷TF9050000.

Stan, H.-J. *et al, Biomed. Mass Spectrom.*, 1982, **9**, 483 (*ms*)
Szeto, S.Y. *et al, J. Agric. Food Chem.*, 1982, **30**, 1082 (*metab, glc*)
Ripley, B.D. *et al, J. Assoc. Off. Anal. Chem.*, 1983, **66**, 1084 (*glc*)
Pesticide Manual, 7th Ed., 164.

(Diacetylmethyl)triphenylphosphonium(1+) D-00022

(2,4-Dioxo-3-pentyl)triphenylphosphonium(1+). (1-Acetyl-2-oxopropyl)triphenylphosphonium(1+)

$Ph_3P^{\oplus}CH(COCH_3)_2$

$C_{23}H_{22}O_2P^{\oplus}$ M 361.399 (ion)

Chloride:

$C_{23}H_{22}ClO_2P$ M 396.852

Solid. Mp 168-170°. pK_a 2.75 (50% EtOH aq.), 11.7 ($MeNO_2$).

Ylide: see 3-(Triphenylphosphoranylidene)-2,4-pentanedione, T-00654

Mastryukova, T.A. *et al, Phosphorus*, 1971, **1**, 159 (*synth, props*)

Dialifos, BSI D-00023

S-[*2-Chloro-1-(1,3-dihydro-1,3-dioxo-2*H*-isoindol-2-yl)ethyl*] O,O-*diethyl phosphorodithioate, 9CI.* S-*2-Chloro-1-phthalimidoethyl* O,O-*diethyl phosphorodithioate. Torak. Dialifor*

[10311-84-9]

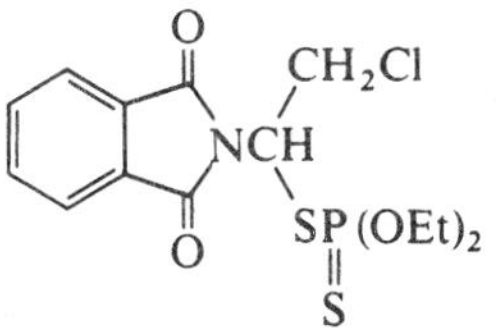

$C_{14}H_{17}ClNO_4PS_2$ M 393.839

Non-systemic agricultural insecticide and acaricide. Cryst. (toluene/hexane). Mp 67-69° (62-64°).

▷Exp. teratogen. TD5165000.

B.P., 1 091 738, (*1964*); *CA*, **65**, 16910d (*synth*)
Freed, V.H. *et al, Pestic. Biochem. Physiol.*, 1979, **10**, 203 (*props*)
Stan, H.-J. *et al, Biomed. Mass. Spectrom.*, 1982, **9**, 483 (*ms*)
Stan, H.-J. *et al, J. Chromatogr.*, 1983, **279**, 173 (*glc*)
Pesticide Manual, 6th Ed., 160; 7th Ed., 168.
The Agrochemicals Handbook, Royal Society of Chemistry, 1983, A129.
Sax, N.I., *Dangerous Properties of Industrial Materials*, 6th Ed., Van Nostrand-Reinhold, 1984, 540.

Diamidafos D-00024

Phenyl N,N′-*dimethylphosphorodiamidate, 9CI*

[1754-58-1]

$(MeNH)_2P(O)OPh$

$C_8H_{13}N_2O_2P$ M 200.177

Insecticide; use now discontinued. Cryst. ($CHCl_3/Et_2O$ or CCl_4). Sl. sol. EtOAc, C_6H_6. Mp 103-104°.

▷TD2800000.

Audrieth, L.F. *et al, J. Am. Chem. Soc.*, 1942, **64**, 1337 (*synth*)
Meikle, R.W., *Bull. Environ. Contam. Toxicol.*, 1978, **19**, 589 (*props*)
Mulliez, M., *Phosphorus Sulfur*, 1980, **8**, 27 (*synth, ir, pmr*)
Roth, H.J. *et al, Arch. Pharm.* (*Weinheim, Ger.*), 1981, **314**, 85 (*synth, pmr*)

2,2-Diamino-2,2,4,4,6,6-hexahydro-4,4,6,6-tetraphenyl-1,3,5-triaza-2,4,6-triphosphorine, 9CI D-00025

2,2-Diamino-2,2,4,4,6,6-hexahydro-4,4,6,6-tetraphenylcyclotriphosphazene. 2,2,4,4,6,6-Hexahydro-4,4,6,6-tetraphenyl-1,3,5-triaza-2,4,6-triphosphorine-2,2-diamine

[50457-19-7]

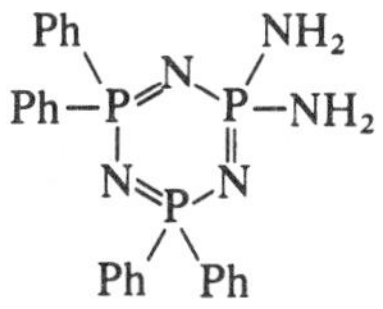

$C_{24}H_{24}N_5P_3$ M 475.408

Needles. Mp 194-196°.

N,N,N′,N′-*Tetra-Me:*

$C_{28}H_{32}N_5P_3$ M 531.516

Solid. Mp 145°.

N,N′-*Di-Ph:* [23669-19-4].

$C_{36}H_{32}N_5P_3$ M 627.604

Solid. Mp 242°.

Schmidpeter, A. *et al, Chem. Ber.*, 1976, **109**, 2340 (*pmr, P nmr, deriv*)
Schmidpeter, A. *et al, Phosphorus*, 1976, **6**, 113 (*synth, pmr, P nmr*)

1,4:3,6-Dianhydro-2,5-deoxy-2,5-bis(diphenylphosphino)iditol D-00026

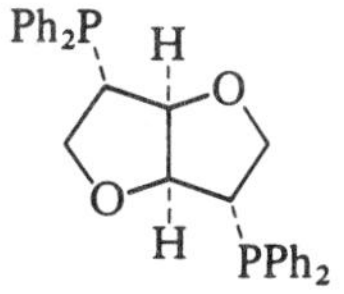

$C_{30}H_{28}O_2P_2$ M 482.498

L-form [88133-77-1]

Chiral ligand for asymm. hydrogenation. Needles. Mp 110-112°. $[\alpha]^{20}_D$ +32.8° (c, 5.0 in $CHCl_3$).

Bakos, J. *et al, J. Organomet. Chem.*, 1983, **253**, 249 (*synth, ms, pmr, P nmr, use*)

Diazinon, BSI **D-00027**

2-Isopropyl-6-methyl-4-pyrimidyl phosphorothioic acid O,O′-diethyl ester. O,O-Diethyl O-[6-methyl-2-(1-methylethyl)-4-pyrimidinyl] phosphorothioate, 8CI. *Basudin. Diazitol. Neocidol. Nucidol. Dimpylate. Spectacide*

[333-41-5]

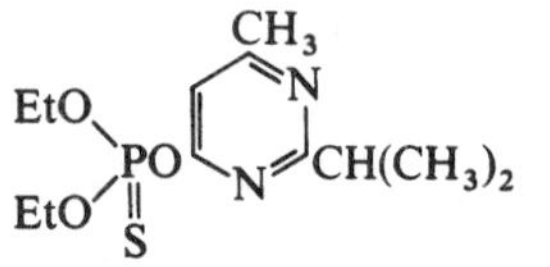

$C_{12}H_{21}N_2O_3PS$ M 304.343

Nonsystemic insecticide for rice and fruit trees. Liq. d^{20} 1.117. $Bp_{0.05}$ 85-90°. n_D^{20} 1.498. Slowly dec. by H_2O and dil. acids.

▷Highly toxic, TLV (skin) 0.1. Exp. teratogen. TF3325000.

Gasser, R., *Z. Naturforsch., B*, 1953, **8**, 225 (*rev, use*)
Gysin, H., *Chimia*, 1954, **8**, 205, 221 (*synth*)
Shvetsova-Shilovskaya, K.D. *et al*, *Zh. Obshch. Khim.*, 1956, **26**, 808; *CA*, **50**, 14769 (*synth*)
Mortland, M.M. *et al*, *J. Agric. Food Chem.*, 1967, **15**, 163 (*uv, complexes*)
Keith, L.H. *et al*, *J. Assoc. Off. Anal. Chem.*, 1968, **51**, 1063 (*pmr*)
Pardue, J. *et al*, *J. Agric. Food Chem.*, 1970, **18**, 405 (*metab, ir, ms, glc*)
Ross, R.T. *et al*, *Anal. Chim. Acta*, 1970, **52**, 139 (*P nmr*)
Gore, R.C. *et al*, *J. Assoc. Off. Anal. Chem.*, 1971, **54**, 1040 (*ir, uv*)
Stan, H.-J. *et al*, *Fresenius' Z. Anal. Chem.*, 1977, **287**, 271 (*glc, ms*)
Stan, H.-J. *et al*, *Biomed. Mass Spectrom.*, 1982, **9**, 483 (*ms*)
Stan, H.-J. *et al*, *J. Chromatogr.*, 1983, **279**, 173 (*glc*)
Pesticide Manual, 6th Ed., 163; 7th Ed., 171.
The Agrochemicals Handbook, Royal Society of Chemistry, 1983, A131.
Sax, N.I., *Dangerous Properties of Industrial Materials*, 6th Ed., Van Nostrand-Reinhold, 1984, 544.

(Diazocyclohexylmethyl)phosphonic acid, 9CI **D-00028**

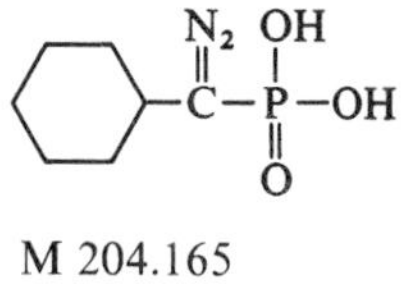

$C_7H_{13}N_2O_3P$ M 204.165

Di-Me ester: Dimethyl (diazocyclohexylmethyl)-phosphonate.
$C_9H_{17}N_2O_3P$ M 232.219
Liq. $Bp_{0.1\text{-}0.15}$ 85-89°.

Marmor, R.S. *et al*, *J. Org. Chem.*, 1977, **36**, 128 (*synth, ir, pmr, props*)

2-Diazo-2-(diphenylphosphino)-1-phenylethanone **D-00029**

(Benzoyldiazomethyl)diphenylphosphine

Ph_2PCN_2COPh

$C_{20}H_{15}N_2OP$ M 330.325

Oxide: [17507-54-9]. *2-Diazo-2-(diphenylphosphinyl)-1-phenylethanone.*
$C_{20}H_{15}N_2O_2P$ M 346.324
Cryst. (C_6H_6/pet. ether). Mp 139°, 152° dec. Photolysis → $Ph_2P(O)CCOPh$:.

Sulfide: [58921-82-7]. *2-Diazo-2-(diphenylphosphinothioyl)-1-phenylethanone.*
$C_{20}H_{15}N_2OPS$ M 362.385
Solid (Et_2O). Mp 123° dec. Photolysis → $Ph_2P(S)CCOPh$:.

Petzgold, G. *et al*, *Naturwissenschaften*, 1967, **54**, 469 (*synth, oxide*)
Illger, W. *et al*, *Justus Liebigs Ann. Chem.*, 1972, **760**, 1.
Cauquis, G. *et al*, *Org. Mass Spectrom.*, 1975, **10**, 1021 (*sulfide, ms*)
Tomioka, H. *et al*, *J. Org. Chem.*, 1977, **42**, 552 (*oxide, props*)
Dirisia, B., *Tetrahedron*, 1979, **35**, 181 (*sulfide, synth, ir, pmr*)
Regitz, M. *et al*, *Justus Liebigs Ann. Chem.*, 1979, 1002 (*oxide*)

Diazo(diphenylphosphinyl)acetic acid, 9CI **D-00030**

$Ph_2P(O)CN_2COOH$

$C_{14}H_{11}N_2O_3P$ M 286.226

Et ester: [17507-55-0]. *Ethyl diazo(diphenylphosphinyl)acetate.*
$C_{16}H_{15}N_2O_3P$ M 314.280
Forms a carbene on photolysis. Cryst. (diisopropyl ether). Mp 99-100°.

Regitz, M. *et al*, *Chem. Ber.*, 1969, **102**, 2216 (*synth, ir*)
Illger, W. *et al*, *Justus Liebigs Ann. Chem.*, 1972, **760**, 1 (*use*)
Regitz, M. *et al*, *Justus Liebigs Ann. Chem.*, 1979, 1002 (*synth*)

1-Diazo-1-(diphenylphosphinyl)-2-propanone, 9CI **D-00031**

[19847-57-5]

$Ph_2P(O)CN_2COCH_3$

$C_{15}H_{13}N_2O_2P$ M 284.254

Cryst. (CCl_4/pet. ether). Mp 83-85°. Forms a carbene on photolysis.

Regitz, M. *et al*, *Chem. Ber.*, 1969, **102**, 2216 (*synth, ir*)
Illger, W. *et al*, *Justus Liebigs Ann. Chem.*, 1972, **760**, 1 (*props*)
Regitz, M. *et al*, *Justus Liebigs Ann. Chem.*, 1979, 1002 (*synth*)

(1-Diazoethyl)phenylphosphinic acid, 9CI **D-00032**

[67515-40-6]

$H_3CCHN_2P(O)(OH)Ph$

$C_8H_{10}N_2O_2P$ M 197.153

Me ester: [56042-55-8]. *Methyl (1-diazoethyl)-phenylphosphinate.*
$C_9H_{12}N_2O_2P$ M 211.180
Yellow liq. d_4^{20} 1.20. n_D^{20} 1.5535.

Et ester: [63469-88-5]. *Ethyl (1-diazoethyl)-phenylphosphinate.*
$C_{10}H_{14}N_2O_2P$ M 225.207
Liq. d_4^{20} 1.16. n_D^{20} 1.5438.

Felcht, U. *et al*, *Chem. Ber.*, 1975, **108**, 2040 (*methyl ester, ir, pmr, uv*)
Gareev, R.D. *et al*, *Zh. Obshch. Khim.*, 1975, **45**, 942 (*Engl. transl.* p. 926) (*methyl ester*)
Gareev, R.D. *et al*, *Zh. Obshch. Khim.*, 1979, **49**, 503 (*Engl. transl.* p. 442) (*ethyl ester*)

(1-Diazoethyl)phosphonic acid, 9CI **D-00033**

$H_3CCN_2P(O)(OH)_2$

$C_2H_5N_2O_3P$ M 136.047
With Cu powder, esters → phosphorylated carbene.
Di-Me ester: [26584-15-6]. *Dimethyl (1-diazoethyl)-phosphonate.*
$C_4H_9N_2O_3P$ M 164.100
Liq. $Bp_{0.2}$ 50-52°. n_D^{20} 1.4583.
Di-Et ester: Diethyl (1-diazoethyl)phosphonate.
$C_6H_{13}N_2O_3P$ M 192.154
Liq. $Bp_{0.14}$ 49°. n_D^{25} 1.4503.

Marmor, R.S. *et al, J. Org. Chem.*, 1971, **36**, 128 (*synth, ir, pmr*)
Seyferth, D. *et al, J. Org. Chem.*, 1971, **36**, 1379 (*props*)
Callot, H.J. *et al, J. Org. Chem.*, 1972, **50**, 1078 (*props*)
Regitz, M. *et al, Chem. Ber.*, 1979, **112**, 2509 (*props*)
Kato, T. *et al, J. Org. Chem.*, 1980, **45**, 2587 (*props*)

(Diazomethyl)diphenylphosphine oxide, 9CI — D-00034

(Diphenylphosphinyl)diazomethane
[5353-66-2]

$Ph_2P(O)CHN_2$

$C_{13}H_{11}N_2OP$ M 242.216
Cryst. (EtOAc/Et_2O). Mp 61-62°.

Kreutzkamp, D.N. *et al, Angew. Chem., Int. Ed. Engl.*, 1965, **4**, 1078 (*synth*)
Regitz, M. *et al, Justus Liebigs Ann. Chem.*, 1971, **748**, 207; 1979, 1002 (*synth, use*)

Diazomethylphosphonic acid — D-00035

$N_2CHP(O)(OH)_2$

$CH_3N_2O_3P$ M 122.020
Esters act as carbene sources. They form anions which convert ketones into alkynes.
▷Esters are potentially carcinogenic and explosive
Di-Me ester: [27491-70-9]. *Dimethyl (diazomethyl)-phosphonate.*
$C_3H_7N_2O_3P$ M 150.074
Yellow oil. $Bp_{0.1}$ 59°.
Di-Et ester: [25411-73-8]. *Diethyl (diazomethyl)-phosphonate.*
$C_5H_{11}N_2O_3P$ M 178.127
Yellow oil. $Bp_{0.35}$ 69-70°. n_D^{20} 1.4534.

Regitz, M. *et al, Justus Liebigs Ann. Chem.*, 1969, **730**, 194 (*synth, ir*)
Colvin, E.W. *et al, J. Chem. Soc., Perkin Trans. 1*, 1977, 869 (*ir, pmr, props*)
Tomioka, H. *et al, J. Org. Chem.*, 1977, **42**, 552 (*ir, pmr*)
Regitz, M. *et al, Justus Liebigs Ann. Chem.*, 1979, 1002 (*props*)
Lahti, P.M. *et al, J. Am. Chem. Soc.*, 1981, **103**, 7011 (*props*)
Gilbert, J.C. *et al, J. Org. Chem.*, 1982, **47**, 1837; 1983, **48**, 448 (*props, use*)
Hauske, J.R. *et al, J. Org. Chem.*, 1982, **47**, 5019 (*use*)

(Diazophenylmethyl)diphenylphosphine oxide, 9CI — D-00036

(α-Diphenylphosphinyl)phenyldiazomethane
[17507-57-2]

$Ph_2P(O)CN_2Ph$

$C_{19}H_{15}N_2OP$ M 318.314
Photolysis gives $Ph_2P(O)CPh$:. Cryst. (EtOH). Mp 160°.

Petzgold, G. *et al, Naturwissenschaften*, 1967, **54**, 469 (*synth*)
Yoshifuji, M. *et al, Tetrahedron Lett.*, 1979, 2415 (*analogues, sulfide*)
Regitz, M. *et al, Tetrahedron*, 1981, **37**, 1039.
Man, G. *et al, Chem. Ber.*, 1982, **115**, 669.

(Diazophenylmethyl)phosphonic acid, 9CI — D-00037

α-Diazobenzylphosphonic acid
[63147-97-7]

$PhCN_2P(O)(OH)_2$

$C_7H_7N_2O_3P$ M 198.118
Esters react with alkenes to give 1-phenylcyclopropane-1-phosphonic acids.
Di-Me ester: [16965-72-3]. *Dimethyl (diazophenylmethyl)phosphonate. Dimethyl α-diazobenzylphosphonate.*
$C_9H_{11}N_2O_3P$ M 226.171
Orange cryst. (Et_2O at −70°). Mp 44-44.5°.
Diisopropyl ester: [38749-18-7]. *Diisopropyl (diazophenylmethyl)phosphonate. Diisopropyl α-diazobenzylphosphonate.*
$C_{13}H_{19}N_2O_3P$ M 282.278
Orange cryst. (Et_2O at −70°). Mp 32-33°.

Seyferth, D. *et al, J. Org. Chem.*, 1971, **36**, 1379 (*synth, ir, pmr, props*)
Scherer, H. *et al, Chem. Ber.*, 1972, **105**, 3357 (*synth, ir, pmr*)
Gurudata, N. *et al, Can. J. Chem.*, 1973, **51**, 1142 (*pmr, cmr*)
Regitz, M. *et al, Chem. Ber.*, 1979, **112**, 2509 (*props*)
Maas, G. *et al, Phosphorus Sulfur*, 1983, **14**, 143 (*use*)

Dibenzo[*d,f*][1,3,2]dioxaphosphepin, 10CI — D-00038

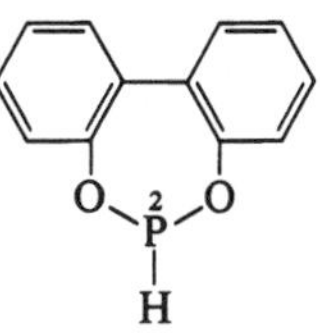

$C_{12}H_9O_2P$ M 216.176
2-Oxide: [53554-11-3]. *2,2'-Biphenylene phosphonate. 2,2'-Biphenylene phosphite.*
$C_{12}H_9O_3P$ M 232.175
Cryst. (toluene). Mp 109-110°. Tautomeric.
2-Ph: see 6-Phenyldibenzo[*d,f*][1,3,2]dioxaphosphepin, P-00107
2-Chloro: see 6-Chlorodibenzo[*d,f*][1,3,2]-dioxaphosphepin, C-00032
2-Methoxy: see 6-Methoxydibenzo[*d,f*][1,3,2]-dioxaphosphepin, M-00032

Ovchinnikov, V.V. *et al, Izv. Akad. Nauk SSSR, Ser. Khim.*, 1978, 689 (*Engl. transl.* p. 595) (*ir, P nmr, props*)
Ovchinnikov, V.V. *et al, Zh. Obshch. Khim.*, 1982, **52**, 707 (*Engl. transl.* p. 615) (*synth*)

6*H*-Dibenzo[*c,e*][1,2]oxaphosphorin, 9CI — D-00039

$C_{12}H_9OP$ M 200.176
6-Oxide: [35948-25-5].
$C_{12}H_9O_2P$ M 216.176

Cryst. (EtOH). Mp 117°.

Chernyshev, E.A. *et al*, *Zh. Obshch. Khim.*, 1972, **42**, 93 (*Engl. transl.* p. 88) (*synth, ir*)

5*H*-Dibenzophosphole, 9CI **D-00040**

*1*H-*Benzo*[b]*phosphindole. 9-Phosphafluorene*

[244-87-1]

$C_{12}H_9P$ M 184.177

Cryst. Mp 47-48°. $Bp_{0.1}$ 65°. Easily oxidized.

5-Me: see 5-Methyl-5H-dibenzophosphole, M-00115

5-Ph: see 5-Phenyl-5H-dibenzophosphole, P-00110

Braye, E.H. *et al*, *Tetrahedron*, 1971, 5523.

9*H*-Dibenzo[2,3:4,5]phospholo[1,2-*f*]-phosphanthridene, 9CI **D-00041**

*17-Phospha-9*H-*fluoreno*[9,9a-b]*phenanthrene*

[40894-65-3]

$C_{24}H_{17}P$ M 336.372

17-Oxide: [40894-65-3].

$C_{24}H_{17}OP$ M 352.371

Solid. Mp 265°.

Hellwinkel, D. *et al*, *Chem. Ber.*, 1972, **105**, 3878 (*synth, ir, uv, pmr*)

Dibenzoyl phosphate **D-00042**

Benzoic-phosphoric (2:1) anhydride, 9CI, 8CI

$(PhCOO)_2P(O)OH$

$C_{14}H_{11}O_6P$ M 306.211

Isol. as Na salt. *N*-Benzoylating agent. Half-life in phosphate buffer pH 7.4 is ca. 45 hr.

Chantrenne, H., *Nature (London)*, 1947, **160**, 603 (*synth, props*)

1,2-Dibenzyl-1,2-dimethyldiphosphine, 8CI **D-00043**

1,2-Dimethyl-1,2-bis(phenylmethyl)diphosphine, 9CI

$PhCH_2PMePMeCH_2Ph$

$C_{16}H_{20}P_2$ M 274.282

(RS,RS)-form

(±)-*form*

Disulfide:

$C_{16}H_{20}P_2S_2$ M 338.402

Cryst. (EtOH). Mp 120-123°.

(RS,SR)-form

meso-*form*

Disulfide: Cryst. (Me_2CO). Mp 188-189°.

Maier, L., *Chem. Ber.*, 1961, **94**, 3043 (*disulfide, synth*)

Lambert, I.B. *et al*, *J. Am. Chem. Soc.*, 1970, **92**, 3093 (*synth, pmr, nmr*)

Keck, H. *et al*, *Org. Mass Spectrom.*, 1979, **14**, 149 (*disulfide, ms*)

***P,P*-Dibenzyl *P′,P′*-diphenyl diphosphate** **D-00044**

P,P-*Diphenyl* P′,P′-*bis(phenymethyl) diphosphate, 9CI.* P,P-*Dibenzyl* P′,P′-*diphenyl pyrophosphate*

$(PhO)_2P(O)OP(O)(OCH_2Ph)_2$

$C_{26}H_{24}O_7P_2$ M 510.419

Viscous, undistillable resin.

Corby, N.S. *et al*, *J. Chem. Soc.*, 1952, 1234 (*synth, props*)

Kenner, G.W. *et al*, *J. Chem. Soc.*, 1956, 1231 (*synth, props*)

1,2-Dibenzyl-1,2-diphenyldiphosphine, 8CI **D-00045**

1,2-Diphenyl-1,2-bis(phenylmethyl)diphosphine, 9CI

$PhCH_2PPhPPhCH_2Ph$

$C_{26}H_{24}P_2$ M 398.423

Cryst. (ligroin). Mp 134°. Bp_2 240-265°.

Disulfide:

$C_{26}H_{24}P_2S_2$ M 462.543

Cryst. (EtOH). Mp 189.5-190.5°.

Steininger, E., *Chem. Ber.*, 1963, **96**, 3184 (*synth*)

Crofts, P.C. *et al*, *J. Chem. Soc.*, 1964, 2486 (*sulfide, cryst struct*)

1,6-Dibenzyl-1,6-diphosphacycloundecane **D-00046**

1,6-Bis(phenylmethyl)-1,6-diphosphacycloundecane, 9CI

[69573-04-2]

$C_{23}H_{32}P_2$ M 370.453

Solid. Mp 155°.

B,2PhCH₂Br: 1,1,6,6-Tetrabenzyl-1,6-diphosphoniacycloundecane dibromide.

$C_{37}H_{46}Br_2P_2$ M 712.526

Cryst. (EtOH/Me_2CO). Mp 285°.

Horner, L. *et al*, *Phosphorus Sulfur*, 1978, **5**, 171.

1,4-Dibenzyl-1,4-diphosphepane, 9CI **D-00047**

1,4-Dibenzyl-1,4-diphosphacycloheptane

[69572-97-0]

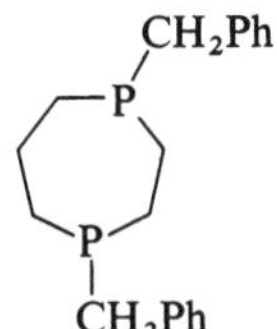

$C_{19}H_{24}P_2$ M 314.346

Solid. Mp 120°.

B,2PhCH₂Br: [69572-97-0]. *1,1,4,4-Tetrabenzyl-1,4-diphosphepanium dibromide.*

$C_{33}H_{38}Br_2P_2$ M 656.419

Cryst. (MeCN). Mp 242°.

Horner, L. *et al*, *Phosphorus Sulfur*, 1978, **5**, 171.

1,4-Dibenzyl-1,4-diphosphocane — D-00048

1,4-Bis(phenylmethyl)-1,4-diphosphocane, 9CI. *1,4-Dibenzyl-1,4-diphosphacyclooctane*

[69572-98-1]

P–CH_2Ph ; P / CH_2Ph

$C_{20}H_{26}P_2$ M 328.373

Solid. Mp 125-130°.

B,2PhCH$_2$Br: [69572-88-9]. *1,1,4,4-Tetrabenzyl-1,4-diphosphocanium dibromide.*

$C_{34}H_{40}Br_2P_2$ M 670.446

Cryst. (EtOH). Mp 290°.

Horner, L. *et al*, *Phosphorus Sulfur*, 1978, **5**, 171.

1,5-Dibenzyl-1,5-diphosphocane — D-00049

1,5-Bis(phenylmethyl)-1,5-diphosphocane, 9CI. *1,5-Dibenzyl-1,5-diphosphacyclooctane*

[69572-99-2]

$C_{20}H_{26}P_2$ M 328.373

Solid. Mp 105°.

B,2PhCH$_2$Br: [69572-88-9]. *1,1,5,5-Tetrabenzyl-1,5-diphosphocanium dibromide.*

$C_{34}H_{40}Br_2P_2$ M 670.446

Cryst. (EtOH). Mp 200°.

Horner, L. *et al*, *Phosphorus Sulfur*, 1978, **5**, 171.

1,4-Dibenzyl-1,4-diphosphonane — D-00050

1,4-Bis(phenylmethyl)-1,4-diphosphonane, 9CI. *1,4-Dibenzyl-1,4-diphosphacyclononane*

[69573-01-9]

$PhCH_2$–P P–CH_2Ph

$C_{21}H_{42}P_2$ M 356.510

Solid. Mp 160°.

B,2PhCH$_2$Br: [69572-90-3]. *1,1,4,4-Tetrabenzyl-1,4-diphosphonanium dibromide.*

$C_{35}H_{56}Br_2P_2$ M 698.583

Cryst. (EtOH). Mp 305°.

Horner, L. *et al*, *Phosphorus Sulfur*, 1978, **5**, 171.

1,5-Dibenzyl-1,5-diphosphonane, 9CI — D-00051

1,5-Dibenzyl-1,5-diphosphacyclononane

[69573-00-8]

$C_{21}H_{42}P_2$ M 356.510

Solid. Mp 127-135°.

B,2PhCH$_2$Br: [69572-89-0]. *1,1,5,5-Tetrabenzyl-1,5-diphosphonanium dibromide.*

$C_{35}H_{56}Br_2P_2$ M 698.583

Cryst. + $1H_2O$ (EtOH/Me_2CO). Mp 195°.

Horner, L. *et al*, *Phosphorus Sulfur*, 1978, **5**, 171.

Dibenzylphenylphosphine — D-00052

Phenylbis(phenylmethyl)phosphine, 9CI

[7650-90-0]

$PhP(CH_2Ph)_2$

$C_{20}H_{19}P$ M 290.344

Ligand for Ni, Pd, and Pt. Foul-smelling cryst. (oxygen-free EtOH). Mp 68-70°.

▷Toxic

Oxide: [24442-45-3].

$C_{20}H_{19}OP$ M 306.343

pK_a 6.12 ($MeNO_2$).

Sulfide: [13298-01-6].

$C_{20}H_{19}PS$ M 322.404

No phys. props. reported.

Verstuyft, A.W. *et al*, *Inorg. Chem.*, 1977, **16**, 2776 (*pmr, cmr*)
Inorg. Synth., 1978, **18**, 169 (*synth*)

Dibenzyl phosphate, 8CI — D-00053

Bis(phenylmethyl) phosphate, 9CI. *Dibenzyl hydrogen phosphate. Dibenzyl phosphoric acid*

[1623-08-1]

$(PhCH_2O)_2P(O)OH$

$C_{14}H_{15}O_4P$ M 278.244

Cryst. as H-bonded dimer and is dimeric or trimeric in soln. Used for phosphorylation of alcohols, esp. sugars. Needles ($CHCl_3$/pet. ether, EtOH, or Et_2O). Sl. sol. H_2O. Mp 78-79°. pK_a 0.71.

Tetramethylammonium salt: Solid. Mp 185°.

Cyclohexylammonium salt: Needles (EtOH). Mp 173°.

Ag salt: [50651-75-7]. Mp 212-216° dec.

Chloride: *see Dibenzyl phosphorochloridate, D-00060*

Amide: [3905-76-8]. *Dibenzyl phosphoramidate. Dibenzyl amidophosphate. Bis(phenylmethyl) phosphoramidate.*

$C_{14}H_{16}NO_3P$ M 277.259

Needles (CCl_4). Mp 103.5-104.5°.

Anilide: [56883-96-6]. *Dibenzyl phenylphosphoramidate. Bis(phenylmethyl) phenylphosphoramidate.*

$C_{20}H_{20}NO_3P$ M 353.357

Cryst. (1,2-dichloroethane/hexane). Mp 91-92.5°.

Atherton, F.R. *et al*, *J. Chem. Soc.*, 1945, 382, 660 (*amide, anilide*)
Baddiley, J. *et al*, *J. Chem. Soc.*, 1949, 815 (*synth, derivs*)
Clark, V.M. *et al*, *J. Chem. Soc.*, 1950, 2023, 2030 (*synth*)
Chabrier, P. *et al*, *C.R. Hebd. Seances Acad. Sci.*, 1957, **244**, 2730 (*synth*)
Dunitz, J.D. *et al*, *Acta Crystallogr.*, 1959, **9**, 327 (*cryst struct*)
Mitsunobu, O. *et al*, *J. Org. Chem.*, 1965, **30**, 1071 (*synth*)
Courtmanche, P. *et al*, *Bull. Soc. Chim. Fr.*, 1967, 3911 (*props*)
Jacobsen, P. *et al*, *Org. Mass Spectrom.*, 1972, **6**, 1303 (*amide, ms*)
Kimura, J., *Bull. Chem. Soc. Jpn.*, 1979, **52**, 1191 (*use*)

Dibenzylphosphine, 8CI — D-00054

Bis(phenylmethyl)phosphine, 9CI

[56522-04-4]

$HP(CH_2Ph)_2$

$C_{14}H_{15}P$ M 214.246

Liq. Bp_3 115-120°.

B,HI: Dibenzylphosphonium iodide.

$C_{14}H_{16}IP$ M 342.159

Cryst. (EtOH). Mp 211° dec.

Oxide: [13238-16-9].
$C_{14}H_{15}OP$ M 230.246
Plates (C_6H_6/hexane). Mp 109-110°. Tautomeric with Dibenzylphosphinous acid, D-00058 .

Miller, R.C. *et al*, *J. Am. Chem. Soc.*, 1956, **78**, 5299 (*oxide*)
Issleib, K. *et al*, *Chem. Ber.*, 1959, **92**, 704 (*synth*)
Sanchez, M. *et al*, *Spectrochim. Acta, Part A*, 1967, **23**, 2617 (*oxide, ir*)
Purdum, W.R. *et al*, *J. Org. Chem.*, 1974, **39**, 2904 (*oxide, use*)
Kostyanovskii, R.G., *Izv. Akad. Nauk SSSR, Ser. Khim.*, 1975, 901 (*Engl. transl.* p. 816) (*synth*)
Petrov, K.A. *et al*, *Zh. Obshch. Khim.*, 1976, **46**, 2494 (*Engl. transl.* p. 2387) (*oxide, use*)

Dibenzylphosphinic acid, 8CI D-00055

Bis(phenylmethyl)phosphinic acid, 9CI
[7369-51-9]

$(PhCH_2)_2P(O)OH$

$C_{14}H_{15}O_2P$ M 246.245
Needles (EtOH). Sol. Me_2CO, dioxan, CCl_4. Mp 192°. pK_a 6.35 (95% EtOH aq.), 7.94 (EtOH).
Me ester: [21713-63-3]. *Methyl dibenzylphosphinate.*
$C_{15}H_{17}O_2P$ M 260.272
Cryst. (hexane). Mp 75-76°.
Et ester: [34679-43-1]. *Ethyl dibenzylphosphinate.*
$C_{16}H_{19}O_2P$ M 274.299
Cryst. (DMF aq.). Mp 65-66°.
Fluoride: [51010-76-5].
$C_{14}H_{14}FOP$ M 248.236
Solid. Mp 108-109°.
Chloride: [18255-62-4].
$C_{14}H_{14}ClOP$ M 264.691
Cryst. or prisms (CCl_4). Mp 124-125°, 173-175°. $Bp_{0.00002}$ 120° subl.
Amide: [34679-31-7]. *Dibenzylphosphinic amide.*
$C_{14}H_{16}NOP$ M 245.260
Solid. Mp 132-133°.
Anilide: P,P-*Dibenzyl*-N-*phenylphosphinic amide.* P,P-*Dibenzylphosphinic anilide.*
$C_{20}H_{20}NOP$ M 321.358
Solid. Mp 170-171°.

Miller, R.C. *et al*, *J. Am. Chem. Soc.*, 1956, **78**, 5299 (*synth*)
Zhuravleva, L.P. *et al*, *Zh. Obshch. Khim.*, 1968, **38**, 342 (*Engl. transl.* p. 341) (*synth, props*)
Zhuravleva, L.P. *et al*, *Zh. Obshch. Khim.*, 1971, **41**, 1950 (*Engl. transl.* p. 1966) (*chloride, ester, amides*)
Markovskii, L.N. *et al*, *Synthesis*, 1973, 787 (*fluoride*)
Golubski, Z.E., *Synthesis*, 1980, 632 (*ester, ir, pmr*)
Heuschmann, M. *et al*, *Chem. Ber.*, 1982, **115**, 3384 (*chloride, pmr*)

Dibenzylphosphinodithioic acid, 8CI D-00056

Bis(phenylmethyl)phosphinodithioic acid, 9CI

$(PhCH_2)_2P(S)SH$

$C_{14}H_{15}PS_2$ M 278.366
Cryst. (C_6H_6/ligroin). Mp 132.5-133.5°.
Benzyl ester: Benzyl dibenzylphosphinodithioate.
$C_{21}H_{21}PS_2$ M 368.491
Solid. Mp 111-112°.
Anhydrosulfide: [22737-45-7].
$C_{28}H_{28}P_2S_3$ M 522.657
Solid. Mp 188°.

Mastryukova, T.A. *et al*, *Zh. Obshch. Khim.*, 1961, **31**, 507 (*Engl. transl.* p. 464) (*synth*)
Almasi, L. *et al*, *Chem. Ber.*, 1969, **102**, 1489 (*anhydrosulfide*)

Dibenzylphosphinothioic acid D-00057

Bis(phenylmethyl)phosphinothioic acid, 9CI
[21187-16-6]

$(PhCH_2)_2P(S)OH \rightleftharpoons (PhCH_2)_2P(O)SH$

$C_{14}H_{15}OPS$ M 262.306
Cryst. (AcOH/pet. ether). Mp 188-191°. pK_a 4.64 (80% EtOH aq.), 8.16 (EtOH), 10.80 ($MeNO_2$).
Na salt: Solid. Mp 232-236°.
Chloride:
$C_{14}H_{14}ClPS$ M 280.751
Solid. Mp 130.5-131.5°.

Mastryukova, T.A. *et al*, *Zh. Obshch. Khim.*, 1961, **31**, 507 (*Engl. transl.* p. 464) (*chloride*)
Crofts, P.C. *et al*, *J. Chem. Soc. (B)*, 1968, 1416 (*synth*)
Mikolajczyk, M. *et al*, *Pol. J. Chem.*, 1979, **53**, 317 (*pmr*)

Dibenzylphosphinous acid D-00058

Bis(phenylmethyl)phosphinous acid, 9CI
[2328-95-2]

$(PhCH_2)_2POH$

$C_{14}H_{15}OP$ M 230.246
Free acid exists as the phosphoryl tautomer. See under Dibenzylphosphine, D-00054 .
Et ester: Ethyl dibenzylphosphinite. Ethyl bis(phenylmethyl)phosphinite.
$C_{16}H_{19}OP$ M 258.299
Liq. d_4^{20} 1.04. Bp_1 120-121°. n_D^{20} 1.5737.
Butyl ester: Butyl dibenzylphosphinite. Butyl bis(phenylmethyl)phosphinite.
$C_{18}H_{23}OP$ M 286.353
Liq. d_4^{20} 1.02. Bp_2 148-149°. n_D^{20} 1.5590.
Ph ester: Phenyl dibenzylphosphinite. Phenyl bis(phenylmethyl)phosphinite.
$C_{20}H_{19}OP$ M 306.343
Liq. $Bp_{2.5}$ 200-203°.
Chloride: [17850-02-1]. *Chlorodibenzylphosphine. Chlorobis(phenylmethyl)phosphine.*
$C_{14}H_{14}ClP$ M 248.691
Solid. Mp 81.5-82°. Bp_{12} 234-236°, $Bp_{1.5}$ 157-163°.
Bromide: [17850-04-3]. *Dibenzylbromophosphine. Bromobis(phenylmethyl)phosphine.*
$C_{14}H_{14}BrP$ M 293.142
Liq. Bp_3 160-165°.
Diethylamide: P,P-*Dibenzyl*-N,N-*diethylphosphinous amide.*
$C_{18}H_{24}NP$ M 285.368
Solid-liq. Sol. Et_2O, C_6H_6, $CHCl_3$, pet. ether. Mp 19-22°. Bp_2 168-170°.

Miller, R.C. *et al*, *J. Am. Chem. Soc.*, 1956, **78**, 5299 (*synth*)
Kabachnik, M.I. *et al*, *Dokl. Akad. Nauk SSSR, Ser. Sci. Khim.*, 1960, **135**, 323 (*Engl. transl.* p. 1267) (*esters*)
Sander, M., *Chem. Ber.*, 1960, **93**, 1220 (*phenyl ester*)
Barabanov, Yu.I. *et al*, *Zh. Obshch. Khim.*, 1969, **39**, 836 (*Engl. transl.* p. 799) (*chloride*)
Z'ola, M.I. *et al*, *Zh. Obshch. Khim.*, 1970, **40**, 1937 (*Engl. transl.* p. 1922) (*chloride*)
Appel, R. *et al*, *Chem. Ber.*, 1971, **104**, 3859 (*chloride, amide, synth*)

Dibenzyl phosphonate, 8CI D-00059

Bis(phenylmethyl) phosphonate, 9CI. Dibenzyl phosphite

[17176-77-1]

$$(PhCH_2O)_2P(O)H \rightleftharpoons (PhCH_2O)_2POH$$

$C_{14}H_{15}O_3P$ M 262.244
Tautomeric. Mp 0-5°, 17°. $Bp_{0.1}$ 165°, $Bp_{0.001}$ 110-120°. n_D^{18} 1.5521.

▷May explode during distillation; violent dec. at 160°

Atherton, F.R. *et al*, *J. Chem. Soc.*, 1945, 382 (*synth*)
Sheehan, J.C. *et al*, *J. Am. Chem. Soc.*, 1950, **72**, 1312 (*synth*)
Bellamy, L.J. *et al*, *J. Chem. Soc.*, 1952, 475 (*ir*)
Oehme, F. *et al*, *Chem. Ber.*, 1957, **90**, 772 (*struct*)
Bretherick, L., *Handbook of Reactive Chemical Hazards*, 2nd Ed., Butterworths, London and Boston, 1979, 727.
Sax, N.I., *Dangerous Properties of Industrial Materials*, 6th Ed., Van Nostrand-Reinhold, 1984, 548.

Dibenzyl phosphorochloridate, 8CI D-00060

Bis(phenylmethyl) phosphorochloridate, 9CI. Dibenzyl chlorophosphonate. Dibenzyl chlorophosphate
[538-37-4]

$$(PhCH_2O)_2P(O)Cl$$

$C_{14}H_{14}ClO_3P$ M 296.690
Phosphorylating agent, esp. for nucleosides, but now largely superseded. Thick oil. Dec. slowly at r.t.

▷Can dec. violently during vacuum dist.

Atherton, F.R. *et al*, *J. Chem. Soc.*, 1945, 382; 1948, 1106 (*synth, use*)
Kenner, G.W. *et al*, *J. Chem. Soc.*, 1952, 3675 (*synth*)
Brown, D.M., *Adv. Org. Chem.*, 1963, **3**, 74 (*rev*)
Sosnovsky, G. *et al*, *J. Org. Chem.*, 1969, **34**, 968 (*pmr*)
Alewood, P.F. *et al*, *Aust. J. Chem.*, 1984, **37**, 429 (*use*)
Sax, N.I., *Dangerous Properties of Industrial Materials*, 6th Ed., Van Nostrand-Reinhold, 1984, 548.

N,N'-Dibenzylphosphorodiamidic acid, 8CI D-00061

N,N'-*Bis(phenylmethyl)phosphorodiamidic acid, 9CI*
[30546-51-1]

$$(PhCH_2NH)_2P(O)OH$$

$C_{14}H_{17}N_2O_2P$ M 276.274
Needles (Me_2CO). Mp 114-116°.
Ph ester: [18995-03-4]. *Phenyl* N,N'-*dibenzylphosphorodiamidate.*
$C_{20}H_{21}N_2O_2P$ M 352.372
Needles. Mp 114°.
Fluoride:
$C_{14}H_{16}FN_2OP$ M 278.265
Cryst. (EtOH aq.). Mp 96°.

▷Highly toxic, LD_{50} 10 mg/kg

Michaelis, A., *Justus Liebigs Ann. Chem.*, 1903, **326**, 129 (*phenyl ester*)
Heap, R. *et al*, *J. Chem. Soc.*, 1948, 1313 (*fluoride, synth, tox*)
Cremyln, R.J.W. *et al*, *J. Chem. Soc. (C)*, 1971, 300 (*synth*)

N,N'-Dibenzylphosphorodiamidothioic acid, 8CI D-00062

N,N'-*Bis(phenylmethyl)phosphorodiamidothioic acid, 9CI. Thiophosphoric acid dibenzylamide*

$$(PhCH_2NH)_2P(S)OH \rightleftharpoons (PhCH_2NH)_2P(O)SH$$

$C_{14}H_{17}N_2OPS$ M 292.335
O-*Ph ester:* O-*Phenyl*-N,N'-*dibenzylphosphorodiamidothioate.* O-*Phenyl* N,N'-*bis(phenylmethyl)-phosphorodiamidothioate.*
$C_{20}H_{21}N_2OPS$ M 368.432
Needles (C_6H_6). Mp 73°.

Michaelis, A., *Justus Liebigs Ann. Chem.*, 1903, **326**, 129, 201.

O,O-Dibenzyl phosphorodithioate, 8CI D-00063

O,O-*Bis(phenylmethyl) phosphorodithioate, 9CI.* O,O-*Dibenzyl hydrogen dithiophosphate.* O,O-*Dibenzyl dithiophosphoric acid*
[2253-62-5]

$$(PhCH_2O)_2P(S)SH$$

$C_{14}H_{15}O_2PS_2$ M 310.365
Unstable liq. n_D^{25} 1.5971.
K salt: [38886-39-4]. Solid. Mp 152-154°.
Pb complex: Solid. Mp 117°.

Hu, P.-F. *et al*, *CA*, 1959, **53**, 3120.
Bolotova, G.L. *et al*, *CA*, 1965, **63**, 6897.

(Dibromomethyl)phosphonic acid, 9CI D-00064

[71778-99-9]

$$Br_2CHP(O)(OH)_2$$

$CH_3Br_2O_3P$ M 253.815
Di-Me ester: [71918-67-7]. *Dimethyl (dibromomethyl)-phosphonate.*
$C_3H_7Br_2O_3P$ M 281.868
Liq. $Bp_{0.02}$ 70°.
Di-Et ester: [58898-18-3]. *Diethyl (dibromomethyl)-phosphonate.*
$C_5H_{11}Br_2O_3P$ M 309.922
Undist. liq.
Difluoride:
$CHBr_2F_2OP$ M 257.797
Liq. $Bp_{0.5}$ 44°.
Dibromide:
$CHBr_4OP$ M 379.608
Liq. $Bp_{0.1}$ 96°.
Bis(dimethylamide): P-*(Dibromomethyl)*-N,N,N',N'-*tetramethylphosphonic diamide.*
$C_5H_{13}Br_2N_2OP$ M 307.952
Cryst. ($CHCl_3$/pentane). Mp 116°.

Savignac, P. *et al*, *Synthesis*, 1976, 197 (*ester, use*)
Elkaim, J.C. *et al*, *Synth. React. Inorg. Met.-Org. Chem.*, 1979, **9**, 479 (*synth, ir, P nmr, pmr*)

(Dibromomethyl)-triphenylphosphonium(1+), 9CI D-00065

[56506-90-2]

$$Ph_3P^{\oplus}CHBr_2$$

$C_{19}H_{16}Br_2P^{\oplus}$ M 435.117 (ion)
Bromide:
$C_{19}H_{16}Br_3P$ M 515.021
Cryst. (MeCN). Mp 144-147°.
Ylide: [42867-45-8]. *(Dibromomethylene)-triphenylphosphorane.*
$C_{19}H_{15}Br_2P$ M 434.109
Reactive Wittig reagent, prepd. *in situ*. Used in prepn. of 1,1-dibromo-1-alkenes, and hence of terminal acetylenes.

Corey, E.J. *et al*, *Tetrahedron Lett.*, 1972, 3769 (*ylide, use*)
Tronchet, J.M.L. *et al*, *Helv. Chim. Acta*, 1974, **57**, 1505 (*ylide, use*)
Wolkoff, P., *Can. J. Chem.*, 1975, **53**, 1333 (*synth*)
Smithers, R.H., *J. Org. Chem.*, 1978, **43**, 2833; 1980, **45**, 173 (*use*)

(2,5-Dibromophenyl)phosphonic acid, 9CI D-00066

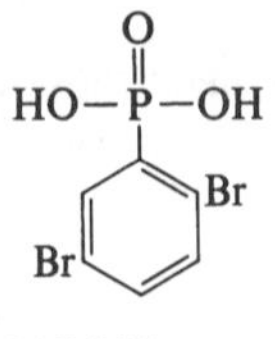

$C_6H_5Br_2O_3P$ M 315.885

Solid. Mp 204-208°.

Di-Et ester: Diethyl (2,5-dibromophenyl)phosphonate.
$C_{10}H_{13}Br_2O_3P$ M 371.993
Liq. $Bp_{0.3}$ 134-139°. n_D^{25} 1.5475.

Dichloride:
$C_6H_3Br_2Cl_2OP$ M 352.777
Solid. Mp 107-114°.

Denham, J.M. *et al*, *J. Org. Chem.*, 1958, **23**, 1298 (*synth, derivs*)

(2,4-Dibromophenyl)phosphoramidic acid, 9CI, 8CI D-00067

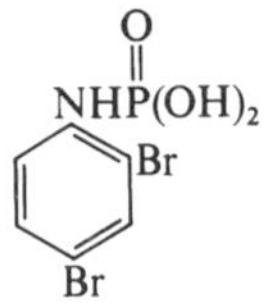

$C_6H_6Br_2NO_3P$ M 330.900

Di-Et ester: Diethyl (2,4-dibromophenyl)-phosphoramidate.
$C_{10}H_{14}Br_2NO_3P$ M 387.007
Oil.

Di-Ph ester: Diphenyl (2,4-dibromophenyl)-phosphoramidate.
$C_{18}H_{14}Br_2NO_3P$ M 483.095
Cryst. (EtOH). Mp 86-88°.

Dichloride:
$C_6H_4Br_2Cl_2NOP$ M 367.791
Solid. Mp 134°.

Dianilide: N-*(2,4-Dibromophenyl)*-N′,N″*-diphenyl-phosphoric triamide.*
$C_{18}H_{16}Br_2N_3OP$ M 481.126
Cryst. (EtOH or AcOH). Mp 228°.

Michaelis, A., *Justus Liebigs Ann. Chem.*, 1903, **326**, 129 (*derivs*)
Zhmurova, I.N. *et al*, *Zh. Obshch. Khim.*, 1961, **31**, 3741 (*Engl. transl.* p. 3495) (*diphenyl ester*)

(1,2-Dibromo-1-phenylpropyl)phosphonic acid, 9CI D-00068

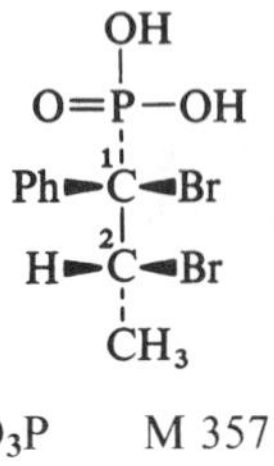

(1*RS*,2*RS*)-*form*

$C_9H_{11}Br_2O_3P$ M 357.966

Monomethyl esters are sources of the monomeric metaphosphate ion intermediate, obtained by the action of base.

(1RS,2RS)-form [67106-81-4]
(±)-*threo-form*
Solid. Mp 180.5-181.5°.

Mono-Me ester: [67106-82-5]. *Methyl hydrogen (1,2-dibromo-1-phenylpropyl)phosphonate.*
$C_{10}H_{13}Br_2O_3P$ M 371.993
Cryst. (C_6H_6/hexane). Mp 148-149.5°.

Di-Me ester: [67106-85-8]. *Dimethyl (1,2-dibromo-1-phenylpropyl)phosphonate.*
$C_{11}H_{15}Br_2O_3P$ M 386.019
No data available.

(1RS,2SR)-form [67106-84-7]
(±)-*erythro-form*
Cryst. (MeCN). Mp 186-187°.

Mono-Me ester: [67106-83-6]. Cryst. ($CHCl_3$/hexane). Mp 156-159°.

Di-Me ester: [67106-86-9]. Oil.

Satterthwait, A.C. *et al*, *J. Am. Chem. Soc.*, 1978, **100**, 3197 (*derivs, synth, props, use*)
Ramirez, F. *et al*, *J. Am. Chem. Soc.*, 1982, **104**, 1345 (*synth, use*)
Calvo, K.C. *et al*, *J. Am. Chem. Soc.*, 1983, **105**, 2827 (*use*)

Dibromotrimethylphosphorane D-00069

Bromotrimethylphosphonium bromide. Trimethylphosphine dibromide
[81492-17-3]

Me_3PBr_2 or $Me_3PBr^{\oplus}$ $Br^{\ominus}$

$C_3H_9Br_2P$ M 235.886

Shown to possess the ionic struct. Converts ROH into RBr. Yellow cryst. Mp 277-278°.

Goubeau, J. *et al*, *Z. Electrochem.*, 1960, **64**, 598 (*synth, props, ir, raman*)
Karsch, H.H., *Phosphorus Sulfur*, 1982, **12**, 217 (*props*)

Dibromotriphenoxyphosphorane, 9CI D-00070

Triphenyl phosphite dibromide
[39943-76-5]

$(PhO)_3PBr_2$ or $(PhO)_3PBr^{\oplus}(PhO)_3PBr_3^{\ominus}$

$C_{18}H_{15}Br_2O_3P$ M 470.096

Probably has ionic struct. Reagent for conversion of alkynic and allenic alcohols to bromides. Air-sensitive pale-yellow solid, becoming liq. on standing.

Br_2 adduct: Mp 105-106°.

Harris, G.S. *et al*, *J. Chem. Soc.*, 1956, 3038 (*synth*)
Rydon, H.N. *et al*, *J. Chem. Soc.*, 1956, 3043 (*synth*)
Kinney, R.S. *et al*, *J. Am. Chem. Soc.*, 1978, **100**, 7902 (*use*)
Tseng, C.K., *J. Org. Chem.*, 1979, **44**, 2793 (*synth, nmr*)
Fieser, M. *et al*, *Reagents for Organic Synthesis*, Wiley, 1967-84, **2**, 446.

Dibromotriphenylphosphorane, 9CI, 8CI D-00071

Triphenylphosphine dibromide. Bromotriphenylphosphonium bromide
[1034-39-5]

Ph_3PBr_2

$C_{18}H_{15}Br_2P$ M 422.098

Ionises in soln. to $Ph_3PBr^{\oplus}Br^{\ominus}$. Brominating agent, used to cleave ethers, esters and lactones to give alkyl bromides or acyl bromides. Converts phenols into acyl bromides, and deoxygenates epoxides. Brominates $\alpha\beta$-unsatd. ketones at the β-posn. Large hygroscopic cryst. Mp 260-270°. Rapidly dec. in air/moisture to give a mixed salt Mp 140-141°.

Br_2 adduct: Orange cryst. Mp 117°. Ionic.
I_2 adduct: Red needles. Mp 79°.

Beveridge, A.D. *et al*, *J. Chem. Soc.* (*A*), 1966, 520 (*synth, struct*)
Anderson, A.G. *et al*, *J. Org. Chem.*, 1972, **37**, 626 (*use*)
De Wit, J. *et al*, *Recl. Trav. Chim. Pays-Bas*, 1973, **92**, 281 (*use*)
Burton, D.J. *et al*, *J. Org. Chem.*, 1975, **40**, 3026 (*use*)
Briggs, E.M. *et al*, *Synthesis*, 1980, 295 (*use*)
Piers, E. *et al*, *Can. J. Chem.*, 1982, **60**, 210 (*use*)
Org. Synth., Coll. Vol. 5, 142, 249 (*use*)
Fieser, M. *et al*, *Reagents for Organic Synthesis*, Wiley, 1967-84, **6**, 645.

Di-3-butenylphenylphosphine, 9CI D-00072

[26681-85-6]

$PhP(CH_2CH_2CH{=}CH_2)_2$

$C_{14}H_{19}P$ M 218.278

Ligand for Ir, Rh and Ru. Liq. Bp_2 120-122°.

Clark, P.W. *et al*, *Can. J. Chem.*, 1974, **52**, 1714 (*synth, pmr, nmr*)
Clark, P.W. *et al*, *Inorg. Chem.*, 1979, **18**, 2067 (*cmr, complexes*)

(Dibutoxyphosphinyl)acetic acid, 9CI D-00073

$(H_3CCH_2CH_2CH_2O)_2P(O)CH_2COOH$

$C_{10}H_{21}O_5P$ M 252.247

Oil. pK_{a1} 3.60 (H_2O, 25°), 4.92 (50% EtOH apK_{a2} *q., 25°).

Et ester: Ethyl (dibutoxyphosphinyl)acetate.
$C_{12}H_{25}O_5P$ M 280.300
Liq. d 1.06. Bp_{10} 176.5-177°. n_D 1.4335.

Butyl ester: Tributyl phosphonoacetate.
$C_{14}H_{29}O_5P$ M 308.354
Liq. d_4^{20} 1.02. Bp_8 182-183°. n_D^{20} 1.4383.

Benzyl ester: Benzyl (dibutoxyphosphinyl)acetate.
$C_{17}H_{27}O_5P$ M 342.371
$Bp_{0.015}$ 156-165°. n_D^{22} 1.4850.

Amide: (Dibutoxyphosphinyl)acetamide.
$C_{10}H_{22}NO_4P$ M 251.262
Solid. Mp 94-96°.

N,N-*Diethylamide: see (2-Diethylamino-2-oxoethyl)-phosphonic acid, D-00268*

Kamai, G. *et al*, *Dokl. Akad. Nauk SSSR, Ser. Sci. Khim.*, 1950, **72**, 301 (*ethyl ester*)
Bodnarchuk, N.D. *et al*, *Zh. Obshch. Khim.*, 1969, **39**, 1707 (*Engl. transl.* p. 1673) (*amide*)
Malevannaya, R.A. *et al*, *Zh. Obshch. Khim.*, 1971, **41**, 1426 (*Engl. transl.* p. 1426) (*butyl ester*)
Tsvetkov, E.N. *et al*, *Zh. Obshch. Khim.*, 1974, **44**, 1225 (*Engl. transl.* p. 1203) (*props, pmr*)
Colvin, E.W. *et al*, *Tetrahedron Lett.*, 1982, **23**, 3835 (*synth, benzyl ester, ir, pmr*)

1,3-Di-*tert*-butyl-2,4-bis(dimethylamino)-1,3,2,4-diazadiphosphetidine D-00074

1,3-Bis(1,1-dimethylethyl)-N,N,N′,N′-*tetramethyl-1,3,2,4-diazadiphosphetidine-2,4-diamine, 10CI.* P,P′-*Bis(dimethylamino)*-N,N′-*di*-tert-*butylcyclodiphosphazane*

[62757-64-6]

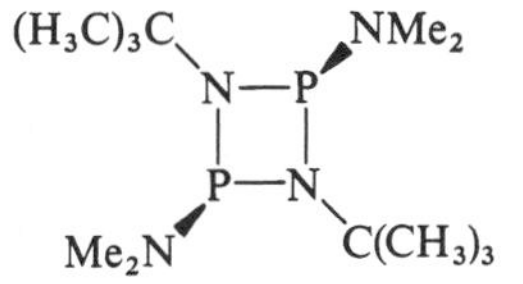

cis-form

$C_{12}H_{30}N_4P_2$ M 292.343

Cis-form [65268-60-2]
Not obtained in pure form.

2,4-Dioxide: [65789-46-0].
$C_{12}H_{30}N_4O_2P_2$ M 324.342
Air-stable needles. Mp 194-196°.

2-Sulfide: [65753-97-1].
$C_{12}H_{30}N_4P_2S$ M 324.403
Needles ($CHCl_3$/pentane). Mp 73-75°.

2,4-Disulfide: [65388-38-7].
$C_{12}H_{30}N_4P_2S_2$ M 356.463
Octahedral cryst. Mp 214°.

2-Selenide: [65213-17-4].
$C_{12}H_{30}N_4P_2Se$ M 371.303
Needles. Mp 102-104°.

2,4-Diselenide: [65753-96-0].
$C_{12}H_{30}N_4P_2Se_2$ M 450.263
Cryst. (C_6H_6). Mp 253-255°.

Trans-form [65213-16-3]
Solid. Mp 55°.

2,4-Dioxide: [65753-94-8]. Needles. Mp 205°.
2-Sulfide: [65753-98-2]. Needles (pet. ether). Mp 127-129°.
2,4-Disulfide: [65789-47-1]. Needles. Mp 255-256°.
2-Selenide: [65268-61-3]. Needles. Mp 126-127°.
2,4-Diselenide: [65753-95-9]. Cryst. (C_6H_6). Mp 264°.
2-Telluride: [65213-18-5].
$C_{12}H_{30}N_4P_2Te$ M 419.943
Pale-yellow cryst. (CH_2Cl_2/pet. ether). Mp 120° dec.

Muir, K.W., *Acta Crystallogr., Sect. B*, 1977, **33**, 3586 (*disulfide, cryst struct*)
Keat, R. *et al*, *Org. Magn. Reson.*, 1979, **12**, 391 (*selenides, nmr*)
Keat, R. *et al*, *J. Chem. Soc., Dalton Trans.*, 1980, 928 (*derivs, pmr, nmr*)
Keat, R. *et al*, *J. Chem. Soc., Dalton Trans.*, 1980, 1858 (*pmr*)

4,4-Dibutyl-1-*tert*-butyl-1,4-dihydro-1,4-phosphastannin, 10CI D-00075

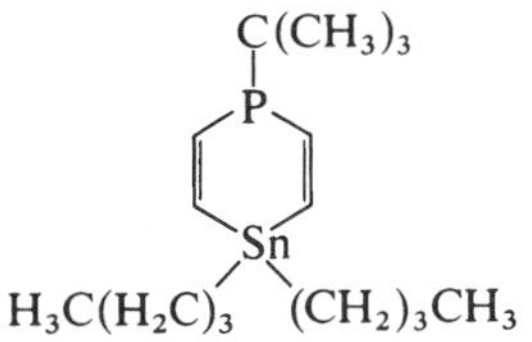

$C_{16}H_{31}PSn$ M 373.085

Liq. $Bp_{0.0006}$ 120-125°.

1-Oxide:
$C_{16}H_{31}OPSn$ M 389.084
Solid. Mp 34-38°. $Bp_{0.0001}$ 180°.

Märkl, G. *et al*, *Tetrahedron Lett.*, 1976, 2599 (*synth, ir, ms, pmr, oxide*)

Märkl, G. *et al*, *Synthesis*, 1977, 842 (*synth*)

Dibutyl butylphosphonate, 9CI, 8CI D-00076

[78-46-6]

$H_3CCH_2CH_2CH_2P(O)(OCH_2CH_2CH_2CH_3)_2$

$C_{12}H_{27}O_3P$ M 250.317

Insect repellant. Extractant for actinides from radioactive wastes; can be used to separate Am(III) from Eu(III). Liq. with pleasant odour. Bp_{20} 160-162°. n_D^{25} 1.4302.

▷SZ7000000.

Kosolapoff, G., *J. Am. Chem. Soc.*, 1945, **67**, 1180 (*synth*)
Doto, P.C. *et al*, *ACS Symp. Ser.*, 1981, **161**, 109 (*use, rev*)
Horwitz, E.P. *et al*, *Sep. Sci. Technol.*, 1981, **16**, 403, 417 (*props, use*)
Kalina, D.G. *et al*, *Sep. Sci. Technol.*, 1981, **16**, 1127 (*props, use*)
Muscatello, A.C. *et al*, *Sep. Sci. Technol.*, 1982, **17**, 859 (*props, use*)

***N,P*-Di-*tert*-butylcarboimidophosphene** D-00077

N,P-*Bis(1,1-dimethylethyl)carboimidophosphene*

$(H_3C)_3CP{=}C{=}NC(CH_3)_3$

$C_9H_{18}NP$ M 171.222

Liq. Bp_{10} 65-66°. n_D^{20} 1.4900.

Kolodiazhnyi, O.I., *Tetrahedron Lett.*, 1982, **23**, 4933 (*synth, ir, ms, pmr, cmr, nmr*)

3,5-Di-*tert*-butyl-2-chloro-2,3-dihydro-1,3,2-oxazaphosphole, 9CI D-00078

[71456-85-4]

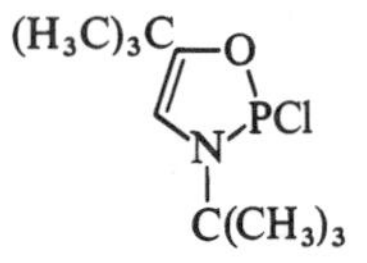

$C_{10}H_{19}ClNOP$ M 235.693

Solid. Mp 63-64°. $Bp_{0.01}$ 67-70°.

2-Oxide: [62591-42-8].

$C_{10}H_{19}ClNO_2P$ M 251.692

Solid. Mp 112-113°. $Bp_{0.05}$ 90-105°. Subl. at 120°/13 mm.

2-Sulfide: [71456-84-3].

$C_{10}H_{19}ClNOPS$ M 267.753

Solid. Mp 91-91.5°. $Bp_{0.05}$ 85-90°. Subl. at 125°/15 mm.

Balitskii, Yu.V. *et al*, *Zh. Obshch. Khim.*, 1977, **47**, 227 (*Engl. transl.* p. 210) (*oxide, ir, pmr*)
Balitskii, Yu.V. *et al*, *Zh. Obshch. Khim.*, 1980, **50**, 291 (*Engl. transl.* p. 231) (*synth*)
Balitskii, Yu.V. *et al*, *Zh. Obshch. Khim.*, 1980, **50**, 2195 (*Engl. transl.* p. 1767) (*derivs, ir, nmr*)

2,4-Di-*tert*-butyl-3-chloro-1,2,4,3-thiadiazaphosphetidine D-00079

3-Chloro-2,4-bis(1,1-dimethylethyl)-1,2,4,3-thiadiazaphosphetidine, 10CI

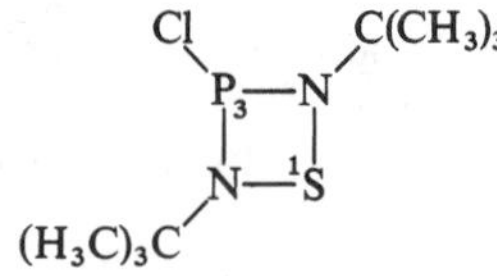

$C_8H_{18}ClN_2PS$ M 240.730

1,1-Dioxide: [76037-02-0].

$C_8H_{18}ClN_2O_2PS$ M 272.729

Cryst. (CH_2Cl_2/pentane). Mp 61-64°. $Bp_{0.5}$ 86-91°.

1-Oxide, 3-Sulfide: [67353-93-9].

$C_8H_{18}ClN_2OPS_2$ M 288.790

Oil. Mp 42-51°. $Bp_{0.01}$ 84-86°. A mixture of stereoisomers.

Scherer, O.J. *et al*, *Z. Naturforsch., B*, 1978, **33**, 652 (*deriv, synth, pmr, P nmr*)
Cowley, A.H. *et al*, *Inorg. Chem.*, 1981, **20**, 712 (*synth, pmr, cmr, P nmr*)

Dibutyl chlorothiophosphonate D-00080

Thiohypochlorous acid anhydrosulfide with dibutyl phosphorothioate, 9CI. S-*Chloro* O,O-*dibutyl phosphorothioate*

[58293-33-7]

$(H_3CCH_2CH_2CH_2O)_2P(O)SCl$

$C_8H_{18}ClO_3PS$ M 260.715

Yellow liq. d_4^{20} 1.14. Bp_2 95°, $Bp_{0.8}$ 71°. n_D^{20} 1.4677.

▷Asphyxiating vapour

Lenard-Borecka, B. *et al*, *Rocz. Chem.*, 1957, **31**, 1167 (*synth*)
Michalski, J. *et al*, *Rocz. Chem.*, 1963, **37**, 1479 (*synth*)
Mel'nik, Ya.I. *et al*, *Zh. Obshch. Khim.*, 1978, **48**, 326 (*Engl. transl.* p. 292) (*synth*)

1,3-Dibutyl-2,3-dihydro-4,5-dimethyl-1*H*-1,3,2-diazaphosphole 2-oxide, 9CI D-00081

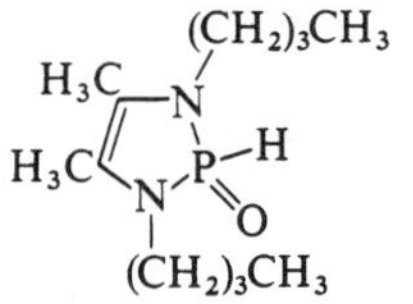

$C_{12}H_{25}N_2OP$ M 244.316

2-Chloro: [75135-40-9].

$C_{12}H_{24}ClN_2OP$ M 278.761

Liq. d_4^{20} 1.10. $Bp_{0.06}$ 111°. n_D^{20} 1.5032.

2-Ethoxy: [75135-39-6].

$C_{14}H_{29}N_2O_2P$ M 288.369

Liq. d_4^{20} 1.01. $Bp_{0.1}$ 114°. n_D^{20} 1.4810.

Kibardin, M.A. *et al*, *Izv. Akad. Nauk SSSR, Ser. Khim.*, 1980, 1452; *CA*, **93**, 168203 (*synth, ir, pmr, P nmr*)

3,5-Di-*tert*-butyl-2,5-dihydro-1,2-oxaphosphole 2-oxide D-00082

*2,5-Dihydro-3,5-bis(1,1-dimethylethyl)-1,2-oxaphosphole 2-oxide, 9CI. 3,5-Di-*tert*-butyl-1,2-oxaphosphol-3-ene, 8CI*

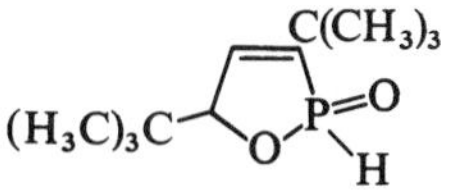

$C_{11}H_{21}O_2P$ M 216.259

2-Chloro: [42087-74-1].
$C_{11}H_{20}ClO_2P$ M 250.705
Cryst. ($CHCl_3$/heptane). Mp 147-148°. $Bp_{0.35}$ 83° subl.

2-Bromo: [30338-49-9].
$C_{11}H_{20}BrO_2P$ M 295.156
Solid. Mp 133-133.5°. Stable indefinitely when pure and dry.

2-Hydroxy: [42087-75-2].
$C_{11}H_{21}O_3P$ M 232.259
Solid. Mp 162-164°.

Macomber, R.S. *et al*, *J. Org. Chem.*, 1971, **36**, 2713 (*bromo, methoxy, synth*)
Elder, R.C. *et al*, *J. Org. Chem.*, 1973, **38**, 4177 (*hydroxy, chloro, synth, ms, ir, pmr, nmr, cryst struct*)
Macomber, R.S. *et al*, *J. Org. Chem.*, 1983, **48**, 1425 (*methoxy, pmr*)

3,5-Di-*tert*-butyl-2,3-dihydro-1,3,2-oxazaphosphole D-00083

*3,5-Bis-(1,1-dimethylethyl)-2,3-dihydro-1,3,2-oxaphosphole, 9CI. 3,5-Di-*tert*-butyl-Δ⁴-1,3,2-oxazaphospholine, 8CI*

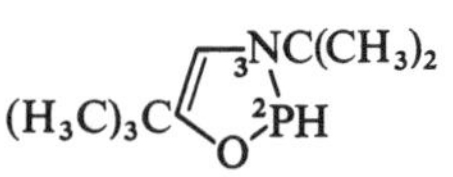

$C_6H_{12}NOP$ M 145.141

2-Methoxy: [70035-95-9].
$C_{11}H_{22}NO_2P$ M 231.274
Cryst. (pet. ether). Mp 39.5-40.5°. $Bp_{0.5}$ 50°.

2-Ethoxy: [73992-81-1].
$C_{12}H_{24}NO_2P$ M 245.301
Liq. $Bp_{0.01}$ 40°. n_D^{20} 1.4648.

2-Ethoxy, 2-oxide: [73992-82-2].
$C_{12}H_{24}NO_3P$ M 261.300
Solid. Mp 70-71°. $Bp_{0.01}$ 93-95°.

Balitskii, Yu.V. *et al*, *Zh. Obshch. Khim.*, 1979, **49**, 42 (*Engl. transl.* p. 34) (*methoxy, synth, ir, pmr, P nmr*)
Boldeskul, I.E. *et al*, *Zh. Prikl. Spektrosk.*, 1979, **31**, (*Engl. transl.* p. 887) (*ir, raman*)
Balitskii, Yu.V. *et al*, *Zh. Obshch. Khim.*, 1980, **50**, 291 (*Engl. transl.* p. 231) (*synth, ir, raman, P nmr, pmr*)

2,6-Di-*tert*-butyl-1,4-dihydro-4-phenyl-1,4-azaphosphorine D-00084

2,6-Bis(1,1-dimethylethyl)-1,4-dihydro-4-phenyl-1,4-azaphosphorine

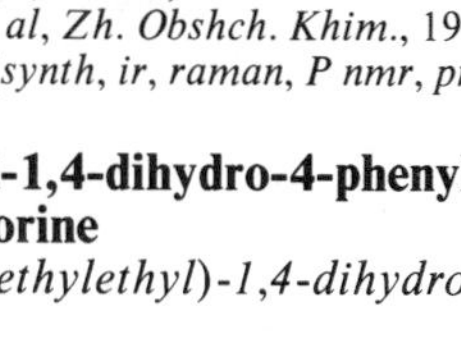

$C_{18}H_{26}NP$ M 287.384

4-Oxide: [34735-03-0].
$C_{18}H_{26}NOP$ M 303.383
Cryst. ($CHCl_3$/cyclohexane or MeOH). Mp 277-278°.

B,PhBr: *2,6-Di-*tert*-butyl-1,4-dihydro-4,4-diphenyl-1,4-azaphosphorinium bromide.*
$C_{24}H_{31}BrNP$ M 444.393
Solid. Mp 263°.

1-Me, 4-oxide:
$C_{19}H_{28}NOP$ M 317.410
Hygroscopic solid. Sol. H_2O, EtOH, C_6H_6. Mp 111-112°.

Williams, J.C. *et al*, *Tetrahedron Lett.*, 1971, 4749 (*oxide, synth, uv, pmr, P nmr*)
Hungerford, L. *et al*, *J. Heterocycl. Chem.*, 1972, **9**, 347 (*oxide, cryst struct*)
Skolimowski, J. *et al*, *Synthesis*, 1979, 109 (*oxide, synth, ir, pmr*)
Yatsimirskii, K.B. *et al*, *Zh. Neorg. Khim.*, 1980, **25**, 2213 (*Engl. transl.* p. 1227) (*deriv, synth, complexes*)
Skolimowski, J. *et al*, *Phosphorus Sulfur*, 1984, **19**, 159 (*derivs, pmr, nmr*)

1,3-Di-*tert*-butyl-2,4-dimethoxy-1,3,2,4-diazadiphosphetidine D-00085

*1,3-Bis(1,1-dimethylethyl)-2,4-dimethoxy-1,3,2,4-diazadiphosphetidine, 9CI. N,N′-Di-*tert*-butyl-P,P′-dimethoxycyclodiphosphazane*

[64899-38-3]

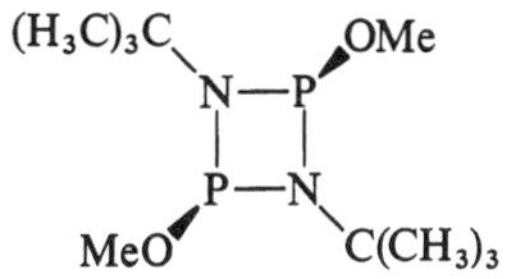

cis-form

$C_{10}H_{24}N_2O_2P_2$ M 266.259
Liq. $Bp_{0.5}$ 66-68°.

2,4-Disulfide: Cryst. (C_6H_6). Mp 124°. Config. of this preparation unclear.

cis-form [67902-07-2]
$Bp_{0.1}$ 66°.

2-Sulfide: [67902-10-7].
$C_{10}H_{24}N_2O_2P_2S$ M 298.319
Liq. $Bp_{0.02}$ 54°.

2-Selenide: [67902-12-9].
$C_{10}H_{24}N_2O_2P_2Se$ M 345.219
Structural assignment uncertain: either $Bp_{0.07}$ 80° *or* $Bp_{0.02}$ 72-6°, Mp 34-6°.

2,4-Disulfide: [71802-06-7].
$C_{10}H_{24}N_2O_2P_2S_2$ M 330.379
Cryst. (CH_2Cl_2/pet. ether). Mp 159-160°.

2,4-Diselenide: [71783-14-7].
$C_{10}H_{24}N_2O_2P_2Se_2$ M 424.179
Solid. Mp 141°.

trans-form [67902-08-3]
Mp 56-60°.

2-Sulfide: [71783-13-6]. Liq. $Bp_{0.03}$ 60°.
2-Selenide: [71783-16-9]. Structural assignment uncertain: either $Bp_{0.02}$ 72-6°, Mp 34-6° *or* $Bp_{0.07}$ 80°.
2,4-Disulfide: [71783-12-5]. Solid. Mp 152-154°.
2,4-Diselenide: [71783-15-8]. Solid. Mp 152-155°.

Colquhoun, I.J. *et al*, *J. Chem. Soc., Dalton Trans.*, 1977, 1674 (*synth, pmr, P nmr*)
Keat, R. *et al*, *J. Chem. Soc., Dalton Trans.*, 1978, 634 (*ir, cmr, pmr, P nmr*)
Bulloch, G. *et al*, *Org. Magn. Reson.*, 1979, **12**, 708 (*pmr, P nmr*)
Keat, R. *et al*, *Org. Magn. Reson.*, 1979, **12**, 391 (*selenides, P nmr*)
Keat, R. *et al*, *J. Chem. Soc., Dalton Trans.*, 1979, 1224 (*synth, derivs, pmr, P nmr*)

1,3-Di-*tert*-butyl-2,4-dimethyl-1,3,2,4-diazadiphosphetidine **D-00086**

1,3-Bis(1,1-dimethylethyl)-2,4-dimethyl-1,3,2,4-diazadiphosphetidine, 9CI. N,N′-*Di-*tert-*butyl-*P,P′-*dimethylcyclodiphosphazane*

[61152-26-9]

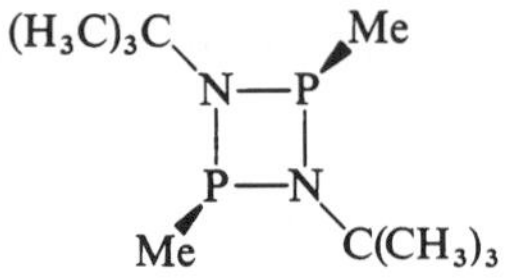

cis-form

$C_{10}H_{24}N_2P_2$ M 234.261

cis-form forms Pt-complexes. Low melting solid. Mp 34-36°. $Bp_{0.5}$ 31-32°.

2,4-Dioxide: [61152-30-5].
$C_{10}H_{24}N_2O_2P$ M 235.286
Cryst. (C_6H_6/pentane). Mp 150-152°. $Bp_{0.001}$ 90-100° subl. 2 Stereoisomers have been characterized, Mps 160° and 230-232°, configs. uncertain.

2,4-Disulfide: [61152-31-6].
$C_{10}H_{24}N_2P_2S_2$ M 298.381
Forms complexes with stannic halides. Cryst. (C_6H_6). Mp 206-208°. $Bp_{0.001}$ 120-145° subl.

2,4-Diselenide:
$C_{10}H_{24}N_2P_2Se_2$ M 392.181
Solid. Mp >180°. $Bp_{0.001}$ 90-95° subl. A mixture of stereoisomers.

2,4-Ditelluride: [61152-33-8].
$C_{10}H_{24}N_2P_2Te_2$ M 489.461
Yellow cryst. Mp ca 60°.

Scherer, O.J. *et al*, *Chem. Ber.*, 1974, **107**, 552 (*disulfide, synth, pmr*)
Scherer, O.J. *et al*, *Chem. Ber.*, 1976, **109**, 2996 (*synth, derivs, pmr, ms, P nmr*)
Keat, R. *et al*, *J. Chem. Soc., Dalton Trans.*, 1979, 1224 (*dioxides, synth, pmr, P nmr*)
Scherer, O.J. *et al*, *Z. Naturforsch, B*, 1982, **37**, 1041 (*complexes*)

1,3-Di-*tert*-butyl-4,4-dimethyl-1,3,2-diazaphosphasiletidine **D-00087**

4,4-Dimethyl-1,3-bis(1,1-dimethylethyl)-1,3-diaza-2-phospha-4-silacyclobutane, 10CI, 9CI

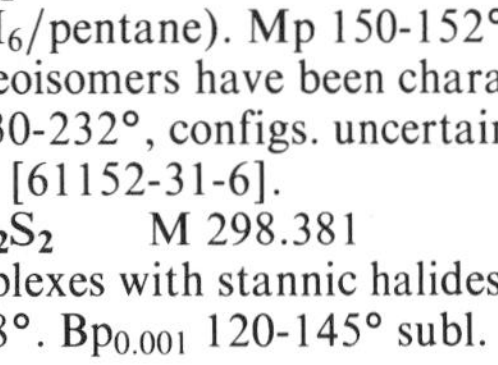

$C_{10}H_{25}N_2PSi$ M 232.380

2-Chloro: [77382-31-1].
$C_{10}H_{24}ClN_2PSi$ M 266.825
Solid. Mp 34-36°. $Bp_{0.001}$ 43-47°.

2-Me: [85414-46-6]. *1,3-Di-*tert-*butyl-2,4,4-trimethyl-1,3-diaza-2-phospha-4-silacyclobutane.*
$C_{11}H_{27}N_2PSi$ M 246.407
Liq. $Bp_{0.1}$ 25-26°.

2-Dimethylamino: [82881-16-1]. N,N,*4,4-Tetramethyl-1,3-bis(1,1-dimethylethyl)-1,3-diaza-2-phosphacyclobutane-2-amine, 9CI.*
$C_{12}H_{30}N_3PSi$ M 275.448
Cryst. (MeCN at low temp.). Mp 26-27°. $Bp_{0.001}$ 20° subl.

2-Dimethylamino, 2-sulfide: [82881-24-1].
$C_{12}H_{30}N_3PSSi$ M 307.508
Solid. Mp 85-87°.

*2-(Di-*tert-*butyl)amino:* [82881-21-8]. N,N,*1,3-Tetrakis(1,1-dimethylethyl)-4,4-dimethyl-1,3-diaza-2-phosphacyclobutane-2-amine, 9CI.*
$C_{18}H_{42}N_3PSi$ M 359.609
Cryst. (MeCN at low temp.). Mp 79-81°. $Bp_{0.0001}$ 70-85°.

*2-(Di-*tert-*butyl)amino, 2-sulfide:* [82881-29-6].
$C_{18}H_{42}N_3PSSi$ M 391.669
Solid. Mp 165-166°.

Scherer, O.J. *et al*, *Chem. Ber.*, 1982, **115**, 2076 (*derivs, synth, P nmr, pmr*)
Scherer, O.J. *et al*, *J. Organomet. Chem.*, 1983, **243**, C33 (*complexes*)

4,8-Di-*tert*-butyl-2,10-dimethyl-12*H*-dibenzo[*d,g*][1,3,2]dioxaphosphocin **D-00088**

*4,8-Bis(1,1-dimethylethyl)-2,10-dimethyl-12*H*-dibenzo*[d,g]*[1,3,2]dioxaphosphocin, 9CI*

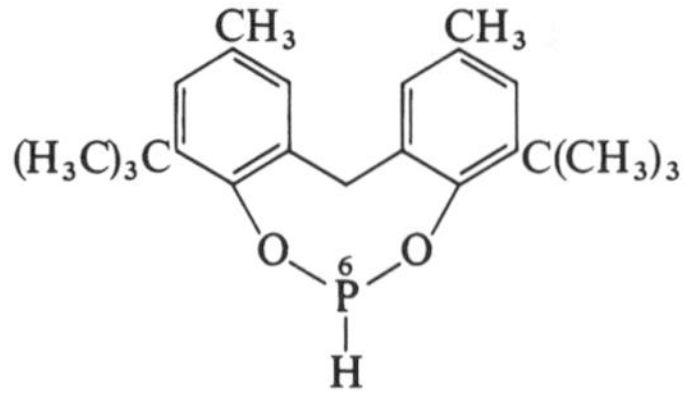

$C_{22}H_{29}O_2P$ M 356.444

Derivs. are particularly thermostable.

6-Ethoxy: [34573-97-2].
$C_{24}H_{31}O_3P$ M 398.481
Cryst. (octane). Mp 192°.

6-Phenoxy:
$C_{28}H_{33}O_3P$ M 448.541
Solid. Mp 159-160°.

6-(1-Naphthoxy):
$C_{32}H_{35}O_3P$ M 498.600
Solid. Mp 180-181°.

6-Oxide: [15233-97-3].
$C_{22}H_{29}O_3P$ M 372.443
Cryst. (EtOH). Mp 153-154°.

Kadyrova, V.Kh. *et al*, *Zh. Obshch. Khim.*, 1971, **41**, 1688 (*Engl. transl.* p. 1696) (*derivs, synth, props*)
Verizhnikov, L.V. *et al*, *Zh. Obshch. Khim.*, 1971, **41**, 2162 (*Engl. transl.* p. 2187) (*P nmr*)

1,2-Di-*tert*-butyl-1,2-dimethyldiphosphine **D-00089**

1,2-Bis(1,1-dimethylethyl)-1,2-dimethyldiphosphine, 10CI

[25196-14-9]

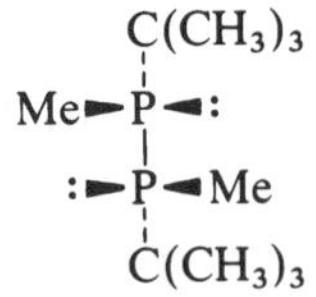

(RS,RS)-form

$C_{10}H_{24}P_2$ M 206.247

(RS,RS)-form [58621-89-9]
$Bp_{0.3}$ 52-53°.
Disulfide: [95837-80-2]. Cryst. (MeOH). Mp 118°.

(RS,SR)-form
No phys. props. reported.
Disulfide: [73719-60-5].
$C_{10}H_{24}P_2S_2$ M 270.367
Cryst. (MeOH at low temp.). Mp 141°. Monosulfides also known, characterised spectroscopically.

Scherer, O.J. *et al*, *Chem. Ber.*, 1970, **103**, 71 (*synth, pmr, nmr*)
Aime, S. *et al*, *J. Chem. Soc., Dalton Trans.*, 1976, 2144 (*pmr, cmr, nmr*)
McFarlane, H.C.E. *et al*, *J. Chem. Soc., Dalton Trans.*, 1980, 240 (*pmr, nmr*)

Alagna, L. *et al*, *Z. Naturforsch.*, *A*, 1981, **36**, 68 (*disulfides, pe*)
Haegele, G. *et al*, *Z. Naturforsch.*, *B*, 1984, **39**, 1574; *J. Chem. Soc., Dalton Trans.*, 2803 (*synth, pmr, P nmr*)

1-(Di-*tert*-butyldiphosphinylidene)-methanediamine D-00090

*1-[Bis(1,1-dimethylethyl)diphosphinylidene]-methanediamine, 11CI. 2-Diaminomethylene-1,1-di-*tert*-butyldiphosphine*

$[(H_3C)_3C]_2P{-}P{=}C(NH_2)_2$

$C_9H_{22}N_2P_2$ M 220.234

N,N,N',N'-*Tetra-Me:* [101409-46-5].
$C_{13}H_{30}N_2P_2$ M 276.341
Yellow-orange liq. $Bp_{0.03}$ 90-92°.

N,N,N',N'-*Tetra-Et:* [101409-48-7].
$C_{17}H_{38}N_2P_2$ M 332.448
Yellow-orange liq. $Bp_{0.03}$ 120-122°.

Romanenko, V.D. *et al*, *Zh. Obshch. Khim.*, 1985, **55**, 1437 (*Engl. transl.* p. 1280) (*synth, pmr, cmr, P nmr*)

1,2-Di-*tert*-butyldiphosphirane D-00091

1,2-Bis(1,1-dimethylethyl)diphosphirane, 9CI
[68969-73-7]

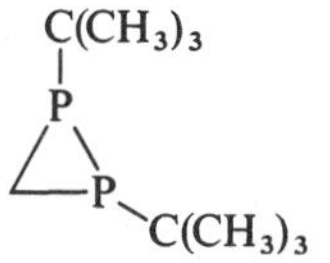

$Bp_{0.005}$ 26-28°. Dimerises readily to give 1,2,4,5-Tetra-*tert*-butyl-1,2,4,5-tetraphosphorinane, T-00027 .

Baudler, M. *et al*, *Z. Naturforsch.*, *B*, 1978, **33**, 1208 (*synth, props, cmr*)
Gleiter, R. *et al*, *Chem. Ber.*, 1981, **114**, 1004 (*pe*)

Dibutyldiphosphonic acid, 8CI D-00092

Dibutylpyrophosphonic acid. Butylphosphonic acid monoanhydride

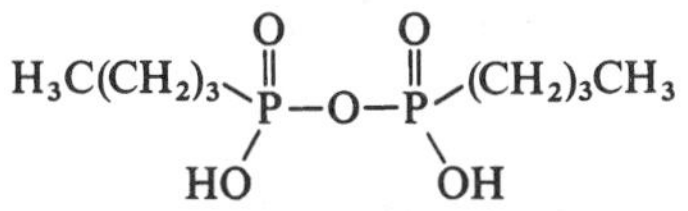

$C_8H_{20}O_5P_2$ M 258.191

Di-Et ester: Diethyl dibutyldiphosphonate.
$C_{12}H_{28}O_5P_2$ M 314.298
Liq. d_4^{20} 1.06. Bp_2 145-146°. n_D^{20} 1.4390.

Pudovik, A.N. *et al*, *Zh. Obshch. Khim.*, 1960, **30**, 2624 (*Engl. transl.* p. 2624) (*synth*)

2,10-Di-*tert*-butyl-1,3,6,9,11,14-hexathia-2,10-diphosphacyclohexadecane, 10CI D-00093

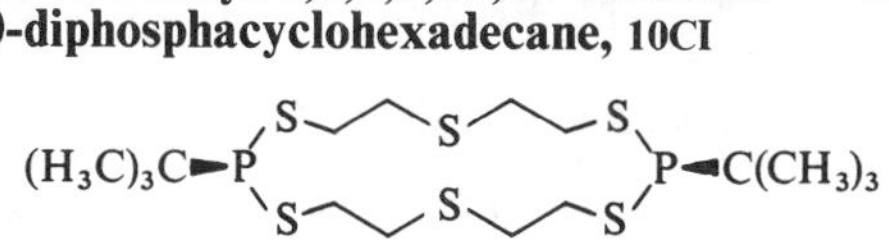

$C_{16}H_{34}P_2S_6$ M 480.752

cis-form [77075-96-8]

2,10-Dioxide: [77123-34-3].
$C_{16}H_{34}O_2P_2S_6$ M 512.751
Solid. Mp 150-152°, Mp 148-150°. Two stereoisomers, stereochemistry not assigned (see below).

2,10-Disulfide: [77075-98-0].
$C_{16}H_{34}P_2S_8$ M 544.872
Solid. Mp 147-148°.

trans-form [77121-94-9]

2,10-Dioxide: [77075-97-9]. Solid. Mp 148-150°, Mp 150-152°. Stereochemistry not assigned (see above).

2,10-Disulfide: [77121-95-0]. Solid. Mp 144-145°.

Martin, J. *et al*, *Nouv. J. Chim.*, 1980, **4**, 515 (*synth, derivs, nmr, cmr, complexes*)
Grand, A. *et al*, *Acta Crystallogr., Sect. B*, 1982, **38**, 3052 (*disulfide, cryst struct*)

(3,5-Di-*tert*-butyl-4-hydroxybenzyl)-phosphonic acid, 8CI D-00094

*[[3,5-Bis(1,1-dimethylethyl)-4-hydroxyphenyl]methyl]-phosphonic acid, 9CI. 2,6-Di-*tert*-butyl-4-(phosphonomethyl)phenol*
[10175-90-3]

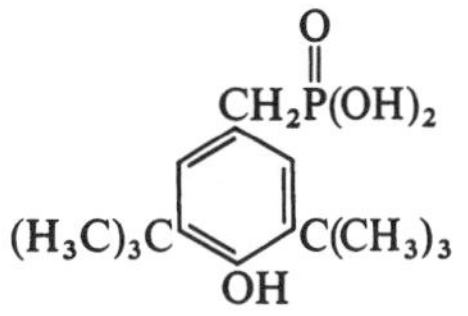

$C_{15}H_{25}O_4P$ M 300.334

Diesters and monoester salts are polymer stabilizers (trade names Irganox 1093, 1425). Cryst. + $2H_2O$. Mp 208-210°.

Di-Et ester: [976-56-7]. *Diethyl [(3,5-di-*tert*-butyl-4-hydroxyphenyl)methyl]phosphonate. Irganox 1222A.* Stabilizer for polymers. Cryst. (pet. ether). Mp 115-116°. $Bp_{0.05}$ 172-176°.

*Bis(trimethylsilyl) ester: Bis(trimethylsilyl)[3,5-di-*tert*-butyl-4-hydroxyphenyl)methyl]phosphonate.*
$C_{21}H_{41}O_4PSi_2$ M 444.697
Solid. Mp 134-136°.

Tavs, P., *Chem. Ber.*, 1970, **103**, 2428 (*ester*)
Gross, H. *et al*, *J. Prakt. Chem.*, 1978, **320**, 344 (*synth, derivs, pmr*)

(Di-*tert*-butylimino)(diisopropylamino)-phosphorane D-00095

[Bis[(1,1-dimethylethyl)imino]][bis(1-methylethyl)-amino]phosphorane

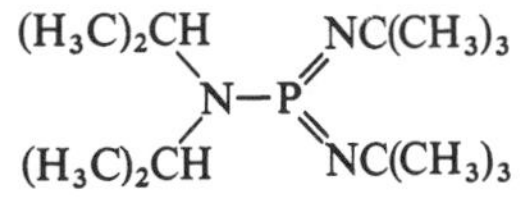

$C_{14}H_{32}N_3P$ M 273.401

Liq. $Bp_{0.1}$ 76°.

Niecke, E. *et al*, *Chem. Ber.*, 1982, **115**, 185 (*synth, cmr, P nmr, pmr, props*)

2,3-Di-*tert*-butyl-1-isopropylphosphadiarsirane D-00096

2,3-Bis(1,1-dimethylethyl)-1-(1-methylethyl)-phosphadiarsirane, 9CI

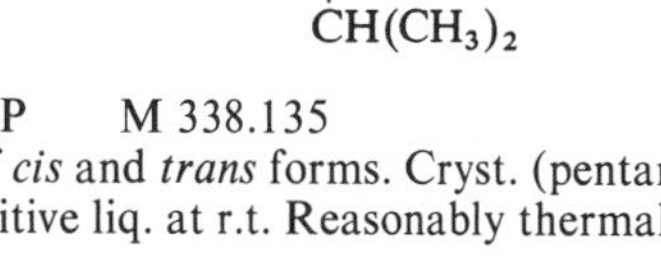

$C_{11}H_{25}As_2P$ M 338.135

Mixture of *cis* and *trans* forms. Cryst. (pentane at −78°). Air-sensitive liq. at r.t. Reasonably thermally stable.

Baudler, M. *et al*, *Angew. Chem.*, *Int. Ed. Engl.*, 1979, **18**, 877 (*synth, P nmr*)

3,4-Di-*tert*-butyl-2-(4-methylphenyl)-1,3,4,2-thiadiazaphospholidin-5-one 2-sulfide D-00097

*3,4-Bis(1,1-dimethylethyl)-2-(4-methylphenyl)-1,3,4,2-thiadiazaphospholidin-5-one 2-sulfide. 3,4-Di-*tert*-butyl-2-*p*-tolyl-1,3,4,2-thiadiazaphospholidin-5-one 2-sulfide*

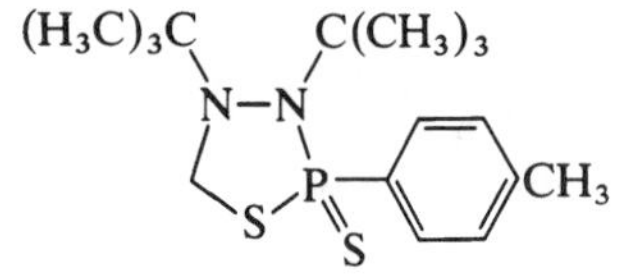

$C_{16}H_{25}N_2O_2PS_2$ M 372.479

(2α,3β,4α)-form [73678-99-6]

Cryst. ($CHCl_3$/hexane). Mp 132-133°.

(2α,3α,4β)-form

Solid. Mp 123-124°. When heated at 70° gives the higher Mp stereoisomer.

L'abbé, G. *et al*, *Bull. Soc. Chim. Belg.*, 1979, **88**, 737 (*synth, ir, pmr, cmr, ms, cryst struct*)

Dibutylmethylphosphine, 9CI D-00098

[33374-48-0]

$MeP(CH_2CH_2CH_2CH_3)_2$

$C_9H_{21}P$ M 160.239

Liq. Bp_{12} 75-79°.

Oxide: [14062-37-4].

$C_9H_{21}OP$ M 176.238

Solid. Mp ca. 35°. Bp_2 181°. Forms tetradeca- and octadecahydrates.

Sulfide: [55482-36-5].

$C_9H_{21}PS$ M 192.299

Solid. Mp 22.5-24°. $Bp_{0.3}$ 85-88°.

Cook, A.G. *et al*, *J. Org. Chem.*, 1972, **37**, 3342 (*oxide*)
Quin, L.D. *et al*, *Org. Magn. Reson.*, 1974, **6**, 503 (*synth, derivs, cmr, nmr*)

Di-*tert*-butylmethylphosphine D-00099

Methylbis(1,1-dimethylethyl)phosphine, 10CI

[6002-40-0]

$MeP[C(CH_3)_3]_2$

$C_9H_{21}P$ M 160.239

Ligand for Cr, Ir, Pd, Pt, Rh and Ru. Liq. which oxidises v. easily in air. Bp_{12} 58°. n_D^{20} 1.4707.

Oxide: [18351-81-0].

$C_9H_{21}OP$ M 176.238

Low-melting solid. Mp 35-37°. $Bp_{0.08}$ 60-62°.

Karsch, H.H. *et al*, *Z. Naturforsch., B*, 1977, **32**, 762.
Kolodyazhnyi, O.I., *Zh. Obshch. Khim.*, 1981, **51**, 2466 (*Engl. transl.* p. 2125) (*synth, nmr*)
Kolodyazhnyi, O.I., *Zh. Obshch. Khim.*, 1982, **52**, 1086 (*Engl. transl.* p. 946) (*oxide, pmr, nmr*)

4,4′-Di-*tert*-butyl-1,1′,2,2′,5,5′,6,6′-octahydro-1,1′-diphenyl-2,2′-biphosphorin D-00100

4,4′-Bis(1,1-dimethylethyl)-1,1′,2,2′,5,5′,6,6′-octahydro-1,1′-diphenyl-2,2′-biphosphinine. 4,4′-Bis(1,1-dimethylethyl)-1,1′,2,2′,5,5′,6,6′-octahydro-1,1′-diphenyl-2,2′-biphosphorin

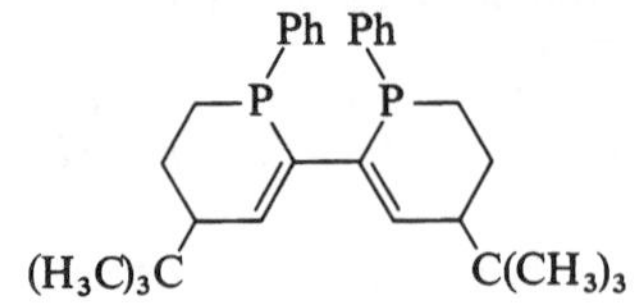

$C_{30}H_{42}P_2$ M 464.609

P,P′-Disulfide: [70641-33-7].

$C_{30}H_{42}P_2S_2$ M 528.729

One of v. few known 2,2′-diphosphorin derivs. Solid. Mp 210°.

Mathey, F. *et al*, *Can. J. Chem.*, 1978, **57**, 723 (*synth, ir, nmr*)

2,7-Di-*tert*-butyl-1-phenylphosphepin D-00101

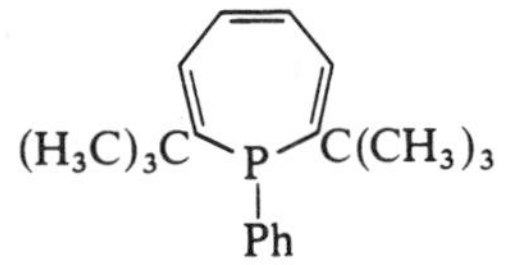

$C_{20}H_{27}P$ M 298.407

First stable monocyclic phosphepin to be synth. Pale-yellow viscous oil. $Bp_{0.01}$ 130-150° (oven). Thermally stable.

Märkl, G. *et al*, *Angew. Chem.*, *Int. Ed. Engl.*, 1984, **23**, 894 (*synth, uv, ir, ms, pmr, cmr, P nmr*)

Dibutylphenylphosphine, 9CI D-00102

[6372-44-7]

$PhP(CH_2CH_2CH_2CH_3)_2$

$C_{14}H_{23}P$ M 222.309

Ligand for metals of Groups VIB and VIII; also Mn, Hg and Ti. Foul-smelling liq. Bp_{50} 185°, $Bp_{0.2}$ 83°. n_D^{20} 1.5249.

▷Toxic. SY8400000.

B,MeI: Dibutylmethylphenylphosphonium iodide.

$C_{15}H_{26}IP$ M 364.249

Rods (H_2O). Mp 137°, Mp 168°.

Oxide: [10557-66-1].

$C_{14}H_{23}OP$ M 238.309

Solid. Mp 50-52°. $Bp_{0.7}$ 147-150°.

Sulfide:

$C_{14}H_{23}PS$ M 254.369

Solid. Mp 47°.

Selenide:

$C_{14}H_{23}PSe$ M 301.269

Solid. Mp 54°.

Davies, W.C. *et al*, *J. Chem. Soc.*, 1929, 23 (*synth, derivs*)
Schindlbauer, H., *Monatsh. Chem.*, 1963, **94**, 99 (*uv*)
Zingaro, R.A. *et al*, *Inorg. Chem.*, 1963, **2**, 192; *J. Chem. Eng. Data*, 1963, **8**, 226 (*sulfide, selenide, ir*)
Wetzel, R.B. *et al*, *J. Org. Chem.*, 1974, **39**, 1531 (*oxide, pmr*)
Kormachev, V.V. *et al*, *Zh. Obshch. Khim.*, 1975, **45**, 307 (*Engl. transl.* p. 293) (*oxide, props*)
Gray, G.A. *et al*, *J. Am. Chem. Soc.*, 1976, **98**, 2109 (*cmr, nmr*)
Inorg. Synth., 1978, **18**, 171 (*synth, props*)

Di-*tert*-butylphenylphosphine **D-00103**

Bis(1,1-dimethylethyl)phenylphosphine, 9CI
[32673-25-9]

$PhP[C(CH_3)_3]_2$

$C_{14}H_{23}P$ M 222.309
Ligand for Ir, Pd, Pt, Rh, and Ru. Liq. $Bp_{0.6}$ 97-100°.
*B,MeI: Di-*tert*-butylmethylphosphonium iodide.*
$C_{15}H_{26}IP$ M 364.249
Solid. Mp 231-233°.

Mann, B.E. *et al, J. Chem. Soc. (A)*, 1971, 1104, 2976 (*synth, nmr, pmr, deriv, complexes*)
Ratovskii, G.V. *et al, Zh. Obshch. Khim.*, 1978, **48**, 1520 (*Engl. transl.* p. 1394) (*uv*)
Ratovskii, G.V. *et al, Zh. Obshch. Khim.*, 1981, **51**, 1504 (*Engl. transl.* p. 1276) (*struct, uv, cmr*)
Panov, A.M. *et al, Zh. Obshch. Khim.*, 1985, **55**, 2243 (*Engl. transl.* p. 1991) (*ir, raman, uv, cmr*)

Dibutyl phosphate, 9CI, 8CI **D-00104**

Dibutyl hydrogen phosphate. Dibutyl phosphoric acid
[107-66-4]

$(H_3CCH_2CH_2CH_2O)_2P(O)OH$

$C_8H_{19}O_4P$ M 210.209
Metal extractant used in regeneration of irradiated nuclear fuels. Metabolite of tributyl phosphate. Oil. d_4^{20} 1.06. $Bp_{0.05}$ 135-138°. pK_a 1.75 (H_2O), pK_a 3.00 (75% EtOH aq.), pK_a 6.97 (EtOH), pK_a 10.49 ($MeNO_2$). n_D^{20} 1.4288.
▷TB9605000.

Fluoride: see Dibutyl phosphorofluoridate, D-00153
Chloride: see Dibutyl phosphorochloridate, D-00140
Anhydride: see Tetrabutyl pyrophosphate, T-00024
Amide: see Dibutyl phosphoramidate, D-00133
Anilide: see Phenylphosphoramidic acid, P-00257

Cook, H.G. *et al, J. Chem. Soc.*, 1949, 635 (*fluoride, synth, tox*)
Grosse-Ruyken, H. *et al, J. Prakt. Chem.*, 1962, **18**, 287 (*synth*)
Mitsunobu, O. *et al, J. Org. Chem.*, 1965, **30**, 1071 (*synth*)
Bliznyuk, N.K. *et al, Zh. Obshch. Khim.*, 1967, **37**, 1119 (*Engl. transl.* p. 1061) (*synth*)
Solovkin, A.S., *Radiokhimiya*, 1982, **24**, 56 (*Engl. transl.* p. 49) (*use, rev*)
Krejzler, J. *et al, J. Radioanal. Nucl. Chem.*, 1984, **85**, 57 (*use*)

Di-*tert*-butyl phosphate, 8CI **D-00105**

*Bis(1,1-dimethylethyl) phosphate, 9CI. Di-*tert*-butyl dihydrogen phosphate. Di-*tert*-butyl phosphoric acid*
[33494-81-4]

$[(H_3C)_3CO]_2P(O)OH$

$C_8H_{19}O_4P$ M 210.209
Solid. Mp 83-84°.

Tetramethylammonium salt: Cryst. (dimethoxyethane). Mp 224.5-225.5°.
Cyclohexylammonium salt: Needles (MeOH), cryst. (dimethoxyethane). Mp 190-191°.
Dicyclohexylammonium salt: Needles (cyclohexane). Mp 177-179°.
Anilinium salt: Cryst. (Me_2CO/pet. ether). Mp 110-112°.
*Chloride: see Di-*tert*-butyl phosphorochloridate, D-00141*
Bromide: see Di-*tert*-butyl phosphorobromidate, D-00138
Amide: [41222-85-9]. *Di-*tert*-butyl phosphoramidate. Bis(1,1-dimethyl) phosphoramidate. Di-*tert*-butyl amidophosphate.*
$C_8H_{20}NO_3P$ M 209.225
Cryst. (pet. ether). Mp 136-137° (122-125°).

Cox, J.R. *et al, J. Org. Chem.*, 1969, **34**, 2600 (*synth, pmr*)
Zwierzak, A. *et al, Tetrahedron*, 1971, **27**, 3163 (*synth, ir, pmr*)
Zawadzki, S. *et al, Tetrahedron*, 1973, **29**, 315 (*amide, synth, ir, pmr, P nmr*)
Mlotkowska, B. *et al, Pol. J. Chem.*, 1979, **53**, 359 (*esters, ir, pmr, P nmr*)
Badet, B. *et al, Synthesis*, 1982, 291 (*esters, ir, pmr, P nmr, cmr, ms*)

Di-*tert*-butylphosphazene **D-00106**

N-[*(1,1-Dimethylethyl)phosphinidene*]*-2-methyl-2-propanamine, 11CI*
[95552-76-4]

$(H_3C)_3CP{=}NC(CH_3)_3$

$C_8H_{18}NP$ M 159.211
Intensely yellow liq. Monomeric. v. sensitive to air and moisture. Stable at −40°.

Niecke, E. *et al, Angew. Chem., Int. Ed. Engl.*, 1981, **20**, 1034 (*synth, ms, pmr, cmr, P nmr*)
Arif, A.M. *et al, J. Am. Chem. Soc.*, 1985, **107**, 2553 (*complexes*)
Elbel, S. *et al, J. Chem. Soc., Dalton Trans.*, 1985, 879 (*ms, pe*)

Dibutylphosphine, 9CI **D-00107**

[1732-72-5]

$HP(CH_2CH_2CH_2CH_3)_2$

$C_8H_{19}P$ M 146.212
Used in prep. of catalysts for hydrogenation of CO. Bp 183-186°, Bp_{14} 70°. pK_a 4.51 (H_2O). n_D^{20} 1.4552. Li deriv. exhibits v. weak blue-green chemiluminescence.
Oxide: [15754-54-8].
$C_8H_{19}OP$ M 162.211
Cryst. (C_6H_6 or hexane). Mp 53-56°, 66°. Tautomeric with Dibutylphosphinous acid, D-00121 .
Sulfide: [689-62-3].
$C_8H_{19}PS$ M 178.272
Liq. Bp_1 120-123°. n_D^{20} 1.5087. Tautomeric with dibutylphosphinothious acid.

Issleib, K. *et al, Chem. Ber.*, 1959, **92**, 704 (*synth*)
Rauhut, M.M. *et al, J. Org. Chem.*, 1963, **28**, 473 (*synth*)
Hewertson, W. *et al, J. Chem. Soc.*, 1964, 1020 (*synth*)
Issleib, K. *et al, Z. Anorg. Allg. Chem.*, 1964, **328**, 21 (*deriv*)
Fritzche, H. *et al, Chem. Ber.*, 1965, **98**, 1681 (*synth*)
Maier, L., *Helv. Chim. Acta*, 1966, **49**, 1249 (*sulfide, synth, ir*)
Fluck, E. *et al, Z. Naturforsch., B*, 1967, **22**, 805 (*oxide, nmr*)
Grayson, M. *et al, Tetrahedron*, 1967, **23**, 1065 (*oxide, synth*)
Strecker, R.A. *et al, J. Am. Chem. Soc.*, 1973, **95**, 210 (*synth*)

Di-*tert*-butylphosphine **D-00108**

Bis(1,1-dimethylethyl)phosphine, 9CI
[819-19-2]

$HP[C(CH_3)_3]_2$

$C_8H_{19}P$ M 146.212
Ligand for metals of Groups VIB, VIIB, IVA, and VIII. Bp_2 34-35°.
Oxide: [684-19-5].
$C_8H_{19}OP$ M 162.211
Hygroscopic cryst. Mp 55-59°. Bp_9 112°. Tautomeric with di-*tert*-butylphosphinous acid.

Hoffmann, H. *et al, Chem. Ber.*, 1966, **99**, 1134; 1967, **100**, 692 (*synth*)
Issleib, K. *et al, J. Organomet. Chem.*, 1968, **13**, 283 (*deriv*)
Crofts, P.C. *et al, J. Chem. Soc. (C)*, 1970, 332 (*oxide, ir, ms*)
Grim, S.O. *et al, J. Chem. Eng. Data*, 1970, **15**, 497 (*nmr*)
Kostyanovskii, R.G. *et al, Org. Mass Spectrom.*, 1972, **6**, 1199; 1976, **11**, 237 (*ms, synth, oxide*)
Lappert, M.F. *et al, J. Chem. Soc., Dalton Trans.*, 1975, 1207 (*pe*)
Dimitriev, V.I. *et al, Zh. Obshch. Khim.*, 1978, **48**, 1533 (*Engl. transl.* p. 1405) (*oxide*)

Dibutylphosphinic acid, 8CI D-00109

[866-32-0]

$$(H_3CCH_2CH_2CH_2)_2P(O)OH$$

$C_8H_{19}O_2P$ M 178.211
Solid. Sol. H_2O, org. solvs. Mp 70-71°. Bp_3 208-210°. pK_a 3.32 (H_2O), 5.63 (80% EtOH aq.), 8.73 (EtOH).
▷SZ4580000.
Me ester: [7163-67-9]. *Methyl dibutylphosphinate.*
$C_9H_{21}O_2P$ M 192.237
Extractant for U and Pu. Liq. Miscible with water <35° and >191°. d_4^{20} 0.95. Bp_5 110°. n_D^{25} 1.4441.
Et ester: [7100-92-7]. *Ethyl dibutylphosphinate.*
$C_{10}H_{23}O_2P$ M 206.264
Extractant for U and Pu. Miscible with water <14° and >226°. d_4^{22} 0.93. Bp_8 114-117°. n_D^{20} 1.4436.
Butyl ester: [2950-47-2]. *Butyl dibutylphosphinate.*
$C_{12}H_{27}O_2P$ M 234.318
Extractant for Fe, Tl, Zr, U and Pu. Liq. Misc. H_2O >253°. d_4^{20} 0.93. $Bp_{0.5}$ 131-133°. n_D^{25} 1.4442.
Fluoride: [666-00-2].
$C_8H_{18}FOP$ M 180.202
Liq. Bp_3 73-74°. n_D^{20} 1.3723.
Chloride: [683-16-9].
$C_8H_{18}ClOP$ M 196.656
Liq. d_4^{20} 1.03. Bp_{22} 163-166°, Bp_2 115°. n_D^{20} 1.4670.
Isocyanate: [15790-45-1].
$C_9H_{18}NO_2P$ M 203.220
Liq. d_4^{20} 1.01. $Bp_{0.2}$ 84-85°. n_D^{20} 1.4520.
Anhydride: [34979-31-2].
$C_{16}H_{36}O_3P_2$ M 338.406
Viscous oil. Bp_2 170-172°. n_D^{20} 1.4628. Reacts rapidly with water.
Amide: see P,P-Dibutylphosphinic amide, D-00111

Kosolapoff, G.M. *et al, J. Am. Chem. Soc.*, 1951, **73**, 4101 (*synth, chloride, anhydride, esters*)
Burger, L.L., *J. Phys. Chem.*, 1958, **62**, 590 (*esters, use*)
Kosolapoff, G.M. *et al, J. Chem. Soc.*, 1959, 3950 (*synth*)
Buckler, S.A., *J. Am. Chem. Soc.*, 1962, **84**, 3093 (*butyl ester*)
Detoni, S. *et al, Spectrochim. Acta*, 1964, **20**, 949 (*ir*)
Blasse, R., *Recl. Trav. Chim. Pays-Bas*, 1965, **84**, 267 (*chloride, ir*)
Neimysheva, A.A. *et al, Zh. Obshch. Khim.*, 1966, **36**, 1090 (*Engl. transl.* p. 1105) (*chloride*)
Haake, P. *et al, Tetrahedron*, 1967, **24**, 565 (*ms*)
Derkach, G.I. *et al, Zh. Obshch. Khim.*, 1968, **38**, 1784 (*Engl. transl.* p. 1739) (*isocyanate*)
Cook, R.D. *et al, J. Am. Chem. Soc.*, 1973, **95**, 8088 (*ester, props*)
Petrov, K.A. *et al, Zh. Obshch. Khim.*, 1982, **52**, 58 (*Engl. transl.* p. 53) (*ester, synth, pmr*)
Nikitin, E.V. *et al, Zh. Obshch. Khim.*, 1982, **52**, 2721 (*Engl. transl.* p. 2400) (*fluoride, synth, ir, P and F nmr*)

Di-*tert*-butylphosphinic acid, 8CI D-00110

Bis(1,1-dimethylethyl)phosphinic acid, 9CI
[677-76-9]

$$[(H_3C)_3C]_2P(O)OH$$

$C_8H_{19}O_2P$ M 178.211
Dimeric in the solid state. Cryst. (H_2O). Mp 210°. pK_a 4.24 (H_2O), 6.26 (75% EtOH aq.).
Me ester: [58156-53-9]. *Methyl di-*tert-*butylphosphinate.*
$C_9H_{21}O_2P$ M 192.237
Liq. $Bp_{0.7}$ 76-77°.
Et ester: [19935-95-6]. *Ethyl di-*tert-*butylphosphinate.*
$C_{10}H_{23}O_2P$ M 206.264
Liq. $Bp_{0.1}$ 41-43°.
Ph ester: [27286-05-1]. *Phenyl di-*tert-*butylphosphinate.*
$C_{14}H_{23}O_2P$ M 254.308
Liq. $Bp_{0.2}$ 130-140° (bath).
Trimethylsilyl ester: [42346-42-9]. *Trimethylsilyl di-*tert-*butylphosphinate.*
$C_{11}H_{27}O_2PSi$ M 250.392
Solid. Mp 66-67°. Bp_1 75°.
Fluoride: [29149-37-9].
$C_8H_{18}FOP$ M 180.202
Oil which solidifies at r.t. Bp_{12} 96-98°.
Chloride: [677-74-7].
$C_8H_{18}ClOP$ M 196.656
Cryst. (pet. ether). Mp 82.5°. Bp_{15} 120°, Bp_5 101-105°. Subl. at 35°/1 mm.
Bromide: [677-72-5].
$C_8H_{18}BrOP$ M 241.107
Solid. Mp 98-99°.
Iodide:
$C_8H_{18}IOP$ M 288.108
Solid. Mp 89.5-90.5°.
Azide: [74722-07-9].
$C_8H_{18}N_3OP$ M 203.223
Liq. $Bp_{0.5}$ 75-80° (oven). Could not be completely freed of H_2O.

Angstadt, H.P., *J. Am. Chem. Soc.*, 1964, **86**, 5040 (*bromide*)
Crofts, P.C. *et al, J. Chem. Soc. (C)*, 1970, 332 (*synth, chloride, ir, pmr, P nmr*)
Fild, M. *et al, J. Chem. Soc. (A)*, 1970, 2359 (*fluoride, pmr, F and P nmr*)
Stewart, A.P. *et al, J. Chem. Soc. (C)*, 1970, 1263 (*phenyl ester, ir, pmr*)
Cook, A.G. *et al, J. Org. Chem.*, 1972, **37**, 3342 (*synth, props*)
Kostyanovskii, R.G. *et al, Org. Mass Spectrom.*, 1972, 1199 (*chloride, ms*)
Kuchen, W. *et al, Z. Anorg. Allg. Chem.*, 1975, **413**, 266 (*silyl ester, ir, pmr, P nmr*)
Druyan, M.E. *et al, J. Am. Chem. Soc.*, 1976, **98**, 4801 (*cryst struct*)
Abbas, K.A. *et al, Can. J. Chem.*, 1977, **55**, 3740 (*synth, ester*)
Dahl, O., *J. Chem. Soc., Perkin Trans. 1*, 1978, 947 (*ester*)
Harger, M.J.P. *et al, J. Chem. Soc., Perkin Trans. 1*, 1981, 736 (*chloride, azide*)

P,P-Dibutylphosphinic amide, 9CI D-00111

[53534-28-4]

$(H_3CCH_2CH_2CH_2)_2P(O)NH_2$

$C_8H_{20}NOP$ M 177.226

N,N-*Di-Me:* [50387-23-0]. P,P-*Dibutyl*-N,N-*dimethylphosphinic amide.*
$C_{10}H_{24}NOP$ M 205.279
Liq. n_D^{25} 1.4561.

N-*Et:* P,P-*Dibutyl*-N-*ethylphosphinic amide.*
$C_{10}H_{24}NOP$ M 205.279
Liq. d_4^{20} 0.94. $Bp_{0.37}$ 157-158°. n_D^{25} 1.4598.

N,N-*Di-Et:* [66379-68-8]. P,P-*Dibutyl*-N,N-*diethylphosphinic amide.*
$C_{12}H_{28}NOP$ M 233.333
Liq. d_4^{20} 0.92. $Bp_{0.2}$ 126-127°. n_D^{20} 1.4581.

Razumov, A.I. *et al, Zh. Obshch. Khim.*, 1957, **27**, 754 (*Engl. transl.* p. 827) (*synth*)
Litvinenko, L.M. *et al, Zh. Obshch. Khim.*, 1973, **43**, 1794 (*Engl. transl.* p. 1777) (*synth*)

Dibutylphosphinodithioic acid, 9CI D-00112

[32435-35-1]

$(H_3CCH_2CH_2CH_2)_2P(S)SH$

$C_8H_{19}PS_2$ M 210.332

Liq. d_4^{20} 1.03. Bp_2 99-99.5°, Bp_2 144-146°. pK_{a1} 1.79 (7% EtOH aq.), pK_{a2} 2.52 (80% EtOH aq.). n_D^{20} 1.5481.

Na salt: [71550-48-6]. Solid. Mp 113-115°.

NH_4 salt: [2573-70-8]. Cryst. (C_6H_6). Mp 97-99°.

Anilinium salt: Solid. Mp 110-111°.

Me ester: [49873-34-9]. *Methyl dibutylphosphinodithioate.*
$C_9H_{21}PS_2$ M 224.359
Liq. d_4^{20} 1.01. Bp_2 123°. n_D^{20} 1.5392.

Et ester: [70111-25-0]. *Ethyl dibutylphosphinodithioate.*
$C_{10}H_{23}PS_2$ M 238.385
Liq. Bp_3 101-103°.

Anhydrosulfide: [22737-09-3].
$C_{16}H_{36}P_2S_3$ M 386.588
Cryst. (C_6H_6/hexane). Mp 170-172°.

Mastryukova, T.A. *et al, Zh. Obshch. Khim.*, 1961, **31**, 507 (*Engl. transl.* p. 464) (*synth, props*)
Rauhut, M.M. *et al, J. Org. Chem.*, 1961, **26**, 5133 (*synth, salts*)
Peters, G., *J. Org. Chem.*, 1962, **27**, 2198 (*synth*)
Almasi, L. *et al, Chem. Ber.*, 1969, **102**, 1489 (*anhydrosulfide*)
Kuramshin, I.Ya. *et al, Zh. Obshch. Khim.*, 1973, **43**, 1456 (*Engl. transl.* p. 1446) (*ester, ir, pmr, P nmr*)
Tsvetkov, E.N. *et al, Izv. Akad. Nauk SSSR, Ser. Khim.*, 1979, 426 (*Engl. transl.* p. 394) (*esters*)
Keck, H. *et al, Phosphorus Sulfur*, 1983, **14**, 225 (*ms*)

Di-*tert*-butylphosphinodithioic acid D-00113

Bis(1,1-dimethylethyl)phosphinodithioic acid, 9CI

$[(H_3C)_3C]_2P(S)SH$

$C_8H_{19}PS_2$ M 210.332

Liq. which fails to cryst.

Anhydrosulfide: [5995-10-8]. *Di*-tert-*butylphosphinodithioic anhydrosulfide.*
$C_{16}H_{36}P_2S_3$ M 386.588
Cryst. (MeOH/Et_2O). Mp 185-186°.

Hoffmann, H. *et al, Chem. Ber.*, 1966, **99**, 1134 (*synth*)
Issleib, K. *et al, Chem. Ber.*, 1966, **99**, 1320 (*anhydrosulfide*)
Haegele, G. *et al, Z. Naturforsch., B*, 1974, **29**, 349 (*anhydrosulfide, ir, pmr, P nmr*)
Harris, R.K. *et al, J. Chem. Soc., Dalton Trans.*, 1978, 9 (*anhydrosulfide, cmr, P nmr*)

Dibutylphosphinoselenoic acid, 9CI D-00114

$[H_3C(CH_2)_3]_2P(Se)OH \rightleftharpoons [H_3C(CH_2)_3]_2P(O)SeH$

$C_8H_{19}OPSe$ M 241.171

Liq. $Bp_{0.05}$ 93-94°. pK_{a1} 2.71 (7% EtOH aq.), pK_{a2} 4.90 (80% EtOH aq.). n_D^{20} 1.5260.

Chloride: [55249-22-4].
$C_8H_{18}ClPSe$ M 259.617
d_4^{20} 1.248. Bp_{18} 153-154°. n_D^{20} 1.5341.

N,N-*Diethylamide:* [54638-48-1]. P,P-*Dibutyl*-N,N-*diethylphosphinoselenoic amide.*
$C_{12}H_{28}NPSe$ M 296.294
Liq. d_4^{20} 1.11. $Bp_{0.5}$ 134°. n_D^{20} 1.5182.

Markowska, A., *Bull. Acad. Pol. Sci., Ser. Sci. Chim.*, 1965, **13**, 149; *CA*, **63**, 9791 (*synth*)
Nuretdinov, I.A. *et al, Zh. Obshch. Khim.*, 1974, **44**, 2588 (*Engl. transl.* p. 2548) (*amide*)
Nuretdinov, I.A. *et al, Izv. Akad. Nauk SSSR, Ser. Khim.*, 1975, 327 (*Engl. transl.* p.263) (*nqr*)
Shagidullin, R.R. *et al, Izv. Akad. Nauk SSSR, Ser. Khim.*, 1976, 184 (*Engl. transl.* p. 174) (*chloride, uv*)
Nuretdinov, I.A. *et al, Zh. Obshch. Khim.*, 1978, **48**, 1071 (*Engl. transl.* p. 975) (*chloride, ir, P nmr*)

***P,P*-Di-*tert*-butylphosphinoselenoic amide** D-00115

P,P-*Bis(1,1-dimethylethyl)phosphinoselenoic amide, 9CI*

$[(H_3C)_3C]_2P(Se)NH_2$

$C_8H_{20}NPSe$ M 240.186

N-*Ph:* [56898-62-5]. P,P-*Di*-tert-*butyl*-N-*phenylphosphinoselenoic amide. Di*-tert-*butylphosphinoselenoic anilide.*
$C_{14}H_{24}NPSe$ M 316.284
Solid. Mp 150-151°.

McFarlane, W. *et al, J. Chem. Soc., Dalton Trans.*, 1976, 2351 (*synth, N and P nmr*)

Di-*tert*-butylphosphinotellurous acid D-00116

Bis(1,1-dimethylethyl)phosphinotellurous acid, 9CI

$[(H_3C)_3C]_2PTeH$

$C_8H_{19}PTe$ M 273.812

Tautomeric. Some derivs. have been characterised spectroscopically.

Du Mont, W.W. *et al, Z. Naturforsch., B*, 1981, **36**, 332 (*derivs, Te nmr*)

Dibutylphosphinothioic acid, 9CI D-00117

$(H_3CCH_2CH_2CH_2)_2P(S)OH \rightleftharpoons (H_3CCH_2CH_2CH_2)_2P(O)SH$

$C_8H_{19}OPS$ M 194.271

V. hygroscopic solid. Mp 24-25°. $Bp_{0.2}$ 79-80°. pK_a 2.91 (7% EtOH aq.) (0.8% thiol form), 5.14 (80% EtOH aq.)(0% thiol form).

Anilinium salt: Cryst. (heptane). Mp 70-71°.

S-*Me ester:* [51049-70-8]. S-*Methyl dibutylphosphinothioate.*
$C_9H_{21}OPS$ M 208.298
Liq. d_4^{20} 1.00. $Bp_{2\text{-}3}$ 164°. n_D^{20} 1.4969.
O-*Et ester:* [24611-05-0]. O-*Ethyl dibutylphosphinothioate.*
$C_{10}H_{23}OPS$ M 222.325
Liq. d_4^{20} 0.94. Bp_9 135-136°. n_D^{20} 1.4670.
O-*Isopropyl ester:* O-*Isopropyl dibutylphosphinothioate.*
$C_{11}H_{25}OPS$ M 236.352
Liq. $Bp_{0.1}$ 84° dec.
O-*Trimethylsilyl ester:* [27502-54-1]. O-*Trimethylsilyl dibutylphosphinothioate.*
$C_{11}H_{27}OPSSi$ M 266.453
Oil. Bp_4 123-125°.
Fluoride: [30779-42-1].
$C_8H_{18}FPS$ M 196.262
Characterised spectroscopically.
Chloride: [23834-60-8].
$C_8H_{18}ClPS$ M 212.717
d_4^{20} 1.04. Bp_{10} 134.5-135°. n_D^{20} 1.5098.
Bromide: [55656-88-7].
$C_8H_{18}BrPS$ M 257.168
Used to derivatize nucleotides, and prepare nucleoside polyphosphates. d_4^{19} 1.24. Bp_6 143-144°.
Iodide: [81373-57-1].
$C_8H_{18}IPS$ M 304.168
Liq. $Bp_{0.06}$ 104-107°.
Anilide: P,P-*Dibutyl*-N-*phenylphosphinothioic amide. Dibutylphosphinothioic anilide.*
$C_{14}H_{24}NPS$ M 269.384
Needles (hexane). Mp 66-67°.
Anhydride: [67003-67-2].
$C_{16}H_{36}OP_2S_2$ M 370.527
Liq. d_4^{20} 1.01. $Bp_{0.001}$ 105°. n_D^{20} 1.5075.

Kabachnik, M.I. *et al, Tetrahedron,* 1960, **9**, 10 (*props*)
Peters, G., *J. Org. Chem.,* 1962, **27**, 2198 (*chloride, anilide*)
Pudovik, A.N. *et al, Zh. Obshch. Khim.,* 1963, **39**, 2231 (*Engl. transl.* p. 2177) (*ester*)
Küchen, W. *et al, Z. Anorg. Allg. Chem.,* 1964, **333**, 71 (*synth*)
Issleib, K. *et al, J. Organomet. Chem.,* 1970, **22**, 375 (*silyl ester*)
Reddy, G.S. *et al, Z. Naturforsch., B,* 1970, **25**, 1199 (*fluoride, F and P nmr*)
Kuramshin, I.Ya. *et al, Zh. Obshch. Khim.,* 1973, **43**, 1456 (*Engl. transl.* p. 1446) (*ester, pmr, ir, P nmr*)
Furusawa, K. *et al, J. Chem. Soc., Perkin Trans. 1,* 1976, 1711 (*bromide, synth, use*)
Hata, T. *et al, Chem. Lett.,* 1976, 987 (*bromide, use*)
Tsvetkov, E.N. *et al, Izv. Akad. Nauk SSSR, Ser. Khim.,* 1979, 426 (*Engl. transl.* p. 394) (*chloride*)
Feshchenko, N.G. *et al, Zh. Obshch. Khim.,* 1982, **52**, 222 (*Engl. transl.* p. 202) (*iodide, P nmr*)

Di-*tert*-butylphosphinothioic acid D-00118

Bis(1,1-dimethylethyl)phosphinothioic acid, 9CI
[53159-03-8]

$$[(H_3C)_3C]_2P(S)OH \rightleftharpoons [CH_3C)_3C]_2P(O)SH$$

$C_8H_{19}OPS$ M 194.271
Completely in thione (thioxo) form. pK_a 3.91 (7% EtOH aq.), 6.09 (80% EtOH aq.), 13.20 ($MeNO_2$).
O-*Me ester:* [62246-59-7]. O-*Methyl di*-tert-*butylphosphinothioate.*
$C_9H_{21}OPS$ M 208.298
No phys. props. reported.
S-*Me ester:* [49873-30-5]. S-*Methyl di*-tert-*butylphosphinothioate.*
$C_9H_{21}OPS$ M 208.298
No phys. props. reported.
O-*Ph ester:* [63027-82-7]. O-*Phenyl di*-tert-*butylphosphinothioate.*
$C_{14}H_{23}OPS$ M 270.369
Cryst. (pet. ether). Mp 53-54°.
O-*Benzyl ester:* O-*Benzyl di*-tert-*butylphosphinothioate.*
$C_{15}H_{25}OPS$ M 284.396
Solid. Mp 87-89°.
O-*Trimethylsilyl ester:* [42346-43-0]. O-*Trimethylsilyl di*-tert-*butylphosphinothioate.* Viscous liq. $Bp_{0.7}$ 64°.
Fluoride: [29149-40-4].
$C_8H_{18}FPS$ M 196.262
$Bp_{0.4}$ 67-70°.
Chloride: [27509-07-5].
$C_8H_{18}ClPS$ M 212.717
Cryst. (pet. ether) with camphoraceous odour. Mp 101-102°. Bp_6 120°.
Bromide: [27509-09-7].
$C_8H_{18}BrPS$ M 257.168
Cryst. Mp 139-140°.
*Amide: see Di-tert-*butylphosphinothioic amide, D-00119
Anhydride: [72170-80-0].
$C_{16}H_{36}OP_2S_2$ M 370.527
Solid. Mp 144-145°.

Kabachnik, M.I. *et al, Tetrahedron,* 1960, **9**, 10 (*synth, props*)
Fild, M. *et al, J. Chem. Soc. (A),* 1970, 2359 (*fluoride, pmr, F and P nmr*)
Küchen, W. *et al, Chem. Ber.,* 1970, **103**, 2114 (*bromide, ir, P nmr, pmr*)
Stewart, A.P. *et al, J. Chem. Soc. (C),* 1970, 1263 (*chloride, phenyl ester, ir, pmr*)
Küchen, W. *et al, Z. Anorg. Allg. Chem.,* 1975, **413**, 266 (*silyl ester, ir*)
Mastryukova, T.A. *et al, Phosphorus Sulfur,* 1976, **1**, 211 (*esters*)
Foss, V.L. *et al, Zh. Obshch. Khim.,* 1979, **49**, 1724 (*Engl. transl.* p. 1510) (*anhydride, P nmr*)
Monkiewicz, J. *et al, Bull. Acad. Pol. Sci., Ser. Sci. Chim.,* 1980, **28**, 351 (*benzyl ester*)

Di-*tert*-butylphosphinothioic amide D-00119

Bis(1,1-dimethylethyl)phosphinothioic amide

$$[(H_3C)_3C]_2P(S)NH_2$$

$C_8H_{20}NPS$ M 193.286
N-*Me:* [55382-99-5]. P,P-*Di*-tert-*butyl*-N-*methylphosphinothioic amide.*
$C_9H_{22}NPS$ M 207.313
Solid. Mp 115-117°.
N-*Ph:* [56898-61-4]. P,P-*Di*-tert-*butyl*-N-*phenylphosphinothioic amide. Di*-tert-*butylphosphinothioic anilide.*
$C_{14}H_{24}NPS$ M 269.384
Solid. Mp 150-151°.

Urata, K. *et al, Bull. Chem. Soc. Jpn.,* 1974, **47**, 2709 (*synth*)
McFarlane, W. *et al, J. Chem. Soc., Dalton Trans.,* 1976, 2351 (*synth, nmr*)

Di-*tert*-butylphosphinothious acid D-00120

Bis(1,1-dimethylethyl)phosphinothious acid, 10CI

$$[(H_3C)_3C]_2PSH$$

$C_8H_{19}PS$ M 178.272
Thought to exist as the thiophosphoryl tautomer; see under Di-*tert*-butylphosphine, D-00108 .

Me ester: [75956-78-4]. *Methyl di*-tert-*butylphosphinothioite.*
$C_9H_{21}PS$ M 192.299
Liq. Bp_{10} 92-93°. n_D^{20} 1.4990.
Anhydrosulfide: [62969-15-7].
$C_{16}H_{36}P_2S$ M 322.468
Cryst. (C_6H_6/hexane). Mp 77-78°.

Patsanovskii, I.I. *et al*, *Dokl. Akad. Nauk SSSR, Ser. Sci. Khim.*, 1980, **254**, 414 (*Engl. transl.* p. 771) (*ester, synth, P nmr*)
Foss, V.L. *et al*, *Zh. Obshch. Khim.*, 1982, **52**, 1054, 1063 (*Engl. transl.* pp. 916, 924) (*anhydrosulfide, synth, pmr, P nmr*)

Dibutylphosphinous acid, 9CI D-00121

[50602-70-5]

$(H_3CCH_2CH_2CH_2)_2POH$

$C_8H_{19}OP$ M 162.211
Free acid exists in the tautomeric phosphoryl form. See under Dibutylphosphine, D-00107 .
Et ester: [56660-55-0]. *Ethyl dibutylphosphinite.*
$C_{10}H_{23}OP$ M 190.265
Liq. Bp_{15} 112-116°.
Butyl ester: [6418-33-7]. *Butyl dibutylphosphinite.*
$C_{12}H_{27}OP$ M 218.318
Liq. $Bp_{1.5}$ 68-69°, $Bp_{0.8}$ 97°. n_D^{20} 1.4520.
Trimethylsilyl ester: [13683-02-8]. *Trimethylsilyl dibutylphosphinite.*
$C_{11}H_{27}OPSi$ M 234.393
Liq. Bp_{12} 99-101°.
Chloride: [4323-64-2]. *Dibutylchlorophosphine.* Fuming liq. Bp_{12} 92-3°. n_D^{20} 1.4743.
Bromide: Bromodibutylphosphine.
$C_8H_{18}BrP$ M 225.108
Liq. Bp_{17} 118-119°.
Iodide: [75271-90-8]. *Dibutyliodophosphine.*
$C_8H_{18}IP$ M 272.108
Hygroscopic, light-yellow cryst. Mp 49-51°. Bp_{13} 120-121°, $Bp_{2.5}$ 93°.
Cyanide: Dibutylcyanophosphine.
$C_9H_{18}NP$ M 171.222
Solid. Mp 89-92°.
Isocyanate:
$C_9H_{18}NOP$ M 187.221
Solid. Mp 94-97°. $Bp_{0.05}$ 58-61°. Exists as a dimer.
Amide: see P,P-Dibutylphosphinous amide, D-00123

Rauhut, M.M. *et al*, *J. Am. Chem. Soc.*, 1958, **80**, 6690 (*synth*)
Issleib, K. *et al*, *Chem. Ber.*, 1959, **92**, 2681 (*halides*)
Kabachnik, M.I. *et al*, *Dokl. Akad. Nauk SSSR, Ser. Sci. Khim.*, 1960, **135**, 323 (*Engl. transl.* p. 1267) (*esters*)
Voigt, D. *et al*, *Bull. Soc. Chim. Fr.*, 1964, 3087 (*butyl ester*)
Kolotilo, M.V. *et al*, *CA*, 1969, **71**, 108528 (*isocyanate*)
Issleib, K. *et al*, *J. Organomet. Chem.*, 1970, **22**, 375 (*trimethylsilyl ester, synth, P nmr*)
Weichmann, H. *et al*, *Z. Anorg. Allg. Chem.*, 1980, **462**, 7 (*iodide, synth, P nmr*)
Mikaya, A.I. *et al*, *Zh. Obshch. Khim.*, 1982, **52**, 1998 (*Engl. transl.* p. 1776) (*ethyl ester, ms*)
Wolfsberger, W., *J. Organomet. Chem.*, 1986, **317**, 167 (*chloride, cmr, P nmr*)

Di-*tert*-butylphosphinous acid, 8CI D-00122

Bis(1,1-dimethylethyl)phosphinous acid, 9CI
[52809-04-8]

$[(H_3C)_3C]_2POH$

$C_8H_{19}OP$ M 162.211
The free acid exists as the phosphoryl tautomer. See under Di-*tert*-butylphosphine, D-00108 .
Me ester: [70073-11-9]. *Methyl di*-tert-*butylphosphinite. Methyl bis(1,1-dimethylethyl)phosphinite.*
$C_9H_{21}OP$ M 176.238
Bp_8 63-64°, $Bp_{2.5}$ 33-34°. Forms Pt complexes.
Et ester: [58309-95-8]. *Ethyl di*-tert-*butylphosphinite. Ethyl bis(1,1-dimethylethyl)phosphinite.*
$C_{10}H_{23}OP$ M 190.265
Liq. Bp_7 59-60°, $Bp_{0.5}$ 41-42°. Forms Pt complexes.
Ph ester: [27286-18-6]. *Phenyl di*-tert-*butylphosphinite. Phenyl bis(1,1-dimethylethyl)phosphinite.*
$C_{14}H_{23}OP$ M 238.309
Liq. $Bp_{0.2}$ 92-95°. Forms stable quaternary adducts with MeI (Mp 218-218.5°) and $PhCH_2Br$ (Mp 189-192°).
Trimethylsilyl ester: [53483-28-6]. *Trimethylsilyl di*-tert-*butylphosphinite. Trimethylsilyl bis(1,1-dimethylethyl)phosphinite.*
$C_{11}H_{27}OPSi$ M 234.393
Liq. Bp_{11} 77-79°. n_D^{20} 1.4472.
Fluoride: [29146-24-5]. *Di*-tert-*butylfluorophosphine.*
$C_8H_{18}FP$ M 164.202
Forms Cr, Mo, W, and Ni complexes. Foul-smelling liq. Bp_{50} 71°. n_D^{20} 1.4340.
▷Toxic
Chloride: see Di-tert-butylphosphinous chloride, D-00125
Bromide: [39106-95-1]. *Bromodi*-tert-*butylphosphine.*
$C_8H_{18}BrP$ M 225.108
Yellowish oil. Bp_5 65-66°.
Iodide: [66783-70-8]. *Di*-tert-*butyliodophosphine.*
$C_8H_{18}IP$ M 272.108
Yellow oil. Bp_{10} 105-107°.
Anhydride: [60714-29-6].
$C_{16}H_{36}OP_2$ M 306.407
Solid. Mp 58-59°. Bp_1 113-115°.
Amide: see P,P-Di-*tert*-butylphosphinous amide, D-00124

Crofts, P.C. *et al*, *J. Chem. Soc.*, 1970, 332 (*synth, ir, ms*)
Fild, M. *et al*, *J. Chem. Soc. (A)*, 1970, 2359 (*fluoride, synth, F and P nmr, pmr*)
Stewart, A.P. *et al*, *J. Chem. Soc. (C)*, 1970, 1263 (*phenyl ester, synth, ir, pmr*)
Lappert, M.F. *et al*, *J. Chem. Soc., Dalton Trans.*, 1975, 1207 (*fluoride, pe*)
Bartsch, R. *et al*, *Chem. Ber.*, 1978, **111**, 1420 (*bromide, iodide, synth, pmr, P nmr*)
Dahl, O., *J. Chem. Soc., Perkin Trans. 1*, 1978, 947 (*synth, esters, pmr, P nmr*)
Oberhammer, H. *et al*, *Inorg. Chem.*, 1978, **17**, 1254 (*fluoride, struct*)
Inorg. Synth., 1978, **18**, 173 (*fluoride, synth*)
Dombek, B.D., *J. Organomet. Chem.*, 1979, **169**, 315 (*esters, synth, pmr, complexes*)
Foss, V.L. *et al*, *Zh. Obshch. Khim.*, 1979, **49**, 1724 (*Engl. transl.* p. 1510) (*anhydride, synth, P nmr*)
Foss, V.L. *et al*, *Zh. Obshch. Khim.*, 1979, **49**, 2418 (*Engl. transl.* p. 2134) (*trimethylsilyl ester*)
Efimova, V.D. *et al*, *Zh. Obshch. Khim.*, 1984, **54**, 1673 (*Engl. transl.* p. 1490) (*esters, synth, P nmr*)

***P,P*-Dibutylphosphinous amide** D-00123

$(H_3CCH_2CH_2CH_2)_2PNH_2$

$C_8H_{20}NP$ M 161.226

N,N-*Di-Me:* P,P-*Dibutyl*-N,N-*dimethylphosphinous amide. Dibutylphosphinous dimethylamide.*
$C_{10}H_{24}NP$ M 189.280
Liq. Bp_{720} 214°, $Bp_{1.5}$ 51.5°. n_D^{20} 1.4592.

N-*Et:* P,P-*Dibutyl*-N-*ethylphosphinous amide. Dibutylphosphinous ethylamide.*
$C_{10}H_{24}NP$ M 189.280
Liq. Bp_2 75°.

N,N-*Di-Et:* P,P-*Dibutyl*-N,N-*diethylphosphinous amide. Dibutylphosphinous diethylamide.*
$C_{12}H_{28}NP$ M 217.334
Liq. Bp_{11} 112-114°. n_D^{20} 1.4642.

Issleib, K. *et al*, *Chem. Ber.*, 1959, **92**, 2681 (*diethyl*)
Voskuil, W. *et al*, *Recl. Trav. Chim. Pays-Bas*, 1962, **81**, 993 (*diethyl*)
Nöth, H. *et al*, *Chem. Ber.*, 1963, **96**, 1109 (*dimethyl*)
Clemens, D.M. *et al*, *Inorg. Chem.*, 1965, **4**, 1222 (*ethyl, synth, ir, nmr*)

***P,P*-Di-*tert*-butylphosphinous amide, 8CI** D-00124

P,P-*Bis(1,1-dimethylethyl)phosphinous amide, 9CI. Aminodi*-tert-*butylphosphine*
[17858-28-5]

$[(H_3C)_3C]_2PNH_2$

$C_8H_{20}NP$ M 161.226
Sol. C_6H_6. Mp −1° to +1°. Bp_2 33-34°.

B,HCl: Solid. Mp 295°. Bp_1 200° subl.

N-*Me:* [21183-89-1]. P,P-*Di*-tert-*butyl*-N-*methylphosphinous amide.* N-*Methyl*-P,P-*bis(1,1-dimethylethyl)phosphinous amide.*
$C_9H_{22}NP$ M 175.253
Solid. Mp 53-55°. $Bp_{0.01}$ 55° subl.

N-*Ph:* [56898-59-0]. P,P-*Di*-tert-*butyl*-N-*phenylphosphinous amide. Di*-tert-*butylphosphinous anilide.*
$C_{14}H_{24}NP$ M 237.324
Liq. Bp_1 80-82°.

N-*Trimethylsilyl:* [17858-29-6]. P,P-*Di*-tert-*butyl*-N-*trimethylsilylphosphinous amide.*
$C_{11}H_{28}NPSi$ M 233.408
Liq. Mp −3°. Bp_1 46°.

Scherer, O.J. *et al*, *Chem. Ber.*, 1968, **101**, 4184 (*synth, derivs, ir, pmr*)
McFarlane, W. *et al*, *J. Chem. Soc., Dalton Trans.*, 1976, 2351 (*phenyl, synth, P nmr*)
Ross, B. *et al*, *Chem. Ber.*, 1979, **112**, 1756 (*synth, P nmr, pmr, cmr*)

Di-*tert*-butylphosphinous chloride D-00125

Bis(1,1-dimethylethyl)phosphinous chloride, 9CI. Di-tert-*butylchlorophosphine*
[13716-10-4]

$[(H_3C)_3C]_2PCl$

$C_8H_{18}ClP$ M 180.657
Liq. Bp_{48} 100-102°, Bp_{13} 70-72°. n_D^{20} 1.4830.

Voskuil, W. *et al*, *Recl. Trav. Chim. Pays-Bas*, 1963, **82**, 302 (*synth*)
Fild, M. *et al*, *J. Chem. Soc. (A)*, 1970, 2359 (*P nmr, pmr, use*)
Kostyanovskii, R.G. *et al*, *Org. Mass Spectrom.*, 1972, **6**, 1199 (*ms*)
Inorg. Synth., 1973, **14**, 4 (*synth, nmr, pmr*)
Labarre, M.-C. *et al*, *J. Mol. Struct.*, 1975, **26**, 17 (*synth, P nmr*)
Lappert, M.F. *et al*, *J. Chem. Soc., Dalton Trans.*, 1975, 1207 (*pe*)
Wolfsberger, W., *J. Organomet. Chem.*, 1986, **317**, 167 (*cmr, P nmr*)

***N,N*-Dibutylphosphonamidic acid** D-00126

N,N-*Dibutylphosphoramidous acid*

$(H_3CCH_2CH_2CH_2)_2NPH(O)OH \rightleftharpoons (H_3CCH_2CH_2CH_2)_2NP(OH)_2$

$C_8H_{20}NO_2P$ M 193.225
Tautomeric.

Et ester: Ethyl N,N-*dibutylphosphonamidate.*
$C_{10}H_{24}NO_2P$ M 221.279
Liq. $Bp_{0.03}$ 68-70°. n_D^{20} 1.4421.

Butyl ester: Butyl N,N-*dibutylphosphonamidate.*
$C_{12}H_{28}NO_2P$ M 249.332
Liq. $Bp_{0.05}$ 72-74°. n_D^{20} 1.4421.

Zwierzak, A. *et al*, *Tetrahedron*, 1967, **23**, 2243 (*synth*)

Dibutyl phosphonate, 9CI D-00127

Dibutyl phosphite
[1809-19-4]

$(H_3CCH_2CH_2CH_2O)_2P(O)H \rightleftharpoons (H_3CCH_2CH_2CH_2O)_2POH$

$C_8H_{19}O_3P$ M 194.210
Tautomeric; almost completely in phosphonate form. Liq. d_4^{20} 0.995. Bp_8 116-117°. pK_a 20.8 (DMSO, 25°). n_D^{20} 1.4240. Na salt v. sol. in hydrocarbons.

▷Mod. toxic. Emits highly toxic fumes on heating or contact with acids. HS6475000.

Cade, J.A. *et al*, *J. Chem. Soc.*, 1954, 2030 (*synth*)
Harless, H.R., *Anal. Chem.*, 1961, **33**, 1387 (*ms*)
Moedritzer, K., *J. Inorg. Nucl. Chem.*, 1962, **22**, 19 (*P nmr*)
Coulson, E.J. *et al*, *J. Chem. Soc.*, 1965, 2364 (*synth*)
Goodell, L.J. *et al*, *Can. J. Chem.*, 1969, **47**, 2461 (*pmr*)
Pudovik, A.N. *et al*, *Zh. Obshch. Khim.*, 1975, **45**, 2123 (*Engl. transl.* p. 2092) (*synth*)
Sax, N.I., *Dangerous Properties of Industrial Materials*, 6th Ed., Van Nostrand-Reinhold, 1984, 554.

Di-*tert*-butyl phosphonate, 8CI D-00128

Bis(1,1-dimethylethyl) phosphonate, 9CI. Di-tert-*butyl phosphite*
[13086-84-5]

$[(H_3C)_3CO]_2P(O)H \rightleftharpoons [(H_3C)_3CO]_2POH$

$C_8H_{19}O_3P$ M 194.210
Tautomeric; almost completely in phosphonate form. Liq. d^{25} 0.96. Bp_{10} 70-72°, $Bp_{0.4}$ 42°. n_D^{25} 1.4168. Dec. vigorously > ca. 70°.

Gerrard, W. *et al*, *J. Chem. Soc.*, 1953, 1920 (*synth*)
Young, R.W., *J. Am. Chem. Soc.*, 1953, **75**, 4620 (*synth, ir*)
Goldwhite, H. *et al*, *J. Chem. Soc.*, 1957, 2409 (*synth, ir*)
Mark, V. *et al*, *J. Org. Chem.*, 1964, **29**, 1006 (*synth, pmr, P nmr*)
Chapman, T.M. *et al*, *J. Org. Chem.*, 1973, **38**, 250 (*synth*)

Dibutyl phosphonite, 9CI D-00129

[30653-71-5]

$(H_3CCH_2CH_2CH_2O)_2PH$

$C_8H_{19}O_2P$ M 178.211

Liq. Bp_1 38-39°. n_D^{20} 1.4360. Stable for 3 months at 20° in inert atmos. Fumes in air: vigorously oxid. to Diethyl phosphonate, D-00338 Addn. of S→ *O,O*-Diethyl phosphonothioate, D-00341 .

Lutsenko, I.F. *et al*, *Dokl. Akad. Nauk SSSR, Ser. Sci. Khim.*, 1970, **193**, 828 (*Engl. transl.* p. 553) (*synth, ir, P nmr, pmr, props*)
Lutsenko, I.F. *et al*, *Organomet. Chem. Synth.*, 1971, **1**, 169 (*synth, ir, pmr, P nmr, props*)
Foss, V.L. *et al*, *Zh. Obshch. Khim.*, 1972, **42**, 954 (*Engl. transl.* p. 944) (*synth*)

Di-*tert*-butylphosphonochloridous acid — D-00130

(1,1-Dimethylethyl)phosphonochloridous acid, 10CI

$(H_3C)_3CPCl(OH)$

$C_4H_{10}ClOP$ M 140.549
Esters of this acid are generally unstable.
Butyl ester: [70446-63-8]. *Butyl* tert-*butylphosphonochloridite.*
$C_8H_{18}ClOP$ M 196.656
Liq. Can be kept at 20° for 3 months or heated for 1 hr. at 150° without change.

Foss, V.L. *et al*, *Zh. Obshch. Khim.*, 1979, **49**, 559.

O,O-Dibutyl phosphonoselenoate, 9CI — D-00131

O,O-*Dibutyl phosphoroselenoite*

$(H_3CCH_2CH_2CH_2O)_2P(Se)H \rightleftharpoons (H_3CCH_2CH_2CH_2O)_2PSeH$

$C_8H_{19}O_2PSe$ M 257.171
Tautomeric. Liq. d_4^{25} 1.20. $Bp_{0.5}$ 76-78°. n_D^{25} 1.4834.

Kuznetsov, E.V. *et al*, *CA*, 1958, **52**, 8938.

O,O-Dibutyl phosphonothioate, 9CI — D-00132

O,O-*Dibutyl thiophosphite*
[17529-47-4]

$(H_3CCH_2CH_2CH_2O)_2P(S)H \rightleftharpoons (H_3CCH_2CH_2CH_2O)_2PSH$

$C_8H_{19}O_2PS$ M 210.271
Tautomeric but exists almost completely in the thiophosphoryl form. Liq. with strong, sickly odour. Bp_3 88-89°. n_D^{20} 1.4535.

Murav'ev, I.V. *et al*, *Zh. Obshch. Khim.*, 1968, **38**, 133 (*Engl. transl.* p. 133) (*synth*)

Dibutyl phosphoramidate, 9CI, 8CI — D-00133

Dibutyl phosphoramide. Dibutyl amidophosphate. Dibutyl phosphoryl amide
[870-52-0]

$(H_3CCH_2CH_2CH_2O)_2P(O)NH_2$

$C_8H_{20}NO_3P$ M 209.225
Liq. d_4^{20} 1.04. Bp_{13} 182-185°, $Bp_{0.1}$ 117°. n_D^{20} 1.4353.
▷TB2802500.

Petrov, K.A. *et al*, *Zh. Obshch. Khim.*, 1960, **30**, 1233 (*Engl. transl.* p. 1256) (*synth*)
Tsolis, A.K. *et al*, *Tetrahedron Lett.*, 1964, 3217 (*synth*)
Jakobsen, P. *et al*, *Org. Mass. Spectrom.*, 1972, **6**, 1303 (*ms*)

Dibutylphosphoramidic acid, 9CI — D-00134

$(H_3CCH_2CH_2CH_2)_2NP(O)(OH)_2$

$C_8H_{20}NO_3P$ M 209.225
Di-Me ester: [74130-08-8]. *Dimethyl dibutylphosphoramidate.*
$C_{10}H_{24}NO_3P$ M 237.278
Liq. Bp_{15} 96°. n_D^{25} 1.4355.
Di-Et ester: [67828-17-5]. *Diethyl dibutylphosphoramidate.*
$C_{12}H_{28}NO_3P$ M 265.332
Liq. Bp_3 115°.
▷TB2801500.
Diisopropyl ester: Diisopropyl dibutylphosphoramidate.
$C_{14}H_{32}NO_3P$ M 293.385
Liq. $Bp_{1.5}$ 96°.
Dibutyl ester: [53796-00-2]. *Dibutyl dibutylphosphoramidate.*
$C_{16}H_{36}NO_3P$ M 321.439
Extractant for rare earth metals and Ga. Liq. Bp_8 166-168°. n_D^{20} 1.4390.
Dichloride: [33876-59-4].
$C_8H_{18}Cl_2NOP$ M 246.116
Liq. $Bp_{1.5}$ 122-124°.

Cheymol, J. *et al*, *C.R. Hebd. Seances Acad. Sci.*, 1959, **249**, 1240 (*dimethyl ester*)
Ger. Pat., 1 033 200, (*1958*); *CA*, **54**, 14124 (*esters*)
Cates, L.A. *et al*, *J. Med. Chem.*, 1971, **14**, 647 (*dichloride*)
Bebikh, G. *et al*, *Zh. Obshch. Khim.*, 1977, **47**, 2196 (*Engl. transl.* p. 2005) (*dibutyl ester, synth, ir*)

Dibutylphosphoramidochloridous acid, 9CI — D-00135

$(H_3CCH_2CH_2CH_2)_2NP(OH)Cl$

$C_8H_{19}ClNOP$ M 211.671
Et ester: Ethyl dibutylphosphoramidochloridite.
$C_{10}H_{23}ClNOP$ M 239.725
Liq. $Bp_{0.05}$ 64-65°. n_D^{20} 1.4672.
Butyl ester: Butyl dibutylphosphoramidochloridite.
$C_{12}H_{27}ClNOP$ M 267.778
Liq. $Bp_{0.08}$ 73-76°. n_D^{20} 1.4675.

Zwierzak, A. *et al*, *Tetrahedron*, 1967, **23**, 2243.

Dibutylphosphoramidous acid, 9CI — D-00136

$(H_3CCH_2CH_2CH_2)_2NP(OH)_2$

$C_8H_{20}NO_2P$ M 193.225
Di-Et ester: [41064-81-7]. *Diethyl dibutylphosphoramidite.*
$C_{12}H_{28}NO_2P$ M 249.332
Liq. Bp_{11} 117-117.5°. n_D^{20} 1.4396.
Dibutyl ester: [32596-67-1]. *Dibutyl dibutylphosphoramidite.*
$C_{16}H_{36}NO_2P$ M 305.440
Liq. Bp_3 130-132°. n_D^{20} 1.4398.

Petrov, K.A. *et al*, *Zh. Obshch. Khim.*, 1962, **32**, 3065 (*Engl. transl.* p. 3015) (*dibutyl ester*)
Kabachnik, M.I. *et al*, *Zh. Obshch. Khim.*, 1982, **52**, 1033 (*Engl. transl.* p. 899) (*diethyl ester*)

Di-*tert*-butylphosphoramidous acid, 8CI — D-00137

Bis(1,1-dimethylethyl)phosphoramidous acid, 9CI

$[(H_3C)_3C]_2NP(OH)_2$

$C_8H_{20}NO_2P$ M 193.225

Difluoride:

$C_8H_{18}F_2NP$ M 197.207

Liq. Bp_{12} 58-60°.

Dichloride: [56607-96-6].

$C_8H_{18}Cl_2NP$ M 230.117

Solid. Mp 30-32°. $Bp_{0.1}$ 59-61°.

Dibromide:

$C_8H_{18}Br_2NP$ M 319.019

Solid. Mp 60-62°.

Gouesnard, J-P. *et al, Can. J. Chem.*, 1980, **58**, 1295 (*esters, N and P nmr, struct*)
Scherer, O.J. *et al, Chem. Ber.*, 1975, **108**, 2478 (*dihalides, synth, pmr, F and P nmr, ms, struct*)

Di-*tert*-butyl phosphorobromidate, 8CI D-00138

*Bis(1,1-dimethylethyl) phosphorobromidate, 9CI. Di-*tert*-butyl bromophosphate. Di-*tert*-butyl phosphoryl bromide*

[59346-65-5]

$[(H_3C)_3CO]_2P(O)Br$

$C_8H_{18}BrO_3P$ M 273.106

Selective phosphorylating agent containing acid-labile protecting groups for primary and secondary alcohols. Activating agent for peptide synthesis superior to the chloride in its reactivity. Unstable liq.

Gajda, T. *et al, Synthesis*, 1976, 243; 1977, 623 (*synth, ir, pmr, nmr, use*)
Gorecka, A. *et al, Synthesis*, 1978, 474 (*use*)

Dibutyl phosphorobromidite, 9CI, 8CI D-00139

Dibutyl bromophosphite

[53764-94-6]

$(H_3CCH_2CH_2CH_2O)_2PBr$

$C_8H_{18}BrO_2P$ M 257.107

Fuming liq. Bp_2 62-64°, $Bp_{0.05}$ 50°. n_D^{20} 1.4642. Easily hydrol.

Gerrard, W. *et al, J. Chem. Soc.*, 1955, 277 (*synth, props*)
Novikova, Z.S. *et al, Zh. Obshch. Khim.*, 1974, **44**, 1857 (*Engl. transl.* p. 1805) (*synth, P nmr*)

Dibutyl phosphorochloridate, 9CI, 8CI D-00140

Dibutyl phosphoryl chloride. Dibutyl chlorophosphate

[819-43-2]

$(H_3CCH_2CH_2CH_2O)_2P(O)Cl$

$C_8H_{18}ClO_3P$ M 228.655

Pungent liq. d_4^{14} 1.08. Bp_{15} 132-135°, $Bp_{0.5}$ 80-81°. n_D^{25} 1.4378.

Gerrard, W., *J. Chem. Soc.*, 1940, 1464 (*synth*)
De Roos, A.M. *et al, Recl. Trav. Chim. Pays-Bas*, 1958, **77**, 946 (*synth*)
Grosse-Ruyken, H. *et al, J. Prakt. Chem.*, 1962, **18**, 287.
Bliznyuk, N.K. *et al, Zh. Obshch. Khim.*, 1967, **37**, 1353 (*Engl. transl.*, p. 1279) (*synth*)
Sosnovsky, G. *et al, J. Org. Chem.*, 1969, **34**, 968.
Sax, N.I., *Dangerous Properties of Industrial Materials*, 6th Ed., Van Nostrand-Reinhold, 1984, 551.

Di-*tert*-butyl phosphorochloridate, 8CI D-00141

*Bis(1,1-dimethylethyl) phosphorochloridate, 9CI. Di-*tert*-butyl phosphoryl chloride. Di-*tert*-butyl chlorophosphate*

[56119-60-9]

$[(H_3C)_3CO]_2P(O)Cl$

$C_8H_{18}ClO_3P$ M 228.655

Phosphorylating possessing acid labile groups, but less effective than Di-*tert*-butyl phosphorobromidate, D-00138 . Liq. n_D^{20} 1.4286. Dec. on attempted dist.

Goldwhite, H. *et al, J. Chem. Soc.*, 1957, 2409 (*synth*)
Gajda, T. *et al, Synthesis*, 1976, 243 (*synth, ir, pmr, P nmr*)

Dibutyl phosphorochloridite, 9CI, 8CI D-00142

Dibutyl chlorophosphite

[4124-92-9]

$(H_3CCH_2CH_2CH_2O)_2PCl$

$C_8H_{18}ClO_2P$ M 212.656

Liq. d_0^{15} 1.01. Bp_{10} 96-98°. n_D^{20} 1.4455.

Gerrard, W., *J. Chem. Soc.*, 1940, 1464; 1953, 1920 (*synth, props*)
Michalski, J. *et al, J. Chem. Soc.*, 1961, 4904 (*synth*)

Di-*tert*-butyl phosphorochloridite, 8CI D-00143

*Bis(1,1-dimethylethyl) phosphorochloridite, 9CI. Di-*tert*-butyl chlorophosphite*

[78543-77-8]

$[(H_3C)_3CO]_2PCl$

$C_8H_{18}ClO_2P$ M 212.656

Liq. V. unstable.

Mark, V. *et al, J. Org. Chem.*, 1964, **29**, 1006 (*P nmr*)

S,S-Dibutyl phosphorochloridodithioate, 9CI, 8CI D-00144

S,S-*Dibutyl chlorodithiophosphate*

[2797-55-9]

$(H_3CCH_2CH_2CH_2S)_2P(O)Cl$

$C_8H_{18}ClOPS_2$ M 260.776

Liq. d_4^{20} 1.12. $Bp_{0.7}$ 121-124°. n_D^{20} 1.5253.

Sorokina, S.F. *et al, Zh. Obshch. Khim.*, 1973, **43**, 750 (*Engl. transl.* p. 748) (*synth, P nmr*)

O,O-Dibutyl phosphorochloridothioate, 9CI, 8CI D-00145

O,O-*Dibutyl thiophosphoryl chloride.* O,O-*Dibutyl chlorothiophosphate*

[2524-07-4]

$(H_3CCH_2CH_2CH_2O)_2P(S)Cl$

$C_8H_{18}ClO_2PS$ M 244.716

Liq. d_4^{20} 1.48. $Bp_{0.5}$ 76°. n_D^{25} 1.4697.

Fletcher, J.H. *et al, J. Am. Chem. Soc.*, 1950, **72**, 2461 (*synth*)
McIvor, R.A. *et al, Can. J. Chem.*, 1956, **34**, 1611 (*ir*)
Popov, E.M. *et al, Zh. Obshch. Khim.*, 1959, **29**, 1998 (*Engl. transl.* p. 1967) (*raman*)

Meinhardt, N.A. *et al*, *J. Org. Chem.*, 1960, **25**, 1991 (*synth*)
Kas'yanova, E.F. *et al*, *Zh. Obshch. Khim.*, 1969, **39**, 365 (*Engl. transl.* p. 342)
Omelanczuk, J. *et al*, *Tetrahedron*, 1975, **31**, 2809 (*synth, P nmr*)

N,N'-Dibutylphosphorodiamidic acid, 9CI, 8CI D-00146

$[H_3C(CH_2)_3NH]_2P(O)OH$

$C_8H_{21}N_2O_2P$ M 208.240

Et ester: [27933-10-4]. *Ethyl* N,N'-*dibutylphosphorodiamidate.*
$C_{10}H_{25}N_2O_2P$ M 236.293
Liq. Bp_{15} 159-160°. n_D^{20} 1.4422.

Butyl ester: [27933-15-9]. *Butyl* N,N'-*dibutylphosphorodiamidate.*
$C_{12}H_{29}N_2O_2P$ M 264.347
Liq. Bp_{18} 192-195°. n_D^{20} 1.4505.

Ph ester: [18995-06-7]. *Phenyl* N,N'-*dibutylphosphorodiamidate.*
$C_{14}H_{25}N_2O_2P$ M 284.337
Solid. Mp 54-55°.

Fluoride: [590-69-2]. *Butafox.*
$C_8H_{20}FN_2OP$ M 210.231
Needles (pet. ether). Mp 59.5°. $Bp_{2.5}$ 177°.

▷Extremely neurotoxic. TD3500000.

Heap, R. *et al*, *J. Chem. Soc.*, 1948, 1313 (*fluoride, synth, tox*)
Davies, D.R. *et al*, *Biochem. Pharmacol.*, 1966, **15**, 1783 (*fluoride, tox*)
Hashimoto, S. *et al*, *CA*, 1968, **68**, 40154 (*phenyl ester*)
Luchkovskaya, O.N. *et al*, *Zh. Obshch. Khim.*, 1970, **40**, 644 (*Engl. transl.* p. 615) (*esters, synth*)
Laskorin, B.N. *et al*, *Izv. Akad. Nauk SSSR, Ser. Khim.*, 1978, 1201 (*Engl. transl.* p. 1045) (*butyl ester, ir*)

N,N'-Dibutylphosphorodiamidodithioic acid, 9CI, 8CI D-00147

Dithiophosphoric acid bis(butylamide)

$(H_3CCH_2CH_2CH_2NH)_2P(S)SH$

$C_8H_{21}N_2PS_2$ M 240.361

Butylammonium salt: [25522-77-4]. Cryst. (C_6H_6). Mp 140-141°.

Becke-Goehring, M. *et al*, *Z. Anorg. Allg. Chem.*, 1969, **369**, 73 (*synth, ir, P nmr, complexes*)

O,O-Dibutyl phosphorodiselenoate D-00148

O,O-*Dibutyl phosphorodiselenoic acid, 9CI, 8CI.* O,O-*Dibutyl hydrogen phosphorodiselenoate.* O,O-*Dibutyl diselenophosphate*

[62920-99-4]

$(H_3CCH_2CH_2CH_2O)_2P(Se)SeH$

$C_8H_{19}O_2PSe_2$ M 336.131

Unstable acid isol. as K salt.

K salt: [19843-47-7]. Cryst. (ligroin/EtOH). Readily sol. Me_2CO, MeCN, EtOH, insol. Et_2O. Mp 93-94°.

Kudchadker, M.V. *et al*, *Can. J. Chem.*, 1968, **46**, 1415 (*synth, ir*)
Zemlyanski, N.I. *et al*, *Zh. Obshch. Khim.*, 1971, **41**, 1691 (*Engl. transl.* p. 1699) (*synth*)
Gorak, R.D. *et al*, *Zh. Obshch. Khim.*, 1972, **42**, 56 (*Engl. transl.* p. 52) (*synth*)

Se,Se-Dibutyl phosphorodiselenothioate, 9CI D-00149

Se,Se-*Dibutyl phosphorodiselenothioic acid.* Se,Se-*Dibutyl hydrogen diselenothiophosphate*

$(H_3CCH_2CH_2CH_2Se)_2P(O)SH \rightleftharpoons (H_3CCH_2CH_2CH_2Se)_2P(S)OH$

$C_8H_{19}OPSSe_2$ M 352.191

O-*Et ester:* [66498-99-5]. Se,Se-*Dibutyl* O-*ethyl phosphorodiselenothioate.*
$C_{10}H_{23}OPSSe_2$ M 380.245
Liq. d_4^{20} 1.40. $Bp_{0.01}$ 95°. n_D^{20} 1.5772.

O-*Propyl ester:* [66499-00-1]. Se,Se-*Dibutyl* O-*propyl phosphorodiselenothioate.*
$C_{11}H_{25}OPSSe_2$ M 394.272
Liq. d_4^{20} 1.37. $Bp_{0.01}$ 100°. n_D^{20} 1.5670.

O-*Butyl ester:* [66499-01-2]. O,Se,Se-*Tributyl phosphorodiselenothioate.*
$C_{12}H_{27}OPSSe_2$ M 408.298
Liq. d_4^{20} 1.33. $Bp_{0.01}$ 110°. n_D^{20} 1.5615.

Kolodii, Ya.I. *et al*, *Zh. Obshch. Khim.*, 1978, **48**, 331 (*Engl. transl.* p. 296) (*synth*)

O,O-Dibutyl phosphorodithioate, 9CI, 8CI D-00150

O,O-*Dibutyl hydrogen dithiophosphate.* O,O-*Dibutyl dithiophosphoric acid*

[2253-44-3]

$(H_3CCH_2CH_2CH_2O)_2P(S)SH$

$C_8H_{19}O_2PS_2$ M 242.331

Extractant for Ni, Zr, U and Ga. Zr complex used as lubricant additive. Oil. d_4^{20} 1.06. Bp_2 99-99.5°, $Bp_{0.8\text{-}1.0}$ 120°. pK_a 1.83 (7% EtOH aq.), pK_a 2.64 (80% EtOH aq.). n_D^{20} 1.4971.

K salt: [3549-51-7]. Flotation agent for Cu and Ni sulfide ores. Solid. Mp 147-147.5°.

McIver, R.A. *et al*, *Can. J. Chem.*, 1958, **36**, 820 (*ir*)
Kabachnik, M.I. *et al*, *Tetrahedron*, 1960, **9**, 10 (*synth*)
Almasi, L. *et al*, *CA*, 1965, **62**, 2729 (*synth*)
Bolotova, G.L. *et al*, *CA*, 1965, **63**, 6897 (*synth*)
Lefferts, J.L. *et al*, *Inorg. Chem.*, 1980, **19**, 1662 (*synth, complexes*)

O,O-Di-*tert*-butyl phosphorodithioate, 8CI D-00151

O,O-*Bis(1,1-dimethyl) phosphorodithioate, 9CI.* O,O-*Di*-tert-*butyl hydrogen dithiophosphate.* O,O-*Di*-tert-*butyl dithiophosphoric acid*

$[(H_3C)_3CO]_2P(S)SH$

$C_8H_{19}O_2PS_2$ M 242.331

Unstable liq. n_D^{25} 1.4761.

Pb salt: Solid. Mp 109° dec.

Hu, P.-F. *et al*, *CA*, 1959, **53**, 3120.

S,S-Dibutyl phosphorodithioate, 9CI, 8CI D-00152

S,S-*Dibutyl hydrogen dithiophosphate.* S,S-*Dibutyl dithiophosphate.* S,S-*Dibutyl dithiophosphoric acid*

$(H_3CCH_2CH_2CH_2S)_2P(O)OH$

$C_8H_{19}O_2PS_2$ M 242.331

Cyclohexylammonium salt: [72284-34-5]. Solid. Mp 151-152°.

Chloride: see S,S-Dibutyl phosphorochloridodithioate, D-00144
Amide: S,S-*Dibutyl phosphoramidodithioate.* S,S-*Dibutyl amidodithiophosphate.*
$C_8H_{20}NOPS_2$ M 241.346
Oil. n_D^{25} 1.5306. Dec. on attempted dist.

Ger. Pat., 2 013 956, (*1971*); *CA*, **74**, 87380 (*amide*)
Yamaguchi, K. *et al*, *Chem. Lett.*, 1979, 1057, (*synth*)

Dibutyl phosphorofluoridate, 9CI, 8CI **D-00153**

Dibutyl phosphoryl fluoride. Dibutyl fluorophosphate. Dibutoxyphosphinyl fluoride
[674-48-6]

$(H_3CCH_2CH_2CH_2O)_2P(O)F$

$C_8H_{18}FO_3P$ M 212.201
Liq. Bp_{12} 102°. Low toxicity, produces negligible miosis.

Cook, H.G. *et al*, *J. Chem. Soc.*, 1949, 635 (*synth, tox*)
Landau, M.A. *et al*, *Zh. Strukt. Khim.*, 1970, **11**, 513 (*Engl. transl.* p. 467) (*struct*)
Ryzhikov, B.D. *et al*, *Zh. Strukt. Khim.*, 1975, **16**, 754 (*Engl. transl.* p. 700) (*F nmr*)
Nikitin, E.V. *et al*, *Dokl. Akad. Nauk SSSR, Ser. Sci. Khim.*, 1980, **252**, 922 (*Engl. transl.* p. 442) (*synth, F and P nmr*)
Gubaidullin, M.G., *Zh. Obshch. Khim.*, 1982, **52**, 2469 (*props*)
Mager, P.P., *Toxicol. Lett.*, 1982, **11**, 67 (*tox*)
Gupta, O.D. *et al*, *J. Chem. Soc., Chem. Commun.*, 1984, 416 (*synth*)

***O,O*-Dibutyl phosphoroselenoate, 9CI, 8CI** **D-00154**

O,O-*Dibutyl phosphoroselenoic acid.* O,O-*Dibutyl hydrogen phosphoroselenoate*
[62558-54-7]

$(H_3CCH_2CH_2CH_2O)_2P(Se)OH \rightleftharpoons (H_3CCH_2CH_2CH_2O)_2P(O)SeH$

$C_8H_{19}O_3PSe$ M 273.170
K salt: [85290-14-8]. Cryst. (CH_3Cl_3/Et_2O or C_6H_6/pet. ether).
Chloride: [55578-33-1]. O,O-*Dibutyl phosphorochloridoselenoate, 9CI, 8CI.* O,O-*Dibutyl chloroselenophosphate.* O,O-*Dibutyl selenophosphoryl chloride.*
$C_8H_{18}ClO_2PSe$ M 291.616
Liq. d_4^{20} 1.30. $Bp_{0.07}$ 80.5°. n_D^{20} 1.4920. Unstable to moisture.

Foss, O., *Acta Chem. Scand.*, 1947, **1**, 8 (*synth*)
Nuretdinov, I.A. *et al*, *Zh. Obshch. Khim.*, 1975, **45**, 533 (*Engl. transl.* p. 526) (*chloride, synth, ir, pmr, P nmr*)

***O,O*-Dibutyl phosphorothioate, 9CI, 8CI** **D-00155**

O,O-*Dibutyl hydrogen phosphorothioate.* O,O-*Dibutylphosphorothioic acid.* O,O-*Dibutyl hydrogen thiophosphate.* O,O-*Dibutyl thiophosphoric acid*
[10163-62-9]

$(H_3CCH_2CH_2CH_2O)_2P(O)SH \rightleftharpoons (H_3CCH_2CH_2CH_2O)_2P(S)OH$

$C_8H_{19}O_3PS$ M 226.270
Conveniently isol. and stored as K or NH_4 salt; 38% thiol form (7% EtOH aq.), 11% thiol form (80% EtOH aq.) at r.t. Oil. d_4^{20} 1.07. $Bp_{0.08}$ 88-89°. n_D^{20} 1.4654.
NH_4 salt: [35329-22-7]. Solid. Mp 147-150°.
K salt: [51825-87-7]. Cryst. ($CHCl_3$).
*Chloride: see O,O-*Dibutyl phosphorochloridothioate, D-00145

Foss, O., *Acta Chem. Scand.*, 1947, **1**, 8 (*synth*)
Kabachnik, M.I. *et al*, *Tetrahedron*, 1960, **9**, 10 (*synth, struct*)
Pesin, V.G. *et al*, *Zh. Obshch. Khim.*, 1961, **31**, 2508 (*Engl. transl.* p. 2337) (*synth*)
Mastryukova, T.A. *et al*, *Zh. Obshch. Khim.*, 1974, **44**, 1001 (*Engl. transl.* p. 963); *Phosphorus Sulfur*, 1976, **1**, 211 (*props*)

***O,O*-Di-*tert*-butyl phosphorothioate, 8CI** **D-00156**

O,O-*Bis(1,1-dimethylethyl) phosphorothioate, 9CI.* O,O-*Di*-tert-*butyl phosphorothioic acid.* O,O-*Di*-tert-*butyl hydrogen phosphorothioate.* O,O-*Di*-tert-*butyl hydrogen thiophosphate*
[45098-72-4]

$[(H_3C)_3CO]_2P(O)SH \rightleftharpoons [(H_3C)_3CO]_2P(S)OH$

$C_8H_{19}O_3PS$ M 226.270
Cryst. (Et_2O). Mp 90-93° dec.
Na salt: [37173-14-1]. Cryst. (2-propanol/pet. ether). Mp 156° dec.
Triethylammonium salt: Reagent for synth. of *S*-alkyl phosphorothioic acids. Cryst. (pet. ether). Mp 80-81°.

Mark, V. *et al*, *J. Org. Chem.*, 1964, **29**, 1006 (*synth, ir, P nmr*)
Zwierzak, A. *et al*, *Z. Naturforsch., B*, 1971, **26**, 386 (*synth, ir, use*)
Chapman, T.M. *et al*, *J. Org. Chem.*, 1973, **38**, 250 (*synth, pmr*)

4,6-Di-*tert*-butylspiro[1,3,2-benzodioxaphosphole-2,1′-[2,8,9]-trioxa[1]-phosphaadamantane], 8CI **D-00157**

4,6-Bis(1,1-dimethylethyl)spiro[1,3,2-benzodioxaphosphole-2,1′-[2,8,9]trioxa[1]phosphatricyclo[3.3.1.1^{3,7}]-decane], 9CI
[75550-55-9]

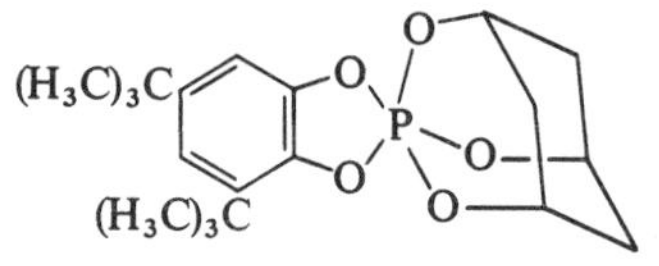

$C_{20}H_{29}O_5P$ M 380.420
Solid. Mp 112-114°.

Navech, J. *et al*, *Tetrahedron Lett.*, 1980, **21**, 1449 (*synth, P nmr*)

2,9-Di-*tert*-butyl-1,3,8,10-tetraoxa-2,9-diphosphacyclotetradecane **D-00158**

2,9-Bis(1,1-dimethylethyl)-1,3,8,10-tetraoxa-2,9-diphosphacyclotetradecane, 10CI

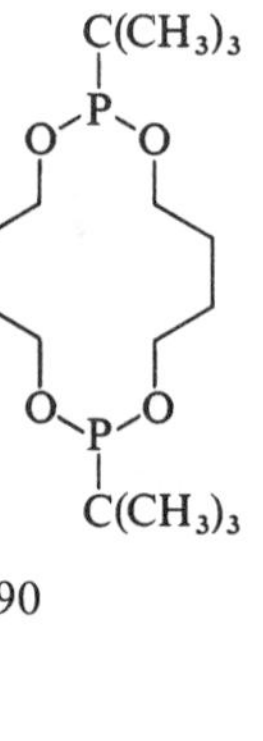

$C_{16}H_{34}O_4P_2$ M 352.390

2,9-Disulfide:
$C_{16}H_{34}O_4P_2S_2$ M 416.510
Obtained as a mixture of stereoisomers.

Dutasta, J.-P. *et al*, *Tetrahedron Lett.*, 1977, 801 (*synth, deriv, pmr, P nmr, cmr*)

2,8-Di-*tert*-butyl-1,3,7,9-tetrathia-2,8-diphosphacyclododecane D-00159

2,8-Bis(1,1-dimethylethyl)-1,3,7,9-tetrathia-2,8-diphosphacyclododecane, 9CI

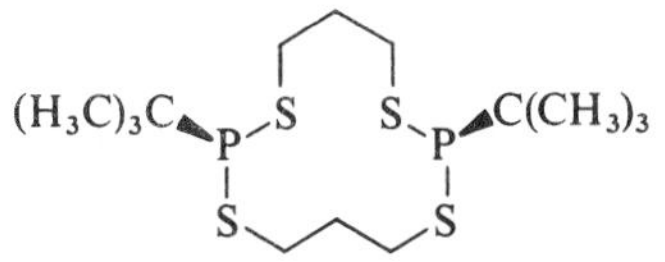

(H₃C)₃C–P(S)(S) ... P–C(CH₃)₃ *cis-form*

$C_{14}H_{30}P_2S_4$ M 388.579
Dimer of 2-*tert*-Butyl-1,3,2-dithiaphosphorinane. Exists as mixt. of *cis*- and *trans*-forms. At 80°, the *trans* form is converted into the *cis* form, and at 160°, the monomer is also produced.

cis-form
2,8-Disulfide:
$C_{14}H_{30}P_2S_6$ M 452.699
Solid. Mp 162-164°.

trans-form
2,8-Disulfide: Solid. Mp 250-252°.

Dutasta, J.-P. *et al*, *J. Org. Chem.*, 1977, **42**, 1662 (*synth, props, derivs, pmr, cmr*)
Martin, J. *et al*, *Org. Magn. Reson.*, 1981, **15**, 87 (*P nmr*)

(1,2-Dicarboxyethyl)-triphenylphosphonium(1+) D-00160

$Ph_3P^{\oplus}CH(COOH)CH_2COOH$

$C_{22}H_{20}O_4P^{\oplus}$ M 379.371 (ion)
2M NaOH aq. causes dec. of salts to Ph_3PO and succinic acid.
Chloride: [64598-21-6].
$C_{22}H_{20}ClO_4P$ M 414.824
Cryst. (MeOH/EtOAc). Mp 125-130° dec. Possibly decarboxylates to some extent on recryst.
Bromide: [64598-17-0].
$C_{22}H_{20}BrO_4P$ M 459.275
Cryst. + $1H_2O$. Mp 140° dec.
Di-Et ester, iodide:
$C_{26}H_{28}O_4IP$ M 562.383
Cryst. + $1H_2O$. Mp 104°.
Ylide: see (Triphenylphosphoranylidene)butanedioic acid, T-00646

Hoffmann, H., *Chem. Ber.*, 1961, **94**, 1331 (*bromide, ester, props*)
Hudson, R.F. *et al*, *Helv. Chim. Acta*, 1963, **46**, 2178 (*chloride, props*)

Dichlofenthion, BSI D-00161

O-*2,4-Dichlorophenyl* O,O-*diethyl phosphorothioate, 9CI, 8CI*
[97-17-6]

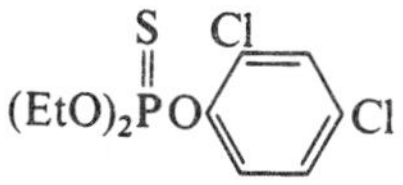

$(EtO)_2P(S)O$–C₆H₃Cl₂ (2,4-dichlorophenyl)

$C_{10}H_{13}Cl_2O_3PS$ M 315.151
Nonsystemic nematocide, soil insecticide. Liq. d_4^{20} 1.31. $Bp_{0.2}$ 126-131°. n_D^{25} 1.5318.
▷TF0350000.

Mandel'baum, Ya.A. *et al*, *Zh. Obshch. Khim.*, 1959, **29**, 1149 (*Engl. transl.* p. 1120) (*synth*)
U.S.P., 3 004 054, (*1961*); *CA*, **56**, 8636 (*synth*)
Ross, R.T. *et al*, *Anal. Chim. Acta*, 1970, **52**, 139 (*P nmr*)
Nuretdinov, I.A. *et al*, *Izv. Akad. Nauk SSSR, Ser. Khim.*, 1971, 1266 (*Engl. transl.* p. 1170) (*synth, P nmr, ir, tox*)
Nicholas, M.L. *et al*, *J. Assoc. Off. Anal. Chem.*, 1976, **59**, 1071 (*raman*)
Busch, K.L. *et al*, *Appl. Spectrosc.*, 1978, **32**, 388 (*ms*)
Ripley, B.D. *et al*, *J. Assoc. Off. Anal. Chem.*, 1983, **66**, 1084 (*glc*)
Pesticide Manual, 6th Ed., 167.

2,4-Dichloro-2,4-bis(dimethylamino)-2,2,4,4,6,6-hexahydro-6,6-diphenyl-1,3,5-triaza-2,4,6-triphosphorine, 9CI D-00162

2,4-Dichloro-2,4-bis(dimethylamino)-2,2,4,4,6,6-hexahydro-6,6-diphenylcyclotriphosphazene
[17242-13-6]

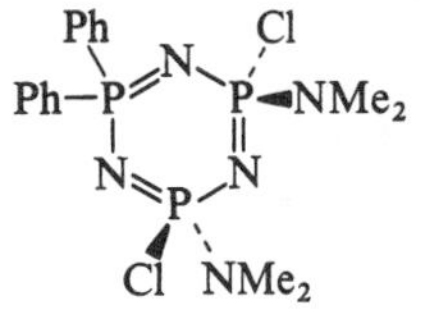

Ph, Ph–P, N, Cl, P–NMe₂, N, N, P, Cl, NMe₂ *(2RS,4RS)-form*

$C_{16}H_{22}Cl_2N_5P_3$ M 448.211
(2RS,4RS)-form [23728-60-1]
trans-*form*
Solid. Mp 144°.
(2RS,4SR)-form [60492-83-3]
cis-*form*
No phys. props. reported.

Hills, K. *et al*, *J. Chem. Soc.*, 1964, 130 (*synth*)
Desai, V.B. *et al*, *J. Chem. Soc. (A)*, 1969, 1977 (*synth, pmr*)

2,4-Dichloro-1,3-di-*tert*-butyl-1,3,2,4-diazadiphosphetidine, 8CI D-00163

2,4-Dichloro-1,3-bis(1,1-dimethylethyl)-1,3,2,4-diazadiphosphetidine, 9CI. P,P'-Dichloro-N,N'-di-tert-butylcyclodiphosphazane
[24335-35-1]

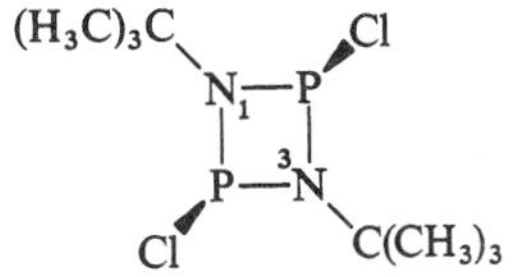

(H₃C)₃C–N¹–P–Cl, P–³N–C(CH₃)₃, Cl *cis-form*

$C_8H_{18}Cl_2N_2P_2$ M 275.097
Dec. in moist atmosphere.

cis-form [35107-68-7]
Mp 42°.
2-Oxide: [49774-26-7].
$C_8H_{18}Cl_2N_2OP_2$ M 291.097
Liq. $Bp_{0.1}$ 90-95°.
2-Sulfide: [49774-27-8].
$C_8H_{18}Cl_2N_2P_2S$ M 307.157
Liq. $Bp_{0.05}$ 86-90°.
2-Oxide, 4-sulfide: [49774-28-9].
$C_8H_{18}Cl_2N_2OP_2S$ M 323.157
Solid. Mp 85-101°.

trans-form
Mp 67-68° (?).
2,4-Dioxide: [42366-28-9].
$C_8H_{18}Cl_2N_2O_2P_2$ M 307.096
Needles. Mp 139-140°. Slowly dec. at r.t.

Jefferson, R. *et al*, *J. Chem. Soc., Dalton Trans.*, 1973, 1414 (*synth, derivs, pmr, P nmr*)
Keat, R. *et al*, *Angew. Chem., Int. Ed. Engl.*, 1973, **12**, 311 (*dioxide*)
Manojlović-Muir, L. *et al*, *J. Chem. Soc., Dalton Trans.*, 1974, 2395 (*dioxide, cryst struct*)
Muir, K.W., *J. Chem. Soc., Dalton Trans.*, 1975, 259 (*cryst struct*)
Bulloch, G. *et al*, *J. Chem. Soc., Dalton Trans.*, 1978, 764 (*pmr, cmr*)
Bulloch, G. *et al*, *Org. Magn. Reson.*, 1979, **12**, 708 (*derivs, pmr, P nmr*)
Kuhn, N. *et al*, *J. Organomet. Chem.*, 1983, **243**, C47 (*complexes*)

5,10-Dichloro-5,10-dihydrophosphanthrene, 9CI, 8CI D-00164

[63586-83-4]

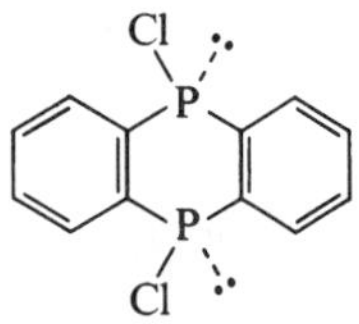

$C_{12}H_8Cl_2P_2$ M 285.049

Cryst. Mp 135-137°.

5,10-Dioxide: [63586-87-8].

$C_{12}H_8Cl_2O_2P_2$ M 317.048

Solid. Mp >300°.

Kovaleva, T.V., *Zh. Obshch. Khim.*, 1977, **47**, 1036 (*Engl. transl.* p. 950)

2,2-Dichloro-2,2-dihydro-4-trichloromethyl-1,3,2-diazaphosphete, 9CI D-00165

[42563-95-1]

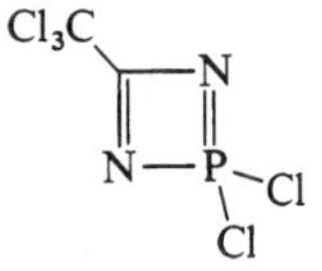

$C_2Cl_5N_2P$ M 260.274

Prisms (C_6H_6). Mp 83-84°. $Bp_{0.05}$ 117-119°.

Kukhar', V.P. *et al*, *Zh. Obshch. Khim.*, 1973, **43**, 743 (*Engl. transl.* p. 741) (*synth, props, ir, nqr*)
Kukhar', V.P. *et al*, *Zh. Obshch. Khim.*, 1976, **46**, 1462 (*Engl. transl.* p. 1436) (*synth*)

2,4-Dichloro-1,3-dimethyl-1,3,2,4-diazadiphosphetidine, 9CI D-00166

[1679-91-0]

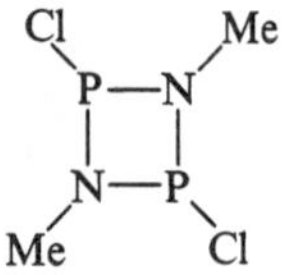

$C_2H_6Cl_2N_2P_2$ M 190.936

2,4-Dioxide: [5944-59-2].

$C_2H_6Cl_2N_2O_2P_2$ M 222.935

Cryst. (cyclohexane). Mp 103°.

2,4-Disulfide: [5944-50-3].

$C_2H_6Cl_2N_2P_2S_2$ M 255.056

Cryst. (pet. ether or cyclohexane). Mp 120-122°.

Latscha, H.P. *et al*, *Z. Anorg. Allg. Chem.*, 1968, **359**, 81 (*derivs, synth*)
Keat, R., *J. Chem. Soc., Dalton Trans.*, 1972, 2189 (*disulfide, synth, pmr, nmr*)
Bulloch, G. *et al*, *J. Chem. Soc., Dalton Trans.*, 1974, 2010 (*derivs, nmr, pmr*)
Dagleish, W.H. *et al*, *J. Magn. Reson.*, 1975, **20**, 359 (*dioxide, nqr*)

2,5-Dichloro-3,4-dimethyl-1,3,4,2,5-thiadiazadiphospholidine, 9CI D-00167

[59725-20-1]

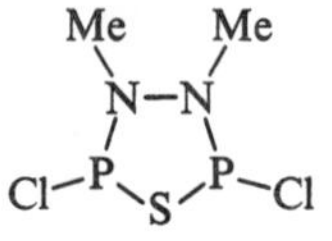

$C_2H_6Cl_2N_2P_2S$ M 222.996

Liq.

▷Reacts explosively with H_2O

Nöth, H. *et al*, *Chem. Ber.*, 1976, **109**, 1942 (*synth, pmr, P nmr, ms, props*)

2,2-Dichloroethenyl phosphorodichloridate, 9CI D-00168

2,2-Dichlorovinyl phosphorodichloridate, 8CI

[20202-72-6]

$Cl_2C{=}CHOP(O)Cl_2$

$C_2HCl_3O_2P$ M 194.361

Liq. Bp_{14} 80-82°. n_D^{20} 1.4962.

Malenko, D.M. *et al*, *Zh. Obshch. Khim.*, 1982, **52**, 2794 (*Engl. transl.*) (*synth, P nmr*)

2,2-Dichloro-2,2,4,4,6,6-hexahydro-4,4,6,6-tetraphenyl-1,3,5-triaza-2,4,6-triphosphorine, 9CI D-00169

2,2-Dichloro-2,2,4,4,6,6-hexahydro-4,4,6,6-tetraphenylcyclotriphosphazene

[3606-94-8]

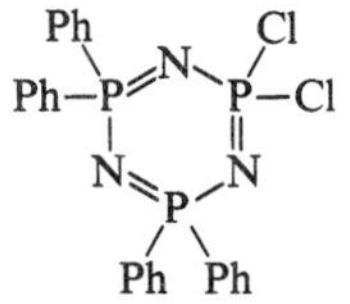

$C_{24}H_{20}Cl_2N_3P_3$ M 514.269

Cryst. (C_6H_6/hexane). Mp 141.5-143°.

McBee, E.T. *et al*, *Inorg. Chem.*, 1965, **4**, 1672 (*synth*)
Wagner, A.J. *et al*, *J. Inorg. Nucl. Chem.*, 1971, **33**, 1307 (*uv*)
Keat, R. *et al*, *J. Chem. Soc., Dalton Trans.*, 1972, 1648 (*nqr*)
Keat, R. *et al*, *J. Chem. Soc., Dalton Trans.*, 1976, 1582 (*P nmr*)
Allen, C.W. *et al*, *J. Chem. Soc., Dalton Trans.*, 1978, 173 (*ms*)
Krishnamurthy, S.S. *et al*, *Org. Magn. Reson.*, 1981, **15**, 205 (*cmr*)

4,6-Dichloro-1,3,5,7,9,10-hexamethyl-1,3,5,7,9,10-hexaaza-4,6-diphospha(4,6-P^V)dispiro[3.1.3.1]decane-2,8-dione, 8CI D-00170

[25674-41-3]

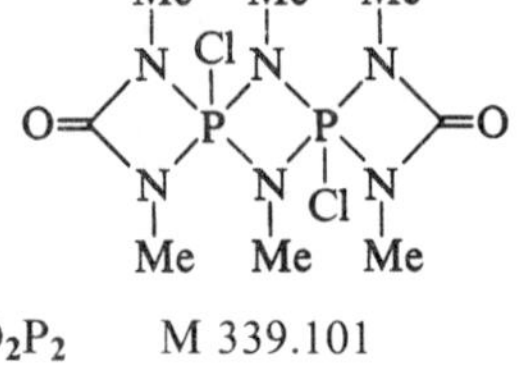

$C_6H_{18}Cl_2N_6O_2P_2$ M 339.101

Cryst. ($CHCl_3$). Mp 178-180° dec.

Becke-Goehring, M. *et al*, *Z. Anorg. Allg. Chem.*, 1970, **372**, 285 (*synth, ir, ms, pmr, P nmr*)

O-(2,5-Dichloro-4-iodophenyl) O-ethyl ethylphosphonothioate, 9CI **D-00171**

CIBA 18244

[25177-27-9]

$C_{10}H_{12}Cl_2IO_2PS$ M 425.048

Soil insecticide. Solid. Mp 60-1°.

Ger. Pat., 1 925 653, (*1969*); *CA*, **72**, 79223 (*synth*)
Harris, C.R., *J. Econ. Entomol.*, 1973, **66**, 216 (*use*)

Dichloromethylenebisphosphonic acid, 9CI **D-00172**

Dichloromethanediphosphonic acid. Clodronic acid

[10596-23-3]

$Cl_2C[P(O)(OH)_2]_2$

$CH_4Cl_2O_6P_2$ M 244.893

Used clinically in the treatment of tumoral bone disease. Prevents mineral formation in bone. pK_{a1} 1.70, pK_{a2} 2.13, pK_{a3} 5.66, pK_{a4} 8.30 (25°, H_2O).

Tetra-Et ester: [19928-97-3]. *Tetraethyl dichloromethylenebisphosphonate.*
$C_9H_{20}Cl_2O_6P_2$ M 357.107
Liq. $Bp_{0.05}$ 119-120°. n_D^{20} 1.4619.

Tetraisopropyl ester: Tetraisopropyl dichloromethylenebisphosphonate.
$C_{11}H_{24}ClO_6P_2$ M 349.708
Solid. Mp 56°.

Quimby, O.T. *et al*, *J. Organomet. Chem.*, 1968, **13**, 199 (*esters, synth, P nmr*)
Francis, M.D. *et al*, *J. Chem. Educ.*, 1978, **55**, 760 (*rev*)
Fleisch, H. *et al*, *Calcif. Tissue Int.*, 1979, **27**, 91 (*props, rev*)
Kukhar, V.P. *et al*, *Zh. Obshch. Khim.*, 1979, **49**, 1470 (*Engl. transl.* p. 1284) (*esters, synth, ir*)
Fonong, T. *et al*, *Anal. Chem.*, 1983, **55**, 1089 (*complexes*)

Dichloromethylenetris(dimethylamino)phosphorane **D-00173**

*1-Dichloromethylene-*N,N,N′,N′,N″,N″*-hexamethylphosphoranetriamine, 10CI, 9CI*

[59578-17-5]

$(Me_2N)_3P{=}CCl_2$

$C_7H_{18}Cl_2N_3P$ M 246.119

Prepared *in situ*. Wittig reagent for the formn. of 1,1-dichloroalkenes and cyclopropanes and for modifications to carbohydrates.

Salmond, W.G., *Tetrahedron Lett.*, 1974, 1237, 1239 (*synth, use*)
Tronchet, J.M.J. *et al*, *Helv. Chim. Acta*, 1976, **59**, 941 (*use*)
European Pat., 7 142, (*1980*); *CA*, **93**, 71110s (*synth, use*)

P-Dichloromethylphosphonamidic acid, 9CI **D-00174**

$CH_4Cl_2NO_2P$ M 163.928

Me ester: [85437-50-9]. *Methyl* P*-dichloromethylphosphonamidate.*
$C_2H_6Cl_2NO_2P$ M 177.955
Solid. Mp 120°.

Et ester: [42003-30-5]. *Ethyl* P*-dichloromethylphosphonamidate.*
$C_3H_8Cl_2NO_2P$ M 191.981
Cryst. (CCl_4). Mp 78-79°.

Shokol, V.A. *et al*, *Zh. Obshch. Khim.*, 1973, **43**, 267 (*Engl. transl.* p. 266) (*synth*)
Khazanchi, R. *et al*, *Agric. Biol. Chem.*, 1983, **47**, 331 (*synth, derivs, pharmacol*)

(Dichloromethyl)phosphonic acid, 9CI **D-00175**

[13113-88-7]

$Cl_2CHP(O)(OH)_2$

$CH_3Cl_2O_3P$ M 164.913

Solid. Mp 123°. pK_{a1} 1.44, pK_{a2} 4.12 (H_2O, 25°).

Di-Me ester: [58993-56-9]. *Dimethyl (dichloromethyl)-phosphonate.*
$C_3H_7Cl_2O_3P$ M 192.966
Liq. Bp_{12} 120°, $Bp_{0.1}$ 76-77°.

Di-Et ester: [3167-62-2]. *Diethyl (dichloromethyl)-phosphonate.*
$C_5H_{11}Cl_2O_3P$ M 221.020
Liq. d_4^{20} 1.30. Bp_{10} 133°, $Bp_{0.8}$ 89.5°. n_D^{20} 1.4537.

Mono-Ph ester: Monophenyl (dichloromethyl)-phosphonate.
$C_7H_8Cl_2O_3P$ M 242.018
Metabolite of the diphenyl ester. Liq. $Bp_{0.8}$ 68-70°.

Di-Ph ester: [40911-36-2]. *Diphenyl (dichloromethyl)-phosphonate.*
$C_{13}H_{11}Cl_2O_3P$ M 317.108
V. effective against rice blast (*P. oryzae*). Cryst. (C_6H_6/pet. ether). Mp 60-62°.

Mono-(4-nitrophenyl) ester: Hydrogen 4-nitrophenyl (dichloromethyl)phosphonate.
$C_7H_7Cl_2NO_5P$ M 287.016
Metabolite of bis(4-nitrophenyl) ester. Solid. Mp 27-28°.

Bis(4-nitrophenyl) ester: Bis(4-nitrophenyl)(dichloromethyl)phosphonate.
$C_{14}H_{10}Cl_2N_2O_8P$ M 436.121
V. effective fungicide against rice blast (*P. oryzae*).

Difluoride: see (Dichloromethyl)phosphonic difluoride, D-00178

Dichloride: see (Dichloromethyl)phosphonic dichloride, D-00177

Diamide: see P-(Dichloromethyl)phosphonic diamide, D-00176

Aksnes, G. *et al*, *Acta Chem. Scand.*, 1960, **14**, 1485 (*ester*)
Bel'skii, V.E. *et al*, *Zh. Obshch. Khim.*, 1972, **42**, 2427 (*Engl. transl.* p. 2421) (*diethyl ester, P nmr*)
Roy, N.K. *et al*, *Indian J. Chem.*, 1972, **10**, 1159 (*derivs, diaryl esters*)
Elkain, J.-C. *et al*, *Tetrahedron Lett.*, 1975, 4409 (*synth, derivs, ir*)

Van der Veken, B.J. *et al*, *J. Mol. Struct.*, 1975, **28**, 371 (*ir, raman*)
Griffiths, W.R. *et al*, *Phosphorus Sulfur*, 1978, **5**, 101 (*esters, ms*)
Bedi, S. *et al*, *J. Environ. Sci. Health, B*, 1979, **14**, 443; 1980, **15**, 259 (*esters, use, metab*)
Hall, C.R. *et al*, *J. Chem. Soc., Perkin Trans. 1*, 1984, 669.

P-(Dichloromethyl)phosphonic diamide, 9CI **D-00176**

$Cl_2CHP(O)(NH_2)_2$

CH_5ClN_2OP M 127.490

N,N,N′,N′-*Tetra-Me:* [58993-59-2]. P-(*Dichloromethyl*)*phosphonic bis*(*dimethylamide*). P-*Dichloromethyl*-N,N,N′,N′-*tetramethylphosphonic diamide.*
$C_5H_{13}Cl_2N_2OP$ M 219.050
Solid. Mp 95°.

N,N,N′,N′-*Tetra-Et:* P-(*Dichloromethyl*)*phosphonic bis*(*diethylamide*). P-*Dichloromethyl*-N,N,N′,N′-*tetraethylphosphonic diamide.*
$C_9H_{21}Cl_2N_2OP$ M 275.157
Cryst. (heptane). Mp 124-125°. Bp_8 157-160°.

N,N′-*Di-Ph:* (*Dichloromethyl*)*phosphonic dianilide.* P-*Dichloromethyl*-N,N′-*diphenylphosphonic diamide.*
$C_{13}H_{13}Cl_2N_2OP$ M 315.138
Solid. Mp 176-177°.

Kinnear, A.M. *et al*, *J. Chem. Soc.*, 1952, 3437 (*dianilide*)
Abramov, V.S. *et al*, *Zh. Obshch. Khim.*, 1971, **41**, 100 (*Engl. transl.* p. 96) (*tetraethyl, synth*)
Elkain, J.-C. *et al*, *Tetrahedron Lett.*, 1975, 4409 (*tetramethyl, synth, ir*)

(Dichloromethyl)phosphonic dichloride, 9CI **D-00177**

[29941-08-0]

$Cl_2CHP(O)Cl_2$

$CHCl_4OP$ M 201.804
Bp_{10} 86°.

Kinnear, A.M. *et al*, *J. Chem. Soc.*, 1952, 3437 (*synth*)
Moedritzer, K. *et al*, *J. Chem. Eng. Data*, 1962, **7**, 307 (*nmr*)
Maier, L., *Helv. Chim. Acta*, 1971, **53**, 1949 (*synth, P nmr, pmr*)
Elkain, J.C. *et al*, *Tetrahedron Lett.*, 1975, 4409 (*synth, ir*)
Griffiths, W.R. *et al*, *Phosphorus Sulfur*, 1978, **4**, 341 (*ms*)

(Dichloromethyl)phosphonic difluoride, 9CI **D-00178**

Cl_2CHPOF_2

$CHCl_2F_2OP$ M 168.895
d_4^{20} 1.66. Bp 126-128°, Bp_{12} 37°. n_D^{20} 1.3987.

Ivanova, Zh.M. *et al*, *Zh. Obshch. Khim.*, 1968, **38**, 1334 (*Engl. transl.* p. 1284) (*synth*)
Elkain, J.-C. *et al*, *Tetrahedron Lett.*, 1975, 4409 (*synth, ir*)

(Dichloromethyl)phosphonic diisocyanate, 9CI **D-00179**

[21783-56-2]

$(OCN)_2P(O)CHCl_2$

$C_3HCl_2N_2O_3P$ M 214.932
Liq. d^{20} 1.62. Bp_{15} 126-127°. n_D^{20} 1.5006.

Derkach, G.I. *et al*, *Zh. Obshch. Khim.*, 1968, **38**, 1784 (*Engl. transl.* p. 1739) (*synth*)

(Dichloromethyl)phosphonothioic acid **D-00180**

$Cl_2CHP(S)(OH)_2 \rightleftharpoons Cl_2CHP(O)(OH)(SH)$

$CH_3Cl_2O_2PS$ M 180.973

Difluoride: [20157-29-3].
$CHCl_2F_2PS$ M 184.955
Liq. d_4^{20} 1.58. Bp 108-109°. n_D^{20} 1.4650.

Dichloride:
$CHCl_4PS$ M 217.865
Liq. d_4^{20} 1.17. Bp_6 70-71°. n_D^{20} 1.5832.

Ivanova, Zh.M. *et al*, *Zh. Obshch. Khim.*, 1968, **38**, 1334 (*Engl. transl.* p. 1284) (*dichloride, difluoride, synth*)
Hall, C.R. *et al*, *J. Chem. Soc., Perkin Trans. 1*, 1984, 669.

(Dichloromethyl)phosphonous acid **D-00181**

$Cl_2CHP(OH)_2$

$CH_3Cl_2O_2P$ M 148.913

Diisopropyl ester: [80920-98-5]. *Diisopropyl* (*dichloromethyl*)*phosphonite.*
$C_7H_{15}Cl_2O_2P$ M 233.074
$Bp_{0.5}$ 65°. n_D^{20} 1.4530.

Difluoride: [55343-32-3]. (*Dichloromethyl*)-*difluorophosphine.*
$CHCl_2F_2P$ M 152.895
Bp 55-56°.

Dichloride: [23415-85-2]. *Dichloro*(*dichloromethyl*)-*phosphine.*
$CHCl_4P$ M 185.805
Liq. d_4^{20} 1.63. Bp 150-151°, Bp_8 36-37°. n_D^{20} 1.5370.

Dibromide: [80920-97-4]. *Dibromo*(*dichloromethyl*)-*phosphine.*
$CHBr_2Cl_2P$ M 274.707
Bp_{36} 108°, $Bp_{0.5}$ 43°. n_D^{20} 1.6263.

Shokol, V.A. *et al*, *Zh. Obshch. Khim.*, 1969, **39**, 856 (*Engl. transl.* p. 820) (*dichloride, synth*)
Harris, R.K. *et al*, *J. Chem. Soc., Dalton Trans.*, 1975, 61 (*dichloride, difluoride, synth, pmr*)
Prischchenko, A.A. *et al*, *Zh. Obshch. Khim.*, 1981, **51**, 2630 (*Engl. transl.* p. 2268) (*derivs, synth, pmr, P nmr*)
Hinke, A. *et al*, *Phosphorus Sulfur*, 1983, **15**,, 93 (*dibromide, synth, P nmr*)

(Dichloromethyl)triphenylphosphonium(1+), 9CI, 8CI **D-00182**

$Ph_3P^{\oplus}CHCl_2$

$C_{19}H_{16}Cl_2P^{\oplus}$ M 346.215 (ion)
Salts are sources of the Wittig reagent, and may also be used in peptide synth.

Chloride: [57212-38-1].
$C_{19}H_{16}Cl_3P$ M 381.668
Cryst. (CH_2Cl_2/Et_2O). Mp 228° dec.

Ylide: [6779-08-4]. (*Dichloromethylene*)-*triphenylphosphorane.*
$C_{19}H_{15}Cl_2P$ M 345.207
Reactive Wittig reagent, prepd. *in situ.*

Speziale, A.J. *et al*, *J. Am. Chem. Soc.*, 1962, **84**, 854 (*ylide, use*)

Appel, R. *et al*, *Chem. Ber.*, 1976, **109**, 58 (*synth, pmr, cmr, nmr*)
Clement, B.A. *et al*, *J. Org. Chem.*, 1976, **41**, 556 (*ylide*)
Appel, R. *et al*, *Synthesis*, 1977, 699 (*synth, pmr, nmr*)
Appel, R. *et al*, *Chem. Ber.*, 1981, **114**, 858 (*use*)

O-2,4-Dichlorophenyl *S*-methyl isopropylphosphoramidothioate D-00183

O-2,4-Dichlorophenyl S-*methyl (1-methylethyl)-phosphoramidothioate*, 9CI
[29373-96-4]

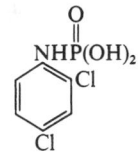

$C_{10}H_{14}Cl_2NO_2PS$ M 314.166

(+)-*form* [22151-84-4]
Cryst. (hexane). Mp 105-106°. $[\alpha]_D$ +22.7° (c, 0.60 in C_6H_6).

(−)-*form* [22260-28-2]
Cryst. (hexane). Mp 104-106°. $[\alpha]_D$ −23.4° (c, 0.56 in MeOH).

(±)-*form* [22151-85-5]
Cryst. (hexane). Mp 85-88° (84-85.5°).

Seiber, J.N. *et al*, *Tetrahedron*, 1969, **25**, 381 (*synth*)

(2,3-Dichlorophenyl)phosphonic acid, 9CI D-00184

$C_6H_5Cl_2O_3P$ M 226.983
Solid. Mp 200-202°.

Denham, J.M. *et al*, *J. Org. Chem.*, 1958, **23**, 1298 (*synth*)

(2,5-Dichlorophenyl)phosphonic acid, 9CI D-00185

$C_6H_5Cl_2O_3P$ M 226.983
Cryst. + $1H_2O$ (H_2O). Mp 194-197°.

Di-Et ester: [68373-17-1]. *Diethyl (2,5-dichlorophenyl)-phosphonate.*
$C_{10}H_{13}Cl_2O_3P$ M 283.091
Predominant metab. of Bromophos-Ethyl in *B. microplus*. Liq. Bp_3 160-164°.

Kosolapoff, G.M. *et al*, *J. Am. Chem. Soc.*, 1947, **69**, 2020 (*synth, ester*)
Freedman, L.D. *et al*, *J. Am. Chem. Soc.*, 1953, **75**, 1379 (*synth*)
Schuntner, C.A. *et al*, *Aust. J. Biol. Sci.*, 1978, **31**, 317 (*ester*)

(3,4-Dichlorophenyl)phosphonic acid, 8CI D-00186

$C_6H_5Cl_2O_3P$ M 226.983
Hygroscopic solid. Mp 153°.

Di-Et ester: Diethyl (3,4-dichlorophenyl)phosphonate.
$C_{10}H_{13}Cl_2O_3P$ M 283.091
Liq. Bp_2 155-157°.

Kosolapoff, G.M. *et al*, *J. Am. Chem. Soc.*, 1947, **69**, 2020 (*synth*)

(3,5-Dichlorophenyl)phosphonic acid, 9CI D-00187

$C_6H_5Cl_2O_3P$ M 226.983
Solid. Mp 188-190°.

Denham, J.M. *et al*, *J. Org. Chem.*, 1958, **23**, 1298 (*synth*)

(2,4-Dichlorophenyl)phosphoramidic acid, 9CI, 8CI D-00188

$C_6H_6Cl_2NO_3P$ M 241.998
Cryst. (EtOH). Mp 167°.

Di-Et ester: Diethyl (2,4-dichlorophenyl)-phosphoramidate.
$C_{10}H_{14}Cl_2NO_3P$ M 298.105
Needles (Et_2O). Mp 106°.

Di-Ph ester: Diphenyl (2,4-dichlorophenyl)-phosphoramidate.
$C_{18}H_{14}Cl_2NO_3P$ M 394.193
Cryst. (EtOH). Mp 71-73°.

Dichloride:
$C_6H_4Cl_4NOP$ M 278.889
Cryst. (C_6H_6/pet. ether). Mp 126°.

Michaelis, A., *Justus Liebigs Ann. Chem.*, 1903, **326**, 129 (*synth, diethyl ester*)
Zhmurova, I.N. *et al*, *Zh. Obshch. Khim.*, 1961, **31**, 3741 (*Engl. transl.* p. 3495) (*diphenyl ester*)

2,4-Dichlorophenyl phosphorodichloridate, 9CI, 8CI D-00189

[19430-75-2]

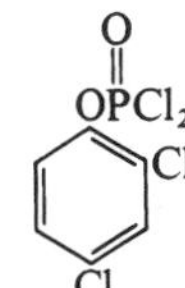

$C_6H_3Cl_4O_2P$ M 279.874
Fuming liq. Bp_5 136°, $Bp_{0.1}$ 70°. n_D^{20} 1.5560.

Maguire, M.H. *et al*, *J. Chem. Soc.*, 1953, 1479 (*synth*)
Cremlyn, R.J.W. *et al*, *J. Chem. Soc., Perkin Trans. 1*, 1972, 583 (*synth, ir*)
Owen, G.R. *et al*, *Synthesis*, 1974, 704 (*synth*)

(Dichlorophosphinyl)acetic acid, 9CI, 8CI D-00190

$Cl_2P(O)CH_2COOH$

$C_2H_3Cl_2O_3P$ M 176.924

Me ester: [31460-11-4]. *Methyl dichlorophosphinylacetate.*
$C_3H_5Cl_2O_3P$ M 190.950
Liq. d_4^{20} 1.49. $Bp_{0.07}$ 78-79°. n_D^{20} 1.4775.

Et ester: [36951-54-7]. *Ethyl dichlorophosphinylacetate.*
$C_4H_7Cl_2O_3P$ M 204.977
Liq. d_4^{20} 1.40. Bp_4 103-104°. n_D^{20} 1.4700.

Chloride:
$C_2H_2Cl_3O_2P$ M 195.369
Liq. $Bp_{1\text{-}2}$ 79-80°.

Nylen, P., *Ber.*, 1924, **57**, 1023 (*chloride*)

Bodnarchuk, N.D. *et al*, *Zh. Obshch. Khim.*, 1970, **40**, 1210 (*Engl. transl.* p. 1201) (*ester*)
Malevannaya, R.A. *et al*, *Zh. Obshch. Khim.*, 1972, **42**, 765 (*Engl. transl.* p. 757) (*ester, synth, ir, pmr*)
Althoff, W. *et al*, *Z. Naturforsch., B*, 1976, **36**, 153 (*chloride, cmr*)
Morita, T. *et al*, *Chem. Lett.*, 1980, 435 (*ester*)

Dichlorophosphinylcarbamic acid, 9CI, 8CI **D-00191**

HOOCNHP(O)Cl_2

$CH_2Cl_2NO_3P$ M 177.911

Me ester: [14557-33-6]. *Methyl dichlorophosphinylcarbamate.*
$C_2H_4Cl_2NO_3P$ M 191.938
Solid. Mp 47-50°.

Et ester: [61670-34-6]. *Ethyl dichlorophosphinylcarbamate.*
$C_3H_6Cl_2NO_3P$ M 205.965
Solid. Mp 23-25°.

Isopropyl ester: Isopropyl dichlorophosphinylcarbamate.
$C_4H_8Cl_2NO_3P$ M 219.992
Solid. Mp 75-77°.

Kirsanov, A.V. *et al*, *Zh. Obshch. Khim.*, 1959, **29**, 2256 (*Engl. transl.* p. 2222) (*esters*)
Egorov, Yu.P. *et al*, *Zh. Prikl. Spektrosk.*, 1968, **9**, 980 (*Engl. transl.* p. 1338) (*methyl ester, ir*)
Nuzhdina, Yu.A. *et al*, *Zh. Strukt. Khim.*, 1972, **13**, 72 (*Engl. transl.* p. 61) (*methyl ester, ir, struct*)

3-(Dichlorophosphinyl)propanoic acid, 9CI, 8CI **D-00192**

$Cl_2P(O)CH_2CH_2COOH$

$C_3H_5Cl_2O_3P$ M 190.950

Me ester: [17391-48-9]. *Methyl 3-dichlorophosphinylpropanoate.*
$C_4H_7Cl_2O_3P$ M 204.977
Liq. d_4^{20} 1.43. $Bp_{0.4}$ 85-87°. n_D^{20} 1.4790.

Et ester: [17391-49-0]. *Ethyl 3-dichlorophosphinylpropanoate.*
$C_5H_9Cl_2O_3P$ M 219.004
Liq. d_4^{20} 1.39. $Bp_{0.4}$ 97-99°. n_D^{20} 1.4712.

Razumova, N.A. *et al*, *Zh. Obshch. Khim.*, 1967, **37**, 1136 (*Engl. transl.* p. 1078) (*synth, ir, pmr*)

3,3-Dichloro-1,1,2,2-tetrahydro-1,1,1,2,2,2-hexaphenyldiphosphirane **D-00193**

[76001-34-8]

Needles (C_6H_6/pet. ether). Mp 166°.

Kansal, N.M. *et al*, *Indian J. Chem., Sect. B*, 1980, **19**, 610, 611 (*synth, props*)
Verma, S. *et al*, *Heterocycles*, 1981, **16**, 1537.

2,4-Dichloro-2,4,6,6-tetrakis(dimethylamino)-2,2,4,4,6,6-hexahydro-1,3,5-triaza-2,4,6-triphosphorine, 9CI **D-00194**

2,4-Dichloro-2,4,6,6-tetrakis(dimethylamino)-2,2,4,4,6,6-hexahydrocyclotriphosphazene

[22514-10-9]

(*4RS,6RS*)*-form*

$C_8H_{24}Cl_2N_7P_3$ M 382.152

cis- and *trans-*forms equilibrate in soln.

(*4RS,6RS*)-*form* [5591-07-1]
trans-*form*
Cryst. (pet. ether). Mp 72°. $Bp_{0.01}$ 40° subl.

(*4RS,6SR*)-*form* [963-05-3]
cis-*form*
Cryst. (pet. ether). Mp 104°.

Keat, R. *et al*, *J. Chem. Soc.*, 1965, 2215 (*synth, struct*)
Strahlberg, R. *et al*, *Spectrochim. Acta, Part A*, 1967, **23**, 2005 (*ir, raman*)
Green, B. *et al*, *J. Inorg. Nucl. Chem.*, 1971, **33**, 3687 (*pmr, ir, glc*)
Dagleish, W.H. *et al*, *J. Chem. Soc., Dalton Trans.*, 1975, 309 (*nqr*)
Keat, R. *et al*, *J. Chem. Soc., Dalton Trans.*, 1976, 1582 (*P nmr*)
Goldschmidt, J.M.E. *et al*, *J. Inorg. Nucl. Chem.*, 1980, **42**, 618 (*synth*)
Friedmann, N. *et al*, *J. Chem. Soc., Dalton Trans.*, 1981, 103 (*props*)

7,7-Dichloro-2,3,5,6-tetrakis(trifluoromethyl)-1,4-diphosphabicyclo[2.2.1]hepta-2,5-diene, 10CI **D-00195**

[65534-45-4]

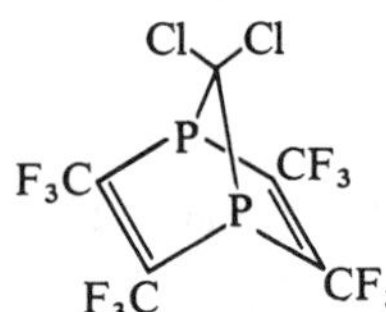

$C_9Cl_2F_{12}P_2$ M 468.933

Yellow needles. Mp 39-39.5° (sealed tube).

Kabayashi, Y. *et al*, *J. Am. Chem. Soc.*, 1980, **102**, 252 (*synth, ir, ms, F nmr*)

3,9-Dichloro-2,4,8,10-tetraoxa-3,9-diphosphaspiro[5.5]undecane, 9CI **D-00196**

[3643-70-7]

$C_5H_8Cl_2O_4P_2$ M 264.969

Cryst. ($CHCl_3$). Mp 121-123°. Bp_8 147-148°. Fairly stable in moist air.

3,9-Dioxide: [714-87-4].
$C_5H_8Cl_2O_6P_2$ M 296.968
Fine needles (AcOH). Mp 233-235°.

3,9-Disulfide: [5305-81-7].
$C_5H_8Cl_2O_4P_2S_2$ M 329.089
Plates (AcOH). Mp 192-193°.

Lucas, H.J. *et al*, *J. Am. Chem. Soc.*, 1950, **72**, 5491 (*synth*)

Rätz, R. *et al*, *J. Heterocycl. Chem.*, 1963, **3**, 14, 20 (*synth, deriv*)
Rätz, R. *et al*, *J. Org. Chem.*, 1963, **28**, 1608 (*dioxide, props*)
Pivawer, P.M. *et al*, *J. Heterocycl. Chem.*, 1967, **4**, 599 (*synth*)
Haegele, G. *et al*, *Z. Naturforsch., B*, 1971, **26**, 1 (*disulfide, pmr*)

2,2-Dichloro-1,1,1-triethyldiphosphinium 2-oxide(1+), 9CI **D-00197**

Triethylphosphoniophosphoryl chloride

$Et_3P^{\oplus}P(O)Cl_2$

$C_6H_{15}Cl_2OP_2^{\oplus}$ M 236.037 (ion)

Chloride: [39124-58-8].

$C_6H_{15}Cl_3OP_2$ M 271.490

Cryst. (Et_2O at −30°). Mp 27-28°. $H_2O \rightarrow Et_3PO$.

Lindner, E. *et al*, *Chem. Ber.*, 1972, **105**, 3261 (*synth, ir*)

Dichlorotrimethylphosphorane, 9CI, 8CI **D-00198**

Chlorotrimethylphosphonium chloride. Trimethylphosphine dichloride

[2725-66-8]

Me_3PCl_2 or $Me_3PCl^{\oplus}Cl^{\ominus}$

$C_3H_9Cl_2P$ M 146.984

Possesses the ionic struct. Solid. Mp 267-268°.

Goubeau, J. *et al*, *Z. Elektrochem.*, 1960, **64**, 598 (*synth, props, ir, raman*)
Appel, R. *et al*, *Z. Anorg. Allg. Chem.*, 1968, **363**, 176 (*synth, props*)
Dillon, K.B. *et al*, *J. Chem. Soc., Dalton Trans.*, 1976, 1243 (*nmr, nqr*)
Appel, R. *et al*, *Chem. Ber.*, 1977, **110**, 2382 (*synth*)

Dichlorotriphenoxyphosphorane, 9CI, 8CI **D-00199**

Triphenyl phosphite dichloride

[15493-07-9]

$(PhO)_3PCl_2$

$C_{18}H_{15}Cl_2O_3P$ M 381.194

Probably ionic struct.; electrically conducting in MeCN. Converts alcohols into alkyl chlorides. Pale-green solid. Mp 80-81°. Hydrolysed to triphenyl phosphate.

Coe, D.G. *et al*, *J. Chem. Soc.*, 1954, 2281 (*synth, struct, use*)
Harris, G.S. *et al*, *J. Chem. Soc.*, 1956, 3038 (*props*)
Ramirez, F. *et al*, *Tetrahedron*, 1968, **24**, 5041.
Tseng, C.K., *J. Org. Chem.*, 1979, **44**, 2793 (*struct*)

Dichlorotriphenylphosphorane, 9CI, 8CI **D-00200**

Triphenylphosphine dichloride

[2526-64-9]

Ph_3PCl_2

$C_{18}H_{15}Cl_2P$ M 333.196

Thought to be unionized in benzene, but ionic in solid state and in HCl(l). Useful chlorinating agent; converts esters and lactones into acyl chlorides and chlorinates $\alpha\beta$-unsat. ketones at the β-position. Solid. Mp 207-210°.

Appel, R. *et al*, *Chem. Ber.*, 1976, **109**, 58; 1977, **110**, 2382 (*synth, cmr, nmr*)
Dillon, K.B. *et al*, *J. Chem. Soc., Dalton Trans.*, 1976, 1243 (*nmr, nqr, struct*)
Org. Synth., 1978, **58**, 64 (*use*)
Sergienko, L.M. *et al*, *Zh. Obshch. Khim.*, 1979, **49**, 317 (*Engl. transl.* p. 275) (*uv*)
Timokhin, B.V. *et al*, *Zh. Obshch. Khim.*, 1981, **51**, 1989 (*Engl. transl.* p. 1711) (*nmr*)
Piers, E. *et al*, *Can. J. Chem.*, 1982, **60**, 210 (*use*)

Dichlorotris(trichloromethyl)phosphorane **D-00201**

$Cl_2P(CCl_3)_3$

$C_3Cl_{11}P$ M 456.990

Prisms (pet. ether). Mp 190° dec. With boiling H_2O, loses only one Cl from the phosphorus atom.

Ginsburg, V.A. *et al*, *Zh. Obshch. Khim.*, 1958, **28**, 728 (*Engl. transl.* p. 810) (*synth, props*)

2,2-Dichlorovinyl methyl octyl phosphate, 8CI **D-00202**

2,3-Dichloroethenyl methyl octyl phosphate, 9CI. Vincofos, USAN. Vingard

[17196-88-2]

$$MeO-P(=O)(OCH=CCl_2)-O(CH_2)_7CH_3$$

$C_{11}H_{21}Cl_2O_4P$ M 319.164

Anthelmintic. $Bp_{0.01}$ 128°.

Ger. Pat., 1 251 745, (*1967*); *CA*, **68**, 39322k (*synth, pharmacol*)

Dichlorvos, BSI **D-00203**

O-(*2,2-Dichloroethenyl*) O,O-*dimethyl phosphate, 9CI. 2,2-Dichlorovinyl dimethyl phosphate, 8CI.* O,O-*Dimethyl* O-(*2,2-dichlorovinyl*) *phosphate. Dichlorophos. D.D.V.P.. Vapona*

[62-73-7]

$(MeO)_2P(O)OCH{=}CCl_2$

$C_4H_7Cl_2O_4P$ M 220.977

Veterinary anthelmintic, insecticide. Cholinesterase inhibitor. Liq. with aromatic odour. Mod. sol. H_2O (1% w/v), misc. EtOH. d_4^{25} 1.42. Bp_{14} 120°, Bp_3 88°. n_D^{20} 1.4550. Hydrolysed rel. slowly by H_2O, more quickly by alkalis.

▷Highly toxic by skin absorption and inhalation, TLV 1, exp. teratogen. TC0350000.

Barthel, W.F. *et al*, *J. Am. Chem. Soc.*, 1955, **77**, 2424 (*synth*)
Keith, L.H. *et al*, *J. Assoc. Off. Anal. Chem.*, 1968, **51**, 1063 (*pmr*)
Ross, R.T. *et al*, *Anal. Chim. Acta*, 1970, **52**, 139 (*P nmr*)
Gore, R.C. *et al*, *J. Assoc. Off. Anal. Chem.*, 1971, **54**, 1040 (*ir, uv*)
Betteridge, D. *et al*, *Anal. Chem.*, 1972, **44**, 2005 (*pe*)
Gillett, J.W. *et al*, *Residue Rev.*, 1972, **44**, 115, 161 (*tox, rev*)
Gaydou, E.M., *Bull. Soc. Chim. Fr.*, 1973, 2275 (*synth, props*)
Gaydou, E.M. *et al*, *Org. Mass Spectrom.*, 1974, **9**, 157 (*ms*)
Nikonorov, K.V. *et al*, *Zh. Obshch. Khim.*, 1974, **44**, 1267 (*Engl. transl.* p. 1245) (*synth*)
Nicholas, M.L. *et al*, *J. Assoc. Off. Anal. Chem.*, 1976, **59**, 1071 (*raman*)
Szalontai, G., *J. Chromatogr.*, 1976, **124**, 9 (*hplc*)
Stan, H.-J. *et al*, *Fresenius' Z. Anal. Chem.*, 1977, **287**, 271 (*glc, ms*)
Szalontai, G. *et al*, *Org. Magn. Reson.*, 1977, **10**, 63 (*cmr*)

Wright, A.S. *et al*, *Arch. Toxicol.*, 1979, **42**, 1 (*metab, props, bibl*)
Pesticide Manual, 6th Ed., 180; 7th Ed., 185.
Agrochemicals Handbook, Royal Society of Chemistry, London, 1983, A141.
Sax, N.I., *Dangerous Properties of Industrial Materials*, 6th Ed., Van Nostrand-Reinhold, 1984, 568.

Dicrotophos **D-00204**

3-(Dimethylamino)-1-methyl-3-oxo-1-propenyl dimethyl phosphate, 9CI. Dimethyl 2-dimethylcarbamoyl-1-methylvinyl phosphate. Carbicron. Ektafos. Bidrin

[3735-78-2]

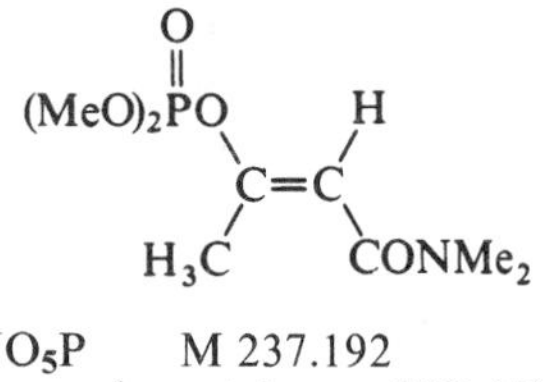

$C_8H_{16}NO_5P$ M 237.192

The tech. prod. contains ca. 85% (*E*)-isomer. Systemic insecticide and acaricide. Amber liq. (mixed isomers). Misc. H_2O, Me_2CO, EtOH. d_{15}^{15} 1.216. $Bp_{0.1}$ 130°, $Bp_{0.0013}$ 90-95°. n_D^{23} 1.4680.

▷Highly toxic, TLV 0.25

(E)-form [141-66-2]

Insecticidal activity > that of the (*Z*)-form.

▷TC3850000.

(Z)-form [18250-63-0]

Less insecticidally active isomer.

Menzer, R.E. *et al*, *J. Agric. Food Chem.*, 1965, **13**, 102 (*ir, metab*)
Keith, L.H. *et al*, *J. Assoc. Off. Anal. Chem.*, 1968, **51**, 1063 (*pmr*)
Ross, R.T. *et al*, *Anal. Chem. Acta*, 1970, **52**, 139 (*P nmr*)
Beynon, K.I. *et al*, *Residue Rev.*, 1973, **47**, 55 (*rev*)
Szalontai, G., *J. Chromatogr.*, 1976, **124**, 9 (*hplc*)
Busch, K.L. *et al*, *Appl. Spectrosc.*, 1978, **32**, 388 (*ms*)
Stan, H.-J. *et al*, *Biomed. Mass Spectrom.*, 1982, **9**, 483 (*ms*)
Stan, H.-J. *et al*, *J. Chromatogr.*, 1983, **279**, 173 (*glc*)
Agrochemicals Handbook, Royal Society of Chemistry, London, 1983, A143.
Pesticide Manual, 6th Ed., 184; 7th Ed., 190.
Sax, N.I., *Dangerous Properties of Industrial Materials*, 6th Ed., Van Nostrand-Reinhold, 1984, 733.

(Dicyanomethylene)triphenylphosphorane **D-00205**

2-(Triphenylphosphoranylidene)-1,3-propanedinitrile. (Triphenylphosphoroanylidene)malononitrile

$Ph_3P{=}C(CN)_2$

$C_{21}H_{15}N_2P$ M 326.337

Stabilised ylide. Cryst. (EtOH or toluene). Mp 188-189°.

Horner, L. *et al*, *Chem. Ber.*, 1958, **91**, 437 (*synth, uv*)
Martin, D. *et al*, *Chem. Ber.*, 1967, **100**, 187 (*synth*)
Cooks, R.G. *et al*, *Tetrahedron*, 1968, **24**, 3289 (*ms*)
Bestmann, H.J. *et al*, *Justus Liebigs Ann. Chem.*, 1974, 1688 (*ir*)

Dicyclohexyldiphosphonic acid, 8CI **D-00206**

Dicyclohexylpyrophosphonic acid. Cyclohexylphosphonic acid monoanhydride

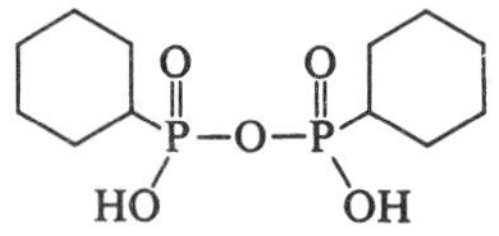

$C_{12}H_{24}O_5P_2$ M 310.266

Biscyclohexylamine salt: Cryst. (H_2O). Mp 242-243°.

Ruveda, M.A. *et al*, *Tetrahedron*, 1972, **28**, 5011 (*synth, ir, pmr*)

2,4-Dicyclohexyl-1,3,2,4-dithiadiphosphetane 2,4-disulfide **D-00207**

2,4-Bis(cyclohexyl)-1,3,2,4-dithiadiphosphetane 2,4-disulfide, 9CI. Cyclic anhydrosulfide from cyclohexylphosphonothioic acid, 8CI

[1024-06-2]

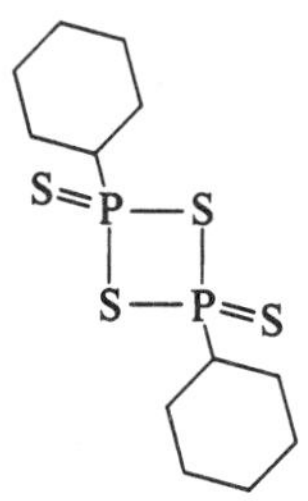

$C_{12}H_{22}P_2S_4$ M 356.493

Reacts with alcohols to give *O*-alkyl esters of Cyclohexylphosphonodithioic acid, C-00261 . Cryst. (Chlorobenzene). Mp 176°.

Newallis, P.E. *et al*, *J. Org. Chem.*, 1961, **27**, 3829 (*synth*)
Grishina, O.N. *et al*, *Zh. Obshch. Khim.*, 1977, **47**, 72 (*Engl. transl.* p. 64) (*synth, P nmr*)
Andreev, N.A. *et al*, *Zh. Obshch. Khim.*, 1982, **52**, 1785 (*Engl. transl.* p. 1581) (*synth, ir, P nmr, struct*)

Dicyclohexylphenylphosphine, 9CI **D-00208**

[6476-37-5]

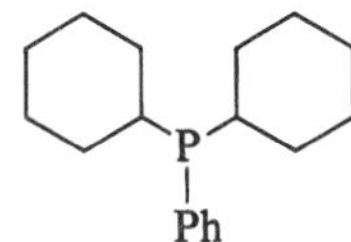

$C_{18}H_{27}P$ M 274.385

Ligand for Cr, Hg, and Group VIII metals. Foul-smelling cryst. (Me_2CO or EtOH). Mp 57-58°. pK_a 3.40.

Oxide:

$C_{18}H_{27}OP$ M 290.384

Cryst. (pet. ether or EtOH). Mp 165°, 187-188°.

Sulfide:

$C_{18}H_{27}PS$ M 306.445

Cryst. (EtOH). Mp 168°.

Issleib, K. *et al*, *Chem. Ber.*, 1961, **94**, 392 (*synth, derivs*)
Screttas, C. *et al*, *J. Org. Chem.*, 1962, **27**, 2573 (*synth, derivs*)
Goetz, H. *et al*, *Justus Liebigs Ann. Chem.*, 1977, 556 (*uv, pe*)
Troy, D. *et al*, *Bull. Soc. Chim. Fr. Part I*, 1979, 241 (*uv*)
Naaktgeboren, A.J. *et al*, *J. Am. Chem. Soc.*, 1980, **102**, 3350 (*nmr*)
Inorg. Synth., **18**, 171 (*synth*)

Dicyclohexyl phosphate, 9CI, 8CI **D-00209**

Dicyclohexyl hydrogen phosphate. Dicyclohexyl phosphoric acid

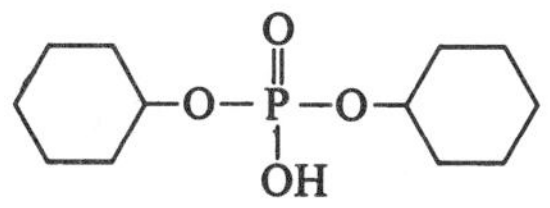

$C_{12}H_{23}O_4P$ M 262.285

Cyclohexylammonium salt: Solid. Mp 212°.

Fluoride: Dicyclohexyl phosphorofluoridate. Dicyclohexyl phosphoryl fluoride. Dicyclohexyl fluorophosphate.
$C_{12}H_{20}FO_3P$ M 262.260
Liq. $Bp_{0.6}$ 125-128°.

▷More toxic than 35850-7. Forms a persistent gas

Chloride: [51672-37-8]. *Dicyclohexyl phosphorochloridate. Dicyclohexyl phosphoryl chloride. Dicyclohexyl chlorophosphate.*
$C_{12}H_{20}ClO_3P$ M 278.715
Liq. d^{20} 1.17. n_D^{20} 1.4772.

Chapman, N.B. *et al, J. Chem. Soc.*, 1948, 1010 (*fluoride, synth, tox*)
Goldwhite, H. *et al, J. Chem. Soc.*, 1955, 2040 (*fluoride*)
Tichý, V. *et al, Chem. Zvesti*, 1958, **12**, 345; *CA*, **52**, 18258 (*chloride*)
Rubinstein, M. *et al, Tetrahedron*, 1975, 2107 (*synth*)

Dicyclohexylphosphine, 9CI **D-00210**

[829-84-5]

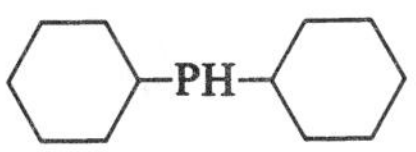

$C_{12}H_{23}P$ M 198.287

Air-sensitive liq. Bp 281-283°, Bp_3 105-108°. pK_a 4.55 (H_2O). n_D^{25} 1.5142.

Li salt: [19966-81-5]. Pale-yellow solid. Insol. Et_2O, C_6H_6, pet. ether, sl. sol. dioxan.

Oxide: [14717-29-4].
$C_{12}H_{23}OP$ M 214.287
Cryst. (hexane or C_6H_6). Mp 72.5-74.5°. Tautomeric with dicyclohexylphosphinous acid.

Sulfide: [14610-03-8].
$C_{12}H_{23}PS$ M 230.347
Cryst. (EtOH or C_6H_6). Mp 107-108°. Tautomeric with dicyclohexylphosphinothious acid.

Rauhut, M.M. *et al, J. Org. Chem.*, 1961, **26**, 5133, 5138 (*sulfide, synth*)
Niebergall, H. *et al, Chem. Ber.*, 1962, **95**, 64 (*synth*)
Issleib, K. *et al, Z. Naturforsch., B*, 1965, **20**, 916 (*synth*)
Sanchez, M. *et al, Spectrochim. Acta, Part A*, 1967, **23**, 2617 (*oxide, sulfide, ir*)
Timokhin, B.V. *et al, Zh. Obshch. Khim.*, 1975, **45**, 2561 (*Engl. transl.* p. 2517) (*sulfide, ir*)

Dicyclohexylphosphinic acid, 9CI **D-00211**

[832-39-3]

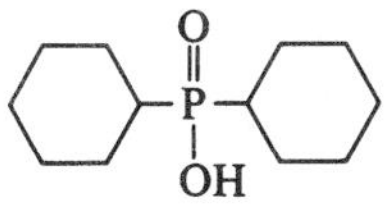

$C_{12}H_{23}O_2P$ M 230.286

May be used to separate Co from Ni in aq. soln. Cryst. (Me_2CO or C_6H_6). Mp 143°. pK_a 5.92 (75% EtOH aq.), 6.64 (95% EtOH aq.), 14.56 ($MeNO_2$).

Me ester: [76905-66-3]. *Methyl dicyclohexylphosphinate.*
$C_{13}H_{25}O_2P$ M 244.313
Cryst. (EtOH). Mp 58-59°.

Et ester: [13788-50-6]. *Ethyl dicyclohexylphosphinate.*
$C_{14}H_{27}O_2P$ M 258.340
Liq. Bp_2 158-160°.

Cyclohexyl ester: Cyclohexyl dicyclohexylphosphinate.
$C_{18}H_{33}O_2P$ M 312.431
Cryst. (pet. ether). Mp 85-86°.

Ph ester: [13689-21-9]. *Phenyl dicyclohexylphosphinate.*
$C_{18}H_{27}O_2P$ M 306.384
Liq. $Bp_{0.7}$ 121-124°. n_D^{20} 1.4438.

Fluoride: [38169-17-4].
$C_{12}H_{22}FOP$ M 232.277
No phys. props. reported.

Chloride: [15873-72-0].
$C_{12}H_{22}ClOP$ M 248.732
Cryst. (C_6H_6). Mp 108.5°.

Stiles, A.R. *et al, J. Am. Chem. Soc.*, 1952, **74**, 3282 (*synth*)
Issleib, K. *et al, Z. Anorg. Allg. Chem.*, 1954, **277**, 258 (*synth, chloride, esters*)
Buckler, S.A., *J. Am. Chem. Soc.*, 1962, **84**, 3093 (*ester*)
Müller, E. *et al, Chem. Ber.*, 1967, **100**, 521 (*synth, chloride*)
Knunyants, I.L. *et al, Dokl. Akad. Nauk SSSR, Ser. Sci. Khim.*, 1971, **201**, 862 (*Engl. transl.* p. 992) (*fluoride, F and P nmr*)
Aslanov, L.A. *et al, Zh. Strukt. Khim.*, 1979, **20**, 758 (*Engl. transl.* p. 646) (*cryst struct*)
Nifant'ev, Ê.E. *et al, Zh. Obshch. Khim.*, 1980, **50**, 1744 (*Engl. transl.* p. 1416) (*synth, P nmr*)

Dicyclohexylphosphinodithioic acid, 9CI **D-00212**

[2512-58-5]

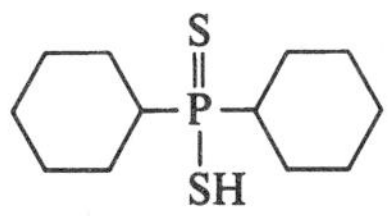

$C_{12}H_{23}PS_2$ M 262.407

Na salt acts as a flotation agent for Cu ores. Cryst. (pet. ether). Mp 103-105°.

Me ester: [65728-06-5]. *Methyl dicyclohexylphosphinodithioate.*
$C_{13}H_{25}PS_2$ M 276.434
Solid. Mp 75-75.5°.

Butyl ester: [7697-75-8]. *Butyl dicyclohexylphosphinodithioate.*
$C_{16}H_{31}PS_2$ M 318.515
Liq. $Bp_{0.15}$ 167-175°.

Cyclohexyl ester: Cyclohexyl dicyclohexylphosphinodithioate.
$C_{18}H_{34}PS_2$ M 345.560
Solid. Mp 82-85°.

Benzyl ester: [65728-16-7]. *Benzyl dicyclohexylphosphinodithioate.*
$C_{19}H_{29}PS_2$ M 352.532
Solid. Mp 107-108°.

Rauhut, M.M. *et al, J. Org. Chem.*, 1961, **26**, 5133 (*synth*)
Grayson, M. *et al, J. Org. Chem.*, 1967, **32**, 236 (*ester, ir*)
U.S.P., 4 064 106 (*1977*); *CA*, **88**, 106270 (*esters*)

Dicyclohexylphosphinothioic acid, 9CI **D-00213**

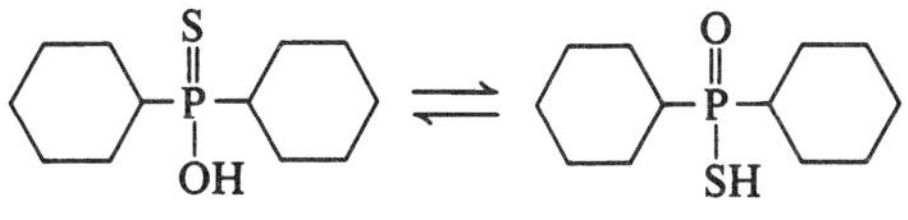

$C_{12}H_{23}OPS$ M 246.347

Cryst. (CH_2Cl_2/MeCN/hexane). Mp 130-137°.
S-Ph ester: [7697-79-2]. *S-Phenyl dicyclohexylphosphinothioate.*
$C_{18}H_{27}OPS$ M 322.444
Cryst. (pet. ether). Mp 91-93.5°.
Fluoride: [38169-29-8].
$C_{12}H_{22}FPS$ M 248.338
No phys. props. reported.
Chloride: [38169-30-1].
$C_{12}H_{22}ClPS$ M 264.793
No phys. props. reported.

Issleib, K. *et al, Z. Kristallogr., Kristallgeom., Kristallphys., Kristallchem.*, 1964, **119**, 472 (*cryst struct*)
Grayson, M. *et al, J. Org. Chem.*, 1967, **32**, 236 (*ester*)
Knunyants, I.L. *et al, Dokl. Akad. Nauk SSSR, Ser. Sci. Khim.*, 1971, **201**, 862 (*Engl. transl.* p. 992) (*halides, P nmr*)
Mattes, R. *et al, Z. Anorg. Allg. Chem.*, 1983, **499**, 67 (*synth, cryst struct*)

Dicyclohexylphosphinous acid, 9CI — D-00214

$C_{12}H_{23}OP$ M 214.287
The free acid exists in the tautomeric phosphoryl form. See Dicyclohexylphosphine, D-00210 .
Et ester: [80413-46-3]. *Ethyl dicyclohexylphosphinite.*
$C_{14}H_{27}OP$ M 242.340
Liq. Bp_1 111-113°. n_D^{20} 1.4950.
Trimethylsilyl ester: Trimethylsilyl dicyclohexylphosphinite.
$C_{15}H_{31}OPSi$ M 286.469
Liq. Bp_1 108-110°. n_D^{20} 1.4919.
Chloride: [16523-54-9]. *Chlorodicyclohexylphosphine.*
$C_{12}H_{22}ClP$ M 232.733
Oily liq. d_4^{20} 1.22. Bp_{10} 164-166°, $Bp_{0.4}$ 50.5-51°. n_D^{20} 1.5330.
Bromide: Bromodicyclohexylphosphine.
$C_{12}H_{22}BrP$ M 277.184
Oil. Bp_{12} 171°.
Iodide: Dicyclohexyliodophosphine.
$C_{12}H_{22}IP$ M 324.184
Yellow oil. Mp −8°. Bp_{13} 190°.

Issleib, K. *et al, Chem. Ber.*, 1959, **92**, 2681 (*halides*)
Grayson, M. *et al, Tetrahedron*, 1967, **23**, 1065 (*synth*)
Kabachnik, M.I. *et al, Izv. Akad. Nauk SSSR, Ser. Khim.*, 1967, 949 (*Engl. transl.* p. 923) (*chloride, ethyl ester*)
Babkina, Ê.I. *et al, Zh. Obshch. Khim.*, 1968, **38**, 1772 (*Engl. transl.* p. 1727) (*chloride, synth, glc*)
Appel, R. *et al, Chem. Ber.*, 1975, **108**, 1783 (*chloride, P nmr*)
Kapoor, P.N. *et al, Helv. Chim. Acta*, 1977, **60**, 2824 (*synth*)
Foss, V.L. *et al, Zh. Obhsch. Khim.*, 1979, **49**, 2418 (*Engl. transl.* p. 2134) (*trimethylsilyl ester*)

Dicyclohexyl phosphonate, 9CI — D-00215

Dicyclohexyl phosphite
[3808-22-8]

$C_{12}H_{23}O_3P$ M 246.286
Tautomeric; almost completely in phosphonate form. Liq. Bp_3 152-153°.

Kuskov, V.K. *et al, Dokl. Akad. Nauk SSSR, Ser. Sci. Khim.*, 1953, **92**, 323; *CA*, **49**, 155 (*synth*)

N,N'-Dicyclohexylphosphorodiamidic acid, 9CI, 8CI — D-00216

[17390-19-1]

$C_{12}H_{25}N_2O_2P$ M 260.315
Solid. Mp 143°.
Et ester: [57673-93-5]. *Ethyl* N,N'-*dicyclohexylphosphorodiamidate.*
$C_{14}H_{29}N_2O_2P$ M 288.369
No phys. props. reported.
Ph ester: Phenyl N,N'*-dicyclohexylphosphorodiamidate.*
$C_{18}H_{29}N_2O_2P$ M 336.413
Cryst. (EtOH aq.). Mp 124-125°.
Fluoride: [350-65-2].
$C_{12}H_{24}FN_2OP$ M 262.307
Cryst. (C_6H_6 or EtOH aq.). Mp 127°.
▷Highly toxic, LD_{50} 9 mg/kg. TD3510000.

Audrieth, L.F. *et al, J. Am. Chem. Soc.*, 1942, **64**, 1337 (*phenyl ester*)
Heap, R. *et al, J. Chem. Soc.*, 1948, 1313 (*fluoride, synth, tox*)
Cremlyn, R.W.J. *et al, J. Chem. Soc. (C)*, 1971, 300 (*synth, ir*)
Mathis, R. *et al, C.R. Hebd. Seances Acad. Sci., Ser. C*, 1975, **281**, 437 (*ethyl ester, ir*)

N,N'-Dicyclohexylphosphorodiamidodithioic acid, 9CI, 8CI — D-00217

Dithiophosphoric acid bis(cyclohexylamide)

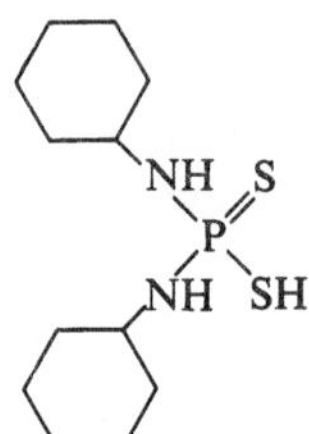

$C_{12}H_{25}N_2PS_2$ M 292.437
Cyclohexylammonium salt: [25522-75-2]. Solid. Mp 197-200° dec.

Becke-Goehring, M. *et al, Z. Anorg. Allg. Chem.*, 1969, **369**, 73 (*synth*)

O,O-Dicyclohexyl phosphorodiselenoate, 9CI, 8CI D-00218

O,O-*Dicyclohexyl hydrogen diselenophosphate.* O,O-*Dicyclohexylphosphorodiselenoic acid*

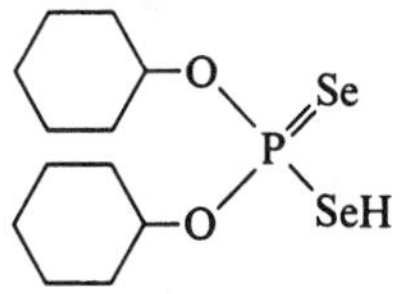

$C_{12}H_{23}O_2PSe_2$ M 388.206

K salt: [19483-48-8]. Cryst. (EtOH/ligroin). Mp 202-208° dec.

Kudchadker, M.V. *et al*, *Can. J. Chem.*, 1968, **46**, 1415 (*synth, ir, pmr*)

Didecylphenylphosphine, 9CI D-00219

[43081-11-4]

$PhP[(CH_2)_9CH_3]_2$

$C_{26}H_{47}P$ M 390.631

Liq. Bp_4 241-243°. n_D^{20} 1.4990.

Oxide: [17262-57-6].

$C_{26}H_{47}OP$ M 406.630

Solid (heptane). Mp 55-56°. Bp_3 254-256°.

Fedorova, G.K. *et al*, *Zh. Obshch. Khim.*, 1964, **34**, 511 (*Engl. transl.* p. 513)

Didecyl phosphate, 9CI D-00220

Didecyl hydrogen phosphate. Didecyl phosphoric acid

[7795-87-1]

$[H_3C(CH_2)_9O]_2P(O)OH$

$C_{20}H_{43}O_4P$ M 378.531

Extractant for Hf. Solid. Mp 37-39°, Mp 50-51°. pK_a 3.28 (75% v/v EtOH aq.).

Tetramethylammonium salt: Solid. Mp 212°.

Chloride: Didecyl phosphorochloridate. Didecyl phosphoryl chloride. Didecyl chlorophosphate.

$C_{20}H_{42}ClO_3P$ M 396.977

Oil. n_D^{25} 1.4552.

Chabrier, P. *et al*, *C.R. Hebd. Seances Acad. Sci.*, 1957, **244**, 2730 (*synth*)
Petrov, K.A. *et al*, *Zh. Obshch. Khim.*, 1961, **31**, 1709 (*Engl. transl.* p. 1596) (*synth*)
Grosse-Ruyken, H. *et al*, *J. Prakt. Chem.*, 1962, **18**, 287 (*synth, chloride*)
Bliznyuk, N.K. *et al*, *Zh. Obshch. Khim.*, 1967, **37**, 1119 (*Engl. transl.* p. 1061) (*synth*)

Didecylphosphinic acid, 9CI D-00221

[7507-08-6]

$[H_3C(CH_2)_9]_2P(O)OH$

$C_{20}H_{43}O_2P$ M 346.532

Cryst. (hexane or hexane/Me_2CO/C_6H_6). Mp 88-88.5°. pK_a 6.1.

Chloride: [33493-44-6].

$C_{20}H_{42}ClOP$ M 364.978

Solid. Mp 50-51°. $Bp_{0.035}$ 174-177°.

Drinkard, W.C. *et al*, *J. Am. Chem. Soc.*, 1952, **74**, 5520 (*salts*)
Williams, R.H. *et al*, *J. Am. Chem. Soc.*, 1955, **77**, 3411 (*synth*)
Feshchenko, N.G. *et al*, *Zh. Obshch. Khim.*, 1970, **40**, 2385 (*Engl. transl.* p. 2373) (*chloride*)
Haynes, J.S. *et al*, *Can. J. Chem.*, 1985, **63**, 1111 (*synth*)

Didecyl phosphonate, 9CI D-00222

Didecyl phosphite

[7000-66-0]

$[H_3C(CH_2)_9O]_2P(O)H \rightleftharpoons [H_3C(CH_2)_9O]_2POH$

$C_{20}H_{42}O_3P$ M 361.524

Tautomeric. Liq. d_4^{20} 0.92. Bp_1 201-202°. n_D^{20} 1.4460.

Petrov, K.A. *et al*, *Zh. Obshch. Khim.*, 1962, **32**, 3723 (*Engl. transl.* p. 3650) (*synth*)
Mandel'baum, Ya.A. *et al*, *CA*, 1968, **69**, 43338 (*synth*)

O,O-Didecyl phosphorochloridothioate, 9CI, 8CI D-00223

O,O-*Didecyl thiophosphoryl chloride.* O,O-*Didecyl chlorothiophosphate*

[42070-34-8]

$[H_3C(CH_2)_9O]_2P(S)Cl$

$C_{20}H_{42}ClO_2PS$ M 413.037

Oil. n_D^{20} 1.4993.

Isaglyants, V.I. *et al*, *Zh. Prikl. Khim.*, 1972, **45**, 2599; 1974, **47**, 615 (*Engl. transl.* pp. 2731, 616) (*synth, props*)

O,O-Didecyl phosphorodithioate, 9CI, 8CI D-00224

O,O-*Didecyl hydrogen phosphorodithioate.* O,O-*Didecyl dithiophosphate.* O,O-*Didecyl dithiophosphoric acid*

[2253-59-0]

$[H_3C(CH_2)_9O]_2P(S)SH$

$C_{20}H_{43}O_2PS_2$ M 410.652

K salt: [3419-35-0]. Solid. Mp 158-158.5°.

Bolotova, G.L. *et al*, *CA*, 1965, **63**, 6897 (*synth*)
Szczepaniak, W. *et al*, *Pol. J. Chem.*, 1979, **53**, 755 (*ir, raman, uv, complexes*)

Didodecyl phosphate D-00225

Didodecyl hydrogen phosphate. Dilauryl phosphate. Didodecyl phosphoric acid

[7057-92-3]

$[H_3C(CH_2)_{11}O]_2P(O)OH$

$C_{24}H_{51}O_4P$ M 434.638

Cryst. (MeOH or Et_2O). Mp 59°.

Brown, D.A. *et al*, *J. Chem. Soc.*, 1955, 1584 (*synth*)
Ramirez, F. *et al*, *Phosphorus Sulfur*, 1978, **4**, 43 (*synth*)

Didodecylphosphinic acid D-00226

[6196-71-0]

$[H_3C(CH_2)_{11}]_2P(O)OH$

$C_{24}H_{51}O_2P$ M 402.639

Cryst. (C_6H_6, C_6H_6/EtOH, or $CHCl_3$/Me_2CO). Mp 94-95°. pK_a 6.2.

Chloride:

$C_{24}H_{50}ClOP$ M 421.085

Solid. Mp 66-67°. $Bp_{0.035}$ 188-191°.

Williams, R.H. *et al*, *J. Am. Chem. Soc.*, 1955, **77**, 3411 (*synth*)
Feshchenko, N.G. *et al*, *Zh. Obshch. Khim.*, 1970, **40**, 2385 (*Engl. transl.* p. 2373) (*chloride*)
Haynes, J.S. *et al*, *Can. J. Chem.*, 1985, **63**, 1111 (*synth*)

Didodecyl phosphonate D-00227

Didodecyl phosphite
[21302-09-0]

$[H_3C(CH_2)_{11}O]_2P(O)H \rightleftharpoons [H_3C(CH_2)_{11}O]_2POH$

$C_{24}H_{50}O_3P$ M 417.631
Tautomeric; almost completely in phosphonate form. Heat stabilizer. Oil additive. Solid. Mp 31.5°. $Bp_{0.6}$ 188-192°.

B.P., 1 298 156, (*1970*); *CA*, **78**, 83841 (*synth*)
U.S.P., 3 725 515, (*1970*); *CA*, **78**, 147344 (*synth*)

4,5-Diethoxy-2*H*-1,3,2-diazaphosphole D-00228

$C_6H_{11}N_2O_2P$ M 174.139
2-Chloro:
$C_6H_{10}ClN_2O_2P$ M 208.584
Solid. Mp 28-32°.
2-Phenoxy:
$C_{12}H_{15}N_2O_3P$ M 266.236
Liq. $Bp_{0.2}$ 137-139°.
2-Me: see 4,5-Diethoxy-2-methyl-2H-1,3,2-diazaphosphole, D-00235
2-Ph: see 4,5-Diethoxy-1,3-dimethyl-2-phenyl-2H-1,3,2-diazaphosphole, D-00230

Dregval', G.F. *et al*, *Zh. Obshch. Khim.*, 1963, **33**, 2952 (*Engl. transl.* p. 2880) (*phenoxy, synth*)
Derkach, G.I. *et al*, *Zh. Obshch. Khim.*, 1966, **36**, 1087 (*Engl. transl.* p. 1101) (*derivs, synth*)

1,1-Diethoxy-2,2-diethyldiphosphine 1-oxide, 9CI D-00229

[66193-16-6]

$(EtO)_2P(O)-PEt_2$

$C_8H_{20}O_3P_2$ M 226.192
Liq. d_4^{20} 1.04. Bp_2 90°. n_D^{20} 1.4709.

Foss, V.L. *et al*, *Phosphorus Sulfur*, 1977, **3**, 299 (*synth, P nmr*)
Kabachnik, M.M. *et al*, *Zh. Obshch. Khim.*, 1979, **49**, 1446 (*Engl. transl.* p. 1264) (*synth*)

4,5-Diethoxy-1,3-dimethyl-2-phenyl-2*H*-1,3,2-diazaphosphole, 8CI D-00230

$C_{12}H_{15}N_2O_2P$ M 250.236
Solid. Mp 62.5°. $Bp_{0.25}$ 149-150°.
2-Sulfide:
$C_{12}H_{15}N_2O_2P$ M 250.236
Prisms (cyclohexane). Mp 141-142°.

Dregval', G.F. *et al*, *Zh. Obshch. Khim.*, 1963, **33**, 2952 (*Engl. transl.* p. 2880)

2,5-Diethoxy-1,3,2,5-dioxadiphosphorinane, 9CI D-00231

2,5-Diethoxy-1,3,2,5-dioxadiphosphinane

$C_6H_{14}O_4P$ M 181.148
5-Oxide:
$C_6H_{14}O_5P$ M 197.147
Liq. d_4^{20} 1.25. $Bp_{0.017}$ 140-150° (bath). n_D^{20} 1.4704. Mixt. of stereoisomers 5:1.
5-Oxide, 2-sulfide:
$C_6H_{14}O_5PS$ M 229.207
Liq. d_4^{20} 1.31. n_D^{20} 1.4940. Mixt. of stereoisomers.

Nifant'ev, É.E. *et al*, *Zh. Obshch. Khim.*, 1979, **49**, 2627 (*Engl. transl.* p. 2331) (*synth, pmr, P nmr*)

(Diethoxyethenylidene)triphenylphosphorane, 9CI D-00232

(Diethoxyvinylidene)triphenylphosphorane. 1,1-Diethoxy-2-(triphenylphosphoranylidene)ethylene
[21882-77-9]

$Ph_3P{=}C{=}C(OEt)_2$

$C_{24}H_{25}O_2P$ M 376.434
Reagent for prep. of ethyl esters of carboxylic acids in aprotic solvents. Cryst. (cyclohexane), reactive to water. Mp 83°.

Bestmann, H.J. *et al*, *Chem. Ber.*, 1973, **106**, 2602 (*synth, pmr, props*)
Bestmann, H.J. *et al*, *Angew. Chem., Int. Ed. Engl.*, 1977, **16**, 349 (*rev*)
Bestmann, H.J. *et al*, *Justus Liebigs Ann. Chem.*, 1977, 276 (*props*)
Bestmann, H.J. *et al*, *J. Chem. Res.* (*S*), 1979, 313 (*props*)
Albright, T.A. *et al*, *Z. Naturforsch.*, *B*, 1980, **35**, 343 (*struct*)
Bestmann, H.J. *et al*, *Synthesis*, 1981, 998 (*use*)
Bestmann, H.J., *Chem. Ber.*, 1982,, **115**, 161 (*synth*)
Houben-Weyls Method. Org. Chem., Band E1, 1982, 759 (*rev*)

(2,2-Diethoxyethylidene)triphenylphosphorane, 9CI D-00233

1,1-Diethoxy-2-(triphenylphosphoranylidene)ethane
[71276-94-3]

$Ph_3P{=}CHCH(OEt)_2$

$C_{24}H_{27}O_2P$ M 378.450
Reagent for prep. of (*Z*)-α,β-unsatd. aldehydes by stereoselective Wittig reaction. Yellow, hygroscopic cryst. Mp 64-66°.

Bestmann, H.J. *et al*, *Justus Liebigs Ann. Chem.*, 1982, 363 (*use*)
Bestmann, H.J. *et al*, *Chem. Ber.*, 1982, **115**, 161 (*synth, pmr, cmr, use*)

(2,2-Diethoxyethyl)phosphonic acid D-00234

$(EtO)_2CHCH_2P(O)(OH)_2$

Di-Et ester: [7598-61-0]. *Diethyl (2,2-diethoxyethyl)-phosphonate.*
$C_{10}H_{23}O_5P$ M 254.262

Liq. d_4^{20} 1.05. $Bp_{0.12}$ 97-98°. n_D^{20} 1.4280.
Di-Ph ester: [72616-61-6]. *Diphenyl (2,2-diethoxyethyl)phosphonate.*
$C_{18}H_{23}O_5P$ M 350.350
Liq. $Bp_{0.15}$ 197-199°. n_D^{20} 1.5275.

Razumov, A.I. *et al*, *Zh. Obshch. Khim.*, 1964, **34**, 2589 (*Engl. transl.* p. 2612) (*synth*)
Maier, L. *et al*, *Z. Anorg. Allg. Chem.*, 1972, **394**, 111 (*synth, pmr, P nmr*)
Halman, M. *et al*, *J. Chem. Soc., Perkin Trans. 2*, 1976, 1210 (*synth, pmr*)
Cates, L.A. *et al*, *J. Med. Chem.*, 1980, **23**, 300 (*synth*)
Yanai, S., *Phosphorus Sulfur*, 1982, **12**, 369 (*ms*)

4,5-Diethoxy-2-methyl-2*H*-1,3,2-diazaphosphole, 8CI — D-00235

EtO, EtO, N, N, P, Me

$C_7H_{13}N_2O_2P$ M 188.166
Liq. d_4^{20} 1.10. $Bp_{0.7}$ 91-93°. n_D^{20} 1.4971.
2-Oxide:
$C_7H_{13}N_2O_3P$ M 204.165
Needles (cyclohexane). Mp 118-120°.
2-Sulfide:
$C_7H_{13}N_2O_2PS$ M 220.226
Prisms (C_6H_6). Mp 85-86°.

Derkach, G.I. *et al*, *Zh. Obshch. Khim.*, 1966, **36**, 1087 (*Engl. transl.* p. 1101)

Diethoxyphosphinecarboxylic acid oxide, 9CI — D-00236

O,O-*Diethylphosphonoformic acid*

$(EtO)_2P(O)COOH$

$C_5H_{11}O_5P$ M 182.113
Esters are plant growth regulators.
Me ester: [41760-84-3]. *Diethyl methyl phosphonoformate.*
$C_6H_{13}O_5P$ M 196.139
Used in heterocyclic synth. Liq. Bp_3 87-93°.
Et ester: [1474-78-8]. *Triethyl phosphonoformate.*
$C_7H_{15}O_5P$ M 210.166
Used in heterocyclic synth. Liq. d_0^0 1.14. $Bp_{12.5}$ 133°.

Reetz, T. *et al*, *J. Am. Chem. Soc.*, 1955, **77**, 3813 (*synth*)
Ogata, J. *et al*, *J. Org. Chem.*, 1970, **35**, 596 (*synth, props*)
Warren, S. *et al*, *J. Chem. Soc. (B)*, 1971, 618 (*synth, props*)
Shotani, S. *et al*, *Chem. Pharm. Bull.*, 1973, **21**, 1160 (*synth, props*)
Gakis, N. *et al*, *Helv. Chim. Acta*, 1975, **58**, 748 (*use*)
Morita, T. *et al*, *Bull. Chem. Soc. Jpn.*, 1981, **54**, 267 (*synth*)

(Diethoxyphosphinothioyl)acetic acid, 9CI — D-00237

$(EtO)_2P(S)CH_2COOH$

$C_6H_{13}O_4PS$ M 212.200
Et ester: Ethyl (diethoxyphosphinothioyl)acetate.
$C_8H_{17}O_4PS$ M 240.254
Liq. d_4^{20} 1.12. Bp_5 105-106°, $Bp_{0.03}$ 80-82°. n_D^{20} 1.4689.

Kabachnik, M.I. *et al*, *Izv. Akad. Nauk SSSR, Ser. Khim.*, 1953, 163 (*Engl. transl.* p. 145)
Moskva, V.V. *et al*, *Zh. Obshch. Khim.*, 1971, **41**, 1495 (*Engl. transl.* p. 1499)

(Diethoxyphosphinothioyl)phosphorimidic acid, 9CI — D-00238

N-*(Diethoxyphosphinothioyl)*-P,P,P-*trimethoxyphosphine imide*

$(EtO)_2P(S)N{=}P(OH)_3$

$C_4H_{13}NO_5P_2S$ M 249.158
Free acid tautomeric with (diethoxyphosphinothioyl)-phosphoramidic acid.
Tri-Me ester: [51945-70-1]. *Trimethyl (diethoxyphosphinothioyl)phosphorimidate.* N-*(Diethoxyphosphinothioyl)*-P,P,P-*trimethoxyphosphazene.* N-*(Diethoxyphosphinothioyl)*-P,P,P-*trimethoxyphosphine imide.*
$C_7H_{19}NO_5P_2S$ M 291.238
Liq. d_4^{20} 1.222. n_D^{20} 1.4758.
Tri-Et ester: [7166-10-1]. *Triethyl (diethoxyphosphinothioyl)phosphorimidate.* P,P,P-*Triethoxy*-N-*(diethoxyphosphinothioyl)phosphazene.* N-*(Diethoxyphosphinothioyl)*-P,P,P-*triethoxyphosphine imide.*
$C_{10}H_{25}NO_5P_2S$ M 333.319
Liq.
▷At 140°, isom. to mixture of N-Et and S-Et isomers
Tri-Ph ester: [51576-47-7]. *Triphenyl (diethoxyphosphinothioyl)phosphorimidate.* N-*(Diethoxyphosphinothioyl)*-P,P,P-*triphenoxyphosphazene.* N-*(Diethoxyphosphinothioyl)*-P,P,P-*triphenoxyphosphine imide.*
$C_{22}H_{25}NO_5P_2S$ M 477.451
Solid. Mp 38-39°.

Khodak, A.A. *et al*, *Zh. Obshch. Khim.*, 1974, **44**, 27, 256 (*Engl. transl.* pp. 24, 241) (*esters, synth, props*)

[(Diethoxyphosphinothioyl)thio]acetic acid, 9CI — D-00239

$(EtO)_2P(S)SCH_2COOH$

$C_6H_{13}O_4PS_2$ M 244.260
pK_a 3.98 (50% dioxan aq., 20°).
Me ester: see Methylacetophos, M-00086
Et ester: see Acetophos, A-00004
Fluoride:
$C_6H_{12}FO_3PS$ M 214.191
Liq. $Bp_{0.1}$ 84-90°.
Chloride:
$C_6H_{12}ClO_3PS$ M 230.646
Liq. d_4^{25} 1.29. $Bp_{0.5}$ 108°. n_D^{25} 1.5185.
Amide: [2047-14-5]. O,O-*Diethyl* S-*(2-amino-2-oxoethyl) phosphorodithioate.* O,O-*Diethyl* S-*(carbamoylmethyl) phosphorodithioate.*
$C_6H_{14}NO_3PS_2$ M 243.275
Cryst. (CCl_4 or diisopropyl ether). Mp 59.5-60.5°.
▷TD8250000.
Isopropylamide: see Prothoate, P-00499

Hoegberg, E.I. *et al*, *J. Am. Chem. Soc.*, 1951, **73**, 557 (*derivs, synth*)
Almasi, L. *et al*, *Chem. Ber.*, 1965, **98**, 613 (*derivs, ir*)
Carter, P.L. *et al*, *J. Appl. Chem.*, 1968, **18**, 257 (*synth, derivs*)
Suplicy, N. *et al*, *J. Econ. Entomol.*, 1972, **65**, 1585 (*derivs, tox*)

[(Diethoxyphosphinothioyl)thio]butanedioic acid, 9CI D-00240

[(Diethoxyphosphinothioyl)thio]succinic acid

$(EtO)_2P(S)SCH(COOH)CH_2COOH$

$C_8H_{15}O_6PS_2$ M 302.297

(±)-form

Di-Me ester: Dimethyl [(diethoxyphosphinothioyl)thio]-butanedioate.
$C_{10}H_{19}O_6PS_2$ M 330.350
Liq. d_4^{20} 1.22. $Bp_{0.025}$ 116-120°. n_D^{20} 1.4985.

Di-Et ester: [3700-86-5]. *Diethyl [(diethoxyphosphinothioyl)thio]butanedioate.*
$C_{12}H_{23}O_6PS_2$ M 358.404
Liq. d_4^{20} 1.17. Bp_3 157-162°. n_D^{20} 1.4910.

▷TD6650000.

Diisopropyl ester: Diisopropyl [(diethoxyphosphinothioyl)thio]butanedioate.
$C_{14}H_{27}O_6PS_2$ M 386.457
Liq. d_4^{20} 1.15. $Bp_{0.025}$ 117-121°. n_D^{20} 1.4815.

Mel'nikov, N.N. *et al*, *Zh. Obshch. Khim.*, 1953, **23**, 1352 (*Engl. transl.* p. 1416)

(Diethoxyphosphinyl)acetic acid, 9CI D-00241

Diethyl (carboxymethyl)phosphonate

[3095-95-2]

$(EtO)_2P(O)CH_2COOH$

$C_6H_{13}O_5P$ M 196.139

Esters are Wittig-Horner reagents for the synth. of esters of $\alpha\beta$-unsat. carboxylic acids. Oil or cryst. d_4^{20} 1.25. Mp <18°, Mp 27-28°. pK_a 3.48 (H_2O), pK_a 4.66 (50% EtOH aq., 25°).

Cyclohexylammonium salt: Cryst. (Me_2CO/Et_2O). Mp 107-108°.

Me ester: [1067-74-9]. *Diethyl (methoxycarbonylmethyl)phosphonate.*
$C_7H_{15}O_5P$ M 210.166
Liq. d_4^{20} 1.14. Bp_9 131.5-132.5°. n_D^{20} 1.4335.

Et ester: [867-13-0]. *Diethyl (ethoxycarbonylmethyl)-phosphonate. Triethyl phosphonoacetate.*
$C_8H_{17}O_5P$ M 224.193
Liq. d 1.13. Bp_{20} 152-153°, Bp_9 142-145°. n_D^{20} 1.4320.

▷AG9800000.

Benzyl ester: [7396-44-3]. *Diethyl (benzyloxycarbonylmethyl)phosphonate.*
$C_{13}H_{19}O_5P$ M 286.264
Liq. $Bp_{1.5}$ 174-179°. n_D^{23} 1.4932.

Trimethylsilyl ester: [66130-90-3]. *Diethyl [(trimethylsilyloxy)carbonyl]phosphonate.*
$C_9H_{21}O_5PSi$ M 268.321
Liq. $Bp_{0.0005}$ 93°.

Marvel, G. *et al*, *J. Chim. Phys.*, 1967, **64**, 1686 (*ethyl ester, nmr*)
Nishikawa, T., *Tetrahedron*, 1967, **23**, 2181 (*esters, ms*)
Malevannaya, R.A. *et al*, *Zh. Obshch. Khim.*, 1971, **41**, 1426 (*Engl. transl.* p. 1432) (*synth, salts*)
Bodnarchuk, N.D. *et al*, *Zh. Obshch. Khim.*, 1971, **41**, 1464 (*Engl. transl.* p. 1470) (*synth*)
Moskva, V.V. *et al*, *Zh. Obshch. Khim.*, 1976, **46**, 534 (*Engl. transl.* p. 529) (*ester, use*)
Lombardo, L. *et al*, *Synthesis*, 1978, 131 (*silyl ester, synth, use, pmr*)
Coutrot, P. *et al*, *Synthesis*, 1978, 133 (*synth, pmr, use*)
Maier, L. *et al*, *Phosphorus Sulfur*, 1978, **5**, 45 (*ester, synth*)
Clayton, J.P. *et al*, *J. Chem. Soc., Perkin Trans. 1*, 1979, 308 (*synth, ir, pmr, use*)
Bottin-Strzalko, T. *et al*, *J. Org. Chem.*, 1980, **45**, 1270 (*pmr, cmr, P nmr*)
Kočovsky, P. *et al*, *Collect. Czech. Chem. Commun.*, 1980, **45**, 921 (*use*)
Rice, K.C., *J. Org. Chem.*, 1982, **47**, 3617 (*ester, use*)

α-(Diethoxyphosphinyl)benzeneacetic acid, 9CI D-00242

α-(Diethoxyphosphinyl)phenylacetic acid

[38654-91-0]

$(EtO)_2P(O)CHPhCOOH$

$C_{12}H_{17}O_5P$ M 272.237

(±)-form

Prisms (Et_2O at −60°). Mp 55°.

Et ester: Ethyl α-(diethoxyphosphinyl)benzeneacetate. Triethyl 2-phosphono-2-phenylacetate.
$C_{14}H_{21}O_5P$ M 300.291
Liq. d_4^{20} 1.11. Bp_1 152-162°, $Bp_{0.3}$ 108-110°. n_D^{20} 1.4810.

Kreutzkamp, N. *et al*, *Arch. Pharm. (Weinheim, Ger.)*, 1962, **295**, 276 (*ester, synth*)
Berry, J.P. *et al*, *J. Org. Chem.*, 1972, **37**, 4396 (*ester, synth*)
Kresze, G. *et al*, *Justus Liebigs Ann. Chem.*, 1972, **756**, 112 (*ester, synth, ir, pmr, use*)
Kirilov, M. *et al*, *Bull. Soc. Chim. Fr., Part 2*, 1973, 3051 (*ester, synth, ir*)
Kolodyazhnyi, O.I. *et al*, *Zh. Obshch. Khim.*, 1979, **49**, 2458 (*Engl. transl. p. 2170*) (*synth, ir, pmr*)

2-(Diethoxyphosphinyl)benzoic acid D-00243

[22537-93-5]

HOOC, P(O)(OEt)$_2$ (ortho-substituted benzene)

$C_{11}H_{15}O_5P$ M 258.210

Me ester: [72974-42-6]. *Methyl 2-(diethoxyphosphinyl)-benzoate.*
$C_{12}H_{17}O_5P$ M 272.237
Liq. $Bp_{0.4}$ 145°.

Et ester: see under 2-Phosphonobenzoic acid, P-00378

Miles, J.A. *et al*, *J. Org. Chem.*, 1978, **43**, 4668 (*props*)
Hall, N. *et al*, *J. Chem. Soc., Perkin Trans. 1*, 1979, 2634 (*ester, synth, pmr*)
Hirao, T. *et al*, *Bull. Chem. Soc. Jpn.*, 1982, **55**, 909 (*ester*)

4-(Diethoxyphosphinyl)benzoic acid D-00244

[1527-34-0]

$C_{11}H_{15}O_5P$ M 258.210

Me ester: Methyl 4-(diethoxyphosphinyl)benzoate.
$C_{12}H_{17}O_5P$ M 272.237
Liq. $Bp_{0.1}$ 144-148°. n_D^{20} 1.4942.

Et ester: see under 4-Phosphonobenzoic acid, P-00380

Tavs, P., *Chem. Ber.*, 1970, **103**, 2428.

(Diethoxyphosphinyl)butanedioic acid D-00245

(Diethoxyphosphinyl)succinic acid

$(EtO)_2P(O)CH(COOH)CH_2COOH$

$C_8H_{15}O_7P$ M 254.176
Esters are useful Wittig-Horner reagents.

(±)-*form*

Di-Me ester: Dimethyl (diethoxyphosphinyl)-butanedioate. Dimethyl (diethoxyphosphinyl)-succinate.
$C_{10}H_{19}O_7P$ M 282.230
Liq. $Bp_{0.4}$ 122-126°. n_D^{22} 1.4420.

Di-Et ester: Tetraethyl phosphonobutanedioate. Diethyl (diethoxyphosphinyl)butanedioate. Tetraethyl phosphonosuccinate.
$C_{12}H_{23}O_7P$ M 310.283
Liq. $Bp_{0.4}$ 127°. n_D^{23} 1.4389.

Grell, W. *et al, Justus Liebigs Ann. Chem.*, 1954, **693**, 134 (*synth, ir, pmr, use*)
Harvey, R.G., *Tetrahedron*, 1966, **22**, 2561 (*synth, pmr*)
Pudovik, A.N. *et al, Zh. Obshch. Khim.*, 1975, **45**, 272 (*Engl. transl.* p. 257) (*synth*)
Linke, S. *et al, Justus Liebigs Ann. Chem.*, 1980, 542 (*synth, use, ir, pmr*)

2-(Diethoxyphosphinyl)butanoic acid, 9CI D-00246

$(EtO)_2P(O)CH(COOH)CH_2CH_3$

$C_8H_{17}O_5P$ M 224.193

(±)-*form*

Et ester: [17145-91-4]. *Ethyl 2-(diethoxyphosphinyl)-butanoate. Triethyl 2-phosphonobutanoate.*
$C_{10}H_{21}O_5P$ M 252.247
Wittig-Horner reagent for the synth. of enoic ethyl esters. Liq. d_4^{20} 1.11. $Bp_{0.6}$ 117-118°. n_D^{20} 1.4310 (1.4530).

Ph ester: [53327-49-4]. *Phenyl 2-(diethoxyphosphinyl)-butanoate.*
$C_{14}H_{21}O_5P$ M 300.291
Cryst. (cyclohexane). Mp 97-98°.

Berry, J.P. *et al, J. Org. Chem.*, 1972, **37**, 4396 (*synth*)
Rachon, J. *et al, Z. Chem.*, 1974, **14**, 152 (*synth, ir, pmr*)
Runge, W. *et al, Justus Liebigs Ann. Chem.*, 1975, 1361 (*use*)
Moskva, V.V. *et al, Zh. Obshch. Khim.*, 1976, **46**, 536 (*Engl. transl.* p. 529) (*synth*)
Huston, R. *et al, Helv. Chim. Acta*, 1982, **65**, 1563 (*use*)

4-(Diethoxyphosphinyl)butanoic acid, 9CI D-00247

[38694-48-3]

$(EtO)_2P(O)CH_2CH_2CH_2COOH$

$C_8H_{17}O_5P$ M 224.193
Cryst. (Et_2O). Mp 46.5-47.5°. pK_a 4.57 (H_2O), pK_a 5.67 (50% EtOH aq., 25°).

Me ester: Methyl 4-(diethoxyphosphinyl)butanoate.
$C_9H_{19}O_5P$ M 238.220
Liq. $Bp_{0.01}$ 104-105°. n_D^{25} 1.4376.

Et ester: [2327-69-7]. *Ethyl 4-(diethoxyphosphinyl)-butanoate. Triethyl 4-phosphonobutanoate.*
$C_{10}H_{21}O_5P$ M 252.247
Liq. d_4^{20} 1.09. Bp_4 144-145°. n_D^{20} 1.4378.

Benzyl ester: Benzyl 4-(diethoxyphosphinyl)butanoate.
$C_{15}H_{23}O_5P$ M 314.317
Liq. $Bp_{0.005}$ 136-140°. n_D^{25} 1.4933.

Falbe, J. *et al, Chem. Ber.*, 1965, **98**, 2312 (*esters*)
Nishiwaki, T., *Tetrahedron*, 1967, **23**, 2181 (*ester, ms*)
Berry, J.P. *et al, J. Org. Chem.*, 1972, **37**, 4396 (*ester*)
Tsvetkov, E.N. *et al, Zh. Obshch. Khim.*, 1975, **45**, 716 (*Engl. transl.* p. 706) (*synth, props*)

4-(Diethoxyphosphinyl)-2-butenoic acid, 9CI D-00248

4-Diethylphosphonocrotonic acid. 4-Diethoxyphosphinylcrotonic acid

[24676-98-0]

$(EtO)_2P(O)CH_2CH{=}CHCOOH$

$C_8H_{15}O_5P$ M 222.177

Me ester: [24676-98-0]. *Methyl 4-(diethoxyphosphinyl)-2-butenoate.*
$C_9H_{17}O_5P$ M 236.204
Wittig-Horner reagent for synth. of Me esters of di- and poly-ene carboxylic acids. Liq. $Bp_{0.3}$ 115-130°.

(*E*)-*form* [79563-56-7]
Solid. Mp 85-86.5°.

Et ester: [42516-28-9]. *Ethyl 4-(diethoxyphosphinyl)-2-butenoate. Triethyl 4-phosphonocrotonate.*
$C_{10}H_{19}O_5P$ M 250.231
A Wittig-Horner reagent for synth. of ethyl esters of di- and poly-ene carboxylic acids. Liq. $Bp_{0.4}$ 121-123°. n_D^{20} 1.4539.

Sato, K. *et al, J. Org. Chem.*, 1967, **32**, 177 (*synth, ir, pmr, use*)
Burden, R.S. *et al, J. Chem. Soc. (C)*, 1969, 2477 (*synth*)
Roush, W.R., *J. Am. Chem. Soc.*, 1978, **100**, 3599 (*use*)
Floyd, D.M. *et al, Tetrahedron Lett.*, 1981, **22**, 2847 (*synth*)
Takeda, K. *et al, Chem. Pharm. Bull.*, 1982, **30**, 4000 (*use*)

N-(Diethoxyphosphinyl)carbamic acid, 9CI, 8CI D-00249

Diethyl carboxyphosphoramidate

$(EtO)_2P(O)NHCOOH$

$C_5H_{12}NO_5P$ M 197.127
Distn. of esters → Diethyl phosphorisocyanatidate, D-00366 .

Me ester: [65289-19-2]. *Methyl* N-*(diethoxyphosphinyl)carbamate.*
$C_6H_{14}NO_5P$ M 211.154
Liq. n_D^{20} 1.4400.

Et ester: [35852-07-4]. *Ethyl* N-*(diethoxyphosphinyl)-carbamate.*
$C_7H_{16}NO_5P$ M 225.181
Liq. n_D^{20} 1.4380.

Shokol, V.A. *et al, Zh. Obshch. Khim.*, 1977, **47**, 2202 (*Engl. transl.* p. 2010) (*synth, props*)

(Diethoxyphosphinyl)fluoroacetic acid, 9CI D-00250

$(EtO)_2P(O)CHFCOOH$

$C_6H_{12}FO_5P$ M 214.130

(±)-*form*

Me ester: Methyl (diethoxyphosphinyl)fluoroacetate.
$C_7H_{14}FO_4P$ M 212.157
Liq. d_4^{20} 1.20. Bp_1 112-115°. n_D^{20} 1.4138.

Et ester: [2536-16-3]. *Ethyl (diethoxyphosphinyl)-fluoroacetate.*
$C_8H_{16}FO_4P$ M 226.184
Wittig-Horner reagent widely used in synth. of fluoroalkenes. Liq. d_4^{20} 1.19. $Bp_{0.7}$ 125°. n_D^{20} 1.4196. Forms carbanion with NaH or $LiN[CH(CH_3)_2]_2$.

N,N-*Diethylamide:* N,N-*Diethyl(diethoxyphosphinyl)-fluoroacetamide.*
$C_{10}H_{21}FNO_4P$ M 269.252
Liq. d_4^{20} 1.12. $Bp_{0.5}$ 131-132°. n_D^{20} 1.4319.

Fokin, A.V. *et al*, *Zh. Org. Khim.*, 1971, **7**, 249 (*Engl. transl.* p. 241) (*derivs, synth*)
Pawson, B.A. *et al*, *J. Med. Chem.*, 1979, **22**, 1059 (*use*)
Liu, R.S.H. *et al*, *J. Am. Chem. Soc.*, 1981, **103**, 7195 (*use*)
Lovey, A.J. *et al*, *J. Med. Chem.*, 1982, **25**, 71 (*use*)
Asato, A.E. *et al*, *J. Am. Chem. Soc.*, 1983, **105**, 2923 (*use*)

4-(Diethoxyphosphinyl)-3-methyl-2-butenoic acid, 9CI — D-00251

$(EtO)_2P(O)CH_2C(CH_3)=CHCOOH$ (*E*)-*form*

$C_9H_{17}O_5P$ M 236.204

(*E*)-form

Me ester: [19945-56-3].
$C_{10}H_{19}O_5P$ M 250.231
Liq. Bp_1 107-108°. n_D^{24} 1.4617.

(*Z*)-form

Me ester: [19945-48-3]. Liq. Bp_1 99-100°. n_D^{24} 1.4590.

Davis, J.B. *et al*, *J. Chem. Soc. (C)*, 1966, 2154 (*synth, use*)
Pattenden, G.L. *et al*, *J. Chem. Soc. (C)*, 1968, 1984 (*synth, uv, ir, pmr*)
Oida, S. *et al*, *Chem. Pharm. Bull.*, 1973, **21**, 528 (*ir, pmr, use*)
Corey, E.J. *et al*, *J. Org. Chem.*, 1974, **39**, 821 (*synth, use*)
Gedye, R.N. *et al*, *Can. J. Chem.*, 1977, **55**, 1218 (*synth, uv, ir, pmr*)
Borowiecki, L. *et al*, *Justus Liebigs Ann. Chem.*, 1982, 1766 (*use*)

(Diethoxyphosphinyl)oxoacetic acid, 9CI — D-00252

$(EtO)_2P(O)COCOOH$

$C_6H_{11}O_6P$ M 210.123

Me ester: [37736-69-9]. *Methyl (diethoxyphosphinyl)-oxoacetate.*
$C_7H_{13}O_6P$ M 224.150
Liq. $Bp_{0.5}$ 145°. n_D^{20} 1.4407.

Et ester: [16540-25-3]. *Ethyl (diethoxyphosphinyl)-oxoacetate.*
$C_8H_{15}O_8P$ M 270.175
Liq. $Bp_{0.05}$ 140°. n_D^{20} 1.4535.

Ger. Pat., 2 111 672, (*1972*); *CA*, **77**, 152349 (*synth*)

2-(Diethoxyphosphinyl)-3-oxobutanoic acid, 9CI — D-00253

[3730-54-9]

$H_3CCOCH(COOH)P(O)(OEt)_2$

$C_8H_{15}O_6P$ M 238.177

▷EK7785000.

Et ester: [3730-54-9]. *Ethyl 2-(diethoxyphosphinyl)-3-oxobutanoate. Triethyl 3-oxo-2-phosphonobutanoate.*
$C_{10}H_{19}O_6P$ M 266.230
Liq. d_4^{20} 1.13. $Bp_{0.5}$ 93-94°. n_D^{20} 1.4430. Exists as 91% enol form in CCl_4, and 54% enol in MeCN.
▷EK7785000.

Kreutzkamp, N., *Chem. Ber.*, 1955, **88**, 195 (*synth*)
Pudovik, A.N. *et al*, *Zh. Obshch. Khim.*, 1975, **45**, 22 (*Engl. transl.* p. 19) (*synth, ir*)
Pudovik, A.N. *et al*, *Zh. Obshch. Khim.*, 1976, **46**, 458 (*Engl. transl.* p. 456) (*ir, pmr, tautom*)

4-(Diethoxyphosphinyl)-3-oxobutanoic acid, 9CI — D-00254

$(EtO)_2P(O)CH_2COCH_2COOH$

$C_8H_{15}O_6P$ M 238.177

Et ester: [65043-08-5]. *Ethyl 4-(diethoxyphosphinyl)-3-oxobutanoate. Triethyl 3-oxo-4-phosphonobutanoate.*
$C_{10}H_{19}O_6P$ M 266.230
Wittig-Horner reagent for the synth. of ethyl esters of β-oxo-γ,δ-unsat. carboxylic acids. Annulates with chalcones → cyclohexenones. $Bp_{0.15}$ 118-120°.

Jagodić, V., *Croat. Chem. Acta*, 1977, **49**, 487 (*synth, ir, pmr*)
Bodalski, R. *et al*, *Tetrahedron Lett.*, 1980, **21**, 2287 (*synth, P nmr, use*)
Bodalski, R. *et al*, *Pol. J. Chem.*, 1983, **57**, 315 (*use*)
Taylor, E.C. *et al*, *Tetrahedron Lett.*, 1983, **24**, 5453 (*use*)
Callant, P. *et al*, *Synth. Commun.*, 1984, **14**, 163 (*use*)

2-[(Diethoxyphosphinyl)oxy]propanoic acid — D-00255

O-*(Diethoxyphosphinyl)lactic acid*

$(EtO)_2P(O)OCH(CH_3)COOH$

$C_7H_{15}O_6P$ M 226.166

(±)-form

Me ester: Methyl 2-[(diethoxyphosphinyl)oxy]-propanoate.
$C_8H_{17}O_6P$ M 240.192
d_4^{20} 1.15. $Bp_{0.5}$ 92°. n_D^{20} 1.4200.

Et ester: Ethyl 2-[(diethoxyphosphinyl)oxy]propanoate. Triethyl O-*phosphonolactate. Triethyl phospholactate.*
$C_9H_{19}O_6P$ M 254.219
Liq. d_4^{20} 1.12-1.13. Bp_8 147°, $Bp_{0.5}$ 105-106°. n_D^{20} 1.4210.

Pudovik, A.N. *et al*, *Zh. Obshch. Khim.*, 1963, **33**, 483; 1969, **39**, 2424; 1972, **42**, 333 (*Engl. transl.* pp 475, 2364, 323) (*synth*)

2-[(Diethoxyphosphinyl)oxy]-2-propenoic acid — D-00256

2-Hydroxyacrylic acid anhydride with diethyl phosphate

$H_2C=C(COOH)OP(O)(OEt)_2$

$C_7H_{13}O_6P$ M 224.150

Me ester: [76179-26-5]. *Methyl 2-[(diethoxyphosphinyl)oxy]-2-propenoate.*
$C_8H_{15}O_6P$ M 238.177
Liq. $Bp_{0.005}$ 80°. n_D^{20} 1.4370.

Et ester: [1991-30-6]. *Ethyl 2-[(diethoxyphosphinyl)-oxy]-2-propenoate. Triethyl 2-phosphonoxy-2-propenoate. Triethyl phosphoenolpyruvate.*
$C_9H_{17}O_6P$ M 252.203
Liq. $Bp_{0.005}$ 79-80°. n_D^{20} 1.4339.

Lichtenthaler, F.W. *et al*, *Chem. Ber.*, 1962, **95**, 1971 (*props*)

Zawadzki, M. *et al*, *Organika*, 1979, 9; *CA*, **94**, 30142 (*esters, synth, pmr, P nmr*)

(Diethoxyphosphinyl)phosphorimidic acid, 9CI, 8CI D-00257

$(EtO)_2P(O)N{=}P(OH)_3$

$C_4H_{13}NO_6P_2$ M 233.097

Free acid tautomeric with (diethoxyphosphinyl)-phosphoramidic acid.

Tri-Me ester: [87992-68-5]. *Trimethyl (diethoxyphosphinyl)phosphorimidate.* N-*(Diethoxyphosphinyl)*-P,P,P-*trimethoxyphosphine imide.* N-*(Diethoxyphosphinyl)*-P,P,P-*trimethoxyphosphazene.*
$C_7H_{19}NO_6P_2$ M 275.178
Liq. Isom. on dist.

Tri-Et ester: [2397-48-0]. *Triethyl (diethoxyphosphinyl)phosphorimidate.* N-*Diethoxyphosphinyl*-P,P,P-*triethoxyphosphine imide.* N-*Diethoxyphosphinyl*-P,P,P-*triethoxyphosphazene.*
$C_{10}H_{25}NO_6P_2$ M 317.258
Liq. d_4^{20} 1.13. $Bp_{0.5}$ 121°. n_D^{20} 1.4345.

Triisopropyl ester: [18313-80-9]. *Triisopropyl (diethoxyphosphinyl)phosphorimidate.* N-*Diethoxyphosphinyl*-P,P,P-*trisisopropoxyphosphine imide.* N-*Diethoxyphosphinyl*-P,P,P-*tris(1-methylethoxy)-phosphazene.*
$C_{13}H_{31}NO_6P_2$ M 359.339
Liq. $Bp_{0.03}$ 91-93°.

Kabachnik, M.I. *et al*, *Izv. Akad. Nauk SSSR, Ser. Khim.*, 1961, 819 (*Engl. transl.* p. 758) (*triethyl ester, synth, ir*)
Gilyarov, V.A. *et al*, *Zh. Obshch. Khim.*, 1972, **42**, 2148 (*Engl. transl.* p. 2145) (*synth*)
Riesel, L. *et al*, *Z. Anorg. Allg. Chem.*, 1977, **435**, 61; 1979, **451**, 5; 1983, **502**, 21 (*synth, P nmr*)

2-(Diethoxyphosphinyl)propanoic acid, 9CI D-00258

[30094-28-1]

$(EtO)_2P(O)CH(CH_3)COOH$

$C_7H_{15}O_5P$ M 210.166

(±)-*form*

Me ester: Methyl 2-(diethoxyphosphinyl)propanoate.
$C_8H_{17}O_5P$ M 224.193
No phys. props. reported.

Et ester: see Ethyl 2-(diethoxyphosphinyl)propanoate, E-00059

Trimethylsilyl ester: Trimethylsilyl 2-(diethoxyphosphinyl)propanoate.
$C_{10}H_{23}O_5PSi$ M 282.348
Liq. Bp_{107} 92-94°.

Ph ester: Phenyl (2-diethoxyphosphinyl)propanoate.
$C_{13}H_{19}O_5P$ M 286.264
Cryst. (pet. ether). Mp 72-73°.

Bottin-Strzalko, T. *et al*, *J. Chem. Soc., Chem. Commun.*, 1976, 905 (*synth, cmr, pmr, P nmr*)
Boehme, H. *et al*, *Arch. Pharm. (Weinheim, Ger.)*, 1979, **312**, 49, 60 (*pmr, cmr, props*)
Katsumuva, S. *et al*, *Chem. Lett.*, 1982, 1689 (*silyl ester*)

3-(Diethoxyphosphinyl)propanoic acid, 9CI D-00259

[3095-96-3]

$(EtO)_2P(O)CH_2CH_2COOH$

$C_7H_{15}O_5P$ M 210.166

Cryst. (Et_2O/pentane). Mp 37.5-38.5°. pK_a 4.28 (H_2O), pK_a 5.26 (50% EtOH aq.).

Me ester: [1112-94-3]. *Methyl 3-(diethoxyphosphinyl)-propanoate.*
$C_8H_{17}O_5P$ M 224.193
Liq. d_4^{20} 1.12. Bp_{14} 154-155°. n_D^{20} 1.4346.

Et ester: [3699-67-0]. *Ethyl 3-(diethoxyphosphinyl)-propanoate. Triethyl 3-phosphonopropanoate.*
$C_9H_{19}O_5P$ M 238.220
d_4^{20} 1.10. Bp_{12} 156-158°, $Bp_{0.2}$ 110-111°. n_D^{20} 1.4315.

Amide:
$C_7H_{16}NO_4P$ M 209.181
Cryst. (C_6H_6). Mp 74.5-76°.

Nitrile: 3-(Diethoxyphosphinyl)propanenitrile.
$C_7H_{14}NO_3P$ M 191.166
Liq. $Bp_{0.1}$ 110°. n_D 1.4382.

Pudovik, A.N. *et al*, *Zh. Obshch. Khim.*, 1952, **22**, 467 (*Engl. transl.* p. 473) (*ester*)
Harvey, R.G., *Tetrahedron*, 1966, **22**, 2561 (*derivs, synth, pmr*)
Gazizov, T.Kh. *et al*, *Zh. Obshch. Khim.*, 1971, **41**, 1957 (*Engl. transl.* p. 1973) (*ester*)
Kreutzkamp, N. *et al*, *Chem. Ber.*, 1967, **100**, 709 (*ester*)
Tsvetkov, E.N. *et al*, *Zh. Obshch. Khim.*, 1975, **45**, 716 (*Engl. transl.* p. 706) (*synth, props*)

2-(Diethoxyphosphinyl)propenoic acid, 9CI D-00260

2-(Diethoxyphosphinyl)acrylic acid

$H_2C{=}C(COOH)P(O)(OEt)_2$

$C_7H_{13}O_5P$ M 208.150

Me ester: [993-88-4]. *Methyl 2-(diethoxyphosphinyl)-propenoate.*
$C_8H_{15}O_5P$ M 222.177
Employed in a synth. of 2,5-dihydrothiophenes. Liq. d_4^{20} 1.14. $Bp_{0.15}$ 95-98°. n_D^{20} 1.4450.

Et ester: [20345-61-3]. *Ethyl 2-(diethoxyphosphinyl)-propenoate. Triethyl 2-phosphonoacrylate.*
$C_9H_{17}O_5P$ M 236.204
Employed in synth. of annelated alkenes, (alkylthiomethyl)acrylates and chromenes. Liq. d_4^{20} 1.12. Bp_1 101-102°. n_D^{20} 1.4440.

Amide:
$C_7H_{14}NO_4P$ M 207.166
Solid. Mp 73°.

Nitrile:
$C_7H_{12}NO_3P$ M 189.150
Liq. d_4^{20} 1.10. Bp_1 82-83°. n_D^{20} 1.4420.

Pudovik, A.N. *et al*, *Zh. Obshch. Khim.*, 1966, **36**, 1232 (*Engl. transl.* p. 1247) (*nitrile*)
Pudovik, A.N. *et al*, *Zh. Obshch. Khim.*, 1967, **37**, 2790 (*Engl. transl.* p. 2660) (*esters*)
Pudovik, A.N. *et al*, *Zh. Obshch. Khim.*, 1972, **42**, 88 (*Engl. transl.* p. 83) (*esters*)
Ide, J. *et al*, *Chem. Lett.*, 1978, 401 (*ester, pmr, ir, use*)
McIntosh, J.M. *et al*, *Can. J. Chem.*, 1978, **56**, 226 (*ester, pmr, use*)
Minami, T. *et al*, *Chem. Lett.*, 1978, 285 (*nitrile, ir, pmr*)
Semmelhack, M.F. *et al*, *J. Org. Chem.*, 1978, **43**, 1259 (*ester, use*)

3-(Diethoxyphosphinyl)propenoic acid, 9CI D-00261

3-(Diethoxyphosphinyl)acrylic acid

$(EtO)_2P(O)CH{=}CHCOOH$

$C_7H_{13}O_5P$ M 208.150

Me ester: [3944-25-0]. *Methyl 3-(diethoxyphosphinyl)-propenoate. Methyl 3-(diethylphosphinyl)acrylate.*

$C_8H_{15}O_5P$ M 222.177

Liq. Bp_1 109-110°. n_D^{20} 1.4483.

Et ester: [995-37-9]. *Ethyl 3-(diethoxyphosphinyl)-propenoate. Triethyl 3-phosphonoacrylate.*

$C_9H_{17}O_5P$ M 236.204

Liq. d_4^{20} 1.14. Bp_5 138-142°. n_D^{20} 1.4565.

Coover, H.W. *et al, J. Am. Chem. Soc.*, 1957, **79**, 1963 (*synth*)
Gareev, R.D. *et al, Zh. Obshch. Khim.*, 1982, **52**, 2637 (*Engl. transl.* p. 2333) (*props*)
Kirillova, L.M. *et al, Zh. Obshch. Khim.*, 1965, **35**, 1146 (*Engl. transl.* p. 1148) (*synth*)

4,4-Diethoxy-2,4,4,5-tetrahydro-2-oxo-5-(4-nitrophenyl)-1,3,4-oxazaphosphole, 10CI — D-00262

[74605-47-3]

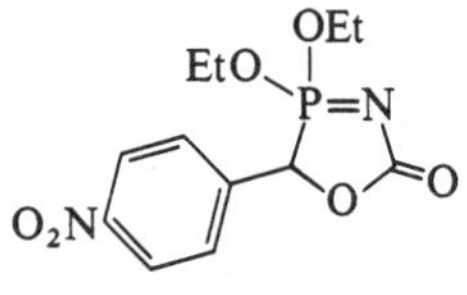

$C_{12}H_{15}N_2O_6P$ M 314.234

Solid, stable in dry air, but easily hydrol. Mp 140°.

Tarasova, R.I. *et al, Zh. Obshch. Khim.*, 1980, **50**, 538 (*Engl. transl.* p. 427) (*synth, props*)
Tarasova, R.I. *et al, Zh. Obshch. Khim.*, 1980, **50**, 757 (*Engl. transl.* p. 600) (*struct, props, P nmr*)

Diethoxytriphenylphosphorane, 9CI — D-00263

[18509-25-6]

$Ph_3P(OEt)_2$

$C_{25}H_{25}O_2P$ M 388.445

Reagent for cyclodehydration of diols and aminoalcohols giving cyclic ethers and cyclic amines. Viscous oil. Dec. in $CHCl_3$, but can be stored in toluene soln. (N_2).

Denny, D.B. *et al, Phosphorus*, 1971, **1**, 151 (*synth, props, use*)
Robinson, P.L. *et al, J. Org. Chem.*, 1983, **48**, 5396; 1985, **50**, 3860 (*props, use*)
von Itzstein, M. *et al, Aust. J. Chem.*, 1983, **36**, 557 (*P nmr*)
Robinson, P.L. *et al, J. Am. Chem. Soc.*, 1985, **107**, 5210 (*synth, pmr, cmr, P nmr, props, use*)
Kelly, J.W. *et al, J. Am. Chem. Soc.*, 1986, **108**, 7681 (*synth, use, props*)
Kelly, J.W. *et al, J. Org. Chem.*, 1986, **51**, 95 (*synth, P nmr, use*)
Robinson, P.L. *et al, Phosphorus Sulfur*, 1986, **26**, 15 (*props, use*)
Murray, W.T. *et al, J. Org. Chem.*, 1987, **52**, 525 (*use*)

(Diethylaminocarbonyl)phosphonic acid, 9CI — D-00264

(Diethylcarbamoyl)phosphonic acid. NN-*Diethylphosphonoformamide*

$(HO)_2P(O)CONEt_2$

$C_5H_{12}NO_4P$ M 181.128

Esters are neutral, chelating, extractants especially for rare earth and actinide elements.

Di-Et ester: Tetraethyl carbamoylphosphonate.

$C_9H_{20}NO_4P$ M 237.235

Liq. d_4^{25} 1.08. Bp_4 117-119°. n_D^{25} 1.4470.

Diisopropyl ester: Diisopropyl (diethylcarbamoyl)-phosphonate.

$C_{11}H_{24}NO_4P$ M 265.289

Liq. d_4^{25} 1.03. $Bp_{1.5}$ 114-115°. n_D^{20} 1.4430.

Dihexyl ester: Dihexyl (diethylcarbamoyl)phosphonate.

$C_{17}H_{36}NO_4P$ M 349.449

Liq. d_4^{25} 0.97. $Bp_{3.5}$ 183-184.5°. n_D^{25} 1.4458.

Arbuzov, B.A. *et al, Izv. Akad. Nauk SSSR, Ser. Khim.*, 1952, 875 (*Engl. transl.* p. 781) (*esters, synth, conformn*)
Yamamoto, M. *et al, Radiochim. Acta*, 1981, **29**, 205 (*props, use*)

2-Diethylamino-1,3,2-dioxaphospholan-4,5-dicarboxylic acid, 9CI — D-00265

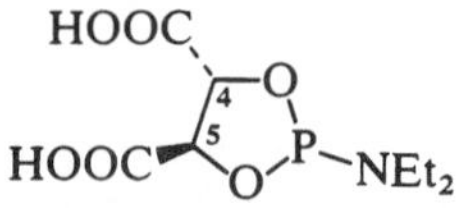

(4*R*,5*R*)-*form*

$C_8H_{14}NO_6P$ M 251.175

(4*R*,5*R*)-form

Di-Me ester: [57090-22-9].

$C_{10}H_{18}NO_6P$ M 279.229

Liq. d_4^{20} 1.21. Bp_1 122-123°. n_D^{20} 1.4717.

Di-Et ester:

$C_{12}H_{22}NO_6P$ M 307.283

Liq. d_4^{20} 1.15. $Bp_{2.5-3}$ 142-144°. n_D^{20} 1.4641.

Di-Et ester, 2-sulfide:

$C_{12}H_{22}NO_6PS$ M 339.343

Solid. Mp 35-36°. Bp_1 165.5-167°.

Diisopropyl ester:

$C_{14}H_{26}NO_6P$ M 335.336

Liq. d_4^{20} 1.10. Bp_1 138-139°. n_D^{20} 1.4572.

(4*RS*,5*RS*)-form

(±)-*form*

Di-Me ester: Liq. d_4^{20} 1.21. Bp_1 121-122°. n_D^{20} 1.4715.

Di-Et ester: Liq. d_4^{20} 1.14. Bp_1 132-133°. n_D^{20} 1.4643.

Di-Et ester, 2-sulfide: Liq. d_4^{20} 1.20. Bp_2 193-195°. n_D^{20} 1.4810.

Diisopropyl ester: Liq. d_4^{20} 1.09. Bp_2 153-154°. n_D^{20} 1.4577.

Grechkin, N.P. *et al, Izv. Akad. Nauk SSSR, Ser. Khim.*, 1967, 1990 (*Engl. transl.* p. 1907)
Samitov, Yu.Yu. *et al, Izv. Akad. Nauk SSSR, Ser. Khim.*, 1975, 1518 (*Engl. transl.* p. 1407)

2-Diethylaminonaphtho[1,2-*d*]-1,3,2-dioxaphosphole — D-00266

1,2-Naphthylene diethylphosphoramidite

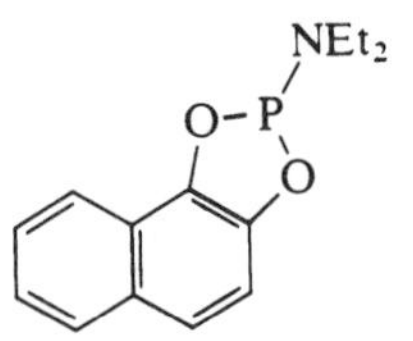

$C_{14}H_{16}NO_2P$ M 261.260

Liq. Bp_1 140-142°.

2-Oxide: 1,2-Naphthylene diethylphosphoramidate.

$C_{14}H_{16}NO_3P$ M 277.259

Solid. Mp 53-54°.

2-Sulfide: O,O-*1,2-Naphthylene diethylphosphoramidothioate.*

$C_{14}H_{16}NO_2PS$ M 293.320

Solid. Mp 75-76°.

Nifant'ev, É.E. *et al*, *Zh. Obshch. Khim.*, 1981, **51**, 1528 (*Engl. transl.* p. 1295) (*synth, P nmr*)
Voropai, L.M. *et al*, *Zh. Obshch. Khim.*, 1985, **55**, 65 (*Engl. transl.* p. 55) (*derivs*)

2-Diethylaminonaphtho[1,8-*de*]-1,3,2-dioxaphosphorin D-00267

N,N-*Diethylnaphtho*[*1,8*-de]-*1,3,2-dioxaphosphorin-2-amine*, *10CI*. *1,8-Naphthylene diethylphosphoramidite*
[72310-32-8]

$C_{14}H_{16}NO_2P$ M 261.260
Oil which cryst. slowly. Mp 49-50°. Bp_2 173-175°. n_D^{20} 1.6080.
2-Oxide: [79750-12-2]. *1,8-Naphthylene diethylphosphoramidate.*
$C_{14}H_{16}NO_3P$ M 277.259
Solid. Mp 90-91°.
2-Sulfide: [79750-11-1]. O,O-*1,8-Naphthylene diethylphosphoramidothioate.*
$C_{14}H_{16}NO_2PS$ M 293.320
Solid. Mp 97-98°.

Nifant'ev, É.E. *et al*, *Zh. Obshch. Khim.*, 1981, **51**, 1528 (*Engl. transl.* p. 1295) (*synth, P nmr*)
Voropai, L.M. *et al*, *Zh. Obshch. Khim.*, 1985, **55**, 65 (*Engl. transl.* p. 55) (*synth, P nmr*)

(2-Diethylamino-2-oxoethyl)phosphonic acid, 9CI D-00268

(*Diethylcarbamoylmethyl*)*phosphonic acid*
[66959-05-5]

$Et_2NCOCH_2P(O)(OH)_2$

$C_6H_{14}NO_4P$ M 195.155
Di-Et ester: [3699-76-1]. *Diethyl* (*2-diethylamino-2-oxoethyl*)*phosphonate. Diethyl* (*diethylcarbamoylmethyl*)*phosphonate. DEDECMP.*
$C_{10}H_{22}NO_4P$ M 251.262
A Wittig-Horner reagent for the synth. of $\alpha\beta$-unsaturated carboxylic acid diethylamides. Liq. $Bp_{1.2}$ 135°. n_D^{25} 1.4560.
Diisopropyl ester: Diisopropyl (*diethylcarbamoylmethyl*)*phosphonate. DPDECMP.*
$C_{12}H_{26}NO_4P$ M 279.315
Forms U, Hg, Sm, and other metal complexes.
Dibutyl ester: [7439-68-1]. *Dibutyl* (*2-diethylamino-2-oxoethyl*)*phosphonate. Dibutyl* (*diethylcarbamoylmethyl*)*phosphonate. DBDECMP.*
$C_{14}H_{30}NO_4P$ M 307.369
Extractant for actinide elements.
Dicyclohexyl ester: [77761-66-1]. *Dicyclohexyl* (*2-diethylamino-2-oxoethyl*)*phosphonate. Dicyclohexyl* (*diethylcarbamoylmethyl*)*phosphonate.*
$C_{18}H_{34}NO_4P$ M 359.445
Extractant for lanthanide and transplutonium elements.

Speziale, A.J. *et al*, *J. Org. Chem.*, 1958, **23**, 1883 (*synth*)
Hejno, K. *et al*, *Collect. Czech. Chem. Commun.*, 1973, **38**, 3511; 1976, **41**, 479 (*use*)
Jarolím, V. *et al*, *Collect. Czech. Chem. Commun.*, 1974, **39**, 587, 596; 1977, **42**, 1894 (*use*)
Landor, P.D. *et al*, *J. Chem. Soc., Perkin Trans. 1*, 1977, 93 (*synth, ir, pmr*)
Horwitz, E.P. *et al*, *Sep. Sci. Technol.*, 1981, **16**, 403, 417 (*use*)
Kem, K.M. *et al*, *J. Org. Chem.*, 1981, **46**, 5188 (*synth, pmr, P nmr*)
Shoun, R.R. *et al*, *Radiochim. Acta*, 1981, **29**, 143 (*use*)
Yamamoto, M. *et al*, *Radiochim. Acta*, 1981, **29**, 205 (*use*)
Bowen, S.M. *et al*, *Inorg. Chim. Acta*, 1982, **59**, 53; **61**, 155 (*ester, complexes*)
Bowen, S.M. *et al*, *Inorg. Chem.*, 1983, **22**, 286 (*ester, complexes, cryst struct*)

2-Diethylamino-3-phenyl-1,3,2-oxazaphospholidine D-00269

N,N-*Diethyl-3-phenyl-1,3,2-oxazaphospholidin-2-amine*, *9CI*. *2-Diethylaminotetrahydro-3-phenyl-1,3,2-oxazaphosphole. 2-Diethylamino-3-phenyl-1,3,2-oxazaphospholane*
[31707-09-2]

$C_{12}H_{19}N_2OP$ M 238.269
Liq. d_4^{20} 1.09. $Bp_{0.07}$ 116-117°. n_D^{20} 1.5625.
2-Oxide: [66850-41-7].
$C_{12}H_{19}N_2O_2P$ M 254.268
Cryst. (C_6H_6). Mp 87-87.5°.
2-Sulfide: [35776-11-5].
$C_{12}H_{19}N_2OPS$ M 270.329
Solid. Mp 104-105°.

Pudovik, A.N. *et al*, *Zh. Obshch. Khim.*, 1970, **40**, 1477 (*Engl. transl.* p. 1463) (*synth, sulfide*)
Pudovik, M.A. *et al*, *Zh. Obshch. Khim.*, 1981, **51**, 518 (*Engl. transl.* p. 402) (*oxide, synth, P nmr*)

O,O-Diethyl *S*-benzoyl phosphorodithioate D-00270

Benzenecarbothioic (O,O-*diethylphosphorodithioic*) *anhydrosulfide*, *9CI*
[1497-32-1]

$(EtO)_2P(S)SCOPh$

$C_{11}H_{15}O_3PS_2$ M 290.331
Selective *N*-benzoylating agent. Liq. Bp_4 100°.

Nair, P.G. *et al*, *Chem. Ind.* (*London*), 1974, 704 (*synth, use*)

Diethyl [(benzylideneamino)methyl]phosphonate D-00271

Diethyl [[(*phenylmethylene*)*amino*]*methyl*]*phosphonate*, *9CI*. *Diethyl* [(*benzylimino*)*methyl*]*phosphonate*
[50917-73-2]

$(EtO)_2P(O)CH_2N{=}CHPh \rightleftharpoons (EtO)_2P(O)CH{=}NCH_2Ph$

$C_{12}H_{18}NO_3P$ M 255.253
Wittig-Horner reagent employed in the synth. of $\alpha\beta$-unsat. imines, and thence branched-chain ketones. Liq.

Ratcliffe, R.W. *et al*, *Tetrahedron Lett.*, 1973, 4645 (*synth, use*)
Dehnel, A. *et al*, *Bull. Soc. Chim. Fr. Part II*, 1978, 95 (*use*)
Martin, S.F. *et al*, *J. Org. Chem.*, 1979, **44**, 3391 (*use*)

Diethyl benzylphosphonate D-00272

Diethyl (phenylmethyl)phosphonate, 9CI

[1080-32-6]

$PhCH_2P(O)(OEt)_2$

$C_{11}H_{17}O_3P$ M 228.227

Important Wittig-Horner reagent for synth. of styrenes and similar phenylethenes. Oily liq. Bp_{25} 172°, $Bp_{0.2}$ 124-126°. pK_a 27.9 (DMSO). n_D^{20} 1.4960.

▷SZ6600000.

Allen, D.W. *et al*, *J. Chem. Soc., Perkin Trans. 2*, 1977, 789 (*pmr, nmr, props*)
Ernst, L., *Org. Magn. Reson.*, 1977, **9**, 35 (*cmr, nmr*)
Bottin-Strzalko, T. *et al*, *J. Org. Chem.*, 1978, **43**, 4346 (*anion, pmr, nmr, cmr*)
Fedoryński, M. *et al*, *J. Org. Chem.*, 1978, **43**, 4682 (*synth*)
Griffiths, W.R., *Phosphorus Sulfur*, 1978, **5**, 101 (*ms*)
Bradamante, S. *et al*, *J. Org. Chem.*, 1980, **45**, 105 (*cmr*)
Fresneda, P.M. *et al*, *Synthesis*, 1981, 222 (*synth*)

Diethyl *tert*-butoxycyanomethylphosphonate D-00273

Diethyl cyano(1,1-dimethylethyl)methylphosphonate, 9CI

[59463-49-9]

$(EtO)_2P(O)CH(CN)OC(CH_3)_3$

$C_{10}H_{20}NO_4P$ M 249.246

Reagent for the homologation of carbonyl compds. to carboxylic acids, ester or amides. Viscous pale-yellow oil. $Bp_{0.5}$ 116-118°.

Dinizo, S.E. *et al*, *J. Org. Chem.*, 1976, **41**, 2846 (*synth, use, ir, pmr, ms*)
Dinizo, S.E. *et al*, *J. Am. Chem. Soc.*, 1977, **99**, 182 (*use*)

Diethyl chlorothiophosphonate D-00274

Thiohypochlorous acid anhydrosulfide with diethyl phosphorothioate, 9CI. S-*Chloro* O,O-*diethyl phosphorothioate*

[1186-08-9]

$(EtO)_2P(O)SCl$

$C_4H_{10}ClO_3PS$ M 204.608

Yellow liq. d_4^{17} 1.28. $Bp_{3.5}$ 90-91°, $Bp_{0.12}$ 45-46°. n_D^{20} 1.4668.

▷Asphyxiating vapour

Lenard-Borecka, B. *et al*, *Rocz. Chem.*, 1957, **31**, 1167 (*synth*)
Petrov, K.A. *et al*, *Zh. Obshch. Khim.*, 1959, **29**, 3030 (*Engl. transl.* p. 2995) (*synth, props*)
Michalski, J. *et al*, *Rocz. Chem.*, 1963, **37**, 1479 (*synth*)
Bochwic, B. *et al*, *Bull. Acad. Pol. Sci., Ser. Sci. Chim.*, 1968, **16**, 463 (*props*)
Krawiecka, B. *et al*, *J. Chem. Soc., Chem. Commun.*, 1974, 630 (*props*)
Lenard-Borecka, B. *et al*, *Bull. Acad. Pol. Sci., Ser. Sci. Chim.*, 1974, **22**, 201 (*props*)

***O,O*-Diethyl chlorothiophosphonothioate, 9CI** D-00275

Thiohypochlorous acid anhydrosulfide with O,O-*diethyl phosphorodithioate, 10CI.* S-*Chloro* O,O-*diethyl phosphorodithioate*

[1639-18-5]

$(EtO)_2P(S)SCl$

$C_4H_{10}ClO_2PS_2$ M 220.669

Golden-yellow liq. d_4^{20} 1.27. $Bp_{0.4}$ 65-66°. n_D^{20} 1.5270.

▷Asphyxiating vapour

Almasi, L. *et al*, *Chem. Ber.*, 1964, **97**, 661; 1965, **98**, 613 (*synth, ir*)

Diethyl 2-(cyclohexylamino)-2-methylethenyl phosphonate D-00276

Diethyl [2-(cyclohexylimino)propyl]phosphonate, 9CI

[52330-21-9]

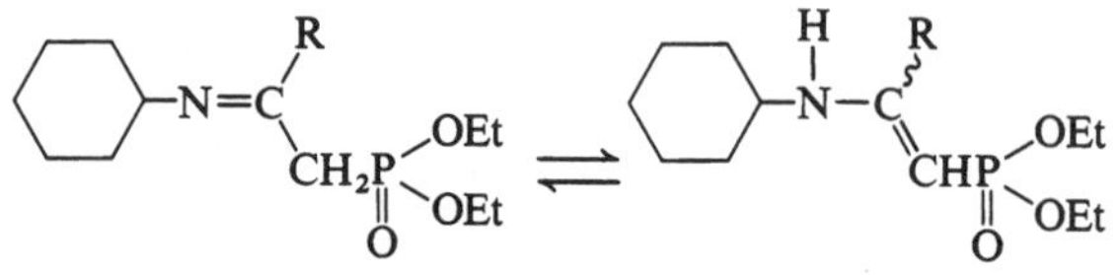

R = CH_3

$C_{13}H_{26}NO_3P$ M 275.327

Tautomeric. Forms an anion which converts aldehydes and ketones into methyl ethenyl ketones *via* the unsat. imine. No phys. props. reported.

Chatta, M.S. *et al*, *Tetrahedron Lett.*, 1971, 1419 (*use*)

Diethyl [2-(cyclohexylamino)-2-phenylethenyl]phosphonate, 9CI D-00277

Diethyl(2-cyclohexylimino-2-phenylethyl)phosphonate

[55365-20-3]

As Diethyl 2-(cyclohexylamino)-2-methylethenyl phosphonate, D-00276 with

R = Ph

$C_{18}H_{28}NO_3P$ M 337.398

Tautomeric. Forms an anion which converts aldehydes and ketones into phenyl ethenyl ketones *via* the unsat. imines.

Chatta, M.S. *et al*, *Tetrahedron Lett.*, 1971, 1419 (*use*)

Diethyl 2-(cyclohexylamino)-vinylphosphonate D-00278

Diethyl 2-(cyclohexylamino)ethenyl phosphonate, 9CI

[20061-84-1]

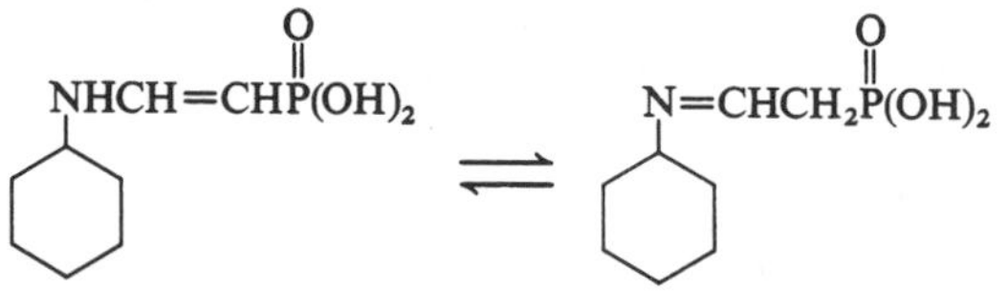

$C_{12}H_{24}NO_3P$ M 261.300

Reagent for conversion of aldehydes and ketones into α,β-unsaturated aldehydes. Cryst. (pentane). Mp 58-61°. $Bp_{0.08}$ 126-141°.

Nagata, W. *et al*, *J. Chem. Soc. (C)*, 1969, 460 (*synth, use, ir, pmr, uv*)
Org. Synth., 1973, **53**, 44, 104 (*synth, ir, pmr, use*)
Rickards, R.W. *et al*, *J. Org. Chem.*, 1980, **45**, 751 (*use*)

Diethyl dibromophosphoramidate, 9CI D-00279

[32755-13-8]

$(EtO)_2P(O)NBr_2$

$C_4H_{10}Br_2NO_3P$ M 310.910

Source of diethoxyphosphinylnitrene: intermediate for synth. of aziridines and 2-bromoalkylamines. Orange oil. n_D^{20} 1.5244.

NaBr adduct: Yellow cryst. Mp 118-119°. Approx. 3:1 stoicheiometry.

Zwierzak, A. *et al, Synthesis*, 1971, 323 (*synth, ir, pmr, props*)
Zawadzki, S. *et al, Tetrahedron*, 1973, **29**, 315 (*synth, ir, pmr, P nmr*)
Zwierzak, A. *et al, Tetrahedron*, 1973, **29**, 3899 (*use*)
Zawadzki, S. *et al, Tetrahedron*, 1981, **37**, 3675 (*props*)

Diethyl dichlorophosphoramidate — D-00280

[18368-11-1]

$(EtO)_2P(O)NCl_2$

$C_4H_{10}Cl_2NO_3P$ M 222.008

Reagt. for synth. of 2-chloroalkylamines and aziridines. Mobile yellow liq. d_4^{25} 1.34. $Bp_{0.01}$ 56-57°. n_D^{20} 1.4620.

▷Explosions may occur in large scale prepns.

Markovskii, L.N. *et al, Zh. Obshch. Khim.*, 1970, **40**, 543 (*Engl. transl.* p. 509) (*synth*)
Zwierzak, A. *et al, Tetrahedron*, 1970, **26**, 3521, 3527 (*synth, ir, props*)
Block, H.D. *et al, Angew. Chem., Int. Ed. Engl.*, 1971, **10**, 491 (*haz*)
Treppendahl, S. *et al, Acta Chem. Scand., Ser. B*, 1974, **28**, 657 (*ms*)
Pinchuk, A.M. *et al, Zh. Obshch. Khim.*, 1975, **45**, 1240 (*Engl. transl.* p. 1219) (*use*)
Egorov, Yu.P. *et al, Zh. Obshch. Khim.*, 1975, **45**, 1716 (*Engl. transl.* p. 1683) (*ir, nmr, conformn*)

Diethyl diethylphosphoramidate, 9CI, 8CI — D-00281

Diethyl diethylamidophosphate

[3167-69-9]

$(EtO)_2P(O)NEt_2$

$C_8H_{20}NO_3P$ M 209.225

Liq. d_4^{20} 1.01-1.04. Bp_{25} 116-119°, Bp_1 60°. n_D^{20} 1.4318 (1.4214).

Petrov, K.A. *et al, Zh. Obshch. Khim.*, 1956, **26**, 3378 (*Engl. transl.* p. 3761) (*synth*)
Abramov, V.S. *et al, Zh. Obshch. Khim.*, 1969, **39**, 2234 (*Engl. transl.* p. 2180) (*synth*)
Appel, R. *et al, Z. Anorg. Allg. Chem.*, 1975, **414**, 241 (*synth*)
Zwierzak, A., *Synthesis*, 1975, 507 (*synth, P nmr*)
Zverev, V.V. *et al, Izv. Akad. Nauk SSSR, Ser. Khim.*, 1979, 84 (*Engl. transl.* p. 74) (*pe*)

5,7-Diethyl-6,7-dihydro-5*H*-dibenzo[*d,f*][1,3]diphosphepin, 8CI — D-00282

$C_{17}H_{20}P_2$ M 286.293

B,2EtBr: [5353-22-0]. *5,5,7,7-Tetraethyl-6,7-dihydro-5*H*-dibenzo*[d,f][*1,3*]*diphosphepinium dibromide.*
$C_{21}H_{30}Br_2P_2$ M 504.224
V. hygroscopic solid. Mp 228-230°.

Allen, D.W. *et al, J. Chem. Soc. (C)*, 1967, 1869.

Diethyl 2-dimethylaminoethyl phosphate — D-00283

[3958-21-2]

$(EtO)_2P(O)OCH_2CH_2NMe_2$

$C_8H_{20}NO_4P$ M 225.224

Liq. d_4^{20} 1.06. Bp_2 98°, $Bp_{0.05}$ 60°. n_D^{20} 1.4160 (n_D^{25} 1.4220).

Hydrogen oxalate: Solid. Mp 112°.

▷Reversible inhibitor of cholinesterases

B,MeI: Solid. Mp 100°.

Tammelin, L.E., *Acta Chem. Scand.*, 1957, **11**, 1340 (*synth*)
Malinovski, M.S. *et al, Zh. Obshch. Khim.*, 1960, **30**, 3454 (*Engl. transl.* p. 3422) (*synth*)
Gramstad, T., *Spectrochim. Acta*, 1963, **19**, 1391 (*ir*)
Aksnes, G. *et al, Acta Chem. Scand.*, 1965, **19**, 888 (*synth, ir*)
Maglothin, J.A. *et al, Biochemistry*, 1974, **13**, 3520 (*props*)

O,O-Diethyl *S*-(2-dimethylaminoethyl) phosphorothioate — D-00284

O,O-*Diethyl* S-(*2-dimethylaminoethyl*) *thiophosphate*

[6736-03-4]

$(EtO)_2P(O)SCH_2CH_2NMe_2$

$C_8H_{20}NO_3PS$ M 241.285

Long acting anticholinesterase. Controls glaucoma. Liq. d_4^{20} 1.08. $Bp_{0.1}$ 85°. n_D^{25} 1.4655.

Hydrogen oxalate: [470-94-0]. Solid. Mp 116°.

▷Highly toxic. TF2025000.

B,MeI: see Ecothiopate, E-00001

Tammelin, L.E. *et al, Acta Chem. Scand.*, 1957, **11**, 1340 (*synth*)
Warren, C. *et al, J. Pharm. Sci.*, 1971, **60**, 1548.
Amitai, G. *et al, J. Med. Chem.*, 1976, **19**, 810 (*synth*)

N,N-Diethyl-2,6-dimethyl-1-phenyl-1,4-phosphaborin-4(1*H*)-amine, 9CI — D-00285

4-Diethylamino-2,6-dimethyl-1-phenyl-1,4(*1*H)-*phosphaborin*

[57590-54-2]

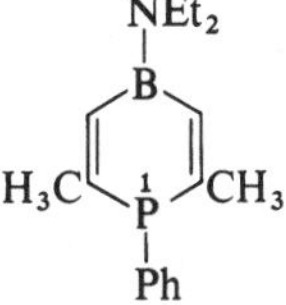

$C_{16}H_{23}BP$ M 257.141

Cryst. (hexane). Mp 65-66°. $Bp_{0.001}$ 77-78°.

1-Oxide: [87367-43-9].
$C_{16}H_{23}BOP$ M 273.141
Solid. Mp 86-88°.

1-Sulfide: [87367-44-0].
$C_{16}H_{23}BPS$ M 289.201
Cryst. Mp 78-80°.

1-Selenide: [87453-82-5].
$C_{16}H_{23}BPSe$ M 336.101
Cryst. Mp 121-124°.

Berger, H.-O. *et al, J. Organomet. Chem.*, 1983, **250**, 33.

Diethyl dimethylphosphoramidate, 9CI — D-00286

Diethyl (*dimethylamido*)*phosphate*

[2404-03-7]

$(EtO)_2P(O)NMe_2$

$C_6H_{16}NO_3P$ M 181.171
Liq. d_4^{20} 1.05. Bp_{20} 95-96°, $Bp_{0.1}$ 54-56°. n_D^{20} 1.4180.

Kamai, G. *et al*, *Zh. Obshch. Khim.*, 1957, **27**, 3064 (*Engl. transl.* p. 3093) (*synth*)
Savignac, P. *et al*, *J. Organomet. Chem.*, 1974, **72**, 361 (*synth*)
Egorov, Yu.P. *et al*, *Zh. Obshch. Khim.*, 1975, **45**, 1716 (*Engl. transl.* p. 1683) (*ir, P nmr*)
Bel'skii, V.E. *et al*, *Zh. Priklad. Spektrosk.*, 1980, **33**, 361; *CA*, **93**, 227847 (*ir*)
Davidowitz, B. *et al*, *S. Afr. J. Chem.*, 1980, **33**, 74 (*synth, props*)

2,3-Diethyl-2,3-diphosphabicyclo[2.2.1]hept-5-ene, 8CI — D-00287

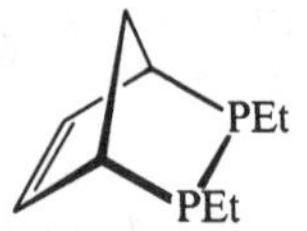

$C_9H_{16}P_2$ M 186.173
$Bp_{0.4}$ 74-78°.
Disulfide: [18005-69-1].
$C_9H_{16}P_2S_2$ M 250.293
Solid. Mp 100°.

Schmidt, U. *et al*, *Chem. Ber.*, 1968, **101**, 1381 (*synth*)

2,3-Diethyl-2,3-diphosphabicyclo[2.2.2]octane, 8CI — D-00288

$C_{10}H_{20}P_2$ M 202.216
Liq. $Bp_{0.5}$ 73-77°.
Disulfide: [18005-70-4].
$C_{10}H_{20}P_2S_2$ M 266.336
Solid. Mp 118°.

Schmidt, U. *et al*, *Chem. Ber.*, 1968, **101**, 1381 (*synth*)

P,P'-Diethyl diphosphate, 9CI — D-00289

P,P'-*Diethyl pyrophosphate*
[1707-71-7]

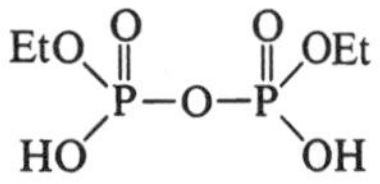

$C_4H_{12}O_7P_2$ M 234.082
Isol. as Ba salt.

Nagasawa, K. *et al*, *Chem. Pharm. Bull.*, 1973, **21**, 2438 (*synth*)
Glanek, T. *et al*, *J. Phys. Chem.*, 1976, **80**, 639 (*P nmr*)

Diethyldiphosphonic acid, 9CI — D-00290

Diethylpyrophosphonic acid. Ethylphosphonic acid monoanhydride
[28397-15-1]

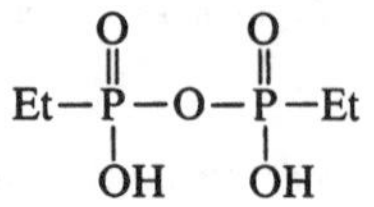

$C_4H_{12}O_5P_2$ M 202.083
Bis(cyclohexylamine) salt: [40203-43-8]. Cryst. (Me_2CO aq.). Mp 210-211°.
Di-Me ester: [34637-96-2]. *Dimethyl diethyldiphosphonate.*
$C_6H_{16}O_5P_2$ M 230.137
No phys. props. reported.
Di-Et ester: [7369-43-9]. *Diethyl diethyldiphosphonate. Diethyl diethylpyrophosphonate.*
$C_8H_{20}O_5P_2$ M 258.191
Liq. d_4^{20} 1.14. Bp_9 154-154.5°, $Bp_{0.0001}$ 72°. n_D^{20} 1.4357.

Pudovik, A.N. *et al*, *Zh. Obshch. Khim.*, 1960, **30**, 2624 (*Engl. transl.* p. 2606) (*ester, synth*)
Turpin, R. *et al*, *Bull. Soc. Chim. Fr.*, 1971, 3878 (*ester, synth*)
Ruveda, M.A. *et al*, *Tetrahedron*, 1972, **28**, 5011 (*synth, ir, pmr*)

P,P'-Diethyldiphosphonic diamide, 9CI — D-00291

P-*Ethylphosphonamidic anhydride*

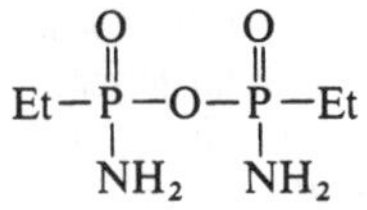

$C_4H_{14}N_2O_3P_2$ M 200.114
N,N,N',N'-*Tetra-Me:* [35412-81-1]. P,P'-*Diethyl-N,N,N',N'-tetramethyldiphosphonic diamide.*
$C_8H_{22}N_2O_3P_2$ M 256.221
Liq. $Bp_{0.3}$ 119°.
▷Cholinesterase inhibitor
N,N,N',N'-*Tetra-Et: Hexaethyldiphosphonic diamide.*
$C_{12}H_{30}N_2O_3P_2$ M 312.328
Liq. d_4^{20} 1.06. Bp_2 163.5-164.5°. n_D^{20} 1.4607.
▷Potential cholinesterase inhibitor

Razumov, A.I. *et al*, *CA*, 1958, **52**, 237.
Joesten, M.D. *et al*, *Inorg. Chem.*, 1972, **11**, 429 (*tetramethyl, synth, pmr, P nmr, tox, complexes*)

Diethyl ethylphosphonate — D-00292

[78-38-6]

$EtP(O)(OEt)_2$

$C_6H_{15}O_3P$ M 166.156
Useful synthetic intermediate. Solvent for synth. of metal complexes using stronger ligands. Pleasant smelling liq. d_{15}^{21} 1.03. Bp_{16} 90-92°, Bp_2 62°. n_D^{18} 1.4172.

Ford-Moore, A.H. *et al*, *J. Chem. Soc.*, 1947, 1465 (*synth*)
Saunders, B.C. *et al*, *J. Chem. Soc.*, 1948, 699 (*synth*)
Org. Synth., Coll. Vol., **4**, 326 (*synth*)
Gerken, T.A. *et al*, *J. Magn. Reson.*, 1976, **24**, 155 (*pmr*)
Ernst, L., *Org. Magn. Reson.*, 1977, **9**, 35 (*nmr, pmr*)
Hall, C.D. *et al*, *J. Chem. Soc., Perkin Trans. 2*, 1977, 1232 (*pmr*)
Griffiths, W.R. *et al*, *Phosphorus Sulfur*, 1978, **5**, 101 (*ms*)
Labintsev, V.B. *et al*, *Zh. Org. Khim.*, 1978, **14**, 1371 (*Engl. transl.* p. 1277) (*ms*)
Sass, S. *et al*, *Org. Mass Spectrom.*, 1979, **14**, 257 (*ms*)
Karayannis, N.M. *et al*, *Inorg. Chim. Acta.*, 1982, **65**, L135 (*use*)
Savignac, P. *et al*, *Synthesis*, 1982, 725 (*anion*)

Diethyl ethylphosphoramidate, 9CI — D-00293

[1946-09-4]

$(EtO)_2P(O)NHEt$

$C_6H_{16}NO_3P$ M 181.171

Liq. d_4^{20} 1.06. Bp_{25} 135°, $Bp_{0.03}$ 79-81°. n_D^{20} 1.4254.

Kabachnik, M.I. *et al*, *Izv. Akad. Nauk SSSR, Ser. Khim.*, 1961, 816 (*Engl. transl.* p. 755) (*synth, ir*)
Nikonorov, K.V. *et al*, *Izv. Akad. Nauk SSSR, Ser. Khim.*, 1968, 587 (*Engl. transl.* p. 569) (*synth*)
Williamson, M.P. *et al*, *J. Phys. Chem.*, 1968, **72**, 4043 (*pmr*)
Appel, R. *et al*, *Z. Anorg. Allg. Chem.*, 1975, **414**, 241 (*synth*)
Mathis, R. *et al*, *C.R. Hebd. Seances Acad. Sci., Ser. C*, 1975, **281**, 437 (*ir*)
Zwierzak, A., *Synthesis*, 1975, 507; 1982, 920; 1984, 223 (*synth, ir, pmr, P nmr*)
Al-Rawi, J.M.B. *et al*, *Org. Magn. Reson.*, 1983, **21**, 75 (*cmr*)

O,O-Diethyl *S*-[2-(ethylthio)ethyl] phosphorothioate, 9CI, 8CI — D-00294

Demeton-S

$(EtO)_2P(O)SCH_2CH_2SEt$

$C_8H_{19}O_3PS_2$ M 258.330

See also Demeton-*O*, D-00019 . Systemic insecticide. Liq. Sol. H_2O. d_4^{21} 1.132. Bp_2 132-134°, $Bp_{0.25}$ 100°. More toxic and more easily hydrol. in base than Demeton-*O*, D-00019 Obt. by heating *O,O*-Diethyl *S*-[2-(ethylthio)ethyl] phosphorothioate, D-00294 .

Fukuto, T.R. *et al*, *J. Am. Chem. Soc.*, 1954, **76**, 5103 (*props*)
Hoffman, F.W. *et al*, *J. Am. Chem. Soc.*, 1958, **80**, 1150 (*synth*)
Stan, H.-J. *et al*, *Fresenius' Z. Anal. Chem.*, 1977, **287**, 271; *Biomed. Mass Spectrom.*, 1982, **9**, 483 (*glc, ms*)
Ripley, B.D. *et al*, *J. Assoc. Off. Anal. Chem.*, 1983, **66**, 1084 (*glc*)
Pesticide Manual, 6th Ed., 154; 7th Ed., 162.
Sax, N.I., *Dangerous Properties of Industrial Materials*, 6th Ed., Van Nostrand-Reinhold, 1984, 538.

N,N-Diethyl-*P*-(methoxymethyl)-phosphonamidic acid, 9CI — D-00295

$MeOCH_2P(O)(OH)NEt_2$

$C_6H_{16}NO_3P$ M 181.171

Et ester: [59375-25-6]. *Ethyl* N,N-*diethyl*-P-*(methoxymethyl)phosphonamidate.*
$C_8H_{20}NO_3P$ M 209.225
Liq. d_4^{20} 1.03. Bp_{15} 143-145°. n_D^{20} 1.4511.

Propyl ester: [59375-25-7]. *Propyl* N,N-*diethyl*-P-*(methoxymethyl)phosphonamidate.*
$C_9H_{22}NO_3P$ M 223.251
Liq. d_4^{20} 1.03. Bp_{15} 150-152°. n_D^{20} 1.4480.

Krutskii, L.N. *et al*, *Zh. Obshch. Khim.*, 1976, **46**, 507 (*Engl. transl.* p. 505) (*synth, pmr*)

N,N-Diethyl-*P*-methylphosphonamidic acid, 9CI — D-00296

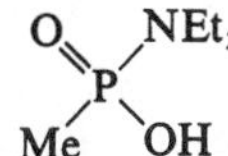

$C_5H_{14}NO_2P$ M 151.145

Me ester: [34605-98-6]. *Methyl* N,N-*diethyl*-P-*methylphosphonamidate.*
$C_6H_{16}NO_2P$ M 165.172
Liq. d_4^{20} 1.02. Bp_{32} 118-120°, $Bp_{0.5}$ 71-74°. n_D^{20} 1.4382.

Et ester: [2404-81-1]. *Ethyl* N,N-*diethyl*-P-*methylphosphonamidate.*
$C_7H_{18}NO_2P$ M 179.198
Liq. d_4^{20} 1.01. Bp_{10} 95-97°, $Bp_{0.008}$ 40-42°. n_D^{20} 1.4360.

Isopropyl ester: Isopropyl N,N-*diethyl*-P-*methylphosphonamidate.*
$C_8H_{20}NO_2P$ M 193.225
Liq. d_4^{20} 1.08. $Bp_{0.1}$ 80-81°. n_D^{20} 1.4395.

Ph ester: [4645-95-8]. *Phenyl* N,N-*diethyl*-P-*methylphosphonamidate.*
$C_{11}H_{18}NO_2P$ M 227.242
Liq. d_4^{20} 1.09. Bp_{10} 160°, Bp_3 165°. n_D^{20} 1.5061.

Chloride: [27930-69-4].
$C_5H_{13}ClNOP$ M 169.591
Liq. d_4^{20} 1.13. Bp_{10} 112°. n_D^{20} 1.4671.

Razumov, A.I. *et al*, *Zh. Obshch. Khim.*, 1957, **27**, 2389 (*Engl. transl.* p. 2450) (*esters*)
Green, M. *et al*, *J. Chem. Soc.*, 1958, 3129 (*chloride*)
Keay, L., *J. Org. Chem.*, 1963, **28**, 329 (*ester, synth*)
Abramov, V.S. *et al*, *Zh. Obshch. Khim.*, 1969, **39**, 2658 (*Engl. transl.* p. 2596) (*esters, synth*)
Nesterov, L.N. *et al*, *Zh. Obshch. Khim.*, 1969, **39**, 2457 (*Engl. transl.* p. 2397) (*derivs*)
Eliseenkev, V.N. *et al*, *Zh. Obshch. Khim.*, 1976, **46**, 23 (*Engl. transl.* p. 23) (*ester*)
Zavlin, P.M. *et al*, *Zh. Obshch. Khim.*, 1977, **47**, 1981 (*Engl. transl.* p. 1812) (*ester*)

N,N-Diethyl-*P*-methylphosphonamidothioic acid, 9CI — D-00297

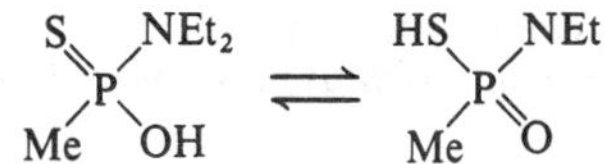

$C_5H_{14}NOPS$ M 167.205

O-*Ph ester:* [31650-60-9]. O-*Phenyl* N,N-*diethyl*-P-*methylphosphonamidothioate.*
$C_{11}H_{18}NOPS$ M 243.303
Liq. d_4^{20} 1.12. $Bp_{0.05}$ 111-112°. n_D^{20} 1.5480.

S-*Ph ester:* [31650-52-9]. S-*Phenyl* N,N-*diethyl*-P-*methylphosphonamidothioate.*
$C_{11}H_{18}NOPS$ M 243.303
Liq. d_4^{20} 1.13. $Bp_{0.03}$ 129-130°. n_D^{20} 1.5625.

Fluoride:
$C_5H_{13}FNPS$ M 169.197
Oil. Bp_2 72°.

Chloride: [3450-35-9].
$C_5H_{13}ClNPS$ M 185.651
Liq. d_4^{20} 1.15. Bp_8 107-109°, Bp_2 68-69°. n_D^{20} 1.5242.

Ger. Pat., 1 099 532, (*1959*); *CA*, **56**, 3515 (*fluoride*)
Godovikov, N.N. *et al*, *Zh. Obshch. Khim.*, 1961, **31**, 1638 (*Engl. transl.* p. 1516) (*chloride*)
Nesterov, L.V. *et al*, *Zh. Obshch. Khim.*, 1970, **40**, 1237 (*Engl. transl.* p. 1228) (*chloride, esters, synth, ir, pmr, P nmr*)

Diethyl methylphosphonate — D-00298

[683-08-9]

$MeP(O)(OEt)_2$

$C_5H_{13}O_3P$ M 152.130

Useful synthetic intermediate. BuLi → anion which may be alkylated or acylated. Pleasant smelling liq. d_0^0 1.07. Bp 192-194°, Bp_{11} 80°, Bp_1 52-53°. n_D^{14} 1.4062 (n_D^{16} 1.4120).

Ford-Moore, A.H. *et al*, *J. Chem. Soc.*, 1947, 1465 (*synth*)
Org. Synth., Coll. Vol., **4**, 326 (*synth*)
Bel'skii, V.E. *et al*, *Zh. Obshch. Khim.*, 1972, **42**, 2427 (*Engl. transl.* p. 2421) (*P nmr*)
Griffiths, W.R. *et al*, *Phosphorus Sulfur*, 1978, **5**, 101 (*ms*)
Labintsev, V.B. *et al*, *Zh. Org. Khim.*, 1978, **14**, 1371 (*Engl. transl.* p. 1277) (*ms*)
Hall, C.D. *et al*, *J. Chem. Soc., Perkin Trans. 2*, 1977, 1232 (*pmr*)

Sass, S. *et al*, *Org. Mass Spectrom.*, 1979, **14**, 257 (*ms*)

Diethyl methylphosphoramidate, 9CI — D-00299

Diethyl (methylamido)phosphate
[6326-73-4]

$(EtO)_2P(O)NHMe$

$C_{15}H_{14}NO_3P$ M 287.254
Reagent for the synth of secy. methylamines. Liq. Bp_{15} 130°, $Bp_{0.01}$ 85°. n_D^{20} 1.4215.

Saunders, B.C. *et al*, *J. Chem. Soc.*, 1948, 699 (*synth*)
Rengaraju, S. *et al*, *J. Org. Chem.*, 1972, **37**, 3304 (*pmr*)
Collins, D.J. *et al*, *Aust. J. Chem.*, 1974, **27**, 1759 (*synth*)
Mathis, R. *et al*, *C.R. Hebd. Seances Acad. Sci., Ser. C*, 1975, **281**, 437; 1977, **284**, 767 (*ir*)
Schlak, O. *et al*, *Z. Anorg. Allg. Chem.*, 1976, **419**, 275 (*ir, pmr, P nmr*)
Corbel, B. *et al*, *Can. J. Chem.*, 1980, **58**, 2183 (*synth, use*)
Mizrahi, V. *et al*, *J. Org. Chem.*, 1982, **47**, 3533 (*synth, ms*)
Al-Rawi, J.M.A. *et al*, *Org. Magn. Reson.*, 1983, **21**, 75 (*cmr*)
Davidowitz, B. *et al*, *Org. Mass. Spectrom.*, 1984, **19**, 128 (*ms*)

O,O-Diethyl *SS*-methyl phosphoro(dithioperoxo)thioate, 9CI — D-00300

O,O-*Diethyl* S-(*methylthio*) *phosphorodithioate*
[50704-30-8]

$(EtO)_2P(S){-}S{-}S{-}Me$

$C_5H_{13}O_2PS_3$ M 232.310
Liq. d_4^{20} 1.21. $Bp_{2.5}$ 100.5-101.5°. n_D^{20} 1.5500.

Almasi, L. *et al*, *Chem. Ber.*, 1962, **95**, 1582 (*synth*)

O,O-Diethyl *Se*-methyl phosphoroselenoite — D-00301

[55987-89-8]

$(EtO)_2PSeMe$

$C_5H_{13}O_2PSe$ M 215.090
Viscous yellow oil. Thermally stable.

Anderson, J.W. *et al*, *Inorg. Nucl. Chem. Lett.*, 1975, **11**, 233 (*synth, pmr*)

O,O-Diethyl *O*-[4-(methylthio)phenyl] phosphorothioate, 9CI — D-00302

[3070-15-3]

$(EtO)_2P(S)O{-}C_6H_4{-}SMe$

$C_{11}H_{17}O_3PS_2$ M 292.347
Insecticide. Liq. d_4^{25} 1.19. $Bp_{0.01}$ 85°. n_D^{25} 1.5462.
▷TF4020000.

S-*oxide: see Fensulfothion, F-00006*
S,S-*Dioxide:* [14255-72-2]. O,O-*Diethyl* O-[4-(*methylsulfonyl*)*phenyl*] *phosphorothioate.*
$C_{11}H_{17}O_5PS_2$ M 324.346
Metab. of Fensulfothion, F-00006 . Solid. Mp 43-44°.
▷TF3935000.

Schrader, G. *et al*, *Angew. Chem.*, 1961, **73**, 331 (*synth, tox*)
Neely, W.B. *et al*, *J. Agric. Food Chem.*, 1970, **18**, 45 (*synth*)
Rainsford, K.D., *Pestic. Biochem. Physiol.*, 1978, **8**, 302 (*tox*)
Hill, A.R.C. *et al*, *Analyst*, 1984, **109**, 483 (*glc*)

Diethyl 2-nitrophenyl phosphate — D-00303

[4532-02-9]

$2\text{-}O_2N{-}C_6H_4{-}OP(O)(OEt)_2$

$C_{10}H_{14}NO_6P$ M 275.197
Liq. $Bp_{0.45}$ 145°. n_D^{25} 1.4967.
▷Powerful inhibitor of acetylcholinesterase

Fukui, K. *et al*, *Bull. Chem. Soc. Jpn.*, 1961, **34**, 1224 (*tox*)
Boter, H.L. *et al*, *Recl. Trav. Chim. Pays-Bas*, 1965, **84**, 1279 (*synth*)
van Hooidonk, C. *et al*, *Recl. Trav. Chim. Pays-Bas*, 1967, **86**, 449 (*props*)

Diethyl 3-nitrophenyl phosphate — D-00304

[4532-06-3]

$C_{10}H_{14}NO_6P$ M 275.197
Liq. $Bp_{0.02}$ 141°. n_D^{25} 1.4994.
▷Powerful inhibitor of acetylcholinesterase

Fukui, K. *et al*, *Bull. Chem. Soc. Jpn.*, 1961, **34**, 1224 (*tox*)
Boter, H.L. *et al*, *Recl. Trav. Chim. Pays-Bas*, 1965, **84**, 1279 (*synth*)
van Hooidonk, C. *et al*, *Recl. Trav. Chim. Pays-Bas*, 1967, **86**, 449 (*synth, props*)
Hansch, C., *J. Org. Chem.*, 1970, **35**, 620 (*props*)

Diethylphenylphosphine, 9CI — D-00305

[1605-53-4]

Et_2PPh

$C_{10}H_{15}P$ M 166.202
Ligand for metals of groups IB, IIB, VIB, VIIB, and VIII. Liq. Bp 221.9°, Bp_8 95°, Bp °.

Oxide: [24323-92-0].
$C_{10}H_{15}OP$ M 182.202
Solid. Mp 56°. Bp_2 150-152° (117-119°).

Sulfide: [14684-35-0].
$C_{10}H_{15}PS$ M 198.262
Pale-yellow viscous oil or solid. Mp 36-38°. Bp_7 165-167°, $Bp_{0.8}$ 127°. n_D^{25} 1.5891.

Selenide:
$C_{10}H_{15}PSe$ M 245.162
$Bp_{0.7}$ 149°. n_D^{25} 1.6086.

Zingaro, R.A. *et al*, *J. Chem. Eng. Data*, 1963, **8**, 226 (*synth, derivs*)
Mann, B.E., *J. Chem. Soc., Perkin Trans. 2*, 1972, 30 (*cmr, nmr*)
Allen, E.A. *et al*, *Spectrochim. Acta, Part A*, 1974, **30**, 1219 (*ir, raman*)
McEwen, W.E. *et al*, *J. Org. Chem.*, 1976, **41**, 1684 (*props*)
Postle, S.R., *Phosphorus Sulfur*, 1977, **3**, 269 (*derivs, cmr*)
Inorg. Synth., 1978, **18**, 170 (*synth*)
Ratovskii, G.V. *et al*, *Zh. Obshch. Khim.*, 1978, **48**, 1520 (*Engl. transl.* p. 1394) (*uv*)
Dorokhova, V.V. *et al*, *Zh. Obshch. Khim.*, 1979, **49**, 83 (*Engl. transl.* p. 68) (*uv, oxide, sulfide, raman*)
Timokhin, B.V. *et al*, *Zh. Obshch. Khim.*, 1979, **49**, 1235 (*Engl. transl.* p. 1083) (*synth, nmr*)
Ratovskii, G.V. *et al*, *Zh. Obshch. Khim.*, 1981, **51**, 1504 (*Engl. transl.* p. 1276) (*uv, cmr, conformn*)
Petrov, K.A. *et al*, *Zh. Obshch. Khim.*, 1982, **52**, 58 (*Engl. transl.* p. 53) (*oxide*)

P,P-Diethyl-*N*-phenylphosphinimidic acid D-00306

P,P-*Diethyl*-N-*phenylphosphinic amide*

$$Et_2P(OH){=}NPh \rightleftharpoons Et_2P(O)NHPh$$

$C_{10}H_{16}NOP$ M 197.216

Acid exists only in tautomeric amide form (see 47883-8).

Et ester: [33345-04-9]. *Ethyl* P,P-*diethyl*-N-*phenylphosphinimidate*.
$C_{12}H_{20}NOP$ M 225.270
pK_a 17.81 ($MeNO_2$).

Anilide: [5586-05-0]. P,P-*Diethyl*-N,N′-*diphenylphosphinimidic amide*.
$C_{16}H_{21}N_2P$ M 272.329
Solid. Mp 161-163°.

Gilyarov, V.A. *et al*, *Zh. Obshch. Khim.*, 1966, **36**, 282 (*Engl. transl.* p. 293) (*anilide*)
Kabachnik, M.I., *Phosphorus*, 1971, **1**, 117 (*ester*)

N,N-Diethyl-*P*-phenylphosphonamidic acid, 9CI D-00307

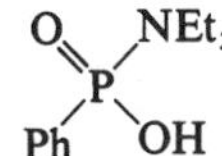

$C_{10}H_{16}NO_2P$ M 213.216

Isol. as diethylammonium salt.

Me ester: [39030-35-8]. *Methyl* N,N-*diethyl*-P-*phenylphosphonamidate*.
$C_{11}H_{18}NO_2P$ M 227.242
Liq. d_4^{20} 1.10. Bp_1 110°. n_D^{20} 1.5144.

Ph ester: [24102-75-8]. *Phenyl* N,N-*diethyl*-P-*phenylphosphonamidate*.
$C_{16}H_{20}NO_2P$ M 289.313
Liq. d_4^{20} 1.22. $Bp_{0.03}$ 172°. n_D^{20} 1.5833.

Fluoride:
$C_{10}H_{15}FNOP$ M 215.207
Liq. d_4^{21} 1.31. Bp_{15} 153-156°. n_D^{21} 1.4970.

Chloride: [17833-40-8].
$C_{10}H_{15}ClNOP$ M 231.661
Liq. d_4^{25} 1.22. $Bp_{0.5}$ 132°. n_D^{25} 1.5243.

Cherbuliez, E. *et al*, *Helv. Chim. Acta*, 1961, **44**, 1820 (*synth*)
Ivanova, Zh.M. *et al*, *Zh. Obshch. Khim.*, 1965, **35**, 1974 (*Engl. transl.* p. 1965) (*fluoride*)
Tolkmith, H. *et al*, *J. Med. Chem.*, 1967, **10**, 1074 (*chloride*)
Pudovik, A.N. *et al*, *Zh. Obshch. Khim.*, 1969, **39**, 1890 (*Engl. transl.* p. 1851) (*ester, synth*)
Kamai, G.Kh. *et al*, *Zh. Obshch. Khim.*, 1972, **42**, 1295 (*Engl. transl.* p. 1290) (*ester, synth, ir*)
Jentsche, R. *et al*, *J. Prakt. Chem.*, 1977, **319**, 871 (*ester, synth, pmr*)

N,N-Diethyl-*P*-phenylphosphonamidodithioic acid, 9CI D-00308

Ph NEt₂ / P / S SH

$C_{10}H_{16}NPS_2$ M 245.337

Et ester: [26990-39-6]. *Ethyl* N,N-*diethyl*-P-*phenylphosphonamidodithioate*.
$C_{12}H_{20}NPS_2$ M 273.390
Liq. d_4^{20} 1.12. $Bp_{0.1}$ 116-118°. n_D^{20} 1.5942.

Isopropyl ester: Isopropyl N,N-*diethyl*-P-*phenylphosphonamidodithioate*.
$C_{13}H_{22}NPS_2$ M 287.417
Liq. d_4^{20} 1.09. $Bp_{0.08}$ 115-117°. n_D^{20} 1.5824.

Rizpolozhenskii, N.I. *et al*, *Izv. Akad. Nauk SSSR, Ser. Khim.*, 1970, 622 (*Engl. transl.* p. 571) (*synth*)
Shagidullin, R.R. *et al*, *Izv. Akad. Nauk SSSR, Ser. Khim.*, 1971, 1024 (*Engl. transl.* p. 940) (*ethyl ester, ir*)

N,N-Diethyl-*P*-phenylphosphonamidothioic acid, 9CI D-00309

[18628-83-6]

S NEt₂ / P / Ph OH ⇌ HS NEt₂ / P / Ph O

$C_{10}H_{16}NOPS$ M 229.276

Cryst. (Et_2O/pet. ether or EtOH). Mp 158-161° (153-154°).

Chloride: [5120-50-3].
$C_{10}H_{15}ClNPS$ M 247.722
Solid. Mp 43-44°. $Bp_{0.4}$ 115°.

O-*Me ester:* [73577-40-9]. O-*Methyl* N,N-*diethyl*-P-*phenylphosphonamidothioate*.
$C_{11}H_{18}NOPS$ M 243.303
Oil. d_4^{20} 1.11. n_D^{20} 1.5592.

S-*Me ester:* [73577-42-1]. S-*Methyl* N,N-*diethyl*-P-*phenylphosphonamidothioate*.
$C_{11}H_{18}NOPS$ M 243.303
Oil. d_4^{20} 1.12. n_D^{20} 1.7572.

S-*Et ester:* [26990-33-0]. S-*Ethyl* N,N-*diethyl*-P-*phenylphosphonamidothioate*.
$C_{12}H_{20}NOPS$ M 257.330
Liq. d_4^{20} 1.10. $Bp_{0.08}$ 105-108°. n_D^{20} 1.5533.

S-*Isopropyl ester:* S-*Isopropyl* N,N-*diethyl*-P-*phenylphosphonamidothioate*.
$C_{13}H_{22}NOPS$ M 271.357
Liq. d_4^{20} 1.08. $Bp_{0.1}$ 111-113°. n_D^{20} 1.5431.

S-*Ph ester:* [791-36-6]. S-*Phenyl* N,N-*diethyl*-P-*phenylphosphonamidothioate*.
$C_{16}H_{20}NOPS$ M 305.374
Liq. d_4^{20} 1.16. $Bp_{0.001}$ 160-162°. n_D^{20} 1.5985.

Rizpolozhenskii, N.I. *et al*, *Izv. Akad. Nauk SSSR, Ser. Khim.*, 1970, 622 (*Engl. transl.* p. 571) (*esters, synth*)
Shagidullin, R.R. *et al*, *Izv. Akad. Nauk SSSR, Ser. Khim.*, 1971, 1024 (*Engl. transl.* p. 940) (*ethyl ester, ir*)
Maier, L., *Phosphorus*, 1975, **5**, 253 (*synth, pmr, nmr*)
Almasi, L. *et al*, *J. Prakt. Chem.*, 1979, **321**, 913 (*synth, methyl esters, pmr, P nmr*)
Andreev, N.A. *et al*, *Zh. Obshch. Khim.*, 1979, **49**, 718 (*Engl. transl.* p. 623) (*chloride, synth, ir, P nmr*)

Diethyl phenylphosphonate D-00310

Diethyl benzenephosphonate

[1754-49-0]

$$(EtO)_2P(O)Ph$$

$C_{10}H_{15}O_3P$ M 214.200

Liq. $Bp_{0.2}$ 96-98°. n_D^{20} 1.4926.

▷TA0377000.

Tavs, P. *et al*, *Chem. Ber.*, 1970, **103**, 2428 (*synth*)
Allen, D.W. *et al*, *J. Chem. Soc., Perkin Trans. 2*, 1972, 63 (*nmr, derivs*)
Chernova, A.V. *et al*, *Izv. Akad. Nauk SSSR, Ser. Khim.*, 1972, 722 (*Engl. transl.* p. 693) (*uv, raman*)
Redmore, D., *J. Org. Chem.*, 1973, **38**, 1306 (*ms*)
Modro, T.A., *Can. J. Chem.*, 1977, **55**, 3681 (*cmr*)
Dorokhova, V.V. *et al*, *Zh. Obshch. Khim.*, 1979, **49**, 83 (*Engl. transl.* p. 68) (*uv*)
Hirao, T. *et al*, *Synthesis*, 1981, 56 (*synth, ir*)
Hollingshaus, J.G. *et al*, *J. Toxicol. Environ. Health*, 1981, **8**, 619 (*tox*)

Diethyl phenylphosphonite, 9CI D-00311

[1638-86-4]

$PhP(OEt)_2$

$C_{10}H_{15}O_2P$ M 198.201

Ligand for metals of Groups IB, VB, VIB, VIIB, and VIII. Liq. d_4^{20} 1.02. Bp 235-237°, Bp_1 63-65°. n_D^{20} 1.5113.

Kabachnik, M.I. *et al, Izv. Akad. Nauk SSSR, Ser. Khim.*, 1960, 133 (*Engl. transl.* p. 122) (*synth*)
Sander, M., *Chem. Ber.*, 1960, **93**, 1220 (*synth*)
Coskran, K.J. *et al, J. Am. Chem. Soc.*, 1968, **90**, 5437 (*uv*)
Williamson, M.P. *et al, J. Phys. Chem.*, 1968, **72**, 4043 (*pmr*)
Tolman, C.A., *J. Am. Chem. Soc.*, 1970, **92**, 2956 (*P nmr*)
Chernova, A.V. *et al, Izv. Akad. Nauk SSSR, Ser. Khim.*, 1972, 722 (*Engl. transl.* p. 693) (*raman*)
Modro, T.A, *Can. J. Chem.*, 1977, **55**, 3681 (*cmr*)
Zverev, V.V. *et al, Dokl. Akad. Nauk SSSR, Ser. Sci. Khim.*, 1981, **256**, 1412 (*Phys. Chem., Eng. transl.* p. 133) (*pe*)

Diethyl phenylphosphoramidate, 9CI D-00312

Diethyl phosphoranilidate. Diethyl phosphoric anilide. N-(*Diethoxyphosphinyl*)*aniline*

[1445-38-1]

$(EtO)_2P(O)NHPh$

$C_{10}H_{16}NO_3P$ M 229.215

Cryst. (EtOH aq.). Mp 95-96°. $Bp_{0.07}$ 116-118°.

N-*Trimethylsilyl: Diethyl (trimethylsilylphenyl)-phosphoramidate.*

$C_{13}H_{24}NO_3PSi$ M 301.397

Liq. d_4^{20} 1.05. Bp_1 99-100°. n_D^{20} 1.4821.

McOmbie, H. *et al, J. Chem. Soc.*, 1945, 380, 921 (*synth*)
McMurray, M.J. *et al, Org. Mass. Spectrom.*, 1970, **3**, 1031 (*ms*)
Zwierzak, A. *et al, Tetrahedron*, 1973, **29**, 3899 (*synth, ir, pmr, P nmr*)
Mathis, R. *et al, C.R. Hebd. Seances Acad. Sci., Ser. C*, 1975, **281**, 437 (*ir*)
Gareev, R.D., *Zh. Obshch. Khim.*, 1976, **46**, 2662 (*Engl. transl.* p. 2543) (*deriv, ir, pmr*)
Ayed, N. *et al, C.R. Hebd. Seances Acad. Sci., Ser. C*, 1977, **285**, 222 (*ir*)
Modro, T.A., *Phosphorus Sulfur*, 1979, **5**, 331 (*cmr*)
Bradamante, S. *et al, J. Org. Chem.*, 1980, **45**, 114 (*pmr, cmr*)
Davidowitz, B. *et al, Org. Mass. Spectrom.*, 1984, **19**, 128 (*ms*)

***N,N*-Diethyl-*N'*-phenylphosphoramidoimidic acid,** 9CI D-00313

$PhN{=}P(OH)_2NEt_2$

$C_{10}H_{17}N_2O_2P$ M 228.230

Free acid tautomeric with *N,N*-diethyl-*N*-phenylphosphorodiamic acid.

Di-Et ester: [33992-85-7]. *Diethyl* N,N-*diethyl*-N'-*phenylphosphoramidimidate.* P,P-*Diethoxy*-P-*diethylamino*-N-*phenylphosphazene.* P,P-*Diethoxy*-P-*diethylamino*-N-*phenylphosphine imide.*

$C_{14}H_{25}N_2O_2P$ M 284.337

Oil. d_4^{20} 1.04. pK_a 17.16 ($MeNO_2$). n_D^{20} 1.5096.

Dichloride: [3185-65-7]. P,P-*Dichloro*-P-*diethylamino*-N-*phenylphosphine imide.* P,P-*Dichloro*-P-*diethylamino*-N-*phenylphosphazene.*

$C_{10}H_{15}Cl_2N_2P$ M 265.122

Liq. $Bp_{0.1}$ 96-98°. n_D^{20} 1.5607.

Gutmann, V. *et al, Monatsh. Chem.*, 1965, **96**, 836 (*synth, props*)
Nifant'ev, E.E. *et al, Zh. Obshch. Khim.*, 1971, **41**, 2011 (*Engl. transl.* p. 2032) (*ester*)

***O,O*-Diethyl *SS*-phenylphosphoro(dithioperoxoate),** 9CI D-00314

Diethoxyphosphinyl phenyl disulfide. O,O-*Diethyl* S-(*phenylthio*) *phosphorothioate*

[7439-48-7]

$(EtO)_2P(O)SSPh$

$C_{10}H_{15}O_3PS_2$ M 278.320

Liq. $Bp_{0.4}$ 112°. n_D^{25} 1.5126.

U.S.P., 3 109 770, (*1963*); *CA*, **60**, 2841 (*synth*)
Miyamoto, T. *et al, Agric. Biol. Chem.*, 1980, **44**, 2581 (*synth, ir, pmr*)

***O,O*-Diethyl *SS*-phenyl phosphoro(dithioperoxo)thioate,** 9CI D-00315

O,O-*Diethyl* S-(*phenylthio*) *phosphorodithioate*

[77748-20-0]

$(EtO)_2P(S){-}S{-}S{-}Ph$

$C_{10}H_{15}O_2PS_3$ M 294.381

Liq. d_4^{20} 1.23. Bp_2 153°. n_D^{20} 1.5993.

Almasi, L. *et al, Chem. Ber.*, 1962, **95**, 1582 (*synth*)
Miyamoto, T. *et al, Agric. Biol. Chem.*, 1980, **44**, 2581 (*synth, ir, pmr*)

***O,O*-Diethyl *Se*-phenyl phosphoroselenoite** D-00316

[64202-92-2]

$(EtO)_2PSePh$

$C_{10}H_{15}O_2PSe$ M 277.161

Liq. d_4^{20} 1.30. $Bp_{0.06}$ 78°. n_D^{20} 1.5740.

Kolodii, Ya.I. *et al, Ukr. Khim. Zh.*, 1977, **43**, 721; *CA*, **87**, 134280.

***O,O*-Diethyl *S*-phenyl phosphorothioate,** 9CI, 8CI D-00317

O,O-*Diethyl* S-*phenyl thiophosphate*

[18852-83-0]

$PhSP(O)(OEt)_2$

$C_{10}H_{15}O_3PS$ M 246.260

Liq. $Bp_{0.02}$ 84-88°.

Torii, S. *et al, J. Org. Chem.*, 1979, **44**, 2938 (*synth, ir, pmr*)

***O,O*-Diethyl *S*-[(phenylthio)methyl] phosphorodithioate,** 9CI D-00318

[25795-00-0]

$(EtO)_2P(S)SCH_2SPh$

$C_{11}H_{17}O_2PS_2$ M 276.348

Liq. d_4^{20} 1.20. $Bp_{0.03}$ 128°. n_D^{20} 1.5909.

Shvetsova-Shilovskaya, K.D. *et al, Zh. Obshch. Khim.*, 1960, **30**, 193 (*Engl. transl.* p. 205) (*synth*)
Miyamoto, T. *et al, Agric. Biol. Chem.*, 1980, **44**, 2581 (*synth, ir, ms, pmr*)

Diethyl[3-phenyl-1-[(trimethylsilyl)oxy]-2-propenyl]phosphonate, 10CI D-00319

[66731-80-4]

$PhCH{=}CHCH(OSiMe_3)PO(OEt)_2$

$C_{16}H_{27}O_4PSi$ M 342.446
Reagent for synth. of β-subst. carboxylic esters and γ-subst. lactones. $Bp_{0.1}$ 134-137°.

Hata, T. *et al*, *Tetrahedron Lett.*, 1979, 2047 (*props*)
Sekine, M. *et al*, *Bull. Chem. Soc. Jpn.*, 1982, **55**, 224 (*nmr, props, synth*)

Diethyl phosphate, 9CI, 8CI — D-00320

Diethyl phosphoric acid. Diethyl hydrogen phosphate
[598-02-7]

$(EtO)_2P(O)OH$

$C_4H_{11}O_4P$ M 154.102
Prod. of metab. or photodegradn. of many commercial agrochemicals contg. the $(EtO)_2P(O)$ or $(EtO)_2P(S)$ groupings, e.g. Chlorpyrifos, C-00194 Parathion, P-00002 Mephosfolan, M-00004 and Dioxathion, D-00974 . Polymerisation catalyst. Syrup. d_4^{20} 1.10-1.18. $Bp_{0.01}$ 116-118°. pK_a 4.66 (95% EtOH, aq.,), pK_a 6.79 (EtOH), pK_a 9.90 ($MeNO_2$). n_D^{25} 1.4148.
Tetramethylammonium salt: Solid. Mp 170°.
Anilinum salt: [7108-85-2]. Solid. Mp 72-75°.
Cyclohexylammonium salt: [60766-82-7]. Solid. Mp 78-80°. Forms monohydrate Mp 65-67°.
Dicyclohexylammonium salt: [34608-90-7]. Cryst. (ligroin). Mp 140-140.5° (132-134°).
Fluoride: see Diethyl phosphorofluoridate, D-00387
Chloride: see Diethyl phosphorochloridate, D-00373
Bromide: see Diethyl phosphorobromidate, D-00371
Cyanide: see Diethyl phosphorocyanidate, D-00381
Azide: see Diethyl phosphorazidate, D-00362
Amide: see Diethyl phosphoramidate, D-00342
Anilide: see Diethyl phenylphosphoramidate, D-00312
Hydrazide: see Diethyl phosphorohydrazidate, D-00391
Anhydride: see Tetraethyl pyrophosphate, T-00069

Toy, A.D.F., *J. Am. Chem. Soc.*, 1948, **70**, 3882 (*synth*)
Mitsunoba, O. *et al*, *J. Org. Chem.*, 1965, **30**, 1071 (*synth*)
Bliznyuk, N.K. *et al*, *Zh. Obshch. Khim.*, 1967, **37**, 1119 (*Engl. transl.* p. 1061) (*synth*)
Sosnovsky, G. *et al*, *Chem. Ind. (London)*, 1967, 1297 (*salts*)
Sosnovsky, G. *et al*, *J. Org. Chem.*, 1972, **37**, 2267 (*derivs*)
Glonek, T. *et al*, *J. Phys. Chem.*, 1976, **80**, 639 (*P nmr*)
Yamaguchi, H. *et al*, *Bull. Chem. Soc. Jpn.*, 1981, **54**, 1891 (*synth, ir, pmr*)
Sax, N.I., *Dangerous Properties of Industrial Materials*, 6th Ed., Van Nostrand-Reinhold, 1984, 581.

Diethylphosphine, 9CI — D-00321

[627-49-6]

Et_2PH

$C_4H_{11}P$ M 90.105
Liq. Bp 85°. n_D^{20} 1.477. Highly air-sensitive.
▷Highly toxic. Ignites in air
Li deriv.: [19093-80-2]. Plates + 1dioxan. Insol. Et_2O, Me_2CO, sol. dioxan.
Oxide: [7215-33-0].
$C_4H_{11}OP$ M 106.104
Liq. $Bp_{0.6}$ 53-55°. n_D^{20} 1.4549. Dec. slowly at 180-200°. Tautomeric with Diethylphosphinous acid, D-00333 .
Sulfide: [6591-06-6].
$C_4H_{11}PS$ M 122.165
Liq. $Bp_{0.3}$ 51-55°. n_D^{20} 1.5350. Tautomeric with diethylphosphonothious acid.

Issleib, K. *et al*, *Chem. Ber.*, 1959, **92**, 1118 (*deriv*)
Maier, L. *et al*, *J. Phys. Chem.*, 1962, **66**, 901 (*nmr*)
Niebergall, H. *et al*, *Chem. Ber.*, 1962, **95**, 64 (*synth*)
Maier, L., *Helv. Chim. Acta*, 1966, **49**, 1249 (*sulfide, ir, nmr*)
Hays, H.R., *J. Org. Chem.*, 1968, **33**, 3690 (*oxide, ir, pmr, nmr*)
Cullen, W.R. *et al*, *J. Fluorine Chem.*, 1971, **1**, 227 (*pe*)
Kostyanovskii, R.G. *et al*, *Org. Mass Spectrom.*, 1972, **6**, 1183 (*ms*)
Rasulev, U.Kh. *et al*, *Zh. Obshch. Khim.*, 1974, **44**, 2198 (*Engl. transl.* p. 2158) (*ms*)
Kabachnik, M.F. *et al*, *Aust. J. Chem.*, 1975, **28**, 755 (*ir*)
Mosbo, J.A. *et al*, *Phosphorus Sulfur*, 1981, **11**, 11 (*struct*)
Bretherick, L., *Handbook of Reactive Chemical Hazards*, 2nd Ed., Butterworths, London and Boston, 1979, 510.
Sax, N.I., *Dangerous Properties of Industrial Materials*, 6th Ed., Van Nostrand-Reinhold, 1984, 581.
Hazards in the Chemical Laboratory, (Bretherick, L., Ed.), 3rd Ed., Royal Society of Chemistry, London, 1981, 290.

Diethylphosphinic acid — D-00322

[813-76-3]

$Et_2P(O)(OH)$

$C_4H_{11}O_2P$ M 122.103
Liq. or cryst. Sol. H_2O, EtOH, C_6H_6, spar. sol. pet. ether. Mp 19°. Bp_{21} 194-195°, $Bp_{1.5}$ 174°, $Bp_{0.08}$ 92°. pK_a 3.29 (H_2O).
Me ester: [5689-41-8]. *Methyl diethylphosphinate.*
$C_5H_{13}O_2P$ M 136.130
Liq. Sol. H_2O, EtOH, Et_2O. d_0^{20} 1.026. Bp_{12} 86°. n_D^{20} 1.4392.
Et ester: [4775-09-1]. *Ethyl diethylphosphinate.*
$C_6H_{15}O_2P$ M 150.157
Liq. with faint but pleasant odour. Sol. H_2O, EtOH. d_4^{20} 1.00. Bp_{12} 87-88°. n_D^{20} 1.4301.
Isopropyl ester: [18632-47-8]. *Isopropyl diethylphosphinate.*
$C_7H_{17}O_2P$ M 164.184
Liq. d_0^{20} 0.97. Bp_{14} 93-95°. n_D^{20} 1.4337.
Ph ester: [63027-79-2]. *Phenyl diethylphosphinate.*
$C_{10}H_{15}O_2P$ M 198.201
Liq. d_4^{20} 1.10. $Bp_{0.5}$ 98°. n_D^{20} 1.5210.
Benzyl ester: [13274-91-4]. *Benzyl diethylphosphinate.*
$C_{11}H_{17}O_2P$ M 212.228
Liq. $Bp_{0.5}$ 133-135°. Dec. at 180°.
Trimethylsilyl ester: [42346-39-4]. *Trimethylsilyl diethylphosphinate.*
$C_7H_{19}O_2PSi$ M 194.285
Bp_1 83-84°, $Bp_{0.06}$ 40-41°.
Fluoride: [756-78-5].
$C_4H_{10}FOP$ M 124.095
No phys. props. reported.
Chloride: [1112-37-4].
$C_4H_{10}ClOP$ M 140.549
Liq. d_4^{20} 1.14. Bp_{20} 120-122°, $Bp_{0.7}$ 62.5-64.5°. n_D^{20} 1.4680.
Azide: [20495-46-9].
$C_4H_{10}N_3OP$ M 147.116
Liq. Bp_8 97°.

Isocyanate: [20434-08-6].
$C_5H_{10}NO_2P$ M 147.113
Liq. d_4^{20} 1.12. $Bp_{0.06}$ 56-58°. n_D^{20} 1.4550.
Isothiocyanate: [20434-13-4].
$C_5H_{10}NOPS$ M 163.174
Liq. $Bp_{0.7}$ 88-90°.
Anhydride: [7495-97-8].
$C_8H_{20}O_3P_2$ M 226.192
Liq. Bp_{14} 186-188°, $Bp_{0.5}$ 124-125°.
▷SZ4725000.
Anhydride with diethylphosphinothioic acid: [4885-53-4]. *Diethylphosphinic diethylphosphinothioic anhydride.*
$C_8H_{20}O_2P_2S$ M 242.252
$Bp_{0.07}$ 92-94°. n_D^{23} 1.5060.
Amide: see P,P-Diethylphosphinic amide, D-00323

Kosolapoff, G.M. *et al, J. Am. Chem. Soc.*, 1951, **73**, 5466 (*anhydride, chloride*)
Crofts, P.C. *et al, J. Am. Chem. Soc.*, 1953, **75**, 3379 (*synth*)
Kosolapoff, G.M. *et al, J. Chem. Soc.*, 1959, 3950 (*synth, chloride*)
Kuchen, W. *et al, Chem. Ber.*, 1962, **95**, 1703 (*synth, anhydride*)
Pollart, K.A. *et al, J. Org. Chem.*, 1962, **27**, 444 (*chloride*)
Blasse, R.G., *Recl. Trav. Chim. Pays-Bas*, 1965, **84**, 267 (*chloride, ir*)
Henning, H.-G. *et al, J. Prakt. Chem.*, 1966, **33**, 188 (*benzyl ester*)
Neimysheva, A.A. *et al, Zh. Obshch. Khim.*, 1966, **36**, 1090 (*Engl. transl.* p. 1105) (*chloride, props*)
Derkach, G.I. *et al, Zh. Obshch. Khim.*, 1968, **38**, 1784 (*Engl. transl.* p. 1738) (*isocyanate*)
Haake, P. *et al, Tetrahedron*, 1968, **24**, 565 (*synth, esters, ms*)
Tomaschewski, G. *et al, Arch. Pharm.* (*Weinheim, Ger.*), 1968, **301**, 520 (*isothiocyanate, isocyanate*)
Knunyants, I.L. *et al, Dokl. Akad. Nauk SSSR, Ser. Sci. Khim.*, 1971, **201**, 862 (*Engl. transl.* p. 992) (*chloride, fluoride, F and P nmr*)
Murray, M. *et al, J. Chem. Soc.* (*B*), 1971, 1714 (*esters, P nmr*)
Cook, R.D. *et al, J. Am. Chem. Soc.*, 1973, **95**, 8088 (*esters, props*)
Appel, R. *et al, Chem. Ber.*, 1974, **107**, 2658; 1975, **108**, 1783 (*ir, pmr, P nmr*)
Kuchen, W. *et al, Z. Anorg. Allg. Chem.*, 1975, **413**, 266 (*silyl ester*)
Schröder, H. *et al, Z. Anorg. Allg. Chem.*, 1975, **418**, 247 (*azide, cmr, P nmr, pmr, ir, raman, ms*)
Kharrasova, F.M. *et al, Zh. Obshch. Khim.*, 1978, **48**, 1046 (*Engl. transl.* p. 953) (*phenyl ester, P nmr*)
Harger, M.J.P. *et al, J. Chem. Soc., Perkin Trans. 1*, 1981, 736 (*synth, chloride, ester, azide, ir, pmr*)

P,P-Diethylphosphinic amide, 9CI D-00323

$Et_2P(O)NH_2$

$C_4H_{12}NOP$ M 121.119
V. hygroscopic cryst. (Et_2O). Mp 62-64°. When heated, the amide loses NH_3 to give the bis(diethylphosphinic) imide.
N-*Me:* [82134-83-6]. P,P-*Diethyl*-N-*methylphosphinic amide.*
$C_5H_{14}NOP$ M 135.145
Liq. $Bp_{0.5}$ 129-131°.
N-*Et:* N,P,P-*Triethylphosphinic amide.*
$C_6H_{16}NOP$ M 149.172
Liq. d_4^{20} 0.99. $Bp_{0.5}$ 120.5-122°. n_D^{20} 1.4603.
N,N-*Di-Et:* [24304-64-1]. *Tetraethylphosphinic amide.*
$C_8H_{20}NOP$ M 177.226
Liq. d_4^{20} 0.94. Bp_{16} 134-135°. pK_a 5.8 ($MeNO_2$). n_D^{20} 1.4564.
N-*Ethoxycarbonyl: Ethyl* N-*(diethylphosphinyl)-carbamate.*
$C_7H_{16}NO_3P$ M 193.182
Solid. Bp 119-121°.
N-*Phenylaminocarbonyl:* N-*(Diethylphosphinyl)*-N′-*phenylurea.*
$C_{11}H_{17}N_2O_2P$ M 240.241
Solid. Mp 127-128° dec.

Razumov, A.I. *et al, Zh. Obshch. Khim.*, 1957, **27**, 754 (*Engl. transl.* p. 827) (*derivs*)
Muckhacheva, O.A. *et al, Zh. Obshch. Khim.*, 1962, **32**, 2696 (*Engl. transl.* p. 2654) (*synth, props*)
Mizrahi, V. *et al, J. Org. Chem.*, 1982, **47**, 3533 (*deriv, ms, pmr*)

Diethylphosphinodiselenoic acid, 9CI D-00324

$Et_2P(Se)SeH$

$C_4H_{11}PSe_2$ M 248.025
Unstable orange oil. pK_a 2.18 (80% 2-propanol).
Na salt: Solid + $2H_2O$. Mp 129-130°.
Et ester: Ethyl diethylphosphinodiselenoate.
$C_6H_{15}PSe_2$ M 276.078
Liq. $Bp_{0.13}$ 105°. n_D^{20} 1.6193.
Propyl ester: Propyl diethylphosphinodiselenoate.
$C_7H_{17}PSe_2$ M 290.105
Liq. $Bp_{0.1}$ 98°. n_D^{20} 1.6037.
Anhydroselenide: [2408-96-0].
$C_8H_{20}P_2Se_3$ M 415.074
Plates (pet. ether). Mp 65°.

Küchen, W. *et al, Chem. Ber.*, 1966, **99**, 1663 (*synth, derivs*)
Hägele, G. *et al, Z. Naturforsch., B*, 1974, **29**, 349 (*anhydroselenide, P nmr*)

Diethylphosphinodithioic acid, 9CI D-00325

[866-54-6]

$Et_2P(S)SH$

$C_4H_{11}PS_2$ M 154.225
Forms salts and complexes with most metals. Alkali salts used as metal ore flotation agents. Liq. d_4^{20} 1.13. Bp_{13} 130-131°, $Bp_{2.5}$ 68-69°. pK_{a1} 1.74 (7% EtOH aq.), pK_{a2} 2.62 (80% EtOH aq.), pK_{a3} 5.06 (EtOH).
Na salt: Solid. Mp 153°. Forms dihydrate Mp 124-7°.
NH_4 salt: [19073-47-3]. Solid. Mp 193°.
Me ester: [49873-27-0]. *Methyl diethylphosphinodithioate.*
$C_5H_{13}PS_2$ M 168.251
Liq. d_4^{20} 1.09. Bp_2 71°. n_D^{20} 1.5669.
Et ester: [5745-32-4]. *Ethyl diethylphosphinodithioate.*
$C_6H_{15}PS_2$ M 182.278
Forms Sn complexes. Liq. d_4^{20} 1.06. $Bp_{0.05}$ 61-62°. n_D^{20} 1.5535.
Isopropyl ester: Isopropyl diethylphosphinodithioate.
$C_7H_{17}PS_2$ M 196.305
Liq. d_4^{25} 1.02. $Bp_{0.04}$ 60-62°. n_D^{25} 1.5385.
Ph ester: Phenyl diethylphosphinodithioate.
$C_{10}H_{15}PS_2$ M 230.322
Liq. $Bp_{0.1}$ 126°.
Trimethylsilyl ester: Trimethylsilyl diethylphosphinodithioate.
$C_7H_{19}PS_2Si$ M 226.406

Liq. $Bp_{0.06}$ 65-67°.
Anhydrosulfide: [22737-43-5].
$C_8H_{20}P_2S_3$ M 274.374
Solid. Mp 52°. $Bp_{0.5}$ 173-176°.

Mastryukova, T.A. *et al, Zh. Obshch. Khim.*, 1961, **31**, 507 (*Engl. transl.* p. 464) (*synth, props*)
Kuchen, W. *et al, Chem. Ber.*, 1963, **96**, 1733 (*synth*)
Akamsin, V.D. *et al, Izv. Akad. Nauk SSSR, Ser. Khim.*, 1966, 493 (*Engl. transl.* p. 463) (*esters*)
Bulgakova, R.A. *et al, Izv. Akad. Nauk SSSR, Ser. Khim.*, 1968, 672 (*Engl. transl.* p. 654) (*synth, ir*)
Almasi, L. *et al, Chem. Ber.*, 1969, **102**, 1489 (*anhydrosulfide*)
Mastryukova, T.A. *et al, Zh. Obshch. Khim.*, 1971, **41**, 1938 (*Engl. transl.* p. 1953) (*props*)
Exner, O. *et al, Collect. Czech. Chem. Commun.*, 1973, **38**, 677 (*synth, esters*)
Kuramshin, I.Ya. *et al, Zh. Obshch. Khim.*, 1973, **43**, 1456 (*Engl. transl.* p. 1446) (*ester*)
Kuchen, W. *et al, Z. Anorg. Allg. Chem.*, 1975, **413**, 266 (*silyl ester, ir*)

2-(Diethylphosphino)ethanethiol, 9CI — D-00326

Diethyl(2-mercaptoethyl)phosphine
[3190-77-0]

$Et_2PCH_2CH_2SH$

$C_6H_{15}PS$ M 150.218
Used in extraction of metals from aq. soln. Ligand for Au (particularly). Liq. $Bp_{0.001}$ 43-44°. pK_a 7.4. n_D^{20} 1.5202. O_2 catalyses isomerization to Et_3PS.

Fr. Pat., 1 401 930, (*1965*); *CA*, **63**, 11615 (*synth*)
Weinstook, J. *et al, J. Med. Chem.*, 1974, **17**, 139.

Diethylphosphinoselenoic acid, 9CI — D-00327

[4002-41-9]

$Et_2P(Se)OH \rightleftharpoons Et_2P(O)SeH$

$C_4H_{11}OPSe$ M 185.064
Solid-liq. d_4^{25} 1.44. Mp 17°. $Bp_{0.05}$ 76-77°. pK_{a1} 2.47 (7% EtOH aq.), pK_{a2} 4.67 (80% EtOH aq.). n_D^{25} 1.5624.
O-*Me ester:* [58722-79-5]. O-*Methyl diethylphosphinoselenoate.*
$C_5H_{13}OPSe$ M 199.091
Liq. d_4^{20} 1.33. Bp_{10} 102-103°. n_D^{20} 1.5312.
O-*Et ester:* [39078-28-9]. O-*Ethyl diethylphosphinoselenoate.*
$C_6H_{15}OPSe$ M 213.118
Liq. d_4^{20} 1.26. Bp_8 104-105°. n_D^{20} 1.5160.
O-*Isopropyl ester:* [59085-24-4]. O-*Isopropyl diethylphosphinoselenoate.*
$C_7H_{17}OPSe$ M 227.144
d_4^{20} 1.21. Bp_{11} 111-112°. n_D^{20} 1.5098.
Se-*Butyl ester:* [60819-73-0]. Se-*Butyl diethylphosphinoselenoate.*
$C_8H_{19}OPSe$ M 241.171
Liq.
O-*Ph ester:* [50351-54-7]. O-*Phenyl diethylphosphinoselenoate.*
$C_8H_{15}OPSe$ M 237.140
Liq. d_4^{20} 1.32. $Bp_{0.1}$ 103-104°. n_D^{20} 1.5838.
Chloride: [6395-52-4].
$C_4H_{10}ClPSe$ M 203.510
d_4^{20} 1.46. Bp_{12} 106-107°. n_D^{20} 1.5682.
Bromide:
$C_4H_{10}BrPSe$ M 247.961
Red oil. $Bp_{12.5}$ 112-115°.
Azide: [70629-41-3].
$C_4H_{10}N_3PSe$ M 210.077
Liq. $Bp_{2.5}$ 97°.
Amide: see P,P-Diethylphosphinoselenoic amide, D-00328

Markowski, A. *et al, Rocz. Chem.*, 1960, **34**, 1675; *CA*, **56**, 7345 (*synth, props, chloride*)
Shagidullin, R.R. *et al, Izv. Akad. Nauk SSSR, Ser. Khim.*, 1975, 197 (*Engl. transl.* p. 190); 1976, 184 (*Engl. transl.* p. 174) (*uv*)
Bayandina, E.V. *et al, Zh. Obshch. Khim.*, 1976, **46**, 288 (*Engl. transl.* p. 285) (*chloride, esters*)
Vandyukova, I.I. *et al, Izv. Akad. Nauk SSSR, Ser. Khim.*, 1976, 1390 (*Engl. transl.* p. 1334) (*ester, ir, raman*)
Nuretdinov, I.A. *et al, Zh. Obshch. Khim.*, 1978, **48**, 1071 (*Engl. transl.* p. 975) (*chloride, P nmr, ir*)
Schröder, H.F. *et al, Z. Anorg. Allg. Chem.*, 1979, **451**, 158 (*azide, ms, P nmr, cmr*)

P,P-Diethylphosphinoselenoic amide, 9CI, 8CI — D-00328

$Et_2P(Se)NH_2$

$C_4H_{11}NPSe$ M 183.071
N-*Me:* [59085-25-5]. P,P-*Diethyl*-N-*methylphosphinoselenoic amide.*
$C_5H_{13}NPSe$ M 197.098
Solid. Mp 59-60°.
N,N-*Di-Me:* [35525-42-9]. P,P-*Diethyl*-N,N-*dimethylphosphinoselenoic amide.*
$C_6H_{15}NPSe$ M 211.125
Liq. d_4^{20} 1.27. Bp_{17} 152-153°. n_D^{20} 1.5529.
N,N-*Di-Et:* [39181-31-2]. *Tetraethylphosphinoselenoic amide.*
$C_8H_{20}NPSe$ M 240.186
Liq. d_4^{20} 1.20. $Bp_{0.2}$ 105-106°. n_D^{20} 1.5350.
N-*Ph:* [59085-26-6]. P,P-*Diethyl*-N-*phenylphosphinoselenoic amide.*
$C_{10}H_{15}NPSe$ M 259.169
Solid. Mp 101-102°.

Osokin, D.J. *et al, Org. Magn. Reson.*, 1972, **4**, 831 (*nqr*)
Stec, W.J. *et al, Phosphorus*, 1972, **2**, 97 (*P nmr*)
Bayandina, E.V. *et al, Zh. Obshch. Khim.*, 1976, **46**, 288 (*Engl. transl.* p. 285) (*synth*)
Shagidullin, R.R. *et al, Izv. Akad. Nauk SSSR, Ser. Khim.*, 1976, 184 (*Engl. transl.* p. 174) (*uv*)

Diethylphosphinoselenothioic acid, 9CI — D-00329

$Et_2P(S)SeH \rightleftharpoons Et_2P(Se)SH$

$C_4H_{11}PSSe$ M 201.125
Unstable yellow oil. pK_a 2.29 (80% 2-propanol aq.).
S-*Et ester:* [39078-29-0]. S-*Ethyl diethylphosphinoselenothioate.*
$C_6H_{15}PSSe$ M 229.178
Oil. d_4^{20} 1.32. Bp_{11} 142-143°. n_D^{20} 1.5879.
Se-*Et ester:* Se-*Ethyl diethylphosphinoselenothioate.*
$C_6H_{15}PSSe$ M 229.178
Oil. $Bp_{0.06}$ 71.5°. n_D^{20} 1.5854.
Se-*Propyl ester:* Se-*Propyl diethylphosphinoselenothioate.*
$C_7H_{17}PSSe$ M 243.205
Oil. $Bp_{0.06}$ 86-87°. n_D^{20} 1.5736.

Küchen, W. *et al, Chem. Ber.*, 1966, **99**, 1663 (*synth, esters*)
Bayandina, E.V. *et al, Zh. Obshch. Khim.*, 1976, **46**, 288 (*Engl. transl.* p. 2851) (*ester*)

Shagidullin, R.R. *et al*, *Izv. Akad. Nauk SSSR, Ser. Khim.*, 1976, 184 (*Engl. transl.* p. 174) (*ester, uv*)

Diethylphosphinothioic acid, 9CI D-00330

[866-53-5]

$Et_2P(S)OH \rightleftharpoons Et_2P(O)SH$

$C_4H_{11}OPS$ M 138.164

Liq. or solid. d_4^{20} 1.11. Mp 11.5-12.5°. Bp_1 103.5-105°, $Bp_{0.08}$ 72.5-73.5°. pK_{a1} 2.80 (7% EtOH aq.) (1% thiol form), pK_{a2} *4.88 (80% EtOH aq.) (0.1% thiol form), pK_{a3} *8.47 (100% EtOH), pK_{a4} *11.94 ($MeNO_2$). n_D^{20} 1.5267.

Na salt: Cryst. + $3H_2O$. Mp 58-59°.

Cyclohexylammonium salt: Solid. Mp 145-147°. Bp_1 62-64°.

O-*Me ester:* [14806-54-3]. O-*Methyl diethylphosphinothioate.*
$C_5H_{13}OPS$ M 152.191
Liq. d_4^{20} 1.04. Bp_7 75-76°. n_D^{20} 1.4953.

S-*Me ester:* [41116-69-2]. S-*Methyl diethylphosphinothioate.*
$C_5H_{13}OPS$ M 152.191
Liq. d_4^{20} 1.08. Bp_2 93° (Bp_7 91-91.5°). n_D^{20} 1.5090, 1.5024.

O-*Et ester:* [4885-49-8]. O-*Ethyl diethylphosphinothioate.*
$C_6H_{15}OPS$ M 166.218
No phys. props. reported.

S-*Et ester:* [4885-47-6]. S-*Ethyl diethylphosphinothioate.*
$C_6H_{15}OPS$ M 166.218
Liq. d_4^{20} 1.04. Bp_4 102°. n_D^{25} 1.5022.

O-*Isopropyl ester:* [51993-67-0]. O-*Isopropyl diethylphosphinothioate.*
$C_7H_{17}OPS$ M 180.244
Liq. Bp_{11} 96.5-97°.

S-*Isopropyl ester:* [63873-24-5]. S-*Isopropyl diethylphosphinothioate.*
$C_7H_{17}OPS$ M 180.244
No phys. props. reported.

S-*Butyl ester:* [18032-97-8]. S-*Butyl diethylphosphinothioate.*
$C_8H_{19}OPS$ M 194.271
Liq. d_4^{20} 1.01. $Bp_{0.03}$ 68.5-69°. n_D^{20} 1.4923.

O-*Ph ester:* [63027-81-6]. O-*Phenyl diethylphosphinothioate.*
$C_{10}H_{15}OPS$ M 214.262
Liq. $Bp_{0.5}$ 160° (oven).

O-*Trimethylsilyl ester:* [4885-54-5]. O-*Trimethylsilyl diethylphosphinothioate.*
$C_7H_{19}OPSSi$ M 210.346
Liq. $Bp_{0.1}$ 45-46°. n_D^{20} 1.4730.

Fluoride: [681-02-7].
$C_4H_{10}FPS$ M 140.155
Liq. Bp_{12} 75°.

Chloride: [3982-89-6].
$C_4H_{10}ClPS$ M 156.610
Liq. d_4^{20} 1.14. Bp_{27} 112°, Bp_4 60-61°. n_D^{20} 1.5292.

Bromide: [3981-46-2].
$C_4H_{10}BrPS$ M 201.061
Liq. Bp_{10} 92-93°.

Iodide: [81373-55-9].
$C_4H_{10}IPS$ M 248.061
Liq. $Bp_{0.02}$ 87-92°.

Azide: [58347-14-1].
$C_4H_{10}N_3PS$ M 163.177
Yellow liq. Bp_1 70-71°.

Anhydride: [4895-48-1].
$C_8H_{20}OP_2S_2$ M 258.313
Solid. Mp 42.5°. n_D^{50} 1.537.

Anhydride with acetic acid: [4885-52-3]. O-*Acetyl diethylphosphinothioate.*
$C_6H_{13}O_2PS$ M 180.201
Liq. $Bp_{0.1}$ 55°. n_D^{20} 1.4992.

Amide: see P,P-Diethylphosphinothioic amide, D-00331

Kabachnik, M.I. *et al*, *Tetrahedron*, 1960, **9**, 10 (*synth, props*)
Mastryukova, T.A. *et al*, *Zh. Obshch. Khim.*, 1961, **31**, 507 (*Engl. transl.* p. 464) (*chloride*)
Kuchen, W. *et al*, *Justus Liebigs Ann. Chem.*, 1962, **652**, 28 (*ester, chloride, bromide*)
Kuchen, W. *et al*, *Z. Anorg. Allg. Chem.*, 1964, **333**, 71 (*synth, props*)
Maier, L., *Helv. Chim. Acta*, 1964, **47**, 27 (*chloride, bromide*)
Akansin, V.D. *et al*, *Izv. Akad. Nauk SSSR, Ser. Khim.*, 1967, 1987 (*Engl. transl.* p. 1904) (*ester*)
Knunyants, I.L. *et al*, *Dokl. Akad. Nauk SSSR, Ser. Sci. Khim.*, 1971, **201**, 862 (*Engl. transl.*, p. 992) (*halides, P nmr*)
Kuramshin, I.Ya. *et al*, *Zh. Obshch. Khim.*, 1973, **43**, 1456 (*Engl. transl.* p. 1446) (*esters, ir, pmr, P nmr*)
Steinberger, H. *et al*, *Z. Naturforsch., B*, 1973, **28**, 44 (*silyl ester, P nmr, use*)
Haegele, R. *et al*, *Z. Naturforsch., B*, 1974, **29**, 349 (*nmr*)
Mastryukova, T.A. *et al*, *Zh. Obshch. Khim.*, 1974, **44**, 1001 (*Engl. transl.* p. 963) (*esters*)
Kuchen, W. *et al*, *Z. Anorg. Allg. Chem.*, 1975, **413**, 266 (*silyl ester, ir, ms*)
Schröder, H.F. *et al*, *Z. Anorg. Allg. Chem.*, 1975, **418**, 247 (*azide, ms, ir, raman, pmr, cmr, P nmr*)
Mastryukova, T.A., *Phosphorus Sulfur*, 1976, **1**, 211 (*esters*)
Bone, S.A. *et al*, *J. Chem. Soc., Perkin Trans. 1*, 1977, 437 (*phenyl ester, pmr*)
Turpin, R. *et al*, *Bull. Soc. Chim. Fr. Part I*, 1977, 999 (*anhydride, P nmr*)
Feshchenko, N.G. *et al*, *Zh. Obshch. Khim.*, 1982, **52**, 222 (*Engl. transl.* p. 202) (*iodide, P nmr*)
Keck, H. *et al*, *Phosphorus Sulfur*, 1985, **24**, 343 (*esters, ms*)

P,P-Diethylphosphinothioic amide, 9CI D-00331

[5022-56-0]

$Et_2P(S)NH_2$

$C_4H_{12}NPS$ M 137.179

N,N-*Di-Me:* [17513-66-5]. P,P-*Diethyl*-N,N-*dimethylphosphinothioic amide.*
$C_6H_{16}NPS$ M 165.233
Forms Sn complexes.

N,N-*Di-Et:* [18032-89-8]. *Tetraethylphosphinothioic amide.*
$C_8H_{20}NPS$ M 193.286
Liq. d_4^{20} 0.99. $Bp_{0.0004}$ 71-72°. n_D^{20} 1.5122.

N-*Ph:* [42847-83-6]. P,P-*Diethyl*-N-*phenylphosphinothioic amide.*
$C_{10}H_{16}NPS$ M 213.277
No phys. props. reported.

N-*Trimethylsilyl:* P,P-*Diethyl*-N-*trimethylsilylphosphinothioic amide.*
$C_7H_{20}NPSSi$ M 209.361
Liq. $Bp_{0.05}$ 117-119°.

N-(*Diethylphosphinothioyl*): [42847-86-7]. N-(*Diethylphosphinothioyl*)-P,P-*diethylphosphinothioic amide.*
$C_8H_{21}NP_2S_2$ M 257.328
pK_a 9.6.

N-*Benzoyl:* N-*Benzoyl*-P,P-*diethylphosphinothioic amide.*
$C_{11}H_{16}NOPS$ M 241.287

Cryst. (cyclohexane). Mp 93-94°. pK_a 11.7.

Akamsin, V.D. *et al*, *Izv. Akad. Nauk SSSR, Ser. Khim.*, 1967, 1983 (*Engl. transl.* p. 1900) (*deriv*)
Boedeker, J., *J. Organomet. Chem.*, 1973, **56**, 255 (*ir*)
Steinberger, H. *et al*, *Z. Naturforsch., B*, 1974, **29**, 611 (*deriv, ir, ms, P nmr*)
Kuramshin, I.Ya. *et al*, *Zh. Obshch. Khim.*, 1975, **45**, 1194 (*Engl. transl.* p. 1177) (*deriv, ir*)
Boedeker, J. *et al*, *J. Prakt. Chem.*, 1976, **318**, 149 (*synth, ir*)
Pudovik, A.N. *et al*, *Zh. Obshch. Khim.*, 1976, **46**, 765 (*Engl. transl.* p. 764) (*deriv, ir, complexes*)

Diethylphosphinothious acid, 9CI — D-00332

Et_2PSH

$C_4H_{11}PS$ M 122.165

Free acid probably exists as diethylphosphine sulfide. See under Diethylphosphine, D-00321 .

Et ester: [6588-00-7]. *Ethyl diethylphosphinothioite.*
$C_6H_{15}PS$ M 150.218
Liq. Bp_7 50-52°. n_D^{20} 1.5047. Forms cryst. phosphonium adducts with MeI, EtI, and $PhCH_2Br$.

Isopropyl ester: Isopropyl diethylphosphinothioite. 1-Methylethyl diethylphosphinothioite.
$C_7H_{17}PS$ M 164.245
Liq. Bp_{20} 77-79°. n_D^{20} 1.4963.

Ph ester: [18788-05-1]. *Phenyl diethylphosphinothioite.*
$C_{10}H_{15}PS$ M 198.262
Liq. d_4^{20} 1.03. Bp_1 107-108°. n_D^{20} 1.5830.

Akamsin, V.D. *et al*, *Dokl. Akad. Nauk SSSR, Ser. Sci. Khim.*, 1966, **168**, 807 (*Engl. transl.* p. 547) (*alkyl esters, synth*)
Shlyk, Yu.N. *et al*, *Zh. Obshch. Khim.*, 1968, **38**, 193 (*Engl. transl.* p. 194) (*phenyl ester, synth*)
Razumov, A.I. *et al*, *Zh. Obshch. Khim.*, 1972, **42**, 1250 (*Engl. transl.* p. 1245) (*ethyl ester, synth, P nmr*)
Chernova, A.V. *et al*, *Zh. Obshch. Khim.*, 1979, **49**, 2002 (*Engl. transl.* p. 1760) (*ethyl ester, uv*)
Zverev, V.V. *et al*, *Zh. Obshch. Khim.*, 1983, **53**, 1968 (*Engl. transl.* p. 1775) (*ethyl ester, pe*)

Diethylphosphinous acid, 9CI — D-00333

Et_2POH

$C_4H_{11}OP$ M 106.104

Free acid exists as the phosphoryl tautomer. See under Diethylphosphine, D-00321 .

Et ester: [2303-77-7]. *Ethyl diethylphosphinite.*
$C_6H_{15}OP$ M 134.158
Air-sensitive liq. Bp 128-131°, Bp_{15} 80-85°. n_D^{20} 1.4328.
▷May ignite in air

Isopropyl ester: [83582-59-6]. *Isopropyl diethylphosphinite.*
$C_7H_{17}OP$ M 148.184
Air-sensitive liq. Bp_{96} 78-79°, Bp_{50} 60°. n_D^{20} 1.4320.
▷May ignite in air

Ph ester: [25781-03-7].
$C_{10}H_{15}OP$ M 182.202
Liq. d_4^{20} 1.00. Bp_{15} 109°. n_D^{20} 1.5290.

Chloride: see Diethylphosphinous chloride, D-00335

Bromide: [20472-46-2]. *Bromodiethylphosphine.*
$C_4H_{10}BrP$ M 169.001
Liq. Bp_{745} 153-154°.

Iodide: [20472-47-3]. *Diethyliodophosphine.*
$C_4H_{10}IP$ M 216.001
Light-yellow cryst. Mp 46-47°. Bp_{15} 65-67°.

Cyanide: [26306-14-9]. *Cyanodiethylphosphine.*
$C_5H_{10}NP$ M 115.114
Liq. Bp_{22} 71-72°.

*Amide: see P,P-*Diethylphosphinous amide, D-00334

Arbuzov, B.A. *et al*, *Dokl. Akad. Nauk SSSR*, 1953, **89**, 291; *CA*, **48**, 7540 (*esters*)
Razumov, A. *et al*, *Zh. Obshch. Khim.*, 1956, **26**, 1436 (*Engl. transl.* p. 1615) (*esters, synth*)
Issleib, K. *et al*, *Chem. Ber.*, 1959, **92**, 2681 (*bromide, iodide, synth*)
Saegusa, T. *et al*, *J. Org. Chem.*, 1970, **35**, 4238 (*cyanide, synth, ir*)
Kharrasova, F.M. *et al*, *Zh. Obshch. Khim.*, 1978, **48**, 1046 (*Engl. transl.* p. 953) (*ester*)
Van Lindhoudt, J.P. *et al*, *Spectrochim. Acta, Part A*, 1979, **35**, 1307 (*halides, pmr, P nmr, cmr*)
Romanenko, V.D. *et al*, *Synthesis*, 1980, 823 (*iodide, synth, P nmr*)
Feshchenko, N.G. *et al*, *Zh. Obshch. Khim.*, 1982, **52**, 222 (*Engl. transl.* p. 202) (*iodide, synth, P nmr*)

P,P-Diethylphosphinous amide, 9CI — D-00334

Aminodiethylphosphine

Et_2PNH_2

$C_4H_{12}NP$ M 105.119
Bp_1 0°. n_D^{20} 1.4765.

N-*Me:* [88834-10-0]. P,P-*Diethyl*-N-*methylphosphinous amide.*
$C_5H_{14}NP$ M 119.146
Liq. Bp_{35} 54-55°. n_D^{20} 1.4690.

N,N-*Di-Me:* [15272-12-5]. P,P-*Diethyl*-N,N-*dimethylphosphinous amide.*
$C_6H_{16}NP$ M 133.173
Liq. Bp_{715} 141-143°. n_D^{20} 1.4550.

N,N-*Di-Et:* [686-20-4]. *Tetraethylphosphinous amide.*
$C_8H_{20}NP$ M 161.226
Liq. Bp_{723} 178-183°. n_D^{20} 1.4678.

N-tert-*Butyl:* [88834-15-5]. N-tert-*Butyl*-P,P-*diethylphosphinous amide.*
$C_8H_{20}NP$ M 161.226
Liq. Bp_7 43-44°. n_D^{20} 1.4580.

N-*Ph:* [5586-04-9]. P,P-*Diethyl*-N-*phenylphosphinous amide. Diethylphosphinous anilide.*
$C_{10}H_{16}NP$ M 181.217
Bp_1 88-90°, $Bp_{0.1}$ 61-62°. n_D^{20} 1.5665.

N,N-*Bis(trimethylsilyl):* [73270-05-0]. P,P-*Diethyl*-N,N-*bis(trimethylsilyl)phosphinous amide.*
$C_{10}H_{28}NPSi_2$ M 249.483
Liq. $Bp_{1.6}$ 68-69°.

Issleib, K. *et al*, *Chem. Ber.*, 1959, **92**, 2681 (*diethyl, synth*)
Maier, L., *Helv. Chim. Acta*, 1964, **47**, 2129 (*derivs, synth, P nmr*)
Genkina, G.K. *et al*, *Zh. Obshch. Khim.*, 1966, **36**, 282 (*Engl. transl.* p. 293) (*phenyl, synth*)
Van Linthoudt, J.P. *et al*, *Spectrochim. Acta, Part A*, 1980, **36**, 315 (*diethyl, cmr, P nmr, pmr*)
Neilson, R.H. *et al*, *Inorg. Chem.*, 1982, **21**, 3568 (*bistrimethylsilyl, synth, P nmr*)
Foss, V.L. *et al*, *Zh. Obshch. Khim.*, 1983, **53**, 2184 (*Engl. transl.* p. 1969) (*synth, derivs, P nmr*)

Diethylphosphinous chloride, 9CI — D-00335

Chlorodiethylphosphine

[686-69-1]

Et_2PCl

$C_4H_{10}ClP$ M 124.550
Air-sensitive liq. d_4^{25} 0.984. Bp_{721} 129-131°, Bp_{100} 70-73°, Bp_{50} 48-52°. n_D^{20} 1.4772.

Issleib, K. *et al*, *Chem. Ber.*, 1959, **92**, 2681 (*synth*)
Maier, L., *J. Inorg. Nucl. Chem.*, 1962, **24**, 1073 (*synth, nmr*)
Maier, L., *Helv. Chim. Acta*, 1964, **47**, 2137 (*synth, P nmr*)
Van Linthoudt, J.P. *et al*, *Spectrochim. Acta, Part A*, 1979, **35**, 1307 (*cmr, P nmr, pmr*)
Wolfsberger, W., *Chem.-Ztg.*, 1983, **107**, 77 (*rev*)
Wolfsberger, W., *J. Organomet. Chem.*, 1986, **317**, 167 (*cmr, P nmr*)
Fox, R.B. *et al*, *J. Am. Chem. Soc.*, 1953, **75**, 3967 (*derivs*)
Harless, H.R., *Anal. Chem.*, 1961, **33**, 1387 (*ms*)
Luz, Z. *et al*, *J. Am. Chem. Soc.*, 1961, **83**, 4513 (*pmr*)
Moedritzer, K., *J. Inorg. Nucl. Chem.*, 1962, **22**, 19 (*P nmr*)
Hayashi, M. *et al*, *Bull. Chem. Soc. Jpn.*, 1977, **50**, 1510 (*props, use*)
Ovchinnikov, V.V. *et al*, *Zh. Obshch. Khim.*, 1978, **48**, 2424 (*Engl. transl.* p. 2199) (*P nmr, tautom*)
Guthrie, P.J. *et al*, *Can. J. Chem.*, 1979, **57**, 236 (*tautom*)
Zverev, V.V. *et al*, *Izv. Akad. Nauk SSSR, Ser. Khim.*, 1979, 84 (*Engl. transl.* p. 74) (*pe*)
Sidky, M.M. *et al*, *Org. Prep. Proceed. Int.*, 1982, **14**, 225 (*use*)

N,P-Diethylphosphonamidic acid, 9CI D-00336

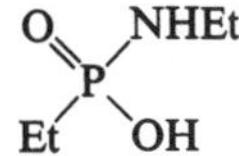

$C_4H_{12}NO_2P$ M 137.118
Et ester: [20395-32-8]. *Ethyl* N,P-*diethylphosphonamidate.*
$C_6H_{16}NO_2P$ M 165.172
Liq. d_4^{20} 1.02. $Bp_{2.5}$ 111.5-112°. n_D^{20} 1.4380.
2-Methylpropyl ester: 2-Methylpropyl N,P-*diethylphosphonamidate. Isobutyl* N,P-*diethylphosphonamidate.*
$C_8H_{20}NO_2P$ M 193.225
Liq. Bp_1 129-129.5°. n_D^{20} 1.4358.

Razumov, A.I. *et al*, *Zh. Obshch. Khim.*, 1957, **27**, 2389 (*Engl. transl.* p. 2389)

N,N-Diethylphosphonamidic acid, 9CI D-00337

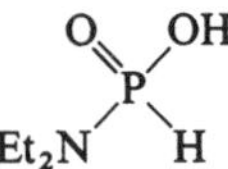

$C_4H_{12}NO_2P$ M 137.118
Me ester: Methyl N,N-*diethylphosphonamidate.*
$C_5H_{14}NO_2P$ M 151.145
Liq. d_4^{20} 1.03. Bp_{12} 86-87°. n_D^{20} 1.4364.
Et ester: [16141-72-3]. *Ethyl* N,N-*diethylphosphonamidate.*
$C_6H_{16}NO_2P$ M 165.172
Liq. $Bp_{0.5}$ 44-45°. n_D^{20} 1.4344.
Butyl ester: [16276-72-5]. *Butyl* N,N-*diethylphosphonamidate.*
$C_8H_{20}NO_2P$ M 193.225
Liq. Bp_1 93-94°, $Bp_{0.05}$ 52-54°. n_D^{20} 1.4395.

Zwierzak, A. *et al*, *Tetrahedron*, 1967, **23**, 2243 (*synth, ir, tlc*)
Abramov, V.S. *et al*, *Zh. Obshch. Khim.*, 1969, **39**, 2237 (*Engl. transl.* p. 2183) (*synth*)
Nifant'ev, E.E. *et al*, *Zh. Obshch. Khim.*, 1972, **42**, 1936 (*Engl. transl.* p. 1929) (*synth*)

Diethyl phosphonate, 9CI D-00338

Diethyl phosphite
[762-04-9]

$(EtO)_2P(O)H \rightleftharpoons (EtO)_2POH$

$C_4H_{11}O_3P$ M 138.103
Tautomeric. Ethylating agent for *N*-heterocyclic compds. *S*-ethylates thioamides. Hydrophosphorylating agent for alkenes and alkynes. Liq. d_0^{20} 1.074. Bp 187-188°, Bp_2 50-51°. n_D^{20} 1.4076.
▷TG7875000.

McCombie, H. *et al*, *J. Chem. Soc.*, 1945, 380 (*synth*)
Meyrick, C.I. *et al*, *J. Chem. Soc.*, 1950, 225 (*ir, raman*)

Diethyl phosphonite, 9CI D-00339

[20502-85-6]

$(EtO)_2PH$

$C_4H_{11}O_2P$ M 122.103
Stable in Et_2O at −20°. Cannot be isol. pure.

Nifant'ev, E.Ê. *et al*, *Zh. Obshch. Khim.*, 1976, **46**, 1184 (*Engl. transl.* p. 1167) (*P nmr*)
Gallagher, M.J. *et al*, *Aust. J. Chem.*, 1980, **33**, 287.

O,O-Diethyl phosphonoselenoate, 9CI D-00340

O,O-*Diethyl phosphoroselenoite*
[23416-28-6]

$(EtO)_2P(Se)H \rightleftharpoons (EtO)_2PSeH$

$C_4H_{11}O_2PSe$ M 201.063
Tautomeric. Liq. d_4^{25} 1.37. $Bp_{1.2}$ 46-47°. n_D^{25} 1.4965.

Kuznetsov, E.V. *et al*, *CA*, 1958, **52**, 8938.
McFarlane, W. *et al*, *J. Chem. Soc., Dalton Trans.*, 1973, 2162 (*pmr, P and Se nmr*)

O,O-Diethyl phosphonothioate, 9CI D-00341

O,O-*Diethyl thiophosphite*
[999-01-9]

$(EtO)_2P(S)H \rightleftharpoons (EtO)_2PSH$

$C_4H_{11}O_2PS$ M 154.163
Tautomeric but exists almost completely in thiophosphoryl form. Liq. with strong, sickly odour. d_4^{20} 1.07. Bp_{20} 83-84°. pK_a 8.65 (50% EtOH aq., 25°). n_D^{20} 1.4594.

Murav'ev, I.V. *et al*, *Zh. Obshch. Khim.*, 1968, **38**, 133 (*Engl. transl.* p. 133) (*synth*)
Nifant'ev, E.Ê. *et al*, *Zh. Obshch. Khim.*, 1983, **53**, 2695 (*Engl. transl.* p. 2429) (*synth, P nmr*)

Diethyl phosphoramidate, 9CI, 8CI D-00342

Diethyl amidophosphate. Diethyl phosphoryl amide. Diethyl phosphoramide
[1068-21-9]

$(EtO)_2P(O)NH_2$

$C_4H_{12}NO_3P$ M 153.117
Fireproofing agent. Reagent used is synth. of pure primary and secondary amines. Needles (CCl_4/cyclohexane). Mp 55.5°. $Bp_{0.2}$ 131-138°. V. hygroscopic.

Saunders, B.C. *et al*, *J. Chem. Soc.*, 1948, 699 (*synth*)
Inorg. Synth., 1953, **4**, 77 (*synth*)
McIvor, R.A. *et al*, *Can. J. Chem.*, 1958, **36**, 820 (*ir*)
Egorov, Yu.P. *et al*, *Zh. Obshch. Khim.*, 1975, **45**, 1716 (*Engl. transl.* p. 1683) (*ir, P nmr*)

Fluck, E. *et al*, *Z. Anorg. Allg. Chem.*, 1975, **412**, 47 (*pe*)
Koziara, A. *et al*, *Synthesis*, 1982, 918 (*use*)
Vafina, N.M. *et al*, *Zh. Obshch. Khim.*, 1982, **52**, 213 (*Engl. transl.* p. 194) (*synth*)
Zwierzak, A., *Synthesis*, 1982, 920 (*synth, ir, pmr, P nmr, use*)

Diethylphosphoramidic acid, 9CI, 8CI — D-00343

Phosphoric acid diethylamide

$Et_2NP(O)(OH)_2$

$C_4H_{12}NO_3P$ M 153.117

Di-Me ester: [65659-19-0]. *Dimethyl diethylphosphoramidate.*
$C_6H_{16}NO_3P$ M 181.171
Liq. d_4^{20} 1.07. Bp_8 82.5-83°. n_D^{20} 1.4265.
Di-Et ester: see Diethyl diethylphosphoramidate, D-00281
Diisopropyl ester: [74124-48-4]. *Diisopropyl diethylphosphoramidate. Bis(1-methylethyl) diethylphosphoramidate.*
$C_{10}H_{24}NO_3P$ M 237.278
Liq. Bp_{20} 115°. n_D^{20} 1.4205.
Dibutyl ester: [6626-39-7]. *Dibutyl diethylphosphoramidate.*
$C_{12}H_{28}NO_3P$ M 265.332
Liq. d_4^{20} 0.97. Bp_2 124-126°. n_D^{20} 1.4380.
Di-Ph ester: [6214-04-6]. *Diphenyl diethylphosphoramidate.*
$C_{16}H_{20}NO_3P$ M 305.313
Cryst. (C_6H_6, pet. ether or EtOH aq.). Mp 61-62°.
Dibenzyl ester: [3881-20-7]. *Dibenzyl diethylphosphoramidate. Bis(phenylmethyl) diethylphosphoramidate.*
$C_{18}H_{24}NO_3P$ M 333.366
Liq. n_D^{20} 1.5317.
Difluoride: see Diethylphosphoramidic difluoride, D-00345
Dichloride: see Diethylphosphoramidic dichloride, D-00344
Diamide: see N,N-Diethylphosphoric triamide, D-00364

Kamai, G. *et al*, *Zh. Obshch. Khim.*, 1957, **27**, 3064 (*Engl. transl.* p. 3093) (*esters*)
Cheymol, J. *et al*, *C.R. Hebd. Seances Acad. Sci.*, 1959, **249**, 1240 (*esters*)
Stock, J.A. *et al*, *J. Chem. Soc. (C)*, 1966, 637 (*diphenyl ester*)
Abramov, V.S. *et al*, *Zh. Obshch. Khim.*, 1969, **39**, 2658 (*Engl. transl.* p. 2596) (*dibutyl ester*)
Zwierzak, A., *Synthesis*, 1975, 507 (*dibenzyl ester*)

Diethylphosphoramidic dichloride, 9CI — D-00344

Diethylamino phosphoryl dichloride. Diethylaminophosphonic dichloride

[1498-54-0]

$Et_2NP(O)Cl_2$

$C_4H_{10}Cl_2NOP$ M 190.009
Liq. Bp 220°, Bp_{15} 110°.

Michaelis, A., *Justus Liebigs Ann. Chem.*, 1903, **326**, 129 (*synth*)
Cowley, A.H. *et al*, *J. Am. Chem. Soc.*, 1965, **87**, 4454 (*pmr*)
Coustures, Y. *et al*, *Bull. Soc. Chim. Fr.*, 1973, 926 (*P nmr, ir*)
Yvernault, T. *et al*, *C.R. Hebd. Seances Acad. Sci., Ser. C*, 1978, **287**, 519 (*synth*)

Kajiwara, M. *et al*, *J. Inorg. Nucl. Chem.*, 1980, **42**, 171 (*synth*)

Diethylphosphoramidic difluoride, 9CI — D-00345

[359-94-4]

$Et_2NP(O)F_2$

$C_4H_{10}F_2NOP$ M 157.100
Liq. with acrid odour. d_4^{20} 1.15. Bp 114°, Bp_{17} 55-56°. n_D^{20} 1.3730.

Olah, G.A. *et al*, *Justus Liebigs Ann. Chem.*, 1959, **625**, 88 (*synth*)
Reddy, G.S. *et al*, *Z. Naturforsch. B.*, 1970, **25**, 1199 (*synth, P nmr*)
Ivanova, Zh.M. *et al*, *Zh. Obshch. Khim.*, 1972, **42**, 2115 (*Engl. transl.* p. 2110) (*synth, ir*)
Coustures, Y. *et al*, *Bull. Soc. Chim. Fr.*, 1973, 926 (*ir, P nmr*)

Diethylphosphoramidochloridic acid, 9CI, 8CI — D-00346

$Et_2NPCl(O)OH$

$C_4H_{11}ClNO_2P$ M 171.563
Derivs. may exist in chiral forms.

(±)-form

Et ester: Ethyl diethylphosphoramidochloridate.
$C_6H_{15}ClNO_2P$ M 199.617
Liq. d_4^{20} 1.10. Bp_{10} 104-105°, Bp_1 60-62°. n_D^{20} 1.4382.
Isopropyl ester: Isopropyl diethylphosphoramidochloridate. 1-Methylethyl diethylphosphoramidochloridate.
$C_7H_{17}ClNO_2P$ M 213.644
Liq. d_4^{20} 1.09. Bp_1 100°. n_D^{20} 1.4276.
Ph ester: Phenyl diethylphosphoramidochloridate.
$C_{10}H_{15}ClNO_2P$ M 247.661
Liq. $Bp_{0.4}$ 118°. n_D^{25} 1.507.
Chloride: see Diethylphosphoramidic dichloride, D-00344

Malatesta, P., *Farmaco, Ed. Sci.*, 1953, **8**, 193; *CA*, **47**, 12077 (*synth*)
Wolff, M.E. *et al*, *J. Am. Chem. Soc.*, 1957, **79**, 1970 (*phenyl ester*)
Abramov, V.S. *et al*, *Zh. Obshch. Khim.*, 1969, **39**, 1003 (*Engl. transl.* p. 974) (*esters*)

Diethylphosphoramidochloridoselenoic acid, 9CI — D-00347

$Et_2N(Cl)P(=Se)OH \rightleftharpoons Et_2N(Cl)P(=O)SeH$

$C_4H_{11}ClNOPSe$ M 234.524

(±)-form

O-Et ester: O-Ethyl diethylphosphoramidochloridoselenoate.
$C_6H_{15}ClNOPSe$ M 262.577
Liq. d_4^{20} 1.31. $Bp_{0.1}$ 70-71°. n_D^{20} 1.5075.
O-Ph ester: O-Phenyl diethylphosphoramidochloridoselenoate.
$C_{10}H_{15}ClNOPSe$ M 310.621
Liq. d_4^{20} 1.37. $Bp_{0.1}$ 123-125°. n_D^{20} 1.5660.

Nuretdinov, I.A. *et al*, *Izv. Akad. Nauk SSSR, Ser. Khim.*, 1969, 1535 (*Engl. transl.* p. 1423) (*esters, synth*)

Shagidullin, R.R. *et al*, *Izv. Akad. Nauk SSSR, Ser. Khim.*, 1976, 184 (*Engl. transl.* p. 174) (*uv*)

Diethylphosphoramidochloridothioic acid, 9CI, 8CI D-00348

S, OH, P, Et_2N, Cl ⇌ HS, O, P, Et_2N, Cl

$C_4H_{11}ClNOPS$ M 187.624

(±)-*form*

O-*Et ester:* O-*Ethyl diethylphosphoramidochloridothioate.*
$C_6H_{15}ClNOPS$ M 215.677
Liq. d_4^{20} 1.14. Bp_{10} 105-106°. n_D^{20} 1.4931.

O-*Ph ester:* O-*Phenyl diethylphosphoramidochloridothioate.*
$C_{10}H_{15}ClNOPS$ M 263.721
Liq. d_4^{20} 1.20. $Bp_{0.25}$ 113-116°. n_D^{20} 1.5523.

Kabachnik, M.I., *Zh. Obshch. Khim.*, 1959, **29**, 2182 (*Engl. transl.* p. 2149) (*ethyl ester*)
Mandel'baum, Ya.A. *et al*, *Zh. Obshch. Khim.*, 1968, **38**, 1754 (*Engl. transl.* p. 1709) (*phenyl ester*)

O,S-Diethyl phosphoramidodithioate, 9CI, 8CI D-00349

O,S-*Diethyl amidodithiophosphate*
[29809-53-8]

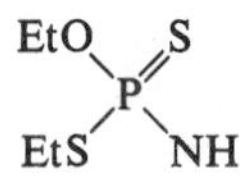

$C_4H_{12}NOPS_2$ M 185.239

(±)-*form*

Oil. $Bp_{0.01}$ 78°. n_D^{24} 1.5530.

U.S.P., 3 787 539, (*1974*); *CA*, **80**, 95248

S,S-Diethyl phosphoramidodithioate, 9CI, 8CI D-00350

S,S-*Diethyl amidodithiophosphate*
[30993-94-3]

$(EtS)_2P(O)NH_2$

$C_4H_{12}NOPS_2$ M 185.239
Possesses insecticidal and acaricidal props. Cryst. Mp 75°. $Bp_{0.03}$ 125°.

Ger. Pat., 2 013 956, (*1971*); *CA*, **74**, 87380

Diethylphosphoramidofluoridic acid, 9CI, 8CI D-00351

$Et_2NPF(O)OH$

$C_4H_{11}FNO_2P$ M 155.109
Derivs. may exist in chiral forms.

(±)-*form*

Et ester: [667-21-0]. *Ethyl diethylphosphoramidofluoridate. Triethyl phosphoramidofluoride.*
$C_6H_{15}FNO_2P$ M 183.162
Liq. d_4^{20} 1.07. Bp_{14} 86-87°. n_D^{20} 1.4045.

Isopropyl ester: Isopropyl diethylphosphoramidofluoridate. 1-Methylethyl diethylphosphoramidofluoridate.
$C_7H_{17}FNO_2P$ M 197.189
Liq. d_4^{20} 1.08. Bp_1 82-83°. n_D^{20} 1.4366.

Fluoride: see Diethylphosphoramidic difluoride, D-00345

Malatesta, P., *Farmaco, Ed. Sci.*, 1953, **8**, 193; *CA*, **47**, 12077 (*synth, derivs*)
Stolzer, C. *et al*, *Chem. Ber.*, 1960, **93**, 1323 (*ethyl ester*)
Gusar', N.I. *et al*, *Zh. Obshch. Khim.*, 1976, **46**, 1981 (*Engl. transl.* p. 1910) (*ethyl ester*)

Diethylphosphoramidofluoridodithioic acid, 9CI, 8CI D-00352

Et_2N, S, P, F, SH

$C_4H_{11}FNPS_2$ M 187.230
Esters are chiral.

Me ester: [20494-69-3]. *Methyl diethylphosphoramidofluoridodithioate. Methyl (dimethylamido)-fluorodithiophosphate.*
$C_5H_{13}FNPS_2$ M 201.257
Liq. $Bp_{0.02}$ 45-47°.

Et ester: [20494-70-6]. *Ethyl diethylphosphoramidofluoridodithioate. Ethyl (dimethylamido)-fluorodithiophosphate.*
$C_6H_{15}FNPS_2$ M 215.283
Liq. $Bp_{0.03}$ 53-56°.

Ph ester: [17620-75-6]. *Phenyl diethylphosphoramidofluoridodithioate. Phenyl (dimethylamido)-fluorodithiophosphate.*
$C_{10}H_{15}FNPS_2$ M 263.327
Liq. $Bp_{0.3}$ 127-128°.

Roesky, H.W., *Chem. Ber.*, 1968, **101**, 636, 2977 (*synth, ir, ms, pmr, F and P nmr*)

Diethylphosphoramidoselenoic acid, 9CI D-00353

Selenophosphoric acid diethylamide

$Et_2NP(Se)(OH)_2 \rightleftharpoons Et_2NP(O)(SeH)(OH)$

$C_4H_{12}NO_2PSe$ M 216.078

O,O-*Di-Me ester:* [24330-37-8]. O,O-*Dimethyl diethylphosphoramidoselenoate.*
$C_6H_{16}NO_2PSe$ M 244.132
Liq. d_4^{20} 1.33. $Bp_{0.02}$ 65.5°. n_D^{20} 1.5000.

O,O-*Di-Et ester:* [24330-38-9]. O,O-*Diethyl diethylphosphoramidoselenoate.*
$C_8H_{20}NO_2PSe$ M 272.185
Liq. d_4^{20} 1.23. $Bp_{0.2}$ 84°. n_D^{20} 1.4837.

O,O-*Diisopropyl ester:* [24330-40-3]. O,O-*Diisopropyl diethylphosphoramidoselenoate.*
$C_{10}H_{24}NO_2PSe$ M 300.239
Liq. d_4^{20} 1.15. $Bp_{0.05}$ 57.5°. n_D^{20} 1.4720.

Difluoride: [53213-43-7].
$C_4H_{10}F_2NO_2PSe$ M 252.059
Liq. Bp_4 44°. Sensitive to moisture and air.

Nikonorova, L.K. *et al*, *Izv. Akad. Nauk SSSR, Ser. Khim.*, 1969, 464 (*Engl. transl.* p. 413) (*synth, ir*)
Roesky, H.W. *et al*, *Z. Naturforsch., B*, 1973, **28**, 697 (*difluoride, synth, ir, ms, pmr, P and F nmr*)
Shagidullin, R.R. *et al*, *Izv. Akad. Nauk SSSR, Ser. Khim.*, 1976, 184 (*Engl. transl.* p. 174) (*diethyl ester, uv*)

O,O-Diethyl phosphoramidothioate, 9CI, 8CI D-00354

O,O-*Diethyl thiophosphoryl amide.* O,O-*Diethyl amidothiophosphate*
[17321-48-1]

$(EtO)_2P(S)NH_2$

$C_4H_{12}NO_2PS$ M 169.178
Liq. d_4^{20} 1.146. Bp_9 112-115°, $Bp_{0.15}$ 67-69°.

Mel'nikov, N.N. *et al*, *Zh. Obshch. Khim.*, 1955, **25**, 828 (*Engl. transl.* p. 793) (*synth*)
Burn, A.J. *et al*, *J. Chem. Soc.*, 1961, 5532 (*synth*)
Nyquist, R.A. *et al*, *Spectrochim. Acta, Part A*, 1967, **23**, 2505 (*ir*)
Quistad, G.B. *et al*, *J. Agric. Food Chem.*, 1970, **18**, 189 (*synth, pharmacol*)

O,S-Diethyl phosphoramidothioate, 9CI, 8CI D-00355
O,S-*Diethyl amidothiophosphate*
[16271-10-6]

EtO(EtS)P(=O)NH2

$C_4H_{12}NO_2PS$ M 169.178
(±)-*form*
Cryst. (C_6H_6/pet. ether). Mp 51°.

Burn, A.J. *et al*, *J. Chem. Soc.*, 1961, 5532 (*synth*)
Quistad, G.B. *et al*, *J. Agric. Food Chem.*, 1970, **18**, 189 (*synth, pharmacol*)

Diethylphosphoramidothioic acid, 9CI, 8CI D-00356
Thiophosphoric diethylamide

$Et_2NP(S)(OH)_2 \rightleftharpoons Et_2NP(O)(OH)(SH)$

$C_4H_{12}NO_2PS$ M 169.178
O,O-*Di-Me ester:* [31599-85-6]. O,O-*Dimethyl diethylphosphoramidothioate.*
$C_6H_{16}NO_2PS$ M 197.232
Liq. d_4^{20} 1.08. Bp_1 62-63°. n_D^{20} 1.4785.
O,O-*Di-Et ester:* [5851-18-3]. O,O-*Diethyl diethylphosphoramidothioate.*
$C_8H_{20}NO_2PS$ M 225.285
Liq. d_0^{15} 1.01. Bp_{20} 110°, Bp_1 69-70°. n_D^{20} 1.4695.
O,O-*Diisopropyl ester:* [55655-35-1]. O,O-*Diisopropyl diethylphosphoramidothioate.*
$C_{10}H_{24}NO_2PS$ M 253.339
Liq. d_4^{20} 1.02. $Bp_{0.9}$ 100°. n_D^{20} 1.4620.
O,O-*Dibutyl ester:* [18628-85-8]. O,O-*Dibutyl diethylphosphoramidothioate.*
$C_{12}H_{28}NO_2PS$ M 281.392
Liq. d_4^{20} 0.98. $Bp_{0.3}$ 91-93°. n_D^{20} 1.4610.
O,O-*Di-Ph ester:* [31860-18-0]. O,O-*Diphenyl diethylphosphoramidothioate.*
$C_{16}H_{20}NO_2PS$ M 321.373
Solid. Mp 114-114.5°.
Difluoride: see Diethylphosphoramidothioic difluoride, D-00358
Dichloride: see Diethylphosphoramidothioic dichloride, D-00357
Diamide: N,N-*Diethylphosphorothioic triamide.*
$C_4H_{14}N_3PS$ M 167.208
Cryst. (CCl_4 or $CHCl_3$). Mp 66-67°.

Michaelis, A., *Justus Liebigs Ann. Chem.*, 1903, **326**, 129, 201 (*esters*)
Mel'nikov, N.N. *et al*, *Zh. Obshch. Khim.*, 1955, **25**, 828 (*Engl. transl.* p. 793) (*esters*)
Mel'nikov, N.N. *et al*, *Zh. Obshch. Khim.*, 1961, **31**, 3605 (*Engl. transl.* p. 3361) (*diphenyl ester, synth, ir*)
Barabanov, V.I. *et al*, *Zh. Obshch. Khim.*, 1970, **40**, 2464 (*Engl. transl.* p. 2451) (*dimethyl ester*)
Gloe, K. *et al*, *J. Prakt. Chem.*, 1975, **317**, 529 (*diamide*)

Diethylphosphoramidothioic dichloride, 9CI, 8CI D-00357
(*Diethylamino*)*phosphonothioic dichloride*
[2523-98-0]

$Et_2NP(S)Cl_2$

$C_4H_{10}Cl_2NPS$ M 206.069
Liq. d_4^{20} 1.27. Bp_{14} 107°, Bp_1 63°. n_D^{20} 1.5262.

Godovikov, N.N. *et al*, *Zh. Obshch. Khim.*, 1961, **31**, 1628 (*Engl. transl.* p. 1516) (*synth*)
Cowley, A.H. *et al*, *J. Am. Chem. Soc.*, 1965, **87**, 4454 (*pmr*)

Diethylphosphoramidothioic difluoride, 9CI, 8CI D-00358
(*Diethylamino*)*phosphonothioic difluoride*
[359-95-5]

$Et_2NP(S)F_2$

$C_4H_{10}F_2NPS$ M 173.160
Liq. Bp_{15} 48°. n_D^{20} 1.4260 (1.4361).

Olah, G.A. *et al*, *Justus Liebigs Ann. Chem.*, 1959, **625**, 88 (*synth*)
Horn, H.-G. *et al*, *Z. Naturforsch, B*, 1966, **21**, 617, 729 (*pmr, F and P nmr, ir*)
Reddy, G.S. *et al*, *Z. Naturforsch, B*, 1970, **25**, 1199 (*synth, P nmr*)

Diethylphosphoramidotrithioic acid, 9CI, 8CI D-00359
Trithiophosphoric acid diethylamide

$Et_2NP(S)(SH)_2$

$C_4H_{12}NPS_3$ M 201.299
Di-Ph ester: Diphenyl diethylphosphoramidotrithioate.
$C_{16}H_{20}NPS_3$ M 353.494
Liq. d_4^{20} 1.19. $Bp_{0.005}$ 190-205° (bath). n_D^{20} 1.6405.

Petrov, K.A. *et al*, *Zh. Obshch. Khim.*, 1962, **32**, 3019 (*Engl. transl.* p. 3070)

Diethylphosphoramidous acid, 9CI D-00360

$Et_2NP(OH)_2$

$C_4H_{12}NO_2P$ M 137.118
Di-Et ester: [20262-87-7]. *Diethyl diethylphosphoramidite.*
$C_8H_{20}NO_2P$ M 193.225
Liq. Bp_{10} 59-60°. n_D^{20} 1.4305.
Di-Ph ester: [14747-96-7]. *Diphenyl diethylphosphoramidite.*
$C_{16}H_{20}NO_2P$ M 289.313
Liq. $Bp_{0.035}$ 107-108°. n_D^{20} 1.5530.
Difluoride: [363-84-8].
$C_4H_{10}F_2NP$ M 141.100
Ligand for Ni, Mo, Fe, Rh and W. Bp_{100} 47°. n_D^{20} 1.3840.
Dichloride: see Diethylphosphoramidous dichloride, D-00361
Dibromide: [4154-80-7].
$C_4H_{10}Br_2NP$ M 262.911

Liq. Bp_{15} 107°. n_D^{20} 1.5720.
Diiodide: [54305-75-8].
$C_4H_{10}I_2NP$ M 356.912
Dark-red liq. Mp −60° to −55°. $Bp_{0.1}$ 83-85°. Stable at r.t. in absence of moisture, light and O_2.

Schmutzler, R., *Inorg. Chem.*, 1964, **3**, 415 (*difluoride, synth, ir, F nmr, complexes*)
Barlow, C.G. *et al, J. Chem. Soc. (A)*, 1966, 228 (*difluoride, P nmr, pmr*)
Gorbatenko, Zh.K. *et al, Zh. Obshch. Khim.*, 1974, **44**, 2375 (*Engl. transl.* p. 2311) (*diiodide, diphenyl ester*)
Pudovik, A.N. *et al, Zh. Obshch. Khim.*, 1975, **45**, 2621 (*Engl. transl.* p. 2582) (*diethyl ester*)
Muratova, A.A. *et al, Zh. Obsch. Khim.*, 1976, **46**, 1729 (*Engl. transl.* p. 1680) (*diethyl ester, ir*)
Romanenko, V.D. *et al, Synthesis*, 1980, 823 (*diiodide, synth, P nmr*)
Romanenko, V.D. *et al, Zh. Obshch. Khim.*, 1980, **50**, 1660; *CA*, **93**, 220856 (*dibromide*)
Chojnowski, J. *et al, J. Organomet. Chem.*, 1981, **215**, 355 (*diphenyl ester*)

Diethylphosphoramidous dichloride, 9CI D-00361

(Diethylamino)dichlorophosphine. Diethylaminophosphorous dichloride
[1069-08-5]

Et_2NPCl_2

$C_4H_{10}Cl_2NP$ M 174.009
Air-sensitive liq. d_0^{15} 1.096. Bp 189°, Bp_{14} 73-75°. n_D^{20} 1.4679.

Michaelis, A., *Justus Liebigs Ann. Chem.*, 1903, **326**, 129 (*synth*)
v. Wazer, J.R. *et al, J. Am. Chem. Soc.*, 1964, **86**, 811 (*P nmr*)
Biryukov, I.P. *et al, Zh. Obshch. Khim.*, 1972, **42**, 1223 (*Engl. transl.* p. 1217) (*nqr*)
Barlos, K. *et al, Z. Naturforsch., B*, 1978, **33**, 515 (*N nmr*)
Van Lindhoudt, J.P. *et al, Spectrochim. Acta, Part A*, 1980, **36**, 315 (*pmr, P nmr, cmr*)

Diethyl phosphorazidate, 9CI, 8CI D-00362

Diethyl phosphoryl azide
[1516-68-3]

$(EtO)_2P(O)N_3$

$C_4H_{10}N_3O_3P$ M 179.115
Reagent for peptide synth and modified Curtius reactions. Liq. d_4^{20} 1.17. Bp_{11} 101-103°, Bp_3 69-70°. n_D^{25} 1.4251.
▷Virulent and powerful anticholinesterase

Kabachnik, M.I. *et al, Izv. Akad. Nauk SSSR, Ser. Khim.*, 1961, 819 (*Engl. transl.* p. 758) (*synth*)
Scott, F.L. *et al, J. Org. Chem.*, 1962, **27**, 4255 (*synth, ir*)
Ninomiya, K. *et al, Tetrahedron*, 1974, **30**, 2151 (*use*)
Shiori, T. *et al, Chem. Pharm. Bull.*, 1974, **22**, 855 (*use*)
Treppendahl, S. *et al, Acta Chem. Scand., Ser. B*, 1974, **28**, 657 (*ms*)
Sisto, A. *et al, Synthesis*, 1985, 294 (*use*)

O,O-Diethyl phosphorazidothioate, 9CI, 8CI D-00363

O,O-*Diethyl thiophosphoryl azide.* O,O-*Diethyl azidothiophosphate*
[1516-67-2]

$(EtO)_2P(S)N_3$

$C_4H_{10}N_3O_2PS$ M 195.176
Liq. d_4^{20} 1.66. Bp_1 57-58°. n_D^{20} 1.4720. Thermally stable.
▷TB6180000.

Gilyarov, V.A. *et al, Zh. Obshch. Khim.*, 1966, **36**, 274 (*Engl. transl.* p. 285) (*synth*)
Khodak, A.A. *et al, Zh. Obshch. Khim.*, 1974, **44**, 27, 256 (*Engl. transl.* pp. 24, 241) (*synth*)

N,N-Diethylphosphoric triamide, 9CI D-00364

[25316-38-5]

$Et_2NP(O)(NH_2)_2$

$C_4H_{14}N_3OP$ M 151.148
Fireproofing agent. Cryst. (CCl_4). Mp 81°.
N′,N″-*Di-Ph:* N,N-*Diethyl-*N′,N″*-diphenylphosphoric triamide. Diethylphosphoramidic dianilide.*
$C_{16}H_{22}N_3OP$ M 303.343
Fine needles (EtOH). Mp 150°.

Michaelis, A., *Justus Liebigs Ann. Chem.*, 1903, **326**, 129, 172 (*deriv*)
Goehring, M. *et al, Chem. Ber.*, 1956, **89**, 1768 (*synth*)
Kojiwara, M. *et al, J. Inorg. Nucl. Chem.*, 1980, **42**, 171 (*props*)

Diethyl phosphoriodidite, 9CI, 8CI D-00365

Diethyl iodophosphite
[20502-50-5]

$(EtO)_2PI$

$C_4H_{10}IO_2P$ M 248.000
Liq. Bp_1 65°.

Kabachnik, M.M. *et al, Zh. Obshch. Khim.*, 1976, **46**, 433 (*Engl. transl.* p. 428) (*synth, P nmr*)

Diethyl phosphorisocyanatidate, 9CI, 8CI D-00366

Diethyl phosphoryl isocyanate
[20039-33-2]

$(EtO)_2P(O)NCO$

$C_5H_{10}NO_4P$ M 179.112
Liq. d_4^{20} 1.18. Bp_{14} 91-93°, $Bp_{0.035}$ 40-43°. n_D^{20} 1.4175 V. reactive to water and other protic compounds.

Samarai, L.I. *et al, Zh. Obshch. Khim.*, 1966, **36**, 1433; 1969, **39**, 1511, 1712 (*Engl. transl.* pp. 1084, 1480, 1678) (*synth*)
Treppendahl, S. *et al, Acta Chem. Scand., Ser. B*, 1974, **28**, 657 (*synth, ms*)
Shokol, V.A. *et al, Zh. Obshch. Khim.*, 1977, **47**, 2202 (*Engl. transl.* p. 2010) (*synth*)
Zwierzak, A. *et al, Synthesis*, 1982, 922 (*synth, P nmr*)

O,O-Diethyl phosphorisocyanatidothioate, 9CI, 8CI D-00367

O,O-*Diethyl thiophosphoryl isocyanate*
[13620-62-7]

$(EtO)_2P(S)NCO$

$C_5H_{10}NO_3PS$ M 195.173
Reactive liq. d_4^{20} 1.16. Bp_{25} 87-89°. n_D^{20} 1.4582. Sensitive to moisture.

Samarai, L.I. *et al, Zh. Obshch. Khim.*, 1966, **36**, 1433 (*Engl. transl.* p. 1439) (*synth*)
Shokol, V.A. *et al, Zh. Obshch. Khim.*, 1969, **39**, 2137 (*Engl. transl.* p. 2088) (*synth*)

_O,S_-Diethyl phosphorisocyanatidothioate, 9CI, 8CI D-00368

O,S-*Diethyl phosphoryl isocyanate*

[25359-61-9]

EtO(EtS)P(O)NCO

$C_5H_{10}NO_3PS$ M 195.173

(±)-*form*

Reactive liq. d_4^{20} 1.22. Bp_2 73-74°. n_D^{20} 1.4787. Sensitive to moisture.

Samarai, L.I. *et al*, *Zh. Obshch. Khim.*, 1969, **39**, 1511 (*Engl. transl.* p. 1480)

Diethyl phosphor(isothiocyanatidate), 9CI, 8CI D-00369

Diethyl phosphoryl isothiocyanate

[6374-26-1]

$(EtO)_2P(O)NCS$

$C_5H_{10}NO_3PS$ M 195.173

Reactive liq. $Bp_{0.08}$ 84°, $Bp_{0.12}$ 58°. n_D^{20} 1.4795.

Treppendahl, S. *et al*, *Acta Chem. Scand., Ser. B*, 1974, **28**, 657 (*synth, ms*)
Lopusinski, A. *et al*, *Justus Liebigs Ann. Chem.*, 1977, 924 (*synth*)
Pudovik, A.N. *et al*, *Zh. Obshch. Khim.*, 1979, **49**, 1425 (*Engl. transl.* p. 1248) (*synth, ir, P nmr*)
Lopusinski, A. *et al*, *Tetrahedron*, 1982, **38**, 679 (*synth, P nmr*)

_O,O_-Diethyl phosphor(isothiocyanatido)-thioate, 9CI, 8CI D-00370

O,O-*Diethyl thiophosphoryl isothiocyanate*

[26190-34-1]

$(EtO)_2P(S)NCS$

$C_5H_{10}NO_2PS_2$ M 211.233

Liq. $Bp_{3.5}$ 90-92°, $Bp_{0.03}$ 53-54°. n_D^{20} 1.5105. Sensitive to moisture.

▷TE6479000.

Kosinskaya, I.M. *et al*, *Zh. Obshch. Khim.*, 1976, **46**, 2227 (*Engl. transl.* p. 2147) (*synth*)
Zimin, M.G. *et al*, *Zh. Obshch. Khim.*, 1982, **52**, 482 (*Engl. transl.* p. 423) (*P nmr*)
Burski, J. *et al*, *Tetrahedron*, 1983, **39**, 4175 (*synth*)

Diethyl phosphorobromidate, 9CI, 8CI D-00371

Diethyl phosphoryl bromide. Diethyl bromophosphate. Diethyl bromophosphonate

[51761-27-4]

$(EtO)_2P(O)Br$

$C_4H_{10}BrO_3P$ M 216.999

Reagent for use in the synth. of amides and peptides in high yield with only slight racemisation. Pungent liq. $Bp_{1.5}$ 75°. n_D^{20} 1.4406.

Goldwhite, H. *et al*, *J. Chem. Soc.*, 1955, 3564 (*synth*)
Górecka, A. *et al*, *Synthesis*, 1978, 474 (*synth, use, ir, pmr, nmr*)
Michalski, J. *et al*, *J. Chem. Soc., Perkin Trans. 1*, 1980, 833 (*P nmr*)

Diethyl phosphorobromidite, 9CI, 8CI D-00372

Diethyl bromophosphite

[20502-48-1]

$(EtO)_2PBr$

$C_4H_{10}BrO_2P$ M 201.000

Fuming liq. Bp_{14} 55-59°. n_D^{20} 1.4758. Easily hydrol.

Novikova, Z.S. *et al*, *Zh. Obshch. Khim.*, 1974, **44**, 1857 (*Engl. transl.* p. 1825) (*synth, P nmr*)
Malenko, D.M. *et al*, *Zh. Obshch. Khim.*, 1979, **49**, 308 (*Engl. transl.* p. 267) (*synth*)

Diethyl phosphorochloridate, 9CI, 8CI D-00373

Diethyl phosphoryl chloride. Diethyl chlorophosphate. Diethyl chlorophosphonate

[814-49-3]

$(EtO)_2P(O)Cl$

$C_4H_{10}ClO_3P$ M 172.548

Phosphorylating agent. Assists reduction of phenols and enols. Reagent for conversion of phenols to amines. Liq. with irritating, unpleasant odour. d_4^{20} 1.21. Bp_{15} 88-89°, $Bp_{0.4}$ 39°. Fumes in moist air.

▷Highly toxic. TD1400000.

Inorg. Synth., 1953, **4**, 78 (*synth, bibl*)
Bliznyaks, N.K. *et al*, *Zh. Obshch. Khim.*, 1967, **37**, 1353 (*Engl. transl.* p. 1279) (*synth*)
Williamson, M.P. *et al*, *J. Phys. Chem.*, 1968, **72**, 4043 (*pmr, bibl*)
Ireland, R. *et al*, *Tetrahedron Lett.*, 1969, 2145 (*use*)
Potenza, J.A. *et al*, *J. Am. Chem. Soc.*, 1969, **91**, 4356 (*epr, nmr*)
Sosnovsky, G. *et al*, *J. Org. Chem.*, 1969, **34**, 968 (*synth*)
Pritchard, J.G., *Org. Mass Spectrom.*, 1970, **3**, 163 (*ms*)
Zverev, V.V. *et al*, *Izv. Akad. Nauk SSSR, Ser. Khim.*, 1979, **28**, 84 (*Engl. transl.* p. 74) (*pe*)
Chattopadhyay, S. *et al*, *J. Electron. Spectrosc. Relat. Phenom.*, 1981, **24**, 27 (*pe*)
Gloede, J., *Z. Anorg. Allg. Chem.*, 1982, **484**, 231 (*synth*)
Araki, S. *et al*, *J. Chem. Soc., Perkin Trans. 1*, 1984, 969 (*use*)
Sax, N.I., *Dangerous Properties of Industrial Materials*, 6th Ed., Van Nostrand-Reinhold, 1984, 574.

Diethyl phosphorochloridite, 9CI, 8CI D-00374

Diethyl chlorophosphite

[589-57-1]

$(EtO)_2PCl$

$C_4H_{10}ClO_2P$ M 156.549

Rgt. for peptide synthesis. Fuming pungent liq. d_0^{20} 1.075. Bp 153-155°, Bp_{30} 63-65°, Bp_{12} 45-50°. n_D^{20} 1.4360.

▷Reacts explosively with H_2O

Young, R.W. *et al*, *J. Am. Chem. Soc.*, 1956, **78**, 2126 (*synth, use*)
Fluck, E. *et al*, *Z. Anorg. Allg. Chem.*, 1961, **307**, 113 (*P nmr*)
Michalski, J. *et al*, *J. Chem. Soc.*, 1961, 4904 (*synth*)
Jones, R.A.Y. *et al*, *Angew. Chem., Int. Ed. Engl.*, 1962, **1**, 32 (*nmr*)
Gloede, J. *et al*, *J. Prakt. Chem.*, 1974, **316**, 703 (*synth, nmr*)
Malenko, D.M. *et al*, *Zh. Obshch. Khim.*, 1979, **49**, 308 (*Engl. transl.* p. 267) (*synth, P nmr*)
Sax, N.I., *Dangerous Properties of Industrial Materials*, 6th Ed., Van Nostrand-Reinhold, 1984, 672.

_S,S_-Diethyl phosphorochloridodithioate, 9CI, 8CI **D-00375**

S,S-*Diethyl chlorodithiophosphate*

[28522-97-6]

$(EtS)_2P(O)Cl$

$C_4H_{10}ClOPS_2$ M 204.669

Liq. d_4^{20} 1.40. Bp_{22} 145-150°, Bp_6 129-131.5°. n_D^{20} 1.5193.

▷Mild vesicant

Mastin, T.W. *et al*, *J. Am. Chem. Soc.*, 1945, **67**, 1662 (*synth*)
Petrov, K.A. *et al*, *Zh. Obshch. Khim.*, 1961, **31**, 1366 (*Engl. transl.* p. 1265)

Diethyl phosphorochloridodithioite, 9CI **D-00376**

Diethyl dithiochlorophosphite

[1486-42-6]

$(EtS)_2PCl$

$C_4H_{10}ClPS_2$ M 188.670

Liq. Bp_5 84-85°, $Bp_{0.02}$ 48-50°. n_D^{20} 1.5793.

Sinyashin, O.G. *et al*, *Zh. Obshch. Khim.*, 1981, **51**, 2410; 1983, **53**, 502, 1706 (*Engl. transl.* pp. 2078, 436, 1535) (*synth, P nmr*)
Kostin, V.P. *et al*, *Zh. Obshch. Khim.*, 1982, **52**, 498 (*Engl. transl.* p. 437) (*synth*)

_O,O_-Diethyl phosphorochloridoselenoate, 9CI, 8CI **D-00377**

O,O-*Diethyl selenophosphoryl chloride*. O,O-*Diethyl chloroselenophosphate*

[13030-37-0]

$(EtO)_2P(Se)Cl$

$C_4H_{10}ClO_2PSe$ M 235.509

Liq. d_4^{25} 1.43-1.44. Bp_{10} 90-91.5°, $Bp_{0.8}$ 52-53°. n_D^{20} 1.4940. Moisture sensitive.

Krawiecki, C. *et al*, *Rocz. Chem.*, 1969, **43**, 869 (*synth*)
Nuretdinov, I.A. *et al*, *Zh. Obshch. Khim.*, 1975, **45**, 533 (*Engl. transl.*, p. 526) (*synth, ir, pmr, P nmr*)

_O,O_-Diethyl phosphorochloridothioate, 9CI, 8CI **D-00378**

Diethyl thiophosphoryl chloride. O,O-*Diethyl chlorothiophosphate*

[2524-04-1]

$(EtO)_2P(S)Cl$

$C_4H_{10}ClO_2PS$ M 188.609

Liq. d_4^{20} 1.202. Bp_{20} 94-96°, Bp_2 60°. n_D^{20} 1.4685.

▷TD1780000.

Fletcher, J.H. *et al*, *J. Am. Chem. Soc.*, 1950, **72**, 2461 (*synth*)
McIvor, R.A. *et al*, *Can. J. Chem.*, 1956, **34**, 1611 (*ir*)
Popov, E.M. *et al*, *Zh. Obshch. Khim.*, 1959, **29**, 1998 (*Engl. transl.* p. 1967) (*raman*)
Williamson, M.P. *et al*, *J. Phys. Chem.*, 1968, **72**, 4043 (*pmr*)
Nyquist, R.A., *Spectrochim. Acta, Part A*, 1969, **25**, 47 (*ir*)
Potenza, J.A. *et al*, *J. Am. Chem. Soc.*, 1969, **91**, 4356 (*P nmr, epr*)
Pritchard, J.G., *Org. Mass Spectrom.*, 1970, **3**, 163 (*ms*)
Whitehead, M.A. *et al*, *J. Chem. Soc. (A)*, 1971, 1738 (*nqr*)
Bentrude, W.G. *et al*, *J. Am. Chem. Soc.*, 1973, **95**, 2286 (*synth*)
Omelanczuk, J. *et al*, *Tetrahedron*, 1975, **31**, 2809 (*synth, P nmr*)

_O,S_-Diethyl phosphorochloridothioate, 9CI, 8CI **D-00379**

O,S-*Diethyl chlorothiophosphate*

[3711-51-1]

EtO O
P
EtS Cl

$C_4H_{10}ClO_2PS$ M 188.609

(±)-*form*

Liq. d_4^{20} 1.22. Bp_5 95-96°. n_D^{20} 1.4844.

Petrov, K.A. *et al*, *Zh. Obshch. Khim.*, 1961, **31**, 1366 (*Engl. transl.* p. 1265) (*synth*)
Lippman, A.E., *J. Org. Chem.*, 1965, **30**, 3217 (*synth, pmr, P nmr*)

Diethyl phosphorochloridotrithioate, 9CI, 8CI **D-00380**

Diethyl chlorotrithiophosphate

[16001-01-7]

$(EtS)_2P(S)Cl$

$C_4H_{10}ClPS_3$ M 220.730

Liq. Bp_2 110-113°.

▷Vesicant

Mastin, T.W. *et al*, *J. Am. Chem. Soc.*, 1945, **67**, 1662 (*synth*)

Diethyl phosphorocyanidate, 9CI, 8CI **D-00381**

Diethyl phosphoryl cyanide. Diethyl cyanophosphonate

$(EtO)_2P(O)CN$

$C_5H_{10}NO_3P$ M 163.113

Reagent for peptide synth. in liq. or solid phase with little racemization. With thiols, carboxylic acids → thioesters; alcohols → esters, and amines → amides. With *tert*-butyl hydroperoxide, alcohols → *tert*-butylperoxy esters. Reagent for *C*-acylation of active methylene compounds. Aldehydes and ketones yield diethyl 1-cyanoalkyl phosphates and thence on hydrol., 1-hydroxyalkyl carboxylic acids. Converts sodium sulfinates to thiocyanates. Liq. d_4^{20} 1.088. Bp_{14} 95-97°. n_D^{20} 1.4047.

Saunders, B.C. *et al*, *J. Chem. Soc.*, 1948, 699 (*synth, tox*)
Shiori, T. *et al*, *Tetrahedron*, 1976, **32**, 2211 (*synth, ir, pmr, use*)
Hamada, Y. *et al*, *Chem. Pharm. Bull.*, 1977, **25**, 224; 1984, **32**, 3683 (*use*)
Yokoyama, Y. *et al*, *Chem. Pharm. Bull.*, 1977, **25**, 2423 (*use*)
Shiori, T. *et al*, *J. Org. Chem.*, 1978, **43**, 3631 (*use*)
Patsanovskii, I.I. *et al*, *Dokl. Akad. Nauk SSSR, Ser. Sci. Khim.*, 1980, **255**, 383 (*synth, P nmr*)
Harusawa, S. *et al*, *Tetrahedron Lett.*, 1982, **23**, 447 (*use*)
Takuma, S. *et al*, *Chem. Pharm. Bull.*, 1982, **30**, 3147 (*use*)
Harusawa, S. *et al*, *Chem. Pharm. Bull.*, 1983, **31**, 2932 (*use*)
Kato, N. *et al*, *Chem. Pharm. Bull.*, 1984, **32**, 1679 (*use*)

_N,N'_-Diethylphosphorodiamidic acid **D-00382**

$(EtNH)_2P(O)OH$

$C_4H_{13}N_2O_2P$ M 152.133

Me ester: Methyl N,N'-*diethylphosphorodiamidate*.
$C_5H_{15}N_2O_2P$ M 166.159

No phys. props. reported.
Et ester: [27933-13-7]. *Ethyl* N,N′-*diethylphosphorodiamidate.*
$C_6H_{17}N_2O_2P$ M 180.186
Fireproofing agent. Solid. Mp 46.5°. Bp_{15} 155-157°.
Ph ester: [7450-70-6]. *Phenyl* N,N′-*diethylphosphorodiamidate.*
$C_{10}H_{17}N_2O_2P$ M 228.230
Solid or liq. d_4^{20} 1.11. Mp 38-39°. $Bp_{0.7}$ 179°. n_D^{20} 1.5152.
Fluoride:
$C_4H_{12}FN_2OP$ M 154.124
No phys. props. reported.
▷Neurotoxic
Chloride:
$C_4H_{12}ClN_2OP$ M 170.578
Solid. Mp 74°.

Michaelis, A., *Justus Liebigs Ann. Chem.*, 1915, **407**, 290 (*chloride*)
Petrov, K.A. *et al*, *Zh. Obshch. Khim.*, 1960, **30**, 4060 (*Engl. transl.* p. 4023) (*phenyl ester*)
Luchkovskaya, O.N. *et al*, *Zh. Obshch. Khim.*, 1970, **40**, 664 (*Engl. transl.* p. 615) (*ethyl ester*)
Roth, H.J. *et al*, *Arch. Pharm.* (*Weinheim, Ger.*), 1981, **314**, 85; 1982, **315**, 581 (*esters, synth, pmr*)
Mager, P.P., *Toxicol. Lett.*, 1982, **11**, 67 (*fluoride*)

O,O-Diethyl phosphorodiselenoate, 9CI, 8CI D-00383

O,O-*Diethyl hydrogen phosphorodiselenoate.* O,O-*Diethyl phosphorodiselenoic acid.* O,O-*Diethyl diselenophosphate*
[39595-20-5]

$(EtO)_2P(Se)SeH$

$C_4H_{11}O_2PSe_2$ M 280.023
Unstable acid; isol. as K salt.
K salt: [19483-45-5]. Cryst. (EtOH/pet. ether). Sol. Me_2CO, MeOH, EtOH, insol. Et_2O, pet. ether. Mp 136-137°.

Kudchadker, M.V. *et al*, *Can. J. Chem.*, 1968, **46**, 1415 (*synth, ir, pmr*)
Zemlyanskii, N.I. *et al*, *Zh. Obshch. Khim.*, 1971, **41**, 1691 (*Engl. transl.* p. 1699) (*synth*)
Gorak, R.D. *et al*, *Zh. Obshch. Khim.*, 1972, **42**, 56 (*Engl. transl.* p. 52) (*synth*)
Stec, W.J. *et al*, *Phosphorus*, 1972, **2**, 57 (*P nmr*)

O,O-Diethyl phosphorodithioate, 9CI, 8CI D-00384

O,O-*Diethyl dithiophosphate.* O,O-*Diethyl dithiophosphoric acid*
[298-06-6]

$(EtO)_2P(S)SH$

$C_4H_{11}O_2PS_2$ M 186.223
Liq. d_4^{20} 1.17. Bp_9 92-94°, Bp_1 60°. pK_a 2.56 (80% EtOH aq.), pK_a 2.84 (EtOH), pK_a 1.62 (7% EtOH aq.). n_D^{20} 1.5120, 1.5070.
▷TD7350000.
NH_4 salt: [1068-22-0]. Employed for determination of Ni in presence of other metals. Cryst. Mp 164-166°.
▷BP8145000.
K salt: [3454-66-8]. Cryst. (EtOH/Et_2O). Mp 152-153°, Mp 195°.
Na salt: [3338-24-7]. *Hostaflot.* Used in flotation of U and Cd ores.
▷TD8550000.
Diethylammonium salt: [39857-88-0]. Cryst. (C_6H_6/pet. ether). Mp 83-85°.
Morpholinium salt: Solid. Mp 79-81°.

Kabachnik, M.I. *et al*, *Tetrahedron*, 1960, **9**, 10 (*synth*)
Zemlyanskii, N.I. *et al*, *Zh. Obshch. Khim.*, 1960, **30**, 4056 (*Engl. transl.* p. 4018) (*raman*)
Nyquist, R.A., *Spectrochim. Acta, Part A*, 1969, **25**, 47 (*ir*)
Jowitt, R.N. *et al*, *J. Chem. Soc.* (*A*), 1970, 1702 (*ir, pmr, complexes*)
Zemlyanskii, N.I., *Zh. Obshch. Khim.*, 1972, **42**, 54 (*Engl. transl.* p. 50) (*synth*)
Khaskin, B.A. *et al*, *Zh. Obshch. Khim.*, 1973, **43**, 1916; 1974, **44**, 95 (*Engl. transl.* pp. 1901, 93) (*salts*)
Olah, G.A. *et al*, *J. Org. Chem.*, 1975, **40**, 2582 (*props, P nmr*)
Lefferts, J.L. *et al*, *Inorg. Chem.*, 1980, **19**, 1662 (*synth, complexes*)
Chattopadhyay, S. *et al*, *J. Electron. Spectrosc. Relat. Phenom.*, 1981, **24**, 27 (*pe*)

O,S-Diethyl phosphorodithioate, 9CI, 8CI D-00385

O,S-*Diethyl hydrogen phosphorodithioate.* O,S-*Diethyl dithiophosphoric acid*

EtO(EtS)P(=O)SH ⇌ EtO(EtS)P(=S)OH

$C_4H_{11}O_2PS_2$ M 186.223
Na salt: [35376-07-9]. Amorph. solid.

Pishchimuka, P., *J. Russ. Phys. Chem. Soc.*, 1912, **44**, 1406; *CA*, **7**, 987.

S,S-Diethyl phosphorodithioate, 9CI, 8CI D-00386

S,S-*Diethyl dithiophosphate.* S,S-*Diethyl dithiophosphoric acid.* S,S-*Diethyl hydrogen dithiophosphate*
[67941-86-0]

$(EtS)_2P(O)OH$

$C_4H_{11}O_2PS_2$ M 186.223
Cyclohexylammonium salt: [72284-31-2]. Solid. Mp 133-135°.
Fluoride: S,S-*Diethyl phosphorofluoridodithioate.* S,S-*Diethyl fluorodithiophosphate.*
$C_4H_{10}FOPS_2$ M 188.215
Liq. Bp_{15} 104-105°.
Chloride: *see S,S-Diethyl phosphorochloridodithioate, D-00375*

Chapman, N.B. *et al*, *J. Chem. Soc.*, 1948, 1010 (*fluoride*)
Yamaguchi, K. *et al*, *Chem. Lett.*, 1979, 1057 (*synth*)

Diethyl phosphorofluoridate, 9CI, 8CI D-00387

Diethyl phosphoryl fluoride. Diethyl fluorophosphate. Diethoxyphosphinyl fluoride
[358-74-7]

$(EtO)_2P(O)F$

$C_4H_{10}FO_3P$ M 156.093
Liq. d_4^{27} 1.14. Bp_{32} 80-80.5°. n_D^{27} 1.3710.
▷Neurotoxic. TE5600000.

Olah, G. *et al*, *J. Org. Chem.*, 1959, **24**, 1568 (*synth*)
Berry, J.P. *et al*, *J. Org. Chem.*, 1968, **33**, 1664 (*synth*)
Landau, M.A. *et al*, *Zh. Strukt. Khim.*, 1970, **11**, 513 (*Engl. transl.* p. 467) (*struct*)

Nikitin, E.V. *et al*, *Dokl. Akad. Nauk SSSR, Ser. Sci. Khim.*, 1980, **252**, 922 (*Engl. transl.* p. 442) (*synth, F and P nmr*)
Gubaidullin, M.G., *Zh. Obshch. Khim.*, 1982, **52**, 2469 (*Engl. transl.* p. 2182) (*props*)
Haas, A. *et al*, *Z. Anorg. Allg. Chem.*, 1983, **501**, 79 (*synth, pmr, F and P nmr*)
Gupta, O.D. *et al*, *J. Chem. Soc., Chem. Commun.*, 1984, 416 (*synth*)
Inorg. Synth., 1986, **24**, 62 (*synth*)
Sax, N.I., *Dangerous Properties of Industrial Materials*, 6th Ed., Van Nostrand-Reinhold, 1984, 577.

O,S-Diethyl phosphorofluoridodithioate D-00388

O,S-*Diethyl fluorodithiophosphate*

[20494-66-0]

EtO(EtS)P(=S)F

$C_4H_{10}FOPS_2$ M 188.215

(±)-*form*

Liq. d_4^{20} 1.20. $Bp_{0.15}$ 28-30°. n_D^{20} 1.4197.

Roesky, H., *Chem. Ber.*, 1968, **101**, 2977 (*synth, ir, ms, F nmr*)

O,O-Diethyl phosphorofluoridothioate D-00389

O,O-*Diethyl thiophosphoryl fluoride*. O,O-*Diethyl fluorothiophosphate*

[358-11-7]

$(EtO)_2P(S)F$

$C_4H_{10}FO_2PS$ M 172.154

Liq. Bp_{10} 79-81°.

Olah, G. *et al*, *Justus Liebigs Ann. Chem.*, 1957, **602**, 118 (*synth*)
Ivanova, Zh.M. *et al*, *Zh. Obshch. Khim.*, 1968, **38**, 551 (*Engl. transl.* p. 538) (*synth*)

Diethyl phosphorofluoridotrithioate, 9CI, 8CI D-00390

Diethyl fluorotrithiophosphate

[20494-72-8]

$(EtS)_2P(S)F$

$C_4H_{10}FPS_3$ M 204.275

Liq. d^{20} 1.25. $Bp_{0.07}$ 62-65°. n_D^{20} 1.5612.

Roesky, H., *Chem. Ber.*, 1968, **101**, 2977 (*synth, ir, ms, F nmr*)

Diethyl phosphorohydrazidate, 9CI, 8CI D-00391

Diethyl phosphoric acid hydrazide. (*Diethoxyphosphinyl*)*hydrazine*

[56183-69-8]

$(EtO)_2P(O)NHNH_2$

$C_4H_{13}N_2O_3P$ M 168.132

Used in a synth. of pyrroles. Liq. $Bp_{0.5}$ 100-102°. n_D^{20} 1.4474.

Zverev, V.V. *et al*, *Izv. Akad. Nauk SSSR, Ser. Khim.*, 1975, 1051 (*Engl. transl.* p. 961) (*pe*)
Zwierzak, A. *et al*, *Synthesis*, 1976, 835 (*synth, ir, pmr, P nmr*)
Baldwin, J.E. *et al*, *J. Chem. Soc., Chem. Commun.*, 1982, 624 (*use*)

O,O-Diethyl phosphorohydrazidothioate, 9CI, 8CI D-00392

O,O-*Diethyl thiophosphoryl hydrazide*. O,O-*Diethyl thiophosphoric hydrazide*

[25005-76-9]

$(EtO)_2P(S)NHNH_2$

$C_4H_{13}N_2O_2PS$ M 184.193

Oil. d_4^{20} 1.19. $Bp_{0.12}$ 93.5-94°. n_D^{20} 1.5009.

Tolkmith, H., *J. Am. Chem. Soc.*, 1962, **84**, 2097 (*synth*)
Lomakina, V.I. *et al*, *Zh. Obshch. Khim.*, 1971, **41**, 1204 (*Engl. transl.* p. 1215) (*synth*)

O,O-Diethyl phosphoroselenoate, 9CI, 8CI D-00393

O,O-*Diethyl phosphoroselenoic acid*. O,O-*Diethyl hydrogen phosphoroselenoate*. O,O-*Diethyl hydrogen selenophosphate*

[7452-28-0]

$(EtO)_2P(Se)OH \rightleftharpoons (EtO)_2P(O)SeH$

$C_4H_{11}O_3PSe$ M 217.063

Tautomeric. Isol. as Na or K salt. Deoxygenates oxiranes to give alkenes.

Na salt: [7452-97-2]. Needles. Mp 146°.

K salt: [41118-97-2]. Solid or cryst. (dioxan/C_6H_6/pet. ether). Mp 170° dec.

Cyclohexylammonium salt: Solid. Mp 98-99° (82°).

Dicyclohexylammonium salt: [80138-63-2]. Solid. Mp 157-158°.

*Chloride: see O,O-*Diethyl phosphorochloridoselenoate, D-00377

Stec, W.J. *et al*, *Phosphorus*, 1972, **2**, 97 (*P nmr*)
Markowska, A. *et al*, *Bull. Acad. Pol. Sci., Ser. Sci. Chim.*, 1973, **21**, 537 (*synth, props*)
Glidewell, C. *et al*, *J. Chem. Soc., Dalton Trans.*, 1977, 527 (*P nmr*)
Mastryukova, T.A. *et al*, *Zh. Obshch. Khim.*, 1978, **48**, 1447 (*Engl. transl.* p. 1329) (*props*)
Bruzik, K. *et al*, *Pol. J. Chem.*, 1980, **54**, 141 (*P nmr*)
Kudelska, W. *et al*, *Tetrahedron*, 1981, **37**, 2989 (*synth, props*)

O,O-Diethyl phosphoroselenothioate, 9CI, 8CI D-00394

O,O-*Diethyl phosphoroselenothioic acid*. O,O-*Diethyl hydrogen phosphoroselenothioate*. O,O-*Diethyl selenothiophosphate*

[10565-43-2]

$(EtO)_2P(Se)SH \rightleftharpoons (EtO)_2P(S)SeH$

$C_4H_{11}O_2PSSe$ M 233.123

Exists preferentially in the thiol form. Unstable acid, isol. as K salt.

K salt: [15887-56-6]. Hygroscopic needles (Me_2CO/C_6H_6/pet. ether). Mp 156° dec. Slowly liberates Se.

Cyclohexylammonium salt: [13106-27-9]. Needles (C_6H_6/pet. ether). Mp 98-99°.

Zemlyanskii, N.I. *et al*, *Zh. Obshch. Khim.*, 1966, **36**, 1240 (*Engl. transl.* p. 1255) (*synth*)
Krawiecki, C. *et al*, *Rocz. Chem.*, 1969, **43**, 869 (*synth*)
Stec, W.J. *et al*, *Phosphorus*, 1972, **2**, 97 (*P nmr*)
Markowska, A. *et al*, *Bull. Acad. Pol. Sci., Ser. Sci. Chim.*, 1973, **21**, 537 (*synth*)
Bruzik, K. *et al*, *Pol. J. Chem.*, 1980, **54**, 141 (*synth, P nmr*)

O,O-Diethyl phosphorotelluroate, 9CI **D-00395**

O,O-*Diethyl hydrogen phosphorotelluroate.* O,O-*Diethyl phosphorotelluroic acid.* O,O-*Diethyl tellurophosphate*

$(EtO)_2P(Te)OH \rightleftharpoons (EtO)_2P(O)TeH$

$C_4H_{11}O_3PTe$ M 265.703

Na salt: [65857-68-3]. Reagent for deoxygenating epoxides (particularly terminal epoxides) to alkenes, and for the dehalogenation of α-halogenoketones. Cryst. solid. Unstable in air.

Clive, D.L.J. *et al, J. Org. Chem.*, 1980, **45**, 2347; 1982, **47**, 1124 (*synth, use, pmr, P nmr*)

O,O-Diethyl phosphorothioate, 9CI, 8CI **D-00396**

O,O-*Diethyl thiophosphate.* O,O-*Diethyl hydrogen thiophosphate.* O,O-*Diethyl thiophosphoric acid*

[2465-65-8]

$(EtO)_2P(S)OH \rightleftharpoons (EtO)_2P(O)SH$

$C_4H_{11}O_3PS$ M 170.163

The thiolo form predominates in aq. solns. and the thiono in alcohols. Corrosion inhibitor. Liq. d_4^{20} 1.181. $Bp_{2.5}$ 106-107°, $Bp_{0.05}$ 76-78°. pK_a 1.49 (H_2O), pK_a 2.84 (EtOH). n_D^{20} 1.4719.

Na salt: [5852-63-1]. Cryst. ($CHCl_3/Et_2O$). Mp 203° (196-198°).

K salt: [5871-17-0]. Cryst. (Me_2CO or EtOH). Mp 197-201°.

NH_4 salt: [5871-16-9]. Cryst. Mp 144-145.5°.

Dicyclohexylammonium salt: [13941-62-3]. Solid. Mp 155-156°.

Fluoride: see O,O-Diethyl phosphorofluoridothioate, D-00389

Chloride: see O,O-Diethyl phosphorochloridothioate, D-00378

▷*TG3675000.*

Amide: see O,O-Diethyl phosphoramidothioate, D-00354

▷*TG3700000.*

Anilide: see Phenylphosphoramidothioic acid, P-00262

Azide: see O,O-Diethyl phosphorazidothioate, D-00363

Hydrazide: see O,O-Diethyl phosphorohydrazidothioate, D-00392

2-Phenylhydrazide: see 2-Phenylphosphorohydrazidothioic acid, P-00296

Isocyanate: see O,O-Diethyl phosphorisocyanatidothioate, D-00367

*Isothiocyanate: see O,O-*Diethyl phosphor(isothiocyanatido)thioate, D-00370

Anhydride: see *O,O,O,O*-Tetraethyl dithiopyrophosphate, T-00045

Mastin, T.W. *et al, J. Am. Chem. Soc.*, 1945, **67**, 1662 (*synth*)
McIvor, R.A. *et al, Can. J. Chem.*, 1958, **36**, 820 (*ir*)
Kabachnik, M.I. *et al, Tetrahedron*, 1960, **9**, 10 (*synth, props*)
Pesin, V.G. *et al, Zh. Obshch. Khim.*, 1961, **31**, 2508 (*Engl. transl.* p. 2337) (*synth*)
Boter, H.L. *et al, Recl. Trav. Chim. Pays-Bas*, 1966, **85**, 1099 (*synth*)
Chapman, T.M. *et al, J. Org. Chem.*, 1973, **38**, 250 (*synth*)
Pogarelyi, V.K. *et al, Dokl. Akad. Nauk SSSR, Ser. Sci. Khim.*, 1974, **214**, 385 (*Engl. transl.* p. 50) (*struct*)
Olah, G.A. *et al, J. Org. Chem.*, 1975, **40**, 2582 (*props, pmr*)
Kudelska, W. *et al, Tetrahedron*, 1986, **42**, 629 (*synth, P nmr*)

O,S-Diethyl phosphorothioate, 9CI, 8CI **D-00397**

O,S-*Diethyl hydrogen phosphorothioate.* O,S-*Diethyl phosphorothioic acid*

[53882-44-5]

EtO(EtS)P(=O)OH

$C_4H_{11}O_3PS$ M 170.163

Oil. n_D^{20} 1.4711.

Dicyclohexylammonium salt: Solid. Mp 136-138°.

*Chloride: see O,S-*Diethyl phosphorochloridothioate, D-00379

Amide: see *O,S*-Diethyl phosphoramidothioate, D-00355

Harvey, R.G. *et al, J. Org. Chem.*, 1963, **28**, 470 (*deriv*)
Zwierzak, A. *et al, Synthesis*, 1975, 270 (*synth, ir, pmr, P nmr*)

N,N-Diethylphosphorothioic triamide, 9CI **D-00398**

[39096-95-2]

$Et_2NP(S)(NH_2)_2$

$C_4H_{14}N_3PS$ M 167.208

Cryst. ($CHCl_3$ or CCl_4). Mp 66-67°.

N′,N″-*Di-Ph:* N,N-*Diethyl*-N′,N″-*diphenylphosphorothioic triamide.*
$C_{16}H_{22}N_3PS$ M 319.404
Needles (EtOH). Mp 192°.

Michaelis, A., *Justus Liebigs Ann. Chem.*, 1903, **326**, 129, 201 (*deriv*)
Goehring, M. *et al, Chem. Ber.*, 1956, **89**, 1768 (*synth*)
Gloe, K. *et al, J. Prakt. Chem.*, 1975, **317**, 529 (*synth, props*)

N,N-Diethyl-P-propadienylphosphonamidic acid, 9CI **D-00399**

P-*Allenyl*-N,N-*diethylphosphonamidic acid*

$H_2C{=}C{=}CH{-}P(=O)(NEt_2)OH$

$C_7H_{14}NO_2P$ M 175.167

Esters isom. to those of *N,N*-Diethyl-*P*-1-propynylphosphonamidic acid, D-00403 when treated with sodium alkoxide.

Et ester: [18629-48-6]. *Ethyl* N,N-*diethyl*-P-*propadienylphosphonamidate.*
$C_9H_{18}NO_2P$ M 203.220
Liq. d_4^{20} 1.03. Bp_2 105-107°. n_D^{20} 1.4800.

Butyl ester: [18629-49-7]. *Butyl* N,N-*diethyl*-P-*propadienylphosphonamidate.*
$C_{11}H_{22}NO_2P$ M 231.274
Liq. Bp_2 122-124°. n_D^{20} 1.4760.

Abramov, V.S. *et al, Zh. Obshch. Khim.*, 1968, **38**, 677 (*Engl. transl.* p. 656)

Diethyl propanoyl phosphate **D-00400**

Propanoic acid anhydride with diethyl phosphate, 9CI, 8CI. *Propanoic-diethyl phosphoric anhydride*

$H_3CCH_2COOP(O)(OEt)_2$

$C_7H_{15}O_5P$ M 210.166

Liq. $Bp_{0.001}$ 65-67°. n_D^{20} 1.4151.

Cramer, F. *et al*, *Chem. Ber.*, 1958, **91**, 704 (*synth*)
Michalski, J. *et al*, *Chem. Ber.*, 1962, **95**, 1629 (*synth*)

N,N-Diethyl-*P*-propylphosphonamidic acid, 9CI — D-00401

$H_3CCH_2CH_2P(O)(NEt_2)OH$

$C_7H_{18}NO_2P$ M 179.198

Fluoride:
$C_7H_{17}FNOP$ M 181.190
Liq. d_4^{20} 1.03. Bp_{12} 110.5°. n_D^{20} 1.4238.

Chloride:
$C_7H_{17}ClNOP$ M 197.644
Liq. d_0^{20} 1.07. Bp_4 112-113°. n_D^{20} 1.4642.

Razumov, A.I. *et al*, *Zh. Obshch. Khim.*, 1958, **28**, 194 (*Engl. transl.* p. 194)

O,O-Diethyl *SS*-propyl phosphoro(dithioperoxo)thioate, 9CI — D-00402

O,O-*Diethyl* S-*(propylthio) phosphorodithioate*

$(EtO)_2P(S){-}S{-}S{-}CH_2CH_2CH_3$

$C_7H_{17}O_2PS_3$ M 260.364
Liq. d_4^{20} 1.15. $Bp_{0.5}$ 106-107°. n_D^{20} 1.5370.

Almasi, L. *et al*, *Chem. Ber.*, 1962, **95**, 1582 (*synth, ir*)

N,N-Diethyl-*P*-1-propynylphosphonamidic acid, 9CI — D-00403

$H_3CC{\equiv}C{-}P(O)(NEt_2)OH$

$C_7H_{14}NO_2P$ M 175.167

Et ester: [21204-52-4]. *Ethyl* N,N-*diethyl*-P-*1-propynylphosphonamidate.*
$C_9H_{18}NO_2P$ M 203.220
Liq. d_4^{20} 1.02. Bp_2 111-113°. n_D^{20} 1.4665.

Butyl ester: [18629-73-7]. *Butyl* N,N-*diethyl*-P-*1-propynylphosphonamidate.*
$C_{11}H_{22}NO_2P$ M 231.274
Liq. Bp_2 135-137°. n_D^{20} 1.4650.

Abramov, V.S. *et al*, *Zh. Obshch. Khim.*, 1968, **38**, 677 (*Engl. transl.* p. 656)

5,8-Diethyl-5,6,7,8-tetrahydrobibenzo[*e,g*][1,4]diphosphocin, 8CI — D-00404

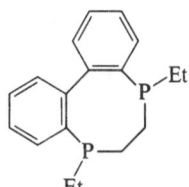

$C_{18}H_{22}P_2$ M 300.319

B2EtBr: [5274-20-4]. *5,5,8,8-Tetraethyl-5,6,7,8-tetrahydrodibenzo*[e,g][*1,4*]*diphosphocinium dibromide. 5,5,8,8-Tetraethyl-5,8-diphosphonia-1,2:3,4-dibenzocyclooctadiene dibromide.*
$C_{22}H_{32}Br_2P_2$ M 518.250
Hygroscopic solid ($CHCl_3/Me_2CO$). Mp 268-272°.

Allen, D.W. *et al*, *J. Chem. Soc.* (*C*), 1967, 1869 (*synth, props*)

Diethylthiodiphosphonic acid — D-00405

Ethylphosphonodithioic acid anhydrosulfide

$Et(HO)P(S){-}S{-}P(S)(OH)Et$

$C_4H_{12}O_2P_2S_3$ M 250.265

O,O′-*Di-Et ester:* [67397-51-7]. O,O′-*Diethyl diethylthiodiphosphonate.* O-*Ethyl ethylphosphonodithioic anhydrosulfide.*
$C_8H_{20}O_2P_2S_3$ M 306.372
Solid. Mp 67-68°.

O,O′-*Diisopropyl ester:* O,O′-*Diisopropyl diethylthiodiphosphonate.* O-*Isopropyl ethylphosphonodithioic anhydrosulfide.*
$C_{10}H_{24}O_2P_2S_3$ M 334.426
Solid. Mp 48°.

Difluoride:
$C_4H_{10}F_2P_2S_3$ M 254.247
Liq. d^{20} 1.35. $Bp_{0.05}$ 75°. n_D 1.5699. Mixture of stereoisomers.

Roesky, H.W., *Chem. Ber.*, 1968, **101**, 3679 (*difluoride, synth, ir, pmr, F nmr*)
Harris, R.K. *et al*, *J. Chem. Soc., Dalton Trans.*, 1972, 1590 (*difluoride, synth, pmr, F and P nmr*)
Hantz, A. *et al*, *J. Prakt. Chem.*, 1978, **320**, 183 (*esters, synth, ms*)
Andreev, N.A. *et al*, *Zh. Obshch. Khim.*, 1982, **52**, 1785 (*Engl. transl.* p. 1581) (*diethyl ester, synth, P nmr*)

Diethyl trifluoroacetyl phosphate — D-00406

Trifluoroacetic acid anhydride with diethyl phosphate, 9CI. Trifluoroacetic-diethyl phosphoric anhydride
[650-09-9]

$(EtO)_2P(O)OCOCF_3$

$C_6H_{10}F_3O_5P$ M 250.111
Liq. Bp_7 82-84°, $Bp_{0.01}$ 30°. n_D^{20} 1.3610.

Lambie, A.J., *Tetrahedron Lett.*, 1966, 3709 (*synth*)
Boev, V.I. *et al*, *Zh. Obshch. Khim.*, 1979, **49**, 2505 (*Engl. transl.* p. 2212) (*synth, ir*)

Diethyl trimethylsilyl phosphate — D-00407

[18306-68-8]

$(EtO)_2P(O)OSiMe_3$

$C_7H_{19}O_4PSi$ M 226.284
Liq. d_4^{20} 1.02. Bp_{12} 97-98°, $Bp_{0.1}$ 48-49°. n_D^{20} 1.4070.

Fehér, F. *et al*, *Chem. Ber*, 1957, **90**, 134 (*synth, raman*)
Kirilov, M. *et al*, *Chem. Ber.*, 1971, **104**, 3073 (*synth*)
Nesterov, L.V. *et al*, *Zh. Obshch. Khim.*, 1978, **48**, 790 (*Engl. transl.* p. 722) (*synth, pmr, P nmr*)

Diethyl trimethylsilyl phosphite, 9CI, 8CI — D-00408

[13716-45-5]

$(EtO)_2POSiMe_3$

$C_7H_{19}O_3PSi$ M 210.285
Reagent for synthesising ketones from aldehydes *via* α-hydroxyphosphonic esters. Liq. d_4^{20} 0.95. Bp_{12} 59-60°. n_D^{20} 1.4114.

Bugerenko, E.F. *et al*, *Zh. Obshch. Khim.*, 1973, **43**, 216 (*Engl. transl.* p. 218) (*synth, P nmr*)

Chernyshev, E.A. *et al, Zh. Obshch. Khim.*, 1975, **45**, 242 (*Engl. transl.* p. 231) (*synth*)
Pudovik, M.A. *et al, Zh. Obshch. Khim.*, 1975, **45**, 700 (*Engl. transl.* p. 682) (*synth*)
Evans, D.A. *et al, J. Am. Chem. Soc.*, 1978, **100**, 3467 (*props*)
Hata, T. *et al, Tetrahedron Lett.*, 1978, 363 (*use*)
Kozlowska, M. *et al, Pol. J. Chem.*, 1978, **52**, 347; *CA*, **89**, 42521.
Sekine, M. *et al, J. Org. Chem.*, 1981, **46**, 2097 (*props*)
Mizhiritskii, M.D. *et al, Zh. Obshch. Khim.*, 1982, **52**, 2089 (*Engl. transl.* p. 1859) (*synth, P nmr*)
Sekine, M. *et al, J. Chem. Soc., Perkin Trans. 1*, 1982, 2509 (*use*)
Azuhata, T. *et al, Synthesis*, 1983, 916 (*use*)
Creary, X. *et al, J. Am. Chem. Soc.*, 1983, **105**, 2851 (*synth, pmr, props, use*)

Diethynylphosphinic acid, 9CI D-00409

[34833-62-0]

$(HC{\equiv}C)_2P(O)OH$

$C_4H_3O_2P$ M 114.040

Dimethylamide: P,P-*Diethynyl*-N,N-*dimethylphosphinic amide.*
C_6H_7NOP M 140.101
Solid. Mp 83°.

Charrier, C. *et al, C.R. Hebd. Seances Acad. Sci., Ser. C*, 1967, **264**, 995 (*synth, pmr*)

Diferrocenylphenylphosphine, 8CI D-00410

1,1″-(Phenylphosphinidene)bisferrocene, 9CI

[12278-69-2]

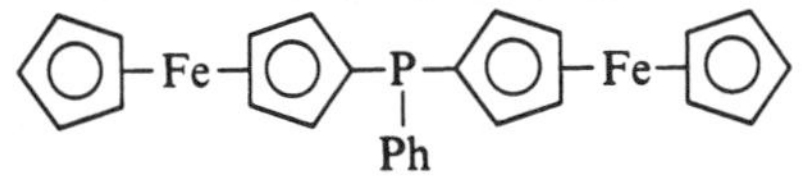

$C_{26}H_{23}Fe_2P$ M 478.135

Orange cryst. (hexane), golden needles (EtOH). Mp 194-195.5°.

B,MeI: Diferrocenylmethylphenylphosphonium iodide.
$C_{27}H_{26}Fe_2IP$ M 620.075
Yellow-orange cryst. (EtOH). Mp 263-265° (sealed tube).

B,PhCH₂Cl: Benzyldiferrocenylphenylphosphonium chloride. Diferrocenylphenyl(phenylmethyl)-phosphonium chloride.
$C_{33}H_{30}ClFe_2P$ M 604.721
Cryst. (EtOH/Et_2O). Mp 205-206°.

Oxide: [12204-30-7]. *1,1″-(Phenylphosphinylidene)-bisferrocene.*
$C_{26}H_{23}Fe_2OP$ M 494.135
Orange powder. Mp 239-241°.

Sollott, G.P. *et al, J. Org. Chem.*, 1963, **28**, 1090 (*synth*)
Neuse, E.W., *J. Organomet. Chem.*, 1967, **7**, 349 (*oxide, uv*)
Neuse, E.W. *et al, J. Macromol. Sci., Part A*, 1967, **1**, 371 (*synth*)
Kotz, J.C. *et al, J. Organomet. Chem.*, 1973, **52**, 387; 1975, **91**, 87 (*complexes*)
Eberhard, L. *et al, J. Organomet. Chem.*, 1974, **80**, 109 (*oxide, props*)
McEwen, W.E. *et al, J. Org. Chem.*, 1976, **41**, 1684 (*synth, deriv*)

Difficidin D-00411

[95152-88-8]

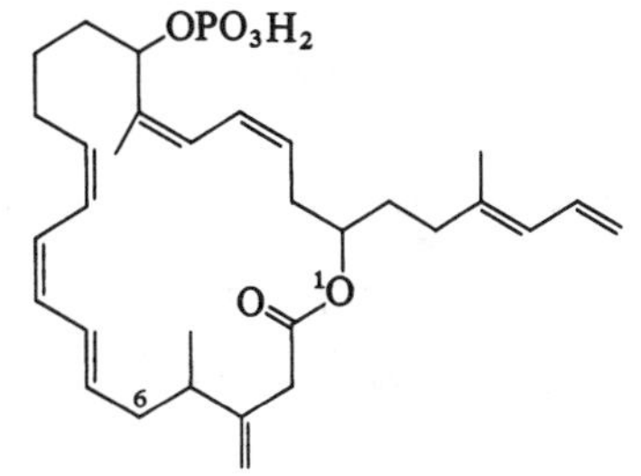

$C_{31}H_{45}O_6P$ M 544.667

Macrolide-type antibiotic. Isol. from *Bacillus subtilis.* Antibacterial agent.

6-Hydroxy: [95152-89-9]. ***Oxydifficidin.***
$C_{31}H_{45}O_7P$ M 560.666
From *B. subtilis.* Antibacterial agent.

16-Dephosphonyl: ***Difficol.***
$C_{31}H_{46}O_3$ M 466.703
From *B. subtilis.*

16-Dephosphonyl, 6-hydroxy: **Oxydifficol.**
$C_{31}H_{46}O_4$ M 482.702
From *B. subtilis.*

Eur. Pat., 128 505, (*1984*); *CA*, **102**, 111467 (*isol*)
Eur. Pat., 190 658, (*1986*); *CA*, **106**, 31376 (*deriv*)

2,2-Difluoro-2,2,4,4,6,6-hexahydro-4,4,6,6-tetraphenyl-1,3,5-triaza-2,4,6-triphosphorine, 9CI D-00412

2,2-Difluoro-2,2,4,4,6,6-hexahydro-4,4,6,6-tetraphenylcyclotriphosphazene

[21050-24-8]

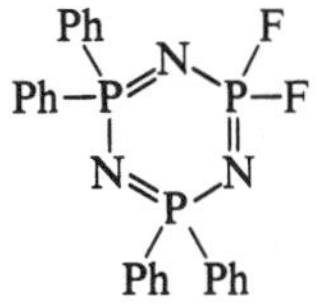

$C_{24}H_{20}F_2N_3P_3$ M 481.360

No phys. props. reported.

Wagner, A.J. *et al, J. Inorg. Nucl. Chem.*, 1971, **33**, 1307 (*uv*)
Stec, W.J. *et al, J. Inorg. Nucl. Chem.*, 1972, **34**, 1100 (*pe*)
Schumann, K. *et al, Phosphorus*, 1973, **3**, 51 (*P nmr*)
Allen, C.W. *et al, J. Chem. Soc., Dalton Trans.*, 1974, 1685 (*ms*)
Allen, C.W., *J. Organomet. Chem.*, 1977, **125**, 215 (*cmr*)
Allen, C.W. *et al, Inorg. Chem.*, 1980, **19**, 1719 (*pe*)

(Difluoromethylene)(trifluoromethyl)-phosphine, 9CI D-00413

[72344-34-4]

$F_3CP{=}CF_2$

C_2F_5P M 149.988

Liq. Stable at −78° readily forms oligomers at higher temps.

Eshtiagh-Hosseini, H. *et al, J. Organomet. Chem.*, 1979, **181**, C1 (*synth, F and P nmr*)
Burg, A.B., *Inorg. Chem.*, 1983, **22**, 2573 (*props, ir, F nmr*)
Grobe, J. *et al, Angew. Chem., Int. Ed. Engl.*, 1984, **23**, 710 (*props, synth, F nmr*)
Binnewies, M. *et al, Phosphorus Sulfur*, 1985, **21**, 349 (*ms*)
Hosseini, H.E. *et al, J. Organomet. Chem.*, 1985, **296**, 351 (*synth, pmr, F and P nmr*)
Steger, B. *et al, Inorg. Chem.*, 1986, **25**, 3177 (*ed*)

(Difluoromethylene)triphenylphosphorane, 9CI **D-00414**

[33558-14-4]

$Ph_3P{=}CF_2$

$C_{19}H_{15}F_2P$ M 312.298

Used in Wittig reactions to prepare 1,1-difluoro-1-alkenes. Reactive ylide.

Fuqua, S.A. *et al*, *J. Org. Chem.*, 1965, **30**, 1027 (*synth, use*)
Naae, D.G. *et al*, *Tetrahedron Lett.*, 1975, 3789 (*synth*)
Wheaton, G.A. *et al*, *J. Org. Chem.*, 1983, **48**, 917 (*synth*)

(Difluoromethyl)phosphonic acid, 9CI **D-00415**

$F_2CHP(O)(OH)_2$

$CH_3F_2O_3P$ M 132.003

Biscyclohexylammonium salt: Solid. Mp 193-195°.

Di-Et ester: Diethyl (difluoromethyl)phosphonate.
$C_5H_{11}F_2O_3P$ M 188.111
Wittig-Horner reagent for the synth. of 1,1-difluoroalkenes. Liq. Bp_{12} 85.5-86.5°.

Soborovskii, L.Z. *et al*, *Zh. Obshch. Khim.*, 1959, **29**, 1144 (*Engl. transl.* p. 1115) (*synth*)
McKenna, C.E. *et al*, *J. Org. Chem.*, 1981, **46**, 4536 (*synth, pmr, F and P nmr*)
Obayashi, M. *et al*, *Tetrahedron Lett.*, 1982, **23**, 2323 (*synth, use*)
Hall, C.R. *et al*, *J. Chem. Soc., Perkin Trans. 1*, 1985, 233 (*esters*)
Blackburn, G.M. *et al*, *J. Chem. Soc., Perkin Trans. 1*, 1987, 181.

Difluorotrimethoxyphosphorane, 9CI, 8CI **D-00416**

Trimethyl phosphite difluoride

[17167-31-6]

$(MeO)_3PF_2$

$C_3H_9F_2O_3P$ M 162.073

Hygroscopic cryst. which when stored at −20°, dec. to $(MeO)_2P(O)F$. Bp_{10} 38°. n_D^{20} 1.3620.

De'ath, N.J. *et al*, *Phosphorus*, 1974, **3**, 205 (*synth*)
Gontar, A.F. *et al*, *Izv. Akad. Nauk SSSR, Ser. Khim.*, 1981, 2632 (*Engl. transl.* p. 2188) (*synth, props*)
Ruppert, I., *Z. Anorg. Allg. Chem.*, 1981, **477**, 59 (*props, cmr, nmr*)

Difluorotrimethylphosphorane, 9CI, 8CI **D-00417**

Trimethylphosphine difluoride

[661-42-7]

Me_3PF_2

$C_3H_9F_2P$ M 114.075

Liq. Mp −31°. Bp_{760} 75°. Weakly ionized in MeCN.

Downs, A.J. *et al*, *Spectrochim. Acta, Part A*, 1967, **23**, 681 (*ir, raman, pmr, nmr*)
Blazer, T.A. *et al*, *Z. Naturforsch., B*, 1969, **24**, 1081 (*ms*)
Appel, R. *et al*, *Chem. Ber.*, 1974, **107**, 2169 (*synth, nmr*)
Harris, G.S., *J. Fluorine Chem.*, 1980, **16**, 293 (*struct*)

Difluorotriphenoxyphosphorane, 9CI, 8CI **D-00418**

Triphenyl phosphite difluoride

[17167-32-7]

$(PhO)_3PF_2$

$C_{18}H_{15}F_2O_3P$ M 348.285

Moisture-sensitive cryst. (MeCN).

Peake, S.C. *et al*, *Inorg. Chem.*, 1971, **10**, 2723 (*synth, nmr, props*)
Ruppert, I., *Z. Anorg. Allg. Chem.*, 1981, **477**, 59 (*synth, cmr, nmr*)
Il'in, E.G. *et al*, *Dokl. Akad. Nauk SSSR, Ser. Sci. Khim.*, 1982, **266**, 123 (*Engl. transl.* p. 300) (*struct, nmr*)

Difluorotriphenylphosphorane, 9CI, 8CI **D-00419**

Triphenylphosphine difluoride

[845-64-7]

Ph_3PF_2

$C_{18}H_{15}F_2P$ M 300.287

Converts primary and secondary alcohols into alkyl fluorides. Cryst. (MeCN or DMF). Mp 141°, 160-162°. Converted by cold aq. alkali into Ph_3PO.

Firth, W.C. *et al*, *Inorg. Chem.*, 1965, **4**, 765 (*synth, ir*)
Kobayashi, Y. *et al*, *Chem. Pharm. Bull.*, 1968, **16**, 1009 (*synth, use*)
Yagupol'skii, L.M. *et al*, *Zh. Obshch. Khim.*, 1968, **38**, 2813 (*Engl. transl.* p. 2714) (*synth*)
Blazer, T.A. *et al*, *Z. Naturforsch., B*, 1969, **24**, 1081 (*ms*)
Appel, R. *et al*, *Chem. Ber.*, 1974, **107**, 2169 (*synth, nmr*)

Difluorotris(trifluoromethyl)phosphorane, 9CI, 8CI **D-00420**

[661-45-0]

$F_2P(CF_3)_3$

$C_3F_{11}P$ M 275.989

Molecule consists of trigonal bipyramid with CF_3 groups in the equatorial positions. Source of F_2C: at 120°. Liq. or gas. Mp −102°. Bp 20°.

Mahler, W., *Inorg. Chem.*, 1963, **2**, 230 (*synth, ir*)
Muetterties, E.L. *et al*, *Inorg. Chem.*, 1963, **2**, 613 (*ir, nmr*)
Nixon, J.F. *et al*, *Spectrochim. Acta*, 1964, **20**, 1835 (*nmr*)
Birchall, J.M. *et al*, *J. Chem. Soc., Perkin Trans. 1*, 1973, 1773 (*use*)
Cavell, R.G. *et al*, *J. Am. Chem. Soc.*, 1977, **99**, 7841 (*cmr*)
Oberhammer, H. *et al*, *Inorg. Chem.*, 1982, **21**, 275 (*ed*)

Di-2-furanylphosphinic acid, 9CI **D-00421**

Di-2-furylphosphinic acid

[65887-64-1]

$C_8H_7O_4P$ M 198.115

Cryst. (hexane/EtOH). Mp 149°.

Et ester: [65887-66-3]. *Ethyl di-2-furanylphosphinate.*
$C_{10}H_{11}O_4P$ M 226.168
Cryst. (hexane). Mp 76°. $Bp_{0.02}$ 116-118°.

Allen, D.W. *et al*, *J. Chem. Soc., Perkin Trans. 2*, 1977, 1705 (*synth*)

Andreae, S. *et al*, *Z. Chem.*, 1979, **19**, 98 (*ester, synth*)

Di-2-furanylphosphinothioic acid D-00422

Di-2-furylphosphinothioic acid

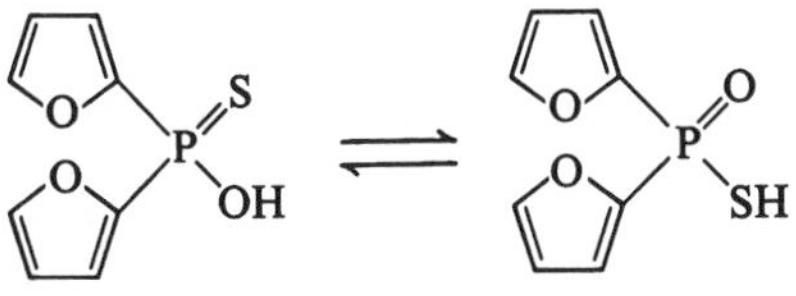

$C_8H_7O_3PS$ M 214.175

Et ester: [70519-56-1]. O-*Ethyl di-2-furanylphosphinothioate.*
$C_{10}H_{11}O_3PS$ M 242.229
Solid. Mp 40-41°. $Bp_{0.02}$ 128-129°.

Andreae, S. *et al*, *Z. Chem.*, 1979, **19**, 98.

P^1,P^4-Diguanosine 5′-tetraphosphate D-00423

Guanosine 5′-(pentahydrogen tetraphosphate) 5′,5′-ester of guanosine, 9CI, 8CI. Diguanosine 5′,5‴-(P^1,P^4-tetraphosphate). GP_4G
[4130-19-2]

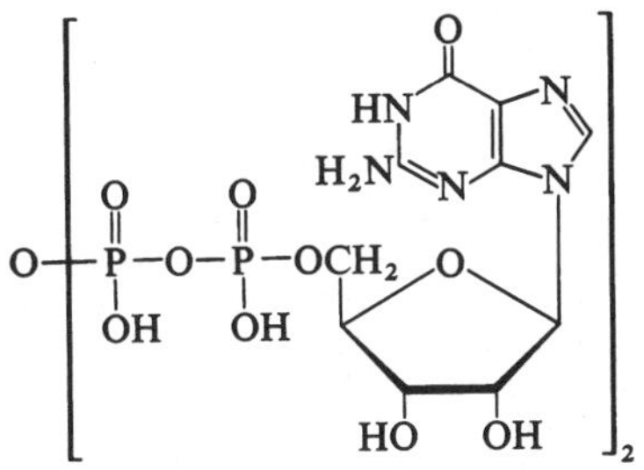

$C_{20}H_{28}N_{10}O_{21}P_4$ M 868.391
Present in brine shrimp (*Artemia salina*) eggs and in rat liver. Is converted into ATP in the metabolism of the brine shrimp. λ_{max} 256-7 nm (ε 21 700) (pH 2 and 10).

Biochem. Prep., 1966, **11**, 27 (*isol, synth*)
van Denbos, G. *et al*, *J. Biol. Chem.*, 1974, **249**, 2816.
Sy, J., *Biochemistry*, 1975, **14**, 970.
Vallejo, C.G. *et al*, *Biochim. Biophys. Acta*, 1976, **438**, 304.

Diheptyl phosphate, 9CI, 8CI D-00424

Diheptyl hydrogen phosphate. Diheptyl phosphoric acid
[3900-12-7]

$$[H_3C(CH_2)_6O]_2P(O)OH$$

$C_{14}H_{31}O_4P$ M 294.370
Oil. d_4^{20} 1.00. n_D^{20} 1.4383.
Chloride: Diheptyl phosphorochloridate. Diheptyl phosphoryl chloride. Diheptyl chlorophosphate.
$C_{14}H_{30}ClO_3P$ M 312.816
Oil. n_D^{25} 1.4410.

Petrov, K.A. *et al*, *Zh. Obshch. Khim.*, 1961, **31**, 1709 (*Engl. transl.* p. 1596) (*synth*)
Grosse-Ruyken, H. *et al*, *J. Prakt. Chem.*, 1962, **18**, 287 (*synth, chloride*)

Diheptylphosphinic acid, 9CI D-00425

[3011-76-5]

$$[H_3C(CH_2)_6]_2P(O)OH$$

$C_{14}H_{31}O_2P$ M 262.371
Useful in extraction of rare earth metals. Solid. Mp 77-78°. pK_a 5.29 (75% v/v EtOH aq.).
Heptyl ester: [3058-19-3]. *Heptyl diheptylphosphinate.*
$C_{21}H_{45}O_2P$ M 360.559
Liq. $Bp_{0.5}$ 164°. n_D^{20} 1.452.

Williams, R.H. *et al*, *J. Am. Chem. Soc.*, 1955, **77**, 3411 (*synth*)
Petrov, K.A. *et al*, *Zh. Obshch. Khim.*, 1960, **30**, 1964 (*Engl. transl.* p. 1943) (*synth*)
Voigt, D. *et al*, *Bull. Soc. Chim. Fr.*, 1964, 3087 (*ester*)
Khramov, V.P. *et al*, *Zh. Neorg. Khim.*, 1973, **18**, 3180; *CA*, **80**, 59101 (*salts, ir*)
Katzin, L.I. *et al*, *Spectrochim. Acta, Part A*, 1978, **34**, 51 (*ir*)

Diheptyl phosphonate, 9CI D-00426

Diheptyl phosphite
[6163-90-2]

$$[H_3C(CH_2)_6O]_2P(O)H \rightleftharpoons [H_3C(CH_2)_6O]_2POH$$

$C_{14}H_{31}O_3P$ M 278.371
Tautomeric. Liq. d_4^{20} 0.93. Bp_1 162-163°. n_D^{20} 1.4364.

Petrov, K.A. *et al*, *Zh. Obshch. Khim.*, 1962, **32**, 1277 (*Engl. transl.* p. 1250) (*synth*)
B.P., 1 298 156, (*1970*); *CA*, **78**, 83841 (*manuf*)

O,O-Diheptyl phosphorodithioate, 9CI, 8CI D-00427

O,O-*Diheptyl hydrogen dithiophosphate.* O,O-*Diheptyl dithiophosphoric acid*

$$[H_3C(CH_2)_6O]_2P(S)SH$$

$C_{14}H_{31}O_2PS_2$ M 326.491
Zn and Pb complexes used as lubricant additives.
K salt: [3287-88-5]. Solid. Mp 154-154.5°.

Bolotova, G.L. *et al*, *CA*, 1965, **63**, 6897 (*synth*)
Mazitova, F.N. *et al*, *Zh. Obshch. Khim.*, 1980, **50**, 1718 (*Engl. transl.* p. 1393)

Dihexadecyl phosphate, 9CI D-00428

Dihexadecyl hydrogen phosphate. Dicetyl phosphate. Dicetyl phosphoric acid. Dihexadecyl phosphoric acid

$$[H_3C(CH_2)_{15}O]_2P(O)OH$$

$C_{32}H_{67}O_4P$ M 546.853
Cryst. (CH_2Cl_2). Mp 74-75°.

Brown, D. *et al*, *J. Chem. Soc.*, 1955, 1584 (*synth, cryst struct*)
Petrov, K.A. *et al*, *Zh. Obshch. Khim.*, 1961, **31**, 1709 (*Engl. transl.* p. 1596) (*synth*)
Ramirez, F. *et al*, *Phosphorus Sulfur*, 1978, **4**, 43 (*synth*)

Dihexadecylphosphinic acid, 9CI D-00429

$$[H_3C(CH_2)_{15}]_2P(O)OH$$

$C_{32}H_{67}O_2P$ M 514.854
Cryst. (C_6H_6 or C_6H_6/EtOH). Mp 102.5-103.5°.

Williams. R.H. *et al*, *J. Am. Chem. Soc.*, 1955, **77**, 3411.

Dihexadecyl phosphonate, 9CI D-00430

Dihexadecyl phosphite

[37032-33-0]

$[H_3C(CH_2)_{15}O]_2P(O)H \rightleftharpoons [H_3C(CH_2)_{15}O]_2POH$

$C_{32}H_{67}O_3P$ M 530.853

Tautomeric; almost completely in phosphonate form. Liq. which slowly solidifies. Cryst. (Me_2CO). Mp 46-47°, 55-57°. $Bp_{0.02}$ 245-250°.

▷TG8200000.

Petrov, K.A. *et al*, *Zh. Obshch. Khim.*, 1962, **32**, 1277 (*Engl. transl.* p. 1250) (*synth*)
B.P., 1 298 156, (*1970*); *CA*, **78**, 83841 (*synth*)
U.S.P., 3 725 515, (*1970*); *CA*, **78**, 147344 (*synth*)

Dihexylphenylphosphine, 9CI D-00431

[18297-98-8]

$PhP[(CH_2)_5CH_3]_2$

$C_{18}H_{31}P$ M 278.417

Liq. Bp_{50} 236°, Bp_1 162-166°.

B,MeI: Dihexylmethylphenylphosphonium iodide.
$C_{19}H_{34}IP$ M 420.356
Needles. Mp 67°.

Oxide: [19259-62-2].
$C_{18}H_{31}OP$ M 294.416
Cryst. (heptane). Mp 55°.

Jackson, I.K. *et al*, *J. Chem. Soc.*, 1931, 2109 (*synth*)
Fedorova, G.K. *et al*, *Zh. Obshch. Khim.*, 1964, **34**, 511 (*Engl. transl.* p. 513) (*oxide*)
McEwen, W.E. *et al*, *J. Am. Chem. Soc.*, 1964, **86**, 2378 (*synth, derivs*)

Dihexyl phosphate, 9CI, 8CI D-00432

Dihexyl hydrogen phosphate. Dihexyl phosphoric acid

[3900-13-8]

$[H_3C(CH_2)_5O]_2P(O)OH$

$C_{12}H_{27}O_4P$ M 266.317

Oil. d_4^{20} 1.02. $Bp_{0.0001}$ 133-138°. pK_a 2.08 (H_2O), pK_a 3.14 (75% v/v EtOH aq.). n_D^{20} 1.4350.

Chloride: Dihexyl phosphorochloridate. Dihexyl phosphoryl chloride. Dihexyl chlorophosphate.
$C_{12}H_{26}ClO_3P$ M 284.762
Oil. d_4^{20} 1.01. Bp_8 165-166°. n_D^{20} 1.4381.

Amide: Dihexyl phosphoramidate. Dihexyl phosphoryl amide. Dihexyl amidophosphate.
$C_{12}H_{28}NO_3P$ M 265.332
Liq. d_4^{20} 1.00. $Bp_{0.23}$ 168-169°. n_D^{20} 1.4452.

Petrov, K.A. *et al*, *Zh. Obshch. Khim.*, 1960, **30**, 1233 (*Engl. transl.* p. 1256) (*chloride, amide*)
Petrov, K.A. *et al*, *Zh. Obshch. Khim.*, 1961, **31**, 1709 (*Engl. transl.* p. 1596) (*synth*)
Grosse-Ruyken, H. *et al*, *J. Prakt. Chem.*, 1962, **18**, 287 (*synth, chloride*)
Bliznyuk, N.K. *et al*, *Zh. Obshch. Khim.*, 1967, **37**, 1119 (*Engl. transl.* p. 1061) (*synth*)

Dihexylphosphine, 9CI D-00433

[24674-43-9]

$HP[(CH_2)_5CH_3]_2$

$C_{12}H_{27}P$ M 202.319

Oxide: [17529-42-9].
$C_{12}H_{27}OP$ M 218.318
Cryst. (hexane). Mp 75-76°. Tautomeric with dihexylphosphinous acid.

Williams, R.H. *et al*, *J. Am. Chem. Soc.*, 1952, **74**, 5418; 1955, **77**, 3411 (*synth*)
Sanchez, M. *et al*, *Spectrochim. Acta, Part A*, 1967, **23**, 2617 (*oxide, synth, ir*)

Dihexylphosphinic acid, 9CI D-00434

[7646-81-3]

$[H_3C(CH_2)_5]_2P(O)OH$

$C_{12}H_{27}O_2P$ M 234.318

Solid. Mp 76-78°.

▷SZ5150000.

Et ester: Ethyl dihexylphosphinate.
$C_{14}H_{31}O_2P$ M 262.371
Extractant for U and Pu. Liq. d_4^{25} 0.91. $Bp_{1.5}$ 143°. n_D^{20} 1.4460.

Hexyl ester: [19259-63-3]. *Hexyl dihexylphosphinate.*
$C_{18}H_{39}O_2P$ M 318.479
Extractant for U and Co.

Fluoride: [84923-96-6].
$C_{12}H_{26}FOP$ M 236.309
Liq. Bp_3 125-127°.

Chloride: [33493-42-4].
$C_{12}H_{26}ClOP$ M 252.764
Liq.-solid. Mp 22-23°. $Bp_{0.03}$ 120-122°.

N*-Butylamide:* N*-Butyl-*P,P*-dihexylphosphinic amide.*
$C_{16}H_{36}NOP$ M 289.440
Liq. $Bp_{0.05}$ 154-156°. n_D^{25} 1.4597.

Burger, L.L., *J. Phys. Chem.*, 1958, **62**, 590 (*ester, use*)
Silver, H.B., *J. Chem. Soc. (C)*, 1967, 1326 (*amide*)
Giancotti, V. *et al*, *J. Chem. Soc. (A)*, 1968, 757 (*cryst struct*)
Feshchenko, N.G. *et al*, *Zh. Obshch. Khim.*, 1970, **40**, 2385 (*Engl. transl.* p. 2373) (*chloride*)
Nikitin, E.V. *et al*, *Zh. Obshch. Khim.*, 1982, **52**, 2721 (*Engl. transl.* p. 2400) (*fluoride, synth, ir, F and P nmr*)

Dihexyl phosphonate, 9CI D-00435

Dihexyl phosphite

[6151-90-2]

$[H_3C(CH_2)_5O]_2P(O)H \rightleftharpoons [H_3C(CH_2)_5O]_2POH$

$C_{12}H_{27}O_3H$ M 220.351

Tautomeric. Liq. d_4^{20} 0.95. $Bp_{1.5}$ 143-144°. pK_a 20.9 (DMSO at 25°). n_D^{20} 1.4395.

Petrov, K.A. *et al*, *Zh. Obshch. Khim.*, 1962, **32**, 3723 (*Engl. transl.* p. 3650) (*synth*)
B.P., 1 298 156, (*1970*); *CA*, **78**, 83841 (*manuf*)
Tsvetkov, E.N. *et al*, *Izv. Akad. Nauk SSSR, Ser. Khim.*, 1978, 1981 (*Engl. transl.* p. 1743) (*props*)

Dihexyl phosphorochloridite, 9CI, 8CI D-00436

Dihexyl chlorophosphite

[66379-66-6]

$[H_3C(CH_2)_5O]_2PCl$

$C_{12}H_{26}ClO_2P$ M 268.763

Pungent, fuming liq. Bp_8 145-147°. n_D^{20} 1.4435.

▷Reacts violently with H_2O

Petrov, K.A. *et al*, *Zh. Obshch. Khim.*, 1961, **31**, 2373 (*Engl. transl.* p. 2211) (*synth, props*)
Razumov, A.I. *et al*, *Zh. Obshch. Khim.*, 1962, **32**, 4063 (*Engl. transl.* p. 3987) (*synth*)

O,O-Dihexyl phosphorodithioate, 9CI, 8CI **D-00437**
O,O-*Dihexyl hydrogen dithiophosphate.* O,O-*Dihexyl dithiophosphoric acid*
[78-64-8]

$[H_3C(CH_2)_5O]_2P(S)SH$

$C_{12}H_{27}O_2PS_2$ M 298.438
Zn and Pb complexes used as lubricant additives.
K salt: [3287-87-4]. Solid. Mp 154-155.5°.

Bolotova, G.L. *et al*, *CA*, 1965, **63**, 6897.

(9,10-Dihydro-9-acridinyl)phosphonic acid, 9CI, 8CI **D-00438**
9-Acridanphosphonic acid. 9,10-Dihydro-9-phosphonoacridine

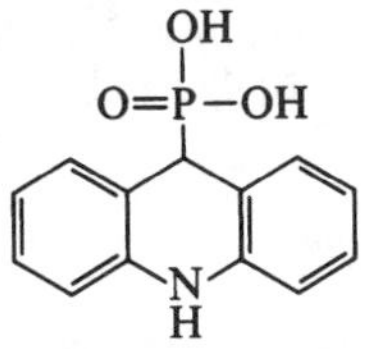

$C_{13}H_{12}NO_3P$ M 261.216
Di-Me ester: [65674-21-7]. *Dimethyl (9,10-dihydro-9-acridinyl)phosphonate.*
$C_{15}H_{16}NO_3P$ M 289.270
Solid. Mp 170-171°.
Di-Et ester: Diethyl (9,10-dihydro-9-acridinyl)-phosphonate.
$C_{17}H_{20}NO_3P$ M 317.324
Cryst. (EtOH or C_6H_6/hexane). Mp 189-190°.
N-*Me, Di-Me ester: Dimethyl (9,10-dihydro-5-methyl-9-acridinyl)phosphonate.*
$C_{15}H_{18}NO_3P$ M 291.286
Solid. Mp 98-99°.
N-*Me, Di-Et ester: Diethyl (9,10-dihydro-5-methyl-9-acridinyl)phosphonate.*
$C_{17}H_{22}NO_3P$ M 319.339
Cryst. (C_6H_6/hexane). Mp 89-91°.
N-*Me, diisopropyl ester: Diisopropyl (9,10-dihydro-5-methyl-9-acridinyl)phosphonate.*
$C_{19}H_{26}NO_3P$ M 347.393
Solid. Mp 124-125.5°.

Redmore, D., *J. Org. Chem.*, 1969, **34**, 1420 (*synth, pmr, ir*)
Sheinkmann, A.K. *et al*, *Zh. Obshch. Khim.*, 1974, **44**, 1472 (*Engl. transl.* p. 1445) (*synth*)
Akiba, K. *et al*, *Synthesis*, 1977, 862 (*synth, pmr*)
Ishikawa, K. *et al*, *Bull. Chem. Soc. Jpn.*, 1978, **51**, 2684 (*ir, props, pmr*)

2,2-Dihydro-1,3,2-benzodioxaphosphole **D-00439**

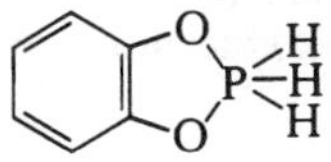

$C_6H_7O_2P$ M 142.094
2,2,2-Trimethoxy: [62785-52-8].
$C_9H_{13}O_5P$ M 232.172
Noncryst.
2,2-Dimethoxy, 2-Ph: [62785-51-7].
$C_{14}H_{15}O_4P$ M 278.244
Noncryst.
2-Methoxy, 2,2-di-Ph:
$C_{19}H_{17}O_3P$ M 324.315
Solid. Mp 84-85°.
2,2,2-Tri-Ph:
$C_{21}H_{19}O_2P$ M 334.354
Solid. Mp 75° dec.
2,2,2-Triphenoxy: see 2,2-Dihydro-2,2,2-triphenoxy-1,3,2-benzodioxaphosphole, D-00602

Antczak, S. *et al*, *J. Chem. Soc., Perkin Trans. 1*, 1977, 278 (*synth, pmr, P nmr*)

7,12-Dihydrobenzo[*e*]phenophosphazine, 9CI **D-00440**

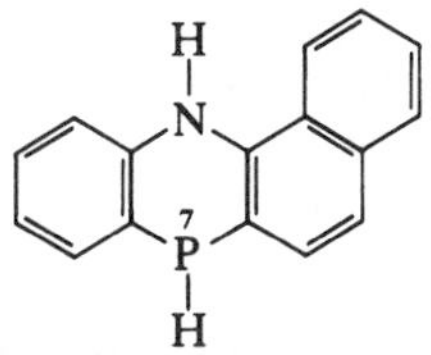

$C_{16}H_{12}NP$ M 249.251
7-Oxide: [53798-59-7].
$C_{16}H_{12}NOP$ M 265.251
Cryst. (EtOH aq.). Mp 238-240°.
7-Hydroxy, 7-oxide: [53798-62-2]. *Benzo*[e]-*phenazaphosphinic acid.*
$C_{16}H_{12}NO_2P$ M 281.250
Cryst. (AcOH). Mp 290-295°.

Jenkins, R.N. *et al*, *J. Org. Chem.*, 1975, **40**, 766 (*synth, ir*)
Freedman, L.D. *et al*, *J. Org. Chem.*, 1975, **40**, 2684 (*uv*)

4,5-Dihydro-3*H*-benzo[*e*]phosphindole, 10CI **D-00441**

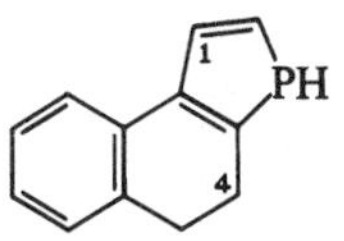

$C_{12}H_{11}P$ M 186.193
*3-Me: 4,5-Dihydro-3-methyl-3*H-*benzo*[e]*phosphindole.*
$C_{13}H_{13}P$ M 200.219
Oil. Forms a dimeric oxide, Mp 263-4° (toluene).
3-Ph: [72054-43-4]. *4,5-Dihydro-3-phenyl-3*H-*benzo*[e]*phosphindole.*
$C_{18}H_{15}P$ M 262.290
Could not be purified. Forms a dimeric oxide, Mp 229-30° (toluene).

Quin, L.D. *et al*, *Org. Magn. Reson.*, 1979, **12**, 442 (*P nmr*)
Quin, L.D. *et al*, *Phosphorus Sulfur*, 1982, **12**, 161 (*synth, derivs*)

2,3-Dihydro-1*H*-benzo[*e*]phosphindole 3-oxide, 9CI **D-00442**

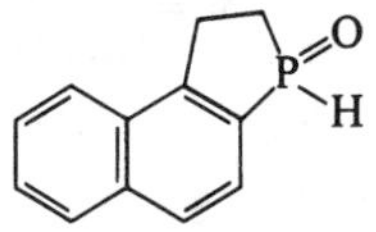

$C_{12}H_{11}OP$ M 202.192
3-Hydroxy: [70610-33-2].
$C_{12}H_{11}O_2P$ M 218.191
Solid. Mp 212-216°.
3-Ph: [70610-32-1].
$C_{18}H_{15}OP$ M 278.290
Solid. Mp 200-202°.

Orton, W.L. *et al*, *Phosphorus Sulfur*, 1979, **5**, 349 (*synth, cmr, P nmr*)

(3,4-Dihydro-1*H*-2-benzopyran-1-yl)-phosphonic acid, 9CI D-00443

*3,4-Dihydro-1-phosphono-1*H*-2-benzopyran*

$C_9H_{11}O_4P$ M 214.157

Di-Et ester: [16259-87-2]. *Diethyl (3,4-dihydro-1*H*-2-benzopyran-1-yl)phosphonate.*
$C_{13}H_{19}O_4P$ M 270.264
Liq. $Bp_{0.1}$ 125°. n_D^{22} 1.5427.

Gross, H. *et al*, *Justus Liebigs Ann. Chem.*, 1967, **707**, 35.

2,2-Dihydro-1,3,2-benzoxazaphosphole D-00444

C_6H_6NOP M 139.093

2,2-Dimethoxy: [36952-33-7].
$C_8H_{10}NO_3P$ M 199.146
Solid. Mp 168-169°.

2,2-Diethoxy: [36952-32-6].
$C_{10}H_{14}NO_3P$ M 227.199
Solid. Mp 104.5-105°.

2,2-Diphenoxy: [36952-34-8].
$C_{18}H_{14}NO_3P$ M 323.287
Solid. Mp 201-202°.

2,2-Di-Ph: [41458-68-8].
$C_{18}H_{14}NOP$ M 291.288
Solid. Dec. at 311°.

2,2-Dichloro: [36952-30-4].
$C_6H_4Cl_2NOP$ M 207.983
Solid. Mp 270°.

2,2-Bis(dimethylamino): [67201-32-5].
$C_{10}H_{16}N_3OP$ M 225.230
Solid. Mp 174-177°.

Kabachnik, M.I. *et al*, *Dokl. Akad. Nauk SSSR, Ser. Sci. Khim.*, 1972, **204**, 1352 (*synth*)
Stegmann, H.B. *et al*, *Synthesis*, 1973, 162 (*diphenyl*)
Cadogan, J.I.G. *et al*, *J. Chem. Soc., Chem. Commun.*, 1978, 182 (*dimethylamino, synth, pmr, P nmr*)

2,2-Dihydro-3,4-bis(methoxycarbonyl)-2,2-dimethyl-5,5-diphenyl-5*H*-1,2-azaphosphole, 9CI, 8CI D-00445

[33536-07-1]

$C_{21}H_{22}NO_4P$ M 383.383
A cyclic phosphine imide.

Schmidpeter, A. *et al*, *Angew. Chem., Int. Ed. Engl.*, 1971, **10**, 396 (*pmr, nmr*)
Schmidpeter, A. *et al*, *Chem. Ber.*, 1978, **111**, 3747 (*props*)
Sheldrick, W.S. *et al*, *Chem. Ber.*, 1980, **113**, 55 (*cryst struct*)

2,2-Dihydro-3,3-bis(trifluoromethyl)-1-[[2,2,2-trifluoro-1-(trifluoromethyl)-ethylidene]amino]azaphosphiridine, 9CI D-00446

$C_6H_3F_{12}N_2P$ M 362.058

2,2,2-Trimethoxy: [49629-23-4].
$C_9H_9F_{12}N_2O_3P$ M 452.136
Liq. Bp_{12} 83°.

2,2,2-Triethoxy: [49629-24-5].
$C_{12}H_{15}F_{12}N_2O_3P$ M 494.217
Liq. Bp_{12} 96°.

2,2,2-Tris(dimethylamino): [49661-82-7].
$C_{12}H_{18}F_{12}N_5P$ M 491.262
$Bp_{0.001}$ 56°.

Burger, K. *et al*, *Angew. Chem., Int. Ed. Engl.*, 1973, **12**, 502 (*synth, ir, F nmr, pmr*)

1,2-Dihydro-1,2-bis(triphenylphosphonio)-benzocyclobutene(2+), 9CI D-00447

Bicyclo[4.2.0]octa-1,3,5-trien-7,8-ylenebis[triphenyl-phosphonium](2+), 8CI

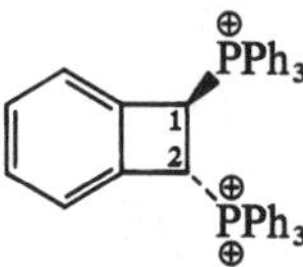

$C_{44}H_{36}P_2^{\oplus\oplus}$ M 626.716 (ion)

(1RS,2RS)-form

(±)-trans-*form*

Dibromide: [1183-90-0].
$C_{44}H_{36}Br_2P_2$ M 786.524
Source of 1,2-Bis(triphosphoranylidene)-benzocyclobutene, B-00460 . Plates (CH_2Cl_2/Et_2O). Mp 230.5-232°.

Diperchlorate: [16522-04-6].
$C_{44}H_{36}Cl_2O_8P_2$ M 825.617
Source of 1,2-Bis(triphosphoranylidene)-benzocyclobutene, B-00460 . Needles. Mp 278-280° dec.

Bis-ylide: see 1,2-Bis(triphosphoranylidene)-benzocyclobutene, B-00460

Blomquist, A.T. *et al*, *J. Am. Chem. Soc.*, 1964, **86**, 5041 (*bromide, pmr*)
Blomquist, A.T. *et al*, *J. Am. Chem. Soc.*, 1967, **89**, 4996 (*synth, ir, uv*)
Sondheimer, F. *et al*, *J. Org. Chem.*, 1981, **46**, 4594 (*use*)

10,11-Dihydro-1*H*-dibenz[*b,e*][1,4]-azaphosphepine 5-oxide, 9CI D-00448

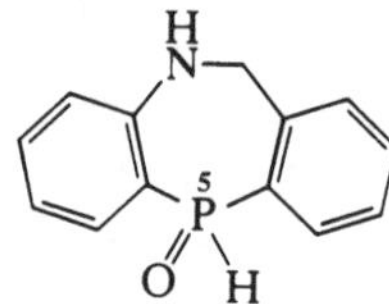

$C_{13}H_{12}NOP$ M 229.218

5-Me: [65884-60-8].
$C_{14}H_{14}NOP$ M 243.244
Cryst. (EtOAc). Mp 178°.

5-Ph: [65884-59-5].
$C_{19}H_{16}NOP$ M 305.315
Cryst. (EtOH). Mp 275-276°.

Segall, Y. *et al, J. Chem. Res. (S)*, 1977, 310 (*synth, uv, pmr, ms*)

5,10-Dihydro-11*H*-dibenz[*b,e*][1,4]-azaphosphepin-11-one 5-oxide, 9CI — D-00449

$C_{13}H_{10}NO_2P$ M 243.201
5-Me: [65884-58-4].
$C_{14}H_{12}NO_2P$ M 257.228
Solid. Mp 260°.
5-Ph: [65884-57-3].
$C_{19}H_{14}NO_2P$ M 319.299
Cryst. (EtOH). Mp 349° (324°).
N-*Me*, *5-Ph:*
$C_{20}H_{16}NO_2P$ M 333.326
Cryst. (C_6H_6). Mp 230-233°.

Segall, Y. *et al, J. Chem. Res. (S)*, 1977, 310 (*synth, ir, uv, pmr, ms*)
Petrov, K.A. *et al, Zh. Obshch. Khim.*, 1978, **48**, 1187 (*Engl. transl.* p. 1087) (*synth, ir, ms*)

6,7-Dihydro-5*H*-dibenzo[*d,f*][1,3,2]-diazaphosphepine, 10CI — D-00450

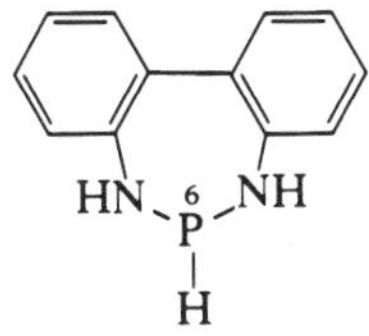

$C_{12}H_{11}N_2P$ M 214.206
6-Me, 6-oxide: [68521-29-9].
$C_{13}H_{13}N_2OP$ M 244.232
Cryst. (MeOH/Et_2O). Mp 252-255°.
6-Ph, 6-oxide: [68521-33-5].
$C_{18}H_{15}N_2OP$ M 306.303
Cryst. (EtOH aq.). Mp 330-333° dec.

Naidu, M.S.R. *et al, Bull. Chem. Soc. Jpn.*, 1978, **51**, 2156 (*synth, ir, ms*)

10,11-Dihydro-5*H*-dibenzo[*b,f*]phosphepin, 9CI — D-00451

1-Phospha-2,3:6,7-dibenzocyclohepta-2,6-diene
Oxide: [59506-93-3].
$C_{14}H_{13}OP$ M 228.230
Air-sensitive cryst. (cyclohexane). Mp 94°.

Segall, Y. *et al, Phosphorus Sulfur*, 1980, **8**, 243 (*synth, ir*)

5,10-Dihydro-5,10-dihydroxyphosphanthrene 5,10-dioxide, 9CI, 8CI — D-00452

9,10-Dihydro-9,10-dihydroxy-9,10-diphosphaanthracene 9,10-dioxide
[63586-90-3]

$C_{12}H_{10}O_4P_2$ M 280.156
Cryst. (EtOH aq.). Mp >300°.

Kovaloveva, T.V., *Zh. Obshch. Khim.*, 1977, **47**, 1036 (*Engl. transl.* p. 950) (*synth*)

1,5-Dihydro-3-dimethylamino-2,4,3-benzodioxaphosphepin — D-00453

N,N-*Dimethyl-1,5-dihydro-2,4,3-benzodioxaphosphepin-3-amine, 9CI*
[69813-53-2]

$C_{10}H_{14}NO_2P$ M 211.200
Liq. $Bp_{0.03}$ 78-80°. n_D^{21} 1.5552.
3-Oxide:
$C_{10}H_{14}NO_3P$ M 227.199
Characterised spectroscopically.
3-Sulfide:
$C_{10}H_{14}NO_2PS$ M 243.260
Cryst. (C_6H_6). Mp 159°.
3-Selenide:
$C_{10}H_{14}NO_2PSe$ M 290.160
Cryst. Mp 176.5°.

Grand, A. *et al, Acta Crystallogr., Sect. B*, 1978, **34**, 199 (*sulfide, cryst struct*)
Guimaraes, A.C. *et al, Org. Magn. Reson.*, 1978, **11**, 411 (*sulfide, pmr, cmr, P nmr*)
Arbuzov, B.A. *et al, Izv. Akad. Nauk SSSR, Ser. Khim.*, 1982, **588**, 1195 (*Engl. transl.* pp. 520, 1069) (*synth, conformn*)
Shagidullin, R.R. *et al, Izv. Akad. Nauk SSSR, Ser. Khim.*, 1984, 1803 (*Engl. transl.* p. 1649) (*derivs, ir, raman, conformn*)
Kadyrov, R.A. *et al, Izv. Akad. Nauk SSSR, Ser. Khim.*, 1985, 799 (*Engl. transl.* p. 724)

4,5-Dihydro-2-dimethylamino-1,3,2-benzodioxaphosphepin — D-00454

N,N-*Dimethyl-4,5-dihydro-1,3,2-benzodioxaphosphepin-2-amine, 10CI*

$C_{10}H_{14}NO_2P$ M 211.200
2-Oxide: [72374-39-1].
$C_{10}H_{14}NO_3P$ M 227.199
Cryst. (cyclohexane). Mp 98°.

Gervais, C. *et al, Tetrahedron*, 1979, **35**, 745 (*synth, pmr*)

2,3-Dihydro-2-dimethylamino-5-methyl-3-phenyl-1,3,4,2-oxadiazaphosphole D-00455

2,3-Dihydro-N,N,5-trimethyl-3-phenyl-1,3,4,2-oxadiazaphosphol-2-amine, 9CI. 2-Dimethylamino-5-methyl-3-phenyl-Δ⁴-1,3,4,2-oxadiazaphospholine, 8CI

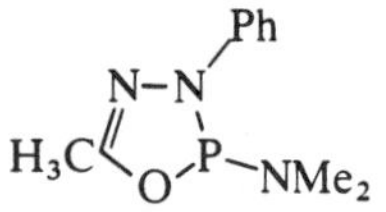

$C_{10}H_{14}N_3OP$ M 223.214

2-Oxide: [31596-41-5].
$C_{10}H_{14}N_3O_2P$ M 239.213
Solid. Mp 78-81°.

2-Sulfide: [21398-21-0].
$C_{10}H_{14}N_3OPS$ M 255.274
Solid. Mp 56-58°.

Italinskaya, T.L. *et al*, *Zh. Obshch. Khim.*, 1968, **38**, 2265 (*Engl. transl.* p. 2192) (*sulfide*)
Shvetsov-Shilovskii, N.I. *et al*, *Zh. Obshch. Khim.*, 1971, **41**, 1200 (*Engl. transl.* p. 1210) (*oxide*)

2,3-Dihydro-1,3-dimethyl-1*H*-1,3,2-benzodiazaphosphole, 9CI D-00456

1,3-Dimethyl-1H-1,3,2-benzodiazaphospholine

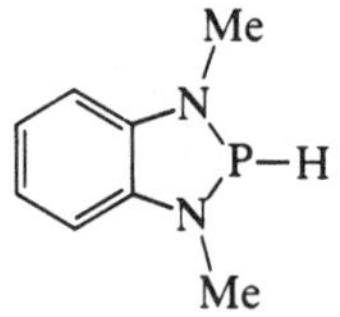

$C_8H_{11}N_2P$ M 166.162

2-Chloro: [59901-85-8].
$C_8H_{10}ClN_2P$ M 200.607
Waxy solid. Dec. in air, and on attempted purification.

2-Ph: *see 2,3-Dihydro-1,3-dimethyl-2-phenyl-1H-benzodiazaphosphole, D-00467*

2-Diisopropylamino:
$C_{14}H_{24}N_3P$ M 265.337
Waxy solid. Mp 60°.

Anisimova, O.S. *et al*, *Zh. Obshch. Khim.*, 1976, **46**, 808 (*Engl. transl.* p. 807) (*ms, uv*)
Jennings, W.B. *et al*, *J. Chem. Soc., Perkin Trans. 2*, 1981, 1411 (*derivs, pmr, pe, cmr, N nmr*)

2,3-Dihydro-1,3-dimethyl-1,3,2-benzodiazaphosphorin-4(1*H*)-one, 9CI D-00457

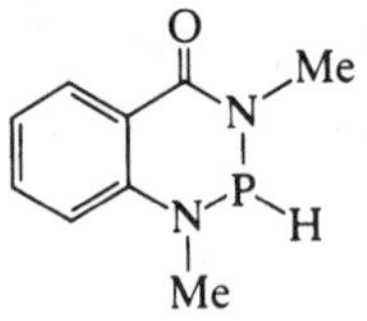

$C_9H_{11}N_2OP$ M 194.172

2-Oxide: [68614-93-7].
$C_9H_{11}N_2O_2P$ M 210.172
Cryst. (CH_2Cl_2/Et_2O). Mp 124-126°.

2-Me, 2-oxide: [71476-14-7].
$C_{10}H_{13}N_2OP$ M 208.199
Cryst. (Et_2O). Mp 123-126°.

Coppola, G.M. *et al*, *J. Heterocycl. Chem.*, 1978, **15**, 1169; 1979, **16**, 897 (*synth, ir, pmr*)

2,3-Dihydro-2,7-dimethyl-1,4,2-benzodithiaphosphorin, 9CI, 8CI D-00458

[50636-30-1]

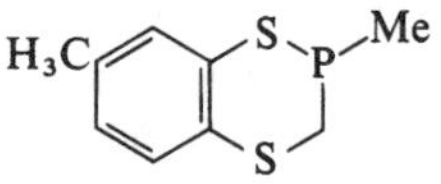

$C_9H_{11}PS_2$ M 214.280
Liq. $Bp_{0.01}$ 96-100°.

2-Selenide: [50636-32-3].
$C_9H_{11}PS_2Se$ M 293.240
Solid. Mp 58-60°. $Bp_{0.01}$ 180-183°.

Wieber, M. *et al*, *Chem. Ber.*, 1973, **106**, 2733 (*synth, pmr*)

2,2-Dihydro-1,3-dimethyl-1,3,2-diazaphosphetidin-4-one, 11CI, 9CI, 8CI D-00459

2,2-Dihydro-1,3-dimethyl-4-oxo-1,3,2-diazaphosphetidine, 10CI

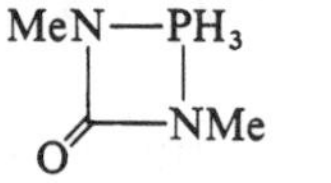

$C_3H_9N_2OP$ M 120.091

2,2,2-Trifluoro: [32707-12-3].
$C_3H_6F_3N_2OP$ M 174.062
Fuming liq. Bp_{20} 39°.

2,2,2-Trichloro: [3576-20-3].
$C_3H_6Cl_3N_2OP$ M 223.426
Fuming liq. $Bp_{1.5}$ 78-79°.

2-Me, 2,2-difluoro: [31053-08-4]. *2,2-Difluoro-2,2-dihydro-1,2,3-trimethyl-1,3,2-diazaphosphetidin-4-one.*
$C_4H_9F_2N_2OP$ M 170.098
Liq. $Bp_{0.1}$ 39°.

2-Ph, 2,2-difluoro: [32707-15-6]. *2,2-Difluoro-2,2-dihydro-1,3-dimethyl-2-phenyl-1,3,2-diazaphosphetidin-4-one.*
$C_9H_{11}F_2N_2OP$ M 232.169
Liq. $Bp_{0.5}$ 130°.

Ulrich, H. *et al*, *Angew. Chem.*, 1964, **76**, 647 (*trichloride, synth, ir, pmr*)
Dunmur, R.E. *et al*, *J. Chem. Soc. (A)*, 1971, 1289 (*derivs, synth, ir, pmr, F and P nmr*)
Baldwin, M.A. *et al*, *Org. Mass Spectrom.*, 1977, **12**, 275 (*ms*)

2,3-Dihydro-4,5-dimethyl-2,7-diphenyl-1*H*-1,2-azaphosphepine, 9CI D-00460

4,5-Dimethyl-2,7-diphenyl-1-aza-2-phospha-4,5-cycloheptadiene

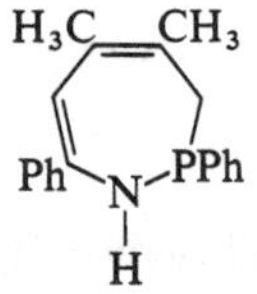

$C_{19}H_{20}NP$ M 293.347

2-Oxide: [38421-29-3].
$C_{19}H_{20}NOP$ M 309.347
Cryst. (methylcyclohexane or $Me_2CO/EtOH$) in 2 forms. Mp 196°, Mp 203° (dimorph.). Forms differ only in details of H-bonding.

2-Sulfide: [40168-00-1].
$C_{19}H_{20}NPS$ M 325.407
Cryst. (EtOH). Mp 131°.

Lampin, J.-P. *et al*, *Tetrahedron*, 1972, **28**, 5367 (*oxide, synth, ir*)
Eberhard, L. *et al*, *Tetrahedron*, 1973, **29**, 2909 (*sulfide, synth, pmr*)

Lampin, J.-P. *et al, Acta Crystallogr., Sect. B*, 1974, **30**, 1626 (*oxide, cryst struct*)
Sheldrick, W.S., *Acta Crystallogr., Sect. B*, 1975, **31**, 2491 (*oxide, cryst struct*)

2,2-Dihydro-2,2-dimethyl-3,5-diphenyl-3*H*-1,2-azaphosphole-4-carboxylic acid, 9CI D-00461

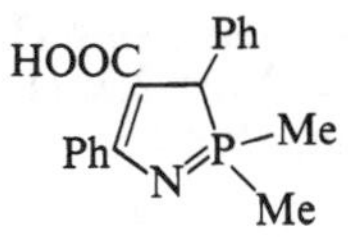

$C_{18}H_{18}NO_2P$ M 311.319

(±)-*form*

Et ester: [37385-44-7]. *2,2-Dihydro-4-ethoxycarbonyl-2,2-dimethyl-3,5-diphenyl-3*H*-1,2-azaphosphole.*
$C_{20}H_{22}NO_2P$ M 339.373
Solid. Mp 129-131°.

Nitrile: [37385-42-5]. *4-Cyano-2,2-dihydro-3,5-diphenyl-3*H*-1,2-azaphosphole.*
$C_{18}H_{17}N_2OP$ M 308.319
Solid. Mp 118-130°.

Schmidpeter, A. *et al, Z. Naturforsch., B*, 1972, **27**, 769 (*synth*)

2,3-Dihydro-4,5-dimethyl-2,7-diphenyl-1,2-oxaphosphepin 2-oxide, 9CI D-00462

[41902-21-0]

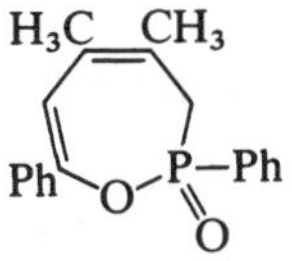

$C_{19}H_{19}O_2P$ M 310.332
Cryst. (methylcyclohexane). Mp 125°.

Mathey, F. *et al, Bull. Soc. Chim. Fr. Part II*, 1973, 2783 (*synth, props*)
Mathey, F., *Tetrahedron*, 1973, **29**, 707 (*synth, ir, pmr*)
Mathey, F. *et al, Can. J. Chem.*, 1975, **53**, 855 (*synth, pmr*)

2,3-Dihydro-4,5-dimethyl-2,2-diphenyl-1,2-thiaphosphepin, 9CI D-00463

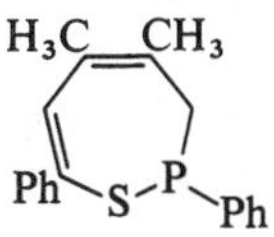

$C_{19}H_{19}PS$ M 310.393

2-Sulfide: [61157-01-5].
$C_{19}H_{19}PS_2$ M 342.453
No phys. props. reported.

Mathey, F. *et al, J. Organomet. Chem.*, 1976, **117**, 377 (*synth, pmr*)

2,5-Dihydro-5,5-dimethyl-1,2-oxaphosphole 2-oxide, 9CI D-00464

5,5-Dimethyl-1,2-oxaphosphol-3-ene 2-oxide, 8CI
[59474-16-7]

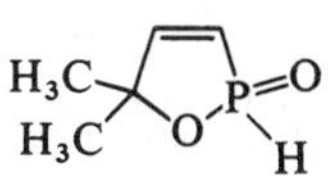

$C_5H_9O_2P$ M 132.099

2-Chloro: [75779-67-8].
$C_5H_8ClO_2P$ M 166.544
Cryst. Mp 83-85.5° (71-72°). Bp_1 72-74°.

2-Hydroxy: [59474-16-7].
$C_5H_9O_3P$ M 148.098
Cryst. (Me_2CO). Mp 156-157.5°. Bp_1 210-212°.

2-Methoxy:
$C_6H_{11}O_3P$ M 162.125
Liq. d_4^{20} 1.16. Bp_1 86-88°. n_D^{20} 1.4608.

2-Ethoxy:
$C_7H_{13}O_3P$ M 176.152
Liq. d_4^{20} 1.12. $Bp_{1.5}$ 102-103°. n_D^{20} 1.4549.

2-Phenoxy: [85152-45-0].
$C_{11}H_{13}O_3P$ M 224.196
Oil which cryst. slowly. Mp 33-35°. $Bp_{0.08}$ 140° (oven).

2-Diethylamino: [85152-46-1].
$C_9H_{18}NO_2P$ M 203.220
Oil which cryst. slowly. Mp 40-42°. $Bp_{0.21}$ 88-91°.

Macomber, R.S. *et al, J. Org. Chem.*, 1976, **41**, 3191 (*synth, ir, P nmr, pmr*)
Mikhailova, T.S. *et al, Zh. Obshch. Khim.*, 1980, **50**, 1690 (*Engl. transl.* p. 1370) (*derivs, synth, ir, P nmr, pmr*)
Macomber, R.S. *et al, J. Org. Chem.*, 1981, **46**, 4038; 1983, **48**, 1420 (*derivs, synth, ir, ms, pmr*)

5,5-Dihydro-4,4-dimethyl-2,3,5,5,5-pentaphenyl-1,2,5-oxazaphospholidine D-00465

[14561-29-6]

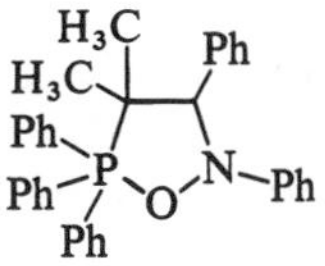

$C_{34}H_{32}NOP$ M 501.607
Solid. Mp 142-143° dec.

Wulff, J. *et al, Angew. Chem., Int. Ed. Engl.*, 1967, **6**, 457 (*synth, P nmr, pmr*)

5,10-Dihydro-5,10-dimethylphenophosphazine, 9CI D-00466

[58943-95-6]

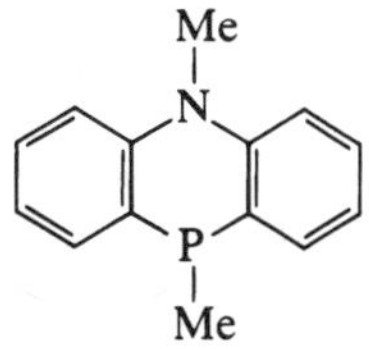

$C_{14}H_{14}NP$ M 227.245
Cryst. (EtOH). Mp 107-108°.

10-Oxide: [58943-97-8].
$C_{14}H_{14}NOP$ M 243.244
Cryst. (C_6H_6). Mp 205-207°.

10-Sulfide: [77205-54-0].
$C_{14}H_{14}NPS$ M 259.305
Cryst. (2-propanol). Mp 183.5-184°.

Smirnov, A.N. *et al, Zh. Obshch. Khim.*, 1977, **47**, 2468 (*Engl. transl.* p. 2256) (*synth, oxide, uv, ms, ir*)
Piskunova, O.G. *et al, Zh. Obshch. Khim.*, 1978, **48**, 1316 (*Engl. transl.* p. 1205) (*oxide, uv*)
Negrebetskii, V.V., *Zh. Strukt. Khim.*, 1979, **20**, 540 (*Engl. transl.* p. 459) (*cmr*)
Demidova, N.I. *et al, Zh. Obshch. Khim.*, 1980, **50**, 2809 (*sulfide, synth, uv*)

2,3-Dihydro-1,3-dimethyl-2-phenyl-1*H*-benzodiazaphosphole, 9CI D-00467

*1,3-Dimethyl-2-phenyl-1*H*-benzodiazaphospholine*

[59901-84-7]

$C_{14}H_{15}N_2P$ M 242.260

Solid. Mp 62-64°. $Bp_{0.0001}$ 95-97°.

2-Sulfide: [4600-12-8].
$C_{14}H_{15}N_2PS$ M 274.320
Cryst. (hexane). Mp 99.5-101.5°.

Anisimova, O.S. *et al*, *Zh. Obshch. Khim.*, 1976, **46**, 808 (*Engl. transl.* p. 807) (*synth, ms, uv*)

2,5-Dihydro-3,4-dimethyl-1-phenyl-1*H*-phosphole, 9CI D-00468

3,4-Dimethyl-1-phenyl-3-phospholene, 8CI

[14409-95-1]

$C_{12}H_{15}P$ M 190.224

Liq. Bp_{20} 158-160°, $Bp_{0.15}$ 74-75°.

*B,MeI: 2,5-Dihydro-1,3,4-trimethyl-1-phenyl-1*H*-phospholium bromide.*
$C_{13}H_{18}IP$ M 332.163
Solid. Mp 210°.

*B,PhCH$_2$Br: 1-Benzyl-2,5-dihydro-3,4-dimethyl-1-phenyl-1*H*-phospholium bromide.*
$C_{19}H_{22}BrP$ M 361.261
Solid.

1-Oxide: [710-89-4].
$C_{12}H_{15}OP$ M 206.224
Liq. which solidifies. $Bp_{0.1}$ 134-135°.

1-Sulfide: [39997-45-0].
$C_{12}H_{15}PS$ M 222.284
Deoxygenates epoxides to alkenes, and also converts epoxides into thiiranes. Cryst. (hexane or EtOH). Mp 78-80°.

1-Selenide: [58608-01-8].
$C_{12}H_{15}PSe$ M 269.184
Converts epoxides into alkenes. Cryst. (MeOH). Mp 88-89°.

Fritzsche, H. *et al*, *Chem. Ber.*, 1964, **97**, 1988 (*synth, derivs*)
Quin, L.D. *et al*, *J. Org. Chem.*, 1964, **29**, 836 (*synth, deriv*)
Lampin, J.-P. *et al*, *Bull. Soc. Chim. Fr.*, 1972, 3494 (*use*)
Chang, L.L. *et al*, *Phosphorus*, 1974, **4**, 265 (*synth, derivs, P nmr*)
Mathey, F., *C.R. Hebd. Seances Acad. Sci., Ser. C*, 1975, **281**, 881 (*sulfide, selenide, props, use*)
Mathey, F. *et al*, *J. Organomet. Chem.*, 1976, **105**, 73 (*props*)

1,3-Dihydro-1,3-dimethylspiro[2*H*-1,3,2-benzodiazaphosphole-2,2′[1,3,2]-benzodioxaphosphole] D-00469

$C_{14}H_{15}N_2O_2P$ M 274.258

2-Me: [79129-17-2]. *1,3-Dihydro-1,2,3-trimethylspiro[2*H*-1,3,2-benzodiazaphosphole-2,2′-[1,3,2]-benzodioxaphosphole].*
$C_{15}H_{17}N_2O_2P$ M 288.285
Cryst. (Et_2O). Mp 122°.

Wieber, M. *et al*, *Z. Anorg. Allg. Chem.*, 1981, **477**, 108 (*synth, pmr, P nmr*)

4′,5′-Dihydro-3′,5′-dimethylspiro[1,3,2-benzoxazaphosph(P^V)ole-2,2′-[1,3,2]-oxazaphospholidin]-4′-one D-00470

[51676-08-5]

(5′*S*)-*form*

$C_{10}H_{13}N_2O_3P$ M 240.198

Tautomeric with monocyclic (P^{III})-form.

(5′S)-form

3-Me: [51676-09-6]. Liq. $Bp_{0.02}$ 118°. n_D^{20} 1.474. Tautomeric.

(±)-form

Liq. $Bp_{0.04}$ 95-97°. n_D^{20} 1.471.

Burgada, R. *et al*, *J. Organomet. Chem.*, 1974, **66**, 255 (*synth, P nmr, props*)

2,3-Dihydro-3,5-dimethyl-1,3,4,2-thiadiazaphosphole 2-sulfide, 9CI D-00471

$C_3H_7N_2PS$ M 134.135

2-Chloro: [35169-11-0].
$C_3H_6ClN_2PS_2$ M 200.641
Liq. $Bp_{0.22}$ 65-66°.

2-Ethoxy: [35169-18-5].
$C_5H_{11}N_2OPS_2$ M 210.248
Liq. $Bp_{0.1}$ 76-77°.

Italinskaya, T.L. *et al*, *Zh. Obshch. Khim.*, 1971, **41**, 1980 (*Engl. transl.* p. 1998) (*synth*)

4,7-Dihydro-1,3,2-dioxaphosphepin, 9CI D-00472

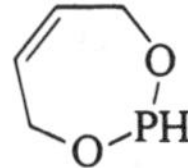

$C_4H_7O_2P$ M 118.072

2-Phenoxy: [59413-42-2].
$C_{10}H_{11}O_3P$ M 210.169
No phys. props. reported.

2-Ph: see 4,7-Dihydro-2-phenyl-1,3,2-dioxaphosphepin, D-00561

2-Dimethylamino: [69813-37-2].
$C_6H_{12}NO_2P$ M 161.140
Liq. $Bp_{0.1}$ 42°. n_D^{22} 1.4946.

Tabony, J., *Spectrochim. Acta, Part A*, 1979, **35**, 217 (*nmr*)
Arbuzov, B.A. *et al*, *Izv. Akad. Nauk SSSR, Ser. Khim.*, 1982, 588 (*Engl. transl.* p. 520) (*dialkylamino, synth, nmr, pmr, conformn*)

4,7-Dihydro-1,3,2-dioxaphosphepin 2-sulfide, 9CI D-00473

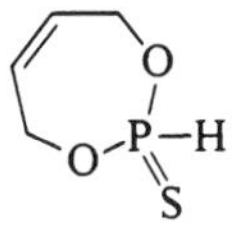

$C_4H_7O_2PS$ M 150.132

2-Chloro: [69813-43-0].
$C_4H_6ClO_2PS$ M 184.577
No phys. props. reported.
2-Phenoxy: [69813-44-1].
$C_{10}H_{11}O_3PS$ M 242.229
No phys. props. reported.
2-Me: [69813-46-3].
$C_5H_9O_2PS$ M 164.159
No phys. props. reported.
2-Ph: see under 4,7-Dihydro-2-phenyl-1,3,2-dioxaphosphepin, D-00561

Guimaraes, A.C. *et al*, *Org. Magn. Reson.*, 1978, **11**, 411 (*pmr, cmr, nmr*)
Tabony, J., *Spectrochim. Acta, Part A*, 1979, **35**, 217 (*nmr*)

[2-(1,3-Dihydro-1,3-dioxo-2*H*-isoindol-2-yl)ethyl]triphenylphosphonium(1+), 9CI D-00474

Triphenyl(2-phthalimidoethyl)phosphonium(1+)

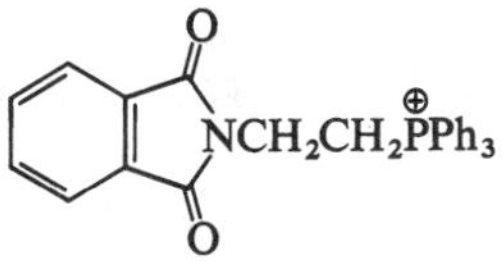

$C_{28}H_{23}NO_2P^{\oplus}$ M 436.469 (ion)
Salts exhibit antileukemia activity.
Chloride: [65273-60-1].
$C_{28}H_{23}ClNO_2P$ M 471.922
Cryst. + $0.5H_2O$. Mp 240-242°.
Bromide: [65273-64-5].
$C_{28}H_{23}BrNO_2P$ M 516.373
Cryst. (2-propanol). Mp 242-243°.

Dubois, R.J. *et al*, *J. Med. Chem.*, 1978, **21**, 303 (*synth*)

(1,3-Dihydro-1,3-dioxo-2*H*-isoindol-2-yl)-phosphonothioic acid, 9CI D-00475

Phthalimide N-*phosphonothioic acid. Phthalimidophosphonothioic acid*

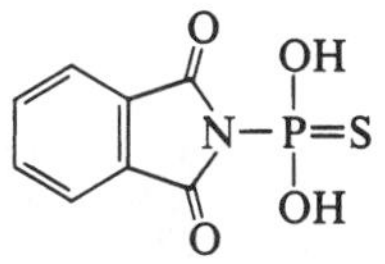

$C_8H_6NO_4PS$ M 243.173

O,O-*Di-Me ester:* [5022-30-0]. O,O-*Dimethyl (1,3-dihydro-1,3-dioxo-2*H*-isoindol-2-yl)phosphonothioate.*
$C_{10}H_{10}NO_4PS$ M 271.227
Solid. Mp 126.5-128°.
O,O-*Di-Et ester: see Ditalimfos, D-01237*
O,O-*Diisopropyl ester:* O,O-*Diisopropyl (1,3-dihydro-1,3-dioxo-2*H*-isoindol-2-yl)phosphonothioate.*
$C_{14}H_{18}NO_4PS$ M 327.334
Solid. Mp 77-81°.

Ger. Pat., 2 256 253, (*1973*); *CA*, **79**, 42140

2,3-Dihydro-1,3-diphenyl-1*H*-1,3-benzodiphosphole, 9CI D-00476

[38234-85-4]

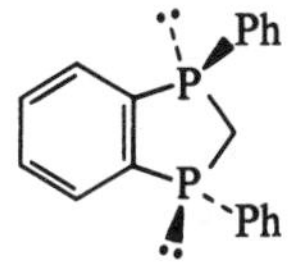

(1*RS*,3*RS*)-*form*

$C_{19}H_{16}P_2$ M 306.283
Cryst. (Me_2CO aq.). Mp 79-81°. $Bp_{0.1}$ 150°. Fails to yield a disulfide or a bismethiodide by direct reaction.

(*1RS,3RS*)-*form*
(±)-trans-*form*
*B,2MeBr: 2,3-Dihydro-1,3-dimethyl-1,3-diphenyl-1*H*-1,3-benzodiphospholium dibromide.*
$C_{21}H_{22}Br_2P_2$ M 496.160
Microcryst. powder + $1H_2O$. Mp 197-200°.

(*1RS,3SR*)-*form*
cis-*form*
B,2MeBr: [72091-04-4]. Cryst. + $1H_2O$. Mp >280°.

Mann, F.G. *et al*, *J. Chem. Soc., Perkin Trans. 1*, 1972, 2548 (*synth, ir, pmr, ms, props*)
Roberts, N.K. *et al*, *J. Am. Chem. Soc.*, 1979, **101**, 6254 (*salts, pmr*)

1,4-Dihydro-1,4-diphenyl-1,4-benzodiphosphorin, 9CI D-00477

[38240-97-0]

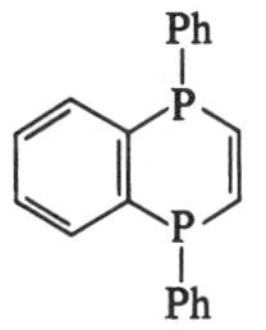

$C_{20}H_{16}P_2$ M 318.294
Needles (EtOH). Mp 108-109°.

Mann, F.G. *et al*, *J. Chem. Soc., Perkin Trans. 1*, 1972, 2548 (*synth, ir, pmr, ms*)

2,3-Dihydro-2,3-diphenyl-1,3,2-benzoxazaphosphole, 9CI D-00478

2,3-Diphenyl-1,3,2-benzoxazaphospholine

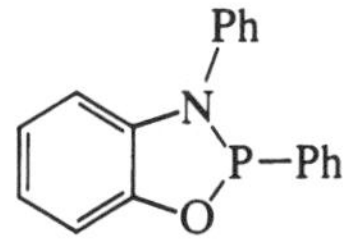

$C_{18}H_{14}NOP$ M 291.288

2-Oxide: [58656-41-0].
$C_{18}H_{14}NO_2P$ M 307.288
Solid. Mp 127-128°.

Cadogan, J.I.G. *et al*, *J. Chem. Soc., Perkin Trans. 1*, 1975, 2376, 2386 (*synth, pmr, P nmr*)
Cadogan, J.I.G. *et al*, *J. Chem. Soc., Perkin Trans. 1*, 1983, 1489 (*synth*)

8,8-Dihydro-8,8-diphenyl-2*H*,6*H*-[1,2]-oxaphospholo[4,3,2-*hi*][2,1]-benzoxaphosphole, 9CI D-00479

[51922-82-8]

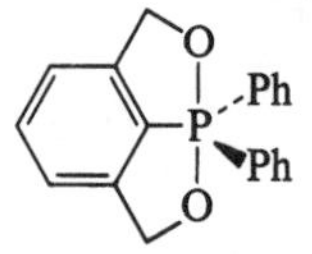

$C_{20}H_{17}O_2P$ M 320.327

Needles (MeOH). Mp 131-134°.

Hellwinkel, D. *et al*, *Chem. Ber.*, 1978, **111**, 13 (*synth, ir, cmr, pmr, P nmr, ms*)

2,2-Dihydro-4,5-diphenyl-1,3,2-oxathiaphosphole, 9CI D-00480

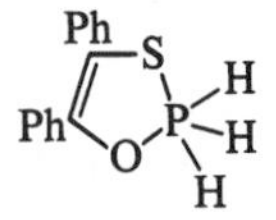

$C_{14}H_{13}OPS$ M 260.290

2,2,2-Trimethoxy: [57084-22-7].

$C_{17}H_{19}O_4PS$ M 350.368

Solid. Mp 98-101°.

Arbuzov, B.A. *et al*, *Izv. Akad. Nauk SSSR, Ser. Khim.*, 1975, 1658 (*Engl. transl.* p. 1548) (*synth, ir, pmr, nmr*)
Arbuzov, B.A. *et al*, *Dokl. Akad. Nauk SSSR, Ser. Sci. Khim.*, 1976, **228**, 865 (*Engl. transl.* p. 515) (*nmr, pmr*)

10,10-Dihydro-2,3-diphenyl-[1,3,2]-oxazaphospholo[2,3-*b*][1,3,2]-benzoxazaphosphole, 9CI D-00481

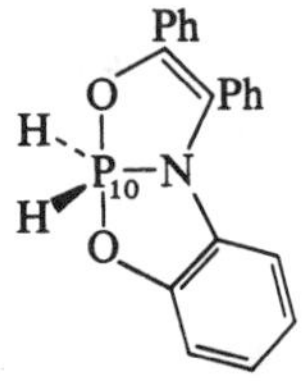

$C_{20}H_{16}NO_2P$ M 333.326

10,10-Dimethoxy: [55590-39-1].

$C_{22}H_{20}NO_4P$ M 393.378

Cryst (MeCN). Mp 148-150°.

10,10-Diphenoxy: [67347-97-1].

$C_{32}H_{24}NO_4P$ M 517.520

Cryst. (MeCN). Mp 100-102°.

10,10-Di-Me: [55590-36-8].

$C_{22}H_{20}NO_2P$ M 361.379

Needles (MeCN). Mp 158-160°.

10,10-Di-Ph: [55590-37-9].

$C_{32}H_{24}NO_2P$ M 485.521

Yellow needles (MeCN). Mp 177-178°.

10,10-Bis(dimethylamino): [55590-38-0].

$C_{24}H_{26}N_3O_2P$ M 419.462

Cryst. (MeCN). Mp 134-135°.

Schmidpeter, A. *et al*, *Angew. Chem., Int. Ed. Engl.*, 1975, **14**, 517.
Sheldrick, W.S. *et al*, *Acta Crystallogr., Sect. B*, 1976, **32**, 925 (*diphenyl, cryst struct*)
Schmidpeter, A. *et al*, *Chem. Ber.*, 1978, **14**, 2086 (*derivs, synth, pmr, P nmr*)

5,10-Dihydro-5,10-diphenylphosphanthrene, 9CI, 8CI D-00482

9,10-Dihydro-9,10-diphenyl-9,10-diphosphaanthracene

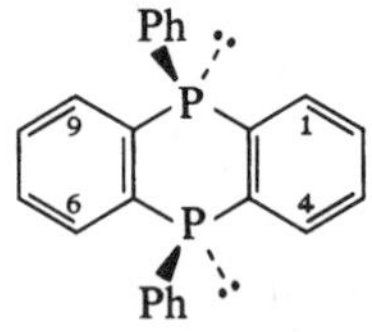

$C_{24}H_{18}P_2$ M 368.354

Can exist in *cis*- and *trans*-forms, each in a "butterfly" conformation.

B,2MeI: 9,10-Dihydro-9,10-dimethyl-9,10-diphenylphosphaanthranium diiodide.

$C_{26}H_{24}I_2P_2$ M 652.232

Cryst. (MeOH/DMF). Mp 385° dec., Mp 330° approx. The two Mp's may refer to *cis*- and *trans*-forms.

Monooxide:

$C_{24}H_{18}OP_2$ M 384.353

Cryst. (Me_2CO aq.). Mp 231-232.5°.

5,10-Dioxide:

$C_{24}H_{18}O_2P_2$ M 400.353

Hygroscopic cryst. (EtOH aq.), needles (EtOH aq.). Mp 276-278°, Mp >400°. Two stereoisomers known, both stable on subl. at 15 mm.

Davis, M. *et al*, *J. Chem. Soc.*, 1964, 3770.

1,1-Dihydro-1,1-diphenyl-3*H*-phosphindole D-00483

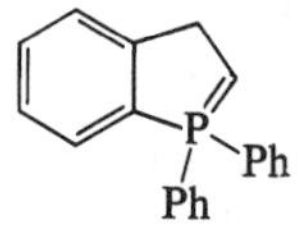

$C_{20}H_{17}P$ M 288.328

Cyclic ylide produced *in situ*.

Märkl, G., *Z. Naturforsch., B*, 1963, **18**, 84.

1,1-Dihydro-1,1-diphenylphosphorin, 9CI, 8CI D-00484

1,1-Dihydro-1,1-diphenylphosphinine

[19108-57-7]

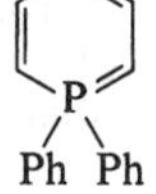

$C_{17}H_{15}P$ M 250.279

Yellow powder.

Märkl, G., *Angew. Chem.*, 1963, **75**, 669 (*synth, uv*)
Vilceanu, R. *et al*, *Rev. Roum. Chim.*, 1968, **13**, 533 (*struct*)
Mracec, M. *et al*, *Rev. Roum. Chim.*, 1971, **16**, 449 (*struct*)
Schafer, W., *J. Am. Chem. Soc.*, 1976, **98**, 4410 (*struct*)

1,1-Dihydro-1,1,3,3,5,5-hexamethyl-1-phospha-3,5-disilacyclohex-1-ene, 9CI D-00485

[58802-35-0]

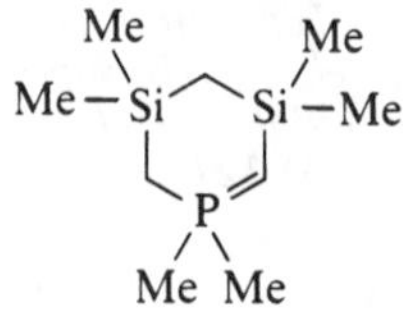

$C_9H_{23}PSi_2$ M 218.425

A cyclic ylide. Forms Ir and Rh complexes. Mp −8°. Bp_2 75°.

B,HCl: 1,1,3,3,5,5-Hexamethyl-1-phosphonia-3,5-disilacyclohexane chloride.
$C_9H_{23}ClPSi_2$ M 253.878
Solid. Mp 225°.

Schmidbaur, H. *et al*, *Chem. Ber.*, 1978, **111**, 2696 (*synth, pmr, cmr, P nmr*)
Murray, B.D. *et al*, *Organometallics*, 1983, **2**, 1700 (*complexes*)

1,1-Dihydro-1,1,3,3,4,4-hexamethyl-1-phospha-3,4-disilacyclopent-1-ene, 9CI D-00486

[39980-64-8]

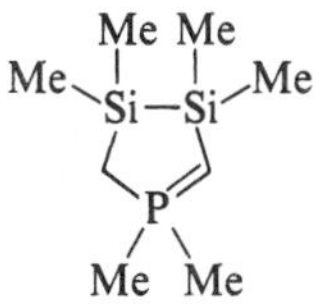

$C_8H_{21}PSi_2$ M 204.399
Cyclic ylide. Liq. Mp −6° to −5°. $Bp_{0.1}$ 53°.

Schmidbaur, H. *et al*, *Chem. Ber.*, 1972, **105**, 3173 (*synth*)

5,10-Dihydro-5-hydroxyacridophosphine 5-oxide, 9CI D-00487

5,10-Dihydro-5-hydroxydibenzo[b,e]*phosphorin 5-oxide. 5,10-Dihydro-5-hydroxydibenzo*[b,e]*phosphinine 5-oxide*

[18593-24-3]

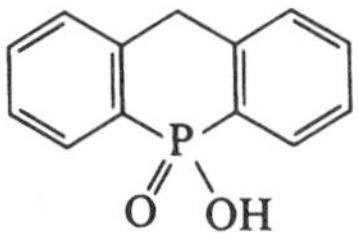

$C_{13}H_{11}O_2P$ M 230.202
Solid. Mp >225° dec.

5-Methoxy (*5-Me ester*)*:* [18593-25-4]. *5,10-Dihydro-5-methoxyacridophosphine 5-oxide.*
$C_{14}H_{13}O_2P$ M 244.229
Characterised spectroscopically.

Doak, G.O. *et al*, *J. Org. Chem.*, 1964, **29**, 2382 (*synth, uv*)
de Koe, P. *et al*, *Angew. Chem., Int. Ed. Engl.*, 1967, **6**, 567 (*synth*)
Haake, P. *et al*, *J. Org. Chem.*, 1969, **34**, 788 (*synth, ester, ms*)

1,3-Dihydro-1-hydroxy-1,2,3-benziodoxaphosphole 3-oxide, 9CI D-00488

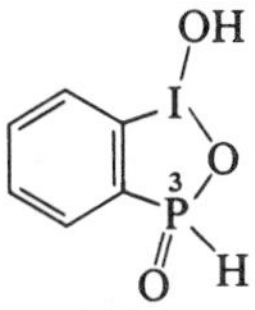

$C_6H_6IO_3P$ M 283.990
Derivs. represent the probable structs. of derivs. of *o*-Iodosophenylphosphonic or *o*-Iodosophenylphosphinic acids.

3-Hydroxy: [54185-80-7]. *1,3-Dihydro-1,3-dihydroxy-1,2,3-benziodoxaphosphole 3-oxide. 2-Iodosophenylphosphonic acid.*
$C_6H_6IO_4P$ M 299.989
Solid. Mp 210-212°.

3-Me: [67673-30-7]. *1,3-Dihydro-1-hydroxy-3-methyl-1,2,3-benzodoxaphosphole 3-oxide.* (*2-Iodosophenyl*)*-methylphosphinic acid.*
$C_7H_8IO_3P$ M 298.017
Solid. Mp 178-185°.

3-Ph: [54185-79-4]. *1,3-Dihydro-1-hydroxy-3-phenyl-1,2,3-benziodoxaphosphole 3-oxide.* (*2-Iodosophenyl*)*phenylphosphinic acid.*
$C_{12}H_{10}IO_3P$ M 360.087
Cryst. (50% EtOH aq.). Mp 209-211°.

Freedman, L.D. *et al*, *Phosphorus*, 1974, **3**, 277 (*synth, struct*)
Balthazor, T.M. *et al*, *J. Org. Chem.*, 1978, **43**, 4538 (*synth, cryst struct, pmr, ms*)

4,5-Dihydro-4-hydroxybenzo[*lmn*]-phosphanthridine 4-oxide, 9CI D-00489

4,5-Dihydro-4-hydroxy-4-phosphapyrene 4-oxide

[38021-51-1]

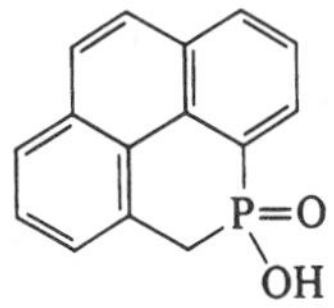

$C_{15}H_{11}O_2P$ M 254.224
Cryst. (H_2O). Mp 255-257°.

Robinson, C.N. *et al*, *J. Heterocycl. Chem.*, 1972, **9**, 735 (*synth*)

1,3-Dihydro-1-hydroxy-2,1-benzoxaphosphole 1-oxide, 9CI D-00490

[75777-32-1]

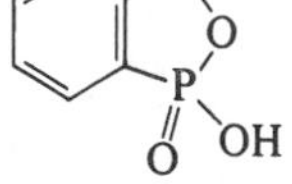

$C_7H_7O_3P$ M 170.104
Cryst. (Me_2CO). Mp 167°.

1-Ethoxy (*Et ester*)*:* [75777-31-0].
$C_9H_{11}O_3P$ M 198.158
Cryst. (C_6H_6/pet. ether). Mp 60°.

Miles, J.A. *et al*, *J. Org. Chem.*, 1982, **47**, 1677 (*synth, deriv, pmr*)

6,7-Dihydro-6-hydroxy-5*H*-dibenzo[*c,e*]-phosphepin 6-oxide, 9CI D-00491

[40964-82-7]

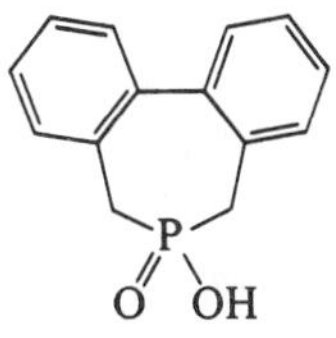

Solid. Mp 246-248°. $Bp_{0.6}$ 290° subl.

Robinson, C.N. *et al*, *Tetrahedron Lett.*, 1972, 4977 (*synth*)

10,11-Dihydro-5-hydroxy-5*H*-dibenzo[*b,f*]phosphepin 5-oxide, 9CI D-00492

1-Hydroxy-1-oxo-1-phospha-2,3:6,7-dibenzocyclohepta-2,6-diene

[30309-73-0]

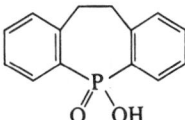

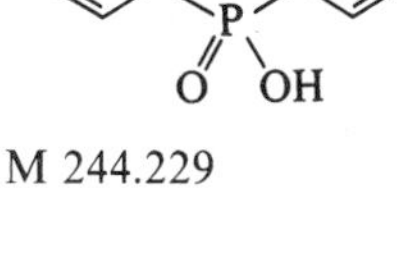

$C_{14}H_{13}O_2P$ M 244.229

Cryst. (EtOH or EtOH aq.). Mp 255-260° (246-251°).
Chloride: [75231-66-2]. *5-Chloro-10,11-dihydro-5*H-*dibenzo*[b,f]*phosphepin 5-oxide.*
$C_{14}H_{12}ClOP$ M 262.675
Solid. Mp 75-80°. Unpurified.
5-Methoxy (*5-Me ester*): [75231-67-3]. *10,11-Dihydro-5-methoxy-5*H-*dibenzo*[b,f]*phosphepin 5-oxide.*
$C_{15}H_{15}O_2P$ M 258.256
Cryst. (cyclohexane). Mp 54°.

Suggs, J.L., *J. Org. Chem.*, 1971, **36**, 2566 (*synth, pmr, uv, ms*)
Segall, Y. *et al*, *Phosphorus Sulfur*, 1980, **8**, 243 (*synth, ir, uv, pmr, ms, derivs*)

2,5-Dihydro-1-hydroxy-3,4-dimethyl-1*H*-phosphole 1-oxide, 9CI **D-00493**
1-Hydroxy-3,4-dimethyl-3-phospholene 1-oxide, 8CI
[695-67-0]

$C_6H_{11}O_2P$ M 146.125
Cryst. (H_2O). Mp 122-123°. $Bp_{0.05}$ 168-169°.
▷SZ6100400.
1-Methoxy (*Me ester*): [697-29-0]. *2,5-Dihydro-1-methoxy-3,4-dimethyl-1*H-*phosphole 1-oxide.*
$C_7H_{13}O_2P$ M 160.152
Liq. d_4^{20} 1.11. Bp_{10} 131°. n_D^{20} 1.4892.
1-Ethoxy (*Et ester*): [1005-95-4]. *1-Ethoxy-2,5-dihydro-3,4-dimethyl-1*H-*phosphole 1-oxide.*
$C_8H_{15}O_2P$ M 174.179
Hygroscopic needles (pentane). Mp 42.5-43°. $Bp_{0.4}$ 100°.
1-Phenoxy (*Ph ester*): *2,5-Dihydro-3,4-dimethyl-1-phenoxy-1*H-*phosphole 1-oxide.*
$C_{12}H_{15}O_2P$ M 222.223
Liq. d_4^{20} 1.15. $Bp_{2.5}$ 146-147°. n_D^{20} 1.5480.
*Chloride: 1-Chloro-2,5-dihydro-3,4-dimethyl-1*H-*phosphole 1-oxide.*
$C_6H_{10}ClOP$ M 164.571
Solid. Mp 83-86°. Bp_{10} 132°.
*Bromide: 1-Bromo-2,5-dihydro-3,4-dimethyl-1*H-*phosphole 1-oxide.*
$C_6H_{10}BrOP$ M 209.022
Solid. Mp 85-87°. $Bp_{0.08}$ 160° ($Bp_{0.55}$ 102-104°).

Arbusov, B.A. *et al*, *Izv. Akad. Nauk SSSR, Ser. Khim.*, 1962, 65 (*Engl. transl.* p. 57) (*ester*)
Aubuzov, B.A. *et al*, *Dokl. Akad. Nauk SSSR, Ser. Sci. Khim.*, 1964, **159**, 582 (*Engl. transl.* p. 1205); *CA*, **62**, 6505 (*synth, derivs, pmr*)
Clarke, F.B. *et al*, *J. Am. Chem. Soc.*, 1971, **93**, 4541 (*esters, synth, props*)
Hammond, P.J. *et al*, *J. Chem. Soc., Perkin Trans. 2*, 1982, 205 (*ester, bromide, pmr, P nmr*)

2,3-Dihydro-2-hydroxy-1*H*-isophosphindole 2-oxide, 9CI **D-00494**
[20148-17-8]

$C_8H_9O_2P$ M 168.132
Cryst. (H_2O). Mp 142-148°, 156-158°.
Me ester: [42104-62-1]. *2,3-Dihydro-2-methoxy-1*H-*isophosphindole 2-oxide.*
$C_9H_{11}O_2P$ M 182.158
Solid. Mp 37-47°. $Bp_{1.3}$ 150°.

Robinson, C.N. *et al*, *J. Heterocycl. Chem.*, 1973, **10**, 395 (*synth, ir, pmr*)
Middlemas, E.D. *et al*, *J. Org. Chem.*, 1979, **44**, 2587 (*synth, P nmr*)

1,2-Dihydro-2-hydroxyisophosphinoline 2-oxide **D-00495**
[57328-54-8]

$C_9H_9O_2P$ M 180.143
Cryst. (THF). Mp 148-150°.

de Graaf, H.G. *et al*, *Tetrahedron*, 1975, **31**, 1097 (*synth, ir*)

2,3-Dihydro-1-hydroxy-4-methyl-1*H*-phosphole 1-oxide, 9CI **D-00496**
1-Hydroxy-3-methyl-2-phospholene 1-oxide, 8CI
[3858-24-0]

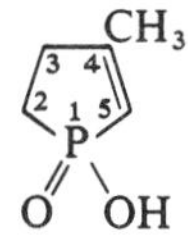

$C_5H_9O_2P$ M 132.099
Cryst. (Et_2O/CH_2Cl_2). Mp 120-121.5°.
▷SZ6102500.
1-Methoxy (*Me ester*): *2,3-Dihydro-1-methoxy-4-methyl-1*H-*phosphole 1-oxide.*
$C_6H_{11}O_2P$ M 146.125
$Bp_{0.4}$ 79°.
1-Ethoxy (*Et ester*): [697-32-5]. *2,3-Dihydro-1-ethoxy-4-methyl-1*H-*phosphole 1-oxide.*
$C_7H_{13}O_2P$ M 160.152
$Bp_{0.07}$ 94-95°. n_D^{20} 1.4842.
1-Phenoxy (*Ph ester*): [2921-71-3]. *2,3-Dihydro-4-methyl-1-phenoxy-1*H-*phosphole 1-oxide.*
$C_{11}H_{13}O_2P$ M 208.196
Cryst. (Et_2O/pentane). Mp 57-59°. $Bp_{0.05}$ 145-155°. n_D^{20} 1.5560.

Hasserodt, U. *et al*, *Tetrahedron*, 1963, **19**, 1563 (*synth, derivs, uv, ir*)
Quin, L.D. *et al*, *J. Org. Chem.*, 1968, **33**, 1034 (*ester*)
Myers, D.K. *et al*, *J. Org. Chem.*, 1971, **36**, 1285 (*synth, pmr*)

2,3-Dihydro-1-hydroxy-5-methyl-1*H*-phosphole 1-oxide, 9CI **D-00497**
2-Methyl-1-hydroxy-2-phospholene 1-oxide
[3858-24-0]
$C_5H_9O_2P$ M 132.099
▷SZ6102500.
1-Ethoxy (*Et ester*): *1-Ethoxy-2,3-dihydro-5-methyl-1*H-*phosphole 1-oxide.*
$C_7H_{13}O_2P$ M 160.152
Liq. $Bp_{0.07}$ 85-91°. n_D^{20} 1.4870.

Hasserodt, U. *et al*, *Tetrahedron*, 1963, **19**, 1563 (*synth, uv*)

2,5-Dihydro-1-hydroxy-2-methyl-1*H*-phosphole-1-oxide, 9CI D-00498

1-Hydroxy-2-methyl-3-phospholene 1-oxide, 8CI

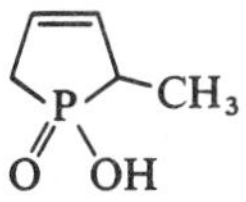

$C_5H_9O_2P$ M 132.099

1-Ethoxy (Et ester): [18874-12-4]. *1-Ethoxy-2,5-dihydro-2-methyl-1*H*-phosphole 1-oxide.*
$C_7H_{13}O_2P$ M 160.152
Liq. d_4^{20} 1.08. $Bp_{1.2}$ 72-73°. n_D^{20} 1.4770.

1-Phenoxy (Ph ester): [18874-12-9]. *2,5-Dihydro-2-methyl-1-phenoxy-1*H*-phosphole 1-oxide.*
$C_{11}H_{13}O_2P$ M 208.196
Liq. d_4^{20} 1.17. Bp_2 145-147°. n_D^{20} 1.5471.

Chloride: see 1-Chloro-2,5-dihydro-2-methyl-1H-phosphole, C-00046

N,N-*Diethylamide:* [18874-05-0]. *1-Diethylamino-2,5-dihydro-2-methyl-1*H*-phosphole 1-oxide.*
$C_9H_{18}NOP$ M 187.221
Liq. d_4^{20} 1.03. $Bp_{1.5}$ 115-118°. n_D^{20} 1.4919.

Hunger, K. *et al*, *Tetrahedron*, 1964, **20**, 1593 (*ir, uv*)
Bliznyuk, N.K. *et al*, *Zh. Obshch. Khim.*, 1967, **37**, 1811 (*Engl. transl.* p. 1726) (*synth*)
Zubtsova, L.I. *et al*, *Zh. Obshch. Khim.*, 1976, **46**, 2209 (*Engl. transl.* p. 2125) (*ester, synth, ir, pmr*)

2,5-Dihydro-1-hydroxy-3-methyl-1*H*-phosphole 1-oxide, 9CI D-00499

1-Hydroxy-3-methyl-3-phospholene 1-oxide, 8CI

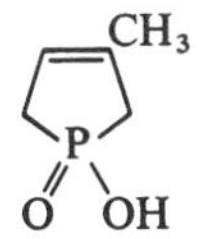

$C_5H_9O_2P$ M 132.099

1-Methoxy (Me ester): [695-59-0]. *2,5-Dihydro-1-methoxy-3-methyl-1*H*-phosphole 1-oxide.*
$C_6H_{11}O_2P$ M 146.125
Liq. d_4^{20} 1.14. Bp_9 121-122°. n_D^{20} 1.4852.
▷SZ6103800.

1-Ethoxy (Et ester): [697-31-4]. *1-Ethoxy-2,5-dihydro-3-methyl-1*H*-phosphole 1-oxide.*
$C_7H_{13}O_2P$ M 160.152
Liq. d_4^{20} 1.09. Bp_9 116-117°. n_D^{20} 1.4792.

1-Phenoxy (Ph ester): [2921-71-3]. *2,5-Dihydro-3-methyl-2-phenoxy-1*H*-phosphole 1-oxide.*
$C_{11}H_{13}O_2P$ M 208.196
Solid. d_4^{20} 1.15. Mp 71°. Bp_2 152-154°. n_D^{20} 1.5420.

Chloride: see 1-Chloro-2,5-dihydro-3-methyl-1H-phosphole, C-00047

Razumova, N.A. *et al*, *Zh. Obshch. Khim.*, 1964, **34**, 2949 (*Engl. transl.* p. 2983) (*esters, ir, pmr*)
Bliznyak, N.K. *et al*, *Zh. Obshch. Khim.*, 1967, **37**, 1811 (*Engl. transl.* p. 1726) (*ester*)

2,5-Dihydro-2-hydroxy-1,2-oxaphosphole 2-oxide, 9CI D-00500

2-Hydroxy-1,2-oxaphosphol-3-ene 2-oxide, 8CI

[68492-55-7]

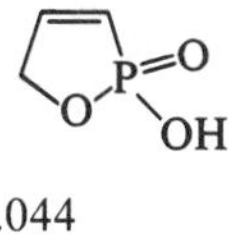

$C_3H_5O_3P$ M 120.044

Cryst. Mp 110-111°.

2-Butoxy: [68492-54-6].
$C_7H_{13}O_3P$ M 176.152
Oil.

Machida, Y. *et al*, *J. Org. Chem.*, 1979, **44**, 865 (*synth, tlc, ir, ms, pmr, cmr*)

3,6-Dihydro-2-hydroxy-2*H*-1,2-oxaphosphorin 2-oxide, 9CI D-00501

*3,6-Dihydro-2-hydroxy-2*H*-1,2-oxaphosphinin 2-oxide.*
2-Butenylphostonic acid

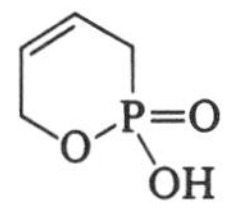

$C_4H_7O_3P$ M 134.071

2-Methoxy (Me ester): [53910-02-4].
$C_5H_9O_3P$ M 148.098
Yields monomeric methyl metaphosphate when heated at 600°. Liq. $Bp_{0.0001}$ 50-60° (bath).

Clapp, C.H. *et al*, *J. Am. Chem. Soc.*, 1974, **96**, 6710 (*synth, use, ir, ms, pmr*)

5,10-Dihydro-10-hydroxyphenophosphazine 10-sulfide, 9CI, 8CI D-00502

[40215-39-2]

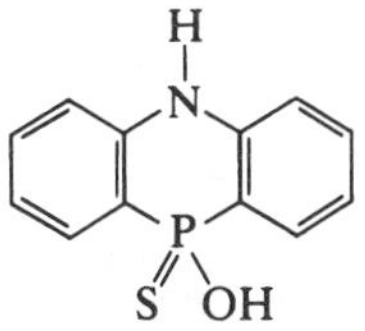

$C_{12}H_{10}NOPS$ M 247.251

Cryst. (EtOH aq.). Mp 213° dec. (block). Tautomeric.

10-Methoxy (O-*Me ester*): [40195-10-6]. *5,10-Dihydro-10-methoxyphenophosphazine 10-sulfide.*
$C_{13}H_{12}NOPS$ M 261.278
Needles (AcOH). Mp 184-185.5°.

10-Methylthio (S-*Me ester*): *5,10-Dihydro-10-methylthiophenophosphazine 10-oxide.*
$C_{13}H_{12}NOPS$ M 261.278
Cryst. (AcOH). Mp 266°.

10-Ethoxy (O-*Et ester*): *5,10-Dihydro-10-ethoxyphenophosphazine 10-sulfide.*
$C_{14}H_{14}NOPS$ M 275.304
Cryst. (EtOH). Mp 211-212°.

10-Ethylthio (S-*Et ester*): *10-Ethylthio-5,10-dihydrophenophosphazine 10-oxide.*
$C_{14}H_{14}NOPS$ M 275.304
Cryst. (EtOAc). Mp 221-222°.

10-Isopropoxy (O-*isopropyl ester*): *5,10-Dihydro-10-isopropoxyphenophosphazine 10-sulfide.*
$C_{15}H_{16}NOPS$ M 289.331
Cryst. (EtOH). Mp 233-234°.

Häring, M., *Helv. Chim. Acta*, 1960, **43**, 1826 (*synth, ir*)
Cheplanova, I.V. *et al*, *Izv. Akad. Nauk SSSR, Ser. Khim.*, 1972, 2283 (*Engl. transl.* p. 2216) (*ester*)

1,2-Dihydro-1-hydroxy-1-phenyl-4*H*-3,1,2-benzothiazaphosphorin-4-one 1-sulfide D-00503

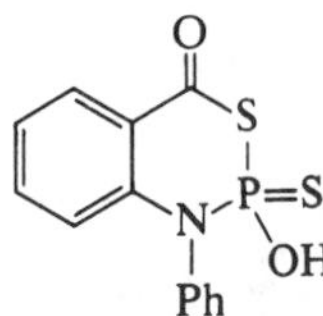

$C_{13}H_{10}NOPS$ M 259.262

2-Methoxy (O-*Me ester*): [23856-98-6].
$C_{14}N_{12}NO_2PS_2$ M 477.334
Solid. Mp 135°.

2-Ethoxy (O-*Et ester*): [23900-29-0].
$C_{15}H_{14}NO_2PS_2$ M 335.375
Solid. Mp 88°.

Legrand, L. *et al*, *Bull. Soc. Chim. Fr.*, 1969, 1173 (*synth*)

5,6-Dihydro-5-hydroxyphosphanthridine 5-oxide, 9CI, 8CI D-00504

5,6-Dihydro-5-hydroxydibenzo[b,d]*phosphorin 5-oxide, 8CI*

[20702-05-0]

$C_{13}H_{11}O_2P$ M 230.202

Monoclinic cryst. (EtOH). Mp 236-238°.

5-Methoxy (*Me ester*): [38033-18-0]. *5-Methoxy-5,6-dihydrophosphanthridene 5-oxide. 5-Methoxy-5,6-dihydrodibenzo*[b,d]*phosphorin 5-oxide.*
$C_{14}H_{13}O_2P$ M 244.229
Cryst. (H_2O or pet. ether). Mp 171.5-172.5°.

5-Phenoxy (*Ph ester*): *5-Phenoxy-5,6-dihydrophosphanthridene 5-oxide. 5-Phenoxy-5,6-dihydrodibenzo*[b,d]*phosphorin 5-oxide.*
$C_{19}H_{15}O_2P$ M 306.300
Solid. Mp 75-77°. $Bp_{0.1}$ 197-198°.

Anhydride:
$C_{26}H_{20}O_3P_2$ M 442.390
Cryst. (EtOH or xylene). Mp 272-276°.

Chloride (*5-Chloro*): *5-Chloro-5,6-dihydrophosphanthridine 5-oxide. 5-Chloro-5,6-dihydrodibenzo*[b,d]-*phosphorin 5-oxide.*
$C_{13}H_{10}ClOP$ M 248.648
Pale-yellow cryst. (C_6H_6/pet. ether). Mp 125-126°.

Anilide: 5-Anilino-5,6-dihydrophosphanthridene 5-oxide. 5-Anilino-5,6-dihydrodibenzo[b,d]*phosphorin 5-oxide.*
$C_{19}H_{16}NOP$ M 305.315
Cryst. (EtOH). Mp 202.5-204°.

Lynch, E.R. *et al*, *J. Chem. Soc.*, 1962, 3729 (*synth, derivs*)
Wheatley, P.J., *J. Chem. Soc.*, 1962, 3733 (*cryst struct*)
Tebby, J.C. *et al*, *J. Chem. Soc.* (*C*), 1971, 1064 (*ester, pmr*)
Robinson, C.N. *et al*, *J. Heterocycl. Chem.*, 1972, **9**, 735 (*ester*)

1,2-Dihydro-1-hydroxyphosphinoline 1-oxide, 9CI D-00505

[52427-76-6]

$C_9H_9O_2P$ M 180.143

Cryst. by subl. Mp 156-157°.

1-Methoxy (*Me ester*): [52427-77-7]. *1,2-Dihydro-1-methoxyphosphinoline 1-oxide.*
$C_{10}H_{11}O_2P$ M 194.169
Hygroscopic liq. $Bp_{0.005}$ 130°.

Collins, D.J. *et al*, *Aust. J. Chem.*, 1974, **27**, 815 (*synth, ir, uv, ms, pmr*)

2,3-Dihydro-1-hydroxy-1*H*-phosphole 1-oxide, 9CI D-00506

1-Hydroxy-2-phospholene 1-oxide, 8CI

[694-24-6]

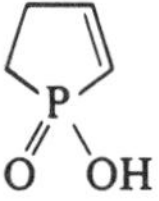

$C_4H_7O_2P$ M 118.072

Cryst. (C_6H_6/$CHCl_3$). Insol. C_6H_6. Mp 107°.

1-Methoxy (*Me ester*): [694-66-6]. *2,3-Dihydro-1-methoxy-1*H*-phosphole 1-oxide.*
$C_5H_9O_2P$ M 132.099
Liq. $Bp_{0.08}$ 143-146°. n_D^{20} 1.4918.
▷SZ6104000.

1-Ethoxy (*Et ester*): [695-63-6]. *1-Ethoxy-2,3-dihydro-1*H*-phosphole 1-oxide.*
$C_6H_{11}O_2P$ M 146.125
Liq. $Bp_{0.1}$ 75-78°. n_D^{20} 1.4836.

1-Phenoxy (*Ph ester*): [14967-93-2]. *2,3-Dihydro-1-phenoxy-1*H*-phosphole 1-oxide.*
$C_{10}H_{11}O_2P$ M 194.169
Cryst. (toluene). Mp 60-61°. $Bp_{0.3}$ 150-155°.

Chloride: see 1-Chloro-2,3-dihydro-1H-phosphole, C-00051

Hasserodt, U. *et al*, *Tetrahedron*, 1963, **19**, 1563 (*esters, uv*)
Moedritzer, K., *Synth. React. Inorg. Metal-Org. Chem.*, 1975, **5**, 45 (*synth, ir, ms, pmr, P nmr*)
Moedritzer, K. *et al*, *Phosphorus Sulfur*, 1981, **9**, 293; **10**, 279 (*derivs, ir, ms, cmr, P nmr, pmr*)

2,5-Dihydro-1-hydroxy-1*H*-phosphole 1-oxide, 9CI D-00507

1-Hydroxy-3-phospholene 1-oxide, 8CI

[39063-70-2]

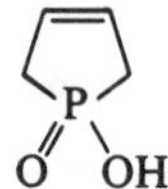

$C_4H_7O_2P$ M 118.072

Cryst. (toluene). Sol. C_6H_6. Mp 88°.

1-Methoxy (*Me ester*): [694-65-5]. *2,5-Dihydro-1-methoxy-1*H*-phosphole 1-oxide.*
$C_5H_9O_2P$ M 132.099
Liq. d_4^{20} 1.19. Bp_9 102-103°, $Bp_{0.06}$ 55-60°. n_D^{20} 1.4890.
▷SZ6104100.

1-Ethoxy (Et ester): [695-62-5]. *1-Ethoxy-2,5-dihydro-1*H*-phosphole 1-oxide.*
$C_6H_{11}O_2P$ M 146.125
Sweet-smelling liq. d_4^{20} 1.13. Bp_9 106-107°, $Bp_{0.2}$ 64-66°. n_D^{20} 1.4806.
▷SZ6101000.
*1-Isopropoxy (Isopropyl ester): 2,5-Dihydro-1-(1-methylethoxy)-1*H*-phosphole 1-oxide.* Liq. $Bp_{0.3}$ 66-67°. n_D^{20} 1.4645.
1-Phenoxy (Ph ester): [5234-91-3]. *2,5-Dihydro-1-phenoxy-1*H*-phosphole 1-oxide.* Cryst. (methylcyclohexane). Mp 49-50°. $Bp_{0.1}$ 110°. n_D^{20} 1.5568.
*Fluoride: 1-Fluoro-2,5-dihydro-1*H*-phosphole 1-oxide.*
C_4H_6FOP M 120.063
d_4^{20} 1.29. Bp_2 58°. n_D^{20} 1.4606.
Chloride: see 1-Chloro-2,5-dihydro-1H-phosphole, C-00052

Bogolybov, G.M. *et al*, *Zh. Obshch. Khim.*, 1971, **41**, 527 (*Engl. transl.* p. 527) (*derivs, ms*)
Moedritzer, K., *Synth. React. Inorg. Metal-Org. Chem.*, 1975, **5**, 45 (*synth, ms, ir, P nmr, pmr*)
Moedritzer, K. *et al*, *Phosphorus Sulfur*, 1981, **10**, 279 (*derivs*)
Moedritzer, K. *et al*, *Phosphorus Sulfur*, 1981, **9**, 293 (*ester, ir, ms, cmr, P nmr, pmr*)
Richardson, L.W. *et al*, *J. Chem. Phys.*, 1980, **73**, 5556 (*ester, props*)

2,3-Dihydro-2-hydroxy-1,3,5-triphenyl-1*H*-1,2-azaphosphole 2-oxide, 9CI — D-00508

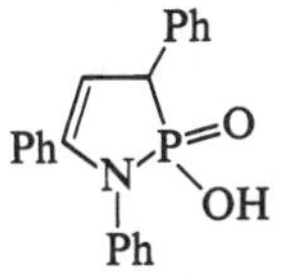

$C_{21}H_{17}NO_2P$ M 346.345
(±)-*form*
2-Ethoxy (Et ester): [60678-61-7]. *2,3-Dihydro-2-ethoxy-1,3,5-triphenyl-1*H*-1,2-azaphosphole 2-oxide.*
$C_{23}H_{21}NO_2P$ M 374.398
Cryst. (Et_2O). Mp 179-180°.

Kawashima, T. *et al*, *Bull. Chem. Soc. Jpn.*, 1976, **49**, 1924 (*synth, ir, ms, pmr*)

2,3-Dihydro-1*H*-isophosphindole, 9CI — D-00509

Isophosphindoline
[4478-20-0]

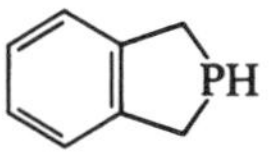

C_8H_9P M 136.133
Liq. $Bp_{2.5}$ 75°.
2-Ph: see 2,3-Dihydro-2-phenyl-1H-isophosphindole, D-00562

Robinson, C.N. *et al*, *J. Heterocycl. Chem.*, 1973, **10**, 395 (*synth, ir, pmr*)

1,2-Dihydroisophosphinoline — D-00510

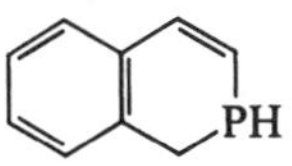

C_9H_9P M 148.144
Liq.

de Graaf, H.G. *et al*, *Tetrahedron*, 1975, **31**, 1097 (*ir, ms, pmr*)

2,3-Dihydro-2-mercapto-1,3,2-benzoxazaphosphole 2-sulfide — D-00511

[45893-59-2]

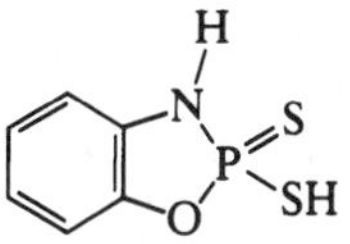

$C_6H_6NOPS_2$ M 203.213
Solid. Mp 127° dec.

Micu-Semeniuc, R. *et al*, *Inorg. Chim. Acta*, 1979, **33**, 281 (*synth, ir*)

2,5-Dihydro-1-mercaptophosphole 1-sulfide — D-00512

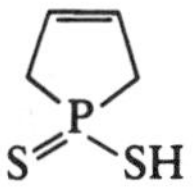

$C_4H_7PS_2$ M 150.193
2-Ethylthio (Et ester):
$C_6H_{11}PS_2$ M 178.247
Solid. Mp 32°. $Bp_{0.5}$ 98-101°.

Razumova, N.A. *et al*, *Zh. Obshch. Khim.*, 1967, **37**, 1919 (*Engl. transl.* p. 1824)

2,3-Dihydro-2-methoxy-5-methyl-3-phenyl-1,3,4,2-oxadiazaphosphole — D-00513

2-Methoxy-5-methyl-3-phenyl-1,3,4,2-oxadiazaphospholine

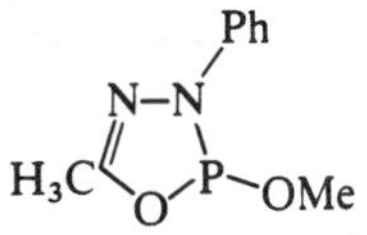

$C_9H_{11}N_2O_2P$ M 210.172
2-Sulfide: [20076-68-0].
$C_9H_{11}N_2O_2PS$ M 242.232
Solid. Mp 71-73°. $Bp_{0.2-0.25}$ 143-150°.

Italinskaya, T.L. *et al*, *Zh. Obshch. Khim.*, 1968, **38**, 2265 (*Engl. transl.* p. 2192)

2,3-Dihydro-2-methoxy-5-methyl-1,2-thiaphosphole, 9CI — D-00514

2-Methoxy-5-methyl-1,2-thiaphosphol-4-ene, 8CI

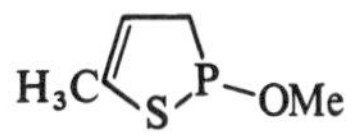

C_5H_9OPS M 148.159
2-Oxide: [23081-71-2].
$C_5H_9O_2PS$ M 164.159
Liq. d_4^{20} 1.10. $Bp_{0.5}$ 43-44°. n_D^{20} 1.5140.

Kovalev, L.S. *et al*, *Zh. Obshch. Khim.*, 1969, **39**, 869 (*Engl. transl.* p. 833) (*synth, pmr, ir*)

5,10-Dihydro-10-methylacridophosphine, 9CI — D-00515

[58751-96-5]

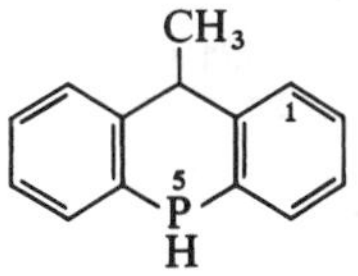

$C_{14}H_{13}P$ M 212.230

Mp 103-106°.
5-Chloro: [58751-94-3]. *5-Chloro-5,10-dihydro-10-methylacridophosphine.*
$C_{14}H_{12}ClP$ M 246.676
Solid. Mp 121-123°. Mixt. of two stereoisomers.
5-Hydroxy, 5-oxide: [58751-95-4]. *5,10-Dihydro-5-hydroxy-10-methylacridophosphine 5-oxide.*
$C_{14}H_{13}O_2P$ M 244.229
Cryst. (EtOH). Mp 298-304°.

Jongsma, C. *et al*, *Tetrahedron*, 1975, **31**, 2931 (*ir, ms, pmr, derivs*)

1,5-Dihydro-3-methyl-2,4,3-benzodioxaphosphepin, 9CI D-00516
[69813-52-1]

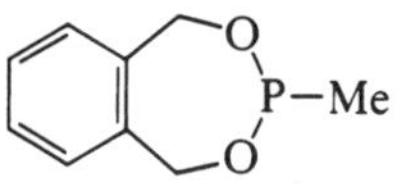

$C_9H_{11}O_2P$ M 182.158
4-Sulfide: [65845-21-8].
$C_9H_{11}O_2PS$ M 214.218
Solid.

Gand, A. *et al*, *Acta Crystallogr., Sect. B*, 1978, **34**, 199 (*sulfide, cryst struct*)
Guimaraes, A.C. *et al*, *Org. Magn. Reson.*, 1978, **11**, 411 (*sulfide, cmr, pmr, P nmr*)
Dutasta, J.P. *et al*, *Chem. Phys. Lett.*, 1981, **77**, 336 (*sulfide, P nmr*)

2,3-Dihydro-2-methyl-1,4,2-benzodioxaphosphorin, 8CI D-00517

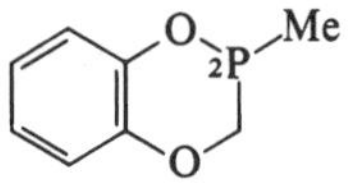

$C_8H_9O_2P$ M 168.132
2-Oxide: [17718-78-4].
$C_8H_9O_3P$ M 184.131
Solid. Mp 52-57°. $Bp_{0.001}$ 114°.

Wieber, M. *et al*, *Monatsh. Chem.*, 1968, **99**, 261 (*synth, pmr*)

2,3-Dihydro-1-methyl-1*H*-benzo[*g*]phosphindole, 9CI D-00518
[81487-80-1]

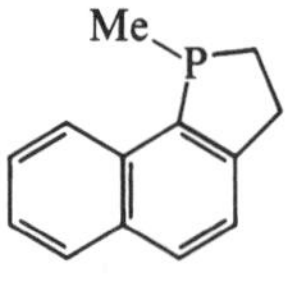

$C_{13}H_{13}P$ M 200.219
Yellow oil.
B,MeI: [81488-03-1]. *2,3-Dihydro-1,1-dimethyl-1*H*-benzo*[g]*phosphindolinium iodide.*
$C_{14}H_{16}IP$ M 342.159
Cryst. (MeOH). Mp 296-297°.
1-Oxide: [81487-79-8].
$C_{13}H_{13}OP$ M 216.219
Solid. Mp 210.5-212.5°.

Quin, L.D. *et al*, *Phosphorus Sulfur*, 1982, **12**, 161 (*synth, derivs, pmr, P nmr*)

2,3-Dihydro-2-methyl-1,3,2-benzothiazaphosphole, 8CI D-00519
[13968-96-2]

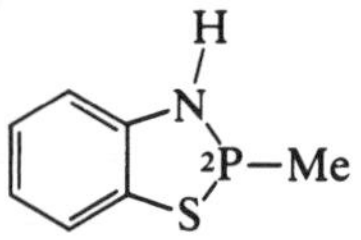

C_7H_8NPS M 169.181
Solid. Mp 120°. Bp_2 122-124°.
2-Sulfide: [13969-07-8].
$C_7H_8NPS_2$ M 201.241
Solid. Mp 88°.
2-Selenide:
C_7H_8NPSSe M 248.141
Oil.

Wieber, M. *et al*, *Chem. Ber.*, 1967, **100**, 974 (*synth, derivs*)

2,3-Dihydro-2-methyl-1*H*-4,1,2-benzothiazaphosphorine, 8CI D-00520

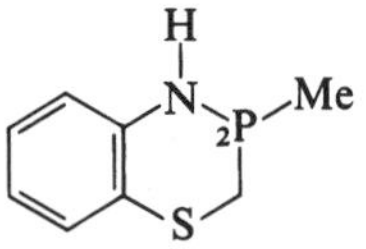

$C_8H_{10}NPS$ M 183.207
2-Oxide: [17718-81-9].
$C_8H_{10}NOPS$ M 199.207
Cryst. (EtOAc). Mp 192-196°.
2-Sulfide: [17718-82-0].
$C_8H_{10}NPS_2$ M 215.267
Solid. Mp 102-109°. $Bp_{0.001}$ 174-178°.

Wieber, M. *et al*, *Monatsh. Chem.*, 1968, **99**, 261 (*synth, pmr*)

3,4-Dihydro-2-methyl-2*H*-1,4,2-benzothiazaphosphorine, 9CI D-00521
[50636-31-2]

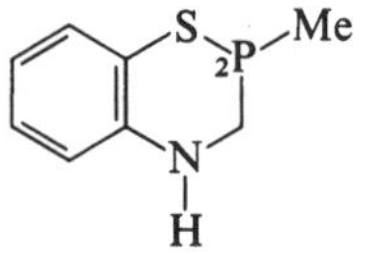

$C_8H_{10}NPS$ M 183.207
Liq. $Bp_{0.01}$ 98-101°.
2-Sulfide: [50636-33-4].
$C_8H_{10}NPS_2$ M 215.267
Solid. Mp 80-82°.
2-Selenide: [50636-34-5].
$C_8H_{10}NPSSe$ M 262.167
Solid. Mp 108-110°.

Wieber M. *et al*, *Chem. Ber.*, 1973, **106**, 2733 (*synth, pmr*)

3,4-Dihydro-2-methyl-2*H*-1,4,2-benzoxazaphosphorine, 8CI D-00522
*3,4-Dihydro-2-methyl-2*H*-1,4,2-benzoxazaphosphinine*

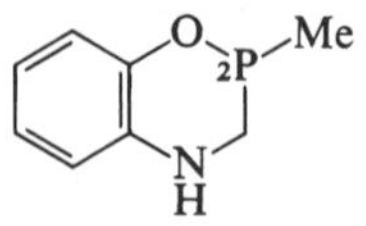

$C_8H_{10}NOP$ M 167.147
2-Oxide: [17718-79-5].
$C_8H_{10}NO_2P$ M 183.146

Solid. Mp 68-74°. $Bp_{0.001}$ 180° dec.
2-Sulfide: [17718-80-8].
$C_8H_{10}NOPS$ M 199.207
Solid. Mp 96-100°. $Bp_{0.001}$ 166-169°.

Wieber, M. *et al*, *Monatsh. Chem.*, 1968, **99**, 261 (*synth, pmr*)

1,2-Dihydro-1-methyl-4*H*-3,1,2-benzoxazaphosphorin-4-one, 11CI D-00523

$C_8H_8NO_2P$ M 181.130
1-Methoxy: [93523-22-9].
$C_9H_{10}NO_3P$ M 211.157
Liq. d_4^{20} 1.30. $Bp_{0.05}$ 118-119°. n_D^{20} 1.5851.
1-Ethoxy: [93523-23-0].
$C_{10}H_{12}NO_3P$ M 225.183
Liq. d_4^{20} 1.24. $Bp_{0.04}$ 128-129°. n_D^{20} 1.5816.
1-Isopropoxy: [93523-24-1].
$C_{11}H_{14}NO_3P$ M 239.210
Liq. d_4^{20} 1.20. $Bp_{0.06}$ 133-135°. n_D^{20} 1.5621.
1-Chloro: [93772-86-2].
$C_8H_7ClNO_2P$ M 215.576
Yellow cryst. Mp 50-51°. $Bp_{0.08}$ 125-126°.

Kuliev, A.K. *et al*, *Zh. Obshch. Khim.*, 1984, **54**, 1671 (*Engl. transl.* p. 1489) (*synth, pmr, P nmr*)

10,11-Dihydro-5-methyl-5*H*-dibenzo[*b,f*]phosphepin, 9CI D-00524

1-Methyl-1-phospha-2,3:6,7-dibenzocyclohepta-2,6-diene
[72978-05-3]

$C_{15}H_{15}P$ M 226.257
Oil. $Bp_{0.02}$ 115-125°.
B,MeI: [72978-08-6]. *10,11-Dihydro-5,5-dimethyl-5*H*-dibenzo*[b,f]*phosphepinium iodide.*
$C_{16}H_{18}IP$ M 368.196
Cryst. (EtOH). Mp 265-266°.
Oxide: [59470-47-2].
$C_{15}H_{15}OP$ M 242.257
Cryst. (EtOH). Mp 108°.

Allen, D.W. *et al*, *J. Chem. Soc., Perkin Trans. 1*, 1979, 2326 (*synth, derivs, pmr*)
Segall, Y. *et al*, *Phosphorus Sulfur*, 1980, **8**, 243 (*oxide*)

2,3-Dihydro-5-methyl-1,2-oxaphosphole 2-oxide, 9CI D-00525

5-Methyl-1,2-oxaphosphol-4-ene 2-oxide, 8CI

$C_4H_7O_2P$ M 118.072
2-Chloro:
$C_4H_6ClO_2P$ M 152.517
Liq. d^{20} 1.34. $Bp_{0.5}$ 73°. n_D^{20} 1.4820.
2-Methoxy:
$C_5H_9O_3P$ M 148.098
Liq. d^{20} 1.22. $Bp_{0.5}$ 74-75°. n_D^{20} 1.4595.
2-Diethylamino:
$C_8H_{16}NO_2P$ M 189.194
Liq. d^{20} 1.08. $Bp_{0.5}$ 85-86°. n_D^{20} 1.4713.

Novitskii, N.I. *et al*, *CA*, 1968, **69**, 43981.

5,10-Dihydro-5-methyl-10-phenylphenophosphazine, 9CI, 8CI D-00526

[896-84-4]

$C_{19}H_{16}NP$ M 289.316
Cryst. (EtOH). Mp 158-160°.
10-Oxide: [58943-96-7].
$C_{19}H_{16}NOP$ M 305.315
Cryst. (C_6H_6/cyclohexane). Mp 162-163°. Forms 1:1 adduct with PhOH.

Kupchik, E. *et al*, *J. Organomet. Chem.*, 1967, **10**, 181 (*synth, uv, ir, pmr*)
Gurkova, S.N. *et al*, *Zh. Strukt. Khim.*, 1977, **18**, 62 (*Engl. transl.* p. 51) (*cryst struct*)
Smirnov, A.N. *et al*, *Zh. Obshch. Khim.*, 1977, **47**, 2468 (*Engl. transl.* p. 2256) (*uv, ir, ms, oxide*)
Gusev, A.I. *et al*, *Zh. Strukt. Khim.*, 1979, **20**, 632 (*Engl. transl.* p. 537) (*oxide, cryst struct*)

2,3-Dihydro-5-methyl-3-phenyl-2-phenylamino-1,3,4,2-oxadiazaphosphole D-00527

*2,3-Dihydro-5-methyl-*N*,3-diphenyl-1,3,4,2-oxadiazaphosphol-2-amine, 9CI. 2-Anilino-5-methyl-3-phenyl-Δ^4-1,3,4,2-oxadiazaphospholine, 8CI*
[19016-52-5]

$C_{14}H_{14}N_3OP$ M 271.258
Solid. Mp 101-103°.
2-Oxide: [33706-30-8].
$C_{14}H_{14}N_3O_2P$ M 287.257
Solid. Mp 146-148°.
2-Sulfide: [21398-17-4].
$C_{14}H_{14}N_3OPS$ M 303.318
Solid. Mp 107-109°.

Italinskaya, T.L. *et al*, *Zh. Obshch. Khim.*, 1968, **38**, 2265 (*Engl. transl.* p. 2192) (*synth*)
Shvetsov-Shilovskii, N.I. *et al*, *Zh. Obshch. Khim.*, 1971, **41**, 1200 (*Engl. transl.* p. 1210) (*oxide*)

2,5-Dihydro-3-methyl-1-phenyl-1*H*-phosphole, 9CI D-00528

1-Phenyl-3-methyl-1-phospha-3-cyclopentene
[15450-93-8]

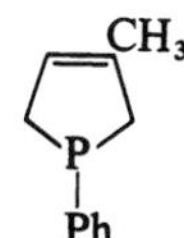

$C_{11}H_{13}P$ M 176.197
Complexes with $PdCl_2$ to give a hydrogenation catalyst. Bp_{16} 133-134°.

B,EtI: 1-Ethyl-2,5-dihydro-3-methyl-1-phenyl-1H-phospholium iodide.
$C_{13}H_{18}PI$ M 332.163
Solid. Mp 84-85.5°.
B,PhCH$_2$Br: 1-Benzyl-2,5-dihydro-3-methyl-1-phenyl-1H-phospholium bromide.
$C_{18}H_{20}PBr$ M 347.234
Solid. Mp 178-180°.
Oxide:
$C_{11}H_{13}OP$ M 192.197
Liq. $Bp_{0.24}$ 133-134°.

Quin, L.D. *et al*, *J. Org. Chem.*, 1964, **29**, 836; 1968, **33**, 1034; 1971, **36**, 1297 (*synth, deriv, ir, P nmr, pmr*)
Breen, J.J. *et al*, *Phosphorus*, 1972, **2**, 55 (*nmr*)
Watanabe, S. *et al*, *Can. J. Chem.*, 1973, **51**, 848 (*synth, use*)
Fieser, M. *et al*, *Reagents for Organic Synthesis*, Wiley, 1967-84, **5**, 503.

2,3-Dihydro-5-methyl-3-phenyl-1,3,4,2-thiadiazaphosphole 2-sulfide, 9CI D-00529

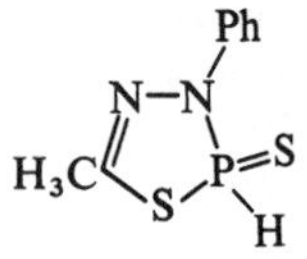

$C_8H_9N_2PS_2$ M 228.266
2-Chloro: [35169-09-6].
$C_8H_8ClN_2PS_2$ M 262.711
Liq. $Bp_{0.2}$ 165-166°. n_D^{20} 1.6598.
2-Ethoxy: [35169-20-7].
$C_{10}H_{13}N_2OPS_2$ M 272.319
Liq. $Bp_{0.2}$ 142-144°.

Italinskaya, T.L. *et al*, *Zh. Obshch. Khim.*, 1971, **41**, 1980 (*Engl. transl.* p. 1998) (*synth*)

2,5-Dihydro-3-methyl-1*H*-phosphole, 9CI D-00530

3-Methyl-3-phospholene, 8CI. 3-Methyl-1-phospha-3-cyclopentene

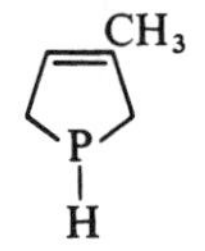

C_5H_9P M 100.100
Liq. d_4^{20} 0.94. Bp_{45} 49.5°. n_D^{20} 1.5182.
1-Methoxy: 2,5-Dihydro-1-methoxy-3-methyl-1H-phosphole.
$C_6H_{11}OP$ M 130.126
Liq. d_4^{20} 0.98. Bp_{10} 54-55°. n_D^{20} 1.4950.
1-Ethoxy: 1-Ethoxy-2,5-dihydro-3-methyl-1H-phosphole.
$C_7H_{13}OP$ M 144.153
Liq. d_4^{20} 0.97. Bp_9 53-56°. n_D^{20} 1.4905.
1-Ethylthio: 1-Ethylthio-1,5-dihydro-3-methyl-1H-phosphole.
$C_7H_{13}PS$ M 160.213
Liq. d_4^{20} 1.02. Bp_{10} 87-88°.
1-Dimethylamino: 2,5-Dihydro-1-dimethylamino-3-methyl-1H-phosphole.
$C_7H_{14}NP$ M 143.168
Liq. d_4^{20} 0.91. Bp_{10} 51-52°. n_D^{20} 1.5045.

Bogolyubov, G.M. *et al*, *Zh. Obshch. Khim.*, 1963, **33**, 2419 (*Engl. transl.* p. 2359) (*synth, ir*)
Vizel', A.O. *et al*, *Zh. Obshch. Khim.*, 1973, **43**, 2137 (*Engl. transl.* p. 2128) (*derivs*)

2,2-Dihydro-4-methyl-2,2,2,3-tetraphenyl-1,2-oxaphosphetane, 9CI D-00531

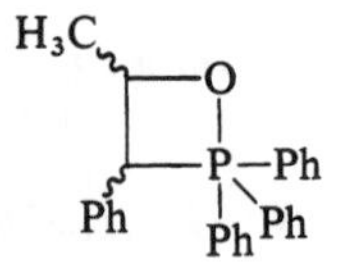

$C_{27}H_{25}OP$ M 396.468
Capable of existence in *cis-* and *trans*-forms. Wittig reaction intermed. Stable at <−20°.

Vedejs, E. *et al*, *J. Am. Chem. Soc.*, 1973, **95**, 5778; 1981, **103**, 2823.

2,3-Dihydro-1*H*-naphtho[1,8-*de*]-1,3,2-diazaphosphorine, 9CI D-00532

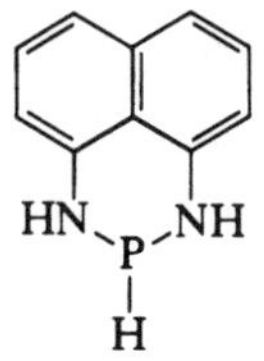

$C_{10}H_9N_2P$ M 188.168
2-(4-Nitrophenoxy), 2-oxide: [75861-17-5].
$C_{16}H_{12}N_3O_4P$ M 341.262
Solid. Mp 235-238°.
2-(4-Chloro-3-methylphenoxy), 2-oxide: [53596-32-0].
$C_{17}H_{14}ClN_2O_2P$ M 344.737
Cryst. ($CHCl_3$). Mp 206-207°.

Naidu, M.S.R. *et al*, *Indian. J. Chem.*, 1974, **12**, 349 (*synth*)
Naidu, M.S.R. *et al*, *Indian J. Chem., Sect. B*, 1977, **15**, 706; 1979, **17**, 458 (*synth, ms*)

2,2-Dihydro-1,2-oxaphosphetane, 9CI D-00533

[40110-50-7]

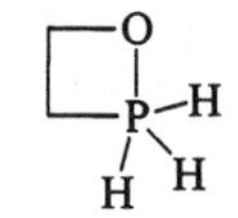

C_2H_7OP M 78.050
Observed only as derivs.
2,2,2-Tributyl: [64470-11-7].
$C_{14}H_{31}OP$ M 246.372
Observed spectroscopically. Probably in equilibrium with acyclic betaine.
2,2,2-Tri-Ph: [57043-19-3].
$C_{20}H_{20}OP$ M 307.351
Observed at low temp.

Schlosser, M. *et al*, *Chimia*, 1977, **31**, 219 (*tributyl*)
Bestmann, H.J. *et al*, *J. Chem. Soc., Chem. Commun.*, 1980, 978 (*struct*)
Höller, R. *et al*, *J. Am. Chem. Soc.*, 1980, **102**, 4632 (*struct, stereochem*)
Vedejs, E. *et al*, *J. Am. Chem. Soc.*, 1981, **103**, 2823 (*triphenyl*)

2,2-Dihydro-1,2-oxaphospholane D-00534

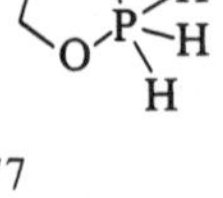

C_3H_9OP M 92.077
Derivs. are fairly unstable.
2,2,2-Tri-Me: [61152-20-3].
$C_6H_{15}OP$ M 134.158

Liq. Mp −28°. Bp_{50} 80°.
2,2,2-Tri-Et: [61152-21-4].
$C_9H_{21}OP$ M 176.238
Liq. Mp −40°. Bp_{12} 95°.
2,2,2-Tri-Ph: [14580-93-9].
$C_{21}H_{21}OP$ M 320.370
Behaves as a Wittig reagent. Cryst., unstable and hygroscopic. Mp 114-115°. At 300° yields (3-phenoxypropyl)diphenylphosphine oxide. With HI yields (3-hydroxypropyl)triphenylphosphonium iodide.

Hands, A.R. *et al, J. Chem. Soc. (C)*, 1967, 1099 (*triphenyl, synth, ir, pmr, props*)
Leppard, D.G. *et al, Helv. Chim. Acta*, 1976, **59**, 695 (*triphenyl, synth, ir, pmr, use*)
Schmidbaur, H. *et al, Chem. Ber.*, 1976, **109**, 3151 (*trimethyl, triethyl, synth, ir, cmr, nmr, pmr*)
Hercouet, A. *et al, Tetrahedron*, 1981, **37**, 2855 (*triphenyl, synth*)

(1,3-Dihydro-3-oxo-1-isobenzofuranyl)-phosphonic acid, 9CI — D-00535

Phthalide-3-phosphonic acid. 3-Phosphonophthalide

$C_8H_7O_5P$ M 214.114
Cryst. + $1H_2O$ (H_2O). Mp 253-254°.
Di-Me ester: [61260-15-9]. *Dimethyl (1,3-dihydro-3-oxo-1-isobenzofuranyl)phosphonate.*
$C_{10}H_{11}O_5P$ M 242.168
Cryst. (C_6H_6/cyclohexane). Mp 96-97°.
Diisopropyl ester: Diisopropyl (1,3-dihydro-3-oxo-1-isobenzofuranyl)phosphonate.
$C_{14}H_{19}O_5P$ M 298.275
Cryst. (C_6H_6/cyclohexane). Mp 98-99°.

Ramirez, F. *et al, J. Am. Chem. Soc.*, 1961, **83**, 173 (*synth, esters, ir, struct*)

2,3-Dihydro-1,2,3,4,5-pentakis(trifluoromethyl)-1*H*-1,2,3-triphosphole — D-00536

1,2,3,4,5-Pentakis(trifluoromethyl)-1,2,3-phosphol-4-ene

$C_7F_{15}P_3$ M 461.974
Liq. Stable up to 200° in absence of air. Bp ca. 160°.
▷Spont. flammable in air

Mahler, W., *J. Am. Chem. Soc.*, 1964, **86**, 2306 (*synth, props, ir, uv, F nmr*)

2,2-Dihydro-2,2,2,4,5-pentaphenyl-1,3,2-dioxaphospholane — D-00537

$C_{32}H_{27}O_2P$ M 474.538
(4RS,5RS)-form [61570-43-2]
(±)-trans-*form*
Dec. at ca. 50° → Ph_3PO and 1,2-diphenyloxirane.

Bartlett, P.D. *et al, J. Org. Chem.*, 1977, **42**, 1661 (*synth, props*)
Kelly, J.W. *et al, J. Am. Chem. Soc.*, 1986, **108**, 7681 (*P nmr*)

1,4-Dihydro-1,2,3,4,9-pentaphenyl-1,4-phosphinidenenaphthalene, 9CI — D-00538

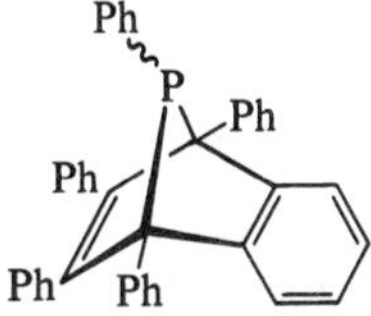

$C_{40}H_{29}P$ M 540.643
Oxide: [37755-73-0]. *2,3-Benzo-1,4,5,6,7-pentaphenyl-7-phosphabicyclo[2.2.1]hept-5-ene 7-oxide.*
$C_{40}H_{29}OP$ M 556.642
Cannot be reduced to the phosphine by $LiAlH_4$ or Si_2Cl_6. At 155°, eliminates $[PhPO]_x$ and gives 1,2,3,4-tetraphenylnaphthalene.

Stille, J.K. *et al, J. Am. Chem. Soc.*, 1972, **94**, 4761 (*synth, props*)

2,3-Dihydro-1,2,3,4,5-pentaphenyl-1*H*-1,2,3-triphosphole, 9CI — D-00539

1,2,3,4,5-Pentaphenyl-1,2,3-phosphol-4-ene
[75600-55-4]

$C_{32}H_{25}P_3$ M 502.471
Solid. Mp 129°.
1,3-Disulfide: [42451-97-8].
$C_{32}H_{25}P_3S_2$ M 566.591
Cryst. Mp 208-210°.

Ecker, A. *et al, Chem. Ber.*, 1973, **106**, 1453 (*disulfide*)
Charrier, C. *et al, J. Org. Chem.*, 1981, **46**, 3 (*synth, P nmr*)

(9,10-Dihydro-9-phenanthreneyl)-triphenylphosphonium(1+) — D-00540

(9,10-Dihydro-9-phenanthryl)-triphenylphosphonium(1+)

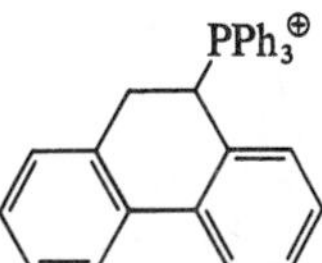

$C_{32}H_{26}P^{\oplus}$ M 441.531 (ion)
Bromide:
$C_{32}H_{26}BrP$ M 521.435

Cryst. Mp 223-224°. With aq. NaOH → 9,10-dihydrophenanthrene; NaOMe → phenanthrene. PhLi or $NaNH_2$ yields ylide.
Ylide: [(*9,10-Dihydro-9-phenanthrenenyl*)*methylene*]-*triphenylphosphorane.*
$C_{32}H_{25}P$ M 440.523

Bestmann, H.J. *et al*, *Chem. Ber.*, 1966, **99**, 28 (*synth, props*)

2,2-Dihydrophenanthro[9,10-*d*]-1,3,2-dioxaphosphole, 10CI, 9CI D-00541

4,5-(2′,2″-Biphenyleno)-2,2-dihydro-1,3,2-dioxaphosphole

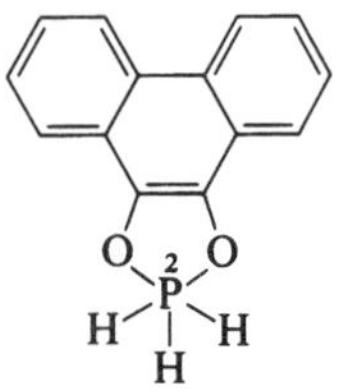

$C_{14}H_{11}O_2P$ M 242.213
2,2,2-Trimethoxy: [4903-06-4].
$C_{17}H_{17}O_5P$ M 332.292
Cryst. (hexane). Mp 74-75°.
2,2,2-Triisopropoxy: [3009-75-4].
$C_{23}H_{29}O_5P$ M 416.453
Cryst. (C_6H_6). Mp 107-108°. Exists in orthorhombic and monoclinic forms.
2,2,2-Triphenoxy: [4850-60-6].
$C_{32}H_{23}O_5P$ M 518.504
Cryst. (C_6H_6/hexane). Mp 147-148°.
2,2,2-Tri-Ph: [6546-78-7].
$C_{32}H_{23}O_2P$ M 470.506
Reagent for conversion of diols, amino alochols and mercaptoalcohols into heterocycles. Cryst. (C_6H_6). Mp 165-166°.

Ramirez, F. *et al*, *J. Am. Chem. Soc.*, 1963, **85**, 3252 (*synth, ir, P nmr*)
Hamilton, W.C. *et al*, *J. Am. Chem. Soc.*, 1967, **89**, 2268 (*cryst struct*)
Spratley, R.D. *et al*, *J. Am. Chem. Soc.*, 1967, **89**, 2272 (*cryst struct*)
Ramirez, F. *et al*, *J. Org. Chem.*, 1968, **33**, 3787 (*triphenyl, ir, P nmr*)
Ogata, Y. *et al*, *J. Chem. Soc., Perkin Trans. 2*, 1972, 493 (*synth*)
Castelijns, M.M.C.F. *et al*, *J. Org. Chem.*, 1981, **46**, 47 (*trimethoxy, pmr, P nmr*)
Denney, D.B. *et al*, *J. Am. Chem. Soc.*, 1981, **103**, 2054 (*trimethoxy, cmr, P nmr*)
Kelly, J.W. *et al*, *J. Org. Chem.*, 1986, **51**, 4473 (*triphenyl, synth, cmr, use*)

5,10-Dihydrophenophosphazine, 9CI, 8CI D-00542

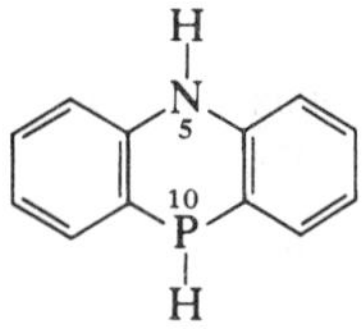

$C_{12}H_{10}NP$ M 199.191
10-Oxide: [53778-28-2].
$C_{12}H_{10}NOP$ M 215.191
Prisms (AcOH). Mp 214-216°.
10-Oxide; B,HCl: Cryst. (EtOH aq.). Mp 146-147°.
10-Oxide, B,HBr: Cryst. (EtOH aq.). Mp 156°.
10-Ph: see 5,10-Dihydro-10-phenylphenophosphazine, D-00563
10-Chloro: see 10-Chloro-5,10-dihydrophenophosphazine, C-00049
5,10-Di-Me: see 5,10-Dihydro-5,10-dimethylphenophosphazine, D-00466
5-Me, 10-Ph: see 5,10-Dihydro-5-methyl-10-phenylphenophosphazine, D-00526

Häring, M., *Helv. Chim. Acta*, 1960, **43**, 1826 (*synth, ir, derivs*)
Jenkins, R.N. *et al*, *J. Org. Chem.*, 1975, **40**, 766 (*synth*)

2,3-Dihydro-2-phenoxy-1*H*-1,3,2-benzodiazaphosphole, 9CI, 8CI D-00543

Phenyl N,N′-o-*phenylenephosphorodiamidite*

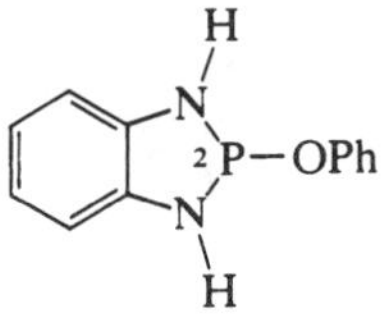

$C_{12}H_{11}N_2OP$ M 230.205
2-Oxide: [41227-61-6]. *Phenyl* N,N′-o-*phenylenephosphorodiamidate.*
$C_{12}H_{11}N_2O_2P$ M 246.205
Cryst. (bromobenzene). Mp 151-153°.
3-Sulfide: [39695-34-6]. O-*Phenyl* N,N′-o-*phenylenephosphorodiamidothioate.*
$C_{12}H_{11}N_2OPS$ M 262.265
Cryst. (C_6H_6). Mp 176°.

Edmundson, R.S. *J. Chem. Soc. (C)*, 1969, 2730 (*sulfide, synth, ir*)
Barthelot, M. *et al*, *Spectrochim. Acta, Part A*, 1973, **29**, 79 (*sulfide, ir*)
Koizumi, T. *et al*, *Chem. Pharm. Bull.*, 1973, **21**, 202 (*oxide, synth, ir, ms, props*)
Koizumi, T. *et al*, *Tetrahedron Lett.*, 1977, 1913 (*oxide, props*)
Edmundson, R.S., *Org. Mass Spectrom.*, 1983, **18**, 150 (*sulfide, ms*)

1,5-Dihydro-2-phenoxy-2,4,3-benzodioxaphosphepin, 9CI D-00544

[59413-43-3]

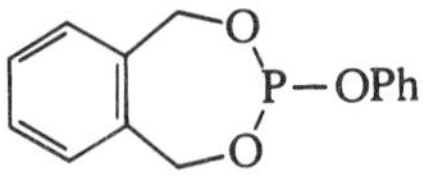

$C_{14}H_{13}O_3P$ M 260.229
3-Oxide: [49785-03-7].
$C_{14}H_{13}O_4P$ M 276.228
Solid. Mp 96-97°.
3-Sulfide: [69813-41-8].
$C_{14}H_{13}O_2PS$ M 276.289
Characterised spectroscopically.

Sato, T. *et al*, *J. Chem. Soc., Chem. Commun.*, 1973, 494 (*oxide, synth, pmr*)
Guimanaes, A.C. *et al*, *Org. Magn. Reson.*, 1978, **11**, 411 (*synth, sulfide, cmr, pmr, P nmr*)

2,3-Dihydro-2-phenoxy-1,3,2-benzothiazaphosphole, 9CI D-00545

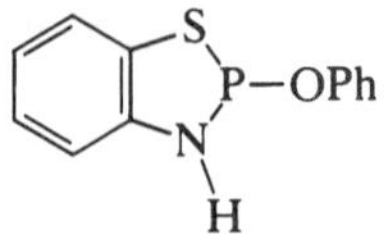

$C_{12}H_{10}NOPS$ M 247.251
2-Oxide:
$C_{12}H_{10}NO_2PS$ M 263.250

Liq. $Bp_{0.2}$ 235°.

Koizumi, T. *et al, Tetrahedron Lett.*, 1977, 1913 (*deriv, synth, props*)

1,3-Dihydro-1-phenoxy-2,1-benzoxaphosphole, 10CI **D-00546**

1-Phenoxy-2-oxa-1-phosphaindane

[79157-80-5]

$C_{13}H_{11}O_2P$ M 230.202

Solid. Mp 37-39°. $Bp_{0.05}$ 104-105°.

1-Sulfide: [79157-84-9].

$C_{13}H_{11}O_2PS$ M 262.262

Cryst. (CCl_4/hexane). Mp 52-53°.

Dahl, B.M. *et al, J. Chem. Soc., Perkin Trans. 1*, 1981, 2239 (*synth, ms, pmr, P nmr*)

5,10-Dihydro-5-phenylacridophosphine, 9CI **D-00547**

9,10-Dihydro-9-phenyl-9-phosphaanthracene

[59273-35-7]

$C_{19}H_{15}P$ M 274.301

Liq. $Bp_{0.01}$ 140°.

B,PhCH₂Br: 5-Benzyl-5,10-dihydro-5-phenylacridophosphonium bromide.

$C_{26}H_{22}BrP$ M 445.338

Solid. Mp 325-326°.

Jongsma, C. *et al*, Tet, 1976, **32**, 121 (*synth, pmr, ms, derivs*)

5,10-Dihydro-10-phenylacridophosphine, 9CI **D-00548**

$C_{19}H_{15}P$ M 274.301

5-Ph, 5-oxide: [80034-71-5]. *5,10-Dihydro-5,10-diphenylacridophosphine 5-oxide.*

$C_{25}H_{19}OP$ M 366.398

Cryst. (EtOH). Mp 230-232°.

5-Ph; B,MeI: 5,10-Dihydro-5-methyl-5,10-diphenylacridophosphonium iodide.

$C_{26}H_{22}IP$ M 492.338

Solid. Mp 226-229°.

5-Hydroxy, 5-oxide: [55364-12-0]. *5,10-Dihydro-5-hydroxy-10-phenylacridophosphine 5-oxide.*

$C_{19}H_{15}O_2P$ M 306.300

Cryst. (EtOH). Mp 246-248°.

5-Chloro: [20995-83-9]. *5-Chloro-5,10-dihydro-10-phenylacridophosphine.*

$C_{19}H_{14}ClP$ M 308.746

Solid. Mp 94-101°. $Bp_{0.001}$ 131°.

Jongsma, C. *et al, Tetrahedron*, 1974, **30**, 3465 (*oxide, ir, ms, pmr*)

Petrov, K.A. *et al, Zh. Obshch. Khim.*, 1977, **47**, 2516 (*Engl. transl.* p. 2299) (*deriv*)

Petrov, K.A. *et al, Zh. Obshch. Khim.*, 1981, **51**, 2142 (*Engl. transl.* p. 1844) (*deriv*)

4,5-Dihydro-3-phenyl-3*H*-1,3-azaphosphole, 9CI **D-00549**

3-Phenyl-Δ¹-1,3-azaphospholine, 8CI

[30596-36-2]

$C_9H_{10}NP$ M 163.158

Liq. $Bp_{0.07}$ 64°.

King, R.B. *et al, J. Am. Chem. Soc.*, 1971, **93**, 564 (*synth, pmr*)

King, R.B. *et al, J. Chem. Soc., Perkin Trans. 1*, 1974, 1371 (*synth, ir, pmr, ms*)

1,5-Dihydro-3-phenyl-2,4,3-benzodioxaphosphepin, 10CI **D-00550**

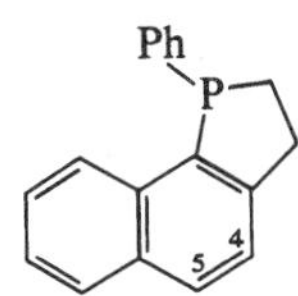

$C_{14}H_{13}O_2P$ M 244.229

Liq. $Bp_{0.1}$ 130°.

Mathey, F. *et al, J. Chem. Soc., Chem. Commun.*, 1980, 191 (*synth, pmr, P nmr, props*)

2,3-Dihydro-1-phenyl-1*H*-benzo[*g*]phosphindole, 10CI **D-00551**

$C_{18}H_{15}P$ M 262.290

1-Oxide: [70610-38-7].

$C_8H_{15}OP$ M 158.180

Solid. Mp 200-202°.

4,5-Dihydro, 1-Oxide: [70610-37-6].

$C_{18}H_{17}OP$ M 280.305

Cryst. (cumene). Mp 151.5-154°.

Orton, W.L. *et al, Phosphorus Sulfur*, 1979, **5**, 349 (*synth, cmr, P nmr*)

Quin, L.D. *et al, Phosphorus Sulfur*, 1982, **12**, 161 (*props*)

1,2-Dihydro-1-phenyl-4*H*-3,1,2-benzothiazaphosphorin-4-thione **D-00552**

$C_{13}H_{10}NPS_2$ M 275.322

2-Methoxy, 2-sulfide: [23857-21-8].

$C_{14}H_{12}NOPS_3$ M 337.409

Solid. Mp 84°.

2-Ethoxy, 2-sulfide: [23857-22-9].

$C_{15}H_{14}NOPS_3$ M 351.435

Solid. Mp 98°.

Legrand, L. *et al, Bull. Soc. Chim. Fr.*, 1969, 1173.

1,3-Dihydro-1-phenyl-2,1-benzoxaphosphole, 9CI **D-00553**

[79157-82-7]

$C_{13}H_{11}OP$ M 214.203

Cryst. (hexane). Mp 62-63°. $Bp_{0.1}$ 100° (oven).

1-Oxide: [75777-29-6].

$C_{13}H_{11}O_2P$ M 230.202

Solid. Mp 99-101°. $Bp_{0.6}$ 180° (oven).

1-Sulfide: [79157-86-1].

$C_{13}H_{11}OPS$ M 246.263

Cryst. (CCl_4). Mp 116.5-117.5°.

Dahl, B.M. *et al, J. Chem. Soc., Perkin Trans. 1*, 1981, 2239 (*synth, sulfide, pmr, P nmr, ms*)

Miles, J.A. *et al, J. Org. Chem.*, 1982, **47**, 1677 (*oxide, pmr, ms*)

3,4-Dihydro-3-phenyl-1*H*-2,3-benzoxaphosphorin, 10CI **D-00554**

$C_{14}H_{13}OP$ M 228.230

Isol. as Ni complex.

3-Oxide: [74722-24-0].

$C_{14}H_{13}O_2P$ M 244.229

Cryst. (C_6H_6). Mp 120°. $Bp_{0.1}$ 210°.

3-Sulfide: [74722-25-1].

$C_{14}H_{13}OPS$ M 260.290

Cryst. (CCl_4/hexane). Mp 77°.

Mathey, F. *et al, J. Chem. Soc., Chem. Commun.*, 1980, 191 (*synth, derivs, complex, pmr, P nmr*)

4,5-Dihydro-2-phenyl-1,3,2-benzoxathiaphosphepin, 8CI **D-00555**

$C_{14}H_{13}OPS$ M 260.290

2-Oxide: [32101-77-2].

$C_{14}H_{13}O_2PS$ M 276.289

Solid. Mp 87°.

Nguyen, H.P. *et al, C.R. Hebd. Seances Acad. Sci., Ser. C*, 1971, **272**, 1145 (*synth, pmr*)

2,3-Dihydro-3-phenyl-4*H*-1,3,2-benzoxazaphosphorin-4-one, 8CI **D-00556**

$C_{13}H_{10}NO_2P$ M 243.201

2-Methoxy: [15494-46-9].

$C_{14}H_{12}NO_3P$ M 273.227

Cryst. (C_6H_6/pet. ether). Mp 75°.

2-Ethoxy: [15494-47-0].

$C_{15}H_{14}NO_3P$ M 287.254

Cryst. (C_6H_6/pet. ether). Mp 93-95°.

2-Phenoxy: [15494-49-2].

$C_{19}H_{14}NO_3P$ M 335.298

Cryst. (C_6H_6/pet. ether). Mp 103-105°.

2-Chloro: see 2-Chloro-2,3-dihydro-3-phenyl-4H-1,3,2-benzoxazaphosphorin-4-one, C-00050

2-Acetoxy:

$C_{15}H_{12}NO_4P$ M 301.238

Cryst. (CCl_4). Mp 76-79°.

Sabirova, R.A. *et al, Zh. Obshch. Khim.*, 1967, **37**, 732 (*Engl. transl.* p. 686)

2,2-Dihydro-3-phenyl-Δ¹-1,3,2-diazaphosphetin-4-one, 8CI **D-00557**

$C_7H_7N_2OP$ M 166.119

2,2-Dichloro: [22403-60-7]. *2,2-Dichloro-2,2-dihydro-3-phenyl-Δ¹-1,3,2-diazaphosphetin-4-one.*

$C_7H_5Cl_2N_2OP$ M 235.009

Solid. Mp 212-214°.

2,2-Diphenoxy: [22403-62-9]. *2,2-Dihydro-2,2-diphenoxy-3-phenyl-Δ¹-1,3,2-diazaphosphetin-4-one.*

$C_{19}H_{15}N_2O_3P$ M 350.313

Liq.

2,2-Di-Ph: [22403-61-8]. *2,2-Dihydro-2,2,3-triphenyl-Δ¹-1,3,2-diazaphosphetin-4-one.*

$C_{19}H_{15}N_2OP$ M 318.314

Solid. Mp 214-216°.

Kolotilo, M.V. *et al, Zh. Obshch. Khim.*, 1969, **39**, 463 (*Engl. transl.* p. 437)

2,3-Dihydro-2-phenyl-1*H*-dibenz[*e,g*]isophosphindole, 9CI **D-00558**

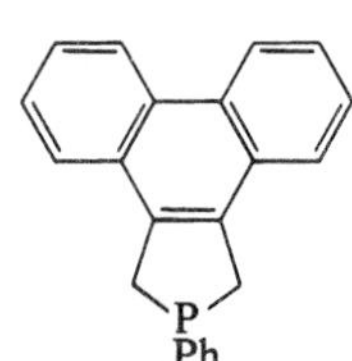

$C_{22}H_{17}P$ M 312.350

2-Oxide: [74078-07-2].

$C_{22}H_{17}OP$ M 328.349

Plates (EtOH/ligroin). Mp 197-198°.

Quin, L.D. *et al, J. Am. Chem. Soc.*, 1982, **104**, 1893 (*synth, pmr, cmr, P nmr*)

10,11-Dihydro-5-phenyl-5*H*-dibenzo[*b,f*]phosphepin, 9CI **D-00559**

1-Phenyl-1-phospha-2,3:6,7-dibenzocyclohepta-2,6-diene

[30309-74-1]

$C_{20}H_{17}P$ M 288.328

Cryst. exhibiting isodimorphism (EtOH). Mp 75-75.5° and 94.5-95°. $Bp_{0.01}$ 170-190°.

*B,MeI: 10,11-Dihydro-5-methyl-5-phenyl-5*H*-dibenzo*[b,f]*phosphepinium iodide.*

$C_{21}H_{20}IP$ M 430.267

Cryst. (EtOH). Mp 251-252°.
Oxide: [30309-75-2].
$C_{20}H_{17}OP$ M 304.327
Needles (EtOH) or plates (C_6H_6). Mp 177° and 185°. $Bp_{0.001}$ 115°.
Sulfide:
$C_{20}H_{17}PS$ M 320.388
Cryst. (EtOH). Mp 157-158°.

Mann, F.G. *et al*, *J. Chem. Soc.*, 1953, 1130 (*synth, derivs, uv*)
Suggs, J.L. *et al*, *J. Org. Chem.*, 1971, **36**, 2566 (*synth, oxide, ms, uv, pmr*)
Allen, D.W. *et al*, *J. Chem. Soc., Perkin Trans. 1*, 1979, 2326 (*synth, derivs*)
Allen, D.W. *et al*, *Z. Naturforsch., B*, 1980, **35**, 133 (*oxide, cryst struct*)
Segall, Y. *et al*, *Phosphorus Sulfur*, 1980, **8**, 243 (*oxide, synth, uv, ir, ms, pmr*)

7,12-Dihydro-12-phenyl-5*H*-dibenz[c,*f*][1,5]oxaphosphocin, 10CI D-00560

$C_{20}H_{17}OP$ M 304.327
B,MeI: [68669-06-7]. *7,12-Dihydro-12-methyl-12-phenyl-5*H*-dibenz*[c,f][*1,5*]*oxaphosphocinium iodide.*
$C_{21}H_{20}IOP$ M 446.267
Solid. Mp 254-256°.
12-Oxide: [68669-00-1].
$C_{20}H_{17}O_2P$ M 320.327
Solid. Mp 217-219°.

Petrov, K.A. *et al*, *Zh. Obshch. Khim.*, 1978, **48**, 2025 (*Engl. transl.* p. 1841) (*synth, ir, ms, pmr, P nmr*)

4,7-Dihydro-2-phenyl-1,3,2-dioxaphosphepin, 11CI D-00561

[69813-38-3]

$C_{10}H_{11}O_2P$ M 194.169
Liq. $Bp_{0.7}$ 94-96°. Yields monomeric phenylmetaphosphonate when pyrolysed.
2-Sulfide: [69813-45-2].
$C_{10}H_{11}O_2PS$ M 226.229
Solid. Mp 69-70°.

Bracker, S. *et al*, *J. Chem. Soc., Chem. Commun.*, 1983, 857.

2,3-Dihydro-2-phenyl-1*H*-isophosphindole, 9CI D-00562

2-Phenylisophosphindoline

$C_{14}H_{13}P$ M 212.230
Bp_{15} 182-186°, $Bp_{0.2}$ 110-113°. Quite stable to air when pure.
*B,MeI: 2,3-Dihydro-2-methyl-2-phenyl-1*H*-isophosphindolium iodide.*
$C_{15}H_{16}IP$ M 354.170
Cryst. (EtOH). Mp 208-210°.
B,PhBr: [34982-17-7]. *2,3-Dihydro-2,2-diphenyl-1*H*-isophosphindolium bromide.*
$C_{20}H_{18}BrP$ M 369.240
Not easily purified.
B,PhClO₄: *2,3-Dihydro-2,2-diphenyl-1*H*-isophosphindolium perchlorate.*
$C_{20}H_{18}ClO_4P$ M 388.787
Solid. Mp 180-181°.
Oxide: [50869-62-0].
$C_{14}H_{13}OP$ M 228.230
Solid + $1H_2O$. Mp 99-102°. Bp_{12} 100° subl.

Mann, F.G. *et al*, *J. Chem. Soc.*, 1958, 2516 (*synth, props, derivs*)
Märkl, G., *Angew. Chem., Int. Ed. Engl.*, 1963, **2**, 620 (*deriv*)
Snider, T.E. *et al*, *Phosphorus*, 1971, **1**, 59 (*salts, pmr*)
Chan, T.H. *et al*, *Phosphorus*, 1974, **3**, 225 (*oxide*)
Chan, T.H. *et al*, *Tetrahedron*, 1975, **31**, 2537 (*oxide, pmr*)

5,10-Dihydro-10-phenylphenophosphazine, 9CI D-00563

[79722-69-3]

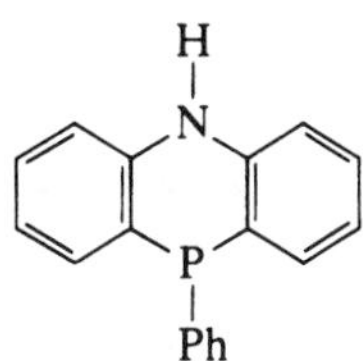

$C_{18}H_{14}NP$ M 275.289
Solid. Mp 158-160°.
10-Oxide: [73785-73-6].
$C_{18}H_{14}NOP$ M 291.288
No phys. props reported.
10-Sulfide: [79691-65-9].
$C_{18}H_{14}NPS$ M 307.349
Cryst. (DMF aq.). Mp 271-272°.
5-Me: see 5,10-Dihydro-5-methyl-10-phenylphenophosphazine, D-00526

Freeman, H.S. *et al*, *J. Org. Chem.*, 1981, **46**, 5373 (*synth, oxide, ir, ms*)
Demidova, N.I. *et al*, *Zh. Obshch. Khim.*, 1982, **52**, 1099 (*Engl. trans.* p. 958) (*sulfide*)

2,7-Dihydro-1-phenyl-1*H*-phosphepin, 9CI D-00564

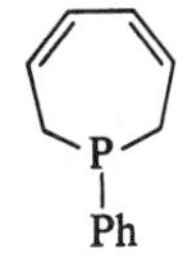

$C_{12}H_{13}P$ M 188.208
Liq. $Bp_{0.01}$ 70-74°.
Oxide: [29634-20-6].
$C_{12}H_{13}OP$ M 204.208
Liq. $Bp_{0.2}$ 160°.

Märkl, G. *et al*, *Tetrahedron Lett.*, 1970, 1273; 1973, 1455 (*synth, pmr, ir*)

4,5-Dihydro-1-phenyl-1*H*-phosphepin, 9CI **D-00565**
[42202-49-3]

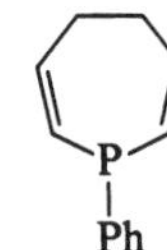

$C_{12}H_{13}P$ M 188.208
Oxide: [42202-57-3].
$C_{12}H_{13}OP$ M 204.208
$Bp_{0.05}$ 135-140°.
B,MeClO₄: [42202-53-9]. *4,5-Dihydro-1-methyl-1-phenyl-1*H*-phosphepinium perchlorate.*
$C_{13}H_{16}ClO_4P$ M 302.694
Solid. Mp 107-108°.
B,PhCH₂Br: [42202-12-0]. *1-Benzyl-4,5-dichloro-1-phenyl-1*H*-phosphepinium bromide.*
$C_{19}H_{20}BrP$ M 359.245
Solid. Mp 209-210°.

Märkl, G. *et al, Tetrahedron Lett.*, 1973, 1455 (*synth, ir, pmr*)

2,3-Dihydro-1-phenyl-1*H*-phosphole, 9CI **D-00566**
1-Phenyl-2-phospholene, 8CI
[28278-55-9]

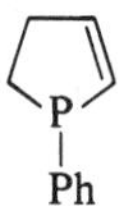

$C_{10}H_{11}P$ M 162.171
Liq. Bp_{14} 120°.
*B,MeI: 2,3-Dihydro-1-methyl-1-phenyl-1*H*-phospholium iodide.*
$C_{11}H_{14}IP$ M 304.110
Solid. Mp 122°.
1-Oxide: [703-03-7].
$C_{10}H_{11}OP$ M 178.170
Solid. Mp 80°.
1-Sulfide: [52988-62-2].
$C_{10}H_{11}PS$ M 194.231
Liq. $Bp_{0.1}$ 135-140°.

Fritzsche, H. *et al, Chem. Ber.*, 1964, **97**, 1988 (*synth, deriv*)
Bundgaard, T. *et al, Tetrahedron Lett.*, 1972, 3353 (*cmr*)
Moedritzer, K., *Synth. React. Inorg. Metal-Org. Chem.*, 1974, **4**, 119 (*sulfide*)
Morris, D.L. *et al, Phosphorus*, 1974, **4**, 69 (*oxide, props*)
Moedritzer, K. *et al, Synth. React. Inorg. Metal-Org. Chem.*, 1977, **7**, 311 (*oxide, ir, raman, cmr, P nmr, pmr, cryst struct*)

2,5-Dihydro-1-phenyl-1*H*-phosphole, 9CI **D-00567**
1-Phenyl-3-phospholene, 8CI
[28278-54-8]

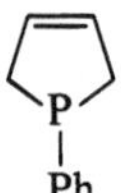

$C_{10}H_{11}P$ M 162.171
1-Oxide: [5186-73-2].
$C_{10}H_{11}OP$ M 178.170
Solid. Mp 85°.

Breen, J.J. *et al, Phosphorus*, 1972, **2**, 55 (*P nmr*)
Moedritzer, K. *et al, Synth. React. Inorg. Metal-Org. Chem.*, 1977, **7**, 311 (*oxide, ms, ir, raman, cmr, pmr, P nmr, cryst struct*)

5,6-Dihydrophosphanthridene, 9CI, 8CI **D-00568**
5,6-Dihydrodibenzo[b,d]*phosphorin, 8CI. 9,10-Dihydro-9-phosphaphenanthrene*
[20335-94-8]

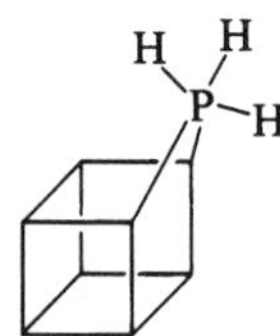

$C_{13}H_{11}P$ M 198.204
Liq. $Bp_{0.01}$ 132°.
5-Me, 5-oxide: [17847-95-9]. *5,6-Dihydro-5-methylphosphanthridene 5-oxide.*
$C_{14}H_{13}OP$ M 228.230
Oil + 0.5 H_2O.
5,5-Di-Me, chloride: [82404-45-3]. *5,6-Dihydro-5,5-dimethylphosphanthridinium chloride. 9,10-Dihydro-9,9-dimethyl-9-phosphoniaphenanthrene bromide.*
$C_{15}H_{16}ClP$ M 262.718
Cryst. (CH_2Cl_2/Et_2O). Mp 170° dec.
5-Ph, 5-oxide: [17847-94-8]. *5,6-Dihydro-5-phenylphosphanthridene 5-oxide.*
$C_{19}H_{15}OP$ M 290.301
Solid. Mp 131-132°.
5,5-Di-Ph, bromide: 5,6-Dihydro-5,5-diphenylphosphanthridinium bromide. 9,10-Dihydro-9,9-diphenyl-9-phosphoniaphenanthrene bromide.
$C_{25}H_{20}BrP$ M 431.311
Cryst. (AcOH or EtOH/EtOAc). Mp 335-337°.

Allen, D.W. *et al, J. Chem. Soc. (C)*, 1969, 252 (*synth, ir, pmr*)
Tebby, J.C. *et al, J. Chem. Soc. (C)*, 1971, 1064 (*pmr*)
Costa, T. *et al, Chem. Ber.*, 1982, **115**, 1367 (*synth, nmr, pmr*)

9,9-Dihydro-9-phosphapentacyclo[4.3.0.0²˒⁵.0³˒⁸.0⁴˒⁷]nonane, 9CI **D-00569**
9,9-Dihydro-9-phosphahomocubane. Homocubylphosphorane

$C_8H_{11}P$ M 138.149
9,9,9-Tri-Me: [33845-96-4]. *Homocubyltrimethylphosphorane.*
$C_{11}H_{17}P$ M 180.229
Oil. $Bp_{0.000001}$ ca. 20°.
9,9,9-Tri-Ph: [30092-40-1]. *Homocubyltriphenylphosphorane.*
$C_{26}H_{23}P$ M 366.441
Cryst. (Me_2CO). Mp 118-119°. Stable to H_2O. At 120°, → Ph_3P.

Katz, T.J. *et al, J. Am. Chem. Soc.*, 1970, **92**, 6701 (*synth, props, pmr*)
Bushweller, C.H. *et al, Tetrahedron Lett.*, 1972, 2401 (*triphenyl, pmr*)
Turnblom, E.W. *et al, J. Am. Chem. Soc.*, 1973, **95**, 4292 (*synth, ir, ms, cmr, P nmr*)

2,3-Dihydro-1*H*-phosphindole 1-oxide, 9CI **D-00570**

C_8H_9OP M 152.132

1-Ph: [31236-96-1].

$C_{14}H_{13}OP$ M 228.230

Sl. hygroscopic cryst. Mp 107-108°.

1-Benzyl:

$C_{15}H_{15}OP$ M 242.257

Cryst. (toluene/cyclohexane). Mp 111°.

Chan, T-H. *et al, Can. J. Chem.*, 1971, **49**, 530 (*deriv, pmr*)
Collins, D.J. *et al, Aust. J. Chem.*, 1974, **27**, 831 (*derivs, uv, ir, ms, pmr*)
Nief, F. *et al, Nouv. J. Chim.*, 1981, **5**, 187 (*deriv, ir, pmr, P nmr*)

2,3-Dihydro-1*H*-phosphole, 9CI **D-00571**

2-Phospholene, 8CI

[1769-52-4]

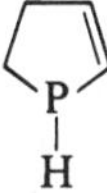

C_4H_7P M 86.073

Liq. Bp_{760} 75°.

1-Chloro: see 1-Chloro-2,3-dihydro-1H-phosphole, C-00051

1-Ph: see 2,3-Dihydro-1-phenyl-1H-phosphole, D-00566

Fritzsche, H. *et al, Chem. Ber.*, 1965, **98**, 1681.

2,5-Dihydro-1*H*-phosphole, 9CI **D-00572**

3-Phospholene, 8CI

[4544-86-9]

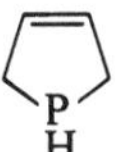

C_4H_7P M 86.073

Liq. Bp 102-104°. Rapidly oxid., but not especially flammable.

1-Chloro: see 1-Chloro-2,5-dihydro-1H-phosphole, C-00052

*1-Bromo: 1-Bromo-2,5-dihydro-1*H*-phosphole.*

C_4H_6BrP M 164.969

Bp_{32} 77-80°.

*1-Benzyl: 1-Benzyl-2,5-dihydro-1*H*-phosphole. 1-Benzyl-3-phospholene.*

$C_{11}H_{13}P$ M 176.197

Solid. Mp 34-34.5°. $Bp_{0.2}$ 71-72°.

Coggon, P. *et al, J. Am. Chem. Soc.*, 1970, **92**, 5779 (*benzyl, uv, P nmr, ms, cryst struct*)
Myers, D.K. *et al, J. Org. Chem.*, 1971, **36**, 1285 (*bromo, ir, P nmr, pmr*)
Breen, J.J. *et al, Phosphorus*, 1972, **2**, 55 (*benzyl*)
Durig, J.R. *et al, J. Chem. Phys.*, 1980, **73**, 5564 (*ir, struct*)
Richardson, L.W. *et al, J. Chem. Phys.*, 1980, **73**, 5556 (*synth, props, pmr, ir, raman*)

1,1-Dihydrophosphorinane, 8CI **D-00573**

1,1-Dihydrophosphinane

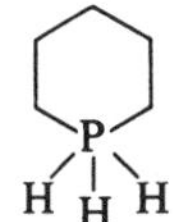

$C_5H_{13}P$ M 104.131

P,P,P-*Trifluoro:* [1478-39-3]. *1,1,1-Trifluoro-1,1-dihydrophosphorinane. 1,1,1-Trifluoro-1,1-dihydrophosphinane.*

$C_5H_{10}F_3P$ M 158.103

Liq. Bp_{40} 64-65°.

Schmutzler, R., *Inorg. Chem.*, 1964, **3**, 421 (*synth, ir, ms*)
Reddy, G.S. *et al, Z. Naturforsch., B*, 1970, **25**, 1199 (*F nmr*)

1,3-Dihydrospiro[2*H*-1,3,2-benzodiazaphosphole-2,2′-[1,3,2]-benzodioxaphosphole], 10CI, 9CI **D-00574**

[50597-14-3]

$C_{12}H_{11}N_2O_2P$ M 246.205

Solid. Mp 108°.

1-Benzyl: [89767-35-1].

$C_{19}H_{17}N_2O_2P$ M 336.329

Solid. Mp 108°.

Charbonnel, Y. *et al, C.R. Hebd. Seances Acad. Sci., Ser. C*, 1973, **277**, 571 (*synth, ir, pmr, P nmr*)
Mathis, R. *et al, C.R. Hebd. Seances Acad. Sci., Ser. C*, 1975, **280**, 809 (*ir*)
Malavaud, C. *et al, Can. J. Chem.*, 1984, **62**, 43 (*deriv, P nmr*)

1,3-Dihydrospiro[2*H*-1,3,2-benzodiazaphosphole-2,2′-[1,3,2]dioxaphosholane], 9CI **D-00575**

$C_8H_{11}N_2O_2P$ M 198.161

2-Ph: [71559-29-0].

$C_{14}H_{15}N_2O_2P$ M 274.258

Solid. Mp 137-141°.

Cadogan, J.I.G. *et al, J. Chem. Soc., Chem. Commun.*, 1979, 189 (*synth, P nmr*)

1,4-Dihydrospiro[2*H*-3,1,2-benzoxazaphosphorine-2,2′-[1,3,2]-dioxaphospholane], 10CI **D-00576**

$C_9H_{12}NO_3P$ M 213.172

2-Ph: [71559-28-9].

$C_{15}H_{15}NO_3P$ M 288.262

Solid. Mp 95-98°.

Cadogan, J.I.G. *et al, J. Chem. Soc., Chem. Commun.*, 1979, 189 (*synth, P nmr*)

6,7-Dihydro-5,5,7,7-tetramethyl-5*H*-dibenzo[*d,f*][1,3]diphosphepinium(2+), 10CI **D-00577**

$C_{17}H_{22}P_2^{\oplus\oplus}$ M 288.308 (ion)

Dibromide: [82404-41-9].
$C_{17}H_{22}Br_2P_2$ M 448.116
Needles ($MeOH/CH_2Cl_2/Et_2O$). Mp 308°.

7,7-Dihydro,monobromide: [82404-42-0]. *5,5,7,7-Tetramethyl*[d,f][*1*,λ^5*3*]*diphosphepinium bromide.*
$C_{17}H_{21}BrP_2$ M 367.204
Needles ($CHCl_3/Et_2O$). Mp 112°.

Costa, T. *et al*, *Chem. Ber.*, 1982, **115**, 1367 (*synth, P nmr, pmr*)

2,2-Dihydro-4,4,5,5-tetramethyl-2,2,2-triphenyl-1,3,2-dioxaphospholane **D-00578**
[49595-63-3]

$C_{24}H_{27}O_2P$ M 378.450
Solid. Dec. at ca. 50°.

Bartlett, P.D. *et al*, *J. Am. Chem. Soc.*, 1973, **95**, 6486 (*synth, ir, P nmr*)
Baumstark, A.L. *et al*, *J. Org. Chem.*, 1980, **45**, 3593 (*synth, props*)

3,4-Dihydro-2,3,4,5-tetraphenyl-2*H*-1,2,3-diazaphosphole, 9CI **D-00579**

(3*RS*,4*RS*)-*form*

$C_{26}H_{21}N_2P$ M 392.439

(3RS,4RS)-form [55520-53-1]
(±)-trans-*form*
Solid. Mp 144-146°.

(3RS,4SR)-form [55520-54-2]
(±)-cis-*form*
Solid. Mp 183-185°.

Baccolini, G. *et al*, *J. Org. Chem.*, 1975, **40**, 2318; 1978, **43**, 216 (*synth, oxides, pmr*)
Baccolini, G. *et al*, *J. Chem. Soc., Perkin Trans. 1*, 1979, 2329; 1983, 535 (*props*)

3,4-Dihydro-2,3,4,5-tetraphenyl-2*H*-1,3,2-diazaphosphole, 10CI **D-00580**
1,2,4,5-Tetraphenyl-Δ^3*-1,3,2-diazaphospholine, 9CI*

$C_{26}H_{21}N_2P$ M 392.439

2-Oxide:
$C_{26}H_{21}N_2OP$ M 408.438
Solid. Stereoisomeric forms, Mp 174-7° and 202-4°, known. Stereochemistries not assigned.

Baccolini, G. *et al*, *J. Org. Chem.*, 1974, **39**, 2650; 1975, **40**, 2318 (*synth, pmr*)

1,2-Dihydro-1,2,3,4-tetraphenyl-1,2-diphosphete, 9CI **D-00581**
[42451-95-6]

$C_{26}H_{20}P_2$ M 394.392
Cryst. Mp 159°. Treatment with H_2O_2, S, or Li causes P-P bond cleavage.

Ecker, A. *et al*, *Chem. Ber.*, 1973, **106**, 1453 (*synth, ms, props*)
Charrier, C. *et al*, *J. Org. Chem.*, 1981, **46**, 3 (*synth, props, cryst struct*)

1,4-Dihydro-1,1,4,4-tetraphenyl-1,4-diphosphorinium(2+), 9CI **D-00582**
1,4-Dihydro-1,1,4,4-tetraphenyl-1,4-diphosphininium(2+)

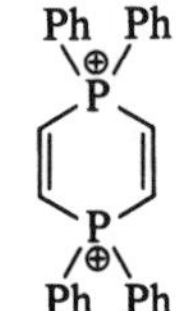

$C_{28}H_{24}P_2^{\oplus\oplus}$ M 422.445 (ion)

Dichloride:
$C_{28}H_{24}Cl_2P_2$ M 493.351
Cryst. + $1H_2O$ (MeOH/EtOAc). Mp 251-253°.

Dibromide: [15924-61-5].
$C_{28}H_{24}Br_2P_2$ M 582.253
Cryst. (MeCN/MeOH). Mp 306-308° dec.

Aguiar, A. *et al*, *J. Am. Chem. Soc.*, 1967, **89**, 4235 (*synth, ir, pmr*)

2,5-Dihydro-2,3,4,5-tetraphenyl-1,2,5-oxadiphosphole 2,5-dioxide, 9CI **D-00583**
[42451-96-7]

$C_{26}H_{20}O_3P_2$ M 442.390
Cryst. Mp 183-185°.

Ecker, A. *et al*, *Chem. Ber.*, 1973, **106**, 1453 (*synth, ms*)

1,2-Dihydro-1,2,4,6-tetraphenylphosphorin, 8CI **D-00584**
1,2-Dihydro-1,2,4,6-tetraphenylphosphinine
[13689-23-1]

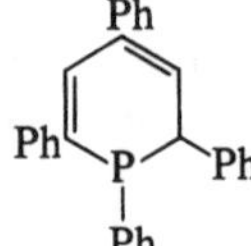

$C_{29}H_{23}P$ M 402.474
Cryst. (EtOH). Mp 144-145°.

Oxide: [13689-26-4].
$C_{29}H_{23}OP$ M 418.474
Needles. Mp 156-158°.
B,MeI: 1,2-Dihydro-1-methyl-1,2,4,6-tetraphenylphosphorinium iodide.
$C_{30}H_{26}IP$ M 544.414
Solid. Mp 160-162°.

Märkl, G. *et al, Angew. Chem., Int. Ed. Engl.*, 1967, **6**, 87 (*synth, pmr, uv, derivs*)

2,5-Dihydro-2,3,4,5-tetraphenyl-1,2,5-thiadiphosphole, 9CI D-00585

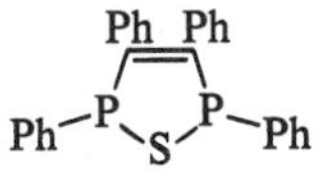

$C_{26}H_{20}P_2S$ M 426.452
2,5-Disulfide: [42451-98-9].
$C_{26}H_{20}P_2S_3$ M 490.572
Cryst. Mp 257-260°.

Ecker, A. *et al, Chem. Ber.*, 1973, **106**, 1453 (*synth, pmr, ms*)

3,6-Dihydro-2*H*-1,2-thiaphosphorin 2-sulfide, 8CI D-00586

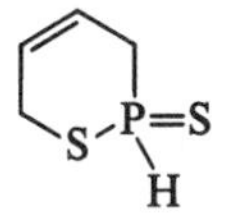

$C_4H_7PS_2$ M 150.193
2-Et: [41274-67-9].
$C_6H_{11}PS_2$ M 178.247
Liq. $Bp_{0.05}$ 114°.
2-Ph: [32309-30-1].
$C_{10}H_{11}PS_2$ M 226.291
Solid. Mp 70°.

Ecker, A. *et al, Monatsh. Chem.*, 1973, **104**, 503 (*synth, pmr*)

N-(4,5-Dihydro-2-thiazolyl)-*P*-methylphosphonamidic acid, 9CI D-00587

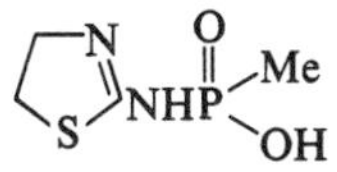

$C_4H_9N_2O_2PS$ M 180.161
Esters appear to exist mainly as the exocyclic imino tautomer.
Et ester: [68236-58-8]. *Ethyl* N-*(4,5-dihydro-2-thiazolyl)*-P-*methylphosphonamidate.*
$C_6H_{13}N_2O_2PS$ M 208.215
Solid. Mp 114-115°.
Ph ester: [68236-47-5]. *Phenyl* N-*(4,5-dihydro-2-thiazolyl)*-P-*methylphosphonamidate.*
$C_{10}H_{13}N_2O_2PS$ M 256.259
Insecticide and acaricide. Solid. Mp 137-138° (130-130.5°).

Ger. Pat., 2 703 363, (*1978*); *CA*, **90**, 6535 (*derivs, synth, use*)
B.P., 1 544 778, (*1979*); *CA*, **92**, 22498 (*derivs, synth, use*)
Razvodovskaya, L.V. *et al, Zh. Obshch. Khim.*, 1980, **50**, 329 (*Engl. transl.* p. 266) (*ester, synth, pmr, ir, struct*)

N-(4,5-Dihydro-2-thiazolyl)-*P*-methylphosphonamidothioic acid D-00588

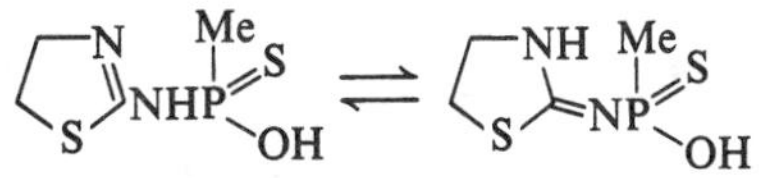

$C_4H_9N_2OPS_2$ M 196.222
SH ⇌ OH Tautomerism also possible. Esters appear to exist mainly in the exocyclic-imino form.

(±)-form

O-*Et ester:* O-*Ethyl* N-*(4,5-dihydro-2-thiazolyl)*-P-*methylphosphonamidothioate.*
$C_6H_{13}N_2OPS_2$ M 224.275
Solid. Mp 103-105°.
O-*Ph ester:* O-*Phenyl* N-*(4,5-dihydro-2-thiazolyl)*-P-*methylphosphonamidothioate.*
$C_{10}H_{13}N_2OPS_2$ M 272.319
Herbicide. Solid. Mp 130-130.5°.

Razvodovskaya, L.V. *et al, Zh. Obshch. Khim.*, 1980, **50**, 329 (*Engl. transl.* p. 266) (*synth, ir, pmr*)

2,2-Dihydro-2,2,2-trichloro-4,4,5,5-tetrakis(trifluoromethyl)-1,3,2-dioxaphospholane D-00589

[70311-67-0]

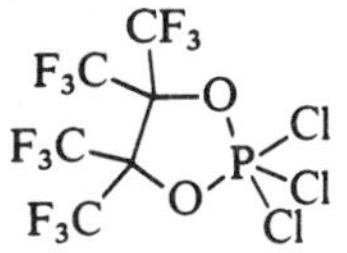

$C_6Cl_3F_{12}O_2P$ M 469.378
Liq. Bp_{90} 97°.

Roeschenthaler, G.-V. *et al, Z. Anorg. Allg. Chem.*, 1979, **450**, 79 (*synth, ir, ms, F and P nmr*)

2,2-Dihydro-2,2,2-triethoxy-4-[(4-methylphenyl)sulfonyl]-1,4,2-oxazaphospholidine, 9CI D-00590

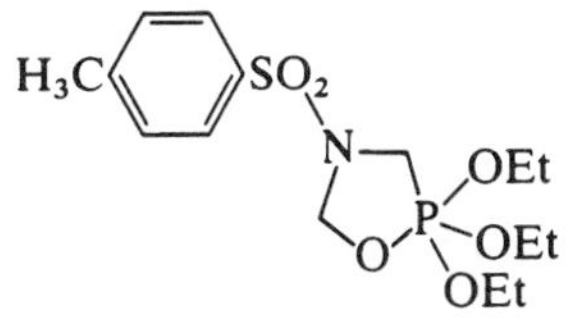

$C_{15}H_{26}NO_6PS$ M 379.407
Needles ($CHCl_3$/hexane). Mp 115-117°.

Scharf, D.J. *et al, J. Org. Chem.*, 1976, **41**, 28 (*synth, ir, pmr*)

2,2-Dihydro-2,2,2-trimethoxy-3,5-bis(pentafluorophenyl)-1,4,2-dioxaphospholane D-00591

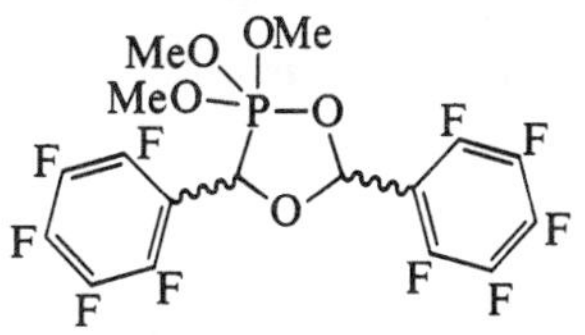

$C_{17}H_{11}F_{10}O_5P$ M 516.229
Cryst. (C_6H_6/pentane). Mp 83-86°. Prepd. as 1:1 mixture of *cis* and *trans* forms.

Ramirez, F. *et al, J. Org. Chem.*, 1969, **34**, 3385 (*synth, pmr, P nmr*)

2,2-Dihydro-2,2,2-trimethoxy-4,5-bis(trifluoromethyl)-1,3,2-dithiaphosphole, 9CI D-00592

[37895-63-9]

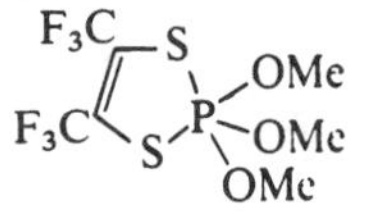

$C_7H_9F_6O_3PS_2$ M 350.230

Dec. readily at r.t.

Death, N.J. *et al, J. Chem. Soc., Chem. Commun.*, 1972, 395 (*synth, P nmr*)

2,2-Dihydro-2,2,2-trimethoxy-4,5-dimethyl-1,3,2-dioxaphosphole, 9CI D-00593

Trimethoxy(dimethylethylenedioxy)phosphorane

[1665-79-8]

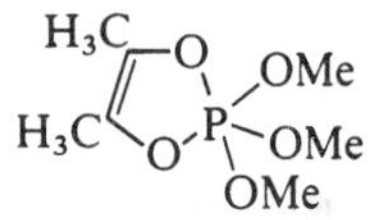

$C_7H_{15}O_5P$ M 210.166

Reaction with acyl chlorides yields phosphates of α-hydroxy-β-diketones and β-keto-α-hydroxycarboxylic acid chlorides. Reagent for conversion of α-diketones into ketones, and iso(thio)cyanates into 2,4-imidazolidine diones, 4-oxazolones and 4-thiazolones. Liq. $Bp_{0.5}$ 57-60°. Easily hydrolysed.

Ramirez, F. *et al, J. Org. Chem.*, 1968, **33**, 1192 (*use*)
Ramirez, F. *et al, Tetrahedron*, 1973, **29**, 3741; 1975, **31**, 2007 (*synth, pmr, props, use*)
Ramirez, F. *et al, J. Am. Chem. Soc.*, 1975, **97**, 3809 (*props*)
Stephenson, L.M. *et al, J. Org. Chem.*, 1976, **41**, 2928 (*use*)
David, S. *et al, J. Chem. Soc., Perkin Trans. 1*, 1980, 1262 (*use*)
Castelijns, M.M.C.F. *et al, J. Org. Chem.*, 1981, **46**, 47 (*pmr, cmr, P nmr, props*)

2,2-Dihydro-2,2,2-trimethoxynaphtho[1,2-*d*]-1,3,2-dioxaphosphole, 8CI D-00594

[15052-46-7]

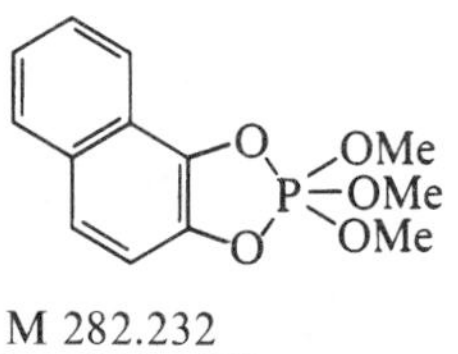

$C_{13}H_{15}O_5P$ M 282.232

Pale-yellow oil. Bp_1 148-150°.

Ramirez, F. *et al, J. Org. Chem.*, 1968, **33**, 20 (*synth, ir, pmr, P nmr*)

2,2-Dihydro-2,2,2-trimethoxy-4,4,5,5-tetrakis(trifluoromethyl)-1,3,2-dioxaphospholane D-00595

[6509-88-2]

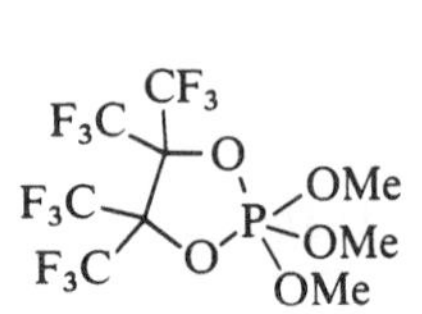

$C_9H_9F_{12}O_5P$ M 456.122

Liq. $Bp_{0.5}$ 61-62°.

Ramirez, F. *et al, J. Org. Chem.*, 1968, **33**, 3787 (*F and P nmr*)

Kibardin, A.M. *et al, Izv. Akad. Nauk SSSR, Ser. Khim.*, 1981, 1095 (*Engl. transl.* p. 855) (*synth*)

2,3-Dihydro-3,3,5-trimethyl-1,2-oxaphosphole 2-oxide, 9CI D-00596

3,3,5-Trimethyl-1,2-oxaphosphol-4-ene 2-oxide, 8CI

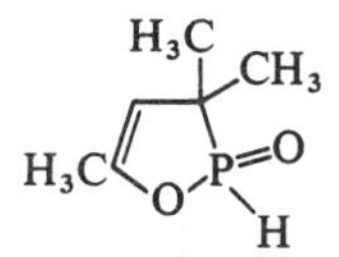

$C_6H_{11}O_2P$ M 146.125

▷Alkoxy and phenoxy derivs are powerful anticholinesterases

2-Chloro: see 2-Chloro-2,3-dihydro-3,3,5-trimethyl-1,2-oxaphosphole, C-00054

2-Methoxy: [4335-89-1].
$C_7H_{13}O_3P$ M 176.152
Liq. d_4^{20} 1.12. Bp_2 87-8°. n_D^{20} 1.4592.
▷RP3450000.

2-Ethoxy: [7614-58-6].
$C_8H_{15}O_3P$ M 190.178
Liq. d_4^{20} 1.08. $Bp_{0.015}$ 72-73°. n_D^{20} 1.4525.

2-Phenoxy: [7563-17-9].
$C_{12}H_{15}O_3P$ M 238.222
Liq. d_4^{20} 1.16. $Bp_{0.01}$ 130-133°. n_D^{20} 1.5190.

2-Me: [32503-56-3].
$C_7H_{13}O_2P$ M 160.152
$Bp_{0.03}$ 73-75°. n_D^{20} 1.4722.

2-Ph: [4529-76-4].
$C_{12}H_{15}O_2P$ M 222.223
Liq. d_4^{20} 1.03. $Bp_{0.09}$ 136°. n_D^{20} 1.5419.

2-Diethylamino: [20335-88-0].
$C_{10}H_{20}NO_2P$ M 217.247
Solid. Mp 54-55°. Bp_3 110-113°.

Arbuzov, B.A. *et al, Izv. Akad. Nauk SSSR, Ser. Khim.*, 1966, 1848 (*Engl. transl.* p. 1786) (*alkoxy, phenoxy, synth, props*)
Novitski, K.I. *et al, Zh. Obshch. Khim.*, 1967, **37**, 2280 (*Engl. transl.* p. 2167) (*phenoxy, synth*)
Nurtdinov, S.Kh. *et al, Zh. Obshch. Khim.*, 1970, **40**, 36 (*Engl. transl.* p. 33) (*phenyl, synth, ir*)
Nurtdinov, S.Kh. *et al, Zh. Obshch. Khim.*, 1970, **40**, 2189 (*Engl. transl.* p. 2176) (*methoxy, synth*)

2,2-Dihydro-2,2,2-trimethyl-1,2-oxaphosphorinane, 9CI D-00597

2,2-Dihydro-2,2,2-trimethyl-1,2-oxaphosphinane

[69783-48-8]

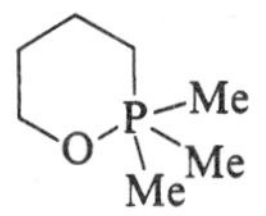

$C_7H_{17}OP$ M 148.184

Liq. Bp_{10} 67°.

Schmidbaur, H. *et al, Chem. Ber.*, 1979, **112**, 501 (*synth, props*)

[(5,6-Dihydro-4,4,6-trimethyl-4*H*-1,3-oxazin-2-yl)methyl]phosphonic acid, 9CI D-00598

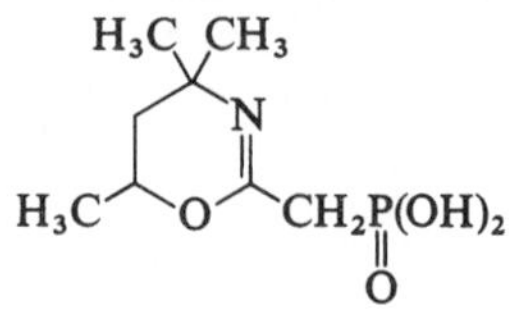

$C_8H_{16}NO_4P$ M 221.192

Esters are Wittig-Horner reagents for synth. of 2-ethenyl-1,3-oxazines and thence ethenyl aldehydes and ketones.

*Di-Me ester: Dimethyl [(5,6-dihydro-4,4,6-trimethyl-4*H*-1,3-oxazin-2-yl]phosphonate.*

$C_{10}H_{20}NO_4P$ M 249.246

Cryst. (pet. ether). Mp 56-57°. Bp_1 102°.

Di-Et ester: [50431-01-1]. *Diethyl [(5,6-dihydro-4,4,6-trimethyl-4*H*-1,3-oxazin-2-yl]phosphonate.*

$C_{12}H_{24}NO_4P$ M 277.300

Pale-yellow liq. $Bp_{0.075}$ 109°.

Malone, G.R. *et al*, *J. Org. Chem.*, 1974, **39**, 623 (*synth, ir, pmr, use*)

[(5,6-Dihydro-4,4,6-trimethyl-4*H*-1,3-oxazin-2-yl)methyl]-triphenylphosphonium(1+), 9CI D-00599

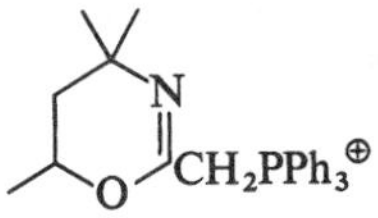

$C_{26}H_{29}NOP^{\oplus}$ M 402.495 (ion)

Chloride: [50259-42-2].

$C_{26}H_{29}ClNOP$ M 437.948

Source of ylide. Cryst. (MeCN/Et_2O). Mp 224°.

*Ylide: [(5,6-Dihydro-4,4,6-trimethyl-4*H*-1,3-oxazin-2-yl)methylene]triphenylphosphorane.*

$C_{26}H_{28}NOP$ M 401.487

Used in Wittig reactions for synth. of vinyloxazines and hence of $\alpha\beta$-unsatd. aldehydes, ketones and acids.

Malone, G.R. *et al*, *J. Org. Chem.*, 1974, **39**, 623 (*synth, uv, ir, pmr, use*)

2,5-Dihydro-1,3,4-trimethyl-1*H*-phosphole, 9CI D-00600

1,3,4-Trimethyl-3-phospholene, 8CI

[14410-05-0]

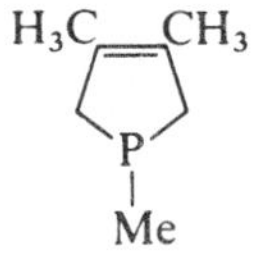

$C_7H_{13}P$ M 128.153

Ligand for Ni. Liq. Bp 160-161°, Bp_{50} 100°, $Bp_{0.5}$ 93-96° (?).

*B,MeBr: 2,5-Dihydro-1,1,3,4-tetramethyl-1*H*-phospholium bromide.*

$C_8H_{16}BrP$ M 223.092

Solid. Mp 186-187°.

B,PhBr: see 2,5-Dihydro-3,4-dimethyl-1-phenyl-1H-phosphole, D-00468

*B,PhCH$_2$Br: 1-Benzyl-2,5-dihydro-1,3,4-trimethyl-1*H*-phospholium bromide.*

$C_{14}H_{20}BrP$ M 299.190

Cryst. (MeOH/EtOAc). Mp 185.5-187.5°.

1-Oxide:

$C_7H_{13}OP$ M 144.153

Liq. or hygroscopic solid. $Bp_{0.05}$ 105°.

Quin, L.D. *et al*, *J. Org. Chem.*, 1968, **33**, 1034 (*synth, derivs, pmr*)

Hammond, P.J. *et al*, *J. Chem. Soc., Perkin Trans. 2*, 1982, 205 (*synth, oxide, pmr, P nmr*)

Buchanan, G.W. *et al*, *Org. Magn. Reson.*, 1983, **21**, 436 (*derivs, cmr*)

2,2-Dihydro-2,2,2-trimethyl-4,4,5,5-tetrakis(trifluoromethyl)-1,3,2-dioxaphospholane D-00601

[17244-54-1]

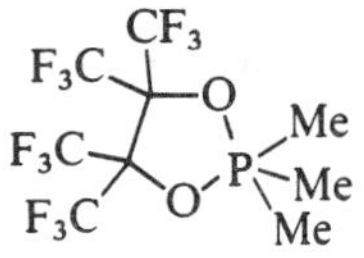

$C_9H_9F_{12}O_2P$ M 408.123

Cryst. (hexane at −20°). Mp 15°.

Ramirez, F. *et al*, *J. Org. Chem.*, 1968, **33**, 3787 (*synth, P nmr*)

2,2-Dihydro-2,2,2-triphenoxy-1,3,2-benzodioxaphosphole D-00602

[19667-83-5]

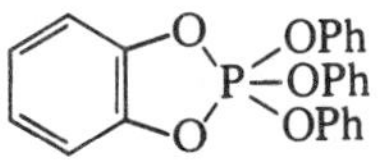

$C_{24}H_{19}O_5P$ M 418.385

Oil. $Bp_{0.04}$ 165-194°.

Ramirez, F. *et al*, *Tetrahedron*, 1968, **24**, 5041 (*synth, props, P nmr*)

Gloede, J. *et al*, *Z. Anorg. Allg. Chem.*, 1980, **471**, 147 (*P nmr*)

Gloede, J. *et al*, *J. Prakt. Chem.*, 1981, **323**, 621 (*synth, P nmr*)

4,4-Dihydro-4,4,4-triphenoxytrioxaphosphetane, 9CI, 8CI D-00603

Triphenylphosphite ozonide

[29833-83-8]

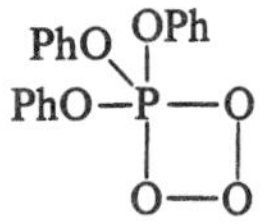

$C_{18}H_{15}O_6P$ M 358.287

Said to be a source of singlet oxygen [$^1O_2$], but this has been questioned. Not isolable; synthesised *in situ*. Dec. >−30°; stable at −78° for 7 days.

Murray, R.W. *et al*, *J. Am. Chem. Soc.*, 1969, **91**, 5358 (*synth, props, use*)

Koch, E. *et al*, *Tetrahedron*, 1970, **26**, 3503 (*synth, props*)

Bartlett, P.D. *et al*, *J. Org. Chem.*, 1980, **45**, 3000, 4269 (*synth, props*)

Stephenson, L.M. *et al*, *J. Am. Chem. Soc.*, 1982, **104**, 5819 (*props*)

Bartlett, P.D. *et al*, *J. Am. Chem. Soc.*, 1983, **105**, 1984 (*props*)

1,4-Dihydro-2,4,6-triphenyl-1,4-azaphosphorine, 9CI D-00604

1,4-Dihydro-2,4,6-triphenyl-1,4-azaphosphinine

[39768-15-5]

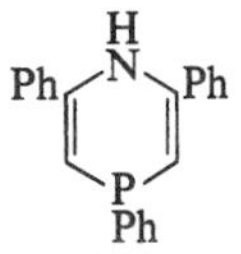

$C_{22}H_{18}NP$ M 327.365

Solid. Mp 145-146°.

4-Oxide: [34735-02-9].

$C_{22}H_{18}NOP$ M 343.364

Solid. Mp 315°.

Märkl, G. *et al*, *Angew. Chem., Int. Ed. Engl.*, 1972, **11**, 1019 (*synth, ms, pmr*)
Moskalevskaya, L.S., *Zh. Obshch. Khim.*, 1983, **53**, 545 (*Engl. transl.* p. 472) (*oxide*)

2,3-Dihydro-1,2,3-triphenyl-1*H*-1,3,2-benzodiphospharsole, 9CI D-00605

$C_{24}H_{19}AsP_2$ M 444.283
(*1α,2β,3α)-form* [38234-83-2]
Needles (Me_2CO aq.). Mp 175-176°.
1,3-Disulfide: [38305-79-2].
$C_{24}H_{19}AsP_2S_2$ M 508.403
Cryst. (Me_2CO). Mp 219-220°.

Mann, F.G. *et al*, *J. Chem. Soc., Perkin Trans. 1*, 1972, 2548 (*synth, ir, ms*)

2,3-Dihydro-1,2,3-triphenyl-1*H*-benzotriphosphole, 9CI D-00606
1,2,3-Triphenyl-1,2,3-triphosphaindane
[57194-41-9]

$C_{24}H_{19}P_3$ M 400.335
(1α,2β,3α)-form
Mp 184-186°.
1-Sulfide:
$C_{24}H_{19}P_3S$ M 432.395
Cryst. (Me_2CO aq.). Mp 122-123°.
1,3-Disulfide: [37913-15-8].
$C_{24}H_{19}P_3S_2$ M 464.455
Solid. Mp 184-186°.
Trisulfide: Prisms (Me_2CO aq.). Mp 121.5-122°.

Daly, J.J., *J. Chem. Soc. (A)*, 1966, 1020 (*cryst struct*)
Mann, F.G. *et al*, *J. Chem. Soc. (C)*, 1966, 916 (*synth*)
Mann, F.G. *et al*, *J. Chem. Soc., Perkin Trans. 1*, 1972, 1631 (*synth, ms, P nmr, ir, pmr*)
Schmidpeter, A. *et al*, *Chem. Ber.*, 1985, **118**, 3849.

2,2-Dihydro-2,2,2-triphenyl-2*H*-1,2-benzoxaphosphorin, 9CI D-00607
[62839-81-0]

$C_{26}H_{21}OP$ M 380.425
Orange solid. Mp 196°. Dec. by MeI.

Bestmann, H.J. *et al*, *Tetrahedron Lett.*, 1977, 79 (*synth, props, P nmr*)

2,2-Dihydro-2,2,2-triphenyl-4,4-bis(trifluoromethyl)-3-(triphenylphosphoranylidene)-1,2-oxaphosphetane, 8CI D-00608
[14181-19-2]

$C_{40}H_{30}F_6OP_2$ M 702.614
Stable Wittig reaction intermed. Cryst. (Et_2O/diglyme). Mp 157-158°.

Birum, G.H. *et al*, *J. Org. Chem.*, 1967, **32**, 3554 (*synth, F and P nmr*)
Chioccola, G. *et al*, *J. Chem. Soc. (A)*, 1968, 568 (*cryst struct*)

2,3-Dihydro-2,4,9-triphenyl-1*H*-1,3,2-diazaphosphonine 2-oxide, 9CI D-00609
[56562-96-0]

$C_{24}H_{21}N_2OP$ M 384.416
Solid. Mp 190°.
1,3-Di-Me: [56562-99-3].
$C_{26}H_{25}N_2OP$ M 412.470
Solid. Mp 160°.

Lampin, J.P. *et al*, *C.R. Hebd. Seances Acad. Sci., Ser. C*, 1975, **280**, 1153 (*synth, ir, pmr*)

5,5-Dihydro-5,5,5-triphenyl-5*H*-dibenzophosphole, 9CI D-00610
Triphenylbiphenylenephosphorane

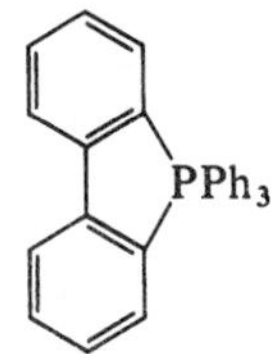

$C_{30}H_{23}P$ M 414.485
Cryst. (cyclohexane). Mp 155.5-156.5°.

Wittig, G. *et al*, *Chem. Ber.*, 1964, **97**, 741 (*synth*)

2,2-Dihydro-2,2,2-triphenyl-1,3,2-dioxaphospholane, 9CI D-00611
Ethylenedioxytriphenylphosphorane
[34736-69-1]

$C_{20}H_{19}O_2P$ M 322.343

Denney, D.B. *et al*, *J. Am. Chem. Soc.*, 1971, **93**, 4004.
von Itzstein, M. *et al*, *J. Chem. Soc., Perkin Trans. 1*, 1986, 437 (*synth, cmr, P nmr*)

Dihydro(triphenylphosphoranylidene)-2,5-furandione, 9CI D-00612

Triphenylphosphoranylidenesuccinic anhydride

[906-65-0]

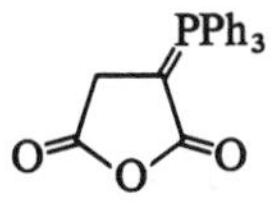

$C_{22}H_{17}O_3P$ M 360.348

Promoter for polymerizations; Wittig reagent for butenolide synth. Mp 162.5-163.5° dec. (157-159°).

Aksnes, G., *Acta Chem. Scand.*, 1961, **15**, 692 (*synth, ir*)
Hedaya, E. *et al*, *Tetrahedron*, 1968, **24**, 2241 (*synth, ir, pmr*)
Cameron, A.F. *et al*, *J. Chem. Soc., Perkin Trans. 2*, 1975, 1030 (*synth, cmr*)
McMurray, J.E. *et al*, *Tetrahedron Lett.*, 1977, 2869 (*use*)

1,1-Dihydro-2,4,6-triphenylphosphorin, 10CI D-00613

1,1-Dihydro-2,4,6-triphenylphosphinin

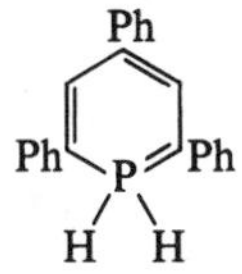

$C_{23}H_{19}P_2$ M 357.351

1,1-Difluoro: [40425-79-4]. *1,1-Difluoro-1,1-dihydro-2,4,6-triphenylphosphorin.*
$C_{23}H_{17}F_2P_2$ M 393.332
Solid. Mp 129-131°.

1,1-Dibromo: 1,1-Dibromo-1,1-dihydro-2,4,6-triphenylphosphorin.
$C_{23}H_{17}Br_2P_2$ M 515.143
Solid. Mp 60° dec.

1,1-Dichloro: [40425-71-6]. *1,1-Dichloro-1,1-dihydro-2,4,6-triphenylphosphorin.*
$C_{23}H_{17}Cl_2P_2$ M 426.241
Cryst. (CCl_4/MeCN). Mp 101-102°.

1,1-Dimethoxy: 1,1-Dihydro-1,1-dimethoxy-2,4,6-triphenylphosphorin.
$C_{25}H_{23}O_2P$ M 386.429
Solid. Mp 111-113°.

1,1-Bis(methylthio): [36240-22-9]. *1,1-Dihydro-1,1-bis(methylthio)-2,4,6-triphenylphosphorin.*
$C_{25}H_{23}PS_2$ M 418.550
Solid. Mp 146-147°.

1,1-Diphenoxy: [20995-73-7]. *1,1-Dihydro-1,1-diphenoxy-2,4,6-triphenylphosphorin.*
$C_{35}H_{27}O_2P$ M 510.571
Solid. Mp 154-156°.

1,1-Bis(phenylthio): 1,1-Dihydro-1,1-bis(phenylthio)-2,4,6-triphenylphosphorin.
$C_{35}H_{27}PS_2$ M 542.692
Solid. Mp 118-119°.

Kanter, H. *et al*, *Chem. Ber.*, 1977, **110**, 395 (*synth, pmr, uv, ms*)

2,2-Dihydro-2,2,2-triphenyl-4,4,5,5-tetrakis(trifluoromethyl)-1,3,2-dioxaphospholane D-00614

[6509-85-9]

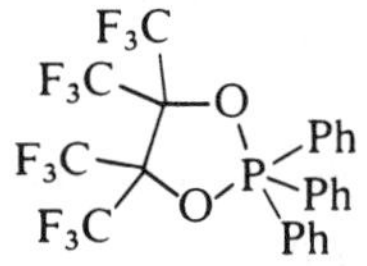

$C_{24}H_{15}F_{12}O_2P$ M 594.336

In situ source of hexafluoroactone. Cryst. (pentane). Mp 105° dec.

Birum, G.H. *et al*, *J. Org. Chem.*, 1967, **32**, 3554 (*synth, F and P nmr, props, use*)
Ramirez, F. *et al*, *J. Org. Chem.*, 1968, **33**, 3787 (*synth*)

2,3-Dihydro-1*H*-1,2,3-triphosphole, 9CI D-00615

1,2,3-Triphosphol-4-ene, 8CI. 4,5-Dehydro-1,2,3-triphospholane

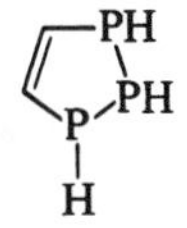

$C_2H_5P_3$ M 121.983

1,2,3-Tri-Me: [68090-74-4]. *2,3-Dihydro-1,2,3-trimethyl-1H-1,2,3-triphosphole.*
$C_5H_{11}P_3$ M 164.063
Liq. Bp_{14} 103-105°.

1,2,3-Tri-Ph: [75600-57-6]. *2,3-Dihydro-1,2,3-triphenyl-1H-1,2,3-triphosphole.*
$C_{20}H_{17}P_3$ M 350.276
Solid. Mp 79°.

Baudler, M. *et al*, *Z. Naturforsch., B*, 1978, **33**, 691 (*synth, ir, ms, P nmr, pmr*)
Charrier, C. *et al*, *J. Org. Chem.*, 1981, **46**, 3 (*synth, P nmr*)

1,3-Dihydroxy-1*H*-1,2,4,3-benziodadioxaphosphorin 3-oxide, 9CI D-00616

2-Iodosophenylphosphoric acid

[40329-00-8]

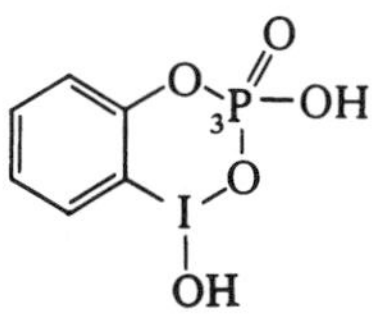

$C_6H_6IO_5P$ M 315.989

Needles. Mp 123-124°. Dec. to red oil.

3-Methoxy (Me ester): [40329-01-9]. *1-Hydroxy-3-methoxy-1H-1,2,4,3-benziodadioxaphosphorine 3-oxide.*
$C_7H_8IO_5P$ M 330.015
Solid. Mp 122-123°. Dec. to red oil.

Leffler, J.E. *et al*, *J. Org. Chem.*, 1973, **38**, 2719 (*synth, deriv, ir, pmr*)
Brody, K.R. *et al*, *J. Nucl. Med.*, 1978, **19**, 848 (*metab*)

2,5-Dihydroxy-3,6-dimethyl-3,6-bis(dihydroxyphosphinyl)-1,4,2,5-dioxadiphosphorinane 2,5-dioxide D-00617

[2,5-Dihydroxy-3,6-dimethyl-2,5-dioxo-1,4,2,5-dioxadiphosphorinan-3,6-diyl]bisphosphonic acid

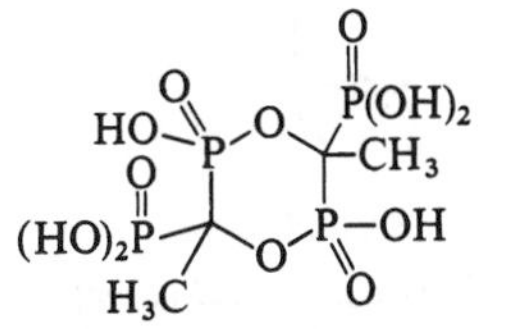

$C_4H_{12}O_{12}P_4$ M 376.027

Cryst. + 3 or 4 H_2O (H_2O). pK_{a1} 2.7, pK_{a2} 3.1, pK_{a3} 9.6.

Hexa-Na salt: Cryst. + $14H_2O$ (MeOH).

Di-Ca salt: Cryst. + $10H_2O$.

Hexa-NH_4 salt: Cryst. + $2H_2O$.
Hexa-Me ester:
$C_{10}H_{24}O_{12}P_4$ M 460.187
Cryst. (Me_2CO/pet. ether). Mp 139-141°.

Blaser, B. *et al, Z. Anorg. Allg. Chem.*, 1971, **381**, 247 (*synth*)
Brun, G. *et al, Rev. Chim. Mineral.*, 1972, **9**, 453 (*synth, ir, raman*)
Philippot, E. *et al, Rev. Chim. Mineral.*, 1972, **9**, 591 (*cryst struct*)
Collins, A.J. *et al, J. Chem. Soc., Dalton Trans.*, 1974, 960 (*pmr, cryst struct*)

2,5-Dihydroxy-1,4,2,5-dioxadiphosphorinane 2,5-dioxide D-00618

2,5-Dihydroxy-1,4,2,5-dioxadiphosphinane 2,5-dioxide

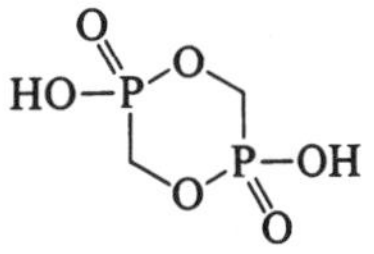

$C_2H_6O_6P_2$ M 188.013
2,5-Diethoxy (di-Et ester): [5021-92-1]. *2,5-Diethoxy-1,4,2,5-dioxaphosphorinane 2,5-dioxide.*
$C_6H_{14}O_6P_2$ M 244.121
Solid. Mp 147-150°.

Pudovik, A.N. *et al, Zh. Obshch. Khim.*, 1976, **46**, 770 (*Engl. transl.* p. 769)
Ramanenko, V.D. *et al, Zh. Obshch. Khim.*, 1978, **48**, 2222 (*Engl. transl.* p. 2018)

1,3-Dihydroxy-1,3-diphosphepane 1,3-dioxide, 10CI D-00619

[65617-68-7]

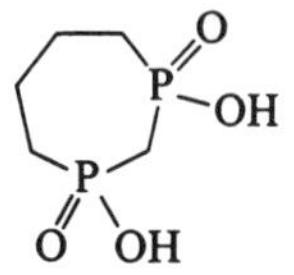

$C_5H_{12}O_4P_2$ M 198.095
Solid. Mp 166-167°.
Diisopropyl ester: 1,3-Diisopropoxy-1,3-diphosphepane 1,3-dioxide.
$C_{11}H_{24}O_4P_2$ M 282.256
Liq. d_4^{20} 1.11. $Bp_{0.1}$ 149°. n_D^{20} 1.4712.

Novikova, Z.S. *et al, Zh. Obshch. Khim.*, 1977, **47**, 2636 (*Engl. transl.* p. 2409) (*synth, deriv*)

1,3-Dihydroxy-1,3-diphosphetane 1,3-dioxide, 10CI D-00620

[66686-22-4]

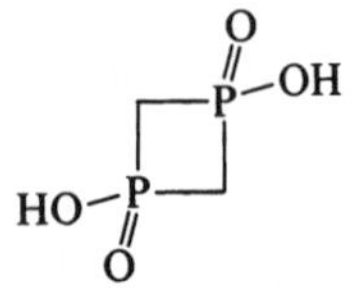

$C_2H_6O_4P_2$ M 156.015
1,3-Diisopropoxy (diisopropyl ester): 1,3-Bis(1-methylethoxy)-1,3-diphosphetane 1,3-dioxide.
$C_8H_{18}O_4P_2$ M 240.175
Mp 103-104°. Hot conc. HCl aq. Causes ring cleavage.

Novikova, Z.S. *et al, Zh. Obshch. Khim.*, 1977, **47**, 2636 (*Engl. transl.* p. 2409) (*synth, pmr, P nmr*)

1,3-Dihydroxy-1,3-diphospholane 1,3-dioxide, 9CI D-00621

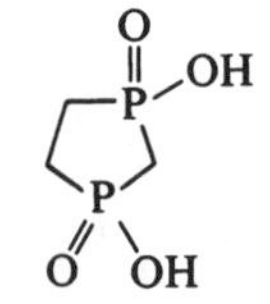

$C_3H_8O_4P_2$ M 170.041
Solid. Mp 147-149°.
1,3-Diisopropoxy (1,3-diisopropyl ester): 1,3-Diisopropoxy-1,3-diphospholane 1,3-dioxide.
$C_9H_{20}O_4P_2$ M 254.202
Solid. Mp 85-87°.

Novikova, Z.S. *et al, Zh. Obshch. Khim.*, 1977, **47**, 2636 (*Engl. transl.* p. 2409) (*synth, pmr, P nmr, deriv*)

1,3-Dihydroxy-1,3-diphosphorinane 1,3-dioxide, 9CI D-00622

1,3-Dihydroxy-1,3-diphosphinane 1,3-dioxide
[65650-33-1]

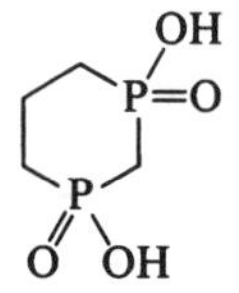

$C_4H_{10}O_4P_2$ M 184.068
Solid. Mp 243-244°.
1,3-Diisopropoxy (1,3-diisopropyl ester): 1,3-Diisopropoxy-1,3-diphosphorinane 1,3-dioxide.
$C_{10}H_{22}O_4P_2$ M 268.229
Solid. Mp 84-85°.

Novikova, Z.S. *et al, Zh. Obshch. Khim.*, 1977, **47**, 2636 (*Engl. transl.* p. 2409) (*synth, pmr, P nmr*)

2,5-Dihydroxy-3-(1-methylpropyl)-1,3,2,5-oxazadiphospholidine 2,5-dioxide, 9CI D-00623

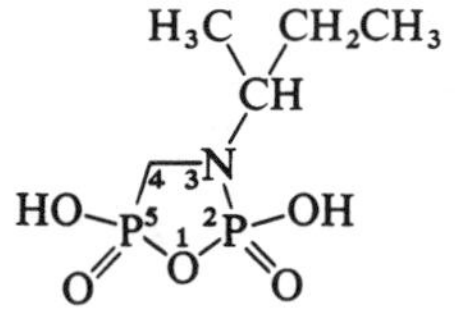

$C_5H_{13}NO_5P$ M 198.135
2,5-Diethoxy (Di-Et ester):
$C_9H_{21}NO_5P$ M 254.242
Liq. d_4^{20} 1.14. $Bp_{0.02}$ 108-112°. n_D^{20} 1.4341.
2,5-Dipropoxy (Dipropyl ester):
$C_{11}H_{25}NO_5P$ M 282.296
Liq. d_4^{20} 1.09. $Bp_{0.6}$ 110-114°. n_D^{20} 1.4555.

Suvalova, E.A. *et al, Zh. Obshch. Khim.*, 1978, **48**, 1281 (*Engl. transl.* p. 1172) (*synth*)

(2,3-Dihydroxypropyl)phosphonic acid, 9CI D-00624

3-Phosphono-1,2-propanediol
[13563-34-3]

$HOCH_2CH(OH)CH_2P(O)(OH)_2$

$C_3H_9O_5P$ M 156.075

(±)-*form*

Di-Et ester: Diethyl (2,3-dihydroxypropyl)phosphonate.
$C_7H_{17}O_5P$ M 212.182
Liq. $Bp_{0.25}$ 142°.

Griffin, C.E. *et al*, *J. Org. Chem.*, 1969, **34**, 1532 (*ester, synth, pmr*)
Bonsen, P.P.M. *et al*, *Chem. Phys. Lipids*, 1972, **8**, 199 (*derivs*)

1,8-Dihydroxy-3a,4,7,7a-tetrahydro-4,7-phosphinidene-1H-phosphindole 1,8-dioxide, 8CI D-00625

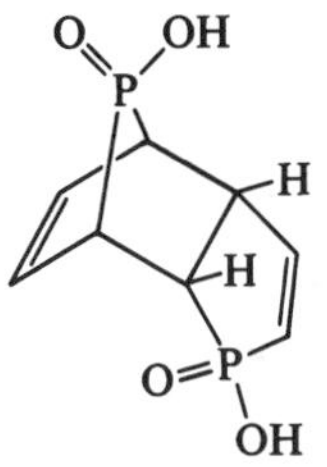

$C_8H_{10}O_4P_2$ M 232.112
Cryst. (Me_2CO aq.). Mp 102-104°.
Tetrahydro: [24576-92-9]. *Octahydro-1,8-dihydroxy-4,7-phosphinidene-1*H*-phosphindole 1,8-dioxide.*
$C_8H_{14}O_4P_2$ M 236.144
Cryst. (Me_2CO aq.). Mp 254-255°.
1,8-Diethoxy (Di-Et ester): [16182-89-1]. *1,8-Diethoxy-3*a*,4,7,7*a*-tetrahydro-4,7-phosphinidene-1*H*-phosphinidole 1,8-dioxide.*
$C_{12}H_{18}O_4P_2$ M 288.219
Rhombs. (EtOAc/hexane). Mp 125-126°.
Tetrahydro, 1,8-diethoxy: [16182-90-4]. *1,8-Diethoxyoctahydro-4,7-phosphinidene-1*H*-phosphindole 1,8-dioxide.*
$C_{12}H_{22}O_4P_2$ M 292.251
Cryst. (EtOAc/hexane). Mp 80-81°.

Kluger, R. *et al*, *J. Am. Chem. Soc.*, 1967, **89**, 3919 (*esters, ir*)
Kluger, R. *et al*, *J. Am. Chem. Soc.*, 1969, **91**, 4143 (*synth, ir*)
Yuan-yuan, H.C. *et al*, *J. Am. Chem. Soc.*, 1969, **91**, 4150 (*cryst struct*)

3,9-Dihydroxy-2,4,8,10-Tetraoxa-3,9-diphosphaspiro[5.5]undecane 3,9-dioxide, 9CI D-00626

[947-28-4]

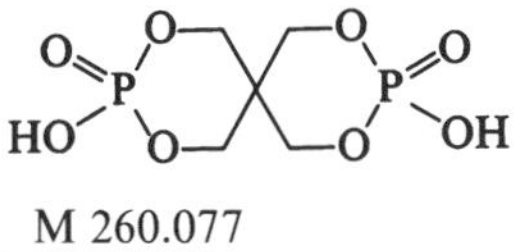

$C_5H_{10}O_8P_2$ M 260.077
Cryst. (AcOH). Mp 314°.
Mono(dimethylammonium) salt: Cryst. (1,4-butanediol). Mp 266°.
Bis(dimethylammonium) salt: Cryst. (EtOH/Me_2CO). Mp 214-215°.
3,9-Dichloride: see 3,9-Dichloro-2,4,8,10-tetraoxa-3,9-diphosphaspiro[*5.5*]*undecane, D-00196*
3,9-Diamide:
$C_5H_{12}N_2O_6P_2$ M 258.107
Cryst. (H_2O). Mp 303-305°.

Rätz, R. *et al*, *J. Org. Chem.*, 1963, **28**, 1608 (*synth*)
Pivawer, P.M. *et al*, *J. Heterocycl. Chem.*, 1967, **4**, 599 (*amide*)
Brault, J.-F. *et al*, *Bull. Soc. Chim. Fr. Part II*, 1974, 677 (*synth*)

1,3-Dihydroxytriphosphetane 1,3-dioxide, 10CI D-00627

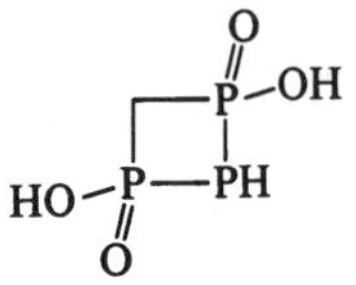

$CH_5O_4P_3$ M 173.969
2-Isopropyl, diisopropyl ester: 1,3-Diisopropoxy-2-isopropyltriphosphetane 1,3-dioxide.
$C_9H_{21}O_4P_3$ M 286.184
Liq. d_4^{20} 1.24. n_D^{20} 1.5173.
*2-*tert*-Butyl, diisopropyl ester:* [73424-19-8]. *2-*tert*-Butyl-1,3-diisopropoxytriphosphetane 1,3-dioxide.*
$C_{11}H_{25}O_4P_3$ M 314.237
Liq. d_4^{20} 1.12. n_D^{20} 1.4995.

Kabachnik, M.M. *et al*, *Zh. Obshch. Khim.*, 1980, **50**, 228; *CA*, **92**, 181288 (*synth, pmr, P nmr*)
Kabachnik, M.M. *et al*, *Zh. Obshch. Khim.*, 1982, **52**, 763; *CA*, **97**, 39017 (*synth, pmr, P nmr*)

Di-1H-imidazol-1-ylphosphinic acid, 9CI D-00628

[68593-85-1]

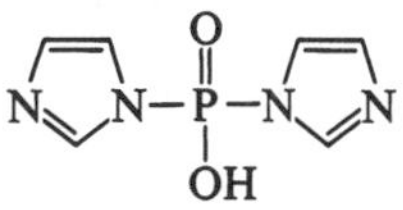

$C_6H_7N_4O_2P$ M 198.121
Tributylammonium salt: [68698-15-7]. Used in phosphorylations. Prepd. *in situ*.
Imidazolium salt: Granular solid. Mp 185-195°.
Ph ester: [15706-68-0]. *Phenyl di-1*H*-imidazol-1-ylphosphinate.*
$C_{12}H_{11}N_4O_2P$ M 274.218
Hygroscopic cryst. Mp 96-97°.
Chloride: [61561-85-1].
$C_6H_6ClN_4OP$ M 216.566
No phys. props. reported.
Imidazolide: see Tri-1-imidazolylphosphine, T-00495

Cramer, F. *et al*, *Chem. Ber.*, 1961, **94**, 1612 (*synth, ester*)
Sergeeva, N.F. *et al*, *Biorg. Khim.*, 1976, **2**, 1056; *CA*, **86**, 43956.
Konieczny, M. *et al*, *Z. Naturforsch.*, *B*, 1978, **33**, 1033 (*ester, ms*)
Kozarich, J.W., *Nucleic Acid Chem.*, 1978, **2**, 853 (*synth, use*)

Diiodotrimethylphosphorane D-00629

Iodotrimethylphosphonium iodide. Trimethylphosphine diiodide

Me_3PI_2 or $Me_3PI^{\oplus}$ $I^{\ominus}$

$C_3H_9I_2P$ M 329.887
Shown to possess the ionic struct. Yellow cryst. Mp 279-280°.

Goubeau, J. *et al*, *Z. Electrochem.*, 1960, **64**, 598 (*synth, props, ir, raman*)

Diiodotriphenoxyphosphorane, 9CI D-00630

Triphenyl phosphite diiodide
[55136-69-1]

$(PhO)_3PI_2$

$C_{18}H_{15}I_2O_3P$ M 564.097

Brown rhombs (chlorobenzene). Mp 68-69°. Also forms higher polyiodides.

I_2 adduct: Dark-red prisms or plates. Mp 76-78°.

Rydon, H.N. *et al*, *J. Chem. Soc.*, 1956, 3043 (*synth*)
Feshchenko, N.G. *et al*, *Zh. Obshch. Khim.*, 1975, **45**, 283 (*Engl. transl.* p. 269) (*synth, props*)
Kostina, V.G., *Zh. Obshch. Khim.*, 1981, **51**, 1962 (*Engl. transl.* p. 1688) (*polyiodides*)

Diiodotriphenylphosphorane, 9CI — D-00631

Triphenylphosphinic diiodide

[6396-07-2]

Ph_3PI_2

$C_{18}H_{15}I_2P$ M 516.099

Thought to possess ionic struct. Iodinates $\alpha\beta$-unsatd. ketones in β-posn. Primary and secondary alcs. (but not tertiary) converted into alkyl iodides. Solid. Mp 172-175° dec.

▷TB6100000.

I_2 adduct: Iodotriphenylphosphonium triiodide.
$C_{18}H_{15}I_4P$ M 769.908
Mp 132°.

Beveridge, A.B. *et al*, *J. Chem. Soc. (A)*, 1966, 520.
Parrett, F.W., *Spectrochim. Acta, Part A*, 1970, **26**, 1271 (*ir, raman*)
Mokovetskii, Yu.P. *et al*, *Zh. Obshch. Khim.*, 1980, **50**, 2436 (*Engl. transl.* p. 1967)
Romanenko, V.D. *et al*, *Synthesis*, 1980, 823 (*synth*)
Haynes, R.K. *et al*, *Aust. J. Chem.*, 1982, **35**, 517 (*use*)
Piers, E. *et al*, *Can. J. Chem.*, 1982, **60**, 210 (*use*)

(Diisopropoxyphosphinothioyl)thioacetic acid — D-00632

[[Bis(1-methylethoxy)phosphinothioyl]thio]acetic acid, 9CI

[69923-44-0]

$[(H_3C)_2CHO]_2P(S)SCH_2COOH$

$C_8H_{17}O_4PS_2$ M 272.314

Et ester: [79691-79-5]. *Ethyl (diisopropoxyphosphinothioyl)thioacetate.*
$C_{10}H_{21}O_4PS_2$ M 300.367
Liq. d_4^{20} 1.13. Bp_2 135-136°. n_D^{20} 1.4908.

Hsun, L. *et al*, *CA*, 1966, **64**, 14896 (*synth*)
Mastryukova, T.A. *et al*, *Zh. Obshch. Khim.*, 1981, **51**, 1475 (*Engl. transl.* p. 1251) (*ester, synth, ms, P nmr*)

(Diisopropoxyphosphinyl)acetic acid — D-00633

[Bis(1-methylethoxy)phosphinyl]acetic acid

$[(H_3C)_2CHO]_2P(O)CH_2COOH$

$C_8H_{17}O_5P$ M 224.193

Et ester: Ethyl (diisopropoxyphosphinyl)acetate.
$C_{10}H_{21}O_5P$ M 252.247
Oil. d_4^{20} 1.14 (1.07). Bp_{11} 142-143°, $Bp_{0.1}$ 100-102°. n_D^{20} 1.4658 (1.4285).

Isopropyl ester: Isopropyl (diisopropoxyphosphinyl)-acetate.
$C_{11}H_{23}O_5P$ M 266.273
Oil. d_4^{20} 1.11. $Bp_{0.1}$ 100-101°. n_D^{20} 1.4668.

Amide: [Bis(1-methylethoxy)phosphinyl]acetamide. (Diisopropoxyphosphinyl)acetamide.
$C_8H_{18}NO_4P$ M 223.208
Solid. Mp 98-99°.

N,N-*Diethylamide: see (2-Diethylamino-2-oxoethyl)-phosphonic acid, D-00268*

Kamai, G. *et al*, *Dokl. Akad. Nauk SSSR, Ser. Sci. Khim.*, 1950, **72**, 301 (*synth*)
Bodnarchuk, N.D. *et al*, *Zh. Obshch. Khim.*, 1969, **39**, 1707 (*Engl. transl.* p. 1673) (*ester, amide*)
Moskva, V.V. *et al*, *Zh. Obshch. Khim.*, 1976, **46**, 534 (*Engl. transl.* p. 529) (*esters*)

Diisopropyl chlorothiophosphonate, 8CI — D-00634

Thiohypochlorous acid anhydrosulfide with bis(1-methylethyl) phosphorothioate, 9CI. Bis(1-methylethyl) chlorothiophosphonate. S-*Chloro* O,O-*diisopropyl phosphorothioate*

[51756-01-5]

$[(H_3C)_2CHO]_2P(O)SCl$

$C_6H_{14}ClO_3PS$ M 232.662

Yellow liq. d_4^{20} 1.18-1.20. Bp_2 86°, $Bp_{0.8}$ 71°. n_D^{20} 1.4658 (1.4587).

▷Asphyxiating vapour

Lenard-Borecka, B. *et al*, *Rocz. Chem.*, 1957, **31**, 1167 (*synth*)
Mel'nik, Ya.I. *et al*, *Zh. Obshch. Khim.*, 1978, **48**, 326 (*Engl. transl.* p. 292) (*synth*)

O,O-Diisopropyl chlorothiophosphonothioate — D-00635

Thiohypochlorous acid, anhydrosulfide with O,O-*bis(1-methylethyl) dithioate, 10CI.* O,O-*Bis(1-methylethyl) chlorothiophosphonothioate, 9CI.* S-*Chloro* O,O-*diisopropyl phosphorodithioate*

[5572-32-7]

$[(H_3C)_2CHO]_2P(S)SCl$

$C_6H_{14}ClO_2PS_2$ M 248.722

Golden-yellow liq. d_4^{20} 1.19. $Bp_{0.4}$ 78°. n_D^{20} 1.5145.

▷Asphyxiating odour

Almasi, L. *et al*, *Chem. Ber.*, 1965, **98**, 3546.

N,N-Diisopropyl-*P*-methylphosphonamidic acid, 8CI — D-00636

P-*Methyl*-N,N-*bis(1-methylethyl)phosphonamidic acid, 9CI*

$MeP(O)(OH)N[CH(CH_3)_2]_2$

$C_7H_{18}NO_2P$ M 179.198

Et ester: Ethyl N,N-*diisopropyl*-P-*methylphosphonamidate.*
$C_9H_{22}NO_2P$ M 207.252
Liq. d_4^{20} 1.08. $Bp_{0.1}$ 80-81°. n_D^{20} 1.4395.

Razumov, A.I. *et al*, *CA*, 1958, **52**, 237 (*synth*)

N,N-Diisopropyl-P-methylphosphonamidothioic acid D-00637

N,N-*Bis(1-methylethyl)*-P-*methylphosphonamidothioic acid*

$$Me(HO)P(=S)N[CH(CH_3)_2]_2 \rightleftharpoons Me(O=)P(SH)N[CH(CH_3)_2]_2$$

$C_7H_{18}NOPS$ M 195.259

Chloride: [51747-25-2].

$C_7H_{17}ClNPS$ M 213.705

Solid. Mp 45°. $Bp_{0.1}$ 82°.

Burdon, J. *et al, J. Chem. Soc., Perkin Trans. 2*, 1976, 1052 (*synth, pmr*)

Diisopropylphenylphosphine D-00638

Bis(1-methylethyl)phenylphosphine, 9CI

[6372-43-6]

$$PhP[CH(CH_3)_2]_2$$

$C_{12}H_{19}P$ M 194.256

Ligand for W, Ni, Pt, Pd, and Rh. Liq. Bp_9 106-107°, Bp_2 72-78°. n_D^{20} 1.5284. Protons of isopropyl CH_3 groups are magnetically nonequivalent.

B,MeI: Diisopropylmethylphenylphosphonium iodide.

$C_{13}H_{22}IP$ M 336.195

Solid. Mp 134-137°.

Oxide: [16543-38-7].

$C_{12}H_{19}OP$ M 210.255

Solid. Mp 49-50°. $Bp_{0.1}$ 94°.

Sulfide: [50538-13-1].

$C_{12}H_{19}PS$ M 226.316

Solid. Mp 76°.

Grim, S.O. *et al, J. Org. Chem.*, 1967, **32**, 781 (*nmr*)
Kosolapoff, G.M. *et al, J. Chem. Soc. (C)*, 1967, 1789 (*oxide*)
McFarlane, W., *J. Chem. Soc., Chem. Commun.*, 1968, 229 (*pmr*)
Krasil'nikova, E.A. *et al, Zh. Obshch. Khim.*, 1973, **43**, 1701 (*Engl. transl.* p. 1685) (*sulfide*)
Grim, S.O. *et al, Phosphorus*, 1974, **4**, 189 (*synth, nmr, deriv*)
Timokhin, B.V. *et al, Zh. Obshch. Khim.*, 1977, **47**, 1267 (*Engl. transl.* p. 1167) (*synth, nmr*)
Ratovskii, G.V. *et al, Zh. Obshch. Khim.*, 1978, **48**, 1520 (*Engl. transl.* p. 1394) (*uv*)
Okano, T. *et al, Bull. Chem. Soc. Jpn.*, 1981, **54**, 3799 (*complexes, use*)
Panov, A.M. *et al, Zh. Obshch. Khim.*, 1985, **55**, 2243; *J. Gen. Chem. USSR (Engl. Transl.)*, 1991 (*uv, cmr*)

Diisopropyl phosphate, 8CI D-00639

Bis(1-methylethyl) phosphate, 9CI. *Diisopropyl hydrogen phosphate. Diisopropyl phosphoric acid*

[1804-93-9]

$$[(H_3C)_2HCO]_2P(O)OH$$

$C_6H_{15}O_4P$ M 182.156

NH_4 salt: Solid. Mp 133°.

Tetramethylammonium salt: Solid. Mp 128°.

Cyclohexylammonium salt: [57734-11-9]. Needles (Me_2CO). Mp 191-193°.

Fluoride: see Diisopropyl phosphorofluoridate, D-00665

Chloride: see Diisopropyl phosphorochloridate, D-00657

Bromide: [80354-12-7]. *Diisopropyl phosphorobromidate. Diisopropyl phosphoryl bromide. Diisopropyl bromophosphate.*

$C_6H_{14}BrO_3P$ M 245.053

Liq. $Bp_{0.4}$ 77-78°. Easily hydrolysed.

▷Acidic vapour

Amide: see Diisopropyl phosphoramidate, D-00653

Anilide: see Phenylphosphoramidic acid, P-00257

Anhydride: see Tetraisopropyl diphosphate, T-00166

Wagner-Jauregg, J. *et al, J. Am. Chem. Soc.*, 1951, **73**, 5202 (*derivs*)
Goldwhite, H. *et al, J. Chem. Soc.*, 1955, 3564 (*bromide*)
Chabrier, P. *et al, C.R. Hebd. Seances Acad. Sci.*, 1957, **244**, 2730 (*synth*)
Blackburn, G.M. *et al, J. Chem. Soc. (C)*, 1966, 239 (*synth*)
Sosnovsky, G. *et al, Chem. Ind. (London)*, 1967, 1297 (*synth*)

Diisopropylphosphine, 8CI D-00640

Bis(1-methylethyl)phosphine, 9CI

[20491-53-6]

$$HP[CH(CH_3)_2]_2$$

$C_6H_{15}P$ M 118.158

Liq. Bp 118-119°.

Oxide: [53753-58-5].

$C_6H_{15}OP$ M 134.158

Liq. $Bp_{1.5}$ 54-55°. n_D^{20} 1.4538. Tautomeric with Diisopropylphosphinous acid, D-00648 .

Sulfide: [68836-92-0].

$C_6H_{15}PS$ M 150.218

Bp_1 70-72°. pK_a 18.9 (DMSO). n_D^{20} 1.5205. Tautomeric with diisopropylphosphinothious acid.

Kabachnik, M.I. *et al, Izv. Akad. Nauk SSSR, Ser. Khim.*, 1963, 1227 (*Engl. transl.* p. 1120) (*oxide*)
Issleib, K. *et al, J. Organomet. Chem.*, 1968, **13**, 283 (*synth, deriv*)
Kostyanovskii, R.G. *et al, Org. Mass Spectrom.*, 1972, **6**, 1199; 1976, **11**, 23 (*synth, ms*)
Grim, S.O. *et al, Phosphorus*, 1974, **4**, 189 (*nmr*)
Mosbo, J.A. *et al, Phosphorus Sulfur*, 1981, **11**, 11 (*struct*)
Foss, V.L. *et al, Zh. Obshch. Khim.*, 1982, **52**, 1034 (*Engl. transl.* p. 916) (*sulfide, synth, nmr, props*)

Diisopropylphosphinic acid, 8CI D-00641

Bis(1-methylethyl)phosphinic acid, 10CI, 9CI

[680-03-5]

$$[(H_3C)_2CH]_2P(O)OH$$

$C_6H_{15}O_2P$ M 150.157

Solid. Mp 48-50°. $Bp_{0.02}$ 84-86°. pK_a 3.56.

Me ester: [18632-46-7]. *Methyl diisopropylphosphinate.*

$C_7H_{17}O_2P$ M 164.184

Liq. $Bp_{0.1}$ 38°.

Et ester: [19935-94-5]. *Ethyl diisopropylphosphinate.*

$C_8H_{19}O_2P$ M 178.211

Liq. $Bp_{0.1}$ 60°.

Isopropyl ester: Isopropyl diisopropylphosphinate.

$C_9H_{21}O_2P$ M 192.237

Liq. $Bp_{0.01}$ 36-38°.

Fluoride: [1536-92-1].

$C_6H_{14}FOP$ M 152.148

Liq. d^{25} 1.06. n_D^{25} 1.3800.

Chloride: [1112-15-8].

$C_6H_{14}ClOP$ M 168.603

Liq. d_4^{20} 1.08. Bp_3 77°, $Bp_{0.12}$ 50°. n_D^{20} 1.4669.

Anhydride: [67949-89-7].

$C_{12}H_{28}O_3P_2$ M 282.299

No phys. props. reported.
Amide: see P,P-Diisopropylphosphinic amide, D-00642
Azide: [78300-89-7].
$C_6H_{14}N_3OP$ M 175.170
Liq. $Bp_{0.5}$ 72-78° (oven).
Isocyanate: [866-68-2].
$C_7H_{14}NO_2P$ M 175.167
Liq. d_4^{20} 1.06. $Bp_{0.02}$ 54-56°. n_D^{20} 1.4605.
Isothiocyanate: [54100-43-5].
$C_7H_{14}NOPS$ M 191.227
No phys. props. reported.

Crofts, P.C. *et al*, *J. Am. Chem. Soc.*, 1953, **75**, 3379 (*synth*)
Larsson, L., *Ark. Kemi*, 1958, **13**, 259; *CA*, **53**, 14919 (*fluoride*)
Neimysheva, A.A. *et al*, *Zh. Obshch. Khim.*, 1966, **36**, 1090 (*Engl. transl.* p. 1105) (*chloride*)
Derkach, G.I. *et al*, *Zh. Obshch. Khim.*, 1968, **38**, 1784 (*Engl. transl.* p. 1739) (*isocyanate*)
Haake, P. *et al*, *Tetrahedron*, 1968, **24**, 565 (*esters, ms*)
Krunyants, I.L. *et al*, *Dokl. Akad. Nauk SSSR, Ser. Sci. Khim.*, 1971, **201**, 862 (*Engl. transl.* p. 952) (*halides, nmr*)
Harger, M.J.P. *et al*, *J. Chem. Soc., Perkin Trans. 1*, 1980, 705; 1981, 736 (*synth, pmr, azide, ir*)
Rahil, J. *et al*, *J. Org. Chem.*, 1981, **46**, 3048 (*chloride, esters*)

P,P-Diisopropylphosphinic amide, 8CI — D-00642

P,P-*Bis(1-methylethyl)phosphinic amide*, *9CI*
[57115-47-6]

$$[(H_3C)_2CH]_2P(O)NH_2$$

$C_6H_{16}NOP$ M 149.172
Cryst. (pet. ether). Mp 135-137.5°.
N,N-*Dichloro:* N,N-*Dichloro*-P,P-*diisopropylphosphinic amide.*
$C_6H_{14}Cl_2NOP$ M 218.062
Cryst. (pet. ether). Mp 32-35°.
N,N-*Di-Me:* [29276-10-6]. P,P-*Diisopropyl*-N,N-*dimethylphosphinic amide.*
$C_8H_{20}NOP$ M 177.226
Liq. Bp_3 100°.
N-tert-*Butyl:* N-tert-*Butyl*-P,P-*diisopropylphosphinic amide.*
$C_{10}H_{24}NOP$ M 205.279
Solid. Mp 182-183°.

Koizumi, T. *et al*, *J. Am. Chem. Soc.*, 1973, **95**, 8073 (*deriv*)
Harger, M.J.P. *et al*, *J. Chem. Soc., Perkin Trans. 1*, 1980, 705 (*synth, deriv, ir, pmr*)
Veits, Yu.A. *et al*, *Zh. Obshch. Khim.*, 1984, **54**, 469 (*Engl. transl.* p. 419) (*deriv, P nmr*)

Diisopropylphosphinodithioic acid — D-00643

Bis(1-methylethyl)phosphinodithioic acid, *9CI*
[32338-34-4]

$$[(H_3C)_2CH]_2P(S)SH$$

$C_6H_{15}PS_2$ M 182.278
Liq. d_4^{20} 1.09. Bp_3 76-76.5°, $Bp_{0.02}$ 92-94°. pK_{a1} 1.64 (7% EtOH aq.), pK_{a2} 2.66 (80% EtOH aq.).
Me ester: [49873-31-6]. *Methyl diisopropylphosphinodithioate.*
$C_7H_{17}PS_2$ M 196.305
Liq. d_4^{20} 1.06. Bp_2 71-72°. n_D^{20} 1.5537.
Et ester: [54565-51-4]. *Ethyl diisopropylphosphinodithioate.*
$C_8H_{19}PS_2$ M 210.332
Liq. d_4^{20} 1.00. Bp_2 105-107°. n_D^{20} 1.5450.
Anhydrosulfide: [66521-47-9].
$C_{12}H_{28}P_2S_3$ M 330.481
Solid. Mp 102-103°.

Kabachnik, M.I. *et al*, *Tetrahedron*, 1960, **9**, 10 (*props*)
Mastryukova, T.A. *et al*, *Zh. Obshch. Khim.*, 1961, **31**, 507 (*Engl. transl.* p. 464) (*synth*)
Kuramshin, I.Ya. *et al*, *Zh. Obshch. Khim.*, 1973, **43**, 1456 (*Engl. transl.* p. 1446) (*ester, ir, pmr, P nmr*)
Ishmaeva, E.A. *et al*, *Zh. Obshch. Khim.*, 1974, **44**, 2625 (*Engl. transl.* p. 2582) (*ester*)
Corbin, J.L. *et al*, *Org. Prep. Proced. Int.*, 1975, **7**, 309 (*synth*)
Harris, R.K. *et al*, *J. Chem. Soc., Dalton Trans.*, 1978, 9 (*anhydrosulfide, cmr, pmr, P nmr*)

Diisopropylphosphinoselenoic acid, 8CI — D-00644

Bis(1-methylethyl)phosphinoselenoic acid, *9CI*

$$[(H_3C)_2CH]_2P(Se)OH \rightleftharpoons [(H_3C)_2CH]_2P(O)SeH$$

$C_6H_{15}OPSe$ M 213.118
Solid. Mp 88°. pK_a 2.87 (7% EtOH aq.), 5.23 (80% EtOH aq.).

Markowska, A., *Bull. Acad. Pol. Sci., Ser. Sci. Chim.*, 1965, **13**, 149; *CA*, **63**, 9791 (*synth*)

Diisopropylphosphinotellurous acid — D-00645

Bis(1-methylethyl)phosphinotellurous acid

$$[(H_3C)_2CH]_2PTeH$$

$C_6H_{15}PTe$ M 245.758
Tautomeric with diisopropylphosphine telluride. Some derivs. have been characterised spectroscopically.

Du Mont, W.W. *et al*, *Z. Naturforsch., B*, 1981, **36**, 332 (*derivs, Te nmr*)

Diisopropylphosphinothioic acid — D-00646

Bis(1-methylethyl)phosphinothioic acid, *9CI*
[22307-83-1]

$$[(H_3C)_2CH]_2P(S)OH \rightleftharpoons [(H_3C)_2CH]_2P(O)SH$$

$C_6H_{15}OPS$ M 166.218
Solid. Mp 76-77.5°. pK_a 3.03 (7% EtOH aq.) (0.5% thiol form), 5.46 (80% EtOH aq.) (0% thiol form), 12.40 ($MeNO_2$).
O-*Me ester:* [62246-59-7]. O-*Methyl diisopropylphosphinothioate.*
$C_7H_{17}OPS$ M 180.244
No phys. props. reported.
S-*Me ester:* [49873-30-5]. S-*Methyl diisopropylphosphinothioate.*
$C_7H_{17}OPS$ M 180.244
d_4^{20} 1.03. Bp_1 88°. n_D^{20} 1.5015.
O-*Ph ester:* [63027-82-7]. O-*Phenyl diisopropylphosphinothioate.*
$C_{12}H_{19}OPS$ M 242.315
Liq. $Bp_{0.3}$ 170° (oven).
Chloride: [23834-61-9].
$C_6H_{14}ClPS$ M 184.663
d_0^{20} 1.10. $Bp_{2.5}$ 58.5-59°. n_D^{20} 1.5232.
Methylamide: P,P-*Diisopropyl*-N-*methylphosphinothioic amide.*
$C_7H_{18}NPS$ M 179.260
Liq. $Bp_{0.2}$ 85-88°.

Kabachnik, M.I. *et al*, *Tetrahedron*, 1960, **9**, 10 (*props*)

Mastryukova, T.A. *et al*, *Zh. Obshch. Khim.*, 1961, **31**, 507 (*Engl. transl.* p. 464) (*chloride*)
Kuramshin, I.Ya. *et al*, *Zh. Obshch. Khim.*, 1973, **43**, 1456 (*Engl. transl.* p. 1446) (*ester, ir, pmr, P nmr*)
Mastryukova, T.A. *et al*, *Phosphorus Sulfur*, 1976, **1**, 211 (*esters*)
Bone, S.A. *et al*, *J. Chem. Soc., Perkin Trans. 1*, 1977, 437 (*phenyl ester*)

Diisopropylphosphinothious acid D-00647

$[(H_3C)_2CH]_2PSH$

$C_6H_{15}PS$ M 150.218
Free acid exists as the thiophosphoryl tautomer. See Diisopropylphosphine, D-00640 .
Anhydrosulfide: [62969-13-5].
$C_{12}H_{28}P_2S$ M 266.361
Exists in tautom. equilibrium with the monosulfide of Tetraisopropyldiphosphine, T-00167 .

Foss, V.L. *et al*, *Zh. Obshch. Khim.*, 1982, **52**, 1054 (*synth, P nmr, pmr*)

Diisopropylphosphinous acid D-00648

Bis(1-methylethyl)phosphinous acid, 9CI

$[(H_3C)_2CH]_2POH$

$C_6H_{15}OP$ M 134.158
The free acid exists as the phosphoryl tautomer. See under Diisopropylphosphine, D-00640 .
Et ester: [6225-99-6]. *Ethyl diisopropylphosphinite.*
$C_8H_{19}OP$ M 162.211
Liq. Bp_{10} 57°. n_D^{20} 1.4425.
Butyl ester: Butyl diisopropylphosphinite.
$C_{10}H_{23}OP$ M 190.265
Liq. Bp_1 36.5-37°. n_D^{20} 1.4452.
Trimethylsilyl ester: [72740-04-6]. *Trimethylsilyl diisopropylphosphinite. Trimethylsilyl bis(1-methylethyl)-phosphinite.*
$C_9H_{23}OPSi$ M 206.339
Liq. Bp_{10} 59-60°. n_D^{20} 1.4392.
Chloride: see Diisopropylphosphinous chloride, D-00650
Bromide: [55383-01-2]. *Bromodiisopropylphosphine.*
$C_6H_{14}BrP$ M 197.054
Liq. Bp_{37} 92°.
Iodide: [59612-04-3]. *Iododiisopropylphosphine.*
$C_6H_{14}IP$ M 244.055
Liq. Bp_2 62°.
Cyanide: Cyanodiisopropylphosphine.
$C_7H_{14}NP$ M 143.168
Liq. Bp_1 34-35°. n_D^{20} 1.4501.
*Amide: see P,P-*Diisopropylphosphinous amide, D-00649

Kabachnik, M.I. *et al*, *Dokl. Akad. Nauk SSSR, Ser. Sci. Khim.*, 1960, **135**, 323 (*Engl. transl.* p. 1267) (*ester*)
Kabachnik, M.I. *et al*, *Izv. Akad. Nauk SSSR, Ser. Khim.*, 1963, 1227 (*Engl. transl.* p. 1120) (*synth*)
Novikova, Z.S. *et al*, *Zh. Obshch. Khim.*, 1976, **46**, 2213 (*Engl. transl.* p. 2128) (*ester*)
Foss, V.L. *et al*, *Zh. Obshch. Khim.*, 1979, **49**, 2418 (*Engl. transl.* p. 2134) (*trimethylsilyl ester, synth, nmr*)
Kabachnik, M.M. *et al*, *Zh. Obshch. Khim.*, 1979, **49**, 1446 (*Engl. transl.* p. 1264) (*iodide, synth, nmr*)
Chervin, I. *et al*, *Izv. Akad. Nauk SSSR, Ser. Khim.*, 1981, 1769 (*Engl. transl.* p. 1438) (*cyanide, synth, ir, pmr*)
Hinke, A. *et al*, *Phosphorus Sulfur*, 1983, **15**, 93 (*bromide*)

P,P-Diisopropylphosphinous amide, 8CI D-00649

P,P-*Bis(1-methylethyl)phosphinous amide, 9CI*
[70609-27-7]

$[(H_3C)_2CH]_2PNH_2$

$C_6H_{16}NP$ M 133.173
Liq. Bp_{15} 55.5°. n_D^{20} 1.4743.
N-*Me:* [88834-11-1]. P,P-*Diisopropyl-*N*-methylphosphinous amide.*
$C_7H_{18}NP$ M 147.200
Liq. Bp_{15} 57.5°. n_D^{20} 1.4660.
N,N-*Di-Me:* [72740-05-7]. P,P-*Diisopropyl-*N,N*-dimethylphosphinous amide.*
$C_8H_{20}NP$ M 161.226
Liq. Bp_8 46.5°. n_D^{20} 1.4624.
N,N-*Di-Et:* [65768-04-9]. N,N-*Diethyl-*P,P*-diisopropylphosphinous amide.*
$C_{10}H_{24}NP$ M 189.280
Liq. Bp_7 73-74°. n_D^{20} 1.4693.
N-tert-*Butyl:* P,P-*Diisopropyl-*N-tert-*butylphosphinous amide.*
$C_{10}H_{24}NP$ M 189.280
Liq. Bp_7 67-68°. n_D^{20} 1.4620.
N-*Ph:* [70589-60-5]. P,P-*Diisopropyl-*N*-phenylphosphinous amide. Diisopropylphosphinous anilide.*
$C_{12}H_{20}NP$ M 209.270
Liq. Bp_1 95-96°. n_D^{20} 1.5500.

Ross, B. *et al*, *Chem. Ber.*, 1979, **112**, 1756 (*synth, cmr*)
Foss, V.L. *et al*, *Zh. Obshch. Khim.*, 1982, **52**, 1054 (*Engl. transl.* p. 916) (*dimethyl, synth, P nmr*)
Foss, V.L. *et al*, *Zh. Obshch. Khim.*, 1983, **53**, 2184 (*Engl. transl.* p. 1969) (*synth, derivs*)

Diisopropylphosphinous chloride, 8CI D-00650

Bis(1-methylethyl)phosphinous chloride, 9CI. Chlorodiisopropylphosphine
[40244-90-4]

$[(H_3C)_2CH]_2PCl$

$C_6H_{14}ClP$ M 152.603
Liq. Bp 155-158°, Bp_{10} 46-47°. n_D^{20} 1.4752.

Voskuil, W. *et al*, *Recl. Trav. Chim. Pays-Bas*, 1963, **82**, 302 (*synth*)
Kostyanovskii, R.G. *et al*, *Org. Mass Spectrom.*, 1972, **6**, 1199 (*ms*)
Kostyanovskii, R.G. *et al*, *Izv. Akad. Nauk SSSR, Ser. Khim.*, 1975, 901 (*Engl. transl.* p. 816) (*synth, pmr*)
Mikaya, A.I. *et al*, *Zh. Obshch. Khim.*, 1982, **52**, 1998 (*Engl. transl.* p. 1776) (*ms*)

Diisopropyl phosphonate, 8CI D-00651

Bis(1-methylethyl) phosphonate, 9CI. Diisopropyl phosphite. Bis(1-methylethyl) phosphite
[1809-20-7]

$[(H_3C)_2CHO]_2P(O)H \rightleftharpoons [(H_3C)_2CHO]_2POH$

$C_6H_{15}O_4P$ M 182.156
Tautomeric; almost completely in phosphonate form. Can act as a reducing agent. Liq. d_4^{20} 1.00. Bp_{53} 106-108°, Bp_8 69-71°. n_D^{20} 1.4080.
▷SZ7660000.

Meyrick, C.I. *et al*, *J. Chem. Soc.*, 1950, 225 (*ir, Raman*)
Harless, H.R., *Anal. Chem.*, 1961, **33**, 1387 (*ms*)
Coulson, E.J. *et al*, *J. Chem. Soc.*, 1965, 2364 (*synth*)
Mark, V. *et al*, *J. Org. Chem.*, 1967, **32**, 1187 (*P nmr*)

Nasakin, O.E. *et al*, *Zh. Prikl. Khim.*, 1982, **55**, 1399 (*Engl. transl.* p. 1286) (*use*)

O,O-Diisopropyl phosphonothioate D-00652

O,O-*Bis(1-methylethyl) phosphonothioate*, *9CI*. O,O-*Diisopropyl thiophosphite*

[14717-27-2]

$$[(H_3C)_2CHO]_2P(S)H \rightleftharpoons [(H_3C)_2CHO]_2PSH$$

$C_6H_{15}O_2PS$ M 182.217

Tautomeric, but exists almost completely in the thiophosphoryl form. Liq. with strong, sickly odour. d_4^{20} 1.02. Bp_{10} 76-78°. n_D^{20} 1.4541.

Murav'ev, I.V. *et al*, *Zh. Obshch. Khim.*, 1968, **38**, 133 (*Engl. transl.* p. 133) (*synth*)
Nifant'ev, E.É. *et al*, *Zh. Obshch. Khim.*, 1983, **53**, 2695 (*Engl. transl.* p. 2429) (*synth, P nmr*)

Diisopropyl phosphoramidate, 8CI D-00653

Bis(1-methylethyl) phosphoramidate, *9CI*. *Diisopropyl amidophosphate*. *Diisopropyl phosphoramide*

[6415-20-9]

$$[(H_3C)_2CHO]_2P(O)NH_2$$

$C_6H_{16}NO_3P$ M 181.171

Cryst. (pet. ether). Mp 56-57°.

Atherton, F.R. *et al*, *J. Chem. Soc.*, 1945, 660 (*synth*)
Jakobsen, P. *et al*, *Org. Mass Spectrom.*, 1972, **6**, 1303 (*ms*)
Cload, P.A. *et al*, *Org. Mass Spectrom.*, 1983, **18**, 57 (*ms*)

Diisopropylphosphoramidochloridous acid D-00654

Bis(1-methylethyl)phosphoramidochloridous acid, *11CI*

$$[(H_3C)_2CH]_2NPClOH$$

$C_6H_{15}ClNOP$ M 183.617

Me ester: [86030-43-5]. *Methyl diisopropylphosphoramidochloridite.*
$C_7H_{17}ClNOP$ M 197.644
Reagent for oligonucleotide synth. in both liq. and solid phases, and used also in glycerophospholipid synth. Liq. $Bp_{0.02}$ 35-36°.

2-Cyanoethyl ester: [89992-70-1]. *2-Cyanoethyl diisopropylphosphoramidochloridite.*
$C_9H_{18}ClN_2OP$ M 236.681
Phosphitylation reagent used in oligonucleotide synth. Liq. $Bp_{0.08}$ 103-105°.

McBride, L.J. *et al*, *Tetrahedron Lett.*, 1983, **24**, 245 (*synth, P nmr*)
Sinha, N.D. *et al*, *Nucleic Acids Res.*, 1984, **12**, 4539 (*synth, ms, pmr, P nmr, use*)
Damha, M.J. *et al*, *Tetrahedron Lett.*, 1985, **26**, 4839 (*use*)
Herdering, W. *et al*, *J. Org. Chem.*, 1985, **50**, 5314 (*use*)
Pon, R.T. *et al*, *Nucleic Acids Res.*, 1985, **13**, 6447 (*use*)
Bruzik, K.S. *et al*, *J. Org. Chem.*, 1986, **51**, 2368 (*use*)

Diisopropylphosphoramidoselenoic acid, 8CI D-00655

Bis(1-methylethyl)phosphoramidoselenoic acid, *9CI*. *Phosphoroselenoic acid diisopropylamide*

$$[(H_3C)_2CH]_2N{-}P(=Se)(OH)_2 \rightleftharpoons [(H_3C)_2CH]_2N{-}P(=O)(SeH)(OH)$$

$C_6H_{16}NO_2PSe$ M 244.132

O,O-*Di-Et ester:* [35060-65-2]. O,O-*Diethyl diisopropylphosphoramidoselenoate.*
$C_{10}H_{24}NO_2PSe$ M 300.239
Liq. d_4^{20} 1.18. $Bp_{0.05}$ 69°.

Nuretdinov, I.A. *et al*, *Izv. Akad. Nauk SSSR, Ser. Khim.*, 1971, 2095 (*Engl. transl.* p. 1989) (*synth, ir, P nmr*)

Diisopropylphosphoramidous acid, 8CI D-00656

Bis(1-methylethyl)phosphoramidous acid, *9CI*

$$[(H_3C)_2CH]_2NP(OH)_2$$

$C_6H_{16}NO_2P$ M 165.172

Di-Ph ester: Diphenyl diisopropylphosphoramidite.
$C_{18}H_{24}NO_2P$ M 317.367
Liq. d^{20} 1.04. $Bp_{0.03}$ 122-123°. n_D^{20} 1.5406.

Difluoride: [921-27-7].
$C_6H_{14}F_2NP$ M 169.154
No phys. props. reported.

Dichloride: [921-26-6].
$C_6H_{14}Cl_2NP$ M 202.063
Useful phosphylating agent.

Koketsu, J. *et al*, *Kogyo Kagaku Zasshi*, 1969, **72**, 2503; *CA*, **72**, 79165 (*ester*)
Gouesnard, J.P. *et al*, *Can. J. Chem.*, 1980, **58**, 1295 (*difluoride, N and P nmr*)
Gynane, M.J.S. *et al*, *J. Chem. Soc., Dalton Trans.*, 1980, 2428 (*dichloride, use*)

Diisopropyl phosphorochloridate, 8CI D-00657

Bis(1-methylethyl) phosphorochloridate, *9CI*. *Diisopropyl phosphoryl chloride*. *Diisopropyl chlorophosphate*

[2574-25-6]

$$[(H_3C)_2CHO]_2P(O)Cl$$

$C_6H_{14}ClO_3P$ M 200.602

Fuming liq. Bp_{14} 95-96°, Bp_1 48°. n_D^{25} 1.4146.

Atherton, F.R. *et al*, *J. Chem. Soc.*, 1948, 1106 (*synth*)
De Roos, A.M. *et al*, *Recl. Trav. Chim. Pays-Bas*, 1958, **77**, 946 (*synth*)
Sosnovsky, G. *et al*, *J. Org. Chem.*, 1969, **34**, 968 (*synth*)
Meyers, A.I. *et al*, *J. Org. Chem.*, 1979, **44**, 2250 (*synth, pmr*)

Diisopropyl phosphorochloridite, 8CI D-00658

Bis(1-methylethyl) phosphorochloridite, *9CI*. *Diisopropyl chlorophosphite*

[41662-51-5]

$$[(H_3C)_2CHO]_2PCl$$

$C_6H_{14}ClO_2P$ M 184.602

Pungent, fuming liq. Bp_{12} 62-64°. n_D^{20} 1.4242. Easily hydrolysed.

Michalski, J. *et al*, *J. Chem. Soc.*, 1961, 4904 (*synth*)
Collins, D.J. *et al*, *Aust. J. Chem.*, 1983, **36**, 2517 (*synth, pmr, P nmr*)

O,O-Diisopropyl phosphorochloridothioate, 8CI D-00659

O,O-*Bis(1-methylethyl) phosphorochloridothioate*, *9CI*. O,O-*Diisopropyl thiophosphoryl chloride*. O,O-*Diisopropyl chlorothiophosphate*

[2524-06-3]

$[(H_3C)_2CHO]_2P(S)Cl$

$C_6H_{14}ClO_2PS$ M 216.662

Oily liq. Bp_1 53-55°. n_D^{20} 1.4588.

▷TD1820000.

Fletcher, J.H. *et al, J. Am. Chem. Soc.*, 1950, **72**, 2461 (*synth*)
Popov, E.M. *et al, Zh. Obshch. Khim.*, 1959, **29**, 1998 (*Engl. transl.* p. 1967) (*raman*)
Meinhardt, N.A. *et al, J. Org. Chem.*, 1960, **25**, 1991 (*synth*)
Whitehead, M.A. *et al, J. Chem. Soc. (A)*, 1971, 1738 (*nqr*)
Omelanczuk, J. *et al, Tetrahedron*, 1975, **31**, 2809 (*synth, P nmr*)

Diisopropyl phosphorochloridothioite **D-00660**

Bis(1-methylethyl) phosphorochloridodithioite. Diisopropyl dithiochlorophosphite

[36696-09-0]

$[(H_3C)_2CHS]_2PCl$

$C_6H_{14}ClPS_2$ M 216.723

Pungent, odorous liq. d_4^{20} 1.13. $Bp_{0.02}$ 63-65°. n_D^{20} 1.5538.

Stepashkina, L.V. *et al, Izv. Akad. Nauk SSSR, Ser. Khim.*, 1972, 380 (*Engl. transl.* p. 330) (*synth*)
Sorokina, S.F. *et al, Zh. Obshch. Khim.*, 1973, **43**, 750 (*Engl. transl.* p. 748) (*synth*)

N,N'-Diisopropylphosphorodiamidic acid, 8CI **D-00661**

N,N'-*Bis(1-methylethyl)phosphorodiamidic acid, 9CI*

$[(H_3C)_2CHNH]_2P(O)OH$

$C_6H_{17}N_2O_2P$ M 180.186

Ph ester: [18995-05-6]. *Phenyl* N,N'-*diisopropylphosphorodiamidate.*
$C_{12}H_{21}N_2O_2P$ M 256.284
Solid. Mp 61-62°.

Fluoride: [371-86-8]. *Mipafox.*
$C_6H_{16}FN_2OP$ M 182.177
Cryst. (pet. ether). Mp 55-56°.
▷Highly neurotoxic. TD3675000.

Chloride: [41273-83-0].
$C_6H_{16}ClN_2OP$ M 198.632
Liq. $Bp_{0.05}$ 138-139°.

Davies, D.R. *et al, Biochem. Pharmacol.*, 1966, **15**, 1783 (*fluoride, tox*)
Hashimoto, S. *et al, CA*, 1968, **68**, 40154 (*phenyl ester*)
Ko, E.C.F. *et al, Can. J. Chem.*, 1973, **51**, 597 (*chloride, synth, props, pmr*)
Soliman, S.A. *et al, J. Environ. Sci. Health, Part B*, 1980, **15**, 207 (*fluoride, synth, tox*)
Chemnitius, J.-M. *et al, Biochem. Pharmacol.*, 1983, **32**, 1693 (*pharmacol*)

O,O-Diisopropyl phosphorodiselenoate **D-00662**

O,O-*Bis(1-methylethyl) phosphorodiselenoic acid, 9CI, 8CI.* O,O-*Bis(1-methylethyl) hydrogen phosphorodiselenoate.* O,O-*Diisopropyl hydrogen phosphorodiselenoate.* O,O-*Bis(1-methylethyl) phosphorodiselenoate, 9CI.* O,O-*Diisopropyl diselenophosphate.* O,O-*Diisopropyl phosphorodiselenoic acid*

[62920-98-3]

$[(H_3C)_2CHO]_2P(Se)SeH$

$C_6H_{15}O_2PSe_2$ M 308.077

Unstable acid isol. as K salt.

K salt: [34585-95-0]. Cryst. (EtOH/pet. ether). Mp 98-99°.

Zemlyanski, N.I. *et al, Zh. Obshch. Khim.*, 1971, **41**, 1691 (*Engl. transl.* p. 1699) (*synth*)
Gorak, R.D. *et al, Zh. Obshch. Khim.*, 1972, **42**, 56 (*Engl. transl.* p. 52) (*synth*)

O,O-Diisopropyl phosphorodithioate, 8CI **D-00663**

O,O-*Bis(1-methylethyl) phosphorodithioate, 9CI.* O,O-*Diisopropyl dithiophosphate.* O,O-*Diisopropyl hydrogen dithiophosphate*

[107-56-2]

$[(H_3C)_2CHO]_2P(S)SH$

$C_6H_{15}O_2PS_2$ M 214.277

Liq. d_4^{20} 1.09. Bp_3 71-72°. pK_a 1.82 (7% EtOH aq.), pK_a 2.65 (80% EtOH aq.). n_D^{20} 1.4918.

K salt: [3419-34-9]. Solid. Mp 193°.

Dimethylammonium salt: [70723-42-1]. Solid.

Anilinium salt: [32997-75-4]. Solid. Mp 102-103.5°.

Cyclohexylammonium salt: [50329-34-5]. Solid. Mp 79-80°.

Kabachnik, M.I. *et al, Tetrahedron*, 1960, **9**, 10 (*synth*)
Zemlyanski, N.I. *et al, Zh. Obshch. Khim.*, 1961, **31**, 880 (*Engl. transl.* p. 811) (*synth*)
Khaskin, B.A. *et al, Zh. Obshch. Khim.*, 1973, **43**, 1916; 1974, **44**, 95 (*Engl. transl.* pp. 1901, 93) (*synth*)
Olah, G.A. *et al, J. Org. Chem.*, 1975, **40**, 2582 (*props, pmr, P nmr*)
Glidewell, C., *Inorg. Chim. Acta*, 1977, **25**, 159 (*salts, P nmr, complexes*)
Zimin, M.G. *et al, Zh. Obshch. Khim.*, 1978, **48**, 1020 (*Engl. transl.* p. 930) (*P nmr*)
Kalinen, A.E. *et al, Izv. Akad. Nauk SSSR, Ser. Khim.*, 1979, 783 (*Engl. transl.* p. 727) (*cryst struct*)

S,S-Diisopropyl phosphorodithioate, 8CI **D-00664**

S,S-*Bis(1-methylethyl) phosphorodithioate, 9CI.* S,S-*Diisopropyldithiophosphate.* S,S-*Diisopropyl hydrogen dithiophosphate.* S,S-*Diisopropyl dithiophosphoric acid*

[42409-57-4]

$[(H_3C)_2CHS]_2P(O)OH$

$C_6H_{15}O_2PS_2$ M 214.277

Cyclohexylammonium salt: [72284-32-3]. Solid. Mp 202-203°.

Yamaguchi, K. *et al, Chem. Lett.*, 1979, 1057 (*synth*)

Diisopropyl phosphorofluoridate, 8CI **D-00665**

Bis(1-methylethyl) phosphorofluoridate, 9CI. Diisopropyl fluorophosphonate. Diisopropyl fluorophosphate. Dyflos, BAN. Isoflurophate. Floropryl. Diflupyl. Fluostigmine

[55-91-4]

$[(H_3C)_2CHO]_2P(O)F$

$C_6H_{14}FO_3P$ M 184.147

Phosphorylating agent, ophthalmic cholinergic used to treat glaucoma, parasympathomimetic agent. Liq. Spar. sol. H_2O. d_4^{25} 1.06. Mp −82°. Bp_{25} 84-85°, Bp_5 46°. n_D^{20} 1.3814. Liq. hydrol. by moisture to HF. Stable in glass container at r.t.

▷Highly toxic, exp. carcinogen. TE5075000.

Saunders, B.C. *et al*, *J. Chem. Soc.*, 1948, 695 (*synth, tox*)
Ford-Moore, A.H. *et al*, *J. Chem. Soc.*, 1953, 1776 (*synth*)
Olah, G.A. *et al*, *J. Org. Chem.*, 1959, **24**, 1568 (*synth*)
Boter, H.L. *et al*, *Recl. Trav. Chim. Pays-Bas*, 1966, **85**, 1099 (*synth*)
Hudson, R.F. *et al*, *J. Chem. Soc. (B)*, 1969, 325.
Landau, M.A. *et al*, *Zh. Struckt. Khim.*, 1970, **11**, 513 (*Engl. transl.* p. 467) (*struct*)
Reddy, G.S. *et al*, *Z. Naturforsch, Part B*, 1970, **25**, 1199 (*F and P nmr*)
Nikitin, E.V. *et al*, *Dokl. Akad. Nauk SSSR, Ser. Sci. Khim.*, 1980, **252**, 922 (*Engl. transl.* p. 442) (*synth, F and P nmr*)
Gordon, J.J. *et al*, *Arch. Toxicol.*, 1983, **52**, 71 (*tox*)
Sax, N.I., *Dangerous Properties of Industrial Materials*, 6th Ed., Van Nostrand-Reinhold, 1984, 592.

O,O-Diisopropyl phosphoroselenoate D-00666

O,O-*Bis(1-methylethyl) phosphoroselenoate*, 9CI. O,O-*Diisopropyl hydrogen phosphoroselenoate*. O,O-*Bis(1-methylethyl) phosphoroselenoic acid*. O,O-*Diisopropyl phosphoroselenoic acid*

[64643-66-9]

$$[(H_3C)_2CHO]_2P(Se)OH \rightleftharpoons [(H_3C)_2CHO]_2P(O)SeH$$

$C_6H_{15}O_3PSe$ M 245.116

Unstable acid isol. as Na or K salt.

Na salt: [58228-29-8]. Cryst. (pet. ether). Sol. Et_2O, C_6H_6.

K salt: [64202-98-8]. Cryst. ($CHCl_3/Et_2O$).

Foss, O., *Acta Chem. Scand.*, 1947, **1**, 8 (*synth*)
Glidewell, C. *et al*, *J. Chem. Soc., Dalton Trans.*, 1977, 527 (*P nmr*)

O,O-Diisopropyl phosphorothioate D-00667

O,O-*Bis(1-methylethyl) phosphorothioate*, 9CI. O,O-*Diisopropyl hydrogen thiophosphate*. O,O-*Diisopropyl thiophosphoric acid*

[4486-44-6]

$$[(H_3C)_2CHO]_2P(O)SH \rightleftharpoons [(H_3C)_2CHO]_2P(S)OH$$

$C_6H_{15}O_3PS$ M 198.216

Conveniently isol. and stored as Na, K or ammonium salt. Contains 54% thiol form (7% EtOH aq.), 29% thiol form (80% EtOH aq.). Metab. and photodegradation product of *S*-Benzyl *O,O*-diisopropyl phosphorothioate, B-00045 . Liq. d_4^{20} 1.09. $Bp_{1.5}$ 89-90°. n_D^{20} 1.4592.

NH_4 salt: [29918-57-8]. Cryst. (C_6H_6/Me_2CO). Mp 157°.

Dicyclohexylammonium salt: Solid. Mp 199-200°.

Chloride: see O,O-Diisopropyl phosphorochloridothioate, D-00659

Anilide: see Phenylphosphoramidothioic acid, P-00262

2-Phenylhydrazide: see 2-Phenylphosphorohydrazidothioic acid, P-00296

Anhydride: see *sym*-Tetraisopropyl dithiopyrophosphate, T-00168

Kabachnik, M.I. *et al*, *Tetrahedron*, 1960, **9**, 10 (*synth, props*)
Zemlyanskii, N.I. *et al*, *Zh. Obshch. Khim.*, 1961, **30**, 4056 (*Engl. transl.* p. 4018) (*raman*)
Boter, H.L. *et al*, *Recl. Trav. Chim. Pays-Bas*, 1966, **85**, 1099 (*synth*)
Yamamoto, H. *et al*, *Agric. Biol. Chem.*, 1973, **37**, 1553.
Murai, T. *et al*, *Agric. Biol. Chem.*, 1977, **41**, 803.
Tolmacheva, N.A. *et al*, *Zh. Obshch. Khim.*, 1978, **48**, 1078 (*Engl. transl.* p. 982) (*synth*)

P,P′-Diisopropyl-*N,N,N′,N′*-tetramethyldiphosphonic diamide D-00668

N,N,N′,N′-*Tetramethyl*-P,P′-*bis(1-methylethyl)-diphosphonic diamide*, 9CI. P-*Isopropyl*-N,N-*dimethylphosphonamidic anhydride*

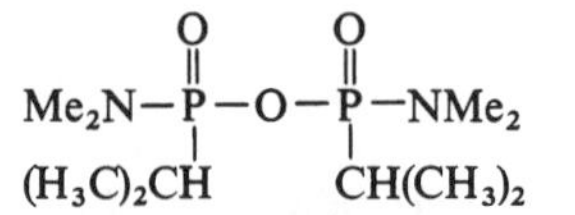

$$Me_2N-\overset{O}{\overset{\|}{\underset{(H_3C)_2CH}{\underset{|}{P}}}}-O-\overset{O}{\overset{\|}{\underset{CH(CH_3)_2}{\underset{|}{P}}}}-NMe_2$$

$C_{10}H_{26}N_2O_3P_2$ M 284.275

Ligand for Mg, Cu, Co, and Ni. Liq. Bp_2 135-140°. Mixt. of racemic and meso forms.

Joestan, M.D. *et al*, *Inorg. Chem.*, 1972, **11**, 429 (*synth, P nmr, complexes, uv*)
Joestan, M.D. *et al*, *J. Chem. Soc., Chem. Commun.*, 1973, 18 (*P nmr*)

Dimefox D-00669

Tetramethylphosphorodiamidic fluoride, 9CI. *Bis(dimethylamido)phosphoryl fluoride*. *Bis(dimethylamino)phosphinic fluoride*. *Pestox XIV*

[115-26-4]

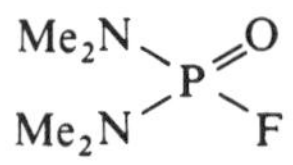

$$(Me_2N)_2P(=O)F$$

$C_4H_{12}FN_2OP$ M 154.124

Pesticide; mostly used as soil treatment to protect hop plants. Now discontinued. Liq. Misc. H_2O, most org. solvs. d^{20} 1.115. Bp_4 67°, Bp 194° (calc.). n_D^{20} 1.4171.

▷Highly toxic; LD_{50} (oral, rats) 5-6 mg/Kg. TD4025000.

Heap, R. *et al*, *J. Chem. Soc.*, 1948, 1313 (*synth, tox*)
Cavell, R.G., *Can. J. Chem.*, 1967, **45**, 1309 (*synth, ir, ms, pmr, F nmr*)
Reddy, G.S. *et al*, *Z. Naturforsch.*, *B*, 1970, **25**, 1199 (*synth, P nmr*)
Robinson, E.A. *et al*, *Spectrochim. Acta, Part A*, 1972, **28**, 1099 (*pmr, P nmr*)
Köttgen, D. *et al*, *Z. Phys. Chem. (Frankfurt u. Main)*, 1974, **92**, 285 (*ir, raman*)
Stan, H.-J. *et al*, *Fresenius' Z. Anal. Chem.*, 1977, **287**, 271 (*ms*)
Light, R.W. *et al*, *Phosphorus Sulfur*, 1980, **8**, 255 (*synth, P nmr*)
Stan, H.-J. *et al*, *J. Chromatogr.*, 1983, **279**, 173 (*glc*)
Pesticide Manual, 6th Ed, 196.

3,9-Dimercapto-2,4,8,10-tetraoxa-3,9-diphosphaspiro[5.5]undecane 3,9-disulfide, 8CI D-00670

Pentaerythritol bis(cyclic phosphorodithioate). *Pentaerythritol bis(cyclic hydrogen dithiophosphate)*

[13824-09-4]

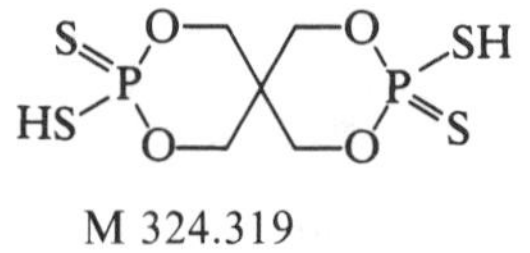

$C_5H_{10}O_4P_2S_4$ M 324.319

Plates (MeOH). Mp 179-181°.

Di-Na salt: Cryst. + $2H_2O$. Mp 208°.

Di-K salt: Cryst. + $1H_2O$. Mp 191-194°.
3,9-Bis(methylthio) (Di-Me ester):
$C_7H_{14}O_4P_2S_2$ M 288.253
Cryst. (EtOH). Mp 212-215°.
3,9-Bis(ethylthio) (Di-Et ester):
$C_9H_{18}O_4P_2S_2$ M 316.306
Solid. Mp 158-159°.

Rätz, R. *et al*, *J. Heterocycl. Chem.*, 1966, **3**, 14 (*ester*)
Delventhal, J. *et al*, *Z. Naturforsch, B*, 1971, **26**, 190 (*synth, derivs*)

Dimethoate, BSI D-00671

O,O-Dimethyl S-(2-methylamino-2-oxoethyl) phosphorodithioate, 9CI. O,O-Dimethyl S-(methylcarbamoylmethyl) phosphorodithioate. O,O-Dimethyl phosphorodithioate S-ester with 2-mercapto-N-methylacetamide, 8CI. Cygon. Rogor. Roxion. Fostion MM
[60-51-5]

$$(MeO)_2P(S)SCH_2CONHMe$$

$C_5H_{12}NO_3PS_2$ M 229.248
Broad range contact and systemic acaricide and insecticide. Cryst. (Et_2O or toluene/hexane). Sol. H_2O, alcohols, $CHCl_3$, C_6H_6, ketones. Mp 51-52°. Stable between pH 2 and 7.
▷TE1750000.

Berkelhammer, G. *et al*, *J. Org. Chem.*, 1961, **26**, 2281 (*synth*)
Keith, L.H. *et al*, *J. Assoc. Off. Anal. Chem.*, 1968, **51**, 1063 (*pmr*)
Gore, R.C. *et al*, *J. Assoc. Off. Anal. Chem.*, 1971, **54**, 1040 (*ir, uv*)
Szalontai, G., *J. Chromatogr.*, 1976, **124**, 9 (*hplc*)
Paasivirta, J. *et al*, *Org. Magn. Reson.*, 1977, **9**, 708 (*pmr, cmr, P nmr*)
Stan, H.-J. *et al*, *Fresenius' Z. Anal. Chem.*, 1977, **287**, 271; *Biomed. Mass Spectrom.*, 1982, **9**, 483 (*glc, ms*)
Daldrup, T. *et al*, *Fresenius' Z. Anal. Chem.*, 1981, **308**, 413 (*tlc, glc, hplc*)
Ripley, B.D. *et al*, *J. Assoc. Off. Anal. Chem.*, 1983, **66**, 1084 (*glc*)
Charalambous, J. *et al*, *Phosphorus Sulfur*, 1984, **19**, 267 (*ms*)
Pesticide Manual, 7th Ed., 205.
Agrochemicals Handbook, Royal Society of Chemistry, London, 1983, A153.
Sax, N.I., *Dangerous Properties of Industrial Materials*, 6th Ed., Van Nostrand-Reinhold, 1984, 608.

Dimethoxyphosphinecarboxylic acid oxide, 9CI D-00672

Dimethoxyphosphinylformic acid. Dimethyl phosphonoformate

$$(MeO)_2P(O)COOH$$

$C_3H_7O_5P$ M 154.059
Me ester: [31142-23-1]. *Trimethyl phosphonoformate.*
$C_4H_9O_5P$ M 168.086
Liq. Bp_1 81-86°. Dec. by NaH, $NaNH_2$ or NaOEt.
Et ester: [1982-13-4]. *Ethyl dimethoxyphosphinylformate.*
$C_5H_{11}O_5P$ M 182.113
Bp_{12} 124-126°.

Takamizawa, A. *et al*, *Chem. Pharm. Bull.*, 1964, **12**, 398.
Shiotani, S. *et al*, *Chem. Pharm. Bull.*, 1973, **21**, 1160.

3-[(Dimethoxyphosphinothioyl)oxy]-2-methyl-2-propenoic acid, 9CI D-00673

3-Hydroxy-2-methacrylic acid O-ester with O,O-dimethyl phosphorothioate, 8CI

$(MeO)_2P(=S)O$–CH=C(CH_3)COOH *(E)-form*

$C_6H_{11}O_5PS$ M 226.184
Ph ester: [38289-60-0]. *Phenyl 3-[(dimethoxyphosphinothioyl)oxy]-2-methyl-2-propenoate.*
$C_{12}H_{15}O_5PS$ M 302.281
Liq. n_D^{23} 1.5277.

(E)-form

Me ester: [62610-77-9]. *Methyl 3-[(dimethoxyphosphinothioyl)oxy]-2-methyl-2-propenoate. Methacrifos.*
$C_7H_{13}O_5PS$ M 240.210
Insecticide and acaricide with vapour, contact, and stomach action. Liq. $Bp_{0.01}$ 90°.

Ger. Pat., 2 147 588, (*1972*); *CA*, **77**, 34698 (*esters*)
Pesticide Manual, 7th Ed., 357.

(Dimethoxyphosphinothioyl)phosphorimidic acid, 9CI D-00674

$$(MeO)_2P(S)N{=}P(OH)_3$$

$C_2H_9NO_5P_2S$ M 221.104
Free acid tautomeric with (dimethoxyphosphinothioyl)-phosphoramidic acid.
Tri-Et ester: [54975-79-0]. *Triethyl (dimethoxyphosphinothioyl)phosphorimidate. P,P,P-Triethoxy-N-(dimethoxyphosphinothioyl)phosphine imide. P,P,P-Triethoxy-N-(dimethoxyphosphinothioyl)phosphazene.*
$C_8H_{21}NO_5P_2S$ M 305.265
Liq. d_4^{20} 1.19. $Bp_{0.004}$ 80-81°. n_D^{20} 1.4712.
Tributyl ester: [51576-53-5]. *Tributyl (dimethoxyphosphinothioyl)phosphorimidate. P,P,P-Tributoxy-N-(dimethoxyphosphinothioyl)phosphine imide. P,P,P-Tributoxy-N-(dimethoxyphosphinothioyl)phosphazene.*
$C_{14}H_{33}NO_5P_2S$ M 389.426
Liq. d_4^{20} 1.08. $Bp_{0.009}$ 92-93°. n_D^{20} 1.4662.
Tri-Ph ester: [51576-54-6]. *Triphenyl (dimethoxyphosphinothioyl)phosphorimidate. N-(Dimethoxyphosphinothioyl)-P,P,P-triphenoxyphosphine imide. N-(Dimethoxyphosphinothioyl)-P,P,P-triphenoxyphosphazene.*
$C_{20}H_{21}NO_5P_2S$ M 449.397
Solid. Mp 42-43°.

Khodak, A.A. *et al*, *Zh. Obshch. Khim.*, 1975, **45**, 262 (*Engl. transl.* p. 248) (*synth, ir, P nmr*)

[(Dimethoxyphosphinothioyl)thio]acetic acid, 9CI D-00675

[1113-01-5]

$$(MeO)_2P(S)SCH_2COOH$$

$C_4H_9O_4PS_2$ M 216.206
Solid. Mp 43-45°.
▷TE0360000.
Me ester: [757-86-8]. *Methyl [(dimethoxyphosphinothioyl)thio]acetate.*
$C_5H_{11}O_4PS_2$ M 230.233

Liq. d_{20}^{20} 1.31. $Bp_{0.1}$ 110°. n_D^{20} 1.5207.
Et ester: [1068-13-9]. *Ethyl [(dimethoxyphosphinothioyl)thio]acetate.*
$C_6H_{13}O_4PS_2$ M 244.260
Liq. d_{20}^{20} 1.31. $Bp_{0.2}$ 110-113°. n_D^{20} 1.5200.
Amide: [3692-87-3]. O,O-*Dimethyl* S-(*2-amino-2-oxoethyl) phosphorodithioate.*
$C_4H_{10}NO_3PS_2$ M 215.222
Cryst. (CCl_4). Mp 62-63°.
Methylamide: see Dimethoate, D-00671

Hoegberg, E.I. *et al*, *J. Am. Chem. Soc.*, 1951, **73**, 557 (*derivs*)
U.S.P., 3 047 459, (*1962*); *CA*, **58**, 1349 (*esters*)
Grimmer, F. *et al*, *Z. Naturforsch.*, *B*, 1968, **23**, 10 (*synth, derivs*)
Sauer, H.H., *J. Agric. Food Chem.*, 1972, **20**, 578 (*tlc, tox*)

α-[(Dimethoxyphosphinothioyl)thio]-benzeneacetic acid, 9CI D-00676

[13376-78-8]

(MeO)$_2$P(S)SCHPhCOOH

$C_{10}H_{13}O_4PS_2$ M 292.304

(±)-*form*
Metab. of Phenthoate, P-00078 . Mp 46-48°.
Et ester: see Phenthoate, P-00078

Takade, D.Y. *et al*, *Arch. Environ. Contam. Toxicol.*, 1976, **5**, 63 (*synth, pmr*)

[(Dimethoxyphosphinothioyl)thio]-butanedioic acid, 9CI D-00677

[(Dimethoxyphosphinothioyl)thio]succinic acid
[1190-28-9]

(MeO)$_2$P(S)SCH(^{1}COOH)CH$_2$^{2}COOH

$C_6H_{11}O_6PS_2$ M 274.243

(±)-*form*
Cryst. ($CHCl_3$). Mp 127-129° (115-118°).
Di-Me ester: [3700-89-8]. *Dimethyl [(dimethoxyphosphinothioyl)thio]butanedioate.*
$C_8H_{15}O_6PS_2$ M 302.297
Liq. d_4^{20} 1.29. $Bp_{0.2}$ 134.5°. n_D^{20} 1.5070.
▷TE0300000.
1-Et ester: β-Et ester. 1-Ethyl [(dimethoxyphosphinothioyl)thio]butanedioate.
$C_8H_{15}O_6PS_2$ M 302.297
Cryst. ($CHCl_3$/hexane). Mp 42-45°.
2-Et ester: α-Et ester. 2-Ethyl [(dimethoxyphosphinothioyl)thio]butanedioate.
$C_8H_{15}O_6PS_2$ M 302.297
Cryst. ($CHCl_3$/hexane). Mp 55-57°.
Di-Et ester: see Malathion, M-00001
Diisopropyl ester: Bis(1-methylethyl) [(dimethoxyphosphinothioyl)thio]butanedioate. Diisopropyl [(dimethoxyphosphinothioyl)thio]succinate.
$C_{12}H_{23}O_6PS_2$ M 358.404
Liq. d_4^{20} 1.18. $Bp_{0.05}$ 122.5-123°. n_D^{20} 1.4810.

Mel'nikov, N.N. *et al*, *Zh. Obshch. Khim.*, 1953, **23**, 1352 (*Engl. transl.* p. 1416) (*esters, synth*)
March, R.B. *et al*, *J. Econ. Entomol.*, 1956, **49**, 185 (*synth*)
Mostafa, I.Y. *et al*, *Z. Naturforsch.*, *B*, 1972, **27**, 1115 (*metab*)
Wolfe, N.L. *et al*, *J. Agric. Food Chem.*, 1975, **23**, 1212 (*synth, derivs*)

(Dimethoxyphosphinyl)acetic acid, 9CI D-00678

Dimethyl (carboxymethyl)phosphonate
[34159-46-1]

(MeO)$_2$P(O)CH$_2$COOH

$C_4H_9O_5P$ M 168.086

Both the acid and its esters form carbanions used in Wittig-Horner reactions for synth. of α,β-unsaturated acids and esters. Hygroscopic cryst. (butanone). Mp 40-41°. pK_a 3.34 (H_2O), pK_a 4.45 (50% EtOH aq., 25°).
K salt: Cryst. (MeOH). Mp 122-123°.
Me ester: [5927-18-4]. *Methyl (dimethoxyphosphinyl)acetate. Trimethyl phosphonoacetate.*
$C_5H_{11}O_5P$ M 182.113
Liq. d_4^{20} 1.264. Bp_5 117-118°. n_D^{20} 1.4373.
Et ester: [311-46-6]. *Ethyl (dimethoxyphosphinyl)acetate.*
$C_6H_{13}O_5P$ M 196.139
Liq. d_4^{20} 1.21. Bp_{10} 134.5-135°, $Bp_{0.4}$ 91-92°. n_D^{20} 1.4348.
Ph ester: [17604-67-0]. *Phenyl (dimethoxyphosphinyl)acetate.*
$C_{10}H_{13}O_5P$ M 244.183
Liq. d_4^{20} 1.34. $Bp_{0.07}$ 136-137°. n_D^{20} 1.5260.

Pudovik, A.N. *et al*, *Zh. Obshch. Khim.*, 1964, **34**, 3942 (*Engl. transl.* p. 4003) (*ester*)
Malevannaya, R.A. *et al*, *Zh. Obshch. Khim.*, 1971, **41**, 1426 (*Engl. transl.* p. 1432) (*ester*)
Moskva, V.V. *et al*, *Zh. Obshch. Khim.*, 1976, **46**, 534 (*Engl. transl.* p. 529) (*esters*)
Nagoka, H. *et al*, *Tetrahedron*, 1981, **37**, 3873 (*ester, use*)
Eschenmoser, W. *et al*, *Helv. Chim. Acta*, 1982, **65**, 353 (*ester, use*)
Roush, W.R. *et al*, *J. Org. Chem.*, 1983, **48**, 758 (*use*)

2-(Dimethoxyphosphinyl)butanoic acid D-00679

(MeO)$_2$P(O)CH(COOH)CH$_2$CH$_3$

$C_6H_{13}O_5P$ M 196.139

(±)-*form*
Me ester: Methyl 2-(dimethoxyphosphinyl)butanoate. Trimethyl 2-phosphonobutanoate.
$C_7H_{15}O_5P$ M 210.166
Liq. d_4^{20} 1.23. $Bp_{0.1}$ 109-110°. n_D^{20} 1.4690.

Moskva, V.V. *et al*, *Zh. Obshch. Khim.*, 1976, **46**, 534 (*Engl. transl.* p. 529)

N-(Dimethoxyphosphinyl)carbamic acid, 9CI D-00680

Dimethyl carboxyphosphoramidate

$C_3H_8NO_5P$ M 169.074

Me ester: [995-17-5]. *Methyl* N-*(dimethoxyphosphinyl)carbamate.* OO-*Dimethyl methylurethane phosphate. Trimethyl phosphonocarbamate.*
$C_4H_{10}NO_5P$ M 183.100
Cryst. (C_6H_6). Mp 63-65°. Dec. on heating giving MeOH and dimethyl phosphorylisocyanate.
Et ester: [65289-20-5]. *Ethyl* N-*(dimethoxyphosphinyl)carbamate.* OO-*Dimethyl ethylurethane phosphate.*
$C_5H_{12}NO_5P$ M 197.127
Liq. n_D^{20} 1.4420. Dec. on heating giving ethanol and dimethyl phosphorylisocyanate.

Kirsanov, A.V. *et al*, *Zh. Obshch. Khim.*, 1956, **26**, 2642; 1957, **27**, 1002 (*Engl. transl.* pp. 2947, 1084) (*esters, synth, props*)
Egorov, Yu.P. *et al*, *Zh. Prikl. Spektrosk.*, 1968, **9**, 980; 1969, **11**, 515 (*Engl. transl.* pp. 1338, 1090) (*ir*)
Nuzhdina, Yu.A. *et al*, *Zh. Prikl. Spektrosk.*, 1970, **13**, 310 (*Engl. transl.* p. 1084) (*ir*)

Shokol, V.A. *et al*, *Zh. Obshch. Khim.*, 1977, **47**, 2202 (*Engl. transl.* p. 2010) (*esters*)

3-[(Dimethoxyphosphinyl)oxy]-2-butenoic acid **D-00681**

3-[(Dimethoxyphosphinyl)oxy]crotonic acid, 8CI

[22716-98-9]

$(MeO)_2P(O)O(CH_3)C{=}CH(COOH)$ (*E*)-form

$C_6H_{11}O_6P$ M 210.123

Me ester: see Mevinphos, M-00440

Et ester: [5675-55-8]. *Ethyl 3-[(dimethoxyphosphinyl)-oxy]crotonate.*

$C_8H_{15}O_6P$ M 238.177

Bp_2 139-140° (typically). Has been obt. as (*E*) + (*Z*) mixt. of various compositions.

1-Phenylethyl ester: see Crotoxyphos, C-00207

Methylamide: see Monocrotophos, M-00459

Dimethylamide: see Dicrotophos, D-00204

(*E*)-form [21300-86-7]

Cis-*form*

Et ester: [26729-68-0]. No data available for pure isomer.

(*Z*)-form

Trans-*form*

Et ester: [78550-26-0]. Liq. Bp_2 139-140°. n_D^{20} 1.2100.

U.S.P., 3 366 713, (*1968*); *CA*, **68**, 95340 (*synth*)
U.S.P., 3 400 177, (*1968*); *CA*, **70**, 87064 (*synth, derivs*)
Ger. Pat., 1 940 003, (*1970*); *CA*, **72**, 89801 (*synth*)
Arbuzov, B.A. *et al*, *Izv. Akad. Nauk SSSR, Ser. Khim.*, 1975, 2763 (*Engl. transl.* p. 2650) (*ethyl ester, synth, ir, P nmr*)

2-[(Dimethoxyphosphinyl)oxy]propanoic acid **D-00682**

O-*(Dimethoxyphosphinyl)lactic acid*

[33313-49-4]

$(MeO)_2P(O)OCH(CH_3)COOH$

$C_5H_{11}O_6P$ M 198.112

(±)-form

Me ester: Methyl 2-[(dimethoxyphosphinyl)oxy]-propanoate. Trimethyl O-*phosphonolactate. Trimethyl phospholactate.*

$C_6H_{13}O_6P$ M 212.139

Liq. d_4^{20} 1.24. $Bp_{0.5}$ 89°. n_D^{20} 1.4200.

Et ester: Ethyl 2-[(dimethoxyphosphinyl)oxy]-propanoate.

$C_7H_{15}O_6P$ M 226.166

Liq. d_4^{20} 1.20. $Bp_{0.5}$ 90°. n_D^{20} 1.4208.

Pudovik, A.N. *et al*, *Zh. Obshch. Khim.*, 1963, **33**, 483; 1969, **39**, 2424 (*Engl. transl.* pp. 475, 2364)

(Dimethoxyphosphinyl)phosphorimidic acid, 9CI **D-00683**

$(MeO)_2P(O)N{=}P(OH)_3$

$C_2H_9NO_6P_2$ M 205.044

Free acid tautomeric with (dimethoxyphosphinyl)-phosphoramidic acid.

Tri-Me ester: [13294-07-0]. *Trimethyl (dimethoxyphosphinyl)phosphorimidate.* P,P,P-*Trimethoxy*-N-*(dimethoxyphosphinyl)phosphazene.* P,P,P-*Trimethoxy*-N-*(dimethoxyphosphinyl)-phosphine imide.*

$C_5H_{15}NO_6P_2$ M 247.124

Liq. d_4^{20} 1.31. $Bp_{0.2}$ 104-106°. n_D^{20} 1.4430.

Tri-Et ester: [18313-77-4]. *Triethyl (dimethoxyphosphinyl)phosphorimidate.* P,P,P-*Triethoxy*-N-*(dimethoxyphosphinyl)phosphine imide.* P,P,P-*Triethoxy*-N-*(dimethoxyphosphinyl)-phosphazene.*

$C_8H_{21}NO_6P_2$ M 289.205

Liq. $Bp_{0.05}$ 100°.

Tris(2-methylpropyl) ester: [39528-26-2]. *Tris(2-methylpropyl) (dimethoxyphosphinyl)phosphorimidate. Triisobutyl (dimethoxyphosphinyl)phosphorimidate.* P,P,P-*Triisobutoxy*-N-*(dimethoxyphosphinyl)-phosphine imide.* P,P,P-*Triisobutoxy*-N-*(dimethoxyphosphinyl)phosphazene.*

$C_{14}H_{33}NO_6P_2$ M 373.365

d_4^{20} 1.07. Bp_{10} 107-109°, $Bp_{0.001}$ 70-71°. n_D^{20} 1.4360.

Kabachnik, M.I. *et al*, *Izv. Akad. Nauk SSSR, Ser. Khim.*, 1961, 819 (*Engl. transl.* p. 758) (*trimethyl ester, synth, ir*)
Gilyarov, V.A. *et al*, *Zh. Obshch. Khim.*, 1972, **42**, 2148 (*Engl. transl.* p. 2145) (*esters*)
Riesel, L. *et al*, *Z. Anorg. Allg. Chem.*, 1983, **502**, 21 (*esters, synth, P nmr*)

2-(Dimethoxyphosphinyl)propanoic acid, 9CI **D-00684**

Dimethyl [1-(methoxycarbonyl)ethyl]phosphonate

$(MeO)_2P(O)CH(CH_3)COOH$

$C_5H_{11}O_5P$ M 182.113

Esters are Wittig-Horner reagents.

(±)-form

Me ester: Methyl 2-(dimethoxyphosphinyl)propanoate. Trimethyl 2-phosphonopropionate.

$C_6H_{13}O_5P$ M 196.139

Liq. $Bp_{0.02}$ 97-105°.

Et ester: Ethyl 2-(dimethoxyphosphinyl)propanoate.

$C_7H_{15}O_5P$ M 210.166

Liq. $Bp_{0.1}$ 100-110°.

Davidson, R.M. *et al*, *J. Org. Chem.*, 1980, **45**, 2698 (*synth, pmr*)
Maier, G. *et al*, *Chem. Ber.*, 1981, **114**, 3959 (*synth, use*)

3-(Dimethoxyphosphinyl)propanoic acid, 9CI **D-00685**

$(MeO)_2P(O)CH_2CH_2COOH$

$C_5H_{11}O_5P$ M 182.113

Liq. d_4^{20} 1.305. n_D^{20} 1.4520.

Me ester: [18733-15-8]. *Methyl 3-(dimethoxyphosphinyl)propanoate. Trimethyl 3-phosphonopropanoate.*

$C_6H_{13}O_5P$ M 196.139

Liq. d_4^{20} 1.22. Bp_{10} 137-138°, $Bp_{0.03}$ 76°. n_D^{20} 1.4390.

Pudovik, A.N. *et al*, *Zh. Obshch. Khim.*, 1952, **22**, 473 (*Engl. transl. p. 537*) (*ester, synth*)
Tsivunin, V.S. *et al*, *Zh. Obshch. Khim.*, 1970, **40**, 1995 (*Engl. transl. p. 1983*) (*synth, ir*)
Evans, D.A. *et al*, *J. Am. Chem. Soc.*, 1978, **100**, 3467 (*ester, synth, ir, pmr*)

3-(Dimethoxyphosphinyl)propenoic acid, 9CI **D-00686**

3-(Dimethoxyphosphinyl)acrylic acid

$(MeO)_2P(O)CH{=}CHCOOH$

$C_5H_9O_5P$ M 180.097
Me ester: Methyl 3-(dimethoxyphosphinyl)propenoate. Trimethyl 3-phosphonoacrylate.
$C_6H_{11}O_5P$ M 194.124
Liq. d_4^{20} 1.24. Bp_{15} 148-151°. n_D^{20} 1.4585.

Kirillova, L.M. *et al, Zh. Obshch. Khim.*, 1965, **35**, 1146 (*Engl. transl.* p. 1148) (*synth*)

(Dimethoxyphosphinyl)thioacetic acid, 9CI D-00687

[41304-05-6]

$(MeO)_2P(O)SCH_2COOH$

$C_4H_9O_5PS$ M 200.146
Et ester: [2088-72-4]. *Ethyl (dimethoxyphosphinyl)-thioacetate.*
$C_6H_{13}O_5PS$ M 228.199
Liq. $Bp_{0.01}$ 76-80°.
▷AI7175000.

Torii, S. *et al, J. Org. Chem.*, 1979, **44**, 2938 (*synth, ir, pmr*)

Dimethylamine-1,1′-diphosphonic acid D-00688

Iminobis(methylene)bisphosphonic acid, 9CI. (Iminodimethylene)diphosphonic acid, 8CI. Iminobis-(methanesphosphonic acid). 1,1′-Diphosphonodimethylamine
[17261-34-6]

$HN[CH_2P(O)(OH)_2]_2$

$C_2H_9NO_6P_2$ M 205.044
Cryst. (EtOH aq.). Mp 237°.
Tetra-Et ester: [3654-42-0]. *Tetraethyl iminobis(methylene)bisphosphonate.*
$C_{10}H_{25}NO_6P_2$ M 317.258
Liq. $Bp_{0.3}$ 150-151°.
▷NJ7035000.

Petrov, K.A. *et al, Zh. Obshch. Khim.*, 1959, **29**, 591 (*synth, ester*)
Szczepaniak, W. *et al, Phosphorus Sulfur*, 1979, **7**, 337 (*deriv*)
Tsirul'nikova, N.V. *et al, Zh. Obshch. Khim.*, 1981, **51**, 1028 (*Engl. transl.* p. 859) (*synth, props, pmr*)

2-Dimethylamino-1,3,2-benzodioxaphosphole, 8CI D-00689

N,N-*Dimethyl-1,3,2-benzodioxaphosphol-2-amine, 9CI.* o-*Phenylene dimethylphosphoramidite*
[18389-60-1]

$C_8H_{10}NO_2P$ M 183.146
Liq. d_4^{20} 1.18. Bp_{20} 118°, Bp_1 76°. n_D^{20} 1.5418. Easily oxidised.
2-Oxide: [56185-09-2]. o-*Phenylene dimethylphosphoramidate.*
$C_8H_{10}NO_3P$ M 199.146
Cryst. (Et_2O/pet. ether). Mp 74°. $Bp_{0.5}$ 123-132°.

Sanchez, M. *et al, Bull. Soc. Chim. Fr.*, 1968, 773 (*synth*)
Simonin, M.-P. *et al, Org. Magn. Reson.*, 1972, **4**, 113 (*pmr*)
Mathis, R. *et al, Spectrochim. Acta, Part A*, 1974, **30**, 357 (*ir*)
Perregaard, J. *et al, Recl. Trav. Chim. Pays-Bas*, 1974, **93**, 252 (*oxide, synth, pmr*)
Gray, G.A. *et al, J. Am. Chem. Soc.*, 1977, **99**, 3243 (*N and O nmr*)
Nuretdinov, I.A. *et al, Zh. Fiz. Khim.*, 1979, **53**, 126 (*Engl. transl.* p. 66) (*nqr*)
Duangthai, S. *et al, Org. Magn. Reson.*, 1982, **20**, 33 (*nmr*)
Kukhar', V.P. *et al, Zh. Obshch. Khim.*, 1982, **52**, 562 (*synth, P nmr*)

[4-(Dimethylamino)butyl]triphenylphosphonium(1+) D-00690

$Ph_3P^{\oplus}CH_2CH_2CH_2CH_2NMe_2$

$C_{24}H_{29}NP^{\oplus}$ M 362.474 (ion)
Bromide: [83299-97-2].
$C_{24}H_{29}BrNP$ M 442.378
Source of ylide. Solid. Mp 199-202°, Mp 212-214°.
Ylide: (4-Dimethylaminobutylidene)-triphenylphosphorane.
$C_{24}H_{28}NP$ M 361.466
Used in Wittig reactions for aminoalkylation of aldehydes and ketones.

De Castro Dantas, T.N. *et al, Phosphorus Sulfur*, 1982, **13**, 97 (*synth, pmr, nmr*)
Maryanoff, B.E. *et al, J. Am. Chem. Soc.*, 1985, **107**, 217 (*synth, props*)

(Dimethylaminocarbonyl)phosphonic acid, 9CI D-00691

(Dimethylcarbamoyl)phosphonic acid

$(HO)_2P(O)CONMe_2$

$C_3H_8NO_4P$ M 153.074
Monoanilinium salt: Cryst. (Me_2CO/MeOH). Mp 148-150°.
Bis(trimethylsilyl) ester: Bis(trimethylsilyl) (dimethylaminocarbonyl)phosphonate. Bis(trimethylsilyl) (N,N-*dimethylcarbamoyl)phosphonate.*
$C_9H_{24}NO_4PSi_2$ M 297.438
Liq. $Bp_{0.1}$ 80-82°.

Morita, T. *et al, Bull. Chem. Soc. Jpn.*, 1978, **51**, 2169 (*synth, ir, pmr*)

4-Dimethylamino-2,2-dihydro-2,2-diphenyl-1,3,5,2-triazaphosphorine, 8CI D-00692

[15309-56-5]

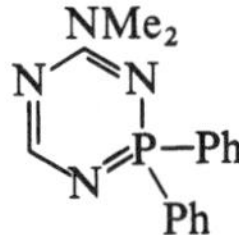

$C_{16}H_{17}N_4P$ M 296.311
Plates (C_6H_6 or Et_2O). Mp 146.5-147°.

Schmidpeter, A. *et al, Angew. Chem., Int. Ed. Engl.*, 1967, **6**, 565 (*synth, P nmr, pmr*)

2-Dimethylamino-5,5-dimethyl-1,3,2-dioxaphosphorinane, 9CI **D-00693**

N,N,*5,5-Tetramethyl-1,3,2-dioxaphosphorin-2-amine*, *9CI*. N,N,*5,5-Tetramethyl-1,3,2-dioxaphosphinin-2-amine*. *Neopentylene dimethylphosphoramidite*

[56465-64-6]

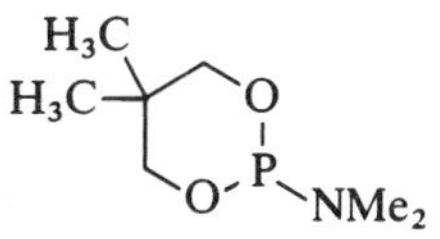

$C_7H_{16}NO_2P$ M 177.183

Pungent liq. Bp_6 64-66°.

2-Oxide: [876-48-2]. *Neopentylene dimethylphosphoramidate.*
$C_7H_{16}NO_3P$ M 193.182
Cryst. (EtOAc/pet. ether). Mp 104.5-106°.

2-Sulfide: [15905-28-9]. *Neopentylene dimethylphosphoramidothioate.*
$C_7H_{16}NO_2PS$ M 209.243
Cryst. (pet. ether). Mp 73-73.5°.

2-Selenide: [91670-70-1]. *Neopentylene dimethylphosphoramidoselenoate.*
$C_7H_{16}NO_2PSe$ M 256.143
Characterised spectroscopically.

Edmundson, R.S., *Tetrahedron*, 1964, **20**, 2781 (*oxide, synth, ir*)
Bartle, K.D. *et al*, *Tetrahedron*, 1967, **23**, 1701 (*synth, sulfide, pmr*)
Majoral, J.-P. *et al*, *Bull. Soc. Chim. Fr.*, 1972, 606 (*oxide, ir, P nmr*)
Majoral, J.-P. *et al*, *Spectrochim. Acta Part A*, 1972, **28**, 2247 (*oxide, ir*)
Francis, G.W. *et al*, *Acta Chem. Scand., Ser. B*, 1976, **30**, 31 (*oxide, ms*)
Grand, A. *et al*, *Acta Crystallogr., Sect. B*, 1978, **34**, 199 (*sulfide, cryst struct*)
Edmundson, R.S., *Phosphorus Sulfur*, 1981, **9**, 307 (*sulfide, ms*)
Schiff, D.E. *et al*, *Inorg. Chem.*, 1984, **23**, 3373 (*cryst struct, pe, cmr, selenide, P nmr*)

2-Dimethylamino-3,4-dimethyl-5-phenyl-1,3,2-oxazaphospholidine **D-00694**

N,N,*3,4-Tetramethyl-5-phenyl-1,3,2-oxazaphospholidin-2-amine*, *9CI*. *2-Dimethylamino-3,4-dimethyl-5-phenyl-1,3,2-oxazaphospholane*

[32018-63-6]

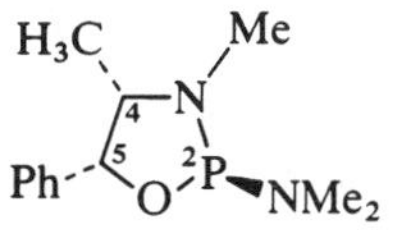

(*2R,4S,5R*)*-form*

$C_{12}H_{19}N_2OP$ M 238.269

(2R,4S,5R)-form [54061-79-9]
$Bp_{0.01}$ 110-120°. $[\alpha]^{2}_{D}$ −27° (or +27°). Chirality at P uncertain.

(2S,4R,5S)-form [54061-80-2]
$Bp_{0.01}$ 110-120°. $[\alpha]^{22}_{D}$ +27° (or −27°). Chirality at P uncertain.

Contreras, R. *et al*, *Synth. Inorg. Met.-Org. Chem.*, 1973, **3**, 37 (*synth, pmr, P nmr*)
Bernard, D. *et al*, *Phosphorus*, 1974, **3**, 187 (*synth, pmr, P nmr*)
Hall, C.R. *et al*, *J. Chem. Soc., Perkin Trans. 1*, 1979, 1646 (*sulfide, synth, props, pmr*)

2-Dimethylamino-1,3,2-dioxaphospholane **D-00695**

N,N-*Dimethyl-1,3,2-dioxaphospholan-2-amine*, *9CI*. *2-Dimethylamino-1,3-dioxa-2-phosphacyclopentane*. *Ethylene dimethylphosphoramidite*

[7114-39-8]

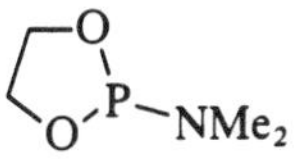

$C_4H_{10}NO_2P$ M 135.102

Pungent liq. Dimethylaminating agent. Bp_{10} 60-62°. n_D^{20} 1.4695.

2-Oxide: [7114-66-1]. *Ethylene dimethylphosphoramidate.*
$C_4H_{10}NO_3P$ M 151.102
Cryst. (Et_2O). Mp 50-52° (47.5-48.5). Bp_1 113-114.5°.

2-Sulfide: [7114-53-6]. *Ethylene dimethylphosphoramidothioate.*
$C_4H_{10}NO_2PS$ M 167.162
Cryst. (EtOH). Mp 46-46.5°. Bp_2 102-103°.

2-Selenide: Ethylene dimethylphosphoramidoselenoate.
$C_4H_{10}NO_2PS_3$ M 231.282
Cryst. (Et_2O). Mp 62-63°.

Abramov, V.S. *et al*, *Zh. Obshch. Khim.*, 1966, **36**, 923 (*Engl. transl.* p. 938) (*derivs*)
Edmundson, R.S. *et al*, *J. Chem. Soc. (C)*, 1966, 1997 (*synth, sulfide*)
Revel, M. *et al*, *Bull. Soc. Chim. Fr., Part I*, 1973, 1195 (*derivs, pmr, P nmr*)
Mathis, R. *et al*, *Spectrochim. Acta, Part A*, 1974, **30**, 357 (*ir*)
Tan, Han-Wan *et al*, *Tetrahedron Lett.*, 1975, 619 (*cmr, P nmr*)
Yamashita, Y., *J. Polym. Sci., Polym. Symp.*, 1976, **56**, 447 (*synth, ir, pmr, props*)
Gatta, F. *et al*, *Synthesis*, 1979, 718 (*use*)
Houalla, D. *et al*, *Nouv. J. Chim.*, 1979, **3**, 507 (*pe*)
Modro, T.A. *et al*, *J. Org. Chem.*, 1981, **46**, 1923 (*oxide, synth, pmr, props*)
Jennings, W.B. *et al*, *J. Chem. Soc., Perkin Trans. 2*, 1984, 1207 (*cmr, conformn*)

2-Dimethylamino-1,3,2-dioxaphosphorinane, 8CI **D-00696**

N,N-*Dimethyl-1,3,2-dioxaphosphorinan-2-amine*, *9CI*. *2-Dimethylamino-1,3,2-dioxaphosphinane*. N,N-*Dimethyl-1,3,2-dioxaphosphinan-2-amine*. *Trimethylene dimethylphosphoramidite*

[17454-25-0]

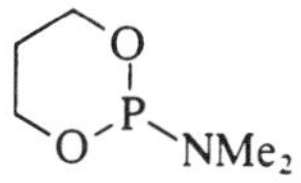

$C_5H_{12}NO_2P$ M 149.129

2-Sulfide: O,O-*Trimethylene dimethylphosphoramidothioate.*
$C_5H_{12}NO_2PS$ M 181.189
Solid. Mp 40°. $Bp_{0.05}$ 80°. n_D^{20} 1.5123.

Nifant'ev, É.E. *et al*, *Dokl. Akad. Nauk SSSR, Ser. Sci. Khim.*, 1973, **208**, 651 (*Phys. Chem., Engl. transl.* p. 100) (*cmr*)
Mosbo, J.A. *et al*, *J. Org. Chem.*, 1977, **42**, 1549 (*oxide, ir, P nmr*)
Nuretdinov, I.A. *et al*, *Zh. Fiz. Khim.*, 1979, **53**, 126 (*Engl. transl.* p. 66) (*nqr*)
Arshinova, R.P. *et al*, *Zh. Obshch. Khim.*, 1981, **51**, 1757 (*pe*)
Edmundson, R.S., *Org. Mass Spectrom.*, 1983, **18**, 150 (*sulfide, ms*)
Shagidullin, R.R. *et al*, *Izv. Akad. Nauk SSSR, Ser. Khim.*, 1984, 2517 (*Engl. transl.* p., 2306) (*ir, raman*)

1-[1-(Dimethylamino)ethyl]-2-(diphenylphosphino)ferrocene, 9CI D-00697

α-[[2-(Diphenylphosphino)ferrocenyl]ethyl]-dimethylamine. PPFA

[60816-98-0]

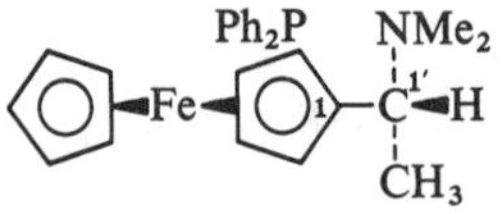

(R_C,R planar)

$C_{26}H_{28}FeNP$ M 441.335

Metal complexes used for asymmetric induction.

(R)$_C$(R)$_{planar}$-form [74311-54-9]

(1S,1'R)-form

Orange oil. $[\alpha]^{25}_D$ +364° (c, 0.4 in $CHCl_3$).

(R)$_C$(S)$_{planar}$-form [55700-44-2]

(1R,1'R)-form

Orange cryst. (EtOH). Mp 139°. $[\alpha]^{25}_D$ −361° (c, 0.6 in EtOH).

$PdCl_2$ complex: [76374-09-9].

$C_{26}H_{28}Cl_2FeNPPd$ M 618.661

Catalyst for asymmetric hydrosilylation of alkenes. Red needles (CH_2Cl_2/hexane).

(S)$_C$(R)$_{planar}$-form [55630-58-3]

(1S,1'S)-form

$NiCl_2$ complex used for asymmetric cross-coupling of Grignard reagents with vinylic halides. Orange cryst. (EtOH). Mp 139°. $[\alpha]^{25}_D$ +361° (c, 0.6 in EtOH).

(RS,RS)-form

(Norbornadiene)rhodiumhexafluorophosphate complex:

$C_{37}H_{36}F_6FeNP_2Rh$ M 829.389

Hydrogenation catalyst. Orange cryst. (EtOH/CH_2Cl_2). Mp 192° dec.

Hayashi, T. *et al*, *J. Am. Chem. Soc.*, 1976, **98**, 3718 (*use*)
Kumada, M. *et al*, *CA*, 1976, **85**, 143275 (*synth*)
Tamao, K. *et al*, *Tetrahedron Lett.*, 1979, **23**, 2155.
Cullen, W.R. *et al*, *J. Am. Chem. Soc.*, 1980, **102**, 988 (*struct*)
Einstein, F.W.B. *et al*, *Acta Crystallogr., Sect. B*, 1980, **36**, 39 (*struct*)
Hayashi, T. *et al*, *Tetrahedron Lett.*, 1980, **21**, 1871.
Hayashi, T. *et al*, *Bull. Chem. Soc. Jpn.*, 1980, **53**, 1138 (*synth, cd, use, uv, pmr, ir*)
Hayashi, T., *ACS Symp. Ser.*, 1982, **185**, 177 (*rev*)
Yamamoto, K. *et al*, *Tetrahedron Lett.*, 1982, **23**, 3089 (*use*)
Van der Steen, F.H. *et al*, *Acta Crystallogr., Sect. C*, 1986, **42**, 547 (*struct, complex*)

(2-Dimethylaminoethyl)-triphenylphosphonium(1+) D-00698

$Ph_3P^{\oplus}CH_2CH_2NMe_2$

$C_{22}H_{25}NP^{\oplus}$ M 334.420 (ion)

Source of ylide.

Bromide: [21331-80-6].

$C_{22}H_{25}BrNP$ M 414.324

Cryst. ($CHCl_3$/EtOAc). Mp 196-199°, 204-206°..

Ylide: (2-Dimethylaminoethylidene)-triphenylphosphorane.

$C_{22}H_{24}NP$ M 333.412

Used in aminoalkylations of aldehydes and ketones.

Marxer, A. *et al*, *Helv. Chim. Acta*, 1978, **61**, 1708 (*synth, use*)
De Castro Dantas, T.N. *et al*, *Phosphorus Sulfur*, 1982, **13**, 97 (*synth, use, nmr*)
Maryanoff, B.E. *et al*, *J. Am. Chem. Soc.*, 1985, **107**, 217 (*synth, props*)

2-Dimethylamino-4-isopropyl-5,5-dimethyl-1,3,2-dioxaphosphorinane D-00699

N,N,5,5-Tetramethyl-4-(1-methylethyl)-1,3,2-dioxaphosphorinan-2-amine, 9CI

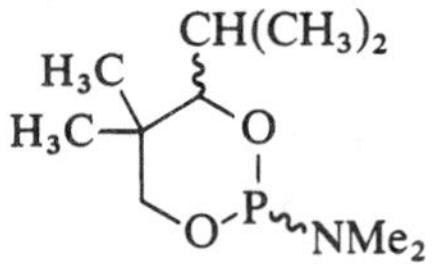

$C_{10}H_{22}NO_2P$ M 219.263

Liq. d^{20}_4 1.00. Bp_9 93-95°. n_D^{20} 1.4703. Mixt. of diastereoisomers.

Nasonovskii, I.S. *et al*, *Zh. Obshch. Khim.*, 1975, **45**, 724 (*Engl. transl.* p. 714) (*synth, P nmr*)

(Dimethylaminomethyl)diphenylphosphine D-00700

1-Diphenylphosphino-N,N-dimethylmethanamine, 9CI

[13119-19-2]

$Ph_2PCH_2NMe_2$

$C_{15}H_{18}NP$ M 243.288

$Bp_{0.1}$ 130-134°.

B,HCl: [66830-60-2]. Mp 215-216°.

B,MeI:

$C_{16}H_{21}INP$ M 385.227

Solid. Mp 118-121°.

Kellner, K. *et al*, *J. Organomet. Chem.*, 1978, **149**, 167 (*synth, derivs, pmr*)
McEwen, W.E. *et al*, *J. Am. Chem. Soc.*, 1980, **102**, 2746 (*synth, derivs, pmr*)

[(Dimethylamino)methylene]bisphosphonic acid D-00701

Trimethylamine-1,1-diphosphonic acid

[29712-30-9]

$Me_2NCH[P(O)(OH)_2]_2$

$C_3H_{11}NO_6P_2$ M 219.071

Probably exists in dipolar form. Complexing agent. Esters are Wittig-Horner reagents. Converts aldehydes into homologous carboxylic acids. Cryst. + $1H_2O$. Mp 237-241° dec. pK_{a1} 2.16, pK_{a2} 5.01, pK_{a3} 8.82, pK_{a4} 11.15.

Tetra-Me ester: Tetramethyl [(dimethylamino)-methylene]bisphosphonate.

$C_7H_{19}NO_6P_2$ M 275.178

Liq. $Bp_{0.03}$ 95-97°. n_D^{23} 1.4611.

Tetra-Et ester: [18855-52-2]. *Tetraethyl [(dimethylamino)methylene]bisphosphonate.*

$C_{11}H_{27}NO_6P_2$ M 331.285

Liq. $Bp_{0.03}$ 114-115°. n_D^{21} 1.4508.

Tetra-Et ester; B,MeI:

$C_{12}H_{27}INO_6P_2$ M 470.200

Solid. Mp 116-118°. With K_2CO_3 aq. forms an ylide Mp 38-40°.

Gross, H. *et al*, *J. Prakt. Chem.*, 1969, **311**, 563, 577, 925 (*derivs, props, pmr*)
Gross, H. *et al*, *J. Prakt. Chem.*, 1972, **314**, 969 (*ester, cmr*)
Gross, H. *et al*, *J. Prakt. Chem.*, 1976, **318**, 116 (*synth, pmr*)
Costisella, B. *et al*, *Tetrahedron*, 1981, **37**, 1227 (*props, use*)
Degenhardt, C.R., *Synth. Commun.*, 1982, **12**, 415 (*use*)

2-Dimethylamino-1,3,2-oxazaphospholidine, 8CI D-00702

N,N-Dimethyl-1,3,2-oxaphospholidin-2-amine, 9CI

[59515-30-9]

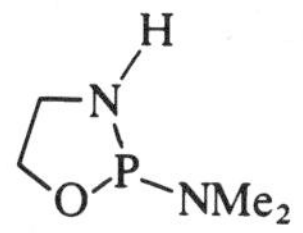

$C_4H_{11}N_2OP$ M 134.117

2-Oxide: [79574-60-0].

$C_4H_{11}N_2O_2P$ M 150.117

Cryst. Mp 80-82°.

Stokes, J.B. *et al, Phosphorus Sulfur*, 1981, **10**, 139 (*oxide, synth, P nmr*)

Kenttamaa, H.I. *et al, Int. J. Mass Spectrom. Ion Phys.*, 1983, **47**, 463 (*oxide, ms*)

(2-Dimethylamino-2-oxoethyl)phosphonic acid, 9CI D-00703

(Dimethylcarbamoylmethyl)phosphonic acid

$(HO)_2P(O)CH_2CONMe_2$

$C_4H_{10}NO_4P$ M 167.101

Di-Me ester: [62285-48-7]. *Dimethyl (dimethylcarbamoylmethyl)phosphonate. Dimethyl (2-dimethylamino-2-oxoethyl)phosphonate.*

$C_6H_{14}NO_4P$ M 195.155

Liq. $Bp_{0.1}$ 120°.

Di-Et ester: [3842-86-2]. *Diethyl (2-dimethylamino-2-oxoethyl)phosphonate. Diethyl (dimethylcarbamoylmethyl)phosphonate.*

$C_8H_{18}NO_4P$ M 223.208

No phys. props. reported.

Dioctyl ester: [83631-23-6]. *Dioctyl (dimethylcarbamoylmethyl)phosphonate. Dioctyl (2-dimethylamino-2-oxoethyl)phosphonate.*

$C_{20}H_{42}NO_4P$ M 391.530

Employed in the recovery of americium from nuclear fuels.

Catsikis, B.D. *et al, J. Inorg. Nucl. Chem.*, 1974, **36**, 1039 (*ir, complexes*)

Kem, K.M. *et al, J. Org. Chem.*, 1981, **46**, 5188 (*synth, pmr, P nmr*)

Partlett, P.A. *et al, J. Org. Chem.*, 1982, **47**, 1284 (*synth, ir, pmr, cmr*)

2-Dimethylamino-4-phenyl-1,3,2-dioxaphosphorinane D-00704

N,N-Dimethyl-4-phenyl-1,3,2-dioxaphosphorinan-2-amine

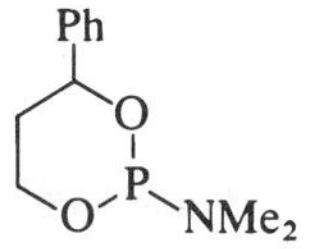

$C_{11}H_{16}NO_2P$ M 225.227

Liq. d_4^{20} 1.13. Bp_1 117-118°. n_D^{20} 1.5398. Readily oxidised.

Nasonovskii, I.S. *et al, Zh. Obshch. Khim.*, 1975, **45**, 724 (*Engl. transl.* p. 714) (*synth, P nmr*)

(4-Dimethylaminophenyl)methylphosphinic acid, 9CI D-00705

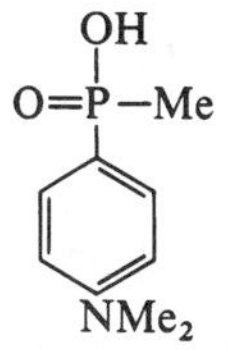

$C_9H_{14}NO_2P$ M 199.189

Me ester: Methyl (4-dimethylaminophenyl)-methylphosphinate.

$C_{10}H_{16}NO_2P$ M 213.216

Liq. $Bp_{0.4}$ 170-172°.

Me ester; B,MeI:

$C_{11}H_{19}INO_2P$ M 355.155

Solid (EtOH). Mp 161° dec. Has been resolved.

Coyne, D.M. *et al, J. Am. Chem. Soc.*, 1956, **78**, 3061.

Green, M. *et al, J. Chem. Soc.*, 1958, 3129.

(4-Dimethylaminophenyl)phosphonic acid, 9CI D-00706

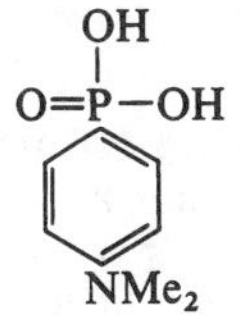

$C_8H_{12}NO_3P$ M 201.161

Cryst. (EtOH). Mp 133°.

Di-Et ester: [1754-43-4]. *Diethyl (4-dimethylaminophenyl)phosphonate.*

$C_{12}H_{20}NO_3P$ M 257.269

Liq. $Bp_{0.01}$ 155-158°.

Michaelis, A. *et al, Justus Liebigs Ann. Chem.*, 1890, **260**, 1 (*synth*)

B.P., 1 200 273, (*1970*); *CA*, **77**, 152351 (*ester*)

Grabiak, R.C. *et al, Phosphorus Sulfur*, 1980, **9**, 197 (*dichloride, pmr, P nmr*)

[3-(Dimethylamino)propyl]-triphenylphosphonium(1+) D-00707

$Ph_3P^{\oplus}CH_2CH_2CH_2NMe_2$

$C_{23}H_{27}NP^{\oplus}$ M 348.447 (ion)

Chloride:

$C_{23}H_{27}ClNP$ M 383.900

Source of ylide. Cryst. (EtOAc). Mp 206-208°. Butyllithium → ylide.

Bromide:

$C_{23}H_{27}BrNP$ M 428.351

Source of ylide. Solid. Mp 270-278° dec.

Ylide: [3-(Dimethylamino)propylidene]-triphenylphosphorane.

$C_{23}H_{26}NP$ M 347.439

Wittig reagent.

De Castro Dantas, T.N. *et al, Phosphorus Sulfur*, 1982, **13**, 97 (*synth*)

Maryanoff, B.E. *et al, J. Am. Chem. Soc.*, 1985, **107**, 217 (*bromide, synth, props*)

(Dimethylamino)tetrafluorophosphorane D-00708

Tetrafluoro-N,N-dimethylphosphoranamine, 9CI. Tetrafluorodimethylaminophosphorane, 8CI

[2353-98-2]

Me_2NPF_4

$C_2H_6F_4NP$ M 151.043

Liq. which fumes on exp. to moist air. Mp −80°. Bp 64°.

Brown, D.H. *et al*, *J. Chem. Soc. (A)*, 1966, 171 (*synth, ir, pmr, nmr*)
Demitras, G.C. *et al*, *Inorg. Chem.*, 1967, **6**, 1903 (*synth, pmr, ir, nmr*)
Cowley, A.H. *et al*, *J. Am. Chem. Soc.*, 1973, **95**, 6506 (*pe*)
Eisenhut, M. *et al*, *J. Am. Chem. Soc.*, 1974, **96**, 5385 (*nmr, struct*)
Inorg. Synth., 1978, **18**, 181 (*synth*)

(Dimethylamino)tetraiodophosphorane **D-00709**

Tetraiodo-N,N-dimethylphosphoranamine, 9CI. *Tetraiododimethylaminophosphorane*, 8CI

[27971-26-2]

Me_2NPI_4

$C_2H_6I_4NP$ M 582.668

Light-orange prisms (C_6H_6). Mp 121-122°. Readily hydrolysed by atm. moisture.

Feshchenko, N.G. *et al*, *Zh. Obshch. Khim.*, 1970, **40**, 500 (*Engl. transl.* p. 464) (*synth*)

2-Dimethylamino-4,4,5,5-tetramethyl-1,3,2-dioxaphospholane, 8CI **D-00710**

Hexamethyl-1,3,2-dioxaphospholan-2-amine, 9CI. *2-Dimethylamino-4,4,5,5-tetramethyl-1,3-dioxa-2-phosphacyclopentane. 4,4,5,5-Tetramethylethylene dimethylphosphoramidite*

[14274-42-1]

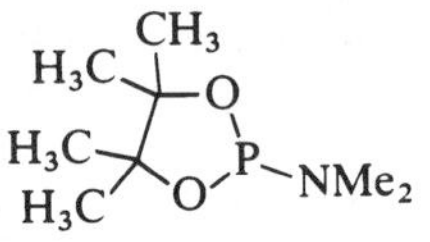

$C_8H_{18}NO_2P$ M 191.209

Pungent liq. d_4^{20} 1.01. Bp_2 63°. n_D^{20} 1.4632. Easily oxidised.

2-Oxide: [16492-44-7]. *4,4,5,5-Tetramethylethylene dimethylphosphoramidate.*
$C_8H_{18}NO_3P$ M 207.209
Deliquescent needles (pet. ether). Mp 51°.

2-Sulfide: [14274-51-2]. *4,4,5,5-Tetramethylethylene dimethylphosphoramidothioate.*
$C_8H_{18}NO_2PS$ M 223.269
Solid. Mp 32-33°.

Edmundson, R.S. *et al*, *J. Chem. Soc. (C)*, 1966, 1997 (*synth, derivs*)
Fontal, B. *et al*, *Tetrahedron*, 1966, **22**, 3275 (*synth, pmr*)
Sanchez, M. *et al*, *Bull. Soc. Chim. Fr.*, 1968, 773 (*synth*)
Al-Rawi, J.M.A. *et al*, *Org. Magn. Reson.*, 1984, **22**, 336 (*cmr*)

[2-(Dimethylarsino)phenyl]-dimethylphosphine, 10CI **D-00711**

o-*(Dimethylphosphinophenyl)dimethylarsine. 1-(Dimethylarsino)-2-(dimethylphosphino)benzene*

[34664-65-8]

$C_{10}H_{16}AsP$ M 242.132

Ligand for Pt. Liq. $Bp_{0.5}$ 92°.

B,MeI: [*2-(Dimethylarsino)phenyl*]-*trimethylphosphonium iodide.*
$C_{11}H_{19}AsIP$ M 384.071
Cryst. ($EtOH/Et_2O$). Mp 210-212° dec.

Levason, W. *et al*, *J. Chem. Soc., Dalton Trans.*, 1979, 1718 (*synth, pmr, nmr, deriv*)
Gulliver, D.J. *et al*, *J. Chem. Soc., Dalton Trans.*, 1981, 2153 (*complexes*)

2,5-Dimethyl-1,3,2-benzodithiaphosphole, 8CI **D-00712**

[13969-32-9]

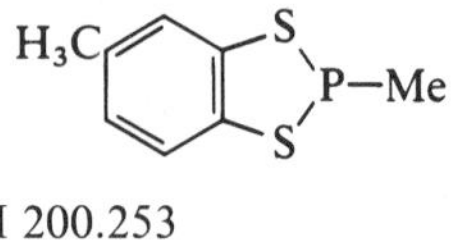

$C_8H_9PS_2$ M 200.253

Liq. Bp_2 110-4°, $Bp_{0.05}$ 80-85°.

Wieber, M. *et al*, *Chem. Ber.*, 1967, **100**, 974 (*pmr, derivs*)
Beer, D.D. *et al*, *Phosphorus Sulfur*, 1983, **77**, 283 (*synth, P nmr*)

***N,N'*-Dimethyl-*N,N'*-bis[[bis(dimethylamino)phosphinyl]-amino]ethane** **D-00713**

HMPT Dehydrodimer

[51833-57-9]

$(Me_2N)_2P(O)NMeCH_2CH_2NMeP(O)(NMe_2)_2$

$C_{12}H_{34}N_6O_2P_2$ M 356.387

Powerful solvent. Liq. $Bp_{0.005}$ 165-175°. Inert to strong bases e.g. BuLi, Ph_3CK.

Nee, G. *et al*, *J. Org. Chem.*, 1983, **48**, 1111 (*synth, pmr, use*)

1,4-Dimethyl-7,9-bis(trifluoromethyl)-1,4,6,8,10-pentaaza-5-phospha(5-P^V)-spiro[4.5]deca-5,7,9-triene, 10CI **D-00714**

1,3-Dimethyl-4',6'-bis(trifluoromethyl)spiro[1,3,2(P^V)-diazaphospholidine-2,2'-[1,3,5,2(P^V)triazaphosphorine]

[64595-14-8]

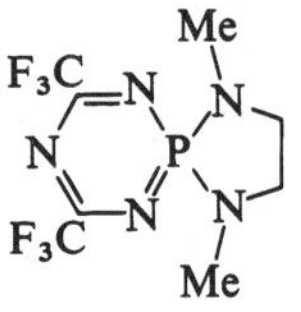

$C_8H_{10}F_6N_5P$ M 321.165

Solid. Mp 178°. $Bp_{0.01}$ 60° subl.

Schöning, G. *et al*, *Chem. Ber.*, 1977, **110**, 3231 (*synth, ir, ms, pmr, F and P nmr*)

Dimethyl chlorophosphite, **D-00715**

[3743-07-5]

$(MeO)_2PCl$

$C_2H_6ClO_2P$ M 128.495

Heat stabiliser for polymers. Ligand for Ni and Pt. Fuming liq. Bp_{56} 40-43°. n_D^{25} 1.4370.

▷Lippman's synth. may give an explosive residue

Lippman, A.E., *J. Org. Chem.*, 1965, **30**, 3217 (*synth, P nmr, pmr*)
Fritzowsky, N. *et al*, *Z. Anorg. Allg. Chem.*, 1971, **386**, 203 (*synth, ir, raman*)
Scherer, O.J. *et al*, *Z. Naturforsch., B*, 1972, **27**, 1429 (*pmr, synth*)

Barlos, K. *et al*, *Chem. Ber.*, 1980, **113**, 3716 (*nqr*)
Sinyashin, O.G. *et al*, *Zh. Obshch. Khim.*, 1981, **51**, 2410 (*Engl. transl.* p. 2078) (*synth*)

Dimethyl chlorothiophosphonate, 8CI **D-00716**

Thiohypochlorous acid anhydrosulfide with O,O*-diethyl phosphorothioate*, 9CI. S*-Chloro-*O,O*-dimethyl phosphorothioate*

[13894-35-4]

$(MeO)_2P(O)SCl$

$C_2H_6ClO_3PS$ M 176.554

Yellow liq. $Bp_{0.6}$ 41-42°. n_D^{20} 1.4820.

▷Asphyxiating vapour

Michalski, J. *et al*, *Rocz. Chem.*, 1963, **37**, 1479 (*synth*)
Mueller, W.H. *et al*, *J. Org. Chem.*, 1966, **31**, 3537 (*synth, props*)
Kutyrev, G.A. *et al*, *Zh. Obshch. Khim.*, 1979, **49**, 524; 1981, **51**, 1003 (*Engl. transl.* pp. 458, 837) (*props*)

1,3-Dimethyl-1,3,2-diazaphosphetidin-4-one, 11CI, 9CI **D-00717**

1,3-Dimethyl-4-oxo-1,3,2-diazaphosphetidine, 10CI

MeN—PH
O=C—NMe

$C_3H_7N_2OP$ M 118.075

2-Methoxy: [29476-24-2].
$C_4H_9N_2O_2P$ M 148.101
Liq. $Bp_{0.05}$ 30°.

2-Dimethylamino: [22298-05-1].
$C_5H_{12}N_3OP$ M 161.143
Liq. $Bp_{0.2}$ 57°.

Devillers, J. *et al*, *Bull. Soc. Chim. Fr.*, 1968, 4670 (*dimethylamino, synth, ir, P nmr*)
Bernard, D. *et al*, *C.R. Hebd. Seances Acad. Sci., Ser. C*, 1970, **271**, 418 (*synth, pmr, P nmr*)
Burgada, R., *Bull. Soc. Chim. Fr.*, 1971, 136 (*methoxy, pmr, P nmr*)
Mathis, R. *et al*, *Spectrochim. Acta, Part A*, 1974, **30**, 357 (*ir*)

1,2-Dimethyl-1,2,4-diazaphospholidine, 10CI **D-00718**

1,2-Dimethyl-1,2-diaza-4-phosphacyclopentane

H, P, N, N—Me, Me

$C_4H_{11}N_2P$ M 118.118

4-Ph: [78211-40-2]. *1,2-Dimethyl-4-phenyl-1,2,4-diazaphospholidine.*
$C_{10}H_{15}N_2P$ M 194.216
Oil. $Bp_{0.01}$ 120°.

4-Cyclohexyl: 4-Cyclohexyl-1,3-dimethyl-1,2,4-diazaphospholidine.
$C_{10}H_{21}N_2P$ M 200.263
Oil. $Bp_{0.01}$ 105-110°.

Märkl, G. *et al*, *Tetrahedron Lett.*, 1981, **22**, 229 (*synth, pmr, ms*)

1,3-Dimethyl-1,3,2-diazaphospholidine, 9CI, 8CI **D-00719**

Me, N, P—H, N, Me

$C_4H_{11}N_2P$ M 118.118

Derivs. are easily oxidised, and are good deoxygenating and desulfurating agents.

2-Fluoro: see 2-Fluoro-1,3-dimethyl-1,3,2-diazaphospholidine, F-00014

2-Chloro: see 2-Chloro-1,3-dimethyl-1,3,2-diazaphospholidine, C-00055

2-Methoxy: [7137-86-2]. *Methyl* N,N′*-ethanediyl-*N,N′*-dimethylphosphorodiamidite.*
$C_5H_{13}N_2OP$ M 148.144
Liq. Bp_{40} 77-78°, Bp_{11} 60°. n_D^{20} 1.475.

2-Ethoxy: see 2-Ethoxy-1,3-dimethyl-1,3,2-diazaphospholidine, E-00030

2-Phenoxy: see 1,3-Dimethyl-2-phenoxy-1,3,2-diazaphospholidine, D-00788

2-Me: [32294-62-5].
$C_5H_{13}N_2P$ M 132.145
Liq. Bp 149-150°.

2-Ph: see 1,3-Dimethyl-2-phenyl-1,3,2-diazaphospholidine, D-00793

2-Dimethylamino: see 1,3-Dimethyl-2-dimethylamino-1,3,2-diazaphospholidine, D-00724

Burgada, R., *Bull. Soc. Chim. Fr.*, 1971, 136 (*methoxy, synth, P nmr, pmr*)
Zuckermann, J.J. *et al*, *Inorg. Chem.*, 1971, **10**, 1028 (*methyl, synth, pmr*)
Worley, S.D. *et al*, *Inorg. Chem.*, 1979, **18**, 3581 (*methoxy, synth, pe*)
Worley, S.D. *et al*, *J. Electron Spectrosc. Relat. Phenom.*, 1982, **25**, 135 (*derivs, pe, struct*)

1,3-Dimethyl-1,3,2-diazaphospholidine 2-oxide, 9CI **D-00720**

N,N′*-Dimethyl-*N,N′*-ethylenephosphonic diamide*

[51452-84-7]

Me, N, P=O, H, N, Me

$C_4H_{11}N_2OP$ M 134.117

Liq. d_4^{20} 1.13. $Bp_{0.004}$ 61-64°. n_D^{20} 1.4873.

2-Ethoxy: see under 2-Ethoxy-1,3-dimethyl-1,3,2-diazaphospholidine, E-00030

2-Phenoxy: see under 1,3-Dimethyl-2-phenoxy-1,3,2-diazaphospholidine, D-00788

2-Ph: see under 1,3-Dimethyl-2-phenyl-1,3,2-diazaphospholidine, D-00793

Pudovik, M.A. *et al*, *Zh. Obshch. Khim.*, 1973, **43**, 2147 (*Engl. transl.* p. 2138) (*synth, ir, pmr*)

1,3-Dimethyl-1,3,2-diazaphospholidin-4-one 2-oxide, 9CI **D-00721**

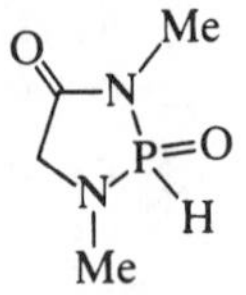

$C_4H_9N_2O_2P$ M 148.101

2-Phenoxy: [64642-31-5]. *1,3-Dimethyl-2-phenoxy-1,3-diazaphospholidin-4-one 2-oxide.*
$C_{10}H_{13}N_2O_3P$ M 240.198
Cryst. (THF/Et_2O). Mp 84-88°.

2-Me: [64642-30-4]. *1,2,3-Trimethyl-1,3-diazaphospholidin-4-one 2-oxide.*
$C_5H_{11}N_2O_2P$ M 162.128
Cryst. (THF/Et_2O). Mp 106-107°.

2-Ph: [64642-25-7]. *1,3-Dimethyl-2-phenyl-1,3-diazaphospholidin-4-one 2-oxide.*
$C_{10}H_{13}N_2O_2P$ M 224.199
Cryst. (Et_2O). Mp 60-63°.

Mulliez, M. *et al, Synthesis*, 1977, 478.

1,4-Dimethyl-1,4,2-diazaphospholidin-5-one 2-oxide, 9CI, 8CI **D-00722**

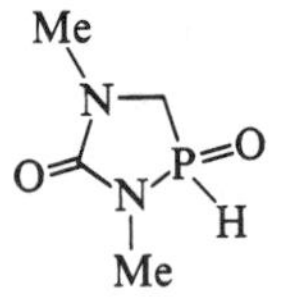

$C_4H_9N_2O_2P$ M 148.101

2-Bromo:
$C_4H_8BrN_2O_2P$ M 226.997
Cryst. (Me_2CO). Mp 88° dec.

2-Methoxy: [59917-79-2]. *1-Methoxy-1,3-dioxo-2,4-dimethyl-1-phospha-2,4-diazacyclopentane.*
$C_5H_{11}N_2O_3P$ M 178.127
Oil. $Bp_{0.7}$ 112-114°.

Petersen, H., *Justus Liebigs Ann. Chem.*, 1969, **726**, 88 (*bromo, synth*)
Kluger, R. *et al, J. Am. Chem. Soc.*, 1979, **101**, 5995 (*methoxy, synth, ir, cmr, P nmr, pmr*)

Dimethyl dichlorophosphoramidate **D-00723**

[29727-86-4]

$(MeO)_2P(O)NCl_2$

$C_2H_6Cl_2NO_3P$ M 193.954
Liq. $Bp_{0.6}$ 65-67°. n_D^{20} 1.4690.

Zwierzak, A. *et al, Tetrahedron*, 1970, **26**, 3521 (*synth, ir*)

1,3-Dimethyl-2-dimethylamino-1,3,2-diazaphospholidine **D-00724**

N,N,*1,3-Tetramethyl-1,3,2-diazaphospholidin-2-amine, 9CI*

[6069-38-1]

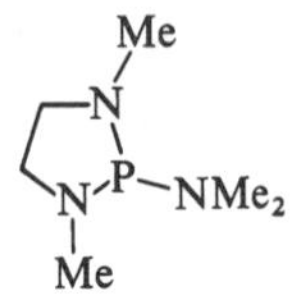

$C_6H_{16}N_3P$ M 161.186

Useful reagent for deoxygenations and desulfurations of e.g. epoxides, peroxides, disulfides. Liq. Bp_{13} 65-68°, $Bp_{1.8}$ 30-32°. n_D^{20} 1.478. Rapidly oxid.

2-Oxide: [7778-06-5].
$C_6H_{16}N_3OP$ M 177.186
$Bp_{0.2}$ 80°.

Burgada, R., *Bull. Soc. Chim. Fr.*, 1971, 136 (*synth, nmr, pmr*)
Mathis, R. *et al, Spectrochim. Acta, Part A*, 1974, **30**, 357 (*ir*)
Worley, S.D. *et al, Inorg. Chem.*, 1979, **18**, 3581 (*synth, pe*)
Gray, G.A. *et al, Org. Magn. Reson.*, 1980, **14**, 8 (*cmr*)
Hargis, J.H. *et al, J. Am. Chem. Soc.*, 1980, **102**, 13 (*cmr, N nmr*)
Gonbeau, D. *et al, Inorg. Chem.*, 1981, **20**, 1966 (*pe, struct*)
Worley, S.D. *et al, J. Electron Spectrosc. Relat. Phenonom.*, 1982, **25**, 135 (*pe, struct*)
Bollinger, J.C. *et al, Org. Mass Spectrom.*, 1985, **20**, 318 (*oxide, ms*)

4,5-Dimethyl-2-dimethylamino-1,3,2-dioxaphospholane, 8CI **D-00725**

N,N,*4,5-Tetramethyl-1,3,2-dioxaphospholan-2-amine, 9CI. 2,3-Butylene dimethylphosphoramidite. 4,5-Dimethyl-2-dimethylamino-1,3-dioxa-2-phosphacyclohexane*

[55666-79-0]

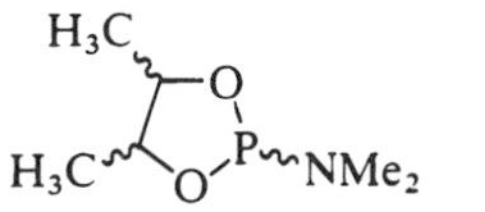

$C_6H_{14}NO_2P$ M 163.156

Stereoisomeric mixt. All 3 possible racemates recognised spectroscopically. Pungent liq. d_4^{20} 1.03. $Bp_{8.5}$ 65.5°. n_D^{24} 1.4619. Easily oxidised.

2-Sulfide: [14274-50-1]. O,O-*2,3-Butylene dimethylphosphoramidothioate.*
$C_6H_{14}NO_2PS$ M 195.216
Liq. $Bp_{0.2}$ 103-104°. n_D^{21} 1.5018. Diastereoisomeric forms recognised.

Edmundson, R.S. *et al, J. Chem. Soc. (C)*, 1966, 1997 (*synth, deriv*)
Sanchez, M. *et al, Bull. Soc. Chim. Fr.*, 1968, 773 (*synth*)
Dahl, O., *Tetrahedron Lett.*, 1982, **23**, 1493 (*P nmr*)
Nielsen, J. *et al, J. Chem. Soc., Perkin Trans. 2*, 1984, 553 (*pmr*)

5,5-Dimethyl-2-dimethylaminotetrahydro-2*H*-1,3,2-oxazaphosphorine **D-00726**

*Tetrahydro-*N,N,*5,5-tetramethyl-2*H*-1,3,2-oxazaphosphorin-2-amine, 9CI. 5,5-Dimethyl-2-dimethylamino-1,3,2-oxazaphosphorinane*

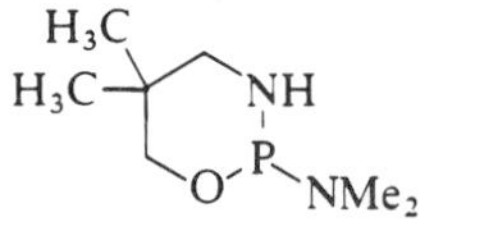

$C_7H_{17}N_2OP$ M 176.198

2-Oxide: [88946-46-7].
$C_7H_{17}N_2O_2P$ M 192.197
Solid. Mp 110-111°.

Holmes, R.R. *et al, J. Am. Chem. Soc.*, 1984, **106**, 2353 (*oxide, ir, ms, pmr, P nmr, cryst struct*)
Setzer, W.N. *et al, J. Am. Chem. Soc.*, 1985, **107**, 2083 (*oxide, pmr, P nmr, conformn*)

Dimethyl dimethylphosphoramidate, 9CI **D-00727**

Dimethyl (dimethylamido)phosphate

[597-07-9]

$(MeO)_2P(O)NMe_2$

$C_4H_{12}NO_3P$ M 153.117
Liq. d_4^{20} 1.13. Bp_{11} 72-72.5°, $Bp_{0.1}$ 46-48°. n_D^{20} 1.4175.

Kamai, G. *et al, Zh. Obshch. Khim.*, 1957, **27**, 3064 (*Engl. transl.* p. 3093) (*synth*)
Cheymol, J. *et al, C.R. Hebd. Seances Acad. Sci.*, 1959, **249**, 1240 (*synth*)
Musina, A.A. *et al, Zh. Strukt. Khim.*, 1976, **17**, 315 (*Engl. transl.* p. 271) (*complexes*)
Schlak, O. *et al, Z. Anorg. Allg. Chem.*, 1976, **419**, 275 (*ir, pmr, P nmr*)
Pressl, K.-D., *Z. Anorg. Allg. Chem.*, 1977, **434**, 171 (*ir, raman*)
Mizrahi, V. *et al, J. Org. Chem.*, 1982, **47**, 3533 (*synth, ms*)
Davidowitz, B. *et al, Org. Mass. Spectrom.*, 1984, **19**, 128 (*ms*)

6,9-Dimethyl-1,3-dioxa-6,9-diaza-2-phosphacycloundecane, 10CI D-00728

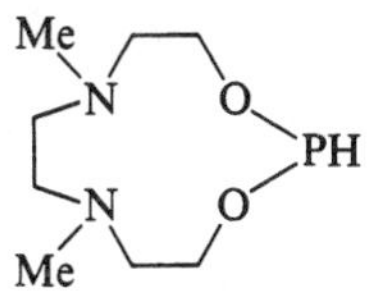

$C_8H_{19}N_2O_2P$ M 206.224
Alkoxy derivs. from alkali-metal complexes.

2-Chloro:
$C_8H_{18}ClN_2O_2P$ M 240.669
No data available.

2-Methoxy: [64762-31-8].
$C_9H_{21}N_2O_3P$ M 236.250
No data available.

2-Ethoxy:
$C_{10}H_{23}N_2O_3P$ M 250.277
No data available.

Sliwa, H. *et al, Tetrahedron Lett.*, 1977, 1583 (*synth, derivs*)
Grandjean, J. *et al, Tetrahedron Lett.*, 1978, 1861 (*complexes*)
Powell, J. *et al, J. Organomet. Chem.*, 1983, **243**, C1 (*complexes*)

4,9-Dimethyl-1,6-dioxa-4,9-diaza-5-phospha(5-P^V)spiro[4.4]nonane, 9CI D-00729

[18389-63-4]

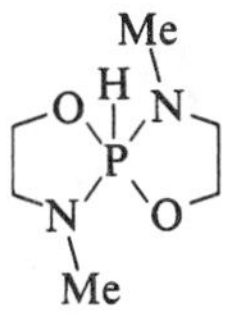

$C_6H_{15}N_2O_2P$ M 178.170
Exists largely (90%) as the monocyclic, tricoordinate tautomer at 20°. Liq. d_4^{20} 1.13. $Bp_{0.005}$ 60°. n_D^{20} 1.4755.

Sanchez, M. *et al, Bull. Soc. Chim. Fr.*, 1968, 773 (*synth, pmr, P nmr*)
Burgada, R. *et al, J. Organomet. Chem.*, 1974, **66**, 255 (*P nmr, struct*)

2,6-Dimethyl-1,3,2,6-dioxadiphosphocane D-00730

2,6-Dimethyl-1,3-dioxa-2,6-diphosphacyclooctane

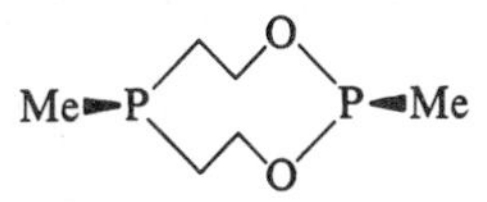

cis-form

$C_6H_{14}O_2P_2$ M 180.123
Stereoisomers separable only as derivs.

cis-form [79251-56-2]

2,6-Disulfide: [79251-54-0].
$C_6H_{14}O_2P_2S_2$ M 244.243
Solid. Mp 174°.

2,6-Diselenide: [79251-52-8].
$C_6H_{14}O_2P_2Se_2$ M 338.043
Solid. Mp 184°.

trans-form [79251-57-3]

2,6-Disulfide: [79251-55-1]. Solid. Mp 116°.
2,6-Diselenide: [79251-53-9]. Solid. Mp 152°.

Dutasta, J.-P. *et al, Tetrahedron Lett.*, 1981, **22**, 2549 (*synth, nmr*)
Piccinni-Leopardi, C. *et al, Acta Crystallogr., Sect. B*, 1982, **38**, 2197 (*disulfide, cryst struct*)
Piccinni-Leopardi, C. *et al, J. Chem. Soc., Perkin Trans. 2*, 1986, 85 (*pmr, cmr, P nmr, cryst struct*)

2,2-Dimethyl-1,3-dioxa-5-phospha-2-silacyclohexane, 9CI D-00731

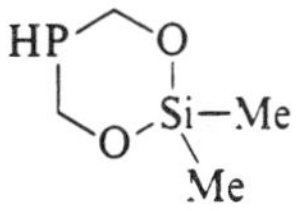

$C_4H_{11}O_2PSi$ M 150.189

5-Me: [68726-16-9].
$C_5H_{13}O_2PSi$ M 164.216
d^{20} 1.05. Bp_{10} 52°. n_D^{20} 1.4802.

5-Me, 5-sulfide: [72617-25-5].
$C_5H_{13}O_2PSSi$ M 196.276
Solid. Mp 93-95°. $Bp_{0.03}$ 30-40° subl.

5-Ph: [71787-50-3].
$C_{10}H_{15}O_2PSi$ M 226.287
Liq. Bp_3 95-97°. n_D^{18} 1.5392.

5-Ph, 5-sulfide: [80202-58-0].
$C_{10}H_{15}O_2PSSi$ M 258.347
Solid. Mp 80-81°.

Glukhikh, V.I. *et al, Dokl. Phys. Chem.*, 1979, **247**, 1179 (*Engl. transl.* p. 682) (*P nmr, pmr, cmr*)
Patsanovskii, I.I. *et al, Zh. Obshch. Khim.*, 1979, **49**, 2399 (*Engl. transl.* p. 2117) (*conformn*)
Voronkov, M.G. *et al, Dokl. Akad. Nauk SSSR, Ser. Sci. Khim.*, 1979, **247**, 609 (*Engl. transl.* p. 355) (*synth, ir, P nmr, conformn*)
Panov, A.M. *et al, Zh. Obshch. Khim.*, 1981, **51**, 2631 (*Engl. transl.* p. 2269) (*phenyl, sulfide, synth, ms, P nmr*)
Voronkov, M.G. *et al, Zh. Obshch. Khim.*, 1981, **51**, 2176 (*Engl. transl.* p. 1872) (*methyl, sulfide, synth, P nmr, ms*)
Gusev, I.A. *et al, Dokl. Akad. Nauk SSSR, Ser. Sci. Khim.*, 1983, **270**, 1398 (*Engl. transl.* p. 208) (*methyl, sulfide, cryst struct*)

2,4-Dimethyl-1,3,2-dioxaphospholane, 9CI D-00732

Propylene methylphosphonite. 2,4-Dimethyl-1,3-dioxa-2-phosphacyclopentane

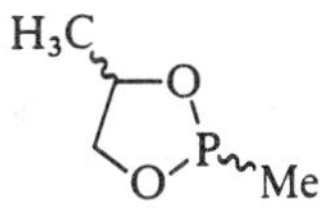

$C_4H_9O_2P$ M 120.088

2-Oxide: Propylene methylphosphonate.
$C_4H_9O_3P$ M 136.087
Liq. d_4^{20} 1.23. $Bp_{0.5}$ 77-79°. n_D^{20} 1.4414. Mixt. of *cis*- and *trans*-forms.

Petrov, K.A. *et al*, *Zh. Obshch. Khim.*, 1965, **35**, 732 (*Engl. transl.* p. 731) (*oxide*)
Evdakov, V.P. *et al*, *Zh. Obshch. Khim.*, 1967, **37**, 441 (*Engl. transl.* p. 412) (*oxide*)

4,5-Dimethyl-1,3,2-dioxaphospholane, 8CI D-00733

4,5-Dimethyl-1,3-dioxa-2-phosphacyclopentane. 2,3-Butylene phosphonite

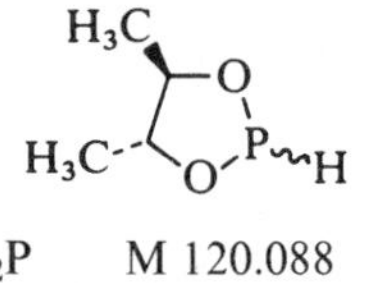

(4*RS*,5*RS*)-*form*

$C_4H_9O_2P$ M 120.088

2-Oxide: [16352-17-3]. *2,3-Butylene phosphonate. 2,3-Butylene phosphite. 2,3-Butylene hydrogen phosphite.*
$C_4H_9O_3P$ M 136.087
pK_a 2.04 (50% EtOH aq.). Tautomeric.

2-Sulfide: [16368-15-3]. O,O-*2,3-Butylene phosphonothioate.* O,O-*2,3-Butylene thiophosphonate.* O,O-*2,3-Butylene hydrogen thiophosphite.*
$C_4H_9O_2PS$ M 152.148
$Bp_{0.03}$ 58-59°. pK_a 2.68 (50% EtOH aq.). n_D^{20} 1.5019. Tautomeric.

(4*RS*,5*RS*)-*form*
(±)-*form*
2-Oxide: [66322-40-5].
$C_4H_9O_3P$ M 136.087
Liq. $Bp_{0.05}$ 60-63°. n_D^{20} 1.4380.
2-Sulfide: [66288-90-2].
$C_4H_9O_2PS$ M 152.148
Liq. $Bp_{0.04}$ 60°. n_D^{20} 1.5001.

(4*RS*,5*SR*)-*form*
cis-*form*. meso-*form*
2-Oxide: [66288-96-8]. Liq. $Bp_{0.05}$ 60-63°. n_D^{22} 1.4369. Diastereoisomers at P characterised spectroscopically.

Zwierzak, A. *et al*, *Can. J. Chem.*, 1967, **45**, 2501 (*synth, ir*)
Mikolajczyk, M. *et al*, *J. Chem. Soc., Perkin Trans. 1*, 1977, 2213 (*oxide, synth, P nmr*)
Ovchinnikov, V.V. *et al*, *Zh. Obshch. Khim.*, 1978, **48**, 2424; 1979, **49**, 1693 (*Engl. transl.* pp. 2199, 1482) (*ir, P nmr, struct*)
Ofitserov, E.N. *et al*, *Zh. Obshch. Khim.*, 1981, **51**, 738 (*Engl. transl.* p. 602) (*sulfide, synth, pmr*)

2,4-Dimethyl-1,3,2-dioxaphosphorinane, 9CI D-00734

1,3-Butylene methylphosphonite. 2,4-Dimethyl-1,3,2-dioxaphosphinane. 2,4-Dimethyl-1,3-dioxa-2-phosphacyclohexane

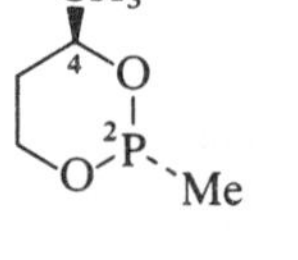

(2*RS*,4*RS*)-*form*

$C_5H_{11}O_2P$ M 134.114

2-Sulfide: [31860-23-8]. O,O-*1,3-Butylene methylphosphonothioate.* O,O-*1,3-Butylene methylthiophosphonate.*
$C_5H_{11}O_2PS$ M 166.174
Liq. d_4^{20} 1.25. $Bp_{0.02}$ 70° (bath). n_D^{20} 1.5215. Stereoisomerically pure, but config. not assigned.

(2*RS*,4*RS*)-*form* [57745-06-9]
(±)-trans-*form*
2-Oxide: [57745-06-9]. *1,3-Butylene methylphosphonate.*
$C_5H_{11}O_3P$ M 150.114
Liq. d_4^{20} 1.20. $Bp_{0.3}$ 93°, $Bp_{0.0001}$ 58-59°, $Bp_{0.01}$ 103-105°. n_D^{20} 1.4520.

(2*RS*,4*SR*)-*form* [57761-68-9]
(±)-cis-*form*
2-Oxide: [57761-68-9]. Liq. d_4^{20} 1.22. $Bp_{0.01}$ 82-85°, $Bp_{0.0001}$ 98-99°. n_D^{20} 1.4575.

Arbuzov, B.A. *et al*, *Izv. Akad. Nauk SSSR, Ser. Khim.*, 1974, 665 (*Engl. transl.* p. 629) (*oxides, synth, P nmr, raman*)
Predvoditelev, D.A. *et al*, *Zh. Obshch. Khim.*, 1974, **44**, 1697 (*Engl. transl.* p. 1667) (*sulfide, synth, tlc, pmr, P nmr*)
Lesiak, K. *et al*, *Phosphorus*, 1975, **6**, 65 (*oxides, ir, P nmr, cmr*)
Mosbo, J.A. *et al*, *J. Org. Chem.*, 1977, **42**, 1549 (*oxides, synth, ir, pmr, P nmr*)
Shakirov, I.Kh. *et al*, *Zh. Obshch. Khim.*, 1978, **48**, 508 (*Engl. transl.* p. 458) (*oxide, ir, raman*)
Nifant'ev, É.E. *et al*, *Tetrahedron*, 1981, **37**, 3183 (*cmr*)

5,5-Dimethyl-1,3,2-dioxaphosphorinane, 9CI D-00735

5,5-Dimethyl-1,3,2-dioxaphosphinane. 5,5-Dimethyl-1,3-dioxa-2-phosphacyclohexane. Neopentylene phosphonite
[25236-29-7]

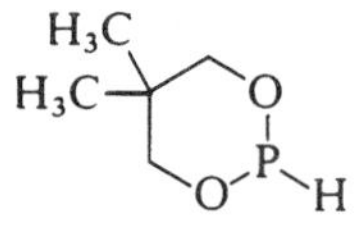

$C_5H_{11}O_2P$ M 134.114
Liq. Bp_1 36-37°. n_D^{20} 1.468.

2-Oxide: [4090-60-2]. *Neopentylene phosphonate. Neopentylene phosphite.*
$C_5H_{11}O_3P$ M 150.114
Cryst. (Et_2O). Mp 56-57°. Bp_1 135-136°, $Bp_{0.05}$ 93-94°. Tautomeric.

2-Sulfide: [2428-07-1]. O,O-*Neopentylene phosphonothioate.* O,O-*Neopentylene thiophosphite.*
$C_5H_{11}O_2PS$ M 166.174
Leaflets (pet. ether or C_6H_6/pet. ether). Mp 83-84.5°. Tautomeric.

Zwierzak, A., *Can. J. Chem.*, 1967, **45**, 2501 (*derivs, synth, ir*)
Nifant'ev, É.E., *Zh. Obshch. Khim.*, 1971, **41**, 2368 (*Engl. transl.* p. 2394) (*oxide*)
Borisenko, A.A. *et al*, *J. Chem. Soc., Chem. Commun.*, 1972, 406 (*oxide, cmr*)
Ovchinnikov, V.V. *et al*, *Zh. Obshch. Khim.*, 1978, **48**, 2424 (*Engl. transl.* p. 2199) (*oxide, P nmr, tautom*)
Nifant'ev, É.E. *et al*, *Tetrahedron*, 1981, **37**, 3183 (*synth, pmr, P nmr*)

1,2-Dimethyl-4,5-diphenyl-1*H*-1,3-azaphosphole D-00736

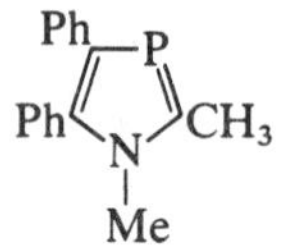

$C_{17}H_{16}NP$ M 265.294
Solid. Mp 115-117°.

Märkl, G. *et al*, *Tetrahedron Lett.*, 1986, **27**, 4419 (*synth, pmr, cmr, P nmr*)

2,4-Dimethyl-9,13-diphenyl-1,5-diaza-9,13-diphosphacyclohexadeca-1,4-diene, 9CI D-00737

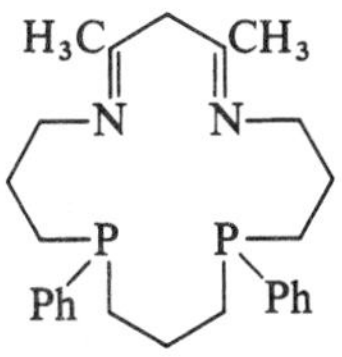

$C_{26}H_{36}N_2P_2$ M 438.531

Isol. only as Ni complex.

Scanlon, L.G. *et al*, *J. Am. Chem. Soc.*, 1980, **102**, 6849 (*cryst struct, synth*)
Scanlon, L.G. *et al*, *Inorg. Chem.*, 1982, **21**, 1215 (*synth, cryst struct*)

12,14-Dimethyl-4,8-diphenyl-1,11-diaza-4,8-diphosphacyclotetradeca-11,14-diene, 10CI D-00738

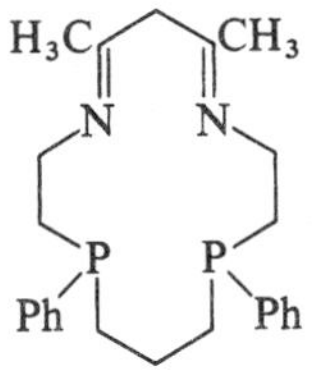

$C_{24}H_{32}N_2P_2$ M 410.478

Isol. only as Ni complex.

Scanlon, L.G. *et al*, *J. Am. Chem. Soc.*, 1980, **102**, 6849.
Scanlon, L.G. *et al*, *Inorg. Chem.*, 1982, **21**, 1215.

1,4-Dimethyl-3,5-diphenyl-1,4,2-diazaphospholidine 2-oxide D-00739

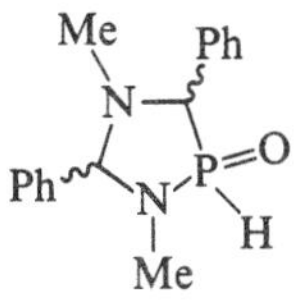

$C_{16}H_{19}N_2OP$ M 286.313

2-Chloro: [85290-04-6].
$C_{16}H_{18}ClN_2OP$ M 320.758
Cryst. (CH_2Cl_2). Mp 143°.

2-Ethoxy: [85290-05-7].
$C_{18}H_{23}N_2O_3P$ M 346.365
Liq. $Bp_{0.06}$ 100°.

Kibardin, M.A. *et al*, *Izv. Akad. Nauk SSSR, Ser. Khim.*, 1983, 432 (*Engl. transl.* p. 390)

3,8-Dimethyl-2,7-diphenyl-1,6-dioxa-4,9-diaza-5-phospha(5-*P*^V^)spiro[4.4]nonane, 9CI D-00740

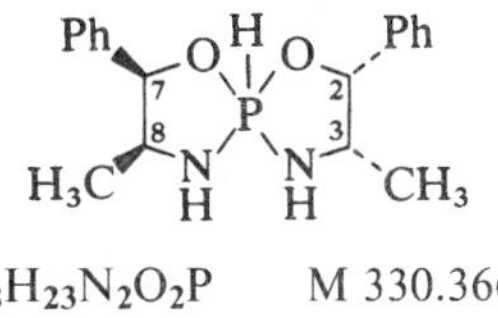

(2*R*,3*S*,7*R*,8*S*)-*form*

$C_{18}H_{23}N_2O_2P$ M 330.366

(2*R*,3*S*,7*R*,8*S*)-*form*
Solid. Mp 134°. $[\alpha]^{30}_{D}$ −197°. At 30° contains 40:60 ratio of helix forms.

(2*S*,3*R*,7*S*,8*R*)-*form*
Solid. Mp 134°. $[\alpha]^{30}_{D}$ +192°.

(2*S*,3*R*,7*R*,8*S*)-*form*
Solid. Mp 168°.

Contreras, R. *et al*, *Phosphorus*, 1972, **2**, 67 (*struct*)
Contreras, R. *et al*, *Synth. Inorg. Met.-Org. Chem.*, 1973, **3**, 37 (*synth, pmr*)
Klaebe, A. *et al*, *Phosphorus Sulfur*, 1977, **3**, 61; 1981, **10**, 53 (*props*)
Klaebe, A. *et al*, *Tetrahedron*, 1982, **38**, 2111 (*synth*)

6,9-Dimethyl-2,3-diphenyl-1,4-dioxa-6,9-diaza-5-phospha(5-*P*^V^)spiro[4.4]non-2-ene, 8CI D-00741

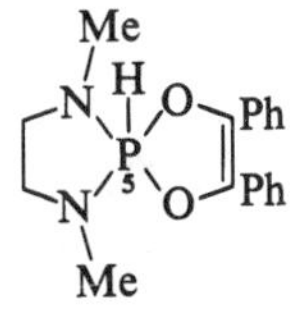

$C_{18}H_{21}N_2O_2P$ M 328.350

5-Methoxy:
$C_{19}H_{23}N_2O_3P$ M 358.376
Gum.

5-Dimethylamino: [7137-89-5]. N,N,*6,9-Tetramethyl-2,3-diphenyl-1,4-dioxa-6,9-diaza-5-phospha*(*5-P*^V^)*-spiro*[*4.4*]*non-2-en-5-amine.*
$C_{20}H_{26}N_3O_2P$ M 371.418
Solid. Mp 98-99°.

Ramirez, F. *et al*, *Tetrahedron*, 1968, **24**, 2275 (*synth, pmr, ir, P nmr*)

1,3-Dimethyl-6,7-diphenyl-5,8-dioxa-1,3-diaza-4-phospha(4-*P*^V^)spiro[3.4]oct-6-en-2-one, 9CI D-00742

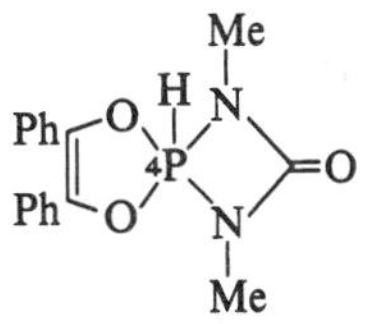

$C_{17}H_{17}N_2O_3P$ M 328.307

4-Methoxy: [29476-27-5].
$C_{18}H_{19}N_2O_4P$ M 358.333
Oil.

4-Dimethylamino: [29476-26-4].
$C_{19}H_{21}N_3O_3P$ M 370.367
Solid. Mp 133-134°.

Bernard, D. *et al*, *Tetrahedron*, 1975, **31**, 797 (*synth, pmr, P nmr*)

9,9-Dimethyl-2,8-diphenyl-3,7-dioxa-1-phosphabicyclo[3.3.1]nonane, 10CI D-00743

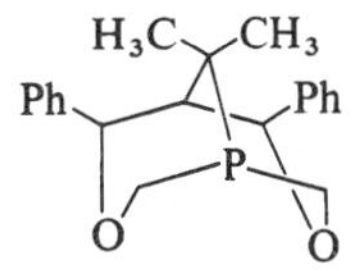

$C_{20}H_{23}O_2P$ M 326.374

B,MeI: 1,9,9-Trimethyl-2,8-diphenyl-3,7-dioxa-1-phosphoniabicyclo[*3.3.1*]*nonane.*
$C_{21}H_{26}IOP$ M 452.314

Solid. Mp 215-217° dec.
1-Sulfide: [77801-12-8].
$C_{20}H_{23}O_2PS$ M 358.434
Solid. Mp 236-238°.

Oehme, H. *et al*, *Z. Anorg. Allg. Chem.*, 1980, **471**, 155 (*synth, pmr, cmr, stereochem, derivs*)
Zschunke, A. *et al*, *Phosphorus Sulfur*, 1980, **9**, 117 (*cmr, nmr*)

1,2-Dimethyl-1,2-diphenyldiphosphine, 9CI **D-00744**
[3676-96-8]

(*RS,RS*)*-form*

$C_{14}H_{16}P_2$ M 246.228
Mp 73-76°. $Bp_{0.5}$ 128-130°. Mixt. of stereoisomers. At >150°, equilibrates between racemic and meso forms.
(*RS,RS*)-*form* [35196-93-1]
(±)*-form*
Monosulfide: [73719-56-9].
$C_{14}H_{16}P_2S$ M 278.288
Characterised spectroscopically.
Disulfide: [13639-75-3].
$C_{14}H_{16}P_2S_2$ M 310.348
Needles (EtOH). Mp 145-146°.
(*RS,SR*)-*form* [35196-92-0]
meso-*form*
Monosulfide: [73719-57-0]. Characterised spectroscopically.
Disulfide: Lozenge-like cryst. ($CHCl_3/Et_2O$ or Me_2CO). Insol. EtOH. Mp 206-208°.

Wheatley, P.J., *J. Chem. Soc.*, 1960, 523 (*disulfides, cryst struct*)
Maier, L., *J. Inorg. Nucl. Chem.*, 1962, **24**, 275 (*synth, nmr*)
Lambert, J.B. *et al*, *J. Am. Chem. Soc.*, 1968, **90**, 6401; 1970, **92**, 3093 (*synth, props, sulfides, pmr, nmr*)
Lee, J.D., *J. Inorg. Nucl. Chem.*, 1970, **32**, 3209 (*struct*)
Aime, S. *et al*, *J. Chem. Soc., Dalton Trans.*, 1976, 2144 (*cmr, nmr*)
Appel, R. *et al*, *Chem. Ber.*, 1977, **110**, 376 (*synth, nmr*)
Keek, H. *et al*, *Org. Mass Spectrom.*, 1979, **14**, 149 (*disulfide, ms*)
McFarlane, H.C.E. *et al*, *J. Chem. Soc., Dalton Trans.*, 1980, 240 (*sulfides, pmr, nmr*)

2,2-Dimethyl-1,3-diphenyl-1,3-diphosphorinane, 9CI **D-00745**
2,2-Dimethyl-1,3-diphenyl-1,3-diphosphinane
[60600-20-6]

$C_{18}H_{22}P_2$ M 300.319
Solid. Mp 97-98°.

Hauser, A. *et al*, *Phosphorus*, 1975, **5**, 261 (*P nmr*)
Issleib, K. *et al*, *Synth. React. Inorg. Metal-Org. Chem.*, 1976, **6**, 179 (*synth*)

2,2-Dimethyl-5,6-diphenyl-1-oxa-5-phospha-2-silacyclohexane **D-00746**

$C_{17}H_{21}OPSi$ M 300.412
Mixt. of stereoisomers, not separated. Liq. $Bp_{0.5}$ 141-145°.

Couret, C. *et al*, *J. Organomet. Chem.*, 1979, **182**, 9 (*synth, P nmr*)

***N,N*-Dimethyl-*P,P*-diphenylphosphinous amide**, 9CI **D-00747**
(*Dimethylamino*)*diphenylphosphine*
[6840-01-3]

Ph_2PNMe_2

$C_{14}H_{16}NP$ M 229.261
Ligand for Cr, Ni, Rh, Sn, W and V. Liq. Bp_3 127-129°. Readily oxidised.

Yoder, C.L. *et al*, *J. Inorg. Nucl. Chem.*, 1970, **32**, 3689 (*synth, nmr*)
Gouesnard, J.-P. *et al*, *Can. J. Chem.*, 1980, **58**, 1295 (*N and P nmr*)
Vaccher, C. *et al*, *J. Mol. Catal.*, 1981, **12**, 329 (*complexes*)

1,3-Dimethyl-5,7-diphenyl-1,3,5,7-tetraaza-4-phospha(4-P^V)spiro[3.3]heptane-2,6-dione **D-00748**

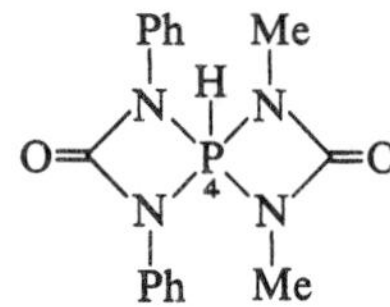

$C_{16}H_{17}N_4O_2P$ M 328.310
4-Chloro: [77655-19-7].
$C_{16}H_{16}ClN_4O_2P$ M 362.755
Cryst. (MeCN). Mp 140-142°.
4-(Diethoxyphosphinyl): [80049-55-4].
$C_{20}H_{26}N_4O_5P_2$ M 464.397
Powder (THF).

Roesky, H.W. *et al*, *Chem. Ber.*, 1981, **114**, 1554 (*chloro, synth, ms, ir, pmr, P nmr, cryst struct*)
Roesky, H.W. *et al*, *Inorg. Chem.*, 1982, **21**, 844 (*synth, ir, ms, P nmr*)

8,8-Dimethyl-2,3-diphenyl-1,4,6,10-tetraoxa-5-phospha(5-P^V)spiro[4.5]dec-2-ene, 8CI **D-00749**

$C_{19}H_{21}O_4P$ M 344.346
5-Phenoxy:
$C_{25}H_{25}O_5P$ M 436.443
Cryst. (C_6H_6). Mp 94°. Unstable.

Ramirez, F. *et al*, *Tetrahedron*, 1968, **24**, 1785 (*synth, ir, P nmr*)

Dimethyldiphosphinogermane **D-00750**
(*Dimethylgermylene*)*bisphosphine*, 9CI. *Bis*(*phosphino*)*-dimethylgermanium*

[20519-93-1]

$Me_2Ge(PH_2)_2$

$C_2H_{10}GeP_2$ M 168.639
Mp −100.2°.

Norman, A.D., *Inorg. Chem.*, 1970, **9**, 870 (*synth, ir, nmr*)
Dahl, A.R., *Inorg. Chem.*, 1975, **14**, 1093, 2562; *J. Am. Chem. Soc.*, 1975, **97**, 6364.

Dimethyldiphosphonic acid, 9CI — D-00751

Dimethylpyrophosphonic acid. Methylphosphonic acid monoanhydride
[1070-90-2]

$C_2H_8O_5P_2$ M 174.030
Hygroscopic cryst. (butanone/$CHCl_3$). Mp 141-142°.
Biscyclohexylamine salt: Cryst. (Me_2CO aq.). Mp 234-236°.
Di-Me ester: Dimethyl dimethyldiphosphonate. Dimethyl dimethylpyrophosphonate.
$C_4H_{12}O_5P_2$ M 202.083
d_0^{17} 1.32. Bp_4 137.5-138°. n_D^{17} 1.4370.
Di-Et ester: [32288-17-8]. *Diethyl dimethyldiphosphonate. Diethyl dimethylpyrophosphonate.*
$C_6H_{16}O_5P_2$ M 230.137
Liq. d_4^{20} 1.20. $Bp_{2.5}$ 122°. n_D^{20} 1.4360.
Bis(dimethylamide): P,P′-*Dimethyl*-N,N,N′,N′-*tetramethyldiphosphonic diamide. Hexamethyldiphosphonamide. Hexamethylpyrophosphonamide.*
$C_6H_{18}N_2O_3P_2$ M 228.167
Liq. Bp_3 135-140°.
▷Chloinesterase inhibitor

Petrov, K.A. *et al*, *Zh. Obshch. Khim.*, 1959, **29**, 1822 (*Engl. transl.* p. 1793) (*ester*)
Evdakov, V.P. *et al*, *Zh. Obshch. Khim.*, 1964, **34**, 3952 (*Engl. transl.* p. 4012) (*synth*)
Pudovik, A.N. *et al*, *Zh. Obshch. Khim.*, 1964, **34**, 2213 (*Engl. transl.* p. 2224) (*ester*)
Joesten, M.D. *et al*, *Inorg. Chem.*, 1972, **11**, 429 (*diamide, synth, pmr, P nmr*)
Ruveda, M.A. *et al*, *Tetrahedron*, 1972, **28**, 5011 (*synth, ir, pmr*)

2,4-Dimethyl-1,3,2,4-dithiadiphosphetane 2,4-disulfide, 9CI — D-00752

Cyclic anhydrosulfide from methylphosphonotrithioic acid, 8CI
[1121-81-9]

Cryst. (chlorobenzene). Mp 206-211°.

Newallis, P.E. *et al*, *J. Org. Chem.*, 1961, **27**, 3829 (*synth, cryst struct, props*)
Baudler, M. *et al*, *Z. Naturforsch., B*, 1967, **22**, 222 (*synth*)
Andreev, N.A. *et al*, *Zh. Obshch. Khim.*, 1982, **52**, 1785 (*Engl. transl.* p. 1581) (*struct, P nmr*)

N-(1,2-Dimethylethenylenedioxyphosphoryl)-imidazole — D-00753

*1-(4,5-Dimethyl-1,3,2-dioxaphosphol-2-yl)-1*H*-imidazole* P*-oxide, 9CI. 4,5-Dimethyl-2-(1′-imidazolyl)-2-oxo-1,3,2-dioxaphosphole*
[57648-76-7]

$C_7H_9N_2O_3P$ M 200.133
Powerful phosphorylating reagent in synth. of unsymm. dialkyl phosphates. Cryst. (C_6H_6/hexane or CH_2Cl_2/hexane). Mp 62-64°.

Ramirez, F. *et al*, *J. Am. Chem. Soc.*, 1975, **97**, 7181 (*use*)
Ramirez, F. *et al*, *Synthesis*, 1975, 637; 1976, 819 (*synth, pmr, P nmr, cmr*)
Ramirez, F. *et al*, *J. Org. Chem.*, 1978, **43**, 3635 (*cryst struct*)
Ramirez, F. *et al*, *Phosphorus Sulphur*, 1978, **4**, 43 (*use*)
Ramirez, F. *et al*, *Tetrahedron*, 1983, **39**, 2197 (*use*)

Dimethyl ethylphosphonate — D-00754

[6163-75-3]

$EtP(O)(OMe)_2$

$C_4H_{11}O_3P$ M 138.103
Useful synthetic intermediate for synth. of homologous phosphonates. Pleasant smelling liq. d_4^{20} 1.15. Bp_{15} 73-75°. n_D^{20} 1.4145.

Evelyn, L. *et al*, *Org. Magn. Reson.*, 1973, **5**, 141 (*pmr*)
Griffiths, W.R. *et al*, *Phosphorus Sulfur*, 1978, **5**, 101 (*ms*)
Bretherick, L., *Handbook of Reactive Chemical Hazards*, 2nd Ed., Butterworth, London, 1979, 510.
Savignac, P. *et al*, *Synthesis*, 1982, 725 (*anion, derivs*)

O,O-Dimethyl *S*-[(ethylthio)methyl] phosphorodithioate — D-00755

$(MeO)_2P(S)SCH_2SEt$

$C_5H_{13}O_2PS_3$ M 232.310
Possesses insecticidal props. Liq. $Bp_{0.01}$ 78°.
B.P., 797 307, (*1958*); *CA*, **53**, 3058

Dimethylgermylphosphine, 9CI, 8CI — D-00756

Dimethyl(phosphino)germane
[26465-28-1]

Me_2GeHPH_2

C_2H_9GeP M 136.657
Liq. Mp −106.5°. Bp 74.2°. The phosphino group undergoes thermal redistribution.

Norman, A.D., *Inorg. Chem.*, 1970, **9**, 870 (*synth*)
Dahl, A.R., *Inorg. Chem.*, 1975, **14**, 1095.

2,10-Dimethyl-1,3,6,9,11,14-hexaoxa-2,10-diphosphacyclohexadecane, 9CI — D-00757

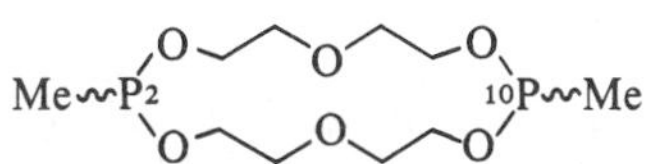

$C_{10}H_{22}O_6P_2$ M 300.228
A mixture of diastereoisomers in equilibrium with the monomer. See 1,3,6,2-Trioxaphosphocane, T-00592
Characterized as the 2,8-disulfide.

Dutasta, J.-P. *et al, J. Am. Chem. Soc.*, 1978, **100**, 1925 (*cmr, P nmr, conformn*)

1,3-Dimethylisophosphinoline, 9CI D-00758

1,3-Dimethyl-2-phosphanaphthalene. 1,3-Dimethylbenzo[c]*phosphorin. 1,3-Dimethylbenzo*[c]-*phosphinine*

[57328-61-7]

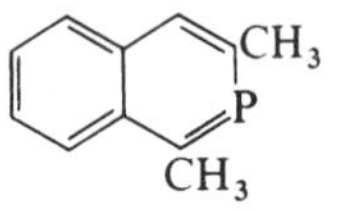

$C_{11}H_{11}P$ M 174.182

Pale-yellow liq. $Bp_{0.01}$ ca. 70°.

de Graaf, H.G. *et al*, Tet, 1975, **31**, 1097 (*synth, uv, pmr*)
Jongsma, C. *et al, Org. Mass Spectrom.*, 1975, **10**, 515 (*ms*)

Dimethyl(2-methylphenyl)phosphine, 10CI, 9CI D-00759

Dimethyl-o-*tolylphosphine*, 8CI. o-(*Dimethylphosphino*)*toluene*

[41845-30-1]

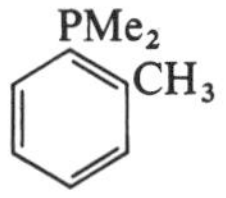

$C_9H_{13}P$ M 152.175

Ligand for Pt and Pd.

Oxide: [79317-54-7].
$C_9H_{13}OP$ M 168.175
Solid. Mp 85-86.5°. Bp_2 150-152°.

Jones, I.W. *et al, J. Chem. Soc., Perkin Trans. 2*, 1979, 501 (*uv*)
Bondarenko, N.A. *et al, Izv. Akad. Nauk SSSR, Ser. Khim.*, 1981, 1596 (*Engl. transl.* p. 1289) (*oxide*)

Dimethyl(3-methylphenyl)phosphine, 9CI D-00760

Dimethyl-m-*tolylphosphine.* m-(*Dimethylphosphino*)-*toluene*

[33733-60-7]

$C_9H_{13}P$ M 152.175

Liq. Bp_{10} 85-86°. n_D^{20} 1.5570.

Oxide: [53888-90-7].
$C_9H_{13}OP$ M 168.175
Cryst. Mp 60-61°. Bp_6 151-152°.

Malakhova, I.G. *et al, Izv. Akad. Nauk SSSR, Ser. Khim.*, 1974, 1842 (*Engl. transl.* p. 1761) (*synth, oxide*)
Jones, I.W. *et al, J. Chem. Soc., Perkin Trans. 2*, 1979, 501 (*uv*)

Dimethyl(4-methylphenyl)phosphine, 9CI D-00761

Dimethyl-p-*tolylphosphine.* p-(*Dimethylphosphino*)-*toluene*

[20676-64-6]

$C_9H_{13}P$ M 152.175

Forms Pd and Mn complexes. Liq. possessing foul odour. Bp 210°, Bp_{13} 90-91°. n_D^{20} 1.5561.

Oxide: [53888-89-4].
$C_9H_{13}OP$ M 168.175
Cryst. (cyclohexane), hygroscopic needles (hexane). Mp 95°. Bp_7 168-170°.

Sulfide: [54844-87-0].
$C_9H_{13}PS$ M 184.235
Cryst. (cyclohexane). Mp 96°.

Mann, F.G. *et al, J. Chem. Soc.*, 1937, 527 (*synth, derivs*)
Raevskii, O.A. *et al, Izv. Akad. Nauk SSSR, Ser. Khim.*, 1974, 453 (*Engl. transl.* p. 422) (*synth, struct*)
Malakhova, I.G. *et al, Izv. Akad. Nauk SSSR, Ser. Khim.*, 1974, 1842 (*Engl. transl.* p. 1761) (*oxide*)
Veistuyft, A.W. *et al, Inorg. Chem.*, 1975, **14**, 1495 (*derivs, pmr, complexes*)
Kaim, W. *et al, Chem. Ber.*, 1978, **111**, 3843 (*esr*)
Jones, I.W. *et al, J. Chem. Soc., Perkin Trans. 2*, 1979, 501 (*uv*)
Gonchareva, L.V. *et al, Zh. Obshch. Khim.*, 1981, **51**, 1450 (*Engl. transl.* p. 1230) (*sulfide, ir*)

Dimethyl methylphosphonate D-00762

[756-79-6]

MeP(O)(OMe)$_2$

$C_3H_9O_3P$ M 124.076

Extractant for rare earths. Useful synth. intermed. Forms an anion which may be alkylated or acylated. Methylates nucleotide bases, and converts carboxylic acids into Me esters. Pleasant-smelling liq. Bp_{10} 66-68°. n_D^{20} 1.4082.

▷SZ9120000.

Nesterov, L.V. *et al, Zh. Obshch. Khim.*, 1967, **37**, 728 (*Engl. transl.* p. 683) (*synth*)
Oae, S. *et al, Tetrahedron*, 1972, **28**, 549 (*synth*)
Vinogradov, L.I. *et al, Zh. Obshch. Khim.*, 1974, **44**, 37 (*Engl. transl.* p. 35) (*ir, pmr*)
Sutter, P. *et al, Phosphorus Sulfur*, 1978, **4**, 335 (*use*)
Sass, S. *et al, Org. Mass Spectrom.*, 1979, **14**, 257 (*ms*)
Zverev, V.V. *et al, Izv. Akad. Nauk SSSR, Ser. Khim.*, 1979, **28**, 84 (*Engl. transl.* p. 74) (*pe*)
Disselnkötter, H. *et al, Justus Liebigs Ann. Chem.*, 1982, 150 (*use*)
Smith, J.G. *et al, J. Org. Chem.*, 1983, **48**, 1110 (*anion, use*)
Van der Veken, B.J. *et al, J. Mol. Struct.*, 1983, **96**, 233 (*ir, conformn*)
Templeton, M.K. *et al, J. Am. Chem. Soc.*, 1985, **107**, 97, 774 (*props*)

N,N-Dimethyl-*P*-(1-methylpropyl)-phosphonamidic chloride, 9CI D-00763

P-sec-*Butyl*-N,N-*dimethylphosphamidic chloride*, 8CI

[23685-56-5]

$H_3CCH_2CH(CH_3)PCl(O)NMe_2$

$C_6H_{15}ClNOP$ M 183.617

Liq. d_4^{25} 1.10. $Bp_{0.2}$ 73-75°. n_D^{25} 1.4638. Presumably a mixt. of diastereoisomers.

Razvodovskaya, L.V. *et al, Zh. Obshch. Khim.*, 1969, **39**, 1260 (*Engl. transl.* p. 1230)

5,5-Dimethyl-2-methylseleno-1,3,2-dioxaphosphorinane, 9CI D-00764

5,5-Dimethyl-2-methylseleno-1,3,2-dioxaphosphinane. Se-*Methyl* O,O-*neopentylene phosphoroselenoite. 5,5-Dimethyl-2-methylseleno-1,3-dioxa-2-phosphacyclohexane*

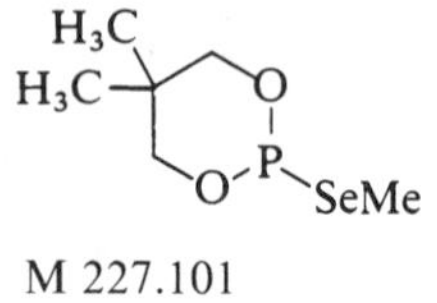

$C_6H_{13}O_2PSe$ M 227.101

2-Oxide: Se-*Methyl* O,O-*neopentylene phosphoroselenoate.*
$C_6H_{13}O_3PSe$ M 243.101
Solid. Mp 90-92°.
2-Sulfide: Se-*Methyl* O,O-*neopentylene phosphoroselenothioate.*
$C_6H_{13}O_2PSSe$ M 259.161
Solid. Mp 71-73°.
2-Selenide: [73098-19-8]. Se-*Methyl* O,O-*neopentylene phosphorodiselenoate.*
$C_6H_{13}O_2PSe_2$ M 306.061
Cryst. (C_6H_6/pet. ether). Mp 91°.

Stec, W.J. *et al, J. Org. Chem.*, 1976, **41**, 227 (*oxide, sulfide, synth, P nmr*)
Lesiak, K. *et al, Pol. J. Chem.*, 1979, **53**, 2041 (*selenide, synth, pmr, P nmr*)

4,5-Dimethyl-2-methylthio-1,3,2-dioxaphospholane, 8CI — D-00765

O,O-*2,3-Butylene* S-*methyl thiophosphite.* O,O-*2,3-Butylene* S-*methyl phosphorothioite. 4,5-Dimethyl-2-methylthio-1,3-dioxa-2-phosphacyclopentane*

(4*RS*,5*RS*)-*form*

$C_5H_{11}O_2PS$ M 166.174

(4RS,5RS)-form
(±)-*form*
2-Oxide: [66242-30-6]. O,O-*2,3-Butylene* S-*methyl thiophosphate.* O,O-*2,3-Butylene* S-*methyl phosphorothioate.*
$C_5H_{11}O_3PS$ M 182.174
Liq. $Bp_{0.05}$ 92-94°. n_D^{20} 1.4850.

(4RS,5SR)-form
meso-*form*
2-Oxide: [66322-41-6]. Liq. $Bp_{0.05}$ 92-94°. n_D^{20} 1.4859. Has the (2α,4β,5β)-config (SMe *trans* to CH_3 groups).

Ishmaeva, E.A. *et al, Zh. Obshch. Khim.*, 1975, **45**, 946 (*Engl. transl.* p. 931) (*sulfide, conformn*)
Mikolajczyk, M. *et al, J. Chem. Soc., Perkin Trans. 1*, 1977, 2213 (*oxides, pmr, P nmr*)

5,5-Dimethyl-2-methylthio-1,3,2-dioxaphosphorinane, 9CI — D-00766

S-*Methyl* O,O-*neopentylene phosphorothioite. 5,5-Dimethyl-2-methylthio-1,3-dioxa-2-phosphacyclohexane. 5,5-Dimethyl-2-methylthio-1,3,2-dioxaphosphinane*
[39846-24-7]

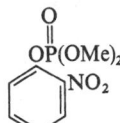

$C_6H_{13}O_2PS$ M 180.201
Liq. with obnoxious odour. Bp_{30} 120°. n_D^{20} 1.5070.
2-Oxide: [1005-98-7]. S-*Methyl* O,O-*neopentylene phosphorothioate.*
$C_6H_{13}O_3PS$ M 196.201
Cryst. (Et_2O). Mp 81-81.5°.
2-Sulfide: [935-59-1]. S-*Methyl* O,O-*neopentylene phosphorodithioate.*
$C_6H_{13}O_2PS_2$ M 212.261
Cryst. (pet. ether). Mp 85.5-86°.

Edmundson, R.S., *Tetrahedron*, 1964, **20**, 2781 (*derivs, synth, ir*)
Zwierzak, A., *Phosphorus*, 1972, **2**, 19 (*synth, ir*)
Francis, G.W. *et al, Acta Chem. Scand., Ser. B*, 1976, **30**, 31 (*oxide, ms*)
Dale, A.J., *Acta Chem. Scand., Ser. B*, 1976, **30**, 255 (*oxide, pmr*)
Stec, W.J. *et al, J. Org. Chem.*, 1976, **41**, 227 (*derivs, synth, P nmr*)
Gramstad, T. *et al, Acta Chem. Scand., Ser. B*, 1977, **31**, 345 (*ir*)
Edmundson, R.S., *Phosphorus Sulfur*, 1981, **9**, 307 (*sulfide, ms*)

Dimethyl 4-(methylthio)phenyl phosphate, 9CI — D-00767

[3254-63-5]

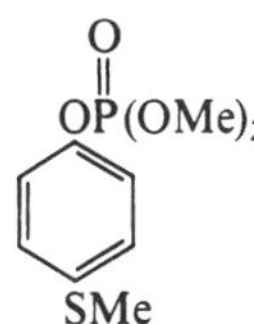

$C_9H_{13}O_4PS$ M 248.233
Agricultural insecticide and acaricide. Liq. Spar. sol. H_2O, sol. Me_2CO, dioxan, CCl_4, EtOH, xylene. d_4^{21} 1.273. n_D^{25} 1.5349.
▷TC5075000.

U.S.P., 3 151 022, (*1962*); *CA*, **60**, 10602a (*synth, ir, use*)
Pesticide Manual, 6th Ed., 203.
Bull, D.L. *et al, J. Agric. Food Chem.*, 1970, **18**, 1134 (*metab*)

Dimethyl 2-nitrophenyl phosphate — D-00768

[4938-89-0]

$C_8H_{10}NO_6P$ M 247.144
Liq. $Bp_{0.10}$ 127°. n_D^{25} 1.5110.
▷Inhibits acetylcholinesterase, trypsin, and chymotrypsin

Boter, H.L. *et al, Recl. Trav. Chim. Pays-Bas*, 1965, **84**, 1279 (*synth, props*)
Jentzsche, R. *et al, J. Prakt. Chem.*, 1977, **319**, 862 (*synth, ir, pmr, P nmr, props*)

Dimethyl 3-nitrophenyl phosphate — D-00769

[4532-05-2]
$C_8H_{10}NO_6P$ M 247.144
Liq. $Bp_{0.14}$ 132°. n_D^{25} 1.5130.
▷Inhibits trypsin and chymotrypsin

Boter, H.L. *et al, Recl. Trav. Chim. Pays-Bas*, 1965, **84**, 1279 (*synth*)

Dimethyl 4-nitrophenyl phosphate — D-00770

Methyl paraoxon
[950-35-6]
$C_8H_{10}NO_6P$ M 247.144
Metab. and product of photodegradation of methyl parathion. Liq. $Bp_{0.1}$ 142°. n_D^{20} 1.5211.
▷ Powerful inhibitor of acetylcholinesterase. Miotic. TC5250000.

de Roos, A.M. *et al, Recl. Trav. Chim. Pays-Bas*, 1958, **77**, 946 (*synth, ir*)

Miyamato, J. *et al*, *CA*, 1968, **69**, 9889.
Jentzsch, R. *et al*, *J. Prakt. Chem.*, 1975, **317**, 721 (*synth, ir, pmr, P nmr, props*)
Abe, T. *et al*, *Bull. Environ. Contam. Toxicol.*, 1979, **22**, 791 (*hplc*)
Mitsuo, N. *et al*, *Chem. Pharm. Bull.*, 1980, **28**, 1327 (*props*)

O,O-Dimethyl *Se,Se*-(2-nitrophenyl) phosphoro(diselenoperoxoate) D-00771

[58228-22-1]

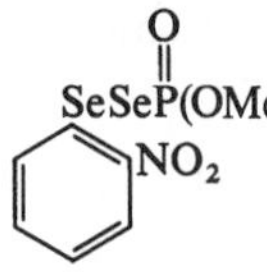

$C_8H_{10}NO_5PSe_2$ M 389.064

Solid. Mp 42°.

Austad, T., *Acta Chem. Scand., Ser. A*, 1975, **29**, 895 (*synth*)

O,O-Dimethyl-*S,Se*-2-nitrophenylphosphoro(selenothioperoxo)-thioate D-00772

[58228-19-6]

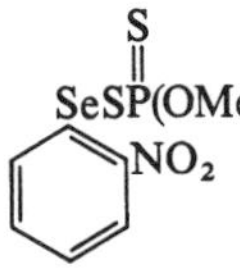

$C_8H_{10}NO_4PSSe$ M 326.165

Cryst. (CS_2 or C_6H_6/pet. ether). Mp 89° (79°).

Foss, O., *J. Am. Chem. Soc.*, 1947, **69**, 2236 (*synth*)
Austad, T., *Acta Chem. Scand., Ser. A*, 1975, **29**, 895 (*synth*)

O,O-Dimethyl *S*-(4-nitrophenyl) phosphorothioate, 9CI D-00773

1-[(Dimethoxyphosphinyl)thio]-4-nitrobenzene

[3820-53-9]

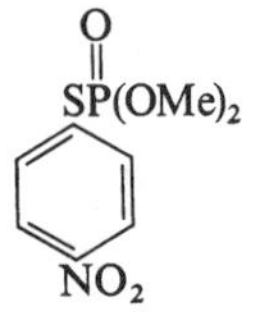

$C_8H_{10}NO_5PS$ M 263.204

Solid. Mp 55-56°.

▷TG0240000.

Pilgram, K. *et al*, *Tetrahedron*, 1965, **21**, 1999 (*synth, props*)
Kataev, E.G. *et al*, *Zh. Obshch. Khim.*, 1967, **37**, 2059 (*synth, ir*)
Meyer, H.J. *et al*, *Bull. Soc. Chim. Belg.*, 1978, **87**, 517 (*ms*)

O,S-Dimethyl *O*-4-nitrophenyl phosphorothioate, 9CI D-00774

O,S-*Dimethyl* O-p-*nitrophenyl thiophosphate*

[597-89-7]

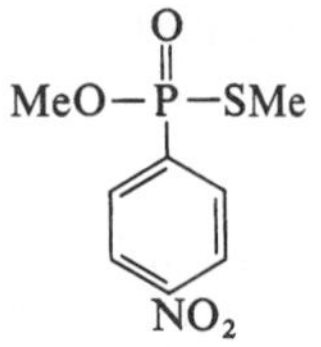

$C_8H_{10}NO_4PS$ M 247.205

▷TG0300000.

(+)-*form*

Cryst. (MeOH). Mp 43-44°. $[\alpha]_D^{21}$ +35.6° (c, 1.25 in MeOH).

(−)-*form*

Cryst. (MeOH). Mp 43°. $[\alpha]_D^{21}$ −35.6° (c, 1.25 in MeOH).

(±)-*form*

Cryst. (MeOH at low temp.) Pale-yellow oil when impure. n_D^{20} 1.5626 (supercooled). Complexes with metal halides.

Hilgetag, G. *et al*, *J. Prakt. Chem.*, 1959, **8**, 207, 224 (*synth*)
Hilgetag, G. *et al*, *Chem. Ber.*, 1965, **98**, 864 (*complexes*)
Lippman, A.E. *et al*, *J. Org. Chem.*, 1965, **30**, 3217 (*synth, pmr, P nmr*)

3,7-Dimethyl-1,6-octadien-3-yl dihydrogen phosphate, 9CI D-00775

Linaloyl phosphate. Linaloyl dihydrogen phosphate

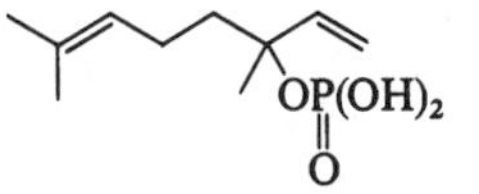

$C_{10}H_{19}O_4P$ M 234.231

Di-NH₄ salt: Solid. Spar. sol. Me_2CO, sl. sol. H_2O, EtOH, MeOH. Mp 148-150° dec.

Cramer, F. *et al*, *Justus Liebigs Ann. Chem.*, 1962, **654**, 180 (*synth*)

3,7-Dimethyl-2,6-octadien-1-yl dihydrogen phosphate, 9CI D-00776

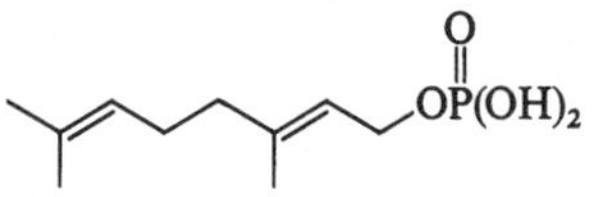

(*E*)-*form*

$C_{10}H_{19}O_4P$ M 234.231

(*E*)-*form*

Geranyl phosphate. Geranyl dihydrogen phosphate

Bis(cyclohexylammonium salt): Cryst. (H_2O). Mp 190-192° dec.

(*Z*)-*form*

Neryl phosphate. Neryl dihydrogen phosphate

Bis(cyclohexylammonium) salt: Solid. Mp 177° dec.

Cramer, F. *et al*, *Justus Liebigs Ann. Chem.*, 1962, **654**, 180 (*synth*)
Julia, M. *et al*, *Tetrahedron*, 1986, **42**, 3841 (*synth, pmr, cmr, P nmr*)

(3,7-Dimethyl-2,6-octadienyl)-triphenylphosphonium(1+), 9CI D-00777

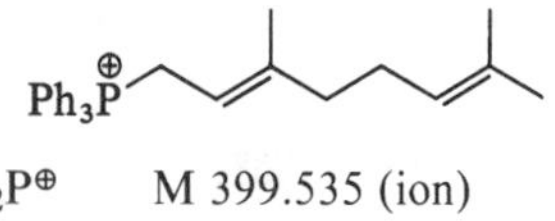

(*E*)-*form*

$C_{28}H_{32}P^{\oplus}$ M 399.535 (ion)

(*E*)-*form*

Geranyltriphenylphosphonium(1+)

Bromide: [41273-34-1].

$C_{28}H_{32}BrP$ M 479.439

Source of ylide. Cryst. (EtOAc/CH_2Cl_2). Mp 189°.

Ylide: [77706-35-5]. *(3,7-Dimethyl-2,6-octadienylidene)triphenylphosphorane.*

$C_{28}H_{31}P$ M 398.527

Used in Wittig synth. of naturally occurring polyenes e.g. allofarnesene, piericidin analogues.

(Z)-form

Neryltriphenylphosphonium(1+)

Bromide: [60346-00-1]. Source of ylide. Cryst. (EtOH). Mp 192-193°.

Ylide: [80388-56-3]. Used in Wittig reactions.

Surmatis, J.D. *et al*, *J. Org. Chem.*, 1963, **28**, 2735 (*synth, use*)
Hjortos, J., *Acta Crystallogr., Sect. B*, 1973, **29**, 767 (*cryst struct*)
Barlow, L. *et al*, *J. Chem. Soc., Perkin Trans. 1*, 1976, 1029 (*synth, ir, pmr, cmr, isom*)
Pfander, H. *et al*, *Helv. Chim. Acta*, 1980, **63**, 1367 (*synth, ir, pmr, cmr*)
Yoshida, S. *et al*, *Agric. Biol. Chem.*, 1980, **44**, 2913 (*use*)
Williams, H.J. *et al*, *Tetrahedron*, 1981, **37**, 2763 (*use*)

(2,7-Dimethyl-2,4,6-octatrien-1,8-diyl)-bis[triphenylphosphonium](2+), 9CI D-00778

1,8-Bis(diphenylphosphonio)-2,7-dimethyl-2,4,6-octatriene(2+)

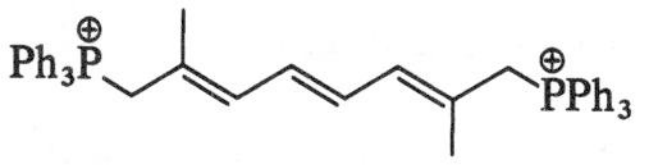

(E,E,E)-form

$C_{46}H_{44}P_2^{\oplus\oplus}$ M 658.801 (ion)

(E,E,E)-form

Dibromide: [57545-56-9].
$C_{46}H_{44}Br_2P_2$ M 818.609
Source of bisylide, obt. by treatment with PhLi. Cryst. (MeOH/EtOAc). Mp 280°.

Ylide: [39776-40-4]. *(2,7-Dimethyl-2,4,6-octatrien-1,8-diylidene)bis[triphenylphosphorane].*
$C_{46}H_{42}P_2$ M 656.785
Used in Wittig synth. of naturally occurring polyenes e.g. Isozeanthin. Violet-red.

Surmatis, J.D. *et al*, *J. Org. Chem.*, 1961, **26**, 1171 (*synth, ylide*)
Surmatis, J.D. *et al*, *Helv. Chim. Acta*, 1970, **53**, 974 (*use*)
Bernhard, K. *et al*, *Helv. Chim. Acta*, 1980, **63**, 1473 (*use*)
Buchecker, R. *et al*, *Helv. Chim. Acta*, 1982, **65**, 896 (*use*)
Haag, A. *et al*, *Helv. Chim. Acta*, 1982, **65**, 1795 (*use*)

2,7-Dimethyl-1,2,7-oxadiphosphepane 2,7-dioxide, 10CI D-00779

[72492-62-7]

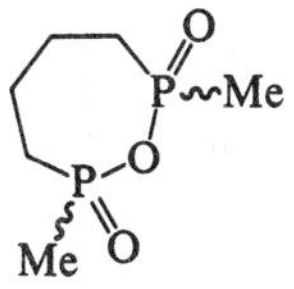

$C_6H_{14}O_3P_2$ M 196.122
Glassy solid. Mp 65°. $Bp_{0.5}$ 185°.

Ger. Pat., 2 811 628, (*1979*); *CA*, **92**, 58976 (*synth*)

2,6-Dimethyl-1,2,6-oxadiphosphorinane 2,6-dioxide, 10CI D-00780

[72492-61-6]

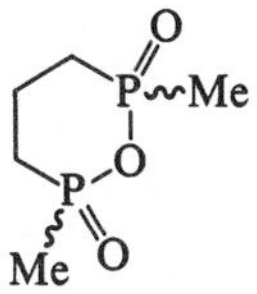

$C_5H_{12}O_3P_2$ M 182.096
Glassy solid. Mp 122-130°. $Bp_{0.3}$ 163°.

Ger. Pat., 2 811 628, (*1979*); *CA*, **92**, 58976 (*synth, pmr*)

4,5-Dimethyl-2-oxa-3-phenyl-1-phosphabicyclo[2.2.1]hept-5-ene, 9CI D-00781

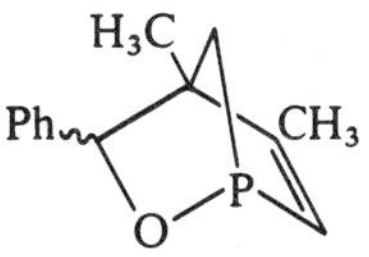

$C_{13}H_{15}OP$ M 218.235

1-Sulfide:
$C_{13}H_{15}OPS$ M 250.295
Solid. Mp 142°.

Lampin, J.-P. *et al*, *Bull. Soc. Chim. Fr.*, 1972, 3494 (*synth*)

(1,1-Dimethyl-3-oxobutyl)ethylphosphinic acid, 9CI D-00782

[19270-90-7]

$H_3CCOCH_2C(CH_3)_2PEt(O)OH$

$C_8H_{17}O_3P$ M 192.194
Cryst. (Et_2O). Mp 112-113°.

Me ester: [19270-91-8]. *Methyl (1,1-dimethyl-3-oxobutyl)ethylphosphinate.*
$C_9H_{19}O_3P$ M 206.221
Liq. d_4^{20} 1.07. Bp_3 112-113°, $Bp_{0.1}$ 79-80°. n_D^{20} 1.4642.

Et ester: Ethyl (1,1-dimethyl-3-oxobutyl)-ethylphosphinate.
$C_{10}H_{21}O_3P$ M 220.248
Liq. d_4^{20} 1.04. Bp_6 128-130°. n_D^{20} 1.4610.

Isopropyl ester: Isopropyl (1,1-dimethyl-3-oxobutyl)-ethylphosphinate.
$C_{11}H_{23}O_3P$ M 234.275
Solid. Mp 82-84°.

Nurtdinov, S.Kh. *et al*, *Zh. Obshch. Khim.*, 1970, **40**, 36 (*Engl. transl.* p. 33) (*synth, esters*)

(3,3-Dimethyl-2-oxobutyl)phosphonic acid, 9CI D-00783

[68084-38-0]

$(H_3C)_3CCOCH_2P(O)(OH)_2$

$C_6H_{13}O_4P$ M 180.140

Dicyclohexylammonium salt: [68064-39-1]. Solid. Mp 175-178.5°.

Di-Et ester: [814-16-4]. *Diethyl (3,3-dimethyl-2-oxobutyl)phosphonate.*
$C_{10}H_{21}O_4P$ M 236.247
Liq. d_4^{20} 1.06. $Bp_{0.6}$ 103-104°. n_D^{20} 1.4410.

Bis(trimethylsilyl) ester: [68064-28-8]. *Bis(trimethylsilyl) (3,3-dimethyl-2-oxobutyl)phosphonate.*
$C_{12}H_{29}O_4PSi_2$ M 324.503
Liq. $Bp_{0.7}$ 85-86°.

Sturtz, G., *Bull. Soc. Chim. Fr.*, 1964, 2349 (*synth, ir, use*)
Fukuyama, Y. *et al*, *J. Am. Chem. Soc.*, 1977, **99**, 646 (*use*)
Mathey, F. *et al*, *Tetrahedron*, 1978, **34**, 649 (*ester*)
Van der Weerdt, A.J.A. *et al*, *J. Chem. Soc., Perkin Trans. 2*, 1978, 155 (*use*)
Zygmunt, J. *et al*, *Synthesis*, 1978, 609 (*silyl ester*)

(3,3-Dimethyl-2-oxobutyl)-triphenylphosphonium(1+), 9CI D-00784

$Ph_3P^{\oplus}CH_2COC(CH_3)_3$

$C_{24}H_{26}OP^{\oplus}$ M 361.443 (ion)

Bromide: [26487-84-3].
$C_{24}H_{26}BrOP$ M 441.347
Aq. NaOH affords the ylide. Cryst. (EtOAc/$CHCl_3$). Mp 217-219°.

Ylide: see 3,3-Dimethyl-1-triphenylphosphoranylidene-2-butanone, D-00934

Schevchuk, M.I. *et al*, *Zh. Obshch. Khim.*, 1970, **40**, 48 (*Engl. transl.* p. 45) (*synth, use*)
Ingham, C.F. *et al*, *Aust. J. Chem.*, 1975, **28**, 2499 (*synth, use*)
Nesmeyanov, N.A. *et al*, *J. Organomet. Chem.*, 1977, **129**, 41 (*pmr, P nmr, tautom*)

(3,3-Dimethyl-2-oxoheptyl)phosphonic acid, 9CI D-00785

$H_3C(CH_2)_3C(CH_3)_2COCH_2P(O)(OH)_2$

$C_9H_{19}O_4P$ M 222.220

Di-Me ester: [39746-15-1]. *Dimethyl (3,3-dimethyl-2-oxoheptyl)phosphonate.*
$C_{11}H_{21}O_4P$ M 248.258
Wittig-Horner reagent widely used in the synth. of prostaglandins. Liq.

Banerjee, A.K. *et al*, *Prostaglandins*, 1978, **16**, 541 (*synth, use*)
Arndt, R.R. *et al*, *S. Afr. J. Chem.*, 1981, **34**, 121 (*synth, use, ir, ms, pmr*)
Muccino, R.R. *et al*, *Prostaglandins*, 1981, **21**, 615 (*use*)

(3,3-Dimethyl-2-oxo-4-phenylbutyl)-phosphonic acid, 9CI D-00786

$PhCH_2C(CH_3)_2COCH_2P(O)(OH)_2$

$C_{12}H_{17}O_4P$ M 256.238

Di-Me ester: [41640-08-8]. *Dimethyl (3,3-dimethyl-2-oxo-4-phenylbutyl)phosphonate.*
$C_{14}H_{21}O_4P$ M 284.291
Widely employed in prostaglandin syntheses. Cryst. (Et_2O). Mp 48-50°.

Ger. Pat., 2 659 215, (*1976*); *CA*, **88**, 6402 (*synth, use*)
U.S.P., 4 032 576, (*1977*); *CA*, **90**, 22413 (*synth, use*)

(3,3-Dimethyl-2-oxopropyl)-triphenylphosphonium(1+) D-00787

$Ph_3P^{\oplus}CH_2COC(CH_3)_3$

$C_{24}H_{26}OP^{\oplus}$ M 361.443 (ion)

Bromide:
$C_{24}H_{26}BrOP$ M 441.347
Cryst. (EtOAc/$CHCl_3$). Mp 217-219°.

Shevchuk, M.I. *et al*, *Zh. Obshch. Khim.*, 1970, **40**, 48 (*Engl. transl.* p. 45) (*synth, use*)
Nesmeyanov, N.A. *et al*, *J. Organomet. Chem.*, 1977, **129**, 41 (*tautom, pmr, nmr*)

1,3-Dimethyl-2-phenoxy-1,3,2-diazaphospholidine, 9CI D-00788

Phenyl N,N′-*1,2-ethanediyl*-N,N′-*dimethylphosphorodiamidite. Phenyl* N,N′-*ethylene*-N,N′-*dimethylphosphorodiamidite*

[54622-61-6]

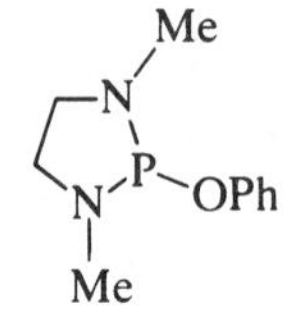

Me, N, P, OPh, N, Me

$C_{10}H_{15}N_2OP$ M 210.215
Solid. Mp 58°. $Bp_{0.03}$ 139-141°.

Revel, M. *et al*, *Bull. Soc. Chim. Fr., Part I*, 1973, 1195 (*pmr, P nmr, sulfide*)
Savignac, P. *et al*, *J. Organomet. Chem.*, 1974, **66**, 63 (*synth, P nmr*)

4,5-Dimethyl-2-phenoxy-1,3,2-dioxaphospholane, 9CI D-00789

2,3-Butylene phenyl phosphite. 4,5-Dimethyl-2-phenoxy-1,3-dioxa-2-phosphacyclopentane

[70870-36-9]

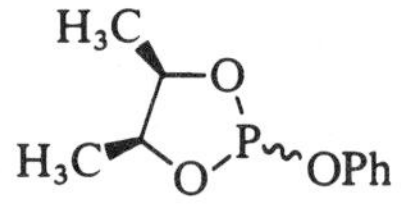

H_3C, O, H_3C, O, P, OPh

$C_{10}H_{13}O_3P$ M 212.185

(4RS,5SR)-form
meso-*form*
No phys. props. reported. Diastereoisomers at P characterised spectroscopically.

2-Oxide:
$C_{10}H_{13}O_4P$ M 228.184
No phys. props. reported. Diastereoisomers at P characterised spectroscopically.

Corriu, R.J.P. *et al*, *Tetrahedron Lett.*, 1983, **24**, 4323 (*oxide, synth, P nmr*)
Nielsen, J. *et al*, *J. Chem. Soc., Perkin Trans. 2*, 1984, 553 (*pmr*)

5,5-Dimethyl-2-phenoxy-1,3,2-dioxaphosphorinane, 9CI D-00790

5,5-Dimethyl-2-phenoxy-1,3,2-dioxaphosphinane. 5,5-Dimethyl-2-phenoxy-1,3-dioxa-2-phosphacyclohexane. Neopentylene phenyl phosphite

[3057-08-7]

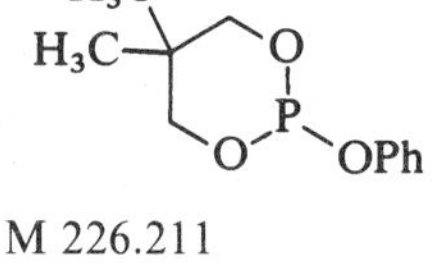

H_3C, H_3C, O, P, O, OPh

$C_{11}H_{15}O_3P$ M 226.211
Liq. Bp_8 130-132°. Easily hydrolysed.

2-Oxide: [884-89-9]. *Neopentylene phenyl phosphate.*
$C_{11}H_{15}O_4P$ M 242.211
Needles (cyclohexane/EtOH or Et_2O). Mp 134-136° (130-131°).

2-Sulfide: [31951-84-5]. O,O-*Neopentylene* O-*phenyl phosphorothioate.*
$C_{11}H_{15}O_3PS$ M 258.271
Cryst. (CCl_4/hexane). Mp 97-99°.

2-Selenide: [95539-24-5]. O,O-*Neopentylene* O-*phenyl phosphoroselenoate.*
$C_{11}H_{15}O_3PSe$ M 305.171

Cryst. (C_6H_6). Mp 119-121°.

Kainosho, M., *J. Phys. Chem.*, 1969, **73**, 3516 (*pmr*)
Hall, L.D. *et al*, *Can. J. Chem.*, 1972, **50**, 2092 (*oxide, pmr, P nmr*)
Majoral, J.-P. *et al*, *Bull. Soc. Chim. Fr.*, 1972, 606 (*oxide, pmr, ir*)
Grand, A. *et al*, *Acta Crystallogr., Sect. B*, 1975, **31**, 2502 (*sulfide, cryst struct*)
Murai, A. *et al*, *Org. Mass Spectrom.*, 1976, **11**, 175 (*oxide, ms*)
Penney, C.L., *Can. J. Chem.*, 1978, **56**, 2396 (*oxide, props, pmr*)
Edmundson, R.S., *Phosphosus Sulfur*, 1981, **9**, 307 (*sulfide, ms*)
Edmundson, R.S., *Org. Mass Spectrom.*, 1983, **18**, 150 (*oxide, pe*)
Arshinova, R.P. *et al*, *Izv. Akad. Nauk SSSR, Ser. Khim.*, 1984, 2512 (*Engl. transl.* p. 2301) (*derivs, uv, pmr, raman*)
El Khatib, F. *et al*, *Phosphorus Sulfur*, 1984, **20**, 55 (*ozonide*)
Rueger, C. *et al*, *J. Prakt. Chem.*, 1984, **326**, 622 (*synth*)
Vlassa, M. *et al*, *J. Prakt. Chem.*, 1984, **326**, 1011 (*sulfide, ir, pmr, P nmr*)
Klochkov, V.V. *et al*, *Izv. Akad. Nauk SSSR, Ser. Khim.*, 1985, 316 (*Engl. transl.* p. 238) (*pmr, conformn*)

3,4-Dimethyl-2-phenoxy-5-phenyl-1,3,2-oxazaphospholidine, 9CI, 8CI — D-00791

3,4-Dimethyl-2-phenoxy-5-phenyl-1,3,2-oxazaphospholane

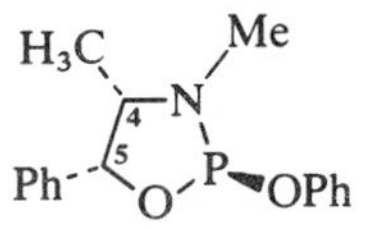

(*2R,4S,5R*)*-form*

$C_{16}H_{18}NO_2P$ M 287.297

(*2R,4S,5R*)-form

2-Oxide:
$C_{16}H_{18}NO_3P$ M 303.297
Cryst. (C_6H_6). Mp 93-94°. $[\alpha]_D^{20}$ −133°, $[\alpha]_D^{20}$ −102° ($CHCl_3$).

(*2S,4S,5R*)-form

2-Oxide: $[\alpha]_D$ −34° ($CHCl_3$).

(*2S,4R,5S*)-form

2-Oxide: Cryst. (C_6H_6). Mp 94-94°. $[\alpha]_D^{20}$ +131.6°.

Devillers, D. *et al*, *C.R. Hebd. Seances Acad. Sci., Ser. C*, 1968, 849 (*synth, P nmr*)
Devillers, J. *et al*, *Bull. Soc. Chim. Fr.*, 1970, 182, 4341 (*pmr, config*)
Devillers, J. *et al*, *Org. Magn. Reson.*, 1974, **6**, 211 (*conformn*)
Cooper, D.B. *et al*, *J. Chem. Soc., Perkin Trans. 1*, 1977, 1969.

5,5-Dimethyl-2-phenylamino-1,3,2-dioxaphosphorinane — D-00792

*5,5-Dimethyl-*N*-phenyl-1,3,2-dioxaphosphorinan-2-amine*, 9CI. *2-Anilino-5,5-dimethyl-1,3,2-dioxaphosphorinane. 2-Anilino-5,5-dimethyl-1,3-dioxa-2-phosphacyclohexane. Neopentylene phenylphosphoramidite*

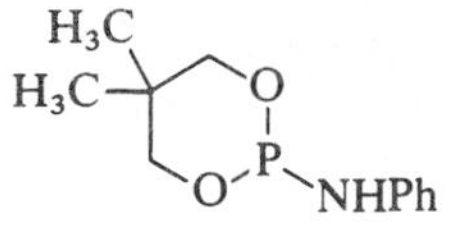

$C_{11}H_{16}NO_2P$ M 225.227
Solid. Mp 65-67°.

2-Oxide: [57237-68-0]. *Neopentylene phenylphosphoramidate.*
$C_{11}H_{16}NO_3P$ M 241.226
Solid. Mp 167°.

2-Sulfide: [57237-69-1]. *Neopentylene phenylphosphoramidothioate.*
$C_{11}H_{16}NO_2PS$ M 257.287
Solid. Mp 175-176°.

2-Selenide: [57237-70-4]. *Neopentylene phenylphosphoramidoselenoate.*
$C_{11}H_{16}NO_2PSe$ M 304.187
Cryst. (EtOH). Mp 175-176°.

Cameron, T.C. *et al*, *Acta Crystallogr., Sect. B*, 1976, **32**, 492 (*oxide, cryst struct*)
Stec, W.J. *et al*, *J. Org. Chem.*, 1976, **41**, 227 (*synth, derivs, P nmr*)
Lesiak, K. *et al*, *Pol. J. Chem.*, 1979, **53**, 2041 (*selenide*)
Edmundson, R.S., *Phosphorus Sulfur*, 1981, **9**, 307 (*oxide, ms*)
Bartczak, T.-J. *et al*, *Acta Crystallogr., Sect. C*, 1983, **39**, 731 (*selenide, cryst struct*)
Van Nuffel, P. *et al*, *J. Mol. Struct.*, 1984, **125**, 1 (*oxide*)

1,3-Dimethyl-2-phenyl-1,3,2-diazaphospholidine, 9CI — D-00793

N,N′*-1,2-Ethanediyl-*N,N′*-dimethyl-*P*-phenylphosphonous diamide*

[22429-12-5]

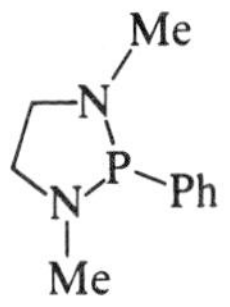

$C_{10}H_{15}N_2P$ M 194.216
Liq. $Bp_{0.7}$ 95°, $Bp_{0.05}$ 78°.

2-Oxide: [6226-05-7]. N,N′*-1,2-Ethanediyl-*N,N′*-dimethyl-*P*-phenylphosphonic diamide.*
$C_{10}H_{15}N_2P$ M 194.216
Liq. $Bp_{0.7}$ 130°.

Ulrich, H. *et al*, *J. Organomet. Chem.*, 1967, **32**, 1360 (*oxide, synth, pmr*)
Zuckermann, J.J. *et al*, *Inorg. Chem.*, 1971, **10**, 1028 (*synth, pmr*)
Albrand, J.P. *et al*, *Tetrahedron*, 1972, **28**, 819 (*synth*)
Gray, G.A. *et al*, *J. Am. Chem. Soc.*, 1977, **99**, 3243 (*N nmr*)
Gray, G.A. *et al*, *Org. Magn. Reson.*, 1980, **14**, 8 (*cmr*)

4,5-Dimethyl-2-phenyl-1,3,2-dioxaphospholane, 9CI — D-00794

2,3-Butylene phenylphosphonite. 4,5-Dimethyl-2-phenyl-1,3-dioxa-2-phosphacyclopentane

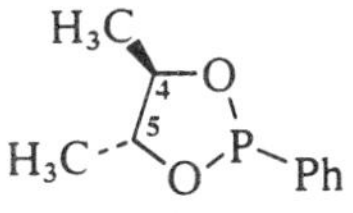

$C_{10}H_{13}O_2P$ M 196.185

(*4R,5R*)-form

Liq. Bp_1 75°. $[\alpha]_D^{20}$ +35.3° (c, 10 in toluene).

2-Oxide: 2,3-Butylene phenylphosphonate.
$C_{10}H_{13}O_3P$ M 212.185
Liq. Bp_1 140°.

Richter, W.J., *Phosphorus Sulfur*, 1981, **10**, 395 (*synth, pmr, P nmr, ms*)

5,5-Dimethyl-2-phenyl-1,3,2-dioxaphosphorinane, 9CI — D-00795

5,5-Dimethyl-2-phenyl-1,3,2-dioxaphosphinane. 5,5-Dimethyl-2-phenyl-1,3-dioxa-2-phosphacyclohexane. Neopentylene phenylphosphonite

[7526-31-0]

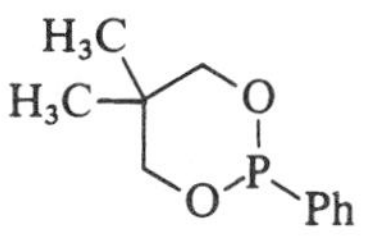

$C_{11}H_{15}O_2P$ M 210.212

Reactive liq. or solid. Bp_6 120°. Deliquescent. Sensitive to air. Sublimes.

B,MeI: 2,5,5-Trimethyl-2-phenyl-1,3,2-dioxaphosphorinanium iodide.

$C_{12}H_{18}IO_2P$ M 352.151

Cryst. ($CHCl_3$/EtOAc). Unstable in air.

2-Oxide: [882-69-9]. *Neopentylene phenylphosphonate.*

$C_{11}H_{15}O_3P$ M 226.211

Solid. Mp 109-111° (103-105°). $Bp_{0.1}$ 150-160°. With CF_3SO_3Me forms a methoxyphosphonium triflate Mp 89-91°.

2-Sulfide: [34284-47-4]. O,O-*Neopentylene phenylphosphonothioate.*

$C_{11}H_{15}O_2PS$ M 242.272

Cryst. (cyclohexane). Mp 55.5-56°.

McConnell, R.L *et al, J. Org. Chem.*, 1959, **24**, 630 (*oxide*)
Tavs, P., *Chem. Ber.*, 1970, **103**, 2428 (*oxide*)
Verkade, J.G. *et al, J. Am. Chem. Soc.*, 1970, **92**, 7125 (*synth, pmr, P nmr, deriv*)
White, D.W., *Phosphorus*, 1971, **1**, 33 (*synth, props*)
White, D.W. *et al, J. Magn. Reson.*, 1971, **4**, 123 (*pmr*)
Hall, L.D. *et al, Can. J. Chem.*, 1972, **50**, 2092 (*oxide, pmr, P nmr*)
Majoral, J.-P. *et al, Bull. Soc. Chim. Fr.*, 1972, 606 (*oxide, ir, pmr*)
Francis, G.W. *et al, Acta Chem. Scand., Ser. B*, 1976, **30**, 31 (*oxide, ms*)
Arshinova, R.P. *et al, Zh. Obshch. Khim.*, 1979, **49**, 2661 (*Engl. transl.* p. 2359) (*struct, derivs*)
Singh, G. *et al, J. Org. Chem.*, 1979, **44**, 1060; 1984, **49**, 5132 (*props, pmr*)
Edmundson, R.S., *Phosphorus Sulfur*, 1981, **9**, 307 (*sulfide, synth, ms*)
Arbuzov, B.A. *et al, Zh. Obshch. Khim.*, 1982, **52**, 2428 (*Engl. transl.* p. 2148) (*uv*)

1,4-Dimethyl-2-phenyl-7-oxa-2-phosphabicyclo[2.2.1]heptane, 10CI D-00796

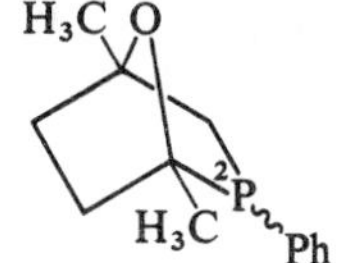

$C_{13}H_{17}OP$ M 220.250

2-Oxide: [79184-79-5].

$C_{13}H_{17}O_2P$ M 236.250

Oil. Stereochemistry unknown.

Kashman, Y. *et al, Tetrahedron Lett.*, 1981, **22**, 2695 (*synth, ms, ir, pmr, cmr*)

2,6-Dimethyl-4-phenyl-1,4-oxaphosphorinane, 9CI D-00797

2,6-Dimethyl-4-phenyl-1,4-oxaphosphinane. 2,6-Dimethyl-1-oxa-4-phenyl-1-phosphacyclohexane

cis-form

$C_{12}H_{17}OP$ M 208.239

cis-form [71761-33-6]

B,PhBr: [66224-11-1]. *2,6-Dimethyl-4,4-diphenyl-1,4-oxaphosphorinanium bromide. 2,6-Dimethyl-4,4-diphenyl-1,4-oxaphosphinanium bromide. 2,6-Dimethyl-4,4-diphenyl-1-oxa-4-phosphoniacyclohexane bromide.* Cryst. (EtOH/Et_2O). Mp 243-245°.

Samaan, S. *et al, Chem. Ber.*, 1978, **111**, 579 (*synth, cmr, nmr*)

2,8-Dimethyl-10-phenyl-10*H*-phenothiaphosphine, 8CI D-00798

$C_{20}H_{17}PS$ M 320.388

Cryst. (AcOH). Mp 147°.

5,5,10-Trioxide:

$C_{20}H_{17}O_3PS$ M 368.386

Cryst. (AcOH aq.). Mp 263°.

Granoth, I. *et al, Tetrahedron*, 1969, **25**, 3919 (*synth, deriv, ir, uv*)

2,8-Dimethyl-10-phenyl-10*H*-phenoxaphosphine, 8CI D-00799

[21981-56-6]

$C_{20}H_{17}OP$ M 304.327

Cryst. (EtOH). Mp 80°.

*B,MeI: 2,8,10-Trimethyl-10-phenyl-10*H*-phenoxaphosphinium iodide.*

$C_{21}H_{20}IOP$ M 446.267

Cryst. (Me_2CO/Et_2O). Mp 240°.

10-Oxide: [21990-63-6].

$C_{20}H_{17}O_2P$ M 320.327

Cryst. (dibutyl ether). Mp 191-192°.

Granoth, I. *et al, Isr. J. Chem.*, 1968, **6**, 651; 1970, **8**, 621 (*synth, derivs, uv, ir, ms*)
Weringa, W.D. *et al, Org. Mass Spectrom.*, 1973, **7**, 459 (*ms*)

5,6-Dimethyl-7-phenyl-7-phosphabicyclo[2.2.1]hept-5-ene-2,3-dicarboxylic acid, 10CI D-00800

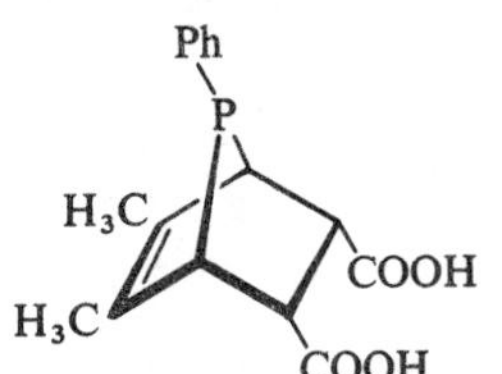

Endo,endo,syn-form

$C_{17}H_{17}O_4P$ M 316.293

endo,endo,syn-form

Di-Me ester: [78177-26-7].

$C_{19}H_{21}O_4P$ M 344.346

Characterised spectroscopically.

7-Sulfide, anhydride:

$C_{17}H_{15}O_3PS$ M 330.337

Cryst. (Et_2O/Me_2CO). Mp 157°.

7-Sulfide, Di-Me ester:

$C_{19}H_{21}O_4PS$ M 376.406

Solid. Mp 163°.

endo,endo,anti-form

Di-Me ester: [78177-25-0]. Characterised spectroscopically.

Kashman, Y. *et al, Tetrahedron*, 1976, **32**, 2427 (*synth, uv, ir, pmr*)

Mathey, F. *et al*, *Tetrahedron Lett.*, 1981, **22**, 319 (*esters, pmr, P nmr*)

4,4-Dimethyl-1-phenyl-1-phospha-4-silacyclohexa-2,5-diene, 9CI **D-00801**

[65736-00-7]

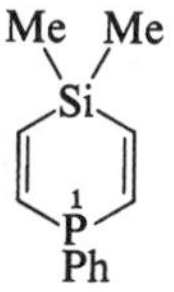

$C_{12}H_{15}PSi$ M 218.310

Liq. $Bp_{0.0004}$ 110-115°.

Märkl, G. *et al*, *Synthesis*, 1977, 842 (*synth, ms, pmr*)

2,2-Dimethyl-1-phenyl-1-phospha-2-silacyclopentane, 9CI **D-00802**

[54770-04-6]

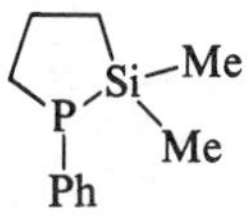

$C_{11}H_{17}PSi$ M 208.315

Liq. $Bp_{0.3}$ 79-80°.

Couret, C. *et al*, *C.R. Hebd. Seances Acad. Sci., Ser. C*, 1974, **279**, 225 (*synth, P nmr*)
Couret, C. *et al*, *Angew. Chem.*, 1976, **88**, 445 (*pmr*)
Couret, C. *et al*, *J. Organomet. Chem.*, 1977, **141**, 35 (*props*)

Dimethylphenylphosphine, 9CI **D-00803**

[672-66-2]

Me_2PPh

$C_8H_{11}P$ M 138.149

Ligand for Ti, V, and metals of groups IB, VIB, VIIB, and VIII. Liq. Bp 192°, Bp_{14} 84°. n_D^{20} 1.5620. Air-sensitive. Forms red adducts with polynitroaryls. Subl. at 100°/0.003 mm.

B,MeBr: Trimethylphenylphosphonium bromide.
$C_9H_{14}BrP$ M 233.087
Cryst. ($EtOH/Et_2O$). Mp 284-286°.

Oxide: [10311-08-7].
$C_8H_{11}OP$ M 154.148
Forms Gd, Fe, Mo and Ni complexes. Cryst. (toluene). Mp 115°.

Sulfide: [1707-00-2].
$C_8H_{11}PS$ M 170.209
Forms Cu, Ir, and Rh complexes. Cryst. (EtOH or C_6H_6/pet. ether). Mp 44°.

Aksnes, G. *et al*, *Acta Chem. Scand.*, 1963, **17**, 1616 (*synth*)
Peterson, D.J. *et al*, *J. Org. Chem.*, 1965, **30**, 1939 (*synth*)
Deeming, A.J. *et al*, *J. Chem. Soc. (A)*, 1969, 597 (*pmr, complexes*)
Mann, B.E., *J. Chem. Soc., Perkin Trans. 2*, 1972, 30 (*cmr, pmr*)
Allen, E.A. *et al*, *Spectrochim. Acta, Part A*, 1974, **30**, 1219 (*ir, raman*)
Henrick, K. *et al*, *Aust. J. Chem.*, 1975, **28**, 1473 (*ms*)
Verstuyft, A.W. *et al*, *Inorg. Chem.*, 1975, **14**, 1495 (*derivs, pmr*)
Inorg. Synth., 1977, **17**, 185 (*synth, ir*)
Ratovskii, G.V. *et al*, *Zh. Obshch. Khim.*, 1978, **48**, 1520 (*Engl. transl.* p. 1394) (*uv*)
Goncharova, L.V. *et al*, *Zh. Obshch. Khim.*, 1981, **51**, 1450 (*Engl. transl.* p. 1230) (*sulfide, ir*)
Mosbo, J.A. *et al*, *Phosphorus Sulfur*, 1981, **11**, 11 (*struct*)
Panov, A.M. *et al*, *Zh. Obshch. Khim.*, 1985, **55**, 2243; *J. Gen. Chem. USSR (Engl. Transl.)*, 1991 (*uv, cmr, ir, raman*)
Ratovskii, G.V. *et al*, *Zh. Obshch. Khim.*, 1985, **55**, 571; *J. Gen. Chem. USSR (Engl. Transl.)*, 505 (*pe*)

(2,4-Dimethylphenyl)phosphinic acid, 8CI **D-00804**

[81592-48-5]

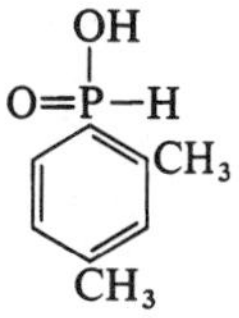

$C_8H_{11}O_2P$ M 170.147

Cryst. Mp 100°.

Michaelis, A., *Justus Liebigs Ann. Chem.*, 1896, **293**, 261.

(2,5-Dimethylphenyl)phosphinic acid, 9CI **D-00805**

p-*Xylylphosphinic acid*

[84372-54-3]

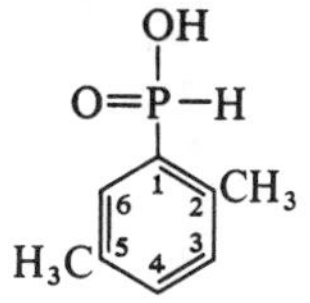

$C_8H_{11}O_2P$ M 170.147

Tautomeric with (2,5-dimethylphenyl)phosphonous acid, but free acid probably exists in phosphinic acid form. Esters possess herbicidal activity against dicotyledoneous weeds. Cryst. only with great difficulty.

Me ester: [16391-15-4]. *Methyl (2,5-dimethylphenyl)-phosphinate.*
$C_9H_{13}O_2P$ M 184.174
Characterised spectroscopically.

Et ester: [16391-16-5]. *Ethyl (2,5-dimethylphenyl)-phosphinate.*
$C_{10}H_{15}O_2P$ M 198.201
Characterised spectroscopically.

Isopropyl ester: Isopropyl (2,5-dimethylphenyl)-phosphinate.
$C_{11}H_{17}O_2P$ M 212.228
Characterised spectroscopically.

Weller, J., *Ber.*, 1888, **21**, 1492 (*synth*)
Wolf, R. *et al*, *Spectrochim. Acta, Part A*, 1967, **23**, 1641 (*esters, ir*)

(3,4-Dimethylphenyl)phosphinic acid **D-00806**

3,4-Xylylphosphinic acid, 8CI

$C_8H_{11}O_2P$ M 170.147

Tautomeric with (3,4-dimethylphenyl)phosphonous acid but free acid exists in phosphinic acid form. Cryst. Mp 43°.

Me ester: [16391-19-8]. *Methyl (3,4-dimethylphenyl)-phosphinate.*
$C_9H_{13}O_2P$ M 184.174
No phys. props. reported.

Et ester: [16391-20-1].
$C_{10}H_{15}O_2P$ M 198.201
No phys. props. reported.

Isopropyl ester: [16391-21-2].
$C_{11}H_{17}O_2P$ M 212.228
No phys. props. reported.

Michaelis, A., *Justus Liebigs Ann. Chem.*, 1896, **293**, 261 (*synth, derivs*)

Wolf, R. *et al*, *Spectrochim. Acta, Part A*, 1967, **23**, 1641 (*esters, ir*)

3,4-Dimethyl-1-phenyl-1*H*-phosphole, 9CI — D-00807

[30540-36-4]

$C_{12}H_{13}P$ M 188.208

Ligand for metals of Groups VIB, VIIB, and VIII. Liq. $Bp_{0.2}$ 71°. n_D^{20} 1.6074. Undergoes cycloaddition reactions. Isomerises at 150°, dimerises at 170°. Oxidn. gives dimeric oxide.

*B,MeI: 1,3,4-Trimethyl-1-phenyl-1*H*-phospholium iodide.*
$C_{13}H_{16}IP$ M 330.148
Solid. Mp 169°. Monomeric.

*B,PhCH₂Br: 1-Benzyl-3,4-dimethyl-1-phenyl-1*H*-phospholium bromide.*
$C_{19}H_{20}BrP$ M 359.245
Solid. Mp 165°. Monomeric.

1-Oxide:
$C_{12}H_{13}OP$ M 204.208
Monomeric oxide unknown.

1-Sulfide:
$C_{12}H_{13}PS$ M 220.268
Solid. Mp 113°. Monomeric.

1-Selenide: [38066-23-8].
$C_{12}H_{13}PSe$ M 267.168
Cryst. (hexane). Mp 103°.

Mathey, F. *et al*, *Bull. Soc. Chim. Fr.*, 1970, 4433 (*synth, pmr, P nmr*)
Mathey, F. *et al*, *Org. Magn. Reson.*, 1972, **4**, 171 (*synth, P nmr, props*)
Mathey, F., *Tetrahedron*, 1972, **28**, 4171 (*selenide, synth, pmr, P nmr*)
Kashman, Y. *et al*, *Tetrahedron*, 1973, **29**, 191; 1976, **32**, 2427 (*sulfide, pmr, props*)
Grey, G.A. *et al*, *Org. Magn. Reson.*, 1980, **14**, 14 (*cmr*)
MacDongall, J.J. *et al*, *Inorg. Chem.*, 1980, **19**, 709 (*pmr, cmr, P nmr, complexes*)
Breque, A. *et al*, *Synthesis*, 1981, 983 (*synth, P nmr*)
Mercier, F. *et al*, *J. Organomet. Chem.*, 1984, **263**, 55 (*props*)

N,N-Dimethyl-*P*-phenylphosphonamidic acid, 9CI — D-00808

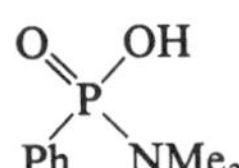

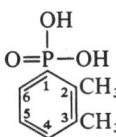

$C_8H_{12}NO_2P$ M 185.162

Tetramethylammonium salt: Cryst. (Me_2CO/Et_2O). Mp 247° dec.

Me ester: [55215-28-6]. *Methyl* N,N*-dimethyl-*P*-phenylphosphonamidate.*
$C_9H_{14}NO_2P$ M 199.189
Liq. $Bp_{0.2}$ 88-92°.

Chloride: [6840-02-4].
$C_8H_{11}ClNOP$ M 203.608
Liq. $Bp_{0.01}$ 90°.

Felcht, U. *et al*, *Justus Liebigs Ann. Chem.*, 1977, 1309 (*synth, ir, pmr*)
Rahil, J. *et al*, *J. Am. Chem. Soc.*, 1981, **103**, 1723 (*ester, chloride, synth, pmr*)

N,N-Dimethyl-*P*-phenylphosphonamidic azide, 9CI — D-00809

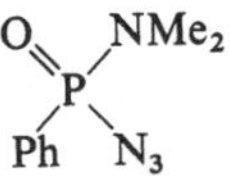

$C_8H_{11}N_4OP$ M 210.175

Liq. $Bp_{0.001}$ 86-88°. n_D^{20} 1.5473. Rel. insensitive to shock.

Baldwin, R.A., *J. Org. Chem.*, 1965, **30**, 3866 (*synth, ir*)

Dimethyl phenylphosphonate — D-00810

Dimethyl benzenephosphonate

[2240-41-7]

$(MeO)_2P(O)Ph$

$C_8H_{11}O_3P$ M 186.147

Liq. d_4^{20} 1.19. Bp_{13} 143.5-144-5°. n_D^{20} 1.5093.

Tavs, P. *et al*, *Tetrahedron*, 1967, **23**, 4677 (*synth*)
Kharrasova, F.M. *et al*, *Zh. Obshch. Khim.*, 1968, **38**, 1262 (*Engl. transl.* p. 1215) (*synth*)
Bunnett, J.F. *et al*, *J. Org. Chem.*, 1974, **39**, 3612 (*synth*)
Harger, M.J.P. *et al*, *J. Chem. Soc., Perkin Trans. 2*, 1980, 1505 (*pmr*)
Hollingshaus, J.G. *et al*, *J. Toxicol. Environ. Health*, 1981, **8**, 619 (*tox*)
Ohms, G. *et al*, *Z. Anorg. Allg. Chem.*, 1982, **486**, 22 (*cmr, P nmr*)

(2,3-Dimethylphenyl)phosphonic acid, 8CI — D-00811

$C_8H_{11}O_3P$ M 186.147

Di-Et ester: [72596-34-0]. *Diethyl (2,3-dimethylphenyl)phosphonate.*
$C_{12}H_{19}O_3P$ M 242.254
Obt. only as mixt. with diethyl ester of (3,4-Dimethylphenyl)phosphonic acid, D-00815 .

Ohmari, H. *et al*, *J. Chem. Soc., Perkin Trans. 1*, 1979, 2023.

(2,4-Dimethylphenyl)phosphonic acid, 9CI — D-00812

$C_8H_{11}O_3P$ M 186.147

Cryst. (H_2O). Mp 194°.

Di-Et ester: [58983-20-3]. *Diethyl (2,4-dimethylphenyl)phosphonate.*
$C_{12}H_{19}O_3P$ M 242.254
Liq. $Bp_{0.0005}$ 89°.

Rudinskas, A.J. *et al*, *J. Org. Chem.*, 1976, **41**, 2411 (*synth, ir, pmr*)

(2,5-Dimethylphenyl)phosphonic acid, 9CI — D-00813

[58983-19-0]

$C_8H_{11}O_3P$ M 186.147

Cryst. (H_2O). Mp 179-180°.

Di-Et ester: [58983-18-9]. *Diethyl (2,5-dimethylphenyl)phosphonate.*
$C_{12}H_{19}O_3P$ M 242.254

Liq. $Bp_{0.025}$ 92-94°.
Dichloride:
$C_8H_9Cl_2OP$ M 223.038
Liq. d^{18} 1.31. Bp 280°.

Weller, J., *Ber.*, 1888, **21**, 1492 (*synth, dichloride*)
Rudinskas, A.J. *et al*, *J. Org. Chem.*, 1976, **41**, 2411 (*synth, ester, ir, pmr, ms*)

(2,6-Dimethylphenyl)phosphonic acid, 9CI **D-00814**

$C_8H_{11}O_3P$ M 186.147
Di-Me ester: Dimethyl (2,6-dimethylphenyl)-phosphonate.
$C_9H_{13}O_3P$ M 200.174
Liq. $Bp_{0.3}$ 92-93°.

Obrycki, R. *et al*, *J. Org. Chem.*, 1968, **33**, 632.

(3,4-Dimethylphenyl)phosphonic acid, 9CI **D-00815**

$C_8H_{11}O_3P$ M 186.147
pK_{a1} 2.03, pK_{a2} 7.69 (H_2O).
Di-Et ester: [72596-30-6]. *Diethyl(3,4-dimethylphenyl)-phosphonate.*
$C_{12}H_{19}O_3P$ M 242.254
Obt. as mixt. with Di-Et ester of (2,3-Dimethylphenyl)phosphonic acid, D-00811 .

Nuallain, C.O., *J. Inorg. Nucl. Chem.*, 1974, **36**, 339 (*props*)
Ohmori, H. *et al*, *J. Chem. Soc., Perkin Trans. 1*, 1979, 2023 (*ester*)

(3,5-Dimethylphenyl)phosphonic acid, 8CI **D-00816**

$C_8H_{11}O_3P$ M 186.147
Di-Et ester: [79238-62-3]. *Diethyl (3,5-dimethylphenyl)phosphonate.*
$C_{12}H_{19}O_3P$ M 242.254
Characterised spectroscopically.

Cooper, D. *et al*, *J. Chem. Soc., Perkin Trans. 1*, 1981, 2127 (*synth, pmr*)

Dimethyl phenylphosphoramidate **D-00817**

Dimethyl phosphoric anilide. N-(Dimethoxyphosphinyl)aniline
[58046-12-1]

$(MeO)_2P(O)NHPh$

$C_8H_{12}NO_3P$ M 201.161
Cryst. (H_2O, hexane, or C_6H_6/hexane). Mp 88-89°.

McOmbie, H. *et al*, *J. Chem. Soc.*, 1945, 921 (*synth*)
Buchanan, G.W. *et al*, *Org. Magn. Reson.*, 1980, **14**, 517 (*cmr*)
Buchanan, G.W. *et al*, *Can. J. Chem.*, 1980, **58**, 2442 (*N nmr*)
Du Plessis, M.P. *et al*, *Acta Crystallogr., Sect. B*, 1982, **38**, 1504 (*cryst struct*)
Modro, T.A. *et al*, *J. Org. Chem.*, 1982, **47**, 3208 (*props*)
Moerat, A. *et al*, *Phosphorus Sulfur*, 1983, **14**, 179 (*synth, props*)

(2,4-Dimethylphenyl)phosphoramidic acid, 9CI, 8CI **D-00818**

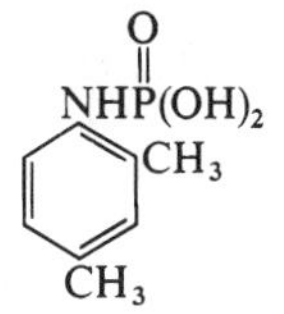

O
NHP(OH)₂ with CH_3 ring substituents

$C_8H_{12}NO_3P$ M 201.161
Di-Et ester: Diethyl (2,4-dimethylphenyl)-phosphoramidate.
$C_{12}H_{20}NO_3P$ M 257.269
Solid. Mp 96°.
Di-Ph ester: Diphenyl (2,4-dimethylphenyl)-phosphoramidate.
$C_{20}H_{20}NO_3P$ M 353.357
Solid. Mp 115°.
Dichloride:
$C_8H_{10}Cl_2NOP$ M 238.053
Needles (pet. ether). Mp 79°.

Michaelis, A., *Justus Liebigs Ann. Chem.*, 1903, **326**, 129.

2,6-Dimethyl-4-phenylphosphorin, 8CI **D-00819**

2,6-Dimethyl-4-phenylphosphabenzene
[22208-82-8]

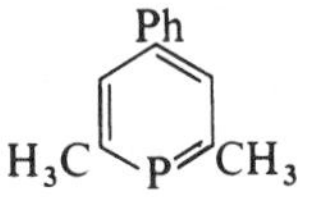

Ph
H_3C P CH_3

$C_{13}H_{13}P$ M 200.219
Orthorhombic cryst. Mp 62-63°.

Bart, J.C.J. *et al*, *Angew. Chem., Int. Ed. Engl.*, 1968, **7**, 811.
Bart, J.C.J. *et al*, *J. Chem. Soc. (A)*, 1970, 567 (*cryst struct*)

2,5-Dimethyl-1-phenylphosphorinan-4-ol, 10CI **D-00820**

4-Hydroxy-2,5-dimethyl-1-phenylphosphacyclohexane
[78721-85-4]

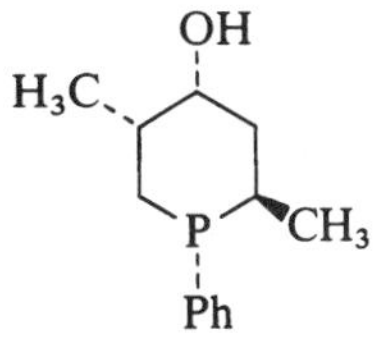

OH
H_3C
P CH_3 *(1RS,2RS,4RS,5RS)-form*
Ph

$C_{13}H_{19}OP$ M 222.266

(1RS,2RS,4RS,5RS)-form
(1α,2β,4α,5α)-form
1-Sulfide: [67424-74-2].
$C_{13}H_{19}OPS$ M 254.326
Solid. Mp 118-119°.
1-Sulfide, 4-O-Ac: [67424-75-3].
$C_{15}H_{21}O_2PS$ M 296.363
Solid. Mp 101-102°.

(1RS,2SR,4RS,5SR)-form
(1α,3α,4α,5β)-form
1-Sulfide: [67424-72-0]. Solid. Mp 172-173°.
1-Sulfide, 4-O-Ac: [67424-73-1]. Solid. Mp 131-132°.

(1RS,2SR,4SR,5SR)-form
(1α,2α,4β,5β)-form
1-Sulfide: [67384-54-7]. Solid. Mp 128-129°.

1-Sulfide, 4-O-Ac: [67384-55-8]. Solid. Mp 134-135°.

(1RS,2RS,4SR,5RS)-form

(1α,2β,4β,5α)-form

1-Sulfide: [67424-76-4]. Solid. Mp 120-121°.

1-Sulfide, 4-O-Ac: [67424-77-5]. Solid. Mp 162-163°.

Bosyakov, Y. *et al, Zh. Obshch. Khim.*, 1978, **48**, 1299 (*Engl. transl.* p. 1189)

2,5-Dimethyl-1-phenylphosphorinan-4-one, 9CI — D-00821

[54877-14-4]

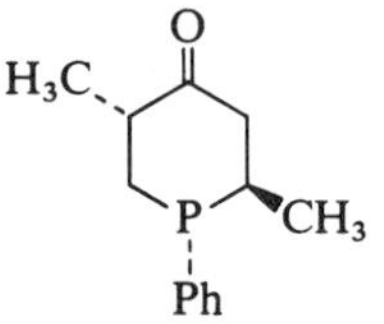

(1*RS*,2*RS*,5*RS*)-*form*

$C_{13}H_{17}OP$ M 220.250

Liq. d_4^{20} 1.08. $Bp_{0.02}$ 108-110°. n_D^{20} 1.5739. Stereochem. composition unknown.

B,MeI: 1,2,5-Trimethyl-4-oxo-1-phenylphosphorinanium iodide.

$C_{14}H_{20}IOP$ M 362.190

Cryst. (EtOH). Mp 185-187°.

Semicarbazone: Cryst. (Me_2CO). Mp 46-48°.

(1RS,2RS,5RS)-form

(1α,2β,5α)-form

1-Oxide: [67424-79-7].

$C_{13}H_{17}O_2P$ M 236.250

Cryst. (C_6H_6). Mp 128-129°.

1-Sulfide: [65831-36-9].

$C_{13}H_{17}OPS$ M 252.310

Cryst. (ligroin). Mp 108-109°.

1-Selenide: [67424-67-3].

$C_{13}H_{17}OPSe$ M 299.210

Solid. Mp 121-122°.

(1SR,2RS,5RS)-form

(1α,2α,5β)-form

1-Oxide: [65831-35-8]. No props. reported.

1-Sulfide: [65831-34-7]. Cryst. (ligroin). Mp 119-120°.

1-Selenide: [67424-66-2]. Solid. Mp 119-120°.

(1RS,2SR,5RS)-form

(1α,2α,5α)-form

1-Selenide: [67424-69-5]. Solid. Mp 137-138°.

Azerbaev, I., *Zh. Obshch. Khim.*, 1975, **45**, 1730 (*Engl. transl.* p. 1696) (*synth, derivs*)

Bosyakov, Yu.G. *et al, Zh. Obshch. Khim.*, 1978, **48**, 1293 (*Engl. transl.* p. 1184) (*synth, derivs*)

Krasnomolova, L.P. *et al, Zh. Fiz. Khim.*, 1980, **54**, 1447; 1981, **55**, 1522 (*Engl. transl.* pp. 827, 852) (*cmr, pmr, conformn*)

3,5-Dimethylphenyl phosphorodichloridate, 9CI — D-00822

3,5-Xylyl phosphorodichloridate. 3,5-Dimethylphenyl dichlorophosphate

[775-08-6]

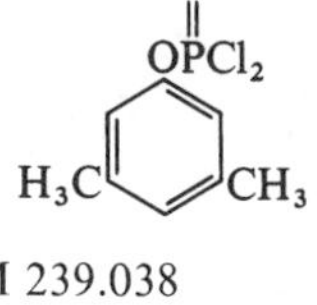

$C_8H_9Cl_2O_2P$ M 239.038

Oil. $Bp_{0.5}$ 82-83°.

Orloff, H.D. *et al, J. Am. Chem. Soc.*, 1958, **80**, 727 (*synth*)

3′,4′-Dimethyl-5′-phenylspiro[1,3,2-benzoxazaphosphole-2(3*H*),2′-[1,3,2]-oxazaphospholidine], 10CI — D-00823

3′,4′-Dimethyl-5′-phenylspiro[1,3,2-benzoxazaphospholine-2,2′-[1,3,2]oxazaphospholidine], 9CI

[30318-19-5]

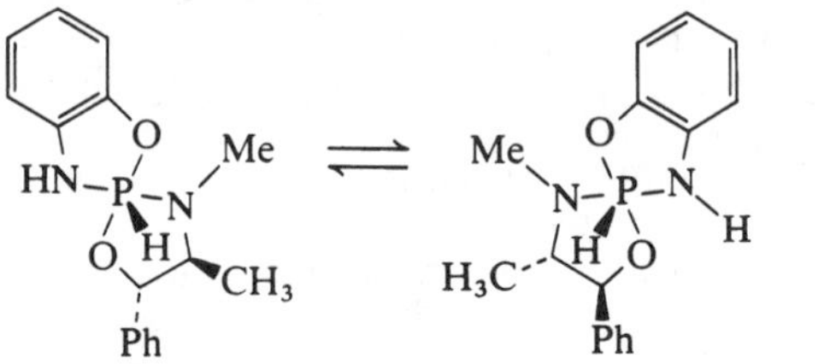

(4′*S*,5′*S*)-(*M*)-*form* (4′*S*,5′*S*)-(*P*)-*form*

$C_{16}H_{19}N_2O_2P$ M 302.312

(4′S,5′S)-(M)-form

Cryst. (C_6H_6). $[\alpha]_{436}^{20}$ −252° (extrap.).

(4′S,5′S)-(P)-form

$[\alpha]_{436}^{20}$ +215° (extrap.).

Klaebe, A. *et al, J. Chem. Soc., Perkin Trans. 2*, 1974, 1668 (*synth, pmr, props*)

Klaebe, A. *et al, Tetrahedron*, 1982, **38**, 2111 (*synth, pmr*)

4,5-Dimethyl-2-phenylthio-1,3,2-dioxaphospholane, 8CI — D-00824

O,O-*2,3-Butylene* S-*phenyl thiophosphite.* O,O-*2,3-Butylene* S-*phenyl phosphorothioite. 4,5-Dimethyl-2-phenylthio-1,3-dioxa-2-phosphacyclopentane*

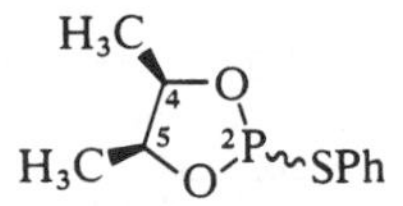

$C_{10}H_{13}O_2PS$ M 228.245

(4RS,5SR)-form

meso-*form*

Liq. $Bp_{0.05}$ 80-84°. n_D^{19} 1.5726. Contains 90% (2α,4β,5β) and 10% (2α,4α,5α)-diastereoisomers.

2-Sulfide: O,O-*2,3-Butylene* S-*phenyl dithiophosphate.* O,O-*2,3-Butylene* S-*phenyl phosphorodithioate.*

$C_{10}H_{13}O_2PS_2$ M 260.305

Liq. $Bp_{0.2}$ 115-120°. n_D^{19} 1.5813. Contains 80% (2α,4β,5β) and 20% (2α,4α,5α)-diastereoisomers.

Mikolajczyk, M. *et al, J. Chem. Soc., Perkin Trans. 1*, 1977, 2213 (*synth, cmr, P nmr*)

1,4-Dimethyl-2-phosphabicyclo[2.2.1]-heptane 2-oxide, 10CI — D-00825

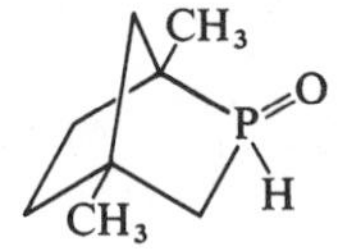

$C_8H_{15}OP$ M 158.180

2-Me: 1,2,4-Trimethyl-2-phosphabicyclo[2.2.1]heptane 2-oxide.

$C_9H_{17}OP$ M 172.206

Liq. Exists in *endo*- and *exo*-forms; struct. assignments not known.

2-Ph: 1,4-Dimethyl-2-phenyl-2-phosphabicyclo[2.2.1]-heptane 2-oxide.
$C_{14}H_{19}OP$ M 234.277
Solid. Mp 93°, 147°. Exists in *endo-* and *exo*-forms; struct. assigments not known.

Kashman, Y. *et al, Tetrahedron Lett.*, 1976, 2819 (*synth, cmr, pmr, ir, ms*)

Dimethyl phosphate, 9CI, 8CI D-00826

Dimethyl hydrogen phosphate. Dimethyl phosphoric acid
[813-78-5]

$(MeO)_2P(O)OH$

$C_2H_7O_4P$ M 126.049
Final prod. of hydrolysis or metab. of many insecticides, and of *O,O*-Dimethyl phosphorodithioate, D-00898 by *Thiobacillus thioflavus*. Fireproofing agent, cross-linking agent in polymerisation. Syrup. d^{20} 1.35. $Bp_{0.0001}$ 78-80°. pK_a 1.25 (H_2O), pK_a 3.01 (80% EtOH aq.), pK_a 4.58 (95% EtOH aq.), pK_a 6.37 (EtOH), pK_a 9.50 ($MeNO_2$). n_D^{20} 1.4080.

Tetramethylammonium salt: Solid. Mp 215°.
Fluoride: see Dimethyl phosphorofluoridate, D-00901
Chloride: see Dimethyl phosphorochloridate, D-00887
Bromide: see Dimethyl phosphorobromidate, D-00886
Cyanide: see Dimethyl phosphorocyanidate, D-00894
Amide: see O,O-Dimethyl phosphoramidate, D-00854
Anilide: see Dimethyl phenylphosphoramidate, D-00817
Azide: see Dimethyl phosphorazidate, D-00878
Hydrazide: see Dimethyl phosphorohydrazidate, D-00904
Anhydride: see Tetramethyl pyrophosphate, T-00229

Chabrier, P. *et al, C. R. Hebd. Seances Acad. Sci.*, 1957, **244**, 2730 (*synth*)
Petrov, K.A. *et al, Zh. Obshch. Khim.*, 1961, **31**, 1709 (*Engl. transl.* p. 1596) (*synth*)
Uhlenhopp, E.L. *et al, J. Org. Chem.*, 1969, **34**, 2237 (*nmr*)
Glonek, T. *et al, J. Phys. Chem.*, 1976, **80**, 639 (*P nmr*)
Gorenstein, D.G. *et al, J. Am. Chem. Soc.*, 1979, **101**, 5869 (*props, struct*)
Lowe, G. *et al, J. Chem. Soc., Chem. Commun.*, 1979, 733 (*O and P nmr*)
Alagona, G. *et al, J. Am. Chem. Soc.*, 1985, **107**, 2229 (*struct*)

Dimethylphosphine, 9CI D-00827

[676-59-5]

Me_2PH

C_2H_7P M 62.051
Liq. with disgusting odour. Bp 25°.
▷Highly toxic. Readily ignites in air
Oxide: see Dimethylphosphine oxide, D-00828
Sulfide: [6591-05-5].
C_2H_7PS M 94.111
Low melting solid. Mp 39°. $Bp_{0.5}$ 48-50°. pK_a 17.6 (DMSO). n_D^{20} 1.5570; tautomeric with dimethylphosphinothious acid.

Bartell, L.S., *J. Chem. Phys.*, 1960, **32**, 832 (*ed*)
Maier, L., *Helv. Chim. Acta*, 1966, **49**, 1249 (*sulfide, synth, nmr*)
Manatt, S.L. *et al, J. Am. Chem. Soc.*, 1966, **88**, 2689 (*nmr, pmr*)
Inorg. Synth., 1968, **11**, 128, 157; 1982, **21**, 180 (*synth*)
Crosbie, K.D. *et al, J. Inorg. Nucl. Chem.*, 1969, **31**, 3684 (*synth*)
Fields, R. *et al, J. Chem. Soc. (C)*, 1970, 197 (*synth, uv*)
Kostyanovskii, R.G. *et al, Org. Mass Spectrom.*, 1972, **6**, 1183 (*ms*)
Weigert, F.J. *et al, Inorg. Chem.*, 1973, **12**, 313 (*cmr*)
Appel, R., *Chem. Ber.*, 1974, **107**, 2658 (*sulfide, synth, pmr, nmr*)
Lappert, M. *et al, J. Chem. Soc., Dalton Trans.*, 1975, 1207 (*pe*)
Wanczek, K-P. *et al, Z. Naturforsch., A*, 1975, **30**, 329 (*ms*)
Elbel, S. *et al, Z. Naturforsch., B*, 1976, **31**, 1472 (*pe*)
Durig, J.R. *et al, J. Phys. Chem.*, 1977, **81**, 1588 (*ir, raman*)
Durig, J.R. *et al, J. Chem. Phys.*, 1977, **67**, 2216 (*struct*)
Aue, D.H. *et al, J. Am. Chem. Soc.*, 1980, **102**, 5151 (*pe*)
Mosbo, J.A. *et al, Phosphorus Sulfur*, 1981, **11**, 11 (*struct*)
Bretherick, L., *Handbook of Reactive Chemical Hazards*, 2nd Ed., Butterworths, London and Boston, 1979, 401.
Sax, N.I., *Dangerous Properties of Industrial Materials*, 6th Ed., Van Nostrand-Reinhold, 1984, 611.
Hazards in the Chemical Laboratory, (Bretherick, L., Ed.), 3rd Ed., Royal Society of Chemistry, London, 1981, 305.

Dimethylphosphine oxide D-00828

[7211-39-4]

$Me_2P(O)H$

C_2H_7OP M 78.050
Tautomeric with Dimethylphosphinous acid, D-00844 . Cryst. Mp 39-41°. Bp_6 65-67°, Bp_1 54°. pK_a 27.1 (DMSO).

Hays, H.R, *J. Org. Chem.*, 1968, **33**, 3690 (*synth, nmr, ir, pmr*)
Seel, F. *et al, Chem. Ber.*, 1972, **105**, 406 (*synth, ms, ir*)
Kleiner, H.J., *Justus Liebigs Ann. Chem.*, 1974, 751 (*synth*)
Elbel, S. *et al, J. Chem. Soc., Dalton Trans.*, 1976, 1762 (*pe*)
Seel, F. *et al, Z. Anorg. Allg. Chem.*, 1976, **423**, 67 (*ms*)

Dimethylphosphinic acid D-00829

[3283-12-3]

$Me_2P(O)(OH)$

$C_2H_7O_2P$ M 94.050
Mp 87.5-88.5°. pK_{a1} 3.08 (H_2O), pK_{a2} 4.72 (75% EtOH), pK_{a3} 13.18 ($MeNO_2$).
NH_4 salt: [51528-33-7]. Cryst. Mp 173° dec.
Me ester: [14337-77-0]. *Methyl dimethylphosphinate.*
$C_3H_9O_2P$ M 108.077
Liq. Bp_{12} 78.5-79.5°. n_D^{22} 1.4299.
Et ester: [2511-19-5]. *Ethyl dimethylphosphinate.*
$C_4H_{11}O_2P$ M 122.103
Liq. d_4^{25} 1.03. Bp_{15} 88-89°. n_D^{25} 1.4261.
Ph ester: [57244-61-8]. *Phenyl dimethylphosphinate.*
$C_8H_{11}O_2P$ M 170.147
Liq. Bp_{27} 164-166°.
Trimethylsilyl ester: [42346-36-1]. *Trimethylsilyl dimethylphosphinate.*
$C_5H_{15}O_2PSi$ M 166.232
Needles (ligroin at −70°). Mp 48-49°. $Bp_{0.5}$ 58-60°.
Fluoride: see Dimethylphosphinic fluoride, D-00832
Chloride: see Dimethylphosphinic chloride, D-00831

Cyanide: [17534-96-2].
C_3H_6NOP M 103.060
Solid. Mp 40-41°, Mp 53-56°. Bp 128-130°, $Bp_{1.4}$ 74°.
Azide: [58347-13-0].
$C_2H_6N_3OP$ M 119.063
Liq.-solid. Mp 14-17°. Bp_1 46-48°.
Isocyanate: [15790-43-9].
$C_3H_6NO_2P$ M 119.060
Solid. Mp 53-56°. $Bp_{0.02}$ 58-60°.
Isothiocyanate: [20443-12-3].
C_3H_6NOPS M 135.120
Solid. Mp 50-52.5°. $Bp_{0.1}$ 73-74°.
Amide: see P,P-Dimethylphosphinic amide, D-00830
Hydrazide:
$C_2H_9N_2OP$ M 108.080
Cryst. (EtOH/pet. ether). Mp 158-161°.
Anhydride: [14337-82-7].
$C_4H_{12}O_3P_2$ M 170.085
Solid. Mp 119°, Mp 132-134°. Bp_{15} 190-192°, $Bp_{0.3}$ 160°, Bp_1 135°.

Reinhardt, H. *et al, Chem. Ber.*, 1957, **90**, 1656 (*synth, derivs*)
Moedritzer, K., *J. Am. Chem. Soc.*, 1961, **83**, 4381 (*anhydride*)
Steininger, E., *Monatsh. Chem.*, 1966, **97**, 383 (*hydrazide*)
Giordano, F. *et al, Acta Crystallogr.*, 1967, **22**, 678 (*cryst struct*)
Derkach, G.I. *et al, Zh. Obshch. Khim.*, 1968, **38**, 1784 (*Engl. transl.* p. 1739) (*isocyanate*)
Haake, P. *et al, Tetrahedron*, 1968, **24**, 565 (*ms, esters*)
Tomaschewski, G. *et al, Arch. Pharm. (Weinheim, Ger.)*, 1968, **301**, 520 (*cyanide, isocyanate, isothiocyanate*)
Lindner, E. *et al, Z. Anorg. Allg. Chem.*, 1970, **376**, 28 (*anhydride, ir*)
Murray, M. *et al, J. Chem. Soc. (B)*, 1971, 1714 (*esters, P nmr*)
Seel, F. *et al, Chem. Ber.*, 1971, **104**, 2972 (*ms, P nmr, pmr, anhydride*)
Appel, R. *et al, Chem. Ber.*, 1974, **107**, 2658 (*synth, sales, P nmr, pmr, ir, ms*)
Kuchen, W. *et al, Z. Anorg. Allg. Chem.*, 1975, **413**, 266 (*silyl ester, ir, pmr, P nmr, ms*)
Schröder, H.F. *et al, Z. Anorg. Allg. Chem.*, 1975, **418**, 247 (*azide, ms, ir, raman, pmr, P nmr, cmr*)
Al'fonsov, V.A. *et al, Zh. Obshch. Khim.*, 1981, **51**, 2657 (*Engl. transl.* p. 2290) (*ester*)
Odeurs, R.L. *et al, J. Mol. Struct.*, 1984, **117**, 235 (*ir*)

P,P-Dimethylphosphinic amide, 9CI — D-00830

[65972-11-4]

$Me_2P(O)NH_2$

C_2H_8NOP M 93.065
N,N-*Di-Me:* [50663-05-3]. *Tetramethylphosphinic amide.*
$C_4H_{12}NOP$ M 121.119
Liq. d^{20} 1.02. Bp_2 89-91°. pK_a 10.0 ($MeNO_2$). $n_D^{22.5}$ 1.435.
N,N-*Di-Et:* [56080-45-6]. N,N-*Diethyl*-P,P-*dimethylphosphinic amide.*
$C_6H_{16}NOP$ M 149.172
d_4^{20} 1.02. Fp ~15°. Bp_{26} 131-142°, $Bp_{1.5}$ 78°. n_D^{23} 1.452.
N-tert-*Butyl:* [68036-30-6]. N-tert-*Butyl*-P,P-*dimethylphosphinic amide.*
$C_6H_{16}NOP$ M 149.172
Cryst. (Et_2O). Mp 90.5-91.5°.
N-(*Dimethylphosphinyl*): P,P-*Dimethyl*-N-(*dimethylphosphinyl*)*phosphinic amide.*
$C_4H_{13}NO_2P_2$ M 169.100
Isol. as K salt; cryst. (CH_2Cl_2/Et_2O). Mp 253-254°.
N-*Methoxycarbonyl: Methyl* N-(*dimethylphosphinyl*)-*carbamate.*
$C_4H_{10}NO_3P$ M 151.102
Solid. Mp 152-154°.
N-*Phenylaminocarbonyl:* N-(*Dimethylphosphinyl*)-N'-*phenylurea.*
$C_9H_{13}N_2O_2P$ M 212.188
Solid. Mp 157-159° dec.

Burg, A.B. *et al, J. Am. Chem. Soc.*, 1960, **82**, 2145 (*synth*)
Derkach, G.I. *et al, Zh. Obshch. Khim.*, 1968, **38**, 1784 (*Engl. transl.* p. 1739) (*derivs*)
Schmidpeter, A. *et al, Chem. Ber.*, 1968, **101**, 815 (*deriv, pmr*)
Pantzer, R. *et al, Z. Anorg. Allg. Chem.*, 1975, **416**, 297 (*deriv, ir, raman*)
Modro, T.A. *et al, J. Org. Chem.*, 1978, **43**, 5000 (*deriv*)

Dimethylphosphinic chloride, 8CI — D-00831

[1111-92-8]

$Me_2P(O)Cl$

C_2H_6ClOP M 112.496
Solid. Mp 67°. Bp 202-204°, Bp_{25} 104°.

Moedritzer, K., *J. Am. Chem. Soc.*, 1961, **83**, 4381 (*synth, P nmr, ir*)
Pollart, K.A. *et al, J. Org. Chem.*, 1962, **27**, 4444 (*synth*)
Durig, J.R. *et al, J. Phys. Chem.*, 1967, **71**, 3815 (*ir, raman*)
Neimysheva, A.A. *et al, Zh. Obshch. Khim.*, 1967, **37**, 2255 (*Engl. transl.* p. 2140) (*nqr, props*)
Seel, F. *et al, Chem. Ber.*, 1971, **104**, 2972 (*ms, pmr, P nmr*)
Fink, M. *et al, Justus Liebigs Ann. Chem.*, 1974, 741 (*synth*)
Appel, R. *et al, Z. Anorg. Allg. Chem.*, 1975, **414**, 236 (*synth*)
Elbel, S. *et al, J. Chem. Soc., Dalton Trans.*, 1976, 1762 (*pe*)

Dimethylphosphinic fluoride, 8CI — D-00832

[753-70-8]

$Me_2P(O)F$

C_2H_6FOP M 96.041
Liq. d^{20} 1.18. Bp 169-171°, Bp_{16} 69-71°. n_D^{20} 1.3920.

Halmann, M., *J. Chem. Soc.*, 1959, 305 (*synth, props*)
Schmutzler, R., *J. Inorg. Nucl. Chem.*, 1963, **25**, 335 (*synth*)
Nixon, J.F. *et al, Spectrochim. Acta*, 1964, **20**, 1835 (*P nmr*)
Kulakova, V.N. *et al, Zh. Obshch. Khim.*, 1969, **39**, 838 (*Engl. transl.* p. 802) (*synth*)
Riess, J.G. *et al, Inorg. Chem.*, 1973, **12**, 2874 (*pmr, F nmr*)
Fokin, A.V. *et al, Izv. Akad. Nauk SSSR, Ser. Khim.*, 1976, 2435 (*Engl. transl.* p. 2271) (*F and P nmr*)

P,P-Dimethylphosphinimidic amide — D-00833

$Me_2P(NH_2){=}NH$

$C_2H_9N_2P$ M 92.080
N,N-*di-Me:* [49778-03-2]. N,N,P,P-*Tetramethylphosphinimidic amide.*
$C_4H_{13}N_2P$ M 120.134
Liq. $Bp_{5.5}$ 73-74°.
N,N'-*di-Me:* [60414-60-0]. N,N',P,P-*Tetramethylphosphinimidic amide.*
$C_4H_{13}N_2P$ M 120.134
Bidentate chelate ligand for Sn and Pt. Solid. $Bp_{0.001}$ 85-90° subl. (bath). Butyllithium → Li deriv.
N,N,N'-*tri-Me:* [49778-06-5]. *Pentamethylphosphinimidic amide.*
$C_5H_{15}N_2P$ M 134.161

Liq. Bp_7 73-74°. Forms MeI adduct, Mp 196-201° dec.

N,N-*di-Me*, N'-*trimethylsilyl:* [73296-42-1]. N,N,P,P-*Tetramethyl*-N'-*trimethylsilylphosphinimidic amide.*
$C_7H_{21}N_2PSi$ M 192.316
Liq. Bp_5 62°.

N,N,N'-*Tris(trimethylsilyl):* [21385-93-3]. P,P-*Dimethyl*-N,N,N'-*tris(trimethylsilyl)phosphinimidic amide.*
$C_{11}H_{33}N_2PSi_3$ M 308.625
Liq. $Bp_{0.4}$ 78-80°.

Issleib, K. *et al, Synth. Inorg. Met.-Org. Chem.*, 1973, **3**, 255 (*deriv, pmr, P nmr*)
Scherer, O.J. *et al, Inorg. Chim. Acta*, 1976, **19**, 38 (*synth, complexes*)
Wilburn, J.C. *et al, Inorg. Chem.*, 1977, **16**, 2519 (*deriv, pmr, cmr, ir*)
Scherer, O.J. *et al, Z. Anorg. Allg. Chem.*, 1979, **449**, 167 (*synth, nmr, pmr*)
Wisian-Neilson, P. *et al, Inorg. Chem.*, 1980, **19**, 1875 (*deriv, pmr, cmr, P nmr*)

Dimethylphosphinodithioic acid, 9CI — D-00834

[16367-68-3]

$Me_2P(S)SH$

$C_2H_7PS_2$ M 126.171
Mp 47-50°.

Na salt: [34669-04-0]. Needles (moist 2-propanol/dioxan). Mp 167-169°. Forms a dihydrate, Mp 210-220°, 223-225° dec.

K salt: Solid. Mp 130°.

Me ester: [29952-68-9]. *Methyl dimethylphosphinodithioate.*
$C_3H_9PS_2$ M 140.198
Solid (Et_2O/pet. ether). Mp 41°. Bp_{14} 112-114°, Bp_1 50-52°.

Et ester: Ethyl dimethylphosphinodithioate.
$C_4H_{11}PS_2$ M 154.225
Liq. d_4^{20} 1.10. $Bp_{7.5}$ 104.5-106°. n_D^{20} 1.5695.

Butyl ester: Butyl dimethylphosphinodithioate.
$C_6H_{15}PS_2$ M 182.278
Liq. d_4^{20} 1.05. Bp_3 69.5-71°. n_D^{20} 1.5484.

Ph ester: [5745-37-9]. *Phenyl dimethylphosphinodithioate.*
$C_8H_{11}PS_2$ M 202.269
Solid. Mp 86-88°. Bp_{11} 173-175°.

Trimethylsilyl ester: Trimethylsilyl dimethylphosphinodithioate.
$C_5H_{15}PS_2Si$ M 198.353
Solid-liq. Mp 19-20°. $Bp_{0.1}$ 78°.

Anhydrosulfide: [23092-11-7]. *Dimethylphosphinodithioic anhydrosulfide.*
$C_4H_{12}P_2S_3$ M 218.266
Needles (Et_2O or CH_2Cl_2/pet. ether). Mp 100-101° (91°).

Haake, P. *et al, J. Am. Chem. Soc.*, 1967, **89**, 2650 (*synth, ir, pmr*)
Almasi, L. *et al, Chem. Ber.*, 1969, **102**, 1489 (*anhydrosulfide*)
Cavell, R.G. *et al, Inorg. Chem.*, 1971, **10**, 2710 (*synth, ir*)
Küchen, W. *et al, Chem. Ber.*, 1972, **105**, 132 (*synth*)
Wheatland, D.A. *et al, Inorg. Chem.*, 1972, **11**, 2340 (*anhydrosulfide, pmr, P nmr*)
Pantzer, R. *et al, Z. Anorg. Allg. Chem.*, 1973, **395**, 262 (*ester, ir, raman*)
Küchen, W. *et al, Z. Anorg. Allg. Chem.*, 1975, **413**, 266 (*silyl ester, ir, ms, pmr*)
Tsvetkov, E.N. *et al, Izv. Akad. Nauk SSSR, Ser. Khim.*, 1979, 426 (*Engl. transl.* p. 394) (*esters*)
Horner, L. *et al, Phosphorus Sulfur*, 1980, **8**, 221 (*synth*)
Fritz, R.H. *et al, Z. Anorg. Allg. Chem.*, 1986, **537**, 17 (*silyl ester, ms, P nmr*)

Dimethylphosphinoselenoic acid, 9CI, 8CI — D-00835

$Me_2P(Se)OH \rightleftharpoons Me_2P(O)SeH$

C_2H_7OPSe M 157.010

O-*Me ester:* [51072-22-1]. O-*Methyl dimethylphosphinoselenoate.*
C_3H_9OPSe M 231.092
No phys. props. reported.

Chloride: [61509-77-1].
C_2H_6ClPSe M 175.456
Liq. Mp 0°, Mp 15-17°. Bp_{11} 86-88°, $Bp_{1.5}$ 52°.

Bromide: [70629-47-9].
C_2H_6BrPSe M 219.907
Yellowish solid. Mp 23-24°. $Bp_{1.5}$ 61°.

Azide: [70442-86-3].
$C_2H_6N_3PSe$ M 182.023
Solid. Mp 69-70°.

Amide: see P,P-Dimethylphosphinoselenoic amide, D-00836

McFarlane, W. *et al, J. Chem. Soc., Dalton Trans.*, 1973, **20**, 2162 (*ester, P and Se nmr, pmr*)
Elbel, S. *et al, J. Chem. Soc., Dalton Trans.*, 1976, 1762 (*chloride, pe*)
Scherer, O.J. *et al, Z. Anorg. Allg. Chem.*, 1979, **449**, 167 (*chloride, pmr, P nmr*)
Schröder, H.F. *et al, Z. Anorg. Allg. Chem.*, 1979, **451**, 158 (*chloride, azide, pmr, P nmr, cmr, ms, ir, raman*)

P,P-Dimethylphosphinoselenoic amide, 9CI, 8CI — D-00836

$Me_2P(Se)NH_2$

C_2H_8NPSe M 156.026

N-*Me:* [70058-99-0]. N,P,P-*Trimethylphosphinoselenoic amide.*
$C_3H_{10}NPSe$ M 170.052
Solid. $Bp_{0.001}$ 70-80° subl.

N,N-*Di-Me:* [51168-02-6]. *Tetramethylphosphinoselenoic amide.*
$C_4H_{12}NPSe$ M 184.079
No phys. props. reported.

N-*Ph:* [56898-58-9]. P,P-*Dimethyl*-N-*phenylphosphinoselenoic amide. Dimethylphosphinoselenoic anilide.*
$C_8H_{12}NPSe$ M 232.123
Needles (toluene). Mp 114-116°.

McFarlane, W. *et al, J. Chem. Soc., Dalton Trans.*, 1973, 2162; 1976, 2351 (*synth, N and P and Se nmr, pmr*)
Scherer, O.J. *et al, Z. Anorg. Allg. Chem.*, 1979, **449**, 167 (*synth, pmr, P nmr*)

Dimethylphosphinoselenothioic acid, 9CI, 8CI — D-00837

$Me_2P(Se)SH \rightleftharpoons Me_2P(S)SeH$

C_2H_7PSSe M 173.071

Se-*Me ester:* [24490-40-2]. Se-*Methyl dimethylphosphinoselenothioate.*
C_3H_9PSSe M 187.098
No phys. props. reported.

McFarlane, W. *et al, J. Chem. Soc., Dalton Trans.*, 1972, 1397 (*Se nmr*)

Dimethylphosphinoselenous acid, 9CI, 8CI D-00838

$Me_2PSeH \rightleftharpoons Me_2PH{=}Se$

C_2H_7PSe M 141.011

Me ester: [24490-39-9]. *Methyl dimethylphosphinoselenoite.*
C_3H_9PSe M 155.038
Pale-yellow liq. with unpleasant odour. Bp 138°. Forms Mo, Cr, and W-containing complexes.

Trifluoromethyl ester: Trifluoromethyl dimethylphosphinoselenoite.
$C_3H_6F_3PSe$ M 209.009
Unstable.

Ph ester: [20626-86-2]. *Phenyl dimethylphosphinoselenoite.*
$C_8H_{11}PSe$ M 217.109
Liq. with unpleasant odour. d_4^{20} 1.32. $Bp_{0.5}$ 67°. n_D^{20} 1.6279.

Shlyk, Yu.N. *et al, Zh. Obshch. Khim.*, 1968, **38**, 193 (*Engl. transl.* p. 194) (*synth*)
Dehnert, P. *et al Z. Naturforsch., B*, 1979, **34**, 1646 (*synth, pmr, nmr*)
Böhm, M.C. *et al, Chem. Ber.*, 1981, **114**, 2300 (*pe*)

Dimethylphosphinothioic acid D-00839

[5761-95-5]

$Me_2P(S)OH \rightleftharpoons Me_2P(O)SH$

C_2H_7OPS M 110.110
Cryst. (C_6H_6/pet. ether). Mp 41.5-43° (s.t.). pK_a 2.5 (7% EtOH aq.), 4.5 (80% EtOH aq.). V. hygroscopic.

Na salt: Solid + $1.5H_2O$. Mp 147-148.5°.

NH_4 salt: [6046-15-7]. Solid. Mp 152°.

O-*Me ester:* [29952-66-7]. O-*Methyl dimethylphosphinothioate.*
C_3H_9OPS M 124.137
Liq. d_4^{20} 1.09. Bp_{10} 64-65°.

S-*Me ester:* [29952-67-8]. S-*Methyl dimethylphosphinothioate.*
C_3H_9OPS M 124.137
Mp 40°. Bp_{12} 106°, $Bp_{2.5}$ 85°.

O-*Ph ester:* [5553-01-5]. O-*Phenyl dimethylphosphinothioate.*
$C_8H_{11}OPS$ M 186.208
Solid. Mp 46-47°.

O-*Trimethylsilyl ester:* [42346-37-2].
$C_5H_{15}OPSSi$ M 182.292
Solid. Mp 45°. $Bp_{0.1}$ 40°.

Fluoride: [811-71-2].
C_2H_6FPS M 112.102
Liq. Bp_{25} 63.5-64.5°. n_D^{24} 1.4875.

Chloride: see Dimethylphosphinothioic chloride, D-00842

Bromide: see Dimethylphosphinothioic bromide, D-00841

Iodide: 81373-54-8
C_2H_6IPS M 220.008
Solid. Mp 63-64°. $Bp_{0.2}$ 85°.

Azide: [27260-90-8].
$C_2H_6N_3PS$ M 135.123
Cryst. (Me_2CO). Mp 69-70°.

*Amide: see P,P-*Dimethylphosphinothioic amide, D-00840

Anhydride: [38055-44-6].
$C_4H_{12}OP_2S_2$ M 202.206
Cryst. (EtOH/toluene). Mp 198-200°. Forms Cd, Cu, and Hg complexes.

Mastryukova, T.A. *et al, Zh. Obshch. Khim.*, 1962, **32**, 3579 (*Engl. transl.* p. 3512) (*synth, props*)
Schmutzler, R., *J. Inorg. Nucl. Chem.*, 1963, **25**, 335 (*fluoride*)
Panteleev, A.R. *et al, Izv. Akad. Nauk SSSR, Ser. Khim.*, 1967, 1644 (*Engl. trans.* p. 1557) (*O-phenyl ester*)
Koch, W. *et al, CA*, 1971, **74**, 125796 (*cyanide*)
Köttgen, D. *et al, Z. Anorg. Allg. Chem.*, 1972, **389**, 269 (*fluoride, ir, Raman*)
Wheatland, D.A. *et al, Inorg. Chem.*, 1972, **11**, 2340 (*anhydride, ir*)
Kuramshin, I.Ya. *et al, Zh. Obshch. Khim.*, 1973, **43**, 1456 (*Engl. transl.* p. 1446) (*S-Me ester, ir, pmr, nmr*)
Steinberger, H. *et al, Z. Naturforsch., B*, 1973, **28**, 44 (*silyl ester, nmr, use*)
Hägele, R. *et al, Z. Naturforsch., B*, 1974, **29**, 349 (*nmr, pmr*)
Küchen, W. *et al, Z. Anorg. Allg. Chem.*, 1975, **413**, 266 (*silyl ester, ir, ms*)
Schröder, H.F. *et al, Z. Anorg. Allg. Chem.*, 1975, **418**, 247 (*azide, ms, ir, raman, pmr, nmr, cmr*)
Müller, J., *Chem. Ber.*, 1977, **110**, 788 (*azide, cryst struct*)
Burkhardt, W.D. *et al, Z. Anorg. Allg. Chem.*, 1978, **442**, 19 (*methyl ester, ir, raman*)
Harris, R.K. *et al, J. Chem. Soc., Dalton Trans.*, 1978, 9 (*anhydride, nmr, cmr*)
Istomin, B.I. *et al, Zh. Obshch. Khim.*, 1981, **51**, 2393 (*Engl. trans.* p. 2063) (*phenyl ester, props*)
Feshchenko, N.G. *et al, Zh. Obshch. Khim.*, 1982, **52**, 222 (*Engl. trans.* p. 202) (*iodide*)
Mattes, R. *et al, Z. Anorg. Allg. Chem.*, 1983, **499**, 67 (*cryst struct*)

P,P-Dimethylphosphinothioic amide, 9CI D-00840

[6851-71-4]

$Me_2P(S)NH_2$

C_2H_8NPS M 109.126
Cryst. (C_6H_6). Mp 99-101°. Bp 273-275° dec.

N,N-*Di-Me:* [6839-95-8]. *Tetramethylphosphinothioic amide.*
$C_4H_{12}NPS$ M 137.179
Solid. Mp 61.5-63°. Bp_5 89-92°.

N-*Ph:* [42847-72-3]. P,P-*Dimethyl-*N-*phenylphosphinothioic amide.* P,P-*Dimethylphosphinothioic anilide.*
$C_8H_{12}NPS$ M 185.223
Cryst. (C_6H_6/pet. ether or pet. ether/Et_2O). V. sol. Me_2CO, Et_2O, less sol. MeCN, C_6H_6. Mp 117°, 128-131°.

N-*Benzyl:* [36190-30-4]. N-*Benzyl-*P,P-*dimethylphosphinothioic amide.*
$C_9H_{14}NPS$ M 199.250
Cryst. (pet. ether). Mp 91-92°.

N-*Benzoyl:* N-*Benzoyl-*P,P-*dimethylphosphinothioic amide.*
$C_8H_{12}NOPS$ M 201.223
Cryst. (H_2O). Mp 123-124°. pK_a 11.3.

N-*Trimethylsilyl:* [42346-45-2]. P,P-*Dimethyl-*N-*trimethylsilylphosphinothioic amide.*
$C_5H_{16}NPSSi$ M 181.307
Mp 75°. $Bp_{0.1}$ 98-99°.

N,N-*Bis(trimethylsilyl):* P,P-*Dimethyl-*N,N-*bis(trimethylsilyl)phosphinothioic amide.*
$C_8H_{24}NPSSi_2$ M 253.489
Solid. Mp 70-75°.

N-*Dimethylphosphinothioyl:* [18509-37-0]. P,P-*Dimethyl-*N-*(dimethylphosphinothioyl)phosphinothioic amide.*
$C_4H_{13}NP_2S_2$ M 201.221
Leaflets. Mp 178°. pK_a 8.7. Forms Na salt (monohydrate, Mp 103-4°, dihydrate Mp 112°).

Schmidpeter, A. *et al*, *Chem. Ber.*, 1968, **101**, 815 (*synth, derivs*)
Ellis, K. *et al*, *J. Chem. Soc., Perkin Trans. 1*, 1972, 1184 (*deriv, pmr*)
Steinberger, H. *et al*, *Z. Naturforsch., B*, 1974, **29**, 611 (*deriv*)
Appel, R. *et al*, *Chem. Ber.*, 1975, **108**, 2349 (*anilide, pmr, P nmr*)
Pantzer, R. *et al*, *Z. Anorg. Allg. Chem.*, 1975, **416**, 297 (*deriv, ir, raman*)
Boedeker, J. *et al*, *J. Prakt. Chem.*, 1976, **318**, 149 (*derivs*)
McFarlane, W. *et al*, *J. Chem. Soc., Dalton Trans.*, 1976, 2351 (*anilide, N and P nmr*)
Wilburn, J.C. *et al*, *Inorg. Chem.*, 1977, **16**, 2519 (*deriv*)

Dimethylphosphinothioic bromide, 9CI, 8CI — D-00841

[6839-93-6]

$Me_2P(S)Br$

C_2H_6BrPS M 173.007
Reagent for protection in peptide synth. Liq. or cryst. Mp 34°. Bp_{718} 205-207°, Bp_{15} 76-78°. n_D^{35} 1.5482.

Maier, L., *Chem. Ber.*, 1961, **94**, 3051 (*synth*)
Durig, J.R. *et al*, *J. Phys. Chem.*, 1967, **71**, 3875 (*ir, raman*)
Inorg. Synth., 1970, **12**, 287 (*synth*)
Elbel, S. *et al*, *J. Chem. Soc., Dalton Trans.*, 1976, 1762 (*pe*)
Fokin, A.V. *et al*, *Izv. Akad. Nauk SSSR, Ser. Khim.*, 1976, **25**, 2435 (*Engl. transl.* p. 2271) (*nmr*)

Dimethylphosphinothioic chloride, 9CI, 8CI — D-00842

[993-12-4]

$Me_2P(S)Cl$

C_2H_6ClPS M 128.556
Reagent for protection in peptide synth. Liq. or cryst. d_4^{20} 1.23-1.27. Mp 24-25°. Bp_{16} 107-107.5°, (Bp_{11} 69-70°). n_D^{20} 1.5460.

Maier, L., *J. Inorg. Nucl. Chem.*, 1962, **24**, 1073 (*synth, nmr*)
Mastryukova, T.A. *et al*, *Zh. Obshch. Khim.*, 1962, **32**, 3579 (*Engl. transl.* p. 3512) (*synth*)
Durig, J.R. *et al*, *J. Phys. Chem.*, 1967, **71**, 3815 (*ir, raman*)
Goubeau, J. *et al*, *Z. Anorg. Allg. Chem.*, 1968, **360**, 182 (*ir, raman*)
Inorg. Synth., 1974, **15**, 191 (*synth*)
Elbel, S. *et al*, *J. Chem. Soc., Dalton Trans.*, 1976, 1762 (*pe*)
Fokin, A.V. *et al*, *Izv. Akad. Nauk SSSR, Ser. Khim.*, 1976, **25**, 2435 (*Engl. transl.* p. 2271) (*nmr*)
Ueki, M. *et al*, *Bull. Chem. Soc. Jpn.*, 1983, **56**, 1187 (*use*)

Dimethylphosphinothious acid — D-00843

Me_2PSH

C_2H_7PS M 94.111
The free acid exists as the thiophosphoryl tautomer. See Dimethylphosphine, D-00827 .
Me ester: Methyl dimethylphosphinothioite.
C_3H_9PS M 108.138
Liq. Bp 121°.
Trifluoromethyl ester: Trifluoromethyl dimethylphosphinothioite.
$C_3H_6F_3PS$ M 162.109
Unstable liq.

Dehnert, P. *et al*, *Z. Naturforsch., B*, 1979, **34**, 1646 (*synth, props, P nmr, pmr*)

Dimethylphosphinous acid, 9CI — D-00844

Me_2POH

C_2H_7OP M 78.050
The free acid exists as the phosphoryl tautomer Dimethylphosphine oxide, D-00828 .
Me ester: [20502-88-9]. *Methyl dimethylphosphinite.*
C_3H_9OP M 92.077
No phys. props. reported.
Butyl ester: Butyl dimethylphosphinite.
$C_6H_{15}OP$ M 134.158
Liq. Bp_{48} 56-58°.
Ph ester: [25781-02-6]. *Phenyl dimethylphosphinite.*
$C_8H_{11}OP$ M 154.148
Liq. $Bp_{0.2}$ 43-45°.
Fluoride: see Dimethylphosphinous fluoride, D-00848
Chloride: see Dimethylphosphinous chloride, D-00846
Bromide: [2240-31-5]. *Bromodimethylphosphine.*
C_2H_6BrP M 140.947
Solid. Mp 93-95°. Bp_{716} 100-105°.
Iodide: [4731-62-8]. *Iododimethylphosphine.*
C_2H_6IP M 187.948
Solid. Mp 105-110° subl.
Cyanide: see Dimethylphosphinous cyanide, D-00847
*Amide: see P,P-*Dimethylphosphinous amide, D-00845

Maier, L., *Helv. Chim. Acta*, 1963, **46**, 2026; 1964, **47**, 2137 (*bromide, synth, P nmr*)
Maier, L. *et al*, *Helv. Chim. Acta*, 1969, **52**, 858 (*esters*)
Seel, F. *et al*, *Chem. Ber.*, 1972, **105**, 406; *Justus Liebigs Ann. Chem.*, 1972, **756**, 181 (*methyl ester, nmr, pmr, ir, ms*)
Kleiner, H.J., *Justus Liebigs Ann. Chem.*, 1974, 751 (*phenyl ester, synth*)
Elbel, S. *et al*, *Z. Naturforsch., B*, 1976, **31**, 178 (*bromide, pe*)
Feshchenko, N.G. *et al*, *Zh. Obshch. Khim.*, 1982, **52**, 222 (*Engl. transl.* p. 202) (*iodide, synth, P nmr*)

P,P-Dimethylphosphinous amide — D-00845

Me_2PNH_2

C_2H_8NP M 77.066
N,N-*Di-Me: see Tetramethylphosphinous amide, T-00213*
N,N-*Di-Ph:* [20626-85-1]. P,P-*Dimethyl*-N,N-*diphenylphosphinous amide.*
$C_{14}H_{16}NP$ M 229.261
Liq. $Bp_{0.5}$ 127°.
N,N-*Bis(trimethylsilyl):* [63744-11-6]. P,P-*Dimethyl*-N,N-*bis(trimethylsilyl)phosphinous amide.*
$C_8H_{24}NPSi_2$ M 221.429
Liq. Bp_4 55-60°.

Shlyk, Yu.N. *et al*, *Zh. Obshch. Khim.*, 1968, **38**, 193 (*Engl. transl.* p. 194) (*diphenyl, synth*)
Neilson, R.H. *et al*, *Inorg. Chem.*, 1982, **21**, 3568 (*trimethylsilyl, synth, P nmr*)

Dimethylphosphinous chloride, 9CI — D-00846

Chlorodimethylphosphine
[811-62-1]

Me_2PCl

C_2H_6ClP M 96.496
Pale-yellow liq. d_4^{25} 1.23. Mp −2°. Bp_{749} 73-74°. n_D^{20} 1.4760.

▷Ignites in air

Maier, L., *Helv. Chim. Acta*, 1963, **46**, 2026; 1964, **47**, 2137 (*synth, pmr, nmr*)
Seel, F. *et al*, *Z. Anorg. Allg. Chem.*, 1968, **363**, 233 (*ir, ms, pmr*)
Fild, M. *et al*, *J. Chem. Soc. (A)*, 1970, 2359 (*synth, P nmr, pmr*)
Inorg. Synth., 1974, **15**, 191 (*synth*)
Durig, J.R. *et al*, *J. Mol. Struct.*, 1975, **27**, 403 (*ir, raman*)
Lappert, M.F. *et al*, *J. Chem. Soc., Dalton Trans.*, 1975, 1207 (*pe*)
Elbel, S. *et al*, *Z. Naturforsch., B*, 1976, **31**, 178 (*pe*)
Barlos, K. *et al*, *Chem. Ber.*, 1980, **113**, 3716 (*nqr*)
Frenking, G. *et al*, *Phosphorus Sulfur*, 1980, **8**, 337, 343 (*pe, struct*)
Weger, E. *et al*, *Org. Mass Spectrom.*, 1983, **18**, 327 (*ms*)
Wolfsberger, W., *J. Organomet. Chem.*, 1986, **317**, 167 (*cmr, P nmr*)

Dimethylphosphinous cyanide, 9CI — D-00847

Cyanodimethylphosphine

[31641-57-3]

Me_2PCN

C_3H_6NP M 87.061

Liq. Mp 0°, −20°. Bp_{10} 40-45°.

▷Spont. ignites in air

Jones, C.F. *et al* *Inorg. Chem.*, 1971, **10**, 1536 (*synth*)
Durig, J.R. *et al*, *Inorg. Chem.*, 1974, **13**, 2302 (*ir, raman, struct*)
Elbel, S. *et al*, *Z. Naturforsch., B*, 1976, **31**, 1472 (*pe*)
Kostyanovskii, R.G. *et al*, *Org. Mass Spectrom.*, 1980, **15**, 397 (*ms*)
Chervin, I.I. *et al*, *Izv. Akad. Nauk SSSR, Ser. Khim.*, 1981, 1769 (*Engl. transl.* p. 1438) (*props*)

Dimethylphosphinous fluoride — D-00848

Fluorodimethylphosphine

[507-15-3]

Me_2PF

C_2H_6FP M 80.042

Mp −109°. Bp 26°. Readily disproportionates at r.t.

▷Spont. ignites in air

Seel, F. *et al*, *Angew. Chem., Int. Ed. Engl.*, 1967, **6**, 708 (*synth, ms, ir, F and P nmr, pmr*)
Seel, F. *et al*, *Z. Anorg. Allg. Chem.*, 1968, **363**, 233 (*ir, pmr, F and P nmr, ms, props*)
Fild, M. *et al*, *J. Chem. Soc. (A)*, 1970, 2359 (*synth, pmr, F and P nmr*)
Lappert, M.F. *et al*, *J. Chem. Soc., Dalton Trans.*, 1975, 1207 (*pe*)

N,P-Dimethylphosphonamidic acid, 9CI — D-00849

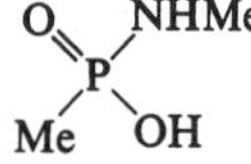

$C_2H_8NO_2P$ M 109.064

Me ester: [7351-35-1]. *Methyl* N,P-*dimethylphosphonamidate.*
$C_3H_{10}NO_2P$ M 123.091
Liq. d_4^{20} 1.13. $Bp_{0.02}$ 72-73°. n_D^{20} 1.4423.

Et ester: [13703-32-7]. *Ethyl* N,P-*dimethylphosphonamidate.*
$C_4H_{12}NO_2P$ M 137.118
Liq. d_4^{20} 1.10. $Bp_{0.02}$ 88°. n_D^{20} 1.4370.

Isopropyl ester: [13703-11-2]. *Isopropyl* N,P-*dimethylphosphonamidate.*
$C_5H_{14}NO_2P$ M 151.145
Liq. d_4^{20} 1.04. $Bp_{0.06}$ 75°. n_D^{20} 1.4350.

Ph ester: [4645-89-0]. *Phenyl* N,P-*dimethylphosphonamidate.*
$C_8H_{12}NO_2P$ M 185.162
Liq. d_4^{20} 1.16. $Bp_{0.2}$ 132°. n_D^{20} 1.5252.

Fluoride:
C_2H_7FNOP M 111.056
Liq. Bp_4 103°.

Petrov, K.A. *et al*, *Zh. Obshch. Khim.*, 1960, **30**, 4060 (*Engl. transl.* p. 4023) (*ester*)
Goldwhite, H. *et al*, *J. Am. Chem. Soc.*, 1966, **88**, 3572 (*pmr*)
Shokol, V.A. *et al*, *Zh. Obshch. Khim.*, 1966, **36**, 1636 (*Engl. transl.* p. 1636) (*synth*)
Grechkin, N.P. *et al*, *Zh. Obshch. Khim.*, 1978, **48**, 1305 (*Engl. transl.* p. 1194) (*synth*)

N,P-Dimethylphosphonamidothioic acid, 9CI — D-00850

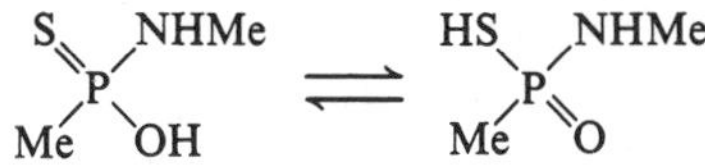

C_2H_8NOPS M 125.125

O-*Et ester:* [21844-11-1]. O-*Ethyl* N,P-*dimethylphosphonamidothioate.*
$C_4H_{12}NOPS$ M 153.179
Liq. d_4^{25} 1.09. $Bp_{0.22}$ 65.5°. n_D^{25} 1.5003.

S-*Me ester:* [67242-35-7]. S-*Methyl* N,P-*dimethylphosphonamidothioate.*
$C_3H_{10}NOPS$ M 139.152
Solid. Mp 40-42°.

Fluoride: [811-99-4]. N,P-*Dimethylphosphonamidothioic fluoride.*
C_2H_7FNPS M 127.116
Oil. Bp_1 58°.
▷V. toxic

Ger. Pat. 1 099 532, (*1959*); *CA*, **56**, 3515 (*fluoride*)
Grapov, A.F. *et al*, *Zh. Obshch. Khim.*, 1968, **38**, 2260 (*Engl. transl.* p. 2187) (*ethyl ester, synth*)
Reddy, G.S. *et al*, *Z. Naturforsch., B*, 1970, **25**, 1199 (*fluoride, F and P nmr*)
Wustner, D.A. *et al*, *J. Agric. Food Chem.*, 1978, **26**, 1104 (*methyl ester, synth, tox*)

Dimethyl phosphonate, 9CI — D-00851

Dimethyl phosphite

[868-85-9]

$(MeO)_2P(O)H \rightleftharpoons (MeO)_2POH$

$C_2H_7O_3P$ M 110.049

Tautomeric. Methylating agent for *N*-heterocyclic compds.; *S*-methylates thioamides.

Hydrophosphorylating reagent for alkenes and alkynes. d_0^{25} 1.191, d_4^{20} 1.20. Bp 170-171°, Bp_8 56.5°. n_D^{20} 1.403.

▷SZ7710000.

Harless, H.R., *Anal. Chem.*, 1961, **33**, 1387 (*ms*)
Moedritzer, K., *J. Inorg. Nucl. Chem.*, 1962, **22**, 19 (*nmr*)
Lavalley, J.C. *et al*, *Spectrochim. Acta, Part A*, 1972, **28**, 2091 (*ir, raman*)
Stec, W.J. *et al*, *J. Chem. Soc., Perkin Trans. 2*, 1972, 463 (*nmr*)
Skvortsov, N.K. *et al*, *Zh. Obshch. Khim.*, 1976, **46**, 521 (*Engl. transl.* p. 518) (*pmr, P nmr*)
Hayashi, M. *et al*, *Bull. Chem. Soc. Jpn.*, 1977, **50**, 1510 (*use*)
Kläm, W. *et al*, *Org. Magn. Reson.*, 1977, **10**, 126 (*pmr, complexes*)

Glidewell, C. *et al*, *J. Chem. Educ.*, 1980, **57**, 740 (*synth, ir, pmr, P nmr*)
Pietro, W.J. *et al*, *J. Am. Chem. Soc.*, 1982, **104**, 3594 (*tautom*)
Sidky, M.M. *et al*, *Org. Prep. Proceed. Int.*, 1982, **14**, 225 (*use*)
Kenttämaa, M.I., *J. Am. Chem. Soc.*, 1985, **107**, 1881 (*ms, tautom*)

Dimethyl phosphonite, 9CI **D-00852**
[20502-63-0]

$(MeO)_2PH$

$C_2H_7O_2P$ M 94.050
Not isol. Unstable.

Stec, W.J. *et al*, *C.R. Hebd. Seances Acad. Sci., Ser. C*, 1975, **281**, 727 (*P nmr*)
Gallagher, M.J. *et al*, *Aust. J. Chem.*, 1980, **33**, 287 (*P nmr*)

***O,O*-Dimethyl phosphonothioate, 9CI** **D-00853**
O,O-Dimethyl thiophosphite
[5930-72-3]

$(MeO)_2P(S)H \rightleftharpoons (MeO)_2PSH$

$C_2H_7O_2PS$ M 126.110
Tautomeric, but exists almost completely in thiophosphoryl form. Insol. H_2O. d^{20} 1.19. $Bp_{16.5}$ 52-53.5°. n_D^{20} 1.4768. Readily oxidized.

Kabachnik, M.I. *et al*, *Izv. Akad. Nauk SSSR, Ser. Khim.*, 1952, 727 (*Engl. transl.* p. 661) (*synth*)
Stec, W.J. *et al*, *J. Chem. Soc., Perkin Trans. 2*, 1972, 463 (*pmr, P nmr*)

***O,O*-Dimethyl phosphoramidate, 9CI, 8CI** **D-00854**
O,O-Dimethyl phosphoric amide. O,O-Dimethyl phosphoramide. O,O-Dimethyl phosphoryl amide
[2697-42-9]

$(MeO)_2P(O)NH_2$

$C_2H_8NO_3P$ M 125.064
Prisms (Et_2O). Sol. H_2O, EtOH, Me_2CO. Spar. sol. Et_2O, C_6H_6, CCl_4. Mp 40-42°. $Bp_{0.5}$ 98°. n_D^{20} 1.4347.

Zhuravleva, L.P. *et al*, *Zh. Obshch. Khim.*, 1965, **35**, 998 (*Engl. transl.* p. 1003) (*synth*)
Öney, I. *et al*, *J. Am. Chem. Soc.*, 1967, **89**, 6972 (*synth, props*)
Zwierzak, A. *et al*, *Tetrahedron*, 1970, **26**, 3521 (*synth, ir*)
Jacobsen, P. *et al*, *Org. Mass Spectrom.*, 1972, **6**, 1303 (*ms*)

Dimethylphosphoramidic acid, 9CI **D-00855**
Phosphoric acid dimethylamide

$Me_2NP(O)(OH)_2$

$C_2H_8NO_3P$ M 125.064
Di-Me ester: see Dimethyl dimethylphosphoramidate, D-00727
Di-Et ester: see Diethyl dimethylphosphoramidate, D-00286
Diisopropyl ester: [2404-04-8]. *Diisopropyl dimethylphosphoramidate. Bis(1-methylethyl) dimethylphosphoramidate.*
$C_8H_{20}NO_3P$ M 209.225
Liq. Bp_{12} 84-84.5°. n_D^{20} 1.4160.
Di-Ph ester: [6415-21-0]. *Diphenyl dimethylphosphoramidate.*
$C_{14}H_{16}NO_3P$ M 277.259
Liq. $Bp_{0.2}$ 154°. n_D^{25} 1.5407.
Difluoride: see Dimethylphosphoramidic difluoride, D-00858
Dichloride: see Dimethylphosphoramidic dichloride, D-00856
Diamide: see N,N-Dimethylphosphoric triamide, D-00880

Kamai, G. *et al*, *Zh. Obshch. Khim.*, 1957, **27**, 3064 (*Engl. transl.* p. 3093) (*esters, synth*)
Keat, R. *et al*, *J. Chem. Soc. (A)*, 1968, 703 (*diphenyl ester, pmr*)
Perregaard, J. *et al*, *Recl. Trav. Chim. Pays-Bas*, 1974, **93**, 252 (*diphenyl ester, synth*)

Dimethylphosphoramidic dichloride, 9CI, 8CI **D-00856**
(Dimethylamino)phosphonic dichloride. Dimethylamino phosphoryl dichloride
[677-43-0]

$Me_2NP(O)Cl_2$

$C_2H_6Cl_2NOP$ M 161.955
Liq. $d_{15.5}^{15.5}$ 1.37. Bp 194-195°, Bp_{10} 76°, Bp_1 45°. n_D^{25} 1.4610.
▷TB4025000.

Michaelis, A., *Justus Liebigs Ann. Chem.*, 1903, **326**, 129, 172 (*synth*)
Cowley, A.H. *et al*, *J. Am. Chem. Soc.*, 1965, **87**, 4454 (*pmr*)
Köttgen, D. *et al*, *Z. Anorg. Allg. Chem.*, 1971, **385**, 56 (*ir, raman*)
Osokin, D.Ya. *et al*, *Org. Magn. Reson.*, 1972, **4**, 831 (*nqr*)
Robinson, E.A. *et al*, *Spectrochim. Acta, Part A*, 1972, **28**, 1099 (*pmr, P nmr*)
Gouesnard, J.P. *et al*, *Nouv. J. Chim.*, 1982, **6**, 143 (*N nmr*)

Dimethylphosphoramidic dicyanide, 9CI, 8CI **D-00857**
Dimethylaminophosphonic dicyanide

$Me_2NP(O)(CN)_2$

$C_4H_6N_3OP$ M 143.085
Liq. Bp_2 106°.

Ger. Pat., 918 603, (*1954*); *CA*, **50**, 4448

Dimethylphosphoramidic difluoride, 9CI **D-00858**
Dimethylamino phosphoryl difluoride
[354-43-8]

$Me_2NP(O)F_2$

$C_2H_6F_2NOP$ M 129.046
Liq. Bp_{12} 47-49°. n_D^{20} 1.4361.

Olah, G.A. *et al*, *Justus Liebigs Ann. Chem.*, 1959, **625**, 88 (*synth*)
Robinson, E.A. *et al*, *Spectrochim Acta, Part A*, 1972, **28**, 1099 (*pmr, P nmr*)
Köttgen, D. *et al*, *Z. Phys. Chem.*, 1974, **92**, 289 (*ir, raman*)
Ryzhikov, B.D. *et al*, *Zh. Strukt. Khim.*, 1975, **16**, 754 (*Engl. transl.* p. 700) (*F nmr*)

Dimethyl phosphoramidite, 9CI **D-00859**
[39230-41-6]

$(MeO)_2PNH_2$

$C_2H_8NO_2P$ M 109.064

Liq. Bp_{30} 57-60°.

Scherer, O.J. *et al*, *Z. Naturforsch., B*, 1972, **27**, 1429 (*synth, pmr*)

Dimethylphosphoramidochloridic acid, 9CI, 8CI **D-00860**

$$Cl-P(=O)(OR)-NMe_2$$

(*S*)-*form of esters*

$C_2H_7ClNO_2P$ M 143.510

(S)-form

Me ester: [71877-82-2]. *Methyl dimethylphosphoramidochloridate.*
$C_3H_9ClNO_2P$ M 157.536
Liq. Bp_9 85°. $[\alpha]_D$ −33° (c, 1.8 in $CHCl_3$).

(±)-form

Et ester: Ethyl dimethylphosphoramido chloridate.
$C_4H_{11}ClNO_2P$ M 171.563
Liq. $Bp_{0.01}$ 51°.

Ph ester: Phenyl dimethylphosphoramidochloridate.
$C_8H_{11}ClNO_2P$ M 219.607
Liq. Bp_{10} 149-150°. n_D^{25} 1.5130.

Greenhalgh, R. *et al*, *Can. J. Chem.*, 1970, **48**, 1351 (*ethyl ester*)
Perregaard, J. *et al*, *Recl. Trav. Chim. Pays-Bas*, 1974, **93**, 252 (*aryl esters*)
Hall, C.R. *et al*, *J. Chem. Soc., Perkin Trans. 1*, 1979, 1646 (*methyl ester, stereochem*)

Dimethylphosphoramidochloridothioic acid, 9CI, 8CI **D-00861**

$$Me_2NPCl(S)OH \rightleftharpoons Me_2NPCl(O)SH$$

$C_2H_7ClNOPS$ M 159.570

(±)-form

O-*Me ester:* O-*Methyl dimethylphosphoramidochloridothioate.* O-*Methyl chlorodimethylamidothiophosphate.*
$C_3H_9ClNOPS$ M 173.597
Liq. $Bp_{1.5}$ 50-52°.

O-*Et ester:* O-*Ethyl dimethylphosphoramidochloridothioate.* O-*Ethyl chlorodimethylamidothiophosphate.*
$C_4H_{11}ClNOPS$ M 187.624
Liq. d_4^{20} 1.20. Bp_3 72-73°. n_D^{20} 1.4972.

O-*Ph ester:* O-*Phenyl dimethylphosphoramidochloridothioate.* O-*Phenyl chlorodimethylamidothiophosphate.*
$C_8H_{11}ClNOPS$ M 235.668
Liq. d_4^{20} 1.27. $Bp_{0.28}$ 104-107°. n_D^{20} 1.5640.

Kabachnik, M.I. *et al*, *Zh. Obshch. Khim.*, 1959, **29**, 2182 (*Engl. transl.* p. 2149) (*ethyl ester*)
Mandel'baum, Ya.A. *et al*, *Zh. Obshch. Khim.*, 1968, **38**, 1754 (*Engl. transl.*, p. 1709) (*phenyl ester*)

Dimethylphosphoramidochloridous acid, 9CI **D-00862**

$$Me_2NP(OH)Cl$$

C_2H_7ClNOP M 127.510

Me ester: [70063-12-6]. *Methyl dimethylphosphoramidochloridite.*
C_3H_9ClNOP M 141.537
Employed in nucleoside phosphitylation. Liq. d^{25} 1.12. Bp_{13} 40-42°.

Et ester: [66442-24-8]. *Ethyl dimethylphosphoramidochloridite.*
$C_4H_{11}ClNOP$ M 155.564
Liq. Bp_{15} 45°. n_D^{20} 1.4730.

2-Cyanoethyl ester: [89992-69-8]. *2-Cyanoethyl dimethylphosphoramidochloridite.*
$C_5H_{10}ClN_2OP$ M 180.574
Employed in nucleoside phosphitylation.

Ph ester: [82237-09-0]. *Phenyl dimethylphosphoramidochloridite.*
$C_8H_{11}ClNOP$ M 203.608
Liq. $Bp_{0.1}$ 64.5-65.5°. n_D^{20} 1.5468.

Beaucage, S.L. *et al*, *Tetrahedron Lett.*, 1981, **22**, 1859 (*methyl ester, synth, P nmr, pmr, use*)
Kukhar', V.P. *et al*, *Zh. Obshch. Khim.*, 1982, **52**, 562 (*Engl. transl.* p. 492) (*ethyl, phenyl esters, synth, P nmr*)
Sinha, N.D. *et al*, *Tetrahedron Lett.*, 1983, **24**, 5843 (*cyanoethyl ester, synth, pmr, P nmr, ms, use*)
Josephson, S. *et al*, *Acta Chem. Scand., Ser. A*, 1984, **38**, 539 (*use*)
Sinha, N.D. *et al*, *Nucleic Acids Res.*, 1984, **12**, 4539 (*use, pmr, ms, P nmr*)

Dimethylphosphoramidocyanidic acid, 9CI, 8CI **D-00863**

$$Me_2N-P(=O)(CN)-OH$$

$C_3H_7N_2O_2P$ M 134.074
Derivs. can exist in chiral forms.

(±)-form

Me ester: Methyl dimethylphosphoramidocyanidate.
$C_4H_9N_2O_2P$ M 148.101
Liq. Bp_{13} 89-91°.

Et ester: see Tabun, T-00001

Isopropyl ester: Isopropyl dimethylphosphoramidocyanidate.
$C_6H_{13}N_2O_2P$ M 176.155
Liq. Bp_2 95°.

Cyclohexyl ester: Cyclohexyl dimethylphosphoramidocyanidate.
$C_9H_{18}N_2O_2P$ M 217.227
Liq. Bp_2 140°.

Ger. Pat., 767 511, (*1937*); *CA*, **49**, 14795
Ger. Pat., 767 830, (*1939*)

Dimethylphosphoramidocyanidothioic acid **D-00864**

$$Me_2N-P(=S)(OH)-CN \rightleftharpoons Me_2N-P(=O)(SH)-CN$$

$C_3H_7N_2OPS$ M 150.135
Derivs. can exist in chiral forms.

(±)-form

O-*Et ester:* O-*Ethyl dimethylphosphoramidocyanidothioate.* O-*Ethyl cyano(dimethylamido)thiophosphate.*
$C_5H_{11}N_2OPS$ M 178.188
Liq. Bp_2 94°.

Ger. Pat., 664 438, (*1935*); 767 511, (*1937*); *CA*, **49**, 14795

_O,S_-Dimethyl phosphoramidodithioate, 9CI, 8CI **D-00865**

O,S-*Dimethyl amidodithiophosphate. Monitor*

[29809-48-1]

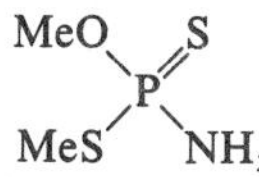

$C_2H_8NOPS_2$ M 157.185

(±)-*form*

Liq. $Bp_{0.2}$ 98° (86°). n_D^{21} 1.5820.

Hamer, N.K. *et al, J. Chem. Soc., Perkin Trans. 2*, 1974, 1184 (*synth, derivs, props*)
U.S.P., 3 787 539, (*1974*); *CA*, **80**, 95248 (*synth*)

_S,S_-Dimethyl phosphoramidodithioate, 9CI, 8CI **D-00866**

S,S-*Dimethyl amidodithiophosphate*

[32979-55-8]

$(MeS)_2P(O)NH_2$

$C_2H_8NOPS_2$ M 157.185

Cryst. (MeOH/toluene). Mp 105-106°.

N-*Ph:* S,S-*Dimethyl phenylphosphoramidodithioate.* S,S-*Dimethyl anilidodithiophosphate.*
$C_8H_{12}NOPS_2$ M 233.283
Solid. Mp 101°.

Fahmy, M.A.H. *et al, J. Org. Chem.*, 1972, **37**, 617 (*synth, props*)
Ger. Pat., 2 013 956, (*1971*); *CA*, **74**, 87380 (*deriv*)

Dimethylphosphoramidofluoridodithioic acid, 9CI, 8CI **D-00867**

$Me_2N(F)P(=S)SH$

$C_2H_7FNPS_2$ M 159.176

Esters are chiral.

Me ester: Methyl dimethylphosphoramidofluoridodithioate. Methyl fluoro(dimethylamido)dithiophosphate.
$C_3H_9FNPS_2$ M 173.203
Liq. $Bp_{0.01}$ 38-41°.

Et ester: [20494-70-6]. *Ethyl dimethylphosphoramidofluoridodithioate. Ethyl fluoro(dimethylamido)-dithiophosphate.*
$C_4H_{11}FNPS_2$ M 187.230
Liq. $Bp_{0.02}$ 33.5-35.5°.

Ph ester: [17620-74-5]. *Phenyl dimethylphosphoramidofluoridodithioate. Ethyl fluoro(dimethylamido)-dithiophosphate.*
$C_8H_{11}FNPS_2$ M 235.274
Liq. $Bp_{0.3}$ 103-104°.

Roesky, H.W., *Chem. Ber.*, 1968, **101**, 636, 2977 (*synth, ir, ms, pmr, F and P nmr*)

Dimethylphosphoramidofluoridothioic acid, 9CI, 8CI **D-00868**

$Me_2NPF(S)OH \rightleftharpoons Me_2NPF(O)SH$

C_2H_7FNOPS M 143.116

(±)-*form*

O-*Et ester:* O-*Ethyl dimethylphosphoramidofluoridothioate.* O-*Ethyl fluorodimethylamidothiophosphate.*
$C_4H_{11}FNOPS$ M 171.169
Liq. Bp_6 41-42°. n_D^{20} 1.4478.

O-*Ph ester:* O-*Phenyl dimethylphosphoramidofluoridothioate.* O-*Phenyl fluorodimethylamidothiophosphate.*
$C_8H_{11}FNOPS$ M 219.213
Liq. Bp_3 93-93.5°. n_D^{20} 1.5268.

Olah, G.A. *et al, Justus Liebigs Ann. Chem.*, 1957, **602**, 123 (*ethyl ester*)
Olah, G.A. *et al, Can. J. Chem.*, 1960, **38**, 2053 (*phenyl ester*)

Dimethylphosphoramidoselenoic acid, 9CI **D-00869**

Selenophosphoric acid dimethylamide

$Me_2NP(Se)(OH)_2 \rightleftharpoons Me_2NP(O)(SeH)(OH)$

$C_2H_8NO_2PSe$ M 188.024

O,O-*Di-Me ester:* [56595-15-4]. O,O-*Dimethyl dimethylphosphoramidoselenoate.*
$C_4H_{12}NO_2PSe$ M 216.078
Liq. Mp −19°. $Bp_{7.5}$ 86-87°.

O,O-*Di-Et ester:* [24330-42-5]. O,O-*Diethyl dimethylphosphoramidoselenoate.*
$C_6H_{16}NO_2PSe$ M 244.132
Liq. d_4^{20} 1.29. Bp_{19} 108°. n_D^{20} 1.4870.

Difluoride: [53213-42-6].
$C_2H_6F_2NPSe$ M 192.007
Liq. Bp_{30} 40°. Sensitive to air and moisture.

Nikonorov, L.K. *et al, Izv. Akad. Nauk SSSR, Ser. Khim.*, 1969, 464 (*Engl. transl.* p. 413) (*diethyl ester, synth, ir*)
Roesky, H.W. *et al, Z. Naturforsch., B*, 1973, **28**, 697 (*difluoride, synth, ir, pmr, ms, P and F nmr*)
Pohl, W. *et al, An. Quim.*, 1974, **70**, 1209 (*dimethyl ester, synth, ir, raman*)

_O,O_-Dimethyl phosphoramidothioate, 9CI, 8CI **D-00870**

O,O-*Dimethyl thiophosphoryl amide.* O,O-*Dimethyl amidothiophosphate*

[17321-47-0]

$(MeO)_2P(S)NH_2$

$C_2H_8NO_2PS$ M 141.124

Liq. d_4^{20} 1.26. Bp_{10} 105-108°, $Bp_{1.2}$ 72-73°. n_D^{20} 1.4982.

Mel'nikov, N.N. *et al, Zh. Obshch. Khim.*, 1955, **25**, 828 (*Engl. transl.* p. 793) (*synth*)
Nyquist, R.A. *et al, Spectrochim. Acta, Part A*, 1967, **23**, 2505 (*ir*)
Quistad, G.B. *et al, J. Agric. Food Chem.*, 1970, **18**, 189 (*synth*)
Fahmy, F.A.H. *et al, J. Org. Chem.*, 1972, **37**, 617 (*props*)
Meyer, H.J. *et al, Bull. Soc. Chim. Belg.*, 1978, **87**, 517 (*ms*)
Schneider, P. *et al, J. Prakt. Chem.*, 1982, **324**, 1063 (*props*)

Dimethylphosphoramidothioic acid, 9CI, 8CI **D-00871**

Thiophosphoric dimethylamide

$Me_2NP(S)(OH)_2 \rightleftharpoons Me_2NP(O)(OH)(SH)$

$C_2H_8NO_2PS$ M 141.124

O,O-*Di-Me ester:* [28167-51-3]. O,O-*Dimethyl dimethylphosphoramidothioate.*
$C_4H_{12}NO_2PS$ M 169.178
Liq. d_4^{20} 1.17. Bp_9 77-78°. n_D^{20} 1.4766.

O,O-*Di-Et ester:* [4167-52-6]. O,O-*Diethyl dimethylphosphoramidothioate.*
$C_6H_{16}NO_2PS$ M 197.232

Liq. d_4^{20} 1.06. Bp_3 97-100°. n_D^{20} 1.4635.
O,O-*Diisopropyl ester:* O,O-*Diisopropyl dimethylphosphoramidothioate.*
$C_8H_{20}NO_2PS$ M 225.285
Liq. d_4^{20} 1.01. $Bp_{0.6}$ 71°. n_D^{20} 1.4585.
O,O-*Di*-tert-*butyl ester:* O,O-*Di*-tert-*butyl dimethylphosphoramidothioate.* O,O-*Bis*(*1,1-dimethylethyl*) *dimethylphosphoramidothioate.*
$C_{10}H_{24}NO_2PS$ M 253.339
Liq. d_{18}^{18} 1.01. $Bp_{0.5}$ 76°. n_D^{19} 1.463.
Difluoride: see Dimethylphosphoramidothioic difluoride, D-00873
Dichloride: see Dimethylphosphoramidothioic dichloride, D-00872
Diamide: NN-*Dimethylphosphorothioic triamide.*
$C_2H_{10}N_3PS$ M 139.155
Solid. Mp 107°.

Mel'nikov, N.N. *et al*, *Zh. Obshch. Khim.*, 1955, **25**, 828 (*Engl. transl.* p. 793) (*esters*)
Goehring, M. *et al*, *Chem. Ber.*, 1956, **89**, 1768 (*diamide*)
Burgada, R. *et al*, *Bull. Soc. Chim. Fr.*, 1963, 2154 (*ester, synth, pmr*)
Quistad, G.B. *et al*, *J. Agric. Food Chem.*, 1970, **18**, 189 (*dimethyl ester*)
Greenhalgh, R. *et al*, *J. Agric. Food Chem.*, 1975, **23**, 325 (*synth*)

Dimethylphosphoramidothioic dichloride, 9CI D-00872

(*Dimethylamino*)*phosphonothioyl dichloride*
[1498-65-3]

$Me_2NP(S)Cl_2$

$C_2H_6Cl_2NPS$ M 178.016
Liq. d_4^{20} 1.377. Bp_{10} 75-76°. n_D^{20} 1.5404.

Godovikov, N.N. *et al*, *Zh. Obshch. Khim.*, 1961, **31**, 1628 (*Engl. transl.* p. 1516) (*synth*)
Keat, R. *et al*, *J. Chem. Soc.* (*A*), 1968, 703 (*pmr*)
Köttgen, D. *et al*, *Z. Anorg. Allg. Chem.*, 1971, **385**, 56 (*ir, raman*)
Osokin, D.Ya. *et al*, *Org. Magn. Reson.*, 1972, **4**, 831 (*nqr*)
Vilesov, F.I. *et al*, *Z. Phys. Chem.* (*Leipzig*), 1974, **255**, 661 (*pe*)
Shagidullin, R.R. *et al*, *Dokl. Akad. Nauk SSSR, Ser. Sci. Khim.*, 1975, **222**, 897 (*Phys. Chem.*, *Engl. transl.* p. 564) (*uv*)

Dimethylphosphoramidothioic difluoride, 9CI, 8CI D-00873

(*Dimethylamino*)*phosphonothioic difluoride*
[812-13-5]

$Me_2NP(S)F_2$

$C_2H_6F_2NPS$ M 145.107
Liq. Bp_{30} 33-34°. n_D^{20} 1.4190.

Olah, G.A. *et al*, *Justus Liebigs Ann. Chem.*, 1959, **625**, 88 (*synth*)
Horn, H.-G. *et al*, *Z. Naturforsch.*, *B*, 1966, **21**, 617, 729 (*pmr, F and P nmr, ir*)
Reddy, G.S. *et al*, *Z. Naturforsch*, *B*, 1970, **25**, 1199 (*synth, pmr, P nmr*)
Köttgen, D. *et al*, *Z. Phys. Chem.* (*Frankfurt am Main*), 1974, **92**, 285 (*ir, raman*)

Dimethylphosphoramidotrithioic acid, 9CI, 8CI D-00874

$Me_2NP(S)(SH)_2$

$C_2H_8NPS_3$ M 173.246

Di-Ph ester: Diphenyl dimethylphosphoramidotrithioate.
$C_{14}H_{16}NPS_3$ M 325.441
Visc. liq. $Bp_{0.05}$ 110°.

Fr. Pat., 1 378 035, (*1964*); *CA*, **62**, 10371

Dimethylphosphoramidous acid, 9CI D-00875

$Me_2NP(OH)_2$

$C_2H_8NO_2P$ M 109.064
Di-Me ester: [20217-54-3]. *Dimethyl dimethylphosphoramidite.*
$C_4H_{12}NO_2P$ M 137.118
Liq. Bp_{190} 88-89°.
Di-tert-*butyl ester: Di*-tert-*butyl dimethylphosphoramidite.*
$C_{10}H_{24}NO_2P$ M 221.279
Liq. Bp_8 60°. n_D^{17} 1.433.
Di-Ph ester: [19620-78-1]. *Diphenyl dimethylphosphoramidite.*
$C_{14}H_{16}NO_2P$ M 261.260
Liq. d^{20} 1.11. $Bp_{0.15}$ 118-120°. n_D^{20} 1.5660.
Difluoride: see Dimethylphosphoramidous difluoride, D-00877
Dichloride: see Dimethylphosphoramidous dichloride, D-00876
Dibromide: [20502-37-8].
$C_2H_6Br_2NP$ M 234.858
Liq. d^{20} 1.96. Mp −29°. $Bp_{0.7}$ 26°. n_D^{20} 1.6012.
Diiodide: [20502-39-0].
$C_2H_6I_2NP$ M 328.859
Dark-red liq. Mp −25° to −20°. $Bp_{0.1}$ 69-70°. Stable at r.t. in absence of oxygen, moisture, and light.
Dicyanide:
$C_4H_6N_3P$ M 127.085
Liq. d^{21} 1.08. Mp 11°. Bp_2 59°. n_D^{20} 1.4780.

Nöth, H. *et al*, *Chem. Ber.*, 1961, **94**, 1505; 1963, **96**, 1109 (*dibromide, dicyanide*)
Burgada, R. *et al*, *Bull. Soc. Chim. Fr.*, 1963, 2154 (*ester, synth, ir, nmr*)
Bentrude, W.G. *et al*, *J. Org. Chem.*, 1972, **37**, 642 (*ester, synth, pmr*)
Osokin, D.Ya. *et al*, *Org. Magn. Reson.*, 1972, **4**, 831 (*nqr*)
Gorbatenko, Zh.K. *et al*, *Zh. Obshch. Khim.*, 1974, **44**, 2357 (*Engl. transl.* p. 2311) (*diiodide, ester*)
Lappert, M.F. *et al*, *J. Chem. Soc., Dalton Trans.*, 1975, 1207 (*pe*)
Gorbatenko, Zh.K. *et al*, *Zh. Obshch. Khim.*, 1977, **47**, 1915 (*Engl. transl.* p. 1752) (*diiodide*)
Kukhar', V.P. *et al*, *Zh. Obshch. Khim.*, 1982, **52**, 562 (*Engl. transl.* p. 492) (*ester, synth, P nmr*)
Worley, S.D. *et al*, *J. Electron. Spectrom. Relat. Phenom.*, 1982, **25**, 135 (*esters, pe, struct*)

Dimethylphosphoramidous dichloride, 9CI D-00876

Dichloro(*dimethylamino*)*phosphine*
[683-85-2]

Me_2NPCl_2

$C_2H_6Cl_2NP$ M 145.956
Ligand for Al, Fe, and Rh. Fuming liq. Bp 150°, Bp_{37} 56°. n_D^{20} 1.5020.
▷V. reactive to oxidizing agents. Reacts violently with H_2O and may inflame

Burg, A.B. *et al*, *J. Am. Chem. Soc.*, 1958, **80**, 1107 (*synth*)
Nöth, H. *et al*, *Chem. Ber.*, 1961, **94**, 1505 (*synth*)

Durig, J.R. *et al, J. Phys. Chem.*, 1971, **75**, 3837 (*ir, raman*)
Biryukov, I.P. *et al, Zh. Obshch. Khim.*, 1972, **42**, 1223 (*Engl. transl.* p. 1217) (*nqr*)
Lappert, M. *et al, J. Chem. Soc., Dalton Trans.*, 1975, 1207 (*pe*)
Barlos, K. *et al, Z. Naturforsch., B*, 1978, **33**, 515 (*N nmr*)
Gray, G.A. *et al, Org. Magn. Reson.*, 1980, **14**, 8 (*cmr*)
Duangthai, S. *et al, Org. Magn. Reson.*, 1982, **20**, 33 (*nmr, struct*)
Kukhar', V.P. *et al, Zh. Obshch. Khim.*, 1982, **52**, 562 (*Engl. transl.* p. 492) (*synth, P nmr*)

Dimethylphosphoramidous difluoride, 9CI D-00877

Difluoro(dimethylamino)phosphine
[814-97-1]

Me_2NPF_2

$C_2H_6F_2NP$ M 113.047
Ligand for metals of groups VIB and VIII. Liq. d_4^{23} 1.10. Mp −87°. Bp_{704} 50°. n_D^{20} 1.3605. Readily oxidized.

Nöth, H. *et al, Chem. Ber.*, 1963, **96**, 1298 (*synth*)
Schmutzler, R., *Inorg. Chem.*, 1964, **3**, 415 (*ir, F and P nmr*)
Morris, E.D. *et al, Inorg. Chem.*, 1969, **8**, 1673 (*cryst struct*)
Cradock, S. *et al, J. Chem. Soc., Faraday Trans. 2*, 1972, **68**, 940 (*pe*)
Osokin, D.Ya. *et al, Org. Magn. Reson.*, 1972, **4**, 831 (*nqr*)
Bach, M.-C. *et al, J. Mol. Struct.*, 1973, **17**, 23 (*struct*)
Forti, P. *et al, J. Am. Chem. Soc.*, 1973, **95**, 756 (*microwave, struct*)
Lappert, M. *et al, J. Chem. Soc., Dalton Trans.*, 1975, 1207 (*pe*)
Barlos, K. *et al, Z. Naturforsch., B*, 1978, **33**, 515 (*N nmr*)
Hill, W.E. *et al, Inorg. Chim. Acta*, 1979, **35**, 135 (*synth, complexes*)

Dimethyl phosphorazidate, 9CI, 8CI D-00878

Dimethyl phosphoryl azide
[57468-68-5]

$(MeO)_2P(O)N_3$

$C_2H_6N_3O_3P$ M 151.061
Liq. d_4^{20} 1.30. Bp_{10} 79.5-81°. n_D^{20} 1.4276.
▷Probably highly toxic

Kabachnik, M.I. *et al, Izv. Akad. Nauk SSSR, Ser. Khim.*, 1961, 819 (*Engl. transl.* p. 758) (*synth*)
Buder, W. *et al, Z. Anorg. Allg. Chem.*, 1975, **418**, 72 (*P nmr, ir, raman*)
Felcht, U. *et al, Justus Liebigs Ann. Chem.*, 1977, 1309 (*synth, ir, pmr*)
Müller, J. *et al, Z. Anorg. Allg. Chem.*, 1979, **450**, 149 (*P nmr, ir, raman*)

O,O-Dimethyl phosphorazidothioate, 9CI, 8CI D-00879

O,O-*Dimethyl thiophosphoryl azide.* O,O-*Dimethyl azidothiophosphate*
[51576-57-9]

$(MeO)_2P(S)N_3$

$C_2H_6N_3O_2PS$ M 167.122
Liq. $Bp_{0.5}$ 38-39°.

Ger. Pat., 880 443, (*1953*); *CA*, **48**, 12167

N,N-Dimethylphosphoric triamide, 9CI D-00880

[19316-37-1]

$Me_2NP(O)(NH_2)_2$

$C_2H_{10}N_3OP$ M 123.094
Efficient plant fertilizer. Cryst. (Me_2CO). Mp 119°. Low stability at pH 4-8.

N′,N″-*Di-Me:* [16853-36-4]. N,N,N′,N″-*Tetramethylphosphoric triamide.*
$C_4H_{14}N_3OP$ M 151.148
Oil which solidifies. Mp 38°. Bp_2 145°.

Goehring, M. *et al, Chem. Ber.*, 1956, **89**, 1768 (*synth*)
Terry, P.H. *et al, J. Agric. Food Chem.*, 1973, **21**, 500 (*deriv*)
Baldwin, M.A. *et al, Org. Mass Spectrom.*, 1977, **12**, 279 (*synth, deriv, ms*)
Roemer, W. *et al, CA*, 1979, **91**, 73618 (*props*)

2,6-Dimethylphosphorin, 9CI D-00881

2,6-Dimethylphosphabenzene. 2,6-Dimethylphosphinine
[56577-95-8]

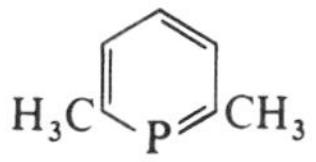

C_7H_9P M 124.122

Ashe, A.J. *et al, Tetrahedron Lett.*, 1975, 1083 (*synth, pmr*)

1,3-Dimethylphosphorinane, 10CI D-00882

1,3-Dimethylphosphacyclohexane. 1,3-Dimethylphosphinane

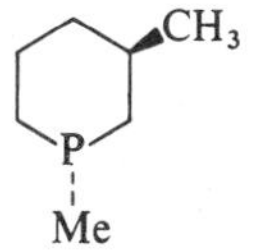

(1*RS*,3*RS*)-*form*

$C_7H_{14}P$ M 129.161
Liq. Bp_{57} 81-85°. Mixt. of *cis*- and *trans*-form.

B,MeI: 1,1,3-Trimethylphosphorinanium iodide. 1,1,3-Trimethylphosphinanium iodide.
$C_8H_{17}IP$ M 271.101
Cryst. (EtOH/EtOAc). Mp 213-214°.

Oxide:
$C_7H_{14}OP$ M 145.161
Liq. Mixt. of stereoisomers.

Sulfide:
$C_7H_{14}PS$ M 161.221
Solid. Mp 43-58°. Mixt. of stereoisomers.

Quin, L.D. *et al, J. Org. Chem.*, 1978, **43**, 1424 (*synth, derivs, cmr, P nmr, pmr*)

1,4-Dimethylphosphorinane, 10CI D-00883

1,4-Dimethylphosphacyclohexane. 1,4-Dimethylphosphinane

$C_7H_{14}P$ M 129.161
Liq. Bp_{45-49} 68-74°. Mixt. of *cis*- and *trans*-forms.

B,MeI: 1,1,4-Trimethylphosphoranium iodide. 1,1,4-Trimethylphosphinanium iodide.
$C_8H_{17}P^{\oplus}$ M 144.196 (ion)
Cryst. ($CHCl_3$/hexane). Mp 313° dec.

Oxide:
$C_7H_{14}OP$ M 145.161
Liq. Mixt. of stereoisomers.

Sulfide:
C_7H_4PS M 151.142
Solid. Mp 71-94°. Mixt. of stereoisomers.

Quin, L.D. *et al, J. Org. Chem.*, 1978, **43**, 1424 (*synth, derivs, cmr, P nmr, pmr*)

_O,O_-Dimethyl phosphorisocyanatidothioate, 9CI, 8CI D-00884

O,O-*Dimethyl phosphoryl isocyanate*

[20039-32-1]

$(MeO)_2P(S)NCO$

$C_3H_6NO_3PS$ M 167.119

Reactive liq. d_4^{20} 1.27. Bp_8 58-59°. n_D^{20} 1.4720. Sensitive to moisture.

Samarai, L.I. *et al, Zh. Obshch. Khim.*, 1969, **39**, 1511, 1712 (*Engl. transl.* pp. 1480, 1678)

_O,S_-Dimethyl phosphorisocyanatidothioate, 9CI, 8CI D-00885

O,S-*Dimethyl phosphoryl isocyanate*

[25359-60-8]

MeO(MeS)P(=O)NCO

$C_3H_6NO_3PS$ M 167.119

(±)-*form*

Reactive liq. d_4^{20} 1.38. Bp_1 56-58°. n_D^{20} 1.4959. Sensitive to moisture.

Samarai, L.I. *et al, Zh. Obshch. Khim.*, 1969, **39**, 1511 (*Engl. transl.* p. 1480)

Dimethyl phosphorobromidate, 9CI, 8CI D-00886

Dimethyl phosphoryl bromide. Dimethyl bromophosphate. Dimethoxyphosphinyl bromide

[24167-74-6]

$(MeO)_2P(O)Br$

$C_2H_6BrO_3P$ M 188.945

Fuming liq. $Bp_{0.8}$ 45-47°.

Goldwhite, H. *et al, J. Chem. Soc.*, 1955, 3564 (*synth*)
Kainosho, M. *et al, J. Phys. Chem.*, 1970, **74**, 2853 (*pmr*)
Satterthwait, A.C. *et al, J. Am. Chem. Soc.*, 1980, **102**, 4464 (*pmr, P nmr*)

Dimethyl phosphorochloridate, 9CI, 8CI D-00887

Dimethyl phosphoryl chloride. Dimethoxyphosphinyl chloride. Dimethyl chlorophosphate

[813-77-4]

$(MeO)_2P(O)Cl$

$C_2H_6ClO_3P$ M 144.494

Fireproofing agent. Synthetic intermediate. Liq. Bp_{25} 80°. n_D^{25} 1.4107.

McIvor, R.A. *et al, Can. J. Chem.*, 1956, **34**, 1819 (*synth*)
Sosnovsky, G. *et al, J. Org. Chem.*, 1969, **34**, 968 (*synth*)
Kainosho, M. *et al, J. Phys. Chem.*, 1970, **74**, 2853 (*pmr*)
Pritchard, J.G., *Org. Mass. Spectrom.*, 1970, **3**, 163 (*ms*)
Hornung, V. *et al, Z. Anorg. Allg. Chem.*, 1971, **380**, 137 (*ir, raman*)
Maier, L., *Helv. Chim. Acta*, 1973, **56**, 492 (*P nmr*)

_S,S_-Dimethyl phosphorochloridodithioate, 9CI, 8CI D-00888

S,S-*Dimethyl chlorodithiophosphate*

[28522-96-5]

$(MeS)_2P(O)Cl$

$C_2H_6ClOPS_2$ M 176.616

Liq. $Bp_{0.8}$ 60-61°.

Wafa, O.A. *et al, Z. Anorg. Allg. Chem.*, 1970, **378**, 273 (*synth, ir, raman*)
Goubeau, J. *et al, Spectrochim. Acta, Part A*, 1971, **27**, 1703 (*ir*)

Dimethyl phosphorochloridodithioite, 9CI D-00889

Dimethyl dithiochlorophosphite

[14061-49-5]

$(MeS)_2PCl$

$C_2H_6ClPS_2$ M 160.616

Liq. Bp_2 65-67°. n_D^{20} 1.6309.

Fritzowsky, N. *et al, Z. Anorg. Allg. Chem.*, 1971, **386**, 67 (*synth, ir, raman*)

Dimethyl phosphorochloridoselenoite, 9CI D-00890

Dimethyl chlorodiselenophosphite

[55776-66-4]

$(MeSe)_2PCl$

$C_2H_6ClPSe_2$ M 254.416

Viscous yellow oil. Thermally stable.

Anderson, J.W. *et al, Inorg. Nucl. Chem. Lett.*, 1975, **11**, 233 (*synth, pmr*)

_O,O_-Dimethyl phosphorochloridothioate, 9CI, 8CI D-00891

OO-*Dimethyl chlorothiophosphate.* OO-*Dimethyl thiophosphoryl chloride*

[2524-03-0]

$(MeO)_2P(S)Cl$

$C_2H_6ClO_2PS$ M 160.555

Synergist for insecticides; synthetic intermediate. Liq. d 1.32. Bp_{12} 68°. n_D^{25} 1.4776. Sensitive to H_2O.

▷TD1830000.

Fletcher, J.H. *et al, J. Am. Chem. Soc.*, 1950, **72**, 2461 (*synth*)
Vasil'ev, A.F., *Zh. Prikl. Spektrosk.*, 1966, **5**, 524 (*Engl. transl.* p. 391) (*ir, struct*)
Cooks, R.G. *et al, J. Chem. Soc. (B)*, 1968, 1327 (*ms*)
Nyquist, R.A. *et al, Spectrochim. Acta, Part A*, 1968, **74**, 187 (*ir, raman*)
Kainosho, M. *et al, Bull. Chem. Soc. Jpn.*, 1969, **42**, 1713 (*pmr*)
Pritchard, J.G., *Org. Mass Spectrom.*, 1970, **3**, 163 (*ms*)
Hornung, V. *et al, Z. Anorg. Allg. Chem.*, 1971, **380**, 137 (*ir, raman*)
Whitehead, M.A. *et al, J. Chem. Soc. (A)*, 1971, 1738 (*nqr*)
Omelańczuk, J. *et al, Tetrahedron*, 1975, **31**, 2809 (*synth, props, P nmr*)
Shagidullin, R.R. *et al, Dokl. Akad. Nauk SSSR, Ser. Sci. Khim.*, 1975, **222**, 897 (*Engl. transl.* p. 564) (*uv*)

O,S-Dimethyl phosphorochloridothioate, 9CI, 8CI **D-00892**

O,S-*Dimethyl chlorothiophosphate*

[3711-50-0]

MeO–P(=O)(–Cl)–SMe (*R*)-*form*

$C_2H_6ClO_2PS$ M 160.555

(***R***)***-form*** [71877-80-0]

Obt. admixed with 32% (*S*)-isomer. Opt. rotn. not reported.

(***S***)***-form*** [71877-81-1]

Obt. admixed with the (*R*)-form. Opt. rotn. not reported.

(±)-***form***

Nucleoside phosphorylating agent. Liq. Bp_3 63-66°. n_D^{20} 1.4928 (impure).

Lippman, A.E., *J. Org. Chem.*, 1965, **30**, 3217 (*synth, pmr, P nmr*)
Hall, C.R. *et al, J. Chem. Soc., Perkin Trans. 1*, 1979, 1646 (*pmr*)
Asseline, U. *et al, Tetrahedron Lett.*, 1981, **22**, 847 (*use*)

Dimethyl phosphorochloridotrithioate, 9CI, 8CI **D-00893**

Dimethyl chlorotrithiophosphate

[16001-03-9]

$(MeS)_2P(S)Cl$

$C_2H_6ClPS_3$ M 192.676

Liq. d_4^{20} 1.409. Bp_2 92-94°. n_D^{20} 1.6535.

Murav'ev, I.V. *et al, Zh. Obshch. Khim.*, 1975, **45**, 1746 (*Engl. transl.* p. 1711) (*synth*)
Wafa, O.A. *et al, Z. Anorg. Allg. Chem.*, 1970, **378**, 273 (*synth, ir, raman*)

Dimethyl phosphorocyanidate **D-00894**

Dimethyl phosphoryl cyanide. Dimethoxyphosphinyl cyanide

[3853-90-2]

$(MeO)_2P(O)CN$

$C_3H_6NO_3P$ M 135.059

Liq. Bp_4 65-66°.

Shiori, T. *et al, Tetrahedron*, 1976, **32**, 2211 (*synth, ir, pmr*)

N,N′-Dimethylphosphorodiamidic acid, 9CI, 8CI **D-00895**

Phosphoric acid bismethylamide

$(MeNH)_2P(O)OH$

$C_2H_9N_2O_2P$ M 124.079

Et ester: [41265-74-1]. *Ethyl* N,N′-*dimethylphosphorodiamidate.*
$C_4H_{13}N_2O_2P$ M 152.133
Liq. or solid. Mp 35-38°. Bp_3 132-136°, $Bp_{0.005}$ 110-111°. n_D^{25} 1.4489.

Ph ester: see Diamidafos, D-00024

Arceneaux, R.L. *et al, J. Org. Chem.*, 1959, **24**, 1419 (*ester*)
Mulliez, M., *Phosphorus Sulfur*, 1980, **8**, 27 (*synth, pmr*)

N,N′-Dimethylphosphorodiamidodithioic acid, 8CI **D-00896**

$(MeNH)_2P(S)SH$

$C_2H_9N_2PS_2$ M 156.200

Methylammonium salt: [20536-00-9]. Solid. Mp 175-178°.

Fluck, E. *et al, Z. Anorg. Allg. Chem.*, 1968, **359**, 102.

O,O-Dimethyl phosphorodiselenoate, 9CI, 8CI **D-00897**

O,O-*Dimethyl phosphorodiselenoic acid.* O,O-*Dimethyl hydrogen diselenophosphate.* O,O-*Dimethyl diselenophosphate*

$(MeO)_2P(Se)SeH$

$C_2H_7O_2PSe_2$ M 251.970

K salt: [34585-92-7]. Soft solid of low Mp. Readily sol. Me_2CO, MeOH, EtOH, insol. Et_2O, pet. ether.

Zemlyanskii, N.I. *et al, Zh. Obshch. Khim.*, 1971, **41**, 1691 (*Engl. transl.* p. 1699) (*synth*)

O,O-Dimethyl phosphorodithioate, 9CI, 8CI **D-00898**

O,O-*Dimethyl dithiophosphate.* O,O-*Dimethyl dithiophosphoric acid*

[756-80-9]

$(MeO)_2P(S)SH$

$C_2H_7O_2PS_2$ M 158.170

Metab. of Malathion, M-00001 . Liq. d_4^{20} 1.29. Bp_4 56-57°, $Bp_{0.15}$ 34-35°. pK_a 1.55 (7% EtOH), pK_a 2.64 (80% EtOH). n_D^{25} 1.5328.

▷TE0525000.

NH_4 salt: [1066-97-3]. Cryst. Mp 145-146° dec.
K salt: Cryst. Mp 171-172° dec.

McIvor, R.A. *et al, Can. J. Chem.*, 1956, **34**, 1819 (*synth*)
Kabachnik, M.I. *et al*, Tet, 1960, **9**, 10 (*synth*)
Zemlyanskii, N.I. *et al, Zh. Obshch. Khim.*, 1961, **30**, 4056 (*Engl. transl.* p. 4018) (*raman*)
Nyquist, R.A. *et al, Spectrochim. Acta, Part A*, 1969, **25**, 47 (*ir*)
Olah, G.A. *et al, J. Org. Chem.*, 1975, **40**, 2582 (*props, pmr, P nmr*)
Wolfe, N.L. *et al, J. Agric. Food Chem.*, 1975, **23**, 1212 (*cmr*)
Paasivirta, J. *et al, Org. Magn. Reson.*, 1977, **9**, 708 (*pmr, cmr, P nmr*)
Lefferts, J.L. *et al, Inorg. Chem.*, 1980, **19**, 1662 (*synth, complexes*)

O,S-Dimethyl phosphorodithioate, 9CI, 8CI **D-00899**

O,S-*Dimethyl hydrogen phosphorodithioate.* O,S-*Dimethyl dithiophosphoric acid.* O,S-*Dimethyl dithiophosphate*

(MeO)(MeS)P(=O)SH ⇌ (MeO)(MeS)P(=S)OH

$C_2H_7O_2PS_2$ M 158.170

(±)-***form*** [68897-36-9]

K salt: [18226-72-7]. Solid or cryst. (Et_2O). Mp 152.5-153.5°.

Cyclohexylammonium salt: [77547-94-5]. Solid. Mp 202°.

Itskova, A.L. *et al, Zh. Obshch. Khim.*, 1968, **38**, 2556 (*Engl. transl.* p. 2471) (*synth*)
Mikolajczyk, M. *et al, J. Am. Chem. Soc.*, 1978, **100**, 7003 (*pmr*)
Kutyrev, G.A. *et al, Zh. Obshch. Khim.*, 1980, **50**, 2738 (*Engl. transl.* p. 2216) (*synth*)

S,S-Dimethyl phosphorodithioate, 9CI, 8CI D-00900

S,S-*Dimethyl hydrogen dithiophosphate*. S,S-*Dimethyl dithiophosphoric acid*

$(MeS)_2P(O)OH$

$C_2H_7O_2PS_2$ M 158.170

K salt: [22608-55-5]. Hygroscopic cryst. Mp 214-215°, Mp >250°.

Fluoride: [25237-22-3]. S,S-*Dimethyl phosphorofluoridodithioate*. S,S-*Dimethyl fluorodithiophosphate*.
$C_2H_6FOPS_2$ M 160.161
Liq. d_4^{20} 1.36. $Bp_{0.03}$ 30-31°. n_D^{20} 1.5223.

Chloride: see S,S-Dimethyl phosphorochloridodithioate, D-00888

Amide: see S,S-Dimethyl phosphoramidodithioate, D-00866

Anilide: see S,S-Dimethyl phosphoramidodithioate, D-00866

U.S.P., 3 309 371, (*1967*); *CA*, **68**, 87013 (*synth*)
Itskova, A.L. *et al*, *Zh. Obshch. Khim.*, 1968, **38**, 2556 (*Engl. transl.* p. 2471) (*synth*)
Roesky, H., *Z. Naturforsch.*, *B*, 1969, **24**, 818 (*fluoride, synth, ir, ms, pmr, F and P nmr*)

Dimethyl phosphorofluoridate, 9CI, 8CI D-00901

Dimethyl phosphoryl fluoride. Dimethyl fluorophosphate. Dimethoxyphosphinyl fluoride

[5954-50-7]

$(MeO)_2P(O)F$

$C_2H_6FO_3P$ M 128.040

Liq. Bp 149°, Bp_{20} 58-59°.

▷Neurotoxic. TE6125000.

Olah, G.A. *et al*, *J. Org. Chem.*, 1959, **24**, 1568 (*synth*)
Landau, M.A. *et al*, *Zh. Strukt. Khim.*, 1970, **11**, 513 (*Engl. transl.* p. 467) (*struct*)
Köttgen, D. *et al*, *Z. Anorg. Allg. Chem.*, 1974, **405**, 275 (*synth, ir, raman*)
Ryzhikov, B.D. *et al*, *Zh. Strukt. Khim.*, 1975, **16**, 754 (*Engl. transl.* p. 700) (*pmr*)
Mager, P.P., *Toxicol. Lett.*, 1982, **11**, 67 (*tox*)
Gubaidullin, M.G., *Zh. Obshch. Khim.*, 1982, **52**, 2469 (*Engl. transl.* p. 2182) (*props*)

O,O-Dimethyl phosphorofluoridothioate, 9CI, 8CI D-00902

O,O-*Dimethyl thiophosphoryl fluoride*. O,O-*Dimethyl fluorothiophosphate*. [(*Dimethoxyphosphino*)*thioyl*]-*fluoride*

[428-67-1]

$(MeO)_2P(S)F$

$C_2H_6FO_2PS$ M 144.100

Liq. Bp 136°, Bp_{20} 75-77°. n_D^{20} 1.4528.

Olah, G. *et al*, *Justus Liebigs Ann. Chem.*, 1957, **602**, 118 (*synth*)
Reddy, G.S. *et al*, *Z. Naturforsch.*, *B*, 1970, **25**, 1199 (*F and P nmr*)
Köttgen, D. *et al*, *Z. Anorg. Allg. Chem.*, 1974, **405**, 275 (*synth, ir, raman*)

Dimethyl phosphorofluoridotrithioate, 9CI, 8CI D-00903

Dimethyl fluorotrithiophosphate

[20494-71-7]

$(MeS)_2P(S)F$

$C_2H_6FPS_3$ M 176.222

Liq. d^{20} 1.36. $Bp_{0.01}$ 44-46°. n_D^{20} 1.5913.

Roesky, H., *Chem. Ber.*, 1968, **101**, 2977 (*synth, ir, ms, F nmr*)

Dimethyl phosphorohydrazidate, 9CI, 8CI D-00904

Dimethoxyphosphinylhydrazine

[58816-61-8]

$(MeO)_2P(O)NH^2NH_2$

$C_2H_9N_2O_3P$ M 140.078

Needles (C_6H_6). Mp 73-74°.

N^2-*Isopropylidene:*
$C_5H_{13}N_2O_3P$ M 180.143
Cryst. (cyclohexane). Mp 84-85°.

N^2-*Benzylidene:*
$C_9H_{13}N_2O_3P$ M 228.187
Needles (cyclohexane). Mp 94-95°.

Felcht, U. *et al*, *Justus Liebigs Ann. Chem.*, 1977, 1309 (*synth, derivs, ir, pmr*)

O,O-Dimethyl phosphorohydrazidothioate, 9CI, 8CI D-00905

O,O-*Dimethyl thiophosphoryl hydrazide*

[59895-94-4]

$(MeO)_2P(S)NHNH_2$

$C_2H_9N_2O_2PS$ M 156.139

Oil. d_4^{25} 1.37. n_D^{25} 1.5428.

Tolkmith, H., *J. Am. Chem. Soc.*, 1962, **84**, 2097 (*synth*)

O,O-Dimethyl phosphoroselenoate, 9CI, 8CI D-00906

O,O-*Dimethyl phosphoroselenoic acid*. O,O-*Dimethyl hydrogen phosphoroselenoate*

$(MeO)_2P(Se)OH \rightleftharpoons (MeO)_2P(O)SeH$

$C_2H_7O_3PSe$ M 189.009

Na salt: [58228-28-7]. Needles (MeOAc). V. sol. Py, insol. Et_2O, $CHCl_3$.

K salt: [85290-15-9]. Cryst. (Py/Et_2O).

Foss, O., *Acta Chem. Scand.*, 1947, **1**, 8 (*synth*)
Glidewell, C. *et al*, *J. Chem. Soc., Dalton Trans.*, 1977, 527 (*P nmr*)

O,O-Dimethyl phosphorothioate, 9CI, 8CI D-00907

O,O-*Dimethyl hydrogen thiophosphate*. O,O-*Dimethyl phosphorothioic acid*. O,O-*Dimethyl thiophosphoric acid*

[1112-38-5]

$(MeO)_2P(O)SH \rightleftharpoons (MeO)_2P(S)OH$

$C_2H_7O_3PS$ M 142.109

The thiol acid form predominates in aq. solns. but the thioic in alcohols. Metab. and photodegradation prod. of various insecticides. Liq. d_4^{20} 1.25. $Bp_{0.5}$ 87-87.5°. pK_a 1.18 (H_2O), pK_a 2.5 (EtOH). n_D^{20} 1.4615.

NH_4 salt: [40633-14-5]. Cryst. Mp 70-78°.

Na salt: [23754-87-2]. Solid. Sol. Py, insol. Et_2O, $CHCl_3$. Mp 195-197°.
Trimethylammonium salt: [16284-88-1]. Solid. Mp 44-45°.
Dicyclohexylammonium salt: [13941-61-2]. Cryst. (C_6H_6/hexane). Mp 182°.
Fluoride: see O,O-Dimethyl phosphorofluoridothioate, D-00902
*Chloride: see O,O-*Dimethyl phosphorochloridothioate, D-00891
Amide: see *O,O*-Dimethyl phosphoramidothioate, D-00870
Anilide: see Phenylphosphoramidothioic acid, P-00262
Azide: see *O,O*-Dimethyl phosphorazidothioate, D-00879
Hydrazide: see *O,O*-Dimethyl phosphorohydrazidothioate, D-00905
2-Phenylhydrazide: see 2-Phenylphosphorohydrazidothioic acid, P-00296
Isocyanate: see *O,O*-Dimethyl phosphorisocyanatidothioate, D-00884
Anhydride: see *sym-O,O,O,O*-Tetramethyl dithiopyrophosphate, T-00203

Kabachnik, M.I. *et al*, *Tetrahedron*, 1960, **9**, 10.
Pesin, V.G. *et al*, *Zh. Obshch. Khim.*, 1961, **31**, 2508 (*Engl. transl.* p. 2337) (*synth*)
Rymareva, T.G. *et al*, *Zh. Obshch. Khim.*, 1973, **43**, 676 (*Engl. transl.* p. 671) (*synth*)
Bruzik, K. *et al*, *J. Org. Chem.*, 1981, **46**, 1625 (*synth, P nmr*)
Nasser, F.A.K. *et al*, *J. Organomet. Chem.*, 1983, **244**, 17 (*synth*)

O,S-Dimethyl phosphorothioate, 9CI, 8CI — D-00908

O,S-*Dimethyl thiophosphate.* O,S-*Dimethyl hydrogen thiophosphate.* O,S-*Dimethyl phosphorothioic acid*
[42576-53-4]

MeO(MeS)P(=O)OH

$C_2H_7O_3PS$ M 142.109
Metab. of *O,S*-Dimethyl phosphoramidodithioate, D-00865 .
NH_4 salt: [10574-82-0]. Solid. Mp 164-165°.
Tetramethylammonium salt: Solid or cryst. (Me_2CO/2-propanol). Mp 110°, Mp 195-201°.

Hilgetag, G. *et al*, *J. Prakt. Chem.*, 1959, **8**, 90 (*synth*)
Voronkova, V.V. *et al*, *Zh. Obshch. Khim.*, 1967, **37**, 1017 (*Engl. transl.* p. 961) (*synth*)
Mel'nikov, N.N. *et al*, *Zh. Obshch. Khim.*, 1967, **37**, 1024 (*Engl. transl.* p. 967) (*synth*)
Werner, R.A., *J. Econ. Entomol.*, 1973, **66**, 867.
Hall, C.R. *et al*, *J. Chem. Soc., Perkin Trans. 1*, 1979, 1646 (*derivs, stereochem*)

N,N-Dimethylphosphorothioic triamide, 9CI — D-00909

$Me_2NP(S)(NH_2)_2$

$C_2H_{10}N_3PS$ M 139.155
Solid. Mp 107°.
N′,N″-*Di-Ph:* [57858-66-9]. N,N-*Dimethyl*-N′,N″-*diphenylphosphorothioic triamide.*
$C_{14}H_{18}N_3PS$ M 291.350
Needles (EtOH). Sol. C_6H_6. Mp 209-210°.

Michaelis, A., *Justus Liebigs Ann. Chem.*, 1903, **326**, 129, 201 (*deriv*)
Goehring, M. *et al*, *Chem. Ber.*, 1956, **89**, 1768 (*synth*)

(2,2-Dimethylpropylidyne)phosphine, 9CI — D-00910

(tert-*Butylmethylidyne*)*phosphine*
[78129-68-7]

$(H_3C)_3CC\equiv P$

C_5H_9P M 100.100
Ligand for Pt.

Becker, G. *et al*, *Z. Naturforsch.*, *B*, 1981, **36**, 16.
Burckett-St. Laurent, J.C.T.R. *et al*, *J. Chem. Soc., Chem. Commun.*, 1981, 1141 (*complex, cryst struct*)
Oberhammer, H. *et al*, *J. Mol. Struct.*, 1981, **75**, 283 (*struct*)
Becker, G. *et al*, *Phosphorus Sulfur*, 1983, **14**, 267 (*rev*)

(1,1-Dimethylpropyl)phosphonic acid, 9CI — D-00911

tert-*Amylphosphonic acid.* (*2-Methyl-2-butyl*)-*phosphonic acid*
[4672-27-9]

$H_3CCH_2C(CH_3)_2P(O)(OH)_2$

$C_5H_{13}O_3P$ M 152.130
Solid. Mp 147-148°. pK_{a1} 2.88, pK_{a2} 8.96 (H_2O, 25°).

Murav'ev, I.V. *et al*, *Zh. Obshch. Khim.*, 1975, **45**, 1746 (*Engl. transl.* p. 1711) (*synth*)

(2,2-Dimethylpropyl)phosphonic acid, 9CI — D-00912

Neopentylphosphonic acid

$(H_3C)_3CCH_2P(O)(OH)_2$

$C_5H_{13}O_3P$ M 152.130
pK_{a1} 2.84, pK_{a2} 8.65 (H_2O, 25°).
Di-Me ester: [54552-76-0]. *Dimethyl* (*2,2-dimethylpropyl*)*phosphonate. Dimethyl neopentylphosphonate.*
$C_7H_{17}O_3P$ M 180.183
Liq. $Bp_{0.65}$ 52°. n_D^{20} 1.4270.
Di-Ph ester: [54552-75-9]. *Diphenyl* (*2,2-dimethylpropyl*)*phosphonate. Diphenyl neopentylphosphonate.*
$C_{17}H_{21}O_3P$ M 304.325
Solid. Mp 57.5-58°.
Dichloride: [54552-71-5].
$C_5H_{11}Cl_2OP$ M 189.021
Solid/liq. Mp 18°. Bp_{15} 96°. n_D^{20} 1.4651.

Hilbert, C. *et al*, *J. Prakt. Chem.*, 1974, **316**, 790 (*derivs, synth, pmr*)

2,2-Dimethylpropyl phosphorodichloridite, 9CI — D-00913

Neopentyl phosphorodichloridite. Neopentyl dichlorophosphite

$(H_3C)_3CCH_2OPCl_2$

$C_5H_{11}Cl_2OP$ M 189.021
Fuming liq. Bp_{10} 47.5-48°.

Gerrard, W. *et al*, *J. Chem. Soc.*, 1950, 2088 (*synth, props*)

(2,2-Dimethylpropyl)-triphenylphosphonium(1+) D-00914

$Ph_3P^{\oplus}CH_2C(CH_3)_3$

$C_{23}H_{26}P^{\oplus}$ M 333.432 (ion)

Iodide:
$C_{23}H_{26}IP$ M 460.337
Source of ylide. Solid. Mp 197°, 209°.

Ylide: [3739-96-6]. *(2,2-Dimethylpropylidene)-triphenylphosphorane.*
$C_{23}H_{25}P$ M 332.424
Wittig reagent prepd. *in situ*. Deep-orange.

Seyferth, D. *et al*, *J. Am. Chem. Soc.*, 1965, **87**, 4156 (*synth, ylide*)
Reith, B.A. *et al*, *J. Org. Chem.*, 1974, **39**, 2728 (*synth, props*)

5,6-Dimethylspiro[1,3,2-benzodioxaphosphole-2,1′(3′H)-[2,1]benzoxaphosphole, 9CI D-00915
[79157-74-7]

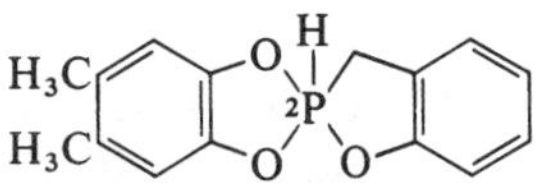

$C_{15}H_{15}O_3P$ M 274.255
Cryst. (hexane). Mp 131-134°.

2-Phenoxy: [79157-76-9].
$C_{21}H_{19}O_4P$ M 366.352
Cryst. (hexane). Mp 105-108°.

2-Ph: [79157-78-1].
$C_{21}H_{19}O_3P$ M 350.353
Cryst. (hexane). Mp 136-138°.

2-Dimethylamino: [79157-77-0]. N,N,*5,6-Tetramethylspiro[1,3,2-benzodioxaphosphole-2,1′(3′*H*)-[2,1]-benzoxaphosphol]-2-amine.*
$C_{17}H_{20}NO_3P$ M 317.324
Cryst. (toluene/hexane). Mp 126-129°.

2-Anilino: [79157-75-8]. *5,6-Dimethyl-*N*-phenylspiro[1,3,2-benzodioxaphosphole-2,1′(3′*H*)-[2,1]-benzoxaphosphol]-2-amine.*
$C_{21}H_{20}NO_3P$ M 365.368
Cryst. (toluene/hexane). Mp 153-155°.

Dahl, B.M. *et al*, *J. Chem. Soc., Perkin Trans. 1*, 1981, 2239 (*synth, pmr*)

3,3′-Dimethyl-2,2′-(3H,3′H)-spirobi[1,3,2-benzothiazaphosphole], 9CI D-00916

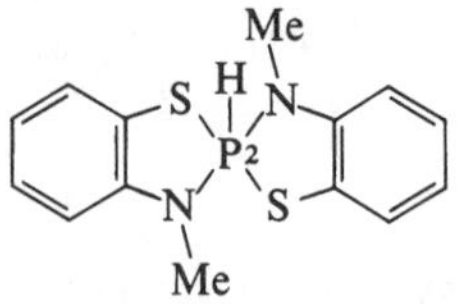

$C_{14}H_{15}N_2PS_2$ M 306.380

2-Chloro: [63503-62-8].
$C_{14}H_{14}ClN_2PS_2$ M 340.825
Solid. Sol. hot MeCN, spar. sol. other org. solvs. Mp 250-252° dec.

Kukhar', V.P. *et al*, *Zh. Obshch. Khim.*, 1977, **47**, 476 (*Engl. transl.* p. 436) (*synth, P nmr*)

3,3′-Dimethyl-2,2(3H,3′H)-spirobi[1,3,2-benzoxazaphosphole], 9CI D-00917
[38057-89-5]

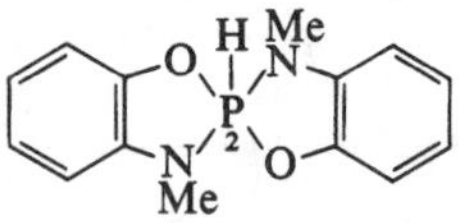

$C_{14}H_{15}N_2O_2P$ M 274.258
Exists completely in pentacoordinate form. Cryst. (C_6H_6). Mp 133°.

2-Me: [79129-18-3]. *2,3,3′-Trimethyl-2,2′(3*H*,3′*H*)-spirobi[1,3,2-benzoxazaphosphole].*
$C_{15}H_{17}N_2O_2P$ M 288.285
Cryst. ($CHCl_3/Et_2O$). Mp 155°.

Burgada, R. *et al*, *J. Organomet. Chem.*, 1974, **66**, 255 (*synth, P nmr, struct*)
Wieber, M. *et al*, *Z. Anorg. Allg. Chem.*, 1981, **477**, 108 (*deriv, synth, pmr, P nmr*)

1,3-Dimethylspiro[1,3,2-diazaphospholidine-2,2′-phenanthro[9,10-d][1,3,2]-dioxaphosphole], 8CI D-00918

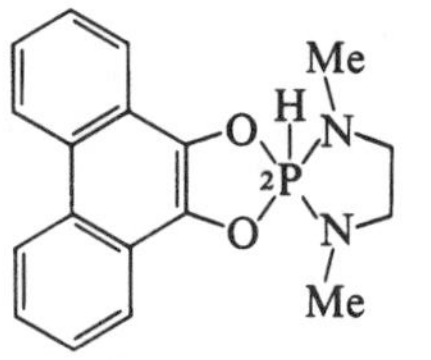

$C_{18}H_{19}N_2O_2P$ M 326.334

2-Methoxy: [15607-06-4].
$C_{19}H_{21}N_2O_3P$ M 356.360
Cryst. (CH_2Cl_2/pentane). Mp 89-91°.

2-Dimethylamino: [15607-05-3]. N,N,*1,3-Tetramethylspiro[1,3,2-diazaphospholidine-2,2′-phenanthro[9,10-*d*][1,3,2]dioxaphosphol]-2-amine.*
$C_{20}H_{24}N_3O_2P$ M 369.402
Cryst. (CH_2Cl_2/pentane). Mp 121-123°.

Ramirez, F. *et al*, *J. Am. Chem. Soc.*, 1967, **89**, 6276 (*synth, ir, pmr, P nmr*)

4,4-Dimethylspiro[1,3,2-dioxaphospholane-2,1′-[2,6,7]trioxa[1]-phosphabicyclo[2.2.1]heptane], 9CI D-00919

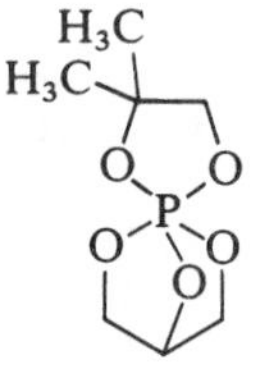

$C_7H_{13}O_5P$ M 208.150
Characterised spectroscopically.

Campbell, B.S. *et al*, *J. Am. Chem. Soc.*, 1976, **98**, 2924 (*synth, pmr, cmr, P nmr*)

O,O-Dimethyl O-4-sulfamoylphenyl phosphorothioate D-00920

O,O-Dimethyl O-(4-aminosulfonylphenyl) phosphorothioate, 9CI. O,O-Dimethyl phosphorothioate, ester with p-hydroxybenzene sulfonamide, 8CI. Cyflee

[115-93-5]

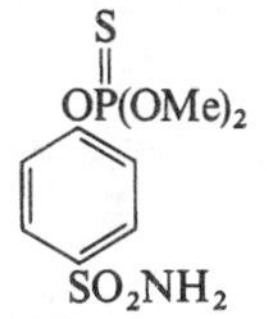

$C_8H_{12}NO_5PS_2$ M 297.280

Vet. insecticide. Solid. Insol. H_2O, sol. C_6H_6, Me_2CO, Et_2O, EtOH. Mp 73-74°.

▷TF7525000.

Pesticide Manual, 7th Ed., 209.

15,16-Dimethyl-1,4,8,11-tetraaza-15,16-diphosphatricyclo[9.3.1.1^{4,8}]hexadecane, 11CI D-00921

[90696-72-3]

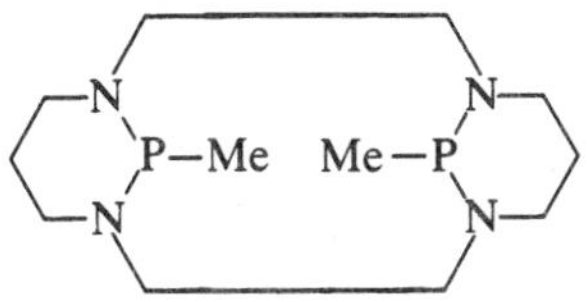

$C_{12}H_{26}N_4P_2$ M 288.312

Potential bidentate ligand, possibly binding *trans* rather than *cis*. Cryst. (toluene/hexane). Mp 157-159°.

Hope, H. *et al*, *Inorg. Chem.*, 1984, **23**, 2550 (*synth, pmr, P nmr, cryst struct*)

6,9-Dimethyl-2,2,3,3-tetrakis(trifluoromethyl)-1,4-dioxa-6,9-diaza-5-phospha(5-P^V)spiro[4.4]nonane, 9CI, 8CI D-00922

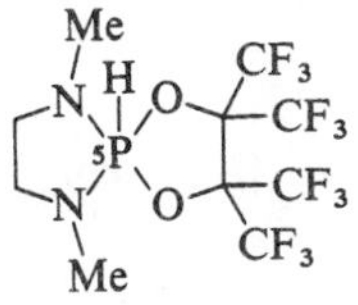

$C_{10}H_{11}F_{12}N_2O_2P$ M 450.164

5-Phenoxy: [53666-65-2]. *6,9-Dimethyl-5-phenoxy-2,2,3,3-tetrakis(trifluoromethyl)-1,4-dioxa-6,9-diaza-5-phospha(5-P^V)spiro[4.4]nonane.*

$C_{16}H_{15}F_{12}N_2O_3P$ M 542.261

Cryst. (pet. ether). Mp 69.5-71°.

5-(Phenylthio): 6,9-Dimethyl-5-phenylthio-2,2,3,3-tetrakis(trifluoromethyl)-1,4-dioxa-6,9-diaza-5-phospha(5-P^V)spiro[4.4]nonane.

$C_{16}H_{15}F_{12}N_2O_2PS$ M 558.321

Cryst. (pet. ether). Mp 46-52°.

5-Dimethylamino: [17244-68-7]. *6,9,N,N-Tetramethyl-2,2,3,3-tetrakis(trifluoromethyl)-1,4-dioxa-6,9-diaza-5-phospha(P^V)spiro[4.4]nonan-5-amine.*

$C_{12}H_{16}F_{12}N_3O_2P$ M 493.232

Cryst. (pet. ether). Mp 86-87° (78-82°).

Ramirez, F. *et al*, *J. Am. Chem. Soc.*, 1967, **89**, 6283 (*synth, ir, pmr*)

Bone, S.A. *et al*, *J. Chem. Soc., Perkin Trans. 1*, 1974, 2125 (*synth, pmr, F and P nmr*)

6,10-Dimethyl-2,2,3,3-tetrakis(trifluoromethyl)-1,4-dioxa-6,10-diaza-5-phospha(5-P^V)spiro[4.5]decane, 10CI D-00923

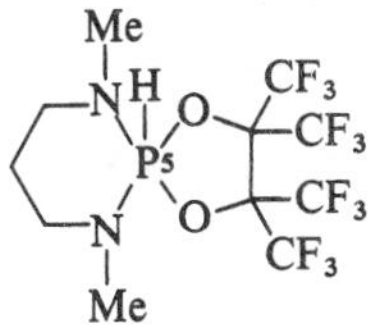

$C_{11}H_{13}F_{12}N_2O_2P$ M 464.190

5-Phenoxy: [62013-28-9].

$C_{17}H_{17}F_{12}N_2O_3P$ M 556.287

Solid. Mp 50-52°.

Bone, S.A. *et al*, *J. Chem. Soc., Perkin Trans. 1*, 1977, 80 (*synth, ms, pmr, F and P nmr*)

2,3-Dimethyl-1,4,6,9-tetraoxa-5-phospha(5-P^V)spiro[4.4]non-2-ene, 9CI, 8CI D-00924

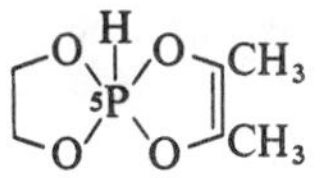

$C_6H_{11}O_4P$ M 178.124

5-Methoxy: [15661-97-9].

$C_7H_{13}O_5P$ M 208.150

Liq. $Bp_{0.1}$ 73°.

5-Dimethylamino: [17454-09-0]. *N,N,2,3-Tetramethyl-1,4,6,9-tetraoxa-5-phospha(5-P^V)spiro[4.4]non-2-en-5-amine.*

$C_8H_{16}NO_4P$ M 221.192

Liq. $Bp_{0.2}$ 94°. n_D^{25} 1.4750.

Ramirez, F. *et al*, *Tetrahedron*, 1968, **24**, 1785 (*synth, ir, pmr, P nmr*)

Dimethylthiodiphosphonic acid, 9CI D-00925

Methylphosphonothioic monoanhydride

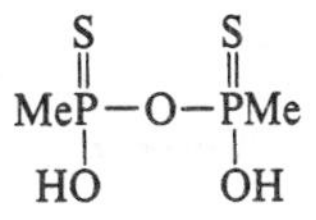

$C_2H_8O_3PS_2$ M 175.177

O,O-Di-Et ester: [34255-82-8]. *O-Ethyl methylphosphonothioic monoanhydride. O-(Ethoxymethylphosphinothioyl) O′-ethyl methylphosphonothioate.*

$C_6H_{16}O_3P_2S_2$ M 262.258

Liq. $Bp_{0.2}$ 93-97°. n_D^{25} 1.5058.

S,S-Di-Et ester: S-Ethyl methylphosphonothioic monoanhydride. O-[(Ethylthio)methylphosphinyl] S-ethyl methylphosphonothioate.

$C_6H_{16}O_3P_2S_2$ M 262.258

Liq. d_4^{20} 1.26. $Bp_{0.04}$ 120-124°. n_D^{20} 1.5200.

O,O-Di-Ph ester: [34255-83-9]. *O-Phenyl ethylphosphonothioic monoanhydride. O-(Phenoxymethylphosphinothioyl) O′-phenyl methylphosphonothioate.*

$C_{14}H_{16}O_3P_2S_3$ M 390.406

Solid. Mp 76-78°.

Difluoride:

$C_2H_6F_2P_2S_3$ M 226.194

Liq. d^{20} 1.44. Bp_2 88-90°. n_D 1.5847. Mixture of stereoisomers.

Roesky, H.W. *et al*, *Chem. Ber.*, 1968, **101**, 3679 (*difluoride, ms, ir, pmr, F nmr, synth*)

Gladshtein, B.M. *et al*, *Zh. Obshch. Khim.*, 1969, **39**, 1001 (*Engl. transl.* p. 972) (*esters*)
Harris, R.K. *et al*, *J. Chem. Soc., Dalton Trans.*, 1972, 1590 (*difluoride, synth, F and P nmr, pmr*)

1,5-Dimethyl-1*H*-1,2,4,3-triazaphosphole, 10CI D-00926

[63148-51-6]

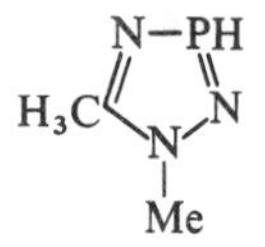

$C_3H_7N_3P$ M 116.082

Forms Cr, Au, Mo complexes. Solid. Mp 39-41°. Bp_{13} 91-93°. Unusual in possessing dicoordinate phosphorus.

Schmidpeter, A. *et al*, *Angew. Chem., Int. Ed. Engl.*, 1977, **16**, 546 (*synth, nmr*)
Pohl, S., *Chem. Ber.*, 1979, **112**, 3159 (*cryst struct*)

Dimethyl 2-(tributylphosphoranylidene)-1,3-dithiol-4,5-dicarboxylate, 9CI D-00927

[68629-93-6]

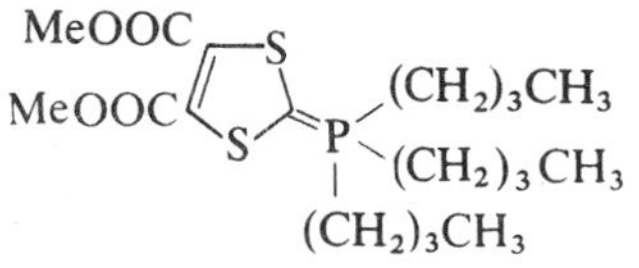

$C_{19}H_{33}O_4PS_2$ M 420.561

Conj. tetrafluoroborate:
$C_{19}H_{34}BF_4O_4PS_2$ M 508.373
Source of ylide. Cryst. ($MeCN/Et_2O$). Mp 120-121°.

Sato, M. *et al*, *J. Org. Chem.*, 1979, **44**, 930 (*synth, props, use*)
Lakshmikantham, M.V. *et al*, *Heterocycles*, 1980, **14**, 271 (*synth*)

***N,N*-Dimethyl-*P*-trifluoromethylphosphonamidothioic fluoride, 9CI** D-00928

[18799-74-1]

$F_3CPF(S)NMe_2$

$C_3H_6F_4NPS$ M 195.114

Volatile liq. V.p. 8.5 mm at 21°, 48 mm at 54°.

Dobbie, R.C. *et al*, *J. Am. Chem. Soc.*, 1968, **90**, 2015 (*ir, pmr, F nmr*)
Cavell, R.G. *et al*, *Inorg. Chem.*, 1978, **17**, 3086 (*synth*)

[3,7-Dimethyl-9-(2,6,6-trimethyl-1-cyclohexen-1-yl)-2,4,6,8-nonatetraenyl]-triphenylphosphonium(1+), 9CI D-00929

Triphenyl(retinyl)phosphonium(1+). Axerophytyltriphenylphosphonium(1+)

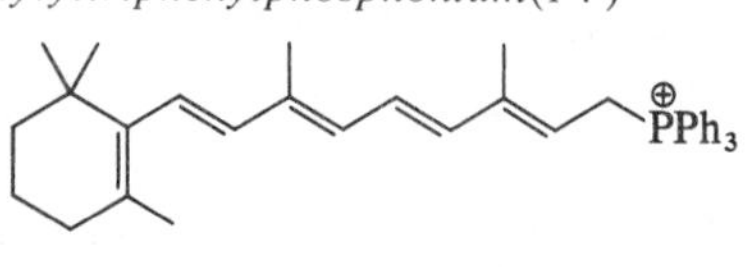

$C_{38}H_{44}P^{\oplus}$ M 531.739 (ion)

***(E,E,E,E)*-form**

Iodide:
$C_{38}H_{44}IP$ M 658.644
Mp 87-89°. $NaNH_2/NH_3(l)$ → ylide.

Periodate: [43207-24-5].
$C_{38}H_{44}IO_4P$ M 722.641
With LiOEt → all *trans*-β-carotene.

Ylide:
$C_{38}H_{43}P$ M 530.731
Wittig reagent used in polyene synth. Violet. Autoxidation → β-carotene.

Fr. Pat., 1 511 772, (*1968*); *CA*, **77**, 88726 (*chloride, synth*)
Bestmann, H.J. *et al*, *Justus Liebigs Ann. Chem.*, 1973, 760 (*periodate, iodide, props*)
Bestmann, H.J. *et al*, *Angew. Chem., Int. Ed. Engl.*, 1976, **15**, 298 (*ylide, props*)
Neidlein, R. *et al*, *Arch. Pharm. (Weinheim, Ger.)*, 1980, **313**, 970 (*ylide, use*)
Byers, J. *et al*, *J. Org. Chem.*, 1983, **48**, 1515 (*chloride, synth, use*)

4,5-Dimethyl-2-trimethylsilyloxy-1,3,2-dioxaphospholane D-00930

2,3-Butylene trimethylsilyl phosphite. 4,5-Dimethyl-2-trimethylsilyloxy-1,3-dioxa-2-phosphacyclopentane

[88576-13-0]

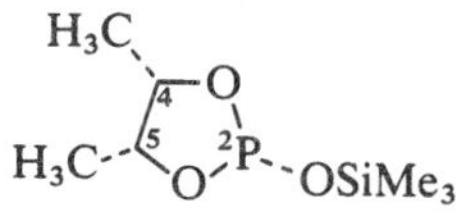

(2α,4α,5α)-*form*

$C_7H_{17}O_3PSi$ M 208.269

(2α,4α,5α)-form [61335-47-5]
cis-*form*

2-Sulfide: O,O-*2,3-Butylene* O-*trimethylsilyl phosphorothioate.* O,O-*2,3-Butylene* O-*trimethylsilyl thiophosphate.*
$C_7H_{17}O_3PSSi$ M 240.329
Liq. $Bp_{0.01}$ 82-84°. n_D^{22} 1.4652.

(2α,4β,5β)-form
trans-*form*
2-Sulfide: Liq. $Bp_{0.01}$ 82-84°. n_D^{22} 1.4650.

(2α,4α,5β)-(±)-form [66288-94-6]
2-Sulfide: Liq. $Bp_{0.01}$ 82-84°. n_D^{20} 1.4655.

Ishmaeva, E.A. *et al*, *Izv. Akad. Nauk SSSR, Ser. Khim.*, 1976, 1880 (*Engl. transl.* p. 1771) (*sulfide*)
Mikolajczyk, M. *et al*, *J. Chem. Soc., Perkin Trans. 1*, 1977, 2213 (*sulfides, synth, pmr, P nmr*)

Dimethyl trimethylsilyl phosphate D-00931

[18135-13-2]

$(MeO)_2P(O)OSiMe_3$

$C_5H_{15}O_4PSi$ M 198.230

Liq. d_4^{20} 1.08. Bp_{12} 87-88°, $Bp_{0.1}$ 39-40°. n_D^{20} 1.4110.

Fehér, F. *et al*, *Chem. Ber.*, 1957, **90**, 134 (*synth, raman*)
Schmidbaur, H. *et al*, *Chem. Ber.*, 1974, **107**, 1731 (*synth*)
Nesterov, L.V. *et al*, *Zh. Obshch. Khim.*, 1978, **48**, 790 (*Engl. transl.* p. 722) (*synth, pmr, P nmr*)
Kolodyazhnyi, Yu.V. *et al*, *Zh. Obshch. Khim.*, 1981, **51**, 806 (*Engl. transl.* p. 665) (*ir, raman, conformn*)

Dimethyl(trimethylsilyl)phosphine, 9CI, 8CI D-00932

(Dimethylphosphino)trimethylsilane

[26464-99-3]

Me_3SiPMe_2

$C_5H_{15}PSi$ M 134.233
Liq. Bp 130°.
▷Spontaneously flammable

Inorg. Synth., 1971, **13**, 26 (*synth*)
Fritz, G. *et al*, *Z. Anorg. Allg. Chem.*, 1974, **409**, 137 (*nmr, pmr*)

8,8-Dimethyl-1,6,10-trioxa-5-phospha(5-P^V)spiro[4.5]dec-2-ene, 10CI — D-00933

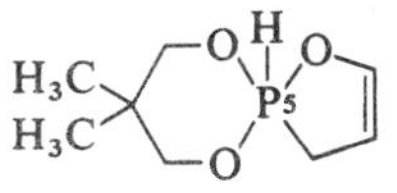

$C_8H_{15}O_3P$ M 190.178

5-Methoxy: [64931-13-1].
$C_9H_{17}O_4P$ M 220.205
Thick liq. d_4^{20} 1.18. $Bp_{0.01}$ 67-68°. n_D^{20} 1.4760. Hydrolyses readily.

Arbuzov, B.A. *et al*, *Izv. Akad. Nauk SSSR, Ser. Khim.*, 1977, 2000 (*Engl. transl.* p. 1852) (*synth, ir, pmr*)

3,3-Dimethyl-1-triphenylphosphoranylidene-2-butanone, 9CI, 8CI — D-00934

(3,3-Dimethyl-2-oxobutylidene)triphenylphosphorane
[26487-93-4]

$Ph_3P{=}CHCOC(CH_3)_3$

$C_{24}H_{25}OP$ M 360.435
Wittig reagent. Useful in heterocycle synth. Cryst. (EtOAc or EtOH aq.). Mp 175-177°.

Gara, A.P. *et al*, *Aust. J. Chem.*, 1970, **23**, 307 (*ms*)
Shevchuk, M.I. *et al*, *Zh. Obshch. Khim.*, 1970, **40**, 48 (*Engl. transl.* p. 45) (*synth, uv*)
Ingham, C.F. *et al*, *Aust. J. Chem.*, 1975, **28**, 2499 (*synth, use*)
Shevchuk, M.I. *et al*, *Zh. Obshch. Khim.*, 1975, **45**, 2609 (*Engl. transl.* p. 2571) (*use*)

Di-4-morpholinylphosphinic acid, 9CI — D-00935

Dimorpholinophosphinic acid, 8CI

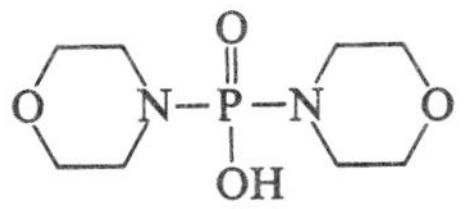

$C_8H_{17}N_2O_4P$ M 236.207

Cyclohexylammonium salt: Solid. Mp 250-252°.
Et ester: [4881-12-3]. *Ethyl di-4-morpholinylphosphinate.*
$C_{10}H_{21}N_2O_4P$ M 264.261
Hygroscopic cryst. (C_6H_6/cyclohexane). Mp 48°. $Bp_{0.003}$ 102-108°.
Cyclohexyl ester: Cyclohexyl di-4-morpholinylphosphinate.
$C_{14}H_{27}N_2O_4P$ M 318.352
Hygroscopic prisms (C_6H_6/pet. ether). Mp 53°.
Ph ester: [4881-17-8]. *Phenyl di-4-morpholinylphosphinate.*
$C_{14}H_{21}N_2O_4P$ M 312.305
Converts benzophenone oximes into benzamidines. Prisms (cyclohexane). Mp 84°.
▷SZ5300000.
Fluoride:
$C_8H_{16}FN_2O_3P$ M 238.198
Cryst. (Et_2O). Mp 40°. Exhibits moderate AChE activity.
Chloride: [7264-90-6].
$C_8H_{16}ClN_2O_3P$ M 254.653
A useful phosphorylating reagent. Cryst. (C_6H_6/cyclohexane or Et_2O). Rather unstable, particularly in a moist atm. Mp 80-82°. $Bp_{0.02}$ 137-140°.
Morpholide: see Trimorpholinophosphine, T-00569

Heap, R. *et al*, *J. Chem. Soc.*, 1948, 1313 (*fluoride, synth, tox*)
Montgomery, H.A.C. *et al*, *J. Chem. Soc.*, 1958, 1963 (*chloride, bromide, ester*)
Cremlyn, R.J.W. *et al*, *J. Chem. Soc. (C)*, 1971, 2028 (*anhydride*)
Ning, R.Y. *et al*, *J. Org. Chem.*, 1976, **41**, 2720 (*chloride, ir, use*)
Jensen, K.G. *et al*, *Acta Chem. Scand., Ser. B*, 1979, **33**, 319 (*ester, use*)
Shevchenko, M.V. *et al*, *Zh. Obshch. Khim.*, 1984, **54**, 1499 (*Engl. transl.* p. 1336) (*synth, nmr*)

Di-4-morpholinylphosphinous acid — D-00936

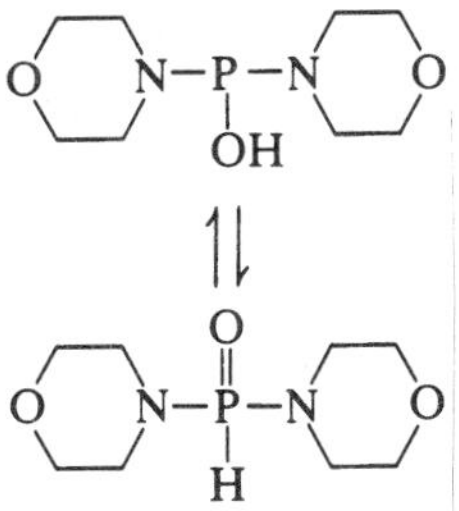

$C_8H_{17}N_2O_3P$ M 220.208

Me ester: [93183-35-8]. *Methyl di-4-morpholinylphosphinite.*
$C_9H_{19}N_2O_3P$ M 234.234
Reagent for phosphitylation of nucleosides and synth. of oligoribonucleotides. Liq. $Bp_{15\mu}$ 82-84°.
Et ester: [4881-11-2]. *Ethyl di-4-morpholinylphosphinite.*
$C_{10}H_{21}N_2O_3P$ M 248.261
Liq. $Bp_{0.08}$ 65-68°. n_D^{20} 1.4934.
Ph ester: [93374-14-2]. *Phenyl di-4-morpholinylphosphinite.*
$C_{14}H_{21}N_2O_3P$ M 296.305
Liq. $Bp_{0.02}$ 99-110°.

Barone, A.D. *et al*, *Nucleic Acids Res.*, 1984, **12**, 4051 (*synth, pmr, P nmr, ms*)
Mitel'man, I.E. *et al*, *Zh. Obshch. Khim.*, 1984, **54**, 1245 (*Engl. transl.* p. 1114) (*synth*)
Shevchenko, M.V. *et al*, *Zh. Obshch. Khim.*, 1984, **54**, 1499 (*Engl. transl.* p. 1336) (*synth, P nmr*)

Di-1-naphthyl phosphate, 8CI — D-00937

Di-1-naphthalenyl phosphate, 9CI. Di-1-naphthyl hydrogen phosphate. Di-1-naphthyl phosphoric acid

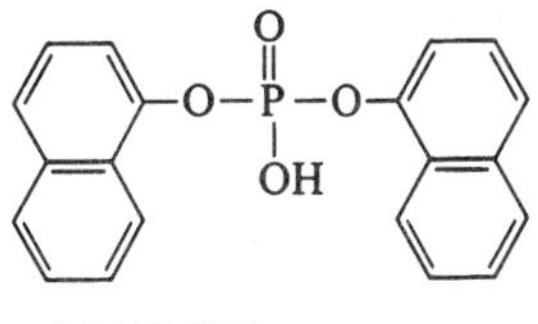

$C_{20}H_{15}O_4P$ M 350.310
Needles (MeOH aq.). Mp 137-139°.
Pyridinium salt: Fluffy needles (C_6H_6). Mp 105-106°.

Chloride: Di-1-naphthyl phosphorochloridate.
$C_{20}H_{14}ClO_3P$ M 368.756
Needles (C_6H_6/pet. ether). Mp 88-90°.

Friedman, O.M. *et al, J. Am. Chem. Soc.*, 1950, **72**, 624 (*synth, chloride*)

Di-2-naphthyl phosphate, 8CI D-00938

Di-2-naphthalenyl phosphate, 9CI. Di-2-naphthyl hydrogen phosphate. Di-2-naphthyl phosphoric acid

$C_{20}H_{15}O_4P$ M 350.310
Needles or prisms. Insol. H_2O, EtOH, pet. ether. Mp 147-148°. pK_{a1} 3.22 (95% EtOH aq.), pK_{a2} 4.68 (EtOH).

Amide: Di-2-naphthyl phosphoramidate.
$C_{20}H_{16}NO_3P$ M 349.325
Solid. Insol. H_2O, sl. sol. hot EtOH, Et_2O, $CHCl_3$. Mp 215°.

Kunz, P., *Ber.*, 1894, **27**, 2559 (*synth*)
Autenrieth, W., *Ber.*, 1897, **30**, 2369 (*synth*)
Spivak, L.I. *et al, Zh. Obshch. Khim.*, 1974, **44**, 870 (*Engl. transl.* p. 838) (*props*)

Di-1-naphthylphosphine D-00939

Di-1-naphthalenylphosphine, 9CI
[39864-75-0]

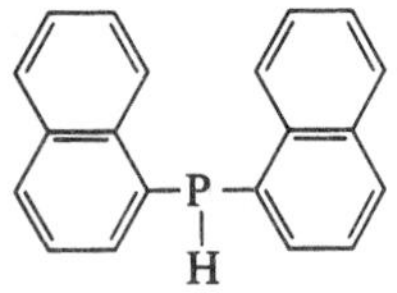

$C_{20}H_{15}P$ M 286.312
Solid. Mp 91-97°. $Bp_{0.1}$ 200-204°. Li deriv. exhibits v. weak orange chemiluminescence.

B,MeI: Methyldi-1-naphthylphosphonium iodide.
$C_{21}H_{18}IP$ M 428.251
Cryst. (EtOH). Mp 229-231°.

Oxide: [13440-07-8].
$C_{20}H_{15}OP$ M 302.312
Cryst. (C_6H_6/hexane). Mp 79°, 166-168°. Tautomeric with di-1-naphthylphosphinous acid.

Strecker, R.A. *et al, J. Am. Chem. Soc.*, 1973, **95**, 210 (*synth*)
Tewari, R.S. *et al, Zh. Obshch. Khim.*, 1973, **43**, 997 (*Engl. transl.* p. 991) (*synth*)
Hobbs, C.F. *et al, J. Org. Chem.*, 1981, **46**, 4422 (*synth, oxide, pmr, nmr*)

Di-2-naphthylphosphine D-00940

Di-2-naphthalenylphosphine, 9CI
[78871-06-4]

$C_{20}H_{15}P$ M 286.312
Cryst. Mp 94-96°. $Bp_{0.3}$ 208-213°.

Oxide: [78871-05-3].
$C_{20}H_{15}OP$ M 302.312
Cryst. (C_6H_6/cyclohexane). Mp 112-113°. Tautomeric with di-2-naphthylphosphinous acid.

Hobbs, C.F. *et al, J. Org. Chem.*, 1981, **46**, 4422 (*synth, oxide, pmr, nmr*)

Di-1-naphthylphosphinic acid D-00941

Di-1-naphthalenylphosphinic acid, 9CI

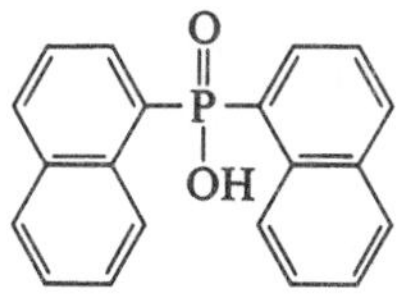

$C_{20}H_{15}O_2P$ M 318.311
Cryst. (EtOH). Mp 219-220°.

Et ester: Ethyl di-1-naphthylphosphinate.
$C_{22}H_{19}O_2P$ M 346.365
Cryst. (MeOH aq.). Mp 138.5-139.5°.

Chloride:
$C_{20}H_{14}ClOP$ M 336.757
Undistillable oil.

Anhydride:
$C_{40}H_{28}O_3P$ M 587.633
Solid. Mp 255-260°.

Anilide: P,P-*Di-1-naphthyl*-N-*phenylphosphinic amide. Di-α-naphthylphosphinic anilide.*
$C_{26}H_{19}NOP$ M 392.416
Cryst. (2-ethoxyethanol). Mp 289-291°.

Issleib, K. *et al, Chem. Ber.*, 1961, **94**, 392 (*synth*)
Crofts, P.C. *et al, J. Chem. Soc.*, 1964, 1240 (*synth, derivs*)

Di-1-naphthyl phosphorochloridite D-00942

Di-1-naphthalenyl phosphorochloridite, 9CI

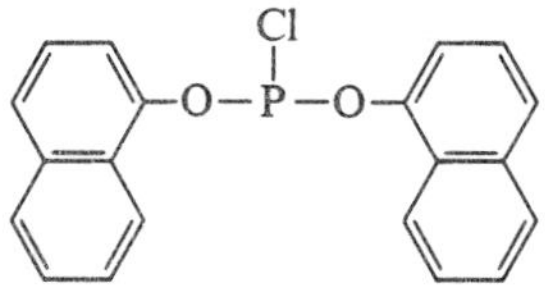

$C_{20}H_{14}ClO_2P$ M 352.756
Oil. $Bp_{0.03}$ 174°. n_D^{20} 1.6706.

Russ. Pat., 189 848, (*1966*); *CA*, **68**, 2706 (*synth*)

Di-2-naphthyl phosphorochloridite D-00943

Di-2-naphthalenyl phosphorochloridite, 9CI

$C_{20}H_{14}ClO_2P$ M 352.756
Liq. $Bp_{0.03}$ 181-183°.

Russ. Pat., 189 848, (*1966*); *CA*, **68**, 2706 (*synth*)

O,O-Di-1-naphthyl phosphorodithioate, 8CI D-00944

O,O-*Di-1-naphthalenyl phosphorodithioate, 9CI.* O,O-*Di-1-naphthyl hydrogen dithiophosphate.* O,O-*Di-1-naphthyl dithiophosphoric acid*
[50566-50-2]

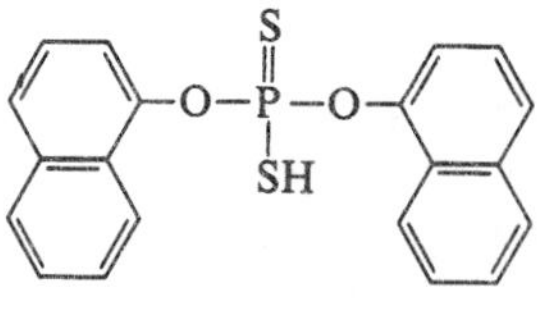

$C_{20}H_{15}O_2PS_2$ M 382.431
Solid. Mp 88-89°.

4-Methylanilinium salt: Solid. Mp 142-144°.

Micu-Semeniuc, R. *et al, Inorg. Chim. Acta*, 1979, **33**, 281 (*synth, ir, complexes*)

O,O-Di-2-naphthyl phosphorodithioate, 8CI D-00945

O,O-*Di-2-naphthalenyl phosphorodithioate, 9CI.* O,O-*Di-2-naphthyl hydrogen dithiophosphate.* O,O-*Di-2-naphthyl dithiophosphoric acid*

$C_{20}H_{15}O_2PS_2$ M 382.431
Solid. Mp 119° (112-113°).
Triethylammonium salt: Solid. Mp 107-108°.

Komkov, I.P. *et al*, *CA*, 1968, **68**, 39241.
Micu-Semeniuc, R. *et al*, *Inorg. Chim. Acta*, 1979, **33**, 281 (*synth, ir, complexes*)

(2,4-Dinitrophenyl)phosphonic acid, 8CI D-00946

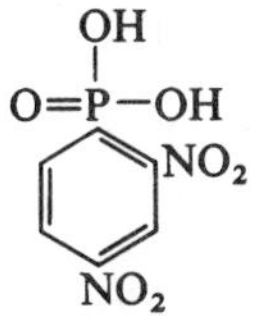

$C_6H_5N_2O_7P$ M 248.088
Di-Me ester: [23081-76-7]. *Dimethyl (2,4-dinitrophenyl)phosphonate.*
$C_8H_9N_2O_7P$ M 276.142
Yellow oil which cryst. (Et_2O/pet. ether). Mp 70-73°. $Bp_{0.1}$ 132-135°.
Di-Et ester: [23081-75-6]. *Diethyl (2,4-dinitrophenyl)-phosphonate.*
$C_{10}H_{13}N_2O_7P$ M 304.196
Liq. $Bp_{0.1}$ 155°.
Diisopropyl ester: [23081-77-8]. *Diisopropyl (2,4-dinitrophenyl)phosphonate.*
$C_{12}H_{17}N_2O_7P$ M 332.249
Solid. Mp 37-38°.

Cadogan, J.I.G. *et al*, *J. Chem. Soc. (C)*, 1969, 1314 (*synth, ir, pmr*)

(2,4-Dinitrophenyl)phosphoramidic acid, 9CI, 8CI D-00947

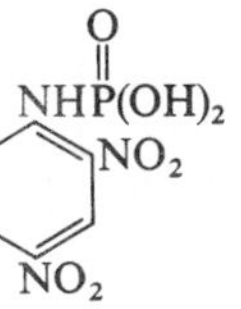

$C_6H_6N_3O_7P$ M 263.103
Di-Ph ester: Diphenyl (2,4-dinitrophenyl)-phosphoramidate.
$C_{18}H_{14}N_3O_7P$ M 415.298
Cryst. (EtOH). Mp 144-146°.

Zhmurova, I.N. *et al*, *Zh. Obshch. Khim.*, 1961, **31**, 3741 (*Engl. transl.* p. 3495)

Dinonyl phosphate, 9CI, 8CI D-00948
Dinonyl hydrogen phosphate. Dinonyl phosphoric acid
[3138-48-0]

$[H_3C(CH_2)_8O]_2P(O)OH$

$C_{18}H_{39}O_4P$ M 350.477
Separates Eu from Hf. Oil or solid. Mp 41-42°.
Chloride: Dinonyl phosphorochloridate. Dinonyl phosphoryl chloride. Dinonyl chlorophosphate.
$C_{18}H_{38}ClO_3P$ M 368.923
Oil. n_D^{25} 1.4527.

Grosse-Ruyken, H. *et al*, *J. Prakt. Chem.*, 1962, **18**, 287 (*synth, chloride*)
Bliznyuk, N.K. *et al*, *Zh. Obshch. Khim.*, 1967, **37**, 1119 (*Engl. transl.* p. 1061) (*synth*)

Dinonylphosphinic acid, 9CI D-00949
[38021-01-1]

$[H_3C(CH_2)_8]_2P(O)OH$

$C_{18}H_{39}O_2P$ M 318.479
Cryst. (hexane or hexane/Me_2CO). Mp 84-85°.

Williams, R.H. *et al*, *J. Am. Chem. Soc.*, 1955, **77**, 3411 (*synth*)
Bello, P., *Gazz. Chim. Ital.*, 1973, **103**, 537 (*cryst struct*)

Dinonyl phosphonate, 9CI D-00950
Dinonyl phosphite
[6838-97-7]

$[H_3C(CH_2)_8O]_2P(O)H \rightleftharpoons [H_3C(CH_2)_8O]_2POH$

$C_{18}H_{39}O_3P$ M 334.478
Tautomeric. Liq. d_4^{20} 0.93. Bp_2 200-202°. n_D^{20} 1.4418-1.4435.

Petrov, K.A. *et al*, *Zh. Obshch. Khim.*, 1962, **32**, 1277, 3723 (*Engl. transl.* pp. 1250, 3650) (*synth*)
Mandel'baum, Ya.A. *et al*, *CA*, 1968, **69**, 43338 (*synth*)

***O,O*-Dinonyl phosphorodithioate, 9CI, 8CI** D-00951
O,O-Dinonyl hydrogen phosphorodithioate. O,O-Dinonyl dithiophosphoric acid. O,O-Dinonyl dithiophosphate
[2253-58-9]

$[H_3C(CH_2)_8O]_2P(S)SH$

$C_{18}H_{39}O_2PS_2$ M 382.599
K salt: [3287-90-9]. Solid. Mp 150-153°.

Bolotova, G.L. *et al*, *CA*, 1965, **63**, 6897.

Dioctadecyl phosphate D-00952
Distearyl phosphate. Dioctadecyl hydrogen phosphate. Dioctadecyl phosphoric acid

$[H_3C(CH_2)_{17}O]_2P(O)OH$

$C_{36}H_{75}O_4P$ M 602.960
Cryst. (MeOH). Mp 82°.

Brown, D.A. *et al*, *J. Chem. Soc.*, 1955, 1584 (*synth*)
Ramirez, F. *et al*, *Phosphorus Sulfur*, 1978, **4**, 43 (*synth*)

Dioctadecylphosphinic acid, 9CI D-00953

$[H_3C(CH_2)_{17}]_2P(O)OH$

$C_{36}H_{75}O_2P$ M 570.961
Cryst. (C_6H_6 or C_6H_6/EtOH). Mp 105.5-106°.

Williams, R.H. *et al*, *J. Am. Chem. Soc.*, 1955, **77**, 3411.

Dioctadecyl phosphonate, 9CI D-00954
Distearyl phosphonate. Dioctadecyl phosphite. Distearyl phosphite
[19047-85-9]

$[H_3C(CH_2)_{17}O]_2P(O)H \rightleftharpoons [H_3C(CH_2)_{17}O]_2POH$

$C_{36}H_{75}O_3P$ M 586.960
Tautomeric; almost completely in phosphonate form. Stabilizer for polymers and lubricating oils. Solid. Mp 54-57°.

U.S.P., 3 036 109, (*1962*); *CA*, **57**, 13612

Dioctylphenylphosphine, 9CI **D-00955**

[14086-46-5]

$PhP[(CH_2)_7CH_3]_2$

$C_{22}H_{39}P$ M 334.524

Used in catalysts for manufacture of polyurethanes. Liq. Bp_{50} 277°.

B,MeI: Methyldioctylphenylphosphonium iodide.
$C_{23}H_{42}IP$ M 476.463
Solid. Mp 81°.

Oxide: [2845-09-2].
$C_{22}H_{39}OP$ M 350.523
Solid. Mp 42-43°. Bp_5 245-247°. pK_a 7.90 (conjugate acid in $MeNO_2$).

Jackson, I.K. *et al*, *J. Chem. Soc.*, 1931, 2109 (*synth, derivs*)
Fedorova, G.K. *et al*, *Zh. Obshch. Khim.*, 1964, **34**, 511 (*Engl. transl.* p. 513) (*oxide*)

Dioctyl phosphate, 9CI **D-00956**

Dioctyl hydrogen phosphate. Dioctyl phosphoric acid

[3115-39-7]

$[H_3C(CH_2)_7O]_2P(O)OH$

$C_{16}H_{35}O_4P$ M 322.424

Extractant for Hf. Oil or cryst. Mp 29-30°. pK_a 3.32 (75% v/v EtOH aq.). n_D^{20} 1.4464 (supercooled).

Tetramethylammonium salt: Solid. Mp 173°.

Chloride: Dioctyl phosphorochloridate. Dioctyl phosphoryl chloride. Dioctyl dichlorophosphate.
$C_{16}H_{34}ClO_3P$ M 340.870
Liq. n_D^{25} 1.4452.

Chabrier, P. *et al*, *C.R. Hebd. Seances Acad. Sci.*, 1957, **244**, 2730 (*synth*)
Petrov, K.A. *et al*, *Zh. Obshch. Khim.*, 1961, **31**, 1709 (*Engl. transl.* p. 1596) (*synth*)
Grosse-Ruyken, H. *et al*, *J. Prakt. Chem.*, 1962, **18**, 287 (*synth, chloride*)
Navratil, O. *et al*, *Hydrometallurgy*, 1981, **7**, 289 (*use*)

Dioctylphosphine, 9CI **D-00957**

[3541-77-3]

$HP[(CH_2)_7CH_3]_2$

$C_{16}H_{35}P$ M 258.426

Sweet-smelling liq. $Bp_{1.1}$ 140-142°. pK_a 4.41 (H_2O). n_D^{25} 1.4629.

Oxide: [3011-82-3].
$C_{16}H_{35}OP$ M 274.426
Medium for sepn. of thorium and uranium from rare earths. Cryst. (hexane). Mp 85-86°. Tautomeric with dioctylphosphinous acid.

▷SZ0180500.

Miller, R.C., *J. Org. Chem.*, 1959, **24**, 2013 (*synth, oxide*)
Pass, F. *et al*, *Monatsh. Chem.*, 1959, **90**, 792 (*synth*)
Henderson, W.A. *et al*, *J. Am. Chem. Soc.*, 1960, **82**, 5794 (*synth, nmr*)
Schindlbauer, H. *et al*, *Monatsh. Chem.*, 1961, **92**, 868 (*ir*)
Muratova, A.A. *et al*, *Zh. Obshch. Khim.*, 1972, **42**, 976 (*Engl. transl.* p. 966) (*oxide, complexes*)

Dioctylphosphinic acid, 9CI **D-00958**

[683-19-2]

$[H_3C(CH_2)_7]_2P(O)OH$

$C_{16}H_{35}O_2P$ M 290.425

Useful extractant for metals of lanthanide and actinide series. Cryst. (pet. ether). Mp 84°. Bp_1 325-327°. pK_a 5.3 (75% EtOH aq.).

Cyclohexylammonium salt: Prisms (MeCN). Mp 141-142°.

Butyl ester: [16644-61-4]. *Butyl dioctylphosphinate.*
$C_{20}H_{43}O_2P$ M 346.532
Liq. $Bp_{0.15}$ 158°. n_D^{25} 1.4515.

Chloride: [5849-35-4].
$C_{16}H_{34}ClOP$ M 308.871
Low-melting solid. Mp 43-44°. $Bp_{0.03}$ 140-143°.

Butylamide: N-*Butyl*-P,P-*dioctylphosphinic amide.*
$C_{22}H_{44}NOP$ M 369.569
Liq. $Bp_{0.1}$ 170°. n_D^{25} 1.4610.

Williams, R.H. *et al*, *J. Am. Chem. Soc.*, 1955, **77**, 3411 (*synth, chloride*)
Petrov, K.A. *et al*, *Zh. Obshch. Khim.*, 1960, **30**, 1964 (*Engl. transl.* p. 1943) (*synth*)
Peppard, D.F. *et al*, *J. Inorg. Nucl. Chem.*, 1965, **27**, 2065 (*synth, props, use*)
Silver, H.B., *J. Chem. Soc. (C)*, 1967, 1326 (*synth, derivs*)
Feshchenko, N.G. *et al*, *Zh. Obshch. Khim.*, 1970, **40**, 2385 (*Engl. transl.* p. 2373) (*chloride*)

Dioctyl phosphonate, 9CI **D-00959**

Dioctyl phosphite

[1809-14-9]

$[H_3C(CH_2)_7O]_2P(O)H \rightleftharpoons [H_3C(CH_2)_7O]_2POH$

$C_{16}H_{35}O_3P$ M 306.424

Tautomeric. Heat and light stabilizer for plastics. Liq. d_0^{20} 0.93. Bp_3 190-191°, $Bp_{0.03}$ 151-153°. n_D^{20} 1.4413.

Carlson, E.J. *et al*, *J. Chem. Soc.*, 1965, 2364 (*synth*)

***O,O*-Dioctyl phosphorodithioate,** 9CI, 8CI **D-00960**

O,O-*Dioctyl hydrogen phosphorodithioate.* O,O-*Dioctyl dithiophosphate.* O,O-*Dioctyl dithiophosphoric acid*

[2253-57-8]

$[H_3C(CH_2)_7O]_2P(S)SH$

$C_{16}H_{35}O_2PS_2$ M 354.545

Extractant for Ni and Zn. Zn complex is an additive for lubricants.

K salt: [3287-89-6]. Solid. Mp 153.5-154.5°.

Bolotova, G.L. *et al*, *CA*, 1965, **63**, 6897.
Szczepaniak, W. *et al*, *Pol. J. Chem.*, 1979, **53**, 755 (*complex, uv*)

2,8-Dioxa-5-aza-1-phosphabicyclo[3.3.0]-octane **D-00961**

[65693-26-7]

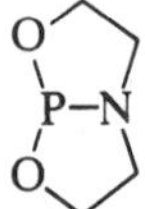

$C_4H_8NO_2P$ M 133.086

No phys. props. reported.

Houalla, D. *et al*, *Nouv. J. Chim.*, 1979, **3**, 507 (*pe*)
Denny, D.B. *et al*, *Phosphorus Sulfur*, 1983, **15**, 281 (*pmr, cmr, P nmr*)

Gupta, O.D. *et al*, *J. Chem. Soc., Chem. Commun.*, 1984, 416 (*props*)

1,4-Dioxa-6-aza-5-phospha(5-P^V)-spiro[4.4]nonan-7-one, 9CI D-00962

$C_5H_{10}NO_3P$ M 163.113

5-Ph: [60890-66-6].

$C_{11}H_{14}NO_3P$ M 239.210

Hygroscopic cryst. (MeCN). Mp 158°. Tautomeric with monocyclic struct.

Saegusa, T. *et al*, *Macromolecules*, 1976, **9**, 724 (*synth, pmr, ir*)

1,6-Dioxa-4,9-diaza-5-phospha(5-P^V)-spiro[4.4]nonane, 8CI D-00963

[1491-22-1]

$C_4H_{11}N_2O_2P$ M 150.117

Tautomeric; 0% monocyclic tricoordinate form at 20°, 25% at 150°.

5-Ph: [55983-41-0].

$C_{10}H_{15}N_2O_2P$ M 226.214

Solid. Mp 167°.

Mathis, R. *et al*, *Spectrochim. Acta, Part A*, 1969, **25**, 1201 (*ir*)
Burgada, R. *et al*, *J. Organomet. Chem.*, 1974, **66**, 255 (*P nmr, struct*)
Malavaud, C. *et al*, *Tetrahedron Lett.*, 1975, 497 (*phenyl, synth, P nmr*)
Meunier, P.L. *et al*, *Inorg. Chem.*, 1976, **15**, 2572 (*cryst struct*)

1,4-Dioxa-6,9-dithia-5-phospha(5-P^V)-spiro[4.4]nonane, 10CI D-00964

[38964-67-9]

$C_4H_9O_2PS_2$ M 184.208

5-Methoxy: [75373-19-2].

$C_5H_{11}O_3PS_2$ M 214.234

Solid. Mp 48-50°.

Majoral, J.-P. *et al*, *Tetrahedron Lett.*, 1980, **21**, 1307 (*synth, P nmr*)

[2-(1,3-Dioxan-2-yl)ethyl]-triphenylphosphonium(1+), 9CI D-00965

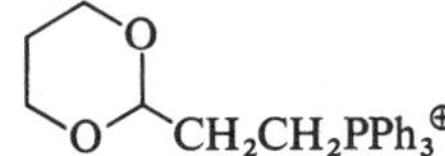

$C_{24}H_{26}O_2P^{\oplus}$ M 377.442 (ion)

Bromide: [69891-92-5].

$C_{24}H_{26}BrO_2P$ M 457.346

With RLi → ylide. Solid. Mp 205-208°.

Ylide: [69891-57-2]. *[2-(1,3-Dioxan-2-yl)ethylidene]-triphenylphosphorane.*

$C_{24}H_{25}O_2P$ M 376.434

Wittig rgt., for chain extension of aldehydes and ketones. Used in leukotriene synth. Orange.

Stowell, J.C. *et al*, *Synthesis*, 1979, 132 (*synth, use*)
Cohen, N. *et al*, *J. Am. Chem. Soc.*, 1983, **105**, 3661 (*use*)

1,3-Dioxa-2-phosphacyclotridecane, 8CI D-00966

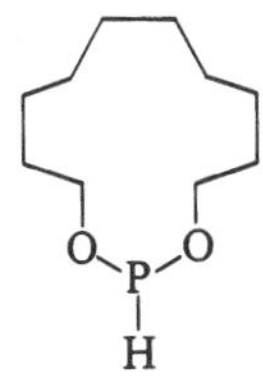

$C_{10}H_{21}O_2P$ M 204.248

2-Fluoro: [19948-00-6]. *2-Fluoro-1,3-dioxa-2-phosphacyclotridecane. Decamethylene phosphorofluoridite.*

$C_{10}H_{20}FO_2P$ M 222.239

Liq. d_4^{20} 1.10. Bp_2 80°. n_D^{20} 1.4798.

2-Chloro: [19948-35-7]. *2-Chloro-1,3-dioxa-2-phosphacyclotridecane. Decamethylene phosphorochloridite.*

$C_{10}H_{20}ClO_2P$ M 238.694

Liq. d_4^{20} 1.10. Bp_2 130°. n_D^{20} 1.4860.

Razumova, N.A. *et al*, *Zh. Obshch. Khim.*, 1968, **38**, 1117 (*Engl. transl.* p. 1072) (*derivs, synth, nmr*)

1,4-Dioxa-5-phospha(5-P^V)spiro[4.4]non-7-ene, 9CI, 8CI D-00967

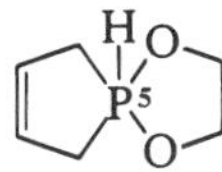

$C_6H_{11}O_2P$ M 146.125

5-Me: [32966-78-2].

$C_7H_{13}O_2P$ M 160.152

Liq. d_4^{20} 1.13. Bp_1 60°. n_D^{20} 1.5010.

5-Ph: [24901-16-4].

$C_{12}H_{15}O_2P$ M 222.223

Solid. Mp 60-62°.

Razumova, N.A. *et al*, *Zh. Obshch. Khim.*, 1969, **39**, 2368 (*Engl. transl.* p. 2305) (*synth, pmr, ir*)
Evtikhov, Zh.L. *et al*, *Zh. Obshch. Khim.*, 1971, **41**, 479 (*Engl. transl.* p. 471) (*synth*)

1,3,2-Dioxaphosphepane, 9CI, 8CI D-00968

1,3-Dioxa-2-phosphacycloheptane

[4757-24-8]

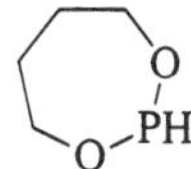

$C_4H_9O_2P$ M 120.088

2-Fluoro: [19948-32-4]. *Tetramethylene phosphorofluoridite.*

$C_4H_8FO_2P$ M 138.078

Liq. d^{20} 1.22. Bp_{16} 38°. n_D^{20} 1.4450.

2-Chloro: [16352-25-3]. *Tetramethylene phosphorochloridite.*

$C_4H_8ClO_2P$ M 154.533

Pungent liq. Bp_8 74.5-75°, $Bp_{0.2}$ 51-53°. n_D^{20} 1.4994.

2-Methoxy: [69576-78-9]. *Methyl tetramethylene phosphite.*

$C_5H_{11}O_3P$ M 150.114

Liq. Bp_9 56-57°, $Bp_{4.5\text{-}5.0}$ 54-55°. n_D^{20} 1.4642.
2-Phenoxy: [59413-41-1]. *Tetramethylene phenyl phosphite.*
$C_{10}H_{13}O_3P$ M 212.185
Liq. $Bp_{0.3}$ 94°. n_D^{22} 1.5361.
2-Me: see 2-Methyl-1,3,2-dioxaphosphepane, M-00125
*2-tert-*Butyl: [63790-94-3]. *Tetramethylene* tert-*butylphosphonite.*
$C_8H_{17}O_2P$ M 176.195
Characterised spectroscopically.
2-Ph: see 2-Phenyl-1,3,2-dioxaphosphepane, P-00114

Arbusov, A.E. *et al*, *Izv. Akad. Nauk SSSR, Ser. Khim.*, 1952, 770; *CA*, **47**, 9900 (*chloro, alkoxy derivs, synth*)
Ayres, D.C. *et al*, *J. Chem. Soc.*, 1957, 1109 (*phenoxy*)
Zwierzak, A., *Can. J. Chem.*, 1967, **45**, 2501 (*chloro*)
Razumova, N.A. *et al*, *Zh. Obshch. Khim.*, 1968, **38**, 1117 (*Engl. transl.* p. 1072) (*fluoro*)
Dutasta, J.P. *et al*, *Tetrahedron Lett.*, 1977, 801 (*tert*-butyl, synth, nmr)
Weiss, R. *et al*, *J. Org. Chem.*, 1979, **44**, 1860 (*methoxy, synth, nmr*)
Shagidullin, R.R. *et al*, *Izv. Akad. Nauk SSSR, Ser. Khim.*, 1981, **30**, 567 (*Engl. transl.* p. 410) (*chloro, raman, conformn*)

1,3,2-Dioxaphosphepane 2-sulfide, 9CI, 8CI D-00969

O,O′-*Tetramethylene phosphonothioate*

$C_4H_9O_2PS$ M 152.148
2-Phenoxy: [69813-49-6]. O,O-*Tetramethylene* O-*phenyl phosphorothioate.*
$C_{10}H_{13}O_3PS$ M 244.245
Characterised spectroscopically.
2-Me: see under 2-Methyl-1,3,2-dioxaphosphepane, M-00125
2-Ph: see under 2-Phenyl-1,3,2-dioxaphosphepane, P-00114

Dutasta, J.P. *et al*, *Tetrahedron Lett.*, 1977, 801 (*tert*-butyl, dimer, nmr)
Guimaraes, A.C. *et al*, *Org. Magn. Reson.*, 1978, **11**, 411 (*derivs, cmr, nmr, pmr*)

1,3,2-Dioxaphospholane, 9CI, 8CI D-00970

1,3-Dioxa-2-phosphacyclopentane. Ethylene phosphonite
[4757-26-0]

$C_2H_5O_2P$ M 92.034
2-Oxide: [1003-11-8]. *2-Hydroxy-1,3,2-dioxaphospholane. Ethylene phosphonate. Ethylene phosphite.*
$C_2H_5O_3P$ M 108.033
Liq. d_4^{20} 1.45. Bp_1 76-78°, $Bp_{0.1}$ 95-99°. n_D^{20} 1.4700 (1.4851). Tautomeric.
▷TG7210000.
2-Sulfide: [16393-52-5]. *2-Mercapto-1,3,2-dioxaphospholane.* O,O-*Ethylene phosphonothioate.* O,O-*Ethylene thiophosphite.*
$C_2H_5O_2PS$ M 124.094
Oil. d_4^{20} 1.41. n_D^{20} 1.5401. Dec. at 100°. Tautomeric.

Krawiecki, Cz. *et al*, *J. Chem. Soc.*, 1960, 881 (*sulfide*)
Nifant'ev, É.E. *et al*, *Zh. Org. Khim.*, 1975, **11**, 2206 (*Engl. transl.* p. 2235) (*oxide, synth, P nmr, props*)

1,3,2-Dioxaphosphonane, 8CI D-00971

1,3-Dioxa-2-phosphacyclononane
[6566-81-0]

$C_6H_{13}O_2P$ M 148.141
2-Fluoro: [19948-34-6]. *2-Fluoro-1,3,2-dioxaphosphonane. Hexamethylene fluorophosphite.*
$C_6H_{12}FO_2P$ M 166.132
Liq. d_4^{20} 1.08. Bp_1 66°. n_D^{20} 1.4270.
2-Chloro: [19948-33-5]. *2-Chloro-1,3,2-dioxaphosphonane. Hexamethylene chlorophosphite.*
$C_6H_{12}ClO_2P$ M 182.586
Liq. d_4^{20} 1.21. $Bp_{0.16}$ 78°. n_D^{20} 1.4880.

Razumova, N.A. *et al*, *Zh. Obshch. Khim.*, 1968, **38**, 1117 (*Engl. transl.* p. 1072) (*synth, pmr*)

1,3,2-Dioxaphosphorinane, 9CI D-00972

1,3,2-Dioxaphosphinane. 1,3-Dioxa-2-phosphacyclohexane. Trimethylene phosphonite
[4757-27-1]

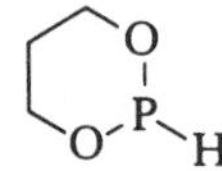

$C_3H_7O_2P$ M 106.061
Obt. only in tributylstannane soln.
2-Oxide: [16352-21-9]. *Trimethylene phosphonate. Trimethylene phosphite.*
$C_3H_7O_3P$ M 122.060
Cryst. (Et_2O) or liq. Mp 30°. Bp_1 120-120.5°, $Bp_{2.5}$ 97-98°. n_D^{20} 1.4570. Tautomeric.
2-Sulfide: [16368-16-4]. O,O-*Trimethylene phosphonothioate.* O,O-*Trimethylene phosphorothioite.* O,O-*Trimethylene thiophosphite.*
$C_3H_7O_2PS$ M 138.121
Solid. Mp 34-35°. $Bp_{0.003}$ 47°. Tautomeric.

Oswald, A.A., *Can. J. Chem.*, 1959, **37**, 1498 (*oxide*)
Krawiecki, Cz. *et al*, *J. Chem. Soc.*, 1960, 881 (*sulfide*)
Nifant'ev, É.E. *et al*, *Zh. Obshch. Khim.*, 1970, **40**, 1248; 1971, **41**, 2368 (*Engl. transl.* pp. 1239, 2394) (*oxide, synth, pmr, P nmr*)
Predvoditelev, D.A. *et al*, *Zh. Obshch. Khim.*, 1973, **43**, 73; 1974, **44**, 748, 1697, 2629 (*Engl. transl.* pp. 70, 720, 1667, 2586) (*sulfide, synth, ir, P nmr, props*)
Mosbo, J.A. *et al*, *J. Org. Chem.*, 1977, **42**, 1549 (*ir, P nmr, conformn*)

1,4,2-Dioxaphosphorinane 2-oxide, 9CI D-00973

1,4,2-Dioxaphosphinane 2-oxide

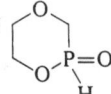

$C_3H_7O_3P$ M 122.060
2-Et:
$C_5H_{11}O_3P$ M 150.114
Liq. d_4^{20} 1.23. Bp_1 115-116°. n_D^{20} 1.4750.

2-Ethoxy:
$C_5H_{11}O_4P$ M 166.113
Liq. d_4^{20} 1.25. $Bp_{0.5}$ 97-97.5°. n_D^{20} 1.4336.

Arbusov, B.A. *et al, Izv. Akad. Nauk SSSR, Ser. Khim.*, 1960, 1767 (*Engl. transl.* p. 1649) (*synth*)

Dioxathion, BSI — D-00974

1,4-Dioxan-2,3-diyl S,S-*di*(O,O-*diethyl phosphorodithioate*), *9CI*. S,S′-(*1,4-Dioxane-2,3-diyl*) O,O,O′,O′-*tetraethyl di*(*phosphorothioate*). *2,3*-p-*Dioxanedithiol* S,S-*bis*(O,O-*diethyl phosphorodithioate*). *Delnav*
[78-34-2]

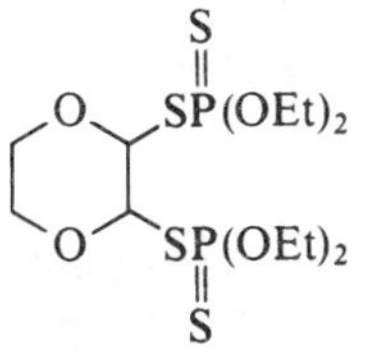

$C_{12}H_{26}O_6P_2S_4$ M 456.521
Tech. prod. is a mixt. of *cis*- and *trans*-isomers plus other compds. Non-systemic insecticide and acaricide.
Brown liq. (tech. grade). d_4^{26} 1.257. $Bp_{0.5}$ 60-68°. n_D^{20} 1.5409 (1.5420). Solns. in C_6H_6/hexane are more stable than pure compd.
▷TE3350000.

Cis-form [16088-56-5]
Liq.
▷More toxic to flies and rats than trans-form. TD7527000.

Trans-form [16270-86-3]
Liq.
▷TD7526000.

Diveley, W.R. *et al, J. Am. Chem. Soc.*, 1959, **81**, 139 (*synth*)
Schuntner, C.A. *et al, J. Chromatogr.*, 1967, **27**, 272 (*isom, ir*)
Babad, H. *et al, Anal. Chim. Acta*, 1968, **41**, 259 (*pmr*)
Harned, W.H. *et al, J. Agric. Food Chem.*, 1976, **24**, 689 (*isom, props, tlc, ir, pmr, ms*)
Stan, H.-J. *et al, Fresenius' Z. Anal. Chem.*, 1977, **287**, 271 (*glc, ms*)
Stan, H.-J. *et al, Biomed. Mass. Spectrom.*, 1982, **9**, 483 (*ms*)
Ripley, B.D. *et al, J. Assoc. Off. Anal. Chem.*, 1983, **66**, 1084 (*glc*)
Pesticide Manual, 6th Ed., 213; 7th Ed., 217.
The Agrochemicals Handbook, Royal Society of Chemistry, 1983, A163.
Sax, N.I., *Dangerous Properties of Industrial Materials*, 6th Ed., Van Nostrand-Reinhold, 1984, 538.

(1,4-Dioxo-1,4-butanediyl)bisphosphonic acid, 9CI — D-00975

$(HO)_2P(O)COCH_2CH_2COP(O)(OH)_2$

$C_4H_8O_8P_2$ M 246.050
Tetra-Me ester: Tetramethyl (1,4-dioxo-1,4-butanediyl)bisphosphonate.
$C_8H_{16}O_8P_2$ M 302.157
Liq. Bp_1 161-163°.

U.S.P., 3 012 054, (*1961*); *CA*, **57**, 4698

[(1,3-Dioxolan-2-yl)methyl]-triphenylphosphonium(1+), 9CI — D-00976

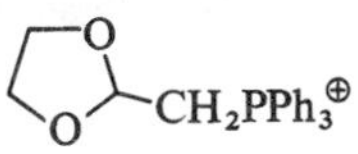

$C_{22}H_{22}O_2P^{\oplus}$ M 349.388 (ion)

Bromide: [52509-14-5].
$C_{22}H_{22}BrO_2P$ M 429.292
Prisms (CH_2Cl_2/Et_2O or Me_2CO/EtOAc). Mp 172-174° (191.5-193° after drying), 199-201°.
Iodide: [69216-75-7].
$C_{22}H_{22}IO_2P$ M 476.293
Cryst. ($CHCl_3$/EtOAc). Mp 212°.
Ylide: [78950-65-9]. [*(1,3-Dioxolan-2-yl)methylene*]-*triphenylphosphorane.*
$C_{22}H_{21}O_2P$ M 348.380
Reagent for vinylogation of aldehydes, RCHO → RCH=CHCHO.

Cresp, T.M. *et al, J. Chem. Soc., Perkin Trans. 1*, 1974, 37 (*bromide, synth, use*)
Cresp, T.M. *et al, J. Am. Chem. Soc.*, 1977, **99**, 194 (*use*)
Darby, N. *et al, J. Org. Chem.*, 1977, **42**, 1960 (*use*)
Cristau, H.-J. *et al, Synthesis*, 1978, 826 (*bromide, synth, use*)
Nishitani, S. *et al, Tetrahedron Lett.*, 1981, **22**, 2099 (*use*)
Cristau, H.-J. *et al, Phosphorus Sulfur*, 1982, **14**, 63 (*iodide, synth*)

Dioxybis[diphenylphosphine oxide], 9CI — D-00977

Bisdiphenylphosphinic peroxide. Bisdiphenylphosphinyl peroxide
[4250-08-2]

$Ph_2P(O)-O-O-P(O)Ph_2$

$C_{24}H_{20}O_4P_2$ M 434.367
Cryst. ($CHCl_3$/hexane at −10°). Mp 88-89°. Store at −80°.

Dannley, R.L. *et al, J. Am. Chem. Soc.*, 1965, **87**, 4805 (*synth, props*)
Dannley, R.L. *et al, J. Org. Chem.*, 1972, **37**, 418 (*props*)
Yaouane, J.J. *et al, Synthesis*, 1985, 807 (*synth, pmr, P nmr*)

Dipentylphenylphosphine, 9CI — D-00978

Diamylphenylphosphine

$PhP[(CH_2)_4CH_3]_2$

$C_{16}H_{27}P$ M 250.363
Liq. Part. misc. with EtOH at 15°. Bp_{50} 210°.
B,MeI: Methyldipentylphenylphosphonium iodide.
$C_{17}H_{30}IP$ M 392.302
Leaflets (H_2O). Mp 90.5°.

Davies, W.C. *et al, J. Chem. Soc.*, 1929, 1262.

Dipentyl phosphate, 9CI, 8CI — D-00979

Dipentyl hydrogen phosphate. Dipentyl phosphoric acid. Diamyl phosphate. Diamyl hydrogen phosphate. Diamyl phosphoric acid
[3138-42-9]

$[H_3C(CH_2)_4O]_2P(O)OH$

$C_{10}H_{23}O_4P$ M 238.263
Extractant for Hf. Oil. n_D^{20} 1.4370.
Tetramethylammonium salt: Solid. Mp 212°.
Fluoride: Dipentyl phosphorofluoridate. Diamyl phosphoryl fluoride. Diamyl fluorophosphate. Dipentyl phosphoryl chloride. Diamyl phosphorofluoridate. Dipentyl fluorophosphate.
$C_{10}H_{22}FO_3P$ M 240.254
Liq. Bp_{30} 143-144°.
▷Toxic
Chloride: [14254-40-1]. *Dipentyl phosphorochloridate. Diamyl phosphoryl chloride. Dipentyl phosphoryl chloride. Diamyl phosphorochloridate.*
$C_{10}H_{22}ClO_3P$ M 256.709

Oil. Bp_1 131-133°, $Bp_{0.04}$ 71-72°. n_D^{25} 1.4373.
Anilide: Dipentyl phenylphosphoramidate. Diamyl phenylphosphoramidate.
$C_{16}H_{28}NO_3P$ M 313.376
Oil. $Bp_{2.5}$ 201-203°. n_D^{20} 1.4810.

Cook, H.G. *et al*, *J. Chem. Soc.*, 1949, 635 (*fluoride, synth, tox*)
Ramaswami, D. *et al*, *J. Am. Chem. Soc.*, 1953, **75**, 1763 (*chloride, anilide*)
Chabrier, P. *et al*, *C.R. Hebd. Seances Acad. Sci.*, 1957, **244**, 2730 (*synth*)
de Roos, A.M. *et al*, *Recl. Trav. Chim. Pays-Bas*, 1958, **77**, 946 (*chloride*)
Grosse-Ruyken, H. *et al*, *J. Prakt. Chem.*, 1962, **18**, 287 (*synth, chloride*)
Navratil, O. *et al*, *Hydrometallurgy*, 1981, **7**, 289 (*use*)

Dipentylphosphine, 9CI **D-00980**
Diamylphosphine
[24674-42-8]

$HP[(CH_2)_4CH_3]_2$

$C_{10}H_{23}P$ M 174.265
Oxide:
$C_{10}H_{23}OP$ M 190.265
Cryst. (hexane). Mp 65-66°. Tautomeric with dipentylphosphinous acid.

Williams, R.H. *et al*, *J. Am. Chem. Soc.*, 1952, **74**, 5418; 1955, **77**, 3410 (*oxide, synth*)
Grim, S.O. *et al*, *Phosphorus*, 1974, **4**, 189 (*nmr*)

Dipentylphosphinic acid, 9CI **D-00981**
Diamylphosphinic acid
[24935-94-2]

$[H_3C(CH_2)_4]_2P(O)OH$

$C_{10}H_{23}O_2P$ M 206.264
Solid. Mp 68-69°.
Pentyl ester: [3058-18-2]. *Pentyl dipentylphosphinate.*
$C_{15}H_{33}O_2P$ M 276.398
Liq. Bp_1 119°. n_D^{20} 1.446.
Fluoride: [84923-95-5].
$C_{10}H_{22}FOP$ M 208.255
Liq. Bp_2 94°.
Chloride: [1604-65-5].
$C_{10}H_{22}ClOP$ M 224.710
Liq. $Bp_{0.06}$ 97°. n_D^{20} 1.4655.

Williams, R.H. *et al*, *J. Am. Chem. Soc.*, 1955, **77**, 3411 (*synth*)
Christen, P.J. *et al*, *Recl. Trav. Chim. Pays-Bas*, 1959, **78**, 543 (*chloride*)
Voigt, D. *et al*, *Bull. Soc. Chim. Fr.*, 1964, 3087 (*ester*)
Blasse, R.G., *Recl. Trav. Chim. Pays-Bas*, 1965, **84**, 267 (*chloride, ir*)
Nikitin, E.V. *et al*, *Zh. Obshch. Khim.*, 1982, **52**, 2721 (*Engl. transl.* p. 2400) (*fluoride, synth, ir, nmr*)

Dipentyl phosphonate, 9CI **D-00982**
Diamyl phosphonate. Dipentyl phosphite. Diamyl phosphite
[1809-17-2]

$$[H_3C(CH_2)_4O]_2P(O)H \rightleftharpoons [H_3C(CH_2)_4O]_2PO$$

$C_{10}H_{23}O_3P$ M 222.264
Tautomeric. Liq. d_4^{20} 0.96. $Bp_{0.05}$ 96-97°. pK_a 21.0 (DMSO at 25°). n_D^{20} 1.4287.

Coulson, E.J. *et al*, *J. Chem. Soc.*, 1965, 2364 (*synth, props*)
Mandel'baum, Ya.A. *et al*, *CA*, 1968, **69**, 43338 (*synth*)
Tsvetkov, E.N. *et al*, *Izv. Akad. Nauk SSSR, Ser. Khim.*, 1978, **27**, 1981 (*Engl. transl.* p. 1743) (*props*)

Dipentyl phosphorochloridite, 9CI, 8CI **D-00983**
Diamyl phosphorochloridite. Diamyl chlorophosphite
[62935-09-5]

$(H_3CCH_2CH_2CH_2CH_2O)_2PCl$

$C_{10}H_{22}ClO_2P$ M 240.709
Pungent liq. Bp_1 91.5-92°. n_D^{20} 1.4415.

Razumov, A.I., *CA*, 1964, **60**, 1571 (*synth*)

***O,O*-Dipentyl phosphorochloridothioate, 9CI, 8CI** **D-00984**
O,O-*Diamyl thiophosphoryl chloride.* O,O-*Diamyl chlorothiophosphate*

$[H_3C(CH_2)_4O]_2P(S)Cl$

$C_{10}H_{22}ClO_2PS$ M 272.769
Liq. d^{20} 1.05. $Bp_{0.9}$ 116-118°. n_D^{20} 1.4700.

Almasi, L. *et al*, *CA*, 1963, **58**, 5556.

***O,O*-Dipentyl phosphorodiselenoate** **D-00985**
O,O-*Dipentyl hydrogen phosphorodiselenoate.* O,O-*Dipentyl phosphorodiselenoic acid.* O,O-*Diamyl phosphorodiselenoate.* O,O-*Diamyl diselenophosphate*

$(H_3CCH_2CH_2CH_2CH_2O)_2P(Se)SeH$

$C_{10}H_{23}O_2PSe_2$ M 364.184
Unstable acid isol. as K salt.
K salt: [19483-49-9]. Cryst. + $1H_2O$ (ligroin). Mp 132-133°.

Kudchadker, M.V. *et al*, *Can. J. Chem.*, 1968, **46**, 1415 (*synth, ir*)

***O,O*-Dipentyl phosphorodithioate, 9CI, 8CI** **D-00986**
O,O-*Dipentyl hydrogen dithiophosphate.* O,O-*Diamyl dithiophosphoric acid.* O,O-*Diamyl phosphorodithioate*
[2253-54-5]

$[H_3C(CH_2)_4O]_2P(S)SH$

$C_{10}H_{23}O_2PS_2$ M 270.384
Zn complex used as lubricant additive. Oil. d^{20} 1.04. $Bp_{0.015}$ 89°. n_D^{20} 1.4920.
K salt: [3287-86-3]. Flotation agent. Solid. Mp 144-145°.

Almasi, L. *et al*, *CA*, 1965, **62**, 2729 (*synth*)
Bolotova, G.L. *et al*, *CA*, 1965, **63**, 6897 (*synth*)
Mazitova, F.N. *et al*, *Zh. Obshch. Khim.*, 1980, **50**, 1718 (*Engl. transl.* p. 1393) (*synth, P nmr*)

(Diphenoxyphosphinothioyl)phosphorimidic acid, 9CI **D-00987**

$(PhO)_2P(S)N{=}P(OH)_3$

$C_{12}H_{13}NO_5P_2S$ M 345.246

Free acid tautomeric with (diphenoxyphosphinothioyl)-phosphoramidic acid.

Tri-Me ester: [63013-65-0]. *Trimethyl (diphenoxyphosphinothioyl)phosphorimidate.* P,P,P-*Trimethoxy*-N-*(diphenoxyphosphinothioyl)phosphine imide.* P,P,P-*Trimethoxy*-N-*(diphenoxyphosphinothioyl)phosphazene.*
$C_{15}H_{19}NO_5P_2S$ M 387.326
Thick liq. n_D^{20} 1.5621. Isom. at 120°.

Tri-Et ester: [51576-55-7]. *Triethyl (diphenoxyphosphinothioyl)phosphorimidate.* P,P,P-*Triethoxy*-N-*(diphenoxyphosphinothioyl)phosphine imide.* P,P,P-*Triethoxy*-N-*(diphenoxyphosphinothioyl)phosphazene.*
$C_{18}H_{25}NO_5P_2S$ M 429.407
No phys. props. reported.

Tri-Ph ester: [79272-36-9]. *Triphenyl (diphenoxyphosphinothioyl)phosphorimidate.* P,P,P-*Triphenoxy*-N-*(diphenoxyphosphinothioyl)phosphine imide.* P,P,P-*Triphenoxy*-N-*(diphenoxyphosphinothioyl)phosphazene.*
$C_{30}H_{25}NO_5P_2S$ M 573.539
Solid or prisms (MeOH). Mp 97-98°.

Trichloride: P,P,P-*Trichloro*-N-*(diphenoxyphosphinylthioyl)phosphine imide.* P,P,P-*Trichloro*-N-*(diphenoxyphosphinothioyl)-phosphazene.*
$C_{12}H_{10}Cl_3NO_2P_2S$ M 400.583
Viscous liq. d_4^{20} 1.45. n_D^{20} 1.6016.

Trianilide: N,N′,N″-*Triphenyl (diphenoxyphosphinothioyl)-phosphorotriamidoimidate.* N-*(Diphenoxyphosphinothioyl)*-P,P,P-*tris(phenylamino)phosphazene.* N-*(Diphenoxyphosphinothioyl)*-P,P,P-*tris(phenylamino)phosphine imide.* P,P,P-*Trianilino*-N-*(diphenoxyphosphinothioyl)phosphine imide.*
$C_{30}H_{28}N_4O_3P$ M 523.550
Plates (EtOH). Mp 195-197°.

Kirsanov, A.V. *et al*, *Zh. Obshch. Khim.*, 1958, **28**, 2478 (*Engl. transl.* p. 2514) (*deriv, props*)
Khodak, A.A. *et al*, *Zh. Obshch. Khim.*, 1974, **44**, 27 (*Engl. transl.* p. 24); 1977, **47**, 273 (*Engl. transl.* p. 251) (*esters, synth, pmr, props*)
Khodak, A.A. *et al*, *Izv. Akad. Nauk SSSR, Ser. Khim.*, 1982, 465 (*Engl. transl.* p. 421) (*trichloride, triphenyl ester, synth, ir, pmr, P nmr*)

(Diphenoxyphosphinyl)acetic acid, 9CI D-00988

Diphenyl (carboxymethyl)phosphonate
[34159-52-9]

$(PhO)_2P(O)CH_2COOH$

$C_{14}H_{13}O_5P$ M 292.227
The dianion is employed in Wittig-Horner reactions. Cryst. (butanone). Mp 143-144°. pK_a 3.24 (H_2O), 4.46 (50% EtOH aq.).

Cyclohexylammonium salt: Cryst. (hexane). Mp 63-64°.

Ph ester: [83037-86-9]. *Phenyl(diphenoxyphosphinyl)-acetate. Triphenyl phosphonoacetate.*
$C_{20}H_{17}O_5P$ M 368.325
Employed in Wittig-Horner reactions.

Malevannaya, R.A. *et al*, *Zh. Obshch. Khim.*, 1971, **41**, 1426 (*Engl. transl.* p. 1432) (*synth*)
Tsvetkov, E.N. *et al*, *Zh. Obshch. Khim.*, 1974, **44**, 1225 (*Engl. transl.* p. 1203) (*props, pmr*)
You, D.A. *et al*, *Tetrahedron Lett.*, 1982, **23**, 2143 (*use*)

N-(Diphenoxyphosphinyl)carbamic acid, 9CI, 8CI D-00989

$(PhO)_2P(O)NHCOOH$

$C_{13}H_{12}NO_5P$ M 293.215

Me ester: Methyl N-*(diphenoxyphosphinyl)carbamate.*
$C_{14}H_{14}NO_5P$ M 307.242
Prisms (C_6H_6/pet. ether). Mp 109-111°.

Et ester: Ethyl N-*(diphenoxyphosphinyl)carbamate.*
$C_{15}H_{16}NO_5P$ M 321.269
Cryst. (C_6H_6/pet. ether). Mp 94-96° dec.

Isopropyl ester: Isopropyl N-*(diphenoxyphosphinyl)-carbamate.*
$C_{16}H_{18}NO_5P$ M 335.296
Cryst. (pet. ether). Mp 79-81° dec.

Ph ester: Phenyl N-*(diphenoxyphosphinyl)carbamate.*
$C_{19}H_{16}NO_5P$ M 369.313
Cryst. powder (dichloroethane). Mp 121-123°.

Benzyl ester: Benzyl N-*(diphenoxyphosphinyl)-carbamate.*
$C_{20}H_{18}NO_5P$ M 383.340
Solid. Mp 79-81°.

Kirsanov, A.V. *et al*, *Zh. Obshch. Khim.*, 1956, **26**, 2642; 1957, **27**, 1002; 1961, **31**, 1607 (*Engl. transl.* pp. 2947, 1084, 1496) (*synth, props*)

[(Diphenoxyphosphinyl)methyl]-triphenylphosphonium(1+) D-00990

$Ph_3P^{\oplus}CH_2P(O)(OPh)_2$

$C_{31}H_{27}OP_2^{\oplus}$ M 477.501 (ion)
Source of Diphenyl [(triphenylphosphoranylidene)-methyl]phosphonate, D-01163 .

Chloride:
$C_{31}H_{27}ClOP_2$ M 512.954
Solid. Mp 210-222°.

Iodide:
$C_{31}H_{27}IOP_2$ M 604.406
Solid. Mp 188°.

Jones, G.H. *et al*, *Tetrahedron Lett.*, 1968, 5731 (*chloride*)
Montgomery, J.A. *et al*, *J. Heterocycl. Chem.*, 1974, **11**, 211 (*iodide, pmr*)

3,9-Diphenoxy-2,4,8,10-tetraoxa-3,9-diphosphaspiro[5.5]undecane, 9CI D-00991

Pentaerythritol diphenyl phosphite
[144-35-4]

PhO—P(O—)(O—) spiro (—O)(—O)P—OPh

$C_{17}H_{18}O_6P_2$ M 380.273
Polymer stabilizer. Cryst. (hexane). Mp 123-124°.

Monooxide: [93701-60-1].
$C_{17}H_{18}O_7P_2$ M 396.273
Phys. props. not described. Characterised spectroscopically.

Dioxide: [55120-33-7].
$C_{17}H_{18}O_8P_2$ M 412.272
Fire-proofing agent. Solid.

Arbuzov, B.A. *et al*, *Izv. Akad. Nauk SSSR, Ser. Khim.*, 1973, 2426 (*Engl. transl.* p. 2372) (*synth*)
Rueger, C. *et al*, *J. Prakt. Chem.*, 1984, **326**, 622 (*synth, derivs, props*)
Pätoprstý, V. *et al*, *Magn. Reson. Chem.*, 1985, **23**, 122 (*cmr*)

Diphenoxytriphenylphosphorane, 10CI **D-00992**

[17663-89-7]

$(PhO)_2PPh_3$

$C_{30}H_{25}O_2P$ M 448.500

Grochowski, E. *et al*, *J. Am. Chem. Soc.*, 1982, **104**, 6876 (*synth, nmr*)

[2-(Diphenylarsino)ethyl]diphenylphosphine, 9CI **D-00993**

1-Diphenylphosphino-2-diphenylarsinoethane. Arphos

[23582-06-1]

$Ph_2PCH_2CH_2AsPh_2$

$C_{26}H_{24}AsP$ M 442.371

Ligand for metals of Groups IIB, VIB, VIIB, VIII and IIIA. Solid (C_6H_6/MeOH or EtOH/CH_2Cl_2). Mp 116-118°.

▷Toxic

King, R.B. *et al*, *J. Am. Chem. Soc.*, 1971, **93**, 4158 (*synth, ir, ms, pmr, nmr*)
Chow, K.-K. *et al*, *J. Organomet. Chem.*, 1973, **59**, 247 (*ms*)
Chow, K.-K. *et al*, *Inorg. Chim. Acta*, 1975, **14**, 5 (*synth, complexes*)
Chow, K.-K. *et al*, *J. Chem. Soc., Dalton Trans.*, 1976, 1429 (*synth*)
Inorg. Synth., 1976, **16**, 188 (*synth*)
Bemi, L. *et al*, *J. Am. Chem. Soc.*, 1982, **104**, 438 (*complexes*)
Goel, A.B. *et al*, *Inorg. Chim. Acta*, 1982, **59**, 237 (*complexes*)

[2-(Diphenylarsino)phenyl]-diphenylphosphine **D-00994**

2-[(Diphenylphosphino)phenyl]diphenylarsine

[55677-75-3]

PPh_2 $AsPh_2$

$C_{30}H_{24}AsP$ M 490.415

Ligand for Co, Ni, Pd, Pt, Mo, Mn, V. White cryst. Mp 198° (190.5-192.5°).

P-*Sulfide:*
$C_{30}H_{24}AsPS$ M 522.475
Cryst. (butanol). Mp 190-192°.

Nicpon, P.E. *et al*, *Inorg. Chem.*, 1967, **6**, 145 (*synth, complexes, sulfide*)
Levason, W. *et al*, *J. Organomet. Chem.*, 1975, **84**, 239 (*ms, complexes*)
Talay, R. *et al*, *Z. Naturforsch., B*, 1981, **36**, 451 (*synth, nmr, ms*)

2,3-Diphenyl-1,3-azaphospholidine, 9CI **D-00995**

2,3-Diphenyl-1,3-azaphosphacyclopentane

[15207-49-5]

Ph P Ph N H

$C_{15}H_{16}NP$ M 241.272

Cryst. Mp 78.5-79.5°.

B,HCl: Solid. Mp 211-213°.

3-Sulfide: [15107-69-4].
$C_{15}H_{16}NPS$ M 273.332
Cryst. (EtOH/Et_2O). Mp 158-159°.

1-Et: 1-Ethyl-2,3-diphenyl-1,3-azaphosphacyclopentane.
$C_{17}H_{20}NP$ M 269.325
Liq. Bp_2 172-174°. pK_a 5.90 (66% EtOH aq., 25°).

1-Ph: [52112-04-6]. *1,2,3-Triphenyl-1,3-azaphosphacyclopentane.*
$C_{21}H_{20}NP$ M 317.369
Cryst. (EtOH). Mp 158-161°.

Issleib, K. *et al*, *Chem. Ber.*, 1967, **100**, 2685; 1968, **101**, 3619 (*synth, derivs*)
Oehme, H. *et al*, *Phosphorus*, 1973, **3**, 159 (*synth*)
Zschunke, A. *et al*, *Z. Chem.*, 1973, **13**, 310 (*pmr*)

2,6-Diphenyl-1,4-azaphosphorine, 9CI **D-00996**

2,6-Diphenyl-1-aza-4-phosphabenzene

[39768-11-1]

P Ph N Ph

$C_{16}H_{12}NP$ M 249.251

Pale-yellow, viscous oil.

Märkl, G. *et al*, *Angew Chem., Int. Ed. Engl.*, 1972, **11**, 1019 (*synth, ms, pmr*)

1,4-Diphenyl-2,3-bis(triphenylphosphoranylidene)-1,4-butanedione, 8CI **D-00997**

1,2-Dibenzoyl-1,2-bis(triphenylphosphoranylidene)-ethane

[25132-69-8]

Ph_3P COPh Ph_3P COPh

$C_{52}H_{40}O_2P_2$ M 758.834

Resonance stabilized ylide. Cryst. ($CHCl_3$/Et_2O). Mp 199-200°.

Shaw, M.A. *et al*, *J. Chem. Soc. (C)*, 1970, 5 (*synth, ir, pmr*)

Diphenyl cyclooctylphosphoramidate, 9CI **D-00998**

PB-1

[86126-37-6]

O H PhO–P–N OPh

$C_{20}H_{26}NO_3P$ M 359.404

Toxin from the red tide dinoflagellate *Ptychodiscus brevis*. Appears to be a metab. of *P. brevis*, although the possibility that it may be derived from an unsuspected contaminant in the artificial culture medium cannot be totally excluded.

DiNovi, M. *et al*, *Tetrahedron Lett.*, 1983, **24**, 855 (*isol, uv, pmr, cmr, struct, synth*)

(2,4-Diphenyl-2,4-cyclopentadien-1-ylidene)triphenylphosphorane, 9CI D-00999

1,4-Diphenyl-5-(triphenylphosphoranylidene)-1,3-cyclopentadiene

[54710-83-7]

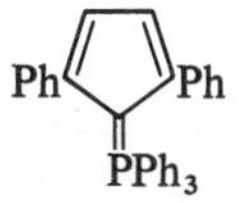

$C_{35}H_{27}P$ M 478.572

Stable ylide. Needles (butanol). Mp 234-235°.

Regitz, M. *et al*, *Tetrahedron*, 1967, **23**, 2701 (*synth, uv*)

3,5-Diphenyl-1*H*-1,2,4-diazaphosphole, 9CI D-01000

[93646-66-3]

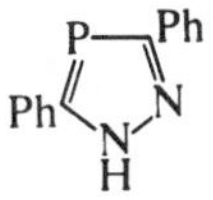

$C_{14}H_{11}N_2P$ M 238.228

Cryst. (Et_2O). Mp 214-217° (200°).

1-Me: [93646-72-1].

$C_{15}H_{13}N_2P$ M 252.255

Pale-yellow cryst. (Et_2O). Mp 105-108° (104-106°).

1-Ph: [93714-99-9].

$C_{20}H_{15}N_2P$ M 314.326

Yellow cryst. (Et_2O). Mp 201-203°.

Märkl, G. *et al*, *Angew. Chem., Int. Ed. Engl.*, 1984, **23**, 901 (*synth, nmr, ir, ms*)

Schmidpeter, A. *et al*, *Angew. Chem., Int. Ed. Engl.*, 1984, **23**, 903 (*synth, P-nmr, derivs*)

5,10-Diphenyldibenzo[1,4]phospharsenin D-01001

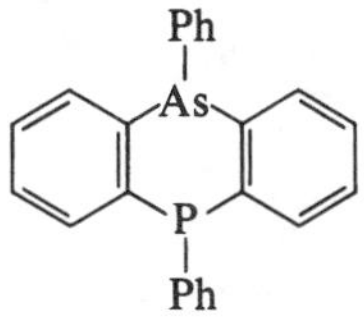

$C_{24}H_{18}AsP$ M 412.302

Prisms (butanone). Mp 189-190°. Quaternises on P.

B,MeI: 5-Methyl-5,10-diphenyldibenzo[*1,4*]*-phosphoniaarsenin.*

$C_{25}H_{21}AsIP$ M 554.241

Cryst. (EtOH). Mp 300-302°.

B,PhCH₂Br: 5-Phenyl-5,10-diphenyldibenzo[*1,4*]*-phosphoniaarsenin.*

$C_{31}H_{25}AsBrP$ M 583.338

Cryst. + $1H_2O$ (H_2O). Mp 318-322°.

P-*Oxide:*

$C_{24}H_{18}AsOP$ M 428.301

Cryst. (EtOH aq.). Mp 214-216°.

As,P-*Dioxide:*

$C_{24}H_{18}AsO_2P$ M 444.300

Cryst. + $2H_2O$ (EtOH aq.). Mp 360°.

Davis, M. *et al*, *J. Chem. Soc.*, 1964, 3770 (*synth, derivs*)

Diphenyl dichlorophosphoramidate D-01002

[27125-51-5]

$(PhO)_2P(O)NCl_2$

$C_{12}H_{10}Cl_2NO_3P$ M 318.096

Cryst. (heptane). Mp 30-32°.

Markovskii, L.N. *et al*, *Zh. Obshch. Khim.*, 1970, **40**, 543 (*Engl. transl.* p. 509) (*synth*)

Drewelies, K. *et al*, *Z. Naturforsch., B*, 1982, **37**, 1402.

2,5-Diphenyl-1,3,5,2-dioxaphosphaborinane, 9CI D-01003

[72729-56-7]

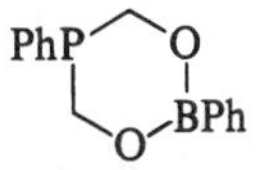

$C_{14}H_{14}BO_2P$ M 256.047

Cryst. (C_6H_6 or MeCN). Mp 105°.

5-Sulfide: [74659-21-5].

$C_{14}H_{14}BO_2PS$ M 288.107

Cryst. (MeCN). Mp 133-134°. Forms a complex with Py.

Arbuzov, B.A. *et al*, *Izv. Akad. Nauk SSSR, Ser. Khim.*, 1979, 2349 (*Engl. transl.* p. 2170) (*synth, pmr*)

Arbuzov, B.A. *et al*, *Izv. Akad. Nauk SSSR, Ser. Khim.*, 1980, 954; *CA*, **93**, 114622 (*conformn, sulfide*)

Arbuzov, B.A. *et al*, *Izv. Akad. Nauk SSSR, Ser. Khim.*, 1983, 1374 (*Engl. transl.* p. 1245) (*synth, P nmr*)

***O,O*-Diphenyl (diphenoxyphosphinothioyl)-phosphoramidothioate** D-01004

O,O,O′,O′-*Tetraphenyl thioimidodiphosphate, 9CI*

[90430-77-6]

$(PhO)_2P(S){-}NH{-}P(S)(OPh)_2$

$C_{24}H_{21}NO_4P_2S_2$ M 513.502

Note 9CI name identical with that for *O,O*-Diphenyl (diphenoxyphosphinyl)phosphoramidothioate, D-01005 . Extractant for Hg, and metals of groups IB and IVB. No phys. props. recorded.

Navrátil, O. *et al*, *Z. Chem.*, 1984, **24**, 30.

***O,O*-Diphenyl (diphenoxyphosphinyl)-phosphoramidothioate** D-01005

O,O,O′,O′-*Tetraphenyl thioimidodiphosphate, 9CI*

$(PhO)_2P(O){-}NH{-}P(S)(OPh)_2$

$C_{24}H_{21}O_5PS$ M 452.461

Note 9CI name identical with that for *O,O*-Diphenyl (diphenoxyphosphinothioyl)phosphoramidothioate, D-01004 . Cryst. (C_6H_6/pet. ether). Mp 100-102°.

Kirsanov, A.V. *et al*, *Zh. Obshch. Khim.*, 1958, **28**, 2478 (*Engl. transl.* p. 2514)

***P,P*-Diphenyl-*N*-(diphenylphosphinothioyl)phosphinimidic acid, 8CI** D-01006

P,P-*Diphenyl*-N-*(diphenylphosphinothioyl)phosphonic amide*

$Ph_2P(OH){=}NP(S)Ph_2 \rightleftharpoons Ph_2P(O)NHP(S)Ph_2$

$C_{24}H_{21}NOP_2S$ M 433.444

Tautomeric. Solid. Mp 268-269°.

O-*Me ester:* [17162-82-2]. O-*Methyl* P,P-*diphenyl*-N-*(diphenylphosphinothioyl)phosphinimidate.* P-(O-*Methoxy*)-P,P-*diphenyl*-N-*(diphenylphosphinothioyl)phosphazene.* P-(O-*Methoxy*)-P,P-*diphenyl*-N-*(diphenylphosphinothioyl)phosphine imide.*

$C_{25}H_{23}NOP_2S$ M 447.470

Solid. Mp 151-152°.

Chloride: [17162-81-1].

$C_{24}H_{20}ClNP_2S$ M 451.889

Solid. Sol. CCl_4, C_6H_6. Mp 114-116°. Easily hydrol.

Amide:

$C_{24}H_{22}N_2P_2S$ M 432.459

Needles. Mp 128-129°.

Schmidpeter, A. *et al*, *Chem. Ber.*, 1967, **100**, 3063, 3979 (*synth, pmr, P nmr, ir*)
Wang, F.T. *et al*, *Synth. React. Inorg. Metal-Org. Chem.*, 1978, **8**, 119 (*synth*)

P,P-Diphenyl-*N*-(diphenylphosphinyl)-phosphinimidic acid D-01007

P,P-*Diphenyl*-N-*diphenylphosphinylphosphinic amide*

[31239-06-2]

$Ph_2P(OH){=}NP(O)Ph_2 \rightleftharpoons Ph_2P(O)NHP(O)Ph_2$

$C_{24}H_{21}NO_2P_2$ M 417.383

Tautomeric. Cryst.

Et ester: Ethyl P,P-*diphenyl*-N-(*diphenylphosphinyl*)-*phosphinimidate.* P-*Ethoxy*-P,P-*diphenyl*-N-(*diphenylphosphinyl*)*phosphazene.* P-*Ethoxy*-P,P-*diphenyl*-N-(*diphenylphosphinyl*)*phosphine imide.*

$C_{26}H_{25}NO_2P$ M 414.463

Cryst. (pet. ether). Mp 98-99°.

Chloride:

$C_{24}H_{20}ClNOP$ M 404.855

Cryst. ($CHCl_3/Et_2O$). Mp 282-284°.

Gilson, I.T. *et al*, *Inorg. Chem.*, 1965, **4**, 273 (*chloride, ir*)
Gilyarov, V.A. *et al*, *Zh. Obshch. Khim.*, 1966, **36**, 274 (*Engl. transl.* p. 285) (*ester*)

1,12-Diphenyl-1,12-diphosphacyclodocosane, 9CI D-01008

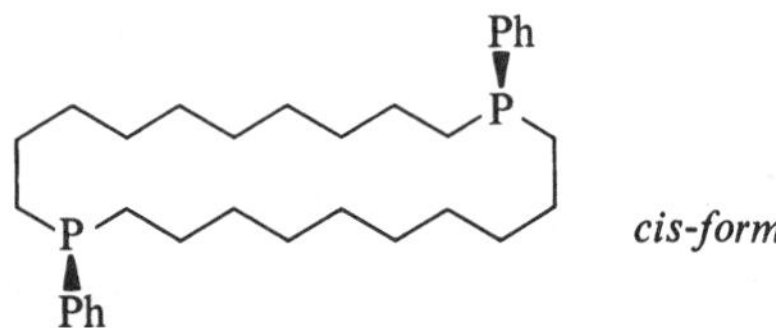

cis-form

$C_{32}H_{50}P_2$ M 496.695

cis-form [51064-83-6]

Solid. Mp 90-91°.

1,12-Dioxide: [51096-68-5].

$C_{32}H_{50}O_2P_2$ M 528.693

Hygroscopic cryst. ($CHCl_3$/hexane). Mp 171-173°.

trans-form [51021-81-9]

Solid. Mp 97-97.5°.

1,12-Dioxide: [51096-67-4]. Hygroscopic cryst. ($CHCl_3$/hexane). Mp 190-191°.

Chan, T.K. *et al*, *J. Org. Chem.*, 1974, **39**, 1748 (*synth, struct*)

1,10-Diphenyl-1,10-diphosphacyclooctadecane, 9CI D-01009

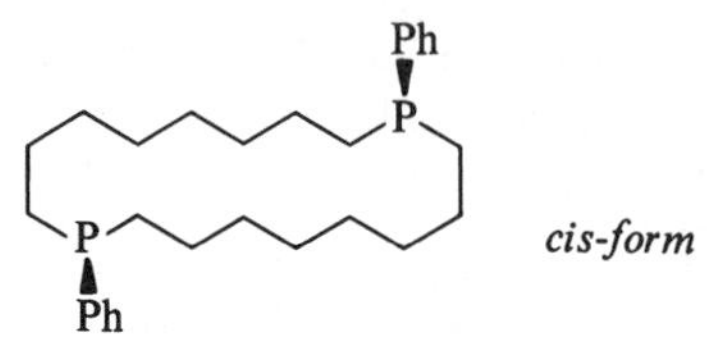

cis-form

$C_{28}H_{42}P_2$ M 440.587

cis-form [51154-91-7]

Mp 55-56°.

1,10-Dioxide: [51096-67-4].

$C_{28}H_{42}O_2P_2$ M 472.586

Hygroscopic cryst. ($CHCl_3$/hexane). Mp 173-174°.

trans-form [51021-80-8]

Solid. Mp 103-104°.

1,10-Dioxide: [51021-83-1]. Hygroscopic cryst. ($CHCl_3$/hexane). Mp 193-195°.

Chan, T.K. *et al*, *J. Org. Chem.*, 1974, **39**, 1748 (*synth, struct*)

1,12-Diphenyl-1,12-diphosphacyclotetracosane, 9CI D-01010

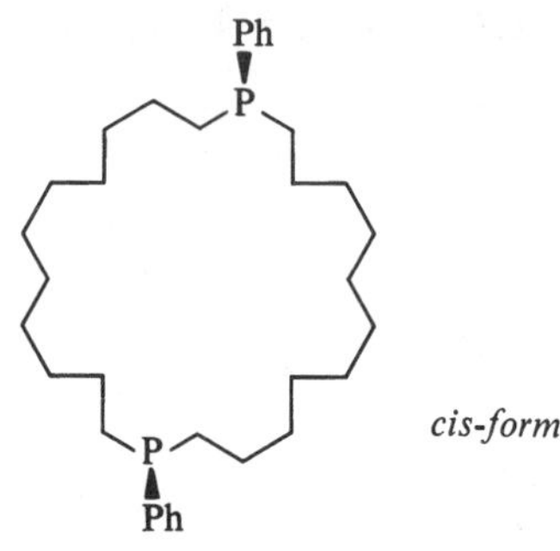

cis-form

$C_{36}H_{58}P_2$ M 552.802

cis-form [51064-84-7]

Solid. Mp 42.5-43°.

1,12-Dioxide: [51021-79-5].

$C_{36}H_{58}O_2P_2$ M 584.801

Hygroscopic cryst. ($CHCl_3$/hexane). Mp 146-147°.

trans-form [51021-82-0]

Solid. Mp 35-36°.

1,12-Dioxide: [51096-69-6]. Hygroscopic cryst. ($CHCl_3$/hexane). Mp 164-166°.

Chan, T.K. *et al*, *J. Org. Chem.*, 1974, **39**, 1748 (*synth, struct*)

P,P'-Diphenyl diphosphate, 9CI D-01011

P,P'-*Diphenyl pyrophosphate*

[1231-03-4]

PhO(HO)P(=O)–O–P(=O)(OPh)OH

$C_{12}H_{12}O_7P_2$ M 330.170

Biscyclohexylammonium salt: [2237-89-0]. Platelets or needles (H_2O). Mp 260-261°.

Kaye, H. *et al*, *J. Chem. Soc.* (*C*), 1967, 1420 (*synth*)
Clarke, V.M. *et al*, *J. Chem. Soc.* (*C*), 1969, 74 (*synth*)
Ruveda, M.A. *et al*, *Tetrahedron*, 1973, **29**, 585 (*synth*)
Fujisawa, T. *et al*, *Bull. Chem. Soc. Jpn.*, 1983, **56**, 3529 (*synth*)

1,2-Diphenyldiphosphine, 9CI D-01012

[34478-62-1]

PhPH–PHPh

$C_{12}H_{12}P_2$ M 218.174

Low melting solid which disproportionates at r.t. (accelerated by acids, bases, or light). Mp 36° (irreversible). Stored at –30° under inert gas.

***(RS,RS)*-form** [39768-63-3]
(±)-*form*
Not separately characterised.
***(RS,SR)*-form** [39768-52-0]
meso-*form*
Not separately characterised.

Albrand, J.P. *et al, J. Chem. Soc., Chem. Commun.*, 1976, 876 (*nmr*)
Baudler, M. *et al, Chem. Ber.*, 1978, **111**, 1210 (*synth, ir, ms, raman, pmr, nmr*)
Baudler, M. *et al, Phosphorus Sulfur*, 1980, **9**, 81.
Albrand, J.P. *et al, ACS Symp. Ser.*, 1981, **171**, 577 (*rev*)
Kawashima, T. *et al, Heterocycles*, 1982, **17**, 341 (*use*)
Newman, T.H. *et al, J. Organomet. Chem.*, 1982, **225**, 211 (*deriv*)

1,3-Diphenyl-1,3-diphospholane, 9CI D-01013

[53088-59-8]

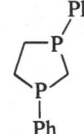

$C_{15}H_{16}P_2$ M 258.239
Cryst. (pentane). Mp 75-79°. $Bp_{0.05}$ 150°. Mixt. of stereoisomers.
B,2MeBr: [18582-75-7]. *1,3-Dimethyl-1,3-diphenyl-1,3-diphospholanium dibromide.*
$C_{17}H_{22}Br_2P_2$ M 448.116
Solid. Mp 328°. Probably possesses the *meso* config.
B,2PhBr: [74144-31-3]. *1,1,3,3-Tetraphenyl-1,3-diphospholanium dibromide.*
$C_{27}H_{26}Br_2P_2$ M 572.258
Source of the diylide 1,1,3,3-Tetrahydro-1,1,3,3-tetraphenyl-1,3-diphosphacyclopenta-1,2-diene, T-00150.

Horner, L. *et al, Tetrahedron Lett.*, 1966, 5783 (*synth*)
Issleib, K. *et al, Z. Anorg. Allg. Chem.*, 1974, **406**, 178 (*synth*)
Schmidbaur, H. *et al, Angew. Chem., Int. Ed. Engl.*, 1980, **19**, 555 (*use*)
Schmidbaur, H. *et al, Chem. Ber.*, 1986, **119**, 2832 (*synth, ms, P nmr*)

Diphenyldiphosphonic acid, 9CI D-01014

Diphenylpyrophosphonic acid. Phenylphosphonic acid monoanhydride

$C_{12}H_{12}O_5P_2$ M 298.171
Cryst. (H_2O). Mp 79.5-80°, 115-118° (dihydrate).
Bisaniline salt: Cryst. (2-propanol). Mp 211-213°.
Bis(morpholine) salt: Cryst. (EtOH/EtOAc). Mp 203-207°.
Di-Ph ester: Diphenyl diphenyldiphosphonate.
$C_{24}H_{20}O_5P_2$ M 450.367
Cryst. (Et_2O). Mp 96°.

Anschütz, L. *et al, Chem. Ber.*, 1956, **89**, 688 (*synth*)
Fox, R.B. *et al, J. Org. Chem.*, 1961, **26**, 2542 (*deriv*)
Dannley, R.L. *et al, J. Am. Chem. Soc.*, 1965, **87**, 4805 (*ester*)
Gallagher, M.J. *et al, J. Chem. Soc. (C)*, 1966, 2176; 1971, 593 (*deriv, ir, synth*)

1,2-Diphenyl-1,2-diphosphorinane, 9CI D-01015

1,2-Diphenyl-1,2-diphosphinane
[6190-11-0]

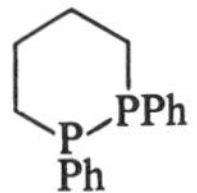

$C_{16}H_{18}P_2$ M 272.266
Solid. Mp 92°.
1,2-Disulfide: [6168-27-0].
$C_{16}H_{18}P_2S_2$ M 336.386
Solid. Mp 171°.

Issleib, K. *et al, Chem. Ber.*, 1966, **99**, 1310 (*synth, disulfide*)
Appel, R. *et al, Chem. Ber.*, 1975, **108**, 1783, 2349 (*props*)
Diemert, K. *et al, Chem. Ber.*, 1982, **115**, 1947 (*synth, P nmr*)

1,3-Diphenyl-1,3-diphosphorinane, 9CI D-01016

1,3-Diphenyl-1,3-diphosphacyclohexane. 1,3-Diphenyl-1,3-diphosphinane
$C_{16}H_{18}P_2$ M 272.266
Liq. $Bp_{0.25}$ 120-160°. Mixt. of stereoisomers.

Schmidbaur, H. *et al, Chem. Ber.*, 1986, **119**, 2832 (*synth, ms, P nmr*)

1,4-Diphenyl-1,4-diphosphorinane, 9CI D-01017

1,4-Diphenyl-1,4-diphosphinane
[18899-86-4]

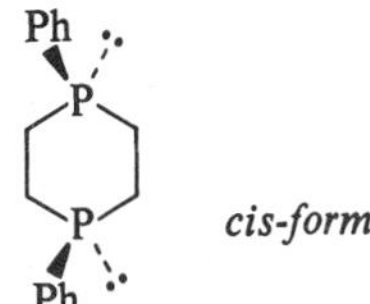

$C_{16}H_{18}P_2$ M 272.266
Solid. Mp 92-95° (evac. tube), 104-105°. Bp_7 300-320°.
B,2PhBr: [2316-28-1]. *1,1,4,4-Tetraphenyl-1,4-diphosphorinanium dibromide. 1,1,4,4-Tetraphenyl-1,4-diphosphinanium dibromide.*
$C_{28}H_{28}Br_2P_2$ M 586.285
Cryst. (MeOH/Me_2CO). Mp 324-325°. Forms a monohydrate.

cis-form

1,4-Dioxide: [27771-03-5].
$C_{16}H_{18}O_2P_2$ M 304.265
Solid. Mp 222-225°.
B,2MeBr: [18582-73-5]. *1,4-Dimethyl-1,4-diphenyl-1,4-diphosphorinanium dibromide. 1,4-Dimethyl-1,4-diphenyl-1,4-diphosphinanium dibromide.*
$C_{18}H_{24}Br_2P_2$ M 462.143
Cryst. + $1H_2O$. Mp 332°.

trans-form

1,4-Dioxide: [27771-04-6]. Solid. Mp 325°.
B,2MeBr: [18582-74-6]. Cryst. + $0.5H_2O$. Mp 350°.

Hinton, R.C. *et al, J. Chem. Soc.*, 1959, 2835 (*synth, derivs*)
Issleib, K. *et al, Chem. Ber.*, 1963, **96**, 279 (*synth, disulfide*)
Aguiar, A.M. *et al, J. Am. Chem. Soc.*, 1965, **85**, 671 (*deriv, pmr*)
Issleib, K. *et al, Chem. Ber.*, 1968, **101**, 2197 (*derivs*)
Driver, G.E. *et al, J. Chem. Soc., Chem. Commun.*, 1970, 1504 (*dioxide*)

Gallagher, M.J. *et al*, *Cryst. Struct. Commun.*, 1979, **8**, 583, 587 (*dioxide, cryst struct*)

1′,3′-Diphenyldispiro[1,3,2-benzodioxaphosphole-2,2′-[1,3,2,4]-diazaphosphetidine-4′,2″-[1,3,2]-benzodioxaphosphole], 9CI D-01018

1,3-Diphenyl-2,2:4,4-bi(o-phenylenedioxy)-1,3,2λ⁵,4λ⁵-diazadiphosphetidine

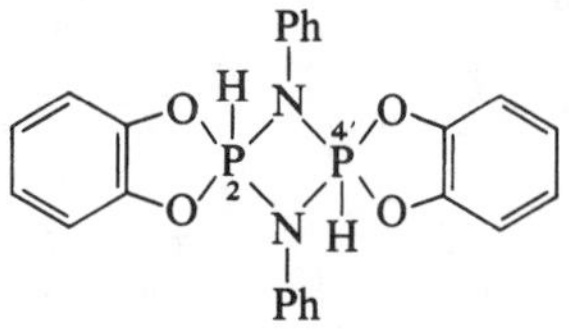

$C_{24}H_{20}N_2O_4P_2$ M 462.381

2,4′-Dimethoxy: [74927-63-2].
$C_{26}H_{24}N_2O_6P_2$ M 522.433
Solid. Mp 213-215°.

2,4′-Diethoxy: [74927-64-3].
$C_{28}H_{28}N_2O_6P_2$ M 550.487
Solid. Mp 225-227°.

2,4′-Diisopropoxy: [84635-70-1].
$C_{30}H_{32}N_2O_6P_2$ M 578.540
Solid. Mp 194-196°.

2,4′-Diphenoxy: [74927-66-5].
$C_{36}H_{28}N_2O_6P_2$ M 646.575
Solid. Mp 246-248°.

2,4′-Difluoro: [84635-68-7].
$C_{24}H_{18}F_2N_2O_4P_2$ M 498.362
Solid. Mp 231-233°.

2,4′-Dichloro: [69695-11-0].
$C_{24}H_{18}Cl_2N_2O_4P_2$ M 531.271
Solid. Mp 240-242°.

2,4′-Dibromo: [84635-69-8].
$C_{24}H_{18}Br_2N_2O_4P_2$ M 620.173
Solid. Mp 229-231°.

2,4′-Di-Ph:
$C_{36}H_{28}N_2O_4P_2$ M 614.576
Solid. Mp 280-282°.

2,4′-Bis(dimethylamino): N,N,N′,N′-*Tetramethyl-1′,3′-diphenyldispiro[1,3,2-benzodioxophosphole-2,2′-[1,3,2,4]diazaphosphetidine-4′,2″-[1,3,2]-benzodioxaphosphol]-2,2′-diamine, 9CI.*
$C_{28}H_{30}N_4O_4P_2$ M 548.517
Cryst. (C_6H_6). Mp 174-175°.

2,4′-Bis(phenylamino):
$C_{36}H_{30}N_4O_4P_2$ M 644.605
Solid. Mp 244-246°.

Kukhar', V.P. *et al*, *Zh. Obshch. Khim.*, 1978, **48**, 2657; 1980, **50**, 1477; 1982, **52**, 2227 (*Engl. transl.* pp. 2410, 1191, 1982) (*synth, P nmr*)

2,4-Diphenyl-1,3,2,4-dithiadiphosphetane 2,4-disulfide, 9CI D-01019

Cyclic dianhydrosulfide from phenylphosphonotrithoic acid, 8CI

[1024-07-3]

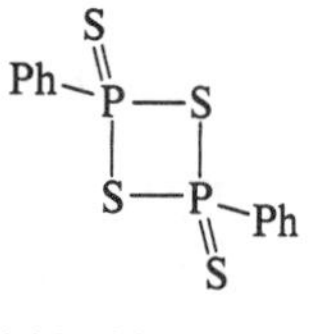

$C_{12}H_{10}P_2S_4$ M 344.399

Useful reagent for sulfurations. Cryst. (PhCl or toluene). Mp 231-236°, 242-243°.

Newallis, P.E. *et al*, *J. Org. Chem.*, 1961, **27**, 3829 (*synth*)
Maier, L., *Helv. Chim. Acta*, 1963, **46**, 1812 (*synth, cryst struct*)
Baudler, M. *et al*, *Z. Naturforsch., B*, 1967, **22**, 222 (*synth, props, P nmr*)
Grishina, O.N. *et al*, *Zh. Obshch. Khim.*, 1977, **47**, 72 (*Engl. transl.* p. 64) (*synth, P nmr*)
Keck, H. *et al*, *Phosphorus Sulfur*, 1978, **4**, 173 (*ms*)
Andreev, N.A. *et al*, *Zh. Obshch. Khim.*, 1982, **52**, 1785 (*Engl. transl.* p. 1581) (*synth, ir, struct*)

Diphenyl hypodiphosphonous acid, 9CI D-01020

PhP(OH)P(OH)Ph ⇌ PhPH(O)PH(O)Ph

$C_{12}H_{12}O_2P_2$ M 250.173

Dibromide: [88017-79-2].
$C_{12}H_{10}Br_2P_2$ M 375.967
Cryst. (C_6H_6). Mp 125.5-127°.

Diiodide:
$C_{12}H_{10}I_2P_2$ M 469.968
Yelow cryst. (C_6H_6, toluene or chlorobenzene). Mp 178-180°.

Hoffmann, H. *et al*, *Chem. Ber.*, 1961, **94**, 186 (*diiodide*)
Feshchenko, M.G. *et al*, *Zh. Obshch. Khim.*, 1972, **42**, 284; 1976, **46**, 252 (*Engl. transl.* pp. 273, 248) (*diiodide*)
Hassler, K. *et al*, *Z. Anorg. Allg. Chem.*, 1978, **443**, 125 (*ir, raman*)
Hinke, A. *et al*, *Chem. Ber.*, 1983, **116**, 3003 (*dibromide, synth, props, P nmr*)

(Diphenylmethylene)(2,4,6-trimethylphenyl)phosphine, 9CI D-01021

(Diphenylmethylene)mesitylphosphine

[67565-91-7]

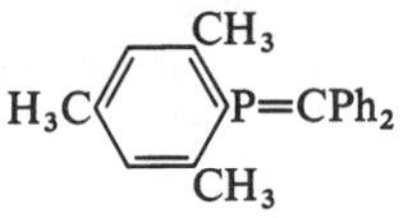

$C_{22}H_{21}P$ M 316.382

Ligand for Cr. Yellow, thermally stable cryst. Mp 51-58°. $Bp_{0.001}$ 140°. Forms adduct with Se (but not with Te).

Klebach, T.C. *et al*, *J. Am. Chem. Soc.*, 1978, **100**, 4886 (*synth, pmr, cmr, nmr, uv*)
Klebach, T.C. *et al*, *J. Organomet. Chem.*, 1981, **210**, 211 (*complexes*)
Van der Knaap, T.A. *et al*, *J. Am. Chem. Soc.*, 1983, **105**, 4026; *J. Organomet. Chem.*, 1983, **244**, 363.

(Diphenylmethylene)triphenylphosphorane, 9CI D-01022

[4214-38-4]

$Ph_3P{=}CPh_2$

$C_{31}H_{25}P$ M 428.512

Stable ylide. Red needles (C_6H_6). Mp 173°. Ozonolysis gives Ph_2CO and Ph_3PO.

Horner, L. *et al*, *Justus Liebigs Ann. Chem.*, 1955, **591**, 135 (*synth*)
Van Dijk, J.M.F. *et al*, *Recl. Trav. Chim. Pays-Bas*, 1974, **93**, 155 (*struct*)
Tewari, R.S. *et al*, *Z. Naturforsch., B*, 1975, **30**, 513 (*synth*)
Casey, C.P. *et al*, *J. Am. Chem. Soc.*, 1977, **99**, 2127 (*synth*)
Fox, M.A., *J. Am. Chem. Soc.*, 1979, **101**, 5339 (*synth*)
Caminade, A.M. *et al*, *Phosphorus Sulfur*, 1983, **14**, 381 (*props*)

(Diphenylmethyl)phenylphosphinic acid, 9CI **D-01023**
Benzhydrylphenylphosphinic acid

$Ph_2CHPPh(O)OH$

$C_{19}H_{17}O_2P$ M 308.316
Cryst. (EtOH aq.). Mp 244-246°.
Et ester: Ethyl(diphenylmethyl)phenylphosphinate. Ethyl benzhydrylphenylphosphinate.
$C_{21}H_{21}O_2P$ M 336.369
Needles (EtOH). Mp 235-237°.

Henning, H-G. *et al*, *J. Prakt. Chem.*, 1965, **29**, 86 (*ester*)
Freeman, K.L. *et al*, *Aust. J. Chem.*, 1966, **19**, 2025, 2159 (*synth, pmr*)

Diphenyl methylphosphonate **D-01024**
[7526-26-3]

$MeP(O)(OPh)_2$

$C_{13}H_{13}O_3P$ M 248.218
Solid. Mp 35-36°. Bp_9 192-192.5°, $Bp_{0.15}$ 138°.

Landauer, S.R. *et al*, *J. Chem. Soc.*, 1953, 2224 (*synth*)
Hudson, H.R. *et al*, *J. Chem. Soc., Perkin Trans. 1*, 1974, 982 (*synth, P nmr*)
Berkova, G.A. *et al*, *Zh. Obshch. Khim.*, 1977, **47**, 1431 (*Engl. transl.* p. 1315) (*pmr*)
Honig, M.L. *et al*, *J. Org. Chem.*, 1977, **42**, 379 (*synth*)
Bel'skii, V.E. *et al*, *Zh. Obshch. Khim.*, 1981, **51**, 2442 (*Engl. transl.* p. 2104) (*props*)

(Diphenylmethyl)-triphenylphosphonium(1+), 9CI **D-01025**

$Ph_3P^{\oplus}CHPh_2$

$C_{31}H_{26}P^{\oplus}$ M 429.520 (ion)
Bromide: [7333-65-5].
$C_{31}H_{26}BrP$ M 509.424
Source of (Diphenylmethylene)triphenylphosphorane, D-01022 . Cryst. ($CHCl_3$/EtOAc or MeOH aq.). Mp 227-228° dec. NaH, NaOEt or $NaN(SiMe_3)_2$ → ylide.
Ylide: see (Diphenylmethylene)triphenylphosphorane, D-01022

Horner, L. *et al*, *Justus Liebigs Ann. Chem.*, 1955, **591**, 135 (*synth, props*)
Tewari, R.S. *et al*, *Z. Naturforsch., B*, 1975, **30**, 513 (*synth*)

2,3-Diphenyl-1,3-oxaphosphepane, 10CI **D-01026**

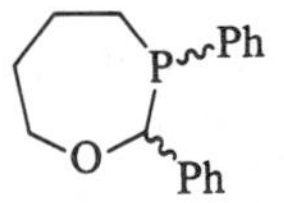

$C_{17}H_{19}OP$ M 270.310
Viscous liq. $Bp_{0.5}$ 168-171°. Mixt. of *cis-* and *trans-* forms.

Oehme, H. *et al*, *Z. Chem.*, 1979, **19**, 57 (*synth, pmr*)

2,3-Diphenyl-1,3-oxaphospholane, 9CI **D-01027**
[37181-95-6]

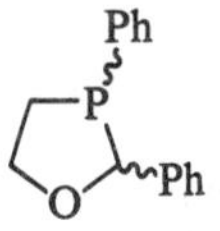

$C_{15}H_{15}OP$ M 242.257
Geom. isomers not separately characterised. Liq. Bp_6 181-184°.
B,MeI: 3-Methyl-2,3-diphenyl-1,3-oxaphospholanium iodide.
$C_{16}H_{18}IOP$ M 384.196
Mp 154-157°.
3-Oxide: [37182-02-8].
$C_{15}H_{15}O_2P$ M 258.256
Solid. Mp 86-90°.
3-Sulfide: [37182-04-0].
$C_{15}H_{15}OPS$ M 274.317
Solid. Mp 134-137°.

Oehme, H. *et al*, *Tetrahedron*, 1972, **28**, 2587 (*synth, derivs*)
Zschunke, A. *et al*, *Org. Magn. Reson.*, 1975, **7**, 470 (*pmr, nmr*)

2,3-Diphenyl-1,3-oxaphosphorinane, 9CI **D-01028**
2,3-Diphenyl-1,3-oxaphosphinane. 2,3-Diphenyl-1-oxa-3-phosphacyclohexane
[38706-86-4]

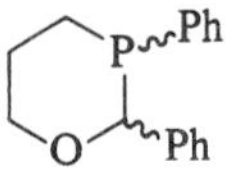

$C_{16}H_{17}OP$ M 256.283
Stereoisomers not separately characterised. Solid. Mp 43°. Bp_2 175-178°. Forms a Pd complex.
B,MeI: [38706-92-2]. *3-Methyl-2,3-diphenyl-1,3-oxaphosphorinanium iodide.*
$C_{17}H_{20}IOP$ M 398.223
Cryst. (EtOH or EtOH/Et_2O). Mp 224-226°.
3-Sulfide:
$C_{16}H_{17}OPS$ M 288.343
Cryst. (EtOH). Mp 94-97°.

Issleib, K. *et al*, *Synth. Inorg. Met.-Org. Chem.*, 1972, **2**, 223 (*synth, derivs*)
Zschunke, A. *et al*, *Phosphorus Sulfur*, 1978, **5**, 81; 1980, **9**, 117 (*cmr, P nmr, stereochem*)
Zschunke, A. *et al*, *J. Organomet. Chem.*, 1981, **222**, 353 (*complex*)

2,3-Diphenyl-1,3,2-oxazaphospholidine, 9CI **D-01029**
2,3-Diphenyl-1,3,2-oxazaphospholane
[5679-73-2]

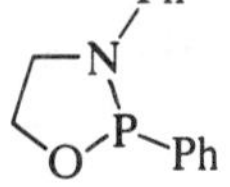

$C_{14}H_{14}NOP$ M 243.244
Solid. Mp 78-81° (75-76°). $Bp_{0.13}$ 150-153°.
2-Sulfide: [59085-32-4].
$C_{14}H_{14}NOPS$ M 275.304
Cryst. (C_6H_6/Et_2O). Mp 142-144°.

Fujisawa, T. *et al*, *Bull. Chem. Soc. Jpn.*, 1967, **40**, 147 (*synth, ir, props*)
Mitsunobu, O. *et al*, *Bull. Chem. Soc. Jpn.*, 1967, **40**, 2964 (*synth*)

Dustmukhmeelov, T.T. *et al*, *Zh. Obshch. Khim.*, 1976, **46**, 300 (*Engl. transl.* p. 294) (*sulfide, synth, ms*)
Beer, P.D. *et al*, *Phosphorus Sulfur*, 1983, **17**, 283 (*synth, pmr, P nmr*)

(1,3-Diphenyl-3-oxopropyl)-phenylphosphinic acid D-01030

$PhCOCH_2CHPhPPh(O)OH$

$C_{21}H_{19}O_3P$ M 350.353

(±)-*form*
Cryst. (EtOH aq.). Mp 243-244°.

Bergesen, K., *Acta Chem. Scand.*, 1965, **19**, 1784 (*synth*)

Diphenyl(4-phenyl-1,3-butadienyl)-phosphine, 9CI D-01031

1-Diphenylphosphino-4-phenyl-1,3-butadiene

$Ph_2PCH{=}CHCH{=}CHPh$

$C_{22}H_{19}P$ M 314.366
Ligand for Fe, Mo, and Ni. Cryst. Mp 90-91°.
Oxide:
$C_{22}H_{19}OP$ M 330.365
Solid. Mp 175-176°.

(*E,E*)-*form* [73210-33-0]
B,PhCH₂Br: Benzyldiphenyl(4-phenyl-1,3-butadienyl)-phosphonium bromide.
$C_{29}H_{26}BrP$ M 485.402
Cryst. Mp 104°.
Sulfide:
$C_{22}H_{19}PS$ M 346.426
Cryst. (Et_2O). Mp 140-145°.

Kabachnik, M.I. *et al*, *Dokl. Akad. Nauk SSSR, Ser. Sci. Khim.*, 1965, **162**, 339 (*Engl. transl.* p. 460) (*oxide*)
Gloyna, D. *et al*, *Angew. Chem.*, 1966, **78**, 907 (*sulfide*)
Mathey, F. *et al*, *Inorg. Chem.*, 1980, **19**, 1813 (*synth, pmr, P nmr, derivs*)

Diphenyl(4-phenyl-1,3-butadiynyl)-phosphine, 9CI D-01032

1-Diphenylphosphino-4-phenyl-1,3-butadiyne

$Ph_2PC{\equiv}CC{\equiv}CPh$

$C_{22}H_{15}P$ M 310.334
Solid. Mp 75°.
Oxide: [7608-32-4].
$C_{22}H_{15}OP$ M 326.334
Cryst. (dimethoxyethane/pet. ether). Mp 131°.
Sulfide:
$C_{22}H_{15}PS$ M 342.394
Solid. Mp 89°.

Charner, C. *et al*, *Bull. Soc. Chim. Fr.*, 1966, 1002 (*synth, derivs, uv, ir*)
Disteldorf, W. *et al*, *Chem. Ber.*, 1976, **109**, 546 (*oxide, ir*)

Diphenyl(2-phenylethenyl)phosphine, 9CI D-01033

Diphenylstyrylphosphine
[794-39-8]

$Ph_2PCH{=}CHPh$

$C_{20}H_{17}P$ M 288.328

(*E*)-*form* [14090-06-3]
Mp 55-56°. $Bp_{0.25}$ 175-180°.
B,MeI: Methyl diphenyl(phenylethenyl)phosphonium iodide.
$C_{21}H_{20}IP$ M 430.267
Solid. Mp 168°.
Oxide: [3582-82-9].
$C_{20}H_{17}OP$ M 304.327
Cryst. (EtOH aq. or octane). Mp 168-169°.
Sulfide: [3582-83-0].
$C_{20}H_{17}PS$ M 320.388
Cryst. (hexane). Mp 106.5-107.5°.

(*Z*)-*form* [14090-07-4]
Cryst. (MeOH). Mp 90-92°.
Oxide: [78045-10-0]. Cryst. (cyclohexane). Mp 103-104°. Converted by PCl_5 into the (*E*)-oxide.

Aguiar, A.M. *et al*, *J. Org. Chem.*, 1967, **30**, 3527 (*synth, derivs*)
Brophy, J.J. *et al*, *J. Chem. Soc. (C)*, 1968, 2760 (*synth, deriv*)
Gloyna, D. *et al*, *J. Prakt. Chem.*, 1976, **318**, 327 (*oxide, pmr*)
Regitz, M. *et al*, *Justus Liebigs Ann. Chem.*, 1979, 1002 (*oxide*)
Berndt, K.-G. *et al*, *J. Prakt. Chem.*, 1981, **323**, 445 (*oxide, ms*)
Duncan, M. *et al*, *Org. Magn. Reson.*, 1981, **15**, 37 (*pmr, cmr, P nmr*)
Gloyna, D. *et al*, *J. Prakt. Chem.*, 1982, **324**, 107 (*spectra*)

Diphenyl(2-phenylethyl)phosphine, 9CI D-01034

Phenethyldiphenylphosphine
[5952-49-8]

$Ph_2PCH_2CH_2Ph$

$C_{20}H_{19}P$ M 290.344
Oil. Bp_{11} 223-224°. n_D^{20} 1.6312.
Oxide: [3582-84-1].
$C_{20}H_{19}OP$ M 306.343
Cryst. (cyclohexane, Et_2O or Me_2CO/hexane). Mp 103-104°.

Aguiar, A.M. *et al*, *J. Org. Chem.*, 1965, **30**, 2826 (*oxide, ir, pmr*)
Schindlbauer, H., *Monatsh. Chem.*, 1965, **96**, 2058 (*synth*)
Grayson, I.J. *et al*, *J. Chem. Soc., Perkin Trans. 1*, 1976, 2556 (*oxide, ir, pmr, ms*)
Sartori, P. *et al*, *Phosphorus*, 1980, **8**, 115 (*oxide, ir, pmr, ms*)

Diphenyl phenylphosphoramidate, 9CI D-01035

Diphenyl phosphoranilidate. N-*(Diphenoxyphosphinyl)-aniline*
[3848-51-9]

$(PhO)_2P(O)NHPh$

$C_{18}H_{16}NO_3P$ M 325.303
Cryst. (EtOH or C_6H_6/hexane). Mp 131-131.5°.

Audrieth, L.F. *et al*, *J. Am. Chem. Soc.*, 1942, **64**, 1337 (*synth*)
Nielsen, M.L., *Inorg. Chem.*, 1964, **3**, 1760 (*synth*)
McMurray, W.J. *et al*, *Org. Mass. Spectrom.*, 1970, **3**, 1031 (*ms*)
Cremlyn, R.J.W., *Aust. J. Chem.*, 1973, **26**, 1591 (*synth, ir, P nmr*)
Ayed, N. *et al*, *C.R. Hebd. Seances Acad. Sci., Ser. C*, 1977, **285**, 221 (*ir*)

Diphenyl phosphate, 9CI, 8CI D-01036

Hydrogen diphenyl phosphate. Diphenyl phosphoric acid. Diphenyl hydrogen phosphate
[838-85-7]

(PhO)$_2$P(O)OH

$C_{12}H_{11}O_4P$ M 250.190

Solid. Sol. H_2O. Mp 68-70°. pK_a 1.85 (H_2O), pK_a 2.90 (50% EtOH aq.) (2.38), pK_a 4.81 (EtOH).

Dihydrate: Mp 51°.

NH_4 salt: Solid. Mp 130°.

Cyclohexylammonium salt: [13798-41-9]. Cryst. (EtOH). Mp 199-200°.

Tetramethylammonium salt: Solid. Mp 210°.

Anilinium salt: Solid. Mp 165-166°.

Fluoride: see Diphenyl phosphorofluoridate, D-01133

Chloride: see Diphenyl phosphorochloridate, D-01120

Bromide: [70677-19-9]. *Diphenyl phosphorobromidate. Diphenylphosphoryl bromide. Diphenyl bromophosphate.*

$C_{12}H_{10}BrO_3P$ M 313.087

Cryst. Mp 46-47°. Bp_3 198-200°.

Cyanide: see Diphenyl phosphorocyanidate, D-01124

Azide: see Diphenyl phosphorazidate, D-01112

Anhydride: see Tetraphenyl diphosphate, T-00256

Anilide: see Diphenyl phenylphosphoramidate, D-01035

Hydrazide: see Diphenyl phosphorohydrazidate, D-01136

Lecocq, J. *et al*, *J. Chem. Soc.*, 1954, 2381 (*synth*)
Chabrier, P. *et al*, *C.R. Hebd. Seances Acad. Sci.*, 1957, **244**, 2730 (*synth*)
Kamai, G. *et al*, *CA*, 1958, **52**, 9946 (*bromide*)
Rubinstein, M. *et al*, *Tetrahedron*, 1975, **31**, 2107 (*synth*)
Cabre-Castellvi, J. *et al*, *Afinidad*, 1982, **39**(382), 508; *CA*, **98**, 160816 (*ir, derivs*)

Diphenylphosphine D-01037

[829-85-6]

Ph_2PH

$C_{12}H_{11}P$ M 186.193

Foul-smelling liq. d[16] 1.07. Bp 280°, Bp_{25} 155-157°, Bp_1 103°. pK_a 0.03 (H_2O). n_D^{20} 1.6279. V. easily oxid. in air.

▷May ignite in air

Oxide: [4559-70-0].

$C_{12}H_{11}OP$ M 202.192

Needles (Et_2O). Mp 53-56°. Deliquescent. Tautomeric with Diphenylphosphinous acid, D-01089 .

Sulfide: [6591-07-7].

$C_{12}H_{11}PS$ M 218.253

Cryst. (MeCN). Mp 95-97°. Tautomeric with Diphenylphosphinothious acid, D-01085 .

Selenide: [5853-64-5].

$C_{12}H_{11}PSe$ M 265.153

Cryst. (MeCN). Mp 111-112°. Tautomeric with Diphenylphosphinoselenous acid, D-01081 .

Li deriv: Yellow cryst. + 1 dioxan; orange soln. in THF.

Schindlbauer, H., *Monatsh. Chem.*, 1963, **94**, 99 (*uv*)
Johnson, A.W. *et al*, *Can. J. Chem.*, 1965, **43**, 1338.
Maier, L., *Helv. Chim. Acta*, 1966, **49**, 1000, 1249 (*selenide, sulfide, ir, nmr*)
Inorg. Synth., 1967, **9**, 19; 1976, **16**, 161 (*synth, pmr*)
Olah, G.A. *et al*, *J. Org. Chem.*, 1969, **34**, 1832 (*pmr, nmr*)
McFarlane, W. *et al*, *J. Chem. Soc., Dalton Trans.*, 1973, 2162 (*selenide, nmr*)
Debies, T.P. *et al*, *Inorg. Chem.*, 1974, **13**, 308 (*pe*)
Goda, K. *et al*, *Bull. Chem. Soc. Jpn.*, 1977, **50**, 545 (*sulfide, raman*)
Petrov, É.S. *et al*, *Zh. Obshch. Khim.*, 1979, **49**, 2410 (*Engl. transl.* p. 2127) (*props*)
Vincent, E. *et al*, *Spectrochim. Acta, Part A*, 1980, **36**, 699 (*pmr, cmr, nmr*)
Bretherick, L., *Handbook of Reactive Chemical Hazards*, 2nd Ed., Butterworths, London and Boston, 1979, 710 (*haz*)

Diphenylphosphine oxide, 9CI D-01038

[4559-70-0]

$Ph_2P(O)H$

$C_{12}H_{11}OP$ M 202.192

Tautomeric with Diphenylphosphinous acid, D-01089 . Hygroscopic needles (Et_2O). Mp 57°. pK_a 20.7 (DMSO), 11.9 (1,2-dimethoxyethane).

Hunt, B.B. *et al*, *J. Chem. Soc.*, 1957, 2413 (*synth, props, ir*)
Grayson, M. *et al*, *Tetrahedron*, 1967, **23**, 1065 (*synth*)
Sanchez, M. *et al*, *Spectrochim. Acta, Part A*, 1967, **23**, 2617 (*ir*)
Emoto, T. *et al*, *Bull. Chem. Soc. Jpn.*, 1974, **47**, 2449.

Diphenylphosphinic acid, 9CI, 8CI D-01039

[1707-03-5]

$Ph_2P(O)OH$

$C_{12}H_{11}O_2P$ M 218.191

Needles (C_6H_6, conc. HNO_3 or EtOH aq.). Mp 195-196°. pK_a 2.32 (7% EtOH). Forms anhydride at 230°.

▷SZ5315000.

Dibenzylammonium salt: [29544-53-4]. Cryst. (EtOH aq.). Mp 133-137°.

Me ester: [1706-90-7]. *Methyl diphenylphosphinate.*

$C_{13}H_{13}O_2P$ M 232.218

Mp 57°. Bp_{29} 232°, $Bp_{0.3}$ 162-163°. Slightly hygroscopic.

Et ester: [1733-55-7]. *Ethyl diphenylphosphinate.*

$C_{14}H_{15}O_2P$ M 246.245

Mp 42°. $Bp_{0.1}$ 178-180°. V. hygroscopic, taking up 0.5 mol H_2O.

Isopropyl ester: [1706-91-8]. *Isopropyl diphenylphosphinate.*

$C_{15}H_{17}O_2P$ M 260.272

Cryst. (C_6H_6/pet. ether or Me_2CO aq.). Mp 97-99°.

tert-*Butyl ester:* [1706-92-9]. tert-*Butyl diphenylphosphinate.*

$C_{16}H_{19}O_2P$ M 274.299

Cryst. (MeOH, aq.). Mp 111.5-112°.

Ph ester: [1706-96-3]. *Phenyl diphenylphosphinate.*

$C_{18}H_{15}O_2P$ M 294.289

Prisms (EtOH). Mp 136-138°.

▷SZ5330000.

Benzyl ester: [5573-42-2]. *Benzyl diphenylphosphinate.*

$C_{19}H_{17}O_2P$ M 308.316

Cryst. Mp 67-77°.

1-Naphthyl ester: 1-Naphthyl diphenylphosphinate.

$C_{22}H_{17}O_2P$ M 344.349

Solid. Mp 122-124°.

Fluoride: see Diphenylphosphinic fluoride, D-01042

Chloride: see Diphenylphosphinic chloride, D-01041

Bromide: [75770-58-0].
$C_{12}H_{10}BrOP$ M 281.088
Liq. $Bp_{0.05}$ 160-166°, $Bp_{0.003}$ 138-142°. n_D^{20} 1.6373; n_D^{25} 1.6287.
Cyanide: [4669-69-6].
$C_{13}H_{10}NOP$ M 227.202
Good protecting reagent for thiol groups. Solid. Mp 54-55°. $Bp_{0.005}$ 125°. n_D^{20} 1.5930.
*Amide: see P,P-*Diphenylphosphinic amide, D-01040
Azide: [4129-17-3].
$C_{12}H_{10}N_3OP$ M 243.204
Oil. $Bp_{0.01}$ 138-140°.
▷SZ5870000.
Isocyanate: [6779-62-0].
$C_{13}H_{10}NO_2P$ M 243.201
Solid. Mp 52-54°. $Bp_{0.5}$ 160°.
Isothiocyanate: [20443-14-5].
$C_{13}H_{10}NOPS$ M 259.262
Solid. Mp 34.5-36.5°. $Bp_{0.05}$ 165-170°.
Anhydride: [5849-36-5].
$C_{24}H_{20}O_3P_2$ M 418.368
Cryst. (C_6H_6 or C_6H_6/hexane). Mp 143.5-144°. $Bp_{0.2}$ 230°.
Anhydride with acetic acid: [7594-78-7]. *Acetic diphenylphosphinic anhydride.*
$C_{14}H_{13}O_3P$ M 260.229
Acetylating agent. Cryst. (Et_2O). Mp 93-97°. Disporportionates when heated.
Anhydride with benzoic acid: [4693-63-4]. *Benzoic diphenylphosphinic anhydride.*
$C_{19}H_{15}O_3P$ M 322.299
Benzoylating agent for *O*, *N*, *S* and *C* nucleophiles. Cryst. (C_6H_6). Mp 108°, 123-124°. Disproportionates in boiling benzene.

Crofts, P.C. *et al, J. Chem. Soc.*, 1964, 1240 (*synth, anhydride, ester*)
Berlin, K.D. *et al, J. Org. Chem.*, 1965, **30**, 1267 (*esters, pmr*)
Inorg. Synth., 1966, **8**, 71 (*synth*)
Harwood, H.J. *et al, J. Org. Chem.*, 1967, **32**, 3882 (*anhydride, ir*)
Venezky, D.L. *et al, J. Org. Chem.*, 1967, **32**, 838 (*anhydrides*)
Sosnovsky, G. *et al, J. Org. Chem.*, 1968, **33**, 2325 (*anhydrides, ir*)
Tomaschewski, G. *et al Arch. Pharm.* (*Weinheim, Ger.*), 1968, **301**, 520 (*isothiocyanate*)
Haake, P. *et al, J. Org. Chem.*, 1969, **34**, 788 (*ms*)
Tomaschewski, G. *et al, J. Prakt. Chem.*, 1969, **311**, 256 (*isocyanate*)
Morgan, W.E. *et al, Inorg. Chem.*, 1971, **10**, 926 (*pe*)
Williams, A. *et al, J. Chem. Soc.* (*B*), 1971, 1967 (*aryl esters*)
Hudson, H.R. *et al, J. Chem. Soc., Perkin Trans. 1*, 1972, 1595 (*bromide*)
Fenske, D. *et al, Chem. Ber.*, 1973, **106**, 1139 (*cryst struct*)
Appel, R. *et al, Chem. Ber.*, 1976, **109**, 805 (*esters, ir, nmr, pmr*)
Felcht, U. *et al, Chem. Ber.*, 1976, **109**, 3675 (*ester, ir, pmr*)
Allen, D.W. *et al, J. Chem. Soc., Perkin Trans. 2*, 1977, 1705 (*ester, pmr, P nmr*)
Modro, T.A., *Phosphorus Sulfur*, 1979, **5**, 331 (*ester, cmr*)
Schröder, H. *et al, Z. Anorg. Allg. Chem.*, 1979, **451**, 158 (*azide, ms, ir, raman, cmr, P nmr, pmr*)
Gloede, J. *et al, Z. Anorg. Allg. Chem.*, 1980, **471** 147 (*bromide, P nmr*)
Horner, L. *et al, Phosphorus Sulfur*, 1981, **11**, 157 (*cyanide, ir, P nmr*)
Lindner, E. *et al, Chem. Ber.*, 1981, **114**, 2272 (*anhydride, ir, raman, ms, pmr*)
Harger, M.J.P. *et al, Tetrahedron*, 1982, **38**, 1511 (*azide*)

P,P-Diphenylphosphinic amide, 9CI **D-01040**

[5994-87-6]

$Ph_2P(O)NH_2$

$C_{12}H_{12}NOP$ M 217.207
Needles (toluene). Mp 165-167°.
N-*Me:* N-*Methyl*-P,P-*diphenylphosphinic amide.*
$C_{13}H_{14}NOP$ M 231.233
Cryst. Mp 112-114°.
N,N-*Di-Me:* [3732-84-1]. N,N-*Dimethyl*-P,P-*diphenylphosphinic amide.*
$C_{14}H_{16}NOP$ M 245.260
Cryst. (C_6H_6/ligroin). Insol. Et_2O, sol. EtOH, C_6H_6, hot CCl_4. Mp 105-106.5°.
N-*Propyl:* P,P-*Diphenyl*-N-*propylphosphinic amide.*
$C_{15}H_{18}NOP$ M 259.287
Cryst. (EtOH). Mp 93°.
N-*Isopropyl:* [71847-21-7]. N-*Isopropyl*-P,P-*diphenylphosphinic amide.*
$C_{15}H_{18}NOP$ M 259.287
Solid. Mp 146-148°.
N-tert-*Butyl:* [68036-31-7]. N-tert-*Butyl*-P,P-*diphenylphosphinic amide.*
$C_{16}H_{20}NOP$ M 273.314
Cryst. (Et_2O). Mp 136-136.5°.
N,N-*Di-Et:* N,N-*Diethyl*-P,P-*diphenylphosphinic amide.*
$C_{16}H_{20}NOP$ M 273.314
Cryst. (EtOH). Mp 142°.
N-*Ph: see N,P,P-*Triphenylphosphinic amide, T-00630
N,N-Di-Ph: [67071-69-6]. *Tetraphenylphosphinic amide.*
$C_{24}H_{20}NOP$ M 369.402
Cryst. (EtOH). Mp 105-106°.
N-*Cyclohexyl:* [6941-20-4]. N-*Cyclohexyl*-P,P-*diphenylphosphinic amide.*
$C_{18}H_{22}NOP$ M 299.352
Solid. Mp 197-197.5°.
N-*Benzyl:* [27127-08-8]. N-*Benzyl*-P,P-*diphenylphosphinic amide.*
$C_{19}H_{18}NOP$ M 307.331
Cryst. (EtOH or cyclohexane). Mp 111-112°.
N,N-*Dibenzyl:* [27127-09-9]. N,N-*Dibenzyl*-P,P-*diphenylphosphinic amide.*
$C_{26}H_{24}NOP$ M 397.455
Cryst. (cyclohexane). Mp 125-126°.
N-*1-Naphthyl:* N-*1-Naphthyl*-P,P-*diphenylphosphinic amide.*
$C_{22}H_{18}NOP$ M 343.364
Cryst. (EtOH). Mp 188-190°.
N-*2-Naphthyl:* N-*2-Naphthyl*-P,P-*diphenylphosphinic amide.*
$C_{22}H_{18}NOP$ M 343.364
Cryst. (EtOH). Mp 264-268°.
N-*Phenylsulfonyl:* P,P-*Diphenyl*-N-*phenylsulfonylphosphinic amide.*
$C_{18}H_{16}NO_3PS$ M 357.363
Prisms (EtOH). Mp 205-206°.
N-*Ac:* [21341-80-0]. N-*Acetyl*-P,P-*diphenylphosphinic amide.*
$C_{14}H_{14}NO_2P$ M 259.244
Solid. Mp 179-181°.
N-*Benzoyl:* [59022-97-8]. N-*Benzoyl*-P,P-*diphenylphosphinic amide.*
$C_{19}H_{16}NO_2P$ M 321.315
Needles (MeOH). Mp 193-195°.
N-*Diphenylphosphino:* [7648-76-2].
$C_{24}H_{21}NOP_2$ M 401.384
Cryst. (C_6H_6/pentane). Mp 171-173°.
N-*Diphenylphosphinyl: see N-(Diphenylphosphinyl)-P,P-diphenylphosphinic amide, D-01096*

N-*Diphenylphosphinothioyl:* [17162-83-3].
$C_{24}H_{21}NOP_2S$ M 433.444
Cryst. (MeOH). Mp 172-174°.

Zhmurova, I.N. *et al*, *Zh. Obshch. Khim.*, 1959, **29**, 2083 (*Engl. transl.* p. 2052) (*synth, derivs*)
Shevchenko, V.I. *et al*, *Zh. Obshch. Khim.*, 1960, **30**, 1566 (*Engl. transl.* p. 1573) (*phenylsulfonyl*)
Smith, N.L. *et al*, *J. Org. Chem.*, 1961, **26**, 5145 (*derivs, ir*)
Gutman, V.V. *et al*, *Monatsh. Chem.*, 1962, **93**, 1114 (*derivs*)
Derkach, G.I. *et al*, *Zh. Obshch. Khim.*, 1964, **34**, 604 (*Engl. transl.* p. 605) (*benzoyl*)
Tomaschewski, G. *et al*, *Z. Chem.*, 1968, **8**, 461 (*derivs*)
Morgan, W.E. *et al*, *Inorg. Chem.*, 1971, **10**, 926 (*pe*)
Koizumi, T. *et al*, *J. Am. Chem. Soc.*, 1973, **95**, 8073 (*synth, derivs, uv, pmr*)
Muzhar-Ul-Hague, *et al*, *J. Chem. Soc., Perkin Trans. 2*, 1976, 1601 (*dimethyl, cryst struct*)
Modro, T.A. *et al*, *J. Org. Chem.*, 1978, **43**, 5000 (*deriv, props*)
Zwerziak, A. *et al*, *Synthesis*, 1979, 691 (*derivs*)
Oliva, G. *et al*, *Acta Crystallogr., Sect. B*, 1981, **37**, 474 (*cryst struct*)

Diphenylphosphinic chloride, 8CI — D-01041

[1499-21-4]

$Ph_2P(O)Cl$

$C_{12}H_{10}ClOP$ M 236.637
Useful protecting reagent for peptide amino groups. Oily liq. d_4^{20} 1.28. $Bp_{0.25}$ 147°. n_D^{20} 1.6091.

Higgins, W.A., *J. Am. Chem. Soc.*, 1955, **77**, 1864 (*synth*)
Crofts, P.C. *et al*, *J. Chem. Soc.*, 1964, 1240 (*synth*)
Neimysheva, A.A. *et al*, *Zh. Obshch. Khim.*, 1966, **36**, 1090 (*Engl. transl.* p. 1105) (*synth*)
Fluck, E. *et al*, *Z. Anorg. Allg. Chem.*, 1967, **354**, 139 (*P nmr*)
Neimysheva, A.A. *et al*, *Zh. Obshch. Khim.*, 1967, **37**, 2255 (*Engl. transl.* p. 2140) (*nqr*)
Sosnovsky, G. *et al*, *J. Org. Chem.*, 1968, **33**, 2325 (*ir*)
Sergienko, L.M. *et al*, *Zh. Obshch. Khim.*, 1979, **49**, 317 (*Engl. transl.* p. 275); 1980, **50**, 1958 (*Engl. transl.* p. 1578) (*uv*)
Horner, L. *et al*, *Phosphorus Sulfur*, 1981, **11**, 157, 349 (*props, use*)
Liorancaite, L. *et al*, *Nucleic Acids Symp. Ser.*, 1981, 215 (*use*)
Harger, M.J.P. *et al*, *Tetrahedron*, 1982, **38**, 1511 (*synth*)

Diphenylphosphinic fluoride, 8CI — D-01042

[1135-98-4]

$Ph_2P(O)F$

$C_{12}H_{10}FOP$ M 220.183
Liq. Bp_3 144°.

Schmutzler, R., *J. Inorg. Nucl. Chem.*, 1963, **25**, 335.
Müller, A. *et al*, *Z. Chem.*, 1970, **10**, 231 (*ir, P nmr*)
Knunyants, I.L. *et al*, *Dokl. Akad. Nauk SSSR, Ser. Sci. Khim.*, 1971, **201**, 862 (*F and P nmr*)
Jones, T.R.B. *et al*, *Org. Mass Spectrom.*, 1977, **12**, 317 (*ms*)
Horner, L. *et al*, *Phosphorus Sulfur*, 1981, **11**, 157 (*props*)
Nikitin, E.V. *et al*, *Zh. Obshch. Khim.*, 1982, **52**, 2721 (*synth, ir, F and P nmr*)

P,P-Diphenylphosphinimidic amide — D-01043

$Ph_2P(NH_2){=}N'H$

$C_{12}H_{13}N_2P$ M 216.222
N,N′-*Di-Ph: see N,P,P*-Triphenylphosphinimidic acid, T-00631
N,N′-Bis(trimethylsilyl): [21955-74-8]. P,P-*Diphenyl*-N,N′-*bis(trimethylsilyl)phosphinimidic amide.*
$C_{18}H_{29}N_2PSi_2$ M 360.585
Liq. $Bp_{0.1}$ 114-116°.
N,N′-*Bis(trimethylsilyl), Na salt:* [61500-29-6]. Dec. at 280°.
N,N′-*Bis(triphenylsilyl):* P,P-*Diphenyl*-N,N′-*bis(triphenylsilyl)phosphinimidic amide.*
$C_{48}H_{41}N_2PSi_2$ M 733.010
Cryst. (heptane). Mp 161-162°.
N-*Triphenylphosphoranylidene: see P,P-Diphenyl-N-(triphenylphosphoranylidene)phosphinimidic amide, D-01164*

Paciorek, K.L. *et al*, *J. Org. Chem.*, 1966, **31**, 2426 (*triphenysilyl, pmr, ir*)
Schmidbaur, H. *et al*, *Chem. Ber.*, 1969, **102**, 564 (*trimethylsilyl, pmr, ir*)

P,P-Diphenylphosphinimidic fluoride — D-01044

P-*Fluoro*-P,P-*diphenylphosphazene.* P-*Fluoro*-P,P-*diphenylphosphine imide*

$Ph_2PF{=}NH$

$C_{12}H_{11}FNP$ M 219.198
N-*Me:* [25959-38-0]. N-*Methyl*-P,P-*diphenylphosphinimidic fluoride.*
$C_{13}H_{13}FNP$ M 233.225
Cryst. (toluene). Poorly sol. org. solvs. Mp 144-146.5°.

Schmutzler, R., *J. Chem. Soc., Dalton Trans.*, 1973, 2687 (*synth, ir, pmr*)

(Diphenylphosphino)acetic acid, 9CI — D-01045

(*Carboxymethyl)diphenylphosphine*
[3064-56-0]

Ph_2PCH_2COOH

$C_{14}H_{13}O_2P$ M 244.229
Ligand for metals of Group VIII. Solid. Mp 120-121°. pK_a 5.49, 6.05 (50% EtOH aq.). pK_{H_2A} 7.6.
Et ester: Ethyl (diphenylphosphino)acetate.
$C_{16}H_{17}O_2P$ M 272.283
Liq. Bp_5 183-185°, $Bp_{0.05}$ 134°.
Amide: 2-(Diphenylphosphino)acetamide.
$C_{14}H_{14}NOP$ M 243.244
Solid. Mp 173-174°.
Oxide: see (Diphenylphosphinyl)acetic acid, D-01093
Sulfide: see (Diphenylphosphinothioyl)acetic acid, D-01086

Issleib, K. *et al*, *Chem. Ber.*, 1960, **93**, 803; 1961, **94**, 2244 (*synth, ester*)
Issleib, K. *et al*, *Z. Anorg. Allg. Chem.*, 1964, **330**, 295 (*pmr, amide, ir*)
Jarolim, T. *et al*, *J. Inorg. Nucl. Chem.*, 1976, **38**, 125 (*synth, ir, uv, complexes*)
Reetz, M.T. *et al*, *Justus Liebigs Ann. Chem.*, 1977, 242 (*ester, props, pmr*)
Stepanov, I.A. *et al*, *Zh. Obshch. Khim.*, 1979, **49**, 2389 (*Engl. transl.* p. 2109) (*ester*)
Braunstein, P. *et al*, *J. Am. Chem. Soc.*, 1981, **103**, 5115 (*ester, pmr, cmr, P nmr, complexes*)

(Diphenylphosphino)acetonitrile, 9CI — D-01046

(*Cyanomethyl)diphenylphosphine*
[31201-88-4]

Ph_2PCH_2CN

$C_{14}H_{12}NP$ M 225.229

Ligand for Ir, Ni, Pd, Pt, Rh, Fe and Au. Liq. $Bp_{0.1}$ 138-140°.

Oxide: [23040-22-4]. *(Diphenylphosphinyl)acetonitrile.*
$C_{14}H_{12}NOP$ M 241.229
Wittig reagent for stereoselective synth. of 2-alkenenitriles. pK_a 17.3 (DMSO).

Sulfide: [69039-10-7]. *(Diphenylphosphinothioyl)-acetonitrile. (Diphenylphosphino)ethanenitrile.*
$C_{14}H_{12}NPS$ M 257.289
Pale-yellow solid. Mp 136-138°.

Raevskii, O.A. *et al*, *Izv. Akad. Nauk. SSSR, Ser. Khim.*, 1973, 1339 (*Engl. transl.* p. 1293) (*oxide, ir*)
Dahl, O., *Acta Chem. Scand., Ser. B*, 1976, **30**, 799 (*synth, nmr*)
Petrov, E.S. *et al*, *Izv. Akad. Nauk. SSSR, Ser. Khim.*, 1976, 782 (*Engl. transl.*, p. 762) (*oxide*)
Loupy, A. *et al*, *Synthesis*, 1977, 126 (*oxide, synth, use*)
Braunstein, P. *et al*, *J. Chem. Res. (S)*, 1978, 232 (*synth, derivs, complexes*)
Braunstein, P. *et al*, *J. Am. Chem. Soc.*, 1981, **103**, 5115 (*pmr, cmr, nmr, complexes*)

2-(Diphenylphosphino)aniline D-01047

2-Diphenylphosphinobenzenamine, 9CI. *(2-Aminophenyl)diphenylphosphine*
[65423-44-1]

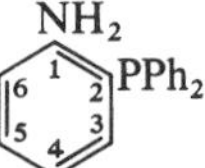

$C_{18}H_{16}NP$ M 277.305
Ligand for Pt. Fluffy needles (EtOH). Mp 82.5-83°.

Oxide: [23081-74-5]. *2-(Diphenylphosphinyl)aniline.*
$C_{18}H_{16}NOP$ M 293.304
Cryst. (EtOH). Mp 166-167°.

Cadogan, J.I.G. *et al*, *J. Chem. Soc. (C)*, 1969, 1314 (*oxide, synth, pmr, ir*)
Cooper, M.K. *et al*, *Inorg. Chem.*, 1978, **17**, 880 (*synth, oxide, complexes*)
Dolzhnikova, E.N. *et al*, *Zh. Obshch. Khim.*, 1978, **48**, 525 (*Engl. transl.* p. 474) (*oxide*)
Poponova, R.V. *et al*, *Zh. Obshch. Khim.*, 1978, **48**, 1956 (*Engl. transl.* p. 1782) (*oxide, ms*)

3-(Diphenylphosphino)aniline D-01048

3-Diphenylphosphinobenzenamine, 9CI. *(3-Aminophenyl)diphenylphosphine*
[36267-33-1]

$C_{18}H_{16}NP$ M 277.305
Cryst. (ligroin). Mp 99-101°.

B,MeI: (3-Aminophenyl)methyldiphenylphosphonium iodide.
$C_{19}H_{19}INP$ M 419.244
Cryst. ($MeOH/Et_2O$). Mp 212-213°.

N-*Ac: 3-(Diphenylphosphino)acetanilide.*
$C_{20}H_{18}NOP$ M 319.342
Cryst. (Me_2CO aq.). Mp 111-112°.

N-*Ac; B,MeI: (3-Acetylaminophenyl)-methyldiphenylphosphonium iodide.*
$C_{21}H_{21}INOP$ M 461.281
Cryst. (CH_2Cl_2/Et_2O). Mp 226-228°.

Oxide: [36357-48-9]. *3-(Diphenylphosphinyl)aniline.*
$C_{18}H_{16}NOP$ M 293.304
Cryst. (ligroin). Mp 160-161°.

Sulfide: [36267-37-5]. *3-(Diphenylphosphinothioyl)-aniline.*
$C_{18}H_{16}NPS$ M 309.365
Cryst. (C_6H_6/pentane). Mp 182-183°.

Sulfide, N-*Ac: 3-(Diphenylphosphinothioyl)acetanilide.*
$C_{20}H_{18}NOPS$ M 351.402
Cryst. (MeOH aq.). Mp 145-146°.

Horner, L. *et al*, *Chem. Ber.*, 1958, **91**, 52 (*oxide*)
Schiemenz, G. *et al*, *Phosphorus*, 1972, **1**, 187 (*synth, pmr, ir, derivs*)

4-(Diphenylphosphino)aniline D-01049

4-Diphenylphosphinobenzenamine, 9CI. *(4-Aminophenyl)diphenylphosphine*
[33834-35-4]

$C_{18}H_{16}NP$ M 277.305
Oil.

B,MeI: (4-Aminophenyl)methyldiphenylphosphonium iodide.
$C_{19}H_{19}INP$ M 419.244
Cryst. ($MeOH/CHCl_3$/pentane). Mp 219-220°.

Oxide: [33834-36-5]. *4-(Diphenylphosphinyl)aniline.*
$C_{18}H_{16}NOP$ M 293.304
Cryst. (EtOAc/MeOH). Mp 238-239° (235°).

Sulfide: [33834-37-6]. *4-(Diphenylphosphinothioyl)-aniline.*
$C_{18}H_{16}NPS$ M 309.365
Cryst. (Me_2CO). Mp 168-173°.

Horner, L. *et al*, *Chem. Ber.*, 1958, **91**, 52 (*oxide*)
Schiemenz, G. *et al*, *Phosphorus*, 1972, **1**, 187 (*synth, ir, pmr, derivs*)

2-(Diphenylphosphino)benzaldehyde, 9CI D-01050

[50777-76-9]

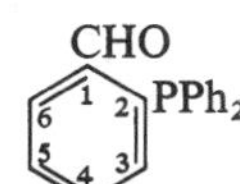

$C_{19}H_{15}OP$ M 290.301
Ligand for metals of Groups VIIB and VIII. Cryst. (MeOH). Mp 116°.

Oxide: [50777-77-0]. *2-(Diphenylphosphinyl)-benzaldehyde.*
$C_{19}H_{15}O_2P$ M 306.300
Cryst. (C_6H_6/pentane). Mp 120-121°.

Sulfide: [50777-78-1]. *2-(Diphenylphosphinothioyl)-benzaldehyde.*
$C_{19}H_{15}OPS$ M 322.361
Cryst. (MeOH). Mp 131-133°.

Schiemenz, G.P. *et al*, *Justus Liebigs Ann. Chem.*, 1973, 1480 (*synth, ir, pmr, derivs*)

3-(Diphenylphosphino)benzaldehyde, 9CI D-01051

[50777-69-0]

$C_{19}H_{15}OP$ M 290.301
Oil.

Oxide: [50777-71-4]. *3-(Diphenylphosphinyl)-benzaldehyde.*
$C_{19}H_{15}O_2P$ M 306.300
Oil.

Sulfide: [50777-70-3]. *3-(Diphenylphosphinothioyl)-benzaldehyde.*
$C_{19}H_{15}OPS$ M 322.361

Oil.

Schiemenz, G.P. *et al*, *Justus Liebigs Ann. Chem.*, 1973, 1480 (*synth, ir, pmr, derivs*)

4-(Diphenylphosphino)benzaldehyde **D-01052**

[5068-18-8]

$C_{19}H_{15}OP$ M 290.301

Solid. Mp 71-72°.

Oxide: [5068-23-5]. *4-(Diphenylphosphinyl)-benzaldehyde.*

$C_{19}H_{15}O_2P$ M 306.300

Solid. Mp 105-106°.

Schiemenz, G.P. *et al*, *Justus Liebigs Ann. Chem.*, 1973, 1480 (*synth, ir, pmr, derivs*)
Zhmurova, I.N. *et al*, *Zh. Obshch. Khim.*, 1977, **47**, 2207 (*Engl. transl.* p. 2015) (*props*)
Gloyna, D. *et al*, *Z. Chem.*, 1980, **20**, 258 (*oxide*)

2-(Diphenylphosphino)benzoic acid, 9CI **D-01053**

(2-Carboxyphenyl)diphenylphosphine

[17261-28-8]

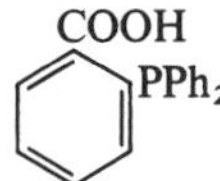

$C_{19}H_{15}O_2P$ M 306.300

Light-yellow needles (EtOH aq.). Mp 187-188° (174-177°).

B,MeI: *(2-Carboxyphenyl)methyldiphenylphosphonium iodide.*

$C_{20}H_{18}IO_2P$ M 448.239

Cryst. (Me_2CO/C_6H_6). Mp 169-173° dec.

Me ester: [79932-99-3].

$C_{20}H_{17}O_2P$ M 320.327

Cryst. Mp 96-97°.

Oxide: [2572-40-9].

$C_{19}H_{15}O_3P$ M 322.299

Cryst. (EtOH, EtOH aq., or AcOH aq.). Mp 252-254°, 262-264°, 274-275°. pK_a 5.37 (50% EtOH aq.).

Oxide, Me ester: [79317-63-8].

$C_{20}H_{17}O_3P$ M 336.326

Cryst. (cyclohexane/butanone). Mp 134-135°.

Issleib, K. *et al*, *Z. Anorg. Allg. Chem.*, 1967, **353**, 197 (*synth, derivs, complexes*)
Luckenbach, R. *et al*, *Z. Naturforsch., B*, 1977, **32**, 1038 (*synth, ir, oxide*)
Petrov, K.A. *et al*, *Zh. Obshch. Khim.*, 1977, **47**, 2516 (*Engl. transl.* p. 2299) (*oxide*)
Segall, Y. *et al*, *J. Chem. Res. (S)*, 1977, 310 (*oxide, synth, ms, ir, uv*)
Rauchfoss, T-B. *et al*, *J. Am. Chem. Soc.*, 1981, **103**, 6769 (*synth, ester*)
Talay, R. *et al*, *Z. Naturforsch., B*, 1981, **36**, 451 (*synth, ir, oxide*)

3-(Diphenylphosphino)benzoic acid, 9CI **D-01054**

(3-Carboxyphenyl)diphenylphosphine

[2129-30-8]

$C_{19}H_{15}O_2P$ M 306.300

Cryst. (EtOH). Mp 157-160°. pK_a 5.51 (50% EtOH aq.).

Me ester: [69209-32-1]. *Methyl 3-(diphenylphosphino)-benzoate.*

$C_{20}H_{17}O_2P$ M 320.327

Cryst. (MeOH). Mp 71.5-72.5°.

Oxide: [2129-29-5]. *3-(Diphenylphosphinyl)benzoic acid.*

$C_{19}H_{15}O_3P$ M 322.299

Cryst. (EtOH). Mp 234-236°. pK_a 5.11 (50% EtOH aq.).

Sulfide: [14378-87-1]. *3-(Diphenylphosphinothioyl)-benzoic acid.*

$C_{19}H_{15}O_2PS$ M 338.360

Solid (EtOAc/hexane). Mp 171-172°. pK_a 5.24 (50% EtOH aq.).

Schindlbauer, H., *Monatsh. Chem.*, 1965, **96**, 1021 (*synth*)
Monagle, J.J. *et al*, *J. Org. Chem.*, 1967, **32**, 2477 (*oxide*)
Tsvetkov, E.N. *et al*, *Tetrahedron*, 1969, **25**, 5623 (*synth, derivs*)
Shvets, A.A. *et al*, *Zh. Obshch. Khim.*, 1978, **48**, 2185 (*Engl. transl.* p. 1986) (*derivs*)
Ratovskii, G.V. *et al*, *Zh. Obshch. Khim.*, 1979, **49**, 548 (*Engl. transl.* p. 479) (*ester, uv*)

4-(Diphenylphosphino)benzoic acid, 9CI **D-01055**

(4-Carboxyphenyl)diphenylphosphine

[2129-31-9]

$C_{19}H_{15}O_2P$ M 306.300

Cryst. (AcOH). Mp 156-158°. pK_a 5.39 (50% EtOH aq.).

Me ester: [5032-51-9]. *Methyl 4-(diphenylphosphino)-benzoate.*

$C_{20}H_{17}O_2P$ M 320.327

Cryst. (MeOH). Mp 98-99°.

Oxide: [2272-04-0]. *4-(Diphenylphosphinyl)benzoic acid.*

$C_{19}H_{15}O_3P$ M 322.299

Cryst. (MeOH aq. or dioxan aq.). Mp 273-274°. pK_a 4.88 (50% EtOH aq.).

Oxide, Me ester: Methyl 4-(diphenylphosphinyl)-benzoate.

$C_{20}H_{17}O_3P$ M 336.326

Cryst. (MeOH aq. or Et_2O/hexane). Mp 113-116°.

Sulfide: [5068-24-6]. p-*(Diphenylphosphinothioyl)-benzoic acid.*

$C_{19}H_{15}O_2PS$ M 338.360

Cryst. (EtOAc/hexane). Mp 177-178.5°. pK_a 4.97 (50% EtOH aq.).

Sulfide, Me ester: Methyl (4-diphenylphosphinothioyl)-benzoate.

$C_{20}H_{17}O_2PS$ M 352.387

Cryst. (MeOH). Mp 84-85°.

Schiemenz, G.P. *et al*, *Chem. Ber.*, 1966, **99**, 504 (*synth, derivs*)
Monagle, J.J. *et al*, *J. Org. Chem.*, 1967, **32**, 2477 (*oxide*)
Tsvetkov, E.N. *et al*, *Tetrahedron*, 1969, **25**, 5623 (*synth, derivs*)
Shvets, A.A. *et al*, *Zh. Obshch. Khim.*, 1978, **48**, 2185 (*Engl. transl.* p. 1986) (*derivs*)
Ratovskii, G.V. *et al*, *Zh. Obshch. Khim.*, 1979, **49**, 548 (*Engl. transl.* p. 479) (*uv*)
Bondarenko, N.A. *et al*, *Izv. Akad. Nauk SSSR, Ser. Khim.*, 1981, **30**, 1596 (*Engl. transl.* p. 1289) (*oxide, props*)

2-(Diphenylphosphino)benzonitrile, 9CI **D-01056**

(2-Cyanophenyl)diphenylphosphine

[34825-99-5]

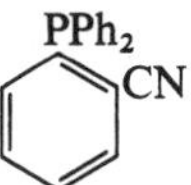

$C_{19}H_{14}NP$ M 287.300

Ligand for Mn, Pt, Pd and Re. Cryst. (2-propanol). Mp 148-149°.

Payne, D.H. *et al*, *Inorg. Nucl. Chem. Lett.*, 1972, **8**, 73 (*complexes*)
Storhoff, B.N. *et al*, *J. Organomet. Chem.*, 1981, **205**, 161 (*synth*)

3-(Diphenylphosphino)benzonitrile, 9CI — D-01057

(3-Cyanophenyl)diphenylphosphine

[14378-89-3]

$C_{19}H_{14}NP$ M 287.300

Cryst. (MeOH). Mp 98-99°.

Oxide: [22836-20-0]. *3-(Diphenylphosphinyl)-benzonitrile.*
$C_{19}H_{14}NOP$ M 303.299
Cryst. (C_6H_6/ligroin). Mp 110-111°.

Sulfide: 3-(Diphenylphosphinothioyl)benzonitrile.
$C_{19}H_{14}NPS$ M 319.360
Cryst. (MeOH). Mp 130-132°.

Schiemenz, G.P. *et al*, *Chem. Ber.*, 1969, **102**, 1883 (*synth, derivs*)
Schiemenz, G.P. *et al*, *Justus Liebigs Ann. Chem.*, 1976, 2126 (*synth*)
Shvets, A.A. *et al*, *Zh. Obshch. Khim.*, 1978, **48**, 2185 (*Engl. transl.* p. 1986) (*synth*)
Ratovskii, G.V. *et al*, *Zh. Obshch. Khim.*, 1979, **49**, 548 (*Engl. transl.* p. 479) (*synth, uv, raman*)

4-(Diphenylphosphino)benzonitrile, 9CI — D-01058

(4-Cyanophenyl)diphenylphosphine

[5068-16-6]

$C_{19}H_{14}NP$ M 287.300

Rods (MeOH). Mp 86-87°.

Oxide: [5032-54-2]. *4-(Diphenylphosphinyl)benzonitrile.*
$C_{19}H_{14}NOP$ M 303.299
Cryst. (MeOH aq. or C_6H_6/pet. ether). Mp 141-143° (132-134°).

Sulfide: [5032-60-0]. *4-(Diphenylphosphinothioyl)-benzonitrile.*
$C_{19}H_{14}NPS$ M 319.360
Cryst. (MeOH). Mp 105-108°.

Schiemenz, G.P., *Chem. Ber.*, 1966, **99**, 504 (*synth, derivs*)
Schiemenz, G.P. *et al*, *Justus Liebigs Ann. Chem.*, 1976, 2126 (*synth*)
Shvets, A.A. *et al*, *Zh. Obshch. Khim.*, 1978, **48**, 2185 (*Engl. transl.* 1986) (*synth*)
Ratovskii, G.V. *et al*, *Zh. Obshch. Khim.*, 1979, **49**, 548 (*Engl. transl.* p. 479) (*uv*)

2-(Diphenylphosphino)-*N*,*N*-dimethylaniline — D-01059

*2-Diphenylphosphino-*N,N*-dimethylbenzenamine*, 9CI. *(2-Dimethylaminophenyl)diphenylphosphine*

[4358-50-3]

NMe$_2$, PPh$_2$

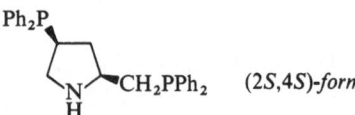

$C_{20}H_{20}NP$ M 305.358

Ligand for Ir, Pd, Pt, Mn, Rh, and Sn. Cryst. (EtOH). Mp 113-114°, 121°.

Oxide: [61102-69-0]. *2-Diphenylphosphinyl-*N,N*-dimethylaniline.*
$C_{20}H_{20}NOP$ M 321.358

Forms Ni, Rh, and Sn complexes. Needles. Mp 132°.

Fritz, H.P. *et al*, *J. Chem. Soc.*, 1965, 5210 (*synth, complexes*)
Rauchfuss, J.B. *et al*, *J. Am. Chem. Soc.*, 1974, **96**, 3098 (*synth, complexes*)
Shvets, A.A. *et al*, *Zh. Obshch. Khim.*, 1976, **46**, 1701 (*Engl. transl.* p. 1654) (*ir*)
Mercer, G.D. *et al*, *J. Am. Chem. Soc.*, 1977, **99**, 6551 (*oxide, ir*)
Roundhill, D.M. *et al*, *Inorg. Chem.*, 1980, **19**, 3365 (*pmr, cmr*)
Horner, L. *et al*, *Phosphorus Sulfur*, 1983, **15**, 165 (*synth*)

3-(Diphenylphosphino)-*N*,*N*-dimethylaniline, 8CI — D-01060

*3-Diphenylphosphino-*N,N*-dimethylbenzenamine*, 9CI. *(3-Dimethylaminophenyl)diphenylphosphine*

[5931-54-4]

$C_{20}H_{20}NP$ M 305.358

Cryst. (MeOH). Mp 78-80°.

Oxide: [61564-31-6]. *3-Diphenylphosphinyl-*N,N*-dimethylaniline.*
$C_{20}H_{20}NOP$ M 321.358
Cryst. (MeOH aq. or C_6H_6/pet. ether). Mp 145-148°.

Sulfide: [61564-32-7]. *3-Diphenylphosphinothioyl-*N,N*-dimethylaniline.*
$C_{20}H_{20}NPS$ M 337.418
Cryst. (MeOH aq.). Mp 125-127°.

Schiemenz, G.P. *et al*, *Justus Liebigs Ann. Chem.*, 1976, 2126 (*synth, derivs*)

4-Diphenylphosphino-*N*,*N*-dimethylaniline — D-01061

*4-Diphenylphosphino-*N,N*-dimethylbenzenamine*, 9CI. *(4-Dimethylaminophenyl)diphenylphosphine*

[739-58-2]

$C_{20}H_{20}NP$ M 305.358

Cryst. (EtOH). Mp 153°.

Oxide: [797-72-8]. *4-Diphenylphosphinyl-*N,N*-dimethylaniline.*
$C_{20}H_{20}NOP$ M 321.358
Cryst. (MeOH aq. or toluene). Mp 182-183° (174-176°).

Sulfide: [1041-99-2]. *4-Diphenylphosphinothioyl-*N,N*-dimethylaniline.*
$C_{20}H_{20}NPS$ M 337.418
Cryst. (MeOH or EtOH). Mp 183°.

Goetz, H. *et al*, *Justus Liebigs Ann. Chem.*, 1967, **665**, 1 (*synth, derivs, ir, uv*)
Dreissig, W. *et al*, *Z. Kristallogr., Kristallgeom., Kristallphys., Kristallchem.*, 1972, **135**, 294 (*cryst struct*)
Schiemenz, G.P. *et al*, *Justus Liebigs Ann. Chem.*, 1976, 2126 (*derivs*)

4-(Diphenylphosphino)-2-(diphenylphosphinomethyl)pyrrolidine, 9CI — D-01062

Ph$_2$P, N, H, CH$_2$PPh$_2$ *(2S,4S)-form*

$C_{29}H_{29}NP_2$ M 453.502

(2S,4S)-form [61478-29-3]

Ligand for Rh in asymmetric hydrogenations and synth. of chiral dipeptides. Solid. Mp 103-104°. $[\alpha]_D^{20}$ −15.75° (c, 1.00 in C_6H_6).

N-*Ac:*
$C_{31}H_{31}NOP_2$ M 495.540
Solid. Mp 118-119°. $[\alpha]^{20}_{D}$ −12° (c, 1.08 in C_6H_6).
N-tert-*Butyloxycarbonyl: see N-tert-*Butoxycarbonyl-4-diphenylphosphino-2-diphenylphosphinomethylpyrrolidine, B-00532

Achiwa, K. *et al*, *Tetrahedron Lett.*, 1978, 1119, 4683 (*derivs, use*)
Ojima, I. *et al*, *J. Org. Chem.*, 1980, **45**, 4728 (*synth, ir, pmr, complex*)
Achiwa, K. *et al*, *J. Organomet. Chem.*, 1981, **218**, 249 (*derivs, complexes*)
Ojima, I. *et al*, *J. Org. Chem.*, 1982, **47**, 1329 (*use*)

N-Diphenylphosphino-*P*,*P*-diphenylphosphinous amide, 9CI D-01063

Bis(diphenylphosphino)amine. Tetraphenyldiphosphazane
[2960-37-4]

$Ph_2P—NH—PPh_2$

$C_{24}H_{21}NP_2$ M 385.384
Cryst. (C_6H_6 or toluene). Mp 144.5-146.5°.

Nöth, H. *et al*, *Z. Anorg. Allg. Chem.*, 1967, **349**, 225 (*synth, P nmr*)
Wang, F.T. *et al*, *Synth. React. Inorg. Met.-Org. Chem.*, 1978, **8**, 119 (*synth, derivs*)
Keat, R. *et al*, *J. Chem. Soc., Dalton Trans.*, 1981, 2192 (*P nmr*)
Nöth, H. *et al*, *Z. Naturforsch., B*, 1984, **39**, 744 (*cryst struct*)

Diphenylphosphinodiselenoic acid, 9CI, 8CI D-01064

Diphenyldiselenophosphinic acid
[25756-78-9]

$Ph_2P(Se)SeH$

$C_{12}H_{11}PSe_2$ M 344.113
Anhydroselenide: [98188-41-1].
$C_{24}H_{20}P_2Se_3$ M 607.250
Solid. Mp 126°.

Colquhoun, I.J. *et al*, *Org. Magn. Reson.*, 1977, **12**, 473 (*P nmr*)
Blake, A.J. *et al*, *J. Mol. Struct.*, 1982, **78**, 265 (*synth, ms*)
Horn, H.-G. *et al*, *Chem. Ztg.*, 1985, **109**, 77 (*synth, pmr, P nmr, ms*)

Diphenylphosphinodithioic acid, 9CI D-01065

[1015-38-9]

$Ph_2P(S)SH$

$C_{12}H_{11}PS_2$ M 250.313
Reacts with nitriles with the formation of thioamides. Forms many metal- and organometal-complexes, and is hence used in flotation technology, and for extraction of the Pt group metals from aq. acidic solns. Pale-yellow cryst. (Et_2O). Mp 55-57°. pK_{a1} 1.77 (7% EtOH aq.), pK_{a2} 2.69 (80% EtOH aq.), pK_{a3} 3.48 (EtOH).
Na salt: [5827-17-8]. May be used for the determination of Ru, Rh, Pt, Pd, Au, Ag, Ir and Os. Cryst. (Me_2CO-/pet. ether). Mp 254-255°.
Me ester: [15288-70-7]. *Methyl diphenylphosphinodithioate.*
$C_{13}H_{13}PS_2$ M 264.339
Cryst. (EtOH or naphtha). Mp 82-83°.
Et ester: [33329-02-1]. *Ethyl diphenylphosphinodithioate.*
$C_{14}H_{15}PS_2$ M 278.366
Liq. d^{20}_{4} 1.19. $Bp_{0.3}$ 166-167°. n_D^{20} 1.6611.
Isopropyl ester: [59568-76-2]. *Isopropyl diphenylphosphinodithioate.*
$C_{15}H_{17}PS_2$ M 292.393
Mp 78-87.5°.
tert-*Butyl ester:* tert-*Butyl diphenylphosphinodithioate.*
$C_{16}H_{19}PS_2$ M 306.420
Needles (pet. ether). Mp 91-92°.
2-Propenyl ester: 2-Propenyl diphenylphosphinodithioate. Allyl diphenylphosphinodithioate.
$C_{15}H_{15}PS_2$ M 290.377
Liq. $Bp_{0.04}$ 160°.
Ph ester: [57644-86-7]. *Phenyl diphenylphosphinodithioate.*
$C_{18}H_{15}PS_2$ M 326.410
Cryst. (EtOH or pet. ether). Mp 57-59°, 123-125°.
Benzyl ester: [57644-87-8]. *Benzyl diphenylphosphinodithioate.*
$C_{19}H_{17}PS_2$ M 340.437
Undist. oil.
Trimethylsilyl ester: Trimethylsilyl diphenylphosphinodithioate.
$C_{15}H_{19}PS_2Si$ M 322.494
Liq. $Bp_{0.02}$ 151-153°.
Anhydrosulfide: [6079-78-3].
$C_{24}H_{20}P_2S_3$ M 466.550
Cryst. (2-propanol). Mp 118-121°.
Anhydride with acetic acid: S-*Acetyl diphenylphosphinodithioate.*
$C_{14}H_{13}OPS_2$ M 292.350
Solid. Mp 86-90°.
Anhydride with benzoic acid: S-*Benzoyl diphenylphosphinodithioate.*
$C_{19}H_{15}OPS_2$ M 354.421
Solid. Mp 120-121°.

Hopkins, T.R. *et al*, *J. Am. Chem. Soc.*, 1956, **78**, 4447 (*anhydrosulfide, esters*)
Küchen, W. *et al*, *Chem. Ber.*, 1968, **101**, 3454 (*synth, P nmr*)
Almasi, L. *et al*, *Chem. Ber.*, 1969, **102**, 1489 (*anhydrosulfide*)
Spence, R.A. *et al*, *Aust. J. Chem.*, 1969, **22**, 2359 (*synth, ir, ms, props*)
Küchen, W. *et al*, *Z. Anorg. Allg. Chem.*, 1975, **413**, 266 (*silyl ester, ir, ms, pmr, P nmr*)
Murav'ev, I.V. *et al*, *Zh. Obshch. Khim.*, 1976, **46**, 789 (*Engl. transl.* p. 787) (*esters*)
Goda, K. *et al*, *Bull. Chem. Soc. Jpn.*, 1978, **51**, 260, 818 (*esters, use*)
Harris, R.K. *et al*, *J. Chem. Soc., Dalton Trans.*, 1978, 9 (*anhydrosulfide, pmr, P nmr, cmr*)
Küchen, W. *et al*, *Chem. Ber.*, 1981, **114**, 3485 (*props, use*)
Keek, H. *et al*, *Phosphorus Sulfur*, 1983, **14**, 225 (*ms*)

2-(Diphenylphosphino)ethylamine D-01066

2-(Diphenylphosphino)ethanamine, 9CI. (2-Aminoethyl)diphenylphosphine
[4848-43-5]

$Ph_2PCH_2CH_2NH_2$

$C_{14}H_{16}NP$ M 229.261
Ligand for Co and Pd. Liq. $Bp_{ca. 60}$ ca. 150°.

Kinoshita, I. *et al*, *Bull. Chem. Soc. Jpn.*, 1980, **53**, 3715; 1981, **54**, 725 (*synth, complexes*)

N-(Diphenylphosphino)-N-ethyl-P,P-diphenylphinous amide, 9CI D-01067

Ethylimidobisdiphenylphosphine. N,N-Bis(diphenylphosphino)ethylamine

[2960-41-0]

$EtN(PPh_2)_2$

$C_{26}H_{25}NP_2$ M 413.438

Cryst. (EtOH aq.). Mp 99°.

Ewart, G. *et al, J. Chem. Soc.*, 1964, 1543 (*synth, derivs*)
Cross, R.J. *et al, J. Chem. Soc., Dalton Trans.*, 1976, 1424 (*pmr, P nmr*)
Keat, R. *et al, J. Chem. Soc., Dalton Trans.*, 1981, 2192 (*P nmr*)

(Diphenylphosphino)ferrocene, 10CI, 9CI D-01068

Ferrocenyldiphenylphosphine

[12098-17-8]

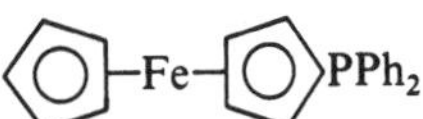

$C_{22}H_{19}FeP$ M 370.213

Orange needles (EtOH). Mp 122-124°.

B,MeI: Ferrocenylmethyldiphenylphosphonium iodide.
$C_{23}H_{22}FeIP$ M 512.152
Dark orange cryst. Sol. hot H_2O. Mp 187-188° dec.

B,PhCH$_2$Cl: Ferrocenyldiphenyl(phenylmethyl)-phosphonium chloride. Benzyl(ferrocenyl)-diphenylphosphonium chloride.
$C_{29}H_{26}ClFeP$ M 496.798
Cryst. ($EtOH/Et_2O$). Mp 178-179°.

Oxide: [54060-24-1]. *Diphenylphosphinylferrocene.*
$C_{22}H_{19}FeOP$ M 386.212
Orange needles (heptane). Mp 163-165°.

Sollott, G.P. *et al, J. Org. Chem.*, 1963, **28**, 1090 (*synth, ir, derivs*)
Kotz, J.C. *et al, J. Organomet. Chem.*, 1973, **52**, 387 (*pmr, ir*)
Eberhard, L. *et al, J. Organomet. Chem.*, 1974, **80**, 109 (*deriv, pmr*)
McEwen, W.E. *et al, J. Org. Chem.*, 1976, **41**, 1684 (*synth, derivs*)
Smalley, A.W., *Org. Prep. Proced. Int.*, 1978, **10**, 195 (*synth*)
Seyferth, D. *et al, Organometallics*, 1982, **1**, 1275 (*cmr*)

(Diphenylphosphino)methanol, 9CI D-01069

Hydroxymethyldiphenylphosphine. Methyloldiphenylphosphine

[5958-44-1]

Ph_2PCH_2OH

$C_{13}H_{13}OP$ M 216.219

Viscous oil. d_4^{20} 0.93. Bp_1 113°. n_D^{20} 1.6137. In the presence of mineral acid, or when heated, isom. to methyldiphenylphosphine oxide.

Oxide: [884-74-2].
$C_{13}H_{13}O_2P$ M 232.218
Cryst. (C_6H_6 or $C_6H_6/EtOAc$). Mp 138-139°, 192-193°.

Sulfide:
$C_{13}H_{13}OPS$ M 248.279
Solid. Mp 71-72°.

Me ether: Methoxymethyldiphenylphosphine.
$C_{14}H_{15}OP$ M 230.246
Liq. $Bp_{0.1}$ 138-139°.

Me ether, oxide: (Diphenylphosphinyl)dimethyl ether.
$C_{14}H_{15}O_2P$ M 246.245
Solid (C_6H_6/pet. ether). Mp 116-117°.

Petrov, K.A. *et al, Zh. Obshch. Khim.*, 1961, **31**, 3417 (*Engl. transl.* p. 3186) (*synth, oxide*)
Trippett, S., *J. Chem. Soc.*, 1961, 2813 (*synth, oxide, uv*)
Hellmann, H. *et al, Justus Liebigs Ann. Chem.*, 1962, **659**, 49 (*oxide, sulfide*)
Matrosov, E.I. *et al, Spectrochim. Acta, Part A*, 1972, **28**, 313 (*ir*)
Tebby, J.C., *Phosphorus*, 1976, **6**, 253 (*oxide, pmr*)
Fieser, M. *et al, Reagents for Organic Synthesis*, Wiley, 1967-84, **7**, 229; **9**, 301 (*deriv, use*)

2-(Diphenylphosphino)phenol, 9CI D-01070

(2-Hydroxyphenyl)diphenylphosphine

[60254-10-6]

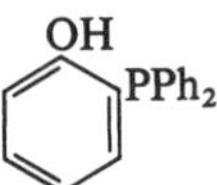

$C_{18}H_{15}OP$ M 278.290

Ligand for Ni, Pd, Pt and Tl. Cryst. (DMF). Mp 164-165°. $Bp_{0.5}$ 135-140° subl.

B,MeI: (2-Hydroxyphenyl)methyldiphenylphosphonium iodide.
$C_{19}H_{18}IOP$ M 420.229
Cryst. ($EtOH/Et_2O$). Mp 226°.

Oxide: [16522-52-4]. *2-(Diphenylphosphinyl)phenol.*
$C_{18}H_{15}O_2P$ M 294.289
Cryst. (EtOH or AcOH). Mp 248.5-249.5° (229-230°).

Sulfide: [16522-54-6]. *2-(Diphenylphosphinothioyl)-phenol.*
$C_{18}H_{15}OPS$ M 310.350
Cryst. (dibutyl ether). Mp 141-142°.

Me ether: see (2-Methoxyphenyl)diphenylphosphine, M-00056

Amarskii, E.G. *et al, Zh. Obshch. Khim.*, 1974, **44**, 461 (*Engl. transl.* p. 443) (*oxide*)
Henning, H.-G. *et al, J. Prakt. Chem.*, 1976, **318**, 69 (*synth, uv, pmr, derivs*)
Rauchfuss, T.B., *Inorg. Chem.*, 1977, **16**, 2966 (*complexes*)
Bondarenko, N.A. *et al, Izv. Akad. Nauk SSSR, Ser. Khim.*, 1979, 432 (*Engl. transl.* p. 399) (*oxide*)
Landvatter, E.F. *et al, Organometallics*, 1982, **1**, 506 (*complexes*)

3-(Diphenylphosphino)phenol, 9CI D-01071

(3-Hydroxyphenyl)diphenylphosphine

[32341-34-7]

$C_{18}H_{15}OP$ M 278.290

Cryst. (MeOH). Mp 55-56°, Mp 74-77°.

Ac:
$C_{20}H_{17}O_2P$ M 320.327
Cryst. (EtOH). Mp 55-56°.

Oxide:
$C_{18}H_{15}O_2P$ M 294.289
Cryst. (C_6H_6 or EtOH aq.). Mp 158-160°, Mp 186-187°°. pK_a 10.2 (50% EtOH aq., 25°).

Oxide, Ac:
$C_{20}H_{17}O_3P$ M 336.326
Needles (EtOH). Mp 158-159°.

Me ether: see (3-Methoxyphenyl)diphenylphosphine, M-00057

Horner, L. *et al*, *Chem. Ber.*, 1958, **91**, 52 (*oxide, synth*)
Lanza, L., *J. Prakt. Chem.*, 1964, **25**, 294 (*synth, oxide*)
Monagle, J.J. *et al*, *J. Org. Chem.*, 1967, **32**, 2477 (*oxide, synth*)
Schiemenz, G.P. *et al*, *Justus Liebigs Ann. Chem.*, 1976, 2126 (*synth, pmr*)
Tkachev, V.V. *et al*, *Izv. Akad. Nauk SSSR, Ser. Khim.*, 1979, **28**, 1159 (*Engl. transl.* p. 1084) (*oxide, cryst struct*)
Bondarenko, N.A. *et al*, *Izv. Akad. Nauk SSSR, Ser. Khim.*, 1980, 106 (*Engl. transl.* p. 92) (*synth, ir*)

4-(Diphenylphosphino)phenol, 9CI **D-01072**

(4-Hydroxyphenyl)diphenylphosphine
[5068-21-3]

$C_{18}H_{15}OP$ M 278.290
Needles (C_6H_6/pet. ether). Mp 94-97°, 105-106°, 114-115°. pK_a 10.46 (50% EtOH aq.).
B,HBr: Solid (MeOH aq.). Mp 202-203°.
Oxide: [793-43-1]. *4-(Diphenylphosphinyl)phenol.*
$C_{18}H_{15}O_2P$ M 294.289
Cryst. (EtOH or MeOH). Mp 250-251°. pK_a 9.38 (50% EtOH aq.).
Sulfide: [20650-56-0]. *4-(Diphenylphosphinothioyl)-phenol.*
$C_{18}H_{15}OPS$ M 310.350
Cryst. (C_6H_6). Mp 176.5-177.5°. pK_a 9.49 (50% EtOH aq.).
Me ether: see (4-Methoxyphenyl)diphenylphosphine, M-00058

Neunhoeffer, O. *et al*, *Chem. Ber.*, 1961, **94**, 2519 (*synth*)
Monagle, J.J. *et al*, *J. Org. Chem.*, 1967, **32**, 2477 (*oxide*)
Tsvetkov, E.N. *et al*, *Tetrahedron*, 1969, **25**, 5623 (*synth, derivs*)
Schiemenz, G.P. *et al*, *Justus Liebigs Ann. Chem.*, 1976, 2126 (*synth, pmr*)
Pudovik, A.N. *et al*, *Zh. Obshch. Khim.*, 1978, **48**, 696, 1455 (*Engl. transl.* p. 638, 1337) (*oxide, ir, nmr*)

3-(Diphenylphosphino)propanenitrile, 9CI **D-01073**

(2-Cyanoethyl)diphenylphosphine

$Ph_2PCH_2CH_2CN$

$C_{15}H_{14}NP$ M 239.256
Cryst. (MeOH). Mp 64-65.5°. $Bp_{0.06}$ 175-178°.
Oxide: [5032-67-7]. *3-(Diphenylphosphinyl)-propanenitrile. 2-Diphenylphosphinylpropionitrile.*
$C_{15}H_{14}NOP$ M 255.255
No phys. props. reported.
Sulfide: 3-(Diphenylphosphinothioyl)propanenitrile. 2-Diphenylphosphinothioylpropionitrile.
$C_{15}H_{14}NPS$ M 271.316
Cryst. (pet. ether). Mp 125-127°.

Mann, F.G. *et al*, *J. Chem. Soc.*, 1952, 4453 (*synth*)
Schindlbauer, H., *Monatsh. Chem.*, 1963, **94**, 99 (*uv*)
Maier, L., *Helv. Chim. Acta*, 1966, **49**, 1249 (*sulfide, ir, nmr*)
Schiemenz, G., *Chem. Ber.*, 1966, **99**, 514 (*synth*)
Pudovik, A.N. *et al*, *Zh. Obshch. Khim.*, 1969, **39**, 334 (*Engl. transl.* p. 314) (*sulfide*)

3-(Diphenylphosphino)-1-propanol **D-01074**

(3-Hydroxypropyl)diphenylphosphine
[2360-09-0]

$Ph_2PCH_2CH_2CH_2OH$

$C_{15}H_{17}OP$ M 244.272
Cryst. (pet. ether). Mp 60-61°. Bp_1 172-173°.
Oxide: [889-57-6].
$C_{15}H_{17}O_2P$ M 260.272
Cryst. (EtOH/Et_2O). Mp 103°.

Aksnes, G. *et al*, *Acta Chem. Scand.*, 1964, **18**, 1586 (*oxide, ir*)

1-(Diphenylphosphino)-2-propanone, 10CI **D-01075**

(Diphenylphosphino)acetone. Acetonyldiphenylphosphine
[17729-74-7]

$Ph_2PCH_2COCH_3$

$C_{15}H_{15}OP$ M 242.257
Oxide: [1733-52-4]. *1-(Diphenylphosphinyl)-2-propanone. Diphenylphosphinylacetone.*
$C_{15}H_{15}O_2P$ M 258.256
Cryst. (C_6H_6/pet. ether or dimethoxyethane/Et_2O). Mp 127-128°.
Oxide, 2,4-dinitrophenylhydrazone: [79543-93-4]. Cryst. Mp 197.5-198°.

Saunders, M. *et al*, *Tetrahedron Lett.*, 1959, 8 (*synth, ir, uv*)
Trippett, S. *et al*, *J. Chem. Soc.*, 1961, 1266.
Reetz, M.T. *et al*, *Justus Liebigs Ann. Chem.*, 1977, 242 (*pmr*)
Mathey, F. *et al*, *Tetrahedron*, 1978, **34**, 649 (*synth*)

2-(Diphenylphosphino)pyridine **D-01076**

Diphenyl-2-pyridylphosphine
[37943-90-1]

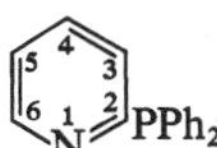

$C_{17}H_{14}NP$ M 263.278
Ligand for Co, Cu, Au, Ni, Pd and Ru. Cryst. (MeOH aq. or pet. ether). Mp 84-85°. $Bp_{0.05}$ 163°.
B,MeI:
$C_{18}H_{17}INP$ M 405.217
Cryst. (MeOH/Et_2O). Mp 141-142°, Mp 157.5-158.5° dec.
P-*Oxide:* [64741-30-6]. *2-(Diphenylphosphinyl)-pyridine.*
$C_{17}H_{14}NOP$ M 279.277
Cryst. (cyclohexane). Mp 109-110°.
Sulfide: 2-(Diphenylphosphinothioyl)pyridine.
$C_{17}H_{14}NPS$ M 295.338
Cryst. (EtOH). Mp 119°.
Sulfide; B,MeI: Yellow cryst. (EtOH). Mp 167.5-168.5°.

Mann, F.G. *et al*, *J. Org. Chem.*, 1948, **13**, 502 (*synth, derivs*)
Schmidbaur, H. *et al*, *Z. Naturforsch., B*, 1980, **35**, 1329 (*synth, deriv, pmr, cmr, nmr*)
Horner, L. *et al*, *Phosphorus Sulfur*, 1983, **15**, 165 (*synth*)

4-(Diphenylphosphino)pyridine, 10CI **D-01077**

Diphenyl-4-pyridylphosphine
[54750-98-0]

$C_{17}H_{14}NP$ M 263.278
Ligand for Co and Ni. Cryst. (EtOH). Mp 66-66.5°. $Bp_{0.02}$ 210° (oven).
B,MeBr: Methyldiphenyl-4-pyridinylphosphonium bromide.
$C_{18}H_{17}BrNP$ M 358.217
Cryst. (Me_2CO). Mp 269-270°.
P-*Oxide:* [54750-99-1]. *4-(Diphenylphosphinyl)-pyridine.*
$C_{17}H_{14}NOP$ M 279.277

Cryst. (EtOAc). Mp 153-155°.
N,P-*Dioxide:*
$C_{17}H_{14}NO_2P$ M 295.277
Solid. Mp 205-207.5°.

Weiner, M.A. *et al*, *Inorg. Chem.*, 1975, **14**, 1714 (*synth, complexes*)
Hassner, A. *et al*, *Tetrahedron*, 1978, **34**, 2069 (*synth, pmr, ms*)
Newkome, G.R. *et al*, *J. Org. Chem.*, 1978, **43**, 947 (*synth*)

Diphenylphosphinoselenoic acid, 9CI, 8CI D-01078

[67998-60-1]

$Ph_2P(Se)OH \rightleftharpoons Ph_2P(O)SeH$

$C_{12}H_{11}OPSe$ M 281.152
Solid. Mp 120-124° dec.
O-*Me ester:* [20180-11-4]. O-*Methyl diphenylphosphinoselenoate.*
$C_{13}H_{13}OPSe$ M 295.179
Solid. Mp 88-89°.
O-*Et ester:* [39181-19-6]. O-*Ethyl diphenylphosphinoselenoate.*
$C_{14}H_{15}OPSe$ M 309.206
Solid. Mp 53-54°, 61-62°.
O-*Ph ester:* O-*Phenyl diphenylphosphinoselenoate.*
$C_{18}H_{15}OPSe$ M 357.250
Solid. Mp 114-115°.
Se-*Ph ester:* [2049-62-9]. Se-*Phenyl diphenylphosphinoselenoate.*
$C_{18}H_{15}OPSe$ M 357.250
Prisms (pet. ether). Mp 78-80°.
Chloride: [55249-23-5].
$C_{12}H_{10}ClPSe$ M 299.598
d_4^{20} 1.49. $Bp_{0.015}$ 140-142°. n_D^{20} 1.6869.
Azide: [70629-46-8].
$C_{12}H_{10}N_3PSe$ M 306.165
Mp 25°. Dec. on heating.
Amide: see P,P-Diphenylphosphinoselenoic amide, D-01079

Michaelis, A. *et al*, *Ber.*, 1885, **18**, 2109 (*ester*)
Petragnani, N. *et al*, *Chem. Ber.*, 1968, **101**, 3070 (*ester, ir*)
Stec, W.J. *et al*, *Phosphorus*, 1972, 97 (*P nmr*)
McFarlane, W. *et al*, *J. Chem. Soc., Dalton Trans.*, 1973, 2162 (*ester, pmr, P and Se nmr*)
Strangeland, L.J. *et al*, *Acta Chem. Scand.*, 1973, **27**, 3919 (*ester*)
Vandyukava, I.I. *et al*, *Izv. Akad. Nauk SSSR, Ser. Khim.*, 1976, 1390 (*Engl. transl.* p. 1334) (*ester, ir, raman*)
Nuretdinov, I.A. *et al*, *Zh. Obshch. Khim.*, 1978, **48**, 1071 (*Engl. transl.* p. 975) (*chloride, ir, P nmr*)
Mastryukova, T.A. *et al*, *Zh. Obshch. Khim.*, 1978, **48**, 1447 (*Engl. transl.* p. 1329) (*synth*)
Bayandina, E.V. *et al*, *Zh. Obshch. Khim.*, 1978, **48**, 2673 (*Engl. transl.* p. 2424) (*ester, chloride, P nmr*)
Schröder, H.F. *et al*, *Z. Anorg. Allg. Chem.*, 1979, **451**, 158 (*chloride, azide, pmr, P nmr, cmr, ms, ir, raman*)

P,P-Diphenylphosphinoselenoic amide, 9CI, 8CI D-01079

$Ph_2P(Se)NH_2$

$C_{12}H_{12}NPSe$ M 280.167
N,N-*Di-Me:* [23486-86-4]. N,N-*Dimethyl*-P,P-*diphenylphosphinoselenoic amide.*
$C_{14}H_{16}NPSe$ M 308.221
Cryst. (Et_2O/hexane). Mp 87-88°.
N-*Ph:* [65438-85-9]. N,P,P-*Triphenylphosphinoselenoic amide.*
$C_{18}H_{16}NPSe$ M 356.265
Solid. Mp 165-167°.

Carlson, R.R., *Inorg. Chem.*, 1974, **13**, 1741 (*deriv*)
McFarlane, W. *et al*, *J. Chem. Soc., Dalton Trans.*, 1973, 2162 (*pmr, nmr*)
El-Borgi, A. *et al*, *C.R. Hebd. Seances Acad. Sci., Ser. C*, 1977, **284**, 983 (*ir*)
Bayandina, E.V. *et al*, *Zh. Obshch. Khim.*, 1978, **48**, 2673 (*Engl. transl.* p. 2424) (*synth*)

Diphenylphosphinoselenothioic acid, 9CI, 8CI D-01080

[66499-12-5]

$Ph_2P(Se)SH \rightleftharpoons Ph_2P(S)SeH$

$C_{12}H_{11}PSSe$ M 297.213
NEt_3 salt: [67998-65-6]. Cryst. (C_6H_6/pet. ether). Mp 114-115°.
S-*Et ester:* [69741-76-0]. S-*Ethyl diphenylphosphinoselenothioate.*
$C_{14}H_{15}PSSe$ M 325.266
Solid. Mp 28-29°.

Bayandina, E.Y. *et al*, *Zh. Obshch. Khim.*, 1978, **48**, 2673 (*Engl. transl.* p. 2424) (*ester*)
Mastryukova, T.A. *et al*, *Zh. Obshch. Khim.*, 1978, **48**, 1447 (*Engl. transl.* p. 1329) (*synth*)

Diphenylphosphinoselenous acid, 9CI, 8CI D-01081

Ph_2PSeH

$C_{12}H_{11}PSe$ M 265.153
The free acid exists as the selenophosphoryl tautomer. See under Diphenylphosphine, D-01037 .
Me ester: [55776-57-3]. *Methyl diphenylphosphinoselenoite.*
$C_{13}H_{13}PSe$ M 279.179
Viscous yellow oil. Thermally stable.
Ph ester: [24091-50-7]. *Phenyl diphenylphosphinoselenoite.*
$C_{18}H_{15}PSe$ M 341.250
Yellow cryst. Mp 54°. When heated, isom. to triphenylphosphine selenide.

McLean, R.A.N., *Inorg. Nucl. Chem. Lett.*, 1969, **5**, 745 (*phenyl ester, synth, props*)
Anderson, J.W. *et al*, *Inorg. Nucl. Chem. Lett.*, 1975, **11**, 233 (*methyl ester, pmr*)

Diphenylphosphinothioic acid, 9CI D-01082

[14278-72-9]

$Ph_2P(S)OH \rightleftharpoons Ph_2P(O)SH$

$C_{12}H_{11}OPS$ M 234.252
O-*Me ester:* [3096-08-0]. O-*Methyl diphenylphosphinothioate.*
$C_{13}H_{13}OPS$ M 248.279
Used in synth. of benzoxazoles. Cryst. (MeOH or Et_2O/pet. ether). Mp 84-86.5° (49-50.5°). Isomorphous with anologous selenide.
S-*Me ester:* [3096-30-5]. S-*Methyl diphenylphosphinothioate.*
$C_{13}H_{13}OPS$ M 248.279
No phys. props. reported.
O-*Et ester:* [3133-27-5]. O-*Ethyl diphenylphosphinothioate.*
$C_{14}H_{15}OPS$ M 262.306
Solid. Mp 42-43°.

S-Et ester: [3096-04-6]. *S-Ethyl diphenylphosphinothioate.*
$C_{14}H_{15}OPS$ M 262.306
Cryst. (Et_2O). Mp 75-76°.
O-Ph ester: [17534-85-9]. *O-Phenyl diphenylphosphinothioate.*
$C_{18}H_{15}OPS$ M 310.350
Solid. Mp 123-125°.
S-Ph ester: [5101-78-1]. *S-Phenyl diphenylphosphinothioate.*
$C_{18}H_{15}OPS$ M 310.350
Cryst. (hexane or C_6H_6). Mp 90-91°.
O-Benzyl ester: O-Benzyl diphenylphosphinothioate.
$C_{19}H_{17}OPS$ M 324.376
Solid. Mp 54-56°.
S-Benzyl ester: [3096-05-7]. *S-Benzyl diphenylphosphinothioate.*
$C_{19}H_{17}OPS$ M 324.376
Cryst. (Me_2CO). Mp 94-95°.
O-Trimethylsilyl ester: [27387-65-1]. *O-Trimethylsilyl diphenylphosphinothioate.*
$C_{15}H_{19}OPSSi$ M 306.434
Cryst. (hexane). Mp 81-82°.
Fluoride: [1648-39-1].
$C_{12}H_{10}FPS$ M 236.243
Liq. $Bp_{1.5}$ 168-172°. n_D^{20} 1.6250.
Chloride: see Diphenylphosphinothioic chloride, D-01084
Bromide: [16534-67-1].
$C_{12}H_{10}BrPS$ M 297.149
Liq. $Bp_{0.01}$ 155°. n_D^{20} 1.6939.
Iodide: [84589-83-3].
$C_{12}H_{10}IPS$ M 344.149
Liq. $Bp_{0.06}$ 131°.
Azide: [4129-20-8].
$C_{12}H_{10}N_3PS$ M 259.265
Oil. Dec. at Bp.
Cyanide: [4569-37-3].
$C_{13}H_{10}NPS$ M 243.262
d_4^{25} 1.20. Mp 50°. $Bp_{0.25}$ 149-151°. n_D^{20} 1.6414.
Isocyanate:
$C_{13}H_{10}NOPS$ M 259.262
Solid. Mp 42-43°.
Isothiocyanate:
$C_{13}H_{10}NPS_2$ M 275.322
Cryst. (pet. ether). Mp 48°. $Bp_{0.01}$ 176-178°.
Anhydride: [3096-09-1]. *Diphenylphosphinothioic anhydride.*
$C_{24}H_{20}OP_2S_2$ M 450.489
Cryst. (C_6H_6). Mp 197-198°.
Amide: see P,P-Diphenylphosphinothioic amide, D-01083
Piperidide: 1-(Diphenylphosphinothioyl)piperidine.
$C_{17}H_{20}NPS$ M 301.385
Solid. Mp 101-102°.

Kabachnik, M.I. *et al*, *Tetrahedron*, 1960, **9**, 10.
Schmutzler, R., *J. Inorg. Nucl. Chem.*, 1963, **25**, 335 (*fluoride*)
Johns, I.B. *et al*, *J. Org. Chem.*, 1964, **29**, 1970 (*cyanide*)
Maier, L., *Helv. Chim. Acta*, 1964, **47**, 120 (*bromide*)
Mastryukova, T.A. *et al*, *Zh. Obshch. Khim.*, 1965, **35**, 1197 (*Engl. transl.* p. 1201) (*synth, anhydride, esters*)
Davidson, R.S., *J. Chem. Soc. (C)*, 1967, 2131 (*esters*)
Schmidpeter, A. *et al*, *Chem. Ber.*, 1967, **100**, 3052 (*isothiocyanate*)
Schindlbauer, H. *et al*, *Monatsh. Chem.*, 1968, **99**, 1792 (*ester*)
Lepicard, G. *et al*, *Acta Crystallogr., Sect. B*, 1969, **25**, 617 (*ester, cryst struct*)
Issleib, K. *et al*, *J. Organomet. Chem.*, 1970, **22**, 375 (*silyl ester*)
Knunyants, I.L. *et al*, *Dokl. Akad. Nauk SSSR, Ser. Sci. Khim.*, 1971, **201**, 862 (*Engl. transl.* p. 992) (*halides, P and F nmr*)
Wheatland, D.A. *et al*, *Inorg. Chem.*, 1972, **11**, 2340 (*anhydride*)
Ojima, I. *et al*, *Bull. Chem. Soc. Jpn.*, 1973, **46**, 2559 (*isocyanate*)
Lindner, E. *et al*, *Chem. Ber.*, 1974, **107**, 135 (*esters, ir, raman*)
Fluck, E. *et al*, *Z. Anorg. Allg. Chem.*, 1975, **412**, 47 (*pe*)
Williams, A. *et al*, *J. Chem. Soc., Perkin Trans. 2*, 1975, 1010 (*ester, uv*)
Mastryukova, T.A., *Phosphorus Sulfur*, 1976, **1**, 211 (*esters*)
Goda, K. *et al*, *Bull. Chem. Soc. Jpn.*, 1977, **50**, 545 (*ester, ir, pmr, P nmr*)
Schröder, H. *et al*, *Z. Anorg. Allg. Chem.*, 1979, **451**, 158 (*azide, ir, raman, ms, pmr, P nmr, cmr*)
Agrawal, S. *et al*, *Indian J. Chem., Sect. B*, 1980, **19**, 132 (*bromide, use*)
Al'fonsov, V.A. *et al*, *Zh. Obshch. Khim.*, 1982, **52**, 2199 (*Engl. transl.* p. 1947) (*iodide, P nmr*)
Yoshifuji, M. *et al*, *Bull. Chem. Soc. Jpn.*, 1982, **55**, 873 (*ester, use*)

P,P-Diphenylphosphinothioic amide, 9CI D-01083

[17366-80-2]

$Ph_2P(S)NH_2$

$C_{12}H_{12}NPS$ M 233.267
Prisms (CH_2Cl_2/Et_2O). Sol. Et_2O, C_6H_6, CCl_4, hot EtOH. Mp 100-102°. Loses NH_3 when heated at 280° to give the imide.
N-Ac: N-Acetyl-P,P-diphenylphosphinothioic amide.
$C_{14}H_{14}NOPS$ M 275.304
Cryst. (H_2O). Mp 250-252°.
N-Benzoyl: N-Benzoyl-P,P-diphenylphosphinothioic amide.
$C_{19}H_{16}NOPS$ M 337.375
Cryst. (EtOH). Mp 190-191°. pK_a 10.3.
N-Me: [16523-61-8]. *N-Methyl-P,P-diphenylphosphinothioic amide.*
$C_{13}H_{14}NPS$ M 247.294
Solid. Mp 129-132°.
N,N-Di-Me: [17513-68-7]. *N,N-Dimethyl-P,P-diphenylphosphinothioic amide.*
$C_{14}H_{16}NPS$ M 261.321
Cryst. (C_6H_6/pet. ether), prisms (EtOH). Sol. Et_2O, CCl_4, C_6H_6. Mp 89-90°.
N-tert-Butyl: [63577-87-7]. *N-tert-Butyl-P,P-diphenylphosphinothioic amide.*
$C_{16}H_{20}NPS$ M 289.374
Solid. Mp 120.5-121.5°.
N-Ph: [4129-41-3]. *N,P,P-Triphenylphosphinothioic amide.*
$C_{18}H_{16}NPS$ M 309.365
Prisms (EtOH). Insol. Et_2O, sol. EtOH, CCl_4, C_6H_6. Mp 165-167°.
N-(Trimethylsilyl): [18796-57-1]. *P,P-Diphenyl-N-trimethylsilylphosphinothioic amide.*
$C_{15}H_{20}NPSSi$ M 305.449
No phys. props. reported.
N-Diphenylphosphinothioyl: see N-Diphenylphosphinothioyl-P,P-diphenylphosphinothioic amide, D-01087
N-Diphenylphosphinyl: see P,P-Diphenylphosphinic amide, D-01040

Zhmurova, I.N. *et al*, *Zh. Obshch. Khim.*, 1959, **29**, 2083 (*Engl. transl.* p. 2052) (*synth*)
Sisler, H.H. *et al*, *J. Org. Chem.*, 1961, **26**, 611 (*deriv, ir*)
Schmidpeter, A. *et al*, *Chem. Ber.*, 1967, **100**, 3052 (*derivs*)
Shaw, R.A. *et al*, *Angew. Chem., Int. Ed. Engl.*, 1967, **6**, 556 (*props*)

Ojima, I. *et al*, *Bull. Chem. Soc. Jpn.*, 1973, **46**, 2559 (*synth*)
Bödeker, J. *et al*, *Z. Chem.*, 1975, **15**, 56 (*deriv, props, use*)
Cauquis, G. *et al*, *Org. Mass Spectrom.*, 1975, **10**, 770 (*ms*)
Kuramshin, I.Ya. *et al*, *Zh. Obshch. Khim.*, 1975, **45**, 1194 (*Engl. transl.* p. 1177) (*deriv, ir*)
Bödeker, J. *et al*, *J. Prakt. Chem.*, 1976, **318**, 149 (*derivs*)
Sridhara, N.S. *et al*, *Z. Naturforsch., B*, 1978, **33**, 212 (*synth, ms, ir, derivs*)

Diphenylphosphinothioic chloride — D-01084

[1015-37-8]

$Ph_2P(S)Cl$

$C_{12}H_{10}ClPS$ M 252.698

Reagent for *N*-protection of amino acids and peptides. Also used in a synth. of benzothiazoles. Liq. $Bp_{0.55}$ 152-155°. n_D^{20} 1.6563.

Maier, L., *Helv. Chim. Acta*, 1964, **47**, 120 (*synth, nmr*)
Spence, R.A. *et al*, *Aust. J. Chem.*, 1969, **22**, 2359 (*use*)
Collins, D.J. *et al*, *Aust. J. Chem.*, 1974, **27**, 841 (*synth, ir, pmr*)
Postle, R.S., *Phosphorus Sulfur*, 1977, **3**, 269 (*cmr*)
Fedorova, G.K. *et al*, *Zh. Obshch. Khim.*, 1982, **52**, 214 (*Engl. transl.* p. 195) (*synth*)
Yoshifuji, M. *et al*, *Bull. Chem. Soc. Jpn.*, 1982, **55**, 870 (*use*)

Diphenylphosphinothious acid, 9CI — D-01085

[55905-05-0]

Ph_2PSH

$C_{12}H_{11}PS$ M 218.253

Free acid probably exists as diphenylphosphine sulfide. See under Diphenylphosphine, D-01037 .

Me ester: [75231-53-7]. *Methyl diphenylphosphinothioite.*
$C_{13}H_{13}PS$ M 232.279
Liq. $Bp_{0.05}$ 125°.

Et ester: [20472-49-5]. *Ethyl diphenylphosphinothioite.*
$C_{14}H_{15}PS$ M 246.306
Liq. d_4^{20} 1.11. $Bp_{0.03}$ 132-134°. n_D^{20} 1.6410. HI cleaves both P-S and S-C bonds; I_2 cleaves S-C bond only.

Butyl ester: [1486-38-0]. *Butyl diphenylphosphinothioite.*
$C_{16}H_{19}PS$ M 274.360
Liq. d_4^{20} 1.12. Bp_1 183-186°. n_D^{20} 1.6122.

Ph ester: [14311-22-9]. *Phenyl diphenylphosphinothioite.*
$C_{18}H_{15}PS$ M 294.350
Ligand for Cu, Ni, Pd. Solid. Mp 52°. AT 100°, isom. to Triphenylphosphine sulfide, T-00629 .

Benzyl ester: [63832-89-3]. *Benzyl diphenylphosphinothioite. Phenylmethyl diphenylphosphinothioite.*
$C_{19}H_{17}PS$ M 308.377
Ligand for Pd. Solid.

Dietsche, W.H., *Tetrahedron*, 1967, **23**, 3049 (*butyl ester*)
McLean, R.A.N., *Inorg. Nucl. Chem. Lett.*, 1969, **5**, 745 (*phenyl ester, synth, props*)
Razumov, A.I. *et al*, *Zh. Obshch. Khim.*, 1972, **42**, 1250 (*Engl. transl.* p. 1245) (*ethyl ester, synth, P nmr*)
Verstuyft, A.W. *et al*, *Inorg. Chem.*, 1977, **16**, 2776 (*benzyl ester, pmr, nmr*)
Giles, J.R.M. *et al*, *J. Chem. Soc., Perkin Trans. 2*, 1981, 1211 (*methyl ester, synth, P nmr*)
Al'fonsov, V.A. *et al*, *Zh. Obshch. Khim.*, 1982, **52**, 2199 (*Engl. transl.* p. 1957) (*ethyl ester, props*)

(Diphenylphosphinothioyl)acetic acid, 9CI — D-01086

[1706-99-6]

$Ph_2P(S)CH_2COOH$

$C_{14}H_{13}O_2PS$ M 276.289

Cryst. (C_6H_6). Insol. H_2O. Mp 190°. pK_a 4.97 (50% EtOH aq.).

Et ester: [69039-09-4]. *Ethyl (diphenylphosphinothioyl)acetate.*
$C_{16}H_{17}O_2PS$ M 304.343
Oil. Bp_{5-6} 225-235°.

Issleib, K. *et al*, *Chem. Ber.*, 1960, **93**, 803 (*synth*)
Tsvetkov, E.N. *et al*, *Zh. Obshch. Khim.*, 1974, **44**, 1225 (*Engl. transl.* p. 1203) (*props*)
Braunstein, P. *et al*, *J. Chem. Res. (S)*, 1978, 232 (*esters*)
Ng, S.W. *et al*, *Organometallics*, 1982, **1**, 714 (*synth*)

N-Diphenylphosphinothioyl-*P*,*P*-diphenylphosphinothioic amide, 9CI — D-01087

Tetraphenyldithioimidophosphinic acid

[6588-07-4]

$Ph_2P(S)NHP(S)Ph_2$

$C_{24}H_{21}NP_2S_2$ M 449.504

Cryst. (CH_2Cl_2). Mp 214°.

K salt: Solid. Mp 363-366°.

NH_4 salt: Solid. Mp 210-212° dec.

Schmidpeter, A. *et al*, *Z. Anorg. Allg. Chem.*, 1966, **345**, 106 (*synth, nmr, props*)
Wang, F.T. *et al*, *Synth. React. Inorg. Met.-Org. Chem.*, 1978, **8**, 119 (*synth*)
Nöth, H. *et al*, *Z. Naturforsch., B*, 1982, **37**, 1491 (*cryst struct*)
Husebye, S. *et al*, *Acta Chem. Scand., Ser. A*, 1983, **37**, 439 (*cryst struct*)

N-Diphenylphosphino-*P*,*P*,*P*-triphenylphosphine imide — D-01088

$Ph_3P{=}NPPh_2$

$C_{30}H_{25}NP_2$ M 461.482

Cryst. (cyclohexane). Mp 124-125°.

Oxide: N-*Diphenylphosphinyl*-P,P,P-*triphenylphosphine imide.*
$C_{30}H_{25}NOP_2$ M 477.481
Cryst. (2-propanol aq.). Mp 152-153°, 170-171°.

Sulfide: N-*Diphenylphosphinothioyl*-P,P,P-*triphenylphosphine imide.*
$C_{30}H_{25}NP_2S$ M 493.542
Cryst. (C_6H_6/pentane). Mp 179-179.5°.

Baldwin, R.A. *et al*, *J. Am. Chem. Soc.*, 1961, **83**, 4466 (*oxide*)
Baldwin, R.A. *et al*, *J. Org. Chem.*, 1965, **30**, 3860 (*derivs*)
Wiegräbe, W. *et al*, *Chem. Ber.*, 1968, **101**, 1414 (*oxide, ir*)
Mardersteig, H.G. *et al*, *Z. Anorg. Allg. Chem.*, 1969, **368**, 254 (*synth, sulfide, P nmr*)

Diphenylphosphinous acid, 9CI — D-01089

[24630-80-6]

Ph_2POH

$C_{12}H_{11}OP$ M 202.192

The free acid exists in the tautomeric phosphoryl form (see under Diphenylphosphine, D-01037 . Cryst. Mp 57°.

Me ester: [4020-99-9]. *Methyl diphenylphosphinite.*
$C_{13}H_{13}OP$ M 216.219

Liq. d_0^{15} 1.040. Bp_{10} 151-152°, Bp_4 118-120°. n_D^{20} 1.6038.

Et ester: [719-80-2]. *Ethyl diphenylphosphinite.*
$C_{14}H_{15}OP$ M 230.246
Liq. d_4^{20} 1.075. Bp_{17} 155-157°, Bp_3 128-130°. n_D^{20} 1.5910.

Butyl ester: [13360-94-6]. *Butyl diphenylphosphinite.*
$C_{16}H_{19}OP$ M 258.299
Liq. d_4^{20} 1.039. Bp_5 174°, Bp_2 127-128°. n_D^{20} 1.5733.

2-Propenyl ester: 2-Propenyl diphenylphosphinite. Allyl diphenylphosphinite.
$C_{15}H_{15}OP$ M 242.257
Liq. $Bp_{0.15}$ 135-149°. When heated strongly, isomerizes to allyldiphenylphosphine oxide (see under Diphenyl-2-propenylphosphine, D-01144 .

Ph ester: [13360-92-4]. *Phenyl diphenylphosphinite.*
$C_{18}H_{15}OP$ M 278.290
Liq. Bp_2 167-170°. n_D^{20} 1.6331.

Fluoride: see Diphenylphosphinous fluoride, D-01092

Chloride: see Diphenylphosphinous chloride, D-01091

Bromide: [1079-65-8]. *Bromodiphenylphosphine.*
$C_{12}H_{10}BrP$ M 265.089
Fuming liq. d^{20} 1.67. Bp_{12} 179-180°, $Bp_{0.02}$ 111°. n_D^{20} 1.4707.

Iodide: [20472-52-0]. *Iododiphenylphosphine.*
$C_{12}H_{10}IP$ M 312.089
Deep red-brown liq. Bp_{23} 215-216°, $Bp_{2\text{-}3}$ 167-168°, $Bp_{0.05}$ 120-123°.

Cyanide: [4791-48-4].
$C_{13}H_{10}NP$ M 211.202
Liq. $Bp_{1.5}$ 180-182°, $Bp_{0.25}$ 110-112°. n_D^{22} 1.6195. Complexes with group VIB metals.

Isothiocyanate: [61582-41-0].
$C_{13}H_{10}NPS$ M 243.262
Liq. $Bp_{0.03}$ 155°.

*Amide: see P,P-*Diphenylphosphinous amide, D-01090

Sander, M., *Chem. Ber.*, 1960, **93**, 1220 (*phenyl ester, synth*)
Petrov, K.A. *et al, Zh. Obshch. Khim.*, 1961, **31**, 3027 (*Engl. transl.* p. 2823) (*bromide, synth*)
Grayson, M. *et al*, Tet, 1967, **23**, 1065 (*synth*)
Dietsche, W., *Justus Liebigs Ann. Chem.*, 1968, **712**, 21 (*ethyl ester, synth*)
Jones, C.E. *et al, Inorg. Chem.*, 1971, **10**, 1536 (*cyanide, synth, ir, pmr, P nmr, ms*)
Uznanski, B. *et al, Synthesis*, 1975, 735 (*cyanide, synth, ir*)
Colle, K.S. *et al, J. Org. Chem.*, 1978, **43**, 571 (*methyl ester, synth, P nmr, pmr*)
Pudovik, A.N. *et al, Zh. Obshch. Khim.*, 1979, **49**, 1425 (*Engl. transl.* p. 1248) (*isothiocyanate, synth, ir, P nmr*)
Romanenko, V.D. *et al, Synthesis*, 1980, 823 (*iodide, synth*)
Vincent, E. *et al, J. Mol. Struct.*, 1980, **65**, 239 (*halides, P nmr*)
Hinke, A. *et al, Phosphorus Sulfur*, 1983, **15**, 93 (*bromide, synth, P nmr*)
Verdonck, L. *et al, Spectrochim. Acta, Part A*, 1984, **40**, 299 (*halides, ir, raman*)

P,P-Diphenylphosphinous amide **D-01090**

Ph_2PNH_2

$C_{12}H_{12}NP$ M 201.207

N,N-*Di-Me: see N,N-Dimethyl-P,P-diphenylphosphinous amide, D-00747*

N,N-*Di-Et:* [1636-15-3]. N,N-*Diethyl*-P,P-*diphenylphosphinous amide. Diphenylphosphinous diethylamide.*
$C_{16}H_{20}NP$ M 257.314
Liq. $Bp_{0.07}$ 125-127°.

N,N-*Diisopropyl:* N,N-*Diisopropyl*-P,P-*diphenylphosphinous amide. Diphenylphosphinous diisopropylamide.*
$C_{18}H_{24}NP$ M 285.368
Solid. Mp 69°. $Bp_{1.5}$ 150-155°.

Cowley, A.H. *et al, J. Am. Chem. Soc.*, 1970, **92**, 5206 (*synth, pmr, stereochem*)
Yoder, C.L. *et al, J. Inorg. Nucl. Chem.*, 1970, **32**, 3689 (*dimethyl, cmr*)
El-Borgi, A. *et al, C.R. Hebd. Seances Acad. Sci., Ser. C*, 1977, **284**, 983 (*ir*)
Cristau, H.J. *et al, Synthesis*, 1980, 551 (*synth, ir, pmr*)
Horner, L. *et al, Phosphorus Sulfur*, 1980, **8**, 215 (*synth*)

Diphenylphosphinous chloride, 9CI **D-01091**

Chlorodiphenylphosphine

[1079-66-9]

Ph_2PCl

$C_{12}H_{10}ClP$ M 220.638

Ligand for metals of groups IB, VIB, VIIB, and VIII. Air-sensitive pale-yellow liq. d 1.229. Bp 320°, Bp_5 174°, $Bp_{0.6}$ 124-126°. n_D^{20} 1.636.

▷Lachrymator

Horner, L. *et al, Chem. Ber.*, 1961, **94**, 2122 (*synth*)
Maier, L., *J. Inorg. Nucl. Chem.*, 1962, **24**, 1073 (*synth*)
Shagidullin, R.R. *et al, Izv. Akad. Nauk SSSR, Ser. Khim.*, 1971, 183 (*Engl. transl.* p. 165) (*uv*)
Petrov, K.A. *et al, Zh. Obshch. Khim.*, 1973, **43**, 37 (*Engl. transl.* p. 34) (*synth*)
Maier, L. *et al, Phosphorus*, 1974, **4**, 41 (*pmr, nmr*)
Appel, R. *et al, Chem. Ber.*, 1975, **108**, 1783; 1977, **110**, 376 (*synth, nmr*)
Weinberg, K.G., *J. Org. Chem.*, 1975, **40**, 3586 (*synth*)
Modro, T.A., *Can. J. Chem.*, 1977, **55**, 3681 (*cmr*)
Vincent, E. *et al, J. Mol. Struct.*, 1980, **65**, 239 (*P nmr*)
Bergrová-Přadná, S., *Collect. Czech. Chem. Commun.*, 1981, **46**, 2289 (*ir*)
Bodner, G.M. *et al, J. Organomet. Chem.*, 1983, **243**, 305 (*cmr*)

Diphenylphosphinous fluoride, 9CI, 8CI **D-01092**

Fluorodiphenylphosphine

[20472-53-1]

Ph_2PF

$C_{12}H_{10}FP$ M 204.183

Disproportionates slowly to Ph_2PF_3 + Ph_2PPPh_2.

Brown, C. *et al, J. Chem. Soc. (C)*, 1970, 878 (*synth, props*)

(Diphenylphosphinyl)acetic acid, 9CI **D-01093**

[1831-63-6]

$Ph_2P(O)CH_2COOH$

$C_{14}H_{13}O_3P$ M 260.229

Cryst. (H_2O, MeCN or butanone/hexane). Mp 144-146°. pK_a 3.74 (H_2O), 4.83 (50% EtOH aq.).

Me ester: [21993-16-8]. *Methyl (diphenylphosphinyl)-acetate.*
$C_{15}H_{15}O_3P$ M 274.255
Cryst. (EtOAc/pet. ether). Mp 116-117°.

Hydrazide: see Diphenylphosphinylacetic acid hydrazide, D-01094

Issleib, K. *et al, Chem. Ber.*, 1961, **94**, 2244 (*synth*)

Richard, J.J. *et al*, *J. Org. Chem.*, 1963, **28**, 123 (*synth, ir*)
Grim, S.O. *et al*, *J. Inorg. Nucl. Chem.*, 1977, **39**, 499 (*synth, nmr, pmr, props*)
Torr, R.S. *et al*, *J. Chem. Soc. Pakistan*, 1979, **1**, 15 (*ester, synth, pmr, ir, ms*)
Ng, S.W. *et al*, *Organometallics*, 1982, **1**, 714 (*ir*)

Diphenylphosphinylacetic acid hydrazide, 9CI D-01094

Fosenazide, INN. *Gidifen. Fosfabenzid*
[16543-10-5]

$Ph_2P(O)CH_2CONHNH_2$

$C_{14}H_{15}N_2O_2P$ M 274.258
Tranquilliser, antiepileptic agent, antitubercular. Cryst. ($EtOH/Et_2O$). Mp 89-90°.

Razumov, A.I. *et al*, *Zh. Obshch. Khim.*, 1967, **37**, 421 (*Engl. transl.* p. 393); 1976, **46**, 1412 (*Engl. transl.* p. 1388) (*synth, pharmacol*)
Zaikonnikova, I.V. *et al*, *CA*, 1980, **93**, 179505 (*pharmacol*)
Razumov, A.I. *et al*, *Zh. Prikl. Khim.* (*Leningrad*), 1983, **56**, 342 (*Engl. transl.* p. 325) (*synth*)

α-(Diphenylphosphinyl)benzeneacetonitrile, 9CI D-01095

α-(Diphenylphosphinyl)benzyl cyanide. (α-Cyanobenzyl)diphenylphosphine oxide
[65164-90-1]

$Ph_2P(O)CH(CN)Ph$

$C_{20}H_{16}NOP$ M 317.326
Mp 151°, 225-226°.

Teichmann, H. *et al*, *Tetrahedron Lett.*, 1977, 2889 (*synth*)
Pudovik, A.N. *et al*, *Zh. Obshch. Khim.*, 1978, **48**, 1001 (*Engl. transl.* p. 913) (*synth*)

N-(Diphenylphosphinyl)-*P,P*-diphenylphosphinic amide, 9CI D-01096

$Ph_2P(O)NHP(O)Ph_2 \rightleftharpoons Ph_2P(O)N{=}P(OH)Ph_2$

$C_{24}H_{21}NO_2P_2$ M 417.383
Thought to exist in imino form in solid state. Cryst. (MeOH or propanol). Mp 272-273° (262-265°).

Fluck, E. *et al*, *Chem. Ber.*, 1963, **96**, 3091 (*synth*)
Gilyarov, V.A. *et al*, *Zh. Obshch. Khim.*, 1966, **36**, 274.
Haubold, W. *et al*, *Z. Anorg. Allg. Chem.*, 1971, **380**, 23 (*synth*)
Wang, F.T. *et al*, *Synth. React. Inorg. Met.-Org. Chem.*, 1978, **8**, 119 (*synth*)
Williams, D.J., *Inorg. Nucl. Chem. Lett.*, 1980, **16**, 189 (*synth, complexes*)
Nöth, H., *Z. Naturforsch., B*, 1982, **37**, 1491 (*cryst struct*)

O-(Diphenylphosphinyl)hydroxylamine, 9CI D-01097

[72804-56-7]

$Ph_2P(O)ONH_2$

$C_{12}H_{12}NO_2P$ M 233.206
Reagent for *N*-aminations, and aminations at carbon of Na, Mg, and Li carbanions. Cryst. (MeOH). Spar. sol. Et_2O, $CHCl_3$, CH_2Cl_2. Mp 130-135° dec. Can be stored at −40° for several months. Dec. by Me_2CO, DMSO.
N-*Ac:*
$C_{14}H_{14}NO_3P$ M 275.243
Cryst. (CH_2Cl_2/Et_2O). Mp 131-133°.
Acetone oxime: 2-[(Diphenylphosphino)oxyimino]-propane.
$C_{15}H_{16}NO_2P$ M 273.271
Cryst. (CH_2Cl_2/Et_2O). Mp 117-119°.

Harger, M.J.P., *J. Chem. Soc., Perkin Trans. 1*, 1981, 3284 (*synth, derivs, props, ir, ms, pmr*)
Boche, G. *et al*, *Tetrahedron Lett.*, 1982, **23**, 5399 (*use*)
Colvin, E.W. *et al*, *Tetrahedron Lett.*, 1982, **23**, 3835 (*use*)
Klöster, W. *et al*, *Synthesis*, 1982, 592 (*use*)

(Diphenylphosphinyl)isocyanoacetic acid, 9CI D-01098

$Ph_2P(O)CH(NC)COOH$

$C_{15}H_{12}NO_3P$ M 285.238
tert-*Butyl ester:* [77891-20-4].
$C_{19}H_{20}NO_3P$ M 341.346
Used in Wittig synth. of αβ-unsatd. isocyanides. Cryst. ($CHCl_3/Et_2O$). Mp 167°.

Rachoń, J., *Justus Liebigs Ann. Chem.*, 1981, 99 (*synth, use*)

(Diphenylphosphinylmethylene)-triphenylphosphorane, 9CI D-01099

[13411-65-9]

$Ph_3P{=}CHP(O)Ph_2$

$C_{31}H_{26}OP_2$ M 476.493
Derived from [(Diphenylphosphinyl)methyl]-triphenylphosphonium(1+), D-01101 . Cryst. (C_6H_6/hexane). Mp 157-158°.

Ramirez, F. *et al*, *J. Am. Chem. Soc.*, 1961, **83**, 3539 (*synth, ir, pmr*)

N-[(Diphenylphosphinyl)methyl]-*N*-methylaniline D-01100

N-*[(Diphenylphosphinyl)methyl]-*N*-methylbenzenamine*, 9CI
[76527-75-8]

$Ph_2P(O)CH_2NMePh$

$C_{20}H_{20}NOP$ M 321.358
Reagent for converting ketones into homologues enamines. Mp 118-119°.

Broekhof, L.N.J.M. *et al*, *Tetrahedron Lett.*, 1980, **21**, 2671 (*synth, use*)
Van den Gen, A. *et al*, *ACS Symp. Ser.*, 1981, **171**, 47 (*rev, use*)

[(Diphenylphosphinyl)methyl]-triphenylphosphonium(1+) D-01101

$Ph_3P^{\oplus}CH_2P(O)Ph_2$

$C_{31}H_{27}OP_2^{\oplus}$ M 477.501 (ion)
Chloride: [16403-36-4].
$C_{31}H_{27}ClOP_2$ M 512.954
Solid. Mp 246°.
Bromide:
$C_{31}H_{27}BrOP_2$ M 557.405
Solid. Mp 186-194°.
Tetraphenylborate: [16389-70-1].
$C_{55}H_{47}BOP_2$ M 796.733
Cryst. (Me_2CO/EtOH). Mp 114-117° (?), 214-217°.
4-Methylbenzenesulfonate: [37627-97-7].
$C_{38}H_{34}O_3P_2S$ M 632.692

Needles ($EtOH/Et_2O$). Mp 198-201°.

Wegener, W. *et al*, *Z. Chem.*, 1972, **12**, 103 (*tetraphenylborate, toluenesulfonate*)
Appel, R. *et al*, *Chem. Ber.*, 1978, **111**, 2054 (*chloride, nmr*)
Gloyna, D., *Z. Chem.*, 1982, **22**, 215 (*bromide, tetraphenylborate*)

4-(Diphenylphosphinyl)-3-oxobutanoic acid D-01102

$Ph_2P(O)CH_2COCH_2COOH$

$C_{16}H_{15}O_4P$ M 302.266

Et ester: [76842-91-6]. *Ethyl 4-(diphenylphosphinyl)-3-oxobutanoate.*
$C_{18}H_{19}O_4P$ M 330.319
Dianion reacts with aldehydes or ketones to give γ,δ-unsatd. β-ketoesters. Cryst. (toluene/Et_2O/hexane). Mp 95°. NaH yields mono- or dianion.

Van der Goorbergh, J.A.M. *et al*, *Tetrahedron Lett.*, 1980, **21**, 3621 (*synth, use*)

N,P-Diphenylphosphonamidic acid, 9CI D-01103

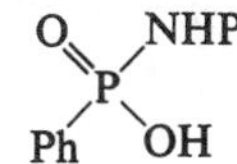

$C_{12}H_{12}NO_2P$ M 233.206

Anilinium salt: Solid. Mp 210-212°.
Me ester: [38938-31-7]. *Methyl* N,P-*diphenylphosphonamidate.*
$C_{13}H_{14}NO_2P$ M 247.233
Cryst. (MeCN or CH_2Cl_2/pet. ether). Mp 123.5-125°.
Ph ester: [57668-23-2]. *Phenyl* N,P-*diphenylphosphonamidate.*
$C_{18}H_{16}NO_2P$ M 309.304
Prisms (EtOH). Mp 145-146°.
▷SZ6225000.
Fluoride:
$C_{12}H_{11}FNOP$ M 235.197
Cryst. (C_6H_6). Readily sol. EtOH, Me_2CO, hot C_6H_6, insol. pet. ether, cold H_2O. Mp 109-112°.
Chloride: [77929-80-7].
$C_{12}H_{11}ClNOP$ M 251.652
Needles. Sol. hot C_6H_6, CCl_4, sl. sol. Et_2O. Mp 139-140°. Dec. in water at r.t.

Hersman, M.F. *et al*, *J. Org. Chem.*, 1958, **23**, 1889 (*ethyl, phenyl esters*)
Zhmurova, I.N. *et al*, *Zh. Obshch. Khim.*, 1963, **33**, 182, 549 (*Engl. transl.* pp. 175, 542) (*chloride, synth, esters*)
Ivanova, Zh.M. *et al*, *Zh. Obshch. Khim.*, 1965, **35**, 1974 (*Engl. transl.* p. 1965) (*fluoride*)
Kamai, G.Kh. *et al*, *Zh. Obshch. Khim.*, 1972, **42**, 1295 (*Engl. transl.* p. 1290) (*methyl ester*)
Williams, A. *et al*, *J. Chem. Soc., Perkin Trans. 2*, 1975, 1010 (*phenyl ester, derivs, uv, hydrolysis*)
Harger, M.J.P., *J. Chem. Soc., Perkin Trans. 1*, 1983, 2699 (*methyl ester, ms, ir, pmr*)

N,P-Diphenylphosphonamidothioic acid, 9CI D-01104

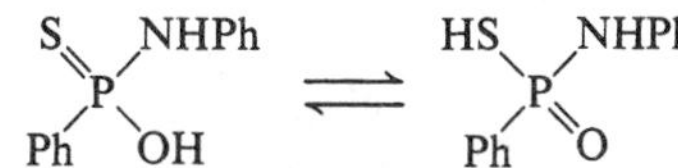

$C_{12}H_{12}NOPS$ M 249.267

O-*Ph ester:* O-*Phenyl* N,P-*diphenylphosphonamidothioate.*
$C_{18}H_{16}NOPS$ M 325.364
Cryst. (EtOH). Mp 103° (93-96°).

Hershmann, M.F. *et al*, *J. Org. Chem.*, 1958, **23**, 1889 (*synth*)
Reist, E.J. *et al*, *J. Org. Chem.*, 1960, **25**, 666 (*synth*)

Diphenyl phosphonate D-01105

Diphenyl phosphite
[4712-55-4]

$(PhO)_2P(O)H \rightleftharpoons (PhO)_2POH$

$C_{12}H_{11}O_3P$ M 234.191

Tautomeric; almost completely in phosphonate form. Reagent for peptide synth. Liq. d^{20} 1.231. Mp 12°, Mp 25°. Bp_{26} 218-219°, $Bp_{0.008}$ 100°. n_D^{20} 1.5575.

Walsh, E.N., *J. Am. Chem. Soc.*, 1959, **81**, 3023 (*synth*)
Vinogradov, L.I. *et al*, *Zh. Obshch. Khim.*, 1972, **42**, 1724 (*Engl. transl.* p. 1712) (*nmr*)
Nifant'ev, É.E. *et al*, *Zh. Obshch. Khim.*, 1972, **42**, 1936 (*Engl. transl.* p. 1929) (*synth*)
Yamazaki, N. *et al*, *Tetrahedron*, 1974, **30**, 1323 (*use*)
Yamazaki, N. *et al*, *Chem. Lett.*, 1977, 185 (*synth, props*)

O,O-Diphenyl phosphonothioate, 9CI D-01106

O,O-*Diphenyl thiophosphite*
[58045-33-3]

$(PhO)_2P(S)H \rightleftharpoons (PhO)_2PSH$

$C_{12}H_{11}O_2PS$ M 250.251

Tautomeric but exists almost completely in the thiophosphoryl form. Viscous liq.

Tashma, Z., *J. Org. Chem.*, 1983, **48**, 3966 (*synth, ir, ms, props*)

Diphenyl phosphoramidate, 9CI D-01107

Diphenyl phosphoramide. Diphenyl amidophosphate. Diphenyl phosphoryl amide
[2015-56-7]

$(PhO)_2P(O)NH_2$

$C_{12}H_{12}NO_3P$ M 249.205

Cryst. ($CHCl_3$/pet. ether). Mp 148-150°. Dec. at 180°.

Atherton, F.R. *et al*, *J. Chem. Soc.*, 1947, 674 (*synth*)
Goehring, M. *et al*, *Chem. Ber.*, 1956, **89**, 1768 (*synth*)
Walsh, E.N., *J. Am. Chem. Soc.*, 1959, **81**, 3023 (*synth*)
Nielsen, M.L. *et al*, *J. Phys. Chem.*, 1964, **68**, 152 (*nmr*)
Jakobsen, P. *et al*, *Org. Mass Spectrom.*, 1972, **6**, 1303 (*ms*)
Cremlyn, R.J.W., *Aust. J. Chem.*, 1973, **26**, 1591 (*synth, ir, P nmr*)
Al-Rawi, J.M.A. *et al*, *Org. Magn. Reson.*, 1983, **21**, 75 (*cmr*)

Diphenylphosphoramidic acid, 9CI D-01108

$Ph_2NP(O)(OH)_2$

$C_{12}H_{12}NO_3P$ M 249.205

Di-Et ester: [34044-22-9]. *Diethyl diphenylphosphoramidate.*
$C_{16}H_{20}NO_3P$ M 305.313
Cryst. Mp 175°.
Di-Ph ester: Diphenyl diphenylphosphoramidate.
$C_{24}H_{20}NO_3P$ M 401.401
Cryst. (heptane or C_6H_6/heptane). Mp 103-104°.
Difluoride: [54981-26-9].
$C_{12}H_{10}F_2NOP$ M 253.188
Low melting solid. Mp ca. 25°. Bp_3 118°.
Dichloride:
$C_{12}H_{10}Cl_2NOP$ M 286.097

Solid. Mp 57°.
Dianilide: N,N′,N″,N″-*Tetraphenylphosphoric triamide.*
$C_{24}H_{22}N_3OP$ M 399.431
Cryst. (EtOAc). Mp 232°.

Otto, P., *Ber.*, 1895, **28**, 613 (*derivs, synth*)
Michaelis, A., *Justus Liebigs Ann. Chem.*, 1903, **326**, 129 (*dichloride*)
Nielsen, M.L., *Inorg. Chem.*, 1964, **3**, 1760 (*diphenyl ester*)
Fluck, E. *et al*, *Z. Anorg. Allg. Chem.*, 1975, **411**, 125; 1975, **412**, 47 (*difluoride, synth, F and P nmr, struct*)

2,2′-Diphenylphosphoramidic dihydrazide, 9CI D-01109

$(PhNHNH)_2P(O)NH_2$

$C_{12}H_{16}N_5OP$ M 277.265
N,N-*Di-Me:* [62729-72-0]. N,N-*Dimethyl-2,2′-diphenylphosphoramidic dihydrazide.*
$C_{14}H_{20}N_5OP$ M 305.319
Cryst. (EtOH). Mp 194-195°.
N,N-*Di-Et:* N,N-*Diethyl-2,2′-diphenylphosphoramidic dihydrazide.*
$C_{16}H_{24}N_5OP$ M 333.372
Solid. Mp 184-185°.

Michaelis, A., *Justus Liebigs Ann. Chem.*, 1903, **326**, 129 (*derivs*)
Majoral, J.P. *et al*, *Tetrahedron*, 1976, **32**, 2633 (*dimethyl deriv, synth, pmr, P nmr*)

Diphenyl phosphoramidite, 9CI D-01110

$(PhO)_2PNH_2$

$C_{12}H_{12}NO_2P$ M 233.206
Solid. Mp 40-42°. Slowly dec. at r.t. giving white P.

Becke-Goehring, M. *et al*, *Chem. Ber.*, 1958, **91**, 1188 (*synth, props*)

Diphenylphosphoramidous acid, 9CI D-01111

$Ph_2NP(OH)_2$

$C_{12}H_{12}NO_2P$ M 233.206
Di-Ph ester: [17506-98-8]. *Diphenyl diphenylphosphoramidite.*
$C_{24}H_{20}NO_2P$ M 385.401
Solid. Mp 86-87°.
Difluoride: [58521-14-5].
$C_{12}H_{10}F_2NP$ M 237.188
Ligand for Mo and Ni. Distillable liq. d_{20} 1.22. Mp 11°. n_D^{20} 1.5667.
Dichloride: [4614-90-8].
$C_{12}H_{10}Cl_2NP$ M 270.097
Distillable solid. Sl. sol. Et_2O, C_6H_6, $CHCl_3$. d_{25} 1.28. Mp 21-23°. n_D^{25} 1.6284.
Dibromide:
$C_{12}H_{10}Br_2NP$ M 358.999
Cryst. Mp 67-68°.
Diiodide: [54305-77-0].
$C_{12}H_{10}I_2NP$ M 453.000
Solid. Mp 83-85°.

Gorbatenko, Zh.K. *et al*, *Zh. Obshch. Khim.*, 1974, **44**, 2357 (*Engl. transl.* p. 2311) (*diphenyl ester, diiodide*)
Falius, H. *et al*, *Z. Anorg. Allg. Chem.*, 1976, **420**, 65 (*dichloride, difluoride, dibromide, synth, F and P nmr, pmr*)

Diphenyl phosphorazidate, 9CI, 8CI D-01112

Diphenyl phosphoryl azide. DPPA
[26386-88-9]

$(PhO)_2P(O)N_3$

$C_{12}H_{10}N_3O_3P$ M 275.203
Reagent for synth. of peptides in both liq. and solid phases with little racemization. Used in Curtius reactions. With RCOOH, alcohols → carbamates, amines → amides, thiols → thioesters. Aminates aromatic and heteroaromatic organometallic reagents. Phosphorylating reagent equally reactive to *O*- and *N*-nucleophiles. With Ph_3P and diethyl azodicarboxylate, converts alcohols into azides. Oil. $Bp_{0.17}$ 157°. Stable, nonexplosive. Dec. at 170°.

Shiori, T. *et al*, *J. Am. Chem. Soc.*, 1972, **94**, 6203 (*synth, use*)
Cremlyn, R.J.W., *Aust. J. Chem.*, 1973, **26**, 1591 (*synth, props*)
Lal, B. *et al*, *Tetrahedron Lett.*, 1977, 1977 (*use*)
Ozawa, K. *et al*, *Chem. Pharm. Bull.*, 1977, **25**, 122 (*use*)
Yokoyama, Y. *et al*, *Chem. Pharm. Bull.*, 1977, **25**, 2423 (*use*)
Brady, S.F. *et al*, *J. Org. Chem.*, 1979, **44**, 3101 (*use*)
Horner, L. *et al*, *Phosphorus Sulfur*, 1981, **11**, 157 (*props, use*)
Olsen, R.H. *et al*, *J. Org. Chem.*, 1982, **47**, 4605 (*use*)
Mori, S. *et al*, *Chem. Pharm. Bull.*, 1986, **34**, 1524 (*use*)

O,O-Diphenyl phosphorazidothioate, 9CI, 8CI D-01113

O,O-*Diphenyl thiophosphoryl azide.* O,O-*Diphenyl azidothiophosphate*
[51576-61-5]

$(PhO)_2P(S)N_3$

$C_{12}H_{10}N_3O_2PS$ M 291.264
Cryst. (EtOH/hexane). Mp 25-27°.

Khodak, A.A. *et al*, *Zh. Obshch. Khim.*, 1974, **44**, 27 (*Engl. transl.* p. 24) (*synth*)

N,N′-Diphenylphosphoric triamide, 9CI D-01114

$(PhNH)_2P(O)NH_2$

$C_{12}H_{14}N_3OP$ M 247.236
Cryst. (EtOH). Sl. sol. C_6H_6, $CHCl_3$, insol. Et_2O, CCl_4, pet. ether. Mp 212-213°.
N″-*Benzoyl: Benzoylphosphoramidic dianilide.*
$C_{19}H_{18}N_3O_2P$ M 351.344
Needles (EtOH). Sol. Me_2CO, spar. sol. C_6H_6, EtOH, $CHCl_3$. Mp 215-216°.

Derkach, G.I. *et al*, *Zh. Obshch. Khim.*, 1961, **31**, 2391 (*Engl. transl.* p. 2228) (*deriv*)
Kirsanov, A.V. *et al*, *Zh. Obshch. Khim.*, 1961, **31**, 598 (*Engl. transl.* p. 557) (*synth*)
Levchenko, E.S. *et al*, *Zh. Obshch. Khim.*, 1964, **34**, 1145 (*Engl. transl.* p. 1136) (*synth*)

Diphenyl phosphoriodidite, 9CI, 8CI D-01115

Diphenyl iodophosphite
[59547-94-3]

$(PhO)_2PI$

$C_{12}H_{10}IO_2P$ M 344.088
Preparable, but disproportionates readily at ambient temp.

Feshchenko, N.G. *et al*, *Zh. Obshch. Khim.*, 1976, **46**, 777 (*Engl. transl.* p. 775)

Diphenyl phosphorisocyanatidate, 9CI, 8CI **D-01116**
Diphenyl phosphoryl isocyanate
[2487-04-9]

$(PhO)_2P(O)NCO$

$C_{13}H_{10}NO_4P$ M 275.200
Liq. d_{20}^{20} 1.28. Bp_5 184-186°, $Bp_{0.1}$ 134-136°. n_D^{20} 1.5470
Reactive to water and other protic compounds.

Kirsanov, A.V. *et al, Zh. Obshch. Khim.*, 1957, **27**, 1002 (*Engl. transl.* p. 1084) (*synth*)
Samarai, L.I. *et al, Zh. Obshch. Khim.*, 1966, **36**, 1433 (*Engl. transl.* p. 1439); 1969, **39**, 1712 (*Engl. transl.* p. 1678) (*synth*)

***O,O*-Diphenyl phosphorisocyanatidothioate,** 9CI, 8CI **D-01117**
O,O-*Diphenyl thiophosphoryl isocyanate*
[13561-75-6]

$(PhO)_2P(S)NCO$

$C_{13}H_{10}NO_3PS$ M 291.261
Reactive solid. Mp 51-52°. $Bp_{0.05}$ 127-129°.

Samarai, L.I. *et al, Zh. Obshch. Khim.*, 1966, **36**, 1433; 1969, **39**, 1712 (*Engl. transl.* pp. 1439, 1678) (*synth*)
Shokol, V.A. *et al, Zh. Obshch. Khim.*, 1969, **39**, 2137 (*Engl. transl.* p. 2088) (*synth*)

Diphenyl phosphorobromidite, 9CI, 8CI **D-01118**
Diphenyl bromophosphite
[70445-76-0]

$(PhO)_2PBr$

$C_{12}H_{10}BrO_2P$ M 297.088
Pungent liq. Bp_{11} 189-192°.

Strecker, W. *et al, Ber.*, 1916, **49**, 63 (*synth*)
Gloede, J. *et al, J. Prakt. Chem.*, 1979, **321**, 1029 (*P nmr*)
Tseng, C.K., *J. Org. Chem.*, 1979, **44**, 2793.
Chojnowski, J. *et al, J. Organomet. Chem.*, 1981, **215**, 355 (*P nmr*)

***O,O*-Diphenyl phosphorobromidothioate,** 9CI, 8CI **D-01119**
O,O-*Diphenyl thiophosphoryl bromide.* O,O-*Diphenyl bromothiophosphate*

$(PhO)_2P(S)Br$

$C_{12}H_{10}BrO_2PS$ M 329.148
Needles (EtOH). Sol. Et_2O, C_6H_6, poorly sol. MeOH, EtOH. Mp 72.5°. Bp_{11} 200°.

Strecker, W. *et al, Ber.*, 1916, **49**, 63 (*synth*)

Diphenyl phosphorochloridate, 9CI, 8CI **D-01120**
Diphenyl chlorophosphate. Diphenyl phosphoryl chloride
[2524-64-3]

$(PhO)_2P(O)Cl$

$C_{12}H_{10}ClO_3P$ M 268.636
Phosphorylating agent for alcohols, amines, and thiols; selective to RNH_2 in presence of ROH. Mild dehydrating agent e.g. converts oximes into nitriles. Activating reagent in the conversion of carboxylic acids into esters and amides. Liq. d_4^{20} 1.296. Bp_{272} 314-316°, Bp_{21} 212-215°, $Bp_{0.1}$ 134°. n_D^{20} 1.5490.

Freeman, H.F., *J. Am. Chem. Soc.*, 1938, **60**, 750 (*synth*)
Walsh, E.N., *J. Am. Chem. Soc.*, 1959, **81**, 3023 (*synth*)
Sosnovsky, G. *et al, J. Org. Chem.*, 1969, **34**, 968 (*synth*)
Just, G. *et al, J. Prakt. Chem.*, 1971, **313**, 69 (*synth*)
Horner, L. *et al, Phosphorus Sulfur*, 1981, **11**, 157 (*props, use*)
Arrieta, A. *et al, Synth. Commun.*, 1983, **13**, 471 (*use*)

Diphenyl phosphorochloridite, 9CI, 8CI **D-01121**
Diphenyl chlorophosphite
[5382-00-3]

$(PhO)_2PCl$

$C_{12}H_{10}ClO_2P$ M 252.637
Liq. d_0^{20} 1.25. Bp_{10} 179-181°. n_D^{25} 1.5789.

Fossman, J.P. *et al, J. Am. Chem. Soc.*, 1953, **75**, 3145 (*synth*)
Jones, R.A.Y. *et al, Angew. Chem., Int. Ed. Engl.*, 1962, **1**, 32 (*P nmr*)

***O,O*-Diphenyl phosphorochloridoselenoate,** 9CI, 8CI **D-01122**
O,O-*Diphenyl selenophosphoryl chloride.* O,O-*Diphenyl chloroselenophosphate*
[39014-78-4]

$(PhO)_2P(Se)Cl$

$C_{12}H_{10}ClO_2PSe$ M 331.597
Cryst. (MeOH). Sol. alcohols. Mp 59-59.5°. Bp_{11} 200°. Becomes pink on exp. to light. At 180° → Se + PCl_3 + *O,O,O*-Triphenyl phosphoroselenoate, T-00667 .

Strecker, W. *et al, Ber.*, 1916, **49**, 63 (*synth*)
Enikeev, K.M. *et al, Zh. Obshch. Khim.*, 1983, **53**, 2143 (*Engl. transl.* p. 1933) (*nmr*)

***O,O*-Diphenyl phosphorochloridothioate,** 9CI, 8CI **D-01123**
O,O-*Diphenyl thiophosphoryl chloride.* O,O-*Diphenyl chlorothiophosphate*
[22077-44-7]

$(PhO)_2P(S)Cl$

$C_{12}H_{10}ClO_2PS$ M 284.697
Cryst. (EtOH or ligroin). Mp 67°. Bp_1 180-183°.

Autenrieth, W. *et al, Ber.*, 1925, **58**, 840 (*synth*)
Gottleib, H., *J. Am. Chem. Soc.*, 1932, **54**, 748 (*synth*)
Miller, B., *J. Am. Chem. Soc.*, 1960, **82**, 3924 (*props*)
Cooks, R.G. *et al, J. Chem. Soc. (B)*, 1968, 1327 (*ms*)
Reimschüssel, W. *et al, Int. J. Chem. Kinet.*, 1980, **12**, 979; 1981, **13**, 417 (*synth, props*)
Mikolajczyk, M. *et al, J. Org. Chem.*, 1982, **47**, 1188 (*props*)

Diphenyl phosphorocyanidate, 9CI, 8CI **D-01124**
Diphenyl phosphoryl cyanide
[51354-18-8]

$(PhO)_2P(O)CN$

$C_{13}H_{10}NO_3P$ M 259.201
Phosphorylating agent; selective for –OH in the presence of $-NH_2$. Peptide coupling reagent. Liq. $Bp_{0.005}$ 140°. n_D^{20} 1.5328.

Horner, L. *et al, Phosphorus Sulfur*, 1981, **11**, 157 (*synth, ir, P nmr, props*)

N,N'-Diphenylphosphorodiamidic acid, 9CI, 8CI D-01125

Phosphoric acid dianilide

[4743-42-4]

$(PhNH)_2P(O)OH$

$C_{12}H_{13}N_2O_2P$ M 248.221

Plates (dioxan). Mp 177-180°, Mp 199-205°, Mp 212-214°. pK_a 4.38 (50% EtOH aq.).

Cyclohexylammonium salt: Cryst. (Me_2CO). Mp 175-178°.

Me ester: Methyl N,N'*-diphenylphosphorodiamidate. Methyl phosphoric dianilide.*
$C_{13}H_{15}N_2O_2P$ M 262.247
Cryst. (EtOH aq.). Mp 108-109°.

Et ester: [5586-09-4]. *Ethyl* N,N'*-diphenylphosphorodiamidate. Ethyl phosphoric dianilide.*
$C_{14}H_{17}N_2O_2P$ M 276.274
Cryst. (EtOH aq.). Mp 119-120°.

Isopropyl ester: Isopropyl N,N'*-diphenylphosphorodiamidate. Isopropyl phosphoric dianilide.*
$C_{15}H_{19}N_2O_2P$ M 290.301
Cryst. (EtOH aq.). Mp 149-150°.

Ph ester: [18995-02-3]. *Phenyl* N,N'*-diphenylphosphorodiamidate. Phenyl phosphoric dianilide.*
$C_{18}H_{17}N_2O_2P$ M 324.318
Cryst. (EtOH aq.). Mp 179-180° (165-168°).

Anhydride:
$C_{24}H_{24}N_4O_3P_2$ M 478.426
Solid. Mp 212-214°.

Fluoride: see N,N'-Diphenylphosphorodiamidic fluoride, D-01127

Chloride: see N,N'-Diphenylphosphorodiamidic chloride, D-01126

Cates, L.A. *et al, J. Am. Pharm. Assoc.*, 1959, **48**, 547 (*esters*)
Ettel, V. *et al, CA*, 1964, **61**, 14561 (*esters*)
Cremlyn, R.J.W. *et al, J. Chem. Soc. (C)*, 1971, 300 (*synth, derivs, ir, P nmr, pmr*)
Williams, A. *et al, J. Chem. Soc., Perkin Trans. 2*, 1972, 1454 (*diphenyl ester, synth, props*)
Nifant'ev, É.E. *et al, Zh. Obshch. Khim.*, 1974, **44**, 108 (*Engl. transl.* p. 106) (*synth*)
Wagner, S. *et al, Z. Chem.*, 1974, **14**, 478 (*synth, derivs*)
Trishin, Yu.G. *et al, Zh. Obshch. Khim.*, 1979, **49**, 48 (*Engl. transl.* p. 39) (*aryl esters*)
Cabré-Castellvi, J. *et al, Afinidad*, 1982, **39**, 508 (*synth, anhydride, ir, use*)

N,N'-Diphenylphosphorodiamidic chloride, 9CI, 8CI D-01126

Dianilido phosphorochloridate. Dianilinochlorophosphine oxide

[5625-99-0]

$(PhNH)_2P(O)Cl$

$C_{12}H_{12}ClN_2OP$ M 266.666

Phosphorylating agent. Cryst. (EtOH). Mp 165-167°.

▷TD3150000.

Cook, H.G. *et al, J. Chem. Soc.*, 1949, 2921 (*synth*)
Cremlyn, R.J.W. *et al, J. Chem. Soc. (C)*, 1971, 300 (*ir, pmr*)
Townsend, L.B. *et al, Nucleic Acid Chemistry*, 1978, Wiley, N.Y. **2**, 801, 809, 993 (*synth, use*)
DeRooij, J.F.M. *et al, Tetrahedron*, 1979, **35**, 2913 (*use*)

N,N'-Diphenylphosphorodiamidic fluoride, 9CI, 8CI D-01127

Dianilido phosphoryl fluoride

[330-08-5]

$(PhNH)_2P(O)F$

$C_{12}H_{12}FN_2OP$ M 250.212

Needles (EtOH aq.). Sol. hot H_2O. Mp 145°.

▷Neurotoxic. TD3680000.

Heap, R. *et al, J. Chem. Soc.*, 1948, 1313 (*synth*)
Cook, H.G. *et al, J. Chem. Soc.*, 1949, 2921 (*synth*)
Davies, D.R. *et al, Biochem. Pharmacol.*, 1966, **15**, 1783 (*pharmacol*)

N,N'-Diphenylphosphorodiamidodithioic acid, 9CI, 8CI D-01128

Dithiophosphoric acid dianilide

[70578-12-0]

$(PhNH)_2P(S)SH$

$C_{12}H_{13}N_2PS_2$ M 280.342

Solid. Mp 161-163°. Stable to ice cold aq. acid and alkali; rapidly hydrolysed at higher temp. and by boiling H_2O.

Buck, A.C. *et al, J. Am. Chem. Soc.*, 1948, **70**, 744 (*synth, props*)
Micu-Semeniuc, R. *et al, Inorg. Chim. Acta*, 1979, **33**, 281 (*synth, ir, complexes*)

N,N'-Diphenylphosphorodiamidothioic acid, 9CI, 8CI D-01129

Thiophosphoric acid dianilide

$(PhNH)_2P(S)OH \rightleftharpoons (PhNH)_2P(O)SH$

$C_{12}H_{13}N_2OPS$ M 264.281

S-*Me ester:* [64853-94-7]. S-*Methyl* N,N'*-diphenylphosphorodiamidothioate.* S-*Ethyl phosphorothioate dianilide.*
$C_{13}H_{15}N_2OPS$ M 278.308
Cryst. (C_6H_6). Mp 168-169°.

O-*Ph ester:* O-*Phenyl* N,N'*-diphenylphosphorodiamidothioate.* O-*Phenyl phosphorothioate dianilide.*
$C_{18}H_{17}N_2OPS$ M 340.379
Solid. Mp 122-123°.

Reist, E.J. *et al, J. Org. Chem.*, 1960, **25**, 666 (*phenyl ester*)
Lesiak, K. *et al, Synth. Commun.*, 1977, **7**, 339 (*methyl ester, synth, P nmr*)

O,O-Diphenyl phosphorodiselenoate, 9CI, 8CI D-01130

O,O-*Diphenyl phosphorodiselenoate.* O,O-*Diphenyl diselenophosphate.* O,O-*Diphenyl hydrogen diselenophosphate*

[34585-91-6]

$(PhO)_2P(Se)SeH$

$C_{12}H_{11}O_2PSe_2$ M 376.111

Cryst. (C_6H_6/pet. ether). Mp 67-68°.

K salt: [34585-97-2]. Solid. Mp 80-81°, Mp 172°.

Zemlyanskii, N.I. *et al, Zh. Obshch. Khim.*, 1971, **41**, 1691 (*Engl. transl.* p. 1699) (*synth*)
Yatsimirskaya, T. *et al, J. Chem. Soc. Pak.*, 1980, **2**, 127 (*uv, ir, props*)
Busev, A.I. *et al, Acta Chim. Acad. Sci. Hung.*, 1981, **108**, 1; *CA*, **96**, 75234 (*uv, props*)

O,O-Diphenyl phosphorodithioate, 9CI, 8CI D-01131

O,O-*Diphenyl hydrogen phosphorodithioate.* O,O-*Diphenyl dithiophosphate.* O,O-*Diphenyl dithiophosphoric acid*

[2253-60-3]

$(PhO)_2P(S)SH$

$C_{12}H_{11}O_2PS_2$ M 282.311
Cryst. (hexane). Mp 61°. pK_a 1.81 (7% EtOH aq.), pK_a 2.66 (80% EtOH aq.).
K salt: [3514-82-7]. Cryst. (Me_2CO/C_6H_6). Mp 191.5-192°.
Diethylammonium salt: [51576-73-9]. Solid. Mp 177-178°.
Triethylammonium salt: Solid. Mp 99-99.5°.
4-Methylanilinium salt: Cryst. (H_2O). Mp 110-112°.

Kabachnik, M.I. *et al, Tetrahedron*, 1960, **9**, 10 (*synth*)
Zemlyanski, N.I. *et al, Zh. Obshch. Khim.*, 1962, **32**, 1962 (*Engl. transl.* p. 1942) (*synth*)
Khaskin, B.A. *et al, Zh. Obshch. Khim.*, 1974, **44**, 95 (*Engl. transl.* p. 93) (*synth*)
Mazitova, F.N. *et al, Zh. Obshch. Khim.*, 1980, **50**, 815 (*Engl. transl.* p. 652) (*synth*)
Rozen, A.M. *et al, Zh. Obshch. Khim.*, 1982, **52**, 1235 (*Engl. transl.* p. 1086) (*props, use*)

***S,S*-Diphenyl phosphorodithioate**, 9CI — D-01132

S,S-*Diphenyl hydrogen dithiophosphate.* S,S-*Diphenyl dithiophosphoric acid*
[23282-01-1]

$(PhS)_2P(O)OH$

$C_{12}H_{11}O_2PS_2$ M 282.311
Phosphorylating agent. Metab. of Edifenphos, E-00002 .
Cyclohexylammonium salt: [67941-87-1].
Phosphorylating agent for nucleosides. Solid. Mp 178-179°.
Fluoride: [25237-23-4]. S,S-*Diphenyl phosphorofluoridodithioate.*
$C_{12}H_{10}FOPS_2$ M 284.303
Solid. Mp 88°.

Roesky, H., *Z. Naturforsch., B*, 1969, **24**, 818 (*fluoride, synth, ir, pmr, F nmr*)
Ueyama, I. *et al, Agric. Biol. Chem.*, 1973, **37**, 1543; 1975, **39**, 1719.
Yamaguchi, K. *et al, Chem. Lett.*, 1979, 1057 (*synth*)
Sekine, M. *et al, Bull. Chem. Soc. Jpn.*, 1981, **54**, 3815 (*synth, ir, pmr, use*)

Diphenyl phosphorofluoridate, 9CI, 8CI — D-01133

Diphenyl phosphoryl fluoride. Diphenyl fluorophosphate. Diphenoxyphosphinyl fluoride
[403-65-6]

$(PhO)_2P(O)F$

$C_{12}H_{10}FO_3P$ M 252.181
Phosphorylating agent; selective for ROH in the presence of RNH_2. $Bp_{0.1}$ 128°. n_D^{19} 1.522.

Chapman, N.B. *et al, J. Chem. Soc.*, 1948, 1010 (*synth, tox*)
Reddy, G.S. *et al, Z. Naturforsch., Part B*, 1970, **25**, 1199 (*synth, F and P nmr*)
Ryzhikov, B.D. *et al, Zh. Strukt. Khim.*, 1975, **16**, 754 (*Engl. transl.* p. 700) (*F nmr*)
Horner, L. *et al, Phosphorus Sulfur*, 1981, **11**, 155 (*synth, P nmr, props, use*)

***O,O*-Diphenyl phosphorofluoridothioate**, 9CI, 8CI — D-01134

O,O-*Diphenyl thiophosphoryl fluoride.* O,O-*Diphenyl fluorodithiophosphate. Diphenoxyphosphinothioyl fluoride*
[1648-37-9]

$(PhO)_2P(S)F$

$C_{12}H_{10}FO_2PS$ M 268.242
Liq. Bp_1 112-113°.

Olah, G.A. *et al, Can. J. Chem.*, 1962, **40**, 1917 (*synth*)
Reddy, G.S. *et al, Z. Naturforsch., B*, 1970, **25**, 1199 (*F and P nmr*)

Diphenyl phosphorofluoridotrithioate, 9CI, 8CI — D-01135

Diphenyl fluorotrithiophosphate
[17620-71-2]

$(PhS)_2P(S)F$

$C_{12}H_{10}FPS_3$ M 300.363
Liq. $Bp_{0.2}$ 168-172°.

Roesky, H.W., *Chem. Ber.*, 1968, **101**, 636 (*synth, ir, F and P nmr*)

Diphenyl phosphorohydrazidate, 9CI, 8CI — D-01136

Diphenyl phosphoryl hydrazine. (Diphenoxyphosphinyl)hydrazine
[33862-44-1]

$(PhO)_2P(O)NHNH_2$

$C_{12}H_{13}N_2O_3P$ M 264.220
Cryst. (EtOH aq.). Mp 116°.
2-Isopropylidene deriv.:
$C_{15}H_{17}N_2O_3P$ M 304.285
Solid. Mp 142°.
2-Benzylidene deriv.:
$C_{19}H_{17}N_2O_3P$ M 352.329
Solid. Mp 114°.

Audrieth, L.F. *et al, J. Org. Chem.*, 1953, **20**, 1288 (*synth, derivs*)
Cremlyn, R.J.W. *et al, J. Chem. Soc. (C)*, 1971, 3011 (*synth, ir, P nmr*)
Jahns, H.-J. *et al, J. Prakt. Chem.*, 1973, **315**, 512 (*derivs*)
El Deek, M., *J. Chem. Eng. Data*, 1980, **25**, 171 (*ms*)

***O,O*-Diphenyl phosphorohydrazidothioate**, 9CI, 8CI — D-01137

O,O-*Diphenyl thiophosphoryl hydrazide.* O,O-*Diphenyl thiophosphoric hydrazide*
[53144-23-3]

$(PhO)_2P(S)NHNH_2$

$C_{12}H_{13}N_2O_2PS$ M 280.281
Mono- or bi-dentate ligand for first row transition metals. Cryst. (MeOH). Mp 63°.

Engelhardt, U. *et al, Z. Naturforsch., B*, 1981, **36**, 791 (*synth, complexes*)

Diphenyl phosphoroisothiocyanatidate — D-01138

$(PhO)_2P(O)NCS$

$C_{13}H_{10}NO_3PS$ M 291.261
Reagent for carboxyl and peptide identification by thiohydantoin formn. d 1.29. $Bp_{0.1}$ 210°.

Kenner, G.W. *et al, J. Chem. Soc.*, 1953, 673 (*synth*)

_O,O_-Diphenyl phosphoroselenoate, 9CI, 8CI **D-01139**

O,O-Diphenyl phosphoroselenoic acid. O,O-Diphenyl hydrogen phosphoroselenoate

[78961-32-7]

$(PhO)_2P(Se)OH \rightleftharpoons (PhO)_2P(O)SeH$

$C_{12}H_{11}O_3PSe$ M 313.151

*Chloride: see O,O-*Diphenyl phosphorochloridoselenoate, D-01122

Bromide: O,O-Diphenyl phosphorobromidoselenoate. O,O-Diphenyl selenophosphoryl bromide. O,O-Diphenyl bromoselenophosphate.

$C_{12}H_{10}BrO_2PSe$ M 376.048

Cryst. (ligroin). Mp 64-65°.

Amide: O,O-Diphenyl phosphoramidoselenoate. O,O-Diphenyl amidoselenophosphate.

$C_{12}H_{12}NO_2PSe$ M 312.166

Cryst. (CCl_4). Mp 78°.

Strecker, W. *et al*, *Ber.*, 1916, **49**, 63 (*synth*)
Arbuzov, B.A. *et al*, *Izv. Akad. Nauk SSSR, Ser. Khim.*, 1983, 675 (*Engl. transl.* p. 613) (*props*)

_O,O_-Diphenyl phosphoroselenothioate, 9CI, 8CI **D-01140**

O,O-Diphenyl hydrogen phosphoroselenothioate. O,O-Diphenyl phosphoroselenothioic acid

$(PhO)_2P(Se)SH \rightleftharpoons (PhO)_2P(S)SeH$

$C_{12}H_{11}O_2PSSe$ M 329.211

Unstable acid isol. as K salt.

K salt: [15899-82-8]. Cryst. (Me_2CO aq.). Mp 190-191°.

Zemlyanskii, N.I. *et al*, *Proc. Acad. Sci. USSR*, 1965, **163**, 808; *CA*, **63**, 16240 (*synth*)

_O,O_-Diphenyl phosphorothioate, 9CI, 8CI **D-01141**

O,O-Diphenyl hydrogen phosphorothioate. O,O-Diphenyl hydrogen thiophosphate. O,O-Diphenyl thiophosphoric acid

[14156-07-1]

$(PhO)_2P(O)SH \rightleftharpoons (PhO)_2P(S)OH$

$C_{12}H_{11}O_3PS$ M 266.251

80% Thiol form (7% EtOH aq.); 79% thiol form (80% EtOH aq.) at r.t. Solid. Mp 86-87°.

K salt: [4836-72-0]. Cryst. (C_6H_6/Me_2CO). Mp 165-167°.

Triethylammonium salt: [85913-94-6]. Oil.

Dicyclohexylammonium salt: Solid. Mp 127-128°.

Cyclohexylammonium salt: Cryst. (EtOH aq.). Mp 193-195°.

Fluoride: see O,O-Diphenyl phosphorofluoridothioate, D-01134

*Chloride: see O,O-*Diphenyl phosphorochloridothioate, D-01123

Bromide: see *O,O*-Diphenyl phosphorobromidothioate, D-01119

Kabachnik, M.I. *et al*, *Tetrahedron*, 1960, **9**, 10 (*synth, struct*)
Miller, B., *J. Am. Chem. Soc.*, 1960, **82**, 3924 (*synth*)
Mastryukova, T.A. *et al*, *Zh. Obshch. Khim.*, 1974, **44**, 1001 (*Engl. transl.* p. 963); *Phosphorus Sulfur*, 1976, **1**, 211 (*props*)
Nasser, F.A.K. *et al*, *J. Organomet. Chem.*, 1983, **244**, 17 (*synth*)
Kudelska, W. *et al*, *Tetrahedron*, 1986, **42**, 629 (*P nmr*)

Diphenyl-1,2-propadienylphosphine, 9CI **D-01142**

Allenyldiphenylphosphine. Diphenylphosphinoallene

$Ph_2PCH{=}C{=}CH_2$

$C_{15}H_{13}P$ M 224.241

Obt. as mixt. with acetylenic isomer, Diphenyl(1-propynyl)phosphine, D-01146 . Liq. $Bp_{0.05}$ 105°.

Oxide: [13172-76-4]. *Diphenylphosphinylpropadiene.*

$C_{15}H_{13}OP$ M 240.241

Solid. Mp 106°.

Simonnin, M.-P. *et al*, *Org. Magn. Reson.*, 1969, **1**, 27 (*synth, oxide, pmr*)
Berlan, J. *et al*, *Bull. Soc. Chim. Fr. Part II*, 1979, 183 (*oxide, props*)

Diphenyl-1-propenylphosphine, 9CI **D-01143**

1-Diphenylphosphinopropene

[69822-97-5]

$Ph_2PCH{=}CHCH_3$

$C_{15}H_{15}P$ M 226.257

(E)-form [72138-53-7]

Liq. $Bp_{0.5}$ 125-127°.

Oxide: [4608-06-4].

$C_{15}H_{15}OP$ M 242.257

Ligand for Zn, Hg, Sb, Sn, Fe, and U. Cryst. (toluene). Mp 124-125°.

Sulfide: [21298-94-2].

$C_{15}H_{15}PS$ M 258.317

No phys. props. reported.

(Z)-form [28691-78-3]

Liq. $Bp_{1.1}$ 142-146°.

Oxide: [4569-35-1].

$C_{15}H_{15}OP$ M 242.257

Cryst. (toluene). Mp 113-116°.

Sulfide: [78045-07-5]. No phys. props. reported.

Welch, F.J. *et al*, *J. Polymer Sci., Part A*, 1965, **3**, 3427, 3429 (*oxide, complexes, ir, pmr*)
Collins, D.J. *et al*, *Aust. J. Chem.*, 1974, **27**, 2365 (*oxide*)
Grim, S.O. *et al*, *J. Org. Chem.*, 1980, **45**, 250 (*synth, pmr, P nmr*)
Duncan, M. *et al*, *Org. Magn. Reson.*, 1981, **15**, 37 (*pmr, cmr, P nmr*)

Diphenyl-2-propenylphosphine, 9CI **D-01144**

Allyldiphenylphosphine, 8CI

[2741-38-0]

$H_2C{=}CHCH_2PPh_2$

$C_{15}H_{15}P$ M 226.257

Ligand for Cr, Mo, Re, Fe and Rh. Liq. Bp_{15} 194-200°, $Bp_{0.4}$ 150°.

B,MeI: Methyldiphenyl-2-propenylphosphonium iodide.

$C_{16}H_{18}IP$ M 368.196

Cryst. (Et_2O/EtOH). Mp 145° dec.

B,PhCH₂Br: Benzyldiphenyl-2-propenylphosphonium bromide.

$C_{22}H_{22}BrP$ M 397.294

Solid. Mp 201-203°.

Oxide: [4141-48-4].

$C_{15}H_{15}PO$ M 242.257

Solid (C_6H_6/pet. ether, cyclohexane or EtOH). Mp 108-109° (99-100°). Bp_2 200-202°, $Bp_{0.01}$ 172-176°.
Sulfide: [10061-87-7].
$C_{15}H_{15}PS$ M 258.317
Solid (hexane). Mp 37-43°, Mp 49-50°. $Bp_{1\text{-}2}$ 184-185°, Bp °.
$C_{15}H_{15}PSe$ M 305.217
Solid. Mp 78-79°.

Browning, M.C. *et al*, *J. Chem. Soc.*, 1962, 693 (*synth, props*)
Issleib, K. *et al*, *Chem. Ber.*, 1963, **96**, 407 (*synth, props*)
Downie, I.M. *et al*, *J. Chem. Soc.*, 1965, 5771 (*oxide*)
Dietsche, W.H., Tet, 1967, **23**, 3049 (*sulfide*)
Clark, P.W. *et al*, *Can. J. Chem.*, 1974, **52**, 1714 (*pmr, nmr*)
Snider, T.E. *et al*, *Org. Prep. Proced. Int.*, 1974, **6**, 221 (*synth, pmr*)
Kormachev, V.V. *et al*, *Zh. Obshch. Khim.*, 1976, **46**, 1264 (*Engl. transl.* p. 1244) (*synth, oxide*)
Postle, S.R., *Phosphorus Sulfur*, 1977, **3**, 269 (*sulfide, cmr*)

Diphenylpropylphosphine, 9CI D-01145

Propyldiphenylphosphine, 8CI
[7650-84-2]

$Ph_2PCH_2CH_2CH_3$

$C_{15}H_{17}P$ M 228.273
Ligand for metals of groups VIB, VIIB, and VII. Liq. Bp_9 195-202°, Bp_4 138-139°. pK_a 2.64. n_D^{20} 1.5929.
Oxide: [4252-88-4].
$C_{15}H_{17}OP$ M 244.272
Cryst. (EtOAc). Mp 100-101°.
Sulfide: [31581-34-7].
$C_{15}H_{17}PS$ M 260.333
Plates. Mp 100-101°.

Trippett, S., *J. Chem. Soc.*, 1961, 2813 (*oxide*)
Goetz, H. *et al*, *Justus Liebigs Ann. Chem.*, 1967, **704**, 1; 1970, **742**, 59 (*synth, sulfide*)
Davidson, A.H. *et al*, *J. Chem. Soc., Perkin Trans. 1*, 1977, 550 (*oxide*)
Goff, S.D. *et al*, *Org. Mass Spectrom.*, 1977, **12**, 33 (*oxide, ms*)
Dmitriev, V.I. *et al*, *Zh. Obshch. Khim.*, 1978, **48**, 52 (*Engl. transl.* p. 42) (*synth, nmr*)
Grim, S.O. *et al*, *J. Org. Chem.*, 1980, **45**, 250 (*synth, nmr*)

Diphenyl(1-propynyl)phosphine, 9CI D-01146

1-(Diphenylphosphino)propyne
[6224-94-8]

$Ph_2PC{\equiv}CCH_3$

$C_{15}H_{13}P$ M 224.241
Cryst. (Et_2O). Mp 33°. $Bp_{0.1}$ 143°. pK_a 3.30 ($MeNO_2$). HBr or HCl give cyclic 1,4-diphosphorinium salts.
B,MeI: Methyldiphenyl-(1-propynyl)phosphonium iodide.
$C_{16}H_{16}IP$ M 366.181
Cryst. (EtOH/EtOAc). Mp 139°.
Oxide: [6224-95-9]. *1-(Diphenylphosphinyl)propyne.*
$C_{15}H_{13}OP$ M 240.241
Cryst. (C_6H_6/pet. ether or Et_2O/pet. ether). Mp 94°.
Sulfide: 1-(Diphenylphosphinothioyl)propyne.
$C_{15}H_{13}PS$ M 256.301
Cryst. (hexane). Mp 88°.

Charrier, C. *et al*, *Bull. Soc. Chim. Fr.*, 1966, 1002 (*synth, uv, ir, pmr, derivs*)
Drenth, W. *et al*, *Recl. Trav. Chim. Pays-Bas*, 1968, **87**, 41 (*props*)
Galishev, V.A. *et al*, *Zh. Obshch. Khim.*, 1973, **43**, 1470 (*Engl. transl.* p. 1460) (*synth*)
Lequan, R.M. *et al*, *Org. Magn. Reson.*, 1975, **7**, 392 (*oxide, pmr, cmr*)
Disteldorf, W. *et al*, *Chem. Ber.*, 1976, **109**, 546 (*oxide, ir*)

2,5′-Diphenylspiro[4*H*-1,3,2-benzodioxaphosphorin-2,2′-[1,3,2]-dioxaphospholane]-4,4′-dione, 10CI D-01147

[70430-51-2]

$C_{21}H_{15}O_6P$ M 394.320
Cryst. (Et_2O). Mp 104-106°.

Kobayashi, S. *et al*, *Chem. Lett.*, 1979, 393 (*synth, ir, pmr, P nmr*)

Diphenyl succinimido phosphate D-01148

1-[(Diphenoxyphosphinyl)oxy]-2,5-pyrrolidinedione, 9CI
[75513-55-2]

$C_{16}H_{14}NO_6P$ M 347.263
Reagent for peptide synth., and for conversion of γ-amino acids into lactams. Leaflets. Mp 88-90°.

Ogura, H. *et al*, *Tetrahedron Lett.*, 1980, **21**, 1467 (*synth, use*)
Ogura, H. *et al*, *Heterocycles*, 1981, **15**, 467 (*use*)

6,12-Diphenyl-1,4,8,11-tetraoxa-6,12-diaza-5,7-diphospha(5,7-P^V)-dispiro[4.1.4.1]dodecane, 10CI D-01149

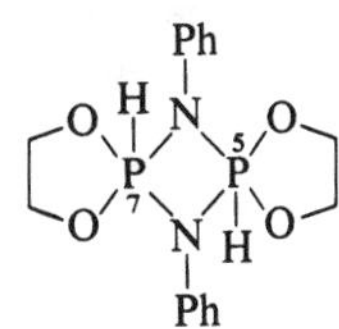

$C_{16}H_{20}N_2O_4P_2$ M 366.293
5,7-Dimethoxy: [67374-21-4].
$C_{18}H_{24}N_2O_6P_2$ M 426.345
Cryst. (C_6H_6). Mp 149-150°.
5,7-Diethoxy:
$C_{20}H_{28}N_2O_6P_2$ M 454.399
Cryst. (C_6H_6). Mp 132-133°.
5,7-Diphenoxy: [67583-31-7].
$C_{28}H_{28}N_2O_6P_2$ M 550.487
Cryst. (C_6H_6/PhCN). Mp 213-214°.
5,7-Bis(ethylthio):
$C_{20}H_{28}N_2O_4P_2S_2$ M 486.520
Cryst. (C_6H_6). Mp 198-199°.
5,7-Difluoro: [67374-25-8].
$C_{16}H_{18}F_2N_2O_4P_2$ M 402.274
Cryst. (MeCN). Mp 209-210°.

Gilyarov, V.A. *et al*, *Zh. Obshch. Khim.*, 1978, **48**, 732 (*Engl. transl.* p. 670) (*derivs*)

2,3-Diphenyl-1,4,6,9-tetraoxa-5-phospha(5-P^V)spiro[4.4]non-2-ene, 8CI D-01150

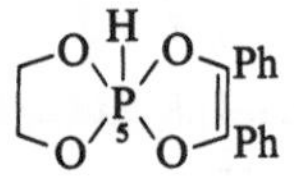

$C_{16}H_{15}O_4P$ M 302.266

5-Methoxy: [15607-00-8].

$C_{17}H_{17}O_5P$ M 332.292

Cryst. (C_6H_6/heptane). Mp 73-76°.

5-Phenoxy:

$C_{22}H_{19}O_5P$ M 394.363

Cryst. (C_6H_6). Mp 129-130°.

5-Dimethylamino: [17454-08-9]. N,N-*Dimethyl-2,3-diphenyl-1,4,6,9-tetraoxa-5-phospha(5-*P^V*)spiro[4.4]-non-2-en-5-amine.*

$C_{18}H_{20}NO_4P$ M 345.334

Pale-yellow cryst. (C_6H_6). Mp 90-92°.

Ramirez, F. *et al*, *Tetrahedron*, 1968, **24**, 1785 (*synth, ir, pmr, P nmr*)

2,3-Diphenyl-1,3-thiaphospholane, 9CI D-01151

[52125-84-5]

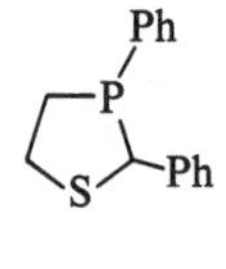

$C_{15}H_{15}PS$ M 258.317

Liq. $Bp_{0.1}$ 170-175°.

3-Oxide: [80959-91-7].

$C_{15}H_{15}OPS$ M 274.317

Characterised by nmr.

Issleib, K. *et al*, *Phosphorus*, 1973, **3**, 113 (*synth, P nmr*)
Barrau, J. *et al*, *J. Organomet. Chem.*, 1981, **221**, 271 (*oxide, P nmr*)

2,3-Diphenyl-1,3-thiaphosphorinane, 9CI D-01152

2,3-Diphenyl-1,3-thiaphosphinane

[51329-19-2]

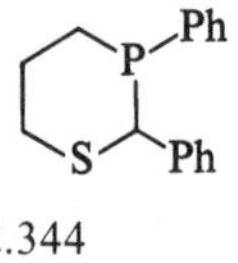

$C_{16}H_{17}PS$ M 272.344

Solid. Mp 78-83°.

Issleib, K. *et al*, *Z. Anorg. Allg. Chem.*, 1973, **402**, 189 (*synth, pmr*)

Diphenylthiohypophosphonic acid, 8CI D-01153

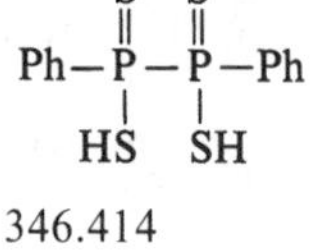

Ph–P(=S)(SH)–P(=S)(SH)–Ph

$C_{12}H_{12}P_2S_4$ M 346.414

Dec. at 225°.

Issleib, K. *et al*, *Chem. Ber.*, 1961, **94**, 107 (*synth*)

Fluck, E. *et al*, *Angew. Chem., Int. Ed. Engl.*, 1967, **6**, 883 (*P nmr*)

Diphenyl(2-trifluoromethylphenyl)phosphine, 9CI D-01154

[25688-44-2]

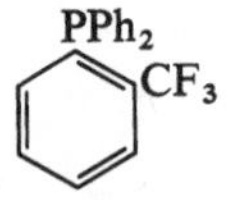

$C_{19}H_{14}F_3P$ M 330.289

Solid. Mp 86-87°.

Grim, S.O. *et al*, *Phosphorus Sulfur*, 1977, **3**, 191.

Diphenyl(3-trifluoromethylphenyl)phosphine, 9CI D-01155

[35099-14-0]

$C_{19}H_{14}F_3P$ M 330.289

Oil which slowly crysts. Mp 44-45°. Bp_1 192°.

B,MeI: Methyldiphenyl(3-trifluoromethylphenyl)-phosphonium iodide.

$C_{20}H_{17}F_3IP$ M 472.228

Cryst. (EtOH/EtOAc) or needles. Mp 180-181° (170-172°).

Oxide:

$C_{19}H_{14}F_3OP$ M 346.288

Cryst. (EtOH/hexane). Mp 115-116°.

Waite, N.E. *et al*, *Phosphorus*, 1971, **1**, 139 (*synth, methiodide*)
Allen, D.W. *et al*, *J. Chem. Soc., Perkin Trans. 1*, 1976, 2529 (*synth, derivs*)
Allen, D.W. *et al*, *J. Chem. Soc., Perkin Trans. 1*, 1979, 1499 (*props*)

Diphenyl(trimethylsilyl)phosphine, 9CI D-01156

Diphenylphosphinotrimethylsilane

[17154-34-6]

Ph_2PSiMe_3

$C_{15}H_{19}PSi$ M 258.374

Valuable intermediate for alkylation or acylation at phosphorus. Malodorous liq. Reactive to air and moisture. Turns yellow on standing. Bp_1 126-127°. n_D^{25} 1.600.

Becher, H.J. *et al*, *Chem. Ber.*, 1973, **106**, 177 (*synth, ir, use*)
Inorg. Synth., 1977, **17**, 186 (*synth*)
Schumann, H. *et al*, *Z. Naturforsch., B*, 1977, **32**, 513 (*nmr*)
Kunze, U. *et al*, *Z. Anorg. Allg. Chem.*, 1979, **456**, 155 (*ir, pmr, props*)
Brunner, H. *et al*, *Z. Naturforsch., B*, 1982, **37**, 404 (*use*)
Lindner, E. *et al*, *Chem. Ber.*, 1982, **115**, 2478 (*use*)

P,P-Diphenyl-*N*-trimethylsilylphosphinimidic acid D-01157

P,P-*Diphenyl*-N-*trimethylsilylphosphinic amide*

$Ph_2P(OH){=}NSiMe_3 \rightleftharpoons Ph_2P(O)NHSiMe_3$

$C_{15}H_{20}NOPSi$ M 289.388

Acid exists only tautomeric amide form.

Trimethylsilyl ester: [66416-58-8]. *Trimethylsilyl* P,P-*diphenyl*-N-*trimethylsilylphosphinimidate.*

$C_{18}H_{28}NOPSi_2$ M 361.570

Solid. Mp 37-39°. $Bp_{0.01}$ 103-104°.

Fluoride: [61701-85-7]. P,P-*Diphenyl*-N-*trimethylsilylphosphinimidic fluoride.* P-*Fluoro*-P,P-*diphenyl*-N-*trimethylsilylphosphazene.* P-*Fluoro*-P,P-*diphenyl*-N-*trimethylsilylphosphine imide.*

$C_{15}H_{19}FNPSi$ M 291.379

Liq. $Bp_{0.1}$ 98°.

Wisian-Neilson, P. *et al*, *Inorg. Chem.*, 1977, **16**, 1460 (*fluoride, F nmr, pmr, ir, ms, props*)
Neilson, R.H. *et al*, *Inorg. Chem.*, 1978, **17**, 1880 (*ester, ir, pmr, cmr*)

2,3-Diphenyl-1,6,10-trioxa-4-aza-5-phospha(5-P^V)spiro[4.5]dec-2-ene, 10CI — D-01158

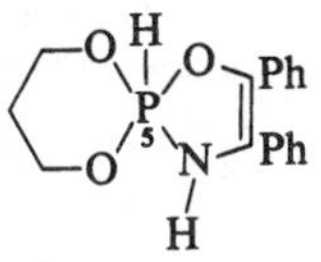

$C_{17}H_{18}NO_3P$ M 315.308
5-Ph: [71559-20-1].
$C_{23}H_{22}NO_3P$ M 391.405
Solid. Mp 94° dec.

Cadogan, J.I.G. *et al*, *J. Chem. Soc., Chem. Commun.*, 1979, 189 (*synth, P nmr*)

2,3-Diphenyl-1,4,6-trioxa-9-aza-5-phospha(5-P^V)spiro[4.4]non-2-ene, 9CI — D-01159

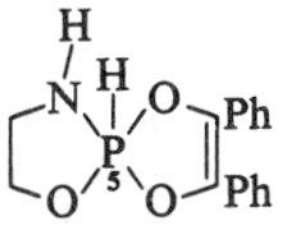

$C_{16}H_{16}NO_3P$ M 301.281
5-Methoxy: [57661-74-2].
$C_{17}H_{18}NO_4P$ M 331.307
Solid. Mp 100°.

Bernard, D. *et al*, *Phosphorus*, 1975, **5**, 285 (*synth, pmr, ir*)
Bernard, D. *et al*, *Tetrahedron*, 1975, **31**, 797 (*synth, pmr, P nmr*)

P,P-Diphenyl-*N*-(triphenylarsoranylidene)phosphinous amide — D-01160

$Ph_3As{=}NPPh_2$

$C_{30}H_{25}AsNP$ M 505.430
P-*Oxide:* [6249-37-2]. P,P-*Diphenyl*-N-(*triphenylarsoranylidene*)*phosphinic amide.*
$C_{30}H_{25}AsNOP$ M 521.429
Cryst. (Et_2O), needles (C_6H_6/hexane). Mp 137-141°, Mp 167-169°.

Froeyen, P., *Acta Chem. Scand.*, 1971, **25**, 983 (*synth*)
Cadogan, J.I.G. *et al*, *J. Chem. Soc., Perkin Trans. 1*, 1974, 466 (*synth, ir, pmr*)

Diphenyl(triphenylgermyl)phosphine, 8CI — D-01161

(Diphenylphosphino)triphenylgermane
[2816-34-4]

Ph_3GePPh_2

$C_{30}H_{25}GeP$ M 489.091
Colourless needles (methylcyclohexane). Mp 159-161°. Oxidises and hydrolyses readily.

Brooks, E.H. *et al*, *J. Chem. Soc.*, 1965, 4283 (*synth*)
Schumann, H. *et al*, *Chem. Ber.*, 1969, **102**, 2900 (*synth, ir*)

1,4-Diphenyl-2-(triphenylphosphoranylidene)-1,4-butanedione — D-01162

Triphenyl[(benzoylphenacyl)methylene]phosphorane. (1,2-Dibenzoylethylidene)triphenylphosphorane

$Ph_3P{=}C(COPh)CH_2COPh$

$C_{34}H_{27}O_2P$ M 498.560
Resonance stabilised ylide. Cryst. Mp 121-122°.

Ramirez, F. *et al*, *Tetrahedron*, 1966, **22**, 567 (*synth, ir, pmr, nmr, props*)

Diphenyl [(triphenylphosphoranylidene)methyl]phosphonate — D-01163

[22400-41-5]

$(PhO)_2P(O)CH{=}PPh_3$

$C_{31}H_{26}O_3P_2$ M 508.492
Wittig reagent for the synth. of diphenyl esters of $\alpha\beta$-unsat. phosphonic acids. Cryst. (EtOAc). Mp 149-150°.

Jones, G.H. *et al*, *Tetrahedron Lett.*, 1968, 5731 (*synth, use*)
Montgomery, J.A. *et al*, *J. Heterocycl. Chem.*, 1974, **11**, 211 (*synth, use*)
Montgomery, J.A. *et al*, *J. Med. Chem.*, 1979, **22**, 109 (*use*)

P,P-Diphenyl-*N*-(triphenylphosphoranylidene)phosphinimidic amide — D-01164

$HN'{=}PPh_2N{=}PPh_3$

$C_{30}H_{26}N_2P_2$ M 476.496
B, HCl: [24033-37-2]. Cryst. (Me_2CO). Mp 245-247°.
N-Ph: [6002-26-2]. N',P,P-*Triphenyl*-N-(*triphenylphosphoranylidene*)*phosphinimidic amide.*
$C_{36}H_{30}N_2P_2$ M 552.594
Needles. Mp 192-193° (184-186°). pK_a 21.42 ($MeNO_2$).
N'-(*4-Nitrophenyl*): [42581-74-8]. N'-(*4-Nitrophenyl*)-P,P-*diphenyl*-N-(*triphenylphosphoranylidene*)-*phosphinimidic amide.*
$C_{36}H_{29}N_3O_2P_2$ M 597.592
Cryst. (C_6H_6/hexane). Mp 186-188°. pK_a 17.76 ($MeNO_2$).
N'-(*4-Fluorophenyl*): [51870-63-4]. N'-(*4-Fluorophenyl*)-P,P-*diphenyl*-N-(*triphenylphosphoranylidene*)*phosphinimidic amide.*
$C_{36}H_{29}FN_2P_2$ M 570.584
Cryst. (C_6H_6/pet. ether). Mp 162-164°. pK_a 20.93 ($MeNO_2$).

Bock, H. *et al*, *Chem. Ber.*, 1966, **99**, 1068 (*synth*)
Zhmurova, I.N. *et al*, *Zh. Obshch. Khim.*, 1974, **44**, 79 (*Engl. transl.* p. 76) (*derivs, synth*)

P,P-Diphenyl-*N*-(triphenylphosphoranylidene)phosphinous amide — D-01165

N-*Diphenylphosphino*-P,P,P-*triphenylphosphazene.* N-*Diphenylphosphino*-P,P,P-*triphenylphosphine imide.* P,P,P-*Triphenyl*-N-*diphenylphosphinoiminophosphorane*

$Ph_2PN{=}PPh_3$

$C_{30}H_{25}NOP_2$ M 477.481
Cryst. (cyclohexane). Mp 125°.
Oxide: [2156-69-6]. P,P-*Diphenyl*-N-(*triphenylphosphoranylidene*)*phosphinic amide.* N-*Diphenylphosphinyl*-P,P,P-*triphenylphosphazene.* N-*Diphenylphosphinyl*-P,P,P-*triphenylphosphine imide.*
$C_{30}H_{25}NO_2P_2$ M 493.481
Cryst. (2-propanol aq.). Mp 152-153°, Mp 170-171°.

Sulfide: [4129-24-2]. P,P-*Diphenyl*-N-(*triphenylphosphoranylidene*)*phosphinothioic amide*. N-*Diphenylphosphinothioyl*-P,P,P-*triphenylphosphazene*. N-*Diphenylphosphinothioyl*-P,P,P-*triphenylphosphine imide*.
$C_{30}H_{25}NOP_2S$ M 509.541
Cryst. (C_6H_6). Mp 173-175°.

Selenide: P,P-*Diphenyl*-N-(*triphenylphosphoranylidene*)*phosphinoselenoic amide*. N-*Diphenylselenoyl*-P,P,P-*triphenylphosphazene*. N-*Diphenylselenoyl*-P,P,P-*triphenylphosphine imide*.
$C_{30}H_{25}NOP_2Se$ M 556.441
Cryst. Mp 193-194°.

Telluride: P,P-*Diphenyl*-N-(*triphenylphosphoranylidene*)*phosphinotelluroic amide*. N-*Diphenyltelluroyl*-P,P,P-*triphenylphosphazene*. N-*Diphenyltelluroyl*-P,P,P-*triphenylphosphine imide*.
$C_{30}H_{25}NOP_2Te$ M 605.081
Cryst. Mp 150° dec.

Baldwin, R.A. *et al*, *J. Am. Chem. Soc.*, 1961, **83**, 4466 (*oxide*)
Wiegraebe, W. *et al*, *Chem. Ber.*, 1968, **101**, 1414 (*oxide, ir*)
Mardersteig, H.G. *et al*, *Z. Anorg. Allg. Chem.*, 1969, **368**, 254; 1970, **375**, 272 (*synth, sulfide, derivs, P nmr*)
Biddlestone, M. *et al*, *J. Chem. Soc., Dalton Trans.*, 1975, 2527 (*synth*)
Cameron, A.F. *et al*, *Acta Crystallogr., Sect. B*, 1979, **35**, 1373 (*oxide, cryst struct*)
Flindt, E.-P., *Z. Anorg. Allg. Chem.*, 1982, **487**, 119 (*synth, ms, ir, pmr, P nmr*)

1,3-Diphenyl-2-(triphenylphosphoranylidene)-1,3-propanedione, 9CI — D-01166

(*Dibenzoylmethylene*)*triphenylphosphorane*. *Dibenzoyl*(*triphenylphosphoranylidene*)*methane*
[1474-24-4]

$Ph_3P{=}C(COPh)_2$

$C_{33}H_{25}O_2P$ M 484.533
Resonance-stabilised ylide. Mp 177-178°, 191-192°. pK_{a1} 2.69 (80% EtOH aq.), pK_{a2} 10.03 ($MeNO_2$) (conj. phosphonium salt). Pyrolysis at 250-80°/0.01 mm gives PhCOC≡CPh.

Chopard, P.A. *et al*, *J. Org. Chem.*, 1965, **30**, 1015 (*synth, ir, props*)
Mastryukova, T.A. *et al*, *Phosphorus*, 1972, **1**, 159 (*props*)
Brittain, J.M. *et al*, *Tetrahedron*, 1979, **35**, 1139 (*nmr*)

Diphenyl(triphenylstannyl)phosphine, 9CI — D-01167

(*Diphenylphosphino*)*triphenyltin*. *Triphenylstannyldiphenylphosphine*
[4632-37-5]

Ph_2PSnPh_3

$C_{30}H_{25}PSn$ M 535.191
Cryst. (C_6H_6). Sol. C_6H_6, toluene. Mp 127-130° (103-105°). Oxidises in air. Does not form a phosphonium salt with MeI. With alkali, gives diphenylphosphinic acid.

Campbell, I.G.M. *et al*, *J. Chem. Soc.*, 1964, 1389 (*synth*)
Schumann, H. *et al*, *Chem. Ber.*, 1964, **97**, 2395; 1969, **102**, 2900 (*synth*)
Engelhardt, G. *et al*, *Z. Naturforsch., B*, 1967, **22**, 352 (*ir, raman, nmr*)
Antoniadis, A. *et al*, *Z. Naturforsch., B*, 1979, **34**, 116 (*synth*)
Schumann, H. *et al*, *J. Organomet. Chem.*, 1980, **190**, 5363 (*synth, pmr, nmr, complexes*)

Diphenylvinylphosphine, 8CI — D-01168

Ethenyldiphenylphosphine, 9CI
[2155-96-6]

$Ph_2PCH{=}CH_2$

$C_{14}H_{13}P$ M 212.230
Ligand for Cr, Hg, Mo, Pd, Ni, Rh and W. Air-sensitive liq. $Bp_{0.5}$ 116-118°. n_D^{25} 1.6215.

B,MeI: [56598-40-4]. *Ethenylmethyldiphenylphosphonium iodide*. *Methyldiphenylvinylphosphonium iodide*.
$C_{15}H_{16}IP$ M 354.170
Cryst. (EtOH/Et_2O or Me_2CO/C_6H_6). Mp 124-125°.

Oxide: [2096-78-8].
$C_{14}H_{13}OP$ M 228.230
Cryst. (toluene/heptane or cyclohexane). Mp 117°.

Sulfide: [21776-15-8].
$C_{14}H_{13}PS$ M 244.290
Cryst. (Et_2O/pet. ether). Mp 53.5-55°.

Berlin, K.D. *et al*, *J. Org. Chem.*, 1961, **26**, 2537 (*synth, oxide, ir*)
Rabinowitz, R. *et al*, *J. Org. Chem.*, 1961, **26**, 4623 (*synth, derivs*)
Wu, C. *et al*, *J. Org. Chem.*, 1965, **30**, 1229 (*synth, ir, pmr, nmr, complexes*)
Peterson, D.J., *J. Org. Chem.*, 1966, **31**, 950 (*prep, sulfide, nmr*)
Collins, D.J. *et al*, *Aust. J. Chem.*, 1974, **27**, 841, 2365 (*oxide, ir, pmr, use, props*)
Albright, T.A. *et al*, *J. Org. Chem.*, 1975, **40**, 3437 (*oxide, cmr*)
King, R.B. *et al*, *J. Am. Chem. Soc.*, 1975, **97**, 46 (*use, sulfide*)
Clark, P.W. *et al*, *J. Organomet. Chem.*, 1977, **139**, 385 (*complexes, use*)

1,4-Diphosphabicyclo[2.2.2]octane — D-01169

[280-61-5]

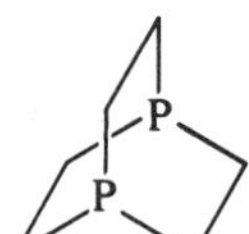

$C_6H_{12}P_2$ M 146.108
Solid. Sol. MeOH, EtOH, Et_2O, C_6H_6. Mp 252° (sealed tube under N_2).

Dioxide:
$C_6H_{12}O_2P_2$ M 178.107
Solid by subl. Not well characterised.

Disulfide:
$C_6H_{12}P_2S_2$ M 210.228
Retains benzene tenaciously. Dec. >400° without melting.

B,2MeI: *1,4-Dimethyl-1,4-diphosphoniabicyclo*[*2.2.2*]-*octane diiodide*.
$C_8H_{18}I_2P_2$ M 429.987
Insol. lower alcs. Mp 375-380° dec.

Hinton, R.C. *et al*, *J. Chem. Soc.*, 1959, 2835 (*synth, derivs, props*)

1,5-Diphosphabicyclo[3.3.0]octane D-01170

*Tetrahydro-1*H,5H-*[1,2]diphospholo[1,2-*a*][1,2]-diphosphole, 9CI*

[66872-86-4]

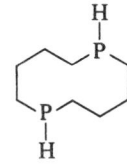

$C_6H_{12}P_2$ M 146.108

Liq. $Bp_{0.1}$ 60-63°.

Monosulfide:

$C_6H_{12}P_2S$ M 178.168

Cryst. (EtOH). Mp 203-205°.

Disulfide: [66872-89-7].

$C_6H_{12}P_2S_2$ M 210.228

Cryst. (butanol/1,2-dichloroethane). Mp 276° dec.

Cryst. in orthorhombic and monoclinic modifications.

B,MeI: [66872-92-2]. *Tetrahydro-4-methyl-1*H,5H-*[1,2]diphospholo[1,2-*a*][1,2]diphospholium iodide.*

$C_7H_{15}IP_2$ M 288.048

Cryst. ($EtOH/Et_2O$). Mp 335-337°.

Issleib, K. *et al*, *Org. Magn. Reson.*, 1977, **10**, 172 (*derivs, cmr, P nmr*)
Issleib, K. *et al*, *Phosphorus Sulfur*, 1978, **4**, 137 (*synth, P nmr, derivs*)
Hartung, H. *et al*, *Z. Anorg. Allg. Chem.*, 1979, **458**, 130 (*disulfide, cryst struct*)

1,6-Diphosphecane, 9CI D-01171

1,6-Diphosphacyclodecane

$C_8H_{18}P_2$ M 176.178

1,6-Di-Ph; B,2PhCH$_2$Br: 1,6-Dibenzyl-1,6-diphenyl-1,6-diphosphecanium dibromide.

$C_{34}H_{40}Br_2P_2$ M 670.446

Cryst. (EtOH). Mp 355-360°.

1,6-Di-Ph, 1,6-dioxide:

$C_{20}H_{26}O_2P_2$ M 360.372

Cryst. + $2H_2O$ (EtOH).

1,6-Dibenzyl: [61142-49-2]. *1,6-Dibenzyl-1,6-diphosphecane.*

$C_{22}H_{26}P_2$ M 352.395

Solid. Mp 110-115°.

1,6-Dibenzyl, 1,6-dioxide:

$C_{22}H_{26}O_2P_2$ M 384.394

Cryst. (MeOH). Mp 335°.

1,1,6,6-Tetrabenzyl, dibromide: 1,1,6,6-Tetrabenzyl-1,6-diphosphecanium dibromide.

$C_{36}H_{40}Br_2P_2$ M 694.468

Solid. Mp 330°.

Dräger, M., *Chem. Ber.*, 1974, **107**, 3246 (*deriv, cryst struct*)
Horner, L. *et al*, *Phosphorus Sulfur*, 1978, **5**, 171 (*synth*)

1,2-Diphosphinobenzene D-01172

1,2-Phenylenebisphosphine, 9CI. 1,2-Diphosphinobenzene

[80510-04-9]

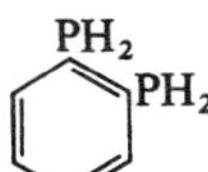

$C_6H_8P_2$ M 142.077

Liq. $Bp_{3.5}$ 71-73°. Air-sensitive.

Issleib, K. *et al*, *Tetrahedron Lett.*, 1981, **22**, 4475 (*synth, cmr, nmr*)
Kyba, E.P. *et al*, *Organometallics*, 1983, **2**, 1877 (*synth, props, nmr*)

1,4-Diphosphinobenzene D-01173

1,4-Phenylenebisphosphine, 9CI

[78550-67-1]

$C_6H_8P_2$ M 142.077

Solid. Mp 69-70.5°. Bp 226°. Sublimes.

Evleth, E.M. *et al*, *J. Org. Chem.*, 1962, **27**, 2192 (*synth, ir*)
Baldwin, R.A. *et al*, *J. Org. Chem.*, 1967, **32**, 2172 (*synth*)
Cabelli, D.E. *et al*, *J. Am. Chem. Soc.*, 1981, **103**, 3286 (*uv, pe, struct*)

1,4-Diphosphinobutane D-01174

1,4-Butanediylbisphosphine, 9CI. Tetramethylenebisphosphine, 8CI

[5518-64-9]

$H_2PCH_2CH_2CH_2CH_2PH_2$

$C_4H_{12}P_2$ M 122.086

Liq. Mp −52.5°. Bp 170-172°, Bp_{13} 64.5°. n_D^{22} 1.5287.

Sander, M., *Chem. Ber.*, 1962, **95**, 473 (*synth*)
Maier, L., *Helv. Chim. Acta*, 1966, **49**, 842 (*synth, ir, nmr*)
Issleib, K. *et al*, *Phosphorus Sulfur*, 1977, **3**, 203 (*derivs, props*)

1,10-Diphosphinodecane D-01175

1,10-Decanediylbisphosphine, 9CI. Decamethylenebisphosphine

$H_2P(CH_2)_{10}PH_2$

$C_{10}H_{24}P_2$ M 206.247

Liq. $Bp_{0.5}$ 75°.

Ger. Pat., 1 126 867, (*1962*); *CA*, **57**, 8619

1,2-Diphosphinoethane D-01176

1,2-Ethanediylbisphosphine, 9CI. Ethylenebisphosphine, 8CI

[5518-62-7]

$H_2PCH_2CH_2PH_2$

$C_2H_8P_2$ M 94.033

Bidentate ligand for Cr, Mo, Pd, Re and W. ^{99}Te complex used for hepatobiliary and myocardial imaging. Foul-smelling liq. Mp −62.5°. Bp_{725} 114-117°. Attacks grease.

▷Ignites in air

Maier, L., *Helv. Chim. Acta*, 1966, **49**, 842 (*synth, ir, pmr, nmr*)
Inorg. Synth., 1973, **14**, 10 (*synth*)
Issleib, K. *et al*, *Phosphorus*, 1977, **3**, 203 (*derivs*)
Bretherick, L., *Handbook of Reactive Chemical Hazards*, 2nd Ed., Butterworths, London and Boston, 1979, 402.

1,6-Diphosphinohexane D-01177

1,6-Hexanediylbisphosphine, 9CI. Hexamethylenebisphosphine

$H_2P(CH_2)_6PH_2$

$C_6H_{16}P_2$ M 150.140

Liq. Bp_1 78°. n_D^{20} 1.5058.

Sander, M., *Chem. Ber.*, 1962, **95**, 473.

Diphosphinomethane **D-01178**

Methylenebisphosphine, 9CI

[5518-61-6]

$$H_2PCH_2PH_2$$

CH_6P_2 M 80.006

Liq. Bp_{726} 76-77°.

Hays, H.R. *et al*, *J. Org. Chem.*, 1966, **31**, 3391 (*synth, ir, nmr*)
Maier, L., *Helv. Chim. Acta*, 1966, **49**, 842 (*synth, nmr*)
Issleib, K., *Z. Chem.*, 1984, **24**, 261 (*synth, nmr*)

1,5-Diphosphinopentane **D-01179**

1,5-Pentanediylbisphosphine, 9CI. Pentamethylenebisphosphine, 8CI

[29936-36-5]

$$H_2P(CH_2)_5PH_2$$

$C_5H_{14}P_2$ M 136.113

Liq. Bp 191-193°, Bp_1 32-35°.

Maier, L., *Helv. Chim. Acta*, 1970, **53**, 1940 (*synth, nmr*)
Issleib, K. *et al*, *Phosphorus Sulfur*, 1977, **3**, 203 (*props, derivs*)

1,3-Diphosphinopropane **D-01180**

1,3-Propanediylbisphosphine, 9CI. Trimethylenebisphosphine, 8CI

[3619-91-8]

$$H_2PCH_2CH_2CH_2PH_2$$

$C_3H_{10}P_2$ M 108.060

Foul-smelling liq. Bp_{725} 129-131°.

Maier, L., *Helv. Chim. Acta*, 1966, **49**, 842 (*synth, ir, pmr, nmr*)
Issleib, K. *et al*, *Phosphorus Sulfur*, 1977, **3**, 203; 1978, **4**, 137 (*props, use*)

1,2-Diphospholane, 10CI **D-01181**

[6680-55-3]

$C_3H_8P_2$ M 106.044

Liq. Bp 160-161°.

1,2-Diisopropyl: [82159-33-9].
$C_9H_{20}P_2$ M 190.205
Liq. $Bp_{0.03}$ 112°.

1,2-Di-Ph:
$C_{15}H_{16}P_2$ M 258.239
Liq. $Bp_{1.5}$ 180°.

1,2-Di-Ph, 1,2-disulfide: [67285-74-9]. *1,2-Diphenyl-1,2-diphospholane 1,2-disulfide.*
$C_{15}H_{16}P_2S_2$ M 322.359
Solid. Mp 178-180°. *Trans*-config.

Issleib, K. *et al*, *Chem. Ber.*, 1966, **99**, 310 (*deriv*)
Lee, L.D., *J. Organomet. Chem.*, 1977, **137**, 193 (*deriv, cryst struct*)
Issleib, K. *et al*, *Phosphorus Sulfur*, 1978, **4**, 137 (*synth, derivs, nmr*)
Diemert, K. *et al*, *Chem. Ber.*, 1982, **115**, 1947 (*deriv*)

2,3-Diphosphonobutanedioic acid, 9CI **D-01182**

2,3-Diphosphonosuccinic acid

[5693-11-8]

HOOCCHP(O)(OH)$_2$
|
HOOCCHP(O)(OH)$_2$

$C_4H_8O_{10}P_2$ M 278.049

Stabilizing agent for photographic colour developers. Corrosion inhibitor. Sodium salts are hypoglycemic agents.

Hexa-Me ester: Hexamethyl 2,3-diphosphonobutanedioate. Dimethyl 1,2-bis(dimethoxyphosphinyl)-butanedioate.
$C_{10}H_{20}O_{10}P_2$ M 362.210
Solid. Mp 89.5-91.5°.

P,P,P′,P′-*Tetraethyl, di-C-Me ester: Dimethyl 1,2-bis(diethoxyphosphinyl)succinate.*
$C_{14}H_{28}O_{10}P_2$ M 418.317
Liq. $Bp_{0.1}$ 172-175°. n_D^{25} 1.4535.

Nicholson, D.A., *Phosphorus*, 1972, **2**, 143 (*synth, esters, P nmr*)

4,4-Diphosphonobutanoic acid, 9CI **D-01183**

[74514-55-9]

$$HOOCCH_2CH_2CH[P(O)(OH)_2]_2$$

$C_4H_{10}O_8P_2$ M 248.066

Complexing agent. Immunosuppressant.

P,P′-*Bis(diethyl), C-Me ester: Methyl 4,4-bis(diethoxyphosphinyl)butanoate.*
$C_{13}H_{28}O_8P_2$ M 374.307
Liq. d_4^{20} 1.17. $Bp_{0.4}$ 180-185°. n_D^{20} 1.4512.

Pudovik, A.N. *et al*, *Zh. Obshch. Khim.*, 1970, **40**, 499 (*synth*)
Worms, K.-H. *et al*, *Z. Anorg. Allg. Chem.*, 1979, **457**, 219 (*synth*)

2,5-Diphosphonohexanedioic acid, 9CI **D-01184**

2,5-Diphosphonoadipic acid

$$(HO)_2P(O)CH(COOH)CH_2CH_2CH(COOH)\text{-}P(O)(OH)_2$$

$C_6H_{12}O_{10}P_2$ M 306.102

Hexa-Et ester: Diethyl 2,5-bis(diethoxyphosphinyl)-hexanedioate.
$C_{18}H_{36}O_{10}P_2$ M 474.424
Useful Wittig-Horner reagent for annulene synth. $Bp_{0.00005}$ 198-205°. Mixt. of stereoisomers.

Vogel, E. *et al*, *Angew. Chem., Int. Ed. Engl.*, 1980, **19**, 919 (*synth, P nmr*)

(Diphosphonomethyl)butanedioic acid, 9CI **D-01185**

(Diphosphonomethyl)succinic acid

[51395-42-7]

$$HOOCCH_2CH(COOH)CH[P(O)(OH)_2]_2$$

$C_5H_{10}O_{10}P_2$ M 292.076

(±)-form

Stabilizer for photographic colour developers. Cryst. + $1H_2O$. Mp 149.5°. Rapidly taken up by rat skeleton.

Hexa-Et ester: Diethyl [bis(diethoxyphosphinyl)-methyl]butanedioate.
$C_{17}H_{34}O_{10}P_2$ M 460.397
Liq. $Bp_{0.05}$ 197-213°.

Worms, K.-H. *et al*, *Z. Anorg. Allg. Chem.*, 1979, **457**, 219 (*synth*)

2,4-Diphosphonopentanedioic acid **D-01186**

2,4-Diphosphonoglutaric acid

$(HO)_2P(O)CH(COOH)CH_2CH(COOH)P(O)(OH)_2$

$C_5H_{10}O_{10}P_2$ M 292.076

Hexa-Et ester: [68539-80-0]. *Diethyl 2,4-bis(diethoxyphosphinyl)pentanedioate.*
$C_{17}H_{34}O_{10}P_2$ M 460.397
Useful Wittig-Horner reagent for annulene synth. Liq. $Bp_{0.00001}$ 185-195°. Mixt. of stereoisomers.

Wagemann, W. *et al*, *Angew. Chem., Int. Ed. Engl.*, 1978, **17**, 956 (*use*)
Vogel, E. *et al*, *Angew. Chem., Int. Ed. Engl.*, 1980, **19**, 41 (*synth, use*)
Savignac, M. *et al*, *Can. J. Chem.*, 1982, **60**, 840 (*synth, pmr*)

3,3-Diphosphonopropanoic acid, 9CI **D-01187**

[4775-92-2]

$HOOCCH_2CH[P(O)(OH)_2]_2$

$C_3H_8O_8P_2$ M 234.039
Complexing agent. Mp 192°.

Penta-Me ester: Methyl 3,3-bis(dimethoxyphosphinyl)-propanoate.
$C_8H_{18}O_8P_2$ M 304.173
d_4^{20} 1.32. $Bp_{3.5}$ 176-178°. n_D^{20} 1.4573.

Penta-Et ester: Ethyl 3,3-bis(diethoxyphosphinyl)-propanoate.
$C_{13}H_{28}O_8P_2$ M 374.307
Liq. d_4^{20} 1.16. $Bp_{0.5}$ 155-156°. n_D^{20} 1.4442.

Pudovik, A.N. *et al*, *Zh. Obshch. Khim.*, 1965, **35**, 354 (*Engl. transl.* p. 354) (*esters*)
Worms, K.-H. *et al*, *Z. Anorg. Allg. Chem.*, 1979, **457**, 219 (*synth*)

***N,O*-Diphosphotyrosine** **D-01188**

*[1-Carboxy-2-(4-phosphonoxyphenyl)ethyl]-phosphoramidic acid. 2-*p*-Phosphonoxybenzyl-1-phosphoramidopropanoic acid*

$C_9H_{13}NO_9P_2$ M 341.151

(S)-form

L-form

Tetrabenzyl ester: Dibenzyl [1-carboxy-2-[4-(dibenzyloxyphosphinyloxy)phenyl]ethyl]-phosphoramidate.
$C_{37}H_{37}NO_9P_2$ M 701.648
Waxy solid. Mp 95-97°.

Li, S.-O. *et al*, *J. Am. Chem. Soc.*, 1955, **77**, 1866.

Dipin **D-01189**

1,4-Bis[bis(1-aziridinyl)phosphinyl]piperazine, 9CI

[738-99-8]

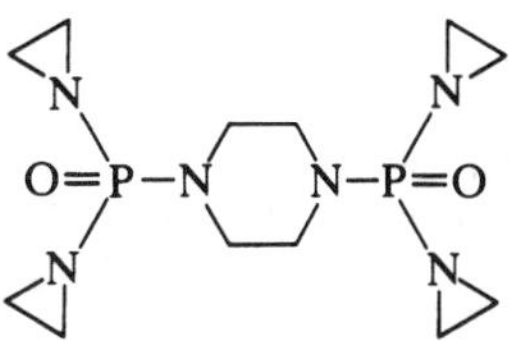

$C_{12}H_{24}N_6O_2P_2$ M 346.308
Insect sterilant. Immunosuppressant. Particularly useful in treatment of cancer of larynx. Cryst. (C_6H_6). Mp 187-189°.

▷SZ1300000.

Savin, Yu.I. *et al*, *Khim.-Farm. Zh.*, 1976, **10**, 49 (*Engl. transl.* p. 1627) (*synth, metab*)
Chistyakov, V.V. *et al*, *Khim.-Farm. Zh.*, 1984, **18**, 1290; 1985, **19**, 599 (*Engl. transl.* pp. 732, 351) (*ms, metab*)

Di-1-piperidinylphosphinic acid, 9CI **D-01190**

Di-1-piperidylphosphinic acid

[78949-15-2]

$C_{10}H_{21}N_2O_2P$ M 232.262

Me ester: Methyl di-1-piperidinylphosphinate.
$C_{11}H_{23}N_2O_2P$ M 246.289
Cryst. (pet. ether). Mp 45-47°. Bp_1 125°. n_D^{25} 1.4872.

Et ester: Ethyl di-1-piperidinylphosphinate.
$C_{12}H_{25}N_2O_2P$ M 260.315
Bp_{10} 160-165°.

Ph ester: Phenyl di-1-piperidinylphosphinate.
$C_{16}H_{25}N_2O_2P$ M 308.359
Liq. Bp_{10} 215-216°.

Fluoride:
$C_{10}H_{20}FN_2OP$ M 234.253
Liq. $Bp_{0.3}$ 145°. Exhibits mod. anticholinesterase activity.

Chloride:
$C_{10}H_{20}ClN_2OP$ M 250.708
Liq. Bp_{12} 184°.

Anilide: N*-Phenyl-*P,P*-di-1-piperidinylphosphinic amide. Di-1-piperidinylphosphinic anilide.*
$C_{16}H_{25}N_3OP$ M 306.367
Hexagonal prisms. Mp 159°.

Piperidide: see Tri-1-piperidinophosphine, T-00704

Michaelis, A., *Justus Liebigs Ann. Chem.*, 1903, **326**, 129 (*derivs*)
Heap, R. *et al*, *J. Chem. Soc.*, 1948, 1313 (*fluoride, tox*)

Di-1-piperidinylphosphinodithioic acid, 9CI **D-01191**

Dipiperidinophosphinodithioic acid, 8CI

$C_{10}H_{21}N_2PS_2$ M 264.383

Piperidinium salt: [25522-78-5]. Cryst. (C_6H_6). Mp 145° (sinters at 138°).

Becke-Goehring, M. *et al*, *Z. Anorg. Allg. Chem.*, 1969, **369**, 73 (*synth, P nmr*)

Di-2-propenyl phosphate, 9CI **D-01192**

Di-2-propenyl hydrogen phosphate. Diallyl phosphate

[7748-09-6]

$(H_2C{=}CHCH_2O)_2P(O)OH$

$C_6H_{11}O_4P$ M 178.124

Fluoride: Di-2-propenyl phosphorofluoridate. Diallyl phosphorofluoridate.

$C_6H_{10}FO_3P$ M 180.115

Liq. Bp_{23} 99-100°.

Chloride: [16383-57-6]. *Di-2-propenyl phosphorochloridate. Diallyl phosphorochloridate.*

$C_6H_{10}ClO_3P$ M 196.570

Liq. $Bp_{0.9}$ 89-90°.

Goldwhite, H. *et al*, *J. Chem. Soc.*, 1955, 2040.

Di-2-propenylphosphinic acid, 9CI **D-01193**

Diallylphosphinic acid, 8CI

[35622-32-3]

$(H_2C{=}CHCH_2)_2P(O)OH$

$C_6H_{11}O_2P$ M 146.125

Esters restore cholinesterase activity after lethal doses of phosphacol. Liq. d_4^{20} 1.09. $Bp_{0.4}$ 158-160°. n_D^{20} 1.4883. Esters undergo isomerization to di-1-propenylphosphinic esters when treated with *tert*-butoxide anion.

Me ester: [24610-83-1]. *Methyl di-2-propenylphosphinate. Methyl diallylphosphinate.*

$C_7H_{13}O_2P$ M 160.152

Liq. d^{20} 1.03. $Bp_{0.07}$ 78-79°. n_D^{20} 1.4710.

Et ester: [757-71-1]. *Ethyl di-2-propenylphosphinate. Ethyl diallylphosphinate.*

$C_8H_{15}O_2P$ M 174.179

Liq. d_4^{20} 1.00. Bp_{15} 118-120°. n_D^{20} 1.4676.

2-Propenyl ester: [5559-92-2]. *2-Propenyl di-2-propenylphosphinate. Allyl diallylphosphinate.*

$C_9H_{15}O_2P$ M 186.190

Liq. d_4^{20} 1.01. $Bp_{0.12}$ 79-81°. n_D^{20} 1.4808.

Chloride: [17620-21-2].

$C_6H_{10}ClOP$ M 164.571

Liq. d^{20} 1.13. Bp_{10} 122-123°. n_D^{20} 1.4971.

Dimethylamide: [24610-81-9]. N,N-*Dimethyl*-P,P-*di-2-propenylphosphinic amide.*

$C_8H_{16}NOP$ M 173.194

Liq. $Bp_{0.05}$ 83-84°. n_D^{20} 1.4880.

Diethylamide: [24610-82-0]. N,N-*Diethyl*-P,P-*di-2-propenylphosphinic amide.*

$C_{10}H_{20}NOP$ M 201.248

Liq. $Bp_{0.07}$ 95-97°. n_D^{20} 1.4830.

Liorber, B.G. *et al*, *Zh. Obshch. Khim.*, 1964, **34**, 1855 (*Engl. transl.* p. 1867) (*ethyl ester*)

Liorber, B.G. *et al*, *Zh. Obshch. Khim.*, 1966, **36**, 314 (*Engl. transl.* p. 323) (*allyl ester, props*)

Razumov, A.I. *et al*, *Khim.-Farm. Zh.*, 1967, **1**, 41; *CA*, **68**, 69080 (*chloride*)

Razumov, A.I. *et al*, *Khim.-Farm. Zh.*, 1969, **3**, 20; *CA*, **72**, 31925 (*methyl ester, amides*)

Razumov, A.I. *et al*, *Zh. Obshch. Khim.*, 1972, **42**, 496 (*Engl. transl.* p. 494) (*synth*)

Razumov, A.I. *et al*, *Zh. Obshch. Khim.*, 1974, **44**, 51; 1975, **45**, 1946 (*Engl. transl.* pp. 48, 1911) (*esters*)

Di-2-propenyl phosphonate, 9CI **D-01194**

Diallyl phosphonate. Di-2-propenyl phosphite. Diallyl phosphite

[23679-20-1]

$(H_2C{=}CHCH_2O)_2P(O)H \rightleftharpoons (H_2C{=}CHCH_2O)_2POH$

$C_6H_{11}O_3P$ M 162.125

Tautomeric. Oil. d_{25}^{25} 1.08. Bp_2 80°, Bp_1 58-61°. n_D^{25} 1.4459. Polymerises to transparent flame-resistant solid.

▷Liable to explode during distillation

Toy, A.D.F. *et al*, *J. Am. Chem. Soc.*, 1954, **76**, 2191 (*synth*)

Harless, H.R., *Anal. Chem.*, 1961, **33**, 1387 (*ms*)

Bretherick, L., *Handbook of Reactive Chemical Hazards*, 2nd Ed., Butterworths, London and Boston, 1979, 598.

Sax, N.I., *Dangerous Properties of Industrial Materials*, 6th Ed., Van Nostrand-Reinhold, 1984, 541.

Hazards in the Chemical Laboratory, (Bretherick, L., Ed.), 3rd Ed., Royal Society of Chemistry, London, 1981, 267.

Di-2-propenylphosphoramidic acid, 9CI **D-01195**

Diallylphosphoramidic acid

$(H_2C{=}CHCH_2)_2NP(O)(OH)_2$

$C_6H_{12}NO_3P$ M 177.139

Di-Et ester: Diethyl di-2-propenylphosphoramidate. Diethyl diallylphosphoramidate.

$C_{10}H_{20}NO_3P$ M 233.247

Liq. Bp_9 118°.

Di-Ph ester: Diphenyl di-2-propenylphosphoramidate. Diphenyl diallylphosphoramidate.

$C_{18}H_{20}NO_3P$ M 329.335

Solid. Mp 37°. Resinifies when heated at 90-110°.

U.S.P., 2 852 550, (*1958*); *CA*, **53**, 4131 (*diethyl ester*)

Arni, P.C. *et al*, *J. Appl. Chem.*, 1964, **14**, 221 (*diphenyl ester, synth, props*)

***O,O*-Di-2-propenyl phosphorochloridothioate,** 9CI **D-01196**

O,O-*Diallyl phosphorochloridothioate,* 8CI. O,O-*Diallyl chlorothiophosphate.* O,O-*Diallyl thiophosphoryl chloride*

[34819-50-6]

$(H_2C{=}CHCH_2O)_2P(S)Cl$

$C_6H_{10}ClO_2PS$ M 212.631

Liq. $Bp_{0.15}$ 72-74°.

U.S.P., 3 833 623, (*1974*); *CA*, **81**, 151560

***O,O*-Di-2-propenyl phosphorodithioate,** 9CI **D-01197**

O,O-*Diallyl phosphorodithioate,* 8CI. O,O-*Diallyl hydrogen dithiophosphate.* O,O-*Diallyl dithiophosphoric acid*

[5851-14-9]

$(H_2C{=}CHCH_2O)_2P(S)SH$

$C_6H_{11}O_2PS_2$ M 210.245

Liq. d_4^{20} 1.17. $Bp_{0.001}$ 64-66°. n_D^{20} 1.5326.

K salt: [20442-45-9]. Cryst. (Me_2CO/Et_2O). Mp 126°.

Mel'nik, Ya.I. *et al*, *Zh. Obshch. Khim.*, 1970, **40**, 791 (*Engl. transl.* p. 768) (*synth*)

Zemlyanski, N.I. *et al*, *Zh. Obshch. Khim.*, 1972, **42**, 54 (*Engl. transl.* p. 50) (*synth*)

O,O-Di-2-propenyl phosphorothioate, 9CI, 8CI D-01198

O,O-Diallyl phosphorothioate, 8CI. O,O-Diallyl hydrogen phosphorothioate. O,O-Diallyl hydrogen thiophosphate. O,O-Diallyl thiophosphoric acid

$(H_2C{=}CHCH_2O)_2P(O)SH \rightleftharpoons (H_2C{=}CHCH_2O)_2P(S)OH$

$C_6H_{11}O_3PS$ M 194.185

NH_4 salt: Solid. Mp 133-140°.

Pesin, V.G. *et al*, *Zh. Obshch. Khim.*, 1961, **31**, 2508 (*Engl. transl.* p. 2337)

(Dipropoxyphosphinyl)acetic acid D-01199

$(H_3CCH_2CH_2O)P(O)CH_2COOH$

$C_8H_{17}O_5P$ M 224.193

Et ester: Ethyl (dipropoxyphosphinyl)acetate.
$C_{10}H_{21}O_5P$ M 252.247
Liq. d 1.08. Bp_{10} 155.5-156°. n_D 1.4292.

Amide: (Dipropoxyphosphinyl)acetamide.
$C_8H_{18}NO_4P$ M 223.208
Solid. Mp 78-80°.

Kamai, G. *et al*, *Dokl. Akad. Nauk SSSR, Ser. Sci. Khim.*, 1950, **72**, 301; *CA*, **45**, 542 (*derivs*)
Bodnarchuk, N.D. *et al*, *Zh. Obshch. Khim.*, 1969, **39**, 1707 (*Engl. transl. p. 1673*) (*amide*)

Dipropyl chlorothiophosphonate D-01200

Thiohypochlorous acid anhydrosulfide with dipropyl phosphorothioate, 9CI
[55655-36-2]

$(H_3CCH_2CH_2O)_2P(O)SCl$

$C_6H_{14}ClO_3PS$ M 232.662
Yellow liq. d_4^{20} 1.20. $Bp_{0.5}$ 84°. n_D^{20} 1.4679.

▷Asphyxiating vapour

Lenard-Borecka, B. *et al*, *Rocz. Chem.*, 1957, **31**, 1167 (*synth*)
Michalski, J. *et al*, *Rocz. Chem.*, 1963, **37**, 1479 (*synth*)

O,O-Dipropyl chlorothiophosphonothioate, 9CI D-01201

Thiohypochlorous acid anhydrosulfide with O,O-dipropyl phosphorodithioate, 10CI. S-Chloro O,O-dipropyl phosphorodithioate
[5187-31-5]

$(H_3CCH_2CH_2O)_2P(S)SCl$

$C_6H_{14}ClO_2PS_2$ M 248.722
Golden-yellow liq. d_4^{20} 1.20. $Bp_{0.6}$ 93°. n_D^{20} 1.5175.

▷Asphyxiating vapour

Almasi, L. *et al*, *Chem. Ber.*, 1965, **98**, 3546.

Dipropyldiphosphonic acid, 8CI D-01202

Dipropylpyrophosphonic acid. Propylphosphonic acid monoanhydride

$$H_3CCH_2CH_2{-}P(=O)(OH){-}O{-}P(=O)(OH){-}CH_2CH_2CH_3$$

$C_6H_{16}O_5P_2$ M 230.137

Di-Et ester: Diethyl dipropyldiphosphonate.
$C_{10}H_{24}O_5P_2$ M 286.244
Liq. d_4^{20} 1.11. $Bp_{0.0001}$ 102-103°. n_D^{20} 1.4380.

Turpin, R. *et al*, *Bull. Soc. Chim. Fr.*, 1971, 3878 (*synth*)

Dipropyl phosphate, 9CI, 8CI D-01203

Dipropyl hydrogen phosphate. Dipropyl phosphoric acid
[1804-93-9]

$(H_3CCH_2CH_2O)_2P(O)OH$

$C_6H_{15}O_4P$ M 182.156
Oil. d_4^{20} 1.10. pK_a 1.59 (H_2O, 25°), pK_a 3.29 (80% EtOH aq.). n_D^{20} 1.4252.

Tetramethylammonium salt: Solid. Mp 132°.
Chloride: see Dipropyl phosphorochloridate, D-01217
Bromide: Dipropyl phosphorobromidate. Dipropyl phosphoryl bromide. Dipropyl bromophosphate. Dipropoxyphosphinyl bromide.
$C_6H_{14}BrO_3P$ M 245.053
Liq. $Bp_{0.4}$ 88-90°.
Amide: see Dipropyl phosphoramidate, D-01213
Anilide: see Phenylphosphoramidic acid, P-00257
Anhydride: see Tetrapropyl diphosphate, T-00285

Ramaswami, D. *et al*, *J. Am. Chem. Soc.*, 1953, **75**, 1763 (*chloride, anilide*)
Goldwhite, H. *et al*, *J. Chem. Soc.*, 1955, 3564 (*bromide*)
Chabrier, P. *et al*, *C.R. Hebd. Seances Acad. Sci.*, 1957, **244**, 2730 (*synth*)
Petrov, K.A. *et al*, *Zh. Obshch. Khim.*, 1961, **31**, 1709 (*Engl. transl.* p. 1596) (*synth*)
Grosse-Ruyken, H. *et al*, *J. Prakt. Chem.*, 1962, **18**, 287 (*synth*)

Dipropylphosphine, 9CI D-01204

[19357-87-0]

$HP(CH_2CH_2CH_3)_2$

$C_6H_{15}P$ M 118.158
Ligand for Cr and Pt. Liq. Bp 136°. n_D^{20} 1.541. Forms Cr and Pt complexes.

Li salt: Needles + $0.5Et_2O$.
Oxide: [27443-18-1].
$C_6H_{15}OP$ M 134.158
Cryst. Mp 48-50°. $Bp_{1.5}$ 71-72°. Tautomeric with Dipropylphosphinous acid, D-01209 .
Sulfide: [54044-00-7].
$C_6H_{15}PS$ M 150.218
Liq. $Bp_{0.1}$ 119°. Tautomeric with dipropylphosphinothious acid.

Horn, P.E. *et al*, *J. Chem. Soc.*, 1963, 1036 (*synth*)
Kabachnik, M.I. *et al*, *Izv. Akad. Nauk SSSR, Ser. Khim.*, 1963, 1227 (*Engl. transl.* p. 1120) (*oxide*)
Issleib, K. *et al*, *Z. Anorg. Allg. Chem.*, 1964, **328**, 21 (*deriv*)
Appel, R. *et al*, *Chem. Ber.*, 1974, **107**, 2658 (*sulfide, pmr, nmr, ir, ms*)

Dipropylphosphinic acid, 9CI D-01205

[867-33-4]

$(H_3CCH_2CH_2)_2P(O)OH$

$C_6H_{15}O_2P$ M 150.157

Solid. Sl. sol. H_2O, EtOH, C_6H_6. Mp 58.5-59.5°. pK_a 3.46 (H_2O).

Et ester: Ethyl dipropylphosphinate.
$C_8H_{19}O_2P$ M 178.211
Bp_{14} 110-112°. n_D^{21} 1.4369.

Fluoride: [665-92-9].
$C_6H_{14}FOP$ M 152.148
No phys. props. reported.

Chloride: [1113-11-7].
$C_6H_{14}ClOP$ M 168.603
Liq. d_4^{20} 1.08. Bp_4 84-88°. n_D^{20} 1.4681.

Trimethylsilyl ester: [53483-29-7]. *Trimethylsilyl dipropylphosphinate.*
$C_9H_{23}O_2PSi$ M 222.339
Liq. $Bp_{0.1}$ 76°.

Anhydride: [34979-30-1].
$C_{12}H_{28}O_3P_2$ M 282.299
Feathery needles. Reacts rapidly with H_2O. d_9^{30} 1.02. Mp 28-30°. Bp_{15} 200-202°. n_D^{30} 1.4570.

Amide: see P,P-Dipropylphosphinic amide, D-01206

Kosolapoff, G.M. *et al*, *J. Am. Chem. Soc.*, 1951, **73**, 4101 (*anhydride, ethyl ester*)
Crofts, P.C. *et al*, *J. Am. Chem. Soc.*, 1953, **75**, 3379 (*synth, props*)
Kuchen, W.K. *et al*, *Chem. Ber.*, 1962, **95**, 1703 (*synth, props*)
Blasse, R.G., *Recl. Trav. Chim. Pays-Bas*, 1965, **84**, 267 (*chloride, ir*)
Neimysheva, A.A. *et al*, *Zh. Obshch. Khim.*, 1966, **36**, 1090 (*Engl. transl.* p. 1105) (*chloride*)
Knunyants, I.L. *et al*, *Dokl. Akad. Nauk SSSR, Ser. Sci. Khim.*, 1971, **201**, 862 (*Engl. transl.* p. 992) (*halides, nmr*)
Turpin, R. *et al*, *Bull. Soc. Chim. Fr.*, 1971, 3878 (*anhydride, props*)
Appel, R. *et al*, *Chem. Ber.*, 1974, **107**, 2658 (*ir, pmr, nmr*)
Kuchen, W. *et al*, *Z. Anorg. Allg. Chem.*, 1975, **413**, 266 (*silyl ester*)

***P,P*-Dipropylphosphinic amide, 9CI** D-01206

$(H_3CCH_2CH_2)_2P(O)NH_2$

$C_6H_{16}NOP$ M 149.172

Hygroscopic scales. Mp 68-69.5° (sealed tube).

N,N-*Di-Et:* N,N-*Diethyl*-P,P-*dipropylphosphinic amide.*
$C_{10}H_{24}NOP$ M 205.279
Liq. $Bp_{0.22}$ 104°. n_D^{20} 1.4566.

Razumov, A.I. *et al*, *Zh. Obshch. Khim.*, 1957, **27**, 754 (*Engl. transl.* p. 827) (*diethylamide*)
Mukhacheva, O.A. *et al*, *Zh. Obshch. Khim.*, 1962, **32**, 2696 (*Engl. transl.* p. 2654) (*synth*)

Dipropylphosphinodithioic acid, 9CI D-01207

[22689-71-0]

$(H_3CCH_2CH_2)_2P(S)SH$

$C_6H_{15}PS_2$ M 182.278

Some esters are lubricating oil additives. d_4^{20} 1.07. Bp_{12} 146-147°, Bp_2 91-91.5°. pK_{a1} 1.84 (7% EtOH aq.), pK_{a2} 2.83 (40% EtOH aq.), pK_{a3} 2.63 (80% EtOH aq.). n_D^{20} 1.5607.

Na salt: [53482-97-6]. Solid. Mp 177-178° (anhyd.). Forms trihydrate Mp 87-9°.

Anhydrosulfide: [22737-44-6].
$C_{12}H_{28}P_2S_3$ M 330.481
Cryst. (C_6H_6/hexane). Mp 47.5°. $Bp_{0.7}$ 193.5-195°.

Kabachnik, M.I. *et al*, *Tetrahedron*, 1960, **9**, 10 (*props*)
Kuchen, W. *et al*, *Chem. Ber.*, 1963, **96**, 1733 (*synth, props*)
Toropova, V.P. *et al*, *Zh. Obshch. Khim.*, 1968, **38**, 2088 (*Engl. transl.* p. 2022) (*props*)
Almasi, L. *et al*, *Chem. Ber.*, 1969, **102**, 1489 (*anhydrosulfide*)
Shagidullin, R.R. *et al*, *Izv. Akad. Nauk SSSR, Ser. Khim.*, 1973, 541 (*Engl. transl.* p. 518) (*ir*)
Keck, H. *et al*, *Phosphorus Sulfur*, 1983, **14**, 225 (*ms*)

Dipropylphosphinothioic acid, 9CI D-01208

[867-34-5]

$(H_3CCH_2CH_2)_2P(S)OH \rightleftharpoons (H_3CCH_2CH_2)_2P(O)SH$

$C_6H_{15}OPS$ M 166.218

Mp 31-32°. Bp_2 98.5-99°. pK_a 2.83 (7% EtOH aq.) (1% thiol form), 5.10 (80% EtOH aq.) (0% thiol form), 8.55 (100% EtOH).

NH_4 salt: Solid. Mp 117-119°.

O-*Me ester:* O-*Methyl dipropylphosphinothioate.*
$C_7H_{17}OPS$ M 180.244
Liq. Bp_{12} 118-119°.

S-*Me ester:* [49873-32-7]. S-*Methyl dipropylphosphinothioate.*
$C_7H_{17}OPS$ M 180.244
Liq. d_4^{20} 1.02. Bp_2 120-120.5°. n_D^{20} 1.4984.

O-*Propyl ester:* [17643-91-3]. O-*Propyl dipropylphosphinothioate.*
$C_9H_{21}OPS$ M 208.298
Liq. $Bp_{0.5}$ 81.2°. n_D^{20} 1.4778.

S-*Butyl ester:* S-*Butyl dipropylphosphinothioate.*
$C_{10}H_{23}OPS$ M 222.325
Liq. Bp_8 131-133°. n_D^{20} 1.4630.

Fluoride: [38169-28-7].
$C_6H_{14}FPS$ M 168.209
Liq.

Chloride: [2524-18-7].
$C_6H_{14}ClPS$ M 184.663
d^{20} 1.09. $Bp_{11.5}$ 120-121°, $Bp_{2.5}$ 73.5-74°. n_D^{20} 1.5182.

Bromide: [55287-83-7].
$C_6H_{14}BrPS$ M 229.114
Liq.

Iodide: [81373-56-0].
$C_6H_{14}IPS$ M 276.115
Liq. $Bp_{0.08}$ 90-92°.

Trimethylsilylamide: [55287-82-6]. P,P-*Dipropyl*-N-*trimethylsilylphosphinothioic amide.*
$C_{29}H_{24}NPSSi$ M 477.635
Liq. $Bp_{0.05}$ 134-135°.

Anhydride: [67003-66-1].
$C_{12}H_{28}OP_2S_2$ M 314.420
Solid. Mp 40°.

Mastryukova, T.A. *et al*, *Zh. Obshch. Khim.*, 1959, **29**, 1450; *CA*, **54**, 9729 (*synth*)
Kabachnik, M.I. *et al*, *Tetrahedron*, 1960, **9**, 10 (*props*)
Mastryukova, T.A. *et al*, *Zh. Obshch. Khim.*, 1961, **31**, 507 (*Engl. transl.* p. 464) (*chloride*)
Petrov, K.A. *et al*, *Zh. Obshch. Khim.*, 1961, **31**, 2889 (*Engl. transl.* p. 2692) (*propyl ester*)
Kuchen, W. *et al*, *Justus Liebigs Ann. Chem.*, 1962, **652**, 28 (*chloride, ester*)
Petrov, K.A. *et al*, *Zh. Obshch. Khim.*, 1962, **32**, 3720 (*Engl. transl.* p. 3647) (*butyl ester*)
Knunyants, L.L. *et al*, *Dokl. Akad. Nauk SSSR, Ser. Sci. Khim.*, 1971, **201**, 862 (*Engl. transl.* p. 992) (*fluoride, F and P nmr*)

Kuramshin, I.Ya. *et al*, *Zh. Obshch. Khim.*, 1973, **43**, 1456 (*Engl. transl.* p. 1446) (*ester*)
Kuchen, W. *et al*, *Z. Anorg. Allg. Chem.*, 1964, **333**, 71 (*synth*)
Steinberger, H. *et al*, *Z. Naturforsch., B*, 1974, **29**, 611 (*silylamide, ir, ms, pmr*)
Turpin, R. *et al*, *Bull. Soc. Chim. Fr. Part I*, 1977, 999 (*anhydride, P nmr*)
Feshchenko, N.G. *et al*, *Zh. Obshch. Khim.*, 1982, **52**, 222 (*Engl. transl.* p. 202) (*iodide, P nmr*)

Dipropylphosphinous acid, 9CI D-01209

[66193-26-8]

$(H_3CCH_2CH_2)_2POH$

$C_6H_{15}OP$ M 134.158

Free acid exists as phosphoryl tautomer. See under Dipropylphosphine, D-01204 .

Butyl ester: Butyl dipropylphosphinite.
$C_{10}H_{23}OP$ M 190.265
Liq. Bp_1 47-48°. n_D^{20} 1.4470.

Chloride: [41157-34-0]. *Chlorodipropylphosphine.*
$C_6H_{14}ClP$ M 152.603
Liq. Bp_{15} 99-101°.

Bromide: Bromodipropylphosphine.
$C_6H_{14}BrP$ M 197.054
Liq. Bp_{15} 143-147°.

Iodide: [81373-58-2]. *Iododipropylphosphine.*
$C_6H_{14}IP$ M 244.055
Solid. Mp 73-76°. Bp_{12} 95°.

Cyanide: Cyanodipropylphosphine.
$C_7H_{14}NP$ M 143.168
Solid. Mp 81-83° dec.

Kabachnik, M.I. *et al*, *Dokl. Akad. Nauk SSSR, Ser. Sci. Khim.*, 1960, **135**, 323 (*Engl. transl.* p. 1267) (*ester*)
Kabachnik, M.I. *et al*, *Izv. Akad. Nauk SSSR, Ser. Khim.*, 1963, 1227 (*Engl. transl.* p. 1120) (*synth*)
Appel, R. *et al*, *Chem. Ber.*, 1975, **108**, 1783 (*chloride, synth, P nmr*)
Feshchenko, N.G. *et al*, *Zh. Obshch. Khim.*, 1982, **52**, 222 (*Engl. transl.* p. 202) (*iodide*)
Wolfsberger, W. *et al*, *J. Organomet. Chem.*, 1986, **317**, 167 (*chloride, cmr, P nmr*)

Dipropyl phosphonate, 9CI D-01210

Dipropyl phosphite

[1809-21-8]

$(H_3CCH_2CH_2O)_2P(O)H \rightleftharpoons (H_3CCH_2CH_2O)_2POH$

$C_6H_{15}O_3P$ M 166.156

Tautomeric; almost completely in phosphonate form. Liq. d_4^{20} 1.02. Bp_{17} 101-102°, Bp_2 43-45°. n_D^{20} 1.4175.

Houalla, D. *et al*, *Bull. Soc. Chim. Fr.*, 1960, 129 (*ir*)
Harless, H.R., *Anal. Chem.*, 1961, **33**, 1387 (*ms*)
Luz, Z. *et al*, *J. Am. Chem. Soc.*, 1961, **83**, 4513 (*pmr*)
Coulson, E.J. *et al*, *J. Chem. Soc.*, 1965, 2364 (*synth*)
Hudson, H.R. *et al*, *J. Chem. Soc., Perkin Trans. 2*, 1974, 1575 (*P nmr*)

O,O-Dipropyl phosphonoselenoate, 9CI D-01211

O,O-*Dipropyl phosphoroselenoite*

$(H_3CCH_2CH_2O)_2P(Se)H \rightleftharpoons (H_3CCH_2CH_2O)_2PSeH$

$C_6H_{15}O_2PSe$ M 229.117

Tautomeric. Liq. d_4^{25} 1.26. $Bp_{0.5}$ 54-55°. n_D^{25} 1.4870.

Kuznetsov, E.V. *et al*, *CA*, 1958, **52**, 8938.

O,O-Dipropyl phosphonothioate, 9CI D-01212

O,O-*Dipropyl thiophosphite*

[14609-96-2]

$(H_3CCH_2CH_2O)_2P(S)H \rightleftharpoons (H_3CCH_2CH_2O)_2PSH$

$C_6H_{15}O_2PS$ M 182.217

Tautomeric, but exists almost completely in the thiophoshoryl form. Liq. with strong, sickly odour. Bp_1 54-55°. n_D^{20} 1.4585.

Nifant'ev, E.É., *et al*, *Zh. Obshch. Khim.*, 1983, **53**, 2695 (*Engl. transl.* p. 2429) (*synth, P nmr*)

Dipropyl phosphoramidate, 9CI, 8CI D-01213

Dipropyl phosphoryl amide. Dipropyl amidophosphate

[17123-09-0]

$(H_3CCH_2CH_2O)_2P(O)NH_2$

$C_6H_{16}NO_3P$ M 181.171

Cryst. (C_6H_6). Mp 39-42°. $Bp_{0.6}$ 115-116°.

Cadogan, J.I.G. *et al*, *J. Chem. Soc. (C)*, 1967, 1356 (*synth*)
Jacobsen, P. *et al*, *Org. Mass Spectrom.*, 1972, **6**, 1303 (*ms*)

Dipropylphosphoramidic acid, 9CI D-01214

$(H_3CCH_2CH_2)_2NP(O)(OH)_2$

$C_6H_{16}NO_3P$ M 181.171

Di-Me ester: Dimethyl dipropylphosphoramidate.
$C_8H_{20}NO_3P$ M 209.225
Liq. Bp_{15} 112°. n_D^{20} 1.4300.

Di-Et ester: [91043-34-4]. *Diethyl dipropylphosphoramidate.*
$C_{10}H_{24}NO_3P$ M 237.278
Liq. $Bp_{0.5}$ 73-75°. n_D^{20} 1.4280.

Dipropyl ester: [17123-10-3]. *Dipropyl dipropylphosphoramidate.*
$C_{12}H_{28}NO_3P$ M 265.332
Liq. $Bp_{0.07}$ 78-79°. n_D^{20} 1.4314 (1.4417).

Di-Ph ester: Diphenyl dipropylphosphoramidate.
$C_{18}H_{24}NO_3P$ M 333.366
Cryst. (EtOH). Mp 72-74°.

Difluoride:
$C_6H_{14}F_2NOP$ M 185.153
Liq. d_4^{20} 1.09. Bp_{27} 93°. n_D^{20} 1.3919.

Dichloride: [40881-98-9].
$C_6H_{14}Cl_2NOP$ M 218.062
Liq. d_4^{20} 1.18. Bp_{14} 127°. n_D^{20} 1.4653.

Audrieth, L.F. *et al*, *J. Prakt. Chem.*, 1959, **8**, 117 (*diphenyl ester*)
Cheymol, J. *et al*, *C.R. Hebd. Seances Acad. Sci.*, 1959, **249**, 1240 (*dimethyl ester*)
Cadogan, J.I.G. *et al*, *J. Chem. Soc.*, 1967, 1356 (*esters*)
Coustures, Y. *et al*, *Bull. Soc. Chim. Fr.*, 1973, 926 (*dihalides, ir, P nmr*)

Dipropylphosphoramidous acid, 9CI D-01215

$(H_3CCH_2CH_2)_2NP(OH)_2$

$C_6H_{16}NO_2P$ M 165.172

Di-Ph ester: [25781-00-4]. *Diphenyl dipropylphosphoramidite.*
$C_{18}H_{24}NO_2P$ M 317.367
Liq. d^{20} 1.06. $Bp_{0.11}$ 155-156°. n_D^{20} 1.5491.

Dichloride: [32597-23-2].
$C_6H_{14}Cl_2NP$ M 202.063

Liq. Bp 220-5°, Bp_{11} 95°.
Diiodide: [64673-58-1].
$C_6H_{14}I_2NP$ M 384.966
Dark-red liq. $Bp_{0.1}$ 100-102°. Stable at r.t. in absence of moisture, O_2, and light.

Michaelis, A., *Justus Liebigs Ann. Chem.*, 1903, **326**, 129 (*dichloride*)
Koketsu, J. *et al*, *Kogyo Kagaku Zasshi*, 1969, **72**, 2503; *CA*, **72**, 79165 (*diphenyl ester*)
Gorbatenko, Zh.K. *et al*, *Zh. Obshch. Khim.*, 1977, **47**, 1915 (*Engl. transl.* p. 1752) (*diiodide*)

O,O-Dipropyl phosphorazidothioate, 9CI, 8CI — D-01216

O,O-*Dipropyl thiophosphoryl azide.* O,O-*Dipropyl azidothiophosphate*
[51576-56-8]

$(H_3CCH_2CH_2O)_2P(S)N_3$

$C_6H_{14}N_3O_2PS$ M 223.229
Liq. Bp_1 77-78°.

Ger. Pat., 880 443, (*1953*); *CA*, **48**, 12167

Dipropyl phosphorochloridate, 9CI — D-01217

Dipropyl phosphoryl chloride. Dipropyl chlorophosphate
[2510-89-6]

$(H_3CCH_2CH_2O)_2P(O)Cl$

$C_6H_{14}ClO_3P$ M 200.602
Liq. Bp_{12} 106-107°, $Bp_{0.3}$ 74°. n_D^{20} 1.4236.

Ramaswami, D. *et al*, *J. Am. Chem. Soc.*, 1953, **75**, 1763 (*synth*)
de Roos, A.M. *et al*, *Recl. Trav. Chim. Pays-Bas*, 1958, **77**, 946 (*synth*)
Sosnovsky, G. *et al*, *J. Org. Chem.*, 1969, **34**, 968 (*synth*)

Dipropyl phosphorochloridite, 9CI, 8CI — D-01218

Dipropyl chlorophosphite
[20003-39-8]

$(H_3CCH_2CH_2O)_2PCl$

$C_6H_{14}ClO_2P$ M 184.602
Pungent liq. d_4^{20} 1.06. Bp_{10} 69-70°. n_D^{20} 1.4419. Easily hydrolysed.

Michalski, J. *et al*, *J. Chem. Soc.*, 1961, 4904 (*synth*)
Gazizov, T.Kh. *et al*, *Zh. Obshch. Khim.*, 1983, **53**, 1954 (*Engl. transl.* p. 1762) (*synth*)

S,S-Dipropyl phosphorochloridodithioate, 9CI, 8CI — D-01219

S,S-*Dipropyl chlorodithiophosphate*
[28522-98-7]

$(H_3CCH_2CH_2S)_2P(O)Cl$

$C_6H_{14}ClOPS_2$ M 232.723
Liq. d_4^{20} 1.18. $Bp_{0.7}$ 108-110°. n_D^{20} 1.5352.

Sorokina, S.F. *et al*, *Zh. Obshch. Khim.*, 1973, **43**, 750 (*Engl. transl.* p. 748) (*synth*)

Dipropyl phosphorochloridodithioite, 9CI — D-01220

Dipropyl dithiochlorophosphite
[4104-04-5]

$(H_3CCH_2CH_2S)_2PCl$

$C_6H_{14}ClPS_2$ M 216.723
Liq. $Bp_{0.02}$ 75-76°. n_D^{20} 1.5642.

Sinyashin, O.G. *et al*, *Zh. Obshch. Khim.*, 1983, **53**, 472, 502 (*Engl. transl.* pp. 415, 436) (*synth, P nmr*)

O,O-Dipropyl phosphorochloridothioate, 9CI, 8CI — D-01221

O,O-*Dipropyl thiophosphoryl chloride.* O,O-*Dipropyl chlorothiophosphate. Dipropoxyphosphinothioyl chloride*
[2524-05-2]

$(H_3CCH_2CH_2O)_2P(S)Cl$

$C_6H_{14}ClO_2PS$ M 216.662
Liq. $Bp_{1.5}$ 68-69°. n_D^{20} 1.4661.

Fletcher, J.H. *et al*, *J. Am. Chem. Soc.*, 1950, **72**, 2461 (*synth*)
McIvor, R.A. *et al*, *Can. J. Chem.*, 1956, **34**, 1611 (*ir*)
Popov, E.M. *et al*, *Zh. Obshch. Khim.*, 1959, **29**, 1998 (*Engl. transl.* p. 1967) (*raman*)
Meinhardt, N.A. *et al*, *J. Org. Chem.*, 1960, **25**, 1991 (*synth*)
Omelanczuk, J. *et al*, *Tetrahedron*, 1975, **31**, 2809 (*synth, P nmr*)

Dipropyl phosphorochloridotrithioate, 9CI, 8CI — D-01222

Dipropyl chlorotrithiophosphate
[55916-42-2]

$(H_3CCH_2CH_2S)_2P(S)Cl$

$C_6H_{14}ClPS_3$ M 248.783
Liq. d_4^{20} 1.18. Bp_1 118-120°. n_D^{20} 1.5690.

Murav'ev, I.V. *et al*, *Zh. Obshch. Khim.*, 1975, **45**, 1746 (*Engl. transl.* p. 1711)

N,N'-Dipropylphosphorodiamidic acid, 9CI, 8CI — D-01223

$(H_3CCH_2CH_2NH)_2P(O)OH$

$C_6H_{17}N_2O_2P$ M 180.186
Plates (Me_2CO). Mp 135-135.5°.
Et ester: Ethyl N,N'-*dipropylphosphorodiamidate.*
$C_8H_{21}N_2O_2P$ M 208.240
Needles (Et_2O). Mp 108°.
Ph ester: [57193-50-7]. *Phenyl* N,N'-*dipropylphosphorodiamidate.*
$C_{12}H_{21}N_2O_2P$ M 256.284
Solid. Mp 66-67°.
Fluoride:
$C_6H_{16}FN_2OP$ M 182.177
No phys. props. reported.
Chloride: [1754-60-5].
$C_6H_{16}ClN_2OP$ M 198.632
Cryst. (C_6H_6/pet. ether or CH_2Cl_2/Et_2O). Mp 88°, Mp 104-105°.

Michaelis, A., *Justus Liebigs Ann. Chem.*, 1915, **407**, 290 (*ethyl ester*)
Traylor, P.S. *et al*, *J. Am. Chem. Soc.*, 1965, **87**, 553 (*synth, chloride*)
Mager, P.P., *Toxicol. Lett.*, 1982, **11**, 67 (*fluoride*)
Roth, H.J. *et al*, *Arch. Pharm.* (*Weinheim, Ger.*), 1981, **314**, 85; 1982, **315**, 581 (*esters, synth, pmr*)

Sax, N.I., *Dangerous Properties of Industrial Materials*, 6th Ed., Van Nostrand-Reinhold, 1984, 420.

O,O-Dipropyl phosphorodiselenoate, 9CI, 8CI D-01224

O,O-*Dipropyl hydrogen phosphorodiselenoate.* O,O-*Dipropyl phosphorodiselenoic acid.* O,O-*Dipropyl diselenophosphate*

[62920-97-2]

$(H_3CCH_2CH_2O)_2P(Se)SeH$

$C_6H_{15}O_2PSe_2$ M 308.077

Unstable, isol. as K salt.

K salt: [19483-46-6]. Cryst. (EtOH/pet. ether). Mp 94-95°.

Kudchadker, M.V. *et al, Can. J. Chem.*, 1968, **46**, 1415 (*synth, pmr, ir*)
Zemlyanski, N.I. *et al, Zh. Obshch. Khim.*, 1971, **41**, 1691 (*Engl. transl.* p. 1699) (*synth*)
Gorak, R.D. *et al, Zh. Obshch. Khim.*, 1972, **42**, 56 (*Engl. transl.* p. 42) (*synth*)

O,O-Dipropyl phosphorodithioate, 9CI, 8CI D-01225

Dipropyl phosphorodithioic acid. O,O-*Dipropyl dithiophosphate.* O,O-*Dipropyl hydrogen dithiophosphate*

[2253-43-2]

$(H_3CCH_2CH_2O)_2P(S)SH$

$C_6H_{15}O_2PS_2$ M 214.277

Liq. d_4^{20} 1.10. Bp_2 80-82°. pK_a 1.75 (7% EtOH aq.), pK_a 2.57 (80% EtOH aq.). n_D^{20} 1.4990.

K salt: [3287-84-1]. Solid. Mp 165°.

Anilinium salt: [67333-92-0]. Solid. Mp 80-81°.

McIvor, R.A. *et al, Can. J. Chem.*, 1958, **36**, 820 (*ir*)
Kabachnik, M.I. *et al, Tetrahedron*, 1960, **9**, 10 (*synth, props*)
Zemlyanskii, N.I. *et al, Zh. Obshch. Khim.*, 1972, **42**, 54 (*Engl. transl.* p. 50)
Zimin, M.G. *et al, Zh. Obshch. Khim.*, 1978, **48**, 1020 (*Engl. transl.* p. 930) (*P nmr*)
Lefferts, J.L. *et al, Inorg. Chem.*, 1980, **19**, 1662 (*synth, complexes*)

O,S-Dipropyl phosphorodithioate, 9CI, 8CI D-01226

O,S-*Dipropyl hydrogen dithiophosphate.* O,S-*Dipropyl dithiophosphoric acid*

$(H_3CCH_2CH_2O)(H_3CCH_2CH_2S)P(=O)SH \rightleftharpoons (H_3CCH_2CH_2O)(H_3CCH_2CH_2S)P(=S)OH$

$C_6H_{15}O_2PS_2$ M 214.277

Some derivs. may exist in chiral modifications.

NH_4 salt: [35329-37-4]. Solid. Mp 142-143°.

Itskova, A.L. *et al, Zh. Obshch. Khim.*, 1971, **41**, 2618 (*Engl. transl.* p. 2651)

S,S-Dipropyl phosphorodithioate, 9CI, 8CI D-01227

S,S-*Dipropyl hydrogen dithiophosphate.* S,S-*Dipropyl dithiophosphoric acid*

$(H_3CCH_2CH_2S)_2P(O)OH$

$C_6H_{15}O_2PS_2$ M 214.277

NH_4 salt: [35329-38-5]. Solid. Mp 197-198°.

Chloride: see S,S-Dipropyl phosphorochloridodithioate, D-01219

Amide: S,S-*Dipropyl phosphoramidodithioate.* S,S-*Dipropyl amidodithiophosphate.*
$C_6H_{16}NOPS_2$ M 213.292
Possesses insecticidal and acaricidal props. Semicryst. Dec. on attempted dist.

Ger. Pat., 2 013 956, (*1971*); *CA*, **74**, 87380 (*amide*)
Itskova, A.L. *et al, Zh. Obshch. Khim.*, 1974, **41**, 2618 (*Engl. transl.* p. 2651) (*synth*)

Dipropyl phosphorofluoridate, 9CI, 8CI D-01228

Dipropyl phosphoryl fluoride. Dipropyl fluorophosphate. Dipropoxyphosphinyl fluoride

[381-45-3]

$(H_3CCH_2CH_2O)_2P(O)F$

$C_6H_{14}FO_3P$ M 184.147

Liq. Bp_{20} 98-100°.

▷Toxic, but less so than 35850-7

Chapman, N.B. *et al, J. Chem. Soc.*, 1948, 1010 (*synth, tox*)
Sheluchenko, V.V. *et al, Dokl. Akad. Nauk SSSR, Ser. Sci. Khim.*, 1967, **177**, 376 (*Engl. transl.* p. 1050) (*F and P nmr*)
Landau, M.A. *et al, Zh. Strukt. Khim.*, 1970, **11**, 513 (*Engl. transl.* p. 467) (*struct*)
Ryzhikov, B.D. *et al, Zh. Strukt. Khim.*, 1975, **16**, 754 (*Engl. transl.* p. 700) (*F nmr*)
Gubaidullin, M.G., *Zh. Obshch. Khim.*, 1982, **52**, 2469 (*Engl. transl.* p. 2182) (*props*)

O,O-Dipropyl phosphoroselenoate, 9CI, 8CI D-01229

O,O-*Dipropyl phosphoroselenoic acid.* O,O-*Dipropyl hydrogen phosphoroselenoate*

$(H_3CCH_2CH_2O)_2P(Se)OH \rightleftharpoons (H_3CCH_2CH_2O)_2P(O)SeH$

$C_6H_{15}O_3PSe$ M 245.116

Unstable oil. d_4^{25} 1.32. n_D^{25} 1.4855.

Na salt: Cryst. (pet. ether). Sol. Et_2O.

K salt: [59745-41-4]. Cryst. ($CHCl_3/Et_2O$).

Chloride: [55578-32-0]. O,O-*Dipropyl phosphorochloridoselenoate.*
$C_6H_{14}ClO_2PSe$ M 263.562
Liq. d_4^{20} 1.33. $Bp_{0.05}$ 61.5°. n_D^{20} 1.4902. Unstable to moisture.

Foss, O., *Acta Chem. Scand.*, 1947, **1**, 8 (*synth*)
Nuretdinov, I.A. *et al, Zh. Obshch. Khim.*, 1975, **45**, 533 (*Engl. transl.* p. 526) (*chloride, synth, ir, pmr, P nmr*)
Markowska, A. *et al, Rocz. Chem.*, 1960, **34**, 1675 (*synth*)

O,O-Dipropyl phosphorothioate, 9CI, 8CI D-01230

O,O-*Dipropyl hydrogen phosphorothioate.* O,O-*Dipropyl thiophosphoric acid.* O,O-*Dipropyl hydrogen thiophosphate*

[4486-43-5]

$(H_3CCH_2CH_2O)_2P(O)SH \rightleftharpoons (H_3CCH_2CH_2O)_2P(S)OH$

$C_6H_{15}O_3PS$ M 198.216

50% thiol form (7% EtOH aq.), 24% thiol form (80% EtOH aq.) at r.t. Conveniently isol. and stored as Na or K salt. Liq. d_4^{20} 1.10. $Bp_{0.09}$ 108.5-109.5°. n_D^{20} 1.4678.

Na salt: Solid. Sol. Et_2O.
K salt: Cryst. (propanol/Et_2O).
Chloride: see O,O-Dipropyl phosphorochloridothioate, D-01221
Anilide: O,O-*Dipropyl phenylphosphoramidothioate.*
$C_{12}H_{20}NO_2PS$ M 273.329
Liq. d_4^{20} 1.11. Bp_2 148-149°. n_D^{20} 1.5388.
Anhydride: see Aspon, A-00144

Foss, O., *Acta Chem. Scand.*, 1947, **1**, 8 (*synth*)
Kabachnik, M.I. *et al*, *Dokl. Akad. Nauk SSSR, Ser. Sci. Khim.*, 1954, **96**, 991; *CA*, **49**, 8842 (*anilide*)
Kabachnik, M.I. *et al*, *Tetrahedron*, 1960, **9**, 10 (*synth, struct*)
Zemlyanskii, N.I. *et al*, *Zh. Obshch. Khim.*, 1960, **30**, 4056 (*Engl. transl.* p. 4018) (*raman*)

Dipropynylphosphinic acid, 9CI — D-01231

$(H_3CC{\equiv}C)_2P(O)OH$

$C_6H_7O_2P$ M 142.094
Dimethylamide: N,N-*Dimethyl*-P,P-*dipropynylphosphinic amide.*
$C_8H_{12}NOP$ M 169.163
Liq. $Bp_{0.3}$ 140°.

Charrier, C. *et al*, *C.R. Hebd. Seances Acad. Sci., Ser. C*, 1967, **264**, 995.

O,O-Di-2-propynyl phosphorodithioate, 9CI — D-01232

O,O-*Di-2-propynyl hydrogen dithiophosphate.* O,O-*Di-2-propynyl dithiophosphoric acid*
[26819-90-9]

$(HC{\equiv}CCH_2O)_2P(S)SH$

$C_6H_7O_2PS_2$ M 206.214
Unstable liq. d_4^{20} 1.27. $Bp_{0.001}$ 76°. n_D^{20} 1.5618.
▷Reaction mixtures may explode if overheated. Ag salt explodes when heated
K salt: Solid. Mp 138°.
Pb complex: Solid. Mp 80°.

Zemlyanskii, N.I. *et al*, *Zh. Obshch. Khim.*, 1969, **39**, 2461; 1972, **42**, 54 (*Engl. transl.* pp. 2401, 50) (*synth*)
Mel'nik, Ya.I. *et al*, *Zh. Obshch. Khim.*, 1970, **40**, 791 (*Engl. transl.* p. 768) (*synth*)

P,P-Di-1-pyrrolidinylphosphinic acid — D-01233

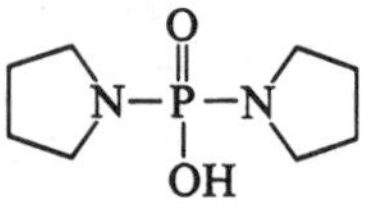

$C_8H_{17}N_2O_2P$ M 204.208
Ph ester: Phenyl bis(1-pyrrolidinyl)phosphinate.
$C_{14}H_{21}N_2O_2P$ M 280.306
Liq. $Bp_{0.15}$ 160-170°. n_D^{20} 1.5362.
Chloride:
$C_8H_{16}ClN_2OP$ M 222.654
Mp 44-48°. $Bp_{0.9}$ 143-145°.
Dimethylamide: [53439-65-9]. N,N-*Dimethyl*-P,P-*di-1-pyrrolidinylphosphinic amide.*
$C_{10}H_{21}N_3OP$ M 230.269
Liq. d_4^{25} 1.09. $Bp_{0.8}$ 126°. n_D^{25} 1.4961.
Diethylamide: [69981-37-9]. N,N-*Diethyl*-P,P-*di-1-pyrrolidinylphosphinic amide.*
$C_{12}H_{25}N_3OP$ M 258.323
Liq. d_4^{25} 1.06. $Bp_{0.6}$ 137°. n_D^{25} 1.4914.
Pyrrolidide: see Tri-1-pyrrolidylphosphine, T-00724

Yvernault, T. *et al*, *C.R. Hebd. Seances Acad. Sci., Ser. C*, 1978, **287**, 519 (*amides*)
Wilson, S.R. *et al*, *Synth. Commun.*, 1982, **12**, 657 (*chloride*)
Bollinger, J.-C. *et al*, *Can. J. Chem.*, 1983, **61**, 328 (*amides*)

2,2′-Diselenobis[5,5-dimethyl-1,3,2-dioxaphosphorinane], 9CI — D-01234

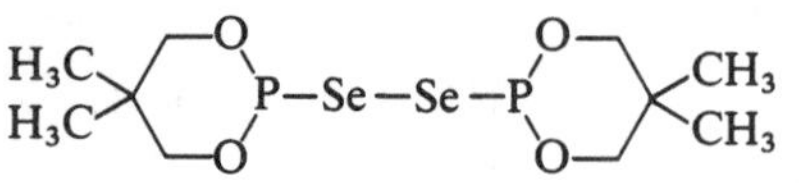

$C_{10}H_{20}O_4P_2Se_2$ M 424.133
2,2′-Disulfide: [26905-08-8]. *2,2′-Diselenobis[5,5-dimethyl-2-thioxo-1,3,2-dioxaphosphorinane].*
$C_{10}H_{20}O_4P_2S_2Se_2$ M 488.253
Needles (EtOH).

Katritzky, A.R. *et al*, *J. Chem. Soc. (B)*, 1970, 140 (*synth, ir, pmr, ms*)
Bruzik, K. *et al*, *Pol. J. Chem.*, 1980, **54**, 141 (*synth, ms, P nmr*)

Diselenoxo(2,4,6-tri-*tert*-butylphenyl)phosphorane — D-01235

*Selenoxo[2,4,6-tris(1,1-dimethylethyl)phenyl]phosphine selenide, 11CI. Diselenoxo[2,4,6-tris(1,1-dimethylethyl)-phenyl]phosphorane. 2,4,6-Tri-*tert*-butylphenylmetadiselenophosphonate*
[90599-66-9]

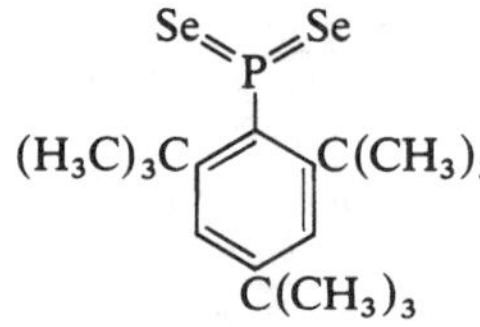

$C_{18}H_{29}PSe_2$ M 434.321
Green cryst. Mp 45-47° dec. Stable at r.t. Sensitive to light.

Yoshifuji, M. *et al*, *Chem. Lett.*, 1984, 603 (*synth, ms, uv, pmr, P nmr*)

Disulfoton, BSI — D-01236

O,O-*Diethyl* S-(*2-ethylthioethyl*) *phosphorodithioate, 8CI.* O,O-*Diethyl* S-*ethylmercaptoethyl dithiophosphate. Disyston. Dithiodemeton. Dithiosystox*
[298-04-4]

$(EtO)_2P(S)SCH_2CH_2SEt$

$C_8H_{19}O_2PS_3$ M 274.391
Insecticide. Oil. Insol. H_2O. d_4^{20} 1.144. $Bp_{0.1}$ 105-106°, $Bp_{0.01}$ 87°. n_D^{20} 1.5330.
▷Highly toxic, TLV 0.1. TD9275000.

U.S.P., 3 041 367, (*1962*); *CA*, **57**, 16399 (*synth*)
U.S.P., 3 082 240, (*1963*); *CA*, **59**, 5077 (*synth*)
Keith, L.H. *et al*, *J. Assoc. Off. Anal. Chem.*, 1968, **51**, 1063 (*pmr*)
Getz, M.E. *et al*, *J. Assoc. Off. Anal. Chem.*, 1968, **51**, 1101 (*tlc*)
Ross, R.T. *et al*, *Anal. Chim. Acta*, 1970, **52**, 139 (*P nmr*)
Stan, H.-J. *et al*, *Fresenius' Z. Anal. Chem.*, 1977, **287**, 271; *Biomed. Mass Spectrom.*, 1982, **9**, 483 (*glc, ms*)
Stan, H.-J. *et al*, *J. Chromatogr.*, 1983, **279**, 173 (*glc*)
Pesticide Manual, 6th Ed., 221; 7th Ed., 224.
Agrochemicals Handbook, Royal Society of Chemistry, London, 1983, A167.

Sax, N.I., *Dangerous Properties of Industrial Materials*, 6th Ed., Van Nostrand-Reinhold, 1984, 517.

Ditalimfos **D-01237**

O,O-*Diethyl (1,3-dihydro-1,3-dioxo-2*H*-isoindol-2-yl)-phosphonothioate, 9CI.* O,O-*Diethyl phthalimidophosphonothioate, 8CI*

[5131-24-8]

$C_{12}H_{14}NO_4PS$ M 299.281

Insecticide: now superseded. Also a fungicide in the control of powdery mildew. Solid. Mp 81.5-83.5°.

▷Mild skin irritant. TB2050000.

Ger. Pat., 2 256 253, (*1973*); *CA*, **79**, 42140 (*synth*)
U.S.P., 4 204 996, (*1980*); *CA*, **93**, 220480 (*synth*)
Morel, J.L., *CA*, 1976, **84**, 145815 (*tox*)
Agrochemicals Handbook, Royal Society of Chemistry, 1st Ed., 168.

Ditetradecylphosphinic acid, 9CI **D-01238**

$[H_3C(CH_2)_{13}]_2P(O)OH$

$C_{28}H_{59}O_2P$ M 458.747

Cryst. (C_6H_6 or C_6H_6/EtOH). Mp 97-98°.

Williams, R.H. *et al*, *J. Am. Chem. Soc.*, 1955, **77**, 3411.

Ditetradecyl phosphonate, 9CI **D-01239**

Ditetradecyl phosphite

[21482-16-6]

$[H_3C(CH_2)_{13}O]_2P(O)H \rightleftharpoons [H_3C(CH_2)_{13}O]_2POH$

$C_{28}H_{59}O_3P$ M 474.746

Tautomeric; almost completely in phosphonate form. Cryst. (pet. ether). Mp 42-44°. $Bp_{0.4}$ 230°.

U.S.P., 3 725 515, (*1970*); *CA*, **78**, 147344 (*synth*)
B.P., 1 298 156, (*1970*); *CA*, **78**, 83841 (*synth*)

1,3-Dithiane-2-phosphonic acid **D-01240**

$C_4H_9O_3PS_2$ M 200.207

Di-Et ester: Diethyl 1,3-dithian-2-ylphosphonate. 2-(Diethoxyphosphinyl)-1,3-dithian.

$C_8H_{17}O_3PS_2$ M 256.314

Wittig-Horner reagent for the synth. of ketene dithioacetals. Liq. n_D^{20} 1.5235.

Mlotkowska, B. *et al*, *J. Prakt. Chem.*, 1977, **319**, 17 (*synth, pmr, P nmr*)
Comins, D.L. *et al*, *Synthesis*, 1978, 309 (*synth*)
Mikolajczyk, M. *et al*, *Tetrahedron*, 1978, **34**, 3081 (*use*)
Lee, T.-J. *et al*, *J. Org. Chem.*, 1982, **47**, 4750 (*use*)

(1,3-Dithian-2-ylmethyl)-triphenylphosphonium(1+), 9CI **D-01241**

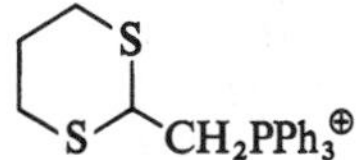

$C_{23}H_{24}PS_2^{\oplus}$ M 395.536 (ion)

Bromide: [69141-63-5].

$C_{23}H_{24}BrPS$ M 443.380

Source of ylide. Cryst. ($CHCl_3$/EtOAc). Mp 238°.

Iodide: [69141-64-6].

$C_{23}H_{24}IPS$ M 490.381

Source of ylide. Cryst. ($CHCl_3$/EtOAc). Mp 222°.

Ylide: (1,3-Dithian-2-ylmethylene)-triphenylphosphorane.

$C_{23}H_{23}PS_2$ M 394.528

Wittig reagent for RCHO → RCH=CHCHO.

Cristau, H.-J. *et al*, *Synthesis*, 1978, 826 (*synth, use*)
Cristau, H.-J. *et al*, *Phosphorus Sulfur*, 1982, **14**, 63 (*synth, ir, pmr, nmr*)

1,3-Dithian-2-yltriphenylphosphonium(1+), 9CI **D-01242**

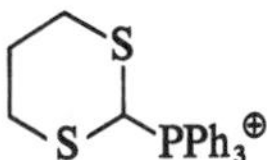

$C_{22}H_{22}PS_2^{\oplus}$ M 381.510 (ion)

Rgt. for prepn. of ketene dithioacetals. With aq. KOH → ylide.

Chloride: [63822-67-3].

$C_{22}H_{22}ClPS_2$ M 416.963

Solid. Mp 182-184°.

Ylide: [63822-68-4]. *1,3-Dithian-2-ylidenetriphenylphosphorane.*

$C_{22}H_{21}PS_2$ M 380.502

Reacts rapidly with aldehydes. Bright-yellow.

Kruse, C.G. *et al*, *Tetrahedron Lett.*, 1977, 885 (*synth, use, pmr*)

1,3,2-Dithiaphospholane, 9CI **D-01243**

[6669-37-0]

$C_2H_5PS_2$ M 124.155

2-Fluoro: [33672-92-3]. *2-Fluoro-1,3,2-dithiaphospholane. Ethylene phosphorofluoridodithioite.*

$C_2H_4FPS_2$ M 142.146

No phys. props. reported.

2-Chloro: see 2-Chloro-1,3,2-dithiaphospholane, C-00071

2-Methoxy: [35437-30-0]. *2-Methoxy-1,3,2-dithiaphospholane.* S,S-*Ethylene*-O-*methylphosphorodithioite.*

$C_3H_7OPS_2$ M 154.181

Liq. Bp_2 74-75°.

2-Methylthio: [35437-31-1]. *2-(Methylthio)-1,3,2-dithiaphospholane. Ethylene methyl phosphorotrithioite.*

$C_3H_7PS_3$ M 170.242

Liq. Bp_2 108-116°.

2-Ethoxy: see 2-Ethoxy-1,3,2-dithiaphospholane, E-00038

2-Me: see 2-Methyl-1,3,2-dithiophospholane, M-00139

2-Ph: see 2-Phenyl-1,3,2-dithiaphospholane, P-00121

2-Dimethylamino: [41855-87-2]. S,S-*Ethylene diethylphosphoramidodithioite.* N,N-*Dimethyl-1,3,2-dithiaphosphetane-2-amine.*
$C_4H_{10}NPS_2$ M 167.223
No phys. props. reported.
2-Diethylamino: N,N-*Diethyl* S,S-*ethylene phosphoramidodithioite.* N,N-*Diethyl-1,3,2-dithiaphospholane-2-amine.*
$C_6H_{14}NPS_2$ M 195.277
Liq. Bp_1 88-90°.
2-Oxide: see 1,3,2-Dithiaphospholane 2-oxide, D-01244
2-Sulfide: see 1,3,2-Dithiaphospholane 2-sulfide, D-01245

Peake, S.C. *et al*,, *J. Chem. Soc., Perkin Trans. 2*, 1972, 380 (*synth, nmr, pmr*)
Albrand, J.-P. *et al*, *Org. Magn. Reson.*, 1973, **5**, 33 (*pmr*)
Davidson, G. *et al*, *Spectrochim. Acta, Part A*, 1983, **39**, 419 (*ir, raman*)
Goubeau, D. *et al*, *J. Mol. Struct.*, 1983, **98**, 109 (*pe, cmr, struct*)

1,3,2-Dithiaphospholane 2-oxide D-01244

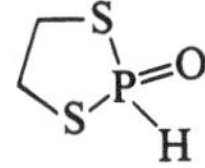

$C_2H_5OPS_2$ M 140.155
2-Methoxy: [35437-36-6]. *2-Methoxy-1,3,2-dithiaphospholane 2-oxide.* S,S-*Ethylene* O-*methyl phosphorodithioate.*
$C_3H_7O_2PS_2$ M 170.181
Involatile solid.
2-Me: see under 2-Methyl-1,3,2-dithiophospholane, M-00139
2-Isopropyl:
$C_5H_{11}OPS_2$ M 182.235
Liq. $Bp_{1.4}$ 120° (oven).
2-Ph: see under 2-Phenyl-1,3,2-dithiaphospholane, P-00121
2-Dimethylamino: [75373-22-7]. S,S-*Ethylene* N,N-*dimethyl phosphoramidodithioate.* N,N-*Dimethyl-1,3,2-dithiaphospholane-2-amine 2-oxide.*
$C_4H_{10}NOPS_2$ M 183.223

Ishmaeva, E.A. *et al*, *Izv. Akad. Nauk SSSR, Ser. Khim.*, 1971, 1317 (*Engl. transl.* p. 1220) (*derivs, struct*)
Peake, S.C. *et al*, *J. Chem. Soc., Perkin Trans. 2*, 1972, 380 (*derivs, synth, nmr, pmr*)
Kim, B.H. *et al*, *Phosphorus Sulfur*, 1982, **13**, 337 (*isopropyl, ir, ms, pmr*)

1,3,2-Dithiaphospholane 2-sulfide D-01245

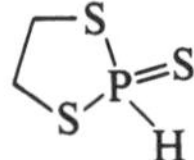

$C_2H_5PS_3$ M 156.215
2-Chloride: see 2-Chloro-1,3,2-dithiaphospholane, C-00071
2-Phenoxy: [42475-16-1]. *2-Phenoxy-1,3,2-dithiaphospholane 2-sulfide.* S,S-*Ethylene* O-*phenyl phosphorotrithioate.*
$C_8H_9OPS_3$ M 248.312
Cryst. (hexane). Mp 81°.
2-Me: see under 2-Methyl-1,3,2-dithiophospholane, M-00139
2-Isopropyl:
$C_5H_{11}PS_3$ M 198.296
Solid. Mp 44-45°.
2-Ph: see under 2-Phenyl-1,3,2-dithiaphospholane, P-00121
2-Diethylamino: S,S-*Ethylene diethylphosphoramidodithioite.* N,N-*Diethyl-1,3,2-dithiaphospholane-2-amine 2-sulfide.*
$C_6H_{14}NPS_3$ M 227.337
Solid. Mp 35°.

Ishmaeva, E.A. *et al*, *Izv. Akad. Nauk SSSR, Ser. Khim.*, 1971, 1317 (*Engl. transl.* p. 1220) (*derivs, struct*)
Peake, S.C. *et al*, *J. Chem. Soc., Perkin Trans. 2*, 1972, 380 (*derivs, synth, nmr, pmr*)
Revel, M. *et al*, *Bull. Soc. Chim. Fr., Part I*, 1973, 1195 (*phenoxy, synth, nmr*)
Kim, B.H. *et al*, *Phosphorus Sulfur*, 1982, **13**, 337 (*isopropyl, ms, pmr, ir*)

1,3,2-Dithiaphosphole, 10CI D-01246

[288-78-8]

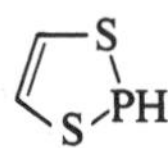

$C_2H_3PS_2$ M 122.139
2-Chloro: [68090-10-8].
$C_2H_2ClPS_2$ M 156.585
Solid. Mp 28.5-30°. $Bp_{0.8}$ 59-61°.
2-Methoxy: [68090-11-9].
$C_3H_5OPS_2$ M 152.166
Liq. $Bp_{0.4}$ 41-42°.

Chen, C.H. *et al*, *Synthesis*, 1978, 667 (*synth, pmr, P nmr, ms*)

1,3,2-Dithiaphosphorinane, 9CI, 8CI D-01247

[6680-67-7]

$C_3H_7PS_2$ M 138.182
Readily oxidisable solid. Mp 40°. Bp_1 50-53°.
2-Chloro: see 2-Choro-1,3,2-dithiaphosphorinane, C-00199
2-Methoxy: [55134-66-2]. *2-Methoxy-1,3,2-dithiaphosphorinane.* O-*Methyl* S,S-*trimethylene phosphorodithioite.*
$C_4H_9OPS_2$ M 168.208
Liq. d_4^{20} 1.30. $Bp_{0.25}$ 74-75°. n_D^{20} 1.6150.
2-Ethoxy: [55157-75-0]. *2-Ethoxy-1,3,2-dithiaphosphorinane.* O-*Ethyl* S,S-*trimethylene phosphorodithioite.*
$C_5H_{11}OPS_2$ M 182.235
Liq. d_4^{20} 1.23. $Bp_{0.25}$ 88-89°. n_D^{20} 1.5908.
2-Me: see 2-Methyl-1,3,2-dithiaphosphorinane, M-00137
2-Ph: [57115-66-9]. *2-Phenyl-1,3,2-dithiaphosphorinane. Trimethylene phenylphosphonodithioite.*
$C_9H_{11}PS_2$ M 214.280
Liq. $Bp_{0.5}$ 100°.

Nifant'ev, É.E. *et al*, *Zh. Obshch. Khim.*, 1974, **44**, 1694 (*Engl. transl.* p. 1664) (*alkoxy derivs, synth*)
Borisenko, A.A. *et al*, *Zh. Obshch. Khim.*, 1978, **48**, 1251 (*Engl. transl.* p. 1144) (*derivs, cmr, nmr, pmr*)
Arshinova, R.P. *et al*, *Zh. Obshch. Khim.*, 1980, **50**, 829 (*Engl. transl.* p. 665) (*phenyl, stereochem*)
Nifant'ev, É.E. *et al*, *Tetrahedron*, 1981, **37**, 3183 (*synth, cmr, nmr, pmr*)

1,3,2-Dithiaphosphorinane 2-oxide **D-01248**

1,3,2-Dithiaphosphinane 2-oxide. S,S′-*Trimethylene phosphonodithioate.* S,S′-*1,3-Propanediyl phosphonodithioate*

[37443-71-3]

$C_3H_7OPS_2$ M 154.181

Solid. Mp 121-122°.

2-Chloro: see 2-Choro-1,3,2-dithiaphosphorinane, C-00199

2-Me: see under 2-Methyl-1,3,2-dithiaphosphorinane, M-00137

2-Diethylamino: S,S′-*1,3-Propanediyl* N,N-*diethylphosphoramidodithioate.*

$C_7H_{16}NOPS_2$ M 225.303

Oil. d_4^{20} 1.23. n_D^{20} 1.5810.

Wieber, M. *et al, Monatsh. Chem.*, 1968, **99**, 1153 (*methyl, synth, pmr*)

Nifant'ev, É.E. *et al, Dokl. Akad. Nauk SSSR, Ser. Sci. Khim.*, 1972, **203**, 593 (*Engl. transl.* p. 262) (*synth, derivs*)

1,3,2-Dithiaphosphorinane 2-sulfide, 9CI **D-01249**

1,3,2-Dithiaphosphinane 2-sulfide. Trimethylene phosphonotrithioate. S,S′-*1,3-Propanediyl phosphonotrithioate*

$C_3H_7PS_3$ M 170.242

2-Chloro: see 2-Choro-1,3,2-dithiaphosphorinane, C-00199

2-Methoxy: [66821-75-8]. O-*Methyl* S,S-*trimethylene phosphonotrithioate.*

$C_4H_9OPS_3$ M 200.268

2-Me: see under 2-Methyl-1,3,2-dithiaphosphorinane, M-00137

2-Ph: [57115-77-2]. *Trimethylene phenylphosphonotrithioate.* S,S′-*1,3-Propanediyl phenylphosphonotrithioate.* No phys. props. reported.

Martin, J. *et al, Org. Magn. Reson.*, 1977, **9**, 637 (*methoxy, pmr, cmr, P nmr*)

Borisenko, A.A. *et al, Zh. Obshch. Khim.*, 1978, **48**, 1251 (*Engl. transl.* p. 1144) (*ethoxy, pmr, cmr, P nmr*)

Martin, J. *et al, Org. Magn. Reson.*, 1981, **15**, 87 (*phenyl, pmr, nmr, cmr*)

Di-2-thienylphosphinic acid, 9CI **D-01250**

Bis(2-thienyl)phosphinic acid, 9CI

[5849-47-8]

$C_8H_7O_2PS_2$ M 230.236

Cryst. (hexane/EtOH). Mp 193°.

Et ester: [65887-65-2]. *Ethyl bis(2-thienyl)phosphinate.*

$C_{10}H_{11}O_2PS_2$ M 258.289

Cryst. (hexane). Mp 61°.

Martin, K.R. *et al, J. Heterocycl. Chem.*, 1966, **3**, 92 (*synth, uv, pmr*)

Allen, D.W. *et al, J. Chem. Soc., Perkin Trans. 2*, 1977, 1705 (*synth, ester, P nmr*)

2,2′-Dithiobis[5,5-dimethyl-1,3,2-dioxaphosphorinane], 9CI **D-01251**

Bis(5,5-dimethyl-1,3,2-dioxaphosphorinan-2-yl) disulfide

$C_{10}H_{20}O_4P_2S_2$ M 330.333

2,2′-Dioxide: [15995-44-5]. *Bis(5,5-dimethyl-2-oxo-1,3,2-dioxaphosphorinan-2-yl) disulfide.*

$C_{10}H_{20}O_6P_2S_2$ M 362.332

Cryst (EtOAc). Mp 139-141° (126-129°).

2,2′-Disulfide: [4073-59-0]. *Bis(5,5-dimethyl-2-thioxo-1,3,2-dioxaphosphorinan-2-yl) disulfide.*

$C_{10}H_{20}O_4P_2S_4$ M 394.453

Cryst. (EtOH). Mp 133.5-134°.

2,2′-Diselenide: [26905-09-9]. *Bis(5,5-dimethyl-2-selenoxo-1,3,2-dioxaphosphorinan-2-yl) disulfide.*

$C_{10}H_{20}O_4P_2S_2Se_2$ M 488.253

Needles (EtOH). Mp 116-117°.

Edmundson, R.S., *Tetrahedron*, 1965, **21**, 2379 (*dioxide, disulfide, ir*)

Katritzky, A.R. *et al, J. Chem. Soc. (B)*, 1970, 140 (*diselenide, ir, ms, pmr*)

Drabowicz, J. *et al, Synthesis*, 1980, 32 (*dioxide, P nmr*)

(1,3-Dithiolan-2-ylmethyl)-triphenylphosphonium(1+), 9CI **D-01252**

$C_{22}H_{22}PS_2^{\oplus}$ M 381.510 (ion)

Bromide: [69141-62-4].

$C_{22}H_{22}BrPS$ M 429.354

Cryst. ($CHCl_3$/EtOAc). Mp 236°.

Ylide: (1,3-Dithiolan-2-ylmethylene)-triphenylphosphorane.

$C_{22}H_{21}PS_2$ M 380.502

Used in prepn. of $\alpha\beta$-unsatd. aldehydes.

Cristau, H.-J. *et al, Synthesis*, 1978, 826 (*synth, use*)

Cristau, H.-J. *et al, Phosphorus Sulfur*, 1982, **14**, 63 (*synth, ir, pmr, nmr*)

1,3-Dithiolan-2-yltriphenylphosphonium(1+), 9CI **D-01253**

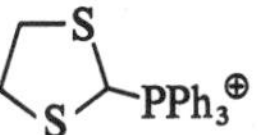

$C_{21}H_{20}PS_2^{\oplus}$ M 367.483 (ion)

With butyllithium, salts give the ylide.

Tetrafluoroborate: [77432-49-6].

$C_{21}H_{20}BF_4PS_2$ M 454.286

Cryst. (EtOH or $CHCl_3$). Mp 237°.

Ylide: 1,3-Dithiol-2-ylidenetriphenylphosphorane.

$C_{21}H_{19}PS_2$ M 366.475

Reacts with aldehydes to give ketone dithioacetals. Yellow.

Tanimoto, S. *et al, Synthesis*, 1981, 53 (*synth, use*)

Dithioxo(2,4,6-tri-*tert*-butylphenyl)-phosphorane **D-01254**

*Thioxo[2,4,6-tris(1,1-dimethylethyl)phenyl]phosphine sulfide, 11CI. 2,4,6-Tri-*tert*-butylphenylmetadithiophosphonate*

[88001-79-0]

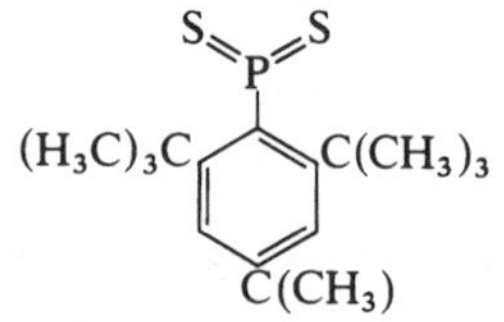

$C_{18}H_{29}PS_2$ M 340.521

Yellow amorph. powder, or yellow or orange cryst. (toluene/MeCN). Mp 92-93°, Mp 150-151°. Solid stable to oxygen and moisture. Dec. above Mp.

Appel, R. *et al*, *Angew. Chem., Int. Ed. Engl.*, 1983, **22**, 1004 (*synth, cryst struct*)

Navech, J. *et al*, *Phosphorus Sulfur*, 1984, **21**, 105; 1986, **26**, 83 (*synth, ir, uv, ms, pmr, P nmr, props, bibl*)

Yoshifuji, M. *et al*, *Chem. Lett.*, 1984, 317 (*synth, ir, uv, pmr, P nmr, props*)

Diundecyl phosphate, 9CI **D-01255**

Diundecyl hydrogen phosphate. Diundecyl phosphoric acid

$[H_3C(CH_2)_{10}O]_2P(O)OH$

$C_{22}H_{47}O_4P$ M 406.585

Cryst. (EtOH aq.). Mp 53.5-54°.

Mathieson, D.W. *et al*, *J. Pharm. Pharmacol.*, 1957, **9**, 612 (*synth, deriv*)

Diundecylphosphinic acid, 9CI **D-01256**

$[H_3C(CH_2)_{10}]_2P(O)OH$

$C_{22}H_{47}O_2P$ M 374.586

Cryst. (hexane or hexane/Me_2CO). Mp 89-90°.

Williams, R.H. *et al*, *J. Am. Chem. Soc.*, 1955, **77**, 3411.

Divinyl phosphate, 8CI **D-01257**

Diethenyl phosphate, 9CI. Diethenyl hydrogen phosphate. Divinyl phosphoric acid. Divinyl hydrogen phosphate

$(H_2C{=}CHO)_2P(O)OH$

$C_4H_7O_4P$ M 150.071

Chloride: Divinyl phosphorochloridate. Diethenyl phosphorochloridate. Divinyl chlorophosphate.
$C_4H_6ClO_3P$ M 168.516
Liq. d_4^{20} 1.24. Bp_{11} 58-59°. n_D^{20} 1.4319.

Bromide: [36219-46-2]. *Diethenyl phosphorobromidate. Divinyl phosphorobromidate. Divinyl bromophosphate.*
$C_4H_6BrO_3P$ M 212.967
Liq. d_D^{20} 1.53. Bp_{12} 75-76°. n_D^{20} 1.4650.

Lutsenko, I.F. *et al*, *Dokl. Akad. Nauk SSSR, Ser. Sci. Khim.*, 1963, **148**, 846; *CA*, **59**, 3759.

Gololobov, Yu.G. *et al*, *Zh. Obshch. Khim.*, 1965, **35**, 1460 (*Engl. transl.* p. 1462) (*chloride*)

Divinylphosphinic acid, 8CI **D-01258**

Diethenylphosphinic acid, 9CI

$(H_2C{=}CH)_2P(O)OH$

$C_4H_7O_2P$ M 118.072

V. hygroscopic liq. d_4^{20} 1.19. $Bp_{0.002}$ 130-132°. pK_a 2.54 (7% EtOH aq.), 4.34 (80% EtOH aq.).

Me ester: [41924-72-5]. *Methyl diethenylphosphinate. Methyl divinylphosphinate.*
$C_5H_9O_2P$ M 132.099
Liq. d_4^{20} 1.08. Bp_{10} 91°. n_D^{20} 1.4685.

Et ester: [30594-15-1]. *Ethyl diethenylphosphinate. Ethyl divinylphosphinate.*
$C_6H_{11}O_2P$ M 146.125
Liq. d_4^{20} 1.03. $Bp_{0.07}$ 46°. n_D^{20} 1.4640.

Ph ester: [30594-16-2]. *Phenyl diethenylphosphinate. Phenyl divinylphosphinate.* Liq. d_4^{20} 1.13. $Bp_{0.1}$ 105°, $Bp_{0.001}$ 73-74°. n_D^{20} 1.5370.

Chloride: [34833-61-9].
C_4H_6ClOP M 136.518
Liq. d_4^{20} 1.20. Bp_{10} 92°. n_D^{20} 1.4971.

Diethylamide: P,P-*Diethenyl*-N,N-*diethylphosphinic amide.*
$C_8H_{16}NOP$ M 173.194
Liq. $Bp_{0.1}$ 80°. n_D^{20} 1.4845.

Kabachnik, M.I. *et al*, *Zh. Obshch. Khim.*, 1963, **33**, 382 (*Engl. transl.* p. 375) (*synth, props*)

Maier, L., *Phosphorus*, 1971, **1**, 111 (*synth, esters, pmr, nmr, chloride*)

Levin, Ya.A. *et al*, *Zh. Obshch. Khim.*, 1973, **43**, 578 (*Engl. transl.* p. 580) (*derivs*)

Naumov, V.A. *et al*, *Zh. Strukt. Khim.*, 1976, **17**, 304 (*Engl. transl.* p. 262) (*chloride, ed, struct*)

Zverev, V.V. *et al*, *Zh. Obshch. Khim.*, 1981, **51**, 303 (*Engl. transl.* p. 242) (*ester, pe*)

Divinylphosphinothioic acid, 8CI **D-01259**

Diethenylphosphinothioic acid, 9CI

$(H_2C{=}CH)_2P(S)OH \rightleftharpoons (H_2C{=}CH)_2P(O)SH$

C_4H_7OPS M 134.132

O-*Et ester:* [30594-17-3]. O-*Ethyl divinylphosphinothioate.*
$C_6H_{11}OPS$ M 162.186
Liq. Bp_1 65-68°. n_D^{20} 1.5147.

S-*Butyl ester:* [41924-81-6]. S-*Butyl divinylphosphinothioate.*
$C_8H_{15}OPS$ M 190.240
Liq. d_4^{20} 1.04. $Bp_{0.07}$ 85°. n_D^{20} 1.5200.

Chloride: [15850-00-7].
C_4H_6ClPS M 152.578
Liq. Bp_{740} 220-221°, $Bp_{0.01}$ 28-30°. n_D^{20} 1.5920.

Maier, L., *Helv. Chim. Acta*, 1971, **54**, 275 (*ester, chloride, P nmr*)

Maier, L., *Phosphorus*, 1971, **1**, 111 (*ester, chloride, P nmr*)

Levin, Ya.A. *et al*, *Zh. Obshch. Khim.*, 1973, **43**, 578 (*Engl. transl.* p. 580) (*ester*)

Zverev, V.V. *et al*, *Zh. Obshch. Khim.*, 1981, **51**, 303 (*Engl. transl.* p. 342) (*pe*)

2,2,4,4,6,6,8,8,9,9,10,10-Dodecaethyl-1,3,5,7-tetraphospha-2,4,6,8,9,10-hexagermatricyclo[3.3.1.1^{3,7}]decane, 9CI **D-01260**

adamanta-*Hexakis(diethylgermyl)tetraphosphane*

[57584-33-5]

As 2,2,4,4,6,6,8,8,9,9,10,10-Dodecamethyl-1,3,5,7-tetraphospha-2,4,6,8,9,10-hexagermatricyclo[3.3.1.1^{3,7}]decane, D-01263 with

R = Et

$C_{24}H_{60}Ge_6P_4$ M 908.173
Solid. Mp 220-222°.

Dahl, A.R. *et al*, *J. Am. Chem. Soc.*, 1975, **97**, 6364.

7,8,10,11,13,14,16,17,19,20,22,27-Dodecahydro-5*H*-dibenzo[*r,u*][1,4,7,10,13,16,20]-hexaoxaphosphacyclotricosin, 9CI **D-01261**

2,3:22,23-Dibenzo-5,8,11,14,17,20-hexaoxa-1-phospha-2,22-cyclotricosadiene

$C_{24}H_{33}O_6P$ M 448.495

27-Ph, 27-oxide: [69928-16-1].
$C_{30}H_{37}O_7P$ M 540.592
Ligand for group I metals. Microcryst. (Et_2O at −20°). Mp 64-66°.

Kaplan, L.J. *et al*, *J. Org. Chem.*, 1979, **44**, 2226 (*synth, pmr, complexes*)

2,3,4,5,6,7,8,9,10,11,12,13-Dodecahydro-2,5,9,12-tetraphenyl-1*H*-2,5,9,12-benzotetraphosphacyclopentadecin, 9CI **D-01262**

[65257-79-6]

$C_{39}H_{42}P_4$ M 634.656
Stereochemistry unknown. Ligand for Ni. Air-sensitive oil.

Del Donno, T.A. *et al*, *J. Am. Chem. Soc.*, 1977, **99**, 8051 (*synth, pmr, ir, complexes*)

2,2,4,4,6,6,8,8,9,9,10,10-Dodecamethyl-1,3,5,7-tetraphospha-2,4,6,8,9,10-hexagermatricyclo[3.3.1.1^3,7^]decane, 10CI, 9CI **D-01263**

adamanta-*Hexakis(dimethylgermyl)tetraphosphane*

[28133-43-9]

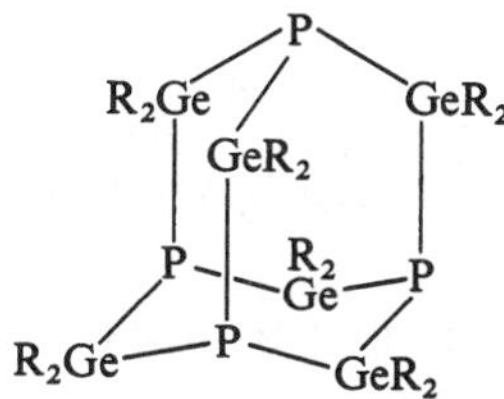

R = Me

$C_{12}H_{36}Ge_6P_4$ M 739.851
Cubic cryst. Mp 300° dec.

Dahl, A. *et al*, *J. Am. Chem. Soc.*, 1975, **97**, 6364.
Hoenle, W. *et al*, *Z. Anorg. Allg. Chem.*, 1978, **442**, 91 (*cryst struct*)

Dodecyldimethylphosphine, 9CI **D-01264**

$H_3C(CH_2)_{11}PMe_2$

$C_{14}H_{31}P$ M 230.373
$Bp_{0.03}$ 80-83°.

B,MeI: Dodecyltrimethylphosphonium iodide.
$C_{15}H_{34}IP$ M 372.312
Cryst. Mp 87-88°.

Oxide: [876-95-4].
$C_{14}H_{31}OP$ M 246.372
Surfactant. Cryst. (hexane). Sol. H_2O. Mp 84-85°. $Bp_{0.5}$ 152-160°.

Laughlin, R.G., *J. Org. Chem.*, 1965, **30**, 1322 (*oxide*)
Hays, H.R., *J. Org. Chem.*, 1966, **31**, 3817 (*synth, derivs, pmr, nmr*)
Kleiner, H.J., *Justus Liebigs Ann. Chem.*, 1974, 751 (*oxide*)
McDonald, M.P. *et al*, *Spectrochim. Acta, Part A*, 1978, **34**, 933 (*oxide, ir*)

Dodecylphosphonous acid, 9CI **D-01265**

$H_3C(CH_2)_{11}P(OH)_2 \rightleftharpoons H_3C(CH_2)_{11}PH(O)OH$

$C_{12}H_{27}O_2P$ M 234.318

Diisopropyl ester: Diisopropyl dodecylphosphonite. Bis(1-methylethyl) dodecylphosphonite.
$C_{18}H_{39}O_2P$ M 318.479
Liq. Bp_8 165-170°.

Dichloride: [2629-35-8]. *Dichlorododecylphosphine.*
$C_{12}H_{25}Cl_2P$ M 271.209
Liq. $Bp_{1.5}$ 135-140°.

Mastalerz, P. *et al*, *Rocz. Chem.*, 1964, **38**, 1529 (*ester, synth*)

Dodecyltriphenylphosphonium(1+), 9CI **D-01266**

$Ph_3P^{\oplus}(CH_2)_{11}CH_3$

$C_{30}H_{40}P^{\oplus}$ M 431.620 (ion)
Topical antifungal agent.

Bromide: [15510-55-1].
$C_{30}H_{40}BrP$ M 511.524
Hygroscopic cryst. (EtOAc/Et_2O). Mp 87-88°, 98-99°. NaH or butyllithium → ylide.

Ylide: [54208-04-7]. *Dodecylidenetriphenylphosphorane.*
$C_{30}H_{39}P$ M 430.612
Used in Wittig reaction. Red.

Jerchel, M. *et al*, *Chem. Ber.*, 1950, **83**, 277.
Chasin, D.G. *et al*, *Chem. Phys. Lipids*, 1971, **6**, 8 (*synth, use*)
Bestmann, H.J. *et al*, *Angew. Chem., Int. Ed. Engl.*, 1976, **15**, 298 (*use*)
Lie Ken Jie, M.S.F. *et al*, *Chem. Phys. Lipids*, 1978, **21**, 275 (*use*)

E

Ecothiopate **E-00001**

2-[(Diethoxyphosphinyl)thio]-N,N,N-*trimethylethanaminium*, 9CI. *Diethoxyphosphinylthiocholine. (2-Mercaptoethyl)trimethylammonium ester with diethyl thiophosphoric acid*

[6736-03-4]

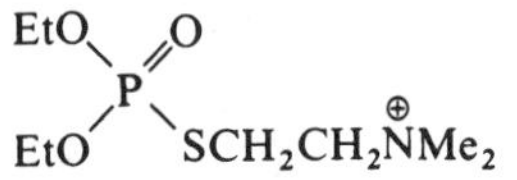

$(EtO)_2P(O)SCH_2CH_2N^{\oplus}Me_3$

$C_9H_{23}NO_3PS^{\oplus}$ M 256.319 (ion)

Iodide: [513-10-0]. Drug used in treatment of glaucoma. Cholinesterase inhibitor. Cryst. Sol. H_2O. Mp 124-124.5° (138°). Unstable in aq. soln.

▷BR6985000.

Hydrogen oxalate: Solid. Mp 116°.

Tammelin, L.E., *Acta Chem. Scand.*, 1957, **11**, 1340 (*synth*)
Chatten, L.G. *et al*, *J. Pharm. Sci.*, 1967, **56**, 834.
Hussain, A. *et al*, *J. Pharm. Sci.*, 1968, **57**, 411 (*props*)
Dale, R.D., *Anal. Prof. Drug Subst.*, 1974, **3**, 233 (*rev, bibl*)
Merck Index, 11th Ed., 3485.

Edifenphos, BSI **E-00002**

O-*Ethyl* S,S-*diphenyl phosphorodithioate*, 9CI, 8CI. *Hinosan*

[17109-49-8]

$(PhS)_2P(O)OEt$

$C_{14}H_{15}O_2PS_2$ M 310.365

Rice fungicide. Yellowish liq. Prac. insol. H_2O. d_4^{20} 1.23. $Bp_{0.01}$ 154°. n_D^{22} 1.6112. Hydrol. in strongly acid or alk. media.

▷TE3850000.

B.P., 1 083 377, (*1965*); *CA*, **68**, 49297v (*synth, use*)
Tomizawa, C. *et al*, *J. Environ. Sci. Health, Part B*, 1976, **11**, 231 (*metab*)
Pesticide Manual, 6th Ed., 229; 7th Ed., 231.
The Agrochemicals Handbook, Royal Society of Chemistry, 1983, A174.

Ensanchomycin **E-00003**

[56748-23-3]

Struct. unknown

$C_{63}H_{110}N_9O_{36}P$ M 1600.574

Isol. from *Streptomyces cinnanonensis* and *S. melanogenes*. Antibiotic active against gram-positive and -negative bacteria.

U.S.P., 3 891 754, (*1975*); *CA*, **83**, 176643q (*isol*)
U.S.P., 3 927 210, (*1975*); *CA*, **84**, 103847v (*isol*)

1-*tert*-Butyl-3,4-dimethyl-1*H*-phosphole, 8CI **E-00004**

*1-(1,1-Dimethylethyl)-3,4-dimethyl-1*H*-phosphole*

[38066-25-0]

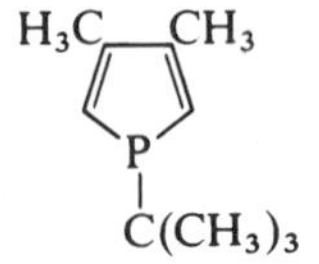

$C_{10}H_{17}P$ M 168.218

Ligand for metals of Groups VIB and VIII. Solid. Mp 45°. $Bp_{0.2}$ 35-36°.

B,MeI: 1-tert-*Butyl-1,3,4-trimethyl-1*H*-phospholium iodide.*

$C_{11}H_{20}IP$ M 310.157

Solid. Mp 252°.

1-Sulfide: [38066-26-1].

$C_{10}H_{17}PS$ M 200.278

Cryst. (hexane). Mp 160°.

Mathey, F., *Tetrahedron*, 1972, **28**, 4171 (*synth, derivs, pmr, P nmr*)
Schaefer, W. *et al*, *J. Am. Chem. Soc.*, 1976, **98**, 407 (*pe*)
Gray, G.A. *et al*, *Org. Magn. Reson.*, 1980, **14**, 14 (*cmr*)
MacDougall, J.J. *et al*, *Inorg. Chem.*, 1980, **19**, 709 (*pmr, cmr, P nmr, complexes*)
Santini, C.C. *et al*, *J. Am. Chem. Soc.*, 1980, **102**, 5809 (*props, complexes*)
Mercier, F. *et al*, *Inorg. Chem.*, 1985, **24**, 4141 (*props, complexes*)

Ethafos **E-00005**

O-*(2,4-Dichlorophenyl)* O-*ethyl* S-*propyl phosphorothioate*

[38527-91-2]

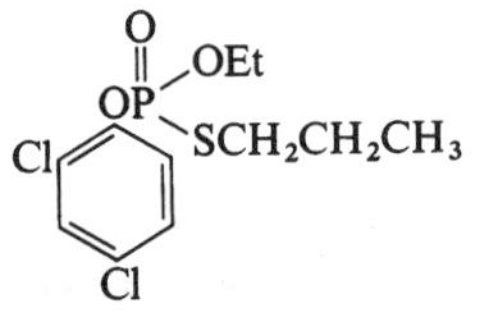

$C_{11}H_{15}Cl_2O_3PS$ M 329.177

Acaricide and insecticide. Nonphytotoxic. Rel. nontoxic.

▷TF0370000.

Takase, I. *et al*, *Nippon Noyaku Gakkaishi*, 1982, **7**, 763; *CA*, **98**, 126247.

1,2-Ethanediphosphonic acid **E-00006**

1,2-Ethanediylbisphosphonic acid, 9CI. *1,2-Ethylenediphosphonic acid*, 8CI. *1,2-Diphosphonoethane*

[6145-31-9]

$(HO)_2P(O)CH_2CH_2P(O)(OH)_2$

$C_2H_8O_6P_2$ M 190.029

Plant growth regulator; flotation agent. Cryst. or needles ($EtOH/Et_2O$). Mp 223-224°. pK_{a1} 1.50, pK_{a2} 2.96, pK_{a3} 7.50, pK_{a4} 9.08 (H_2O, 25°).

Tetra-Me ester: [5927-50-4]. *Tetramethyl 1,2-ethanediylbisphosphonate. 1,2-Bis(dimethoxyphosphinyl)ethane.*

$C_6H_{16}O_6P_2$ M 246.136

Liq. $Bp_{1.5}$ 166-168°. n_D^{15} 1.4428.
Tetra-Et ester: [995-32-4]. *Tetraethyl 1,2-ethanediylbisphosphonate. 1,2-Bis(diethoxyphosphinyl)ethane.*
$C_{10}H_{24}O_6P_2$ M 302.244
Deodorant; extractant for transuranic elements. Liq. Bp_{14} 200-202°, Bp_1 167°. $n_D^{16.5}$ 1.4425.
Tetrabutyl ester: [919-48-2]. *Tetrabutyl 1,2-ethanediylbisphosphonate. 1,2-Bis(dibutoxyphosphinyl)ethane.*
$C_{18}H_{40}O_6P_2$ M 414.458
Used in purification of Ga.
Tetra-Ph ester: [42451-26-3]. *Tetraphenyl 1,2-ethanediylbisphosphonate. 1,2-Bis(diphenoxyphosphinyl)-ethane.*
$C_{26}H_{24}O_6P_2$ M 494.420
Solid. Mp 155-155.5°.
Tetrachloride: [1499-30-5]. *1,2-Ethanediylbis[phosphonic dichloride]. 1,2-Bis(dichlorophosphinyl)-ethane.*
$C_2H_4Cl_4O_2P_2$ M 263.812
Solid. Mp 167-170°.

Ford-Moore, A.H. *et al*, *J. Chem. Soc.*, 1947, 1465 (*tetraethyl ester*)
Moedritzer, K. *et al*, *J. Inorg. Nucl. Chem.*, 1961, **22**, 297 (*ester, synth, ir, P nmr*)
Maier, L., *Helv. Chim. Acta*, 1965, **48**, 133 (*tetrachloride*)
Kosolapoff, G.M. *et al*, *J. Chem. Soc.*, 1966, 757 (*tetrachloride, ir*)
Brophy, J.J. *et al*, *Aust. J. Chem.*, 1967, **20**, 503 (*esters, pmr*)
Shner, S.M. *et al*, *Zh. Obshch. Khim.*, 1967, **37**, 418 (*Engl. transl.* p. 390) (*synth, tetrakis(2-chloroethyl)* ester)
Sommer, K., *Z. Anorg. Allg. Chem.*, 1970, **376**, 37 (*synth, P nmr, tetrachloride*)
Peterson, S.W. *et al*, *J. Phys. Chem.*, 1977, **81**, 466 (*cryst struct*)
Yatmirski, K.B. *et al*, *Zh. Neorg. Khim.*, 1977, **22**, 435 (*Engl. transl.* p. 236) (*props, complexes*)

1,2-Ethanediphosphonothioic acid E-00007

Ethanediylbis(phosphonothioic acid), 9CI

$$(HO)_2P(S)CH_2CH_2P(S)(OH)_2$$

$C_2H_8O_4P_2S_2$ M 222.150
Tetrachloride: [1661-12-7].
$C_2H_4Cl_4P_2S_2$ M 295.933
Cryst. (Et_2O). Mp 99.5-100.5°. $Bp_{0.2}$ 150°.
Tetrabromide: [82159-30-6].
$C_2H_4Br_4P_2S_2$ M 473.737
Cryst. (CH_2Cl_2 at −80°). Mp 128-129°.

Maier, L., *Helv. Chim. Acta*, 1965, **48**, 133 (*chloride, synth, ir, P nmr*)
Diemert, K. *et al*, *Chem. Ber.*, 1982, **115**, 1947 (*bromide, synth, P nmr*)

1,2-Ethanediphosphonous acid E-00008

1,2-Ethanediylbis(phosphonous acid). 1,2-Ethylenediphosphonous acid. 1,2-Ethanediylbis(phosphonic acid)

$$(HO)_2PCH_2CH_2P(OH)_2 \rightleftharpoons HO(O)PHCH_2CH_2PH(O)OH$$

$C_2H_8O_4P_2$ M 158.030
Tetra-Ph ester: [78819-34-8]. *Tetraphenyl 1,2-ethanediylbisphosphonite.*
$C_{26}H_{24}O_4P_2$ M 462.421
Cryst. Forms an Fe complex.
Bis(dichloride): [28240-69-9].
$C_2H_4Cl_4P_2$ M 231.813
Bp_2 81-82°.
Bis(dibromide): [39063-58-6].
$C_2H_4Br_4P_2$ M 409.617
Liq. $Bp_{0.1}$ 105°.
Bis[bis(diethylamide)]: [86926-28-5]. P,P′-*1,2-Ethanediylbis[*N,N,N′,N′*-tetraethylphosphonous diamide].*
$C_{18}H_{44}N_4P_2$ M 378.520
Liq. $Bp_{0.1}$ 110°, $Bp_{0.005}$ 135-140°.

Sommer, K., *Z. Anorg. Allg. Chem.*, 1970, **376**, 37 (*tetrachloride, synth, P nmr*)
Diemert, K. *et al*, *Chem. Ber.*, 1982, **115**, 1947 (*tetrabromide, P nmr*)
Day, V.W. *et al*, *Organometallics*, 1983, **2**, 494 (*ester, pmr, cmr, P nmr, complex*)
King, R.B. *et al*, *J. Org. Chem.*, 1984, **49**, 1784 (*amide, synth, ms, pmr, cmr, P nmr, ir*)

1,2-Ethanediphosphoramidic acid E-00009

1,2-Ethanediylbis(phosphoramidic acid), 9CI

$$(HO)_2P(O)NHCH_2CH_2NHP(O)(OH)_2$$

$C_2H_{10}N_2O_6P_2$ M 220.058
Tetrapropyl ester: Tetrapropyl 1,2-ethanediylbisphosphoramidate. 1,2-Bis(dipropoxyphosphinylamino)-ethane.
$C_{14}H_{34}N_2O_6P_2$ M 388.380
Cryst. (hexane). Mp 81°.
Tetraisopropyl ester: [56045-72-8]. *Tetraisopropyl 1,2-ethanediylbisphosphoramidate. 1,2-Bis(diisopropoxyphosphinylamino)ethane.*
$C_{14}H_{34}N_2O_6P_2$ M 388.380
Cryst. (hexane). Mp 84°.
Tetra-Ph ester: [34670-52-5]. *Tetraphenyl 1,2-ethanediylbisphosphoramidate. 1,2-Bis(diphenoxyphosphinylamino)ethane.*
$C_{26}H_{26}N_2O_6P_2$ M 524.449
Cryst. (C_6H_6 or EtOAc). Mp 131-132°.

Ger. Pat., 1 034 173, (*1956*); *CA*, **54**, 10865 (*esters, synth*)
Edmundson, R.S., *J. Chem. Soc.* (*C*), 1971, 3614 (*phenyl ester*)

1,2-Ethanediphosphoramidothioic acid E-00010

1,2-Ethanediylbisphosphoramidothioic acid, 9CI

$$(HO)_2P(S)NHCH_2CH_2NHP(S)(OH)_2$$

$C_2H_{10}N_2O_4P_2S_2$ M 252.180
O,O,O,O-*Tetrapropyl ester:* O,O,O,O-*Tetrapropyl 1,2-ethanediylbisphosphoramidothioate. 1,2-Bis[(dipropoxyphosphinothioyl)amino]ethane.*
$C_{14}H_{34}N_2O_4P_2S_2$ M 420.501
Oil. n_D^{20} 1.4821.
O,O,O,O-*Tetra-Ph ester:* [34670-59-2]. O,O,O,O-*Tetraphenyl 1,2-ethanediylbisphosphoramidothioate. 1,2-Bis[(diphenoxyphosphinothioyl)amino]ethane.*
$C_{26}H_{26}N_2O_4P_2S_2$ M 556.570
Cryst. (EtOAc/cyclohexane). Mp 86°.

Ger. Pat., 1 033 201, (*1958*); *CA*, **54**, 11994
Edmundson, R.S., *J. Chem. Soc.* (*C*), 1971, 3614.

1,2-Ethanediylbis[(2-methoxyphenyl)-phenylphosphine], 9CI E-00011

1,2-Bis[(2-methoxyphenyl)phenylphosphino]ethane. DiPAMP

OMe Ph, P–CH_2CH_2–P, MeO, Ph *(R*,R*)-form*

$C_{28}H_{28}O_2P_2$ M 458.476

Rh complexes used as catalysts in reductions with high enantioselectivity.

(R*,R*)-form

Cryst. (MeOH). Mp 102-104°. $[\alpha]^{20}_{D}$ −85.0° (MeOH).

Dioxide:

$C_{28}H_{28}O_4P_2$ M 490.474

Mp 205-207°. $[\alpha]^{20}_{D}$ −46° (MeOH).

(RS,SR)-form

meso-*form*

Dec. on heating.

Dioxide: Solid. Mp 205-207°.

Vineyard, B.D. *et al*, *J. Am. Chem. Soc.*, 1977, **99**, 5946 (*synth, use*)
Solodar, J., *J. Org. Chem.*, 1978, **43**, 1789 (*use*)
Brown, J.M. *et al*, *J. Am. Chem. Soc.*, 1980, **102**, 3040 (*use*)
Scott, J.W. *et al*, *J. Org. Chem.*, 1981, **46**, 5086 (*use*)
Wife, R.L. *et al*, *Synthesis*, 1983, 71 (*synth*)

1,2-Ethanediylbis[methylphenylphosphine], 9CI E-00012

1,2-Bis(methylphenylphosphino)ethane

[23808-01-7]

Me(Ph)P–CH_2CH_2–P(Ph)Me *(RS,RS)-form*

$C_{16}H_{20}P_2$ M 274.282

Ligand for Os and Ru.

(RS,RS)-form [17325-54-1]

(±)-*form*

Oil. Bp 250-260°.

B,2MeI: 1,2-Ethanediylbis[dimethylphenylphosphonium] dibromide.

$C_{18}H_{26}Br_2P_2$ M 464.159

Solid. Mp 234-235°.

B,2PhCH₂Br:

$C_{30}H_{34}Br_2P_2$ M 616.354

Solid. Mp 293°.

Dioxide: [38234-90-1].

$C_{16}H_{20}O_2P_2$ M 306.280

Cryst. (diethyl carbonate). Mp 190-191° or 246-247°. Stereochemical assignment to the oxides is uncertain (see also below).

(RS,SR)-form [17325-53-0]

meso-*form*

Solid. Mp 90°. $Bp_{0.2}$ 130°.

B,2PhCH₂Br: Cryst. + $1H_2O$. Mp 278°.

Dioxide: [38234-89-8]. Cryst. (diethyl carbonate). Mp 246-247° or 190-191°. Stereochemical assignment to the oxides is uncertain (see above).

Issleib, K. *et al*, *Chem. Ber.*, 1963, **96**, 279 (*synth, derivs*)
Horner, L. *et al*, *Tetrahedron Lett.*, 1966, 5783 (*synth, derivs, props*)
Mann, F.G. *et al*, *J. Chem. Soc., Perkin Trans. 1*, 1972, 2548 (*oxides*)

1,2-Ethanediylbis[methylphosphinic acid], 9CI E-00013

Ethylenebis(methylphosphinic acid)

[51807-11-5]

Me(HO)P(O)–CH_2CH_2–P(O)(OH)Me

$C_4H_{12}O_4P_2$ M 186.084

Cryst. (MeOH). Mp 189-191°.

Diisopropyl ester:

$C_{10}H_{24}O_4P_2$ M 270.245

Cryst. (cyclohexane). Sol. H_2O. Mp 84-86° and 115-117°. Low-melting form probably the racemic stereoisomer; high-melting form the meso-.

Mastalerz, P., *Rocz. Chem.*, 1964, **38**, 61.

[1,2-Ethanediylbis[nitrilobis[methylene]]]-tetrakisphosphonic acid, 9CI E-00014

N,N,N′,N′-*Tetrakis(phosphonomethyl)-1,2-ethanediamine*

[1429-50-1]

$[(HO)_2P(O)CH_2]_2NCH_2CH_2N[CH_2P(O)(OH)_2]_2$

$C_6H_{20}N_2O_{12}P_4$ M 436.125

Behaves as a hexabasic acid, existing in a zwitterionic form. Sequestering agent, used as a corrosion inhibitor, in electroplating, in the extraction of americium and europium, and as an antidote to Be poisoning. Mp 214° dec. pK_{a1} 2.53, pK_{a2} 3.64, pK_{a3} 4.75, pK_{a4} 5.85, pK_{a5} 6.98, pK_{a6} 8.10, pK_{a7} 9.21, pK_{a8} 10.32.

Monohydrate: Mp 250° dec.

Moedritzer, K. *et al*, *J. Org. Chem.*, 1966, **31**, 1603 (*synth, P nmr, props*)
Motekaitis, R.J. *et al*, *Inorg. Chem.*, 1976, **15**, 2303 (*synth, complexes*)
Kurochkina, L.V. *et al*, *Zh. Neorg. Khim.*, 1978, **23**, 2676 (*Engl. transl.* p. 1481) (*complexes*)
Levin, V.I. *et al*, *Zh. Neorg. Khim.*, 1981, **26**, 1180 (*Engl. transl.* p. 637) (*complexes, props*)
Rizkalla, E.N. *et al*, *Inorg. Chem.*, 1983, **22**, 1478 (*synth, pmr, P nmr, complexes*)

1,2-Ethanediylbis[phenylphosphinic acid], 9CI E-00015

1,2-Bis(hydroxyphenylphosphinyl)ethane

[1089-77-6]

Ph–P(O)(OH)CH_2CH_2P(O)(OH)–Ph

$C_{14}H_{16}O_4P_2$ M 310.226

Cryst. (EtOH). Mp 276-278°.

Di-Et ester: Diethyl 1,2-ethanediylbis(phenylphosphinate). 1,2-Bis(ethoxyphenylphosphinyl)ethane.

$C_{18}H_{24}O_4P_2$ M 366.333

Cryst. in 2 forms, prob. diastereoisomers. Mp 68-69°, 101-102°.

Diisopropyl ester: Diisopropyl 1,2-ethanediylbis(phenylphosphinate).

$C_{20}H_{28}O_4P_2$ M 394.386

Cryst. in 2 forms, prob. diastereoisomers. Mp 113-114°, 146-147°.

Dichloride: [2259-81-6].

$C_{14}H_{14}Cl_2O_2P_2$ M 347.117

Cryst. (dioxane or dibutyl ether). Mp 179-181°. Stereochem. unknown.

Diamide:

$C_{14}H_{18}N_2O_2P_2$ M 308.256

Cryst. (MeOH). Mp 202-204° dec. When heated cyclises with loss of NH_3. Stereochem. unknown.

Dianilide: N,N-*Diphenyl*-P,P-*1,2-ethanediylbis[phenylphosphinic amide]*.

$C_{26}H_{26}N_2O_2P_2$ M 460.451

Cryst. (EtOH). Mp 240-246°. Stereochem. unknown.

Mastalerz, P., *Rocz. Chem.*, 1965, **39**, 33 (*chloride, amides*)
Abramov, V.S. *et al*, *Zh. Obshch. Khim.*, 1968, **38**, 1794 (*Engl. transl.* p. 1748) (*esters*)
Lindner, E. *et al*, *Z. Naturforsch., B*, 1978, **33**, 1457 (*synth, nmr*)
Garst, M.E., *Synth. Commun.*, 1979, **9**, 261 (*synth*)

1,2-Ethanediylbis[phenylphosphinodithioic acid] E-00016

HSP(S)PhCH$_2$CH$_2$PPh(S)SH

$C_{14}H_{16}P_2S_2$ M 310.348
Cryst. (C_6H_6). Mp 142-144°.

Issleib, K. *et al*, *Chem. Ber.*, 1968, **101**, 2197.

1,2-Ethanediylbis[triphenylphosphonium](2+), 9CI E-00017

1,2-Ethylenebis[triphenylphosphonium](2+). 1,2-Bis(triphenylphosphonio)ethane
[13275-02-0]

$Ph_3P^{\oplus}CH_2CH_2P^{\oplus}Ph_3$

$C_{38}H_{34}P_2^{\oplus\oplus}$ M 552.634 (ion)
Dibromide: [1519-45-5].
$C_{38}H_{34}Br_2P_2$ M 712.442
Cryst. ($CHCl_3$/EtOAc). Mp 297-300°, 313-314°.
▷Exhibits anticholinesterase activity towards vertebrates and schistosomes. Can explode at 80-90°

Pattenden, G. *et al*, *J. Chem. Soc. (C)*, 1969, 531 (*synth, ir, nmr*)
Schweizer, E.E. *et al*, *J. Org. Chem.*, 1973, **38**, 3069 (*synth, use*)
Swartz, W.E. *et al*, *Spectrochim. Acta, Part A*, 1974, **30**, 1561 (*pe*)
Wood, G.W. *et al*, *J. Org. Chem.*, 1975, **40**, 636 (*ms*)
Leddy, B.P. *et al*, *Tetrahedron Lett.*, 1980, **21**, 2261 (*props*)
Willcockson, W.S. *et al*, *Comp. Biochem. Physiol. C*, 1982, **72**, 101 (*tox*)

1,2-Ethenediylbis(phenylphosphinic acid), 9CI E-00018

Ethylenebis(phenylphosphinic acid)

(Structure: H(Ph-P(=O)(OH))C=C(H)P(=O)(OH)Ph) (*E*)-*form*

$C_{14}H_{14}O_4P$ M 277.236
(*E*)-form
Di-Me ester: [40612-16-6]. *Dimethyl 1,2-ethenediylbis(phenylphosphinate).*
$C_{16}H_{20}O_4P$ M 307.305
No phys. props. reported.
Di-Et ester: [20408-23-5]. *Diethyl 1,2-ethenediylbis(phenylphosphinate).*
$C_{18}H_{22}O_4P_2$ M 364.317
Solid. Mp 135-136°. May be a mixt. of (*E*−) and (*Z*−)-forms.

Kataev, E.G. *et al*, *Dokl. Akad. Nauk SSSR, Ser. Sci. Khim.*, 1968, **179**, 862 (*Engl. transl.* p. 292) (*synth*)
Samitov, Yu.Yu. *et al*, *Zh. Obshch. Khim.*, 1975, **45**, 2130 (*Engl. transl.* p. 2097) (*pmr, nmr, stereochem*)

1,2-Ethenediylbis[triphenylphosphonium](2+), 9CI E-00019

1,2-Bis(triphenylphosphonio)ethylene(2+)
[54854-20-5]

$Ph_3P^{\oplus}CH{=}CHPPh_3^{\oplus}$

$C_{38}H_{32}P_2^{\oplus\oplus}$ M 550.618 (ion)
Reacts with nucleophilic active H compds. to give 2 subst. vinyl phosphonium salts ("phosphoniovinylation") or 2,2-disubst. ethylphosphonium salts ("phosphonioethylation").
(*E*)-form
Dibromide: [53332-51-7].
$C_{38}H_{32}Br_2P_2$ M 710.426
Solid. Mp 268-271° dec.
Diiodide: [53332-52-8].
$C_{38}H_{32}I_2P_2$ M 804.427
Orange cryst. Mp 256-258°.
(*Z*)-form
Dibromide: [64518-03-2]. No phys. props. reported.

Kataev, E.G. *et al*, *Zh. Org. Khim.*, 1974, **10**, 1050 (*Engl. transl.* p. 1059) (*synth*)
Samitov, Yu.Yu. *et al*, *Zh. Obshch. Khim.*, 1975, **45**, 2130 (*Engl. transl.* p. 2097) (*pmr*)
Christau, H.-J. *et al*, *Phosphorus Sulfur*, 1982, **14**, 63, 73 (*use*)
Flitsch, W. *et al*, *Chem. Ber.*, 1982, **115**, 1547 (*props*)

Ethenylidenebisphosphonic acid, 9CI E-00020

Vinylidenebisphosphonic acid. Ethene-1,1-diphosphonic acid. 1,1-Diphosphonoethene. Ethylene-1,1-diphosphonic acid

$H_2C{=}C[P(O)(OH)_2]_2$

$C_2H_6O_6P_2$ M 188.013
Cryst. (EtOH). Mp 232-234°.
Tetra-Me ester: Tetramethyl ethenylidenebisphosphonate.
$C_6H_{14}O_6P_2$ M 244.121
Liq. $Bp_{0.05}$ 113-118°.
Tetra-Et ester: Tetraethyl ethenylidenebisphosphonate.
$C_{10}H_{22}O_6P_2$ M 300.228
Liq. $Bp_{0.05}$ 115-116°.
Tetraisopropyl ester: Tetraisopropyl ethenylidenebisphosphonate.
$C_{14}H_{30}O_6P_2$ M 356.335
Liq.

Berdnikov, E.A. *et al*, *Zh. Obshch. Khim.*, 1985, **55**, 579 (*Engl. transl.* p 512) (*synth, pmr, cmr, P nmr*)
Degenhardt, C.R. *et al*, *J. Org. Chem.*, 1986, **51**, 3488 (*esters, synth, pmr, P nmr*)

Ethiofos, USAN E-00021

2-[(3-Aminopropyl)amino]ethanethiol dihydrogenphosphate, 8CI. Gammaphos. NSC-296961. WR-2721
[20537-88-6]

$H_2NCH_2CH_2NHCH_2CH_2SP(O)(OH)_2$

$C_5H_{15}N_2O_3PS$ M 214.219
Radioprotective agent. Mp 160-161°.
▷TE6491000.

Piper, J.R. *et al*, *J. Med. Chem.*, 1969, **12**, 236 (*synth, pharmacol*)
Rozman, R.S., *Anal. Chem.*, 1976, **48**, 989 (*tlc*)

Ethion, BSI, ISO **E-00022**

O,O,O′,O′-Tetraethyl S,S′-methylene bis(phosphorodithioate), 9CI. *S,S′-Methylene-O,O,O′,O′-tetraethyl di(phosphorodithioate)*, 8CI. *Bis[S-(diethoxyphosphinothioyl)thio]methane. Niagara 1240. Nialate*

[563-12-2]

$(EtO)_2P(S)SCH_2SP(S)(OEt)_2$

$C_9H_{22}O_4P_2S_4$ M 384.458

Pesticide with cholinesterase inhibitory activity. Oily liq. Sol. Me_2CO, C_6H_6, $CHCl_3$, EtOH, Et_2O, sl. sol. H_2O, mod. sol. ligroin. d_4^{20} 1.31 (1.23). Fp −13° (−12°). $Bp_{0.3}$ 164-165°. n_D^{20} 1.5478.

▷Highly toxic, TLV 0.4. TE4550000.

Shvetsova-Shilovskaya, K.D. *et al*, *Zh. Obshch. Khim.*, 1959, **29**, 3593 (*synth*)
Keith, L.H. *et al*, *Anal. Chim. Acta*, 1969, **44**, 447 (*pmr*)
Ross, R.T. *et al*, *Anal. Chim. Acta*, 1970, **52**, 139 (*P nmr*)
Gore, R.C. *et al*, *J. Assoc. Off. Anal. Chem.*, 1971, **54**, 1040 (*ir, uv*)
Nicholas, M.L. *et al*, *J. Assoc. Off. Anal. Chem.*, 1976, **59**, 1071 (*raman*)
Stan, H.J. *et al*, *Fresenius' Z. Anal. Chem.*, 1977, **287**, 271 (*glc, ms*)
Daldrup, T. *et al*, *Fresenius' Z. Anal. Chem.*, 1981, **308**, 413 (*tlc, hplc, glc*)
Stan, H.J. *et al*, *Biomed. Mass Spectrom.*, 1982, **9**, 483 (*ms*)
Pesticide Manual, 6th Ed., 244; 7th Ed., 244.
Agrochemicals Handbook, The Royal Society of Chemistry, London, 1983, A182.
Sax, N.I., *Dangerous Properties of Industrial Materials*, 6th Ed., Van Nostrand-Reinhold, 1984, 646.

Ethoprophos, BSI **E-00023**

O-Ethyl S,S-dipropyl phosphorodithioate, 9CI, 8CI. *Prophos. Mocap*

[13194-48-4]

$(H_3CCH_2CH_2S)_2P(O)OEt$

$C_8H_{19}O_2PS_2$ M 242.331

Fungicide, nematocide and soil insecticide. Yellow liq. Spar. sol. H_2O, v. sol. most org. solvs. d_4^{20} 1.094. $Bp_{0.2}$ 86-91°.

▷Exhibits high oral tox. TE4025000.

B.P., 1 155 078, (*1961*); *CA*, **65**, 14362d
Krijgsman, W. *et al*, *J. Chromatogr.*, 1976, **117**, 201 (*glc*)
Verschoyle, V.D. *et al*, *Arch. Toxicol.*, 1982, **51**, 221 (*tox*)
Ding, X.D. *et al*, *J. Agric. Food Chem.*, 1984, **32**, 622 (*hplc*)
The Agrochemical Handbook, Royal Society of Chemistry, 1983, A185.
Pesticide Manual, 6th Ed., 248; 7th Ed., 247.
Sax, N.I., *Dangerous Properties of Industrial Materials*, 6th Ed., Van Nostrand-Reinhold, 1984, 836.

2-Ethoxy-1,3,2-benzodioxaphosphole, 9CI **E-00024**

Ethyl o-phenylene phosphite

[10072-02-3]

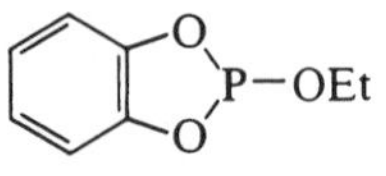

$C_8H_9O_3P$ M 184.131

Bp_{17} 99-100°, $Bp_{0.06}$ 91°. n_D^{20} 1.5073.

2-Oxide: [10508-76-6]. *Ethyl o-phenylene phosphate.*
$C_8H_9O_4P$ M 200.130
Liq. Bp_{11} 149.5-151°, $Bp_{0.05}$ 80-83°. n_D^{20} 1.4950.

2-Sulfide: [58661-92-0]. *O-Ethyl O,O-o-phenylene thiophosphate.*
$C_8H_9O_3PS$ M 216.191
Characterised spectroscopically.

Anschütz, L. *et al*, *J. Prakt. Chem.*, 1932, **133**, 65 (*synth*)
Gross, H. *et al*, *Chem. Ber.*, 1966, **99**, 2631 (*oxide*)
Gloede, J. *et al*, *Z. Anorg. Allg. Chem.*, 1980, **471**, 147 (*oxide, P nmr*)
Budilova, I.Yu. *et al*, *Zh. Obshch. Khim.*, 1983, **53**, 285 (*oxide, synth, ir, pmr*)

1-(Ethoxycarbonyl)-cyclopropyltriphenylphosphonium(1+), 9CI **E-00025**

Carbethoxycyclopropyltriphenylphosphonium(1+)

[52186-87-7]

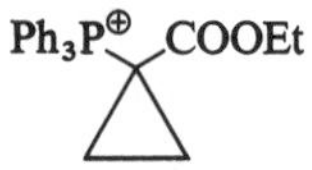

$C_{24}H_{24}O_2P^{\oplus}$ M 375.426 (ion)

Tetrafluoroborate: [52186-89-7]. Reagent for the cycloalkenylation of β-keto esters or 1,3-diketones, spiroannelation reactions with enolates of α-formylcycloalkanones, and for synth. of heterocyclic compds., e.g. 2,3-dihydrofurans. Cryst. ($CHCl_3/Et_2O$). Mp 179-181°.

Fuchs, P.L., *J. Am. Chem. Soc.*, 1974, **96**, 1607 (*synth, use*)
Dauben, W.G. *et al*, *J. Am. Chem. Soc.*, 1975, **97**, 1622 (*use*)
Dauben, W.G. *et al*, *Tetrahedron Lett.*, 1975, 4353 (*use*)
Hendrick, C.A. *et al*, *Bioorg. Chem.*, 1978, **7**, 235 (*use*)
Muchowski, J.M. *et al*, *Tetrahedron Lett.*, 1980, **21**, 4585 (*use*)
Flitsch, W. *et al*, *Justus Liebigs Ann. Chem.*, 1983, 521 (*use*)

2-Ethoxy-2,3-dihydro-1*H*-1,3,2-benzodiazaphosphole, 9CI **E-00026**

Ethyl N,N′-o-phenylenephosphorodiamidite

[58825-45-9]

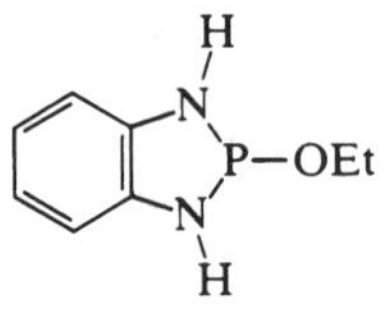

$C_8H_{11}N_2OP$ M 182.161

Liq. d_4^{20} 1.21. $Bp_{0.003}$ 135°. n_D^{20} 1.6015.

2-Sulfide: [58825-46-0]. *O-Ethyl N,N′-o-phenylenephosphorodiamidothioate.*
$C_8H_{11}N_2OPS$ M 214.221
Cryst. ($C_6H_6/CHCl_3$). Mp 149-150°.

Pudovik, M.A. *et al*, *Zh. Obshch. Khim.*, 1976, **46**, 230 (*Engl. transl.* p. 227) (*synth, deriv*)
Shagidullin, R.R. *et al*, *Zh. Obshch. Khim.*, 1984, **54**, 1283 (*Engl. transl.* p. 1148) (*ir, props*)

2-Ethoxy-2,3-dihydro-1,3,2-benzoxazaphosphole, 9CI **E-00027**

2-Ethoxy-1,3,2-benzoxazaphospholine

[7051-22-1]

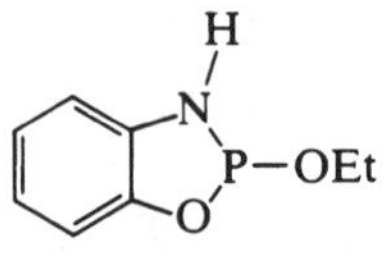

$C_8H_{10}NO_2P$ M 183.146

Liq. d_4^{20} 1.20. Bp_3 108-109°, $Bp_{0.03}$ 77-78°. n_D^{20} 1.5556. Reactive to air.

Kirpichnikov, P.A. *et al*, *Zh. Obshch. Khim.*, 1966, **36**, 1147 (*Engl. transl.* p. 1161) (*synth*)
Pudovik, A.N. *et al*, *Zh. Obshch. Khim.*, 1972, **42**, 1901 (*Engl. transl.* p. 1895) (*synth*)
Pudovik, M.A. *et al*, *Zh. Obshch. Khim.*, 1975, **45**, 266; 1982, **52**, 1302; 1983, **53**, 2468 (*Engl. transl.* pp. 252, 1144, 2226) (*synth, props*)
Shagidullin, R.R. *et al*, *Zh. Obshch. Khim.*, 1984, **54**, 1283 (*Engl. transl.* p. 1148) (*ir*)

3-Ethoxy-2,3-dihydro-2-methyl-5-phenyl-1*H*-1,2,4,3-triazaphosphole, 9CI E-00028

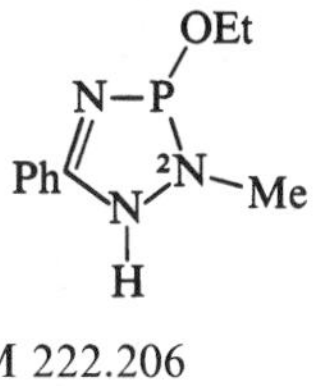

$C_{10}H_{13}N_3OP$ M 222.206

3-Oxide: [79117-07-0].
$C_{10}H_{14}N_3O_2P$ M 239.213
Cryst. ($CHCl_3$). Mp 90-92°.

3-Sulfide: [59007-36-2].
$C_{10}H_{14}N_3OPS$ M 255.274
No phys. props. reported.

Mathis, R. *et al*, *C.R. Hebd. Seances Acad. Sci., Ser. B*, 1976, **282**, 67 (*sulfide, ir*)
Schmidpeter, A., *Z. Anorg. Allg. Chem.*, 1981, **475**, 211 (*oxide, synth*)

4-Ethoxy-1,4-dihydro-1,3,5-triphenyl-1,2,4-diazaphosphorine 4-oxide, 9CI E-00029

[75482-41-6]

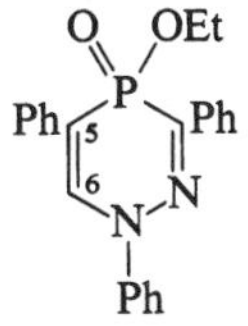

$C_{23}H_{21}N_2O_2P$ M 388.405
Mp 138-139°.

5,6-Dihydro: [37628-15-2]. *4-Ethoxy-1,4,5,6-tetrahydro-1,3,5-triphenyl-1,2,4-diazaphosphorine 4-oxide, 9CI.*
$C_{23}H_{23}N_2O_2P$ M 390.421
Cryst. (THF/hexane). Mp 150-151°.

Kosovtsev, V.V. *et al*, *Zh. Obshch. Khim.*, 1971, **41** 2649 (*Engl. transl.* p. 2682) (*synth, ir, pmr*)
Platonov, A.Yu. *et al*, *Zh. Obshch. Khim.*, 1980, **50**, 1269 (*Engl. transl.* p. 1026) (*deriv, synth, pmr, P nmr*)

2-Ethoxy-1,3-dimethyl-1,3,2-diazaphospholidine E-00030

Ethyl N,N′-*1,2-ethanediyl*-N,N′-*dimethylphosphorodiamidite*

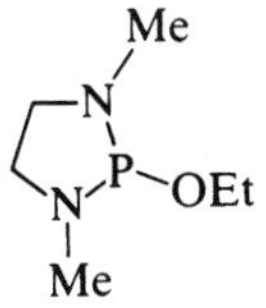

$C_6H_{15}N_2OP$ M 162.171
Liq. $Bp_{0.3}$ 85-86°. Sensitive to oxygen and moisture.

2-Oxide: [10026-27-4]. *Ethyl* N,N′-*1,2-ethanediyl*-N,N′-*dimethylphosphorodiamidate.*
$C_6H_{15}N_2O_2P$ M 178.170
Liq. $Bp_{0.3}$ 82-84°.

Ulrich, H. *et al*, *J. Org. Chem.*, 1967, **32**, 1360 (*oxide, synth, pmr*)
Savignac, P. *et al*, *J. Organomet. Chem.*, 1974, **66**, 63, 81; **72**, 361 (*synth, P nmr*)
Worley, S.D. *et al*, *J. Electron. Spectrosc. Rel. Phenom.*, 1982, **25**, 135 (*pe, struct*)

2-Ethoxy-4,5-dimethyl-1,3,2-dioxaphospholane, 9CI E-00031

Ethyl dimethylethylene phosphite. 2,3-Butylene ethyl phosphite. 2-Ethoxy-4,5-dimethyl-1,3-dioxa-2-phosphacyclopentane

[53255-91-7]

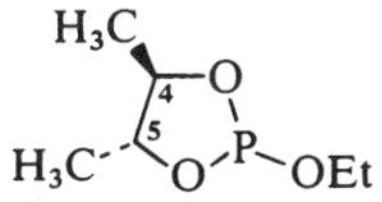

(4*R*,5*R*)-*form*

$C_6H_{13}O_3P$ M 164.141
Liq. with characteristic phosphite odour. Bp_{16} 76°. n_D^{20} 1.4422. Mixt. of diastereoisomers.

2-Oxide: [16492-17-4]. *2,3-Butylene ethyl phosphate.*
$C_6H_{13}O_4P$ M 180.140
Liq. $Bp_{0.001}$ 90°. Mixt. of diastereoisomers.

2-Sulfide: O,O-*Butylene* O-*ethyl thiophosphate.* O,O-*Butylene* O-*ethyl phosphorothioate.*
$C_6H_{13}O_3PS$ M 196.201
Liq. $Bp_{0.06}$ 66°. n_D^{19} 1.4785. Mixt. of diastereoisomers.

(4R,5R)-form

d_4^{25} 1.04. Bp_{10} 54°. $[\alpha]_D^{25}$ +49.95°. n_D^{25} 1.4297.

(4RS,5RS)-form

2-Oxide: Liq. $Bp_{0.5}$ 105-110°. n_D^{20} 1.4291.

Garner, H.K. *et al*, *J. Am. Chem. Soc.*, 1950, **72**, 5497 (*synth*)
Aksnes, G. *et al*, *Acta Chem. Scand.*, 1966, **20**, 2463 (*synth, props*)
Edmundson, R.S. *et al*, *J. Chem. Soc. (C)*, 1966, 1997; *J. Chem. Soc. (B)*, 1967, 577 (*derivs*)

2-Ethoxy-5,5-dimethyl-1,3,2-dioxaphosphorinane, 9CI E-00032

Ethyl neopentylene phosphite. 2-Ethoxy-5,5-dimethyl-1,3,2-dioxaphosphinane. 2-Ethoxy-5,5-dimethyl-1,3-dioxa-2-phosphacyclohexane

[1007-57-4]

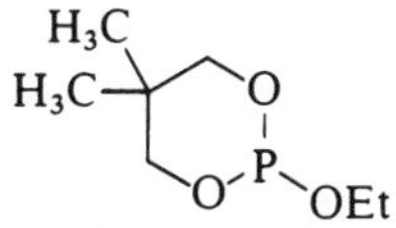

$C_7H_{15}O_3P$ M 178.167
Liq. with characteristic phosphite odour. Bp_{16} 76-77°, Bp_2 40°. n_D^{20} 1.4435.

2-Oxide: [1007-80-3]. *Ethyl neopentylene phosphate.*
$C_7H_{15}O_4P$ M 194.167
Liq. $Bp_{0.3}$ 94-95°. $n_D^{24.5}$ 1.4401.

2-Sulfide: [1076-00-2]. O-*Ethyl* O,O-*neopentylene phosphorothioate.*
$C_7H_{15}O_3PS$ M 210.227
Cryst. (Et_2O). Mp 62-63°.

Edmundson, R.S., *Tetrahedron*, 1964, **20**, 2781 (*synth, derivs, ir*)
Bartle, K.D. *et al*, *Tetrahedron*, 1967, **23**, 1701 (*oxide, pmr*)
Majoral, J.-P. *et al*, *Bull. Soc. Chim. Fr.*, 1972, 606 (*oxide, ir*)
Majoral, J.-P. *et al*, *Spectrochim. Acta, Part A*, 1972, **28**, 2247 (*ir*)
Efremov, Yu.Ya. *et al*, *Khim. Geterotsikl. Soedin.*, 1974, 1620 (*Engl. transl.* p. 1424) (*ms*)
Fazliev, D.F. *et al*, *Zh. Obshch. Khim.*, 1976, **46**, 1832 (*Engl. transl.* p. 1776) (*ir*)
Vogt, W. *et al*, *Makromol. Chem.*, 1976, **177**, 1779 (*oxide, synth, props*)

2-Ethoxy-1,3,2-dioxaphospholan-4,5-dicarboxylic acid, 9CI **E-00033**

(*4R,5R*)-*form*

$C_6H_9O_7P$ M 224.107

(*4R,5R*)-*form*

Di-Et ester: [41773-68-6].
$C_{10}H_{17}O_7P$ M 280.214
Liq. $Bp_{0.07}$ 115°. n_D^{20} 1.4502.

(*4RS,5RS*)-*form*

(±)-*form*

Di-Et ester: [41773-69-7]. Liq. $Bp_{0.09}$ 122°. n_D^{22} 1.4488.

(*4RS,5SR*)-*form*

meso-*form*

Di-Et ester: [41773-70-0]. Liq. $Bp_{0.05}$ 122°. n_D^{20} 1.4512.

Makarova, N.A. *et al*, *Izv. Akad. Nauk SSSR, Ser. Khim.*, 1973, 653 (*Engl. transl.* p. 627) (*synth, P nmr*)

2-Ethoxy-1,3,2-dioxaphospholane, 9CI **E-00034**

Ethyl ethylene phosphite

[695-11-4]

$C_4H_9O_3P$ M 136.087

Liq. d_4^{25} 1.23. Bp_{21} 60-61°. n_D^{25} 1.4390.

▷TG7230000.

2-Oxide: see 2-Ethoxy-1,3,2-dioxaphospholane 2-oxide, E-00035

2-Sulfide: see 2-Ethoxy-1,3,2-dioxaphospholane 2-sulfide, E-00036

Lucas, H.J. *et al*, *J. Am. Chem. Soc.*, 1950, **72**, 5491 (*synth*)
Cason, J. *et al*, *J. Org. Chem.*, 1959, **24**, 247 (*synth*)
Jones, R.A.Y. *et al*, *J. Chem. Soc.*, 1960, 4376 (*ir, pmr*)
Efremov, Yu.Ya. *et al*, *Khim. Geterotsikl. Soedin.*, 1972, 1329 (*Engl. transl.* p. 1202) (*ms*)
Armour, M.A. *et al*, *J. Chem. Soc., Perkin Trans. 2*, 1975, 1185 (*P nmr*)
Besserre, D. *et al*, *Org. Magn. Reson.*, 1980, **13**, 235, 313 (*cmr, conformn*)
Worley, S.D. *et al*, *J. Electron. Spectrosc. Relat. Phenom.*, 1982, **25**, 135 (*pe*)

2-Ethoxy-1,3,2-dioxaphospholane 2-oxide, 9CI **E-00035**

Ethyl ethylene phosphate

[823-31-4]

Liq. $Bp_{0.7-0.8}$ 106-108°. Easily hydrolysed.

Keay, L. *et al*, *J. Chem. Soc.*, 1961, 710 (*synth*)
Libiszowski, J. *et al*, *J. Polym. Sci., Polym. Chem. Ed.*, 1978, **16**, 1275 (*synth, pmr, P nmr, props*)
Gorenstein, D.G. *et al*, *J. Am. Chem. Soc.*, 1982, **104**, 6130 (*props*)
Taira, K. *et al*, *J. Org. Chem.*, 1984, **49**, 4531 (*synth, pmr, P nmr, props*)
Gorenstein, D.G. *et al*, *Tetrahedron*, 1987, **43**, 469 (*props*)

2-Ethoxy-1,3,2-dioxaphospholane 2-sulfide, 9CI **E-00036**

O-*Ethyl* O,O-*ethylene phosphorothioate*, *8CI*. *Ethyl ethylene thiophosphate*

[14295-11-5]

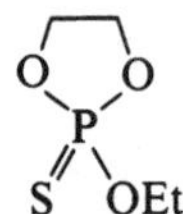

$C_4H_9O_3PS$ M 168.147

Liq. d^{21} 1.27. Bp_3 98-99°. n_D^{25} 1.4849. Very sensitive to heat undergoing isomerisation.

Yamasaki, T. *et al*, *CA*, 1956, **50**, 314 (*synth*)
Cason, J. *et al*, *J. Org. Chem.*, 1959, **24**, 247 (*synth*)
Edmundson, R.S. *et al*, *J. Chem. Soc. (C)*, 1966, 1997 (*synth, ir*)

2-Ethoxy-1,3,2-dioxaphosphorinane, 9CI **E-00037**

Ethyl trimethylene phosphite. 2-Ethoxy-1,3,2-dioxaphosphinane. 2-Ethoxy-1,3-dioxa-2-phosphacyclohexane

[696-58-2]

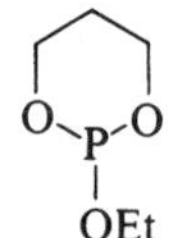

$C_5H_{11}O_3P$ M 150.114

Oil. d_4^{25} 1.12. Bp_{25} 77.1°.

2-Oxide: [697-39-2]. *Ethyl trimethylene phosphate.*
$C_5H_{11}O_4P$ M 166.113
Oil. d_4^{25} 1.26. Mp 8°. $Bp_{0.7}$ 111°.

2-Sulfide: [33148-58-2]. O-*Ethyl* O,O-*trimethylene thiophosphate.* O-*Ethyl* O,O-*trimethylene phosphorothioate.*
$C_5H_{11}O_3PS$ M 182.174
$Bp_{0.01}$ 95-100° (bath). n_D^{20} 1.4930.

Lucas, H.J. *et al*, *J. Am. Chem. Soc.*, 1950, **72**, 5491 (*synth*)
Jones, R.A.Y. *et al*, *J. Chem. Soc.*, 1960, 4376 (*ir, pmr*)
Efremov, Yu.Ya. *et al*, *Khim. Geterotsikl. Soedin.*, 1974, 1620 (*Engl. transl.* p. 1424) (*ms*)
Predvoditelev, D.A. *et al*, *Zh. Obshch. Khim.*, 1976, **46**, 40 (*Engl. transl.* p. 39) (*sulfide*)
Vogt, W. *et al*, *Makromol. Chem.*, 1976, **177**, 1779 (*oxide, synth, props*)
Lapiensis, G. *et al*, *Macromolecules*, 1977, **10**, 130 (*oxide, synth, pmr, P nmr*)

2-Ethoxy-1,3,2-dithiaphospholane **E-00038**

O-Ethyl ethylene dithiophosphite. O-Ethyl ethylene phosphorodithioite

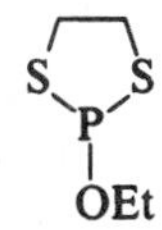

$C_4H_9OPS_2$ M 168.208

Liq. d_0^{20} 1.26. Bp_5 98-99°.

Cu_2Cl_2 addn. compd.: Cryst., dec. rapidly in air. Mp 122-124°.

Cu_2I_2 addn. compd.: Cryst., dec. rapidly in air. Mp 134°.

Arbuzov, A.E. *et al*, *Izv. Akad. Nauk SSSR, Otd. Khim. Nauk*, 1952, 453; *CA*, 1953, **47**, 4833 (*synth, derivs*)

(2-Ethoxyethenyl)-triphenylphosphonium(1+), 9CI **E-00039**

(2-Ethoxyvinyl)triphenylphosphonium(1+)

$Ph_3P^{\oplus}CH{=}CHOEt$

$C_{22}H_{22}OP^{\oplus}$ M 333.389 (ion)

Has been used in the synth. of furans.

Bromide: [71276-93-2].
$C_{22}H_{22}BrOP$ M 413.293
Cryst. (Me_2CO/EtOAc). Mp 115°.

Tetrafluoroborate:
$C_{22}H_{22}BF_4OP$ M 420.193
Cryst. (EtOAc). Mp 124°.

Tetraphenylborate:
$C_{46}H_{42}BOP$ M 652.621
Cryst. ($MeCN/Et_2O$). Mp 168-170°.

(Z)-form

Iodide: [29749-84-6].
$C_{22}H_{22}IOP$ M 460.293
Solid. Mp 147-148°.

Swan, J.M. *et al*, *Aust. J. Chem.*, 1971, **24**, 777 (*synth*)
Devlin, C.J. *et al*, *Tetrahedron*, 1972, **28**, 3501 (*iodide*)
Garst, M. *et al*, *J. Org. Chem.*, 1974, **39**, 584 (*use*)
Bestmann, H.J., *Justus Liebigs Ann. Chem.*, 1982, 1359 (*use*)
Cristau, H-J *et al*, *Phosphorus Sulfur*, 1982, **14**, 63, 73 (*synth, props*)

2-Ethoxy-4-methyl-1,3,2-dioxaphospholane, 9CI **E-00040**

2-Ethoxy-4-methyl-1,3-dioxa-2-phosphacyclopentane. Ethyl propylene phosphite

[696-50-4]

$C_5H_{11}O_3P$ M 150.114

Liq. with characteristic phosphite odour. Bp_{10} 68-69°. n_D^{20} 1.4460. Mixt. of diastereoisomers.

2-Oxide: Ethyl propylene phosphate.
$C_5H_{11}O_4P$ M 166.113
Mixt. of *cis-* and *trans-*forms.

2-Sulfide: O-*Ethyl* O,O-*propylene thiophosphate.* O-*Ethyl* O,O-*propylene phosphorothioate.*
$C_5H_{11}O_3PS$ M 182.174
Oil. n_D^{21} 1.4770. Mixt. of *cis-* and *trans-*forms.

Aksnes, G. *et al*, *Acta Chem. Scand.*, 1966, **20**, 2463 (*synth, props*)
Edmundson, R.S. *et al*, *J. Chem. Soc. (C)*, 1966, 1997 (*sulfide*)
Arbuzov, B.A. *et al*, *Dokl. Akad. Nauk SSSR, Ser. Sci. Khim.*, 1972, **204**, 1349 (*Engl. transl.* p. 523)
Efremov, Yu.Ya. *et al*, *Khim. Geterotsikl. Soedin.*, 1972, 1329 (*Engl. transl.* p. 1202) (*ms*)

2-Ethoxy-4-methyl-1,3,2-dioxaphosphorinane, 9CI **E-00041**

2-Ethoxy-4-methyl-1,3,2-dioxaphosphinane. 1,3-Butylene ethyl phosphite. 2-Ethoxy-4-methyl-1,3-dioxa-2-phosphacyclohexane

[874-65-7]

$C_6H_{13}O_3P$ M 164.141

Liq. with characteristic phosphite odour. Bp_{11} 56-57°, Bp_{10} 68-69°. n_D^{20} 1.4435. *Cis-* and *trans-*forms characterised spectroscopically.

2-Sulfide: [58995-93-0]. O,O-*1,3-Butylene* O-*ethyl thiophosphate.*
$C_6H_{13}O_3PS$ M 196.201
Liq. $Bp_{0.01}$ 95-100° (bath). n_D^{20} 1.4930.

Aksnes, G. *et al*, *Acta Chem. Scand.*, 1966, **20**, 2463; 1967, **21**, 1028 (*synth, glc*)
Bodkin, C. *et al*, *J. Chem. Soc., Perkin Trans. 2*, 1972, 2049.
Efremov, Yu.Ya. *et al*, *Khim. Geterotsikl. Soedin.* 1974, 1620 (*ms*)
Predvoditelev, D.A. *et al*, *Zh. Obshch. Khim.*, 1976, **46**, 40 (*Engl. transl.* p. 39) (*sulfide*)
Fazliev, D.F. *et al*, *Zh. Obshch. Khim.*, 1976, **46**, 1832 (*Engl. transl.* p. 1776) (*ir, raman*)

2-Ethoxy-4-methyl-1,3,2-dithiaphosphorinane, 9CI **E-00042**

2-Ethoxy-4-methyl-1,3,2-dithiaphosphinane. S,S-*1,3-Butylene* O-*ethyl phosphorodithioite*

$C_6H_{13}OPS_2$ M 196.262

Liq. d_4^{20} 1.19. Bp_1 78°. n_D^{20} 1.5750. Mixt. of diastereoisomers.

2-Sulfide: S,S-*1,3-Butylene* O-*ethyl phosphorotrithioate.*
$C_6H_{13}OPS_3$ M 228.322
Solid. Mp 37-38°. Mixt. of diastereoisomers.

Nifant'ev, E.É. *et al*, *Zh. Obshch. Khim.*, 1974, **44**, 1694 (*Engl. transl.* p. 1664) (*synth, P nmr*)
Borisenko, A.A. *et al*, *Zh. Obshch. Khim.*, 1978, **48**, 1251 (*Engl. transl.* p. 1144) (*pmr, cmr, P nmr*)

2-Ethoxy-3-methyl-1,3,2-thiazaphospholidine **E-00043**

$C_5H_{12}NOPS$ M 165.190

2-Oxide: [21124-99-2].
$C_5H_{12}NO_2PS$ M 181.189

Oil. $Bp_{0.05}$ 100-102°. n_D^{20} 1.5070.
2-Sulfide:
$C_5H_{12}NOPS_2$ M 197.250
Oil. $Bp_{0.05}$ 100-102°.

Savignac, P. *et al*, *C.R. Hebd. Seances Acad. Sci., Ser. C*, 1968, **267**, 183; 1970, **270**, 2086.

2-Ethoxynaphtho[1,2-*d*]-1,3,2-dioxaphosphole E-00044
Ethyl 1,2-naphthylene phosphite
[23988-42-3]

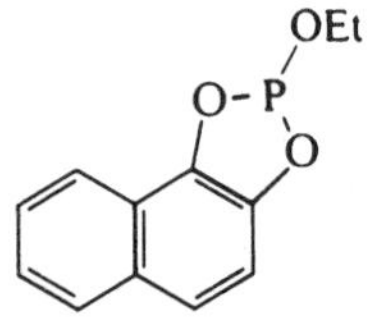

$C_{12}H_{11}O_3P$ M 234.191
Liq. $Bp_{0.11}$ 111-116°.
2-Oxide: Ethyl 1,2-naphthylene phosphate.
$C_{12}H_{11}O_4P$ M 250.190
Solid. Mp 41-42°.
2-Sulfide: O-*Ethyl* O,O-*1,2-naphthylene thiophosphate.*
$C_{12}H_{11}O_3PS$ M 266.251
Solid. Mp 185-186°.

Denny, D.B. *et al*, *J. Am. Chem. Soc.*, 1969, **91**, 5821 (*synth, pmr, P nmr*)
Voropai, L.M. *et al*, *Zh. Obshch. Khim.*, 1985, **55**, 65 (*Engl. transl.* p. 55) (*synth, derivs*)

2-Ethoxynaphtho[1,8-*de*]-1,3,2-dioxaphosphorin, 10CI E-00045
Ethyl 1,8-naphthylene phosphite
[72310-30-6]

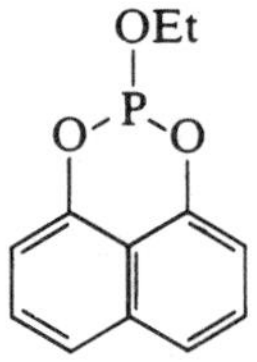

$C_{12}H_{11}O_3P$ M 234.191
Liq. Bp_2 132-134°. n_D^{20} 1.6029.
2-Oxide: Ethyl 1,8-naphthylene phosphate.
$C_{12}H_{11}O_4P$ M 250.190
Solid. Mp 66-67°.
2-Sulfide: [74536-91-7]. O-*Ethyl* O,O-*1,8-naphthylene thiophosphate.*
$C_{12}H_{11}O_3PS$ M 266.251
Solid. Mp 91-92°.

Nifant'ev, É.E. *et al*, *Zh. Obshch. Khim.*, 1981, **51**, 1528 (*Engl. transl.* p. 1295) (*synth, P nmr*)
Voropai, L.M. *et al*, *Zh. Obshch. Khim.*, 1985, **55**, 65 (*Engl. transl.* p. 55) (*derivs*)

(2-Ethoxy-2-oxoethyl)-triphenylphosphonium, 9CI E-00047
(Carboxymethyl)triphenylphosphonium ethyl ester, 8CI. Carbethoxymethyltriphenylphosphonium
[42809-80-3]

$Ph_3P^{\oplus}CH_2COOEt$

$C_{22}H_{22}O_2P^{\oplus}$ M 349.388 (ion)
pK_a 8.95, 9.2.
Chloride: [17577-28-5].
$C_{22}H_{22}ClO_2P$ M 384.841
Cryst. (CH_2Cl_2/CCl_4). Mp 87° dec., 144° dec. (after intensive drying). Forms a dihydrate.
Bromide: [1530-45-6].
$C_{22}H_{22}BrO_2P$ M 429.292
Cryst. (C_6H_6). Mp 158° dec. pK_a 8.81.

Isler, O. *et al*, *Helv. Chim. Acta*, 1957, **40**, 1242 (*bromide*)
Considine, W., *J. Org. Chem.*, 1962, **27**, 647 (*chloride, ir*)
Quemeneur, F. *et al*, *J. Chem. Res. (S)*, 1979, 187 (*bromide*)

2-Ethoxy-3-phenyl-1,3,2-oxazaphospholidine, 9CI E-00048
2-Ethoxytetrahydro-3-phenyl-1,3,2-oxazophosphole. 2-Ethoxy-3-phenyl-1,3,2-oxazaphospholane
[5679-72-1]

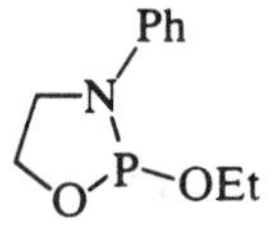

$C_{10}H_{14}NO_2P$ M 211.200
Liq. d_4^{20} 1.16. $Bp_{0.06}$ 100-101°. n_D^{20} 1.5619.
2-Sulfide: [31700-79-5].
$C_{10}H_{14}NO_2PS$ M 243.260
Solid. Mp 75-77° (64-65°).

Mitsumobu, O. *et al*, *Bull. Chem. Soc. Jpn.*, 1967, **40**, 2964 (*synth*)
Pudovik, A.N. *et al*, *Zh. Obshch. Khim.*, 1970, **40**, 1477; 1971, **41**, 2407 (*Engl. transl.* pp. 1463, 2434) (*synth, sulfide*)
Efremov, Yu.Yu., *Khim. Geterotsikl. Soedin.*, 1973, 894 (*Engl. transl.* p. 824) (*ms*)

(2-Ethoxy-2-propenylidene)-triphenylphosphorane, 9CI E-00049

2-Ethoxyallylidenetriphenylphosphorane

[62639-98-9]

$Ph_3P{=}CHCH(OEt){=}CH_2$

$C_{23}H_{23}OP$ M 346.408

Reagent for the cyclohexenone annelation of enones. Mp >60° dec.

Ramirez, F. *et al*, *J. Org. Chem.*, 1957, **22**, 41 (*synth, use*)
Martin, S.F. *et al*, *J. Org. Chem.*, 1977, **42**, 1664 (*synth, use*)
Bestmann, H.J. *et al*, *Angew. Chem., Int. Ed. Engl.*, 1981, **20**, 575 (*synth, props, use*)

(2-Ethoxypropenyl)-triphenylphosphonium(1+) E-00050

$Ph_3P^{\oplus}CH_2C(OEt){=}CH_2$

$C_{23}H_{24}OP^{\oplus}$ M 347.416 (ion)

Salts dec. by alkali → Me_2CO.

Iodide:
$C_{23}H_{24}IOP$ M 474.320
Cryst. (MeOH/EtOAc). Mp 163-165°. With butyllithium forms the ylide.

Ylide: see (2-Ethoxy-2-propenylidene)-triphenylphosphorane, E-00049

Ramirez, F. *et al*, *J. Org. Chem.*, 1957, **22**, 41 (*synth, props, uv, ir*)

2-Ethoxytetrahydro-2*H*-1,3,2-oxazaphosphorine, 9CI E-00051

*2-Ethoxytetrahydro-2*H*-1,3,2-oxazaphosphinine. 2-Ethoxy-1,3,2-oxazaphosphorinane*

[35726-77-3]

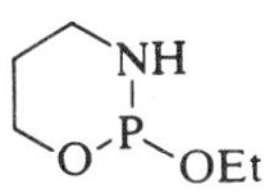

$C_5H_{12}NO_2P$ M 149.129

Liq. d_4^{20} 1.10. $Bp_{0.4}$ 47-48°. n_D^{20} 1.4670.

2-Sulfide: [19733-71-2].
$C_5H_{12}NO_2PS$ M 181.189
Liq. Bp_2 151°. n_D^{20} 1.5180.

Kil'desheva, O.V. *et al*, *Izv. Akad. Nauk SSSR, Ser. Khim.*, 1968, 398 (*Engl. transl.* p. 386) (*sulfide*)
Pudovik, A.N. *et al*, *Zh. Obshch. Khim.*, 1971, **41**, 2180 (*Engl. transl.* p. 2205) (*synth*)
Shagidullin, R.R. *et al*, *Zh. Obshch. Khim.*, 1984, **54**, 1283 (*Engl. transl.* p. 1168) (*ir*)

2-Ethoxy-4,4,5,5-tetramethyl-1,3,2-dioxaphospholane, 9CI E-00052

2-Ethoxy-4,4,5,5-tetramethyl-1,3-dioxa-2-phosphacyclopentane. Ethyl tetramethylethylene phosphite. Ethyl pinacolyl phosphite

[38206-24-5]

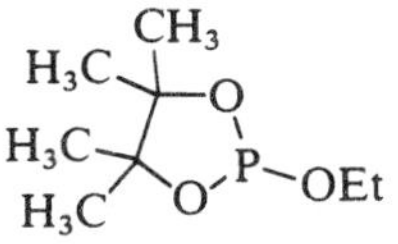

$C_8H_{17}O_3P$ M 192.194

Liq. with characteristic phosphite odour. Bp_{14} 75-76°. n_D^{20} 1.4392.

2-Sulfide: [14274-45-4]. *O-Ethyl O,O-tetramethylethylene phosphorothioate.*
$C_8H_{17}O_3PS$ M 224.254
Oil. n_D^{19} 1.4775.

Edmundson, R.S. *et al*, *J. Chem. Soc. (C)*, 1966, 1997 (*sulfide*)
Arbuzov, B.A. *et al*, *Dokl. Akad. Nauk SSSR, Ser. Sci. Khim.*, 1972, **204**, 1349 (*Engl. transl.* p. 523)
Efremov, Yu.Ya. *et al*, *Khim. Geterosikl. Soedin.*, 1972, 1329 (*Engl. transl.* p. 1202) (*ms*)

2-Ethoxy-1,3,2-thiazaphospholidine, 8CI E-00053

2-Ethoxy-1,3,2-thiazaphospholane

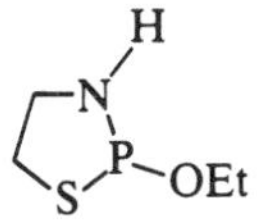

$C_4H_{10}NOPS$ M 151.163

2-Oxide: [13346-80-0].
$C_4H_{10}NO_2PS$ M 167.162
Liq. d_4^{20} 1.31. $Bp_{0.02}$ 139°. n_D^{20} 1.5230.

2-Sulfide: [13346-70-8].
$C_4H_{10}NOPS_2$ M 183.223
Oil. d_4^{20} 1.30. n_D^{20} 1.5874. Unstable to distn.

U.S.P., 3 285 999, (*1966*); *CA*, **66**, 28778 (*derivs*)
Savignac, P. *et al*, *C.R. Hebd. Seances Acad. Sci., Ser. C*, 1968, **266**, 1791 (*oxide*)

***O*-Ethyl *O,O*-bis(4-nitrophenyl) phosphorothioate, 9CI** E-00054

[7508-73-8]

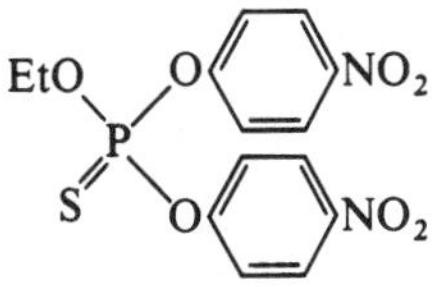

$C_{14}H_{13}N_2O_7PS$ M 384.300

Powerful insecticide. Cryst. (EtOH). Mp 125°.

▷TE6650000.

Mandel'baum, Y.A. *et al*, *Dokl. Akad. Nauk SSSR*, 1955, **100**, 77; *CA*, 1956, **50**, 1650 (*synth*)
Hanic, F. *et al*, *Acta Crystallogr.*, 1958, **11**, 127 (*cryst struct*)
Ahmed, M.K. *et al*, *J. Agric. Food Chem.*, 1958, **6**, 740 (*metab*)
Gersmann, H.R. *et al*, *Recl. Trav. Chim. Pays-Bas*, 1958, **77**, 1018 (*uv*)
Ketelaar, J.A.A. *et al*, *Recl. Trav. Chim. Pays-Bas*, 1959, **78**, 190 (*ir*)
Joiner, R.L. *et al*, *Pestic. Biochem. Physiol.*, 1973, **2**, 371 (*pharmacol, use*)
Markowska, A. *et al*, *Nouv. J. Chim.*, 1979, **3**, 409 (*pmr, P nmr, props*)

Ethyl carbonate-diethyl phosphate anhydride E-00055

Ethoxycarbonyl diethyl phosphate

[4456-15-9]

$EtOOCOP(O)(OEt)_2$

$C_7H_{15}O_6P$ M 226.166

Liq. d_{20}^{20} 1.17. Bp_1 85-86°. n_D^{20} 1.4139. Easily hydrolysed.

Shamshurin, A.A. *et al*, *Zh. Obshch. Khim.*, 1965, **35**, 1877 (*Engl. transl.* p. 1871)
Lambie, A.J., *Tetrahedron Lett.*, 1966, 3709.

Ethyl (chlorosulfonyl)-phosphorodichlorimidate, 9CI E-00056

[32755-55-8]

$ClSO_2N{=}PCl_2OEt$

$C_2H_5Cl_3NO_3PS$ M 260.459
Liq. Isom. in presence of Et_2O.

Roesky, H.W. *et al*, *Chem. Ber.*, 1971, **104**, 3204 (*synth, pmr, ir, P nmr, props*)

Ethyl cyano(diethoxyphosphinyl)acetate, 9CI E-00057

Triethyl cyanophosphonoacetate, 8CI. *Triethyl 2-phosphono-2-cyanoacetate*
[13504-83-1]

$(EtO)_2P(O)CH(CN)COOEt$

$C_9H_{16}NO_5P$ M 249.203
Liq. $Bp_{0.01}$ 90-118°.

Martin, D., *Chem. Ber.*, 1967, **100**, 187 (*synth, ir*)

Ethyl (dichlorophosphinyl)-phosphorodichloroimidate, 10CI E-00058

P,P-*Dichloro*-N-*dichlorophosphinyl*-P-*ethoxyphosphazene*. P,P-*Dichloro*-N-*dichlorophosphinyl*-P-*ethoxyphosphine imide*
[74308-16-0]

$Cl_2P(O)N{=}PCl_2OEt$

$C_2H_5Cl_4NO_2P_2$ M 278.827
Liq. Readily isomerises.

Riesel, L., *Z. Chem.*, 1980, **20**, 98 (*synth*)
Thomas, B. *et al*, *Z. Chem.*, 1980, **20**, 335 (*synth, P nmr, props*)

Ethyl 2-(diethoxyphosphinyl)propanoate, 9CI E-00059

Triethyl 2-phosphonopropionate, 8CI
[3699-66-9]

$(EtO)_2P(O)CH(CH_3)COOEt$

$C_9H_{19}O_5P$ M 238.220

(±)-form

Reagent for Wittig-Horner reactions. Liq. d_4^{20} 1.11. Bp_{12} 143-144°, $Bp_{0.6}$ 95-98°. n_D^{20} 1.4320.

Arbuzov, A.E. *et al*, *Ber.*, 1927, **60**, 291 (*synth*)
Nishiwaki, T., Tet, 1967, **23**, 2181 (*ms*)
Kresze, G. *et al*, *Justus Liebigs Ann. Chem.*, 1972, **756**, 112 (*synth*)
Gallagher, G. *et al*, *Synthesis*, 1974, 122 (*use*)
Schaumann, E. *et al*, *Justus Liebigs Ann. Chem.*, 1979, 1715 (*synth, use*)
Minami, T. *et al*, *Synthesis*, 1982, 231 (*use*)

Ethyl diethylphosphoramidochloridite, 9CI E-00060

[14114-77-3]

$Et_2NPClOEt$

$C_6H_{15}ClNOP$ M 183.617
Pungent liq. $Bp_{1.3}$ 44-45°. n_D^{20} 1.4680. Dec. to EtCl at 170-190°.

Zwierzak, A. *et al*, *Tetrahedron*, 1967, **23**, 2243 (*synth, props*)
Butkova, O.L. *et al*, *Izv. Akad. Nauk SSSR, Ser. Khim.*, 1982, 2390 (*Engl. transl.* p. 2106)

4′-Ethyl-1,3-dihydro-5-phenyl-3,3-bis(trifluoromethyl)spiro[2*H*-1,4,2-diazaphosphole-2,1′-[2,6,7]trioxa[1]-phosphabicyclo[2.2.2]octane], 9CI E-00061

$C_{16}H_{17}F_6N_2O_3P$ M 430.286
1-(2,6-Dimethylphenyl):
$C_{24}H_{25}F_6N_2O_3P$ M 534.437
Solid. Mp 146° dec.
1-(2,4,6-Trimethylphenyl): [75619-23-7].
$C_{25}H_{27}F_6N_2O_3P$ M 548.464
Solid. Mp 138° dec.

Burger, K. *et al*, *Z. Naturforsch., B*, 1980, **35**, 749 (*synth, ir, pmr, F and P nmr*)

***O*-Ethyl *S*-[2-(diisopropylamino)ethyl] methylphosphonothioate** E-00062

S-[*2-[Bis(1-methylethyl)amino]ethyl*] O-*ethyl methylphosphonothioate*, 9CI. *VX agent*
[50782-69-9]

$C_{11}H_{26}NO_2PS$ M 267.366
Acetylcholinesterase inhibitor.
▷Extremely toxic, LD_{50} (rats, percutaneous) 0.012 mg/Kg. (*S*)-(−)-form more toxic than (*R*)-(+)-form. TB1090000.

B.P., 1 346 409, (*1974*); *CA*, **81**, 4068 (*synth*)
Epstein, E. *et al*, *Phosphorus*, 1974, **4**, 157 (*props*)
U.S.P., 3 911 059, (*1975*); *CA*, **84**, 31243 (*synth*)
Hall, C.R. *et al*, *J. Pharm. Pharmacol.*, 1977, **29**, 574 (*tox*)
Sass, S. *et al*, *Org. Mass Spectrom.*, 1979, **14**, 257 (*ms*)

***P*-Ethyl-*N*,*N*-dimethylphosphonamidodithioic acid, 9CI** E-00063

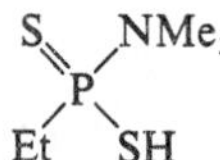

$C_4H_{12}NPS_2$ M 169.239
Me ester: [67242-51-7]. *Methyl* P-*methyl*-N,N-*dimethylphosphonamidodithioate*.
$C_5H_{14}NPS_2$ M 183.266
Liq. $Bp_{0.025}$ 54.5°. n_D^{25} 1.5636.
Ph ester: *Phenyl* P-*ethyl*-N,N-*dimethylphosphonamidodithioate*.
$C_{10}H_{16}NPS_2$ M 245.337
Liq. $Bp_{0.01}$ 104°.

Ger. Pat., 1 139 119, (*1962*); *CA*, **58**, 12601 (*phenyl ester*)
Wustner, D.A. *et al*, *J. Agric. Food Chem.*, 1978, **26**, 1104 (*methyl ester*)

Ethyl dimethylphosphoramidofluoridate, 9CI, 8CI E-00064

[358-29-2]

$Me_2NPF(O)OEt$

$C_4H_{11}FNO_2P$ M 155.109
▷TB4710000.

(±)-form
Liq. Bp_{18} 76-78°.
▷V. toxic. Miotic

Cook, H.G. *et al, J. Chem. Soc.*, 1949, 2921 (*synth, tox*)
Mager, P.P., *Toxicol. Lett.*, 1982, **11**, 67 (*tox*)

O-Ethyl *O,S*-dimethyl phosphorothioate, 10CI, 9CI — E-00065

O-Ethyl O,S-dimethyl thiophosphate

MeO–P(=O)(–SMe)(OEt) *(R)-form*

$C_4H_{11}O_3PS$ M 170.163
***(R)*-form** [57557-26-3]
Bp_{15} 115°, $Bp_{0.2}$ 75° (bath). $[\alpha]_D$ +1.0° (c, 1.0 in $CHCl_3$).
***(S)*-form** [57557-25-2]
$[\alpha]_D$ −0.9° (c, 1.0 in $CHCl_3$).

Hall, C.R. *et al, J. Chem. Soc., Perkin Trans. 1*, 1979, 1104, 1646 (*synth, stereochem*)
Hall, C.R. *et al, J. Chem. Soc., Perkin Trans. 1*, 1981, 2368 (*synth*)

Ethyldiphenylphosphine, 9CI — E-00066

[607-01-2]

Ph_2PEt

$C_{14}H_{15}P$ M 214.246
Ligand for metals of Groups IB, VB, VIB, and VIII, and also Al, Ga. Foul-smelling liq. Easily oxid. Bp_{22} 184°, $Bp_{1.4}$ 112°. pK_a 2.62.
B,MeI: Ethylmethyldiphenylphosphonium iodide.
$C_{15}H_{18}IP$ M 356.185
Cryst. (EtOH). Mp 186-187°.
Oxide: [1733-57-9].
$C_{14}H_{15}OP$ M 230.246
Cryst. (Et_2O). Mp 113°, Mp 123.5-124°.
Sulfide: [1017-98-7].
$C_{14}H_{15}PS$ M 246.306
Solid. Mp 67°.
Selenide:
$C_{14}H_{15}PSe$ M 293.206
Solid. Mp 49°.

Zingaro, R.A. *et al, J. Chem. Eng. Data*, 1963, **8**, 226 (*synth, derivs*)
Mann, B.E., *J. Chem. Soc., Perkin Trans. 2*, 1972, 30 (*cmr, P nmr*)
Allen, E.A. *et al, Spectrochim. Acta, Part A*, 1974, **30**, 1219 (*ir, raman*)
Mandel, F.S. *et al, J. Magn. Reson.*, 1974, **14**, 235 (*pmr, P nmr*)
Albright, T.A. *et al, J. Org. Chem.*, 1975, **40**, 3437 (*oxide, cmr, P nmr*)
Inorg. Synth., 1976, **16**, 155 (*synth*)
Postle, S.R., *Phosphorus Sulfur*, 1977, **3**, 269 (*sulfide, cmr*)
Dmitriev, V.I. *et al, Zh. Obshch. Khim.*, 1978, **48**, 52 (*Engl. transl.* p. 42) (*synth, nmr*)
Vincent, E. *et al, Spectrochim. Acta, Part A*, 1980, **36**, 699 (*synth, pmr, cmr, P nmr, complexes*)
Ginsberg, A.P. *et al, Inorg. Chem.*, 1982, **21**, 3666 (*complexes*)
Sax, N.I., *Dangerous Properties of Industrial Materials*, 6th Ed., Van Nostrand-Reinhold, 1984, 656.

P-Ethyl-*N,N*-diphenylphosphonamidic acid, 9CI — E-00067

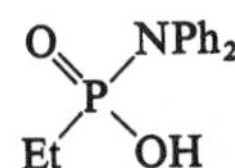

$C_{14}H_{16}NO_2P$ M 261.260
Et ester: [24102-76-9]. *Ethyl* P-*ethyl*-N,N-*diphenylphosphonamidate.*
$C_{16}H_{20}NO_2P$ M 289.313
Liq. d_0^{20} 1.11. $Bp_{0.03}$ 100-101°. n_D^{20} 1.5698.

Pudovik, A.N. *et al, Zh. Obshch. Khim.*, 1969, **39**, 1890 (*Engl. transl.* p. 1851)

Ethyl ethylphosphinate — E-00068

[998-80-1]

EtOPH(O)Et

$C_4H_{11}O_2P$ M 122.103
(+)-form [31355-98-3]
$[\alpha]_D^{20}$ +12.04°. n_D^{20} 1.4208.
(−)-form [59624-92-9]
$[\alpha]_D^{20}$ −5.4°. n_D^{20} 1.4190.
(±)-form
Liq. d_4^{20} 1.02. Bp_{16} 80-81°. n_D^{25} 1.4238.

Michalski, J. *et al, Rocz. Chem.*, 1956, **30**, 799; *CA*, **54**, 10832 (*synth*)
Gonçalves, H. *et al, Bull. Soc. Chim. Fr.*, 1961, 1595 (*synth, ir*)
Sanchez, M. *et al, Spectrochim. Acta, Part A*, 1967, **23**, 2617 (*ir*)
Vinogradov, L.I. *et al, Zh. Obshch. Khim.*, 1972, **42**, 1724 (*Engl. transl.* p. 1712) (*pmr, nmr*)
Andreev, N.A. *et al, Zh. Obshch. Khim.*, 1979, **49**, 2230; 1980, **50**, 803 (*Engl. transl.* pp. 641, 1959) (*synth, ir, nmr*)
Buina, N.A. *et al, Izv. Akad. Nauk SSSR, Ser. Khim.*, 1979, 2362 (*Engl. transl.* p. 2184) (*synth*)

O-Ethyl ethylphosphonochloridothioate, 9CI — E-00069

[1497-68-3]

EtO–P(=S)(Cl)(Et) *(R)-form*

$C_4H_{10}ClOPS$ M 172.609
***(R)*-form** [4789-37-1]
Liq. $Bp_{0.3}$ 26-27°. $[\alpha]_D$ −82.50° (neat). n_D^{24} 1.4923.
***(S)*-form** [13547-42-7]
Liq. $Bp_{0.05}$ 20°. $[\alpha]_D^{20}$ +95° (c, 1 in diisopropyl ether), $[\alpha]_D^{20}$ +81.2° (neat). n_D^{20} 1.4912.
(±)-form [13547-40-5]
Pugent liq. d_4^{20} 1.15. $Bp_{0.7}$ 34-36°. n_D^{24} 1.4921.

Hoffmann, F.W. *et al, J. Am. Chem. Soc.*, 1958, **80**, 3945 (*synth*)
Mikolajczyk, M. *et al, Tetrahedron*, 1972, **28**, 4357 (*config, uv, cd*)
Stec, W.J. *et al, J. Org. Chem.*, 1976, **41**, 227, 233 (*enantiomers, synth, P nmr*)
Allahyari, R. *et al, J. Agric. Food Chem.*, 1977, **25**, 471 (*enantiomers, synth, config*)
Hirashima, A. *et al, Agric. Biol. Chem.*, 1983, **47**, 829 (*synth*)

O-Ethyl ethylphosphonodithioate, 9CI **E-00070**

O-Ethyl hydrogen ethylphosphonodithioate

[995-79-9]

(*R*)-*form* of esters

$C_4H_{11}OPS_2$ M 170.224

(*R*)-*form*

Et ester: Liq. $Bp_{0.5}$ 49°. $[\alpha]_D$ −72.0° (neat).

Ph ester: see Fonofos, F-00052

4-Bromophenyl ester: S-*4-Bromophenyl* O-*ethyl ethylphosphonodithioate. Bromofonofos.*

$C_{10}H_{14}BrOPS_2$ M 325.218

Solid. Mp 31.5°. $[\alpha]_D^{24}$ +258.1° (neat), −129.5° (c, 9.65 in cyclohexane).

(*S*)-*form*

Et ester: [23124-79-0]. Liq. $Bp_{0.2}$ 45-47°. $[\alpha]_D$ +62.4° (neat). n_D^{22} 1.5223.

Ph ester: see Fonofos, F-00052

4-Bromophenyl ester: Mp 31.5°. $[\alpha]_D^{24}$ −257.9° (neat), $[\alpha]_D$ +130.7° (c, 17.0 in cyclohexane).

(±)-*form*

Liq. $Bp_{0.09}$ 65-67°. n_D^{20} 1.5412.

Et ester: [3347-32-8]. O,S-*Diethyl ethylphosphonodithioate.*

$C_6H_{15}OPS_2$ M 198.278

Liq. d_4^{20} 1.09. Bp_2 56-7°. n_D^{20} 1.5200.

Ph ester: see Fonofos, F-00052

4-Chlorophenyl ester: [2984-64-7]. O-*Ethyl* S-*4-chlorophenyl ethylphosphonodithioate.* N-*2596.*

$C_{10}H_{14}ClOPS_2$ M 280.767

Soil insecticide.

▷TA5425000.

Chupp, J.P. *et al, J. Org. Chem.*, 1962, **27**, 3832 (*synth*)

Omelanczuk, J. *et al, Tetrahedron*, 1971, **27**, 5587 (*ethyl ester, enantiomers*)

Mikolajczyk, M. *et al, Tetrahedron*, 1972, **28**, 4357 (*ethyl ester, config, uv, cd*)

Ishmaeva, E.A. *et al, Zh. Obshch. Khim.*, 1974, **44**, 2625 (*Engl. transl.* p. 2582) (*ethyl ester, synth*)

Allahyari, R. *et al, J. Agric. Food Chem.*, 1977, **25**, 47 (*bromophenyl ester, config, cryst struct*)

Miaullis, J.B. *et al, J. Agric. Food Chem.*, 1977, **25**, 501 (*chlorophenyl ester*)

Andreev, N.A. *et al, Zh. Obshch. Khim.*, 1982, **52**, 1785 (*Engl. transl.* p. 1581) (*synth, ir, P nmr*)

O-Ethyl ethylphosphonofluoridothioate, 9CI **E-00071**

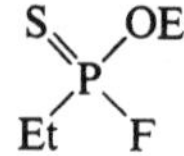

$C_4H_{10}FOPS$ M 156.155

(+)-*form* [24680-53-3]

$[\alpha]_D$ +18.75° (pure).

(−)-*form* [24680-57-7]

$[\alpha]_D$ −13.9° (pure).

(±)-*form*

Liq. Bp_{45} 65°. n_D^{20} 1.4422.

Pliszka-Krawiecka, B. *et al, Bull. Acad. Polon. Sci., Ser. Sci. Chim.*, 1969, **17**, 75 (*synth*)

O-Ethyl ethylphosphonoselenoate, 9CI **E-00072**

O-*Monoethyl ethylphosphonoselenoate.* O-*Ethyl hydrogen ethylphosphonoselenoate*

[3958-00-7]

$C_4H_{11}O_2PSe$ M 201.063

The phosphoryl struct. predominates.

(*R*)-*form* [53228-53-8]

Liq. d_4^{20} 1.44. $Bp_{0.008}$ 70-70.5°. $[\alpha]_D$ +3.72°, +11.36° (neat). n_D^{20} 1.5235.

O-*Trimethylsilyl ester:* O-*Ethyl* O-*trimethylsilyl ethylphosphonoselenoate.*

$C_7H_{19}O_2PSeSi$ M 273.245

d_4^{20} 1.18. Bp_1 54-55°. $[\alpha]_D^{20}$ −4.03° (neat). n_D^{20} 1.4785.

(*S*)-*form* [53228-52-7]

Liq. d_4^{20} 1.44. $Bp_{0.01}$ 78°. $[\alpha]_D$ −17.54° (neat). n_D^{20} 1.5220.

Se-*Me ester:* [65426-09-7]. O-*Ethyl* Se-*methyl ethylphosphonoselenoate.*

$C_5H_{13}O_2PSe$ M 215.090

d_4^{20} 1.38. $Bp_{0.08}$ 58-59°. $[\alpha]_D$ −37.8°, $[\alpha]_D^{20}$ −72.84°. n_D^{20} 1.5040.

O-*Trimethylsilyl ester:* d_4^{20} 1.19. Bp_2 58°. $[\alpha]_D$ +5.16° (neat). n_D^{20} 1.4794.

(±)-*form* [53228-54-9]

Liq. d_4^{20} 1.44. $Bp_{0.03}$ 80-82°. n_D^{20} 1.5235.

Se-*Me ester:* d_4^{20} 1.38. $Bp_{0.07}$ 55-56°. n_D^{20} 1.5028.

O-*Trimethylsilyl ester:* Liq. d_4^{20} 1.18. $Bp_{0.075}$ 42-43°. n_D^{20} 1.4790.

Nuretdinov, I.A. *et al, Izv. Akad. Nauk SSSR, Ser. Khim.*, 1974, 483 (*Engl. transl.* p. 455) (*resoln, P and Se nmr*)

Nuretdinov, I.A. *et al, Izv. Akad. Nauk SSSR, Ser. Khim.*, 1977, 2635 (*Engl. transl.* p. 2441) (*derivs*)

Nuretdinov, I.A. *et al, Izv. Akad. Nauk SSSR, Ser. Khim.*, 1978, 437 (*Engl. transl.* p. 378) (*synth, ir, P nmr*)

Buina, N.A. *et al, Izv. Akad. Nauk SSSR, Ser. Khim.*, 1979, 2362 (*Engl. transl.* p. 2184)

Nuretdinov, I.A. *et al, Zh. Obshch. Khim.*, 1980, **50**, 1429; *CA*, **93**, 203996 (*esters, synth, ir, nmr*)

O-Ethyl ethylphosphonoselenothioate, 9CI **E-00073**

[66499-13-6]

$C_4H_{11}OPSSe$ M 217.124

(±)-*form*

Dicyclohexylammonium salt: Solid. Mp 113-115°.

Mastryukova, T.A. *et al, Zh. Obshch. Khim.*, 1978, **48**, 463, 1447 (*Engl. transl.* pp. 412, 1329)

O-Ethyl ethylphosphonothioate, 9CI E-00074

[7776-66-1]

$$S{=}\overset{\text{OEt}}{\underset{\text{Et}}{P}}{-}OH \quad (R)\text{-form}$$

$C_4H_{11}O_2PS$ M 154.163

Tautomeric with thiol form. Metab. of Fonofos.

(*R*)-form [4789-36-0]

Oil. $Bp_{0.1}$ 64-65°. $[\alpha]_D^{24}$ +14.82° (neat). n_D^{24} 1.4882.

Dicyclohexylammonium salt: Mp 158-160°. $[\alpha]_D$ +6.85° (c, 3.23 in MeOH).

S-*Me ester:* O-*Ethyl* S-*methyl ethylphosphonothioate.*

$C_5H_{13}O_2PS$ M 168.190

Liq. Bp_2 62°. $[\alpha]_D^{20}$ +72.8° (C_6H_6). n_D^{20} 1.4782.

S-*Et ester:* [7348-85-8]. O,S-*Diethyl ethylphosphonothioate.*

$C_6H_{15}O_2PS$ M 182.217

No phys. props. reported.

S-*Ph ester:* O-*Ethyl* S-*phenyl ethylphosphonothioate.*

$C_{10}H_{15}O_2PS$ M 230.261

$[\alpha]_D^{24}$ +112.6° (c, 0.533 in cyclohexane).

Chloride: see O-Ethyl ethylphosphonochloridothioate, E-00069

(*S*)-form [5152-74-9]

Oil. $Bp_{1\text{-}2}$ 92-94°, $Bp_{0.08}$ 61-62°. $[\alpha]_D^{24}$ −15.45° (neat). n_D^{24} 1.4887.

Dicyclohexylammonium salt: Mp 160-161°. $[\alpha]_D$ −7.11° (c, 2.15 in MeOH).

S-*Me ester:* Bp_1 55°. $[\alpha]_D^{20}$ −75.8° (C_6H_6). n_D^{20} 1.4782.

S-*Et ester:* Liq. $Bp_{0.6}$ 55°. $[\alpha]_D$ −70.9° (neat). n_D^{20} 1.4727.

S-*Ph ester:* $[\alpha]_D$ −121.7° (c, 0.636 in cyclohexane). More toxic to mice and flies than the (*R*)-form.

Chloride: see O-Ethyl ethylphosphonochloridothioate, E-00069

(±)-form [36585-29-2]

Oil. $Bp_{0.1}$ 64-65°, $Bp_{0.01}$ 49-53°. n_D^{22} 1.4837.

Na salt: Solid (pet. ether). Mp 185-186°.

Dicyclohexylammonium salt: Cryst. (Me_2CO/pet. ether). Mp 166-168°.

S-*Et ester:* Liq. d_4^{20} 1.07. Bp_{12} 103-104°, $Bp_{0.9}$ 62°. n_D^{22} 1.4720.

S-*Ph ester:* [944-21-8]. *Fonofos oxon.*

▷Toxic

O-(*2,4,5-Trichlorophenyl*) *ester: see Trichloronate, T-00419*

O-(*2,5-Dichloro-4-iodophenyl*) *ester: see O-(2,5-Dichloro-4-iodophenyl) O-ethyl ethylphosphonothioate, D-00171*

Chloride: see O-Ethyl ethylphosphonochloridothioate, E-00069

Kabachnik, M.I. *et al, Zh. Obshch. Khim.*, 1956, **26**, 2228 (*Engl. transl.* p. 2491) (*ir, esters, synth*)
Kabachnik, M.I. *et al, Izv. Akad. Nauk SSSR, Ser. Khim.*, 1956, 193 (*Engl. transl.* p. 185) (*esters, synth*)
Aaron, H.S. *et al, J. Am. Chem. Soc.*, 1958, **80**, 107 (*resoln, derivs*)
Mikolajczyk, M., *Tetrahedron*, 1967, **23**, 1543 (*synth*)
Omelanczuk, J. *et al, Tetrahedron*, 1971, **27**, 5587 (*diethyl ester*)
Mikolajczyk, M. *et al, Tetrahedron*, 1972, **28**, 4357 (*cd, config*)
Stec, W.J. *et al, J. Org. Chem.*, 1976, **41**, 227, 233 (*methyl ester, synth, P nmr*)
Allahyari, R. *et al, J. Agric. Food Chem.*, 1977, **25**, 471 (*synth, resoln, props, cryst struct*)
Lee, P.W. *et al, Pestic. Biochem. Physiol.*, 1978, **8**, 146, 158 (*phenyl ester, metab*)
Mikolajczyk, M. *et al, J. Am. Chem. Soc.*, 1978, **100**, 7003 (*P nmr*)

2-Ethylhexyl diphenyl phosphate, 9CI, 8CI E-00075

Octicizer, USAN

[1241-94-7]

$$(PhO)_2P(=O)OCH_2CH(CH_2CH_3)(CH_2)_3CH_3$$

$C_{20}H_{27}O_4P$ M 362.405

Plasticiser, fireproofer.

▷TC6125000.

Hummel, D., *Kunststoffe*, 1965, **55**, 102 (*ir*)
Ger. Pat., 2 538 091, (*1976*); *CA*, **85**, 22502x (*synth, use*)
Ger. Pat., 2 833 341, (*1980*); *CA*, **93**, 26101 (*synth*)

Ethylidenetriphenylphosphorane, 9CI E-00076

[1754-88-7]

$$Ph_3P{=}CHCH_3$$

$C_{20}H_{19}P$ M 290.344

A Wittig reagent. Obt. as salt-free, red cryst.

Grim, S.O. *et al, J. Chem. Soc., Chem. Commun.*, 1967, 1191 (*P nmr*)
Grim, S.O. *et al, J. Org. Chem.*, 1968, **33**, 2993 (*synth, uv*)
Yamamoto, Y. *et al, J. Organomet. Chem.*, 1975, **97**, 479 (*complexes*)
Albright, T.A. *et al, J. Am. Chem. Soc.*, 1976, **98**, 6249 (*cmr, P nmr*)
Ostoja Starzewski, K.A. *et al, J. Am. Chem. Soc.*, 1976, **98**, 8486 (*synth, pe*)
Ostoja Starzewski, K.A. *et al, Phosphorus*, 1976, **6**, 177 (*cmr*)
Petragnani, N. *et al, J. Organomet. Chem.*, 1976, **114**, 281 (*props*)
Schlosser, M. *et al, Chimia*, 1982, **36**, 396 (*synth*)

Ethylidynephosphine, 10CI E-00077

1-Phosphapropyne

[67517-97-9]

$$P{\equiv}CCH_3$$

C_2H_3P M 58.019

Gas. Stable to condensation (N_2 liq.) and revaporization.

Kroto, H.W. *et al, J. Mol. Spectrosc.*, 1979, **77**, 270 (*synth, microwave*)
Minh Tho Nguyen, *Z. Naturforsch., A*, 1984, **39**, 169 (*struct*)
Ohno, K. *et al, Chem. Lett.*, 1984, 413 (*ir*)
Pellerin, B. *et al, Tetrahedron Lett.*, 1986, **27**, 5723 (*synth, pmr, cmr, P nmr*)

Ethylimidodiphosphoric acid, 9CI E-00078

Ethyliminodiphosphoric acid. Ethylimidodiphosphoric acid

$$EtN[P(O)(OH)_2]_2$$

$C_2H_9NO_6P_2$ M 205.044

Tetra-Et ester: [3654-42-0]. *Pentaethyl imidodiphosphate. Pentaethyl iminodiphosphate. Tetraethyl ethylimidodiphosphate.*

$C_{10}H_{25}NO_6P_2$ M 317.258

Liq. d_4^{20} 1.15. $Bp_{0.08}$ 100-102°. n_D^{20} 1.4350.

▷NJ7035000.

Tetraisopropyl ester: Tetraisopropyl ethyliminodiphosphate. Tetrakis(1-methylethyl) ethylimidodiphosphate.

$C_{14}H_{33}NO_6P_2$ M 373.365

Liq. d_4^{20} 1.10. $Bp_{0.02}$ 98-99°. n_D^{20} 1.4319.

Bisdifluoride: Ethylimidobisphosphoryl fluoride.

$C_2H_5F_4NO_2P_2$ M 213.008

$Bp_{0.01}$ 30°.

Shokol, V.A. *et al*, *CA*, 1968, **69**, 10065.

Pinchuk, A.M. *et al*, *Zh. Obshch. Khim.*, 1975, **45**, 2394 (*Engl. transl.* p. 2352)

Riesch, L. *et al*, *Zh. Anorg. Allg. Chem.*, 1986, **539**, 183 (*fluoride*)

Ethylimidodi(phosphorous acid), 10CI — E-00079

Ethyliminodi(phosphorous acid), 9CI

$EtN[P(OH)_2]_2$

$C_2H_9NO_4P$ M 142.071

Tetra-Et ester: [65395-34-8]. *Tetraethyl ethylimidodiphosphite.*

$C_{10}H_{25}NO_4P$ M 254.286

Bidentate ligand for Pd and Pt.

Tetra-Ph ester: [57857-79-1]. *Tetraphenyl ethylimidodiphosphite.*

$C_{26}H_{25}NO_4P_2$ M 477.435

Ligand for Cr, Mo, Pd, Pt, Rh, W and Fe. Liq. $Bp_{0.03}$ 200-202°. n_D^{20} 1.5882.

Tetrachloride: see Ethylimidodiphosphorous tetrachloride, E-00080

Tetrakis(dimethylamide): [73551-19-6]. N‴*-Ethyl-*N,N,N′,N′,N″,N″,N‴,N‴*-octamethylimidodiphosphorous tetraamide.*

$C_{10}H_{29}N_5P_2$ M 281.320

Liq. $Bp_{0.2}$ 102°. Over 10 weeks at ambient temps.→ Hexamethylphosphorous triamide, H-00080 and 1,3-dimethyl-2,4-bis(dimethylamino)-1,3,2,4-diazadiphosphetidine.

Gorbatenko, Zh.K. *et al*, *Zh. Obshch. Khim.*, 1975, **45**, 2367 (*Engl. transl.* p. 2325) (*phenyl ester, synth*)

DuPreez, A.L. *et al*, *J. Organomet. Chem.*, 1977, **141**, C10 (*ethyl ester, complexes*)

Keat, R. *et al*, *J. Chem. Soc., Dalton Trans.*, 1980, 321; 1981, 2192 (*amide, synth, P nmr*)

Ethylimidodiphosphorous tetrachloride, 10CI — E-00080

Ethyliminodiphosphorous tetrachloride, 9CI. Ethylaminobis(dichlorophosphine)

[17648-17-8]

$EtN(PCl_2)_2$

$C_2H_5Cl_4NP_2$ M 246.828

Liq. $Bp_{0.4}$ 62°.

Nixon, J.F., *J. Chem. Soc. (A)*, 1968, 2689 (*synth, ir, pmr, raman*)

Jefferson, R. *et al*, *J. Chem. Soc., Dalton Trans.*, 1973, 1414 (*synth, pmr, P nmr*)

Gorbatenko, Zh.K. *et al*, *Zh. Obshch. Khim.*, 1975, **45**, 2367 (*Engl. transl.* p. 2325) (*synth*)

Colquhoun, I.J. *et al*, *J. Chem. Soc., Dalton Trans.*, 1977, 1674 (*pmr, P nmr*)

[(Ethylimino)methylethenyl]phosphonic acid, 9CI — E-00081

$EtN{=}C{=}C(CH_3)P(O)(OH)_2$

$C_5H_{10}NO_3P$ M 163.113

(±)-*form*

Di-Et ester: [73473-51-5]. *Diethyl [(ethylimino)-methylethenyl]phosphonate, 9CI.*

$C_9H_{18}NO_3P$ M 219.220

Reacts with Na salts of hydroxyaryl ketones to give dihydropyrans. Stable liq. Bp_1 86-90°.

Motiyoshiya, J. *et al*, *J. Org. Chem.*, 1980, **45**, 5385 (*synth, use, ir, pmr*)

O-Ethyl isopropylphosphonothioate, 8CI — E-00082

O*-Ethyl (1-methylethyl)phosphonothioate, 9CI.* O*-Ethyl hydrogen isopropylphosphonothioate*

[37912-98-4]

HO–P(=S)(–OEt)–CH(CH₃)₂ ⇌ O=P(–SH)(–OEt)–CH(CH₃)₂ (*R*)*-form*

$C_5H_{13}O_2PS$ M 168.190

(*R*)-*form* [38315-75-2]

Liq. $[\alpha]_D$ +7.30° (neat).

O*-Me ester: see O-Methyl isopropylphosphonothioate, M-00166*

Chloride: see O-Methyl isopropylphosphonochloridothioate, M-00165

(*S*)-*form* [38607-71-5]

Liq. $[\alpha]_D$ −7.10° (neat). n_D^{20} 1.4921.

O*-Me ester: see O-Methyl isopropylphosphonothioate, M-00166*

Chloride: see O-Methyl isopropylphosphonochloridothioate, M-00165

(±)-*form* [36585-75-8]

Chloride: see O-Methyl isopropylphosphonochloridothioate, M-00165

Mikolajczyk, M. *et al*, *Pol. J. Chem.*, 1979, **53**, 317 (*esters, pmr*)

1-Ethyl-2-methyl-1,2,3-azaphospharsolidine, 9CI — E-00083

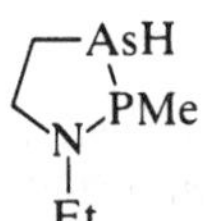

$C_5H_{13}AsNP$ M 193.060

3-Et: [60680-84-4].

$C_7H_{17}AsNP$ M 221.113

Liq. $Bp_{1.0}$ 55-57°.

3-Ph: [60680-85-5].

$C_{11}H_{17}AsNP$ M 269.157

Liq. $Bp_{0.2}$ 104-108°.

Tzschach, A. *et al*, *Z. Chem.*, 1976, **16**, 278 (*synth*)

Ethylmethylphenylphosphine, 9CI E-00084

[15849-84-0]

Me / Ph–P–: / Et (R)-form

$C_9H_{13}P$ M 152.175

(R)-form [52119-19-4]

B,PhCH₂I: Benzylethylmethylphenylphosphonium iodide.

$C_{16}H_{20}IP$ M 370.212

$[\alpha]_D$ −15.8°. (R)-config.

Oxide: [26515-05-9].

$C_9H_{13}OP$ M 168.175

Solid. Mp 47-51°. $Bp_{0.15}$ 95-105°. $[\alpha]_D^{25}$ −22.8° (c, 2.168 in H_2O). Has (S)-config.

Sulfide: [62621-05-0].

$C_9H_{13}PS$ M 184.235

$[\alpha]_D$ −1.1° (MeOH) (4% o.p.). Has (S)-config.

(S)-form [72974-35-7]

$[\alpha]_D$ +3.4° (toluene).

Oxide: [17045-47-5]. Mp 47-48°. $[\alpha]_D^{25}$ +22.4° (c, 2.184 in H_2O). Has (R)-config.

Sulfide: [41899-40-5]. $[\alpha]_D$ +22.3° (MeOH). (R)-config.

(±)-form

Liq. Bp_{14} 93-94°. n_D^{25} 1.5524.

B,MeI: Dimethylethylphenylphosphonium iodide.

$C_{10}H_{16}IP$ M 294.115

Cryst. (2-propanol). Mp 148.5-150° (138-140°).

Oxide: [7309-49-1]. Liq. $Bp_{0.3}$ 110-112°.

Sulfide: Cryst. (pet. ether). Mp 41-41.5° (33-34°). $Bp_{0.2}$ 108-110°. n_D^{20} 1.5996.

Maier, L., *Helv. Chim. Acta*, 1964, **47**, 120 (*sulfide, nmr*)
McEwen, W.E. *et al*, *J. Am. Chem. Soc.*, 1964, **86**, 2378 (*oxide*)
Snider, T.E. *et al*, *Org. Prep. Proced. Int.*, 1974, **6**, 221 (*synth*)
Henrick, K. *et al*, *Aust. J. Chem.*, 1975, **28**, 1473 (*ms*)
Jones, I.W. *et al*, *J. Chem. Soc., Perkin Trans. 2*, 1979, 501 (*uv*)
Omelańczuk, J. *et al*, *J. Chem. Soc., Chem. Commun.*, 1980, 24 (*synth, sulfide*)
Payne, N.C. *et al*, *Can. J. Chem.*, 1980, **58**, 15 (*synth, deriv*)

Ethylmethylphosphinic acid, 9CI E-00085

[51528-32-6]

EtMeP(O)OH

$C_3H_9O_2P$ M 108.077

Liq. Mp 7-8°. Bp_{11} 170-172°, $Bp_{1.5}$ 122°. n_D^{20} 1.4514.

Et ester: [19229-33-0]. *Ethyl ethylmethylphosphinate.*

$C_5H_{13}O_2P$ M 136.130

Liq. Bp_{17} 94-98°. n_D^{20} 1.4337.

2-Methylpropyl ester: 2-Methylpropyl ethylmethylphosphinate. Isobutyl ethylmethylphosphinate.

$C_7H_{17}O_2P$ M 164.184

Liq. Bp_3 85°. n_D^{20} 1.4308.

Chloride: [13213-38-2].

C_3H_8ClOP M 126.522

Liq. d_4^{20} 1.19. Bp_{15} 99-102°, Bp_1 55°. n_D^{20} 1.4702.

Anhydride: [51528-37-1].

$C_6H_{16}O_3P_2$ M 198.138

Liq. Bp_{15} 189-193°, $Bp_{0.2}$ 115-119°. n_D^{20} 1.4701.

Crofts, P.C. *et al*, *J. Chem. Soc.*, 1958, 2995 (*synth*)
Maier, L. *et al*, *Chem. Ber.*, 1961, **94**, 3051, 3056 (*synth, chloride*)
Maier, L., *Helv. Chim. Acta*, 1964, **47**, 1448 (*anhydride, ester*)
Zinov'ev, Yu.M. *et al*, *Zh. Obshch. Khim.*, 1964, **34**, 929 (*Engl. transl.* p. 923) (*chloride, esters*)
Neimysheva, A.A. *et al*, *Zh. Obshch. Khim.*, 1966, **36**, 1090 (*Engl. transl.* p. 1105) (*chloride*)
Knunyants, I.L. *et al*, *Dokl. Akad. Nauk SSSR, Ser. Sci. Khim.*, 1971, **201**, 862 (*Engl. transl.*, p. 992) (*halides, nmr*)
Finke, M. *et al*, *Justus Liebigs Ann. Chem.*, 1974, 741 (*anhydride, chloride, ester*)

Ethylmethylphosphinodithioic acid, 9CI E-00086

[66220-54-0]

MeEtP(S)SH

$C_3H_9PS_2$ M 140.198

NH₄ salt: Cryst. ($EtOH/Et_2O$). Mp 183-186°.

Et ester: [13298-37-8]. *Ethyl ethylmethylphosphinodithioate.*

$C_5H_{13}PS_2$ M 168.251

Liq. d_4^{20} 1.10. $Bp_{0.03}$ 64-66°. n_D^{20} 1.5650.

Krasil'nikova, E.A. *et al*, *Zh. Obshch. Khim.*, 1968, **38**, 609 (*Engl. transl.* p. 587) (*ester*)
Diemert, K. *et al*, *Phosphorus Sulfur*, 1977, **3**, 131 (*synth, P nmr*)

Ethylmethylphosphinous acid, 9CI E-00087

EtMePOH

C_3H_9OP M 92.077

Free acid exists as phosphoryl tautomer.

Me ester: [58910-86-4]. *Methyl ethylmethylphosphinite.*

$C_4H_{11}OP$ M 106.104

Liq. d_4^{20} 1.00. Bp_7 87.5-88.5°. n_D^{20} 1.4485.

Et ester: Ethyl ethylmethylphosphinite.

$C_5H_{13}OP$ M 120.131

Liq. Bp_{15} 67-70°. n_D^{20} 1.4275.

Chloride: [2240-32-6]. *Chloroethylmethylphosphine.*

C_3H_8ClP M 110.523

d_4^{20} 1.03 (1.05). Bp_{720} 100-3°, Bp_{73} 40-1°. n_D^{20} 1.4728.

Bromide: [2240-33-7]. *Bromoethylmethylphosphine.*

C_3H_8BrP M 154.974

Liq. Bp_{720} 128-129°.

Petrov, K.A. *et al*, *Zh. Obshch. Khim.*, 1961, **31**, 2889 (*Engl. transl.* p. 2692) (*ester*)
Petrov, K.A. *et al*, *Zh. Obshch. Khim.*, 1961, **31**, 3085 (*Engl. transl.* p. 2876) (*ester*)
Maier, L., *Helv. Chim. Acta*, 1964, **47**, 2137 (*chloride, bromide, synth, P nmr*)
Zinovev, Yu.M. *et al*, *Zh. Obshch. Khim.*, 1964, **34**, 929 (*Engl. transl.* p. 923) (*chloride*)
Sommer, K., *Z. Anorg. Allg. Chem.*, 1970, **379**, 56 (*chloride, synth, P nmr*)

N-Ethyl-P-methylphosphonamidic acid, 9CI E-00088

O=P(NHEt)(Me)(OH)

$C_3H_{10}NO_2P$ M 123.091

Me ester: [13703-06-5]. *Methyl* N-*ethyl*-P-*methylphosphonamidate.*

$C_4H_{12}NO_2P$ M 137.118

Liq. d_4^{20} 1.08. $Bp_{0.02}$ 78-79°. n_D^{20} 1.4402.
Et ester: [13703-09-8]. *Ethyl* N-*ethyl*-P-*methylphosphonamidate.*
$C_5H_{14}NO_2P$ M 151.145
Liq. d_4^{20} 1.05. $Bp_{0.4}$ 91-93°. n_D^{20} 1.4372.
Isopropyl ester: [13703-12-3]. *Isopropyl* N-*ethyl*-P-*methylphosphonamidate.*
$C_6H_{16}NO_2P$ M 165.172
Liq. d_4^{20} 1.01. $Bp_{0.03}$ 69-71°. n_D^{20} 1.4338.
Ph ester: Phenyl N-*ethyl*-P-*methylphosphonamidate.*
$C_9H_{14}NO_2P$ M 199.189
Liq. d_4^{20} 1.13. $Bp_{0.2}$ 142°. n_D^{20} 1.5169.

Petrov, K.A. *et al*, *Zh. Obshch. Khim.*, 1960, **30**, 4060 (*Engl. transl.* p. 4023) (*synth*)
Shokol, V.A. *et al*, *Zh. Obshch. Khim.*, 1966, **36**, 1636 (*Engl. transl.* p. 1636) (*synth*)

Ethyl methylphosphonate E-00089

Monoethyl methylphosphonate, 9CI. *Ethyl hydrogen methylphosphonate*
[1832-53-7]

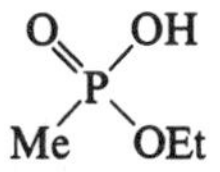

$C_3H_9O_3P$ M 124.076
Liq. d_4^{20} 1.19. $Bp_{0.1}$ 108-110°. n_D^{20} 1.4245.
Fluoride: see Methylphosphonofluoridic acid, M-00303
Chloride: see Methylphosphonochloridic acid, M-00296
Ph ester: see Ethyl phenyl methylphosphonate, E-00100

Keay, L., *Can. J. Chem.*, 1965, **43**, 2637 (*synth*)
Petrov, K.A. *et al*, *Zh. Obshch. Khim.*, 1965, **35**, 723 (*Engl. transl.* p. 723) (*synth*)
Cadogan, J.I.G. *et al*, *J. Chem. Soc. (B)*, 1971, 1988 (*synth*)
Kryuchkov, A.A. *et al*, *Izv. Akad. Nauk SSSR, Ser. Khim.*, 1978, 1985 (*Engl. transl.* p. 1746) (*props*)

O-Ethyl methylphosphonochloridothioate, 9CI E-00090

[2524-16-5]

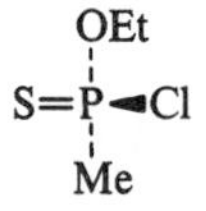

(*R*)-*form*

C_3H_8ClOPS M 158.582
(*R*)-*form* [38315-81-0]
Liq. $[\alpha]_D$ +78.60° (neat).
(±)-*form*
Liq. d_4^{25} 1.18. Bp_{10} 55-56°, Bp_4 41°. n_D^{25} 1.4950.

Hoffmann, F.W. *et al*, *J. Am. Chem. Soc.*, 1958, **80**, 3945 (*synth*)
Nesterov, L.V. *et al*, *Zh. Obshch. Khim.*, 1970, **40**, 1237 (*Engl. transl.* p. 1228) (*synth*)
Mikolajczyk, M. *et al*, *Tetrahedron*, 1972, **28**, 3855 (*props*)

O-Ethyl methylphosphonodithioate E-00091

[999-83-7]

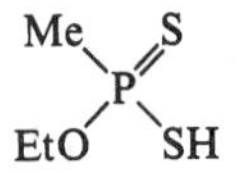

$C_3H_9OPS_2$ M 156.197
Liq. $Bp_{0.2}$ 45-48°. pK_a 3.16 (EtOH). $n_D^{22.5}$ 1.5440.
Et ester: see Methylphosphonodithioic acid, M-00301

Chupp, J.P. *et al*, *J. Org. Chem.*, 1962, **27**, 3832 (*synth*)

O-Ethyl methylphosphonoselenoate E-00092

O-*Ethyl hydrogen methylphosphonoselenoate*
[66376-30-5]

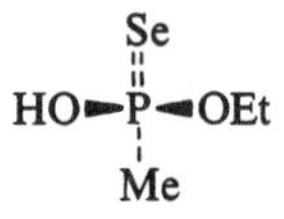

(*R*)-*form*

$C_3H_9O_2PSe$ M 187.037
(*R*)-*form*
Se-*Me ester:* O-*Ethyl* Se-*methyl methylphosphonoselenoate.* Liq. $[\alpha]_D$ +81° (c, 0.4 in $CHCl_3$).
(*S*)-*form* [61739-43-3]
Se-*Me ester:*
$C_4H_{11}O_2PSe$ M 201.063
Liq. $[\alpha]_D$ −80° (c, 0.4 in $CHCl_3$).
(±)-*form*
Liq. d_4^{20} 1.52. $Bp_{0.01}$ 76-77°. n_D 1.5289.

Hall, L.R. *et al*, *Tetrahedron Lett.*, 1976, 3645 (*deriv, pmr*)
Cooper, D.B. *et al*, *J. Chem. Soc., Perkin Trans. 1*, 1977, 1969 (*deriv, pmr*)
Nuretdinov, I.A. *et al*, *Izv. Akad. Nauk SSSR, Ser. Khim.*, 1978, 437 (*Engl. transl.* p. 378) (*synth, nmr*)

O-Ethyl methylphosphonothioate, 9CI E-00093

O-*Ethyl hydrogen methylphosphonothioate*
[18005-40-8]

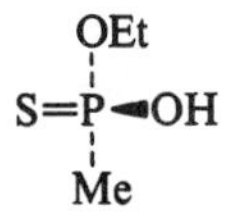

(*R*)-*form*

$C_3H_9O_2PS$ M 140.137
(*R*)-*form* [38315-72-9]
Liq. $[\alpha]_D$ +10.35° (neat).
(R)-*1-Phenylethylammonium salt:* [53518-61-9]. Solid. Mp 139-140°. $[\alpha]_D^{25}$ +10.64° (c, 3.1 in MeOH).
(S)-*1-Phenylethylammonium salt:* Solid. Mp 107.5-109°. $[\alpha]_D^{20}$ +5.33°.
Dicyclohexylammonium salt: Cryst. (EtOAc). Mp 127°. $[\alpha]_D^{25}$ +8.47° (c, 5.0 in MeOH).
(*S*)-*form* [38315-77-4]
Liq. $[\alpha]_D$ −10.5° (neat).
(R)-*1-Phenylethylammonium salt:* Mp 108-110°. $[\alpha]_D^{20}$ −5.00°.
(S)-*1-Phenylethylammonium salt:* [53518-62-0]. Mp 139-140°. $[\alpha]_D^{25}$ −10.56° (c, 3.1 in MeOH).
Dicyclohexylammonium salt: Cryst. (EtOAc). Mp 127°. $[\alpha]_D^{25}$ −8.47° (c, 5.0 in MeOH).
(±)-*form* [36585-70-3]
d_4^{20} 1.18. $Bp_{0.55\text{-}0.6}$ 87-93°. n_D^{28} 1.4907.
Na salt: Solid. Mp 216-218°.

Dicyclohexylammonium salt: [73790-51-9]. Cryst. (EtOH). Mp 161-163°.

▷TB1103500.

Benzylammonium salt: [61372-42-7]. Solid. Mp 98-99.5°.

S-*Et ester:* O,S-*Diethyl methylphosphonothioate.*
$C_5H_{13}O_2PS$ M 168.190
Liq. d_4^{20} 1.10. Bp_{10} 96-98°. n_D^{20} 1.4742.

S-*Butyl ester:* S-*Butyl* O-*ethyl methylphosphonothioate.*
$C_7H_{17}O_2PS$ M 196.244
Liq. d_4^{20} 1.05. Bp_9 112-114°. n_D^{20} 1.4815.

▷Cholinesterase inhibitor

Chloride: see O-Ethyl methylphosphonochloridothioate, E-00090

Petrov, K.A. *et al*, *Zh. Obshch. Khim.*, 1961, **31**, 179 (*Engl. transl.* p. 168) (*synth, esters*)
Pelchowicz, Z. *et al*, *J. Chem. Soc.*, 1962, 3824 (*synth*)
Nesterov, L.V. *et al*, *Zh. Obshch. Khim.*, 1970, **40**, 1237 (*Engl. transl.* p. 1228) (*ethyl ester, synth, pmr, P nmr*)
Mikolajczyk, M. *et al*, *Tetrahedron*, 1972, **28**, 3855, 4357 (*config, cd, props*)
Tashma, Z. *et al*, *Org. Mass Spectrom.*, 1973, **7**, 955; *Anal. Chem.*, 1982, **54**, 2130 (*ms*)
Kabachnik, M.I., *Tetrahedron*, 1976, **32**, 1719 (*synth, pmr, P nmr*)
Hall, C.D. *et al*, *J. Chem. Soc., Perkin Trans. 2*, 1977, 1232 (*esters, pmr*)
Sass, S. *et al*, *Org. Mass Spectrom.*, 1979, **14**, 357 (*ethyl ester, ms*)

O-Ethyl Se-methyl phosphorochloridoselenoite, 9CI — E-00094

O-*Ethyl* Se-*methyl chloroselenophosphite*
[55776-64-2]

EtOPClSeMe

$C_3H_8ClOPSe$ M 205.482
Viscous yellow oil. Thermally stable.

Anderson, J.W. *et al*, *Inorg. Nucl. Chem. Lett.*, 1975, **11**, 233 (*synth, pmr*)

O-Ethyl O-methyl phosphorodithioate, 9CI, 8CI — E-00095

O-*Ethyl* O-*methyl dithiophosphoric acid.* O-*Ethyl* O-*methyl hydrogen dithiophosphate*
[20115-21-3]

MeO(EtO)P(=S)SH

$C_3H_9O_2PS_2$ M 172.197
Liq. d_4^{20} 1.25. Bp_2 52-54°. n_D^{20} 1.5260.
K salt: Solid. Mp 172-173°.

Kotovich, B.P. *et al*, *Zh. Obshch. Khim.*, 1968, **38**, 1282 (*Engl. transl.* p. 1235) (*synth*)

O-Ethyl S-methyl phosphorodithioate, 9CI, 8CI — E-00096

O-*Ethyl* S-*methyl hydrogen phosphorodithioate.* O-*Ethyl* S-*methyl dithiophosphate.* O-*Ethyl* S-*methyl dithiophosphoric acid*

MeS(EtO)P(=O)SH ⇌ MeS(EtO)P(=S)OH

$C_3H_9O_2PS_2$ M 172.197

(±)-form
Dimethylammonium salt: Solid.

Ger. Pat., 2 506 618, (*1976*); *CA*, **85**, 176867 (*synth, pmr*)

S-Ethyl O-methyl phosphorodithioate, 9CI, 8CI — E-00097

S-*Ethyl* O-*methyl dithiophosphate.* S-*Ethyl* O-*methyl hydrogen dithiophosphate.* S-*Ethyl* O-*methyl dithiophosphoric acid*

MeO(EtS)P(=O)SH ⇌ MeO(EtS)P(=S)OH

$C_3H_9O_2PS_2$ M 172.197

(±)-form
Dimethylammonium salt: Solid.
Diethylmethylammonium salt: Cryst. (pet. ether). Mp 61-62°.

Mel'nikov, N.N. *et al*, *Zh. Obshch. Khim.*, 1963, **33**, 2456 (*Engl. transl.* p. 2394)
Ger. Pat., 2 506 618, (*1976*); *CA*, **85**, 176867 (*pmr*)

O-Ethyl O-methyl phosphorothioate, 10CI — E-00098

O-*Ethyl* O-*methyl hydrogen phosphorothioate.* O-*Ethyl* O-*methyl thiophosphoric acid*

MeO–P(=O)(SH)–OEt (*R*)-*form*

$C_3H_9O_3PS$ M 156.136

(R)-form [71348-05-5]
Liq. $Bp_{0.5}$ 112°. $[\alpha]_D$ +1.5° (c, 3.0 in $CHCl_3$).

(S)-form [71348-18-0]
$[\alpha]_D$ −1.4° (c, 3.0 in $CHCl_3$).

Hall, C.R. *et al*, *J. Chem. Soc., Perkin Trans. 1*, 1979, 1104 (*synth, config*)

O-Ethyl O-4-nitrophenyl phenylphosphonothionate, 9CI — E-00099

O-*Ethyl* O-p-*nitrophenyl benzenethiophosphonate. EPN*
[2104-64-5]

4-O_2NC$_6$H$_4$OP(=S)(OEt)Ph

$C_{14}H_{14}NO_4PS$ M 323.303

▷TB1925000.

(+)-form [65580-79-2]
$[\alpha]_D^{20}$ +29.7° (c, 7.9 in $CHCl_3$).
▷More toxic to flies than (−)-form; exhibits no delayed neurotoxicity

(−)-form [65580-80-5]
$[\alpha]_D^{20}$ −27.7° (c, 7.0 in $CHCl_3$).
▷Exhibits delayed neurotoxicity

(±)-form
Nonsystemic insecticide, cholinesterase inhibitor. Off-white cryst. Insol. H_2O, sol. most org. solvs. Mp 36°. $Bp_{2.5}$ 175-180°. Coml. product is viscous oil d^{25} 1.27. Stable in neutral and acidic soln.

▷V. highly toxic, TLV 0.5. Exhibits delayed neurotoxicity

Schrader, G., *Z. Naturforsch., B*, 1963, **18**, 965 (*pharmacol*)
Keith, L.H. *et al, J. Assoc. Off. Anal. Chem.*, 1968, **51**, 1063 (*pmr*)
Ross, R.T. *et al, Anal. Chim. Acta*, 1970, **52**, 139 (*P nmr*)
Gore, R.C. *et al, J. Assoc. Off. Anal. Chem.*, 1971, **54**, 1040 (*ir, uv*)
Holmstead, R.L. *et al, J. Assoc. Off. Anal. Chem.*, 1974, **57**, 1050 (*ms*)
Krijgsman, W. *et al, J. Chromatogr.*, 1976, **117**, 201 (*glc*)
Nicholas, M.L. *et al, J. Assoc. Off. Anal. Chem.*, 1976, **59**, 1071 (*raman*)
Ohkawa, H. *et al, Bull. Environ. Contam. Toxicol.*, 1977, **18**, 534 (*tox*)
Gifkins, M.R. *et al, Cryst. Struct. Commun.*, 1980, **9**, 571 (*cryst struct*)
Chrzanowski, R.L. *et al, J. Agric. Food Chem.*, 1981, **29**, 580; 1982, **30**, 155 (*metab*)
Purnanand, R.K.D., *Synthesis*, 1983, 731 (*synth, ir, pmr*)
Pesticide Manual, 6th Ed., 235; 7th Ed., 236.
Sax, N.I., *Dangerous Properties of Industrial Materials*, 6th Ed., Van Nostrand-Reinhold, 1984, 641.

Ethyl phenyl methylphosphonate **E-00100**

[38074-88-3]

OPh / O=P◄OEt / Me *(R)-form*

$C_9H_{13}O_3P$ M 200.174

(R)-form [70741-64-9]

Liq. $Bp_{0.02}$ 77-78°. $[\alpha]_D$ +11.7° (c, 0.80 in $CHCl_3$).

(S)-form [70741-65-0]

Liq. $Bp_{0.02}$ 77-78°. $[\alpha]_D$ −10.9° (c, 1.36 in $CHCl_3$).

(±)-form

Liq. d_4^{20} 1.15. Bp_2 134°. n_D^{20} 1.5010.

Zavlin, P.M. *et al, Zh. Obshch. Khim.*, 1972, **42**, 1257 (*Engl. transl.* p. 1253) (*synth*)
Otsuki, T. *et al, Synthesis*, 1981, 811 (*synth, pmr*)

Ethylphenylphosphinic acid, 9CI, 8CI **E-00101**

[13317-44-7]

OR / Et►P◄Ph / O *(R)*$_P$-form of esters

$C_8H_{11}O_2P$ M 170.147

(R)-form

Me ester: [69423-57-0]. *Methyl ethylphenylphosphinate.*
$C_9H_{13}O_2P$ M 184.174
Liq. $[\alpha]_D^{19}$ +41.8° (c, 1.7 in MeOH).

(S)-form

Me ester: [65665-33-0]. $[\alpha]_D^{18}$ −40.1° (c, 1.7 in MeOH).

(±)-form

Solid. Mp 79.5-80.5°.

Dicyclohexylammonium salt: Solid. Mp 148-149°.
Me ester: Liq. $Bp_{0.8}$ 109°. n_D^{23} 1.5218.
Et ester: Ethyl ethylphenylphosphinate.
$C_{10}H_{15}O_2P$ M 198.201
Ethylating agent for *N*-heterocyclic compds. Liq. d_4^{20} 1.08. Bp_1 108-109°. n_D^{20} 1.5065, 1.5125.
Isopropyl ester: Isopropyl ethylphenylphosphinate.
$C_{11}H_{17}O_2P$ M 212.228
Liq. d_4^{20} 1.06. Bp_9 146.5°, $Bp_{0.1}$ 89-90°. n_D^{20} 1.5078.
Fluoride:
$C_8H_{10}FOP$ M 172.139
No phys. props. reported.
Chloride:
$C_8H_{10}ClOP$ M 188.593
Liq. d_4^{20} 1.21. Bp_3 123-125°. n_D^{20} 1.5494.
Azide:
$C_8H_{10}N_3OP$ M 195.160
Liq. $Bp_{0.1}$ 130-135° (oven).

Budzikiewicz, H. *et al, Monatsh. Chem.*, 1965, **96**, 1739 (*methyl ester, ms*)
Neimysheva, A.A. *et al, Zh. Obshch. Khim.*, 1966, **36**, 1090 (*Engl. transl.* p. 1105) (*chloride*)
Knunyants, I.L. *et al, Dokl. Akad. Nauk SSSR, Ser. Sci. Khim.*, 1971, **201**, 862 (*Engl. transl.* p. 992) (*halides, nmr*)
Legin, G.Ya., *Zh. Obshch. Khim.*, 1973, **43**, 2202 (*Engl. transl.* p. 2194) (*esters*)
Appel, R. *et al, Chem. Ber.*, 1976, **109**, 805 (*esters, ir, pmr, nmr*)
Hayashi, M. *et al, Bull. Chem. Soc. Jpn.*, 1976, **49**, 283 (*ethyl ester, use*)
Kharrasova, F.M. *et al, Zh. Obshch. Khim.*, 1976, **46**, 2237 (*Engl. transl.*, p. 2150) (*synth, derivs*)
Garst, M.E., *Synth. Commun.*, 1979, **9**, 261 (*synth*)
Harger, M.J.P., *J. Chem. Soc., Perkin Trans. 2*, 1980, 1505 (*methyl ester, pmr*)
Harger, M.J.P. *et al, Tetrahedron*, 1982, **38**, 3073 (*azide, pmr*)

P-Ethyl-*P*-phenylphosphinic amide, 9CI **E-00102**

[51028-14-9]

$PhEtP(O)NH_2$

$C_8H_{12}NOP$ M 169.163
Cryst. (C_6H_6). Mp 105-109°.
N-*Ph:* [51028-07-0]. P-*Ethyl*-N,P-*diphenylphosphinic amide.*
$C_{14}H_{16}NOP$ M 245.260
Cryst. (EtOAc). Mp 167-170°.

Harger, M.J.P., *J. Chem. Soc., Perkin Trans. 1*, 1975, 514 (*synth, pmr, ir*)
Harger, M.J.P., *J. Chem. Soc., Perkin Trans. 1*, 1977, 605 (*deriv, ir, pmr*)

Ethylphenylphosphinodithioic acid, 9CI **E-00103**

[20384-85-4]

PhEtP(S)SH

$C_8H_{11}PS_2$ M 202.269
Cryst. (EtOH). Mp 64.5°.
NH_4 salt: Solid. Mp 166-170°.
Et ester: [6588-35-8]. *Ethyl ethylphenylphosphinodithioate.*
$C_{10}H_{15}PS_2$ M 230.322
Liq. d_4^{20} 1.14. $Bp_{0.1}$ 112-115°. n_D^{20} 1.6140.
Isopropyl ester: Isopropyl ethylmethylphosphinodithioate.
$C_{11}H_{17}PS_2$ M 244.349
Liq. d_4^{20} 1.11. $Bp_{0.05}$ 97-98°. n_D^{20} 1.5993.

Newallis, P.E. *et al, J. Org. Chem.*, 1962, **27**, 3829 (*synth, P nmr*)
Akamsin, V.D. *et al, Izv. Akad. Nauk SSSR, Ser. Khim.*, 1966, 493 (*Engl. transl.* p. 463) (*esters*)

Ethylphenylphosphinoselenoic acid, 9CI E-00104

PhEtP(Se)OH ⇌ PhEtP(O)SeH

$C_8H_{11}OPSe$ M 233.108

(±)-*form*

O-*Me ester:* O-*Methyl ethylphenylphosphinoselenoate.*
$C_9H_{13}OPSe$ M 247.135
Liq. d_4^{20} 1.37. $Bp_{0.015}$ 94-95°. n_D^{20} 1.5938.

O-*Et ester:* O-*Ethyl ethylphenylphosphinoselenoate.*
$C_{10}H_{15}OPSe$ M 261.162
Liq. d_4^{20} 1.33. $Bp_{0.05}$ 94-96°. n_D^{20} 1.5806.

Chloride:
$C_8H_{10}ClPSe$ M 251.554
Liq. d_4^{20} 1.47. $Bp_{0.001}$ 94-96°. n_D^{20} 1.6385.

Bayandina, E.V. *et al*, *Zh. Obshch. Khim.*, 1978, **48**, 2673 (*Engl. transl.* p. 2424) (*derivs, synth, P nmr*)

Ethylphenylphosphinothioic acid, 9CI E-00105

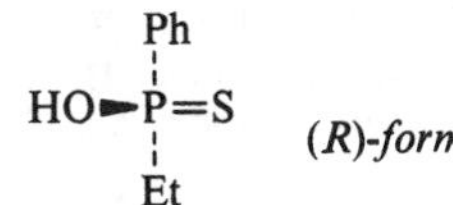

$C_8H_{11}OPS$ M 186.208

(*R*)-*form* [73176-32-6]
$[\alpha]_D$ +24.4° (c, 0.6 in $CHCl_3$) (98% opt. pure).

N,N-*Dimethylamide:* P-*Ethyl*-N,N-*dimethyl*-P-*phenylphosphinothioic amide.*
$C_{10}H_{16}NPS$ M 213.277
$[\alpha]_D$ −1.3° (c, 3.72 in C_6H_6).

(*S*)-*form* [38607-74-8]

Dicyclohexylammonium salt: [72974-38-0]. Solid. $[\alpha]_D$ −9.8° (c, 2.4 in MeOH).

O-*Me ester:* [38605-14-0]. O-*Methyl ethylphenylphosphinothioate.*
$C_9H_{13}OPS$ M 200.235
Liq. $Bp_{0.05}$ 70°. $[\alpha]_D$ +25.1° (neat). n_D^{28} 1.5405.

S-*Me ester:* [57322-10-8]. S-*Methyl ethylphenylphosphinothioate.*
$C_9H_{13}OPS$ M 200.235
$[\alpha]_D$ −104.4° (c, 2.93 in C_6H_6).

O-*Menthyl ester:* (S)$_P$-O-*Menthyl ethylphenylphosphinothioate.*
$C_{18}H_{29}OPS$ M 324.460
Solid. Mp 108-110°. $[\alpha]_D$ −49.5° (c, 2.83 in C_6H_6).

(±)-*form* [68908-02-1]
$Bp_{0.06}$ 111-112°. n_D^{20} 1.5972. Loses H_2S on distillation.

Dicyclohexylammonium salt: Solid or cryst. (Me_2CO). Mp 158°.

S-*Me ester:* [76420-34-3]. Liq. $Bp_{0.3}$ 122° (oven). n_D^{20} 1.5755.

S-*Et ester:* [33626-88-9]. S-*Ethyl ethylphenylphosphinothioate.*
$C_{10}H_{15}OPS$ M 214.262
Liq. d_4^{20} 1.13. $Bp_{0.05}$ 93°. n_D^{20} 1.5690.

S-*Propyl ester:* S-*Propyl ethylphenylphosphinothioate.*
$C_{11}H_{17}OPS$ M 228.288
d_4^{20} 1.10. $Bp_{0.08}$ 102-104°. n_D^{20} 1.5594.

Fluoride: [38169-27-6].
$C_8H_{10}FPS$ M 188.199
No phys. props. reported.

Chloride: [5075-15-0].
$C_8H_{10}ClPS$ M 204.654
Oil. Bp_1 125-127°. $Bp_{0.1}$ 100-103°.

N,N-*Dimethylamide:* [83587-40-0]. P-*Ethyl*-N,N-*dimethyl*-P-*phenylphosphinothioic amide.* Liq. $Bp_{0.05}$ 116°.

Green, M. *et al*, *J. Chem. Soc.*, 1958, 3129 (*ester, resoln*)
Pollart, K.A. *et al*, *J. Org. Chem.*, 1962, **27**, 4444 (*chloride, P nmr*)
Ratajczak, A., *Rocz. Chem.*, 1962, **36**, 175 (*synth*)
Knunyants, I.L. *et al*, *Dokl. Akad. Nauk SSSR, Ser. Sci. Khim.*, 1971, **201**, 862 (*fluoride, F and P nmr*)
Mikolajczyk, M. *et al*, *Tetrahedron*, 1972, **28**, 4357 (*esters, uv, cd*)
Mikolajczyk, M. *et al*, *J. Am. Chem. Soc.*, 1978, **100**, 7003 (*salts, P nmr*)
Mikolajczyk, M. *et al*, *Tetrahedron*, 1979, **35**, 1531 (*ester, use, amide*)
Harger, M.J.P., *J. Chem. Soc., Perkin Trans. 2*, 1980, 1505 (*synth, esters*)
Kaushik, M.P. *et al*, *Indian J. Chem., Sect. B*, 1981, **20**, 932 (*chloride*)
Johnson, C.R. *et al*, *J. Am. Chem. Soc.*, 1982, **104**, 7041 (*dimethylamide, cmr, P nmr, ir, use*)

Ethylphenylphosphinothioselenoic acid, 10CI E-00106

PhEtP(Se)SH ⇌ PhEtP(S)SeH

$C_8H_{11}PSSe$ M 249.169
Rather unstable in free form.

S-*Et ester:* [69741-80-6]. S-*Ethyl ethylphenylphosphinothioselenoate.*
$C_{10}H_{15}PSSe$ M 277.222
Liq. d_4^{20} 1.35. $Bp_{0.015}$ 120-121°. n_D^{20} 1.6321.

Bayandina, E.V. *et al*, *Zh. Obshch. Khim.*, 1978, **48**, 2673 (*Engl. transl.* p. 2424) (*synth*)

Ethylphenylphosphinothious acid, 9CI E-00107

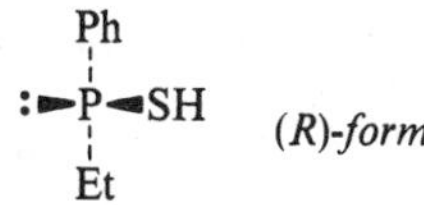

$C_8H_{11}PS$ M 170.209
Free acid tautomeric with ethylphenylphosphine sulfide.

(*R*)-*form*

Et ester: [62621-04-9]. *Ethyl ethylphenylphosphinothioite.*
$C_{10}H_{15}PS$ M 198.262
Liq. $[\alpha]_D$ +9.7° (neat: ca. 10% o.p.).

(*S*)-*form*

Et ester: [62621-07-2]. $[\alpha]_D$ −9.0° (neat).

(±)-*form*

Et ester: [6587-98-0]. Liq. d_4^{20} 1.04. $Bp_{0.05}$ 73-75°. n_D^{20} 1.5832. Forms a stable adduct with MeI.

Isopropyl ester: [6588-46-1]. *Isopropyl ethylphenylphosphinothioite.*
$C_{11}H_{17}PS$ M 212.289
Liq. d_4^{20} 1.01. $Bp_{0.07}$ 71-73°. n_D^{20} 1.5680.

Butyl ester: Butyl ethylphenylphosphinothioite.
$C_{12}H_{19}PS$ M 226.316
Liq. with unpleasant odour. d_4^{20} 1.01. $Bp_{0.05}$ 86-88°. n_D^{20} 1.5665.

Akamsin, V.D. *et al*, *Izv. Akad. Nauk SSSR, Ser. Khim.*, 1966, 493 (*Engl. transl.* p. 463)
Omelanczuk, J. *et al*, *J. Chem. Soc., Chem. Commun.*, 1976, 1025.

Hygroscopic liq. $Bp_{0.1}$ 150-160° (oven).

Felcht, U. *et al*, *Justus Liebigs Ann. Chem.*, 1977, 1309 (*synth, ir, pmr*)
Harger, M.J.P. *et al*, *Tetrahedron*, 1982, **38**, 3073 (*synth, ir, pmr, ms*)

Ethylphenylphosphinous acid, 9CI E-00108

Ph
:—P—OH (*R*)-*form*
Et

$C_8H_{11}OP$ M 154.148

Free acid exists in the tautomeric phosphoryl form.

(***R***)-***form***

Me ester: [57322-08-4]. *Methyl ethylphenylphosphinite.*
$C_9H_{13}OP$ M 168.175
Liq. $[\alpha]_D$ +35.2° (neat).

Diethylamide: [71162-24-8]. N,N,P-*Triethyl*-P-*phenylphosphinous amide.* $[\alpha]_D$ −86.7°.

(***S***)-***form***

Diethylamide: [71162-25-9]. $[\alpha]_D$ +82.7°.

(±)-***form***

Me ester: [20337-70-6]. Liq. d_4^{20} 1.00. $Bp_{3.5}$ 60.5-61.5°. n_D^{20} 1.5357.

Et ester: [61388-10-1]. *Ethyl ethylphenylphosphinite.*
$C_{10}H_{15}OP$ M 182.202
Liq. Bp_4 81.5-82.5°. n_D^{20} 1.5190.

Ph ester: [20337-75-1]. *Phenyl ethylphenylphosphinite.*
d_4^{20} 1.08. Bp_{11} 173.5-174.5°. n_D^{20} 1.5880.

Chloride: [15849-83-9]. *Chloroethylphenylphosphine.*
$C_8H_{10}ClP$ M 172.594
Liq. d_4^{20} 1.12. Bp_{11} 100-101°, Bp_2 73-75°. n_D^{20} 1.5707.

Bromide: [22880-68-8]. *Bromoethylphenylphosphine.*
$C_8H_{10}BrP$ M 217.045
Liq. d_4^{20} 1.37. $Bp_{0.03}$ 58-59°. n_D^{20} 1.6005.

Cyanide: Cyanoethylphenylphosphine.
$C_9H_{10}NP$ M 163.158
Solid. Mp 72-73°.

Diethylamide: [75281-59-3]. Liq. $Bp_{0.05}$ 84-87°. n_D^{23} 1.5373.

Kamai, G. *et al*, *Zh. Obshch. Khim.*, 1962, **32**, 2848 (*Engl. transl.* p. 2806) (*chloride*)
Maier, L., *J. Inorg. Nucl. Chem.*, 1962, **24**, 1073 (*chloride, synth, P nmr*)
Steininger, E., *Chem. Ber.*, 1963, **96**, 3184 (*chloride*)
Kharrasova, F.M. *et al*, *Zh. Obshch. Khim.*, 1968, **38**, 359 (*Engl. transl.* p. 358) (*esters*)
Rizpolozhenskii, N.I. *et al*, *Izv. Akad. Nauk SSSR, Ser. Khim.*, 1969, 370 (*Engl. transl.* p. 321) (*bromide*)
Horner, L. *et al*, *Phosphorus Sulfur*, 1980, **8**, 209, 215, 225, 235 (*diethylamide, synth, props*)
Wolfsberger, W., *J. Organomet. Chem.*, 1986, **317**, 167 (*chloride, cmr, P nmr*)

N-Ethyl-P-phenylphosphonamidic acid, 9CI E-00109

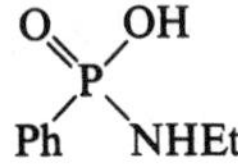

$C_8H_{12}NO_2P$ M 185.162

Me ester: [65111-16-2]. *Methyl* N-*ethyl*-P-*phenylphosphonamidate.*
$C_9H_{14}NO_2P$ M 199.189

P-Ethyl-N-phenylphosphonamidic acid, 9CI E-00110

NHPh
O=P—OR (*R*)-*form* of esters
Et

$C_8H_{12}NO_2P$ M 185.162

(***R***)-***form***

Et ester: [57237-62-4]. *Ethyl* P-*ethyl*-N-*phenylphosphonamidate.*
$C_{10}H_{16}NO_2P$ M 213.216
Oil. $[\alpha]_D^{20}$ +77.0° (C_6H_6).

(***S***)-***form***

Et ester: [57237-61-3]. Oil. $[\alpha]_D^{20}$ −40.2° (C_6H_6).

(±)-***form***

Me ester: Methyl P-*ethyl*-N-*phenylphosphonamidate.*
$C_9H_{14}NO_2P$ M 199.189
Cryst. (pet. ether). Mp 71-72°. $Bp_{0.2}$ 175-180° (oven).

Stec, W.J. *et al*, *J. Org. Chem.*, 1976, **41**, 227 (*synth, ir, P nmr*)
Harger, M.J.P. *et al*, *Tetrahedron*, 1982, **38**, 3073 (*synth, ms, ir, pmr*)

Ethyl phenyl phosphonate, 9CI E-00111

Ethyl phenyl phosphite
[20442-56-2]

PhOPH(O)OEt ⇌ PhOP(OH)OEt

$C_8H_{11}O_3P$ M 186.147
Tautomeric. Liq. d_4^{20} 1.15-1.17. $Bp_{0.8}$ 112-114°. n_D^{20} 1.4995.

Wolf, R. *et al*, *Bull. Soc. Chim. Fr.*, 1960, 124 (*synth, ir*)
Houalla, D. *et al*, *Bull. Soc. Chim. Fr.*, 1960, 129 (*ir*)
Maklyaev, V.K. *et al*, *Zh. Obshch. Khim.*, 1962, **32**, 3421 (*Engl. trans.* p. 3357) (*synth*)
Mandel'baum, Ya.A. *et al*, *Zh. Obshch. Khim.*, 1972, **42**, 502 (*Engl. transl.* p. 500) (*synth*)

O-Ethyl phenylphosphonochloridothioate, 9CI E-00112

[5075-13-8]

S
Cl—P—OEt (*R*)-*form*
Ph

$C_8H_{10}ClOPS$ M 220.653
▷TA3765000.

(***R***)-***form***
$[\alpha]_D^{20}$ +50.9° (neat).

(***S***)-***form*** [68280-30-8]
$[\alpha]_D^{20}$ −51.2° (neat), $[\alpha]_D$ −79.2° (c, 0.8 in CCl_4).

(±)-***form***
Oil. d_4^{20} 1.34. n_D^{20} 1.5741, 1.6049.

Bliznyuk, N.K. *et al*, *Zh. Obshch. Khim.*, 1967, **37**, 1122 (*Engl. transl.* p. 1064) (*synth*)

Ohkawa, H. *et al*, *Agric. Biol. Chem.*, 1977, **41**, 369 (*synth*)
Ohkawa, H. *et al*, *Bull. Environ. Contam. Toxicol.*, 1977, **18**, 534 (*use*)
Nomeir, A.A. *et al*, *Pestic. Biochem. Physiol.*, 1979, **10**, 121 (*synth*)
Corriu, R.J.P. *et al*, *Tetrahedron*, 1980, **36**, 1617 (*synth, P nmr*)
Yoshikawa, H. *et al*, *Agric. Biol. Chem.*, 1980, **44**, 1447 (*synth*)
Hirashima, A. *et al*, *Agric. Biol. Chem.*, 1983, **47**, 829 (*synth, config*)

O-Ethyl phenylphosphonodithioate, 9CI **E-00113**

[1007-94-9]

$C_8H_{11}OPS_2$ M 218.268

Liq. d_4^{20} 1.23. $Bp_{0.003}$ 83-85°. n_D^{20} 1.6020.

Dicyclohexylammonium salt: Solid. Mp 105-107°.

Chupp, J.P. *et al*, *J. Org. Chem.*, 1962, **27**, 3832 (*salts*)
Mizrakh, L.I. *et al*, *Zh. Obshch. Khim.*, 1966, **36**, 469 (*Engl. transl.* p. 487) (*synth, derivs*)

O-Ethyl phenylphosphonothioate, 9CI **E-00114**

[6230-93-9]

(*S*)-*form*

$C_8H_{11}O_2PS$ M 202.207

Tautomeric. Metabolite of *O*-Ethyl *O*-4-nitrophenyl phenylphosphonothionate, E-00099 .

(*S*)-***form*** [67964-17-4]

1-Phenylethylammonium salt: Mp 126°. $[\alpha]_D$ −6.3° (c, 1 in MeOH).

Dicyclohexylammonium salt: Mp 146°. $[\alpha]_D$ −6.8° (MeOH).

Chloride: see O-Ethyl phenylphosphonochloridothioate, E-00112

(±)-***form*** [67964-16-3]

d_4^{20} 1.22. $Bp_{0.2}$ 120-7°, $Bp_{0.04}$ 93-4°. n_D^{20} 1.5732, 1.5643.

Na salt: Cryst. (C_6H_6/pet. ether). Mp 223-224° dec.

NH_4 salt: Solid. Mp 141-144°.

Dicyclohexylammonium salt: Solid. Mp 148.5-150.5°.

S-*Me ester: see Phenylphosphonothioic acid, P-00244*

S-*Et ester: see Phenylphosphonothioic acid, P-00244*

O-*4-Nitrophenyl ester: see O-Ethyl O-4-nitrophenyl phenylphosphonothionate, E-00099*

O-*4-Cyanophenyl ester: see Cyanofenphos, C-00213*

S-*Benzyl ester: see Inezin, I-00009*

O-*4-Bromo-2,5-dichlorophenyl ester: Leptophos-Ethyl.*

$C_{14}H_{12}BrCl_2O_2PS$ M 426.091

Insecticide. Cryst. (EtOH). Mp 62-64°.

O-*8-Quinolinyl ester: see Quintiofos, Q-00006*

Chloride: see O-Ethyl phenylphosphonochloridothioate, E-00112

O-*2,4-Dichlorophenyl ester: see S-Seven, S-00004*

Mizrakh, L.I. *et al*, *Zh. Obshch. Khim.*, 1966, **36**, 469 (*Engl. transl.* p. 483) (*synth, deriv*)

Daly, J.J. *et al*, *Helv. Chim. Acta*, 1972, **55**, 1991 (*derivs, P nmr*)
Mikolajczyk, M. *et al*, *Pol. J. Chem.*, 1979, **53**, 317 (*esters, pmr*)
Corriu, R.J. *et al*, *Tetrahedron*, 1980, **36**, 1617 (*resoln, derivs*)
Chrzanowski, R.L. *et al*, *J. Agric. Food Chem.*, 1981, **29**, 580.
Andreev, N.A. *et al*, *Zh. Obshch. Khim.*, 1982, **52**, 1530 (*Engl. transl.* p. 1352) (*synth, P nmr*)

Ethyl phenyl phosphoramidate **E-00115**

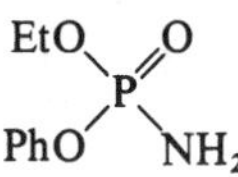

$C_8H_{12}NO_3P$ M 201.161

Reagent for synth. of chiral phosphate triesters.

(+)-***form*** [72163-96-3]

Cryst. (diisopropyl ether/hexane). Mp 63-64°. $[\alpha]_D^{27}$ +6.5° (c, 1.61 in $CHCl_3$).

(−)-***form*** [72163-97-4]

Cryst. (diisopropyl ether/hexane). Mp 64-65°. $[\alpha]_D^{27}$ −6.2° (c, 1.48 in $CHCl_3$).

Kobayashi, Y. *et al*, *Chem. Pharm. Bull.*, 1979, **27**, 1641 (*synth, ir, pmr, use*)

Ethylphenylphosphoramidic acid, 9CI **E-00116**

PhNEtP(O)$(OH)_2$

$C_8H_{12}NO_3P$ M 201.161

Di-Et ester: [25626-89-5]. *Diethyl ethylphenylphosphoramidate.*

$C_{12}H_{20}NO_3P$ M 257.269

Liq. d_4^{20} 1.10. Mp −24° to −23°. $Bp_{0.6}$ 90-91°. n_D^{20} 1.4948.

Ching Yee Cheng, *et al*, *J. Chem. Soc., Perkin Trans. 1*, 1976, 1739.
Gilyarov, V.A. *et al*, *Izv. Akad. Nauk SSSR, Ser. Khim.*, 1980, 151 (*Engl. transl.* p. 136)

O-Ethyl *O*-phenyl phosphorochloridothioate, 9CI **E-00117**

O-*Ethyl* O-*phenyl chlorothiophosphate.* O-*Ethyl* O-*phenyl thiophosphoryl chloride*

[38052-05-0]

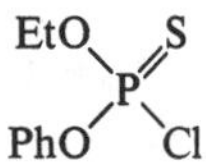

$C_8H_{10}ClO_2PS$ M 236.653

(±)-***form***

Liq. d_4^{20} 1.28. $Bp_{0.5}$ 117-119°. n_D^{20} 1.5494.

Mandel'baum, Ya.A. *et al*, *Zh. Obshch. Khim.*, 1960, **30**, 194 (*Engl. synth.* p. 207) (*synth*)
Bliznyuk, N.K. *et al*, *Zh. Obshch. Khim.*, 1967, **37**, 1122 (*Engl. transl.* p. 1064) (*synth*)

O-Ethyl *S*-phenyl phosphorochloridothioate, 9CI, 8CI **E-00118**

O-*Ethyl* S-*phenyl chlorothiophosphate*

[18351-20-7]

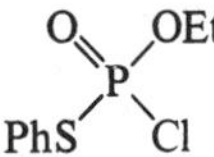

$C_8H_{10}ClO_2PS$ M 236.653

(±)-form
Liq. d_4^{20} 1.34. Bp_1 102-103°. n_D^{20} 1.6049.

Bliznyuk, N.K. *et al*, *Zh. Obhsch. Khim.*, 1967, **37**, 1122 (*Engl. transl.* p. 1064) (*synth*)

S-Ethyl *O*-phenyl phosphorochloridothioate, 9CI, 8CI — E-00119

S-Ethyl O-phenyl chlorothiophosphate
[60249-40-3]

$C_8H_{10}ClO_2PS$ M 236.653

(±)-form
Liq. d_4^{20} 1.29. Bp_5 144-145°. n_D^{20} 1.5508.

Petrov, K.A. *et al*, *Zh. Obshch. Khim.*, 1961, **31**, 1366 (*Engl. transl.* p. 1265) (*synth*)

O-Ethyl *O*-phenyl phosphorodithioate, 9CI, 8CI — E-00120

O-Ethyl O-phenyl hydrogen dithiophosphate. O-Ethyl O-phenyl dithiophosphoric acid

$C_8H_{11}O_2PS_2$ M 234.267

K salt: [20868-28-4]. Solid. Mp 166-168°.

Kotovich, B.P. *et al*, *Zh. Obshch. Khim.*, 1968, **38**, 1763 (*Engl. transl.* p. 1718)

O-Ethyl *S*-phenyl phosphorodithioate, 9CI, 8CI — E-00121

O-Ethyl S-phenyl dithiophosphate. O-Ethyl S-phenyl dithiophosphoric acid
[57350-23-9]

$C_8H_{11}O_2PS_2$ M 234.267

No phys. props. reported; characterised as Me ester. Metab. of Edifenphos, E-00002 .

Ueyama, S. *et al*, *Agric. Biol. Chem.*, 1975, **39**, 1719.

Ethyl phosphinate, 9CI — E-00122

Ethyl hypophosphite
[14684-32-3]

$EtOP(O)H_2 \rightleftharpoons EtOP(OH)H$

$C_2H_7O_2P$ M 94.050

Liq. which cryst. at −20°. d_4^{20} 1.11. Bp_2 31-32°. n_D^{20} 1.4250. Dec. rapidly at r.t. giving yellow solid; stable only at <0° in inert atm. Oxid. in air. Stabilised by trace of EtOH.

▷Spont. flammable

Kabachnik, M.I. *et al*, *Izv. Akad. Nauk SSSR*, *Ser. Khim.*, 1960, 146 (*Engl. transl.* p. 138) (*synth, props*)
Fitch, S.J., *J. Am. Chem. Soc.*, 1964, **86**, 61 (*synth, pmr, P nmr*)
Pinnick, H.W. *et al*, *Synth. Commun.*, 1979, **9**, 535 (*synth, pmr, cmr, P nmr, ir, props*)

Ethylphosphine — E-00123

[593-68-0]

$EtPH_2$

C_2H_7P M 62.051

Volatile liq. with unpleasant odour. Bp 25°. Dec. by H_2O. Forms cryst. salts with HCl, HI, stable only in conc. acid soln.

▷Ignites in air, explodes in contact with Cl_2, Br_2 or fuming HNO_3

Kreutzkamp, N., *Chem. Ber.*, 1954, **87**, 919.
Wymore, C.E. *et al*, *J. Inorg. Nucl. Chem.*, 1960, **14**, 42.
Fluck, E. *et al*, *Chem. Ber.*, 1965, **98**, 2674 (*nmr*)
Bogolyubov, G.M. *et al*, *Zh. Obshch. Khim.*, 1969, **39**, 1808 (*Engl. transl.* p. 1772) (*ms*)
Durig, J.R. *et al*, *J. Chem. Phys.*, 1975, **63**, 2303; 1976, **64**, 1930 (*ir, raman, microwave*)
Goetz, H. *et al*, *Phosphorus Sulfur*, 1978, **4**, 309 (*struct*)
Mosbo, J.A. *et al*, *Phosphorus Sulfur*, 1981, **11**, 11 (*struct*)
Bretherick, L., *Handbook of Reactive Chemical Hazards*, 2nd Ed., Butterworths, London and Boston, 1979, 401.
Sax, N.I., *Dangerous Properties of Industrial Materials*, 6th Ed., Van Nostrand-Reinhold, 1984, 672.
Hazards in the Chemical Laboratory, (Bretherick, L., Ed.), 3rd Ed., Royal Society of Chemistry, London, 1981, 334.

Ethylphosphinic acid, 9CI — E-00124

[4363-06-8]

EtP(O)(OH)H

$C_2H_7O_2P$ M 94.050

Tautomeric with Ethylphosphonous acid, E-00153 free acid exists in phosphinic acid form. Oil. d_4^{20} 1.28. n_D^{20} 1.4880.

Me ester: [27852-48-8]. *Methyl ethylphosphinate.*
$C_3H_9O_2P$ M 108.077
Liq. d_4^{20} 1.08. Bp_{11} 65-66°. n_D^{20} 1.4273.

Et ester: see Ethyl ethylphosphinate, E-00068

Isopropyl ester: [27852-49-9]. *Isopropyl ethylphosphinate.*
$C_5H_{13}O_2P$ M 136.130
Liq. Bp_{10} 69-70°. $[\alpha]_{589}$+5.0°, −5.5°. n_D^{20} 1.4233. Has been obt. in partially resolved forms.

tert-*Butyl ester:* [27852-53-5]. tert-*Butyl ethylphosphinate.*
$C_6H_{15}O_2P$ M 150.157
Liq. $Bp_{0.1}$ 35°. n_D^{20} 1.4328.

Ph ester: [21162-06-1]. *Phenyl ethylphosphinate.*
$C_8H_{11}O_2P$ M 170.147
Liq. $Bp_{0.08}$ 62-64°. n_D^{20} 1.5190.

Diethylamide: [70403-02-0]. N,N,P-*Triethylphosphinic amide.*
$C_6H_{16}NOP$ M 149.172
Liq. n_D^{20} 1.4590.

Arbuzov, B.A. *et al*, *Izv. Akad. Nauk SSSR*, *Ser. Khim.*, 1954, 622 (*Engl. transl.* p. 531) (*esters*)
Michalski, J. *et al*, *Rocz. Chem.*, 1956, **30**, 799; *CA*, **54**, 10832 (*esters*)
Pudovik, A.N. *et al*, *Zh. Obshch. Khim.*, 1968, **38**, 305 (*Engl. transl.* p. 304) (*ester*)
Benschop, H.P. *et al*, *J. Chem. Soc.*, *Chem. Commun.*, 1970, 1431 (*ester, resoln*)
Andreev, N.A. *et al*, *Zh. Obshch. Khim.*, 1979, **49**, 332, 2230 (*Engl. transl.* pp. 288, 1959) (*synth, amide, nmr, ir*)

Ethylphosphinothioic acid — E-00125

$EtP(S)OH \rightleftharpoons Et(O)SH \rightleftharpoons EtP(OH)(SH)$

C_2H_7OPS M 110.110

Exists in tautomeric equilibrium between thione and thiol forms, in further tautomeric equilibrium with ethylphosphonothious acid (minor component).

O-Et ester: O-*Ethyl ethylphosphinothioate.*
$C_4H_{11}OPS$ M 138.164
Liq. Bp_{19} 84°. n_D^{20} 1.4894.
O-Isopropyl ester: O-*Isopropyl ethylphosphinothioate.*
$C_5H_{13}OPS$ M 152.191
Liq. Bp_{12} 79°. n_D^{25} 1.4742.

Michalski, J. *et al*, *Rocz. Chem.*, 1962, **36**, 1781; *CA*, **59**, 10109.
Pudovik, A.N. *et al*, *Zh. Obshch. Khim.*, 1968, **38**, 1533 (*Engl. transl.* p. 1483)
Hantz, H. *et al*, *J. Prakt. Chem.*, 1978, **320**, 183.
Quistad, G.B. *et al*, *J. Agric. Food Chem.*, 1970, **18**, 189 (*esters, synth, props*)
Wustner, D.A. *et al*, *J. Agric. Food Chem.*, 1978, **26**, 1104 (*esters, synth, tox*)

P-Ethylphosphonamidic acid, 9CI — E-00126

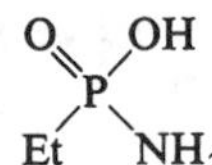

$C_2H_8NO_2P$ M 109.064
Et ester: [62992-28-3]. *Ethyl* P-*ethylphosphonamidate.*
$C_4H_{12}NO_2P$ M 137.118
Needles (toluene) or cryst. (Et_2O/pet. ether). Mp 32°, 62.5-63.5°.
Fluoride: [29070-42-6]. P-*Ethylphosphonamidic fluoride.*
C_2H_7FNOP M 111.056
Solid. Mp 47°.

Roesky, H.W. *et al*, *Z. Anorg. Allg. Chem.*, 1970, **375**, 140 (*fluoride, synth, ir, ms, pmr, F nmr*)
Issleib, K. *et al*, *Z. Anorg. Allg. Chem.*, 1977, **428**, 16 (*ester, synth*)
Feuer, H. *et al*, *J. Org. Chem.*, 1978, **43**, 4676 (*ester, synth, ir, pmr*)

P-Ethylphosphonamidodithioic acid, 9CI — E-00127

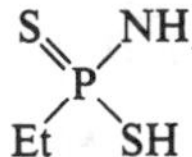

$C_2H_8NPS_2$ M 141.186
Me ester: [26350-31-2]. *Methyl* P-*ethylphosphonamidodithioate.*
$C_3H_{10}NPS_2$ M 155.212
Liq. $Bp_{0.05}$ 90-91°. n_D^{25} 1.6086.

Wustner, D.A. *et al*, *J. Agric. Food Chem.*, 1978, **26**, 1104 (*synth, tox*)

P-Ethylphosphonamidothioic acid, 9CI — E-00128

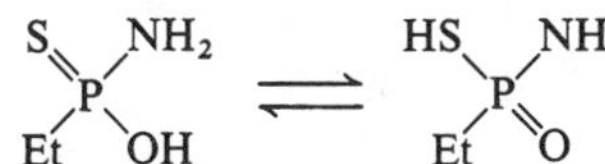

C_2H_8NOPS M 125.125
S-Me ester: [26350-29-8]. S-*Methyl* P-*ethylphosphonamidothioate.*
$C_3H_{10}NOPS$ M 139.152
Solid. Mp 65-67°.
S-Et ester: [26350-28-7]. S-*Ethyl* P-*ethylphosphonamidothioate.*
$C_4H_{12}NOPS$ M 153.179
Solid. Mp 46-48°.
Fluoride: [21693-82-3].
C_2H_7FNPS M 127.116
Liq. d_4^{20} 1.25. $Bp_{0.01}$ 46°. n_D^{20} 1.5048.

Roesky, H.W., *Z. Naturforsch.*, *B*, 1969,, **24**, 5 (*fluoride, synth, ir, pmr, F nmr*)

Ethylphosphonic acid, 9CI, 8CI — E-00129

[6779-09-5]

$EtP(O)(OH)_2$

$C_2H_7O_3P$ M 110.049
Hygroscopic cryst. Mp 44°, 61.5-62.5°. Bp_8 330-340°. pK_{a1} 2.43 (H_2O, 25°), pK_{a2} 8.05 (H_2O, 25°).
Anilinium salt: Cryst. (2-propanol). Mp 143-144°.
Di-Me ester: see Dimethyl ethylphosphonate, D-00754
Mono-Et ester: [7305-61-5]. *Ethyl hydrogen ethylphosphonate. Monoethyl ethylphosphonate.*
$C_4H_{11}O_3P$ M 138.103
Metab. of Fonofos, F-00052 . $Bp_{0.04}$ 92°. n_D^{25} 1.4281.
Di-Et ester: see Diethyl ethylphosphonate, D-00292
Monobutyl ester: Monobutyl ethylphosphonate. Butyl hydrogen ethylphosphonate.
$C_6H_{15}O_3P$ M 166.156
Liq. d_4^{20} 1.06. Bp_1 140-142°. n_D^{20} 1.4300.
Dibutyl ester: [2404-58-2]. *Dibutyl ethylphosphonate.*
$C_{10}H_{23}O_3P$ M 222.264
Liq. Bp_{15} 129.5°. n_D^{20} 1.4273.
Di-Ph ester: [7526-29-6]. *Diphenyl ethylphosphonate.*
$C_{14}H_{15}O_3P$ M 262.244
Liq. d_4^{20} 1.19. Bp_{13} 202°, $Bp_{0.5}$ 140-142°. n_D^{20} 1.5500.
Dibenzyl ester: Dibenzyl ethylphosphonate.
$C_{16}H_{19}O_3P$ M 290.298
$Bp_{0.6}$ 168°.
Bis(trimethylsilyl) ester: [1641-57-2]. *Bis(trimethylsilyl) ethylphosphonate.*
$C_8H_{23}O_3PSi_2$ M 254.413
Liq. Bp_{25} 104-106°. n_D^{20} 1.4111.
Difluoride: see Ethylphosphonic difluoride, E-00132
Dichloride: see Ethylphosphonic dichloride, E-00131
Diisocyanate: [2736-46-1].
$C_4H_5N_2O_3P$ M 160.069
Liq. Bp_7 80-81°, $Bp_{0.7}$ 58-59°.
Diamide: see P-Ethylphosphonic diamide, E-00130

Kosolapoff, G.M. *et al*, *J. Am. Chem. Soc.*, 1945, **67**, 1180 (*synth, dibutyl ester*)
Voronkov, M.G. *et al*, *Zh. Obshch. Khim.*, 1965, **35**, 106 (*Engl. transl.* p. 105) (*silyl ester*)
Petrov, K.A. *et al*, *Zh. Obshch. Khim.*, 1965, **35**, 732 (*Engl. transl.* p. 731) (*monobutyl ester*)
Derkach, G.I. *et al*, *Zh. Obshch. Khim.*, 1965, **35**, 1881 (*Engl. transl.* p. 1876) (*diisocyanate*)
Henning, H.-G. *et al*, *J. Prakt. Chem.*, 1966, **33**, 188 (*dibenzyl ester*)
Cadogan, J.I.G. *et al*, *J. Chem. Soc. (B)*, 1971, 1988 (*monoethyl ester*)
Barabanov, V.P. *et al*, *Zh. Obshch. Khim.*, 1972, **42**, 2431 (*Engl. transl.* p. 2425) (*props*)
Harvey, D.J. *et al*, *Org. Mass Spectrom.*, 1974, **9**, 111 (*silyl ester, ms*)
Kharrasova, F.M. *et al*, *Zh. Obshch. Khim.*, 1978, **48**, 1041 (*Engl. transl.* p. 948) (*diphenyl ester*)

Andreev, N.A. *et al, Zh. Obshch. Khim.*, 1979, **49**, 2230 (*Engl. transl.* p. 1959) (*synth, ir, P nmr*)

P-Ethylphosphonic diamide, 9CI E-00130

EtP(O)(NH$_2$)$_2$

$C_2H_9N_2OP$ M 108.080

N,N,N′,N′-*Tetra-Me:* [14655-69-7]. P-*Ethyl*-N,N,N′,N′-*tetramethylphosphonic diamide. Ethylphosphonic bis(dimethylamide).*
$C_6H_{17}N_2OP$ M 164.187
Liq. d_4^{26} 1.00. Bp_{31} 142°, $Bp_{0.5}$ 63-65°. n_D^{23} 1.4567.

N,N′-*Di-Ph:* [56530-33-7]. P-*Ethyl*-N,N′-*diphenylphosphonic diamide. Ethylphosphonic dianilide.*
$C_{14}H_{17}N_2OP$ M 260.275
Solid. Mp 148-149°.

N,N′-*Dicyclohexyl:* N,N′-*Dicyclohexyl*-P-*ethylphosphonic diamide.*
$C_{14}H_{29}N_2OP$ M 272.370
Solid. Mp 161-162°.

N,N,N′,N′-*Tetra-Et: Pentaethylphosphonic diamide.*
$C_{10}H_{25}N_2OP$ M 220.294
Liq. Bp_{26} 183°. n_D^{30} 1.4591.

Kinnear, A.M. *et al, J. Chem. Soc.*, 1952, 3437 (*dianilide*)
Kosolapoff, G.M. *et al, J. Org. Chem.*, 1956, **21**, 413 (*bisdialkylamides*)
Normant, H., *C.R. Hebd. Seances Acad. Sci., Ser. C*, 1967, **264**, 707 (*bisdialkylamides*)

Ethylphosphonic dichloride, 9CI E-00131

[1066-50-8]

EtP(O)Cl$_2$

$C_2H_5Cl_2OP$ M 146.941
Liq. d_4^{20} 1.37. Bp 174-175°, Bp_{15} 66°. n_D^{20} 1.4650. Easily hydrolysed at r.t. but not at 0°.

▷TA1780000.

Clay, J.P., *J. Org. Chem.*, 1951, **16**, 892 (*synth, props*)
Kinnear, A.M. *et al, J. Chem. Soc.*, 1952, 3437 (*synth*)
Geiseler, G. *et al, Ber. Bunsenges. Phys. Chem.*, 1967, **71**, 478 (*ir, raman*)
Neimysheva, A.A. *et al, Zh. Obshch. Khim.*, 1967, **37**, 2255 (*Engl. transl.* p. 2140) (*nqr*)
Steger, E. *et al, Spectrochim. Acta, Part A*, 1967, **23**, 2189 (*ir, raman*)
Maier, L., *Helv. Chim. Acta*, 1973, **56**, 492 (*synth, pmr, P nmr*)
Moedritzer, K. *et al, Synth. React. Inorg. Met.-Org. Chem.*, 1974, **4**, 417 (*synth, pmr, P nmr*)
Shagidullin, R.R. *et al, Izv. Akad. Nauk SSSR, Ser. Khim.*, 1976, **25**, 801 (*Engl. transl.* p. 781) (*ir, raman, struct*)
Griffiths, W.R. *et al, Phosphorus Sulfur*, 1978, **4**, 341 (*ms*)
Morita, T. *et al, Chem. Lett.*, 1980, 435 (*synth*)

Ethylphosphonic difluoride, 9CI E-00132

[753-98-0]

EtP(O)F$_2$

$C_2H_5F_2OP$ M 114.031
Liq. Bp 110-111°. n_D^{20} 1.3376.

Schmutzler, R., *J. Chem. Soc.*, 1964, 4551 (*synth, P nmr*)
Fluck, E. *et al, Z. Anorg. Allg. Chem.*, 1972, **393**, 126 (*F and P nmr*)
Shagidullin, R.R. *et al, Zh. Obshch. Khim.*, 1983, **53**, 84 (*Engl. transl.* p. 68) (*ir, raman*)

Ethylphosphonisocyanatidic acid, 9CI E-00133

Et–P(=O)(OH)–NCO

$C_3H_6NO_3P$ M 135.059

Et ester: [13534-39-9]. *Ethyl ethylphosphonisocyanatidate.*
$C_5H_{10}NO_3P$ M 163.113
Liq. d_4^{20} 1.13. Bp_{15} 93-95°, $Bp_{0.35\text{-}30.4}$ 55-56°. n_D^{20} 1.4320.

Derkach, G.I. *et al, Zh. Obshch. Khim.*, 1966, **36**, 2215 (*Engl. transl.* p. 2210) (*props*)
Gubnitskaya, E.S. *et al, Zh. Obshch. Khim.*, 1968, **38**, 1530 (*Engl. transl.* p. 1479) (*synth*)

Ethylphosphonobromidothious acid, 9CI E-00134

EtPBrSH

C_2H_6BrPS M 173.007

Et ester: [17151-43-8]. *Ethyl ethylphosphonobromidothioite.*
$C_4H_{10}BrPS$ M 201.061
Liq. d_4^{20} 1.40. Bp_{10} 80-82°. n_D^{20} 1.5620.

Butyl ester: [20626-67-9]. *Butyl ethylphosphonobromidothioite.*
$C_6H_{14}BrPS$ M 229.114
Liq. d_4^{20} 1.32. Bp_9 118-119°. n_D^{20} 1.5560.

Benzyl ester: [27127-15-7]. *Benzyl ethylphosphonobromidothioite. Phenylmethyl ethylphosphonobromidothioite.*
$C_9H_{12}BrPS$ M 263.132
Liq. d_4^{20} 1.342. Bp_6 154-156°. n_D^{20} 1.5934.

Krasil'nikova, E.A. *et al, Zh. Obshch. Khim.*, 1968, **38**, 1098 (*Engl. transl.* p. 1101) (*synth*)
Shitov, L.N. *et al, Zh. Obshch. Khim.*, 1969, **39**, 1251 (*Engl. transl.* p. 1220) (*synth, P nmr*)
Potapov, A.M. *et al, Zh. Obshch. Khim.*, 1970, **40**, 566 (*Engl. transl.* p. 534) (*synth*)

Ethylphosphonochloridic acid, 9CI E-00135

RO–P(=O)(Et)–Cl (***R***)*-form* of esters

$C_2H_6ClO_2P$ M 128.495

(*R*)-*form*

Et ester: Ethyl ethylphosphonochloridate.
$C_4H_{10}ClO_2P$ M 156.549
Liq. $[\alpha]_D^{27}$ +39.0° (c, 3.9 in $CHCl_3$), $[\alpha]_D$ +31.5° (c, 3.3 in C_6H_6).

(*S*)-*form*

Et ester: Liq. $Bp_{0.2}$ 28-29°. $[\alpha]_D^{20}$ −15.05° (neat).

(±)-*form*

Et ester: Liq. $Bp_{0.4}$ 34-35°. n_D^{26} 1.4347.

Isopropyl ester: Isopropyl ethylphosphonochloridate.
$C_5H_{12}ClO_2P$ M 170.575
d_4^{20} 1.10. $Bp_{9.5}$ 76.5-78.5°, Bp_1 38°. n_D^{25} 1.4322.

2-Methylpropyl ester: 2-Methylpropyl ethylphosphonochloridate. Isobutyl ethylphosphonochloridate.
$C_6H_{14}ClO_2P$ M 184.602

Liq. $Bp_{0.1}$ 56-58°.
Benzyl ester: Benzyl ethylphosphonochloridate.
$C_9H_{12}ClO_2P$ M 218.619
Unstable oil.

Anand, N. *et al*, *J. Chem. Soc.*, 1951, 1867 (*esters, synth*)
Coe, D.H. *et al*, *J. Chem. Soc.*, 1957, 3607 (*esters, synth*)
Razumov, A.I. *et al*, *Zh. Obshch. Khim.*, 1957, **27**, 2389 (*Engl. transl.* p. 2450) (*esters, synth*)
Aaron, H.S. *et al*, *J. Am. Chem. Soc.*, 1962, **84**, 617 (*synth*)
Michalski, J. *et al*, *J. Am. Chem. Soc.*, 1978, **100**, 5386 (*synth, P nmr*)

Ethylphosphonochloridodithioic acid, 9CI E-00136

EtPCl(S)SH

$C_2H_6ClPS_2$ M 160.616
Et ester: [17162-55-9]. *Ethyl ethylphosphonochloridodithioate.*
$C_4H_{10}ClPS_2$ M 188.670
Solid. d_4^{20} 1.21. Bp_9 99-101°. n_D^{20} 1.5697.
Isopropyl ester: [17162-57-1]. *Isopropyl ethylphosphonochloridodithioate.*
$C_5H_{10}ClPS_2$ M 200.681
Liq. d_4^{20} 1.17. $Bp_{0.02}$ 52-52.5°. n_D^{20} 1.5555.
tert-*Butyl ester:* [78570-13-5]. tert-*Butyl ethylphosphonochloridodithioate.*
$C_6H_{14}ClPS_2$ M 216.723
Liq. $Bp_{0.7}$ 75-78°.

Akamsin, V.D. *et al*, *Izv. Akad. Nauk SSSR, Ser. Khim.*, 1967, 1976 (*Engl. transl.* p. 1894) (*synth*)
European Pat., 25 270, (*1981*); *CA*, **95**, 150886 (*tert*-butyl ester)

Ethylphosphonochloridothioic acid, 9CI E-00137

C_2H_6ClOPS M 144.556
O-*Me ester: see O-Methyl ethylphosphonochloridothioate, M-00154*
S-*Me ester:* [13113-92-3]. S-*Methyl ethylphosphonochloridothioate.*
C_3H_8ClOPS M 158.582
Liq. d_4^{20} 1.29. $Bp_{2.5}$ 81°. n_D^{20} 1.5162.
O-*Et ester: see O-Ethyl ethylphosphonochloridothioate, E-00069*
S-*Et ester:* [13113-93-4]. S-*Ethyl ethylphosphonochloridothioate.*
$C_4H_{10}ClOPS$ M 172.609
Liq. d_4^{20} 1.22. $Bp_{2.5}$ 84-85°. n_D^{20} 1.5085.
O-*Propyl ester:* [1497-67-2]. O-*Propyl ethylphosphonochloridothioate.*
$C_5H_{12}ClOPS$ M 186.636
Liq. d_4^{25} 1.12. $Bp_{1.5}$ 54°. n_D^{25} 1.4874.
S-*Propyl ester:* [13113-94-5]. S-*Propyl ethylphosphonochloridothioate.*
$C_5H_{12}ClOPS$ M 186.636
Liq. d_4^{20} 1.18. Bp_2 98°. n_D^{20} 1.5045.
S-*Isopropyl ester:* [13113-99-0]. S-*Isopropyl ethylphosphonochloridothioate.*
$C_5H_{12}ClOPS$ M 186.636
Liq. d_4^{20} 1.17. Bp_3 98°. n_D^{20} 1.4997.
O-*Ph ester:* [53056-49-8]. O-*Phenyl ethylphosphonochloridothioate.*
$C_8H_{10}ClOPS$ M 220.653
Liq. $Bp_{0.24}$ 99°. n_D^{25} 1.5642.

Chupp, J.P. *et al*, *J. Org. Chem.*, 1962, **27**, 3832 (*synth*)
Neimysheva, A.A. *et al*, *Zh. Obshch. Khim.*, 1966, **36**, 500 (*Engl. transl.* p. 520) (*synth*)
Gubaidullin, M.G., *Zh. Obshch. Khim.*, 1977, **47**, 2659 (*Engl. transl.* p. 2428) (*props*)

Ethylphosphonochloridothious acid, 9CI E-00138

EtPCl(SH)

C_2H_6ClPS M 128.556
Et ester: [6588-28-9]. *Ethyl ethylphosphonochloridothioite.*
$C_4H_{10}ClPS$ M 156.610
Liq. d_4^{20} 1.08. Bp_{14} 64°. n_D^{20} 1.5280.
Isopropyl ester: [16543-45-6]. *Isopropyl ethylphosphonochloridothioite.*
$C_5H_{12}ClPS$ M 170.637
Liq. d_4^{20} 1.06. Bp_{10} 65-67°. n_D^{20} 1.5202.
Butyl ester: [16182-83-5]. *Butyl ethylphosphonochloridothioite.*
$C_6H_{14}ClPS$ M 184.663
Liq. d_4^{20} 1.06. Bp_9 96-97°. n_D^{20} 1.5160.

Akamsin, V.D. *et al*, *Izv. Akad. Nauk SSSR, Ser. Khim.*, 1967, 825 (*synth*)
Krasil'nikova, E.A. *et al*, *Zh. Obshch. Khim.*, 1967, **37**, 1409 (*Engl. transl.* p. 1339) (*synth*)
Shitov, L.N. *et al*, *Zh. Obshch. Khim.*, 1969, **39**, 1251 (*Engl. transl.* p. 1220) (*synth, P nmr*)

Ethylphosphonochloridous acid, 9CI E-00139

EtPCl(OH)

C_2H_6ClOP M 112.496
Et ester: [69805-25-0]. *Ethyl ethylphosphonochloridite.*
$C_4H_{10}ClOP$ M 140.549
Liq. which fumes in air. Bp_{18} 35-36°. n_D^{20} 1.4554.
Easily hydrol.

Steininger, E., *Chem. Ber.*, 1962, **95**, 2993.

Ethylphosphonodithioic acid, 9CI E-00140

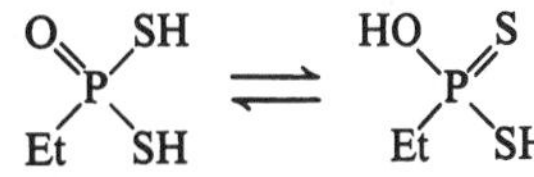

$C_2H_7OPS_2$ M 142.170
O-*Me*, S-*Ph ester:* O-*Methyl* S-*phenyl ethylphosphonodithioate.*
$C_9H_{13}OPS_2$ M 232.295
Liq. $Bp_{0.3}$ 132°. n_D^{20} 1.6060.
O,S-*Di-Et ester: see O-Ethyl ethylphosphonodithioate, E-00070*
S,S-*Di-Et ester:* [18032-95-6].
$C_6H_{15}OPS_2$ M 198.278
Liq. d_4^{20} 1.11. $Bp_{0.04}$ 69-70°. n_D^{20} 1.5357.
▷TA5700000.
S,S-*Dibutyl ester:*
$C_{10}H_{23}OPS_2$ M 254.385
Liq. d_4^{20} 1.04. $Bp_{0.04}$ 103-104.5°. n_D^{20} 1.5153.
O-*Et ester: see O-Ethyl ethylphosphonodithioate, E-00070*
O-*Et*, S-*Ph ester: see Fonofos, F-00052*
Anhydrosulfide: see Diethylthiodiphosphonic acid, D-00405

Menn, J.J. *et al*, *J. Econ. Entomol.*, 1965, **58**, 734 (*synth, tox*)

Akamsin, V.D. *et al*, *Izv. Akad. Nauk SSSR, Ser. Khim.*, 1967, 1987 (*Engl. transl.* p. 1904) (*synth*)
Horner, L. *et al*, *Phosphorus Sulfur*, 1985, **22**, 13 (*derivs, synth, pmr*)

Ethylphosphonodithious acid, 9CI **E-00141**

EtP(SH)$_2$

$C_2H_7PS_2$ M 126.171

Di-Et ester: [15714-21-3]. *Diethyl ethylphosphonodithioite.*
$C_6H_{15}PS_2$ M 182.278
Liq. d_4^{20} 1.02. Bp_{10} 98-100°. n_D^{20} 1.5490.
Dibutyl ester: [6588-02-9]. *Dibutyl ethylphosphonodithioite.*
$C_8H_{23}PS_2$ M 214.363
Liq. n_D^{20} 1.5258.
Di-Ph ester: Diphenyl ethylphosphonodithioite.
$C_{14}H_{15}PS_2$ M 278.366
Liq. d_4^{20} 1.17. $Bp_{0.03}$ 148-152°. n_D^{20} 1.6597.
Dibenzyl ester: [20354-83-0]. *Dibenzyl ethylphosphonodithioite. Bis(phenylmethyl) ethylphosphonodithioite.*
$C_{16}H_{19}PS_2$ M 306.420
Liq. d_4^{20} 1.15. $Bp_{0.03}$ 162-163°. n_D^{20} 1.6330.

Krasil'nikova, E.A. *et al*, *Zh. Obshch. Khim.*, 1968, **38**, 609 (*Engl. transl.* p. 587) (*synth*)
Nifant'ev, É.E. *et al*, *Zh. Obshch. Khim.*, 1970, **40**, 1420 (*Engl. transl.* p. 1406) (*dibutyl ester, synth*)
Razumov, A.I. *et al*, *Zh. Obshch. Khim.*, 1972, **42**, 1250 (*Engl. transl.* p. 1245) (*synth, nmr*)
Chernova, A.V. *et al*, *Zh. Obshch. Khim.*, 1979, **49**, 2002 (*Engl. transl.* p. 1760) (*diethyl ester, uv*)

Ethylphosphonofluoridic acid, 9CI **E-00142**

[2171-87-1]

EtPF(O)OH

$C_2H_6FO_2P$ M 112.040
Liq. $Bp_{0.2}$ 72°.
Me ester: [665-03-2]. *Methyl ethylphosphonofluoridate.*
$C_3H_8FO_2P$ M 126.067
Liq. Bp_{12} 46-48°. n_D^{20} 1.3835.
Et ester: [650-20-4]. *Ethyl ethylphosphonofluoridate.*
$C_4H_{10}FO_2P$ M 140.094
Liq. d_{20}^{20} 1.13. Bp_{15} 54°. n_D^{25} 1.3860.
Isopropyl ester: [1189-87-3]. *Isopropyl ethylphosphonofluoridate.*
$C_5H_{10}FO_2P$ M 152.105
Liq. Bp_{16} 74°.
▷TA8110000.

Yakubovich, A.Ya. *et al*, *Zh. Obshch. Khim.*, 1961, **31**, 1517 (*Engl. transl.* p. 1405) (*ester, synth*)
Bebbington, A. *et al*, *J. Chem. Soc. (C)*, 1966, 1410 (*ester, synth*)
Landau, M.A. *et al*, *Zh. Strukt. Khim.*, 1970, **11**, 513 (*Engl. transl.* p. 467) (*esters, struct*)
Bender, R. *et al*, *Phosphorus*, 1974, **4**, 183 (*synth, pmr, nmr*)
Kozhushko, B.N. *et al*, *Zh. Obshch. Khim.*, 1980, **50**, 1273 (*Engl. transl.* p. 1029) (*esters, synth*)
Gubaidullin, M.G., *Zh. Obshch. Khim.*, 1982, **52**, 2469 (*Engl. transl.* p. 2182) (*esters, props*)
Mager, P.P., *Toxicol. Lett.*, 1982, **11**, 67 (*tox*)

Ethylphosphonofluoridodithioic acid, 9CI **E-00143**

[21207-78-3]

EtPF(S)SH

$C_2H_6FPS_2$ M 144.162
Liq. d^{20} 1.28. Bp_{12-13} 64-67°. n_D 1.5433.
Anhydrosulfide: see Diethylthiodiphosphonic acid, D-00405

Roesky, H.W., *Chem. Ber.*, 1968, **101**, 3679 (*synth, ir, pmr*)

Ethylphosphonoperoxoic acid, 9CI **E-00144**

EtP(O)(OH)OOH

$C_2H_7O_4P$ M 126.049
OO-tert-*Butyl,* O-*Me ester:* [6795-02-4]. OO-tert-*Butyl* O-*methyl ethylphosphonoperoxoate.*
$C_7H_{17}O_4P$ M 196.183
Liq. d_4^{20} 1.05. $Bp_{0.02}$ 56-57°. n_D^{20} 1.4282.
OO-tert-*Butyl,* O-*Et ester:* [34109-34-7]. OO-tert-*Butyl* O-*ethyl ethylphosphonoperoxoate.*
$C_8H_{19}O_4P$ M 210.209
No phys. props. reported.
OO-tert-*Butyl,* O-*isopropyl ester:* [31238-31-0]. OO-tert-*Butyl* O-*isopropyl ethylphosphonoperoxoate.*
$C_9H_{21}O_4P$ M 224.236
Liq. $Bp_{0.075}$ 59-60°. n_D^{20} 1.4243.
OO-tert-*Butyl,* O-*butyl ester:* [31246-33-0]. O-*Butyl* OO-tert-*butyl ethylphosphonoperoxoate.*
$C_{10}H_{23}O_4P$ M 238.263
Liq. $Bp_{0.15}$ 74-75°. n_D^{20} 1.4297.

Sosnovsky, G. *et al*, *Synthesis*, 1971, 144 (*esters, synth*)
Sosnovsky, G. *et al*, *J. Org. Chem.*, 1972, **37**, 2267 (*props*)
Maslennikov, V.P. *et al*, *Dokl. Akad. Nauk SSSR, Ser. Sci. Khim.*, 1973, **209**, 1109 (*Engl. transl.* p. 302) (*esters, props*)

Ethylphosphonoselenoic acid, 9CI, 8CI **E-00145**

[13992-06-8]

Se=P(Et)(OH)OH ⇌ HSe–P(=O)(Et)OH

$C_2H_7O_2PSe$ M 173.010
Mono-O-Et ester: see O-Ethyl ethylphosphonoselenoate, E-00072
O,O-*Di-Et ester:* [39181-15-2].
$C_6H_{15}O_2PSe$ M 229.117
Liq. d_0^{20} 1.27. $Bp_{11.5}$ 97.5°, $Bp_{0.1}$ 52-53°. n_D^{20} 1.4872.
O, *Se-Di-Et ester:* [65426-08-6]. O,Se-*Diethyl ethylphosphonoselenoate.*
$C_6H_{15}O_2PSe$ M 229.117
Liq. d_4^{20} 1.31. $Bp_{0.1}$ 61-62°. n_D^{20} 1.4973.
O-*Monopropyl ester:* [55927-32-7]. O-*Propyl hydrogen phosphonoselenoate.*
$C_5H_{13}O_2PSe$ M 215.090
Liq. d_4^{20} 1.37. $Bp_{0.015}$ 110-111°. n_D^{20} 1.5136.
O,O-*Dipropyl ester:* O,O-*Dipropyl phosphonoselenoate.*
$C_8H_{19}O_2PSe$ M 257.171
d_4^{20} 1.20. $Bp_{0.1}$ 72-73°. n_D^{20} 1.4827.
O-*Monoisopropyl ester:* O-*Isopropyl hydrogen ethylphosphonoselenoate.* O-*1-Methylethyl ethylphosphonoselenoate.*
$C_5H_{13}O_2PSe$ M 215.090
d_4^{20} 1.36. $Bp_{0.015}$ 86-88°. n_D^{20} 1.5090.

Gryszkiewicz-Trochimowski, E. *et al*, *Bull. Soc. Chim. Fr.*, 1960, 1794 (*O,O-dialkyl esters, synth*)
Stec, W.J. *et al*, *Phosphorus*, 1972, **2**, 97 (*diethyl ester, pmr*)

Shagidullin, R.R. *et al*, *Izv. Akad. Nauk SSSR, Ser. Khim.*, 1975, 197 (*Engl. transl.* p. 190) (*diethyl ester, uv*)
Nuretdinov, I.A., *Izv. Akad. Nauk SSSR, Ser. Khim.*, 1977, 2635 (*Engl. transl.* p. 2441) (*OSe-diethyl ester, synth, P nmr*)
Nuretdinov, I.A., *Izv. Akad. Nauk SSSR, Ser. Khim.*, 1978, 437 (*Engl. transl.* p. 378) (*monoalkyl esters, synth, P nmr*)

P-Ethylphosphonoselenoic diamide, 9CI — E-00146

EtP(Se)$(NH_2)_2$

$C_2H_9N_2PSe$ M 171.040

N,N,N′,N′-*Tetra-Me:* [55205-96-4]. *Ethylphosphonoselenoic bis(dimethylamide).* P-*Ethyl*-N,N,N′,N′-*tetramethylphosphonoselenoic diamide.*
$C_6H_{17}N_2PSe$ M 227.147
No phys. props. reported.

N,N,N′,N′-*Tetra-Et:* [39181-30-1]. *Ethylphosphonoselenoic bis(diethylamide). Pentaethylphosphonoselenoic diamide.*
$C_{10}H_{25}N_2PSe$ M 283.255
No phys. props. reported.

N,N′-*Di-Ph:* P-*Ethyl*-N,N′-*diphenylphosphonoselenoic diamide. Ethylphosphonoselenoic dianilide.*
$C_8H_{17}N_2PSe$ M 251.169
Cryst. (MeOH). Mp 146-147°.

Gryszkiewicz-Trochimowski, E. *et al*, *Bull. Soc. Chim. Fr.*, 1960, 1794 (*diphenyl*)
Stec, W.J. *et al*, *Phosphorus*, 1972, **2**, 97 (*tetraethyl, P nmr*)
Shagidullin, R.R. *et al*, *Izv. Akad. Nauk SSSR, Ser. Khim.*, 1975, **24**, 197; 1976, **25**, 184 (*Engl. transl.* pp. 190, 174) (*tetramethyl, uv*)

Ethylphosphonoselenoic dichloride, 9CI, 8CI — E-00147

[14705-46-5]

EtP(Se)Cl_2

$C_2H_5Cl_2PSe$ M 209.901
Liq. d_4^{20} 1.68. Bp_{25} 92-94°, $Bp_{1.5}$ 38-39°. n_D^{20} 1.5850.

Gryszkiewicz-Trochimowski, E. *et al*, *Bull. Soc. Chim. Fr.*, 1960, 1794 (*synth*)
Ivin, S.Z. *et al*, *Zh. Obshch. Khim.*, 1961, **31**, 4052 (*Engl. transl.* p. 3780) (*synth*)
Roesky, H.W. *et al*, *Z. Naturforsch., B*, 1973, **28**, 697 (*synth, ir, pmr, raman, P nmr*)
Nuretdinov, I.A. *et al*, *Izv. Akad. Nauk SSSR, Ser. Khim.*, 1975, 327 (*Engl. transl.* p. 263) (*nqr*)
Shagidullin, R.R. *et al*, *Izv. Akad. Nauk SSSR, Ser. Khim.*, 1976, 184 (*Engl. transl.* p. 174) (*uv*)

Ethylphosphonothioic acid, 9CI — E-00148

[13991-98-5]

Et(S=)P(OH)OH ⇌ Et(HS)P(=O)OH

$C_2H_7O_2PS$ M 126.110

Mono-O-Me ester: see O-Methyl ethylphosphonothioate, M-00155

O,O-*Di-Me ester:* O,O-*Dimethyl ethylphosphonothioate.*
$C_4H_{11}O_2PS$ M 154.163
Liq. d_4^{25} 1.10. $Bp_{2.2}$ 40°. n_D^{25} 1.4722.

O-*Mono-Et ester: see O-Ethyl ethylphosphonothioate, E-00074*

O-*Me*, O-*Et diester: see under O-Methyl ethylphosphonothioate, M-00155*

S-*Me*, O-*Et diester: see under O-Ethyl ethylphosphonothioate, E-00074*

O,O-*Di-Et ester:* [2455-45-0]. O,O-*Diethyl ethylphosphonothioate.*
$C_6H_{15}O_2PS$ M 182.217
Liq. d_4^{25} 1.03. Bp_{10} 80-82°, Bp_1 42°. n_D^{25} 1.4620.

O,S-*Di-Et ester: see O-Ethyl ethylphosphonothioate, E-00074*

O,O-*Diisopropyl ester:* O,O-*Diisopropyl ethylphosphonothioate.*
$C_8H_{19}O_2PS$ M 210.271
Liq. $Bp_{0.5}$ 48°. n_D^{25} 1.4521.

O-*Et*, S-*Ph diester: see O-Ethyl ethylphosphonothioate, E-00074*

O-*Et*, O-*(2,4,5-trichlorophenyl)diester: see Trichloronate, T-00419*

O-*Et*, O-*(2,5-dichloro-4-iodophenyl)diester: see O-(2,5-Dichloro-4-iodophenyl) O-ethyl ethylphosphonothioate, D-00171*

O,O-*Di-Ph ester:* [74206-36-3]. O,O-*Diphenyl ethylphosphonothioate.*
$C_{14}H_{15}O_2PS$ M 278.305
Liq. $Bp_{0.06}$ 150-155°. n_D^{20} 1.5786.

Difluoride: see Ethylphosphonothioic difluoride, E-00151

Dichloride: see Ethylphosphonothioic dichloride, E-00150

Dibromide:
$C_2H_5Br_2PS$ M 251.903
Liq. Bp_{10} 82-6°.

Diamide: see P-Ethylphosphonothioic diamide, E-00149

Hoffmann, F.W. *et al*, *J. Am. Chem. Soc.*, 1958, **80**, 3945 (*O,O-dialkyl esters, synth*)
Pudovik, M.A. *et al*, *Zh. Obshch. Khim.*, 1980, **50**, 740 (*diphenyl ester, synth, P nmr*)
Maier, L., *Helv. Chim. Acta*, 1964, **47**, 27 (*dibromide, synth, P nmr*)

P-Ethylphosphonothioic diamide, 9CI — E-00149

Ethylthiophosphonic diamide

[30043-15-3]

EtP(S)$(NH_2)_2$

$C_2H_9N_2PS$ M 124.140
Cryst. (hexane). Mp 137°.

N,N,N′,N′-*Tetra-Me:* [3981-45-1]. P-*Ethyl*-N,N,N′,N′-*tetramethylphosphonothioic diamide. Ethylphosphonothioic bis(dimethylamide).*
$C_6H_{17}N_2PS$ M 180.247
Liq. Bp_{12} 120-125°. n_D^{20} 1.5145.

N,N,N′,N′-*Tetra-Et:* [30610-36-7]. *Pentaethylphosphonothioic diamide.*
$C_{10}H_{25}N_2PS$ M 236.355
Liq. d_4^{20} 1.01. $Bp_{0.0001}$ 70-75°. n_D^{20} 1.5096.

N,N′-*Di-Ph:* [61173-01-1]. P-*Ethyl*-N,N′-*diphenylphosphonothioic diamide. Ethylphosphonothioic dianilide.*
$C_{14}H_{17}N_2PS$ M 276.335
Solid. Mp 112-114°.

N,N′-*Dibenzyl:* [62992-30-7]. N,N′-*Dibenzyl*-P-*ethylphosphonothioic diamide.*
$C_{16}H_{21}N_2PS$ M 304.389
Needles (heptane). Mp 52°.

Kabachnik, M.I. *et al*, *Dokl. Akad. Nauk SSSR, Ser. Sci. Khim.*, 1956, **110**, 217; *CA*, **51**, 4982 (*diphenyl, synth*)
Maier, L., *Helv. Chim. Acta*, 1964, **47**, 27 (*tetramethyl, synth, P nmr*)

Issleib, K. *et al*, *Z. Anorg. Allg. Chem.*, 1977, **428**, 16 (*synth, dibenzyl*)
Nifant'ev, É.E. *et al*, *Zh. Obshch. Khim.*, 1977, **47**, 299 (*Engl. transl.* p. 276) (*tetraethyl, synth*)

Ethylphosphonothioic dichloride, 9CI **E-00150**

Ethylthiophosphonic dichloride

[993-43-1]

$EtP(S)Cl_2$

$C_2H_5Cl_2PS$ M 163.001

Liq. with pungent odour. d_4^{20} 1.35. Bp_{740} 177-181°, Bp_{50} 80-82°. n_D^{20} 1.5428.

Hoffmann, F.W. *et al*, *J. Am. Chem. Soc.*, 1958, **80**, 3945 (*synth*)
Maier, L., *Helv. Chim. Acta*, 1964, **47**, 27 (*synth, nmr*)
Christol, C. *et al*, *J. Chim. Phys.*, 1965, **62**, 246 (*ir, raman*)
Kavavanov, K.V. *et al*, *Zh. Obshch. Khim.*, 1965, **35**, 78 (*Engl. transl.* p. 76) (*synth*)
Shagidullin, R.R. *et al*, *Dokl. Akad. Nauk SSSR, Ser. Sci. Khim.*, 1975, **222**, 897 (*Engl. transl.* p. 564) (*uv*)

Ethylphosphonothioic difluoride, 9CI **E-00151**

Ethylthiophosphonic difluoride

[753-99-1]

$EtP(S)F_2$

$C_2H_5F_2PS$ M 130.092

Liq. d_4^{23} 1.21. Bp 89-90°. n_D 1.4076.

Roesky, H.W., *Chem. Ber.*, 1968, **101**, 3679 (*synth, ir, pmr, F nmr*)
Reddy, G.S. *et al*, *Z. Naturforsch., B*, 1970, **25**, 1199 (*pmr, F and P nmr*)
Köttgen, D. *et al*, *Z. Anorg. Allg. Chem.*, 1972, **389**, 269 (*ir*)
Shagidullin, R.R. *et al*, *Zh. Obshch. Khim.*, 1983, **53**, 84 (*Engl. transl.* p. 68) (*ir, raman*)

Ethylphosphonotrithioic acid, 9CI **E-00152**

$EtP(S)(SH)_2$

$C_2H_7PS_3$ M 158.231

Di-Et ester: [13297-95-5]. *Diethyl ethylphosphonotrithioate.*
$C_6H_{15}PS_3$ M 214.338
Liq. d_4^{20} 1.13. Bp_{14} 152°. n_D^{20} 1.5845.

Dibutyl ester: Dibutyl ethylphosphonotrithioate.
$C_{10}H_{23}PS_3$ M 270.445
Liq. d_4^{20} 1.06. $Bp_{0.04}$ 106°. n_D^{20} 1.5535.

Di-Ph ester: Diphenyl ethylphosphonotrithioate.
$C_{14}H_{15}PS_3$ M 310.426
Liq. $Bp_{0.01}$ 144°.

Cyclic bis(anhydrosulfide): [1124-70-5]. *2,4-Diethyl-1,3,2,4-dithiadiphosphetane 2,4-disulfide.*
$C_4H_{10}P_2S_4$ M 248.311
Cryst. (C_6H_6 or chlorobenzene). Mp 148°.

Newallis, P.E. *et al*, *J. Org. Chem.*, 1962, **27**, 3829 (*anhydrosulfide*)
Krasil'nikova, E.A. *et al*, *Zh. Obshch. Khim.*, 1967, **37**, 1409, 2365 (*Engl. transl.* pp. 1339, 2254) (*esters*)
Andreev, N.A. *et al*, *Zh. Obshch. Khim.*, 1982, **52**, 1785 (*Engl. transl.* p. 1581) (*anhydrosulfide, struct, ir, nmr*)

Ethylphosphonous acid **E-00153**

[4736-97-4]

$EtP(OH)_2$

$C_2H_7O_2P$ M 94.050

Free acid exists as the phosphoryl tautomer Ethylphosphinic acid, E-00124 .

Mono-Me ester: see Ethylphosphinic acid, E-00124

Di-Me ester: [15715-42-1]. *Dimethyl ethylphosphonite.*
$C_4H_{11}O_2P$ M 122.103
Liq. d_4^{20} 0.952. Bp_{225} 73.5-74.5°. n_D^{20} 1.4210.
▷Spont. flammable

Mono-Et ester: see Ethylphosphinic acid, E-00124

Di-Et ester: [2651-85-6]. *Diethyl ethylphosphonite.*
$C_6H_{15}O_2P$ M 150.157
Liq. Bp 137-139°, Bp_{10} 35.5-36.5°. n_D^{20} 1.4230.

Monoisopropyl ester: see Ethylphosphinic acid, E-00124

Diisopropyl ester: [76509-66-5]. *Diisopropyl ethylphosphonite.*
$C_8H_{19}O_2P$ M 178.211
Liq. Bp_{12} 50-52°. n_D^{20} 1.4169.

*Mono-*tert*-butyl ester: see Ethylphosphinic acid, E-00124*

Mono-Ph ester: see Ethylphosphinic acid, E-00124

Di-Ph ester: [14744-12-8]. *Diphenyl ethylphosphonite.*
$C_{14}H_{15}O_2P$ M 246.245
Liq. d_4^{20} 1.11. Bp_{11} 129-130°. n_D^{20} 1.5658.

Dibenzyl ester: Dibenzyl ethylphosphonite. Bis(phenylmethyl) ethyphosphonite.
$C_{16}H_{19}O_2P$ M 274.299
Liq. d_4^{20} 1.08. $Bp_{1.5}$ 140-141°. n_D^{20} 1.5499.

Difluoride: [430-78-4]. *Ethyldifluorophosphine.*
$C_2H_5F_2P$ M 98.032
Gas-liq. Bp_{760} 6-7°.

Dichloride: see Ethylphosphonous dichloride, E-00155

Dibromide: [1068-59-3]. *Dibromoethylphosphine.*
$C_2H_5Br_2P$ M 219.843
Liq. d_{20}^{20} 1.82. Bp 161°. n_D^{15} 1.5451.

Diiodide: [13868-72-9]. *Ethyldiiodophosphine.*
$C_2H_5I_2P$ M 313.844
Liq. Bp_3 70-71°.

Diamide: see Ethylphosphonous diamide, E-00154

Arbuzov, B.A. *et al*, *Izv. Akad. Nauk SSSR, Ser. Khim.*, 1952, 854 (*Engl. transl.* p. 765) (*esters*)
Rizpolozhenskii, N.I. *et al*, *Izv. Akad. Nauk SSSR, Ser. Khim.*, 1959, 358 (*Engl. transl.* p. 335) (*esters*)
Maier, L., *Helv. Chim. Acta*, 1963, **46**, 2026 (*dibromide, synth, nmr*)
Mastalerz, P., *Rocz. Chem.*, 1964, **38**, 61 (*diisopropyl ester*)
Drozd, G.I. *et al*, *Zh. Obshch. Khim.*, 1967, **37**, 1343 (*Engl. transl.* p. 1269) (*difluoride, synth, F and P nmr*)
Quin, L.D. *et al*, *Org. Magn. Reson.*, 1974, **6**, 503 (*dimethyl ester, cmr*)
Kharrasova, F.M. *et al*, *Zh. Obshch. Khim.*, 1978, **48**, 1041 (*Engl. transl.* p. 948) (*diphenyl ester, synth, P nmr*)
Van Linthoudt, J.P. *et al*, *Spectrochim. Acta, Part A*, 1979, **35**, 1307 (*dibromide, diiodide, pmr, cmr, P nmr*)
Romanenko, V.D. *et al*, *Synthesis*, 1980, 823 (*diiodide, synth, P nmr*)
Hinke, A. *et al*, *Phosphorus Sulfur*, 1983, **15**, 93 (*dibromide, synth, P nmr*)

Ethylphosphonous diamide **E-00154**

$EtP(NH_2)_2$

$C_2H_9N_2P$ M 92.080

Tetra-Et: [26546-68-9]. *Pentaethylphosphonous diamide. Ethylphosphonous bis(diethylamide).*
$C_{10}H_{25}N_2P$ M 204.295
Liq. Bp_{10} 88-89°. n_D^{20} 1.4680.

Tetrakis(trimethylsilyl): [84050-72-6]. P-*Ethyl-N,N,N',N'-tetrakis(trimethylsilyl)phosphonous diamide.*
$C_{14}H_{41}N_2PSi_4$ M 380.807
Liq. $Bp_{0.05}$ 98-99°.

Zentil, M. *et al, Bull. Soc. Chim. Fr.*, 1971, 376 (*tetraethyl, synth*)
Andreev, N.A. *et al, Zh. Obshch. Khim.*, 1979, **49**, 2230 (*Engl. transl.* p. 1959) (*tetraethyl, synth, ir*)
Van Linthoudt, J.P. *et al, Spectrochim. Acta, Part A*, 1980, **36**, 315 (*tetraethyl, cmr, P nmr, pmr*)
Bei-Li Li *et al, Inorg. Chem.*, 1983, **22**, 575 (*trimethylsilyl, synth, pmr, cmr, P nmr*)

Ethylphosphonous dichloride, 9CI — E-00155

Dichloroethylphosphine
[1498-40-4]

$EtPCl_2$

$C_2H_5Cl_2P$ M 130.941
Liq. with pungent, foul, odour. d_4^{20} 1.26. Bp_{722} 111-112°. n_D^{20} 1.4943. Air-sensitive.
▷TB2465000.

Maier, L., *Helv. Chim. Acta*, 1964, **47**, 2137 (*synth, P nmr*)
Remizov, A.B. *et al, Zh. Obshch. Khim.*, 1973, **43**, 1406 (*Engl. transl.* p. 1393) (*pmr, conformn*)
Quin, L.D. *et al, Org. Magn. Reson.*, 1974, **6**, 503 (*cmr*)
Fishman, A.I. *et al, Spectrochim. Acta, Part A*, 1976, **32**, 651 (*ir, raman*)
Naumova, V.A. *et al, Zh. Strukt. Khim.*, 1977, **18**, 67 (*Engl. transl.* p. 54) (*ed, struct*)
van Linthoudt, J.P. *et al, Spectrochim. Acta, Part A*, 1979, **35**, 1307 (*cmr, pmr, P nmr*)

Ethylphosphoramidic acid, 9CI — E-00156

$EtNHP(O)(OH)_2$

$C_2H_8NO_3P$ M 125.064
Di-Me ester: [20464-99-7]. *Dimethyl ethylphosphoramidate.*
$C_4H_{12}NO_3P$ M 153.117
Liq. d_4^{20} 1.14. Bp_{10} 122-124°. n_D^{20} 1.4275.
Di-Et ester: see Diethyl ethylphosphoramidate, D-00293
Dipropyl ester: [20465-04-7]. *Dipropyl ethylphosphoramidate.*
$C_8H_{20}NO_3P$ M 209.225
Liq. d_4^{20} 1.01. Bp_{20} 160-161°, $Bp_{0.03}$ 85-89°. n_D^{20} 1.4319.
Dibutyl ester: [20465-06-9]. *Dibutyl ethylphosphoramidate.*
$C_{10}H_{24}NO_3P$ M 237.278
Liq. d_4^{20} 0.99. Bp_{10} 165-167°, $Bp_{0.03}$ 98-100°. n_D^{20} 1.4335.
Di-Ph ester: [5756-02-5]. *Diphenyl ethylphosphoramidate.*
$C_{14}H_{16}NO_3P$ M 277.259
Cryst. (pet. ether). Mp 49-50°. $Bp_{0.9}$ 186-188°.
Difluoride: [819-04-5].
$C_2H_6F_2NOP$ M 129.046
Liq. Bp_8 35-36°. n_D^{20} 1.3622.
Dichloride: [61056-26-6].
$C_2H_6Cl_2NOP$ M 161.955
Liq. Bp_{12} 129°.

Olah, G.A. *et al, Justus Liebigs Ann. Chem.*, 1959, **625**, 88 (*difluride*)
Nikonorov, K.V. *et al, Izv. Akad. Nauk SSSR, Ser. Khim.*, 1968, 587 (*Engl. transl.* p. 567) (*esters, synth*)
Luchkovskaya, O.N. *et al, Zh. Obshch. Khim.*, 1970, **40**, 644 (*Engl. transl.* p. 615) (*esters, synth*)
Stock, J.A. *et al, J. Chem. Soc. (C)*, 1966, 637 (*diphenyl ester, synth, props*)
Roth, H.J. *et al, Arch. Pharm. (Weinheim, Ger.)*, 1981, **314**, 85 (*dichloride, synth, pmr*)

Ethylphosphoramidochloridothioic acid, 9CI, 8CI — E-00157

S=P(OH)(EtNH)(Cl) ⇌ HS–P(=O)(EtNH)(Cl)

$C_2H_7ClNOPS$ M 159.570
(±)-***form***
O-*Me ester:* O-*Methyl ethylphosphoramidochloridothioate.*
$C_3H_9ClNOPS$ M 173.597
Liq. d_4^{20} 1.28. $Bp_{0.5}$ 68°. n_D^{20} 1.5090.
O-*Ph ester:* O-*Phenyl ethylphosphoramidochloridothioate.*
$C_8H_{11}ClNOPS$ M 235.668
Liq. d_4^{20} 1.26. $Bp_{0.26}$ 123-125°. n_D^{20} 1.5670.

Mandel'baum, Ya.A. *et al, Zh. Obshch. Khim.*, 1968, **38**, 1754 (*Engl. transl.* p. 1709) (*synth*)

O-Ethyl phosphoramidofluoridothioate, 9CI, 8CI — E-00158

Thiophosphoric acid O-*ethyl ester-fluoride-amide.* O-*Ethyl amidofluorothiophosphate*
[25246-49-5]

$EtOPF(S)NH_2$

C_2H_7FNOPS M 143.116
(±)-***form***
Liq. $Bp_{0.02}$ 39°. n_D^{20} 1.4666.

Roesky, H.W. *et al, Chem. Ber.*, 1969, **102**, 2588 (*synth, ir, ms, pmr, F nmr*)

Ethylphosphoramidothioic acid, 9CI, 8CI — E-00159

$EtNHP(S)(OH)_2 \rightleftharpoons EtNHP(O)(OH)(SH)$

$C_2H_8NO_2PS$ M 141.124
O,O-*Di-Et ester:* [36592-32-2]. O,O-*Diethyl ethylphosphoramidothioate.*
$C_6H_{16}NO_2PS$ M 197.232
Liq. $Bp_{0.25}$ 67-70°. n_D^{20} 1.4716.
O,O-*Diisopropyl ester:* O,O-*Diisopropyl ethylphosphoramidothioate.* O,O-*Bis(1-methylethyl) ethylphosphoramidothioate.*
$C_8H_{20}NO_2PS$ M 225.285
Liq. $Bp_{0.25}$ 72-75°. n_D^{20} 1.4617.
O,O-*Di-*tert-*butyl ester:* O,O-*Di-*tert-*butyl ethylphosphoramidothioate.* O,O-*Bis(1,1-dimethylethyl) ethylphosphoramidothioate.*
$C_{10}H_{24}NO_2PS$ M 253.339
$Bp_{0.25}$ 78-81°. n_D^{20} 1.4662.
O,O-*Di-Ph ester:* O,O-*Diphenyl ethylphosphoramidothioate.*
$C_{14}H_{16}NO_2PS$ M 293.320

Needles (EtOH aq.). Mp 64-64.5°.
Dichloride:
$C_2H_6Cl_2NPS$ M 178.016
Liq. Bp_8 98-102°. n_D^{20} 1.5432.

Klee, F.C. *et al, J. Pharm. Sci.*, 1962, **51**, 423 (*synth*)

Ethylphosphoramidous acid, 9CI E-00160

EtNHP(OH)₂

$C_2H_8NO_2P$ M 109.064
Difluoride:
$C_2H_6F_2NP$ M 113.047
Liq. V.p. 69 mm at 27°.
Dichloride:
$C_2H_6Cl_2NP$ M 145.956
Liq. Bp 222-225° dec., Bp_{11} 92°.

Michaelis, A., *Justus Liebigs Ann. Chem.*, 1903, **326**, 129 (*dichloride*)
Harman, J.S. *et al, J. Chem. Soc. (A)*, 1970, 1935 (*difluoride, synth, ir, pmr*)

Ethylphosphorimidic acid, 9CI E-00161

EtN=P(OH)₃

$C_2H_8NO_3P$ M 125.064
Free acid tautomeric with Ethylphosphoramidic acid, E-00156 .
Tri-Me ester: [55215-24-2]. *Trimethyl ethylphosphorimidate.* N-*Ethyl*-P,P,P-*trimethoxyphosphine imide.* N-*Ethyl*-P,P,P-*trimethoxyphosphazene.*
$C_5H_{14}NO_3P$ M 167.144
Liq. Bp_1 36°.
Tri-Et ester: [37983-59-8]. *Triethyl ethylphosphorimidate.* P,P,P-*Triethoxy*-N-*ethylphosphine imide.* P,P,P-*Triethoxy*-N-*ethylphosphazene.*
$C_8H_{20}NO_3P$ M 209.225
Liq. d_4^{20} 0.99. Bp_8 73-76°. n_D^{20} 1.4194.
Tri-Ph ester: Triphenyl ethylphosphorimidate. N-*Ethyl*-P,P,P-*triphenoxyphosphine imide.* N-*Ethyl*-P,P,P-*triphenoxyphosphazene.*
$C_{20}H_{20}NO_3P$ M 353.357
Greenish cryst. Mp 33-34°. Bp_2 171-173°.

Petrov, K.A. *et al, Zh. Obshch. Khim.*, 1956, **26**, 3378 (*Engl. transl.* p. 3763) (*triphenyl ester*)
Goldwhite, H. *et al, J. Chem. Soc., Dalton Trans.*, 1975, 12 (*trimethyl ester, ir, pmr, P nmr*)
Khodak, A.A. *et al, Zh. Obshch. Khim.*, 1976, **46**, 2482 (*Engl. transl.* p. 2376) (*triethyl ester, ir, P nmr*)
Martin, G.J. *et al, Tetrahedron Lett.*, 1983, **24**, 4989 (*N and P nmr*)

Ethyl phosphorobromidofluoridate, 9CI, 8CI E-00162

Ethyl bromofluorophosphate

EtOP(O)BrF

$C_2H_5BrFO_2P$ M 190.936
Liq. d_{20}^{20} 1.70. Bp_{15} 40-42°. n_D^{20} 1.413. Easily hydrolysed.

Stölzer, C. *et al, Chem. Ber.*, 1960, **93**, 1323.

Ethyl phosphorochloridisocyanatidate, 9CI, 8CI E-00163

[995-87-9]

EtOPCl(O)NCO

$C_3H_5ClNO_3P$ M 169.504
(±)-*form*
Unstable liq. with unpleasant odour. d_4^{20} 1.36. Bp_{55} 100-102°, $Bp_{0.3}$ 29-30°. n_D^{20} 1.4320.

Narbut, A.V. *et al, Zh. Obshch. Khim.*, 1968, **38**, 1321 (*Engl. transl.* p. 1272) (*synth*)
Gubnitskaya, E.A. *et al, Zh. Obshch. Khim.*, 1970, **40**, 1205 (*Engl. transl.* p. 1197) (*synth*)
Shokol, V.A. *et al, Zh. Obshch. Khim.*, 1974, **44**, 2660 (*Engl. transl.* p. 2615) (*synth*)

Ethyl phosphorochloridofluoridate, 9CI, 8CI E-00164

Ethyl chlorofluorophosphate
[762-77-6]

EtOP(O)ClF

$C_2H_5ClFO_2P$ M 146.485
Liq. d_{20}^{20} 1.32. Bp_{50} 50°. n_D^{20} 1.3750. Easily hydrolysed.

Stölzer, C. *et al, Chem. Ber.*, 1960, **93**, 1323 (*synth*)
Sheluchenko, V.V. *et al, Dokl. Akad. Nauk SSSR, Ser. Sci. Khim.*, 1967, **177**, 376 (*Engl. transl.* p. 1050) (*F and P nmr*)
Ivanova, Zh.M. *et al, Zh. Obshch. Khim.*, 1968, **38**, 551 (*Engl. transl.* p. 538) (*synth*)
Neimysheva, A.A. *et al, Zh. Obshch. Khim.*, 1968, **38**, 595 (*Engl. transl.* p. 575) (*props*)

Ethyl phosphorodiamidite, 9CI E-00165

EtOP(NH₂)₂

$C_2H_9N_2OP$ M 108.080
Yellowish amorph. solid. Mp 140-142° dec.

Becke-Goehring, M. *et al, Chem. Ber.*, 1958, **91**, 1188 (*synth, props*)

Ethyl phosphorodibromidite, 9CI, 8CI E-00166

Ethyl dibromophosphite
[20502-49-2]

EtOPBr₂

$C_2H_5Br_2OP$ M 235.843
Pungent liq. d_4^{20} 2.0. Bp_{10} 49°. n_D^{20} 1.5648. Not obt. pure, admixed with 20% PBr_3.

Nuretdinova, O.N. *et al, Izv. Akad. Nauk SSSR, Ser. Khim.*, 1975, 694 (*Engl. transl.* p. 620) (*synth, P nmr*)

Ethyl phosphorodichloridate, 9CI, 8CI E-00167

Ethyl phosphoryl dichloride. Ethyl dichlorophosphate
[1498-51-7]

EtOP(O)Cl₂

$C_2H_5Cl_2O_2P$ M 162.940
Liq. d_{25} 1.38. Bp 167°, Bp_{13} 58°. n_D^{20} 1.4342.
▷TD4390000.

Saunders, B.C. *et al, J. Chem. Soc.*, 1948, 699 (*synth, tox*)
Grunze, H., *Chem. Ber.*, 1959, **92**, 850 (*synth, props*)
Williamson, M.P. *et al, J. Phys. Chem.*, 1968, **72**, 4043 (*pmr, bibl*)
Furlani, C. *et al, J. Chem. Soc., Dalton Trans.*, 1977, 673 (*pe*)

Zverev, V.V. *et al, Zh. Obshch. Khim.*, 1979, **49**, 1737 (*Engl. transl.* p. 1522) (*pe*)
Bel'skii, V.E. *et al, Zh. Prikl. Spektrosk.*, 1980, **33**, 361; *CA*, **93**, 227847 (*ir*)

Popov, E.M. *et al, Zh. Obshch. Khim.*, 1959, **29**, 1998 (*Engl. transl.* p. 1967) (*ir, raman*)
Petrov, K.A. *et al, Zh. Obshch. Khim.*, 1961, **31**, 1366 (*Engl. transl.* p. 1265) (*synth*)

Ethyl phosphorodichloridite, 9CI, 8CI **E-00168**

Ethyl dichlorophosphite

[1498-42-6]

$EtOPCl_2$

$C_2H_5Cl_2OP$ M 146.941

Fuming liq. d_4^{20} 1.29. Bp 116-118°, Bp_{18} 28-30°. n_D^{20} 1.4730.

▷Violently hydrol. by H_2O. Highly irritant and corrosive

Cook, H.G. *et al, J. Chem. Soc.*, 1949, 2921 (*synth, props*)
Schwarz, R. *et al, Chem. Ber.*, 1957, **90**, 952 (*synth*)
Fluck, E. *et al, Z. Anorg. Allg. Chem.*, 1961, **307**, 113 (*P nmr*)
Steyermark, P.R., *J. Org. Chem.*, 1963, **28**, 586.
Armour, M.A. *et al, J. Chem. Soc., Perkin Trans. 2*, 1975, 1185 (*synth*)
Gazizov, T.Kh. *et al, Zh. Obshch. Khim.*, 1977, **47**, 1234 (*Engl. transl.* p. 1137) (*synth*)
Sax, N.I., *Dangerous Properties of Industrial Materials*, 6th Ed., Van Nostrand-Reinhold, 1984, 672.

Ethyl phosphorodichloridodithioate, 9CI, 8CI **E-00169**

Ethyl dichlorodithiophosphate

[16001-05-1]

$EtSP(S)Cl_2$

$C_2H_5Cl_2PS_2$ M 195.061

Liq. d_4^{20} 1.42. Bp_3 61-62°. n_D^{20} 1.6904.

Godovikov, N.N. *et al, Zh. Obshch. Khim.*, 1961, **31**, 1628 (*Engl. transl.* p. 1516) (*synth*)

O-Ethyl phosphorodichloridothioate, 9CI, 8CI **E-00170**

O-*Ethyl dichlorothiophosphate*

[1498-64-2]

$EtOP(S)Cl_2$

$C_2H_5Cl_2OPS$ M 179.001

Liq. d^{20} 1.397. Bp_{20} 68°. n_D^{25} 1.5026.

Inorg. Synth., 1953, **4**, 75.
McIvor, R.A. *et al, Can. J. Chem.*, 1956, **34**, 1611, 1819, 1826 (*ir, synth*)
Williamson, M.P. *et al, J. Phys. Chem.*, 1968, **72**, 4043 (*pmr*)
Nyquist, R.A. *et al, Spectrochim. Acta, Part A*, 1970, **26**, 769 (*ir, raman*)
Whitehead, M.A. *et al, J. Chem. Soc. (A)*, 1971, 1738 (*nqr*)
Vilesov, F.I. *et al, Z. Phys. Chem. (Leipzig)*, 1974, **255**, 661 (*pe*)
Lebedev, N.N. *et al, Zh. Obshch. Khim.*, 1985, **55**, 814 (*Engl. transl.* p. 725) (*synth*)

S-Ethyl phosphorodichloridothioate, 9CI, 8CI **E-00171**

S-*Ethyl dichlorothiophosphate*

[1486-40-4]

$EtSP(O)Cl_2$

$C_2H_5Cl_2OPS$ M 179.001

Liq. d_4^{20} 1.40. Bp_{15} 87-88°. n_D^{20} 1.5279.

Ethyl phosphorodichloridothioite, 9CI **E-00172**

Ethyl dichlorothiophosphite

[1486-43-7]

$EtSPCl_2$

$C_2H_5Cl_2PS$ M 163.001

Pungent, odorous liq. d_4^{20} 1.34. Bp_9 51-53°. n_D^{20} 1.5597. Slowly disproportionates into PCl_3 and Diethyl phosphorochloridodithioite, D-00376 .

Stepashkina, L.V. *et al, Izv. Akad. Nauk SSSR, Ser. Khim.*, 1972, 380 (*Engl. transl.* p. 330)

Ethyl phosphorodifluoridate, 9CI, 8CI **E-00173**

Ethyl phosphoryl difluoride. Ethyl difluorophosphate

[460-52-6]

$EtOP(O)F_2$

$C_2H_5F_2O_2P$ M 130.031

Liq. Bp 85-86°.

▷V. toxic

Gold, A.M., *J. Org. Chem.*, 1961, **26**, 3991 (*synth*)
Roesky, H.W., *Chem. Ber.*, 1967, **100**, 2147 (*synth, pmr, nmr*)
Brown, D.H. *et al, J. Chem. Soc. (A)*, 1969, 872 (*F nmr, tox*)

Ethyl phosphorodifluoridite, 9CI, 8CI **E-00174**

Ethyl difluorophosphite

[24933-27-5]

$EtOPF_2$

$C_2H_5F_2OP$ M 114.031

Ligand for Fe, Ni, and Ir. Gas or liq. d_4^{15} 1.09. Bp 23-24°. n_D^{25} 1.3280.

Binder, H. *et al, Z. Naturforsch., B*, 1972, **27**, 753 (*synth, P nmr*)
Ivanova, Zh.M. *et al, Zh. Obshch. Khim.*, 1972, **42**, 2115 (*Engl. transl.* p. 2110) (*synth*)
Lines, E.L. *et al, Inorg. Chem.*, 1973, **12**, 2111 (*synth, ir, ms, pmr, F and P nmr*)

O-Ethyl phosphorodifluoridothioate, 9CI, 8CI **E-00175**

O-*Ethyl thiophosphoryl difluoride.* O-*Ethyl difluorothiophosphate*

[460-53-7]

$EtOP(S)F_2$

$C_2H_5F_2OPS$ M 146.091

Liq. d_4^{20} 1.23. Mp −124°. Bp 78-79°. n_D^{20} 1.3755.

Olah, G. *et al, J. Org. Chem.*, 1960, **25**, 603 (*synth*)
Seel, F. *et al, Chem. Ber.*, 1962, **95**, 199 (*synth*)
Ivanova, Zh.M. *et al, Zh. Obshch. Khim.*, 1972, **42**, 2115 (*Engl. transl.* p. 2110) (*synth, ir*)
Gusar', N.I. *et al, Zh. Obshch. Khim.*, 1976, **46**, 1981 (*Engl. transl.* p. 1910) (*synth*)

S-Ethyl phosphorodifluoridothioate, 9CI, 8CI E-00176

S-*Ethyl difluorothiophosphate*

[25237-38-1]

$EtSP(O)F_2$

$C_2H_5F_2OPS$ M 146.091

Liq. d_4^{20} 1.32. Bp 111-112°. n_D^{20} 1.4076.

Roesky, H., *Z. Naturforsch.*, *B*, 1969, **24**, 818 (*synth, ir, pmr, F and P nmr*)

Ethyl phosphorodiisocyanatidate, 9CI, 8CI E-00177

Ethylphosphoryl diisocyanate

[24958-80-3]

$EtOP(O)(NCO)_2$

$C_4H_5N_2O_4P$ M 176.068

V. reactive, moisture-sensitive liq. d_4^{20} 1.34. $Bp_{0.5}$ 59-60°. n_D^{20} 1.4442.

Shokol, V.A. *et al*, *Zh. Obshch. Khim.*, 1968, **39**, 1041 (*Engl. transl.* p. 1012) (*esters, synth, ir*)

Borovikov, Yu.Ya. *et al*, *Zh. Obshch. Khim.*, 1977, **47**, 328 (*Engl. transl.* p. 304) (*conformn*)

O-Ethyl phosphorodiisocyanatidothioate, 9CI, 8CI E-00178

O-*Ethyl thiophosphoryl diisocyanate*

[26074-17-9]

$EtOP(S)(NCO)_2$

$C_4H_5N_2O_3PS$ M 192.129

Reactive liq. d_4^{20} 1.30. $Bp_{0.25}$ 36-38°. n_D^{20} 1.4840. Sensitive to moisture.

Shokol, V.A. *et al*, *Zh. Obshch. Khim.*, 1971, **41**, 2380 (*Engl. transl.* p. 2407) (*synth, props*)

Borovikov, Yu.Ya. *et al*, *Zh. Obshch. Khim.*, 1977, **47**, 328 (*Engl. transl.* p. 304) (*conformn*)

Ethyl phosphorofluoridate E-00179

[371-68-6]

EtOPF(O)OH

$C_2H_6FO_3P$ M 128.040

Derivatives may exist in chiral forms. Liq. d_{20}^{20} 1.31. $Bp_{0.0005}$ 50-55°. n_D^{20} 1.3680.

Anilinium salt: Needles ($MeOH/Et_2O$). Mp 84-85°.

Hood, A. *et al*, *J. Am. Chem. Soc.*, 1950, **72**, 4956 (*synth, props*)

Stölzer, C. *et al*, *Chem. Ber.*, 1960, **93**, 1323 (*synth, deriv*)

Sheluchenko, V.V. *et al*, *Dokl. Akad. Nauk SSSR, Ser. Sci. Khim.*, 1967, **177**, 376 (*Engl. transl.* p. 1050) (*F and P nmr*)

Ethyl phosphorofluoridisocyanatidate, 9CI, 8CI E-00180

[20001-73-4]

EtOPF(O)NCO

$C_3H_5FNO_3P$ M 153.050

(±)-form

Reactive liq. d_4^{20} 1.32. Bp_{12} 42-43°. n_D^{20} 1.3902.

Ivanova, Zh.M. *et al*, *Zh. Obshch. Khim.*, 1968, **38**, 551 (*Engl. transl.* p. 538) (*synth*)

Shokol, V.A. *et al*, *Zh. Obshch. Khim.*, 1974, **44**, 2660 (*Engl. transl.* p. 2615) (*synth, props*)

O-Ethyl phosphorofluoridoisocyanatidothioate, 9CI, 8CI E-00181

[54377-81-0]

EtO(F)P(=S)NCO

$C_3H_5FNO_2PS$ M 169.110

(±)-form

Liq. d_4^{20} 1.27. Bp_{20} 25-27°. n_D^{20} 1.4380. Reacts with $ArNH_2$ initially at the C=O gp.

Shokol, V.A. *et al*, *Zh. Obshch. Khim.*, 1974, **44**, 2660 (*Engl. transl.* p. 2615) (*synth, props*)

O-Ethyl phosphorothioate, 9CI, 8CI E-00182

O-*Ethyl hydrogen phosphorothioate.* O-*Ethyl phosphorothioic acid.* O-*Ethyl thiophosphoric acid*

EtO(HO)P(=S)OH ⇌ EtO(HS)P(=O)OH

$C_2H_7O_3PS$ M 142.109

Di-Na salt: Cryst. ($MeOH/Me_2CO$).

Difluoride: see O-Ethyl phosphorodifluoridothioate, E-00175

Dichloride: see O-Ethyl phosphorodichloridothioate, E-00170

Diamide: [40334-49-4]. O-*Ethyl phosphorodiamidothioate.* O-*Ethyl thiophosphoryl diamide.* O-*Ethyl diamidothiophosphate.*

$C_2H_9N_2OPS$ M 140.140

Oil. n_D^{24} 1.5464.

Gay, D.C. *et al*, *J. Chem. Soc.* (*B*), 1970, 1123 (*synth*)

Ger. Pat., 2 135 349, (*1973*); *CA*, **78**, 97806 (*diamide*)

Bäuerlein, E. *et al*, *Phosphorus Sulfur*, 1978, **5**, 53 (*synth, P nmr*)

S-Ethyl phosphorothioate, 9CI, 8CI E-00183

S-*Ethyl dihydrogen thiophosphate.* S-*Ethyl thiophosphoric acid*

$EtSP(O)(OH)_2$

$C_2H_7O_3PS$ M 142.109

Di-Li salt: Cryst. (EtOH aq.).

Monodicyclohexylammonium salt: Solid. Mp 157-158°.

Difluoride: see S-Ethyl phosphorodifluoridothioate, E-00176

Dichloride: see S-Ethyl phosphorodichloridothioate, E-00171

Akerfeldt, S., *Acta Chem. Scand.*, 1962, **16**, 1897 (*synth*)

Zwierzak, A. *et al*, *Z. Naturforsch.*, *B*, 1971, **26**, 386 (*synth, ir, pmr*)

Bäuerlein, E. *et al*, *Phosphorus Sulfur*, 1978, **5**, 53 (*synth, P nmr*)

O-Ethyl phosphorotriselenoate, 8CI E-00184

O-*Ethyl triselenophosphate.* O-*Ethyl hydrogen phosphorotriselenoate.* O-*Ethyl phosphorotriselenoic acid*

$EtOP(Se)(SeH)_2$

$C_2H_7OPSe_3$ M 314.930

Unstable acid, isol. as di-K salt.

Di-K salt: [37443-16-6]. Solid. Mp 82°.

Gorak, R.D. *et al, Zh. Obshch. Khim.*, 1972, **42**, 56 (*Engl. transl.* p. 52)

Ethylpropylphosphinodithioic acid, 9CI **E-00185**

$H_3CCH_2CH_2PEt(S)SH$

$C_5H_{13}PS_2$ M 168.251

Et ester: [13297-93-3]. *Ethyl ethylpropylphosphinodithioate.*

$C_7H_{17}PS_2$ M 196.305

Liq. d_4^{20} 1.07. $Bp_{0.1}$ 77-80°. n_D^{20} 1.5515.

Krasil'nikova, E.A. *et al, Zh. Obshch. Khim.*, 1968, **38**, 609 (*Engl. transl.* p. 587)

***O*-Ethyl *S*-propyl phosphorodithioate** **E-00186**

O-*Ethyl hydrogen* S-*propyl phosphorodithioate.* O-*Ethyl* S-*propyl dithiophosphate*

[50669-78-8]

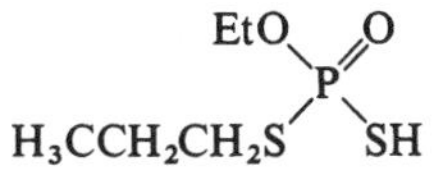

$C_5H_{13}OPS_2$ M 184.251

Metab. of Prothiofos, P-00498 .

Ueyama, I. *et al, Pestic. Biochem. Physiol.*, 1980, **14**, 98.

4-Ethylspiro[2,8,9-trioxa-1-phosphatricyclo[3.3.1.1^{3,7}]decane-1,4′-trioxaphosphetane] **E-00187**

Ethyl bicyclic phosphite ozonide

[58594-17-5]

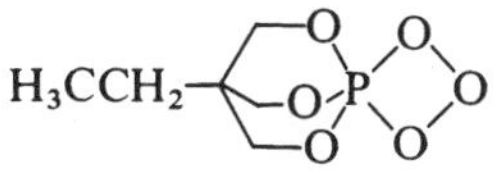

$C_6H_{11}O_6P$ M 210.123

Source of singlet oxygen. Unstable; prep. *in situ.*

Stephenson, L.M. *et al, J. Am. Chem. Soc.*, 1973, **95**, 3074 (*props*)
Bartlett, P.D. *et al, J. Org. Chem.*, 1980, **45**, 3000 (*synth, props*)
Mendenhall, G.D. *et al, J. Photochem.*, 1984, **25**, 227 (*synth, cmr, P nmr, props, use*)

2-Ethyl-2,3,4,5-tetrahydro-2-methyl-1*H*-1,3-benzazaphosphepine, 9CI **E-00188**

[60680-68-4]

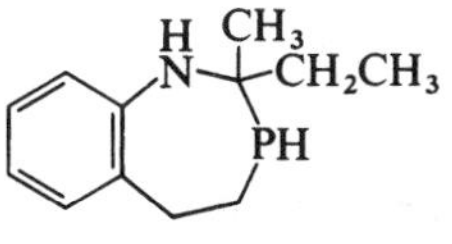

$C_{12}H_{18}NP$ M 207.255

B,HCl: [60680-72-0]. Solid. Mp 155°.

B,MeI:

$C_{13}H_{21}INP$ M 349.194

Cryst. ($EtOH/Et_2O$). Mp 209°.

Issleib, K. *et al, Z. Anorg. Allg. Chem.*, 1976, **424**, 97 (*synth, P nmr*)

2-Ethylthio-4,5-dimethyl-1,3,2-dioxaphospholane, 8CI **E-00189**

O,O-*2,3-Butylene* S-*ethyl thiophosphite. 2-Ethylthio-4,5-dimethyl-1,3-dioxa-2-phosphacyclopentane*

[14274-39-6]

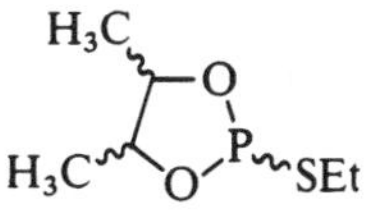

$C_6H_{13}O_2PS$ M 180.201

Malodorous liq. $Bp_{1.4}$ 65°. $n_D^{22.5}$ 1.5015. Mixt. of diasteroisomers.

▷Readily oxidises exothermically

2-Sulfide: [14274-48-7].

$C_6H_{13}O_2PS_2$ M 212.261

Malodorous liq. d_4^{20} 1.20. $Bp_{0.25}$ 106°. n_D^{23} 1.5327. Mixt. of diastereoisomers.

Edmundson, R.S. *et al, J. Chem. Soc. (C)*, 1966, 1997 (*synth, deriv*)
Pudovik, A.N. *et al, Zh. Obshch. Khim.*, 1968, **38**, 307; 1972, **42**, 2638 (*Engl. transl.* pp. 308, 2629) (*sulfide*)
Ishmaeva, E.A. *et al, Zh. Obshch. Khim.*, 1975, **45**, 946 (*Engl. transl.* p. 931) (*sulfide, conformn*)

2-Ethylthio-5,5-dimethyl-1,3,2-dioxaphosphorinane, 9CI **E-00190**

S-*Ethyl* O,O-*neopentylene phosphorothioite. 2-Ethylthio-5,5-dimethyl-1,3-dioxa-2-phosphacyclohexane*

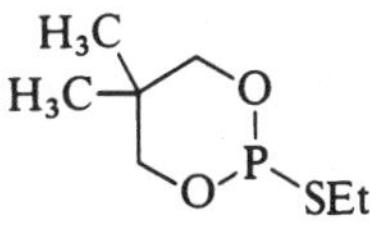

$C_7H_{15}O_2PS$ M 194.228

2-Oxide: S-*Ethyl* O,O-*neopentylene phosphorothioate.*

$C_7H_{15}O_3PS$ M 210.227

Cryst. (Et_2O/pet. ether). Mp 42-44°.

2-Sulfide: S-*Ethyl* O,O-*neopentylene phosphorodithioate.*

$C_7H_{15}O_2PS_2$ M 226.288

Cryst. (pet. ether). Mp 69.5-70.5°.

Edmundson, R.S., *Tetrahedron*, 1964, **20**, 2781 (*synth, ir*)
Edmundson, R.S., *Phosphorus Sulfur*, 1981, 307 (*ms*)

2-Ethylthio-1,3,2-dioxaphospholane, 9CI **E-00191**

S-*Ethyl ethylene thiophosphite.* S-*Ethyl ethylene phosphorothioite*

[14274-25-0]

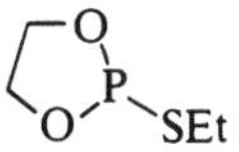

$C_4H_9O_2PS$ M 152.148

Liq. with foul odour. $Bp_{1.5}$ 53-56°. $n_D^{21.5}$ 1.5222.

▷Oxidises exothermically in air

2-Sulfide: [14274-46-5]. S-*Ethyl* O,O-*ethylene dithiophosphate.* S-*Ethyl* O,O-*ethylene phosphorodithioate.*

$C_4H_9O_2PS_2$ M 184.208

Liq. with foul odour. $Bp_{0.4}$ 114°. n_D^{21} 1.5540.

Edmundson, R.S. *et al, J. Chem. Soc. (C)*, 1966, 1997 (*synth*)

[(Ethylthio)methyl]phosphonic acid, 9CI E-00192

[69639-74-3]

$EtSCH_2P(O)(OH)_2$

$C_3H_9O_3PS$ M 156.136

Esters are Wittig-Horner reagents. Solid.

Di-Et ester: [54091-78-0]. *Diethyl [(ethylthio)methyl]-phosphonate.*

$C_7H_{17}O_3PS$ M 212.243

Oil. Bp_{10} 131°, Bp_1 127°. n_D^{24} 1.4671.

Kreutzkamp, N. *et al, Arch. Pharm. (Weinheim, Ger.)*, 1959, **292**, 159 (*ester, synth*)
Mikolajczyk, M. *et al, J. Org. Chem.*, 1979, **44**, 2967 (*synth, pmr, use*)
Wozniak, M. *et al, J. Chem. Soc., Dalton Trans.*, 1981, 2423 (*synth*)
Rozen, A.M. *et al, Zh. Obshch. Khim.*, 1982, **52**, 1232 (*Engl. transl.* p. 1083) (*props*)

2-Ethylthio-4,4,5,5-tetramethyl-1,3,2-dioxaphospholane, 9CI E-00193

S-*Ethyl* O,O-*tetramethylethylene phosphorothioite*

[14274-40-9]

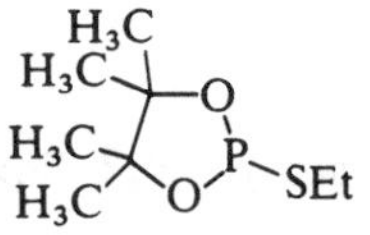

$C_8H_{17}O_2PS$ M 208.255

Odorous liq. $Bp_{2.4-3.0}$ 81-94°. $n_D^{23.5}$ 1.4968. Easily oxidised in air with evolution of heat.

2-Sulfide: [14274-49-8]. S-*Ethyl* O,O-*tetramethylethylene phosphorodithioate.*

$C_8H_{17}O_2PS$ M 208.255

Odorous liq. d_4^{20} 1.16. $Bp_{0.15}$ 99-106°. n_D^{20} 1.5222.

Edmundson, R.S. *et al, J. Chem. Soc. (C)*, 1966, 1997 (*synth*)
Pudovik, A.N. *et al, Zh. Obhsch. Khim.*, 1972, **42**, 2638 (*Engl. transl.* p. 2629) (*sulfide*)
Ishmaeva, E.A. *et al, Zh. Obshch. Khim.*, 1975, **45**, 946 (*Engl. transl.* p. 931) (*sulfide, conformn*)

3-(Ethylthio)-3-thioxo-2-(triphenylphosphoranylidene)propanoic acid, 9CI E-00194

2-(Triphenylphosphoranylidene)-1,1-dithiomalonic acid

$Ph_3P{=}C(COOH)CSSH$

$C_{21}H_{17}O_2PS_2$ M 396.458

Di-Et ester: [73144-64-6]. *Ethyl 3-(ethylthio)-3-thioxo-2-(triphenylphosphoranylidene)propanoate. Ethyl [(ethoxycarbonyl)-(ethylthiothiocarbonyl)-methylene]triphenylphosphorane.*

$C_{25}H_{25}O_2PS_2$ M 452.565

Golden-yellow cryst. Mp 175°.

Bestmann, H.J. *et al, J. Chem. Res. (S)*, 1979, 313 (*synth, props*)

4-Ethyl-2,6,7-trimethyl-2,6,7-triaza-1-phosphabicyclo[2.2.2]octane, 9CI E-00195

[67590-60-7]

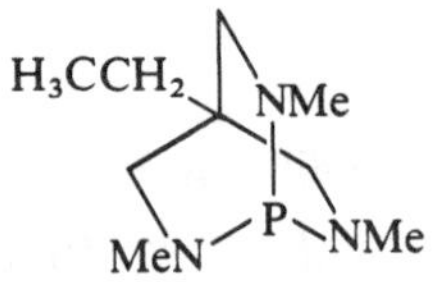

$C_9H_{20}N_3P$ M 201.251

Viscous liq. Bp_{18} 120°.

1-Sulfide: [67590-61-8].

$C_9H_{20}N_3PS$ M 233.311

Solid. Mp 49-51°.

▷XX8700000.

Cooper, G.H. *et al, Eur. J. Med. Chem.-Chim. Ther.*, 1978, **13**, 207 (*synth, tox*)

4-Ethyl-2,6,7-trioxa-1-phosphabicyclo[2.2.2]octane, 9CI E-00196

Trimethylolpropane phosphite. Ethyl bicyclic phosphite. ETPB

[824-11-3]

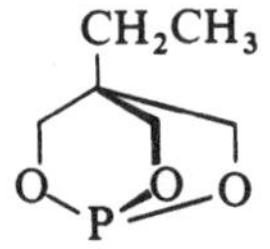

$C_6H_{11}O_3P$ M 162.125

Cryst. Mp 56°. Bp_8 100°, $Bp_{0.1}$ 86° subl.

▷TY6650000.

Oxide: [1005-93-2]. *Ethyl bicyclic phosphate. Trimethylolpropane phosphate.*

$C_6H_{11}O_4P$ M 178.124

Cryst. (H_2O or Me_2CO). Mp 207-208°.

▷TY6475000.

Sulfide: [935-52-4]. *Ethyl bicyclic thiophosphate. Trimethylolpropane thiophosphate.*

$C_6H_{11}O_3PS$ M 194.185

Cryst. (EtOH). Mp 176-178°.

▷TY6655000.

Selenide: [3883-97-4]. *Ethyl bicyclic selenophosphate. Trimethylolpropane selenophosphate.*

$C_6H_{11}O_3PSe$ M 241.085

Cryst. (2-methoxyethanol). Mp 207-210°.

Wadsworth, W.S. *et al, J. Am. Chem. Soc.*, 1962, **84**, 610 (*synth, derivs*)
Chang, W.-H., *J. Org. Chem.*, 1964, **29**, 3711 (*synth, derivs*)
Mosbo, J.A. *et al, J. Magn. Reson.*, 1972, **8**, 243, 250 (*pmr*)
Bellet, E.M. *et al, Science*, 1973, **182**, 1135 (*tox*)
Voorhees, K.J. *et al, Org. Mass Spectrom.*, 1979, **14**, 459 (*ms*)
Kenttamaa, H. *et al, Org. Mass Spectrom.*, 1980, **15**, 520 (*ms*)
Ozoe, Y. *et al, Agric. Biol. Chem.*, 1982, **46**, 555, 2521, 2527 (*oxide, props, tox*)
Samitov, Yu.Yu. *et al, Zh. Obshch. Khim.*, 1984, **54**, 805 (*Engl. transl.* p. 714) (*cmr*)

Ethyltriphenylphosphonium(1+), 9CI E-00197

[39895-79-9]

$Ph_3P^{\oplus}Et$

$C_{20}H_{20}P^{\oplus}$ M 291.352 (ion)

▷Salts. Exhibit AChE activity in verterbrates and schistosomes

Chloride: [896-33-3].

$C_{20}H_{20}ClP$ M 326.805

Cryst. (CH_2Cl_2/EtOAc). Mp 240-241° dec.
Bromide: [1530-32-1].
$C_{20}H_{20}BrP$ M 371.256
Hygroscopic cryst. (EtOH/Et_2O). Mp 206.5-207.5°.
Iodide: [4736-60-1].
$C_{20}H_{20}IP$ M 418.256
Cryst. Mp 167° dec.
▷TA2312000.
Tetrafluoroborate: [2994-53-8].
$C_{20}H_{20}BF_4P$ M 378.155
Cryst. (CH_2Cl_2/Et_2O). Mp 125-127°.
Tetraphenylborate:
$C_{44}H_{40}BP$ M 610.584
Solid. Mp 204-205° dec.

Wittig, G. *et al*, *Justus Liebigs Ann. Chem.*, 1957, **606**, 1 (*synth*)
Horner, L. *et al*, *Chem. Ber.*, 1958, **91**, 67; 1966, **99**, 2789 (*synth*)
Hendrickson, J. *et al*, Tet, 1964, **20**, 449 (*pmr*)
Schlosser, M. *et al*, *Justus Liebigs Ann. Chem.*, 1967, **708**, 1 (*synth*)
Gray, G.A., *J. Am. Chem. Soc.*, 1973, **95**, 7736 (*cmr, nmr*)
Wood, G.W. *et al*, *J. Org. Chem.*, 1975, **40**, 636 (*ms*)
Albright, T.A. *et al*, *J. Am. Chem. Soc.*, 1976, **98**, 6249 (*cmr, nmr*)
McAllister, P.R. *et al*, *J. Med. Chem.*, 1980, **23**, 862 (*pharmacol*)

Ethyl triphenylphosphoranylideneacetate, 9CI, 8CI — E-00198

Carbethoxymethylidenetriphenylphosphorane
[1099-45-2]

$Ph_3P{=}CHCOOEt$

$C_{22}H_{21}O_2P$ M 348.380
Reagent used in Wittig reactions giving alkenes and also in reactions leading to acetylenic compds. Leaflets (EtOAc/pet. ether or C_6H_6/pet. ether). Mp 116-117°, 129-130°.

Issler, O. *et al*, *Helv. Chim. Acta*, 1957, **40**, 1242 (*synth, uv, use*)
Taylor, P.J., *Spectrochim. Acta, Part A*, 1978, **34**, 115 (*ir*)
Brittain, J.M. *et al*, *Tetrahedron*, 1979, **35**, 1139 (*pmr*)
Cassy, J. *et al*, *Bull. Soc. Chim. Fr., Part II*, 1979, 559 (*use*)
Endo, T. *et al*, *Chem. Pharm. Bull.*, 1979, **27**, 2807 (*synth, use*)
Gramenitskaya, V.N. *et al*, *Zh. Org. Khim.*, 1979, **15**, 2090 (*Engl. transl.*, p. 1889) (*use*)
Lang, R.W. *et al*, *Helv. Chim. Acta*, 1980, **63**, 438 (*synth, pmr, use*)
Hanack, H. *et al*, *J. Am. Chem. Soc.*, 1981, **103**, 2356 (*synth, use*)
Uesato, S. *et al*, *Chem. Pharm. Bull.*, 1982, **30**, 927 (*props*)

Ethyl 2-triphenylphosphoranylidenepropanoate, 9CI — E-00199

(1-Carbethoxyethylidene)triphenylphosphorane
[5717-37-3]

$Ph_3P{=}C(CH_3)COOEt$

$C_{23}H_{23}O_2P$ M 362.407
Wittig reagent. Needles (EtOAc/pet. ether). Mp 156-157°, 165-167°.

Issler, O. *et al*, *Helv. Chim. Acta*, 1957, **40**, 1242 (*synth, uv, use*)
Cooks, R.G. *et al*, *Tetrahedron*, 1968, **24**, 3289 (*ms*)
Gray, G.A., *J. Am. Chem. Soc.*, 1973, **95**, 5092, 7736 (*cmr, nmr*)
Kuchař, M. *et al*, *Collect. Czech. Chem. Commun.*, 1973, **38**, 447 (*synth, pmr, props*)
Lumbroso, H. *et al*, *Bull. Soc. Chim. Fr.*, 1974, 819 (*struct*)
Plieninger, H. *et al*, *Justus Liebigs Ann. Chem.*, 1976, 1475 (*synth, use*)
Bartlett, P.A. *et al*, *J. Am. Chem. Soc.*, 1980, **102**, 337 (*use*)
Uesato, S. *et al*, *Chem. Pharm. Bull.*, 1982, **30**, 927 (*use*)

P,P'-1,2-Ethynediylbis[*N,N,N',N'*-tetramethylphosphonous diamide] — E-00200

Bis[(dimethylamino)phosphino]acetylene
[29936-20-7]

$(Me_2N)_2PC{\equiv}CP(NMe_2)_2$

$C_{10}H_{24}N_4P_2$ M 262.274
Solid or liq. Mp 36°. $Bp_{0.3}$ 98-100°.

Kuchen, W. *et al*, *Z. Naturforsch., B*, 1970, **25**, 1189 (*synth, derivs*)

Ethynyldiphenylphosphine, 9CI — E-00201

Diphenylphosphinoacetylene. Diphenylphosphinoethyne
[6104-47-8]

$Ph_2PC{\equiv}CH$

$C_{14}H_{11}P$ M 210.215
Cryst. (EtOH). Mp 35°. pK_a 1.57 ($MeNO_2$).
Oxide: [6104-48-9]. *Diphenylphosphinylacetylene. Diphenylphosphinylethyne.*
$C_{14}H_{11}OP$ M 226.214
Cryst. (C_6H_6/pet. ether). Mp 68°.

Charrier, C. *et al*, *Bull. Soc. Chim. Fr.*, 1966, 1002 (*synth, ir, pmr*)
Lequan, R.M. *et al*, *Org. Magn. Reson.*, 1975, **7**, 392 (*oxide, pmr, cmr, P nmr*)

Ethynylphosphinic acid, 9CI — E-00202

$HC{\equiv}CPH(O)OH$

$C_2H_3O_2P$ M 90.018
Tautomeric with Ethynylphosphonous acid, E-00204 .
Butyl ester: Butyl ethynylphosphinate.
$C_6H_{11}O_2P$ M 146.125
Liq. d_4^{20} 1.03. $Bp_{1.5}$ 65-66°. n_D^{20} 1.4492.

Kabachnik, M.I. *et al*, *Zh. Obshch. Khim.*, 1962, **32**, 3351 (*Engl. transl.* p. 3288)

Ethynylphosphonic acid, 9CI — E-00203

Phosphonoacetylene. Acetylenephosphonic acid
[69310-55-0]

$HC{\equiv}CP(O)(OH)_2$

$C_2H_3O_3P$ M 106.018
Di-Et ester: [4851-51-8]. *Diethyl ethynylphosphonate.*
$C_6H_{11}O_3P$ M 162.125
Liq. $Bp_{0.1}$ 75-77°.
Diisopropyl ester: Diisopropyl ethynylphosphonate.
$C_8H_{15}O_3P$ M 190.178
Liq. $Bp_{0.1}$ 50-60°.
*Bis(*N,N*-dimethylamide):* [16400-19-4]. P-*Ethynyl-*N,N,N',N'*-tetramethylphosphonic diamide.*
$C_6H_{13}N_2OP$ M 160.155
Solid. Mp 66°.

Saunders, B.C. *et al*, *J. Chem. Soc.*, 1963, 3351 (*esters*)

Sturtz, G. *et al*, *Bull. Soc. Chim. Fr.*, 1966, 1707 (*ester*)
Charrier, C. *et al*, *C.R. Hebd. Seances Acad. Sci., Ser. C*, 1967, **264**, 995 (*bisdimethylamide*)
Lequan, R.M. *et al*, *Org. Magn. Reson.*, 1975, **7**, 392 (*ester, pmr, cmr*)
Zakharov, V. *et al*, *Zh. Obshch. Khim.*, 1976, **46**, 1415 (*Engl. transl.* p. 1391) (*P nmr*)
Blackburn, G.M. *et al*, *J. Chem. Soc., Chem. Commun.*, 1978, 870 (*synth*)

Ethynylphosphonous acid, 9CI E-00204

HC≡CP(OH)$_2$ ⇌ HC≡CPH(O)OH

$C_2H_3O_2P$ M 90.018
Free acid exists in the tautomeric form, Ethynylphosphinic acid, E-00202 .
Di-Et ester: [20505-16-2]. *Diethyl ethynylphosphonite.*
$C_6H_{11}O_2P$ M 146.125
Liq. Bp_9 58°. n_D^{20} 1.4470.
Monobutyl ester: see Ethynylphosphinic acid, E-00202
Dibutyl ester: [39222-25-8]. *Dibutyl ethynylphosphonite.*
$C_{10}H_{19}O_2P$ M 202.233
Liq. Bp_2 59-60°. n_D^{20} 1.4520.

Kabachnik, M.I. *et al*, *Zh. Obshch. Khim.*, 1962, **32**, 3351 (*Engl. transl.* p. 3288) (*dibutyl ester*)
Aguiar, A.M. *et al*, *J. Org. Chem.*, 1969, **34**, 2684 (*ir, pmr*)
Tamm, L.A. *et al*, *Zh. Obshch. Khim.*, 1973, **43**, 2178 (*Engl. transl.* p. 2170) (*synth, ir, props*)

Etrimfos E-00205

O-(*6-Ethoxy-2-ethyl-4-pyrimidinyl*) O,O-*dimethyl phosphorothioate*, 9CI. *Ekamet*
[38260-54-7]

S
‖
OP(OMe)$_2$
N
EtO N CH$_2$CH$_3$

$C_{10}H_{17}N_2O_4PS$ M 292.289
Agricultural insecticide. Oil. d_4^{20} 1.20. n_D^{20} 1.5068. Unstable when pure: stable in soln. in non-polar solvents.
▷TF8350000.

B.P., 1 387 661, (*1971*); *CA*, **77**, 152218e (*synth*)
Ioannou, Y.M. *et al*, *Pestic. Biochem. Physiol.*, 1978, **9**, 190 (*metab*)
Stan, H.-J. *et al*, *J. Chromatogr.*, 1983, **279**, 173 (*glc*)
Bottomley, P. *et al*, *Analyst (London)*, 1984, **109**, 85 (*hplc*)
Pesticide Manual, 6th Ed., 255; 7th Ed., 253.

F

Famphur **F-00001**

O-*[4-(Dimethylamino)sulfonyl]phenyl* O,O-*dimethyl phosphorothioate*, 9CI. O,O-*Dimethyl phosphorothioate*, O-*ester with* p-*hydroxy*-N,N-*dimethylbenzenesulfonamide*, 8CI. *Varbex*

[52-85-7]

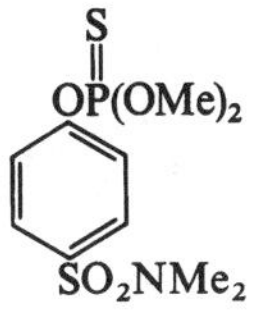

$C_{10}H_{16}NO_5PS_2$ M 325.334

Insecticide and acaricide. Cryst. (toluene/cyclohexane). Spar. sol. H_2O, aliphatic hydrocarbons.

▷TF7650000.

U.S.P., 3 179 560, (*1965*); *CA*, **63**, 1738 (*synth*)
Ding, X.D. *et al*, *J. Agric. Food Chem.*, 1984, **32**, 622 (*hplc*)
Pesticide Manual, 7th Ed., 254.
Sax, N.I., *Dangerous Properties of Industrial Materials*, 6th Ed., Van Nostrand-Reinhold, 1984, 604.

Fenamiphos, BSI **F-00002**

Ethyl 3-methyl-4-(methylthio)phenyl 1-methylethylphosphoramidate, 9CI. *Ethyl 4-methylthio*-m-*tolyl isopropylphosphoramidate*, 8CI. *Nemacur. Phenamiphos*

[22224-92-6]

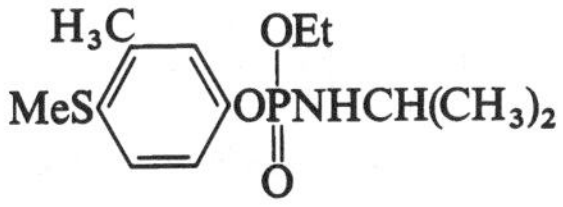

$C_{13}H_{22}NO_3PS$ M 303.355

Systemic agricultural nematocide. Cholinesterase inhibitor. Solid. Spar. sol. H_2O. Mp 49°.

▷TB3675000.

Waggoner, T.B. *et al*, *J. Agric. Food Chem.*, 1972, **20**, 157 (*metab*)
Waggoner, T.B. *et al*, *Residue Rev.*, 1974, **53**, 79 (*rev, pharmacol*)
Bowman, B.T. *et al*, *J. Environ. Sci. Health, Part B*, 1983, **18**, 221 (*props*)
Ripley, B.D. *et al*, *J. Assoc. Off. Anal. Chem.*, 1983, **66**, 1084 (*glc*)
Pesticide Manual, 6th Ed., 257; 7th Ed., 256.

Fenchlorphos, BAN, BSI **F-00003**

O,O-*Dimethyl* O-*(2,4,5-trichlorophenyl) phosphorothioate*, 9CI, 8CI. *Dermaphos. Nankor. Ronnel. Trolene. Korlan. Ectoral*

[299-84-3]

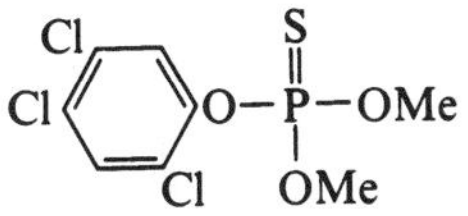

$C_8H_8Cl_3O_3PS$ M 321.542

Insecticide, now superseded. Cryst. powder. Mp 40-42°. $Bp_{0.01}$ 97°. n_D^{35} 1.5597.

▷Mod. toxic, TLV 10. TG0525000.

U.S.P., 2 887 505, (*1960*); *CA*, **54**, 2257 (*synth*)
Keith, L.H. *et al*, *J. Assoc. Off. Anal. Chem.*, 1968, **51**, 1063 (*pmr*)
Ross, R.T. *et al*, *Anal. Chim. Acta*, 1970, **53**, 139 (*P nmr*)
Gore, R.L. *et al*, *J. Assoc. Off. Anal. Chem.*, 1971, **54**, 1040 (*ir, uv*)
U.S.P., 3 862 273, (*1975*); *CA*, **82**, 125074 (*synth*)
Stan, H.-J. *et al*, *Fresenius' Z. Anal. Chem.*, 1977, **287**, 271; *Biomed. Mass Spectrom.*, 1982, **9**, 483 (*glc, ms*)
Daldrup, T. *et al*, *Fresenius' Z. Anal. Chem.*, 1981, **308**, 413 (*tlc, glc, hplc*)
Pesticide Manual, 6th Ed., 260.
The Agrochemicals Handbook, Royal Society of Chemistry, 1983, A197.
Sax, N.I., *Dangerous Properties of Industrial Materials*, 6th Ed., Van Nostrand-Reinhold, 1984, 957.

Fenitrooxon **F-00004**

Dimethyl 3-methyl-4-nitrophenyl phosphate, 9CI, 8CI. *Dimethyl 4-nitro*-m-*tolyl phosphate. Sumioxon*

[2255-17-6]

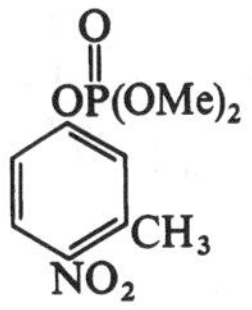

$C_9H_{12}NO_6P$ M 261.171

Liq. Product of photodegradation and metab. of 91414-4. $Bp_{0.07}$ 134-136°. n_D^{20} 1.5165.

▷TC5260000.

Miyamoto, J. *et al*, *Agric. Biol. Chem.*, 1963, **27**, 669 (*synth*)
Ohkawa, H. *et al*, *Agric. Biol. Chem.*, 1974, **38**, 2247 (*tlc*)

Fenitrothion, BAN, BSI **F-00005**

O,O-*Dimethyl* O-*(3-methyl-4-nitrophenyl) phosphorothioate*, 9CI. O,O-*Dimethyl* O-*4-nitro*-m-*tolyl phosphorothioate*, 8CI. O,O-*Dimethyl* O-*(3-methyl-4-nitrophenyl) thiophosphate. Folithion. Sumithion. Nitrophos*

[122-14-5]

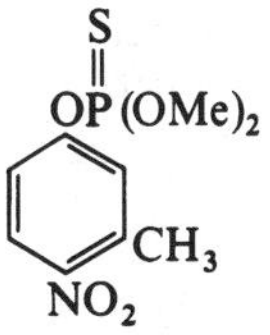

$C_9H_{12}NO_5PS$ M 277.231

Contact insecticide. Yellow oil. Insol. H_2O, spar. sol. ligroin. d_4^{25} 1.323. $Bp_{0.1}$ 140-145°. n_D^{25} 1.5528. Stable to acidic but not basic hydrol. *S*-Me isomer is formed at high temp.

▷TG0350000.

Schrader, G., *Angew. Chem.*, 1961, **73**, 331 (*synth, props*)
Nishizawa, Y. *et al*, *Residue Rev.*, 1970, **60** (*bibl, rev*)
Stan, H.-J. *et al*, *Fresenius' Z. Anal. Chem.*, 1977, **287**, 271 (*glc, ms*)
Volpé, G. *et al*, *Chromatographia*, 1981, **14**, 333 (*hplc*)
Stan, H.-J. *et al*, *Biomed. Mass Spectrom.*, 1982, **9**, 483 (*ms*)
Pesticide Manual, 6th Ed., 262; 7th Ed., 261.

The Agrochemicals Handbook, Royal Society of Chemistry, 1983, A199.
Sax, N.I., *Dangerous Properties of Industrial Materials*, 6th Ed., Van Nostrand-Reinhold, 1984, 683.

Fensulfothion, BSI — F-00006

O,O-*Diethyl* O-[*4-(methylsulfinyl)phenyl*] *phosphorothioate*, *9CI*. *Dasanit*. *Terracur* P
[115-90-2]

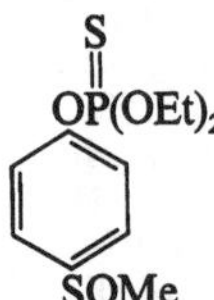

$C_{11}H_{17}O_4PS_2$ M 308.347
Agricultural insecticide and nematocide. Yellow oil. Sol. most org. solvs., spar. sol. H_2O. d_4^{20} 1.202. $Bp_{0.01}$ 138-141°. n_D^{25} 1.540.
▷Highly toxic, TLV 0.1. TF3850000.

B.P., 819 689, (*1956*); *CA*, **55**, 11751h (*synth*)
Stan, H.-J. *et al*, *Biomed. Mass Spectrom.*, 1982, **9**, 483 (*ms*)
Ripley, B.D. *et al*, *J. Assoc. Off. Anal. Chem.*, 1983, **66**, 1084 (*glc*)
Stan, H.-J. *et al*, *J. Chromatog.*, 1983, **279**, 173 (*glc*)
Pesticide Manual, 6th Ed., 264; 7th Ed., 267.
Sax, N.I., *Dangerous Properties of Industrial Materials*, 6th Ed., Van Nostrand-Reinhold, 1984, 580.

Fenthion, BSI — F-00007

O,O-*Dimethyl* O-[(*3-methyl-4-methylthio*)*phenyl*] *phosphorothioate*, *9CI*. O,O-*Dimethyl*-O-(*4-methylthio*-m-*tolyl*) *phosphorothioate*, *8CI*. *4-Methylmercapto-3-methylphenyl dimethyl thiophosphate*. *Mercaptofos*. *Baytex*. *Lebaycid*. *Numerous other synonyms*
[55-38-9]

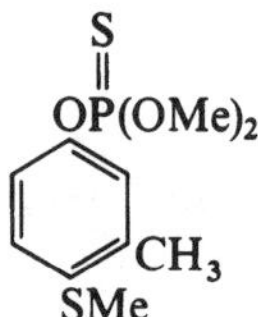

$C_{10}H_{15}O_3PS_2$ M 278.320
Insecticide with low mammalian toxicity. Used against mosquito larvae in tropical fresh waters. Liq. with slight garlic odour. V. spar. sol. H_2O. d_4^{20} 1.25. $Bp_{0.01}$ 87°. n_D^{20} 1.5698. Thermally stable to 210°, alkali-resistant to pH 9. Metab. *via* sulfoxide and sulfone.
▷Highly toxic, TLV 0.1. TF9625000.

S-*Oxide:* O,O-*Dimethyl* O-[(*3-methyl-4-methylsulfinyl*)*phenyl*] *phosphorothioate*. *Fenthion sulfoxide*.
$C_{10}H_{15}O_4PS_2$ M 294.320
Metab. of Fenthion.

S,S-*Dioxide:* O,O-*Dimethyl* O-[(*3-methyl-4-methylsulfonyl*)*phenyl*]*phosphorothioate*. *Fenthion sulfone*.
$C_{10}H_{15}O_5PS_2$ M 310.319
Metab. of Fenthion.

Ger. Pat., 1 116 656, (*1961*); *CA*, **56**, 14170 (*synth, props*)
Ibrahim, F.B. *et al*, *J. Agric. Food Chem.*, 1966, **14**, 369 (*uv*)
Keith, L.H. *et al*, *J. Assoc. Off. Anal. Chem.*, 1968, **51**, 1063 (*pmr*)
Getz, M.E. *et al*, *J. Assoc. Off. Anal. Chem.*, 1968, **51**, 1101 (*tlc*)
Ross, R.T. *et al*, *Anal. Chim. Acta*, 1970, **52**, 139 (*P nmr*)
Stan, H.-J. *et al*, *Fresenius' Z. Anal. Chem.*, 1977, **287**, 271; *Biomed. Mass Spectrom.*, 1982, **9**, 483 (*glc, ms*)
Rainsford, K.D., *Pestic. Biochem. Physiol.*, 1978, **8**, 302 (*tox, metab*)
Ripley, B.D., *J. Assoc. Off. Anal. Chem.*, 1983, **66**, 1084 (*derivs, glc*)
Stan, H.-J. *et al*, *J. Chromatogr.*, 1983, **279**, 173 (*glc*)
Pesticide Manual, 6th Ed., 265; 7th Ed., 269.
The Agrochemicals Handbook, Royal Society of Chemistry, 1983, A203.
Sax, N.I., *Dangerous Properties of Industrial Materials*, 6th Ed., Van Nostrand-Reinhold, 1984, 609.

(Ferrocenylmethyl)-triphenylphosphonium(1+), 9CI — F-00008

$C_{29}H_{26}FeP^{\oplus}$ M 461.345 (ion)
Iodide: [32914-67-3].
$C_{29}H_{26}FeIP$ M 588.250
Dark-yellow leaflets (EtOH). Mp 254-256° dec.
Ylide: (*Ferrocenylmethylene*)*triphenylphosphorane*.
$C_{29}H_{25}FeP$ M 460.337
Used in Wittig reaction. Blood-red.

Helling, J.F. *et al*, *Justus Liebigs Ann. Chem.*, 1961, **640**, 79 (*synth*)
Pauson, P.L. *et al*, *J. Chem. Soc.*, 1963, 2990 (*synth, use*)
Woods, T.A. *et al*, *J. Org. Chem.*, 1975, **40**, 2416 (*ylide, use*)
Koridze, A.A. *et al*, *J. Organomet. Chem.*, 1977, **136**, 57 (*cmr*)
Boev, V.I. *et al*, *Zh. Obshch. Khim.*, 1982, **52**, 1693; *J. Gen. Chem. USSR* (*Engl. Transl.*), 1497 (*synth*)

(9*H*-Fluoren-9-ylideneethenylidene)-triphenylphosphorane, 9CI — F-00009

2-(9-Fluorenylidene)-1-(triphenylphosphoranylidene)-ethene
[57674-98-3]

$C_{33}H_{23}P$ M 450.518
Golden plates (toluene). Mp 188-189°.

Bestmann, H.J. *et al*, *Chem. Ber.*, 1985, **118**, 1709 (*synth, ir, P nmr, props, use*)
Burzlaff, H. *et al*, *Chem. Ber.*, 1985, **118**, 1720 (*cryst struct*)

9*H*-Fluoren-9-ylidenetriphenylphosphorane, 9CI — F-00010

9-(Triphenylphosphoranylidene)fluorene
[4756-25-6]

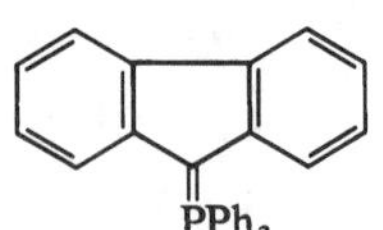

$C_{31}H_{23}P$ M 426.496
V. stable ylide. Yellow cryst. (C_6H_6/pet. ether). Mp 258-260°, 278°. Dissolves reversibly in dil. acid.

Horner, L. *et al*, *Justus Liebigs Ann. Chem.*, 1956, **627**, 142 (*ir*)

Johnson, A.W., *J. Org. Chem.*, 1959, **24**, 282 (*synth, uv*)
Gray, G.A. *et al, J. Am. Chem. Soc.*, 1973, **95**, 5092, 7736 (*cmr, nmr*)
Neidlein, R. *et al, Chem. Ztg.*, 1982, **106**, 233 (*use*)

(9*H*-Fluoren-9-yl)phosphonic acid, 9CI F-00011

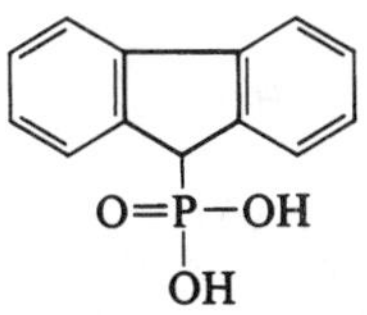

$C_{13}H_{11}O_3P$ M 246.202
Cryst. (C_6H_6/Me_2CO). Mp 255-259°.
Di-Et ester: [7142-76-9]. *Diethyl (9H-fluoren-9-yl)-phosphonate.*
$C_{17}H_{19}O_3P$ M 302.309
Solid. Mp 67.5-70°. Bp_2 148.5-149° subl.

Poshkus, A.C. *et al, J. Org. Chem.*, 1964, **29**, 2567.
Chiusoli, G.P. *et al, J. Chem. Soc., Chem. Commun.*, 1977, 216 (*ester*)

9*H*-Fluoren-9-yltriphenylphosphonium(1+), 9CI F-00012

9-(Triphenylphosphonio)fluorene(1+)

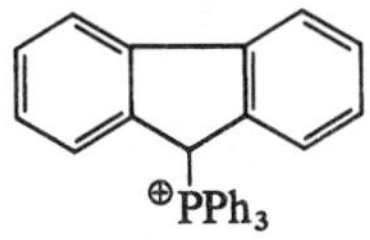

$C_{31}H_{24}P^{\oplus}$ M 427.504 (ion)
Bromide: [7253-07-8].
$C_{31}H_{24}BrP$ M 507.408
Source of 9*H*-Fluoren-9-ylidenetriphenylphosphorane, F-00010 obt. with NaOH aq. Solid. Mp 289-291°, 310° dec.
Tetrafluoroborate: [57945-51-4].
$C_{31}H_{24}BF_4P$ M 514.308
No phys. props. reported.
Ylide: see 9H-Fluoren-9-ylidenetriphenylphosphorane, F-00010

Horner, L. *et al, Justus Liebigs Ann. Chem.*, 1955, **591**, 135 (*synth*)
Johnson, A.W., *J. Org. Chem.*, 1959, **24**, 282 (*synth, use*)

2-Fluoro-1,3,2-benzodioxaphosphole, 9CI F-00013

o-Phenylene fluorophosphite. o-Phenylene phosphorofluoridite
[1526-24-5]

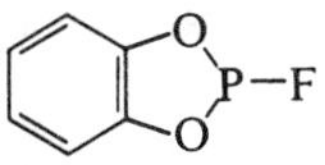

$C_6H_4FO_2P$ M 158.069
Liq. d_4^{20} 1.36. Bp_6 38°. n_D^{20} 1.5160.

Schmutzler, R., *Chem. Ber.*, 1963, **96**, 2435 (*synth*)
Razumov, N.A. *et al, Zh. Obshch. Khim.*, 1968, **38**, 1117 (*Engl. transl.* p. 1072) (*synth, pmr*)

2-Fluoro-1,3-dimethyl-1,3,2-diazaphospholidine F-00014

[33672-91-2]

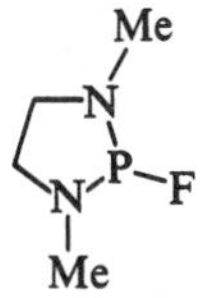

$C_4H_{10}FN_2P$ M 136.109
Liq. Bp 145-150°.
2-Oxide: [76457-37-9].
$C_4H_{10}FN_2OP$ M 152.108
No phys. props. reported.
2-Sulfide: [69284-88-4].
$C_4H_{10}FN_2PS$ M 168.169
Solid. Mp 40°.

Albrand, J.-P. *et al, Org. Magn. Reson.*, 1971, **3**, 75 (*pmr*)
Fleming, S. *et al, Inorg. Chem.*, 1972, **11**, 2534 (*synth, ir, pmr, F and P nmr, complex*)
Light, R.W. *et al, Acta Crystallogr., Sect. B*, 1978, **34**, 3671 (*sulfide, cryst struct*)
Light, R.W. *et al, Phosphorus Sulfur*, 1980, **8**, 255 (*oxide, sulfide, ms, ir, nmr*)

2-Fluoro-4,5-dimethyl-1,3,2-dioxaphospholane, 8CI F-00015

2,3-Butylene fluorophosphite. 2,3-Butylene phosphorofluoridite. 2-Fluoro-4,5-dimethyl-1,3-dioxa-2-phosphacyclopentane
[19952-57-9]

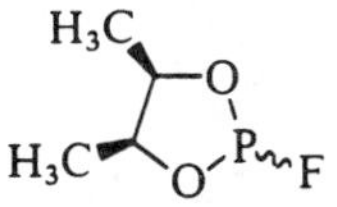

$C_4H_8FO_2P$ M 138.078
(4RS,5SR)-form
meso-*form*
Liq. d_4^{20} 1.16. Bp_{16} 28°. n_D^{20} 1.4020. Mixt. containing 25% ($2\alpha,4\alpha,5\alpha$)- and 75% ($2\beta,4\alpha,5\alpha$)-forms, characterised spectroscopically.
2-Sulfide: O,O-*2,3-Butylene fluorothiophosphate.* O,O-*2,3-Butylene phosphorofluoridothioate.*
$C_4H_8FO_2PS$ M 170.138
Liq. Bp_2 56-60°. Mixt. containing 25% ($2\alpha,4\alpha,5\alpha$) and 75% ($2\beta,4\alpha,5\alpha$)-forms, characterised spectroscopically.

Razumova, N.A. *et al, Zh. Obshch. Khim.*, 1968, **38**, 1117 (*Engl. transl.* p. 1072) (*synth*)
Mikolajczyk, M. *et al, J. Chem. Soc., Perkin Trans. 1*, 1977, 2213 (*sulfide, synth, props, cmr, P nmr*)

2-Fluoro-5,5-dimethyl-1,3,2-dioxaphosphorinane, 9CI F-00016

Neopentylene phosphorofluoridite. 2-Fluoro-5,5-dimethyl-1,3,2-dioxaphosphinane. 2-Fluoro-5,5-dimethyl-1,3-dioxa-2-phosphacyclohexane
[21458-74-2]

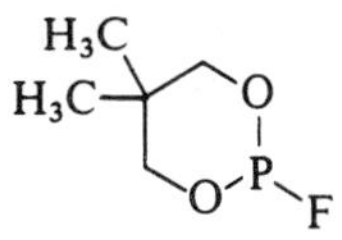

$C_5H_{10}FO_2P$ M 152.105
Liq. Bp_{30} 59-60°.
2-Oxide: [39846-28-1]. *Neopentylene phosphorofluoridate.*
$C_5H_{10}FO_3P$ M 168.104

Cryst. (ligroin). Mp 41-42°. $Bp_{0.1}$ 50° subl.

White, D.W. *et al*, *J. Am. Chem. Soc.*, 1970, **92**, 7125 (*synth, ir*)
Zwierzak, A., *Phosphorus*, 1972, **2**, 19 (*oxide, synth, ir, pmr*)
Hart, G.J. *et al*, *Pestic. Biochem. Physiol.*, 1976, **6**, 464 (*synth, oxide, pmr, F and P nmr, props*)

2-Fluoro-1,3,2-dioxaphospholane, 9CI **F-00017**

Ethylene fluorophosphite. Ethylene phosphorofluoridite. 2-Fluoro-1,3-dioxa-2-phosphacyclopentane

[765-40-2]

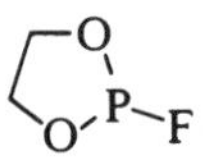

$C_2H_4FO_2P$ M 110.025

Liq. Bp_{170} 48°. $n_D^{23.5}$ 1.4003.

Schmutzler, R., *Chem. Ber.*, 1963, **96**, 2435 (*synth*)
Albrand, J.P. *et al*, *Org. Magn. Reson.*, 1971, **3**, 75 (*pmr*)
Mingaleva, K.S. *et al*, *Zh. Obshch. Khim.*, 1971, **41**, 2431 (*Engl. transl.* p. 2456) (*struct*)

2-Fluoro-1,3,2-dioxaphosphorinane, 9CI **F-00018**

2-Fluoro-1,3,2-dioxaphosphinane. Trimethylene fluorophosphite. Trimethylene phosphorofluoridite. 2-Fluoro-1,3-dioxa-2-phosphacyclohexane

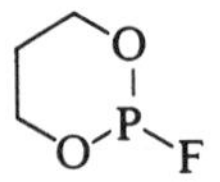

$C_3H_6FO_2P$ M 124.051

Liq. Bp 62°.

2-Oxide: [695-31-8]. *Trimethylene fluorophosphate. Trimethylene phosphorofluoridate.*

$C_3H_6FO_3P$ M 140.051

Liq. $Bp_{0.5}$ 96-98°. n_D^{25} 1.4087.

▷Cholinesterase inhibitor

Fukuto, T.R. *et al*, *J. Med. Chem.*, 1965, **8**, 759 (*oxide, synth, props*)
Ashani, Y. *et al*, *J. Med. Chem.*, 1973, **16**, 446 (*props*)
Coult, D.B. *et al*, *Biochem. J.*, 1976, **155**, 717 (*props*)
Hart, G.J. *et al*, *Pest. Biochem. Physiol.*, 1976, **6**, 464 (*props*)
Margalit, Y. *et al*, *Phosphorus Sulfur*, 1977, **3**, 315 (*pmr, F nmr, props*)
Hacklin, H. *et al*, *Phosphorus Sulfur*, 1985, **25**, 79 (*synth, ms*)

***O*-(2-Fluoroethyl) phosphorodichloridothioate,** 9CI, 8CI **F-00019**

O-(*2-Fluoroethyl*) *dichlorothiophosphate.* O-(*2-Fluoroethyl*) *thiophosphoryl dichloride*

$FCH_2CH_2OP(S)Cl_2$

$C_2H_4Cl_2FOPS$ M 196.991

Liq. d_4^{20} 1.51. Bp_8 67-68°. n_D^{20} 1.5041.

Kabachnik, M.I. *et al*, *Zh. Obshch. Khim.*, 1959, **29**, 1671 (*Engl. transl.* p. 1647) (*synth, props*)

2-Fluoro-4-methyl-1,3,2-dioxaphospholane, 9CI **F-00020**

Propylene fluorophosphite. 1,2-Propylene phosphorofluoridite. 2-Fluoro-4-methyl-1,3-dioxa-2-phosphacyclopentane

[16415-09-1]

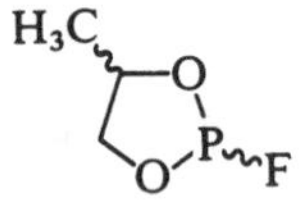

$C_3H_6FO_2P$ M 124.051

Liq. with disagreeable smell. d_4^{20} 1.22. Bp_{100} 44°. n_D^{20} 1.4035. Mixt. of *cis-* and *trans-*forms.

2-Sulfide: Propylene fluorothiophosphate. 1,2-Propylene phosphorofluoridothioate.

$C_3H_6FO_2PS$ M 156.111

Liq. d_4^{20} 1.34. $Bp_{1.0}$ 84°. n_D^{20} 1.4550. Mixt. of *cis-* and *trans-*forms.

Razumova, N.A. *et al*, *Zh. Obshch. Khim.*, 1968, **38**, 1117 (*Engl. transl.* p. 1072)

2-Fluoro-4-methyl-1,3,2-dioxaphosphorinane, 9CI **F-00021**

1,3-Butylene fluorophosphite. 2-Fluoro-4-methyl-1,3,2-dioxaphosphinane. 2-Fluoro-4-methyl-1,3-dioxa-2-phosphacyclohexane

[19952-56-8]

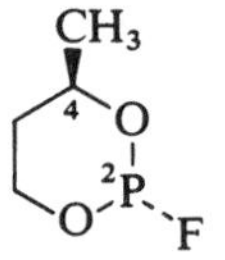

$C_4H_8FO_2P$ M 138.078

Liq. d_4^{20} 1.19. Bp_{16} 30°. n_D^{20} 1.4160. Mixt. of diastereoisomers.

(2RS,4RS)-form [56888-24-5]

trans-*form*

Liq. Bp_{30} 40-42°. n_D^{22} 1.4205 (94% diastereomeric purity).

2-Oxide: [57756-98-6]. Liq. $Bp_{0.1}$ 81-82°. n_D^{22} 1.4135 (90% diastereomeric purity).

2-Sulfide: [56888-26-7]. O,O-*1,3-Butylene fluorothiophosphate.* O,O-*1,3-Butylene phosphorofluoridothioate.*

$C_4H_8FO_2PS$ M 170.138

Cryst. (pet. ether). Mp 33-34°. $Bp_{0.1}$ 61°.

2-Selenide: [59319-64-1]. O,O-*1,3-Butylene fluoroselenophosphate.* O,O-*1,3-Butylene phosphorofluoridoselenoate.*

$C_4H_8FO_2PSe$ M 217.038

Liq. $Bp_{0.2}$ 85°. n_D^{22} 1.5100 (80% diastereomeric purity).

Razumova, N.A. *et al*, *Zh. Obshch. Khim.*, 1968, **38**, 1117 (*Engl. transl.* p. 1072) (*synth*)
Amos, J. *et al*, *Phosphorus*, 1975, **6**, 35 (*synth, ir, pmr, F nmr, props*)
Mikolajczyk, M. *et al*, *Tetrahedron Lett.*, 1975, 1607 (*sulfide, pmr, F and P nmr*)
Hart, G.J. *et al*, *Pestic. Biochem. Physiol.*, 1976, **6**, 464 (*synth, oxide, pmr, F and P nmr, props*)
Okruszek, A. *et al*, *Zh. Naturforsch., B*, 1976, **31**, 354 (*derivs, ms, pmr, F and P nmr*)
Margerlit, Y. *et al*, *Phosphorus Sulfur*, 1977, **3**, 315 (*pmr, F nmr*)
Miller, A. *et al*, *Acta Crystallogr., Sect. B*, 1981, **37**, 1951 (*sulfide, cryst struct*)
Nifant'ev, É.E. *et al*, *Tetrahedron*, 1981, **37**, 3183 (*cmr*)
Lopusinski, A. *et al*, *J. Am. Chem. Soc.*, 1982, **104**, 290 (*oxide, synth, F and P nmr*)

(Fluoromethyl)phenylphosphinic acid, 9CI **F-00022**

$FCH_2PPh(O)OH$

$C_7H_8FO_2P$ M 174.111
Solid. Sol. H_2O, EtOH, C_6H_6. Mp 94-95°.

Yagupol'skii, L.M. *et al*, *Zh. Obshch. Khim.*, 1960, **30**, 4026 (*Engl. transl.* p. 3986)

(Fluoromethyl)phosphonic acid, 9CI **F-00023**

$FCH_2P(O)(OH)_2$

CH_4FO_3P M 114.013
V. hygroscopic solid. Mp 79-85°.
Anilinium salt: Solid. Mp 168-170° dec.
Biscyclohexylammonium salt: Solid. Mp 185-186°.
Diisopropyl ester: Diisopropyl fluoromethylphosphonate.
$C_7H_{16}O_3P$ M 179.175
Liq. d_0^{20} 1.05. Bp_{13} 97-98°. n_D^{20} 1.4080.
Dichloride:
CH_2Cl_2FOP M 150.904
Liq. Bp_{14} 54-55°. n_D^{20} 1.4500.

Gryszkiewicz-Trochimowski, E., *Bull. Soc. Chim. Fr.*, 1967, 4289 (*synth, derivs, pmr*)
Hall, C.R. *et al*, *J. Chem. Soc., Perkin Trans. 1*, 1985, 233 (*esters, synth, pmr, P nmr*)
Blackburn, G.M. *et al*, *J. Chem. Soc., Perkin Trans. 1*, 1987, 181 (*derivs, ir, pmr, F and P nmr*)

(Fluoromethyl)phosphonochloridic acid, 9CI **F-00024**

$FCH_2PCl(O)OH$

CH_3ClFO_2P M 132.459
Isopropyl ester: Isopropyl (fluoromethyl)-phosphonochloridate.
$C_4H_9ClFO_2P$ M 174.539
Liq. Bp_{11} 75-76°. n_D^{20} 1.4158.

Gryszkiewicz-Trochimowska, E., *Bull. Soc. Chim. Fr.*, 1967, 4289 (*synth*)

(Fluoromethyl)triphenylphosponium(1+), 9CI **F-00025**

$Ph_3P^{\oplus}CH_2F$

$C_{19}H_{17}FP^{\oplus}$ M 295.315 (ion)
Iodide: [28096-32-4].
$C_{19}H_{17}FIP$ M 422.220
Poor source of ylide, obt. with RLi. Cryst. (Me_2CO). Mp 158-162°, 168.5-169.5° dec.
Ylide: [28096-33-5]. *(Fluoromethylene)-triphenylphosphorane.*
$C_{19}H_{16}FP$ M 294.308
Wittig reagent for prep. of 1-fluoro-1-alkenes.

Schlosser, M. *et al*, *Synthesis*, 1969, **1**, 75 (*synth, use*)
Burton, D.J. *et al*, *J. Fluorine Chem.*, 1973, **3**, 447 (*use*)
Burton, D.J. *et al*, *J. Org. Chem.*, 1975, **40**, 2796 (*synth, use*)

(2-Fluorophenyl)diphenylphosphine, 9CI **F-00026**

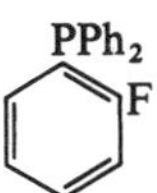

$C_{18}H_{14}FP$ M 280.281
Cryst. (EtOH). Mp 90°.
Oxide:
$C_{18}H_{14}FOP$ M 296.280
Cryst. (toluene). Mp 133°.

McEwen, W.E. *et al*, *J. Am. Chem. Soc.*, 1978, **100**, 7304 (*synth*)
Stegman, H.B. *et al*, *Phosphorus Sulfur*, 1982, **13**, 331 (*synth, cmr, P nmr, ms, oxide*)

(3-Fluorophenyl)diphenylphosphine, 9CI **F-00027**

[21388-29-4]
$C_{18}H_{14}FP$ M 280.281
Cryst. Mp 56-58°.
Oxide: [54300-35-5].
$C_{18}H_{14}FOP$ M 296.280
Cryst. (C_6H_6/pet. ether). Mp 118°, 139° (135°). Dimorphous.
Sulfide: [23588-06-9].
$C_{18}H_{14}FPS$ M 312.341
Cryst. (EtOH). Mp 94°.
Selenide:
$C_{18}H_{14}FPSe$ M 359.241
Solid. Mp 135°.

Rakshys, J.W. *et al*, *J. Am. Chem. Soc.*, 1968, **90**, 5236 (*synth, F nmr*)
Schindlbauer, H. *et al*, *Chem. Ber.*, 1969, **102**, 2914 (*synth, derivs, F nmr*)
De Ketelaere, R.F. *et al*, *J. Mol. Struct.*, 1974, **23**, 233; 1975, **27**, 363 (*derivs, ir, raman, P nmr*)
De Ketelaere, R.F. *et al*, *Phosphorus*, 1974, **5**, 43 (*derivs, ms*)

(4-Fluorophenyl)diphenylphosphine, 9CI **F-00028**

[18437-72-4]
$C_{18}H_{14}FP$ M 280.281
Low-melting solid. Mp 37-39°. Bp_9 207-208°.
Oxide: [18437-73-5].
$C_{18}H_{14}FOP$ M 296.280
Cryst. (pet. ether). Mp 133-135°.
Sulfide: [18437-76-8].
$C_{18}H_{14}FPS$ M 312.341
Cryst. (EtOH). Mp 121-122.5°.
Selenide:
$C_{18}H_{14}FPS$ M 312.341
Solid. Mp 137°.

Schindlbauer, H., *Chem. Ber.*, 1967, **100**, 3432 (*synth, F nmr*)
Rakshys, J.W. *et al*, *J. Am. Chem. Soc.*, 1968, **90**, 5236 (*synth, F nmr*)
Prikoszovich, W. *et al*, *Chem. Ber.*, 1969, **102**, 1969 (*deriv, F and P nmr*)
De Ketelaere, R.F. *et al*, *J. Mol. Struct.*, 1974, **23**, 233; 1975, **27**, 363 (*derivs, ir, raman, nmr*)
De Ketelaere, R.F. *et al*, *Phosphorus*, 1974, **5**, 43 (*derivs, ms*)

[(2-Fluorophenyl)methyl]phosphonic acid, 9CI F-00029

o-Fluorobenzylphosphonic acid

[80395-15-9]

$C_7H_8FO_3P$ M 190.111

Di-Et ester: [63909-54-6]. *Diethyl [(2-fluorophenyl)-methyl]phosphonic acid.*

$C_{11}H_{16}FO_3P$ M 246.218

Wittig-Horner reagent. Liq. Bp_6 146-149°. n_D^{25} 1.483.

Ernst, L., *Org. Magn. Reson.*, 1977, **9**, 35 (*synth, cmr, P nmr*)

[(3-Fluorophenyl)methyl]phosphonic acid, 9CI F-00030

m-Fluorobenzylphosphonic acid

[80395-16-0]

$C_7H_8FO_3P$ M 190.111

Di-Et ester: [63909-57-9]. *Diethyl [(3-fluorophenyl)-methyl]phosphonate.*

$C_{11}H_{16}FO_3P$ M 246.218

Wittig-Horner reagent. Liq. Bp_6 145°. n_D^{25} 1.481.

Ernst, L., *Org. Magn. Reson.*, 1977, **9**, 35 (*synth, P nmr, cmr*)

[(4-Fluorophenyl)methyl]phosphonic acid, 9CI F-00031

p-Fluorobenzylphosphonic acid

[80395-14-8]

$C_7H_8FO_3P$ M 190.111

Di-Et ester: [63909-58-0]. *Diethyl [(4-fluorophenyl)-methyl]phosphonate.*

$C_{11}H_{16}FO_3P$ M 246.218

Wittig-Horner reagent. Liq. Bp_6 148-150°. n_D^{20} 1.478.

Ernst, L., *Org. Magn. Reson.*, 1977, **9**, 35 (*synth, cmr, P nmr*)

[(4-Fluorophenyl)methyl]-triphenylphosphonium(1+), 9CI F-00032

(p-Fluorobenzyl)triphenylphosphonium(1+)

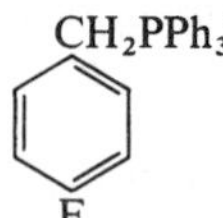

$C_{25}H_{21}FP^{\oplus}$ M 371.413 (ion)

Chloride: [3462-95-1].

$C_{25}H_{21}ClFP$ M 406.866

Source of ylide with MeOLi. Solid. Mp 315-316°.

Bromide: [51044-11-2].

$C_{25}H_{21}BrFP$ M 451.317

Source of ylide with butyllithium. Solid. Mp 300°.

Ylide: [59625-60-4]. *[(4-Fluorophenyl)methylene]-triphenylphosphorane. (p-Fluorobenzylidene)-triphenylphosphorane.*

$C_{25}H_{20}FP$ M 370.405

A Wittig reagent prepd. *in situ.*

Jones, R.A. *et al*, *Aust. J. Chem.*, 1965, **18**, 903 (*chloride*)

Leznoff, C.C. *et al*, *Can. J. Chem.*, 1972, **50**, 582 (*chloride, use*)

Yamato, M. *et al*, *Chem. Pharm. Bull.*, 1977, **25**, 706 (*bromide, use*)

(3-Fluorophenyl)phosphine, 9CI F-00033

[23588-09-2]

C_6H_6FP M 128.086

Liq. Bp_{21} 52-53°. n_D^{20} 1.5366.

Schindlbauer, H. *et al*, *Chem. Ber.*, 1969, **102**, 2914 (*synth, nmr*)

(4-Fluorophenyl)phosphine, 9CI F-00034

[18437-71-3]

C_6H_6FP M 128.086

Liq. Bp 154-156°, Bp_{11} 38-40°. n_D^{20} 1.5338.

Schindlbauer, H., *Chem. Ber.*, 1967, **100**, 3432 (*synth, nmr*)

Maier, L., *Phosphorus*, 1974, **4**, 41 (*synth, pmr, nmr*)

(3-Fluorophenyl)phosphinic acid, 9CI F-00035

[27003-03-8]

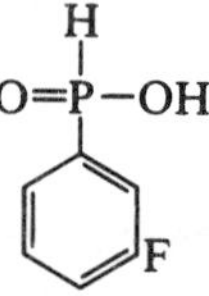

$C_6H_6FO_2P$ M 160.084

Tautomeric with 3-fluorophenylphosphonous acid, but free acid probably exists in phosphinic acid form. Cryst. (C_6H_6/EtOH). Mp 60°.

Schindlbauer, H., *Chem. Ber.*, 1969, **102**, 2914 (*synth*)

(2-Fluorophenyl)phosphonic acid, 9CI F-00036

[700-24-3]

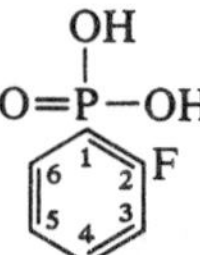

$C_6H_6FO_3P$ M 176.084

Solid. Mp 146-149°. pK_{a1} 1.64 (H_2O), 2.84 (50% EtOH aq.), pK_{a2} 6.80 (H_2O), 7.99 (50% EtOH aq.).

Di-Et ester: [312-05-0]. *Diethyl (2-fluorophenyl)-phosphonate.*

$C_{10}H_{14}FO_3P$ M 232.191

Liq. $Bp_{0.02}$ 90-92°.

Bis(trimethylsilyl) ester: [99136-00-2]. *Bis(trimethylsilyl)(2-fluorophenyl)phosphonate.*

$C_{12}H_{22}FO_3PSi_2$ M 320.447

Liq. $Bp_{0.001}$ 78-80°.

Freedman, L.D. *et al*, *J. Am. Chem. Soc.*, 1953, **75**, 1379 (*synth*)

Bard, R.R. *et al*, *J. Org. Chem.*, 1979, **44**, 4918 (*ester, ir, pmr, ms*)

Issleib, K. *et al*, *Z. Anorg. Allg. Chem.*, 1985, **529**, 151 (*silyl ester, synth, P nmr*)

(3-Fluorophenyl)phosphonic acid, 9CI **F-00037**

[2369-26-8]

$C_6H_6FO_3P$ M 176.084

Di-Me ester: Dimethyl (3-fluorophenyl)phosphonate.
$C_8H_{10}FO_3P$ M 204.137
Liq. Bp_{13} 136-138°. n_D^{20} 1.4776.

Di-Et ester: [23588-07-0]. *Diethyl (3-fluorophenyl)-phosphonate.*
$C_{10}H_{14}FO_3P$ M 232.191
Liq. Bp_{10} 137-138°. n_D^{20} 1.4770.

Dichloride: [23588-13-8].
$C_6H_4Cl_2FOP$ M 212.975
Liq. Bp_{18} 126-130°. n_D^{20} 1.5384.

Schindlbauer, H. *et al*, *Chem. Ber.*, 1969, **102**, 2914 (*derivs, F nmr*)
Grabiak, R.C. *et al*, *Phosphorus Sulfur*, 1980, **9**, 197 (*derivs, P nmr*)

(4-Fluorophenyl)phosphonic acid, 9CI **F-00038**

[349-87-1]

$C_6H_6FO_3P$ M 176.084

Solid. Mp 125-127°.

Di-Me ester: [15704-45-7]. *Dimethyl (4-fluorophenyl)-phosphonate.*
$C_8H_{10}FO_3P$ M 204.137
Liq. $Bp_{0.2}$ 80-81°.

Di-Et ester: [310-40-7]. *Diethyl (4-fluorophenyl)-phosphonate.*
$C_{10}H_{14}FO_3P$ M 232.191
Liq. Bp_8 136-137°, $Bp_{0.65}$ 105°. n_D^{20} 1.4788.

Bis(trimethylsilyl) ester: [104412-67-1]. *Bis(trimethyl-silyl)(4-fluorophenyl)phosphonate.*
$C_{12}H_{22}FO_3PSi_2$ M 320.447
Liq. $Bp_{0.001}$ 85°.

Dichloride: see (4-Fluorophenyl)phosphonic dichloride, F-00039

Bost, R.W. *et al*, *J. Org. Chem.*, 1953, **18**, 362 (*synth*)
Obrycki, R. *et al*, *Tetrahedron Lett.*, 1966, 5049 (*dimethyl ester*)
Schindlbauer, H., *Chem. Ber.*, 1967, **100**, 3432 (*diethyl ester, pmr, F nmr*)
Bunnett, J.F. *et al*, *J. Org. Chem.*, 1978, **43**, 1867 (*diethyl ester, synth, ir, pmr, ms*)
Grabiak, R.C. *et al*, *Phosphorus Sulfur*, 1980, **9**, 197 (*synth, derivs, nmr*)
Issleib, K. *et al*, *Z. Anorg. Allg. Chem.*, 1985, **529**, 151 (*silyl ester, synth, P nmr*)

(4-Fluorophenyl)phosphonic dichloride, 9CI **F-00039**

[657-81-8]

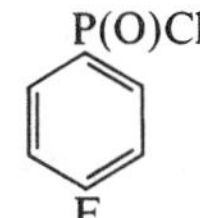

$C_6H_4Cl_2FOP$ M 212.975

Liq. Bp_{17} 124-125°. n_D^{20} 1.5386.

Cherbuliez, E. *et al*, *Helv. Chim. Acta*, 1962, **45**, 2665 (*synth*)
Gutmann, V. *et al*, *Monatsh. Chem.*, 1966, **97**, 1265 (*synth, ir, pmr, F nmr*)
Schindlbauer, H., *Chem. Ber.*, 1967, **100**, 3432 (*synth, F nmr*)
Sergienko, L.M. *et al*, *Zh. Obshch. Khim.*, 1980, **50**, 1958 (*Engl. transl.* p. 1578) (*uv*)
Ratovskii, G.V. *et al*, *Zh. Obshch. Khim.*, 1981, **51**, 714 (*Engl. transl.* p. 580) (*raman*)

(2-Fluorophenyl)phosphonous acid **F-00040**

$C_6H_6FO_2P$ M 160.084

Dichloride: [5592-65-4]. *Dichloro(2-fluorophenyl)-phosphine.*
$C_6H_4Cl_2FP$ M 196.976
Liq. Bp_{15} 105-6°.

Furin, G.G. *et al*, *Zh. Obshch. Khim.*, 1975, **45**, 1473 (*Engl. transl.* p. 1441) (*dichloride, synth, uv*)

(3-Fluorophenyl)phosphonous acid **F-00041**

$C_6H_6FO_2P$ M 160.084

Free acid exists as the phosphoryl tautomer (3-Fluorophenyl)phosphinic acid, F-00035 .

Di-Me ester: [21388-28-3]. *Dimethyl (3-fluorophenyl)-phosphonite.*
$C_8H_{10}FO_2P$ M 188.138
Liq. $Bp_{0.7}$ 45°.

Difluoride: [21388-30-7]. *Difluoro(3-fluorophenyl)-phosphine.*
$C_6H_4F_3P$ M 164.067
Impure liq. Bp_{22} 47°. n_D^{22} 1.4648.

Dichloride: [5510-94-1]. *Dichloro(3-fluorophenyl)-phosphine.*
$C_6H_4Cl_2FP$ M 196.976
Liq. Bp_{15} 90-91°, $Bp_{1.3}$ 68-70°. n_D^{25} 1.5721.

Dicyanide: [21388-33-0].
$C_8H_4FN_2P$ M 178.105
Oil.

Bis(dimethylamide): [21388-27-2]. P-*3-Fluorophenyl-N,N,N′,N′-tetramethylphosphonous diamide.*
$C_{10}H_{16}FN_2P$ M 214.222
Liq. $Bp_{0.2}$ 87°.

Schmidbauer, H., *Monatsh. Chem.*, 1965, **96**, 1936 (*dichloride, synth, ir*)
Rayshys, J.W. *et al*, *J. Am. Chem. Soc.*, 1968, **90**, 5236 (*derivs, synth, nmr*)
De Ketelaere, R. *et al*, *Bull. Soc. Chim. Belg.*, 1969, **78**, 219 (*dichloride, ir, pmr, F and P nmr*)
Muylle, E. *et al*, *Spectrochim. Acta, Part A*, 1976, **32**, 599 (*P nmr*)

(4-Fluorophenyl)phosphonous acid, 9CI **F-00042**

$C_6H_6FO_2P$ M 160.084

Free acid probably exists as phosphoryl tautomer.

Di-Me ester: [21388-36-3]. *Dimethyl (4-fluorophenyl)-phosphonite.*
$C_8H_{10}FO_2P$ M 188.138
Liq. $Bp_{2.7}$ 75°.

Difluoride: [21388-38-5]. *Difluoro(4-fluorophenyl)-phosphine.*
$C_6H_4F_3P$ M 164.067
Impure liq. Bp_{40} 58°.

Dichloride: [5510-93-0]. *Dichloro(4-fluorophenyl)-phosphine.*
$C_6H_4Cl_2FP$ M 196.976
Liq. Bp_{10} 82°. n_D^{20} 1.5690.

Dibromide: [86012-31-9]. *Dibromo(4-fluorophenyl)-phosphine.*
$C_6H_4Br_2FP$ M 285.878
Liq. Bp_5 105-106°.

Diiodide: [24901-24-4]. *(4-Fluorophenyl)-diiodophosphine.*
$C_6H_4FI_2P$ M 379.879
Liq. $Bp_{0.07}$ 104-105°.
Dicyanide: [21388-41-0].
$C_8H_4FN_2P$ M 178.105
Oil.
Bis(dimethylamide): [21388-35-2]. P-*4-Fluorophenyl-N,N,N′,N′-tetraethylphosphonous diamide.*
$C_{10}H_{16}FN_2P$ M 214.222
Liq. $Bp_{1.3}$ 83°.

Rayshys, J.W. *et al*, *J. Am. Chem. Soc.*, 1968, **90**, 5236 (*derivs, synth, F nmr*)
Feshchenko, N.G. *et al*, *Zh. Obshch. Khim.*, 1969, **39**, 2184 (*Engl. transl.* p. 2133) (*diiodide*)
De Ketelaere, R. *et al*, *Bull. Soc. Chim. Belg.*, 1969, **78**, 219 (*dichloride, ir, pmr F and P nmr*)
Muylle, E. *et al*, *Spectrochim. Acta, Part A*, 1976, **32**, 599 (*dichloride, P nmr*)
Hinke, A. *et al*, *Phosphorus Sulfur*, 1983, **15**, 93 (*dibromide, synth, nmr*)

(3-Fluorophenyl)phosphoramidic acid, 9CI F-00043

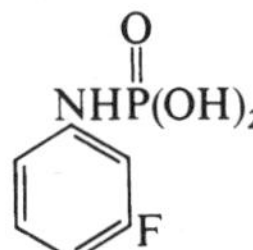

$C_6H_7FNO_3P$ M 191.098
Di-Me ester: [79639-92-2]. *Dimethyl (3-fluorophenyl)-phosphoramidate.*
$C_8H_{11}FNO_3P$ M 219.152
No phys. props. reported.
Di-Et ester: [50672-16-7]. *Diethyl (3-fluorophenyl)-phosphoramidate.*
$C_{10}H_{15}FNO_3P$ M 247.206
Solid. Mp 73-76°.
Diisopropyl ester: Diisopropyl (3-fluorophenyl)-phosphoramidate.
$C_{12}H_{19}FNO_3P$ M 275.259
Solid. Mp 99.5-100°.

Joshi, K.C. *et al*, *Pestic. Sci.*, 1973, **4**, 701 (*esters, synth*)
Foulds, G.A. *et al*, *S. Afr. J. Chem.*, 1981, **34**, 72 (*dimethyl ester, ir*)

(4-Fluorophenyl)phosphoramidic acid, 9CI F-00044

$C_6H_7FNO_3P$ M 191.098
Di-Me ester: [79639-93-3]. *Dimethyl (4-fluorophenyl)-phosphoramidate.*
$C_8H_{11}FNO_3P$ M 219.152
No phys. props. reported.
Di-Et ester: [50672-18-9]. *Diethyl (4-fluorophenyl)-phosphoramidate.*
$C_{10}H_{15}FNO_3P$ M 247.206
Liq. Bp_{11} 61° (?), $Bp_{0.2}$ 138-140°. n_D^{20} 1.4812.
Diisopropyl ester: [50672-29-2]. *Diisopropyl (4-fluorophenyl)phosphoramidate.*
$C_{12}H_{19}FNO_3P$ M 275.259
Solid. Mp 119.5°.

Joshi, K.C. *et al*, *Pestic. Sci.*, 1973, **4**, 701 (*esters, synth*)
Bradamante, S. *et al*, *J. Org. Chem.*, 1980, **45**, 114 (*diethyl ester, pmr, F nmr, cmr*)
Foulds, G.A. *et al*, *S. Afr. J. Chem.*, 1981, **34**, 72 (*dimethyl ester, ir*)
Davidowitz, B. *et al*, *Org. Mass Spectrom.*, 1984, **19**, 128 (*dimethyl ester, ms*)

(3-Fluorophenyl)phosphoramidothioic acid, 9CI F-00045

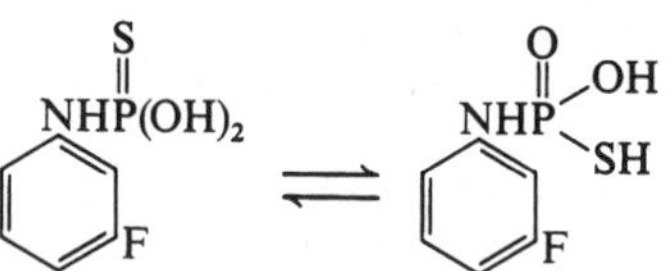

$C_6H_7FNO_2PS$ M 207.159
O,O-*Di-Et ester:* [50672-17-8]. O,O-*Diethyl (3-fluorophenyl)phosphoramidothioate.*
$C_{10}H_{15}FNO_2PS$ M 263.266
Solid. Mp 159°.

Joshi, K.C. *et al*, *Pestic. Sci.*, 1973, **4**, 701.

(4-Fluorophenyl)phosphoramidothioic acid, 9CI F-00046

$C_6H_7FNO_2PS$ M 207.159
O,O-*Di-Et ester:* [50672-19-0]. O,O-*Diethyl (4-fluorophenyl)phosphoramidothioate.*
$C_{10}H_{15}FNO_2PS$ M 263.266
Solid. Mp 157°.
O,O-*Diisopropyl ester:* [50672-30-5]. O,O-*Diisopropyl (4-fluorophenyl)phosphoroamidothioate.*
$C_{12}H_{19}FNO_2PS$ M 291.320
Solid. Mp 199°.

Joshi, K.C. *et al*, *Pestic. Sci.*, 1973, **4**, 701.

1-Fluorophosphorinane, 8CI F-00047

1-Fluorophosphacyclohexane. 1-Fluorophosphinane

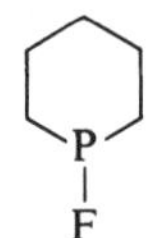

$C_5H_{10}FP$ M 120.106
Oxide: [30903-12-9].
$C_5H_{10}FOP$ M 136.106
Characterised spectroscopically.
Sulfide: [30779-27-2].
$C_5H_{10}FPS$ M 152.166
Characterised spectroscopically.

Reddy, G.S. *et al*, *Z. Naturforsch., B*, 1970, **25**, 1199 (*synth, F and P nmr*)

2-Fluoro-2,2′-spirobi[1,3,2-benzodioxaphosphole], 9CI F-00048

[21229-79-8]

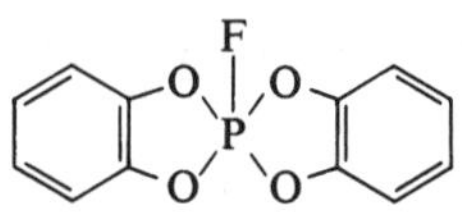

$C_{12}H_8FO_4P$ M 266.165
Trigonal bipyramidal struct. with P-F bond equatorial. Cryst. (Et_2O). Mp 118-119°.

Doak, G.O. *et al*, *J. Chem. Soc. (A)*, 1971, 1295 (*synth, pmr, F and P nmr*)
Wunderlich, H. *et al*, *Acta Crystallogr., Sect. B*, 1974, **30**, 935 (*cryst struct*)
Holmes, R.R. *et al*, *J. Am. Chem. Soc.*, 1977, **99**, 3318 (*struct*)

Fluorotetramethylphosphorane, 9CI F-00049

[36121-18-3]

Me_4PF

$C_4H_{12}FP$ M 110.111

Probably has an ionic struct. Cryst. $Bp_{0.1}$ 20° subl.

Ramaswamy, K. *et al*, *Z. Phys. Chem.*, 1969, **242**, 215 (*struct*)
Schmidbaur, H. *et al*, *Angew. Chem., Int. Ed. Engl.*, 1972, **11**, 144 (*synth, ir, raman, ms, pmr, struct*)

3-Fluoro-1,3-thiaphosphetane, 9CI F-00050

Thiodimethylenephosphinous fluoride

C_2H_4FPS M 110.086

3-Sulfide: [51500-74-4].

$C_2H_4FPS_2$ M 142.146

Liq. d_4^{20} 1.48. Bp_{10} 76-8°. n_D^{20} 1.6008.

Arshinova, R.P. *et al*, *Izv. Akad. Nauk SSSR, Ser. Khim.*, 1973, 2240 (*Engl. transl.* p. 2185) (*sulfide, conformn*)
Shakirov, I.Kh. *et al*, *Dokl. Akad. Nauk SSSR, Ser. Sci. Khim.*, 1974, **219**, 917 (*Engl. transl.* p. 1155 (*sulfide, ir, raman, struct*)
Shagidullin, R.R. *et al*, *Zh. Obshch. Khim.*, 1975, **45**, 536 (*Engl. transl.* p. 530) (*sulfide, synth, ir, raman*)

Flurofamide, INN, USAN F-00051

N-(*Diaminophosphinyl*)-*4-fluorobenzamide*, *9CI*. *Flurfamide*

[70788-28-2]

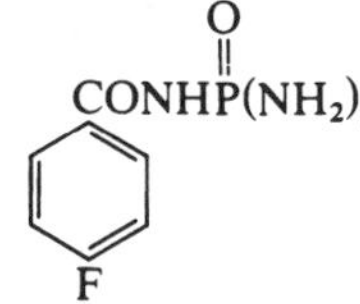

$C_7H_9FN_3O_2P$ M 217.139

Urease inhibitor.

▷CV3458100.

U.S.P., 4 182 881, (*1980*); *CA*, **92**, 146458 (*synth*)

Fonofos, BSI F-00052

O-*Ethyl* S-*phenyl ethylphosphonodithioate*, *9CI*. *Dyfonate*

[944-22-9]

(*R*)-*form* Absolute configuration

$C_{10}H_{15}OPS_2$ M 246.322

▷Highly toxic, TLV (skin) 0.1. TA5950000.

(*R*)-*form* [62705-71-9]

Liq. $[\alpha]_D^{24}$ +156° (neat), $[\alpha]_D^{24}$ −114.6° (c, 0.31 in cyclohexane). n_D^{24} 1.5895.

(*S*)-*form* [62680-03-9]

Liq. $[\alpha]_D^{24}$ −181.5° (neat), $[\alpha]_D^{24}$ +138.4° (c, 0.30 in cyclohexane). n_D^{24} 1.5896.

▷Less toxic (flies, mice) than (*R*)-form

(±)-*form* [66767-39-3]

Soil insecticide. Pale-yellow oil with aromatic odour. V. spar. sol. H_2O. $Bp_{0.1}$ 130°.

Menn, J.J. *et al*, *J. Econ. Entomol.*, 1965, **58**, 734 (*synth, tox*)
Babad, H. *et al*, *Anal. Chim. Acta*, 1968, **41**, 259 (*pmr*)
Krijgsman, W. *et al*, *J. Chromatogr.*, 1976, **117**, 201 (*glc*)
Allahyari, R. *et al*, *J. Agric. Food Chem.*, 1977, **25**, 471 (*resoln, abs config*)
Lee, P.W. *et al*, *Pestic. Biochem. Physiol.*, 1978, **8**, 146, 158 (*tox, metab*)
Stan, H.-J. *et al*, *Biomed. Mass Spectrom.*, 1982, **9**, 483 (*ms*)
Stan, H.-J. *et al*, *J. Chromatogr.*, 1983, **279**, 173 (*glc*)
Pesticide Manual, 6th Ed., 282; 7th Ed., 289.

Fopurine F-00053

P,P-*Bis*(*1-aziridinyl*)-N-[*2-*(*dimethylamino*)-*7-methyl-7*H-*purin-6-yl*]*phosphonic amide*, *9CI*. *Phorpurine*

[42061-52-9]

$C_{12}H_{18}N_8OP$ M 321.301

Antitumour agent with immunosuppressive activity. Solid.

▷SZ5827000.

Slapsite, G., *CA*, 1981, **94**, 202467 (*pharmacol*)
Gol'dberg, E.D. *et al*, *CA*, 1982, **97**, 16813 (*pharmacol*)
Vasil'chenko, V.N. *et al*, *Zh. Strukt. Khim.*, 1982, **23**, 107 (*Engl. transl.* p. 257) (*cryst struct*)
Verkin, B.I. *et al*, *Dokl. Akad. Nauk SSSR, Ser. Sci. Khim.*, 1982, **265**, 115; *CA*, **97**, 181536 (*ms, struct*)

Formaldehyde (triphenylphosphoranylidene)hydrazone, 9CI F-00054

(*Methylenehydrazono*)*triphenylphosphorane*, *8CI*

[15990-54-2]

$Ph_3P{=}NN{=}CH_2$

$C_{19}H_{17}N_2P$ M 304.330

Solid. Mp 145-146° dec. (139-141°).

Wittig, G. *et al*, *Chem. Ber.*, 1955, **88**, 1654 (*synth*)
Zeeh, B. *et al*, *Org. Mass Spectrom.*, 1968, **1**, 791 (*ms*)
Bock, H. *et al*, *Chem. Ber.*, 1969, **102**, 1363 (*synth, ir*)
Albright, T.A. *et al*, *J. Org. Chem.*, 1976, **41**, 2716 (*cmr, P nmr*)
Lumbroso, H. *et al*, *J. Organomet. Chem.*, 1978, **161**, 347 (*struct*)
Bestmann, H.J. *et al*, *J. Organomet. Chem.*, 1980, **192**, 177 (*deriv*)

Formothion, BSI, ISO F-00055

S-[*2-*(*Formylmethylamino*)-*2-oxoethyl*] O,O-*dimethyl phosphorodithioate*, *9CI*. S-(N-*Formyl*-N-*methylcarbamoylmethyl*) O,O-*dimethyl phosphorodithioate*. *Anthio*. *Aflix*

[2540-82-1]

$(MeO)_2P(S)SCH_2CONMeCHO$

$C_6H_{12}NO_4PS_2$ M 257.259

Contact and systemic insecticide. Yellow viscous oil or cryst. mass. Sl. sol. H_2O, misc. most org. solvs. d_4^{20} 1.361. Mp 25-26°. n_D^{20} 1.5541. Dec. on dist. Metab. *via* Dimethoate, D-00671 .

▷TE1050000.

U.S.P., 3 176 035, (*1959*); *CA*, **57**, 16404g (*synth, tox*)
Stan, H.-J. *et al*, *Fresenius' Z. Anal. Chem.*, 1977, **287**, 271; *Biomed. Mass Spectrom.*, 1982, **9**, 483 (*glc, ms*)
Stan, H.-J. *et al*, *J. Chromatogr.*, 1983, **279**, 173 (*glc*)
Pesticide Manual, 6th Ed., 285; 7th Ed., 292.

The Agrochemicals Handbook, Royal Society of Chemistry, 1983, A217.
Sax, N.I., *Dangerous Properties of Industrial Materials*, 6th Ed., Van Nostrand-Reinhold, 1984, 695.

(α-Formylbenzyl)triphenylphosphonium(1+) **F-00056**
[α-Formyl(phenylmethyl)]triphenylphosphonium(1+)

$Ph_3P^{\oplus}CHPhCHO$

$C_{26}H_{22}OP^{\oplus}$ M 381.433 (ion)
Chloride:
$C_{26}H_{22}ClOP$ M 416.886
Solid. Mp 223-225°.
Bromide:
$C_{26}H_{22}BrOP$ M 461.337
Cryst. ($CHCl_3$/EtOH or $MeNO_2$/EtOAc). Mp 235-237°.

Märkl, G., *Tetrahedron Lett.*, 1962, 1027 (*chloride, use*)
Devlin, C.J. *et al*, *J. Chem. Soc., Perkin Trans. 1*, 1974, 453 (*bromide*)

Fosazepam, BAN, USAN **F-00057**
7-Chloro-1-[(dimethylphosphinyl)methyl]-1,3-dihydro-5-phenyl-2H-1,4-benzodiazepin-2-one, 9CI
[35322-07-7]

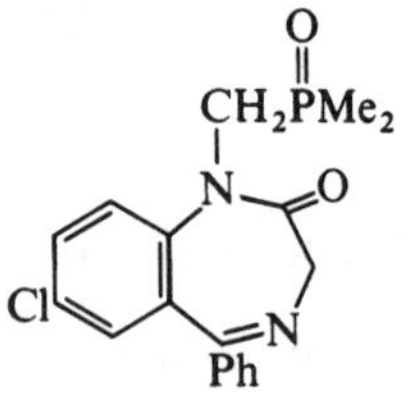

$C_{18}H_{18}ClN_2O_2P$ M 360.779
Hypnotic.

Ger. Pat., 2 022 503, (*1971*); *CA*, **76**, 72570c (*synth*)
Allen, S. *et al*, *Br. J. Clin. Pharmacol.*, 1976, **3**, 165 (*pharmacol*)

Fosfazinomycin *A* **F-00058**
AM 630. Antibiotic AM 630
[87423-10-7]

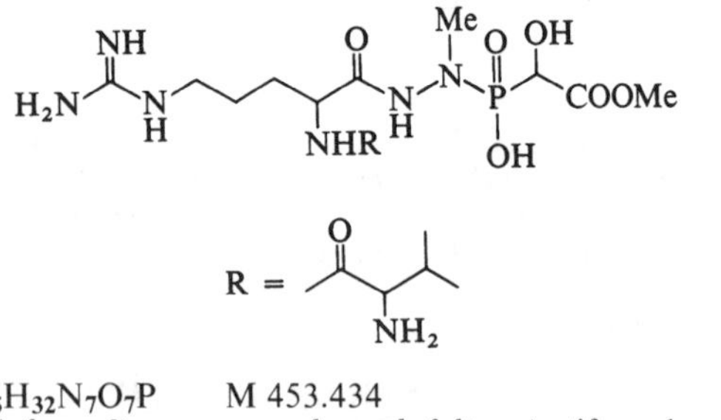

$C_{15}H_{32}N_7O_7P$ M 453.434
Isol. from *Streptomyces lavendofoliae*. Antifungal antibiotic. Hygroscopic powder. Mp 157-161° dec. $[\alpha]^{25}_{D}$ +14.7° (c, 1 in H_2O).

Japan. Pat., 80 118 500, (*1980*); *CA*, **94**, 28884 (*isol*)
Gungi, S. *et al*, *Agric. Biol. Chem.*, 1983, **47**, 2061 (*isol*)
Ogita, T. *et al*, *Tetrahedron Lett.*, 1983, **24**, 2283 (*struct, ir, pmr, cmr, ms*)

Fosfazinomycin *B* **F-00059**
AM 630B. Antibiotic AM 630B
[87423-11-8]
As Fosfazinomycin *A*, F-00058 with

R = H

$C_{10}H_{23}N_6O_6P$ M 354.302
Isol. from *Streptomyces lavendofoliae*. Antifungal antibiotic.
B,H₂CO₃ (?): Mp 148-150° dec. $[\alpha]^{25}_{D}$ +17.2° (c, 1.0 in H_2O). Not clear from the lit. whether this physical data refers to the free base or to the carbonate salt.

Japan. Pat., 81 154 494, (*1981*); *CA*, **96**, 179425 (*isol*)
Gunji, S. *et al*, *Agric. Biol. Chem.*, 1983, **47**, 2061.
Ogita, T. *et al*, *Tetrahedron Lett.*, 1983, **24**, 2283 (*struct, ir, pmr, cmr, ms*)

Fosfocreatinine, INN **F-00060**
(1-Methyl-4-oxo-2-imidazolidinylene)phosphoramidic acid, 9CI
[5786-71-0]

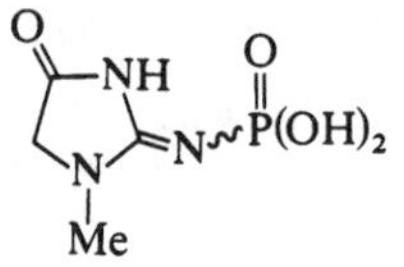

$C_4H_8N_3O_4P$ M 193.099
Tonic.
Di-Na salt: [19604-05-8]. Mp 131-132°.

Ger. Pat., 1 260 473, (*1968*); *CA*, **69**, 36192n (*synth, pharmacol*)
Fr. Pat., 1 536 402, (*1968*); *CA*, **71**, 91475p (*synth*)

Fosmenic acid, INN **F-00061**
(3-Cyclohexen-1-ylhydroxymethyl)phosphinic acid, 8CI
[13237-70-2]

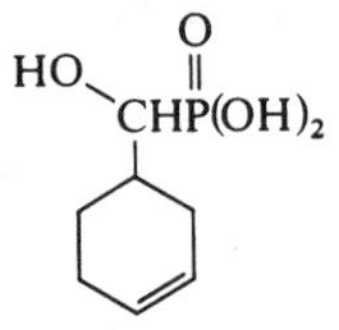

$C_7H_{13}O_3P$ M 176.152
Drug for treatment of atherosclerosis.

Romero, R.L. *et al*, *Med. Monatsschr.*, 1965, **19**, 61 (*pharmacol*)

Fosmidomycin **F-00062**
[3-(Formylhydroxyamino)propyl]phosphonic acid, 9CI
[66508-53-0]

$(HO)_2P(O)CH_2CH_2CH_2N(OH)CHO$

$C_4H_{10}NO_5P$ M 183.100
Produced by *Streptomyces lavendulae* and isol. as Na salt. Clinically useful antibiotic active against gram-negative bacteria.
Mono-Na salt: [66508-37-0]. Cryst. (MeOH/EtOH). Mp 189-191°.

Kojo, Y. *et al*, *J. Antibiot.*, 1980, **33**, 44 (*pharmacol*)
Kuroda, Y. *et al*, *J. Antibiot.*, 1980, **33**, 29 (*struct, ir, pmr*)
Mine, Y. *et al*, *J. Antibiot.*, 1980, **33**, 36 (*pharmacol*)
Okuhara, M. *et al*, *J. Antibiot.*, 1980, **33**, 24 (*isol, struct, ir, pmr, tlc*)
Hemmi, K. *et al*, *Chem. Pharm. Bull.*, 1982, **30**, 111 (*synth, props*)
Murakawa, T. *et al*, *Antimicrob. Agents Chemother.*, 1982, **21**, 224 (*pharmacol*)

Fospirate, USAN **F-00063**

Dimethyl 3,5,6-trichloro-2-pyridinyl phosphate, 9CI, 8CI. Torelle. Dowco 217

[5598-52-7]

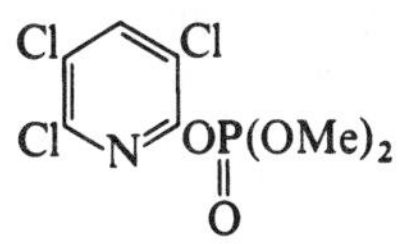

$C_7H_7Cl_3NO_4P$ M 306.469

Oxidn. product from Chlorpyrifos, C-00194 . Veterinary anthelmintic of low mammalian toxy. Cryst. (pet. ether). Mp 86.5-88°.

▷TC5450000.

Rigterink, R.H. *et al*, *J. Agric. Food Chem.*, 1966, **14**, 394 (*synth, pharmacol*)
Whitney, W.K. *et al*, *J. Econ. Entomol.*, 1969, **62**, 567 (*use*)
Baughman, R.G. *et al*, *J. Agric. Food Chem.*, 1977, **25**, 582 (*cryst struct*)

Fostedil, USAN **F-00064**

Diethyl[[4-(2-benzothiazolyl)phenyl]methyl]-phosphonate, 9CI. A-53986. KB-944

[75889-62-2]

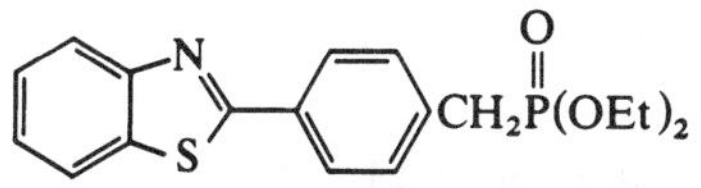

$C_{18}H_{20}NO_3PS$ M 361.395

Vasodilator, Cachannel blocker.

Eur. Pat., 10 120, (*1980*); *CA*, **94**, 4119j (*synth, pharmacol*)
Morita, T. *et al*, *Arzneim.-Forsch.*, 1982, 1047 (*pharmacol*)
Hirakawa, K. *et al*, *Arzneim.-Forsch.*, 1982, 1071 (*tox*)
Thomas, E.W., *J. Chromatogr.*, 1984, **305**, 233 (*hplc*)

Fosthietan, BSI **F-00065**

Diethyl 1,3-dithietan-2-ylidenephosphoramidate, 9CI. Nem-a-tak. Acconem. Geofos

[21548-32-3]

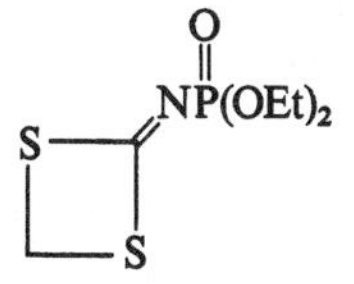

$C_6H_{12}NO_3PS_2$ M 241.259

Nematocide and soil insecticide. Yellow liq. with mercaptan odour. Sol. Me_2CO, $CHCl_3$, MeOH, toluene, mod. sol. H_2O. d^{25} 1.3.

▷V. toxic. NJ6490000.

U.S.P., 4 070 372, (*1978*); *CA*, **88**, 170129 (*synth*)
Stout, S.J. *et al*, *Biomed. Mass Spectrom.*, 1984, **11**, 207 (*glc, ms*)
Pesticide Manual, 6th Ed., 287; 7th Ed., 295.

Fotrin **F-00066**

2,2,4,4,6-Pentakis(1-aziridinyl)-2,2,4,4,6,6-hexahydro-6-(4-morpholinyl)-1,3,5-triaza-2,4,6-triphosphorine, 9CI

[37132-72-2]

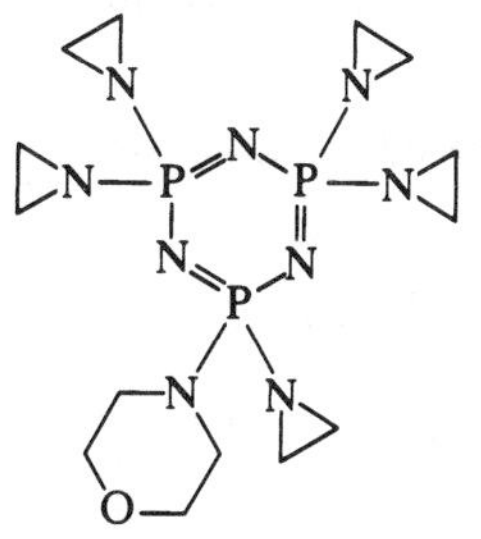

$C_{14}H_{28}N_9OP_3$ M 431.356

Cytostatic, immunosuppressant. Cryst. ($EtOH/Et_2O/EtOAc$). Mp 120.5-121.5°.

▷XX9470000.

Ger. Pat., 2 043 128, (*1972*); *CA*, **77**, 48480 (*synth*)
Ovchinnikova, V.A. *et al*, *Khim.-Farm. Zh.*, 1977, **11**, 18 (*Engl. transl.* p. 1601) (*metab*)
Van der Huizen, A.A. *et al*, *Inorg. Chim. Acta*, 1983, **78**, 239.

Fructose 1,6-bis(dihydrogen phosphate), 9CI, 8CI **F-00067**

Harden-Young ester. Hexose diphosphate. 1,6-Fructose diphosphoric acid

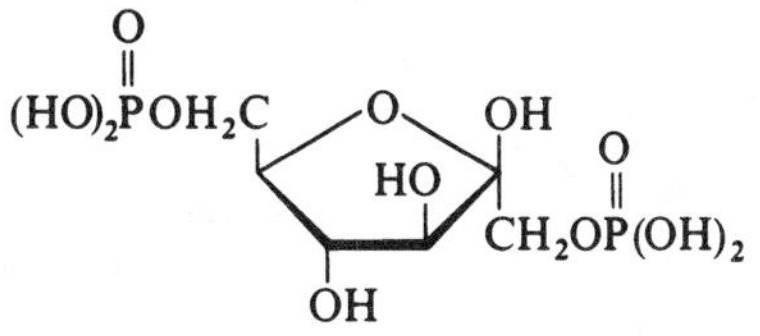

β-Furanose

$C_6H_{14}O_{12}P_2$ M 340.117

D-form [488-69-7]

Prepared by action of yeasts on Glucose, Sucrose and Fructose; formed from Fructose 6-phosphate in the presence of Mg^{++}, ATP and the enzyme phosphohexokinase. Metabolic intermed. $[\alpha]_D^{17}$ +4.04° (c, 13.6 in H_2O). pK_{a1} 1.48, pK_{a2} 6.29. Reversibly cleaved in the presence of Aldolase to give 3-Phosphoglyceraldehyde and 1-Phosphodihydroxyacetone.

Di-Ca salt monohydrate: [34378-77-3]. *Candiolin.* Powder.

Di-Na salt: [26177-85-5]. Constit. of tonics.

Neuberg, C. *et al*, *Arch. Biochem. Biophys.*, 1944, **3**, 33 (*synth*)
Gray, G.R., *Biochemistry*, 1971, **10**, 4705 (*nmr*)
MacDonald, D.L., *The Carbohydrates*, Academic Press, 1972, 2nd Ed., **1A**, 253 (*rev*)
Koerner, T.A.W., *CA*, 1976, **85**, 124249h (*synth, pmr*)

Fructose 1-dihydrogen phosphate, 9CI, 8CI **F-00068**

Fructosyl phosphate

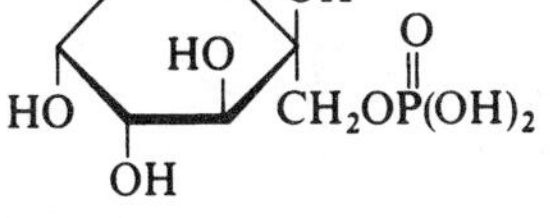

β-Pyranose-*form*

$C_6H_{13}O_9P$ M 260.137

D-form [15978-08-2]

$[\alpha]_D$ −64.2° (H_2O).

Ba salt: $[\alpha]_D$ −39° (H_2O).

Brucine salt: $[\alpha]_D$ −52.1° (H_2O).
Hydrazone: Mp 96-97°. $[\alpha]_D$ −33.6° (1:1 Py/MeOH).

Tanko, B. *et al*, *Biochem. J.*, 1935, **29**, 961.
Graham, D. *et al*, *Nature (London)*, 1965, **208**, 88
Harvey, D.J. *et al*, *J. Chromatogr.*, 1973, **76**, 51 (*glc, ms*)

Fructose 2-dihydrogen phosphate F-00069

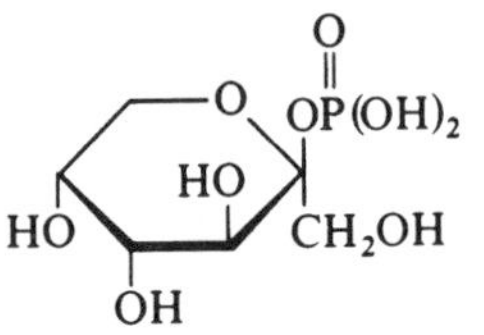

β-Pyranose-*form*

$C_6H_{13}O_9P$ M 260.137

D-form

Ba salt: $[\alpha]_D^{20}$ −83.3° (H_2O).
Na salt: $[\alpha]_D^{20}$ −53.6° (H_2O).
Dicyclohexylammonium salt: $[\alpha]_D^{24}$ −78° (c, 1.0 in H_2O).

Pontis, H.G. *et al*, *Biochem. J.*, 1963, **89**, 452 (*synth*)
MacDonald, D.L., *J. Org. Chem.*, 1966, **31**, 513 (*synth*)

Fructose 6-dihydrogen phosphate, 9CI, 8CI F-00070

Neuberg ester. Fructose-6-phosphoric acid. Hexose monophosphate

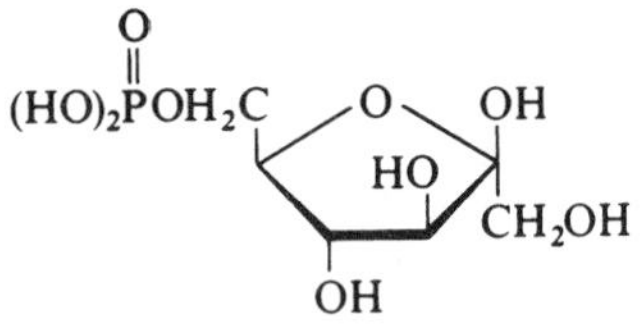

β-Furanose-*form*

$C_6H_{13}O_9P$ M 260.137

D-form [643-13-0]

Present in animal tissues as an equilib. mixt. with Glucose 6-phosphate. Formed by the action of the enzyme phosphohexose isomerase on Glucose 6-phosphate. $[\alpha]_D$ +1.2° (c, 0.9 in H_2O). pK_{a1} 0.97, pK_{a2} 6.11.

Ba salt: $[\alpha]_D$ +3.6° (H_2O).

Neuberg, C. *et al*, *Arch. Biochem. Biophys.*, 1944, **3**, 33 (*synth*)
MacDonald, D.L., *The Carbohydrates*, Academic Press, 1972, 2nd Ed., **IA**, 253 (*rev*)
Beucamp, K. *et al*, *Methoden. Enzym. Anal.* (*3. Aufl.*), 1974, **1**, 558 (*rev*)

(2-Furanylhydroxymethyl)phosphonic acid, 9CI F-00071

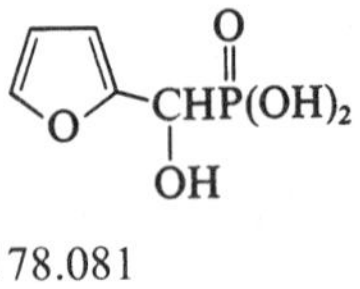

$C_5H_7O_5P$ M 178.081

(±)-*form*

Di-Me ester: Dimethyl (2-furanylhydroxymethyl)-phosphonate. Dimethyl (α-hydroxy-2-furanylmethyl)phosphonate.
$C_7H_{11}O_5P$ M 206.135
Cryst. (EtOH/cyclohexane). Mp 47-48°, Mp 62-63°.

Di-Et ester: [20627-09-2]. *Diethyl (2-furanylhydroxymethyl)phosphonate. Diethyl (α-hydroxy-2-furanylmethyl)phosphonate.*
$C_9H_{15}O_5P$ M 234.188
Liq. d_4^{20} 1.22. Bp_3 179-180°. n_D^{20} 1.4760.

Diisopropyl ester: Diisopropyl α-hydroxy-2-furanylmethylphosphonate.
$C_{11}H_{19}O_5P$ M 262.242
Cryst. (EtOH/cyclohexane). Mp 61-62°.

Abramov, V.S. *et al*, *Zh. Obshch. Khim.*, 1957, **27**, 173 (*Engl. transl.* p. 193)

(2-Furanylmethyl)phosphonic acid, 10CI F-00072

2-(Phosphonomethyl)furan

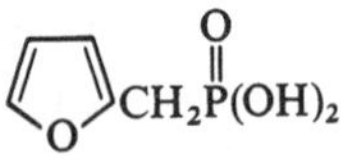

$C_5H_7O_4P$ M 162.082
Solid. Mp 112-120°.

Di-Et ester: Diethyl 2-furanylmethylphosphonate.
$C_9H_{15}O_4P$ M 218.189
Liq. $Bp_{0.7}$ 104-108°.

Bistrimethylsilyl ester: Bis(trimethylsilyl) 2-furanylmethylphosphonate.
$C_{11}H_{23}O_4PSi$ M 278.360
Liq. $Bp_{0.04}$ 73°.

Allen, D.W. *et al*, *J. Chem. Soc., Perkin Trans. 2*, 1977, 789 (*ester, synth, pmr, P nmr*)
Andreae, S. *et al*, *Z. Chem.*, 1980, **20**, 338 (*synth, silyl ester*)

2-Furanylphosphonic acid, 9CI F-00073

2-Furylphosphonic acid. 2-Phosphonofuran
[77113-17-8]

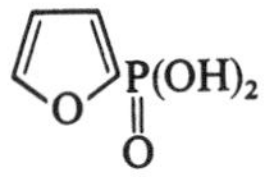

$C_4H_5O_4P$ M 148.055
Oil.

Di-Me ester: [13640-97-6]. *Dimethyl 2-furanylphosphonate.*
$C_6H_9O_4P$ M 176.108
Liq. $Bp_{0.4}$ 111-114°.

Di-Et ester: [36366-55-9]. *Diethyl 2-furanylphosphonate.*
$C_8H_{13}O_4P$ M 204.162
Liq. $Bp_{0.8}$ 98-108°.

Di-Ph ester: [63818-45-1]. *Diphenyl 2-furanylphosphonate.*
$C_{16}H_{13}O_4P$ M 300.250
Needles (hexane). Mp 50°.

Bis(trimethylsilyl) ester: [77113-29-2]. *Bis(trimethylsilyl) 2-furanylphosphonate.*
$C_{10}H_{21}O_4PSi_2$ M 292.418
Liq. $Bp_{0.04}$ 70°.

Obrycki, R. *et al*, *J. Org. Chem.*, 1968, **33**, 632 (*ester, synth, ir, pmr*)
Allen, D.W. *et al*, *J. Chem. Soc., Perkin Trans. 2*, 1972, 63; 1977, 789 (*esters, synth, pmr, P nmr*)
Andreae, S. *et al*, *Z. Chem.*, 1980, **20**, 338 (*synth, derivs*)

3-Furanylphosphonic acid, 9CI F-00074

3-Furylphosphonic acid. 3-Phosphonofuran

$C_4H_5O_4P$ M 148.055

Di-Me ester: [37632-32-9]. *Dimethyl 3-furanylphosphonate.*
$C_6H_9O_4P$ M 176.108
Liq. d_4^{20} 1.27. Bp_3 110°. n_D^{20} 1.4691.

Di-Et ester: [37632-33-0]. *Diethyl 3-furanylphosphonate.*
$C_8H_{13}O_4P$ M 204.162
Liq. d_4^{20} 1.18. Bp_2 111°. n_D^{20} 1.4630.
Dichloride:
$C_4H_3Cl_2O_2P$ M 184.946
Liq. d_4^{20} 1.60. Bp_5 91°. n_D^{20} 1.5281.

Fridland, S.V. *et al*, *Zh. Obshch. Khim.*, 1972, **42**, 121 (*Engl. transl.* p. 115) (*synth*)

G

Galactose 1-dihydrogen phosphate, 8CI — G-00001

α-Pyranose-*form*

$C_6H_{13}O_9P$ M 260.137

α-D-Pyranose-form [2255-14-3]

Occurs in liver, milk, and yeasts. $[\alpha]_D^{25}$ +143° (H_2O). Yeast or bacterial preparations reversibly convert α-Galactose 1-phosphate into Glucose 6-phosphate *via* α-Glucose 1-phosphate.

Ba salt: $[\alpha]_D$ +92° (H_2O).

Di-K salt: [19046-60-7]. $[\alpha]_D^{22}$ +100° (c, 1.57 in H_2O).

Dicyclohexylammonium salt: Mp 147-153°. $[\alpha]_D^{26}$ +78.5° (H_2O).

β-D-Pyranose-form [2520-52-7]

Ba salt: $[\alpha]_D$ +31.2° (H_2O).

Dicyclohexylammonium salt: Mp 145-151°. $[\alpha]_D^{26}$ +21.0° (H_2O).

Biochem. Prep., 1955, **4**, 1 (*synth*)
Putman, E.W. *et al*, *J. Am. Chem. Soc.*, 1957, **79**, 5057.
MacDonald, D.L., *The Carbohydrates*, Academic Press, 1972, 2nd Ed., **1A**, 253 (*rev*)
Lee, C. *et al*, *Biochemistry*, 1976, **15**, 697 (*conformn, pmr*)

Galactose 3-dihydrogen phosphate — G-00002

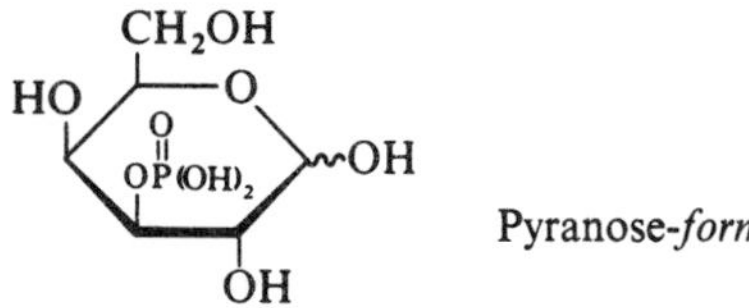

Pyranose-*form*

$C_6H_{13}O_9P$ M 260.137

D-form

Ba salt: [21063-51-4]. $[\alpha]_D^{22}$ +43.3° (c, 0.46 in H_2O).

Di-K salt: $[\alpha]_D^{18}$ +25.2° (c, 0.81 in H_2O).

α-1,2-O-Isopropylidene, 4,6-O-ethylidene: $C_{11}H_{19}O_9P$ M 326.239

$[\alpha]_D^{18}$ −41.5° (c, 0.77 in H_2O).

α-1,2-O-Isopropylidene, 4,6-O-ethylidene, Ba salt: $[\alpha]_D^{20}$ −24° (c, 1.4 in H_2O).

Foster, A.B. *et al*, *J. Chem. Soc.*, 1951, 980.
Chittenden, G.J.F. *et al*, *Biochem. J.*, 1968, **109**, 597(synth).
MacDonald, D.L., *The Carbohydrates*, Academic Press, 1972, 2nd Ed., **1A**, 253 (*rev*)

Galactose 6-dihydrogen phosphate, 9CI, 8CI — G-00003

Pyranose-*form*

$C_6H_{13}O_9P$ M 260.137

D-form

$[\alpha]_D^{20}$ +36.5° (c, 0.6 in H_2O).

Ba salt: $[\alpha]_D^{25}$ +24.5° (H_2O).

α-D-pyranose-form

1,2:3,4-Di-O-isopropylidene: 1,2:3,4-Di-O-isopropylidene-α-D-galactopyranose 6-dihydrogen phosphate. $C_{12}H_{21}O_9P$ M 340.266

$[\alpha]_D^{20}$ −17.9° (c, 1.6 in H_2O).

Foster, A.B. *et al*, *J. Chem. Soc.*, 1951, 980 (*synth*)
Todd, A.R. *et al*, *CA*, 1953, **47**, 6436 (*synth*)
MacDonald, D.L., *The Carbohydrates*, Academic Press, 1972, 2nd Ed., **1A**, 253 (*rev*)
Costello, A.J.R. *et al*, *Carbohydr. Res.*, 1975, **42**, 23(nmr).

Glucose 2-dihydrogen phosphate — G-00004

Pyranose-*form*

$C_6H_{13}O_9P$ M 260.137

D-form

Di-K salt: $[\alpha]_D$ +15° (H_2O).

Farrar, K.R., *J. Chem. Soc.*, 1949, 3131.
MacDonald, D.L., *The Carbohydrates*, Academic Press, 1972, 2nd Ed., **1A**, 253 (*rev*)

Glucose 3-dihydrogen phosphate, 9CI, 8CI — G-00005

Pyranose-*form*

$C_6H_{13}O_9P$ M 260.137

D-form [20701-41-1]

$[\alpha]_D$ +39.5° (H_2O). pK_{a1} 0.84, pK_{a2} 5.67.

Ba salt: $[\alpha]_D$ +26.5° (H_2O).

Brucine salt: $[\alpha]_D$ −14.5° (Py aq.).

Levene, P.A. *et al*, *J. Biol. Chem.*, 1930, **89**, 479 (*synth*)
MacDonald, D.L., *The Carbohydrates*, Academic Press, 1972, 2nd Ed., **1A**, 253 (*rev*)

Tabata, S. *et al*, *CA*, 1972, **76**, 26622y.

Glucose 4-dihydrogen phosphate G-00006

$C_6H_{13}O_9P$ M 260.137

D-form

Brucine salt: $[\alpha]_D$ −45.3° (Py).

Raymond, A.L., *J. Biol. Chem.*, 1936, **113**, 375.
MacDonald, D.L., *The Carbohydrates*, Academic Press, 1972, 2nd Ed., **1A**, 253 (*rev*)

Glucose 6-dihydrogen phosphate, 9CI, 8CI G-00007

Pyranose-*form*

$C_6H_{13}O_9P$ M 260.137

D-form [56-73-5]

Robison ester

Constit. of resting muscle and of the crude mixt. of hexose phosphates obt. by yeast fermentation. $[\alpha]_D$ +35.1° (H_2O). pK_{a1} 0.94, pK_{a2} 6.11.

Ba salt: [58823-95-3]. $[\alpha]_D^{24}$ +17.9° (H_2O).

Di-K salt: [5996-17-8]. $[\alpha]_D^{24}$ +21.2° (c, 1.3 in H_2O).

Robison, R. *et al*, *Biochem. J.*, 1931, **25**, 323 (*isol*)
Biochem. Prep., 1952, **2**, 39 (*synth*)
MacDonald, D.L., *The Carbohydrates*, Academic Press, 1972, 2nd Ed., **1A**, 253 (*rev*)
Harvey, D.J. *et al*, *J. Chromatogr.*, 1973, **76**, 51 (*glc, ms*)
Lis, T., *Carbohydr. Res.*, 1985, **135**, 187 (*cryst struct*)

Glycerol 1,2-didodecanoate 3-phosphocholine G-00008

$CH_2OOC(CH_2)_{10}CH_3$
$CHOOC(CH_2)_{10}CH_3$
$CH_2OP(O)(O^{\ominus})OCH_2CH_2\overset{\oplus}{N}Me_3$

$C_{32}H_{64}NO_8P$ M 621.833

(*R*)-*L-form* [18194-25-7]

Dilauroylphosphatidylcholine

$[\alpha]_D^{26}$ +6.1° (c, 1.5 in 1:1 $CHCl_3$/MeOH).

Baer, E. *et al*, *Biochemistry*, 1962, **1**, 518 (*synth*)
Brandt, A.E. *et al*, *Biochim. Biophys. Acta*, 1967, **144**, 605 (*synth*)

Glycerol 1,2-didodecanoate 3-phosphoethanolamine G-00009

Dodecanoic acid 1-[[[(2-aminoethoxy)-hydroxyphosphinyl]oxy]methyl]-1,2-ethanediyl ester, 9CI. Dilauroylphosphatidylethanolamine

$CH_2OOC(CH_2)_{10}CH_3$
$CHOOC(CH_2)_{10}CH_3$
$CH_2OP(O)(OH)OCH_2CH_2NH_2$

$C_{29}H_{58}NO_8P$ M 579.753

(±)-*form* [42436-56-6]

Mp 208°.

Bevan, T.H. *et al*, *J. Chem. Soc.*, 1951, 2667.
Chapman, D. *et al*, *Trans. Faraday Soc.*, 1966, **62**, 2607 (*pmr, props*)
v. Dijck, P.W.M., *Biochim. Biophys. Acta*, 1976, **455**, 576.
Wood, G.W. *et al*, *Chem. Phys. Lipids*, 1977, **18**, 316 (*ms*)
Cullis, P.R. *et al*, *Biochim. Biophys. Acta*, 1978, **513**, 31 (*nmr*)
Blume, A., *Biochim. Biophys. Acta*, 1979, **557**, 32.

Glycerol 1,2-diheptadecanoate 3-phosphocholine G-00010

$CH_2OOC(CH_2)_{15}CH_3$
$CHOOC(CH_2)_{15}CH_3$
$CH_2OP(O)(O^{\ominus})OCH_2CH_2\overset{\oplus}{N}Me_3$

$C_{42}H_{84}NO_8P$ M 762.101

(*R*)-*L-form* [70897-27-7]

Diheptadecanoylphosphatidylcholine

Sylvius, J.R. *et al*, *Biochim. Biophys. Acta*, 1979, **555**, 175.

Glycerol 1,2-dihexadecanoate 3-phosphate G-00011

Hexadecanoic acid 1-[(phosphonoxy)methyl]-1,2-ethanediyl ester, 9CI. Dipalmitoylphosphatidic acid

[19698-29-4]

$C_{35}H_{69}O_8P$ M 648.899

(*R*)-*form* [7091-44-3]

1,2-Dihexadecanoyl-sn-glycerol 3-phosphate

Mp 70-71°. $[\alpha]_D^{26}$ +4.0° (c, 9.6 in $CHCl_3$).

Di-Me ester: Mp 42.5-43.5°. $[\alpha]_D$ +2.0° (c, 7.6 in $CHCl_3$).

Diphenyl ester: Mp 47-48°. $[\alpha]_D$ +2.4° (c, 2.4 in $CHCl_3$).

(±)-*form* [5129-68-0]

Di-Me ester: Mp 46°.

Dibenzyl ester: Mp 50-50.5°.

Monophenyl ester: Mp 54.5-55.5°.

Baer, E., *J. Biol. Chem.*, 1951, **189**, 235.
Hessel, L.W. *et al*, *Recl. Trav. Chim. Pays-Bas*, 1954, **73**, 150.
Baer, E. *et al*, *J. Biol. Chem.*, 1955, **212**, 39.
Baer, E., *Prog. Chem. Fats Other Lipids*, 1963, **6**, 33 (*rev*)
Aneja, R. *et al*, *Chem. Phys. Lipids*, 1974, **12**, 39 (*pmr*)
Avenado, R. *et al*, *CA*, 1975, **83**, 113590.

Glycerol 1,2-dihexadecanoate 3-phosphocholine **G-00012**

Dipalmitoylphosphatidylcholine. 1,2-Dihexadecanoylglycero-3-phosphocholine

$$\begin{array}{l} CH_2OOC(CH_2)_{14}CH_3 \\ | \\ CHOOC(CH_2)_{14}CH_3 \\ |\quad\ O \\ |\quad\ \| \\ CH_2OPOCH_2CH_2\overset{\oplus}{N}Me_3 \\ \quad\ \ | \\ \quad\ \ O^{\ominus} \end{array}$$

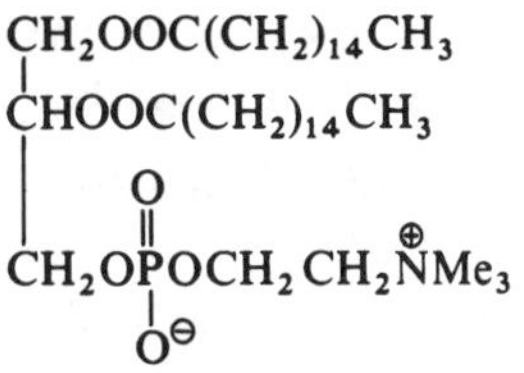

$C_{40}H_{80}NO_8P$ M 734.048

(R)-L-form [63-89-8]

Mp 235-236°. $[\alpha]^{25}_{D}$ +7.0° (c, 5.6 in $CHCl_3$).

Baer, E. *et al*, *J. Am. Chem. Soc.*, 1952, **74**, 158 (*synth*)
Brandt, A.E. *et al*, *Biochim. Biophys. Acta*, 1967, **144**, 605 (*synth*)
Klein, R.A., *J. Lipid Res.*, 1971, **12**, 123 (*ms*)
Ghosh, D. *et al*, *Biochim. Biophys. Acta*, 1972, **266**, 41.
Gordon, D.T. *et al*, *Lipids*, 1972, **7**, 261 (*synth*)
Bunow, M.R. *et al*, *Biochim. Biophys. Acta*, 1977, **489**, 191 (*raman*)

Glycerol 1,2-dihexadecanoate 3-phosphoethanolamine **G-00013**

Hexadecanoic acid 1-[[[(2-aminoethoxy)hydroxyphosphinyl]oxy]methyl]-1,2-ethanediyl ester, 9CI. Dipalmitoylphosphatidylethanolamine. 1,2-Dihexadecanoylglycero-3-phosphoethanolamine

[3026-45-7]

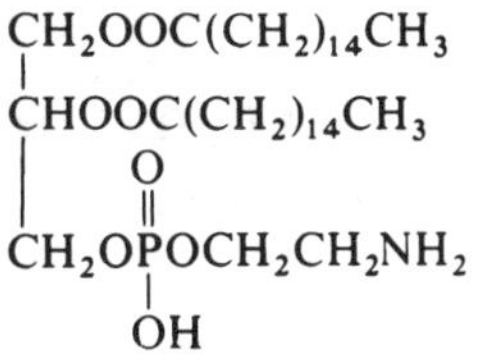

$$\begin{array}{l} CH_2OOC(CH_2)_{14}CH_3 \\ | \\ CHOOC(CH_2)_{14}CH_3 \\ |\quad\ O \\ |\quad\ \| \\ CH_2OPOCH_2CH_2NH_2 \\ \quad\ \ | \\ \quad\ \ OH \end{array}$$

$C_{37}H_{74}NO_8P$ M 691.967

(R)-form [923-61-5]

Mp 185-186°, 210-211°. $[\alpha]_D$ +6.55°.

(±)-form [5681-36-7]

Mp 196°.

v. Veen, A. *et al*, *Recl. Trav. Chim. Pays-Bas*, 1960, **79**, 1085 (*synth*)
Chapman, D. *et al*, *Trans. Faraday Soc.*, 1966, **62**, 2607 (*pmr, props*)
Billimoria, J.D. *et al*, *J. Chem. Soc. (C)*, 1968, 1404 (*synth*)
Billimoria, J.D. *et al*, *Chem. Phys. Lipids*, 1974, **12**, 327 (*synth*)
Wood, G.W. *et al*, *Chem. Phys. Lipids*, 1977, **18**, 316 (*ms*)
Mendelsohn, R., *Biochemistry*, 1978, 3944 (*raman*)
Skarjune, R. *et al*, *Biochemistry*, 1979, **18**, 5903 (*conformn*)

Glycerol 1,2-dihexadecanoate 3-phospho-1′-glycerol **G-00014**

Hexadecanoic acid 1-[[[(2,3-dihydroxypropoxy)hydroxyphosphinyl]oxy]methyl]-1,2-ethanediyl ester, 9CI. Dipalmitoylphosphatidylglycerol

[4537-77-3]

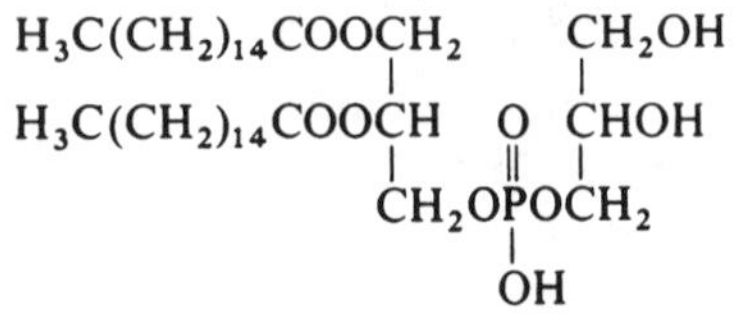

$$\begin{array}{lll} H_3C(CH_2)_{14}COOCH_2 & & CH_2OH \\ \qquad\qquad\quad | & & | \\ H_3C(CH_2)_{14}COOCH & O & CHOH \\ \qquad\qquad\quad | & \| & | \\ \qquad\qquad CH_2O & P & OCH_2 \\ & | & \\ & OH & \end{array}$$

$C_{38}H_{75}O_{10}P$ M 722.978

Lammers, J.G. *et al*, *Recl. Trav. Chim. Pays-Bas*, 1979, **98**, 243 (*synth, cmr*)

Glycerol 1,2-dihexadecanoate 3-phosphoinositol **G-00015**

Dipalmitoylphosphatidylinositol

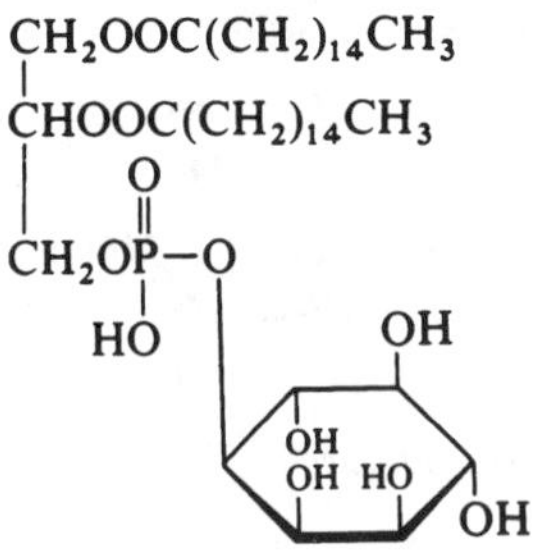

$$\begin{array}{l} CH_2OOC(CH_2)_{14}CH_3 \\ | \\ CHOOC(CH_2)_{14}CH_3 \\ |\quad\ O \\ |\quad\ \| \\ CH_2OP{-}O{-}\text{inositol (OH)}_5 \\ \quad\ \ | \\ \quad\ \ HO \end{array}$$

$C_{41}H_{79}O_{13}P$ M 811.041

Myoinositol-form

myo-*Inositol mono[[2,3-bis(1-oxohexadecyl)oxy]propyl] 1-(dihydrogen phosphate) ester*, 9CI

NH_4 salt: Mp 169-172°. $[\alpha]^{20}_{D}$ +7.48° (c, 0.3 in $CHCl_3$).

Penta-O-benzyl: Mp 53-54°.

Klyashehitskii, B.A. *et al*, *Tetrahedron Lett.*, 1970, 587 (*synth*)
Zhelvakova, E.G., *CA*, 1970, **72**, 133119 (*synth*)
Lyutik, A.Y. *et al*, *CA*, 1975, **82**, 86519 (*synth*)
Sadovnikova, M.S. *et al*, *CA*, 1975, **83**, 131855 (*synth*)
Sukhanov, V.A. *et al*, *CA*, 1978, **88**, 62556 (*synth*)

Glycerol 1,2-dihexadecanoate 3-phosphoserine **G-00016**

Serine 2,3-bis[(1-oxohexadecyl)oxy]propyl hydrogen phosphate ester, 9CI. Dipalmitoylphosphatidylserine

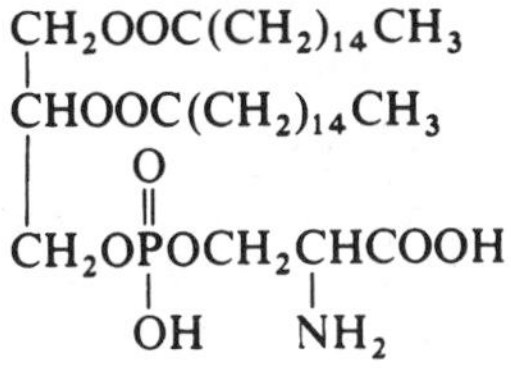

$$\begin{array}{l} CH_2OOC(CH_2)_{14}CH_3 \\ | \\ CHOOC(CH_2)_{14}CH_3 \\ |\quad\ O \\ |\quad\ \| \\ CH_2OPOCH_2CHCOOH \\ \quad\ \ |\qquad\quad | \\ \quad\ \ OH\qquad NH_2 \end{array}$$

$C_{38}H_{74}NO_{10}P$ M 735.977

(R)-L-form [40290-42-4]

*1,2-Dipalmitoyl-*sn*-glycero-L-serine*

Mp 155-158°. $[\alpha]_D$ +5.5° ($CHCl_3$, $MeOH/H_2O$).

Browning, J. *et al*, *Chem. Phys. Lipids*, 1979, **24**, 103 (*pmr*)
Browning, J. *et al*, *Biochemistry*, 1980, **19**, 1262 (*nmr*)

Glycerol 1,2-di-5,8,11,14-icosatetraenoate 3-phosphocholine **G-00017**

1,2-Diicosa-5,8,11,14-tetraenoylglycero-3-phosphocholine

$$\begin{array}{l} CH_2OOC(CH_2)_3(CH{=}CHCH_2)_4(CH_2)_3CH_3 \\ | \\ CHOOC(CH_2)_3(CH{=}CHCH_2)_4(CH_2)_3CH_3 \\ |\quad\ O \\ |\quad\ \| \\ CH_2OPOCH_2CH_2\overset{\oplus}{N}Me_3 \\ \quad\ \ | \\ \quad\ \ O^{\ominus} \end{array}$$

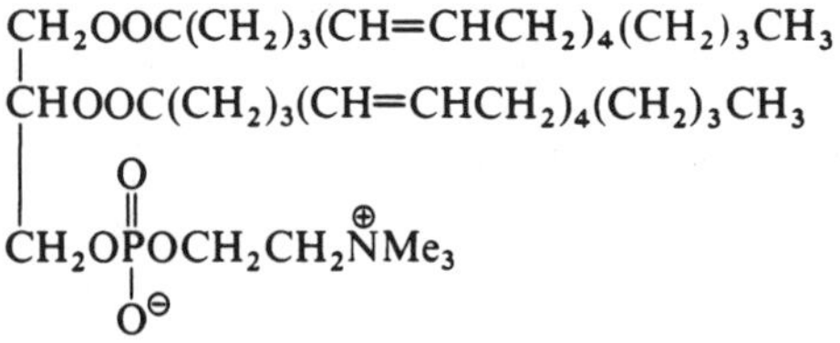

$C_{48}H_{80}NO_8P$ M 830.136

Incorrectly named in 9CI.

(R)-(all-Z)-form [56782-48-0]

Diarachidonylphosphatidylcholine

Wood, G.W., *Chem. Phys. Lipids*, 1977, **18**, 316 (*ms*)

Glycerol 1,2-di-5,8,11,14-icosatetraenoate 3-phosphoethanolamine G-00018

1,2-Diicosa-5,8,11,14-tetraenoylglycero-3-phosphoethanolamine

[14994-07-1]

$CH_2OOC(CH_2)_3(CH{=}CHCH_2)_4(CH_2)_3CH_3$

$CHOOC(CH_2)_3(CH{=}CHCH_2)_4(CH_2)_3CH_3$

$CH_2OP(O)(OH)OCH_2CH_2NH_2$

$C_{45}H_{74}NO_8P$ M 788.055

(R)-(all-Z)-form

Diarachidonoylphosphatidylethanolamine

Renin inhibitor. $[\alpha]^{20}_{D}$ +6.0°.

Bova, L.M. *et al*, *CA*, 1966, **64**, 3343 (*synth, ir*)

Miyazaki, M. *et al*, *Proc. Soc. Exp. Biol. Med.*, 1977, **155**, 468.

Glycerol 1,2-di-11-methyldodecanoate 3-phosphocholine G-00019

$CH_2OOC(CH_2)_9CH(CH_3)_2$

$CHOOC(CH_2)_9CH(CH_3)_2$

$CH_2OP(O)(O^{\ominus})OCH_2CH_2\overset{\oplus}{N}Me_3$

$C_{34}H_{68}NO_8P$ M 649.887

(R)-L-form [71368-25-7]

1,2-Diisotridecanoylphosphatidylcholine

Sylvius, J.R. *et al*, *Chem. Phys. Lipids*, 1979, **24**, 287 (*synth*)

Glycerol 1,2-di-16-methylheptadecanoate 3-phosphocholine G-00020

$CH_2OOC(CH_2)_{14}CH(CH_3)_2$

$CHOOC(CH_2)_{14}CH(CH_3)_2$

$CH_2OP(O)(O^{\ominus})OCH_2CH_2\overset{\oplus}{N}Me_3$

$C_{44}H_{88}NO_8P$ M 790.155

(R)-L-form [60683-79-6]

1,2-Diisooctadecanoylphosphatidylcholine

Sylvius, J.R. *et al*, *Chem. Phys. Lipids*, 1979, **24**, 287.

Glycerol 1,2-di-15-methylhexadecanoate 3-phosphocholine G-00021

$CH_2OOC(CH_2)_{13}CH(CH_3)_2$

$CHOOC(CH_2)_{13}CH(CH_3)_2$

$CH_2OP(O)(O^{\ominus})OCH_2CH_2\overset{\oplus}{N}Me_3$

$C_{42}H_{84}NO_8P$ M 762.101

(R)-L-form [71368-21-3]

Diisoheptadecanoylphosphatidylcholine

Sylvius, J.R. *et al*, *Chem. Phys. Lipids*, 1979, **24**, 287.

Glycerol 1,2-di-14-methylpentadecanoate 3-phosphocholine G-00022

$CH_2OOC(CH_2)_{12}CH(CH_3)_2$

$CHOOC(CH_2)_{12}CH(CH_3)_2$

$CH_2OP(O)(O^{\ominus})OCH_2CH_2\overset{\oplus}{N}Me_3$

$C_{40}H_{80}NO_8P$ M 734.048

(R)-L-form [71368-22-4]

Diisohexadecanoylphosphatidylcholine

Sylvius, J.R. *et al*, *Chem. Phys. Lipids*, 1979, **24**, 287.

Glycerol 1,2-di-13-methyltetradecanoate 3-phosphocholine G-00023

$CH_2OOC(CH_2)_{11}CH(CH_3)_2$

$CHOOC(CH_2)_{11}CH(CH_3)_3$

$CH_2OP(O)(O^{\ominus})OCH_2CH_2\overset{\oplus}{N}Me_3$

$C_{38}H_{76}NO_8P$ M 705.994

(R)-L-form [71368-23-5]

1,2-Diisopentadecanoylphosphatidylcholine

Sylvius, J.R. *et al*, *Chem. Phys. Lipids*, 1979, **24**, 287.

Glycerol 1,2-di-12-methyltridecanoate 3-phosphocholine G-00024

$CH_2OOC(CH_2)_{10}CH(CH_3)_2$

$CHOOC(CH_2)_{10}CH(CH_3)_2$

$CH_2OP(O)(O^{\ominus})OCH_2CH_2\overset{\oplus}{N}Me_3$

$C_{36}H_{72}NO_8P$ M 677.940

(R)-L-form [71368-24-6]

1,2-Diisotetradecanoylphosphatidylcholine

Sylvius, J.R. *et al*, *Chem. Phys. Lipids*, 1979, **24**, 287 (*synth*)

Glycerol 1,2-di-10-methylundecanoate 3-phosphocholine G-00025

$CH_2OOC(CH_2)_8CH(CH_3)_2$

$CHOOC(CH_2)_8CH(CH_3)_2$

$CH_2OP(O)(O^{\ominus})OCH_2CH_2\overset{\oplus}{N}Me_3$

$C_{32}H_{64}NO_8P$ M 621.833

(R)-L-form [71368-26-8]

Diisododecanoylphosphatidylcholine

Sylvius, J.R. *et al*, *Chem. Phys. Lipids*, 1979, **24**, 287 (*synth*)

Glycerol 1,2-dinonadecanoate 3-phosphocholine G-00026

$CH_2OOC(CH_2)_{17}CH_3$
$CHOOC(CH_2)_{17}CH_3$
$CH_2OP(=O)(O^{\ominus})OCH_2CH_2\overset{\oplus}{N}Me_3$

$C_{46}H_{92}NO_8P$ M 818.208

***(R)*-L-form**

Dinonadecanoylphosphatidylcholine

Sylvius, J.R. *et al*, *Biochim. Biophys. Acta*, 1979, **555**, 175.

Glycerol 1,2-di-9,12-octadecadienoate 3-phosphate G-00027

$C_{39}H_{69}O_8P$ M 696.943

Shvets, V.I. *et al*, *CA*, 1966, **65**, 20000 (*synth*)

Glycerol 1,2-di-9,12-octadecadienoate 3-phosphocholine G-00028

1,2-Dioctadeca-9,12-dienoylglycero-3-phospohocholine

$C_{44}H_{80}NO_8P$ M 782.092

***(R)*-L-*(all-Z)*-form** [998-06-1]

Dilinoleoylphosphatidylcholine

Gelatinous. Mp 232°. $[\alpha]^{20}_{D}$ +6.02°.

Demel, R.A. *et al*, *Biochim. Biophys. Acta*, 1972, **266**, 26.
Gordon, D.T. *et al*, *Lipids*, 1972, **7**, 261 (*synth*)
Warner, T.G. *et al*, *J. Lipid Res.*, 1977, **18**, 548 (*synth*)
Wood, G.W. *et al*, *Chem. Phys. Lipids*, 1977, **18**, 316 (*ms*)
Lammers, J.G. *et al*, *Chem. Phys. Lipids*, 1978, **22**, 293 (*nmr, ir*)

Glycerol 1,2-di-9,12-octadecadienoate 3-phosphoethanolamine G-00029

9,12-Octadecadienoic acid 1-[[[(2-aminoethoxy)-hydroxyphosphinyl]oxy]methyl]-1,2-ethanediyl ester, 9CI. 1,2-Di-9,12-octodecadienoylphosphatidylcholine

$C_{41}H_{74}NO_8P$ M 740.011

***(R)*-*(all-Z)*-form** [20707-71-5]

Dilinoleoylphosphatidylethanolamine

Mp 25°. $[\alpha]^{25}_{D}$ +6.2° (c, 1 in $CHCl_3$).

(±)-*(all-Z)*-form

Waxy solid.

Shvets, V.I. *et al*, *CA*, 1962, **56**, 4611 (*synth*)
Pfeiffer, F.R. *et al*, *J. Org. Chem.*, 1969, **34**, 2795 (*synth*)
Billimoria, J.D. *et al*, *Chem. Phys. Lipids*, 1974, **12**, 327 (*synth*)

Glycerol 1,2-di-9,12-octadecadienoate 3-phosphoinositol G-00030

Dilinoloylphosphatidylinositol

$CH_2OOC(CH_2)_7(CH{=}CHCH_2)_2(CH_2)_3CH_3$
$CHOOC(CH_2)_7(CH{=}CHCH_2)_2(CH_2)_3CH_3$
$CH_2OP(=O)(OH)-O-$inositol (OH, OH, HO, OH, OH)

$C_{45}H_{79}O_{13}P$ M 859.085

Shvets, V.I. *et al*, *CA*, 1966, **64**, 14261 (*synth*)
Petrova, M.K. *et al*, *CA*, 1966, **65**, 20207 (*synth*)

Glycerol 1,2-di-9,12-octadecadienoate 3-phosphoserine G-00031

$CH_2OOC(CH_2)_{16}CH_3$
$CHOOC(CH_2)_{16}CH_3$
$CH_2OP(=O)(HO)-O-$inositol (OH, OH, OH, HO, OH)

$C_{42}H_{74}NO_{10}P$ M 784.021

(±)-*(all-Z)*-form

Dilinoleoylphosphatidylserine

Turner, D.L. *et al*, *Lipids*, 1966, **1**, 439 (*synth*)

Glycerol 1,2-dioctadecanoate 3-phosphate G-00032

Octadecanoic acid 1-[(phosphonoxy)methyl]-1,2-ethanediyl ester, 9CI. Distearoylphosphatidic acid

$C_{39}H_{77}O_8P$ M 705.006

***(R)*-form** [17966-16-4]

*1,2-Dioctadecanoyl-*sn*-glycerol 3-phosphate*

Mp 75.5-76.5°. $[\alpha]^{26}_{D}$ +3.8° (c, 9.3 in $CHCl_3$).

Di-Me ester: Mp 52-53°. $[\alpha]_D$ +1.9° (c, 9.9 in $CHCl_3$).

Diphenyl ester: Mp 54.5-55°. $[\alpha]_D$ +2.0° (c, 10.2 in $CHCl_3$).

(±)-form [13563-93-4]

Mp 70-71°.

Phenyl ester: Mp 58-59°.

Dibenzyl ester: Mp 56-57°.

Baer, E., *J. Biol. Chem.*, 1951, **189**, 235.
Hessel, L.W. *et al*, *Recl.Trav. Chim. Pays-Bas*, 1954, **73**, 150.
Baer, E. *et al*, *J. Biol. Chem.*, 1955, **212**, 39.
Baer, E. *et al*, *Arch. Biochem. Biophys.*, 1958, **78**, 294.
Baer, E., *Prog. Chem. Fats Other Lipids*, 1963, **6**, 33 (*rev*)
Avenado, R. *et al*, *CA*, 1975, **83**, 113590.

Glycerol 1,2-dioctadecanoate 3-phospho-2-amino-1-propanol G-00033

Distearoyl-α-glycerylphosphoryl-2-amino-1-propanol. Dioctadecanoyl-α-glycerylphosphoryl-2-amino-1-propanol

$H_3C(CH_2)_{16}COOCH_2$
$H_3C(CH_2)_{16}COO{\blacktriangleright}C{\blacktriangleleft}H$
$CH_2OP(=O)(OH)OCH_2{\blacktriangleright}C(NH_2)(H){\blacktriangleleft}CH_3$

$C_{42}H_{84}NO_8P$ M 762.101

***(R)*-L-form**

Mp 182-183°. $[\alpha]^{23}_{D}$ +7° (c, 2.7 in $CHCl_3$).

Baer, E. *et al*, *J. Biol. Chem.*, 1963, **238**, 3591.

Glycerol 1,2-dioctadecanoate 3-phosphocholine G-00034

Distearoylphosphatidylcholine. 1,2-Dioctadecanoylglycero-3-phosphocholine

[4539-70-2]

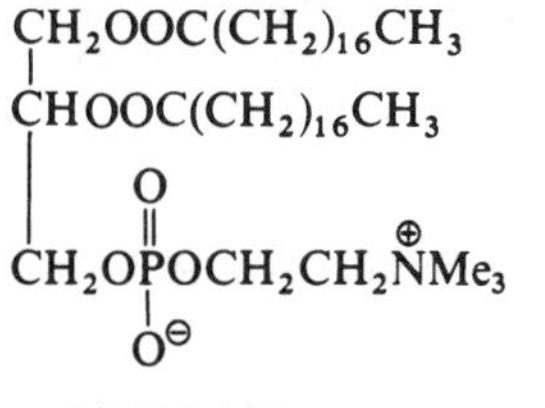

$$\begin{array}{l} CH_2OOC(CH_2)_{16}CH_3 \\ | \\ CHOOC(CH_2)_{16}CH_3 \\ | \quad O \\ | \quad \| \qquad\qquad \oplus \\ CH_2OPOCH_2CH_2NMe_3 \\ \quad\;\; | \\ \quad\;\; O^{\ominus} \end{array}$$

$C_{44}H_{88}NO_8P$ M 790.155

(R)-form [816-94-4]

Mp 231-232°. $[\alpha]_D^{29}$ +6.2° (c, 10 in $CHCl_3$/MeOH). $[\alpha]_D^{20}$ +7.1° (c, 2.2 in $CHCl_3$/MeOH).

Baer, E. *et al*, *Can. J. Biochem. Physiol.*, 1959, **37**, 453 (*synth*)
Chapman, D. *et al*, *J. Biol. Chem.*, 1966, 241, 5044 (*nmr*)
Aneja, R. *et al*, *Biochim. Biophys. Acta*, 1971, **248**, 455 (*synth*)
Ghosh, D. *et al*, *Biochim. Biophys. Acta*, 1972, **266**, 41.
Klein, R.A., *J. Lipid Res.*, 1972, **13**, 672 (*ms*)
Stoffel, W. *et al*, *Hoppe-Seyler's Z. Physiol. Chem.*, 1972, **253**, 1962 (*cmr*)
Wood, G.W., *Chem. Phys. Lipids*, 1977, **18**, 316 (*ms*)

Glycerol 1,2-dioctadecanoate 3-phospho-1′-glycerol G-00035

Octadecanoic acid 1-[[[(2,3-dihydroxypropoxy)-hydroxyphosphinyl]oxy]methyl]-1,2-ethanediyl ester, 9CI. Distearoylphosphatidylglycerol

[4537-78-4]

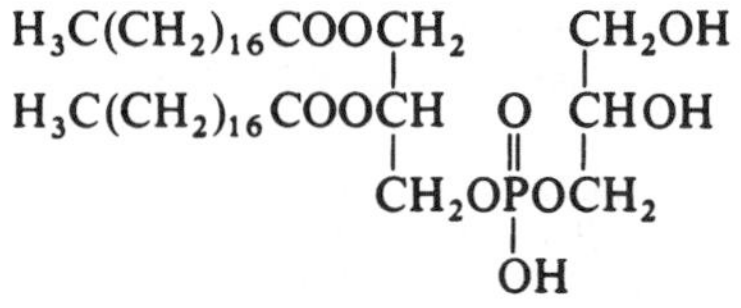

$$\begin{array}{l} H_3C(CH_2)_{16}COOCH_2 \qquad CH_2OH \\ H_3C(CH_2)_{16}COOCH \quad O \quad CHOH \\ \qquad\qquad\qquad\quad | \quad\; \| \quad\; | \\ \qquad\qquad\qquad\quad CH_2OPOCH_2 \\ \qquad\qquad\qquad\qquad\quad\;\; | \\ \qquad\qquad\qquad\qquad\quad\; OH \end{array}$$

$C_{42}H_{83}O_{10}P$ M 779.085

Mp 66.5-67°. $[\alpha]_D^{22}$ +2.0° (c, 10 in $CHCl_3$).

Ba salt: Mp 166°. $[\alpha]_D^{25}$ +9.2° (c, 1. in Py).

Baer, E. *et al*, *J. Biol. Chem.*, 1958, **232**, 895 (*synth*)
Baer, E., *Prog. Chem. Fats Other Lipids*, 1963, **6**, 33 (*rev*)
Saunders, R.M. *et al*, *J. Am. Chem. Soc.*, 1966, **88**, 3844 (*synth*)
Lammers, J.G. *et al*, *Recl. Trav. Chim. Pays-Bas*, 1977, **96**, 216 (*synth*)
Lammers, J.G. *et al*, *Recl. Trav. Chim. Pays-Bas*, 1979, **98**, 243 (*synth, cmr*)

Glycerol 1,2-dioctadecanoate 3-phosphoinositol G-00036

Distearoylphosphatidylinositol

$C_{45}H_{87}O_{13}P$ M 867.148

Myoinositol-form

Mp 135-137°. Other stereoisomers have been synthesised.

Penta-Ac: Mp 119-122°.

Davies, J.M. *et al*, *Chem. Ind.* (*London*), 1959, 1155 (*synth*)
Shevchenko, V.P. *et al*, *CA*, 1977, **86**, 190385 (*synth*)
Krylova, V.N. *et al*, *CA*, 1979, **91**, 19354 (*synth*)

Glycerol 1,2-dioctadecanoate 3-phosphonoethanolamine G-00037

Octadecanoic acid 1-[[[(2-aminoethoxy)-hydroxyphosphinyl]oxy]methyl]-1,2-ethanediyl ester, 9CI. Distearoylphosphatidylethanolamine

[4537-76-2]

$C_{41}H_{82}NO_8P$ M 748.074

(R)-form

Cryst. Mp 185°, 196°. $[\alpha]_D$ +6.2° (c, 3.4 in 9:1 $CHCl_3$/AcOH).

(±)-form

Cryst. Mp 180-181°, 193°.

Hoefnagel, M.A. *et al*, *Recl. Trav. Chim. Pays-Bas*, 1960, **79**, 330 (*synth*)
Chapman, D. *et al*, *Proc. R. Soc. London., Ser. A*, 1966, **290**, 1151 (*ir, pmr, props*)
Chapman, D. *et al*, *Trans. Faraday Soc.*, 1966, **62**, 2607 (*pmr, props*)
Aneja, R. *et al*, *Biochim. Biophys. Acta*, 1969, **187**, 439; 1970, **218**, 102 (*synth*)
Wood, G.W. *et al*, *Chem. Phys. Lipids*, 1977, **18**, 316 (*ms*)
Lammers, J.G. *et al*, *Chem. Phys. Lipids*, 1978, **22**, 293 (*ir, pmr*)

Glycerol 1,2-dioctadecanoate 3-phosphoserine G-00038

Serine 2,3-bis[(1-oxooctadecyl)oxy]propyl hydrogen phosphate ester, 9CI. Distearoylphosphatidylserine

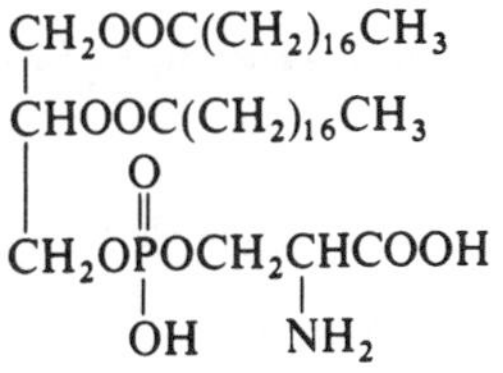

$$\begin{array}{l} CH_2OOC(CH_2)_{16}CH_3 \\ | \\ CHOOC(CH_2)_{16}CH_3 \\ | \quad O \\ | \quad \| \\ CH_2OPOCH_2CHCOOH \\ \quad\;\; | \qquad\quad | \\ \quad\; OH \qquad NH_2 \end{array}$$

$C_{42}H_{82}NO_{10}P$ M 792.084

(R)-L-form [51446-62-9]

*1,2-Distearoyl-*sn*-glycero-3-phospo-L-serine*

Mp 159-161°.

(R)-DL-form

Mp 162°. $[\alpha]_D^{55}$ +7.4° (c, 0.6 in 9:1 $CHCl_3/Me_2CO$).

Baer, E. *et al*, *J. Biol. Chem.*, 1955, **212**, 25.
Baer, E., *Prog. Chem. Fats Other Lipids*, 1963, **6**, 33 (*rev*)
Aneja, R. *et al*, *Biochim. Biophys. Acta*, 1970, **218**, 102.

Glycerol 1,2-di-9,12,15-octadecatrienoate 3-phosphate G-00039

9,12,15-Octadecatrienoic acid 1-[(phosphonoxy)-methyl]-1,2-ethanediyl ester, 9CI

$C_{39}H_{65}O_8P$ M 692.911

(R)-(all-Z)-form

*1,2-Dilinolenoyl-*sn*-glycerol 3-phosphate*

$[\alpha]_D^{20}$ +31° (c, 3.5 in $CHCl_3$).

Rakhit, S. *et al*, *Can. J. Chem.*, 1969, **47**, 2906 (*synth*)

Glycerol 1,2-di-9,12,15-octadecatrienoate 3-phosphocholine G-00040

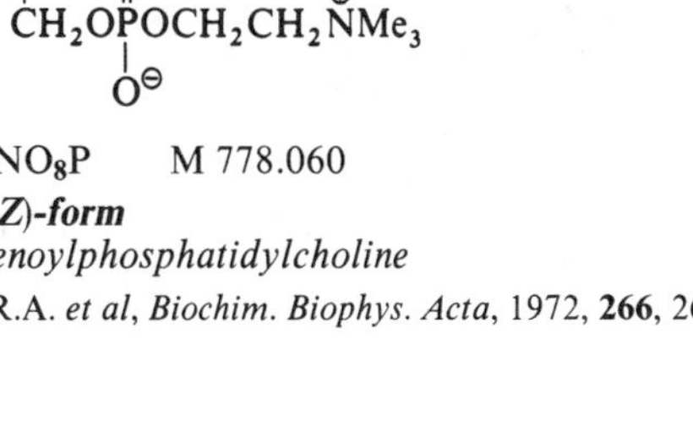

$$\begin{array}{l} CH_2OOC(CH_2)_7(CH{=}CHCH_2)_3CH_3 \\ | \\ CHOOC(CH_2)_7(CH{=}CHCH_2)_3CH_3 \\ | \quad O \\ | \quad \| \qquad\qquad \oplus \\ CH_2OPOCH_2CH_2NMe_3 \\ \quad\;\; | \\ \quad\;\; O^{\ominus} \end{array}$$

$C_{44}H_{76}NO_8P$ M 778.060

(R)-L-(all-Z)-form

Dilinolenoylphosphatidylcholine

Demel, R.A. *et al*, *Biochim. Biophys. Acta*, 1972, **266**, 26.

Warner, T.G. *et al*, *J. Lipid Res.*, 1977, **18**, 548 (*nmr*)
Wood, G.W. *et al*, *Chem. Phys. Lipids*, 1977, **18**, 316 (*ms*)

Glycerol 1,2-di-9,12,15-octadecatrienoate 3-phosphoethanolamine G-00041

9,12,15-Octadecatrienoic acid 1-[[[(2-aminoethoxy)-hydroxyphosphinyl]oxy]methyl]-1,2-ethanediyl ester, 9CI

$$\begin{array}{l} CH_2OOC(CH_2)_7(CH{=}CHCH_2)_3CH_3 \\ | \\ CHOOC(CH_2)_7(CH{=}CHCH_2)_3CH_3 \\ | \quad\ O \\ | \quad\ \| \\ CH_2OPOCH_2CH_2NH_2 \\ \qquad | \\ \qquad OH \end{array}$$

$C_{41}H_{70}NO_8P$ M 735.980

(*R*)-(*all-Z*)-*form* [34813-40-6]
Dilinolenoylphosphatidylethanolamine
$[\alpha]_D$ +6.1° (c, 2 in $CHCl_3$).

Shvets, V.I., *CA*, 1964, **60**, 1579 (*synth*)
Rakhit, S. *et al*, *Can. J. Chem.*, 1969, **47**, 2906 (*synth*)

Glycerol 1,2-di-9-octadecenoate 3-phosphocholine G-00042

1,2-Di-9-octadecenoylglycerol-3-phosphocholine

$$\begin{array}{l} CH_2OOC(CH_2)_7CH{=}CH(CH_2)_7CH_3 \\ | \\ CHOOC(CH_2)_7CH{=}CH(CH_2)_7CH_3 \\ | \quad\ O \\ | \quad\ \| \qquad\qquad\quad \oplus \\ CH_2OPOCH_2CH_2NMe_3 \\ \qquad | \\ \qquad O^{\ominus} \end{array}$$

$C_{44}H_{84}NO_8P$ M 786.123

(*R*)-(*Z,Z*)-*form* [4235-95-4]
Dioleoylphosphatidylcholine
$[\alpha]_D$ +6.1° (c, 10 in $CHCl_3$/MeOH).

Baer, E. *et al*, *Can. J. Biochem. Physiol.*, 1959, **37**, 953 (*synth*)
Stoffel, W. *et al*, *Hoppe-Seyler's Z. Physiol. Chem.*, 1969, **350**, 1385 (*props*)
Klein, R.A., *J. Lipid Res.*, 1971, **12**, 123, 628; 1972, **13**, 672 (*ms*)
Lippert, J., *Biochem. Biophys. Acta*, 1972, **282**, 8 (*raman*)
Barton, P.G. *et al*, *J. Biol. Chem.*, 1975, **250**, 4470 (*isomers*)
Wood, G.W. *et al*, *Chem. Phys. Lipids*, 1977, **18**, 316 (*ms*)
Seelig, J., *Biochemistry*, 1978, **17**, 3310 (*nmr*)

Glycerol 1,2-di-9-octadecenoate 3-phosphoethanolamine G-00043

9-Octadecenoic acid 1-[[[(2-aminoethoxy)-hydroxyphosphinyl]oxy]methyl]-1,2-ethanediyl ester, 9CI. 1,2-Di-9-octadecenoylphosphatidylethanolamine

$C_{41}H_{78}NO_8P$ M 744.043

(*R*)-(*all-E*)-*form*
Dielaidylphosphatidylethanolamine
Cryst. ($CHCl_3$/MeOH). Mp 193°. $[\alpha]_D^{22}$ +6.1° ($CHCl_3$).

(±)-(*all-E*)-*form* [16777-83-6]
Cryst. Mp 188°. Exhibits polymorphism.

(*R*)-(*all-Z*)-*form* [4004-05-1]
Dioleoylphosphatidylethanolamine
Cryst. (MeOH). Mp 195-200°. $[\alpha]_D^{20}$ +6.0° ($CHCl_3$).

(±)-(*all-Z*)-*form* [2462-63-7]
Exhibits polymorphism.

Chapman, D. *et al*, *Biochim. Biophys. Acta*, 1966, **120**, 148.
Chapman, D. *et al*, *Proc. R. Soc. London, Ser. A*, 1966, **290**, 1151 (*ir, nmr, props*)
Chapman, D. *et al*, *Trans. Faraday Soc.*, 1966, **62**, 607.
Billimoria, J.D. *et al*, *J. Chem. Soc. (C)*, 1968, 1404 (*synth*)
Stoffel, W. *et al*, *Hoppe-Seyler's Z. Physiol. Chem.*, 1969, **350**, 1385.
v. Dijck, P.W.M., *Biochim. Biophys. Acta*, 1976, **455**, 576.
Wood, G.W. *et al*, *Chem. Phys. Lipids*, 1977, **18**, 316 (*ms*)
Cullis, P.R. *et al*, *Biochim. Biophys. Acta*, 1978, **513**, 31 (*nmr*)
Lammers, J.G. *et al*, *Recl. Trav. Chim. Pays-Bas*, 1979, **98**, 243 (*synth*)

Glycerol 1,2-di-9-octadecenoate 3-phospho-1′-glycerol G-00044

$$\begin{array}{l} H_3C(CH_2)_7CH{=}CH(CH_2)_7COOCH_2 \qquad CH_2OH \\ H_3C(CH_2)_7CH{=}CH(CH_2)_7COOCH \quad O \quad CHOH \\ \qquad\qquad\qquad\qquad\qquad\qquad\quad CH_2OPOCH_2 \\ \qquad\qquad\qquad\qquad\qquad\qquad\qquad\quad OH \end{array}$$

$C_{42}H_{78}O_{10}P$ M 774.046

(*all-Z*)-*form*
*1,2-Dioleoyl-*sn*-glycerol-3-phospho-1′-*sn*-glycerol*
$[\alpha]_D^{22}$ +2.35° (c, 11 in EtOH), $[\alpha]_D^{21}$ +2.0° (c, 10 in $CHCl_3$).

Baer, E. *et al*, *J. Biol. Chem.*, 1958, **232**, 895 (*synth*)
Baer, E., *Prog. Chem. Fats Other Lipids*, 1963, **6**, 33 (*rev*)

Glycerol 1,2-di-9-octadecenoate 3-phosphoserine G-00045

$C_{42}H_{78}NO_{10}P$ M 788.053

(*R*)-D-*form*
*1,2-Dioleoyl-*sn*-glycero-3-phospho-D-serine*
$[\alpha]_D^{22}$ +21.3° (C_6H_6), +13.4° ($CHCl_3$).

(*R*)-L-*form*
*1,2-Dioleoyl-*sn*-glycero-3-phospho-L-serine*
$[\alpha]_D^{22}$ −7.8° ($CHCl_3$), +5.1° (65:25:4 $CHCl_3$/MeOH/H_2O), −17.8° (C_6H_6).
Ba salt: Mp 165-169°.

(*R*)-DL-*form*
*1,2-Dioleoyl-*sn*-glycero-3-phospho-DL-serine*
$[\alpha]_D^{22}$ +4.7° ($CHCl_3$), +7.0° (65:25:4 ($CHCl_3$/MeOH/H_2O), +4.9° (C_6H_6).

Browning, J. *et al*, *Chem. Phys. Lipids*, 1979, **24**, 103 (*pmr*)
Browning, J. *et al*, *Biochemistry*, 1980, **19**, 1262 (*cmr*)

Glycerol 1,2-dipentadecanoate 3-phosphocholine G-00046

$$\begin{array}{l} CH_2OOC(CH_2)_{13}CH_3 \\ | \\ CHOOC(CH_2)_{13}CH_3 \\ | \quad\ O \\ | \quad\ \| \qquad\qquad\quad \oplus \\ CH_2OPOCH_2CH_2NMe_3 \\ \qquad | \\ \qquad O^{\ominus} \end{array}$$

$C_{38}H_{76}O_8P$ M 691.987

(*R*)-L-*form* [3355-27-9]
Dipentadecanoylphosphatidylcholine

Sylvius, J.R. *et al*, *Biochim. Biophys. Acta*, 1979, **555**, 175.

Glycerol 1,2-ditetradecanoate 3-phosphate G-00047

Tetradecanoic acid 1-[(phosphonoxy)methyl]-1,2-ethanediyl ester, 9CI. Dimyristoylphosphatidic acid

[30170-00-4]
$C_{31}H_{61}O_8P$ M 592.792
(***R***)-***form*** [28874-52-4]
*1,2-Ditetradecanoyl-*sn*-glycerol 3-phosphate*
Mp 61.5-62.5°. $[\alpha]_D^{24}$ +4.4° (c, 11 in $CHCl_3$).
Ba salt: Mp 213-215°.
Di-Me ester: Mp 32-33°. $[\alpha]_D$ +2.3°.
Diphenyl ester: Mp 38-39°. $[\alpha]_D$ +2.6° (c, 10 in $CHCl_3$).
(±)-***form***
Mp 54-55°.
Phenyl ester: Mp 48.5-49.5°.
Dibenzyl ester: Mp 40-41°.

Baer, E., *J. Biol. Chem.*, 1951, **189**, 235.
Hessel, L.W. *et al*, *Recl. Trav. Chim. Pays-Bas*, 1954, **73**, 150.
Baer, E. *et al*, *J. Biol. Chem.*, 1955, **212**, 39.
Baer, E., *Prog. Chem. Fats Other Lipids*, 1963, **6**, 33 (*rev*)
Browning, J. *et al*, *Chem. Phys. Lipids*, 1979, **24**, 103.

Glycerol 1,2-ditetradecanoate 3-phosphocholine G-00048

Dimyristoylphosphatidylcholine. 1,2-Ditetradecanoylglycero-3-phosphocholine

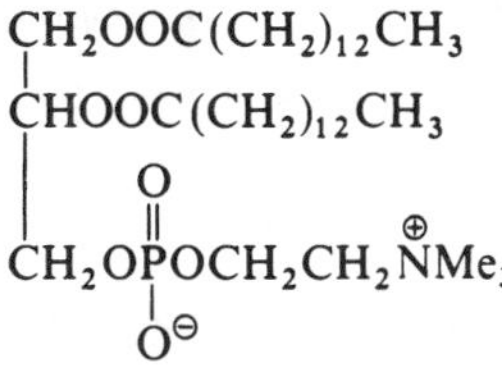

$$\begin{array}{l} CH_2OOC(CH_2)_{12}CH_3 \\ CHOOC(CH_2)_{12}CH_3 \\ CH_2OP(=O)(O^{\ominus})OCH_2CH_2\overset{\oplus}{N}Me_3 \end{array}$$

$C_{36}H_{72}NO_8P$ M 677.940
(***R***)-***form*** [18194-24-6]
Mp 236-237°. $[\alpha]_D^{22}$ +7.0° (c, 10 in 1:1 $CHCl_3$/MeOH).

Baer, E. *et al*, *Can. J. Biochem.*, 1959, **37**, 953.
Klein, R.A., *J. Lipid Res.*, 1972, **13**, 672 (*ms*)
Dorset, D.L., *Biochim. Biophys. Acta*, 1975, **380**, 257.
Koyama, Y. *et al*, *Chem. Phys. Lipids*, 1977, **19**, 74 (*ir, raman*)
Wood, G.W. *et al*, *Chem. Phys. Lipids*, 1977, **18**, 316 (*ms*)

Glycerol 1,2-ditetradecanoate 3-phosphoethanolamine G-00049

Tetradecanoic acid 1-[[[(2-aminoethoxy)-hydroxyphosphinyl]oxy]methyl]-1,2-ethanediyl ester, 9CI. Dimyristoylphosphatidylethanolamine. 1,2-Ditetradecanoylglycero-3-phosphoethanolamine
[20255-95-2]

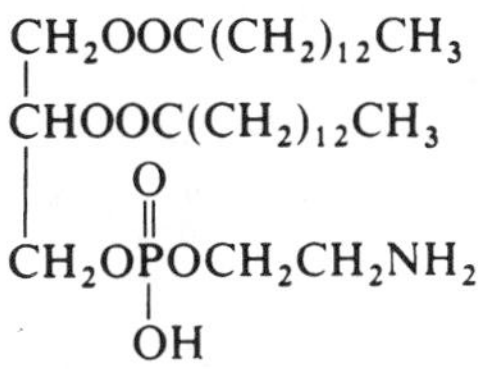

$$\begin{array}{l} CH_2OOC(CH_2)_{12}CH_3 \\ CHOOC(CH_2)_{12}CH_3 \\ CH_2OP(=O)(OH)OCH_2CH_2NH_2 \end{array}$$

$C_{33}H_{66}NO_8P$ M 635.860
(***R***)-***form*** [998-07-2]
Mp 175-177°. $[\alpha]_D^{26}$ +6.7° (c, 8.4 in $CHCl_3$).
(±)-***form*** [5683-46-5]
Cryst. (EtOH). Mp 200°.

Baer, E. *et al*, *J. Am. Chem. Soc.*, 1952, **74**, 152 (*synth*)
Hoefnagel, M.A. *et al*, *Recl. Trav. Chim. Pays-Bas*, 1960, **79**, 330 (*synth*)
Chapman, D. *et al*, *Proc. R. Soc. London, Ser. A*, 1966, **290**, 1151 (*pmr, ir, props*)
Klein, R.A, *J. Lipid Res.*, 1972, **13**, 672 (*ms*)
v. Dijck, P.W.M., *Biochim. Biophys. Acta*, 1976, **455**, 576.
Eibl, H. *et al*, *Chem. Phys. Lipids*, 1978, **22**, 1 (*synth*)
Blume, A., *Biochim. Biophys. Acta*, 1979, **557**, 32.

Glycerol 1,2-ditetradecanoate 3-phosphono-2-amino-1-propanol G-00050

Ditetradecanoyl-α-glycerylphosphoryl-2-amino-1-propanol. Dimyristoyl-α-glycerylphosphoryl-2-amino-1-propanol

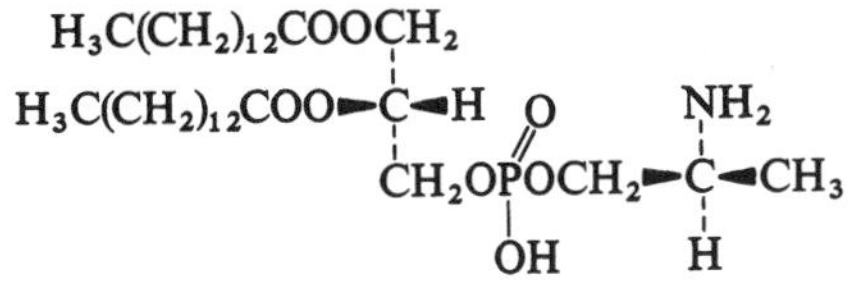

$$\begin{array}{l} H_3C(CH_2)_{12}COOCH_2 \\ H_3C(CH_2)_{12}COO\blacktriangleright C \blacktriangleleft H \\ CH_2OP(=O)(OH)OCH_2\blacktriangleright C(NH_2)(H) \blacktriangleleft CH_3 \end{array}$$

$C_{34}H_{68}NO_8P$ M 649.887
(***R,L***)-***form***
Mp 181-182°. $[\alpha]_D^{24}$ +7.4° (c, 5 in $CHCl_3$).

Baer, E. *et al*, *J. Biol. Chem.*, 1963, **238**, 3591.

Glycerol 1,2-ditetradecanoate 3-phosphoserine G-00051

Dimyristoylphosphatidylserine

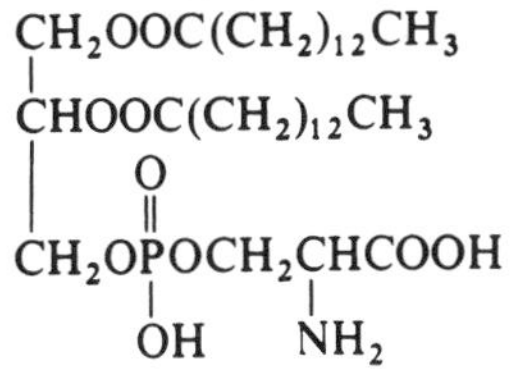

$$\begin{array}{l} CH_2OOC(CH_2)_{12}CH_3 \\ CHOOC(CH_2)_{12}CH_3 \\ CH_2OP(=O)(OH)OCH_2CH(NH_2)COOH \end{array}$$

$C_{34}H_{66}NO_{10}P$ M 679.870
(***R***)-***D-form***
*1,2-Dimyristoyl-*sn*-glycero-3-phospho-D-serine*
$[\alpha]_D^{22}$ +11.5° ($CHCl_3$), +6.3° ($CHCl_3$/MeOH/H_2O), +21.5° (C_6H_6).
(***R***)-***L-form***
*1,2-Dimyristoyl-*sn*-glycero-3-phospho-L-serine*
$[\alpha]_D^{22}$ −6.0° ($CHCl_3$), +6.2° ($CHCl_3$/MeOH/H_2O), −13.5° ($CHCl_3$).
(***R***)-***DL-form***
*1,2-Dimyristoyl-*sn*-glycero-3-phospho-DL-serine*
$[\alpha]_D^{22}$ +3.6° ($CHCl_3$), +7.3° ($CHCl_3$/MeOH/H_2O), +4.2° (C_6H_6).

Browning, J. *et al*, *Chem. Phys. Lipids*, 1979, **24**, 103 (*synth*)
Browning, J. *et al*, *Biochemistry*, 1980, **19**, 1262 (*nmr*)

Glycerol 1,2-ditridecanoate 3-phosphocholine G-00052

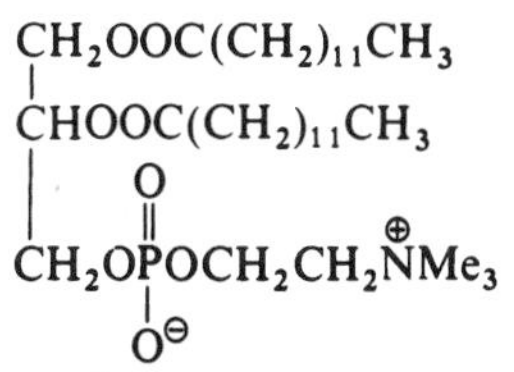

$$\begin{array}{l} CH_2OOC(CH_2)_{11}CH_3 \\ CHOOC(CH_2)_{11}CH_3 \\ CH_2OP(=O)(O^{\ominus})OCH_2CH_2\overset{\oplus}{N}Me_3 \end{array}$$

$C_{34}H_{68}NO_8P$ M 649.887
(***R***)-***L-form*** [71242-28-9]
Ditridecanoylphosphatidylcholine

Sylvius, J.R. *et al*, *Biochim. Biophys. Acta*, 1979, **555**, 175.

Glycerol 1-dodecanoate 2-octadecanoate 3-phosphocholine G-00053

$$\begin{array}{l} CH_2OOC(CH_2)_{10}CH_3 \\ CHOOC(CH_2)_{16}CH_3 \\ CH_2OP(=O)(O^{\ominus})OCH_2CH_2\overset{\oplus}{N}Me_3 \end{array}$$

$C_{38}H_{76}NO_8P$ M 705.994

(±)-***form***

Lauroylstearoylphosphatidylcholine

Mp 225°. The (*R*)-form has also been synthesised.

de Haas, G.H. *et al*, *Tetrahedron Lett.*, 1960, No. 9, 1 (*synth*)
de Haas, G.H. *et al*, *Recl. Trav. Chim. Pays-Bas*, 1961, **80**, 951 (*synth, ir*)

Glycerol 1-dodecanoate 2-9-octadecenoate 3-phosphoinositol G-00054

myo-*Inositol 1-[2-[(1-oxo-9-octadecenyl)oxy]-3-[(1-oxododecyl)oxy]propyl hydrogen phosphate, 9CI*

$$\begin{array}{l} CH_2OOC(CH_2)_{10}CH_3 \\ CHOOC(CH_2)_7CH{=}CH(CH_2)_7CH_3 \\ CH_2OP(=O)(OH)\text{–O–inositol (OH)}_5 \end{array}$$

$C_{39}H_{73}O_{13}P$ M 780.972

Molotkovskii, Y.G. *et al*, *CA*, 1971, **75**, 88864 (*synth*)

Glycerol 1-hexadecanoate 2-5,8,11,14-icosatetraenoate 3-phosphocholine G-00055

1-Hexadecanoyl-2-icosa-5,8,11,14-tetraenoylglycero-3-phosphocholine

$$\begin{array}{l} CH_2OOC(CH_2)_{14}CH_3 \\ CHOOC(CH_2)_3(CH{=}CHCH_2)_4(CH_2)_3CH_3 \\ CH_2OP(=O)(O^{\ominus})OCH_2CH_2\overset{\oplus}{N}Me_3 \end{array}$$

$C_{44}H_{80}NO_8P$ M 782.092

(***R***)-L-(***all-Z***)-***form*** [35418-58-7]

Palmitoylarachidonoylphosphatidylcholine

Evans, R.W. *et al*, *Chem. Phys. Lipids*, 1978, **22**, 207.

Glycerol 1-hexadecanoate 2-9,12-octadecadienoate 3-phosphocholine G-00056

1-Hexadecanoyl-2-octadeca-9,12-dienoylglycero-3-phosphocholine

$$\begin{array}{l} CH_2OOC(CH_2)_{14}CH_3 \\ CHOOC(CH_2)_7CH{=}CHCH_2CH{=}CH(CH_2)_4CH_3 \\ CH_2OP(=O)(O^{\ominus})OCH_2CH_2\overset{\oplus}{N}Me_3 \end{array}$$

$C_{42}H_{80}NO_8P$ M 758.070

(***R***)-L-(***Z,Z***)-***form*** [17708-90-6]

Palmitoyllinoleoylphosphatidylcholine

Demel, R.A. *et al*, *Biochim. Biophys. Acta*, 1972, **266**, 26.
Ghosh, D. *et al*, *Biochim. Biophys. Acta*, 1972, **266**, 41.
Wood, G.W. *et al*, *Chem. Phys. Lipids*, 1977, **18**, 316 (*ms*)
Porter, N.A. *et al*, *Lipids*, 1979, **14**, 20 (*hplc*)

Glycerol 1-hexadecanoate 2-octadecanoate 3-phosphocholine G-00057

Palmitoylstearoylphosphatidylcholine. 1-Hexadecanoyl-2-octadecanoylglycero-3-phosphocholine

$$\begin{array}{l} CH_2OOC(CH_2)_{14}CH_3 \\ CHOOC(CH_2)_{16}CH_3 \\ CH_2OP(=O)(O^{\ominus})OCH_2CH_2\overset{\oplus}{N}Me_3 \end{array}$$

$C_{42}H_{84}NO_8P$ M 762.101

Keough, K.M.W. *et al*, *Biochemistry*, 1979, **18**, 1453.
Porter, N.A. *et al*, *Lipids*, 1979, **14**, 20 (*hplc*)

Glycerol 1-hexadecanoate 2-9,12,15-octadecatrienoate 3-phosphocholine G-00058

[17118-61-5]

$$\begin{array}{l} CH_2OOC(CH_2)_{14}CH_3 \\ CHOOC(CH_2)_7(CH{=}CHCH_2)_3CH_3 \\ CH_2OP(=O)(O^{\ominus})OCH_2CH_2\overset{\oplus}{N}Me_3 \end{array}$$

$C_{42}H_{78}NO_8P$ M 756.054

(***R***)-L-(***all-Z***)-***form***

Palmitoyllinolenoylphosphatidylcholine

Ghosh, D. *et al*, *Chem. Phys. Lipids*, 1971, **7**, 173.
Demel, R.A. *et al*, *Biochim. Biophys. Acta*, 1972, **266**, 26.
Ghosh, D. *et al*, *Biochim. Biophys. Acta*, 1972, **266**, 41.

Glycerol 1-hexadecanoate 2-9,12,15-octadecatrienoate 3-phosphoethanolamine G-00059

$$\begin{array}{l} CH_2OOC(CH_2)_{14}CH_3 \\ CHO(CH_2)_7(CH{=}CHCH_2)_3CH_3 \\ CH_2OP(=O)(OH)OCH_2CH_2NH_2 \end{array}$$

$C_{39}H_{72}NO_8P$ M 713.973

(R)-(all-Z)-form
Palmitoyllinoleoylphosphatidylethanolamine
$[\alpha]^{20}_{D}$ +6.0° (c, 18 in $CHCl_3$).

de Haas, G.H. *et al*, *Biochim. Biophys. Acta*, 1962, **65**, 260.

Glycerol 1-hexadecanoate 2-9-octadecenoate 3-phosphate G-00060

$C_{37}H_{71}O_8P$ M 674.937

(R)-(Z)-form
*1-Palmitoyl-2-oleoyl-*sn*-glycerol 3-phosphate*
Di-NH_4 salt: Mp 192-194°. $[\alpha]_D$ +4.5° (c, 3 in $CHCl_3$).

Molotkovsky, J.G. *et al*, *Chem. Phys. Lipids*, 1976, **17**, 108 (*synth*)

Glycerol 1-hexadecanoate 2-9-octadecenoate 3-phosphocholine G-00061

1-Hexadecanoyl-2-octadec-9-enoylglycero-3-phosphocholine

$$\begin{array}{l} CH_2OOC(CH_2)_{14}CH_3 \\ | \\ CHOOC(CH_2)_7CH{=}CH(CH_2)_7CH_3 \\ | \quad\ O \\ | \quad\ \| \qquad\qquad\quad \oplus \\ CH_2OPOCH_2CH_2NMe_3 \\ \quad\ \ | \\ \quad\ \ O^{\ominus} \end{array}$$

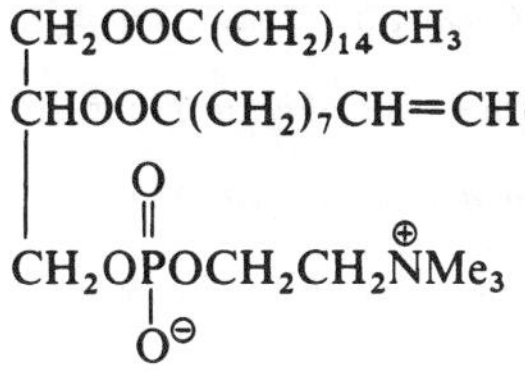

$C_{42}H_{82}NO_8P$ M 760.085

(R)-L-(Z)-form [26853-31-6]
Palmitoyloleoylphosphatidylcholine

Slotboom, A.J. *et al*, *Chem. Phys. Lipids*, 1970, **4**, 15.
Ghosh, D. *et al*, *Biochim. Biophys. Acta*, 1972, **266**, 41.
Wood, G.W. *et al*, *Chem. Phys. Lipids*, 1977, **18**, 316 (*ms*)
Seelig, J. *et al*, *Biochemistry*, 1978, **17**, 3310 (*nmr*)
Porter, N.A., *Lipids*, 1979, **14**, 20 (*hplc*)
Guyer, W. *et al*, *Chem. Phys. Lipids*, 1983, **33**, 313 (*synth*)

Glycerol 1-hexadecanoate 2-9-octadecenoate 3-phosphoethanolamine G-00062

9-Octadecenoic acid 1-[[[(2-aminoethoxy)-hydroxyphosphinyl]oxy]methyl]-2-[(1-oxohexadecyloxy)ethyl ester, 9CI. *1-Hexadecanoyl-2-octadec-9-enoylglycero-3-phosphoethanolamine*

$$\begin{array}{l} CH_2OOC(CH_2)_{14}CH_3 \\ | \\ CHOOC(CH_2)_7CH{=}CH(CH_2)_7CH_3 \\ | \quad\ O \\ | \quad\ \| \\ CH_2OPOCH_2CH_2NH_2 \\ \quad\ \ | \\ \quad\ \ OH \end{array}$$

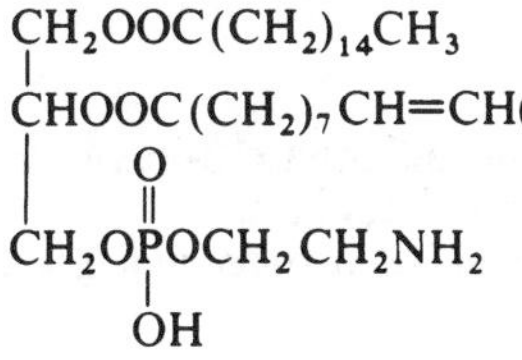

$C_{39}H_{76}NO_8P$ M 718.005

(R)-(Z)-form [39749-44-5]
Palmitoyloleoylphosphatidylethanolamine
Mp 184-185°, 194-196°. $[\alpha]^{22}_{D}$ +6.35° (c, 3 in $CHCl_3$).

Daemen, F.J.M. *et al*, *Chem. Phys. Lipids*, 1967, **1**, 476 (*synth*)
Billimoria, J.D. *et al*, *Chem. Phys. Lipids*, 1974, **12**, 327 (*synth*)
Molotkovksy, I.G. *et al*, *Chem. Phys. Lipids*, 1976, **17**, 108 (*synth*)
Eibl, H. *et al*, *Chem. Phys. Lipids*, 1978, **22**, 1 (*synth*)

Glycerol-1-hexadecanoate 2-9-octadecenoate 3-phosphoinositol G-00063

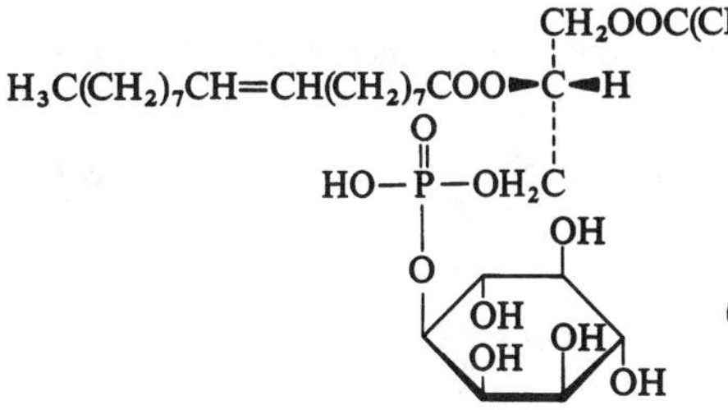

$$\begin{array}{l} \qquad\qquad\qquad\qquad\qquad\quad CH_2OOC(CH_2)_{14}CH_3 \\ H_3C(CH_2)_7CH{=}CH(CH_2)_7COO\blacktriangleright C \blacktriangleleft H \\ \qquad\qquad\quad O \\ \qquad\qquad\quad \| \\ \qquad HO-P-OH_2C \\ \qquad\qquad\quad | \\ \qquad\qquad\quad O-\text{inositol (OH)}_5 \end{array}$$

(R)-D-form

$C_{43}H_{81}O_{13}P$ M 837.079

The (*R*)-symbol designates the chirality of the glyceride moeity (≡ 1-phosphate acc. to the *sn*-system) and the D or L symbol refers to the abs. config. of the inositol ring system.

(R)-D-form
*1-D-1-O-(1′-Palmitoyl-2′-oleoyl-*sn*-glycero-3′-phosphate)myoinositol*
Cryst. ($CHCl_3$/MeOH). Mp 172-174°. $[\alpha]_D$ +6.1° (c, 12 in $CHCl_3$). Natural config.

(R)-L-form
*1-L-1-O-(1′-Palmitoyl-2′-oleoyl-*sn*-glycero-3′-phosphate)myoinositol*
Cryst. ($CHCl_3$/MeOH). Mp 169-172°. $[\alpha]_D$ +9.2° (c, 0.4 in $CHCl_3$).

(R)-DL-form
*DL-1-O-(1′-Palmitoyl-2′-oleoyl-*sn*-glycero-3′-phosphate)myoinositol*
4-Phosphate triammonium salt: Cryst. ($CHCl_3$/MeOH). Mp 187°. $[\alpha]_D$ +1.8° (c, 0.3 in $CHCl_3$).

Molotkovsky, J.G. *et al*, *Chem. Phys. Lipids*, 1973, **11**, 135.
Shevchenko, V.P. *et al*, *Chem. Phys. Lipids*, 1975, **15**, 95.
Sadovnika, M.S. *et al*, *CA*, 1976, **85**, 17835.

Glycerol 1-hexadecanoate 2-tetradecanoate 3-phosphocholine G-00064

Palmitoylmyristoylphosphatidylcholine. 1-Hexadecanoyl-2-tetradecanoylglycero-3-phosphocholine

$$\begin{array}{l} CH_2OOC(CH_2)_{14}CH_3 \\ | \\ CHOOC(CH_2)_{12}CH_3 \\ | \quad\ O \\ | \quad\ \| \qquad\qquad\quad \oplus \\ CH_2OPOCH_2CH_2NMe_3 \\ \quad\ \ | \\ \quad\ \ O^{\ominus} \end{array}$$

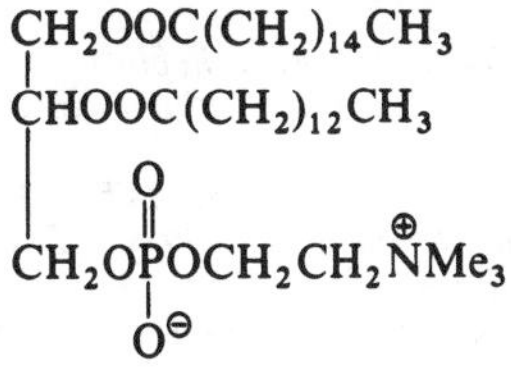

$C_{38}H_{76}NO_8P$ M 705.994

Wood, G.W. *et al*, *Chem. Phys. Lipids*, 1977, **18**, 316 (*ms*)
Keough, K.M.W. *et al*, *Biochemistry*, 1979, **18**, 1453.

Glycerol 1-5,8,11,14-icosatetraenoate 2-octadecanoate 3-phosphoethanolamine G-00065

5,8,11,14-Icosatetraenoic acid 3-[[(2-aminoethoxy)-hydroxyphosphinyl]oxy]-2-[(1-oxooctadecyl)oxy]-propyl ester, 9CI. *1-Icosa-5,8,11,14-tetraenoyl-2-octadecanoylglycero-3-phosphoethanolamine*
[22259-33-2]

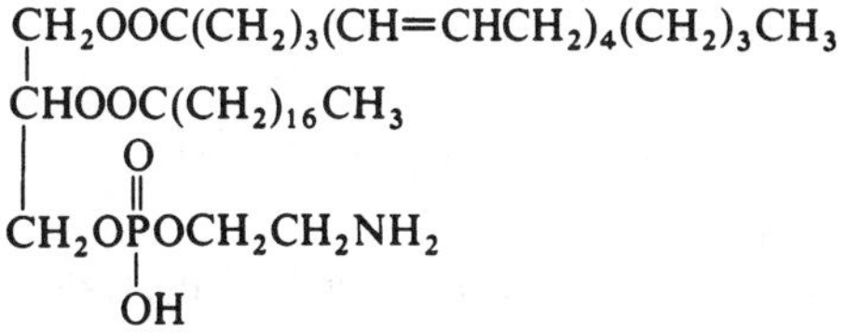

$$\begin{array}{l} CH_2OOC(CH_2)_3(CH{=}CHCH_2)_4(CH_2)_3CH_3 \\ | \\ CHOOC(CH_2)_{16}CH_3 \\ | \quad\ O \\ | \quad\ \| \\ CH_2OPOCH_2CH_2NH_2 \\ \quad\ \ | \\ \quad\ \ OH \end{array}$$

$C_{43}H_{78}NO_8P$ M 768.065

***(R)*-(*all-Z*)-form**
Arachidonoylstearoylphosphatidylethanolamine
Renin inhibitor.

Miyazaki, M. *et al*, *Proc. Soc. Exp. Biol. Med.*, 1977, **155**, 468 (*synth*)

Glycerol 1-monophosphate G-00066

α-Glycerophosphoric acid

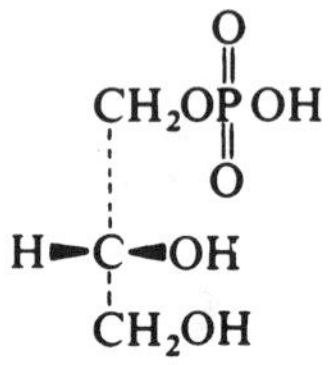

$$\begin{array}{c} CH_2OP(O)(O)OH \\ H \blacktriangleright C \blacktriangleleft OH \\ CH_2OH \end{array} \quad (R)\text{-}form$$

$C_3H_9O_6P$ M 172.074
Natural glycerophosphoric acid obt. from biol. sources is a mixt. of ca. 25% Glycerol 1-monophosphate, G-00066 and 75% Glycerol 2-monophosphate, G-00067 $[\alpha]_D$ ca. −0.5°. Synthetic glycerophosphoric acid prepd. by phosphorylation of glycerol has varying isomeric composition.

***(R)*-form** [5746-57-6]
L-form
Syrup. Dec. on dist. Slowly hydrol. by H_2O.
Ba salt: $[\alpha]_D$ −1.45° (c, 10.3 in 2N HCl).
Di-Me ether, Na salt:
$C_5H_{13}O_6P$ M 200.128
$[\alpha]_D$ −7.2°.
Di-Me ether, di-Me ester:
$C_7H_{17}O_6P$ M 228.181
$[\alpha]_D$ −4.78° (EtOH).

***(S)*-form** [17989-41-2]
D-form
Syrup. Dec. on dist. Slowly hydrol. by H_2O.
Li salt: $[\alpha]_D^{28}$ +3.51° (H_2O).
Di-Me ether, di-Me ester: $[\alpha]_D$ +5.1°.

(±)-form [1509-81-5]
Syrup. Dec. on dist. Slowly hydrol. by H_2O.
Quinine salt: Mp 155°.

Biochem. Prep., 1952, **2**, 31 (*synth, bibl*)
Hossel, L.W. *et al*, *Recl. Trav. Chim. Pays-Bas*, 1954, **73**, 150 (*derivs*)
Baer, E. *et al*, *J. Biol. Chem.*, 1955, **212**, 39 (*synth*)
Harvey, D.J. *et al*, *J. Chem. Soc., Perkin Trans. 1*, 1972, 1074 (*trimethylsilyl, ms*)
Taga, T. *et al*, *J. Chem. Soc., Chem. Commun.*, 1972, 465 (*cryst struct*)
Taguchi, Y. *et al*, *Chem. Pharm. Bull.*, 1975, **23**, 1586 (*synth*)
Yeagle, P.L. *et al*, *J. Am. Chem. Soc.*, 1975, **97**, 7175 (*cmr*)
Rios-Mercadillo, V.M. *et al*, *J. Am. Chem. Soc.*, 1979, **101**, 5828 (*enzymic synth*)
McAlister, J. *et al*, *Acta Crystallogr., Sect. B*, 1980, **36**, 1652 (*cryst struct*)
Gerothanassis, I.P. *et al*, *J. Magn. Reson.*, 1982, **46**, 423 (*O nmr*)
Weissman, J.D. *et al*, *J. Biol. Chem*, 1982, **257**, 3618 (*synth*)
Crans, D.C. *et al*, *J. Am. Chem. Soc.*, 1985, **107**, 7019 (*synth*)

Glycerol 2-monophosphate G-00067

β-Glycerophosphoric acid
[17181-54-3]

$$(HOCH_2)_2CHOP(O)(OH)_2$$

$C_3H_9O_6P$ M 172.074
See note under Glycerol 1-monophosphate, G-00066 . Syrup. Hygroscopic, dec. on dist.
Mono-Na salt: Cariostatic: use in mouthwashes.
Ca salt: [55701-23-0]. Cryst. + $1H_2O$.
Ba salt: Mod. sol. H_2O.
Di-Ph ether:
$C_{15}H_{17}O_6P$ M 324.269
Cryst. (EtOH/pet. ether). Mp 137-137.5°.
Di-Ph ether, Na salt: Cryst. + $10H_2O$. Mp 54°.
1-Dodecanoyl:
$C_{15}H_{31}O_7P$ M 354.379
Needles (as Ba salt). Mp 245-255° (Ba salt).
1-Hexanoyl:
$C_9H_{19}O_7P$ M 270.219
Mp 261-263° (Ba salt).
1,3-Ditetradecanoyl:
$C_{31}H_{61}O_8P$ M 592.792
Cryst. + 0.5 quinoline (EtOAc). Mp 96.5-97.5°.
1,3-Dioctadecanoyl:
$C_{39}H_{77}O_8P$ M 705.006
Cryst. Mp 68.5-69.5°.

Baer, G. *et al*, *J. Biol. Chem.*, 1940, **135**, 321 (*synth*)
Cherbuliez, E. *et al*, *Helv. Chim. Acta*, 1946, **29**, 2006 (*synth*)
Mazhar-Ul-Haque, *et al*, *J. Chem. Soc., Chem. Commun.*, 1966, 214 (*cryst struct*)
Greenwald, J. *et al*, *J. Chem. Soc., Perkin Trans. 2*, 1972, 1095 (*uv, props*)
Harvey, D.J. *et al*, *J. Chem. Soc., Perkin Trans. 1*, 1972, 1074 (*trimethylsilyl, ms*)
Nakagaki, M. *et al*, *Chem. Pharm. Bull.*, 1979, **27**, 1887 (*conformn*)
Inoue, M. *et al*, *Chem. Pharm. Bull.*, 1980, **28**, 1491 (*struct*)

Glycerol 1-9,12-octadecadienoate 2-hexadecanoate 3-phosphocholine G-00068

1-Octadeca-9,12-dienoyl-2-hexadecanoylglycero-3-phosphocholine

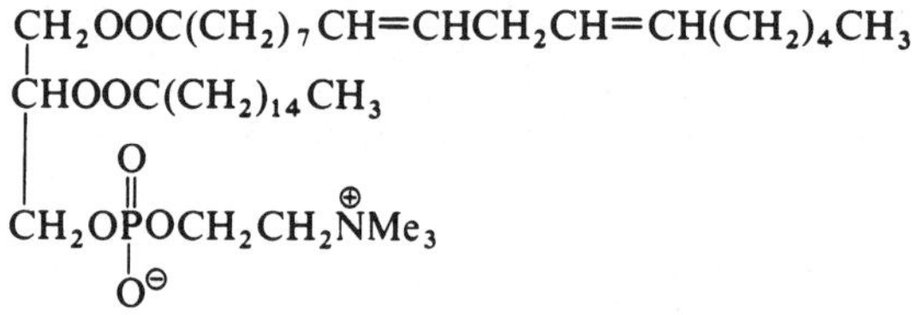

$$\begin{array}{l} CH_2OOC(CH_2)_7CH{=}CHCH_2CH{=}CH(CH_2)_4CH_3 \\ CHOOC(CH_2)_{14}CH_3 \\ CH_2OP(O)(O^{\ominus})OCH_2CH_2\overset{\oplus}{N}Me_3 \end{array}$$

$C_{42}H_{80}NO_8P$ M 758.070

***(R)*-L-(*Z,Z*)-form**
Linoleoylpalmitoylphosphatidylcholine

Demel, R.A. *et al*, *Biochem. Biophys. Acta*, 1972, **266**, 26.

Glycerol 1-9,12-octadecadienoate 2-hexadecanoate 3-phosphoethanolamine G-00069

1-Octadeca-9,12-dienoyl-2-hexadecanoylglycero-3-phosphoethanolamine

$$\begin{array}{l} CH_2OOC(CH_2)_7CH{=}CHCH_2CH{=}CH(CH_2)_4CH_3 \\ CHOOC(CH_2)_{14}CH_3 \\ CH_2OP(O)(OH)OCH_2CH_2NH_2 \end{array}$$

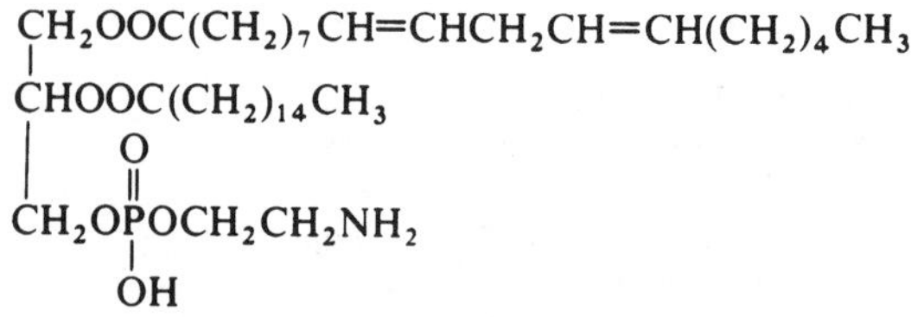

$C_{39}H_{74}NO_8P$ M 715.989

***(R)*-(*all-Z*)-form** [26662-95-3]
Linoleoylpalmitoylphosphatidylethanolamine

Finean, J.B., *Biochim. Biophys. Acta*, 1953, **10**, 371.

Glycerol 1-9,12-octadecadienoate 2-octadecanoate 3-phosphocholine G-00070

1,9,12-Octadecadienoyl-2-octadecanoylglycero-3-phosphocholine

$CH_2OOC(CH_2)_7CH{=}CHCH_2CH{=}CH(CH_2)_4CH_3$
$CHOOC(CH_2)_{16}CH_3$
$CH_2OP(=O)(O^{\ominus})OCH_2CH_2\overset{\oplus}{N}Me_3$

$C_{44}H_{84}NO_8P$ M 786.123

(±)-(*Z,Z*)-*form*

Hygroscopic powder (EtOH/Me_2CO).

Bogolovskii, N.A. *et al*, *CA*, 1963, **58**, 8895 (*synth, ir*)

Glycerol 1-9,12-octadecadienoate 2-octadecanoate 3-phosphoethanolamine G-00071

1-Octadec-9,12-dienoyl-2-octadecanoylglycero-3-phosphocholine

$CH_2OOC(CH_2)_7CH{=}CHCH_2CH{=}CH(CH_2)_4CH_3$
$CHOOC(CH_2)_{16}CH_3$
$CH_2OP(=O)(OH)OCH_2CH_2NH_2$

$C_{41}H_{78}NO_8P$ M 744.043

(±)-(*all-Z*)-*form* [18468-68-3]

Linoleoylstearoylphosphatidylethanolamine

Mp 181-183°, 194-195°.

Shvets, V.I. *et al*, *CA*, 1962, **56**, 4611 (*synth*)
Bogoslovskii, N.A. *et al*, *CA*, 1962, **57**, 14151 (*synth*)
Dorofeeva, L.T. *et al*, *CA*, 1965, **63**, 6852 (*synth*)
Chapman, D. *et al*, *Proc. R. Soc. London, Ser. A*, 1966, **290**, 1151 (*ir, nmr, props*)

Glycerol 1-9,12-octadecadienoate 2-9-octadecenoate 3-phosphocholine G-00072

1-Octadeca-9,12-dienoyl-2-octadec-9-enoylglycero-3-phosphocholine

$CH_2OOC(CH_2)_7CH{=}CHCH_2CH{=}CH(CH_2)_4CH_3$
$CHOOC(CH_2)_7CH{=}CH(CH_2)_7CH_3$
$CH_2OP(=O)(O^{\ominus})OCH_2CH_2\overset{\oplus}{N}Me_3$

$C_{44}H_{82}NO_8P$ M 784.107

(*R*)-L-(*all-Z*)-*form*

Linoleoyloleoylphosphatdiylcholine

Gelatinous. $[\alpha]_D^{20}$ +6.12°.

Shvets, V.I., *CA*, 1965, **62**, 9001 (*synth*)

Glycerol 1-9,12-octadecadienoate 2,9-octadecenoate 3-phosphoinositol G-00073

$CH_2OOC(CH_2)_7(CH{=}CHCH_2)(CH_2)_3CH_3$
$CHOOC(CH_2)_7CH{=}CH(CH_2)_7CH_3$
$CH_2OP(=O)(OH)O$–inositol (cyclohexane bearing OH groups)

$C_{45}H_{81}O_3P$ M 701.107

Shvets, V.I. *et al*, *CA*, 1966, **64**, 14261 (*synth*)

Glycerol 1-octadecanoate 2-dodecanoate 3-phosphocholine G-00074

[15071-88-2]

$CH_2OOC(CH_2)_{16}CH_3$
$CHOOC(CH_2)_{10}CH_3$
$CH_2OP(=O)(O^{\ominus})OCH_2CH_2\overset{\oplus}{N}Me_3$

$C_{38}H_{76}NO_8P$ M 705.994

(*R*)-L-*form*

Stearoyllauroylphosphatidylcholine

Powder (Et_2O). Mp 211-212°.

de Haas, G.H. *et al*, *Tetrahedron Lett.*, 1960, No. 9, 1 (*synth*)

Glycerol 1-octadecanoate 2-hexadecanoate 3-phosphocholine G-00075

Stearoylpalmitoylphosphatidylcholine. 1-Octadecyl-2-hexadecanoylglycerol-3-phosphocholine

$CH_2OOC(CH_2)_{16}CH_3$
$CHOOC(CH_2)_{14}CH_3$
$CH_2OP(=O)(O^{\ominus})OCH_2CH_2\overset{\oplus}{N}Me_3$

$C_{42}H_{84}NO_8P$ M 762.101

Nakajima, T. *et al*, *CA*, 1976, **85**, 176859 (*synth*)
Keough, K.M.W. *et al*, *Biochemistry*, 1979, **18**, 1453.
Porter, N.A. *et al*, *Lipids*, 1979, **14**, 20 (*hplc*)

Glycerol 1-octadecanoate 2-5,8,11,14-icosatetraenoate 3-phosphocholine G-00076

2-Icosa-1-octadecanoyl-5,8,11,14-tetraenoylglycero-3-phosphocholine

$CH_2OOC(CH_2)_{16}CH_3$
$CHOOC(CH_2)_3(CH{=}CHCH_2)_4(CH_2)_3CH_3$
$CH_2OP(=O)(O^{\ominus})OCH_2CH_2\overset{\oplus}{N}Me_3$

$C_{46}H_{84}NO_8P$ M 810.145

(R)-(all-Z)-form [35418-59-8]
Stearoylarachidonoylphosphatidylcholine

Ghosh, D. *et al, Biochim. Biophys. Acta*, 1972, **266**, 41.
Porter, N.A. *et al, Lipids*, 1979, **14**, 20 (*hplc*)

Glycerol 1-octadecanoate 2-5,8,11,14-icosatetraenoate 3-phosphoethanolamine — G-00077

5,8,11,14-Icosatetraenoic acid 1-[[[(2-aminoethoxy)-hydroxyphosphinyl]oxy]methyl]-2-[(1-oxooctadecyl)-oxy]ethyl ester. 1-Octadecanoyl-2-icosa-5,8,11,14-tetraenoylglycero-3-phosphoethanolamine

$CH_2OOC(CH_2)_{16}CH_3$
$CHOOC(CH_2)_3(CH{=}CHCH_2)_4(CH_2)_3CH_3$
$CH_2OP(=O)(OH)OCH_2CH_2NH_2$

$C_{43}H_{78}NO_8P$ M 768.065

(R)-(all-Z)-form
Stearoylarachidonylphosphatidylethanolamine
Renin inhibitor.

Stoffel, W. *et al, Hoppe-Seyler's Z. Physiol. Chem.*, 1969, **350**, 1385.
Miyazaki, M. *et al, Proc. Soc. Exp. Biol. Med.*, 1977, **155**, 468 (*synth*)

Glycerol 1-octadecanoate 2-9,12-octadecadienoate 3-phosphocholine — G-00078

1-Octadecanoyl-2,9,12-octadecadienoylglycero-3-phosphocholine

$CH_2OOC(CH_2)_{16}CH_3$
$CHOOC(CH_2)_7CH{=}CHCH_2CH{=}CH(CH_2)_4CH_3$
$CH_2OP(=O)(O^{\ominus})OCH_2CH_2\overset{\oplus}{N}Me_3$

$C_{44}H_{84}NO_8P$ M 786.123

(R)-L-(all-Z)-form [27098-24-4]
Stearoyllinoleoylphosphatidylcholine
Gelatinous. $[\alpha]_D^{20}$ +5.02°.

Shvets, V.I., *CA*, 1965, **62**, 9001 (*synth*)
Stoffel, W. *et al, Hoppe-Seyler's Z. Physiol. Chem.*, 1969, **350**, 1385 (*props*)
Ghosh, D. *et al, Biochim. Biophys. Acta*, 1972, **266**, 41.
Stoffel, W. *et al, Hoppe-Seyler's Z. Physiol. Chem.*, 1972, **353**, 1962 (*cmr*)

Glycerol 1-octadecanoate 2-9,12-octadecadienoate 3-phosphoethanolamine — G-00079

9,12-Octadecadienoic acid 1-[[[(2-aminoethoxy)-hydroxyphosphinyl]oxy]methyl]-2-[(1-oxooctadecyl)-oxy]ethyl ester, 9CI. *1-Octadecanoyl-2-octadeca-9,12-dienoylglycero-3-phosphoethanolamine*

$CH_2OOC(CH_2)_{16}CH_3$
$CHOOC(CH_2)_7CH{=}CHCH_2CH{=}CH(CH_2)_4CH_3$
$CH_2OP(=O)(OH)OCH_2CH_2NH_2$

$C_{41}H_{78}NO_8P$ M 744.043

(R)-(all-Z)-form [7266-53-7]
Stearoyllinoleoylphosphatidylethanolamine

(±)-(all-Z)-form
Mp 193-194°.

Dorofeeva, L.T. *et al, CA*, 1965, **63**, 6851 (*synth*)
Chapman, D. *et al, Proc. R. Soc. London, Ser. A*, 1966, **290**, 1151 (*ir, nmr, props*)
Stoffel, W. *et al, Hoppe-Seyler's Z. Physiol. Chem.*, 1969, **350**, 1385.

Glycerol 1-octadecanoate 2-9,12,15-octadecatrienoate 3-phosphocholine — G-00080

$CH_2OOC(CH_2)_{16}CH_3$
$CHOOC(CH_2)_7(CH{=}CHCH_2)_3CH_3$
$CH_2OP(=O)(O^{\ominus})OCH_2CH_2\overset{\oplus}{N}Me_3$

$C_{44}H_{82}NO_8P$ M 784.107

(R)-L-(all-Z)-form [35418-57-6]
Stearoyllinolenoylphosphatidylcholine

Stoffel, W. *et al, Hoppe-Seyler's Z. Physiol. Chem.*, 1969, **350**, 1385.
Ghosh, D. *et al, Chem. Phys. Lipids*, 1971, **7**, 173.
Ghosh, D. *et al, Biochim. Biophys. Acta*, 1972, **266**, 41.

Glycerol 1-octadecanoate 2-9,12,15-octadecatrienoate 3-phosphoethanolamine — G-00081

9,12,15-Octadecatrienoic acid 1-[[[(2-aminoethoxy)-hydroxyphosphinyl]oxy]methyl]-2-[(1-oxooctadecyl)-oxy]ethyl ester, 9CI

$CH_2OOC(CH_2)_{16}CH_3$
$CHOOC(CH_2)_7(CH{=}CHCH_2)_3CH_3$
$CH_2OP(=O)(OH)OCH_2CH_2NH_2$

$C_{41}H_{76}NO_8P$ M 742.027

(±)-(all-Z)-form [55822-08-7]
Stearoyllinolenoylphosphatidylethanolamine
Renin inhibitor.

Shvets, V.I., *CA*, 1964, **60**, 179 (*synth*)
Stoffel, W. *et al, Hoppe-Seyler's Z. Physiol. Chem.*, 1969, **350**, 1385.
Miyazaki, M., *Proc. Soc. Exp. Biol. Med.*, 1977, **155**, 468.

Glycerol 1-octadecanoate 2-9-octadecenoate 3-phosphocholine — G-00082

1-Octadecanoyl-2,9-octadecenoylglycero-3-phosphocholine
[6753-56-6]

$CH_2OOC(CH_2)_{16}CH_3$
$CHOOC(CH_2)_7CH{=}CH(CH_2)_7CH_3$
$CH_2OP(=O)(O^{\ominus})OCH_2CH_2\overset{\oplus}{N}Me_3$

$C_{44}H_{86}NO_8P$ M 788.139

(R)-form [56421-10-4]

Stearoyloleoylphosphatidylcholine

Mp 230-231°. $[\alpha]_D^{20}$ +6.0° (c, 9 in $CHCl_3$).

de Haas, G.H. *et al*, *Recl. Trav. Chim. Pays-Bas*, 1961, **80**, 951 (*synth, ir*)
Chapman, D. *et al*, *J. Biol. Chem.*, 1966, **241**, 5044 (*nmr*)
Stoffel, W. *et al*, *Hoppe-Seyler's Z. Physiol. Chem.*, 1969, **350**, 1385.
Klein, R.A., *J. Lipid Res.*, 1971, **12**, 123, 628; 1972, **13**, 672 (*ms*)
Demel, R.A., *Biochem. Biophys. Acta*, 1972, **266**, 26.

Glycerol 1-octadecanoate 2-9-octadecenoate 3-phosphoethanolamine G-00083

Octadecenoic acid 1-[[[(2-aminoethoxy)-hydroxyphosphinyl]oxy]methyl]-2-[(1-oxooctadecyl)-oxy]ethyl ester, 9CI. 1-Octadecanoyl-2-octadec-9-enoylglycero-3-phosphoethanolamine

$CH_2OOC(CH_2)_{16}CH_3$
$CHOOC(CH_2)_7CH{=}CH(CH_2)_7CH_3$
$CH_2OP(=O)(OH)OCH_2CH_2NH_2$

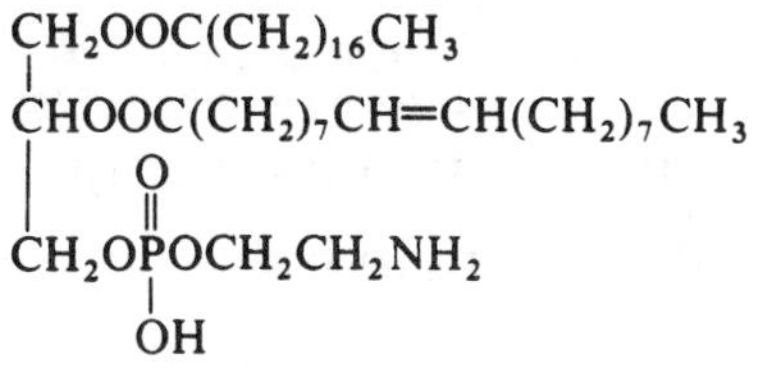

$C_{41}H_{80}NO_8P$ M 746.059

(R)-(Z)-form [6418-95-7]

Stearoyloleoylphosphatidylcholine

Mp 182-183°. $[\alpha]_D$ +6.0° (c, 10 in $CHCl_3$).

(±)-(E)-form

Stearoylelaidoylphosphatidylethanolamine

Mp 190-192°.

(±)-(Z)-form

Mp 189-190°.

Baer, E. *et al*, *Can. J. Biochem.*, 1961, **39**, 1471 (*synth*)
Chapman, D. *et al*, *Biochim. Biophys. Acta*, 1966, **120**, 148.
Chapman, D. *et al*, *Proc. R. Soc. London, Ser. A*, 1966, **290**, 1151 (*ir, nmr, props*)
Chapman, D. *et al*, *Trans. Faraday Soc.*, 1966, **62**, 2607.
Stoffel, W. *et al*, *Hoppe-Seyler's Z. Physiol. Chem.*, 1969, **350**, 1385.

Glycerol 1-octadecanoate 2-9-octadecenoate 3-phosphoinositol G-00084

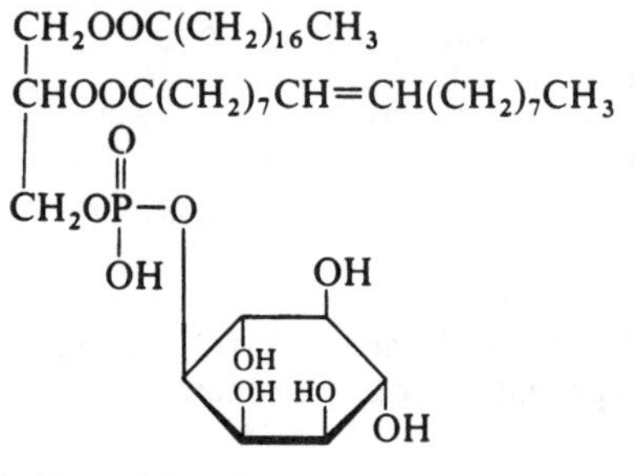

$CH_2OOC(CH_2)_{16}CH_3$
$CHOOC(CH_2)_7CH{=}CH(CH_2)_7CH_3$
$CH_2OP(=O)(OH)$–O–inositol (OH, OH, OH HO, OH)

$C_{45}H_{85}O_{13}P$ M 865.132

(R)-(Z)-form [58116-17-9]

Lyutik, A.Y. *et al*, *CA*, 1975, **82**, 86520; 1976, **84**, 59900 (*synth*)

Glycerol 1-9,12,15-octadecatrienoate 2-hexadecanoate 3-phosphoethanolamine G-00085

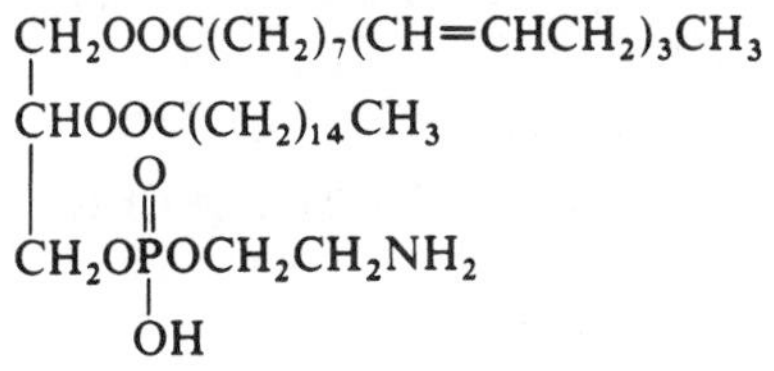

$CH_2OOC(CH_2)_7(CH{=}CHCH_2)_3CH_3$
$CHOOC(CH_2)_{14}CH_3$
$CH_2OP(=O)(OH)OCH_2CH_2NH_2$

$C_{39}H_{72}NO_8P$ M 713.973

(R)-(all-Z)-form

Linoleoylpalmitoylphosphatidylethanolamine

$[\alpha]_D^{20}$ +6.2° (c, 8 in $CHCl_3$).

de Haas, G.H. *et al*, *Biochim. Biophys. Acta*, 1962, **65**, 260.

Glycerol 1-9-octadecenoate 2-hexadecanoate 3-phosphate G-00086

9-Octadecenoic acid 1-[[(1-oxohexadecyl)oxy]methyl]-2-(phosphonoxy)ethyl ester, 9CI

[10015-87-9]

$C_{37}H_{71}O_8P$ M 674.937

(±)-form

Bonsen, P.P.M. *et al*, *Chem. Phys. Lipids*, 1966, **1**, 100 (*synth*)

Glycerol 1-9-octadecenoate 2-hexadecanoate 3-phosphocholine G-00087

1-Octadec-9-enyl-2-hexadecanoylglycero-3-phosphocholine

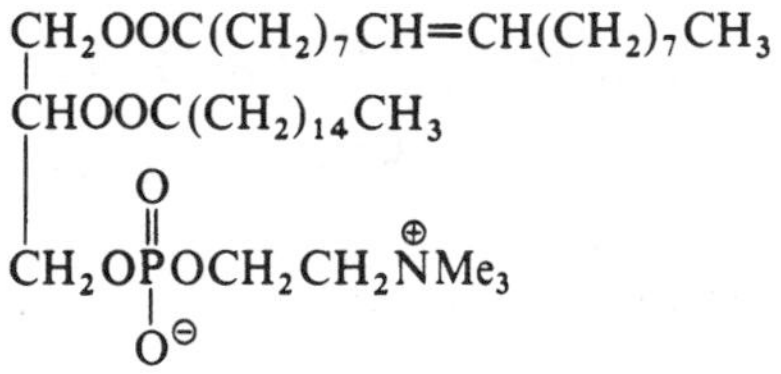

$CH_2OOC(CH_2)_7CH{=}CH(CH_2)_7CH_3$
$CHOOC(CH_2)_{14}CH_3$
$CH_2OP(=O)(O^{\ominus})OCH_2CH_2\overset{\oplus}{N}Me_3$

$C_{42}H_{82}NO_8P$ M 760.085

(R)-L-(Z)-form [59491-62-2]

Oleoylpalmitoylphosphatidylcholine

Wood, G.W. *et al*, *Chem. Phys. Lipids*, 1977, **18**, 316 (*ms*)

Glycerol 1-9-octadecenoate 2-hexadecanoate 3-phospho-1′-glycerol G-00088

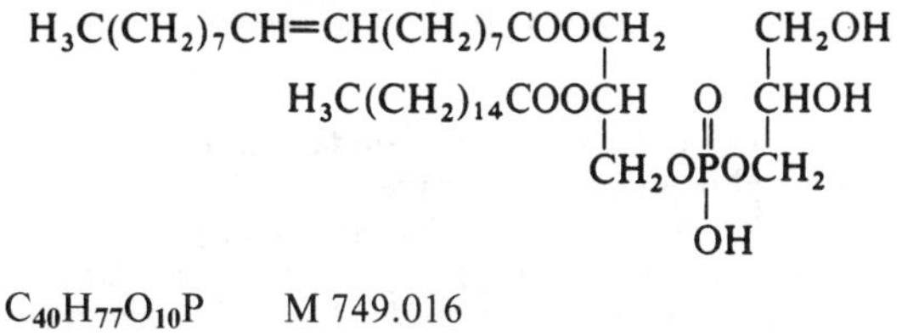

$H_3C(CH_2)_7CH{=}CH(CH_2)_7COOCH_2$
$H_3C(CH_2)_{14}COOCH$
$CH_2OP(=O)(OH)OCH_2$–CHOH–CH_2OH

$C_{40}H_{77}O_{10}P$ M 749.016

(Z)-form

Olein 2-palmito-1-(2,3-dihydroxypropyl)hydrogen phosphate, 8CI

Na salt: [13879-80-6]. Mp 176-179°. $[\alpha]_D^{20}$ +1.02° (c, 10 in $CHCl_3$).

Bonsen, P.P.M. *et al*, *Chem. Phys. Lipids*, 1966, **1**, 33 (*synth*)

Glycerol 1-9-octadecenoate 2-9,12-octadecadienoate 3-phosphoethanolamine G-00089

1-Octadec-9-enoyl-2-octadeca-9,12-dienoylglycero-3-phosphoethanolamine

$$\begin{array}{l} CH_2OOC(CH_2)_7CH{=}CH(CH_2)_7CH_3 \\ CHOOC(CH_2)_7CH{=}CHCH_2CH{=}CH(CH_2)_4CH_3 \\ CH_2OP(O)(OH)OCH_2CH_2NH_2 \end{array}$$

$C_{41}H_{76}NO_8P$ M 742.027

(R)-(all-Z)-form

Oleoyllinoleoylphosphatidylethanolamine

Mp 155°. $[\alpha]_D^{20}$ +6.0° (c, 5 in $CHCl_3$).

Daemen, F.J.M. *et al*, *Recl. Trav. Chim. Pays-Bas*, 1963, **82**, 487 (*synth*)
Volkova, L.V. *et al*, *CA*, 1965, **62**, 13039 (*synth*)
Shvets, V.I. *et al*, *CA*, 1966, **65**, 13534 (*synth*)

Glycerol 1-9-octadecenoate 2-9,12-octadecadienoate 3-phosphoserine G-00090

$C_{42}H_{76}NO_{10}P$ M 786.037

(±)-form

Oleoyllinoleoylphosphatidylserine

Shvets, V.I. *et al*, *CA*, 1968, **68**, 3169 (*synth*)

Glycerol 1-9-octadecenoate 2-octadecanoate 3-phosphocholine G-00091

2-Octadecanoyl-1,9-octadecenoylglycero-3-phosphocholine

$$\begin{array}{l} CH_2OOC(CH_2)_7CH{=}CH(CH_2)_7CH_3 \\ CHOOC(CH_2)_{16}CH_3 \\ CH_2OP(O)(O^{\ominus})OCH_2CH_2\overset{\oplus}{N}Me_3 \end{array}$$

$C_{44}H_{86}NO_8P$ M 788.139

(R)-(Z)-form [7319-55-3]

Oleoylstearoylphosphatidylcholine

Mp 235°. $[\alpha]_D^{20}$ +6.2° (c, 10 in $CHCl_3$).

de Haas, G.H. *et al*, *Recl. Trav. Chim. Pays-Bas*, 1961, **80**, 951 (*synth, ir*)
Infante, R. *et al*, *Biochim. Biophys. Acta*, 1968, **164**, 436 (*synth*)
Demel, R.A., *Biochim. Biophys. Acta*, 1972, **266**, 26 (*props*)
Klein, R.A., *J. Lipid Res.*, 1972, **13**, 672 (*ms*)

Glycerol 1-9-octadecenoate 2-octadecanoate 3-phosphoethanolamine G-00092

9-Octadecenoic acid 3-[[[(2-Aminoethoxy)hydroxyphosphinyl]oxy]methyl]-2-[(1-oxooctadecyl)oxy]propyl ester, 9CI. *1-Octadec-9-enoyl-2-octadecanoylglycero-3-phosphoethanolamine*

$$\begin{array}{l} CH_2OOC(CH_2)_7CH{=}CH(CH_2)_7CH_3 \\ CHOOC(CH_2)_{16}CH_3 \\ CH_2OP(O)(OH)OCH_2CH_2NH_2 \end{array}$$

$C_{41}H_{80}NO_8P$ M 746.059

(R)-(Z)-form

Oleoylstearoylphosphatidylethanolamine

Cryst. Mp 181-182°. $[\alpha]_D$ +6.0° (c, 10 in $CHCl_3$).

(±)-(E)-form

Elaidoylstearoylphosphatidylethanolamine

Mp 190-192°.

(±)-(Z)-form [55176-55-1]

Mp 188-190°.

Baer, E. *et al*, *Can. J. Biochem.*, 1961, **39**, 1477 (*synth*)
Daemen, F.J.M. *et al*, *Recl. Trav. Chim. Pays-Bas*, 1962, **81**, 348 (*synth*)
Chapman, D. *et al*, *Biochim. Biophys. Acta*, 1966, **120**, 148.
Chapman, D. *et al*, *Proc. R. Soc. London, Ser. A*, 1966, **290**, 1151 (*ir, nmr, props*)
Chapman, D. *et al*, *Trans. Faraday Soc.*, 1966, **62**, 2607 (*pmr, props*)

Glycerol 1-9-octadecenoate 2-octadecanoate 3-phosphoserine G-00093

Serine 2-[(1-oxooctadecenyl)oxy]-3-[(1-oxooctadecyl)oxy]propyl ester, dihydrogen phosphate ester, 9CI

$$\begin{array}{l} CH_2OOC(CH_2)_7CH{=}CH(CH_2)_7CH_3 \\ CHOOC(CH_2)_{16}CH_3 \\ CH_2OP(O)(OH)OCH_2CH(NH_2)COOH \end{array}$$

$C_{42}H_{80}NO_{10}P$ M 790.068

(±)-form

Oleoylpalmitoylphosphatidylserine

Turner, D.L. *et al*, *J. Lipid Res.*, 1964, **5**, 616 (*synth*)
Turner, D.L. *et al*, *J. Med. Chem.*, 1966, **9**, 771 (*synth*)

Glycerol 1-9-octadecenoate 2-tetradecanoate 3-phosphate G-00094

$C_{35}H_{67}O_8P$ M 646.883

(R)-form

*1-Oleoyl-2-palmitoyl-*sn*-glycerol 3-phosphate*

Cryst. (Me_2CO). Mp 61-63°. $[\alpha]_D^{18}$ +2.05° (c, 6 in $CHCl_3$).

Ba salt: Mp >360°. $[\alpha]_D^{20}$ +12.4° (c, 8 in $CHCl_3/MeOH$).

Di-Na salt: Mp 208-210°. $[\alpha]_D^{20}$ +5.70° (c, 11 in $CHCl_3$).

Bonsen, P.P.M. *et al*, *Chem. Phys. Lipids*, 1966, **1**, 100.

Glycerol 1-9-octadecenoate 2-tetradecanoate 3-phosphoglycerophosphate G-00095

$$\begin{array}{l} H_3C(CH_2)_7CH{=}CH(CH_2)_7COOCH_2 \\ H_3C(CH_2)_{12}COOCH \\ CH_2OP(O)(OH)OCH_2CH(OH)CH_2OP(O)(OH)-OH \end{array}$$

$C_{38}H_{74}O_{13}P_2$ M 800.942

(Z)-form

Olein 2-myristo-1,2,3-dihydroxypropyl hydrogen phosphate 3-dihydrogen phosphate, 8CI. *1-Oleoyl-2-myristoylglycerol-3-phospho-*rac*-1′-glycerol-3′-phosphate*

Ba salt: [17708-99-5]. Mp >300°. $[\alpha]_D^{20}$ +4.70° (c, 7 in $CHCl_3$/MeOH).

Bonson, P.P.M. *et al*, *Chem. Phys. Lipids*, 1966, **1**, 100.

Glycerol 1-tetradecanoate 2-hexadecanoate 3-phosphocholine **G-00096**

Myristoylpalmitoylphosphatidylcholine. 1-Tetradecanoyl-2-hexadecanoylglycero-3-phosphocholine

[10589-50-1]

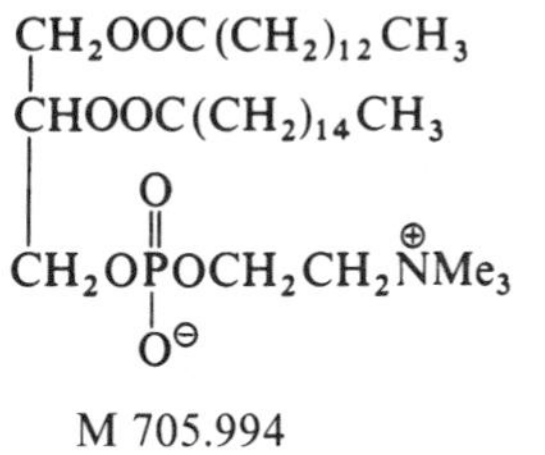

$CH_2OOC(CH_2)_{12}CH_3$
$CHOOC(CH_2)_{14}CH_3$
$CH_2OP(O)(O^{\ominus})OCH_2CH_2\overset{\oplus}{N}Me_3$

$C_{38}H_{76}NO_8P$ M 705.994

Keough, K.M.W. *et al*, *Biochemistry*, 1978, **18**, 1453.

Glyphosate, BSI **G-00097**

N-*(Phosphonomethyl)glycine*, 9CI

[1071-83-6]

$(HO)_2P(O)CH_2NHCH_2COOH$

$C_3H_8NO_5P$ M 169.074

Total post-emergence herbicide. Solid. Sl. sol. H_2O, insol. most org. solvs. Mp ca. 200° dec. pK_{a1} 2.32, pK_{a2} 5.86, pK_{a3} 10.86 (H_2O, 25°).

▷MC1075000.

Isopropylammonium salt: [38641-94-0]. *Roundup*. Total post-emergence herbicide. V. sol. H_2O.

▷MC1080000.

U.S.P., 3 799 758, *(1971)*; *CA*, **77**, 165079l (*synth*)
Rueppel, M.L., *Org. Magn. Reson.*, 1976, **8**, 19 (*pmr, P nmr*)
Rueppel, M.L., *J. Agric. Food Chem.*, 1977, **25**, 517 (*metab, cmr*)
Madsen, H.E.L. *et al*, *Acta Chem. Scand., Ser. A*, 1978, **32**, 79 (*complexes*)
Burns, A.J. *et al*, *J. Chromatogr. Sci.*, 1979, **17**, 333 (*hplc*)
Knuuttila, P. *et al*, *Acta Chem. Scand., Ser. B*, 1979, **33**, 623 (*cryst struct*)
The Agrochemicals Handbook, Royal Society of Chemistry, 1983, A222.
Pesticide Manual, 6th Ed., 292; 7th Ed., 303.

Glyphosine, BSI **G-00098**

N,N-*Bis(phosphonomethyl)glycine*, 9CI. *Polaris*

[2439-99-8]

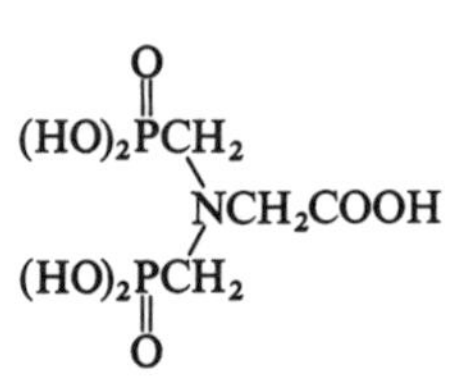

$(HO)_2P(O)CH_2$
NCH_2COOH
$(HO)_2P(O)CH_2$

$C_4H_{11}NO_8P_2$ M 263.080

Plant growth regulator, used particularly in the sugar cane industry. Ligand for In, Mn, Ga, Ca, Mg, Fe, and Cu. Cryst. (EtOH aq.). V. sol. H_2O, sl. sol. EtOH, insol. C_6H_6. Mp 200° dec. pK_{a1} 1.7, pK_{a2} 2.0, pK_{a3} 5.1, pK_{a4} 6.45, pK_{a5} 10.98 (H_2O, 25°). Light-stable.

▷MB9120000.

Westerback, S. *et al*, *J. Am. Chem. Soc.*, 1965, **87**, 2567 (*synth, props, complexes*)
Kireeva, A.Yu. *et al*, *Zh. Obshch. Khim.*, 1973, **43**, 2508 (*Engl. transl.* p. 2494) (*synth, ir, pmr*)
Tsirul'nikova, N.V. *et al*, *Zh. Obshch. Khim.*, 1981, **51**, 1028 (*Engl. transl.* p. 859) (*synth*)
Pesticide Manual, 6th Ed., 293; 7th Ed., 304.

Guanosine 5′-diphosphate **G-00099**

Guanosine 5′-(trihydrogen diphosphate), 9CI. *Guanosine 5′-(trihydrogen pyrophosphate)*, 8CI. *GDP*

[146-91-8]

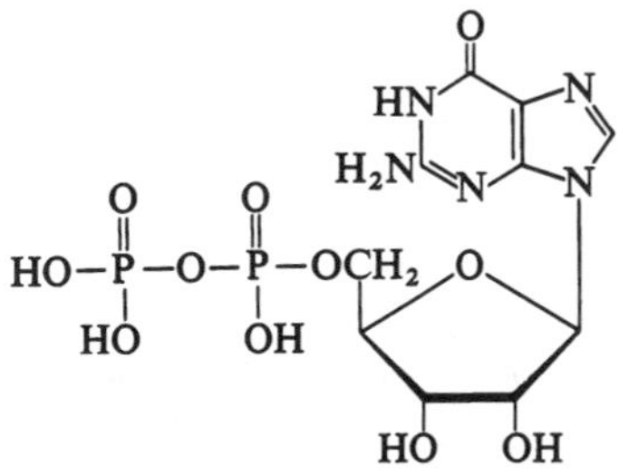

$C_{10}H_{15}N_5O_{11}P_2$ M 443.203

Constit. of many plant and animal tissues. Intermediate in biosynth. of RNA by polynucleotide phosphorylase. Amorph. λ_{max} 256 nm (ϵ 11 800) (pH 2).

Ayengar, P. *et al*, *J. Biol. Chem.*, 1956, **218**, 521 (*isol*)
Michelson, A.M., *Biochim. Biophys. Acta*, 1964, **91**, 1 (*synth*)
Furusawa, K. *et al*, *J. Chem. Soc., Perkin Trans. 1*, 1976, 1711.
Labotka, R.J. *et al*, *J. Am. Chem. Soc.*, 1976, **98**, 3699 (*nmr*)
Lee, C. *et al*, *Biochemistry*, 1976, **15**, 697 (*conformn, pmr*)

Guanosine diphosphate mannose **G-00100**

Guanosine 5′-(trihydrogen diphosphate)mono-α-D-mannopyranosyl ester, 9CI. *Guanosine 5′-(trihydrogen pyrophosphate)mono-α-D-mannopyranosyl ester*, 8CI

[3123-67-9]

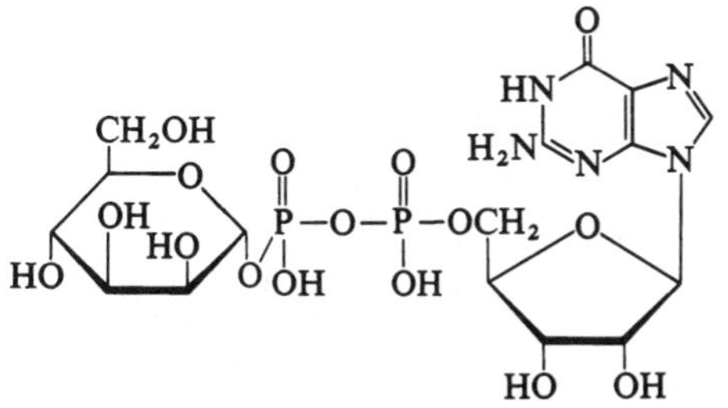

$C_{16}H_{25}N_5O_{16}P_2$ M 605.345

Present in egg white. Presumed intermed. in formation of mannan in yeast. λ_{max} 280, 260 nm (H_2O).

K salt: λ_{max} 252 nm (ϵ 13 700) (H_2O).

Cabib, E. *et al*, *J. Biol. Chem.*, 1954, **206**, 779.
Munch-Petersen, A., *Arch. Biochem. Biophys.*, 1955, **55**, 592 (*synth*)
Donovan, J.W. *et al*, *Arch. Biochem. Biophys.*, 1967, **122**, 17.
Braell, W.A. *et al*, *Anal. Biochem.*, 1976, **74**, 484 (*synth*)
Lee, C. *et al*, *Biochemistry*, 1976, **15**, 697 (*conformn, pmr*)

Guanosine 5′-triphosphate G-00101

Guanosine 5′-(tetrahydrogen triphosphate), 9CI, 8CI. *GTP*

[86-01-1]

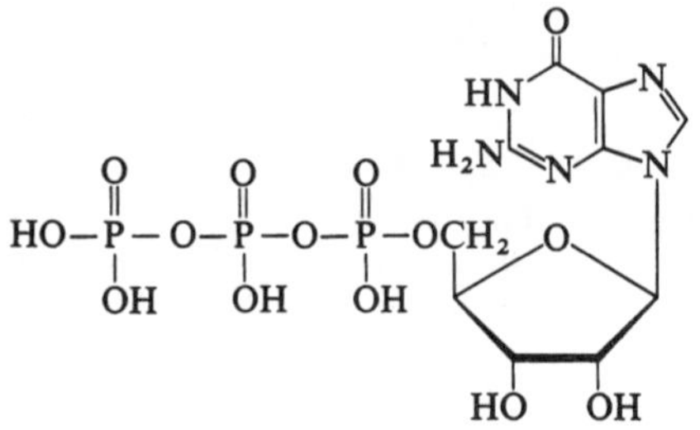

$C_{10}H_{16}N_5O_{14}P_3$ M 523.183

Occurs in many animal and plant tissues. Isol. from yeast. Amorph. λ_{max} 256 nm (pH 1).

▷MF8793000.

Ayengar, P. *et al*, *J. Biol. Chem.*, 1956, **218**, 521 (*isol*)
Kawaguchi, K. *et al*, *Agric. Biol. Chem.*, 1970, **34**, 908 (*synth*)
Chan, S.I. *et al*, *Jerusalem Symp. Quantum Chem. Biochem.*, 1972, **4**, 277; *CA*, **80**, 129365d (*rev, pmr*)
Furusawa, K. *et al*, *J. Chem. Soc., Perkin Trans. 1*, 1976, 1711 (*synth*)
Labotka, R.J. *et al*, *J. Am. Chem. Soc.*, 1976, **98**, 3699 (*nmr*)

3′-Guanylic acid, 9CI, 8CI G-00102

Guanosine 3′-(dihydrogen phosphate), 9CI. *Guanylic acid* b

[117-68-0]

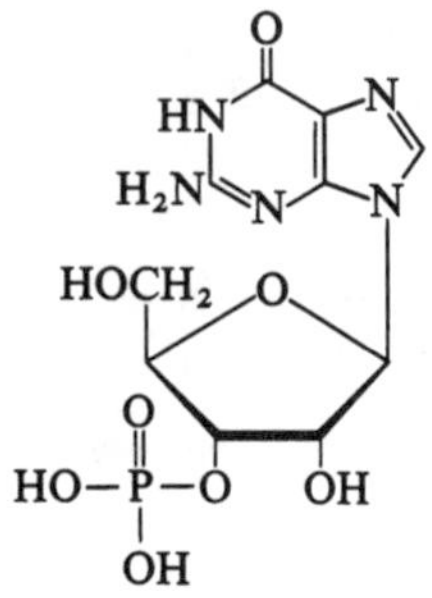

$C_{10}H_{14}N_5O_8P$ M 363.223

Occurs in yeast nucleic acid. Cryst. + $2H_2O$. Mp 180° dec. (208°). $[\alpha]_D^{25}$ −8° (c, 2.0 in H_2O).

Brucine salt: Cryst. + $7H_2O$. Mp 233-240° dec. (anhyd.). $[\alpha]_D^{20}$ −26° (EtOH).

Levene, P.A. *et al*, *J. Biol. Chem.*, 1932, **98**, 9 (*struct*)
Cohn, W.E. *et al*, *Biol. Prep.*, 1957, **5**, 40 (*isol*)
Ts'O, P.O.P. *et al*, *Biochemistry*, 1969, **8**, 997.
Japan. Pat., 71 34 195, (*1971*); *CA*, **76**, 2550x (*synth*)
Tran, D., *Nucleic Acids Res.*, 1975, **2**, 873 (*pmr, conformn*)

5′-Guanylic acid, 9CI, 8CI G-00103

Guanosine 5′-(dihydrogen phosphate). *GMP*

[85-32-5]

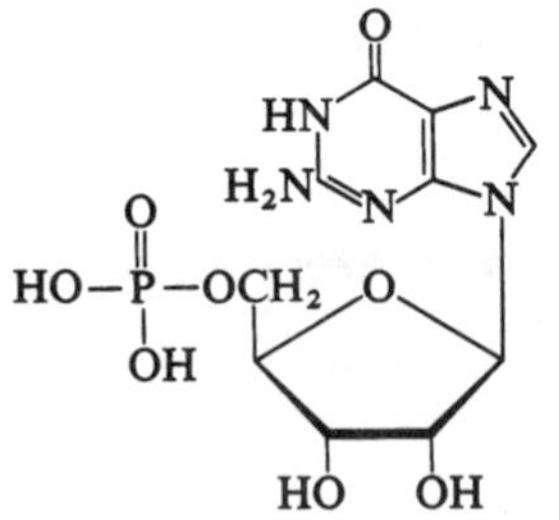

$C_{10}H_{14}N_5O_8P$ M 363.223

Widely distributed in plants and animals, occurs in the hydrolysates of RNA. Mp 190-200° dec. λ_{max} 256 (ϵ 12 400) (pH 2), 260 nm (12 100) (pH 12).

▷MF9283000.

Di-Na salt: Flavour intensifier. Hygroscopic monohydrate. Mp 250° dec. λ_{max} 252.5 nm (ϵ 13 700) (pH 7).

Ba salt: Powder + $8H_2O$. λ_{max} 256 (ϵ 12 400) (pH 2), 260 nm (12 100) (pH 12).

2′,3′-O-Isopropylidene, 5′-(4-nitrophenyl hydrogen phosphate): Hexahydrate. λ_{max} 276 (ϵ 14 000), 258 nm (15 100) (pH 2).

2′,3′-O-Isopropylidene, 5′-(bis-4-nitrophenyl)phosphate: Monohydrate. Mp 263-264°.

Chambers, R.W. *et al*, *J. Am. Chem. Soc.*, 1957, **79**, 3747 (*synth*)
Tener, G.M., *J. Am. Chem. Soc.*, 1961, **83**, 159.
Simoncsits, A. *et al*, *Biochem. Biophys. Acta*, 1975, **395**, 74 (*synth*)
Takaku, O. *et al*, *Agric. Biol. Chem.*, 1975, **39**, 2373 (*pmr*)
Norton, R.S. *et al*, *J. Am. Chem. Soc.*, 1976, **98**, 1007 (*cmr*)
Barnes, C.L. *et al*, *Acta Crystallogr., Sect. B*, 1982, **38**, 812 (*cryst struct*)

H

N,N,N',N',N'',N'',N'''-Heptamethylphosphorimidic triamide, 9CI, 8CI — H-00001

P,P,P-*Tris(dimethylamino)*-N-*methylphosphine imide*. P,P,P-*Tris(dimethylamino)*-N-*methylphosphazene*. P,P,P-*Tris(dimethylamino)*-N-*methyliminophosphorane*

[49778-04-3]

$(Me_2N)_3P{=}NMe$

$C_7H_{21}N_4P$ M 192.243

Liq. Bp_2 64°.

B, HI: Tris(dimethylamino)(methylamino)phosphonium iodide. Heptamethyltetraaminophosphonium iodide.
$C_7H_{22}IN_4P$ M 320.156
Solid. Mp 208-211° dec.

Issleib, K. *et al*, *Synth. Inorg. Metal. Org. Chem.*, 1973, **3**, 255 (*synth, pmr, P nmr*)
Haasemann, P. *et al*, *Z. Anorg. Allg. Chem.*, 1974, **408**, 293 (*synth, ir*)
Goldwhite, H. *et al*, *J. Chem. Soc., Dalton Trans.*, 1975, 12 (*synth, ir, pmr, P nmr*)
Presel, K.D. *et al*, *Z. Anorg. Allg. Chem.*, 1977, **435**, 69 (*complexes*)
Egorov, Yu.P. *et al*, *Teor. Eksp. Khim.*, 1982, **18**, 58 (*Engl. transl.* p. 45) (*P nmr, struct*)

1,7-Heptanediphosphonic acid — H-00002

1,7-Heptanediylbis(phosphonic acid), *9CI*. *Heptamethylenediphosphonic acid*, *8CI*

$(HO)_2P(O)(CH_2)_7P(O)(OH)_2$

$C_7H_{18}O_6P_2$ M 260.163

Solid. Mp 153.5-155°.

Tetra-Et ester: Tetraethyl 1,7-heptanediylbisphosphonate.
$C_{15}H_{34}O_6P_2$ M 372.378
Liq. d_4^{25} 1.07. $Bp_{0.7}$ 188°. n_D^{25} 1.4469.

Bis-dichloride: 1,7-Heptanediylbis(phosphonic dichloride).
$C_7H_{14}Cl_4O_2P_2$ M 333.946
V. hygroscopic solid. Mp 32-33.5°. $Bp_{0.2}$ 142-143° (177-178°).

Kosolapoff, G.M. *et al*, *J. Chem. Soc. (C)*, 1966, 757 (*synth, chloride, ester, ir*)
Raskina, L.P. *et al*, *Zh. Prikl. Khim.*, 1968, **41**, 1544 (*Engl. transl.* p. 1470) (*esters, chloride*)

2,4,6,7,7,8,8-Heptaphenyl-1-phosphabicyclo[4.2.0]octa-2,4-diene, 9CI — H-00003

[58159-86-7]

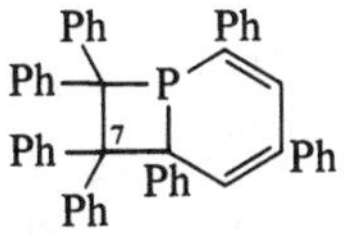

$C_{49}H_{37}P$ M 656.805

Cryst. (AcOH). Mp 189-190°.

Kieselack, P. *et al*, *Chem. Ber.*, 1975, **108**, 3656 (*synth, uv, ms, P nmr, pmr*)

Heptenophos, BSI — H-00004

7-Chlorobicyclo[3.2.0]hepta-2,6-dien-6-yl dimethyl phosphate, *9CI, 8CI*. *Hostaquick*. *Ragadan*

[23560-59-0]

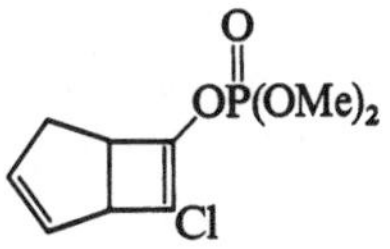

$C_9H_{12}ClO_4P$ M 250.618

Agricultural and veterinary insecticide. Pale-amber liq. Misc. most org. solvs. d_4^{20} 1.294. $Bp_{0.001}$ 94-95°.

▷TB8545000.

B.P., 1 194 603, (*1966*); *CA*, **70**, 106068p (*synth, use*)
Stan, H.-J. *et al*, *J. Chromatogr.*, 1983, **279**, 173 (*glc*)
Pesticide Manual, 6th Ed., 297; 7th Ed., 307.
The Agrochemicals Handbook, Royal Society of Chemistry, 1983, A226.

Heptylphosphine, 9CI — H-00005

$H_3C(CH_2)_6PH_2$

$C_7H_{17}P$ M 132.185

Liq. Bp 150°, Bp 169.5°. n_D^{20} 1.4517.

Watt, G.W. *et al*, *J. Am. Chem. Soc.*, 1948, **70**, 2295 (*synth*)
Schindlbauer, H. *et al*, *Montash. Chem.*, 1961, **92**, 868 (*synth, ir*)

Heptylphosphonic acid, 9CI — H-00006

1-Phosphonoheptane. *1-Heptanephosphonic acid*

[4721-16-8]

$H_3C(CH_2)_6P(O)(OH)_2$

$C_7H_{17}O_3P$ M 180.183

Cryst. (hexane). Mp 105-106°.

Di-Me ester: Dimethyl heptylphosphonate.
$C_9H_{21}O_3P$ M 208.237
Liq. d_4^{20} 1.00. $Bp_{15.5}$ 144°. n_D^{20} 1.4330.

Di-Et ester: [17195-46-9]. *Diethyl heptylphosphonate.*
$C_{11}H_{25}O_3P$ M 236.290
Liq. Bp_1 106-107°. n_D^{20} 1.4270.

Diisopropyl ester: Diisopropyl heptylphosphonate.
$C_{13}H_{29}O_3P$ M 264.344
Liq. Bp_1 100°. n_D^{20} 1.4300.

Diheptyl ester: Diheptyl heptylphosphonate.
$C_{21}H_{45}O_3P$ M 376.558
Liq. d_4^{20} 0.91. Bp_1 198-199°. n_D^{20} 1.4450.

Difluoride: [14576-61-5].
$C_7H_{15}F_2OP$ M 184.165
Liq. d_4^{20} 1.05. Bp 196-197°, Bp_6 74-76°.

Dichloride: [764-11-4].
$C_7H_{15}Cl_2OP$ M 217.075
Liq. d_4^{20} 1.14. $Bp_{4.5}$ 103-104°, Bp_1 87-88°. n_D^{20} 1.4675.

Bis(dimethylamide): [5277-12-3]. P-*Heptyl*-N,N,N',N'-*tetramethylphosphonic diamide. Heptylphosphonic bis(dimethylamide).*
$C_{11}H_{27}N_2OP$ M 234.321

Liq. $Bp_{0.4}$ 113°. n_D^{22} 1.4601.

Kosolapoff, G.M., *J. Am. Chem. Soc.*, 1945, **67**, 1180 (*synth, esters*)
Ford-Moore, A.H. *et al*, *J. Chem. Soc.*, 1947, 1465 (*ester*)
Pudovik, A.N. *et al*, *Zh. Obshch. Khim.*, 1959, **29**, 3342 (*Engl. transl.* p. 3305) (*ester*)
Feshchenko, N.G. *et al*, *Zh. Obshch. Khim.*, 1967, **37**, 473 (*Engl. transl.* p. 441) (*dihalides, esters, synth*)
Geiseler, G. *et al*, *Ber. Bunsenges. Phys. Chem.*, 1967, **71**, 478 (*dichloride, ir, raman*)
Normant, H. *et al*, *C.R. Hebd. Seances Acad. Sci., Ser. C*, 1967, **264**, 707 (*bisdimethylamide*)
Dietze, U., *J. Prakt. Chem.*, 1974, **316**, 293 (*ir*)
Nefant'ev, É.E. *et al*, *Zh. Obshch. Khim.*, 1979, **49**, 1905 (*Engl. transl.* p. 1678) (*synth, dichloride*)

Heptylphosphonothioic acid, 9CI — H-00007

$H_3C(CH_2)_6P(S)(OH)_2 \rightleftharpoons H_3C(CH_2)_6P(O)(OH)(SH)$

$C_7H_{17}O_2PS$ M 196.244
The thione struct. is the preferred one. Syrup.
Monoanilinium salt: [77339-68-5]. Solid. Mp 125.5-125°.
O,O-*Di-Et ester:* O,O-*Diethyl heptylphosphonothioate.*
$C_{11}H_{25}O_2PS$ M 252.351
Liq. Bp_{13} 146-148°. n_D^{20} 1.4640.
O,O-*Dibutyl ester:* O,O-*Dibutyl heptylphosphonothioate.*
$C_{15}H_{33}O_2PS$ M 308.458
Liq. Bp_3 161°. n_D^{20} 1.4622.
O,O-*Bis(trimethylsilyl) ester:* [77339-66-3].
$C_{13}H_{33}O_2PSSi_2$ M 340.607
Liq. Bp_2 113-115°. n_D^{20} 1.4527.

Pudovik, A.N. *et al*, *Zh. Obshch. Khim.*, 1960, **30**, 2348 (*Engl. transl.* p. 2328)
Nifant'ev, E.E. *et al*, *Zh. Obshch. Khim.*, 1980, **50**, 2676 (*Engl. transl.* p. 2159)

Heptylphosphonous acid, 9CI — H-00008

$H_3C(CH_2)_6P(OH)_2 \rightleftharpoons H_3C(CH_2)_6PH(O)OH$

$C_7H_{17}O_2P$ M 164.184
The free acid probably exists in the phosphoryl tautomer.
Diheptyl ester: [3058-21-7]. *Diheptyl heptylphosphonite.*
$C_{21}H_{45}O_2P$ M 360.559
Liq. $Bp_{0.8}$ 155°. n_D^{20} 1.451.
Dichloride: [15573-33-8]. *Dichloroheptylphosphine.*
$C_7H_{15}Cl_2P$ M 201.075
Liq. d_4^{27} 1.06. Bp 228.5°. n_D^{25} 1.4788.
Bis(diethylamide): [20417-42-9]. N,N,N′,N′-*Tetraethyl*-P-*heptylphosphonous diamide.* Liq. Bp_1 145-148°. n_D^{20} 1.4620.

Fox, R.B., *J. Am. Chem. Soc.*, 1950, **72**, 4147 (*dichloride*)
Petrov, K.A. *et al*, *CA*, 1968, **69**, 67487 (*amide*)
Voigt, D. *et al*, *Bull. Soc. Chim. Fr.*, 1964, 3087 (*ester*)

Heptylphosphoramidic acid, 9CI — H-00009

$H_3C(CH_2)_6NHP(O)(OH)_2$

$C_7H_{18}NO_3P$ M 195.198
Dichloride: [34492-20-1].
$C_7H_{16}Cl_2NOP$ M 232.089
Liq. $Bp_{0.004}$ 125-128°.

Mizuma, T. *et al*, *J. Pharm. Soc. Jpn.*, 1961, **81**, 48; *CA*, **55**, 13403.
Dorn, H. *et al*, *J. Prakt. Chem.*, 1971, **313**, 218.

Heptyl phosphorodichloridite, 9CI, 8CI — H-00010

Heptyl dichlorophosphite

$H_3C(CH_2)_6OPCl_2$

$C_7H_{15}Cl_2OP$ M 217.075
Liq. d_0^0 1.11. $Bp_{12.5}$ 107°. n_D^{14} 1.4720.

Razumov, A.I., *Zh. Obshch. Khim.*, 1944, **14**, 464; *CA*, **39**, 4586 (*synth*)

Heptyltriphenylphosphonium(1+), 9CI — H-00011

$Ph_3P^{\oplus}(CH_2)_6CH_3$

$C_{25}H_{30}P^{\oplus}$ M 361.486 (ion)
Bromide: [13423-48-8].
$C_{25}H_{30}BrP$ M 441.390
With $NaN(SiMe_3)_2$ gives the ylide. No phys. props. reported.
Iodide: [59378-88-0].
$C_{25}H_{30}IP$ M 488.390
NaOMe or MeLi yields the ylide. Solid. Mp 128-131°.
Ylide: [55367-56-1]. *Heptylidenetriphenylphosphorane.*
$C_{25}H_{29}P$ M 360.478
Wittig reagent employed in fatty acid synth. Prepd. *in situ.*

Hauser, C.F. *et al*, *J. Org. Chem.*, 1963, **38**, 372 (*bromide*)
Bestmann, H.J. *et al*, *Chem. Ber.*, 1979, **112**, 1923 (*bromide, ylide, use*)
Adlof, R.O. *et al*, *J. Labelled Compd. Radiopharm.*, 1981, **18**, 419 (*iodide, ylide, use*)
Wheaton, G.A. *et al*, *J. Org. Chem.*, 1983, **48**, 917 (*ylide, use*)

Hexabenzylphosphorous triamide — H-00012

Hexakis(phenylmethyl)phosphorous triamide, 9CI.
Tris(dibenzylamino)phosphine
[59758-28-0]

$P[N(CH_2Ph)_2]_3$

$C_{42}H_{42}N_3P$ M 619.788
Forms Pt and Pd complexes. Solid. Mp 99°.

Verstuyft, A.W. *et al*, *Inorg. Chem.*, 1977, **16**, 2776 (*synth, pmr, nmr, complexes*)
Gray, G.A. *et al*, *Org. Magn. Reson.*, 1980, **14**, 8 (*cmr, nmr*)

Hexabutylphosphorous triamide, 9CI — H-00013

*Tris(*N,N*-dibutylamino)phosphine*
[5848-65-7]

$P[N(CH_2CH_2CH_2CH_3)_2]_3$

$C_{24}H_{54}N_3P$ M 415.684
Liq. $Bp_{0.1}$ 140-141°. n_D^{20} 1.4700.
Oxide: [22421-85-8]. *Hexabutylphosphoric triamide. Tris(*N,N*-dibutylamino)phosphine oxide.*
$C_{24}H_{54}N_3PO$ M 431.684
Has been employed in the extraction of the lanthanide elements. Liq. $Bp_{0.01}$ 130-132°. pK_{a1} 12.49, pK_{a2} 5.81 ($MeNO_2$). n_D^{25} 1.4618.
*Sulfide: Hexabutylphosphorothioic triamide. Tris(*N,N*-dibutylamino)phosphine sulfide.*
$C_{24}H_{54}N_3PS$ M 447.744
Low-melting solid. Mp 19-20°. n_D^{20} 1.4869.

Stuebe, C. *et al*, *J. Am. Chem. Soc.*, 1956, **78**, 976 (*synth, derivs*)
Gonnet, C. *et al*, *Anal. Chim. Acta*, 1972, **62**, 227 (*oxide, use*)
Gonnet, C. *et al*, *Bull. Soc. Chim. Fr. Part I*, 1973, 45 (*oxide, use*)
Buchikhin, E.P. *et al*, *Zh. Obshch. Khim.*, 1974, **44**, 1354 (*Engl. transl.* p. 1330) (*oxide, props*)
Marchenko, A.P. *et al*, *Zh. Obshch. Khim.*, 1978, **48**, 551 (*Engl. transl.* p. 501) (*oxide*)
Yakshin, V.V. *et al*, *Dokl. Akad. Nauk SSSR, Ser. Sci. Khim.*, 1979, **247**, 128 (*Engl. transl.* p. 344) (*oxide, props*)

Hexadecyltriphenylphosphonium(1+), 9CI, 8CI — H-00014

$Ph_3P^{\oplus}(CH_2)_{15}CH_3$

$C_{34}H_{52}P^{\oplus}$ M 491.759 (ion)

Bromide: [14866-43-4].
$C_{34}H_{52}BrP$ M 571.663
No phys. props. reported.

Veith, H.J., *Org. Mass Spectrom.*, 1978, **13**, 280 (*ms*)

Hexaethylphosphorous triamide, 8CI — H-00015

Tris(diethylamino)phosphine
[2283-11-6]

$P(NEt_2)_3$

$C_{12}H_{30}N_3P$ M 247.363

Desulphurising agent. Uses similar to those for Hexamethylphosphorous triamide, H-00080 . Liq. Bp 245-246°, $Bp_{0.6}$ 72-74°.

Oxide: [2622-07-3]. *Hexaethylphosphoric triamide, 9CI. Hexaethylphosphoramide.*
$C_{12}H_{30}N_3OP$ M 263.362
Liq. d_4^{20} 0.97. Bp_{12} 135°, $Bp_{1.3}$ 111°. pK_{a1} 11.12, pK_{a2} 7.20 ($MeNO_2$). n_D^{20} 1.4633.

Sulfide: [4154-77-2]. *Hexaethylphosphorothioic triamide, 9CI.*
$C_{12}H_{30}N_3PS$ M 279.423
Liq. $Bp_{0.12}$ 121-122°. n_D^{23} 1.5040.

Selenide: [39181-29-8]. *Hexaethylphosphoroselenoic triamide.*
$C_{12}H_{30}N_3PSe$ M 326.323
Liq. d_4^{20} 1.15. $Bp_{0.003}$ 118°. n_D^{20} 1.5246.

Telluride: [50351-61-1]. No phys. props. reported.

Stuebe, C. *et al*, *J. Am. Chem. Soc.*, 1956, **78**, 976 (*synth, oxide, sulfide*)
Cowley, A.H. *et al*, *J. Am. Chem. Soc.*, 1965, **87**, 4454 (*pmr*)
Coustures, Y. *et al*, *Bull. Soc. Chim. Fr., Part I*, 1973, 926 (*oxide, ir, P nmr*)
Nuretdinov, I.A. *et al*, *Zh. Obshch. Khim.*, 1974, **44**, 2588 (*Engl. transl.* p. 2548) (*selenide, P nmr*)
Hargis, J.H. *et al*, *Inorg. Chem.*, 1977, **16**, 1686 (*pe*)
Barlos, K. *et al*, *Z. Naturforsch., B*, 1978, **33**, 515 (*N nmr*)
Marchenko, A.P. *et al*, *Zh. Obshch. Khim.*, 1978, **48**, 551 (*Engl. transl.* p. 501) (*oxide*)
Davidson, G. *et al*, *Spectrochim. Acta, Part A*, 1979, **35**, 141 (*ir, raman*)
van Linthoudt, J.P. *et al*, *Spectrochim. Acta, Part A*, 1980, **36**, 315 (*cmr, P nmr, pmr*)
Bergesen, K. *et al*, *Acta Chem. Scand., Ser. A*, 1981, **35**, 147 (*cmr, selenide*)
Duangthai, S. *et al*, *Org. Magn. Reson.*, 1982, **20**, 33 (*nmr, struct*)
Rastetter, W.H. *et al*, *J. Org. Chem.*, 1982, **47**, 2785 (*use*)
Bollinger, J.C. *et al*, *Org. Mass Spectrom.*, 1985, **20**, 318 (*ms*)
Fieser, M. *et al*, *Reagents for Organic Synthesis*, Wiley, 1967-84, **3**, 148; **4**, 242; **5**, 323 (*use*)

3,3,4,4,5,5-Hexafluoro-1,2-bis(diphenylphosphino)cyclopentene — H-00016

(3,3,4,4,5,5-Hexafluoro-1-cyclopentene-1,2-diyl)-bis[diphenylphosphine], 9CI

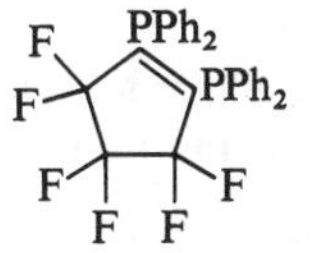

$C_{29}H_{20}F_6P_2$ M 544.415

Ligand for Fe. Cryst. Mp 97-99°.

Dioxide: [4545-95-3]. *1,2-Bis(diphenylphosphinyl)-3,3,4,4,5,5-hexafluorocyclopentene.*
$C_{29}H_{20}F_6O_2P_2$ M 576.414
Cryst. (EtOAc). Mp 176-177°.

Frank, A.W., *J. Org. Chem.*, 1965, **30**, 3663 (*oxide*)
Cullen, W.R. *et al*, *Inorg. Chem.*, 1969, **8**, 95 (*synth, complexes, nmr*)

(Hexafluorocyclobutylidene)-triphenylphosphorane, 9CI — H-00017

1,1,2,2,3,3-Hexafluoro-4-(triphenylphosphoranylidene)-cyclobutane
[50615-91-3]

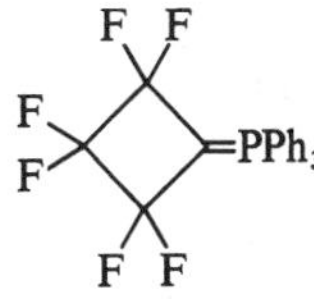

$C_{22}H_{15}F_6P$ M 424.325

Fluorine-stabilised ylide. Cryst.

Stockel, R.F. *et al*, *J. Org. Chem.*, 1968, **33**, 4395 (*synth, pmr*)
Howells, M.A. *et al*, *J. Am. Chem. Soc.*, 1973, **95**, 5366 (*uv, cryst struct*)

3,3,4,4,5,5-Hexafluoro-1,2-cyclopentenediphosphonic acid — H-00018

(3,3,4,4,5,5-Hexafluoro-1-cyclopentene-1,2-diyl)-bisphosphonic acid, 9CI. *3,3,4,4,5,5-Hexafluoro-1,2-diphosphonocyclopentene*
[5942-80-3]

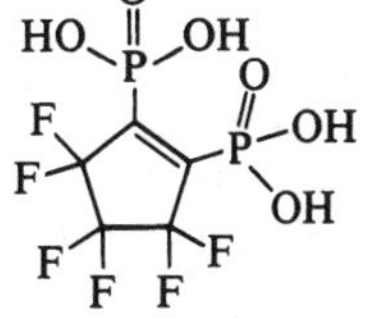

$C_5H_4F_6O_6P_2$ M 336.021

V. hygroscopic solid. Sol. polar solvents, insol. hydrocarbons, chlorinated hydrocarbons. Mp 194-199° dec. pK_{a1} 4.24, pK_{a2} 11.98.

Dianilinium salt: Cryst. (EtOH). Mp 267° dec.

Tetra-Me ester: [4545-88-4]. *Tetramethyl (3,3,4,4,5,5-hexafluoro-1-cyclopentene-1,2-diyl)bisphosphonate.*
$C_9H_{12}F_6O_6P_2$ M 392.128
Liq. $Bp_{0.4}$ 132-134°. n_D^{20} 1.4159.

Tetra-Et ester: [4655-86-1]. *Tetraethyl (3,3,4,4,5,5-hexafluoro-1-cyclopentene-1,2-diyl)bisphosphonate.*
$C_{13}H_{20}F_6O_6P_2$ M 448.235
Liq. $Bp_{0.1}$ 111-112°. n_D^{20} 1.4167.

Bis(dichloride):
$C_5Cl_4F_6O_2P_2$ M 409.804

Solid. Mp 78.5-81.5°. $Bp_{0.45}$ 89-90°.

Frank, A.W., *J. Org. Chem.*, 1965, **30**, 3663 (*esters*)
Frank, A.W., *J. Org. Chem.*, 1966, **31**, 1521 (*synth, ir, props, chloride*)

Hexahydro-1,5-bis(1-phenylethyl)-1*H*-1,5,3-diazaphosphepine, 10CI — H-00019

(*R,R*)-*form*

$C_{20}H_{27}N_2P$ M 326.420

(*R,R*)-*form*

3-Me:
$C_{21}H_{29}N_2P$ M 340.447
Liq. $Bp_{0.00007}$ 200-210°. $[\alpha]^{25}_{D}$ +11.6° (c, 2.3 in $CHCl_3$).

3-Ph:
$C_{26}H_{31}N_2P$ M 402.518
Liq. $Bp_{0.00005}$ 250° (oven). $[\alpha]^{25}_{D}$ +18.0° (c, 1.13 in $CHCl_3$).

(*S,S*)-*form*

3-Et:
$C_{22}H_{31}N_2P$ M 354.474
Liq. $Bp_{0.00004}$ 190-195°. $[\alpha]^{25}_{D}$ −13.7° (c, 1.14 in $CHCl_3$).

3-Cyclohexyl:
$C_{26}H_{37}N_2P$ M 408.565
Liq. $Bp_{0.00005}$ 230-250°. $[\alpha]^{25}_{D}$ −15.4° (c, 0.8 in $CHCl_3$).

Märkl, G. *et al*, *Tetrahedron Lett.*, 1980, **21**, 3467 (*synth*)

Hexahydro-1,3,2-diazaphosphorine 2-oxide, 8CI — H-00020

N,N′-*Trimethylenephosphonic diamide. Hexahydro-1,3,2-diazaphosphinine 2-oxide*

$C_3H_9N_2OP$ M 120.091

2-Methoxy: [16456-53-4]. *Methyl* N,N′-*trimethylenephosphorodiamidate.*
$C_4H_{11}N_2O_2P$ M 150.117
Cryst. (toluene). Mp 86-89°.

2-Phenoxy: see Hexahydro-2-phenoxy-1,3,2-diazaphosphorine, H-00039

2-Me: [16456-51-2]. P-*Methyl*-N,N′-*trimethylenephosphonic diamide.*
$C_4H_{11}N_2OP$ M 134.117
Solid. Mp 116-121°.

2-Ph: [16456-50-1]. N,N′-*Trimethylene*-P-*phenylphosphonic diamide.*
$C_9H_{13}N_2OP$ M 196.188
Solid. Mp 138-141°.

2-Dimethylamino: [16498-14-9]. N″,N″-*Dimethyl*-N,N′-*trimethylenephosphoric triamide.*
$C_5H_{14}N_3OP$ M 163.159
Solid. Mp 127-130°.

U.S.P. 3 463 813, (*1959*); *CA*, **71**, 81439

5,6,7,8,9,10-Hexahydrodibenzo[*b,d*][1,6]-diphosphecin, 8CI — H-00021

5,10-Diphospha-1,2:3,4-dibenzocyclodecadiene

$C_{16}H_{18}P_2$ M 272.266

5,5,10,10-Tetra-Et, dibromide: [16523-75-4]. *5,5,10,10-Tetraethyl-5,6,7,8,9,10-hexahydrodibenzo*[b,d][*1,6*]-*diphosphecinium dibromide. 5,5,10,10-Tetraethyl-5,10-diphosphonia-1,2:3,4-dibenzocyclodecadiene dibromide.*
$C_{24}H_{36}Br_2P_2$ M 546.304
Cryst. + $4H_2O$. Mp 151-152°.

Allen, D.W. *et al*, *J. Chem. Soc.* (*C*), 1967, 1869.

Hexahydro-1,3-dimethyl-1,3,2-diazaphosphorine, 9CI, 8CI — H-00022

Hexahydro-1,3-dimethyl-1,3,2-diazaphosphinine

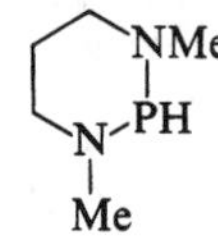

$C_5H_{13}N_2P$ M 132.145

2-Chloro: [40201-85-2]. N,N′-*Dimethyl*-N,N′-*trimethylene phosphorodiamidous chloride.*
$C_5H_{12}ClN_2P$ M 166.590
Pungent liq. d^{20}_{4} 1.16. Bp_{10} 96-98°, $Bp_{0.2}$ 45-47°. n_D^{20} 1.5276.

2-Methoxy: [40201-86-3]. *Methyl* N,N′-*dimethyl*-N,N′-*trimethylene phosphorodiamidite.*
$C_6H_{15}N_2OP$ M 162.171
Liq. Bp_{11} 66°. n_D^{20} 1.4755.

2-Ethoxy: [55666-86-9]. *Ethyl* N,N′-*dimethyl*-N,N′-*trimethylene phosphorodiamidite.*
$C_7H_{17}N_2OP$ M 176.198
Liq. Bp_{13} 80°.

*2-*tert*-Butoxy:* tert-*Butyl* N,N′-*dimethyl*-N,N′-*trimethylene phosphorodiamidite.*
$C_9H_{21}N_2OP$ M 204.251
Liq. Bp_1 48-49°. n_D^{20} 1.4642.

2-Me: N,N′-*Dimethyl*-N,N′-*trimethylene*-P-*methylphosphonous diamide.*
$C_7H_{15}N_2P$ M 158.183
Liq. Bp_{13} 64°. n_D^{22} 1.4927.

2-Ph: [40201-80-7]. N,N′-*Dimethyl*-N,N′-*trimethylene*-P-*phenylphosphonous diamide.*
$C_{11}H_{17}N_2P$ M 208.242
Liq. $Bp_{0.30}$ 119-123°. n_D^{26} 1.5669.

2-Oxide: see Hexahydro-1,3-dimethyl-1,3,2-diazaphosphorine 2-oxide, H-00023

Hutchinson, R.O. *et al*, *J. Am. Chem. Soc.*, 1972, **94**, 9151 (*derivs, synth, nmr, pmr*)
Maryanoff, B.E. *et al*, *J. Org. Chem.*, 1972, **37**, 3475 (*derivs, ms*)
Dennis, R.W. *et al*, *J. Chem. Soc., Perkin Trans. 2*, 1975, 140 (*ethoxy, synth*)
Nuretdinov, I.A. *et al*, *Izv. Akad. Nauk SSSR, Ser. Khim.*, 1978, 950 (*Engl. transl.* p. 824) (*chloro, synth, nmr*)
Nifant'ev, Ê.E. *et al*, *Zh. Obshch. Khim.*, 1979, **49**, 64 (*Engl. transl.* p. 50) (*ethoxy, diethylamino, cmr*)

Hexahydro-1,3-dimethyl-1,3,2-diazaphosphorine 2-oxide, 9CI, 8CI H-00023

N,N'-Dimethyl-N,N'-trimethylenephosphonic diamide

[67364-34-5]

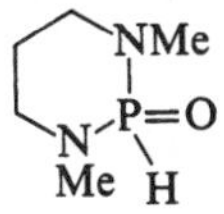

$C_5H_{13}N_2OP$ M 148.144

Liq. d_4^{20} 1.09. $Bp_{0.0001}$ 123-125° (bath). n_D^{20} 1.4963.

2-Ph: [7784-90-9]. N,N'-*Dimethyl*-N,N'-*trimethylene*-P-*phenylphosphonic diamide.*
$C_{11}H_{17}N_2OP$ M 224.242
Liq. $Bp_{0.3}$ 140-143°.

2-Dimethylamino: [7778-09-8]. N,N',N'',N''-*Tetramethyl*-N,N'-*trimethylenephosphoric triamide.*
$C_7H_{18}N_3OP$ M 191.212
Liq. $Bp_{0.1}$ 90-91°.

Ulrich, H. *et al, J. Org. Chem.*, 1967, **32**, 1360 (*dimethylamino, phenyl, synth, pmr*)
Nifant'ev, É.E. *et al, Zh. Obshch. Khim.*, 1978, **48**, 1419 (*Engl. transl.* p. 1302) (*synth, nmr*)

Hexahydro-2,4-dimethyl-3-phenoxy-1,2,4,5,3-tetrazaphosphorine H-00024

Hexahydro-2,4-dimethyl-3-phenoxy-1,2,4,5,3-tetrazaphosphinine

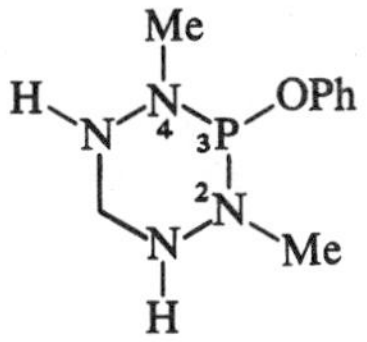

$C_9H_{15}N_4OP$ M 226.217

3-Oxide: [74737-35-2].
$C_9H_{15}N_4O_2P$ M 242.217
Solid. Mp 121-122°.

3-Sulfide: [56634-29-8].
$C_9H_{15}N_4OPS$ M 258.277
Solid. Mp 48-49°.

Majoral, J.P. *et al, Tetrahedron*, 1976, **32**, 2633 (*sulfide, synth, P nmr, pmr*)
Majoral, J.P. *et al, J. Chem. Res. (S)*, 1980, 129 (*oxide, synth, P nmr, pmr*)

2,2,4,4,6,6-Hexahydro-2,2-dimethyl-4,4,6,6-tetraphenyl-1,3,5-triaza-2,4,6-triphosphorine, 9CI H-00025

2,2,4,4,6,6-Hexahydro-2,2-dimethyl-4,4,6,6-tetraphenylcyclotriphosphazene

[50964-72-2]

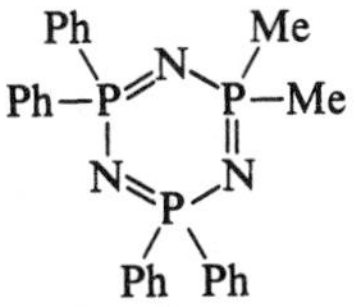

$C_{26}H_{26}N_3P_3$ M 473.433

Cryst. (MeCN or Et_2O/hexane). Mp 142-144°.
Treatment with butyllithium yields a carbanion.

B, HCl: [22880-57-5]. Solid. Mp 180-183°.

Bermann, M. *et al, J. Inorg. Nucl. Chem.*, 1969, **31**, 271 (*ir, pmr, P nmr*)
Appel, R. *et al, Chem. Ber.*, 1973, **106**, 3455 (*synth, ir, ms, pmr*)
Schmidpeter, A. *et al, Chem. Ber.*, 1976, **109**, 2340 (*pmr, P nmr*)
Gallicano, K.D. *et al, J. Inorg. Nucl. Chem.*, 1980, **42**, 923 (*synth*)

7,8,15,16,17,18-Hexahydro-14,18-diphenyl-6*H*,18*H*-1,6*H*-dibenzo[*b,i*][1,11,4,8]-dithiadiphosphacyclotetradecin, 9CI H-00026

13,17-Diphenyl-13,17-diphospha-2,6-dithiatricyclo[16.4.0.0^{7,12}]docosa-7(12),8,10,1(18),19,21-hexaene

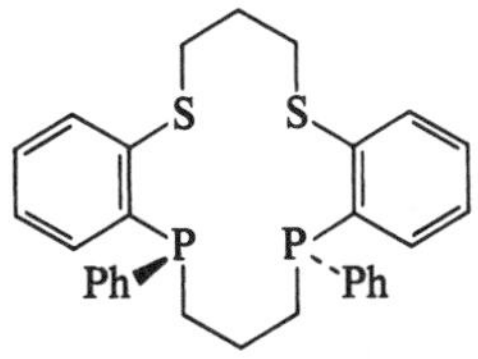

(14*RS*,18*RS*)-*form*

$C_{30}H_{30}P_2S_2$ M 516.635

(14RS,18RS)-form
trans-*form*
Hexagonal plates (EtOAc). Mp 155-157°.

(14RS,18SR)-form
cis-*form*
Cryst. (EtOAc). Mp 127-129°.

Kyba, E.P. *et al, J. Am. Chem. Soc.*, 1985, **107**, 2141 (*synth, pmr, P nmr, cryst struct*)

2,2,4,4,6,6-Hexahydro-2,2,4,4,6,6-hexakis(methylamino)-1,3,5-triaza-2,4,6-triphosphorine, 9CI H-00027

2,2,4,4,6,6-Hexahydro-2,2,4,4,6,6-hexakis(methylamino)cyclotriphosphazene

[1635-63-8]

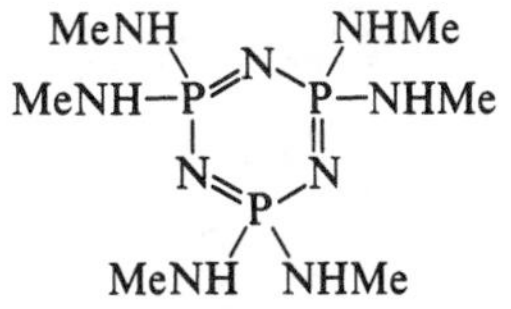

$C_6H_{24}N_9P_3$ M 315.237

Cryst. ($CHCl_3$/pet. ether). Mp 258-259°.

Ray, S.K. *et al, J. Chem. Soc.*, 1961, 872 (*synth*)
Allcock, H.R. *et al, Inorg. Chem.*, 1977, **16**, 3362 (*props*)
Allcock, H.R. *et al, J. Am. Chem. Soc.*, 1981, **103**, 2250; *J. Org. Chem.*, 1981, **46**, 13 (*P nmr*)
Thomas, B. *et al, Phosphorus Sulfur*, 1981, **10**, 375 (*N and P nmr*)
Allcock, H.R. *et al, Inorg. Chem.*, 1982, **21**, 515 (*P nmr, props*)

2,2,4,4,6,6-Hexahydro-2,2,4,4,6,6-hexamethyl-1,3,5-triaza-2,4,6-triphosphorine, 9CI H-00028

2,2,4,4,6,6-Hexahydro-2,2,4,4,6,6-hexamethylcyclotriphosphazene

[6607-30-3]

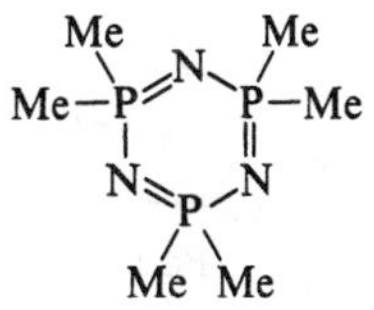

$C_6H_{18}N_3P_3$ M 225.150

Cryst. (pet. ether). Sol. H_2O, polar solvs. Mp 195-196°.
Forms 1:1 adduct with I_2.

Branton, G.R. *et al, J. Chem. Soc. (A)*, 1970, 151 (*pe*)
Markila, P.L. *et al, Can. J. Chem.*, 1974, **52**, 2197 (*adduct, cryst struct*)
Searle, H.T. *et al, J. Chem. Soc., Dalton Trans.*, 1975, 203 (*synth, pmr, ir, P nmr, props, derivs*)
Allcock, H.R. *et al, Inorg. Chem.*, 1977, **16**, 197 (*synth, props*)
Oakley, R.T. *et al, Can. J. Chem.*, 1977, **55**, 4206 (*cryst struct*)
Thomas, B. *et al, Phosphorus Sulfur*, 1981, **10**, 375 (*N and P nmr*)

Hexahydro-2,2,4,4,6,6-hexamethyl-1,3,5-triphenyl-1,3,5,2,4,6-triphosphatrigermanin, 10CI **H-00029**

1,1,3,3,5,5-Hexamethyl-2,4,6-triphenyl-1,3,5-trigerma-2,4,6-triphosphane

[30404-86-5]

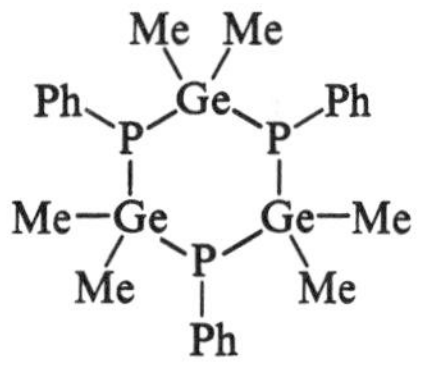

$C_{24}H_{33}Ge_3P_3$ M 632.216

Viscous oil.

Schumann, H. *et al, Chem. Ber.*, 1971, **104**, 333.
Couret, C. *et al, J. Organomet. Chem.*, 1978, **157**, C35.

2,2,4,4,6,6-Hexahydro-2,2,4,4,6,6-hexaphenoxy-1,3,5-triaza-2,4,6-triphosphorine, 9CI **H-00030**

Hexaphenyl cyclophosphonitrilate. 2,2,4,4,6,6-Hexahydro-2,2,4,4,6,6-hexaphenoxycyclotriphosphazene

[1184-10-7]

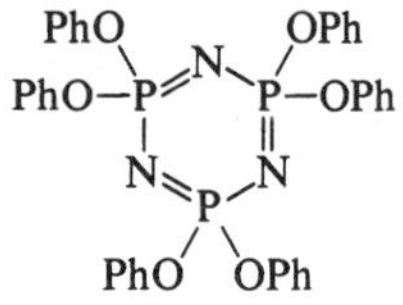

$C_{36}H_{30}N_3O_6P_3$ M 693.571

Solid. Mp 112-113°. Possesses slightly nonplanar ring. Undergoes electrophilic aromatic substitution.

Branton, G.R. *et al, J. Chem. Soc. (A)*, 1970, 151 (*pe*)
Marsh, W.C. *et al, J. Chem. Soc. (A)*, 1971, 169 (*cryst struct*)
Telkova, I.B. *et al, Zh. Obshch. Khim.*, 1973, **43**, 1257 (*Engl. transl.* p. 1247) (*synth, P nmr*)
Gitel', P.O. *et al, Zh. Obshch. Khim.*, 1975, **45**, 1749 (*Engl. transl.* p. 1714) (*synth, uv, ir, P nmr, props*)
Thomas, B. *et al, Phosphorus Sulfur*, 1981, **10**, 375 (*N and P nmr*)

1,3*a*,3*b*,4,6*a*,6*b*-Hexahydro-1,2,3*b*,4,5,6*b*-hexaphenylcyclobuta[1,2-*b*:3,4-*b'*]diphosphole, 8CI **H-00031**

[28124-22-3]

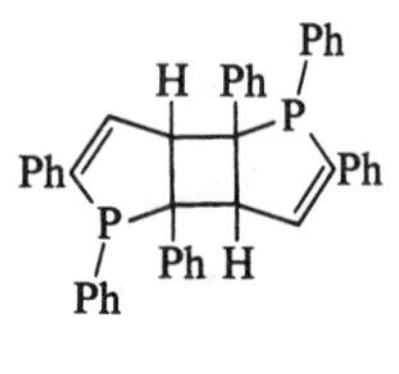

$C_{44}H_{34}P_2$ M 624.700

Stereochemistry unknown. Solid. Mp 229-230°.

Barton, T.J. *et al, Tetrahedron Lett.*, 1969, 5037 (*synth, pmr, P nmr*)

2,2,4,4,6,6-Hexahydro-2,2,4,4,6,6-hexaphenyl-1,3,5-triaza-2,4,6-triphosphorine, 9CI **H-00032**

2,2,4,4,6,6-Hexahydro-2,2,4,4,6,6-hexaphenylcyclotriphosphazene

[1110-78-7]

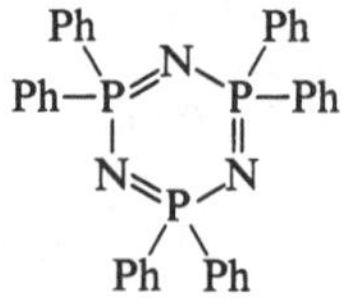

$C_{36}H_{30}N_3P_3$ M 597.574

Prisms. Mp 232° (222°). Ring shows slight distortion to chair form.

Latscha, H.-P., *Z. Anorg. Allg. Chem.*, 1968, **362**, 7 (*P nmr*)
Ahmed, F.R. *et al, Acta Crystallogr., Sect. B*, 1969, **25**, 316 (*cryst struct, rev*)
Biddlestone, M. *et al, J. Chem. Soc. (A)*, 1969, 178; 1971, 2715 (*synth*)
Wagner, A.J. *et al, J. Inorg. Nucl. Chem.*, 1971, **33**, 1307 (*uv*)
Allen, C.W. *et al, J. Chem. Soc., Dalton Trans.*, 1978, 173 (*ms*)
Krishnamurthy, S.S. *et al, Org. Magn. Reson.*, 1981, **15**, 205 (*cmr*)

2,2,4,4,6,6-Hexahydro-2,2,4,4,6,6-hexapropoxy-1,3,5-triaza-2,4,6-triphosphorine, 9CI **H-00033**

2,2,4,4,6,6-Hexahydro-2,2,4,4,6,6-hexapropoxycyclotriphosphazene

[5116-77-8]

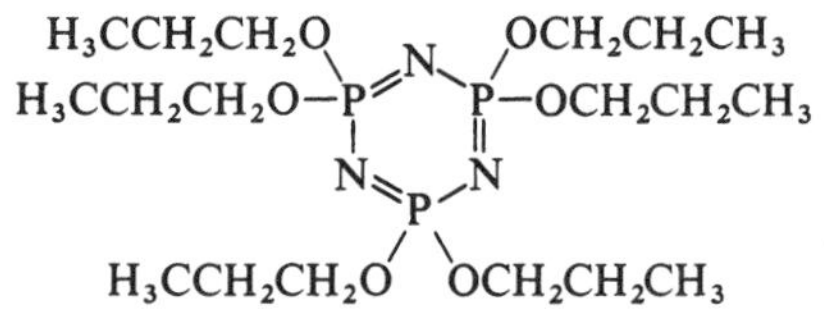

$C_{18}H_{42}N_3O_6P_3$ M 489.468

Fireproofing agent. Oil.

Zeleneva, T.P. *et al, Zh. Obshch. Khim.*, 1973, **43**, 1007 (*Engl. transl.* p. 1000) (*pmr*)
Kajiwara, M. *et al, Polymer*, 1976, **17**, 898 (*synth, P nmr, uv, props*)

Hexahydro-2-hydroxycyclopenta[*d*]-1,3,2-dioxaphosphorin 2-oxide, 9CI **H-00034**

2-Hydroxy-5,6-trimethylene-1,3,2-dioxaphosphorin 2-oxide. Cyclopentatrimethylene phosphoric acid. 2-Hydroxy-2,4-dioxa-3-phosphabicyclo[4.3.0]nonane 3-oxide

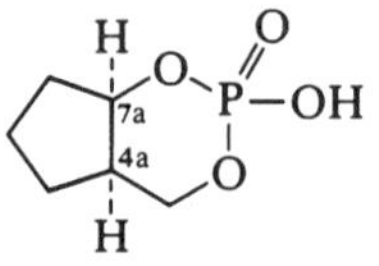

(4a*RS*,7a*RS*)-*form*

$C_6H_{11}O_4P$ M 178.124

(4aRS,7aRS)-form [73424-01-8]

(±)-cis-*form*

Lustrous plates (MeOH/Et_2O). Mp 170-172°.

(4aRS,7aSR)-form [73410-59-0]

(±)-trans-*form*

Cryst. (EtOH/Et_2O). Mp 138.5-139.5°.

Benzylammonium salt: Solid. Mp 153-155°.

2-Methoxy (*Me ester*):
$C_7H_{13}O_4P$ M 192.151
Liq. ca. 95% pure. Stereochem. at P unknown.
2-Phenoxy (*Ph ester*): [68755-19-1].
$C_{12}H_{15}O_4P$ M 254.222
Cryst. Mp 83-85°.
Chloride: [34384-99-1].
$C_6H_{10}ClO_3P$ M 196.570
Liq. unstable to distn. Stereochem. at P unknown.

Ramirez, F. *et al*, *Phosphorus*, 1975, **5**, 73 (*derivs, pmr, P nmr*)
Penny, C.L. *et al*, *Can. J. Chem.*, 1978, **56**, 2396 (*phenyl ester*)
March, F.J. *et al*, *J. Am. Chem. Soc.*, 1980, **102**, 1660 (*props*)

2,3,4,5,6,7-Hexahydro-1*H*-isophosphindole 2-oxide, 10CI H-00035

$C_8H_{13}OP$ M 156.164
2-Me: [65482-10-2].
$C_9H_{15}OP$ M 170.191
Hygroscopic solid. Mp 75-78°. $Bp_{0.03}$ 106-107°.
2-Ph: [80754-56-9].
$C_{14}H_{17}OP$ M 232.261
Cryst. (C_6H_6/ligroin). Mp 75-79°. $Bp_{0.02}$ 153-155°.
2-Hydroxy: [65482-11-3].
$C_8H_{13}O_2P$ M 172.163
Cryst. (Me_2CO). Mp 141-144°.

Quin, L.D. *et al*, *J. Am. Chem. Soc.*, 1977, **99**, 8370; 1982, **104**, 1893 (*synth, pmr, P nmr*)

Hexahydro-1-methyl-4-phenyl-1,4-azaphosphorine, 9CI, 8CI H-00036

Hexahydro-1-methyl-4-phenyl-1,4-azaphosphinine
[52427-44-8]

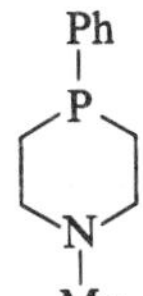

$C_{11}H_{16}NP$ M 193.228
Liq. with phosphine-like odour. $Bp_{0.4}$ 101-102°.
4-Oxide: [52427-40-4].
$C_{11}H_{16}NOP$ M 209.227
Deliquescent cryst. (C_6H_6/cyclohexane). Mp 115-116°.
4-Sulfide: [52427-45-9].
$C_{11}H_{16}NPS$ M 225.288
Needles. Mp 103-104°.

Collins, D.J. *et al*, *Aust. J. Chem.*, 1974, **27**, 841 (*synth, derivs, ir, pmr, ms*)
Gatehouse, B.M. *et al*, *Acta Crystallogr., Sect. B*, 1974, **30**, 2112 (*sulfide, cryst struct*)

Hexahydro-2*H*-[1,2]oxaphosphorino[2,3-*b*][1,2]oxaphosphorin, 9CI H-00037

1-Phospha-2,10-dioxabicyclo[*4.4.0*]*decane*

$C_7H_{13}O_2P$ M 160.152
9-Oxide: [66341-36-4].
$C_7H_{13}O_3P$ M 176.152
Cryst. (Me_2CO). Mp 60-61°. *cis*-geometry.

Bellard, S. *et al*, *Acta Crystallogr., Sect. B*, 1978, **34**, 1032 (*cryst struct*)

Hexahydro-2-phenoxy-1*H*-1,3,2-diazaphosphepine, 8CI H-00038

$C_{10}H_{15}N_2OP$ M 210.215
2-Sulfide: [26387-48-4].
$C_{10}H_{15}N_2OPS$ M 242.275
Cryst. (Et_2O). Mp 86-86.5°.

Edmundson, R.S., *J. Chem. Soc.* (*C*), 1969, 2730 (*synth, ir*)
Edmundson, R.S., *Org. Mass Spectrom.*, 1983, **18**, 150 (*ms, pe*)

Hexahydro-2-phenoxy-1,3,2-diazaphosphorine, 8CI H-00039

Hexahydro-2-phenoxy-1,3,2-diazaphosphinine

$C_9H_{13}N_2OP$ M 196.188
2-Oxide: [16456-52-3]. *Phenyl* N,N′-*trimethylenephosphorodiamidate.*
$C_9H_{13}N_2O_2P$ M 212.188
Cryst. (EtOH). Mp 144.5-145.5°.
2-Sulfide: [26387-47-3]. O-*Phenyl* N,N′-*trimethylenephosphorodiamidothioate.*
$C_9H_{13}N_2OPS$ M 228.248
Cryst. (2-propanol). Mp 51-52°.

Edmundson, R.S., *J. Chem. Soc.* (*C*), 1969, 2730 (*derivs, synth, ir*)
Kraemer, R. *et al*, *Bull. Soc. Chim. Fr.*, 1971, 3580 (*derivs, synth, ir, P nmr*)
Mathis, R. *et al*, *Spectrochim. Acta, Part A*, 1973, **29**, 63 (*oxide, ir*)
Edmundson, R.S., *Org. Mass Spectrom.*, 1983, **18**, 150 (*derivs, ms, pe*)

Hexahydro-3-phenyl-1*H*-1,3-azaphosphepine, 9CI H-00040

[42451-87-6]

$C_{11}H_{16}NP$ M 193.228
Liq. Bp_2 135-136°.

Issleib, K. *et al*, *Z. Chem.*, 1973, **13**, 139 (*synth, derivs*)

Hexahydro-3-phenyl-1,3-azaphosphorine, 8CI H-00041

Hexahydro-3-phenyl-1,3-azaphosphinine

[21200-06-6]

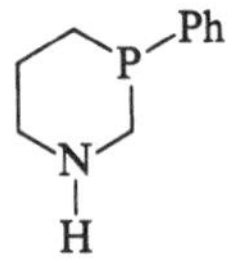

$C_{10}H_{14}NP$ M 179.201

Liq. $Bp_{1.5}$ 127-130°.

1-Ph: [49789-34-6]. *Hexahydro-1,3-diphenyl-1,3-azaphosphorine.*

$C_{16}H_{18}NP$ M 255.299

Liq. $Bp_{0.01}$ 148-153°.

1-Ph, 3-sulfide: [49789-43-7].

$C_{16}H_{18}NPS$ M 287.359

Cryst. ($EtOH/Et_2O$). Sol. C_6H_6, alcohols, Me_2CO. Mp 114-116°.

Issleib, K. *et al*, *Chem. Ber.*, 1968, **101**, 4032 (*synth*)
Oehme, H. *et al*, *J. Prakt. Chem.*, 1973, **315**, 526 (*deriv*)

Hexahydro-4-phenyl-1,4-azaphosphorine, 9CI, 8CI H-00042

Hexahydro-4-phenyl-1,4-azaphosphinine

[15916-61-7]

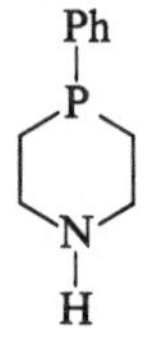

$C_{10}H_{14}NP$ M 179.201

Liq. Bp_3 145-147°. pK_a 8.45 (50% EtOH aq., 25°).

4-Sulfide: [15916-62-8].

$C_{10}H_{14}NPS$ M 211.261

Cryst. (C_6H_6/pet. ether). Insol. Et_2O, H_2O. Mp 114-115°.

1-Ph: Hexahydro-1,4-diphenyl-1,4-azaphosphorine.

$C_{16}H_{18}NP$ M 255.299

Cryst. (EtOH). Mp 89-90°.

1-Ph, 4-oxide:

$C_{16}H_{18}NOP$ M 271.298

Cryst. (H_2O). Mp 145-147°.

1-Me: see Hexahydro-1-methyl-4-phenyl-1,4-azaphosphorine, H-00036

Mann, F.G. *et al*, *J. Chem. Soc.*, 1952, 3039 (*deriv, uv*)
Issleib, K. *et al*, *Chem. Ber.*, 1967, **100**, 2685 (*synth*)

6,6*a*,7,8,9,9*a*-Hexahydro-5-phenyl-1*H*-benzo[*h*]cyclopenta[*c*]phosphinoline, 9CI H-00043

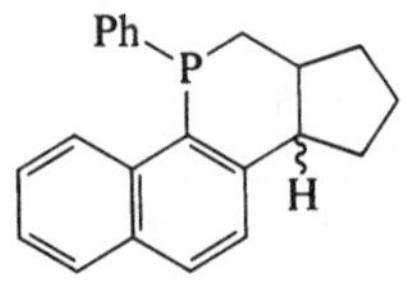

$C_{22}H_{21}P$ M 316.382

B,EtClO₄: [35456-77-0]. *5-Ethyl-6,6*a*,7,8,9,9*a*-hexahydro-5-phenyl-1*H*-benzo*[h]*cyclopenta*[c]*-phosphinolinium perchlorate.*

$C_{24}H_{26}ClO_4P$ M 444.894

Cryst. (2-propanol). Mp 183.5-187°. Possible stereoisomeric mixt.

Chen, C.H. *et al*, *Phosphorus*, 1971, **1**, 49 (*synth, pmr, ir*)

3,4,5,6,7,8-Hexahydro-5-phenyl-2*H*-1,9,5-benzodithiaphosphacycloundecin, 10CI H-00044

6-Phenyl-6-phospha-2,10-dithiabicyclo[9.4.0]-pentadeca-11(1),12,14-triene

[68351-49-5]

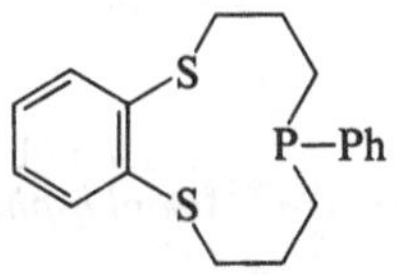

$C_{18}H_{21}PS_2$ M 332.458

Rods (cyclohexane). Mp 109.5-110°.

Kyba, E.P. *et al*, *J. Am. Chem. Soc.*, 1980, **102**, 139 (*synth, pmr, cmr, P nmr, cryst struct*)

4*b*,5,6,10*b*,11,12-Hexahydro-5-phenylbenzo[*c*]phosphanthridine, 10CI H-00045

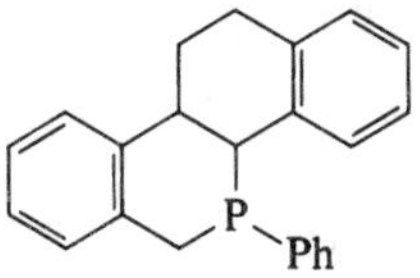

$C_{23}H_{21}P$ M 328.393

B,PhPF₆: [76062-43-6]. *4*b*,5,6,10*b*,11,12-Hexahydro-5,5-diphenylbenzo*[c]*phosphanthridinium hexafluorophosphate.*

$C_{29}H_{26}F_6P_2$ M 550.462

Solid. Mp 130-132.5°.

Radhakrishna, A.S. *et al*, *Pol. J. Chem.*, 1980, **54**, 495 (*synth, pmr*)

1,2,3,4,7,8-Hexahydro-1-phenyl-5,6-bis[(trimethylsilyl)oxy]phosphocin, 9CI H-00046

[40438-07-1]

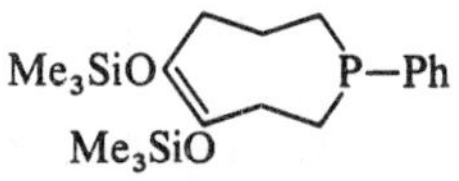

$C_{19}H_{33}O_2PSi_2$ M 380.613

Liq. $Bp_{0.04}$ 124-125°.

van Reijendam, J. *et al*, *Tetrahedron Lett.*, 1972, 5181.

2,3,4,7,8,9-Hexahydro-1-phenyl-5,6-bis[(trimethylsilyl)oxy]-1*H*-phosphonine, 9CI H-00047

[40334-15-4]

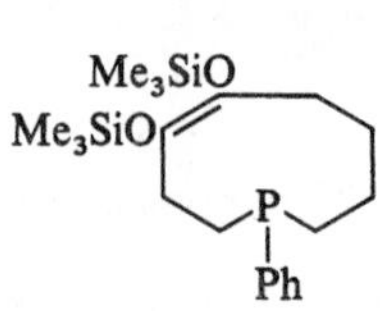

$C_{20}H_{35}O_2PSi_2$ M 394.640

Liq. $Bp_{0.09}$ 134-137°.

Sulfide:

$C_{20}H_{35}O_2PSSi_2$ M 426.700

Solid. Mp 108-110°.

Van Reijendam, J.W. *et al*, *Tetrahedron Lett.*, 1972, 5181.

2,3,3*a*,4,5,9*b*-Hexahydro-4-phenyl-1*H*-cyclopent[*c*]isophosphinoline, 9CI **H-00048**

2,3,3a,4,5,9b-Hexahydro-4,4-diphenyl-1H-cyclopent[c]-isophosphinolinum hexafluorophosphate

$C_{18}H_{19}P$ M 266.322

B, PhBr: [54230-05-6]. *2,3,3a,4,5,9b-Hexahydro-4,4-diphenyl-1H-cyclopent[c]isophosphinolinium bromide.*

$C_{24}H_{24}BrP$ M 423.331

Solid. Mp 326-328°.

B, PhPF₆: *2,3,3a,4,5,9b-Hexahydro-4,4-diphenyl-1H-cyclopent[c]isophosphinolinium hexafluorophosphate.*

$C_{24}H_{24}F_6P_2$ M 488.392

Solid. Mp 264-266°.

Dilbeck, G.A. *et al*, *J. Org. Chem.*, 1975, **40**, 1150 (*synth, ir, pmr*)

2,3,5,6,8,8-Hexahydro-8-phenyl-[1,3,2]-oxazaphospholo[2,3-*b*][1,3,2]-oxazaphosphole, 9CI **H-00049**

[57680-64-5]

$C_{10}H_{14}NO_2P$ M 211.200

Tautomeric with monocyclic form. Forms Fe, Mo, and Rh complexes. Liq. $Bp_{0.001}$ 108°.

Houalla, D. *et al*, *Phosphorus*, 1975, **5**, 229 (*synth, pmr, P nmr*)
Devillers, J. *et al*, *Org. Magn. Reson.*, 1976, **8**, 500 (*pmr*)
Wachter, J. *et al*, *Organometallics*, 1984, **3**, 714 (*complex*)

1,4,4*a*,9,9*a*,10-Hexahydro-11-phenyl-9,10-phosphinideneanthracene, 9CI **H-00050**

$C_{20}H_{19}P$ M 290.344

11-Oxide: [58499-02-8].

$C_{20}H_{19}OP$ M 306.343

Cryst. ($CHCl_3$/hexane). Mp 185-187°.

Chan, T.H. *et al*, *Tetrahedron*, 1975, **31**, 2537 (*synth, ms, ir, pmr*)

Hexahydro-3-phenyl-1,2,4,3-triazaphosphorine, 9CI **H-00051**

Hexahydro-3-phenyl-1,2,4,3-triazaphosphinine

$C_8H_{12}N_3P$ M 181.177

1,4-Di-Me: [55054-73-4].

$C_{10}H_{16}N_3P$ M 209.230

Solid. Mp 58°. Bp_1 119-123°.

1,4-Diisopropyl:

$C_{14}H_{24}N_3P$ M 265.337

Solid. Mp 59°. $Bp_{1.5}$ 134-138°.

1,4-Dibutyl:

$C_{16}H_{28}N_3P$ M 293.391

Liq. Bp_1 150-155°. n_D^{20} 1.5172.

Gol'din, G.S. *et al*, *Zh. Obshch. Khim.*, 1974, **44**, 2668 (*Engl. transl.* p. 2623) (*synth, ms, ir*)

Hexahydrospiro[1,3,2-benzoxazaphosphole-2(3*H*),1′-phospholane], 10CI **H-00052**

$C_{10}H_{20}NOP$ M 201.248

2-Ph: [71934-28-6].

$C_{16}H_{24}NOP$ M 277.345

Solid. Mp 37-39°. Cyclohexane stereochemistry not specified.

Cadogan, J.I.G. *et al*, *J. Chem. Soc., Chem. Commun.*, 1979, 189 (*synth, P nmr*)

1,2,3,4,5,6-Hexahydro-2,2,5,5-tetraphenyl-2,5-benzodiphosphocinium(2+), 8CI **H-00053**

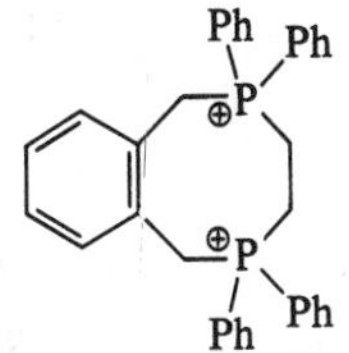

$C_{34}H_{32}P_2^{\oplus\oplus}$ M 502.574 (ion)

System undergoes phosphorus-ring cleavage when treated with NaOH aq.

Dibromide: [15352-69-9]. *Tetraphenyl-1,4-diphosphonia-6,7-benzocyclooct-6-ene.*

$C_{34}H_{32}Br_2P_2$ M 662.382

Cryst. (MeOH). Mp 345-347°.

5,6-Didehydro, dibromide: [15303-25-0]. *1,2,5,6-Tetrahydro-2,2,5,5-tetraphenyl-2,5-benzodiphocinium dibromide. Tetraphenyl-1,4-diphosphonia-6,7-benzocycloocta-2,6-diene.*

$C_{34}H_{30}Br_2P_2$ M 660.367

Cryst. (MeOH). Mp 331-334°.

Aguiar, A.M. *et al*, *J. Org. Chem.*, 1968, **33**, 579 (*synth, props, pmr*)

Hexahydro-3*H*,6*H*-2*a*,5*a*,8*a*-triaza-8*b*-phosphaacenaphthylene, 10CI **H-00054**

[62051-25-6]

$C_8H_{16}N_3P$ M 185.208

Liq. $Bp_{0.3}$ 73-75°.

Atkins, T.J., *Tetrahedron Lett.*, 1978, 4331 (*synth, cmr, P nmr*)

Hexahydro-2a,4a,6a-triaza-6b-phosphacyclopenta[cd]pentalene H-00055

10-Phospha-1,4,7-triazatricyclo[5.2.1.0^{4,10}]decane

[71771-36-3]

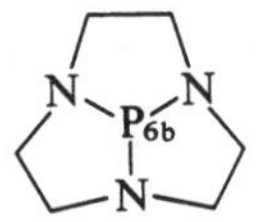

$C_6H_{12}N_3P$ M 157.155

Unstable.

6b-Oxide: [71771-37-4].

$C_6H_{12}N_3OP$ M 173.154

Cryst. (C_6H_6/hexane). Mp 215° dec.

6b-Sulfide: [71771-38-5].

$C_6H_{12}N_3PS$ M 189.215

Cryst. ($CHCl_3/C_6H_6$). Mp 180° dec. $Bp_{0.1}$ 100° subl.

6b-Selenide: [71771-39-6].

$C_6H_{12}N_3PSe$ M 236.115

Cryst. ($CHCl_3/Et_2O$). Mp 218.5-222°. $Bp_{0.1}$ 100° subl.

White, D.W. *et al*, *J. Am. Chem. Soc.*, 1979, **101**, 4921 (*derivs*)
Cowley, A.H. *et al*, *Inorg. Chem.*, 1982, **21**, 543 (*derivs*, *pe*)

2,2,2,3,4,5-Hexahydro-2,2,2-trimethoxy-3-phenylnaphth[2,1-*d*]-1,2-oxaphosphole H-00056

[61704-76-5]

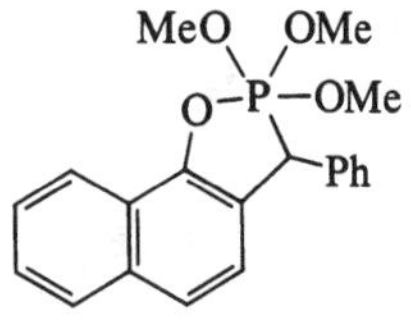

$C_{20}H_{21}O_4P$ M 356.357

Cryst. Mp 63-64.5°.

Arbuzov, B.A. *et al*, *Izv. Akad. Nauk SSSR, Ser. Khim.*, 1976, 2369 (*Engl. transl.* p. 2212)

Hexahydro-1,3,4-trimethyl-1,3,2-diazaphosphorine, 9CI H-00057

Hexahydro-1,3,4-trimethyl-1,3,2-diazaphosphinine

[67364-36-7]

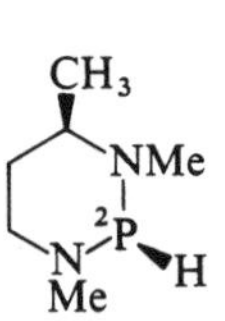

cis-form

$C_6H_{15}N_2P$ M 146.172

Liq., oxid. v. rapidly in air. Bp_8 63-64°. Consists of a mixt. of *cis-* and *trans-*forms.

Oxide: N,N′-*1,3-Butylene* N,N′-*dimethylphosphonic diamide.*

$C_6H_{15}N_2OP$ M 162.171

Liq. d_4^{20} 1.12. $Bp_{0.0001}$ 130-135° (bath). n_D^{20} 1.4930. Consists of a mixt. of *cis-* and *trans-*forms.

2-Chloro: [67364-37-8]. N,N′-*1,3-Butylene*-N,N′-*dimethylphosphorodiamidous chloride.*

$C_6H_{14}ClN_2P$ M 180.617

Liq. $Bp_{1.5}$ 84-85°. n_D^{20} 1.5261. Consists of a mixt. of *cis-* and *trans-*forms.

2-Methoxy: Methyl N,N′-*1,3-butylene*-N,N′-*dimethylphosphorodiamidite.*

$C_7H_{17}N_2OP$ M 176.198

Liq., v. easily oxid. d_4^{20} 1.00. Bp_5 66-67°. n_D^{20} 1.4781. Consists of a mixt. of *cis-* and *trans-*forms.

2-Phenoxy: Phenyl N,N′-*1,3-butylene*-N,N′-*dimethylphosphorodiamidite.*

$C_{12}H_{19}N_2OP$ M 238.269

Liq. d_4^{20} 1.07. Bp_1 116-117°. n_D^{20} 1.5380. Consists of a mixt. of *cis-* and *trans-*forms.

2-Diethylamino:

$C_{10}H_{24}N_3P$ M 217.293

Liq. $Bp_{1.5}$ 87-88°. n_D^{20} 1.4834. Consists of a mixt. of *cis-* and *trans-*forms.

Nifant'ev, É.E. *et al*, *Zh. Obshch. Khim.*, 1978, **48**, 1419 (*Engl. transl.* p. 1302) (*synth*, *oxide*)
Nifant'ev, É.E. *et al*, *Zh. Obshch. Khim.*, 1979, **49**, 64 (*Engl. transl.* p. 53) (*P nmr*, *cmr*, *derivs*)
Nifant'ev, É.E. *et al*, *Tetrahedron*, 1981, **37**, 3183 (*derivs*, *cmr*, *nmr*)

Hexahydro-1,3,5-triphenyl-1,3,5-diazaphosphorine H-00058

[72897-05-3]

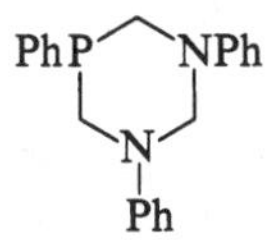

$C_{21}H_{21}N_2P$ M 332.384

Needles (MeCN). Mp 115°.

5-Oxide: [74607-62-8].

$C_{21}H_{21}N_2OP$ M 348.383

Cryst. (MeCN). Mp 147-148°.

5-Sulfide: [74607-65-1].

$C_{21}H_{21}N_2PS$ M 364.444

Solid. Mp 109-112°.

Arbuzov, B.A. *et al*, *Izv. Akad. Nauk SSSR, Ser. Khim.*, 1980, 721; *CA*, **93**, 95334 (*derivs*, *synth*, *P nmr*)
Arbuzov, B.A. *et al*, *Izv. Akad. Nauk SSSR, Ser. Khim.*, 1980, 1571 (*Engl. transl.* p. 1115) (*synth*, *pmr*, *P nmr*)
Märkl, G. *et al*, *Tetrahedron Lett.*, 1981, **22**, 229 (*synth*)

2,2,2,3,3a,11b-Hexahydro-2,2,2-triphenylphenanthro[9,10-*d*]-1,3,2-oxazaphosphole, 9CI H-00059

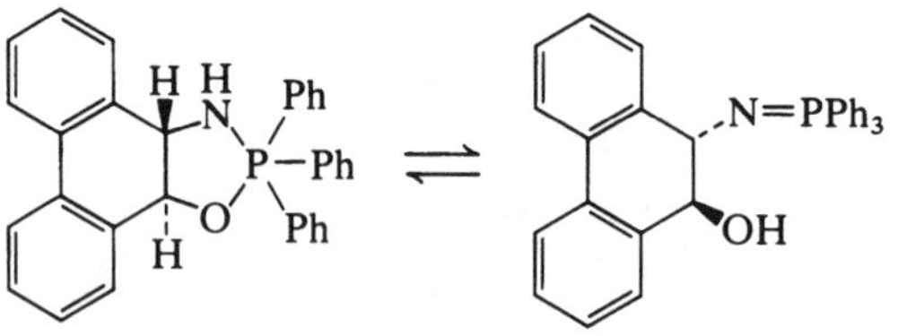

$C_{32}H_{26}NOP$ M 471.537

Tautomeric mixture.

(9RS,10RS)-form [93214-19-8]

(±)-trans-*form*

Solid. Mp 110° dec.

Pöchlauer, P. *et al*, *Helv. Chim. Acta*, 1984, **67**, 1238 (*synth*, *props*, *ir*, *pmr*, *cmr*, *P nmr*)

2,2,4,4,6,6-Hexakis(1-aziridinyl)-2,2,4,4,6,6-hexahydro-1,3,5-triaza-2,4,6-triphosphorine, 9CI H-00060

Apholate. Myco 63. Myko 63

[52-46-0]

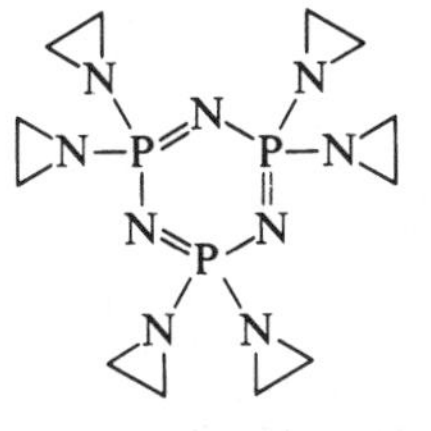

$C_{12}H_{24}N_9P_3$ M 387.303

Insect sterilant. Powerful anticancer agent used clinically against lympholeukaemia. Orthorhombic needles (*m*-xylene) monoclinic cryst. (CS_2). V. sol. H_2O. Mp 150°. Forms solvates with C_6H_6 (2:1) and with CCl_4 (1:3).

▷XX9450000.

Spell, H.L., *Anal. Chem.*, 1967, **39**, 185 (*ir*)
Labarre, J.-F. *et al*, *J. Mol. Struct.*, 1980, **63**, 127 (*struct*)
Monsarrat, B. *et al*, *Biomed. Mass. Spectrom.*, 1980, **7**, 405 (*ms*)
Alix, A.J.P. *et al*, *J. Chim. Phys., Phys. Chim. Biol.*, 1982, **79**, 129 (*struct*)
Cameron, T.S. *et al*, *Acta Crystallogr., Sect. B*, 1982, **38**, 168, 2000 (*cryst struct*)
Lahana, R. *et al*, *THEOCHEM.*, 1982, **4**, 283 (*struct*)
Guerch, G. *et al*, *THEOCHEM.*, 1982, **5**, 317 (*struct*)
Manfait, M. *et al*, *J. Raman Spectrosc.*, 1982, **12**, 212 (*P nmr, ir, raman*)

2,2,4,4,6,6-Hexakis(dimethylamino)-2,2,4,4,6,6-hexahydro-1,3,5-triaza-2,4,6-triphosphorine, 9CI H-00061

2,2,4,4,6,6-Hexakis(dimethylamino)-2,2,4,4,6,6-hexahydrocyclophosphazene

[974-68-5]

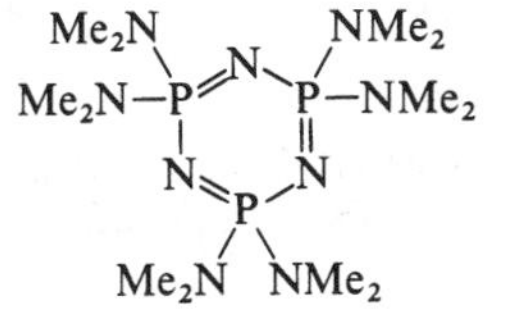

$C_{12}H_{36}N_9P_3$ M 399.398

Plates (pet. ether). Mp 104°. Sl. distorted planar ring with boat conformation.

▷XX9430000.

B, HCl: [1097-84-3]. Cryst. (C_6H_6/pet. ether). Mp 198-200°.

Keat, R. *et al*, *J. Chem. Soc.*, 1965, 2215 (*synth*)
Keat, R. *et al*, *J. Chem. Soc. (A)*, 1968, 703 (*pmr*)
Branton, G.R. *et al*, *J. Chem. Soc. (A)*, 1970, 151 (*pe*)
Green, B. *et al*, *J. Chem. Soc., Dalton Trans.*, 1973, 1042 (*pe*)
Rettig, S.J. *et al*, *Can. J. Chem.*, 1973, **51**, 1295 (*cryst struct*)
Nabi, S.N. *et al*, *J. Chem. Soc., Dalton Trans.*, 1974, 1618; 1975, 588, 2634 (*props*)
Keat, R. *et al*, *J. Chem. Soc., Dalton Trans.*, 1976, 1582 (*P nmr*)

Hexakis(dimethylamino)-μ-oxodiphosphorus(2+) H-00062

*Hexakis(*N*-methylmethanaminato)-μ-oxodiphosphorus(2+), 9CI*

$[(Me_2N)_3POP(NMe_2)_3]^{\oplus\oplus}$

$C_{12}H_{36}N_6OP_2^{\oplus\oplus}$ M 342.404 (ion)

Bis(tetrafluoroborate): [55881-03-3].
$C_{12}H_{36}B_2F_8N_6OP_2$ M 516.011
Used in peptide bond formation. In conjunction with (1-Hydroxy-1*H*-benzotriazolato-*O*)-tris(dimethylamino)phosphorus(1+), H-00111 the degree of racemization is reduced. Solid. Mp 194-204°.

Bis(trifluoromethanesulfonate): [72450-49-8].
$C_{14}H_{36}F_6N_6O_7P_2S_2$ M 640.532
Cryst. (EtOH/2-propanol). Mp 230° dec.

Bis-(4-methylbenzenesulfonate): [55878-19-8].
$C_{26}H_{50}N_6O_{13}P_2S_2$ M 780.781
Hygroscopic solid. Mp 115-117°.

Bates, A.J. *et al*, *Helv. Chim. Acta*, 1975, **58**, 688 (*synth, pmr, nmr, use*)
Galpin, I.J. *et al*, *Tetrahedron*, 1976, **32**, 2417 (*use*)
Aaberg, A. *et al*, *Acta Chem. Scand., Ser. A*, 1980, **34**, 717 (*trifluoromethanesulfonate, cryst struct*)

Hexakis(2-methylpropyl)phosphorous triamide, 9CI H-00063

Hexaisobutylphosphorous triamide, 8CI.
Tris(diisobutylamino)phosphine

$P[N[CH_2CH(CH_3)_2]_2]_3$

$C_{24}H_{54}N_3P$ M 415.684

Oxide: Hexakis(2-methylpropyl)phosphoric triamide. Hexaisobutylphosphoric triamide.
$C_{24}H_{54}N_3OP$ M 431.684
Liq. d_4^{25} 0.91. $Bp_{0.015}$ 116-118°. pK_{a1} 10.13, pK_{a2} 7.81 ($MeNO_2$). n_D^{25} 1.4686.

Marchenko, A.P. *et al*, *Zh. Obshch. Khim.*, 1978, **48**, 551 (*Engl. transl.* p. 501) (*oxide, methiodide*)
Yakshin, V.V. *et al*, *Dokl. Akad. Nauk SSSR, Ser. Sci. Khim.*, 1979, **247**, 128 (*Engl. transl.* p. 344) (*oxide, props*)

2,3,5,6,7,8-Hexakis(trifluoromethyl)-1,4-diphosphabicyclo[2.2.2]octa-2,5,7-triene, 9CI H-00064

2,3,5,6,7,8-Hexakis(trifluoromethyl)-1,4-diphosphabarrelene

[2925-91-9]

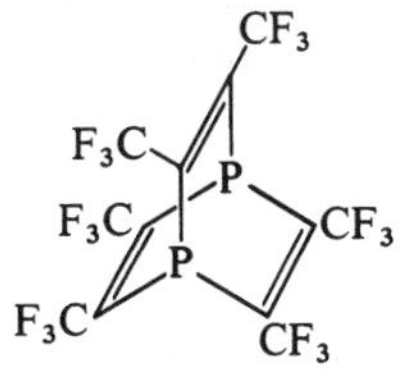

$C_{12}F_{18}P_2$ M 548.051

Cryst. (AcOH). Insol. c.H_2SO_4. Mp 119-120°. Subl. at 100°/100 mm. Does not react with O_2, Br_2, MeI or $PhCH_2Cl$, even at 100°.

Krespan, C.G. *et al*, *J. Am. Chem. Soc.*, 1960, **82**, 1515; 1961, **83**, 3432 (*synth, uv, ir, F nmr*)
Kobayashi, Y. *et al*, *Tetrahedron Lett.*, 1977, 867 (*props*)

2,5,7,10,11,12-Hexakis(trifluoromethyl)-1,6-diphosphahexacyclo-[4.4.2.0^{2,5},0^{3,9},0^{4,8},0^{7,10}]dodec-11-ene H-00065

*1,1*a*,2,3,6*a*,6*b*-Hexahydro-3,4,5,6*a*,6*b*,7-hexakis(trifluoromethyl)-1,2,3,6-ethanediylidene-6*H*-3*a*,6-diphosphacyclobut*[c,d]indene, 9CI

[62839-78-5]

$C_{16}H_4F_{18}P_2$ M 600.126

Plates (Me_2CO). Mp 201-203° (sealed tube).

Kobayashi, Y. *et al*, *J. Org. Chem.*, 1979, **44**, 4930 (*synth, ir, pmr, F nmr*)
Schomburg, D. *et al*, *Phosphorus Sulfur*, 1981, **10**, 17 (*cryst struct*)

4,5,9,10,11,12-Hexakis(trifluoromethyl)-1,8-diphosphatetracyclo[6.2.2.0^{2,7}.0^{3,6}]-dodeca-4,9,11-triene, 9CI H-00066

$C_{16}H_4F_{18}P_2$ M 600.126

(2α,3β,6β,7α)-form [71901-72-9]
exo-*form*
Solid. Mp 175-176° (sealed tube).

Kobayashi, Y. *et al*, *J. Org. Chem.*, 1979, **44**, 4930 (*synth, ir, ms, pmr*)

2,7,9,10,11,12-Hexakis(trifluoromethyl)-1,8-diphosphatetracyclo[6.2.2.0^{2,7}.0^{3,6}]-dodeca-4,9,11-triene, 10CI H-00067

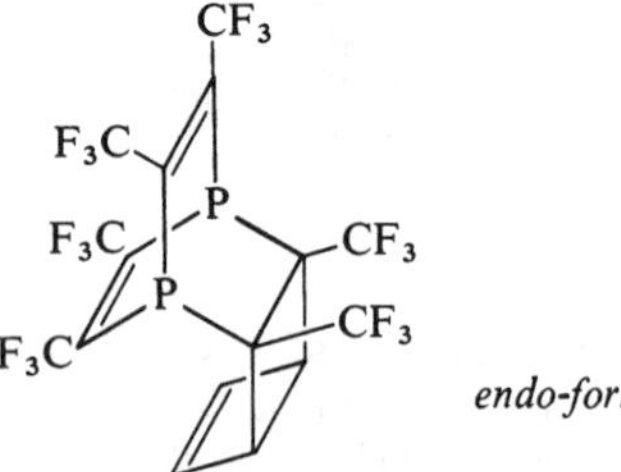

endo-form

$C_{16}H_4F_{18}P_2$ M 600.126

(2α,3α,6α,7α)-form [62929-10-6]
endo-*form*
Plates (pentane). Mp 111-112° (sealed tube).

(2α,3β,6β,7α)-form [62839-77-4]
exo-*form*
Plates (pentane). Mp 97-98° (sealed tube).

Kobayashi, Y. *et al*, *J. Org. Chem.*, 1979, **44**, 4930 (*synth, ir, ms, pmr, F nmr*)

***N,N,N',N'*,6,7-Hexamethyl-1,3-bis(trimethylsilyl)-5,8-dioxa-1,3-diaza-2,4-diphospha(4-*P*^{V})spiro[3.4]oct-6-ene-2,4-diamine, 10CI, 9CI** H-00068

[59992-10-8]

$C_{14}H_{36}N_4O_2P_2Si_2$ M 410.583

Cryst. (MeCN). Mp 52°.

2-Sulfide: [59992-11-9].
$C_{14}H_{36}N_4O_2P_2SSi_2$ M 442.643
Solid. Mp 93-94°.

Zeiss, W., *Angew. Chem., Int. Ed. Engl.*, 1976, **15**, 554 (*synth, deriv, pmr, P nmr*)
Lux, D. *et al*, *Z. Naturforsch., B*, 1980, **35**, 369 (*cryst struct*)

***N,N,N',N',N'',N''*-Hexamethyl-*N'''*-diphenylphosphinophosphorimidic triamide, 9CI** H-00069

P,P,P-*Tris(dimethylamino)*-N-*(diphenylphosphino)*-*phosphazene*. P,P,P-*Tris(dimethylamino)*-N-*(diphenylphosphino)phosphine imide*

[83318-64-3]

$(Me_2N)_3P{=}NPPh_2$

$C_{18}H_{28}N_4P_2$ M 362.394

Cryst. (Et_2O). Mp 48-49°.

Oxide: [56727-72-1].
$C_{18}H_{28}N_4O_2P_2$ M 394.392
Cryst. Mp 77-78°.

Kroshefsky, R.D. *et al*, *Inorg. Chem.*, 1975, **14**, 3090 (*oxide, synth, ir, P nmr, pmr*)
Flindt, E.-P., *Z. Anorg. Allg. Chem.*, 1982, **487**, 119 (*synth, ir, P nmr*)

2,3,5,6,7,8-Hexamethyl-2,3,5,6,7,8-hexaaza-1,4-diphosphabicyclo[2.2.2]octane, 9CI, 8CI H-00070

[3478-74-8]

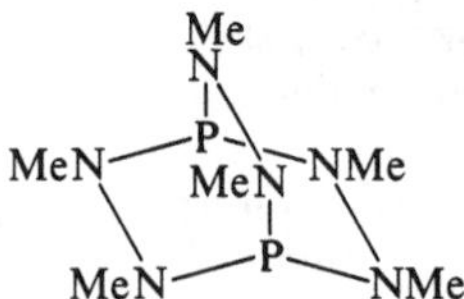

$C_6H_{18}N_6P_2$ M 236.196

Solid. Mp 120-121°. $Bp_{0.001}$ 70° subl.

1,4-Dioxide: [3182-66-9].
$C_6H_{18}N_6O_2P_2$ M 268.195
Cryst. (toluene). Mp 316° (320-325°).

1,4-Disulfide: [3335-14-6].
$C_6H_{18}N_6P_2S_2$ M 300.316
Cryst. (toluene). Mp 340°. Sinters at 270°.

1,4-Diselenide: [38167-29-2].
$C_6H_{18}N_6P_2Se_2$ M 394.116
Solid. Mp 340°.

van Doone, W. *et al*, *Inorg. Chem.*, 1971, **10**, 2591 (*cryst struct*)
Havlicek, M.D. *et al*, *Inorg. Chem.*, 1972, **11**, 1624 (*synth, ir, pmr, props*)
Gilje, J.W. *et al*, *Inorg. Chem.*, 1972, **11**, 1643 (*dioxide, cryst struct*)

Goetze, R. *et al*, *Chem. Ber.*, 1972, **105**, 2637 (*synth, derivs, ms, ir, pmr, P nmr*)
Nöth, H. *et al*, *Chem. Ber.*, 1974, **107**, 1019; 1976, **109**, 1942 (*synth, pmr, P and N nmr*)
Cowley, A.H. *et al*, *Inorg. Chem.*, 1977, **16**, 854 (*pe*)
Bulloch, G. *et al*, *J. Chem. Soc., Dalton Trans.*, 1978, 764 (*pmr, cmr*)

2,4,6,8,9,10-Hexamethyl-2,4,6,8,9,10-hexaaza-1,3,5,7-tetraphosphatricyclo[3.3.1.1^3,7]decane, 10CI, 9CI — H-00071

closo-*Tetraphosphorus hexakis(methylimide).*
2,4,6,8,9,10-Hexamethyl-2,4,6,8,9,10-hexaaza-1,3,5,7-tetraphosphaadamantane
[10369-17-2]

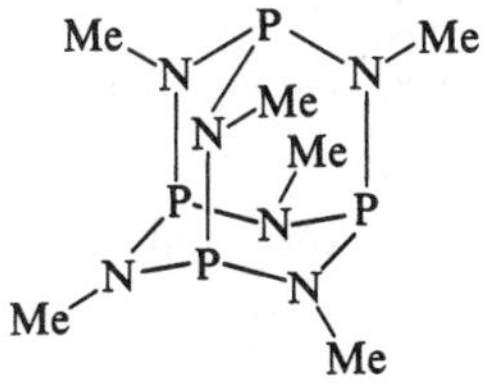

$C_6H_{18}N_6P_4$ M 298.143
Solid. Mp 122-123°. Bp_{737} 303-304°.
1,3,5,7-Tetroxide: [58979-11-6].
$C_6H_{18}N_6O_4P_4$ M 362.141
Solid. Mp 179° dec. $Bp_{0.001}$ 180° subl.
1-Sulfide: [38448-57-6].
$C_6H_{18}N_6P_4S$ M 330.203
Cryst. (pentane). Mp 94°. Polymorphous.
1,3-Disulfide: [38448-56-5].
$C_6H_{18}N_6P_4S_2$ M 362.263
Cryst. (pentane). Mp 125-126°.
1,3,5-Trisulfide: [38448-55-4].
$C_6H_{18}N_6P_4S_3$ M 394.323
Cryst. (pentane). Mp 174°.
1,3,5,7-Tetrasulfide: [37747-07-2].
$C_6H_{18}N_6P_4S_4$ M 426.383
Cryst. ($CHCl_3$/pentane). Mp 246°.
1-Oxide, 3,5,7-Trisulfide:
$C_6H_{18}N_6OP_4S_3$ M 410.323
Solid. Mp 215° dec.
1,3-Dioxide, 5,7-Disulfide:
$C_6H_{18}N_6O_2P_4S_2$ M 394.262
Solid. Mp 185° dec.
1,3,5-Trioxide, 7-Sulfide:
$C_6H_{18}N_6O_3PS$ M 285.280
Solid. Mp 157° dec.

Holmes, R.R. *et al*, *J. Am. Chem. Soc.*, 1961, **83**, 1334 (*synth*)
Elkain, J.C. *et al*, *Phosphorus*, 1973, **2**, 249 (*derivs, P nmr*)
Wolff, A. *et al*, *Bull. Soc. Chim. Fr. Part I*, 1973 1587.
Wolff, A. *et al*, *Org. Mass Spectrom.*, 1974, **9**, 594 (*ms*)
Riess, J.G. *et al*, *Inorg. Chim. Acta*, 1976, **17**, L27 (*tetraoxide, pmr, P nmr*)
Casabianca, F. *et al*, *Inorg. Chem.*, 1977, **16**, 864 (*derivs, synth, ir, ms, pmr*)
Bulloch, G. *et al*, *J. Chem. Soc., Dalton Trans.*, 1978, 764 (*pmr, cmr*)
Casabianca, F. *et al*, *Inorg. Chem.*, 1978, **17**, 3232 (*tetraoxide, tetrasulfide, cryst structs*)
Cotton, F.A. *et al*, *Inorg. Chem.*, 1978, **17**, 3521 (*monosulfide, cryst struct*)
Lee, T.H. *et al*, *J. Am. Chem. Soc.*, 1980, **102**, 2631 (*pe*)
Cotton, F.A. *et al*, *Inorg. Chem.*, 1982, **21**, 3123 (*sulfides, cryst struct*)
Cotton, F.A. *et al*, *Inorg. Chem.*, 1983, **22**, 133 (*pe, cmr*)

N,N,N',N',N'',N''-Hexamethyl-1-methylenephosphoranetriamide, 9CI — H-00072

Tris(dimethylamino)methylenephosphorane
[28706-85-6]

$(Me_2N)_3P{=}CH_2$

$C_7H_{20}N_3P$ M 177.229
Liq. Bp_{14} 87-88°. Forms Au complexes.

Issleib, K. *et al*, *J. Prakt. Chem.*, 1970, **312**, 135 (*synth*)
Issleib, K. *et al*, *Org. Magn. Reson.*, 1973, **5**, 401 (*pmr, nmr*)
Ibrahim, E.H.M. *et al*, *Egypt. J. Chem.*, 1979, **22**, 393; *CA*, **94**, 208940 (*props*)

N,N,N',N',N'',N''-Hexamethyl-*N'''*-phenylphosphorimidic triamide — H-00073

P,P,P-*Tris(dimethylamino)*-N-*phenylphosphine imide.*
P,P,P-*Tris(dimethylamino)*-N-*phenylphosphazene.*
P,P,P-*Tris(dimethylamino)*-N-*phenyliminophosphorane*
[35989-04-9]

$(Me_2N)_3P{=}NPh$

$C_{12}H_{23}N_4P$ M 254.314
Liq. (or solid). d_4^{22} 1.05. $Bp_{0.4}$ 127°, $Bp_{0.5}$ 100°. n_D^{20} 1.5537. Dec. slowly at r.t., rapidly at 100°.

Vetter, H.-J. *et al*, *Chem. Ber.*, 1963, **96**, 1308 (*synth*)
Tarasevich, A.S. *et al*, *Teor. Eksp. Khim.*, 1971, **7**, 828 (*Engl. transl.* p. 676) (*P nmr, struct*)
Goldwhite, H. *et al*, *J. Chem. Soc., Dalton Trans.*, 1975, 1216 (*synth, pmr, cmr, uv*)
Kroshefsky, R.D. *et al*, *Inorg. Chem.*, 1975, **14**, 3090 (*ir, uv, pmr, P nmr*)

1,2,2,3,4,4-Hexamethylphosphetane, 9CI, 8CI — H-00074

[16109-84-5]

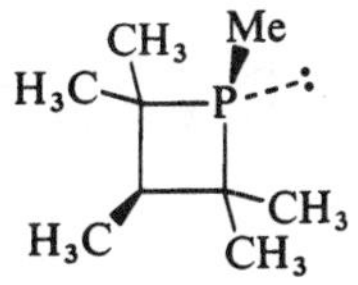

cis-form

$C_9H_{19}P$ M 158.223
B,MeBr: [16084-01-8]. *1,1,2,2,3,4,4-Heptamethylphosphetanium bromide.*
$C_{10}H_{22}BrP$ M 253.162
Solid. Mp >305°.
B,MeI: [28772-00-1]. *1,1,2,2,3,4,4-Heptamethylphosphetanium iodide.*
$C_{10}H_{22}IP$ M 300.162
Cryst. (H_2O). Mp >335°.
Oxide: [16083-94-6].
$C_9H_{19}OP$ M 174.222
Cryst. (Et_2O/pet. ether). Mp 160-162°.

cis-form [35622-00-5]
Oxide: [33530-51-7]. Cryst. (pet. ether). Mp 171-172°.

trans-form [35621-97-7]
B,PhCH$_2$Br: 1-Benzyl-1,2,2,3,4,4-hexamethylphosphetanium bromide.
$C_{16}H_{26}BrP$ M 329.259
Solid. Mp 212-218°.
Oxide: [33530-51-7]. No phys. props. recorded for this isomer; see above for undefined stereoisomer Mp 160-162°.

Cremer, S.E. *et al*, *J. Org. Chem.*, 1967, **32**, 4066 (*synth, oxide*)
Corfield, J.R. *et al*, *J. Chem. Soc. (C)*, 1970, 1855 (*synth, derivs, pmr*)

Gray, G. *et al*, *J. Org. Chem.*, 1972, **37**, 3458, 3470 (*cmr, derivs*)
Gray, G. *et al*, *J. Magn. Reson.*, 1973, **12**, 5 (*cmr*)

Hexamethylphosphoric triamide, 9CI H-00075

Hexamethylphosphoramide. HMPA. HMPT
[680-31-9]

$OP(NMe_2)_3$

$C_6H_{18}N_3OP$ M 179.201
Widely used as polar, aprotic solvent which activates the synth. and use of organometallic compds. (e.g. Li, Mg derivs.). Widespread use in organic synthesis: converts benzamides into dimethylaminoquinolines, dehydrating agent for alcohols and amides. Used in combination with $SOCl_2$ for synth. of alkyl chlorides. Misc. H_2O. d_4^{20} 1.025. Mp 7°. Bp 235°, Bp_{15} 115°, Bp_1 68-70°. n_D^{20} 1.4582.
▷Potent animal carcinogen; suspected human carcinogen. Irritant. TD0875000.

Normant, H., *Angew. Chem., Int. Ed. Engl.*, 1967, **6**, 1046 (*rev*)
Diggle, J.W. *et al*, *J. Phys. Chem.*, 1974, **78**, 1018 (*props*)
Gloe, K. *et al*, *J. Prakt. Chem.*, 1975, **317**, 529 (*synth*)
Pantzer, R. *et al*, *Z. Anorg. Allg. Chem.*, 1975, **416**, 297 (*ir, raman*)
London, A.G. *et al*, *Org. Mass Spectrom.*, 1977, **12**, 283 (*ms*)
Itaya, K. *et al*, *J. Am. Chem. Soc.*, 1978, **100**, 5996 (*purifn*)
Pederson, E.B., *CA*, 1978, **89**, 89752e (*rev*)
Spencer, H., *Chem. Ind. (London)*, 1979, 728 (*tox*)
Bergesen, K. *et al*, *Acta Chem. Scand., Ser. A*, 1981, **35**, 147 (*cmr*)
Cowley, A.H. *et al*, *Inorg. Chem.*, 1982, **21**, 543 (*pe*)
Duangthai, S. *et al*, *Org. Magn. Reson.*, 1982, **20**, 33 (*struct*)
Fujinaga, T. *et al*, *Recom. Methods Purific. Solvents. Tests Impurities*, J.F. Coetzee Ed., Pergamon 1982, 38 (*rev*)
Koidan, G.N. *et al*, *Zh. Obshch. Khim.*, 1982, **52**, 2001 (*Engl. transl.* p. 1779) (*synth, props, derivs*)
Worley, S.D. *et al*, *J. Electron Spectrosc. Relat. Phenom.*, 1982, **25**, 135 (*pe, struct*)
Bollinger, J.C. *et al*, *Org. Mass Spectrom.*, 1985, **20**, 318 (*ms*)
Fieser, M. *et al*, *Reagents for Organic Synthesis*, Wiley, 1967-84, **5**, 323; **6**, 273; **7**, 168; **8**, 240; **9**, 235; **10**, 196.
Sax, N.I., *Dangerous Properties of Industrial Materials*, 6th Ed., Van Nostrand-Reinhold, 1984, 721.
Hazards in the Chemical Laboratory, (Bretherick, L., Ed.), 3rd Ed., Royal Society of Chemistry, London, 1981, 346.

2,2,2′,2′,2″,2″-Hexamethylphosphoric trihydrazide H-00076

$(Me_2NNH)_3P{=}O$

$C_6H_{21}N_6OP$ M 224.245
Solid. Mp 193.5-195°. $Bp_{0.1}$ 105° subl.

Nielson, R.P. *et al*, *Inorg. Chem.*, 1963, **2**, 753 (*synth, pmr*)

N,N,N′,N′,N″,N″-Hexamethylphosphorimidic triamide, 9CI H-00077

P,P,P-*Tris(dimethylamino)phosphine imide*. P,P,P-*Tris(dimethylamino)phosphazene*. P,P,P-*Tris(dimethylamino)iminophosphorane*
[49778-01-0]

$(Me_2N)_3P{=}NH$

$C_6H_{19}N_4P$ M 178.217
Liq. Bp_{10} 96-97°, $Bp_{0.1}$ 60-62°.
B, HCl: [87863-65-8]. N,N,N′,N′,N″,N″-*Hexamethyltetraaminophosphonium chloride. Aminotris(dimethylamino)phosphonium chloride.*
$C_6H_{20}ClN_4P$ M 214.678
Cryst. (CH_2Cl_2/Et_2O). Mp 146-148°.
B, HBr: N,N,N′,N′,N″,N″-*Hexamethyltetraaminophosphonium bromide. Aminotris(dimethylamino)phosphonium bromide.*
$C_6H_{20}BrN_4P$ M 259.129
Cryst. (CH_2Cl_2/Et_2O). Mp 185-186°.
N‴-*Chloro:* [78050-96-1].
$C_6H_{18}ClN_4P$ M 212.662
Undistillable, yellow mobile liq. Slowly dec. at 20°.

Issleib, K. *et al*, *Synth. Inorg. Metal.-Org. Chem.*, 1973, **3**, 255 (*synth, pmr, P nmr*)
Lorberth, J. *et al*, *J. Organomet. Chem.*, 1974, **71**, 159 (*synth, deriv*)
Egorov, Yu.P. *et al*, *Teor. Eksp. Khim.*, 1982, **18**, 58 (*Engl. transl.* p. 45) (*P nmr, struct*)

Hexamethylphosphoroselenoic triamide H-00078

[7422-73-3]

$(Me_2N)_3PSe$

$C_6H_{18}N_3PSe$ M 242.162
Solid.

Osokin, D.Ya. *et al*, *Org. Magn. Reson.*, 1972, **4**, 831 (*nqr*)
Dean, P.A.W. *et al*, *Can. J. Chem.*, 1979, **57**, 754 (*P and Se nmr*)
Rømming, C. *et al*, *Acta Chem. Scand., Ser. A*, 1979, **33**, 187 (*cryst struct*)
Bergesen, K. *et al*, *Acta Chem. Scand., Ser. A*, 1981, **35**, 147 (*cmr*)
Cowley, A.H. *et al*, *Inorg. Chem.*, 1982, **21**, 543 (*pe*)

Hexamethylphosphorothioic triamide H-00079

Tris(dimethylamino)phosphine sulfide
[3732-82-9]

$(Me_2N)_3PS$

$C_6H_{18}N_3PS$ M 195.262
Low-melting solid. d^{30} 1.04. Mp 29°. $Bp_{1.5}$ 94°. n_D^{30} 1.5070.
▷Carcinogenic but less so than 36939-5. TG4390000.

Vetter, H.J. *et al*, *Chem. Ber.*, 1963, **96**, 1308 (*synth*)
Osokin, D.Ya. *et al*, *Org. Magn. Reson.*, 1972, **4**, 831 (*nqr*)
Diggle, J.W. *et al*, *J. Phys. Chem.*, 1974, **78**, 1018 (*synth, props*)
Dorschner, R. *et al*, *Inorg. Chim. Acta*, 1975, **15**, 71 (*struct*)
Pantzer, R. *et al*, *Z. Anorg. Allg. Chem.*, 1975, **416**, 297 (*ir, raman*)
Gray, G.A. *et al*, *J. Am. Chem. Soc.*, 1976, **98**, 857 (*nmr*)
Skvortsov, N.K. *et al*, *Zh. Obshch. Khim.*, 1976, **46**, 521 (*Engl. transl. p. 518*) (*pmr, P nmr*)
Light, R.W. *et al*, *Phosphorus Sulfur*, 1980, **8**, 255 (*ms*)
Cowley, A.H. *et al*, *Inorg. Chem.*, 1982, **21**, 543 (*pe*)

Hexamethylphosphorous triamide, 9CI H-00080

Tris(dimethylamino)phosphine
[1608-26-0]

$P(NMe_2)_3$

$C_6H_{18}N_3P$ M 163.202
Versatile synthetic reagent. Employed in Wittig reactions: Deoxygenates peroxides, sulfoxides, and azoxy compds., and desulfurizes disulfides. Used to selectively functionalize diols, and to convert carboxylic acids into their *N,N*-dimethylamides. Yellow oil. Mp −44°. Bp 162-164°, Bp_{12} 49-51°. n_D^{20} 1.4660.
▷TH3390000.

Oxide: see Hexamethylphosphoric triamide, H-00075
Sulfide: see Hexamethylphosphorothioic triamide, H-00079
Selenide: see Hexamethylphosphoroselenoic triamide, H-00078

Burgada, R., *Bull. Soc. Chim. Fr.*, 1972, 4161 (*props, use*)
Osokin, D.Ya. *et al, Org. Magn. Reson.*, 1972, **4**, 831 (*N nmr*)
Dorschner, R. *et al, Inorg. Chim. Acta*, 1975, **15**, 71 (*struct*)
Lappert, M.F. *et al, J. Chem. Soc., Dalton Trans.*, 1975, 1207 (*pe*)
Gray, G.A. *et al, J. Am. Chem. Soc.*, 1976, **98**, 3857 (*nmr*)
Mason, J. *et al, J. Chem. Soc., Dalton Trans.*, 1977, 2337 (*N nmr*)
Thomas, M.G. *et al, Inorg. Chem.*, 1977, **16**, 994 (*ir, pmr, P nmr*)
Org. Synth., Coll. Vol., **5**, 602 (*synth*)
Davidson, G. *et al, Spectrochim. Acta, Part A*, 1979, **35**, 141 (*ir, raman*)
Chretien, F. *et al, J. Chem. Soc., Perkin Trans. 1*, 1980, 381 (*use*)
Bergesen, K. *et al, Acta Chem. Scand., Ser. A*, 1981, **35**, 147 (*cmr*)
Goubeau, D. *et al, Inorg. Chem.*, 1981, **20**, 1966 (*pe, struct*)
Cowley, A.H. *et al, Inorg. Chem.*, 1982, **21**, 543 (*pe*)
Worley, S.D. *et al, J. Electron. Spectrosc. Relat. Phenomena*, 1982, **25**, 135 (*pe, struct*)
Fieser, M. *et al, Reagents for Organic Synthesis*, Wiley, 1967-84, **4**, 247; **6**, 279.

2,5,5,8,11,11-Hexamethyl-1,3,7,9-tetraoxa-2,8-diphosphacyclododecane — H-00081

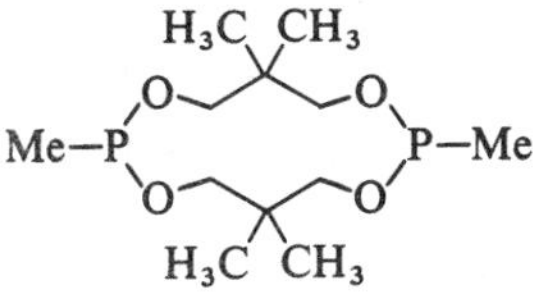

$C_{12}H_{26}O_4P_2$ M 296.283

A dimeric cyclic phosphonite in equilibrium with monomeric and trimeric forms.

2,8-Disulfide:
$C_{12}H_{26}O_4P_2S_2$ M 360.403
Cryst. (C_6H_6 or 1,2-dichlorobenzene). Mp 250°, Mp 280°. Two stereoisomers known.

Albrand, J.P. *et al, J. Am. Chem. Soc.*, 1974, **96**, 4584 (*synth, pmr, P nmr*)

N,N,N′,N′,N″,N″-Hexamethyl-N‴-trimethylsilylphosphorimidic triamide — H-00082

P,P,P-*Tris(dimethylamino)*-N-*(trimethylsilyl)-phosphazene.* P,P,P-*Tris(dimethylamino)*-N-*(trimethylsilyl)phosphine imide*

[53167-50-3]

$(Me_2N)_3P{=}NSiMe_3$

$C_9H_{27}N_4PSi$ M 250.398

Useful synth. intermediate. Liq. $Bp_{1.3}$ 69.5-72°, $Bp_{0.1}$ 52-55°. $n_D^{20.5}$ 1.4548.

Schlak, O. *et al, Z. Anorg. Allg. Chem.*, 1976, **419**, 275 (*synth, ir, pmr, P and Si nmr*)
Flindt, E.-P., *Z. Anorg. Allg. Chem.*, 1978, **447**, 97 (*synth, ir, pmr, P nmr*)

2,2,3,3,7,7-Hexamethyl-1,4,6-trioxa-9-aza-5-phospha(5-P^V)spiro[4.4]nonane — H-00083

[51777-80-1]

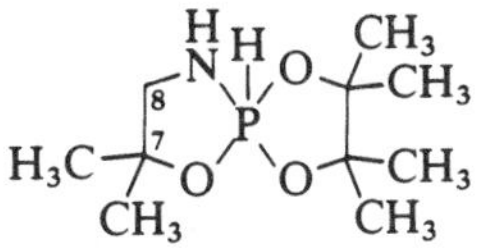

$C_{10}H_{22}NO_3P$ M 235.262

Potentially tautomeric with monocyclic stricture, but at 20-100° exists completely in bicyclic form. Cryst. (pet. ether). Mp 52°.

Burgada, R. *et al, J. Organomet. Chem.*, 1974, **66**, 255 (*synth, struct*)

2,2,3,3,8,8-Hexamethyl-1,4,6-trioxa-9-aza-5-phospha(5P^V)spiro[4.4]nonane, 9CI — H-00084

[33312-92-4]

$C_{10}H_{22}NO_3P$ M 235.262

Tautomeric with monocyclic P^{III} compd. (0% at r.t., 18% at 100°). No phys. props. reported.

Burgada, R. *et al, J. Organomet. Chem.*, 1974, **66**, 255 (*synth, struct*)

1,6-Hexanediphosphonic acid — H-00085

1,6-Hexanediylbisphosphonic acid, 9CI. Hexamethylenediphosphonic acid, 8CI. 1,6-Diphosphonohexane

[4721-22-6]

$(HO)_2P(O)(CH_2)_6P(O)(OH)_2$

$C_6H_{16}O_6P_2$ M 246.136

Corrosion inhibitor. Cryst. (H_2O). Mp 206-208°.

Tetra-Et ester: [5391-92-4]. *Tetraethyl 1,6-hexanediylbisphosphonate.*
$C_{14}H_{32}O_6P_2$ M 358.351
Liq. $Bp_{0.1}$ 185°. n_D^{25} 1.4462.

Bisdichloride:
$C_6H_{12}Cl_4O_2P_2$ M 319.919
Mod. air-stable solid. Mp 60-62°. $Bp_{0.2}$ 166°.

Moedritzer, K. *et al, J. Inorg. Nucl. Chem.*, 1961, **22**, 297 (*synth, ester, ir, P nmr*)
Kosolapoff, G.M. *et al, J. Chem. Soc. (C)*, 1966, 757 (*synth, chloride, ir*)
Van Haverbeke, L. *et al, Bull. Soc. Chim. Belg.*, 1972, **81**, 547; *CA*, **78**, 49963 (*ir, raman*)

1,6-Hexanediphosphonous acid — H-00086

1,6-Hexanediylbis(phosphonous acid), 9CI. Hexamethylenediphosphonous acid

$(HO)_2P(CH_2)_6P(OH)_2 \rightleftharpoons HO(O)PH(CH_2)_6PH(O)OH$

$C_6H_{16}O_4P_2$ M 214.138

Tetra-Me ester: Tetramethyl 1,6-hexanediylbisphosphonite. Tetramethyl hexamethylenediphosphonite.
$C_{10}H_{24}O_4P_2$ M 270.245
Liq. $Bp_{1.5}$ 128-132°. n_D^{21} 1.4704.

Tetra-Et ester: Tetraethyl 1,6-hexanediylbisphosphonite. Tetraethyl hexamethylenediphosphonite.
$C_{14}H_{32}O_4P_2$ M 326.352
Liq. $Bp_{0.02}$ 115-120°. n_D^{20} 1.4590.

Tetraisopropyl ester: Tetraisopropyl 1,6-hexanediylbisphosphonite. Tetraisopropyl hexamethylenebisphosphonite.
$C_{18}H_{40}O_4P_2$ M 382.459

Liq. Bp_2 155-157°. n_D^{22} 1.4515.
Bis(dichloride): [24110-33-6].
$C_6H_{12}Cl_4P_2$ M 287.920
Liq. d_4^{20} 1.32. $Bp_{0.5}$ 111-112°. n_D^{20} 1.5402.
Bis(dibromide): [82159-18-0].
$C_6H_{12}Br_4P_2$ M 465.724
Liq. $Bp_{0.1}$ 148°.
Bis[bis(diethylamide)]: [82159-40-8]. P,P′-*1,6-Hexanediylbis[bis* N,N-*diethylphosphonous diamide]*.
$C_{22}H_{52}N_4P_2$ M 434.627
Liq. $Bp_{0.1}$ 140°.

Sander, M., *Chem. Ber.*, 1962, **95**, 473 (*tetrachloride, esters, synth*)
Babkina, É.I. *et al*, *Zh. Obshch. Khim.*, 1968, **38**, 1772 (*Engl. transl.* p. 1727) (*tetrachloride*)
Diemert, K. *et al*, *Chem. Ber.*, 1982, **115**, 1947 (*tetrabromide, synth, P nmr*)
Diemert, K. *et al*, *Phosphorus Sulfur*, 1983, **15**, 155 (*diethylamide, synth, P nmr*)

Hexaphenyl-1,3-diphospha-2,4-digermacyclobutane H-00087

1,2,2,3,4,4-Hexaphenyl-1,3,2,4-diphosphdigermetane, 8CI
[30404-88-7]

Ph
P
Ph Ph
Ge Ge
Ph Ph
P
Ph

$C_{36}H_{30}Ge_2P_2$ M 669.761
White solid. Mp 40-42°. Oxidises and hydrolyses readily.

Schumann, H. *et al*, *Chem. Ber.*, 1971, **104**, 333.

1,2,2,3,4,4-Hexaphenyl-1,3-diphosphetane, 9CI H-00088

Ph Ph
Ph P
P Ph
Ph Ph

$C_{38}H_{30}P_2$ M 548.603
Cryst. (1,2-dimethoxyethane or THF). Mp 149° dec. Subl. *in vacuo*.

Becker, G. *et al*, *Z. Anorg. Allg. Chem.*, 1981, **479**, 41 (*synth, ms, ir, cmr, P nmr, pmr*)

Hexaphenylhexaphosphorinane, 9CI H-00089

Hexaphenylhexaphosphinane. Phosphobenzene B. *Hexaphenylcyclohexaphosphine*
[4552-71-0]

Ph
P
PhP PPh
PhP PPh
P
Ph

$C_{36}H_{30}P_6$ M 648.476
4 forms known.

Monoclinic-form
Cryst. (THF). Mp 189-193° (open tube), 194-198° (*in vacuo*).

Rhombohedral-form
Cryst. (C_6H_6). Mp 183-186° (open tube), 229-235° (*in vacuo*). On resolidification Mp 130-55°.

Triclinic-form
Cryst. (THF). Mp 185-189° (open tube), 193-198° (*in vacuo*).

Trigonal-form
Cryst. (THF). Mp 190-195° (open tube), 236-240° (*in vacuo*).

Amster, R.L. *et al*, *Can. J. Chem.*, 1964, **42**, 2577 (*ir, raman*)
Daly, J.J., *J. Chem. Soc.*, 1965, 4789 (*cryst struct*)
Daly, J.J., *J. Chem. Soc.* (*A*), 1966, 428 (*cryst struct*)
Maier, L., *Helv. Chim. Acta*, 1966, **49**, 1119 (*synth*)
Dupont, T.J. *et al*, *Inorg. Chem.*, 1973, **12**, 2487 (*synth, uv, P nmr, pmr*)
Baudler, M. *et al*, *Z. Naturforsch., B*, 1976, **31**, 558 (*pmr, struct*)
Hassler, K. *et al*, *Monatsh. Chem.*, 1979, **110**, 919 (*ir, raman*)

Hexaphenylphosphorous triamide, 9CI H-00090

Tris(N,N-*diphenylamino*)*phosphine*
[51528-98-4]

$(Ph_2N)_3P$

$C_{36}H_{30}N_3P$ M 535.627
Cryst. (Et_2O at low temp.). Sl. sol. $CHCl_3$, C_6H_6. Mp 63-65°.
Oxide: [7422-64-2]. *Hexaphenylphosphoric triamide.*
$C_{36}H_{30}N_3OP$ M 551.626
No phys. props. reported.

Nielsen, M.L. *et al*, *J. Phys. Chem.*, 1964, **68**, 152 (*oxide, nmr*)
Badin, M.J., *Z. Anorg. Allg. Chem.*, 1980, **467**, 218 (*synth, ms, ir, P nmr, pmr, cryst struct, sulfide*)

1,1,4,7,10,10-Hexaphenyl-1,4,7,10-tetraphosphadecane, 9CI H-00091

Hexaphenyltriethylenetetraphosphine. Tetraphos
[23582-04-9]

$Ph_2PCH_2CH_2PPhCH_2CH_2PPhCH_2CH_2PPh_2$

$C_{42}H_{42}P_4$ M 670.689
2 Diastereoisomers have been characterised.

Low-melting-form
Cryst. (CH_2Cl_2/MeOH). Mp 99-101°.

High-melting-form
Cryst. (THF/MeOH). Mp 169-171°.
Tetraoxide: [36156-14-6].
$C_{42}H_{42}O_4P_4$ M 734.686
Solid. Mp 344°.

King, R.B. *et al*, *J. Am. Chem. Soc.*, 1971, **93**, 4158 (*synth, ir, pmr, nmr*)
Medved', T.Ya. *et al*, *Izv. Akad. Nauk SSSR, Ser. Khim.*, 1971, 2839 (*Engl. transl.* p. 2709) (*oxide*)
Butler, I.S. *et al*, *J. Organomet. Chem.*, 1974, **66**, 111 (*complexes*)
King, R.B. *et al*, *Z. Naturforsch., B*, 1974, **29**, 574 (*isom*)
Moedritzer, K., *Thermochim. Acta*, 1976, **16**, 173; *CA*, **86**, 42900 (*oxide*)

Hexapropylphosphorous triamide, 9CI H-00092

Tris(N,N-*dipropylamino*)*phosphine*
[5848-64-6]

$P[N(CH_2CH_2CH_3)_2]_3$

$C_{18}H_{42}N_3P$ M 331.524
Liq. $Bp_{0.15}$ 101-103°. n_D^{20} 1.4721.
Oxide: [13987-57-0]. *Hexapropylphosphoric triamide. Tris*(N,N-*dipropylamino*)*phosphine oxide.*
$C_{18}H_{42}N_3OP$ M 347.523

Liq. $Bp_{0.2}$ 125°. pK_{a1} 7.15, pK_{a2} 11.07 ($MeNO_2$). n_D^{20} 1.4630.

Sulfide: [33712-82-2]. *Hexapropylphosphorothioic triamide. Tris*(N,N*-dipropylamino)phosphine sulfide.*
$C_{18}H_{42}N_3PS$ M 363.584
Liq. $Bp_{0.01}$ 108-112°.

Stuebe, C. *et al*, *J. Am. Chem. Soc.*, 1956, **78**, 976 (*synth, derivs*)
Coustures, Y. *et al*, *Bull. Soc. Chim. Fr. Part I*, 1973, 926 (*oxide, ir, P nmr*)
Ando, F. *et al*, *Bull. Chem. Soc. Jpn.*, 1978, **51**, 1481 (*sulfide, pmr, ir*)
Marchenko, A.P. *et al*, *Zh. Obshch. Khim.*, 1978, **48**, 551 (*Engl. transl.* p. 501) (*oxide*)
Hargis, J.K. *et al*, *Inorg. Chem.*, 1977, **16**, 1686 (*pe*)
Bergesen, K. *et al*, *Acta Chem. Scand., Ser. A*, 1981, **35**, 147 (*cmr*)

Hexyldiphenylphosphine, 9CI — H-00093

1-Diphenylphosphinohexane
[18298-00-5]

$Ph_2P(CH_2)_5CH_3$

$C_{18}H_{23}P$ M 270.353
Liq. d_4^{20} 1.00. $Bp_{0.09}$ 138°. n_D^{20} 1.5760.

Oxide: [19259-70-2]. *1-(Diphenylphosphinyl)hexane.*
$C_{18}H_{23}OP$ M 286.353
Cryst. (naphtha). Mp 61-62°.

Sulfide: [6591-16-8]. *1-(Diphenylphosphinothioyl)-hexane.*
$C_{18}H_{23}PS$ M 302.413
Cryst. (EtOH). Mp 51-52°.

Stuebe, C. *et al*, *J. Am. Chem. Soc.*, 1955, **77**, 3526 (*synth, derivs*)
Peterson, D.J., *J. Org. Chem.*, 1966, **31**, 950 (*synth, sulfide, nmr*)
Postle, S.R., *Phosphorus Sulfur*, 1977, **3**, 269 (*sulfide, cmr*)

Hexylmethylphosphinic acid, 9CI — H-00094

$H_3C(CH_2)_5P(O)(OH)Me$

$C_7H_{17}O_2P$ M 164.184
Liq. Bp_1 216-218°.

2-Methylpropyl ester: 2-Methylpropyl hexylmethylphosphinate. Isobutyl hexylmethylphosphinate.
$C_{11}H_{25}O_2P$ M 220.291
Liq. $Bp_{0.25}$ 89°.

Chloride: [51528-28-0].
$C_7H_{16}ClOP$ M 182.630
Liq. $Bp_{0.2}$ 82°.

Anhydride:
$C_{14}H_{32}O_3P_2$ M 310.353
Liq. $Bp_{0.2}$ 118°.

Petrov, K.A. *et al*, *Zh. Obshch. Khim.*, 1960, **30**, 2995 (*Engl. transl.* p. 2967) (*synth*)
Finke, M. *et al*, *Justus Liebigs Ann. Chem.*, 1974, 741 (*derivs*)
Ryzhikov, B.D. *et al*, *Zh. Strukt. Khim.*, 1975, **16**, 754 (*Engl. transl.* p. 700) (*fluoride, nmr*)

Hexyl phosphinate, 9CI — H-00095

$H_3C(CH_2)_5OP(O)H_2 \rightleftharpoons H_3C(CH_2)_5OP(OH)H$

$C_6H_{15}O_2P$ M 150.157
Pentavalent form predominates. Liq. $Bp_{0.2}$ 105-110°. n_D^{20} 1.4355.

Ivanov, B.E. *et al*, *Izv. Akad. Nauk SSSR, Ser. Khim.*, 1967, 1498 (*Engl. transl.* p. 1447) (*synth*)

Hexylphosphine, 9CI — H-00096

[2502-20-7]

$H_3C(CH_2)_5PH_2$

$C_6H_{15}P$ M 118.158
Liq. Bp 127.5-128°. n_D^{20} 1.4482, 1.4527.

Pass, F. *et al*, *Monatsh. Chem.*, 1959, **90**, 148, 792 (*synth*)
Schindlbauer, H. *et al*, *Monatsh. Chem.*, 1961, **92**, 868 (*ir*)

Hexylphosphonic acid, 9CI — H-00097

1-Phosphonohexane. 1-Hexanephosphonic acid
[4721-24-8]

$H_3C(CH_2)_5P(O)(OH)_2$

$C_6H_{15}O_3P$ M 166.156
Cryst. (pet. ether). Mp 105-106°.

Di-Me ester: [6172-92-5]. *Dimethyl hexylphosphonate.*
$C_8H_{19}O_3P$ M 194.210
Liq. Bp_{10} 121-123°. n_D^{20} 1.4276.

Di-Et ester: [16165-66-5]. *Diethyl hexylphosphonate.*
$C_{10}H_{23}O_3P$ M 222.264
Bp_2 103°. n_D^{17} 1.4311.
▷SZ8561000.

Diisopropyl ester: Diisopropyl hexylphosphonate.
$C_{12}H_{27}O_3P$ M 250.317
Bp_2 111.5-113°. n_D^{20} 1.4328.

Difluoride: [14576-59-1].
$C_6H_{13}F_2OP$ M 170.139
Liq. d_4^{20} 1.07. Bp 176-177°, Bp_7 58-60°.

Dichloride: *see Hexylphosphonic dichloride, H-00098*

Ford-Moore, A.H. *et al*, *J. Chem. Soc.*, 1947, 1465 (*ester*)
Griffin, C.E. *et al*, *J. Org. Chem.*, 1960, **25**, 665 (*synth*)
Canavan, A.E. *et al*, *J. Chem. Soc.*, 1962, 331 (*synth, esters*)
Dietze, U., *J. Prakt. Chem.*, 1974, **316**, 293 (*ir*)
Ernst, L., *Org. Magn. Reson.*, 1977, **9**, 35 (*ester, cmr, P nmr*)
Nefant'ev, É.E. *et al*, *Zh. Obshch. Khim.*, 1979, **49**, 1905 (*Engl. transl.* p. 1678) (*synth, P nmr*)

Hexylphosphonic dichloride, 9CI — H-00098

[928-64-3]

$H_3C(CH_2)_5P(O)Cl_2$

$C_6H_{13}Cl_2OP$ M 203.048
Liq. d_4^{20} 1.16. Bp_{15} 124-126°. n_D^{20} 1.4670.

Feshchenko, N.G. *et al*, *Zh. Obshch. Khim.*, 1967, **37**, 473 (*Engl. transl.* p. 441) (*synth*)
Geiseler, G. *et al*, *Ber. Bunsenges. Phys. Chem.*, 1967, **71**, 478 (*ir, raman*)
Bel'skii, V.E. *et al*, *Zh. Obshch. Khim.*, 1974, **44**, 2657 (*Engl. transl.* p. 2612) (*P nmr*)

Hexylphosphonodithioic acid, 9CI — H-00099

$H_3C(CH_2)_5P(O)(SH)_2 \rightleftharpoons H_3C(CH_2)_5P(S)(OH)(SH)$

$C_6H_{15}OPS_2$ M 198.278

O*-Mono-Me ester:* [18788-99-3]. O*-Methyl hydrogen hexylphosphonodithioate.*
$C_7H_{17}OPS_2$ M 212.304

Liq. d^{20} 1.09. $Bp_{0.02}$ 68-71.5°. n_D^{20} 1.5314.
O-*Mono-Et ester:* [13685-85-3]. O-*Ethyl hydrogen hexylphosphonodithioate.*
$C_8H_{19}OPS_2$ M 226.331
Liq. d^{20} 1.06. $Bp_{0.03}$ 87-91°. n_D^{20} 1.5199.

Grishina, O.N. *et al, Neftekhimiya*, 1968, **8**, 111; *CA*, **69**, 2980.

Hexylphosphonothioic acid, 9CI — H-00100

1-Hexanephosphonothioic acid
[77326-10-4]

$$H_3C(CH_2)_5P(S)(OH)_2 \rightleftharpoons H_3C(CH_2)_5P(O)(SH)(OH)$$

$C_6H_{15}O_2PS$ M 182.217
Syrup. n_D^{20} 1.4988.
Dianilinium salt: Solid. Mp 124.5-125°.
Dichloride: [18351-11-6].
$C_6H_{13}Cl_2PS$ M 219.108
Liq. d_4^{20} 1.19. Bp_{10} 108-110°. n_D^{20} 1.5249.
O,O-*Bis(trimethylsilyl) ester:* [77339-65-2]. O,O-*Bis-(trimethylsilyl) hexylphosphonothioate.*
$C_{12}H_{31}O_2PSSi_2$ M 326.580
Liq. Bp_2 94-96°. n_D^{20} 1.4535.

Grishina, O.N. *et al*, *CA*, 1968, **69**, 2980 (*dichloride*)
Nifant'ev, E.É. *et al*, *Zh. Obshch. Khim.*, 1980, **50**, 2676 (*synth, P nmr, derivs*)

Hexylphosphonous acid, 9CI — H-00101

$$H_3C(CH_2)_5P(OH)_2 \rightleftharpoons H_3C(CH_2)_5PH(O)OH$$

$C_6H_{15}O_2P$ M 150.157
The free acid probably exists as the phosphoryl tautomer.
Diisopropyl ester: Diisopropyl hexylphosphonite.
$C_{12}H_{27}O_2P$ M 234.318
Liq. Bp_{15} 112-116°.
Dichloride: [6460-28-2]. *Dichlorohexylphosphine.*
$C_6H_{13}Cl_2P$ M 187.048
d_4^{20} 1.10, 1.19. Bp_{25} 104-107°, Bp_{11} 91-92°. n_D^{20} 1.4850, 1.5010.

Fox, R.B., *J. Am. Chem. Soc.*, 1950, **72**, 4147 (*dichloride*)
Mastalerz, P. *et al*, *Rocz. Chem.*, 1964, **38**, 1529 (*ester*)
Bliznyuk, N.U. *et al*, *Zh. Obshch. Khim.*, 1967, **37**, 890 (*Engl. transl.* p. 840) (*dichloride*)
Nifant'ev, É.E. *et al*, *Zh. Obshch. Khim.*, 1967, **37**, 1366 (*Engl. transl.* p. 1293) (*dichloride*)

Hexylphosphoramidic acid, 9CI — H-00102

$$H_3C(CH_2)_5NHP(O)(OH)_2$$

$C_6H_{16}NO_3P$ M 181.171
Di-Me ester: Dimethyl hexylphosphoramidate.
$C_8H_{20}NO_3P$ M 209.225
Liq. $Bp_{0.5}$ 119°. n_D^{25} 1.4337.
Di-Ph ester: Diphenyl hexylphosphoramidate.
$C_{18}H_{24}NO_3P$ M 333.366
Solid. Mp 48°.
Dichloride:
$C_6H_{14}Cl_2NOP$ M 218.062
Liq. $Bp_{3\text{-}4}$ 135-139°.

Baumgarten, H.E. *et al*, *J. Am. Chem. Soc.*, 1959, **81**, 2132 (*dimethyl ester*)
Baumgarten, H.E. *et al*, *J. Org. Chem.*, 1961, **26**, 1533 (*diphenyl ester*)
Mizuma, T. *et al*, *J. Pharm. Soc. Jpn.*, 1961, **81**, 48; *CA*, **55**, 13403 (*dichloride*)

Hexyl phosphorodichloridite, 9CI, 8CI — H-00103

Hexyl dichlorophosphite
[10496-14-7]

$$H_3C(CH_2)_5OPCl_2$$

$C_6H_{13}Cl_2OP$ M 203.048
Liq. d_0^0 1.14. Bp_{24} 104°. $n_D^{26.5}$ 1.4669.

Razumov, A.I., *Zh. Obshch. Khim.*, 1944, **14**, 464; *CA*, **39**, 4586 (*synth*)

Hexyltriphenylphosphonium(1+), 9CI — H-00104

$$Ph_3P^{\oplus}(CH_2)_5CH_3$$

$C_{24}H_{28}P^{\oplus}$ M 347.459 (ion)
Treatment of salts with butyllithium yields the ylide, widely used in synth. of prostaglandins and pheromones.
▷Salts exhibit anti-acetylcholinesterase activity in vertebrates and schistosomes
Bromide: [4762-26-9].
$C_{24}H_{28}BrP$ M 427.363
Solid. Mp 201-204°.
Iodide: [60106-53-8].
$C_{24}H_{28}IP$ M 474.363
Solid. Mp 130-131°.
Ylide: [1666-79-8]. *Hexylidenetriphenylphosphorane.*
$C_{24}H_{27}P$ M 346.451
Reactive Wittig reagent used in prostaglandin synth. Prepd. *in situ.*

Hauser, C.F. *et al*, *J. Org. Chem.*, 1963, **28**, 372 (*bromide*)
Ohloff, G. *et al*, *Helv. Chim. Acta*, 1973, **56**, 1176 (*ylide, use*)
Müller-Schwartz, D. *et al*, *J. Chem. Ecol.*, 1976, **2**, 389 (*bromide, ir, ylide, use*)
Kowal, R. *et al*, *Pol. J. Chem.*, 1979, **53**, 673 (*iodide*)
Bestmann, H.J. *et al*, *Justus Liebigs Ann. Chem.*, 1981, 2117 (*ylide, use*)
Wilcockson, W.S. *et al*, *Comp. Biochem. Physiol.*, *C*, 1982, **72**, 101 (*pharmacol*)

Hydrazobis[methylphosphinothioic acid], 9CI — H-00105

$$HO-\overset{S}{\overset{\|}{\underset{Me}{\underset{|}{P}}}}-NHNH-\overset{S}{\overset{\|}{\underset{Me}{\underset{|}{P}}}}-OH \rightleftharpoons O=\overset{SH}{\overset{|}{\underset{Me}{\underset{|}{P}}}}-NHNH-\overset{SH}{\overset{|}{\underset{Me}{\underset{|}{P}}}}=O$$

$C_2H_{10}N_2O_2P_2S_2$ M 220.181
O,O′-*Di-Et ester:* O,O′-*Diethyl hydrazobismethylphosphinothioate.*
$C_6H_{18}N_2O_2P_2S_2$ M 276.288
Solid. Mp 159°.

Petrov, K.A. *et al*, *Zh. Obshch. Khim.*, 1970, **40**, 1234 (*Engl. transl.* p. 1225)

4-Hydroperoxycyclophosphamide H-00106

N,N-*Di(2-chloroethyl)tetrahydro-4-hydroperoxy-2*H-*1,3,2-oxazaphosphorin-2-amine 2-oxide*, 9CI. *2-[Bis(2-chloroethyl)amino]tetrahydro-4-hydroperoxy-2*H-*1,3,2-oxazaphosphorine 2-oxide*

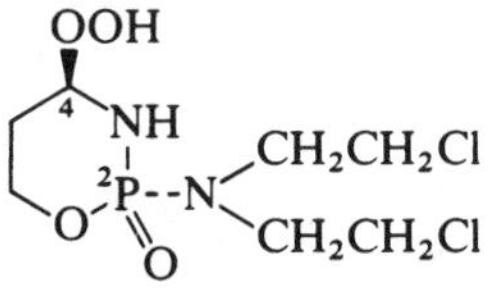

(2RS,4RS)-form

$C_7H_{15}Cl_2N_2O_4P$ M 293.086

Oxidation product of Cyclophosphamide, C-00283 .

(2RS,4RS)-form

(±)-cis-*form*

Microcryst. solid (Me_2CO). Mp 107-108° (103°).

(2RS,4SR)-form

(±)-trans-*form*

Waxy solid.

Van der Steen, J. *et al*, *J. Am. Chem. Soc.*, 1973, **95**, 7535 (*ir, pmr*)
Struck, R.F. *et al*, *J. Am. Chem. Soc.*, 1974, **96**, 313 (*cmr*)
Voelker, G. *et al*, *Arzneim.-Forsch.*, 1974, **24**, 1172 (*tlc*)
Takamizawa, A. *et al*, *J. Med. Chem.*, 1975, **18**, 376 (*synth, ir, pmr, metab*)
Peter, G. *et al*, *Cancer Treat. Rep.*, 1976, **60**, 429 (*synth, props*)
Przybylski, M. *et al*, *Cancer Treat. Rep.*, 1976, **60**, 509 (*ms*)
Camerman, A. *et al*, *Cancer Treat. Rep.*, 1976, **60**, 517 (*cryst struct*)
Camerman, A. *et al*, *Acta Crystallogr., Sect. B*, 1977, **33**, 678 (*cryst struct*)
Takamizawa, A. *et al*, *Heterocycles*, 1977, **7**, 1091 (*rev*)
Borch, R.F. *et al*, *J. Med. Chem.*, 1984, **27**, 485, 490 (*synth, pmr, P nmr*)

2-Hydroxy-1,2-azaphosphetidine 2-oxide H-00107

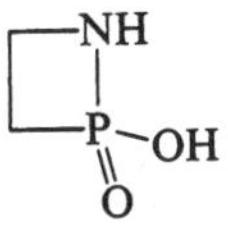

$C_2H_6NO_2P$ M 107.049

Et ester, N-*(2-phenylethyl)*: [69412-49-3]. *2-Ethoxy-1-(2-phenylethyl)-1,2-azaphosphetidine 2-oxide.*

$C_{12}H_{18}NO_2P$ M 239.253

Liq. d^{20} 1.13. $Bp_{0.02}$ 133°. n_D^{20} 1.5215.

Gubnitskaya, E.S. *et al*, *Zh. Obshch. Khim.*, 1978, **48**, 2624; 1980, **50**, 2171 (*Engl. transl.*, pp 2624, 1746) (*synth, ir*)

2-Hydroxy-1,3,2-benzodioxaphosphole 2-oxide H-00108

o-*Phenylenedioxy phosphate.* o-*Phenylenedioxy hydrogen phosphate. Catechol cyclic phosphate*

[4846-23-5]

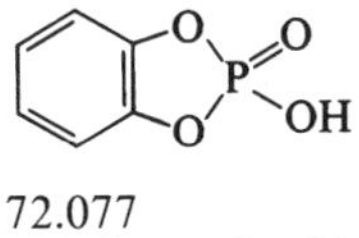

$C_6H_5O_4P$ M 172.077

Phosphorylating agent for nucleosides. Viscous liq. Bp_1 237-238°.

2-Methoxy (Me ester):

$C_7H_7O_4P$ M 186.104

Solid. Mp 59-60°. Bp_{18} 148.5-149°.

2-Bromo (bromide): see 2-Bromo-1,3,2-benzodioxaphosphole, B-00464

Anhydride: see 2,2′-Oxybis[1,3,2-benzodioxaphosphole], O-00087

Cherbuliez, E. *et al*, *Helv. Chim. Acta*, 1951, **34**, 841 (*synth, deriv*)
Kaiser, E.T. *et al*, *J. Am. Chem. Soc.*, 1967, **89**, 6725 (*derivs*)
Khwaja, A.T. *et al*, *Tetrahedron*, 1971, **27**, 6189 (*use*)
Boer, F.P., *Acta Crystallogr., Sect. B*, 1972, **28**, 1201 (*cryst struct*)
Humphris, K.J. *et al*, *J. Chem. Soc., Perkin Trans. 2*, 1973, 831 (*synth*)
Atwood, L. *et al*, *Bioorg. Chem.*, 1976, **5**, 373 (*props*)

2-Hydroxy-4*H*-1,3,2-benzodioxaphosphorin 2-oxide H-00109

Desmethyl salioxon

[40156-84-1]

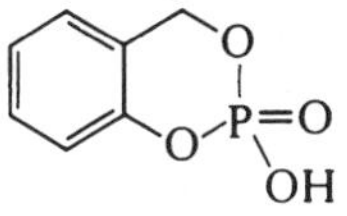

$C_7H_7O_4P$ M 186.104

Solid. Mp 144°.

2-Methoxy (Me ester): [3735-80-6]. *Salioxon.*

$C_8H_9O_4P$ M 200.130

Metab. of Salithion, S-00001 . Liq. $Bp_{0.05}$ 110-112°. n_D^{25} 1.5155.

2-Ethoxy (Et ester):

$C_9H_{11}O_4P$ M 214.157

Liq. $Bp_{0.05}$ 137-139°.

2-(2-Methylphenoxy) (o-*Tolyl ester*):

$C_{14}H_{13}O_4P$ M 276.228

Metab. of Tris(2-methylphenyl) phosphate, T-00803 . Oil. $Bp_{0.09\text{-}0.1}$ 159-161°. $n_D^{24.5}$ 1.5584.

Eto, M. *et al*, *Agric. Biol. Chem.*, 1962, **26**, 630 (*esters, props*)
Eto, M. *et al*, *Biochem. Pharmacol.*, 1962, **11**, 337 (*tolyl ester, synth, ir*)
Eto, M. *et al*, *Agric. Biol. Chem.*, 1963, **27**, 789 (*Me ester, synth, ir, uv*)
Eto, M. *et al*, *Biochem. Toxicol. Insect.*, 1970, 73 (*ester, ms*)
Mihara, K. *et al*, *Agric. Biol. Chem.*, 1974, **38**, 1913.
Eto, M. *et al*, *Agric. Biol. Chem.*, 1981, **45**, 915 (*esters, pmr*)

2-Hydroxy-4*H*-1,3,2-benzodioxaphosphorin 2-sulfide H-00110

Saligenin cyclic thiophosphoric acid

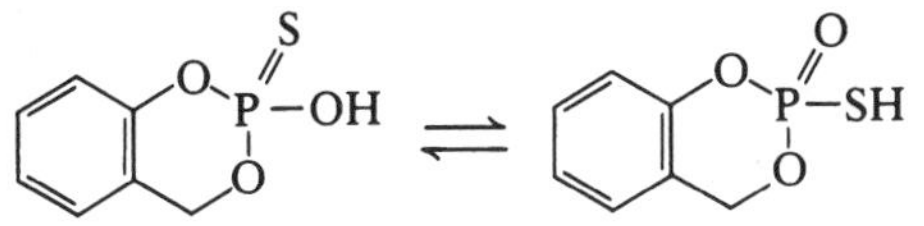

$C_7H_7O_3PS$ M 202.164

Tautomeric.

K salt: Solid. Mp 218°.

2-Methoxy (O-*Me ester*): *see Salithion, S-00001*

2-Methylthio (S-*Me ester*): *see 2-Methylthio-4H-1,3,2-benzodioxaphosphorin, M-00411*

2-Ethoxy (O-*Et ester*): O-*Ethyl saligenin cyclic thiophosphate.*

$C_9H_{11}O_3PS$ M 230.218

Oil. n_D^{25} 1.5495.

2-Ethylthio (S-*Et ester*): S-*Ethyl saligenin cyclic thiophosphate.*

$C_9H_{11}O_3PS$ M 230.218

Oil. $Bp_{0.4}$ 140-145°.

2-Phenoxy (O-*Ph ester*): *see 2-Phenoxy-4H-1,3,2-benzodioxaphosphorin, P-00064*

2-Phenylthio (S-*Ph ester*): S-*Phenyl saligenin cyclic thiophosphate.*

$C_{13}H_{11}O_3PS$ M 278.262

Solid. Mp 88-89°.

Eto, M. *et al*, *Agric. Biol. Chem.*, 1963, **27**, 789 (*esters, synth, ir*)
Kobayashi, K. *et al*, *CA*, 1970, **72**, 100196 (*esters, synth, props*)

(1-Hydroxy-1*H*-benzotriazolato-*O*)-tris(dimethylamino)phosphorus(1+) H-00111

*(1-Hydroxy-1*H*-benzotriazolato-*O*)tris(*N*-methylmethanaminato)phosphorus(1+)*, 9CI. *Le BOP reagent. Castro reagent*

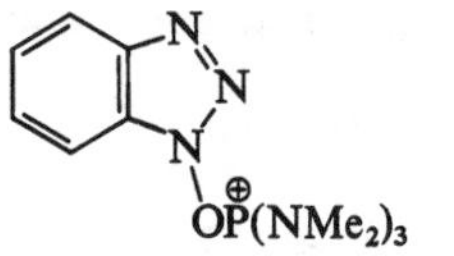

$C_{12}H_{22}N_6OP^{\oplus}$ M 297.319 (ion)

Chloride: [62157-09-9].
$C_{12}H_{22}ClN_6OP$ M 332.772
No phys. props. reported.

Tetrafluoroborate:
$C_{12}H_{22}BF_4N_6OP$ M 384.123
Cryst. (MeCN/pet. ether). Mp 129-131°.

Hexafluorophosphate: [55602-33-6].
$C_{12}H_{22}F_6N_6OP_2$ M 442.283
Coupling agent for peptides (sometimes in conjunction with 49577-2) and also nucleotides; and for esterification of carboxylic acids by phenols. Allows selective esterification in carbohydrates. Cryst. (Me_2CO/Et_2O). Mp 147-149°.

Castro, B. *et al*, *Synthesis*, 1976, 751 (*synth, pmr, nmr, ir, use*)
Galpin, I.J. *et al*, *Tetrahedron*, 1976, **32**, 2417 (*use*)
Castro, B. *et al*, *Synthesis*, 1977, 413 (*use*)
Chapleur, Y. *et al*, *J. Chem. Soc., Perkin Trans. 1*, 1980, 1940, 2683 (*use*)
Appel, R. *et al*, *Chem. Ber.*, 1981, **114**, 2649 (*use*)

2-Hydroxy-2*H*-1,2-benzoxaphosphorin 2-oxide, 9CI H-00112

[76164-17-5]

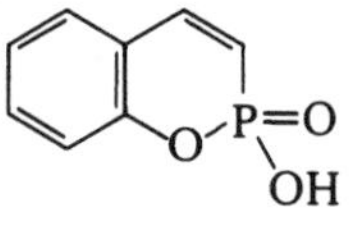

$C_8H_7O_3P$ M 182.115

2-Ethoxy (Et ester): [69750-16-9].
$C_{10}H_{11}O_3P$ M 210.169
Liq. Bp_2 165-167°. n_D^{20} 1.5568.

2-Phenoxy (Ph ester): [69750-15-8].
$C_{14}H_{11}O_3P$ M 258.213
Solid. Mp 96-99°.

Petrov, K.A. *et al*, *Zh. Obshch. Khim.*, 1978, **48**, 2667 (*Engl. transl.* p. 2419) (*derivs, synth, ir, pmr*)

1-Hydroxy-1,1-butanediphosphonic acid H-00113

(1-Hydroxybutylidene)bisphosphonic acid, 9CI. *1,1-Diphosphono-1-butanol*

[16856-53-4]

$H_3CCH_2CH_2C(OH)[P(O)(OH)_2]_2$

$C_4H_{12}O_7P_2$ M 234.082

Complexing agent. Isol. as tetra-Na salt, heptadecahydrate.

Worms, K.-H. *et al*, *Z. Anorg. Allg. Chem.*, 1979, **457**, 209 (*synth*)

(1-Hydroxybutyl)phosphonic acid, 9CI H-00114

1-Phosphono-1-butanol

[26245-90-9]

$H_3CCH_2CH_2CH(OH)P(O)(OH)_2$

$C_4H_{11}O_4P$ M 154.102

(±)-*form*

Plates (C_6H_6/AcOH). Mp 159-161°.

Di-Me ester: Dimethyl (1-hydroxybutyl)phosphonate. Liq. d_4^{20} 1.14. Bp_{15} 143-145°, $Bp_{0.03}$ 99-101°. n_D^{20} 1.4379.

Di-Et ester: Diethyl (1-hydroxybutyl)phosphonate. Liq. d_4^{20} 1.07. Bp_{18} 165-167°, $Bp_{0.03}$ 90-92°. n_D^{20} 1.4371.

McConnell, R.L. *et al*, *J. Am. Chem. Soc.*, 1957, **79**, 1961 (*ester, synth, props*)
Kharasch, M.S. *et al*, *J. Org. Chem.*, 1960, **25**, 1000 (*synth, derivs*)
Nesterov, L.V. *et al*, *Zh. Obshch. Khim.*, 1976, **46**, 1974 (*Engl. transl.* p. 1904) (*esters, synth*)
Nurtdinov, S.Kh. *et al*, *Zh. Obshch. Khim.*, 1979, **49**, 2446 (*Engl. transl.* p. 2159) (*synth, ir, P nmr*)
Baraldi, P.G. *et al*, *Synthesis*, 1982, 653 (*ester, synth, ir, pmr*)

4-Hydroxycyclophosphamide H-00115

*2-[Bis(2-chloroethyl)amino]tetrahydro-2*H*-1,3,2-oxazaphosphorin-4-ol 2-oxide*, 9CI. *2-[Bis(2-chloroethyl)-amino]-1,3,2-oxazaphosphorinan-4-ol 2-oxide*

[40277-05-2]

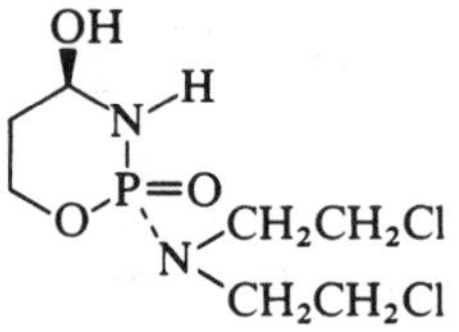

cis-form

$C_7H_{15}Cl_2N_2O_3P$ M 277.087

In aq. soln., *trans*-form exists in tautom. equilibrium with Aldophosphamide, A-00050 . Metab. of Cyclophosphamide, C-00283 and 4-Hydroperoxycyclophosphamide, H-00106 . Cytostatic agent. Labile cryst or needles (CH_2Cl_2/Et_2O). Mp 47.5-48.5°.

Voelker, G. *et al*, *Arzneim.-Forsch.*, 1974, **24**, 1172 (*tlc*)
Takamizawa, A. *et al*, *J. Med. Chem.*, 1975, **18**, 376 (*synth, ir, pmr, metab*)
Przybylski, M. *et al*, *Cancer Treat. Rep.*, 1976, **60**, 509 (*ms*)
Takamizawa, A. *et al*, *Heterocycles*, 1977, **7**, 1091 (*rev*)
Low, J.E. *et al*, *Cancer Res.*, 1982, **42**, 830 (*metab*)
Voelcker, G. *et al*, *Arzneim.-Forsch.*, 1982, **32**, 639 (*pharmacol*)
Borch, R.F. *et al*, *J. Med. Chem.*, 1984, **27**, 490; 1987, **30**, 427 (*props, metab*)

2-Hydroxydibenzo[*d,f*][1,3,2]-dioxaphosphepin 2-oxide, 10CI H-00116

[35227-84-0]

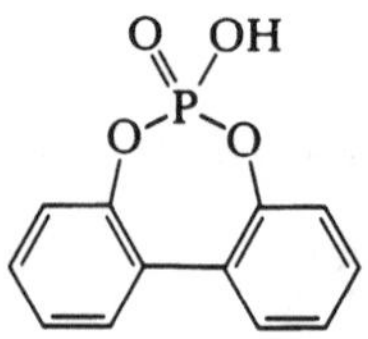

$C_{12}H_9O_4P$ M 248.174

Solid. Mp 262-263° dec.

NH_4 salt: Solid. Mp ca. 200° dec.

Me ester: see 6-Methoxydibenzo[d,f][1,3,2]-dioxaphosphepin, M-00032

Keck, H. *et al*, *Org. Mass Spectrom.*, 1980, **15**, 591 (*ms*)
Kuchen, W. *et al*, *Phosphorus Sulfur*, 1980, **8**, 139 (*synth, derivs, pmr, P nmr*)

5-Hydroxy-5*H*-dibenzo[*b,f*]phosphepin 5-oxide, 9CI — H-00117

1-Hydroxy-1-oxo-1-phospho-2,3:6,7-dibenzocyclohepta-2,4,6-triene

[75231-73-1]

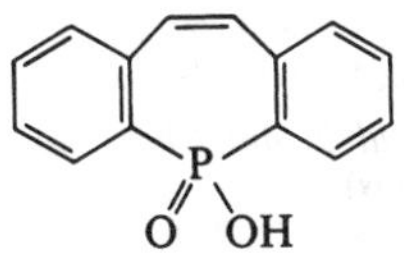

$C_{14}H_{11}O_2P$ M 242.213

Cryst. (EtOH). Mp 310-314°.

Chloride: [75231-74-2]. *5-Chloro-5H-dibenzo[b,f]-phosphepin 5-oxide.*
$C_{14}H_{10}ClOP$ M 260.659
Characterised spectroscopically.

5-Methoxy (5-Me ester): [75231-75-3]. *5-Methoxy-5H-dibenzo[b,f]phosphepin 5-oxide.*
$C_{15}H_{13}O_2P$ M 256.240
Cryst. (pentane). Mp 67-68°.

Segall, Y. *et al*, *Phosphorus Sulfur*, 1980, **8**, 243 (*synth, uv, ir, ms, pmr*)

5-Hydroxy-5*H*-dibenzophosphole 5-oxide, 9CI — H-00118

1-Hydroxy-1H-benzo[b]phosphindole 1-oxide. Phosphafluorinic acid

[524-49-2]

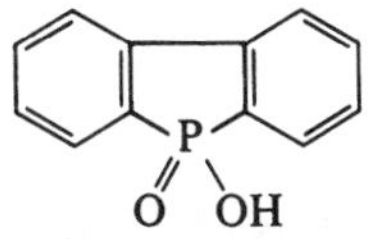

$C_{12}H_9O_2P$ M 216.176

Needles (EtOH). Mp 251-253°. pK_a 3.2 (10% DMSO aq.).

Anhydride: [84530-50-7]. *5,5'-Oxybis[5H-benzophosphindole] 5,5'-dioxide.*
$C_{24}H_{16}O_3P_2$ M 414.336
Insol. hot EtOAc. Mp 262-264°.

5-Methoxy (5-Me ester): [40932-00-1]. *5-Methoxy-5H-dibenzophosphole 5-oxide. 1-Methoxy-1H-benzo[b]-phosphindole 1-oxide.*
$C_{13}H_{11}O_2P$ M 230.202
Cryst. (C_6H_6/pet. ether). Mp 114-115°.

5-Ethoxy (5-Et ester): [55277-64-0]. *5-Ethoxy-5H-dibenzophosphole 5-oxide. 1-Ethoxy-1H-benzo[b]-phosphindole 1-oxide.*
$C_{14}H_{13}O_2P$ M 244.229
No phys. props. reported.

Dodecahydro:
$C_{12}H_{21}O_2P$ M 228.270
Cryst. (EtOH aq.). Mp 153-154.5°.

Freedman, L.D. *et al*, *J. Org. Chem.*, 1956, **21**, 238 (*synth, uv*)
Doak, G.O. *et al*, *J. Org. Chem.*, 1964, **29**, 2382 (*synth*)
Alexander, R.G. *et al*, *J. Chem. Soc., Perkin Trans. 2*, 1974, 1836 (*ms*)
De Boer, J.J. *et al*, *Acta Crystallogr., Sect. B*, 1974, **30**, 797 (*cryst struct*)
Cornforth, J. *et al*, *J. Chem. Soc., Perkin Trans. 1*, 1982, 2289 (*synth, derivs*)

2-Hydroxy-2,3-dihydro-1,2-benzoxaphosphole 2-oxide, 9CI, 8CI — H-00119

2-Hydroxy-1-oxa-2-oxo-2-phosphaindane

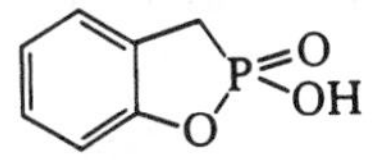

$C_7H_7O_3P$ M 170.104

2-Methoxy (Me ester): [14707-40-5]. *2,3-Dihydro-2-methoxy-1,2-benzoxaphosphole 2-oxide.*
$C_8H_9O_3P$ M 184.131
Liq. d_4^{20} 1.27. $Bp_{0.04}$ 123-124°. n_D^{20} 1.5220.

2-Ethoxy (Et ester): [14707-41-6]. *2-Ethoxy-2,3-dihydro-1,2-benzoxaphosphole 2-oxide.*
$C_9H_{11}O_3P$ M 198.158
Solid. Mp 60°. $Bp_{0.08}$ 145°, $Bp_{0.3}$ 116-118°.

Ageeva, A.B. *et al*, *Izv. Akad. Nauk SSSR, Ser. Khim.*, 1967, 1494 (*Engl. transl.* p. 1443) (*synth*)
Ivanov, B.E. *et al*, *Izv. Akad. Nauk SSSR, Ser. Khim.*, 1967, 226 (*Engl. transl.* p. 228) (*synth, ir*)
Chasor, D.W., *J. Org. Chem.*, 1983, **48**, 4768 (*synth, pmr, cmr, P nmr*)

1-Hydroxy-2,3-dihydro-1*H*-phosphindole 1-oxide — H-00120

[52427-49-3]

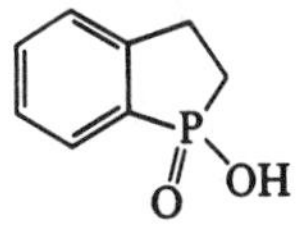

$C_8H_9O_2P$ M 168.132

Cryst. (C_6H_6/cyclohexane). Mp 144-145°.

1-Methoxy (Me ester): [52427-50-7]. *2,3-Dihydro-1-methoxy-1H-phosphindole 1-oxide.*
$C_9H_{11}O_2P$ M 182.158
Hygroscopic oil.

Collins, D.J. *et al*, *Aust. J. Chem.*, 1974, **27**, 831 (*synth, ir, uv, pmr, ms, derivs*)

2-Hydroxy-4,5-dimethyl-1,3,2-dioxaphospholane 2-oxide, 8CI — H-00121

2,3-Butylene cyclic phosphoric acid. 2,3-Butylene hydrogen phosphate. 2-Hydroxy-4,5-dimethyl-1,3-dioxa-2-oxo-2-phosphacyclopentane

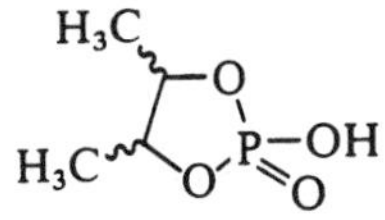

$C_4H_9O_4P$ M 152.086

Trimethylammonium salt: [41821-75-4]. Solid. Dec. at 275°.

Anilinium salt: [31481-91-1]. Solid. Mp 134°.

2-Methoxy (Me ester): see under 2-Methoxy-4,5-dimethyl-1,3,2-dioxaphospholane, M-00033

2-Ethoxy (Et ester): see under 2-Ethoxy-4,5-dimethyl-1,3,2-dioxaphospholane, E-00031

Chloride: see under 2-Chloro-4,4,5,5-tetramethyl-1,3,2-dioxaphospholane, C-00185

Revel, M. *et al*, *Bull. Soc. Chim. Fr.*, 1971, 105 (*P nmr*)
Brault, J.-F. *et al*, *Bull. Soc. Chim. Fr., Part 2*, 1974, 677 (*derivs*)
Zbaida, S. *et al*, *J. Org. Chem.*, 1982, **47**, 1073 (*ir, pmr, P nmr*)

2-Hydroxy-4,5-dimethyl-1,3,2-dioxaphospholane 2-sulfide, 9CI **H-00122**

2-Mercapto-4,5-dimethyl-1,3,2-dioxaphospholane 2-oxide. O,O-*2,3-Butylene hydrogen thiophosphate.* O,O-*2,3-Butylene cyclic phosphorothioic acid*

[61617-39-8]

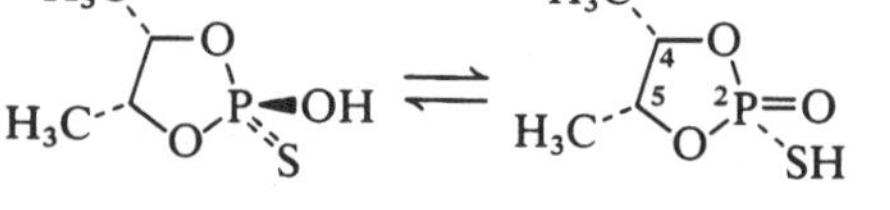

(2α.4α.5α)-*form*

$C_4H_9O_3PS$ M 168.147

Tautomeric. Oil. d_4^{20} 1.277. $Bp_{0.05}$ 210°. pK_a 2.77 (80% EtOH aq.) (97% thiol form).

2-Methoxy (O-*Me ester*)*: see under 2-Methoxy-4,5-dimethyl-1,3,2-dioxaphospholane, M-00033*

2-Methylthio (S-*Me ester*)*: see under 4,5-Dimethyl-2-methylthio-1,3,2-dioxaphospholane, D-00765*

2-Ethoxy (O-*Et ester*)*: see under 2-Ethoxy-4,5-dimethyl-1,3,2-dioxaphospholane, E-00031*

Fluoride: see under 2-Fluoro-4,5-dimethyl-1,3,2-dioxaphospholane, F-00015

Bromide: see under 2-Bromo-4,5-dimethyl-1,3,2-dioxaphospholane, B-00470

Dimethylamide: see under 4,5-Dimethyl-2-dimethylamino-1,3,2-dioxaphospholane, D-00725

(2α,4α,5α)-form

meso-cis-*form*

OH *trans*- to Me group.

Tetramethylammonium salt: Cryst. (propanol/Et_2O). Mp 174-176°.

1H-Imidazolium salt: Cryst. (propanol/Et_2O). Mp 114-116°.

(2α,4α,5β)-form

(±)-*form*

Tetramethylammonium salt: Cryst. (propanol/Et_2O). Mp 155-158°.

1H-Imidazolium salt: [61361-04-4]. Cryst. (propanol/Et_2O). Mp 117-118°.

(2α,4β,5β)-form

meso-trans-*form*

OH *cis*- to Me group.

Tetramethylammonium salt: Cryst. (propanol/Et_2O). Mp 178-180°.

1H-Imidazolium salt: [74560-09-1]. Cryst. (propanol/Et_2O). Mp 103-106°.

Wieczorek, M.W. *et al*, *Cryst. Struct. Commun.*, 1976, **5**, 739 (*cryst struct*)

Mikolajczyk, M. *et al*, *J. Chem. Soc., Perkin Trans. 1*, 1977, 2213 (*synth, ir, P nmr*)

Ovchinnikov, V.V. *et al*, *Zh. Obshch. Khim.*, 1977, **47**, 290 (*Engl. transl.* p. 267) (*synth, props, ir, P nmr*)

Wieczorek, M.W. *et al*, *Acta Crystallogr., Sect. B*, 1980, **36**, 1452 (*cryst struct*)

Wieczorek, M.W. *et al*, *Phosphorus Sulfur*, 1980, **9**, 137 (*cryst struct, bibl*)

2-Hydroxy-4,5-dimethyl-1,3,2-dioxaphosphole 2-oxide, 9CI **H-00123**

Dimethylvinylene phosphoric acid. Dimethylvinylene hydrogen phosphate

[20682-72-8]

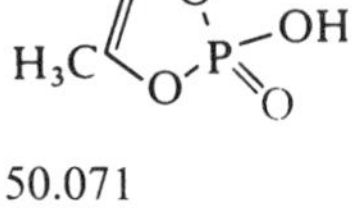

$C_4H_7O_4P$ M 150.071

Cryst. (CH_2Cl_2/Et_2O). Mp 108-110°.

N-*Methylpyridinium salt:* Solid. Mp 105-107°.

2-Methoxy (*Me ester*)*:* [933-43-7]. *Methyl dimethylvinylene phosphate.*

$C_5H_9O_4P$ M 164.097

Liq. which slowly cryst. Mp 42-43°. $Bp_{0.2}$ 64-65°.

2-Ethoxy (*Et ester*)*:* [16764-06-0]. *Ethyl dimethylvinylene phosphate.*

$C_6H_{11}O_4P$ M 178.124

Liq. d_4^{20} 1.18. Bp_{10} 120°. n_D^{20} 1.4325.

*2-*tert-*Butoxy* (tert-*Butyl ester*)*:* tert-*Butyl dimethylvinylene phosphate.*

$C_8H_{15}O_4P$ M 206.178

Solid. Mp 58-60°.

2-Phenoxy (*Ph ester*)*:* [55895-03-9]. *Dimethylvinylene phenyl phosphate.*

$C_{10}H_{11}O_4P$ M 226.168

Liq. $Bp_{0.1}$ 100° (bath).

2-Trimethylsilyloxy (*Trimethylsilyl ester*)*: Dimethylvinylene trimethylsilyl phosphate.*

$C_7H_{15}O_4PSi$ M 222.252

Liq. $Bp_{0.5}$ 80-81°.

Anhydride: see under 2,2′-Oxybis[4,5-dimethyl-1,3,2-dioxaphosphole], O-00088

Chloride: see under 2-Chloro-4,5-dimethyl-1,3,2-dioxaphosphole, C-00057

Imidazolide: see N-(1,2-Dimethylethenylenedioxyphosphoryl)imidazole, D-00753

Gaydou, E.M. *et al*, *Bull. Soc. Chim. Fr., Part 2*, 1973, 2279 (*ethyl ester, pmr, P nmr*)

Pudovik, A.N. *et al*, *Zh. Obshch. Khim.*, 1974, **44**, 1411 (*Engl. transl.* p. 1383) (*ethyl ester, synth, ir, P nmr*)

Ramirez, F. *et al*, *J. Am. Chem. Soc.*, 1975, **97**, 3809 (*esters, synth, pmr, P nmr*)

Ramirez, F. *et al*, *Synthesis*, 1975, 99 (*synth, esters, pmr, P nmr*)

Ramirez, F. *et al*, *Phosphorus Sulfur*, 1978, **4**, 325 (*esters*)

Konovalova, I.V. *et al*, *Zh. Obshch. Khim.*, 1982, **52**, 1965 (*Engl. transl.* p. 1745) (*ethyl ester*)

2-Hydroxy-5,5-dimethyl-1,3,2-dioxaphosphorinane 2-oxide **H-00124**

2-Hydroxy-5,5-dimethyl-1,3-dioxa-2-oxo-2-phosphacyclohexane. 2-Hydroxy-5,5-dimethyl-1,3,2-dioxaphosphinane 2-oxide. Neopentylene cyclic phosphoric acid

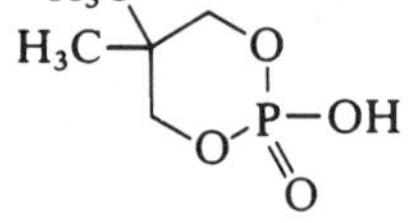

$C_5H_{11}O_4P$ M 166.113

Cryst. + $1H_2O$ (H_2O). Mp 174-176° (170-171°) (anhyd.). pK_a 1.67 (H_2O), pK_a 2.76 (50% EtOH aq.), pK_a 4.20 (95% EtOH aq.), pK_a 9.14 ($MeNO_2$). Exists in orthorhombic and monoclinic modifications.

Tetramethylammonium salt: [41821-80-1]. Solid. Mp 320° dec.

Cyclohexylammonium salt: Needles (EtOH/EtOAc). Mp 240-245°.

Anhydride: see under 2,2′-Oxybis[5,5-dimethyl-1,3,2-dioxaphosphorinane], O-00089

Fluoride: see under 2-Fluoro-5,5-dimethyl-1,3,2-dioxaphosphorinane, F-00016

Chloride: see 2-Chloro-5,5-dimethyl-1,3,2-dioxaphosphorinane 2-oxide, C-00059

Bromide: see under 2-Bromo-5,5-dimethyl-1,3,2-dioxaphosphorinane, B-00471

2-Methoxy (*Me ester*)*: see under 2-Methoxy-5,5-dimethyl-1,3,2-dioxaphosphorinane, M-00034*
2-Ethoxy (*Et ester*)*: see under 2-Ethoxy-5,5-dimethyl-1,3,2-dioxaphosphorinane, E-00032*
*2-*tert*-Butoxy* (tert*-Butyl ester*)*: see under 2-tert-*Butoxy-5,5-dimethyl-1,3,2-dioxaphosphorinane, B-00534
2-Phenoxy (Ph ester): see under 5,5-Dimethyl-2-phenoxy-1,3,2-dioxaphosphorinane, D-00790
2-N,N-Dimethylamide: see under 2-Dimethylamino-5,5-dimethyl-1,3,2-dioxaphosphorinane, D-00693
2-N-tert-Butylamide: see under 2-*tert*-Butylamino-5,5-dimethyl-1,3,2-dioxaphosphorinane, B-00538
Anilide: see under 5,5-Dimethyl-2-phenylamino-1,3,2-dioxaphosphorinane, D-00792

Edmundson, R.S., *Tetrahedron*, 1965, **21**, 2379 (*synth*)
Murayama, W. *et al*, *Bull. Chem. Soc. Jpn.*, 1969, **42**, 1819 (*cryst struct*)
Hall, L.D. *et al*, *Can. J. Chem.*, 1972, **50**, 2092 (*pmr, P nmr*)
Brault, J.-F. *et al*, *Bull. Soc. Chim. Fr., Part II*, 1973, 3149; 1974, 677 (*synth, pmr, P nmr*)
Forrest, G. *et al*, *J. Raman. Spectrosc.*, 1977, **6**, 32 (*ir*)
Ramirez, F. *et al*, *Tetrahedron Lett.*, 1982, **23**, 5375 (*synth, P nmr*)

2-Hydroxy-5,5-dimethyl-1,3,2-dioxaphosphorinane 2-selenide — H-00125

2-Hydroxy-5,5-dimethyl-1,3,2-dioxaphosphinane 2-selenide. O,O*-Neopentylene cyclic phosphoroselenoic acid.* O,O*-Neopentylene hydrogen phosphoroselenoate. 2-Hydroxy-5,5-dimethyl-1,3-dioxa-2-seleno-2-phosphacyclohexane*

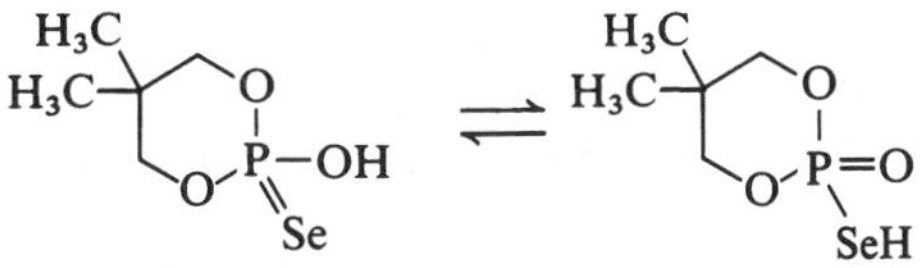

$C_5H_{11}O_3PSe$ M 229.074
Tautomeric.

Triethylammonium salt: Reagent for deoxygenation of carbohydrate epoxides.
2-Methylseleno (Se*-Me ester*)*: see under 5,5-Dimethyl-2-methylseleno-1,3,2-dioxaphosphorinane, D-00764*
2-Phenoxy (O*-Ph ester*)*: see under 5,5-Dimethyl-2-phenoxy-1,3,2-dioxaphosphorinane, D-00790*
Chloride: see under 2-Chloro-5,5-dimethyl-1,3,2-dioxaphosphorinane, C-00058
N,N*-Dimethylamide: see under 2-Dimethylamino-5,5-dimethyl-1,3,2-dioxaphosphorinane, D-00693*
Anilide: see under 5,5-Dimethyl-2-phenylamino-1,3,2-dioxaphosphorinane, D-00792

Kudelska, W. *et al*, *Tetrahedron*, 1981, **37**, 2989 (*P nmr, use*)

3-Hydroxy-2,2-dimethyl-1,3-oxaphosphetane 3-oxide, 9CI — H-00126

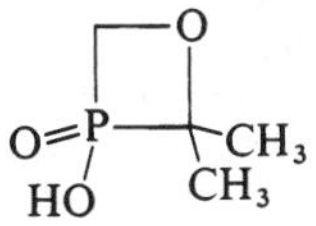

$C_4H_9O_3P$ M 136.087
Isolable only as Na salt, stable in alkaline soln.
Na salt: [57788-18-8]. No phys. props. reported.

Zyablikova, T.A. *et al*, *Zh. Obshch. Khim.*, 1975, **45**, 1984 (*Engl. transl.* p. 1950) (*synth, props, ir, pmr*)

10-Hydroxy-2,8-dimethyl-10*H*-phenothiaphosphine 10-oxide, 9CI, 8CI — H-00127

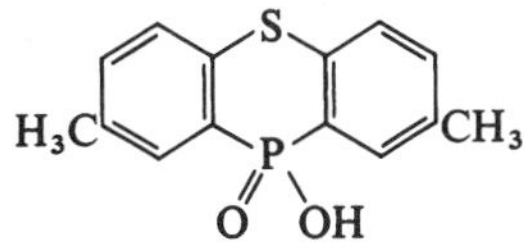

$C_{14}H_{13}O_2PS$ M 276.289
Cryst. (EtOH or DMF). Mp 315°.
Dicyclohexylammonium salt: Cryst. (EtOH). Mp 254°.
10-Methoxy (*Me ester*)*:* [23855-75-6].
$C_{15}H_{15}O_2PS$ M 290.316
Solid. Mp 176°.

Granoth, I. *et al*, *Tetrahedron*, 1969, **25**, 3919 (*synth, deriv*)
Granoth, I. *et al*, *Org. Mass Spectrom.*, 1970, **3**, 1359 (*ms*)

10-Hydroxy-2,8-dimethyl-10*H*-phenoxaphosphine 10-oxide, 8CI — H-00128

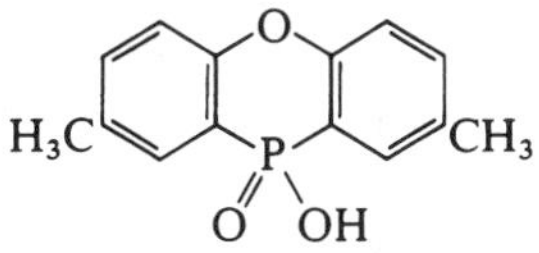

$C_{14}H_{13}O_3P$ M 260.229
Cryst. (EtOH). Mp 307°.
Dicyclohexylammonium salt: Cryst. (EtOH). Mp 227°.

Granoth, I. *et al*, *Isr. J. Chem.*, 1968, **6**, 651 (*synth, ir*)
Hellwinkel, D. *et al*, *Chem. Ber.*, 1978, **111**, 13 (*synth*)

2-Hydroxy-5,5-dimethyl-4-phenyl-1,3,2-dioxaphoshorinane 2-oxide, 9CI — H-00129

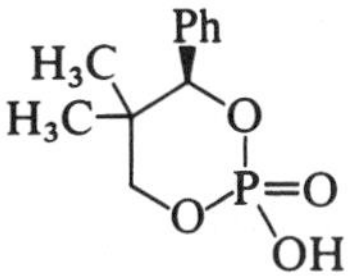

(*R*)-*form*

$C_{11}H_{15}O_4P$ M 242.211

(*R*)-*form*
Solid. $[\alpha]_D$ −60°.
Chloride: 2-Chloro-5,5-dimethyl-4-phenyl-1,3,2-dioxaphosphorinane 2-oxide. Useful for det. of enantiomeric purity of chiral amines. Solid. Mp 162-164.5°. $[\alpha]_{578}$−82.4°.

(*S*)-*form*
Solid. Mp 230-231°. $[\alpha]_D$ +62.5°.

(±)-*form*
Solid. Mp 224-224.5°.
Chloride:
$C_{11}H_{14}ClO_3P$ M 260.657
Solid. Mp 127.5-129.5°.

ten Hoeve, W. *et al*, *J. Org. Chem.*, 1985, **50**, 4508 (*synth, resoln, chloride, pmr, use*)

1-Hydroxy-3,4-dimethyl-1*H*-phosphole 1-oxide, 9CI — H-00130

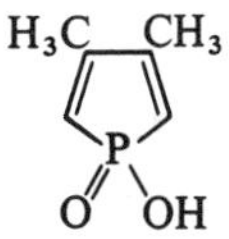

$C_6H_9O_2P$ M 144.110

1-Methoxy (*Me ester*): [34422-49-6]. *1-Methoxy-3,4-dimethyl-1*H*-phosphole 1-oxide.*
$C_7H_{11}O_2P$ M 158.136
Unstable, rapidly dimerises; prepd. and used in soln.
1-Phenoxy (*Ph ester*):
$C_{12}H_{13}O_2P$ M 220.207
Unstable, rapidly dimerises; prepd. and used in soln.

Clarke, F.B. *et al*, *J. Am. Chem. Soc.*, 1971, **93**, 4541 (*synth, uv, props, pmr*)

4-Hydroxydinaphtho[2,1-*d*:1',2'-*f*][1,3,2]-dioxaphosphepin 4-oxide, 9CI — H-00131

1,1'-Binaphthyl 2,2'-cyclic phosphoric acid
[35193-63-6]

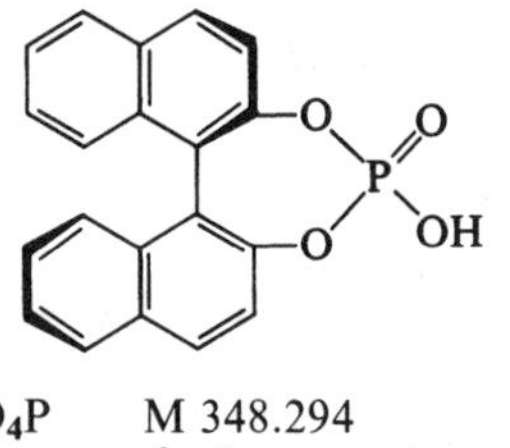

(*R*)-*form*

$C_{20}H_{13}O_4P$ M 348.294
Resolving agent for bases, amino acids, and helicenes. Resolved *via* strychnine, cinchonine, or cinchonidine salts.

(*R*)-*form* [39648-67-4]
Solid. Mp 335-337°. $[\alpha]_D$ −609° (c, 0.24 in MeOH).
Strychnine salt: Solid. Mp 245-247°.
4-Methoxy (*Me ester*): [86334-02-3]. *4-Methoxydinaphtho[2,1-*d*,1',2'-*f*][1,3,2]dioxaphosphepin 4-oxide. 2,2'-(1,1'-Binaphthyl) methyl phosphate.*
$C_{21}H_{15}O_4P$ M 362.321
Cryst. (C_6H_6/pet. ether). $[\alpha]_D$ −490° (c, 19.4 in MeOH).

(*S*)-*form* [35193-64-7]
$[\alpha]_D^{22}$ +530° (c, 1.35 in MeOH). Rotn. quoted in the paper as $[\alpha]^{22}{}_J$, which sems to be a misprint.
2-Methoxy (*Me ester*): [35193-65-8]. Solid. Mp 216°. $[\alpha]_D^{22}$ +544° (c, 0.22 in MeOH).

(±)-*form*
Cryst. (MeOH). Mp 344-346°.
2-Methoxy (*Me ester*): [35193-66-9]. Solid. Mp 211°.

Jacques, J. *et al*, *Tetrahedron Lett.*, 1971, 4617 (*synth, resoln, use*)
Mikeš, F. *et al*, *J. Chromatogr.*, 1978, **149**, 455 (*use*)
Hoyano, Y.Y. *et al*, *Can. J. Chem.*, 1980, **58**, 134 (*synth, resoln, derivs*)
Tetreau, C. *et al*, *Nouv. J. Chim.*, 1980, **4**, 423 (*cd*)

2-Hydroxy-1,3,2-dioxaphosphepane 2-oxide, 9CI, 8CI — H-00132

Tetramethylene phosphoric acid. Tetramethylene hydrogen phosphate
[51374-71-1]

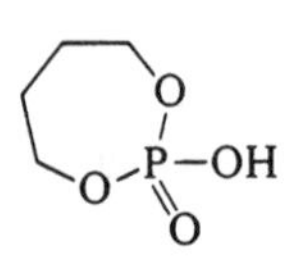

$C_4H_9O_4P$ M 152.086
Cryst. (CH_2Cl_2), plates (C_6H_6). Mp 129-131°.
2-Ethoxy (*Et ester*): [18748-21-5]. *2-Ethoxy-1,3,2-dioxaphosphepane 2-oxide. Ethyl tetramethylene phosphate.*
$C_6H_{13}O_4P$ M 180.140
Liq. $Bp_{0.001}$ 120-130°. n_D^{20} 1.4478.
2-Phenoxy (*Ph ester*): [7191-26-6]. *2-Phenoxy-1,3,2-dioxaphosphepane 2-oxide. Phenyl tetramethylene phosphate.*
$C_{10}H_{13}O_4P$ M 228.184
Solid. Mp 62-64°.

Munoz, A. *et al*, *Bull. Soc. Chim. Fr.*, 1967, 3343 (*ethyl ester*)
Coulter, C.L., *J. Am. Chem. Soc.*, 1975, **97**, 4084 (*cryst struct*)
Penny, C.L. *et al*, *Can. J. Chem.*, 1978, **56**, 2396 (*phenyl ester, pmr*)
Ramirez, F. *et al*, *Tetrahedron Lett.*, 1982, **23**, 5375 (*synth*)

2-Hydroxy-1,3,2-dioxaphospholane — H-00133

Phosphorous acid cyclic ethylene ester, 8CI. Ethylene glycol hydrogen phosphite
[1003-11-8]

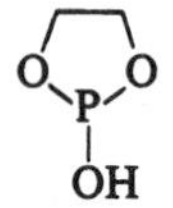

$C_2H_5O_3P$ M 108.033
Tautomeric with Liq., unstable to dist.
▷TG7210000.
Me ester: see 2-Methoxy-1,3,2-dioxaphospholane, M-00036
▷*TG7250000.*
Et ester: see 2-Ethoxy-1,3,2-dioxaphospholane, E-00034
Chloride: see 2-Chloro-1,3,2-dioxaphospholane, C-00063
2-Oxide: see 2-Hydroxy-1,3,2-dioxaphospholane 2-oxide, H-00134
2-Sulfide: see 2-Hydroxy-1,3,2-dioxaphospholane 2-sulfide, H-00135

Arbuzov, A.E. *et al*, *Izv. Akad. Nauk SSSR, Otd. Khim. Nauk*, 1952, 770.

2-Hydroxy-1,3,2-dioxaphospholane 2-oxide, 9CI — H-00134

Ethylene phosphoric acid. Ethylene hydrogen phosphate
[6711-47-3]

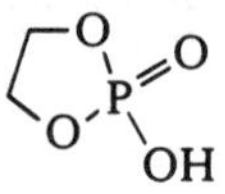

$C_2H_5O_4P$ M 124.033
Solid. Mp 133-135°. Also isol. as Ba or cyclohexylammonium salt. Readily hydrolysed.
Cyclohexylammonium salt: [92136-94-2]. Needles (propanol/Et_2O). Mp 168°.
2-Methoxy (*Me ester*): *see 2-Methoxy-1,3,2-dioxaphospholane 2-oxide, M-00037*
2-Ethoxy (*Et ester*): *see under 2-Ethoxy-1,3,2-dioxaphospholane 2-oxide, E-00035*
2-Phenoxy (*Ph ester*): *see under 2-Phenoxy-1,3,2-dioxaphospholane, P-00067*
Anhydride: see under 2,2'-Oxybis[1,3,2-dioxaphospholane], O-00091
Chloride: see 2-Chloro-1,3,2-dioxaphospholane 2-oxide, C-00064

Anilide: see under 1-Phenylamino-1,3,2-dioxaphospholane, P-00083

Kumamoto, J. *et al*, *J. Am. Chem. Soc.*, 1956, **78**, 4858 (*synth*)
Keay, L. *et al*, *J. Chem. Soc.*, 1961, 710 (*synth*)
Wilcox, R.D. *et al*, *Tetrahedron Lett.*, 1968, 6001 (*synth, deriv*)
Sturtevan, J.M. *et al*, *J. Am. Chem. Soc.*, 1973, **95**, 8168 (*props*)
Nguyen Thanh Thuong, *et al*, *Bull. Soc. Chim. Fr., Part II*, 1974, 667 (*pmr*)
Gerlt, J.A. *et al*, *J. Biol. Chem.*, 1975, **250**, 5059 (*props*)
Forrest, G. *et al*, *J. Raman Spectrosc.*, 1977, **6**, 32 (*ir*)
Taira, K. *et al*, *J. Org. Chem.*, 1984, **49**, 4531 (*props*)

2-Hydroxy-1,3,2-dioxaphospholane 2-sulfide, 9CI — H-00135

O,O-*Ethylene hydrogen thiophosphate.* O,O-*Ethylene phosphorothioate.* O,O-*Ethylene thiophosphoric acid*

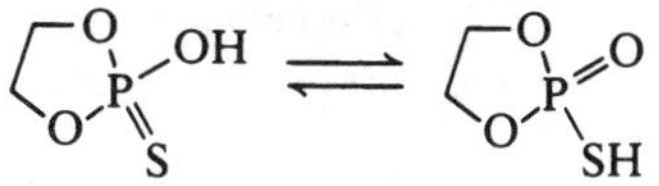

$C_2H_5O_3PS$ M 140.093

Tautomeric.

1H-*Imidazolium salt:* [23233-31-0]. Cryst. (propanol/Et_2O). Mp 117-119°.

2-Ethoxy (O-*Et ester*)*: see 2-Ethoxy-1,3,2-dioxaphospholane 2-sulfide, E-00036*

2-Phenoxy (O-*Ph ester*)*: see 2-Phenoxy-1,3,2-dioxaphospholane, P-00067*

Chloride: see 2-Chloro-1,3,2-dioxaphospholane 2-sulfide, C-00065

Dimethylamide: see 2-Dimethylamino-1,3,2-dioxaphospholane, D-00695

Anilide: see 1-Phenylamino-1,3,2-dioxaphospholane, P-00083

Mikolajczyk, M. *et al*, *J. Chem. Soc., Perkin Trans. 1*, 1976, 371 (*synth, pmr, P nmr*)
Wieczorek, M.W. *et al*, *Acta Crystallogr., Sect. B*, 1978, **34**, 3138 (*cryst struct*)
Galdecki, Z. *et al*, *Cryst. Struct. Commun.*, 1981, **10**, 137 (*cryst struct*)

2-Hydroxy-1,3,2-dioxaphosphole 2-oxide, 9CI — H-00136

2-Hydroxy-1,3-dioxa-2-oxo-2-phosphacyclopentene. Ethylenedioxyphosphinic acid. Vinylenephosphoric acid

[59892-97-6]

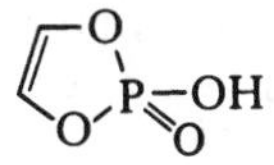

$C_2H_3O_4P$ M 122.017

2-Methoxy (*Me ester*)*:* [39902-05-1]. *Methyl vinylene phosphate.*
$C_3H_5O_4P$ M 136.044
Liq. $Bp_{0.05}$ 45°.

Gaydou, E.M. *et al*, *Bull. Soc. Chim. Fr., Part 2*, 1973, 2279 (*synth, pmr, P nmr*)

2-Hydroxy-1,3,2-dioxaphosphonane 2-oxide, 10CI — H-00137

Hexamethylene cyclic phosphoric acid. Hexamethylene hydrogen phosphate

[68755-20-4]

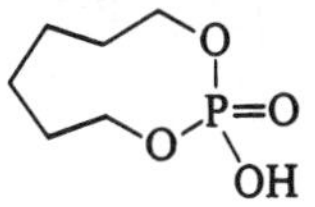

$C_6H_{13}O_4P$ M 180.140

Gum.

Ph ester: [68755-15-7]. *2-Phenoxy-1,3,2-dioxaphosphonane 2-oxide. Hexamethylene phenyl phosphate.*
$C_{12}H_{17}O_4P$ M 256.238
Oil.

Penney, C.L. *et al*, *Can. J. Chem.*, 1978, **56**, 2396 (*synth, pmr*)

2-Hydroxy-1,3,2-dioxaphosphorinane 2-oxide, 9CI — H-00138

Trimethylene phosphoric acid. Trimethylene hydrogen phosphate. 2-Hydroxy-1,3,2-dioxaphosphinane 2-oxide. 2-Hydroxy-1,3-dioxa-2-oxo-2-phosphacyclohexane

[13507-10-3]

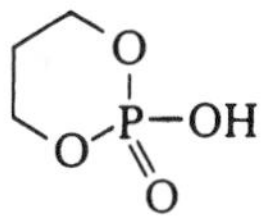

$C_3H_7O_4P$ M 138.060

Cryst. ($MeCN/Et_2O$). Mp 102-102.5°. pK_a 1.87 (H_2O), pK_a 2.84 (50% EtOH aq.), pK_a 4.30 (95% EtOH aq.), pK_a 9.27 ($MeNO_2$).

Tetramethylammonium salt: [41821-77-6]. Solid. Mp 293-295° dec.

Anhydride: see 2,2'-Oxybis[1,3,2-dioxaphosphorinane], O-00092

2-Methoxy (*Me ester*)*: see 2-Methoxy-1,3,2-dioxaphosphorinane, M-00038*

2-Ethoxy (*Et ester*)*: see 2-Ethoxy-1,3,2-dioxaphosphorinane, E-00037*

2-Phenoxy (*Ph ester*)*: see 2-Phenoxy-1,3,2-dioxaphosphorinane, P-00068*

Fluoride: see 2-Fluoro-1,3,2-dioxaphosphorinane, F-00018

Chloride: see 2-Chloro-1,3,2-dioxaphosphorinane 2-oxide, C-00067

N,N-*Dimethylamide: see 2-Dimethylamino-1,3,2-dioxaphosphorinane, D-00696*

Khorana, H.G. *et al*, *J. Am. Chem. Soc.*, 1957, **79**, 430 (*synth*)
Brault, J.-F. *et al*, *Bull. Soc. Chim. Fr., Part II*, 1973, 3149; 1974, 677 (*synth, pmr*)
Sturtevant, J.M. *et al*, *J. Am. Chem. Soc.*, 1973, **95**, 8168 (*props*)
Forrest, G. *et al*, *J. Raman Spectrosc.*, 1977, **6**, 32 (*ir, raman*)
Lapiensis, G. *et al*, *Macromolecules*, 1977, **10**, 1301 (*esters*)
Ramirez, F. *et al*, *Tetrahedron Lett.*, 1982, **23**, 5375 (*synth, P nmr*)
Jankowska, J. *et al*, *Synthesis*, 1984, 408 (*synth, cmr*)

2-Hydroxy-4,5-diphenyl-1,3,2-dioxaphosphole 2-oxide, 9CI H-00139

Diphenylethylenedioxyphosphoric acid

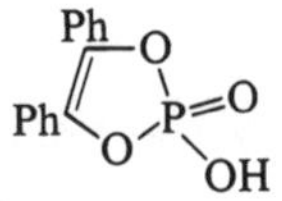

$C_{14}H_{11}O_4P$ M 274.212

2-Methoxy (*Me ester*): [84382-98-9].
$C_{15}H_{13}O_4P$ M 288.239
Solid. Mp 76°. $Bp_{0.07}$ 80°.

2-Ethoxy (*Et ester*): [16830-69-6].
$C_{16}H_{15}O_4P$ M 302.266
Liq. $Bp_{0.005}$ 150-152°.

Gozman, I.P. *et al*, *Zh. Obshch. Khim.*, 1967, **37**, 881 (*Engl. transl.* p. 831) (*ethoxy*)
Komovalova, I.V. *et al*, *Zh. Obshch. Khim.*, 1982, **52**, 1965 (*Engl. transl.* p. 1745) (*methoxy*)

4-Hydroxy-2,6-diphenyl-4*H*-1,4-oxaphosphorin 4-oxide, 10CI H-00140

4-Hydroxy-2,6-diphenyl-1,4-oxaphosphacyclohexa-2,5-diene 4-oxide

[56153-45-8]

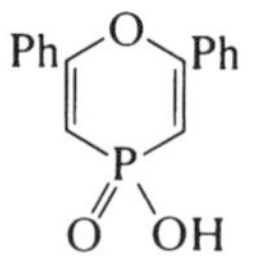

$C_{16}H_{13}O_3P$ M 284.251
Cryst. (EtOH). Mp 238-239°. pK_a 4.57 (MeOH).

Triethylammonium salt: [70886-00-9]. Cryst. (C_6H_6/Et_2O). Mp 115-120°.

4-Ethoxy (*Et ester*): [61183-55-9]. *4-Ethoxy-2,6-diphenyl-4*H*-1,4-oxaphosphorin 4-oxide.*
$C_{18}H_{17}O_3P$ M 312.304
Flakes (Et_2O/pentane). Mp 141-142°.

Anhydride: [70886-04-3]. *Oxybis(2,6-diphenyl-4*H*-1,4-oxaphosphorin)-4,4′-dioxide.*
$C_{32}H_{24}O_5P_2$ M 550.486
Cryst. (dioxan/Et_2O). Mp 290-292°.

Chloride: [70886-01-0]. *4-Chloro-2,6-diphenyl-4*H*-1,4-oxaphosphorin 4-oxide.*
$C_{16}H_{12}ClO_2P$ M 302.696
Cryst. (C_6H_6). Mp 197-199°.

Amide: [70886-02-1]. *2,6-Diphenyl-4*H*-1,4-oxaphosphorin-4-amine 4-oxide.*
$C_{16}H_{14}NO_2P$ M 283.266
Cryst. (chlorobenzene). Mp 280°.

Moskalevskaya, L.S. *et al*, *Zh. Obshch. Khim.*, 1975, **45**, 950; 1979, **49**, 1015 (*Engl. transl.* pp. 937, 880) (*synth, derivs, ir, pmr*)
Chattha, M.S., *Chem. Ind.* (*London*), 1976, 484 (*ester, ir, pmr*)
Fedorova, G.K. *et al*, *Zh. Obshch. Khim.*, 1984, **54**, 1481 (*Engl. transl.* p. 1321) (*synth*)

***N*-Hydroxy-*P,P*-diphenylphosphinic amide, 9CI** H-00141

[73452-52-5]

$Ph_2P(O)NHOH$

$C_{12}H_{12}NO_2P$ M 233.206
Solid. Mp 145-146° dec.

O-*Ac:*
$C_{14}H_{14}NO_3P$ M 275.243
Cryst. (CH_2Cl_2/Et_2O). Mp 164-165°.

O-(*4-Methylbenzenesulfonyl*): Cryst. (MeOH aq.). Mp 133-134°.

Harger, M.J.P., *J. Chem. Soc., Perkin Trans. 1*, 1983, 2699 (*synth, ms, ir*)
Harger, M.J.P., *Tetrahedron Lett.*, 1983, **24**, 3115 (*props*)

1-Hydroxy-3,4-diphenyl-1*H*-phosphole 1-oxide H-00142

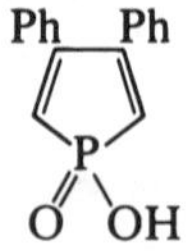

$C_{16}H_{13}O_2P$ M 268.251
Microcryst. powder + $1H_2O$. Mp 181-182°. Dimerizes on cryst. from 1,2-dichloroethane.

1-Ethoxy (*Et ester*): *1-Ethoxy-3,4-diphenyl-1*H*-phosphole 1-oxide.*
$C_{18}H_{17}O_2P$ M 296.305
No phys. props. reported. Undergoes cycloaddition with dimethyl 2-butynedioate.

1-Phenoxy (*Ph ester*): *1-Phenoxy-3,4-diphenyl-1*H*-phosphole 1-oxide.*
$C_{22}H_{17}O_2P$ M 344.349
Needles (Et_2O/C_6H_6 or MeCN). Mp 156.5-157°.

Clarke, F.C. *et al*, *J. Am. Chem. Soc.*, 1971, **93**, 4541 (*synth, derivs, pmr, uv, P nmr, props*)

4-Hydroxy-2,6-diphenyl-4*H*-1,4-thiaphosphorine 2-oxide H-00143

4-Hydroxy-2,6-diphenyl-1,4-thiaphospha-2,5-cyclohexadiene 2-oxide

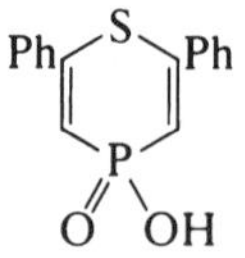

$C_{16}H_{13}O_2PS$ M 300.311
Cryst. (MeOH aq.). Mp 197-198°.

4-Methoxy (*Me ester*): *4-Methoxy-2,6-diphenyl-4*H*-1,4-thiaphosphorin 4-oxide.*
$C_{17}H_{14}O_2PS$ M 313.330
Cryst. (hexane). Mp 117-118°.

4-Ethoxy (*Et ester*): *4-Ethoxy-2,6-diphenyl-4*H*-1,4-thiaphosphorin 4-oxide.*
$C_{18}H_{16}O_2PS$ M 327.357
Cryst. (hexane). Mp 116-117°.

Chloride (*4-Chloro*): *4-Chloro-2,6-diphenyl-4*H*-1,4-thiaphosphorin 4-oxide.*
$C_{16}H_{12}ClOPS$ M 318.757
Solid. Mp 130-142°.

Proklina, N.V. *et al*, *Zh. Obshch. Khim.*, 1985, **55**, 2147 (*Engl. transl.* p. 1905) (*synth, derivs, ir, pmr, P nmr*)

(1-Hydroxyethylidene)bisphosphonic acid, 9CI H-00144

Hydroxyethane-1,1-diphosphonic acid. Etidronic acid, *BAN, USAN, INN*

[2809-21-4]

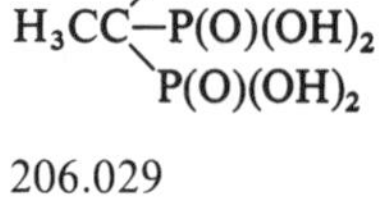

$H_3CC(OH)(P(O)(OH)_2)_2$

$C_2H_8O_7P_2$ M 206.029

Calcification inhibitor for Paget's disease. Syrup, cryst. + $1H_2O$ (AcOH aq.). Sol. H_2O, EtOH, MeOH. Mp 105°. pK_{a1} 2.01 (1.7), pK_{a2} 3.08 (2.47), pK_{a3} 7.60 (7.28), pK_{a4} 11.96 (10.29), pK_{a5} 13.63 (11.13) (H_2O, 25°).

Di-Na salt: [7414-83-7]. *HEDSPA*. Used in dentifrices.
▷JL5950000.

Tri-Na salt: [2666-14-0]. Used in dentifrices.
▷JL6650000.

Tetra-Me ester: [15207-88-2]. *Tetramethyl (1-hydroxyethylidene)bisphosphonate.*
$C_6H_{16}O_7P_2$ M 262.136
Cryst. (C_6H_6/hexane). Mp 68-71°.

Tetra-Et ester: [20427-93-4]. *Tetraethyl (1-hydroxyethylidene)bisphosphonate.*
$C_{10}H_{24}O_7P_2$ M 318.243
Liq. which isomerises on distn.

Nicholson, D.A. *et al*, *J. Org. Chem.*, 1970, **35**, 3149; 1971, **36**, 3843 (*ester, synth, P nmr*)
Blaser, B. *et al*, *Z. Anorg. Allg. Chem.*, 1971, **381**, 247 (*synth, props, P nmr*)
Wiers, B.H., *Inorg. Chem.*, 1971, **10**, 2581 (*salts*)
Bikhman, B.I. *et al*, *Zh. Neorg. Khim.*, 1973, **18**, 2406 (*Engl. transl.* p. 1273) (*ir, salts*)
Maier, L., *Helv. Chim. Acta*, 1973, **56**, 1257 (*ester, synth, pmr, P nmr, props*)
Collins, A.J. *et al*, *J. Appl. Chem. Biotechnol.*, 1977, **27**, 651 (*props, salts*)
Worms, K.H. *et al*, *Z. Anorg. Allg. Chem.*, 1979, **457**, 209 (*synth*)
Motekaitis, R.J. *et al*, *Inorg. Chem.*, 1980, **19**, 1646 (*props, complexes*)
Rizhalla, E.N. *et al*, *Talanta*, 1980, **27**, 715 (*complexes*)
Goeva, L.V., *Radiokhimiya*, 1982, **22**, 591 (*Engl. transl.* p. 489) (*props*)
Fonong, T. *et al*, *Anal. Chem.*, 1983, **55**, 1089 (*props*)

(1-Hydroxyethyl)phosphonic acid, 9CI — H-00145

1-Phosphonoethanol
[20188-02-7]

$$H_3CCH(OH)P(O)(OH)_2$$

$C_2H_7O_4P$ M 126.049

(±)-*form*
Oil which slowly cryst.

Monoanilinium salt: [75502-79-3]. Solid. Mp 168-169°.

Di-Me ester: [10184-66-4]. *Dimethyl (1-hydroxyethyl)-phosphonate.*
$C_4H_{11}O_4P$ M 154.102
Liq. d_4^{20} 1.24. $Bp_{0.08}$ 95°. n_D^{20} 1.4469.

Di-Et ester: [15336-73-9]. *Diethyl (1-hydroxyethyl)-phosphonate.*
$C_6H_{15}O_4P$ M 182.156
d_4^{20} 1.12. $Bp_{0.1}$ 120°. n_D^{20} 1.4328.

O-*Trimethylsilyl ether:* see [*1-[(Trimethylsilyl)oxy]-ethyl]phosphonic acid, T-00555*

Kharasch, M.S. *et al*, *J. Org. Chem.*, 1960, **25**, 1000 (*synth, derivs*)
Bel'skii, V.E. *et al*, *Zh. Obshch. Khim.*, 1972, **42**, 2427 (*Engl. transl.* p. 2421) (*ester, P nmr*)
Evelyn, L. *et al*, *Org. Magn. Reson.*, 1973, **5**, 141 (*ester, pmr*)
Gazizov, T.Kh. *et al*, *Zh. Obshch. Khim.*, 1977, **47**, 1234 (*Engl. transl.* p. 1137) (*ester, synth*)
Nurtdinov, S.Kh. *et al*, *Zh. Obshch. Khim.*, 1979, **49**, 2446 (*Engl. transl.* p. 2159) (*ester, synth*)
Baraldi, P.G. *et al*, *Synthesis*, 1982, 653 (*ester, ir, pmr*)

(2-Hydroxyethyl)phosphonic acid, 9CI — H-00146

2-Phosphonoethanol
[22987-21-9]

$$HOCH_2CH_2P(O)(OH)_2$$

$C_2H_7O_4P$ M 126.049
Major metab. of Ethepon in *Hevea brasiliensis*. Syrup.

Monoanilinium salt: [75502-79-3]. Solid. Mp 104-109°.

Mono-dicyclohexylammonium salt: [75502-80-6]. Solid. Mp 165-169°.

Di-Me ester: [54731-72-5]. *Dimethyl (2-hydroxyethyl)-phosphonate.*
$C_4H_{11}O_4P$ M 154.102
Liq. d_4^{20} 1.23. Bp_{1-2} 98-101°. n_D^{20} 1.4439.

Di-Me ester, Ac: see *(2-Acetoxyethyl)phosphonic acid, A-00006*

Di-Et ester: [39997-40-5]. *Diethyl (2-hydroxyethyl)-phosphonate.*
$C_6H_{15}O_4P$ M 182.156
Liq. d_4^{20} 1.12. Bp_2 112-115°. n_D^{20} 1.4372.

Di-Et ester, Ac: see *(2-Acetoxyethyl)phosphonic acid, A-00006*

Di-Et ester, Et ether: [5191-39-5]. *Diethyl (2-ethoxyethyl)phosphonate.*
$C_8H_{19}O_4P$ M 210.209
Liq. d_4^{20} 1.04. Bp_3 87°. n_D^{20} 1.4258.

Bis(trimethylsilyl) ester: [75502-78-2]. *Bis(trimethylsilyl) (2-hydroxyethyl)phosphonate.*
$C_8H_{23}O_4PSi_2$ M 270.412
Liq. $Bp_{0.01}$ 84°.

Pudovik, A.N. *et al*, *Zh. Obshch. Khim.*, 1963, **33**, 2755 (*Engl. transl.* p. 2684) (*ether*)
Songstadt, J., *Acta Chem. Scand.*, 1967, **21**, 1681 (*esters, synth, ir*)
Brel', A.K. *et al*, *Zh. Obshch. Khim.*, 1980, **50**, 2134 (*CA*, **94**, 139888) (*synth, ir, pmr*)
Gloede, J. *et al*, *J. Prakt. Chem.*, 1980, **322**, 327 (*synth, derivs, P nmr*)

(2-Hydroxyethyl)phosphonothioic acid, 9CI — H-00147

[65253-76-1]

$$HOCH_2CH_2P(S)(OH)_2 \rightleftharpoons HOCH_2CH_2P(O)(OH)(SH)$$

$C_2H_7O_3PS$ M 142.109
Grease-like substance.

Monoanilinium salt: [68669-55-6]. Cryst. (EtOH). Mp 153-154°.

Fedorova, G.K. *et al*, *Zh. Obshch. Khim.*, 1978, **78**, 2015 (*Engl. transl.* p. 1833) (*synth*)

(2-Hydroxyethyl)phosphoramidic acid — H-00148

[23545-86-0]

$$HOCH_2CH_2NHP(O)(OH)_2$$

$C_2H_8NO_4P$ M 141.063
Solid. Mp 242°.

Di-Et ester: [14662-78-3]. *Diethyl (2-hydroxyethyl)-phosphoramidate.*
$C_6H_{16}NO_4P$ M 197.170
Liq. d_4^{20} 1.16. $Bp_{0.008}$ 110-112°. n_D^{20} 1.4455.

Diisopropyl ester: Diisopropyl (2-hydroxyethyl)-phosphoramidate. Bis(1-methylethyl) (2-hydroxyethyl)phosphoramidate.
$C_8H_{20}NO_4P$ M 225.224

Liq. $Bp_{0.2}$ 156-157°.

Plapinger, R.E. *et al, J. Am. Chem. Soc.*, 1953, **75**, 5757 (*synth*)
Greenhalgh, R. *et al, Can. J. Chem.*, 1967, **45**, 495 (*esters, synth, ir, pmr*)
Pudovik, M.A. *et al, Zh. Obshch. Khim.*, 1976, **46**, 21 (*Engl. transl.* p. 20) (*diethyl ester, synth, props*)
Zwierzak, A. *et al, Synthesis*, 1984, 223 (*synth*)

O-(2-Hydroxyethyl) phosphorodichloridothioate H-00149

O-(*2-Hydroxyethyl) dichlorothiophosphate*

$HOCH_2CH_2OP(S)Cl_2$

$C_2H_5Cl_2O_2PS$ M 195.000

Me ether: [24108-60-9]. O-(*2-Methoxyethyl) phosphorodichloridothioate*, 9CI, 8CI. O-(*2-Methoxyethyl) dichlorothiophosphate.*
$C_3H_7Cl_2O_2PS$ M 209.027
Precursor to several potent insecticides. Liq. Bp_8 81-82°.

Et ether: [13891-64-0]. O-(*2-Ethoxyethyl) phosphorodichloridothioate*, 9CI, 8CI. O-(*2-Ethoxyethyl) dichlorothiophosphate.* O-(*2-Ethoxyethyl) thiophosphoryl dichloride.*
$C_4H_9Cl_2O_2PS$ M 223.054
Liq. d_4^{20} 1.30. n_D^{20} 1.4868.

Butyl ether: [18351-08-1]. O-(*2-Butoxyethyl) phosphorodichloridothioate*, 9CI, 8CI. O-(*2-Butoxyethyl) thiophosphoryl dichloride.* O-(*2-Butoxyethyl) dichlorothiophosphate.*
$C_{16}H_{13}Cl_2O_2PS$ M 371.217
Liq. d_4^{20} 1.22. Bp_1 95-96°. n_D^{20} 1.4787.

Srivastava, K.C., *Aust. J. Chem.*, 1966, **19**, 2397.
Bliznyuk, N.K. *et al, Zh. Obshch. Khim.*, 1967, **37**, 1122 (*Engl. transl.* p. 1064)
Ger. Pat., 1 924 972, (*1969*); *CA*, **72**, 90042

(2-Hydroxyethyl)-trimethylphosphonium(1+), 9CI H-00150

2-(Trimethylphosphonio)ethanol. Phosphacholine
[59738-68-0]

$Me_3P^{\oplus}CH_2CH_2OH$

$C_5H_{14}OP^{\oplus}$ M 121.139 (ion)
Phosphorus analogue of choline. Employed as a ^{31}P probe in biological membrane studies.

Chloride: [58887-04-0].
$C_5H_{14}ClOP$ M 156.592
V. deliquescent needles. Mp >250° dec.

Iodide: [59694-43-8].
$C_5H_{14}IOP$ M 248.043
Needles. Less hygroscopic than chloride. Dec. >250° without melting.

Sim, E. *et al, Biochem. J.*, 1975, **151**, 555; 1976, **154**, 105 (*use*)
Edwards, R.G. *et al, Biochem. Biophys. Acta*, 1976, **431**, 303 (*synth, pmr, nmr*)

(2-Hydroxyethyl)-triphenylphosphonium(1+), 9CI H-00151

2-(Triphenylphosphonio)ethanol(1+)

$Ph_3P^{\oplus}CH_2CH_2OH$

$C_{20}H_{20}OP^{\oplus}$ M 307.351 (ion)

Chloride: [58887-04-0].
$C_{20}H_{20}ClOP$ M 342.804
Solid. Mp 240-242°.

Bromide: [53710-27-3].
$C_{20}H_{20}BrOP$ M 387.255
Solid. Mp 217-218.5°.

Iodide: [59694-43-8].
$C_{20}H_{20}IOP$ M 434.256
Solid. Mp 185-186°.

Seyferth, D. *et al, J. Organomet. Chem.*, 1966, **6**, 205 (*synth*)
Kunz, H., *Phosphorus*, 1974, **3**, 273 (*deriv*)

2-Hydroxyhexahydro-4*H*-1,3,2-benzodioxaphosphorin 2-oxide H-00152

2-Hydroxy-1,3-dioxa-2-phosphadecalin 2-oxide. 2-Hydroxy-5,6-tetramethylene-1,3,2-dioxaphosphorinane 2-oxide

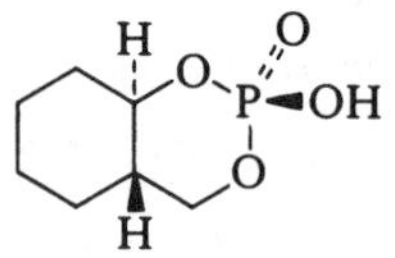

(2α,4α,5α)-*form*

$C_7H_{13}O_4P$ M 192.151

(2α,4aβ,8aα)-form
trans-axial-*form*

2-Phenoxy (2-Ph ester): [74378-77-1].
$C_{13}H_{17}O_4P$ M 268.249
Cryst. (MeCN). Mp 157-159°.
2-(4-Methoxyphenoxy) (2-Methoxyphenyl ester): [74378-76-0]. Cryst. (Et_2O). Mp 142-144°.
2-(4-Nitrophenoxy) (2-(4-Nitrophenyl)ester): [71771-33-0]. Cryst. (MeCN). Mp 122-125°.
2-(2,4-Dinitrophenoxy) (2-(2,4-Dinitrophenyl)ester): [74431-09-7]. Cryst. (MeCN). Mp 127-130°.

(2α,4aα,8aβ)-form
trans-eq.-*form*

2-Phenoxy: [74410-66-5]. Cryst. (toluene). Mp 65-68°.
2-(4-Methoxyphenoxy): [74410-65-4]. Cryst. (Et_2O). Mp 111-114°.
2-(4-Nitrophenoxy): [71806-76-3]. Cryst. (CCl_4). Mp 94°.
2-(2,4-Dinitrophenoxy): [74378-78-2]. Cryst. (toluene). Mp 110-113°.

Gorenstein, D.G. *et al, J. Am. Chem. Soc.*, 1979, **101**, 4925, 1980, **102**, 5077 (*synth, props, cmr, P nmr, ir, pmr*)
Rowell, R. *et al, J. Am. Chem. Soc.*, 1981, **103**, 5894 (*props*)
Van Nuffel, P. *et al, J. Mol. Struct.*, 1984, **125**, 1 (*struct*)

1-Hydroxy-1,1-hexanediphosphonic acid H-00153

1-Hydroxyhexylidenebisphosphonic acid, 9CI. *1,1-Diphosphono-1-hexanol*
[76254-55-2]

$H_3C(CH_2)_4C(OH)[P(O)(OH)_2]_2$

$C_6H_{16}O_7P_2$ M 262.136
Isol. as as hydrated sodium salt.

Blaser, B. *et al, Z. Anorg. Allg. Chem.*, 1971, **381**, 247 (*synth*)

Hydroxymethanediphosphonic acid H-00154

(Hydroxymethylene)bis(phosphonic acid), 9CI. *Oxidronic acid*, USAN, INN. *HMDP*
[15468-10-7]

$HOCH[P(O)(OH)_2]_2$

$CH_6O_7P_2$ M 192.002

Appears to be known only as salts. Weak inhibitor of AMV reverse transcriptase and DNA-polymerase. Calcium regulator.

Di-Na salt: [14255-61-9]. Mp 244-249°.

Eriksson, B. *et al*, *Biochem. Biophys. Acta*, 1982, **696**, 115 (*props*)
Eubank, W.B. *et al*, *J. Parasitol.*, 1982, **68**, 599 (*props*)
Shinoda, H. *et al*, *Calcif. Tissue Int.*, 1983, **35**, 87 (*props*)

2-Hydroxy-4-methyl-1,3,2-dioxaphospholane 2-oxide, 8CI — H-00155

Propylene cyclic phosphate. 1,2-Propylene hydrogen phosphate

[20636-79-7]

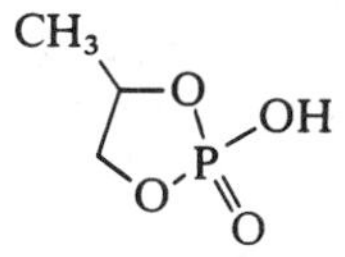

$C_3H_7O_4P$ M 138.060

(±)-*form*

Tetramethylammonium salt: Solid. Mp 222-226°.

Anilinium salt: Solid. Mp 153°.

2-Methoxy (Me ester): see 2-Methoxy-4-methyl-1,3,2-dioxaphospholane, M-00043

2-Ethoxy (Et ester): see 2-Ethoxy-4-methyl-1,3,2-dioxaphospholane, E-00040

2-tert-*Butoxy* (tert-*Butyl ester*)*: see 2-tert-*Butoxy-1,3,2-dioxaphospholane, B-00535

2-Phenoxy (Ph ester): see 4-Methyl-2-phenoxy-1,3,2-dioxaphospholane, M-00190

Chloride: see 2-Chloro-4-methyl-1,3,2-dioxaphospholane, C-00092

Dimethylamide: see 4-Methyl-2-dimethylamino-1,3,2-dioxaphospholane, M-00116

Revel, M. *et al*, *Bull. Soc. Chim. Fr.*, 1971, 105 (*synth, P nmr*)
Brault, J.-F. *et al*, *Bull. Soc. Chim. Fr., Part II*, 1974, 677 (*synth*)
Buchwald, S.L. *et al*, *J. Am. Chem. Soc.*, 1984, **106**, 4916.

2-Hydroxy-4-methyl-1,3,2-dioxaphosphorinane 2-oxide, 9CI — H-00156

1,3-Butylene hydrogen phosphate. 1,3-Butylene cyclic phosphoric acid. 2-Hydroxy-4-methyl-1,3,2-dioxaphosphinane 2-oxide

[16727-61-0]

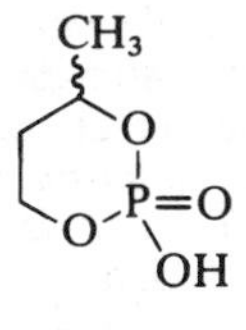

$C_4H_9O_4P$ M 152.086

Solid. Mp 72-73.5°. pK_a 1.75 (H_2O), pK_a 2.85 (50% EtOH aq.), pK_a 3.21 (80% EtOH aq.), pK_a 4.54 (95% EtOH aq.), pK_a 9.40 ($MeNO_2$).

Tetramethylammonium salt: Solid. Mp 250°.

Benzylammonium salt: [68755-21-5]. Solid. Mp 160-161°.

2-Methoxy (Me ester): see 2-Methoxy-4-methyl-1,3,2-dioxaphosphorinane 2-oxide, M-00045

2-tert-*Butoxy* (tert-*Butyl ester*)*: see 2-tert-*Butoxy-4-methyl-1,3,2-dioxaphosphorinane, B-00537

Fluoride: see under 2-Fluoro-4-methyl-1,3,2-dioxaphosphorinane, F-00021

Chloride: see 2-Chloro-4-methyl-1,3,2-dioxaphosphorinane 2-oxide, C-00094

Bromide: see under 2-Bromo-4-methyl-1,3,2-dioxaphosphorinane, B-00474

2-Dimethylamide: see 4-Methyl-2-dimethylamino-1,3,2-dioxaphosphorinane 2-oxide, M-00118

2-tert-Butylamide: see 2-*tert*-Butylamino-4-methyl-1,3,2-dioxaphosphorinane, B-00541

2-Anilide: see under 4-Methyl-2-phenylamino-1,3,2-dioxaphosphorinane, M-00193

Brault, J.-F. *et al*, *Bull. Soc. Chim. Fr., Part 2*, 1973, 3149; 1974, 677 (*synth, pmr, P nmr*)
Kryuchkov, A.A. *et al*, *Izv. Akad. Nauk SSSR, Ser. Khim.*, 1978, 1985 (*Engl. transl.* p. 1746)
Penney, C.L. *et al*, *Can. J. Chem.*, 1978, **56**, 2396 (*synth, derivs, pmr*)
Jankowska, J. *et al*, *Synthesis*, 1984, 408 (*synth, cmr*)

2-Hydroxy-4-methyl-1,3,2-dioxaphosphorinane 2-selenide — H-00157

O,O-*1,3-Butylene phosphoroselenoic acid.* O,O-*1,3-Butylene hydrogen phosphoroselenoate*

[86343-95-5]

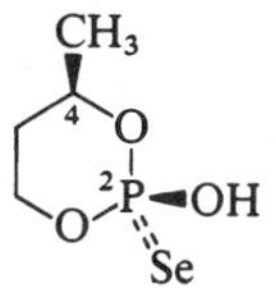

(2*RS*,4*RS*)-*form*

$C_4H_9O_3PSe$ M 215.047

Tautomeric.

(2*RS*,4*RS*)-*form* [39826-73-8]

(±)-cis-*form*

Dicyclohexylammonium salt: [26349-88-2]. Cryst. (propanol/pet. ether). Mp 199-202°.

2-Methoxy (O-Me ester): see 2-Methoxy-4-methyl-1,3,2-dioxaphosphorinane 2-selenide, M-00046

(2*RS*,4*SR*)-*form* [39826-70-5]

(±)-trans-*form*

Tetramethylammonium salt: V. hygroscopic solid. Mp 114-117°.

Dicyclohexylammonium salt: [39826-69-2]. Cryst. ($MeOH/Et_2O$). Mp 174-176°.

2-Methoxy (O-Me ester): see 2-Methoxy-4-methyl-1,3,2-dioxaphosphorinane 2-selenide, M-00046

Mikolajczyk, M. *et al*, *Tetrahedron*, 1972, **28**, 5411 (*synth, pmr, P nmr*)

2-Hydroxy-4-methyl-1,3,2-dioxaphosphorinane 2-sulfide, 9CI — H-00158

O,O-*1,3-Butylene hydrogen phosphorothioate.* O,O-*1,3-Butylene phosphorothioic acid*

[39826-54-5]

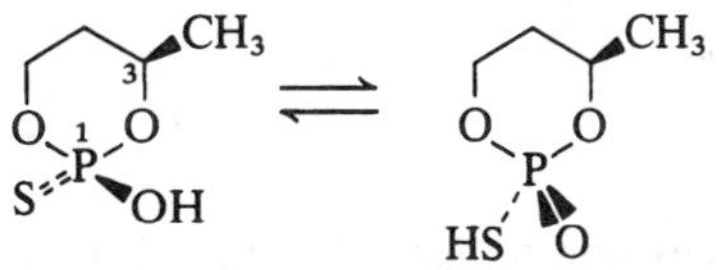

(2*RS*,4*RS*)-*form*

$C_4H_9O_3PS$ M 168.147

Tautomeric, thiol form predominates in soln. Oil. d_4^{20} 1.32. $Bp_{0.05}$ 170° (oven). pK_a 1.47 (7% EtOH aq.) (95% thiol), pK_a 2.67 (80% EtOH aq.) (99% thiol).

2-Methoxy (O-Me ester): see 2-Methoxy-4-methyl-1,3,2-dioxaphosphorinane 2-sulfide, M-00047

2-Methylthio (S-Me ester): see under 4-Methyl-2-methylthio-1,3,2-dioxaphosphorinane, M-00174

2-Chloride: see 2-Chloro-4-methyl-1,3,2-dioxaphosphorinane 2-sulfide, C-00095
Anhydride: see 2,2′-Oxybis[4-methyl-1,3,2-dioxaphosphorinane], O-00096

(2RS,4RS)-form
(±)-trans-*form*
NH_4 salt: Cryst. (propanol). Mp 144-147°.
Tetramethylammonium salt: Solid. Mp 122-126°.
Diethylammonium salt: Cryst. (C_6H_6/Et_2O). Mp 117-120°.
Dicyclohexylammonium salt: [86919-76-8]. Cryst. (propanol/pet. ether). Mp 194-196°.

(2RS,4SR)-form
(±)-cis-*form*
NH_4 salt: Cryst. (propanol/Et_2O). Mp 188-192°.
Tetramethylammonium salt: Cryst. (Me_2CO/propanol). Mp 203-206°.
Diethylammonium salt: Cryst. (C_6H_6/Et_2O). Mp 122-126°.
Dicyclohexylammonium salt: Cryst. (propanol/pet. ether). Mp 208-211°.

Mikolajczyk, M. *et al, Tetrahedron*, 1972, **28**, 5411 (*synth, pmr, P nmr*)
Stec, W.J. *et al, J. Inorg. Nucl. Chem.*, 1972, **34**, 1100 (*pe*)
Ovchinnikov, V.V. *et al, Zh. Obshch. Khim.*, 1977, **47**, 290 (*Engl. transl.* p. 267) (*synth, ir, pmr, P nmr*)
Bartczak, T.J. *et al, Acta Crystallogr., Sect. C*, 1983, **39**, 1059, 1467 (*cryst struct*)

(1-Hydroxy-1-methylethyl)phosphonic acid, 9CI — H-00159

2-Phosphono-2-propanol

$(H_3C)_2C(OH)P(O)(OH)_2$

$C_3H_9O_4P$ M 140.075
Cryst. (AcOH). Mp 167-169°.
Monoanilinium salt: Solid. Mp 176-178°.
Di-Et ester: [6632-88-8]. *Diethyl (1-hydroxy-2-methylethyl)phosphonate.*
$C_7H_{17}O_4P$ M 196.183
Liq. d_4^{20} 1.09. Bp_{10} 130-131°. n_D^{20} 1.4318.
Di-Ph ester: Diphenyl (1-hydroxy-1-methylethyl)phosphonate.
$C_{15}H_{17}O_4P$ M 292.271
Cryst. (pet. ether). Mp 113-114°.
Bis(trimethylsilyl) ester, O-trimethylsilyl ether:
$C_{12}H_{33}O_4PSi_3$ M 356.621
Liq. $Bp_{0.7}$ 86-90°.

Conant, J.B. *et al, J. Am. Chem. Soc.*, 1921, **43**, 1928 (*synth*)
Conant, J.B. *et al, J. Am. Chem. Soc.*, 1923, **45**, 762 (*ester*)
Abramov, V.S., *Zh. Obshch. Khim.*, 1952, **22**, 647 (*Engl. transl.* p. 709) (*ester, synth*)
Bel'skii, V.E. *et al, Zh. Obshch. Khim.*, 1972, **42**, 2427 (*Engl. transl.* p. 2421) (*ester, P nmr*)
Sekine, M. *et al, Chem. Lett.*, 1977, 485 (*synth*)

(Hydroxymethyl)phenylphosphinic acid, 9CI — H-00160

$HOCH_2PPh(O)OH$

$C_7H_9O_3P$ M 172.120
Solid. Mp 127-131°, 218-219°.
Me ester: [64128-97-8]. *Methyl (hydroxymethyl)phenylphosphinate.*
$C_8H_{11}O_3P$ M 186.147
Solid. Mp 68-70°.

Tebby, J.C., *Phosphorus*, 1976, **6**, 253 (*synth, pmr*)
Grapov, A.F. *et al, Zh. Obshch. Khim.*, 1977, **47**, 1465 (*Engl. transl.* p. 1347) (*synth, ester*)

3-(Hydroxymethylphosphinyl)propanoic acid, 9CI — H-00161

(2-Carboxyethyl)methylphosphinic acid
[15090-23-0]

$HOOCCH_2CH_2PMe(O)OH$

$C_4H_9O_4P$ M 152.086
Solid. Mp 94°.
Di-Me ester: Methyl 3-(methoxymethylphosphinyl)propanoate. Methyl [2-(methoxycarbonyl)ethyl]methylphosphinate.
$C_6H_{13}O_2P$ M 148.141
Liq. d_4^{20} 1.18. $Bp_{0.22}$ 113-114°. n_D^{20} 1.4520.
Di-Et ester: Ethyl 3-(ethoxymethylphosphinyl)propanoate. Ethyl [2-(ethoxycarbonyl)ethyl]methylphosphinate.
$C_8H_{17}O_2P$ M 176.195
d_4^{20} 1.10. $Bp_{0.005}$ 92-93°. n_D^{20} 1.4470.
Dichloride: 3-(Chloromethylphosphinyl)propanoyl chloride. [2-(Chloroformyl)ethyl]methylphosphinic chloride.
$C_4H_7Cl_2O_2P$ M 188.978
Liq. d_4^{20} 1.40. $Bp_{0.01}$ 114-115°. n_D^{20} 1.5005.

Khairullin, V.K. *et al, Zh. Obshch. Khim.*, 1967, **37**, 710 (*Engl. transl.* p. 666) (*synth, ir, derivs*)

(Hydroxymethyl)phosphonic acid, 9CI — H-00162

Phosphonomethanol
[2617-47-2]

$HOCH_2P(O)(OH)_2$

CH_5O_4P M 112.022
Cryst. (EtOH/EtOAc). Mp 88-90°, 98.5-100.5°. pK_{a1} 2.21, pK_{a2} 5.65.
Monoanilinium salt: Cryst. (EtOH). Mp 167-168°.
Di-Me ester: [24630-67-9]. *Dimethyl (hydroxymethyl)phosphonate.*
$C_3H_9O_4P$ M 140.075
Liq.
Di-Et ester: [3084-40-0]. *Diethyl (hydroxymethyl)phosphonate.*
$C_5H_{13}O_4P$ M 168.129
Liq. $Bp_{0.2}$ 103-105°. n_D^{20} 1.4342.
Bis(triethylsilyl) ester: Bis(triethylsilyl) (hydroxymethyl)phosphonate.
$C_{13}H_{33}O_4PSi_2$ M 340.546
Liq. n_D^{20} 1.4472. Dec. on attempted distn.
Ac: see (Acetoxymethyl)phosphonic acid, A-00007
Trimethylsilyl: see [(Trimethylsilyloxy)methyl]phosphonic acid, T-00556

Kharasch, M.S. *et al, J. Org. Chem.*, 1960, **25**, 1000 (*diethyl ester, synth*)
Orlov, N.F. *et al, Zh. Obshch. Khim.*, 1966, **36**, 578 (*Engl. transl.* p. 537) (*silyl ester*)
Maier, L., *Z. Anorg. Allg. Chem.*, 1972, **394**, 117 (*synth, P nmr, salts*)
Brun, G. *et al, Bull. Soc. Fr. Mineral. Cristallogr.*, 1974, **97**, 79; *CA*, **81**, 83219 (*cryst struct*)
Tebby, J.C. *et al, Phosphorus*, 1975, **5**, 273 (*ms*)
Griffiths, W.R. *et al, Phosphorus*, 1976, **6**, 223 (*synth*)
Rueppel, M.L. *et al, Org. Magn. Reson.*, 1976, **8**, 19 (*pmr, P nmr*)
Tebby, J.C. *et al, Phosphorus*, 1976, **6**, 253 (*pmr*)
Baraldi, P.G. *et al, Synthesis*, 1982, 653 (*ester, synth, ir, pmr, props*)

Holý, A. *et al*, *Collect. Czech. Chem. Commun.*, 1982, **47**, 3447 (*esters, synth, derivs*)

Hydroxymethyltriphenylphosphonium(1+), 9CI H-00163

(Triphenylphosphonio)methanol

$Ph_3P^{\oplus}CH_2OH$

$C_{19}H_{18}OP^{\oplus}$ M 293.324 (ion)

Chloride: [5293-84-4].
$C_{19}H_{18}ClOP$ M 328.777
Solid. Mp 190-192°.

Iodide:
$C_{19}H_{18}IOP$ M 420.229
Yellow, malodorous oil. n_D^{20} 1.5948.

Hellmann, H. *et al*, *Justus Liebigs Ann. Chem.*, 1962, **659**, 49 (*chloride*)
Tebby, J.C., *Phosphorus*, 1976, **6**, 253 (*iodide, pmr*)
Frank, A.W. *et al*, *J. Org. Chem.*, 1977, **42**, 4040 (*chloride, pmr*)
Frank, A.W. *et al*, *Phosphorus Sulfur*, 1978, **5**, 19 (*iodide, ir, pmr, props*)

2-Hydroxynaphtho[2,3-*d*]-1,3,2-dioxaphosphole 2-oxide, 9CI, 8CI H-00164

2,3-Naphthyl cyclic phosphoric acid. 2,3-Naphthylene phosphoric acid

[73749-41-4]

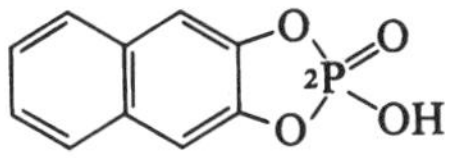

$C_{10}H_7O_4P$ M 222.137
Solid. Mp 160° dec.

2-Methoxy (Me ester): [86507-04-2].
$C_{11}H_9O_4P$ M 236.163
Powder.

2-Ethoxy (Et ester): [62290-27-1].
$C_{12}H_{11}O_4P$ M 250.190
Cryst. (DMSO). Mp >300°.

2-Chloride (chloro): see 2-Chloronaphtho[2,3-d]-1,3,2-dioxaphosphole, C-00119

Bhatia, M.S. *et al*, *Ann. Chim. (Paris)*, 1976, **1**, 239 (*ester, synth, ir*)
B.P, 2 022 276, (*1979*); *CA*, **92**, 224274 (*synth*)
Eur. Pat., 74 611, (*1983*); *CA*, **99**, 53476 (*ester, synth, pmr*)

2-Hydroxy-1,2-oxaphosphepane 2-oxide, 8CI H-00165

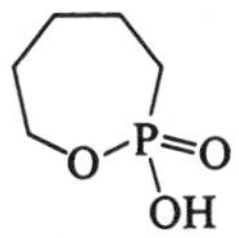

$C_5H_{11}O_3P$ M 150.114

2-Ethoxy (Et ester): [14923-49-0]. *2-Ethoxy-1,2-oxaphosphepane 2-oxide.*
$C_7H_{15}O_3P$ M 178.167
Liq. $Bp_{0.7}$ 93-97°. n_D^{20} 1.4596.

Songstadt, J., *Acta Chem. Scand.*, 1967, **21**, 1681 (*synth*)

2-Hydroxy-1,2-oxaphospholane 2-oxide, 9CI H-00166

Propylphostonic acid

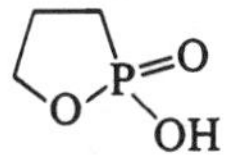

$C_3H_7O_3P$ M 122.060

2-Ethoxy (Et ester): [1193-39-1]. *2-Ethoxy-1,2-oxaphospholane 2-oxide. Ethyl propylphostonate.*
$C_5H_{11}O_3P$ M 150.114
Liq. d_4^{20} 1.11. Bp_{10} 128°, $Bp_{0.55}$ 74°. n_D^{20} 1.4498.

Pudovik, A.N. *et al*, *Zh. Obshch. Khim.*, 1964, **34**, 2582 (*Engl. transl.* p. 2604) (*synth, ir*)
Eberhard, A. *et al*, *J. Am. Chem. Soc.*, 1965, **87**, 253 (*synth, ir*)
Aksnes, G. *et al*, *Acta Chem. Scand.*, 1966, **20**, 2518 (*synth, props*)

2-Hydroxy-1,2-oxaphosphorinane 2-oxide, 9CI H-00167

2-Hydroxy-1,2-oxaphosphinane 2-oxide. Butylphostonic acid

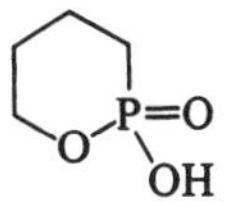

$C_4H_9O_3P$ M 136.087
Solid. Mp 104-106°.

2-Ethoxy (Et ester): [1194-41-8]. *2-Ethoxy-1,2-oxaphosphorinane 2-oxide. Ethyl butylphostonate.*
$C_6H_{13}O_3P$ M 164.141
Liq. d_4^{20} 1.07. Bp_{10} 130°, $Bp_{0.8}$ 88-90°. n_D^{20} 1.4581.

2-Phenoxy (Ph ester): [55549-60-5]. *2-Phenoxy-1,2-oxaphosphorinane 2-oxide. Phenyl butylphostonate.*
$C_{10}H_{13}O_3P$ M 212.185
Liq. $Bp_{0.3}$ 151-155°.

2-Chloro (chloride): [55549-63-8]. *2-Chloro-1,2-oxaphosphorinane 2-oxide. Butylphostonic chloride.*
$C_4H_8ClO_2P$ M 154.533
Liq. $Bp_{0.1}$ 96-102°.

Eberhard, A. *et al*, *J. Am. Chem. Soc.*, 1965, **87**, 253 (*synth, ethyl ester, ir*)
Aksnes, G. *et al*, *Acta. Chem. Scand.*, 1966, **20**, 2508 (*ethyl ester*)
Songstadt, J., *Acta. Chem. Scand.*, 1967, **21**, 1681 (*ethyl ester, ir*)
Stutz, H. *et al*, *Z. Chem.*, 1975, **15**, 52 (*derivs, synth*)
Matrosov, E.I. *et al*, *Izv. Akad. Nauk SSSR, Ser. Khim.*, 1977, **26**, 791 (*Engl. transl.* p. 719) (*ethyl ester, synth, nmr*)

1-Hydroxy-2,2,3,4,4-pentamethylphosphetane 1-oxide, 9CI, 8CI H-00168

1,1,2,3,3-Pentamethyltrimethylenephosphinic acid

[35210-25-4]

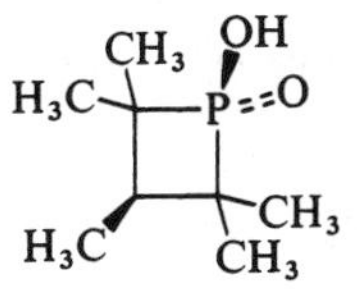

cis-form

$C_8H_{17}O_2P$ M 176.195

The reported *cis*- and *trans*-isomers are doubtful. Cryst. (heptane). Mp 72-74°. pK_a 4.66 (75% EtOH aq.). Forms a dihydrate, Mp 54-58°.

Benzyl ester: Benzyl 1,1,2,3,3-pentamethyltrimethylenephosphinate. 1-Phenylmethoxy-2,2,3,4,4-pentamethylphosphetane 1-oxide.
$C_{15}H_{21}O_2P$ M 264.303
Liq. $Bp_{3.0}$ 170-173°. Mixture of *cis*- and *trans*-isomers.
Fluoride: 1-Fluoro-2,2,3,4,4-pentamethylphosphetane 1-oxide. 1,1,2,3,3-Pentamethyltrimethylenephosphinic fluoride.
$C_{25}H_{23}FO_2P$ M 405.428
Solid. Mp 50°. $Bp_{0.3}$ 52°.
Anhydride:
$C_{16}H_{32}O_3P_2$ M 334.375
Air-stable solid. Mp 150-151°. Single stereoisomer.

cis-form [17405-94-6]
Mp 74.1°. pK_a 2.69.
Me ester: Methyl 1,1,2,3,3-pentamethyltrimethylenephosphinate. 1-Methoxy-1,2,3,4,4-pentamethylphosphetane 1-oxide.
$C_9H_{19}O_2P$ M 190.222
Solid-liq. Mp ca. 20°.
Et ester: [17405-96-8]. *Ethyl 1,1,2,3,3-pentamethyltrimethylenephosphinate. 1-Ethoxy-2,2,3,4,4-pentamethylphosphetane 1-oxide.*
$C_{10}H_{21}O_2P$ M 204.248
d_4^{20} 1.10. Bp_{10} 112°. n_D^{20} 1.4620.
Chloride: see 1-Chloro-2,2,3,4,4-pentamethylphosphetane, C-00124

trans-form [17405-93-5]
Mp 74-76°. pK_a 2.95.
Me ester: [26490-21-1]. Liq. $Bp_{0.4}$ 58-59°.
Et ester: [17405-95-7]. Liq. d_4^{20} 1.05. Bp_{10} 112°, $Bp_{0.2}$ 53-55°. n_D^{20} 1.4570.
Isopropyl ester: Isopropyl 1,1,2,3,3-pentamethyltrimethylenephosphinate. 1-Methylethoxy-2,2,3,4,4-pentamethylphosphetane 1-oxide.
$C_{11}H_{23}O_2P$ M 218.275
Liq. $Bp_{0.1}$ 53-55°.
Ph ester: Phenyl 1,1,2,3,3-pentamethyltrimethylenephosphinate. 1-Phenoxy-2,2,3,4,4-pentamethylphosphetane 1-oxide.
$C_{14}H_{21}O_2P$ M 252.292
Solid. Mp 54-55°. $Bp_{0.1}$ 102-104°.
Chloride: see 1-Chloro-2,2,3,4,4-pentamethylphosphetane, C-00124
Bromide:
$C_8H_{16}BrOP$ M 239.092
Cryst. (pet. ether). Mp 76-78°.
Azide: 1,1,2,3,3-Pentamethyltrimethylenephosphinic azide. 1-Azido-2,2,3,4,4-pentamethylphosphetane 1-oxide.
$C_8H_{16}N_3OP$ M 201.208
Cryst. (Et_2O). Mp 32-33°. $Bp_{0.1}$ 65°.

Bergesen, K., *Acta Chem. Scand.*, 1967, **21**, 1587 (*synth, pmr, ester*)
Gray, G. *et al*, *J. Org. Chem.*, 1972, **37**, 3458 (*cmr, nmr, esters*)
Cooke, R. *et al*, *J. Am. Chem. Soc.*, 1973, **95**, 8088.
DeBruin, K. *et al*, *J. Am. Chem. Soc.*, 1973, **95**, 4681 (*derivs, pmr*)
Emsley, J. *et al*, *J. Chem. Soc., Dalton Trans.*, 1974, 633 (*esters, pmr, nmr*)
Wiseman, J. *et al*, *J. Am. Chem. Soc.*, 1974, **96**, 4262 (*anhydride, azides*)

1-Hydroxy-2,2,3,4,4-pentamethylphosphetane 1-sulfide, 9CI H-00169

1,1,2,3,3-Pentamethyltrimethylenephosphinothioic acid

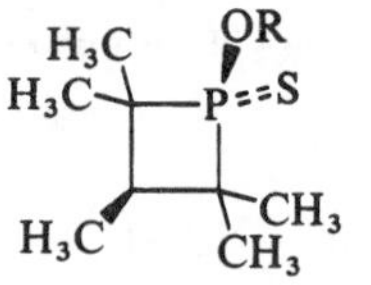

cis-form

$C_8H_{17}OPS$ M 192.255
Thione (R = H) form in equilibrium with thiol form.

cis-form
Chloride: See 1-Chloro-2,2,3,4,4-pentamethylphosphetane, C-00124
1-Methylthio (S-*Me ester*): *1-Methylthio-2,2,3,4,4-pentamethylphosphetane 1-oxide.* S-*Methyl 1,1,2,3,3-pentamethyltrimethylenephosphinothioate.*
$C_9H_{19}OPS$ M 206.282
Characterised spectroscopically.
1-Ethoxy (O-*Et ester*): *1-Ethoxy-2,2,3,4,4-pentamethylphosphetane 1-sulfide.* O-*Ethyl 1,1,2,3,3-pentamethyltrimethylenephosphinothioate.*
$C_{10}H_{21}OPS$ M 220.309
Liq. $Bp_{0.2}$ 100° (bath).
1-Ethylthio (S-*Et ester*): *1-Ethylthio-2,2,3,4,4-pentamethylphosphetane 1-oxide.* S-*Ethyl 1,1,2,3,4-pentamethyltrimethylenephosphinothioate.*
$C_{10}H_{21}OPS$ M 220.309
Liq. $Bp_{0.25}$ 80-82°.

trans-form
Chloride: See 1-Chloro-2,2,3,4,4-pentamethylphosphetane, C-00124
S-*Me ester:* Characterised spectroscopically.
O-*Et ester:* Liq. $Bp_{0.2}$ 61°.
S-*Et ester:* Liq. $Bp_{0.2}$ 110° (bath).

Corfield, J.R. *et al*, *J. Chem. Soc., Perkin Trans. 1*, 1972, 713 (*synth, pmr, nmr*)
Ardrey, R. *et al*, *J. Chem. Soc., Dalton Trans.*, 1973, 2641 (*pmr, ms*)
DeBruin, K.E. *et al*, *J. Am. Chem. Soc.*, 1973, **95**, 4681 (*pmr*)

10-Hydroxy-10*H*-phenoxaphosphine 10-oxide, 9CI, 8CI H-00170

Phenoxaphosphinic acid
[15042-79-2]

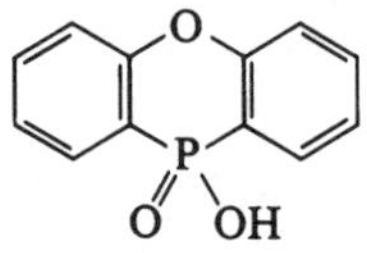

$C_{12}H_9O_3P$ M 232.175
Cryst. (EtOH). Mp 231-234°.

Doak, G.O. *et al*, *J. Org. Chem.*, 1964, **29**, 2382 (*synth, uv*)
Levy, J.B. *et al*, *J. Org. Chem.*, 1968, **33**, 474 (*synth*)
Granoth, I. *et al*, *J. Chem. Soc. (B)*, 1971, 2391 (*ms*)

2-Hydroxy-4-phenyl-1,3,2-dioxaphosphorinane 2-oxide, 9CI H-00171

[59556-98-8]

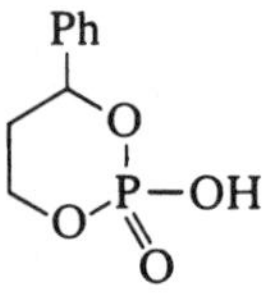

$C_9H_{11}O_4P$ M 214.157

Cryst. (EtOH). Mp 129-129.5°. pK_a 1.67 (H_2O), pK_a 2.75 (50% EtOH aq.), pK_a 2.92 (80% EtOH aq.), pK_a 4.13 (95% EtOH aq.), pK_a 8.94 ($MeNO_2$).

Tsuboi, M. *et al*, *Bull. Chem. Soc. Jpn.*, 1967, **40**, 1813 (*pmr*)
Matrosov, E.I. *et al*, *Izv. Akad. Nauk SSSR, Ser. Khim.*, 1976, 530 (*Engl. transl.* p. 512) (*synth*)
Kryuchkov, A.A. *et al*, *Izv. Akad. Nauk SSSR, Ser. Khim.*, 1978, 1985 (*Engl. transl.* p. 1746) (*props*)

(Hydroxyphenylmethyl)bisphosphonic acid, 9CI H-00172

[2809-26-9]

$PhC(OH)[P(O)(OH)_2]_2$

$C_7H_{10}O_7P_2$ M 268.099
Cryst. + $2H_2O$. Mp 191°.

Tetra-Me ester: [32249-59-5]. *Tetramethyl (hydroxyphenylmethyl)bisphosphonate.*
$C_{11}H_{18}O_7P_2$ M 324.207
Cryst. (Et_2O). Mp 130-133°.

Nicholson, D.A. *et al*, *J. Org. Chem.*, 1971, **36**, 3843 (*ester, P nmr*)
Worms, K.-H. *et al*, *Z. Anorg. Allg. Chem.*, 1979, **457**, 203 (*synth*)

(Hydroxyphenylmethyl)phenylphosphinic acid, 9CI H-00173

(*α-Hydroxybenzyl)phenylphosphinic acid*
[2409-69-8]

PhCH(OH)PPh(O)OH

$C_{13}H_{13}O_3P$ M 248.218
Solid. Mp 200-202°.

Et ester: Ethyl (α-hydroxybenzyl)phenylphosphinate.
$C_{15}H_{17}O_3P$ M 276.271
Solid. Mp 75°.

Pudovik, A.N. *et al*, *Zh. Obshch. Khim.*, 1969, **39**, 1715 (*Engl. transl.* p. 1681) (*synth*)
Campbell, I.G.M. *et al*, *J. Chem. Soc. (C)*, 1971, 1836 (*synth*)

(Hydroxyphenylmethyl)phosphonic acid, 9CI H-00174

(*α-Hydroxybenzyl)phosphonic acid*
[1127-41-9]

$PhCH(OH)P(O)(OH)_2$

$C_7H_9O_4P$ M 188.119

(±)-*form*
Cryst. (C_6H_6/AcOH). Mp 173°.

Monoanilinium salt: [1212-80-2]. Solid. Mp 183-184°, Mp 201°.

Mono-Me ester: [79296-50-7]. *Methyl hydrogen (α-hydroxybenzyl)phosphonate.*
$C_8H_{11}O_4P$ M 202.146
Solid. Mp 149°. Resolved giving (−)-enantiomer, mp 110°, $[\alpha]^{25}_{578}$ −51° (c, 1 in MeOH).

Di-Me ester: [6329-46-0]. *Dimethyl (hydroxyphenylmethyl)phosphonate. Dimethyl (α-hydroxybenzyl)phosphonate.*
$C_9H_{13}O_4P$ M 216.173
Solid, cryst. (Me_2CO). Mp 102°.

Di-Et ester: Diethyl (hydroxyphenylmethyl)-phosphonate. Diethyl (α-hydroxybenzyl)phosphonate.
$C_{11}H_{17}O_4P$ M 244.227
Solid. Mp 73°, Mp 83°.

Diisopropyl ester: Bis(1-methylethyl) (hydroxyphenylmethyl)phosphonate. Diisopropyl (α-hydroxybenzyl)phosphonate.
$C_{13}H_{21}O_4P$ M 272.280
Solid. Mp 92°.

Kharasch, M.S. *et al*, *J. Org. Chem.*, 1960, **25**, 1000 (*synth, derivs*)
Kreutzkamp, N. *et al*, *Arch. Pharm. (Weinheim, Ger.)*, 1962, **295**, 188 (*synth*)
Nesterov, L.V. *et al*, *Zh. Obshch. Khim.*, 1976, **46**, 1974 (*Engl. transl.* p. 1904) (*diethyl ester, synth*)
Jacques, J. *et al*, *Tetrahedron*, 1981, **37**, 1727 (*methyl esters, synth, resoln*)
Pirkle, W.H. *et al*, *J. Am. Chem. Soc.*, 1981, **103**, 3964 (*diethyl ester, resoln*)
Texier-Boullet, F. *et al*, *Synthesis*, 1982, **165**, 916 (*esters, synth, ir, pmr*)

[(2-Hydroxyphenyl)methyl]phosphonic acid, 9CI H-00175

o-*Hydroxybenzylphosphonic acid*

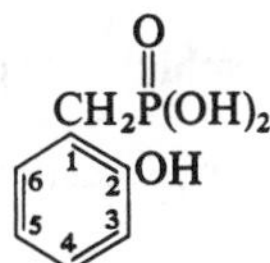

$C_7H_9O_4P$ M 188.119

Di-Me ester: [68997-87-5]. *Dimethyl [(2-hydroxyphenyl)methyl]phosphonate. Dimethyl* o-*hydroxybenzylphosphonate.*
$C_9H_{13}O_4P$ M 216.173
Cryst. (Et_2O).

Di-Et ester: [50375-72-9]. *Diethyl [(2-hydroxyphenyl)-methyl]phosphonate. Diethyl* o-*hydroxybenzylphosphonate.*
$C_{11}H_{17}O_4P$ M 244.227
Liq. n_D^{20} 1.5150.

Ivanov, B.É. *et al*, *Izv. Akad. Nauk SSSR, Ser. Khim.*, 1973, 1825 (*props*)
U.S.P., 4 069 340, (*1976*); *CA*, **88**, 120766 (*synth*)
Vogt, W., *Phosphorus Sulfur*, 1978, **5**, 123 (*synth, pmr*)

[(4-Hydroxyphenyl)methyl]phosphonic acid, 9CI H-00176

p-*Hydroxybenzylphosphonic acid*

$C_7H_9O_4P$ M 188.119

Di-Me ester: [68997-88-6]. *Dimethyl [(4-hydroxyphenyl)methyl]phosphonate. Dimethyl* p-*hydroxybenzylphosphonate.*
$C_9H_{13}O_4P$ M 216.173
No phys. props. reported.

Di-Et ester: [3173-38-4]. *Diethyl [(4-hydroxyphenyl)-methyl]phosphonate. Diethyl* p-*hydroxybenzylphosphonate.*
$C_{11}H_{17}O_4P$ M 244.227
Solid. Mp 89-91°.

U.S.P., 4 069 340, (*1976*); *CA*, **88**, 120766
Vogt, W., *Phosphorus Sulfur*, 1978, **5**, 123.

[(2-Hydroxyphenyl)methyl]-triphenylphosphonium(1+), 9CI H-00177

(o-*Hydroxybenzyl)triphenylphosphonium*(1+)

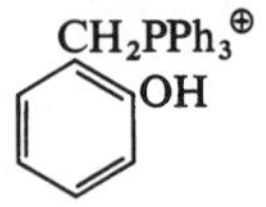

$C_{25}H_{22}OP^{\oplus}$ M 369.422 (ion)

Precursor for benzofuran synth.

Bromide: [70340-04-4].

$C_{25}H_{22}BrOP$ M 449.326

Source of ylide. Solid. Mp 246-248°.

Ylide: [(*2-Hydroxyphenyl)methyl*]-*triphenylphosphorane*. (o-*Hydroxybenzylidene*)-*triphenylphosphorane*.

$C_{25}H_{21}OP$ M 368.414

Used in Wittig reactions.

Hercouet, A. *et al*, *Tetrahedron Lett.*, 1979, 2145 (*synth, use*)
Begasse, B. *et al*, *Tetrahedron*, 1980, **36**, 3409 (*use*)
Hercouet, A. *et al*, *Tetrahedron*, 1981, **37**, 2867 (*use*)

3-(Hydroxyphenylphosphinyl)propanoic acid, 9CI H-00178

(*2-Carboxyethyl)phenylphosphinic acid*

[14657-64-8]

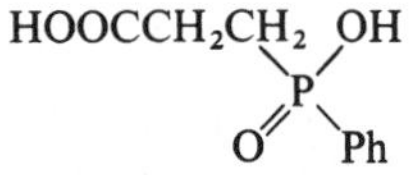

$C_9H_{11}O_4P$ M 214.157

Solid. Mp 156-157°.

Di-Me ester: Methyl 3-(methoxyphenylphosphinyl)-propanoate.

$C_{11}H_{15}O_4P$ M 242.211

Liq. d_4^{20} 1.21. $Bp_{0.0011}$ 143°. n_D^{20} 1.5782.

Di-Et ester: Ethyl 3-(ethoxyphenylphosphinyl)-propanoate.

$C_{13}H_{19}O_4P$ M 270.264

Liq. d_4^{20} 1.14. $Bp_{0.002}$ 145-146°. n_D^{20} 1.5048.

Dichloride: 3-(Chlorophenylphosphinyl)propanoyl chloride.

$C_9H_9Cl_2O_2P$ M 251.049

d_4^{20} 1.37. $Bp_{0.001}$ 230°. n_D^{20} 1.5600.

Bochwic, B. *et al*, *Rocz. Chem.*, 1952, **26**, 593; *CA*, **49**, 2345.
Pudovik, A.N. *et al*, *Zh. Obshch. Khim.*, 1967, **37**, 455 (*Engl. transl.* p. 423)

(2-Hydroxyphenyl)phosphonic acid, 9CI H-00179

[53104-46-4]

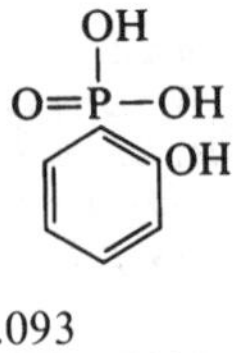

$C_6H_7O_4P$ M 174.093

Cryst. Mp 124-127°, 178-179°. pK_{a1} 1.66, pK_{a2} 6.46, pK_{a3} 15.40.

Di-Me ester: Dimethyl (2-hydroxyphenyl)phosphonate.

$C_8H_{11}O_4P$ M 202.146

Solid. Mp 97-98°.

Di-Et ester: [69646-14-6]. *Diethyl (2-hydroxyphenyl)-phosphonate.*

$C_{10}H_{15}O_4P$ M 230.200

Liq. $Bp_{0.3}$ 128-130°, $Bp_{0.05}$ 92-97°.

Benzyl ether: [2-(Benzyloxy)phenyl]phosphonic acid. [2-(Phenylmethoxy)phenyl]phosphonic acid.

$C_{13}H_{13}O_4P$ M 264.217

Cryst. (Me_2CO aq.). Mp 156-157°.

Freedman, L.D. *et al*, *J. Org. Chem.*, 1960, **25**, 140 (*synth*)
Lukin, A.M. *et al*, *Zh. Obshch. Khim.*, 1960, **30**, 1597 (*Engl. transl.* p. 1600) (*synth*)
Obrycki, R. *et al*, *J. Org. Chem.*, 1968, **33**, 632 (*ester, ir*)
Nuallain, C.O., *J. Inorg. Nucl. Chem.*, 1974, **36**, 339, 1420 (*props*)
Melvin, L.S., *Tetrahedron Lett.*, 1981, **22**, 3375 (*diethyl ester*)
Cambie, R.C. *et al*, *Aust. J. Chem.*, 1982, **35**, 827 (*diethyl ester, ir, pmr, cmr, ms*)
Heinicke, J. *et al*, *J. Organomet. Chem.*, 1986, **317**, 11 (*ester, synth, P nmr*)

(3-Hydroxyphenyl)phosphonic acid, 9CI H-00180

[33733-31-2]

$C_6H_7O_4P$ M 174.093

Solid. Mp 149-151°. pK_{a1} 2.08 (H_2O), 3.40 (50% EtOH aq.), pK_{a2} 5.53 (H_2O), 6.35 (50% EtOH aq.).

▷SZ8566100.

Di-Me ester: Dimethyl (3-hydroxyphenyl)phosphonate.

$C_8H_{11}O_4P$ M 202.146

Solid. Mp 91-92°.

Doak, G.O. *et al*, *J. Am. Chem. Soc.*, 1952, **74**, 753 (*synth*)
Obrycki, R. *et al*, *J. Org. Chem.*, 1968, **33**, 632 (*ester*)
Nuallain, C.O., *J. Inorg. Nucl. Chem.*, 1974, **36**, 339 (*props*)

(4-Hydroxyphenyl)phosphonic acid, 9CI H-00181

[33795-18-5]

$C_6H_7O_4P$ M 174.093

Solid. Mp 177°. pK_{a1} 2.29 (H_2O), 5.75 (50% EtOH aq.), pK_{a2} 3.74 (H_2O), 6.61 (50% EtOH aq.).

▷SZ8566300.

Di-Me ester: Dimethyl (4-hydroxyphenyl)phosphonate.

$C_8H_{11}O_4P$ M 202.146

Solid. Mp 77-78°.

Mono-Et ester: [77173-37-6]. *Ethyl hydrogen (4-hydroxyphenyl)phosphonate.*

$C_8H_{11}O_4P$ M 202.146

Metab. of EPN.

Di-Et ester: [28255-39-2]. *Diethyl (4-hydroxyphenyl)-phosphonate.*

$C_{10}H_{15}O_4P$ M 230.200

Cryst. (C_6H_6). Mp 31-33°.

Doak, G.O. *et al*, *J. Am. Chem. Soc.*, 1952, **74**, 753 (*synth*)
Obrycki, R. *et al*, *J. Org. Chem.*, 1968, **33**, 632 (*ester*)
Tavs, P., *Chem. Ber.*, 1970, **103**, 2428 (*ester*)
Nuallain, C.O., *J. Inorg. Nucl. Chem.*, 1974, **36**, 339 (*props*)

(2-Hydroxyphenyl)-triphenylphosphonium(1+) H-00182

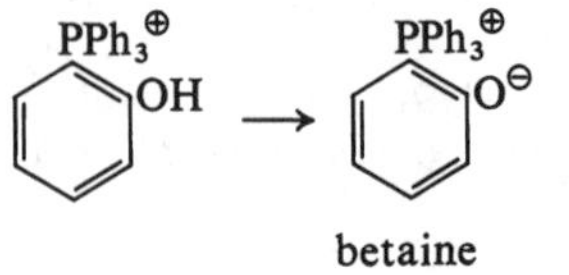

betaine

$C_{24}H_{20}OP^{\oplus}$ M 355.395 (ion)

With $NaHCO_3$ aq., salts form the betaine. At 350-400°, gives (2-phenoxyphenyl)diphenylphosphine.

Bromide:
$C_{24}H_{20}BrOP$ M 435.299
Solid. Mp 260-266°.
Iodide: [21230-91-1].
$C_{24}H_{20}IOP$ M 482.300
Cryst. (EtOH/EtOAc). Mp 284-286°.

Bestmann, H.J. *et al, Justus Liebigs Ann. Chem.*, 1968, **716**, 98 (*synth, props*)

(4-Hydroxyphenyl)-triphenylphosphonium(1+), 9CI, 8CI **H-00183**

$C_{24}H_{20}OP^{\oplus}$ M 355.395 (ion)
Salts are corrosion inhibitors. With NaOH aq. the salts give the anion (betaine), mesomeric with 4-Triphenylphosphoranylidene-2,5-cyclohexadien-1-one, T-00649 .
Bromide: [22883-70-1].
$C_{24}H_{20}BrOP$ M 435.299
Cryst. (H_2O). Mp 254-262° (sinters at 248°).
Ylide: see 4-Triphenylphosphoranylidene-2,5-cyclohexadien-1-one, T-00649

Horner, L. *et al, Chem. Ber.*, 1958, **91**, 52 (*synth, props*)

1-Hydroxy-3-phosphetanecarboxylic acid 1-oxide **H-00184**

3-Carboxy-1-hydroxyphosphetane 1-oxide

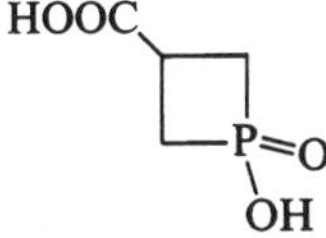

$C_4H_7O_4P$ M 150.071
Cryst. (Me_2CO). Mp 136-138°.
Di-Me ester: 1-Methoxy-3-methoxycarbonylphosphetane 1-oxide.
$C_6H_{11}O_4P$ M 178.124
Liq. d_4^{20} 1.27. $Bp_{0.03}$ 85°. n_D^{20} 1.4670.
C-Et ester: 3-Ethoxycarbonyl-1-hydroxyphosphetane 1-oxide.
$C_6H_{11}O_4P$ M 178.124
Cryst. (EtOH). Mp 101-102°.

Zyablikova, T.A. *et al, Izv. Akad. Nauk SSSR, Ser. Khim.*, 1969, 373 (*Engl. transl.* p. 324) (*synth, derivs, pmr*)

1-Hydroxyphospholane 1-oxide, 9CI, 8CI **H-00185**

Tetramethylenephosphinic acid
[6787-46-8]

Cryst. (hexane). Mp 53-54.5°.
1-Methoxy (Me ester): Methyl tetramethylenephosphinate. 2-Methoxyphospholane 1-oxide.
$C_5H_{11}O_2P$ M 134.114
Liq. $Bp_{0.6}$ 74°. n_D^{20} 1.4702.
1-Ethoxy (Et ester): [10545-61-6]. *Ethyl tetramethylenephosphinate. 2-Ethoxyphospholane 1-oxide.*
$C_6H_{13}O_2P$ M 148.141
Liq. Bp_{10} 108°. n_D^{20} 1.4620.
Chloride: see 1-Chlorophospholane, C-00167

Helferich, B. *et al, Justus Liebigs Ann. Chem.*, 1962, **658**, 100 (*synth, ester*)
Hunger, K. *et al, Tetrahedron*, 1964, **20**, 1593 (*esters*)
Alver, E. *et al, Acta Chem. Scand.*, 1969, **23**, 1101 (*cryst struct*)
Sommer, K., *Z. Anorg. Allg. Chem.*, 1970, **379**, 56 (*synth*)
Blackburn, G.M. *et al, Tetrahedron*, 1971, **27**, 2903 (*ester, P nmr*)
Asubiojo, O.I. *et al, J. Am. Chem. Soc.*, 1977, **99**, 7707 (*ester, props*)

1-Hydroxy-1*H*-phosphole 1-oxide **H-00186**

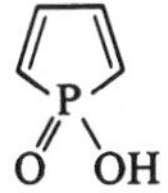

$C_4H_5O_2P$ M 116.056
Et ester:
$C_6H_9O_2P$ M 144.110
Dimerises rapidly; see 1,8-Dihydroxy-3*a*,4,7,7*a*-tetrahydro-4,7-phosphinidene-1*H*-phosphindole 1,8-dioxide, D-00625 .

Usher, D.A. *et al, J. Am. Chem. Soc.*, 1964, **86**, 4732.
Kluger, R. *et al, J. Am. Chem. Soc.*, 1967, **89**, 3919.

2-Hydroxy-3-(phosphonoxy)propanal, 9CI **H-00187**

Glyceraldehyde-3-phosphate. 3-Phosphoglyceraldehyde. Glyceraldehyde 3-dihydrogen phosphate
[142-10-9]

CHO
H–C–OH
$CH_2OP(O)(OH)_2$
(*R*)-*form*

$C_3H_7O_6P$ M 170.058
Bactericidal and inhibits phosphoglyceride synth. Biosynthetic intermed.

Nisselbaum, J.S. *et al, Adv. Enzyme Regul.*, 1972, **10**, 273 (*rev*)
Fahey, R.C. *et al, Anal. Biochem.*, 1974, **57**, 547 (*cd*)
Hofer, H.W., *Anal. Biochem.*, 1974, **61**, 54.
Hall, A. *et al, Biochemistry*, 1975, **14**, 4348 (*props*)
Tang, C.-T. *et al, Antimicrob. Agents Chemother.*, 1977, **11**, 147 (*props*)
Triantophylides, C. *et al, J. Chem. Soc., Perkin Trans. 2*, 1977, 1719 (*deriv, glc*)
Seriani, A. *et al, Biochemistry*, 1979, **18**, 1192 (*synth, cmr*)
Duke, C.C. *et al, Carbohydr. Res.*, 1981, **95**, 1 (*pmr, cmr*)
Rendina, A.R. *et al, Anal. Biochem.*, 1981, **117**, 213 (*purifn*)
Wong, C.-H. *et al, J. Org. Chem.*, 1983, **48**, 3199 (*synth*)

1-Hydroxyphosphorinane 1-oxide, 9CI, 8CI **H-00188**

Pentamethylenephosphinic acid. 1-Hydroxyphosphinane 1-oxide

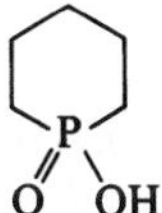

$C_5H_{11}O_2P$ M 134.114
Cryst. (Et_2O or C_6H_6/hexane). Mp 129°. pK_a 2.73.
1-Ethoxy (Et ester): [14923-50-3]. *Ethyl pentamethylenephosphinate. 1-Ethoxyphosphorinane 1-oxide.*
$C_7H_{15}O_2P$ M 162.168
Liq. Bp_{10} 109°, $Bp_{0.2}$ 71-72°.
Chloride: see 1-Chlorophosphorinane, C-00168

Aksnes, G. *et al, Acta Chem. Scand.*, 1966, **20**, 2508 (*synth, ester*)
Sommer, K., *Z. Anorg. Allg. Chem.*, 1970, **379**, 56 (*synth*)

Wetzel, R.B. *et al*, *J. Am. Chem. Soc.*, 1974, **96**, 5789 (*ester, pmr*)

1-Hydroxy-1,1-propanediphosphonic acid H-00189

(1-Hydroxypropylidene)bisphosphonic acid, 9CI. *1,1-Diphosphono-1-propanol*

[21089-13-4]

$[(HO)_2P(O)]_2C(OH)CH_2CH_3$

$C_3H_{10}O_7P_2$ M 220.055

Prevents mineral formn. in bone. Isolated as a hydrated sodium salt.

Blaser, R. *et al*, *Z. Anorg. Allg. Chem.*, 1971, **381**, 247 (*synth, salts*)
Shinoda, H. *et al*, *Calcif. Tissue Int.*, 1983, **35**, 87 (*props*)

(1-Hydroxypropyl)phosphonic acid, 9CI H-00190

1-Phosphono-1-propanol

[53621-85-5]

$H_3CCH_2CH(OH)P(O)(OH)_2$

$C_3H_9O_4P$ M 140.075

(±)-*form*

Solid. Mp 159°.

Di-Me ester: Dimethyl (1-hydroxypropyl)phosphonate.
$C_5H_{13}O_4P$ M 168.129
Liq. d_4^{20} 1.18. $Bp_{0.04}$ 90-93°. n_D^{20} 1.4418.

Di-Et ester: Diethyl (1-hydroxypropyl)phosphonate.
$C_7H_{17}O_4P$ M 196.183
Liq. d_4^{20} 1.09. $Bp_{0.03}$ 89-91°. n_D^{20} 1.4370.

O-*Ac: see (1-Acetoxypropyl)phosphonic acid, A-00008*

Bis(trimethylsilyl)ester, O-*Trimethylsilyl ether:*
$C_{12}H_{33}O_4PSi_3$ M 356.621
Liq. $Bp_{0.2}$ 91-95°.

Ramirez, F. *et al*, *J. Am. Chem. Soc.*, 1964, **86**, 514 (*ester, synth, ir, P nmr*)
Bel'skii, V.E. *et al*, *Zh. Obshch. Khim.*, 1972, **42**, 2427 (*Engl. transl.* p. 2421) (*ester, P nmr*)
Nesterov, L.V. *et al*, *Zh. Obshch. Khim.*, 1976, **46**, 1974 (*Engl. transl.* p. 1904) (*esters, synth*)
Texier-Boullet, F. *et al*, *Synthesis*, 1982, 165 (*ester, synth, ir, pmr*)

(2-Hydroxypropyl)phosphonic acid, 9CI H-00191

1-Phosphono-2-propanol

[76274-77-6]

$H_3CCH(OH)CH_2P(O)(OH)_2$

$C_3H_9O_4P$ M 140.075

(±)-*form*

Cyclohexylammonium salt: Lustrous needles (EtOH/THF). Mp 194-196° dec.

Di-Me ester: Dimethyl (2-hydroxypropyl)phosphonate. 1-(Dimethoxyphosphinyl)-2-propanol.
$C_5H_{13}O_4P$ M 168.129
Liq. $Bp_{0.03}$ 92°.

Di-Et ester: Diethyl (2-hydroxypropyl)phosphonate. 1-(Diethoxyphosphinyl)-2-propanol.
$C_7H_{17}O_4P$ M 196.183
Liq. $Bp_{0.03}$ 72°.

Di-Et ester, Ac: see (2-Acetoxypropyl)phosphonic acid, A-00009

Diisopropyl ester: Diisopropyl (2-hydroxypropyl)-phosphonate. 1-(Diisopropoxyphosphinyl)-2-propanol.
$C_9H_{21}O_4P$ M 224.236
Liq. $Bp_{0.05}$ 82°.

Baboulene, M. *et al*, *Phosphorus Sulfur*, 1979, **7**, 101 (*synth, ir, pmr*)

(3-Hydroxypropyl)phosphonic acid, 9CI H-00192

3-Phosphonopropanol

[53054-21-0]

$HOCH_2CH_2CH_2P(O)(CH)_2$

$C_3H_9O_4P$ M 140.075

Oil or cryst. solid.

Di-Me ester: [54731-74-7]. *Dimethyl (3-hydroxypropyl)phosphonate.*
$C_5H_{13}O_4P$ M 168.129
Liq. d_4^{20} 1.20. $Bp_{1.5-2.0}$ 111-112°. n_D^{20} 1.4481.

*Di-Me ester, Ac: see 1-Trimethylsilyl-2-trimethylsilyloxy-2-(2,4,6-tri-tert-*butylphenyl)-methylenephosphine, T-00560

Di-Et ester: [55849-69-9]. *Diethyl (3-hydroxypropyl)-phosphonate.*
$C_7H_{17}O_4P$ M 196.183
Liq. d_4^{20} 1.10. Bp_1 94-96°. n_D^{20} 1.4418.

*Di-Et ester, Ac: see 1-Trimethylsilyl-2-trimethylsilyloxy-2-(2,4,6-tri-tert-*butylphenyl)-methylenephosphine, T-00560

Stiles, A.R. *et al*, *J. Am. Chem. Soc.*, 1952, **74**, 3282 (*synth*)
Brun, G. *et al*, *Bull. Soc. Fr. Mineral Cristalogr.*, 1974, **97**, 79; *CA*, **81**, 83219 (*cryst struct*)
Brel', A.K. *et al*, *Zh. Obshch. Khim.*, 1980, **50**, 2134 (*CA*, **94**, 139888) (*esters, synth, ir, pmr*)

(3-Hydroxypropyl)-triphenylphosphonium(1+), 9CI H-00193

3-(Triphenylphosphonio)-1-propanol

$Ph_3^{\oplus}PCH_2CH_2CH_2OH$

$C_{21}H_{22}OP^{\oplus}$ M 321.378 (ion)

Chloride: [54674-84-9].
$C_{21}H_{22}ClOP$ M 356.831
Cryst. (Et_2O/EtOH). Mp 224°.

Bromide: [51860-45-8].
$C_{21}H_{22}BrOP$ M 401.282
Cryst. ($CHCl_3$/AcOH). Mp 222-224°.

Iodide:
$C_{21}H_{22}IOP$ M 448.282
Solid. Mp 210-211°.

Ylide: see 2,2-Dihydro-1,2-oxaphospholane, D-00534

Hands, A.R. *et al*, *J. Chem. Soc. (C)*, 1967, 1099; 1968, 2448 (*props*)
Kunz, H. *Justus Liebigs Ann. Chem.*, 1973, 2001 (*bromide*)
Aksnes, G. *et al*, *Phosphorus Sulfur*, 1977, **3**, 157 (*chloride, props*)

1-Hydroxy-1,1′(3*H*,3′*H*)-spirobi[2,1-benzoxaphosphole]-3,3′-dione H-00194

[67754-89-6]

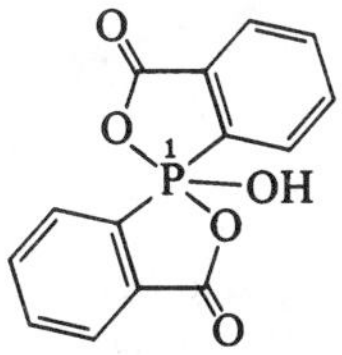

$C_{14}H_9O_5P$ M 288.196

Cryst. (C_6H_6). Mp 228-230°.

Triethylammonium salt: Solid. Mp 118-120°.

1-Methoxy: [71182-73-5].

$C_{15}H_{11}O_5P$ M 302.223

Characterised spectroscopically. Easily hydrolysed.

Segall, Y. *et al*, *J. Am. Chem. Soc.*, 1978, **100**, 5130 (*synth, ms, P nmr*)

Segall, Y. *et al*, *J. Am. Chem. Soc.*, 1979, **101**, 3687 (*deriv, pmr, P nmr*)

1-Hydroxy-1,2,3,4-tetrahydrophosphinoline 1-oxide H-00195

[52293-07-9]

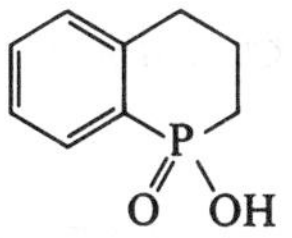

$C_9H_{11}O_2P$ M 182.158

Cryst. (EtOH). Mp 146-147°. $Bp_{0.001}$ 135° subl.

1-Methoxy (Me ester): [52221-86-0]. *1,2,3,4-Tetrahydro-1-methoxyphosphinoline 1-oxide.*

$C_{10}H_{13}O_2P$ M 196.185

Oil. $Bp_{0.005}$ 130° (oven).

1-Ethoxy (Et ester): [52221-87-1]. *1-Ethoxy-1,2,3,4-tetrahydrophosphinoline 1-oxide.*

$C_{11}H_{15}O_2P$ M 210.212

Hygroscopic oil. $Bp_{0.003}$ 140° (oven).

Rowley, L.E. *et al*, *Aust. J. Chem.*, 1974, **27**, 801 (*synth, uv, ir, ms, pmr*)

2-Hydroxy-4,4,5,5-tetramethyl-1,3,2-dioxaphospholane 2-oxide, 8CI H-00196

Pinacol cyclic phosphoric acid

[41821-76-5]

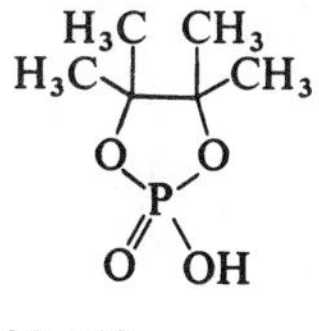

$C_6H_{13}O_4P$ M 180.140

Solid. Mp 296°.

Anilinium salt: Solid. Mp 138-139.5°.

Tetramethylammonium salt: Solid. Mp 296°.

Cyclohexylammonium salt: Solid. Mp 234-236°.

Chabrier, P. *et al*, *C.R. Hebd. Seances Acad. Sci., Ser. C*, 1973, **276**, 1135 (*synth*)

Brault, J.-F. *et al*, *Bull. Soc. Chim. Fr. Part 2*, 1974, 677 (*synth*)

Leroux, Y. *et al*, *Tetrahedron Lett.*, 1981, **22**, 3393.

2-Hydroxy-4,4,5,5-tetramethyl-1,3,2-dioxaphospholane 2-sulfide, 8CI H-00197

2-Mercapto-4,4,5,5-tetramethyl-1,3,2-dioxaphospholane 2-oxide. O,O-*Tetramethylethylene hydrogen phosphorothioate.* O,O-*Tetramethylethylene thiophosphate*

[45840-49-1]

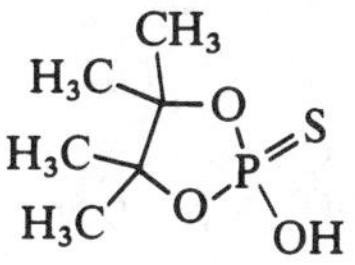

$C_6H_{13}O_3PS$ M 196.201

Tautomeric. pK_a 4.71 (propanol).

Imidazolium salt: [23233-32-1]. Cryst. (propanol/Et_2O). Mp 175-179.5°.

Dicyclohexylammonium salt: [59523-66-9]. Cryst. (propanol/Et_2O). Mp 212-215°.

2-Ethoxy (O-*Et ester*)*: see 2-Ethoxy-4,4,5,5-tetramethyl-1,3,2-dioxaphospholane, E-00052*

Chloride: see 2-Chloro-4,4,5,5-tetramethyl-1,3,2-dioxaphospholane, C-00185

N,N-*Dimethylamide: see 2-Dimethylamino-4,4,5,5-tetramethyl-1,3,2-dioxaphospholane, D-00710*

Mikolajczyk, M. *et al*, *J. Chem. Soc., Perkin Trans. 1*, 1976, 371 (*pmr*)

Wieczorek, M.W. *et al*, *Acta Crystallogr., Sect. B*, 1978, **34**, 3414 (*cryst struct*)

4-Hydroxy-2,2,6,6-tetramethyl-1-oxa-4-phospha-2,6-disilacyclohexane 4-oxide, 9CI H-00198

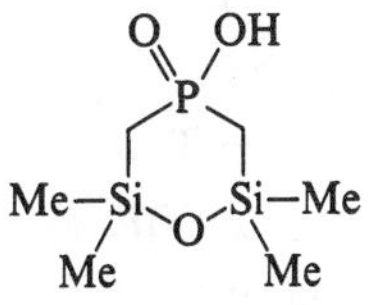

$C_6H_{17}O_3PSi_2$ M 224.343

4-Butoxy (butyl ester): [63382-79-6].

$C_{10}H_{25}O_3PSi_2$ M 280.450

Liq. n_D^{20} 1.4357.

Dvořak, M. *et al*, *Chem. Prum.*, 1977, **27**, 234; *CA*, **87**, 135546 (*synth, props*)

1-Hydroxy-3,3,3′,3′-tetramethyl-1,1′(3*H*,3′*H*)-spirobi[2,1-benzoxaphosphole], 9CI H-00199

[67759-42-6]

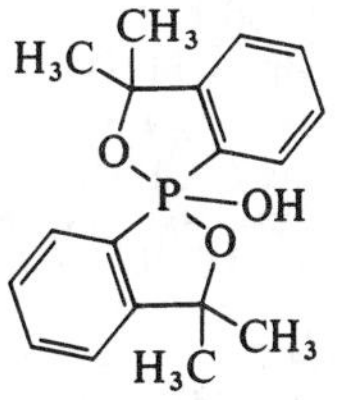

$C_{18}H_{21}O_3P$ M 316.336

Tautom. with monocyclic struct. Cryst. (EtOH). Mp 181°.

Na salt: [67759-43-7]. Mp >280°.

1-Methoxy:

$C_{19}H_{23}O_2P$ M 314.363

Solid. Mp 115°.

Granoth, I. *et al*, *J. Am. Chem. Soc.*, 1979, **101**, 4618 (*synth, derivs, ir, ms, pmr, P nmr*)

1-Hydroxy-2,3,4,5-tetraphenyl-1*H*-phosphole 1-oxide H-00200

[34674-30-1]

$C_{28}H_{21}O_2P$ M 420.446

Orange cryst. (EtOH). Mp 260-262°, Mp 287-289°.

1-Methoxy (Me ester): [74676-25-8]. *1-Methoxy-2,3,4,5-tetraphenyl-1*H*-phosphole 1-oxide.*
$C_{29}H_{23}O_2P$ M 434.473
Yellow solid. Mp 208-210°.

1-Ethoxy (Et ester): [74676-24-7]. *1-Ethoxy-2,3,4,5-tetraphenyl-1*H*-phosphole 1-oxide.*
$C_{30}H_{25}O_2P$ M 448.500
Yellow solid. Mp 196-197°.

1-Phenoxy (Ph ester): [74676-26-9]. *1-Phenoxy-2,3,4,5-tetraphenyl-1*H*-phosphole 1-oxide.*
$C_{34}H_{25}O_2P$ M 496.544
Yellow solid. Mp 188-189°.

Braye, E.H. *et al*, *Tetrahedron*, 1971, **27**, 5523 (*synth, ir*)
Freedman, L.D. *et al*, *Phosphorus*, 1974, **4**, 199 (*synth, ms*)
Yasufuku, K. *et al*, *J. Am. Chem. Soc.*, 1980, **102**, 4363 (*esters, pmr*)

3-Hydroxy-1,3-thiaphosphetane 3-oxide, 9CI, 8CI H-00201

Thiodimethylenephosphinic acid

[22585-74-6]

$C_2H_5O_2PS$ M 124.094

Cryst. (C_6H_6). Mp 142°.

Chloride: see 3-Chloro-1,3-thiaphosphetane, C-00188

3-Methoxy (Me ester): [28459-97-4]. *Methyl thiodimethylenephosphinate.*
$C_3H_7O_2PS$ M 138.121
Liq. d_4^{20} 1.36. $Bp_{0.028}$ 58-59°. n_D^{20} 1.5312.

3-Ethoxy (Et ester): [22585-75-7]. *Ethyl thiodimethylenephosphinate.*
$C_4H_9O_2PS$ M 152.148
Liq. d_4^{20} 1.28. $Bp_{0.03}$ 68°. n_D^{20} 1.5170.

2-Phenoxy (Ph ester): Phenyl thiodimethylenephosphinate.
$C_8H_9O_2PS$ M 200.192
Solid. Mp 44-6°. $Bp_{0.05}$ 115-117°. n_D^{20} 1.5868.

Gilyazov, M.M. *et al*, *Izv. Akad. Nauk SSSR, Ser. Khim.*, 1970, 1177 (*Engl. transl.* p. 1117) (*derivs, synth*)
Arshinova, R.P. *et al*, *Dokl. Akad. Nauk SSSR, Ser. Sci. Khim.*, 1972, **204**, 1118 (*Engl. transl.* p. 504) (*derivs, synth, pmr*)
Arbuzov, B.A. *et al*, *Izv. Akad. Nauk SSSR, Ser. Khim.*, 1973, 1964 (*Engl. transl.* p. 1913) (*derivs, struct*)
Shagidullin, R.R. *et al*, *Izv. Akad. Nauk SSSR, Ser. Khim.*, 1973, 458 (*Engl. transl.* p. 442) (*derivs, ir*)
Shakirov, I.Kh. *et al*, *Dokl. Akad. Nauk SSSR, Ser. Sci. Khim.*, 1974, **219**, 917 (*Engl. transl.* p. 1155) (*derivs, ir, raman, struct*)

[5-(4-Hydroxy-2,6,6-trimethyl-1-cyclohexen-1-yl)-3-methyl-2,4-pentadienyl]-triphenylphosphonium(1+), 9CI H-00202

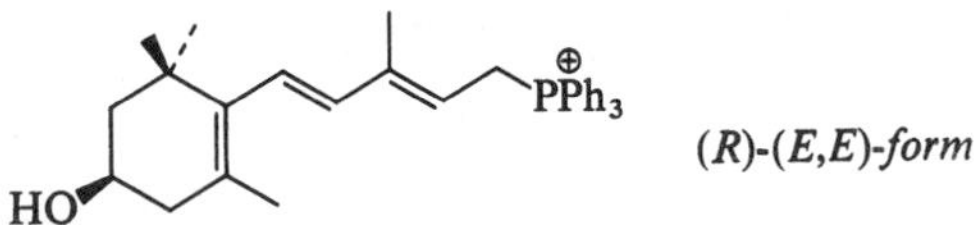

(R)-(E,E)-form

$C_{33}H_{38}OP^{\oplus}$ M 481.636 (ion)

Salts used in polyene syntheses *via* Wittig reactions.

(R)-(E,E)-form

Chloride: [76686-31-2].
$C_{33}H_{38}ClOP$ M 517.089
Cryst. Mp 209-210°. $[\alpha]_D^{25}$ −57.2° (c, 1 in $CHCl_3$).

Bromide: [76682-26-3].
$C_{33}H_{38}BrOP$ M 561.540
Cryst. (Me_2CO/Et_2O). Mp 188°.

(S)-(E,E)-form

Chloride: [76686-23-2]. Cryst. Mp 207-209°. $[\alpha]_D^{25}$ +55.2° (c, 1 in $CHCl_3$).

(±)-(E,E)-form

Bromide: [79734-41-1]. Pale-yellow solid. Mp 97-108°.

Loeber, D.E. *et al*, *J. Chem. Soc. (C)*, 1971, 404 (*synth, use*)
Pfander, H. *et al*, *Helv. Chim. Acta*, 1980, **63**, 1377 (*synth, use*)
Rüttimann, A. *et al*, *Helv. Chim. Acta*, 1980, **63**, 1456 (*synth, pmr*)
Mayer, H. *et al*, *Helv. Chim. Acta*, 1980, **63**, 1467 (*synth, use*)

1-Hydroxy-2,4,6-trimethyl-1,3,5-dioxaphosphorinane 5-oxide H-00203

[20540-86-7]

$C_6H_{13}O_4P$ M 180.140

Cryst. (Me_2CO). Mp 163-165°.

5-Methoxy (Me ester): [20540-88-9].
$C_7H_{15}O_4P$ M 194.167
Liq. d_4^{20} 1.21. $Bp_{0.003}$ 72°. n_D^{20} 1.4663.

Zyablikova, T.A. *et al*, *Izv. Akad. Nauk SSSR, Ser. Khim.*, 1968, 397 (*Engl. transl.* p. 384) (*synth, deriv*)

5-Hydroxy-2,4,6-triphenyl-1,3,5-dioxaphosphorinane 5-oxide, 10CI H-00204

[25154-95-4]

$C_{21}H_{19}O_4P$ M 366.352

Cryst. (MeOH). Mp 203°.

5-Methoxy (Me ester): [81048-09-1].
$C_{22}H_{21}O_4P$ M 380.379
Cryst. (MeOH). Mp 194°.

Arbuzov, B.A. *et al*, *Izv. Akad. Nauk SSSR, Ser. Khim.*, 1981, 2776, 2803 (*Engl. transl.* p. 2336) (*synth, P nmr, pmr*)

I

Ifosfamide, BAN, USAN **I-00001**

N,3-Bis(2-chloroethyl)tetrahydro-2H-1,3,2-oxazaphosphorin-2-amine 2-oxide, 9CI. *3-(2-Chloroethyl)-2-[(2-chloroethyl)amino]tetrahydro-2H-1,3,2-oxazaphosphorine 2-oxide,* 8CI. *Mitoxana. Isophosphamide. Isoendoxan*

[3778-73-2]

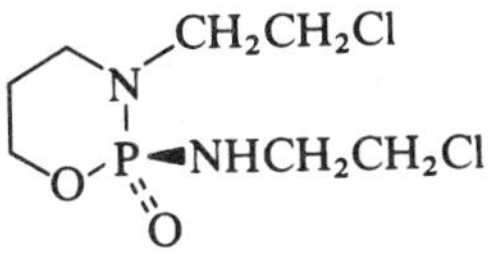

(*R*)-*form*

$C_7H_{15}Cl_2N_2O_2P$ M 261.087

▷RP6050000.

(*R*)-*form* [66849-34-1]

Cryst. (Et_2O). Mp 61-63°. $[\alpha]^{25}_D$ +39° (c, 4.0 in MeOH).

(*S*)-*form* [66849-33-0]

Cryst. (Et_2O). Mp 62-63°. $[\alpha]^{25}_D$ −38.8° (c, 4.2 in MeOH).

(±)-*form* [84711-20-6]

Antileukemic, antineoplastic, immunosupressant. Cryst. (Et_2O).

Schulten, H.-R., *Biomed. Mass Spectrom.*, 1974, **1**, 223 (*ms*)
Perales, A. *et al*, *Acta Crystallogr., Sect. B*, 1977, **33**, 1935 (*cryst struct*)
Takamizawa, A. *et al*, *Chem. Pharm. Bull.*, 1977, **25**, 2900 (*metab*)
Kinas, R.W. *et al*, *Bull. Acad. Pol. Sci. Ser. Sci. Chim.*, 1978, **26**, 39 (*resoln*)
Ludeman, S.M. *et al*, *J. Org. Chem.*, 1979, **44**, 1163 (*synth*)
Pankiewicz, K. *et al*, *J. Am. Chem. Soc.*, 1979, **101**, 7712 (*config*)
Wroblewski, A.E. *et al*, *Inorg. Chem.*, 1980, **19**, 3713 (*complexes, resoln*)
Su, C.N. *et al*, *J. Am. Chem. Soc.*, 1982, **104**, 7343 (*cd*)
Misiura, K. *et al*, *J. Med. Chem.*, 1983, **26**, 674 (*metab*)
Buess, M.L. *et al*, *Org. Magn. Reson.*, 1984, **22**, 67 (*nqr*)
Brade, W.P. *et al*, *Cancer Treat. Rev.*, 1985, **12**, 1 (*pharmacol, tox, use, rev*)

Imcarbofos, USAN **I-00002**

[(2-Methoxy-1,4-phenylene)bis(iminocarbonothioyl)]-bis(phosphoramidic acid) tetraethyl ester, 9CI

[66608-32-0]

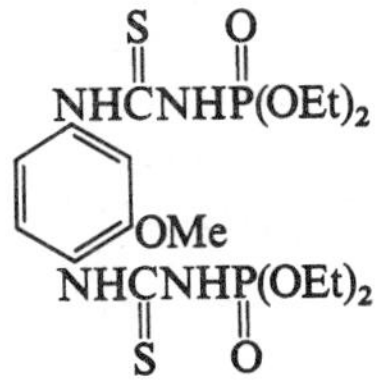

$C_{17}H_{30}N_4O_7P_2S_2$ M 528.514

Veterinary anthelmintic. Solid. Mp 143-144°.

Ger. Pat., 2 739 215, (*1978*); *CA*, **89**, 24005 (*synth*)
U.S.P., 4 137 310, (*1979*); *CA*, **90**, 186634w (*synth, pharmacol*)

1*H*-Imidazol-1-ylphosphonic acid, 9CI **I-00003**

1-Imidazolephosphonic acid. 1-Phosphonoimidazole

[15496-31-8]

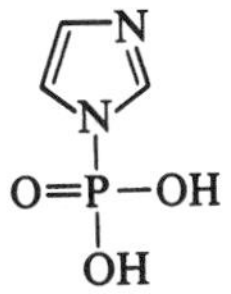

$C_3H_5N_2O_3P$ M 148.058

Di-Me ester: [67723-07-3]. *Dimethyl 1H-imidazol-1-ylphosphonate. N-Dimethylphosphorylimidazole. N-Dimethoxyphosphinylimidazole.*

$C_5H_9N_2O_3P$ M 176.111

Liq. n_D^{20} 1.4705.

Di-Et ester: [16913-98-7]. *Diethyl 1H-imidazol-1-ylphosphonate. N-Diethylphosphorylimidazole. N-Diethoxyphosphinylimidazole.*

$C_7H_{13}N_2O_3P$ M 204.165

Liq. $Bp_{0.3}$ 79-80°. n_D^{22} 1.4488.

Diisopropyl ester: [67711-52-8]. *Diisopropyl 1H-imidazol-1-ylphosphonate. N-Diisopropylphosphorylimidazole. N-Diisopropoxyphosphinylimidazole.*

$C_9H_{17}N_2O_3P$ M 232.219

Liq.

Di-Ph ester: [66778-06-1]. *Diphenyl 1H-imidazol-1-ylphosphonate. N-Diphenylphosphorylimidazole. N-Diphenoxyphosphinylimidazole.*

$C_{15}H_{13}N_2O_3P$ M 300.253

Good reagent for dehydrating aldoximes to nitriles. Solid. Mp 92-94° dec.

Bisdimethylamide: [28003-13-6]. *P-Imidazol-1-yl-N,N,N',N'-tetramethylphosphonic diamide.*

$C_7H_{15}N_4OP$ M 202.195

Liq. $Bp_{0.6}$ 135°.

Nikolenko, L.N. *et al*, *Zh. Obshch. Khim.*, 1967, **37**, 1350 (*Engl. transl.* p. 1276) (*ester*)
Degterev, E.V. *et al*, *Zh. Obshch. Khim.*, 1970, **40**, 2262 (*Engl. transl.* p. 2250) (*use*)
Ranganathan, N. *et al*, *J. Org. Chem.*, 1978, **43**, 4853 (*esters, synth, pmr*)
Sosnovsky, G. *et al*, *Z. Naturforsch., B*, 1978, **33**, 1165 (*ester, use*)
Etemad-Moghadam, G. *et al*, *Phosphorus Sulfur*, 1981, **12**, 61 (*ester, use*)
Weber, A.L., *J. Mol. Evol.*, 1981, **18**, 24.
Dabkowski, W. *et al*, *Chem. Ber.*, 1982, **115**, 1636 (*ester, pmr, P nmr*)

Iminocyclophosphamide **I-00004**

N,N-Bis(2-chloroethyl)-5,6-dihydro-2H-1,3,2-oxazaphosphorin-2-amine 2-oxide. 2-Bis(2-chloroethyl)-amino-5,6-dihydro-2H-1,3,2-oxazaphosphorine 2-oxide

[84489-09-8]

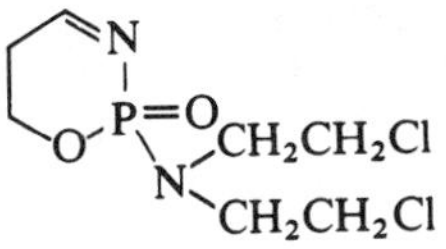

$C_7H_{13}Cl_2N_2O_2P$ M 259.072

Isol. as HCN adduct. Intermediate in metab. of Cyclophosphamide, C-00283 and in aq. chemistry of 4-Hydroperoxycyclophosphamide, H-00106 .

Fenselau, C. *et al*, *Drug. Metab. Dispos.*, 1982, **10**, 636.
Borch, R.F. *et al*, *J. Med. Chem.*, 1984, **27**, 485, 490.
Boyd, V.L. *et al*, *J. Med. Chem.*, 1987, **30**, 366.

(1-Iminoethyl)phosphoramidothioic acid, 9CI — I-00005

HN=C(CH_3)NHP(S)$(OH)_2$

$C_2H_7N_2O_2PS$ M 154.123

O,S-*Di-Me ester:* O,S-*Dimethyl* (*1-iminoethyl*)-*phosphoramidothioate.*
$C_4H_{11}N_2O_2PS$ M 182.177
Solid. Mp 130-135°.

O,O-*Di-Et ester:* [67777-11-1]. O,O-*Diethyl* (*1-iminoethyl*)*phosphoramidothioate.*
$C_6H_{15}N_2O_2PS$ M 210.230
Solid. Mp 34-35°.

O,S-*Di-Et ester:* O,S-*Diethyl* (*1-iminoethyl*)-*phosphoramidothioate.*
$C_6H_{15}N_2O_2PS$ M 210.230
Semicryst.

O,O-*Bis*(*4-chlorophenyl*) *ester:* [4104-14-7]. O,O-*Bis*(*4-chlorophenyl*) (*1-iminoethyl*)*phosphoramidothioate. Phosacetim.*
$C_{14}H_{13}Cl_2N_2O_2PS$ M 375.209
Pesticide.

▷TB4725000.

Ger. Pat., 2 758 173, (*1978*); *CA*, **89**, 146448

1*H*-Indene-1-phosphonous acid — I-00006

1H-*1-Indenylphosphonous acid*, 9CI

$C_9H_9O_2P$ M 180.143

Di-Et ester: Diethyl 1H-*1-indenylphosphonite.*
$C_{13}H_{17}O_2P$ M 236.250
Liq. d_4^{20} 1.07. $Bp_{1.5}$ 99-99.5°. n_D^{20} 1.5491.

Kabachnik, M.I. *et al*, *Izv. Akad. Nauk SSSR, Ser. Khim.*, 1960, 133 (*Engl. transl.* p. 122)

1*H*-Inden-2-ylphosphonic acid, 9CI — I-00007

$C_9H_9O_3P$ M 196.142

Cryst. (AcOH) or plates (H_2O). Mp 184°.

Di-Et ester: Diethyl 1H-*inden-2-ylphosphonate.*
$C_{13}H_{17}O_3P$ M 252.249
Liq. Bp_1 150°.

Bergmann, E. *et al*, *Ber.*, 1930, **63**, 1158 (*synth*)
Anisimov, K.N. *et al*, *Izv. Akad. Nauk SSSR, Ser. Khim.*, 1956, 16; *CA*, **50**, 13784 (*ester*)
Kenyon, G.L. *et al*, *J. Am. Chem. Soc.*, 1966, **88**, 3557 (*synth, pmr*)

1-Indenyltriphenylphosphonium(1+) — I-00008

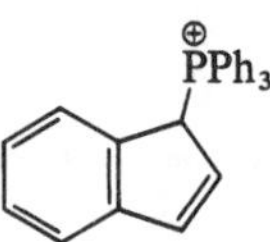

$C_{27}H_{22}P^{\oplus}$ M 377.445 (ion)

Bromide:
$C_{27}H_{22}BrP$ M 457.349
Hygroscopic cryst. Mp 190-195°. 2M NH_4OH yields the ylide.

Ylide: [13125-82-1]. *1-Indenylidenetriphenylphosphorane.*
$C_{27}H_{21}P$ M 376.437
Resonance-stabilised ylide. Green-yellow needles (C_6H_6/pet. ether). Mp 218-220°. Can be purified by reversible soln. in dil. aq. acid.

Crofts, P.C. *et al*, *J. Chem. Soc.* (*C*), 1967, 1093 (*synth, use*)
Ford, J.A., *Tetrahedron Lett.*, 1968, 815 (*ylide*)

Inezin — I-00009

O-*Ethyl* S-(*phenylmethyl*) *phenylphosphonothioate*, 9CI.
S-*Benzyl* O-*ethyl phenylphosphonothioate*, 8CI
[21722-85-0]

$C_{15}H_{17}O_2PS$ M 292.332

Insecticide. Now superseded. Liq. Bp_3 195-200°. Relatively nontoxic, exhibits no delayed neurotoxicity.

Uesugi, Y. *et al*, *Agric. Biol. Chem.*, 1972, **36**, 313 (*metab*)
Soliman, S.A., *J. Toxicol. Environ. Health*, 1982, **10**, 907 (*synth, tox*)

2-Iodo-5,5-dimethyl-1,3,2-dioxaphosphorinane, 9CI — I-00010

Neopentylene phosphoroiodidite. 2-Iodo-5,5-dimethyl-1,3,2-dioxaphosphinane. 2-Iodo-5,5-dimethyl-1,3-dioxa-2-phosphacyclohexane
[82568-46-5]

$C_5H_{10}IO_2P$ M 260.011

2-Sulfide: Neopentylene phosphoroiodidothioate.
$C_5H_{10}IO_2PS$ M 292.071
Cryst. (EtOH). Mp 142-143°.

Edmundson, R.S., *Chem. Ind.* (*London*), 1965, 1220 (*sulfide, ir*)

Iodofenphos, BAN, BSI **I-00011**

O-*(2,5-Dichloro-4-iodophenyl)* O,O-*dimethyl phosphorothioate, 9CI. Nuvanol* N

[18181-70-9]

$C_8H_8Cl_2IO_3PS$ M 412.994

Stored product and public health insecticide. Cryst. Mp 76°.

▷Mod. toxic orally; potential skin irritant. TF0175000.

B.P., 1 057 609, *(1964)*; *CA*, **65**, 13761h

Baughman, R.G. *et al*, *J. Agric. Food Chem.*, 1982, **30**, 293 *(cryst struct)*

Fogy, I. *et al*, *Int. J. Mass Spectrom. Ion Phys.*, 1983, **48**, 319.

Agrochemical Handbook, Royal Society of Chemistry, London, 1983, A239.

Pesticide Manual, 6th Ed., 304; 7th Ed., 319.

Sax, N.I., *Dangerous Properties of Industrial Materials*, 6th Ed., Van Nostrand-Reinhold, 1984, 745.

(Iodomethyl)phenylphosphinic acid, 9CI **I-00012**

$ICH_2PPh(O)OH$

$C_7H_8IO_2P$ M 282.017

Et ester: Ethyl (iodomethyl)phenylphosphinate.

$C_9H_{12}IO_2P$ M 310.071

Liq. $Bp_{0.5}$ 153°.

Henning, H-G., *J. Prakt. Chem.*, 1965, **29**, 93.

(Iodomethyl)phosphonic acid, 9CI **I-00013**

[13298-02-7]

$ICH_2P(O)(OH)_2$

CH_4IO_3P M 221.919

Hygroscopic needles (dichloroethane). Mp 89-91°. pK_a 1.6 (H_2O, 25).

Di-Et ester: [10419-77-9]. *Diethyl (iodomethyl)-phosphonate.*

$C_5H_{12}IO_3P$ M 278.026

Liq. $Bp_{0.7}$ 101°. n_D^{17} 1.4975.

Di-Ph ester: [84441-54-3]. *Diphenyl (iodomethyl)-phosphonate.*

$C_{13}H_{12}IO_3P$ M 374.114

Cryst. (Et_2O). Mp 76°.

Ford-Moore, A.W. *et al*, *J. Chem. Soc.*, 1947, 1465 *(ester)*

Pitrè, D. *et al*, *J. Prakt. Chem.*, 1966, **32**, 317 *(synth)*

Paul, G., *Z. Chem.*, 1982, **22**, 307 *(ester)*

[(1-Iodo-2-oxo-2-phenyl)ethylidene]-triphenylphosphorane **I-00014**

2-Iodo-2-(triphenylphosphoranylidene)acetophenone

$Ph_3P{=}CICOPh$

$C_{26}H_{20}IOP$ M 506.322

Stabilised ylide. Orange plates (EtOH). Mp 156.5-157.5°, 186-187°.

Speziale, A.J. *et al*, *J. Am. Chem. Soc.*, 1963, **85**, 2790 *(ir, nmr)*

Speziale, A.J. *et al*, *J. Org. Chem.*, 1963, **28**, 465 *(synth, ir, uv)*

Speziale, A.J. *et al*, *J. Am. Chem. Soc.*, 1965, **87**, 5603 *(cryst struct)*

Stephens, F.S. *et al*, *J. Chem. Soc.*, 1965, 5640 *(cryst struct)*

Grigarenko, A.A. *et al*, *Zh. Obshch. Khim.*, 1966, **36**, 1121 *(Engl. transl.* p. 1134) *(synth)*

[(2-Iodophenyl)methyl]phosphonic acid, 9CI **I-00015**

o-*Iodobenzylphosphonic acid*

$C_7H_8IO_3P$ M 298.017

Di-Et ester: [62680-68-6]. *Diethyl [(2-iodophenyl)-methyl]phosphonate.*

$C_{11}H_{16}IO_3P$ M 354.124

Oil. Bp_{10} 193°, $Bp_{0.0025}$ 134-138°. n_D^{25} 1.550.

Ernst, L., *Org. Magn. Reson.*, 1977, **9**, 35 *(synth, cmr, P nmr)*

Staab, H.A. *et al*, *Chem. Ber.*, 1977, **110**, 619 *(synth)*

(2-Iodophenyl)phosphonic acid, 9CI **I-00016**

[54185-82-9]

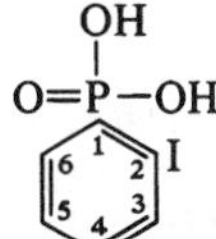

$C_6H_6IO_3P$ M 283.990

Solid. Mp 219-222°. pK_{a1} 1.74 (H_2O), 3.06 (50% EtOH aq.), pK_{a2} 7.06 (H_2O), 8.40 (50% EtOH aq.).

I-*Oxide: see 1,3-Dihydro-1-hydroxy-1,2,3-benziodoxaphosphole 3-oxide, D-00488*

Freedman, L.D. *et al*, *J. Am. Chem. Soc.*, 1953, **75**, 1379 *(synth)*

Freedman, L.D. *et al*, *Phosphorus*, 1974, **3**, 277 *(uv, oxide)*

(3-Iodophenyl)phosphonic acid, 9CI **I-00017**

$C_6H_6IO_3P$ M 283.990

Solid. Mp 183-184°.

Kosolapoff, G.M., *J. Am. Chem. Soc.*, 1948, **70**, 3465 *(synth)*

Denham, J.M. *et al*, *J. Org. Chem.*, 1958, **23**, 1298 *(synth)*

(4-Iodophenyl)phosphonic acid, 9CI **I-00018**

[4042-59-5]

$C_6H_6IO_3P$ M 283.990

Cryst. + $1H_2O$ (HCl aq.). Mp 228-229°.

I-*Oxide: (4-Iodosophenyl)phosphonic acid.*

$C_6H_6IO_4P$ M 299.989

Mp 218-220°.

Kosolapoff, G.M., *J. Am. Chem. Soc.*, 1948, **70**, 3465 *(synth)*

Freedman, L.D. *et al*, *J. Med. Chem.*, 1965, **8**, 891 *(synth)*

Freedman, L.D. *et al*, *Phosphorus*, 1974, **3**, 277 *(uv, deriv)*

O-(4-Iodophenyl) phosphorodichloridothioate, 9CI, 8CI I-00019

O-(4-Iodophenyl) dichlorothiophosphate. O-(4-Iodophenyl) thiophosphoryl dichloride

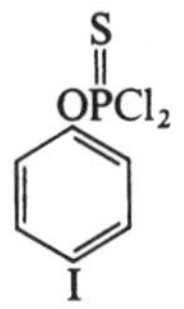

$C_6H_4Cl_2IOPS$ M 352.941

Low melting solid. Mp 30-31°. Bp_5 145-147°.

Protsenko, L.D. *et al*, *Zh. Obshch. Khim.*, 1964, **34**, 2233 (*Engl. transl.* p. 2244)

Iodotetraphenoxyphosphorane, 8CI I-00020

Tetraphenoxyphosphonium iodide

$(PhO)_4PI$ or $(PhO)_4P^{\oplus}I^{\ominus}$

$C_{24}H_{20}IO_4P$ M 530.298

Possesses the ionic struct. Mp 87-90°.

Nesterov, L.V. *et al*, *Zh. Obshch. Khim.*, 1967, **37**, 1843 (*Engl. transl.* p. 1756) (*props*)
Nesterov, L.V. *et al*, *Izv. Akad. Nauk SSSR, Ser. Khim.*, 1971, 414 (*Engl. transl.* p. 346) (*nmr*)

Isazofos, BSI I-00021

O-[5-Chloro-1-(1-methylethyl)-1H-1,2,4-triazol-3-yl] O,O-diethyl phosphorothioate, 9CI. O-(5-Chloro-1-isopropyl-1H-1,2,4-triazol-3-yl) O,O-diethyl phosphorothioate

[42509-80-8]

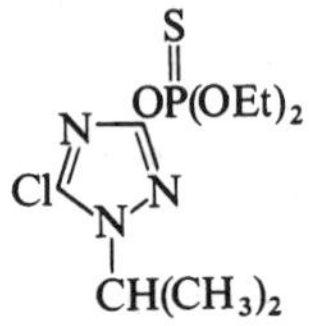

$C_9H_{17}ClN_3O_3PS$ M 313.738

Nematocide. Yellow liq. $Bp_{0.0001}$ 100°.

▷TE7750000.

Egli, H., *J. Agric. Food Chem.*, 1982, **30**, 861.
Pesticide Manual, 7th Ed., 323.

(Isocyanatomethyl)phosphonic acid, 8CI I-00022

$(HO)_2P(O)CH_2NCO$

$C_2H_4NO_4P$ M 137.032

Di-Me ester: [70525-48-3]. *Dimethyl (isocyanatomethyl)phosphonate.*
$C_4H_8NO_4P$ M 165.085
Liq. d_4^{20} 1.21. $Bp_{0.05}$ 68-70°. n_D^{20} 1.4195.

Di-Et ester: [21955-18-0]. *Diethyl (isocyanatomethyl)-phosphonate.*
$C_6H_{12}NO_4P$ M 193.139
Liq. d_4^{20} 1.17. $Bp_{0.02}$ 69-70°. n_D^{20} 1.4361.

Diisopropyl ester: Diisopropyl (isocyanatomethyl)-phosphonate.
$C_8H_{16}NO_4P$ M 221.192
Liq. d_4^{20} 1.09. $Bp_{0.03}$ 65-66°. n_D^{20} 1.4315.

Difluoride:
$C_2H_2F_2NO_2P$ M 141.014
Liq. d_4^{20} 1.55. Bp_{15} 68-69°. n_D^{20} 1.4085.

Dichloride: [23041-18-1].
$C_2H_2Cl_2NO_2P$ M 173.923
Liq. d_4^{20} 1.57. $Bp_{0.15}$ 59-60°. n_D^{20} 1.4980.

Shokol, V.A. *et al*, *Zh. Obshch. Khim.*, 1969, **39**, 938 (*Engl. transl.* p. 908) (*dichloride*)
Shokol, V.A. *et al*, *Zh. Obshch. Khim.*, 1970, **40**, 535 (*Engl. transl.* p. 502) (*derivs, synth, ir*)
Shokol, V.A. *et al*, *Zh. Obshch. Khim.*, 1970, **40**, 1458 (*Engl. transl.* p. 1445) (*esters, synth, ir*)
Shokol, V.A. *et al*, *Zh. Obshch. Khim.*, 1979, **49**, 312 (*Engl. transl.* p. 271) (*esters, synth*)

2-Isocyano-5,5-dimethyl-1,3,2-dioxaphosphorinane, 9CI I-00023

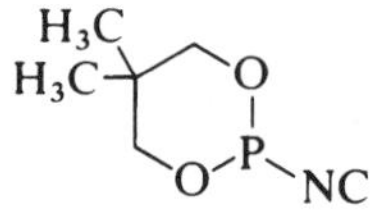

$C_6H_{10}NO_2P$ M 159.124

2-Oxide: [55379-58-3].
$C_6H_{10}NO_3P$ M 175.124
First known phosphoryl isocyanide. Cryst. (Et_2O/heptane). Mp 70.5-72°. Unstable and isomerises when heated.

Stec, W.J. *et al*, *J. Chem. Soc., Chem. Commun.*, 1974, 923 (*deriv, ir, P nmr, props*)

(Isocyanomethyl)phosphonic acid, 9CI I-00024

$(HO)_2P(O)CH_2NC$

$C_2H_4NO_3P$ M 121.032

Di-Et ester: [41003-94-5]. *Diethyl (isocyanomethyl)-phosphonate.*
$C_6H_{12}NO_3P$ M 177.139
A Wittig-Horner reagt. for synth. of 1-isocyano-1-alkenes, (1-aminoalkyl)phosphonic acids and oxazolinyl- and thiazolinylphosphonic acids. Liq. $Bp_{0.2}$ 90°.

Schöllkopf, U. *et al*, *Justus Liebigs Ann. Chem.*, 1974, 44 (*synth, ir, pmr, use*)
Rachoń, J. *et al*, *Justus Liebigs Ann. Chem.*, 1981, 709, 186, 1693 (*props, use*)

Isofenphos, BSI I-00025

1-Methylethyl 2-[[ethoxy[(1-methylethyl)amino]-phosphinothioyl]oxy]benzoate, 9CI. O-Ethyl O-2-isopropoxycarbonylphenyl isopropylphosphoramidothioate. Oftanol

[25311-71-1]

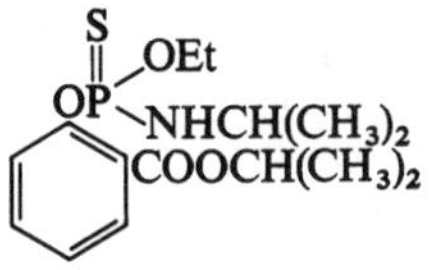

$C_{15}H_{24}NO_4PS$ M 345.393

Agricultural insecticide with contact and stomach action. Oil. V. spar. sol. H_2O. d_4^{20} 1.13. $Bp_{0.01}$ 120°.

▷VO4395500.

B.P., 1 224 323, (*1967*); *CA*, **74**, 125186z (*synth*)
Ripley, B.D. *et al*, *J. Assoc. Off. Anal. Chem.*, 1983, **66**, 1084 (*glc*)
Pesticide Manual, 6th Ed., 309; 7th Ed., 325.
The Agrochemicals Handbook, Royal Society of Chemistry, 1983, A236.

1*H*-Isophosphindole-1,3(2*H*)-dione, 9CI — I-00026

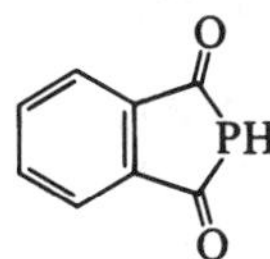

$C_8H_5O_2P$ M 164.100

2-Ph:

$C_{14}H_9O_2P$ M 240.198

Needles. Mp 73-74°. Yields $PhPH_2$ when heated with aniline.

2-Ph, 2-sulfide: [54552-79-3].

$C_{14}H_9O_2PS$ M 272.258

Solid. Mp 78-79°.

Enol-form

Trimethylsilyl ether:

$C_{11}H_{13}O_2PSi$ M 236.282

Red-orange liq. Stable below −30°. In 1,2-dimethoxyethane or THF, solns. stable at −5° for a few days.

Issleib, K. *et al*, *Z. Anorg. Allg. Chem.*, 1974, **408**, 266 (*synth, deriv*)
Fenske, D. *et al*, *Chem. Ber.*, 1976, **109**, 359 (*synth, ir*)
Markowski, L.N. *et al*, *Zh. Obshch. Khim.*, 1982, **52**, 2796 (*Engl. transl.* p. 2465) (*silyl ether, ir, P nmr, pmr*)

Isophosphinoline, 9CI — I-00027

2-Phosphanaphthalene. Benzo[c]*phosphorin. Benzo*[c]-*phosphinine*

[253-37-2]

C_9H_7P M 146.128

Mp 82-84°. Subl. at 55°/0.0001 mm.

de Graaf, H.G. *et al*, Tet, 1975, **31**, 1097 (*synth, uv, ms, P nmr, pmr*)
Jongsma, C. *et al*, *Org. Mass Spectrom.*, 1975, **10**, 575 (*ms*)
Galasso, V., *J. Magn. Reson.*, 1979, **34**, 199 (*struct, nmr*)

4-Isopropyl-5,5-dimethyl-1,3,2-dioxaphosphorinane — I-00028

5,5-Dimethyl-4-(1-methylethyl)-1,3,2-dioxaphosphorinane, 9CI

[69220-13-9]

(2*RS*,4*RS*)-*form*

$C_8H_{17}O_2P$ M 176.195

Liq. Bp_1 37-39°. n_D^{20} 1.468. Mixt. of *cis*- and *trans*-forms.

(*2RS,4RS*)-*form*

(±)-trans-*form*

2-Oxide: Oil. d_4^{20} 1.11. Bp_1 103-105°. n_D^{20} 1.4620.

(*2RS,4SR*)-*form*

(±)-cis-*form*

2-Oxide:

$C_8H_{17}O_3P$ M 192.194

Oil. d_4^{20} 1.11. Bp_1 120-122°. n_D^{20} 1.4635.

Nifant'ev, É.E. *et al*, *Tetrahedron*, 1981, **37**, 3183 (*synth, pmr, P nmr*)
Matrosov, E.I. *et al*, *Izv. Akad. Nauk SSSR, Ser. Khim.*, 1976, 530 (*Engl. transl.* p. 512) (*oxide*)

5-Isopropyl-2,6-dimethyl-6-phosphabicyclo[3.1.1]hept-2-ene 6-oxide — I-00029

2,6-Dimethyl-5-(1-methylethyl)-6-phosphabicyclo[3.1.1]*hept-2-ene 6-oxide, 9CI*

[67173-22-3]

$C_{11}H_{19}OP$ M 198.244

(*1S,5S,6R*)-*form*

Prod. of reacn. of $MePCl_2/AlCl_3$ with (−)-α-pinene. Cryst. (Et_2O). Mp 124-125°. $[\alpha]_D^{25}$ −76° ($CHCl_3$).

Vilkas, E. *et al*, *J. Chem. Soc., Chem. Commun.*, 1978, 125 (*synth, ir, ms, pmr, cmr, P nmr, cryst struct*)

O-Isopropyl *O,S*-dimethyl phosphorothioate, 8CI — I-00030

O,S-*Dimethyl* O-(*1-methylethyl*) *phosphorothioate, 9CI.* O-*Isopropyl* O,S-*dimethyl thiophosphate*

[75511-33-0]

(*R*)-*form*

$C_5H_{13}O_3PS$ M 184.190

(*R*)-*form* [57557-28-5]

$[\alpha]_D$ −3.0° (c, 0.6 in $CHCl_3$).

(*S*)-*form* [57557-27-4]

$[\alpha]_D^{21}$ +3.1° (c, 0.4 in $CHCl_3$).

Hall, C.R. *et al*, *J. Chem. Soc., Perkin Trans. 1*, 1979, 1646 (*props*)
Cooper, D.B. *et al*, *J. Chem. Soc., Perkin Trans. 1*, 1977, 1969 (*synth, pmr, stereochem*)

Isopropyldiphenylphosphine, 8CI — I-00031

(*1-Methylethyl*)*diphenylphosphine, 9CI*

[6372-40-3]

$Ph_2PCH(CH_3)_2$

$C_{15}H_{17}P$ M 228.273

Ligand for Ni, Rh, Ir, and Group VIB metals. Cryst. or liq. Mp 33°. Bp_{13} 165°, $Bp_{0.5}$ 145-147°. n_D^{20} 1.6027. Air-sensitive.

B,MeI: Isopropylmethyldiphenylphosphonium iodide.

$C_{16}H_{20}IP$ M 370.212

Cryst. ($CHCl_3$/pet. ether). Mp 225-226°.

Oxide: [2959-75-3].

$C_{15}H_{17}OP$ M 244.272

Cryst. (EtOAc). Mp 145-146°.

Sulfide: [66295-79-2].

$C_{15}H_{17}PS$ M 260.333

Cryst. Mp 93°.

Gough, S.D.T., *J. Chem. Soc.*, 1961, 4263 (*synth, deriv*)
Horner, L. *et al*, *Justus Liebigs Ann. Chem.*, 1961, **646**, 65 (*synth*)
Albright, T.A. *et al*, *J. Org. Chem.*, 1975, **40**, 3437 (*oxide, pmr, cmr*)
Inorg. Synth., 1976, **16**, 192 (*synth*)

Goff, S.D. *et al*, *Org. Mass Spectrom.*, 1977, **12**, 33 (*oxide, ms*)
Postle, S.R., *Phosphorus Sulfur*, 1977, **3**, 269 (*sulfide, synth, cmr*)
Dmitriev, V.I. *et al*, *Zh. Obshch. Khim.*, 1978, **48**, 52 (*Engl. transl.* p. 42) (*synth*)
Vincent, E. *et al*, *Spectrochim. Acta, Part A*, 1980, **36**, 699 (*pmr, cmr, nmr*)
Bertz, S.H. *et al*, *J. Am. Chem. Soc.*, 1981, **103**, 5932 (*oxide*)
Okano, T. *et al*, *Bull. Chem. Soc. Jpn.*, 1981, **54**, 3799 (*complex, use*)

7-Isopropylidene-4,8,8-trimethyl-1,6-dioxa-4-aza-5-phospha(5-P^V)spiro[4.4]-nonan-9-one I-00032

4,8,8-Trimethyl-7-(1-methylethylidene)-1,6-dioxa-4-aza-5-phospha(5-P^V)spiro[4.4]nonan-9-one, 9CI

$C_{10}H_{20}NO_3P$ M 233.247

5-Dimethylamino: [35854-55-8].
$C_{12}H_{25}N_2O_3P$ M 276.315
Solid. Mp 63-65°. Dec. on attempted distn.

Bentrude, W.G. *et al*, *J. Am. Chem. Soc.*, 1972, **94**, 923 (*synth, ir, uv, pmr, P nmr, props*)

4-Isopropyl-2-methoxy-5,5-dimethyl-1,3,2-dioxaphosphorinane I-00033

2-Methoxy-5,5-dimethyl-4-(1-methylethyl)-1,3,2-dioxaphosphorinane, 9CI

(*2RS,4RS*)*-form*

$C_9H_{19}O_3P$ M 206.221

(*2RS,4RS*)-*form* [95115-83-6]
(±)-trans-*form*
$Bp_{0.5}$ 65-67°. n_D^{18} 1.4552.
2-Selenide: [95115-84-7]. Cryst. (pet. ether). Mp 72-73°.

(*2RS,4SR*)-*form* [95115-82-5]
(±)-cis-*form*
Liq. Bp_3 68.5°. n_D^{18} 1.4583. Stereomutates slowly to the trans phosphite.
2-Selenide:
$C_9H_{19}O_3PSe$ M 285.181
Oil.

Nifant'ev, E.E. *et al*, *Tetrahedron*, 1981, **37**, 3183 (*P nmr*)
Edmundson, R.S. *et al*, *J. Chem. Soc., Perkin Trans. 2*, 1985, 69 (*synth, deriv, ir, pmr, P nmr, props*)

Isopropyl methylphosphinate I-00034

1-Methylethyl methylphosphinate
[21204-48-8]

(*R*)*-form*

$C_4H_{11}O_2P$ M 122.103

(*R*)-*form*
$[\alpha]_D$ +32.25° (c, 4.31 in EtOH) (calc).

(*S*)-*form*
$[\alpha]_D$ −17.4° (EtOH) 58% o.p., −30° (EtOH). Rapidly racemized by MeO^-.

(±)-*form*
d_4^{20} 1.01. Bp_{11} 69-70°. n_D^{20} 1.4209.

Petrov, K.A. *et al*, *Zh. Obshch. Khim.*, 1961, **31**, 179 (*Engl. transl.* p. 168) (*synth*)
Gladshtein, B.M. *et al*, *Zh. Obshch. Khim.*, 1969, **39**, 1951 (*Engl. transl.* p. 1913) (*synth*)
Benschop, H.P. *et al*, *J. Chem. Soc., Chem. Commun.*, 1970, 1431 (*resoln*)
Reiff, L.P. *et al*, *J. Am. Chem. Soc.*, 1970, **92**, 5275.
Szafraniec, L.J. *et al*, *J. Org. Chem.*, 1982, **47**, 1936 (*config*)

Isopropylmethylphosphinic acid I-00035

Methyl(1-methylethyl)phosphinic acid, 9CI

$(H_3C)_2CHP(O)(OH)Me$

$C_4H_{11}O_2P$ M 122.103
$Bp_{0.05}$ 96-98°. n_D^{24} 1.4502.
Chloride: [3393-57-5].
$C_4H_{10}ClOP$ M 140.549
Liq. d_4^{20} 1.16. Bp_4 71°. n_D^{20} 1.4670.

Crofts, P.C. *et al*, *J. Chem. Soc.*, 1958, 2995 (*synth*)
Neimysheva, A.A. *et al*, *Zh. Obshch. Khim.*, 1966, **36**, 1090 (*Engl. transl.* p. 1105) (*chloride*)
Neimysheva, A.A. *et al*, *Zh. Obshch. Khim.*, 1967, **37**, 2255 (*Engl. transl.*, p. 2140) (*nqr*)
Knunyants, I.L. *et al*, *Dokl. Akad. Nauk SSSR, Ser. Sci. Khim.*, 1971, **201**, 862 (*Engl. transl.* p. 992) (*halides, nmr*)

N-Isopropyl-*P*-methylphosphonamidic acid, 8CI I-00036

P-*Methyl*-N-*(1-methylethyl)phosphonamidic acid*, 9CI

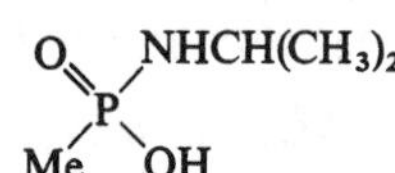

$C_4H_{12}NO_2P$ M 137.118

Me ester: [13703-07-6]. *Methyl* N-*isopropyl*-P-*methylphosphonamidate.*
$C_5H_{14}NO_2P$ M 151.145
Liq. d_4^{20} 1.04. $Bp_{0.03}$ 81-83°. n_D^{20} 1.4373.

Et ester: [13792-55-7]. *Ethyl* N-*isopropyl*-P-*methylphosphonamidate.*
$C_6H_{16}NO_2P$ M 165.172
Liq. Bp_3 118°, $Bp_{0.03}$ 66-67°. n_D^{20} 1.4350.

Isopropyl ester: [13703-13-4]. *Isopropyl* N-*isopropyl*-P-*methylphosphonamidate.*
$C_7H_{18}NO_2P$ M 179.198
Liq. Bp_3 120°, $Bp_{0.07}$ 85-87°. n_D^{20} 1.4318.

Ph ester: [4645-91-4]. *Phenyl* N-*isopropyl*-P-*methylphosphonamidate.*
$C_{10}H_{16}NO_2P$ M 213.216
Liq. Bp_2 192°.

Shokol, V.A. *et al*, *Zh. Obshch. Khim.*, 1966, **36**, 1636 (*Engl. transl.* p. 1636) (*synth*)
Tomchina, L.F. *et al*, *Zh. Obshch. Khim.*, 1968, **38**, 564 (*Engl. transl.* p. 549) (*synth*)
Zavlin, P.M. *et al*, *Zh. Obshch. Khim.*, 1975, **45**, 239 (*Engl. transl.* p. 226) (*synth*)
Zavlin, P.M. *et al*, *Zh. Obshch. Khim.*, 1977, **47**, 1981 (*Engl. transl.* p. 1812) (*synth*)

N-Isopropyl-*P*-methylphosphonamidothioic acid I-00037

P-*Methyl*-N-*(1-methylethyl)phosphonamidothioic acid*, *9CI*

S=P(Me)(OH)NHCH(CH$_3$)$_2$ ⇌ O=P(Me)(SH)NHCH(CH$_3$)$_2$

$C_4H_{12}NOPS$ M 153.179

O-*Et ester:* [21843-98-1]. O-*Ethyl* N-*isopropyl*-P-*methylphosphonamidothioate.*
$C_6H_{16}NOPS$ M 181.232
Liq. d_4^{20} 1.03. $Bp_{0.2}$ 69-70°. n_D^{25} 1.4878.

Grapov, A.F. *et al*, *Zh. Obshch. Khim.*, 1968, **38**, 2260 (*Engl. transl.* p. 2187)

Isopropyl methylphosphonate I-00038

1-Methylethyl methylphosphonate. Isopropyl hydrogen methylphosphonate
[1832-54-8]

O=P(OH)(Me)OCH(CH$_3$)$_2$

$C_4H_{11}O_3P$ M 138.103

Oil. d_4^{20} 1.10. $Bp_{0.1}$ 102-104°. n_D^{20} 1.4210. Non-toxic.

Ph ester: Isopropyl phenyl methylphosphonate. 1-Methylethyl phenyl methylphosphonate.
$C_7H_{17}O_3P$ M 180.183
d_4^{20} 1.16. Bp_4 174°. n_D^{20} 1.5043.

Fluoride: see Sarin, S-00002
Chloride: see Methylphosphonochloridic acid, M-00296

Keay, L, *Can. J. Chem.*, 1965, **43**, 2637 (*synth*)
Petrov, K.A. *et al*, *Zh. Obshch. Khim.*, 1965, **35**, 723 (*Engl. transl.* p. 723) (*synth*)
Cadogan, J.I.G. *et al*, *J. Chem. Soc. (B)*, 1971, 1988 (*synth*)
Zavlin, P.M. *et al*, *Zh. Obshch. Khim.*, 1972, **42**, 1257 (*Engl. transl.* p. 1253) (*phenyl ester*)

O-Isopropyl methylphosphonochloridothioate I-00039

O-*(1-Methylethyl) methylphosphonochloridothioate*, *9CI*
[2524-17-6]

S=P(Cl)(Me)OCH(CH$_3$)$_2$ (*R*)-*form*

$C_4H_{10}ClOPS$ M 172.609

(*R*)-*form* [80799-69-5]
Liq. $Bp_{0.05}$ 22°. $[\alpha]_D$ −85.80° (neat). n_D^{20} 1.4844.

(*S*)-*form*
Liq. $[\alpha]_D$ +80.60° (neat).

(±)-*form*
Liq. d_4^{20} 1.13. Bp_2 45-46°. n_D^{20} 1.4822.

Hoffmann, F.W. *et al*, *J. Am. Chem. Soc.*, 1958, **80**, 3945 (*synth*)
Boter, H.L. *et al*, *Recl. Trav. Chim. Pays-Bas*, 1966, **85**, 27 (*synth*)
Omelanczuk, J. *et al*, *Tetrahedron*, 1971, **27**, 5587 (*synth*)
Mikolajczyk, M. *et al*, *Tetrahedron*, 1972, **28**, 3835 (*props*)
Szafraniec, L.J. *et al*, *J. Org. Chem.*, 1982, **47**, 1936.

O-Isopropyl methylphosphonodithioate, 8CI I-00040

O-*(1-Methylethyl) methylphosphonodithioate*, *9CI*. O-*Isopropyl hydrogen methylphosphonodithioate*

S=P(SH)(Me)OCH(CH$_3$)$_2$ (*R*)-*form*

$C_4H_{11}OPS_2$ M 170.224

(*R*)-*form*
Et ester: [34666-63-2]. S-*Ethyl* O-*isopropyl methylphosphonothioate.*
$C_6H_{15}OPS_2$ M 198.278
Liq. $[\alpha]_D$ −45.60° (c, 3.61 in C_6H_6).

(*S*)-*form*
Et ester: Liq. $Bp_{0.7}$ 57°. $[\alpha]_D$ +45.9° (c, 2.93 in C_6H_6). n_D^{20} 1.5158.

(±)-*form* [999-87-1]
Liq. $Bp_{0.01}$ 38°.

Omelanczuk, J. *et al*, *Tetrahedron*, 1971, **27**, 5587 (*synth*)
Mikolajczyk, M. *et al*, *Tetrahedron*, 1972, **28**, 4357 (*uv, cd, config*)

O-Isopropyl methylphosphonothioate I-00041

O-*(1-Methylethyl) methylphosphonothioate*, *9CI*. O-*Isopropyl hydrogen methylphosphonothioate*
[20627-00-3]

S=P(OH)(Me)OCH(CH$_3$)$_2$ (*R*)-*form*

$C_4H_{11}O_2PS$ M 154.163

Tautomeric with thiol form.

(*R*)-*form* [26547-89-7]
(R)-*1-Phenylethylammonium salt:* Solid. Mp 158-158.5°. $[\alpha]_D^{24}$ +10.3° (c, 2.98 in MeOH).
Dicyclohexylammonium salt: Solid. Mp 124-125°. $[\alpha]_D^{27}$ +7.30° (c, 2.2 in MeOH).
S-*Me ester:* O-*Isopropyl* S-*methyl methylphosphonothioate.*
$C_5H_{13}O_2PS$ M 168.190
Liq. Bp_4 73-75°. $[\alpha]_D^{28}$ +87.8° (c, 3.3 in C_6H_6). n_D^{25} 1.4712.
S-*Et ester:* S-*Ethyl* O-*isopropyl methylphosphonothioate.*
$C_6H_{15}O_2PS$ M 182.217
Liq. $Bp_{0.35}$ 46°. $[\alpha]_D^{25}$ +48.8° (c, 2.66 in C_6H_6). n_D^{25} 1.4661.

(*S*)-*form* [44657-29-6]
Liq. $Bp_{0.01}$ 46°. $[\alpha]_D$ −12.25° (neat). n_D^{20} 1.4828.
(S)-*1-Phenylethylammonium salt:* Solid. Mp 157.5-158°. $[\alpha]_D^{24}$ −10.6° (c, 2.09 in MeOH).
Dicyclohexylammonium salt: Solid. Mp 124-126°. $[\alpha]_D^{25}$ −7.7° (c, 5 in MeOH).
S-*Me ester:* [44657-29-6]. Liq. $Bp_{4.5}$ 76°, $Bp_{0.7}$ 47-49°. $[\alpha]_D^{25}$ −86.8° (c, 1.87 in C_6H_6). n_D^{25} 1.4705.
S-*Et ester:* Liq. Bp_9 94-95°, Bp_1 54°. $[\alpha]_D$ −14.10° (neat), $[\alpha]_D$ −63.9° (c, 2.29 in C_6H_6). n_D^{20} 1.4668.

(±)-*form* [36585-72-5]
Liq. $Bp_{0.6}$ 84-85°. n_D^{23} 1.4795.
Dicyclohexylammonium salt: Cryst. (EtOH). Mp 171-173°.
S-*Et ester:* Liq. Bp_9 94-95°.

S-Ph ester: O-*Isopropyl* S-*phenyl methylphosphonothioate.*
$C_{10}H_{15}O_2PS$ M 230.261
Liq. $Bp_{0.04}$ 113-114°.
Chloride: see O-Isopropyl methylphosphonochloridothioate, I-00039

Petrov, K.A. *et al*, *Zh. Obshch. Khim.*, 1961, **31**, 176 (*Engl. transl.* p. 168) (*ester*)
Aaron, H.S. *et al*, *J. Am. Chem. Soc.*, 1962, **84**, 617 (*resoln*)
Pelchowicz, Z. *et al*, *J. Chem. Soc.*, 1962, 3824 (*synth*)
Boter, H.L. *et al*, *Recl. Trav. Chim. Pays-Bas*, 1967, **86**, 399 (*resoln*)
Omelańczuk, J. *et al*, *Tetrahedron*, 1971, **27**, 5587 (*props*)
Mikolajczyk, M. *et al*, *Tetrahedron*, 1972, **28**, 3855 (*esters, ord*)
Mikolajczyk, M. *et al*, *Tetrahedron*, 1972, **28**, 4357 (*cd, config*)
Moriyama, M. *et al*, *J. Am. Chem. Soc.*, 1983, **105**, 4727 (*synth, resoln*)

2-Isopropyl 1,3,2-oxathiaphospholane, 9CI, 8CI — I-00042

O,S-*Ethylene isopropylphosphonothioate*

[structure: five-membered ring S–P–O with P–$CH(CH_3)_2$]

$C_5H_{11}OPS$ M 150.175

2-Oxide: [84549-28-0]. *2-Isopropyl-1,3,2-oxathiaphospholane 2-oxide.* O,S-*Ethylene isopropylphosphonothioate.*
$C_5H_{11}O_2PS$ M 166.174
Liq. $Bp_{1.2}$ 99-101°.
2-Sulfide: [84549-29-1]. *2-Isopropyl-1,3,2-oxathiaphospholane 2-sulfide.* O,S-*Ethylene isopropylphosphonodithioate.*
$C_5H_{11}OPS_2$ M 182.235
Liq. Bp_2 110° (oven).

Kim, B.H. *et al*, *Phosphorus Sulfur*, 1982, **13**, 337 (*synth, ir, ms, pmr*)

Isopropylphenylphosphinic acid, 8CI — I-00043

(*1-Methylethyl*)*phenylphosphinic acid, 9CI*
[16543-43-4]

$(H_3C)_2CHPPh(O)OH$

$C_9H_{13}O_2P$ M 184.174

(±)-form

Me ester: [76420-29-6]. *Methyl isopropylphenylphosphinate.*
$C_{10}H_{15}O_2P$ M 198.201
Liq. $Bp_{1.5}$ 95-100°. Known in opt. active forms.
Isopropyl ester: Isopropyl isopropylphenylphosphinate.
$C_{12}H_{19}O_2P$ M 226.255
d_0^{17} 1.08. Bp_{11} 146-147°. n_D^{20} 1.4929.
Chloride: [13213-43-9].
$C_9H_{12}ClOP$ M 202.620
Liq. d_4^{20} 1.16. Bp_3 116-121°. n_D^{20} 1.5430.
Azide: [85656-05-9].
$C_9H_{12}N_3OP$ M 209.187
Liq. $Bp_{0.1}$ 135-140° (oven).
Amide: see P-Isopropyl-P-phenylphosphinic amide, I-00044

Arbusov, A.E. *et al*, *Zh. Obshch. Khim.*, 1945, **15**, 766; *CA*, **41**, 105 (*ester*)
Neimysheva, A.A. *et al*, *Zh. Obshch. Khim.*, 1966, **36**, 1090 (*Engl. transl.* p. 1105) (*chloride*)
Kosolapoff, G.M. *et al*, *J. Chem. Soc.* (*C*), 1967, 1789.
Legin, G.Ya., *Zh. Obshch. Khim.*, 1973, **43**, 2202 (*Engl. transl.* p. 2194) (*esters*)
Harger, M.J.P., *J. Chem. Soc., Perkin Trans. 1*, 1975, 514 (*chloride*)
Harger, M.J.P., *Tetrahedron Lett.*, 1978, 2927 (*ester, pmr*)
Harger, M.J.P, *J. Chem. Soc., Perkin Trans. 2*, 1980, 1505 (*ester*)
Harger, M.J.P. *et al*, *Tetrahedron*, 1982, **38**, 3073 (*azide, pmr*)

P-Isopropyl-*P*-phenylphosphinic amide, 9CI — I-00044

P-(*1-Methylethyl*)-P-*phenylphosphinic amide*
[51028-16-1]

[structure: $(H_3C)_2CH$, O, Ph, NH_2 on P]

$C_9H_{14}NOP$ M 183.189

(±)-form

Cryst. (C_6H_6). Mp 129-131°.
B,HCl: Solid. Mp 74-76°.
N-*Ph:* P-*Isopropyl*-N,P-*diphenylphosphinic amide.* Cryst. (EtOAc/$CHCl_3$). Mp 168-171°.

Harger, M.J.P., *J. Chem. Soc., Perkin Trans. 1*, 1975, 514; 1977, 605 (*synth, ir, pmr, deriv*)

Isopropylphenylphosphinodithioic acid — I-00045

(*1-Methylethyl*)*phenylphosphinodithioic acid, 9CI*
[20384-86-5]

$(H_3C)_2CHPPh(S)SH$

$C_9H_{13}PS_2$ M 216.295
Liq. $Bp_{0.001}$ 92°. n_D^{22} 1.6354.
NH_4 salt: Solid. Mp 148-158°.

Newallis, P.E. *et al*, *J. Org. Chem.*, 1962, **27**, 3829 (*synth, P nmr*)
Dietsche, W.H., *Tetrahedron*, 1967, **23**, 3049.

Isopropylphenylphosphinothioic acid, 8CI — I-00046

(*1-Methylethyl*)*phenylphosphinothioic acid*
[53159-02-7]

$(H_3C)_2CHPPh(S)OH \rightleftharpoons (H_3C)_2CHPPh(O)SH$

$C_9H_{13}OPS$ M 200.235

(+)-form [76380-92-2]

Dicyclohexylammonium salt: [76380-91-1]. $[\alpha]_D$ +11.96° (c, 2.2 in MeOH).
S-*Me ester:* [62246-61-1]. S-*Methyl isopropylphenylphosphinothioate.* $[\alpha]_D$ +141.4° (c, 1.91 in C_6H_6).

(±)-form [76380-88-6]

Solid.
Dicyclohexylammonium salt: [76380-89-7]. Cryst. (CH_2Cl_2/Et_2O). Mp 160-162°.
S-*Me ester:* [76380-85-3].
$C_{10}H_{15}OPS$ M 214.262
Liq. $Bp_{0.3}$ 120-125° (oven).
Chloride:
$C_9H_{12}ClPS$ M 218.681
Liq. Bp_3 123°.
N,N-*Dimethylamide:* [83587-43-3]. P-*Isopropyl*-N,N-*dimethyl*-P-*phenylphosphinothioic amide.*
$C_{11}H_{18}NPS$ M 227.304

Liq. $Bp_{0.05}$ 111°.

Mastryukova, T.A. *et al*, *Phosphorus Sulfur*, 1976, **1**, 211 (*esters*)
Harger, M.J.P., *J. Chem. Soc., Perkin Trans. 2*, 1980, 1505 (*acid, ester*)
Kaushik, M.P. *et al*, *Indian J. Chem., Sec. B*, 1981, **20**, 932 (*chloride*)
Johnson, C.R. *et al*, *J. Am. Chem. Soc.*, 1982, **104**, 7041 (*amide, use, pmr, cmr, ir*)

Isopropyl phenyl phosphonate — I-00047

(1-Methylethyl) phenyl phosphonate, 9CI. *Isopropyl phenyl phosphite*. *(1-Methylethyl) phenyl phosphite*
[14609-90-6]

$(H_3C)_2CHOPH(O)OPh \rightleftharpoons (H_3C)_2CHOP(OH)OPh$

$C_9H_{13}O_3P$ M 200.174
Tautomeric. Liq. d_4^{20} 1.16. $Bp_{1.5}$ 106-108°. n_D^{20} 1.4970.

Wolf, R. *et al*, *Bull. Soc. Chim. Fr.*, 1960, 124 (*synth, ir*)
Houalla, D. *et al*, *Bull. Soc. Chim. Fr.*, 1960, 129 (*ir*)
Mandel'baum, Ya.A. *et al*, *Zh. Obshch. Khim.*, 1972, **42**, 502 (*Engl. transl.* p. 500) (*synth*)

Isopropylphosphine, 8CI — I-00048

(1-Methylethyl)phosphine, 9CI
[4538-29-8]

$(H_3C)_2CHPH_2$

C_3H_9P M 76.078
Liq. Bp 41°.

Hays, H.R. *et al*, *J. Org. Chem.*, 1966, **31**, 3391 (*synth, nmr*)
Durig, J.R. *et al*, *J. Phys. Chem.*, 1976, **80**, 2493 (*ir, raman*)
Kostyanovskii, R.G. *et al*, *Org. Mass Spectrom.*, 1976, **11**, 237 (*ms*)
Durig, J.R. *et al*, *J. Mol. Spectrosc.*, 1978, **70**, 27 (*microwave*)
Mosbo, J.A. *et al*, *Phosphorus Sulfur*, 1981, **11**, 11 (*struct*)

Isopropylphosphinic acid — I-00049

(1-Methylethyl)phosphinic acid, 9CI

$(H_3C)_2CHPH(O)OH$

$C_3H_9O_2P$ M 108.077
Tautomeric with isopropylphosphonous acid. Free acid exists in phosphinic acid form. Liq. d^{13} 1.19.
Butyl ester: [70446-71-8]. *Butyl isopropylphosphinate*.
$C_7H_{17}O_2P$ M 164.184
Liq. d_4^{20} 0.96. $Bp_{2.5}$ 58.5-60°. n_D^{20} 1.4321.

Guichard, F., *Ber.*, 1899, **32**, 1572 (*synth*)
Kabachnik, M.I. *et al*, *Dokl. Akad. Nauk SSSR, Ser. Sci. Khim.*, 1959, **125**, 1260 (*Engl. transl.* p. 309) (*ester*)

Isopropylphosphonic acid — I-00050

(1-Methylethyl)phosphonic acid, 9CI. *2-Phosphonopropane*
[4721-37-3]

$(H_3C)_2CHP(O)(OH)_2$

$C_3H_9O_3P$ M 124.076
Several esters are used in the extraction and sepn. of lanthamides. Cryst. (C_6H_6). Mp 74-75°. pK_{a1} 2.66, pK_{a2} 8.44 (H_2O, 25°).
Monoanilinium salt: Cryst. (MeCN). Mp 175-177°.
Di-Me ester: [54552-77-1]. *Dimethyl isopropylphosphonate*.
$C_5H_{13}O_3P$ M 152.130
Liq. Bp_{13} 73-74°. n_D^{20} 1.4203.
Di-Et ester: [1538-69-8]. *Diethyl isopropylphosphonate*.
$C_7H_{17}O_3P$ M 180.183
Liq. Bp_{19} 83°. n_D^{20} 1.4149.
Di-Ph ester: [1538-72-3]. *Diphenyl isopropylphosphonate*.
$C_{15}H_{17}O_3P$ M 276.271
Cryst. (pet. ether). Mp 42.5-43°.
Difluoride: [677-42-9].
$C_3H_7F_2OP$ M 128.058
Liq. Bp 114-115°.
Dichloride: see Isopropylphosphonic dichloride, I-00052
Diisocyanate: [2736-47-2].
$C_5H_7N_2O_3P$ M 174.096
Liq. d_4^{20} 1.26. Bp_4 60-61°. n_D^{20} 1.4628.

Crofts, P.C. *et al*, *J. Am. Chem. Soc.*, 1953, **75**, 3379 (*synth, props*)
Myers, T.C. *et al*, *J. Am. Chem. Soc.*, 1954, **76**, 4172 (*ester*)
Haven, A.C., *J. Am. Chem. Soc.*, 1956, **78**, 842 (*diisocyanate*)
Ivanova, Zh.M. *et al*, *Zh. Obshch. Khim.*, 1968, **38**, 1334 (*Engl. transl.* p. 1284) (*difluoride*)
Ivanova, Zh.M. *et al*, *Zh. Obshch. Khim.*, 1969, **39**, 1037 (*Engl. transl.* p. 1008) (*diisocyanate*)
Griffiths, W.R. *et al*, *Phosphorus*, 1975, **5**, 273 (*ms*)
Griffiths, W.R. *et al*, *Phosphorus Sulfur*, 1978, **5**, 101 (*esters, ms*)
Villieras, J. *et al*, *J. Organomet. Chem.*, 1978, **144**, 17 (*ester, synth, ir, pmr*)

P-Isopropylphosphonic diamide, 8CI — I-00051

P-*(1-Methylethyl)phosphonic diamide*, 9CI

$(H_3C)_2HCP(O)(NH_2)_2$

$C_3H_{11}N_2OP$ M 122.106
N,N,N′,N′-*Tetra-Me:* P-*Isopropyl*-N,N,N′,N′-*tetramethylphosphonic diamide*. *Isopropylphosphonic bis(dimethylamide)*.
$C_7H_{19}N_2OP$ M 178.214
Liq. d_4^{20} 0.98. $Bp_{0.15}$ 74-76°. n_D^{25} 1.4548.
N,N′-*Di-Ph:* N,N′-*Diphenyl*-P-*propylphosphonic diamide*. *Isopropylphosphonic dianilide*.
$C_{15}H_{19}N_2OP$ M 274.302
Solid. Mp 199-201°.

Kinnear, A.M. *et al*, *J. Chem. Soc.*, 1952, 3437 (*dianilide*)
Razvodovskaya, L.V. *et al*, *Zh. Obshch. Khim.*, 1969, **39**, 1260 (*Engl. transl.* p. 1230) (*bisdimethylamide*)

Isopropylphosphonic dichloride, 8CI — I-00052

(1-Methylethyl)phosphonic dichloride, 9CI
[1498-46-0]

$(H_3C)_2CHP(O)Cl_2$

$C_3H_7Cl_2OP$ M 160.967
Pungent liq. d_4^{20} 1.30. Bp_{746} 189°, Bp_{20} 80°. n_D^{25} 1.4646.

Kinnear, A.M. *et al*, *J. Chem. Soc.*, 1952, 3437 (*synth*)
Christol, C. *et al*, *J. Chim. Phys.*, 1965, **62**, 246 (*ir, raman*)
Geiseler, G. *et al*, *Ber. Bunsenges. Phys. Chem.*, 1967, **71**, 478 (*ir, raman*)
Griffiths, W.R. *et al*, *Phosphorus Sulfur*, 1978, **4**, 341 (*ms*)

Isopropylphosphonochloridothioic acid, 8CI — I-00053

(1-Methylethyl)phosphonochloridothioic acid, 9CI

$(H_3C)_2CHPCl(S)OH \rightleftharpoons (H_3C)_2CHPCl(O)SH$

C_3H_8ClOPS M 158.582

O-Me ester: see O-Methyl isopropylphosphonochloridothioate, M-00165

O-Et ester: see O-Ethyl isopropylphosphonothioate, E-00082

S-Et ester: [19057-06-8]. *S-Ethyl isopropylphosphonochloridothioate.*
$C_5H_{12}ClOPS$ M 186.636
Liq. d_4^{20} 1.18. $Bp_{2.0}$ 88°. n_D^{20} 1.5055.

O-Propyl ester: O-Propyl isopropylphosphonochloridothioate.
$C_6H_{14}ClOPS$ M 200.663
Liq. $Bp_{0.7}$ 58-59°. n_D^{25} 1.4865.

Neimysheva, A.A. *et al*, *Zh. Obshch. Khim.*, 1967, **37**, 1822 (*Engl. transl.* p. 1736) (*S-Ethyl ester, synth, hydrol*)
Chupp, J.P. *et al*, *J. Org. Chem.*, 1962, **27**, 3832 (*O-propyl ester, synth*)

Isopropylphosphonothioic acid, 8CI I-00054

(1-Methylethyl)phosphonothioic acid, 9CI

$(H_3C)_2CHP(S)(OH)_2 \rightleftharpoons (H_3C)_2CHP(O)(OH)(SH)$

$C_3H_9O_2PS$ M 140.137

O-Me ester: see O-Methyl isopropylphosphonothioate, M-00166

O-Et ester: see O-Ethyl isopropylphosphonothioate, E-00082

O-Me, O′-Et-ester: see O-Methyl isopropylphosphonothioate, M-00166

O,O-Di-Et ester: [52038-87-6]. *O,O-Diethyl isopropylphosphonothioate.*
$C_7H_{17}O_2PS$ M 196.244
Liq. Bp_5 75°.

O-Me ester, chloride: see O-Methyl isopropylphosphonochloridothioate, M-00165

O-Et ester, chloride: see Isopropylphosphonochloridothioic acid, I-00053

Dichloride: see Isopropylphosphonothioic dichloride, I-00055

Kaushik, M.P. *et al*, *J. Org. Chem.*, 1980, **45**, 2250 (*diethyl ester, synth, ir, pmr*)

Isopropylphosphonothioic dichloride I-00055

(1-Methylethyl)phosphonothioic dichloride, 9CI. Isopropylthiophosphonic dichloride

[1498-60-8]

$(H_3C)_2CHP(S)Cl_2$

$C_3H_7Cl_2PS$ M 177.028

Liq. d^{20} 1.30. Bp_{15} 77°. n_D^{17} 1.5391.

Christol, C. *et al*, *J. Chim. Phys.*, 1965, **62**, 246 (*ir, raman*)
Kuchen, W. *et al*, *Chem. Ber.*, 1970, **103**, 2114 (*synth, ir, pmr, P nmr*)
Kaushik, M.P. *et al*, *J. Org. Chem.*, 1980, **45**, 2270 (*synth, ir, pmr*)

Isopropylphosphonous acid I-00056

(1-Methylethyl)phosphonous acid, 9CI

$(H_3C)_2CHP(OH)_2 \rightleftharpoons (H_3C)_3CHPH(O)OH$

$C_3H_9O_2P$ M 108.077

Free acid exists as the phosphoryl tautomer, Isopropylphosphinic acid, I-00049 .

Diisopropyl ester: Diisopropyl isopropylphosphonite. Bis(1-methylethyl) (1-methylethyl)phosphonite.
$C_9H_{21}O_2P$ M 192.237
Liq. $Bp_{12.5}$ 105.5-107.5°, Bp_{14} 59.5-60.5°. n_D^{20} 1.4150.

Monobutyl ester: see Isopropylphosphinic acid, I-00049

Dibutyl ester: [70446-65-0]. *Dibutyl isopropylphosphonite. Dibutyl (1-methylethyl)phosphonite.*
$C_{11}H_{25}O_2P$ M 220.291
Liq. Bp_9 94-98°. n_D^{20} 1.4352.

Diphenyl ester: Diphenyl isopropylphosphonite.
$C_{15}H_{17}O_2P$ M 260.272
Liq. d_0^{17} 1.16. Bp_{11} 212-214°. n_D^{17} 1.5782.

Dichloride: see Isopropylphosphonous dichloride, I-00057

Dibromide: Dibromoisopropylphosphine.
$C_3H_7Br_2P$ M 233.870
Liq. Bp_{40} 80°.

Razumov, A.I. *et al*, *Izv. Akad. Nauk SSSR, Ser. Khim.*, 1952, 894; *CA*, **47**, 10466 (*ester*)
Razumov, A.I. *et al*, *CA*, 1957, **51**, 6503 (*ester*)
Kamai, G. *et al*, *CA*, 1957, **51**, 11273 (*ester*)
Foss, V.L. *et al*, *Zh. Obshch. Khim.*, 1979, **49**, 559 (*Engl. transl.* p. 489) (*dibutyl ester, synth, P nmr*)
Hinke, A. *et al*, *Phosphorus Sulfur*, 1983, **15**, 93 (*dibromide, synth, nmr*)

Isopropylphosphonous dichloride I-00057

(1-Methylethyl)phosphonous dichloride, 9CI. Dichloroisopropylphosphine

[25235-15-8]

$(H_3C)_2CHPCl_2$

$C_3H_7Cl_2P$ M 144.968

Pungent liq. Bp_{745} 130°, Bp_{202} 88°. n_D^{25} 1.4868.

Perry, B.J. *et al*, *Can. J. Chem.*, 1963, **41**, 2299 (*synth*)
Dutasta, J.P. *et al*, *J. Chem. Soc., Chem. Commun.*, 1975, 747 (*conformn*)
Kostyanovskii, R.G. *et al*, *Izv. Akad. Nauk SSSR, Ser. Khim.*, 1975, 901 (*Engl. transl.* p. 816) (*synth, pmr*)

Isopropylphosphoramidic acid, 8CI I-00058

(1-Methylethyl)phosphoramidic acid, 9CI

[33876-53-8]

$(H_3C)_2CHNHP(O)(OH)_2$

$C_3H_{10}NO_3P$ M 139.091

Di-Me ester: [74124-43-9]. *Dimethyl isopropylphosphoramidate.*
$C_5H_{14}NO_3P$ M 167.144
No phys. props. reported.

Di-Et ester: [22685-19-4]. *Diethyl isopropylphosphoramidate.*
$C_7H_{18}NO_3P$ M 195.198
No phys. props. reported.

Diisopropyl ester: [74124-46-2]. *Diisopropyl isopropylphosphoramidate.*
$C_9H_{22}NO_3P$ M 223.251
Liq. Bp_1 86-89° (impure).

Di-Ph ester: [5756-04-7]. *Diphenyl isopropylphosphoramidate.*
$C_{15}H_{18}NO_3P$ M 291.286
Cryst. (pet. ether). Mp 75-76°.

Difluoride:
$C_3H_8F_2NOP$ M 143.073
Liq. Bp_{10} 62-63°. n_D^{20} 1.3689.

Dichloride: [33876-58-3].
$C_3H_8Cl_2NOP$ M 175.982
Cryst. (Et_2O). Mp 51-52°.

Kabachnik, M.I. *et al*, *Izv. Akad. Nauk SSSR, Ser. Khim.*, 1961, 816 (*Engl. transl.* p. 755) (*diisopropyl ester*)
Garrison, A.W. *et al*, *Spectrochim. Acta, Part A*, 1969, **25**, 77 (*diethyl ester, ir, nmr*)
Cates, L.A. *et al*, *J. Med. Chem.*, 1971, **14**, 647 (*dichloride*)
Modro, T.A. *et al*, *J. Org. Chem.*, 1978, **43**, 5000 (*diphenyl ester, synth, props*)
Al-Rawi, J.M.A. *et al*, *Org. Magn. Reson.*, 1983, **21**, 75 (*diethyl ester, cmr*)
Davidowitz, B. *et al*, *Org. Mass. Spectrom.*, 1984, **19**, 128 (*dimethyl ester, ms*)

Isopropylphosphoramidothioic acid, 8CI **I-00059**

(1-Methylethyl)phosphoramidothioic acid, 9CI

$(H_3C)_2CHNHP(S)(OH)_2 \rightleftharpoons$
$(H_3C)_2CHNHP(O)(OH)(SH)$

$C_3H_{10}NO_2PS$ M 155.151

O,O-*Di-Et ester:* [6737-25-3]. O,O-*Diethyl isopropylphosphoramidothioate.*
$C_7H_{18}NO_2PS$ M 211.258
No data reported.

O,O-*Di-Ph ester:* [63577-86-6]. O,O-*Diphenyl isopropylphosphoramidothioate.*
$C_{15}H_{18}NO_2PS$ M 307.346
No data reported.

O-*(4-Methyl-2-nitrophenyl) ester: see* O-*(4-Methyl-2-nitrophenyl) isopropylphosphoramidothioate, M-00180*

O-*(2,4-Dichlorophenyl)* O-*methyl ester: see Zytron, Z-00002*

O-*(2,4-Dichlorophenyl)* S-*methyl ester: see O-2,4-Dichlorophenyl S-methyl isopropylphosphoramidothioate, D-00183*

O-*Ethyl* O-*(4-methyl-2-nitrophenyl) ester: see Amiprophos M, A-00131*

O-*Ethyl* O-*(2-isopropoxycarbonylphenyl) ester: see Isofenphos, I-00025*

Ayed, N. *et al*, *C.R. Hebd. Seances Acad. Sci., Ser. C*, 1977, **285**, 221 (*ir*)
El-Borgi, A. *et al*, *C.R. Hebd. Seances Acad. Sci., Ser. C*, 1977, **284**, 983 (*ir*)
Mathis, R. *et al*, *C.R. Hebd. Seances Acad. Sci., Ser. C*, 1977, **281**, 437; 1977, **284**, 767 (*ir*)
El-Rawi, J.M.A. *et al*, *Org. Magn. Reson.*, 1983, **21**, 75 (*cmr*)

Isopropyl phosphorodichloridate, 8CI **I-00060**

1-Methylethyl phosphorodichloridate, 9CI. Isopropyl dichlorophosphate. Isopropyl phosphoryl dichloride

[56376-11-5]

$(H_3C)_2CHOP(O)Cl_2$

$C_3H_7Cl_2O_2P$ M 176.967

Pungent liq. d^{25} 1.28. Bp_{13} 60°. Easily hydrolysed.

Grunze, H., *Chem. Ber.*, 1959, **92**, 850 (*synth*)
Jauhiainen, T.P. *et al*, *Finn. Chem. Lett.*, 1976, 185 (*synth, props*)

Isopropyl phosphorodichloridite, 8CI **I-00061**

1-Methylethyl phosphorodichloridite, 9CI. Isopropyl dichlorophosphite

[31430-66-7]

$(H_3C)_2CHOPCl_2$

$C_3H_7Cl_2OP$ M 160.967

Pungent liq. Bp_{20} 40°.

Ger. Pat., 1 175 659, (*1964*); *CA*, **62**, 1566

O-Isopropyl phosphorodichloridothioate, 8CI **I-00062**

O-*(1-Methylethyl) phosphorodichloridothioate, 9CI.* O-*Isopropyl thiophosphoryl dichloride.* O-*Isopropyl dichlorothiophosphate*

[19021-61-5]

$(H_3C)_2CHOP(S)Cl_2$

$C_3H_7Cl_2OPS$ M 193.027

Liq. Bp_{10} 59-60°.

U.S.P., 3 005 005, (*1961*); *CA*, **56**, 5836 (*manuf*)
U.S.P., 3 365 532, (*1968*); *CA*, **68**, 77731 (*synth*)
Shagidullin, R.R. *et al*, *Dokl. Akad. Nauk SSSR, Ser. Sci. Khim.*, 1975, **222**, 897 (*Engl. transl.* p. 564) (*uv*)
Shagidullin, R.R. *et al*, *Izv. Akad. Nauk SSSR, Ser. Khim.*, 1975, 1527 (*Engl. transl.* p. 1414) (*ir, raman, conformn*)

Isopropyl phosphorodichloridothioite, 8CI **I-00063**

1-Methylethyl phosphorodichloridothioite, 9CI. Isopropyl dichlorothiophosphite

[36696-25-0]

$(H_3C)_2CHSPCl_2$

$C_3H_7Cl_2PS$ M 177.028

Pungent, odorous liq. d_4^{20} 1.26. Bp_9 57-58°. n_D^{20} 1.5425. Slowly disproportionates to Diisopropyl phosphorochloridothioite, D-00660 and PCl_3.

Stepashkina, L.V. *et al*, *Izv. Akad. Nauk SSSR, Ser. Khim.*, 1972, 300 (*Engl. transl.* p. 330) (*synth, props*)

S-Isopropyl phosphorothioate, 8CI **I-00064**

S-*(1-Methylethyl) phosphorothioate, 9CI.* S-*Isopropyl dihydrogen thiophosphate.* S-*Isopropyl thiophosphoric acid*

$(H_3C)_2CHSP(O)(OH)_2$

$C_3H_9O_3PS$ M 156.136

Monodicyclohexylammonium salt: Solid. Mp 156°.

Dichloride: [26121-97-1]. S-*Isopropyl phosphorodichloridothioate.* S-*Isopropyl dichlorothiophosphate.*
$C_3H_7Cl_2OPS$ M 193.027
Liq. Bp_5 70-80°.

Kobayashi, K. *et al*, *CA*, 1970, **72**, 100196 (*dichloride*)
Zwierzak, A. *et al*, *Z. Naturforsch., B*, 1971, **26**, 386 (*synth, ir, pmr*)

Isopropyltriphenylphosphonium(1+), 8CI **I-00065**

(1-Methylethyl)triphenylphosphonium(1+), 9CI

$Ph_3P^{\oplus}CH(CH_3)_2$

$C_{21}H_{22}P^{\oplus}$ M 305.379 (ion)

With butyllithium, salts give the ylide.

Bromide: [1530-33-2].
$C_{21}H_{22}BrP$ M 385.283
Solid. Mp 239-240°.

Iodide: [24470-78-8].
$C_{21}H_{22}IP$ M 432.283

Cryst. ($EtOH/Et_2O$). Mp 195-196°.
Perchlorate:
$C_{21}H_{22}ClO_4P$ M 404.829
Cryst. (EtOH). Mp 182-183°.
Ylide: see (1-Methylethylidene)triphenylphosphorane, M-00153

Hendrickson, J. *et al*, Tet, 1964, **20**, 449 (*iodide, synth, pmr*)
Jaenicke, L. *et al*, *Justus Liebigs Ann. Chem.*, 1973, 1252 (*bromide*)
Albright, T.A. *et al*, *J. Am. Chem. Soc.*, 1975, **97**, 940, 2942; 1976, **98**, 6249 (*cmr, nmr*)
Reid, D.H. *et al*, *J. Chem. Soc., Perkin Trans. 1*, 1979, 2334 (*synth*)
Smith, K.M. *et al*, *J. Am. Chem. Soc.*, 1980, **102**, 2437 (*iodide*)

2-Isoselenocyanato-5,5-dimethyl-1,3,2-dioxaphosphorinane I-00066

[57436-71-2]

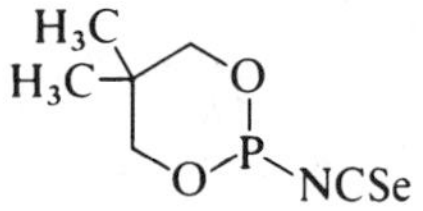

$C_6H_{10}NO_2PSe$ M 238.084
Unstable and readily isomerises on warming.
2-Oxide: [55379-57-2].
$C_6H_{10}NO_3PSe$ M 254.084
Cryst. (C_6H_6/heptane). Mp 64-67°.

Stec, W.J. *et al*, *J. Chem. Soc., Chem. Commun.*, 1974, 923 (*synth, deriv, ir, P nmr, props*)

Isothioate, BSI I-00067

O,O-*Dimethyl* S-[*2-[(1-methylethyl)thio]ethyl*] *phosphorodithioate*, 9CI. O,O-*Dimethyl* S-[*(2-isopropylthio)ethyl*] *phosphorodithioate*, 8CI
[36614-38-7]

$(MeO)_2P(S)SCH_2CH_2SCH(CH_3)_2$

$C_7H_{17}O_2PS_3$ M 260.364
Systemic insecticide. Light-yellow-brown liq. with sl. aromatic odour. d_4^{20} 1.18. $Bp_{0.01}$ 53-56°. n_D^{21} 1.5360.
▷Mod. toxic. TE4465000.

Nagasawa, T. *et al*, *Anal. Methods. Pestic. Plant Growth Regul.*, 1978, **10**, 75 (*rev*)
Pesticide Manual, 7th Ed., 330.

Isoxathion I-00068

O,O-*Diethyl* O-(*5-phenyl-3-isoxazolyl*) *phosphorothioate*, 9CI. *Karphos*
[18854-01-8]

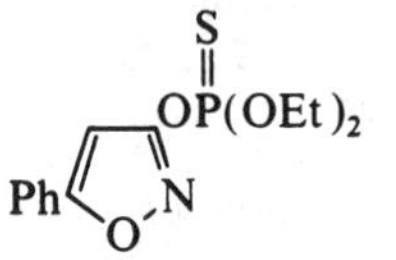

$C_{13}H_{16}NO_4PS$ M 313.307
Broad spectrum non-systemic contact insecticide. Molluscicide. Yellowish liq. $Bp_{0.2}$ 160° (bath). n_D^{18} 1.5497. Dec. by alkali.
▷Moderately toxic. TF5600000.

B.P., 1 115 585, (*1965*); *CA*, **69**, 36110z (*synth*)
Ando, K., *Jpn. Pestic. Inf.*, 1978, 27 (*rev*)
Nakamura, T. *et al*, *Anal. Methods Pestic. Plant Growth Regul.*, 1978, **10**, 83 (*rev*)
Pesticide Manual, 6th Ed., 188; 7th Ed., 331.

K

Kitazin **K-00001**

O,O-*Diethyl* S-*(phenylmethyl) phosphorothioate*, 9CI. S-*Benzyl* O,O-*diethyl phosphorothioate*, 8CI

[13286-32-3]

$$PhCH_2SP(=O)(OEt)_2$$

$C_{11}H_{17}O_3PS$ M 260.287

Contact insecticide and fungicide. d_4^{20} 1.16. Bp_3 140-141°, $Bp_{0.02}$ 110-114°. n_D^{20} 1.5275.

▷TE6500000.

Kabachnik, M.I. *et al*, *Zh. Obshch. Khim.*, 1955, **25**, 1924 (*Engl. transl.* p. 1867) (*synth*)

Cadogan, J.I.G. *et al*, *J. Chem. Soc.*, 1961, 5524 (*synth*)

Meyer, H.J. *et al*, *Bull. Soc. Chim. Belg.*, 1978, **87**, 517 (*ms*)

Torii, S. *et al*, *J. Org. Chem.*, 1979, **44**, 2938 (*synth, ir, pmr*)

L

Leptophos, BSI — L-00001

O-4-Bromo-2,5-dichlorophenyl O-methyl phenylphosphonothioate, 9CI. *Phosvel. Abar*
[21609-90-5]

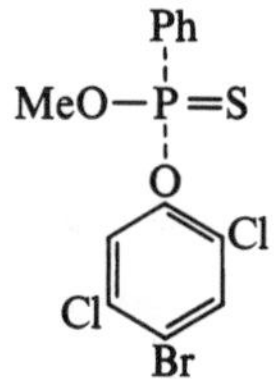

(R)-form

$C_{13}H_{10}BrCl_2O_2PS$ M 412.065

▷Highly toxic. TB1720000.

(R)-form
Mp 88-90°. $[\alpha]_D^{24}$ +46.13° (c, 10.6 in C_6H_6). More toxic (housefly, mouse) than (*S*)-form.

(S)-form
Mp 88-90°. $[\alpha]_D^{24}$ −46.26° (c, 10.0 in C_6H_6) (98% o.p.). Shows more delayed neurotoxic effect (hen) than (*R*)-form.

(±)-form
Agricultural insecticide, now superseded. Cryst. (EtOH). V. sol. hexane, Me_2CO, v. spar. sol. H_2O. Mp 71-72°. Stable at pH 1-7, dec. at pH >7. Degraded by uv.

▷Exhibits delayed neurotoxicity; causes irreversible ataxia

B.P., 1 197 111, (*1965*); *CA*, **71**, 81519s
Holmstead, R.L. *et al*, *Arch. Environ. Contam. Toxicol.*, 1973, **1**, 133 (*metabolism, tlc, ms*)
Steurbaut, W. *et al*, *Bull. Soc. Chim. Belges*, 1975, **84**, 791 (*synth, ir, pmr*)
Lee, R.W. *et al*, *Arch. Environ. Contam. Toxicol.*, 1976, **4**, 443 (*metab*)
Busch, K.L. *et al*, *Appl. Spectrosc.*, 1978, **32**, 388 (*ms*)
Hollingshaus, J.G. *et al*, *J. Agric. Food Chem.*, 1979, **27**, 1197 (*tox*)
Allahyari, R. *et al*, *J. Agric. Food Chem.*, 1980, **28**, 594 (*enantiomers, synth, tox*)
Lapp, R.L. *et al*, *Cryst. Struct. Commun.*, 1980, **9**, 65 (*cryst struct*)
Zayad, S.M.A.D. *et al*, *J. Labelled Comp. Radiopharm.*, 1981, **18**, 521 (*synth*)
Francis, B.M. *et al*, *J. Environ. Sci. Health*, 1982, **17**, 611 (*tox*)
Pesticide Manual, 6th Ed., 318.
Sax, N.I., *Dangerous Properties of Industrial Materials*, 6th Ed., Van Nostrand-Reinhold, 1984, 558.

Lombricine — L-00002

Serine 2-[(aminoiminomethyl)amino]ethyl hydrogen phosphate (ester), 9CI. *O^3-(2-Guanidinoethoxyphosphinicoserine)*

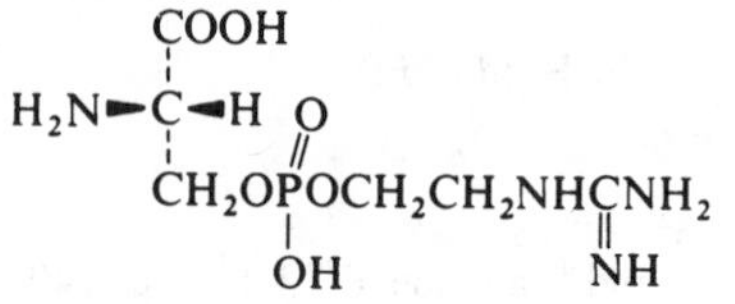

(S)-form

$C_6H_{15}N_4O_6P$ M 270.181

(R)-form
Constit. of *Ophelia bicornis* muscle. Mp 229-230°. $[\alpha]_D^{25.4}$ +14.4° (c, 0.917 in H_2O).

(S)-form [18416-85-8]
Constit. of the body wall muscle of the echiuroid worm *Thalassema neptuni*. Cryst. (EtOH aq.). Mp 233°. $[\alpha]_D^{25}$ −11.3° (c, 0.380 in H_2O).

α,N,N-*Di-Me:* [40524-74-1]. *Thalassemine.*
$C_8H_9N_4O_6P$ M 288.156
Constit. of *T.* viscera. Cryst. + $1H_2O$ (H_2O). Mp 184°. $[\alpha]_D^{26}$ −11.3° (c, 0.919 in H_2O).

Thoai, N.-V. *et al*, *Biochim. Biophys. Acta*, 1954, **14**, 76 (*isol*)
Pant, R., *Biochem. J.*, 1959, **73**, 30 (*isol*)
Dubey, S.S., *Indian J. Chem.*, 1963, **1**, 453 (*isol*)
Thoai, N.-V. *et al*, *Biochemistry*, 1972, **11**, 3890 (*isol*)

M

Malathion, BSI, ISO **M-00001**

[(Dimethoxyphosphinothioyl)thio]butanedioic acid diethyl ester, 9CI. S-*(1,2-Dicarbethoxyethyl)* O,O-*dimethyl dithiophosphate*

[121-75-5]

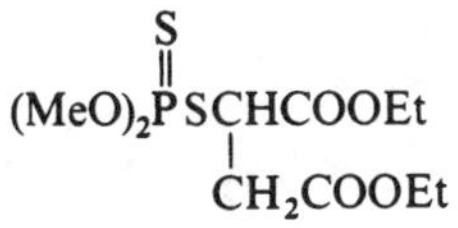

$$(MeO)_2\overset{S}{\overset{\|}{P}}SCH(COOEt)CH_2COOEt$$

$C_{10}H_{19}O_6PS_2$ M 330.350

▷WM8400000.

(±)-***form***

Wide range insecticide. Yellow liq. with characteristic odour. V. spar. sol. H_2O. d_4^{25} 1.23. Mp −7°. $Bp_{0.7}$ 156-157° sl. dec. n_D^{25} 1.4985. Low acute toxicity.

▷Exp. carcinogen, TLV 10

Cassaday, J.T. *et al*, *J. Am. Chem. Soc.*, 1951, **73**, 557 (*synth*)
Melnikov, N.N. *et al*, *Zh. Obshch. Khim.*, 1953, **23**, 1352 (*synth*)
Keith, L.H. *et al*, *J. Assoc. Off. Anal. Chem.*, 1968, **51**, 1063 (*pmr*)
Ackerman, H., *J. Chromatogr.*, 1969, **44**, 414 (*tlc*)
Gaines, T.B., *Toxicol. Appl. Pharmacol.*, 1969, **14**, 515 (*tox*)
Ross, R.T. *et al*, *Anal. Chim. Acta*, 1970, **52**, 139 (*P nmr*)
Gore, R.C. *et al*, *J. Assoc. Off. Anal. Chem.*, 1971, **54**, 1040 (*ir, uv*)
Holmstead, R.L. *et al*, *J. Assoc. Off. Anal. Chem.*, 1974, **57**, 1050 (*ms, metab*)
Nicholas, M.L. *et al*, *J. Assoc. Off. Anal. Chem.*, 1976, **59**, 1071 (*raman*)
Szalontai, G., *J. Chromatogr.*, 1976, **124**, 9 (*hplc*)
Stan, H.-J. *et al*, *Fresenius' Z. Anal. Chem.*, 1977, **287**, 271 (*glc, ms, metab*)
Busch, K.L. *et al*, *Appl. Spectrosc.*, 1978, **32**, 388 (*ms*)
Mulla, M.S. *et al*, *Residue Rev.*, 1981, **81**, 1 (*metab, rev*)
Stan, H.-J. *et al*, *J. Chromatogr.*, 1983, **279**, 173 (*glc*)
Agrochemicals Handbook, Royal Society of Chemistry, London, 1983, A249.
Ding, X.D. *et al*, *J. Agric. Food Chem.*, 1984, **32**, 622 (*hplc*)
Pesticide Manual, 6th Ed., 321; 7th Ed., 337.
Sax, N.I., *Dangerous Properties of Industrial Materials*, 6th Ed., Van Nostrand-Reinhold, 1984, 604.

Mecarbam, BSI **M-00002**

Ethyl 6-ethoxy-2-methyl-7-oxa-5-thia-2-aza-9-phosphanonanoate 6-sulfide, 9CI. S-(N-*Ethoxycarbonyl*-N-*methylcarbamoylmethyl)* O,O-*diethyl phosphorodithioate. Ethyl [[(diethoxyphosphinothioyl)thio]acetyl] methylcarbamate. Murfotox. Pestan*

[2595-54-2]

$$(EtO)_2P(S)SCH_2CONMeCOOEt$$

$C_{10}H_{20}NO_5PS_2$ M 329.365

Agricultural insecticide and acaricide with sl. systemic props. Pale-yellow oil. Spar. sol. H_2O. d_{20}^{20} 1.223. $Bp_{0.02}$ 144°. n_D^{20} 1.5138.

▷High oral tox. FB3850000.

B.P., 867 780, (*1957*); *CA*, **55**, 25756h (*synth, use*)
Krijgsman, W. *et al*, *J. Chromatogr.*, 1976, **117**, 201 (*glc*)
Lynch, V.P. *et al*, *Pestic. Sci.*, 1981, **12**, 65 (*synth, props*)
Charalambous, J. *et al*, *Phosphorus Sulfur*, 1984, **19**, 267 (*ms*)
Pesticide Manual, 6th Ed., 328; 7th Ed., 344.

Menazon, BSI **M-00003**

O,O′-*Dimethyl* S-*[(4,6-diamino-1,3,5-triazin-2-yl)methyl] phosphorodithioate, 9CI.* S-*[(4,6-Diamino-1,3,5-triazin-2-yl)methyl]* O,O-*dimethyl phosphorodithioate.* S-*[(4,6-Diamino-*sym*-triazin-2-yl)methyl]* O,O′-*dimethyl dithiophosphate. PP 175. Saphicol. Azadithion*

[78-57-9]

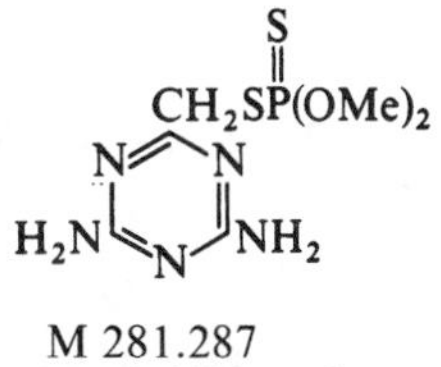

$C_6H_{12}N_5O_2PS_2$ M 281.287

Pesticide; cholinesterase inhibitor. Systemic aphicide, now superseded. Cryst. (MeOH). Spar. sol. H_2O, most org. solvs., sol. THF, 2-methoxyethanol. Mp 160-162° dec. Dec. in aq. acid soln.

▷TD5600000.

Calderbank, A. *et al*, *Chem. Ind. (London)*, 1961, 630 (*pharmacol, uv, ir*)
Calderbank, A., *J. Chem. Soc. (C)*, 1966, 56 (*synth*)
Krijgsman, W. *et al*, *J. Chromatogr.*, 1976, **117**, 201 (*glc*)
Pesticide Manual, 6th Ed., 331.

Mephosfolan, BSI **M-00004**

Diethyl (4-methyl-1,3-dithiolan-2-ylidene)phosphoramidate, 9CI. PP-*Diethyl cyclic propylene phosphonodithioimidocarbonate, 8CI. Cytrolane*

[950-10-7]

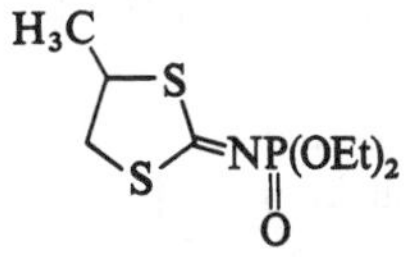

$C_8H_{16}NO_3PS_2$ M 269.313

▷JP1050000.

(±)-***form***

Agricultural insecticide with contact and stomach action. Yellow liq. Sol. Me_2CO, EtOH, C_6H_6, CH_2Cl_2, sl. sol. H_2O. $Bp_{0.001}$ 120°. n_D^{26} 1.539.

▷V. toxic on oral admin.

B.P., 974 138, (*1961*); *CA*, **59**, 10066f (*synth*)
Ku, C.C. *et al*, *J. Agric. Food Chem.*, 1979, **27**, 1046 (*props*)
Ku, C.C. *et al*, *ACS Symp. Ser.*, 1981, **158**, 97 (*metab*)
Abou-Donia, S.A. *et al*, *J. Chromatogr.*, 1982, **240**, 532 (*hplc*)
Pesticide Manual, 6th Ed., 332; 7th Ed., 347.

2-Mercapto-1,3,2-benzodioxaphosphole 2-sulfide, 9CI **M-00005**

O,O-o-*Phenylene dithiophosphoric acid.* O,O-o-*Phenylene hydrogen dithiophosphate.* O,O-o-*Phenylene phosphorothioic acid*

[50577-97-4]

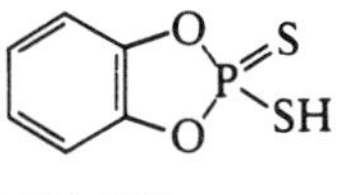

$C_6H_5O_2PS_2$ M 204.198

Oil.

Triethylammonium salt: [31202-14-9]. Solid. Mp 137-138°.

2-Methylthio (Me ester):
$C_7H_7O_2PS_2$ M 218.225
Solid. Mp 49°.

2-Ethylthio (Et ester):
$C_8H_9O_2PS_2$ M 232.252
Solid. Mp 28-29°.

Zemlyanskii, N.I. *et al*, *Zh. Obshch. Khim.*, 1969, **39**, 1591 (*Engl. transl.* p. 1559) (*ester*)
Kalashnikov, V.P., *Zh. Obshch. Khim.*, 1970, **40**, 1954 (*Engl. transl.* p. 1939) (*derivs, ir*)
Micu-Semeniac, R. *et al*, *Inorg. Chim. Acta*, 1979, **33**, 281 (*synth, ir, complex*)

2-Mercapto-4*H*-1,3,2-benzodioxaphosphorin 2-sulfide M-00006

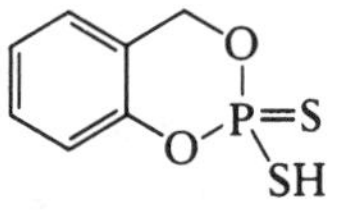

$C_7H_7O_2PS_2$ M 218.225

2-Methylthio (2-Me ester): see under 2-Methylthio-4H-1,3,2-benzodioxaphosphorin, M-00411

2-Ethylthio (2-Et ester):
$C_9H_{11}O_2PS_2$ M 246.278
Oil. $Bp_{0.2}$ 145-147°. n_D^{25} 1.6221.

2-Isopropylthio (2-Isopropyl ester):
$C_{10}H_{13}O_2PS_2$ M 260.305
Oil. $Bp_{0.1}$ 140-143°. n_D^{25} 1.5920.

2-Phenylthio (2-Ph ester):
$C_{13}H_{11}O_2PS_2$ M 294.322
Solid. Mp 79-80°.

Kobayashi, K. *et al*, *J. Agric. Chem. Soc. Jpn.*, 1966, **40**, 315; *CA*, **66**, 10883 (*synth*)
Netherlands Pat., 6 615 246, (*1967*); *CA*, **67**, 116710

2-Mercaptodibenzo[*d,f*][1,3,2]dioxaphosphepin 2-oxide, 10CI M-00007

2-Hydroxydibenzo[d,f] [*1,3,2*]*dioxaphosphepin 2-sulfide.* O,O-*(2,2′-Biphenylene) cyclic phosphorothioate.* O,O-*(2,2′-Biphenylene) hydrogen thiophosphate*

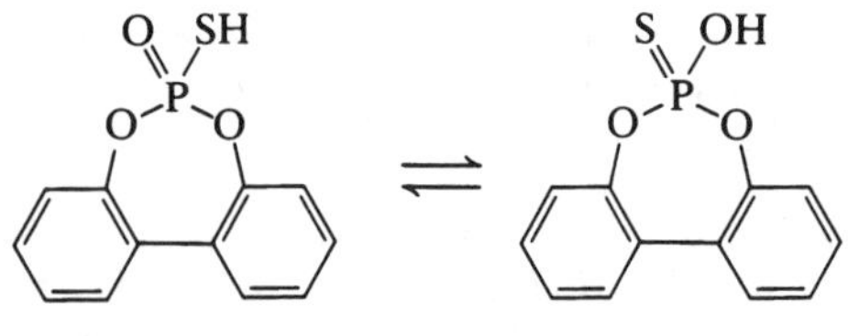

$C_{12}H_9O_3PS$ M 264.235

Tautomeric. Viscous, hygroscopic oil.

Na salt: Cryst. + $1H_2O$. Mp 250-252°.

Ag salt: Solid. Mp 309°.

Triethylammonium salt: Needles (EtOH). Mp 197-198°.

NH_4 salt: [76045-08-4]. Solid. Sol. H_2O, alcohols. Mp 224-225°.

Keck, H. *et al*, *Org. Mass Spectrom.*, 1980, **15**, 591 (*ms*)
Kuchen, W. *et al*, *Phosphorus Sulfur*, 1980, **8**, 139 (*synth, derivs, pmr, P nmr*)

6-Mercaptodibenzo[*d,f*][1,3,2]dioxaphosphepin 6-sulfide, 9CI M-00008

O,O-*2,2′-Biphenylyl cyclic dithiophosphate.* O,O-*2,2′-Biphenylyl hydrogen phosphorodithioate*

[68560-86-1]

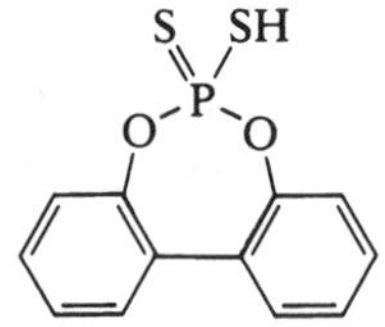

$C_{12}H_9O_2PS_2$ M 280.296

Solid. Sol. H_2O. Mp 104-105°. Forms Ni complex.

NH_4 salt: [68560-85-0]. Cryst. ($MeOH/C_6H_6$/pet. ether). Mp 275-277°.

Kuchen, W. *et al*, *Z. Naturforsch, B*, 1978, **33**, 1049 (*synth, P nmr, complex*)
Keck, H. *et al*, *Org. Mass. Spectrom.*, 1980, **15**, 591 (*ms*)

2-Mercapto-4,5-dimethyl-1,3,2-dioxaphospholane 2-sulfide, 9CI M-00009

O,O-*2,3-Butylene hydrogen dithiophosphate.* O,O-*2,3-Butylene dithiophosphoric acid.* O,O-*2,3-Butylene phosphorodithioate*

[695-68-1]

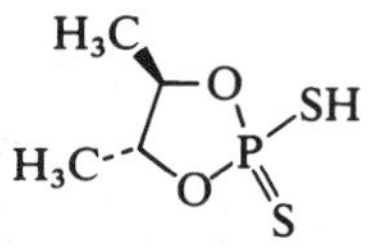

(4*R*,5*R*)*-form*

$C_4H_9O_2PS_2$ M 184.208

Oil, malodorous when impure. d_4^{20} 1.29. $Bp_{0.2}$ 97-98°. pK_a 2.85 (C_6H_6). n_D^{20} 1.5481. Mixt. of diastereoisomers.

NH_4 salt: Solid which could not be cryst.

(4R,5R)-form

(−)*-4,5-*trans*-form*

K salt: Solid. Mp 288-292°. $[\alpha]_D$ −43.3° (c 0.6 in MeCN).

Tetraethylammonium salt: Solid. Mp 178-192°. $[\alpha]_D$ −4.4° (c, 1.2 in MeCN).

(4RS,5RS)-form

(±)*-4,5-*trans*-form*

K salt: Solid. Mp 288-290°.

Tetraethylammonium salt: Solid. Mp 170-174°.

Edmundson, R.S. *et al*, *J. Chem. Soc. (C)*, 1966, 1997 (*synth*)
Pudovik, A.N. *et al*, *Zh. Obshch. Khim.*, 1968, **38**, 307 (*Engl. transl.* p. 308) (*synth*)
Ovchinnikov, V.V. *et al*, *Zh. Obshch. Khim.*, 1979, **49**, 1693 (*Engl. transl.* p. 1482) (*synth*)
Biscarini, P. *et al*, *Inorg. Chim. Acta*, 1983, **74**, 65 (*synth, ir, pmr, P nmr, complexes*)
Chauhan, H.P.S. *et al*, *Phosphorus Sulfur*, 1983, **15**, 99 (*synth, ir, pmr*)

2-Mercapto-5,5-dimethyl-1,3,2-dioxaphosphorinane 2-oxide M-00010

O,O-*Neopentylene cyclic phosphorothioic acid. 2-Mercapto-5,5-dimethyl-1,3-dioxa-2-oxo-2-phosphacyclohexane.* O,O-*Neopentylene hydrogen thiophosphate*

[45734-11-0]

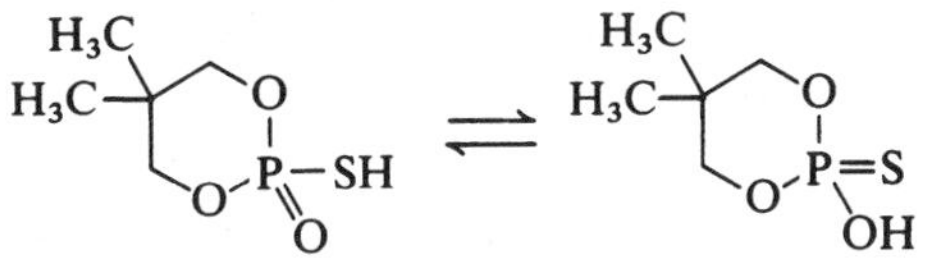

$C_5H_{11}O_3PS$ M 182.174

Tautomeric. Cryst. (C_6H_6/pet. ether). Mp 74-75°. pK_a 1.43 (7% EtOH aq., 99.5% thiol form), 2.49 (80% EtOH aq., 99.8% thiol form).

Triethylammonium salt: [4090-61-3]. Cryst. (EtOAc). Mp 70.5-72.5°.

Cyclohexylammonium salt: [4073-55-6]. Needles (2-propanol/pet. ether). Mp 238-242°.

Methyltriethylammonium salt: [75716-68-6]. Cryst. Mp 74.5-75°.

Anhydride: see 2,2′-Oxybis[5,5-dimethyl-1,3,2-dioxaphosphorinane], O-00089

Anhydrosulfide: see 2,2′-Thiobis[5,5-dimethyl-1,3,2-dioxaphosphorinane] 2,2′-dioxide, T-00310

2-Methoxy (O-*Me ester*)*: see under 2-Methoxy-5,5-dimethyl-1,3,2-dioxaphosphorinane, M-00034*

2-Methylthio (S-*Me ester*)*: see under 5,5-Dimethyl-2-methylthio-1,3,2-dioxaphosphorinane, D-00766*

2-Ethoxy (O-*Et ester*)*: see under 2-Ethoxy-5,5-dimethyl-1,3,2-dioxaphosphorinane, E-00032*

2-Ethylthio (S-*Et ester*)*: see under 2-Ethylthio-5,5-dimethyl-1,3,2-dioxaphosphorinane, E-00190*

2-Phenoxy (O-*Ph ester*)*: see under 5,5-Dimethyl-2-phenoxy-1,3,2-dioxaphosphorinane, D-00790*

Bromide: see under 2-Bromo-5,5-dimethyl-1,3,2-dioxaphosphorinane, B-00471

Iodide: see under 2-Iodo-5,5-dimethyl-1,3,2-dioxaphosphorinane, I-00010

Dimethylamide: see 2-Dimethylamino-5,5-dimethyl-1,3,2-dioxaphosphorinane, D-00693

tert-*Butylamide: see under 2-tert-*Butylamino-5,5-dimethyl-1,3,2-dioxaphosphorinane, B-00538

Anilide: see under 5,5-Dimethyl-2-phenylamino-1,3,2-dioxaphosphorinane, D-00792

Edmundson, R.S., *Tetrahedron*, 1965, **21**, 2379 (*synth*)
Stec, W.J. *et al*, *J. Phys. Chem.*, 1971, **75**, 3975 (*pe, P nmr*)
Ovchinnikov, V.V. *et al*, *Zh. Obshch. Khim.*, 1977, **47**, 290 (*Engl. transl.* p. 267) (*synth, ir, P nmr, props*)
Bruzik, K. *et al*, *J. Org. Chem.*, 1981, **46**, 1618, 1625 (*synth, P nmr*)
Frey, P.A. *et al*, *J. Am. Chem. Soc.*, 1986, **108**, 1720 (*P nmr, struct*)

2-Mercapto-5,5-dimethyl-1,3,2-dioxaphosphorinane 2-sulfide, 9CI M-00011

2-Mercapto-5,5-dimethyl-1,3,2-dioxaphosphinane 2-sulfide. O,O-*Neopentylene phosphorodithioate.* O,O-*Neopentylene hydrogen dithiophosphate. 2-Mercapto-5,5-dimethyl-2-thioxo-1,3-dioxa-2-phosphacyclohexane*

[697-45-0]

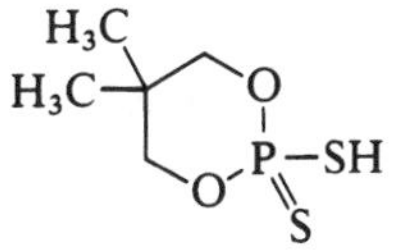

$C_5H_{11}O_2PS_2$ M 198.234

Cryst. (Et_2O or pet. ether). Mp 81-82°.

Na salt: Solid. Mp 103-105°.

Cyclohexylammonium salt: Cryst. ($CHCl_3$/pet. ether). Mp 216-217°.

Anhydrosulfide: see under 2,2′-Thiobis[5,5-dimethyl-1,3,2-dioxaphosphorinane], T-00309

2-Methylthio (S-*Me ester*)*: see under 5,5-Dimethyl-2-methylthio-1,3,2-dioxaphosphorinane, D-00766*

2-Ethylthio (S-*Et ester*)*: see under 2-Ethylthio-5,5-dimethyl-1,3,2-dioxaphosphorinane, E-00190*

Edmundson, R.S., *Tetrahedron*, 1965, **21**, 2379 (*synth*)
Bartle, K.D. *et al*, *Tetrahedron*, 1967, **23**, 1701 (*pmr*)
Brault, J.-F. *et al*, *Bull. Soc. Chim. Fr., Part 2*, 1973, 3149 (*pmr, P nmr*)
Edmundson, R.S., *Phosphorus Sulfur*, 1981, **9**, 307 (*ms*)
Samitov, Yu.Yu. *et al*, *Zh. Obshch. Khim.*, 1981, **51**, 711 (*Engl. transl.* p. 577) (*cmr, P nmr*)

4-Mercaptodinaphtho[2,1-*d*:1′,2′-*f*][1,3,2]dioxaphosphepin 4-sulfide, 9CI M-00012

O,O-*1,1′-Binaphthylene hydrogen dithiophosphate.* O,O-*1,1′-Binaphthylene phosphorodithioate*

[188-36-3]

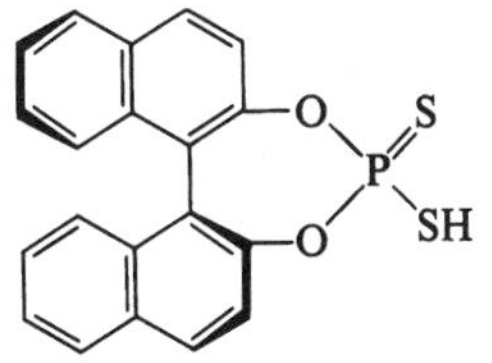

(*S*)-*form*

$C_{20}H_{13}O_2PS_2$ M 380.415

(*S*)-*form* [70144-34-2]

Forms a (−)-Ni complex.

(+)-*1-Phenylethylammonium salt:* [70144-36-4]. Cryst. (2-propanol, aq.). $[\alpha]^{25}_{378}$+1404° (c, 0.01 in EtOH).

(±)-*form* [70101-69-8]

Cryst. (toluene/hexane). Mp 242-243°.

Hoffmann, E.W. *et al*, *Angew. Chem., Int. Ed. Engl.*, 1979, **18**, 415 (*synth, resoln*)
Poll, W. *et al*, *Acta Crystallogr., Sect. B*, 1980, **36**, 1191 (*complex, cryst struct*)

2-Mercapto-1,3,2-dioxaphosphorinane 2-sulfide, 9CI **M-00013**

2-Mercapto-1,3,2-dioxaphosphinane 2-sulfide. O,O-*Trimethylene phosphorodithioate.* O,O-*Trimethylene hydrogen dithiophosphate.* O,O-*Trimethylene dithiophosphoric acid*

[55055-14-6]

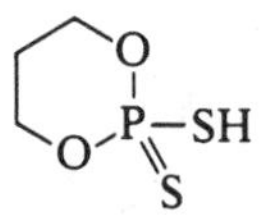

$C_3H_7O_2PS_2$ M 170.181

K salt: [85556-99-6]. Solid. Dec. >200°.

Triethylammonium salt: [55055-15-7]. Cryst. (Me_2CO or dioxan). Mp 81-82°, Mp 99-100°.

2-Methylthio (2-Me ester): 2-Methylthio-1,3,2-dioxaphosphorinane 2-sulfide. O,O-*Trimethylene* S-*methyl phosphorodithioate.*
$C_4H_9O_2PS_2$ M 184.208
Solid. Mp 44-45°. $Bp_{0.008}$ 85-90° (bath).

Zemlyanskii, N.I. *et al*, *Zh. Obshch. Khim.*, 1972, **42**, 1647 (*Engl. transl.* p. 1639) (*synth*)
Predvoditelev, D.A. *et al*, *Zh. Obshch. Khim.*, 1974, **44**, 2629 (*Engl. transl.* p. 2586) (*derivs, synth, ir, pmr, P nmr*)
Gilbert, B.C. *et al*, *J. Chem. Soc., Perkin Trans. 2*, 1984, 629 (*esr*)

2-Mercapto-1,3,2-dithiaphosphorinane 2-sulfide, 9CI **M-00014**

Trimethylene phosphorotetrathioate

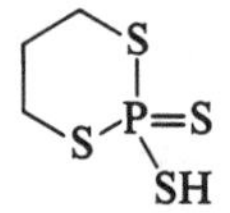

$C_3H_7PS_4$ M 202.302

Triethylammonium salt: [37505-84-3]. Cryst. (Me_2CO). Mp 99-100°.

2-Methylthio (Me ester): [37505-86-5]. *Methyl trimethylene phosphorotetrathioate.*
$C_4H_9PS_4$ M 216.329
Cryst. (pet. ether). Mp 64-65°.

Zemlyanskii, N.I. *et al*, *Zh. Obshch. Khim.*, 1972, **42**, 1647 (*Engl. transl.* p. 1639)

2-Mercapto-4-methyl-1,3,2-dioxaphospholane 2-oxide, 9CI **M-00015**

2-Hydroxy-4-methyl-1,3,2-dioxaphospholane 2-sulfide. Propylene cyclic thiophosphoric acid. O,O-*1,2-Propylene hydrogen thiophosphate*

[78928-19-5]

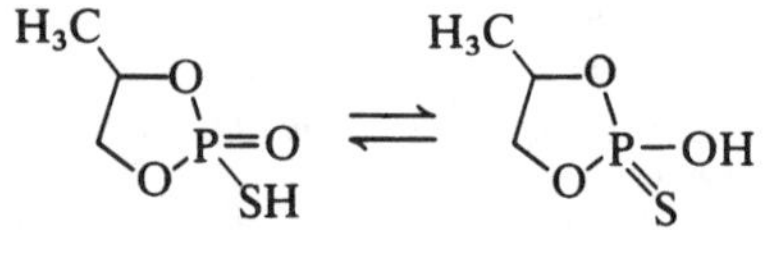

$C_3H_7O_3PS$ M 154.120

Tautomeric.

Tetramethylammonium salt: Cryst. (propanol/Et_2O). Mp 177-180°. Characterised spectroscopically as *cis/trans* mixt.

1H-Imidazolium salt: Characterised spectroscopically as *cis/trans* mixt.

2-Methoxy (O-Me ester): see 2-Methoxy-4-methyl-1,3,2-dioxaphospholane, M-00043

2-Methylthio (S-Me ester): see 4-Methyl-2-methylthio-1,3,2-dioxaphospholane, M-00173

2-Ethoxy (O-Et ester): see 2-Ethoxy-1,3,2-dioxaphospholane 2-sulfide, E-00036

Chloride: see 2-Chloro-4-methyl-1,3,2-dioxaphospholane, C-00092

Dimethylamide: see 4-Methyl-2-dimethylamino-1,3,2-dioxaphospholane, M-00116

Mikolajczyk, M. *et al*, *J. Chem. Soc., Perkin Trans. 1*, 1976, 371 (*synth, pmr, P nmr*)

2-Mercapto-4-methyl-1,3,2-dioxaphospholane 2-sulfide **M-00016**

O,O-*Propylene dithiophosphate.* O,O-*Propylene hydrogen dithiophosphate*

[17043-52-6]

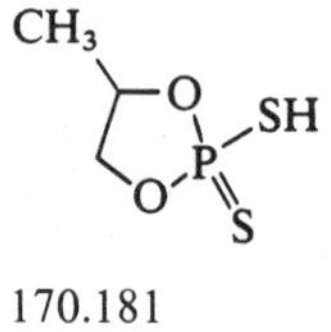

$C_3H_7O_2PS_2$ M 170.181

(±)-form

Liq., malodorous if impure. d_4^{20} 1.38. $Bp_{0.4}$ 96-96.5°. n_D^{20} 1.5610.

Na salt: Solid. Mp 85-87°.

2-Methylthio (2-Me ester): see 4-Methyl-2-methylthio-1,3,2-dioxaphospholane, M-00173

2-Benzylthio (2-Benzyl ester): S-*Benzyl* O,O-*propylene phosphorodithioate.* S-*Benzyl* O,O-*propylene dithiophosphate.*
$C_{10}H_{13}O_2PS_2$ M 260.305
Liq. d_4^{20} 1.28. $Bp_{0.4}$ 157-158°. n_D^{20} 1.5974.

Pudovik, A.N. *et al*, *Zh. Obshch. Khim.*, 1968, **38**, 307 (*Engl. transl.* p. 308) (*synth*)
Cherkasov, R.A. *et al*, *Zh. Obshch. Khim.*, 1976, **46**, 957 (*Engl. transl.* p. 956) (*synth*)

2-Mercapto-4-methyl-1,3,2-dioxaphosphorinane 2-sulfide **M-00017**

O,O-*1,3-Butylene hydrogen dithiophosphate.* O,O-*1,3-Butylene dithiophosphoric acid. 2-Mercapto-4-methyl-1,3,2-dioxaphosphinane 2-sulfide. 2-Mercapto-4-methyl-1,3-dioxa-2-thioxo-2-phosphacyclohexane*

[20639-78-5]

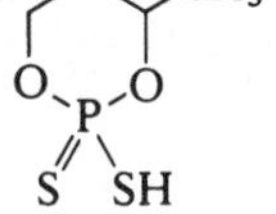

$C_4H_9O_2PS_2$ M 184.208

Oil. d_4^{20} 1.33. $Bp_{0.4}$ 130-131°. n_D^{20} 1.5669.

Diethylammonium salt: Cryst. (EtOH). Mp 130-132°.

Triethylammonium salt: [55055-16-8]. Solid. Mp 85-86°.

2-Methylthio (Me ester): see under 4-Methyl-2-methylthio-1,3,2-dioxaphosphorinane, M-00174

Pudovik, A.N. *et al*, *Zh. Obshch. Khim.*, 1968, **38**, 307 (*Engl. transl.* p. 308) (*synth*)
Predvoditelev, D.A. *et al*, *Zh. Obshch. Khim.*, 1974, **44**, 2629 (*Engl. transl.* p. 2586) (*derivs, synth, tlc, pmr*)

(Mercaptomethyl)phosphonic acid, 9CI **M-00018**

[36120-86-2]

$HSCH_2P(O)(OH)_2$

CH_5O_3PS M 128.082

Di-Et ester: [70660-05-8]. *Diethyl (mercaptomethyl)-phosphonate.*

$C_5H_{13}O_3PS$ M 184.190

Liq. n_D^{24} 1.4658.

S-*Et: see* [(*Ethylthio*)*methyl*]*phosphonic acid, E-00192*

S-*Ph: see* [(*Phenylthio*)*methyl*]*phosphonic acid, P-00326*

S-*Benzyl: see* [(*Benzylthio*)*methyl*]*phosphonic acid, B-00072*

Ivasyuk, N.V. *et al, Zh. Obshch. Khim.*, 1971, **41**, 2199 (*Engl. transl.* p. 2224) (*synth*)
Mikolajczyk, M. *et al, J. Org. Chem.*, 1979, **44**, 2967 (*synth, pmr, P nmr*)

2-Mercapto-4,4,5,5-tetramethyl-1,3,2-dioxaphospholane 2-sulfide, 9CI M-00019

O,O-*Tetramethylethylene hydrogen phosphorodithioate.* O,O-*Tetramethylethylene dithiophosphate. Pinacol cyclic dithiophosphoric acid*

[699-36-5]

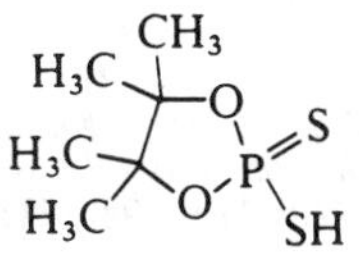

$C_6H_{13}O_2PS_2$ M 212.261

Cryst. (2-propanol). Mp 67.5-68.5°. pK_a 2.63 (80% EtOH aq.).

Na salt: [59825-34-2]. Solid. Mp 110-111°.

NH_4 *salt:* [86428-81-1]. Cryst. (EtOAc). Mp 229.5-230°.

Et ester: see N,N′,N″,N‴- Tetrakis(dihydroxyphosphinylmethyl)-1,4,7,10-tetraazacyclododecane, T-00175

Edmundson, R.S. *et al, J. Chem. Soc.* (*C*), 1966, 1997 (*synth*)
Pudovik, A.N. *et al, Zh. Obshch. Khim.*, 1972, **42**, 2638 (*Engl. transl.* p. 2629) (*ir, props*)
Cherkasov, R.A. *et al, Zh. Obshch. Khim.*, 1976, **46**, 957 (*Engl. transl.* p. 956)
Ovchinnikov, V.V. *et al, Zh. Obshch. Khim.*, 1979, **49**, 1693 (*Engl. transl.* p. 1482) (*synth*)

2-Mercapto-1,3,5,2-triazaphosphorine-4,6(1*H*,5*H*)-dione 2-sulfide, 9CI M-00020

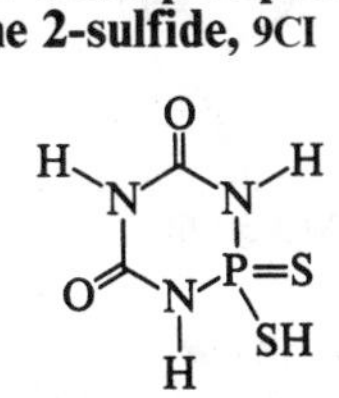

$C_2H_4N_3O_2PS_2$ M 197.166

Aurivillius, B. *et al, Acta Chem. Scand., Ser. A*, 1975, **29**, 717 (*cryst struct*)

Methamidophos, BSI M-00021

O,S-*Dimethyl phosphoramidothioate, 9CI.* O,S-*Dimethyl amidothiophosphate. Tamaron*

[10265-92-6]

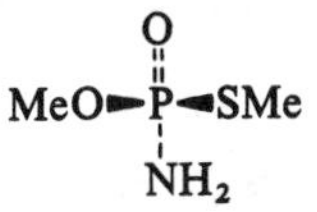

(*R*)-*form*

$C_2H_8NO_2PS$ M 141.124

▷Exhibits high oral tox. Cholinesterase inhibitor. TB4970000.

(*R*)-*form* [65960-95-4]

$[\alpha]_D$ +24° (c, 0.5 in $CHCl_3$).

(±)-*form*

Agricultural systemic insecticide and acaricide. Metab. of Acephate, A-00003 . Needles (ligroin). Sol. H_2O, EtOH, mod. sol. $CHCl_3$, C_6H_6, Et_2O. Mp 44.5°.

Holmstead, R.L. *et al, J. Assoc. Off. Anal. Chem.*, 1974, **57**, 1050 (*ms*)
Lubkowitz, J.A. *et al, J. Agric. Food Chem.*, 1974, **22**, 151 (*synth, ms, pmr*)
Hall, C.R. *et al, J. Chem. Soc., Perkin Trans. 1*, 1979, 1646 (*stereochem*)
Robinson, C.P. *et al, J. Appl. Toxicol.*, 1982, **2**, 217 (*tox*)
Schneider, P. *et al, J. Prakt. Chem.*, 1982, **324**, 1063 (*props*)
Thompson, C.M. *et al, J. Agric. Food Chem.*, 1982, **30**, 282; 1983, **31**, 696 (*synth, pmr, cmr, tox*)
Ripley, B.D. *et al, J. Assoc. Off. Anal. Chem.*, 1983, **66**, 1084 (*glc*)
Agrochemicals Handbook, Royal Society of Chemistry, London, 1983, A265.
Pesticide Manual, 6th Ed., 340; 7th Ed., 359.

Methanediphosphonic acid M-00022

Methylenebisphosphonic acid, 9CI. Methylenediphosphonic acid, 8CI. Medronic acid, BAN, USAN

[1984-15-2]

$(HO)_2P(O)CH_2P(O)(OH)_2$

$CH_6O_6P_2$ M 176.002

Complexing agent, pharmaceutical aid. Used for prevention of dental caries. Cryst. (2-methyl-2-propanol or AcOH). Mp 201°. pK_{a2} 3.05, pK_{a3} 7.35, pK_{a4} 10.96.

Tetra-Me ester: [16001-93-7]. *Tetramethyl methylenebisphosphonate.*

$C_5H_{14}O_6P_2$ M 232.110

Liq. $Bp_{0.05}$ 87-90°. n_D^{20} 1.4523.

Tetra-Et ester: [1660-94-2]. *Tetraethyl methylenebisphosphonate.*

$C_9H_{22}O_6P_2$ M 288.217

Liq. $Bp_{0.1}$ 90-94°. n_D^{20} 1.4412 (1.4300).

▷SZ9150000.

Tetrakis(2-propenyl) ester: Tetraallyl methylenediphosphonate.

$C_{13}H_{22}O_6P_2$ M 336.261

Liq.

▷Dec. violently on attempted distn. SZ9150000.

Tetraisopropyl ester: [1660-95-3]. *Tetrakis(1-methylethyl) methylenebisphosphonate. Tetraisopropyl methylenediphosphonate.*

$C_{13}H_{30}O_6P_2$ M 344.324

Liq. $Bp_{0.1}$ 96-100°. n_D^{20} 1.4317.

Tetra-Ph ester: [84441-53-2]. *Tetraphenyl methylenebisphosphonate.*

$C_{25}H_{22}O_6P_2$ M 480.393

Needles (Et_2O). Mp 81-82°.

Difluoride: see Methylenebisphosphonic difluoride, M-00146
Dichloride: see Methylenebis(phosphonic dichloride), M-00145
Tetraamide: see P,P′-Methylenebis(phosphonic diamide), M-00144

Schwarzenbach, G. *et al*, *Monatsh. Chem.*, 1950, **81**, 202 (*synth*)
Moedritzer, K. *et al*, *J. Inorg. Nucl. Chem.*, 1961, **22**, 297 (*synth, tetraethyl ester, ir, P nmr*)
Grabenstetter, R.J. *et al*, *J. Phys. Chem.*, 1967, **71**, 4194 (*P nmr, props*)
Quimby, O.T. *et al*, *J. Organomet. Chem.*, 1968, **13**, 199 (*esters, derivs*)
Nicholson, D.A. *et al*, *J. Org. Chem.*, 1970, **35**, 3149 (*esters, synth, P nmr*)
Sommer, K., *Z. Anorg. Allg. Chem.*, 1970, **376**, 37 (*synth, P nmr*)
Althoff, W. *et al*, *Z. Naturforsch., B*, 1976, **31**, 153 (*cmr*)
Peterson, S.W. *et al*, *J. Phys. Chem.*, 1977, **81**, 466 (*cryst struct*)
Griffiths, W.R. *et al*, *Phosphorus Sulfur*, 1978, **5**, 101 (*ester, ms*)
Browning, J. *et al*, *Inorg. Chem.*, 1981, **20**, 3912 (*complexes*)
Menge, M. *et al*, *Arch. Pharm. (Weinheim, Ger.)*, 1981, **314**, 218 (*ester, synth, pmr*)
Czekarski, T. *et al*, *J. Prakt. Chem.*, 1982, **324**, 537 (*esters, synth*)
Paul, G. *et al*, *Z. Chem.*, 1982, **22**, 307 (*ester, synth, pmr, cmr, nmr*)
Bottin-Strzalko, T. *et al*, *Phosphorus Sulfur*, 1985, **22**, 217 (*ir, raman, pmr, cmr, P nmr*)
Lang, G. *et al*, *Z. Anorg. Allg. Chem.*, 1986, **536**, 187 (*synth, ester*)

Methanediphosphonothioic acid **M-00023**

Methylenebisphosphonothioic acid, 9CI

$$(HO)_2P(S)CH_2P(S)(OH)_2$$

$CH_6O_4P_2S_2$ M 208.124

Tetra-Et ester: [63366-59-6]. O,O,O,O-*Tetraethyl methylenebisphosphonothioate.*
$C_9H_{22}O_4P_2S_2$ M 320.338
Liq. d_4^{20} 1.16. Bp_1 140°. n_D^{20} 1.4910.

Tetrafluoride: [84534-45-2].
$CH_2F_4P_2S_2$ M 216.088
No phys. props. reported.

Tetrachloride: [1499-32-7].
$CH_2Cl_4P_2S_2$ M 281.906
Cryst. (hexane). Mp 30.5°. $Bp_{0.06}$ 96-100°. n_D^{20} 1.6358.

Tetrabromide:
$CH_2Br_4P_2S_2$ M 459.710
Cryst. (C_6H_6). Mp 75°.

Bis[bis(dimethylamide)]: [63366-61-0]. P,P′-*Methylenebis*[(N,N,N′,N′-*tetramethyl*)-*phosphonodiamidothioate*].
$C_9H_{14}N_4P_2S_2$ M 304.304
Solid. Mp 115°.

Maier, L., *Helv. Chim. Acta*, 1965, **48**, 133 (*tetrachloride, synth, ir, P nmr*)
Nonikova, Z.S. *et al*, *Zh. Obshch. Khim.*, 1977, **47**, 775 (*Engl. transl.* p. 707) (*derivs, synth, pmr, P nmr*)
Czekański, T. *et al*, *J. Prakt. Chem.*, 1982, **324**, 537 (*tetraethyl ester, synth, pmr*)
Rankin, D.W.H. *et al*, *J. Chem. Soc., Dalton Trans.*, 1982, 2079 (*amide, struct, ed*)

Methanediphosphonous acid **M-00024**

Methylenediphosphonous acid, 8CI. *Methylenebisphosphonous acid*

$$(HO)_2PCH_2P(OH)_2$$

$CH_6O_4P_2$ M 144.004

Free acid exists as phosphoryl tautomer.

Tetra-Me ester: Tetramethyl methylenebisphosphonite.
$C_5H_{14}O_4P_2$ M 200.111
Liq. d_4^{20} 1.11. $Bp_{1.5}$ 65°. n_D^{20} 1.4695.

Tetra-Et ester: Tetraethyl methylenebisphosphonite.
$C_9H_{22}O_4P_2$ M 256.218
Liq. d_4^{20} 1.03. Bp_1 90°. n_D^{20} 1.4595.

Tetra-Ph ester: [87648-02-0]. *Tetraphenyl methylenebisphosphonite.*
$C_{25}H_{22}O_4P_2$ M 448.394
Oil.

Bis(difluoride): [60839-30-7].
$CH_2F_4P_2$ M 151.968
Liq. Bp 44° (est.).

Bis(dichloride): [28240-68-8].
$CH_2Cl_4P_2$ M 217.786
Liq. d_4^{20} 1.63. Bp_1 48°. n_D^{20} 1.5944.

Bis(dibromide): [63366-48-3].
$CH_2Br_4P_2$ M 395.590
Liq. d_4^{20} 2.69. Bp_1 108°, $Bp_{0.1}$ 80°.

Bis(diethylamide): [64151-43-5]. *Methylenebis*[N,N,N′,N′-*tetraethylphosphonous diamide*].
$C_{17}H_{42}N_4P_2$ M 364.493
Liq. $Bp_{0.1}$ 97°.

Sommer, K. *et al*, *Z. Anorg. Allg. Chem.*, 1970, **376**, 37 (*tetrachloride, nmr*)
Fild, M. *et al*, *Chem.-Ztg.*, 1977, **101**, 259 (*derivs, synth, P nmr*)
Novikova, Z.S. *et al*, *Zh. Obshch. Khim.*, 1977, **47**, 775 (*Engl. transl.* p. 707) (*chloride, bromide, esters, synth, pmr, P nmr*)
Diemert, K. *et al*, *Chem. Ber.*, 1982, **115**, 1947 (*tetrabromide, synth, P nmr*)
Diemert, K. *et al*, *Phosphorus Sulfur*, 1983, **15**, 155 (*tetraamide, synth, nmr*)
Karsch, H.H., *Z. Naturforsch., B*, 1983, **38**, 1027 (*tetraphenyl ester, synth, pmr, cmr, P nmr*)
Novikov, V.P. *et al*, *Izv. Akad. Nauk SSSR, Ser. Khim.*, 1983, 2252 (*Engl. transl.* p. 2033) (*chloride, ir, raman*)

Methidathion, BSI, ISO **M-00025**

O,O-*Dimethyl* S-[(*5-methoxy-2-oxo-1,3,4-thiadiazol-3(2H)-yl*)-*methyl*] *phosphorodithioate*, 9CI. O,O-*Dimethyl* S-(*4-mercaptomethyl-2-methoxy-Δ²-1,3,4-thiadiazolin-5-one*) *phosphorodithioate*, 8CI. *Supracide. Ultracide*

[950-37-8]

S
‖
N—NCH₂SP(OMe)₂
MeO⟨S⟩=O

$C_6H_{11}N_2O_4PS_3$ M 302.318

Insecticide, nonsystemic acaricide. Cryst. (MeOH). V. spar. sol. H_2O. Mp 39-40°.

▷TE2100000.

Esser, H.O. *et al*, *Helv. Chim. Acta*, 1968, **51**, 513 (*metab*)
Rüfenacht, K., *Helv. Chim. Acta*, 1974, **57**, 1658 (*synth*)
Stan, H.-J. *et al*, *Fresenius' Z. Anal. Chem.*, 1977, **287**, 271; *Biomed. Mass Spectrom.*, 1982, **9**, 483 (*glc, ms*)
Daldrup, T. *et al*, *Fresenius' Z. Anal. Chem.*, 1981, **308**, 413 (*tlc, glc, hplc*)
Ripley, B.D. *et al*, *J. Assoc. Off. Anal. Chem.*, 1983, **64**, 1084 (*glc*)
Pesticide Manual, 6th Ed., 343; 7th Ed., 361.
The Agrochemicals Handbook, The Royal Society of Chemistry, 1983, A268.

2-Methoxy-1,3,2-benzodioxaphosphole, 9CI **M-00026**
Methyl o-*phenylene phosphite*
[20570-25-6]

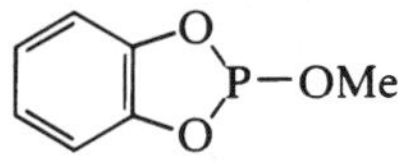

$C_7H_7O_3P$ M 170.104
Liq. Bp_{15} 76-77°, Bp_3 73°. n_D^{19} 1.5209.
2-Oxide: [702-86-3]. *Methyl* o-*phenylene phosphate.*
$C_7H_7O_4P$ M 186.104
Oil. Bp_{10} 139.5-140°. n_D^{19} 1.5060.

Anschutz, L. *et al*, *J. Prakt. Chem.*, 1932, **133**, 65 (*synth*)
Gross, H. *et al*, *Chem. Ber.*, 1966, **99**, 2631 (*oxide*)
Gloede, J. *et al*, *Z. Anorg. Allg. Chem.*, 1980, **471**, 147 (*oxide, P nmr*)

(1-Methoxycarbonylethyl)-triphenylphosphonium(1+) **M-00027**

$Ph_3P^{\oplus}CH(CH_3)COOMe$

$C_{22}H_{22}O_2P^{\oplus}$ M 349.388 (ion)
Iodide:
$C_{22}H_{22}IO_2P$ M 476.293
Solid. Mp 138-146°, 152-154°. With $NaNH_2$ yields the ylide.
Ylide: see Methyl (2-triphenylphosphoranylidene)-propanoate, M-00434

Bestmann, H.J. *et al*, *Chem. Ber.*, 1962, **95**, 2921 (*synth, use*)
Lang, R.W. *et al*, *Helv. Chim. Acta*, 1980, **63**, 438 (*synth, use, pmr*)

[(Methoxycarbonyl)methyl]phosphonamidic acid **M-00028**
Aminohydroxyphosphinylacetic acid methyl ester. Methyl aminohydroxyphosphinylacetate

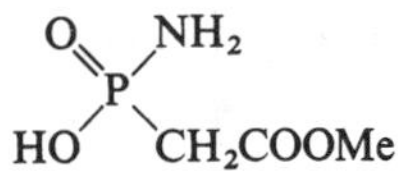

$C_3H_8NO_4P$ M 153.074
Et ester: [31460-21-6]. *Ethyl [(methoxycarbonyl)-methyl]phosphonamidate. Methyl (aminoethoxyphosphinyl)acetate.*
$C_5H_{12}NO_4P$ M 181.128
Solid. Mp 76-77°.
Ph ester: [31460-22-7]. *Phenyl [(methoxycarbonyl)-methyl]phosphonamidate.*
$C_9H_{12}NO_4P$ M 229.172
Solid. Mp 75-76°.

Bodnarchuk, N.D. *et al*, *Zh. Obshch. Khim.*, 1970, **40**, 1210 (*Engl. transl.* p. 1201)

[[2-(Methoxycarbonyl)phenyl]methyl]-triphenylphosphonium(1+), 9CI **M-00029**
(o-*Methoxycarbonylbenzyl*)*triphenylphosphonium*(*1+*)

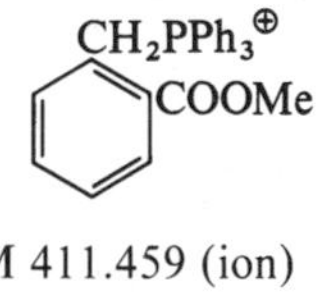

$C_{27}H_{24}O_2P^{\oplus}$ M 411.459 (ion)
Bromide: [60494-73-1].
$C_{27}H_{24}BrO_2P$ M 491.363
Source of ylide. Solid. Mp 239-242°. LiOMe or 1,5-diazabicyclo[4.3.0]non-5-ene → ylide.
Ylide: [[*2-(Methoxycarbonyl)phenyl*]*methylene*]-*triphenylphosphorane.* (o-*Methoxycarbonylbenzylidene*)*triphenylphosphorane.*
$C_{27}H_{23}O_2P$ M 410.451
Used in Wittig reactions.

Dunn, J.P. *et al*, *J. Med. Chem.*, 1977, **20**, 1557 (*bromide*)
Begasse, B. *et al*, *Tetrahedron*, 1980, **36**, 3409.
Bellinger, G.C.A. *et al*, *J. Chem. Soc., Perkin Trans. 1*, 1982, 2819 (*use*)

[[3-(Methoxycarbonyl)phenyl]methyl]-triphenylphosphonium(1+), 9CI **M-00030**
(m-*Methoxycarbonylbenzyl*)*triphenylphosphonium*(*1+*)
$C_{27}H_{24}O_2P^{\oplus}$ M 411.459 (ion)
Bromide: [56981-97-6].
$C_{27}H_{24}BrO_2P$ M 491.363
Source of ylide. Solid. Mp 234-236°.
Ylide: [[*3-(3-Methoxycarbonyl)phenyl*]*methylene*]-*triphenylphosphorane.* (m-*Methoxycarbonylbenzylidene*)*triphenylphosphorane.*
$C_{27}H_{23}O_2P$ M 410.451
Used in Wittig reactions.

Yamato, M. *et al*, *Chem. Pharm. Bull.*, 1977, **25**, 706 (*synth, use*)
Sankaran, V. *et al*, *J. Polym. Sci., Polym. Chem. Ed.*, 1980, **18**, 1821 (*synth, ir, pmr, use*)

[[4-(Methoxycarbonyl)phenyl]methyl]-triphenylphosphonium(1+), 9CI **M-00031**
(p-*Methoxycarbonylbenzyl*)*triphenylphosphonium*(*1+*)
$C_{27}H_{24}O_2P^{\oplus}$ M 411.459 (ion)
Bromide: [1253-46-9].
$C_{27}H_{24}BrO_2P$ M 491.363
Source of ylide. Prisms (CH_2Cl_2/EtOAc). Mp 232-235°, 258-260°. Na_2CO_3 aq. → ylide.
Ylide: [[*4-(Methoxycarbonyl)phenyl*]*methylene*]-*triphenylphosphorane.* (p-*Methoxycarbonylbenzylidene*)*triphenylphosphorane.*
$C_{27}H_{23}O_2P$ M 410.451
Used in Wittig reactions.

Cresp, T.M. *et al*, *J. Chem. Soc., Perkin Trans. 1*, 1974, 2435 (*synth*)
Yamato, M. *et al*, *Chem. Pharm. Bull.*, 1977, **25**, 706 (*synth, use*)
Sankaran, V. *et al*, *J. Polym. Sci., Polym. Chem. Ed.*, 1979, **17**, 3949 (*synth, ir, use*)
Knoppová, V. *et al*, *Collect. Czech. Chem. Commun.*, 1981, **46**, 515 (*synth, use*)
Lee, B.H. *et al*, *J. Polym. Sci., Polym. Chem. Ed.*, 1982, **20**, 393 (*use*)

6-Methoxydibenzo[*d,f*][1,3,2]-dioxaphosphepin, 8CI **M-00032**
2,2′-Biphenyldiyl methyl phosphite
[14074-89-6]

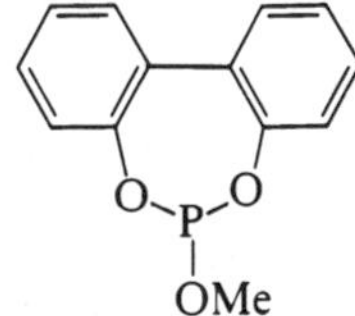

$C_{13}H_{11}O_3P$ M 246.202

d_4^{20} 1.26. $Bp_{0.004}$ 110-111°. n_D^{20} 1.6085.

6-Oxide: [76045-15-3]. *2,2′-Biphenyldiyl methyl phosphate.*

$C_{13}H_{11}O_4P$ M 262.201

Cryst. (C_6H_6/ligroin). Mp 104-106°.

Verizhnikov, L.V. *et al*, *Zh. Obshch. Khim.*, 1967, **37**, 1355 (*Engl. transl.* p. 1287) (*synth*)
Keck, H. *et al*, *Org. Mass. Spectrom.*, 1980, **15**, 591 (*ms*)
Kuchen, W. *et al*, *Phosphorus Sulfur*, 1980, **8**, 139 (*oxide, pmr, P nmr*)

2-Methoxy-4,5-dimethyl-1,3,2-dioxaphospholane, 9CI M-00033

2,3-Butylene methyl phosphite. 2-Methoxy-4,5-dimethyl-1,3-dioxa-2-phosphacyclopentane

[41821-87-8]

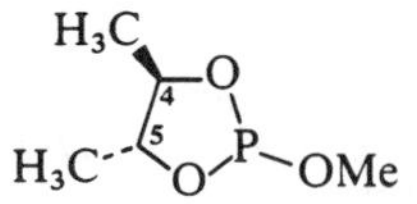

(4R,5R)-*form*

$C_5H_{11}O_3P$ M 150.114

2-Oxide: [27303-86-2]. *2,3-Butylene methyl phosphate.*

$C_5H_{11}O_4P$ M 166.113

Liq. $Bp_{0.65}$ 62-64°. Mixt. of diastereoisomers.

2-Sulfide: [66242-28-2]. O,O-*2,3-Butylene* O-*methyl thiophosphate.* O,O-*2,3-Butylene* O-*methyl phosphorothioate.*

$C_5H_{11}O_3PS$ M 182.174

Oil. n_D^{21} 1.4810. Mixt. of diastereoisomers.

(4R,5R)-form

Liq. with characteristic phosphite odour. d_4^{25} 1.20. Bp_{10} 46°. $[\alpha]_D^{25}$ +53.63°. n_D^{25} 1.4318.

(4RS,5RS)-form [27303-84-0]

Liq. with characteristic phosphite odour. Bp_{18} 62-64°.

(4RS,5SR)-form

meso-*form*. *4,5*-cis-*form*

Liq. with characteristic phosphite odour. Bp_{15} 70-72°. Diastereoisomers at P characterised spectroscopically.

2-Oxide: Liq. $Bp_{0.35}$ 84°. Epimers at P characterised spectroscopically.

Garner, H.K. *et al*, *J. Am. Chem. Soc.*, 1950, **72**, 5497 (*synth*)
Denney, D.Z. *et al*, *J. Am. Chem. Soc.*, 1969, **91**, 6838 (*synth, oxide, pmr, P nmr*)
Pouchoulin, G. *et al*, *Org. Magn. Reson.*, 1976, **8**, 518 (*pmr, cmr, P nmr*)
Mikolajczyk, M. *et al*, *J. Chem. Soc., Perkin Trans. 1*, 1977, 2213 (*sulfide, cmr, pmr, P nmr*)
Corriu, R.J.P. *et al*, *Tetrahedron Lett.*, 1983, **24**, 4323 (*oxide, P nmr*)

2-Methoxy-5,5-dimethyl-1,3,2-dioxaphosphorinane, 9CI M-00034

Methyl neopentylene phosphite. 2-Methoxy-5,5-dimethyl-1,3,2-dioxaphosphinane. 2-Methoxy-5,5-dimethyl-1,3-dioxa-2-phosphacyclohexane

[1005-69-2]

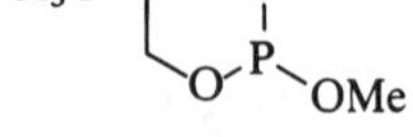

$C_6H_{13}O_3P$ M 164.141

Liq. with characteristic phosphite odour. Bp_{18} 66°.

2-Oxide: [1005-96-5]. *Methyl neopentylene phosphate.*

$C_6H_{13}O_4P$ M 180.140

Solid. Mp 94°. $Bp_{0.5}$ 118-121°.

2-Sulfide: [1005-97-6]. O-*Methyl* O,O-*neopentylene phosphorothioate.*

$C_6H_{13}O_3PS$ M 196.201

Cryst. (Et_2O). Mp 93.5-94.5°.

Edmundson, R.S., *Tetrahedron*, 1964, **20**, 2781 (*synth, derivs, ir*)
Bartle, K.D. *et al*, *Tetrahedron*, 1967, **23**, 1701 (*derivs, pmr*)
Gagnaire, D. *et al*, *Bull. Soc. Chim. Fr.*, 1967, 2240 (*pmr*)
Majoral, J.-P. *et al*, *Bull. Soc. Chim. Fr.*, 1972, 606 (*oxide, ir, P nmr*)
Haemers, M. *et al*, *Tetrahedron*, 1973, **29**, 3539 (*cmr, P nmr*)
Stec, W.J., *Z. Naturforsch, B*, 1974, **29**, 109 (*selenide, P nmr*)
Grand, A. *et al*, *Acta Crystallogr., Sect. B*, 1975, **31**, 2523 (*selenide, cryst struct*)
Francis, G.W. *et al*, *Acta Chem. Scand., Ser. B*, 1976, **30**, 31 (*oxide, ms*)
Van Nuffel, P. *et al*, *Cryst. Struct. Commun.*, 1980, **9**, 733 (*oxide, cryst struct*)
Edmundson, R.S., *Phosphorus Sulfur*, 1981, **9**, 307 (*sulfide, ms*)
Van Nuffel, P. *et al*, *Acta Crystallogr., Sect. B*, 1981, **37**, 133 (*sulfide, cryst struct*)

2-Methoxy-3,4-dimethyl-5-phenyl-1,3,2-oxazaphospholidine, 9CI M-00035

2-Methoxy-3,4-dimethyl-5-phenyl-1,3,2-oxazaphospholane

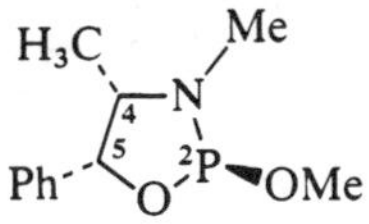

(2R,4S,5R)-*form*

$C_{11}H_{16}NO_2P$ M 225.227

Derivs. are reagents for synth. of chiral methyl phosphoric and methyl phosphorothioic esters.

(2R,4S,5R)-form [53648-95-6]

Liq. $Bp_{0.7}$ 102-104°. Mixt. with (2*S*,4*S*,5*R*)-form.

2-Oxide: [54655-13-9]. Syrup. $[\alpha]_D$ −110° ($CHCl_3$). Has 2*S*-config.

2-Sulfide: Liq. $Bp_{0.1}$ 140° (both). $[\alpha]_D$ −140° ($CHCl_3$). Has 2*S*-config.

(2S,4S,5R)-form [53648-94-5]

Liq. $Bp_{0.7}$ 102-104°. Mixt. with (2*R*,4*S*,5*R*)-form.

2-Oxide: Syrup. $[\alpha]_D$ −37°. Has 2*R*-config.

2-Sulfide: Cryst. (diisopropyl ether). Mp 88-90°. $[\alpha]_D$ +2° ($CHCl_3$). Has 2*R*-config.

(2R,4R,5S)-form [53648-97-8]

Liq. $Bp_{0.7}$ 102-104°. Mixt. with (2*S*,4*R*,5*S*)-form.

(2S,4R,5S)-form [53648-96-7]

Liq. $Bp_{0.7}$ 102-104°. Mixt. with (2*R*,4*R*,5*S*)-form.

Bernard, D. *et al*, *Phosphorus*, 1974, **3**, 187 (*synth, P nmr*)
Cooper, D.B. *et al*, *J. Chem. Soc., Perkin Trans. 1*, 1977, 1969 (*derivs, tlc, pmr*)

2-Methoxy-1,3,2-dioxaphospholane, 9CI M-00036

2-Methoxy-1,3-dioxa-2-phosphacyclopentane. Ethylene methyl phosphite

[3741-36-4]

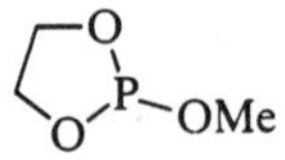

$C_3H_7O_3P$ M 122.060

Liq. with characteristic phosphite odour. d_4^{25} 1.225. Bp_{35} 60-62°. n_D^{25} 1.4406.

▷TG7250000.

2-Oxide: see 2-Methoxy-1,3,2-dioxaphospholane 2-oxide, M-00037

2-Selenide: [67761-25-5]. O,O-*Ethylene* O-*methyl phosphoroselenoate.*

$C_3H_7O_3PSe$ M 201.020

Characterised spectroscopically.

Lucas, H.J. *et al, J. Am. Chem. Soc.*, 1950, **72**, 5491 (*synth*)

Burgada, R. *et al, Bull. Soc. Chim. Fr.*, 1971, 136; 1974 (*Part II*), 341 (*pmr, P nmr*)

Pouchoulin, G. *et al, Org. Magn. Reson.*, 1976, **8**, 518 (*pmr, cmr, P nmr*)

Labintsev, L.V. *et al, Zh. Org. Khim.*, 1978, **14**, 1134 (*Engl. transl.* p. 1058) (*ms*)

Besserre, D. *et al, Org. Magn. Reson.*, 1980, **13**, 235, 313 (*cmr, conformn*)

Hodges, R.V. *et al, J. Am. Chem. Soc.*, 1980, **102**, 932 (*props*)

Gonbeau, D. *et al, J. Mol. Struct.*, 1983, **98**, 109 (*pe, struct*)

2-Methoxy-1,3,2-dioxaphospholane 2-oxide, 9CI M-00037

Ethylene methyl phosphate. 2-Methoxy-1,3-dioxa-2-oxo-2-phosphacyclopentane

[2196-04-5]

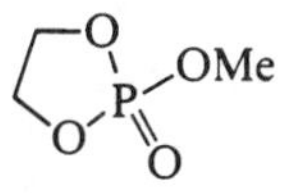

$C_3H_7O_4P$ M 138.060

Liq. $Bp_{0.09}$ 80°.

Revel, M. *et al, Bull. Soc. Chim. Fr., Pt. 1*, 1973, 1195 (*pmr, P nmr*)

Shagidullin, R.R. *et al, Zh. Obshch. Khim.*, 1976, **46**, 1021 (*Engl. transl.* p. 1017) (*ir, raman*)

Libiszowski, J. *et al, J. Polym. Sci., Polym. Chem. Ed.*, 1978, **16**, 1275 (*synth, pmr, P nmr, props*)

Yasuda, H. *et al, Macromolecules*, 1982, **15**, 1231 (*synth, props, ir, pmr, cmr*)

Taira, K. *et al, J. Am. Chem. Soc.*, 1984, **106**, 1521 (*props*)

Taira, K. *et al, J. Org. Chem.*, 1984, **49**, 4531 (*synth, pmr, P nmr, props*)

Kluger, R. *et al, J. Am. Chem. Soc.*, 1985, **107**, 6006 (*props*)

Kluger, R. *et al, J. Org. Chem.*, 1986, **51**, 207 (*props*)

Gorenstein, D.G. *et al, Tetrahedron*, 1987, **43**, 469 (*props*)

2-Methoxy-1,3,2-dioxaphosphorinane, 9CI M-00038

2-Methoxy-1,3,2-dioxaphosphinane. 2-Methoxy-1,3-dioxa-2-phosphacyclohexane. Methyl trimethylene phosphite

[31121-06-9]

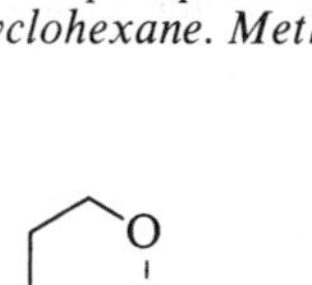

$C_4H_9O_3P$ M 136.087

Liq. with characteristic phosphite odour. Bp_{15} 70°.

▷Exposure may produce headache

2-Oxide: [33554-05-1]. *Methyl trimethylene phosphate.*

$C_4H_9O_4P$ M 152.086

No phys. props. reported.

2-Sulfide: [33148-57-1]. O-*Methyl* O,O-*trimethylene phosphorothioate.*

$C_4H_9O_3PS$ M 168.147

No phys. props. reported.

2-Selenide: [52912-90-0]. O-*Methyl* O,O-*trimethylene phosphoroselenoate.*

$C_4H_9O_3PSe$ M 215.047

Prob. undergoes easy rearrangement on distn.

Bergesen, K. *et al, Acta Chem. Scand.*, 1971, **25**, 2257 (*synth, pmr*)

Haemers, M. *et al, Tetrahedron*, 1973, **29**, 3539 (*cmr, P nmr*)

Stec, W.J., *Z. Naturforsch., B*, 1974, **29**, 109 (*selenide, P nmr*)

Mosbo, J.A. *et al, J. Org. Chem.*, 1977, **42**, 1549 (*oxide, ir, P nmr*)

Arshinova, R.P., *Phosphorus Sulfur*, 1978, **5**, 131 (*conformn*)

Nifant'ev, É.E. *et al, Tetrahedron*, 1981, **37**, 3183 (*pmr*)

Eliel, E.L. *et al, J. Am. Chem. Soc.*, 1986, **108**, 6651 (*cmr, O and P nmr*)

2-Methoxy-4,5-diphenyl-1,3,2-dioxaphospholane, 9CI M-00039

2-Methoxy-4,5-diphenyl-1,3-dioxa-2-phosphacyclopentane

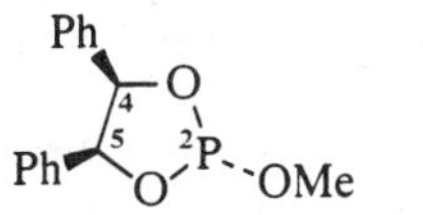

$C_{15}H_{15}O_3P$ M 274.255

(2α,4β,5β)-form [54054-03-4]

Cryst. solid. Mp 56-57°. $Bp_{0.05}$ 130-135°, $Bp_{0.1}$ 70° subl. Unstable in moist air.

2-Oxide: [54017-36-6].

$C_{15}H_{15}O_4P$ M 290.255

Cryst. (MeOH/hexane). Mp 74-75°. $Bp_{0.05}$ 90° subl.

2-Sulfide: [66289-05-2].

$C_{15}H_{15}O_3PS$ M 306.315

Cryst. (Et_2O/pet. ether at −70°). Mp 93-97°.

Newton, M.G. *et al, J. Am. Chem. Soc.*, 1974, **96**, 7790 (*synth, cryst struct*)

Mikolajczyk, M. *et al, J. Chem. Soc., Perkin Trans. 1*, 1977, 2213 (*synth, sulfide, P nmr*)

Cullis, P.M. *et al, J. Chem. Soc., Perkin Trans. 1*, 1981, 2317 (*oxide, P nmr*)

Nielsen, J. *et al, J. Chem. Soc., Perkin Trans. 2*, 1984, 553 (*pmr, P nmr*)

[(2-Methoxyethoxy)methyl]phosphonic acid M-00040

$MeOCH_2OCH_2P(O)(OH)_2$

$C_6H_{15}O_5P$ M 198.155

Di-Et ester: Diethyl[(2-methoxyethoxy)methyl]-phosphonate, 9CI.

$C_8H_{19}O_5P$ M 226.209

Reagent for the conversion of aldehydes and ketones into ethenyl-1-(2-methoxyethoxy) ethers. Liq. $Bp_{0.8}$ 104-105°. Yields an anion with $LiN[CH(CH_3)_2]_2$.

Kluge, A.F. *et al, J. Org. Chem.*, 1979, **44**, 4847 (*synth, use*)

(2-Methoxyethyl)phosphoramidothioic acid, 9CI, 8CI M-00041

$MeOCH_2CH_2NHP(S)(OH)_2 \rightleftharpoons MeOCH_2CH_2NH-P(O)(OH)(SH)$

$C_3H_{10}NO_3PS$ M 171.151

O,O-*Di-Me ester:* [35812-40-9]. O,O-*Dimethyl (2-methoxyethyl)phosphoramidothioate.*

$C_5H_{14}NO_3PS$ M 199.204

Liq. $Bp_{0.12}$ 78-80°. n_D^{20} 1.4853.

O,O-*Di-Et ester:* [35812-41-0]. O,O-*Diethyl (2-methoxyethyl)phosphoramidothioate.*

$C_7H_{18}NO_3PS$ M 227.258

Liq. $Bp_{0.06}$ 75-77°. n_D^{20} 1.4733.

Cates, L.A. *et al, J. Med. Chem.*, 1971, **14**, 1022 (*synth, tox*)

7-Methoxy-2,3,5,6,7,8-hexakis(trifluoromethyl)-1,4-diphosphabicyclo[2.2.2]octa-2,5-diene, 10CI **M-00042**

[62218-18-2]

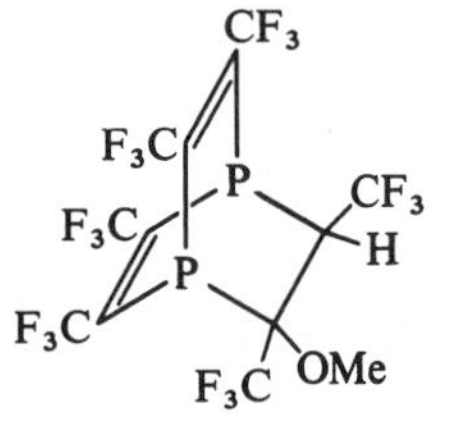

$C_{13}H_3F_{18}OP_2$ M 579.085

Source of 2,3,5,6-tetrakis(trifluoromethyl)-1,4-diphosphorine *in situ*. Plates (pet. ether). Mp 68-70° (sealed tube).

Kobayashi, Y. *et al, J. Org. Chem.*, 1980, **45**, 4683 (*props, use*)
Kobayashi, Y. *et al, J. Am. Chem. Soc.*, 1980, **102**, 252 (*synth, ir, pmr, ms, F and P nmr*)
Kobayashi, Y. *et al, J. Am. Chem. Soc.*, 1981, **103**, 2465 (*props, use*)

2-Methoxy-4-methyl-1,3,2-dioxaphospholane, 9CI **M-00043**

2-Methoxy-4-methyl-1,3-dioxa-2-phosphacyclopentane. Methyl propylene phosphite

[6156-15-6]

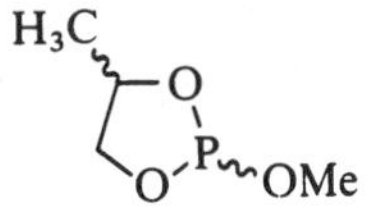

$C_4H_9O_3P$ M 136.087

Liq. with characteristic phosphite odour. d_4^{25} 1.14. Bp_{25} 60.8°. n_D^{25} 1.4354. *Cis* and *trans*-forms characterised spectroscopically.

2-Oxide: Methyl propylene phosphate.
$C_4H_9O_4P$ M 152.086
Liq. Bp_{70} 250°, $Bp_{1.1}$ 87-93°. *Cis* and *trans* forms characterised spectroscopically.

2-Sulfide: O-*Methyl* O,O-*propylene thiophosphate.* O-*Methyl* O,O-*propylene phosphorothioate.*
$C_4H_9O_3PS$ M 168.147
Oil. n_D^{20} 1.4810. *Cis* and *trans* forms characterised spectroscopically.

Lucas, H.J. *et al, J. Am. Chem. Soc.*, 1950, **72**, 5491 (*synth*)
Denney, D.Z. *et al, J. Am. Chem. Soc.*, 1969, **91**, 6838 (*pmr, P nmr*)
Bentrude, W.G. *et al, J. Am. Chem. Soc.*, 1976, **98**, 1850 (*cmr, pmr, P nmr*)
Bentrude, W.G. *et al, J. Am. Chem. Soc.*, 1976, **98**, 5348 (*oxide, pmr*)
Mikolajczyk, M. *et al, J. Chem. Soc., Perkin Trans. 1*, 1976, 371 (*sulfide, pmr, P nmr*)
Abbott, S.J. *et al, J. Am. Chem. Soc.*, 1979, **101**, 4323 (*oxide, synth, ms*)
Hansen, D.E. *et al, J. Biol. Chem.*, 1981, **256**, 5967 (*oxide, P nmr*)
Yasuda, H. *et al, Macromolecules*, 1982, **15**, 1231 (*synth, oxide, pmr, cmr, ir, props*)
Gonbeau, D. *et al, J. Mol. Struct.*, 1983, **98**, 109 (*struct*)

2-Methoxy-4-methyl-1,3,2-dioxaphosphorinane, 9CI **M-00044**

2-Methoxy-4-methyl-1,3,2-dioxaphosphinane. 2-Methoxy-4-methyl-1,3-dioxa-2-phosphacyclohexane. 1,3-Butylene methyl phosphite

[33892-95-4]

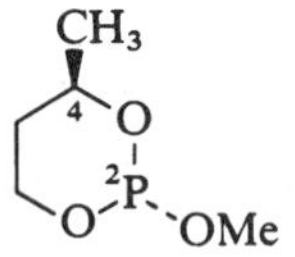

(2*RS*,4*RS*)-*form*

$C_5H_{11}O_3P$ M 150.114

(2*RS*,4*RS*)-*form* [7735-81-1]

(±)-trans-*form*

Liq. with characteristic phosphite odour. d_4^{20} 1.11. Bp_{60} 90-92°, Bp_7 48-50°. n_D^{20} 1.4420 (1.4481).

CuCl adduct (*1:1*): Solid. Mp 137-138°.

2-Oxide: see 2-Methoxy-4-methyl-1,3,2-dioxaphosphorinane 2-oxide, M-00045

2-Sulfide: see 2-Methoxy-4-methyl-1,3,2-dioxaphosphorinane 2-sulfide, M-00047

2-Selenide: see 2-Methoxy-4-methyl-1,3,2-dioxaphosphorinane 2-selenide, M-00046

(2*RS*,4*SR*)-*form* [7735-85-5]

(±)-cis-*form*

Liq. with characteristic phosphite odour. d_4^{20} 1.12. $Bp_{0.01}$ 30-32°. n_D^{20} 1.4515.

CuCl adduct (*1:1*): Solid. Mp 128-129°.

2-Oxide: see 2-Methoxy-4-methyl-1,3,2-dioxaphosphorinane 2-oxide, M-00045

2-Sulfide: see 2-Methoxy-4-methyl-1,3,2-dioxaphosphorinane 2-sulfide, M-00047

Bodkin, C.L. *et al, J. Chem. Soc. (B)*, 1971, 1137 (*synth, config*)
Mikolajczyk, M. *et al, Tetrahedron*, 1972, **28**, 5411 (*synth*)
Haemers, M. *et al, Tetrahedron Lett.*, 1973, 2241 (*cmr*)
Arbuzov, B.A. *et al, Izv. Akad. Nauk SSSR, Ser. Khim.*, 1974, 665 (*Engl. transl.* p. 629) (*synth, derivs, conformn*)
Stec, W.J. *et al, J. Org. Chem.*, 1976, **41**, 233 (*synth, P nmr*)
Arshinova, R.P. *et al, Zh. Obshch. Khim.*, 1981, **51**, 1757 (*Engl. transl.* p. 1503) (*pe*)
Nifant'ev, E.Ê. *et al, Tetrahedron*, 1981, **37**, 3183 (*cmr*)
Eliel, E.L. *et al, J. Am. Chem. Soc.*, 1986, **108**, 665 (*cmr, O and P nmr*)

2-Methoxy-4-methyl-1,3,2-dioxaphosphorinane 2-oxide, 9CI **M-00045**

2-Methoxy-4-methyl-1,3,2-dioxaphosphinane 2-oxide. 1,3-Butylene methyl phosphate. 2-Methoxy-4-methyl-1,3-dioxa-2-oxo-2-phosphacyclohexane

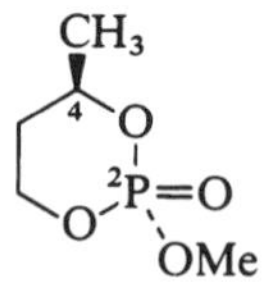

(2*RS*,4*RS*)-*form*

$C_5H_{11}O_4P$ M 166.113

(2*RS*,4*RS*)-*form*

(±)-trans-*form*

Liq. d_4^{20} 1.26. $Bp_{0.0001}$ 84-84.5°. n_D^{20} 1.4390.

(2*RS*,4*SR*)-*form* [33996-04-2]

(±)-cis-*form*

Liq. d_4^{20} 1.25. $Bp_{0.0001}$ 97-98°. n_D^{20} 1.4375.

Mosbo, J.A. *et al, J. Am. Chem. Soc.*, 1972, **94**, 8224 (*P nmr, stereochem*)
Stec, W.J. *et al, Tetrahedron*, 1973, **29**, 539 (*synth, ir, pmr, P nmr*)

Arbuzov, B.A. *et al*, *Izv. Akad. Nauk SSSR, Ser. Khim.*, 1974, 665 (*Engl. transl.* p. 629) (*synth, raman, P nmr, conformn*)
Mosbo, J.A. *et al*, *J. Org. Chem.*, 1977, **42**, 1549 (*ir, P nmr, conformn*)
Mosbo, J.A., *Org. Magn. Reson.*, 1978, **11**, 281 (*oxide, conformn*)

2-Methoxy-4-methyl-1,3,2-dioxaphosphorinane 2-selenide, 9CI M-00046

2-Methoxy-4-methyl-1,3,2-dioxaphosphinane 2-selenide. O,O-*1,3-Butylene* O-*methyl selenophosphate*. O,O-*1,3-Butylene* O-*methyl phosphoroselenoate*. *2-Methoxy-4-methyl-1,3-dioxa-2-selenoxo-2-phosphacyclohexane*

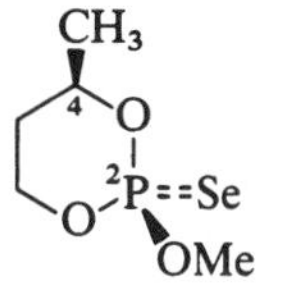

(2*RS*,4*RS*)-*form*

$C_5H_{11}O_3PSe$ M 229.074

(2RS,4RS)-form [33996-01-9]
(±)-cis-*form*
Liq. $Bp_{0.1}$ 85-87°. n_D^{20} 1.5216 (88% diastereomerically pure).

(2RS,4SR)-form [33996-02-0]
(±)-trans-*form*
Liq. $Bp_{0.2}$ 100-102°. n_D^{20} 1.5268 (96% diastereomerically pure).

Stec, W.J., *Z. Naturforsch., B*, 1974, **29**, 109 (*P nmr*)
Stec, W.J. *et al*, *J. Org. Chem.*, 1976, **41**, 233 (*synth, pmr, P nmr*)

2-Methoxy-4-methyl-1,3,2-dioxaphosphorinane 2-sulfide, 9CI M-00047

2-Methoxy-4-methyl-1,3,2-dioxaphosphinane 2-sulfide. O,O-*1,3-Butylene* O-*methyl thiophosphate*. O,O-*1,3-Butylene* O-*methyl phosphorothioate*. *2-Methoxy-4-methyl-1,3-dioxa-2-thioxo-2-phosphacyclohexane*

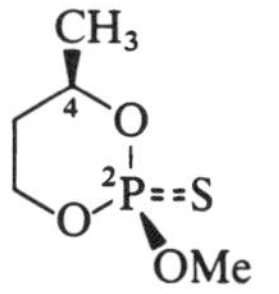

(2*RS*,4*RS*)-*form*

$C_5H_{11}O_3PS$ M 182.174

With trifluoroacetic acid, or MeI, rearranges to *S*-Me compd. (see under 4-Methyl-2-methylthio-1,3,2-dioxaphosphorinane, M-00174 .

(2RS,4RS)-form [23168-88-9]
(±)-cis-*form*
Liq. d_4^{20} 1.25. $Bp_{0.0001}$ 83-84°. n_D^{20} 1.4958.

(2RS,4SR)-form [23168-89-0]
(±)-trans-*form*
Liq. d_4^{20} 1.24. $Bp_{0.0001}$ 74°. n_D^{20} 1.4920.

Bodkin, C.L. *et al*, *J. Chem. Soc. (B)*, 1971, 1137 (*synth, config*)
Mikolajczyk, M. *et al*, *Tetrahedron*, 1972, **28**, 5411 (*synth, pmr, P nmr*)
Arbuzov, B.A. *et al*, *Izv. Akad. Nauk SSSR, Ser. Khim.*, 1974, 665 (*Engl. transl.*, p. 629) (*P nmr, raman, conformn*)
Stec, W.J. *et al*, *J. Org. Chem.*, 1976, **41**, 233, 1291 (*synth, P nmr, props*)
Mikolajczyk, M. *et al*, *J. Org. Chem.*, 1977, **42**, 190 (*props*)
Łopusinski, A. *et al*, *J. Am. Chem. Soc.*, 1982, **104**, 290 (*props*)

3-Methoxy-15-methyl-18-nor-15-phosphaestra-1,3,5(10),6,8,13-hexaene M-00048

*2,3,10,11-Tetrahydro-7-methoxy-3-methyl-1*H-*naphtho[1,2-g]phosphindole, 10CI*

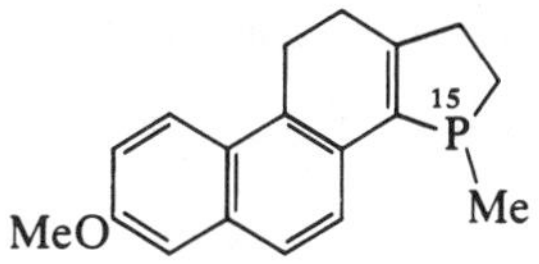

$C_{18}H_{19}OP$ M 282.321

15-Oxide: [63347-76-2].
$C_{18}H_{19}O_2P$ M 298.321
Solid. Mp 194° dec.

Symmes, C. *et al*, *J. Org. Chem.*, 1979, **44**, 1048 (*synth, pmr*)

2-Methoxy-4-methyl-1,2-oxaphospholane M-00049

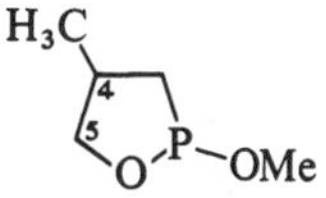

$C_5H_{11}OP$ M 118.115

2-Oxide: [27503-04-4].
$C_5H_{11}O_2P$ M 134.114
Liq. Bp_1 129°. n_D^{21} 1.4540.

Bergesen, K. *et al*, *Acta Chem. Scand.*, 1970, **24**, 1122 (*synth, pmr*)

(6-Methoxy-3-methyl-6-oxo-2,4-hexadienyl)triphenylphosphonium(1+) M-00050

(6-Methoxycarbonyl-3-methyl-2,4-hexadienyl)-triphenylphosphonium(1+)

$Ph_3P^{\oplus}CH_2CH{=}C(CH_3)CH{=}CHCOOMe$

$C_{26}H_{26}O_2P^{\oplus}$ M 401.464 (ion)

Used in synth. of Me natural bixin. Glassy solid.

Pattenden, G. *et al*, *J. Chem. Soc. (C)*, 1970, 235 (*synth, use*)

3-Methoxy-17-methyl-15-phenyl-15-phosphaestra-1,3,5(10),8,16-pentaene, 9CI M-00051

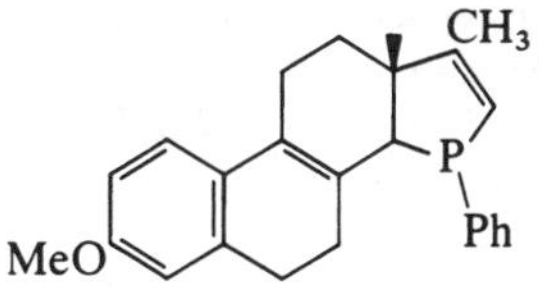

$C_{25}H_{27}OP$ M 374.461

(±)-form

15-Oxide: [62241-69-4].
$C_{25}H_{27}O_2P$ M 390.461
Isomerises when heated.

15-Sulfide: [62241-68-3].
$C_{25}H_{27}OPS$ M 406.521
Cryst. (EtOAc). Mp 158°.

Kashman, Y. *et al*, *Tetrahedron*, 1976, **32**, 2427 (*synth, uv, pmr, ms, ir*)

(Methoxymethyl)phenylphosphinic acid, 9CI M-00052

$MeOCH_2PPh(O)OH$

$C_8H_{11}O_3P$ M 186.147

Et ester: Ethyl (methoxymethyl)phenylphosphinate.

$C_{10}H_{15}O_3P$ M 214.200

Liq. d_0^0 1.15. Bp_2 138-139°. n_D^{20} 1.4891.

Arbuzov, A.E., *Izv. Akad. Nauk SSSR, Ser. Khim.*, 1945, 167; *CA*, **40**, 3411.

(Methoxymethyl)-triphenylphosphonium(1+), 9CI M-00053

$Ph_3P^{\oplus}CH_2OMe$

$C_{20}H_{20}OP$ M 307.351

Salts yield the ylide when treated with NaH, $NaNH_2$ or Bu^tOH.

Chloride: [4009-98-7].

$C_{20}H_{20}ClOP$ M 342.804

Cryst. ($CHCl_3$/EtOAc). Mp 201-202° dec. (192-194°).

Bromide: [33670-32-5].

$C_{20}H_{20}BrOP$ M 387.255

Cryst. ($CHCl_3$/EtOAc).

Iodide: [68305-75-9].

$C_{20}H_{20}IOP$ M 434.256

Cryst. (CCl_4/CH_2Cl_2). Mp 178-180°.

Tetraphenylborate:

$C_{44}H_{40}BOP$ M 626.583

Cryst. (Me_2CO/EtOH). Mp 198-199°.

Ylide: [20763-19-3].

Methoxymethylenetriphenylphosphorane.

$C_{20}H_{19}OP$ M 306.343

Used in carbohydrate and prostaglandin chemistry.

Wittig, G. *et al*, *Chem. Ber.*, 1961, **94**, 1373 (*synth*)
Hendrickson, J. *et al*, Tet, 1964, **20**, 449 (*synth, pmr*)
Gray, G.A., *J. Am. Chem. Soc.*, 1973, **95**, 7736 (*cmr, nmr*)
Eklind, K. *et al*, *Acta Chem. Scand., Ser. B*, 1974, **28**, 260 (*ylide, use*)
Albright, T.A. *et al*, *J. Am. Chem. Soc.*, 1975, **97**, 2946 (*nmr*)
Corey, E.J. *et al*, *J. Am. Chem. Soc.*, 1976, **98**, 6417 (*ylide, use*)
Sukenik, C.N. *et al*, *J. Am. Chem. Soc.*, 1976, **98**, 6613 (*ylide, use*)
Jung, M.E. *et al*, *Synthesis*, 1978, 588 (*synth, ylide, use*)
Yamamoto, Y. *et al*, *Bull. Chem. Soc. Jpn.*, 1980, **53**, 3436 (*synth, pmr, cmr, ylide*)

3-Methoxy-18-nor-17-phosphaestra-1,3,5(10),9(11)-tetraen-15-one M-00054

1,2,3a,3b,4,5,11,11a-Octahydro-7-methoxy-3H-naphtho[2,1-c]phosphindol-3-one, 9CI

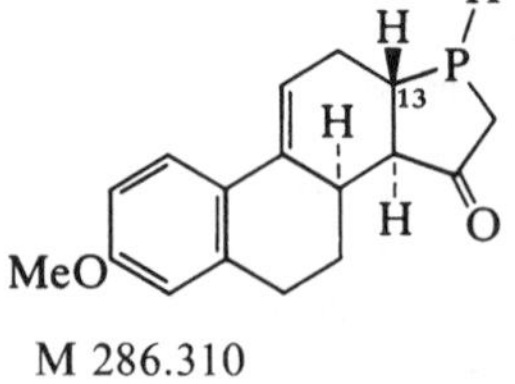

$C_{17}H_{19}O_2P$ M 286.310

13α-form

17-Me, 17-oxide: [50731-88-9].

$C_{18}H_{21}O_3P$ M 316.336

Solid. Mp 148°.

17-Ph, 17-oxide: [50731-87-8]. Solid. Mp 162-164°.

13β-form

17-Ph, 17-oxide: [50897-31-9].

$C_{23}H_{23}O_3P$ M 378.407

Solid. Mp 232-234°.

Kashman, Y. *et al*, *Tetrahedron Lett.*, 1973, 3217 (*synth, props, ir, ms, pmr*)

(2-Methoxy-2-oxoethyl)-triphenylphosphonium(1+), 9CI M-00055

(Carboxymethyl)triphenylphosphonium methyl ester, 8CI. *Carbomethoxymethyltriphenylphosphonium(1+)*

$Ph_3P^{\oplus}CH_2COOMe$

$C_{21}H_{20}O_2P^{\oplus}$ M 335.362 (ion)

Treatment of salts with 2M NaOH yields the ylide.

Chloride: [2181-97-7].

$C_{21}H_{20}ClO_2P$ M 370.815

Solid. Mp 155° dec.

Bromide: [1779-58-4].

$C_{21}H_{20}BrO_2P$ M 415.266

Cryst. (H_2O). Mp 164-165° dec.

Iodide: [39720-65-5].

$C_{21}H_{20}IO_2P$ M 462.266

Cryst. (CH_2Cl_2/Et_2O). Mp 162-163°.

Tetrafluoroborate:

$C_{21}H_{20}BF_4O_2P$ M 422.165

Cryst. (CH_2Cl_2/Et_2O). Mp 130-131°.

Isler, O. *et al*, *Helv. Chim. Acta*, 1957, **40**, 1242 (*synth*)
Grim, S.O. *et al*, *J. Phys. Chem.*, 1966, **70**, 581 (*nmr*)
Gallagher, M.J. *et al*, *Aust. J. Chem.*, 1968, **21**, 1197 (*pmr*)
Albright, T.A. *et al*, *J. Am. Chem. Soc.*, 1976, **98**, 6249 (*cmr*)
Nesterov, L.V. *et al*, *Zh. Obshch. Khim.*, 1977, **47**, 1259 (*Engl. transl.* p. 1161) (*bromide, iodide, nmr*)
Ayrey, G. *et al*, *J. Labelled Compd. Radiopharm.*, 1978, **14**, 935 (*bromide, ylide*)
Lang, R.W. *et al*, *Helv. Chim. Acta*, 1980, **63**, 438 (*bromide, use*)

(2-Methoxyphenyl)diphenylphosphine, 9CI M-00056

o-*(Diphenylphosphino)methoxybenzene*. o-*(Diphenylphosphino)anisole*. o-*Anisyldiphenylphosphine*

[53111-20-9]

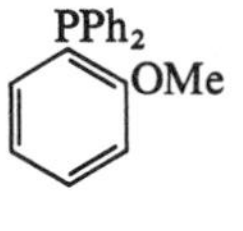

$C_{19}H_{17}OP$ M 292.316

Ligand for Ir, Ni, Pt, Rh and Ru. Prisms (C_6H_6/EtOH or MeOH). Mp 124-126°.

B,MeI: (2-Methoxyphenyl)methyldiphenylphosphonium iodide.

$C_{20}H_{20}IOP$ M 434.256

Microcryst. (Me_2CO). Mp 164-166°.

B,PhCH₂Br: Benzyl(2-methoxyphenyl)-diphenylphosphonium bromide.

$C_{26}H_{24}BrOP$ M 463.353

Cryst. (EtOH/EtOAc). Mp 229-231°.

Oxide: [51986-54-0].

$C_{19}H_{17}O_2P$ M 308.316

Cryst. (EtOH). Mp 163-165°. Forms Sn complex.

Sulfide: [60254-11-7].

$C_{19}H_{17}OPS$ M 324.376

Cryst. (C_6H_6). Mp 159-160°.

Amarskii, E.G. *et al*, *Zh. Obshch. Khim.*, 1974, **44**, 461 (*Engl. transl.* p. 443) (*oxide, synth*)
Jones, C.E. *et al*, *J. Chem. Soc., Dalton Trans.*, 1974, 992 (*synth, pmr, complexes*)
McEwen, W.E. *et al*, *J. Am. Chem. Soc.*, 1975, **97**, 1787 (*ms, uv, pmr, nmr*)

Henning, H.-G. *et al*, *J. Prakt. Chem.*, 1976, **318**, 69 (*synth, deriv*)
Shvets, A.A. *et al*, *Zh. Obshch. Khim.*, 1976, **46**, 1701 (*Engl. transl.* p. 1654) (*oxide, ir*)
Grim, S.O. *et al*, *Phosphorus Sulfur*, 1977, **3**, 191 (*synth, nmr*)
Roundhill, D.M. *et al*, *Inorg. Chem.*, 1980, **19**, 3365 (*cmr*)

(3-Methoxyphenyl)diphenylphosphine, 9CI **M-00057**

m-*(Diphenylphosphino)methoxybenzene*. m-*(Diphenylphosphino)anisole*. m-*Anisyldiphenylphosphine*

[13145-84-1]

$C_{19}H_{17}OP$ M 292.316

Cryst. (EtOH). Mp 60-61°. Bp_{15} 220°.

B,PhCH₂Cl: Benzyl(3-methoxyphenyl)-diphenylphosphonium chloride.
$C_{26}H_{24}ClOP$ M 418.902
Cryst. (EtOAc/$CHCl_3$). Mp 262-263°.

Oxide: m-*(Diphenylphosphinyl)anisole.*
$C_{19}H_{17}O_2P$ M 308.316
Cryst. (pet. ether). Mp 112-113°.

Horner, L. *et al*, *Chem. Ber.*, 1958, **91**, 52 (*oxide*)
Lamza, L., *J. Prakt. Chem.*, 1964, **25**, 294 (*synth, derivs*)
McEwen, W.E. *et al*, *J. Am. Chem. Soc.*, 1975, **97**, 1787 (*synth, ms, pmr, nmr*)

(4-Methoxyphenyl)diphenylphosphine, 9CI **M-00058**

p-*Anisyldiphenylphosphine*. p-*(Diphenylphosphino)-anisole*

[896-89-9]

$C_{19}H_{17}OP$ M 292.316

Cryst. solid (MeOH). Mp 64-65°, 78-79°.

B,MeI: (4-Methoxyphenyl)methyldiphenylphosphonium iodide.
$C_{20}H_{20}IOP$ M 434.256
Solid. Mp 132-133°.

B,PhCH₂Cl: Benzyl(4-methoxyphenyl)-diphenylphosphonium chloride.
$C_{26}H_{24}ClOP$ M 418.902
Cryst. (EtOH/EtOAc). Mp 202-203°.

Oxide: [795-44-8]. p-*(Diphenylphosphinyl)anisole.*
$C_{19}H_{17}O_2P$ M 308.316
Cryst. (C_6H_6/pet. ether). Mp 116-117° (105-108°).

Sulfide: [14180-49-5]. p-*(Diphenylphosphinothioyl)-anisole.*
$C_{19}H_{17}OPS$ M 324.376
Solid (EtOH). Mp 128-129°.

Goetz, H. *et al*, *Justus Liebigs Ann. Chem.*, 1963, **665**, 1 (*synth, uv, ir, derivs*)
Schiemenz, G.P., *Justus Liebigs Ann. Chem.*, 1971, **752**, 30 (*synth, derivs*)
McEwen, W.E. *et al*, *J. Am. Chem. Soc.*, 1975, **97**, 1787 (*synth, pmr, uv, ms, nmr, derivs*)

(2-Methoxyphenyl)methylphenylphosphine, 9CI **M-00059**

Methyl(2-methoxyphenyl)phenylphosphine. o-*Anisylmethylphenylphosphine*. *PAMP*

[1485-88-7]

(*S*)-*form*

$C_{14}H_{15}OP$ M 230.246

One of the most effective non-chelating ligands for use in catalysts for asymmetric homogeneous hydrogenations.

(*S*)-*form* [65337-14-6]
$[\alpha]_D$ +13.6° (MeOH).

Oxide:
$C_{14}H_{15}O_2P$ M 246.245
Solid. Mp 70-75°. $[\alpha]_D^{20}$ +25.9° (c, 1.0 in MeOH). Has (*R*)-config.

Vineyard, B.D. *et al*, *J. Am. Chem. Soc.*, 1977, **99**, 5946 (*synth, oxide*)
Solodar, J., *J. Org. Chem.*, 1978, **43**, 1787 (*complex, use*)
Chodkiewicz, W. *et al*, *Tetrahedron Lett.*, 1979, 1069, 3573 (*synth*)
Brown, J. *et al*, *J. Am. Chem. Soc.*, 1980, **102**, 3040 (*complexes*)
Trost, B.M. *et al*, *Tetrahedron Lett.*, 1981, **22**, 4929 (*use*)

[(3-Methoxyphenyl)methyl]phosphonic acid, 9CI **M-00060**

m-*Methoxybenzylphosphonic acid*

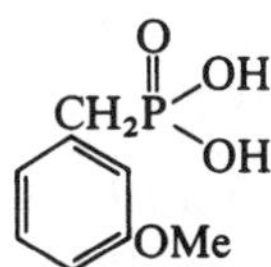

$C_8H_{11}O_4P$ M 202.146

Di-Et ester: [60815-18-2]. *Diethyl [(3-methoxyphenyl)-methyl]phosphonate.*
$C_{12}H_{19}O_4P$ M 258.253
Wittig-Horner reagent. Liq. $Bp_{0.05}$ 121-124°.

Zimmerman, H.E. *et al*, *J. Am. Chem. Soc.*, 1976, **98**, 5574 (*synth, ir, pmr*)

[(4-Methoxyphenyl)methyl]phosphonic acid, 9CI **M-00061**

p-*Methoxybenzylphosphonic acid*

[40299-61-4]

$C_8H_{11}O_4P$ M 202.146

Cryst. (H_2O). Mp 204-206°.

Di-Me ester: [17105-65-6]. *Dimethyl [(4-methoxyphenyl)methyl]phosphonate.*
$C_{10}H_{15}O_4P$ M 230.200
Wittig-Horner reagent. Oil. d_4^{20} 1.23. $Bp_{0.21}$ 125-128°. n_D^{20} 1.5110.

Di-Et ester: [1145-93-3]. *Diethyl [(4-methoxyphenyl)-methyl]phosphonate.*
$C_{12}H_{19}O_4P$ M 258.253
Wittig-Horner reagent. Liq. d_4^{20} 1.13. Bp_3 176-178°, $Bp_{0.25}$ 123-125°. n_D^{20} 1.5040.

Prokof'eva, A.F. *et al*, *Zh. Obshch. Khim.*, 1971, **41**, 1702 (*Engl. transl.* p. 1710) (*ester*)
Williams, A. *et al*, *J. Chem. Soc., Perkin Trans. 2*, 1973, 25 (*synth, props*)

Franke, A. *et al*, *Synthesis*, 1979, 712 (*ester, synth, use*)
Fresneda, P.M. *et al*, *Synthesis*, 1981, 222 (*ester, synth*)

[(2-Methoxyphenyl)methyl]-triphenylphosphonium(1+), 10CI, 9CI M-00062

(o-*Methoxybenzyl)triphenylphosphonium(1+)*, *8CI*

$C_{26}H_{24}OP^{\oplus}$ M 383.449 (ion)

Salts are source of ylide, used in Wittig reactions.

Chloride: [52045-25-7].
$C_{26}H_{24}ClOP$ M 418.902
Prisms (CH_2Cl_2/Et_2O). Mp 249-251°.

Bromide: [64820-07-1].
$C_{26}H_{24}BrOP$ M 463.353
Solid. Mp 229-231°.

Ylide: [59659-68-6]. [*(2-Methoxyphenyl)methylene]-triphenylphosphorane*, *9CI*. (*2-Methoxybenzylidene)-triphenylphosphorane*, *8CI*.
$C_{26}H_{23}OP$ M 382.441
Used in Wittig reactions.

Cresp, T.M. *et al*, *J. Chem. Soc., Perkin Trans. 1*, 1974, 2435 (*chloride, synth*)
Yamoto, M. *et al*, *Chem. Pharm. Bull.*, 1977, **25**, 706 (*chloride, synth, use*)
Reimann, E., *Arch. Pharm.* (*Weinheim, Ger.*), 1979, **312**, 772 (*bromide, synth, use*)
Brown, C. *et al*, *J. Chem. Soc., Perkin Trans. 1*, 1982, 3007 (*use*)

[(3-Methoxyphenyl)methyl]-triphenylphosphonium(1+), 9CI M-00063

(m-*Methoxybenzyl)triphenylphosphonium(1+)*, *8CI*

$C_{26}H_{24}OP^{\oplus}$ M 383.449 (ion)

Salts are source of ylide, used in Wittig reactions.

Chloride: [18880-05-2].
$C_{26}H_{24}ClOP$ M 418.902
Solid. Mp 278-279°.

Bromide: [72311-12-7].
$C_{26}H_{24}BrOP$ M 463.353
Solid. Mp 265-267°.

Ylide: [64970-91-8]. [*(3-Methoxyphenyl)methylene]-triphenylphosphorane*, *9CI*. (*3-Methoxybenzylidene)-triphenylphosphorane*, *8CI*.
$C_{26}H_{23}OP$ M 382.441
Used in Wittig reactions.

Confalone, P.N. *et al*, *J. Am. Chem. Soc.*, 1977, **99**, 7020 (*use*)
Yamoto, M. *et al*, *Chem. Pharm. Bull.*, 1977, **25**, 706 (*chloride, synth, use*)
Matsumoto, T. *et al*, *Bull. Chem. Soc. Jpn.*, 1979, **52**, 212 (*chloride, use*)
Reimann, E., *Arch. Pharm.* (*Weinheim, Ger.*), 1979, **312**, 772 (*bromide, synth, use*)

[(4-Methoxyphenyl)methyl]-triphenylphosphonium(1+), 9CI M-00064

(p-*Methoxybenzyl)triphenylphosphonium(1+)*, *8CI*

$C_{26}H_{24}OP^{\oplus}$ M 383.449 (ion)

Chloride: [3462-97-3].
$C_{26}H_{24}ClOP$ M 418.902
Cryst. ($CHCl_3$/pet. ether). Mp 245-247°.

Bromide: [1530-38-7].
$C_{26}H_{24}BrOP$ M 463.353
Solid. Mp 242-243°.

Ylide: [21960-26-9]. [*(4-Methoxyphenyl)methylene]-triphenylphosphorane*, *10CI*, *9CI*. *4-Methoxybenzylidenetriphenylphosphorane*, *8CI*.
$C_{26}H_{23}OP$ M 382.441
Used in Wittig reactions.

Ketcham, R. *et al*, *J. Org. Chem.*, 1962, **27**, 4666 (*chloride, synth, use*)
Griffin, C.E. *et al*, *J. Organomet. Chem.*, 1965, **3**, 414 (*pmr*)
Leznoff, C.C. *et al*, *Can. J. Chem.*, 1972, **50**, 528 (*bromide, synth, use*)
Yamato, M. *et al*, *Chem. Pharm. Bull.*, 1977, **25**, 706 (*chloride, synth, use*)
Reimann, E., *Arch. Pharm.* (*Weinheim, Ger.*), 1979, **312**, 772 (*bromide, synth, use*)

(2-Methoxyphenyl)-2-naphthalenylphenyl-phosphine, 9CI M-00065

o-*Anisyl-β-naphthylphenylphosphine*

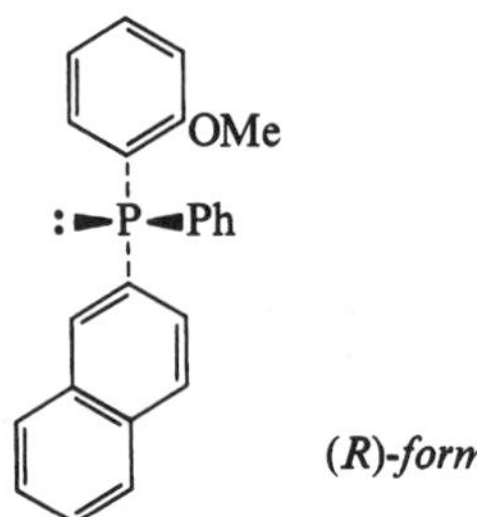

$C_{23}H_{19}OP$ M 342.376

(R)-form

$[\alpha]_D$ +8.1° (C_6H_6), $[\alpha]_{400}$+35.5° (C_6H_6).

Oxide:
$C_{23}H_{19}O_2P$ M 358.376
Glass. $Bp_{0.05}$ 220° (oven). $[\alpha]_D$ −1.81° (c, 1-3 in MeOH) (44% o.p.), $[\alpha]_{400}$−28.5° (MeOH). Oxide has (*S*)-config.

(S)-form

$[\alpha]_D$ −10.0° (C_6H_6), $[\alpha]_{400}$−43.8° (C_6H_6).

Oxide: $[\alpha]_{400}$+9.0° (C_6H_6).

(±)-form

Oxide: Cryst. (CH_2Cl_2/hexane). Mp 139-142°. $Bp_{0.1}$ 210-220° (kugelrohr). Subl. at 160°/0.5 mm.

Lewis, R.A. *et al*, *J. Am. Chem. Soc.*, 1969, **91**, 7009 (*oxide, synth, pmr*)
Naumann, K. *et al*, *J. Am. Chem. Soc.*, 1969, **91**, 7012 (*synth*)

(4-Methoxyphenyl)-1-naphthalenylphenyl-phosphine, 9CI M-00066

p-*Anisyl-α-naphthylphenylphosphine*

$C_{23}H_{19}OP$ M 342.376

(+)-form [64822-31-7]

Mp 120°. $[\alpha]_D$ +5° (c, 0.09 in toluene).

B,PhCH₂Cl: *(4-Methoxyphenyl)methyl-1-naphthalenylphenylphosphonium chloride*. p-*Anisylmethyl-α-naphthylphenylphosphonium chloride*. Solid. Mp 253°. $[\alpha]_D$ +15.7° (c, 1.49 in MeOH).

B,PhCH₂Br: *(4-Methoxyphenyl)methyl-1-naphthalenylphenylphosphonium bromide*. p-*Anisylmethyl-α-naphthylphenylphosphonium bromide*. $[\alpha]_D$ +11.0° (c, 0.137 in $PhNO_2$/DMF).

(±)-***form*** [64822-27-1]
Mp 124°.
B,PhCH₂Cl:
$C_{24}H_{22}ClOP$ M 392.864
Solid. Mp 273° dec.
B,PhCH₂Br:
$C_{24}H_{22}BrOP$ M 437.315
Solid. Mp 248° dec.
Oxide:
$C_{23}H_{19}O_2P$ M 358.376
$Bp_{0.5}$ 260°.

Horner, L. *et al, Tetrahedron Lett.*, 1964, 1421 (*synth, resoln*)
Lewis, R.A. *et al, J. Am. Chem. Soc.*, 1969, **91**, 7009 (*synth, oxide*)
Tani, K. *et al, J. Am. Chem. Soc.*, 1977, **99**, 7876 (*resoln, complexes*)

(4-Methoxyphenyl)phosphine, 9CI M-00067

p-*Anisylphosphine*
[13153-12-3]

C_7H_9OP M 140.121
Liq. Bp_{11} 89-91°. Forms As, Ta, Os, and Ru complexes.

Maier, L., *Phosphorus*, 1974, **4**, 41 (*synth, pmr, nmr*)
Huttner, G. *et al, J. Organomet. Chem.*, 1980, **191**, 161 (*complexes*)

(2-Methoxyphenyl)phosphinic acid, 9CI M-00068

o-*Anisylphosphinic acid*

$C_7H_9O_3P$ M 172.120
Tautomeric with 2-methoxyphenylphosphonous acid, but free acid probably exists in phosphinic acid form. Cryst. (CCl_4/EtOH). Mp 100-102°.

Quin, L.D. *et al, J. Am. Chem. Soc.*, 1961, **83**, 4124.

(4-Methoxyphenyl)phosphinic acid, 9CI M-00069

p-*Anisylphosphinic acid*
[53534-65-9]
$C_7H_9O_3P$ M 172.120
Tautomeric with (4-methoxyphenyl)phosphonous acid, but free acid probably exists in phosphinic acid form. Solid. Mp 110.5-112.5°. pK_a 1.75 (H_2O), ca. 3.5 (EtOH).
Anilinium salt: Solid. Mp 97-98°.
Et ester: [38766-21-1]. *Ethyl (4-methoxyphenyl)-phosphinate.*
$C_9H_{13}O_3P$ M 200.174
Characterised spectroscopically.

Vinogradov, L.I. *et al, Zh. Obshch. Khim.*, 1972, **42**, 1724 (*Engl. transl.* p. 1712) (*ester, ir*)
Herrin, T.R. *et al, J. Med. Chem.*, 1977, **20**, 660 (*synth*)

(2-Methoxyphenyl)phosphonic acid, 9CI M-00070

o-*Anisylphosphonic acid*
[7506-85-6]

$C_7H_9O_4P$ M 188.119
Cryst. (6M HCl). Mp 179-182°, Mp 204-206°. pK_{a1} 2.16 (H_2O), 3.62 (50% EtOH aq.), pK_{a2} 7.77 (H_2O), 8.87 (50% EtOH aq.).
Di-Me ester: [15286-16-5]. *Dimethyl (2-methoxyphenyl)phosphonate.*
$C_9H_{13}O_4P$ M 216.173
Liq. $Bp_{0.1}$ 110-112°.
Di-Et ester: [15286-17-6]. *Diethyl (2-methoxyphenyl)-phosphonate.*
$C_{11}H_{17}O_4P$ M 244.227
Liq. $Bp_{0.14}$ 122-124°.
Bis(trimethylsilyl) ester: [99136-02-4]. *Bis(trimethylsilyl)(2-methoxyphenyl)phosphonate.*
$C_{13}H_{25}O_4PSi_2$ M 332.483
Liq. $Bp_{0.001}$ 95°.

Dawson, N.D. *et al, J. Org. Chem.*, 1953, **18**, 207 (*synth, ester*)
Obrycki, R. *et al, J. Org. Chem.*, 1968, **33**, 632 (*esters, ir*)
Nuallain, C.O., *J. Inorg. Nucl. Chem.*, 1974, **36**, 339 (*props*)
Bunnett, J.F. *et al, J. Org. Chem.*, 1978, **43**, 1867 (*ester, ir, pmr, ms*)
Grabiak, R.C. *et al, Phosphorus Sulfur*, 1980, **9**, 197 (*synth, derivs, pmr, P nmr*)
Issleib, K. *et al, Z. Anorg. Allg. Chem.*, 1985, **529**, 151 (*silyl ester, synth, P nmr*)

(3-Methoxyphenyl)phosphonic acid, 9CI M-00071

m-*Anisylphosphonic acid*
[77918-49-1]
$C_7H_9O_4P$ M 188.119
Solid. Mp 139-142°.
Di-Me ester: Dimethyl (3-methoxyphenyl)phosphonate.
$C_9H_{13}O_4P$ M 216.173
Liq. $Bp_{0.1}$ 116-117°.
Di-Et ester: [65442-22-0]. *Diethyl (3-methoxyphenyl)-phosphonate.*
$C_{11}H_{17}O_4P$ M 244.227
Liq. $Bp_{1.5}$ 143°.

Obrycki, R. *et al, J. Org. Chem.*, 1968, **33**, 632 (*ester, synth, ir*)
Bunnett, J.F. *et al, J. Org. Chem.*, 1978, **43**, 1867 (*ester, synth, ir, pmr, ms*)
Grabiak, R.C. *et al, Phosphorus Sulfur*, 1980, **9**, 197 (*synth, derivs, P nmr*)

(4-Methoxyphenyl)phosphonic acid, 9CI M-00072

p-*Anisylphosphonic acid*
[21778-19-8]
$C_7H_9O_4P$ M 188.119
Cryst. (H_2O). Mp 179-179.5° (168-169°). pK_{a1} 2.17 (2.32) (H_2O), pK_{a2} 7.90 (5.6) (H_2O).
Di-Me ester: [15286-19-8]. *Dimethyl (4-methoxyphenyl)phosphonate.*
$C_9H_{13}O_4P$ M 216.173

Liq. $Bp_{0.1}$ 113-114°.
Di-Et ester: [3762-33-2]. *Diethyl (4-methoxyphenyl)-phosphonate.*
$C_{11}H_{17}O_4P$ M 244.227
Liq. d_4^{20} 1.20. Bp_2 152-154°, $Bp_{0.4}$ 117-118°. n_D^{20} 1.5109.
Di-Ph ester: [85599-22-0]. *Diphenyl (4-methoxyphenyl)phosphonate.*
$C_{19}H_{17}O_4P$ M 340.315
Liq. $Bp_{2.5}$ 233-235°.
Dichloride: see (4-Methoxyphenyl)phosphonic dichloride, M-00073
Dianilide: P-*(4-Methoxyphenyl)*-N,N′-*diphenylphosphonic diamide.* p-*Anisylphosphonic dianilide.*
$C_{19}H_{19}N_2O_2P$ M 338.345
Cryst. (Me_2CO/Et_2O). Mp 210°.

Morrison, D.C., *J. Am. Chem. Soc.*, 1951, **73**, 5896 (*dianilide*)
Obrycki, R. *et al*, *J. Org. Chem.*, 1968, **33**, 632 (*esters, synth, pmr*)
Barabanov, V.P. *et al*, *Zh. Obshch. Khim.*, 1972, **42**, 2431 (*Engl. transl.* p. 2425) (*synth, props*)
Kharrasova, F.M. *et al*, *Zh. Obshch. Khim.*, 1973, **43**, 2642 (*Engl. transl.* p. 2621) (*esters, P nmr*)
Williams, A. *et al*, *J. Chem. Soc., Perkin Trans. 2*, 1973, 25 (*synth*)
Daasch, L.W. *et al*, *Phosphorus*, 1975, **5**, 189 (*difluoride, ms*)
Allen, D.W. *et al*, *J. Chem. Soc., Perkin Trans. 2*, 1977, 789 (*ester, synth, pmr, P nmr*)
Hirao, T. *et al*, *Synthesis*, 1981, 56; *Bull. Chem. Soc. Jpn.*, 1982, **55**, 909 (*diethyl ester, ir*)
Osuka, A. *et al*, *Synthesis*, 1983, 69 (*diethyl, diphenyl esters, ir, pmr*)

(4-Methoxyphenyl)phosphonic dichloride, 8CI — M-00073

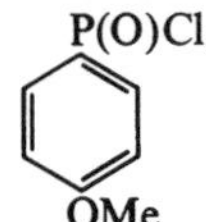

$C_7H_7Cl_2O_2P$ M 225.011
Liq. $Bp_{6\text{-}7}$ 149-153°.

Lecher, H.Z. *et al*, *J. Am. Chem. Soc.*, 1956, **78**, 5018 (*synth*)
Kharrasova, F.M. *et al*, *Zh. Obshch. Khim.*, 1973, **43**, 2642 (*Engl. transl.* p. 2621) (*P nmr*)
Dorokhova, V.V. *et al*, *Zh. Obshch. Khim.*, 1979, **49**, 83 (*Engl. transl.* p. 68) (*raman, uv*)
Sergienko, L.M. *et al*, *Zh. Obshch. Khim.*, 1980, **50**, 1958 (*Engl. transl.* p. 1978) (*uv*)
Ratovskii, G.V. *et al*, *Zh. Obshch. Khim.*, 1981, **51**, 714 (*Engl. transl.* p. 580) (*raman*)

(2-Methoxyphenyl)phosphoramidic acid, 9CI — M-00074

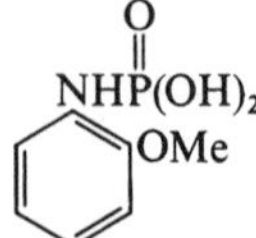

$C_7H_{10}NO_4P$ M 203.134
Di-Me ester: Dimethyl (2-methoxyphenyl)-phosphoramidate.
$C_9H_{14}NO_4P$ M 231.188
Solid. Mp 112-113°.
Difluoride:
$C_7H_8F_2O_2P$ M 193.110
Solid. Mp 61°. Bp_6 105-106°.

Olah, G.A. *et al*, *J. Org. Chem.*, 1959, **24**, 1443 (*difluoride*)
Cadogan, J.I.G. *et al*, *J. Chem. Soc. (C)*, 1969, 2813 (*ester*)

(3-Methoxyphenyl)phosphoramidic acid, 9CI — M-00075

$C_7H_{10}NO_4P$ M 203.134
Bis(cyclohexylammonium) salt: Solid. Indefinite Mp.
Di-Me ester: Dimethyl (3-methoxyphenyl)-phosphoramidate.
$C_9H_{14}NO_4P$ M 231.188
No phys. props. reported.

Williams, A. *et al*, *J. Chem. Soc. (B)*, 1971, 1973 (*synth, uv*)
Foulds, G.A. *et al*, *S. Afr. J. Chem.*, 1981, **34**, 72 (*dimethyl ester, ir*)

(4-Methoxyphenyl)phosphoramidic acid, 9CI — M-00076

$C_7H_{10}NO_4P$ M 203.134
Di-Me ester: [25627-05-8]. *Dimethyl (4-methoxyphenyl)phosphoramidate.*
$C_9H_{14}NO_4P$ M 231.188
Solid. Mp 109-110°.
Di-Et ester: Diethyl (4-methoxyphenyl)-phosphoramidate.
$C_{11}H_{18}NO_4P$ M 259.241
Solid. Mp 59-60°.
Di-Ph ester: [62753-76-8]. *Diphenyl (4-methoxyphenyl)phosphoramidate.*
$C_{19}H_{18}NO_4P$ M 355.329
Cryst. (EtOH). Mp 139-141°.
Dichloride:
$C_7H_8Cl_2NO_2P$ M 240.025
Cryst. (Et_2O). Mp 71-72°.

Kropacheva, A.A. *et al*, *Zh. Obshch. Khim.*, 1959, **29**, 556 (*Engl. transl.* p. 553) (*dichloride*)
Zhmurova, I.N. *et al*, *Zh. Obshch. Khim.*, 1961, **31**, 3741 (*Engl. transl.* p. 3495) (*diphenyl ester*)
Cadogan, J.I.G. *et al*, *J. Chem. Soc. (C)*, 1969, 2813 (*esters*)
Williams, A. *et al*, *J. Chem. Soc. (B)*, 1971, 1973 (*synth, uv*)
Buchanan, G.W. *et al*, *Can. J. Chem.*, 1980, **58**, 2442 (*N nmr*)
Buchanan, G.W. *et al*, *Org. Magn. Reson.*, 1980, **14**, 517 (*cmr*)
du Plessis, M.P. *et al*, *S. Afr. J. Chem.*, 1980, **33**, 124 (*dimethyl ester, cryst struct*)
Davidowitz, B. *et al*, *Org. Mass Spectrom.*, 1984, **19**, 128 (*dimethyl ester, ms*)

(2-Methoxyphenyl)phosphoramidothioic acid, 9CI — M-00077

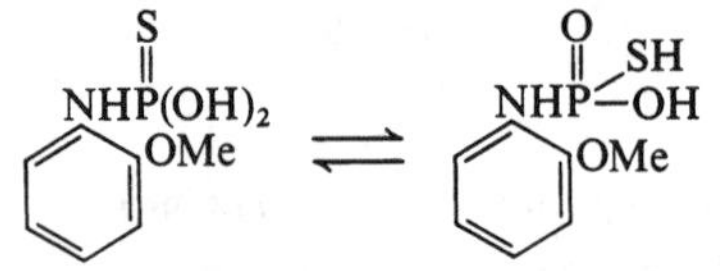

$C_7H_{10}NO_3PS$ M 219.195
O,O-*Di-Et ester:* [15832-95-8]. O,O-*Diethyl (2-methoxyphenyl)phosphoramidothioate.*
$C_{11}H_{18}NO_3PS$ M 275.302
Liq. n_D^{25} 1.5213.
Difluoride:
$C_7H_8F_2NOPS$ M 223.177
Solid. Mp 46-47°. Bp_1 93°.

Olah, G.A. *et al*, *J. Org. Chem.*, 1959, **24**, 1443 (*difluoride*)
Japan. Pat., 62 4 458, (*1962*); *CA*, **67**, 73347

(3-Methoxyphenyl)phosphoramidothioic acid, 9CI, 8CI — M-00078

$C_7H_{10}NO_3PS$ M 219.195

O,O-*Di-Et ester:* [15833-50-8]. O,O-*Diethyl (3-methoxyphenyl)phosphoramidothioate.*
$C_{11}H_{18}NO_3PS$ M 275.302
Liq. n_D^{23} 1.5344.

Japan. Pat., 62 4 458, (*1962*); *CA*, **67**, 73347

(4-Methoxyphenyl)phosphoramidothioic acid, 9CI **M-00079**

$C_7H_{10}NO_3PS$ M 219.195

O,O-*Di-Et ester:* [15832-94-7]. O,O-*Diethyl (4-methoxyphenyl)phosphoramidothioate.*
$C_{11}H_{18}NO_3PS$ M 275.302
Liq. n_D^{24} 1.5370.

O,O-*Di-Ph ester:* O,O-*Diphenyl (4-methoxyphenyl)-phosphoramidothioate.*
$C_{19}H_{18}NO_3PS$ M 371.390
Cryst. (EtOH/3-methylbutanol). Mp 83-85°.

Borke, M.L. *et al*, *J. Am. Pharm. Assoc.*, 1958, **47**, 461.

3-Methoxyphenyl phosphorodichloridate, 9CI **M-00080**

3-Methoxyphenyl phosphoryl dichloride. 3-Methoxyphenyl dichlorophosphate

[20464-67-9]

$C_7H_7Cl_2O_3P$ M 241.010

Liq. d_4^{19} 1.44. Bp_5 132-133°. n_D^{19} 1.5330.

Katyshkina, V.V. *et al*, *Zh. Obshch. Khim.*, 1956, **26**, 3060 (*Engl. transl.* p. 3407)
Kuz'menko, I.I. *et al*, *Zh. Obshch. Khim.*, 1968, **38**, 158 (*Engl. transl.* p. 156)

4-Methoxyphenyl phosphorodichloridate, 9CI **M-00081**

4-Methoxyphenyl phosphoryl dichloride. 4-Methoxyphenyl dichlorophosphate

[20464-68-0]

$C_7H_7Cl_2O_3P$ M 241.010

Liq. d_4^{23} 1.43. Bp_6 140°. n_D^{23} 1.5305.

Katyshkina, V.V. *et al*, *Zh. Obshch. Khim.*, 1956, **26**, 3060 (*Engl. transl.* p. 3407)
Kuz'menko, I.I. *et al*, *Zh. Obshch. Khim.*, 1968, **38**, 158 (*Engl. transl.* p. 156)

***O*-(4-Methoxyphenyl) phosphorodichloridothioate, 9CI** **M-00082**

O-*(4-Methoxyphenyl) thiophosphoryl dichloride.* O-*(4-Methoxyphenyl) dichlorothiophosphate*

[18961-94-9]

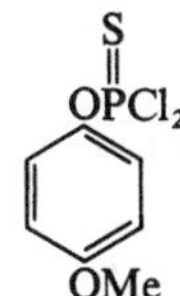

$C_7H_7Cl_2O_2PS$ M 257.071

Liq. Bp_3 127-134°.

U.S.P., 3 022 329, (*1962*); *CA*, **57**, 9743

2-(4-Methoxyphenyl)-1,3,2-thiazaphosphetidine 2-sulfide, 9CI **M-00083**

$C_8H_{10}NOPS_2$ M 231.267

Oil.

N-Me: [79922-17-1].
$C_9H_{12}NOPS_2$ M 245.294
Oil.

Joergensen, K.A. *et al*, *Bull. Soc. Chim. Belg.*, 1980, **89**, 247 (*synth*)

(3-Methoxy-2-propenyl)-triphenylphosphonium(1+) **M-00084**

$Ph_3P^{\oplus}CH_2CH{=}CHOMe$

$C_{22}H_{20}OP^{\oplus}$ M 331.373 (ion)

Bromide:
$C_{22}H_{20}BrOP$ M 411.277
Source of ylide. Solid. Mp 209-210°. Butyllithium → ylide.

Ylide: [82784-34-7]. *(3-Methoxy-2-propenylidene)-triphenylphosphorane. 3-Methoxyallylidenetriphenylphosphorane.*
$C_{22}H_{21}OP$ M 332.381
Used for 3-carbon homologation.

Martin, S.F. *et al*, *Tetrahedron Lett.*, 1977, 3875 (*synth, use*)

2-Methoxy-4,4,5,5-tetramethyl-1,3,2-dioxaphospholane, 8CI **M-00085**

Methyl tetramethylethylene phosphite. Methyl pinacolyl phosphite

[14812-60-3]

$C_7H_{15}O_3P$ M 178.167

Bp_{48} 91-92.5°, Bp_{11} 63°, Bp_1 38-39°. n_D^{20} 1.4417.

2-Oxide: [7443-26-7]. *Methyl tetramethylethylene phosphate.*
$C_7H_{15}O_4P$ M 194.167
Solid. Mp 101-103°. Forms adducts with Et_3N.

Fontal, B. *et al*, *Tetrahedron*, 1966, **22**, 3275 (*synth, pmr*)
Denny, D.Z. *et al*, *J. Am. Chem. Soc.*, 1969, **91**, 6838 (*synth, oxide*)
Burgada, R., *Bull. Soc. Chim. Fr.*, 1971, 136 (*synth*)
Pouchoulin, G. *et al*, *Org. Magn. Reson.*, 1976, **8**, 518 (*pmr, cmr, P nmr*)
Besserre, D. *et al*, *Org. Magn. Reson.*, 1980, **13**, 235 (*cmr*)
Yasuda, H. *et al*, *Macromolecules*, 1982, **15**, 1231 (*oxide, pmr, ir*)

Methylacetophos **M-00086**

Methyl [(diethoxyphosphinothioyl)thio]acetate, 9CI

$(EtO)_2P(S)SCH_2COOMe$

$C_7H_{15}O_4PS_2$ M 258.287

Liq. d_{20}^{20} 1.19. $Bp_{0.15}$ 122°. n_D^{20} 1.4994. Less toxic than ethyl ester. Low mammalian toxicity claimed.

U.S.P., 3 047 459, (*1962*); *CA*, **58**, 1349 (*synth, use*)

10-Methylacridophosphine, 9CI M-00087

10-Methyl-9-phosphaanthracene. 10-Methyldibenzo[b,e]*phosphorin. 10-Methyldibenzo*[be]-*phosphinine*

[57422-79-4]

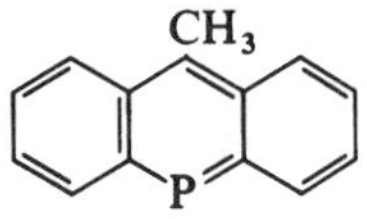

$C_{14}H_{11}P$ M 210.215

Yellow cryst. (cyclohexane). Mp 120° dec. With maleic anhydride, forms a Diels-Alder adduct. Mp 271-4° ($CHCl_3$).

Jongsma, C. *et al*, *Org. Mass Spectrom.*, 1975, **10**, 515 (*ms*)
Jongsma, C. *et al*, *Tetrahedron*, 1975, **31**, 2931 (*uv, ms, synth*)

(Methylaminocarbonyl)phosphonic acid, 9CI M-00088

N-*Methylcarbamoylphosphonic acid*

MeNHCOP(O)(OH)$_2$

$C_2H_6NO_4P$ M 139.047

Monoesters are plant growth regulators.

Di-Me ester: Dimethyl (methylaminocarbonyl)-phosphonate.

$C_4H_{10}NO_4P$ M 167.101

Liq. d_4^{20} 1.29. Bp_2 135-137°. n_D^{20} 1.4585.

Di-Et ester: [59682-40-5]. *Diethyl (methylaminocarbonyl)phosphonate.*

$C_6H_{14}NO_4P$ M 195.155

Liq. d_4^{20} 1.18. Bp_2 142-143°. n_D^{20} 1.4525.

Pudovik, A.N. *et al*, *Zh. Obshch. Khim.*, 1955, **25**, 1369 (*Engl. transl.* p. 1317) (*synth, esters*)

1-Methyl-1,2,5-azadiphospholidine, 9CI M-00089

1-Methyl-1-aza-2,5-diphosphacyclopentane

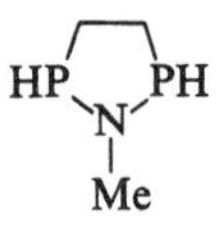

$C_3H_9NP_2$ M 121.058

2,5-Difluoro: [52810-61-4]. *2,5-Difluoro-1-methyl-1,2,5-azadiphospholidine.*

$C_3H_7F_2NP_2$ M 157.039

Fairly unstable volatile liq. $Bp_{4.5}$ 26°.

2,5-Di-Ph: [60600-14-8]. *1-Methyl-2,5-diphenyl-1,2,5-azadiphospholidine.*

$C_{15}H_{17}NP_2$ M 273.254

Solid. Mp 112-114°.

Falardeau, E.R. *et al*, *Inorg. Chem.*, 1975, **14**, 132 (*difluoro, synth, ir, ms, pmr, F nmr*)
Issleib, K. *et al*, *Synth. React. Inorg. Met.-Org. Chem.*, 1976, **6**, 179 (*diphenyl*)

2-Methyl-1,3,2-benzodioxaphosphole, 9CI M-00090

o-*Phenylene methylphosphonite*

[13968-98-4]

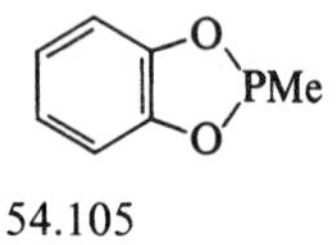

$C_7H_7O_2P$ M 154.105

Liq. Mp 2°. Bp_{10} 76°.

2-Oxide: [13969-05-6]. o-*Phenylene methylphosphonate.*

$C_7H_7O_3P$ M 170.104

Solid. Mp 84°. Bp_2 120°.

2-Sulfide: [15004-44-1]. O,O-o-*Phenylene thiophosphonate.*

$C_7H_7O_2PS$ M 186.165

Solid. Mp 50°. Bp_2 88°.

2-Selenide: O,O-o-*Phenylene selenophosphonate.*

$C_7H_7O_2PSe$ M 233.065

Solid. Mp 46°. Bp_2 90°.

Wieber, M. *et al*, *Chem. Ber.*, 1967, **100**, 974 (*synth, pmr, derivs*)
Wieber, M. *et al*, *Monatsh. Chem.*, 1970, **101**, 776 (*synth, pmr*)
Gloede, J., *Z. Anorg. Allg. Chem.*, 1983, **500**, 59 (*oxide, P nmr*)

Methyl 2,3-bis-*O*-diphenylphosphino-4,6-*O*-benzylideneglucopyranoside M-00091

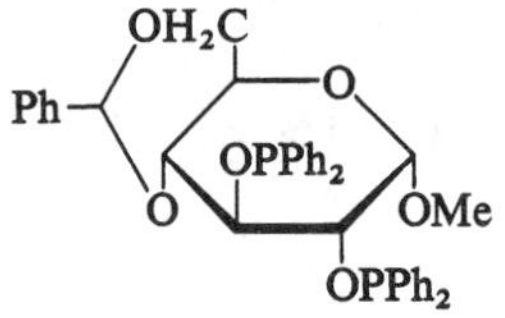

$C_{38}H_{36}O_6P_2$ M 650.646

α-D-form [37605-43-9]

Reagent for asymmetric hydrogenations. Cryst. ($CHCl_3$/pet. ether). Mp 130-131°. $[\alpha]_D^{25}$ −9° (c, 8.8 in $CHCl_3$).

Cullen, W.R. *et al*, *Tetrahedron Lett.*, 1978, 1635 (*synth, use*)

4′-Methyl-4,5-bis(trifluoromethyl)-spiro[1,3,2-dithiaphosphole-2,1′-[2,6,7]-trioxa[1]phosphabicyclo[2.2.2]octane], 9CI M-00092

[60049-49-2]

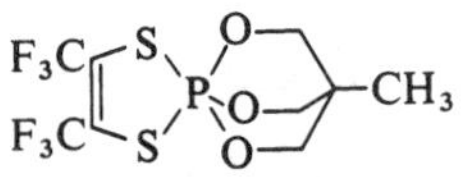

$C_9H_9F_6O_3PS_2$ M 374.252

Solid. Mp 65-67°. $Bp_{0.3}$ 123-125°.

Campbell, B.S. *et al*, *J. Am. Chem. Soc.*, 1976, **98**, 2924 (*synth, pmr, P and F nmr*)

(2-Methyl-1,3-butadienyl)phosphonic acid, 9CI M-00093

[24590-63-4]

$H_3C{=}CHC(CH_3){=}CHP(O)(OH)_2$

$C_5H_9O_3P$ M 148.098

Derivs. obt. as mixts. of (*E*)- and (*Z*)-forms. Viscous light-yellow liq.

Monoanilinium salt: Cryst. (Me_2CO). Mp 149-150°.

Di-Me ester: [4037-12-1]. *Dimethyl (2-methyl-1,3-butadienyl)phosphonate.*

$C_7H_{13}O_3P$ M 176.152

Liq. Bp_1 86-87°. n_D^{20} 1.4870.

Di-Et ester: [7158-34-1]. *Diethyl (2-methyl-1,3-butadienyl)phosphonate.*

$C_9H_{17}O_3P$ M 204.205

Liq. Bp_1 92-93°.

Di-Ph ester: [59611-97-1]. *Diphenyl (2-methyl-1,3-butadienyl)phosphonate.*

$C_{17}H_{17}O_3P$ M 300.293

Liq. d_4^{20} 1.16. n_D^{20} 1.5688.
Dichloride: [4981-27-5].
$C_5H_7Cl_2OP$ M 184.989
Liq. d_4^{20} 1.29. Bp_2 77.5-78°. n_D^{20} 1.5400.
Diamide: see (2-Methyl-1,3-butadienyl)phosphonic diamide, M-00096

Mashlyakovskii, L.N. *et al, Zh. Obshch. Khim.*, 1965, **35**, 1577 (*Engl. transl.* p. 1582) (*dichloride, ir, pmr*)
Mashlyakovskii, L.N. *et al, Zh. Obshch. Khim.*, 1967, **37**, 1307 (*Engl. transl.* p. 1237) (*dichloride, ester, ir, pmr*)
Timofeeva, T.N. *et al, Zh. Obshch. Khim.*, 1969, **39**, 354, 1048 (*Engl. transl.* pp. 332, 1019) (*derivs, pmr*)
Mashlyakovskii, L.N. *et al, Vysokomol. Soedin., Sect. A*, 1976, **18**, 308 (*Engl. transl.* p. 354) (*polymers, pmr, P nmr, ir*)
Shtengel', K.Kh. *et al, Zh. Obshch. Khim.*, 1976, **46**, 434 (*Engl. transl.* p. 430) (*esters, synth*)

(3-Methyl-1,2-butadienyl)phosphonic acid M-00094

[1831-37-4]

$(H_3C)_2C{=}C{=}CHP(O)(OH)_2$

$C_5H_9O_3P$ M 148.098
Di-Me ester: [17166-43-7]. *Dimethyl (3-methyl-1,2-butadienyl)phosphonate.*
$C_7H_{13}O_3P$ M 176.152
Liq. d_4^{20} 1.06. Bp_2 94-95°. n_D^{20} 1.4733.
Di-Et ester: [3201-84-1]. *Diethyl (3-methyl-1,2-butadienyl)phosphonate.*
$C_9H_{17}O_3P$ M 204.205
Liq. d_4^{20} 1.02. Bp_{10} 122-123°. n_D^{20} 1.4642.
Difluoride:
$C_5H_7F_2OP$ M 152.080
Liq. d_4^{20} 1.16. $Bp_{1.5}$ 24°. n_D^{20} 1.4215.
Dichloride:
$C_5H_7Cl_2OP$ M 184.989
Liq. d_4^{20} 1.26. $Bp_{1.5}$ 79°. n_D^{20} 1.5740.

Mark, V., *Tetrahedron Lett.*, 1962, 281 (*ester*)
Pudovik, A.N. *et al, Zh. Obshch. Khim.*, 1965, **35**, 1210; 1969, **39**, 1646 (*Engl. transl.* pp. 1214, 1614) (*esters*)
Cherbuliez, E. *et al, Helv. Chim. Acta*, 1966, **49**, 2395 (*ms*)
Ignat'ev, V.M. *et al, Zh. Obshch. Khim.*, 1967, **37**, 1898; *CA*, **68**, 29796 (*derivs*)
Krudy, G.A. *et al, J. Org. Chem.*, 1978, **43**, 4656 (*cmr*)
Altenbach, H.J. *et al, Tetrahedron Lett.*, 1981, **22**, 5175 (*ester, props, use*)

(3-Methyl-1,3-butadienyl)phosphonic acid M-00095

3-Methyl-1-phosphono-1,3-butadiene

$H_2C{=}C(CH_3)CH{=}CHP(O)(OH)_2$

$C_5H_9O_3P$ M 148.098
Di-Me ester: [65670-22-6]. *Dimethyl (3-methyl-1,3-butadienyl)phosphonate.*
$C_7H_{13}O_3P$ M 176.152
Liq. $Bp_{0.1}$ 80° (oven).
Di-Et ester: [79521-54-3]. *Diethyl (3-methyl-1,3-butadienyl)phosphonate.*
$C_9H_{17}O_3P$ M 204.205
No phys. props. reported.
Difluoride:
$C_5H_7F_2OP$ M 152.080
Liq. d_4^{20} 1.16. Bp_2 29°. n_D^{20} 1.4445.

(E)-form

Dichloride: [70589-76-3].
$C_5H_7Cl_2OP$ M 184.989
Mp 45-47°. Bp_3 93-94°.

(Z)-form

Dichloride: [75779-68-9]. Liq. d_4^{20} 1.41. $Bp_{1-1.5}$ 102-107°. n_D^{20} 1.5458.

Slovokhotova, N.A. *et al, Izv. Akad. Nauk SSSR, Ser. Khim.*, 1961, 71 (*Engl. transl.*, p. 62) (*diethyl ester, ir*)
Welter, W. *et al, Chem. Ber.*, 1978, **111**, 3068 (*dimethyl ester, ir, pmr*)
Shekhade, A.M. *et al, Zh. Obshch. Khim.*, 1979, **49**, 564 (*Engl. transl.* p. 493) (*dichloride, ir, pmr*)
Mikhailova, T.-S. *et al, Zh. Obshch. Khim.*, 1980, **50**, 1690 (*Engl. transl.* p. 1370) (*dichloride, ir, pmr*)

(2-Methyl-1,3-butadienyl)phosphonic diamide M-00096

$H_2C{=}CHC(CH_3){=}CHP(O)(NH_2)_2$

$C_5H_{11}N_2OP$ M 146.128
N,N′-*Di-Me: (2-Methyl-1,3-butadienyl)phosphonic bis(methylamide).* N,N′-*Dimethyl*-P-*(2-methyl-1,3-butadienyl)phosphonic diamide.*
$C_7H_{15}N_2OP$ M 174.182
Cryst. (C_6H_6). Mp 100-101°. Mixture of (*E*)- and (*Z*)-forms.
N,N′-*Diisopropyl: (2-Methyl-1,3-butadienyl)phosphonic bis(isopropylamide).* P-*(2-Methyl-1,3-butadienyl)*-N,N′-*bis(1-methylethyl)phosphonic diamide.*
$C_{11}H_{23}N_2OP$ M 230.289
Cryst. (Et_2O/pet. ether). Mp 49.5-50°. Mixture of (*E*)- and (*Z*)-forms.
N,N,N′,N′-*Tetra-Me: (2-Methyl-1,3-butadienyl)phosphonic bis(dimethylamide).* N,N,N′,N′-*Tetramethyl*-P-*(2-methyl-1,3-butadienyl)phosphonic diamide.*
$C_9H_{19}N_2OP$ M 202.236
Liq. d_4^{20} 1.00. $Bp_{0.5}$ 86-87.5°. n_D^{20} 1.5078. Mixture of (*E*)- and (*Z*)-forms.
N,N,N′,N′-*Tetra-Et: (2-Methyl-1,3-butadienyl)phosphonic bis(diethylamide).* N,N,N′,N′-*Tetraethyl*-P-*(2-methyl-1,3-butadienyl)phosphonic diamide.*
$C_{13}H_{26}N_2OP$ M 257.335
Liq. Bp_3 133-134°. n_D^{20} 1.5028. Mixture of (*E*) and (*Z*)-forms.

Timofeeva, T.N. *et al, Zh. Obshch. Khim.*, 1969, **39**, 1048 (*Engl. transl.* p. 1019) (*synth, pmr*)
Prorubshchikov, A.Yu. *et al, Zh. Obshch. Khim.*, 1982, **52**, 1793 (*Engl. transl.* p. 1587) (*synth, pmr, cmr*)

(2-Methyl-1,3-butadienyl)phosphonothioic acid M-00097

$H_2C{=}CH(H_3C)C{=}CHP(S)(OH)_2 \rightleftharpoons H_2C{=}CH(H_3C)C{=}CHP(O)(SH)(OH)$

(*E*)-*form*

$C_5H_9O_2PS$ M 164.159
O,O-*Di-Me ester:* O,O-*Dimethyl (2-methyl-1,3-butadienyl)phosphonothioate.*
$C_7H_{13}O_2PS$ M 192.212
Liq. d_4^{20} 1.11. Bp_3 75-76°. n_D^{20} 1.5394. Mixture of (*E*) and (*Z*) stereoisomers.
O,O-*Di-Et ester:*
$C_9H_{17}O_2PS$ M 220.266
Liq. d_4^{20} 1.08. Bp_1 93-94°. n_D^{20} 1.5234. Mixture of (*E*) and (*Z*) stereoisomers.
Dichloride: [4981-29-7].
$C_5H_7Cl_2PS$ M 201.050

Liq. d_4^{20} 1.30. $Bp_{0.5\text{-}1.0}$ 68-70°. n_D^{20} 1.5978. Mixture of (*E*) and (*Z*) stereoisomers.
Bis(diethylamide): N,N,N',N'-*Tetraethyl*-P-(*2-methyl-1,3-butadienyl)phosphonothioic diamide.*
$C_{13}H_{27}N_2PS$ M 274.403
Liq. d_4^{20} 1.10. $Bp_{1.0}$ 110-112°. n_D^{20} 1.5600.

(*E*)-form

Bis(dimethylamide): [83630-54-0]. N,N,N',N'-*Tetramethyl*-P-(*2-methyl-1,3-butadienyl)phosphonothioic diamide.*
$C_9H_{19}N_2PS$ M 218.296
Cryst. (hexane). Mp 52.7-53.2°.

Mashlyakovskii, L.N. *et al*, *Zh. Obshch. Khim.*, 1972, **42**, 2648 (*Engl. transl.* p. 2639) (*esters, dichloride, synth, ir, pmr*)
Prorubshchikov, A.Yu. *et al*, *Zh. Obshch. Khim.*, 1982, **52**, 1793 (*Engl. transl* p. 1587) (*esters, amides, synth, ir, P nmr, cmr, pmr*)

3-Methyl-2-butenylidenetriphenylphosphorane, 9CI M-00098

[31188-53-1]

$Ph_3P{=}CHCH{=}C(CH_3)_2$

$C_{23}H_{23}P$ M 330.408
Has been obt. as red cryst. and salt-free.

Bogdanović, B. *et al*, *Synthesis*, 1972, 481 (*synth, pmr, use*)
Barnett, B.L. *et al*, *Cryst. Struct. Commun.*, 1973, **2**, 427 (*cryst struct*)
Spangler, C.W. *et al*, *J. Chem. Soc., Perkin Trans. 2*, 1979, 810 (*use*)
Baba, K. *et al*, *Chem. Pharm. Bull.*, 1981, **29**, 2182 (*use*)

(1-Methyl-3-butenyl)phosphonic acid M-00099

4-Pentene-2-phosphonic acid. 4-Phosphono-1-pentene

$H_2C{=}CHCH_2CH(CH_3)P(O)(OH)_2$

$C_5H_{11}O_3P$ M 150.114

(±)-form

Di-Et ester: Diethyl (1-methyl-3-butenyl)phosphonate.
$C_9H_{19}O_3P$ M 206.221
Liq. $Bp_{0.3}$ 60-62°.

Savignac, P. *et al*, *Synth. Commun.*, 1979, **9**, 487 (*synth, pmr*)

(2-Methyl-1-butenyl)phosphonic acid M-00100

2-Methyl-1-phosphono-1-butene

$H_3CCH_2C(CH_3){=}CHP(O)(OH)_2$

$C_5H_{11}O_3P$ M 150.114
Dichloride: [17051-68-2].
$C_5H_9Cl_2OP$ M 187.005
Liq. d_4^{20} 1.23. Bp_{10} 89-91°. n_D^{20} 1.5000.

Mashlyakovskii, L.N. *et al*, *Zh. Obshch. Khim.*, 1967, **37**, 1307 (*Engl. transl.* p. 1237) (*synth, ir, pmr*)
Moskva, V.V. *et al*, *Zh. Obshch. Khim.*, 1974, **44**, 2621 (*Engl. transl.* p. 2578) (*synth*)

(3-Methyl-1-butenyl)phosphonic acid M-00101

$(H_3C)_2CHCH{=}CHP(O)(OH)_2$

$C_5H_{11}O_3P$ M 150.114
Dichloride: [17051-69-3].
$C_5H_9Cl_2OP$ M 187.005
Liq. (mixture of E and Z isomers). d_4^{20} 1.29. Bp_2 77.5-78°. n_D^{20} 1.5400.

(*E*)-form

Di-Et ester: [33536-50-4]. *Diethyl(3-methyl-1-butenyl)-phosphonate.*
$C_9H_9O_3P$ M 196.142
$Bp_{0.025}$ 45°. n_D^{25} 1.4372.

(*Z*)-form

Di-Et ester: [18689-34-4]. No phys. props. reported.

Ignat'ev, V.M. *et al*, *Zh. Obshch. Khim.*, 1969, **39**, 2433 (*Engl. transl. p. 2374*) (*glc*)
Carey, F.A. *et al*, *J. Org. Chem.*, 1972, **37**, 939 (*synth, pmr*)
Gilmore, W.F. *et al*, *J. Org. Chem.*, 1973, **38**, 1423 (*synth, ir*)

(3-Methyl-2-butenyl)phosphonic acid, 9CI M-00102

(*γ,γ-Dimethylallyl)phosphonic acid. Isopentenylphosphonic acid*

$(H_3C)_2C{=}CHCH_2P(O)(OH)_2$

$C_5H_{11}O_3P$ M 150.114
Di-Me ester: Dimethyl (3-methyl-2-butenyl)-phosphonate.
$C_7H_{15}O_3P$ M 178.167
Liq. $Bp_{0.2}$ 51-52°.
Di-Et ester: [51795-72-3]. *Diethyl (3-methyl-2-butenyl)phosphonate.*
$C_9H_{19}O_3P$ M 206.221
No phys. props. reported.

Martin, D.J. *et al*, *Tetrahedron*, 1967, **23**, 1831 (*synth, ir, pmr*)
Hirao, T. *et al*, *Bull. Chem. Soc. Jpn.*, 1982, **55**, 909 (*synth*)

(3-Methyl-3-butenyl)phosphonic acid M-00103

2-Methyl-4-phosphono-1-butene

$H_2C{=}C(CH_3)CH_2CH_2P(O)(OH)_2$

$C_5H_{11}O_3P$ M 150.114
Di-Et ester: [71071-55-1]. *Diethyl (3-methyl-3-butenyl)phosphonate.*
$C_9H_{19}O_3P$ M 206.221
Liq. $Bp_{0.1}$ 72-75°.

Savignac, P. *et al*, *Synth. Commun.*, 1979, **9**, 487 (*synth, pmr*)

(3-Methyl-2-butenyl)-triphenylphosphonium(1+), 9CI M-00104

Isopentenyltriphenylphosphonium(1+)
[49868-41-9]

$Ph_3P^{\oplus}CH_2CH{=}C(CH_3)_2$

$C_{23}H_{24}P^{\oplus}$ M 331.416 (ion)
Chloride: [52750-95-5].
$C_{23}H_{24}ClP$ M 366.869
No phys. props. reported.
Bromide: [1530-34-3].
$C_{23}H_{24}BrP$ M 411.320
Converted into ylide with Et_3N, NaOMe, or butyllithium. Cryst. (CH_2Cl_2/EtOAc). Mp 239-240° (228-229°).

Ylide: see 3-Methyl-2-butenylidenetriphenylphosphorane, M-00098

Schweizer, E.E. *et al*, *J. Org. Chem.*, 1971, **36**, 4033 (*bromide, ir, pmr, use*)
McIntosh, J.M. *et al*, *J. Org. Chem.*, 1975, **40**, 1294 (*bromide, pmr*)
Spangler, C.W. *et al*, *J. Chem. Soc., Perkin Trans. 2*, 1979, 810 (*bromide, pmr, ylide, use*)
Baba, K. *et al*, *Chem. Pharm. Bull.*, 1981, **29**, 2182 (*bromide, ylide, use*)

(1-Methylbutyl)phosphonic acid, 9CI — M-00105

2-Pentylphosphonic acid

$H_3CCH_2CH_2CH(CH_3)P(O)(OH)_2$

$C_5H_{13}O_3P$ M 152.130

(±)-*form*

Monoanilinium salt: Cryst. (MeCN). Mp 145-146°.
Di-Et ester: Diethyl (1-methylbutyl)phosphonate. Diethyl 2-pentylphosphonate.
$C_9H_{21}O_3P$ M 208.237
Liq. Bp_1 59-62°.
Dichloride:
$C_5H_{11}Cl_2OP$ M 189.021
Liq. d_4^{25} 1.21. Bp_2 70°. n_D^{25} 1.4670.

Buckler, S.A. *et al*, *Tetrahedron*, 1962, **18**, 1211 (*deriv*)
Geiseler, G. *et al*, *Ber. Bunsenges. Phys. Chem.*, 1967, **71**, 478 (*dichloride, synth, ir, raman*)
Grishina, O.N. *et al*, *Zh. Prikl. Khim.*, 1969, **42**, 2289 (*Engl. transl.* p. 2149) (*dichloride*)
Inokawa, S. *et al*, *Synthesis*, 1973, 364 (*ester*)

(3-Methylbutyl)phosphonic acid, 9CI — M-00106

Isopentylphosphonic acid. 1-Phosphono-3-methylbutane

$(H_3C)_2CHCH_2CH_2P(O)(OH)_2$

$C_5H_{13}O_3P$ M 152.130
Solid. Mp 166°.

Di-Et ester: Diethyl (3-methylbutyl)phosphonate.
$C_9H_{21}O_3P$ M 208.237
$Bp_{2.5}$ 78-79°. n_D^{20} 1.4262.
Dichloride: [16474-65-0].
$C_5H_{11}Cl_2OP$ M 189.021
Liq. d^{20} 1.19. Bp_{55} 122-125°.

Guichard, F., *Chem. Ber.*, 1899, **32**, 1572 (*synth, dichloride*)
Ford-Moore, A.H. *et al*, *J. Chem. Soc.*, 1947, 1465 (*ester*)
Ionin, B.I. *et al*, *Zh. Obshch. Khim.*, 1963, **33**, 2863 (*Engl. transl.* p. 2791) (*ester*)

O-Methyl *tert*-butylphosphonothioate, 8CI — M-00107

O-Methyl (1,1-dimethylethyl)phosphonothioate, 9CI

MeO–P(OH)(=S)–C(CH₃)₃ ⇌ MeO–P(=O)(SH)–C(CH₃)₃ (*R*)-*form*

$C_5H_{13}O_2PS$ M 168.190

(*R*)-*form* [34239-27-5]
Liq. $Bp_{0.05}$ 48°. $[\alpha]_D$ +13.68° (pure).
S-*Me ester:* O,S-*Dimethyl* tert-*butylphosphonothioate.*
$[\alpha]_D$ +66.84° (pure).

(*S*)-*form* [34208-79-2]
Liq. $Bp_{0.0007}$ 54°. $[\alpha]_D$ −11.27° (C_6H_6).

(±)-*form* [34260-92-9]
$Bp_{0.05}$ 50°. n_D^{25} 1.4894.
S-*Me ester:*
$C_6H_{15}O_2PS$ M 182.217
Liq. $Bp_{0.5}$ 43°. n_D^{25} 1.4767.
O-*Trimethylsilyl ester:* O-*Methyl* O-*trimethylsilyl* tert-*butylphosphonothioate.*
$C_8H_{21}O_2PSSi$ M 240.372
Liq. $Bp_{0.01}$ 34°. n_D^{20} 1.4582.
Chloride:
$C_5H_{12}ClOPS$ M 186.636
Solid. Mp 61-63°.

Krawiecka, B. *et al*, *Bull. Acad. Pol. Sci., Ser. Sci. Chim.*, 1971, **19**, 377 (*synth, resoln, derivs, ir, pmr*)
Mikolajczyk, M. *et al*, *Tetrahedron*, 1972, **28**, 4357 (*config, cd, uv*)
Mikolajczyk, M. *et al*, *J. Am. Chem. Soc.*, 1978, **100**, 7003 (*pmr*)
Michalski, J. *et al*, *Phosphorus Sulfur*, 1980, **8**, 263 (*trimethylsilyl ester*)

3-Methylbutyl phosphorodichloridite, 9CI — M-00108

Isopentyl phosphorodichloridite. Isoamyl dichlorophosphite

$(H_3C)_2CHCH_2CH_2OPCl_2$

$C_5H_{11}Cl_2OP$ M 189.021
Fuming liq. Bp_{750} 175-176°.

Schwarz, R. *et al*, *Chem. Ber.*, 1957, **90**, 952 (*synth*)

O-(3-Methylbutyl) phosphorodichloridothioate, 9CI — M-00109

O-*Isopentyl phosphorodichloridothioate*, 8CI. O-*Isopentyl thiophosphoryl dichloride.* O-*Isopentyl dichlorothiophosphate.* O-*(3-Methylbutyl) dichlorothiophosphate*
[18351-13-8]

$(H_3C)_2CHCH_2CH_2OP(S)Cl_2$

$C_5H_{11}Cl_2OPS$ M 221.081
Liq. d_4^{20} 1.19. Bp_1 67-70°. n_D^{20} 1.4811.

Bliznyuk, N.K. *et al*, *Zh. Obshch. Khim.*, 1967, **37**, 1122 (*Engl. transl.* p. 1064) (*synth*)

S-(3-Methylbutyl) phosphorodichloridothioate, 9CI — M-00110

S-*Isopentyl phosphorodichloridothioate*, 8CI. S-*Isopentyl dichlorothiophosphate.* S-*(3-Methylbutyl) dichlorothiophosphate*

$(H_3C)_2CHCH_2CH_2SP(O)Cl_2$

$C_5H_{11}Cl_2OPS$ M 221.081
Liq. d_4^{20} 1.20. Bp_8 114-115°. n_D^{20} 1.5044.

Petrov, K.A. *et al*, *Zh. Obshch. Khim.*, 1961, **31**, 1366 (*Engl. transl.* p. 1265) (*synth*)

(3-Methylbutyl)triphenylphosphonium(1+), 9CI — M-00111

Isopentyltriphenylphosphonium(1+). Isoamyltriphenylphosphonium(1+)

$Ph_3P^{\oplus}CH_2CH_2CH(CH_3)_2$

$C_{23}H_{26}P^{\oplus}$ M 333.432 (ion)
Bromide: [28322-40-9].
$C_{23}H_{26}BrP$ M 413.336
Treatment with butyllithium yields the ylide. Cryst. (MeOH/Et$_2$O or 2-propanol/EtOAc). Mp 152-153°.
Iodide: [52710-37-9].
$C_{23}H_{26}IP$ M 460.337
Ylide source.
Ylide: [39110-24-2]. *3-Methylbutylidenetriphenylphosphorane. Isopentylidenetriphenylphosphorane.*
$C_{23}H_{25}P$ M 332.424
Reactive Wittig reagent. Used in synth. of vitamin D$_3$.

Inhoffen, H.H. *et al, Chem. Ber.*, 1959, **92**, 1959 (*bromide, ylide, use*)
Georghiou, P.E. *et al, J. Chem. Soc., Perkin Trans. 1*, 1973, 888 (*ylide, use*)
Bogoslovskii, N.A. *et al, Zh. Obshch. Khim.*, 1978, **48**, 908 (*Engl. transl.* p. 823) (*bromide, ylide, use*)
Fürst, A. *et al, Helv. Chim. Acta*, 1982, **65**, 1499 (*ylide, use*)

Methyl 6-deoxy-2,3-di-*O*-methyl-6-methylaminoglucopyranoside 4,6-cyclic phosphonamide M-00112

Octahydro-3-methylpyrano[2,3-e]-1,3,2-oxazaphosphorine 2-oxide. 8,9,10-Trimethoxy-4-methyl-2,7,4-dioxaza-3-phosphabicyclo[4.4.0]decane 3-oxide

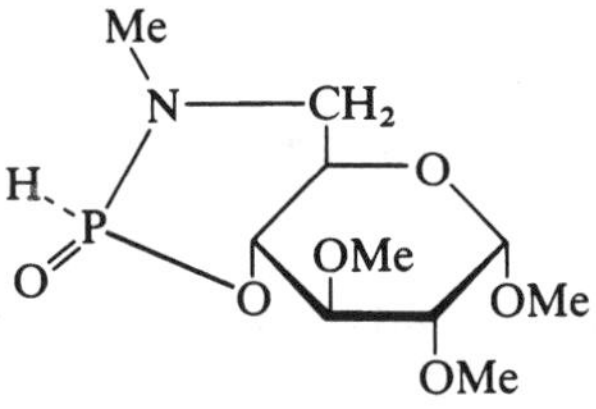

α-D-(*R*)$_p$-*form*

$C_{10}H_{20}NO_6P$ M 281.245

α-D-(R)$_P$-form

P-*Me:*
$C_{11}H_{22}NO_6P$ M 295.272
Syrup.
P-*Ph:*
$C_{16}H_{24}NO_6P$ M 357.342
Syrup. $[\alpha]_D$ +50° (c, 1.5 in CHCl$_3$).
P-*Methoxy:*
$C_{11}H_{21}NO_7P$ M 310.263
Syrup. $[\alpha]_D$ +80° (c, 4 in CHCl$_3$). Has (*S*)P-config.
P-*Ethoxy:*
$C_{12}H_{23}NO_7P$ M 324.290
Syrup. $[\alpha]_D$ +67° (CHCl$_3$). Has (*S*)P-config.
P-(*4-Nitrophenoxy*):
$C_{16}H_{23}N_2O_9P$ M 418.339
Cryst. (diisopropyl ether). Mp 98-101°. $[\alpha]_D$ +59° (c, 1 in CHCl$_3$).

α-D-(S)$_P$-form

P-*Me:* Cryst. (diisopropyl ether/Me$_2$CO). Mp 173-176°. $[\alpha]_D$ +96° (c, 2 in CHCl$_3$).
P-*Ph:* Cryst. (pet. ether). Mp 101-103°. $[\alpha]_D$ +73° (c, 2 in CHCl$_3$).
P-*Chloro:*
$C_{10}H_{19}ClNO_6P$ M 315.690
Cryst. (pet. ether). Mp 123-125°. $[\alpha]_D$ +86° (c, 1.5 in CHCl$_3$). Has (*R*)P-config.
P-*Methoxy:* Cryst. (diisopropyl ether). Mp 110-120°. $[\alpha]_D$ +45° (c, 0.7 in CHCl$_3$). Has (*R*)P-config.
P-*Ethoxy:* $[\alpha]_D$ +53° (c, 0.5 in CHCl$_3$). Has (*R*)P-config.
2-(*4-Nitrophenoxy*): Solid. Mp 120-124°. $[\alpha]_D$ +102° (c, 0.8 in CHCl$_3$).

Harrison, J.M. *et al, J. Chem. Soc., Perkin Trans. 1*, 1975, 1892 (*synth, ir, pmr, P nmr*)
Inch, T.D. *et al, Carbohydrate Res.*, 1975, **45**, 65 (*props*)

Methyl 6-deoxy-2,3-di-*O*-methyl-6-methylaminoglucopyranoside 4,6-cyclic thiophosphonamide M-00113

Octahydro-3-methylpyrano[2,3-e]-1,3,2-oxazaphosphorine 2-sulfide. 8,9,10-Trimethoxy-4-methyl-2,7,4-dioxaza-3-phosphabicyclo[4.4.0]decane 3-sulfide

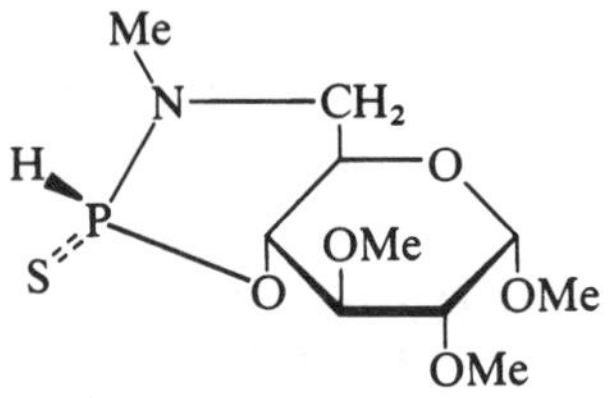

α-D-(*R*)$_p$-*form*

$C_{10}H_{20}NO_5PS$ M 297.305

α-D-(R)$_P$-form

P-*Me:*
$C_{11}H_{22}NO_5PS$ M 311.332
Cryst. (pet. ether). Mp 181-183°. $[\alpha]_D$ +57° (c, 1.8 in CHCl$_3$).
P-*Chloro:*
$C_{10}H_{19}ClNO_5PS$ M 331.751
Cryst. (pet. ether). Mp 123-127°. $[\alpha]_D$ +57° (c, 1.2 in CHCl$_3$). Has (*S*)P-config.

α-D-(S)$_P$-form

P-*Me:* Cryst. (pet. ether). Mp 123°. $[\alpha]_D$ +142° (c, 1.2 in CHCl$_3$).
P-*Methoxy:*
$C_{11}H_{22}NO_6PS$ M 327.332
Cryst. (pet. ether). Mp 70-73°. $[\alpha]_D$ +153° (c, 1 in CHCl$_3$). Has (*R*)P-config.

Harrison, J.M. *et al, J. Chem. Soc., Perkin Trans. 1*, 1975, 1892 (*synth, pmr, P nmr*)

5-Methyl-1*H*-1,2,3-diazaphosphole, 10CI M-00114

[63139-09-3]

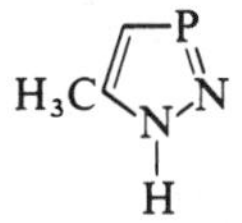

$C_3H_5N_2P$ M 100.060
Solid. Mp 45-47°.
1-Me: [69991-36-2]. *1,5-Dimethyl-1*H*-1,2,3-diazaphosphole.*
$C_4H_7N_2P$ M 114.086
Liq. Bp 140-149°.

Bobkova, R.G. *et al, Zh. Obshch. Khim.*, 1977, **47**, 576 (*Engl. transl.* p. 527) (*synth*)
Negrebetskii, V.V. *et al, Zh. Strukt. Khim.*, 1978, 535 (*Engl. transl.* p. 462) (*pmr, N nmr, tautom*)
Weinmaier, J.H. *et al, Chem. Ber.*, 1980, **113**, 2278 (*deriv, synth, P nmr, pmr*)

5-Methyl-5*H*-dibenzophosphole, 9CI **M-00115**

*1-Methyl-1*H*-benzo*[b]*phosphindole. 9-Methyl-9-phosphafluorene*

[16546-79-5]

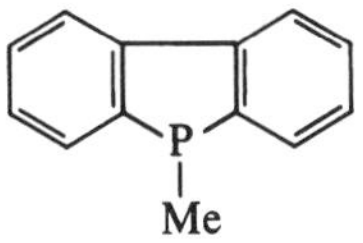

$C_{13}H_{11}P$ M 198.204

Liq. d_4^{20} 1.15. Bp_5 134-135°. n_D^{20} 1.6805.

B, MeI: [5274-22-6]. *5,5-Dimethyl-5*H*-dibenzophospholium iodide.*

$C_{14}H_{14}IP$ M 340.143

Cryst. (EtOH aq.). Mp 281-282°.

5-Oxide: [19190-40-0].

$C_{13}H_{11}OP$ M 214.203

Cryst. (C_6H_6/pet. ether). Mp 114-116°.

5-Sulfide: [33771-53-8].

$C_{13}H_{11}PS$ M 230.264

Cryst. (EtOH). Mp 124-125°.

5-Selenide: [79133-84-9].

$C_{13}H_{11}PSe$ M 277.164

Cryst. (EtOH). Mp 142°.

Allen, D.W. *et al*, *J. Chem. Soc. (C)*, 1967, 1869 (*synth, deriv*)

Allen, D.W. *et al*, *J. Chem. Soc. (C)*, 1969, 252 (*synth, ir, uv, pmr*)

Ezzell, B.R. *et al*, *J. Org. Chem.*, 1969, **34**, 1777 (*synth, oxide*)

Cheryshev, E.A. *et al*, *Zh. Obshch. Khim.*, 1971, **41**, 800 (*Engl. transl.* p. 806) (*synth, derivs*)

Bochkarev, V.N. *et al*, *Zh. Obshch. Khim.*, 1974, **44**, 1273 (*Engl. transl.* p. 1251) (*ms*)

Gray, G.A. *et al*, *Org. Magn. Reson.*, 1980, **14**, 14 (*cmr*)

Allen, D.W. *et al*, *J. Chem. Res. (S)*, 1981, 220 (*selenide, P nmr*)

4-Methyl-2-dimethylamino-1,3,2-dioxaphospholane, 8CI **M-00116**

N,N,*4-Trimethyl-1,3,2-dioxaphospholan-2-amine,* 9CI. *Propylene dimethylphosphoramidite. 4-Methyl-2-dimethylamino-1,3-dioxa-2-phosphacyclopentane*

[7114-42-3]

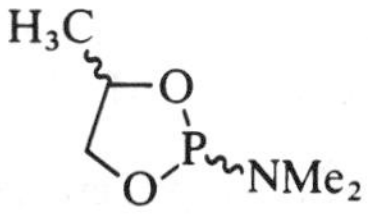

$C_5H_{12}NO_2P$ M 149.129

Liq. with pungent odour. d_4^{20} 1.02. $Bp_{5.5}$ 56°. $n_D^{22.5}$ 1.4632. Characterised spectroscopically as *cis/trans* mixture.

2-Oxide: Propylene dimethylphosphoramidate.

$C_5H_{12}NO_3P$ M 165.128

Liq. d_4^{20} 1.19. Bp_3 110-111°. n_D^{20} 1.4512. *Cis* and *trans* forms characterised spectroscopically.

2-Sulfide: O,O-*Propylene dimethylphosphoramidothioate.*

$C_5H_{12}NO_2PS$ M 181.189

Liq. $Bp_{0.2}$ 98-102°. n_D^{21} 1.5066.

2-Selenide: O,O-*Propylene dimethylphosphoramidoselenoate.*

$C_5H_{12}NO_2PSe$ M 228.089

Cryst. (Et_2O). Mp 53-54°.

Abramov, V.S. *et al*, *Zh. Obshch. Khim.*, 1966, **36**, 923 (*Engl. transl.* p. 938) (*synth, derivs*)

Edmundson, R.S. *et al*, *J. Chem. Soc. (C)*, 1966, 1997 (*synth, deriv*)

Germa, H. *et al*, *Bull. Soc. Chim. Fr.*, 1970, 612 (*pmr, P nmr*)

Amos, J. *et al*, *Phosphorus*, 1975, **6**, 35 (*oxide, synth, ir, pmr*)

Bentrude, W.G. *et al*, *J. Am. Chem. Soc.*, 1976, **98**, 1850 (*cmr, P nmr*)

Bentrude, W.G. *et al*, *J. Am. Chem. Soc.*, 1977, **99**, 4383 (*oxide*)

4-Methyl-2-dimethylamino-1,3,2-dioxaphosphorinane, 8CI **M-00117**

N,N,*4-Trimethyl-1,3,2-dioxaphosphorinan-2-amine,* 9CI. N,N,*4-Trimethyl-1,3,2-dioxaphosphinan-2-amine. 4-Methyl-2-dimethylamino-1,3-dioxa-2-phosphacyclohexane*

[13041-11-7]

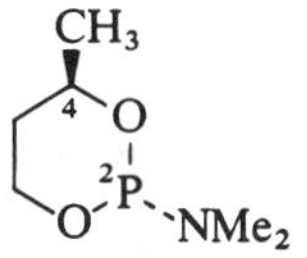

(2*RS*,4*RS*)-*form*

$C_6H_{14}NO_2P$ M 163.156

Phosphorylating agent. Pungent liq. Bp_{20} 75°. n_D^{20} 1.4652. Mixt. of *cis-* and *trans*-diastereoisomers in ratio 1:4.

(2RS,4RS)-form

(±)-trans-*form*

2-Oxide: see 4-Methyl-2-dimethylamino-1,3,2-dioxaphosphorinane 2-oxide, M-00118

2-Sulfide: [67057-63-0]. Solid. Mp 51-52°.

2-Selenide: [67123-93-7]. Solid. Mp 76-77°.

(2RS,4SR)-form

(±)-cis-*form*

2-Oxide: see 4-Methyl-2-dimethylamino-1,3,2-dioxaphosphorinane 2-oxide, M-00118

2-Sulfide: [67057-64-1]. O,O-*1,3-Butylene dimethylphosphoramidothioate.*

$C_6H_{14}NO_2PS$ M 195.216

Solid. Mp 37-38°.

2-Selenide: [67057-65-2]. O,O-*1,3-Butylene dimethylphosphoramidoselenoate.*

$C_6H_{14}NO_2PSe$ M 242.116

Cryst. (Et_2O/hexane). Mp 51-52°.

Kochetkov, N.K. *et al*, *Zh. Obshch. Khim.*, 1970, **40**, 2340 (*Engl. transl.* p. 2330) (*use*)

Nifant'ev, É.E. *et al*, *Zh. Obshch. Khim.*, 1973, **43**, 71 (*Engl. transl.* p. 68) (*synth, stereochem, P nmr*)

Stec, W.J. *et al*, *Phosphorus*, 1973, **2**, 235 (*sulfide, selenide*)

Stec. W.J. *et al*, *J. Org. Chem.*, 1976, **41**, 233 (*synth, derivs, P nmr*)

Zielinska, B. *et al*, *Org. Mass Spectrom.*, 1978, **13**, 65 (*sulfide, selenide, ms*)

Stec, W.J. *et al*, *Org. Mass Spectrom.*, 1980, **15**, 105 (*sulfide, ms*)

Nifant'ev, É.E. *et al*, *Tetrahedron*, 1981, **37**, 3183 (*cmr*)

4-Methyl-2-dimethylamino-1,3,2-dioxaphosphorinane 2-oxide **M-00118**

N,N,*4-Trimethyl-1,3,2-dioxaphosphorinan-2-amine 2-oxide,* 9CI. *4-Methyl-2-dimethylamino-1,3-dioxa-2-oxo-2-phosphacyclohexane. 1,3-Butylene dimethylphosphoramidate*

[21857-33-0]

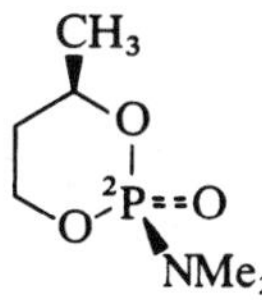

(2*RS*,4*RS*)-*form*

$C_6H_{14}NO_3P$ M 179.155

(2RS,4RS)-form [67057-62-9]
(±)-cis-*form*
Cryst. (Et_2O/hexane). Mp 37-38°. $Bp_{0.8}$ 90° (96% diastereomeric purity). n_D^{20} 1.4522.
(2RS,4SR)-form [67057-61-8]
(±)-trans-*form*
Liq. $Bp_{1.2}$ 97-103° (85% diastereomeric purity). n_D^{20} 1.4555.

Mosbo, J.A. *et al*, *J. Am. Chem. Soc.*, 1972, **94**, 8224; 1973, **95**, 4659 (*stereochem, pmr, P nmr*)
Stec, W. *et al*, *Tetrahedron*, 1973, **29**, 539 (*synth, ir, pmr, P nmr*)
Stec, W. *et al*, *J. Org. Chem.*, 1976, **41**, 233 (*synth*)
Mosbo, J.A., *Org. Magn. Reson.*, 1978, **11**, 281 (*pmr*)
Zielinska, B. *et al*, *Org. Mass Spectrom.*, 1978, **13**, 65 (*P nmr, ms*)

Methyl 2,3-di-*O*-methylglucopyranoside 4,6-cyclic phosphonate M-00119

Hexahydro-6,7,8-trimethoxypyrano[*2,3*-e]*-1,3,2-dioxaphosphorin 2-oxide. 8,9,10-Trimethoxy-2,4,7-trioxa-3-phosphabicyclo*[*4.4.0*]*decane 3-oxide*

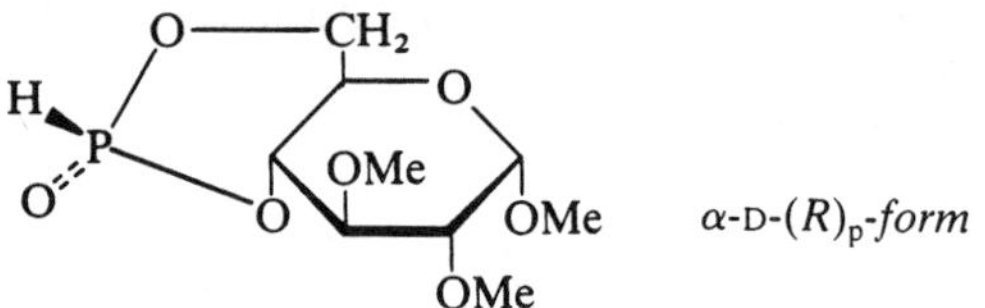

$C_9H_{17}O_7P$ M 268.203

α-D-(R)$_P$-form
P-*Me:*
$C_{10}H_{19}O_7P$ M 282.230
Cryst. (diisopropyl ether). Mp 96-99°. $[\alpha]_D$ +72° (c, 2 in $CHCl_3$).
P-*Ph:*
$C_{15}H_{21}O_7P$ M 344.300
Cryst. (diisopropyl ether). Mp 107-109°. $[\alpha]_D$ +18° (c, 0.5 in $CHCl_3$).
P-*Ethoxy:*
$C_{11}H_{21}O_8P$ M 312.256
$[\alpha]_D$ +90° ($CHCl_3$).
P-*Dimethylamino:*
$C_{11}H_{22}NO_7P$ M 311.271
Cryst. (diisopropyl ether). Mp 145-147°. $[\alpha]_D$ +105° (c, 1.6 in $CHCl_3$).
P-(*4-Nitrophenoxy*)*:*
$C_{15}H_{20}NO_{10}P$ M 405.297
Cryst. (diisopropyl ether). Mp 112-115°. $[\alpha]_D$ +43° (c, 1.1 in $CHCl_3$).
P-(*Propylthio*)*:*
$C_{12}H_{23}O_7PS$ M 342.343
$[\alpha]_D$ +53° (c, 2 in $CHCl_3$). Has $(S)_P$-config.

α-D-(S)$_P$-form
P-*Me:* Cryst. (Me_2CO/pet. ether). Mp 205°. $[\alpha]_D$ +114° (c, 2 in $CHCl_3$).
P-*Ph:* Cryst. (diisopropyl ether). Mp 173°. $[\alpha]_D$ +156° (c, 1.0 in $CHCl_3$).
P-*Ethoxy:* Cryst. (diisopropyl ether). Mp 115°. $[\alpha]_D$ +114° (c, 2 in $CHCl_3$).
P-(*4-Nitrophenoxy*)*:* Cryst. Mp 91-93°. $[\alpha]_D$ +80° (c, 1.3 in $CHCl_3$).
P-*Fluoro:*
$C_9H_{16}FO_7P$ M 286.193
Cryst. (diisopropyl ether). Mp 126°. $[\alpha]_D$ +88° (c, 1.3 in $CHCl_3$).
P-*Chloro:*
$C_9H_{16}ClO_7P$ M 302.648
Cryst. (diisopropyl ether). Mp 127-129°. $[\alpha]_D$ +114.5° (c, 1.8 in $CHCl_3$). Has $(R)_P$-config.
P-*Dimethylamino:* Cryst. (diisopropyl ether). Mp 105-110°. $[\alpha]_D$ +83° (c, 1.1 in $CHCl_3$).
P-(*Propylthio*)*:* $[\alpha]_D$ +29° (c, 0.5 in $CHCl_3$). Has $(R)_P$-config.

Cooper, D.B. *et al*, *J. Chem. Soc., Perkin Trans. 1*, 1974, 1043, 1049 (*synth, ir, pmr, P nmr*)
Harrison, J.M. *et al*, *J. Chem. Soc., Perkin Trans. 1*, 1974, 1053 (*synth, ir, pmr, P nmr*)
Cooper, D.B. *et al*, *J. Chem. Soc., Perkin Trans. 1*, 1974, 1058 (*props*)

Methyl 2,3-di-*O*-methylglucopyranoside 4,6-cyclic phosphonothioate M-00120

Hexahydro-6,7,8-trimethoxypyrano[*2,3*-e]*-1,3,2-dioxaphosphorin 2-sulfide. 8,9,10-Trimethoxy-2,4,7-trioxa-3-phosphabicyclo*[*4.4.0*]*decane 3-sulfide*

α-D-(*R*)$_P$-*form*

$C_9H_{17}O_6P$ M 252.203

α-D-(R)$_P$-form
P-*Me:*
$C_{10}H_{19}O_6PS$ M 298.290
Cryst. (Me_2CO/hexane). Mp 167-168°. $[\alpha]_D$ +75° (c, 1.0 in $CHCl_3$).
P-*Ph:*
$C_{15}H_{21}O_6PS$ M 360.361
Cryst. (Me_2CO/cyclohexane). Mp 158-161°. $[\alpha]_D$ +126° (c, 1.0 in $CHCl_3$).
P-*Ethoxy:*
$C_{11}H_{21}O_7PS$ M 328.316
Cryst. (EtOAc/cyclohexane). Mp 81-82°. $[\alpha]_D$ +136° (c, 1.0 in $CHCl_3$).
P-*Chloro:*
$C_9H_{16}ClO_6PS$ M 318.709
Cryst. (Et_2O/pet. ether). Mp 71-74°. $[\alpha]_D$ +96° (c, 0.8 in $CHCl_3$). Has $(S)_P$-config.

α-D-(S)$_P$-form
P-*Me:* $[\alpha]_D$ +135° (c, 1.0 in $CHCl_3$).
P-*Ph:* Cryst. (cyclohexane). Mp 116-117°. $[\alpha]_D$ +98° (c, 1.0 in $CHCl_3$).
P-*Ethoxy:* Cryst. (EtOH/pet. ether). Mp 140-141°. $[\alpha]_D$ +89° (c, 0.1 in $CHCl_3$).

Cooper, D.B. *et al*, *J. Chem. Soc., Perkin Trans. 1*, 1974, 1049 (*synth, P nmr*)
Harrison, J.M. *et al*, *J. Chem. Soc., Perkin Trans. 1*, 1974, 1053 (*chloride*)

Methyl 2,3-di-*O*-methyl-4-thioglucopyranoside 4,6-cyclic phosphonate M-00121

Hexahydro-6,7,8-trimethoxypyrano[*3,2*-d][*1,3,2*]*-oxathiaphosphorine 2-oxide. 8,9,10-Trimethoxy-4,7-dioxa-2-thia-3-phosphabicyclo*[*4.4.0*]*decane 3-oxide*

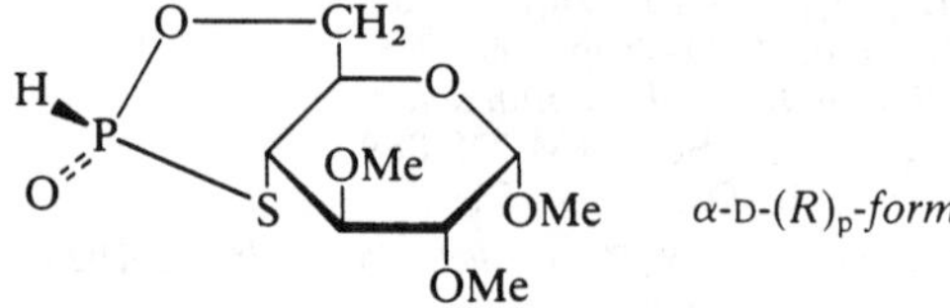

$C_9H_{17}O_6PS$ M 284.263

α-D-(R)$_P$-form

P-*Me:*

$C_{10}H_{19}O_6PS$ M 298.290

Cryst. (EtOH). Mp 183°. $[\alpha]_D$ +73° (c, 1 in $CHCl_3$).

P-*Ethoxy:*

$C_{11}H_{21}O_7PS$ M 328.316

Syrup. $[\alpha]_D$ +46° (c, 2 in $CHCl_3$). Has $(S)_P$-config.

P-*Chloro:*

$C_9H_{16}ClO_6PS$ M 318.709

Cryst. (diisopropyl ether). Mp 127°. $[\alpha]_D$ +41° (c, 0.4 in $CHCl_3$). Has $(S)_P$-config.

α-D-(S)$_P$-form

P-*Me:* Cryst. (Et_2O/pet. ether). Mp 124°. $[\alpha]_D$ +42.5° (c, 1.2 in $CHCl_3$).

P-*Ethoxy:* Cryst. Mp 98°. $[\alpha]_D$ +7° (c, 1.0 in $CHCl_3$). Has $(R)_P$-config.

Cooper, D.B. *et al*, *J. Chem. Soc., Perkin Trans. 1*, 1974, 1049 (*synth, ir, pmr*)

Harrison, J.M. *et al*, *J. Chem. Soc., Perkin Trans. 1*, 1974, 1053 (*chloride, props*)

Methyl 2,3-di-*O*-methyl-6-thioglucopyranoside 4,6-cyclic phosphonate **M-00122**

*8,9,10-Trimethoxy-2,7-dioxa-4-thia-3-phosphabicyclo[4.4.0]decane 3-oxide. Hexahydro-6,7,8-trimethoxypyrano[2,3-*e*]-1,3,2-oxathiaphosphorin 2-oxide*

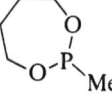

α-D-$(R)_P$-*form*

$C_9H_{17}O_6PS$ M 284.263

α-D-(R)$_P$-form

P-*Me:*

$C_{10}H_{19}O_6PS$ M 298.290

Cryst. (diisopropyl ether). Mp 203-207°. $[\alpha]_D$ +160° (c, 2 in $CHCl_3$).

P-*Ethoxy:*

$C_{11}H_{21}O_7PS$ M 328.316

Syrup. $[\alpha]_D$ +88° (0.7 in $CHCl_3$). Has $(S)_P$-config.

P-*Chloro:*

$C_9H_{16}ClO_6PS$ M 318.709

Cryst. (diisopropyl ether). Mp 129-131°. $[\alpha]_D$ +206° (c, 1.7 in $CHCl_3$). Has $(S)_P$-config.

α-D-(S)$_P$-form

P-*Me:* Cryst. (diisopropyl ether). Mp 159-161°. $[\alpha]_D$ +190° (1.7 in $CHCl_3$).

P-*Ethoxy:* Cryst. (diisopropyl ether). Mp 135-138°. $[\alpha]_D$ +207° (c, 0.8 in $CHCl_3$). Has $(R)_P$-config.

Cooper, D.B. *et al*, *J. Chem. Soc., Perkin Trans. 1*, 1974, 1049, 1058 (*synth, ir, pmr, props*)

Harrison, J.M. *et al*, *J. Chem. Soc., Perkin Trans. 1*, 1974, 1053 (*props*)

Methyl 2,3-di-*O*-methyl-4-thioglucopyranoside 4,6-cyclic phosphonothioate **M-00123**

*Hexahydro-6,7,8-trimethoxypyrano[3,2-*d*][1,3,2]-oxathiaphosphorin 2-sulfide. 8,9,10-Trimethoxy-4,7-dioxa-2-thia-3-phosphabicyclo[4.4.0]decane 3-sulfide*

α-D-$(R)_P$-*form*

$C_9H_{17}O_5PS_2$ M 300.324

α-D-(R)$_P$-form

P-*Ethoxy:*

$C_9H_{21}O_6PS_2$ M 320.355

Cryst. (EtOH/pet. ether). Mp 81-82°.

α-D-(S)$_P$-form

P-*Ethoxy:* Cryst. (EtOAc/pet. ether). Mp 140-141°.

P-*Chloro:*

$C_9H_{16}ClO_5PS_2$ M 334.769

Cryst. (Et_2O/pet. ether). Mp 71-74°. $[\alpha]_D$ +96° (c, 0.8 in $CHCl_3$). Has $(R)_P$-config.

Harrison, J.M. *et al*, *J. Chem. Soc., Perkin Trans. 1*, 1974, 1053 (*synth*)

4-Methyl-2,6-dioxa-7-aza-1-phosphabicyclo[2.2.2]octane 1-oxide **M-00124**

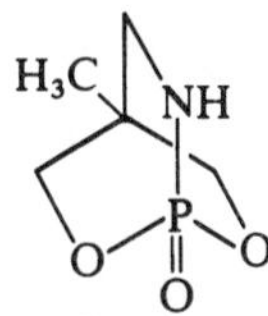

$C_5H_{10}NO_3P$ M 163.113

7-tert-Butyl: [86947-48-0].

$C_9H_{18}NO_3P$ M 219.220

Cryst. (EtOH/pet. ether). Mp 120-120.5°.

7-Ph: [86947-47-9].

$C_{11}H_{14}NO_3P$ M 239.210

Cryst. (EtOAc). Mp 195°.

7-Benzyl: [86947-46-8].

$C_{12}H_{16}NO_3P$ M 253.237

Cryst. (CCl_4). Mp 122°.

Edmundson, R.S., *Synthesis*, 1983, 445 (*synth, ir, pmr*)

2-Methyl-1,3,2-dioxaphosphepane, 9CI **M-00125**

Tetramethylene methylphosphonite

[66295-39-4]

$C_5H_{11}O_2P$ M 134.114

No phys. props. reported.

2-Oxide: [63299-52-5]. *Tetramethylene methylphosphonate.*

$C_5H_{11}O_3P$ M 150.114

Solid. Mp 68.5-70°. $Bp_{0.5}$ 72-76°.

2-Sulfide: [66295-41-8]. *O,O-Tetramethylene methylphosphonothioate.*

$C_5H_{11}O_2PS$ M 166.174

No phys. props. reported.

McKay, A.F. *et al*, *J. Am. Chem. Soc.*, 1954, **76**, 3546 (*oxide*)
Dutasta, J.P. *et al*, *J. Am. Chem. Soc.*, 1978, **100**, 1925 (*nmr*)
Guimaraes, A.C. *et al*, *Org. Magn. Reson.*, 1978, **11**, 411 (*sulfide, pmr, cmr, P nmr*)
Tabony, J. *et al*, *Spectrochim. Acta, Part A*, 1979, **35**, 217 (*sulfide, P nmr*)
Shagidullin, R.R. *et al*, *Izv. Akad. Nauk SSSR, Ser. Khim.*, 1981, 2253 (*Engl. transl.* p. 1849) (*oxide, ir, raman*)

2-Methyl-1,3,2-dioxaphospholane, 8CI — M-00126

2-Methyl-1,3-dioxa-2-phosphacyclopentane. Ethylene methylphosphonite

[32564-13-9]

$C_3H_7O_2P$ M 106.061

2-Oxide: [1831-25-0]. *Ethylene methylphosphonate.*
$C_3H_7O_3P$ M 122.060
Liq. d_4^{20} 1.33. $Bp_{0.5}$ 84-86°. n_D^{20} 1.4466.

2-Sulfide: [18906-32-6]. *Ethylene methylphosphonothioate.*
$C_3H_7O_2PS$ M 138.121
Phys. props. not reported.

Petrov, K.A. *et al*, *Zh. Obshch. Khim.*, 1965, **35**, 732 (*Engl. transl.* p. 731) (*oxide, synth*)
Evdakov, V.P. *et al*, *Zh. Obshch. Khim.*, 1967, **37**, 441 (*Engl. transl.* p. 412) (*oxide, synth*)
Arbusov, B.A. *et al*, *Zh. Obshch. Khim.*, 1973, **43**, 2134 (*Engl. transl.* p. 2125) (*oxide, P nmr*)
Dutasta, J.P. *et al*, *Tetrahedron*, 1979, **35**, 197 (*sulfide, synth, cmr, pmr, P nmr*)

4-Methyl-1,3,2-dioxaphospholane — M-00127

Propylene phosphonite

$C_3H_7O_2P$ M 106.061

(±)-*form*

2-Oxide: 2-Hydroxy-4-methyl-1,3,2-dioxaphospholane. Propylene phosphite. Propylene phosphonate.
$C_3H_7O_3P$ M 122.060
Liq. d_4^{20} 1.27. $Bp_{0.1}$ 91-92°, $Bp_{0.01}$ 66-67°. n_D^{20} 1.4690. Tautomeric.

2-Sulfide: 2-Mercapto-4-methyl-1,3,2-dioxophospholane. O,O-*Propylene thiophosphite.* O,O-*Propylene thiophosphonate.*
$C_3H_7O_2PS$ M 138.121
Oil. $Bp_{0.2}$ 67-68°. n_D^{20} 1.515. Tautomeric.

Zwierzak, A. *et al*, *Can. J. Chem.*, 1967, **45**, 2501 (*oxide, synth, ir*)
Pudovik, A.N. *et al*, *Zh. Obshch. Khim.*, 1969, **39**, 1719 (*Engl. transl.* p. 1685) (*oxide*)
Nifant'ev, E.E. *et al*, *Zh. Org. Khim.*, 1975, **11**, 2206 (*Engl. transl.* p. 2235) (*oxide, synth, P nmr*)
Ovchinnikov, V.V. *et al*, *Zh. Obshch. Khim.*, 1978, **48**, 2424 (*Engl. transl.* p. 2199) (*oxide, P nmr, struct*)
U.S.P., 4 115 559, (*1978*); *CA*, **90**, 72242 (*sulfide*)

2-Methyl-1,3,2-dioxaphosphorinane, 9CI — M-00128

2-Methyl-1,3,2-dioxaphosphinane. 2-Methyl-1,3-dioxa-2-phosphacyclohexane. Trimethylene methylphosphonite

[61558-36-9]

$C_4H_9O_2P$ M 120.088

2-Oxide: [13407-03-9]. *Trimethylene methylphosphonate.*
$C_4H_9O_3P$ M 136.087
Cryst. (CCl_4). Mp 98-99.5°. Bp_3 110-112°.

2-Sulfide: [18882-27-4]. O,O-*Trimethylene methylphosphonothioate.*
$C_4H_9O_2PS$ M 152.148
Solid. Mp 78-80°.

2-Selenide: [18882-26-3]. O,O-*Trimethylene methylphosphonoselenoate.*
$C_4H_9O_2PSe$ M 199.048
Solid. Mp 94-95°.

Mosbo, J.A. *et al*, *J. Org. Chem.*, 1977, **42**, 1549 (*oxide, synth, ir, P nmr*)
Nifant'ev, É.E. *et al*, *Tetrahedron*, 1981, **37**, 3183 (*pmr*)
Shagidullin, R.R. *et al*, *Izv. Akad. Nauk SSSR, Ser. Khim.*, 1981, 1156, 1890, 2253, 2505 (*Engl. transl.* pp. 916, 1557, 1849, 2071) (*derivs, ir, raman, P nmr*)
Shagidullin, R.R. *et al*, *Zh. Obshch. Khim.*, 1982, **52**, 2218 (*Engl. transl.* p. 1973) (*derivs, conformn*)

4-Methyl-1,3,2-dioxaphosphorinane, 9CI — M-00129

4-Methyl-1,3,2-dioxaphosphinane. 4-Methyl-1,3-dioxa-2-phosphacyclohexane. 1,3-Butylene phosphonite

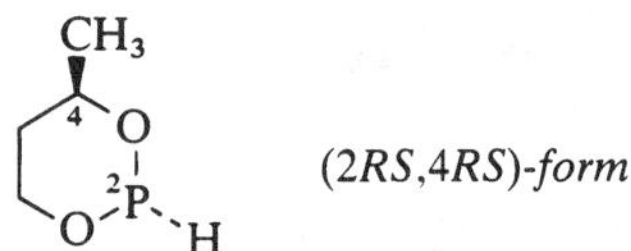

$C_4H_9O_2P$ M 120.088
Bp_{15} 45-46°. n_D^{20} 1.459.

2-Oxide: see 4-Methyl-1,3,2-dioxaphosphorinane 2-oxide, M-00130

2-Sulfide: see 4-Methyl-1,3,2-dioxaphosphorinane 2-sulfide, M-00131

2-Selenide: Characterised spectroscopically as *cis-trans* mixture.

(2RS,4RS)-form [40986-03-6]
(±)-trans-*form*
Liq. Bp_{50} 60-63°. n_D^{20} 1.4630.

(2RS,4SR)-form [80812-96-0]
(±)-cis-*form*
No phys. props. reported.

Foss, V.L. *et al*, *Zh. Obshch. Khim.*, 1973, **43**, 1000 (*Engl. transl.* p. 994) (*synth*)
Stec, W.J. *et al*, *Phosphorus*, 1973, **2**, 237 (*synth, selenide, P nmr*)
Nifant'ev, É.E. *et al*, *Tetrahedron*, 1981, **37**, 3183 (*pmr, cmr, P nmr*)

4-Methyl-1,3,2-dioxaphosphorinane 2-oxide, 9CI M-00130

4-Methyl-1,3,2-dioxaphosphinane 2-oxide. 1,3-Butylene phosphonate. 1,3-Butylene phosphite

[16368-20-0]

(2RS,4RS)-form

$C_4H_9O_3P$ M 136.087

Tautomeric.

(2RS,4RS)-form [26339-67-3]

(±)-cis-*form*

Cryst. (Et_2O). d_4^{20} 1.26. Mp 24-25°. Bp_1 97-97.5°. n_D^{20} 1.4550.

(2RS,4SR)-form [26339-68-4]

(±)-trans-*form*

Cryst. (Et_2O, heptane, or C_6H_6/Et_2O). Mp 56-58° (52-53°).

Zwierzak, A., *Can. J. Chem.*, 1967, **45**, 2501 (*synth, ir*)
Nifant'ev, E.É. *et al*, *Zh. Obshch. Khim.*, 1971, **41**, 2368 (*Engl. transl.* p. 2394) (*synth, P nmr*)
Bodkin, C.L. *et al*, *J. Chem. Soc., Perkin Trans. 2*, 1973, 673 (*synth, config*)
Mosbo, J.A. *et al*, *J. Am. Chem. Soc.*, 1973, **95**, 204 (*synth, ir, pmr*)
Saenger, W. *et al*, *Chem. Ber.*, 1973, **106**, 3519 (*P nmr, cryst struct*)
Mosbo, J.A. *et al*, *J. Org. Chem.*, 1977, **42**, 1549 (*ir, pmr, P nmr, conformn*)
Nifant'ev, E.É. *et al*, *Zh. Obshch. Khim.*, 1981, **51**, 2428 (*Engl. transl.* p. 2092) (*props, ir, raman*)

4-Methyl-1,3,2-dioxaphosphorinane 2-sulfide, 9CI M-00131

4-Methyl-1,3,2-dioxaphosphinane 2-sulfide. 4-Methyl-1,3-dioxa-2-thioxo-2-phosphacyclohexane. O,O-1,3-Butylene thiophosphonate. O,O-1,3-Butylene phosphonothioate

[16368-17-5]

$C_4H_9O_2PS$ M 152.148

Tautomeric. Liq. $Bp_{0.1}$ 84-86°, $Bp_{0.0002}$ 60-65°. pK_a 8.0 (50% EtOH aq.). n_D^{20} 1.5225. Mixt. of diastereoisomers.

Zwierzak, A., *Can. J. Chem.*, 1967, **45**, 2501 (*synth, ir*)
Predvoditelev, D.A. *et al*, *Zh. Obshch. Khim.*, 1973, **43**, 73 (*Engl. transl.* p. 70) (*synth*)
Stec, W.J. *et al*, *Phosphorus*, 1973, **2**, 235, 237 (*synth, pmr, P nmr*)
Ovchinnikov, V.V. *et al*, *Zh. Obshch. Khim.*, 1979, **49**, 1693 (*props*)

[3-(2-Methyl-1,3-dioxolan-2-yl)propyl]-triphenylphosphonium, 8CI M-00132

$C_{25}H_{28}O_2P^{\oplus}$ M 391.469 (ion)

A protected-carbonyl-contg. reagent widely employed in Wittig reactions, e.g. in synth. of geranylacetone and in steroid modification.

Bromide: [5944-33-2].

$C_{25}H_{28}BrO_2P$ M 471.373

Source of ylide. Pale-yellow, electrostatically active cryst. ($MeCN/CH_2Cl_2/EtOAc$). Mp 210-211°, 220-221°. Me_3COK or $NaNH_2/NH_3$ → ylide.

Iodide: [21955-58-8].

$C_{25}H_{28}IO_2P$ M 518.373

Source of ylide. Solid. Mp 167-170°, 204-206°. $NaNH_2/NH_3$ → ylide.

Ylide: [3054-93-1]. *[3-(2-Methyl-1,3-dioxolan-2-yl)-propylidene]triphenylphosphorane.*

$C_{25}H_{27}O_2P$ M 390.461

Widely used in Wittig reactions.

Obol'nikova, E.A. *et al*, *Zh. Obshch. Khim.*, 1964, **34**, 1499 (*Engl. transl.* p. 1506) (*bromide, use*)
Davydova, L.P. *et al*, *Zh. Obshch. Khim.*, 1968, **38**, 2091 (*Engl. transl.* p. 2025) (*iodide*)
Crombie, L. *et al*, *J. Chem. Soc.* (*C*), 1969, 1016 (*bromide, use*)
Uijttowaae, A.P. *et al*, *J. Org. Chem.*, 1979, **44**, 3157 (*iodide, use*)
Fürst, A. *et al*, *Helv. Chim. Acta*, 1982, **65**, 1499 (*use*)
Salinaro, R.F. *et al*, *J. Am. Chem. Soc.*, 1982, **104**, 2228 (*bromide, use*)

5-Methyl-2,3-diphenyl-1,3-oxaphospholane, 9CI M-00133

[37181-98-9]

$C_{16}H_{17}OP$ M 256.283

Liq. Bp_{10} 195-198°. A mixture of four geometric isomers.

Oehme, H. *et al*, *Tetrahedron*, 1972, **28**, 2587 (*synth*)
Zschunke, A. *et al*, *Org. Magn. Reson.*, 1974, **6**, 568 (*conformn, P nmr, pmr*)

6-Methyl-1,5-diphenyl-6-phosphabicyclo[3.2.1]octane M-00134

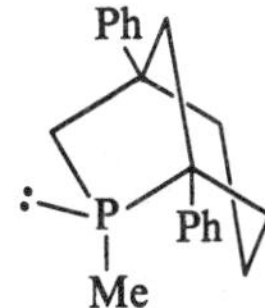

$C_{19}H_{23}P$ M 282.364

Oxide: [68305-42-0].

$C_{19}H_{23}OP$ M 298.364

Oil. Epimers septd. but configs. not assigned.

Rudi, A. *et al*, *Tetrahedron Lett.*, 1978, 2209 (*synth, ir, ms, pmr, cmr*)

Methyldiphenylphosphine, 9CI M-00135

[1486-28-8]

Ph_2PMe

$C_{13}H_{13}P$ M 200.219

Air-sensitive liq. d_4^{20} 1.08. Bp 284°, $Bp_{2.2}$ 118-120°, Bp °.

B,MeI: [1017-88-5]. *Dimethyldiphenylphosphonium iodide.*

$C_{14}H_{16}IP$ M 342.159

Cryst. (EtOH). Mp 253-255°.
Oxide: [2129-89-7].
$C_{13}H_{13}OP$ M 216.219
Complexes with Gd, Mn, Mo, and Sn. Cryst. (hexane). Mp 111-112°. $Bp_{0.5}$ 103-110°.
Sulfide: [18639-74-2].
$C_{13}H_{13}PS$ M 232.279
Complexes with Ir, Cr, and W. Liq. $Bp_{1.5}$ 181°. n_D^{25} 1.6515.
Selenide:
$C_{13}H_{13}PSe$ M 279.179
Liq. $Bp_{1.35}$ 200°. n_D^{25} 1.6780.

Mallion, K.B. *et al*, *J. Chem. Soc.*, 1964, 5716 (*synth*)
Gallagher, M.J., *Aust. J. Chem.*, 1968, **21**, 1197 (*pmr*)
Mann, B.E., *J. Chem. Soc., Perkin Trans. 2*, 1972, 30 (*nmr, cmr*)
Wiley, R.A. *et al*, *Biochem. Pharmacol.*, 1972, **21**, 3235 (*metab*)
Skvortsov, N.K. *et al*, *Zh. Obshch. Khim.*, 1973, **43**, 981 (*Engl. transl.* p. 976) (*pmr, nmr, props*)
Allen, E.A. *et al*, *Spectrochim. Acta, Part A*, 1974, **30**, 1219 (*ir, raman*)
Schott, G. *et al*, *Z. Anorg. Allg. Chem.*, 1974, **404**, 204 (*derivs, use*)
Cauguis, G. *et al*, *Org. Mass Spectrom.*, 1975, **10**, 770 (*ms*)
Henrick, K., *Aust. J. Chem.*, 1975, **28**, 1473 (*ms*)
Verstuyft, A.W. *et al*, *Inorg. Chem.*, 1975, **14**, 1495 (*derivs, pmr, complexes*)
Inorg. Synth., 1976, **16**, 155 (*synth*)
Inorg. Synth., 1977, **17**, 183 (*oxide*)
Goff, S.D. *et al*, *Org. Mass Spectrom.*, 1977, **12**, 33 (*oxide, ms*)
Grim, S.O. *et al*, *Inorg. Chem.*, 1977, **16**, 1770 (*deriv, use*)
Postle, S.R., *Phosphorus Sulfur*, 1977, **3**, 269 (*sulfide, cmr*)
Mathey, F. *et al*, *Tetrahedron*, 1978, **34**, 649 (*oxide, deriv, use*)
Grim, S.O. *et al*, *Inorg. Chem.*, 1980, **19**, 3195 (*sulfide, deriv, use*)
Vincent, E. *et al*, *Spectrochim. Acta, Part A*, 1980, **36**, 699 (*pmr, cmr, nmr*)
Schore, N.E. *et al*, *J. Org. Chem.*, 1981, **46**, 2306 (*deriv, use*)
Mercier, F. *et al*, *J. Organomet. Chem.*, 1982, **231**, 237 (*sulfide, props*)

4-Methyl-1,3,2-dithiaphospholane, 9CI M-00136

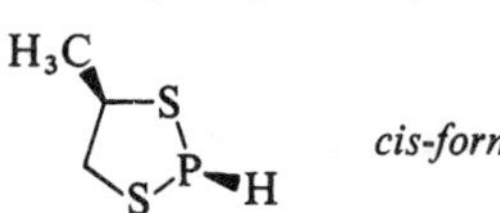

$C_3H_7PS_2$ M 138.182
2-Chloro: [57351-96-9]. *2-Chloro-4-methyl-1,3,2-dithiaphospholane. 1,2-Propylene phosphorochloridodithioite.*
$C_3H_6ClPS_2$ M 172.627
Pungent liq. Bp_1 104°.
2-Methoxy: [57093-51-3]. *2-Methoxy-4-methyl-1,3,2-dithiaphospholane. S,S-1,2-Propylene* O*-methyl phosphorodithioite.*
$C_4H_9OPS_2$ M 168.208
Liq. Bp_9 105°. n_D^{22} 1.585.
2-Ph: 4-Methyl-2-phenyl-1,3,2-dithiaphospholane. 1,2-Propylene phenylphosphonodithioite.
$C_9H_{11}PS_2$ M 214.280
Liq. $Bp_{0.2}$ 118°. Stereoisomer mixt.
2-Dimethylamino: [53827-16-0]. S,S-*1,2-Propylene* N,N-*dimethylphosphoramidodithioite.* N,N,*4-Trimethyl-1,3,2-dithiaphospholan-2-amine.*
$C_5H_{12}NPS_2$ M 181.250
Liq. $Bp_{0.01}$ 76°.

Albrand, J.P. *et al*, *Org. Magn. Reson.*, 1973, **5**, 33 (*phenyl, pmr*)
Bergesen, K. *et al*, *Acta Chem. Scand.*, 1973, **27**, 1103 (*chloro, phenyl, synth, pmr*)
Bondjebel, H. *et al*, *Bull. Soc. Chim. Fr. Part II*, 1975, 628 (*methoxy, dimethylamino, synth*)
Revel, M. *et al*, *Org. Magn. Reson.*, 1976, **8**, 399 (*methoxy, dimethylamino, pmr, P nmr, sulfide, derivs*)
Gonbeau, D. *et al*, *J. Mol. Struct.*, 1983, **98**, 109 (*chloro, methoxy, pe, struct*)

2-Methyl-1,3,2-dithiaphosphorinane, 9CI M-00137

S,S'*-Propanediyl methylphosphonodithioite*
[60991-59-5]

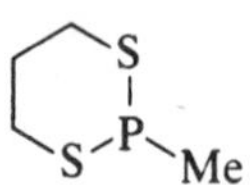

$C_4H_9PS_2$ M 152.209
Foul-smelling liq. Bp_5 102°.
2-Oxide: [18882-25-2]. S,S'*-Propanediyl methylphosphonodithioate.*
$C_4H_9OPS_2$ M 168.208
Cryst. (C_6H_6). Mp 114°.
2-Sulfide: [18882-24-1]. *Trimethylene methylphosphonotrithioate. 1,3-Propanediyl methylphosphonotrithioate.*
$C_4H_9PS_3$ M 184.269
Cryst. (C_6H_6). Mp 123°.
2-Selenide: S,S'*-Propanediyl methylphosphonoselenodithioate.*
$C_4H_9PS_2Se$ M 231.169
Cryst. (C_6H_6). Mp 127-128°.

Wieber, M. *et al*, *Angew. Chem.*, 1964, **76**, 648 (*synth*)
Wieber, M. *et al*, *Monatsh. Chem.*, 1968, **99**, 1153 (*derivs, synth, pmr*)
Grand, A. *et al*, *Acta Crystallogr., Sect. B*, 1976, **32**, 1244 (*sulfide, cryst struct*)
Arshinova, R.P. *et al*, *Zh. Obshch. Khim.*, 1980, **50**, 829 (*Engl. transl.* p. 665) (*sulfide, stereochem*)
Martin, J. *et al*, *Org. Magn. Reson.*, 1981, **15**, 85 (*sulfide, pmr, cmr, P nmr*)

4-Methyl-1,3,2-dithiaphosphorinane, 9CI M-00138

4-Methyl-1,3,2-dithiaphosphinane. S,S'*-1,3-Butylene phosphonodithioite.* S,S'*-1,3-Butanediyl phosphonodithioite*

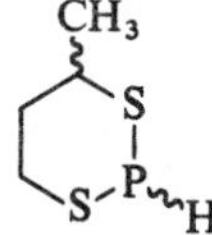

$C_4H_9PS_2$ M 152.209
Low melting solid. Mp 40°. Bp_1 50-53°. Mixture of diastereoisomers.
2-Chloro: [30382-05-9]. S,S'*-1,3-Butylene phosphorochloridodithioite.* S,S'*-1,3-Butanediyl phosphorochloridodithioite.*
$C_4H_8ClPS_2$ M 186.654
Pungent liq. d_4^{20} 1.35. Bp_3 103-104°. n_D^{20} 1.6302. A mixture of *cis* and *trans* stereoisomers.
2-Methoxy: S,S'*-1,3-Butylene* O*-methyl phosphorodithioite.* S,S'*-1,3-Butanediyl* O*-methyl phosphorodithioite.*
$C_5H_{11}OPS_2$ M 182.235
Liq. d_4^{20} 1.23. Bp_1 62°. n_D^{20} 1.5940. A mixture of *cis* and *trans* stereoisomers.
2-Ethoxy: see 2-Ethoxy-4-methyl-1,3,2-dithiaphosphorinane, E-00042
2-Diethylamino: S,S*-1,3-Butylene diethylphosphoramidodithioate.* N,N*-Diethyl-4-methyl-1,3,2-dithiaphosphorinan-2-amine.* Liq. d_4^{20} 1.10. Bp_1 105-107°. n_D^{20} 1.5779. A mixture of *cis* and *trans* stereoisomers.

Nifant'ev, É.E. *et al*, *Zh. Obshch. Khim.*, 1974, **44**, 1694 (*Engl. transl.* p. 1664) (*synth, P nmr*)
Borisenko, A.A. *et al*, *Zh. Obshch. Khim.*, 1978, **48**, 1251 (*Engl. transl.* p. 1144) (*cmr, P nmr, pmr*)
Nifant'ev, É.E. *et al*, *Tetrahedron*, 1981, **37**, 3183 (*synth, derivs, stereochem, P nmr, cmr, pmr*)

2-Methyl-1,3,2-dithiaphospholane M-00139

S,S-*Ethylene methylphosphonodithioite*. *2-Methyl-1,3-dithia-2-phosphacyclopentane*
[35437-28-6]

$C_3H_7PS_2$ M 138.182
Liq. Bp_5 90°, $Bp_{0.7}$ 66-68°.
2-Oxide: [18906-31-5]. S,S-*Ethylene methylphosphonodithioate.*
$C_3H_7OPS_2$ M 154.181
Solid. Mp 67-68°. $Bp_{0.7}$ 141°.
2-Sulfide: [18789-43-0]. *Ethylene methylphosphonotrithioate.*
$C_3H_7PS_3$ M 170.242
Solid. Mp 74-75°. Bp_2 132-134°.
2-Selenide: S,S-*Ethylene methylphosphonoselenodithioate.*
$C_3H_7PS_2Se$ M 217.142
Solid. Mp 68°. Bp_2 148°.

Wieber, M. *et al*, *Monatsh. Chem.*, 1968, **99**, 1153 (*derivs*)
Peake, S.C. *et al*, *J. Chem. Soc., Perkin Trans. 2*, 1972, 380 (*synth, derivs, pmr, P nmr*)
Albrand, J.-P. *et al*, *Org. Magn. Reson.*, 1973, **5**, 33 (*pmr*)
Martin, J. *et al*, *Org. Magn. Reson.*, 1981, **15**, 87 (*derivs, P nmr*)

Methylenebis[methylphosphinic acid], 9CI M-00140

[19014-66-5]

MeP(O)(OH)CH₂P(O)(OH)Me

$C_3H_{10}O_4P_2$ M 172.057
Esters were obt. as mixt. of meso and racemic forms in approx. equal proportions.
Di-Me ester: [78463-56-6]. *Dimethyl methylenebis(methylphosphinate).*
$C_5H_{14}O_4P_2$ M 200.111
No phys. props. reported.
Di-Et ester: [65747-62-8]. *Diethyl methylenebis[methylphosphinate].*
$C_7H_{18}O_4P_2$ M 228.164
d_4^{20} 1.17. $Bp_{0.02}$ 120°. n_D^{20} 1.4645.
Diisopropyl ester: [65747-63-9]. *Diisopropyl methylenebis[methylphosphinate].*
$C_9H_{22}O_4P_2$ M 256.218
d_4^{20} 1.10. $Bp_{0.02}$ 141°. n_D^{20} 1.4574.

Prishchenko, A.A. *et al*, *Zh. Obshch. Khim.*, 1977, **47**, 2689 (*Engl. transl.* p. 2451) (*synth, pmr, nmr*)
Prishchenko, A.A. *et al*, *Zh. Obshch. Khim.*, 1981, **51**, 480; *CA*, **95**, 62313.

[Methylenebis(2,1-phenylenemethylene)]-bis[triphenylphosphonium](2+), 9CI M-00141

2,2'-Bis(triphenylphosphoniomethyl)bibenzyl(2+)

$C_{51}H_{44}P_2^{\oplus\oplus}$ M 718.856 (ion)
Dibromide:
$C_{51}H_{44}Br_2P_2$ M 878.664
Source of ylide.
Diperiodate:
$C_{51}H_{44}I_2O_8P_2$ M 1100.660
Solid. Mp 156° dec. LiOEt yields the bis-ylide and 5*H*-dibenzo[*a,d*]cycloheptene.
Bis-ylide: [*Methylenebis(2,1-phenylenemethylene)*]-*bis*[*triphenylphosphorane*].
$C_{51}H_{42}P_2$ M 716.840
Used in Wittig reactions for synth. of benzannelated benzannulenes.

Bestmann, H.J. *et al*, *Angew. Chem., Int. Ed. Engl.*, 1969, **4**, 583, 830 (*bromide*)
Bestmann, H.J. *et al*, *Chem. Ber.*, 1969, **102**, 2259 (*periodate*)
Gamliel, A. *et al*, *Synthesis*, 1977, 410 (*use*)
Rabinowitz, M. *et al*, *Tetrahedron*, 1979, **35**, 667 (*use*)

Methylenebis[phenylphosphinic acid], 9CI M-00142

[4851-69-8]

Ph–P(O)(OH)CH₂P(O)(OH)–Ph

$C_{13}H_{14}O_4P_2$ M 296.199
Cryst. (EtOH). Mp 228°, 265-268°.

Sommer, K. *et al*, *Z. Anorg. Allg. Chem.*, 1970, **376**, 37 (*synth, nmr*)
Garst, M.E., *Synth. Commun.*, 1979, **9**, 261 (*synth*)

Methylenebis[phosphinic acid], 9CI M-00143

Methanediphosphinic acid
[81050-37-5]

HOPH(O)CH₂PH(O)OH

$CH_6O_4P_2$ M 144.004
Diisopropyl ester: [65747-71-9]. *Diisopropyl methylenebisphosphinate.*
$C_7H_{18}O_4P_2$ M 228.164
Liq. d_4^{20} 1.12. n_D^{20} 1.4571.

Prishchenko, A.A. *et al*, *Zh. Obshch. Khim.*, 1977, **47**, 2689 (*Engl. transl.* p. 2451) (*synth, ir, pmr, nmr*)
King, C. *et al*, *Inorg. Chem.*, 1986, **25**, 1290 (*synth, ir, pmr, P nmr, cryst struct*)

P,P'-Methylenebis(phosphonic diamide), 9CI M-00144

[25598-80-5]

(H₂N)₂P(O)CH₂P(O)(NH₂)₂

$CH_{10}N_4O_2P_2$ M 172.063
Octa-Me: [25598-80-5].
$C_9H_{26}N_4O_2P_2$ M 284.278
Ligand for Mg, Co, Ni, Zn and Cu. Liq. which solidifies to v. hydroscopic solid. Mp 47-60°. $Bp_{0.25}$ 147°. n_D^{20} 1.4888.

Lannert, K.P. *et al*, *Inorg. Chem.*, 1969, **8**, 1775 (*synth, pmr, complexes*)

Methylenebis(phosphonic dichloride), 9CI **M-00145**

Bis(dichlorophosphinyl)methane. Methanediphosphonic tetrachloride

[1499-29-2]

$Cl_2P(O)CH_2P(O)Cl_2$

$CH_2Cl_4O_2P_2$ M 249.785

Cryst. (C_6H_6/pet. ether). Mp 104-105° (98-100°).

Maier, L., *Helv. Chim. Acta*, 1965, **48**, 133 (*synth, P nmr*)
Kosolapoff, G.M. *et al*, *J. Chem. Soc. (C)*, 1966, 757 (*synth*)
Sheldrick, W.S., *J. Chem. Soc., Dalton Trans.*, 1975, 943 (*cryst struct*)
Althoff, W. *et al*, *Z. Naturforsch., B*, 1976, **31**, 153 (*cmr*)
Dagliesh, W.H. *et al*, *J. Chem. Soc., Dalton Trans.*, 1977, 1505 (*nqr*)
Althoff, W. *et al*, *Chem. Ber.*, 1981, **114**, 1082 (*synth*)

Methylenebisphosphonic difluoride, 9CI **M-00146**

Bis(difluorophosphinyl)methane. Methanediphosphonic tetrachloride

[42148-25-4]

$F_2P(O)CH_2P(O)F_2$

$CH_2F_4O_2P_2$ M 183.967

Cryst. Mp 50-52°.

Althoff, W. *et al*, *Z. Naturforsch., B*, 1973, **28**, 98 (*synth, nmr*)
Althoff, W. *et al*, *Z. Naturforsch., B*, 1976, **31**, 153 (*cmr*)
Althoff, W. *et al*, *Chem. Ber.*, 1981, **114**, 1082 (*pmr, F and P nmr*)

Methylenebis[triphenylphosphonium](2+), 9CI **M-00147**

Bis(triphenylphosphonio)methane

[35823-41-7]

$Ph_3P^{\oplus}CH_2P^{\oplus}Ph_3$

$C_{37}H_{32}P_2^{\oplus\oplus}$ M 538.607 (ion)

Dichloride:
$C_{37}H_{32}Cl_2P_2$ M 609.513
Cryst. + $1H_2O$. Mp 239-246°. Drying at 100°/0.1 mm causes loss of HCl.

Dibromide: [14529-09-0].
$C_{37}H_{32}Br_2P_2$ M 698.415
Controls helminths in sheep. Source of both Triphenyl[(triphenylphosphoranylidene)methyl]-phosphonium(1+), T-00697 and Bis(triphenylphosphoranylidene)methane, B-00459 . Cryst. (MeOH/EtOAc). Mp 310-310.5°. pK_a 5.4.

Diiodide: [67260-61-1].
$C_{37}H_{32}I_2P_2$ M 792.416
Cryst. (EtOH). Mp 296-300° dec., 306-311° dec.

Ramirez, F. *et al*, *J. Am. Chem. Soc.*, 1961, **83**, 3539 (*dibromide, prep, ir*)
Driscoll, J.B. *et al*, *J. Org. Chem.*, 1964, **29**, 2427 (*dichloride*)
Mastryukova, T.A. *et al*, *Zh. Obshch. Khim.*, 1978, **48**, 991 (*Engl. transl. p. 904*) (*diiodide*)

Methylenetriphenylphosphorane **M-00148**

[3487-44-3]

$H_2C{=}PPh_3$

$C_{19}H_{17}P$ M 276.317

Reagent for methyleneation of aldehydes and ketones. Yellow cryst. turning white in air. Strong solns. are yellow to red. Forms Li deriv. with *tert*-butyllithium.

Lüttke, W. *et al*, *Angew. Chem.*, 1965, **77**, 867 (*ir*)
Grim, S.O. *et al*, *J. Org. Chem.*, 1968, **33**, 2993 (*uv*)
Williams, D.H. *et al*, *J. Am. Chem. Soc.*, 1968, **90**, 966 (*ms*)
Bart, J.C.J., *J. Chem. Soc. (B)*, 1969, 350 (*cryst struct*)
Albright, T.A. *et al*, *J. Am. Chem. Soc.*, 1975, **97**, 940 (*cmr*)
Ostoja Starzewski, K.A. *et al*, *J. Am. Chem. Soc.*, 1976, **98**, 8486 (*pe*)
Ostoja Starzewski, K.A. *et al*, *Phosphorus*, 1976, **6**, 177 (*pmr, cmr, nmr*)
Seyferth, D. *et al*, *J. Organomet. Chem.*, 1979, **179**, 25 (*synth, props*)
Corey, E.J. *et al*, *J. Am. Chem. Soc.*, 1982, **104**, 4724 (*deriv*)
Schlosser, M. *et al*, *Chimia*, 1982, **36**, 396 (*synth*)

(1-Methylethenyl)phosphonic acid, 9CI **M-00149**

Isopropenylphosphonic acid. 2-Phosphonopropene. (1-Methylvinyl)phosphonic acid

$H_2C{=}C(CH_3)P(O)(OH)_2$

$C_3H_7O_3P$ M 122.060

Di-Me ester: [38281-98-0]. *Dimethyl (1-methylethenyl)-phosphonate.*
$C_5H_{11}O_3P$ M 150.114
Liq. d_4^{20} 1.11. Bp_8 66.5-67°, Bp_1 38-41°. n_D^{20} 1.4318.

Di-Et ester: [20170-34-7]. *Diethyl (1-methylethenyl)-phosphonate.*
$C_7H_{15}O_3P$ M 178.167
Liq. d_4^{20} 1.03. Bp_{10} 95-96°, Bp_1 44-47°. n_D^{20} 1.4291.

Di-Ph ester: [37037-60-8]. *Diphenyl (1-methylethenyl)-phosphonate.*
$C_{15}H_{15}O_3P$ M 274.255
d_4^{20} 1.16. Bp_2 175-178°. n_D^{20} 1.5580.

Dichloride: see (1-Methylethenyl)phosphonic dichloride, M-00150

Pudovik, A.N. *et al*, *Zh. Obshch. Khim.*, 1963, **33**, 3201 (*Engl. transl.* p. 3128) (*synth*)
Nurtdinov, S.Kh. *et al*, *Zh. Obshch. Khim.*, 1972, **42**, 123 (*Engl. transl.* p. 118) (*synth, pmr*)
Pudovik, A.N. *et al*, *Zh. Obshch. Khim.*, 1972, **42**, 1862 (*Engl. transl.* p. 1847) (*P nmr, ir*)
Borisova, E.E. *et al*, *Dokl. Akad. Nauk SSSR, Ser. Sci. Khim.*, 1976, **226**, 1330 (*Engl. transl.* p. 142) (*synth*)
Gareev, R.D. *et al*, *Zh. Obshch. Khim.*, 1978, **48**, 269 (*Engl. transl.* p. 238) (*pmr*)
Kojima, M. *et al*, *Synthesis*, 1979, 147 (*synth*)
Yamashita, M. *et al*, *Bull. Chem. Soc. Jpn.*, 1980, **53**, 1625 (*synth*)

(1-Methylethenyl)phosphonic dichloride **M-00150**

Isopropenylphosphonic dichloride. (2-Propenyl)-phosphinic dichloride

[3944-27-2]

$H_2C{=}C(CH_3)P(O)Cl_2$

$C_3H_5Cl_2OP$ M 158.952

Liq. d_4^{20} 1.34. Bp_{13} 64-66°. n_D^{20} 1.4818.

Bolle, J. *et al*, *C.R. Hebd. Seances Acad. Sci.*, 1965, **261**, 1852 (*synth*)
Remizov, A.B. *et al*, *Zh. Obshch. Khim.*, 1974, **44**, 1863 (*Engl. transl.* p. 1831) (*ir*)

Gareev, R.D. *et al*, *Zh. Obshch. Khim.*, 1978, **48**, 1688 (*Engl. transl.* p. 1544) (*synth, pmr*)
Zverev, V.V. *et al*, *Dokl. Akad. Nauk SSSR, Ser. Sci. Khim.*, 1979, **246**, 1368 (*Engl. transl.* p. 312) (*pe*)

1-Methylethenyl phosphorodichloridate, 9CI **M-00151**

1-Methylvinyl phosphorodichloridate. 1-Methylvinyl dichlorophosphate. 1-Methylvinyl phosphoryl dichloride. Isopropenyl phosphorodichloridate. Isopropenyl dichlorophosphate. Isopropenyl phosphoryl dichloride
[4188-33-4]

$H_2C{=}C(CH_3)OP(O)Cl_2$

$C_3H_5Cl_2O_2P$ M 174.951
Pungent liq. d_4^{20} 1.33. Bp_{30} 69-70°, Bp_9 46-47°. n_D^{20} 1.4500. Sensitive to moisture.

Goldosov, Y.G. *et al*, *Zh. Obshch. Khim.*, 1965, **35**, 1460 (*Engl. transl.* p. 1462) (*synth*)
Novikova, Z.S. *et al*, *Zh. Obshch. Khim.*, 1967, **37**, 2080 (*Engl. transl.* p. 1972) (*synth*)

2-(1-Methylethoxy)-1,3,2-dioxaphospholane, 9CI **M-00152**

Ethylene isopropyl phosphite. 2-Isopropoxy-1,3,2-dioxaphospholane
[40928-00-5]

$POCH(CH_3)_2$

$C_5H_{11}O_3P$ M 150.114
Liq. d_4^{25} 1.13. Bp_{20} 64-66°. n_D^{25} 1.4347. Rapidly hydrolysed.

Lucas, H.J. *et al*, *J. Am. Chem. Soc.*, 1950, **72**, 5491 (*synth*)
Arbuzov, B.A. *et al*, *Izv. Akad. Nauk SSSR, Otd. Khim. Nauk*, 1957, 1046 (*synth*)
Besserre, D. *et al*, *Bull. Soc. Chim. Fr.*, 1974, 845 (*pmr*)

(1-Methylethylidene)triphenylphosphorane, 9CI **M-00153**

Isopropylidenetriphenylphosphorane
[16666-80-1]

$Ph_3P{=}C(CH_3)_2$

$C_{21}H_{21}P$ M 304.371
A reactive Wittig regent for prepn. of 1,1-dimethyl-1-alkenes. Obt. as salt-free, red cryst.

Grim, S.O. *et al*, *J. Org. Chem.*, 1968, **33**, 2993 (*synth, uv*)
Albright, T.A. *et al*, *J. Am. Chem. Soc.*, 1975, **97**, 940; 1976, **98**, 6249 (*cmr, P nmr*)
Ostoja Starzewski, K.A. *et al*, *J. Am. Chem. Soc.*, 1976, **98**, 8486 (*synth, pe*)
Vedejs, E. *et al*, *J. Am. Chem. Soc.*, 1981, **103**, 2823 (*props*)
Belletire, J.L. *et al*, *Synth. Commun.*, 1982, **12**, 469 (*use*)
Schlosser, M. *et al*, *Chimia*, 1982, **36**, 396 (*synth*)

***O*-Methyl ethylphosphonochloridothioate,** 9CI **M-00154**

[1497-69-4]

[Structure: Cl above, MeO–P=S, Et below] (*R*)-*form*

C_3H_8ClOPS M 158.582

(***R***)-***form*** [38315-84-3]
Liq. $Bp_{0.05}$ 20°. $[\alpha]_D$ −57.25°. n_D^{20} 1.5020.

Mikolajczyk, M. *et al*, *Tetrahedron*, 1972, **28**, 3855 (*synth*)

***O*-Methyl ethylphosphonothioate,** 9CI **M-00155**

[Structure: OH above, S=P–OMe, Et below] (*S*)-*form*

$C_3H_9O_2PS$ M 140.137
Tautomeric with thiol form.

(***S***)-***form*** [38344-09-1]
$[\alpha]_D$ −5.25°.
O-*Et ester:* [38344-09-1]. O-*Ethyl* O-*methyl ethylphosphonothioate.*
$C_5H_{13}O_2PS$ M 168.190
Liq. Bp_3 46°. $[\alpha]_D^{20}$ −2.71° (neat). n_D^{20} 1.4662.

(±)-***form*** [36585-74-7]
Chloride: see O-Methyl ethylphosphonochloridothioate, M-00154

Mikolajczyk, M. *et al*, *Tetrahedron*, 1972, **28**, 4357 (*config, uv, cd*)
Stec, W.J. *et al*, *J. Org. Chem.*, 1974, **41**, 233 (*synth, P nmr*)
Michalski, J. *et al*, *J. Am. Chem. Soc.*, 1978, **100**, 5386 (*synth, P nmr*)
Mikolajczyk, M. *et al*, *J. Am. Chem. Soc.*, 1978, **100**, 7003 (*pmr, nmr, config*)

Methylimidodiphosphoric acid, 9CI **M-00156**

Methyliminodiphosphoric acid. Methylimidobisphosphoric acid

$MeN[P(O)(OH)_2]_2$

$CH_7NO_6P_2$ M 191.017

Tetra-Me ester: [18313-83-2]. *Tetramethyl methylimidodiphosphate. Tetramethyl methyliminodiphosphate.*
$C_5H_{15}NO_6P_2$ M 247.124
Liq. d_4^{20} 1.32. $Bp_{0.002}$ 78-79°. n_D^{20} 1.4426.

Tetra-Et ester: [13294-09-2]. *Tetraethyl methylimidodiphosphate. Tetraethyl methyliminodiphosphate.*
$C_9H_{23}NO_6P_2$ M 303.231
Liq. d_4^{20} 1.16. $Bp_{0.03}$ 65.5-66°. n_D^{20} 1.4345. Wide ranging Bps have been recorded.

Tetraisopropyl ester: Tetraisopropyl methyliminodiphosphate. Tetrakis(1-methylethyl) methylimidodiphosphate.
$C_{13}H_{31}NO_6P_2$ M 359.339
Liq. d_4^{20} 1.16. Bp_1 130-131°, $Bp_{0.002}$ 83-85°. n_D^{20} 1.4358.

Tetra-Ph ester: Tetraphenyl methylimidodiphosphate. Tetraphenyl methyliminodiphosphate.
$C_{25}H_{23}NO_6P_2$ M 495.407
Solid. Mp 92-92.5°.

Bisdifluoride: Methylimidobisphosphoryl fluoride.
$CH_3F_4NO_2P_2$ M 198.981
Liq. $Bp_{0.04}$ 71°.

Nielsen, M.L., *Inorg. Chem.*, 1964, **3**, 1760 (*tetraphenyl ester, synth, ir, pmr, P nmr*)
Shokol, V.A. *et al*, *CA*, 1968, **69**, 10065 (*esters, synth*)
Gilyarov, V.A. *et al*, *Zh. Obshch. Khim.*, 1971, **41**, 2355 (*Engl. transl.* p. 2380) (*tetraethyl ester, synth*)
Gololobov, Yu.G. *et al*, *Zh. Obshch. Khim.*, 1976, **46**, 1268 (*Engl. transl.* p. 1248) (*tetraethyl ester, synth*)
Riesch, L. *et al*, *Z. Anorg. Allg. Chem.*, 1986, **539**, 183 (*fluoride, synth, pmr, F nmr, P nmr*)

Methylimidodi(phosphorous acid), 10CI **M-00157**
Methyliminodi(phosphorous acid), 9CI

MeN[P(OH)$_2$]$_2$

$CH_7NO_4P_2$ M 159.018
Tetra-Me ester: [34244-05-8]. *Tetramethyl methylimidodiphosphite.*
$C_5H_{15}NO_4P_2$ M 215.125
Ligand for Cr, Mo, and Fe. Liq. Bp_{12} 92-94°.
Tetra-Et ester: [34244-07-0]. *Tetraethyl methylimidodiphosphite.*
$C_9H_{23}NO_4P_2$ M 271.233
Liq. Bp_3 68-69°.
Tetra-Ph ester: Tetraphenyl methylimidodiphosphite.
$C_{25}H_{23}NO_4P_2$ M 463.409
Liq. $Bp_{0.03}$ 190-191°. n_D^{20} 1.5991.
Tetrafluoride: see Methylimidodiphosphorous tetrafluoride, M-00159
Tetrachloride: see Methylimidodiphosphorous tetrachloride, M-00158
Tetraiodide:
$CH_3I_4NP_2$ M 598.607
Solid. Mp 91-92°.
Tetrakis(dimethylamide): [34244-04-7]. *Nonamethylimidodiphosphorous tetraamide.*
$C_9H_{27}N_5P_2$ M 267.293
Mp 33-35°. $Bp_{0.1}$ 94°.

Metzinger, H.G., *Org. Magn. Reson.*, 1971, **3**, 485 (*esters, amide, pmr*)
Binder, H. *et al*, *Chem. Ber.*, 1974, **107**, 205 (*esters, synth*)
Gorbatenko, Zh.K. *et al*, *Zh. Obshch. Khim.*, 1975, **45**, 2367 (*Engl. transl.* p. 2325) (*phenyl ester, iodide, synth*)
Colquhoun, I.J. *et al*, *J. Chem. Soc., Dalton Trans.*, 1977, 1674 (*halides, amide, methyl ester, pmr, P nmr*)
Zeiss, W. *et al*, *Z. Naturforsch., B*, 1979, **34**, 423 (*bromide, P nmr, pmr, props*)
Keat, R. *et al*, *J. Chem. Soc., Dalton Trans.*, 1980, 321 (*dimethylamide, synth*)
Brown, G.M. *et al*, *Inorg. Chem.*, 1982, **21**, 2139, 3790 (*methyl ester, synth, pmr, P nmr, ms, complexes*)

Methylimidodiphosphorous tetrachloride, 10CI **M-00158**
Methyliminodiphosphorous tetrachloride, 9CI. Methylaminobis(dichlorophosphine)
[17648-16-7]

MeN(PCl$_2$)$_2$

$CH_3Cl_4NP_2$ M 232.801
Liq. or solid. Mp 30°. $Bp_{0.5}$ 72° (47-52°), $Bp_{0.3}$ 80-90°. n_D^{20} 1.5555.

Nixon, J.F., *J. Chem. Soc. (A)*, 1968, 2689 (*synth, pmr, ir, raman*)
Jefferson, R. *et al*, *J. Chem. Soc., Dalton Trans.*, 1973, 1414 (*synth*)
Wannagat, U. *et al*, *Z. Anorg. Allg. Chem.*, 1976, **420**, 119 (*synth, pmr*)
Barlos, K. *et al*, *Z. Naturforsch., B*, 1978, **33**, 515 (*N nmr*)
Bulloch, G. *et al*, *J. Chem. Soc., Dalton Trans.*, 1978, 764 (*cmr, pmr*)
King, R.B. *et al*, *Inorg. Chem.*, 1978, **17**, 2390 (*synth*)

Methylimidodiphosphorous tetrafluoride, 10CI **M-00159**
Methyliminodiphosphorous tetrafluoride, 9CI. Methylaminobis(difluorophosphine)
[17648-18-9]

MeN(PF$_2$)$_2$

$CH_3F_4NP_2$ M 166.983
Ligand for metals of groups VB, VIB and VIII. Liq. Bp 45°.

Nixon, J.F., *J. Chem. Soc. (A)*, 1968, 2689 (*synth, pmr, P nmr, raman*)
Hedberg, E. *et al*, *J. Am. Chem. Soc.*, 1974, **96**, 4417 (*ed*)
Bulloch, G. *et al*, *J. Chem. Soc., Dalton Trans.*, 1978, 764 (*cmr, pmr*)
King, R.B. *et al*, *Inorg. Chem.*, 1978, **17**, 2390 (*synth, P nmr, complexes*)
Andreocci, M.V. *et al*, *Inorg. Chem.*, 1979, **18**, 954 (*pe, complexes*)
Keat, R. *et al*, *J. Chem. Soc., Dalton Trans.*, 1981, 2192 (*P nmr*)

Methylimidodiphosphoryl chloride, 9CI **M-00160**
Methyliminodiphosphoryl tetrachloride
[29777-24-0]

MeN[P(O)Cl$_2$]$_2$

$CH_3Cl_4NO_2P_2$ M 264.800
Solid. Mp 53-54°. Bp_{1-2} 98-102°, $Bp_{0.01}$ 65-70°.

Keat, R., *J. Chem. Soc. (A)*, 1970, 2732 (*synth, ir, pmr, P nmr*)
Hägele, G. *et al*, *J. Chem. Soc., Dalton Trans.*, 1974, 1985 (*pmr, cmr, P nmr*)
Riesel, L. *et al*, *Z. Anorg. Allg. Chem.*, 1974, **404**, 219; 1977, **435**, 61 (*synth*)
Dagleish, W.H. *et al*, *J. Chem. Soc., Dalton Trans.*, 1977, 1505 (*nqr*)
Thomas, B. *et al*, *Z. Anorg. Allg. Chem.*, 1985, **525**, 7 (*N nmr*)

Methylimidothiodiphosphoric acid **M-00161**
Methylthioimidodiphosphoric acid, 9CI. Methyliminothiodiphosphoric acid. Methylimidodiphosphorothioic acid

MeN[P(S)(OH)$_2$]$_2$

$CH_7NO_4P_2S_2$ M 223.138
Note 9CI name same as that for Methylthioimidodiphosphoric acid, M-00413 .
O,O,O,O-*Tetra-Me ester:* [34244-10-5]. O,O,O,O-*Tetramethyl methylthioimidodiphosphate.* O,O,O,O-*Tetramethyl methylimidodiphosphorothioate.*
$C_5H_{15}NO_4PS_2$ M 248.272
Liq. d_4^{20} 1.30. $Bp_{0.02}$ 80°. n_D^{20} 1.5170.
O,O,O,O-*Tetra-Et ester:* O,O,O,O-*Tetraethyl methylthioimidodiphosphate.* O,O,O,O-*Tetraethyl methylimidodiphosphorothioate.*
$C_9H_{23}NO_4PS_2$ M 304.379
Liq. d_4^{20} 1.17. $Bp_{1.5}$ 134-135°. n_D^{20} 1.4940.

Arbuzov, B.A. *et al*, *Izv. Akad. Nauk SSSR, Ser. Khim.*, 1956, 932 (*Engl. transl.* p. 953) (*synth*)
Grechkin, N.P. *et al*, *Zh. Obshch. Khim.*, 1976, **46**, 1753 (*Engl. transl.* p. 1703) (*synth, P nmr*)

1-Methylisophosphinoline, 9CI **M-00162**
1-Methyl-2-phosphanaphthalene
[57328-60-6]

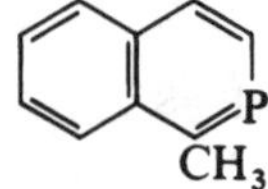

$C_{10}H_9P$ M 160.155

$Bp_{0.0001}$ 70°.

de Graaf, H.G., *Tetrahedron*, 1975, **31**, 1097 (*synth, ir, pmr, uv, P nmr, ms*)

3-Methylisophosphinoline, 9CI **M-00163**

3-Methyl-2-phosphanaphthalene. 3-Methylbenzo[c]*-phosphorin. 3-Methylbenzo*[c]*phosphinine*

[49622-63-1]

$C_{10}H_9P$ M 160.155

Mp 72-74° (64.5-69°).

de Graaf, H.G. *et al*, *Tetrahedron*, 1975, **31**, 1097 (*synth, pmr, cmr, ms*)
Jongsma, C. *et al*, *Org. Mass Spectrom.*, 1975, **10**, 515 (*ms*)

(5-Methyl-2-isopropylcyclohexyl)-diphenylphosphine **M-00164**

[*5-Methyl-2-(1-methylethyl)cyclohexyl*]*-diphenylphosphine, 9CI. (2-Diphenylphosphino)-1-isopropyl-4-methylcyclohexane. 3-(Diphenylphosphino)-*p*-menthane*

[32511-22-1]

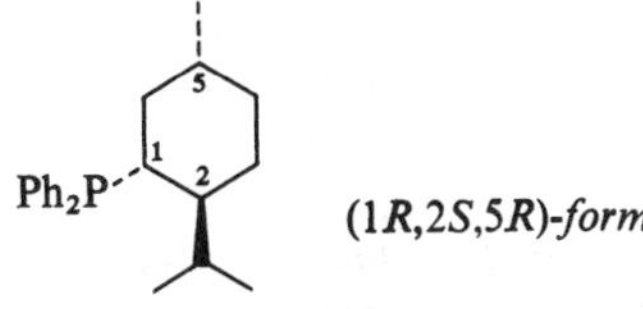

(1*R*,2*S*,5*R*)-*form*

$C_{22}H_{29}P$ M 324.445

Used in conjunction with Rh or Pd, as catalyst for asymmetric hydroformylations, cyclizations, dehydrogenations, hydrogenations and hydrocarboalkoxylations.

(1*R*,2*S*,5*R*)-*form* [43077-31-2]

Menthyldiphenylphosphine. MDPP

Mp 57.5-58.5°. $[\alpha]_D^{20}$ −95.7° (c, 1.07 in CH_2Cl_2).

Oxide:

$C_{22}H_{29}OP$ M 340.444

Solid. Mp 183.5-185°. $[\alpha]_D^{25}$ −87.4° (c, 1.74 in EtOH).

(1*S*,2*S*,5*R*)-*form* [43077-29-8]

Neomenthydiphenylphosphine. NMDPP

Cryst. (pet. ether). Mp 96-99°. $Bp_{0.2}$ 165-170°. $[\alpha]_D^{23}$ +94.4° (c, 1.26 in CH_2Cl_2).

B,PhCH₂Br: Benzylneomenthyldiphenylphosphonium bromide.

$C_{29}H_{36}BrP$ M 495.481

Cryst. (C_6H_6/MeOH). Mp 220-223°.

Oxide: Cryst. (C_6H_6/hexane). Mp 217-221°. $[\alpha]_D^{29}$ +54.5° (c, 1.40 in EtOH).

Kiso, Y. *et al*, *J. Am. Chem. Soc.*, 1972, **94**, 4373 (*use*)
Morrison, J.D. *et al*, *J. Org. Chem.*, 1974, **39**, 270 (*synth, oxide*)
Tanaka, M. *et al*, *Bull. Chem. Soc. Jpn.*, 1974, **47**, 1698 (*use*)
Aguiar, A.M. *et al*, *J. Org. Chem.*, 1976, **41**, 1545 (*synth, cmr, ir*)
Yamamoto, K. *et al*, *J. Organomet. Chem.*, 1981, **210**, 9 (*use*)
Hidao, M. *et al*, *J. Organomet. Chem.*, 1982, **232**, 89 (*use*)

O-Methyl isopropylphosphonochloridothioate **M-00165**

O*-Methyl (1-methylethyl)phosphonochloridothioate, 9CI*

[50636-74-3]

OMe
S=P–Cl (*R*)-*form*
$CH(CH_3)_2$

$C_4H_{10}ClOPS$ M 172.609

(*R*)-*form* [38315-90-1]

Liq. Bp_5 72°. $[\alpha]_D$ −91.80° (neat). n_D^{20} 1.4967.

(*S*)-*form* [38315-86-5]

Liq. Bp_5 72°. $[\alpha]_D$ +93.90° (neat). n_D^{20} 1.4972 (1.4889).

(±)-*form*

Liq. Bp_2 63°. n_D^{20} 1.4948.

Mikolajczyk, M. *et al*, *Tetrahedron*, 1972, **28**, 3855 (*synth, hydrol, config*)
Mikolajczyk, M. *et al*, *Phosphorus*, 1973, **3**, 47 (*synth*)
Omelanczuk, J. *et al*, *Tetrahedron*, 1975, **31**, 2809 (*synth*)

O-Methyl isopropylphosphonothioate **M-00166**

[50636-75-4]

OMe OMe
S=P–OH ⇌ HS–P=O (*R*)-*form*
$CH(CH_3)_2$ $CH(CH_3)_2$

$C_4H_{11}O_2PS$ M 154.163

(*R*)-*form*

Liq. $[\alpha]_D$ +14.67° (neat).

(R)*-1-Phenylethylammonium salt:* Solid. Mp 149-151°. $[\alpha]_D$ +10.7° (c, 1.1 in MeOH).

Dicyclohexylammonium salt: Mp 179-181°. $[\alpha]_D$ +4.60° (c, 1.17 in C_6H_6).

O*-Et ester:* O*-Ethyl* O*-methyl isopropylphosphonothioate.*

$C_6H_{15}O_2PS$ M 182.217

Liq. $Bp_{1.5}$ 44°. $[\alpha]_{589}$−4.14° (neat). n_D^{20} 1.4649. Has the (*S*)-config.

Chloride: see O-Methyl isopropylphosphonochloridothioate, M-00165

(*S*)-*form*

Liq. $[\alpha]_D$ −14.30° (neat).

(S)*-1-Phenylethylammonium salt:* Solid. Mp 150-151°. $[\alpha]_D$ −11.3° (c, 1.63 in MeOH).

Dicyclohexylammonium salt: Solid. Mp 179-181°. $[\alpha]_D$ −4.49° (c, 1.07 in C_6H_6).

O*-Et ester:* Liq. $Bp_{1.5}$ 44°. $[\alpha]_{589}$+1.71° (neat). n_D^{20} 1.4645. Has the (*R*)-config.

Chloride: see O-Methyl isopropylphosphonochloridothioate, M-00165

(±)-*form*

Dicyclohexylammonium salt: Cryst. (C_6H_6/ligroin). Mp 174-176°.

Chloride: see O-Methyl isopropylphosphonochloridothioate, M-00165

Mikolajczyk, M. *et al*, *Tetrahedron*, 1972, **28**, 3855 (*ester, config*)
Mikolajczyk, M. *et al*, *Phosphorus*, 1973, **3**, 47 (*synth, resoln, salts*)
Mikolajczyk, M. *et al*, *J. Am. Chem. Soc.*, 1978, **100**, 7003 (*pmr, P nmr*)

6-Methyl-4-methylene-4*H*-1,3,2-dioxaphosphorin, 9CI **M-00167**

*6-Methyl-4-methylene-4*H*-1,3,2-dioxaphosphinine*

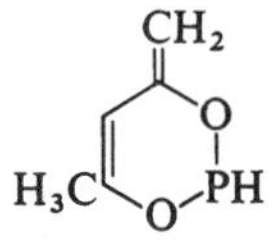

$C_5H_7O_2P$ M 130.083

2-Ethoxy: [87363-91-5].
$C_7H_{11}O_3P$ M 174.136
Liq. d_4^{20} 1.00. Bp_{11} 66-68°. n_D^{20} 1.4880.

2-Ethoxy, 2-oxide:
$C_7H_{11}O_4P$ M 190.135
Liq. d_4^{20} 1.19. $Bp_{0.02}$ 77-78°. n_D^{20} 1.4778.

2-Me, 2-oxide: [87363-94-8].
$C_6H_9O_3P$ M 160.109
Solid. Mp 58°.

2-Ph: [76795-74-9].
$C_{11}H_{11}O_2P$ M 206.180
Solid. Mp 40°.

2-Diethylamino:
$C_9H_{16}NO_2P$ M 201.205
Liq. $Bp_{0.022}$ 58-60°. n_D^{20} 1.5021.

Van Seyerl, J. *et al*, *Z. Naturforsch.*, *B*, 1980, **35**, 1373 (*phenyl, synth, cryst struct, pmr, cmr, P nmr*)
Mukhametov, F.S. *et al*, *Zh. Obshch. Khim.*, 1983, **53**, 1274 (*Engl. transl.* p. 1137) (*synth, ir, pmr*)
Mukhametov, F.S. *et al*, *Zh. Obshch. Khim.*, 1983, **53**, 2703 (*Engl. transl.* p. 2436) (*synth, pmr, P nmr*)

Methyl methylphosphonate **M-00168**

Monomethyl methylphosphonate. Methyl hydrogen methylphosphonate

[1066-53-1]

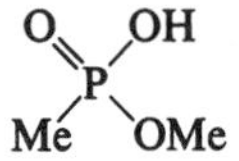

$C_2H_7O_3P$ M 110.049

Oil. d_4^{20} 1.307. Bp_1 121-123°. n_D^{20} 1.4222.

Isopropyl ester: [690-64-2]. *Isopropyl methyl methylphosphonate. Methyl 1-methylethyl methylphosphonate.*
$C_5H_{13}O_3P$ M 152.130
Liq. Bp_{13} 69°. n_D^{20} 1.4125.

Benzyl ester: [710-09-8]. *Benzyl methyl methylphosphonate. Methyl phenylmethyl methylphosphonate.*
$C_9H_{13}O_3P$ M 200.174
Liq. Bp_{13} 159°. n_D^{20} 1.5095.

Ph ester: see Methyl phenyl methylphosphonate, M-00205

Fluoride: see Methylphosphonofluoridic acid, M-00303

Chloride: see Methylphosphonochloridic acid, M-00296

Keay, L., *Can. J. Chem.*, 1965, **43**, 2637 (*synth*)
Petrov, K.A. *et al*, *Zh. Obshch. Khim.*, 1965, **35**, 723 (*Engl. transl.* p. 723) (*synth*)
Thuong, N.T. *et al*, *Bull. Soc. Chim. Fr.*, 1965, 1925 (*mixed esters, synth, pmr*)
Mikulski, C.M. *et al*, *Inorg. Chem.*, 1970, **9**, 2053 (*complexes*)
Cadogan, J.I.G. *et al*, *J. Chem. Soc.* (*B*), 1971, 1988 (*synth*)
Vinogradov, L.I. *et al*, *Zh. Obshch. Khim.*, 1974, **44**, 37 (*Engl. transl.* p. 35) (*pmr, ir, complexes*)
Kluger, R. *et al*, *J. Am. Chem. Soc.*, 1975, **97**, 4298 (*complexes*)

***O*-Methyl methylphosphonochloridothioate, 9CI** **M-00169**

[2524-15-4]

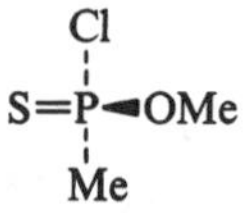

(*S*)-*form*

C_2H_6ClOPS M 144.556

(*S*)-*form* [38315-80-9]
Liq. $[\alpha]_D$ +25.8° (neat).

Mikolajczyk, M. *et al*, *Tetrahedron*, 1972, **28**, 3855 (*synth*)
Pudovik, A.N. *et al*, *Zh. Obshch. Khim.*, 1972, **42**, 317 (*Engl. transl.* p. 308) (*ir*)

***O*-Methyl methylphosphonodithioate** **M-00170**

[993-44-2]

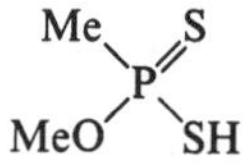

$C_2H_7OPS_2$ M 142.170

Liq. $Bp_{0.01}$ 28°. $n_D^{22.5}$ 1.5659.

Me ester: see Methylphosphonodithioic acid, M-00301

Chupp, J.P. *et al*, *J. Org. Chem.*, 1962, **27**, 3832 (*synth*)

***O*-Methyl methylphosphonothioate, 9CI** **M-00171**

O-*Methyl hydrogen methylphosphonothioate*

[18005-39-5]

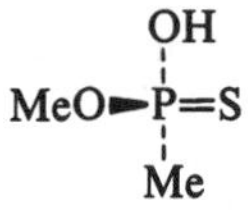

(*R*)-*form*

$C_2H_7O_2PS$ M 126.110

(*R*)-*form* [38315-71-8]

Dicyclohexylammonium salt: Cryst. (EtOAc). Mp 180°. $[\alpha]_D^{25}$ +6.37° (c, 5.0 in MeOH).

(R)-*1-Phenylethylammonium salt:* Solid. Mp 147°. $[\alpha]_D^{25}$ +10.04° (c, 3.1 in MeOH).

O-*Isopropyl ester:* [38315-92-3]. O-*Isopropyl* O-*methyl methylphosphonothioate.*
$C_5H_{13}O_2PS$ M 168.190
Liq. Bp_3 61°. $[\alpha]_D$ −0.50° (neat). n_D^{20} 1.4608. Has the (*S*)-config.

(*S*)-*form* [44252-11-1]

Dicyclohexylammonium salt: Cryst. (EtOAc). Mp 180°. $[\alpha]_D^{25}$ −6.35° (c, 5.0 in MeOH).

(S)-*1-Phenylethylammonium salt:* Solid. Mp 147°. $[\alpha]_D^{25}$ −10.04° (c, 3.1 in MeOH).

O-*Isopropyl ester:* [38315-91-2]. Liq. Bp_3 61°. n_D^{20} 1.4602. Has the (*R*)-config.

(±)-*form* [36585-69-0]

$Bp_{0.3\text{-}0.5}$ 75-78°. n_D^{27} 1.5008.

Tetramethylammonium salt: Solid. Mp 105°.

Dicyclohexylammonium salt: Cryst. (EtOH). Mp 179-180°.

O-*Me,* S-*butyl ester:* S-*Butyl* O-*methyl methylphosphonothioate.*
$C_6H_{15}O_2PS$ M 182.217
Inhibits trypsin and cholinesterases. Liq. Bp_1 64°.

O-*Me,* S-*Ph ester:* O-*Methyl* S-*phenyl methylphosphonothioate.*
$C_8H_{11}O_2PS$ M 202.207

Liq. $Bp_{0.04}$ 101-102°.
Chloride: see O-Methyl methylphosphonochloridothioate, M-00169

Pelchowicz, Z. *et al, J. Chem. Soc.*, 1962, 3824 (*synth, salts*)
Boter, H.L. *et al, Recl. Trav. Chim. Pays-Bas*, 1967, **86**, 399 (*resoln*)
Mikolajczyk, M.M. *et al, Tetrahedron*, 1972, **28**, 3855, 4357 (*esters, synth, cd, config*)
Mikolajczyk, M.M. *et al, J. Am. Chem. Soc.*, 1978, **100**, 7003 (*pmr, P nmr, cmr*)

Methyl(2-methylpropyl)phosphinic acid, 9CI M-00172

Isobutylmethylphosphinic acid

$(H_3C)_2CHCH_2P(O)(OH)Me$

$C_5H_{13}O_2P$ M 136.130

2-Methylpropyl ester: [53314-51-5]. *2-Methylpropyl methyl(2-methylpropyl)phosphinate. Isobutyl isobutylmethylphosphinate.* Liq. $Bp_{0.2}$ 66°.
Fluoride: [2708-92-1]. Liq. $Bp_{3.5}$ 71-72°.
Chloride: [13213-42-8]. Liq. d_4^{20} 1.08. Bp_1 68°. n_D^{20} 1.4614.

Neimysheva, A.A. *et al, Zh. Obshch. Khim.*, 1966, **36**, 1090 (*Engl. transl.* p. 1105) (*chloride*)
Dawson, T.P. *et al, J. Org. Chem.*, 1967, **22**, 1671 (*chloride, fluoride*)
Knunyants, I.L. *et al, Dokl. Akad. Nauk SSSR, Ser. Sci. Khim.*, 1971, **201**, 862 (*Engl. transl.*, p. 992) (*halides, nmr*)
Finke, M. *et al, Justus Liebigs Ann. Chem.*, 1974, 741 (*ester, synth, pmr*)

4-Methyl-2-methylthio-1,3,2-dioxaphospholane, 9CI M-00173

S-*Methyl* O,O-*propylene phosphorothioite.* S-*Methyl* O,O-*propylene thiophosphite. 4-Methyl-2-methylthio-1,3-dioxa-2-phosphacyclopentane*

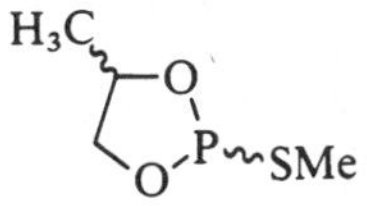

$C_4H_9O_2PS$ M 152.148

2-Oxide: S-*Methyl* O,O-*propylene phosphorothioate.* S-*Methyl* O,O-*propylene thiophosphate.*
$C_4H_9O_3PS$ M 168.147
Liq. $Bp_{0.05}$ 91-92°. n_D^{23} 1.4902. *Cis-* and *trans*-forms characterised spectroscopically.
2-Sulfide: S-*Methyl* O,O-*propylene phosphorodithioate.* S-*Methyl* O,O-*propylene dithiophosphate.*
$C_4H_9O_2PS_2$ M 184.208
No phys. props. reported.

Ishmaeva, E.A. *et al, Zh. Obshch. Khim.*, 1975, **45**, 946 (*Engl. transl.* p. 931) (*sulfide, conformn*)
Mikolajczyk, M. *et al, J. Chem. Soc., Perkin Trans. 1*, 1976, 371 (*oxide, synth, pmr, P nmr*)

4-Methyl-2-methylthio-1,3,2-dioxaphosphorinane, 9CI M-00174

4-Methyl-2-methylthio-1,3,2-dioxaphosphinane. O,O-*1,3-Butylene* S-*methyl phosphorothioite. 4-Methyl-2-methylthio-1,3-dioxa-2-phosphacyclohexane*

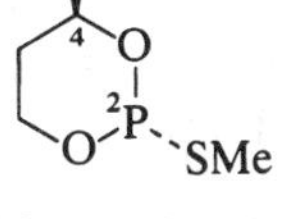

(2*RS*,4*RS*)-*form*

$C_5H_{11}O_2PS$ M 166.174

(2RS,4RS)-form [57270-64-1]
(±)-trans-*form*
Liq. Bp_2 64°. n_D^{25} 1.5142.
2-Oxide: [50902-84-6]. O,O-*1,3-Butylene* S-*methyl thiophosphate.*
$C_5H_{11}O_3PS$ M 182.174
Cryst. (C_6H_6/cyclohexane or Et_2O). Mp 76-78°.
2-Sulfide: [55102-62-0]. O,O-*1,3-Butylene* S-*methyl dithiophosphate.*
$C_5H_{11}O_2PS_2$ M 198.234
Cryst. (EtOH). Mp 62-63°.
2-Selenide: [57270-66-3]. O,O-*1,3-Butylene* S-*methyl phosphoroselenothioate.*
$C_5H_{11}O_2PSSe$ M 245.134
Cryst. (EtOH). Mp 77-78°.

(2RS,4SR)-form [57321-50-3]
(±)-cis-*form*
2-Oxide: [50902-83-5]. Oil. $Bp_{0.05}$ 90°. n_D^{20} 1.4992.
2-Sulfide: [55102-61-9]. Liq. d_4^{20} 1.27. $Bp_{0.008}$ 55-60° (bath). n_D^{20} 1.5548.
2-Selenide: [57270-65-2]. No phys. props. reported.

Predvoditelev, D.A. *et al, Zh. Obshch. Khim.*, 1974, **44**, 2629 (*Engl. transl.* p. 2586) (*sulfide, synth, ir, P nmr*)
Ishmaeva, E.A. *et al, Dokl. Akad. Nauk SSSR, Ser. Sci. Khim.*, 1975, **223**, 351 (*Engl. transl.* p. 410) (*oxide, ir, conformn*)
Okruszek, A. *et al, Z. Naturforsch., B*, 1975, **30**, 430 (*derivs, cmr, P nmr*)
Stec, W.J. *et al, J. Org. Chem.*, 1976, **41**, 1291 (*oxides, synth, P nmr*)
Mikolajczyk, M. *et al, J. Org. Chem.*, 1977, **42**, 190 (*oxide, synth, ir, pmr, P nmr*)

P-Methyl-4-morpholinylphosphonamidic acid, 9CI M-00175

Methyl-4-morpholinylphosphinic acid

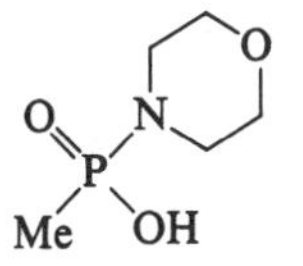

$C_5H_{12}NO_3P$ M 165.128

Isopropyl ester: Isopropyl P-*methyl-4-morpholinylphosphonamidate. Isopropyl methyl-4-morpholinylphosphinate.*
$C_8H_{18}NO_3P$ M 207.209
Liq. Bp_1 86°. n_D^{18} 1.4596.

Coe, D.G. *et al, J. Org. Chem.*, 1959, **24**, 1018 (*synth*)

O-Methyl *O*-1-naphthalenyl phosphorothioate, 9CI M-00176

O-*Methyl* O-*1-naphthyl hydrogen phosphorothioate.* O-*Methyl* O-*1-naphthyl phosphorothioic acid*

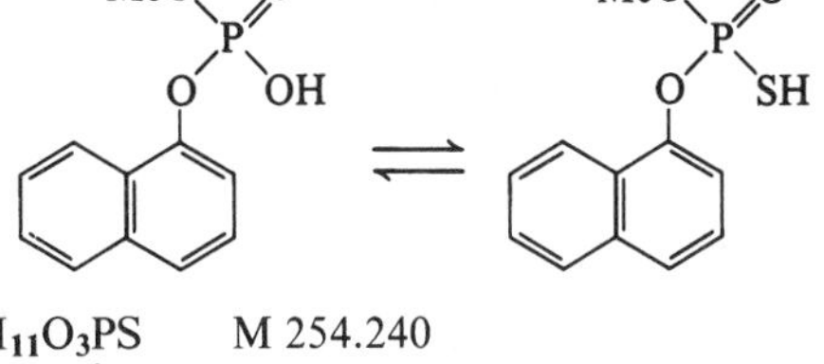

$C_{11}H_{11}O_3PS$ M 254.240
Tautomeric.

(+)-form [48154-37-6]
$[\alpha]_{365}$+57.1°, $[\alpha]_D$ +14.2° (c, 1.10 in $CHCl_3$).
Dicyclohexylammonium salt: [58293-37-1]. Cryst. (EtOH aq.). Mp 150-152°. $[\alpha]_D$ +16.3° (c, 2.14 in $CHCl_3$).

Ephedrine salt: Cryst. (MeOH aq.). Mp 168°. $[\alpha]_{365}$+136.4°, $[\alpha]_D$ +31.2° (c, 1.21 in $CHCl_3$).

(−)-***form*** [46501-98-8]

$[\alpha]_{365}$−53.6°, $[\alpha]_D$ −17.0° (c, 14.3 in $CHCl_3$).

Dicyclohexylammonium salt: [58293-38-2]. Cryst. Mp 132-137°. $[\alpha]_D$ −14.7° (c, 2.17 in $CHCl_3$).

(−)-*Ephedrine salt:* Oil. $[\alpha]_{365}$−134.9°, $[\alpha]_D$ −35.3° (c, 12.32 in $CHCl_3$).

(±)-***form*** [22601-77-0]

Tetramethylammonium salt: Solid. Mp 108-110°.

Omelańczuk, J. *et al*, *Tetrahedron*, 1975, **31**, 2809 (*synth, resoln, props, derivs*)
Akintonwa, D.A.A., *Tetrahedron*, 1978, **34**, 959 (*synth, resoln, props, derivs*)
Mikolajczyk, M. *et al*, *J. Am. Chem. Soc.*, 1978, **100**, 7003 (*P nmr*)

Methyl-2-naphthylphenylphosphine M-00177

Methyl-2-naphthalenylphenylphosphine, 9CI

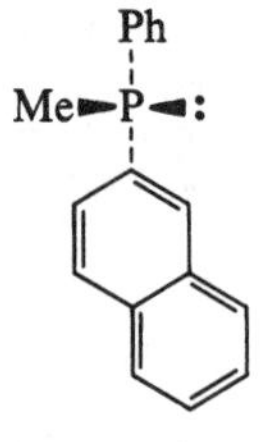

(*R*)-*form*

$C_{17}H_{15}P$ M 250.279

(***R***)-***form***

Oxide:

$C_{17}H_{15}OP$ M 266.279

$[\alpha]_D$ −12° (MeOH). Has (*S*)-config.

(***S***)-***form*** [29415-54-1]

Liq. $[\alpha]_{350}$+144° (c, 1 in C_6H_6).

B,PhCH₂Br: Methyl-2-naphthalenylphenyl(phenylmethyl)phosphonium bromide, 9CI. Benzylmethyl-2-naphthylphenylphosphonium bromide.

$C_{24}H_{22}BrP$ M 421.316

Mp 204-205°. $[\alpha]_D$ +21.5° (c, 4.17 in MeOH). Has (*R*)-config.

Oxide: [20663-39-2]. Mp 125-130°. $[\alpha]_D$ +4.4° (c, 8.11 in MeOH) (37% o.p.). Has (*R*)-config.

Korpium, O. *et al*, *J. Am. Chem. Soc.*, 1968, **90**, 4842 (*oxide*)
Lewis, R.A. *et al*, *J. Am. Chem. Soc.*, 1969, **91**, 7009 (*config, oxide*)
Luckenbach, R., *Phosphorus*, 1971, **1**, 77 (*oxide, salt*)

Methyl-1-naphthylphosphinic acid M-00178

Methyl(1-naphthalenyl)phosphinic acid, 9CI

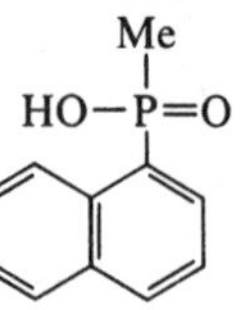

$C_{11}H_{11}O_2P$ M 206.180

Me ester: Methyl methyl-1-naphthylphosphinate.

$C_{12}H_{13}O_2P$ M 220.207

Liq. $Bp_{0.1}$ 128-130°.

Green, M. *et al*, *J. Chem. Soc.*, 1958, 3129.

O-Methyl *O*-4-nitrophenyl cyclohexylphosphoramidothioate, 9CI, 8CI M-00179

[22077-39-0]

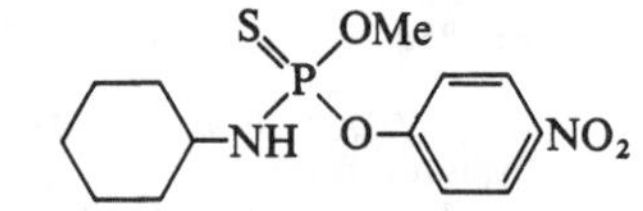

$C_{13}H_{19}N_2O_4PS$ M 330.338

(+)-***form*** [18457-65-3]

Cryst. Mp 57.5-58.5°. $[\alpha]_D$ +18.5°.

(±)-***form*** [13351-39-8]

Pale-yellow needles (MeOH). Mp 83-83.5°.

Gerrard, A.F. *et al*, *J. Chem. Soc.* (*B*), 1967, 1122 (*synth, pmr, ir, props*)
Cooks, R.G. *et al*, *J. Chem. Soc.* (*B*), 1968, 1327 (*ms*)

O-(4-Methyl-2-nitrophenyl) isopropylphosphoramidothioate M-00180

O-(*4-Methyl-2-nitrophenyl*) (*1-methylethyl*)-*phosphoramidothioate, 9CI*

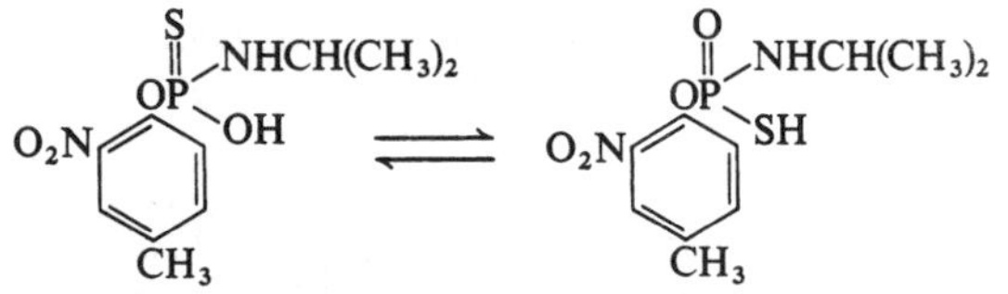

$C_{10}H_{15}N_2O_4PS$ M 290.273

(±)-***form***

O-*Me ester: see Amiprophos-Methyl, A-00132*

S-*Me ester:* S-*Methyl* O-(*4-methyl-2-nitrophenyl*) *isopropylphosphoramidothioate.*

$C_{11}H_{17}N_2O_4PS$ M 304.300

Solid. Mp 82-84°.

O-*Et ester: see Amiprophos M, A-00131*

S-*Et ester:* S-*Ethyl* O-(*4-methyl-2-nitrophenyl*) *isopropylphosphoramidothioate.*

$C_{12}H_{19}N_2O_4PS$ M 318.327

Solid. Mp 73-75°.

Ger. Pat., 2 049 693, (*1971*); *CA*, **75**, 35391

(1-Methylnonyl)phosphonic acid, 8CI M-00181

2-Decylphosphonic acid

$H_3C(CH_2)_7CH(CH_3)P(O)(OH)_2$

$C_{10}H_{23}O_3P$ M 222.264

(±)-***form***

Dichloride: [16474-69-4].

$C_{10}H_{21}Cl_2OP$ M 259.155

Liq. $Bp_{0.02}$ 47°. n_D^{25} 1.4649.

Geiseler, G. *et al*, *Ber. Bunsenges. Phys. Chem.*, 1967, **71**, 478 (*synth, ir, raman*)

2-Methyl-1,2-oxaphospholane M-00182

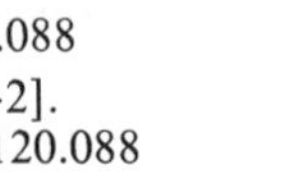

C_4H_9OP M 104.088

2-Oxide: [53314-66-2].

$C_4H_9O_2P$ M 120.088

Acted upon by amines to give (3-aminopropyl)-methylphosphinic acids. Liq. $Bp_{0.15}$ 87°.

2-Sulfide: [73627-86-8].

C_4H_9OPS M 136.148

Liq. $Bp_{0.5}$ 103-104°.

Finke, M. *et al, Justus Liebigs Ann. Chem.*, 1974, 741 (*oxide, synth, pmr*)
Kleiner, H.J. *et al, Justus Liebigs Ann. Chem.*, 1980, 324 (*oxide, use, sulfide, synth*)

5-Methyl-1,2-oxaphospholane 2-oxide, 9CI — M-00183

$C_4H_9O_2P$ M 120.088

2-Methoxy: [27230-13-3].
$C_5H_{11}O_2P$ M 134.114
Liq. Bp_{10} 132°. n_D^{21} 1.4480.

2-Ph: [41392-17-0].
$C_{10}H_{13}O_2P$ M 196.185
Obtained impure.

Bergesen, K. *et al, Acta. Chem. Scand.*, 1970, **24**, 1122 (*methoxy, synth, pmr*)
Arbuzov, B.A. *et al, Izv. Akad. Nauk SSSR, Ser. Khim.*, 1973, 648 (*Engl. transl.* p. 621) (*phenyl, synth*)

5-Methyl-1,3,2-oxathiaphosphole 2-oxide, 9CI — M-00184

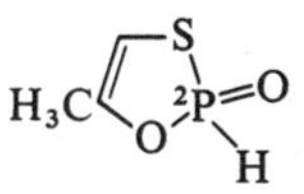

$C_3H_5O_2PS$ M 136.105

2-Ethoxy: [69174-43-2].
$C_5H_9O_3PS$ M 180.158
Liq.; cryst. (Et_2O, low temp.). d_4^{20} 1.266. $Bp_{0.03}$ 69-70°. n_D^{20} 1.4958.

2-Diethylamino:
$C_7H_{14}NO_2PS$ M 207.227
Liq. d_4^{20} 1.17. $Bp_{0.02}$ 86-88°. n_D^{20} 1.5060.

Ivanova, Zh.M. *et al, Zh. Obshch. Khim.*, 1978, **48**, 2376 (*Engl. transl.* p. 2156) (*synth, pmr, nmr*)
Ivanova, Zh.M. *et al, Zh. Obshch. Khim.*, 1979, **49**, 1464 (*Engl. transl.* p. 1279) (*synth*)

(3-Methyloxiranyl)phosphonic acid, 9CI — M-00185

(1,2-Epoxypropyl)phosphonic acid, 8CI
[23112-90-5]

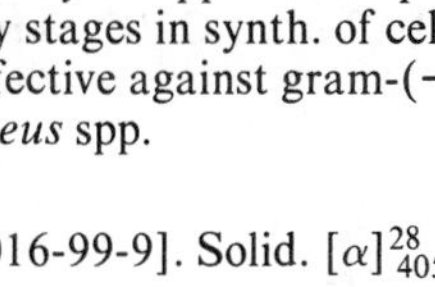

$C_3H_7O_4P$ M 138.060

(2*R*,3*S*)-*form* [23155-02-4]
(−)-cis-*form*. ***Fosfomycin***. *Phosphonomycin. MK 955. Antibiotic MK 955. Antibiotic 833A*
Isol. from *Streptomyces* spp. Broad spectrum bactericide acting on early stages in synth. of cell wall. Low toxicity. V. effective against gram-(−)ve bacteria, incl. resistant *Saureus* spp.
▷SZ7890000.

Di-Na salt: [26016-99-9]. Solid. $[\alpha]_{405}^{28}$ −14.0° (c, 5 in H_2O).
▷SZ7902000.

Ca salt: [26472-47-9]. Solid. Mp >250°. $[\alpha]_{405}^{28}$ −12° (in 0.4M ethylenediamine aq.), $[\alpha]_D$ +3.8° (c, 0.7 in H_2O).

Monobenzylammonium salt: [24349-90-4]. Solid. Mp 170-174° (sealed tube). $[\alpha]_{405}$ −9.1° (c, 5 in H_2O).

Mono-(+)-1-phenylethylammonium salt: [25383-07-7]. Cryst. (EtOH). Mp 169-171° (131-132°, 139-140°). $[\alpha]_{405}^{28}$ −2.6°.

Monoamphetamine salt: Cryst. (EtOH). Mp 163-165°.

Di-Me ester: [25484-41-7]. *Dimethyl (3-methyloxiranyl)phosphonate.*
$C_5H_{11}O_4P$ M 166.113
Liq. Bp_2 55-56°. $[\alpha]_D$ +6.11° (c, 4.33 in MeOH).

(2*S*,3*R*)-*form* [26017-03-8]
(+)-cis-*form*

Mono-(−)-α-phenylethylamine salt: [26016-89-7]. Cryst. (MeOH). Mp 125-126°. $[\alpha]_D^{20}$ −19.4° (DMF).

Di-Me ester: [25460-63-3]. Liq. $[\alpha]_D$ −6.0° (c, 2.75 in MeOH).

(2*RS*,3*SR*)-*form*
(±)-cis-*form*

Mono(benzylammonium) salt: [25489-13-8]. Cryst. (EtOH). Mp 155-157° (152-155°).

Bis(ethylammonium) salt: Cryst. (EtOH). Mp 154-156°.

Dipiperazine salt: Solid. Mp 195°.

Di-Me ester: [25030-78-8]. Liq. $Bp_{0.5}$ 70-71°. n_D^{27} 1.4375.

Di-Et ester: [25030-57-3]. *Diethyl (3-methyl-2-oxiranyl)phosphonate.*
$C_7H_{15}O_4P$ M 194.167
Liq. $Bp_{0.5}$ 78-82°. n_D^{27} 1.4327.

Di-2-propenyl ester: Di-2-propenyl (3-methyl-2-oxiranyl)phosphonate.
$C_9H_{15}O_4P$ M 218.189
Liq. $Bp_{0.5}$ 105-115°.

Dipropyl ester: Dipropyl (3-methyl-2-oxiranyl)-phosphonate.
$C_9H_{19}O_4P$ M 222.220
Liq. $Bp_{0.5}$ 110-111°.

Di-Ph ester: [25992-25-0]. *Diphenyl (3-methyl-2-oxiranyl)phosphonate.*
$C_{15}H_{15}O_4P$ M 290.255
Oil which slowly cryst. Mp 46-48°.

(2*RS*,3*RS*)-*form* [27357-50-2]
(±)-trans-*form*
Few derivs. have been synthesised, and none characterised.

Christensen, B.G. *et al, Science*, 1969, **166**, 123 (*struct, config, synth*)
Girotra, N.N. *et al, Tetrahedron Lett.*, 1969, 4647 (*ester, synth*)
Glannkowski, E.J. *et al, J. Org. Chem.*, 1970, **35**, 3510 (*synth, resoln*)
Ger. Pats., 1 924 231; 1 924 085; 1 924 118; 1 924 135; 2 003 850; (*1970*); *CA*, **72**, 132972; 90634; 132953; 43870; **74**, 3725 (*synth, esters, derivs*)
Rogers, T.O. *et al, Antimicrob. Agents Chemother.*, 1974, **5**, 121 (*biosynth*)
Woodruff, B.H. *et al, Chemotherapy (Basel)*, 1976, **22**, Suppl. 2, 1; 1977, **23**, Suppl. 1, 1 (*rev, pharmacol*)
Gallego, A. *et al, Fosfomycin*, Wiley, 1977.
U.S.P., 4 222 970, (*1980*); *CA*, **94**, 15872 (*synth*)
Mertel, H.E. *et al, J. Labelled Compd. Radiopharm.*, 1982, **19**, 405 (*synth*)
Von Carstenn-Lichterfelde, C. *et al, J. Chem. Soc., Perkin Trans. 2*, 1983, 943 (*uv, ir, pmr*)
Imai, S. *et al, Agric. Biol. Chem.*, 1985, **49**, 873 (*biosynth*)
Yudelevich, V.I. *et al, Khim.-Farm. Zh.*, 1986, **20**, 238 (*CA*, **104**, 192958) (*rev*)

(3-Methyl-2-oxo-2(5*H*)-furanyl)-triphenylphosphonium(1+) M-00186

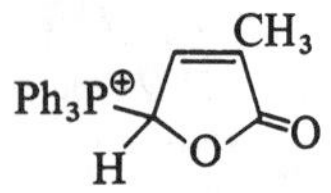

$C_{23}H_{20}O_2P^{\oplus}$ M 359.384 (ion)

Bromide:

$C_{23}H_{20}BrO_2P$ M 439.288

Gum. Dilute alkali yields the ylide.

Ylide: [35298-44-3]. *3-Methyl-5-(triphenylphosphoranylidene)-2(5*H*)-furanone.*

$C_{23}H_{19}O_2P$ M 358.376

Bright-yellow cryst. Mp 186-189° (180-183°). Can be stored for several days at −15°.

Corrie, J.E.T., *Tetrahedron Lett.*, 1971, 4873 (*use*)
Ingham, C.F. *et al*, *Aust. J. Chem.*, 1974, **27**, 1491 (*synth, use*)
Knight, D.W. *et al*, *J. Chem. Soc., Chem. Commun.*, 1974, 188 (*use*)

(3-Methyl-2-oxoheptyl)phosphonic acid, 9CI M-00187

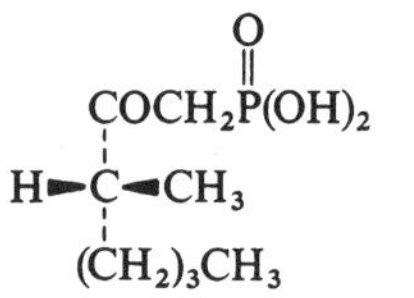

(*R*)-*form*

$C_8H_{17}O_4P$ M 208.194

(*R*)-*form*

Me ester: [39850-00-5]. *Dimethyl (3-methyl-2-oxoheptyl)phosphonate.*

$C_{10}H_{21}O_4P$ M 236.247

Important intermed. for Wittig-Horner synth. of prostaglandins and prostaglandin analogues. Liq. $Bp_{0.5}$ 115-137°. $[\alpha]_D^{25}$ −11.6° (c, 8.6 in Et_2O).

(*S*)-*form*

Me ester: [39850-01-6]. Wittig-Horner intermed. Liq. $[\alpha]_D^{25}$ +15.1° (c, 6.11 in Et_2O).

(±)-*form*

Me ester: [39746-02-6]. Wittig-Horner intermed. Liq. Bp_1 126-129°.

Grieco, P.A. *et al*, *J. Am. Chem. Soc.*, 1973, **95**, 3071 (*synth*)
Hayashi, M. *et al*, *J. Org. Chem.*, 1973, **38**, 1250 (*synth, use*)
Banerjee, A.K. *et al*, *Prostaglandins*, 1978, **16**, 541 (*synth, use*)

(1-Methyl-2-oxo-2-phenylethyl)-triphenylphosphonium(1+), 9CI M-00188

$Ph_3P^{\oplus}CH(CH_3)COPh$

$C_{27}H_{24}OP^{\oplus}$ M 395.460 (ion)

Salts are sources of 1-Phenyl-2-(triphenylphosphoranylidene)-1-propanone, P-00335 .

(±)-*form*

Bromide:

$C_{27}H_{24}BrOP$ M 475.364

Cryst. (MeOH aq.). Mp 290°. MeONa → ylide.

Iodide:

$C_{27}H_{24}IOP$ M 522.364

Cryst. (MeOH/EtOAc). Mp 198°.

Ylide: see (1-Methyl-2-oxo-2-phenylethyl)-triphenylphosphonium(1+), M-00188

Blank, B. *et al*, *J. Med. Chem.*, 1975, **18**, 952 (*bromide*)
Nesmeyanov, N.A. *et al*, *Zh. Org. Khim.*, 1977, **13**, 2465 (*iodide, pmr, nmr*)

2-(1-Methyl-2-oxopropylidene)-phosphorohydrazidothioate oxime M-00189

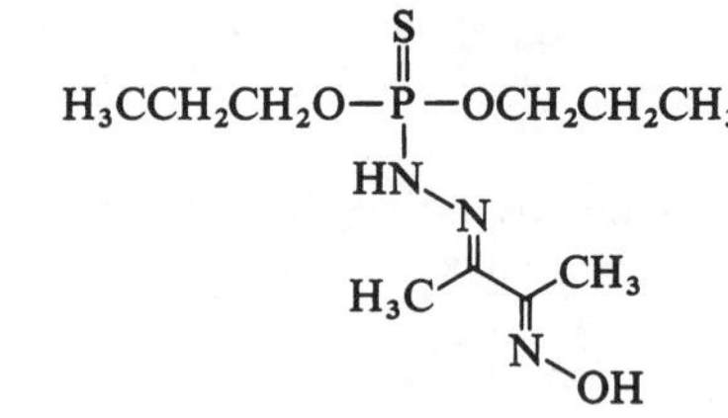

$C_{10}H_{22}N_3O_3PS$ M 295.336

(*E,E*)-*form* [82638-81-1]

Isol. from the Florida red tide dinoflagellate *Gymnodinium breve* (*Ptychodiscus brevis*). Ichthyotoxin. Needles (C_6H_6). Mp 82-83°.

Alam, M. *et al*, *J. Am. Chem. Soc.*, 1982, **104**, 5232 (*isol, cryst struct, ms, ir, pmr*)

4-Methyl-2-phenoxy-1,3,2-dioxaphospholane, 9CI M-00190

Phenyl propylene phosphite. 4-Methyl-2-phenoxy-1,3-dioxa-2-phosphacyclopentane

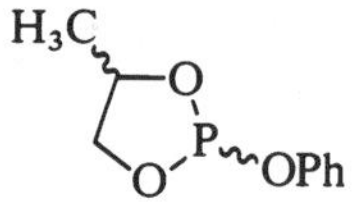

$C_9H_{11}O_3P$ M 198.158

2-Oxide: [22227-09-4]. *Phenyl propylene phosphate.*

$C_9H_{11}O_4P$ M 214.157

Cis- and *trans*-forms characterised spectroscopically.

Maria, P.C. *et al*, *Thermochimica Acta*, 1972, **4**, 505 (*props*)
Revel, M. *et al*, *Org. Magn. Reson.*, 1976, **8**, 399 (*oxide, pmr, P nmr*)

4-Methyl-2-phenoxy-1,3,2-dioxaphosphorinane, 9CI M-00191

4-Methyl-2-phenoxy-1,3,2-dioxaphosphinane. 1,3-Butylene phenyl phosphite. 4-Methyl-2-phenoxy-1,3-dioxa-2-phosphacyclohexane

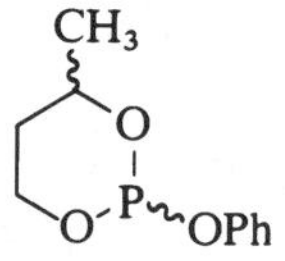

$C_{10}H_{13}O_3P$ M 212.185

Liq. d_4^{20} 1.16. Bp_7 132-133°. n_D^{20} 1.5145. Probably a mixt. of stereoisomers.

2-Oxide: [19219-95-5]. *1,3-Butylene phenyl phosphate.*

$C_{10}H_{13}O_4P$ M 228.184

Cryst. Mp 79-80°.

2-Sulfide: [31951-89-0]. O,O-*1,3-Butylene* O-*phenyl thiophosphate.*

$C_{10}H_{13}O_3PS$ M 244.245

Solid. Mp 71-72°.

Nifant'ev, É.E. *et al*, *Zh. Obshch. Khim.*, 1968, **38**, 1295 (*Engl. transl.* p. 1247) (*synth*)
Penney, C.L. *et al*, *Can. J. Chem.*, 1978, **56**, 2396 (*oxide, pmr*)
Arshinova, R.P. *et al*, *Izv. Akad. Nauk SSSR, Ser. Khim.*, 1979, 2242 (*Engl. transl.* p. 2065) (*sulfide, P nmr*)

3-Methyl-2-phenoxy-1,3,2-oxazaphospholidine, 9CI M-00192

3-Methyl-2-phenoxy-1,3,2-oxazaphospholane

[57301-49-2]

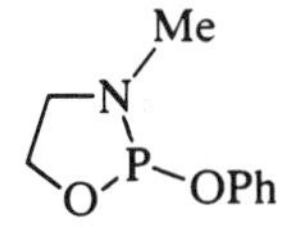

$C_9H_{12}NO_2P$ M 197.173

2-Oxide: [21047-79-0].

$C_9H_{12}NO_3P$ M 213.172

Liq. $Bp_{0.1}$ 138-140°. n_D^{20} 1.5230.

Devillers, J. *et al, Bull. Soc. Chim. Fr.*, 1970, 4341 (*oxide, synth, ir, pmr, P nmr*)
Devillers, J. *et al, Org. Magn. Reson.*, 1971, **3**, 177; 1973, **5**, 511 (*pmr*)
Brown, C. *et al, J. Chem. Soc., Perkin Trans. 2*, 1976, 888 (*oxide, synth, pmr, P nmr, props*)

4-Methyl-2-phenylamino-1,3,2-dioxaphosphorinane M-00193

*4-Methyl-*N*-phenyl-1,3,2-dioxaphosphorinan-2-amine, 9CI. 2-Anilino-4-methyl-1,3,2-dioxaphosphorinane. 1,3-Butylene phenylphosphoramidite. 2-Anilino-4-methyl-1,3-dioxa-2-phosphacyclohexane*

[34875-40-6]

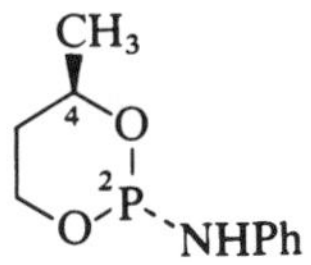

(2*RS*,4*RS*)-*form*

$C_{10}H_{14}NO_2P$ M 211.200

Liq. d_4^{20} 1.14. $Bp_{0.3}$ 115-129°. Mixt. of diastereoisomers.

(2RS,4RS)-form

(±)-trans-*form*

Liq. $Bp_{0.02}$ 99-101°. n_D^{20} 1.5575 (90% diastereomerically pure).

2-Oxide: [67057-49-2]. *1,3-Butylene phenylphosphoramidate.*

$C_{10}H_{14}NO_3P$ M 227.199

Cryst. (EtOAc). Mp 154-156°.

2-Sulfide: [67057-51-6]. O,O-*1,3-Butylene phenylphosphoramidothioate.*

$C_{10}H_{14}NO_2PS$ M 243.260

Cryst. (MeOH). Mp 171-172°.

2-Selenide: [67057-53-8]. O,O-*1,3-Butylene phenylphosphoramidoselenoate.*

$C_{10}H_{14}NO_2PSe$ M 290.160

Cryst. (C_6H_6 or MeOH). Mp 166-167°.

(2RS,4SR)-form

(±)-cis-*form*

Liq.

2-Oxide: [67057-50-5]. Cryst. (EtOAc, C_6H_6, or MeOH). Mp 174-176°.

2-Sulfide: [67057-52-7]. Cryst. (C_6H_6/hexane). Mp 91-92°.

2-Selenide: [67057-54-9]. Cryst. (C_6H_6/hexane). Mp 95-96°.

Stec, W.J. *et al, Tetrahedron*, 1973, **29**, 547 (*oxides, ir, ms, pmr, P nmr*)
Stec, W.J. *et al, J. Chem. Soc., Perkin Trans. 1*, 1975, 1828 (*synth, derivs, pmr, cmr, P nmr, ms*)
Zielinska, B. *et al, Org. Mass Spectrom.*, 1978, **13**, 65 (*derivs, P nmr, ms*)
Gombler, W. *et al, Z. Naturforsch., B*, 1983, **38**, 815 (*N and P nmr*)
Edmundson, R.S. *et al, J. Chem. Soc., Perkin Trans. 1*, 1984, 1943 (*derivs, stereochem*)

(Methylphenylamino)-triphenylphosphonium(1+) M-00194

(N-*Methylbenzenaminato)triphenylphosphorus(1+)*

$Ph_3P^{\oplus}NMePh$

$C_{25}H_{23}NP^{\oplus}$ M 368.437 (ion)

Bromide: [76195-19-2]. *1-Bromo-*N*-methyl-*N*,1,1,1-tetraphenylphosphoranamine.*

$C_{25}H_{23}BrNP$ M 448.341

Mp 157°.

Iodide: [34257-63-1]. *1-Iodo-*N*-methyl-*N*,1,1,1-tetraphenylphosphoranamine, 9CI.*

$C_{25}H_{23}INP$ M 495.342

Salts are used in regio- and stereo-selective alkylations of allylic alcs. by Murahashi method. Cryst. ($CHCl_3$/EtOAc or H_2O). Mp 238-238.5°.

Tanigawa, Y. *et al, J. Am. Chem. Soc.*, 1977, **99**, 2361 (*synth, use*)
Briggs, E.M. *et al, Org. Magn. Reson.*, 1980, **13**, 306 (*nmr*)
Briggs, E.M. *et al, Synthesis*, 1980, 295 (*props, use*)
Horner, L. *et al, Phosphorus Sulfur*, 1980, **8**, 209 (*bromide*)
Goering, H.L. *et al, J. Org. Chem.*, 1981, **46**, 2144 (*synth, use*)

5-Methyl-2-phenyl-1,3,2-benzodithiaphosphole, 10CI M-00195

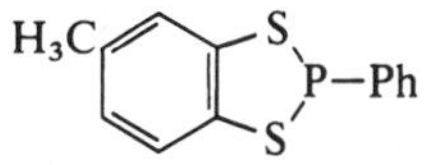

$C_{13}H_{11}PS_2$ M 262.324

Cryst. (hexane). Mp 59-60°. $Bp_{0.1}$ 65° subl.

2-Oxide:

$C_{13}H_{11}OPS_2$ M 278.323

Solid. Mp 98°.

Sau, A.C. *et al, J. Organomet. Chem.*, 1978, **156**, 253 (*synth, pmr, P nmr*)
Acher, F. *et al, Bull. Chem. Soc. Jpn.*, 1982, **55**, 3675 (*oxide, ms, P nmr*)

3-Methyl-2-phenyl-1,3-diaza-2-phosphabicyclo[3.3.1]nonane, 9CI M-00196

[35661-65-5]

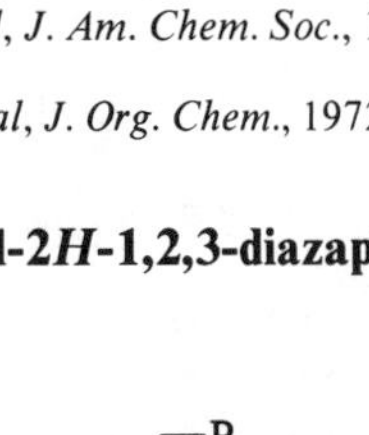

$C_{13}H_{19}N_2P$ M 234.280

$Bp_{0.01}$ 120°. Mixt. of stereoisomers.

Hutchins, R.O. *et al, J. Am. Chem. Soc.*, 1972, **94**, 9151 (*synth, P nmr*)
Maryanoff, B.E. *et al, J. Org. Chem.*, 1972, **37**, 3475 (*ms*)

5-Methyl-2-phenyl-2*H*-1,2,3-diazaphosphole, 9CI, 8CI M-00197

[18108-37-7]

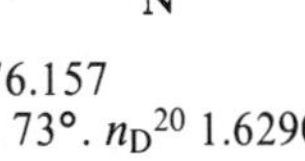

$C_9H_9N_2P$ M 176.157

Liq. d_4^{20} 1.30. $Bp_{0.05}$ 73°. n_D^{20} 1.6290. Forms a yellow Cu_2I_2 adduct.

B,HCl: Cryst. → powder in air. Mp 90°.

Ignatova, N.P. *et al, Khim. Geterosikl. Soedin.*, 1967, 753 (*Engl. transl.* p. 601) (*synth, derivs, props*)
Negrebetskii, V.V. *et al, Zh. Strukt. Khim.*, 1970, **11**, 633 (*Engl. transl.* p. 589) (*P nmr, pmr, struct*)
Bobkova, R.G. *et al, Zh. Obshch. Khim.*, 1977, **47**, 576 (*Engl. transl.* p. 527) (*synth*)
Negrebetskii, V.V. *et al, Zh. Strukt. Khim.*, 1978, **19**, 64 (*Engl. transl.* p. 52) (*cmr, P nmr, struct*)

4-Methyl-2-phenyl-1,3,2-dioxaphospholane, 9CI **M-00198**

Propylene phenylphosphonite. 4-Methyl-2-phenyl-1,3-dioxa-2-phosphacyclopentane

[33327-82-1]

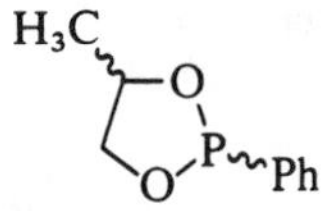

$C_9H_{11}O_2P$ M 182.158

Liq. d_4^{20} 1.15. Bp_3 94-95°. n_D^{20} 1.5455. *Cis-* and *trans-* forms characterised spectroscopically.

2-Oxide: [1831-31-8].
$C_9H_{11}O_3P$ M 198.158
Liq. d_4^{20} 1.24. $Bp_{0.015}$ 94-95°. n_D^{20} 1.5282. *Cis-* and *trans-*forms characterised spectroscopically.

Evdakov, V.P. *et al, Zh. Obshch. Khim.*, 1967, **37**, 441 (*Engl. transl.* p. 412) (*oxide*)
Bagrov, F.V. *et al, Zh. Obshch. Khim.*, 1974, **40**, 2565 (*Engl. transl.* p. 2557) (*synth*)
Bentrude, W.G. *et al, J. Am. Chem. Soc.*, 1976, **98**, 1850 (*cmr, P nmr*)
Revel, M. *et al, Org. Magn. Reson.*, 1976, **8**, 399 (*oxide, pmr, P nmr*)
Bentrude, W.G. *et al, J. Am. Chem. Soc.*, 1977, **99**, 4383 (*oxide, synth, pmr*)
Kwan, B.M. *et al, Phosphorus Sulfur*, 1983, **16**, 271 (*oxide, ms*)

4-Methyl-2-phenyl-1,3,2-dioxaphosphorinane, 9CI **M-00199**

4-Methyl-2-phenyl-1,3,2-dioxaphosphinane. 1,3-Butylene phenylphosphonite. 2,4-Methyl-2-phenyl-1,3-dioxa-2-phosphacyclohexane

[13223-91-1]

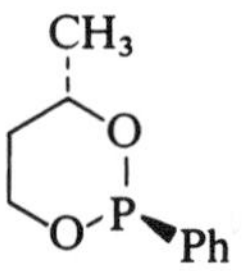

(2*S*,4*S*)-*form*

$C_{10}H_{13}O_2P$ M 196.185

(2*S*,4*S*)-*form*
Liq. $Bp_{0.05}$ 59-60°. $[\alpha]_D^{25}$ +51.6° (c, 5.5 in C_6H_6). Config. not certain.

(±)-*form*
Liq. d_4^{20} 1.13. Bp_3 101-102°. n_D^{21} 1.5420. Mixt. of stereoisomers.

2-Oxide: [57454-28-1]. *1,3-Butylene phenylphosphonate.*
$C_{10}H_{13}O_3P$ M 212.185
Solid. Mp 64°. $Bp_{0.3}$ 158-160°.

Bagrov, F.V. *et al, Zh. Obshch. Khim.*, 1970, **40**, 2565 (*Engl. transl.* p. 2557) (*synth*)
Corfield, G.C. *et al, J. Macromol. Sci., Part A*, 1975, **9**, 1113 (*oxide, P nmr*)
Segi, M. *et al, Chem. Lett.*, 1983, 913 (*pmr, props, use*)

(2-Methylphenyl)diphenylphosphine, 9CI **M-00200**

Diphenyl-o-tolylphosphine, 8CI

[5931-53-3]

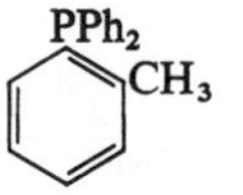

$C_{19}H_{17}P$ M 276.317

Ligand for metals of Groups IB, VIB, and VIII. Air-stable cryst. (EtOH). Mp 73° (67-68°).

Oxide: [6840-26-2].
$C_{19}H_{17}OP$ M 292.316
Cryst. (cyclohexane or Me_2CO). Mp 124-125°.

Sulfide:
$C_{19}H_{17}PS$ M 308.377
Forms Cd complexes.

Selenide: [75492-61-4].
$C_{19}H_{17}PSe$ M 355.277
Cryst. (butanol). Mp 149-150°. Forms Cd complexes.

Schindlbauer, H. *et al, Monatsh. Chem.*, 1965, **96**, 2051 (*synth, ir*)
Schiemenz, G.P. *et al, Angew. Chem.*, 1968, **80**, 558 (*pmr*)
Bennett, M.A. *et al, J. Am. Chem. Soc.*, 1969, **91**, 6266 (*synth, complexes*)
Henrick, K. *et al, Aust. J. Chem.*, 1975, **28**, 1455 (*ms*)
Grim, S.O. *et al, Phosphorus Sulphur*, 1977, **3**, 191 (*nmr*)
Segall, Y. *et al, J. Chem. Res. (S)*, 1977, 310 (*synth, ir, ms, pmr*)
Dean, P.A.W. *et al, Can. J. Chem.*, 1980, **58**, 1627 (*sulfide, selenide, complexes*)
Allen, D.W. *et al, J. Chem. Soc., Dalton Trans.*, 1982, 51 (*selenide, nmr*)

(3-Methylphenyl)diphenylphosphine, 9CI **M-00201**

Diphenyl-m-tolylphosphine

[7579-70-6]

$C_{19}H_{17}P$ M 276.317

Cryst. (MeOH). Mp 51-52°.

Oxide: [6840-27-3].
$C_{19}H_{17}OP$ M 292.316
Cryst. (C_6H_6/pet. ether or Et_2O/pet. ether). Mp 123-124°.

Monagle, J.J. *et al, J. Org. Chem.*, 1967, **32**, 2477 (*synth, oxide*)
Shvets, A.A. *et al, Zh. Obshch. Khim.*, 1978, **48**, 2185 (*Engl. transl.* p. 1986) (*oxide, struct*)
Schwartz, J.E. *et al, J. Org. Chem.*, 1979, **44**, 340 (*oxide, synth, ms, ir*)
Christau, H.J. *et al, J. Organomet. Chem.*, 1980, **185**, 283 (*synth*)

(4-Methylphenyl)diphenylphosphine, 9CI **M-00202**

Diphenyl-p-tolylphosphine

[1031-93-2]

$C_{19}H_{17}P$ M 276.317

Ligand for Cd, Cu, and Ag, and the metals of Group VIII. Prisms (EtOH or EtOH/Et_2O). Mp 67-68°. Bp_{14} 250°.

Oxide: [6840-28-4].
$C_{19}H_{17}OP$ M 292.316
Cryst. (Me_2CO/cyclohexane or C_6H_6/hexane). Mp 132.5-133.5°.

Sulfide: [5587-39-3].
$C_{19}H_{17}PS$ M 308.377
Needles (EtOH). Insol. Et_2O.

Geotz, H. *et al*, *Justus Liebigs Ann. Chem.*, 1963, **665**, 1 (*synth, ir*)
Dmitriev, V.I. *et al*, *Zh. Obshch. Khim.*, 1978, **48**, 52 (*Engl. transl.* p. 42) (*synth*)
Shvets, A.A. *et al*, *Zh. Obshch. Khim.*, 1978, **48**, 2185 (*Engl. transl.* p. 1986) (*oxide, synth, struct*)
Schwartz, J.E. *et al*, *J. Org. Chem.*, 1979, **44**, 340.
Streitweiser, A. *et al*, *J. Org. Chem.*, 1982, **47**, 768 (*props, oxide*)

(2-Methylphenyl)methylphosphinic acid, 9CI M-00203

*Methyl-*o*-tolylphosphinic acid, 8CI*
[61820-24-4]

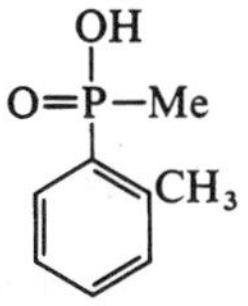

$C_8H_{11}O_2P$ M 170.147
Chloride: [78089-64-2].
$C_8H_{10}ClOP$ M 188.593
Liq. $Bp_{0.1}$ 100°.

Miles, J.A. *et al*, *J. Org. Chem.*, 1981, **46**, 3486 (*chloride, pmr*)
Miles, J.A. *et al*, *J. Org. Chem.*, 1982, **47**, 1677 (*synth*)

(4-Methylphenyl)methylphosphinic acid, 9CI M-00204

*Methyl-*p*-tolylphosphinic acid, 8CI*
$C_8H_{11}O_2P$ M 170.147

(S)$_p$-form
(−)-Menthyl ester: [77145-22-5].
$C_{18}H_{30}O_2P$ M 309.408
Cryst. (pentane). Mp 72-73.5°. $[\alpha]_D^{25}$ −103.2° (c, 22 in C_6H_6).

(±)-form
Cryst. (EtOH). Mp 119-120°.
Me ester: [39013-59-7]. *Methyl (4-methylphenyl)-methylphosphinate. Methyl methyl-*p*-tolylphosphinate.* Liq. d_0^{15} 1.12. Bp_{11} 148°, $Bp_{0.1}$ 89-90°. n_D^{25} 1.5190.
Propyl ester: Propyl (4-methylphenyl)-phenylphosphinate.
$C_{11}H_{17}O_2P$ M 212.228
Liq. d_0^{16} 1.07. Bp_{12} 167°. n_D^{16} 1.5185.
Chloride: [77145-23-4].
$C_8H_{10}ClOP$ M 188.593
Liq. $Bp_{0.4}$ 125°.

Khisamova, Z.L. *et al*, *Zh. Obshch. Khim.*, 1950, **20**, 1162 (*Engl. transl.* p. 1207) (*synth, esters*)
Bunnett, J.F. *et al*, *J. Org. Chem.*, 1973, **38**, 2703 (*synth, props*)
Curci, R. *et al*, *J. Chem. Soc., Perkin Trans. 2*, 1973, 531 (*ester, pmr*)
Sybert, P.D., *Macromolecules*, 1981, **14**, 502 (*chloride, ester, ir, pmr, cmr*)

Methyl phenyl methylphosphonate M-00205

[7526-25-2]

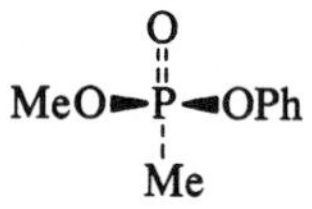

(R)-form

$C_8H_{11}O_3P$ M 186.147

(R)-form [70741-68-3]
Liq. $Bp_{0.02}$ 73-78°. $[\alpha]_D$ +22.5° (c, 1.30 in $CHCl_3$).
(S)-form [70741-69-4]
Liq. $Bp_{0.02}$ 73-78°. $[\alpha]_D$ −21.8° (c, 1.07 in $CHCl_3$).

Otsuki, T. *et al*, *Synthesis*, 1981, 811 (*synth, pmr*)

[(2-Methylphenyl)methyl]phosphonic acid, 9CI M-00206

o-*Methylbenzylphosphonic acid*
[18896-56-5]

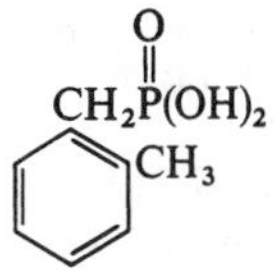

$C_8H_{11}O_3P$ M 186.147
Di-Et ester: [62278-16-9]. *Diethyl [(2-methylphenyl)-methyl]phosphonate.*
$C_{12}H_{19}O_4P$ M 258.253
Wittig-Horner reagent. Liq. $Bp_{0.2}$ 101-105°.

Ernst, L., *Org. Magn. Reson.*, 1977, **9**, 35 (*cmr, nmr*)
Franke, A. *et al*, *Synthesis*, 1979, 712 (*synth, pmr, use*)

[(3-Methylphenyl)methyl]phosphonic acid, 9CI M-00207

m-*Methylbenzylphosphonic acid*
[18945-65-8]
$C_8H_{11}O_3P$ M 186.147
Di-Et ester: [63909-50-2]. *Diethyl [(3-methylphenyl)-methyl]phosphonate.*
$C_{12}H_{19}O_3P$ M 242.254
Wittig-Horner reagent. Liq. d_4^{20} 1.08. $Bp_{0.7}$ 126-127°. n_D^{20} 1.4450.

Mel'nikov, N.N. *et al*, *Zh. Obshch. Khim.*, 1961, **31**, 3953 (*Engl. transl.* p. 3687) (*synth*)
Ernst, L., *Org. Magn. Reson.*, 1977, **9**, 35 (*nmr, cmr*)
Tominaga, Y. *et al*, *J. Heterocycl. Chem.*, 1982, **19**, 871 (*use*)

[(4-Methylphenyl)methyl]phosphonic acid, 9CI M-00208

p-*Methylbenzylphosphonic acid*
[13081-74-8]
$C_8H_{11}O_3P$ M 186.147
Plates or needles (H_2O). Mp 189°.
Di-Et ester: [3762-25-2]. *Diethyl [(4-methylphenyl)-methyl]phosphonate.*
$C_{12}H_{19}O_3P$ M 242.254
Wittig-Horner reagent. Pleasant smelling liq. Bp_{11} 157-160°, $Bp_{0.7}$ 132-135°. n_D^{20} 1.4958.
Dichloride:
$C_8H_9Cl_2OP$ M 223.038
Liq. d_{20}^{20} 1.26. Bp_2 130°. n_D^{20} 1.5430.

Kosolapoff, G.M., *J. Am. Chem. Soc.*, 1945, **67**, 2259 (*synth, ester*)
Mel'nikov, N.N. *et al*, *Zh. Obshch. Khim.*, 1961, **31**, 3953 (*Engl. transl.* p. 3687) (*synth*)
Rafikov, S.R. *et al*, *Zh. Obshch. Khim.*, 1964, **34**, 2230 (*Engl. transl.* p. 2241) (*synth, dichloride, ester*)
Williams, A. *et al*, *J. Chem. Soc., Perkin Trans. 2*, 1973, 25 (*synth*)
Ernst, L., *Org. Magn. Reson.*, 1977, **9**, 35 (*ester, cmr, nmr*)
Fresneda, P.M. *et al*, *Synthesis*, 1981, 222 (*ester, synth*)

[(2-Methylphenyl)methyl]-triphenylphosphonium(1+), 9CI M-00209

*(2-Methylbenzyl)triphenylphosphonium(1+). Triphenyl-*o-*xylylphosphonium(1+)*

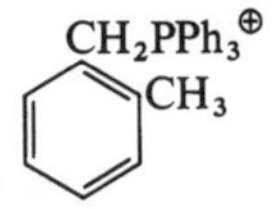

$C_{26}H_{24}P^{\oplus}$ M 367.449 (ion)

Chloride: [63368-36-5].
$C_{26}H_{24}ClP$ M 402.902
Exhibits antitrypanosomal activity. Source of ylide. Cryst. (Me_2CO or EtOH). Mp 279-281°.

Bromide: [1530-36-5].
$C_{26}H_{24}BrP$ M 447.353
Source of ylide. Solid. Mp 253-255° dec.
▷TA2370000.

Ylide: *[(2-Methylphenyl)methylene]-triphenylphosphorane.* o-*Methylbenzylidenetriphenylphosphorane.*
$C_{26}H_{23}P$ M 366.441
Used in Wittig reactions.

Griffin, C.E. *et al, J. Organomet. Chem.*, 1965, **3**, 414 (*bromide, pmr*)
Francis, G.W., *Acta Chem. Scand.*, 1972, **26**, 2969 (*use*)
James, B.G. *et al, J. Chem. Soc., Perkin Trans. 1*, 1974, 1195 (*use*)
Seyferth, D. *et al, J. Am. Chem. Soc.*, 1975, **97**, 7417 (*ylide*)
Yamato, M. *et al, Chem. Pharm. Bull.*, 1977, **25**, 706 (*chloride, use*)
Lapouyade, R. *et al, J. Org. Chem.*, 1982, **47**, 1361 (*use*)

[(3-Methylphenyl)methyl]-triphenylphosphonium(1+), 9CI M-00210

*(3-Methylbenzyl)triphenylphosphonium(1+). Triphenyl-*m-*xylylphosphonium(1+)*

$C_{26}H_{24}P^{\oplus}$ M 367.449 (ion)

Chloride: [63368-37-6].
$C_{26}H_{24}ClP$ M 402.902
Source of ylide. Cryst. (Me_2CO or EtOH). Mp 300°.

Bromide: [1702-41-6].
$C_{26}H_{24}BrP$ M 447.353
Source of ylide. Solid. Mp 271-272° dec.

Ylide: [59625-55-7]. *[(3-Methylphenyl)methylene]-triphenylphosphorane.* m-*Methylbenzylidenetriphenylphosphorane.*
$C_{26}H_{23}P$ M 366.441
Used in Wittig reactions.

Griffin, C.E. *et al, J. Organomet. Chem.*, 1965, **3**, 414 (*bromide, pmr*)
Seyferth, D. *et al, J. Am. Chem. Soc.*, 1975, **97**, 7417 (*use*)
Knorr, U. *et al, Chem. Ber.*, 1976, **109**, 3869 (*use, props*)
Yamato, M. *et al, Chem. Pharm. Bull.*, 1977, **25**, 706 (*chloride, use*)
Knorr, H. *et al, Justus Liebigs Ann. Chem.*, 1978, 1266 (*props*)

[(4-Methylphenyl)methyl]-triphenylphosphonium(1+), 9CI M-00211

*(4-Methylbenzyl)triphenylphosphonium(1+). Triphenyl-*p-*xylylphosphonium(1+)*
[18583-38-5]

$C_{26}H_{24}P^{\oplus}$ M 367.449 (ion)

Chloride: [1530-37-6].
$C_{26}H_{24}ClP$ M 402.902
Exhibits antitrypanosomal activity. Source of ylide. Cryst. (Me_2CO or EtOH). Mp 248°, 262-265°.

Bromide: [2378-86-1].
$C_{26}H_{24}BrP$ M 447.353
Source of ylide.

Ylide: [39110-21-9]. *[(4-Methylphenyl)methylene]-triphenylphosphorane.* p-*Methylbenzylidenetriphenylphosphorane.*
$C_{26}H_{23}P$ M 366.441
Used in Wittig reactions.

Griffin, C.E. *et al, J. Organomet. Chem.*, 1965, **3**, 414 (*pmr*)
Alper, H. *et al, J. Organomet. Chem.*, 1972, **44**, 371 (*ylide, complexes*)
Leznoff, C.C. *et al, Can. J. Chem.*, 1972, **50**, 528 (*use*)
Yamato, M. *et al, Chem. Pharm. Bull.*, 1977, **25**, 706 (*chloride, use*)
Knoppová, V. *et al, Collect. Czech. Chem. Commun.*, 1981, **46**, 515 (*use*)

2-Methyl-3-phenyl-1,3,2-oxazaphospholidine, 9CI M-00212

2-Methyl-3-phenyl-1,3,2-oxazaphospholane
[32287-51-7]

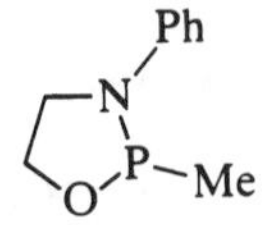

$C_9H_{12}NOP$ M 181.174
Liq. d_4^{20} 1.15. $Bp_{0.1}$ 85°. n_D^{20} 1.5812.

2-Oxide: [32287-54-0].
$C_9H_{12}NO_2P$ M 197.173
Solid. Mp 100.5-102°.

2-Sulfide: [78259-08-2].
$C_9H_{12}NOPS$ M 213.234
Cryst. (C_6H_6). Mp 120-122°.

Ishmaeva, É.A. *et al, Dokl. Akad. Nauk SSSR, Ser. Sci. Khim.*, 1971, **196**, 630 (*Engl. transl. Phys. Chem.* p. 63) (*synth, struct, oxide*)
Pudovik, M.A. *et al, Zh. Obshch. Khim.*, 1981, **51**, 518 (*Engl. transl.* p. 402) (*synth, P nmr, props*)

3-Methyl-2-phenyl-1,3,2-oxazaphospholidine, 9CI M-00213

3-Methyl-2-phenyl-1,3,2-oxazaphospholane
[1885-79-6]

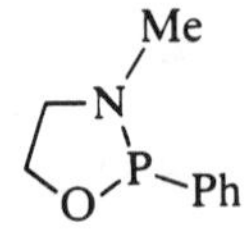

$C_9H_{12}NOP$ M 181.174
Mobile liq. when freshly dist. $Bp_{0.05}$ 75-78°. Gradually dimerises, becoming viscous.

2-Sulfide: [1885-80-9].
$C_9H_{12}NOPS$ M 213.234
Cryst. (cyclohexane). Mp 67°.

Dimer: see Octahydro-3,8-dimethyl-2,7-diphenyl-1,6,3,8,2,7-dioxadiazadiphosphecine, O-00006

Greenhalgh, R. *et al, Phosphorus*, 1972, **2**, 1 (*synth, pmr, P nmr, props*)
Devillers, J. *et al, Org. Magn. Reson.*, 1973, **5**, 511 (*oxide, pmr*)
Robert, J.B. *et al, J. Org. Chem.*, 1978, **43**, 3031 (*pmr, cmr, P nmr, props, sulfide*)

P-(4-Methylphenyl)-N-phenylphosphonamidic acid, 9CI **M-00214**

N-*Phenyl*-P-p-*tolylphosphonamidic acid*

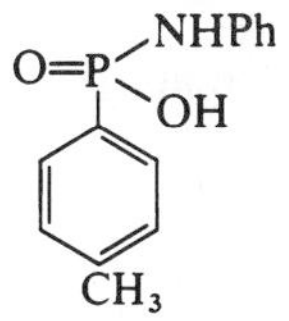

$C_{13}H_{14}NO_2P$ M 247.233

Solid. Mp 150°.

Me ester: Methyl P-(*4-methylphenyl*)-N-*phenylphosphonamidate.*
$C_{14}H_{16}NO_2P$ M 261.260
Solid. Mp 65°.

Et ester: Ethyl P-(*4-methylphenyl*)-N-*phenylphosphonamidate.*
$C_{15}H_{18}NO_2P$ M 275.286
Solid. Mp 53°.

Ph ester: Phenyl P-(*4-methylphenyl*)-N-*phenylphosphonamidate.*
$C_{19}H_{18}NO_2P$ M 323.330
Oil which cryst. Mp 59°. Bp_{48} 283°.

Michaelis, A., *Justus Liebigs Ann. Chem.*, 1896, **293**, 261; 1915, **407**, 316.

Methylphenylphosphine, 9CI **M-00215**

[6372-48-1]

PhPHMe

C_7H_9P M 124.122

Ligand for Co, Fe, and Ru. Highly air-sensitive liq. Bp_{30} 85°, Bp_{10} 59-60°. $n_D{}^{20}$ 1.5695.

Oxide: [19315-13-0].
C_7H_9OP M 140.121
Liq. $Bp_{0.2}$ 93-102°. Tautomeric with Methylphenylphosphinous acid, M-00227 .

Pass, F. *et al*, *Monatsh. Chem.*, 1959, **90**, 792 (*synth*)
Maier, L. *et al*, *Chem. Ber.*, 1961, **94**, 3056 (*synth, nmr*)
Razumov, A.I. *et al*, *Zh. Obshch. Khim.*, 1972, **42**, 1250 (*Engl. transl.* p. 1245) (*nmr*)
Wetzel, R.B. *et al*, *J. Org. Chem.*, 1974, **39**, 1531 (*oxide, synth, pmr*)
Henson, P.D. *et al*, *J. Org. Chem.*, 1974, **39**, 2296 (*synth, ir, pmr, oxide*)
Jones, I.W. *et al*, *J. Chem. Soc., Perkin Trans. 2*, 1979, 501 (*uv*)

(2-Methylphenyl)phosphine, 9CI **M-00216**

o-*Tolylphosphine*

[53772-59-1]

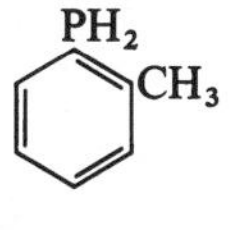

C_7H_9P M 124.122

Liq. Mp 4°. Bp 178°, $Bp_{0.5}$ 52-54°.

Maier, L., *Helv. Chim. Acta*, 1967, **50**, 1747 (*synth*)
Maier, L., *Phosphorus*, 1974, **4**, 41 (*pmr, nmr*)
Jones, I.W. *et al*, *J. Chem. Soc., Perkin Trans. 2*, 1979, 501 (*uv*)
De Santo, J.T. *et al*, *Inorg. Chem.*, 1980, **19**, 3086 (*struct*)

(3-Methylphenyl)phosphine, 9CI **M-00217**

m-*Tolylphosphine, 8CI*

[71337-04-7]

C_7H_9P M 124.122

Jones, I.W. *et al*, *J. Chem. Soc., Perkin Trans. 2*, 1979, 501 (*uv*)

(4-Methylphenyl)phosphine, 9CI **M-00218**

p-*Tolylphosphine, 8CI*

[53772-54-6]

C_7H_9P M 124.122

Bp_{11} 59-61°. pK_a 22.8. Forms W and Fe complexes.

Schindlbauer, H., *Monatsh. Chem.*, 1963, **94**, 99 (*uv*)
Maier, L., *Phosphorus*, 1974, **4**, 41 (*nmr*)
Jones, I.W. *et al*, *J. Chem. Soc., Perkin Trans. 2*, 1979, 501 (*uv*)
Terekhova, M.I. *et al*, *Zh. Obshch. Khim.*, 1982, **52**, 516 (*Engl. transl.* p. 452) (*props*)

Methylphenylphosphinic acid, 9CI, 8CI **M-00219**

[4271-13-0]

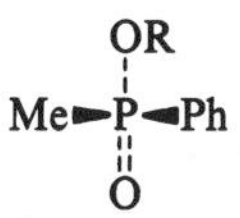

(*R*)$_P$-form of esters

$C_7H_9O_2P$ M 156.121

(R)-form

Me ester: [34647-07-9]. *Methyl methylphenylphosphinate.*
$C_8H_{11}O_2P$ M 170.147
Liq. $Bp_{0.2}$ 75°. $[\alpha]_D$ +56° (c, 4 in C_6H_6) (100% o.p.), $[\alpha]^6_D$ +49° (c, 1.7 in MeOH), $[\alpha]_D$ +57° ($CHCl_3$).

Et ester: [34638-79-4]. *Ethyl methylphenylphosphinate.*
$C_9H_{13}O_2P$ M 184.174
Liq. $Bp_{0.02}$ 76°. $[\alpha]_D$ +49° (c, 4 in C_6H_6) (100% o.p.).

Isopropyl ester: Isopropyl methylphenylphosphinate.
$C_{10}H_{15}O_2P$ M 198.201
Liq. $Bp_{0.04}$ 75°. $[\alpha]^{11}_D$ +46° (c, 0.87 in C_6H_6).

(−)-*Menthyl ester:* [16934-92-2]. *Menthyl methylphenylphosphinate.*
$C_{17}H_{27}O_2P$ M 294.373
Solid. Mp 87-88°. $[\alpha]^{20}_D$ −17.2° (c, 1.0 in C_6H_6).

(S)-form

Me ester: [34647-06-8]. Liq. $Bp_{0.04}$ 90-95°. $[\alpha]^{17}_D$ −49° (c, 1.7 in MeOH), $[\alpha]_D$ −57.5° (C_6H_6), $[\alpha]_D$ −57° ($CHCl_3$).

Et ester: [33642-98-7]. Liq. $Bp_{0.03}$ 76°. $[\alpha]^{11}_D$ −43° (c, 1.7 in C_6H_6).

Isopropyl ester: Liq. $Bp_{0.04}$ 72°. $[\alpha]^{11}_D$ −46° (c, 0.54 in C_6H_6).

(−)-*Menthyl ester:* [16934-93-3]. Solid. Mp 76-78°. $[\alpha]_D$ −92.7° (c, 2.1 in C_6H_6).

(±)-form

Cryst. (EtOH). Mp 135.5-136°. pK_a 2.96.

Me ester: Liq. d^{20}_4 1.16. Bp_{14} 142°, $Bp_{0.1}$ 83-85°. $n_D{}^{20}$ 1.5244.

Et ester: Liq. d^{20}_4 1.10. Bp_{10} 135°. $n_D{}^{20}$ 1.5169.

Isopropyl ester: d^{20}_4 1.08. Bp_{10} 146°. $n_D{}^{20}$ 1.5097.

tert-*Butyl ester:* tert-*Butyl methylphenylphosphinate.*
$C_{11}H_{17}O_2P$ M 212.228

Liq. $Bp_{0.3}$ 70° (oven).
Ph ester: [76420-32-1]. *Phenyl methylphenylphosphinate.*
$C_{13}H_{13}O_2P$ M 232.218
Liq. or solid at r.t. $Bp_{0.5}$ 140-150° (oven).
4-Nitrophenyl ester: 4-Nitrophenyl methylphenylphosphinate.
$C_{13}H_{12}NO_4P$ M 277.216
Used in peptide synth. Cryst. (Et_2O/pet. ether). Mp 85°.
Fluoride:
C_7H_8FOP M 158.112
Liq. $Bp_{0.5}$ 80-82°.
Chloride:
C_7H_8ClOP M 174.566
Solid. d_4^{20} 1.27. Mp 36-38°. $Bp_{0.8}$ 103-105°. n_D^{20} 1.5603.
Azide:
$C_7H_8N_3OP$ M 181.133
Liq. $Bp_{0.2}$ 123-128° (oven). No shock sensitivity; burns slowly in air.
Amide: see P-Methyl-P-phenylphosphinic amide, M-00223
Anhydride:
$C_{14}H_{16}O_3P_2$ M 294.226
Solid. Mp 92-94°. Bp_2 200-210°.

Moedritzer, K., *J. Am. Chem. Soc.*, 1961, **83**, 4381 (*synth, anhydride, chloride, nmr*)
Budzikiewicz, H. *et al*, *Monatsh. Chem.*, 1965, **96**, 1739 (*methyl ester, ms*)
Neimysheva, A.A. *et al*, *Zh. Obshch. Khim.*, 1966, **36**, 1090 (*Engl. transl.* p. 1105) (*chloride*)
Pudovik, A.N. *et al*, *Zh. Obshch. Khim.*, 1966, **36**, 1467; 1968, **38**, 1287 (*Engl. transl.* pp. 1471, 1239) (*esters*)
Baldwin, A. *et al*, *J. Org. Chem.*, 1967, **32**, 2172 (*azide*)
Nudelman, A. *et al*, *J. Am. Chem. Soc.*, 1968, **90**, 3869 (*menthyl ester*)
Lewis, R.A. *et al*, *J. Am. Chem. Soc.*, 1968, **90**, 4847 (*menthyl ester, pmr*)
DeBruin, K.E. *et al*, *J. Org. Chem.*, 1972, **37**, 2272; 1975, **40**, 1523 (*methyl esters*)
Legin, G.Ya., *Zh. Obshch. Khim.*, 1973, **43**, 2202 (*Engl. transl.* p. 2194) (*esters*)
Brooks, R.J. *et al*, *J. Org. Chem.*, 1975, **40**, 2059 (*synth, halides, pmr, props*)
Harger, M.J.P., *J. Chem. Soc., Perkin Trans. 1*, 1977, 2057; *J. Chem. Soc., Perkin Trans. 2*, 1882 (*menthyl esters*)
Garst, M.E., *Synth. Commun.*, 1979, **9**, 261 (*synth*)
Koizumi, T. *et al*, *Synthesis*, 1979, 110 (*esters*)
Harger, M.J.P., *J. Chem. Soc., Perkin Trans. 2*, 1980, 1505 (*esters*)
Mukaiyama, T. *et al*, *Chem. Lett.*, 1981, 1367 (*nitrophenyl ester, use*)
Harger, M.J.P. *et al*, *Tetrahedron*, 1982, **38**, 3073 (*azide, props*)

(2-Methylphenyl)phosphinic acid, 9CI M-00220

o-*Tolylphosphinic acid, 8CI*

O
‖
HO—P—H
(2-CH_3-phenyl)

$C_7H_9O_2P$ M 156.121
Tautomeric with (2-methylphenyl)phosphonous acid, but free acid exists in phosphinic acid form. Solid. Mp 115°.
Anilinium salt (1:1): Solid. Mp 94°.

Plets, V.M., *J. Gen. Chem. USSR*, 1937, **7**, 84, 90; *CA*, **31**, 4965 (*synth*)
Weil, T. *et al*, *Helv. Chim. Acta*, 1953, **36**, 1314 (*synth*)

(3-Methylphenyl)phosphinic acid, 9CI M-00221

m-*Tolylphosphinic acid*
$C_7H_9O_2P$ M 156.121
Syrup.
Phenylhydrazinium salt: Golden needles. Sl. sol. EtOH. Mp 131° dec.

Michaelis, A., *Justus Liebigs Ann. Chem.*, 1896, **293**, 261.

(4-Methylphenyl)phosphinic acid, 9CI M-00222

p-*Tolylphosphinic acid*
[20783-50-0]
$C_7H_9O_2P$ M 156.121
Tautomeric with 49086-x; free acid exists in the phosphinic acid form. Displays some herbicide activity against dicotyledoneous weeds. Solid. Mp 104-106°. pK_a 1.83 (H_2O), ca. 3.3 (EtOH).
▷SZ5775000.
Me ester: [13336-55-5]. *Methyl (4-methylphenyl)-phosphinate.*
$C_8H_{11}O_2P$ M 170.147
Liq. d_4^{20} 1.10. Bp_9 146°. n_D^{20} 1.5210.
Isopropyl ester: [13336-57-7]. *Isopropyl (4-methylphenyl)phosphinate.*
$C_{10}H_{15}O_2P$ M 198.201
d_4^{20} 1.07. Bp_{10} 156°. n_D^{20} 1.5152.
Octyl ester: Octyl (4-methylphenyl)phosphinate.
$C_{15}H_{25}O_2P$ M 268.335
d_4^{20} 1.00. Bp_1 145-146°. n_D^{20} 1.5006.

Weil, T. *et al*, *Helv. Chim. Acta*, 1953, **36**, 1314 (*synth, uv*)
Pudovik, A.N. *et al*, *Zh. Obshch. Khim.*, 1966, **36**, 1467 (*Engl. transl.* p. 1471) (*esters, synth, props*)
Gallagher, M.J. *et al*, *J. Chem. Soc. (C)*, 1971, 593 (*synth, pmr*)
Vinogradov, L.I. *et al*, *Zh. Obshch. Khim.*, 1972, **42**, 1724 (*Engl. transl.* p. 1712) (*ester, pmr, P nmr*)

P-Methyl-*P*-phenylphosphinic amide, 9CI M-00223

NH_2
Ph—P—Me (*R*)-*form*
‖
O

$C_7H_{10}NOP$ M 155.136
(*R*)-*form* [61185-81-7]
N-Ph: [61217-78-5]. P-*Methyl*-N,P-*diphenylphosphinic amide.*
$C_{13}H_{14}NOP$ M 231.233
Needles. $[\alpha]_D$ +29.0° (c, 2.8 in MeOH).
(*S*)-*form* [54835-97-1]
Solid. Mp 98-104°. $[\alpha]_D$ +9.2° (c, 2.8 in MeOH) (≥ 99% o.p.).
N-Ph: [50682-94-5]. Solid. Mp 162-164°. $[\alpha]_D$ −28.6° (c, 2.8 in MeOH) (>95.5% o.p.).
(±)-*form* [65634-93-7]
Cryst. (EtOAc). Mp 101-103°.
N-Ph: [51703-89-0]. Cryst. (Me_2CO/hexane or EtOAc). Mp 125-128°, 144-146°.

Nudelman, A. *et al*, *J. Am. Chem. Soc.*, 1968, **90**, 3869 (*anilide*)
Henson, P.D. *et al*, *J. Org. Chem.*, 1974, **39**, 2296 (*anilides*)

Harger, M.J.P., *J. Chem. Soc., Perkin Trans. 1*, 1977, 2057; 1979, 1294 (*synth, pmr, derivs*)
Harger, M.J.P., *J. Chem. Soc., Perkin Trans. 2*, 1978, 326 (*pmr*)

Methylphenylphosphinodithioic acid, 9CI — M-00224

[20384-84-3]

PhMeP(S)SH

$C_7H_9PS_2$ M 188.242
Liq. n_D^{22} 1.6787. Dec. on attempted distn.
NH_4 salt: [66220-57-3]. Solid. Mp 161-163°.
Butyl ester: Butyl methylphenylphosphinodithioate.
$C_{11}H_{17}PS_2$ M 244.349
Liq. $Bp_{0.015}$ 125°.

Newallis, P.E. *et al, J. Org. Chem.*, 1962, **27**, 3829 (*synth, P nmr*)
Shagidullin, R.R. *et al, Izv. Akad. Nauk SSSR, Ser. Khim.*, 1973, 541 (*Engl. transl.* p. 518) (*ir*)
Diemert, K. *et al, Phosphorus Sulfur*, 1977, **3**, 131 (*synth, complexes*)
Horner, L. *et al, Phosphorus Sulfur*, 1983, **14**, 245 (*ester*)

Methylphenylphosphinothioic acid, 9CI — M-00225

Ph
HO–P=S (*R*)-*form*
Me

C_7H_9OPS M 172.181
(*R*)-*form* [38607-73-7]
Chiral nmr shift reagent. $[\alpha]_D$ +19.5° (c, 0.5 in $CHCl_3$) (95% opt. pure).
Dicyclohexylammonium salt: [18117-87-8]. Solid. Mp 157°. $[\alpha]_D$ +9.25° (MeOH).
O-*Me ester:* [78037-58-8]. O-*Methyl methylphenylphosphinothioate.*
$C_8H_{11}OPS$ M 186.208
$[\alpha]_D$ +8.88° (28% opt. pure).
S-*Me ester:* [38605-11-7]. S-*Methyl methylphenylphosphinothioate.*
$C_8H_{11}OPS$ M 186.208
$[\alpha]_D$ +158° (100% opt. pure).
(*S*)-*form* [35556-64-0]
Chiral nmr shift reagent. Mp 91-93.5°. $[\alpha]_D$ −22.3° (c, 1.95 in MeOH).
Dicyclohexylammonium salt: [18117-86-7]. Cryst. Mp 157-159°. $[\alpha]_D^{25}$ −9.25° (c, 2.5 in MeOH) (83% opt. pure).
O-*Me ester:* [38605-13-9].
$C_8H_{11}OPS$ M 186.208
$Bp_{0.4}$ 80-85°. $[\alpha]_D$ +24.55° (neat). n_D^{19} 1.5770.
S-*Me ester:* [18872-30-5]. $Bp_{0.3}$ 115°. $[\alpha]_D$ −159.1° (c, 2.2 in C_6H_6). n_D^{25} 1.5874.
(±)-*form* [18962-89-5]
Mp 43-46°. $Bp_{0.1}$ 115-125° (oven). n_D^{23} 1.5708. Loses H_2S on distillation.
Dicyclohexylammonium salt: [18117-85-6]. Cryst. (EtOAc). Mp 157°.
O-*Me ester:* [76420-36-5]. Liq. $Bp_{0.1}$ 80-100°.
S-*Me ester:* [76420-33-2]. Liq. $Bp_{0.1}$ 100°.
O-*Et ester:* [42295-71-6]. O-*Ethyl methylphenylphosphinothioate.*
$C_9H_{13}OPS$ M 200.235
No phys. props. reported.
S-*Et ester:* [42295-74-9]. S-*Ethyl methylphenylphosphinothioate.*
$C_9H_{13}OPS$ M 200.235
No phys. props. reported.
O-*Isopropyl ester:* [42295-72-7]. O-*Isopropyl methylphenylphosphinothioate.*
$C_{10}H_{15}OPS$ M 214.262
No phys. props. reported.
S-*Isopropyl ester:* [42295-75-0]. S-*Isopropyl methylphenylhosphinothioate.*
$C_{10}H_{15}OPS$ M 214.262
No phys. props. reported.
O-*Ph ester:* [76380-87-5]. O-*Phenyl methylphenylphosphinothioate.*
$C_{13}H_{13}OPS$ M 248.279
Liq. $Bp_{0.3}$ 135-140° (oven).
S-*Ph ester:* [76420-35-4]. S-*Phenyl methylphenylphosphinothioate.*
$C_{13}H_{13}OPS$ M 248.279
Liq. $Bp_{0.5}$ 160° (oven).
Fluoride: [657-38-5].
C_7H_8FPS M 174.172
No phys. props. reported.
Chloride: [13639-67-8].
C_7H_8ClPS M 190.627
Liq. d_4^{20} 1.25. Bp_3 124-126°, $Bp_{0.3}$ 93°. n_D^{20} 1.6175.
Bromide:
C_7H_8BrPS M 235.078
Liq. $Bp_{0.5}$ 113-116°. n_D^{20} 1.6512.
Amide: see Methylphenylphosphinothioic amide, M-00226

Maier, L., *Chem. Ber.*, 1961, **94**, 3051 (*chloride, bromide*)
Maier, L., *Helv. Chim. Acta*, 1964, **47**, 120 (*chloride, P nmr*)
Benschop, H.P. *et al, Recl. Trav. Chim. Pays-Bas*, 1968, **87**, 362, 387 (*synth, resoln, ester*)
Mikolajczyk, M. *et al, Tetrahedron*, 1972, **28**, 4357 (*esters, uv, cd*)
DeBruin, K.E. *et al, J. Am. Chem. Soc.*, 1973, **95**, 4675 (*esters*)
Skvortsov, N.K. *et al, Zh. Obshch. Khim.*, 1976, **46**, 521 (*Engl. transl.* p. 518) (*ester, pmr, P nmr*)
Harger, M.J.P., *J. Chem. Soc., Perkin Trans. 2*, 1978, 326 (*synth, resoln, pmr, use*)
Mikolajczyk, M. *et al, J. Am. Chem. Soc.*, 1978, **100**, 7003 (*salts, pmr, P nmr*)
Hall, C.R. *et al, Pol. J. Chem.*, 1980, **54**, 489 (*resoln*)
Harger, M.J.P., *J. Chem. Soc., Perkin Trans. 2*, 1980, 1505 (*esters*)
Istomin, B.I. *et al, Zh. Obshch. Khim.*, 1981, **51**, 2393 (*Engl. transl.* p. 2063) (*ester, props*)

Methylphenylphosphinothioic amide, 8CI — M-00226

MePhP(S)NH$_2$

$C_7H_{10}NPS$ M 171.196
N,N-*Di-Me:* N,N,P-*Trimethyl*-P-*phenylphosphinothioic amide.*
$C_9H_{14}NPS$ M 199.250
Light-yellow liq. $Bp_{0.1}$ 110°.
N-*Butyl:* N-*Butyl*-P-*methyl*-P-*phenylphosphinothioic amide.*
$C_{11}H_{18}NPS$ M 227.304
Cryst. (Et_2O/pet. ether). Mp 40°.
N-*Benzyl:* N-*Benzyl*-P-*methyl*-P-*phenylphosphinothioic amide.*
$C_{14}H_{16}NPS$ M 261.321
Cryst. (CH_2Cl_2/pet. ether). Mp 73-74°.

Ellis, K. *et al, J. Chem. Soc., Perkin Trans. 1*, 1972, 1184 (*benzyl, pmr, ms*)

Johnson, C.R. *et al*, *J. Am. Chem. Soc.*, 1982, **104**, 1041 (*dimethyl, synth, use*)
Horner, L. *et al*, *Phosphorus Sulfur*, 1983, **14**, 245 (*butyl, synth, props*)

Methylphenylphosphinous acid, 9CI M-00227

Me
HO–P–:
Ph
(*R*)-*form*

C_7H_9OP M 140.121
Free acid exists in the tautomeric phosphonyl form. See under Methylphenylphosphine, M-00215 .

(*R*)-*form*
Me ester: [57322-08-4]. *Methyl methylphenylphosphinite.*
$C_8H_{11}OP$ M 154.148
Liq. $[\alpha]_D$ +15.8° (neat).

(±)-*form*
Me ester: Liq. Bp_{15} 125-130°.
Chloride: see Methylphenylphosphinous chloride, M-00228
Bromide: Bromomethylphenylphosphine.
C_7H_8BrP M 203.018
Liq. which fumes in air. $Bp_{0.5}$ 65°.
Diethylamide: N,N-*Diethyl*-P-*methyl*-P-*phenylphosphinous amide.*
$C_{11}H_{18}NP$ M 195.244
Solid. Mp 70-74°. $Bp_{0.2}$ 102-110°.
Cyanide:
C_8H_8NP M 149.132
Solid. Mp 61°.

Maier, L., *J. Inorg. Nucl. Chem.*, 1962, **24**, 275 (*bromide, synth, nmr*)
Maier, L., *Helv. Chim. Acta*, 1964, **47**, 2129 (*diethylamide*)
Mikolajczyk, M., *J. Chem. Soc., Chem. Commun.*, 1975, 382 (*ester*)

Methylphenylphosphinous chloride, 9CI M-00228

Chloromethylphenylphosphine
[15849-86-2]

PhPMeCl

C_7H_8ClP M 158.567
Air-sensitive liq. Bp_{30} 110°, Bp_2 66-67°.

Maier, L., *J. Inorg. Nucl. Chem.*, 1962, **24**, 1073 (*synth, P nmr*)
Duff, J.M. *et al*, *J. Chem. Soc., Dalton Trans.*, 1972, 2219 (*synth, pmr*)
Appel, R. *et al*, *Chem. Ber.*, 1977, **110**, 376 (*synth, P nmr*)
Jore, D. *et al*, *J. Organomet. Chem.*, 1978, **149**, C7 (*synth, pmr*)
Bestmann, H.J. *et al*, *Chem. Ber.*, 1982, **115**, 3875 (*synth*)
Wolfsberger, W., *J. Organomet. Chem.*, 1986, **317**, 167 (*cmr, P nmr*)

3-Methyl-1-phenylphospholane, 9CI M-00229

[24901-29-9]

CH_3
P
Ph
(1*R*,3*S*)-*form*

$C_{11}H_{15}P$ M 178.213

(*1R,3S*)-*form*
B,PhCH$_2$I: 1-Benzyl-3-methyl-1-phenylphospholanium iodide.
$C_{18}H_{22}IP$ M 396.250
Solid. Mp 184-185°. $[\alpha]_D^{22}$ −2.14° (c, 16.22 in $CDCl_3$).

(*1S,3R*)-*form* [54932-28-4]
Liq. $Bp_{0.5}$ 90° (oven). $[\alpha]_D^{22}$ +22.18° (c, 6.71 in $CDCl_3$).
B,PhCH$_2$I: Solid. Mp 184.5-185.5°. $[\alpha]_D$ +2.16° (c, 15.6 in $CDCl_3$).
1-Oxide:
$C_{11}H_{15}OP$ M 194.213
Liq. $Bp_{0.5}$ 130° (oven). $[\alpha]_D^{22}$ +23.52° (c, 7.59 in $CDCl_3$).

(*1RS,3RS*)-*form*
(±)-trans-*form*
Liq. $Bp_{0.05}$ 51°.
B,MeBr: 1,3-Dimethyl-1-phenylphospholanium bromide.
$C_{12}H_{18}BrP$ M 273.152
Solid. Mp 157.5-159°.
B,PhCH$_2$Br: 1-Benzyl-3-methyl-1-phenylphospholanium bromide.
$C_{18}H_{22}BrP$ M 349.250
Cryst. (EtOH/EtOAc). Mp 179.5-180°.
1-Oxide: [55043-98-6]. Liq. $Bp_{0.01}$ 120°.

(*1RS,3SR*)-*form* [57664-94-5]
(±)-cis-*form*
$Bp_{0.01}$ 49-51°.
B,MeBr: Mp ca. 100°.
B,PhCH$_2$Br: Cryst. (EtOAc). Mp 171.5-172°.
1-Oxide: [55043-99-7]. Hygroscopic solid. Mp 60-61°. $Bp_{0.05}$ 115-125°.

Marsi, K.L. *et al*, *J. Org. Chem.*, 1972, **37**, 238 (*synth, derivs, pmr*)
Marsi, K.L. *et al*, *J. Org. Chem.*, 1975, **40**, 1843 (*resoln, stereochem*)
Fitzgerald, A. *et al*, *J. Org. Chem.*, 1976, **41**, 1155 (*cryst struct, stereochem, derivs*)
Day, R.O. *et al*, *J. Am. Chem. Soc.*, 1980, **102**, 4387 (*deriv, cryst struct*)

N-Methyl-*P*-phenylphosphonamidic acid, 9CI M-00230

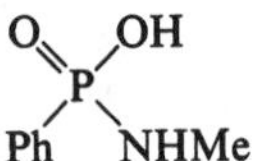

$C_7H_{10}NO_2P$ M 171.135
Me ester: [65111-15-1]. *Methyl* N-*methyl*-P-*phenylphosphonamidate.*
$C_8H_{12}NO_2P$ M 185.162
Cryst. (C_6H_6/Et_2O). Mp 43°.

Felcht, U. *et al*, *Justus Liebigs Ann. Chem.*, 1977, 1309 (*synth, ir, pmr*)
Harger, M.J.P. *et al*, *Tetrahedron*, 1982, **38**, 3073 (*ester, ir, pmr*)

P-Methyl-*N*-phenylphosphonamidic acid, 9CI M-00231

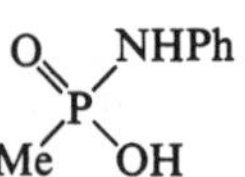

$C_7H_{10}NO_2P$ M 171.135
Me ester: [85656-06-0]. *Methyl* P-*methyl*-N-*phenylphosphonamidate.*
$C_8H_{12}NO_2P$ M 185.162

Cryst. (toluene). Mp 76-77°.
Et ester: [20341-75-7]. *Ethyl* P-*methyl*-N-*phenylphosphonamidate.*
$C_9H_{14}NO_2P$ M 199.189
Liq. d_4^{20} 1.15. Bp_{20} 160-165°, Bp_7 190-192°, $Bp_{0.08}$ 140-142°. n_D^{20} 1.5270.
Isopropyl ester: Isopropyl P-*methyl*-N-*phenylphosphonamidate.*
$C_{10}H_{16}NO_2P$ M 213.216
Cryst. (pet. ether). Mp 95-98°. Known in enantiomeric forms.
Ph ester: [20341-81-5]. *Phenyl* P-*methyl*-N-*phenylphosphonamidate.*
$C_{13}H_{14}NO_2P$ M 247.233
Solid. Mp 94°.

Aaron, H.S. *et al*, *J. Am. Chem. Soc.*, 1962, **84**, 617 (*isopropyl ester*)
Keay, L., *J. Org. Chem.*, 1963, **28**, 329 (*ethyl ester*)
Tomchina, L.F. *et al*, *Zh. Obshch. Khim.*, 1968, **38**, 564 (*Engl. transl.* p. 549) (*ethyl, phenyl esters, synth*)
Reiff, L.P. *et al*, *J. Am. Chem. Soc.*, 1970, **92**, 5275 (*isopropyl ester*)
Ashkinazi, L.A. *et al*, *Zh. Obshch. Khim.*, 1976, **46**, 1749 (*Engl. transl.* p. 1699) (*phenyl ester, conformn, pmr, P nmr*)
Harger, M.J.P. *et al*, *Tetrahedron*, 1982, **38**, 3073 (*methyl ester, synth, ir, pmr*)

P-Methyl-*N*-phenylphosphonamidodithioic acid, 9CI — M-00232

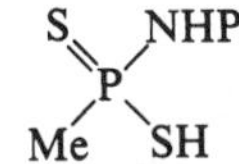

$C_7H_{10}NPS_2$ M 203.256
Me ester: [16284-72-3]. *Methyl* P-*methyl*-N-*phenylphosphonamidodithioate.*
$C_8H_{12}NPS_2$ M 217.283
Solid. Mp 116.5-117°.

Gryszkiewicz-Trochimowski, E., *Bull. Soc. Chim. Fr.*, 1967, 2232.

P-Methyl-*N*-phenylphosphonamidothioic acid, 9CI — M-00233

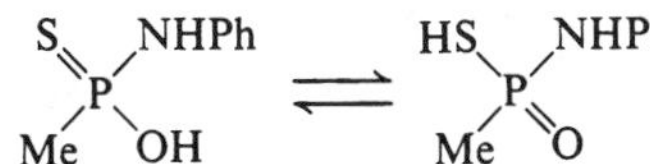

$C_7H_{10}NOPS$ M 187.196
S-*Et ester:* S-*Ethyl* P-*methyl*-N-*phenylphosphonamidothioate.*
$C_9H_{14}NOPS$ M 215.249
Needles (MeOH/C_6H_6). Mp 155°.
S-*Propyl ester:* S-*Propyl* P-*methyl*-N-*phenylphosphonamidothioate.*
$C_{10}H_{16}NOPS$ M 229.276
Cryst. (C_6H_6/pet. ether). Mp 98-100°.

Cadogan, J.I.G., *J. Chem. Soc.*, 1961, 3067 (*synth, ir*)

Methyl phenyl phosphonate, 9CI — M-00234

Methyl phenyl phosphite

PhOPH(O)OMe ⇌ PhOP(OH)OMe

$C_7H_9O_3P$ M 172.120
Tautomeric. Liq. d^{20} 1.22. n_D^{20} 1.5085.

Wolf, R. *et al*, *Bull. Soc. Chim. Fr.*, 1960, 124 (*synth, ir*)
Houalla, D. *et al*, *Bull. Soc. Chim. Fr.*, 1960, 129 (*ir*)

(2-Methylphenyl)phosphonic acid, 9CI — M-00235

o-*Tolylphosphonic acid*

$C_7H_9O_3P$ M 172.120
Solid. Mp 141°.
Di-Me ester: [6840-23-9]. *Dimethyl (2-methylphenyl)-phosphate. Dimethyl* o-*tolylphosphonate.*
$C_9H_{13}O_3P$ M 200.174
Liq. $Bp_{0.25}$ 96-97°.
Di-Et ester: [15286-11-0]. *Diethyl (2-methylphenyl)-phosphonate. Diethyl* o-*tolylphosphonate.*
$C_{11}H_{17}O_3P$ M 228.227
Liq. Bp_{14} 148-150°, $Bp_{0.01}$ 117-118°.
Dichloride: [62386-52-1].
$C_7H_7Cl_2OP$ M 209.011
Liq. $d^{18.5}$ 1.39. Bp 273°.
Dianilide: P-*(2-Methylphenyl)*-N,N′-*diphenylphosphonic diamide.*
$C_{19}H_{19}N_2OP$ M 322.346
Needles. Mp 234°.

Michaelis, A., *Justus Liebigs Ann. Chem.*, 1896, **293**, 261 (*synth, dianilide, dichloride*)
Morrison, D.C., *J. Am. Chem. Soc.*, 1951, **73**, 5896 (*dianilide*)
Obrycki, R. *et al*, *J. Org. Chem.*, 1968, **33**, 632 (*esters, synth, ir*)
Tavs, P., *Chem. Ber.*, 1970, **103**, 2428 (*diethyl ester, synth*)
Bunnett, J.F. *et al*, *J. Org. Chem.*, 1978, **43**, 1867 (*esters, synth, ir, ms, pmr*)
Grabiak, R.C. *et al*, *Phosphorus Sulfur*, 1980, **9**, 197 (*derivs, synth, P nmr*)
Hirao, T. *et al*, *Synthesis*, 1981, 56; *Bull. Chem. Soc. Jpn.*, 1982, **55**, 909 (*diethyl ester, synth*)

(3-Methylphenyl)phosphonic acid, 9CI — M-00236

m-*Tolylphosphonic acid*

$C_7H_9O_3P$ M 172.120
Solid. Mp 121°. pK_{a1} 1.88 (H_2O), pK_{a2} 7.44 (H_2O).
Di-Me ester: Dimethyl (3-methylphenyl)phosphonate. Dimethyl m-*tolylphosphonate.*
$C_9H_{13}O_3P$ M 200.174
Liq. Bp_4 141°, $Bp_{0.45}$ 106-107°.
Di-Et ester: [15286-13-2]. *Diethyl (3-methylphenyl)-phosphonate. Diethyl* m-*tolylphosphonate.*
$C_{11}H_{17}O_3P$ M 228.227
Liq. $Bp_{0.02}$ 119-120°, $Bp_{0.4}$ 108°. n_D^{20} 1.4932.
Bis(trimethylsilyl) ester: [99136-05-7]. *Bis(trimethylsilyl)(3-methylphenyl)phosphonate.*
$C_{13}H_{25}O_3PSi_2$ M 316.483
Liq. $Bp_{0.001}$ 84-86°.
Difluoride: [769-91-5].
$C_7H_7F_2OP$ M 176.102
Characterised spectroscopically.
Dichloride: [63392-65-4].
$C_7H_7Cl_2OP$ M 209.011
Liq. d^{18} 1.35. Bp 275°.

Ashby, E.C. *et al*, *J. Am. Chem. Soc.*, 1953, **75**, 4903 (*props*)
Griffin, C.E. *et al*, *Tetrahedron*, 1966, **22**, 561 (*dimethyl ester, ir, pmr*)
Obrycki, R. *et al*, *J. Org. Chem.*, 1968, **33**, 632 (*esters, ir*)
Szafraniec, L.L., *Org. Mass Spectrom.*, 1974, **6**, 565 (*difluoride, F and P nmr*)

Allen, D.W. *et al, J. Chem. Soc., Perkin Trans. 2*, 1977, 789 (*esters, pmr, P nmr*)
Bunnett, J.F. *et al, J. Org. Chem.*, 1978, **43**, 1867 (*diethyl ester, ir, pmr, ms*)
Grabiak, R.C. *et al, Phosphorus Sulfur*, 1980, **9**, 197 (*synth, derivs, P nmr*)
Issleib, K. *et al, Z. Anorg. Allg. Chem.*, 1985, **529**, 151 (*silyl ester, synth, P nmr*)

(4-Methylphenyl)phosphonic acid, 9CI   M-00237

p-*Tolylphosphonic acid*
[3366-72-1]

$C_7H_9O_3P$ M 172.120
Solid. Mp 198-199° (188-189°). pK_{a1} 2.29 (1.84) (H_2O); 3.62 (50% EtOH aq.pK_{a2} *), pK_{a3} *pK_{a2} 5.74 (7.33) (H_2O)pK_{a4} *; 6.50 (50% EtOH aq.).
▷TA0470000.

Di-Me ester: [6840-25-1]. *Dimethyl (4-methylphenyl)-phosphonate. Dimethyl* p-*tolylphosphonate.*
$C_9H_{13}O_3P$ M 200.174
Liq. Bp_1 106-106.5°.

Di-Et ester: [1754-46-7]. *Diethyl (4-methylphenyl)-phosphonate. Diethyl* p-*tolylphosphonate.*
$C_{11}H_{17}O_3P$ M 228.227
Liq. Bp_{20} 156-162°, $Bp_{0.05}$ 118-119°. n_D^{20} 1.4930.

Di-Ph ester: Diphenyl (4-methylphenyl)phosphonate. Diphenyl p-*tolylphosphonate.*
$C_{19}H_{17}O_4P$ M 340.315
Liq. Bp_{25} 225-227°.

Bis(trimethylsilyl) ester: [104412-66-0]. *Bis(trimethylsilyl)(4-methylphenyl)phosphonate.*
$C_{13}H_{25}O_3PSi_2$ M 316.483
Liq. $Bp_{0.001}$ 42°.

Difluoride: [701-32-6].
$C_7H_7F_2OP$ M 176.102
No phys. props. reported.

Dichloride: see (4-Methylphenyl)phosphonic dichloride, M-00239

Diamide: see P-(4-Methylphenyl)phosphonic diamide, M-00238

Griffin, C.E. *et al, Tetrahedron*, 1966, **22**, 561 (*dimethyl ester, synth, ir, pmr*)
Obrycki, R. *et al, J. Org. Chem.*, 1968, **33**, 632 (*esters, synth*)
Tavs, P., *Chem. Ber.*, 1970, **103**, 2428 (*diethyl ester, synth*)
Williams, A. *et al, J. Chem. Soc., Perkin Trans. 2*, 1973, 25 (*synth, diethyl ester*)
Kharrasova, F.M. *et al, Zh. Obshch. Khim.*, 1973, **43**, 2642 (*Engl. transl.* p. 2621) (*esters, P nmr*)
Szafraniec, L.L., *Org. Magn. Reson.*, 1974, **6**, 565 (*difluoride, F and P nmr*)
Grabiak, R.C. *et al, Phosphorus Sulfur*, 1980, **9**, 197 (*synth, derivs, P nmr, pmr*)
Hirao, T. *et al, Synthesis*, 1981, 56; *Bull. Chem. Soc. Jpn.*, 1982, **55**, 909 (*diethyl ester*)
Osuka, A. *et al, Synthesis*, 1983, 69 (*diethyl, diphenyl esters, ir, pmr*)
Issleib, K. *et al, Z. Anorg. Allg. Chem.*, 1985, **529**, 151 (*silyl ester, synth, P nmr*)

P-(4-Methylphenyl)phosphonic diamide, 9CI   M-00238

P-p-*Tolylphosphonic diamide, 8CI*
[18799-59-2]

NH_2
O=P–NH_2
CH_3

$C_7H_{11}N_2OP$ M 170.150
Cryst. (EtOH). Mp 176°.

N,N′-*Di-Ph:* [52222-65-8]. P-*(4-Methylphenyl)*-N,N′-*diphenylphosphonic diamide.* P-*Tolylphosphonic dianilide.*
$C_{19}H_{19}N_2OP$ M 322.346
Needles (EtOH). Mp 209°.

Michaelis, A., *Justus Liebigs Ann. Chem.*, 1896, **293**, 261 (*synth*)
Michaelis, A. *et al, Justus Liebigs Ann. Chem.*, 1915, **407**, 316 (*synth*)
Tomaschewski, G. *et al, Chem. Ber.*, 1968, **101**, 2037 (*uv*)

(4-Methylphenyl)phosphonic dichloride, 9CI   M-00239

p-*Tolylphosphonic dichloride*
[17566-84-6]

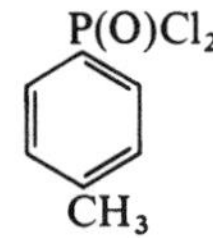

$C_7H_7Cl_2OP$ M 209.011
Low-melting solid. d_4^{20} 1.34. Mp 26.5-27°. Bp_{14} 149.5-150.5°. n_D^{20} 1.5577.

Kharrasova, F.M. *et al, Zh. Obshch. Khim.*, 1968, **38**, 1262 (*Engl. transl.* p. 1215) (*synth*)
Kharrasova, F.M. *et al, Zh. Obshch. Khim.*, 1973, **43**, 2642 (*Engl. transl.* p. 2621) (*P nmr*)
Petrov, K.A. *et al, Zh. Obshch. Khim.*, 1975, **45**, 2428 (*Engl. transl.* p. 2386) (*synth*)
Zakirova, D.U. *et al, Zh. Obshch. Khim.*, 1977, **47**, 1661 (*Engl. transl.* p. 1522) (*nqr*)
Ratovskii, G.V. *et al, Zh. Obshch. Khim.*, 1981, **51**, 714 (*Engl. transl.* p. 580) (*raman*)
Sergienko, L.M. *et al, Zh. Obshch. Khim.*, 1980, **50**, 1958 (*Engl. transl.* p. 1584) (*uv*)

N-Methyl-*P*-phenylphosphonimidic acid, 9CI, 8CI   M-00240

MeN=PPh$(OH)_2$

$C_7H_{10}NO_2P$ M 171.135
Tautomeric with *N*-Methyl-*P*-phenylphosphonamidic acid, M-00230 .

Di-Me ester: [55215-26-4]. *Dimethyl* N-*methyl*-P-*phenylphosphonimidate.* P,P-*Dimethoxy*-N-*methyl*-P-*phenylphosphine imide.*
$C_9H_{14}NO_2P$ M 199.189
Liq. $Bp_{0.3}$ 74°.

Goldwhite, H. *et al, J. Chem. Soc., Dalton Trans.*, 1975, 12 (*synth, ir, pmr, P nmr*)

P-Methyl-N^2-phenylphosphonohydrazidic acid, 9CI M-00241

HO, O / P / Me NHNHPh

$C_7H_{11}N_2O_2P$ M 186.150

Ph ester: [34170-54-2]. *Phenyl* P-*methyl*-N^2-*phenylphosphonohydrazidate.*
$C_{13}H_{15}N_2O_2P$ M 262.247
Cryst. (EtOH aq.). Mp 118-119°.

Grapov, A.F. *et al*, *Zh. Obshch. Khim.*, 1971, **41**, 1441 (*Engl. transl.* p. 1447)

P-Methyl-N^2-phenylphosphonohydrazidothioic acid, 9CI M-00242

S, NHNHPh / P / Me OH ⇌ HS, NHNHPh / P / Me O

$C_7H_{11}N_2OPS$ M 202.210

O-*Ph ester:* [34170-57-5]. O-*Phenyl* P-*methyl*-N^2-*phenylphosphonohydrazidothioate.*
$C_{13}H_{15}N_2OPS$ M 278.308
Cryst. (EtOH). Mp 73-74°.

Grapov, A.F. *et al*, *Zh. Obshch. Khim.*, 1971, **41**, 1441 (*Engl. transl.* p. 1447)

O-Methyl phenylphosphonothioate, 9CI M-00243

[42976-67-0]

OH / MeO–P=S / Ph (*R*)-*form*

$C_7H_9O_2PS$ M 188.181

Tautomeric. Metab. of Leptophos.

(*R*)-*form* [67253-70-7]

$[\alpha]_D^{24}$ +21.72° (neat).

1-Phenylethylammonium salt: Cryst. (EtOAc or EtOAc/hexane). Mp 142-144°. $[\alpha]_D^{24}$ +17.83° (c, 10.75 in MeOH).

S-*Me ester:* O,S-*Dimethyl phenylphosphonothioate.*
$C_8H_{11}O_2PS$ M 202.207
Liq. $[\alpha]_D$ +81° (C_6H_6).

O-*Et ester:* O-*Ethyl* O-*methyl phenylphosphonothioate.*
$C_9H_{13}O_2PS$ M 216.234
Liq. $Bp_{0.03}$ 73°. $[\alpha]_D^{12}$ −7.36° (c, 2.8 in CCl_4).

(*S*)-*form* [67253-71-8]

$[\alpha]_D^{24}$ −21.0° (neat).

1-Phenylethylammonium salt: [73270-51-6]. Cryst. (EtOAc or EtOAc/hexane). Mp 142-144°. $[\alpha]_D^{24}$ −17.68° (c, 14.5 in MeOH).

S-*Me ester:* [77326-05-7]. Liq. $Bp_{0.5}$ 110°. $[\alpha]_D^{22}$ −100.3° (c, 0.43 in C_6H_6).

(±)-*form* [67213-42-7]

Oil. $Bp_{0.02}$ 100°. n_D^{23} 1.5772.

Tetramethylammonium salt: Hygroscopic solid. Sol. $CHCl_3$, v. sol. H_2O, alcohols, insol. Et_2O, C_6H_6. Mp 135°.

1-Phenylethylammonium salt: Cryst. (EtOAc or EtOAc/hexane). Mp 142-144°.

Dicyclohexylammonium salt: Cryst. (2-propanol). Mp 155-156°.

O-*4-Bromo-2,5-dichlorophenyl ester: see Leptophos, L-00001*

O-*2,5-Dichlorophenyl ester:* O-*2,5-Dichlorophenyl* O-*methyl phenylphosphonothioate. Desbromoleptophos.*
$C_{13}H_{11}Cl_2O_2PS$ M 333.168
Insecticide. Cryst. (EtOH). Mp 45-46°.
▷Possesses delayed neurotoxicity

O-*4-Nitrophenyl ester: EPN-Methyl.* O-*Methyl* O-*4-nitrophenyl phenylphosphonothioate.*
$C_{13}H_{12}NO_4PS$ M 309.276
Insecticide. Solid. Mp 31-32°.
▷Neurotoxin

Chloride: see Phenylphosphonochloridothioic acid, P-00232

Thuong, N.T. *et al*, *Bull. Soc. Chim. Fr.*, 1970, 780 (*synth*)
De Bruin, K.F. *et al*, *J. Chem. Soc., Chem. Commun.*, 1975, 753 (*methyl ester, synth*)
Harger, M.J.P. *et al*, *J. Chem. Soc., Perkin Trans. 2*, 1978, 326 (*synth, pmr*)
Mikolajczyk, M. *et al*, *J. Am. Chem. Soc.*, 1978, **100**, 7003 (*pmr*)
Allahyari, R. *et al*, *J. Agric. Food Chem.*, 1980, **28**, 594 (*synth, resoln, derivs, abs config, cryst struct*)
Koizumi, T. *et al*, *Chem. Lett.*, 1980, 1403 (*methyl ester, synth, abs config*)

S-Methyl phenylphosphonothioate M-00244

[77173-41-2]

SMe / HO–P=O / Ph (*S*)-*form*

$C_7H_9O_2PS$ M 188.181

(*S*)-*form*

O-*Menthyl ester:* [33625-82-0]. Cryst. (2,2,4-trimethylpentane). Mp 76.5°. $[\alpha]_D$ −141° (C_6H_6).

(±)-*form*

Oil.

Dicyclohexylammonium salt: Solid. Mp 176.5-178.5°.

O-*Me ester: see O-Methyl phenylphosphonothioate, M-00243*

O-*Et ester:* O-*Ethyl* S-*methyl phenylphosphonothioate.*
$C_9H_{13}O_2PS$ M 216.234
Liq. d_0^{25} 1.17. Bp_3 125-135°, Bp_1 84-85°. n_D^{20} 1.5532.
▷Toxic

Chloride: see Phenylphosphonochloridothioic acid, P-00232

Morrison, D.C., *J. Org. Chem.*, 1956, **21**, 705 (*ethyl ester, synth*)
Donohue, J. *et al*, *J. Am. Chem. Soc.*, 1971, **93**, 3792 (*menthyl ester, synth, cryst struct*)
Ol'khoaya, G.C. *et al*, *Izv. Akad. Nauk SSSR, Ser. Khim.*, 1975, 1837 (*Engl. transl.* p. 1719) (*ethyl ester, synth*)
Chrzanowski, R.L. *et al*, *J. Agric. Food Chem.*, 1981, **29**, 580 (*deriv, ms*)

P-Methyl-*N*-phenylphosphonothioic diamide, 9CI M-00245

S, NHPh / P / Me NH$_2$

$C_7H_{11}N_2PS$ M 186.211

N′-*Me:* [5994-75-2]. N,P-*Dimethyl*-N′-*phenylphosphonothioic diamide.*
$C_8H_{13}N_2PS$ M 200.238
Weak fungicide. Solid. Mp 96.5-97°.
N′-*Isopropyl:* [5994-74-1]. P-*Methyl*-N-*isopropyl*-N′-*phenylphosphonothioic diamide.*
$C_{10}H_{17}N_2PS$ M 228.291
Weak fungicide. Cryst. (Me_2CO). Mp 94-96°.

Mel'nikov, N.V. *et al*, *Zh. Obshch. Khim.*, 1966, **36**, 269 (*Engl. transl.* p. 279)

(2-Methylphenyl)phosphonous acid, 9CI M-00246

o-*Tolylphosphonous acid*

$C_7H_9O_2P$ M 156.121
The free acid exists as the phosphoryl tautomer (2-Methylphenyl)phosphinic acid, M-00220 .
Dichloride: [5310-87-2]. *Dichloro*-o-*tolylphosphine.*
$C_7H_7Cl_2P$ M 193.012
Liq. Bp_{10} 108-109°, $Bp_{0.4}$ 67-68°. n_D^{20} 1.5914.
Dibromide: [6460-26-0]. *Dibromo*-o-*tolylphosphine.*
$C_7H_7Br_2P$ M 281.914
Liq. d_4^{20} 1.80. Bp_1 99-102°.
Bis(diethylamide): N,N,N′,N′-*Tetraethyl*-P-(*2-methylphenyl*)*phosphonous diamide.*
$C_{11}H_{26}N_2P$ M 217.314
Liq. Bp_{10} 155-157°.

Schindlbauer, H., *Monatsh. Chem.*, 1965, **96**, 1936 (*dichloride, diethylamide, synth, ir*)
Bliznyuk, N.K. *et al*, *Zh. Obshch. Khim.*, 1967, **37**, 890 (*Engl. transl.* p. 840) (*dibromide*)
Weinberg, K.G., *J. Org. Chem.*, 1975, **40**, 3586 (*dichloride*)
Clark, P.W. *et al*, *J. Organomet. Chem.*, 1981, **217**, 51 (*dichloride*)

(3-Methylphenyl)phosphonous acid, 9CI M-00247

m-*Tolylphosphonous acid*
$C_7H_9O_2P$ M 156.121
Free acid exists as the phosphoryl tautomer (3-Methylphenyl)phosphinic acid, M-00221 .
Dichloride: [5510-88-3]. *Dichloro*-m-*tolylphosphine.*
$C_7H_7Cl_2P$ M 193.012
Liq. Bp_9 102-104°, $Bp_{0.5}$ 58-56°.
Dibromide: Dibromo-m-*tolylphosphine.*
$C_7H_7Br_2P$ M 281.914
Liq. Bp_2 110°.
Bis(diethylamide): N,N,N′,N′-*Tetraethyl*-P-(*3-methylphenyl*)*phosphonous diamide.*
$C_{11}H_{26}N_2P$ M 217.314
Liq. Bp_{12} 159-61°.

Petrov, K.A. *et al*, *Zh. Obshch. Khim.*, 1961, **31**, 3027 (*Engl. transl.* p. 2823) (*dibromide, synth*)
Schindlbauer, H., *Monatsh. Chem.*, 1965, **96**, 1936 (*dichloride, diethylamide, synth, ir*)
Weinberg, K.G., *J. Org. Chem.*, 1975, **40**, 3586 (*dichloride, synth*)

(4-Methylphenyl)phosphonous acid, 9CI M-00248

p-*Tolylphosphonous acid*
[20676-63-5]
$C_7H_9O_2P$ M 156.121
Free acid exists as the phosphoryl tautomer (4-Methylphenyl)phosphinic acid, M-00222 .
Mono-Me ester: see under (4-Methylphenyl)phosphinic acid, M-00222
Di-Me ester: [63507-03-9]. *Dimethyl (4-methylphenyl)phosphonite.*
$C_9H_{13}O_2P$ M 184.174
Liq. d_0^{10} 1.04. Bp_{14} 107-109°, $Bp_{0.35}$ 68-70°. n_D^{20} 1.5325.
Di-Et ester: Diethyl (4-methylphenyl)phosphonite.
$C_{11}H_{17}O_2P$ M 212.228
Liq. d_4^{20} 1.03. Bp_9 134-136°. n_D^{20} 1.5138.
Monoisopropyl ester: see under (4-Methylphenyl)phosphinic acid, M-00222
Diisopropyl ester: [51303-82-3]. *Diisopropyl* p-*tolylphosphonite.*
$C_{13}H_{21}O_2P$ M 240.281
Liq. Bp_{12} 129-131°. n_D^{20} 1.5003.
Monooctyl ester: see under (4-Methylphenyl)phosphinic acid, M-00222
Di-Ph ester: [36838-17-2]. *Diphenyl (4-methylphenyl)phosphonite.*
$C_{19}H_{17}O_2P$ M 308.316
Liq. d_4^{20} 1.15. $Bp_{0.2}$ 174-180°, $Bp_{0.08}$ 143-145°. n_D^{20} 1.6069 (1.624).
Difluoride: [10568-62-4]. *Difluoro*-p-*tolylphosphine.*
$C_7H_7F_2P$ M 160.103
Liq. d_4^{20} 1.20. Bp_{50} 70-72°. n_D^{20} 1.5040.
Dichloride: see (4-Methylphenyl)phosphonous dichloride, M-00249
Dibromide: Dibromo-p-*tolylphosphine.*
$C_7H_7Br_2P$ M 281.914
Solid. Mp 160-161°.
Diiodide: [24901-23-3]. *Diiodo*-p-*tolylphosphine.*
$C_7H_7I_2P$ M 375.915
Liq. Bp_4 176-177°.
Bis(diethylamide): N,N,N′,N′-*Tetraethyl*-P-(*4-methylphenyl*)*phosphonous diamide.*
$C_{11}H_{26}N_2P$ M 217.314
Liq. Bp_{12} 154-156°.

Lindner, J. *et al*, *Monatsh. Chem.*, 1929, **53**, 263 (*dibromide*)
Kamai, G. *et al*, *Zh. Obshch. Khim.*, 1964, **34**, 439 (*Engl. transl.* p. 442) (*ester*)
Drozd, G.I. *et al*, *Zh. Obshch. Khim.*, 1967, **37**, 958, 1343 (*Engl. transl.* pp. 906, 1269) (*difluoride, synth, F and P nmr*)
Feshchenko, N.G. *et al*, *Zh. Obshch. Khim.*, 1969, **39**, 2184 (*Engl. transl.* p. 2133) (*diiodide*)
Kharrasova, F.M. *et al*, *Zh. Obshch. Khim.*, 1973, **43**, 1930 (*Engl. transl.* p. 1914) (*esters*)
Kharrasova, F.M. *et al*, *Zh. Obshch. Khim.*, 1978, **48**, 1041 (*Engl. transl.* p. 948) (*diphenyl ester, synth, nmr*)
Sybert, P.D. *et al*, *Macromolecules*, 1981, **14**, 502 (*dimethyl ester, synth, ir, pmr, nmr, cmr*)

(4-Methylphenyl)phosphonous dichloride, 9CI M-00249

Dichloro-p-*tolylphosphine*
[1005-32-9]

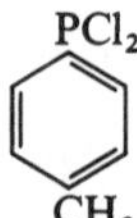

$C_7H_7Cl_2P$ M 193.012

Oil which solidifies on standing. d_4^{20} 1.27. Mp 24°. Bp_{21} 125-127°. n_D^{20} 1.5910.

Schindlbauer, H., *Monatsh. Chem.*, 1965, **96**, 1936 (*synth, ir*)
Kharrasova, F.M. *et al*, *Zh. Obshch. Khim.*, 1966, **36**, 1987 (*Engl. transl.* p. 1979) (*synth*)
Malakhova, I.G. *et al*, *Izv. Akad. Nauk SSSR, Ser. Khim.*, 1974, 1842 (*Engl. transl.* p. 1761) (*synth*)
Weinberg, K.G., *J. Org. Chem.*, 1975, **40**, 3586 (*synth*)
Zakirov, D.U. *et al*, *Zh. Obshch. Khim.*, 1977, **47**, 1661 (*Engl. transl.* p. 1522) (*nqr*)

Methylphenylphosphoramidic acid, 9CI M-00250

PhNMeP(O)(OH)$_2$

$C_7H_{10}NO_3P$ M 187.135

Di-Me ester: [7006-95-3]. *Dimethyl methylphenylphosphoramidate.*
$C_9H_{14}NO_3P$ M 215.188
Liq. d_4^{20} 1.20. $Bp_{0.25}$ 92°. n_D^{20} 1.5130.

Di-Et ester: [52670-78-7]. *Diethyl methylphenylphosphoramidate.*
$C_{11}H_{18}NO_3P$ M 243.242
Liq. d_4^{20} 1.12. Bp_1 91-92°. n_D^{20} 1.5030.

Di-Ph ester: [52670-92-5]. *Diphenyl methylphenylphosphoramidate.*
$C_{19}H_{18}NO_3P$ M 339.330
Needles. Mp 50°.

Dibenzyl ester: Dibenzyl methylphenylphosphoramidate.
$C_{21}H_{22}NO_3P$ M 367.383
Prisms (pet. ether). Mp 86-87°.

Atherton, F.R. *et al*, *J. Chem. Soc.*, 1947, 674 (*dibenzyl ester*)
Kabachnik, M.I. *et al*, *Izv. Akad. Nauk SSSR, Ser. Khim.*, 1956, 790 (*Engl. transl.* p. 809) (*diethyl ester*)
Gilyarov, V.A. *et al*, *Zh. Obshch. Khim.*, 1966, **36**, 708 (*Engl. transl.* p. 722) (*dimethyl ester, derivs*)
Uesugi, Y. *et al*, *Agric. Biol. Chem.*, 1974, **38**, 907 (*esters, synth, ir, pmr*)
Ching Yee Cheng, *et al*, *J. Chem. Soc., Perkin Trans. 1*, 1976, 1739 (*diethyl ester, synth, pmr*)
Buchanan, G.W. *et al*, *Org. Magn. Reson.*, 1980, **14**, 517 (*dimethyl ester, cmr*)
Buchanan, G.W. *et al*, *Can. J. Chem.*, 1980, **58**, 2442 (*dimethyl ester, N nmr*)

(2-Methylphenyl)phosphoramidic acid, 9CI M-00251

o-*Tolylphosphoramidic acid, 8CI*

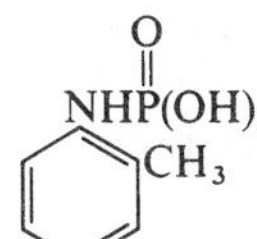

$C_7H_{10}NO_3P$ M 187.135

Di-Me ester: [25626-98-6]. *Dimethyl (2-methylphenyl)-phosphoramidate.*
$C_9H_{14}NO_3P$ M 215.188
Cryst. (C_6H_6/pet. ether). Mp 110-111°.

Di-Et ester: [78504-35-5]. *Diethyl (2-methylphenyl)-phosphoramidate.*
$C_{11}H_{18}NO_3P$ M 243.242
Solid. Mp 95°.

Di-Ph ester: [56168-00-8]. *Diphenyl (2-methylphenyl)-phosphoramidate.*
$C_{19}H_{18}NO_3P$ M 339.330
Cryst. (EtOH). Mp 121-123°.

Difluoride:
$C_7H_8F_2NOP$ M 191.117
Low melting solid. Mp 40°. Bp_6 98-100°.

Dichloride:
$C_7H_8Cl_2NOP$ M 224.026
Cryst. (pet. ether). Mp 91°.

Michaelis, A. *et al*, *Ber.*, 1894, **27**, 2572 (*derivs*)
Olah, G.A. *et al*, *J. Org. Chem.*, 1959,**24**, 1443 (*difluoride*)
Zhmurova, I.N. *et al*, *Zh. Obshch. Khim.*, 1961, **31**, 3741 (*Engl. transl.* p. 3495) (*diphenyl ester*)
Cadogan, J.I.G. *et al*, *J. Chem. Soc. (C)*, 1969, 2813 (*dimethyl ester*)
Buchanan, G.W. *et al*, *Can. J. Chem.*, 1980, **58**, 2442 (*dimethyl ester, N nmr*)
Buchanan, G.W. *et al*, *Org. Magn. Reson.*, 1980, **14**, 517 (*dimethyl ester, cmr*)
Foulds, G.A. *et al*, *S. Afr. J. Chem.*, 1981, **34**, 72 (*dimethyl ester, ir*)
Modro, T.A. *et al*, *J. Org. Chem.*, 1982, **47**, 3208 (*dimethyl ester, synth, props*)
Davidowitz, B. *et al*, *Org. Mass Spectrom.*, 1984, **19**, 128 (*dimethyl ester, ms*)

(3-Methylphenyl)phosphoramidic acid, 9CI M-00252

m-*Tolylphosphoramidic acid*

$C_7H_{10}NO_3P$ M 187.135

Di-Me ester: [25626-99-7]. *Dimethyl (3-methylphenyl)-phosphoramidate.*
$C_9H_{14}NO_3P$ M 215.188
Cryst. (C_6H_6/pet. ether). Mp 97°.

Di-Et ester: [25626-00-3]. *Diethyl (3-methylphenyl)-phosphoramidate.*
$C_{11}H_{18}NO_3P$ M 243.242
Solid. Mp 98°.

Di-Ph ester: [76168-01-9]. *Diphenyl (3-methylphenyl)-phosphoramidate.*
$C_{19}H_{18}NO_3P$ M 339.330
Cryst. (EtOH). Mp 122-123°.

Difluoride:
$C_7H_8F_2NOP$ M 191.117
Liq. Bp_6 124-125°.

Olah, G.A. *et al*, *J. Org. Chem.*, 1959, **24**, 1443 (*difluoride*)
Zhmurova, I.N. *et al*, *Zh. Obshch. Khim.*, 1961, **31**, 3741 (*Engl. transl.* p. 3495) (*diphenyl ester*)
Cadogan, J.I.G. *et al*, *J. Chem. Soc. (C)*, 1969, 2813 (*esters*)
Foulds, G.A. *et al*, *S. Afr. J. Chem.*, 1981, **34**, 72 (*dimethyl ester, ir*)
Modro, T.A. *et al*, *J. Org. Chem.*, 1982, **47**, 3208 (*dimethyl ester, props*)
Davidowitz, B. *et al*, *Org. Mass Spectrom.*, 1984, **19**, 128 (*dimethyl ester, ms*)

(4-Methylphenyl)phosphoramidic acid, 9CI M-00253

p-*Tolylphosphoramidic acid*

$C_7H_{10}NO_3P$ M 187.135
Solid. Mp 270-272° dec.

Bis(cyclohexylammonium) salt: Solid. Mp 160-165°.

Di-Me ester: [25627-01-4]. *Dimethyl (4-methylphenyl)-phosphoramidate.*
$C_9H_{14}NO_3P$ M 215.188
Cryst. (C_6H_6/pet. ether). Mp 110-112°.

Di-Et ester: [20809-97-6]. *Diethyl (4-methylphenyl)-phosphoramidate.*
$C_{11}H_{18}NO_3P$ M 243.242
Solid. Mp 95°.

Di-Ph ester: [62569-07-7]. *Diphenyl (4-methylphenyl)-phosphoramidate.*
$C_{19}H_{18}NO_3P$ M 339.330

Cryst. (EtOH). Mp 138-140°.
Dibenzyl ester: Dibenzyl (4-methylphenyl)-phosphoramidate. Bis(phenylmethyl)(4-methylphenyl)phosphoramidate.
$C_{21}H_{22}NO_3P$ M 367.383
Prisms (pet. ether). Mp 89.5-90.5°.
Difluoride:
$C_7H_8F_2NOP$ M 191.117
Solid. Mp 78-80°. Bp_6 118°.
Dichloride:
$C_7H_8Cl_2NOP$ M 224.026
Cryst. Mp 98°.
Dianilide: N-*(4-Methylphenyl)*-N′,N″*-diphenylphosphoric triamide.*
$C_{19}H_{20}N_3OP$ M 337.360
Cryst. (EtOH). Mp 168°.

Michaelis, A. *et al*, *Ber.*, 1894, **27**, 2572 (*derivs*)
Atherton, F.R. *et al*, *J. Chem. Soc.*, 1947, 674 (*dibenzyl ester*)
Goldwhite, H. *et al*, *J. Chem. Soc.*, 1957, 2409 (*synth*)
Olah, G.A. *et al*, *J. Org. Chem.*, 1959, **24**, 1443 (*difluoride*)
Zhmurova, I.N. *et al*, *Zh. Obshch. Khim.*, 1961, **31**, 3741 (*Engl. transl.* p. 3495) (*diphenyl ester*)
Cadogan, J.I.G. *et al*, *J. Chem. Soc. (C)*, 1969, 2813 (*esters*)
Williams, A. *et al*, *J. Chem. Soc. (B)*, 1971, 1973 (*synth, uv*)
Foulds, G.A. *et al*, *S. Afr. J. Chem.*, 1981, **34**, 72 (*dimethyl ester, ir*)
Davidowitz, B. *et al*, *Org. Mass Spectrom.*, 1984, **19**, 128 (*dimethyl ester, ms*)

(2-Methylphenyl)phosphoramidothioic acid, 9CI **M-00254**

o-*Tolylphosphoramidothioic acid, 8CI*

$C_7H_{10}NO_2PS$ M 203.195
O,O-*Di-Et ester:* [15832-91-4]. O,O-*Diethyl (2-methylphenyl)phosphoramidothioate.*
$C_{11}H_{18}NO_2PS$ M 259.302
Liq. n_D^{25} 1.5302.
O,O-*Di-Ph ester:* O,O-*Diphenyl (2-methylphenyl)-phosphoramidothioate.*
$C_{19}H_{18}NO_2PS$ M 355.390
Cryst. (MeOH or 3-methylbutanol). Mp 68-70°.

Borke, M.L. *et al*, *J. Am. Pharm. Assoc.*, 1958, **47**, 461.

(3-Methylphenyl)phosphoramidothioic acid, 9CI **M-00255**

m-*Tolylphosphoramidothioic acid, 8CI*
$C_7H_{10}NO_2PS$ M 203.195
O,O-*Di-Et ester:* [15832-92-5]. O,O-*Diethyl (3-methylphenyl)phosphoramidothioate.*
$C_{11}H_{18}NO_2PS$ M 259.302
Liq. n_D^{20} 1.5342.
O,O-*Di-Ph ester:* O,O-*Diphenyl (3-methylphenyl)-phosphoramidothioate.*
$C_{19}H_{18}NO_2PS$ M 355.390
Cryst. (EtOH). Mp 86-88°.
Difluoride:
$C_7H_8F_2NPS$ M 207.177
Liq. $Bp_{0.9}$ 85.5°. n_D^{20} 1.5361.

Borke, M.L. *et al*, *J. Am. Pharm. Assoc.*, 1958, **47**, 461 (*diphenyl ester*)
Olah, G.A. *et al*, *J. Org. Chem.*, 1959, **24**, 1443 (*difluoride*)
Japan. Pat., 62 4 458, (*1962*); *CA*, **67**, 73347

(4-Methylphenyl)phosphoramidothioic acid, 9CI **M-00256**

p-*Tolylphosphoramidothioic acid, 8CI*
$C_7H_{10}NO_2PS$ M 203.195
O,O-*Di-Et ester:* [15832-93-6]. O,O-*Diethyl (4-methylphenyl)phosphoramidothioate.*
$C_{11}H_{18}NO_2PS$ M 259.302
Cryst. (pet. ether). Mp 38°.
O,O-*Di-Ph ester:* O,O-*Diphenyl (4-methylphenyl)-phosphoramidothioate.*
$C_{19}H_{18}NO_2PS$ M 355.390
Cryst. (EtOH or 3-methylbutanol). Mp 94-96°.

Borke, M.L. *et al*, *J. Am. Pharm. Assoc.*, 1958, **47**, 461 (*diphenyl ester*)
Burn, A.J. *et al*, *J. Chem. Soc.*, 1961, 5532 (*diethyl ester*)

4-Methylphenyl phosphorodichloridate, 9CI **M-00257**

p-*Tolyl phosphorodichloridate.* p-*Tolyl phosphoryl dichloride.* p-*Cresyl dichlorophosphate*
[878-17-1]

$C_7H_7Cl_2O_2P$ M 225.011
Bp_{12} 145-150°, Bp_1 90-91°.

Orloff, H.D. *et al*, *J. Am. Chem. Soc.*, 1958, **80**, 727 (*synth*)
Kosolapoff, G.M. *et al*, *J. Chem. Soc. (C)*, 1968, 815 (*synth, pmr*)
Just, G. *et al*, *J. Prakt. Chem.*, 1971, **313**, 69 (*synth*)

2-Methylphenyl phosphorodichloridite, 9CI **M-00258**

o-*Tolyl phosphorodichloridite, 8CI.* o-*Tolyl dichlorophosphite*
[31860-10-3]

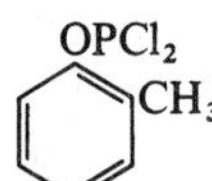

$C_7H_7Cl_2OP$ M 209.011
Pungent oil. Bp_{11} 116°.

Strecker, W. *et al*, *Ber.*, 1916, **49**, 63 (*synth*)

3-Methylphenyl phosphorodichloridite, 9CI **M-00259**

m-*Tolyl phosphorodichloridite, 8CI.* m-*Tolyl dichlorophosphite*
[21720-16-1]
$C_7H_7Cl_2OP$ M 209.011
Pungent liq. Bp_{12} 114°.

Broeker, W., *J. Prakt. Chem.*, 1928, **118**, 287 (*synth*)

4-Methylphenyl phosphorodichloridite, 9CI **M-00260**

p-*Tolyl phosphorodichloridite, 8CI.* p-*Tolyl dichlorophosphite*

[21719-85-7]
$C_7H_7Cl_2OP$ M 209.011
Pungent liq. Bp_{11} 118°.

Strecker, W. *et al*, *Chem. Ber.*, 1916, **49**, 63 (*synth*)

O-(2-Methylphenyl) phosphorodichloridothioate, 9CI M-00261

O-o-*Tolyl phosphorodichloridothioate*, 8CI. O-o-*Tolyl dichlorothiophosphate*. O-o-*Tolyl thiophosphoryl dichloride*

S
‖
$OPCl_2$
CH_3

$C_7H_7Cl_2OPS$ M 241.071
Liq. Bp_6 95-96°. n_D^{20} 1.5672.

Strecker, W. *et al*, *Ber.*, 1916, **49**, 63 (*synth*)
Protsenko, L.D. *et al*, *Zh. Obshch. Khim.*, 1964, **34**, 2233 (*Engl. transl.* p. 2244) (*synth*)

O-(4-Methylphenyl) phosphorodichloridothioate, 9CI M-00262

O-p-*Tolyl phosphorodichloridothioate*, 8CI. O-p-*Tolyl dichlorothiophosphate*. O-p-*Tolyl thiophosphoryl dichloride*

[18961-95-0]
$C_7H_7Cl_2OPS$ M 241.071
Oil. d_4^{20} 1.36. Bp_{12} 138°, Bp_3 91-92°. n_D^{20} 1.5615.

Autenrieth, W. *et al*, *Ber.*, 1925, **58**, 840 (*synth*)
Godovikov, N.N. *et al*, *Zh. Obshch. Khim.*, 1961, **31**, 1628 (*Engl. transl.* p. 1516) (*synth*)
Protsenko, L.D. *et al*, *Zh. Obshch. Khim.*, 1964, **34**, 2233 (*Engl. transl.* p. 2244) (*synth*)

O-Methyl *O*-phenyl phosphorodithioate, 9CI, 8CI M-00263

O-Methyl O-*phenyl hydrogen dithiophosphate*. O-*Methyl* O-*phenyl dithiophosphoric acid*

MeO S
P
PhO SH

$C_7H_9O_2PS_2$ M 220.241
K salt: [20868-30-8]. Solid. Mp 176-178°.

Kotovich, B.P. *et al*, *Zh. Obshch. Khim.*, 1968, **38**, 1763 (*Engl. transl.* p. 1718)

Methylphenyl-2-propenylphosphine, 9CI M-00264

Allylmethylphenylphosphine, 8CI
[54807-86-2]

Ph
:►P◄Me (*R*)-*form*
$CH_2CH{=}CH_2$

$C_{10}H_{13}P$ M 164.186

(*R*)-*form*
Liq. $Bp_{0.05}$ 40°.
B,PhCH₂Br: Methylphenyl(phenylmethyl)-2-propenylphosphonium bromide. Allylbenzylmethylphenylphosphonium bromide.
$C_{17}H_{20}BrP$ M 335.223
Solid. Mp 134-137°. $[\alpha]_D$ −17.0° (c, 1-3 in MeOH). Has (*S*)-config.
Sulfide: [26343-70-4].
$C_{10}H_{13}PS$ M 196.246
Needles (C_6H_6/hexane). Mp 62-63°. $[\alpha]_D$ −18.8° (c, 1-3 in MeOH) (74% o.p.). Has (*S*)-config.

(*S*)-*form*
Liq. $Bp_{0.1}$ 45°.
B,PhCH₂Br: Solid. Mp 159-161°. $[\alpha]_D$ +16.0° (c, 1.50 in MeOH) (100% o.p.). Has (*R*)-config.
Oxide:
$C_{10}H_{13}OP$ M 180.186
$[\alpha]_D$ +21° (c, 1-3 in MeOH) (100% o.p.). Has (*R*)-config.

(±)-*form*
Liq. $Bp_{0.1}$ 50°.
B,PhCH₂Br: Solid. Mp 140-141°.
Sulfide: Needles (C_6H_6/hexane). Mp 62-63°.

Baechler, R.D. *et al*, *J. Am. Chem. Soc.*, 1969, **91**, 5686; 1970, **92**, 3090 (*synth*)
Naumann, N. *et al*, *J. Am. Chem. Soc.*, 1969, **91**, 7012 (*oxide, props*)
Zon, G. *et al*, *J. Am. Chem. Soc.*, 1969, **91**, 7023 (*derivs, props*)
Horner, L. *et al*, *Phosphorus*, 1971, **1**, 73 (*salts*)
Snider, T.E. *et al*, *Org. Prep. Proc. Int.*, 1974, **6**, 221 (*synth, salts*)

Methylphenylpropylphosphine, 9CI M-00265

[4653-62-7]

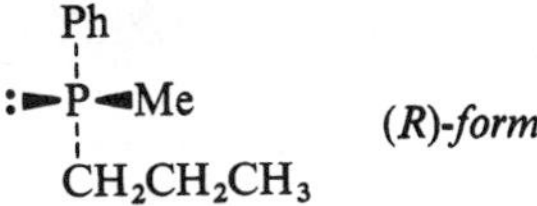

(*R*)-*form*

$C_{10}H_{15}P$ M 166.202

(*R*)-*form* [13153-89-4]
$[\alpha]_D$ −20.6° (MeOH), −18.4° (toluene).
Oxide: [1515-99-7].
$C_{10}H_{15}OP$ M 182.202
$Bp_{0.03\text{-}00.05}$ 110°. $[\alpha]_D^{22}$ −16.0° (c, 1.1-2.0, MeOH) (91% o.p.). Oxide has (*S*)-config.
Sulfide: [13153-91-8].
$C_{10}H_{15}PS$ M 198.262
$[\alpha]_D$ −20.6°. Has (*S*)-config.
Selenide:
$C_{10}H_{15}PSe$ M 245.162
$[\alpha]_D$ −19.48° (100% o.p.). Has (*S*)-config.

(*S*)-*form* [701-03-1]
Liq. $Bp_{0.25}$ 70°. $[\alpha]_D^{22}$ +16.1° (c, 1.1-2.0 in MeOH) (99% o.p.).
B,C₃H₅Br: Methylphenyl(2-propenyl)-propylphosphonium bromide. Allylmethylphenylpropylphosphonium bromide.
$C_{13}H_{20}BrP$ M 287.179
Solid. Mp 105-106°. $[\alpha]_D$ +16.2° (c, 3.60 in MeOH) (93% o.p.).
B,PhCH₂Br: see Benzylmethylphenylpropylphosphonium(1+), B-00052
Oxide: [17170-48-8]. Mp 57-58°. $Bp_{0.1}$ 125°. $[\alpha]_D$ +17.5° (C_6H_6), $[\alpha]_D$ +16.1° (MeOH) (93% o.p.). Has (*R*)-config.
Sulfide: Cryst. Mp 80°. $[\alpha]_D$ +19.5° (c, 4.02 in MeOH) (70% o.p.). Has (*R*)-config.

(±)-*form* [20108-75-2]
Liq. Bp_{20} 116-118°.
B,MeI: Dimethylphenylpropylphosphonium iodide.
$C_{11}H_{18}IP$ M 308.141

Mp 75-79°.
B,PhCH$_2$Br: see Benzylmethylphenylpropylphosphonium(1+), B-00052
Oxide: [2328-23-6]. Solid. Mp 40°. $Bp_{0.2}$ 140°.
Sulfide: [34666-66-5]. No phys. props. reported.

Horner, L. *et al, Chem. Ber.*, 1969, **102**, 3542 (*uv, cd*)
Horner, L. *et al, Phosphorus*, 1971, **1**, 73 (*oxides, salts*)
Marsi, K.L., *J. Org. Chem.*, 1974, **39**, 265 (*salts*)
Luckenbach, R. *et al, Chem. Ber.*, 1975, **108**, 3533 (*sulfide, oxide*)
Luckenbach, R. *et al, Justus Liebigs Ann. Chem.*, 1976, 2305 (*use, oxide*)
Sakaki, K. *et al, Chem. Lett.*, 1977, 1003 (*selenide, props*)
Mikolajczyk, M. *et al, J. Org. Chem.*, 1978, **43**, 2132 (*sulfide, selenide, oxide*)
Omelańczuk, J. *et al, J. Am. Chem. Soc.*, 1979, **101**, 7292 (*sulfide*)
Payne, N.C. *et al, Can. J. Chem.*, 1980, **58**, 15 (*synth, pmr, nmr, methiodide*)
Bestmann, H.J. *et al, Chem. Ber.*, 1982, **115**, 3875 (*config, salts*)

[[(4-Methylphenyl)sulfinyl]methyl]phosphonic acid, 9CI — M-00266

[(p-*Tolylsulfinyl)methyl]phosphonic acid*

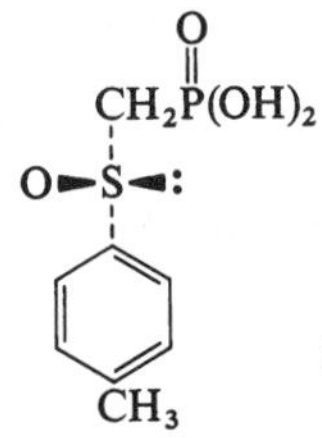

(*R*)-form

$C_8H_{11}PO_4S$ M 234.206

(*R*)-form

Mono-Me ester: Methyl hydrogen [[(4-methylphenyl)sulfinyl]methyl]phosphonate.
$C_9H_{13}O_4PS$ M 248.233
$[\alpha]_D$ −142° (c, 1.2 in $CHCl_3$).
Di-Me ester: [61187-71-1]. *Dimethyl [[(4-methylphenyl)sulfonyl]methyl]phosphonate.* Wittig-Horner reagent for the synth. of opt. active vinyl sulfoxides. $[\alpha]_D$ −149° (c, 1.16 in Me_2CO).

(*S*)-form

Mono-Me ester: $[\alpha]_D$ +103° (c, 1.31 in $CHCl_3$), $[\alpha]_D$ +143°.
Di-Me ester: [63268-43-9]. Wittig-Horner reagent. $[\alpha]^{25}_D$ +176.3°, $[\alpha]_D$ +106° (c, 1.7 in Me_2CO).

(±)-form

Mono-Me ester, quinine salt: Mp 52-59°. $[\alpha]_D$ −78.4° (c, 1.55 in $CHCl_3$).
Di-Me ester: [63231-19-6].
$C_{10}H_{15}O_4PS$ M 262.260
Liq. n_D^{20} 1.5226 (1.5319).

Drabowicz, J. *et al, Synthesis*, 1978, 758; 1979, 39 (*synth, P nmr*)
Mikolajczyk, M. *et al, J. Org. Chem.*, 1978, **43**, 473 (*synth, resoln, pmr, cmr, P nmr, use*)
Goldmann, S. *et al, Chem. Ber.*, 1980, **113**, 831 (*config, use*)
Hoffmann, R.W. *et al, Chem. Ber.*, 1980, **113**, 845, 856 (*use*)
Akkerman, J.M. *et al, Heterocycles*, 1981, **15**, 797 (*use*)

[[(4-Methylphenyl)sulfonyl]methyl]triphenylphosphonium(1+) — M-00267

Triphenyl(tosylmethyl)phosphonium(1+)

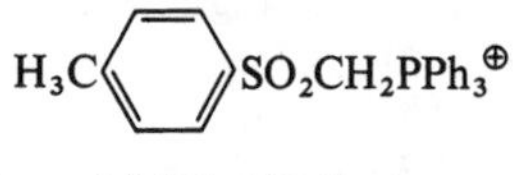

$C_{26}H_{24}O_2PS^{\oplus}$ M 431.508 (ion)

Bromide: [5681-42-5].
$C_{26}H_{24}BrO_2PS$ M 511.412
Cryst. (CH_2Cl_2/C_6H_6). Mp 267-270°. With NaH gives the ylide.
Ylide: see 4-Methylphenyl triphenylphosphoranylidenemethyl sulfone, M-00271

Speziale, A.I. *et al, J. Am. Chem. Soc.*, 1965, **87**, 5603 (*synth, pmr, ir, ylide*)
van Leusen, A.M. *et al, Recl. Trav. Chim. Pays-Bas*, 1972, **91**, 37 (*synth, ylide*)

[(4-Methylphenyl)sulfonyl]phosphorimidic acid, 9CI — M-00268

p-*Toluenesulfonylphosphorimidic acid. Tosylphosphorimidic acid*

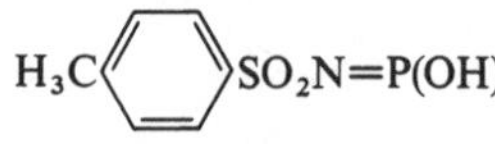

$C_7H_{10}NO_5PS$ M 251.193
Free acid is tautomeric with [(4-methylphenyl)sulfonyl]-phosphoramidic acid.

Tri-Me ester: [17986-07-1]. *Trimethyl [(4-methylphenyl)sulfonyl]phosphorimidate.* P,P,P-*Trimethoxy*-N-p-*toluenesulfonylphosphazene.* P,P,P-*Trimethoxy*-N-p-*toluenesulfonylphosphine imide.*
$C_{10}H_{16}NO_5PS$ M 293.274
No phys. props. reported.
Tri-Et ester: [4779-09-3]. *Triethyl [(4-methylphenyl)-sulfonyl]phosphorimidate.* P,P,P-*Triethoxy*-N-p-*toluenesulfonylphosphazene.* P,P,P-*Triethoxy*-N-p-*toluenesulfonylphosphine imide.*
$C_{13}H_{22}NO_5PS$ M 335.354
Oil. $Bp_{0.01}$ 159-170°. n_D^{25} 1.5041.
Tri-Ph ester: [17436-27-0]. *Triphenyl [(4-methylphenyl)sulfonyl]phosphorimidate.* N-p-*Toluenesulfonyl*-P,P,P-*triphenoxyphosphazene.* N-p-*Toluenesulfonyl*-P,P,P-*triphenoxyphosphine imide.*
$C_{25}H_{22}NO_5PS$ M 479.486
No phys. props. recorded.
Trichloride: [3576-26-9]. P,P,P-*Trichloro*-N-p-*toluenesulfonylphosphazene.* P,P,P-*Trichloro*-N-p-*toluenesulfonylphosphine imide.*
$C_7H_7Cl_3NO_2PS$ M 306.531
Solid. Mp 103-104°.
Tribromide: P,P,P-*Tribromo*-N-p-*toluenesulfonylphosphazene.* P,P,P-*Tribromo*-N-p-*toluenesulfonylphosphine imide.*
$C_7H_7Br_3NO_2PS$ M 439.884
Solid. Mp 140-141°.

Cadogan, J.I.G. *et al, J. Chem. Soc.*, 1961, 3079 (*triethyl ester*)
Wiegräbe, W. *et al, Chem. Ber.*, 1968, **101**, 1414 (*trichloride, synth, ir*)
Markovskii, L.N. *et al, Zh. Org. Khim.*, 1972, **8**, 2057 (*Engl. transl.* p. 2104) (*tribromide*)
Glidewell, C., *Inorg. Chim. Acta*, 1976, **18**, 51 (*trichloride, ms*)
Glidewell, C., *J. Organomet. Chem.*, 1976, **108**, 335 (*esters, ir, pmr, P nmr*)
Kovenya, V.A. *et al, Zh. Obshch. Khim.*, 1976, **46**, 2679 (*Engl. transl.* p. 2557) (*trichloride, synth*)

1-Methyl-5-phenyl-1*H*-1,2,4,3-triazaphosphole, 9CI — M-00269

[52718-98-1]

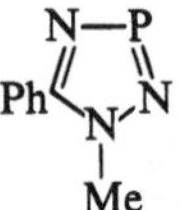

$C_8H_8N_3P$ M 177.145

Liq. $Bp_{0.1}$ 105-110°.

Charbonnel, Y. *et al*, *Tetrahedron*, 1976, **32**, 2039 (*synth*, *nmr*)
Majoral, J.-P. *et al*, *Tetrahedron Lett.*, 1980, **21**, 1307 (*nmr*)

2-Methyl-5-phenyl-2*H*-1,2,4,3-triazaphosphole, 9CI — M-00270

[52713-97-0]

$C_8H_9N_3P$ M 178.153
Solid. Mp 50°. Bp_{15} 163°. Unusual in possessing dicoordinate phosphorus.

Charbonnel, Y. *et al*, *Tetrahedron*, 1976, **32**, 2039 (*synth*, *P nmr*)
Legros, J.P. *et al*, *C.R. Hebd. Seances Acad. Sci., Ser. C*, 1980, **291**, 271 (*cryst struct*)
Schmidpeter, A. *et al*, *Z. Naturforsch., B*, 1983, **38**, 1484 (*props*)

4-Methylphenyl triphenylphosphoranylidenemethyl sulfone — M-00271

[5554-81-4]

$C_{26}H_{23}O_2PS$ M 430.500
Wittig reagent. Cryst. (CH_2Cl_2/hexane). Mp 185-186°.

Speziale, A.J. *et al*, *J. Am. Chem. Soc.*, 1965, **87**, 5603 (*synth*, *ir*, *uv*, *cryst struct*)
Wheatley, P.J., *J. Chem. Soc.*, 1965, 5785 (*cryst struct*)
v. Leusen, A.M. *et al*, *Recl. Trav. Chim. Pays-Bas*, 1972, **91**, 37 (*synth*, *pmr*, *ir*)

8-Methyl-8-phosphabicyclo[3.2.1]octane — M-00272

anti-form

$C_8H_{15}P$ M 142.180

syn-form

Oxide: [55816-41-6].
$C_8H_{15}OP$ M 158.180
V. hygroscopic solid. Mp 150°. $Bp_{0.7}$ 100° subl.

anti-form

Oxide: [55816-42-7]. V. hygroscopic cryst. (cyclohexane). Mp 135°. $Bp_{0.1}$ 90° subl.

Awerbauch, O. *et al*, *Tetrahedron*, 1975, **31**, 33 (*synth*, *ir*, *pmr*)

8-Methyl-8-phosphabicyclo[3.2.1]oct-6-ene 8-oxide — M-00273

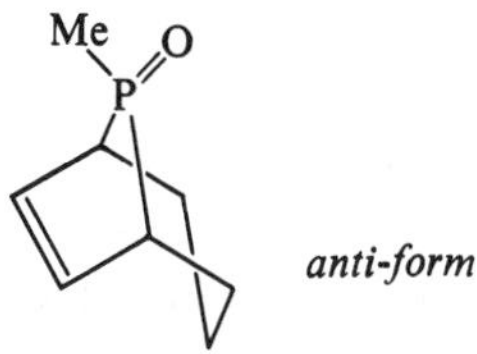

anti-form

$C_8H_{13}OP$ M 156.164
Mp 97-99°.

syn-form [55816-32-5]
Solid. Mp 128-129°. $Bp_{0.05}$ 100° subl.

anti-form [55816-33-6]
Cryst. (cyclohexane). Mp 97-99°. $Bp_{0.15}$ 80° subl.

Awerbauch, O. *et al*, *Tetrahedron*, 1975, **31**, 33 (*synth*, *ir*, *pmr*)

5-Methyl-1-phospha-5-silabicyclo[3.3.1]nonane, 9CI — M-00274

[83622-76-8]

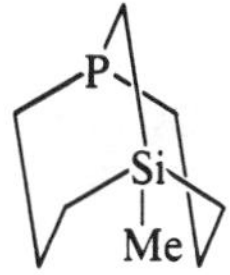

$C_8H_{17}PSi$ M 172.282
Liq. $Bp_{0.05}$ 42°.

B,MeI: 1,5-Dimethyl-1-phosphonia-5-silabicyclo[3.3.1]nonane iodide.
$C_9H_{20}IPSi$ M 314.221
Cryst. (MeCN). Mp 191°.

1-Sulfide: [83622-80-4].
$C_8H_{17}PSSi$ M 204.342
Cryst. (MeOH). Mp 114-116°.

Kühne, U. *et al*, *Phosphorus Sulfur*, 1982, **13**, 153 (*synth*, *derivs*, *ir*, *pmr*, *cmr*, *P nmr*)

4-Methyl-4-phosphatetracyclo[3.3.0.0^{2,8}.0^{3,6}]octane, 10CI, 9CI — M-00275

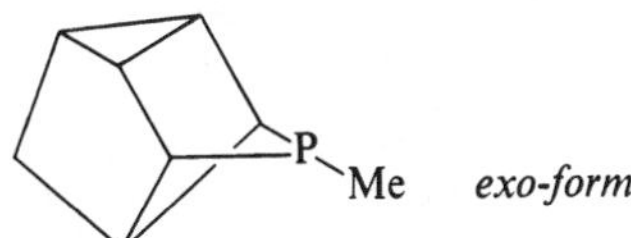

exo-form

$C_8H_{11}P$ M 138.149
Obt. only as exo-endo mixt.

endo-form

4-Oxide:
$C_8H_{11}OP$ M 154.148
Solid. Mp 71-74°.

4-Sulfide:
$C_8H_{11}PS$ M 170.209
Solid. Mp 129-133°.

exo-form

4-Oxide: Cryst. (CCl_4). Mp 156-157°.
4-Sulfide: Solid. Mp 111-113.5°.

Green, M., *J. Chem. Soc.*, 1965, 541 (*oxide*, *ir*, *pmr*, *ms*)
Cremer, S.E. *et al*, *J. Chem. Soc., Chem. Commun.*, 1975, 374 (*oxides*, *cmr*)
Mazhar-ul-Haque, *et al*, *J. Chem. Soc., Perkin Trans. 2*, 1978, 1115 (*cryst struct*)
Quin, L.D. *et al*, *J. Org. Chem.*, 1983, **48**, 4466 (*oxides*, *props*)

Quin, L.D. *et al*, *J. Am. Chem. Soc.*, 1984, **106**, 7021 (*synth, P nmr*)

Methyl phosphinate, 9CI M-00276

Methyl hypophosphite

[14684-31-2]

$$MeOP(O)H_2 \rightleftharpoons MeOP(OH)H$$

CH_5OP M 64.024

Liq. unstable at r.t. d_4^{20} 1.22. $Bp_{2.5}$ 25-25.5°. n_D^{20} 1.4275.

Kabachnik, M.I. *et al*, *Izv. Akad. Nauk SSSR, Ser. Khim.*, 1960, 146 (*Engl. transl.* p. 138) (*synth*)
Fitch, S.J., *J. Am. Chem. Soc.*, 1964, **86**, 61 (*synth, ir, pmr, P nmr*)
Gallagher, M.J. *et al*, *J. Chem. Soc., Chem. Commun.*, 1978, 54 (*use, props*)

Methylphosphine M-00277

[593-54-4]

$$MePH_2$$

CH_5P M 48.024

Ligand for Fe, Mo, Ti, Os, Pt and W. Gas. Almost insol. H_2O. Bp −17.1°. Forms salts which are dec. by H_2O.

▷Highly toxic. Ignites in air.

Inorg. Synth., 1968, **11**, 124 (*synth*)
Crosbie, K.D. *et al*, *J. Inorg. Nucl. Chem.*, 1969, **31**, 3684 (*synth*)
Kostyanovskii, R.G. *et al*, *Org. Mass Spectrom.*, 1972, **6**, 1183 (*ms*)
Quin, L.D. *et al*, *Org. Magn. Reson.*, 1974, **6**, 503 (*cmr*)
Lappert, M.F. *et al*, *J. Chem. Soc., Dalton Trans.*, 1975, 1207 (*pe*)
Albright, T.A., *Org. Mass Reson.*, 1976, **8**, 489 (*pmr, nmr, cmr, struct*)
McKean, C.D. *et al*, *J. Mol. Struct.*, 1978, **49**, 275 (*ir*)
Newman, T.H. *et al*, *J. Organomet. Chem.*, 1980, **197**, 159 (*derivs*)
Mitchell, D.J. *et al*, *Can. J. Chem.*, 1981, **59**, 3280 (*struct*)
Mosbo, J.A. *et al*, *Phosphorus Sulfur*, 1981, **11**, 11 (*struct*)
Cowley, A.A. *et al*, *Inorg. Chem.*, 1982, **21**, 85 (*pe*)
Fluck, E. *et al*, *Z. Anorg. Allg. Chem.*, 1986, **536**, 129 (*ms, cryst struct*)
Bretherick, L., *Handbook of Reactive Chemical Hazards*, 2nd Ed., Butterworths, London and Boston, 1979, 316.
Sax, N.I., *Dangerous Properties of Industrial Materials*, 6th Ed., Van Nostrand-Reinhold, 1984, 830.
Hazards in the Chemical Laboratory, (Bretherick, L., Ed.), 3rd Ed., Royal Society of Chemistry, London, 1981, 402.

Methylphosphinic acid, 9CI M-00278

[4206-94-4]

$$MeP(O)(OH)H$$

CH_5O_2P M 80.023

Tautomeric with Methylphosphonous acid, M-00319 . Oil. pK_a 2.3. Free acid exists in phosphinic acid form. Acid and Na salt in soln. are stable at 100° for several minutes.

Me ester: [16391-06-3]. *Methyl methylphosphinate.*
$C_2H_7O_2P$ M 94.050
d_4^{20} 1.12. n_D^{20} 1.422.

Et ester: [16391-07-4]. *Ethyl methylphosphinate.*
$C_3H_9O_2P$ M 108.077
Liq. d_4^{20} 1.05. Bp_6 58-59°. n_D^{20} 1.4220.

Isopropyl ester: see Isopropyl methylphosphinate, I-00034

Ph ester: Phenyl methylphosphinate.
$C_7H_9O_2P$ M 156.121
Liq. d_4^{20} 1.43. Bp_9 115-116°. n_D^{20} 1.5645.

Petrov, K.A. *et al*, *Zh. Obshch. Khim.*, 1961, **31**, 179 (*Engl. transl.* p. 168) (*esters*)
Fiat, D. *et al*, *J. Chem. Soc.*, 1962, 3837 (*synth, pmr, nmr, props*)
Barabanov, V.I. *et al*, *Zh. Obshch. Khim.*, 1965, **35**, 2225 (*Engl. transl.* p. 2215) (*ester*)
Fontal, B. *et al*, *J. Org. Chem.*, 1966, **31**, 2424 (*synth, pmr*)
Wolf, R. *et al*, *Spectrochim. Acta, Part A*, 1967, **23**, 1641 (*ir, pmr*)
Sanchez, M. *et al*, *Spectrochim. Acta, Part A*, 1967, **23**, 2617 (*ir*)
Daugherty, K.E. *et al*, *Appl. Spectrosc.*, 1968, **22**, 95 (*pmr*)
Grudzev, V.G. *et al*, *Zh. Obshch. Khim.*, 1968, **38**, 1548 (*Engl. transl.* p. 1499) (*ester*)
Marty, R. *et al*, *Org. Magn. Reson.*, 1970, **2**, 141 (*ester, pmr*)

Methylphosphinothioic acid, 9CI, 8CI M-00279

$$MePH(S)OH \rightleftharpoons MePH(O)SH \rightleftharpoons MeP(OH)SH$$

CH_5OPS M 96.084

Exists as tautomeric equilibrium between thione and thiol forms in further equilibrium with methylphosphonothious acid (minor component).

(S)-form

O-*Isopropyl ester:* [30983-71-2]. O-*Isopropyl methylphosphinothioate.*
$C_4H_{11}OPS$ M 138.164
Liq. $[\alpha]_D$ −16.2° (C_6H_6), $[\alpha]_D$ −17.4° (CCl_4), $[\alpha]_D$ −20.4° (EtOH) (68% o.p.). Optically stable but immediately racemised by MeO^-.

(±)-form

O-*Et ester:* O-*Ethyl methylphosphinothioate.*
C_3H_9OPS M 124.137
Liq. d_4^{20} 1.07. n_D^{20} 1.4908.

O-*Propyl ester:* O-*Propyl methylphosphinothioate.*
$C_4H_{11}OPS$ M 138.164
Liq. d_4^{20} 1.03. Bp_5 70-71°. n_D^{20} 1.4864.

O-*Butyl ester:* O-*Butyl methylphosphinothioate.*
$C_5H_{13}OPS$ M 152.191
Bp_9 96-98°. n_D^{20} 1.4840.

Petrov, K.A. *et al*, *Zh. Obshch. Khim.*, 1964, **34**, 2226 (*Engl. transl.* p. 2236)
Sanchez, M. *et al*, *Spectrochim. Acta, Part A*, 1967, **23**, 2617.
Aaron, H.S. *et al*, *J. Am. Chem. Soc.*, 1970, **92**, 6391.

1-Methylphospholane, 9CI, 8CI M-00280

Methyltetramethylenephosphine

[39834-55-4]

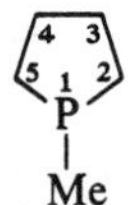

$C_5H_{11}P$ M 102.116

Liq. Bp 122-124°.

B,PhCH₂Br: 1-Benzyl-1-methylphospholanium bromide.
$C_{12}H_{18}BrP$ M 273.152
Solid. Mp 184-184.5°.

Oxide: [5794-87-6].
$C_5H_{11}OP$ M 118.115
Liq. $Bp_{0.65}$ 93-94°.

Sulfide: [1661-17-2].
$C_5H_{11}PS$ M 134.176

Liq. $Bp_{0.01}$ 73-75°. n_D^{20} 1.5573.

Maier, L., *Helv. Chim. Acta*, 1965, **48**, 133 (*sulfide, ir*)
Marsi, K.L., *J. Am. Chem. Soc.*, 1969, **91**, 4724 (*synth, oxide, pmr, P nmr*)
Quin, L.D. *et al*, *J. Org. Chem.*, 1969, **34**, 3700 (*synth, deriv*)
Sommer, K., *Z. Anorg. Allg. Chem.*, 1970, **379**, 56 (*synth, sulfide*)

3-Methylphospholane, 8CI — M-00281

$C_5H_{10}P$ M 101.108

(±)-*form*

Liq. d_4^{20} 0.90. Bp_{70} 119°. n_D^{20} 1.4964.

Bogolyubov, G.M. *et al*, *Zh. Obshch. Khim.*, 1963, **33**, 2419 (*Engl. transl.* p. 2359)

1-Methyl-1*H*-phosphole, 9CI — M-00282

[17167-23-6]

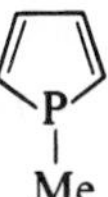

C_5H_7P M 98.084

Unstable liq. with strong phosphine-like odour, not obt. completely pure. Bp_{306} 79°. pK_a 0.5.

B,MeI: Cryst. (MeOH/EtOAc). Mp 190-194°. Struct. not clear.

Quin, L.D. *et al*, *J. Am. Chem. Soc.*, 1967, **89**, 5984; 1969, **91**, 3308 (*synth, uv, P nmr, pmr, ms*)
Quin, L.D. *et al*, *Org. Magn. Reson.*, 1973, **5**, 161 (*cmr*)
Kaufmann, G. *et al*, *Phosphorus*, 1974, **4**, 231 (*aromaticity*)
Schäfer, W. *et al*, *J. Am. Chem. Soc.*, 1976, **98**, 407 (*pe*)

P-Methylphosphonamidic acid, 9CI — M-00283

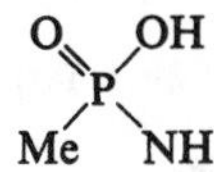

CH_6NO_2P M 95.038

Et ester: [19280-65-0]. *Ethyl* P-*methylphosphonamidate.*
$C_3H_{10}NO_2P$ M 123.091
Solid. Mp 50-53°. $Bp_{0.15}$ 112-114°.

Ph ester: [19280-64-9]. *Phenyl* P-*methylphosphonamidate.*
$C_7H_{10}NO_2P$ M 171.135
Solid. Mp 98°.

Fluoride: [29070-43-7]. P-*Methylphosphonamidic fluoride.*
CH_5FNOP M 97.029
Solid. Mp 38°.

N-*Me: see N,P*-Dimethylphosphonamidic acid, D-00849
N,N-Di-Me: see *N,N,P*-Trimethylphosphonamidic acid, T-00537
N-Et: see *N*-Ethyl-*P*-methylphosphonamidic acid, E-00088
N,N-Di-Et: see *N,N*-Diethyl-*P*-methylphosphonamidic acid, D-00296

Petrov, K.A. *et al*, *Zh. Obshch. Khim.*, 1960, **30**, 4060 (*Engl. transl.* p. 4023) (*phenyl ester, synth*)
Shokol, V.A. *et al*, *Zh. Obshch. Khim.*, 1968, **38**, 871, 1867; 1969, **39**, 1485 (*Engl. transl.* pp. 836, 1815, 1455) (*esters, derivs*)
Roesky, H.W. *et al*, *Z. Anorg. Allg. Chem.*, 1970, **375**, 140 (*fluoride, ir, ms, pmr, F nmr*)

P-Methylphosphonamidothioic acid, 9CI — M-00284

CH_6NOPS M 111.098

S-*Me ester:* [67242-52-8]. S-*Methyl* P-*methylphosphonamidothioate.*
C_2H_8NOPS M 125.125
Solid. Mp 76-78°.

O-*Me ester:* [40334-37-0]. O-*Methyl* P-*methylphosphonamidothioate.*
C_2H_8NOPS M 125.125
No phys. props. reported.

S-*Et ester:* [65331-55-7]. S-*Ethyl* P-*methylphosphonamidothioate.*
$C_3H_{10}NOPS$ M 139.152
Solid. Mp 44-45°.

O-*Et ester:* [34255-89-5]. O-*Ethyl* P-*methylphosphonamidothioate.*
$C_3H_{10}NOPS$ M 139.152
Liq. d_4^{25} 1.14. $Bp_{0.4}$ 90-91°. n_D^{25} 1.5170.

Fluoride: [21693-81-2].
CH_5FNPS M 113.089
Liq. d_4^{20} 1.33. $Bp_{0.02}$ 44-46°. n_D^{20} 1.5110.

Roesky, H.W., *Z. Naturforsch., B*, 1969, **24**, 5 (*fluoride, synth, ir, pmr, F nmr*)
Razvodovskaja, L.V. *et al*, *Zh. Obshch. Khim.*, 1971, **41**, 1446 (*Engl. transl.* p. 1452) (*ester, synth, props*)
Hammock, B.D. *et al*, *Pestic. Biochem. Physiol.*, 1977, **7**, 517 (*esters, tox*)
Wustner, D.A. *et al*, *J. Agric. Food Chem.*, 1978, **26**, 1104 (*S-alkyl esters, synth, tox*)

1-Methylphosphonane, 10CI — M-00285

1-Methylphosphacyclononane

[80461-84-3]

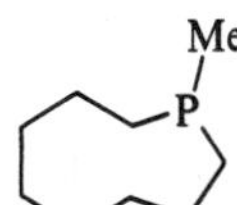

$C_9H_{19}P$ M 158.223

Liq.

B,MeI: [75401-44-4]. *1,1-Dimethylphosphonanium iodide.*
$C_{10}H_{22}IP$ M 300.162
Cryst. (EtOH). Mp 265-267°.

1-Oxide: [75401-31-9].
$C_9H_{19}OP$ M 174.222
Hygroscopic solid by subl. Mp 117-120°.

Quin, L.D. *et al*, *J. Org. Chem.*, 1982, **47**, 905 (*synth, derivs, pmr, cmr, P nmr*)

Methylphosphonazidic acid — M-00286

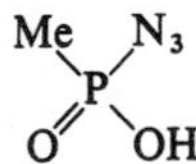

$CH_4N_3O_2P$ M 121.035

Et ester: [31650-79-0]. *Ethyl methylphosphonazidate.*
$C_3H_8N_3O_2P$ M 149.089
Liq. d_4^{20} 1.19. $Bp_{0.1}$ 39°. n_D^{20} 1.4480.

Ph ester: [31683-58-6]. *Phenyl methylphosphonazidate.*
$C_7H_8N_3O_2P$ M 197.133

Liq. d_4^{20} 1.26. $Bp_{0.04}$ 80°. n_D^{20} 1.5289.
Chloride: [33185-66-9].
CH_3ClN_3OP M 139.481
Liq. d_4^{20} 1.41. $Bp_{0.03}$ 46-47°. n_D^{20} 1.4850.

Shokol, V.A. *et al, Zh. Obshch. Khim.*, 1970, **40**, 1680; 1971, **41**, 545 (*Engl. transl.* pp. 1668, 539) (*synth, P nmr*)
Gilyarov, V.A. *et al, Zh. Obshch. Khim.*, 1972, **42**, 2148 (*Engl. transl.* p. 2145) (*synth*)

P-Methylphosphonazidic amide, 9CI   M-00287

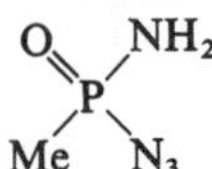

CH_5N_4OP M 120.050
N,N-*Di-Et:* [33078-26-1]. N,N-*Diethyl*-P-*methylphosphonazidic amide.*
$C_5H_{13}N_4OP$ M 176.158
Liq. d_4^{20} 1.09. $Bp_{0.03}$ 76-77°. n_D^{20} 1.4680.
N-*Ph:* P-*Methyl*-N-*phenylphosphonazidic amide.* P-*Methylphosphonazidic anilide.*
$C_7H_9N_4OP$ M 196.148
Viscous liq. n_D^{20} 1.5740.

Shokol, V.A. *et al, Zh. Obshch. Khim.*, 1971, **41**, 545 (*Engl. transl.* p. 539) (*synth, P nmr*)

Methylphosphonic acid, 9CI   M-00288

Methanephosphonic acid
[993-13-5]

MeP(O)(OH)₂

CH_5O_3P M 96.022
Hygroscopic cryst. Sol. H_2O, EtOH, Et_2O. Mp 105°. pK_a 2.68, 6.32 (H_2O).
▷Highly irritant. Corrosive
Anilinium salt: Cryst. (propanol). Mp 149-150°.
Mono-Me ester: see Methyl methylphosphonate, M-00168
Di-Me ester: see Dimethyl methylphosphonate, D-00762
Mono-Et ester: see Ethyl methylphosphonate, E-00089
Di-Et ester: see Diethyl methylphosphonate, D-00298
Monopropyl ester: [4546-11-6]. *Hydrogen propyl methylphosphonate.*
$C_4H_{11}O_3P$ M 138.103
Liq. $Bp_{0.05}$ 106°. n_D^{25} 1.4259.
Dipropyl ester: [6410-56-6]. *Dipropyl methylphosphonate.*
$C_7H_{17}O_3P$ M 180.183
Liq. d_4^{20} 1.01. Bp_9 93-96°, $Bp_{0.25}$ 115.5°. n_D^{20} 1.4210.
Di-2-propenyl ester: Di-2-propenyl methylphosphonate. Diallyl methylphosphonate.
$C_7H_{13}O_3P$ M 176.152
Liq. d_{25}^{25} 1.04. $Bp_{0.5}$ 77-85°. n_D^{25} 1.4468.
Monoisopropyl ester: see Isopropyl methylphosphonate, I-00038
Diisopropyl ester: [1445-75-6]. *Diisopropyl methylphosphonate.*
$C_7H_{17}O_3P$ M 180.183
Forms complexes with many metal salts. Bp_3 66°. $n_D^{16.5}$ 1.4120. Nontoxic; nonmutagenic when pure.
▷SZ9090000.
Dibutyl ester: [2404-73-1]. *Dibutyl methylphosphonate.*
$C_9H_{22}O_3P$ M 209.245
Extractant for metals. Liq. d_4^{20} 0.95. Bp_{15} 132-135°, $Bp_{1.5}$ 86-91°. n_D^{20} 1.4240.
Di-tert-*butyl ester:* [17123-05-6]. *Di*-tert-*butyl methylphosphonate.*
$C_9H_{22}O_3P$ M 209.245
Oil.
Monocyclohexyl ester: [1932-60-1]. *Cyclohexyl hydrogen methylphosphonate.*
$C_7H_{15}O_3P$ M 178.167
Mp 45-48°.
Mono-Ph ester: see Phenyl methylphosphonate, P-00171
Di-Ph ester: see Diphenyl methylphosphonate, D-01024
Bis(trimethylsilyl) ester: [18279-83-9].
$C_7H_{21}O_3PSi_2$ M 240.386
Liq. Bp_{27} 105-107.5°.
Difluoride: see Methylphosphonic difluoride, M-00291
Dichloride: see Methylphosphonic dichloride, M-00290
Dibromide: [19430-64-9].
CH_3Br_2OP M 221.816
Liq. d_4^{20} 2.43. Bp_{728} 191-195°, $Bp_{0.5}$ 58°. n_D^{20} 1.5829.
Dicyanide: [31641-59-5].
$C_3H_3N_2OP$ M 114.043
No phys. props. reported.
Diisocyanate: [1068-20-8].
$C_3H_3N_2O_3P$ M 146.042
Liq. d_4^{20} 1.33. Bp_2 71°. n_D^{20} 1.4680.
Diisothiocyanate: [4519-65-7].
$C_3H_3N_2OPS_2$ M 178.163
Liq. d^{20} 1.37. Bp_{18} 148°, $Bp_{0.05}$ 99-102°. n_D^{20} 1.6215.
Diazide: [33078-24-9].
CH_3N_6OP M 146.048
Liq. d_4^{20} 1.38. $Bp_{0.03}$ 49-50°. n_D^{20} 1.4980.
Diamide: see P-Methylphosphonic diamide, M-00289

Ford-Moore, A.H. *et al, J. Chem. Soc.*, 1947, 1465 (*synth, esters*)
Crofts, P.C. *et al, J. Am. Chem. Soc.*, 1953, **75**, 3379 (*synth, esters*)
Maier, L., *Helv. Chim. Acta*, 1963, **46**, 2667 (*dibromide, ir, P nmr*)
Rabinowitz, R., *J. Org. Chem.*, 1963, **28**, 2975 (*silyl ester, synth*)
Petrov, K.A. *et al, Zh. Obshch. Khim.*, 1964, **34**, 2586 (*Engl. transl.* p. 2608) (*dipropyl ester*)
Bai, L.I. *et al, Zh. Obshch. Khim.*, 1964, **34**, 3609 (*Engl. transl.* p. 3656) (*diisocyanate*)
Mark, V. *et al, J. Org. Chem.*, 1964, **29**, 1006 (*di-tert*-butyl ester, synth, pmr, ir)
Petrov, K.A. *et al, Zh. Obshch. Khim.*, 1965, **35**, 723 (*Engl. transl.* p. 723) (*monocyclohexyl ester*)
Cadogan, J.I.G. *et al, J. Chem. Soc.* (*B*), 1971, 1988 (*monoalkyl esters*)
Shokol, V.A. *et al, Zh. Obshch. Khim.*, 1971, **41**, 545 (*Engl. transl.* p. 539) (*diazide, synth, ir*)
Riess, J.G. *et al, Bull. Soc. Chim. Fr.*, 1972, 3700 (*dibromide, diisothiocyanate, synth, ir, pmr*)
Harvey, D.J. *et al, Org. Mass Spectrom.*, 1974, **9**, 111 (*silyl ester, ms*)
Tebby, J.C. *et al, Phosphorus*, 1975, **5**, 273 (*ms*)
Rueppel, M.L. *et al, Org. Magn. Reson.*, 1976, **8**, 19 (*P nmr, pmr*)
Kodolov, V.I. *et al, Izv. Akad. Nauk SSSR, Ser. Khim.*, 1977, 165 (*Engl. transl.* p. 142) (*pe*)
Pudovik, A.N. *et al, Zh. Obshch. Khim.*, 1979, **49**, 1425 (*Engl. transl.* p. 1248) (*diisothiocyanate*)
Sass, S. *et al, Org. Mass Spectrom.*, 1979, **14**, 257 (*diisopropyl ester, ms*)
Holtzclaw, J.R. *et al, Org. Mass Spectrom.*, 1985, **20**, 90 (*dimethyl ester, ms, struct*)

Sax, N.I., *Dangerous Properties of Industrial Materials*, 6th Ed., Van Nostrand-Reinhold, 1984, 830.

P-Methylphosphonic diamide, 9CI M-00289

[4759-30-2]

$MeP(O)(NH_2)_2$

CH_7N_2OP M 94.053

Employed, as are many of its derivs., in fire-proofing preparations. Plates (MeOH). Mp 128-129° (sealed tube). Hydrolysed in damp atmos. to the diammonium salt of methylphosphonic acid.

N,N′-*Di-Me:* [67704-60-3]. N,N′,P-*Trimethylphosphonic diamide. Methylphosphonic bis(methylamide).*
$C_3H_{11}N_2OP$ M 122.106
Plates (C_6H_6). Mp 63-65°.

N,N,N′,N′-*Tetra-Me:* [2511-17-3]. *Pentamethylphosphonic diamide. Methylphosphonic bis(dimethylamide).*
$C_5H_{15}N_2OP$ M 150.160
Liq. d_4^{30} 1.02. Bp_{32} 138°. n_D^{30} 1.4539.

N,N,N′,N′-*Tetra-Et:* N,N,N′,N′-*Tetraethyl*-P-*methylphosphonic diamide. Methylphosphonic bis(diethylamide).*
$C_9H_{23}N_2OP$ M 206.267
Liq. Bp_{33} 166°. n_D^{30} 1.4565.

N,N′-*Di*-tert-*butyl:* N,N′-*Di*-tert-*butyl*-P-*methylphosphonic diamide. Methylphosphonic bis*(tert-*butylamide*).
$C_9H_{23}N_2OP$ M 206.267
Cryst. (ligroin). Mp 99-100°.

N,N′-*Di-Ph:* [4653-50-3]. P-*Methyl*-N,N′-*diphenylphosphonic diamide. Methylphosphonic dianilide.*
$C_{13}H_{15}N_2O_5$ M 279.272
Cryst. ($C_6H_6/CHCl_3$). Mp 156.5-158.5°.

Kinnear, A.M. *et al, J. Chem. Soc.*, 1952, 3437 (*dianilide*)
Rätz, R., *J. Am. Chem. Soc.*, 1955, **77**, 4170 (*synth*)
Kosolapoff, G.M. *et al, J. Org. Chem.*, 1956, **21**, 413 (*derivs*)
Helferich, B. *et al, Justus Liebigs Ann. Chem.*, 1963, **670**, 48 (*deriv*)
Mel'nikov, N.N. *et al, Zh. Obshch. Khim.*, 1965, **35**, 1771 (*Engl. transl.* p. 1769) (*dianilide*)
Quast, H. *et al, Justus Liebigs Ann. Chem.*, 1981, 943 (*deriv, synth, ir, pmr, P nmr, ms*)

Methylphosphonic dichloride, 9CI M-00290

[676-97-1]

$MeP(O)Cl_2$

CH_3Cl_2OP M 132.914

Dehydrating agent, synthetic intermediate. Reagent for detn. of enantiomeric excess in chiral thiols. Low melting solid with pungent odour. Easily hydrolysed. Mp 32°. Bp 162°, Bp_{53} 98°, Bp_{17} 55°.

▷TA1840000.

Durig, J.R. *et al, Spectrochim. Acta*, 1965, **21**, 1105 (*ir, raman*)
Geiseler, G. *et al, Ber. Bunsenges. Phys. Chem.*, 1967, **71**, 478 (*ir, raman*)
Maier, L., *Helv. Chim. Acta*, 1973, **56**, 492 (*synth, pmr, P nmr*)
Bel'skii, V.E. *et al, Dokl. Akad. Nauk SSSR, Ser. Sci. Khim.*, 1974, **215**, 355 (*Engl. transl.* p. 260) (*ir, nqr*)
Moedritzer, K. *et al, Synth. React. Inorg. Metal-Org. Chem.*, 1974, **4**, 417 (*synth, pmr, P nmr*)
Quast, H. *et al, Synthesis*, 1974, 490 (*synth*)
Elbel, S. *et al, J. Chem. Soc., Dalton Trans.*, 1976, 1762 (*pe*)
Griffiths, W.R. *et al, Phosphorus Sulfur*, 1978, **4**, 341 (*ms*)
Zverev, V.V. *et al, Zh. Obshch. Khim.*, 1979, **49**, 1737; 1980, **50**, 2690 (*Engl. transl.* pp. 1522, 2172) (*pe, struct*)
Morita, T. *et al, Chem. Lett.*, 1980, 435 (*synth*)
Stritjtven, B. *et al, Tetrahedron*, 1987, **43**, 123 (*use*)

Methylphosphonic difluoride, 9CI M-00291

[676-99-3]

$MeP(O)F_2$

CH_3F_2OP M 100.005

Liq. d_4^{20} 1.33 (1.38). Bp 98°, Bp_{27} 22°. n_D^{20} 1.3165.

▷TA1840700.

Lysenko, V.V. *et al, Zh. Obshch. Khim.*, 1966, **36**, 1507 (*Engl. transl.* p. 1512) (*synth*)
Köttgen, D. *et al, Z. Anorg. Allg. Chem.*, 1972, **389**, 269 (*ir, raman*)
Durig, J.R. *et al, J. Mol. Struct.*, 1976, **34**, 9 (*ir, raman, microwave*)
Fokin, A.V. *et al, Izv. Akad. Nauk SSSR, Ser. Khim.*, 1976, 2435 (*Engl. transl.* p. 2271) (*F and P nmr*)
Zverev, V.V. *et al, Zh. Obshch. Khim.*, 1984, **54**, 1265 (*Engl. transl.* p. 1131) (*pe*)

Methylphosphonisocyanatidic acid, 9CI M-00292

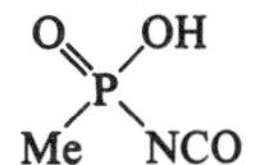

$C_2H_4NO_3P$ M 121.032

Et ester: [19081-02-8]. *Ethyl methylphosphonisocyanatidate.*
$C_4H_8NO_3P$ M 149.086
Liq. d^{20} 1.19. Bp_{10} 80°. n_D^{20} 1.4345.

Ph ester: [17848-02-1]. *Phenyl methylphosphonisocyanatidate.*
$C_8H_8NO_3P$ M 197.130
Liq. d_4^{20} 1.26. $Bp_{0.07}$ 105-108°. n_D^{20} 1.5192.

Amide: see P-Methylphosphonisocyanatidic amide, M-00293

Azide: [33078-25-0].
$C_2H_3N_4O_2P$ M 146.045
Liq. d_4^{20} 1.40. $Bp_{0.04}$ 49-50°. n_D^{20} 1.4835.

Derkach, G.I. *et al, Zh. Obshch. Khim.*, 1967, **37**, 2069 (*Engl. transl.* p. 1961) (*synth*)
Gubnitskaya, E.S. *et al, Zh. Obshch. Khim.*, 1968, **38**, 1530 (*Engl. transl.* p. 1479) (*synth*)
Shokol, V.A. *et al, Zh. Obshch. Khim.*, 1969, **39**, 2197 (*Engl. transl.* p. 2146) (*synth*)
Shokol, V.A. *et al, Zh. Obshch. Khim.*, 1971, **41**, 545 (*Engl. transl.* p. 539) (*synth, nmr, azide*)

P-Methylphosphonisocyanatidic amide, 9CI M-00293

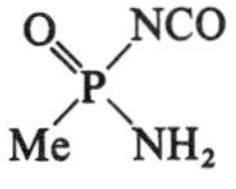

$C_2H_5N_2O_2P$ M 120.047

N,N-*Di-Me:* [17848-04-3]. N,N,P-*Trimethylphosphonisocyanatidic amide.*
$C_4H_9N_2O_2P$ M 148.101
Liq. d_4^{20} 1.17. $Bp_{0.05}$ 49-51°. n_D^{20} 1.4690.

N,N-*Di-Et:* [17848-05-4]. N,N-*Diethyl*-P-*methylphosphonisocyanatidic amide.*
$C_5H_{13}N_2O_2P$ M 164.144
Liq. d_4^{20} 1.11. $Bp_{0.04}$ 76-78°. n_D^{20} 1.4555.

Derkach, G.I. *et al, Zh. Obshch. Khim.*, 1967, **37**, 2069 (*Engl. transl.* p. 1961) (*synth, ir*)

Methylphosphonisothiocyanatidic acid, 9CI M-00294

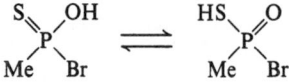

$C_2H_4NO_2PS$ M 137.093

Me ester: [13298-14-1]. *Methyl methylphosphonisothiocyanatidate.*
$C_3H_6NO_2PS$ M 151.120
Liq. d_4^{20} 1.27. $Bp_{0.3}$ 70°. n_D^{20} 1.5192.

Et ester: [13298-15-2]. *Ethyl methylphosphonisothiocyanatidate.*
$C_4H_8NO_2PS$ M 165.146
Liq. d_4^{20} 1.20. $Bp_{0.5}$ 57-60°. n_D^{20} 1.5065.

Isopropyl ester: [13298-19-4]. *Isopropyl methylphosphonisothiocyanatidate.*
$C_5H_{10}NO_2PS$ M 179.173
Liq. d_4^{20} 1.14. $Bp_{0.5}$ 80°. n_D^{20} 1.4978.

Ph ester: [21100-88-9]. *Phenyl methylphosphonisothiocyanatidate.*
$C_8H_8NO_2PS$ M 213.190
Liq. d_4^{20} 1.26. $Bp_{0.1}$ 98-100°. n_D^{20} 1.5734.

Ivanova, Zh.M. *et al*, *Zh. Obshch. Khim.*, 1966, **36**, 162 (*Engl. transl.* p. 169) (*synth*)
Derkach, G.I. *et al*, *Zh. Obshch. Khim.*, 1968, **38**, 1779 (*Engl. transl.* p. 1734) (*synth*)

Methylphosphonobromidothioic acid, 9CI M-00295

CH_4BrOPS M 174.980

O-*Et ester:* O-*Ethyl methylphosphonobromidothioate.*
C_3H_8BrOPS M 203.033
Liq. d^{20} 1.47. Bp_{20} 86-87°. n_D^{20} 1.5290.

O-*Isopropyl ester:* O-*Isopropyl methylphosphonobromidothioate.*
$C_4H_{10}BrOPS$ M 217.060
Liq. Bp_5 73-74°.

Maier, L., *Helv. Chim. Acta*, 1964, **47**, 1448 (*synth, nmr*)

Methylphosphonochloridic acid, 9CI M-00296

[42408-72-0]

(*R*)-*form of esters*

CH_4ClO_2P

(*R*)-form

Isopropyl ester: Isopropyl methylphosphonochloridate.
$C_4H_{10}ClO_2P$
Liq. $Bp_{1.2}$ 37°. $[\alpha]_D^{27}$ −52.3° (c, 2.8 in C_6H_6). n_D^{25} 1.4281.

(*S*)-form

Isopropyl ester: $Bp_{3.5}$ 48-49°. $[\alpha]_D^{31}$ +52.8° (C_6H_6). n_D^{25} 1.4289.

(±)-form

Me ester: [1066-52-0]. *Methyl methylphosphonochloridate.*
$C_2H_6ClO_2P$
Liq. d_4^{16} 1.30. Bp_{22} 73°. n_D^{18} 1.4395.

Et ester: [5284-09-3]. *Ethyl methylphosphonochloridate.*
$C_3H_8ClO_2P$
Liq. d_4^{20} 1.21. Bp_{21} 83°. n_D^{25} 1.4385.

Isopropyl ester: [1445-76-7]. Liq. d_4^{21} 1.15. Bp_{22} 83°, Bp_1 40°. n_D^{23} 1.4285.

Ph ester: [14235-74-6]. *Phenyl methylphosphonochloridate.*
$C_7H_8ClO_2P$ M 190.566
Liq. d_4^{20} 1.28. Bp_{12} 125-127°. n_D^{20} 1.5280.

Gefter, E.L., *Zh. Obshch. Khim.*, 1961, **31**, 3316 (*Engl. transl.* p. 3093) (*phenyl ester*)
Pelchowitz, Z., *J. Chem. Soc.*, 1961, 238.
Aaron, H.S. *et al*, *J. Am. Chem. Soc.*, 1962, **84**, 617 (*isopropyl ester*)
Pudovik, A.N. *et al*, *Zh. Obshch. Khim.*, 1964, **34**, 2213.
Shokol, V.A. *et al*, *Zh. Obshch. Khim.*, 1969, **39**, 2197 (*Engl. transl.* p. 2146) (*phenyl ester*)
Gladshtein, B.M. *et al*, *Zh. Obshch. Khim.*, 1970, **40**, 1245 (*Engl. transl.* p. 1236) (*isopropyl ester*)

Methylphosphonochloridodithioic acid, 9CI M-00297

MePCl(S)SH

CH_4ClPS_2 M 146.589

Me ester: [16284-71-2]. *Methyl methylphosphonochloridodithioate.*
$C_2H_6ClPS_2$ M 160.616
Liq. d_4^{20} 1.32. Bp_{23} 106.5-107°, $Bp_{0.4}$ 58-62°. n_D^{20} 1.6013.

▷Lachrymator

tert-Butyl ester: [79220-12-5]. S-tert-*Butyl methylphosphonochloridodithioate.*
$C_5H_{12}ClPS_2$ M 202.697
Liq. $Bp_{0.2}$ 72-75°.

Ph ester: [22740-38-1]. *Phenyl methylphosphonochloridodithioate.*
$C_7H_8ClPS_2$ M 222.687
Solid. Mp 43-44°. Bp_4 153-156°, $Bp_{0.04}$ 106-107°.

Moedritzer, K. *et al*, *Inorg. Chem.*, 1963, **2**, 1152 (*synth, pmr, P nmr*)
Gryszkiewicz-Trochimowski, E., *Bull. Soc. Chim. Fr.*, 1967, 2232 (*methyl ester*)
Nesterov, L.V. *et al*, *Zh. Obshch. Khim.*, 1970, **40**, 1237 (*Engl. transl.* p. 1228) (*phenyl ester, ir, pmr, P nmr*)
Ger. Pat., 1 956 187, (*1971*); *CA*, **75**, 77028 (*synth*)
Ger. Pat., 2 714 771, (*1978*); *CA*, **90**, 23254 (*synth*)
European Pat. 25 270, (*1981*); *CA*, **95**, 150886 (*tert-butyl ester, synth, use*)

Methylphosphonochloridothioic acid, 9CI M-00298

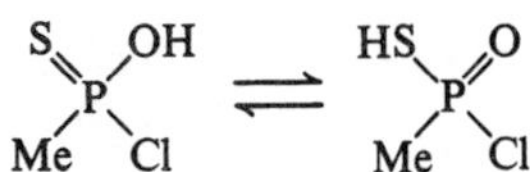

CH_4ClOPS M 130.529

O-*Me ester: see O-Methyl methylphosphonochloridothioate, M-00169*

S-*Me ester:* [13113-89-8]. S-*Methyl methylphosphonochloridothioate.*
C_2H_6ClOPS M 144.556
Liq. d_4^{20} 1.34. Bp_2 70-1°. n_D^{20} 1.5215.

O-*Et ester: see O-Ethyl methylphosphonochloridothioate, E-00090*

S-*Et ester:* [13113-90-1]. S-*Ethyl methylphosphonochloridothioate.*
C_3H_8ClOPS M 158.582

Liq. d_4^{20} 1.26. Bp_{10} 93-95°. n_D^{20} 1.5302 (1.5127).
O-*Propyl ester: see O-Propyl methylphosphonochloridothioate, P-00468*
S-*Propyl ester:* [13113-91-2]. S-*Propyl methylphosphonochloridothioate.*
$C_4H_{10}ClOPS$ M 172.609
Liq. d_4^{20} 1.21. Bp_5 97°. n_D^{20} 1.5039.
O-*Isopropyl ester: see O-Isopropyl methylphosphonochloridothioate, I-00039*
S-*Isopropyl ester:* [13113-98-9]. S-*Isopropyl methylphosphonochloridothioate.*
$C_4H_{10}ClOPS$ M 172.609
Liq. d_4^{20} 1.20. Bp_5 78-9°. n_D^{20} 1.5024.
O-*Butyl ester: see O-Butyl methylphosphonochloridothioate, B-00573*
S-*Butyl ester:* [38751-14-3]. S-*Butyl methylphosphonochloridothioate.*
$C_5H_{12}ClOPS$ M 186.636
Liq. $Bp_{0.25}$ 69.5-70°. n_D^{25} 1.4997.
O-*Ph ester:* [14410-07-2]. O-*Phenyl methylphosphonochloridothioate.*
C_7H_8ClOPS M 206.626
Liq. d_4^{20} 1.30. Bp_{10} 121-123°, $Bp_{0.16}$ 97°. n_D^{20} 1.5740.
S-*Ph ester:* [20433-63-0]. S-*Phenyl methylphosphonochloridothioate.*
C_7H_8ClOPS M 206.626
Liq. d_4^{20} 1.34. $Bp_{0.04}$ 116-118°. n_D^{20} 1.5892.

Chupp, J.P. *et al*, *J. Org. Chem.*, 1962, **27**, 3832 (*synth*)
Moedritzer, K. *et al*, *Inorg. Chem.*, 1963, **2**, 1152 (*synth, pmr, P nmr*)
Neimysheva, A.A. *et al*, *Zh. Obshch. Khim.*, 1966, **36**, 500 (*Engl. transl.* p. 520) (*synth*)
Shitov, L.N. *et al*, *Zh. Obshch. Khim.*, 1968, **38**, 2340 (*Engl. transl.* p. 2268) (*phenyl ester*)
Nesterov, L.V. *et al*, *Zh. Obshch. Khim.*, 1970, **40**, 1237 (*Engl. transl.*, p. 1228) (*synth, ir, pmr, P nmr*)
Gubaidullin, M.G., *Zh. Obshch. Khim.*, 1977, **47**, 2659 (*Engl. transl.* p. 2428) (*props*)

Methylphosphonochloridothious acid, 9CI M-00299

MePCl(SH)

CH_4ClPS M 114.529
Me ester: Methyl methylphosphonochloridothioite.
C_2H_6ClPS M 128.556
Liq. Bp_{12} 45°.
Et ester: Ethyl methylphosphonochloridothioite.
C_3H_8ClPS M 142.583
Liq. d_4^{20} 1.13. $Bp_{0.8}$ 40-41°. n_D^{20} 1.5321.
Butyl ester: Butyl methylphosphonochloridothioite.
$C_5H_{12}ClPS$ M 170.637
Liq. d_4^{20} 1.09. Bp_1 58-59°. n_D^{20} 1.5198.
Ph ester: [23588-03-6]. *Phenyl methylphosphonochloridothioite.*
C_7H_8ClPS M 190.627
Liq. d_4^{20} 1.23. Bp_2 111-112°. n_D^{20} 1.6164.

Shitov, L.N. *et al*, *Zh. Obshch. Khim.*, 1969, **39**, 1251 (*Engl. transl.* p. 1220) (*esters, synth, P nmr*)

Methylphosphonochloridous acid, 9CI M-00300

MePCl(OH)

CH_4ClOP M 98.469
Esters are highly unstable.
Me ester: [51934-48-6]. *Methyl methylphosphonochloridite.*
C_2H_6ClOP M 112.496
No phys. data reported.

Abraham, K.M. *et al*, *Inorg. Chem.*, 1974, **13**, 2346.

Methylphosphonodithioic acid, 9CI M-00301

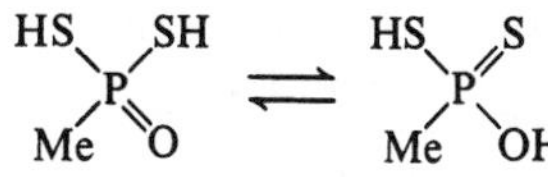

CH_5OPS_2 M 128.144

(*R*)-*form*

O,S-*Di-Et ester:* Liq. $Bp_{0.5}$ 49°. $[\alpha]_D$ −72.0° (neat). n_D^{21} 1.5238.

(*S*)-*form*

O,S-*Di-Et ester:* Liq. $[\alpha]_D$ +62.00° (neat).

(±)-*form*

O-*Me ester: see O-Methyl methylphosphonodithioate, M-00170*
O,S-*Di-Me ester:* [54565-44-5]. O,S-*Dimethyl methylphosphonodithioate.*
$C_3H_9OPS_2$ M 156.197
Liq. d_4^{20} 1.20. Bp_1 35-8°. n_D^{20} 1.5568.
S,S-*Di-Me ester:* [40145-83-3]. S,S-*Dimethyl methylphosphonodithioate.*
$C_3H_9OPS_2$ M 156.197
Liq. $Bp_{2.5}$ 85°.
O-*Et ester: see O-Ethyl methylphosphonodithioate, E-00091*
O,S-*Di-Et ester:* [3347-31-7]. O,S-*Diethyl methylphosphonodithioate.*
$C_5H_{13}OPS_2$ M 184.251
Liq. d_4^{20} 1.11. Bp_3 85-7°. n_D^{20} 1.5259.
S,S-*Di-Et ester:* [995-88-0]. S,S-*Diethyl methylphosphonodithioate.*
$C_5H_{13}OPS_2$ M 184.251
Liq. Bp_{12} 120-3°. n_D^{20} 1.5442.
O-*Propyl ester: see O-Propyl methylphosphonodithioate, P-00469*
O-*Isopropyl ester: see O-Isopropyl methylphosphonodithioate, I-00040*
S-*Et,* O-*isopropyl ester: see under O-Isopropyl methylphosphonodithioate, I-00040*
O-*Ph ester: see O-Phenyl methylphosphonodithioate, P-00172*
S,S-*Di-Ph ester:* [31650-50-7]. S,S-*Diphenyl methylphosphonodithioate.*
$C_{13}H_{13}OPS_2$ M 280.339
Cryst. (CH_2Cl_2). Mp 82-83°.

Nesterov, L.V. *et al*, *Zh. Obshch. Khim.*, 1970, **40**, 1237 (*Engl. transl.* p. 1228) (*S,S-diethyl and S,S-diphenyl esters, synth, ir, P nmr*)

Omelańczuk, J. *et al*, *Tetrahedron*, 1971, **27**, 5587 (*O,S-diethyl ester, config*)
Pantzer, R. *et al*, *Z. Anorg.-Allg. Chem.*, 1973, **395**, 262 (*S,S-dimethyl ester, synth, ir, raman*)
Ishmaeva, É.A. *et al*, *Zh. Obshch. Khim.*, 1974, **44**, 2625 (*Engl. transl.* p. 2582) (*O,S-dimethyl and O,S-diethyl esters, synth, conformn*)
Hall, C.D. *et al*, *J. Chem. Soc., Perkin Trans. 2*, 1977, 1232 (*esters, nmr*)
Bo Long, P., *Aust. J. Chem.*, 1983, **36**, 1027 (*diisopropyl ester, config*)

Methylphosphonodithious acid, 9CI **M-00302**

MeP(SH)$_2$

CH_5PS_2 M 112.144

Di-Me ester: [75956-77-3]. *Dimethyl methylphosphonodithioite.*
$C_3H_9PS_2$ M 140.198
No phys. props. reported.

Di-Et ester: [999-02-0]. *Diethyl methylphosphonodithioite.* Liq. with unpleasant odour. d_4^{20} 1.04 (1.06). Bp_9 85-87°. n_D^{20} 1.5575.

Di-Ph ester: [38476-65-2]. *Diphenyl methylphosphonodithioite.*
$C_{13}H_{13}PS_2$ M 264.339
Liq. d_4^{20} 1.20. $Bp_{0.007}$ 115-120° (bath). n_D^{20} 1.6701.

Dibenzyl ester: Dibenzyl methylphosphonodithioite. Bis-(phenylmethyl) methylphosphonodithioite.
$C_{15}H_{17}PS_2$ M 292.393
Liq. d_4^{20} 1.16. $Bp_{0.002}$ 150-150° (bath). n_D^{20} 1.6328.

Petrov, K.A. *et al*, *Zh. Obshch. Khim.*, 1962, **32**, 3070 (*Engl. transl.* p. 3019) (*diphenyl, dibenzyl esters, synth*)
Razumov, A.I. *et al*, *Zh. Obshch. Khim.*, 1972, **42**, 1250 (*Engl. transl.* p. 1245) (*diethyl, diphenyl esters, synth, P nmr*)
Patsanovskii, I.I. *et al*, *Dokl. Akad. Nauk SSSR, Ser. Sci. Khim.*, 1980, **254**, 414 (*Engl. transl.* p. 771) (*dimethyl ester, ir, raman, conformn*)
Zverev, V.V. *et al*, *Zh. Obshch. Khim.*, 1983, **53**, 1968 (*Engl. transl.* p. 1775) (*dimethyl ester, pe*)

Methylphosphonofluoridic acid, 9CI **M-00303**

[1511-67-7]

MePF(O)OH

CH_4FO_2P M 98.014
Liq. $Bp_{0.15}$ 48°.

▷Esters are neurotoxic

Anilinium salt: Cryst. (dioxane). Mp 135° (softens at 125°).

Me ester: [353-88-8]. *Methyl methylphosphonofluoridate.*
$C_2H_6FO_2P$ M 112.040
Liq. Bp_{55} 71-72°.

Et ester: [673-97-2]. *Ethyl methylphosphonofluoridate.*
$C_3H_8FO_2P$ M 126.067
Liq. Bp_{20} 49°. n_D^{25} 1.3778.

Isopropyl ester: see Sarin, S-00002

1,2,2-Trimethylpropyl ester: see Soman, S-00005

Bebbington, A. *et al*, *J. Chem. Soc. (C)*, 1966, 1410 (*ester, synth*)
Landau, M.A. *et al*, *Zh. Strukt. Khim.*, 1970, **11**, 513 (*Engl. transl.* p. 467) (*esters, struct, P nmr*)
Reddy, G.S. *et al*, *Z. Naturforsch, B*, 1970, **25**, 1199 (*synth, pmr, ester, ir, F and P nmr*)
Bender, R. *et al*, *Phosphorus*, 1974, **4**, 183 (*synth, ir, nmr*)
Ryzhikov, B.D. *et al*, *Zh. Strukt. Khim.*, 1975, **16**, 754 (*Engl. transl.* p. 700) (*esters, F nmr*)
Mager, P.P., *Toxicol. Lett.*, 1982, **11**, 67 (*tox*)

Methylphosphonofluoridodithioic acid, 9CI **M-00304**

[21207-77-2]

MePF(S)SH

CH_4FPS_2 M 130.135
Liq. d^{20} 1.35. Bp_{12-14} 49-50°. n_D 1.5557.

Anhydrosulfide: [21207-81-8]. *Dimethylthiodiphosphonic acid difluoride.*
$C_2H_6F_2P_2S_3$ M 226.194
Liq. d^{20} 1.44. $Bp_{0.01}$ 55-56°. n_D^{20} 1.5847.

Roesky, H.W., *Chem. Ber.*, 1968, **101**, 3679 (*synth, ms, ir, pmr, F nmr*)
Harris, R.K. *et al*, *J. Chem. Soc., Dalton Trans.*, 1972, 1590 (*F and P nmr*)

Methylphosphonofluoridothioic acid, 9CI **M-00305**

Me–P(=S)(OH)F ⇌ Me–P(=O)(SH)F

CH_4FOPS M 114.074

O-*Me ester:* [20518-03-0]. O-*Methyl methylphosphonofluoridothioate.*
C_2H_6FOPS M 128.101
Liq. d_4^{20} 1.20. Bp_{23} 46-49°. n_D^{25} 1.4368.

O-*Et ester:* O-*Ethyl methylphosphonofluoridothioate.*
C_3H_8FOPS M 142.128
Liq. d_4^{20} 1.15. Bp_{15} 38-39°. n_D^{20} 1.4380.

S-*Et ester:* [673-98-3]. S-*Ethyl methylphosphonofluoridothioate.*
C_3H_8FOPS M 142.128
Characterised spectroscopically.

O-*Isopropyl ester:* [4241-37-6]. O-*Isopropyl methylphosphonofluoridothioate.*
$C_4H_{10}FOPS$ M 156.155
Liq. d_4^{20} 1.09. Bp_{14} 45°. n_D^{25} 1.4326.

O-*Butyl ester:* O-*Butyl methylphosphonofluoridothioate.*
$C_5H_{12}FOPS$ M 170.181
Bp_{10} 63°. n_D^{25} 1.4414.

Boter, H.L. *et al*, *Recl. Trav. Chim. Pays-Bas*, 1966, **85**, 27 (*O-alkyl esters, synth*)
Ivanova, Zh.M. *et al*, *Zh. Obshch. Khim.*, 1968, **38**, 1334 (*Engl. transl.* p. 1284) (*O-alkyl esters, synth*)
Landau, M.A. *et al*, *Zh. Strukt. Khim.*, 1970, **11**, 513 (*Engl. transl.* p. 467) (*S-ester, struct*)
Ryzhikov, B.D. *et al*, *Zh. Strukt. Khim.*, 1975, **16**, 754 (*Engl. transl.* p. 700) (*S-ester, F nmr, struct*)

Methylphosphonofluoridothious acid, 9CI **M-00306**

MePF(SH)

CH_4FPS M 98.075

Butyl ester: [22491-81-2]. *Butyl methylphosphonofluoridothioite.*
$C_5H_{12}FPS$ M 154.182
Liq. d_4^{20} 1.07. Bp_{55} 72-74°. n_D^{20} 1.4745.

Sheluchenko, V.V. *et al*, *Zh. Strukt. Khim.*, 1968, **9**, 909 (*Engl. transl.* p. 805) (*butyl, phenyl esters, nmr*)
Drozd, G.I. *et al*, *Zh. Obshch. Khim.*, 1969, **39**, 1417 (*Engl. transl.* p. 1385) (*butyl ester, synth, F and P nmr*)

Methylphosphonofluoridous acid, 9CI — M-00307

MePF(OH)

CH_4FOP M 82.014

Me ester: [22794-04-3]. *Methyl methylphosphonofluoridite.*
C_2H_6FOP M 96.041
Liq. d_4^{20} 1.02. Bp_{750} 56-58°. Readily oxid. in air.

Et ester: [20337-67-1]. *Ethyl methylphosphonofluoridite.*
C_3H_8FOP M 110.068
Liq. d_4^{20} 1.04. Bp_{750} 60-62°. n_D^{20} 1.3935.

Isopropyl ester: [22794-06-5]. *Isopropyl methylphosphonofluoridite.*
$C_4H_{10}FOP$ M 124.095
Liq. Bp_{750} 67-69°. n_D^{20} 1.3820.

Ph ester: [22788-91-6]. *Phenyl methylphosphonofluoridite.*
C_7H_8FOP M 158.112
Liq. d_4^{20} 1.13. Bp_{14} 58-59°. n_D^{20} 1.5020.

Kulakova, V.N. *et al*, *Zh. Obshch. Khim.*, 1969, **39**, 579 (*Engl. transl.* p. 547)

P-Methylphosphonohydrazidic acid, 9CI — M-00308

O=P(NHNH₂)(Me)(OH)

$CH_7N_2O_2P$ M 110.052

Et ester: [19233-71-7]. *Ethyl* P-*methylphosphonohydrazidate.*
$C_3H_{11}N_2O_2P$ M 138.106
Oil.

Isopropyl ester: [19233-70-6]. *Isopropyl* P-*methylphosphonohydrazidate.*
$C_4H_{13}N_2O_2P$ M 152.133
Cryst. Mp 62°.

2-Methylpropyl ester: [19233-68-2]. *2-Methylpropyl* P-*methylphosphonohydrazidate. Isobutyl* P-*methylphosphonohydrazidate.*
$C_5H_{15}N_2O_2P$ M 166.159
Oil. Bp_3 140°. n_D^{20} 1.4560.

Cyclohexyl ester: [19233-65-9]. *Cyclohexyl* P-*methylphosphonohydrazidate.*
$C_7H_{17}N_2O_2P$ M 192.197
Yellow cryst. Mp 77°.

N^2-*Ph: see P-Methyl-N^2-phenylphosphonohydrazidic acid, M-00241*

Énglin, M.A. *et al*, *Zh. Obshch. Khim.*, 1968, **38**, 869 (*Engl. transl.* p. 833) (*synth*)

P-Methylphosphonohydrazidothioic acid, 9CI — M-00309

S=P(NHNH₂)(Me)(OH) ⇌ HS–P(NHNH₂)(Me)=O

CH_7N_2OPS M 126.113

O-*Et ester:* [19233-61-5]. O-*Ethyl* P-*methylphosphonohydrazidothioate.*
$C_3H_{11}N_2OPS$ M 154.166
Oil. $Bp_{0.06}$ 72-76°. n_D^{20} 1.5344.

O-*Ph ester:* [34421-23-3]. O-*Phenyl* P-*methylphosphonohydrazidothioate.*
$C_7H_{11}N_2OPS$ M 202.210
Cryst. (EtOH). Mp 45-47°.

Petrov, K.A. *et al*, *Zh. Obshch. Khim.*, 1970, **40**, 1234 (*Engl. transl.* p. 1225) (*ethyl ester*)
Grapov, A.F. *et al*, *Zh. Obshch. Khim.*, 1971, **41**, 1441 (*Engl. transl.* p. 1447) (*phenyl ester*)

Methylphosphonoperoxoic acid, 9CI — M-00310

MeP(O)(OH)OOH

CH_5O_4P M 112.022

OO-tert-*Butyl*, O-*Me ester:* [6795-01-3]. OO-tert-*Butyl* O-*methyl methylphosphonoperoxoate.*
$C_6H_{15}O_4P$ M 182.156
Liq. d_4^{20} 1.06. $Bp_{0.02}$ 50-51°. n_D^{20} 1.4265.

OO-tert-*Butyl*, O-*Et ester:* [31238-29-6]. OO-tert-*Butyl* O-*ethyl methylphosphonoperoxoate.*
$C_7H_{17}O_4P$ M 196.183
Liq. $Bp_{0.15}$ 51.5-52°. n_D^{20} 1.4248.

OO-tert-*Butyl*, O-*isopropyl ester:* [31238-30-9]. OO-tert-*Butyl* O-*isopropyl methylphosphonoperoxoate.*
$C_8H_{19}O_4P$ M 210.209
Liq. $Bp_{0.015}$ 65-66°. n_D^{20} 1.4224.

OO-tert-*Butyl*, O-*butyl ester:* [31238-32-1]. O-*Butyl* OO-tert-*butyl methylphosphonoperoxoate.*
$C_9H_{21}O_4P$ M 224.236
Liq. $Bp_{0.085}$ 72-74°. n_D^{20} 1.4227.

Sosnovsky, G. *et al*, *Synthesis*, 1971, 144 (*synth*)
Sosnovsky, G. *et al*, *J. Org. Chem.*, 1972, **37**, 2267 (*props*)
Sosnovsky, G. *et al*, *Phosphorus*, 1974, **4**, 255 (*props*)

Methylphosphonoselenoic acid, 9CI, 8CI — M-00311

Se=P(OH)(Me)(OH) ⇌ HSe–P(=O)(Me)(OH)

CH_5O_2PSe M 158.983
Tautomeric.

O,O-*Di-Me ester:* [51072-23-2]. O,O-*Dimethyl methylphosphonoselenoate.*
$C_3H_9O_2PSe$ M 187.037
Liq. Bp_{35} 97-98°.

Mono-O-Et ester: see O-Ethyl methylphosphonoselenoate, E-00092

O,O-*Di-Et ester:* O,O-*Diethyl methylphosphonoselenoate.*
$C_5H_{13}O_2PSe$ M 215.090
Liq. d_4^{20} 1.31. $Bp_{0.1}$ 41-42°. n_D^{20} 1.4891.

Mono-O-propyl ester: O-*Propyl methylphosphonoselenoate.* O-*Propyl hydrogen methylphosphonoselenoate.*
$C_4H_{11}O_2PSe$ M 201.063
Liq. d_4^{20} 1.43. $Bp_{0.02}$ 79-81°. n_D^{20} 1.5208.

O,O-*Dibutyl ester:* O,O-*Dibutyl methylphosphonoselenoate.*
$C_9H_{21}O_2PSe$ M 271.197
Liq. d_4^{20} 1.17. $Bp_{0.1}$ 78-79°. n_D^{20} 1.4830.

Difluoride:
CH_3F_2PSe M 162.965
Liq. Bp_{130} 34°.

Dichloride: see Methylphosphonoselenoic dichloride, M-00312

Dibromide:
CH_3Br_2PSe M 284.776
Liq. d_4^{20} 2.26. Bp_{730} 174-176°, Bp_8 63-65°. n_D^{20} 1.6349.

Gryszkiewicz-Trochimowski, E. *et al*, *Bull. Soc. Chim. Fr.*, 1960, 1794 (*esters*)

Maier, L., *Helv. Chim. Acta*, 1963, **46**, 2667 (*dibromide*)
McFarlane, W. *et al*, *J. Chem. Soc., Dalton Trans.*, 1973, 2162 (*ester, pmr, P and Se nmr*)
Roesky, H.W. *et al*, *Z. Naturforsch., B*, 1973, **28**, 697 (*difluoride, ir, raman, pmr, nmr*)
Shagidullin, R.R., *Izv. Akad. Nauk SSSR, Ser. Khim.*, 1976, **25**, 184 (*Engl. transl.* p. 174) (*ester, uv*)
Vandyukova, I.I. *et al*, *Izv. Akad. Nauk SSSR, Ser. Khim.*, 1976, **25**, 1390 (*Engl. transl.* p. 1334) (*ester, synth*)
Nuretdinov, I.A. *et al*, *Izv. Akad. Nauk SSSR, Ser. Khim.*, 1978, **28**, 437 (*Engl. transl.* p. 378) (*esters*)

Methylphosphonoselenoic dichloride, 9CI, 8CI — M-00312

MeP(Se)Cl_2

CH_3Cl_2PSe M 195.874
Liq. d^{18} 1.75-1.79. Bp_{25} 74-75°, $Bp_{0.1}$ 20-21°. n_D^{18} 1.5970.

Gryszkiewicz-Trochimowski, E. *et al*, *Bull. Soc. Chim. Fr.*, 1960, 1794 (*synth*)
Ivin, S.Z. *et al*, *Zh. Obshch. Khim.*, 1961, **31**, 4052 (*Engl. transl.* p. 3780) (*synth*)
Roesky, H.W. *et al*, *Z. Naturforsch., B*, 1973, **28**, 697 (*synth, ir, P nmr*)
Nuretdinov, I.A. *et al*, *Izv. Akad. Nauk SSSR, Ser. Khim.*, 1975, 327 (*Engl. transl.* p. 263) (*nqr*)
Elbel, S. *et al*, *J. Chem. Soc., Dalton Trans.*, 1976, 1762 (*pe*)
Shagidullin, R.R. *et al*, *Izv. Akad. Nauk SSSR, Ser. Khim.*, 1976, 184 (*Engl. transl.* p. 174) (*uv*)

Methylphosphonothioic acid — M-00313

MeP(S)$(OH)_2$ ⇌ MeP(O)(OH)(SH)

CH_5O_2PS M 112.083
Has not been isol. pure.

O-*Me ester: see O-Methyl methylphosphonothioate, M-00171*
O-*Me, chloride: see O-Propyl methylphosphonochloridothioate, P-00468*
O,O-*Di-Me ester:* [681-06-1]. O,O-*Dimethyl methylphosphonothioate.*
$C_3H_9O_2PS$ M 140.137
Li deriv employed in synth. of substituted methylphosphonothioate esters. Liq. d_4^{25} 1.14. Bp_4 40°. n_D^{25} 1.4744.
O-*Et ester: see O-Ethyl methylphosphonothioate, E-00093*
O-*Et, chloride: see O-Ethyl methylphosphonochloridothioate, E-00090*
O,O-*Di-Et ester:* [6996-81-2]. O,O-*Diethyl methylphosphonothioate.*
$C_5H_{13}O_2PS$ M 168.190
Li deriv. employed in synth. of substd. phosphonothioate esters. Liq. d_4^{25} 1.05. Bp_{14} 77-79°. n_D^{25} 1.4619.
O-*Propyl ester, chloride: see O-Propyl methylphosphonochloridothioate, P-00468*
O-*Isopropyl ester: see O-Isopropyl methylphosphonothioate, I-00041*
O-*Isopropyl ester, chloride: see O-Isopropyl methylphosphonochloridothioate, I-00039*
O,O-*Diisopropyl ester:* O,O-*Diisopropyl methylphosphonothioate.* O,O-*Bis(1-methylethyl)-methylphosphonothioate.*
$C_7H_{17}O_2PS$ M 196.244
Liq. d_4^{25} 0.99. Bp_1 42°. n_D^{25} 1.4512.
O-*Butyl ester, chloride: see O-Butyl methylphosphonochloridothioate, B-00573*
O,S-*Dibutyl ester:* [15536-25-1]. O,S-*Dibutyl methylphosphonothioate.*
$C_9H_{21}O_2PS$ M 224.297
Liq. d_4^{20} 1.01. Bp_{13} 143-146°. n_D^{20} 1.4681.
▷Inhibitor of trypsin and cholinesterases
O,O-*Di-Ph ester:* [27976-73-4]. O,O-*Diphenyl methylphosphonothioate.*
$C_{13}H_{13}O_2PS$ M 264.278
Solid. Mp 35-36°. $Bp_{0.04}$ 131°.
O,S-*Di-Ph ester:* [31650-51-8]. O,S-*Diphenyl methylphosphonothioate.*
$C_{13}H_{13}O_2PS$ M 264.278
Liq. d_4^{20} 1.23. $Bp_{0.03}$ 137-138°. n_D^{20} 1.5986.
Difluoride: see Methylphosphonothioic difluoride, M-00317
Dichloride: see Methylphosphonothioic dichloride, M-00316
Dibromide: see Methylphosphonothioic dibromide, M-00315
Diamide: see P-Methylphosphonothioic diamide, M-00314

Hoffmann, F.W. *et al*, *J. Am. Chem. Soc.*, 1958, **80**, 3945 (*esters*)
Pelchowicz, Z. *et al*, *J. Chem. Soc.*, 1962, 3824 (*esters*)
Nesterov, L.V. *et al*, *Zh. Obshch. Khim.*, 1970, **40**, 1237 (*Engl. transl.* p. 1228) (*esters, synth, ir, pmr, P nmr*)
Burkhardt, W.D. *et al*, *Z. Anorg. Allg. Chem.*, 1978, **442**, 19 (*dimethyl ester, ir, raman*)
Mathey, F. *et al*, *Tetrahedron*, 1978, **34**, 649 (*dimethyl ester, use*)
Sass, S. *et al*, *Org. Mass. Spectrom.*, 1979, **14**, 257 (*diethyl ester, ms*)
Fieser, M. *et al*, *Reagents for Organic Synthesis*, Wiley, 1967-84, **2**, 155.

P-Methylphosphonothioic diamide, 9CI — M-00314

MeP(S)$(NH_2)_2$

CH_7N_2PS M 110.113

N,N'-*Diisopropyl:* [13789-68-9]. P-*Methyl*-N,N'-*diisopropylphosphonothioic diamide.*
$C_7H_{19}N_2PS$ M 194.274
Cryst. (pet. ether). Mp 75°.
N,N'-*Di-Ph:* [13789-63-4]. P-*Methyl*-N,N'-*diphenylphosphonothioic diamide. Methylphosphonothioic dianilide.*
$C_{13}H_{15}N_2PS$ M 262.309
Solid. Mp 177-178°.
N,N,N',N'-*Tetra-Me:* [15849-93-1]. *Pentamethylphosphonothioic diamide. Methylphosphonothioic bis(dimethylamide).*
$C_5H_{15}N_2PS$ M 166.221
Liq. Bp_{725} 245-247°.
N,N,N',N'-*Tetra-Et:* [31650-61-0]. *Methylphosphonothioic bis(diethylamide).* N,N,N',N'-*Tetraethyl*-P-*methylphosphonothioic diamide.*
$C_9H_{23}N_2PS$ M 222.328
Liq. d_4^{20} 1.01. Bp_{15} 149-151°, $Bp_{0.0001}$ 60-65°. n_D^{20} 1.5080.

Gryszkiewicz-Trochimowski, E., *Bull. Soc. Chim. Fr.*, 1967, 2232 (*anilide*)
Mel'nikov, N.N. *et al*, *Zh. Obshch. Khim.*, 1967, **37**, 239 (*Engl. transl.* p. 222) (*diisopropyl, synth, uv*)
Nesterov, L.V. *et al*, *Zh. Obshch. Khim.*, 1970, **40**, 1237 (*Engl. transl.* p. 1228) (*tetraethyl, synth, ir, pmr, P nmr*)

Methylphosphonothioic dibromide, 9CI — M-00315

Methylthiophosphonic dibromide

[5827-24-7]

MeP(S)Br_2

CH_3Br_2PS M 237.876

Low-melting solid. Mp 31-32°. Bp_{720} 203-204°, Bp_{17} 85-87°.

Maier, L., *Helv. Chim. Acta*, 1963, **46**, 2667 (*synth, ir, P nmr*)
Kuchen, W. *et al*, *Chem. Ber.*, 1970, **103**, 2114 (*synth, ir, pmr*)
Elbel, S. *et al*, *J. Chem. Soc., Dalton Trans.*, 1976, 1762 (*pe*)

Methylphosphonothioic dichloride, 9CI M-00316

Methylthiophosphonic dichloride

[676-98-2]

MeP(S)Cl_2

CH_3Cl_2PS M 148.974

Reagent for detn. of ee in chiral amines. Liq. with pungent odour. d_4^{20} 1.35-1.43. Bp 177-178°, Bp_{19} 49-51°. n_D^{20} 1.5428.

▷TB2100000.

Hoffmann, F.W. *et al*, *J. Am. Chem. Soc.*, 1958, **80**, 3945 (*synth*)
Christol, C. *et al*, *J. Chim. Phys.*, 1965, **62**, 246 (*ir, raman*)
Kavavanov, K.V. *et al*, *Zh. Obshch. Khim.*, 1965, **35**, 78 (*Engl. transl.* p. 76) (*synth*)
Nesterov, L.V. *et al*, *Zh. Obshch. Khim.*, 1970, **40**, 1237 (*Engl. transl.* p. 1228) (*ir, pmr, P nmr*)
Shagidullin, R.R. *et al*, *Dokl. Akad. Nauk SSSR, Ser. Sci. Khim.*, 1975, **222**, 897 (*Engl. transl.* p. 564) (*uv*)
Elbel, S. *et al*, *J. Chem. Soc., Dalton Trans.*, 1976, 1762 (*pe*)
Khairullin, V.K. *et al*, *Zh. Obshch. Khim.*, 1978, **48**, 1993 (*Engl. transl.* p. 1813) (*synth*)
Feringa, B.L. *et al*, *J. Org. Chem.*, 1986, **51**, 5484 (*use*)
Sax, N.I., *Dangerous Properties of Industrial Materials*, 6th Ed., Van Nostrand-Reinhold, 1984, 830.

Methylphosphonothioic difluoride, 9CI M-00317

Methylthiophosphonic difluoride

[753-72-0]

MeP(S)F_2

CH_3F_2PS M 116.065

Liq. d_4^{20} 1.28. Bp_{758} 61-62°. n_D^{20} 1.4012.

Aleksandrov, V.N. *et al*, *Zh. Obshch. Khim.*, 1967, **37**, 2714 (*Engl. transl.* p. 2584) (*synth*)
Roesky, H.W., *Chem. Ber.*, 1968, **101**, 3679 (*synth, ir, pmr, F nmr*)
Reddy, G.S. *et al*, *Z. Naturforsch., B*, 1970, **25**, 1199 (*pmr, F and P nmr*)
Köttgen, D. *et al*, *Z. Anorg. Allg. Chem.*, 1972, **389**, 269 (*ir, raman*)
Durig, J.R. *et al*, *Inorg. Chem.*, 1983, **22**, 4134 (*ir, raman, microwave*)

Methylphosphonotrithioic acid, 9CI M-00318

[55453-25-3]

MeP(S)$(SH)_2$

CH_5PS_3 M 144.204

Bimolecular bis(anhydrosulfide): see 2,4-Dimethyl-1,3,2,4-dithiadiphosphetane 2,4-disulfide, D-00752

Di-Me ester: [14806-66-7]. *Dimethyl methylphosphonotrithioate.*

$C_3H_9PS_3$ M 172.258

Liq. d_4^{20} 1.25. Bp_{14} 131-134°. n_D^{20} 1.6386.

Di-Et ester: [31650-57-4]. *Diethyl methylphosphonotrithioate.*

$C_5H_{13}PS_3$ M 200.311

Liq. d^{20} 1.16. Bp_{10} 131-133°. n_D^{20} 1.5965.

Di-Ph ester: [31650-59-6]. *Diphenyl methylphosphonotrithioate.*

$C_{13}H_{13}PS_3$ M 296.399

Solid. Mp 49-50°. $Bp_{0.03}$ 169-171°.

Bis(trimethylsilyl) ester:

$C_7H_{21}PS_3Si_2$ M 288.568

Cryst. (pentane).

Gryszkiewicz-Trochimowskii, E., *Bull. Soc. Chim. Fr.*, 1967, 2232 (*ester, synth*)
Nesterov, L.V. *et al*, *Zh. Obshch. Khim.*, 1970, **40**, 1237 (*Engl. transl.* p. 1228) (*esters, synth*)
Pantzer, R. *et al*, *Z. Anorg. Allg. Chem.*, 1973, **395**, 262 (*ester, synth, ir, raman, P nmr*)
Seel, F. *et al*, *Chem. Ber.*, 1980, **113**, 1837 (*synth*)
Martin, J. *et al*, *Org. Magn. Reson.*, 1981, **15**, 87 (*P nmr*)
Fritz, G. *et al*, *Z. Anorg. Allg. Chem.*, 1986, **537**, 17 (*silyl ester, ms*)
Hahn, J. *et al*, *Z. Anorg. Allg. Chem.*, 1986, **543**, 7 (*silyl ester, synth, pmr, cmr, Si nmr*)

Methylphosphonous acid M-00319

MeP$(OH)_2$

CH_5O_2P M 80.023

Free acid exists as the phosphoryl tautomer Methylphosphinic acid, M-00278 .

Mono-Me ester: see Methylphosphinic acid, M-00278

Di-Me ester: [20278-51-7]. *Dimethyl methylphosphonite.*

$C_3H_9O_2P$ M 108.077

Liq. Bp_{300} 62-65°. n_D^{20} 1.4172.

Mono-Et ester: see Methylphosphinic acid, M-00278

Di-Et ester: [15715-41-0]. *Diethyl methylphosphonite.*

$C_5H_{13}O_2P$ M 136.130

Noxious pungent liq. Bp 120-122° (much dec.), Bp_{50} 47°. n_D^{20} 1.4155.

▷Mod. irritant, flammable

Monoisopropyl ester: see Methylphosphinic acid, M-00278

Diisopropyl ester: [66295-44-1]. *Diisopropyl methylphosphonite.*

$C_7H_{17}O_2P$ M 164.184

Liq. Bp_{40} 57-58°. n_D^{20} 1.4168.

Mono-Ph ester: see Methylphosphinic acid, M-00278

Di-Ph ester: [38316-42-6]. *Diphenyl methylphosphonite.*

$C_{13}H_{13}O_2P$ M 232.218

Liq. d_4^{20} 1.14. Bp_9 144-148°, Bp_1 105-106°. n_D^{20} 1.5560.

Difluoride: see Methylphosphonous difluoride, M-00324

Dichloride: see Methylphosphonous dichloride, M-00322

Dibromide: see Methylphosphonous dibromide, M-00321

Diiodide: see Methylphosphonous diiodide, M-00325

Dicyanide: see Methylphosphonous dicyanide, M-00323

Diisocyanate:

$C_3H_3N_2O_2P$ M 130.043

Liq. Bp_6 33-36°.
Diisothiocyanate:
$C_3H_3N_2PS_2$ M 162.164
Liq. d_4^{20} 1.36. $Bp_{0.1}$ 55-57°. n_D^{20} 1.6741.
Diamide: see Methylphosphonous diamide, M-00320

Hoffmann, F.W. *et al*, *J. Am. Chem. Soc.*, 1958, **80**, 1150 (*esters*)
Petrov, K.A. *et al*, *Zh. Obshch. Khim.*, 1962, **32**, 3065 (*Engl. transl.* p. 3015) (*diphenyl ester*)
Maier, L., *Helv. Chim. Acta*, 1963, **46**, 2667 (*derivs, synth, ir, P nmr*)
Gladshtein, B.M. *et al*, *Zh. Obshch. Khim.*, 1969, **39**, 1951 (*Engl. transl.* p. 1913) (*esters*)
Quin, L.D. *et al*, *Org. Magn. Reson.*, 1974, **6**, 503 (*dimethyl ester, cmr*)
Miles, J.A. *et al*, *Org. Prep. Proceed, Int.*, 1979, **11**, 11 (*diethyl ester, synth, pmr, P nmr*)
Nifant'ev, É.E. *et al*, *Tetrahedron*, 1981, **37**, 3183 (*diethyl ester, cmr*)
Collins, D.J. *et al*, *Aust. J. Chem.*, 1983, **36**, 2517 (*diethyl ester, pmr, P nmr*)

Methylphosphonous diamide **M-00320**

$MeP(NH_2)_2$

CH_7N_2P M 78.053
Tetra-Me: [14937-39-4]. *Pentamethylphosphonous diamide. Methylphosphonous bis(dimethylamide).*
$C_5H_{15}N_2P$ M 134.161
Liq. Bp_{720} 137-141°, Bp_{45} 62-63°. n_D^{20} 1.4630.
N,N,N′,N′-*Tetra-Et:* N,N,N′,N′-*Tetraethyl*-P-*methylphosphonous diamide. Methylphosphonous bis(diethylamide).*
$C_9H_{23}N_2P$ M 190.268
Liq. Bp_{11} 82-85°. n_D^{20} 1.4658.
N,N′-*Di*-tert-*butyl:* [61152-24-7]. N,N′-*Di*-tert-*butyl*-P-*methylphosphonous diamide.*
$C_9H_{23}N_2P$ M 190.268
Liq. Bp_5 50-53°.
N,N,N′,N′-*Tetrakis(trimethylsilyl):* [82581-87-1]. P-*Methyl*-N,N,N′,N′-*tetrakis(trimethylsilyl)-phosphonous diamide.*
$C_{13}H_{39}N_2PSi_4$ M 366.780
Liq. $Bp_{0.05}$ 97-100°.

Petrov, K.A. *et al*, *Zh. Obshch. Khim.*, 1961, **31**, 2377 (*Engl. transl.* p. 2214) (*tetraethyl, synth*)
Maier, L., *Helv. Chim. Acta*, 1963, **46**, 2667 (*derivs, synth, P nmr*)
Scherer, O.J. *et al Chem. Ber.*, 1976, **109**, 2996 (*di-tert*-butyl, synth, pmr, P nmr, ms)
Barlos, K. *et al*, *Z. Naturforsch., B*, 1978, **33**, 515 (*N nmr*)
Gonbeau, D. *et al*, *Inorg. Chem.*, 1981, **20**, 1966 (*tetramethyl, pe*)
Bei-Li Li *et al*, *Inorg. Chem.*, 1983, **22**, 575 (*trimethylsilyl, synth, pmr, cmr, P nmr*)

Methylphosphonous dibromide **M-00321**

Dibromomethylphosphine
[1066-34-8]

$MePBr_2$

CH_3Br_2P M 205.816
Fuming liq. d_4^{20} 2.19. Mp −58°. Bp_{720} 138.5°. n_D^{20} 1.6104.

Maier, L., *Helv. Chim. Acta*, 1963, **46**, 2026 (*synth, ir, P nmr*)
Kuchen, W. *et al*, *Chem. Ber.*, 1965, **98**, 480 (*synth*)
Elbel, S. *et al*, *Z. Naturforsch., B*, 1976, **31**, 178 (*pe*)
Hinke, A. *et al*, *Phosphorus Sulfur*, 1983, **15**, 93 (*synth, P nmr*)

Methylphosphonous dichloride, 9CI **M-00322**

Dichloromethylphosphine
[676-83-5]

$MePCl_2$

CH_3Cl_2P M 116.914
Liq. with pungent, foul odour. d_4^{20} 1.30. Bp_{729} 80-81°. n_D^{20} 1.4960.
▷Pyrophoric, corrosive; reacts violently with H_2O. TB2475000.

Maier, L., *Helv. Chim. Acta*, 1963, **46**, 2026; 1964, **47**, 2137 (*synth, ir, P nmr*)
Geiseler, G. *et al*, *Ber. Bunsenges. Phys. Chem.*, 1967, **71**, 478 (*ir, raman*)
Fild, M. *et al*, *J. Chem. Soc. (A)*, 1970, 2359 (*pmr, F nmr*)
Durig, J.R. *et al*, *J. Phys. Chem.*, 1971, **75**, 1956 (*ir*)
Quin, L.D. *et al*, *Org. Magn. Reson.*, 1974, **6**, 503 (*cmr*)
Lappert, M.F. *et al*, *J. Chem. Soc., Dalton Trans.*, 1975, 1207 (*pe*)
Elbel, S. *et al*, *Z. Naturforsch., B*, 1976, **31**, 178 (*pe*)
Soroka, M. *et al*, *Synthesis*, 1977, 450 (*synth*)
Barlos, K. *et al*, *Chem. Ber.*, 1980, **113**, 3716 (*nqr*)
Frenking, G. *et al*, *Phosphorus Sulfur*, 1980, **8**, 337, 343 (*pe, struct*)
Naumov, V.A. *et al*, *Zh. Strukt. Khim.*, 1983, **24**, 160 (*Engl. transl.* p. 312) (*struct*)

Methylphosphonous dicyanide **M-00323**

Dicyanomethylphosphine
[27388-03-0]

$MeP(CN)_2$

$C_3H_3N_2P$ M 98.044
Solid. Mp 72-73°. $Bp_{0.1}$ 50-56° subl.

Maier, L., *Helv. Chim. Acta*, 1963, **46**, 2667 (*synth, P nmr, ir, cryst struct*)
Jones, C.E. *et al*, *Inorg. Chem.*, 1971, **10**, 1536 (*synth, ir, ms, P nmr*)
Edwards, H.G.M. *et al*, *Spectrochim. Acta, Part A*, 1976, **32**, 739 (*raman*)
Elbel, S. *et al*, *Z. Naturforsch., B*, 1976, **31**, 1472 (*pe*)

Methylphosphonous difluoride, 9CI **M-00324**

Difluoromethylphosphine
[753-59-3]

$MePF_2$

CH_3F_2P M 84.005
Gas. Mp −110°. Bp −28°.

Seel, F. *et al*, *Z. Anorg. Allg. Chem.*, 1965, **341**, 196 (*synth, pmr, F and P nmr, ir, ms*)
Drozd, G.I. *et al*, *Zh. Obshch. Khim.*, 1967, **37**, 1343 (*Engl. transl.* p. 1269) (*synth, ir, P nmr*)
Fild, M. *et al*, *J. Chem. Soc. (A)*, 1970, 2359 (*pmr, F and P nmr*)
Codding, E.G. *et al*, *Inorg. Chem.*, 1974, **13**, 856 (*struct*)
Lappert, M.F. *et al*, *J. Chem. Soc., Dalton Trans.*, 1975, 1207 (*pe*)
Elbel, S. *et al*, *Z. Naturforsch., B*, 1976, **31**, 178 (*pe*)
Durig, J.R. *et al*, *J. Raman Spectrosc.*, 1981, **10**, 44 (*ir, raman*)

Methylphosphonous diiodide, 9CI **M-00325**

Diiodomethylphosphine
[13868-71-8]

$MePI_2$

CH_3I_2P M 299.817
Solid. Mp 33-36°. Bp_7 82-85°.

Maier, L., *Helv. Chim. Acta*, 1963, **46**, 2026 (*synth, ir, P nmr*)
Elbel, S. *et al*, *Z. Naturforsch., B*, 1976, **31**, 178 (*pe*)
Mel'nichuk, E.A. *et al*, *Zh. Obshch. Khim.*, 1979, **49**, 1668 (*Engl. transl.* p. 1457) (*synth, props*)
Cavell, R.G. *et al*, *J. Am. Chem. Soc.*, 1971, **93**, 1130 (*synth, pmr, ir, nmr, ms*)
Robinson, E.A., *et al*, *Spectrochim. Acta, Part A*, 1972, **28**, 1099 (*P nmr, pmr*)

Methylphosphoramidic acid, 9CI — M-00326

Phosphoric acid methylamide

MeNHP(O)(OH)$_2$

CH_6NO_3P M 111.037

Di-Me ester: [52420-88-9]. *Dimethyl methylphosphoramidate.*
$C_3H_{10}NO_3P$ M 139.091
Liq. d_4^{20} 1.20. Bp_{10} 121-122°, Bp_1 81°. n_D^{20} 1.4258.

Di-Et ester: see Diethyl methylphosphoramidate, D-00299

Dipropyl ester: Dipropyl methylphosphoramidate.
$C_7H_{18}NO_3P$ M 195.198
Liq. d_4^{20} 1.04. $Bp_{0.06}$ 94.5-95°. n_D^{20} 1.4350.

Dibutyl ester: [2014-81-5]. *Dibutyl methylphosphoramidate.*
$C_9H_{22}NO_3P$ M 223.251
Liq. $Bp_{0.25}$ 126-128°. n_D^{27} 1.4315.

Di-Ph ester: [2014-79-1]. *Diphenyl methylphosphoramidate.*
$C_{13}H_{14}NO_3P$ M 263.232
Cryst. (pet. ether). Mp 95-96°.

Difluoride: see Methylphosphoramidic difluoride, M-00328

Dichloride: see Methylphosphoramidic dichloride, M-00327

Stock, J.A. *et al*, *J. Chem. Soc. (C)*, 1966, 637 (*esters, synth, props*)
Keat, R., *J. Chem. Soc., Dalton Trans.*, 1974, 876 (*dimethyl ester, P nmr, pmr*)
Nikonorov, L.K. *et al*, *Zh. Obshch. Khim.*, 1975, **45**, 1008 (*Engl. transl.* p. 995) (*esters, synth*)
Mizrahi, V. *et al*, *J. Org. Chem.*, 1982, **47**, 3533 (*dimethyl ester, ms*)
Seel, F. *et al*, *Z. Naturforsch., B*, 1983, **38**, 804.
Davidowitz, B. *et al*, *Org. Mass Spectrom.*, 1984, **19**, 128 (*esters, ir*)

Methylphosphoramidic dichloride, 9CI — M-00327

[36598-86-4]

MeNHP(O)Cl$_2$

CH_4Cl_2NOP M 147.928

Liq. Bp_{20} 130°. Largely dimeric.

Michaelis, A., *Justus Liebigs Ann. Chem.*, 1903, **326**, 129, 172 (*synth*)
Robinson, E.A. *et al*, *Spectrochim. Acta, Part A*, 1972, **28**, 1099 (*pmr, P nmr*)
Shihada, A.-F., *Z. Anorg. Allg. Chem.*, 1975, **411**, 135 (*ir, raman, complexes*)
Roth, H.J. *et al*, *Arch. Pharm. (Weinheim, Ger.)*, 1981, **314**, 85 (*synth, pmr*)

Methylphosphoramidic difluoride, 9CI — M-00328

[3824-48-4]

MeNHP(O)F$_2$

CH_4F_2NOP M 115.019

Roesky, H.W. *et al*, *Z. Anorg. Allg. Chem.*, 1970, **375**, 140 (*synth, ir, pmr, F nmr*)

Methylphosphoramidochloridothioic acid, 9CI, 8CI — M-00329

MeNHPCl(S)OH ⇌ MeNHPCl(O)SH

$CH_5ClNOPS$ M 145.543

(±)-*form*

O-Ph ester: *O-Phenyl methylphosphoramidochloridothioate. O-Phenyl chloro(methylamido)thiophosphate.*
$C_7H_9ClNOPS$ M 221.641
Low melting solid. Mp 22-23°. $Bp_{0.2}$ 105-107°.

Mandel'baum, Ya.A. *et al*, *Zh. Obshch. Khim.*, 1968, **38**, 1754 (*Engl. transl.* p. 1709)

Methylphosphoramidodithioic acid, 9CI, 8CI — M-00330

Dithiophosphoric acid methylamide

MeNHP(O)(SH)$_2$⇌MeNHP(S)(OH)(SH)

CH_6NOPS_2 M 143.158

O-Et S-propyl ester: *O-Ethyl S-propyl methylphosphoramidodithioate.*
$C_6H_{16}NOPS_2$ M 213.292
Liq. d_4^{20} 1.11. $Bp_{0.15}$ 126-127°. n_D^{20} 1.5280.

O-Me S-Ph ester: *O-Methyl-S-phenyl methylphosphoramidodithioate.*
$C_8H_{12}NOPS_2$ M 233.283
Liq. d_4^{20} 1.24. $Bp_{0.25}$ 127-128°. n_D^{20} 1.6110.

O-Et S-Ph ester: *O-Ethyl-S-phenyl methylphosphoramidodithioate.*
$C_9H_{14}NOPS_2$ M 247.309
Liq. d_4^{20} 1.22. $Bp_{0.2}$ 131-132°. n_D^{20} 1.6050.

O-Isopropyl S-Ph ester: *O-Isopropyl S-phenyl methylphosphoramidodithioate.*
$C_{10}H_{16}NOPS_2$ M 261.336
Liq. d_4^{20} 1.16. $Bp_{0.5}$ 149-150°. n_D^{20} 1.5765.

Mandel'baum, Ya.A., *et al*, *Zh. Obshch. Khim.*, 1967, **37**, 2540 (*Engl. transl.*, p. 2417) (*esters, synth*)
Hamer, N.K. *et al*, *J. Chem. Soc., Perkin Trans. 2*, 1974, 1184 (*synth, pmr*)

O-Methyl phosphoramidofluoridothioate, 9CI, 8CI — M-00331

Thiophosphoric acid O-methyl ester-fluoride-amide. O-Methyl amidofluorothiophosphate

[25246-48-4]

MeOPF(S)NH$_2$

CH_5FNOPS M 129.089

(±)-*form*

Liq. $Bp_{0.01}$ 28°. n_D^{20} 1.4738.

Roesky, H.W. *et al*, *Chem. Ber.*, 1969, **102**, 2588 (*synth, ir, ms, pmr, F nmr*)

Methylphosphoramidoselenoic acid, 9CI — M-00332

Phosphoroselenoic acid methylamide

MeNHP(Se)(OH)$_2$ ⇌ MeNHP(O)(SeH)(OH)

CH_6NO_2PSe M 173.998

O,O-*Di-Et ester:* [56341-79-8]. O,O-*Diethyl methylphosphoramidoselenoate.* O,O-*Diethyl selenophosphoryl methylamide.*
$C_5H_{14}NO_2PSe$ M 230.105
Liq. d_4^{20} 1.36. $Bp_{0.04}$ 61.5°. n_D^{20} 1.5035.
O,O-*Diisopropyl ester:* [56341-80-1]. O,O-*Diisopropyl methylphosphoramidoselenoate.* O,O-*Diisopropyl selenophosphoryl methylamide.*
$C_7H_{18}NO_2PSe$ M 258.158
Liq. d_4^{20} 1.23. $Bp_{0.04}$ 57°. n_D^{20} 1.4830.

Nikonorova, L.K., *Zh. Obshch. Khim.*, 1975, **45**, 1008 (*Engl. transl.* p. 995) (*esters, synth, ir*)

Methylphosphoramidothioic acid, 9CI M-00333

$MeNHP(S)(OH)_2 \rightleftharpoons MeNHP(O)(OH)(SH)$

CH_6NO_2PS M 127.098
O,O-*Di-Me ester:* [31464-99-0]. O,O-*Dimethyl methylphosphoramidothioate.*
$C_3H_{10}NO_2PS$ M 155.151
Liq. d_4^{20} 1.19. Bp_1 99.5-100.5°, Bp_1 68-69°. n_D^{20} 1.4920.
O,O-*Di-Et ester:* [17321-49-2]. O,O-*Diethyl methylphosphoramidothioate.*
$C_5H_{14}NO_2PS$ M 183.205
Liq. d_4^{20} 1.10. Bp_{14} 107-109°. n_D^{20} 1.4753.
Dichloride: see Methylphosphoramidothioic dichloride, M-00334

Popov, E.M. *et al*, *Zh. Obshch. Khim.*, 1959, **29**, 1998 (*Engl. transl.* p. 1967) (*esters, ir, raman*)
Barabanov, V.I. *et al*, *Zh. Obshch. Khim.*, 1970, **40**, 2464 (*Engl. transl.* p. 2451) (*dimethyl ester*)
Nikonorova, L.K. *et al*, *Zh. Obshch. Khim.*, 1975, **45**, 1008 (*Engl. transl.* p. 995) (*esters, synth, ir*)
Schneider, P. *et al*, *J. Prakt. Chem.*, 1982, **324**, 1063 (*dimethyl ester, props*)

Methylphosphoramidothioic dichloride, 9CI, 8CI M-00334

[20665-23-0]

$MeNHP(S)Cl_2$

CH_4Cl_2NPS M 163.989
Liq. Bp_{33} 115°, $Bp_{0.1}$ 43°.

Nyquist, R.A. *et al*, *Spectrochim. Acta, Part A*, 1970, **26**, 611 (*ir, raman*)
Binder, H. *et al*, *Z. Anorg. Allg. Chem.*, 1971, **381**, 116 (*synth, P nmr*)
Keat, R., *J. Chem. Soc., Dalton Trans.*, 1972, 2189 (*pmr, P nmr*)

Methylphosphoramidous acid, 9CI M-00335

$MeNHP(OH)_2$

CH_6NO_2P M 95.038
Di-Me ester: [39480-85-8]. *Dimethyl methylphosphoramidite.*
$C_3H_{10}NO_2P$ M 123.091
Liq. Bp_{16} 43-46°.
Di-Et ester: [60030-06-0]. *Diethyl methylphosphoramidite.*
$C_5H_{14}NO_2P$ M 151.145
Liq. Bp_{10} 53-5°. n_D^{20} 1.4295.
Diisopropyl ester: Diisopropyl methylphosphoramidite.
$C_7H_{18}NO_2P$ M 179.198
Liq. Bp_{12} 63.5°. n_D^{20} 1.4288.

Difluoride: [2851-75-4].
CH_4F_2NP M 99.020
Ligand for Co, Fe, W, and Mo. Liq. Bp 52-53° dec.

Barlow, C.G. *et al*, *J. Chem. Soc. (A)*, 1968, 2692 (*difluoride, synth, props, ir, pmr, F and P nmr*)
Harman, J.S. *et al*, *J. Chem. Soc. (A)*, 1970, 1935 (*difluoride*)
Scherer, O.J. *et al*, *Z. Naturforsch, B*, 1972, **27**, 1429 (*dimethyl ester, synth, pmr*)
Nikonorova, L.K. *et al*, *Zh. Obshch. Khim.*, 1976, **46**, 1015 (*Engl. transl.* p. 1012) (*esters*)
Laurenson, G.S. *et al*, *J. Mol. Struct.*, 1979, **54**, 111 (*difluoride, ed*)

Methylphosphorimidic acid, 9CI M-00336

$MeN{=}P(OH)_3$

CH_6NO_3P M 111.037
Free acid tautomeric with Methylphosphoramidic acid, M-00326 .
Tri-Me ester: [54000-84-9]. *Trimethyl methylphosphorimidate. Methylimino trimethyl phosphate.* P,P,P-*Trimethoxy*-N-*methylphosphine imide.*
$C_4H_{12}NO_3P$ M 153.117
Liq. Mp 17-18°. $Bp_{1.5}$ 27°.
Tri-Et ester: [71867-67-9]. *Triethyl methylphosphorimidate. Triethyl methylimido phosphate.* P,P,P-*Triethoxy*-N-*methylphosphine imide.*
$C_7H_{18}NO_3P$ M 195.198
Liq. Bp_1 56-57°.
Trifluoride: P,P,P-*Trifluoro*-N-*methylphosphazene.* P,P,P-*Trifluoro*-N-*methylphosphine imide.*
CH_3F_3P M 103.004
Exists only as a dimer.
Trichloride: [23453-30-7]. P,P,P-*Trichloro*-N-*methylphosphazene.* P,P,P-*Trichloro*-N-*methylphosphine imide.*
CH_3Cl_3P M 152.367
Exists only as a dimer.

Haasemann, P. *et al*, *Z. Anorg. Allg. Chem.*, 1974, **408**, 293 (*trimethyl ester, synth, ir, raman*)
Goldwhite, H., *et al*, *J. Chem. Soc., Dalton Trans.*, 1975, 12 (*trimethyl ester, synth, ir, pmr, P nmr*)
Hay, R.S. *et al*, *J. Chem. Soc., Perkin Trans. 2*, 1979, 756 (*triethyl ester, synth, P nmr, esr*)

2-Methylphosphorin, 9CI M-00337

2-Methylphosphabenzene. 2-Methylphosphinine
[56577-92-5]

P CH₃

C_6H_7P M 110.095

Ashe, A.J. *et al*, *Tetrahedron Lett.*, 1975, 1083 (*synth, pmr, ms*)
Ashe, A.J. *et al*, *J. Am. Chem. Soc.*, 1976, **98**, 545 (*cmr*)
Galasso, V., *J. Magn. Reson.*, 1979, **36**, 181 (*struct*)

4-Methylphosphorin, 9CI M-00338

4-Methylphosphinine
[57242-06-5]
C_6H_7P M 110.095
No phys. data reported.

Ashe, A.J. *et al*, *Tetrahedron Lett.*, 1975, 2749 (*synth, pmr, struct*)

1-Methylphosphorinane, 9CI, 8CI **M-00339**

1-Methylphosphacyclohexane. Methylpentamethylenephosphine. 1-Methylphosphinane

[39763-50-3]

$C_6H_{13}P$ M 116.142

Bp 146°.

B,HI: 1-Methylphosphorinanium iodide. 1-Methylphosphinanium iodide.

$C_6H_{14}IP$ M 244.055

Cryst. Mp 220-225° dec.

B,MeBr: 1,1-Dimethylphosphorinanium bromide. 1,1-Dimethylphosphinanium bromide.

$C_7H_{16}BrP$ M 211.081

Solid. Mp >300°.

Oxide: [39763-49-0].

$C_6H_{13}OP$ M 132.142

Solid. Mp 120-122°. Bp_{10} 143°, $Bp_{0.1}$ 60°. Subl. at Bp_{20} 100°.

Sulfide: [1661-16-1].

$C_6H_{13}PS$ M 148.202

Solid. Mp 49-50°. $Bp_{0.1}$ 104-108°.

Sommer, K., *Z. Anorg. Allg. Chem.*, 1970, **379**, 56 (*sulfide*)
Lambert, J. *et al*, *Tetrahedron*, 1971, **27**, 4245 (*salt, nmr*)
Marsi, K.L. *et al*, *J. Am. Chem. Soc.*, 1973, **95**, 200 (*synth, oxide*)
Featherman, S.I. *et al*, *J. Org. Chem.*, 1974, **39**, 2899 (*synth, sulfide, cmr*)
Featherman, S.I. *et al*, *J. Am. Chem. Soc.*, 1975, **97**, 4349 (*P nmr, conformn*)
Schmidbaur, H. *et al*, *Chem. Ber.*, 1977, **110**, 1576 (*derivs*)

1-Methyl-4-phosphorinanone, 9CI, 8CI **M-00340**

1-Methyl-4-phosphinanone

[16327-48-3]

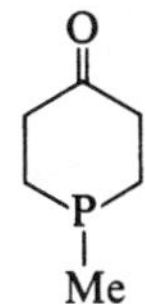

$C_6H_{11}OP$ M 130.126

Liq. $Bp_{1.2}$ 55-57°.

Oxime:

$C_6H_{12}NOP$ M 145.141

Cryst. (MeOH aq.). Mp 88-89°.

Oxide: [54662-09-8].

$C_6H_{11}O_2P$ M 146.125

Cryst. (EtOAc or CH_2Cl_2/pet. ether). Mp 143-144°.

Oxide, oxime:

$C_6H_{12}NO_2P$ M 161.140

Solid. Mp 179-181°.

Shook, H.E. *et al*, *J. Am. Chem. Soc.*, 1967, **89**, 1841 (*synth, ir, pmr*)
Quin, L.D. *et al*, *J. Chem. Soc. (B)*, 1971, 832 (*ms*)
Shalaby, S.W. *et al*, *J. Polym. Sci., Polym. Chem. Ed.*, 1974, **12**, 2917 (*synth, derivs, ir, ms, pmr*)
Breen, J.J. *et al*, *J. Org. Chem.*, 1975, **40**, 2245 (*cmr*)

Methyl phosphorochloridofluoridate, 9CI, 8CI **M-00341**

Methyl chlorofluorophosphate

[754-01-8]

MeOP(O)ClF

CH_3ClFO_2P M 132.459

Liq. d_{20}^{20} 1.44. Bp_{75} 55°. n_D^{20} 1.3671. Easily hydrolysed.

Stölzer, C. *et al*, *Chem. Ber.*, 1960, **93**, 1323 (*synth*)
Sheluchenko, V.V. *et al*, *Dokl. Akad. Nauk SSSR, Ser. Sci. Khim.*, 1967, **177**, 376 (*Engl. transl.* p. 1050) (*F and P nmr*)
Neimysheva, A.A. *et al*, *Zh. Obshch. Khim.*, 1968, **38**, 595 (*Engl. transl.* p. 575) (*props*)

Methyl phosphorodiamidite, 9CI **M-00342**

$MeOP(NH_2)_2$

CH_7N_2OP M 94.053

Yellowish solid. Mp 130-132° dec. Hygroscopic.

Becke-Goehring, M. *et al*, *Chem. Ber.*, 1958, **91**, 1188 (*synth, props*)

***S*-Methyl phosphorodiamidothioate**, 9CI, 8CI **M-00343**

S-Methyl diamidothiophosphate

[20217-86-1]

$MeSP(O)(NH_2)_2$

CH_7N_2OPS M 126.113

Synthetic intermediate. Solid. Mp 133-135.5°.

Quistad, G.B. *et al*, *J. Agric. Food Chem.*, 1970, **18**, 189 (*synth*)

Methyl phosphorodibromidite, 9CI, 8CI **M-00344**

Methyl dibromophosphite

[20502-44-7]

$MeOPBr_2$

CH_3Br_2OP M 221.816

Fuming liq. d^{24} 2.43. Bp_{20} 68-70°. n_D^{24} 1.6233. Unstable.

Drozd, G.I. *et al*, *Zh. Obshch. Khim.*, 1969, **39**, 937 (*Engl. transl.* p. 907) (*synth, ir, P nmr*)

Methyl phosphorodichloridate, 9CI, 8CI **M-00345**

Methyl dichlorophosphate. Methoxydichlorophosphine oxide

[677-24-7]

$MeOP(O)Cl_2$

$CH_3Cl_2O_2P$ M 148.913

Liq. d_4^{25} 1.49. Bp_{15} 62-64°. n_D^{20} 1.4359.

Grunze, H., *Chem. Ber.*, 1959, **92**, 850 (*synth*)
McFarlane, W., *Proc. R. Soc. London, Ser. A*, 1968, **306**, 185 (*nmr*)
Kainosho, M., *J. Phys. Chem.*, 1970, **74**, 2853 (*pmr*)
Gonbeau, J. *et al*, *Spectrochim. Acta, Part A*, 1971, **27**, 1703 (*ir*)
Hornung, V. *et al*, *Z. Anorg. Allg. Chem.*, 1971, **380**, 137 (*synth, ir, raman*)
Oberhammer, H., *J. Mol. Struct.*, 1975, **29**, 370 (*ed*)
Furlani, C. *et al*, *J. Chem. Soc., Dalton Trans.*, 1977, 673 (*pe*)
Pieters, G.H. *et al*, *J. Mol. Struct.*, 1983, **102**, 27, 221; 1984, **114**, 413 (*ir, raman, struct*)

Methyl phosphorodichloridite, 9CI, 8CI **M-00346**

Methyl dichlorophosphite. Methoxydichlorophosphine

[3279-26-3]

$MeOPCl_2$

CH_3Cl_2OP M 132.914

Air-sensitive liq. d_4^{25} 1.43. Mp −91°. Bp 91.5°, 95-96°. n_D^{20} 1.4772. Pyrolysis gives ClP=CH_2.

▷Reacts violently with water

Martin, D.R. *et al*, *J. Am. Chem. Soc.*, 1950, **72**, 4584 (*synth*)
Biryukov, I.P. *et al*, *Zh. Obshch. Khim.*, 1972, **42**, 1223 (*Engl. transl.* p. 1217) (*nqr*)
Hutchins, R.O. *et al*, *J. Am. Chem. Soc.*, 1972, **94**, 9151 (*synth*)
Gordon, M.D. *et al*, *J. Magn. Reson.*, 1976, **22**, 149 (*P nmr*)
Barlos, K. *et al*, *Chem. Ber.*, 1980, **113**, 3716 (*nqr*)
Ogilvie, K.K. *et al*, *Can. J. Chem.*, 1980, **58**, 2686 (*synth, use*)
Vasil'ev, V.V. *et al*, *Zh. Obshch. Khim.*, 1981, **51**, 2134 (*Engl. transl.* p. 1836) (*O nmr*)

Methyl phosphorodichloridodithioate, 9CI, 8CI **M-00347**

Methyl dichlorodithiophosphate

[5390-60-3]

$MeSP(S)Cl_2$

$CH_3Cl_2PS_2$ M 181.034

Liq. Bp_{10} 83-84°.

Wafa, O.A. *et al*, *Z. Anorg. Allg. Chem.*, 1970, **378**, 273 (*synth, ir, raman*)
Khokhlov, P.S. *et al*, *Zh. Obshch. Khim.*, 1980, **50**, 1214 (*Engl. transl.* p. 981) (*props*)

Methyl phosphorodichloridoselenoite, 9CI **M-00348**

Methyl dichloroselenophosphite

[55776-63-1]

$MeSePCl_2$

CH_3Cl_2PSe M 195.874

Viscous yellow oil. Thermally stable.

Anderson, J.W. *et al*, *Inorg. Nucl. Chem. Lett.*, 1975, **11**, 233 (*synth, pmr*)

***O*-Methyl phosphorodichloridothioate**, 9CI, 8CI **M-00349**

O-*Methyl thiophosphoryl dichloride.* O-*Methyl dichlorothiophosphate*

[2523-94-6]

$MeOP(S)Cl_2$

CH_3Cl_2OPS M 164.974

Pungent, malodorous liq. Bp_{11} 45°. n_D^{25} 1.5124. Isom. to *S*-Methyl phosphorodichlorodothioate, M-00351 by heat (autocatalytic, complete in 5 mins. at 130°) or in pyridine.

McIver, R.A. *et al*, *Can. J. Chem.*, 1956, **34**, 1819 (*synth*)
Vasil'ev, A.F. *et al*, *Zh. Prikl. Spektrosk.*, 1966, **5**, 524; 1967, **6**, 485; 1968, **8**, 102 (*Engl. transl.* pp. 391, 319, 64) (*ir, struct*)
Nyquist, R.A., *Spectrochim. Acta, Part A*, 1967, **23**, 1499 (*ir, raman*)
Whitehead, M.A. *et al*, *J. Chem. Soc. (A)*, 1971, 1738 (*nqr*)
Shagidullin, R.R. *et al*, *Dokl. Akad. Nauk SSSR, Ser. Sci. Khim.*, 1975, **222**, 897 (*Engl. transl.* p. 564) (*uv*)
Bezzubov, V.M. *et al*, *Zh. Strukt. Khim.*, 1976, **17**, 98 (*Engl. transl.* p. 79) (*ed*)
Furlani, C. *et al*, *J. Chem. Soc., Dalton Trans.*, 1977, 673 (*pe*)
Shermergorn, I.M. *et al*, *Zh. Obshch. Khim.*, 1983, **53**, 81 (*Engl. transl.* p. 65) (*props*)

Methyl phosphorodichloridothioite, 9CI **M-00350**

Methyl dichlorothiophosphite

[14684-24-3]

$MeSPCl_2$

CH_3Cl_2PS M 148.974

Liq. Bp_8 40°. n_D^{20} 1.5752.

Fritzowsky, N. *et al*, *Z. Anorg. Allg. Chem.*, 1971, **386**, 67 (*synth, ir, raman*)
Zverev, V.V. *et al*, *Zh. Obshch. Khim.*, 1983, **53**, 1968 (*Engl. transl.* p. 1775) (*pe*)

***S*-Methyl phosphorodichlorodothioate**, 9CI, 8CI **M-00351**

S-*Methyl dichlorothiophosphate*

[18281-76-0]

$MeSP(O)Cl_2$

CH_3Cl_2OPS M 164.974

Pungent malodorous liq. Bp_{10} 74.5-75.5°. n_D^{20} 1.5294.

Hilgetag, G. *et al*, *J. Prakt. Chem.*, 1960, **12**, 1 (*synth*)
Wafa, O.A. *et al*, *Z. Anorg. Allg. Chem.*, 1970, **378**, 273 (*synth, ir, raman*)
Nyquist, R.A., *Spectrochim. Acta, Part A*, 1971, **27**, 697 (*ir, raman*)
Naumov, V.A. *et al*, *Dokl. Akad. Nauk SSSR, Ser. Sci. Khim.*, 1976, **228**, 888 (*Engl. transl.* p. 535) (*ed*)
Zverev, V.V. *et al*, *Zh. Obshch. Khim.*, 1979, **49**, 1737 (*Engl. transl.* p. 1522) (*pe*)
Filippova, E.A. *et al*, *Zh. Strukt. Khim.*, 1985, **26**, 55 (*ir, raman*)

Methyl phosphorodifluoridate, 9CI, 8CI **M-00352**

Methyl phosphoryl difluoride. Methyl difluorophosphate

[22382-13-4]

$MeOP(O)F_2$

$CH_3F_2O_2P$ M 116.004

Liq. Bp 65°.

▷V. toxic

Brown, D.H. *et al*, *J. Chem. Soc. (A)*, 1969, 872 (*synth, pmr, tox*)
Roesky, H.W. *et al*, *Z. Anorg. Allg. Chem.*, 1970, **375**, 140 (*synth, ir*)
Köttgen, D. *et al*, *Z. Anorg. Allg. Chem.*, 1970, **405**, 275 (*synth, ir, raman*)
Reddy, G.S. *et al*, *Z. Naturforsch., B*, 1970, **25**, 1199 (*synth, ir, pmr, P nmr*)
Furlani, C. *et al*, *J. Chem. Soc., Dalton Trans.*, 1977, 673 (*pe*)

Methyl phosphorodifluoridite, 9CI, 8CI **M-00353**

Methyl difluorophosphite

[381-65-7]

$MeOPF_2$

CH_3F_2OP M 100.005
Gas. Mp −117°. Bp_{740} −15.5°.

Martin, D.R. *et al*, *J. Am. Chem. Soc.*, 1950, **72**, 4584 (*synth, props*)
Binder, H. *et al*, *Z. Naturforsch.*, *B*, 1972, **27**, 753 (*synth, P nmr*)
Lines, E.L. *et al*, *Inorg. Chem.*, 1973, **12**, 2111 (*synth, ir*)
Codding, E.G. *et al*, *Inorg. Chem.*, 1974, **13**, 178 (*synth, microwave, struct*)
Robinet, E.G. *et al*, *Chem. Phys. Lett.*, 1974, **29**, 449 (*conformn*)
Durig, J.R. *et al*, *Appl. Spectrosc.*, 1980, **34**, 65 (*ir, raman*)

O-Methyl phosphorodifluoridothioate, 9CI, 8CI M-00354

O-Methyl thiophosphoryl difluoride. O-Methyl difluorothiophosphate
[21348-12-9]

$MeOP(S)F_2$

CH_3F_2OPS M 132.065
Liq. Bp 54°.

Charlton, T.L. *et al*, *Inorg. Chem.*, 1968, **7**, 2195 (*synth, ir, ms, pmr, F nmr*)
Durig, J.R. *et al*, *J. Chem. Phys.*, 1969, **50**, 107 (*synth, ir, raman*)
Reddy, G.S. *et al*, *Z. Naturforsch.*, *B*, 1970, **25**, 1199 (*synth, pmr, F and P nmr*)
Köttgen, D. *et al*, *Z. Anorg. Allg. Chem.*, 1974, **405**, 275 (*synth, ir, raman*)

S-Methyl phosphorodifluoridothioate, 9CI, 8CI M-00355

S-Methyl difluorothiophosphate
[25237-37-0]

$MeSP(O)F_2$

CH_3F_2OPS M 132.065
Liq. d_4^{20} 1.42. Bp 105-106°. n_D^{20} 1.3979.

Roesky, H., *Z. Naturforsch.*, *B*, 1969, **24**, 818 (*synth, ir, ms, pmr, F nmr*)

Methyl phosphorodiisocyanatidate, 9CI, 8CI M-00356

Methyl phosphoryl diisocyanate
[24958-79-0]

$MeOP(O)(NCO)_2$

$C_3H_3N_2O_4P$ M 162.041
Liq. d_4^{20} 1.42. $Bp_{0.1}$ 54-55°. n_D^{20} 1.4513. Highly reactive to protic substances.

Shokol, V.A. *et al*, *Zh. Obshch. Khim.*, 1969, **39**, 1041 (*Engl. transl.* p. 1012) (*synth, ir*)

O-Methyl phosphorodiisocyanatidothioate, 9CI, 8CI M-00357

O-Methyl thiophosphoryl diisocyanate
[36384-93-7]

$MeOP(S)(NCO)_2$

$C_3H_3N_2O_3PS$ M 178.102
Reactive liq. d_4^{20} 1.39. $Bp_{0.2}$ 34-36°. n_D^{20} 1.4936. Sensitive to moisture.

Shokol, V.A. *et al*, *Zh. Obshch. Khim.*, 1971, **41**, 2380 (*Engl. transl.* p. 2407) (*synth, props*)

O-Methyl phosphorofluoridoisothiocyanatidothioate, 8CI M-00358

[16365-56-3]

$C_2H_3FNOPS_2$ M 171.144

(±)-*form*
Liq. d_4^{20} 1.36. Bp_{15} 53-54°. n_D^{20} 1.5320. Reacts with $ArNH_2$ initially at the C=S gp.

Ivanova, Zh.M. *et al*, *Zh. Obshch. Khim.*, 1967, **37**, 1144 (*Engl. transl.* p. 1085) (*synth, props*)

O-Methyl phosphorothioate, 9CI, 8CI M-00359

O-Methyl dihydrogen thiophosphate. O-Methyl thiophosphoric acid

CH_5O_3PS M 128.082

Difluoride: see O-Methyl phosphorodifluoridothioate, M-00354
Dichloride: see O-Methyl phosphorodichloridothioate, M-00349
Diamide: [40334-50-7]. *O-Methyl phosphorodiamidothioate. O-Methyl thiophosphoryl diamide. O-Methyl diamidothiophosphate.*
CH_7N_2OPS M 126.113
Solid. Mp 55°.

Ger. Pat., 2 135 349, (*1973*); *CA*, **78**, 97806 (*diamide*)

S-Methyl phosphorothioate, 9CI, 8CI M-00360

S-Methyl dihydrogen thiophosphate. S-Methyl thiophosphoric acid
[21302-85-2]

$MeSP(O)(OH)_2$

CH_5O_3PS M 128.082
Metabolite of 44086-5.

Monodicyclohexylammonium salt: Solid. Mp 159-160°.
Difluoride: see S-Methyl phosphorodifluoridothioate, M-00355
Dichloride: see S-Methyl phosphorodichlorodothioate, M-00351
Diamide: see S-Methyl phosphorodiamidothioate, M-00343

Zwierzak, A. *et al*, *Z. Naturforsch.*, *B*, 1971, **26**, 386 (*synth*)

P-Methyl-1-piperidinylphosphonamidic acid, 9CI M-00361

Methyl-1-piperidinylphosphinic acid

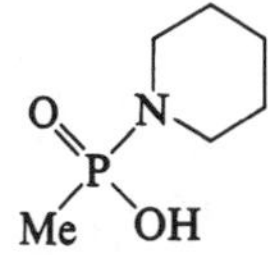

$C_6H_{14}NO_2P$ M 163.156

Isopropyl ester: Isopropyl P-methyl-1-piperidinylphosphonamidate. Isopropyl methyl-1-piperidinylphosphinate.
$C_9H_{20}NO_2P$ M 205.236
Liq. Bp_1 90°. n_D^{16} 1.4590.

Coe, D.G. *et al*, *J. Org. Chem.*, 1959, **24**, 1018 (*synth*)

(2-Methyl-1-propenyl)diphenylphosphine, 9CI **M-00362**

(2,2-Dimethylvinyl)diphenylphosphine. Isobutenyldiphenylphosphine. 1-(Diphenylphosphino)-2-methylpropene

[34193-25-4]

$Ph_2PCH{=}C(CH_3)_2$

$C_{16}H_{17}P$ M 240.284

d_4^{20} 1.04. Bp_4 171-177°, Bp_1 119-121°. n_D^{20} 1.6165.

Oxide: [34295-11-9]. *1-Diphenylphosphinyl-2-methylpropene.*

$C_{16}H_{17}OP$ M 256.283

Cryst. (pet. ether). Mp 149-150°.

Kobovtsev, V.V. *et al*, *Zh. Obshch. Khim.*, 1971, **41**, 2638 (*Engl. transl.* p. 2671) (*synth, chloride, pmr*)
Cann, P.F. *et al*, *J. Chem. Soc., Perkin Trans. 2*, 1972, 304 (*oxide, ir, pmr*)
Koosha, K. *et al*, *Bull. Soc. Chim. Fr.*, 1975, 1284 (*oxide, pmr*)
Grim, S.O. *et al*, *J. Org. Chem.*, 1980, **45**, 250 (*synth, pmr, P nmr*)

(2-Methyl-2-propenylidene)-triphenylphosphorane **M-00363**

[29219-35-0]

$H_2C{=}C(CH_3)CH{=}PPh_3$

$C_{22}H_{21}P$ M 316.382

Stabilised ylide used in Wittig reactions. Solid. Mp 115°.

Köster, R. *et al*, *Justus Liebigs Ann. Chem.*, 1970, **739**, 211 (*synth, pmr, P nmr*)
Bohlmann, F. *et al*, *Chem. Ber.*, 1974, **107**, 1780 (*use*)
Elliott, M. *et al*, *J. Chem. Soc., Perkin Trans. 1*, 1974, 2470 (*use*)
Hirai, M.F. *et al*, *J. Organomet. Chem.*, 1978, **160**, 25 (*complexes*)

Methyl(2-propenyl)phosphinic acid, 9CI **M-00364**

$H_3CCH{=}CHPMe(O)OH$

$C_4H_9O_2P$ M 120.088

Chloride: [56583-26-7].

C_4H_8ClOP M 138.533

Liq. Bp_{27} 99°.

Ger. Pat., 2 357 678 (*1975*); *CA*, **83**, 97567

(2-Methyl-1-propenyl)phosphonic dichloride, 9CI **M-00365**

[4708-01-4]

$(H_3C)_2C{=}CHP(O)Cl_2$

$C_4H_7Cl_2OP$ M 172.978

Liq. d_4^{20} 1.29. Bp_8 84-86°, Bp_1 50-51°. n_D^{20} 1.4970.

Mashlyakovskii, L.N. *et al*, *Zh. Obshch. Khim.*, 1967, **37**, 1307 (*Engl. transl.* p. 1237) (*synth, pmr*)
Moskva, V.V. *et al*, *Zh. Obshch. Khim.*, 1971, **41**, 2577 (*Engl. transl.* p. 2609) (*synth*)
Rybkina, V.V. *et al*, *Zh. Obshch. Khim.*, 1982, **52**, 548 (*Engl. transl.* p. 480) (*synth, pmr, P nmr, props*)

(2-Methyl-1-propenyl)phosphonothioic acid, 9CI **M-00366**

Isobutenylphosphonothioic acid

$(H_3C)_2C{=}CHP(S)(OH)_2 \rightleftharpoons (H_3C)_2C{=}CHP(S)(OH)(SH)$

$C_4H_9O_2PS$ M 152.148

O,O-*Di-Et ester:* O,O-*Diethyl (2-methyl-1-propenyl)-phosphonothioate.*

$C_8H_{17}O_2PS$ M 208.255

Liq. d_4^{20} 1.03. Bp_{11} 117-119°. n_D^{20} 1.4830.

Kamai, G. *et al*, *Zh. Obshch. Khim.*, 1965, **35**, 1817 (*Engl. transl.*, p. 1812)

(2-Methyl-2-propenyl)-triphenylphosphonium(1+), 9CI **M-00367**

[47252-15-3]

$H_2C{=}C(CH_3)CH_2PPh_3^{\oplus}$

$C_{22}H_{22}P^{\oplus}$ M 317.390 (ion)

Chloride: [4303-59-7].

$C_{22}H_{22}ClP$ M 352.843

Source of ylide on treatment with butyllithium or $NaNH_2$. Solid. Mp 191°.

Ylide: see (2-Methyl-2-propenylidene)-triphenylphosphorane, M-00363

Heimgartner, H. *et al*, *Helv. Chim. Acta*, 1971, **54**, 2313 (*synth, use*)
Hirai, M.F. *et al*, *J. Organomet. Chem.*, 1978, **160**, 25 (*complexes*)

2-(2-Methylpropoxy)-1,3,2-dioxaphospholane, 9CI **M-00368**

Ethylene isobutyl phosphite. 2-Isobutoxy-1,3,2-dioxaphospholan

[40925-11-9]

(cyclic OCH_2CH_2O)$POCH_2CH(CH_3)_2$

$C_6H_{13}O_3P$ M 164.141

Liq. d_4^{25} 1.06. Bp_{25} 87.1-87.2°. n_D^{25} 1.4401. Easily hydrolysed.

Lucas, H.J. *et al*, *J. Am. Chem. Soc.*, 1950, **72**, 5491 (*synth*)
Arbuzov, B.A. *et al*, *Izv. Akad. Nauk SSSR, Otd. Khim. Nauk*, 1957, 1046 (*synth*)
Ovchinnikova, N.K. *et al*, *Zh. Org. Khim.*, 1975, **11**, 1839 (*Engl. transl.* p. 1849) (*synth*)

(1-Methylpropyl)diphenylphosphine, 9CI **M-00369**

sec-*Butyldiphenylphosphine, 8CI*

[7650-79-5]

PPh₂ / H–C–CH₃ / CH₂CH₃ (*R*)-*form*

$C_{16}H_{19}P$ M 242.300

(*R*)-*form* [80173-16-6]

Liq. $Bp_{0.01}$ 95°. $[\alpha]_D^{25}$ −1.8° (c, 18.6 in CH_2Cl_2).

B,MeI: Methyl(1-methylpropyl)diphenylphosphonium iodide.

$C_{17}H_{22}IP$ M 384.239

Solid. Mp 210-212°. $[\alpha]_D^{25}$ −4.4° (c, 8.61 in CH_2Cl_2).

Oxide: [80173-17-2].
$C_{16}H_{19}OP$ M 258.299
Solid. Mp 95-96°. $[\alpha]_D^{25}$ −2.8° (c, 8.30 in CH_2Cl_2).
(±)-***form*** [80226-61-5]
Liq. $Bp_{1.5}$ 141-145°, $Bp_{0.5}$ 113-115°.
Oxide: [4252-61-3]. Cryst. (EtOAc). Mp 102-105°.

Grim, S.O. *et al, J. Org. Chem.*, 1967, **32**, 781 (*synth, nmr*)
Grim, S.O. *et al, Phosphorus*, 1974, **4**, 189 (*synth*)
Salvadori, P. *et al, Chim. Ind. (Milan)*, 1981, **63**, 492 (*derivs, pmr, nmr, ms*)

(2-Methylpropyl)diphenylphosphine, 9CI M-00370

Isobutyldiphenylphosphine, 8CI
[5952-47-6]

$Ph_2PCH_2CH(CH_3)_2$

$C_{16}H_{19}P$ M 242.300
Ligand for Fe and Rh. Liq. Bp_{50} 202°, $Bp_{0.5}$ 113-115°. n_D^{20} 1.5906.
Oxide: [63103-76-4].
$C_{16}H_{19}OP$ M 258.299
Cryst. Mp 137.5-138°.
Sulfide:
$C_{16}H_{19}PS$ M 274.360
Rhombic cryst. Mp 80-81°.

Arbuzov, A., *J. Russ. Phys. Chem. Soc.*, 1910, **42**, 549; *CA*, **5**, 1397 (*oxide*)
Schindlbauer, H., *Monatsh. Chem.*, 1965, **96**, 2058 (*synth, nmr*)
Grim, S.O. *et al, J. Org. Chem.*, 1967, **32**, 781 (*synth, nmr*)
Goff, S.D. *et al, Org. Mass Spectrom.*, 1977, **12**, 33 (*oxide, ms*)

(2-Methylpropylidene)-triphenylphosphorane, 9CI M-00371

Isobutylidenetriphenylphosphorane, 8CI
[21960-27-0]

$Ph_3P{=}CHCH(CH_3)_2$

$C_{22}H_{23}P$ M 318.397
Wittig reagent, ligand for Cu, Ag, Sn and Pb. Orange cryst. (THF/pentane). Mp 90-92°.

Yamamoto, Y., *J. Organomet. Chem.*, 1975, **96**, 133 (*synth, complexes*)
Skreekumar, C. *et al, J. Org. Chem.*, 1980, **45**, 4260 (*use*)
Kozikowski, A.P. *et al, Tetrahedron Lett.*, 1982, **23**, 2005 (*use*)
Yamamoto, Y. *et al, Bull. Chem. Soc. Jpn.*, 1982, **55**, 3025 (*complexes*)

(1-Methylpropyl)phenylphosphinic acid, 9CI M-00372

sec-*Butylphenylphosphinic acid*
[82224-12-2]

$H_3CCH_2CH(CH_3)P(O)(OH)Ph$

$C_{10}H_{15}O_2P$ M 198.201
(±)-***form***
Pentyl ester: Pentyl (1-methylpropyl)phenylphosphinate. Amyl sec-*butylphenylphosphinate.*
$C_{15}H_{25}O_2P$ M 268.335
Liq. d_4^{20} 1.01. $Bp_{0.2}$ 128-130°. n_D^{20} 1.4988.
Chloride:
$C_{10}H_{14}ClOP$ M 216.647
Liq. d_4^{20} 1.17. $Bp_{0.6}$ 118-120°. n_D^{20} 1.5348.

Siddall, T.H. *et al, J. Am. Chem. Soc.*, 1962, **84**, 2502 (*chloride, pmr*)
Legin, G.Ya., *Zh. Obshch. Khim.*, 1973, **43**, 2202 (*Engl. transl.* p. 2195) (*chloride, ester*)

(2-Methylpropyl)phenylphosphinic acid, 9CI M-00373

Isobutylphenylphosphinic acid

$(H_3C)_2CHCH_2P(O)(OH)Ph$

$C_{10}H_{15}O_2P$ M 198.201
Solid. Mp 64-65°.
2-Methylpropyl ester: 2-Methylpropyl (2-methylpropyl)phenylphosphinate. Isobutyl isobutylphenylphosphinate.
$C_{14}H_{23}O_2P$ M 254.308
Liq. $Bp_{0.5}$ 115-117°.

Arbuzov, A.E. *et al, J. Russ. Phys. Chem. Soc.*, 1929, **61**, 1905; *CA*, **24**, 5289 (*synth*)
Siddall, T.H. *et al, J. Am. Chem. Soc.*, 1962, **84**, 2502 (*chloride, pmr*)
Henning, H.-G., *J. Prakt Chem.*, 1965, **29**, 93 (*ester*)

(2-Methylpropyl)phosphine, 9CI M-00374

Isobutylphosphine, 8CI
[4023-52-3]

$(H_3C)_2CHCH_2PH_2$

$C_4H_{11}P$ M 90.105
Liq. Bp 60°, Bp 77°. pK_a −0.02 (H_2O). n_D^{25} 1.4308.

Kreutzkampf, N., *Chem. Ber.*, 1954, **87**, 919 (*synth*)
Rauhut, M.M. *et al, J. Org. Chem.*, 1961, **26**, 5138 (*synth*)
Fritzsche, H. *et al, Chem. Ber.*, 1965, **98**, 1681 (*synth*)

Methylpropylphosphinic acid, 9CI M-00375

[73342-45-7]

$H_3CCH_2CH_2P(O)(OH)Me$

$C_4H_{11}O_2P$ M 122.103
Liq. $Bp_{0.0001}$ 122°. n_D^{20} 1.4518.
2-Methylpropyl ester: 2-Methylpropyl methylpropylphosphinate. Isobutyl methylpropylphosphinate.
$C_8H_{19}O_2P$ M 178.211
Liq. $Bp_{0.1}$ 74°.
Chloride: [13213-39-3].
$C_4H_{10}ClOP$ M 140.549
Liq. d_4^{20} 1.13. Bp_{10} 88°. n_D^{20} 1.4635.

Maier, L., *Chem. Ber.*, 1961, **94**, 3051, 3056 (*synth, chloride*)
Neimysheva, A.A. *et al, Zh. Obshch. Khim.*, 1966, **36**, 1090 (*Engl. transl.* p. 1105) (*chloride*)
Knunyants, I.L. *et al, Dokl. Akad. Nauk SSSR, Ser. Sci. Khim.*, 1971, **201**, 862 (*Engl. transl.* p. 992) (*halides, nmr*)
Finke, M. *et al, Justus Liebigs Ann. Chem.*, 1974, 741 (*ester*)

(2-Methylpropyl)phosphinic acid, 9CI M-00376

Isobutylphosphinic acid, 8CI

$[(H_3C)_2CHCH_2]_2PH(O)OH$

$C_4H_{11}O_2P$ M 122.103
Tautomeric with (2-Methylpropyl)phosphonous acid, M-00383 .
Et ester: Ethyl (2-methylpropyl)phosphinate. Ethyl isobutylphosphinate.
$C_6H_{15}O_2P$ M 150.157

Liq. d_4^{20} 0.95. Bp_9 78-79°. n_D^{20} 1.4320.

Nifant'ev, É.E. *et al*, *Zh. Obshch. Khim.*, 1967, **37**, 1366 (*Engl. transl.* p. 1293)

(1-Methylpropyl)phosphonic acid, 9CI **M-00377**

sec-*Butylphosphonic acid*

[6778-87-6]

$H_3CCH_2CH(CH_3)P(O)(OH)_2$

$C_4H_{11}O_3P$ M 138.103

(±)-*form*

Deliquescent plates (C_6H_6/pet. ether). Mp 54-56°. pK_{a1} 2.74, pK_{a2} 8.48 (H_2O, 25°).

Di-Me ester: Dimethyl (1-methylpropyl)phosphonate. Dimethyl sec-*butylphosphonate.*

$C_6H_{15}O_3P$ M 166.156

Liq. Bp_{22} 70-72°, Bp_{11} 83-84°. n_D^{20} 1.4269.

Di-Et ester: Diethyl (1-methylpropyl)phosphonate. Diethyl sec-*butylphosphonate.*

$C_8H_{19}O_3P$ M 194.210

Liq. Bp_1 49-52°. n_D^{20} 1.4279.

Dibutyl ester: Dibutyl (1-methylpropyl)phosphonate. Dibutyl sec-*butylphosphonate.*

$C_{12}H_{27}O_3P$ M 250.317

Liq. Bp_{13} 143-146°. n_D^{20} 1.4322.

Dichloride: see (1-Methylpropyl)phosphonic dichloride, M-00379

Dianilide: P-*(1-Methylpropyl)*-N,N'-*diphenylphosphonic diamide.* sec-*Butylphosphonic dianilide.*

$C_{16}H_{21}N_2OP$ M 288.328

Solid. Mp 152-153°.

Kinnear, A.M. *et al*, *J. Chem. Soc.*, 1952, 3437 (*dianilide*)
Crofts, P.C. *et al*, *J. Am. Chem. Soc.*, 1953, **75**, 3379 (*synth, props*)
Stiles, A.R. *et al*, *J. Am. Chem. Soc.*, 1958, **80**, 714 (*ester*)
Bel'skii, V.E. *et al*, *Zh. Obshch. Khim.*, 1972, **42**, 2427 (*Engl. transl.* p. 2421) (*ester, P nmr*)
Inokawa, S. *et al*, *Synthesis*, 1973, 364; 1977, 179 (*esters, synth, pmr*)
Villieras, J. *et al*, *J. Organomet. Chem.*, 1978, **144**, 263 (*ester, synth, ir, pmr*)

(2-Methylpropyl)phosphonic acid, 9CI **M-00378**

Isobutylphosphonic acid

[4721-34-0]

$(H_3C)_2CHCH_2P(O)(OH)_2$

$C_4H_{11}O_3P$ M 138.103

Cryst. (cyclohexane). Mp 125-126°. pK_{a1} 2.70, pK_{a2} 8.43 (H_2O, 25°).

Di-Et ester: [50655-63-5]. *Diethyl (2-methylpropyl)-phosphonate. Diethyl isobutylphosphonate.*

$C_8H_{19}O_3P$ M 194.210

Liq. Bp_{13} 86-93°, Bp_3 83°.

Bis(2-methylpropyl) ester: [52928-43-5]. *Bis(2-methylpropyl) (2-methylpropyl)phosphonate. Diisobutyl isobutylphosphonate.*

$C_{12}H_{27}O_3P$ M 250.317

Extractant for Cd and Zn from aq. soln. Liq. Bp_{10} 133-134°.

Di-Ph ester: [53235-71-5]. *Diphenyl (2-methylpropyl)-phosphonate. Diphenyl isobutylphosphonate.*

$C_{16}H_{19}O_3P$ M 290.298

Characterised spectroscopically.

Dichloride: [5021-98-7].

$C_4H_9Cl_2OP$ M 174.994

Liq. d_4^{20} 1.25. Bp_{20} 90-92°. n_D^{20} 1.4641.

Arbuzov, A.E. *et al*, *Zh. Fiz. Khim.*, 1913, **45**, 690; *CA*, 7, 3599 (*synth, deriv*)
Fields, E.K. *et al*, *Chem. Ind. (London)*, 1960, 999 (*ester*)
Jason, E.F. *et al*, *J. Org. Chem.*, 1962, **27**, 1402 (*ester*)
Tsvetkov, E.N. *et al*, *Izv. Akad. Nauk SSSR, Ser. Khim.*, 1967, 2375 (*Engl. transl.* p. 2267) (*dichloride*)
Hudson, H.R. *et al*, *J. Chem. Soc., Perkin Trans. 1*, 1974, 982 (*ester, P nmr*)
Ernst, L., *Org. Magn. Reson.*, 1977, **9**, 35 (*ester, cmr, P nmr*)

(1-Methylpropyl)phosphonic dichloride, 9CI **M-00379**

sec-*Butylphosphonic dichloride,* 8CI

[4707-94-2]

$H_3CCH_2CH(CH_3)P(O)Cl_2$

$C_4H_9Cl_2OP$ M 174.994

(±)-*form*

Pungent liq. d_4^{20} 1.25. Bp_{20} 92°, Bp_7 75°. n_D^{20} 1.4701.

Grishina, O.N. *et al*, *Izv. Akad. Nauk SSSR, Ser. Khim.*, 1965, 2140 (*synth*)
Geiseler, G. *et al*, *Ber. Bunsenges. Phys. Chem.*, 1967, **71**, 478 (*ir, raman*)
Grishina, O.N. *et al*, *Zh. Prikl. Khim.*, 1969, **42**, 2289 (*Engl. transl.* p. 2149) (*synth*)
Razvodovskaya, L.V. *et al*, *Zh. Obshch. Khim.*, 1969, **39**, 1260 (*Engl. transl.* p. 1230) (*synth*)

(1-Methylpropyl)phosphonodithioic acid, 9CI **M-00380**

sec-*Butylphosphonodithioic acid,* 8CI

$H_3CCH_2CH(CH_3)P(O)(SH)_2 \rightleftharpoons H_3CCH_2CH(CH_3)P(O)(OH)(SH)$

$C_4H_{11}OPS_2$ M 170.224

(±)-*form*

O-*Et ester:* [55549-40-1]. O-*Ethyl hydrogen (1-methylpropyl)phosphonodithioate.*

$C_6H_{15}OPS_2$ M 198.278

Isol. as tri- or diethylammonium salt.

O-*Et ester, diethylammonium salt:* Solid. Mp 105-107°.

O-*Isopropyl ester:* [13953-50-9]. O-*Isopropyl hydrogen (1-methylpropyl)phosphonodithioate.*

$C_7H_{17}OPS_2$ M 212.304

Liq. d_4^{20} 1.06. $Bp_{0.15}$ 62°. n_D^{20} 1.5180.

Grishina, O.N. *et al*, *Izv. Akad. Nauk SSSR, Ser. Khim.*, 1966, 1617 (*Engl. transl.* p. 1558) (*isopropyl ester*)
Grishina, O.N. *et al*, *Zh. Obshch. Khim.*, 1970, **40**, 579 (*Engl. transl.* p. 548) (*ester, salts*)

(2-Methylpropyl)phosphonothioic acid, 9CI **M-00381**

Isobutylphosphonothioic acid, 8CI

$(H_3C)_2CHCH_2P(S)(OH)_2 \rightleftharpoons (H_3C)_2CHCH_2P(S)(OH)(SH)$

$C_4H_{11}O_2PS$ M 154.163

Waxy solid.

O,O-*Bis(2-methylpropyl)ester:* O,O-*Bis(2-methylpropyl) (2-methylpropyl)phosphonothioate.* O,O-*Diisobutyl isobutylphosphonothioate.*

$C_{12}H_{27}O_2PS$ M 266.378

Bp_{12} 136-136.5°. n_D^{20} 1.4420.

Dichloride: [6588-21-2].

$C_4H_9Cl_2PS$ M 191.055

Liq. d^{20} 1.25. Bp_{50} 110-113°.

Guichard, F., *Ber.*, 1899, **32**, 1572 (*dichloride*)
Razumov, A.I. *et al*, *Izv. Akad. Nauk SSSR, Ser. Khim.*, 1952, 894 (*Engl. transl.* p. 797) (*ester*)

(1-Methylpropyl)phosphonotrithioic acid, 9CI — M-00382

sec-*Butylphosphonotrithioic acid, 8CI*

$H_3CCH_2CH(CH_3)P(S)(SH)_2$

$C_4H_{11}PS_3$ M 186.285

(±)-*form*

Cyclic bis(anhydrosulfide): 2,4-Bis(1-methylpropyl)-1,3,2,4-dithiodiphosphetane 2,4-disulfide.
$C_8H_{18}P_2S_4$ M 304.418
Solid. Mp 92-93°.

Et-Ph diester: Ethyl phenyl (1-methylpropyl)-phosphonotrithioate.
$C_{12}H_{19}PS_3$ M 290.436
Liq. d_4^{20} 1.15. $Bp_{0.01}$ 147-148°. n_D^{20} 1.6230.

Grishina, O.N. *et al*, *Izv. Akad. Nauk SSSR, Ser. Khim.*, 1965, 1619 (*Engl. transl.* p. 1581) (*anhydrosulfide*)
Grishina, O.N. *et al*, *Zh. Obshch. Khim.*, 1967, **37**, 2276 (*Engl. transl.* p. 2162) (*esters, synth*)

(2-Methylpropyl)phosphonous acid, 9CI — M-00383

Isobutylphosphonous acid, 8CI

$(H_3C)_2CHCH_2P(OH)_2 \rightleftharpoons (H_3C)_2CHCH_2PH(O)OH$

$C_4H_{11}O_2P$ M 122.103

The free acid exists as the phosphoryl tautomer, (2-Methylpropyl)phosphinic acid, M-00376 . Bp_{100} 97-99°.

Mono-Et ester: see (2-Methylpropyl)phosphinic acid, M-00376

Bis(2-methylpropyl) ester: Bis(2-methylpropyl) (2-methylpropyl)phosphonite. Diisobutyl isobutylphosphonite.
$C_{12}H_{27}O_2P$ M 234.318
Liq. $Bp_{12.5}$ 105.5-107.5°. n_D^{20} 1.4290.

Dichloride: [17045-33-9]. *Dichloroisobutylphosphine.*
$C_4H_9Cl_2P$ M 158.995
Liq. d_4^{20} 1.20. Bp 149-151°, Bp_{100} 97-99°. n_D^{20} 1.4810.

Razumov, A.I. *et al*, *Izv. Akad. Nauk SSSR, Ser. Khim.*, 1952, 894 (*Engl. transl.* p. 797) (*dichloride, ester, synth*)
Henderson, W.A. *et al*, *J. Org. Chem.*, 1961, **26**, 4770 (*dichloride*)
Nifant'ev, É.E. *et al*, *Zh. Obshch. Khim.*, 1967, **37**, 1366 (*Engl. transl.* p. 1293) (*dichloride*)

(2-Methylpropyl)phosphoramidic acid, 9CI — M-00384

Isobutylphosphoramidic acid, 8CI

$(H_3C)_2CHCH_2NHP(O)(OH)_2$

$C_4H_{12}NO_3P$ M 153.117

Di-Et ester: Diethyl (2-methylpropyl)phosphoramidate. Diethyl isobutylphosphoramidate.
$C_8H_{20}NO_3P$ M 209.225
Liq. Bp_{14} 146°. n_D^{20} 1.4346.

Dibutyl ester: Dibutyl (2-methylpropyl)-phosphoramidate. Dibutyl isobutylphosphoramidate.
$C_{12}H_{28}NO_3P$ M 265.332
Liq. $Bp_{0.7}$ 139-140°. n_D^{25} 1.4351.

Di-Ph ester: Diphenyl (2-methylpropyl)-phosphoramidate. Diphenyl isobutylphosphoramidate.
$C_{16}H_{20}NO_3P$ M 305.313
Needles (EtOH). Mp 58°, Mp 69-70°. Bp_{11} 218°.

Dichloride:
$C_4H_{10}Cl_2NOP$ M 190.009
Liq. Bp_{14} 141°.

Dianilide: N-*2-Methylpropyl-*N′,N″*-diphenylphosphoric triamide.* N-*Isobutyl-*N′,N″*-diphenylphosphoric triamide.*
$C_{16}H_{22}N_3OP$ M 303.343
Solid. Mp 207°.

Michaelis, A., *Justus Liebigs Ann. Chem.*, 1903, **326**, 129; 1915, **407**, 290 (*derivs*)
Stock, J.A. *et al*, *J. Chem. Soc. (C)*, 1966, 637 (*dibutyl ester*)
Zwierzak, A., *Synthesis*, 1982, 920 (*diethyl ester, synth, ir, pmr, P nmr*)

2-Methylpropyl phosphorodichloridite, 9CI, 8CI — M-00385

Isobutyl dichlorophosphite

[52057-41-7]

$(H_3C)_2CHCH_2OPCl_2$

$C_4H_9Cl_2OP$ M 174.994

Pungent fuming liq. d_4^{22} 1.17. Bp_{12} 42-45°. n_D^{20} 1.4642.

Gerrard, W. *et al*, *J. Chem. Soc.*, 1953, 1920 (*synth, props*)

(1-Methylpropyl)-triphenylphosphonium(1+), 9CI — M-00386

sec-*Butyltriphenylphosphonium(1+)*

$Ph_3P^{\oplus}CH(CH_3)CH_2CH_3$

$C_{22}H_{24}P^{\oplus}$ M 319.405 (ion)

Treatment of salts with $NaNH_2/NH_3(l)$ yields the ylide.

Bromide: [3968-92-1].
$C_{22}H_{24}BrP$ M 399.309
Solid or cryst. (Et_2O/diisopropyl ether). Mp 230-232°.

Iodide: [4762-30-5].
$C_{22}H_{24}IP$ M 446.310
Cryst. (H_2O). Mp 213-216° (204-205.5°).

Ylide: [21481-98-1]. *(1-Methylpropylidene)-triphenylphosphorane.* sec-*Butylidenetriphenylphosphorane.*
$C_{22}H_{23}P$ M 318.397
Reactive Wittig reagent, prepd. *in situ*.

Eyles, C.T. *et al*, *J. Chem. Soc. (C)*, 1966, 67 (*bromide*)
Seyferth, D. *et al*, *J. Organomet. Chem.*, 1966, **6**, 205 (*bromide, pmr*)
Grim, S.O. *et al*, *J. Chem. Soc., Chem. Commun.*, 1967, 1191 (*ylide, nmr*)
Tokunaga, H. *et al*, *Bull. Chem. Soc. Jpn.*, 1972, **45**, 506 (*iodide, ylide*)
Albright, T.A. *et al*, *J. Am. Chem. Soc.*, 1975, **97**, 2942 (*bromide, cmr, nmr*)
Schaumann, E. *et al*, *Justus Liebigs Ann. Chem.*, 1979, 1702 (*iodide, pmr, ylide, props*)

(2-Methylpropyl)-triphenylphosphonium(1+), 9CI — M-00387

Isobutyltriphenylphosphonium(1+), 8CI

$Ph_3P^{\oplus}CH_2CH(CH_3)_2$

$C_{22}H_{24}P^{\oplus}$ M 319.405 (ion)
Bromide: [22884-29-3].
$C_{22}H_{24}BrP$ M 399.309
Cryst. (C_6H_6). Mp 188-190°.
Iodide: [60610-05-1].
$C_{22}H_{24}IP$ M 446.310
Solid. Mp 186-188°.

Hendrickson, J.B. *et al, Tetrahedron*, 1964, **20**, 449 (*iodide*)
Jaenicke, L. *et al, Justus Liebigs Ann. Chem.*, 1973, 1252 (*bromide*)
Effenberger, F. *et al, Chem. Ber.*, 1974, **107**, 278 (*bromide*)

(2-Methyl-4-pyridinyl)phosphonic acid, 9CI M-00388

2-Methyl-4-phosphonopyridine

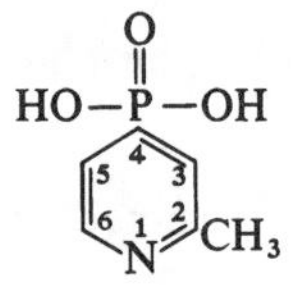

$C_6H_8NO_3P$ M 173.108
Di-Me ester: [78133-47-8]. *Dimethyl 2-methyl-4-pyridinephosphonate.*
$C_8H_{12}NO_3P$ M 201.161
Plates (EtOH). Mp 112-114°.

Katritzky, A.R. *et al, J. Chem. Soc., Perkin Trans. 1*, 1981, 668 (*synth, ir, pmr*)

(3-Methyl-2-pyridinyl)phosphonic acid, 9CI M-00389

3-Methyl-2-phosphonopyridine
[26384-91-8]
$C_6H_8NO_3P$ M 173.108
Cryst. (EtOH aq.). Mp 279-280°.
Di-Et ester: [26384-81-6]. *Diethyl (3-methyl-2-pyridinyl)phosphonate. 2-Diethylphosphoryl-3-methylpyridine. 2-Diethoxyphosphinyl-3-methylpyridine.*
$C_{10}H_{16}NO_3P$ M 229.215
Hygroscopic, absorbing $1H_2O$. $Bp_{0.07}$ 109-110°.

Redmore, D., *J. Org. Chem.*, 1970, **35**, 4114 (*synth, ester, pmr*)
Redmore, D., *Phosphorus Sulfur*, 1979, **5**, 271 (*cmr, P nmr*)

(3-Methyl-4-pyridinyl)phosphonic acid, 9CI M-00390

3-Methyl-4-phosphonopyridine
[58816-02-7]
$C_6H_8NO_3P$ M 173.108
Cryst. + $1H_2O$; cryst. (EtOH aq.). Mp 255-261°, 296°. pK_{a1} 4.85, pK_{a2} 6.90.
Di-Me ester: [78133-49-0]. *Dimethyl (3-methyl-4-pyridinyl)phosphonate. 3-Methyl-4-dimethylphosphorylpyridine. 3-Methyl-4-dimethoxyphosphinylpyridine.*
$C_8H_{12}NO_3P$ M 201.161
Plates (EtOH). Mp 93-94°.
Diisopropyl ester: [58815-97-7]. *Diisopropyl (3-methyl-4-pyridinyl)phosphonate. 3-Methyl-4-(diisopropylphosphoryl)pyridine. 3-Methyl-4-(diisopropoxyphosphinyl)pyridine.*
$C_{12}H_{20}NO_3P$ M 257.269
Liq. $Bp_{0.1}$ 103-105°.

Redmore, D., *J. Org. Chem.*, 1976, **41**, 2148 (*synth, props, cmr, ester*)

Boduszek, B. *et al, Synthesis*, 1979, 452 (*synth, ir, ms, pmr*)
Redmore, D., *Phosphorus Sulfur*, 1979, **5**, 271 (*cmr, P nmr*)
Katritzky, A.R. *et al, J. Chem. Soc., Perkin Trans. 1*, 1981, 668 (*ester, ir, pmr*)

(4-Methyl-2-pyridinyl)phosphonic acid, 9CI M-00391

4-Methyl-2-phosphonopyridine
[26384-90-7]
$C_6H_8NO_3P$ M 173.108
Cryst. (EtOH aq.). Mp 272-276°.
Di-Et ester: [26384-89-4]. *Diethyl (4-methyl-2-pyridinyl)phosphonate. 2-Diethylphosphoryl-4-methylpyridine. 2-Diethoxyphosphinyl-4-methylpyridine.*
$C_{10}H_{16}NO_3P$ M 229.215
Hygroscopic liq., absorbing $1H_2O$. $Bp_{0.05}$ 109-112°.

Redmore, D., *J. Org. Chem.*, 1970, **35**, 4114 (*synth, ester, ir, pmr*)
Redmore, D., *Phosphorus Sulfur*, 1979, **5**, 271 (*cmr, P nmr*)

(6-Methyl-2-pyridinyl)phosphonic acid, 9CI M-00392

2-Methyl-6-phosphonopyridine
[26384-88-3]
$C_6H_8NO_3P$ M 173.108
Cryst. (EtOH aq.). Mp 277-280°.
Di-Et ester: [26384-87-2]. *Diethyl (6-methyl-2-pyridinyl)phosphonate. 2-Diethylphosphoryl-6-methylpyridine. 2-Diethoxyphosphinyl-6-methylpyridine.*
$C_{10}H_{16}NO_3P$ M 229.215
Hygroscopic liq. absorbing $1H_2O$. d_4^{20} 1.14. $Bp_{1.5}$ 140°. n_D^{20} 1.4931.
Dichloride:
$C_6H_4Cl_2NOP$ M 207.983
Solid. Mp 75-81°. $Bp_{0.01}$ 100-110°.

Redmore, D., *J. Org. Chem.*, 1970, **35**, 4114 (*synth, ester, pmr*)
Redmore, D., *Phosphorus Sulfur*, 1979, **5**, 271 (*cmr, P nmr*)
Eliseenkov, V.N. *et al, Khim. Geterosikl. Soedin.*, 1974, **10**, 1354 (*Engl. transl.* p. 1182) (*synth*)

[2-(Methylseleno)phenyl]diphenylphosphine, 9CI M-00393

[16566-17-9]

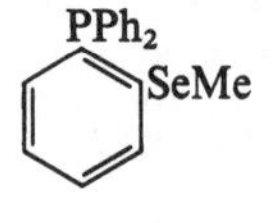

$C_{19}H_{17}PSe$ M 355.277
Ligand for Pd and Ru.

Roundhill, D.M. *et al, J. Am. Chem. Soc.*, 1979, **101**, 5428; *Inorg. Chem.*, 1980, **19**, 3365 (*complexes, cmr, pmr*)

3′-Methylspiro[1,3,2-benzodioxaphosphole-2,2(3′*H*)-[1,3,2]benzothiazaphosphole], 10CI M-00394

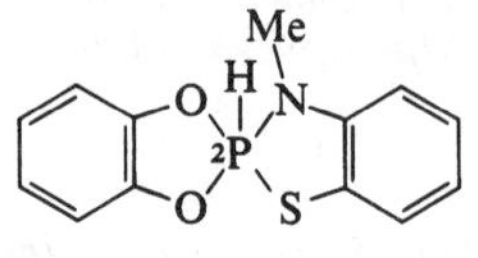

$C_{13}H_{12}NO_2PS$ M 277.277

2-Phenoxy: [69774-94-3].
$C_{19}H_{16}NO_3PS$ M 369.374
Cryst. (CH_2Cl_2/pet. ether or EtOAc/pet. ether). Mp 82-83°.
2-Dimethylamino: [69774-96-5]. *3′,N,N-Trimethylspiro[1,3,2-benzodioxaphosphole-2,2′(3′H)-[1,3,2]-benzothiazaphosphol]-2-amine.*
$C_{15}H_{17}N_2O_2PS$ M 320.345
Cryst. (CH_2Cl_2/pet. ether or EtOAc/pet. ether). Mp 78-79°.

Singh, S. *et al, J. Chem. Soc., Perkin Trans. 1,* 1978, 1438 (*synth, P nmr*)

3′-Methylspiro[1,3,2-benzodioxaphosphole-2,2′(3′H)-[1,3,2]benzoxazaphosphole] M-00395

[51675-92-4]

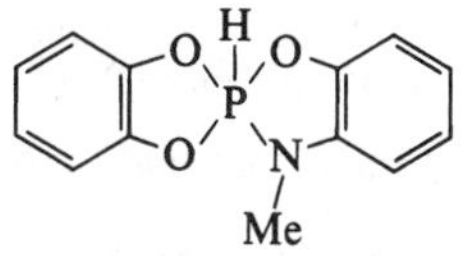

$C_{13}H_{12}NO_3P$ M 261.216
Undistorted trigonal bipyramidal struct. with equatorial P-H bond at r.t. At 150°, undergoes tautom. with monocyclic form. Cryst. (C_6H_6 or Et_2O). Mp 103-105° (99°).
2-Me: [79129-16-1]. *2,3′-Dimethylspiro[1,3,2-benzodioxaphosphole-2,2′(3′H)-[1,3,2]-benzoxazaphosphole].*
$C_{14}H_{14}NO_3P$ M 275.243
Yellow liq. $Bp_{0.1}$ 136-140°.

Burgada, R. *et al, J. Organomet. Chem.,* 1974, **66,** 255 (*synth, P nmr, struct*)
Clark, T.E. *et al, Inorg. Chem.,* 1979, **18,** 1653 (*synth, cryst struct*)
Wieber, M. *et al, Z. Anorg. Allg. Chem.,* 1981, **477,** 108 (*deriv, synth, pmr, P nmr*)

5′-Methylspiro[1,3,2-benzodioxaphosphole-2,2′(3H)-[1,2]oxaphosphole] M-00396

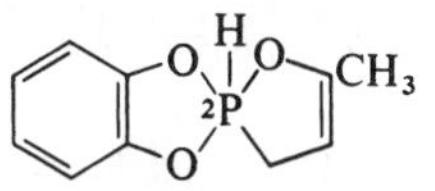

$C_{10}H_{11}O_3P$ M 210.169
2-Methoxy: [55055-20-4].
$C_{11}H_{13}O_4P$ M 240.195
Liq. d_4^{20} 1.28. Bp_1 110-112°. n_D^{20} 1.5400.
2-Ethoxy: [55055-21-5].
$C_{12}H_{15}O_4P$ M 254.222
Liq. d_4^{20} 1.24. Bp_1 131-132°. n_D^{20} 1.5340.
2-Ph: [39055-19-1].
$C_{16}H_{15}O_3P$ M 286.266
Liq. Bp_1 158-160°.

Razumova, N.A. *et al, Zh. Obshch. Khim.,* 1972, **42,** 2114 (*Engl. transl.* p. 2109) (*synth, pmr, P nmr*)
Gruk, M.P. *et al, Zh. Obshch. Khim.,* 1974, **44,** 2645 (*Engl. transl.* p. 2601) (*synth, pmr, P nmr*)
Vasil'ev, V.V. *et al, Zh. Obshch. Khim.,* 1976, **46,** 463, 1739 (*Engl. transl.* pp. 461, 1690) (*synth, P nmr*)

2-Methyl-2,2′-spirobi[1,3,2-benzodioxaphosphole], 9CI M-00397

[21229-04-9]

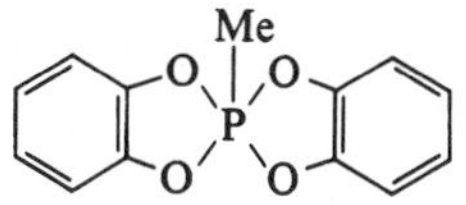

$C_{13}H_{11}O_4P$ M 262.201
Cryst. (toluene). Mp 72-74°. $Bp_{0.1}$ 135-140°.

Wieber, M. *et al, Monatsh. Chem.,* 1970, **101,** 776 (*synth*)
Doak, G.O. *et al, J. Chem. Soc. (A),* 1971, 1295 (*synth, pmr, P nmr*)
Wunderlich, H., *Acta Crystallogr., Sect. B,* 1974, **30,** 939 (*cryst struct*)
Gloede, J., *Z. Anorg. Allg. Chem.,* 1983, **500,** 59 (*synth*)

[(Methylsulfinyl)methyl]phosphonic acid, 9CI M-00398

$MeSOCH_2P(O)(OH)_2$

$C_2H_7O_4PS$ M 158.109
Esters are Wittig-Horner reagents for the synth. of methyl vinyl sulfoxides.

(±)-form

Di-Me ester: [65915-23-3]. *Dimethyl [(methylsulfinyl)-methyl]phosphonate.*
$C_4H_{11}O_4PS$ M 186.162
Cryst. (Me_2CO/C_6H_6). Mp 52-54°.
Di-Et ester: [65915-24-4]. *Diethyl [(methylsulfinyl)-methyl]phosphonate.*
$C_6H_{15}O_4PS$ M 214.216
Oil. n_D^{20} 1.4768.

Mikolajczyk, M. *et al, Synthesis,* 1973, 669 (*synth*)
Mikolajczyk, M. *et al, J. Org. Chem.,* 1975, **40,** 1979 (*use*)
Mikolajczyk, M. *et al, Synthesis,* 1975, 278 (*use*)
Drabowicz, J. *et al, Synthesis,* 1978, 758 (*synth, nmr*)
Mikolajczyk, M. *et al, J. Org. Chem.,* 1978, **43,** 2518 (*props*)

[(Methylsulfonyl)methylene]triphenylphosphorane, 9CI M-00399

[5554-83-6]

$Ph_3P{=}CHSO_2Me$

$C_{20}H_{19}O_2PS$ M 354.403
Cryst. ($CHCl_3$/pet. ether). Sulfone-stabilized ylide. Mp 200-203°.

Speziale, A.J. *et al, J. Am. Chem. Soc.,* 1965, **87,** 5603 (*synth, ir, uv*)
Van Leusen, A.M. *et al, Recl. Trav. Chim. Pays-Bas,* 1972, **91,** 37 (*synth, ir, pmr*)
Reith, B.A. *et al, J. Org. Chem.,* 1974, **39,** 2728 (*pmr, ir*)

[(Methylsulfonyl)methyl]phosphonic acid, 9CI M-00400

$MeSO_2CH_2P(O)(OH)_2$

$C_2H_7O_5PS$ M 174.108
Di-Me ester: [25508-33-2]. *Dimethyl [(methylsulfonyl)-methyl]phosphonate.*
$C_4H_{11}O_5PS$ M 202.162
Cryst. (C_6H_6). Mp 82°.
Di-Et ester: [40137-11-9]. *Diethyl [(methylsulfonyl)-methyl]phosphonate.*
$C_6H_{15}O_5PS$ M 230.215
Cryst. (C_6H_6). Mp 96°.

Shahak, I. *et al*, *Synthesis*, 1969, 170 (*synth*)
Posner, G.H. *et al*, *J. Org. Chem.*, 1972, **37**, 3547 (*synth, use*)
Mikolajczyk, M. *et al*, *Synthesis*, 1975, 278 (*use*)
Fillion, H. *et al*, *J. Heterocycl. Chem.*, 1978, **15**, 753 (*synth, use*)
De Jong, B.E. *et al*, *Recl. Trav. Chim. Pays-Bas*, 1981, **100**, 410 (*synth, pmr, use*)

Methylsulfonylphosphoramidic acid, 9CI **M-00401**

$MeSO_2NHP(O)(OH)_2$

CH_6NO_5PS M 175.096

Di-Me ester: [7109-15-1]. *Dimethyl methylsulfonylphosphoramidate.*
$C_3H_{10}NO_5PS$ M 203.149
Needles. Mp 111-112°.

Di-Et ester: Diethyl methylsulfonylphosphoramidate.
$C_5H_{14}NO_5PS$ M 231.203
Solid. Mp 96-96.5°. Exists largely in the tautomeric imino form.

Dichloride: [29651-27-2].
$CH_4Cl_2NO_3PS$ M 211.987
Cryst. (CCl_4 or ligroin). Mp 78-81°.

Kirsanov, A.V. *et al*, *Zh. Obshch. Khim.*, 1955, **25**, 1140 (*Engl. transl.* p. 1093) (*dichloride*)
Goldstein, J.A., *J. Org. Chem.*, 1977, **42**, 2466 (*dimethyl ester, synth, ir, pmr*)
Kabachnik, M.I. *et al*, *Izv. Akad. Nauk SSSR, Ser. Khim.*, 1961, 819, 1022 (*Engl. transl.* pp. 758, 945) (*diethyl ester, ir, struct*)

(Methylsulfonyl)phosphorimidic acid, 9CI **M-00402**

$MeSO_2N{=}P(OH)_3$

CH_6NO_5PS M 175.096
Tautomeric with Methylsulfonylphosphoramidic acid, M-00401 .

Tri-Me ester: [7109-06-0]. *Trimethyl methylsulfonylphosphorimidate.* P,P,P-*Trimethoxy*-N-*methylsulfonylphosphazene.* P,P,P-*Trimethoxy*-N-*methylsulfonylphosphine imide.*
$C_4H_{12}NO_5PS$ M 217.176
Solid.

Tri-Et ester: Triethyl methylsulfonylphosphorimidate. P,P,P-*Triethoxy*-N-*methylsulfonylphosphazene.* P,P,P-*Triethoxy*-N-*methylsulfonylphosphine imide.*
$C_6H_{18}NO_5PS$ M 247.246
Oil. d_4^{20} 1.22. $Bp_{0.05}$ 102-104°. n_D^{25} 1.4446.

Tri-Ph ester: Triphenyl methylsulfonylphosphorimidate. N-*Methylsulfonyl*-P,P,P-*triphenoxyphosphazene.* N-*Methylsulfonyl*-P,P,P-*triphenoxyphosphine imide.*
$C_{19}H_{18}NO_5PS$ M 403.389
Needles (C_6H_6), cubes (Et_2O), cryst. (MeOH). Mp 90-92°.

Trichloride: [29651-24-9]. P,P,P-*Trichloro*-N-*methylsulfonylphosphazene.*
$CH_3Cl_2NO_2PS$ M 194.980
Needles. Mp 47-50°.

Kirsanov, A.V. *et al*, *Zh. Obshch. Khim.*, 1955, **25**, 187 (*Engl. transl.* p. 171) (*trichloride*)
Kirsanov, A.V. *et al*, *Zh. Obshch. Khim.*, 1958, **28**, 1052 (*Engl. transl.* p. 1023) (*triphenyl ester*)
Goerdelier, J. *et al*, *Chem. Ber.*, 1961, **94**, 1067 (*esters*)
Goldstein, J.A. *et al*, *J. Org. Chem.*, 1977, **42**, 2466 (*trimethyl ester, synth, pmr*)

17-Methyl-6,7,9,10-tetrahydrodibenzo[*d,m*][1,3,6,9,12,2]-pentaoxaphosphacyclotetradecin **M-00403**

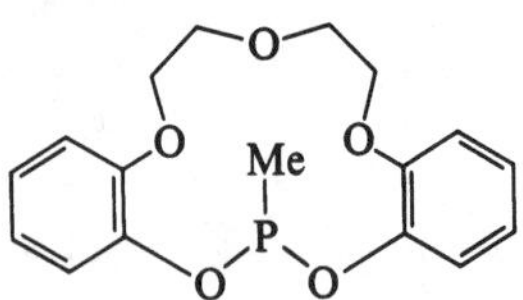

$C_{17}H_{19}O_5P$ M 334.308

17-Oxide: [60331-23-9].
$C_{17}H_{19}O_6P$ M 350.307
Ligand for alkali and alkaline-earth metals. Needles ($Me_2CO/CHCl_3$). Mp 104-106°, Mp 128-130°.

17-Sulfide: [71787-57-0].
$C_{17}H_{19}O_5PS$ M 366.368
Cryst. (pet. ether or EtOAc). Mp 92-93°.

Yatsimirskii, K.B. *et al*, *Teor. Eksp. Khim.*, 1976, **12**, 421; 1983, **19**, 500 (*Engl. transl.* pp. 326, 465) (*synth, ir, pmr, complexes*)
Kudrya, T.N. *et al*, *Zh. Obshch. Khim.*, 1978, **48**, 927 (*Engl. transl.* p. 844) (*synth*)
Yatsimirskii, K.B. *et al*, *Zh. Neorg. Khim.*, 1980, **25**, 63 (*Engl. transl.* p. 32) (*pmr*)
Golovatyi, V.G. *et al*, *Teor. Eksp. Khim.*, 1981, **17**, 849 (*Engl. transl.* p. 671) (*ms*)
Malinovskii, T.S. *et al*, *Zh. Strukt. Khim.*, 1984, **25**, 130 (*sulfide, cryst struct*)
Raevskii, O.A. *et al*, *Izv. Akad. Nauk SSSR, Ser. Khim.*, 1984, 797 (*Engl. transl.* p. 732) (*derivs, ir, complexes*)

3-Methyltetrahydro-2-phenoxy-2*H*-1,3,2-oxazaphosphorine, 10CI **M-00404**

3-Methyl-2-phenoxy-1,3,2-oxazaphosphorinane

$C_{10}H_{14}NO_2P$ M 211.200

2-Sulfide: [71093-78-2].
$C_{10}H_{14}NO_2PS$ M 243.260
Cryst. (pet. ether). Mp 65-66°.

Karolek-Wojciechowska, J. *et al*, *J. Chem. Soc., Perkin Trans. 1*, 1979, 146 (*synth, ir, pmr, cryst struct*)

4′-Methyl-4,4,5,5-tetrakis(trifluoromethyl)-spiro[1,3,2-dioxaphospholane-2,1′-[2,6,7]trioxa[1]phosphabicyclo[2.2.2]-octane] **M-00405**

[36296-79-4]

$C_{11}H_9F_{12}O_5P$ M 480.144
Cryst. (C_6H_6/CH_2Cl_2). Mp 225-228° dec.

Ramirez, F. *et al*, *Phosphorus*, 1971, **1**, 1 (*synth, pmr, P nmr*)

Methyl tetramethylphosphorodiamidate **M-00406**

[7393-11-5]

$(Me_2N)_2P(O)OMe$

$C_5H_{15}N_2O_2P$ M 166.159
Liq. Bp_{15} 95°, Bp_1 49-50°. n_D^{25} 1.4368.

Cheymol, J. *et al*, *C.R. Hebd. Seances Acad. Sci.*, 1959, **249**, 1240 (*synth*)
Terry, P.H. *et al*, *J. Agric. Food Chem.*, 1973, **21**, 500 (*synth*)
Laskorin, B.N. *et al*, *Dokl. Akad. Nauk SSSR, Ser. Sci. Khim.*, 1974, **215**, 595 (*Engl. transl.* p. 184) (*ir, pmr, P nmr*)
Pressl, K.-D. *et al*, *Z. Anorg. Allg. Chem.*, 1977, **434**, 171 (*ir, raman*)

5-Methyl-1,2-thiaphospholane 2-sulfide M-00407

$C_4H_9PS_2$ M 152.209

2-Et: [41392-23-8].
$C_6H_{13}PS_2$ M 180.262
Liq. d_4^{20} 1.15. $Bp_{0.2}$ 118°. n_D^{20} 1.5937.
2-Ph: [41392-21-6].
$C_{10}H_{13}PS_2$ M 228.306
Liq. d_4^{20} 1.25. $Bp_{0.3}$ 156°. n_D^{20} 1.6531.
2-Chloro: [76442-64-3].
$C_4H_8ClPS_2$ M 186.654
Liq. $Bp_{0.01}$ 82°. n_D^{25} 1.6133. Desulfurized by Ph_3P to tervalent chloride.

Arbuzov, B.A. *et al*, *Izv. Akad. Nauk SSSR, Ser. Khim.*, 1973, 648 (*Engl. transl.* p. 621) (*synth*)
U.S.P., 4 231 970, (*1980*); *CA*, **94**, 84303 (*synth, props*)

3-Methyl-1,3,2-thiazaphospholidine M-00408

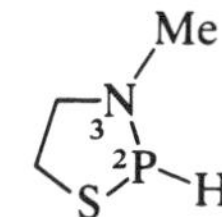

C_3H_8NPS M 121.137

2-Chloro, 2-oxide: [28004-49-1].
$C_3H_7ClNOPS$ M 171.581
Oil. $Bp_{0.003}$ 120°.
2-Chloro, 2-sulfide: [28004-51-5].
$C_3H_7ClNPS_2$ M 187.642
Solid. Mp 30-32°.
2-Methoxy, 2-oxide: [21124-98-1].
$C_4H_{10}NO_2PS$ M 167.162
Oil. $Bp_{0.05}$ 95-97°. n_D 1.5180.
2-Methoxy, 2-sulfide:
$C_4H_{10}NOPS_2$ M 183.223
Oil. $Bp_{0.05}$ 95-97°. n_D^{20} 1.5180.
2-Ethoxy, 2-oxide: [21124-99-2].
$C_5H_{12}NO_2PS$ M 181.189
Oil. $Bp_{0.05}$ 100-102°. n_D^{20} 1.5070.
2-Ethoxy, 2-sulfide:
$C_5H_{12}NOPS_2$ M 197.250
Oil. $Bp_{0.05}$ 100-102°. n_D^{20} 1.5070.

Fr. Pat., 1 537 175, (*1968*); *CA*, **71**, 30585
Savignac, P. *et al*, *C.R. Hebd. Seances Acad. Sci., Ser. C*, 1968, **267**, 183; 1970, **270**, 2086.

P-Methyl-*N*-2-thiazolylphosphonamidic acid, 9CI M-00409

$C_4H_7N_2O_2PS$ M 178.145

Ph ester: [68236-57-7]. *Phenyl* P-*methyl*-N-*2-thiazolylphosphonamidate.*
$C_{10}H_{11}N_2O_2PS$ M 254.243
Insecticide and acaricide. Solid. Mp 83-85°.

Ger. Pat., 2 703 363, (*1978*); *CA*, **90**, 6535 (*derivs, synth, use*)
B.P., 1 544 778, (*1979*); *CA*, **92**, 22498 (*derivs, synth, use*)

2-Methylthioadenosine 5′-(dihydrogen phosphate), 8CI M-00410

2-Methylthio-AMP
[22140-20-1]

$C_{11}H_{16}N_5O_7PS$ M 393.310
Specific inhibitor of platelet aggregation. Mp 192-195° dec.

Tener, G.M., *J. Am. Chem. Soc.*, 1961, **83**, 159.
Michael, F. *et al*, *Nature (London)*, 1969, **222**, 1073 (*synth*)
B.P., 1 226 699, (*1971*); *CA*, **75**, 20922d

2-Methylthio-4*H*-1,3,2-benzodioxaphosphorin M-00411

$C_8H_9O_2PS$ M 200.192

2-Oxide: [18865-25-3].
$C_8H_9O_3PS$ M 216.191
Phosphorylating and alkylating agent. Cryst. (Et_2O). Mp 44°. $Bp_{0.1}$ 144-145°.
2-Sulfide: [7234-22-2].
$C_8H_9O_2PS_2$ M 232.252
Solid. Mp 69-70°.

Kobayashi, K. *et al*, *J. Agric. Chem. Soc. Jpn.*, 1966, **40**, 315; *CA*, **66**, 10883 (*sulfide*)
Eto, M. *et al*, *Agric. Biol. Chem.*, 1968, **32**, 1056; 1974, **38**, 2081; 1981, **45**, 915 (*oxide, use, pmr*)
Iio, M. *et al*, *Agric. Biol. Chem.*, 1973, **37**, 115 (*oxide, synth, use*)

[1-(Methylthio)ethyl]phosphonic acid, 9CI M-00412

$MeSCH(CH_3)P(O)(OH)_2$

$C_3H_9O_3PS$ M 156.136

(±)-*form*

Di-Et ester: Diethyl [1-(methylthio)ethyl]phosphonate.
$C_7H_{17}O_3PS$ M 212.243
Wittig-Horner reagent for the conversion of aldehydes and ketones into higher methyl ketones. Liq. $Bp_{0.2}$ 68-70°. n_D^{13} 1.4620.

Corey, E.J. *et al*, *J. Org. Chem.*, 1970, **35**, 777 (*synth, ir, pmr, use*)
McGuire, H.M. *et al*, *J. Chem. Soc., Perkin Trans. 1*, 1974, 1879 (*use*)
Mikolajczyk, M. *et al*, *J. Org. Chem.*, 1979, **44**, 2967 (*synth, use*)

Methylthioimidodiphosphoric acid, 9CI **M-00413**

$$(HO)_2P(O){-}NMe{-}P(S)(OH)_2 \rightleftharpoons (HO)_2P(O){-}NMe{-}P(O)(OH)(SH)$$

$CH_7NO_5P_2S$ M 207.078

O,O,O,O-*Tetra-Me ester:* O,O-*Dimethyl (dimethoxyphosphinyl)methylphosphoramidothioate.*
$C_5H_{15}NO_5P_2S$ M 263.185
Liq. d_4^{20} 1.31. $Bp_{0.03}$ 86°. n_D^{20} 1.4788.

O,O,O,O-*Tetra-Et ester:* [53227-37-5]. O,O-*Diethyl (diethoxyphosphinyl)methylphosphoramidothioate.*
$C_9H_{23}NO_5P_2S$ M 319.292
Liq. d_4^{20} 1.04. $Bp_{0.03}$ 91°. n_D^{20} 1.4650.

Gusar, N.I. *et al*, *Zh. Obshch. Khim.*, 1974, **44**, 1456 (*Engl. transl.* p. 1430) (*tetra-Et ester*)
Grechkin, N.P. *et al*, *Zh. Obshch. Khim.*, 1976, **46**, 1753 (*Engl. transl.* p. 1703) (*tetra-Me ester, synth, P nmr*)

(Methylthio)imidodiphosphoryl chloride, 9CI **M-00414**
(Methylthio)iminodiphosphoryl tetrachloride
[38568-67-1]

$MeN[P(S)Cl_2]_2$

$CH_3Cl_4NP_2S_2$ M 296.921
Cryst. (pet. ether). Mp 61-62°.

Keat, R., *J. Chem. Soc., Dalton Trans.*, 1972, 2189 (*synth, ir, pmr, P nmr*)
Hägele, G. *et al*, *J. Chem. Soc., Dalton Trans.*, 1974, 1985 (*pmr, P nmr*)
Dagleish, W.H. *et al*, *J. Chem. Soc., Dalton Trans.*, 1977, 1505 (*nqr*)

[(Methylthio)methyl]phosphonic acid, 8CI **M-00415**

$MeSCH_2P(O)(OH)_2$

$C_2H_7O_3PS$ M 142.109
Esters form carbanions which may be alkylated or acylated, and which undergo Wittig-Horner reaction with aldehydes or ketones to give 1-methylthioalkenes, and thence methyl ketones.

Di-Me ester: [25508-32-1]. *Dimethyl [(methylthio)-methyl]phosphonate.*
$C_4H_{11}O_3PS$ M 170.163
Bp_{30} 138-140°, $Bp_{0.02}$ 55°. n_D^{20} 1.4780.

Di-Et ester: [28460-01-7]. *Diethyl [(methylthio)-methyl]phosphonate.*
$C_6H_{17}O_3PS$ M 200.232
Liq. $Bp_{0.2}$ 70-72°. n_D^{20} 1.4635.

S-*oxide: see [(Methylsulfinyl)methyl]phosphonic acid, M-00398*

S-*dioxide: see [(Methylsulfonyl)methyl]phosphonic acid, M-00400*

Green, M., *J. Chem. Soc.*, 1963, 1324 (*synth, ir, pmr, use*)
Shahak, I. *et al*, *Synthesis*, 1969, 170 (*synth, pmr, use*)
Mikolajczyk, M. *et al*, *Synthesis*, 1973, 669 (*synth*)
Mikolajczyk, M. *et al*, *J. Org. Chem.*, 1979, **44**, 2967 (*synth, pmr, P nmr*)
Mikolajczyk, M. *et al*, *Synthesis*, 1980, 127 (*use*)
Smith, J.G. *et al*, *J. Org. Chem.*, 1983, **48**, 1110 (*ir, pmr, ms*)

[(Methylthio)methyl]-triphenylphosphonium(1+), 9CI **M-00416**

$Ph_3P^{\oplus}CH_2SMe$

$C_{20}H_{20}PS^{\oplus}$ M 323.412 (ion)

Chloride: [1779-54-0].
$C_{20}H_{20}ClPS$ M 358.865
Source of ylide. Exhibits herbicidal props. Cryst. ($MeNO_2$). Mp 219.5-220.5°.

Tetraphenylborate:
$C_{44}H_{40}BPS$ M 642.644
Cryst. (Me_2CO/EtOH). Mp 210.5-211°.

Ylide: [23462-73-9]. *[(Methylthio)methylene]-triphenylphosphorane.*
$C_{20}H_{19}PS$ M 322.404
Wittig reagent.

Wittig, G. *et al*, *Chem. Ber.*, 1961, **94**, 1373 (*synth, props*)
Speziale, A.J. *et al*, *J. Am. Chem. Soc.*, 1965, **87**, 5607 (*chloride*)
Raasch, M.S., *J. Org. Chem.*, 1972, **37**, 1347 (*synth, props*)
Schlosser, M. *et al*, *Chimica*, 1982, **36**, 396 (*ylide*)
Cameron, A.G. *et al*, *J. Chem. Soc., Perkin Trans. 1*, 1983, 2979 (*chloride, ir, pmr*)

[2-(Methylthio)phenyl]diphenylphosphine, 10CI, 9CI **M-00417**
o-*(Diphenylphosphino)thioanisole*
[14791-94-7]

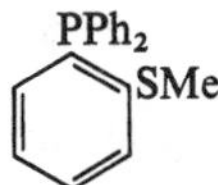

$C_{19}H_{17}PS$ M 308.377
Ligand for Pd and Ru. Cryst. (EtOH). Mp 104.5-105°.

B,PhCH₂Cl: Benzyl(2-methylthiophenyl)-diphenylphosphonium chloride.
$C_{26}H_{24}ClPS$ M 434.962
Cryst. (EtOH/EtOAc). Mp 220-222°.

McEwen, W.E. *et al*, *J. Org. Chem.*, 1976, **41**, 1684 (*synth, pmr, deriv*)
Inorg. Synth., 1976, **16**, 168 (*synth*)
Roundhill, D.M. *et al*, *Inorg. Chem.*, 1980, **19**, 3365 (*cmr, complexes*)

(4-Methylthiophenyl)phosphonic acid, 9CI **M-00418**
[46061-42-1]

$$4\text{-}MeS{-}C_6H_4{-}P(O)(OH)_2$$

$C_7H_9O_3PS$ M 204.180
Solid. Mp 168-169°, 186-189°.

Di-Et ester: Diethyl (4-methylthiophenyl)phosphonate.
$C_{11}H_{17}O_3PS$ M 260.287
Liq. $Bp_{0.1\text{-}0.4}$ 150° (oven).

Miles, J.A. *et al*, *J. Org. Chem.*, 1975, **40**, 343 (*synth, pmr*)
Grabiak, R.C. *et al*, *Phosphorus Sulfur*, 1980, **9**, 197 (*synth, derivs, pmr, nmr*)

Methyl(trichloromethyl)phosphinic acid, 9CI **M-00419**
[69404-21-3]

$Cl_3CPMe(O)OH$

$C_2H_4Cl_3O_2P$ M 197.385
Cryst. (EtOH or C_6H_6). Mp 161-161.5°.
Et ester: [20543-88-8]. *Ethyl methyl(trichloromethyl)-phosphinate.*
$C_4H_8Cl_3O_2P$ M 225.439
Solid. d_4^{20} 1.41. Mp 40°. Bp_5 97-98.5°. n_D^{20} 1.4801.
Isopropyl ester: Isopropyl methyl(trichloromethyl)-phosphinate.
$C_5H_{10}Cl_3O_2P$ M 239.466
Solid. Mp 45°.
Chloride: [20543-86-6].
$C_2H_3Cl_4OP$ M 215.831
Solid. Mp 84°.

Reinhardt, H. *et al*, *Chem. Ber.*, 1957, **90**, 1656 (*synth*)
Dmitrieva, L.E. *et al*, *Zh. Obshch. Khim.*, 1968, **38**, 157 (*Engl. transl.* p. 154) (*chloride, esters*)
Efimova, V.D. *et al*, *Zh. Obshch. Khim.*, 1974, **44**, 55 (*Engl. transl.* p. 51) (*ester*)

Methyl(trifluoromethyl)phosphinic acid M-00420

$Me(F_3C)P(O)OH$

$C_2H_4F_3O_2P$ M 148.021
Me ester: [82403-26-7]. *Methyl methyl(trifluoromethyl)phosphinate.*
$C_3H_6F_3O_2P$ M 162.048
Liq. Bp_{50} 71°.
Butyl ester: [82403-28-9]. *Butyl methyl(trifluoromethyl)phosphinate.*
$C_6H_{12}F_3O_2P$ M 204.129
Liq. Bp_{30} 93-95°.

Maslennikov, I.G. *et al*, *Zh. Obshch. Khim.*, 1982, **52**, 935 (*Engl. transl.* p. 816) (*synth, pmr, nmr*)

[3-Methyl-5-(2,6,6-trimethyl-1-cyclohexen-1-yl)-2,4-pentadienyl]-triphenylphosphonium(1+), 9CI M-00421

(β-Ionylideneethyl)triphenylphosphonium(1+)

(E,E)-form

$C_{33}H_{38}P^{\oplus}$ M 465.637 (ion)

(E,E)-form

Chloride: [53282-28-3].
$C_{33}H_{38}ClP$ M 501.090
Source of ylide, obt. with RLi.
Bromide: [62285-98-7].
$C_{33}H_{38}BrP$ M 545.541
Source of ylide.
Ylide: [71987-74-1]. *[3-Methyl-5-(2,6,6-trimethyl-1-cyclohexen-1-yl)-2,4-pentadienylidene]-triphenylphosphorane.*
$C_{33}H_{37}P$ M 464.629
Used in Wittig reactions.

Buddrus, J. *et al*, *Chem. Ber.*, 1974, **107**, 2050 (*chloride*)
Davalian, D. *et al*, *J. Org. Chem.*, 1979, **44**, 4988 (*bromide*)
Dawson, M.I. *et al*, *J. Med. Chem.*, 1981, **24**, 583 (*use*)
Pardini, V.L. *et al*, *J. Chem. Soc., Perkin Trans. 2*, 1981, 1520 (*iodide*)
Mehta, R.R. *et al*, *J. Chem. Soc., Perkin Trans. 1*, 1982, 2921 (*ylide, synth*)
Broek, A.D. *et al*, *Recl. Trav. Chim. Pays-Bas*, 1983, **102**, 46 (*use*)

[1-Methyl-3-(2,6,6-trimethyl-1-cyclohexen-1-yl)-2-propenyl]-triphenylphosphonium(1+), 9CI M-00422

β-Ionylidenetriphenylphosphonium(1+). 4-[(2,6,6-Trimethyl-1-cyclohexen-1-yl)-3-buten-2-yl]-triphenylphosphonium(1+)

(E)-form

$C_{31}H_{36}P^{\oplus}$ M 439.599 (ion)
With RLi or KOH → ylide.

(E)-form

Chloride: [77837-61-7].
$C_{31}H_{36}ClP$ M 475.052
No phys. props. reported.
Bromide: [66556-69-2].
$C_{31}H_{36}BrP$ M 519.503
Cryst. (C_6H_6 or EtOAc). Mp 99°.
Ylide: β-Ionylidenetriphenylphosphorane.
$C_{31}H_{35}P$ M 438.591
Wittig reagent widely used in polyene, particularly retinoid, synth.

Olivé, J.-L. *et al*, *Bull. Soc. Chim. Fr.*, 1969, 3247 (*bromide, use*)
Eschenmoser, W. *et al*, *Helv. Chim. Acta*, 1978, **61**, 822 (*bromide, use*)
Sueiras, J. *et al*, *J. Am. Chem. Soc.*, 1980, **102**, 6255 (*bromide, use*)
Byers, J. *et al*, *J. Org. Chem.*, 1983, **48**, 1515 (*chloride, use*)

4-Methyl-2,6,7-trioxa-1-phosphabicyclo[2.2.1]heptane, 9CI M-00423

Methylglycerol phosphite
[61580-09-4]

$C_4H_7O_3P$ M 134.071
Liq. Bp_{5-6} 46-48°.
1-Oxide: Methylglycerol bicyclic phosphate.
$C_4H_7O_4P$ M 150.071
Cryst. (CH_2Cl_2/Et_2O at low. temp.). Mp 92-94°.
1-Selenide: [68378-98-3]. *Methylglycerol bicyclic selenophosphate.*
$C_4H_7O_3PSe$ M 213.031
Solid. Mp 74-76°. V. unstable to moisture.

Albright, J.O. *et al*, *J. Coord. Chem.*, 1976, **5**, 225 (*complexes*)
Milbrath, D.S. *et al*, *J. Am. Chem. Soc.*, 1976, **98**, 5493 (*oxide, cryst struct*)
Griend, L.J.V. *et al*, *J. Am. Chem. Soc.*, 1977, **99**, 2459 (*synth, pmr, P nmr, struct, oxide*)
Kroshefsky, R.D. *et al*, *Inorg. Chem.*, 1979, **18**, 469 (*selenide, P nmr*)

4-Methyl-2,6,7-trioxa-1-phosphabicyclo[2.2.2]octane, 9CI M-00424

Methyl bicyclic phosphite. Trimethylolethane phosphite
[1449-91-8]

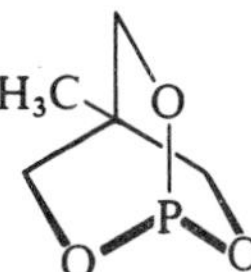

$C_5H_9O_3P$ M 148.098

Ligand for metals of groups IB, VIB, VIIB, and VIII. Solid. Mp 94°. $Bp_{0.02}$ 50° subl.

▷V. toxic. TY7200000.

B,Ph₃CClO₄: 4-Methyl-4-triphenylmethyl-2,6,7-trioxa-1-phosphoniabicyclo[2.2.2]octane perchlorate.
$C_{24}H_{24}ClO_7P$ M 490.876
Cryst. (MeCN). Mp 198°.

B,Ph₃CBF₄: 4-Methyl-4-triphenylmethyl-2,6,7-trioxa-1-phosphoniabicyclo[2.2.2]octane tetrafluoroborate.
$C_{24}H_{24}BF_4O_3P$ M 478.229
Cryst. (Me_2CO). Mp 183°.

1-Oxide: [1449-89-4]. *Trimethylolethane phosphate. Methyl bicyclic phosphate.*
$C_5H_9O_4P$ M 164.097
Mp 249-250°.

▷Toxic. LD_{50} 32 mg/kg (mouse, ip). YK0711700.

1-Sulfide: [3196-56-3]. *Trimethylolethane thiophosphate. Methyl bicyclic thiophosphate.*
$C_5H_9O_3PS$ M 180.158
Cryst. (1,2-dimethoxyethane). Mp 216-218°.

▷Toxic. LD_{50} 34 mg/kg (mouse, ip). TE9450000.

1-Selenide: [67471-54-9]. *Trimethylol selenophosphate. Methyl bicyclic selenophosphate.*
$C_5H_9O_3PSe$ M 227.058
Cryst. (toluene). Mp 246-248°.

Wadsworth, W.S. *et al*, *J. Am. Chem. Soc.*, 1962, **84**, 610.
Verkade, J.G. *et al*, *Inorg. Chem.*, 1965, **4**, 83 (*synth, pmr*)
Nimrod, D.M. *et al*, *J. Am. Chem. Soc.*, 1968, **90**, 2780 (*oxide, cryst struct*)
Keiter, R.L *et al*, *Inorg. Chem.*, 1970, **9**, 404 (*complexes*)
Bellet, E.M. *et al*, *Science*, 1973, **182**, 1135 (*tox*)
Cowley, A.H. *et al*, *Inorg. Chem.*, 1977, **16**, 854; 1984, **23**, 3378 (*pe*)
Kroshefsky, R.D. *et al*, *Inorg. Chem.*, 1979, **18**, 469 (*selenide, synth, P nmr*)
Voorhees, K.J. *et al*, *Org. Mass. Spectrom.*, 1979, **14**, 459 (*ms*)
Kenttamaa, H. *et al*, *Org. Mass. Spectrom.*, 1980, **15**, 520 (*derivs, ms*)
Fanni, T. *et al*, *J. Am. Chem. Soc.*, 1986, **108**, 6311 (*props*)

4-Methyl-3,5,8-trioxa-1-phosphabicyclo[2.2.2]octane, 9CI — M-00425

[18620-05-8]

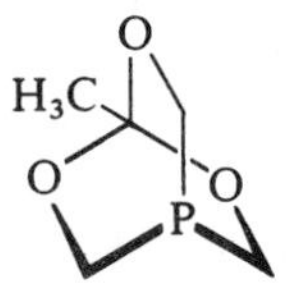

$C_5H_9O_3P$ M 148.098
Ligand for Fe, Pd. Mp 81-83°.

Boros, E.J. *et al*, *Inorg. Chem.*, 1968, **7**, 165 (*synth, derivs, ir, pmr, complexes*)
Ogilvie, F.B. *et al*, *J. Am. Chem. Soc.*, 1970, **92**, 1916 (*pmr, derivs, complexes*)
Allison, D.A. *et al*, *Phosphorus*, 1973, **2**, 257 (*pmr, P nmr*)
Cowley, A.H. *et al*, *Inorg. Chem.*, 1977, **16**, 854 (*pe*)
Kozlov, É.S. *et al*, *Zh. Obshch. Khim.*, 1980, **50**, 1499 (*Engl. transl.* p. 1210) (*synth*)

1-Methyl-2,6,7-trioxa-4-phospha-1-silabicyclo[2.2.2]octane, 9CI — M-00426

[24647-32-3]

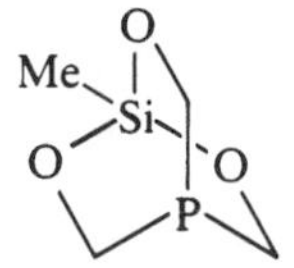

$C_4H_9O_3PSi$ M 164.173
V. hygroscopic cryst. Mp 61-62°.

Rathke, J. *et al*, *J. Org. Chem.*, 1970, **35**, 2310 (*synth, pmr*)
Bertrand, R.D. *et al*, *Phosphorus*, 1973, **3**, 1 (*pmr, P nmr*)

7-Methyl-1,4,6-trioxa-5-phospha(5-P^V)-spiro[4.4]nonan-8-one, 8CI — M-00427

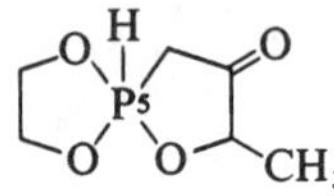

$C_6H_{11}O_4P$ M 178.124

5-Hydroxy: [20553-55-3].
$C_6H_{11}O_5P$ M 194.124
Cryst. Mp 112-115°.

5-Chloro: [20544-49-4].
$C_6H_{10}ClO_4P$ M 212.569
Fuming, hygroscopic solid. Unstable at r.t.

Voznesenskaya, A.Kh. *et al*, *Zh. Obshch. Khim.*, 1968, **38**, 1553 (*Engl. transl.* p. 1504) (*synth, pmr, ir*)

9-Methyl-1,4,6-trioxa-5-phospha(5-P^V)-spiro[4.4]non-7-ene, 10CI — M-00428

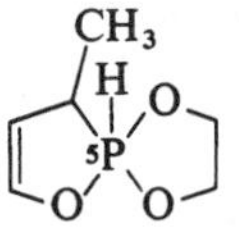

$C_6H_{11}O_3P$ M 162.125

5-Methoxy:
$C_7H_{13}O_4P$ M 192.151
Liq. d_4^{20} 1.23. $Bp_{0.2}$ 75°. n_D^{20} 1.4795.

5-Ethoxy: [66918-48-7].
$C_8H_{15}O_4P$ M 206.178
Liq. At 130°, dec. into ethyl ethylene phosphite.

5-Ph: [67504-68-1].
$C_{12}H_{15}O_3P$ M 238.222
Liq. d_4^{20} 1.24. $Bp_{0.1}$ 105°. n_D^{20} 1.5569.

Razumova, N.A. *et al*, *Zh. Obshch. Khim.*, 1977, **47**, 312 (*Engl. transl.* p. 289) (*synth, pmr, P nmr*)
Ragulin, V.V. *et al*, *Zh. Obshch. Khim.*, 1978, **48**, 707 (*Engl. transl.* p. 650) (*props*)

Methyltriphenoxyphosphonium(1+) — M-00429

Methyltriphenoxyphosphorus(1+), 9CI

$(PhO)_3PMe^{\oplus}$

$C_{19}H_{18}O_3P^{\oplus}$ M 325.323 (ion)

Iodide: [17579-99-6]. *Triphenylphosphite methiodide.*
$C_{19}H_{18}IO_3P$ M 452.228
Reagent for conversion of alcohols to alkyl iodides and epoxides to alkenes. Also used for dehydrations and dehydrohalogenations under mild conditions. Amber cryst. Mp 146°.

Trifluoromethanesulfonate: Used in synth. of ethers, esters, nitriles and diols, and in connection with redn. of alkenes to alkanes. Cryst. Mp 96.5-98.5°.

Verheyden, H. *et al*, *J. Org. Chem.*, 1970, **35**, 2319 (*synth, nmr*)
Phillips, D.I. *et al*, *J. Am. Chem. Soc.*, 1976, **98**, 184 (*trifluoromethanesulfonate, synth, pmr, nmr, ms*)
Lewis, E.S. *et al*, *J. Chem. Soc., Chem. Commun.*, 1978, 424 (*trifluoromethanesulfonate, use*)
Yamada, K. *et al*, *J. Org. Chem.*, 1978, **43**, 2076 (*iodide, use*)
Spangler, C.W. *et al*, *J. Chem. Soc., Perkin Trans. 1*, 1981, 2287 (*iodide, use*)
Fieser, M. *et al*, *Reagents for Organic Synthesis*, Wiley, 1967-84, **1**, 1249; **4**, 559; **6**, 649; **8**, 354 (*use*)

6-Methyl-2,3,4-triphenyl-2*H*-1,5,2-dioxaphosphorin 2-oxide, 10CI — M-00430

6-Methyl-2,3,4-triphenyl-2H-1,5,2-dioxaphosphinine 2-oxide

(2*RS*,6*RS*)-*form*

$C_{22}H_{19}O_3P$ M 362.364

(2RS,6RS)-form [66407-89-4]
(±)-cis-*form*
Cryst. (Et_2O). Mp 199-200°.

(2RS,6SR)-form [66407-90-7]
(±)-trans-*form*
Cryst. (Et_2O or EtOAc/pet. ether). Mp 160-161°.

Regitz, M. *et al*, *Chem. Ber.*, 1978, **111**, 705 (*synth, ir, ms, pmr*)
Maas, G. *et al*, *Chem. Ber.*, 1978, **111**, 726 (*cryst struct*)

6-Methyl-3,5,7-triphenyl-6-phosphatricyclo[3.3.1.0^{2,7}]non-3-ene, 9CI — M-00431

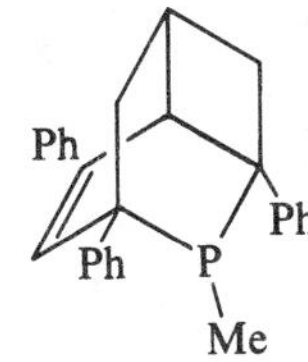

$C_{27}H_{25}P$ M 380.468

6-Oxide: [58525-84-1].
$C_{27}H_{25}OP$ M 396.468
Solid. Mp 189-191°.

Siefert, W.J. *et al*, *Angew. Chem., Int. Ed. Engl.*, 1976, **15**, 238 (*cryst struct*)
Dimroth, K. *et al*, *Chem. Ber.*, 1981, **114**, 1752 (*synth, ir, ms, uv, pmr*)

Methyltriphenylphosphonium(1+), 9CI — M-00432

[15912-74-0]

$Ph_3P^{\oplus}Me$

$C_{19}H_{18}P^{\oplus}$ M 277.325 (ion)

With PhLi, butyllithium, $NaNH_2$, NaH or NaN($SiMe_3$)$_2$, salts yield the ylide.

Chloride: [1031-15-8].
$C_{19}H_{18}ClP$ M 312.778
Cryst. (EtOH/Et_2O). Mp 221-223°.

Bromide: [1779-49-3].
$C_{19}H_{18}BrP$ M 357.229
Cryst. (MeOH/Et_2O). Mp 234-235°.

Iodide: [2065-66-9].
$C_{19}H_{18}IP$ M 404.229
Cryst. (H_2O). Mp 185-187°.

Perchlorate: [20920-23-4].
$C_{19}H_{18}ClO_4P$ M 376.776
Cryst. (CH_2Cl_2/Et_2O). Mp 157-158°.

Tetrafluoroborate:
$C_{19}H_{18}BF_4P$ M 364.129
Cryst. (CH_2Cl_2/Et_2O). Mp 122-123°.

Permanganate:
$C_{19}H_{18}MnO_4P$ M 396.261
Violet powder.

▷Explodes at 70°

Tetraphenylborate:
$C_{43}H_{38}BP$ M 596.557
Mp 195-196°.

Ylide: see Methylenetriphenylphosphorane, M-00148

Wittig, G. *et al*, *Justus Liebigs Ann. Chem.*, 1958, **619**, 10 (*iodide*)
Horner, L. *et al*, *Justus Liebigs Ann. Chem.*, 1961, **646**, 65 (*bromide*)
Mallion, K. *et al*, *J. Chem. Soc.*, 1963, 1327 (*iodide, ir*)
Hendrickson, J. *et al*, Tet, 1964, **20**, 449 (*iodide, pmr*)
Albright, T.A. *et al*, *J. Am. Chem. Soc.*, 1975, **97**, 940, 2942 (*cmr, nmr*)
Starzewski, K.A.O. *et al*, *Phosphorus*, 1976, **6**, 177 (*cmr*)
Grim, S.O. *et al*, *J. Org. Chem.*, 1977, **42**, 1236 (*nmr*)
Nesterov, L.V. *et al*, *Zh. Obshch. Khim.*, 1977, **47**, 1259 (*Engl. transl.* p. 1161) (*perchlorate, tetrafluoroborate*)
Richter, W. *et al*, *Chem. Ber.*, 1977, **110**, 1312 (*chloride*)
Reischl, W. *et al*, *Tetrahedron*, 1979, **35**, 1109 (*permanganate, use*)
Doleschall, G., *Synthesis*, 1981, 478 (*iodide, synth, use*)
Vedejs, E. *et al*, *J. Am. Chem. Soc.*, 1981, **103**, 2823 (*bromide, synth, props*)
Willcockson, W.S. *et al*, *Comp. Biochem. Phys. C*, 1982, **72**, 101 (*pharmacol*)

Methyl triphenylphosphoranylideneacetate, 9CI, 8CI — M-00433

Carbomethoxymethylidenetriphenylphosphorane

[2605-67-6]

$Ph_3P{=}CHCOOMe$

$C_{21}H_{19}O_2P$ M 334.354

Wittig reagent. Thick prisms (EtOAc/pet. ether). Mp 169-169.5° (159-160°). Can be subl.

Issler, O. *et al*, *Helv. Chim. Acta*, 1957, **40**, 1242 (*synth, uv, use*)
Bestmann, H.J. *et al*, *Chem. Ber.*, 1962, **95**, 2921 (*use*)
Matthews, C.N. *et al*, *Tetrahedron Lett.*, 1966, 5707 (*nmr*)
Bestmann, H.J. *et al*, *J. Am. Chem. Soc.*, 1967, **89**, 3936 (*pmr*)
Cooks, R.G. *et al*, *Tetrahedron*, 1968, **24**, 3289 (*ms*)
Ayrey, G. *et al*, *J. Labelled Compd. Radiopharm.*, 1978, **14**, 935 (*synth*)
Lang, R.W. *et al*, *Helv. Chim. Acta*, 1979, **62**, 1025; 1980, **63**, 438 (*synth, use*)
Doleschall, G., *Synthesis*, 1981, 478 (*use*)

Methyl (2-triphenylphosphoranylidene)-propanoate — M-00434

[2605-68-7]

$Ph_3P{=}C(CH_3)COOMe$

$C_{22}H_{21}O_2P$ M 348.380

Wittig reagent used in polyene acid synth. Solid. Mp 151-153° (145°).

Bestmann, H.J. *et al*, *Chem. Ber.*, 1962, **95**, 2921 (*synth*)
Zeliger, H.I. *et al*, *Tetrahedron Lett.*, 1969, 2199 (*pmr*)
Dale, A.J. *et al*, *Acta Chem. Scand.*, 1970, **24**, 2681 (*ir, pmr*)
Lang, R.W. *et al*, *Helv. Chim. Acta*, 1980, **63**, 438 (*synth, pmr, use*)

Methyltris(triphenylphosphine)copper, 9CI **M-00435**

Methylcoppertris(triphenylphosphine)

[38704-10-8]

$MeCu(PPh_3)_3$

$C_{55}H_{48}CuP_3$ M 865.451

Unstable yellow powder.

Toluene solvate: Light-sensitive yellow needles.

Et_2O solvate: Can be stored under N_2 in dark at r.t.

Costa, G. *et al*, *J. Inorg. Nucl. Chem.*, 1964, **26**, 961 (*synth*)
Yamamoto, A. *et al*, *Bull. Chem. Soc. Jpn.*, 1972, **45**, 1583 (*synth*)
Miyashita, A. *et al*, *Bull. Chem. Soc. Jpn.*, 1977, **50**, 1109 (*synth, props*)

4-Methyl-2,6,7-trithia-1-phosphabicyclo[2.2.2]octane, 9CI **M-00436**

[18818-32-1]

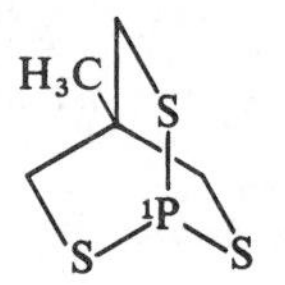

$C_5H_9PS_3$ M 196.280

Needles. Mp 187-189°. Subl. *in vacuo.*

1-Oxide: [18818-34-3].
$C_5H_9OPS_3$ M 212.279
Solid. Mp 236-239°. Subl. *in vacuo.*

1-Sulfide: [18818-33-2].
$C_5H_9PS_4$ M 228.340
Needles. Mp 286-290°. Subl. *in vacuo.*

▷Fairly toxic. YL8850000.

Vandenbroucke, A.C. *et al*, *Inorg. Chem.*, 1968, **7**, 1469 (*synth, derivs, pmr, ir*)
Casida, J.E. *et al*, *Toxicol. Appl. Pharmacol.*, 1976, **36**, 261 (*sulfide, tox*)
Cowley, A.H. *et al*, *Inorg. Chem.*, 1984, **23**, 3378 (*pe*)

Methyl trithion **M-00437**

S-[[(*4-Chlorophenyl)thio*]*methyl*] O,O-*dimethyl phosphorodithioate*, 9CI. *Tri-Me*

[953-17-3]

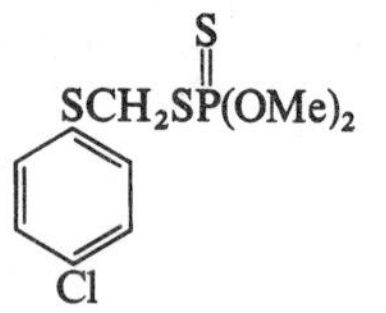

$C_9H_{12}ClO_2PS_3$ M 314.799

Acaricide. Light-yellow to amber liq. V. spar. sol. H_2O, misc. most org. solvs. d_{20}^{20} 1.34-1.35. Mp −18°. n_D^{30} 1.6130.

▷TD5425000.

Babad, H. *et al*, *Anal. Chim. Acta*, 1968, **41**, 259 (*pmr*)
Ackerman, H., *J. Chromatogr.*, 1969, **44**, 414 (*tlc*)
Ross, R.T. *et al*, *Anal. Chim. Acta*, 1970, **52**, 139 (*P nmr*)
Stan, H.-J. *et al*, *Fresenius' Z. Anal. Chem.*, 1977, **287**, 271 (*glc, ms*)
Pesticide Manual, 6th Ed., 111.

Methylvinylphosphinic acid, 8CI **M-00438**

Ethenylmethylphosphinic acid, 9CI

[53314-64-0]

$H_2C{=}CHPMe(O)OH$

$C_3H_7O_2P$ M 106.061

Liq. $Bp_{0.6}$ 128-130°.

Me ester: [63314-88-5]. *Methyl methylvinylphosphinate. Methyl ethenylmethylphosphinate.*
$C_4H_9O_2P$ M 120.088
Liq. Bp_3 48-50°.

Chloride: [36120-75-9].
C_3H_6ClOP M 124.507
Liq. d_4^{20} 1.23. $Bp_{2\text{-}3}$ 65-67°. n_D^{20} 1.4830.

Rogacheva, I.A. *et al*, *Zh. Obshch. Khim.*, 1971, **41**, 2634 (*Engl. transl.* p. 2666) (*chloride*)
Finke, M. *et al*, *Justus Liebigs Ann. Chem.*, 1974, 741 (*synth*)

Meturedepa, USAN **M-00439**

Ethyl [bis(2,2-dimethyl-1-aziridinyl)phosphinyl]-carbamate, 9CI, 8CI. *Turloc*

[1661-29-6]

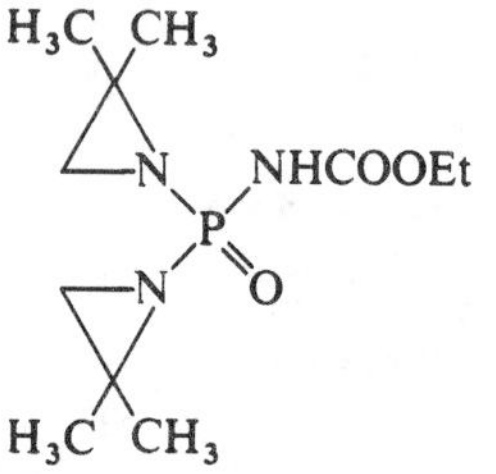

$C_{11}H_{22}N_3O_3P$ M 275.287

Antineoplastic drug. Cryst. (Et_2O). Mp 57-58°.

Bardos, T.J. *et al*, *J. Pharm. Sci.*, 1965, **54**, 187 (*synth, ir, pmr*)
Bořkovec, A.B. *et al*, *J. Med. Chem.*, 1966, **9**, 522 (*pharmacol*)
Bardos, T.J. *et al*, *Ann. N.Y. Acad. Sci.*, 1969, **163**, 1006 (*pharmacol*)
Lalko, D. *et al*, *J. Pharm. Sci.*, 1975, **64**, 230 (*pharmacol*)

Mevinphos, BSI **M-00440**

Methyl 3-[(dimethoxyphosphinyl)oxy]-2-butenoate, 9CI. *3-Hydroxy-2-butenoic acid, methyl ester, dimethyl phosphate*, 8CI. *2-Methoxycarbonyl-1-methylvinyl dimethyl phosphate. Phosdrin*

[7786-34-7]

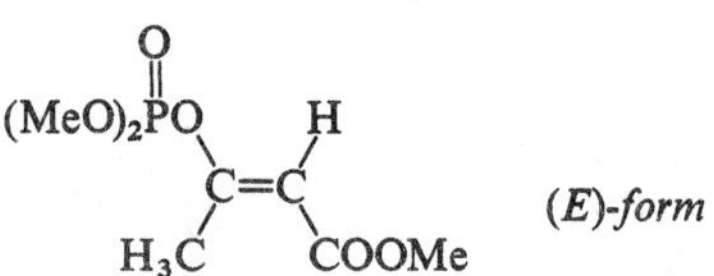

$C_7H_{13}O_6P$ M 224.150

Coml. samples contain ca. 60% (*E*)-form. Contact and systemic insecticide and acaricide. Misc. H_2O and org. solvs. except pet. ether. Rapidly degraded.

▷Highly toxic by inhalation or skin absorption. TLV (skin) 0.1. GQ5250000.

(*E*)-form [298-01-1]
d^{20} 1.23. Mp 21°. n_D^{20} 1.4452.

(*Z*)-form [338-45-4]
d^{20} 1.245. Mp 6.9°. n_D^{20} 1.4524.

U.S.P., 2 685 552, (*1952*); *CA*, **48**, 12365c
Getz, M.E. *et al*, *J. Assoc. Off. Anal. Chem.*, 1968, **51**, 1101 (*tlc*)
Ross, R.T. *et al*, *Anal. Chim. Acta*, 1970, **52**, 139 (*P nmr*)
Beynon, K.I. *et al*, *Residue Rev.*, 1973, **47**, 55 (*rev*)
Gaydou, E.M., *Can. J. Chem.*, 1973, **51**, 3412 (*pmr*)
Nicholas, N.L. *et al*, *J. Assoc. Off. Anal. Chem.*, 1976, **59**, 1071 (*raman*)
Szalontai, G., *J. Chromatogr.*, 1976, **124**, 9 (*hplc*)

Stan, H.-J. *et al*, *Fresenius' Z. Anal. Chem.*, 1977, **287**, 271; *Biomed. Mass Spectrom.*, 1982, **9**, 483 (*glc, ms*)
Szalontai, G., *Org. Mass Spectrom.*, 1977, **10**, 63 (*cmr, P nmr*)
Ripley, B.D. *et al*, *J. Assoc. Off. Anal. Chem.*, 1983, **66**, 1084 (*glc*)
Pesticide Manual, 6th Ed., 363; 7th Ed., 381.
The Agrochemicals Handbook, Royal Society of Chemistry, 1983, A281.
Sax, N.I., *Dangerous Properties of Industrial Materials*, 6th Ed., Van Nostrand-Reinhold, 1984, 601.

Moenomycin M-00441

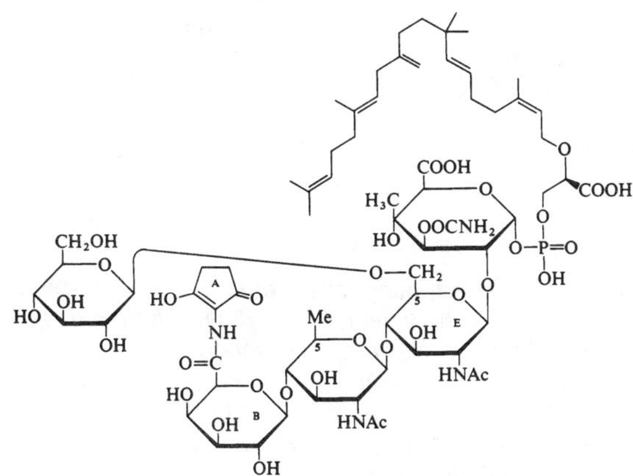

Phospoglycolipid antibiotic complex. Strain also produces Moenomycins B_1 and B_2, later sepd. into Moenomycins *D-H*. Prod. by *Streptomyces bambergensis*.

Moenomycin A [76095-39-1]
$C_{69}H_{107}N_4O_{35}P$ M 1583.584

Chief constit. of Flavomycin, used in animal nutrition. Active against gram-positive bacteria. Mp 184-185°.
Na salt: $[\alpha]^{23}_{D}$ +4.0° (c, 1 in H_2O).

Moenomycin C
$C_{75}H_{135}N_7O_{42}P$ M 1837.887

From *S*. sp. Active against gram-positive and weakly against gram-negative bacteria. Mp 178-179°. $[\alpha]^{23}_{D}$ +4.0° (c, 1 in H_2O).

Huber, G., *Antibiotics*, 1979, Hahn, E.E. Ed., Springer, Berlin, **5.1**, 135 (*rev*)
Welzel, P. *et al*, *Angew. Chem., Int. Ed. Engl.*, 1981, **20**, 121 (*isol, pmr, struct*)
Welzel, P. *et al*, *Tetrahedron*, 1983, **39**, 1583, 2219 (*isol, uv, cmr*)

Monobenzoyl phosphate M-00442

Benzoic acid-phosphoric acid anhydride, 9CI, 8CI
[6659-26-3]

$PhCOOP(O)(OH)_2$

$C_7H_7O_5P$ M 202.103
Substrate for acyl phosphatase assay.
Difluoride: [67598-47-4]. *Benzoyl phosphorodifluoridate. Benzoyl difluorophosphate.*
$C_7H_5F_2O_3P$ M 206.085
Liq. $Bp_{0.01}$ 52°.
Dichloride: [67598-46-3]. *Benzoyl phosphorodichloridate. Benzoyl dichlorophosphate.*
$C_7H_5Cl_2O_3P$ M 238.994
Liq. $Bp_{0.001}$ 77°.

Satchell, D.P.N. *et al*, *Biochem. Biophys. Acta*, 1972, **268**, 233 (*props*)
Camici, G. *et al*, *Experientia*, 1976, **32**, 535 (*synth*)
Yamaguchi, K. *et al*, *J. Am. Chem. Soc.*, 1980, **102**, 4534 (*derivs, synth, ir*)
Effenberger, F. *et al*, *Chem. Ber.*, 1981, **114**, 916, 926 (*halides, synth, ir, P nmr, cmr*)

Monobenzyl phosphate, 8CI M-00443

Mono(phenylmethyl) phosphate, 9CI. *Benzyl dihydrogen phosphate. Benzyl phosphoric acid*
[1623-07-0]

$PhCH_2OP(O)(OH)_2$

$C_7H_9O_4P$ M 188.119
Solid or cryst. (Et_2O/pet. ether). Mp 104.5-105.5° (94-96°).
Monoanilinium salt: [23562-59-6]. Cryst. (EtOH). Mp 158-160° (150-153°).
Bisanilinium salt: Solid. Mp 152-154°.
Bis(cyclohexylammonium) salt: Solid. Mp 214°, Mp 232-234°.
Dichloride: [52692-02-1]. *Benzyl phosphorodichloridate. Phenylmethyl phosphorodichloridate. Benzyl dichlorophosphate.* Oil. Dec. on attempted distn.

Moffatt, J.G. *et al*, *J. Am. Chem. Soc.*, 1957, **79**, 3741 (*synth*)
Williams, A. *et al*, *J. Chem. Soc. (B)*, 1971, 1973 (*synth*)
Zwierzak, A. *et al*, *Tetrahedron*, 1971, **27**, 3163 (*synth, ir, pmr*)
Yamaguchi, H. *et al*, *Bull. Chem. Soc. Jpn.*, 1981, **54**, 1891 (*synth, pmr, ir*)
Misiura, K. *et al*, *J. Med. Chem.*, 1983, **26**, 674 (*dichloride, synth, ms, P nmr*)
Okamoto, Y., *Bull. Chem. Soc. Jpn.*, 1985, **58**, 3393 (*synth*)

Monobenzyl phosphonate M-00444

Monobenzyl phosphite. Benzyl phosphite. Benzyl dihydrogen phosphite
[10542-07-1]

$PhCH_2OPH(O)OH \rightleftharpoons PhCH_2OP(OH)_2$

$C_7H_9O_3P$ M 172.120
Tautomeric. Pale-yellow oil. n_D^{18} 1.5521. Dec. slowly in air or on dist. Generally isol. as NH_4 salt.
NH_4 salt: [56317-74-9]. Needles (2-methoxyethanol/dioxan or EtOAc). Mp 153-155°.

Baddiley, J. *et al*, *J. Chem. Soc.*, 1949, 815 (*synth*)
Michalski, J. *et al*, *J. Chem. Soc.*, 1961, 4904 (*synth*)

Mono(2-biphenylyl) phosphate M-00445

[1,1'-Biphenyl]-2-yl dihydrogen phosphate. [1,1'-Biphenyl]-2-yl phosphoric acid

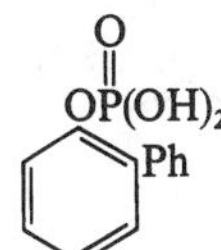

$C_{12}H_{11}O_4P$ M 250.190
Solid. Mp 136-137°.
Dichloride: [36240-28-5]. *[1,1'-Biphenyl]-2-yl phosphorodichloridate. [1,1'-Biphenyl]-2-yl dichlorophosphate.*
$C_{12}H_9Cl_2O_2P$ M 287.082
Liq. d_4^{20} 1.36. Bp_4 164°, $Bp_{0.08}$ 142°. n_D^{20} 1.5883.
Dianilide: *[1,1'-Biphenyl]-2-yl* N,N'-*diphenylphosphorodiamidate.*
$C_{24}H_{21}N_2O_2P$ M 400.416

Solid. Mp 151-152°.

Chernyshev, E.A. *et al*, *Zh. Obshch. Khim.*, 1971, **41**, 2189 (*Engl. transl.* p. 2214) (*dichloride, synth, props*)
Cremlyn, R.J.W. *et al*, *Aust. J. Chem.*, 1974, **27**, 1065 (*synth, ir, derivs*)

Mono-(1,1′-biphenyl)-4-yl phosphate, 9CI **M-00446**

(1,1′-Biphenyl)-4-yl dihydrogen phosphate
[55231-79-3]

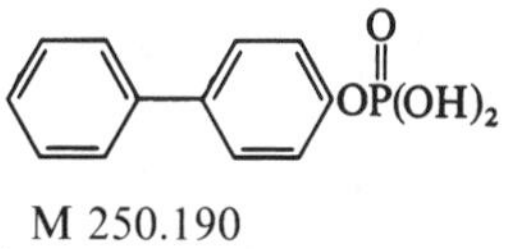

$C_{12}H_{11}O_4P$ M 250.190
Cryst. ($CHCl_3$). Mp 127-129°.
Dichloride: [55231-79-3]. *(1,1′-Biphenyl)-4-yl phosphorodichloridate. (1,1′-Biphenyl)-4-yl dichlorophosphate.*
$C_{12}H_9Cl_2O_2P$ M 287.082
Solid. Mp 53°, Mp 83°. Bp_{11} 206-209°.

U.S.P., 1 960 184, (*1934*) (*synth*)
U.S.P., 2 117 291 (*1938*) (*synth*)

Mono(2-bromoethyl) phosphate, 9CI **M-00447**

2-Bromoethyl dihydrogen phosphate. 2-Bromoethyl phosphoric acid

$BrCH_2CH_2OP(O)(OH)_2$

$C_2H_6BrO_4P$ M 204.945
Isolated as Ba salt.
Dichloride: [4167-02-6]. *2-Bromoethyl phosphorodichloridate.*
$C_2H_4BrCl_2O_2P$ M 241.836
Used in synth. of phosphorylcholine esters and phospholipids. Liq. Bp_2 70-71°.

Zetsche, F. *et al*, *Helv. Chim. Acta*, 1925, **8**, 943 (*synth*)
Jean, H., *Bull. Soc. Chim. Fr.*, 1957, 783 (*chloride*)
Eibl, H. *et al*, *Chem. Phys. Lipids*, 1978, **22**, 1 (*chloride, synth, use*)
Spande, T.F. *et al*, *J. Org. Chem.*, 1980, **45**, 3081 (*chloride, synth, use*)
Tsushima, S. *et al*, *Chem. Pharm. Bull.*, 1982, **30**, 3260 (*chloride, synth, use*)

Mono(4-bromophenyl) phosphate, 9CI **M-00448**

4-Bromophenyl dihydrogen phosphate. 4-Bromophenyl phosphoric acid

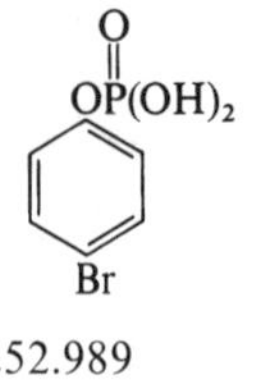

$C_6H_6BrO_4P$ M 252.989
Plates ($CHCl_3$). Mp 161°.

Zetsche, F. *et al*, *Helv. Chim. Acta*, 1926, **9**, 420.

Mono-2-*tert*-butylphenyl phosphate **M-00449**

*Mono 2-(1,1-dimethylethyl)phenyl phosphate, 9CI. 2-*tert-*Butylphenyl phosphate. 2-*tert-*Butylphenyl dihydrogen phosphate*

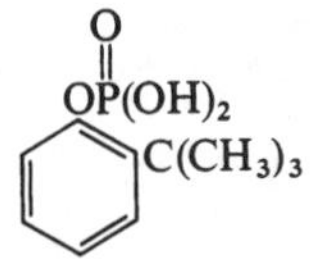

$C_{10}H_{15}O_4P$ M 230.200
Solid. Mp 180-180.5°.
Dichloride: [18351-33-2]. *2-*tert-*Butylphenyl phosphorodichloridate.*
$C_{10}H_{13}Cl_2O_2P$ M 267.091
Liq. $Bp_{0.1}$ 180-180.5°.

Kosolapoff, G.M. *et al*, *J. Chem. Soc. (C)*, 1968, 815 (*synth, pmr, chloride*)

Mono-4-*tert*-butylphenyl phosphate **M-00450**

*Mono-4-(1,1-dimethylethyl)phenyl phosphate, 9CI. 4-*tert-*Butylphenyl dihydrogen phosphate*

$C_{10}H_{15}O_4P$ M 230.200
Cryst. (Me_2CO/hexane). Mp 185-186°.
Dichloride: [18351-36-5]. *4-*tert-*Butylphenyl phosphorodichloridate.*
$C_{10}H_{13}Cl_2O_2P$ M 267.091
Liq. d_4^{20} 1.25. Bp_4 130°, $Bp_{0.01}$ 86-88°. n_D^{20} 1.5147.

Kosolapoff, G.M. *et al*, *J. Chem. Soc. (C)*, 1968, 815 (*synth, derivs, pmr*)
Tacke, R. *et al*, *Justus Liebigs Ann. Chem.*, 1981, 387 (*derivs, synth, ms, pmr*)

Monobutyl phosphate, 9CI, 8CI **M-00451**

Butyl dihydrogen phosphate. Butyl phosphoric acid
[1623-15-0]

$H_3C(CH_2)_3OP(O)(OH)_2$

$C_4H_{11}O_4P$ M 154.102
Isol. as amine salt. Extractant for transuranic and rare earth elements. Metab. of Tributyl phosphate, T-00370 . pK_{a1} 1.89, pK_{a2} 6.84 (H_2O, 25°).
Monoanilinium salt: Cryst. (EtOH). Mp 144-148° (138-140°).
Monocyclohexylammonium salt: [14703-66-3]. Solid. Mp 181-182°.

Obata, T. *et al*, *J. Org. Chem.*, 1967, **32**, 1063 (*synth*)
Zwierzak, A. *et al*, *Tetrahedron*, 1971, **27**, 3163 (*synth, ir, pmr*)
Markowska, A. *et al*, *Phosphorus Sulfur*, 1981, **10**, 245 (*synth*)
Solovkin, A.S., *Radiokhimiya*, 1982, **24**, 56 (*Engl. transl.* p. 49) (*props, use, rev*)
Suzuki, T. *et al*, *J. Agric. Food Chem.*, 1984, **32**, 603.

Mono-*tert*-butyl phosphate, 8CI **M-00452**

Mono(1,1-dimethylethyl) phosphate, 9CI. tert-*Butyl dihydrogen phosphate.* tert-*Butyl phosphoric acid*
[2382-75-4]

$(H_3C)_3COP(O)(OH)_2$

$C_4H_{11}O_4P$ M 154.102
Monoanilinium salt: [78944-60-2]. Solid. Mp 139-143° dec.

Monocyclohexylammonium salt: Cryst. (MeOH/EtOAc). Mp 205-206° dec.
Biscyclohexylammonium salt: Needles (EtOH). Mp 168-170°, Mp 191-193° dec.

Cramer, F. *et al*, *Justus Liebigs Ann. Chem.*, 1962, **654**, 180 (*synth*)
Maynard, S.A. *et al*, *Aust. J. Chem.*, 1963, **16**, 596 (*synth, pmr*)
Williams, A. *et al*, *J. Chem. Soc.* (*B*), 1971, 1973 (*synth*)
Yamaguchi, H. *et al*, *Bull. Chem. Soc. Jpn.*, 1981, **54**, 1891 (*synth*)
Ramirez, F. *et al*, *J. Am. Chem. Soc.*, 1982, **104**, 1345 (*P nmr*)

Monobutyl phosphonate, 9CI — M-00453

Monobutyl phosphite
[16456-56-7]

$$H_3C(CH_2)_3OPH(O)OH \rightleftharpoons H_3C(CH_2)_3OP(OH)_2$$

$C_4H_{11}O_3P$ M 138.103
Tautomeric; almost completely in the phosphonate form. Thermally unstable liq. n_D^{20} 1.4292 (undist.).
▷TG6720000.
Na salt: [56317-58-9]. Hygroscopic needles (EtOH). Mp 177.5-178.5°.
p-*Chlorobenzylisothiouronium salt:* Solid. Mp 150-152°.

Michalski, J. *et al*, *J. Chem. Soc.*, 1961, 4904 (*synth*)
Zwierzak, A. *et al*, *Tetrahedron*, 1973, **29**, 1089 (*synth*)
Kluba, M. *et al*, *Synthesis*, 1978, 134 (*synth, ir, pmr*)
Norén, J.O. *et al*, *J. Med. Chem.*, 1983, **26**, 264 (*ir, pmr*)

Mono(2-chloroethyl) phosphate, 9CI — M-00454

2-Chloroethyl dihydrogen phosphate. 2-Chloroethyl phosphoric acid

$$ClCH_2CH_2OP(O)(OH)_2$$

$C_2H_6ClO_4P$ M 160.494
Isol. as Ba or amine salts.
Monocyclohexylammonium salt: Solid. Mp 161-162°.
Bis(cyclohexylammonium) salt: Cryst. (EtOH/Et_2O). Mp 205-208°.
Dichloride: [1455-05-5]. *2-Chloroethyl phosphorodichloridate. 2-Chloroethyl phosphoryl dichloride.*
$C_2H_4Cl_3O_2P$ M 197.385
Liq. of pungent odour. d_4^{20} 1.54. Bp_{20} 101°, Bp_2 71.5°. n_D^{20} 1.4694 (1.4964).

Renshaw, R.R. *et al*, *J. Am. Chem. Soc.*, 1929, **51**, 953 (*dichloride*)
Cherbuliez, E. *et al*, *Helv. Chim. Acta*, 1958, **41**, 1693 (*synth*)
Maynard, S.A. *et al*, *Aust. J. Chem.*, 1963, **16**, 596 (*synth*)
Pudovik, A.N. *et al*, *Zh. Obshch. Khim.*, 1966, **36**, 1454 (*Engl. transl.* p. 1461) (*dichloride, ester*)
Markowska, A. *et al*, *Bull. Acad. Pol. Sci., Ser. Sci. Chim.*, 1979, **27**, 115 (*synth*)

Mono(2-chloromethyl-4-nitrophenyl) phosphate — M-00455

(2-Chloromethyl-4-nitrophenyl) dihydrogen phosphate. (2-Chloromethyl-4-nitrophenyl) phosphoric acid

O ‖ OP(OH)$_2$, CH_2Cl, NO_2

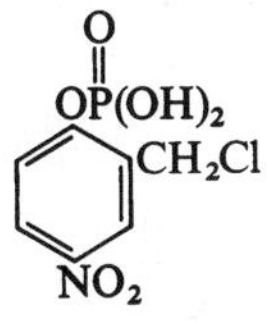

$C_7H_7ClNO_6P$ M 267.562
Powder (H_2O). Mp 137-139°.
Dichloride: [23561-36-6]. *(2-Chloromethyl-4-nitrophenyl) phosphorodichloridate. (2-Chloromethyl-4-nitrophenyl) phosphoryl dichloride. (2-Chloromethyl-4-nitrophenyl) dichlorophosphate.*
$C_7H_5Cl_3NO_4P$ M 304.454
Phosphorylating agent. Pale-yellow viscous liq. which slowly solidifies. Mp 43-44°. $Bp_{0.2}$ 165-167°.

Hata, T. *et al*, *J. Am. Chem. Soc.*, 1969, **91**, 4532 (*synth, uv, derivs*)
Mushika, Y. *et al*, *Chem. Pharm. Bull.*, 1971, **19**, 696 (*use*)

Mono(4-chlorophenyl) mono(5-chloro-8-quinolinyl) phosphate — M-00456

4-Chlorophenyl 5-chloro-8-quinolinyl hydrogen phosphate
[81366-72-5]

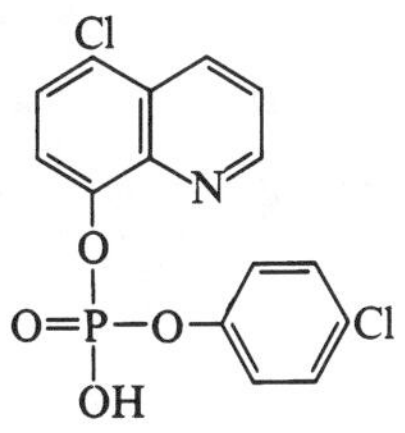

Cl, N, O, O=P–O–, Cl, OH

$C_{15}H_{10}Cl_2NO_4P$ M 370.128
Reagent for oligoribonucleotide synth, as are its derivs. Cryst. (MeCN aq.). Mp 108-110°.
Cyclohexylammonium salt: Solid. Mp 148-150°.
Chloride: [77181-80-7]. *4-Chlorophenyl 5-chloro-8-quinolinyl phosphorochloridate.*
$C_{15}H_9Cl_3NO_3P$ M 388.574
Phosphorylating agent for 5′-OH and 3′-OH groups in partially protected nucleosides.

Takaku, H. *et al*, *Chem. Lett.*, 1981, 543; 1982, 197 (*synth, derivs, use*)

Mono(2-chlorophenyl) phosphate, 9CI — M-00457

2-Chlorophenyl dihydrogen phosphate. 2-Chlorophenyl phosphoric acid
[13428-19-8]

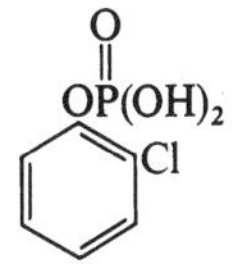

O ‖ OP(OH)$_2$, Cl

$C_6H_6ClO_4P$ M 208.538
Dipentylammonium salt: Plates (MeOH/Et_2O). Mp 179-180°.
Dichloride: *see 2-Chlorophenyl phosphorodichloridate, C-00161*

Rosenmund, K.W. *et al*, *Arch. Pharm.* (*Weinheim, Ger.*), 1943, **281**, 317
Maguire, M.H. *et al*, *J. Chem. Soc.*, 1953, 1479 (*synth*)

Mono(4-chlorophenyl) phosphate, 9CI — M-00458

4-Chlorophenyl dihydrogen phosphate. 4-Chlorophenyl phosphoric acid
[13388-88-0]

$C_6H_6ClO_4P$ M 208.538
Prisms (C_6H_6 or toluene). Mp 93°. pK_{a1} 0.40, pK_{a2} 5.42 (H_2O, 25°). Forms a hemihydrate, Mp 125-125.5°.

Monocyclohexylammonium salt: [79227-87-5]. Solid. Mp 206-208°.
Bis(cyclohexylammonium) salt: [88766-69-2]. Cryst. (MeOH/EtOAc). Mp 204-205°.
Difluoride: 4-Chlorophenyl phosphorodifluoridate. 4-Chlorophenyl phosphoryl difluoride. 4-Chlorophenyl difluorophosphate.
$C_6H_4ClF_2O_2P$ M 212.520
Liq. Bp_{10} 82°.
Dichloride: see 4-Chlorophenyl phosphorodichloridate, C-00162
Dianilide: 4-Chlorophenyl N,N'-*diphenylphosphorodiamidate.*
$C_{18}H_{16}ClN_2O_2P$ M 358.763
Solid. Mp 169-171°.

Maguire, M.H. *et al*, *J. Chem. Soc.*, 1953, 1479 (*synth*)
Maynard, S.A. *et al*, *Aust. J. Chem.*, 1963, **16**, 596 (*synth*)
Williams, A. *et al*, *J. Chem. Soc., Perkin Trans. 2*, 1972, 1454 (*dianilide*)
Blackburn, G.M. *et al*, *J. Chem. Soc., Perkin Trans. 1*, 1980, 1150 (*synth, pmr*)
Effenberger, F. *et al*, *Synthesis*, 1981, 70 (*difluoride, synth, ir, P nmr*)
Markowska, A. *et al*, *Phosphorus Sulfur*, 1981, **10**, 143 (*nmr*)
Bourne, N. *et al*, *J. Org. Chem.*, 1984, **49**, 1200 (*props*)

Monocrotophos, BSI — M-00459

*Dimethyl 1-methyl-3-(methylamino)-3-oxo-1-propenyl phosphate, 9CI. Dimethyl phosphate, ester with 3-hydroxy-*N*-methylcrotonamide, 8CI. Dimethyl 1-methyl-2-methylcarbamoylvinyl phosphate. Nuvacron. Azodrin*
[2157-98-4]

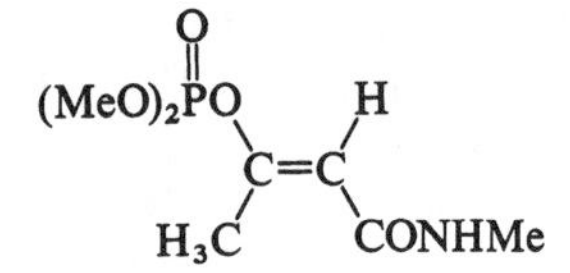

$(MeO)_2P(O)O-C(CH_3)=CH-CONHMe$

$C_7H_{14}NO_5P$ M 223.165
The (*E*)-isomer is the major constit. of the technical prod. Metab. of bidrin.
(*E*)-***form*** [6923-22-4]
Agricultural insecticide with both systemic and contact action. Cryst. Sol. H_2O, Me_2CO, insol. hexane. Mp 54-55°.
▷Highly toxic, TLV 0.25. TC4375000.
(*Z*)-***form*** [919-44-8]
No phys. props. reported.

U.S.P., 3 400 177, (*1965*); *CA*, **70**, 87064r (*synth*)
Menzer, R.E. *et al*, *J. Agric. Food Chem.*, 1965, **13**, 102 (*ir*)
Babad, H. *et al*, *Anal. Chim. Acta*, 1968, **41**, 259 (*pmr*)
Ross, R.T. *et al*, *Anal. Chim. Acta*, 1970, **52**, 139 (*P nmr*)
Beynon, K.I. *et al*, *Residue Rev.*, 1973, **47**, 55 (*metab, rev*)
Szalontai, G., *J. Chromatogr.*, 1976, **124**, 9 (*hplc*)
Busch, K.L. *et al*, *Appl. Spectrosc.*, 1978, **32**, 388 (*ms*)
Stan, H.-J. *et al*, *Biomed. Mass Spectrom.*, 1982, **9**, 483 (*ms*)
Stan, H.-J. *et al*, *J. Chromatogr.*, 1983, **279**, 173 (*glc*)
Pesticide Manual, 6th Ed., 366; 7th Ed., 384.
The Agrochemicals Handbook, Royal Society of Chemistry, 1983, A284.
Sax, N.I., *Dangerous Properties of Industrial Materials*, 6th Ed., Van Nostrand-Reinhold, 1984, 394.

Mono(2-cyanophenyl) phosphate — M-00460

2-Cyanophenyl dihydrogen phosphate. 2-Cyanophenyl phosphoric acid

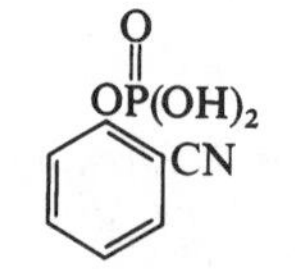

$C_7H_6NO_4P$ M 199.102
Isolated as Ba salt + $4H_2O$.

Cherbuliez, E. *et al*, *Helv. Chim. Acta*, 1956, **39**, 1461, 1844.

Monocyclohexyl phosphate, 9CI, 8CI — M-00461

Cyclohexyl dihydrogen phosphate. Cyclohexyl phosphoric acid

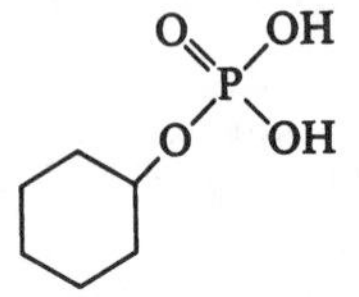

$C_6H_{13}O_4P$ M 180.140
Cryst. (C_6H_6/cyclohexane or $CHCl_3$/cyclohexane). Mp 86°.
Monoanilinium salt: Cryst. (EtOH). Mp 168-169° (160.5-163°).
Bis(cyclohexylammonium) salt: Cryst. (EtOH, MeOH/EtOAc, or C_6H_6/EtOH). Mp 218-219° (208-210°).

Moffatt, J.G. *et al*, *J. Am. Chem. Soc.*, 1957, **79**, 3741 (*synth*)
Montgomery, H.A.C. *et al*, *J. Chem. Soc.*, 1958, 1963 (*synth*)
Obata, T. *et al*, *J. Org. Chem.*, 1967, **32**, 1063 (*synth*)
Gajda, T. *et al*, *Synthesis*, 1977, 623 (*synth, ir, pmr*)
Yamaguchi, H. *et al*, *Bull. Chem. Soc. Jpn.*, 1981, **54**, 1891 (*synth, ir, pmr*)

Monodecyl phosphate, 9CI, 8CI — M-00462

Decyl dihydrogen phosphate. Decyl phosphoric acid
[3921-30-0]

$H_3C(CH_2)_9OP(O)(OH)_2$

$C_{10}H_{23}O_4P$ M 238.263
Di-Na salt used as wetting agent in the textile industry. Cryst. (hexane). Mp 45°. pK_a 4.93 (90% 2-propanol aq., 20°).

Nelson, A.K. *et al*, *Inorg. Chem.*, 1963, **2**, 775 (*synth*)
Okamoto, Y., *Bull. Chem. Soc. Jpn.*, 1985, **58**, 3393 (*synth*)

Mono(2,4-dichlorophenyl) phosphate — M-00463

2,4-Dichlorophenyl dihydrogen phosphate. 2,4-Dichlorophenyl phosphoric acid

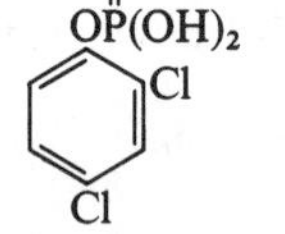

$C_6H_5Cl_2O_4P$ M 242.983
Cryst. (C_6H_6). Mp 138-139°.
Monocyclohexylammonium salt: Solid. Mp 187-188°.
Dianilide: 2,4-Dichlorophenyl N,N'-*diphenylphosphorodiamidate.*
$C_{18}H_{15}Cl_2N_2O_2P$ M 393.208
Solid. Mp 165-167°.

Cremlyn, R.J.W. *et al*, *J. Chem. Soc., Perkin Trans. 1*, 1972, 583 (*synth, derivs*)
Owen, G.R. *et al*, *Synthesis*, 1974, 704 (*synth, deriv*)
Markowska, A. *et al*, *Phosphorus Sulfur*, 1981, **10**, 143 (*synth, P nmr*)

Mono[2-(dimethylamino)-4-nitrophenyl] phosphate, 9CI — M-00464

$C_8H_{11}N_2O_6P$ M 262.158

Selective phosphorylating agent for 5-OH groups in unprotected nucleosides.

Mono-NH_4 salt: Pale-yellow prisms + $0.5H_2O$ (MeOH/Et_2O). Mp 171-172.5°.

Taguchi, Y. *et al*, *Chem. Pharm. Bull.*, 1975, **23**, 1586 (*use*)
Taguchi, Y. *et al*, *J. Org. Chem.*, 1975, **40**, 2310 (*synth, ir, pmr, uv, use*)

Mono(3,7-dimethyl-2,6-octadienyl) diphosphate — M-00465

3,7-Dimethyl-2,6-octadien-1-ol pyrophosphate. 3,7-Dimethyl-2,6-octadien-1-ol trihydrogen pyrophosphate

[16751-02-3]

(*E*)-*form*

$C_{10}H_{20}O_7P_2$ M 314.211

Intermed. in biosynth. of terpenes.

(*E*)-form [763-10-0]

Geranyl pyrophosphate

Tri-NH_4 salt: Platelets (Me_2CO aq.). Mp 120°.

(*Z*)-form [16751-02-3]

Neryl pyrophosphate

Syrup.

Cramer, F. *et al*, *Angew. Chem.*, 1959, **71**, 775 (*synth*)
Eggerer, H., *Chem. Ber.*, 1961, **94**, 174 (*synth, props*)
Cramer, F. *et al*, *Tetrahedron*, 1967, **23**, 3015 (*props*)
Holoway, P.W. *et al*, *Biochem. J.*, 1967, **104**, 57.
Bunton, C.A. *et al*, *J. Org. Chem.*, 1972, **37**, 4036; 1979, **44**, 3238 (*props*)
Croteau, R. *et al*, *Arch. Biochem. Biophys.*, 1976, **176**, 734; 1981, **207**, 460.
Banthorpe, D.V. *et al*, *Phytochemistry*, 1977, **16**, 355 (*props*)
Dixit, V.M. *et al*, *J. Org. Chem.*, 1981, **46**, 1967 (*synth, pmr, P nmr*)

Mono-2,6-dimethylphenyl phosphate, 9CI — M-00466

2,6-Xylyl phosphate, 8CI. *2,6-Dimethylphenyl dihydrogen phosphate*

$C_8H_{11}O_4P$ M 202.146

Needles. Mp 185-186.5°.

Dichloride: [18350-98-6]. *2,6-Dimethylphenyl phosphorodichloridate. 2,6-Xylyl dichlorophosphate.*
$C_8H_9Cl_2O_2P$ M 239.038
Oil. Bp_{12} 139-140°, $Bp_{1.5}$ 110°.

Kosolapoff, G.M. *et al*, *J. Chem. Soc. (C)*, 1968, 815 (*synth, derivs*)

Mono(2,4-dinitrophenyl) phosphate, 9CI, 8CI — M-00467

2,4-Dinitrophenyl dihydrogen phosphate. 2,4-Dinitrophenyl phosphoric acid

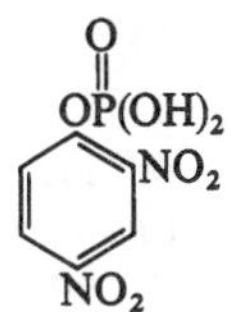

$C_6H_5N_2O_8P$ M 264.088

Sterically hindered amine (1:1) salts act as source of metaphosphate. Cryst. (Et_2O/C_6H_6). Mp 126-127°. pK_a 4.55 (H_2O).

Pyridinium salt: Solid. Mp 156-157°.

Dichloride: [20056-44-4]. *2,4-Dinitrophenyl phosphorodichloridate. 2,4-Dinitrophenyl phosphoryl dichloride. 2,4-Dinitrophenyl dichlorophosphate.*
$C_6H_3Cl_2N_2O_6P$ M 300.979
Syrup which cryst. (C_6H_6/pet. ether). Mp 75-78°. $Bp_{0.05}$ 170-180° (bath). Can be subl.

Azerad, R. *et al*, *Bull. Soc. Chim. Fr.*, 1963, 2078 (*derivs, synth*)
Varvoglis, A.G., *Chim. Chron.*, 1968, **33**, 54; *CA*, **68**, 106075 (*dichloride*)
Ramirez, F. *et al*, *Synthesis*, 1978, 601 (*synth, pmr, P nmr,props, derivs*)
Ramirez, F. *et al*, *Organophosphorus Chemistry*, 1981, **12**, 142 (*rev*)
Bunton, C.A. *et al*, *J. Org. Chem.*, 1985, **50**, 3230 (*props*)

Monododecyl phosphate, 9CI, 8CI — M-00468

Dodecyl dihydrogen phosphate. Dodecyl phosphoric acid

[2627-35-2]

$H_3C(CH_2)_{11}OP(O)(OH)_2$

$C_{12}H_{27}O_4P$ M 266.317

Na salts used in cleansing creams. Cryst. Mp 58°. pK_a 4.96 (90% 2-propanol aq., 20°).

Brown, D.A. *et al*, *J. Chem. Soc.*, 1955, 1584 (*synth*)
Nelson, A.K. *et al*, *Inorg. Chem.*, 1963, **2**, 775 (*synth*)

Monoethyl (aminocarbonyl)phosphonate, 9CI — M-00469

Ethyl carbamoylphosphonate. Ethyl hydrogen carbamoylphosphonate. Ethyl hydrogen (aminocarbonyl)-phosphonate

$C_3H_8NO_4P$ M 153.074

NH_4 salt: [25954-13-6]. *Fosamine ammonium,* BSI. *Krenite.* Contact herbicide against woody plants; used in forestry. Cryst. V. sol. H_2O, MeOH, sl. sol., EtOH, v. spar. sol. Me_2CO, C_6H_6. Mp 175°. Dec. in acid media.

▷BQ4112000.

B.P., 1 243 857, (*1968*); *CA*, **72**, 90627y
Chrzanowski, R.L. *et al*, *J. Agric. Food. Chem.*, 1979, **27**, 550 (*metab*)
Pesticide Manual, 6th Ed., 286; 7th Ed., 293.
Allen, D.W. *et al*, *J. Chem. Soc., Perkin Trans. 2*, 1977, 789 (*deriv, synth*)
Chrzanowski, R.L. *et al*, *J. Agric. Food Chem.*, 1981, **29**, 580 (*derivs, ms*)
Otsuki, T. *et al*, *Synthesis*, 1981, 811 (*esters, synth*)

Mono(2-ethylhexyl) phosphate, 9CI, 8CI **M-00470**

2-Ethylhexyl dihydrogen phosphate. 2-Ethylhexyl phosphoric acid

[1070-03-7]

$H_3CCH_2CH_2CH_2CH(CH_2CH_3)CH_2OP(O)(OH)_2$

$C_8H_{19}O_4P$ M 210.209

Extractant for many metals incl. U, rare earths, and transplutonium. Liq.

Levin, I.S. *et al*, *Dokl. Akad. Nauk SSSR, Ser. Sci. Khim.*, 1961, **139**, 158; *CA*, **56**, 11285 (*use*)
Kuzin, I.A. *et al*, *Zh. Anal. Khim.*, 1969, **24**, 800; *CA*, **71**, 56464 (*purifn*)
Lee, Te-Wei. *et al*, *Sep. Sci. Technol.*, 1981, **16**, 943 (*purifn, use*)

Monoethyl phenylphosphonate **M-00471**

Ethyl hydrogen phenylphosphonate

[4546-19-4]

OR / O=P◄OEt / Ph — (*R*)-*form* of esters

$C_8H_{11}O_3P$ M 186.147

Metab. of Cyanofenphos, C-00213.

(*R*)-*form*

Ph ester: Ethyl phenyl phenylphosphonate.

$C_{14}H_{15}O_3P$ M 262.244

Liq. $Bp_{0.02}$ 110-115°. $[\alpha]^{20}_D$ +34.4° (c, 2.16 in $CHCl_3$).

Benzyl ester: Benzyl phenyl phenylphosphonate.

$C_{15}H_{17}O_3P$ M 276.271

Liq. $Bp_{0.02}$ 112-115°. $[\alpha]^{20}_D$ +17.5° (c, 1.45 in $CHCl_3$).

(*S*)-*form*

Ph ester: Liq. $Bp_{0.02}$ 110-115°. $[\alpha]^{20}_D$ −33.6° (c, 1.26 in $CHCl_3$).

Benzyl ester: Liq. $Bp_{0.02}$ 112-115°. $[\alpha]^{20}_D$ −18.1° (c, 1.66 in $CHCl_3$).

(±)-*form*

Syrup. d^{20}_4 1.23. n_D^{25} 1.5223.

Dicyclohexylammonium salt: [77173-40-1]. Cryst. (hexane). Mp 141°.

Et ester: see Diethyl phenylphosphonate, D-00310

4-Nitrophenyl ester: [2012-00-2]. *Ethyl 4-nitrophenyl phenylphosphonate.*

$C_{14}H_{14}NO_5P$ M 307.242

Metab. and photodecomp. product of *O*-Ethyl *O*-4-nitrophenyl phenylphosphonothionate, E-00099 .

▷TA0380000.

4-Cyanophenyl ester: [62266-03-9]. *4-Cyanophenyl ethyl phenylphosphonate.*

$C_{15}H_{14}NO_3P$ M 287.254

Metab. and photodecomp. product from Cyanofenphos, C-00213 .

Rubinowitz, R., *J. Am. Chem. Soc.*, 1960, **82**, 4564 (*synth*)
Cherbuliez, E. *et al*, *Helv. Chim. Acta*, 1961, **44**, 1812, 1817 (*synth*)

Monoethyl phosphate, 9CI, 8CI **M-00472**

Ethyl dihydrogen phosphate. Ethyl phosphoric acid

[1623-14-9]

$EtOP(O)(OH)_2$

$C_2H_7O_4P$ M 126.049

Isol. as metal (Li, Ca, or Ba) or amine salt. Product of the photodecomp. of some insecticides e.g. Edifenphos, E-00002 and Mephosfolan, M-00004 and of the metab. of others e.g. Prothiofos, P-00498 . Syrup. pK_{a1} 1.60, pK_{a2} 6.62 (H_2O, 25°).

Monoanilinium salt: [2180-42-9]. Cryst. (EtOH). Mp 164-165° (145-147°).

Bis(cyclohexylammonium) salt: [4902-59-4]. Cryst. (Me_2CO aq.). Mp 205-206° (188°).

Difluoride: see Ethyl phosphorodifluoridate, E-00173

Dichloride: see Ethyl phosphorodichloridate, E-00167

Miyano, M. *et al*, *J. Am. Chem. Soc.*, 1955, **77**, 3522, 3524 (*synth*)
Montgomery, H.A.C. *et al*, *J. Chem. Soc.*, 1958, 1963 (*synth*)
Obata, T. *et al*, *J. Org. Chem.*, 1967, **32**, 1063 (*synth*)
Zwierzak, A. *et al*, *Tetrahedron*, 1971, **27**, 3163 (*synth, ir, pmr*)
Kerr, K.A. *et al*, *Acta Crystallogr., Sect. B*, 1979, **35**, 2749 (*cryst struct*)
Yamaguchi, H. *et al*, *Bull. Chem. Soc. Jpn.*, 1981, **54**, 1891 (*synth, ir, pmr*)

Monoethyl phosphonate, 9CI **M-00473**

Monoethyl phosphite

[15845-66-6]

$EtOPH(O)OH \rightleftharpoons EtOP(OH)_2$

$C_2H_7O_3P$ M 110.049

Tautomeric. Thermally unstable liq. pK_a 0.9 (estimated). n_D^{20} 1.4230 (undist.).

Na salt: [39148-16-8]. Needles (EtOH). Mp 183°.

Al salt: [39148-24-8]. *Fosetyl*, BSI. *Phosethyl AL. LS 74 783. Aliette.* Fungicide having widespread use against zoospores and sporangics, but not mycelial growths.

p-*Chlorobenzylisothiouranium salt:* Solid. Mp 140-142°.

NH_4 salt: Cryst. (EtOH/Me_2CO). Mp 99-100°.

Michalski, J. *et al*, *J. Chem. Soc.*, 1961, 4905 (*synth*)
Hammond, P.R., *J. Chem. Soc.*, 1962, 2521 (*synth*)
Zwierzak, A. *et al*, *Tetrahedron*, 1973, **29**, 1089 (*synth*)
Kluba, M. *et al*, *Synthesis*, 1978, 134 (*synth, salts, ir, pmr*)
Davis, R.M., *Plant. Dis.*, 1982, **66**, 218 (*use*)
Pesticide Manual, 7th Ed., 294.

Monoheptyl phosphate, 9CI, 8CI **M-00474**

Heptyl dihydrogen phosphate. Heptyl phosphoric acid

[3900-03-6]

$H_3C(CH_2)_6OP(O)(OH)_2$

$C_7H_{17}O_4P$ M 196.183

Oil.

U.S.P., 3 146 255, (*1964*); *CA*, **61**, 14530

Monohexadecyl phosphate, 9CI, 8CI M-00475

Hexadecyl dihydrogen phosphate. Hexadecyl phosphoric acid

[3539-43-3]

$H_3C(CH_2)_{15}OP(O)(OH)_2$

$C_{16}H_{35}O_4P$ M 322.424

Isol. as Ba salt. Cryst. Mp 73°. pK_a 4.93 (90% 2-propanol aq., 25°).

Plimmer, R.H.A. *et al*, *J. Chem. Soc.*, 1929, 292 (*synth*)
Brown, D.A. *et al*, *J. Chem. Soc.*, 1955, 1584 (*synth*)

Monohexyl phosphate, 9CI M-00476

Hexyl dihydrogen phosphate. Hexyl phosphoric acid

[3900-04-7]

$H_3C(CH_2)_5OP(O)(OH)_2$

$C_6H_{15}O_4P$ M 182.156

Monoanilinium salt: [79312-13-3]. Cryst. (EtOH). Mp 133-134°.

Bis(cyclohexylammonium) salt: Cryst. (EtOH). Mp 196°.

Moffatt, J.G. *et al*, *J. Am. Chem. Soc.*, 1957, **79**, 3741 (*synth*)
Jacob, L. *et al*, *Synthesis*, 1983, 451 (*synth, ir, pmr, cmr*)

Monoisopropyl phosphate, 8CI M-00477

Mono(1-methylethyl) phosphate, 9CI. Isopropyl dihydrogen phosphate. Isopropyl phosphoric acid

[1623-24-1]

$(H_3C)_2CHOP(O)(OH)_2$

$C_3H_9O_4P$ M 140.075

Isolated as Ca or amine salt.

▷TC6320000.

Monoanilinium salt: [1992-41-2]. Cryst. (EtOH, aq.). Mp 159-160°, Mp 167-168°, Mp 177-178°.

Bis(cyclohexylammonium) salt: Solid. Mp 209-212°.

Obata, T. *et al*, *J. Org. Chem.*, 1967, **32**, 1063 (*synth*)
Williams, A. *et al*, *J. Chem. Soc. (B)*, 1971, 1973 (*synth*)
Zwierzak, A. *et al*, *Tetrahedron*, 1971, **27**, 3163 (*synth, ir, pmr*)
Mlotkowska, B. *et al*, *Pol. J. Chem.*, 1979, **53**, 359 (*synth*)
Yamaguchi, H. *et al*, *Bull. Chem. Soc. Jpn.*, 1981, **54**, 1891 (*synth, ir, pmr*)

Monoisopropyl phosphonate M-00478

Mono(1-methylethyl) phosphonate, 9CI. Monoisopropyl phosphite

[42800-31-7]

$(H_3C)_2CHOPH(O)OH \rightleftharpoons (H_3C)_2CHOP(OH)_2$

$C_3H_9O_3P$ M 124.076

Tautomeric; almost completely in phosphonate form. Thermally unstable liq. n_D^{20} 1.4211 (undist.).

Na salt: Hygroscopic needles (EtOH/Et_2O). Mp 132-133°.

Monocyclohexylammonium salt: [77547-92-3]. Solid. Mp 150°.

p-Chlorobenzylisothiouronium salt: Solid. Mp 166-168°.

Al salt: [56318-16-2]. Possesses fungicidal props.

Michalski, J. *et al*, *J. Chem. Soc.*, 1961, 4904 (*synth*)
Zwierzak, A. *et al*, *Tetrahedron*, 1973, **29**, 1089 (*synth*)
Kluba, M. *et al*, *Synthesis*, 1978, 134 (*salts, ir, pmr*)
Troev, K. *et al*, *Phosphorus Sulfur*, 1981, **11**, 363 (*synth*)
Noren, J.O. *et al*, *J. Med. Chem.*, 1983, **26**, 264 (*ir, pmr*)

Monomethoxycarbonyl phosphate M-00479

Methyl carbonate-phosphoric acid anhydride. Methoxycarbonyl dihydrogen phosphate

[4456-11-5]

$MeOOCOP(O)(OH)_2$

CH_5O_6P M 144.021

Isolated as di-Li salt. Substrate for acetate kinase from *E. coli* and of carbamate kinase of *S. faecalis*. Useful for the regeneration of ATP from ADP in enzyme-catalysed reacns. pK_a 4.4 (H_2O). Hydrolysed more rapidly than phosphoenolpyruvic acid.

Shamshurin, A.A. *et al*, *Zh. Obshch. Khim.*, 1965, **35**, 1877 *Engl. transl.* p. 1871) (*derivs, synth, props*)
Kazlauskas, R.J. *et al*, *J. Org. Chem.*, 1985, **50**, 1069 (*synth, pmr, cmr, P nmr, props, use*)

Mono(2-methoxyphenyl) phosphate, 9CI M-00480

2-Methoxyphenyl dihydrogen phosphate. 2-Methoxyphenyl phosphoric acid. Mono-o-anisyl phosphate. o-Anisyl phosphoric acid

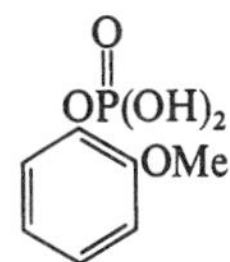

$C_7H_9O_5P$ M 204.119

Needles. Mp 94°.

Tetramethylammonium salt: Solid. Mp 179-181°.

Dichloride: [33965-78-5]. *2-Methoxyphenyl phosphorodichloridate. 2-Methoxyphenyl phosphoryl dichloride. 2-Methoxyphenyl dichlorophosphate.*
$C_7H_7Cl_2O_3P$ M 241.010
Liq. $Bp_{0.1}$ 111°.

Katyshkina, V.V. *et al*, *Zh. Obshch. Khim.*, 1956, **26**, 3060 (*Engl. transl.* p. 3407) (*dichloride*)
Just, G. *et al*, *J. Prakt. Chem.*, 1971, **313**, 69 (*dichloride*)
Jentzsch, R. *et al*, *J. Prakt. Chem.*, 1975, **317**, 721 (*synth, derivs*)

Mono(4-methoxyphenyl) phosphate, 9CI M-00481

4-Methoxyphenyl dihydrogen phosphate. 4-Methoxyphenyl phosphoric acid

$C_7H_9O_5P$ M 204.119

Tetramethylammonium salt: Solid. Mp 136-138°.

Bis(cyclohexylammonium) salt: Prisms (EtOH aq.). Mp 188-189°.

Difluoride: 4-Methoxyphenyl phosphorodifluoridate. 4-Methoxyphenyl phosphoryl difluoride. 4-Methoxyphenyl difluorophosphate.
$C_7H_7F_2O_3P$ M 208.101
Liq. Bp_{10} 103°.

Dichloride: *see 4-Methoxyphenyl phosphorodichloridate, M-00081*

Dianilide: 4-Methoxyphenyl N,N′-*diphenylphosphorodiamidate.*
$C_{19}H_{19}N_2O_3P$ M 354.344
Solid. Mp 163-164°.

Williams, A. *et al, J. Chem. Soc., Perkin Trans. 2*, 1972, 1454 (*dianilide*)
Jentzsch, R. *et al, J. Prakt. Chem.*, 1975, **317**, 721 (*synth*)
Blackburn, G.M. *et al, J. Chem. Soc., Perkin Trans. 1*, 1980, 1150 (*synth, pmr*)
Effenberger, F. *et al, Synthesis*, 1981, 70 (*difluoride, synth, ir, P nmr*)

Mono(3-methyl-2-butenyl) diphosphate, 9CI **M-00482**
3-Methyl-2-butenyl pyrophosphate. γ,γ-Dimethylallyl pyrophosphate
[358-72-5]

$(H_3C)_2C{=}CHCH_2OP(O)(OH)OP(O)(OH)_2$

$C_5H_{12}O_7P_2$ M 246.093
Important intermed. in biosynth. of long chain polyenes incl. rubber and steroidal compds. Cryst. (MeOH).
Tri-NH₄ salt: Cryst. + $1H_2O$ (Me_2CO aq.). Mp 107-109°.
Tris(cyclohexylammonium) salt: Cryst. + $1H_2O$. Mp 118-119°.

Archer, B.L. *et al, Biochem. J.*, 1963, **89**, 565 (*synth*)
Pleininger, H. *et al, Chem. Ber.*, 1965, **98**, 414 (*synth*)
Cornforth, J.W. *et al, J. Biol. Chem*, 1966, **241**, 3970.
Immel, H., *Chem. Ber.*, 1966, **99**, 2409.
Holloway, P.W. *et al, Biochem. J.*, 1967, **104**, 57 (*synth*)
Dugan, R.E. *et al, Anal. Biochem.*, 1968, **22**, 249.
Reed, B.C. *et al, Biochemistry*, 1976, **15**, 3739.
Koyama, T. *et al, J. Am. Chem. Soc.*, 1980, **102**, 3614 (*props*)
Dixit, V.M. *et al, J. Org. Chem.*, 1981, **46**, 1967 (*synth, pmr, P nmr*)

Mono(3-methyl-3-butenyl) diphosphate **M-00483**
Δ^3-Isopentenyl pyrophosphate. 3-Isopentenyl trihydrogen pyrophosphate
[358-71-4]

$CH_2{=}C(CH_3)CH_2CH_2O{-}P(O)(OH){-}O{-}P(O)(OH){-}OH$

$C_5H_{12}O_7P_2$ M 246.093
Important biosynthetic intermed. in formn. of terpenes and steroids.

Lindberg, M. *et al, Biochemistry*, 1962, **1**, 182 (*biosynth*)
Foote, C.D. *et al, Biochemistry*, 1963, **2**, 1254 (*synth*)
Chesterton, C.J. *et al, Arch. Biochem. Biophys.*, 1968, **125**, 76 (*biosynth*)
Dugan, R.E. *et al, Anal. Biochem.*, 1968, **22**, 249.
Tidd, B.K., *J. Chem. Soc. (B)*, 1971, 1168 (*props, derivs*)
Davisson, V.J. *et al, J. Org. Chem.*, 1986, **51**, 4768 (*synth, pmr, cmr, P nmr*)

Mono(3-methylbutyl) phosphate, 9CI **M-00484**
Monoisopentyl phosphate, 8CI. *Isoamyl phosphate. Isopentyl dihydrogen phosphate. Isopentyl phosphoric acid. Isoamyl phosphoric acid*

$(H_3C)_2CHCH_2CH_2OP(O)(OH)_2$

$C_5H_{13}O_4P$ M 168.129
Isolated as an amine salt.
Monoanilinium salt: Cryst. (EtOH). Mp 149-151°.

Obata, T. *et al, J. Org. Chem.*, 1967, **32**, 1063 (*synth*)

Mono(2-methylphenyl) phosphate, 9CI **M-00485**
*Mono-*o*-tolyl phosphate,* 8CI. o*-Tolyl dihydrogen phosphate. 2-Methylphenyl phosphoric acid. Mono-*o*-cresyl phosphate*

$2\text{-}CH_3C_6H_4OP(O)(OH)_2$

$C_7H_9O_4P$ M 188.119
Difluoride: 2-Methylphenyl phosphorodifluoridate. o*-Tolyl phosphodifluoridate.* o*-Tolyl difluorophosphate.*
$C_7H_7F_2O_2P$ M 192.102
Liq. Bp_{10} 68°.
Dichloride: [6964-36-9]. *2-Methylphenyl phosphorodichloridate.* o*-Tolyl phosphorodichloridate.* o*-Tolyl dichlorophosphate.*
$C_7H_7Cl_2O_2P$ M 225.011
Liq. Bp_8 118°, $Bp_{0.2}$ 75°. Hydrolysed in moist air.

Katyshkina, V.V. *et al, Zh. Obshch. Khim.*, 1956, **26**, 3060 (*Engl. transl.* p. 3407) (*dichloride*)
Just, G. *et al, J. Prakt. Chem.*, 1971, **313**, 69 (*dichloride*)
Effenberger, F. *et al, Synthesis*, 1981, 70 (*difluoride, synth, ir, P nmr*)

Mono(3-methylphenyl) phosphate, 9CI **M-00486**
*Mono-*m*-tolyl phosphate,* 8CI. m*-Tolyl dihydrogen phosphate.* m*-Cresyl phosphate*
[22987-28-6]
$C_7H_9O_4P$ M 188.119
Oil.
Dichloride: [940-18-1]. *3-Methylphenyl phosphorodichloridate.* m*-Tolyl phosphorodichloridate.* m*-Cresyl dichlorophosphate.* m*-Tolyl dichlorophosphate.*
$C_7H_7Cl_2O_2P$ M 225.011
Liq. Bp_{10} 124.5-125.5°, Bp_1 89-91°, $Bp_{0.02}$ 58°.

Orloff, H.D. *et al, J. Am. Chem. Soc.*, 1958, **80**, 727 (*dichloride*)
Beever, W.H. *et al, J. Polym. Sci., Polym. Symp.*, 1978, **65**, 41 (*synth, pmr, P nmr*)

Mono(4-methylphenyl) phosphate, 9CI **M-00487**
*Mono-*p*-tolyl phosphate,* 8CI. p*-Tolyl dihydrogen phosphate.* p*-Cresyl phosphate. 4-Methylphenyl phosphoric acid*
[6729-45-9]
$C_7H_9O_4P$ M 188.119
Plates. Mp 116°. pK_{a1} 0.56, pK_{a2} 5.80 (H_2O, 25°).
Monocyclohexylammonium salt: [79227-86-4]. Solid. Mp 191-192°.
Difluoride: 4-Methylphenyl phosphorodifluoridate. 4-Methylphenyl phosphoryl difluoride. 4-Methylphenyl difluorophosphate. p*-Tolyl phosphorodifluoridate.* p*-Tolyl phosphoryl difluoride.* p*-Tolyl difluorophosphate.*
$C_7H_7F_2O_2P$ M 192.102
Liq. Bp_{10} 73-75°.

Dichloride: see 4-Methylphenyl phosphorodichloridate, M-00257

Rapp, M., *Justus Liebigs Ann. Chem.*, 1884, **224**, 156 (*synth*)
Effenberger, F. *et al*, *Synthesis*, 1981, 70 (*difluoride, synth, ir, P nmr*)
Markowska, A. *et al*, *Phosphorus Sulfur*, 1981, **10**, 143 (*P nmr*)
Bourne, N. *et al*, *J. Org. Chem.*, 1984, **49**, 1200 (*uv, props*)

Monomethyl phenylphosphonate M-00488

Methyl hydrogen phenylphosphonate
[10088-45-6]

RO–P(=O)(OMe)–Ph (*R*)-*form* of esters

$C_7H_9O_3P$ M 172.120
Metab. of Leptophos and Leptophos Oxon.

(*R*)-*form*

Ph ester: Methyl phenyl phenylphosphonate.
$C_{13}H_{13}O_3P$ M 248.218
Liq. $Bp_{0.02}$ 105-108°. $[\alpha]^{20}_{D}$ +38.1° (c, 1.45 in $CHCl_3$).
Benzyl ester: Benzyl methyl phenylphosphonate.
$C_{14}H_{15}O_3P$ M 262.244
Liq. $Bp_{0.02}$ 110-111°. $[\alpha]^{20}_{D}$ +15.5° (c, 2.15 in $CHCl_3$).

(*S*)-*form*

Ph ester: Liq. $Bp_{0.02}$ 105-108°. $[\alpha]^{20}_{D}$ −38.8° (c, 1.24 in $CHCl_3$).
Benzyl ester: Liq. $Bp_{0.02}$ 110-111°. $[\alpha]^{20}_{D}$ −17.3° (c, 0.52 in $CHCl_3$).

(±)-*form*

Oil. pK_a 2.97. Heat-sensitive, isol. as Ba or cyclohexylammonium salt.
Mono-cyclohexylammonium salt: [38555-73-6]. Cryst. (Me_2CO). Mp 156-158°.
Mono-dicyclohexylammonium salt: [75620-16-5]. Cryst. (hexane). Mp 132°.
Me ester: see Dimethyl phenylphosphonate, D-00810
(4-Bromo-2,5-dichlorophenyl) ester: [25006-32-0]. *(4-Bromo-2,5-dichlorophenyl)methyl phenylphosphonate. Leptophos Oxon.*
$C_{13}H_{10}BrCl_2O_3P$ M 396.004
Metab. of, and a more powerful neurotoxic agent than, Leptophos.
Chloride: see Phenylphosphonochloridic acid, P-00229

Cherbuliez, E. *et al*, *Helv. Chim. Acta*, 1961, **44**, 1812 (*synth*)
Brooks, R.J. *et al*, *J. Org. Chem.*, 1973, **38**, 1614 (*synth, pmr*)
Holmstead, R.L. *et al*, *Arch. Environ. Contam. Toxicol.*, 1973, **1**, 133; *CA*, **80**, 44526 (*ms*)
Felcht, U. *et al*, *Chem. Ber.*, 1976, **109**, 3675 (*synth, ir*)
Lee, P.W. *et al*, *Arch. Environ. Contam. Toxicol.*, 1976, **4**, 443; *CA*, **86**, 66794 (*deriv*)
Otsuki, T. *et al*, *Synthesis*, 1981, 811 (*ester*)

Monomethyl phosphate, 9CI, 8CI M-00489

Methyl dihydrogen phosphate. Methyl phosphoric acid
[812-00-0]

$MeOP(O)(OH)_2$

CH_5O_4P M 112.022
Isol. as metal (Na, Ba, Ca) or amine salts. Prod. of the photodec. of many pesticides e.g. Iodofenphos, I-00011 and of metab. of others e.g. Trichlorphon, T-00423 and Tetrachlorvinphos, T-00038 . Oil. pK_{a1} 1.54, pK_{a2} 6.31 (H_2O, 25°).
▷TC6594000.
Di-NH_4 salt: [36989-61-4]. Cryst. + $2H_2O$ (EtOH aq.).
Monoanilinium salt: [7704-44-1]. Solid. Mp 167-168°.
Bis(cyclohexylammonium) salt: Solid. Mp 169-172°.
Difluoride: see Methyl phosphorodifluoridate, M-00352
Dichloride: see Methyl phosphorodichloridate, M-00345

Cramer, F. *et al*, *Chem. Ber.*, 1958, **91**, 1181 (*synth*)
Obata, T. *et al*, *J. Org. Chem.*, 1967, **32**, 1063 (*synth, tlc*)
Zwierzak, A. *et al*, *Tetrahedron*, 1971, **27**, 3163 (*synth, ir, pmr*)
Garbassi, F. *et al*, *Acta Crystallogr., Sect. B*, 1972, **28**, 1665 (*cryst struct*)
Yamaguchi, H. *et al*, *Bull. Chem. Soc. Jpn.*, 1981, **54**, 1891 (*synth, ir, pmr*)
Calvo, K.C. *et al*, *J. Am. Chem. Soc.*, 1983, **105**, 2827 (*P nmr*)

Monomethyl phosphonate, 9CI M-00490

Methyl hydrogen phosphonate. Methyl phosphite. Monomethyl phosphite
[13590-71-1]

$MeOPH(O)OH \rightleftharpoons MeOP(OH)_2$

CH_5O_3P M 96.022
Tautomeric; almost completely in phosphonate form. Unstable oil.
▷SZ9650000.
NH_4 salt: Cryst. (EtOH/Et_2O). Mp 110-111°.

Zwierzak, A. *et al*, *Tetrahedron*, 1973, **29**, 1089 (*synth*)
Miyake, C. *et al*, *J. Inorg. Nucl. Chem.*, 1981, **43**, 2407 (*complexes*)

Mono(2-methyl-1-propenyl) phosphate, 9CI M-00491

2-Methyl-1-propenyl dihydrogen phosphate. (2,2-Dimethylvinyl) phosphoric acid. Isobutenyl dihydrogen phosphate. Monoisobutenyl phosphate. Isobutenyl phosphoric acid

$(H_3C)_2C{=}CHOP(O)(OH)_2$

$C_4H_9O_4P$ M 152.086
Bis(cyclohexylammonium) salt: Solid. Mp 189-190°.

Vakulova, A. *et al*, *Dokl. Akad. Nauk SSSR, Ser. Sci. Khim.*, 1962, **147**, 103; *CA*, **58**, 8919 (*synth*)

Mono(1-methylpropyl) phosphate, 9CI M-00492

sec-*Butyl dihydrogen phosphate.* sec-*Butyl phosphoric acid*
[2382-77-6]

$H_3CCH_2CH(CH_3)OP(O)(OH)_2$

$C_4H_{11}O_4P$ M 154.102

(±)-*form*

Monoanilinium salt: [33494-77-8]. Cryst. (EtOH). Mp 158-159°.
Monocyclohexylammonium salt: [79138-43-5]. Solid. Mp 187-189°.

Zwierzak, A. *et al*, *Tetrahedron*, 1971, **27**, 3163 (*synth, ir, pmr*)
Gajda, T. *et al*, *Synthesis*, 1977, 623 (*synth, ir, pmr*)
Markowska, A. *et al*, *Phosphorus Sulfur*, 1981, **10**, 245 (*synth*)
Yamaguchi, H. *et al*, *Bull. Chem. Soc. Jpn.*, 1981, **54**, 1891 (*synth, ir, pmr*)

Mono(2-methylpropyl) phosphate, 9CI **M-00493**

Monoisobutyl phosphate, 8CI. Isobutyl dihydrogen phosphate. Isobutyl phosphoric acid

[2466-73-1]

$(H_3C)_2CHCH_2OP(O)(OH)_2$

$C_4H_{11}O_4P$ M 154.102

Isolated as amine salt.

Monoanilinium salt: [7704-50-9]. Cryst. (EtOH). Mp 158-159.5° (147-148°).

Obata, T. *et al*, *J. Org. Chem.*, 1967, **32**, 1063 (*synth*)
Zwierzak, A. *et al*, *Tetrahedron*, 1971, **27**, 3163 (*synth, ir, pmr*)
Gajda, T. *et al*, *Synthesis*, 1977, 623 (*synth, ir, pmr*)

Mono-1-naphthyl phosphate, 8CI **M-00494**

Mono-1-naphthalenyl phosphate, 9CI. 1-Naphthyl dihydrogen phosphate. 1-Naphthyl phosphoric acid

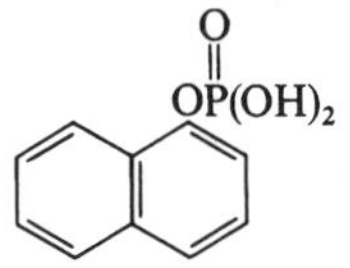

$C_{10}H_9O_4P$ M 224.152

Cryst. Mp 142°, Mp 155-157°. pK_{a1} 1.10 (H_2O), pK_{a2} 4.89 (95% EtOH aq.), pK_{a3} 6.60 (EtOH).

Difluoride: 1-Naphthyl phosphorodifluoridate. 1-Naphthyl difluorophosphate.
$C_{10}H_7F_2O_2P$ M 228.135
Liq. $Bp_{0.1}$ 80°.

Dichloride: [31651-76-0]. *1-Naphthyl phosphorodichloridate. 1-Naphthyl dichlorophosphate.*
$C_{10}H_7Cl_2O_2P$ M 261.044
Liq. Bp 325-327°, Bp_{20} 199-201°. n_D^{22} 1.596.

Friedman, O.M. *et al*, *J. Am. Chem. Soc.*, 1950, **72**, 624 (*synth, dichloride*)
Effenberger, F. *et al*, *Synthesis*, 1981, 70 (*synth, ir, P nmr*)

Mono-2-naphthyl phosphate, 8CI **M-00495**

Mono-2-naphthalenyl phosphate, 9CI. 2-Naphthyl dihydrogen phosphate. 2-Naphthylphosphoric acid

$C_{10}H_9O_4P$ M 224.152

Plates ($CHCl_3$/EtOH or H_2O). Mp 176-177°. pK_{a1} 1.17 (H_2O), pK_{a2} 4.95 (95% EtOH aq.), pK_{a3} 6.61 (EtOH).

Dichloride: [31651-74-8]. *2-Naphthyl phosphorodichloridate. 2-Naphthyl dichlorophosphate.*
$C_{10}H_7Cl_2O_2P$ M 261.044
Oil which solidifies. Mp 39°. Bp_{20} 204-205°, Bp_1 150-155°.

Atherton, F.R. *et al*, *J. Chem. Soc.*, 1945, 382 (*synth*)
Friedman, O.M. *et al*, *J. Am. Chem. Soc.*, 1950, **72**, 624; 1951, **73**, 5292 (*synth, derivs*)

Mono-1-naphthyl phosphonate **M-00496**

1-Naphthalenyl hydrogen phosphonate. Mono-1-naphthyl phosphite

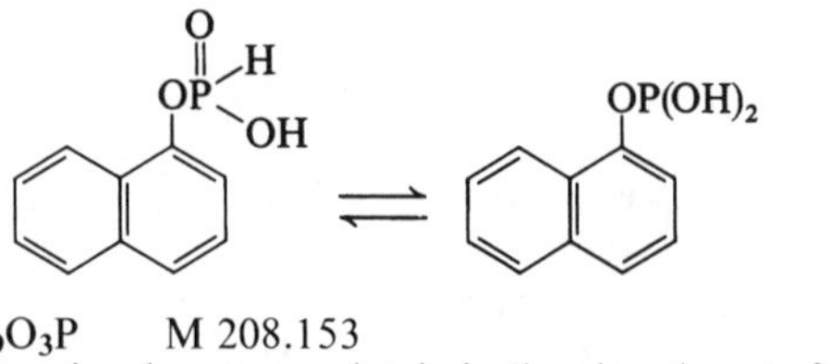

$C_{10}H_9O_3P$ M 208.153

Tautomeric; almost completely in the phosphonate form. Cryst. powder. Mp 82°. Rapidly hydrolysed to 1-naphthol and H_3PO_3.

Kunz, P., *Ber.*, 1894, **27**, 2559 (*synth*)

Mono-2-naphthyl phosphonate **M-00497**

2-Naphthalenyl hydrogen phosphonate. Mono-2-naphthyl phosphite

$C_{10}H_9O_3P$ M 208.153

Tautomeric; almost completely in phosphonate form. Cryst. Mp 111°. Rapidly hydrolysed to H_3PO_3 and 2-naphthol.

Kunz, P., *Ber.*, 1894, **27**, 2559 (*synth*)

Mono(2-nitrophenyl) phosphate, 9CI **M-00498**

2-Nitrophenyl dihydrogen phosphate. 2-Nitrophenyl phosphoric acid

[6064-84-2]

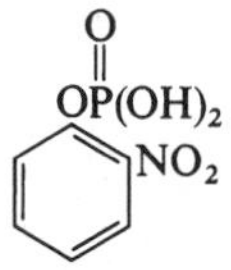

$C_6H_6NO_6P$ M 219.090

Difluoride: 2-Nitrophenyl phosphorodifluoridate. 2-Nitrophenyl phosphoryl difluoride. 2-Nitrophenyl difluorophosphate.
$C_6H_4F_2NO_4P$ M 223.072
Liq. $Bp_{0.01}$ 67-70°.

Dichloride: [20056-38-6]. *2-Nitrophenyl phosphorodichloridate. 2-Nitrophenyl phosphoryl dichloride. 2-Nitrophenyl dichlorophosphate.*
$C_6H_4Cl_2NO_4P$ M 255.982
Oil. $Bp_{2.8}$ 150-152°, $Bp_{0.3}$ 105°.

Katyshkina, V.V. *et al*, *Zh. Obshch. Khim.*, 1956, **26**, 3060 (*Engl. transl.* p. 3407) (*dichloride*)
Orloff, H.D. *et al*, *J. Am. Chem. Soc.*, 1958, **80**, 727 (*dichloride*)
Effenberger, F. *et al*, *Synthesis*, 1981, 70 (*difluoride, synth, ir, P nmr*)

Mono(3-nitrophenyl) phosphate, 9CI, 8CI **M-00499**

3-Nitrophenyl dihydrogen phosphate. 3-Nitrophenyl phosphoric acid

[13388-91-5]

$C_6H_6NO_6P$ M 219.090

pK_a 5.14 (H_2O, 25°).

Dichloride: [38319-29-8]. *3-Nitrophenyl phosphorodichloridate. 3-Nitrophenyl phosphoryl dichloride. 3-Nitrophenyl dichlorophosphate.*
$C_6H_4Cl_2NO_4P$ M 255.982
Oil. $Bp_{0.07}$ 126°.

Dianilide: 3-Nitrophenyl N,N′-*diphenylphosphorodiamidate.*
$C_{18}H_{16}N_3O_4P$ M 369.316
Solid. Mp 171-173°.

Katyshkina, V.V. *et al*, *Zh. Obshch. Khim.*, 1956, **26**, 3060 (*Engl. transl.* p. 3407) (*dichloride*)
Williams, A. *et al*, *J. Chem. Soc. (B)*, 1971, 1973 (*uv, dianilide*)
Lazarus, R.A. *et al*, *J. Chem. Soc., Perkin Trans. 2*, 1980, 373 (*dichloride*)
Bourne, N. *et al*, *J. Org. Chem.*, 1984, **49**, 1200 (*props*)

Mono(4-nitrophenyl) phosphate, 9CI **M-00500**

4-Nitrophenyl dihydrogen phosphate. 4-Nitrophenyl phosphoric acid

[330-13-2]

$C_6H_6NO_6P$ M 219.090

Substrate for phosphatase enzymes. Solid. Mp 154-156°. pK_{a1} 0.30, pK_{a2} 4.96 (H_2O, 25°).

Monoanilinium salt: Solid. Mp 180-182°.

Bis(cyclohexylammonium) salt: [52483-84-8]. Cryst. (Me_2CO). Mp 165-167°. Forms a dihydrate.

Difluoride: 4-Nitrophenyl phosphorodifluoridate. 4-Nitrophenyl phosphoryl difluoride. 4-Nitrophenyl difluorophosphate.
$C_6H_4F_2NO_4P$ M 223.072
Liq. $Bp_{0.01}$ 72-75°.

Dichloride: see 4-Nitrophenyl phosphorodichloridate, N-00064

Dianilide: 4-Nitrophenyl N,N′-*diphenylphosphorodiamidate.*
$C_{18}H_{16}N_3O_4P$ M 369.316
Solid. Mp 169-170°.

Mitsunobu, O. *et al*, *Bull. Chem. Soc. Jpn.*, 1965, **38**, 2100 (*synth*)
Williams, A. *et al*, *J. Chem. Soc. (B)*, 1971, 1973 (*synth, derivs, uv, ms*)
Bel'skii, V.E. *et al*, *Izv. Akad. Nauk SSSR, Ser. Khim.*, 1979, 1633 (*Engl. transl.* p. 1510) (*derivs*)
Effenberger, F. *et al*, *Synthesis*, 1981, 70 (*difluoride, synth, ir, nmr*)
Bourne, N. *et al*, *J. Org. Chem.*, 1984, **49**, 1200 (*props*)
Jones, P.G. *et al*, *Acta Crystallogr., Sect. C*, 1984, **40**, 550 (*cryst struct*)
Beltran, A.M. *et al*, *Tetrahedron Lett.*, 1985, **26**, 1711 (*props*)

Monooctyl phosphate, 9CI, 8CI — M-00501

Dihydrogen octyl phosphate. Octyl phosphoric acid. Octyl dihydrogen phosphate

[3991-73-9]

$H_3C(CH_2)_7OP(O)(OH)_2$

$C_8H_{19}O_4P$ M 210.209

Monoanilinium salt: [7704-54-3]. Cryst. (EtOH). Mp 129-130°.

Bis(cyclohexylammonium) salt: Solid. Mp 153-155°.

Obata, T. *et al*, *J. Org. Chem.*, 1967, **32**, 1063 (*synth, tlc*)
Jacob, L. *et al*, *Synthesis*, 1983, 451 (*synth, ir, pmr*)
Okamoto, Y., *Bull. Chem. Soc. Jpn.*, 1985, **58**, 3393 (*synth*)

Mono(pentachlorophenyl) phosphate, 9CI — M-00502

Pentachlorophenyl dihydrogen phosphate. Pentachlorophenyl phosphoric acid

$C_6Cl_5OP(O)(OH)_2$

$C_6H_2Cl_5O_4P$ M 346.318

Plates. Mp 208-210° (203°).

Monohydrate: Mp 224°.

Dichloride: see Pentachlorophenyl phosphorodichloridate, P-00006

Zincke, T. *et al*, *Ber.*, 1891, **24**, 927; *Justus Liebigs Ann. Chem.*, 1892, **267**, 1 (*synth*)
Cremlyn, R.J.W. *et al*, *J. Chem. Soc., Perkin Trans. 1*, 1972, 583 (*synth, ir, derivs*)

Mono(pentafluorophenyl) phosphate, 9CI — M-00503

Pentafluorophenyl dihydrogen phosphate. Pentafluorophenyl phosphoric acid

$C_6F_5OP(O)(OH)_2$

$C_6H_2F_5O_4P$ M 264.045

Dichloride: Pentafluorophenyl phosphorodichloridate. Pentafluorophenyl dichlorophosphate. Pentafluorophenyl phosphoryl dichloride.
$C_6Cl_2F_5O_2P$ M 300.937
Liq. Bp_{20} 112-115°.

U.S.P., 3 341 630, (*1967*); *CA*, **68**, 77961

Monopentyl phosphate, 9CI, 8CI — M-00504

Monoamyl phosphate. Pentyl dihydrogen phosphate. Pentyl phosphoric acid. Amyl phosphoric acid. Amyl dihydrogen phosphate

[2382-76-5]

$H_3C(CH_2)_4OP(O)(OH)_2$

$C_5H_{13}O_4P$ M 168.129

Monoanilinium salt: Cryst. (EtOH). Mp 139-141°.

Obata, T. *et al*, *J. Org. Chem.*, 1967, **32**, 1063 (*synth*)
Takaku, H. *et al*, *Tetrahedron Lett.*, 1972, 411 (*synth*)

Mono(2-phenoxyphenyl) phosphate, 9CI, 8CI — M-00505

2-Phenoxyphenyl dihydrogen phosphate. 2-Phenoxyphenyl phosphoric acid

O
‖
$OP(OH)_2$
OPh

$C_{12}H_{11}O_5P$ M 266.190

Cryst. (CCl_4). Mp 121-123°.

Dichloride: 2-Phenoxyphenyl phosphorodichloridate. 2-Phenoxyphenyl phosphoryl dichloride. 2-Phenoxyphenyl dichlorophosphate.
$C_{12}H_9Cl_2O_3P$ M 303.081
Liq. Bp_{11} 195-198°.

U.S.P., 1 960 184, (*1934*) (*synth*)

Mono(2-phenylethyl) phosphate, 9CI, 8CI — M-00506

2-Phenylethyl dihydrogen phosphate. 2-Phenylethylphosphoric acid. Phenethyl dihydrogen phosphate. Phenethyl phosphoric acid

[18110-43-5]

$PhCH_2CH_2OP(O)(OH)_2$

$C_8H_{11}O_4P$ M 202.146

Oil or syrup.

Monocyclohexylammonium salt: Cryst. (Me_2CO aq.). Mp 177°.

Dichloride: 2-Phenylethyl phosphorodichloridate. 2-Phenylethyl dichlorophosphate. Phenethyl phosphorodichloridate. Phenethyl dichlorophosphate.
$C_8H_9Cl_2O_2P$ M 239.038
Liq. Bp_{3-5} 128-132°. n_D^{20} 1.5200.

Montgomery, H.A.C. *et al*, *J. Chem. Soc.*, 1958, 1963 (*synth*)
Bauer, S. *et al*, *Isr. J. Chem.*, 1967, **5**, 171 (*synth, props, uv*)
Petrov, K.A. *et al*, *Zh. Obshch. Khim.*, 1970, **40**, 2192 (*Engl. transl.* p. 2179) (*dichloride*)

Monophenyl phosphate, 9CI, 8CI — M-00507

Phenyl dihydrogen phosphate. Phenyl phosphoric acid. Dihydrogen phenyl phosphate

[701-64-4]

$PhOP(O)(OH)_2$

$C_6H_7O_4P$ M 174.093

Cryst. ($CHCl_3$ or H_2O). Mp 99.5-100°. pK_{a1} 0.48 (H_2O, 25°); 3.23 (50% pK_{a2} *EtOH aq.), pK_{a3} *pK_{a2} 5.70 (H_2O, pK_{a4} *25°), pK_{a5} *6.23 (50% EtOH aq.).

Di-NH_4 salt: Solid. Mp 164-165° dec.

Monoanilinium salt: [1992-39-8]. Solid. Mp 169-170°.

Bis(cyclohexylammonium) salt: [13798-39-5]. Prisms (EtOH aq. or H_2O). Mp 190°, Mp 214-215°.

Difluoride: see Phenyl phosphorodifluoridate, P-00285

Dichloride: see Phenyl phosphorodichloridate, P-00279

Dianilide: see N,N′-Diphenylphosphorodiamidic acid, D-01125

Miyano, M., *J. Am. Chem. Soc.*, 1955, **77**, 3524 (*synth*)
Montgomery, H.A.C. *et al*, *J. Chem. Soc.*, 1958, 1963 (*synth*)
Owen, G.R. *et al*, *Synthesis*, 1974, 704 (*synth*)
Kodolov, V.I. *et al*, *Izv. Akad. Nauk SSSR, Ser. Khim.*, 1977, 165 (*Engl. transl.* p. 142) (*pmr, pe*)
Blackburn, G.M. *et al*, *J. Chem. Soc., Perkin Trans. 1*, 1980, 1150 (*pmr*)
Yamaguchi, H. *et al*, *Bull. Chem. Soc. Jpn.*, 1981, **54**, 1891 (*synth, ir, pmr*)
Bourne, N. *et al*, *J. Org. Chem.*, 1984, **49**, 1200 (*uv*)

Monophenyl phosphonate, 9CI **M-00508**

Monophenyl phosphite

[2310-89-6]

$PhOPH(O)OH \rightleftharpoons PhOP(OH)_2$

$C_6H_7O_3P$ M 158.093

Tautomeric. Reagent for amidations, esterifications, and peptide formation without racemization. Liq.

NH_4 salt: [54921-72-1]. Cryst. (EtOH/Et_2O). Mp 172-173°.

Hammond, P.R., *J. Chem. Soc.*, 1962, 2521 (*synth*)
Yamazaki, N. *et al*, *Synthesis*, 1974, 436 (*use*)
Yamazaki, N. *et al*, *Chem. Lett.*, 1977, 185 (*synth*)

Mono-2-propenyl phosphate, 9CI **M-00509**

Allyl phosphate. Monoallyl phosphate. Allyl dihydrogen phosphate

[25022-72-4]

$H_2C{=}CHCH_2OPO_3H_2$

$C_3H_7O_4P$ M 138.060

Prod. during acid fermentation of *Aspergillus niger*. Used to improve adhesives. Triose phosphate isomerase inhibitor. Syrup.

▷Highly toxic. Can explode on distillation

NH_4 salt: Cryst. Mp 93°, 107-112°.

Biscyclohexylammonium salt: [7264-92-8]. Cryst. + $1H_2O$ (Me_2CO aq.). Mp 175°, 200° dec.

Montgomery, H.A.C. *et al*, *J. Chem. Soc.*, 1958, 1963 (*synth*)
Maynard, J.A. *et al*, *Aust. J. Chem.*, 1963, **16**, 596 (*synth*)
Khwaja, T.A. *et al*, *J. Chem. Soc. (C)*, 1970, 2092 (*synth*)
Zwerziak, A. *et al*, *Tetrahedron*, 1971, **27**, 3163 (*synth*)
Sax, N.I., *Dangerous Properties of Industrial Materials*, 6th Ed., Van Nostrand-Reinhold, 1984, 351.

Mono-2-propenyl phosphonate, 9CI **M-00510**

Monoallyl phosphonate. Mono-2-propenyl phosphite. Monoallyl phosphite. Mono-2-propenyl phosphite

[42023-33-6]

$H_2C{=}CHCH_2OPH(O)OH \rightleftharpoons H_2C{=}CHCH_2OP(OH)_2$

$C_3H_7O_3P$ M 122.060

Tautomeric. Liq. n_D^{20} 1.4390. Dec. on dist.

Na salt: Mp 180-185°.

p-*Chlorobenzylisothiouronium salt:* Solid. Mp 145-147°.

Michalski, J. *et al*, *J. Chem. Soc.*, 1961, 4904 (*synth*)
Zwierzak, A. *et al*, *Tetrahedron*, 1973, **29**, 1089 (*synth*)

Monopropyl phenylphosphonate **M-00511**

Propyl hydrogen phenylphosphonate

[7670-93-1]

OR
O=P◀OCH₂CH₂CH₃ (with Ph below P)

(*R*)-*form* of phenyl and benzyl esters

$C_9H_{13}O_3P$ M 200.174

Propyl ester: see under Phenylphosphonic acid, P-00221

(*R*)-*form*

Ph ester: Phenyl propyl phenylphosphonate.

$C_{15}H_{17}O_3P$ M 276.271

$Bp_{0.02}$ 113-117°. $[\alpha]_D^{20}$ +29.5° (c, 2.36 in $CHCl_3$).

Benzyl ester: Benzyl propyl phenylphosphonate. Phenylmethyl propyl phenylphosphonate.

$C_{16}H_{19}O_3P$ M 290.298

Liq. $Bp_{0.02}$ 130-133°. $[\alpha]_D$ +11.4° (c, 1.14 in $CHCl_3$).

(*S*)-*form*

Ph ester: Liq. $Bp_{0.02}$ 113-117°. $[\alpha]_D^{20}$ −29.1° (c, 1.08 in $CHCl_3$).

Benzyl ester: Liq. $Bp_{0.02}$ 130-133°. $[\alpha]_D^{20}$ −13.2° (c, 1.12 in $CHCl_3$).

Cherbuliez, E. *et al*, *Helv. Chim. Acta*, 1961, **44**, 1812, 1817 (*synth*)
Otsuki, T. *et al*, *Synthesis*, 1981, 811 (*esters, synth, pmr*)

Monopropyl phosphate, 9CI, 8CI **M-00512**

Propyl dihydrogen phosphate. Propyl phosphoric acid

[1623-06-9]

$H_3CCH_2CH_2OP(O)(OH)_2$

$C_3H_9O_4P$ M 140.075

Isolated as metal (Ca) or amine salt. pK_{a1} 1.88, pK_{a2} 6.67 (H_2O, 25°).

Monoanilinium salt: [2180-41-8]. Cryst. (EtOH). Mp 137-139°, Mp 150-152°.

Monocyclohexylammonium salt: [79138-42-4]. Solid. Mp 176-177°.

Dichloride: see Propyl phosphorodichloridate, P-00487

Obata, T. *et al*, *J. Org. Chem.*, 1967, **32**, 1063 (*synth*)
Zwierzak, A. *et al*, *Tetrahedron*, 1971, **27**, 3163 (*synth, ir, pmr*)
Markowska, A. *et al*, *Phosphorus Sulfur*, 1981, **10**, 245 (*synth*)
Yamaguchi, H. *et al*, *Bull. Chem. Soc. Jpn.*, 1981, **54**, 1891 (*synth, ir, pmr*)

Monopropyl phosphonate, 9CI **M-00513**

Monopropyl phosphite

[42023-31-4]

$$H_3CCH_2CH_2OPH(O)OH \rightleftharpoons H_3CCH_2CH_2OP(OH)_2$$

$C_3H_9O_3P$ M 124.076

Tautomeric. Thermally unstable liq. n_D^{20} 1.4257 (undist.).

Na salt: Cryst. (EtOH). Mp 195-196°.

p-*Chlorobenzylisothiouronium salt:* Solid. Mp 139-141°.

NH_4 salt: Cryst. (EtOH/Me_2CO). Mp 91-92°.

Michalski, J. *et al*, *J. Chem. Soc.*, 1961, 4904 (*synth*)
Hammond, P.R., *J. Chem. Soc.*, 1962, 2521 (*synth*)
Zwierzak, A. *et al*, *Tetrahedron*, 1973, **29**, 1089 (*synth*)

Monotetradecyl phosphate, 9CI, 8CI **M-00514**

Tetradecyl dihydrogen phosphate. Tetradecyl phosphoric acid

$$H_3C(CH_2)_{13}OP(O)(OH)_2$$

$C_{14}H_{31}O_4P$ M 294.370

Cryst. (hexane). Mp 68°. pK_a 4.94 (90% 2-propanol aq., 20°).

Brown, D.A. *et al*, *J. Chem. Soc.*, 1955, 1584 (*synth*)
Nelson, A.K. *et al*, *Inorg. Chem.*, 1963, **2**, 775 (*synth*)

Mono(3,7,11,15-tetramethyl-2,6,10,14-hexadecatetraenyl) diphosphate **M-00515**

3,7,11,15-Tetramethyl-2,6,10,14-hexadecatetren-1-ol trihydrogen pyrophosphate

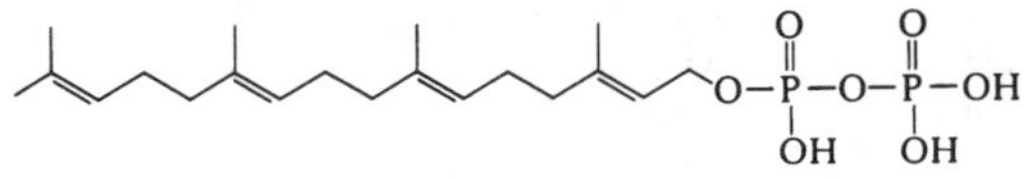

O–P(=O)(OH)–O–P(=O)(OH)–OH

(*E,E,E*)-*form*

$C_{20}H_{36}O_7P_2$ M 450.448

Biosynthetic precursor to terpenes.

(*E,E,E*)-*form* [6699-20-3]

Geranylgeranyl pyrophosphate

Nandi, D.L. *et al*, *Arch. Biochem. Biophys.*, 1964, **105**, 7 (*enzymic synth*)
Upper, C.D. *et al*, *J. Biol. Chem*, 1967, **242**, 3285 (*synth*)
Lee, T.-C. *et al*, *Phytochemistry*, 1972, **11**, 681.
Gregonis, D.E. *et al*, *Biochemistry*, 1974, **13**, 1538 (*synth*)
Moore, T.C. *et al*, *Phytochemistry*, 1976, **15**, 1241 (*biosynth*)

Mono(2,2,2-tribromoethyl) phosphate **M-00516**

Dihydrogen tribromoethyl phosphate. Tribromoethyl phosphoric acid

$$Br_3CCH_2OP(O)(OH)_2$$

$C_2H_4Br_3O_4P$ M 362.737

Solid. Mp 147°.

Dichloride: [53676-22-5]. *2,2,2-Tribromoethyl phosphorodichloridate. 2,2,2-Tribromoethyl phosphoryl dichloride. 2,2,2-Tribromoethyl dichlorophosphate.*
$C_2H_2Br_3Cl_2O_2P$ M 399.628
Precursor to reagents used in phospholipid synth. Fuming oil or solid. Mp ca. 20°. $Bp_{0.4}$ 90-92°.

Owen, G.R. *et al*, *Synthesis*, 1974, 704 (*synth, dichloride*)
Van Boeckel, C.A.A. *et al*, *Tetrahedron*, 1981, **37**, 3751 (*use*)

Mono(3,7,11-trimethyl-2,6,10-dodecatrienyl) diphosphate **M-00517**

3,7,11-Trimethyl-2,6,10-dodecatrien-1-ol trihydrogen phosphate. 3,7,11-Trimethyl-2,6,10-dodecatrienyl pyrophosphate

[13058-04-3]

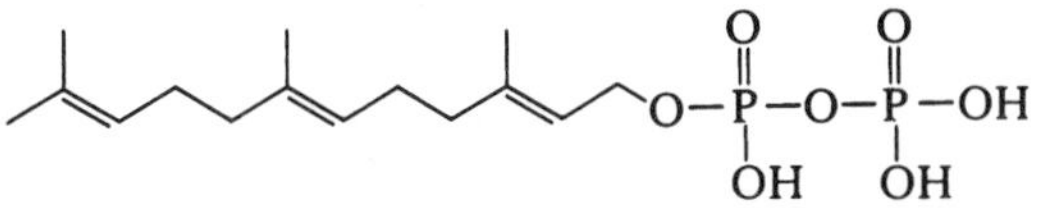

O–P(=O)(OH)–O–P(=O)(OH)–OH

$C_{15}H_{28}O_7P_2$ M 382.330

Intermed. in biosynth. of terpenes.

(*E,E*)-*form* [372-97-4]

Farnesyl pyrophosphate

Krishna, G. *et al*, *Arch. Biochem. Biophys.*, 1966, **114**, 200 (*synth*)
Holoway, P.W. *et al*, *Biochem. J.*, 1967, **104**, 57 (*biosynth*)
Popják, G. *et al*, *J. Biol. Chem*, 1969, **244**, 1897.
Popják, G., *Methods Enzymol*, 1969, **15**, 393 (*enzymic synth*)
Dixit, V.M. *et al*, *J. Org. Chem.*, 1981, **46**, 1967 (*synth, pmr, P nmr*)

Monovinyl phosphate, 8CI **M-00518**

Monoethenyl phosphate, 9CI. *Vinyl phosphate. Ethenyl dihydrogen phoshate. Vinyl phosphoric acid. Vinyl dihydrogen phosphate*

[36885-49-1]

$$H_2C{=}CHOPO(OH)_2$$

$C_2H_5O_4P$ M 124.033

Insecticide. Very unstable oil, usually isol. as the di-Li salt. Sol. Et_2O, dioxane, insol. ligroin. Rapidly hydrolysed by H_2O at room temp.

▷Toxic by oral admin. Inhibits cholinesterases. TC6593000.

Dichloride: [2035-84-9]. *Ethenyl phosphorodichloridate. Vinyl phosphorodichloridate. Vinyl dichlorophosphate.*
$C_2H_3Cl_2O_2P$ M 160.924
Pungent liq. d_4^{20} 1.43. Bp_{30} 36-40°. n_D^{20} 1.4429.

Baer, E. *et al*, *J. Biol. Chem.*, 1959, **234**, 1 (*synth*)
Gololobov, Y.G. *et al*, *Zh. Obshch. Khim.*, 1965, **35**, 1460 (*Engl. transl.* p. 1462) (*dichloride*)

(Morpholinomethyl)diphenylphosphine oxide **M-00519**

4-[(Diphenylphosphinyl)methyl]morpholine

[20684-76-8]

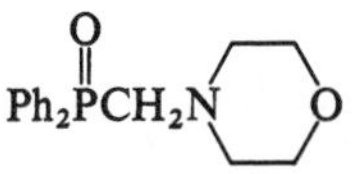

$Ph_2P(O)CH_2N$ (morpholine ring)

$C_{17}H_{20}NO_2P$ M 301.324

Reagent, used as Li deriv., for converting ketones into morpholinoenamines, and aldehydes into α-aminomethyl ketones. Cryst. (methylcyclohexane or EtOAc). Mp 163-165°.

Burger, A. *et al*, *J. Med. Chem.*, 1961, **4**, 225 (*synth, use*)
Gallagher, M.J., *Aust. J. Chem.*, 1968, **21**, 1197 (*synth, pmr, use*)
Broekhof, N.L.J.M. *et al*, *Tetrahedron Lett.*, 1979, 2433 (*synth, use*)
Van den Gen, A. *et al*, *ACS Symp. Ser.*, 1981, **171**, 47 (*rev, use*)

(4-Morpholinylmethylene)bisphosphonic acid, 9CI **M-00520**

4-Morpholinomethanediphosphonic acid. 4-(Diphosphonomethyl)morpholine

[32545-75-8]

$C_5H_{13}NO_7P_2$ M 261.108

Complexing agent. Cryst. + $1H_2O$. Mp 248-250°. pK_{a1} 2.66 (25°), pK_{a2} 4.91, pK_{a3} 8.81, pK_{a4} 11.28.

Tetra-Et ester: [59646-46-7]. *Tetraethyl (4-morpholinylmethylene)bisphosphonate.*

$C_{13}H_{29}NO_7P_2$ M 373.322

Used as a Wittig-Horner reagent in synth. of phosphonoenamines. Liq. $Bp_{0.1}$ 140-145°. n_D^{22} 1.4652.

Plöger, W. *et al*, *Z. Anorg. Allg. Chem.*, 1972, **389**, 119 (*synth, use*)
Gross, H. *et al*, *J. Prakt. Chem.*, 1976, **318**, 116 (*synth, pmr*)
Gross, H. *et al*, *Zh. Obshch. Khim.*, 1978, **48**, 1914 (*Engl. transl.* p. 1746) (*props*)
Costisella, B. *et al*, *Tetrahedron*, 1981, **37**, 1227 (*ester, use*)

(4-Morpholinylmethyl)phosphonic acid **M-00521**

[4730-75-0]

$C_5H_{12}NO_4P$ M 181.128

Cryst. (EtOH aq.). Mp 261-263° dec.

Di-Et ester: [27353-29-3]. *Diethyl (4-morpholinylmethyl)phosphonate.*

$C_9H_{20}NO_4P$ M 237.235

Li deriv. used in synth. of δ-ketoaldehydes and the spiroannelation of cyclohexenones. Liq. Bp_3 137°. pK_a 4.00 (H_2O), 2.18 (EtOH, 25°). n_D^{25} 1.4549.

Fields, E.K., *J. Am. Chem. Soc.*, 1952, **74**, 1528 (*ester, synth*)
Bel'skii, V.E. *et al*, *Izv. Akad. Nauk SSSR, Ser. Khim.*, 1975, **24**, 1624 (*Engl. transl.* p. 1511) (*ester, ir, P nmr*)
Martin, S.F., *J. Org. Chem.*, 1976, **41**, 3337 (*ester, use*)
Fredericks, P.M. *et al*, *Z. Naturforsch., C*, 1981, **36**, 242 (*synth*)

4-Morpholinylphosphinic acid **M-00522**

4-Morpholinephosphinic acid. 4-Morpholinophosphinic acid

$C_4H_{10}NO_3P$ M 151.102

Tautomeric with 4-Morpholinylphosphonous acid, M-00529 .

Et ester: [16276-75-8]. *Ethyl 4-morpholinylphosphinate.*

$C_6H_{14}NO_3P$ M 179.155

Liq. $Bp_{0.01}$ 66-67°. n_D^{20} 1.4630. Tautomeric.

Zwierzak, A. *et al*, *Tetrahedron*, 1967, **23**, 2243 (*synth, ir, tlc*)

4-Morpholinylphosphonic acid, 9CI **M-00523**

Morpholinophosphonic acid, 8CI. 4-Morpholinephosphonic acid. 4-Phosphonomorpholine

[4730-74-9]

$C_4H_8NO_4P$ M 165.085

Di-Me ester: [597-25-1]. *Dimethyl 4-morpholinylphosphonate.*

$C_6H_{14}NO_4P$ M 195.155

Liq. Bp_1 96°. n_D^{25} 1.4530.

▷SZ9660000.

Di-Et ester: [37097-43-1]. *Diethyl 4-morpholinylphosphonate.*

$C_8H_{18}NO_4P$ M 223.208

Liq. Bp_{11} 137°.

Di-Ph ester: [7412-25-1]. *Diphenyl 4-morpholinylphosphonate.*

$C_{16}H_{18}NO_4P$ M 319.296

Cryst. (EtOH or pet. ether). Mp 72.5-73.5°, 87-88°. Bp_5 240-250°.

▷QE8925000.

Dichloride: [1498-57-3].

$C_4H_8Cl_2NO_2P$ M 203.992

Phosphorylating agent. Liq. d_4^{20} 1.43. Mp 6-7.5°. $Bp_{0.1}$ 65-68°. n_D^{20} 1.4975.

Diamide: 4-Morpholinylphosphonic diamide.

$C_4H_{12}N_3O_2P$ M 165.131

Leaflets. Mp 158-161°.

Monoanhydride: Di-4-morpholinyldiphosphonic acid.

$C_8H_{18}N_2O_6$ M 238.240

Useful phosphorylating agent.

Monoanhydride, diammonium salt: Needles. Mp 138-141°.

Saunders, B.C. *et al*, *J. Chem. Soc.*, 1948, 699 (*ester*)
Cheymol, J. *et al*, *C.R. Hebd. Seances Acad. Sci., Ser. C*, 1959, **249**, 1240 (*synth*)
Meise, W. *et al*, *Justus Liebigs Ann. Chem.*, 1966, **693**, 76 (*derivs*)
Felcht, U. *et al*, *Justus Liebigs Ann. Chem.*, 1977, 1309 (*ester, ir, pmr*)
Kasparek, F. *et al*, *Collect. Czech. Chem. Commun.*, 1980, **45**, 386 (*ester, props*)
Dawson, H.P. *et al*, *Org. Mass Spectrom.*, 1982, **17**, 212 (*ester, ms*)

4-Morpholinylphosphonochloridic acid **M-00524**

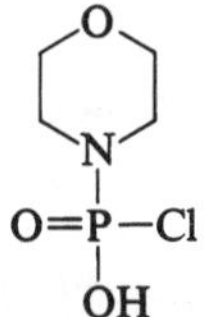

$C_4H_9ClNO_3P$ M 185.547

2,2,2-Tribromoethyl ester: see 2,2,2-Tribromoethyl 4-morpholinylphosphonochloridate, T-00357

Ph ester: [74700-10-0]. *Phenyl 4-morpholinylphosphonochloridate.*

$C_{10}H_{13}ClNO_3P$ M 261.644

Phosphorylating agent. Oil. n_D^{25} 1.5322.

4-Nitrophenyl ester: [79838-05-4]. *4-Nitrophenyl 4-morpholinylphosphonochloridate.*

$C_{10}H_{12}ClN_2O_5P$ M 306.642

Phosphorylating agent. Cryst. (Et_2O/cyclohexane). Mp 55°.

2,4-Dichlorophenyl ester: [58809-22-6]. *2,4-Dichlorophenyl 4-morpholinylphosphonochloridate.*

$C_{10}H_{11}Cl_3NO_3P$ M 330.535

Phosphorylating agent. Cryst. (EtOH at −20°). Mp 46°.

Anhydride with diphenyl phosphate: [69803-89-4].

$C_{16}H_{18}ClNO_6P$ M 386.748

Phosphorylating agent.

Van Boom, J.H. *et al, Tetrahedron*, 1975, **31**, 2953 (*synth*)
Cremlyn, R.J. *et al, Phosphorus Sulfur*, 1979, **7**, 247 (*synth*)
Lysenkova, A.V. *et al, Zhim.-Farm. Zh.*, 1979, **13**, 69 (*Engl. transl.* p. 64) (*use*)
Den Hartog, J.A.J. *et al, Recl. Trav. Chim. Pays-Bas*, 1981, **100**, 285 (*pmr, ms, P nmr, use*)

4-Morpholinylphosphonochloridothioic acid **M-00525**

$C_4H_9ClNO_2PS$ M 201.607

O-*Ph ester:* [74700-13-3]. O-*Phenyl 4-morpholinylphosphonochloridothioate.*

$C_{10}H_{13}ClNO_2PS$ M 277.705

Low-melting solid. Mp 15-17°. Bp_3 108°.

Cremlyn, R. *et al, Phosphorus Sulfur*, 1979, **7**, 247 (*synth, tlc, ir, ms*)

4-Morpholinylphosphonochloridous acid, 9CI **M-00526**

$C_4H_9ClNO_2P$ M 169.547

Me ester: [86030-42-4]. *Methyl 4-morpholinylphosphonochloridite.*

$C_5H_{11}ClNO_2P$ M 183.574

Used in oligonucleotide synth. Liq. $Bp_{0.1}$ 52-54°.

Et ester: [16276-74-7]. *Ethyl 4-morpholinylphosphonochloridite.*

$C_6H_{13}ClNO_2P$ M 197.601

Phosphitylating agent used in oligonucleotide synth. Liq. $Bp_{0.03}$ 51-54°. n_D^{20} 1.4990.

Zwierzak, A. *et al, Tetrahedron*, 1967, **23**, 2243 (*ethyl ester, synth, props*)
Döper, T. *et al, Nucleic Acids Res.*, 1983, **11**, 2575 (*synth, pmr, P nmr, use*)
McBride, L.J. *et al, Tetrahedron Lett.*, 1983, **24**, 245 (*methyl ester, synth, use*)
Sinha, N.D. *et al, Nucleic Acids Res.*, 1984, **12**, 4539 (*ester, synth, ms, pmr, P nmr, use*)

4-Morpholinylphosphonothioic acid, 9CI **M-00527**

Morpholinophosphonothioic acid, 8CI

$C_4H_{10}NO_3PS$ M 183.162

*Mono-*O*-Me ester:* [18370-12-2]. *Monomethyl 4-morpholinylphosphonothioate. Methyl hydrogen phosphoromorpholidothioate.*

$C_5H_{12}NO_3PS$ M 197.188

Deliquescent plates ($CHCl_3$/pet. ether). Mp 111-112°. Has been resolved *via* its quinine salt.

O,O*-Di-Me ester:* [32743-80-9]. O,O*-Dimethyl 4-morpholinylphosphonothioate.*

$C_6H_{14}NO_3PS$ M 211.215

Liq. $Bp_{0.1}$ 80°. n_D^{20} 1.504.

O,O*-Di-Ph ester:* [22077-43-6]. O,O*-Diphenyl 4-morpholinylphosphonothioate.*

$C_{16}H_{18}NO_3PS$ M 335.357

Needles (MeOH). Mp 97-98°.

Dichloride: [74700-09-7].

$C_4H_8Cl_2NOPS$ M 220.053

Cryst. (C_6H_6/pet. ether). Mp 29-31°. $Bp_{1.5}$ 100-102°.

Diamide: see P-4-Morpholinylphosphonothioic diamide, M-00528

Thuong, N.T. *et al, Bull. Soc. Chim. Fr.*, 1964, 1407 (*ester, synth, props*)
Gerrard, A.F. *et al, J. Chem. Soc. (B)*, 1967, 1122 (*monomethyl ester, synth, resoln*)
Cooks, R.G. *et al, J. Chem. Soc. (B)*, 1968, 1327 (*diphenyl ester, synth, ms*)
Cremlyn, R. *et al, Phosphorus Sulfur*, 1979, **7**, 247 (*dichloride, synth, ir*)
Cremlyn, R. *et al, Chem. Ind. (London)*, 1983, 354 (*derivs, synth*)
Prasmickienc, G. *et al, Izv. Akad. Nauk SSSR, Ser. Khim.*, 1983, 2373 (*Engl. transl.* p. 2138) (*dichloride*)

P-4-Morpholinylphosphonothioic diamide, 9CI **M-00528**

$C_4H_{12}N_3OPS$ M 181.192

Oil.

Tetra-Me: N,N,N',N'*-Tetramethyl-*P*-4-morpholinylphosphonothioic diamide.*

$C_8H_{20}N_3OPS$ M 237.299

Oil.

N,N'*-Di-Ph:* P*-Morpholinyl-*N,N'*-diphenylphosphonothioic diamide.* P*-Morpholinylphosphonothioic dianilide.*

$C_{16}H_{20}N_3OPS$ M 333.387

Solid. Mp 213-215°.

Prasmickiene, G. *et al, Izv. Akad. Nauk SSSR, Ser. Khim.*, 1983, 2373 (*Engl. transl.* p. 2138) (*tetramethyl*)
Cremlyn, R. *et al, J. Heterocycl. Chem.*, 1984, **21**, 1457 (*diphenyl, synth, ms*)

4-Morpholinylphosphonous acid, 9CI **M-00529**

Morpholinophosphonous acid, 8CI. 4-Morpholinephosphonous acid

$C_4H_{10}NO_3P$ M 151.102

Free acid probably exists as the phosphoryl tautomer 4-Morpholinylphosphinic acid, M-00522 .

Mono-Et ester: see 4-Morpholinylphosphinic acid, M-00522

Di-Et ester: Diethyl 4-morpholinylphosphonite.

$C_8H_{14}NO_3P$ M 203.177

Liq. $Bp_{0.1}$ 92-94°. n_D^{20} 1.4572.

Dibutyl ester: Dibutyl 4-morpholinylphosphonite.
$C_{12}H_{26}NO_3P$ M 263.316
Liq. $Bp_{0.06}$ 62°. n_D^{20} 1.4530.
Di-Ph ester: [19620-83-8]. *Diphenyl 4-morpholinylphosphonite.*
$C_{16}H_{18}NO_3P$ M 303.297
Liq. d^{20} 1.19. $Bp_{0.02}$ 138-140°. n_D^{20} 1.5763.
Dichloride: [932-74-1].
$C_4H_8Cl_2NOP$ M 187.993
Liq. Bp_{18} 98-100°, $Bp_{0.03}$ 34°. n_D^{20} 1.5360.
Dibromide:
$C_4H_8Br_2NOP$ M 276.895
$Bp_{0.1}$ 76-77°. n_D^{20} 1.6110.
Bis(dimethylamide): N,N,N′,N′-*Tetramethyl*-P-*4-morpholinylphosphonous diamide.*
$C_8H_{20}N_3OP$ M 205.239
Liq. d_4^{20} 1.01. $Bp_{0.2}$ 79-82°. n_D^{20} 1.483.
Dimorpholide: see Trimorpholinophosphine, T-00569

Petrov, K.A. *et al, Zh. Obshch. Khim.*, 1962, **32**, 3065 (*Engl. transl.* p. 3015) (*dibutyl ester*)
Houlla, D. *et al, Bull. Soc. Chim. Fr.*, 1965, 2368 (*dimethylamide*)
Grechkin, N.P. *et al, Izv. Akad. Nauk SSSR, Ser. Khim.*, 1968, 1141 (*Engl. transl.* p. 1090) (*diphenyl ester*)
Lazukina, L.A. *et al, Zh. Obshch. Khim.*, 1974, **44**, 2355 (*Engl. transl.* p. 2309) (*dichloride*)
Kukhar', V.P. *et al, Zh. Obshch. Khim.*, 1982, **52**, 562 (*Engl. transl.* p. 492) (*dichloride, synth, nmr*)
Shevchenko, M.V. *et al, Zh. Obshch. Khim.*, 1984, **54**, 1499 (*Engl. transl.* p. 1336) (*derivs, synth, P nmr*)

N

Naled, BSI **N-00001**

1,2-Dibromo-2,2-dichloroethyl dimethyl phosphate, 9CI. *Dibrom. Bromchlophos*

[300-76-5]

$(MeO)_2P(O)OCHBrCBrCl_2$

$C_4H_7Br_2Cl_2O_4P$ M 380.785

▷TB9450000.

(±)-***form***

Agricultural insecticide, acaricide, bactericide and fungicide. V. spar. sol. H_2O. d_4^{26} 1.96. Mp 26°. $Bp_{0.5}$ 110°. n_D^{25} 1.5108. Rapidly hydrolysed and degraded in sunlight.

▷Highly toxic, TLV 3

B.P., 855 157, (*1957*); *CA*, **55**, 8748f (*manuf, tox*)
Ger. Pat., 1 190 246, (*1965*); *CA*, **63**, 4162 (*synth, tox*)
Getz, M.E. *et al*, *J. Assoc. Off. Anal. Chem.*, 1968, **51**, 1101 (*tlc*)
Keith, L.H. *et al*, *J. Assoc. Off. Anal. Chem.*, 1968, **51**, 1063 (*pmr*)
Szalontai, G. *et al*, *J. Chromatogr.*, 1976, **124**, 9 (*hplc*)
Stan, H.-J. *et al*, *Fresenius' Z. Anal. Chem.*, 1977, **287**, 271; *Biomed. Mass Spectrom.*, 1982, **9**, 483 (*glc, ms*)
Ripley, B.D. *et al*, *J. Assoc. Off. Anal. Chem.*, 1983, **66**, 1084 (*glc*)
Pesticide Manual, 6th Ed., 371; 7th Ed., 389.
Sax, N.I., *Dangerous Properties of Industrial Materials*, 6th Ed., Van Nostrand-Reinhold, 1984, 603.

[1,5-Naphthalenediylbis[methylene]]-bis[phosphonic acid], 9CI **N-00002**

1,5-Bis(phosphonomethyl)naphthalene

$CH_2P(O)(OH)_2$ (1-position); $(HO)_2P(O)CH_2$ (5-position)

$C_{12}H_{14}O_6P_2$ M 316.187

Tetra-Et ester: [25075-80-3]. *Diethyl [1,5-naphthalenediylbis[methylene]]bisphosphonate. 1,5-Bis(diethoxyphosphinyl)naphthalene.*

$C_{20}H_{30}O_6P_2$ M 428.401

Wittig-Horner reagent. No phys. props. reported.

Ernst, L., *J. Chem. Soc., Chem. Commun.*, 1977, 375 (*P nmr*)

[1,6-Naphthalenediylbis[methylene]]-bis[phosphonic acid], 9CI **N-00003**

1,6-Bis(phosphonomethyl)naphthalene

$C_{12}H_{14}O_6P_2$ M 316.187

Tetra-Et ester: [64649-14-5]. *Diethyl [1,6-naphthalenediylbis[methylene]]bisphosphonate. 1,6-Bis(diethoxyphosphinyl)naphthalene.*

$C_{20}H_{30}O_6P_2$ M 428.401

Wittig-Horner reagent. No phys. props. reported.

Ernst, L., *J. Chem. Soc., Chem. Commun.*, 1977, 375 (*P nmr*)

[1,8-Naphthalenediylbis[methylene]]-bis[phosphonic acid] **N-00004**

1,8-Bis(phosphonomethyl)naphthalene

$C_{12}H_{14}O_6P_2$ M 316.187

Tetra-Me ester: Dimethyl[1,8-naphthalenediylbis[methylene]]bisphosphonate. 1,8-Bis(dimethoxyphosphinyl)naphthalene.

$C_{16}H_{22}O_6P_2$ M 372.294

Wittig-Horner reagent. Cryst. ($CHCl_3$/hexane). Mp 171-172°.

Meinwald, J. *et al*, *J. Am. Chem. Soc.*, 1971, **93**, 725 (*synth, ir, ms, pmr, use*)

[2,3-Naphthalenediylbis[methylene]]-bis[phosphonic acid] **N-00005**

2,3-Bis(phosphonomethyl)naphthalene

$C_{12}H_{14}O_6P_2$ M 316.187

Tetra-Et ester: [64649-16-7]. *Tetraethyl [2,3-naphthalenediylbis[methylene]]bisphosphonate. 2,3-Bis(diethoxyphosphinylmethyl)naphthalene.*

$C_{20}H_{30}O_6P_2$ M 428.401

A Wittig-Horner reagent. No phys. props. reported.

Ernst, L., *J. Chem. Soc., Chem. Commun.*, 1977, 375 (*P nmr*)

[2,6-Naphthalenediylbis[methylene]]-bis[phosphonic acid], 9CI **N-00006**

2,6-Bis(phosphonomethyl)naphthalene

$C_{12}H_{14}O_6P_2$ M 316.187

Tetra-Et ester: [23973-60-6]. *Tetraethyl [2,6-naphthalenediylbis[methylene]]bisphosphonate. 2,6-Bis(diethylphosphinylmethyl)naphthalene.*

$C_{20}H_{30}O_6P_2$ M 428.401

Wittig-Horner reagent. No phys. props. reported.

Ernst, L., *J. Chem. Soc., Chem. Commun.*, 1977, 375 (*P nmr*)

[2,7-Naphthalenediylbis[methylene]]-bis[phosphonic acid], 9CI **N-00007**

2,7-Bis(phosphonomethyl)naphthalene

$C_{12}H_{14}O_6P_2$ M 316.187

Tetra-Et ester: [64649-15-6]. *Tetraethyl [2,7-naphthalenediylbis[methylene]]bisphosphonate. 2,7-Bis(diethoxyphosphinylmethyl)naphthalene.*

$C_{20}H_{30}O_6P_2$ M 428.401

A Wittig-Horner reagent. No phys. props. reported.

Ernst, L., *J. Chem. Soc., Chem. Commun.*, 1977, 375 (*synth, nmr*)

Naphthalophos, BAN **N-00008**

2-[(Diethoxyphosphinyl)oxy]-1H-benz[de]isoquinoline-1,3(2H)-dione, 9CI. *N-Hydroxynaphthalimide diethyl phosphate*, 8CI. *Naftalophos*, USAN. *Maretin*. *Rametin*

[1491-41-4]

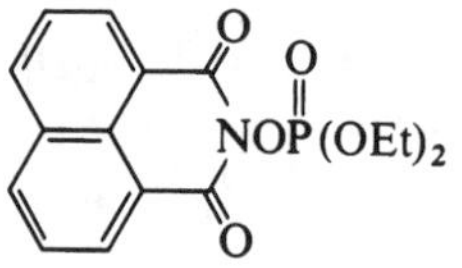

$C_{16}H_{16}NO_6P$ M 349.279

Veterinary anthelmintic and nematocide. Also metab. of corresponding thiophosphate, Bayer 22408. Cryst. (EtOH). Mp 115-116°.

▷QK5775000.

Ger. Pat., 962 608, (*1957*); *CA*, **51**, 15588 (*synth*)
Gatterdan, P.E. *et al*, *J. Econ. Entomol.*, 1962, **55**, 326.
Werbel, L.M. *et al*, *J. Med. Chem.*, 1966, **10**, 32 (*synth*)

Naphtho[1,2-*d*]-1,3,2-dioxaphosphole **N-00009**

1,2-Naphthylene phosphonite

[234-23-1]

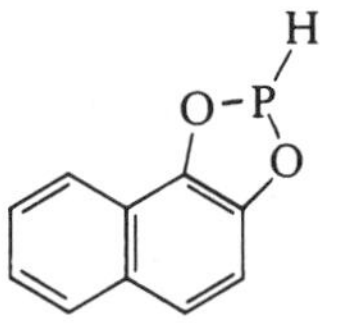

$C_{10}H_7O_2P$ M 190.138

2-Oxide: 1,2-Naphthylene phosphonate.
$C_{10}H_7O_3P$ M 206.137
Solid. Mp 95-97°. Bp_{1-2} 132°. Tautomeric; phosphoryl form predominates.

2-Ethoxy: see 2-Ethoxynaphtho[1,2-d]-1,3,2-dioxaphosphole, E-00044

2-Diethylamino: see 2-Diethylaminonaphtho[1,2-d]-1,3,2-dioxaphosphole, D-00266

Voropai, L.M. *et al*, *Zh. Obshch. Khim.*, 1985, **55**, 65 (*Engl. transl.* p. 55)

Naphtho[1,8-*de*]-1,3,2-dioxaphosphorin, 10CI **N-00010**

1,8-Naphthylene phosphonite

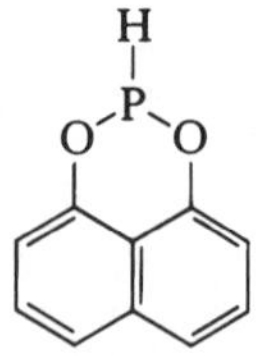

$C_{10}H_7O_2P$ M 190.138

2-Oxide: [79772-19-3]. *1,8-Naphthylene phosphonate.*
$C_{10}H_7O_3P$ M 206.137
Cryst. (pet. ether). Mp 106-107°. Bp_2 186-188°.

Nifant'ev, É.E. *et al*, *Zh. Obshch. Khim.*, 1981, **51**, 1528 (*Engl. transl.* p. 1295) (*deriv, synth, P nmr*)
Voropai, L.M. *et al*, *Zh. Obshch. Khim.*, 1985, **55**, 65 (*Engl. transl.* p. 55) (*deriv, synth, P nmr*)

1-Naphthyldiphenylphosphine **N-00011**

1-Naphthalenyldiphenylphosphine, 9CI. *1-(Diphenylphosphino)naphthalene*

[1162-90-9]

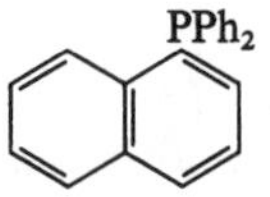

$C_{22}H_{17}P$ M 312.350

Cryst. (EtOH). Sl. sol. pet. ether, Et_2O, EtOH. Mp 124°.

Oxide: [3095-33-8]. *1-(Diphenylphosphinyl)-naphthalene.*
$C_{22}H_{17}OP$ M 328.349
Solid. Mp 178-179°.

Issleib, K. *et al*, *Chem. Ber.*, 1961, **94**, 392 (*synth*)
Schindlbauer, H. *et al*, *Monatsh. Chem.*, 1965, **96**, 285 (*uv*)
Zorn, H. *et al*, *Chem. Ber.*, 1965, **98**, 2431 (*synth, ir*)

2-Naphthyldiphenylphosphine **N-00012**

2-Naphthalenyldiphenylphosphine, 9CI. *2-(Diphenylphosphino)naphthalene*

$C_{22}H_{17}P$ M 312.350

Mp 118°. Dec. at 387°.

Schindlbauer, H. *et al*, *Monatsh. Chem.*, 1965, **96**, 285 (*uv*)
Zorn, H. *et al*, *Chem. Ber.*, 1965, **98**, 2431 (*synth, ir*)

(1-Naphthylmethyl)phosphonic acid **N-00013**

1-Naphthalenylmethylphosphonic acid, 9CI. *1-(Phosphonomethyl)naphthalene*

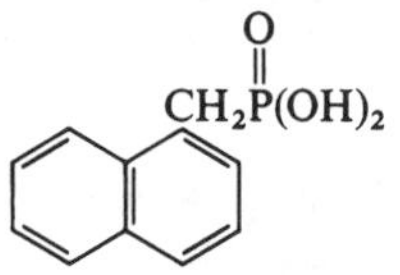

$C_{11}H_{11}O_3P$ M 222.180

Lustrous plates (H_2O). Mp 212-212.5°.

Di-Et ester: [53575-08-9]. *Diethyl 1-naphthylmethylphosphonate.*
$C_{15}H_{19}O_3P$ M 278.287
Wittig-Horner reagent. Oil. Bp_5 205-206°, $Bp_{0.2}$ 202-205°. n_D^{25} 1.5610.

▷SZ9852000.

Kosolapoff, G.M., *J. Am. Chem. Soc.*, 1945, **67**, 2259 (*synth, ester*)
Ernst, L., *Org. Magn. Reson.*, 1977, **9**, 35 (*ester, cmr*)
Franke, A. *et al*, *Synthesis*, 1979, 712 (*ester, synth*)
Tominaga, Y. *et al*, *J. Heterocycl. Chem.*, 1982, **19**, 1125 (*use*)

(2-Naphthylmethyl)phosphonic acid **N-00014**

2-Naphthalenylmethylphosphonic acid, 9CI. *2-(Phosphonomethyl)naphthalene*

$C_{11}H_{11}O_3P$ M 222.180

Solid. Mp 229-230°.

Di-Et ester: [57277-25-5]. *Diethyl 2-naphthylmethylphosphonate.*
$C_{15}H_{19}O_3P$ M 278.287
Wittig-Horner reagent. $Bp_{0.05}$ 148-153°.

Arbuzov, B.A. *et al*, *Zh. Obshch. Khim.*, 1950, **20**, 1249; *CA*, **45**, 1567 (*synth*)
Zimmerman, H.E. *et al*, *J. Am. Chem. Soc.*, 1976, **98**, 5574 (*ester, ir, pmr*)
Ernst, L., *Org. Magn. Reson.*, 1977, **9**, 35 (*ester, cmr, P nmr*)

(1-Naphthylmethyl)-triphenylphosphonium(1+) N-00015

(1-Naphthalenylmethyl)triphenylphosphonium(1+), 9CI

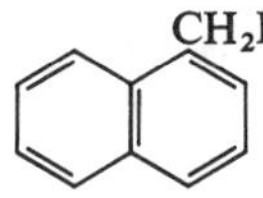

$C_{29}H_{24}P^{\oplus}$ M 403.482 (ion)

Chloride: [23277-00-1].
$C_{29}H_{24}ClP$ M 438.935
Source of ylide. Plates (EtOH). Mp 291-301°.
NaOMe → ylide.

Bromide: [39171-65-8].
$C_{29}H_{24}BrP$ M 483.386
Source of ylide. Cryst. Mp 286-288°. NaOMe → ylide.

Ylide: [60824-80-8]. *1-Naphthalenylmethylenetriphenylphosphorane. (1-Naphthylmethylene)-triphenylphosphorane.*
$C_{29}H_{23}P$ M 402.474
Used in Wittig reactions.

Francis, G.W., *Acta Chem. Scand.*, 1972, **26**, 2969 (*ylide, use*)
Yasuhara, A. *et al*, *Bull. Chem. Soc. Jpn.*, 1972, **45**, 3638 (*chloride, synth, ylide, use*)
Reid, W. *et al*, *Justus Liebigs Ann. Chem.*, 1976, 1415 (*ylide, use*)
Tinnemans, A.H.A. *et al*, *J. Chem. Soc., Perkin Trans. 2*, 1976, 1104 (*bromide, ylide, use*)

(2-Naphthylmethyl)-triphenylphosphonium(1+) N-00016

(2-Naphthalenylmethyl)triphenylphosphonium(1+), 9CI

$C_{29}H_{24}P^{\oplus}$ M 403.482 (ion)

Bromide: [35160-95-3].
$C_{29}H_{24}BrP$ M 483.386
Source of ylide. Cryst. (EtOH). Mp 251-252°.
Treatment with bases, e.g. LiOMe, generates ylide.

Ylide: [18792-78-4]. *2-Naphthalenylmethylenetriphenylphosphorane. (2-Naphthylmethylene)-triphenylphosphorane.*
$C_{29}H_{23}P$ M 402.474
Used in Wittig reactions.

Yasuhara, A. *et al*, *Bull. Chem. Soc. Jpn.*, 1972, **45**, 3638 (*bromide, synth, ylide, use*)
Reid, W. *et al*, *Chem. Ber.*, 1976, **109**, 1506 (*ylide, use*)
Tinnemans, A.H.A. *et al*, *J. Chem. Soc., Perkin Trans. 2*, 1976, 1104 (*bromide, synth, ylide, use*)
Lapouyade, R. *et al*, *J. Org. Chem.*, 1982, **47**, 1361 (*ylide, use*)

1-Naphthylphenylphosphinic acid N-00017

(1-Naphthalenyl)phenylphosphinic acid

[25944-84-7]

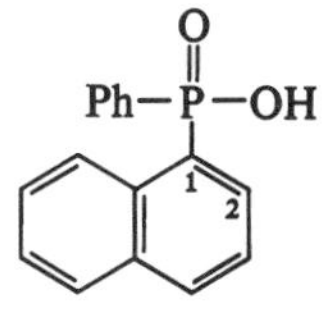

$C_{16}H_{13}O_2P$ M 268.251

(±)-form

Cryst. (MeOH or EtOH). Mp 188-189° (180-183°).

Me ester: [76380-84-2]. *Methyl 1-naphthylphenylphosphinate.*
$C_{17}H_{15}O_2P$ M 282.278
Cryst. (C_6H_6/pet. ether). Mp 108-109°.

Horner, L. *et al*, *Chem. Ber.*, 1958, **91**, 64 (*synth*)
Tars, P., *Chem. Ber.*, 1970, **103**, 2428 (*synth*)
Chauzov, V.A. *et al*, *Zh. Obshch. Khim.*, 1973, **43**, 69 (*Engl. transl.* p. 66) (*synth, chloride*)
Harger, M.J.P., *J. Chem. Soc., Perkin Trans. 2*, 1980, 1505 (*synth, ester*)

2-Naphthylphenylphosphinic acid N-00018

(2-Naphthalenyl)phenylphosphinic acid

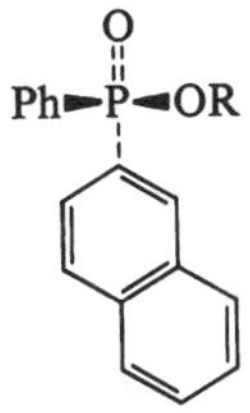

(R)$_P$-form of esters

$C_{16}H_{12}O_2P$ M 267.243

***(R)*$_P$-form**

(−)-Menthyl ester: [21232-91-7].
$C_{26}H_{30}O_2P$ M 405.496
Solid. Mp 87-88°. $[\alpha]_D^{23.6}$ −14° (c, 1.3 in C_6H_6).

***(S)*$_P$-form**

Me ester: [54632-62-1]. *Methyl 2-naphthylphenylphosphinate.*
$C_{17}H_{14}O_2P$ M 281.270
$[\alpha]_D$ −17° (c, 4 in C_6H_6) (60% o.p.).

Et ester: [54632-63-2]. *Ethyl 2-naphthylphenylphosphinate.*
$C_{18}H_{16}O_2P$ M 295.297
$[\alpha]_D$ −22° (c, 4 in C_6H_6) (60% o.p.).

(−)-Menthyl ester: [21232-92-8]. Cryst. (hexane). Mp 103-104°. $[\alpha]_D^{23.6}$ −90° (c, 1.3 in C_6H_6).

(±)-form

Cryst. (EtOH). Mp 165-166°.

Chloride:
$C_{16}H_{11}ClOP$ M 285.689
Liq. $Bp_{0.05}$ 195-198°.

Korpium, O. *et al*, *J. Am. Chem. Soc.*, 1968, **90**, 4842 (*synth*)
Červinka, O. *et al*, *J. Chem. Soc., Chem. Commun.*, 1970, 562 (*synth, chloride, menthyl ester*)
DeBruin, K.E. *et al*, *J. Org. Chem.*, 1975, **40**, 1523 (*esters*)

1-Naphthylphosphinic acid, 8CI N-00019

1-Naphthalenylphosphinic acid, 9CI

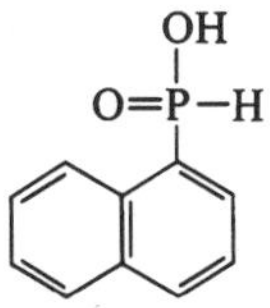

$C_{10}H_9O_2P$ M 192.154

Tautomeric with 1-Naphthylphosphonous acid, N-00025 but free acid probably exists in the phosphinic acid form. Solid. Mp 122-124°.

Weil, T. *et al*, *Helv. Chim. Acta*, 1953, **36**, 1314 (*synth, uv*)

2-Naphthylphosphinic acid, 8CI N-00020

2-Naphthalenylphosphinic acid, 9CI

[61260-18-2]

$C_{10}H_9O_2P$ M 192.154

Tautomeric with 2-Naphthylphosphonous acid, N-00026 but free acid probably exists in the phosphinic acid form. Cryst. (H_2O or C_6H_6). Mp 137°.

Weil, T. *et al*, *Helv. Chim. Acta*, 1953, **36**, 1314 (*synth, uv*)

1-Naphthylphosphonic acid, 8CI N-00021

1-Naphthalenylphosphonic acid, 9CI

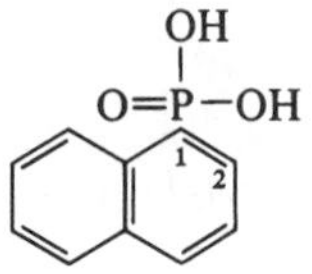

$C_{10}H_9O_3P$ M 208.153

Cryst. (H_2O). Mp 189°.

Di-Me ester: Dimethyl 1-naphthylphosphonate.

$C_{12}H_{13}O_3P$ M 236.207

Liq. $Bp_{0.15}$ 129-131°.

Dichloride:

$C_{10}H_7Cl_2OP$ M 245.044

Solid. Mp 60°.

Lindner, J. *et al*, *Monatsh. Chem.*, 1929, **53**, 274 (*synth, dichloride*)
Obrycki, R. *et al*, *J. Org. Chem.*, 1968, **33**, 632 (*ester*)

2-Naphthylphosphonic acid, 8CI N-00022

2-Naphthalenylphosphonic acid, 9CI

$C_{10}H_9O_3P$ M 208.153

Di-Et ester: Diethyl 2-naphthylphosphonate.

$C_{14}H_{17}O_3P$ M 264.260

Liq. $Bp_{0.05}$ 153-155°.

Dichloride: [40299-92-1].

$C_{10}H_7Cl_2OP$ M 245.044

Solid. Mp 55° (46°). Bp_3 173°, $Bp_{0.048}$ 142-143.5°.

U.S.P., 2 814 645, (*1957*); **A**, **52**, 5464 (*dichloride*)
Tavs, P., *Chem. Ber.*, 1970, **103**, 2428 (*ester*)

1-Naphthylphosphonothioic acid, 8CI N-00023

1-Naphthalenylphosphonothioic acid, 9CI

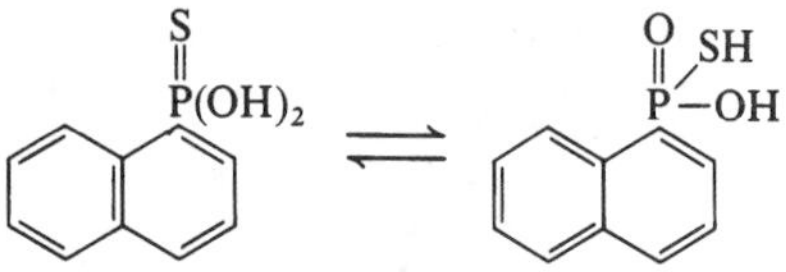

$C_{10}H_9O_2PS$ M 224.214

O-*Me ester:* O-*Methyl naphthylphosphonothioate.*

$C_{11}H_{11}O_2PS$ M 238.240

Solid (as tetramethylammonium salt). Mp 109-113°.

Dichloride: [33446-89-8].

$C_{10}H_7Cl_2PS$ M 261.105

Oil. $Bp_{0.001}$ 125-130°.

Ger. Pat., 2 227 939, (*1972*); *CA*, **78**, 72360 (*dichloride*)

2-Naphthylphosphonothioic acid N-00024

2-Naphthalenylphosphonothioic acid, 9CI

$C_{10}H_9O_2PS$ M 224.214

Dichloride:

$C_{10}H_8Cl_2PS$ M 262.113

Liq. Bp_4 173-174°.

Lecher, H.Z. *et al*, *J. Am. Chem. Soc.*, 1956, **78**, 5018.

1-Naphthylphosphonous acid N-00025

1-Naphthalenylphosphonous acid, 9CI

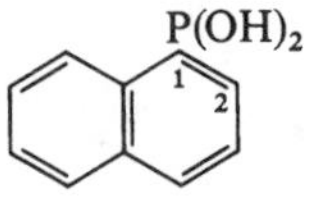

$C_{10}H_9O_2P$ M 192.154

Free acid exists as the phosphoryl tautomer 1-Naphthylphosphinic acid, N-00019 .

Di-Me ester: Dimethyl 1-naphthylphosphonite.

$C_{12}H_{13}O_2P$ M 220.207

Liq. d_0^{20} 1.16. n_D^{20} 1.6096.

Di-Et ester: [84372-47-4]. *Diethyl 1-naphthylphosphonite.*

$C_{14}H_{17}O_2P$ M 248.261

Liq. d_0^{20} 1.10. Bp_{10} 167-168°. n_D^{20} 1.5848.

Diisopropyl ester: Diisopropyl 1-naphthylphosphonite.

$C_{16}H_{21}O_2P$ M 276.314

Liq. d_0^{20} 1.07. Bp_{12} 176-178°. n_D^{20} 1.5648.

Di-Ph ester: Diphenyl 1-naphthylphosphonite.

$C_{22}H_{17}O_2P$ M 344.349

Liq. d_0^{15} 1.19. Bp_{10} 245-247°. n_D^{15} 1.6178.

Dichloride: [36043-00-2]. *Dichloro-1-naphthylphosphine.*

$C_{10}H_7Cl_2P$ M 229.045

Solid. Mp 55°. Bp_{10} 180°, $Bp_{0.5}$ 135-137°.

Dibromide: [32186-90-6]. *Dibromo-1-naphthylphosphine.*

$C_{10}H_7Br_2P$ M 317.947

Solid. Mp 65-68°.

Kamai, G. *et al*, *CA*, 1957, **51**, 11273 (*ester*)
Green, M. *et al*, *J. Chem. Soc.*, 1958, 3129 (*dichloride, ester, synth*)
Kamai, G. *et al*, *Zh. Obshch. Khim.*, 1961, **31**, 3550 (*Engl. transl.* p. 3311) (*esters*)
Duff, J.M. *et al*, *J. Chem. Soc., Dalton Trans.*, 1972, 2219 (*dichloride, synth, pmr*)
Weinberg, K.G., *J. Org. Chem.*, 1975, **40**, 3586 (*dichloride*)

2-Naphthylphosphonous acid N-00026

2-Naphthalenylphosphonous acid, 9CI

$C_{10}H_9O_2P$ M 192.154

Free acid exists as the phosphoryl tautomer 2-Naphthylphosphinic acid, N-00020 .

Dichloride: [57150-67-1]. *Dichloro-2-naphthylphosphine.*

$C_{10}H_7Cl_2P$ M 229.045

Solid. Mp 41-55°, Mp 50-60°. $Bp_{0.2}$ 110°.

Weil, T. *et al*, *Helv. Chim. Acta*, 1953, **36**, 1314.

O-2-Naphthyl phosphorodiamidothioate, 8CI N-00027

O-*2-Naphthalenyl phosphorodiamidothioate, 9CI.* O-*2-Naphthyl diamidothiophosphate*

[15323-49-6]

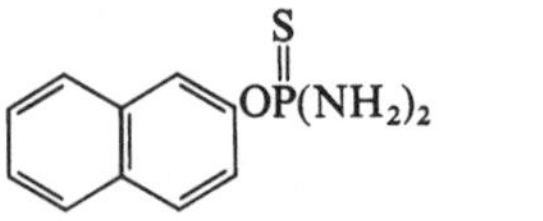

$C_{10}H_{11}N_2OPS$ M 238.243

Solid. Mp 184-185°.

N,N,N′,N′-*Tetra-Me:* O-*2-Naphthyl tetramethylphosphorodiamidothioate.*

$C_{14}H_{19}N_2OPS$ M 294.351

Cryst. (MeOH). Mp 93-94°.

U.S.P., 3 328 494, (*1967*); *CA*, **67**, 73440

1-Naphthylphosphoramidic acid, 8CI **N-00028**
1-Naphthalenylphosphoramidic acid, *9CI*
[36097-61-7]

$C_{10}H_{10}NO_3P$ M 223.168
Solid. Mp 230-235° dec.
Di-Et ester: [49802-17-7]. *Diethyl 1-naphthylphosphoramidate*.
$C_{14}H_{18}NO_3P$ M 279.275
Cryst. (C_6H_6/hexane). Mp 110-113°.
Dichloride:
$C_{10}H_8Cl_2NOP$ M 260.059
Solid. Mp 183° dec.

Li, S.-O. *et al*, *Acta Chem. Sinica*, 1957, **23**, 99; *CA*, **52**, 14552 (*synth*, *dichloride*)
Warner, V.D. *et al*, *J. Med. Chem.*, 1973, **16**, 1185 (*diethyl ester*)

2-Naphthylphosphoramidic acid **N-00029**
2-Naphthalenylphosphoramidic acid, *9CI*
[36097-62-8]
$C_{10}H_{10}NO_3P$ M 223.168
Di-Et ester: *Diethyl 2-naphthylphosphoramidate*.
$C_{14}H_{18}NO_3P$ M 279.275
Liq. $Bp_{0.3}$ 126-129°.
Dibenzyl ester: *Dibenzyl 2-naphthylphosphoramidate*.
$C_{24}H_{32}NO_3P$ M 413.495
Prisms (Et_2O/cyclohexane). Mp 75.5-76.5°.
Dichloride:
$C_{10}H_8Cl_2NOP$ M 260.059
Cryst. ($CHCl_3/Et_2O$). Mp 115-116°.

Atherton, F.R., *J. Chem. Soc.*, 1947, 674 (*dibenzyl ester*)
Cook, H.G. *et al*, *J. Chem. Soc.*, 1949, 2921 (*diethyl ester*)
Kropacheva, A.A. *et al*, *Zh. Obshch. Khim.*, 1959, **29**, 556 (*Engl. transl.* p. 553) (*dichloride*)

1-Naphthyl phosphorodichloridite, 8CI **N-00030**
1-Naphthalenyl phosphorodichloridite, *9CI*. *1-Naphthyl dichlorophosphite*

$C_{10}H_7Cl_2OP$ M 245.044
Liq. d_0^{15} 1.08. Bp_{15} 174-176°.

Kunz, P., *Ber.*, 1894, **27**, 2559 (*synth*)

2-Naphthyl phosphorodichloridite, 8CI **N-00031**
2-Naphthalenyl phosphorodichloridite, *9CI*. *2-Naphthyl dichlorophosphite*
[14966-25-7]
$C_{10}H_7Cl_2OP$ M 245.044
Liq. d_0^{15} 1.08. Bp_{15} 179-181°.

Kunz, P., *Ber.*, 1894, **27**, 2559 (*synth*)

N″-Bis(dimethylamino)phosphinyl-*N,N,N′,N′,N″*-pentamethylphosphorotriamidothioate **N-00032**
Nonamethyl thioimidodiphosphoramide, *9CI*. N″-*Bis(dimethylamino)phosphinothioyl*-N,N,N′,N′,N″-*pentamethylphosphoric triamide*
[55042-06-3]

$$(Me_2N)_2P(O){-}NMe{-}P(S)(NMe_2)_2$$

$C_9H_{27}N_5OP_2S$ M 315.353
Note 9CI name is identical with that of Bis(dimethylaminophosphinothioyl)-pentamethylthiophosphoric triamide, B-00191 . Liq. $Bp_{0.1}$ 115°.

Bulloch, G. *et al*, *J. Chem. Soc., Dalton Trans.*, 1974, 2329 (*synth*, *pmr*, *P nmr*)

[Nitrilotris(methylene)]trisphosphonic acid, 9CI **N-00033**
Nitrilotris(methylenephosphonic acid). sym-*Trimethylaminetriphosphonic acid*
[6419-19-8]

$$N[CH_2P(O)(OH)_2]_3$$

$C_3H_{12}NO_9P_3$ M 299.050
A pentabasic acid with betaine struct. Corrosion inhibitor, complexing agent, e.g. for scale control, and demetallization of wines. Glassy or free-flowing solid. Sol. H_2O, insol. org. solvs. pK_{a1} 0.3, pK_{a2} 1.5, pK_{a3} 4.64, pK_{a4} 5.86, pK_{a5} 7.3, pK_{a6} 12.1 (H_2O, 25°).
▷Irritant
Hexa-Et ester: *Hexaethyl [nitrilotris(methylene)]-trisphosphonate*.
$C_{15}H_{36}NO_9P_3$ M 467.372
d_4^{20} 1.18. $Bp_{0.8}$ 202-204°. n_D^{20} 1.4534.

Petrov, K.A. *et al*, *Zh. Obshch. Khim.*, 1959, **29**, 591 (*Engl. transl.* p. 587) (*ethyl ester*)
Moedritzer, K. *et al*, *J. Org. Chem.*, 1966, **31**, 1603 (*synth*, *props*, *P nmr*)
Carter, R.P. *et al*, *Inorg. Chem.*, 1967, **6**, 639 (*props*)
Krüger, F. *et al*, *Chem.-Ztg.*, 1972, **96**, 691 (*synth*)
Grigor'ev, A.I. *et al*, *Zh. Neorg. Khim.*, 1974, **19**, 1970 (*Engl. transl.* p. 1079) (*ir*, *complexes*)
Nikitina, L.V. *et al*, *Zh. Obshch. Khim.*, 1974, **44**, 1598, 1669 (*Engl. transl.* pp. 1568, 1641) (*props*, *ir*, *complexes*)
U.S.P., 3 832 393, (*1974*); *CA*, **82**, 112160 (*use*)
Nikitina, L.V. *et al*, *Zh. Obshch. Khim.*, 1975, **45**, 552 (*Engl. transl.* p. 546) (*complexes*)
Goeva, L.V. *et al*, *Radiokhimiya*, 1982, **24**, 591 (*Engl. transl.* p. 489) (*complexes*)
Popov, K.I. *et al*, *Zh. Neorg. Khim.*, 1982, **27**, 2756 (*Engl. transl.* p. 1562) (*P nmr*, *props*, *struct*)
Rudomino, M.V. *et al*, *Izv. Akad. Nauk SSSR, Ser. Khim.*, 1984, 2768 (*synth*, *nmr*)
Kaslina, N.A. *et al*, *Zh. Obshch. Khim.*, 1985, **55**, 534 (*props*)

(2-Nitro-2,4-cyclopentadienylidene)-triphenylphosphorane, 9CI **N-00034**
1-Nitro-5-(triphenylphosphoranylidene)-1,3-cyclopentadiene
[41172-19-4]

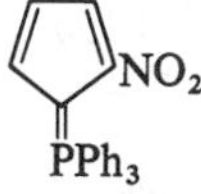

$C_{23}H_{18}NO_2P$ M 371.374
Cryst. ($EtOH/CH_2Cl_2$). Mp 243-244°.

Yoshida, Z. *et al*, *J. Org. Chem.*, 1973, **38**, 3537 (*synth*, *ir*, *uv*, *pmr*)

Yoneda, S. *et al*, *Bull. Chem. Soc. Jpn.*, 1978, **51**, 2605 (*props*)

Berkova, G.A. *et al*, *Zh. Obshch. Khim.*, 1981, **51**, 757 (*Engl. transl.* p. 619) (*P nmr*)

(1-Nitroethyl)phosphonic acid, 10CI **N-00035**

$H_3CCH(NO_2)P(O)(OH)_2$

$C_2H_6NO_5P$ M 155.047

(±)-***form***

Di-Me ester: [60593-25-1]. *Dimethyl (1-nitroethyl)-phosphonate.*

$C_4H_{10}NO_5P$ M 183.100

Liq. Bp_2 102°. pK_a 4.98 (H_2O), 6.30 (50% EtOH aq.), 8.18 (EtOH). n_D^{20} 1.4453.

Di-Et ester: [60593-26-2]. *Diethyl (1-nitroethyl)-phosphonate.*

$C_6H_{14}NO_5P$ M 211.154

Liq. Bp_1 100°. n_D^{20} 1.4413.

Petrov, K.A. *et al*, *Zh. Obshch. Khim.*, 1976, **46**, 1250 (*Engl. transl.* p. 1230); 1979, **49**, 90 (*Engl. transl.* p. 75) (*synth*)

(2-Nitroethyl)phosphonic acid, 10CI **N-00036**

$O_2NCH_2CH_2P(O)(OH)_2$

$C_2H_6NO_5P$ M 155.047

Di-Me ester: [59344-69-3]. *Dimethyl (2-nitroethyl)-phosphonate.*

$C_4H_{10}NO_5P$ M 183.100

Liq. d_4^{20} 1.32. $Bp_{0.05}$ 101-102°. n_D^{20} 1.4458.

Di-Ph ester: Diphenyl (2-nitroethyl)phosphonate.

$C_{14}H_{14}NO_5P$ M 307.242

Solid. Mp 62-63°.

Borisova, E.E. *et al*, *Dokl. Akad. Nauk SSSR, Ser. Sci. Khim.*, 1976, **226**, 1330 (*Engl. transl.* p. 142)
Ranganathan, D. *et al*, *J. Chem. Soc., Chem. Commun.*, 1979, 975.

(Nitromethyl)phosphonic acid, 9CI **N-00037**

$O_2NCH_2P(O)(OH)_2$

CH_4NO_5P M 141.020

Esters can be titrated by alkali in aq. media as weak monobasic acids.

Di-Me ester: [53753-40-5]. *Dimethyl (nitromethyl)-phosphonate.*

$C_3H_8NO_5P$ M 169.074

Wittig-Horner reagent. Liq. Bp_5 127°. pK_a 5.33 (H_2O), 6.56 (50% EtOH aq.), 8.08 (EtOH, 25°). n_D^{20} 1.4434.

Di-Et ester: [53753-37-0]. *Diethyl (nitromethyl)-phosphonate.*

$C_5H_{12}NO_5P$ M 197.127

Wittig-Horner reagent. Liq. d_4^{20} 1.22. $Bp_{0.2}$ 104°. pK_a 5.74 (H_2O), 6.76 (50% EtOH aq.), 8.39 (EtOH, 25°). n_D^{20} 1.4388. Conc. HCl aq. → $NH_2OH.HCl$.

Dibutyl ester: Dibutyl (nitromethyl)phosphonate.

$C_9H_{22}NO_5P$ M 255.250

Wittig-Horner reagent. Liq. d_4^{20} 1.07. $Bp_{0.05}$ 107-109°. n_D^{20} 1.4409.

Petrov, K.A. *et al*, *Zh. Obshch. Khim.*, 1974, **44**, 1649 (*Engl. transl.* p. 1617) (*synth, props*)
Neimysheva, A.A. *et al*, *Zh. Obshch. Khim.*, 1976, **46**, 940 (*Engl. transl.* p. 942) (*synth*)
Petrov, K.A. *et al*, *Zh. Obshch. Khim.*, 1976, **46**, 1242, 1499 (*Engl. transl.* pp. 1222, 1468) (*synth, props*)

(Nitromethyl)triphenylphosphonium(1+) **N-00038**

$Ph_3P^{\oplus}CH_2NO_2$

$C_{19}H_{17}NO_2P^{\oplus}$ M 322.323 (ion)

With aq. alkali → Ph_3PO.

Bromide:

$C_{19}H_{17}BrNO_2P$ M 402.227

Cryst. ($MeNO_2$). Mp 166° dec.

Trippett, S. *et al*, *J. Chem. Soc.*, 1959, 3874 (*synth, props*)

(3-Nitrophenyl)diphenylphosphine, 9CI, 8CI **N-00039**

[31634-79-4]

PPh_2 / NO_2

$C_{18}H_{14}NO_2P$ M 307.288

Yellowish oil.

B,MeI: Methyl(3-nitrophenyl)diphenylphosphonium iodide.

$C_{19}H_{17}INO_2P$ M 449.227

Yellow cryst. (Me_2CO/EtOAc). Mp 192-194°.

Oxide:

$C_{18}H_{14}NO_3P$ M 323.287

Cryst. (cyclohexane). Mp 100-120°.

Sulfide:

$C_{18}H_{14}NO_2PS$ M 339.348

Cryst. (AcOH aq.). Mp 115-116°.

Schiemenz, G.P. *et al*, *Chem. Ber.*, 1971, **104**, 1219 (*synth, derivs, pmr, ir*)

(4-Nitrophenyl)diphenylphosphine, 9CI, 8CI **N-00040**

[5032-63-3]

$C_{18}H_{14}NO_2P$ M 307.288

Cryst. (MeOH). Mp 79°, 96-97°.

B,MeI: Methyl(4-nitrophenyl)diphenylphosphonium iodide.

$C_{19}H_{17}INO_2P$ M 449.227

Red cryst. (Me_2CO/AcOH). Mp 82-86°.

Oxide:

$C_{18}H_{14}NO_3P$ M 323.287

Cryst. (MeOH or MeOH aq.). Mp 152-154°.

Sulfide:

$C_{18}H_{14}NO_2PS$ M 339.348

Cryst. (MeOH). Mp 86-88°.

Schiemenz, G.P. *et al*, *Chem. Ber.*, 1966, **99**, 514; 1971, **104**, 1219 (*synth, derivs*)
Édel'man, T.G. *et al*, *Zh. Obshch. Khim.*, 1972, **42**, 1477 (*Engl. transl.* p. 1469) (*uv, oxide*)
Schiemenz, G.P. *et al*, *Justus Liebigs Ann. Chem.*, 1976, 2126 (*synth, ir*)

[(4-Nitrophenyl)methylene]-triphenylphosphorane **N-00041**

p-*Nitrobenzylidenetriphenylphosphorane*

[6933-17-1]

$Ph_3P{=}CH$ NO_2

$C_{25}H_{20}NO_2P$ M 397.412

A Wittig reagent. Forms complexes containing Group IV elements. Vermilion. Mod. stable.

Butcher, M. *et al*, *Aust. J. Chem.*, 1973, **26**, 2067 (*synth, use*)
Yurchenko, R.I. *et al*, *Zh. Obshch. Khim.*, 1975, **45**, 1735 (*Engl. transl.*, p. 1700) (*uv*)
Bombieri, G. *et al*, *J. Chem. Soc., Perkin Trans. 2*, 1976, 1404 (*props*)

(2-Nitrophenyl)methylphosphinic acid, 9CI **N-00042**

[23081-50-7]

OH
O=P—Me
NO_2

$C_7H_8NO_4P$ M 201.118

Cryst. ($CHCl_3$). Mp 154°.

Et ester: [23081-49-4]. *Ethyl (2-nitrophenyl)-methylphosphinate.*
$C_9H_{12}NO_4P$ M 229.172
Liq. $Bp_{0.2}$ 122°.

(–)-Menthyl ester: Oil. $[\alpha]_D$ –62.8° (c, 0.82 in MeOH).

Chloride:
$C_7H_7ClNO_3P$ M 219.564
Liq. $Bp_{0.4}$ 135°.

Cadogan, J.I.G. *et al*, *J. Chem. Soc. (C)*, 1969, 1314 (*synth, ester, ir, pmr*)
Horner, L. *et al*, *Phosphorus Sulfur*, 1978, **4**, 155 (*derivs*)

(4-Nitrophenyl)methylphosphinic acid, 9CI **N-00043**

[81349-02-2]

$C_7H_8NO_4P$ M 201.118

Solid. Mp 190-192°.

Mastalerz, P., *Rocz. Chem.*, 1963, **37**, 187; *CA*, **59**, 6435.

[(2-Nitrophenyl)methyl]phosphonic acid, 9CI **N-00044**

o-*Nitrobenzylphosphonic acid, 8CI*

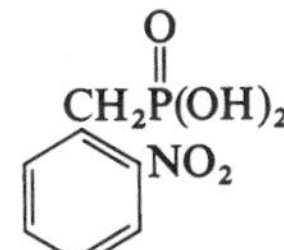

$C_7H_8NO_5P$ M 217.118

Cryst. (H_2O). Mp 190-191° (183-185°).

Di-Me ester: [54006-08-5]. *Dimethyl [(2-nitrophenyl)-methyl]phosphonate.*
$C_9H_{12}NO_5P$ M 245.171
Cryst. (Et_2O/pet. ether). Mp 64°.

Di-Et ester: Diethyl[(2-nitrophenyl)methyl]-phosphonate.
$C_{11}H_{16}NO_5P$ M 273.225
Liq. $Bp_{0.1}$ 160-165°.

Dibutyl ester: Dibutyl [(2-nitrophenyl)methyl]-phosphonate.
$C_{15}H_{24}NO_5P$ M 329.332
Oil. $Bp_{0.05}$ 161°.

Bratkowski, T. *et al*, *Rocz. Chem.*, 1967, **41**, 421 (*synth*)
Issleib, K. *et al*, *Synth. React. Inorg. Met. Org. Chem.*, 1974, **4**, 191 (*esters, synth, use*)
Okamoto, Y. *et al*, *Bull. Chem. Soc. Jpn.*, 1987, **60**, 277 (*synth, uv, pmr*)

[(3-Nitrophenyl)methyl]phosphonic acid, 9CI **N-00045**

m-*Nitrobenzylphosphonic acid*

[40299-58-9]

$C_7H_8NO_5P$ M 217.118

Cryst. (C_6H_6/EtOH or H_2O). Mp 186-187° (177-179°).

Di-Et ester: Diethyl [(4-nitrophenyl)methyl]-phosphonate.
$C_{11}H_{16}NO_5P$ M 273.225
Orange oil. $Bp_{0.1}$ 179-180°. n_D^{22} 1.4975.

Kreutzkamp, N. *et al*, *Arch. Pharm. (Weinheim, Ger.)*, 1961, **294**, 49 (*synth, ester*)
Williams, A. *et al*, *J. Chem. Soc., Perkin Trans. 2*, 1973, 25 (*synth, ester*)
Okamoto, Y. *et al*, *Bull. Chem. Soc. Jpn.*, 1987, **60**, 277 (*synth, uv, pmr*)

[(4-Nitrophenyl)methyl]phosphonic acid, 9CI **N-00046**

p-*Nitrobenzylphosphonic acid*

[1205-62-5]

$C_7H_8NO_5P$ M 217.118

Antiviral agent, alkaline phosphatase inhibitor. Cryst. (H_2O or 2-propanol/methylcyclohexane). Mp 232-234° (225-226° dec.). $Bp_{0.5}$ 183-187°.

Di-Me ester: [39980-20-6].
$C_9H_{12}NO_5P$ M 245.171
Reagent for Wittig-Horner reactions. Cryst. (2-propanol). Mp 75°.

Di-Et ester: [2609-49-6].
$C_{11}H_{16}NO_5P$ M 273.225
Stereospecific Wittig reagent. Liq. Bp_3 199-201°, $Bp_{0.1}$ 148-153°. n_D^{25} 1.5220.

Kosolapoff, G.M., *J. Am. Chem. Soc.*, 1949, **71**, 1876 (*ester*)
Kagan, F. *et al*, *J. Am. Chem. Soc.*, 1959, **81**, 3026 (*synth, props*)
Inglot, A.D. *et al*, *Nature (London)*, 1965, **207**, 784
Scherer, H. *et al*, *Chem. Ber.*, 1972, **105**, 3357 (*ester, ir, pmr*)
Williams, A. *et al*, *J. Chem. Soc., Perkin Trans. 2*, 1973, 25 (*synth*)
Ernst, L., *Org. Magn. Reson.*, 1977, **9**, 35 (*cmr, nmr*)
Okamoto, Y. *et al*, *Bull. Chem. Soc. Jpn.*, 1987, **60**, 277 (*synth, uv, pmr, P nmr*)

[(2-Nitrophenyl)methyl]-triphenylphosphonium(1+), 10CI, 9CI **N-00047**

(o-*Nitrobenzyl)triphenylphosphonium(1+), 8CI*

$CH_2PPh_3^{\oplus}$
NO_2

$C_{25}H_{21}NO_2P^{\oplus}$ M 398.420 (ion)

With Na_2CO_3 aq. or Et_3N → ylide.

Chloride: [52513-28-7].
$C_{25}H_{21}ClNO_2P$ M 433.873

Source of ylide. Cryst. + $1H_2O$. Mp 230° dec.
Bromide: [23308-83-0].
$C_{25}H_{21}BrNO_2P$ M 478.324
Source of ylide. Cryst. (EtOH). Mp 161-162°.
Ylide: [42546-50-9]. *[(2-Nitrophenyl)methylene]-triphenylphosphorane, 10CI, 9CI.* o-*Nitrobenzylidene-triphenylphosphorane, 8CI.*
$C_{25}H_{20}NO_2P$ M 397.412
Used in Wittig reactions. Purple.

Kröhnke, F., *Chem. Ber.*, 1950, **83**, 291 (*synth, props*)
Butcher, M. *et al*, *Aust. J. Chem.*, 1973, **26**, 2067 (*use*)
Ardakani, A.A. *et al*, *J. Org. Chem.*, 1978, **43**, 4128 (*synth, pmr, use*)

[(3-Nitrophenyl)methyl]-triphenylphosphonium(1+) N-00048

(m-*Nitrobenzyl)triphenylphosphonium(1+)*

$C_{25}H_{21}NO_2P^{\oplus}$ M 398.420 (ion)
Chloride: [34906-44-0].
$C_{25}H_{21}ClNO_2P$ M 433.873
Source of ylide on treatment with Na_2CO_3 aq.
Bromide: [1530-41-2].
$C_{25}H_{21}BrNO_2P$ M 478.324
Source of ylide on treatment with Na_2CO_3 aq. Needles (EtOH), rhombs ($CHCl_3/Et_2O$). Mp 269° dec.
Ylide: [61110-97-2]. *[(3-Nitrophenyl)methylene]-triphenylphosphorane.* m-*Nitrobenzylidenetriphenylphosphorane.*
$C_{25}H_{20}NO_2P$ M 397.412
Used in Wittig reactions. Violet ylide, prepd. *in situ*.

Kröhnke, F., *Chem. Ber.*, 1950, **83**, 291 (*bromide, synth, props*)
Schiemenz, G.P., *Tetrahedron*, 1971, **27**, 5723 (*uv*)
Butcher, M. *et al*, *Aust. J. Chem.*, 1973, **26**, 2067 (*ylide*)

[(4-Nitrophenyl)methyl]-triphenylphosphonium(1+), 9CI N-00049

(p-*Nitrobenzyl)triphenylphosphonium(1+)*

$Ph_3\overset{\oplus}{P}CH_2C_6H_4NO_2$

$C_{25}H_{21}NO_2P^{\oplus}$ M 398.420 (ion)
Salts are sources of [(4-Nitrophenyl)methylene]-triphenylphosphorane, N-00041 .
Chloride: [1530-42-3].
$C_{25}H_{21}ClNO_2P$ M 433.873
Needles (CCl_4/pet. ether). Mp 278-280°.
Bromide: [2767-70-6].
$C_{25}H_{21}BrNO_2P$ M 478.324
Prisms (H_2O). Mp 261° dec., 287-290°.
▷TA2390000.

Friedrich, K. *et al*, *Chem. Ber.*, 1959, **92**, 2756 (*bromide*)
Ketcham, R. *et al*, *J. Org. Chem.*, 1962, **27**, 4666 (*chloride*)
Schiemenz, G.P., *Tetrahedron*, 1971, **27**, 5723 (*chloride, uv*)
Kroppová, V. *et al*, *Collect. Czech. Chem. Commun.*, 1981, **46**, 515 (*bromide, use*)

(4-Nitrophenyl-2-oxoethylidene)-triphenylphosphorane N-00050

[1439-43-6]

$Ph_3P{=}CHCOC_6H_4NO_2$

$C_{26}H_{20}NO_3P$ M 425.423
Stabilised Wittig reagent. Cryst. (MeOH aq. or EtOH aq.). Mp 160°, 196-197°.

Kuchar, M. *et al*, *Collect. Czech. Chem. Commun.*, 1972, **37**, 3950 (*synth*)
Wilson, I.F. *et al*, *J. Chem. Soc., Perkin Trans. 1*, 1972, 31 (*uv, pmr, struct*)
Buddrus, J., *Chem. Ber.*, 1974, **107**, 2050 (*synth*)
Seno, M. *et al*, *Bull. Chem. Soc. Jpn.*, 1975, **48**, 2001 (*synth, ir, nmr, pe*)
Nesmeyanov, N.A. *et al*, *J. Organomet. Chem.*, 1977, **129**, 41 (*synth, tautom, ir, pmr, nmr*)

[2-(4-Nitrophenyl)-2-oxoethyl]-triphenylphosphonium(1+), 9CI N-00051

[(4-Nitrobenzoyl)methyl]triphenylphosphonium(1+)
[35497-01-9]

$Ph_3\overset{\oplus}{P}CH_2COC_6H_4NO_2$

$C_{26}H_{21}NO_3P^{\oplus}$ M 426.431 (ion)
When treated with aq. K_2CO_3, salts give the ylide. pK_a 9.95.
Bromide: [17730-93-7].
$C_{26}H_{21}BrNO_3P$ M 506.335
Solid. Mp 157-158° (149-150°).
Iodide: [53282-27-2].
$C_{26}H_{21}INO_3P$ M 553.335
Ylide source.
Ylide: see (4-Nitrophenyl-2-oxoethylidene)-triphenylphosphorane, N-00050

Kuchar, M. *et al*, *Collect. Czech. Chem. Commun.*, 1972, **37**, 3950 (*synth, ir, pmr, ylide*)
Buddrus, J., *Chem. Ber.*, 1974, **107**, 2050 (*synth, ylide*)
Seno, M. *et al*, *Bull. Chem. Soc. Jpn.*, 1975, **48**, 2001 (*ir, P nmr, cryst struct*)
Nesmeyanov, N.A. *et al*, *J. Organomet. Chem.*, 1977, **129**, 41 (*synth, tautom, ir, pmr, P nmr*)

4-Nitrophenyl phenyl phosphate N-00052

4-Nitrophenyl phenyl hydrogen phosphate. 4-Nitrophenyl phenyl phosphoric acid

$O_2NC_6H_4{-}O{-}P(=O)(OH){-}OPh$

$C_{12}H_{10}NO_6P$ M 295.188
Cryst. (C_6H_6). Mp 101-102°.
Chloride: [793-10-2]. *4-Nitrophenyl phenyl phosphorochloridate. 4-Nitrophenyl phenyl chlorophosphate.*
$C_{12}H_9ClNO_5P$ M 313.634
Phosphorylating agent used in synth. of oligonucleotides. Needles (Et_2O/pet. ether). Mp 78-80°. Bp_1 203-209°, $Bp_{0.06}$ 170-174°.

Dilaris, I. *et al*, *J. Org. Chem.*, 1965, **30**, 686 (*synth, derivs*)
Reese, C.B. *et al*, *J. Chem. Soc., Chem. Commun.*, 1977, 802 (*synth, use*)
Lazarus, R.A. *et al*, *J. Am. Chem. Soc.*, 1979, **101**, 4300 (*chloride, synth, ir, P nmr*)

(2-Nitrophenyl)phosphinic acid, 9CI **N-00053**

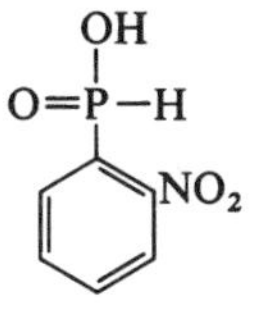

$C_6H_6NO_4P$ M 187.091

Tautomeric with (2-nitrophenyl)phosphonous acid, but free acid exists in phosphinic acid form. Solid. Mp 157°.

Plets, V.M., *J. Gen. Chem. USSR*, 1937, **7**, 84, 90; *CA*, **31**, 4965.

(4-Nitrophenyl)phosphinic acid, 9CI **N-00054**

$C_6H_6NO_4P$ M 187.091

Tautomeric with (4-nitrophenyl)phosphonous acid, but free acid exists in the phosphinic acid form. Solid. Mp 134°.

Plets, V.M., *J. Gen. Chem. USSR*, 1937, **7**, 84, 90; *CA*, **31**, 4965.

(2-Nitrophenyl)phosphonic acid, 9CI **N-00055**

[38696-09-2]

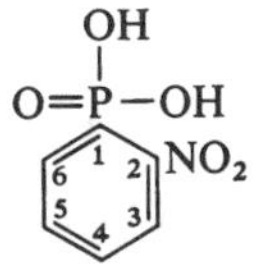

$C_6H_6NO_5P$ M 203.091

Cryst. ($Me_2CO/CHCl_3$). Mp 200-203°.

Mono-Me ester: Methyl hydrogen (2-nitrophenyl)-phosphonate.
$C_7H_8NO_5P$ M 217.118
Cryst. (2-propanol). Mp 153-154°.

Di-Me ester: [23081-47-2]. *Dimethyl (2-nitrophenyl)-phosphonate.*
$C_8H_{10}NO_5P$ M 231.144
Cryst. (pet. ether). Mp 79.5-80.5°.

Mono-Et ester: Ethyl hydrogen (2-nitrophenyl)-phosphonate. Cryst. (EtOAc). Mp 78-79°.

Di-Et ester: [13294-40-1]. *Diethyl (2-nitrophenyl)-phosphonate.*
$C_{10}H_{14}NO_5P$ M 259.198
Needles (pet. ether). Mp 55-56°. $Bp_{0.5}$ 158°.

Diisopropyl ester: Diisopropyl (2-nitrophenyl)-phosphonate.
$C_{12}H_{18}NO_5P$ M 287.252
Liq. $Bp_{0.02}$ 113°.

Cadogan, J.I.G. *et al*, *J. Chem. Soc. (C)*, 1969, 1314 (*synth, esters, ir, pmr, P nmr*)
Modro, T.A. *et al*, *Tetrahedron*, 1972, **28**, 3867 (*uv*)
Dolzhnikova, E.N. *et al*, *Zh. Obshch. Khim.*, 1976, **46**, 1902 (*Engl. transl.* p. 1839) (*esters*)
Naylor, R.A. *et al*, *J. Chem. Soc., Perkin Trans. 2*, 1976, 1908 (*ester*)

(3-Nitrophenyl)phosphonic acid, 9CI **N-00056**

[5337-19-3]

$C_6H_6NO_5P$ M 203.091

Cryst. (Et_2O/C_6H_6). Mp 141-143°, 155-156°. pK_{a1} 1.14 (H_2O), 2.67 (50% EtOH aq.), pK_{a2} 6.53 (H_2O), 5.55 (50% EtOH aq.).

Di-Me ester: Dimethyl (3-nitrophenyl)phosphonate.
$C_8H_{10}NO_5P$ M 231.144
Liq. Bp_1 148-150°.

Diisopropyl ester: Diisopropyl (3-nitrophenyl)-phosphonate.
$C_{12}H_{18}NO_5P$ M 287.252
Liq. Bp_1 156-160°.

Dichloride: [34909-17-6].
$C_6H_4Cl_2NO_3P$ M 239.982
Solid. Mp 88-90°. Bp_3 161-163°.

Dianilide: P-(*3-Nitrophenyl*)-N,N′-*diphenylphosphonic diamide.*
$C_{18}H_{16}N_3O_3P$ M 353.316
Cryst. (MeOH). Mp 254-255°.

Cates, A.L. *et al*, *J. Pharm. Sci.*, 1964, **53**, 969 (*esters*)
Zhmurova, I.N. *et al*, *Zh. Obshch. Khim.*, 1964, **34**, 1171 (*Engl. transl.* p. 1162) (*dichloride, dianilide*)
Modro, T.A. *et al*, *Tetrahedron*, 1972, **28**, 3867 (*uv*)
Williams, A. *et al*, *J. Chem. Soc., Perkin Trans. 2*, 1973, 25 (*synth*)
Nuallain, C.O., *J. Inorg. Nucl. Chem.*, 1974, **36**, 339 (*props*)

(4-Nitrophenyl)phosphonic acid, 9CI **N-00057**

[2175-86-2]

$C_6H_6NO_5P$ M 203.091

Cryst. (EtOAc or EtOH/hexane). Mp 198°. pK_{a1} 1.54 (H_2O), 2.60 (50%, EtOH aq.), pK_{a2} 4.73 (H_2O), 5.54 (50%, EtOH aq.).

▷SZ9430300.

Mono-Me ester: Methyl hydrogen (4-nitrophenyl)-phosphonate.
$C_7H_8NO_5P$ M 217.118
Solid. Mp 129-131°.

Mono-Et ester: [85501-46-8]. *Ethyl hydrogen (4-nitrophenyl)phosphonate.*
$C_8H_{10}NO_5P$ M 231.144
Solid. Mp 111-112.5°. pK_a 2.5.

Di-Et ester: [1754-42-3]. *Diethyl (4-nitrophenyl)-phosphonate.*
$C_{10}H_{14}NO_5P$ M 259.198
Liq. $Bp_{0.2}$ 170°.

Mono-Ph ester: Phenyl hydrogen (4-nitrophenyl)-phosphonate.
$C_{12}H_{10}NO_5P$ M 279.188
Solid. Mp 139-141°. pK_a 2.5.

Dichloride: [34909-18-7].
$C_6H_4Cl_2NO_3P$ M 239.982
Solid. Mp 98-99°. $Bp_{0.1}$ 121-134°.

Dianilide: P-(*4-Nitrophenyl*)-N,N′-*diphenylphosphonic diamide.*
$C_{18}H_{16}N_3O_3P$ M 353.316
Cryst. (EtOH aq.). Mp 195-197.5°.

Doak, G.O. *et al*, *J. Am. Chem. Soc.*, 1954, **76**, 1621 (*anilides, dichloride*)
Burger, A. *et al*, *J. Am. Chem. Soc.*, 1957, **79**, 3575 (*monoesters*)
Modro, T.A. *et al*, *Tetrahedron*, 1972, **28**, 3867 (*uv*)
Williams, A. *et al*, *J. Chem. Soc., Perkin Trans. 2*, 1973, 25 (*synth*)
Mukhtarov, A.S. *et al*, *Zh. Strukt. Khim.*, 1976, **17**, 76 (*Engl. transl.* p. 60) (*synth, esters, esr*)
Hirao, T. *et al*, *Synthesis*, 1981, 56; *Bull. Chem. Soc. Jpn.*, 1982, **55**, 909 (*ester*)

(2-Nitrophenyl)phosphoramidic acid, 9CI **N-00058**

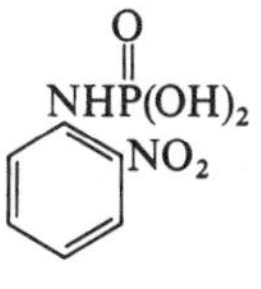

$C_6H_7N_2O_5P$ M 218.105

Di-Ph ester: Diphenyl (2-nitrophenyl)phosphoramidate.
$C_{18}H_{15}N_2O_5P$ M 370.301
Cryst. (EtOH). Mp 108-109°.

Zhmurova, I.N. *et al*, *Zh. Obshch. Khim.*, 1961, **31**, 3741 (*Engl. transl.* p. 3495)
Buchanan, G.W. *et al*, *Org. Magn. Reson.*, 1980, **14**, 517 (*cmr*)
Foulds, G.A. *et al*, *S. Afr. J. Chem.*, 1981, **34**, 72 (*dimethyl ester, ir*)
du Plessis, M.P. *et al*, *Acta Crystallogr., Sect. B*, 1982, **38**, 1504 (*dimethyl ester, cryst struct*)
Rewcastle, G.W. *et al*, *J. Med. Chem.*, 1982, **25**, 1231; 1984, **27**, 1053 (*ester*)

(3-Nitrophenyl)phosphoramidic acid, 9CI **N-00059**

$C_6H_7N_2O_5P$ M 218.105
Di-Me ester: [79639-95-5]. *Dimethyl (3-nitrophenyl)-phosphoramidate.*
$C_8H_{11}N_2O_5P$ M 246.159
Cryst. (MeOH). Mp 144°.
Di-Et ester: [83646-38-2]. *Diethyl (3-nitrophenyl)-phosphoramidate.*
$C_{10}H_{15}N_2O_5P$ M 274.213
Solid. Mp 120°.
Di-Ph ester: [62569-10-2]. *Diphenyl (3-nitrophenyl)-phosphoramidate.*
$C_{18}H_{15}N_2O_5P$ M 370.301
Cryst. (EtOH). Mp 127-128°.
Dichloride:
$C_6H_5Cl_2N_2O_3P$ M 254.997
Cryst. (C_6H_6). Mp 85-86°.
Dianilide: N-*(3-Nitrophenyl)*-N′,N″-*diphenylphosphoric triamide.*
$C_{18}H_{17}N_4O_3P$ M 368.331
Solid. Mp 177°.

Michaelis, A., *Justus Liebigs Ann. Chem.*, 1903, **326**, 129 (*derivs*)
Kropacheva, A.A. *et al*, *Zh. Obshch. Khim.*, 1959, **29**, 556 (*Engl. transl.* p. 553) (*dichloride*)
Zhmurova, I.N. *et al*, *Zh. Obshch. Khim.*, 1961, **31**, 3741 (*Engl. transl.* p. 3495) (*diphenyl ester*)
Foulds, G.A. *et al*, *S. Afr. J. Chem.*, 1981, **34**, 72 (*dimethyl ester, ir*)
Rewcastle, G.W. *et al*, *J. Med. Chem.*, 1982, **25**, 1231 (*dimethyl ester*)
Davidowitz, B. *et al*, *Org. Mass Spectrom.*, 1984, **19**, 128 (*dimethyl ester, ms*)

(4-Nitrophenyl)phosphoramidic acid, 9CI **N-00060**

$C_6H_7N_2O_5P$ M 218.105
Di-Me ester: [78258-13-6]. *Dimethyl (4-nitrophenyl)-phosphoramidate.*
$C_8H_{11}N_2O_5P$ M 246.159
Cryst. (MeOH aq.). Mp 163-164°.
Di-Et ester: [63542-12-1]. *Diethyl (4-nitrophenyl)-phosphoramidate.*
$C_{10}H_{15}N_2O_5P$ M 274.213
Cryst. (MeOH aq.). Mp 139-140°.
Di-Ph ester: Diphenyl (4-nitrophenyl)phosphoramidate.
$C_{18}H_{15}N_2O_5P$ M 370.301
Cryst. (EtOH). Mp 146-148°.
Dichloride:
$C_6H_5Cl_2N_2O_3P$ M 254.997
Solid. Mp 156°.
Dianilide: N-*(4-Nitrophenyl)*-N′,N″-*diphenylphosphoric triamide.*
$C_{18}H_{17}N_4O_3P$ M 368.331
Solid. Mp 242°.

Michaelis, A., *Justus Liebigs Ann. Chem.*, 1903, **326**, 129 (*derivs*)
Zhmurova, I.N. *et al*, *Zh. Obshch. Khim.*, 1961, **31**, 3741 (*Engl. transl.* p. 3495) (*diphenyl ester*)

(2-Nitrophenyl)phosphorimidic acid, 9CI **N-00061**

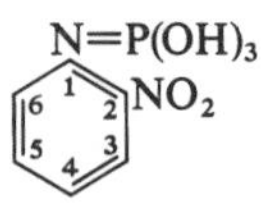

$C_6H_7N_2O_5P$ M 218.105
Free acid tautomeric with (2-Nitrophenyl)-phosphoramidic acid, N-00058 .
Tri-Et ester: [54948-89-9]. *Triethyl (2-nitrophenyl)-phosphorimidate.* P,P,P-*Triethoxy*-N-*(2-nitrophenyl)phosphazene.* P,P,P-*Triethoxy*-N-*(2-nitrophenyl)phosphine imide.*
$C_{12}H_{19}N_2O_5P$ M 302.266
Oil.
Tri-Ph ester: Triphenyl (2-nitrophenyl)-phosphorimidate. N-*(2-Nitrophenyl)*-P,P,P-*triphenoxyphosphazene.* N-*(2-Nitrophenyl)*-P,P,P-*triphenoxyphosphine imide.*
$C_{24}H_{19}N_2O_5P$ M 446.398
Thick liq.
Trichloride: [54998-94-6]. P,P,P-*Trichloro*-N-*(2-nitrophenyl)phosphazene.* P,P,P-*Trichloro*-N-*(2-nitrophenyl)phosphine imide.* P,P,P-*Trichloro*-N-*(2-nitrophenyl)iminophosphorane.*
$C_6H_4Cl_3N_2O_2P$ M 273.443
Solid. Mp 73-75°.

Kirsanov, A.V. *et al*, *Zh. Obshch. Khim.*, 1957, **27**, 2817 (*Engl. transl.* p. 2852) (*trichloride, synth*)
Zhmurova, I.N. *et al*, *Zh. Obshch. Khim.*, 1961, **31**, 3741 (*Engl. transl.* p. 3495) (*triphenyl ester*)
Cadogan, J.I.G. *et al*, *J. Chem. Soc., Perkin Trans. 1*, 1974, 1694 (*trichloride, triethyl ester, pmr, ms*)
Glidewell, C., *Inorg. Chim. Acta*, 1976, **18**, 51 (*trichloride, ms*)

(3-Nitrophenyl)phosphorimidic acid, 9CI **N-00062**

$C_6H_7N_2O_5P$ M 218.105
Free acid tautomieric with (3-Nitrophenyl)-phosphoramidic acid, N-00059 .
Tri-Ph ester: Triphenyl (3-nitrophenyl)-phosphorimidate. P,P,P-*Triphenoxy*-N-*(3-nitrophenyl)phosphazene.* P,P,P-*Triphenoxy*-N-*(3-nitrophenyl)phosphine imide.*
$C_{24}H_{19}N_2O_5P$ M 446.398
Thick liq. d_4^{20} 1.27. n_D^{20} 1.6032.
Trichloride: [53379-57-0]. P,P,P-*Trichloro*-N-*(3-nitrophenyl)phosphazene.* P,P,P-*Trichloro*-N-*(3-nitrophenyl)phosphine imide.* P,P,P-*Trichloro*-N-*(3-nitrophenyl)iminophosphorane.*
$C_6H_4Cl_3N_2O_5P$ M 321.441
Solid. Mp 82-84°.

Kirsanov, A.V. *et al*, *Zh. Obshch. Khim.*, 1957, **27**, 2817 (*Engl. transl.* p. 2852) (*trichloride*)
Zhmurova, I.N. *et al*, *Zh. Obshch. Khim.*, 1961, **31**, 3741 (*Engl. transl.* p. 3495) (*ester*)

(4-Nitrophenyl)phosphorimidic acid, 9CI **N-00063**

$C_6H_7N_2O_5P$ M 218.105

Free acid tautomeric with (4-Nitrophenyl)-phosphoramidic acid, N-00060 .

Tri-Me ester: [22864-69-3]. *Trimethyl (4-nitrophenyl)-phosphorimidate.* P,P,P-*Trimethoxy*-N-*(4-nitrophenyl)phosphine imide.* P,P,P-*Trimethoxy*-N-*(4-nitrophenyl)phosphazene.*
$C_9H_{13}N_2O_5P$ M 260.186
Phys. props. unrecorded.

Tri-Et ester: [56031-25-5]. *Triethyl (4-nitrophenyl)-phosphorimidate.* P,P,P-*Triethoxy*-N-*(4-nitrophenyl)phosphine imide.* P,P,P-*Triethoxy*-N-*(4-nitrophenyl)phosphazene.*
$C_{12}H_{19}N_2O_5P$ M 302.266
Cryst. (pet. ether). Mp 75-77°.

Tri-Ph ester: Triphenyl (4-nitrophenyl)-phosphorimidate. N-*(4-Nitrophenyl*-P,P,P-*triphenoxyphosphine imide.* N-*(4-Nitrophenyl)*-P,P,P-*triphenoxyphosphazene.*
$C_{24}H_{19}N_2O_5P$ M 446.398
Light-yellow needles (EtOH). Mp 76-78°.

Trichloride: [46338-54-9]. P,P,P-*Trichloro*-N-*(4-nitrophenyl)phosphine imide.* P,P,P-*Trichloro*-N-*(4-nitrophenyl)phosphazene.*
$C_6H_4Cl_3N_2O_5P$ M 321.441
Solid. Mp 118-119°.

Kirsanov, A.V. *et al*, *Zh. Obshch. Khim.*, 1957, **27**, 2817 (*Engl. transl.* p. 2852) (*trichloride*)
Zhmurova, I.N. *et al*, *Zh. Obshch. Khim.*, 1961, **31**, 3741 (*Engl. transl.* p. 3495) (*triphenyl ester*)
Kozlov, É.S. *et al*, *Zh. Obshch. Khim.*, 1975, **45**, 767 (*Engl. transl.* p. 753) (*triethyl ester, synth, uv*)
Buchanan, G.W. *et al*, *Org. Magn. Reson.*, 1980, **14**, 517 (*trimethyl ester, cmr*)

4-Nitrophenyl phosphorodichloridate, 9CI N-00064

4-Nitrophenyl phosphoryl dichloride. 4-Nitrophenyl dichlorophosphate
[777-52-6]

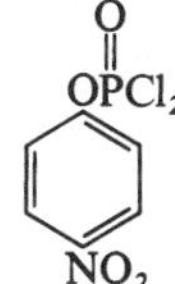

$C_6H_4Cl_2NO_4P$ M 255.982
Phosphorylating agent. Solid. Mp 42-43°. $Bp_{0.55}$ 140°.

Katyshkina, V.V. *et al*, *Zh. Obshch. Khim.*, 1956, **26**, 3060 (*Engl. transl.* p. 3407) (*synth*)
Raevskii, O.A. *et al*, *J. Mol. Struct.*, 1973, **19**, 275 (*ir, conformn*)
Lazarus, R.A. *et al*, *J. Chem. Soc., Perkin Trans. 2*, 1980, 273 (*synth*)
Roth, H.J. *et al*, *Arch. Pharm. (Weinheim, Ger.)*, 1982, **315**, 577 (*synth, haz*)

O-(4-Nitrophenyl) phosphorodichloridothioate, 9CI N-00065

O-(p-*Nitrophenyl) dichlorothiophosphate.* O-(p-*Nitrophenyl) thiophosphoryl dichloride*
[4225-51-8]

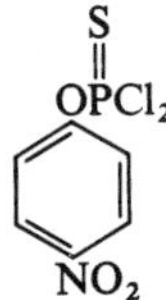

$C_6H_4Cl_2NO_3PS$ M 272.042

Thiophosphorylating agent. Solid. Mp 54°. Bp_4 150°, $Bp_{0.15}$ 130-135°.

Mandel'baum, Ya.A. *et al*, *Zh. Obshch. Khim.*, 1959, **29**, 1149 (*Engl. transl.* p. 1121) (*synth*)
Protsenko, L.D. *et al*, *Zh. Obshch. Khim.*, 1964, **34**, 2233 (*Engl. transl.* p. 2244) (*synth*)

(1-Nitroso-1-methylethyl)phosphonic acid N-00066

$(H_3C)_2C(NO)P(O)(OH)_2$

$C_3H_8NO_4P$ M 153.074

Di-Et ester: [59274-19-0]. *Diethyl (1-nitroso-1-methylethyl)phosphonate.*
$C_7H_{16}NO_4P$ M 209.181
Bright-blue mobile liq. d_4^{20} 1.09. $Bp_{0.04}$ 55-57°. n_D^{20} 1.4280. Monomeric.

Diisopropyl ester: [70448-73-6]. *Diisopropyl (1-nitroso-1-methylethyl)phosphonate.*
$C_9H_{20}NO_4P$ M 237.235
Bright-blue liq. d_4^{20} 1.02. $Bp_{0.05}$ 54.5-58°. n_D^{20} 1.4240.

Levin, Ya.A. *et al*, *Izv. Akad. Nauk SSSR, Ser. Khim.*, 1976, 477 (*Engl. transl.* p. 465) (*ethyl ester, synth, uv, ir, pmr, P nmr*)
Petrov, K.A. *et al*, *Zh. Obhsch. Khim.*, 1979, **49**, 590 (*Engl. transl.* p. 516) (*esters*)

2,5,5,8,11,11,14,17,17-Nonamethyl-1,3,7,9,13,15-hexaoxa-2,8,14-triphosphacyclooctadecane, 9CI N-00067

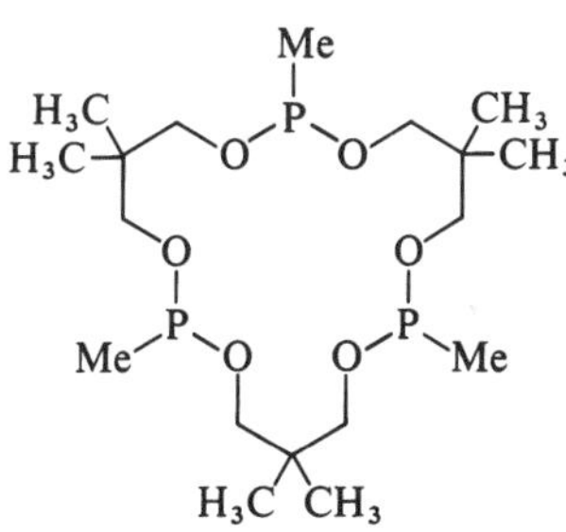

$C_{18}H_{39}O_6P_3$ M 444.424
A trimeric cyclic phosphonite in equilibrium, through the dimer (2,5,5,8,11,11-Hexamethyl-1,3,7,9-tetraoxa-2,8-diphosphacyclododecane, H-00081 , with 5,5-Dimethyl-1,3,2-dioxaphosphorinane, D-00735 .

2,8,14-Trisulfide:
$C_{18}H_{39}O_6P_3S_3$ M 540.604
Viscous oil. Consists of mixture of stereoisomers.

Albrand, J.P. *et al*, *J. Am. Chem. Soc.*, 1974, **96**, 4584 (*form, derivs, pmr, P nmr*)

Nonamethyl imidodiphosphoramide, 9CI N-00068

Nonamethyl iminodiphosphoramide, 8CI. Methylimido bis(phosphoryl dimethylamide)
[34834-03-2]

$(Me_2N)_2P(O)NMeP(O)(NMe_2)_2$

$C_9H_{27}N_5O_2P_2$ M 299.292
Complexes with metals of Groups IIA, IIIA, IB, IIB, VIB, VIIB, and VIII. Hygroscopic cryst. Mp 56°. Bp_{1-2} 158-164°.

Irvine, I. *et al*, *J. Chem. Soc., Dalton Trans.*, 1972, 17 (*pmr, P nmr, ir*)
Haegele, G. *et al*, *J. Chem. Soc., Dalton Trans.*, 1974, 1985 (*cmr, pmr, P nmr*)
Riesel, L. *et al*, *Z. Anorg. Allg. Chem.*, 1974, **404**, 219 (*synth*)

1,9-Nonanediphosphonic acid N-00069

1,9-Nonanediylbisphosphonic acid, 9CI. Nonamethylenediphosphonic acid

$(HO)_2P(O)(CH_2)_9P(O)(OH)_2$

$C_9H_{22}O_6P_2$ M 288.217
Solid. Mp 172-173°.

Tetra-Et ester: Tetraethyl 1,9-nonanediylbisphosphonate. 1,9-Bis(diethoxyphosphinyl)nonane.
$C_{17}H_{38}O_6P_2$ M 400.431
Liq. d_4^{25} 1.05. $Bp_{0.7}$ 203°. n_D^{25} 1.4485.

Tetrachloride:
$C_9H_{18}Cl_4O_2P_2$ M 362.000
V. hygroscopic solid. Mp 38.5-40°. $Bp_{0.2}$ 191°.

Kosolapoff, G.M. *et al*, *J. Chem. Soc. (C)*, 1966, 757 (*synth, derivs, ir*)

Nonaphenyl-1,3,5-triphospha-2,4,6-trigermacyclohexane N-00070

Hexahydro-1,2,2,3,4,4,5,6,6-nonaphenyl-s-triphosphatrigermanin, 8CI
[30404-89-8]

```
            Ph
      Ph    |    Ph
        \   P   /
    Ph—Ge      Ge—Ph
        |       |
        P       P
    Ph/  \ Ge /  \Ph
           / \
          Ph  Ph
```

$C_{54}H_{45}Ge_3P_3$ M 1004.641
White cryst. (C_6H_6/methylcyclohexane). Mp 112-114°. Oxidises and hydrolyses readily.

Schumann, H. *et al*, *Chem. Ber.*, 1971, **104**, 333.

2-Nonenylidenetriphenylphosphorane, 9CI N-00071

[50376-02-8]

$Ph_3P{=}CHCH{=}CH(CH_2)_5CH_3$

$C_{27}H_{31}P$ M 386.516
Prepared as mixture of (*E*) and (*Z*)-forms. Cryst. (Et_2O at −75°).

(***E***)-***form*** [56374-64-2]
Not separately characterised.

(***Z***)-***form*** [56374-63-1]
Not separately characterised.

Bogdanovich, B. *et al*, *Justus Liebigs Ann. Chem.*, 1975, 692 (*synth, pmr*)

3-Nonenyltriphenylphosphonium(1+) N-00072

$Ph_3P^{\oplus}CH_2CH_2CH{=}CH(CH_2)_4CH_3$

$C_{27}H_{32}P^{\oplus}$ M 387.524 (ion)

(***Z***)-***form***

Chloride: [75415-25-7].
$C_{27}H_{32}ClP$ M 422.977
Source of ylide.

Iodide: [72297-07-5].
$C_{27}H_{32}IP$ M 514.428
Solid. Mp 89-90°. Treatment with butyllithium gives the ylide.

Ylide: [73958-09-5]. *3-Nonenylidenetriphenylphosphorane.*
$C_{27}H_{31}P$ M 386.516
Prepd. *in situ*. Widely employed in synth. of leukotrienes and other eicosapolyene compds.

Corey, E.J. *et al*, *J. Am. Chem. Soc.*, 1979, **101**, 6748; 1980, **102**, 1436 (*iodide, ylide, use*)
Rokach, J. *et al*, *Tetrahedron Lett.*, 1980, 1485; 1981, **22**, 979 (*chloride, use*)
Corey, E.J. *et al*, *Tetrahedron Lett.*, 1982, **23**, 3463 (*ylide, use*)
Arai, Y. *et al*, *J. Med. Chem.*, 1983, **26**, 72 (*ylide, use*)
Cohen, N. *et al*, *J. Am. Chem. Soc.*, 1983, **105**, 3661 (*ylide, use*)

4-Nonenyltriphenylphosphonium(1+) N-00073

$Ph_3P^{\oplus}(CH_2)_3CH{=}CH(CH_2)_3CH_3$

$C_{27}H_{32}P^{\oplus}$ M 387.524 (ion)

(***E***)-***form***

Bromide: [59512-90-2].
$C_{27}H_{32}BrP$ M 467.428
Oil. Ylide formed by action of $NaN(SiMe_3)_2$.

Ylide: [59499-32-0]. *4-Nonenylidenetriphenylphosphorane.*
$C_{27}H_{31}P$ M 386.516
Prepd. *in situ* for synth. of pheromones.

(***Z***)-***form***

Bromide: [59512-89-9]. Cryst. (CH_2Cl_2/Et_2O). Mp 106°.

Ylide: [59499-31-9]. Prepd. *in situ* for synth. of pheromones.

Hammoud, A. *et al*, *Bull. Soc. Chim. Fr. Part 2*, 1978, 299 (*synth, ylide, use*)
Bestmann, H.J. *et al*, *Justus Liebigs Ann. Chem.*, 1981, 1705; 1982, 1478 (*synth, ylide, use*)

Nonylphenylphosphinic acid, 9CI N-00074

$H_3C(CH_2)_8P(O)(OH)Ph$

$C_{15}H_{25}O_2P$ M 268.335
Cryst. (Et_2O). Mp 39°.

Me ester: Methyl nonylphenylphosphinate.
$C_{16}H_{27}O_2P$ M 282.362
Liq. d_4^{20} 1.02. Bp_2 180-182°. n_D^{20} 1.5000.

Et ester: Ethyl nonylphenylphosphinate.
$C_{17}H_{29}O_2P$ M 296.389
Liq. Bp_2 187-188.5°. n_D^{20} 1.4970.

Pudovik, A.N. *et al*, *Zh. Obshch. Khim.*, 1960, **30**, 2348 (*Engl. transl.* p. 2328)

Nonylphosphine, 9CI N-00075

[15573-37-2]

$H_3C(CH_2)_8PH_2$

$C_9H_{21}P$ M 160.239
Liq. Bp 187°. n_D^{20} 1.4571.

Schindlbauer, H. *et al*, *Monatsh. Chem.*, 1961, **92**, 868 (*ir*)
Pass, F. *et al*, *Monatsh. Chem.*, 1962, **93**, 230 (*synth*)

Nonylphosphonic acid N-00076

[4730-79-4]

$H_3C(CH_2)_8P(O)(OH)_2$

$C_9H_{21}O_3P$ M 208.237

Di-Me ester: [55339-99-6]. *Dimethyl nonylphosphonate.*
$C_{11}H_{25}O_3P$ M 236.290
Liq. $Bp_{0.2}$ 115-117°.

Diisopropyl ester: Diisopropyl nonylphosphonate.
$C_{15}H_{33}O_3P$ M 292.398
Liq. Bp_1 126-127°. n_D^{20} 1.4315.

Difluoride:
$C_9H_{19}F_2OP$ M 212.219
d_4^{20} 1.02. Bp 226-227°, Bp_6 96-97°.

Dichloride: [14576-64-8].
$C_9H_{19}Cl_2OP$ M 245.128
Liq. d_4^{20} 1.09. Bp_1 115-116°. n_D^{20} 1.4680.

Feshchenko, N.G. *et al*, *Zh. Obshch. Khim.*, 1967, **37**, 473 (*Engl. transl.* p. 441) (*dihalides, esters, synth*)
Geiseler, G. *et al*, *Ber. Bunsenges. Phys. Chem.*, 1967, **71**, 478 (*dichloride, ir, raman*)
Dietze, U., *J. Prakt. Chem.*, 1974, **316**, 293 (*ir*)
Rosini, G. *et al*, *Synthesis*, 1975, 44 (*ester, synth, ir, pmr*)

Nonylphosphoramidic acid, 9CI N-00077

$H_3C(CH_2)_8NHP(O)(OH)_2$

$C_9H_{22}NO_3P$ M 223.251

Dichloride:
$C_9H_{20}Cl_2NOP$ M 260.143
Liq. $Bp_{1.5}$ 133-137°.

Mizuma, T. *et al*, *J. Pharm. Soc. Jpn.*, 1961, **81**, 48; *CA*, **55**, 13403.

Nonyltriphenylphosphonium(1+), 9CI N-00078

$Ph_3P^{\oplus}(CH_2)_8CH_3$

$C_{27}H_{34}P^{\oplus}$ M 389.539 (ion)

Bromide: [60902-45-2].
$C_{27}H_{34}BrP$ M 469.443
Source of ylide, obt. with RLi. Viscous oil.

Iodide: [39589-66-7].
$C_{27}H_{34}IP$ M 516.444
Source of ylide, obt. with RLi. Yellow prisms (Me_2CO/Et_2O). Mp 80-80.5°.

Ylide: [54208-05-8]. *Nonylidenetriphenylphosphorane.*
$C_{27}H_{33}P$ M 388.531
Used in synth. of insect pheromones and natural acetylenes.

Barley, G.C. *et al*, *J. Chem. Soc., Perkin Trans. 1*, 1973, 151 (*iodide, use*)
Ohloff, G. *et al*, *Helv. Chim. Acta*, 1977, **60**, 1161 (*bromide*)
Spur, B. *et al*, *Arch. Pharm. (Weinheim, Ger.)*, 1983, **316**, 789 (*ylide*)

O

Octaethyldiphosphoramide, 9CI **O-00001**

Tetraethylphosphorodiamidic acid anhydride

[38450-57-6]

$(Et_2N)_2P(O)OP(O)(NEt_2)_2$

$C_{16}H_{40}N_4O_3P_2$ M 398.465

Extractant for U. Pale-yellow viscous liq. $Bp_{0.05}$ 130-132°. n_D^{18} 1.4691.

Jankowska, M. *et al*, *J. Radioanal. Chem.*, 1976, **31**, 9 (*synth, props, use*)

3,3,4,4,5,5,6,6-Octafluoro-1,2-bis(diphenylphosphino)cyclohexene **O-00002**

(3,3,4,4,5,5,6,6-Octafluoro-1-cyclohexene-1,2-diyl)-bis[diphenylphosphine], 9CI

$C_{30}H_{20}F_8P_2$ M 594.423

Ligand for Cr, Mo and W. Orange cryst. (hexane). Stable to air and light in both solid and soln. states. Mp 135°.

Cullen, W.R. *et al*, *Can. J. Chem.*, 1969, **47**, 4671 (*synth, ir, pmr, nmr*)

3,3,4,4,5,5,6,6-Octafluoro-1,2-cyclohexenediphosphonic acid **O-00003**

(3,3,4,4,5,5,6,6-Octafluoro-1-cyclohexene-1,2-diyl)-bisphosphonic acid, 9CI

[36206-23-2]

$C_6H_4F_8O_6P_2$ M 386.029

May be useful in reducing bone decalcification.

Tetra-Et ester: [4545-93-1]. *Tetraethyl (3,3,4,4,5,5,6,6,-octafluoro-1-cyclohexene-1,2-diyl)bisphosphonate.*
$C_{14}H_{20}F_8O_6P_2$ M 498.243
Liq. $Bp_{0.5}$ 120-135°. n_D^{25} 1.4140.

Frank, A.W., *J. Org. Chem.*, 1965, **30**, 3663 (*synth*)

Octahydro-1*H*-cyclopropa[*c*]phosphindole 3-oxide, 9CI **O-00004**

2-Phosphatricyclo[4.4.0.0^{4,6}]decane 2-oxide

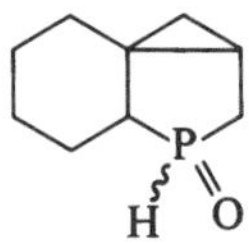

$C_9H_{15}OP$ M 170.191

*3-Me: Octahydro-3-methyl-1*H*-cyclopropa*[c]*-phosphindole 3-oxide.*
$C_{10}H_{17}OP$ M 184.217
Cryst. (CH_2Cl_2/cyclohexane). Mp 153°.

3-Methoxy: [53173-36-7]. *Octahydro-3-methoxy-1*H*-cyclopropa*[c]*phosphindole 3-oxide.*
$C_{10}H_{17}O_2P$ M 200.217
Oil.

3-Ph: [53097-98-6]. *Octahydro-3-phenyl-1*H*-cyclopropa*[c]*phosphindole 3-oxide.*
$C_{15}H_{19}OP$ M 246.288
Cryst. (CH_2Cl_2/cyclohexane). Mp 94-95° and 153-155°. 2 Stereoisomeric forms known.

Kashman, Y. *et al*, *Tetrahedron*, 1973, **29**, 4279 (*synth, ir, ms, pmr*)

Octahydro-1,9-dimethyl-[1,3,2]-diazaphosphorino[1,2-*a*][1,3,2]-diazaphosphorine, 10CI **O-00005**

2,10-Dimethyl-2,6,10-triaza-1-phosphabicyclo[4.4.0]-decane

[75403-51-9]

$C_8H_{18}N_3P$ M 187.224

Liq. $Bp_{0.15}$ 59°.

Denney, D.B. *et al*, *J. Am. Chem. Soc.*, 1980, **102**, 7072 (*synth, props, pmr, cmr, P nmr*)

Octahydro-3,8-dimethyl-2,7-diphenyl-1,6,3,8,2,7-dioxadiazadiphosphecine, 9CI **O-00006**

[75918-60-4]

$C_{18}H_{24}N_2O_2P_2$ M 362.347

Stereochemistry unknown. Dimer of 3-Methyl-2-phenyl-1,3,2-oxazaphospholidine, M-00213 . Cryst. (C_6H_6). Mp 191-193°.

2,7-Disulfide (isomer 1):
$C_{18}H_{24}N_2O_2P_2S_2$ M 426.467
Cryst. (Py). Insol. most solvs. Mp 274-275°. Config. unknown.

2,7-Disulfide (isomer 2):
$C_{18}H_{24}N_2O_2P_2S_2$ M 426.467

Solid. Mp 186-187°. Config. unknown.

Robert, J.B. *et al*, *J. Org. Chem.*, 1978, **43**, 3031 (*synth, pmr, cmr, P nmr*)

2,3,4,5,6,7,8,9-Octahydro-1,9-dimethyl-5-phenyl-1*H*-5,1,9-benzophosphadiarsacycloundecin, 10CI O-00007

2,10-Dimethyl-2,10-diarsa-6-phenyl-6-phosphabicyclo[9.4.0]pentadeca-11(1),12,14-triene

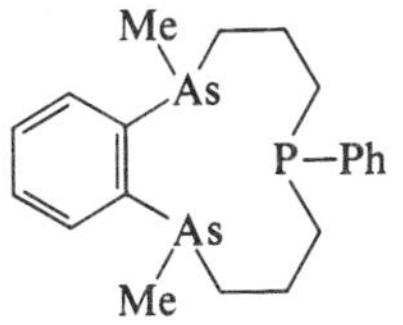

$C_{20}H_{27}As_2P$ M 448.250

Stereochemistry unknown. Ligand for Mo. Viscous liq.

Kyba, E.P. *et al*, *J. Am. Chem. Soc.*, 1980, **102**, 7012 (*synth, pmr, cmr*)

13,14,16,17,31,32,34,35-Octahydro-6,24-dimethyltetrabenzo-[*d,m,r,a'*][1,3,6,9,12,15,20,23,26,2,16]-decaoxadiphosphacyclooctacosin, 10CI O-00008

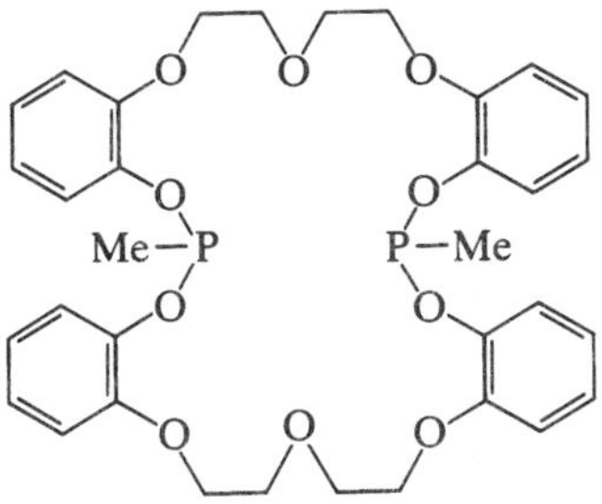

$C_{34}H_{38}O_{10}P_2$ M 668.616

6,24-Dioxide: [65894-65-7].

$C_{34}H_{38}O_{12}P_2$ M 700.615

Cryst. (dichloroethane). Mp 135°, Mp 218°. The Mp's represent possible steroisomers. There is possibly a third form, Mp 127-129°.

6,24-Disulfide:

$C_{34}H_{38}O_{10}P_2S_2$ M 732.736

Solid. Mp 180°.

Kudrya, T.N. *et al*, *Zh. Obshch. Khim.*, 1978, **48**, 927 (*Engl. transl.* p. 844) (*synth*)

Yatsimirskii, K.B. *et al*, *Dokl. Akad. Nauk SSSR, Ser. Sci. Khim.*, 1979, **244**, 1142, 1359 (*Engl. transl.* pp. 78, 89) (*pmr, ms*)

Kirsanov, A.V. *et al*, *Dokl. Akad. Nauk SSSR, Ser. Sci. Khim.*, 1979, **247**, 613 (*Engl. transl.* p. 358) (*synth*)

Golovatyi, V.G. *et al*, *Teor. Eksp. Khim.*, 1981, **17**, 849 (*Engl. transl.* p. 671) (*ms*)

2,3,4,5,6,7,8,9-Octahydro-1,9-diphenyl-1*H*-5,1,9-benzazadiphosphacycloundecine, 10CI O-00009

2,10-Diphenyl-6-aza-2,10-diphosphabicyclo[9.4.0]-pentadeca-11(1),12,14-triene

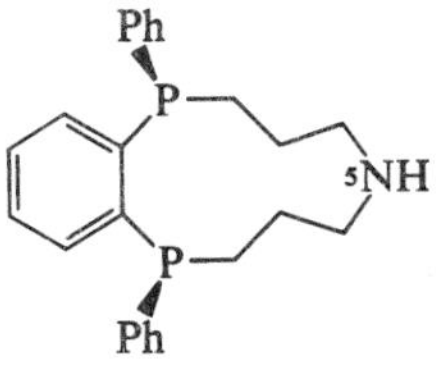

$C_{24}H_{27}NP_2$ M 391.432

(1RS,9SR)-form

meso-*form*

Cryst. (Me_2CO). Mp 112-112.5°.

5-Me: [72110-14-6].

$C_{25}H_{29}NP_2$ M 405.458

Ligand for Co, Cr, Mo and W. Cryst. (Me_2CO). Mp 145-147°.

5-Ph: [72110-15-7].

$C_{30}H_{31}NP_2$ M 467.529

Ligand for Co, Cr, Mo and W. Cryst. (Me_2CO). Mp 147-149°. Forms a solvate, Mp 71-73°.

Kyba, E.P. *et al*, *J. Am. Chem. Soc.*, 1980, **102**, 139 (*synth, pmr, cmr, P nmr, cryst struct*)

Kyba, E.P. *et al*, *Organometallics*, 1982, **1**, 1619 (*synth, pmr, cmr, P nmr, complexes*)

2,3,4,5,6,7,8,9-Octahydro-1,9-diphenyl-1*H*-1,9-benzodiphosphacycloundecin, 9CI O-00010

2,10-Diphenyl-2,10-diphosphabicyclo[9.4.0]pentadeca-11(1),12,14-triene

[72110-16-8]

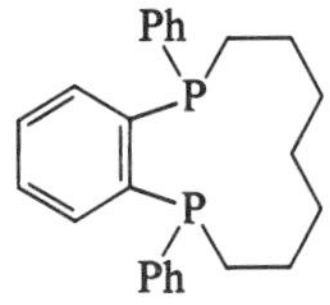

$C_{25}H_{28}P_2$ M 390.444

Stereochemistry uncertain: probably *cis* phenyl groups. Cryst. (Me_2CO).

Kyba, E.P. *et al*, *J. Am. Chem. Soc.*, 1980, **102**, 139 (*synth, pmr, cmr, P nmr*)

1,2,3,4,6,7,8,9-Octahydro-1,9-diphenyl-5,9-benzothiadiphosphacycloundecin, 9CI O-00011

2,10-Diphenyl-2,10-diphospha-6-thiabicyclo[9.4.0]-pentadeca-11(1),12,14-triene

[72110-12-4]

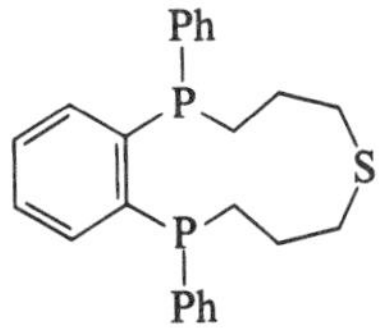

$C_{24}H_{26}P_2S$ M 408.477

Stereochemistry uncertain: probably *cis* phenyl groups. Ligand for Cu, Cr, Mo and W. Cryst. (Me_2CO). Mp 112-112.5°.

Kyba, E.P. *et al*, *J. Am. Chem. Soc.*, 1980, **102**, 139 (*synth, pmr, cmr, P nmr*)

1,2,3,4,6,7,8,9-Octahydro-1,9-diphenyl-5,1,9-benzoxadiphosphacycloundecin, 9CI O-00012

2,10-Diphenyl-6-oxa-2,10-diphosphabicyclo[9.4.0]-pentadeca-11(1),12,14-triene

[72110-13-5]

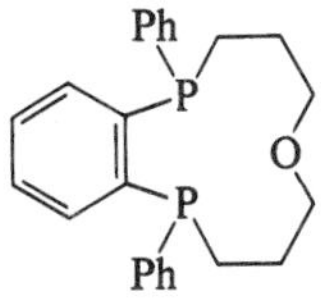

$C_{24}H_{26}OP_2$ M 392.416

Stereochemistry uncertain; probably *cis* phenyl groups. Ligand for Co. Glass.

Kyba, E.P. *et al*, *J. Am. Chem. Soc.*, 1980, **102**, 139 (*synth*, *pmr*, *cmr*, *P nmr*)

Kyba, E.P. *et al*, *Organometallics*, 1982, **1**, 1619.

6,7,8,9,14,15,16,17-Octahydro-9,14-diphenyldibenzo[*b,i*][1,4,8,11]-dithiadiphosphacyclotetradecin, 10CI **O-00013**

6,13-Diphenyl-6,13-diphospha-2,17-dithiatricyclo[16.4.0.0^{7,12}]docosa-7(12),8,10,1(18),19,21-hexaene

[68351-50-8]

$C_{30}H_{30}P_2S_2$ M 516.635

(9RS,14SR)-form [78339-70-5]

cis-*form*

Needles (Me_2CO). Mp 111-112°.

Kyba, E.P. *et al*, *J. Am. Chem. Soc.*, 1981, **103**, 3868 (*synth*, *pmr*, *P nmr*, *cryst struct*)

6,7,8,9,15,16,17,18-Octahydro-9,18-diphenyldibenzo[*b,i*][1,8,4,11]-dithiadiphosphacyclotetradecin, 9CI **O-00014**

6,17-Diphenyl-6,17-diphospha-2,13-dithiatricyclo[16.4.0.0^{7,12}]docosa-7(12),8,10,1(18),19,21-hexaene

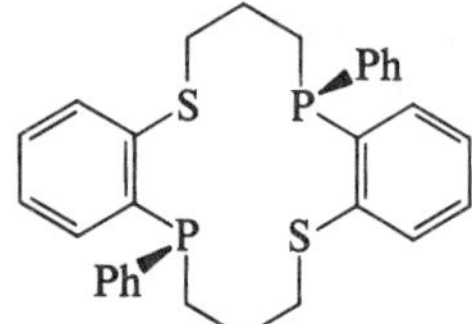

(*9RS,18RS*)*-form*

$C_{30}H_{30}P_2S_2$ M 516.635

(9RS,18RS)-form [94890-89-8]

cis-*form*

Ligand for Pt. Microcryst. (THF/EtOH). Mp 201-206°.

(9RS,18SR)-form

trans-*form*

Cryst. ($CHCl_3$). Mp 202-206°.

Kyba, E.P. *et al*, *J. Am. Chem. Soc.*, 1985, **107**, 2141 (*synth*, *pmr*, *P nmr*, *complexes*, *cryst struct*)

6,7,9,10,16,17,18,19-Octahydro-16,19-diphenyldibenzo[*h,n*][1,4,7,10,13]-trioxadiphosphacyclopentadecin, 10CI **O-00015**

1,2,7,8-Dibenzo-9,12,15-trioxa-3,6-diphospha-3,6-diphenyl-15-crown-5

$C_{30}H_{30}O_3P_2$ M 500.513

Stereochemistry unknown.

16,19-Dioxide: [68402-79-9].

$C_{30}H_{30}O_5P_2$ M 532.512

Complexes with alkali and alkaline-earth metals. Solid. Mp 224-225°. pK_a 6.85 ($MeNO_2$, 25°).

Bodrin, G.V. *et al*, *Izv. Akad. Nauk SSSR, Ser. Khim.*, 1978, 1930 (*Engl. transl.* p. 1700) (*synth*, *P nmr*)

Yatsimirskii, K.B. *et al*, *Zh. Neorg. Khim.*, 1980, **25**, 1788, 2355 (*Engl. transl.* pp. 992, 1302) (*complexes*)

Octahydro-2,4-diphenyl-1,3,5-metheno-1*H*-2,4-diphosphacyclobuta[*c,d*]-pentalene, 9CI **O-00016**

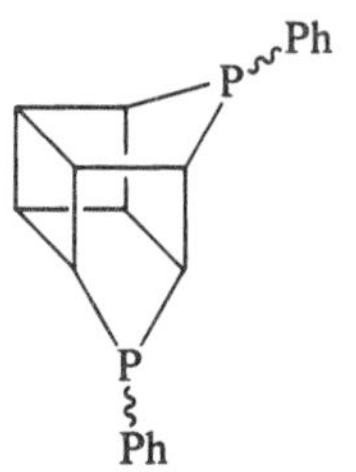

$C_{20}H_{18}P_2$ M 320.310

2,4-Dioxide: [55523-58-5].

$C_{20}H_{18}O_2P_2$ M 352.309

Cryst. ($CHCl_3$/pet. ether). Mp 249-250°. Stereochem. uncertain.

Tomioka, H. *et al*, *Tetrahedron Lett.*, 1974, 4477 (*synth*, *ir*, *pmr*)

Octahydro-1,2,3,4,5,6-hexamethyl-[1,2,3]-triphospholo[4,5-*d*]-1,2,3-triphosphole, 9CI **O-00017**

2,3,4,6,7,8-Hexamethyl-2,3,4,6,7,8-hexaphosphabicyclo[3.3.0]octane

cis-form

$C_8H_{20}P_6$ M 302.089

(3aα,6aα)-form

cis-*form*

Only obt. admixed with 20% *trans*-form.

(3aα,6aβ)-form [67860-71-3]

trans-*form*

Cryst. (C_6H_6). Mp 155°. The Me groups are placed 1*α*,2*β*,3*α*,4*β*,5*α*,6*β*.

Baudler, M. *et al*, *Z. Naturforsch., B*, 1978, **33**, 691 (*synth*, *ir*, *raman*, *pmr*, *P nmr*)

Von Schnering, H.G. *et al*, *Z. Naturforsch., B*, 1978, **33**, 698 (*cryst struct*)

8,9,10,11,12,13,14,15-Octahydro-4-hydroxydinaphtho[2,1-*d*:1′,2′-*f*][1,3,2]-dioxaphosphepin 4-oxide, 9CI **O-00018**

[39648-63-0]

$C_{20}H_{33}O_4P$ M 368.452

Cryst. (EtOH).

4-Methoxy (*Me ester*): [39648-64-1].

$C_{21}H_{35}O_4P$ M 382.479

Solid. Mp 210-211°.

Ger. Pat., 2 212 660, (*1972*); *CA*, **78**, 43129

6,7,11,12,19,20,22,23-Octahydro-9-methyldibenzo[*g,p*][1,3,6,9,12,15,18,21]-heptaoxaphosphacycloeicosin, 10CI **O-00019**

Dibenzo(methylphosphinidene)-20-crown-7

$C_{21}H_{27}O_7P$ M 422.414

9-Oxide: [71817-08-8].

$C_{21}H_{27}O_8P$ M 438.413

Forms complexes with alkali and alkaline-earth metals. Cryst. (Me_2CO or butanone). Sol. $CHCl_3$, DMF. Mp 143-144°.

Kirsanov, A.V. *et al*, *Zh. Obshch. Khim.*, 1980, **50**, 2452 (*Engl. transl.* p. 1980) (*synth, ir*)

Yatsimirskii, K.B. *et al*, *Zh. Neorg. Khim.*, 1980, **25**, 63 (*pmr, complexes*)

Golovatyi, V.G. *et al*, *Teor. Eksp. Khim.*, 1981, **17**, 849 (*Engl. transl.* p. 671) (*ms*)

Bidzilya, V.A. *et al*, *Teor. Eksp. Khim.*, 1982, **18**, 65 (*Engl. transl.* p. 52) (*complexes, pmr*)

Tkachev, V.V. *et al*, *Izv. Akad. Nauk SSSR, Ser. Khim.*, 1985, 1775 (*Engl. transl.* p. 1625) (*cryst struct, ir*)

Octahydro-2-(4-nitrophenoxy)-2*H*-1,3,2-benzoxazaphosphorine, 9CI **O-00020**

5,6-Tetramethylene-2-(4-nitrophenoxy)-1,3,2-oxazaphosphorinane. Tetrahydro-5,6-tetramethylene-2-(4-nitrophenoxy)-2H-1,3,2-oxazaphosphorine

(2*RS*,4a*RS*,8a*SR*)-*form*

$C_{13}H_{17}N_2O_4P$ M 296.262

(*2RS,4aRS,8aSR*)-*form*

(*2α,4aα,8aα*)-*form*

2-Oxide: [71806-77-4]. Cryst. (CCl_4). Mp 124-126°. Has (2*SR*)-config.

(*2RS,4aSR,8aRS*)-*form*

(*2α,4aβ,8aα*)-*form*

2-Oxide: [71771-35-2].

$C_{13}H_{17}N_2O_5P$ M 312.261

Cryst. ($CHCl_3/CCl_4$). Mp 183-186°. Has (2*SR*)-config.

Gorenstein, D.G. *et al*, *J. Am. Chem. Soc.*, 1979, **101**, 4925; 1980, **102**, 5077 (*synth, ir, pmr, P nmr, conformn*)

Rowell, R. *et al*, *J. Am. Chem. Soc.*, 1981, **103**, 5894 (*props*)

1,2,3,4,5,8,9,10-Octahydro-1-phenyl-6,7-bis[(trimethylsilyl)oxy]phosphecine **O-00021**

[40438-08-2]

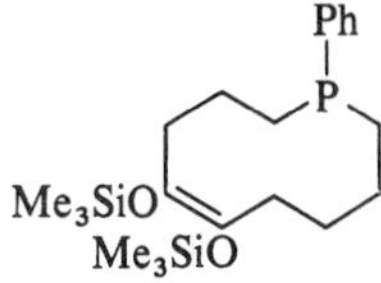

$C_{21}H_{37}O_2PSi_2$ M 408.667

Liq. $Bp_{0.05}$ 142-152°.

van Reijendam, J.W. *et al*, *Tetrahedron Lett.*, 1972, 5181.

Octahydro-3*H*,8*H*,10*bH*-2*a*,5*a*,7*a*,10*a*-tetraaza-10*b*-phosphaindeno[7,1-*cd*]-indene, 10CI **O-00022**

[64317-99-3]

$C_{10}H_{21}N_4P$ M 228.276

Tautomeric with (five membered) ring-opened, trivalent phosphorus form. Solid. $Bp_{0.3}$ 60-70° subl.

*10*b-*Fluoro:* [78725-80-1].

$C_{10}H_{20}FN_4P$ M 246.267

Solid. Mp 63.5-65.5°. $Bp_{0.1}$ 80°. Covalent.

Atkins, T.J. *et al*, *Tetrahedron Lett.*, 1978, 5149 (*synth, struct, P nmr*)

Richman, J.E. *et al*, *J. Am. Chem. Soc.*, 1981, **103**, 5265 (*derivs, F and P nmr*)

Dupart, J.-M. *et al*, *J. Am. Chem. Soc.*, 1982, **104**, 2316; 1983, **105**, 1051 (*complexes*)

Octahydro-7*H*,9*bH*-2*a*,4*a*,6*a*,9*a*-tetraaza-9*b*-phosphapentaleno[1,6-*cd*]-indene, 10CI **O-00023**

[64317-98-2]

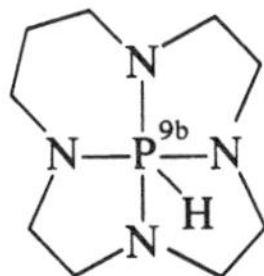

$C_9H_{19}N_4P$ M 214.250

Liq. $Bp_{0.3}$ 92°.

9b-*Fluoro:* [78725-78-7].

$C_9H_{18}FN_4P$ M 232.240

Liq. $Bp_{0.05}$ 70-78°. Covalent.

9b-*Chloro:*

$C_9H_{18}ClN_4P$ M 248.695

Ionic, possibly in equilibrium with dimer.

Atkins, T.J. *et al*, *Tetrahedron Lett.*, 1978, 5149 (*synth, struct, cmr, P nmr*)

Richman, R.E. *et al*, *J. Am. Chem. Soc.*, 1981, **103**, 1291, 5265 (*derivs, struct, F and P nmr*)

Octahydro-5*H*,8*H*,10*bH*-2*a*,4*a*,7*a*,10*a*-tetraaza-10*b*-phosphapentaleno[1,6-*de*]-naphthalene, 10CI O-00024

[71089-75-3]

$C_{10}H_{21}N_4P$ M 228.276

In equilib. with tricyclic, tervalent-phosphorus tautomer. Liq. $Bp_{0.3}$ 106-110°.

10b-Fluoro: [78725-79-8].

$C_{10}H_{20}FN_4P$ M 246.267

Solid. Mp 51.5-53.5°. $Bp_{0.05}$ 80°. Covalent.

10b-Chloro:

$C_{10}H_{20}ClN_4P$ M 262.722

Ionic.

Atkins, T.J. *et al, Tetrahedron Lett.*, 1978, 5149 (*synth, struct, P nmr*)
Richman, J.E. *et al, J. Am. Chem. Soc.*, 1981, **103**, 5265 (*derivs, synth, F and P nmr*)

Octahydro-8*bH*-2*a*,4*a*,6*a*,8*a*-tetraaza-8*b*-phosphapentaleno[1,6-*cd*]pentalene, 10CI O-00025

[64317-97-1]

$C_8H_{17}N_4P$ M 200.223

No tautom. with tervalent phosphorus form. Cryst. Mp 112-115°. Fairly air-stable.

8b-Hydroxy:

$C_8H_{17}N_4OP$ M 216.222

Solid. Mp 210-214°. $Bp_{0.1}$ 170°. Tautomeric with tricyclic phosphine oxide.

8b-Methoxy:

$C_9H_{19}N_4OP$ M 230.249

Oil.

8b-Me: [61754-88-9].

$C_9H_{19}N_4P$ M 214.250

Solid. Mp 35°. Bp_1 120-140° subl.

8b-Ph: [61754-89-0].

$C_{14}H_{21}N_4P$ M 276.320

Solid. Mp 140-142°. $Bp_{0.02}$ 100° subl.

8b-Fluoro: [61754-87-8].

$C_8H_{16}FNP$ M 176.193

Solid. Mp 87-92°. $Bp_{0.1}$ 100° subl. Covalent.

8b-Chloro:

$C_8H_{16}ClN_4P$ M 234.668

Ionic; forms dimer.

Richman, J.E. *et al, Tetrahedron Lett.*, 1977, 559 (*derivs*)
Atkins, T.J. *et al, Tetrahedron Lett.*, 1978, 5149 (*synth, struct, cmr, P nmr*)
Richman, J.E. *et al, J. Am. Chem. Soc.*, 1981, **103**, 1291, 5265 (*derivs, F and P nmr*)
Hamerlinck, J.H.H. *et al, J. Am. Chem. Soc.*, 1983, **105**, 385 (*cryst struct*)
Richman, J.E. *et al, J. Am. Chem. Soc.*, 1983, **105**, 749 (*derivs, cmr, pmr, P nmr*)

Octahydro-1,3,5,7-tetraphenyl-1,5,3,7-diazadiphosphocine, 9CI O-00026

[75593-74-7]

$C_{28}H_{28}N_2P_2$ M 454.490

Mp 190-191° (178-180°).

3,7-Disulfide: [85684-40-8].

$C_{28}H_{28}N_2P_2S_2$ M 518.610

Cryst. (MeCN). Mp 222°.

B,2MeI: [85684-42-0]. *Octahydro-3,7-dimethyl-1,3,5,7-tetraphenyl-1,5,3,7-diazadiphosphocinium iodide.*

$C_{30}H_{34}I_2N_2P_2$ M 738.369

Cryst. (MeCN). Mp 210°.

Märkl, G. *et al, Tetrahedron Lett.*, 1980, **21**, 1409 (*synth, pmr, cmr, complex*)
Arbuzov, B.A. *et al, Dokl. Akad. Nauk SSSR, Ser. Sci. Khim.*, 1981, **257**, 127 (*Engl. transl.* p. 157) (*cryst struct*)
Arbuzov, B.A. *et al, Izv. Akad. Nauk SSSR, Ser. Khim.*, 1983, 1846 (*Engl. transl.* p. 1672) (*synth, P nmr, ms, struct*)
Erastov, O.A. *et al, Izv. Akad. Nauk SSSR, Ser. Khim.*, 1983, 1379 (*Engl. transl.* p. 1250) (*salts, P nmr*)

2,3,4,5,6,7,8,9-Octahydro-1,5,9-triphenyl-1*H*-1,9,5-benzodiphosphaarsacycloundecin, 10CI O-00027

2,6,10-Triphenyl-6-arsa-2,10-diphosphabicyclo[9.4.0]-pentadeca-1(11),12,14-triene

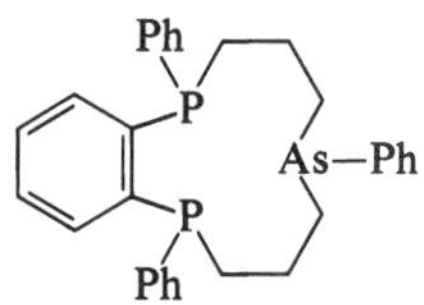

$C_{30}H_{31}AsP_3$ M 559.418

Stereochemistry uncertain (probably *cis* phenyl groups). 2 Meso and one racemic forms possible. Cryst. (Me_2CO). Mp 136.5-138°.

Kyba, E.P. *et al, J. Am. Chem. Soc.*, 1980, **102**, 7012 (*synth, pmr, cmr, P nmr*)

2,3,4,5,6,7,8,9-Octahydro-1,5,9-triphenyl-1*H*-1,5,9-benzotriphosphacycloundecin, 9CI O-00028

2,6,10-Triphenyl-2,6,10-triphosphabicyclo[9.4.0]-pentadeca-1(11),12,14-triene

[65113-29-3]

$C_{30}H_{31}P_3$ M 484.496

Can exist in two meso and one racemic forms. Only one stereoisomer known (illus.) with *cis*-1,9-config. Ligand for Cr, Mo and W. Cryst. (Me_2CO). Mp 163-166°.

Kyba, E.P. *et al, J. Am. Chem. Soc.*, 1977, **99**, 8053 (*synth, pmr, cmr, P nmr*)

Kyba, E.P. *et al*, *J. Am. Chem. Soc.*, 1980, **102**, 139 (*cryst struct*)

Octamethyldiphosphoramide, 9CI — O-00029

*Octamethylpyrophosphoramide, 8CI. Bis-*N,N,N′,N′*-tetramethylphosphorodiamidic anhydride. Schradan, BSI. OMPA. Pestox III. Sytam*

[152-16-9]

$(Me_2N)_2P(O)OP(O)(NMe_2)_2$

$C_8H_{24}N_4O_3P_2$ M 286.250

Systemic insecticide; no longer in use. Viscous liq. Misc. H_2O, insol. pet. ether. d_4^{25} 1.134 (1.09). Bp_2 154°, $Bp_{0.5}$ 120-122°. Hydrol. by acid. Forms complexes with U, Th, rare earth elements, and metals of Groups IB, IIB, VIB, and VIII.

▷Highly toxic. UX5950000.

Gardiner, J.E. *et al*, *J. Chem. Soc.*, 1950, 1769 (*synth*)
Inorg. Synth., 1963, **7**, 73 (*synth*)
Popp, C.J., *Inorg. Chem.*, 1965, **4**, 1418 (*ir, complexes*)
Babad, H. *et al*, *Anal. Chim. Acta*, 1968, **41**, 259 (*pmr*)
Lovins, R.E. *et al*, *J. Agric. Food Chem.*, 1969, **17**, 663 (*ms*)
Pesticide Manual, 6th Ed., 470.
Sax, N.I., *Dangerous Properties of Industrial Materials*, 6th Ed., Van Nostrand-Reinhold, 1984, 874.

sym-Octamethylmonothiopyrophosphoramide — O-00030

Octamethyl thiodiphosphoramide, 9CI. sym-*Octamethylmonothiopyrophosphoric tetraamide*

$(Me_2N)_2P(O){-}S{-}P(O)(NMe_2)_2$

$C_8H_{24}N_4O_2P_2S$ M 302.311

Name 9CI name identical with that for *unsym*-Octamethylmonothiopyrophosphoramide, O-00031 . Liq. d_4^{25} 1.14. $Bp_{0.03}$ 94°. n_D^{25} 1.4675.

Michalski, J., *Rocz. Chem.*, 1955, **29**, 960 (*synth*)

unsym-Octamethylmonothiopyrophosphoramide — O-00031

Octamethyl thiodiphosphoramide, 9CI. O-*[Bis(dimethylamino)phosphinyl] tetramethylphosphorodiamidothioate.* O-*[Bis(dimethylamino)phosphinyl] bis(dimethylamino)phosphinothioate.* unsym-*Octamethylmonothiopyrophosphoric tetraamide*

[51833-60-4]

$(Me_2N)_2P(O){-}O{-}P(S)(NMe_2)_2$

$C_8H_{24}N_4O_2P_2S$ M 302.311

Prob. identical to *sym*-Octamethylmonothiopyrophosphoramide, O-00030
Note 9CI name identical with that of *sym*-Octamethylmonothiopyrophosphoramide, O-00030 . Liq. Bp_3 125-130°.

Ger. Pat., 918 603, (*1941*); *Chem. Zentralbl.*, 1955, 4197

2,2,3,3,7,7,8,8-Octamethyl-1,4,6,9-tetraoxa-5-phospha(5-P^V)spiro[4.4]nonane, 9CI, 8CI — O-00032

[18389-61-2]

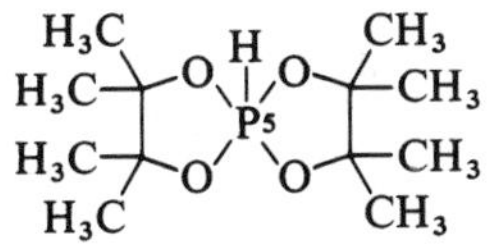

$C_{12}H_{25}O_4P$ M 264.301

Exists completely in the pentacoordinate form. Cryst. (pet. ether). Mp 90°.

5-Me: [56705-27-2].
$C_{13}H_{27}O_4P$ M 278.328
Solid. Mp 60°.

5-Ph: [55983-40-9].
$C_{18}H_{29}O_4P$ M 340.398
Solid. Mp 136°.

5-Benzyl:
$C_{19}H_{31}O_4P$ M 354.425
Solid. Mp 131°.

5-Phenoxy: [57301-52-7].
$C_{18}H_{29}O_5P$ M 356.398
Oil.

5-Phenylselenoxy:
$C_{18}H_{29}O_4PSe$ M 419.358
Cryst. (pet. ether). Mp 112-113°.

5-Benzoyl: [57244-57-2].
$C_{19}H_{29}O_5P$ M 368.409
Cryst. (pet. ether). Mp 158-160°.

Sanchez, M. *et al*, *Bull. Soc. Chim. Fr.*, 1968, 773 (*synth, P nmr*)
Bernard, D. *et al*, *J. Organomet. Chem.*, 1973, **47**, 113 (*pmr, P nmr, struct*)
Malavaud, C. *et al*, *Tetrahedron Lett.*, 1975, 497 (*phenyl, P nmr*)
Savignac, P. *et al*, *J. Organomet. Chem.*, 1975, **93**, 331 (*props*)
Trippett, S. *et al*, *J. Chem. Soc., Perkin Trans. 1*, 1975, 1220 (*benzoyl, ms, pmr, P nmr*)
Bone, S.A. *et al*, *J. Chem. Soc., Perkin Trans. 1*, 1976, 156 (*phenoxy, pmr, P nmr*)
Gonçalves, H. *et al*, *Phosphorus Sulfur*, 1978, **4**, 351 (*derivs, P nmr*)
Johnson, M.P. *et al*, *J. Chem. Soc., Perkin Trans. 1*, 1981, 3074 (*phenylselenoxy, pmr, P nmr*)

1,2,7,8,9,10,15,16-Octamethyl-3,6,11,14-tetraphenyltetrakisphospholo[1,2-*b*:2′,1′-*d*:1″,2″-*f*:2‴,1‴-*h*][1,2,5,6]-tetraphosphocin — O-00033

[80737-80-0]

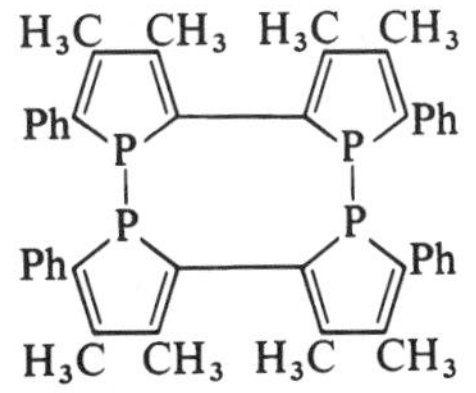

$C_{48}H_{44}P_4$ M 744.771

Red cryst. Mp >260°. Forms a disulfide.

Mathey, F. *et al*, *J. Am. Chem. Soc.*, 1982, **104**, 2077 (*synth, deriv, ms, pmr, P nmr, complexes*)

Octamethyl thiodiphosphoramide, 9CI — O-00034

Tetramethyl phosphorodiamidothioic anhydride

[61614-67-3]

$(Me_2N)_2P(S){-}O{-}P(S)(NMe_2)_2$

$C_8H_{24}N_4OP_2S_2$ M 318.371

Solid. Mp 102-103°.

U.S.P., 4 148 782, (*1979*); *CA*, **91**, 21858

1,8-Octanediphosphonic acid — O-00035

1,8-Octanediylbisphosphonic acid, 9CI. Octamethylenediphosphonic acid

$(HO)_2P(O)(CH_2)_8P(O)(OH)_2$

$C_8H_{20}O_6P_2$ M 274.190

Tetra-Et ester: Tetraethyl 1,8-octanediylbisphosphonate. 1,8-Bis(diethoxyphosphinyl)octane.
$C_{16}H_{36}O_6P_2$ M 386.404
Liq. d_4^{25} 1.05. $Bp_{0.2}$ 186°. n_D^{25} 1.4472.

Tetrachloride:
$C_8H_{16}Cl_4O_2P_2$ M 347.973
Solid, mod. stable in air. Mp 63-65°. $Bp_{0.25}$ 183-184°.

Kosolapoff, G.M. *et al, J. Chem. Soc. (C)*, 1966, 757 (*synth, derivs, ir*)

2,7,9,14,15,16,17,18-Octaoxa-1,8-diphospha(1,8-P^V)-pentacyclo[10.2.1.1^{1,4}.1^{5,8}.1^{8,11}]-octadecane-3,6,10,13-tetrone O-00036

Bis[tartrato(4−)]phosphorane
[80454-84-8]

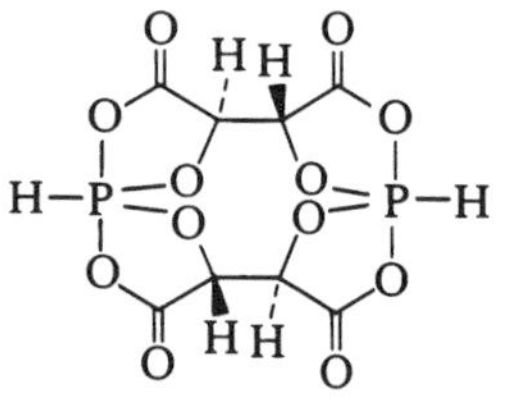

$C_8H_6O_{12}P_2$ M 356.076

(R,R,R,R)-form

Bis(triethylammonium) salt: [80454-59-7]. Sol. $CHCl_3$, Et_2O, DMF.

1,8-Dihydroxy, bistriethylammonium salt: [80483-64-3]. V. sol. DMF, DMSO; sol. MeCN, sp. sol. CH_2Cl_2, Et_2O.

Munoz, A. *et al, Phosphorus Sulfur*, 1981, **11**, 71 (*synth, pmr, P nmr, derivs*)
Dubourg, A. *et al, Phosphorus Sulfur*, 1983, **17**, 97 (*cryst struct*)
Garrigues, B. *et al, Can. J. Chem.*, 1984, **62**, 2170, 2179 (*props, P nmr*)

1,2,3,4,9,10,11,12-Octaphenyl-5,8-diphosphoniadispiro[4.2.4.2]deca-1,3,9,11-tetraene(2+), 9CI O-00037

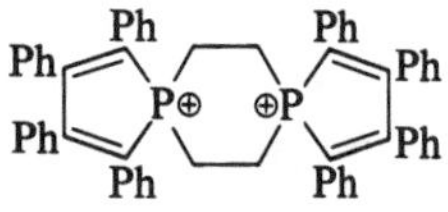

$C_{60}H_{48}P_2^{\oplus\oplus}$ M 830.987 (ion)

Dibromide: [34667-32-8].
$C_{60}H_{48}Br_2P_2$ M 990.795
Cryst. (MeOH). Mp 325-330° dec.

Braye, E.H. *et al, Tetrahedron*, 1971, **27**, 5523 (*synth*)

Octicidin O-00038

N^5-*[Amino(sulfoamino)phosphinyl]-L-ornithine*
[93289-64-6]

$HO_3SNHP(O)(NH_2)NH(CH_2)_3CH(NH_2)COOH$

$C_5H_{15}N_4O_6PS$ M 290.230

Amino acid antibiotic. Major phytotoxin in leaves infected with *Pseudomonas syringae phaseolicola*; formed by degrdn. of Phaseolotoxin. Inhibits ornithine carbamoyl transferase.

Moore, R.E. *et al, Tetrahedron Lett.*, 1984, **25**, 3931 (*isol, struct, nmr*)
Templeton, M.D. *et al, Biochem. J.*, 1985, **228**, 347 (*props*)

Octyldiphenylphosphine, 9CI O-00039

1-Diphenylphosphinooctane
[6737-43-5]

$Ph_2P(CH_2)_7CH_3$

$C_{20}H_{27}P$ M 298.407
Liq. d_4^{20} 0.98. $Bp_{1.4}$ 178-179°. n_D^{20} 1.5655.

Oxide: [29701-85-7]. *1-(Diphenylphosphinyl)octane.*
$C_{20}H_{27}OP$ M 314.406
Solid. Mp 64-65°. pK_a 6.64 ($MeNO_2$) (of conj. acid).

Sulfide: 1-(Diphenylphosphinothioyl)octane.
$C_{20}H_{27}PS$ M 330.467
d_4^{20} 1.05. Mp 14-16°. n_D^{20} 1.5862.

Stuebe, C. *et al, J. Am. Chem. Soc.*, 1955, **77**, 3526 (*synth, derivs*)

Octylphenylphosphinic acid, 9CI O-00040

[31066-81-6]

$H_3C(CH_2)_7P(O)(OH)Ph$

$C_{14}H_{23}O_2P$ M 254.308

Useful extractant for lanthanides and actinides. Solid. Mp 40°. pK_a 4.60 (75% v/v EtOH aq.).

Et ester: Ethyl octylphenylphosphinate.
$C_{16}H_{27}O_2P$ M 282.362
Liq. d_4^{20} 1.01. Bp_2 174-175°. n_D^{20} 1.5000.

Pudovik, A.N. *et al, Zh. Obshch. Khim.*, 1960, **30**, 2348 (*Engl. transl.* p. 2328)

Octylphosphine, 9CI O-00041

[3095-90-7]

$H_3C(CH_2)_7PH_2$

$C_8H_{19}P$ M 146.212
Liq. Bp_{70} 169°,Bp_5 58-60°. pK_a 0.43 (H_2O). n_D^{20} 1.4548. Adds S to give a tetrameric sulfide.

Oxide:
$C_8H_{19}OP$ M 162.211
Cryst. solid. Mp 46-48°. Stable at −75°, poorly stable at r.t. Tautomeric with octylphosphinous acid.

Rauhut, M.M. *et al, J. Org. Chem.*, 1961, **26**, 5138 (*synth*)
Buckler, S.A. *et al, Tetrahedron*, 1962, **18**, 1221 (*oxide, ir, nmr*)
Pass, F. *et al, Monatsh. Chem.*, 1962, **93**, 230 (*synth*)
Maier, L., *Helv. Chim. Acta*, 1965, **48**, 1190 (*synth, ir, nmr*)

Octylphosphonic acid, 9CI O-00042

1-Octanephosphonic acid
[4724-48-5]

$H_3C(CH_2)_7P(O)(OH)_2$

$C_8H_{19}O_3P$ M 194.210
Plates (pet. ether). Mp 102-102.5°.

Mono-anilinium salt: Cryst. (H_2O). Mp 143-145°.
Mono-cyclohexylammonium salt: Solid. Mp 187-188°.
Di-Me ester: [6172-97-0]. *Dimethyl octylphosphonate.*
$C_{10}H_{23}O_3P$ M 222.264

Liq. d_4^{20} 0.98. Bp_{14} 156°. n_D^{20} 1.4350.
Di-Et ester: [1068-07-1]. *Diethyl octylphosphonate.*
$C_{12}H_{27}O_3P$ M 250.317
Liq. d_4^{20} 0.95. Bp_1 119°. n_D^{16} 1.4360.
Dibutyl ester: Dibutyl octylphosphonate.
$C_{16}H_{35}O_3P$ M 306.424
Liq. d_4^{20} 0.93. Bp_2 159-161°. n_D^{25} 1.4391.
Di-Ph ester: [21612-70-4]. *Diphenyl octylphosphonate.*
$C_{20}H_{27}O_3P$ M 346.405
Liq. $Bp_{0.006}$ 148°. n_D^{25} 1.5221.
Difluoride: [14576-63-7].
$C_8H_{17}F_2OP$ M 198.192
Liq. d_4^{20} 1.02. Bp 210-211°, Bp_6 88-90°.
Dichloride: see Octylphosphonic dichloride, O-00043

Kosolapoff, G.M., *J. Am. Chem. Soc.*, 1945, **67**, 1180 (*synthesis, esters*)
Ford-Moore, A.H. *et al*, *J. Chem. Soc.*, 1947, 1465 (*ester*)
Griffin, C.E. *et al*, *J. Org. Chem.*, 1959, **24**, 2049 (*synth*)
Pudovik, A.N. *et al*, *Zh. Obshch. Khim.*, 1959, **29**, 3342 (*Engl. transl.* p. 3305) (*esters*)
Kharasch, M.S. *et al*, *J. Org. Chem.*, 1960, **25**, 100 (*synth, derivs*)
Laughlin, R.G. *et al*, *J. Org. Chem.*, 1962, **27**, 3644 (*ester*)
Feshchenko, N.G. *et al*, *Zh. Obshch. Khim.*, 1967, **37**, 473 (*Engl. transl.* p. 441) (*difluoride*)
Dietze, U., *J. Prakt. Chem.*, 1974, **316**, 293 (*ir*)

Octylphosphonic dichloride, 9CI — O-00043

[3095-94-1]

$H_3C(CH_2)_7P(O)Cl_2$

$C_8H_{17}Cl_2OP$ M 231.101
Liq. d_4^{20} 1.12. Bp_{12} 143-145°, $Bp_{0.01}$ 76-78°. n_D^{20} 1.4680.

Maier, L., *Helv. Chim. Acta*, 1965, **48**, 1190 (*synth, ir, P nmr*)
Feshchenko, N.G. *et al*, *Zh. Obshch. Khim.*, 1967, **37**, 473 (*Engl. transl.* p. 441) (*synth*)
Geiseler, G. *et al*, *Ber. Bunsenges. Phys. Chem.*, 1967, **71**, 478 (*ir, raman*)

Octylphosphonous acid, 9CI — O-00044

$H_3C(CH_2)_7P(OH)_2 \rightleftharpoons H_3C(CH_2)_7PH(O)OH$

$C_8H_{19}O_2P$ M 178.211
Di-Et ester: Diethyl octylphosphonite.
$C_{12}H_{27}O_2P$ M 234.318
Liq. $Bp_{0.12}$ 58-60°. n_D^{25} 1.4384.
Diisopropyl ester: Diisopropyl octylphosphonite.
$C_{14}H_{31}O_2P$ M 262.371
Liq. Bp_4 107-112°.
Dichloride: [15573-34-9]. *Dichlorooctylphosphine.*
$C_8H_{17}Cl_2P$ M 215.102
Liq. Bp 247°, Bp_8 106-110°. n_D^{25} 1.4778.
Dibromide: Dibromooctylphosphine.
$C_8H_{17}Br_2P$ M 304.004
Liq. d_4^{20} 1.4370. Bp_1 114-115°. n_D^{20} 1.5250.

Fox, R.B., *J. Am. Chem. Soc.*, 1950, **72**, 4147 (*dichloride*)
Henderson, W.A. *et al*, *J. Org. Chem.*, 1961, **26**, 4770 (*dichloride*)
Petrov, K.A. *et al*, *Zh. Obshch. Khim.*, 1961, **31**, 3027 (*Engl. transl.* p. 2823) (*dibromide*)
Mastalerz, P. *et al*, *Rocz. Chem.*, 1964, **38**, 1529 (*ester*)
Khokhlov, P.S. *et al*, *Zh. Obshch. Khim.*, 1984, **54**, 2703 (*Engl. transl.* p. 2416) (*dibromide, synth, P nmr*)

Octylphosphoramidic acid, 9CI — O-00045

$H_3C(CH_2)_7NHP(O)(OH)_2$

$C_8H_{20}NO_3P$ M 209.225
Di-Me ester: Dimethyl octylphosphoramidate.
$C_{10}H_{24}NO_3P$ M 237.278
Liq. Bp_1 127°. n_D^{25} 1.4283 (n_D^{20} 1.4392).
Di-Et ester: [53246-95-0]. *Diethyl octylphosphoramidate.*
$C_{12}H_{28}NO_3P$ M 265.332
Liq. n_D^{20} 1.4417.
Dichloride:
$C_8H_{18}Cl_2NOP$ M 246.116
Liq. Bp_3 150-154°.

Baumgarten, H.E. *et al*, *J. Am. Chem. Soc.*, 1959, **81**, 2132 (*dimethyl ester, synth*)
Mizuma, T. *et al*, *J. Pharm. Soc. Jpn.*, 1961, **81**, 48; *CA*, **55**, 13403 (*dichloride*)
Bebikh, G. *et al*, *Zh. Obshch. Khim.*, 1977, **47**, 2196 (*Engl. transl.* p. 2005) (*dimethyl ester, synth, ir*)
Zwierzak, A., *Synthesis*, 1982, 920 (*diethyl ester, synth, ir, pmr, P nmr*)

Octyl phosphorodichloridite, 9CI, 8CI — O-00046

Octyl dichlorophosphite
[56148-84-6]

$H_3C(CH_2)_7OPCl_2$

$C_8H_{17}Cl_2OP$ M 231.101
Liq. d_0^0 1.09. Bp_{12} 123-124°. n_D^{21} 1.4682.

Razumov, A.I., *Zh. Obshch. Khim.*, 1944, **14**, 464; *CA*, 1945, **39**, 4586 (*synth*)

Octyltriphenylphosphonium(1+), 9CI — O-00047

$Ph_3P^{\oplus}(CH_2)_7CH_3$

$C_{26}H_{32}P^{\oplus}$ M 375.513 (ion)
Chloride: [84834-74-2].
$C_{26}H_{32}ClP$ M 410.966
Effective catalyst for dechlorinations of 1,2-dichloroalkanes, etc., by metals.
Bromide: [42036-78-2].
$C_{26}H_{32}BrP$ M 455.417
Ylide source. Solid. Mp 91-98°.
Ylide: [80625-78-1]. *Octylidenetriphenylphosphorane.*
$C_{26}H_{30}P$ M 373.497
Employed in Wittig reactions used in synth. of natural products e.g. insect pheromones.

Jaenicke, L. *et al*, *Justus Liebigs Ann. Chem.*, 1973, 1252 (*bromide, use*)
Kauffmann, T. *et al*, *Chem. Ber.*, 1973, **106**, 1612 (*bromide, use*)
Bestmann, H.J. *et al*, *Justus Liebigs Ann. Chem.*, 1981, 2117 (*use*)
Chukhadzhyan, G.A. *et al*, *CA*, 1983, **98**, 106743 (*chloride, use*)

Omethoate, BSI — O-00048

O,O-*Dimethyl* S-*[2-(methylamino)-2-oxoethyl] phosphorothioate, 9CI.* O,O-*Dimethyl phosphorothioate,* S-*ester with 2-mercapto-*N*-methylacetamide, 8CI.* O,O-*Dimethyl* S-*methylcarbamoylmethyl phosphorothioate. Folimat*
[1113-02-6]

$(MeO)_2P(O)SCH_2CONHMe$

$C_5H_{12}NO_4PS$ M 213.188

Systemic insecticide and acaricide. Oil. Misc. H_2O, almost insol. hexane. d_4^{20} 1.32. Dec. ca. 135°. n_D^{20} 1.4987.

▷TF8050000.

B.P., 983 734, (*1962*); *CA*, **61**, 579c (*synth*)
Paasivirta, J. *et al*, *Org. Magn. Reson.*, 1977, **9**, 708 (*pmr, cmr, P nmr*)
Stan, H.-J. *et al*, *Fresenius' Z. Anal. Chem.*, 1977, **287**, 271 (*glc, ms*)
Daldrup, T. *et al*, *Fresenius' Z. Anal. Chem.*, 1981, **308**, 413 (*tlc, hplc, glc*)
Stan, H.-J. *et al*, *Biomed. Mass Spectrom.*, 1982, **9**, 483 (*ms*)
Charalambous, J. *et al*, *Phosphorus Sulfur*, 1984, **19**, 267 (*ms*)
Pesticide Manual, 6th Ed., 392; 7th Ed., 410.

1-Oxa-5-phospha(5-P^V)spiro[4.5]decane, 10CI — O-00049

[70225-02-4]

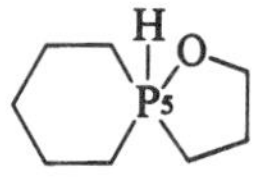

$C_8H_{17}OP$ M 160.195

5-Me: [69783-47-7].
$C_9H_{19}OP$ M 174.222
Liq. $Bp_{0.1}$ 54°.

Schmidbaur, H. *et al*, *Chem. Ber.*, 1979, **112**, 501 (*synth, pmr, cmr*)

6-Oxa-5-phospha(5-P^V)spiro[4.5]decane, 10CI — O-00050

[70225-00-2]

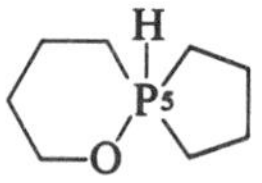

$C_8H_{17}OP$ M 160.195

5-Me: [69783-50-2].
$C_9H_{12}OP$ M 167.167
Liq. $Bp_{0.1}$ 46°. Possesses trigonal bipyramidal struct.

Schmidbaur, H. *et al*, *Chem. Ber.*, 1979, **112**, 501 (*synth, pmr, cmr*)

1-Oxa-5-phospha(5-P^V)spiro[4.4]nonane, 10CI — O-00051

[70225-03-5]

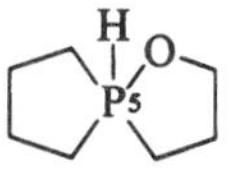

$C_7H_{15}OP$ M 146.169

5-Me: [69783-46-6].
$C_8H_{17}OP$ M 160.195
Liq. Bp_{20} 106°. Possesses trigonal bipyramidal struct.

Schmidbaur, H. *et al*, *Chem. Ber.*, 1979, **112**, 501 (*synth, pmr, cmr*)

1-Oxa-6-phospha(6-P^V)spiro[5.5]undecane, 10CI — O-00052

[70236-64-5]

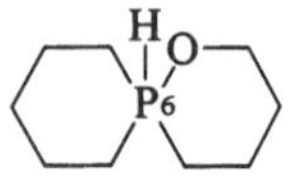

$C_9H_{19}OP$ M 174.222

6-Me: [69783-51-3].
$C_{10}H_{21}OP$ M 188.249
Liq. $Bp_{0.1}$ 53°. Possesses trigonal bipyramidal struct.

Schmidbaur, H. *et al*, *Chem. Ber.*, 1979, **112**, 501 (*synth, pmr, cmr*)

1,2-Oxaphospholan-5-one 2-oxide — O-00053

(2-Carboxyethyl)phosphinic cyclic anhydride

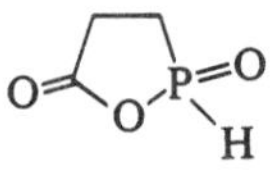

$C_3H_5O_3P$ M 120.044

2-Me: [15171-48-9]. *2-Methyl-1,2-oxaphospholan-5-one 2-oxide. (2-Carboxyethyl)methylphosphinic cyclic anhydride.*
$C_4H_7O_3P$ M 134.071
Solid. Sol. Me_2CO, C_6H_6, insol. CCl_4. Mp 106° (97-98°). $Bp_{0.5}$ 149-150°.

2-Ph: [14651-62-8]. *2-Phenyl-1,2-oxaphospholan-2-one 2-oxide. (2-Carboxyethyl)phenylphosphinic cyclic anhydride.*
$C_9H_9O_3P$ M 196.142
Cryst. (Me_2CO/Et_2O). Mp 74-75°, Mp 87°. $Bp_{0.6}$ 185°, $Bp_{0.003}$ 158-160°. Slowly hydrol. in air.

Pudovik, A.N. *et al*, *Zh. Obshch. Khim.*, 1967, **37**, 411, 423 (*Engl. transl.* pp. 385, 423) (*phenyl*)
Khairullin, V.K. *et al*, *Zh. Obshch. Khim.*, 1967, **37**, 710 (*Engl. transl.* p. 666) (*methyl*)
Ger. Pat., 2 526 689, (*1976*); 2 531 920, (*1976*); *CA*, **86**, 140245, 121154

1,2-Oxaphosphorinane, 9CI — O-00054

1,2-Oxaphosphinane. 1-Oxa-2-phosphacyclohexane

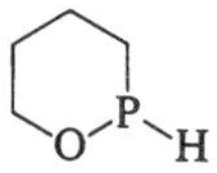

C_4H_9OP M 104.088

2-Chloro, 2-oxide: see 2-Hydroxy-1,2-oxaphosphorinane 2-oxide, H-00167

2-Ph: [87079-91-2].
$C_{10}H_{13}OP$ M 180.186
Liq. $Bp_{0.55}$ 76-79°.

2-Ph, 2-oxide: [55549-39-8].
$C_{10}H_{13}O_2P$ M 196.185
Yields monomeric phenylmetaphosphonate when pyrolysed. Cryst. (Me_2CO). Mp 87-88°. $Bp_{0.7}$ 165°.

Stutz, H. *et al*, *Z. Chem.*, 1975, **15**, 52 (*synth*)
Cadogen, J.I.G. *et al*, *Phosphorus Sulfur*, 1983, **18**, 229.
Kobayashi, S. *et al*, *Bull. Chem. Soc. Jpn.*, 1985, **58**, 2153 (*synth, pmr, cmr, P nmr*)
Kobayashi, S. *et al*, *Macromolecules*, 1986, **19**, 466 (*oxide, pmr, P nmr*)

1,4-Oxaphosphorinane, 9CI, 8CI — O-00055

1,4-Oxaphosphinane. 1-Oxa-4-phosphacyclohexane

[6569-77-3]

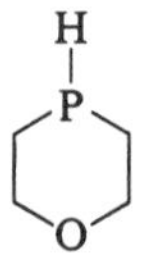

C_4H_9OP M 104.088
Liq. Bp_{16} 45-48°.

4-Ph: see 4-Phenyl-1,4-oxaphosphorinane, P-00176

Tavs, P., *Angew. Chem.*, 1969, **81**, 742 (*synth, ir, ms, pmr, P nmr*)

1,3,2-Oxathiaphospholane, 9CI, 8CI **O-00056**

O,S-*Ethylene phosphonothioite*

[7092-52-6]

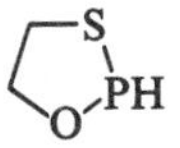

C_2H_5OPS M 108.095

2-Chloro: [20354-32-9]. *2-Chloro-1,3,2-oxathiaphospholane.* O,S-*Ethylene phosphorochloridothioate.*

C_2H_4ClOPS M 142.540

Pungent liq. which is easily hydrolysed. $Bp_{0.4}$ 57°.

2-Methoxy: [23081-69-8]. O,S-*Ethylene* O-*methyl phosphorothioite.*

$C_3H_7O_2PS$ M 138.121

Liq. d_4^{20} 1.37. $Bp_{0.5}$ 42-43°. n_D^{20} 1.5588.

2-Phenoxy: see 2-Phenoxy-1,3,2-oxathiaphospholane, P-00072

2-Methylthio: [41012-80-0]. O,S-*Ethylene* S-*methyl phosphorodithioite.*

$C_3H_7OPS_2$ M 154.181

Liq. $Bp_{0.1}$ 62°.

2-Ph: [38443-57-1]. O,S-*Ethylene phenylphosphonothioite.*

C_8H_9OPS M 184.192

$Bp_{0.5}$ 106°.

2-Dimethylamino: N,N-*Dimethyl-1,3,2-oxathiaphospholan-2-amine,* 9CI. *2-Dimethylamino-1,3,2-oxathiaphospholane.*

$C_4H_{10}NOPS$ M 151.163

Liq. $Bp_{0.5}$ 64°. n_D^{20} 1.550.

Kovalev, L.S. *et al*, *Zh. Obshch. Khim.*, 1969, **39**, 869 (*Engl. transl.* p. 833) (*methoxy, synth, ir, pmr*)
Bergesen, K. *et al*, *Acta Chem. Scand.*, 1972, **26**, 2156 (*chloro, phenyl, synth, pmr*)
Bergesen, K. *et al*, *Acta Chem. Scand.*, 1973, **27**, 357 (*methoxy, methylthio, synth, pmr*)
Willson, M. *et al*, *Bull. Soc. Chim. Fr., Part 2*, 1975, 615 (*dimethylamino, methoxy, synth, pmr, nmr*)
Nuretdinov, I.A. *et al*, *Izv. Akad. Nauk SSSR, Ser. Khim.*, 1978, 950 (*Engl. transl.* p. 824) (*chloro, nqr*)

1,3,2-Oxathiaphosphorinane 2-oxide, 8CI **O-00057**

1,3,2-Oxathiaphosphinane 2-oxide

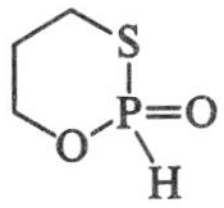

$C_3H_7O_2PS$ M 138.121

2-Isopropoxy: [30342-93-9].

$C_6H_{13}O_3PS$ M 196.201

Liq. n_D^{20} 1.4860.

2-Phenoxy: [30342-92-8].

$C_9H_{11}O_3PS$ M 230.218

Solid. Mp 84-85°.

2-Ph: [30342-95-1].

$C_9H_{11}O_2PS$ M 214.218

Liq. n_D^{20} 1.5803.

Hoang Phuong Nguyen, *et al*, *C.R. Hebd. Seances Acad. Sci., Ser. C*, 1970, **271**, 1465 (*derivs, ir, pmr*)
Arbuzov, B.A. *et al*, *Izv. Akad. Nauk SSSR, Ser. Khim.*, 1983, **32**, 675 (*Engl. transl.* p. 613) (*phenoxy, synth*)
Arbuzov, B.A. *et al*, *Izv. Akad. Nauk SSSR, Ser. Khim.*, 1985, 570 (*Engl. transl.* p. 520) (*uv, pmr, cmr*)

4-Oxocyclophosphamide **O-00058**

*2-[Bis(2-chloroethyl)amino]tetrahydro-2*H*-1,3,2-oxazaphosphorin-4-one 2-oxide,* 9CI. *2-[Bis(2-chloroethyl)-amino]-1,3,2-oxazaphosphorinan-4-one 2-oxide. 4-Ketocyclophosphamide*

[27046-19-1]

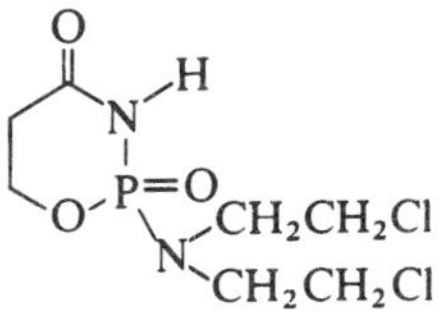

$C_7H_{13}Cl_2N_2O_3P$ M 275.071

Metab. of Cyclophosphamide, C-00283 . Cryst. (H_2O).

▷RQ5857500.

Camerman, N. *et al*, *J. Am. Chem. Soc.*, 1973, **95**, 5038 (*cryst struct*)
Schulten, H.R. *et al*, *Biomed. Mass Spectrom.*, 1974, **1**, 223 (*ms*)
Struck, R.F. *et al*, *J. Am. Chem. Soc.*, 1974, **96**, 313 (*cmr*)
Voelcker, G. *et al*, *Arzneim.-Forsch.*, 1974, **24**, 1172; 1982, **32**, 639 (*tlc*)
Cox, P.J. *et al*, *Biomed. Mass Spectrom.*, 1977, **4**, 371 (*ms*)
Struck, R.F. *et al*, *Cancer Treat Rep.*, 1984, **68**, 765.

(2-Oxoethyl)phosphonic acid, 9CI **O-00059**

Phosphonoacetaldehyde. 2-Phosphonoethanal. (Formylmethyl)phosphonic acid

[16051-76-6]

$(HO)_2P(O)CH_2CHO$

$C_2H_5O_4P$ M 124.033

Thought to lie on the catabolic pathway leading to (2-aminoethyl)phosphonic acid in *T. pyriformis*. Esters form acetals, dinitrophenylhydrazones, and semicarbazones; also enol derivs. Esters may distil only with difficulty.

Di-Me ester: [10038-65-0]. *Dimethyl (2-oxoethyl)-phosphonate. 2-(Dimethoxyphosphinyl)ethanal.*

$C_4H_9O_4P$ M 152.086

Liq. d_4^{20} 1.26. Bp_1 87-88°. n_D^{20} 1.4410.

Di-Et ester: [1606-75-3]. *Diethyl (2-oxoethyl)-phosphonate. 2-(Diethoxyphosphinyl)ethanal.*

$C_6H_{13}O_4P$ M 180.140

Used in synth. of heterocyclic compds. Liq. d_4^{20} 1.15. Bp_3 104-105°. n_D^{20} 1.4390.

Lutsenko, I.F. *et al*, *Zh. Obshch. Khim.*, 1962, **32**, 263 (*Engl. transl.* p. 257) (*synth*)
Tavs, P., *Chem. Ber.*, 1967, **100**, 1571 (*synth, ir, derivs*)
Razumov, A.I. *et al*, *Zh. Obshch. Khim.*, 1971, **41**, 1954 (*Engl. transl.* p. 1970) (*ir, pmr, struct*)
Razumov, A.I. *et al*, *Zh. Obshch. Khim.*, 1978, **48**, 51 (*Engl. transl.* p. 41) (*use*)
Horigane, A. *et al*, *Biochem. Biophys. Acta*, 1979, **572**, 385.
Aboujaoude, E.E. *et al*, *Synthesis*, 1983, 634 (*synth, nmr*)

(2-Oxoethyl)triphenylphosphonium(1+), 9CI **O-00060**

(Formylmethyl)triphenylphosphonium(1+), 8CI

$Ph_3P^{\oplus}CH_2CHO$

$C_{20}H_{18}OP^{\oplus}$ M 305.335 (ion)

Chloride: [62942-43-2].

$C_{20}H_{18}ClOP$ M 340.788

Source of Triphenylphosphoranylideneacetaldehyde, T-00639 . Cryst. ($CHCl_3$/EtOAc). Mp 212-213° dec. pK_a 6.15 (25°, EtOH aq.). At pH10 in 10% EtOH aq., $t_{1/2}$ = 391 ± 21 hr. Exists mostly as *trans*-enol form in $CHCl_3$, but as keto-form (100%) in MeOH.

Bromide: [19753-63-0].
$C_{20}H_{18}BrOP$ M 385.239
Source of Triphenylphosphoranylideneacetaldehyde, T-00639 . Cryst. ($CHCl_3$/EtOAc). Mp 191-192°.
Tetrafluoroborate:
$C_{20}H_{18}BF_4OP$ M 392.139
Solid. Mp 76-77°.
Ylide: see Triphenylphosphoranylideneacetaldehyde, T-00639

Trippett, S. *et al*, *J. Chem. Soc.*, 1961, 1266 (*chloride, synth*)
Issleib, K. *et al*, *Justus Liebigs Ann. Chem.*, 1968, **713**, 12 (*props*)
Devlin, C.J. *et al*, *Tetrahedron*, 1972, **28**, 3501 (*bromide, synth, pmr, ir, ms, nmr*)
Nesmeyanov, N.A. *et al*, *J. Organomet. Chem.*, 1977, **129**, 41 (*derivs, synth, tautom, ir, pmr*)
Wittig, G. *et al*, *Justus Liebigs Ann. Chem.*, 1978, 362 (*use*)
Brittain, J.M. *et al*, *Tetrahedron*, 1979, **35**, 1139 (*nmr*)

(2-Oxoheptyl)phosphonic acid, 9CI — O-00061

1-Phosphono-2-heptanone
[69639-73-2]

$$H_3C(CH_2)_4COCH_2P(O)(OH)_2$$

$C_7H_{15}O_4P$ M 194.167
Esters are Wittig-Horner reagents widely used in the synth. of prostaglandins and related substances. $Bp_{0.5}$ 108-110°, $Bp_{0.15}$ 130°.
Di-Me ester: [36969-89-8]. *Dimethyl (2-oxoheptyl)-phosphonate.*
$C_9H_{19}O_4P$ M 222.220
Liq. $Bp_{0.1}$ 97-101°. n_D^{22} 1.4745.
Di-Et ester: [3450-65-5]. *Diethyl (2-oxoheptyl)-phosphonate.*
$C_{11}H_{23}O_4P$ M 250.274
Liq. $Bp_{0.5}$ 108-110°, $Bp_{0.15}$ 130°. n_D^{22} 1.4667.

Chattha, M. *et al*, *J. Org. Chem.*, 1973, **38**, 2908 (*synth*)
Mathey, F. *et al*, *Tetrahedron*, 1978, **34**, 649 (*synth*)
Corey, E.J. *et al*, *Tetrahedron Lett.*, 1980, **21**, 137 (*use*)
Ansell, M.F. *et al*, *Tetrahedron Lett.*, 1981, **22**, 1141 (*use*)
Greene, A.E. *et al*, *J. Org. Chem.*, 1982, **47**, 2553 (*use*)
Schwartz, S. *et al*, *Tetrahedron*, 1982, **38**, 1261 (*use*)

(2-Oxoheptyl)triphenylphosphonium(1+) — O-00062

$$Ph_3P^{\oplus}CH_2CO(CH_2)_4CH_3$$

$C_{25}H_{28}OP^{\oplus}$ M 375.469 (ion)
Chloride:
$C_{25}H_{28}ClOP$ M 410.922
Cryst. (Me_2CO). Mp 172°. Converted by aq. K_2CO_3 into the ylide.
Bromide:
$C_{25}H_{28}BrOP$ M 455.373
Cryst. (EtOAc/pet. ether). Mp 198-199°. Converted by alkali into the ylide.
Ylide: [33803-58-6]. *1-(Triphenylphosphoranylidene)-2-heptanone.*
$C_{25}H_{27}OP$ M 374.461
Reactive Wittig reagent, used extensively in the synth. of prostaglandins. Solid. Mp 73-74°.

Pailer, M. *et al*, *Monatsh. Chem.*, 1977, **108**, 1059 (*bromide, ylide*)
Pirillo, D. *et al*, *Farmaco, Ed. Sci.*, 1977, **32**, 747; *CA*, **88**, 23049 (*chloride, ylide*)
Kienzle, F. *et al*, *Helv. Chim. Acta*, 1980, **63**, 1425 (*bromide, ylide*)
Pirillo, D. *et al*, *Farmaco, Ed. Sci.*, 1982, **37**, 328; *CA*, **97**, 127314 (*ylide, use*)

2-Oxononylphosphonic acid, 9CI — O-00063

1-Phosphono-2-nonanone

$$H_3C(CH_2)_6COCH_2P(O)(OH)_2$$

$C_9H_{19}O_4P$ M 222.220
Esters are Wittig-Horner reagents used in the synthesis of prostaglandins and their analogues.
Di-Me ester: [37497-25-9]. *Dimethyl (2-oxononyl)-phosphonate.*
$C_{11}H_{23}O_4P$ M 250.274
No phys. props. reported.
Di-Et ester: [40601-40-9]. *Diethyl (2-oxononyl)-phosphonate.*
$C_{13}H_{27}O_4P$ M 278.328
Liq. $Bp_{0.5}$ 125-129°. n_D^{22} 1.4420.

Chattha, M. *et al*, *J. Org. Chem.*, 1973, **38**, 2908 (*synth*)
Dauben, W.G. *et al*, *J. Am. Chem. Soc.*, 1975, **97**, 4973 (*use*)
Mathey, F. *et al*, *Tetrahedron*, 1978, **34**, 649 (*synth*)

(2-Oxooctyl)phosphonic acid, 9CI — O-00064

1-Phosphono-2-octanone

$$H_3C(CH_2)_5COCH_2P(O)(OH)_2$$

$C_8H_{17}O_4P$ M 208.194
Esters are Wittig-Horner reagents for the synth. of prostaglandins and their analogues.
Di-Me ester: [61408-88-6]. *Dimethyl (2-oxooctyl)-phosphonate.*
$C_{10}H_{21}O_4P$ M 236.247
No phys. props. reported.
Di-Et ester: [3452-99-1]. *Diethyl (2-oxooctyl)-phosphonate.*
$C_{12}H_{25}O_4P$ M 264.301
Liq. $Bp_{0.4}$ 123°. n_D^{22} 1.4402.

Sturtz, G., *Bull. Soc. Chim. Fr.*, 1966, 1707 (*synth*)
Chattha, M. *et al*, *J. Org. Chem.*, 1973, **38**, 2908 (*synth*)
Saijo, S. *et al*, *Chem. Pharm. Bull.*, 1980, **28**, 1449 (*use*)
Temesvári-Major, E. *et al*, *Tetrahedron Lett.*, 1980, **21**, 4035 (*use*)

(2-Oxo-4-phenylbutyl)phosphonic acid, 9CI — O-00065

[51008-95-8]

$$PhCH_2CH_2COCH_2P(O)(OH)_2$$

$C_{10}H_{13}O_4P$ M 228.184
Esters are reagents widely employed in prostaglandin synth., and in the synth. of $\alpha\beta$-cyclohexenones. Solid. Mp 153-154°.
Di-Me ester: [41162-19-0]. *Dimethyl (2-oxo-4-phenylbutyl)phosphonate.*
$C_{12}H_{17}O_4P$ M 256.238
Used in cyclohexenone synth. Liq. $Bp_{0.5}$ 120-122°.
Di-Et ester: [40601-45-4]. *Diethyl (2-oxo-4-phenylbutyl)phosphonate.*
$C_{14}H_{21}O_4P$ M 284.291
Liq. $Bp_{0.12}$ 162°.

Chattha, M. *et al*, *J. Org. Chem.*, 1973, **38**, 2908 (*ester*)
Göhring, G. *et al*, *Chem. Ber.*, 1973, **106**, 2460 (*synth, ester, ir, pmr, ms*)
Grieco, P.A. *et al*, *J. Am. Chem. Soc.*, 1973, **95**, 3071 (*ester*)
Grieco, P.A. *et al*, *Synthesis*, 1973, 425 (*use*)

Magerlein, B.J. *et al*, *Prostaglandins*, 1975, **9**, 5 (*use*)

(2-Oxo-2-phenylethyl)phosphonic acid, 9CI **O-00066**

Phenacylphosphonic acid

[4724-50-9]

$PhCOCH_2P(O)(OH)_2$

$C_8H_9O_4P$ M 200.130

Sl. yellow needles. Mp 139°.

Monoanilinium salt: Cryst. (Me_2CO/MeOH). Mp 103-104°.

Di-Me ester: [1015-28-7]. *Dimethyl (2-oxo-2-phenylethyl)phosphonate.*

$C_{10}H_{13}O_4P$ M 228.184

Employed in the synth. of azaprostaglandins.

Di-Et ester: [3453-00-7]. *Diethyl (2-oxo-2-phenylethyl)phosphonate.*

$C_{12}H_{17}O_4P$ M 256.238

Liq. $Bp_{0.1}$ 135°.

Dibutyl ester: [1034-94-2]. *Dibutyl (2-oxo-2-phenylethyl)phosphonate. HDBPP.*

$C_{16}H_{25}O_4P$ M 312.345

Forms rare earth complexes. Employed in extr. of U from acid soln. Liq. $Bp_{0.3}$ 155°.

Bis(dimethylamide): N,N,N′,N′-*Tetramethyl*-P-(*2-oxo-2-phenylethyl)phosphonic diamide.*

$C_{12}H_{19}N_2O_2P$ M 254.268

Liq. Dec. on attempted dist.

Chattha, M. *et al*, *J. Org. Chem.*, 1973, **38**, 2908 (*diethyl ester*)
Maier, L., *Synth. Inorg. Met.-Org. Chem.*, 1973, **3**, 329 (*synth, ir, pmr*)
Martin, J.L. *et al*, *J. Chem. Res. (S)*, 1978, 88 (*dibutyl ester, synth, purifn, props, use*)
Savignac, P. *et al*, *Synthesis*, 1978, 682 (*ester, amide, synth, pmr*)
Pichon, R. *et al*, *Bull. Soc. Chim. Fr., Part II*, 1980, 449 (*dimethyl pmr, ir, uv*)
Burgard, M. *et al*, *J. Phys. Chem.*, 1982, **86**, 4817 (*dibutyl ester, pmr, nmr, cmr, complexes*)

(2-Oxo-2-phenylethyl)-triphenylphosphonium(1+), 9CI **O-00067**

(Benzoylmethyl)triphenylphosphonium(1+). Phenacyltriphenylphosphonium, 8CI

[17638-55-0]

$Ph_3P^{\oplus}CH_2COPh$

$C_{26}H_{22}OP^{\oplus}$ M 381.433 (ion)

Chloride: [1678-18-8].

$C_{26}H_{22}ClOP$ M 416.886

Plates. Forms Pd complex.

Bromide: [6048-29-9].

$C_{26}H_{22}BrOP$ M 461.337

Na_2CO_3→ylide. Cryst. (EtOH/Et_2O or H_2O). Possibly exists in two cryst. forms. Mp 265-268° dec., 284-286° dec. pK_a 6.03 (25°, 66% EtOH aq.). Exists largely as keto tautomer in soln.

▷TA2400000.

Iodide: [6230-82-6].

$C_{26}H_{22}IOP$ M 508.337

NaOMe or Na_2CO_3 aq.→ylide. Cryst. (H_2O). Mp 240-242°, 259-260°.

▷TA2410000.

Ylide: see 1-Phenyl-2-(triphenylphosphoranylidene)-ethanone, P-00333

Ramirez, F. *et al*, *J. Org. Chem.*, 1957, **22**, 41 (*synth, props*)
Horner, L. *et al*, *Justus Liebigs Ann. Chem.*, 1961, **646**, 65 (*synth*)
Albright, T.A. *et al*, *J. Am. Chem. Soc.*, 1975, **97**, 2942; 1976, **98**, 6249 (*cmr, nmr*)
Nesmeyanov, N.A. *et al*, *J. Organomet. Chem.*, 1977, **129**, 41 (*tautom, ir, pmr, nmr*)
Nesmeyanov, N.A. *et al*, *Zh. Org. Khim.*, 1977, **13**, 2465 (*Engl. transl.* p. 2293) (*iodide, ir, nmr, props*)
Fronza, G. *et al*, *J. Organomet. Chem.*, 1978, **157**, 299 (*bromide, ms*)
Brittain, J.M. *et al*, *Tetrahedron*, 1979, **35**, 1139 (*chloride, nmr*)
Doleschall, G., *Synthesis*, 1981, 478 (*iodide, use*)
Antipin, M.Yu. *et al*, *Zh. Strukt. Khim.*, 1984, **25**, 122 (*Engl. transl.* p. 106) (*bromide, cryst struct*)

(2-Oxo-3-phenylpropyl)phosphonic acid, 9CI **O-00068**

$PhCH_2COCH_2P(O)(OH)_2$

$C_9H_{11}O_4P$ M 214.157

Di-Me ester: [52343-38-1]. *Dimethyl(2-oxo-3-phenylpropyl)phosphonate.*

$C_{11}H_{15}O_4P$ M 242.211

Intermed. for synth. of prostaglandin analogues via Wittig reactions. Liq. $Bp_{0.2}$ 142-4°.

U.S.P., 4 029 814, (*1977*); *CA*, **87**, 84618 (*synth*)
U.S.P., 4 035 360, (*1977*); *CA*, **87**, 184071 (*synth, pmr*)
Eur. Pat., 79 694, (*1983*); *CA*, **100**, 6190 (*synth*)

(2-Oxo-1-phenylpropyl)-triphenylphosphonium(1+) **O-00069**

$Ph_3P^{\oplus}CHPhCOCH_3$

$C_{27}H_{24}OP^{\oplus}$ M 395.460 (ion)

Bromide:

$C_{27}H_{24}BrOP$ M 475.364

Cryst. ($CHCl_3$/pet. ether). Mp 251°. When treated with aq. NaOH yields the ylide.

Ylide: see 1-Phenyl-1-(triphenylphosphoranylidene)-2-propanone, P-00334

Gough, S.T.D. *et al*, *J. Chem. Soc.*, 1962, 2333 (*synth, props*)

(3-Oxo-3-phenylpropyl)-triphenylphosphonium(1+) **O-00070**

$Ph_3P^{\oplus}CH_2COCH_2Ph$

$C_{27}H_{24}OP^{\oplus}$ M 395.460 (ion)

Bromide: [19874-73-8].

$C_{27}H_{24}BrOP$ M 475.364

Solid. Mp 243-245°. pK_a 6.19 (67% EtOH aq., 25°).

Iodide:

$C_{27}H_{24}IOP$ M 522.364

Cryst. (MeOH). Mp >140° dec.

Ylide: see 1-Phenyl-3-(triphenylphosphoranylidene)-2-propanone, P-00336

Issleib, K. *et al*, *Justus Liebigs Ann. Chem.*, 1968, **713**, 12 (*bromide*)
Doleschall, G., *Synthesis*, 1981, 478 (*iodide*)

1-Oxo-1,3-propanediphosphonic acid **O-00071**

(1-Oxo-1,3-propanediyl)bisphosphonic acid, 9CI. *1,3-Diphosphono-1-propanone*

$(HO)_2P(O)CH_2CH_2COP(O)(OH)_2$

$C_3H_8O_7P_2$ M 218.040

Tetra-Et ester: [34666-29-0]. *Tetraethyl (1-oxo-1,3-propanediyl)bisphosphonate.*

$C_{11}H_{24}O_7P_2$ M 330.254

Liq. d_4^{20} 1.22. $Bp_{0.05}$ 165-168°. n_D^{20} 1.4612.

Gazizov, T.Kh. *et al*, *Zh. Obshch. Khim.*, 1971, **41**, 1957 (*Engl. transl.* p. 1973) (*synth*)

2-Oxo-1,3-propanediphosphonic acid O-00072

(2-Oxo-1,3-propanediyl)bisphosphonic acid, 9CI. Acetone 1,3-diphosphonic acid

$(HO)_2P(O)CH_2COCH_2P(O)(OH)_2$

$C_3H_8O_7P_2$ M 218.040

Tetra-Me ester: [73125-47-0]. *Tetramethyl (2-oxo-1,3-propanediyl)bisphosphonate.*

$C_7H_{16}O_7P_2$ M 274.147

Employed in Wittig-Horner reactions for the synth. of macrocyclic ketones. Liq. $Bp_{0.08}$ 156-158°.

Tetra-Et ester: Tetraethyl (2-oxo-1,3-propanediyl)-bisphosphonate.

$C_{11}H_{24}O_7P_2$ M 330.254

Liq. d_4^{24} 1.19. $Bp_{0.6}$ 175°. n_D^{24} 1.4490.

Sturtz, G., *Bull. Soc. Chim. Fr.*, 1967, 1345 (*synth, pmr, P nmr*)
Büchi, G. *et al*, *Helv. Chim. Acta*, 1979, **62**, 2661 (*synth, ir, pmr, use*)

(2-Oxo-1,3-propanediyl)-bis(triphenylphosphonium)(2+), 10CI O-00073

Acetone-1,3-bis(triphenylphosphonium)(2+). 1,3-Bis-(triphenylphosphonio)-2-propanone(2+)

$Ph_3P^{\oplus}CH_2COCH_2P^{\oplus}Ph_3$

$C_{39}H_{34}OP_2^{\oplus\oplus}$ M 580.645 (ion)

Dichloride: [4885-57-8].

$C_{39}H_{34}Cl_2OP_2$ M 651.551

Cryst. ($CHCl_3$/MeCN). Mp 266-267°.

Dibromide: [70307-60-7].

$C_{39}H_{34}Br_2OP_2$ M 740.453

Cryst. (CH_2Cl_2/EtOAc). Mp 305°. NaOEt yields the bis-ylide.

Bis-ylide: [6472-98-6]. *1,3-Bis(triphenylphosphoranylidene)-2-propanone.*

$C_{39}H_{32}OP_2$ M 578.629

Wittig reagent prepd. *in situ.*

Ford, J.A. *et al*, *J. Org. Chem.*, 1961, **26**, 1433 (*chloride, use*)
Denney, D.B. *et al*, *J. Org. Chem.*, 1964, **29**, 495 (*chloride, use*)
Stringer, M.B. *et al*, *Aust. J. Chem.*, 1978, **31**, 1607 (*use*)
Bestmann, H.J. *et al*, *Synthesis*, 1979, 201 (*bromide, synth, use*)
Bradbury, R.H. *et al*, *J. Chem. Soc., Perkin Trans. 1*, 1981, 3234 (*use*)

(1-Oxopropyl)phosphonic acid, 9CI O-00074

Propanoylphosphonic acid

[67472-30-4]

$H_3CCH_2COP(O)(OH)_2$

$C_3H_7O_4P$ M 138.060

Esters act as acylating agents to alcohols, enols and primary amines.

Mono-dicyclohexylammonium salt: Solid. Mp 209-212°.

Di-Me ester: [51463-65-1]. *Dimethyl (1-oxopropyl)-phosphonate. Dimethyl propanoylphosphonate.*

$C_5H_{11}O_4P$ M 166.113

Liq. $Bp_{0.5}$ 58-61.5°.

Di-Et ester: [1523-68-8]. *Diethyl (1-oxopropyl)-phosphonate. Diethyl propanoylphosphonate.*

$C_7H_{15}O_4P$ M 194.167

Liq. Bp_7 105°.

Di-Et ester, 2,4-dinitrophenylhydrazone: Mp 93-94°.

Bis(trimethylsilyl) ester: [67472-25-7]. *Bis(trimethylsilyl) (1-oxopropyl)phosphonate. Bis(trimethylsilyl) propanoylphosphonate.*

$C_9H_{23}O_4PSi_2$ M 282.423

Liq. $Bp_{0.5}$ 103-106°. $Bp_{0.3}$ 56-57°.

Berlin, K.D. *et al*, *J. Org. Chem.*, 1965, **30**, 1265 (*synth, ir, pmr*)
Cohen, H. *et al*, *Can. J. Chem.*, 1974, **52**, 66 (*synth, ir*)
Zygmunt, J. *et al*, *Synthesis*, 1978, 609 (*silyl ester, ir, pmr*)
Yamashita, M. *et al*, *Bull. Chem. Soc. Jpn.*, 1980, **53**, 1625 (*props, use*)
Sekine, M. *et al*, *Tetrahedron Lett.*, 1981, **22**, 3617 (*props, use*)
Sekine, M. *et al*, *Chem. Lett.*, 1981, 1087 (*props, use*)

(2-Oxopropyl)phosphonic acid, 9CI O-00075

Acetonylphosphonic acid. Phosphonoacetone

[6913-02-6]

$H_3CCOCH_2P(O)(OH)_2$

$C_3H_7O_4P$ M 138.060

Esters form anions which may be alkylated, and which act as Wittig-Horner reagents. When heated, the mono-Na salt → Me_2CO. Esters exist almost entirely in the keto form but form Mg enolates. Ester anions exist as *Z*- and *E*-forms in soln.

Mono-dicyclohexylammonium salt: [68064-33-5]. Solid. Mp 173-175°.

Di-Me ester: [4202-14-6]. *Dimethyl (2-oxopropyl)-phosphonate. Dimethyl acetonylphosphonate.*

$C_5H_{11}O_4P$ M 166.113

Liq. Bp_{10} 123-124°, $Bp_{0.5}$ 85-88°. n_D^{20} 1.4338.

Di-Et ester: [1067-71-6]. *Diethyl (2-oxopropyl)-phosphonate. Diethyl acetonylphosphonate.*

$C_7H_{15}O_4P$ M 194.167

$Bp_{0.5}$ 88-90°. n_D^{22} 1.4335. Aq. $Ba(OH)_2$ → Me_2CO.

Diisopropyl ester: [67257-36-7]. *Bis(1-methylethyl) (2-oxopropyl)phosphonate. Diisopropyl acetonylphosphonate.*

$C_9H_{19}O_4P$ M 222.220

Liq. $Bp_{0.5}$ 92-96°. n_D^{22} 1.4301.

Di-Ph ester: Diphenyl (2-oxopropyl)phosphonate. Diphenyl acetonylphosphonate.

$C_{15}H_{15}O_4P$ M 290.255

Solid. Mp 127-128°.

Bis(trimethylsilyl) ester: [68064-25-5]. *Bis(trimethylsilyl) (2-oxopropyl)phosphonate. Bis(trimethylsilyl) acetonylphosphonate.*

$C_9H_{23}O_4PSi_2$ M 282.423

Liq. $Bp_{0.4}$ 66-67°.

Dichloride:

$C_3H_5Cl_2O_2P$ M 174.951

Solid. Mp 39-40°. $Bp_{1.5}$ 89.5-90°.

Lutsenko, I.F. *et al*, *Dokl. Akad. Nauk SSSR, Ser. Sci. Khim.*, 1960, **132**, 842; *CA*, **54**, 20842 (*dichloride*)
Cotton, F.A. *et al*, *J. Am. Chem. Soc.*, 1963, **85**, 2394 (*esters, ir, pmr, complexes*)
Kirilov, M. *et al*, *Chem. Ber.*, 1967, **100**, 3139 (*ester, complexes*)

Laskorin, B.N. *et al*, *Zh. Obshch. Khim.*, 1972, **42**, 1261 (*Engl. transl.* p. 1256) (*ester, synth, ir, uv, pmr, P nmr*)
Grieco, P.A. *et al*, *J. Am. Chem. Soc.*, 1973, **95**, 3071; *J. Org. Chem.*, 1973, **38**, 2909 (*ester, dianion, props*)
Kluger, R.K. *et al*, *J. Am. Chem. Soc.*, 1975, **97**, 4298 (*complexes*)
Tsuboi, S. *et al*, *Bull. Chem. Soc. Jpn.*, 1975, **48**, 1331 (*ester, use*)
Laskorin, B.N. *et al*, *Zh. Obshch. Khim.*, 1976, **46**, 2545 (*Engl. transl.* p. 2434) (*ester, ms*)
Mathey, F. *et al*, *Tetrahedron*, 1978, **34**, 649 (*esters*)
Zygmunt, J. *et al*, *Synthesis*, 1978, 609 (*synth*)
Bottin-Strzalko, T. *et al*, *Org. Magn. Reson.*, 1982, **19**, 69 (*ester, anion, pmr, P nmr, cmr*)
Villieras, J. *et al*, *Phosphorus Sulfur*, 1983, **14**, 385 (*esters, use*)

(3-Oxopropyl)phosphonic acid, 9CI — O-00076

3-Phosphonopropanal. (2-Formylethyl)phosphonic acid
[60158-23-8]

$(HO)_2P(O)CH_2CH_2CHO$

$C_3H_7O_4P$ M 138.060
Forms acetals, 2,4-dinitrophenylhydrazone and enol derivs.
Di-Et ester: [3956-95-6]. *Diethyl (3-oxopropyl)-phosphonate.*
$C_7H_{15}O_4P$ M 194.167
Liq. d_4^{20} 1.13. $Bp_{0.05}$ 91-97°. n_D^{20} 1.4400.

Razumov, A.I. *et al*, *Zh. Obshch. Khim.*, 1964, **34**, 2589 (*Engl. transl.* p. 2612) (*synth, derivs*)
Cates, L.A. *et al*, *J. Med. Chem.*, 1980, **23**, 300 (*synth, derivs*)
Varlet, J.M. *et al*, *Tetrahedron*, 1981, **37**, 1377 (*synth, pmr, props*)

P-(2-Oxopropyl)phosphonic diamide, 9CI — O-00077

P-*Acetonylphosphonic diamide, 8CI*

$H_3CCOCH_2P(O)(NH_2)_2$

$C_3H_9N_2O_2P$ M 136.090
N,N,N′,N′-*Tetra-Et:* [19930-80-4]. N,N,N′,N′-*Tetraethyl*-P-*(2-oxopropyl)phosphonic diamide.*
$C_{11}H_{25}N_2O_2P$ M 248.304
Liq. d_4^{20} 1.04. Bp_8 149-150°. n_D^{20} 1.4709.

Pudovik, A.N. *et al*, *Zh. Obshch. Khim.*, 1958, **28**, 2492 (*Engl. transl.* p. 2527)

(1-Oxopropyl)phosphonothioic acid, 9CI — O-00078

Propanoylphosphonothioic acid. Propionylphosphonothioic acid. Propionylthiophosphonic acid

$H_3CCH_2COP(S)(OH)_2 \rightleftharpoons H_3CCH_2COP(O)(OH)(SH)$

$C_3H_7O_3PS$ M 154.120
O,O-*Di-Et ester:* O,O-*Diethyl (1-oxopropyl)-phosphonothioate.*
$C_7H_{15}O_3PS$ M 210.227
Liq. d_4^{20} 1.09. Bp_3 70-72°. n_D^{20} 1.4690.

Pudovik, A.N. *et al*, *Zh. Obshch. Khim.*, 1964, **34**, 3946 (*Engl. transl.* p. 4007)

(2-Oxopropyl)phosphonothioic acid, 9CI — O-00079

Acetonylphosphonothioic acid, 8CI

$H_3CCOCH_2P(S)(OH)_2 \rightleftharpoons H_3CCOCH_2P(O)(OH)(SH)$

$C_3H_7O_3PS_2$ M 186.180
O,O-*Di-Et ester:* [1067-72-7]. O,O-*Diethyl (2-oxopropyl)phosphonothioate.* O,O-*Diethyl acetonylphosphonothioate.*
$C_7H_{15}O_3PS$ M 210.227
Liq. d_4^{21} 1.11. Bp_3 111°. n_D^{21} 1.4782.

Sturtz, G., *Bull. Soc. Chim. Fr.*, 1964, 2340 (*synth, ir*)
Chattha, M. *et al*, *J. Org. Chem.*, 1972, **37**, 1845 (*synth, ir, pmr*)

(2-Oxopropyl)triphenylphosphonium(1+), 9CI — O-00080

Acetonyltriphenylphosphonium(1+), 8CI. Triphenylphosphonoacetone(1+)

$H_3CCOCH_2P^{\oplus}Ph_3$

$C_{21}H_{20}OP^{\oplus}$ M 319.362 (ion)
Chloride: [1235-21-8].
$C_{21}H_{20}ClOP$ M 354.815
Hygroscopic prisms ($EtOH/Et_2O$) or cryst. ($CHCl_3/Et_2O$). Mp 243-245° (235-236°). pK_a 6.6 (80% EtOH aq.), 17.8 ($MeNO_2$). Na_2CO_3 aq. yields the ylide.
▷TA1841000.
Bromide: [2236-01-3].
$C_{21}H_{20}BrOP$ M 399.266
Cryst. ($CHCl_3/Et_2O$). Mp 229°.
Iodide:
$C_{21}H_{20}IOP$ M 446.267
Cryst. (H_2O). Mp 207-209°.
Tetrafluoroborate:
$C_{21}H_{20}BF_4OP$ M 406.166
Cryst. ($CHCl_3/Et_2O$). Mp 154°.
Ylide: see 1-Triphenylphosphoranylidene-2-propanone, T-00662

Ramirez, F. *et al*, *J. Org. Chem.*, 1957, **22**, 41 (*ir, uv*)
Horner, L. *et al*, *Justus Liebigs Ann. Chem.*, 1961, **646**, 65 (*synth*)
Trippett, S. *et al*, *J. Chem. Soc.*, 1961, 1266 (*synth*)
Hendrickson, J. *et al*, Tet, 1964, **20**, 449 (*pmr*)
Albright, T.A. *et al*, *J. Am. Chem. Soc.*, 1975, **97**, 2946; 1976, **98**, 6249 (*cmr, nmr*)
Nesmeyanov, N.A. *et al*, *J. Organomet. Chem.*, 1977, **129**, 41 (*synth, tautom, pmr, nmr, ir*)
Brittain, J.M. *et al*, *Tetrahedron*, 1979, **35**, 1139 (*nmr*)

5-Oxo-1,1,3,3-tetraphenyl-1,3-diphosphorinanium(2+), 9CI — O-00081

5-Oxo-1,1,3,3-tetraphenyl-1,3-diphosphinanium(2+)

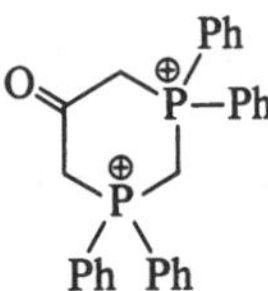

$C_{28}H_{26}OP_2^{\oplus\oplus}$ M 440.460 (ion)
All salts exist in the enol form. pK_{a1} 4.10 (EtOH), 15.05 ($MeNO_2$), pK_{a2} 9.45 (EtOH), 18.10 ($MeNO_2$). With Et_3N in benzene, the salts yield a mesomeric enol monocation.
Dichloride: [35497-20-2].
$C_{28}H_{26}Cl_2OP_2$ M 511.366
Solid. Mp 265-267°.
Dibromide: [35497-19-9].
$C_{28}H_{26}Br_2OP_2$ M 600.268
Solid. Mp 230-250°.

Mastryukova, T.A. *et al*, *Phosphorus*, 1972, **1**, 159 (*synth*)
Mastryukova, T.A. *et al*, *Zh. Obshch. Khim.*, 1972, **42**, 2620 (*Engl. transl.*, p. 2612) (*ir, P nmr*)
Mastryukova, T.A. *et al*, *Zh. Obshch. Khim.*, 1973, **43**, 2613 (*Engl. transl.* p. 2593) (*tautom*)

γ-Oxo-β-(triphenylphosphoranylidene)-benzenebutanoic acid, 9CI **O-00082**

3-Benzoyl-3-(triphenylphosphoranylidene)propanoic acid

$Ph_3P{=}C(COPh)CH_2COOH$

$C_{28}H_{23}O_3P$ M 438.462

Me ester: [17615-05-3]. *Methyl γ-oxo-β-(triphenylphosphoranylidene)benzenebutanoate, 9CI. Methyl 3-benzoyl-3-(triphenylphosphoranylidene)-propionate, 8CI. [(1-Benzoyl-2-methoxycarbonyl)-ethylidene]triphenylphosphorane.*
$C_{29}H_{25}O_3P$ M 452.488
Resonance-stabilised ylide. Cryst. (AcOH). Mp 177-178°.

Bestmann, H.J. *et al*, *Justus Liebigs Ann. Chem.*, 1967, **706**, 68 (*synth, pmr*)

β-Oxo-α-(triphenylphosphoranylidene)-benzenepropanoic acid, 9CI **O-00083**

Benzoyl(triphenylphosphoranylidene)acetic acid

$Ph_3P{=}C(COPh)COOH$

$C_{27}H_{21}O_3P$ M 424.435

Et ester: [1474-31-3]. *Ethyl β-oxo-α-(triphenylphosphoranylidene)benzenepropanoate, 9CI. (α-Ethoxycarbonylphenacylidene)-triphenylphosphorane.*
$C_{29}H_{25}O_3P$ M 452.488
Resonance-stablised ylide for Wittig reacns. Cryst. (EtOAc). Mp 142-143°. At 280°/10 mm yields ethyl 2-phenylpropiolate.

Nitrile: [5032-98-4].
Benzoyl(triphenylphosphoranylidene)acetonitrile, 8CI. α-Cyanophenacylidenetriphenylphosphorane.
$C_{27}H_{20}NOP$ M 405.435
Resonance-stabilised ylide. Prisms (EtOH). Mp 208°. Pyrolysis at 280°/10 mm yields PhC≡CCN.

Gough, S.T.D. *et al*, *J. Chem. Soc.*, 1962, 2333 (*synth, uv*)
Chopard, P.A. *et al*, *J. Org. Chem.*, 1965, **30**, 1015 (*synth, ir*)
Martin, D. *et al*, *Chem. Ber.*, 1967, **100**, 187 (*nitrile*)
Shevchuk, M.I. *et al*, *Zh. Obshch. Khim.*, 1970, **40**, 57 (*Engl. transl.* p. 54) (*uv, props*)
Brittain, J.M. *et al*, *Tetrahedron*, 1979, **35**, 1139 (*nmr*)

3-Oxo-2-(triphenylphosphoranylidene)-butanoic acid, 9CI **O-00084**

$Ph_3P{=}C(COOH)COCH_3$

$C_{22}H_{19}O_3P$ M 362.364

Me ester: [1743-62-0]. *Methyl 3-oxo-2-(triphenylphosphoranylidene)butanoate, 10CI.*
$C_{23}H_{21}O_3P$ M 376.391
Carbonyl-stabilised ylide. Cryst. Mp 153-155°. Stable under neutral or basic conds., with HCl aq. gives the phosphonium chloride.

Märkl, G., *Chem. Ber.*, 1961, **94**, 3005 (*synth, use, ir*)
Chopard, P.A. *et al*, *J. Org. Chem.*, 1965, **30**, 1015 (*synth, ir*)
Bestmann, H.J. *et al*, *Justus Liebigs Ann. Chem.*, 1977, 282 (*synth, ir, pmr*)
Doleschall, G., *Synthesis*, 1981, 478 (*use*)

4-Oxo-5-(triphenylphosphoranylidene)-pentanoic acid **O-00085**

5-(Triphenylphosphoranylidene)levulinic acid

$Ph_3P{=}CHCOCH_2CH_2COOH$

$C_{23}H_{21}O_3P$ M 376.391

Me ester: Methyl 4-oxo-5-(triphenylphosphoranylidene)pentanoate.
$C_{24}H_{23}O_3P$ M 390.418
Obt. by action of Na_2CO_3 on crude phosphonium salt. Reagent for stereoselective conversion of aldehydes into 4-ene-3-keto esters by Wittig reacn. Stable cryst. Sol. DMF.

Ronald, R.C., *J. Org. Chem.*, 1983, **48**, 138 (*synth, use*)

[2-Oxo-2-(triphenylphosphoranylidene)-propyl]phosphonic acid, 9CI **O-00086**

$Ph_3P{=}CHCOCH_2P(O)(OH)_2$

$C_{21}H_{20}O_4P_2$ M 398.334

Di-Et ester: [59925-00-7]. *Diethyl [2-oxo-2-(triphenylphosphoranylidene)propyl]phosphonate.*
$C_{25}H_{28}O_4P_2$ M 454.441
A Wittig and Wittig-Horner reagent for the synth. of unsatd. aldehydes and ketones, and capable of reacting at either reactive site. No phys. props. reported.

Hercouet, A. *et al*, *Tetrahedron Lett.*, 1976, 825.

2,2′-Oxybis[1,3,2-benzodioxaphosphole] **O-00087**

[16421-86-6]

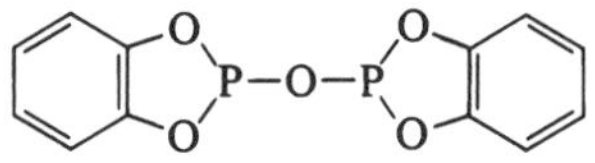

$C_{12}H_8O_5P_2$ M 294.140

Cryst. (pet. ether). Mp 72°. $Bp_{0.25}$ 156°. n_D^{20} 1.5090 (supercooled liq.).

Dioxide: [27255-30-7]. *Catechol cyclic phosphoric acid anhydride.*
$C_{12}H_8O_7P_2$ M 326.139
Cryst. (Et_2O). Mp 136-138°.

Cherbuliez, E. *et al*, *Helv. Chim. Acta*, 1951, **34**, 841 (*dioxide*)
Crofts, P.C. *et al*, *J. Chem. Soc.*, 1958, 4250 (*synth*)
Abraham, K.M. *et al*, *Inorg. Chem.*, 1975, **14**, 1099 (*synth, P nmr*)
Willson, M. *et al*, *Bull. Soc. Chim. Fr.*, 1975, 615 (*synth*)

2,2′-Oxybis[4,5-dimethyl-1,3,2-dioxaphosphole], 9CI **O-00088**

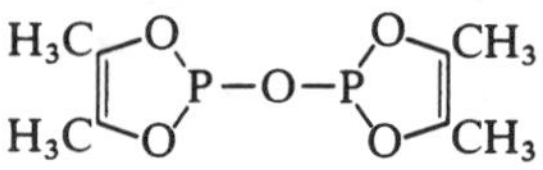

$C_8H_{12}O_5P_2$ M 250.127

Liq. Bp_1 62-64°.

2,2′-Dioxide: [55894-94-5]. *Bis(dimethylvinylene) pyrophosphate. Acetoin diol cyclic pyrophosphate. Acetoinenediol cyclic pyrophosphate.*
$C_8H_{12}O_7P_2$ M 282.126
Powerful phosphorylating agent for synth. of unsym. dialkyl phosphates, phospholipids, phospholiponucleosides. Cryst. (CH_2Cl_2/hexane). Mp 84-86°.

Ramirez, F. *et al*, *J. Am. Chem. Soc.*, 1975, **97**, 3809; 1976, **98**, 5310; 1982, **104**, 5483 (*dioxide, synth, ir, pmr, P nmr, props, use*)

Ricci, J.S. *et al*, *J. Am. Chem. Soc.*, 1975, **97**, 5457 (*dioxide, cryst struct*)
Ramirez, F. *et al*, *Tetrahedron*, 1977, **33**, 599; 1983, **39**, 2157 (*dioxide, use*)
Karlstedt, N.B. *et al*, *Zh. Obshch. Khim.*, 1982, **52**, 1974 (*Engl. transl.*, p. 1754) (*synth*)
Ramirez, F. *et al*, *J. Org. Chem.*, 1983, **48**, 2008 (*dioxide, use*)

2,2′-Oxybis[5,5-dimethyl-1,3,2-dioxaphosphorinane], 9CI **O-00089**

Bis(5,5-dimethyl-1,3,2-dioxaphosphorinan-2-yl) oxide

$C_{10}H_{20}O_5P_2$ M 282.213

2-Oxide: [16368-09-5].
$C_{10}H_{20}O_6P_2$ M 298.212
Cryst. (C_6H_6/Me_2CO). Mp 88-90°.

2,2′-Dioxide: [4090-52-2].
$C_{10}H_{20}O_7P_2$ M 314.211
Cryst. (EtOAc). Mp 193-195°.

2,2′-Disulfide: [4090-51-1]. *Sandoflam.*
$C_{10}H_{20}O_5P_2S_2$ M 346.333
Flameproofer for nylon. Needles (C_6H_6). Mp 233°.

2-Oxide, 2′-Sulfide: see 2,2′-Oxybis[5,5-dimethyl-1,3,2-dioxaphosphorinane] 2-oxide 2′-sulfide, O-00090

Edmundson, R.S., *Tetrahedron*, 1965, **21**, 2379 (*derivs, synth, ir*)
Bartle, K.D. *et al*, *Tetrahedron*, 1967, **23**, 1701 (*derivs, pmr*)
Stec, W.J. *et al*, *Can. J. Chem.*, 1967, **45**, 2513 (*derivs, synth*)
Stec, W.J. *et al*, *J. Phys. Chem.*, 1971, **75**, 3975 (*derivs, pe, P nmr*)
Simpson, P. *et al*, *J. Chem. Soc., Perkin Trans. 1*, 1975, 201 (*dioxide*)
Cook, D.S. *et al*, *J. Chem. Soc., Dalton Trans.*, 1976, 2212 (*dioxide, cryst struct*)
Bukowska-Stryzyzewska, M. *et al*, *Acta Crystallogr., Sect. B*, 1978, **34**, 1357 (*dioxide, cryst struct*)
Edmundson, R.S., *Phosphorus Sulfur*, 1981, **9**, 307 (*derivs, ms*)
Van Nuffel, P. *et al*, *J. Mol. Struct.*, 1984, **125**, 1 (*dioxide, struct*)
Cullis, P.M. *et al*, *J. Chem. Soc., Chem. Commun.*, 1985, 1329 (*dioxide, synth*)

2,2′-Oxybis[5,5-dimethyl-1,3,2-dioxaphosphorinane] 2-oxide 2′-sulfide **O-00090**

2-[(5,5-Dimethyl-1,3,2-dioxaphosphorinan-2-yl)oxy]-5,5-dimethyl-1,3,2-dioxaphosphorinane P*-oxide 2-sulfide, 10CI*

[15762-04-6]

$C_{10}H_{20}O_6P_2S$ M 330.272
Cryst. (Me_2CO/pet. ether). Mp 154-154.5°.

Bartle, K.D. *et al*, *Tetrahedron*, 1967, **23**, 1701 (*pmr*)
Edmundson, R.S., *J. Chem. Soc. (C)*, 1967, 1635 (*synth*)
Stec, W.J. *et al*, *J. Phys. Chem.*, 1971, **75**, 3975 (*pe, P nmr*)
Bukowska-Strzyzewski, M. *et al*, *Acta Crystallogr., Sect. B*, 1980, **36**, 3169 (*cryst struct*)
Reimschussel, W. *et al*, *Org. Mass Spectrom.*, 1980, **15**, 302 (*ms*)
Van Nuffel, P. *et al*, *J. Mol. Struct.*, 1984, **125**, 1 (*struct*)
Paneth, P. *et al*, *J. Am. Chem. Soc.*, 1985, **107**, 1407.
Roeske, C. *et al*, *J. Am. Chem. Soc.*, 1985, **107**, 1409 (*P nmr*)

2,2′-Oxybis[1,3,2-dioxaphospholane], 9CI **O-00091**

Ethylene glycol cyclic P,P:P′,P′*-pyrophosphite, 8CI. Ethylene pyrophosphite*

[3348-43-4]

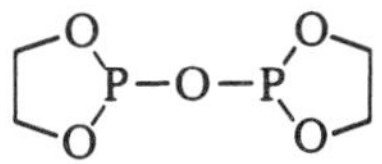

$C_4H_8O_5P_2$ M 198.052
Liq. d_4^{20} 1.43. Bp$_4$ 100-101°. n_D^{20} 1.4900.
▷Reacts violently with H_2O

2,2′-Dioxide: [22063-07-6]. *Ethylene glycol cyclic* P,P:P′,P′*-pyrophosphate. Ethylene pyrophosphate.*
$C_4H_8O_7P_2$ M 230.051
Solid. Mp 120-124°.

Arbuzov, B.A. *et al*, *Dokl. Akad. Nauk SSSR, Ser. Sci. Khim.*, 1953, **91**, 817; *CA*, **48**, 10539 (*synth*)
Houalla, D. *et al*, *Bull. Soc. Chim. Fr.*, 1965, 2368 (*synth, P nmr*)
Wilcox, R.D. *et al*, *Tetrahedron Lett.*, 1968, 6001 (*synth, dioxide*)
Munoz, A. *et al*, *C.R. Hebd. Seances Acad. Sci., Ser. C*, 1971, **273**, 152, 677 (*props*)

2,2′-Oxybis[1,3,2-dioxaphosphorinane], 9CI **O-00092**

2,2′-Oxybis[1,3,2-dioxaphosphinane]. Bis(1,3,2-dioxaphosphorinan-2-yl) oxide. Bis(1,3-dioxa-2-phosphacyclohexan-2-yl) ether

$C_6H_{12}O_5P_2$ M 226.105

2,2′-Dioxide: [21758-82-7].
$C_6H_{12}O_7P_2$ M 258.104
Cryst. (MeCN/Et_2O). Mp 137-137.5°. Reacts easily with H_2O.

Khorana, H.G. *et al*, *J. Am. Chem. Soc.*, 1957, **79**, 430 (*synth*)
Brault, J.-F. *et al*, *Bull. Soc. Chim. Fr., Part II*, 1973, 3149 (*synth*)
Ashani, Y. *et al*, *Biochem. J.*, 1979, **177**, 781 (*P nmr*)

Oxybis[2-(diphenylphosphino)ethane] **O-00093**

Oxy(di-2,1-ethanediyl)bis[diphenylphosphine]. Bis[(2-diphenylphosphino)ethyl]ether

$O(CH_2CH_2PPh_2)_2$

$C_{28}H_{28}OP_2$ M 442.476
Ligand for Ni and Co. Yellow oil.

Dioxide: [65534-50-1]. *Bis[(2-diphenylphosphinyl)-ethyl]ether. Oxybis[2-(diphenylphosphinyl)ethane].*
$C_{28}H_{28}O_3P_2$ M 474.475
Cryst. (EtOAc). Mp 81-82°.

Sacconi, L. *et al*, *Inorg. Chem.*, 1968, **7**, 291 (*synth*)
Postle, S.R. *et al*, *J. Chem. Soc., Perkin Trans. 1*, 1977, 2084 (*oxide, synth, ms, pmr, cmr*)

2,2′-Oxybis[ethanephosphonic acid] **O-00094**

Oxybis(1,2′-ethanediyl)bis(phosphonic acid), 9CI

$(HO)_2(O)PCH_2CH_2OCH_2CH_2P(O)(OH)_2$

$C_4H_{12}O_7P_2$ M 234.082

Cryst. Mp 98-105°.

Tetra-Et ester: Tetraethyl oxybis(1,2′-ethanediyl)-bisphosphonate.
$C_{12}H_{28}O_7P_2$ M 346.297
Liq. $Bp_{0.002}$ 150-161°. n_D^{20} 1.4458.

Tetraisopropyl ester: Tetraisopropyl oxybis(1,2′-ethanediyl)bisphosphonate.
$C_{16}H_{36}O_7P_2$ M 402.404
Liq. d_4^{20} 1.05. $Bp_{0.7}$ 154-157°. n_D^{20} 1.4330.

Petrov, K.A. *et al*, *Zh. Obshch. Khim.*, 1960, **30**, 1960 (*Engl. transl.* p. 1939) (*esters, synth*)
Makljaev, F.I. *et al*, *Zh. Obshch. Khim.*, 1960, **30**, 4053 (*Engl. transl.* p. 4015) (*esters, synth*)
Maier, L. *et al*, *Phosphorus Sulfur*, 1978, **5**, 45 (*synth, ester*)

Oxybis[methanephosphonic acid] O-00095

[Oxybis(methylene)]bisphosphonic acid, 9CI. *Dimethyl ether 1,1′-diphosphonic acid*
[44991-95-9]

$(HO)_2(O)PCH_2OCH_2P(O)(OH)_2$

$C_2H_8O_7P_2$ M 206.029
Viscous mass which cryst. slowly. Mp 94°.

Tetra-Et ester: [33921-08-3]. *Tetraethyl [oxybis(methylene)]bisphosphonate.*
$C_{10}H_{24}O_7P_2$ M 318.243
Liq. $Bp_{0.005}$ 143-145°. n_D^{20} 1.4450.

Tetrabutyl ester: [65824-64-8]. *Tetrabutyl [oxybis(methylene)]bisphosphonate.*
$C_{18}H_{40}O_7P_2$ M 430.457
Liq. d_4^{20} 1.06. $Bp_{0.4}$ 200-202°. n_D^{20} 1.4425.

Petrov, K.A. *et al*, *Zh. Obshch. Khim.*, 1960, **30**, 1960 (*Engl. transl.* p. 1939) (*ester*)
Bel'skii, V.E. *et al*, *Zh. Obshch. Khim.*, 1972, **42**, 2427 (*Engl. transl.* p. 2421) (*ester, P nmr*)
Petrov, K.A. *et al*, *Zh. Obshch. Khim.*, 1977, **47**, 2741 (*Engl. transl.* p. 2494) (*esters*)
Maier, L. *et al*, *Phosphorus Sulfur*, 1978, **5**, 45 (*synth, ester*)

2,2′-Oxybis[4-methyl-1,3,2-dioxaphosphorinane], 9CI O-00096

Bis(4-methyl-1,3,2-dioxaphosphorinan-2-yl) oxide. Bis(4-methyl-1,3,2-dioxaphosphorinan-2-yl) ether. Bis(1,3-butylene) pyrophosphite

$C_8H_{16}O_5P_2$ M 254.159
$Bp_{0.6}$ 102-104°. n_D^{20} 1.4560. Mixt. of diastereoisomers.

2,2′-Disulfide: see Bis(4-methyl-2-thioxo-1,3,2-dioxaphosphorinan-2-yl) oxide, B-00365

Mikolajczyk, M. *et al*, *J. Chem. Soc., Perkin Trans. 2*, 1983, 501 (*synth, ir, pmr, P nmr*)

[Oxybis[methylene]]-bis[triphenylphosphonium](2+), 9CI O-00097

Dimethyl ether bis(triphenylphosphonium)(2+)

$Ph_3P^{\oplus}CH_2OCH_2P^{\oplus}Ph_3$

$C_{38}H_{34}OP_2^{\oplus\oplus}$ M 568.634 (ion)

Dichloride: [55338-02-8].
$C_{38}H_{34}Cl_2OP_2$ M 639.540
Cryst. ($CHCl_3$). Mp 294-296°.

Dibromide: [5368-60-5].
$C_{38}H_{34}Br_2OP_2$ M 728.442
Cryst. ($CHCl_3$ or HBr aq.). Mp 290-294°.

Bis-ylide:
$C_{38}H_{32}OP_2$ M 566.618
Reactive Wittig reagent used in prepn. of oxa-annulenes. Prepd. *in situ.*

Dimroth, K. *et al*, *Chem. Ber.*, 1966, **99**, 634, 642 (*bromide, use*)
Ogawa, H. *et al*, *Tetrahedron Lett.*, 1972, 4129 (*ylide, use*)
Wife, R.L. *et al*, *J. Am. Chem. Soc.*, 1975, **97**, 640 (*use*)
Ojima, J. *et al*, *Bull. Chem. Soc. Jpn.*, 1976, **49**, 3709 (*synth, ylide, use*)

Oxydemeton-methyl, BSI, ISO O-00098

S-*[2-(Ethylsulfinyl)ethyl]* O,O-*dimethyl phosphorothioate*, 9CI, 8CI. S-*2-Ethylsulfinylethyl* O,O-*dimethyl phosphorothioate. Metasystox*-R
[301-12-2]

$(MeO)_2P(O)SCH_2CH_2S(O)Et$

$C_6H_{15}O_4PS_2$ M 246.276
Systemic and contact insecticide. Yellow liq. Sol. H_2O, most org. solvs., insol. pet. ether. d_4^{20} 1.289. Mp <−10°. $Bp_{0.01}$ 106°. n_D^{20} 1.5216.

▷TG1420000.

Schrader, G., *Die Entwichlung neuerinsektizider Phosphorsäure-Ester*, 1963, Verlag Chemie, Weinheim, 416.
Daldrup, T. *et al*, *Fresenius' Z. Anal. Chem.*, 1981, **308**, 413 (*tlc, glc, hplc*)
Pesticide Manual, 6th Ed., 398; 7th Ed., 416.
The Agrochemicals Handbook, Royal Society of Chemistry, 1983, A309.

Oxydeprofos, BSI O-00099

S-*[2-(Ethylsulfinyl)-1-methylethyl]* O,O-*dimethyl phosphorothioate*, 9CI, 8CI. *Metasystox* S. *Estox*
[2674-91-1]

$(MeO)_2P(O)SCH(CH_3)CH_2SOEt$

$C_7H_{17}O_4PS_2$ M 260.303

▷TG1575000.

(±)-form

Agricultural insecticide and acaricide. Yellow oil. Sol. H_2O. d_4^{20} 1.257. $Bp_{0.02}$ 115°. n_D^{25} 1.5149.

B.P., 823 732, (*1955*); *CA*, **54**, 10860d (*synth*)
Keith, L.H. *et al*, *J. Assoc. Off. Anal. Chem.*, 1968, **51**, 1063 (*pmr*)
Nagasawa, K. *et al*, *J. Chromatogr.*, 1969, **39**, 282.
Pesticide Manual, 6th Ed., 253; 7th Ed., 417.

P

Paraoxon **P-00001**

O,O-*Diethyl* O-(*4-nitrophenyl*) *phosphate*, 9CI. *Eticol. Phosphacol. Mintacol. Miotisal* A. *Solugliacit*

[311-45-5]

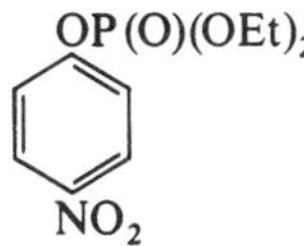

$C_{10}H_{14}NO_6P$ M 275.197

Cholinergic, miotic, insecticide, metab. and product of photodegradative oxidn. of Parathion. Reddish-yellow oily liq. with slight odour. Spar. sol. H_2O. d_4^{20} 1.274. $Bp_{0.5}$ 160-162°.

▷TC2275000.

Fr. Pat., 1 372 446, (*1964*); *CA*, **61**, 16016 (*synth*)
v. Hooidonk, C. *et al*, *Recl. Trav. Chim. Pays-Bas*, 1967, **86**, 449 (*props*)
Keith, L.H. *et al*, *J. Assoc. Off. Anal. Chem.*, 1968, **51**, 1063 (*pmr*)
Boyd, G.R., *J. Agric. Food Chem.*, 1970, **18**, 742 (*synth*)
Ross, R.T. *et al*, *Anal. Chim. Acta*, 1970, **52**, 139 (*P nmr*)
Ashani, Y. *et al*, *J. Med. Chem.*, 1973, **16**, 446 (*props*)
Holmstead, R.L. *et al*, *J. Assoc. Off. Anal. Chem.*, 1974, **57**, 1050 (*ms*)
Kráčmar, J. *et al*, *Pharmazie*, 1976, **31**, 614 (*uv*)
Horner, L. *et al*, *Justus Liebigs Ann. Chem.*, 1977, 61 (*props*)
Stan, H.-J. *et al*, *Fresenius' Z. Anal. Chem.*, 1977, **287**, 271 (*glc, ms*)
Ripley, B.D. *et al*, *J. Assoc. Off. Anal. Chem.*, 1983, **66**, 1084 (*glc*)
Sax, N.I., *Dangerous Properties of Industrial Materials*, 6th Ed., Van Nostrand-Reinhold, 1984, 580.

Parathion, BSI **P-00002**

O,O-*Diethyl* O-(*4-nitrophenyl*) *phosphorothioate*, 9CI. *Diethyl 4-nitrophenyl thiophosphate. Numerous other synonyms*

[56-38-2]

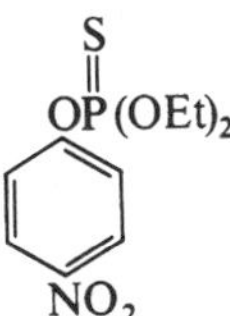

$C_{10}H_{14}NO_5PS$ M 291.258

Nonsystemic agricultural insecticide and acaricide with contact and stomach action. Pale-yellow oil. Insol. H_2O, pet. ether. d_4^{25} 1.26. Mp 6°. Bp 375°, $Bp_{0.6}$ 157-162°. Hydrol. v. slowly under natural conditions. Rapidly hydrolysed under alkaline conditions. n_D^{20} 1.5420. Isom. when heated.

▷Highly toxic, cumulative poison. Absorbed through skin. TF4920000.

Fletcher, J.H. *et al*, *J. Am. Chem. Soc.*, 1948, **70**, 3943 (*synth, props*)
Mandel'baum, Ya.A. *et al*, *Zh. Obshch. Khim.*, 1960, **30**, 194 (*Engl. transl.* p. 207) (*synth*)
Keith, L.H. *et al*, *J. Assoc. Off. Anal. Chem.*, 1968, **51**, 1063 (*pmr*)
Getz, M.E. *et al*, *J. Assoc. Off. Anal. Chem.*, 1968, **51**, 1101 (*tlc*)
Ross, R.T. *et al*, *Anal. Chim. Acta*, 1970, **52**, 139 (*P nmr*)
Gore, R.G. *et al*, *J. Assoc. Off. Anal. Chem.*, 1971, **54**, 1040 (*ir, uv*)
Holmstead, R.L. *et al*, *J. Assoc. Off. Anal. Chem.*, 1974, **57**, 1050 (*ms*)
Nicholas, M.L. *et al*, *J. Assoc. Off. Anal. Chem.*, 1976, **59**, 1071 (*raman*)
Adhya, T.K. *et al*, *Pestic. Biochem. Physiol.*, 1981, **16**, 14 (*metab*)
Mulla, M.S. *et al*, *Residue Rev.*, 1981, **81**, 1 (*metab, rev*)
Stan, H.-J. *et al*, *Biomed. Mass Spectrom.*, 1982, **9**, 483 (*ms*)
Stan, H.-J. *et al*, *J. Chromatogr.*, 1983, **279**, 173 (*glc*)
Rosenberg, A. *et al*, *J. Chromatogr.*, 1984, **294**, 436 (*metab, hplc*)
Pesticide Manual, 6th Ed., 401; 7th Ed., 419.
The Agrochemicals Handbook, The Royal Society of Chemistry, London, 1983, A311.
Sax, N.I., *Dangerous Properties of Industrial Materials*, 6th Ed., Van Nostrand-Reinhold, 1984, 886.

Parathion methyl **P-00003**

OO-*Dimethyl* O-(*4-nitrophenyl*) *phosphorothioate*. *1-[(Dimethoxyphosphinothioyl)oxy]-4-nitrobenzene. Methyl parathion*

[298-00-0]

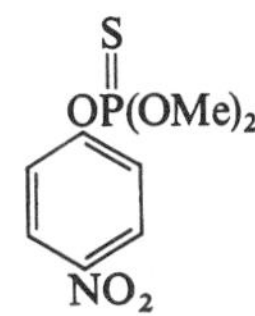

$C_8H_{10}NO_5PS$ M 263.204

Insecticide. Solid. Mp 35-36°. Bp_2 158°.

▷TG0246000.

Mandel'baum, Ya.A. *et al*, *Zh. Obshch. Khim.*, 1959, **29**, 1149 (*Engl. transl.* p. 1121) (*synth*)
Getz, M.E. *et al*, *J. Assoc. Off. Anal. Chem.*, 1968, **51**, 1101 (*tlc*)
Keith, L.H. *et al*, *J. Assoc. Off. Anal. Chem.*, 1968, **51**, 1063 (*pmr*)
Ross, R.T. *et al*, *Anal. Chim. Acta*, 1970, **52**, 139 (*P nmr*)
Holmstead, R.L. *et al*, *J. Assoc. Off. Anal. Chem.*, 1974, **57**, 1050 (*ms*)
Nicholas, M.L. *et al*, *J. Assoc. Off. Anal. Chem.*, 1976, **59**, 1071 (*Raman*)
Szalontai, G., *J. Chromatogr.*, 1976, **124**, 9 (*hplc*)
Stan, H.-J. *et al Fresenius' Z. Anal. Chem.*, 1977, **287**, 271 (*glc, ms*)
Rainsford, K.D., *Pestic. Biochem. Physiol.*, 1978, **8**, 302 (*tlc, tox*)
Adhya, T.K. *et al*, *Pestic. Biochem. Physiol.*, 1981, **16**, 14 (*metab*)
Ripley, B.D. *et al*, *J. Assoc. Off. Anal. Chem.*, 1983, **66**, 1084 (*glc*)
Pesticide Manual, 7th Ed., 420.
Sax, N.I., *Dangerous Properties of Industrial Materials*, 6th Ed., Van Nostrand-Reinhold, 1984, 610.

Pentabutylphosphorodiamidimidic acid, 9CI **P-00004**

Pentabutylphosphoric triamide

$$H_3C(CH_2)_3N{=}P(OH)[N(CH_2CH_2CH_2CH_3)_2]_2 \rightleftharpoons H_3C(CH_2)NHP(O)[N(CH_2CH_2CH_2CH_3)_2]_2$$

$C_{20}H_{46}N_3OP$ M 375.577

Free acid is tautomeric.

Fluoride: [86601-01-6]. N-*Butyl*-P,P-*bis*(*dibutylamino*)-P-*fluorophosphazene*. N-*Butyl*-P,P-*bis*(*dibutylamino*)-P-*fluorophosphine imide*.
$C_{20}H_{45}FN_3P$ M 377.568

Liq. $Bp_{0.01}$ 99-100°. n_D^{20} 1.4507.
Chloride: [66911-82-8]. N-*Butyl*-P-*chloro*-P,P-*bis(dibutylamino)phosphazene*. N-*Butyl*-P-*chloro*-P,P-*bis(dibutylamino)phosphine imide.*
$C_{20}H_{45}ClN_3P$ M 394.022
Liq. d_4^{25} 0.94. $Bp_{0.015}$ 114-115°. n_D^{25} 1.4710.
Chloride; B,HCl: [84591-39-9]. *(Butylamino)-chlorobis(dibutylamino)phosphonium chloride. Pentabutylchlorotriaminophosphonium chloride.*
$C_{20}H_{46}Cl_2N_3P$ M 430.483
Liq. $Bp_{0.05}$ 148-149°.
Bromide: [81675-82-3]. P-*Bromo*-N-*butyl*-P,P-*bis(dibutylamino)phosphazene*. P-*Bromo*-N-*butyl*-P,P-*bis(dibutylamino)phosphine imide.*
$C_{20}H_{45}BrN_3P$ M 438.473
Liq. $Bp_{0.02}$ 128-130°. n_D^{20} 1.4852.
Bromide; B,HBr: [84591-36-6]. *Bromo(butylamino)-bis(dibutylamino)phosphonium bromide. Bromopentabutyltriaminophosphonium bromide.*
$C_{20}H_{46}Br_2N_3P$ M 519.385
Liq. $Bp_{0.05}$ 200-202°.

Marchenko, A.P. *et al*, *Zh. Obshch. Khim.*, 1978, **48**, 551 (*Engl. transl.* p. 501) (*chloride*)
Egorov, Yu.P. *et al*, *Teor. Eksp. Khim.*, 1982, **18**, 58 (*Engl. transl.* p. 45) (*P nmr, struct*)
Miroshnichenko, V.V. *et al*, *Zh. Obshch. Khim.*, 1982, **52**, 2517 (*Engl. transl.* p. 2222) (*ir, P nmr, bromide*)
Marchenko, A.P. *et al*, *Zh. Obshch. Khim.*, 1983, **53**, 698 (*fluoride*)

(Pentachloroethyl)phosphorimidic trichloride, 9CI P-00005

P,P,P-*Trichloro*-N-*(pentachloroethyl)phosphine imide.*
P,P,P-*Trichloro*-N-*(pentachloroethyl)phosphazene.*
P,P,P-*Trichloro*-N-*(pentachloroethyl)-iminophosphorane*
[1067-01-2]

$Cl_3CCCl_2N{=}PCl_3$

C_2Cl_8NP M 352.626
Solid or liq. Mp 21-23°. Bp_5 117-118°.

Shevchenko, V.I. *et al*, *Zh. Obshch. Khim.*, 1963, **33**, 1342, 1591; 1966, **36**, 1645 (*Engl. transl.* pp. 1312, 1553) (*synth, ir*)
Tarasevich, A.S. *et al*, *Teor. Eksp. Khim.*, 1971, **7**, 828 (*Engl. transl.* p. 676) (*P nmr, struct*)
Fluck, E. *et al*, *Z. Anorg. Allg. Chem.*, 1972, **387**, 349 (*P nmr*)
Kyuntzel, I.A. *et al*, *J. Magn. Reson.*, 1975, **20**, 394 (*nqr*)

Pentachlorophenyl phosphorodichloridate, 9CI, 8CI P-00006

Pentachlorophenyl phosphoryl dichloride. Pentachlorophenyl dichlorophosphate
[17725-01-8]

$C_6Cl_5OP(O)Cl_2$

$C_6Cl_7O_2P$ M 383.210
Cryst. (heptane). Mp 98° (91-92°).

Cremlyn, R.J.W. *et al*, *J. Chem. Soc., Perkin Trans. 1*, 1972, 583 (*synth*)
Lazarus, R.A. *et al*, *J. Chem. Soc., Perkin Trans. 2*, 1980, 373 (*synth*)

O-(Pentachlorophenyl) phosphorodichloridothioate, 9CI, 8CI P-00007

O-*(Pentachlorophenyl) dichlorothiophosphate*
[16167-31-0]

$(C_6Cl_5)OP(S)Cl_2$

C_6Cl_7OPS M 399.270
Cryst. (pet. ether). Mp 99-101°.

Yarmukhametova, D.Kh. *et al*, *Izv. Akad. Nauk SSSR, Ser. Khim.*, 1967, 602 (*Engl. transl.* p. 579)

S-(Pentachlorophenyl) phosphorodichloridothioate, 9CI, 8CI P-00008

S-*(Pentachlorophenyl) dichlorothiophosphate*

$C_6Cl_5SP(O)Cl_2$

C_6Cl_7OPS M 399.270
Cryst. (pet. ether). Mp 98°.

Ger. Pat., 1 159 935, (*1963*); *CA*, **60**, 11945

1,3-Pentadienylphosphonic acid, 9CI P-00009

1-Phosphono-1,3-pentadiene

$H_3CCH{=}CHCH{=}CHP(O)(OH)_2$

$C_5H_9O_3P$ M 148.098
Di-Me ester: Dimethyl 1,3-pentadienylphosphonate.
$C_7H_{13}O_3P$ M 176.152
Liq. d_4^{20} 1.07. Bp_2 101-103°. n_D^{20} 1.4932.
Difluoride:
$C_5H_7F_2OP$ M 152.080
Liq. d_4^{20} 1.15. Bp_7 60-62°. n_D^{20} 1.4532.
Dichloride: [4981-25-3].
$C_5H_7Cl_2OP$ M 184.989
Solid. Mp 46-48°. $Bp_{3\text{-}3.5}$ 97-100°.

(E,E)-form

Di-Et ester: [67221-20-9]. *Diethyl 1,3-pentadienylphosphonate.*
$C_9H_{17}O_3P$ M 204.205
Liq. $Bp_{0.01}$ 82-84°. $n_D^{18.5}$ 1.4812.

Mashlyakovskii, L.N. *et al*, *Zh. Obshch. Khim.*, 1965, **35**, 1577 (*Engl. transl.* p. 1582) (*dichloride, ester, synth, ir, pmr*)
Sturtz, G. *et al*, *J. Chem. Res. S.*, 1978, 89 (*ester, ir, pmr*)
Fokin, A.V. *et al*, *Dokl. Akad. Nauk SSSR, Ser. Sci. Khim.*, 1978, **240**, 1131 (*Engl. transl.* p. 292) (*dichloride*)

1,3-Pentadiyn-1-ylphosphonic acid, 9CI P-00010

1-Phosphono-1,3-pentadiyne

$H_3CC{\equiv}CC{\equiv}CP(O)(OH)_2$

$C_5H_5O_3P$ M 144.066
Di-Et ester: Diethyl 1,3-pentadiyn-1-ylphosphonate.
$C_9H_{13}O_3P$ M 200.174
Liq. d_4^{20} 1.08. $Bp_{1.5}$ 134.5-6.5°. n_D^{20} 1.4930.

Ionin, B.I. *et al*, *Dokl. Akad. Nauk SSSR, Ser. Sci. Khim.*, 1963, **152**, 1354 (*Engl. transl.* p. 831) (*synth, ir, pmr, P nmr*)
Turbanova, E.S. *et al*, *Zh. Org. Khim.*, 1982, **18**, 1137 (*Engl. transl.* p. 982) (*struct*)

Pentaethoxyphosphorane, 9CI, 8CI P-00011

[7735-87-7]

$(EtO)_5P$

$C_{10}H_{25}O_5P$ M 256.278

Powerful ethylating agent, reacting with carboxylic acids, phenols, and enols. Liq. Rapidly hydrolysed under neutral conditions.

Denney, D.B. *et al*, *J. Am. Chem. Soc.*, 1969, **91**, 5821 (*nmr*)
Chang, L.L. *et al*, *J. Am. Chem. Soc.*, 1977, **99**, 2293 (*synth, pmr, cmr, nmr*)
Guthrie, J.P., *J. Am. Chem. Soc.*, 1977, **99**, 3991 (*synth, props*)
Fieser, M. *et al*, *Reagents for Organic Synthesis*, Wiley, 1967-84, **2**, 305 (*use*)

Pentaethylphosphorodiamidimidic acid, 9CI — P-00012

EtN=P(OH)(NEt$_2$)$_2$

$C_{10}H_{26}N_3OP$ M 235.309

Free acid tautomeric with pentasilylphosphoric triamide.

Fluoride: [86600-99-9]. P,P-*Bis*(*diethylamino*)-N-*ethyl*-P-*fluorophosphine imide*. P,P-*Bis*(*diethylamino*)-N-*ethyl*-P-*fluorophosphazene*.
$C_{10}H_{25}FN_3P$ M 237.300
Liq. Bp_{14} 104-105°. n_D^{20} 1.4440.

Chloride: [66911-81-7]. P-*Chloro*-P,P-*bis*(*diethylamino*)-N-*ethylphosphine imide*. P-*Chloro*-P,P-*bis*-(*diethylamino*)-N-*ethylphosphazene*.
$C_{10}H_{25}ClN_3P$ M 253.754
Liq. d_4^{25} 1.00. $Bp_{0.015}$ 55-56°. n_D^{25} 1.4783.

Chloride; B,HCl: [84591-37-7]. *Chlorobis*(*diethylamino*)(*ethylamino*)*phosphonium chloride*. *Chloropentaethyltriaminophosphonium chloride*.
$C_{10}H_{26}Cl_2N_3P$ M 290.215
Liq. $Bp_{0.03}$ 125-127°.

Bromide: [73954-60-6]. P-*Bromo*-P,P-*bis*(*diethylamino*)-N-*ethylphosphine imide*. P-*Bromo*-P,P-*bis*-(*diethylamino*)-N-*ethylphosphazene*.
$C_{10}H_{25}BrN_3P$ M 298.205
Liq. $Bp_{0.04}$ 75-76°. n_D^{25} 1.4982.

Bromide; B,HBr: [84591-34-4]. *Bromobis*(*diethylamino*)(*ethylamino*)*phosphonium bromide*. *Bromopentaethyltriaminophosphonium bromide*.
$C_{10}H_{26}Br_2N_3P$ M 379.117
Cryst. (THF). Mp 78-82°.

Marchenko, A.P. *et al*, *Zh. Obshch. Khim.*, 1978, **48**, 551 (*Engl. transl.* p. 501) (*chloride*)
Marchenko, A.P. *et al*, *Zh. Obshch. Khim.*, 1980, **50**, 951 (*bromide, synth, ir, P nmr*)
Egorov, Yu.P. *et al*, *Teor. Eksp. Khim.*, 1982, **18**, 58 (*Engl. transl.* p. 45) (*P nmr, struct*)
Miroshnichenko, V.V. *et al*, *Zh. Obshch. Khim.*, 1982, **52**, 2517 (*Engl. transl.* p. 2222) (*salts, P nmr*)
Marchenko, A.P. *et al*, *Zh. Obshch. Khim.*, 1983, **53**, 698 (*Engl. transl.* p. 608) (*fluoride, synth, ir, P and F nmr*)

(Pentafluoroethyl)phosphorimidic trichloride, 9CI — P-00013

P,P,P-*Trichloro*-N-(*pentafluoroethyl*)*phosphine imide*.
P,P,P-*Trichloro*-N-(*pentafluoroethyl*)*phosphazene*.
P,P,P-*Trichloro*-N-(*pentafluoroethyl*)-*iminophosphorane*
[80248-48-2]

$F_3CCF_2N{=}PCl_3$

$C_2Cl_3F_5NP$ M 270.353

Liq. Hydrol. v. rapidly. Dec. >100°.

Leidinger, W. *et al*, *J. Fluorine Chem.*, 1981, **19**, 85 (*synth, ir, ms, F and P nmr*)

(Pentafluorophenyl)diphenylphosphine, 9CI — P-00014

[5525-95-1]

$Ph_2PC_6F_5$

$C_{18}H_{10}F_5P$ M 352.243

Air-stable tan solid. Mp 68-70°.

Oxide: [5074-74-8].
$C_{18}H_{10}F_5OP$ M 368.242
Cryst. (EtOH). Mp 122-123°.

Sulfide:
$C_{18}H_{10}F_5PS$ M 384.303
Cryst. (C_6H_6). Mp 132°.

Fild, M. *et al*, *Z. Naturforsch.*, *B*, 1967, **22**, 253 (*nmr*)
Hogben, M.G. *et al*, *J. Am. Chem. Soc.*, 1966, **88**, 3457; 1969, **91**, 283 (*nmr*)
Fild, M., *Z. Anorg. Allg. Chem.*, 1968, **358**, 257 (*sulfide, ir, nmr*)
Kemmitt, R.D. *et al*, *J. Chem. Soc.* (*A*), 1968, 2149 (*synth, complexes*)
Burdon, J. *et al*, *J. Chem. Soc.* (*C*), 1969, 2615 (*synth, oxide, sulfide*)
Naae, D.G. *et al*, *J. Fluorine Chem.*, 1979, **13**, 473 (*oxide, cryst struct*)

Pentafluorophenylphosphonic acid, 8CI — P-00015

OH
O=P—OH
F F
F F
F

$C_6H_2F_5O_3P$ M 248.046

Solid. Mp 141-142°.

Di-Me ester: Dimethyl (*pentafluorophenyl*)*phosphonate*.
$C_8H_{10}F_5O_3P$ M 280.131
Liq. Bp_{10} 46-47°. n_D^{23} 1.4443.

Di-Et ester: Diethyl (*pentafluorophenyl*)*phosphonate*.
$C_{10}H_{10}F_5O_3P$ M 304.153
Liq. Bp_8 146-156°.

Difluoride: [22474-68-6].
C_6F_7OP M 252.028
Liq. Bp_{14} 73°.

Fild, M. *et al*, *J. Chem. Soc.* (*A*), 1969, 840 (*fluoride, F and P nmr*)
Furin, G.G. *et al*, *Zh. Obshch. Khim.*, 1975, **45**, 1473 (*Engl. transl.* p. 1441) (*synth, uv, F and P nmr*)
Fild, M. *et al*, *Z. Anorg. Allg. Chem.*, 1978, **439**, 145 (*fluoride, synth*)
Burton, D.J. *et al*, *Synthesis*, 1979, 615 (*ester, synth, ir, F and P nmr, pmr*)

(Pentafluorophenyl)phosphorimidic acid, 9CI — P-00016

(*Pentafluorophenyl*)*phosphoramidic acid*

$(C_6F_5)N{=}P(OH)_3$

$C_6H_3F_5NO_3P$ M 263.060

Free acid is tautomeric.

Tri-Et ester: [88515-16-6]. *Triethyl pentafluorophenylphosphorimidate*. P,P,P-*Triethoxy*-N-*pentafluorophenylphosphazine*. P,P,P-*Triethoxy*-N-(*pentafluorophenyl*)*phosphine imide*.
$C_{12}H_{15}F_5NO_3P$ M 347.221
Liq. $Bp_{0.1}$ 104-106°.

Trichloride: [69162-08-9]. P,P,P-*Trichloro*-N-*(pentafluorophenyl)phosphazene.* P,P,P-*Trichloro*-N-*(pentafluorophenyl)phosphine imide.*
$C_6Cl_3F_5NP$ M 318.397
Liq. which cryst. at r.t. Mp 101.5-103°. $Bp_{0.1}$ 73-97°. Dimeric in solid state: monomeric in non-polar solvents.

Petrova, T.-O. *et al*, *Izv. Akad. Nauk SSSR, Ser. Khim.*, 1978, 2635 (*Engl. transl.* p. 2358) (*trichloride, ir, P and F nmr, props*)
Shermolovich, Yu.G. *et al*, *Zh. Obshch. Khim.*, 1983, **53**, 2150 (*Engl. transl.* p. 1940) (*ester, chloride, synth, P nmr*)

Pentafluorophenyl phosphorodichloridate, 9CI — P-00017

Pentafluorophenyl phosphoryl dichloride. Pentafluorophenyl dichlorophosphate
[17788-07-7]

$C_6F_5OP(O)Cl_2$

$C_6Cl_2F_5O_2P$ M 300.937
Liq. Bp_{20} 112-125°.

U.S.P., 3 341 630, (*1967*); *CA*, **68**, 77961

Pentamethoxyphosphorane, 9CI — P-00018

[1455-07-8]

$(MeO)_5P$

$C_5H_{15}O_5P$ M 186.144
Acts as *O*-, *N*-, and *S*-methylating agent for carboxylic acids, phenols, thiophenols, and amides. Liq. V. rapidly hydrolyzed under neutral conditions.

Chang, L.L. *et al*, *J. Am. Chem. Soc.*, 1977, **99**, 2293 (*pmr, cmr, nmr*)
Denney, D.B. *et al*, *J. Org. Chem.*, 1978, **43**, 4672 (*synth, use*)

Pentamethylpentaphospholane, 9CI — P-00019

Pentamethylcyclopentaphosphine
[1073-98-9]

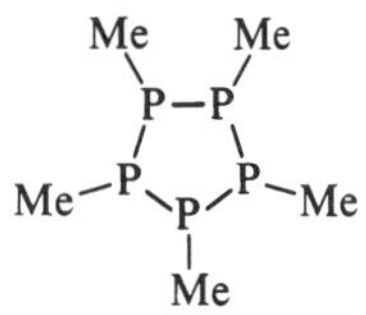

$C_5H_{15}P_5$ M 230.042
Bp_3 135-136°, Bp_1 110°.

Henderson, W.A. *et al*, *J. Am. Chem. Soc.*, 1963, **85**, 2462 (*synth, uv, P nmr*)
Baudler, M. *et al*, *Z. Naturforsch., B*, 1965, **20**, 810 (*ir, raman*)
Cowley, A.H. *et al*, *Inorg. Chem.*, 1966, **5**, 1459 (*ms*)
Issleib, K. *et al*, *Z. Anorg. Allg. Chem.*, 1968, **360**, 77 (*synth, pmr*)
Smith, L.R. *et al*, *J. Am. Chem. Soc.*, 1976, **98**, 3852 (*synth, cmr, P nmr, pmr*)

2,2,3,4,4-Pentamethyl-1-phenylphosphetane, 9CI, 8CI — P-00020

[23041-38-5]

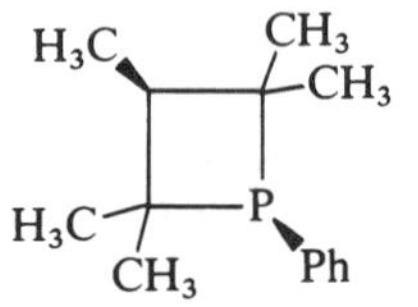

cis-form

$C_{14}H_{21}P$ M 220.294

cis-form [22434-51-1]
Known only admixed with *trans*-form.
Oxide: [16083-91-3].
$C_{14}H_{21}OP$ M 236.293
Cryst. Mp 117-118°.
Sulfide: [30664-59-6].
$C_{14}H_{21}PS$ M 252.354
Cryst. (pet. ether). Mp 92-93°.

trans-form [16083-95-7]
Cryst. Mp ca. 50°. $Bp_{1.5}$ 111°.
B,MeBr: 1,2,2,3,4,4-Hexamethyl-1-phenylphosphetanium bromide.
$C_{15}H_{24}BrP$ M 315.232
Solid. Mp 226-228°.
B,PhCH₂Br: 1-Benzyl-2,2,3,4,4-pentamethyl-1-phenylphosphetanium bromide.
$C_{21}H_{28}BrP$ M 391.330
Cryst. ($CHCl_3$/EtOAc). Mp 220-221°.
Oxide: [20047-46-5]. Solid. Mp 126-127°.
Sulfide: [30664-60-9]. Cryst. (pet. ether). Mp 102-103°.

Cremer, S. *et al*, *J. Org. Chem.*, 1967, **32**, 4066 (*synth, derivs, pmr*)
Hawes, W. *et al*, *J. Chem. Soc. (C)*, 1969, 1465 (*synth, derivs*)
Moret, C., *J. Am. Chem. Soc.*, 1969, **91**, 2235 (*cryst struct*)
Mazhar-ul-Haque, *J. Chem. Soc. (B)*, 1970, 938; 1971, 117 (*oxides, cryst struct*)
Trippett, S. *et al*, *J. Chem. Soc. (C)*, 1971, 334 (*sulfides, ir, pmr, ms*)
Grey, G. *et al*, *J. Org. Chem.*, 1972, **37**, 3470, 3458 (*cmr*)
Grey, G. *et al*, *J. Am. Chem. Soc.*, 1976, **98**, 2109 (*cmr*)

2,2,3,4,4-Pentamethylphosphetanamine 1-oxide, 9CI — P-00021

1-Amino-2,2,3,4,4-Pentamethylphosphetane 1-oxide

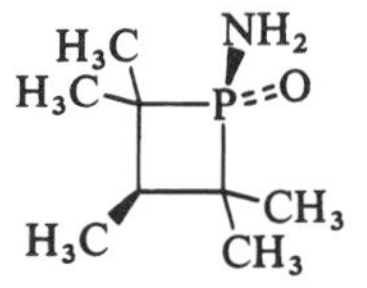

cis-form

$C_8H_{18}NOP$ M 175.210

cis-form
N-*Benzyl:* N-*Benzyl*-P,P-*(1,1,2,3,3-pentamethyltrimethylene)phosphinic amide.*
$C_{15}H_{24}NOP$ M 265.334
Cryst. (CH_2Cl_2/pet. ether). Mp 146°.

trans-form
Cryst. (C_6H_6). Mp 162-163°.
N-*Me:* N,*2,2,3,4,4-Hexamethylphosphetanamine 1-oxide.* N-*Methyl*-P,P-*(1,1,2,3,3-pentamethyltrimethylene)phosphinic amide.*
$C_9H_{20}NOP$ M 189.237
Cryst. (C_6H_6). Mp 146-148°.
N,N-*Di-Me:* N,N,*2,2,3,4,4-Heptamethylphosphetanamine 1-oxide.* N,N-*Dimethyl*-P,P-*(1,1,2,3,3-pentamethyltrimethylene)phosphinic amide.*
$C_{10}H_{22}NOP$ M 203.264
Solid. Mp 94-95°.
N-*Ph:* N-*Phenyl-2,2,3,4,4-pentamethylphosphetanamine 1-oxide.* N-*Phenyl*-P,P-*(1,1,2,3,3-pentamethyltrimethylene)phosphinic amide.*
$C_{14}H_{22}NOP$ M 251.308
Oil.
N-*Benzyl:* Solid. Mp 160-161°.

Corfield, J.R. *et al*, *J. Chem. Soc., Perkin Trans. 1*, 1972, 713 (*synth, pmr*)

DeBruin, K.E. *et al*, *J. Am. Chem. Soc.*, 1973, **95**, 4681 (*synth, pmr*)
Emsley, J. *et al*, *J. Chem. Soc., Dalton Trans.*, 1973, 1576 (*synth, ir, pmr*)
Harger, M.J.P., *J. Chem. Soc., Perkin Trans. 1*, 1974, 2604; 1980, 705 (*synth, ir, pmr, P nmr*)

2,2,3,4,4-Pentamethylphosphetane, 9CI, 8CI — P-00022

[36044-89-0]

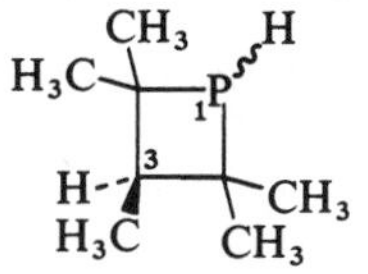

$C_8H_{17}P$ M 144.196
Bp_{80} 80-81°, Bp_6 32°.

1-Chloro: see 1-Chloro-2,2,3,4,4-pentamethylphosphetane, C-00124
1-Me: see 1,2,2,3,4,4-Hexamethylphosphetane, H-00074
1-Hydroxy, 1-oxide: see 1-Hydroxy-2,2,3,4,4-pentamethylphosphetane 1-oxide, H-00168
1-Hydroxy, 1-sulfide: see 1-Hydroxy-2,2,3,4,4-pentamethylphosphetane 1-sulfide, H-00169
1-Amino, 1-oxide: see 2,2,3,4,4-Pentamethylphosphetanamine 1-oxide, P-00021

Corfield, J.R. *et al*, *J. Chem. Soc., Perkin Trans. 1*, 1972, 713 (*synth, ir, P nmr, pmr*)

N,N,N′,N′,P-Pentamethylphosphonimidic diamide, 9CI — P-00023

Methylbis(dimethylamino)iminophosphorane. P,P-*Bis(dimethylamino)*-P-*methylphosphine imide*. P,P-*Bis(dimethylamino)*-P-*methylphosphazene*
[49778-02-1]

HN=PMe$(NMe_2)_2$

$C_5H_{16}N_3P$ M 149.175
Liq. Bp_3 63-64°.

N″-*Me:* [49778-05-4]. *Hexamethylphosphonimidic diamide*. N,P-*Dimethyl*-P,P-*bis(dimethylamino)*-*phosphazene*.
$C_6H_{18}N_3P$ M 163.202
Liq. Bp_3 63-64°.

Issleib, K. *et al*, *Synth. Inorg. Met.-Org. Chem.*, 1973, **3**, 255 (*synth, pmr, P nmr*)

Pentamethylphosphonoselenoic diamide, 9CI — P-00024

[35525-41-8]

MeP(Se)$(NMe_2)_2$

$C_5H_{15}N_2PSe$ M 213.121
No phys. props. reported.

Osokin, D.Ya. *et al*, *Org. Magn. Reson.*, 1972, **4**, 831 (*nqr*)
McFarlane, W. *et al*, *J. Chem. Soc., Dalton Trans.*, 1973, **20**, 2162 (*pmr, P and Se nmr*)

Pentamethylphosphonotelluroic diamide, 9CI — P-00025

[51461-98-4]

MeP(Te)$(NMe_2)_2$

$C_5H_{15}N_2PTe$ M 261.761

Nunetdinova, I.A. *et al*, *Izv. Akad. Nauk SSSR, Ser. Khim.*, 1973, **22**, 2827 (*Engl. transl.* p. 2765) (*nmr*)

Pentamethylphosphoric triamide, 9CI, 8CI — P-00026

[10159-46-3]

$(Me_2N)_2$P(O)NHMe

$C_5H_{16}N_3OP$ M 165.175
Phase transfer catalyst. Anion employed in synth. of aldehydes from allylamines. Liq. $Bp_{0.05}$ 114°. n_D^{23} 1.4622.
▷TD0897000.

Arceneaux, R.L. *et al*, *J. Org. Chem.*, 1959, **24**, 1419 (*synth*)
Baldwin, M.A. *et al*, *Org. Mass Spectrom.*, 1977, **12**, 279 (*ms*)
Coutrot, P. *et al*, *J. Chem. Res. (S)*, 1977, 308 (*use*)
Corbel, B. *et al*, *Can. J. Chem.*, 1980, **58**, 2183 (*synth*)

2′,2′,3′,4′,4′-Pentamethylspiro[1,3,2-benzodioxaphosphole-2,1′-phosphetane], 9CI — P-00027

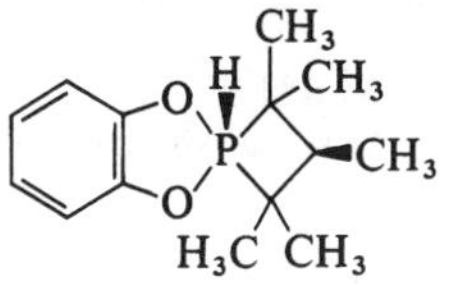

cis-form

$C_{14}H_{21}O_2P$ M 252.292

cis-form

2-Ph:
$C_{20}H_{25}O_2P$ M 328.390
Solid. Mp 65-67°.

trans-form

2-Ph: Solid. Mp 124-125°.
2-Benzyl:
$C_{21}H_{27}O_2P$ M 342.417
Solid. Mp 140-141°.

Antczak, S. *et al*, *J. Chem. Soc., Perkin Trans. 1*, 1977, 278; 1978, 1326 (*synth, pmr, P nmr*)

2,7,10,15,17-Pentamethyl-17*H*-tetrabenzo[*b,d,f,h*]phosphonin, 9CI — P-00028

[59313-80-3]

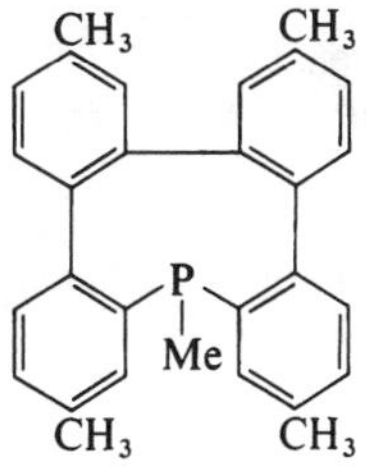

$C_{29}H_{27}P$ M 406.506
Plates (EtOH). Mp 155-157°.

B, MeI: [59325-05-2]. *2,7,10,15,17,17-Hexamethyl-17*H*-tetrabenzo*[b,d,f,h]*phosphoninium iodide*.
$C_{30}H_{30}IP$ M 548.445
Cryst. (EtOH). Mp 317-320° dec.

*B, PhCH₂I: 17-Benzyl-2,7,10,15,17-pentamethyl-17*H*-tetrabenzo*[b,d,f,h]*phosphoninium iodide*.
$C_{36}H_{34}IP$ M 624.543
Needles. Mp 275-281°.

Hellwinkel, D. *et al*, *Chem. Ber.*, 1976, **109**, 1497 (*synth, derivs, pmr, P nmr, struct*)

1,1,2,3,3-Pentamethyl-6,6,7,7-tetrakis(trifluoromethyl)-5,8-dioxa-4-phospha(4-P^V)spiro[3.4]octane, 9CI **P-00029**

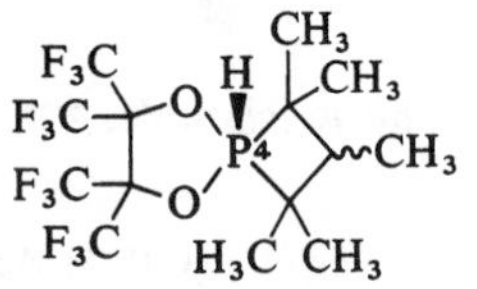

$C_{14}H_{17}F_{12}O_2P$ M 476.242

4-Phenoxy:
$C_{20}H_{21}F_{12}O_3P$ M 568.339
Solid. Mp 40-60°. $Bp_{0.5}$ 125°. Stereoisomeric mixt.

*4-*tert*-Butyl:*
$C_{18}H_{25}F_{12}O_2P$ M 532.349
Cryst. (pentane). Mp 60-75°. Stereoisomeric mixt.

4-Ph (cis-)*:*
$C_{20}H_{21}F_{12}O_2P$ M 552.339
Cryst. (MeOH). Mp 45-50°.

4-Ph (trans-)*:*
$C_{20}H_{21}F_{12}O_2P$ M 552.339
Cryst. (MeOH). Mp 95-97°.

4-(4-Bromophenyl):
$C_{20}H_{20}BrF_{12}O_2P$ M 631.235
Cryst. (MeOH aq.). Possesses square pyramidal geometry at phosphorus.

4-Dimethylamino:
$C_{16}H_{22}F_{12}NO_2P$ M 519.310
Cryst. (pentane). Mp 65-75°. $Bp_{0.5}$ 90-95°. Stereoisomeric mixt.

Howard, J.A. *et al, J. Chem. Soc., Chem. Commun.*, 1973, 856 (*cryst struct*)
Oram, R.K. *et al, J. Chem. Soc., Perkin Trans. 1*, 1973, 1300 (*synth, pmr, F and P nmr*)

1,1,2,3,3-Pentamethyl-6,6,8,8-tetrakis(trifluoromethyl)-5,7-dioxa-4-phospha(4-P^V)spiro[3.4]octane, 9CI **P-00030**

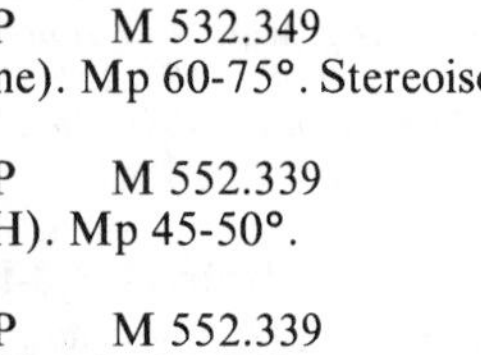

$C_{14}H_{17}F_{12}O_2P$ M 476.242

4-Chloro: [42282-06-4].
$C_{14}H_{16}ClF_{12}O_2P$ M 510.687
Cryst. (pentane at −78°). Mp ca. 20°. Stereochemistry not specified.

Oram, R.K. *et al, J. Chem. Soc., Perkin Trans. 1*, 1973, 1300 (*synth, pmr, F and P nmr*)

1,1-Pentanediphosphonic acid **P-00031**

Pentylidenebisphosphonic acid, 9CI
[4672-28-0]

$$H_3CCH_2CH_2CH_2CH[P(O)(OH)_2]_2$$

$C_5H_{14}O_6P_2$ M 232.110
Anticariogenic agent. V. hygroscopic cryst. (C_6H_6/AcOH). Mp 163-165° (151-152°).

Tetra-Et ester: [53459-47-5]. *Tetraethyl pentylidenebisphosphonate. 1,1-Bis(diethoxyphosphinyl)pentane.*
$C_{13}H_{30}O_6P_2$ M 344.324
Liq. $Bp_{0.3}$ 147-149°. n_D^{20} 1.4428.

Tetraisopropyl ester: Tetraisopropyl pentylidenebisphosphonate. 1,1-Bis(diisopropoxyphosphinyl)pentane.
$C_{17}H_{38}O_6P_2$ M 400.431
Liq. $Bp_{0.012}$ 90-95°.

Bis(dichloride):
$C_5H_{10}Cl_4O_2P_2$ M 305.892
Liq. $Bp_{0.028}$ 115-120°.

Kosolapoff, G.M., *J. Am. Chem. Soc.*, 1953, **75**, 1500 (*synth, ester*)
Hays, H.R. *et al, J. Org. Chem.*, 1966, **31**, 3391 (*synth, derivs, P nmr*)
Quimby, O.T. *et al, J. Organomet. Chem.*, 1968, **13**, 199 (*synth, P nmr*)

1,5-Pentanediphosphonic acid **P-00032**

1,5-Pentanediylbisphosphonic acid, 9CI. Pentamethylenediphosphonic acid. 1,5-Diphosphonopentane
[4672-25-7]

$$(HO)_2P(O)(CH_2)_5P(O)(OH)_2$$

$C_5H_{14}O_6P_2$ M 232.110
Needles (AcOH), cryst. (H_2O). Mp 160°.

Tetra-Et ester: Tetraethyl 1,5-pentanediylbisphosphonate. 1,5-Bis(diethoxyphosphinyl)pentane.
$C_{13}H_{30}O_6P_2$ M 344.324
Liq. $Bp_{0.1}$ 180-181°. n_D^{25} 1.4464.

Bis(dichloride):
$C_5H_{10}Cl_4O_2P_2$ M 305.892
Solid. Mp 60-62°. $Bp_{0.3}$ 165°.

Bride, M.H. *et al, J. Appl. Chem.*, 1961, **11**, 352 (*synth*)
Moedritzer, K. *et al, J. Inorg. Nucl. Chem.*, 1961, **22**, 297 (*synth, ir, nmr*)
Kosolapoff, G.M. *et al, J. Chem. Soc. (C)*, 1966, 757 (*synth, chloride*)

1,5-Pentanediphosphonous acid **P-00033**

1,5-Pentanediylbis(phosphonous acid), 9CI. Pentamethylenediphosphonous acid, 8CI

$$(HO)_2P(CH_2)_5P(OH)_2 \rightleftharpoons HO(O)PH(CH_2)_5PH(O)OH$$

$C_5H_{14}O_4P_2$ M 200.111

Tetra-Et ester: [29936-33-2]. *Tetraethyl 1,5-pentanediylbisphosphonite. Tetraethyl pentamethylenediphosphonite.*
$C_{13}H_{30}O_4P_2$ M 312.325
Liq. Bp_2 135-142°. n_D^{20} 1.4596.

Tetra-Ph ester: Tetraphenyl 1,5-pentanediylbisphosphonite. Tetraphenyl pentamethylenediphosphonite.
$C_{29}H_{30}O_4P_2$ M 504.501
Liq. $Bp_{0.02}$ 158-160°.

Bis(dichloride): [29936-35-4].
$C_5H_{10}Cl_4P_2$ M 273.894
Liq. Bp_2 165-170°.

Bis(dibromide): [82159-17-9].
$C_5H_{10}Br_4P_2$ M 451.698
Liq. $Bp_{0.1}$ 136°.

Bis[bis(diethylamide)]: [29936-34-3]. P,P'*-1,5-Pentanediylbis[bis-*N,N*-diethylphosphonous diamide].*
$C_{21}H_{50}N_4P_2$ M 420.600
Liq. Bp_2 80-93°, $Bp_{0.1}$ 133°. n_D^{20} 1.4717.

Sander, M., *Chem. Ber.*, 1962, **95**, 473 (*tetraethyl ester*)
Adrova, N.A. *et al, Izv. Akad. Nauk SSSR, Ser. Khim.*, 1966, 1824 (*Engl. transl.* p. 1758) (*tetraphenyl ester*)
Maier, L., *Helv. Chim. Acta*, 1970, **53**, 1940 (*derivs, synth, P nmr*)

Diemert, K. *et al*, *Chem. Ber.*, 1982, **115**, 1947 (*tetrabromide, synth, P nmr*)
Diemert, K. *et al*, *Phosphorus Sulfur*, 1983, **15**, 155 (*diethylamide, synth, P nmr*)

1,5-Pentanediylbis[triphenylphosphonium](2+), 9CI — P-00034

Pentamethylenebis[triphenylphosphonium](2+). 1,5-Bis(triphenylphosphonio)pentane(2+)

$Ph_3P^{\oplus}(CH_2)_5PPh_3^{\oplus}$

$C_{41}H_{40}P_2^{\oplus\oplus}$ M 594.715 (ion)

▷Salts exhibit AChE activity in vertebrate and schistosome spp.

Dibromide: [22884-31-7].
$C_{41}H_{40}Br_2P_2$ M 754.523
Source of bisylide. Prisms (DMF). Mp 255-256°.

Bisylide: [38451-19-3]. *1,5-Pentanediylidenebis[triphenylphosphorane].*
$C_{41}H_{38}P_2$ M 592.699
Used in Wittig reactions.

Horner, L. *et al*, *Chem. Ber.*, 1962, **95**, 581 (*synth*)
Bestmann, H.J. *et al*, *Angew. Chem., Int. Ed. Engl.*, 1972, **11**, 508 (*props*)
Booth, B.L. *et al*, *J. Organomet. Chem.*, 1981, **220**, 229 (*complexes*)
Willcockson, W.S. *et al*, *Comp. Biochem. Physiol., C*, 1982, **72**, 101 (*tox*)

Pentaphenoxyphosphorane, 9CI, 8CI — P-00035

[19613-06-0]

$(PhO)_5P$

$C_{30}H_{25}O_5P$ M 496.498
Cryst. (hexane). Mp 103-104°.

Ramirez, F. *et al*, *J. Am. Chem. Soc.*, 1968, **90**, 3507 (*synth, pmr, ir, props*)
Ramirez, F. *et al*, *Tetrahedron*, 1968, **24**, 5041 (*props*)
Archie, W.C. *et al*, *J. Am. Chem. Soc.*, 1973, **95**, 5955 (*uv, props*)
Sarma, R. *et al*, *J. Am. Chem. Soc.*, 1976, **98**, 581 (*cryst struct*)
Deiters, J.A. *et al*, *J. Am. Chem. Soc.*, 1977, **99**, 5461 (*struct*)
Denning, L.W. *et al*, *J. Am. Chem. Soc.*, 1982, **104**, 230 (*nmr*)

2,2,3,4,6-Pentaphenyl-2*H*-1,5,2-dioxaphosphorin 2-oxide, 10CI — P-00036

*2,2,3,4,6-Pentaphenyl-2*H*-1,5,2-dioxaphosphinin 2-oxide*

[52364-40-6]

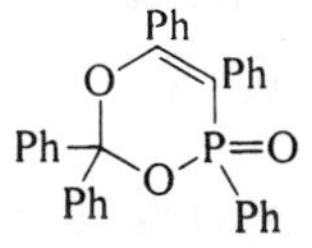

$C_{33}H_{25}O_3P$ M 500.532
Cryst. (EtOAc/Et_2O). Mp 182-183°.

Regitz, M. *et al*, *Chem. Ber.*, 1978, **111**, 705 (*synth*)

Pentaphenylpentaphospholane, 9CI — P-00037

Phosphobenzene A. Pentaphenylcyclopentaphosphine

[3376-52-1]

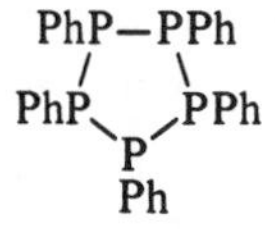

$C_{30}H_{25}P_5$ M 540.396

In soln., the pentaphosphole is evidently in equilibrium with Hexaphenylhexaphosphorinane, H-00089 and Tetraphenyltetraphosphetane, T-00278 . Monoclinic cryst. (MeCN or THF). Mp 154-156° (159.5-161 sealed tube). Pyrolysis → PhP:.

Henderson, W.A. *et al*, *J. Am. Chem. Soc.*, 1963, **85**, 2462.
Daly, J.J., *J. Chem. Soc.*, 1964, 6147 (*cryst struct*)
Maier, L., *Helv. Chim. Acta*, 1966, **49**, 1119 (*synth*)
Schmidt, V. *et al*, *Chem. Ber.*, 1968, **101**, 1381 (*ms*)
DuPont, T.J. *et al*, *Inorg. Chem.*, 1973, **12**, 2487 (*synth, uv, pmr, P nmr*)
Kaska, W.C. *et al*, *Helv. Chim. Acta*, 1974, **57**, 2550 (*synth, props*)
Hoffman, P.R. *et al*, *Inorg. Chem.*, 1975, **14**, 1997 (*synth, conformn, P nmr*)
Baudler, M. *et al*, *Z. Naturforsch., B*, 1976, **31**, 558 (*pmr, conformn*)
Smith, L.R. *et al*, *J. Am. Chem. Soc.*, 1976, **98**, 3852 (*synth, P nmr*)
Hassler, K. *et al*, *Monatsh. Chem.*, 1979, **110**, 919 (*ir*)

Pentaphenyl-1*H*-phosphole, 9CI, 8CI — P-00038

Pentaphenylphosphacyclopentadiene

[1181-62-0]

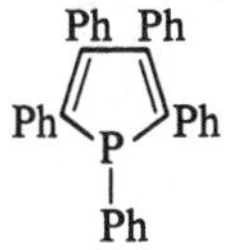

$C_{34}H_{25}P$ M 464.545

Yellowish-green fluor. needles (toluene or CH_2Cl_2). Mp 255-256°. O_3 → Phenylphosphonic acid.

Maleic anhydride adduct:
$C_{36}H_{27}O_2P$ M 522.582
Mp 260-264°.

Oxide: [1641-63-0].
$C_{34}H_{25}OP$ M 480.545
Cryst. (EtOH). Mp 284-285° (270-271°).

Sulfide: [21996-99-6].
$C_{34}H_{25}PS$ M 496.605
Yellow plates (CH_2Cl_2/EtOH). Mp 197-198°.

Braye, E.H. *et al*, *J. Am. Chem. Soc.*, 1961, **83**, 4406 (*synth, uv, derivs*)
Campbell, I.G.M. *et al*, *J. Chem. Soc.*, 1965, 2184 (*synth, oxide*)
Arägar, M. *et al*, *Chem. Ber.*, 1976, **109**, 877 (*oxide, cryst struct*)
Ivanova, N.P. *et al*, *Zh. Obshch. Khim.*, 1977, **47**, 763 (*Engl. transl.* p. 696) (*oxide*)
Yasufuku, K. *et al*, *J. Am. Chem. Soc.*, 1980, **102**, 4363 (*oxide, synth, pmr*)

Pentaphenylphosphorane, 9CI, 8CI — P-00039

[2588-88-7]

Ph_5P

$C_{30}H_{25}P$ M 416.501

Molecule consists of distorted trigonal bipyramid. Cryst. (C_6H_6/cyclohexane) which are photosensitive turning black. Mp 123-124° dec. Acted on by $KNH_2/NH_3(l)$ with P-C cleavage.

Wheatley, P.J., *J. Chem. Soc.*, 1964, 2206 (*cryst struct*)
Wittig, G. *et al*, *Chem. Ber.*, 1964, **97**, 741 (*synth*)
Latscha, H.P., *Z. Naturforsch., B*, 1968, **23**, 139 (*nmr*)
Angelelli, J.M. *et al*, *J. Am. Chem. Soc.*, 1969, **91**, 4500 (*ir*)
Hellwinkel, D. *et al*, *Phosphorus*, 1973, **2**, 167 (*ms*)
Brock, C.P., *Acta Crystallogr., Sect. A*, 1977, **33**, 193 (*struct*)
Sergienko, L.M. *et al*, *Zh. Obshch. Khim.*, 1979, **49**, 317 (*Engl. transl.* p. 275) (*uv*)

2,3,4,5,6-Pentaphenylphosphorin, 8CI **P-00040**

2,3,4,5,6-Pentaphenylphosphinine

[14657-87-5]

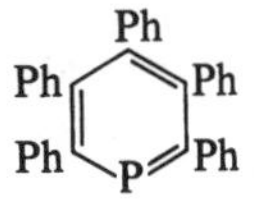

$C_{35}H_{25}P$ M 476.556

Mp 216-217°, 253-255°.

Dimroth, K. *et al*, *Angew. Chem., Int. Ed. Engl.*, 1967, **6**, 711 (*synth, uv, esr*)
Märkl, G. *et al*, *Angew. Chem., Int. Ed. Engl.*, 1967, **6**, 458, 944 (*synth, uv*)

3,3,3,6,6-Pentaphenyl-3,3,3,6-tetrahydro-4,5-bis(methoxycarbonyl)-1,2,3-diazaphosphorine, 9CI **P-00041**

3,3,3,6,6-Pentaphenyl-3,3,3,6-tetrahydro-4,5-bis(methoxycarbonyl)-1,2,3-diazaphosphinine

[53813-58-4]

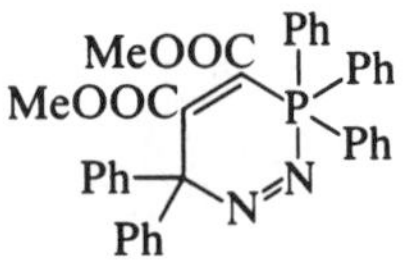

$C_{37}H_{31}N_2O_4P$ M 598.637

Solid. Mp 275-276°.

Shakel, I. *et al*, *Isr. J. Chem.*, 1973, **11**, 729 (*synth, pmr, uv*)

1,2,3,4,9-Pentaphenyl-1*H*-tribenzo[*b,d,f*]phosphepin, 9CI **P-00042**

[60686-35-3]

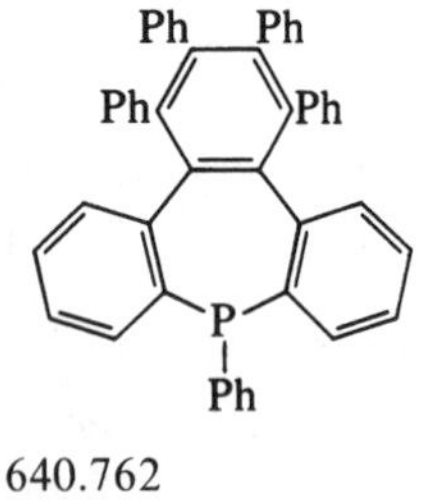

$C_{48}H_{33}P$ M 640.762

Cryst. (C_6H_6/pet. ether). Mp 355-357°, Mp 353-355° (double Mp). Stereoisomer ratio 1.75:1 (130°, PhBr).

*B,MeI: 9-Methyl-1,2,3,4,9-pentaphenyl-1*H*-tribenzo*[b,d,f]*phosphepinium iodide.*
$C_{49}H_{36}IP$ M 782.702
Solid. Mp 354-365° dec., Mp 370° dec. Stereoisomer ratio 20:1 (120°, Py).

9-Oxide:
$C_{48}H_{33}OP$ M 656.762
Solid. Mp 349-350°, Mp 346-347°. Stereoisomer ratio 30:1 (130°, PhBr).

Winter, W., *Chem. Ber.*, 1976, **109**, 2405; 1978, **111**, 2942 (*synth, derivs, ir, ms, P nmr, pmr, cryst struct*)

Pentapropylphosphorodiamidimidic acid, 9CI **P-00043**

Pentapropylphosphoric triamide

$H_3CCH_2CH_2N{=}P(OH)[N(CH_2CH_2CH_3)_2]_2 \rightleftharpoons H_3CCH_2CH_2NHP(O)[N(CH_2CH_2CH_3)_2]_2$

$C_{15}H_{36}N_3OP$ M 305.443

Free acid is tautomeric.

Fluoride: [86601-00-5]. P,P-*Bis(dipropylamino)*-P-*fluoro*-N-*propylphosphazene.* P,P-*Bis(dipropylamino)*-P-*fluoro*-N-*propylphosphine imide.*
$C_{15}H_{35}FN_3P$ M 307.434
Liq. $Bp_{0.02}$ 75-76°. n_D^{20} 1.4470.

Chloride: [66964-03-2]. P-*Chloro*-P,P-*bis(dipropylamino)*-N-*propylphosphazene.* P-*Chloro*-P,P-*bis(dipropylamino)*-N-*propylphosphine imide.*
$C_{15}H_{35}ClN_3P$ M 323.888
Liq. d_4^{20} 0.96. $Bp_{0.02}$ 108-110°. n_D^{25} 1.4740.

Bromide: [73954-61-7]. P-*Bromo*-P,P-*bis(dipropylamino)*-N-*propylphosphazene.* P-*Bromo*-P,P-*bis(dipropylamino)*-N-*propylphosphine imide.*
$C_{15}H_{35}BrN_3P$ M 368.339
Liq. $Bp_{0.015}$ 83-84°. n_D^{20} 1.4896.

Bromide; BHBr: [84591-35-5].
Bromobis(dipropylamino)(propylamino)phosphonium bromide. Triaminobromopentapropylphosphonium bromide.
$C_{15}H_{36}Br_2N_3P$ M 449.251
Cryst. (THF/pet. ether). Mp 52-56°.

Marchenko, A.P. *et al*, *Zh. Obshch. Khim.*, 1978, **48**, 551 (*Engl. transl.* p. 503) (*chloride*)
Egorov, Yu.P. *et al*, *Teor. Eksp. Khim.*, 1982, **18**, 58 (*Engl. transl.* p. 45) (*chloride, P nmr, struct*)
Miroshnichenko, V.V. *et al*, *Zh. Obshch. Khim.*, 1982, **52**, 2517 (*Engl. transl.* p. 2222) (*fluoride, bromide, salts*)
Marchenko, A.P. *et al*, *Zh. Obshch. Khim.*, 1983, **53**, 698 (*Engl. transl.* p. 608) (*fluoride, synth, P nmr*)

4-Pentenyltriphenylphosphonium(1+) **P-00044**

$Ph_3P^{\oplus}(CH_2)_3CH{=}CH_2$

$C_{23}H_{24}P^{\oplus}$ M 331.416 (ion)

Bromide: [56771-29-0].
$C_{23}H_{24}BrP$ M 411.320
Treatment with butyllithium or $NaN(SiMe_3)_2$ gives the ylide. Cryst. (CH_2Cl_2/Et_2O). Mp 175°.

Ylide: [67773-68-6]. *4-Pentenylidenetriphenylphosphorane.*
$C_{23}H_{23}P$ M 330.408
Reactive Wittig reagent prepd. *in situ*.

Hauser, C.F. *et al*, *J. Org. Chem.*, 1963, **28**, 372 (*bromide*)
Bestmann, H.J. *et al*, *Justus Liebigs Ann. Chem.*, 1981, 1705 (*synth, ylide, use*)
Bestmann, H.J. *et al*, *Justus Liebigs Ann. Chem.*, 1982, 1478 (*use*)
Eberle, M.K. *et al*, *J. Org. Chem.*, 1982, **47**, 2210 (*use*)

(3-Penten-1-ynyl)phosphonic acid, 9CI **P-00045**

$H_3CCH{=}CHC{\equiv}CP(O)(OH)_2$

$C_5H_7O_3P$ M 146.082

Di-Et ester: [20537-00-2]. *Diethyl (3-penten-1-ynyl)phosphonate.*
$C_9H_{15}O_3P$ M 202.189
Liq. d_4^{20} 1.06. Bp_2 112-115°. n_D^{20} 1.4790.

Ionin, B.E. *et al*, *Zh. Obshch. Khim.*, 1963, **33**, 2863 (*Engl. transl.* p. 2791) (*synth, ir, pmr*)

(4-Penten-2-yn-1-yl)phosphonic acid, 9CI P-00046

$H_2C{=}CHC{\equiv}CCH_2P(O)(OH)_2$

$C_5H_7O_3P$ M 146.082

Di-Et ester: Diethyl (4-penten-2-yn-1-yl)phosphonate.
$C_9H_{15}O_3P$ M 202.189
Liq. d_4^{20} 1.06. Bp_2 107-107.5°. n_D^{20} 1.4752. EtO^- causes partial conversion to the diethyl ester of (3-Penten-1-ynyl)phosphonic acid, P-00045 .

Ionin, B.I. *et al*, *Zh. Obshch. Khim.*, 1964, **34**, 1174 (*Engl. transl.* p. 1165) (*synth, ir*)

2-Penten-4-ynyltriphenylphosphonium(1+), 10CI P-00047

$HC{\equiv}CCH{=}CHCH_2PPh_3^{\oplus}$

$C_{23}H_{20}P^{\oplus}$ M 327.385 (ion)
Rgt. for Wittig reactions, used in synth. of flexirubin.

Bromide: [61665-57-4].
$C_{23}H_{20}BrP$ M 407.289
Source of ylide.

Ylide: 2-Penten-4-ynylidenetriphenylphosphorane.
$C_{23}H_{19}P$ M 326.377
Used in Wittig reactions.

Achenbach, H. *et al*, *Angew. Chem., Int. Ed. Engl.*, 1977, **16**, 191 (*synth, use*)

Pentyl phosphinate, 9CI P-00048

Amyl hypophosphite
[18108-11-7]

$H_3C(CH_2)_4OP(O)H_2 \rightleftharpoons H_3C(CH_2)_4OP(OH)H$

$C_5H_{13}O_2P$ M 136.130
Liq. d_4^{20} 1.01. $Bp_{0.2}$ 70-80°. n_D^{20} 1.4332.

Ivanov, B.E. *et al*, *Izv. Akad. Nauk SSSR, Ser. Khim.*, 1967, 1498 (*Engl. transl.* p. 1447); *CA*, **68**, 78359 (*synth*)

Pentylphosphine, 9CI P-00049

Amylphosphine

$H_3C(CH_2)_4PH_2$

$C_5H_{13}P$ M 104.131
Liq. Bp 104°. n_D^{20} 1.4431.

Pass, F. *et al*, *Monatsh. Chem.*, 1959, **90**, 148; 1962, **93**, 230 (*synth*)
Hays, H.R. *et al*, *J. Org. Chem.*, 1966, **31**, 3391 (*ms, nmr*)

Pentylphosphonic acid, 9CI P-00050

Amylphosphonic acid
[4672-26-8]

$H_3C(CH_2)_4P(O)(OH)_2$

$C_5H_{13}O_3P$ M 152.130
Cryst. (ligroin). Mp 120.5-121°.

Di-Me ester: [6619-48-3]. *Dimethyl pentylphosphonate.*
$C_7H_{17}O_3P$ M 180.183
Liq. $Bp_{0.2}$ 67-68°.

Di-Et ester: [1186-17-0]. *Diethyl pentylphosphonate.*
$C_9H_{21}O_3P$ M 208.237
Liq. Bp_{17} 167-169°, $Bp_{1.5}$ 86°. $n_D^{16.5}$ 1.4282.

Dipentyl ester: [6418-56-0]. *Dipentyl pentylphosphonate. Diamyl amylphosphonate.*
$C_{15}H_{33}O_3P$ M 292.398
Extractant for Zr(II) from NaCl solns. Liq. Bp_2 150-151°. n_D^{20} 1.4378.

Dichloride: [926-46-5].
$C_5H_{11}Cl_2OP$ M 189.021
Liq. d_4^{20} 1.22. Bp_{13} 101-108°, Bp_5 76°. n_D^{25} 1.4637, n_D^{20} 1.4703.

Kosolapoff, G.M., *J. Am. Chem. Soc.*, 1945, **67**, 1180 (*synth, esters*)
Ford-Moore, A.H. *et al*, *J. Chem. Soc.*, 1947, 1465 (*ester, synth*)
Bogonosceva, N.P. *et al*, *Zh. Obshch. Khim.*, 1963, **33**, 1363 (*Engl. transl.* p. 1332) (*ester, synth*)
Geiseler, G. *et al*, *Ber. Bunsenges. Phys. Chem.*, 1967, **71**, 478 (*dichloride, ir, raman*)
Ernst, L., *Org. Magn. Reson.*, 1973, **9**, 35 (*diethyl ester, nmr, cmr*)
Dietze, U., *J. Prakt. Chem.*, 1974, **316**, 293 (*ir*)
Rosini, G. *et al*, *Synthesis*, 1975, 44 (*dimethyl ester, synth, ir, pmr*)
Nagueira, E.D. *et al*, *Hydrometallurgy*, 1983, **9**, 333 (*dipentyl ester, use*)

Pentylphosphonodithioic acid, 9CI P-00051

Amylphosphonodithioic acid

$H_3C(CH_2)_4P(O)(SH)_2 \rightleftharpoons H_3C(CH_2)_4P(S)(OH)(SH)$

$C_5H_{13}OPS_2$ M 184.251

O-*Mono-Me ester:* [13685-79-5]. O-*Methyl hydrogen pentylphosphonodithioate.*
$C_6H_{15}OPS_2$ M 198.278
Liq. d^{20} 1.11. $Bp_{0.11}$ 55-56°. n_D^{20} 1.5358.

O-*Mono-Et ester:* [18789-05-4]. O-*Ethyl hydrogen pentylphosphonodithioate.*
$C_7H_{17}OPS_2$ M 212.304
Liq. d^{20} 1.07. $Bp_{0.08}$ 81-84°. n_D^{20} 1.5241.

O-*Monoisopropyl ester:* [18789-04-3]. O-*Isopropyl hydrogen pentylphosphonodithioate.*
$C_8H_{19}OPS_2$ M 226.331
Liq. d^{20} 1.04. $Bp_{0.02}$ 72-82°. n_D^{20} 1.5153.

Grishina, O.N. *et al*, *Izv. Akad. Nauk SSSR, Ser. Khim.*, 1966, 1617 (*Engl. transl.* p. 1558) (*synth*)
Grishina, O.N. *et al*, *Neftekhimiya*, 1968, **8**, 111; *CA*, **69**, 2980 (*synth*)
Grishina, O.N. *et al*, *CA*, 1969, **71**, 39085 (*synth*)

Pentylphosphonous acid, 9CI P-00052

Amylphosphonous acid

$H_3C(CH_2)_4P(OH)_2 \rightleftharpoons H_3C(CH_2)_4PH(O)OH$

$C_5H_{13}O_2P$ M 136.130
Free acid probably exists as the phosphoryl tautomer.

Diisopropyl ester: Diisopropyl pentylphosphonite.
$C_{11}H_{25}O_2P$ M 220.291
Liq. Bp_{15} 98-101°.

Dipentyl ester: [3245-78-1]. *Dipentyl pentylphosphonite. Diamyl amylphosphonite.*
$C_{15}H_{33}O_2P$ M 276.398
Bp_1 119°. n_D^{20} 1.446.

Dichloride: [15573-32-7]. *Dichloropentylphosphine.*
$C_5H_{11}Cl_2P$ M 173.022
d_4^{27} 1.10. Bp 184°. n_D^{25} 1.4815.

Fox, R.B., *J. Am. Chem. Soc.*, 1950, **72**, 4147 (*dichloride*)
Mastalerz, P. *et al*, *Rocz. Chem.*, 1964,, **38**, 1529 (*diisopropyl ester*)

Voigt, D. *et al*, *Bull. Soc. Chim. Fr.*, 1964, 3087 (*ester*)

Pentylphosphoramidic acid, 9CI P-00053

Amylphosphoramidic acid

$H_3C(CH_2)_4NHP(O)(OH)_2$

$C_5H_{14}NO_3P$ M 167.144

Di-Et ester: [85231-81-8]. *Diethyl pentylphosphoramidate.*
$C_9H_{22}NO_3P$ M 223.251
Liq. Bp_{25} 185°. n_D^{20} 1.4357.
Dichloride:
$C_5H_{12}Cl_2NOP$ M 204.036
Liq. Bp_{17} 159°, Bp_3 127-128°.
Dianilide: N-*Pentyl*-N′,N″-*diphenylphosphoric triamide.*
$C_{17}H_{24}N_3OP$ M 317.370
Solid. Mp 117°.

Michaelis, A., *Justus Liebigs Ann. Chem.*, 1903, **326**, 129; 1915, **407**, 290 (*derivs*)
Mizuma, T. *et al*, *J. Pharm. Soc. Jpn.*, 1961, **81**, 48; *CA*, **55**, 13403 (*dichloride*)
Zwierzak, A., *Synthesis*, 1982, 920 (*ester, synth, ir, pmr, P nmr*)

Pentyl phosphorodichloridite, 9CI, 8CI P-00054

Amyl phosphorodichloridite. Amyl dichlorophosphite
[6068-99-1]

$H_3C(CH_2)_4OPCl_2$

$C_5H_{11}Cl_2OP$ M 189.021
Pungent liq. d^{20} 1.13. Bp_{14} 71-73°. n_D^{20} 1.4650. Easily hydrolysed.

Nesterov, L.V. *et al*, *Zh. Obshch. Khim.*, 1965, **35**, 2050 (*Engl. transl.* p. 2041) (*synth*)

O-Pentyl phosphorodichloridothioate, 9CI, 8CI P-00055

O-*Amyl dichlorothiophosphate.* O-*Pentyl thiophosphoryl dichloride.* O-*Pentyl dichlorothiophosphate*
[18351-14-9]

$H_3C(CH_2)_4OP(S)Cl_2$

$C_5H_{11}Cl_2OPS$ M 221.081
Liq. d_4^{20} 1.19. Bp_1 74-75°. n_D^{20} 1.4845.

Bliznyuk, N.U. *et al*, *Zh. Obshch. Khim.*, 1967, **37**, 1122 (*Engl. transl.* p. 1064) (*synth*)

Pentyltriphenylphosphonium(1+), 9CI P-00056

Amyltriphenylphosphonium(*1*+)

$Ph_3P^{\oplus}(CH_2)_4CH_3$

$C_{23}H_{26}P^{\oplus}$ M 333.432 (ion)
Bromide: [21406-61-1].
$C_{23}H_{26}BrP$ M 413.336
Cryst. (H_2O or CH_2Cl_2/Et_2O). Mp 167-168°.
Iodide: [35171-55-2].
$C_{23}H_{26}IP$ M 460.337
Solid. Mp 170°.
Ylide: [29541-98-8]. *Pentylidenetriphenylphosphorane.*
$C_{23}H_{25}P$ M 332.424
Reactive Wittig reagent, prepd. *in situ.*

Jaenicke, J. *et al*, *Justus Liebigs Ann. Chem.*, 1973, 1252 (*bromide, use*)
Bestmann, H.J. *et al*, *Chem. Ber.*, 1976, **109**, 1694 (*ylide*)
Kowal, R. *et al*, *Pol. J. Chem.*, 1979, **53**, 673 (*iodide*)
Bestmann, H.J. *et al*, *Justus Liebigs Ann. Chem.*, 1982, 1359 (*ylide, use*)

Peroxycyclophosphamide P-00057

4,4′-Dioxybis[N,N-*bis*(*2-chloroethyl*)*amino*]-*tetrahydro-2*H-*1,3,2-oxazaphosphorine 2,2′-dioxide, 9CI*
[51274-71-6]

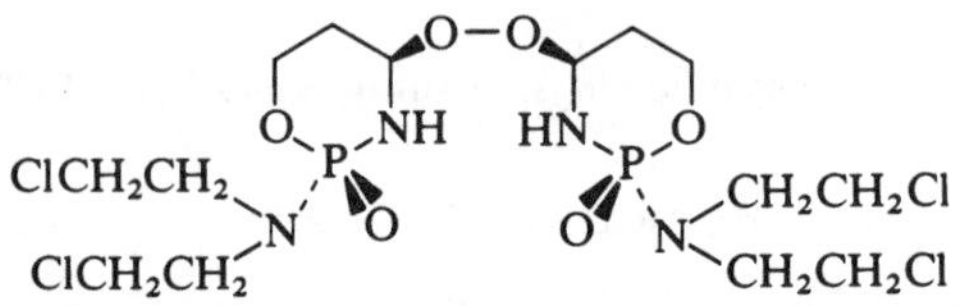

$C_{14}H_{28}Cl_4N_4O_6P_2$ M 552.158
Oxidn. prod. of Cyclophosphamide, C-00283 . Cytotoxic.

Sternglanz, H. *et al*, *J. Am. Chem. Soc.*, 1974, **96**, 4014 (*cryst struct*)
Struck, R.F. *et al*, *J. Am. Chem. Soc.*, 1974, **96**, 313 (*synth, ir, ms, pmr, cmr*)
Takamizawa, A. *et al*, *Tetrahedron Lett.*, 1974, 517 (*synth, ir, pmr*)
Montgomery, J.A. *et al*, *Cancer Treat. Repts.*, 1976, **60**, 381.

Phaseolotoxin P-00058

[62249-77-8]

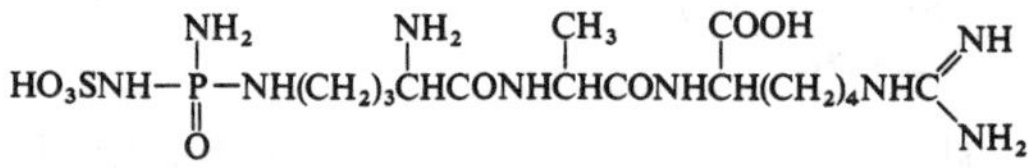

$C_{15}H_{34}N_9O_8PS$ M 531.523
Prod. by *Pseudomonas syringae* pv. *phaseolicola.* Phytotoxin.
▷OL5610000.

Moore, R.E. *et al*, *Tetrahedron Lett.*, 1984, **25**, 3931 (*struct, nmr, bibl*)

Phellanphos P-00059

[*5-Methyl-7-*(*1-methylethyl*)*bicyclo*[*2.2.2*]*oct-5-ene-2,3-diyl*]*bis*[*diphenylphosphine*], *10CI. 5,6-Bis*(*diphenylphosphino*)*-8-isopropyl-2-methylbicyclo*[*2.2.2*]*oct-2-ene*
[72021-47-7]

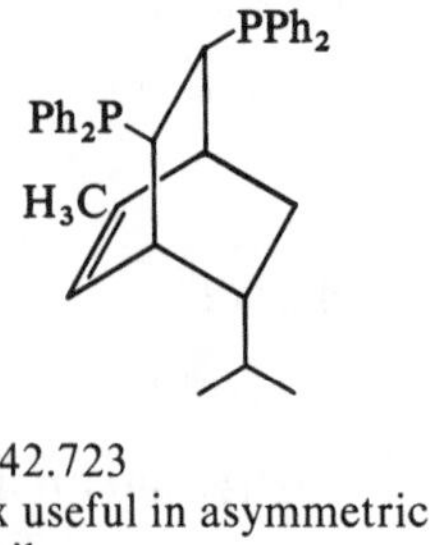

$C_{36}H_{48}P_2$ M 542.723
Forms Rh complex useful in asymmetric reductions. Easily oxidised oil.
Disulfide:
$C_{36}H_{48}P_2S_2$ M 606.843
Cryst. (CH_2Cl_2/hexane). Mp >300°. $[\alpha]_D^{20}$ +138° (c, 1.0 in $CHCl_3$).

Lauer, M. *et al*, *J. Organomet. Chem.*, 1979, **177**, 309 (*synth, pmr, cmr*)

Samuel, O. *et al, Nouv. J. Chim.*, 1981, **5**, 15 (*synth, pmr, cmr, nmr*)
Scott, J.W. *et al, J. Org. Chem.*, 1981, **46**, 5086 (*use*)
Samuel, O. *et al, Phosphorus Sulfur*, 1984, **21**, 145 (*disulfide, cryst struct*)

6-Phenanthridinephosphonic acid **P-00060**

6-Phenanthridinylphosphonic acid, 9CI. 6-Phosphonophenanthridine

[56583-34-7]

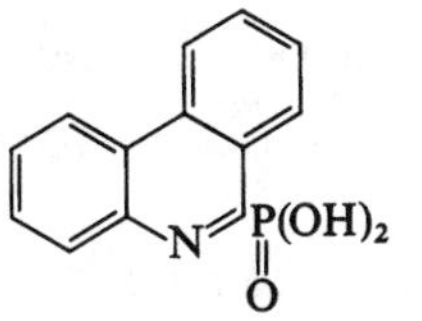

$C_{13}H_{10}NO_3P$ M 259.201

Cryst. (H_2O).

Di-Et ester: [56583-33-6]. *Diethyl 6-phenanthradinylphosphonate.*
$C_{17}H_{18}NO_3P$ M 315.308
Cryst. (C_6H_6/hexane).

U.S.P., 3 888 626, (*1975*); *CA*, **83**, 97569

Phenazaphosphinic acid **P-00061**

5,10-Dihydro-10-hydroxyphenophosphazine 10-oxide, 9CI

[472-43-5]

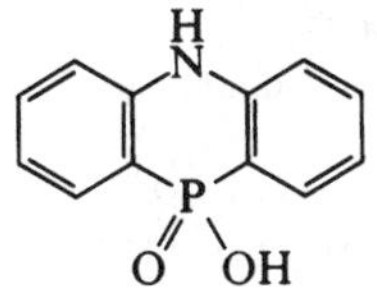

$C_{12}H_{10}NO_2P$ M 231.190

Prisms (EtOH). Mp 277° (preheated block).

▷SZ5380000.

Me ester (10-Methoxy): [64117-64-2]. *Methyl phenazaphosphinate. 5,10-Dihydro-10-methoxyphenophosphazine 10-oxide.*
$C_{13}H_{12}NO_2P$ M 245.217
Prisms (2-propanol). Mp 223-224°.

Häring, M., *Helv. Chim. Acta*, 1960, **43**, 1826 (*synth, ir*)
Freedman, L.D. *et al, J. Org. Chem.*, 1975, **40**, 2684 (*uv*)

Phenkapton, BSI, ISO **P-00062**

S-[[(*2,5-Dichlorophenyl*)*thio*]*methyl*] O,O-*diethyl phosphorodithioate, 9CI*

[2275-14-1]

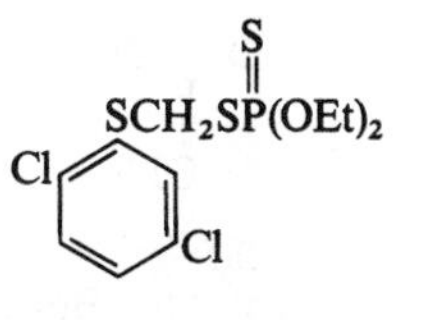

$C_{11}H_{15}Cl_2O_2PS_3$ M 377.298

Insecticide, now superseded.

▷TD5775000.

Keith, L.H. *et al, J. Assoc. Off. Anal. Chem.*, 1968, **51**, 1063 (*pmr*)
Stan, H.-J. *et al, Biomed. Mass Spectrom.*, 1982, **9**, 483 (*ms*)
Stan, H.-J. *et al, J. Chromatogr.*, 1983, **279**, 173 (*glc*)
Agrochemicals Handbook, Royal Society of Chemistry, London, 1983, A318.

2-Phenoxy-1,3,2-benzodioxaphosphole, 9CI **P-00063**

Phenyl o-*phenylene phosphite*

[4591-40-6]

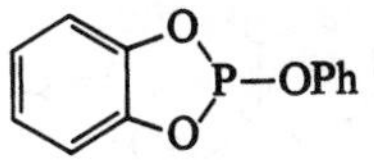

$C_{12}H_9O_3P$ M 232.175

Used for the conversion of sulfoxides to sulfides in high yield. Liq. Bp_{12} 155°, $Bp_{0.7}$ 125-130°. n_D^{21} 1.5760.

Oxide: [52961-95-2]. *Phenyl* o-*phenylene phosphate.*
$C_{12}H_9O_4P$ M 248.174
Viscous oil. $Bp_{0.02}$ 133-147°.

Sulfide: O-*Phenyl* o-*phenylene thiophosphate.*
$C_{12}H_9O_3PS$ M 264.235
Cryst. (Et_2O). Mp 71-2°. Bp_{10} 186°.

Anschütz, L. *et al, J. Prakt. Chem.*, 1932, **133**, 65 (*synth, deriv*)
Humphris, K.J. *et al, J. Chem. Soc., Perkin Trans. 2*, 1973, 826 (*synth*)
Dreux, M. *et al, Synthesis*, 1974, 506 (*synth, pmr, P nmr*)
Ramirez, F. *et al, J. Am. Chem. Soc.*, 1974, **96**, 7269; 1976, **98**, 4330 (*P nmr*)
Ramirez, F. *et al, Phosphorus*, 1976, **6**, 215 (*oxide, synth, pmr, props*)
Gloede, J. *et al, Z. Anorg. Allg. Chem.*, 1983, **500**, 59 (*oxide, synth, P nmr*)

2-Phenoxy-4H-1,3,2-benzodioxaphosphorin, 9CI **P-00064**

*2-Phenoxy-4*H-*1,3,2-benzodioxaphosphinin. Phenyl saligenin cyclic phosphite*

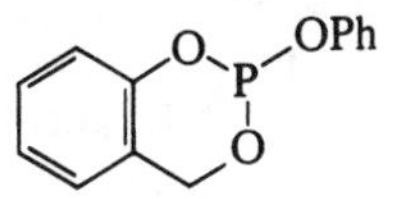

$C_{13}H_{11}O_3P$ M 246.202

2-Oxide: [4081-23-6]. *Phenyl saligenin cyclic phosphate.*
$C_{13}H_{11}O_4P$ M 262.201
Inhibits chymotrypsin and kynurenine formamidase. Cryst. ($CHCl_3$ or C_6H_6/ligroin). Mp 79.5-81°.
▷Neurotoxic

2-Sulfide: [4575-86-4]. O-*Phenyl saligenin cyclic thiophosphate.*
$C_{13}H_{11}O_3PS$ M 278.262
Solid. Mp 36°.

Eto, M. *et al, Agric. Biol. Chem.*, 1962, **26**, 452; 1963, **27**, 723, 789 (*oxide, sulfide, tlc, ir*)
Hall, L.D. *et al, Can. J. Chem.*, 1972, **50**, 2092 (*oxide*)
Galdecki, Z. *et al, Acta Crystallogr., Sect. B*, 1978, **34**, 160 (*oxide, cryst struct*)

6-Phenoxy-6H-dibenzo[c,e][1,2]-oxaphosphorin, 9CI, 8CI **P-00065**

[35948-27-7]

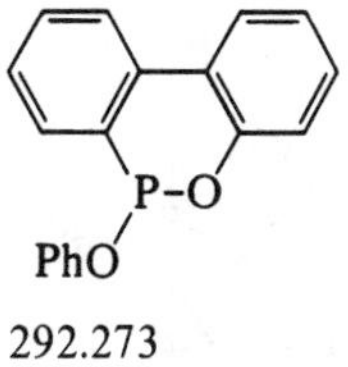

$C_{18}H_{13}O_2P$ M 292.273

Oil. Bp_5 210°.

U.S.P., 3 702 878, (*1972*); *CA*, **78**, 43708 (*synth, sulfide*)
Ger. Pat., 2 034 887, (*1972*); *CA*, **76**, 99823 (*synth, sulfide*)

2-Phenoxy-1,5-dihydro-2,4,3-benzodioxaphosphepin P-00066

2-Phenoxybenzo[d]*[1,3,2]dioxaphosphacyclohept-4-ene*

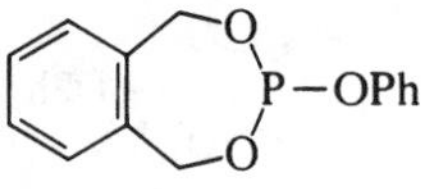

$C_{14}H_{13}O_3P$ M 260.229

Liq. $Bp_{0.02}$ 153°.

Oxide:

$C_{14}H_{13}O_4P$ M 276.228

Cryst. (C_6H_6/cyclohexane). Mp 91-92°.

Sulfide:

$C_{14}H_{13}O_3PS$ M 292.289

Cryst. (C_6H_6/Et_2O). Mp 86-88°.

Selenide:

$C_{14}H_{13}O_3PSe$ M 339.189

Cryst. (C_6H_6/Et_2O). Mp 105-106°.

Arbuzov, B.A. *et al*, *Izv. Akad. Nauk SSSR, Ser. Khim.*, 1985, 1762 (*Engl. transl.* p. 1612) (*synth, derivs, P nmr, pmr, conformn*)

2-Phenoxy-1,3,2-dioxaphospholane, 9CI P-00067

2-Phenoxy-1,3-dioxa-2-phosphacyclopentane. Ethylene phenyl phosphite

[1077-05-0]

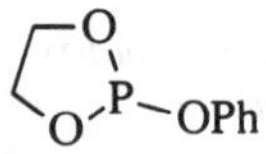

$C_8H_9O_3P$ M 184.131

Liq. $Bp_{0.3}$ 73°. n_D^{20} 1.5342. Readily liberates phenol in moist air.

2-Oxide: [16492-16-3]. *Ethylene phenyl phosphate.*

$C_8H_9O_4P$ M 200.130

Low melting solid. Mp 28-29°. $Bp_{0.3}$ 152-159°.

2-Sulfide: [24453-83-6]. O,O-*Ethylene* O-*phenyl phosphorothioate.* O,O-*Ethylene* O-*phenyl thiophosphate.*

$C_8H_9O_3PS$ M 216.191

Characterised spectroscopically.

Ozonide: see 4-Phenyl-1,2,3,5,8-pentaoxa-4-phospha(4-P^V)spiro[3.4]octane, P-00181

Ayres, D.C. *et al*, *J. Chem. Soc.*, 1957, 1109 (*synth, uv*)
Edmundson, R.S. *et al*, *J. Chem. Soc. (B)*, 1967, 577 (*oxide, synth, ir, props*)
Revel, M. *et al*, *Bull. Soc. Chim. Fr., Part I*, 1973, 1195 (*derivs, pmr, P nmr*)
Rueger, C. *et al*, *J. Prakt. Chem.*, 1984, **326**, 622 (*synth, P nmr*)

2-Phenoxy-1,3,2-dioxaphosphorinane, 9CI P-00068

2-Phenoxy-1,3,2-dioxaphosphinane. 2-Phenoxy-1,3-dioxa-2-phosphacyclohexane. Trimethylene phenyl phosphite

[1078-57-5]

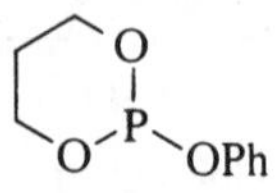

$C_9H_{11}O_3P$ M 198.158

Cryst. (Me_2CO/hexane). Mp 44-46°. $Bp_{0.4}$ 88°.

2-Oxide: [711-07-9]. *Trimethylene phenyl phosphate.*

$C_9H_{11}O_4P$ M 214.157

Needles ($CHCl_3/CCl_4$ or Et_2O/pet. ether). Sl. sol. Et_2O, CCl_4; sol. unchanged or hot H_2O. Mp 76-77° (69.5-71°).

Ayres, D.C. *et al*, *J. Chem. Soc.*, 1957, 1109 (*synth*)
Bergesen, K. *et al*, *Acta Chem. Scand.*, 1971, **25**, 2257 (*synth, pmr*)
Majoral, J.-P. *et al*, *Bull. Soc. Chim. Fr.*, 1971, 95 (*oxide, ir, P nmr*)
Hall, L.D. *et al*, *Can. J. Chem.*, 1972, **50**, 2092 (*oxide, synth, pmr, P nmr*)
Majoral, J.-P. *et al*, *Spectrochim. Acta, Part A*, 1972, **28**, 2247 (*ir*)
Efremov, Yu.Ya. *et al*, *Khim. Geterosikl. Soedin.*, 1974, 1620 (*Engl. transl.* p. 1424) (*ms*)
Arbuzov, B.A. *et al*, *Izv. Akad. Nauk SSSR, Ser. Khim.*, 1977, 2006 (*Engl. transl.* p. 1856) (*synth, raman, ir, conformn, P nmr*)
Arbuzov, B.A. *et al*, *Zh. Obshch. Khim.*, 1982, **52**, 2176 (*Engl. transl.* p. 1937) (*derivs, uv*)
Edmundson, R.S., *Org. Mass Spectrom.*, 1983, **18**, 150 (*oxide, ms*)
Jones, P.G. *et al*, *Acta Crystallogr., Sect. C*, 1984, **40**, 1061 (*oxide, cryst struct*)
Van Lier, J.J.C. *et al*, *Phosphorus Sulfur*, 1984, **19**, 173 (*pmr, cmr, P nmr*)

2-Phenoxy-4,5-diphenyl-1,3,2-dioxaphospholane, 9CI P-00069

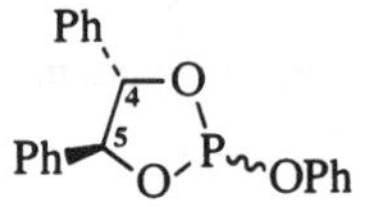

(4*RS*,5*RS*)-*form*

$C_{20}H_{17}O_3P$ M 336.326

(4RS,5RS)-form [91740-93-1]

2-Oxide:

$C_{20}H_{17}O_4P$ M 352.326

Cryst. (EtOH). Mp 116-117°.

(4RS,5SR)-form

meso-*form*

Epimers at P characterised spectroscopically.

2-Oxide: Cryst. (EtOH). Mp 107-109°. Config. at P uncertain.

Modro, T.A. *et al*, *Bull. Acad. Pol. Sci., Ser. Sci. Chim.*, 1972, **20**, 399 (*oxide, synth, ir, pmr*)
Khatib, F.El. *et al*, *Phosphorus Sulfur*, 1984, **20**, 55 (*synth*)

2-Phenoxynaphtho[1,8-*de*]-1,3,2-dioxaphosphorine, 9CI P-00070

1,8-Naphthylene phenyl phosphite

[72310-31-7]

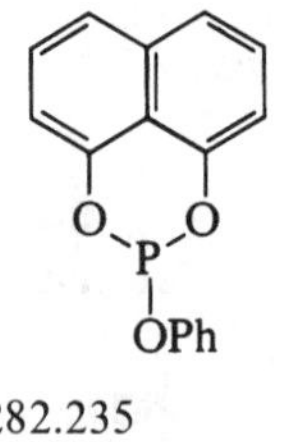

$C_{16}H_{11}O_3P$ M 282.235

Cryst. (pet. ether). Mp 54-55°. Bp_1 178-180°.

2-Oxide: 1,8-Naphthylene phenyl phosphate.

$C_{16}H_{11}O_4P$ M 298.234

Solid. Mp 115-116°.

Nifant'ev, É.E. *et al*, *Zh. Obshch. Khim.*, 1981, **51**, 1528 (*Engl. transl.* p. 1295) (*synth, P nmr*)
Voropai, L.M. *et al*, *Zh. Obshch. Khim.*, 1985, **55**, 65 (*Engl. transl.* p. 55) (*oxide, P nmr*)

2-Phenoxy-1,3,2-oxaselenaphosphorinane 2-oxide, 8CI P-00071

2-Phenoxy-1,3,2-oxaselenaphosphinane 2-oxide

[32102-01-5]

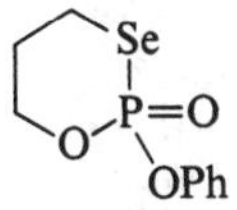

$C_9H_{11}O_3PSe$ M 277.118

Cryst. (CCl_4). Mp 96-98°.

Arbuzov, B.A. *et al, Izv. Akad. Nauk SSSR, Ser. Khim.*, 1983, **32**, 675 (*Engl. transl.* p. 613) (*synth*)
Arshinova, R.P. *et al, Izv. Akad. Nauk SSSR, Ser. Khim.*, 1985, 575 (*Engl. transl.* p. 525) (*pmr, cmr, uv*)

2-Phenoxy-1,3,2-oxathiaphospholane, 9CI P-00072

O,S-*Ethylene* O-*phenyl phosphorothioite*

[38580-66-4]

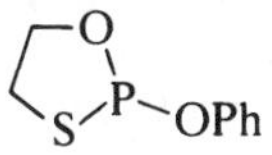

$C_8H_9O_2PS$ M 200.192

Liq. $Bp_{0.5}$ 123°.

2-Oxide: [78961-27-0]. O,S-*Ethylene* O-*phenyl phosphorothioate.*
$C_8H_9O_3PS$ M 216.191
Liq. d_4^{20} 1.37. $Bp_{0.09}$ 136°. n_D^{20} 1.5685.

Bergesen, K. *et al, Acta Chem. Scand.*, 1972, **26**, 2156 (*synth, pmr*)
Nuretdinova, O.N. *et al, Izv. Akad. Nauk SSSR, Ser. Khim.*, 1981, 1130 (*Engl. transl.* p. 890) (*oxide, synth, P nmr*)

2-Phenoxy-3-phenyl-1,3,2-oxazaphospholidine, 9CI P-00073

2-Phenoxy-3-phenyl-1,3,2-oxazaphospholane

[82564-88-3]

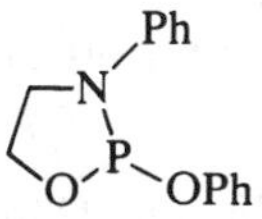

$C_{14}H_{14}NO_2P$ M 259.244

Liq. d_4^{20} 1.62. $Bp_{0.06}$ 126°. n_D^{20} 1.6060.

2-Oxide: [60749-96-4].
$C_{14}H_{14}NO_3P$ M 275.243
Cryst. (CCl_4/hexane). Mp 64-66°.

Brown, C. *et al, J. Chem. Soc., Perkin Trans. 2*, 1976, 888 (*oxide, synth, pmr, P nmr, props*)
Pudovik, M.A. *et al, Zh. Obshch. Khim.*, 1982, **52**, 1481 (*Engl. transl.* p. 1310) (*synth*)

2-Phenoxy-2,2′-spirobi[1,3,2-benzodioxaphosphole], 9CI P-00074

[19579-02-3]

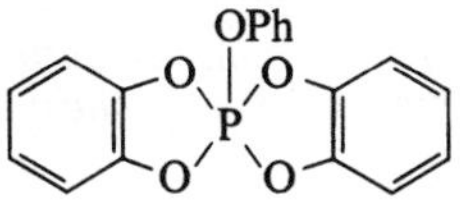

$C_{18}H_{13}O_5P$ M 340.271

Cryst. (C_6H_6/hexane). Mp 110-112° (107-109°). $Bp_{0.02}$ 170-185°.

Ramirez, F. *et al, Tetrahedron*, 1968, **24**, 5041 (*synth, P nmr*)
Sarma, R. *et al, J. Org. Chem.*, 1976, **41**, 473 (*synth, P nmr, cryst struct*)
Gloede, J., *J. Prakt. Chem.*, 1981, **323**, 621 (*synth, P nmr*)
Budilova, I.Yu. *et al, Zh. Obshch. Khim.*, 1984, **54**, 1985 (*Engl. transl.* p. 1771) (*synth, ir, P nmr*)

2-Phenoxy-4,4,5,5-tetramethyl-1,3,2-dioxaphospholane P-00075

4,4,5,5-Tetramethylethylene phenyl phosphite. 4,4,5,5-Tetramethyl-1,3-dioxa-2-phenoxy-2-phosphacyclopentane

[14812-61-4]

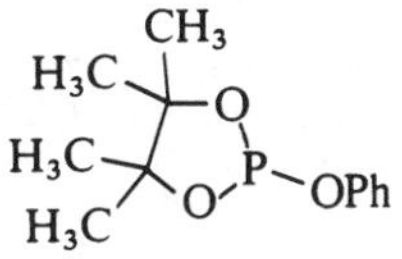

$C_{12}H_{17}O_3P$ M 240.238

Liq. Bp_2 88-90°.

2-Oxide: [22345-03-5]. *4,4,5,5-Tetramethylethylene phenyl phosphate.*
$C_{12}H_{17}O_4P$ M 256.238
No phys. props. recorded. Characterised spectroscopically.

Fontal, B. *et al, Tetrahedron*, 1966, **22**, 3275 (*synth, pmr*)
Kubayashi, S. *et al, Polym. Bull. (Berlin)*, 1980, **3**, 585 (*props, P nmr*)

Phenoxytetraphenylphosphorane, 9CI P-00076

Tetraphenylphenoxyphosphorane

[37084-64-3]

$PhOPPh_4$

$C_{30}H_{25}OP$ M 432.501

Solid, stable in air. Mp 138°. At 200°, dec. to Ph_2O, Ph_3P, and PhCOCOPh.

Razuvaeva, G.A. *et al, Izv. Akad. Nauk SSSR, Ser. Khim.*, 1969, 2234 (*Engl. transl.* p. 2083) (*synth, props*)
Razuvaeva, G.A. *et al, J. Organomet. Chem.*, 1972, **38**, 77 (*props*)

Phenoxytriphenylphosphorus(1+), 9CI P-00077

$Ph_3P^{\oplus}OPh$

$C_{24}H_{20}OP^{\oplus}$ M 355.395 (ion)

Perchlorate: [74289-41-1].
$C_{24}H_{20}ClO_5P$ M 454.846
Solid. Mp >290°.

Hexachloroantimonate:
$C_{24}H_{20}Cl_6OPSb$ M 689.863
Cryst. Mp 203-204°.

Teichmann, H. *et al, J. Prakt. Chem.*, 1972, **314**, 129 (*hexachloroantimonate, ir*)
Ohmori, H. *et al, Chem. Pharm. Bull.*, 1980, **28**, 910 (*perchlorate, synth*)

Phenthoate, BSI P-00078

Ethyl α-[(dimethoxyphosphinothioyl)thio]-benzeneacetate, 9CI. S-*α-Ethoxycarbonylbenzyl* O,O-*dimethyl phosphorodithioate. Cidial. Elsan. Papthion*

[2597-03-7]

$(MeO)_2P(S)SCHPhCOOEt$

$C_{12}H_{17}O_4PS_2$ M 320.358

▷AI7875000.

(+)-***form*** [61391-87-5]
  $[\alpha]^{20}_{D}$ +135.5° (c, 8.0 in $CHCl_3$).
(−)-***form*** [61362-00-3]
  $[\alpha]^{20}_{D}$ −132.3° (c, 4.5 in $CHCl_3$).
(±)-***form*** [61361-99-7]
  Nonsystemic agricultural insecticide. Yellowish oil. V. spar. sol. H_2O, sol. hexane. d_4^{20} 1.23-1.26. $Bp_{0.01}$ 122-125°. n_D^{20} 1.5449. Stable between pH 3.9 and 7.8.

B.P., 803 441, (*1958*); *CA*, **54**, 4502 (*synth*)
B.P., 834 814, (*1959*); *CA*, **55**, 457b
Ohkawa, H. *et al*, *Agric. Biol. Chem.*, 1976, **40**, 1857 (*synth, tox*)
Szalontai, G. *et al*, *J. Chromatsgr.*, 1976, **124**, 9 (*hplc*)
Pesticide Manual, 6th Ed., 415; 7th Ed., 432.
The Agrochemicals Handbook, Royal Society of Chemistry, 1983, A320.

10-Phenylacridophosphine, 10CI — P-00079

10-Phenyldibenzo[b,e]*phosphorin, 9CI. 10-Phenyl-9-phosphaanthracene. 10-Phenyldibenzo*[b,e]*phosphinine*
[20995-81-7]

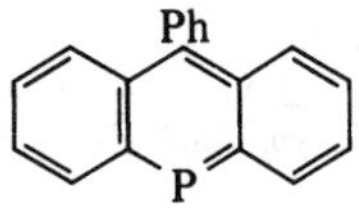

$C_{19}H_{13}P$ M 272.285
Yellow cryst. (toluene or by subl.). Mp 173-176°. Readily reacts with O_2. With maleic anhydride gives a Diels-Alder adduct Mp 259-62°.

de Koe, P. *et al*, *Angew. Chem., Int. Ed. Engl.*, 1968, **7**, 889 (*synth, uv*)
Schäfer, W. *et al*, *Angew. Chem., Int. Ed. Engl.*, 1972, **11**, 924 (*pe*)
Jongsma, C. *et al*, *Tetrahedron*, 1976, **32**, 121 (*props*)

5-Phenyl-10(5*H*)-acridophosphinone, 9CI — P-00080

9,10-Dihydro-9-phenyl-9-phosphaanthracen-10-one
[54086-39-4]

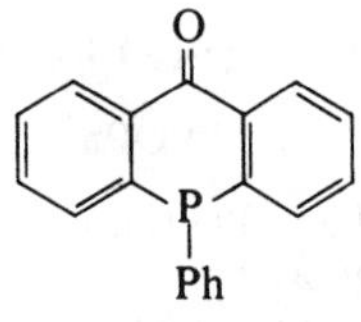

$C_{19}H_{13}OP$ M 288.285
Cryst. (EtOH under N_2). Mp 141-142° (136-137°).
5-Oxide: [54086-38-3].
  $C_{19}H_{13}O_2P$ M 304.284
  Cryst. (EtOH). Mp 220-222°.
B,PhCH₂Cl: 9,10-Dihydro-10-oxo-9-phenyl-9-phosphoniaanthracene chloride.
  $C_{26}H_{20}ClOP$ M 414.870
  Solid. Mp 289°.

Segall, Y. *et al*, *J. Chem. Soc., Chem. Commun.*, 1974, 501 (*synth*)
Jongsma, C. *et al*, Tet, 1976, **32**, 121 (*synth, pmr, ms, ir, oxide*)
Petrov, K.A., *Zh. Obshch. Khim.*, 1977, **47**, 2516 (*Engl. transl.* p. 2299) (*synth, deriv*)

(Phenylamino)carbonylphosphoramidic acid, 9CI — P-00081

Phenylcarbamoylphosphoramidic acid, 8CI. N-*Phenylurea-*N′-*phosphonic acid*

$$PhNHCONHP(O)(OH)_2$$

$C_7H_9N_2O_4P$ M 216.133
Di-Me ester: Dimethyl (*phenylamino*)-*carbonylphosphoramidate.*
  $C_9H_{13}N_2O_4P$ M 244.186
  Cryst. (H_2O). Mp 136-137°.
Di-Et ester: [33202-91-4]. *Diethyl* (*phenylamino*)-*carbonylphosphoramidate.*
  $C_{11}H_{17}N_2O_4P$ M 272.240
  Needles (MeOH). Mp 125-127°.
Diisopropyl ester: Diisopropyl (*phenylamino*)-*carbonylphosphoramidate.*
  $C_{13}H_{21}N_2O_4P$ M 300.294
  Needles (MeOH). Mp 138-140°.
Di-Ph ester: [1817-86-3]. *Diphenyl* (*phenylamino*)-*carbonylphosphoramidate.*
  $C_{19}H_{17}N_2O_4P$ M 368.328
  Needles (EtOH). Mp 155-156°.
Dichloride:
  $C_7H_7Cl_2N_2O_2P$ M 253.024
  Prisms. Mp 124-125°.

Kirsanov, A.V. *et al*, *Zh. Obshch. Khim.*, 1956, **26**, 2285; 1957, **27**, 1002 (*Engl. transl.* pp. 2555, 1084) (*synth*)
Matyusha, A.G. *et al*, *Zh. Obshch. Khim.*, 1971, **41**, 996 (*Engl. transl.* p. 1001)

[(Phenylamino)carbonyl]-phosphoramidothioic acid — P-00082

Phenylcarbamoylphosphoramidothioic acid, 8CI

$$PhNHCONHP(S)(OH)_2 \rightleftharpoons PhNHCONHP(O)(OH)(SH)$$

$C_7H_9N_2O_3PS$ M 232.193
Esters possess fungicidal properties.
O,S-*Di-Me ester:* O,S-*Dimethyl* [(*phenylamino*)-*carbonyl*]*phosphoramidothioate.*
  $C_9H_{13}N_2O_3PS$ M 260.247
  Prisms. Mp 76-78°.
O,O-*Di-Et ester:* [13557-07-8]. O,O-*Diethyl* [(*phenylamino*)*carbonyl*]*phosphoramidothioate.*
  $C_{11}H_{17}N_2O_3PS$ M 288.301
  Needles (Et_2O). Mp 121-122°.
O,S-*Di-Et ester:* [25359-63-1]. O,S-*Diethyl* [(*phenylamino*)*carbonyl*]*phosphoramidothioate.*
  $C_{11}H_{17}N_2O_3PS$ M 288.301
  Prisms. Mp 132-133°.
O,O-*Di-Ph ester:* O,O-*Diphenyl* [(*phenylamino*)-*carbonyl*]*phosphoramidothioate.*
  $C_{19}H_{17}N_2O_3PS$ M 384.389
  Needles (C_6H_6). Mp 166-168°.

Samarai, L.I. *et al*, *Zh. Obshch. Khim.*, 1966, **36**, 1433 (*Engl. transl.* p. 1439); 1969, **39**, 1511 (*Engl. transl.* p. 1480)

1-Phenylamino-1,3,2-dioxaphospholane — P-00083

N-*Phenyl-1,3,2-dioxaphospholan-2-amine, 9CI. 2-Anilino-1,3,2-dioxaphospholane. 2-Anilino-1,3-dioxa-2-phosphacyclopentane. Ethylene phenylphosphoramidite*
[34875-42-8]

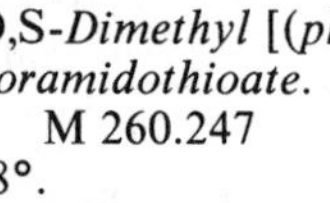

$C_8H_{10}NO_2P$ M 183.146
2-Oxide: [6587-25-3]. *Ethylene phenylphosphoramidate.*
  $C_8H_{10}NO_3P$ M 199.146

Cryst. ($CHCl_3$ or C_6H_6/pet. ether). Mp 123° (108-111°).

2-Sulfide: [65438-83-7]. *Ethylene phenylphosphoramidothioate.*
$C_8H_{10}NO_2PS$ M 215.206
Characterised spectroscopically.

Nifant'ev, É.E. *et al*, *Zh. Obshch. Khim.*, 1971, **41**, 2011 (*Engl. transl.* p. 2032) (*synth*)
Nguyen Thanh Thuong, *et al*, *Bull. Soc. Chim. Fr., Part II*, 1975, 2083 (*oxide*)
Ayed, N. *et al*, *C.R. Hebd. Seances Acad. Sci., Ser. C*, 1977, **285**, 221 (*sulfide, ir*)
Moerat, A. *et al*, *Phosphorus Sulfur*, 1983, **4**, 179 (*oxide, synth, props*)

[(Phenylamino)methyl]phosphonic acid, 9CI P-00084

(Anilinomethyl)phosphonic acid
[60558-59-0]

$PhNHCH_2P(O)(OH)_2$

$C_7H_{10}NO_3P$ M 187.135
Cryst. (H_2O). Mp 197-199°, 221°.

Di-Et ester: [56875-30-0]. *Diethyl [(phenylamino)-methyl]phosphonate.*
$C_{11}H_{18}NO_3P$ M 243.242
Liq. d_4^{20} 1.40. $Bp_{0.5}$ 135-137°. n_D^{20} 1.5225.

Diisopropyl ester: Diisopropyl [(phenylamino)methyl]-phosphonate. Bis(1-methylethyl) [(phenylamino)-methyl]phosphonate.
$C_{13}H_{22}NO_3P$ M 271.295
Cryst. (hexane). Mp 65-67°. $Bp_{0.6\text{-}0.8}$ 145-152°. n_D^{20} 1.5098.

Kreutzkamp, N. *et al*, *Arch. Pharm. (Weinheim, Ger.)*, 1962, **295**, 773 (*synth*)
Islamov, R.G. *et al*, *Zh. Obshch. Khim.*, 1975, **45**, 1444 (*Engl. transl.* p. 1414) (*ester, ir*)
Petrov, K.A. *et al*, *Zh. Obshch. Khim.*, 1977, **47**, 2741 (*Engl. transl.* p. 2494) (*synth, esters*)
Antokhina, L.A. *et al*, *Zh. Obshch. Khim.*, 1983, **53**, 2650 (*Engl. transl.* p. 2390) (*esters, synth, ir, P nmr*)

[3-(Phenylamino)propyl]phosphonic acid, 9CI P-00085

(3-Anilinopropyl)phosphonic acid

$PhNH(CH_2)_3P(O)(OH)_2$

$C_9H_{14}NO_3P$ M 215.188
Prisms (H_2O). Mp 129-130°. pK_{a1} 2.1, pK_{a2} 4.25, pK_{a3} 7.15.

Kosolapoff, G.M. *J. Am. Chem. Soc.*, 1944, **66**, 1511 (*synth*)
Rumpf, P. *et al*, *C.R. Hebd. Seances Acad. Sci.*, 1967, **224**, 919 (*props*)

7-Phenyl-6-aza-7-phosphabicyclo[3.2.1]-octane, 9CI P-00086

$C_{12}H_{16}NP$ M 205.239

6-Cyclohexyl, 7-sulfide: [35638-72-3].
$C_{18}H_{27}NPS$ M 320.452
Solid. Mp not reported.

Healy, J.D. *et al*, *Phosphorus*, 1971, **1**, 157 (*P nmr, cryst struct*)

3-Phenyl-1,3-azaphosphetidine, 10CI P-00087

[78227-61-9]

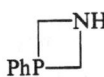

$C_8H_{10}NP$ M 151.147

1-Ph: [74633-13-9]. *1,3-Diphenyl-1,3-azaphosphetidine.*
$C_{14}H_{14}NP$ M 227.245
Solid. Mp 190-191°.

1-Benzyl: [74633-16-2]. *1-Benzyl-3-phenyl-1,3-aza-phosphetidine. 3-Phenyl-1-phenylmethyl-1,3-azaphosphetidine.*
$C_{15}H_{16}NP$ M 241.272
Solid. Mp 141-143°.

Arbuzov, B.A. *et al*, *Izv. Akad. Nauk SSSR, Ser. Khim.*, 1980, 735; *CA*, **93**, 95335 (*synth, P nmr*)
Arbuzov, B.A. *et al*, *Izv. Akad. Nauk SSSR, Ser. Khim.*, 1980, 2129; *CA*, **94**, 30620 (*synth, oxide, P nmr*)

2-Phenyl-1*H*-1,3-azaphosphole P-00088

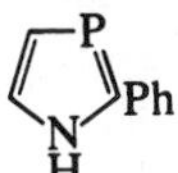

C_9H_8NP M 161.143
Solid. Mp 47°. Bp_1 122-124°.

Heinricke, J. *et al*, *Tetrahedron Lett.*, 1986, **27**, 5699 (*synth, uv, pmr, cmr, P nmr*)

2-Phenyl-1,2-azaphospholidine, 9CI P-00089

2-Phenyl-1-aza-2-phosphacyclopentane
[33730-82-4]

$C_9H_{12}NP$ M 165.174
Liq. $Bp_{0.01}$ 65-70°.

N-*Ph:* [33795-32-3]. *1,2-Diphenyl-1,2-azaphospholidine.*
$C_{15}H_{16}NP$ M 241.272
Solid. Mp 55-57°. $Bp_{0.01}$ 126-128°.

Oehme, H. *et al*, *J. Prakt. Chem.*, 1973, **315**, 526 (*synth*)

2-Phenyl-1,3-azaphospholidine, 9CI, 8CI P-00090

[20477-51-4]

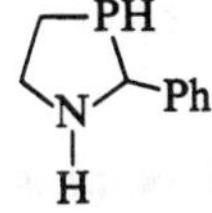

$C_9H_{12}NP$ M 165.174

(±)-*form*
Liq. Bp_2 108-111°. pK_a 6.93 (66% EtOH aq., 25°).

1-Et: 1-Ethyl-2-phenyl-1,3-azaphospholidine.
$C_{11}H_{16}NP$ M 193.228
Liq. $Bp_{0.5}$ 88-90°.

3-Ph: see 2,3-Diphenyl-1,3-azaphospholidine, D-00995

Issleib, K. *et al*, *Chem. Ber.*, 1968, **101**, 3619 (*derivs, synth*)

3-Phenyl-1,3-azaphospholidine, 9CI, 8CI **P-00091**
3-Phenyl-1-aza-3-phosphacyclopentane
[15096-33-0]

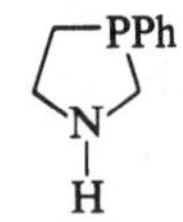

$C_9H_{12}NP$ M 165.174
Liq. Bp_2 109-112°.
1-Et: [20490-50-0]. *1-Ethyl-3-phenyl-1,3-azaphospholidine.*
$C_{11}H_{16}NP$ M 193.228
Liq. $Bp_{3.5}$ 117-118°. pK_a 6.91 (66% EtOH aq., 25°).

Issleib, K. *et al*, *Chem. Ber.*, 1967, **100**, 2685 (*synth*)
Issleib, K. *et al*, *Chem. Ber.*, 1968, **101**, 3619 (*deriv*)

1-Phenyl-1,2-azaphospholidin-5-one, 9CI, 8CI **P-00092**

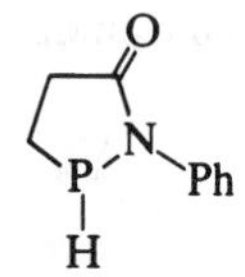

$C_9H_{10}NOP$ M 179.158
2-Et, 2-oxide:
$C_{11}H_{14}NO_2P$ M 223.211
Liq. d_4^{20} 1.23. $Bp_{0.006}$ 169-172°. n_D^{20} 1.5650.
2-Ph, 2-oxide:
$C_{15}H_{14}NO_2P$ M 271.255
Solid. Mp 115-116°. $Bp_{0.06}$ 234°.

Khairullin, V.K. *et al*, *Izv. Akad. Nauk SSSR, Ser. Khim.*, 1968, 1375 (*Engl. transl.* p. 1298)
Pudovik, A.N. *et al*, *Izv. Akad. Nauk SSSR, Ser. Khim.*, 1969, 2076 (*Engl. transl.* p. 1937)

3-Phenyl-1,3-azaphospholidin-2-thione, 10CI **P-00093**
3-Phenyl-1-aza-3-phosphacyclopentane-2-thione
[68286-92-0]

$C_9H_{10}NPS$ M 195.218
Solid. Mp 124-126°.
3-Sulfide: [68287-03-6].
$C_9H_{10}NPS_2$ M 227.278
Cryst. (EtOH). Mp 157-159°.

Oehme, H. *et al*, *J. Prakt. Chem.*, 1978, **320**, 600 (*synth, ir, ms*)

2-Phenyl-1,3,2-benzodioxaphosphole, 8CI **P-00094**
o-*Phenylene phenylphosphonite*
[4759-37-9]

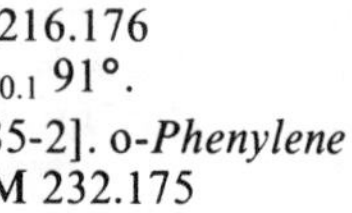

$C_{12}H_9O_2P$ M 216.176
Solid. Mp 28°. $Bp_{0.1}$ 91°.
2-Oxide: [40156-85-2]. o-*Phenylene phenylphosphonate.*
$C_{12}H_9O_3P$ M 232.175
Cryst. (C_6H_6). Mp 124-125°. Bp_9 206°.

Anschütz, L. *et al*, *J. Prakt. Chem.*, 1932, **133**, 65 (*oxide*)
Wieber, M. *et al*, *Monatsh. Chem.*, 1970, **101**, 776 (*synth*)
Vasil'ev, V.V. *et al*, *Zh. Obshch. Khim.*, 1976, **46**, 1739 (*Engl. transl.* p. 1690) (*P nmr*)
Gloede, J. *et al*, *Z. Anorg. Allg. Chem.*, 1983, **500**, 59 (*oxide, props*)

2-Phenyl-4*H*-1,3,2-benzodioxaphosphorin, 8CI **P-00095**

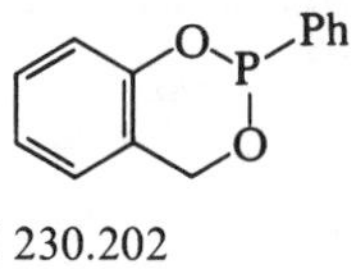

$C_{13}H_{11}O_2P$ M 230.202
2-Oxide: [4242-21-1].
$C_{13}H_{11}O_3P$ M 246.202
Chymotrypsin inhibitor. Solid. Mp 146-148°.
2-Sulfide: [4242-25-5].
$C_{13}H_{11}O_2PS$ M 262.262
Solid. Mp 37°.

Eto, M. *et al*, *Agric. Biol. Chem.*, 1962, **26**, 452, 630; 1963, **27**, 723, 789 (*derivs, ir, tlc*)

3-Phenyl-3*H*-benzophosphepin **P-00096**

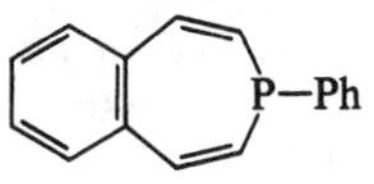

$C_{16}H_{13}P$ M 236.252
Cryst. (MeCN at low temp.). Mp 86-88°. Thermolabile, forming naphthalene and $(PhP)_x$ *via* a valence tautomer.
3-Oxide:
$C_{16}H_{13}OP$ M 252.252
Thermally stable solid. Mp 155-156°.

Märkl, G. *et al*, *Tetrahedron Lett.*, 1983, **24**, 2545 (*synth, cmr, P nmr, pmr, uv, ms*)

2-Phenyl-4*H*-1,3,2-benzoxathiaphosphorin, 8CI **P-00097**

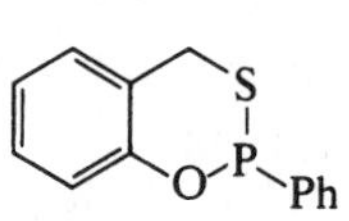

$C_{13}H_{11}OPS$ M 246.263
2-Oxide: [30337-25-8].
$C_{13}H_{11}O_2PS$ M 262.262
Solid. Mp 87°.

Nguyen, H.P. *et al*, *C.R. Hebd. Seances Acad. Sci., Ser. C*, 1970, **271**, 1465 (*synth, pmr*)

Phenylbis[2-(phenylethynyl)phenyl]phosphine, 10CI **P-00098**
[54100-67-3]

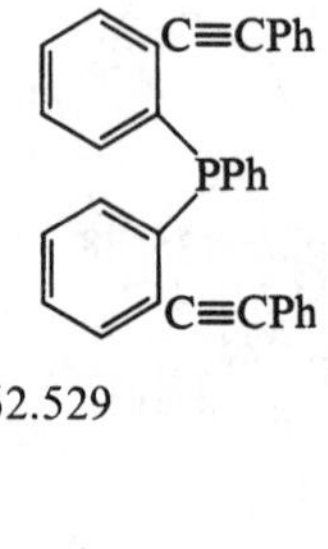

$C_{34}H_{23}P$ M 462.529

Ligand for Fe. Cryst. (xylene). Mp 150-151°. Reacts with H_2O in EtOH aq. to give the oxide of a dihydro deriv.

Oxide:
$C_{34}H_{23}OP$ M 478.529
Cryst. (C_6H_6/pet. ether). Mp 159-160°.

Winter, W., *Chem. Ber.*, 1976, **109**, 2405 (*synth, ms, ir, raman, nmr, deriv*)
Winter, W., *Angew. Chem., Int. Ed. Engl.*, 1978, **17**, 947 (*props*)
Luppold, E. *et al*, *Chem. Ber.*, 1983, **116**, 1923 (*complexes*)

Phenylbis(phenylethynyl)phosphine, 9CI — P-00099

[27258-73-7]

$PhP(C{\equiv}CPh)_2$

$C_{22}H_{15}P$ M 310.334
Ligand for Fe. Cryst. (pet. ether). Mp 53-55°.

Oxide: [31398-95-5].
$C_{22}H_{15}OP$ M 326.334
Cryst. (Me_2CO or EtOH). Mp 150°.

Sulfide: [1883-26-7].
$C_{22}H_{15}PS$ M 342.394
No phys. props. reported.

Mootz, D. *et al*, *Z. Kristallogr., Kristallgeom., Kristallphys., Kristallchem.*, 1969, **130**, 239 (*cryst struct*)
Chekunina, L.I. *et al*, *Zh. Obshch. Khim.*, 1972, **42**, 995 (*Engl. transl.* p. 985) (*synth, ir, uv, nmr*)
Fluck, E. *et al*, *Z. Anorg. Allg. Chem.*, 1972, **393**, 126 (*sulfide, nmr*)
Maumy, M., *Bull. Soc. Chim. Fr.*, 1972, 1600 (*oxide, ir, props*)
Fluck, E. *et al*, *Z. Anorg. Allg. Chem.*, 1975, **412**, 47 (*sulfide, pe*)
Smith, W.F. *et al*, *Inorg. Chem.*, 1977, **16**, 1593 (*synth, ir, complexes*)

1-Phenyl-6,7-bis[(trimethylsilyl)oxy]-phosphacycloundec-6-ene, 9CI — P-00100

[40334-16-5]

$C_{22}H_{34}O_2PSi_2$ M 417.654
Liq. $Bp_{0.04}$ 152-156°.

Sulfide:
$C_{22}H_{34}O_2PSSi_2$ M 449.714
Solid. Mp 100-102°.

van Reijendam, J.W. *et al*, *Tetrahedron Lett.*, 1972, 5181.

Phenylbis(triphenylgermyl)phosphine, 8CI — P-00101

[13371-29-4]

$(Ph_3Ge)_2PPh$

$C_{42}H_{35}Ge_2P$ M 715.892
White cryst. Mp 110°. Oxidises and hydrolyses readily.

Schumann, H. *et al*, *Chem. Ber.*, 1969, **102**, 2900.

P-Phenyl-*N,N′*-bis(triphenylphosphoranylidene)-phosphonous diamide, 9CI — P-00102

N,N′-(*Phenylphosphinidene*)*bis*[P,P,P-*triphenylphosphine imide*]. N,N′-(*Phenylphosphinidene*)*bis*[P,P,P-*triphenylphosphazene*]

$PhP(N{=}PPh_3)_2$

$C_{42}H_{35}N_2P_3$ M 660.673

Oxide: [4129-47-9]. P-*Phenyl*-N,N′-*bis*(*triphenylphosphoranylidene*)*phosphonic diamide*. N,N′-(*Phenylphosphinylidene*)*bis*[P,P,P-*triphenylphosphine imide*].
$C_{42}H_{35}N_2OP_3$ M 676.673
Cryst. (dimethoxyethane). Mp 192-193°.

Sulfide: [4042-71-1]. P-*Phenyl*-N,N′-*bis*(*triphenylphosphoranylidene*)*phosphonothioic diamide*. N,N′-(*Phenylphosphinothioylidene*)*bis*[P,P,P-*triphenylphosphine imide*].
$C_{42}H_{35}N_2P_3S$ M 692.733
Solid. Mp 200°.

Baldwin, R.A., *J. Org. Chem.*, 1965, **30**, 3866 (*oxide*)
Biddlestone, M. *et al*, *J. Chem. Soc., Dalton Trans.*, 1975, 2527 (*sulfide*)
Shtepanek, A.S. *et al*, *Zh. Obshch. Khim.*, 1975, **45**, 1012 (*Engl. transl.* p. 999) (*oxide*)

Phenylbis[3-(triphenylphosphoranylidene)-propenyl]phosphine, 8CI — P-00103

[24442-23-7]

$(Ph_3P{=}CHCH{=}CH)_2PPh$

$C_{48}H_{41}P_3$ M 710.773
Orange-red solid. Sol. C_6H_6, THF, spar. sol. Et_2O. Mp 108-110° (under Ar).

Issleib, K. *et al*, *J. Prakt. Chem.*, 1969, **311**, 857 (*synth*)

Phenyl cyclohexylphosphoramidochloridate, 9CI — P-00104

[58809-19-1]

$C_{12}H_{17}ClNO_2P$ M 273.699

(±)-form

Nucleoside phosphorylating agent. Cryst. (C_6H_6/pentane). Mp 94°.

Van Boom, J.H. *et al*, *Tetrahedron*, 1975, **31**, 2953 (*synth, use*)

9-Phenyl-1,5-diaza-9-phosphabicyclo[3.3.1]nonane, 9CI — P-00105

[35661-66-6]

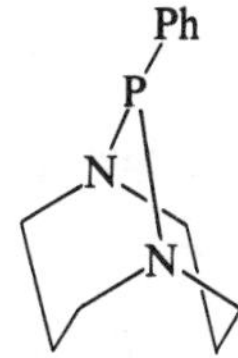

$C_{12}H_{17}N_2P$ M 220.253
Solid. $Bp_{0.01}$ 98-101°. V. air-sensitive.

Hutchins, R.O. *et al*, *J. Am. Chem. Soc.*, 1972, **94**, 9151 (*synth, props, P nmr*)
Maryanoff, B.E. *et al*, *J. Org. Chem.*, 1972, **37**, 3475 (*ms*)

9-Phenyl-1,5-diaza-9-phosphatricyclo[3.3.1.1^3,7^]decane, 9CI P-00106
9-Phenyl-1,5-diaza-9-phosphaadamantane
[35820-68-9]

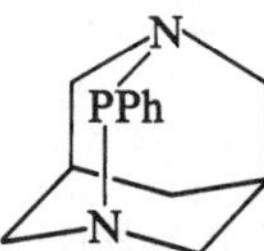

$C_{13}H_{17}N_2P$ M 232.264
Needles. Becomes liq. rapidly on exp. to air.

Hutchins, R.O. *et al*, *J. Am. Chem. Soc.*, 1972, **94**, 9151 (*synth, props, P nmr*)
Maryanoff, B.E. *et al*, *J. Org. Chem.*, 1972, **37**, 3475 (*ms*)

6-Phenyldibenzo[*d,f*][1,3,2]-dioxaphosphepin, 9CI, 8CI P-00107
2,2′-Biphenylene phenylphosphonite

$C_{18}H_{13}O_2P$ M 292.273
Oil which solidifies. Bp_1 215°.
Oxide:
$C_{18}H_{13}O_3P$ M 308.273
Solid. Mp 116-117°. Bp_2 93-95°.

Silcox, C.M. *et al*, *J. Am. Chem. Soc.*, 1966, **88**, 168.

5-Phenyl-5*H*-dibenzo[*b,f*]phosphepin, 9CI P-00108
1-Phenyl-1-phospha-2,3:6,7-dibenzocyclohepta-2,4,6-triene

$C_{20}H_{15}P$ M 286.312
Cryst. (EtOH). Mp 135-136°.
Sulfide: [75238-07-2].
$C_{20}H_{15}PS$ M 318.372
Cryst. (EtOH). Mp 218-219°.
10,11-Dihydro: se 10,11-Dihydro-5-phenyl-5H-dibenzo[b,f]phosphepin, D-00559

Segall, Y. *et al*, *Phosphorus Sulfur*, 1980, **8**, 243 (*synth, uv, pmr, ms*)

6-Phenyl-5*H*-dibenzo[*c,e*]phosphepin-5,7(6*H*)-dione, 9CI P-00109
[58529-35-4]

$C_{20}H_{13}O_2P$ M 316.295
Solid. Mp 166-168°.

Fenske, D. *et al*, *Chem. Ber.*, 1976, **109**, 359 (*synth, ms, ir*)

5-Phenyl-5*H*-dibenzophosphole, 9CI P-00110
*1-Phenyl-1*H*-benzo*[b]*phosphindole. 9-Phenyl-9-phosphafluorene*
[1088-00-2]

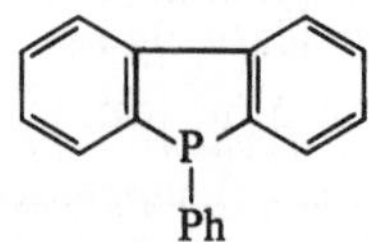

$C_{18}H_{13}P$ M 260.274
Ligand for group VIII metals. Mp 93.5-94°.
*B, MeI: 5-Methyl-5-phenyl-5*H*-dibenzophospholium iodide. 1-Methyl-1-phenyl-1*H*-benzo*[b]*phosphindolium iodide.*
$C_{19}H_{16}IP$ M 402.214
Cryst. (EtOH aq.). Mp 208-209°.
B, PhI: [53530-44-2]. *5,5-Diphenyl-5*H*-dibenzophospholium iodide. 1,1-Diphenyl-1*H*-benzo*[b]*phosphindolium iodide.*
$C_{24}H_{18}IP$ M 464.284
Cryst. (EtOH aq.). Mp 264° dec.
*B, PhCH₂Cl: 5-Benzyl-5-phenyl-5*H*-dibenzophospholium chloride. 1-Benzyl-1-phenyl-1*H*-benzo*[b]*phosphindolium chloride.*
$C_{25}H_{20}ClP$ M 386.860
Cryst. (EtOH aq.). Mp 294-296°.
5-Oxide: [1031-13-6].
$C_{18}H_{13}OP$ M 276.274
Cryst. (C_6H_6 or EtOH aq.). Mp 167-168°.
5-Sulfide: [33771-54-9].
$C_{18}H_{13}PS$ M 292.334
Cryst. (EtOH). Mp 179-180°.
5-Selenide:
$C_{18}H_{13}PSe$ M 339.234
Cryst. (C_6H_6). Mp 162-164°.

Campbell, I.G.M. *et al*, *J. Chem. Soc.*, 1961, 2133; 1965, 2184 (*oxide, selenide*)
Wittig, G. *et al*, *Chem. Ber.*, 1964, **97**, 747 (*synth, deriv*)
Chernyshev, E.A. *et al*, *Zh. Obshch. Khim.*, 1971, **41**, 800 (*synth, derivs*)
Ecker, A. *et al*, *Monatsh. Chem.*, 1971, **102**, 1851 (*sulfide*)
Alexander, R.G. *et al*, *J. Chem. Soc., Perkin Trans. 2*, 1974, 1836 (*ms*)
Nesmeyanov, N.A. *et al*, *J. Organomet. Chem.*, 1976, **110**, 49 (*synth*)
Gray, G.A. *et al*, *Org. Magn. Reson.*, 1980, **14**, 14 (*cmr*)

7-Phenyl-7*H*-dibenzo[*d,f*]phosphonin, 10CI P-00111
[75401-43-3]

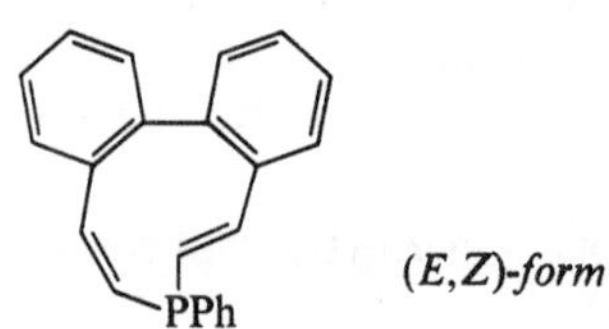

(E,Z)-form

$C_{16}H_{17}P$ M 240.284
***(E,Z)*-form** [74078-11-8]
Solid. Mp 68-73°.
B,MeI: [74078-12-9]. *7-Methyl-7-phenyl-7*H*-dibenzo*[d,f]*phosphoninium iodide.*
$C_{17}H_{20}IP$ M 382.223
Solid. Mp 208-209°.
7-Oxide: [74078-10-7].
$C_{16}H_{17}OP$ M 256.283
Cryst. + $0.5H_2O$ (pet. ether). Mp 166-167°.

Middlemass, E.D. *et al*, *J. Am. Chem. Soc.*, 1980, **102**, 4838 (*synth, uv, P nmr, pmr, oxide*)

Quin, L.D. *et al*, *J. Org. Chem.*, 1982, **47**, 905 (*synth, oxide, uv, P nmr, pmr*)

Phenyl (dichlorophosphinothioyl)-phosphorodichlorimidate, 10CI **P-00112**

P,P-*Dichloro*-P-*phenoxy*-N-(*dichlorophosphinothioyl*)-*phosphine imide*. P,P-*Dichloro*-P-*phenoxy*-N-(*dichlorophosphinothioyl*)*phosphazene*. P,P-*Dichloro*-P-*phenoxy*-N-(*dichlorophosphinothioyl*)-*iminophosphorane*

[79272-33-6]

$Cl_2P(S)N{=}PCl_2OPh$

$C_6H_5Cl_4NOPS$ M 311.957

Liq. d_4^{20} 1.56. $Bp_{0.3}$ 148-149°. n_D^{20} 1.5924.

Khodak, A.A. *et al*, *Izv. Akad. Nauk SSSR, Ser. Khim.*, 1981, 1117 (*Engl. transl.* p. 877) (*synth, ir, P nmr*)

4-Phenyl-3,5-dioxa-1-phosphabicyclo[2.2.1]heptane, 8CI **P-00113**

[21088-66-4]

$C_{10}H_{11}O_2P$ M 194.169

Cryst. (Et_2O). Mp 72-74°. Dec. by H_2O.

1-Oxide: [31149-52-7].

$C_{10}H_{11}O_3P$ M 210.169

Plates (C_6H_6/pet. ether). Sol. $MeNO_2$, MeCN, CH_2Cl_2. Mp 105-106°.

Kozlov, É.S. *et al*, *Zh. Obshch. Khim.*, 1968, **38**, 1881 (*Engl. transl.* p. 1828); 1970, **40**, 1673 (*Engl. transl.* p. 1661) (*synth, props, oxide*)

2-Phenyl-1,3,2-dioxaphosphepane, 9CI **P-00114**

Tetramethylene phenylphosphonite

[7526-37-6]

$C_{10}H_{13}O_2P$ M 196.185

Liq. $Bp_{0.3}$ 80°.

2-Oxide: [7191-20-0]. *Tetramethylene phenylphosphonate*.

$C_{10}H_{13}O_3P$ M 212.185

Cryst. (pet. ether). Mp 76-77.5°. Bp_3 125-127°.

2-Sulfide: [69813-50-9]. O,O-*Tetramethylene phenylphosphonothioate*.

$C_{10}H_{13}O_2PS$ M 228.245

No phys. props. reported.

Korshak, V.V. *et al*, *Izv. Akad. Nauk SSSR, Ser. Khim.*, 1957, 631 (*Engl. transl.* p. 641) (*oxide*)
Felcht, U. *et al*, *Chem. Ber.*, 1976, **109**, 3675.
Guimaraes, A.C. *et al*, *Org. Magn. Reson.*, 1978, **11**, 411 (*sulfide, pmr, cmr, P nmr*)
Petrov, A.A. *et al*, *Dokl. Akad. Nauk SSSR, Ser. Sci. Khim.*, 1981, **259**, 379; *CA*, **95**, 169298.
Kobayashi, S. *et al*, *Macromolecules*, 1986, **19**, 466 (*synth, P nmr, props*)

2-Phenyl-1,3,2-dioxaphospholane, 9CI **P-00115**

2-Phenyl-1,3-dioxa-2-phosphacyclopentane. *Ethylene phenylphosphonite*

[1006-83-3]

$C_8H_9O_2P$ M 168.132

Mobile liq. Bp_{20} 123°, $Bp_{0.4}$ 65°.

2-Oxide: [13468-89-8]. *Ethylene phenylphosphonate*.

$C_8H_9O_3P$ M 184.131

Liq. Mp 56-57°. $Bp_{0.3}$ 182-185°, $Bp_{0.015}$ 81-82°.

2-Sulfide: [36103-10-3]. O,O-*Ethylene phenylphosphonothioate*.

$C_8H_9O_2PS$ M 200.192

Characterised spectroscopically.

Edmundson, R.S. *et al*, *Tetrahedron*, 1967, **23**, 283 (*oxide, synth, ir, props*)
Evdakov, V.P. *et al*, *Zh. Obshch. Khim.*, 1967, **37**, 441 (*Engl. transl.* p. 412) (*oxide, synth*)
Revel, M. *et al*, *Bull. Soc. Chim. Fr., Part I*, 1973, 1195 (*oxide, pmr, P nmr*)
Dousse, G. *et al*, *J. Organomet. Chem.*, 1975, **88**, C35 (*synth*)
Tan, Han-Wan *et al*, *Tetrahedron Lett.*, 1975, 619 (*cmr, P nmr*)
Dutasta, J.P. *et al*, *Tetrahedron*, 1979, **35**, 197 (*synth, sulfide, pmr, cmr, P nmr, props*)

2-Phenyl-1,3,2-dioxaphosphorinane, 9CI **P-00116**

2-Phenyl-1,3,2-dioxaphosphinane. *Trimethylene phenylphosphonite*

[7526-32-1]

$C_9H_{11}O_2P$ M 182.158

Viscous liq. $Bp_{0.15}$ 72-74°. Undergoes ring-opening polymerisation.

2-Oxide: [7191-13-1]. *Trimethylene phenylphosphonate*.

$C_9H_{11}O_3P$ M 198.158

Oil which slowly cryst. Mp 33°. $Bp_{0.6}$ 172°. n_D^{21} 1.5369.

2-Sulfide: [30342-94-0]. O,O-*Trimethylene phenylphosphonothioate*.

$C_9H_{11}O_2PS$ M 214.218

Oil. n_D^{20} 1.5903.

Mukaiyama, T. *et al*, *J. Org. Chem.*, 1964, **29**, 2572 (*synth*)
Edmundson, R.S. *et al*, *Tetrahedron*, 1967, **23**, 283 (*oxide, synth, ir*)
Nguyen Hoang Phuong *et al*, *C.R. Hebd. Seances Acad. Sci., Ser. C*, 1970, **271**, 1465 (*sulfide, ir*)
Majoral, J.P. *et al*, *Spectrochim. Acta, Part A*, 1972, **28**, 2247 (*oxide, ir*)
Singh, G., *J. Org. Chem.*, 1979, **44**, 1060 (*synth, props*)

2-Phenyl-1,3,6,2-dioxathiaphosphocane, 10CI **P-00117**

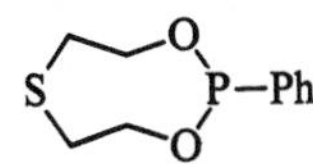

$C_{10}H_{13}O_2PS$ M 228.245

2-Oxide: [74720-40-4].
$C_{10}H_{13}O_3PS$ M 244.245
Solid. Mp 76-77°.
2-Sulfide: [74720-44-8].
$C_{10}H_{13}O_2PS_2$ M 260.305
Liq. Bp_2 155°.

Sharma, R.K. *et al*, *J. Chem. Res. (S)*, 1980, 12 (*synth, ir, ms, pmr*)

Phenyl di-2-propenyl phosphate, 9CI — P-00118

Diallyl phenyl phosphate
[1623-12-7]

$(H_2C{=}CHCH_2O)_2P(O)OPh$

$C_{12}H_{15}O_4P$ M 254.222
Liq. d_{25}^{25} 1.142. Bp_2 151°, $Bp_{0.5}$ 102°. n_D^{25} 1.4957.
Polymerises to transparent flame-resistant solid.

Toy, A.D.F. *et al*, *J. Am. Chem. Soc.*, 1954, **76**, 2191 (*synth*)
Faizullin, I.N. *et al*, *CA*, 1969, **71**, 112543 (*synth, ir*)

Phenyldi-2-propenylphosphine, 9CI — P-00119

Diallylphenylphosphine, 8CI
[29949-75-5]

$PhP(CH_2CH{=}CH_2)_2$

$C_{12}H_{15}P$ M 190.224
Air-sensitive pale-yellow liq. Bp_{14} 127°, $Bp_{0.7}$ 83°. n_D^{25} 1.567. Darkens on long standing.
Oxide:
$C_{12}H_{15}OP$ M 206.224
Hygroscopic solid. Mp 54°. $Bp_{1.3}$ 169° (Bp_2 135-137°).
Sulfide: [54970-71-7].
$C_{12}H_{15}PS$ M 222.284
No phys. props. reported.

Jones, W.J. *et al*, *J. Chem. Soc.*, 1947, 1446 (*synth*)
Berlin, K.D. *et al*, *J. Am. Chem. Soc.*, 1960, **82**, 2712 (*oxide, ir*)
Grim, S.O. *et al*, *J. Chem. Eng. Data*, 1970, **15**, 497 (*nmr*)
Samaan, S., *Tetrahedron Lett.*, 1974, 3927 (*sulfide, use*)

Phenyldipropylphosphine, 9CI — P-00120

[7650-83-1]

$PhP(CH_2CH_2CH_3)_2$

$C_{12}H_{19}P$ M 194.256
Ligand for Mn and metals of Groups VIB and VIII. Liq. Bp_9 103-104°. n_D^{20} 1.5284.

Davies, W.C. *et al*, *J. Chem. Soc.*, 1929, 1262 (*synth*)
Goetz, H. *et al*, *Justus Liebigs Ann. Chem.*, 1968, **715**, 1 (*synth*)
Mann, B.E., *J. Chem. Soc., Perkin Trans. 2*, 1972, 30 (*cmr, nmr*)
Timokhin, B.V. *et al*, *Zh. Obshch. Khim.*, 1977, **47**, 1267 (*Engl. transl.* p. 1167) (*synth, nmr*)

2-Phenyl-1,3,2-dithiaphospholane, 9CI — P-00121

S,S-Ethylene phenylphosphonodithioite. 2-Phenyl-1,3-dithia-2-phosphacyclopentane
[4669-54-9]

$C_8H_9PS_2$ M 200.253
Liq. $Bp_{0.7}$ 130°, $Bp_{0.5}$ 106°.
2-Oxide: [4669-52-7]. S,S-*Ethylene phenylphosphonodithioate.*
$C_8H_9OPS_2$ M 216.252
Liq. $Bp_{0.05}$ 220°.
2-Sulfide: [29021-62-3]. *Ethylene phenylphosphonotrithioate.*
$C_8H_9PS_3$ M 232.313
Solid. Mp 68°.

Bergesen, K. *et al*, *Acta Chem. Scand.*, 1972, **26**, 3037 (*synth, pmr*)
Peake, S.C. *et al*, *J. Chem. Soc., Perkin Trans. 2*, 1972, 380 (*synth, derivs, pmr, P nmr*)
Albrand, J.-P. *et al*, *Org. Magn. Reson.*, 1973, **5**, 33 (*pmr*)
Davidson, G. *et al*, *Spectrochim. Acta, Part A*, 1983, **39**, 419 (*ir, raman*)

Phenyldivinylphosphine — P-00122

Diethenylphenylphosphine, 9CI
[26681-88-9]

$PhP(CH{=}CH_2)_2$

$C_{10}H_{11}P$ M 162.171
Ligand for Cr and Ni. Viscous yellow liq. $Bp_{0.5}$ 55°.
Oxide: [13815-96-8].
$C_{10}H_{11}OP$ M 178.170
Cryst. (cyclohexane). Mp 50-51°. Bp_2 130-130.5°.

Maier, L. *et al*, *J. Am. Chem. Soc.*, 1957, **79**, 5884 (*synth*)
Bundgaard, T. *et al*, *Tetrahedron Lett.*, 1972, 3353 (*cmr*)
Levin, Ya.A. *et al*, *Zh. Obshch. Khim.*, 1972, **42**, 1166 (*Engl. transl.* p. 1156) (*oxide*)
Collins, D.J. *et al*, *Aust. J. Chem.*, 1974, **27**, 841 (*oxide, use*)

1,2-Phenylenebis[dimethylphosphine], 9CI — P-00123

1,2-Bis(dimethylphosphino)benzene. Diphos
[7237-07-2]

$C_{10}H_{16}P_2$ M 198.184
Ligand for metals of Groups VB, VIB and VIII. Oily liq. Bp_{15} 139-141°, $Bp_{0.5}$ 80-83° ($Bp_{0.3}$ 98-101°).
B,MeI: Trimethyl(2-dimethylphosphinophenyl)-phosphonium iodide.
$C_{11}H_{19}IP_2$ M 340.123
Cryst. (EtOH). Mp 275° dec., 298° dec.

Hart, F.A., *J. Chem. Soc.*, 1960, 3324 (*synth*)
Warren, L.F. *et al*, *Inorg. Chem.*, 1976, **15**, 3126 (*synth, pmr, nmr, complexes*)
Levason, W. *et al*, *J. Chem. Soc., Dalton Trans.*, 1979, 1718 (*synth, pmr, nmr, ms, complexes*)
Kyba, E.P. *et al*, *Organometallics*, 1983, **2**, 1877 (*synth, nmr*)

1,3-Phenylenebis[dimethylphosphine], 9CI **P-00124**

1,3-Bis(dimethylphosphino)benzene

$C_{10}H_{16}P_2$ M 198.184

Liq. $Bp_{0.3}$ 103°.

Dioxide: 1,3-Bis(dimethylphosphinyl)benzene.
$C_{10}H_{16}O_2P_2$ M 230.183
Mp 132°.

Disulfide: [82340-10-1]. *1,3-Bis(dimethylphosphinothioyl)benzene.*
$C_{10}H_{16}P_2S_2$ M 262.304
Mp 139-140°.

Kaim, W. *et al*, *Chem. Ber.*, 1981, **114**, 1576; 1982, **115**, 1265 (*synth, derivs, esr, pmr*)

1,4-Phenylenebis[dimethylphosphine], 9CI **P-00125**

1,4-Bis(dimethylphosphino)benzene

[10498-57-4]

$C_{10}H_{16}P_2$ M 198.184

Mp 29°. Bp_{20} 152-155°.

Dioxide: [77876-82-5]. *1,4-Bis(dimethylphosphinyl)-benzene.*
$C_{10}H_{16}O_2P_2$ M 230.183
Spar. sol. Et_2O. Mp 323-325°. $Bp_{0.01}$ 250° subl.

Disulfide: [69220-11-7]. *1,4-Bis(dimethylphosphinothioyl)benzene.*
$C_{10}H_{16}P_2S_2$ M 262.304
Cryst. ($EtOH/CHCl_3$). Mp 249°. Desulfurized by Na.

Diselenide: 1,4-Bis(dimethylphosphinoselenoyl)benzene.
$C_{10}H_{16}P_2Se_2$ M 356.104
Leaflets. Mp 242°.

Kaim, W. *et al*, *Chem. Ber.*, 1978, **111**, 3846 (*synth, pe, pmr*)
Kaim, W., *Z. Naturforsch., B*, 1981, **36**, 150 (*oxide, pe, pmr, esr, selenide*)
Kaim, W. *et al*, *Chem. Ber.*, 1982, **115**, 1265 (*pmr, esr, disulfide*)

1,2-Phenylenebis[diphenylphosphine], 9CI **P-00126**

1,2-Bis(diphenylphosphino)benzene

[13991-08-7]

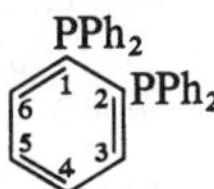

$C_{30}H_{24}P_2$ M 446.467

Cryst. (EtOH/toluene or DMF). Mp 183-185°. $Bp_{0.5}$ 150° subl.

Hart, F.A., *J. Chem. Soc.*, 1960, 3324 (*synth*)
Levason, W. *et al*, *J. Organomet. Chem.*, 1975, **84**, 239 (*ms*)
Talarj, R. *et al*, *Z. Naturforsch., B*, 1981, **36**, 451 (*synth, nmr, ms*)
McFarlane, H.C.E. *et al*, *Polyhedron*, 1983, **2**, 303 (*synth*)

1,3-Phenylenebis[diphenylphosphine], 9CI **P-00127**

1,3-Bis(diphenylphosphino)benzene

[1179-05-1]

$C_{30}H_{24}P_2$ M 446.467

Oil.

B,2MeI: 1,3-Phenylenebis(methyldiphenylphosphonium)diiodide.
Yellow cryst. (EtOH). Mp 199-200°, 282-284°.

Schindlbauer, H. *et al*, *Monatsh. Chem.*, 1965, **96**, 285, 1793 (*uv, derivs*)
Zorn, H. *et al*, *Chem. Ber.*, 1965, **98**, 1965 (*ir, deriv*)
Baldwin, R.A. *et al*, *J. Org. Chem.*, 1967, **32**, 1572 (*synth*)

1,4-Phenylenebis[diphenylphosphine], 9CI **P-00128**

1,4-Bis(diphenylphosphino)benzene

[1179-06-2]

$C_{30}H_{24}P_2$ M 446.467

Cryst. (propanol). Mp 170-171°. Dec. at ca. 360°.

Dioxide: 1,4-Bis(diphenylphosphinyl)benzene.
$C_{30}H_{24}O_2P_2$ M 478.466
Solid. Mp 298-300°.

Zorn, H. *et al*, *Chem. Ber.*, 1965, **98**, 2431 (*dioxide, ir*)
Baldwin, R.A. *et al*, *J. Org. Chem.*, 1967, **32**, 1572 (*synth*)
Zorn, H. *et al*, *Monatsh. Chem.*, 1967, **98**, 731 (*uv, ir*)
Schindlbauer, H. *et al*, *Monatsh. Chem.*, 1967, **98**, 1196 (*synth, uv*)

[1,2-Phenylenebis(methylene)]bisphosphonic acid, 9CI **P-00129**

o-*Xylylenediphosphonic acid. 1,2-Bis(phosphonomethyl)benzene*

[42104-58-5]

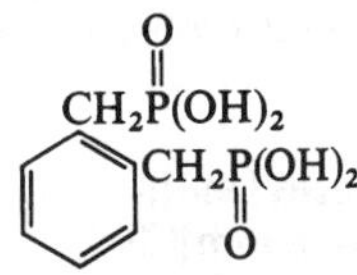

$C_8H_{12}O_6P_2$ M 266.127

Sl. yellow powder. Mp 252° dec. pK_{a1} 5.3, pK_{a2} 10.3, pK_{a3} 10.3, pK_{a4} 10.3.

Tetra-Et ester: [42092-05-7]. *Tetraethyl [1,2-phenylenebis(methylene)]bisphosphonate.*
$C_{16}H_{28}O_6P_2$ M 378.341
Oil.

Ernst, L., *Org. Magn. Reson.*, 1977, **9**, 35 (*cmr, P nmr*)
Maier, L. *et al*, *Phosphorus Sulfur*, 1978, **5**, 45 (*synth*)

[1,3-Phenylenebis(methylene)]bisphosphonic acid, 9CI **P-00130**

m-*Xylylenediphosphonic acid. 1,3-Bis(phosphonomethyl)benzene*

$C_8H_{12}O_6P_2$ M 266.127

Tetra-Et ester: [56875-38-5]. *Tetraethyl [1,2-phenylenebis(methylene)]bisphosphonate.*
$C_{16}H_{28}O_6P_2$ M 378.341
Liq. $Bp_{0.05}$ 190-195°.

Ernst, L., *Org. Magn. Reson.*, 1977, **9**, 35 (*synth, cmr, P nmr*)

[1,4-Phenylenebis(methylene)]bisphosphonic acid, 9CI **P-00131**

p-*Xylylenediphosphonic acid. 1,4-Bis(phosphonomethyl)benzene*

[4546-06-9]

$C_8H_{12}O_6P_2$ M 266.127

Solid or cryst. (H_2O). Mp 275-280° dec., Mp 321-323° dec.

Tetra-Me ester: [52577-04-5]. *Tetramethyl [1,4-phenylenebis(methylene)]bisphosphonate.*
$C_{12}H_{20}O_6P_2$ M 322.234
Used in Wittig-Horner reactions. Liq.

Tetra-Et ester: [4546-04-7]. *Tetraethyl [1,4-phenylenebis(methylene)]bisphosphonate.*
$C_{16}H_{28}O_6P_2$ M 378.341
Cryst. (C_6H_6/pet. ether). Mp 73-74° (70-72°). Bp_7 220°, $Bp_{0.1}$ 203-206°.
Tetrapropyl ester: Tetrapropyl [1,4-phenylenebis(methylene)]bisphosphonate.
$C_{20}H_{36}O_6P_2$ M 434.448
Solid. Mp 76-77°. Bp_1 240-245°.
Tetra-Ph ester: Tetraphenyl [1,4-phenylenebis(methylene)]bisphosphonate.
$C_{32}H_{28}O_6P_2$ M 570.517
Cryst. (C_6H_6/Me_2CO). Mp 172°.
Tetra-chloride:
$C_8H_8Cl_4O_2P_2$ M 339.910
Cryst. (toluene). Mp 171-173°. $Bp_{0.4}$ 190-200°.
Tetra-amide:
$C_8H_{16}N_4O_2P_2$ M 262.188
Solid. Sol. H_2O.

Mel'nikov, N.N. *et al*, *Zh. Obshch. Khim.*, 1961, **31**, 3953 (*Engl. transl.* p. 3681) (*ester*)
Chantrell, P.G. *et al*, *J. Appl. Chem.*, 1965, **15**, 460 (*synth, derivs*)
Abramov, V.S. *et al*, *Zh. Obshch. Khim.*, 1967, **37**, 2243 (*Engl. transl.* p. 2129) (*esters*)
Ernst, L., *Org. Magn. Reson.*, 1977, **9**, 35 (*cmr, nmr*)
Tewari, R.S. *et al*, *Indian J. Chem., Sect. B*, 1977, **15**, 753 (*ester, synth, use*)

1,2-[Phenylenebis[methylene]]-bis[triphenylphosphonium](2+), 9CI — P-00132

(o-*Phenylenedimethyl)bis[triphenylphosphonium](2+)*, *8CI*. *2-Xylylenebis[triphenylphosphonium](2+)*

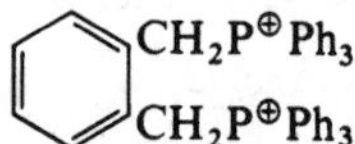

$C_{44}H_{38}P_2^{\oplus\oplus}$ M 628.732 (ion)
Dibromide: [1519-46-6].
$C_{44}H_{38}Br_2P_2$ M 788.540
Cryst. ($CHCl_3$). Mp >340°. LiOEt or butyllithium→bis-ylide.
Bis-ylide: see [1,2-Phenylenebis[methylene]]-bis[triphenylphosphorane], P-00135

Griffin, C.E. *et al*, *J. Org. Chem.*, 1962, **27**, 1627; 1963, **28**, 1715 (*synth*)
Ojima, J. *et al*, *Bull. Chem. Soc. Jpn.*, 1976, **49**, 2840; 1977, **50**, 933 (*use*)
Mitchell, R.H. *et al*, *Can. J. Chem.*, 1977, **55**, 210 (*use*)

[1,3-Phenylenebis[methylene]]-bis[triphenylphosphonium](2+), 9CI — P-00133

1,3-Bis(diphenylphosphiniomethyl)benzene(2+)
$C_{44}H_{38}P_2^{\oplus\oplus}$ M 628.732 (ion)
LiOEt → bisylide.
Dichloride: [66726-75-8].
$C_{44}H_{38}Cl_2P_2$ M 699.638
No phys. props. reported.
Dibromide: [10273-74-2].
$C_{44}H_{38}Br_2P_2$ M 788.540
Used in Wittig reactions leading to polycyclic hydrocarbons. Solid ($CHCl_3$/C_6H_6), prisms (propanol). Mp 246°, 320-325°, 340°.
Bisylide: [1,3-Phenylenebis[methylene]]-bis[triphenylphosphorane], 9CI. (m-*Phenylenedimethylidyne)bis[triphenylphosphorane]*, *8CI*.
$C_{44}H_{36}P_2$ M 626.716
Employed in Wittig reactions; prepd. *in situ.*

Friedrich, F. *et al*, *Chem. Ber.*, 1959, **92**, 2756 (*bromide, synth*)
Shubina, L.V. *et al*, *Zh. Obshch. Khim.*, 1972, **42**, 969 (*Engl. transl.* p. 958) (*bromide, synth, use*)
Storck, W. *et al*, *Makromol. Chem.*, 1975, **176**, 97 (*bromide, synth, use*)
Thulin, B. *et al*, *Acta Chem. Scand., Ser. B*, 1978, **32**, 109 (*chloride, use*)
Ghose, B.N., *J. Prakt. Chem.*, 1982, **324**, 1052 (*bromide, synth, use*)

1,4-[Phenylenebis[methylene]]-bis[triphenylphosphonium](2+), 9CI — P-00134

(p-*Phenylenedimethylene)-bis[triphenylphosphonium](2+)*, *8CI*. *4-Xylylenebis[triphenylphosphonium](2+)*
$C_{44}H_{38}P_2^{\oplus\oplus}$ M 628.732 (ion)
Reacts with NaOEt or LiOEt to give bis-ylide.
Dichloride: [1519-47-7].
$C_{44}H_{38}Cl_2P_2$ M 699.638
Cryst. (DMF or EtOH), cryst. + $2H_2O$(H_2O). Mp 414° (400°).
▷TA3580000.
Dibromide: [40817-03-6].
$C_{44}H_{38}Br_2P_2$ M 788.540
Cryst. (EtOH). Mp >340° dec.

Campbell, T.W. *et al*, *J. Org. Chem.*, 1959, **24**, 730, 1246 (*chloride*)
Friedrich, K. *et al*, *Chem. Ber.*, 1959, **92**, 2756 (*bromide*)

[1,2-Phenylenebis[methylene]]-bis[triphenylphosphorane], 9CI — P-00135

(1,2-Phenylenediyldimethylidyne)-bis[triphenylphosphorane]. 1,2-Bis(triphenylphosphoniomethyl)benzene
[31574-96-6]

$Ph_3P{=}CH$ $CH{=}PPh_3$

$C_{44}H_{36}P_2$ M 626.716
Prepd. *in situ*. Used in Wittig reactions.

Griffin, C.E. *et al*, *J. Org. Chem.*, 1962, **27**, 1627.
Ojima, J. *et al*, *Bull. Chem. Soc. Jpn.*, 1976, **49**, 2840; 1977, **50**, 933 (*use*)
Mitchell, R.H. *et al*, *Can. J. Chem.*, 1977, **55**, 210 (*use*)
Ehrensperger, C.-P. *et al*, *Helv. Chim. Acta*, 1978, **61**, 2813 (*use*)
Staab, H.A. *et al*, *Chem. Ber.*, 1979, **112**, 3895 (*use*)

1,4-Phenylenebis[methylphosphinic acid], 8CI P-00137

[59384-67-7]

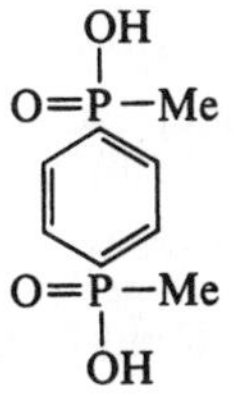

$C_8H_{12}O_4P_2$ M 234.128

Cryst. (EtOH). Mp 233-235°.

Di-Me ester: [10580-44-6]. *Dimethyl 1,4-phenylenebis[methylphosphinate].*

$C_{10}H_{16}O_4P_2$ M 262.182

Solid. Mp 107-109°.

Evleth, E.M. *et al, J. Org. Chem.*, 1962, **27**, 2192 (*synth, ir*)
Baldwin, R.A. *et al, J. Org. Chem.*, 1967, **32**, 2172 (*synth*)

1,4-Phenylenebis[methylphosphinodithioic acid], 9CI P-00138

[66055-07-0]

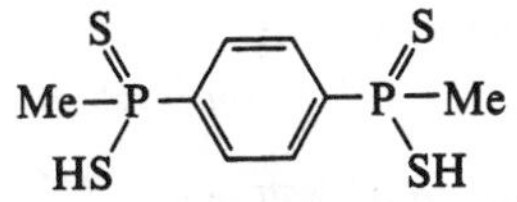

$C_8H_{12}P_2S_4$ M 298.370

Cryst. (xylene). Mp 176-183°.

Di-Me ester: [66055-08-1]. *Dimethyl 1,4-phenylenebis[methylphosphinodithioate].*

$C_{10}H_{16}P_2S_4$ M 326.424

Cryst. (xylene/ligroin). Mp 128-132°.

Diemert, K. *et al, Chem. Ber.*, 1978, **111**, 629 (*synth, nmr*)

1,2-Phenylenebis[phenylphosphinic acid], 9CI P-00139

[5994-56-9]

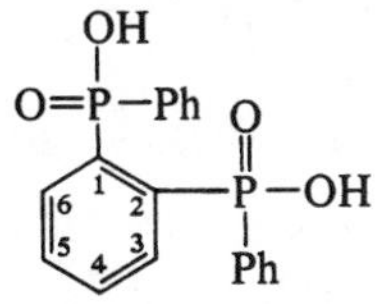

$C_{18}H_{16}O_4P_2$ M 358.270

Cryst. (EtOH aq.). Mp 207-208° (197-198°).

Benzylisothiouronium salt: Cryst. (EtOH aq.). Mp 201-202°.

Di-Me ester: [37909-36-7]. *Dimethyl 1,2-phenylenebis(phenylphosphinate).*

$C_{20}H_{20}O_4P_2$ M 386.323

$Bp_{0.5}$ 260°. On cooling becomes glassy and cannot be recryst.

Di-Et ester: [78605-34-2]. *Diethyl 1,2-phenylenebis(phenylphosphinate).*

$C_{22}H_{24}O_4P_2$ M 414.377

Characterised spectroscopically.

Mann, F.G. *et al, J. Chem. Soc. (C)*, 1966, 916 (*synth, derivs*)
Mann, F.G. *et al, J. Chem. Soc., Perkin Trans. 1*, 1972, 1631 (*synth, ir*)
Kyba, E.P. *et al, Tetrahedron Lett.*, 1981, **22**, 1875 (*ester, P nmr*)

1,4-Phenylenebis[phenylphosphinic acid], 9CI P-00140

[10212-05-2]

$C_{18}H_{16}O_4P_2$ M 358.270

Solid. Mp 330-333°.

Baldwin, R.A. *et al, J. Org. Chem.*, 1967, **32**, 1572.

1,3-Phenylenebis[phosphorodichloridite], 9CI P-00141

1,3-Phenylenebis(dichlorophosphite)

$C_6H_4Cl_4O_2P_2$ M 311.856

Pungent liq. d_0^{18} 1.570. Bp_{56} 240°.

Knauer, W., *Ber.*, 1894, **27**, 2565 (*synth*)

1,4-Phenylenebis[phosphorodichloridite], 9CI P-00142

1,4-Phenylenebis(dichlorophosphite)

[41105-12-8]

$C_6H_4Cl_4O_2P_2$ M 311.856

Solid. Mp 65°. Bp_{65} 200°.

Knauer, W., *Ber.*, 1894, **27**, 2565 (*synth*)

***N,N'*-1,2-Phenylenebis[*P,P,P*-triphenylphosphine imide], 8CI** P-00143

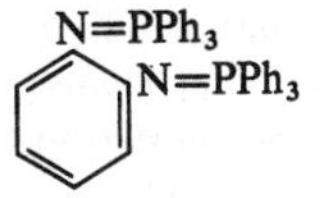

$C_{42}H_{34}N_2P_2$ M 628.692

Yellow cryst. (dioxan). Mp 206°.

Horner, L. *et al, Justus Liebigs Ann. Chem.*, 1956, **627**, 142.

***N,N'*-1,3-Phenylenebis[*P,P,P*-triphenylphosphine imide], 8CI** P-00144

[20339-23-5]

$C_{42}H_{34}N_2P_2$ M 628.692

Cryst. (C_6H_6/pet. ether). Mp 213-214°. pK_a 5.84, 8.61 (EtOH, 95% aq.), pK_a 14.94, 18.01 ($MeNO_2$).

Zhmurova, I.N. *et al, Zh. Obshch. Khim.*, 1968, **38**, 613 (*Engl. transl.* p. 592) (*synth*)
Kukhar', V.P. *et al, Zh. Obshch. Khim.*, 1970, **40**, 1696 (*Engl. transl.* p. 1682) (*props*)

***N,N'*-1,4-Phenylenebis[*P,P,P*-triphenylphosphine imide], 8CI** P-00145

[3356-46-5]

$C_{42}H_{34}N_2P_2$ M 628.692

Yellow cryst. (C_6H_6). Mp 255-257°. pK_a 6.18, 8.85 (EtOH, 95% aq.), pK_a 18.72, 15.66 ($MeNO_2$).

Herring, D.L., *J. Org. Chem.*, 1961, **26**, 3998 (*synth*)
Kukhar', V.P., *et al, Zh. Obshch. Khim.*, 1970, **40**, 1696 (*Engl. transl.* p. 1682) (*props*)

***P,P'*-1,4-Phenylenebis[*N,P,P*-triphenylphosphine imide], 9CI** P-00146

[42003-78-1]

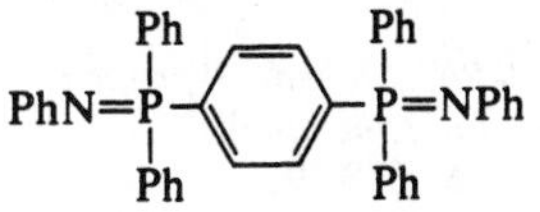

$C_{42}H_{34}N_2P_2$ M 628.692

Yellow cryst. (toluene). Mp 119-120°.

Herring, D.L., *J. Org. Chem.*, 1961, **26**, 3998 (*synth*)
Kireev, V.V. *et al*, *Zh. Obshch. Khim.*, 1973, **43**, 434 (*Engl. transl.* p. 430) (*synth, ir, P nmr*)

1,3-Phenylene diphosphoric acid — P-00147

1,3-Benzenediol bis(dihydrogen phosphate), 9CI. *Resorcinol bis(phosphoric acid). Resorcinol diphosphate*

[36011-88-8]

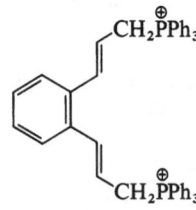

O
‖
OP(OH)₂ ... OP(OH)₂
‖
O

$C_6H_8O_8P_2$ M 270.072

Tetrachloride: [38135-34-1]. *1,3-Phenylene bisphosphorodichloridate.*
$C_6H_4Cl_4O_4P_2$ M 343.855
Liq. d^{15} 1.64. Bp_1 157°.

Katyshkina, V.V. *et al*, *Zh. Obshch. Khim.*, 1956, **26**, 3060 (*Engl. transl.* p. 3407) (*tetrachloride*)
Suh, B. *et al*, *J. Biol. Chem*, 1971, **246**, 7041 (*uv*)

1,4-Phenylenediphosphoric acid — P-00148

1,4-Benzenediol bis(dihydrogen phosphate), 9CI

$C_6H_8O_8P_2$ M 270.072

Tetrafluoride: 1,4-Phenylenebis(phosphorodifluoridate). 1,4-Bis(difluorophosphinyloxy)benzene.
$C_6H_4F_4O_4P_2$ M 278.036
Solid. Mp 36-37°. $Bp_{0.01}$ 70-72°.

Tetrachloride: [41240-73-7]. *1,4-Phenylenebis(phosphorodichloridate). 1,4-Bis(dichlorophosphinyloxy)-benzene.*
$C_6H_4Cl_4O_4P_2$ M 343.855
Liq. Bp_{20} 212°.

Katyshkina, V.V. *et al*, *Zh. Obshch. Khim.*, 1956, **26**, 3060 (*Engl. transl.* p. 3407) (*tetrachloride*)
Effenberger, F. *et al*, *Synthesis*, 1981, 70 (*tetrafluoride, ir, F nmr*)

(1,2-Phenylenedi-2-propene-3,1-diyl)-bis[triphenylphosphonium](2+), 10CI — P-00149

$CH_2\overset{\oplus}{P}Ph_3$ ... $CH_2\overset{\oplus}{P}Ph_3$

$C_{48}H_{42}P_2^{\oplus\oplus}$ M 680.807 (ion)

(E,E)-form

Dibromide: [62761-27-7].
$C_{48}H_{42}Br_2P_2$ M 840.615
Source of ylide. Solid. Mp 264-267°.

Bis(tetrafluoroborate): [73326-54-2].
$C_{48}H_{42}B_2F_8P_2$ M 854.415
Cryst. (MeOH). Mp 208-210°. Yields bis-ylide with RLi.

Bis-ylide: (1,2-Phenylenedi-2-propene-3,1-diylidene)-bis[triphenylphosphorane].
$C_{48}H_{40}P_2$ M 678.792

Used in synth. of benzo[18]annulene by Wittig reaction. Red.

Staab, H.A. *et al*, *Chem. Ber.*, 1979, **112**, 3907 (*synth, use*)

(1-Phenyl-1,2-ethanediyl)-bis[diphenylphosphine], 10CI — P-00150

1,2-Bis(diphenylphosphino)-1-phenylethane. Phenphos

CH_2PPh_2
H–C–PPh_2 (*R*)-*form*
Ph

$C_{32}H_{28}P_2$ M 474.521

Ligand.

(R)-form [69381-91-5]
Needles (CH_2Cl_2/MeOH). Mp 155-158°, 165-175° dec.
$[\alpha]_D^{20}$ −33.2° (c, 0.762 in $CHCl_3$).

(S)-form [69381-90-4]
Cryst. Mp 155-158°. $[\alpha]_D^{20}$ +10.1° (c, 0.99 in CH_2Cl_2).
Dioxide: [82631-91-2].
$C_{32}H_{28}O_2P_2$ M 506.520
Hygroscopic needles ($CHCl_3$/hexane). Mp 315-316°.
$[\alpha]_D^{20}$ −177.7° (c, 0.28 in $CHCl_3$).

(±)-form [79814-92-9]
Dioxide: [82660-49-9]. Cryst. Mp 290-291°.

Aguiar, A.M. *et al*, *J. Org. Chem.*, 1965, **30**, 352 (*dioxide, synth, pmr*)
King, R.B. *et al*, *J. Org. Chem.*, 1979, **44**, 1729 (*synth, pmr, cmr, nmr*)
Brown, J.M. *et al*, *J. Organomet. Chem.*, 1981, **216**, 263 (*pmr, ir, nmr*)
Brown, J.M. *et al*, *J. Chem. Soc., Perkin Trans. 2*, 1982, 489 (*use*)
Morandini, F. *et al*, *Inorg. Chim. Acta*, 1982, **57**, 15 (*complexes*)
Yoshikuni, T. *et al*, *Inorg. Chem.*, 1982, **21**, 2129 (*complexes*)

12-Phenyl-10*H*-5,10-ethenoacridophosphine, 9CI — P-00151

[55523-19-8]

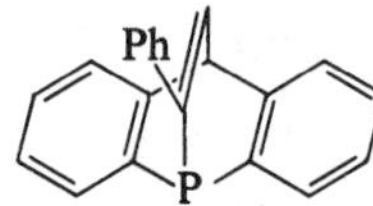

$C_{21}H_{15}P$ M 298.323

Needles. Mp 164-165°.

Maerkl, G. *et al*, *Tetrahedron Lett.*, 1974, 4369 (*synth, ir, ms, uv*)

(2-Phenylethenyl)phosphine, 9CI — P-00152

(2-Phenylvinyl)phosphine. Styrylphosphine

[65094-30-6]

PhCH=CHPH₂

C_8H_9P M 136.133

Briefly storable liq. Polym. readily. d_4^{20} 1.02. Bp_2 82°. n_D^{20} 1.6180.

Bogolyubov, G.M. *et al*, *Zh. Obshch. Khim.*, 1963, **33**, 3774 (*Engl. transl.* p. 3710) (*synth, ir, pmr*)

(2-Phenylethenyl)phosphinic acid, 9CI — P-00153

Styrylphosphinic acid, 8CI

[63263-75-2]

PhCH=CHPH(O)OH

$C_8H_9O_2P$ M 168.132
Cryst. (H_2O). Mp 74-75°.

Me ester: [18788-86-8]. *Methyl (2-phenylethenyl)-phosphinate. Methyl styrylphosphinate.*
$C_9H_{11}O_2P$ M 182.158
d_4^{20} 1.15. $Bp_{0.04}$ 138°. n_D^{20} 1.5886.

Et ester: [18788-85-7]. *Ethyl (2-phenylethenyl)-phosphinate. Ethyl styrylphosphinate.*
$C_{10}H_{13}O_2P$ M 196.185
Liq. d_4^{20} 1.12. $Bp_{0.03}$ 137-138°. n_D^{20} 1.5665.

Isopropyl ester: [18788-83-5]. *Isopropyl styrylphosphinate.*
$C_{11}H_{15}O_2P$ M 210.212
d_4^{20} 1.09. $Bp_{0.03}$ 126-127°. n_D^{20} 1.5545.

Walsh, E.N. *et al, J. Am. Chem. Soc.*, 1955, **77**, 929 (*synth*)
Levin, Ya.A. *et al, Zh. Obshch. Khim.*, 1967, **37**, 2736 (*Engl. transl.* p. 2604) (*esters, ir*)
Shagidullin, R.R. *et al, Izv. Akad. Nauk SSSR, Ser. Khim.*, 1971, 1168 (*Engl. transl.* p. 1082) (*ir, raman*)
Fridland, S.V. *et al, Zh. Obshch. Khim.*, 1978, **48**, 319 (*Engl. transl.* p. 285) (*esters*)

(1-Phenylethenyl)phosphonic acid, 9CI P-00154

α-Styrylphosphonic acid. 1-Phenylvinylphosphonic acid. 1-Phenyl-1-phosphonoethylene
[3220-50-6]

H_2C=CPhP(O)(OH)$_2$

$C_8H_9O_3P$ M 184.131
Solid. V. sol. H_2O, mod. sol. EtOH, sl. sol. C_6H_6. Mp 112°. Decolourises Br_2-H_2O.

Monoanilinium salt: Solid. Mp 86.5-87.5°, 180-181°.

Di-Me ester: [4844-39-7]. *Dimethyl (1-phenylethenyl)-phosphonate.*
$C_{10}H_{13}O_3P$ M 212.185
Liq. d_4^{20} 1.18. $Bp_{0.015}$ 90°. n_D^{20} 1.5270.

Di-Et ester: [25944-64-3]. *Diethyl (1-phenylethenyl)-phosphonate.*
$C_{12}H_{17}O_3P$ M 240.238
Liq. $Bp_{0.1}$ 101-103°. n_D^{20} 1.5158.

Di-Ph ester: [32187-40-9]. *Diphenyl (1-phenylethenyl)-phosphonate.*
$C_{20}H_{17}O_3P$ M 336.326
Liq. d_4^{20} 1.20. $Bp_{0.09}$ 195-196°. n_D^{20} 1.5916.

Conant, J.B. *et al, J. Am. Chem. Soc.*, 1921, **43**, 1928 (*synth, props*)
Tavs, P. *et al, Tetrahedron*, 1970, **26**, 5529 (*ester, synth, ir, uv*)
Maas, G. *et al, Chem. Ber.*, 1976, **109**, 2039 (*ester, synth, ir, pmr*)
Krueger, W.E. *et al, J. Org. Chem.*, 1978, **43**, 2877 (*esters, ir, pmr, ms*)
Yamashita, M. *et al, Bull. Chem. Soc. Jpn.*, 1980, **53**, 1625 (*esters*)
Axelrad, G. *et al, J. Org. Chem.*, 1981, **46**, 5200 (*esters*)
Satterthwait, A.C. *et al, J. Am. Chem. Soc.*, 1981, **103**, 1177 (*deriv, props*)
Hirao, T. *et al, Bull. Chem. Soc. Jpn.*, 1982, **55**, 909 (*ester*)

2-Phenylethenylphosphonic acid, 9CI P-00155

β-Styrylphosphonic acid. 2-Phenylvinylphosphonic acid. 1-Phenyl-2-phosphonoethylene
[1707-08-0]

PhCH=CHP(O)(OH)$_2$

$C_8H_9O_3P$ M 184.131

Lustrous plates (H_2O). Mp 142°, 154-155°. pK_{a1} 2.22, pK_{a2} 6.69 (H_2O, 25°).

Mono-Me ester: Methyl hydrogen (2-phenylethenyl)-phosphonate. Monomethyl styrylphosphonate.
$C_9H_{11}O_3P$ M 198.158
Solid. Mp 83-86°.

Mono-Et ester: Monoethyl (2-phenylethenyl)-phosphonate. Monoethyl styrylphosphonate.
$C_{10}H_{13}O_3P$ M 212.185
Solid. Mp 75-76°. pK_a 1.80 (H_2O, 25°).

Di-Et ester: [1018-24-2]. *Diethyl (2-phenylethenyl)-phosphonate. Diethyl styrylphosphonate.*
$C_{12}H_{17}O_3P$ M 240.238
Liq. d_4^{20} 1.09. $Bp_{1.5}$ 134°. n_D^{20} 1.5298.

Monobenzyl ester: Mono(phenylmethyl) (2-phenylethenyl)phosphonate. Benzyl styrylphosphonate.
$C_{15}H_{15}O_3P$ M 274.255
Cryst. (pet. ether). Mp 91-93°.

Difluoride: see (2-Phenylethenyl)phosphonic difluoride, P-00158

Dichloride: see (2-Phenylethenyl)phosphonic dichloride, P-00157

Di-amide: see P-(2-Phenylethenyl)phosphonic diamide, P-00156

(E)-form

Di-Me ester: [60190-89-8]. Cryst. (Et_2O). Mp 29-30°. $Bp_{0.1}$ 127°.

Di-Et ester: [20408-33-7]. Liq. $Bp_{0.3}$ 136-138°.

Maynard, J.A. *et al, Aust. J. Chem.*, 1963, **16**, 609 (*esters, synth, uv*)
Ionin, B.I. *et al, Zh. Obshch. Khim.*, 1964, **34**, 2630 (*Engl. transl.* p. 2651) (*diethyl ester*)
Griffin, C.E. *et al, J. Org. Chem.*, 1965, **30**, 1935 (*synth, esters*)
Chernova, A.V. *et al, Izv. Akad. Nauk SSSR, Ser. Khim.*, 1972, 722 (*Engl. transl.* p. 693) (*ester, uv, raman*)
Maas, G. *et al, Chem. Ber.*, 1976, **109**, 2039 (*ester, synth, ir, pmr*)
Berkova, G.A. *et al, Zh. Obshch. Khim.*, 1978, **48**, 66 (*Engl. transl.* p. 54) (*ester, pmr*)
Axelrad, G. *et al, J. Org. Chem.*, 1981, **46**, 5200 (*esters, synth*)
Biryulina, V.N. *et al, Zh. Obshch. Khim.*, 1981, **51**, 976 (*Engl. transl.* p. 815) (*complexes*)
Hirao, T. *et al, Bull. Chem. Soc. Jpn.*, 1982, **55**, 909 (*ester, synth*)
Sekine, M. *et al, Bull. Chem. Soc. Jpn.*, 1982, **55**, 224 (*esters, synth, pmr*)
Turkhina, L.A. *et al, Zh. Neorg. Khim.*, 1982, **27**, 523 (*Engl. transl.* p. 297) (*monoalkyl esters*)
Xu, Y. *et al, Synthesis*, 1983, 556 (*ester, synth, ir, pmr*)

P-(2-Phenylethenyl)phosphonic diamide, 9CI P-00156

P-*Styrylphosphonic diamide.* P-*(2-Phenylvinyl)-phosphonic diamide*

PhCH=CHP(O)(NH$_2$)$_2$

$C_8H_{11}N_2OP$ M 182.161

N,N,N′,N′-*Tetra-Me: (2-Phenylethenyl)phosphonic bis-(dimethylamide).* N,N,N′,N′-*Tetramethyl-*P-*(2-phenylethenyl)phosphonic diamide.*
$C_{12}H_{19}N_2OP$ M 238.269
Liq. d_4^{20} 1.08. Bp_2 180°. n_D^{20} 1.5665.

N,N,N′,N′-*Tetra-Et: (2-Phenylethenyl)phosphonic bis-(diethylamide).* N,N,N′,N′-*Tetraethyl-*P-*(2-phenylethenyl)phosphonic diamide.*
$C_{16}H_{27}N_2OP$ M 294.376
Solid. Mp 103.5°.

N,N,N′,N′-*Tetrapropyl: (2-Phenylethenyl)phosphonic bis(dipropylamide).* P-*(2-Phenylethenyl)*-N,N,N′,N′-*tetrapropylphosphonic diamide.*
$C_{20}H_{35}N_2OP$ M 350.483

Liq. d_4^{20} 1.01. Bp_2 192°. n_D^{20} 1.5290.
N,N,N′,N′-*Tetrabutyl: (2-Phenylethenyl)phosphonic bis(dibutylamide)*. N,N,N′,N′-*Tetrabutyl*-P-*(2-phenylethenyl)phosphonic diamide.*
$C_{24}H_{43}N_2OP$ M 406.590
Solid. Mp 39-40°.

Anisimov, K.N. *et al, Izv. Akad. Nauk SSSR, Ser. Khim.*, 1956, 19 (*Engl. transl.* p. 17) (*synth*)
Benkova, G.A. *et al, Zh. Obshch. Khim.*, 1977, **47**, 1431 (*Engl. transl.* p. 1315) (*pmr*)

(2-Phenylethenyl)phosphonic dichloride, 9CI **P-00157**
Styrylphosphonic dichloride
[4708-07-0]

PhCH=CHP(O)Cl_2

$C_8H_7Cl_2OP$ M 221.022
Solid. Mp 70°. Bp_{18} 182-184°, Bp_2 133-134°.
(E)-form [20408-31-5]
Characterised spectroscopically.
(Z)-form [25362-02-1]
Characterised spectroscopically.

Dogadina, A.V. *et al, Zh. Obshch. Khim.*, 1971, **41**, 1662 (*Engl. transl.* p. 1670) (*pmr*)
Gareev, V.S. *et al, Zh. Obshch. Khim.*, 1972, **42**, 1496, 1714 (*Engl. transl.* pp. 1487, 1703) (*synth, ir*)
Dorokhova, V.V. *et al, Zh. Obshch. Khim.*, 1973, **43**, 2172 (*Engl. transl.* p. 2164) (*uv*)
Org. Synth., Coll. Vol., **5**, 1005 (*synth*)
Dorokhova, V.V. *et al, Zh. Obshch. Khim.*, 1979, **49**, 83 (*Engl. transl.* p. 68) (*raman*)
Glukhikh, V.I. *et al, Dokl. Akad. Nauk SSSR, Ser. Sci. Khim.*, 1979, **247**, 1405 (*Phys. Chem., Engl. transl.* p. 710) (*cmr*)
Komarev, V.Ya. *et al, Zh. Obshch. Khim.*, 1980, **50**, 1262 (*Engl. transl.* p. 1020) (*P nmr*)
Timokhin, B.V. *et al, Zh. Obshch. Khim.*, 1981, **51**, 2808 (*Engl. transl.* p. 2420) (*synth, nmr*)

(2-Phenylethenyl)phosphonic difluoride, 9CI **P-00158**
β-Styrylphosphonic difluoride
[705-90-8]

PhCH=CHP(O)F_2

$C_8H_7F_2OP$ M 188.113
Liq. d_4^{20} 1.29 (1.34). Bp_{10} 115-117°. n_D^{20} 1.5340 (1.5140).

Timofeeva, T.N. *et al, Zh. Obshch. Khim.*, 1968, **38**, 1255 (*Engl. transl.* p. 1208) (*synth, pmr*)
Reddy, G.S. *et al, Z. Naturforsch., B*, 1970, **25**, 1199 (*F and P nmr*)
Zakharov, V.I. *et al, Dokl. Akad. Nauk SSSR, Ser. Sci. Khim.*, 1973, **209**, 1343 (*Engl. transl.* p. 329) (*pmr, nmr*)
Fridland, S.V. *et al, Zh. Obshch. Khim.*, 1980, **50**, 784 (*Engl. transl.* p. 624) (*synth*)
Krolevets, A.A. *et al, Izv. Akad. Nauk SSSR, Ser. Khim.*, 1980, 1454; *CA*, **94**, 30843.

(2-Phenylethenyl)phosphonodithioic acid, 9CI **P-00159**
Styrylphosphonodithioic acid, 8CI

PhCH=CHP(O)$(SH)_2$ ⇌ PhCH=CHP(S)(OH)(SH)

$C_8H_9OPS_2$ M 216.252
S,S-*Di-Et ester:* S,S-*Diethyl (2-phenylethenyl)-phosphonodithioate.*
$C_{12}H_{17}OPS_2$ M 272.359
Liq. d_4^{20} 1.15. Bp_1 168°. n_D^{20} 1.6248.
S,S-*Di-Ph ester:* S,S-*Diphenyl (2-phenylethenyl)-phosphonodithioate.*
$C_{20}H_{17}OPS_2$ M 368.447
Solid. Mp 117°.

Anisimov, K.N. *et al, Izv. Akad. Nauk SSSR, Ser. Khim.*, 1956, 19 (*Engl. transl.* p. 17)

(2-Phenylethenyl)phosphonothioic acid, 9CI **P-00160**
β-Styrylphosphonothioic acid, 8CI. (2-Phenylvinyl)-phosphonothioic acid

PhCH=CHP(S)$(OH)_2$ ⇌ PhCH=CHP(O)(OH)(SH)

$C_8H_9O_2PS$ M 200.192
O,O-*Di-Et ester:* O,O-*Diethyl (2-phenylethenyl)-phosphonothioate.*
$C_{12}H_{17}O_2PS$ M 256.299
Liq. d_4^{20} 1.12. Bp_3 146°. n_D^{20} 1.5450.
Dichloride: [26756-19-4].
$C_8H_7Cl_2PS$ M 237.083
Liq. d_4^{20} 1.35. $Bp_{0.08}$ 122-124°. n_D^{20} 1.6573.

Walsh, E.N. *et al, J. Am. Chem. Soc.*, 1955, **77**, 929 (*dichloride*)
Pudovik, A.N. *et al, Zh. Obshch. Khim.*, 1961, **31**, 2656 (*Engl. transl.* p. 2480) (*ester, synth*)
Shagidullin, R.R. *et al, Izv. Akad. Nauk SSSR, Ser. Khim.*, 1971, **20**, 1168 (*Engl. transl.* p. 1082) (*dichloride, ester, uv, raman*)
Galeev, V.S. *et al, Zh. Obshch. Khim.*, 1972, **42**, 1714 (*Engl. transl.* p. 1703) (*dichloride*)
Dorokhova, V.V. *et al, Zh. Obshch. Khim.*, 1973, **43**, 2172 (*Engl. transl.*, p. 2164) (*dichloride, uv, raman*)
Komarov, V.Ya. *et al, Zh. Obshch. Khim.*, 1980, **50**, 1262 (*Engl. transl.* p. 1020) (*ester, dichloride, nmr*)

(2-Phenylethenyl)phosphonous acid, 9CI **P-00161**
Styrylphosphonous acid. β-Phenylvinylphosphonous acid

PhCH=CHP$(OH)_2$

$C_8H_9O_2P$ M 168.132
The free acid exists as the phosphoryl tautomer, (2-Phenylethenyl)phosphinic acid, P-00153 .
Mono-Me ester: see (2-Phenylethenyl)phosphinic acid, P-00153
Di-Me ester: [17316-54-0]. *Dimethyl (2-phenylethenyl)-phosphonite.*
$C_{10}H_{13}O_2P$ M 196.185
Liq. with pleasant odour. d_4^{20} 1.09. $Bp_{0.06}$ 88-90°. n_D^{20} 1.5692.
Mono-Et ester: see (2-Phenylethenyl)phosphinic acid, P-00153
Di-Et ester: [17316-55-1]. *Diethyl (2-phenylethenyl)-phosphonite.*
$C_{12}H_{17}O_2P$ M 224.239
Liq. d_4^{20} 1.03. $Bp_{0.1}$ 106-107°. n_D^{20} 1.5480.
Monoisopropyl ester: see (2-Phenylethenyl)phosphinic acid, P-00153
Diisopropyl ester: [17316-57-3]. *Diisopropyl (2-phenylethenyl)phosphonite.*
$C_{14}H_{21}O_2P$ M 252.292
Liq. $Bp_{0.04}$ 92-94°. n_D^{20} 1.5310.
Bis(diethylamide): [86936-48-3]. N,N,N′,N′-*Tetraethyl*-P-*(2-phenylethenyl)phosphonous diamide.*
$C_{16}H_{27}N_2P$ M 278.376
Liq. $Bp_{0.012}$ 123°. n_D^{20} 1.5610.

Dichloride: see (2-Phenylethenyl)phosphonous dichloride, P-00162

Levin, Yu.A. *et al*, *Zh. Obshch. Khim.*, 1967, **37**, 1327 (*Engl. transl.* p. 1255) (*esters, synth, ir*)
Shagidullin, R.R. *et al*, *Izv. Akad. Nauk SSSR, Ser. Khim.*, 1971, 183, 1168 (*Engl. transl.* pp. 165, 1082) (*uv, raman*)
Fridland, S.V. *et al*, *Zh. Obshch. Khim.*, 1978, **48**, 319 (*Engl. transl.* p. 285) (*ester, synth*)
Nurtdinov, S.Kh. *et al*, *Zh. Obshch. Khim.*, 1983, **53**, 785 (*Engl. transl.* p. 685) (*amides, synth, ir, pmr*)
Kosovtsev, V.V. *et al*, *Zh. Obshch. Khim.*, 1971, **41**, 2638 (*Engl. transl.* p. 2671) (*ester, dimethylamide, pmr*)

(2-Phenylethenyl)phosphonous dichloride, 9CI — P-00162

Styrylphosphonous dichloride. β-Phenylvinylphosphonous dichloride. Dichloro-β-phenylvinylphosphine
[17391-53-6]

$PhCH{=}CHPCl_2$

$C_8H_7Cl_2P$ M 205.023
Liq. d_4^{20} 1.28. Bp_7 142-144°, $Bp_{0.05}$ 82-84°. n_D^{20} 1.6360 (1.5950).

Walsh, E.N. *et al*, *J. Am. Chem. Soc.*, 1955, **77**, 929 (*synth*)
Shagidullin, R.R. *et al*, *Izv. Akad. Nauk SSSR, Ser. Khim.*, 1971, **183**, 1168 (*Engl. transl.* pp. 165, 1082) (*uv, raman*)
Dorokhova, V.V. *et al*, *Zh. Obshch. Khim.*, 1973, **43**, 2172 (*Engl. transl.* p. 2164) (*uv*)
Krolevets, A.A. *et al*, *Izv. Akad. Nauk SSSR, Ser. Khim.*, 1980, 897 (*Engl. transl.* p. 644) (*ir, ms, P nmr*)
Rozinov, V.G. *et al*, *Zh. Obshch. Khim.*, 1981, **51**, 1498; 1982, **52**, 1994 (*Engl. transl.* pp. 1271, 1772) (*synth, P nmr*)
Timokhin, B.V. *et al*, *Zh. Obshch. Khim.*, 1983, **53**, 291 (*Engl. transl.* p. 252) (*synth, P nmr*)

(2-Phenylethyl)phosphine, 9CI — P-00163

2-Phenethylphosphine

$PhCH_2CH_2PH_2$

$C_8H_{11}P$ M 138.149
Liq. Bp_8 75°. n_D^{25} 1.5494.

Rauhut, M.M. *et al*, *J. Org. Chem.*, 1961, **26**, 5138 (*synth*)
Bogolyubov, G.M. *et al*, *Zh. Obshch. Khim.*, 1963, **33**, 3774 (*Engl. transl.* p. 3710) (*synth*)

1-Phenylethylphosphonic acid, 9CI — P-00164

[61470-40-4]

$PhCH(CH_3)P(O)(OH)_2$

$C_8H_{11}O_3P$ M 186.147

(±)-form
Cryst. (EtOAc/methylcyclohexane). Mp 153-154.5°.
Di-Me ester: [69914-34-7]. *Dimethyl 1-phenylethylphosphonate.*
$C_{10}H_{15}O_3P$ M 214.200
Liq. Bp_2 150-152°.
Di-Et ester: Diethyl 1-phenylethylphosphonate.
$C_{12}H_{19}O_3P$ M 242.254
Liq. $Bp_{0.4}$ 103°. n_D^{20} 1.4915.
▷SZ9065000.

Kagan, F. *et al*, *J. Am. Chem. Soc.*, 1959, **81**, 3026 (*synth, ester*)
Inokawa, S. *et al*, *Synthesis*, 1977, 179 (*ester, synth, pmr*)
Villieras, J. *et al*, *J. Organomet. Chem.*, 1978, **144**, 17, 263 (*ester, synth, ir, pmr*)
Yamashita, M. *et al*, *Bull. Chem. Soc. Jpn.*, 1979, **52**, 466 (*ester, synth, pmr*)

(2-Phenylethyl)phosphonic acid, 9CI — P-00165

Phenethylphosphonic acid
[4672-30-4]

$PhCH_2CH_2P(O)(OH)_2$

$C_8H_{11}O_3P$ M 186.147
Cryst. (EtOH/ligroin, or 2-propanol/methylcyclohexane). Mp 138-140°.
Di-Me ester: [69404-44-0]. *Dimethyl 2-phenylethylphosphonate.*
$C_{10}H_{15}O_3P$ M 214.200
No phys. props. reported.
Di-Et ester: [54553-21-8]. *Diethyl 2-phenylethylphosphonate.*
$C_{12}H_{19}O_3P$ M 242.254
Liq. Bp_{14} 173-177°. n_D^{20} 1.4910.
Bis(2-phenylethyl) ester: Bis(2-phenylethyl)(2-phenylethyl)phosphonate.
$C_{24}H_{27}O_3P$ M 394.449
Liq. d_4^{20} 1.13. $Bp_{0.018}$ 185°. n_D^{20} 1.5594.

Bergmann, E. *et al*, *Chem. Ber.*, 1930, **63**, 1158 (*synth*)
Kagan, F. *et al*, *J. Am. Chem. Soc.*, 1959, **81**, 3026 (*synth, esters*)
Petrov, K.A. *et al*, *Zh. Obshch. Khim.*, 1970, **40**, 2192 (*Engl. transl.* p. 2179) (*synth*)
Griffiths, W.R. *et al*, *Phosphorus Sulfur*, 1978, **5**, 101 (*esters, ms*)

Phenylethynylphosphonic acid, 9CI — P-00166

1-Phenyl-2-phosphonoacetylene

$PhC{\equiv}CP(O)(OH)_2$

$C_8H_7O_3P$ M 182.115
Plates. Mp 142°.
Di-Me ester: [33802-53-8]. *Dimethyl phenylethynylphosphonate.*
$C_{10}H_{11}O_3P$ M 210.169
Liq. d_4^{20} 1.18. Bp_1 138-139°. n_D^{20} 1.5410.
Di-Et ester: [3450-67-7]. *Diethyl phenylethynylphosphonate.*
$C_{12}H_{15}O_3P$ M 238.222
Liq. d_4^{20} 1.09. Bp_2 154-155°. n_D^{20} 1.5290.
Dichloride: [4981-32-2].
$C_8H_5Cl_2OP$ M 219.007
Liq. d_4^{20} 1.37. $Bp_{0.5}$ 121-124°. n_D^{20} 1.5938.

Bergmann, E. *et al*, *Ber.*, 1933, **66**, 278 (*synth*)
Mashlyakovskii, L.N. *et al*, *Zh. Obshch. Khim.*, 1965, **35**, 1577 (*Engl. transl.* p. 1582) (*dichloride, pmr, ir*)
Dogadina, A.V. *et al*, *Zh. Obshch. Khim.*, 1971, **41**, 1662 (*Engl. transl.* p. 1670) (*ester, synth*)
Fujii, A. *et al*, *J. Am. Chem. Soc.*, 1971, **93**, 3694 (*ester, synth, ir, pmr, ms*)
Berkova, G.A. *et al*, *Zh. Obshch. Khim.*, 1977, **47**, 1431 (*Engl. transl.* p. 1315) (*bisdimethylamide, P nmr, pmr*)

(Phenylethynyl)phosphonothioic acid, 9CI — P-00167

$PhC{\equiv}CP(S)(OH)_2 \rightleftharpoons PhC{\equiv}CP(O)(OH)(SH)$

$C_8H_7O_2PS$ M 198.176
O,O-*Di-Me ester:* [30238-13-2]. O,O-*Dimethyl (phenylethynyl)phosphonothioate.*
$C_{10}H_{11}O_2PS$ M 226.229
Liq. $Bp_{0.2}$ 119°.

O,O-Di-Et ester: [30238-12-1]. *O,O-Diethyl (phenylethynyl)phosphonothioate.*
$C_{12}H_{15}O_2PS$ M 254.283
Liq. $Bp_{0.15}$ 134-135°.
Dichloride: [13891-84-4].
$C_8H_5Cl_2PS$ M 235.067
Liq. d_4^{20} 1.32. Bp_2 125°. n_D^{20} 1.6510.

Bogolyubov, G.M. *et al*, *Zh. Obshch. Khim.*, 1967, **37**, 229 (*Engl. transl.* p. 211) (*dichloride*)
Chattha, M.S. *et al*, *J. Org. Chem.*, 1971, **36**, 2720 (*esters, synth, ir*)

Phenylimidodiphosphoric acid, 9CI **P-00168**

Phenyliminodiphosphoric acid, 8CI. Anilinodiphosphoric acid

PhN[P(O)(OH)$_2$]$_2$

$C_6H_9NO_6P_2$ M 253.088
Tetra-Et ester: [24949-50-6]. *Tetraethyl phenylimidodiphosphate. Tetraethyl phenyliminodiphosphate.*
$C_{14}H_{25}NO_6P_2$ M 365.302
Liq. d_4^{20} 1.21. $Bp_{0.01}$ 130-132°. n_D^{20} 1.4850.

Gilyarov, V.A. *et al*, *Zh. Obshch. Khim.*, 1971, **41**, 2355 (*Engl. transl.* p. 2380) (*synth*)

2-Phenyl-2*H*-isophosphindole 2-oxide, 9CI **P-00169**

[51998-39-1]

$C_{14}H_{11}OP$ M 226.214
Reactive intermed. Reacts as a diene. Readily forms a dimer unless alternative dienophile present.

Chan, T.H. *et al*, *Tetrahedron*, 1975, **31**, 2537 (*formn, props*)

(Phenylmethylidyne)phosphine, 9CI **P-00170**

[76684-21-4]

PhC≡P

C_7H_5P M 120.090
Condensed at −196°.

Appel, R. *et al*, *Angew. Chem., Int. Ed. Engl.*, 1981, **20**, 197 (*synth*)
Burckett-St. Laurent, J.C.T.R. *et al*, *J. Mol. Spectrosc.*, 1982, **92**, 158 (*microwave*)

Phenyl methylphosphonate **P-00171**

Monophenyl methylphosphonate, 9CI. Phenyl hydrogen methylphosphonate
[13091-13-9]

$C_7H_9O_3P$ M 172.120
Synthetic intermediate. Solid or oil. d_4^{20} 1.26. Mp 32-33°. n_D^{20} 1.5196.
Dicyclohexylammonium salt: Solid. Mp 105-106°.
Me ester: see Methyl phenyl methylphosphonate, M-00205
Et ester: see Ethyl phenyl methylphosphonate, E-00100
Chloride: see Methylphosphonochloridic acid, M-00296

Petrov, K.A. *et al*, *Zh. Obshch. Khim.*, 1961, **31**, 1705 (*Engl. transl.* p. 1592) (*synth*)
Evdakov, V.P. *et al*, *Zh. Obshch. Khim.*, 1964, **34**, 1848 (*Engl. transl.* p. 1860) (*synth*)

O-Phenyl methylphosphonodithioate **P-00172**

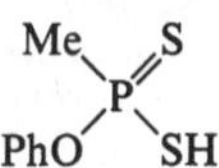

$C_7H_9OPS_2$ M 204.241
Ni salt: Mp 200-202°.
Ph ester: see Methylphosphonodithioic acid, M-00301

Chupp, J.P. *et al*, *J. Org. Chem.*, 1962, **27**, 3832 (*synth*)

2-Phenyl-1-oxa-4-aza-5-phospha(5-P^V)-spiro[4.4]nonane, 10CI **P-00173**

$C_{12}H_{18}NOP$ M 223.254
5-Ph: [71559-18-7].
$C_{18}H_{22}NOP$ M 299.352
Solid. Mp 45-47°.

Cadogan, J.I.G. *et al*, *J. Chem. Soc., Chem. Commun.*, 1979, 189 (*synth, P nmr*)

2-Phenyl-1,2-oxaphospholane, 9CI, 8CI **P-00174**

2-Phenyl-1-oxa-2-phosphacyclopentane
[16324-17-7]

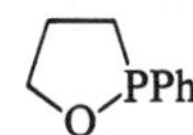

$C_9H_{11}OP$ M 166.159
Liq. $Bp_{0.3}$ 120°, $Bp_{0.1}$ 62-65°.
2-Oxide: [16324-19-9].
$C_9H_{11}O_2P$ M 182.158
$Bp_{0.7}$ 157°.
2-Sulfide: [16324-18-8].
$C_9H_{11}OPS$ M 198.219
Liq. $Bp_{0.01}$ 138°.

Grayson, M. *et al*, *J. Chem. Soc., Chem. Commun.*, 1967, 830 (*synth, derivs, ir, pmr*)
Wetzel, R.B. *et al*, *J. Org. Chem.*, 1974, **39**, 1531 (*oxide, synth, pmr*)
Mathey, F. *et al*, *J. Organomet. Chem.*, 1976, **117**, 377 (*derivs, synth*)
Antczak, S. *et al*, *J. Chem. Soc., Perkin Trans. 1*, 1978, 1326 (*synth, nmr*)
Kobayashi, S. *et al*, *Macromolecules*, 1984, **17**, 107 (*props*)
Singh, G. *et al*, *J. Org. Chem.*, 1984, **49**, 5132 (*oxide, cmr*)
Kobayashi, S. *et al*, *Bull. Chem. Soc. Jpn.*, 1985, **58**, 2153 (*synth, pmr, cmr, P nmr*)

4-Phenyl-4*H*-1,4-oxaphosphorin, 9CI **P-00175**

*4-Phenyl-4-phospha-4*H*-pyran*

$C_{10}H_9OP$ M 176.154

4-Oxide: [37755-70-7].

$C_{10}H_9O_2P$ M 192.154

Cryst. ($CHCl_3$/pentane). Mp 143°.

Maumy, M., *Bull. Soc. Chim. Fr.*, 1972, 1600 (*synth, ir, pmr*)

4-Phenyl-1,4-oxaphosphorinane, 9CI, 8CI **P-00176**

4-Phenyl-1,4-oxaphosphinane. 1-Oxa-4-phenylphosphacyclohexane

[31614-36-5]

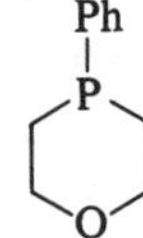

$C_{10}H_{13}OP$ M 180.186

Liq. Sol. Et_2O, THF, EtOH. Bp_4 116-118°.

B,MeI: 4-Methyl-4-phenyl-1,4-oxaphosphorinanium iodide. 4-Methyl-4-phenyl-1,4-oxaphosphinanium iodide. 4-Methyl-4-phenyl-1-oxa-4-phosphoniacyclohexane iodide.

$C_{11}H_{16}IOP$ M 322.125

Cryst. (EtOH/Et_2O). Mp 149-150°.

Oxide: [52427-43-7].

$C_{10}H_{13}O_2P$ M 196.185

Solid. Mp 117-118°.

Issleib, K. *et al*, *J. Prakt. Chem.*, 1970, **312**, 578 (*synth, deriv*)

Collins, D.J. *et al*, *Aust. J. Chem.*, 1974, **27**, 841 (*oxide, ir, ms, pmr*)

3-Phenyl-1,3-oxaphosphorinan-2-one **P-00177**

3-Phenyl-1,3-oxaphosphinan-2-one

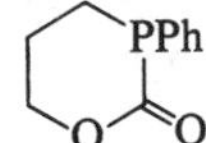

$C_9H_{11}O_2P$ M 182.158

Liq. $Bp_{0.6}$ 124-126°.

B,MeI: 3-Methyl-2-oxo-3-phenyl-1,3-oxaphosphorinanium iodide.

$C_{10}H_{14}IO_2P$ M 324.098

Cryst. (H_2O or CH_2Cl_2). Mp 178-181° dec.

3-Sulfide: [84441-06-5].

$C_9H_{11}O_2PS$ M 214.218

Solid. Insol. H_2O, Et_2O, sol. CH_2Cl_2, EtOH, Me_2CO. Mp ~75°.

Thamm, R. *et al*, *Z. Naturforsch. B.*, 1982, **37**, 965 (*synth, derivs, ms, P nmr*)

2-Phenyl-1,3,2-oxathiaphosphepane, 9CI **P-00178**

2-Phenyl-1-oxa-2-phospha-3-thiacycloheptane

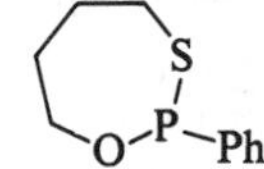

$C_{10}H_{13}OPS$ M 212.246

2-Sulfide: [55781-98-1].

$C_{10}H_{13}OPS_2$ M 244.306

Cryst. (hexane). Mp 62-63.5°.

Nakayama, S. *et al*, *Bull. Chem. Soc. Jpn.*, 1975, **48**, 546 (*synth, ir, pmr*)

2-Phenyl-1,3,2-oxazaphospholidine, 9CI **P-00179**

[19858-95-8]

$C_8H_{10}NOP$ M 167.147

Liq. d_4^{20} 1.18. $Bp_{0.08}$ 89-91°. n_D^{20} 1.5840.

Mitsunobu, O. *et al*, *Bull. Chem. Soc. Jpn.*, 1967, **40**, 2964 (*synth*)

Pudovik, M.A. *et al*, *Zh. Obshch. Khim.*, 1982, **52**, 491 (*Engl. transl.* p. 431) (*synth, P nmr*)

3-Phenyl-1,3,2-oxazaphospholidine, 9CI **P-00180**

3-Phenyl-1,3,2-oxazaphospholane

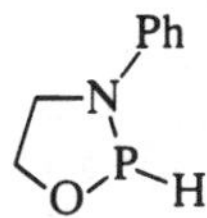

$C_8H_{10}NOP$ M 167.147

Oxide: [51104-85-9].

$C_8H_{10}NO_2P$ M 183.146

Cryst. (C_6H_6). Mp 86-87°.

2-Phenoxy: see 2-Phenoxy-3-phenyl-1,3,2-oxazaphospholidine, P-00073

2-Me: see 2-Methyl-3-phenyl-1,3,2-oxazaphospholidine, M-00212

2-Ph: see 2,3-Diphenyl-1,3,2-oxazaphospholidine, D-01029

2-Chloro: see 2-Chloro-3-phenyl-1,3,2-oxazaphospholidine, C-00138

Pudovik, M.A. *et al*, *Zh. Obshch. Khim.*, 1973, **43**, 2144 (*Engl. transl.* p. 2135) (*oxide, pmr*)

4-Phenyl-1,2,3,5,8-pentaoxa-4-phospha(4-*P*^V)spiro[3.4]octane, 9CI **P-00181**

Ethylene phenyl phosphite ozonide

$C_8H_9O_6P$ M 232.129

Possible source of singlet oxygen. Unstable even at low temp.

Stephenson, L.M. *et al*, *J. Am. Chem. Soc.*, 1973, **95**, 3074 (*synth, props*)

Mendenhall, G.D. *et al*, *J. Photochem.*, 1984, **25**, 227 (*synth, P nmr, props*)

10-Phenyl-10*H*-phenothiaphosphine, 9CI, 8CI **P-00182**

[23844-90-8]

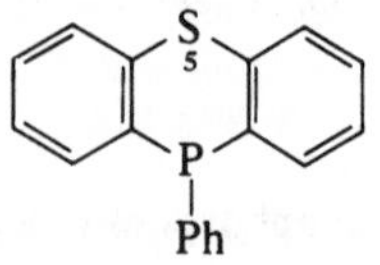

$C_{18}H_{13}PS$ M 292.334

Cryst. (MeOH). Mp 74-76°, Mp 92-93°.

B,MeI: [23781-24-0]. *10-Methyl-10-phenyl-10*H*-phenothiaphosphinium iodide.*

$C_{19}H_{16}IPS$ M 434.274

Pale-yellow plates. Mp 250-260°.

5,5-Dioxide; B,MeI: [23781-25-1]. *10-Methyl-10-phenyl-10*H*-phenothiaphosphinium 5,5-dioxide iodide.*

$C_9H_{16}IO_2PS$ M 346.162

Yellow cryst. Mp 285-290°.

10-Oxide:

$C_8H_{13}OPS$ M 188.224

Solid. Mp 112-113°.

10-Sulfide: [25317-88-8].

$C_8H_{13}PS_2$ M 204.284

Solid. Mp 200-201°.

5,5-Dioxide: [23844-89-5].

$C_8H_{13}O_2PS$ M 204.223

Cryst. (AcOH). Mp 155.5-156.5°.

5,5,10-Trioxide: [23781-30-8].

$C_{18}H_{13}O_3PS$ M 340.333

Solid. Mp 246-247°.

5,5-Dioxide, 10-sulfide: [23781-31-9].

$C_{18}H_{13}O_2PS_2$ M 356.393

Cryst. (xylene). Mp 215-216°.

U.S.P., 3 449 426, (*1969*); *CA*, **71**, 91649 (*synth, derivs*)

10-Phenyl-10*H*-phenoxaphosphine, 9CI, 8CI **P-00183**

[1225-16-7]

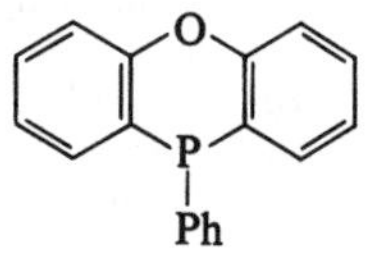

$C_{18}H_{13}OP$ M 276.274

Ligand for Pd. Cryst. (EtOH). Mp 97.5-98°. $Bp_{0.1}$ 170-180°.

B,MeI: [15040-65-0]. *10-Methyl-10-phenyl-10*H*-phenoxaphosphinium iodide.*

$C_{19}H_{16}IOP$ M 418.213

Solid or cryst. (EtOH). Mp 244-245° (236-237°).

*B,PhI: 10,10-Diphenyl-10*H*-phenoxaphosphonium iodide.*

$C_{24}H_{18}IOP$ M 480.284

Cryst. + $1H_2O$ (H_2O). Mp 195-197°.

10-Oxide: [1091-27-6].

$C_{18}H_{13}O_2P$ M 292.273

Cryst. (EtOH/6M HCl or Me_2CO aq.). Mp 177-178°.

10-Sulfide: [36712-24-0].

$C_{18}H_{13}OPS$ M 308.334

Cryst. (C_6H_6/pet. ether). Mp 169-170°.

Mann, F.G. *et al*, *J. Chem. Soc.*, 1953, 3746 (*synth, derivs, uv, complexes*)
Levy, J.B. *et al*, *J. Org. Chem.*, 1965, **30**, 660; 1968, **33**, 474 (*synth, derivs*)
Granoth, I. *et al*, *J. Chem. Soc., Perkin Trans. 2*, 1972, 697 (*synth, derivs, ms*)
Mann, F.G. *et al*, *J. Chem. Soc., Perkin Trans. 2*, 1976, 1383 (*cryst struct*)
Allen, D.W. *et al*, *Phosphorus Sulfur*, 1979, **7**, 309 (*cryst struct*)
Gray, G.A. *et al*, *Org. Magn. Reson.*, 1980, **14**, 14 (*cmr*)

Phenyl(4-phenyl-1,3-butadienyl)phosphinic acid, 9CI **P-00184**

PhCH=CHCH=CHPPh(O)OH

$C_{16}H_{15}O_2P$ M 270.267

Prisms. Mp 151-152°.

Chloride:

$C_{16}H_{14}ClOP$ M 288.713

Viscous liq.

Ph ester: Phenyl phenyl(4-phenyl-1,3-butadienyl)-phosphinate.

$C_{22}H_{19}O_2P$ M 346.365

Microcryst. powder. Mp 98-101°.

Anilide: N,P*-Diphenyl-*P*-(4-phenyl-1,3-butadienyl)-phosphinic amide.*

$C_{22}H_{20}NOP$ M 345.380

Prisms. Mp 195-196°.

Fedorova, G.K. *et al*, *Zh. Obshch. Khim.*, 1966, **36**, 1262 (*Engl. transl.* p. 1278) (*synth*)

Phenyl(phenylbutadiynyl)phosphinic acid, 9CI **P-00185**

PhC≡CC≡CCPh(O)OH

$C_{16}H_{11}O_2P$ M 266.235

Cryst. (MeOH aq. or Et_2O/pet. ether). Mp 141-141.5°.

Fedorova, G.K. *et al*, *Zh. Obshch. Khim.*, 1966, **36**, 1262 (*Engl. transl.* p. 1278) (*synth, ir*)

Phenyl(2-phenylethenyl)phosphinic acid **P-00186**

Phenylstyrylphosphinic acid. Phenyl(2-phenylvinyl)-phosphinic acid

[4895-67-4]

PhCH=CHP(O)(OH)Ph

$C_{14}H_{13}O_2P$ M 244.229

Solid. Mp 150-151°.

Me ester: Methyl phenyl(2-phenylethenyl)phosphinate.

$C_{15}H_{15}O_2P$ M 258.256

Solid. Mp 114-115°.

(E)-form

Et ester: [82943-05-3]. *Ethyl phenyl(2-phenylethenyl)-phosphinate. Ethyl phenylstyrylphosphinate.*

$C_{16}H_{17}O_2P$ M 272.283

Cryst. (pentane). Mp 58.5-62°. $Bp_{0.2}$ 178°.

Fedorova, G.K. *et al*, *CA*, 1966, **64**, 8228 (*synth, esters*)
Gloyna, D. *et al*, *J. Prakt. Chem.*, 1977, **319**, 451; 1982, **324**, 107 (*ester, synth, pmr, ir*)

Phenyl(phenylethynyl)phosphinic acid, 9CI **P-00187**

[4895-54-9]

PhC≡CP(O)(OH)Ph

$C_{14}H_{11}O_2P$ M 242.213

Cryst. (C_6H_6). Mp 126-127°.

Fedorova, G.K. *et al*, *Zh. Obshch. Khim.*, 1967, **37**, 2686 (*Engl. transl.* p. 2557); 1974, **44**, 85 (*Engl. transl.* p. 83)

[Phenyl[(phenylmethyl)amino]methyl]-phosphonic acid, 9CI **P-00188**

[α-(Benzylamino)benzyl]phosphonic acid, 8CI

[25881-35-0]

$PhCH_2NHCHPhP(O)(OH)_2$

$C_{14}H_{16}NO_3P$ M 277.259

(±)-***form***

Cryst. (EtOH aq.). Mp 180-181°, Mp 236-237°.

Di-Et ester: [68374-69-6].

$C_{18}H_{24}NO_3P$ M 333.366

No phys. props. reported.

Tyka, R., *Tetrahedron Lett.*, 1970, 677 (*synth*)
Taube, D.O. *et al*, *Zh. Obshch. Khim.*, 1972, **42**, 351 (*Engl. transl.* p. 341) (*synth*)
Ružić-Toroš, Ž. *et al*, *Acta Crystallogr., Sect. B*, 1978, **34**, 3110 (*ester, cryst struct*)

Phenyl phenylphosphoramidic acid **P-00189**

Phenyl phosphoric monoanilide. Phenyl hydrogen phenylphosphoramidate

[6254-02-0]

O NHPh
P
PhO OH

$C_{12}H_{12}NO_3P$ M 249.205

Cryst. (MeCN). Mp 140-140.5°.

Cyclohexylammonium salt: [2152-00-3]. Solid. Mp 200-203°.

Chloride:

$C_{12}H_{11}ClNO_2P$ M 267.651

Cryst. (MeCN). Mp 129-130°.

Schallen, H. *et al*, *Chem. Ber.*, 1961, **94**, 1621 (*synth*)
Orey, I. *et al*, *J. Am. Chem. Soc.*, 1967, **89**, 6972 (*synth*)
Mestres, R. *et al*, *Synthesis*, 1981, 218; 1982, 288 (*synth*)

Phenyl phenylphosphoramidochloridate, 9CI **P-00190**

[51766-21-3]

Cl O
P
PhNH OPh

$C_{12}H_{11}ClNO_2P$ M 267.651

(±)-***form***

Phosphorylating agent for nucleosides. Activating agent for RCOOH in the synth. of amides and anhydrides. Cryst. (MeCN). Mp 129-133°.

Zieliński, W.S. *et al*, *Synthesis*, 1976, 185 (*synth, ms, P nmr, tlc, use*)
Mestres, R. *et al*, *Synthesis*, 1982, 288 (*use*)
Arrieta, A. *et al*, *Synth. Commun.*, 1983, **13**, 471 (*use*)

***P*-Phenyl-*N*-phenylsulfonylphosphonamidic acid** **P-00191**

N-*Benzenesulfonyl*-P-*phenylphosphonamidic acid*

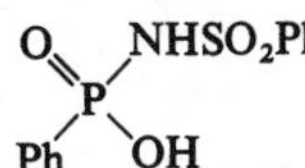

$C_{12}H_{12}NO_4PS$ M 297.265

Me ester: Methyl P-*phenyl*-N-*phenylsulfonylphosphonamidate.*

$C_{13}H_{14}NO_4PS$ M 311.292

Solid. Mp 172-173°.

Et ester: Ethyl P-*phenyl*-N-*phenylsulfonylphosphonamidate.*

$C_{14}H_{16}NO_4PS$ M 325.318

Solid. Mp 141-142°.

Ph ester: Phenyl P-*phenyl*-N-*phenylsulfonylphosphonamidate.*

$C_{18}H_{16}NO_4PS$ M 373.362

Solid. Mp 144-146°.

Shevchenko, V.I. *et al*, *Zh. Obshch. Khim.*, 1960, **30**, 1561 (*Engl. transl.* p. 1568)

***P*-Phenyl-*N*-phenylsulfonylphosphonimidic acid** **P-00192**

$PhSO_2N{=}PPh(OH)_2$

$C_{12}H_{12}NO_4PS$ M 297.265

Tautomeric with *P*-Phenyl-*N*-phenylsulfonylphosphonamidic acid, P-00191 .

Di-Me ester: Dimethyl P-*phenyl*-N-*phenylsulfonylphosphonimidate.* P,P-*Dimethoxy*-P-*phenyl*-N-*phenylsulfonylphosphazene.* P,P-*Dimethoxy*-P-*phenyl*-N-*phenylsulfonylphosphine imide.*

$C_{14}H_{16}NO_4PS$ M 325.318

Cryst. (EtOH). Mp 48°.

Di-Ph ester: Diphenyl P-*phenyl*-N-*phenylsulfonylphosphonimidate.* P,P-*Diphenoxy*-P-*phenyl*-N-*phenylsulfonylphosphazene.* P,P-*Diphenoxy*-P-*phenyl*-N-*phenylsulfonylphosphine imide.*

$C_{24}H_{20}NO_4PS$ M 449.460

Cryst. (EtOH). Mp 64-65°.

Shevchenko, V.I. *et al*, *Zh. Obshch. Khim.*, 1960, **30**, 1561 (*Engl. transl.* p. 1568) (*esters, synth, props*)
Babyak, A.G. *et al*, *Zh. Obshch. Khim.*, 1974, **44**, 469 (*Engl. transl.* p. 453) (*dichloride*)

9-Phenyl-9-phosphabicyclo[3.3.1]nonane, 9CI, 8CI **P-00193**

[14109-35-4]

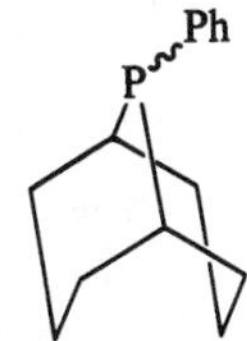

$C_{14}H_{19}P$ M 218.278

$Bp_{0.3}$ 134-135°. Mixed with 9-Phenyl-9-phosphabicyclo[4.2.1]nonane, P-00194 .

Oxide: [57458-74-9].

$C_{14}H_{19}OP$ M 234.277

Cryst. (C_6H_6/cyclohexane). Mp 181-183°. $Bp_{0.015}$ 175-178° subl.

Netherlands Pat., 6 604 094, (*1966*); *CA*, **66**, 65101 (*synth*)
Wiseman, J.R. *et al*, *J. Org. Chem.*, 1976, **41**, 589 (*oxide, ir, pmr*)

9-Phenyl-9-phosphabicyclo[4.2.1]nonane, 9CI, 8CI **P-00194**

[13945-86-3]

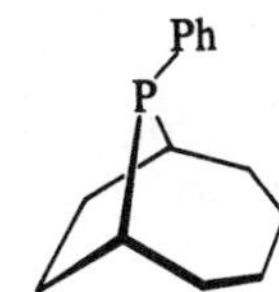

$C_{14}H_{19}P$ M 218.278

Solid. Mp 58-60°. $Bp_{0.05}$ 60° subl.

B,MeI: 9-Methyl-9-phenyl-9-phosphoniabicyclo[4.2.1]-nonane iodide.
$C_{15}H_{22}IP$ M 360.217
Needles (EtOH). Mp 244.5-245°.

B,PhBr: 9,9-Diphenyl-9-phosphoniabicyclo[4.2.1]-nonane bromide.
$C_{20}H_{24}BrP$ M 375.287
Solid. Mp 244-246°. With PhLi, yields one of v. few known bridgehead ylides.

Oxide: [40468-93-7].
$C_{14}H_{19}OP$ M 234.277
Cryst. (C_6H_6/cyclohexane). Mp 157-159°. $Bp_{0.1}$ 150° subl.

Katz, T.J. *et al, J. Am. Chem. Soc.*, 1966, **88**, 3832 (*deriv, pmr*)
Turnblom, E.W. *et al, J. Am. Chem. Soc.*, 1973, **95**, 4292 (*synth, derivs, ir, uv, pmr*)

9-Phenyl-9-phosphabicyclo[3.3.1]nonan-3-one, 9CI — P-00195

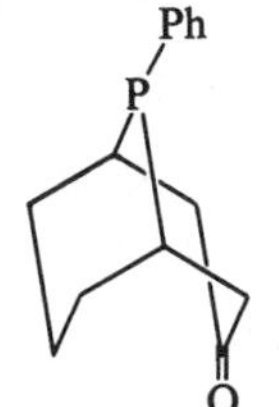

syn-form

$C_{14}H_{17}OP$ M 232.261

syn-form

Oxide:
$C_{14}H_{17}O_2P$ M 248.261
Cryst. (C_6H_6/cyclohexane). Mp 201.5-204°. This is referred to as the *anti*-oxide (O atom *anti* to C=O group).

anti-form

Oxide: Cryst. ($MeCN/C_6H_6$). Mp 305-306° (250° dec.). This is referred to as the *syn*-oxide.

Kashman, Y. *et al, Tetrahedron*, 1972, **28**, 4091 (*synth, derivs, ir, ms*)
Wiseman, J.R. *et al, J. Org. Chem.*, 1976, **41**, 589 (*oxide, synth, ir, pmr, cmr, nmr*)

9-Phenyl-9-phosphabicyclo[4.2.1]nona-2,4,7-triene, 9CI — P-00196

[13887-07-5]

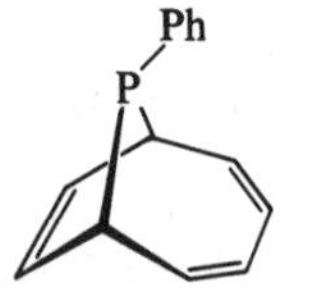

syn-form

$C_{14}H_{13}P$ M 212.230

syn-form

Needle-like cryst. Mp 85.5-86.5°. $Bp_{0.4}$ 160-170°, $Bp_{0.1}$ 80° subl. Darkens rapidly on exp. to air. When heated, yields the *anti*-form.

B,MeI: 9-Methyl-9-phenyl-9-phosphoniabicyclo[4.2.1]-nona-2,4,7-triene iodide.
$C_{15}H_{16}IP$ M 354.170
Pale-yellow, rod-like cryst. (EtOH). Mp 239-240°.

Oxide:
$C_{14}H_{13}OP$ M 228.230
Solid. Mp 182.8-183.4°.

anti-form

Solid. Mp 84.5-85.5°. $Bp_{0.1}$ 80° subl.

B,MeI: Cryst. (EtOH). Mp 219-220°.

Oxide: Needles (Me_2CO). Mp 197.5-199.7°.

Katz, T.J. *et al, J. Am. Chem. Soc.*, 1966, **88**, 3832 (*synth, props, derivs, uv, pmr*)
Turnblom, E.W. *et al, J. Am. Chem. Soc.*, 1973, **95**, 4292 (*synth*)
Evans, W.J. *et al, J. Org. Chem.*, 1981, **46**, 3925 (*synth*)

9-Phenyl-9-phosphabicyclo[6.1.0]none-2,4,6-triene — P-00197

[43017-14-7]

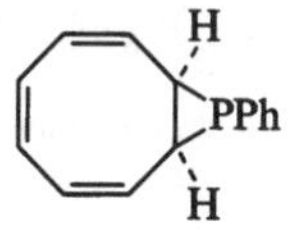

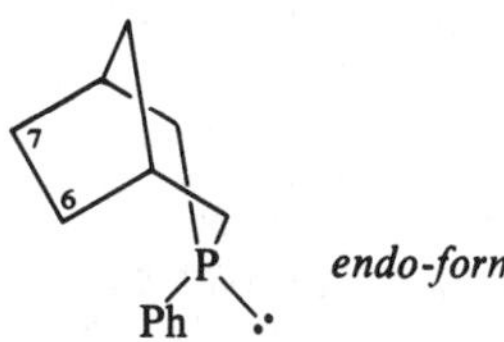

$C_{14}H_{13}P$ M 212.230

Oily at r.t. Cryst. at −78°. Darkens rapidly on exp. to air. On lengthy standing or when heated at 70°, yields 9-Phenyl-9-phosphabicyclo[4.2.1]nona-2,4,7-triene, P-00196 .

Katz, T.J. *et al, J. Am. Chem. Soc.*, 1966, **88**, 3832 (*synth, uv, P nmr, pmr, props*)
Märkl, G. *et al, Tetrahedron Lett.*, 1982, **23**, 4915 (*props*)

3-Phenyl-3-phosphabicyclo[3.2.1]octane, 9CI — P-00198

endo-form

$C_{13}H_{17}P$ M 204.251

B,PhBr: 3,3-Diphenyl-3-phosphoniabicyclo[3.2.1]octane bromide.
$C_{19}H_{22}BrP$ M 361.261
Cryst. (H_2O). Mp 228-230°.

endo-form [76152-73-0]

3-Oxide: [76152-72-2].
$C_{13}H_{17}OP$ M 220.250
Cryst. (cyclohexane). Mp 92-94°.

6,7-Didehydro, 3-oxide: [79150-96-2]. *3-Phenyl-3-phosphabicyclo[3.2.1]oct-6-ene 3-oxide.*
$C_{13}H_{15}OP$ M 218.235
Solid.

exo-form

6,7-Didehydro, 3-oxide: Solid.

Mazhur-ul-Haque, *et al, J. Chem. Soc., Perkin Trans. 2*, 1980, 1467; 1981, 1000 (*synth, derivs, cmr, pmr, cryst struct*)
Mazhur-ul-Haque, *et al, Acta Crystallogr., Sect. C*, 1983, **39**, 383 (*cryst struct*)

8-Phenyl-8-phosphabicyclo[3.2.1]octan-3-one, 9CI, 8CI — P-00199

[29259-72-1]

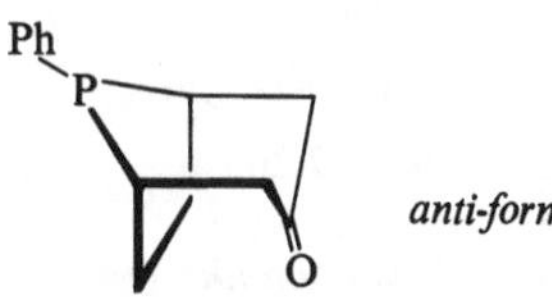

anti-form

$C_{13}H_{15}OP$ M 218.235

syn-form [67580-24-9]

Oxide: [29259-76-5].

$C_{13}H_{15}O_2P$ M 234.234

Solid. Mp 247-248°.

anti-form [29259-72-1]

Cryst. (Me_2CO). Mp 144-146°. $Bp_{0.4}$ 150° subl.

B,MeI: 8-Methyl-8-phenyl-8-phosphoniabicyclo[3.2.1]-octan-3-one iodide.

$C_{14}H_{18}IOP$ M 360.174

Cryst. (H_2O). Mp 239-240°.

B,PhCH₂Cl: 8-Benzyl-8-phenyl-8-phosphoniabicyclo[3.2.1]octan-3-one chloride.

$C_{20}H_{22}ClOP$ M 344.820

Cryst. (EtOH/EtOAc). Mp 267-269°.

Oxide: [29259-77-6]. Cryst. (MeCN). Mp 235-237°.

Kashman, Y. *et al*, *Tetrahedron*, 1970, **26**, 4213 (*synth, ir, uv*)
Rudi, A. *et al*, *Org. Magn. Reson.*, 1977, **10**, 245 (*cmr*)

8-Phenyl-8-phosphabicyclo[3.2.1]oct-6-ene, 9CI — P-00200

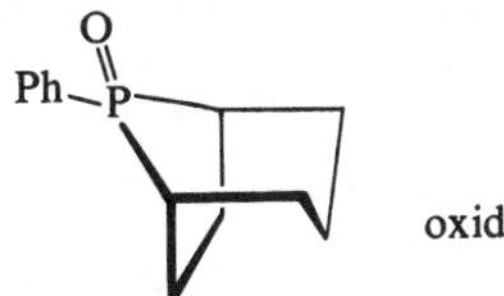

oxide

$C_{13}H_{15}P$ M 202.235

Oxide:

$C_{13}H_{15}OP$ M 218.235

Cryst. (EtOH/EtOAc). Mp 126-127°.

Awerbouch, O. *et al*, *Tetrahedron*, 1975, **31**, 33 (*synth, ir, pmr*)

9-Phenyl-9-phosphapentacyclo[4.3.0.0^{2,5}.0^{3,8}.0^{4,7}]-nonane, 9CI — P-00201

9-Phenyl-9-phosphahomocubane

[25881-31-6]

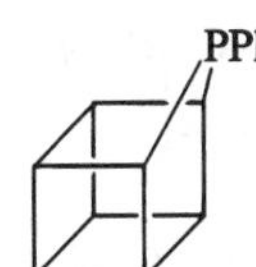

$C_{14}H_{13}P$ M 212.230

Stereochem. at phosphorus uncertain. Solid. Mp 58-59° (54.5-56.5°). $Bp_{0.1}$ 50° subl.

B, MeI: 9-Methyl-9-phenyl-9-phosphoniahomocubane iodide.

$C_{15}H_{16}IP$ M 354.170

Cryst. (EtOH aq.). Mp 195-196°.

B, PhBr: 9,9-Diphenyl-9-phosphoniahomocubane bromide.

$C_{20}H_{18}BrP$ M 369.240

Cryst. (MeOH/Et_2O). Mp 310-311°.

9-Oxide: [28051-32-3].

$C_{14}H_{13}OP$ M 228.230

Mp 122-123° (115-118°) (sealed tube).

9-Sulfide: [67452-76-0].

$C_{14}H_{13}PS$ M 244.290

Solid. Mp 162-162.5°.

Katz, T.J. *et al*, *J. Am. Chem. Soc.*, 1970, **92**, 734 (*synth, oxide, pmr*)
Turnblom, E.W. *et al*, *J. Am. Chem. Soc.*, 1973, **95**, 4292 (*synth, derivs, ir*)
Albarella, J.P. *et al*, *J. Org. Chem.*, 1978, **43**, 4338 (*derivs, ir, ms, uv, pmr*)

9-Phenyl-9-phosphatricyclo[3.3.1.0^{2,8}]-nona-3,6-diene, 9CI, 8CI — P-00202

[25881-29-2]

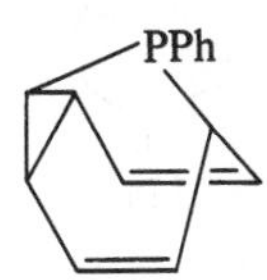

$C_{14}H_{13}P$ M 212.230

Solid. Mp 43-45°.

9-Oxide: [25881-30-5].

$C_{14}H_{13}OP$ M 228.230

Solid. Mp 101-102.5°.

Katz, T.J. *et al*, *J. Am. Chem. Soc.*, 1970, **92**, 734 (*synth, oxide, pmr*)
Turnblom, E. *et al*, *J. Am. Chem. Soc.*, 1973, **95**, 4292 (*props*)

9-Phenyl-9-phosphatricyclo[4.2.1.0^{2,5}]-nona-3,7-diene, 9CI, 8CI — P-00203

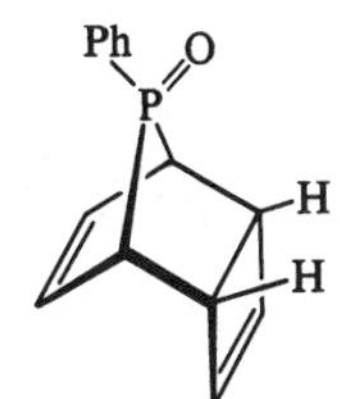

(*Endo-syn*)-*form* of oxide

$C_{14}H_{11}P$ M 210.215

Undergoes destruction when heated in MeOH soln., yielding $PhPH_2$, cyclooctatetraene and dimethyl phenylphosphonite.

endo-syn-*9-Oxide:* [26003-21-9]. Mp 107.5-109°.

endo-anti-*9-Oxide:* [26003-22-5].

$C_{14}H_{11}OP$ M 226.214

Cryst. (C_6H_6/cyclohexane). Mp 178-179°.

Katz, T.J. *et al*, *J. Am. Chem. Soc.*, 1970, **92**, 734 (*synth*)
Quin, L.D. *et al*, *Org. Magn. Reson.*, 1979, **12**, 442 (*nmr*)
Mesch, M.A. *et al*, *Tetrahedron Lett.*, 1980, **21**, 4791 (*props*)

9-Phenyl-9-phosphatricyclo[4.2.1.0^{2,5}]-nonane, 9CI — P-00204

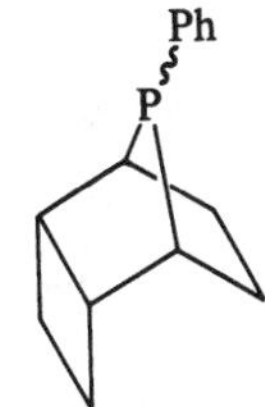

$C_{14}H_{17}P$ M 216.262

($1\alpha,2\alpha,5\alpha,6\alpha$)-*form*

B,PhBr: 9,9-Diphenyl-9-phosphoniatricyclo[4.2.1.0^{2,5}]-nonane bromide.

$C_{20}H_{22}BrP$ M 373.272

Cryst. (MeOH/Et_2O). Mp 264-265°.

9-Oxide: [43017-17-0].

$C_{14}H_{17}OP$ M 232.261

Solid. Mp 52-54°. $Bp_{0.1}$ 135° subl.

Turnblom, E.W. *et al*, *J. Am. Chem. Soc.*, 1973, **95**, 4292 (*synth, ir, ms, uv, pmr, props*)

1-Phenylphosphepane, 9CI, 8CI **P-00205**

[4963-96-6]

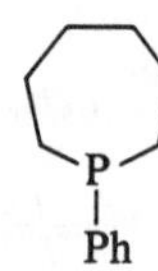

$C_{12}H_{17}P$ M 192.240

Liq. $Bp_{0.035}$ 110-120° (oven).

B,PhBr: [59386-55-9]. *1,1-Diphenylphosphepanium bromide.*

$C_{18}H_{22}BrP$ M 349.250

Cryst. (EtOH/EtOAc). Mp 248-250°.

B,PhCH₂Br: [59432-49-4]. *1-Benzyl-1-phenylphosphepanium bromide.*

$C_{19}H_{24}BrP$ M 363.276

Cryst. (EtOH/EtOAc). Mp 164.5-165.5°.

1-Oxide: [36126-86-0].

$C_{12}H_{17}OP$ M 208.239

Needles (octane). Mp 119-120°.

Sulfide:

$C_{12}H_{17}PS$ M 224.300

Cryst. (CH_2Cl_2). Mp 90.5-91.5°.

Davies, J.H. *et al, J. Chem. Soc. (C)*, 1966, 245 (*synth, sulfide*)
Derkach, N.Ya. *et al, Zh. Obshch. Khim.*, 1971, **41**, 2806 (*Engl. transl.* p. 2836) (*oxide*)
Gray, G.A. *et al, J. Am. Chem. Soc.*, 1976, **98**, 2109 (*salts, oxide, cmr, nmr, pmr*)

1-Phenyl-1*H*-phosphepin, 8CI **P-00206**

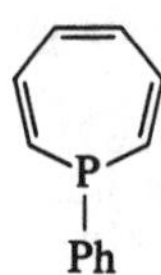

Oxide: [29634-03-5].

$C_{12}H_{11}OP$ M 202.192

Clusters of needles. Mp 91-92°.

Märkl, G. *et al, Tetrahedron Lett.*, 1970, 1273 (*synth, pmr, uv*)

1-Phenyl-1*H*-phosphindole, 8CI **P-00207**

[31236-98-3]

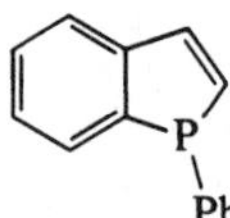

$C_{14}H_{11}P$ M 210.215

Cryst. Mp 66°.

1-Oxide: [31236-97-2].

$C_{14}H_{11}OP$ M 226.214

Cryst. (Et_2O). Mp 92°.

Chan, T-H. *et al, Can. J. Chem.*, 1971, **49**, 530 (*synth, deriv, uv*)
Nief, F. *et al, Phosphorus Sulfur*, 1982, **13**, 259 (*synth, deriv, pmr, P nmr*)

Phenylphosphine, 9CI **P-00208**

Phosphaaniline

[638-21-1]

$PhPH_2$

C_6H_7P M 110.095

Foul-smelling liq. d^{15} 1.001. Bp 160-161°, Bp_{50} 72°. pK_{a1} 24.5 (referred to MeOH), pK_{a2} 22.4 (DMSO). n_D^{20} 1.5796. V. rapidly oxid. in air.

▷Highly toxic, TLV 0.25. Fire hazard. SZ2100000.

Oxide: [10052-96-7]. Tautomeric with Phenylphosphinic acid, P-00210 .

Sulfide: Tautomeric with Phenylphosphinothioic acid, P-00213 .

Horner, L. *et al, Chem. Ber.*, 1959, **92**, 2088.
Sander, M., *Chem. Ber.*, 1960, **93**, 1220 (*synth*)
Issleib, K. *et al, Chem. Ber.*, 1965, **98**, 2091 (*derivs*)
Taylor, R.C. *et al, Synth. React. Inorg. Metal-Org. Chem.*, 1973, **3**, 175 (*synth*)
Kabachnik, M.I. *et al, Aust. J. Chem.*, 1975, **28**, 755 (*ir*)
Becker, G. *et al, Z. Anorg. Allg. Chem.*, 1978, **443**, 42 (*synth, derivs*)
Parr, W.J.E., *J. Chem. Soc., Faraday Trans. 2*, 1978, **74**, 933 (*pmr, cmr, struct*)
Ratovskii, G.V. *et al, Zh. Obshch. Khim.*, 1978, **48**, 1520 (*Engl. transl.* p. 1394) (*uv*)
Vincent, E. *et al, Spectrochim. Acta, Part A*, 1980, **36**, 699 (*pmr, cmr, nmr*)
Zverev, V.V. *et al, Zh. Obshch. Khim.*, 1981, **51**, 303 (*Engl. transl.* p. 242) (*pe*)
Batchelor, R. *et al, J. Am. Chem. Soc.*, 1982, **104**, 674 (*pmr, cmr, nmr*)
Sinyashin, O.G. *et al, Zh. Obshch. Khim.*, 1984, **54**, 1917 (*Engl. transl.* p. 1708) (*synth, nmr*)
Sax, N.I., *Dangerous Properties of Industrial Materials*, 6th Ed., Van Nostrand-Reinhold, 1984, 906.

Phenylphosphinedicarboxylic acid, 9CI **P-00209**

$PhP(COOH)_2$

$C_8H_7O_4P$ M 198.115

Esters dec. by aq. alkali to give phenylphosphine.

Di-Me ester: [79564-52-6]. *Bis(methoxycarbonyl)-phenylphosphine.*

$C_{10}H_{11}O_4P$ M 226.168

Liq. $Bp_{0.4}$ 126-130°.

Di-Et ester: [79564-54-8]. *Bis(ethoxycarbonyl)-phenylphosphine.*

$C_{12}H_{15}O_4P$ M 254.222

Liq. $Bp_{0.6}$ 112-115°.

Di-Ph ester: [79564-58-2]. *Bis(phenoxycarbonyl)-phenylphosphine.*

$C_{20}H_{15}O_4P$ M 350.310

Dec. on heating.

Thamm, R. *et al, Z. Naturforsch., B*, 1981, **36**, 910 (*synth, ir, ms, pmr, nmr*)

Phenylphosphinic acid, 9CI **P-00210**

[1779-48-2]

PhPH(O)(OH)

$C_6H_7O_2P$ M 142.094

Tautomeric with Phenylphosphonous acid, P-00251 The free acid exists in the phosphinic acid form. Cryst. Mp 86-87° (78-80°). pK_{a1} 1.53 (H_2O), pK_{a2} 2.1 (40% MeOH aq.).

Cyclohexylammonium salt: Solid. Mp 198-200°.

Anilinium salt: Solid. Mp 101°.

Me ester: [7162-15-4]. *Methyl phenylphosphinate.*

$C_7H_9O_2P$ M 156.121

Liq. Bp_3 103-104°.

Et ester: [2511-09-3]. *Ethyl phenylphosphinate.*

$C_8H_{11}O_2P$ M 170.147

Prod. of uv irrad. of Inezin. Liq. d_4^{20} 1.13. $Bp_{0.2}$ 102-103°. n_D^{20} 1.5180 (1.5220).
▷SZ5600000.
Isopropyl ester: [13336-50-6]. *Isopropyl phenylphosphinate. 1-Methylethyl phenylphosphinate.*
$C_9H_{13}O_2P$ M 184.174
Liq. d_4^{20} 1.09. Bp_{10} 146°, $Bp_{0.5}$ 80°. n_D^{20} 1.5075, 1.5154.
Ph ester: [52744-21-5]. *Phenyl phenylphosphinate.*
$C_{12}H_{11}O_2P$ M 218.191
Liq. n_D^{25} 1.5924.
(−)-*Menthyl ester: Menthyl phenylphosphinate.*
$C_{16}H_{25}O_2P$ M 280.346
Liq. $Bp_{0.1}$ 165° (bath).
Trimethylsilyl ester: [27262-80-2]. *Trimethylsilyl phenylphosphinate.*
$C_9H_{15}O_2PSi$ M 214.276
Liq. d_4^{20} 1.05. $Bp_{0.07}$ 91°. n_D^{20} 1.4902.
N,N-*Diethylamide:* [70403-04-2]. N,N-*Diethyl*-P-*phenylphosphinic amide.* No phys. props. reported.
Anhydride: [5054-42-2].
$C_{12}H_{12}O_3P_2$ M 266.173
No phys. props. reported.

Frank, A.W., *J. Org. Chem.*, 1961, **26**, 850 (*synth*)
Goncalves, H. *et al*, *Bull. Soc. Chim. Fr.*, 1961, 1595 (*synth, ir, esters*)
Budzikiewicz, H. *et al*, *Monatsh. Chem.*, 1965, **96**, 1739 (*ester, ms*)
Wolf, R. *et al*, *Spectrochim. Acta, Part A*, 1967, **23**, 1641 (*esters, ir*)
Emmick, T.L. *et al*, *J. Am. Chem. Soc.*, 1968, **90**, 3459 (*esters*)
Pudovik, A.N. *et al*, *Zh. Obshch. Khim.*, 1968, **38**, 305 (*Engl. transl.*, p. 306) (*ester*)
Pudovik, A.N. *et al*, *Zh. Obshch. Khim.*, 1969, **39**, 1715 (*Engl. transl.* p. 1681) (*ir, props*)
Benschop, H.P. *et al*, *J. Chem. Soc., Dalton Trans.*, 1970, 1431 (*ester*)
Brazier, J-F. *et al*, *Bull. Soc. Chim. Fr.*, 1970, 1089 (*silyl ester, ir, pmr, nmr*)
Gallagher, M.J. *et al*, *J. Chem. Soc. (C)*, 1971, 593 (*anhydride*)
Wiley, R.H., *Org. Mass Spectrom.*, 1971, **5**, 675 (*ms*)
Andreev, N.A. *et al*, *Zh. Obshch. Khim.*, 1979, **49**, 2230 (*Engl. transl.* p 1959) (*amide*)
Hewitt, D.G., *Aust. J. Chem.*, 1979, **32**, 463 (*ester*)

1,1′-(Phenylphosphinidene)ferrocene, 10CI — P-00211

1-Phenyl-1-phospha[1]ferrocenophane. 1,1′-(Phenylphosphinidene)diylferrocene. (1,1′-Ferrocenediyl)phenylphosphine
[72954-06-4]

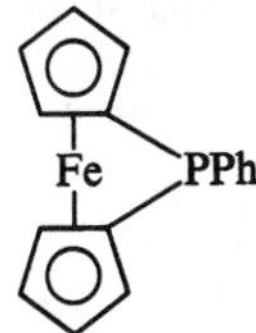

$C_{16}H_{13}FeP$ M 292.099
Forms Fe and W complexes. Air-sensitive, brick-red nuggets (hexane) or red-purple cryst. (hexane at −30°). Mp 100-102°, Mp 104-107° dec. Mod. air-sensitive. Undergoes ring-opening with RLi.
Sulfide: [72954-09-7].
$C_{16}H_{13}FePS$ M 324.159
Mod. air-stable red-brown microcryst. Mp >280°.

Osborne, A.G. *et al*, *J. Organomet. Chem.*, 1980, **193**, 345 (*synth, pmr, cmr, uv, complexes*)
Seyferth, D. *et al*, *J. Organomet. Chem.*, 1980, **185**, C1 (*sulfide*)
Stoeckli-Evans, H. *et al*, *J. Organomet. Chem.*, 1980, **194**, 91 (*struct*)
Seyferth, D. *et al*, *Organometallics*, 1982, **1**, 1275 (*synth, ir, uv, cmr, pmr, ms*)
Withers, H.P. *et al*, *Organometallics*, 1982, **1**, 1283.

2-Phenylphosphinoline, 9CI — P-00212

2-Phenyl-1-phosphanaphthalene. 2-Phenylbenzo[b]*-phosphorin. 2-Phenylbenzo*[b]*phosphinine*
[39768-04-2]

$C_{15}H_{11}P$ M 222.226
Pale-yellow cryst. (EtOH). Mp 101-102°.

Märkl, G. *et al*, *Angew. Chem., Int. Ed. Engl.*, 1972, **11**, 1017 (*synth*)
Schäfer, W. *et al*, *Tetrahedron Lett.*, 1973, 3743 (*pe*)
Daly, J.H. *et al*, *J. Chem. Soc., Dalton Trans.*, 1974, 2388 (*cryst struct*)
Märkl, G. *et al*, *Tetrahedron Lett.*, 1974, 4501, 4369 (*props*)

Phenylphosphinothioic acid, 9CI, 8CI — P-00213

PhPH(S)OH ⇌ PhPH(O)SH ⇌ PhP(SH)(OH)

C_6H_7OPS M 158.154
Exists as tautomeric equilibrium between thione and thiol forms in further equilibrium with phenylphosphonothious acid (minor component).
O-*Et ester:* [6591-08-8]. O-*Ethyl phenylphosphinothioate.*
$C_8H_{11}OPS$ M 186.208
Liq. d_4^{20} 1.14. $Bp_{0.1}$ 105-107°. n_D^{20} 1.5716.
O-*Propyl ester:* O-*Propyl phenylphosphinothioate.*
$C_9H_{13}OPS$ M 200.235
Liq. $Bp_{0.03}$ 100-140° (bath). n_D^{20} 1.5613.

Michalski, J. *et al*, *Rocz. Chem.*, 1962, **36**, 1781; *CA*, **59**, 10109 (*esters*)
Petrov, K.A. *et al*, *Zh. Obshch. Khim.*, 1964, **34**, 2226 (*Engl. transl.* p. 2236) (*esters*)
Sanchez, M. *et al*, *Spectrochim. Acta, Part A*, 1967, **23**, 2617 (*ester, ir*)
Marty, R. *et al*, *Org. Magn. Reson.*, 1970, **2**, 141 (*ester, pmr, nmr*)
Andreev, N.A. *et al*, *Zh. Obshch. Khim.*, 1982, **52**, 1530 (*Engl. transl.* p. 1352) (*ester*)

1-Phenylphosphirane, 9CI, 8CI — P-00214

[22846-16-8]

C_8H_9P M 136.133
Liq. $Bp_{1.5}$ 44-48°.

Chan, S. *et al*, *Tetrahedron*, 1969, **25**, 1097 (*synth, ir, P nmr, pmr*)
Denney, D. *et al*, *J. Am. Chem. Soc.*, 1974, **96**, 317 (*synth, props*)
Gray, G. *et al*, *J. Am. Chem. Soc.*, 1976, **98**, 2109 (*pmr, cmr, P nmr*)

1-Phenylphosphocane, 9CI **P-00215**

1-Phenylphosphacyclooctane

[59386-52-6]

$C_{13}H_{18}P$ M 205.259

Liq. Bp_{15} 120-130° (oven).

B,PhBr: 1,1-Diphenylphosphocanium bromide.

$C_{19}H_{23}BrP$ M 362.268

Cryst. (EtOH/EtOAc). Mp 222.5-223.5°.

B,PhCH₂Br: 1-Benzyl-1-phenylphosphocanium bromide.

$C_{20}H_{25}BrP$ M 376.295

Cryst. (EtOH/EtOAc). Mp 182-183°.

Oxide: [59386-61-7].

$C_{13}H_{18}OP$ M 221.258

Solid by subl. Mp 86-88°.

Gray, G.A. *et al*, *J. Am. Chem. Soc.*, 1976, **98**, 2109 (*synth, derivs, cmr, P nmr, pmr*)

1-Phenylphospholane, 9CI **P-00216**

Phenyltetramethylenephosphine

[3302-87-2]

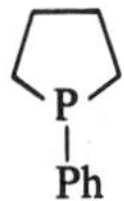

$C_{10}H_{13}P$ M 164.186

Liq. Bp_{14} 125°. n_D^{20} 1.5918.

B,MeI: [3302-93-0]. *1-Methyl-1-phenylphospholanium iodide.*

$C_{11}H_{16}IP$ M 306.126

Cryst. (EtOH). Mp 130°.

B,PhBr: [43017-36-3]. *1,1-Diphenylphospholanium bromide.*

$C_{16}H_{18}BrP$ M 321.196

Solid. Mp 163-164°.

B,PhCH₂Br: [31082-04-9]. *1-Benzyl-1-phenylphospholanium bromide.*

$C_{17}H_{20}BrP$ M 335.223

Cryst. (MeCN). Mp 166-167°.

1-Oxide: [4963-91-1].

$C_{10}H_{13}OP$ M 180.186

Solid. Mp 56-57°. $Bp_{0.2}$ 108-112°.

1-Sulfide:

$C_{10}H_{13}PS$ M 196.246

Cryst. (cyclohexane). Mp 77°.

Issleib, K. *et al*, *Chem. Ber.*, 1961, **94**, 113 (*synth, derivs*)
Davies, J.H. *et al*, *J. Chem. Soc. (C)*, 1966, 245 (*synth, derivs*)
Alver, E. *et al*, *Acta Chem. Scand.*, 1967, **21**, 359 (*cryst struct*)
Sommer, K., *Z. Anorg. Allg. Chem.*, 1970, **379**, 56 (*synth, P nmr*)
Schäfer, W. *et al*, *Angew. Chem., Int. Ed. Engl.*, 1973, **12**, 145 (*pe*)
Fell, B. *et al*, *Synthesis*, 1974, 119 (*oxide*)
Wetzel, R.B. *et al*, *J. Org. Chem.*, 1974, **39**, 1531 (*synth, pmr*)
Gray, G.A. *et al*, *J. Am. Chem. Soc.*, 1976, **98**, 2109 (*cmr, nmr, pmr*)
Koch, C.W. *et al*, *Org. Mass Spectrom.*, 1977, **12**, 624 (*oxide, ms*)
Muchowski, J.M. *et al*, *J. Org. Chem.*, 1981, **46**, 459 (*deriv, pmr, ir*)

1-Phenyl-1*H*-phosphole, 9CI **P-00217**

[20342-00-1]

$C_{10}H_9P$ M 160.155

Forms complexes with Mo and metals of Group VIII. Oil. $Bp_{0.4}$ 64-65°. Yields a dimeric oxide and a dimeric sulfide, but no monomeric derivs.

Dimer: see 3a,4,7,7a-Tetrahydro-1,8-diphenyl-4,7-phosphinidene-1H-phosphindole, T-00105

Märkl, G. *et al*, *Tetrahedron Lett.*, 1968, 1755 (*synth, pmr*)
Bundgaard, T. *et al*, *Tetrahedron Lett.*, 1972, 3353 (*cmr*)
Mathey, F. *et al*, *Org. Magn. Reson.*, 1972, **4**, 171 (*P nmr, pmr*)
Breque, A. *et al*, *Synthesis*, 1981, 983 (*synth, P nmr*)

***P*-Phenylphosphonamidic acid, 9CI** **P-00218**

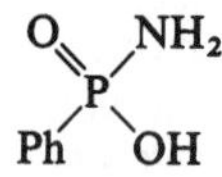

$C_6H_8NO_2P$ M 157.108

Isol. as Na salt.

Me ester: [40334-32-5]. *Methyl* P-*phenylphosphonamidate.*

$C_7H_{10}NO_2P$ M 171.135

Cryst. (C_6H_6). Mp 110-111°.

Et ester: [5326-06-7]. *Ethyl* P-*phenylphosphonamidate.*

$C_8H_{12}NO_2P$ M 185.162

Cryst. (xylene). Sol. $CHCl_3$, Me_2CO, hot H_2O, sl. sol. Et_2O, cold H_2O. Mp 127°.

Ph ester: [5467-82-3]. *Phenyl* P-*phenylphosphonamidate.*

$C_{12}H_{12}NO_2P$ M 233.206

Solid. Mp 127°.

Benzyl ester: [14572-73-7]. *Benzyl* P-*phenylphosphonamidate.*

$C_{13}H_{14}NO_2P$ M 247.233

Cryst. (CCl_4/EtOH). Mp 121-123°.

Fluoride: [29070-40-4].

C_6H_7FNOP M 159.100

Solid. Mp 86°.

N,N-*Di-Me: see N,N*-Dimethyl-*P*-phenylphosphonamidic acid, D-00808

N,N-Di-Et: see *N,N*-Diethyl-*P*-phenylphosphonamidic acid, D-00307

N-Ph: see *N,P*-Diphenylphosphonamidic acid, D-01103

Smith, W.C. *et al*, *J. Org. Chem.*, 1957, **22**, 265 (*esters, synth*)
Shevchenko, V.I. *et al*, *Zh. Obshch. Khim.*, 1959, **29**, 3757 (*synth*)
Petrov, K.A. *et al*, *Zh. Obshch. Khim.*, 1960, **30**, 4060 (*Engl. transl.* p. 4023) (*phenyl ester*)
Fanshawe, W.J. *et al*, *J. Med. Chem.*, 1967, **10**, 16 (*benzyl ester*)
Roesky, H.W. *et al*, *Z. Anorg. Allg. Chem.*, 1970, **375**, 140 (*fluoride, synth, ir, ms, pmr, F nmr*)
Felcht, U. *et al*, *Justus Liebigs Ann. Chem.*, 1977, 1309 (*derivs, synth, ir, pmr*)
Harger, M.J.P. *et al*, *Tetrahedron*, 1982, **38**, 3073 (*derivs*)

P-Phenylphosphonamidothioic acid, 9CI **P-00219**

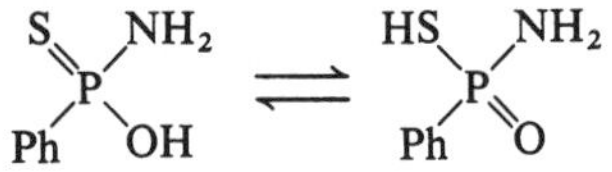

C_6H_8NOPS M 173.169

Fluoride: [24623-72-1].
C_6H_7FNPS M 175.160
Oil. $Bp_{0.01}$ 110°.

Roesky, H.W. *et al*, *Z. Naturforsch.*, *B*, 1969, **24**, 1250 (*synth, ir, pmr, F nmr*)

Phenylphosphonazidic acid **P-00220**

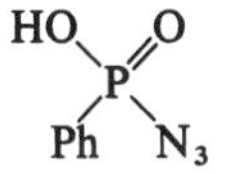

$C_6H_6N_3O_2P$ M 183.106

Me ester: [58816-62-9]. *Methyl phenylphosphonazidate.*
$C_7H_8N_3O_2P$ M 197.133
Oil. $Bp_{0.01}$ 95° (oven).

Felcht, U. *et al*, *Justus Liebigs Ann. Chem.*, 1977, 1309 (*synth, ir, pmr*)

Phenylphosphonic acid, 9CI, 8CI **P-00221**

Benzenephosphonic acid
[1571-33-1]

$PhP(O)(OH)_2$

$C_6H_7O_3P$ M 158.093
Fire retardant. Plates (H_2O or EtOAc). Mp 162-164°. pK_{a1} 3.45, pK_{a2} 6.35 (50% EtOH aq.), pK_{a3} *pK_{a1} 2.05, pK_{a4} *pK_{a2} 5.51 (H_2O, 25°). Titrates as a monobasic acid in Me_2CO or ROH but as dibasic in DMF.
▷TA0350000.

Bis-triethylammonium salt: Cryst. (EtOH/EtOAc). Mp 112-114°.
Mono-Me ester: see Monomethyl phenylphosphonate, M-00488
Di-Me ester: see Dimethyl phenylphosphonate, D-00810
Mono-Et ester: see Monoethyl phenylphosphonate, M-00471
Di-Et ester: see Diethyl phenylphosphonate, D-00310
Monopropyl ester: see Monopropyl phenylphosphonate, M-00511
Dipropyl ester: [20677-03-6]. *Dipropyl phenylphosphonate.*
$C_{12}H_{19}O_3P$ M 242.254
Liq. Bp_1 106-108°. n_D^{20} 1.4895.
Diisopropyl ester: [7237-16-3]. *Diisopropyl phenylphosphonate.*
$C_{12}H_{19}O_3P$ M 242.254
d_4^{20} 1.05. Bp_7 130.5-131.5°, $Bp_{0.1}$ 94°. n_D^{20} 1.4830.
Dibutyl ester: Dibutyl phenylphosphonate.
$C_{14}H_{23}O_3P$ M 270.308
d_4^{20} 1.04. Bp_4 150-151°. n_D^{20} 1.4860.
Bis(2-methylpropyl) ester: [2783-48-4]. *Bis(2-methylpropyl) phenylphosphonate. Diisobutyl phenylphosphonate.*
$C_{14}H_{23}O_3P$ M 270.308
Used for extraction of Fe from aq. media, and of U and Th from spent reactor fuel elements.
Di 2-propenyl ester: [2948-89-2]. *Di-2-propenyl phenylphosphonate. Diallyl phenylphosphonate.*
$C_{12}H_{15}O_3P$ M 238.222
Liq. d_4^{20} 1.12. Bp_{14} 149-150°. n_D^{20} 1.4976.
Dioctyl ester: [1754-47-8]. *Dioctyl phenylphosphonate.*
$C_{22}H_{39}O_3P$ M 382.522
Liq. Bp_4 204-207°, $Bp_{0.001}$ 200°. n_D^{25} 1.4765.
▷TA0379000.
Mono-Ph ester: [2310-87-4]. *Phenyl hydrogen phenylphosphonate.*
$C_{12}H_{11}O_3P$ M 234.191
Needles (EtOH aq. or hexane). Mp 78-79°. $Bp_{0.001}$ 170-178°.
Di-Ph ester: [3049-24-9]. *Diphenyl phenylphosphonate.*
$C_{18}H_{15}O_3P$ M 310.288
Cryst. (pet. ether). Mp 75-75.5°. $Bp_{0.45}$ 163-165°.
Bis(4-nitrophenyl) ester: [38873-91-5].
$C_{18}H_{13}N_2O_7P$ M 400.284
Peptide coupling agent causing little racemization. Cryst. (CCl_4). Mp 94-96°.
Bis(trimethylsilyl) ester: [42449-24-1].
$C_{12}H_{23}O_3PSi_2$ M 302.457
Liq. d_4^{20} 1.02. $Bp_{0.06}$ 77°. n_D^{20} 1.4703.
Difluoride: see Phenylphosphonic difluoride, P-00224
Dichloride: see Phenylphosphonic dichloride, P-00223
Dibromide:
$C_6H_5Br_2OP$ M 283.887
Liq. d_4^{20} 1.95. Bp_{12} 156-158°. n_D^{20} 1.6177.
Diisocyanate: see Phenylphosphonic diisocyanate, P-00226
Diisothiocyanate: [20443-19-0].
$C_8H_5N_2OPS_2$ M 240.234
$Bp_{0.3}$ 146-148°.
Diazide:
$C_6H_5N_6OP$ M 208.119
Liq. $Bp_{0.1}$ 72-74°. n_D^{20} 1.5960.
▷Burns vigorously. Explodes on impact, but stable when in soln.
Dicyanide: [22122-79-8].
$C_8H_5N_2OP$ M 176.114
Liq. $Bp_{0.2\text{-}0.4}$ 70-72°.
Diamide: see P-Phenylphosphonic diamide, P-00222
Dihydrazide: see P-Phenylphosphonic dihydrazide, P-00225
Diimidazolide:
$C_{12}H_{11}N_4OP$ M 258.219
Cryst. (toluene). Mp 99-104°, Mp 115.5-117°.
Di-1-pyrrolide:
$C_{14}H_{13}N_2OP$ M 256.243
Cryst. (cyclohexane). Mp 107-108°.

Burger, A. *et al*, *J. Org. Chem.*, 1951, **16**, 1250 (*monophenyl ester*)
Doak, G.O. *et al*, *J. Am. Chem. Soc.*, 1951, **73**, 5658 (*synth, props*)
Green, B.S. *et al*, *Chem. Ind.* (*London*), 1960, 1306 (*diisothiocyanate*)
Petrov, K.A. *et al*, *Zh. Obshch. Khim.*, 1961, **31**, 3027 (*Engl. transl.* p. 2823) (*dibromide*)
Greenley, R.Z. *et al*, *J. Org. Chem.*, 1964, **29**, 1009 (*diimidazolide, dipyrrolide, P nmr*)
Baldwin, R.A., *J. Org. Chem.*, 1965, **30**, 3866 (*diazide, ir*)
Tavs, P. *et al*, *Tetrahedron*, 1967, **23**, 4677 (*esters, synth*)
Kharrasova, F.M. *et al*, *Zh. Obshch. Khim.*, 1968, **38**, 1262 (*Engl. transl.* p. 1215) (*esters*)
Carraher, C.E. *et al*, *Makromol. Chem.*, 1969, **123**, 144 (*dicyanide, synth, ir, props*)

Dietze, U., *J. Prakt. Chem.*, 1974, **316**, 485 (*ir*)
Harvey, D.J. *et al*, *Org. Mass Spectrom.*, 1974, **9**, 111 (*silyl ester, ms*)
Griffiths, W.R. *et al*, *Phosphorus*, 1975, **5**, 273 (*ms*)
Weakley, T.J.R., *Acta Crystallogr., Sect. B*, 1976, **32**, 2889 (*cryst struct*)
Allen, D.W. *et al*, *J. Chem. Soc., Perkin Trans. 2*, 1977, 789 (*ester, synth, pmr, P nmr*)
Kodolov, V.I. *et al*, *Izv. Akad. Nauk SSSR, Ser. Khim.*, 1977, 165 (*Engl. transl.* p. 142) (*pe*)
Modro, T.A. *et al*, *J. Chem. Soc., Perkin Trans. 2*, 1977, 1479 (*cmr*)
Loran, J.S. *et al*, *J. Chem. Soc., Perkin Trans. 2*, 1977, 418 (*nitrophenyl ester, synth, props*)
Craggs, A. *et al*, *J. Inorg. Nucl. Chem.*, 1978, **40**, 1943 (*dioctyl ester, synth, ir, pmr, use*)
Hirao, T. *et al*, *Synthesis*, 1981, 56 (*esters, synth, ir*)
Osuka, A. *et al*, *Synthesis*, 1983, 69 (*diphenyl ester, synth, ir, pmr*)

P-Phenylphosphonic diamide, 9CI — P-00222

[4707-88-4]

$PhP(O)(NH_2)_2$

$C_6H_9N_2OP$ M 156.124

Cryst. (EtOH). Sol. H_2O (1%), EtOH (0.4%) at r.t. Mp 191°.

▷TA1575000.

N,N′-*Di-Et:* [14360-85-1]. N,N′-*Diethyl*-P-*phenylphosphonic diamide. Phenylphosphonic di(ethylamide).*
$C_{10}H_{17}N_2OP$ M 212.231
Needles. Mp 72-74°.

N,N′-*Diisopropyl:* N,N′-*Diisopropyl*-P-*phenylphosphonic diamide.* N,N′-*Bis(1-methylethyl)*-P-*phenylphosphonic diamide.*
$C_{12}H_{21}N_2OP$ M 240.284
Solid. Mp 159-160°.

N,N-*Di*-tert-*butyl:* [15916-99-1]. N,N′-*Di*-tert-*butyl*-P-*phenylphosphonic diamide.*
$C_{14}H_{25}N_2OP$ M 268.338
Needles (toluene). Mp 191-192°.

N-*Ph:* N,P-*Diphenylphosphonic diamide.*
$C_{12}H_{13}N_2OP$ M 232.221
Needles (EtOH). Mp 170-171°.

N,N′-*Di-Ph:* [4707-91-9]. N,N′,P-*Triphenylphosphonic diamide. Phenylphosphonic dianilide.*
$C_{18}H_{17}N_2OP$ M 308.319
Cryst. ($Me_2CO/CHCl_3$, DMF aq., or EtOH). Mp 213-214°, 224-225°.
▷DA7175000.

N,N′-*Dicyclohexyl:* N,N′-*Dicyclohexyl*-P-*phenylphosphonic diamide. Phenylphosphonic bis(cyclohexylamide).*
$C_{18}H_{29}N_2OP$ M 320.414
Solid. Mp 166-167°.

Morrison, D.C., *J. Am. Chem. Soc.*, 1951, **73**, 5896 (*diphenyl deriv*)
Smith, W.C. *et al*, *J. Org. Chem.*, 1957, **22**, 265 (*synth, props*)
Guttman, V. *et al*, *Monatsh. Chem.*, 1960, **91**, 836 (*derivs, synth*)
Greenley, R.Z. *et al*, *J. Org. Chem.*, 1964, **29**, 1009 (*derivs, synth*)
Lane, A.P. *et al*, *J. Chem. Soc. (A)*, 1967, 1492 (*dialkyl derivs, synth, ir, pmr*)
Quast, H. *et al*, *Justus Liebigs Ann. Chem.*, 1981, 943 (*di-tert-*butyl, diphenyl derivs, ir, pmr, P nmr, ms)

Phenylphosphonic dichloride — P-00223

[824-72-6]

$PhP(O)Cl_2$

$C_6H_5Cl_2OP$ M 194.985

Pungent liq. d_4^{20} 1.40. Bp_{12} 126-127°. n_D^{20} 1.5595.

Kharrasova, F.M. *et al*, *Zh. Obshch. Khim.*, 1968, **38**, 1262 (*Engl. transl.* p. 1215) (*synth*)
Modro, T.A., *Can. J. Chem.*, 1977, **55**, 3681 (*cmr*)
Dorokhova, V.V. *et al*, *Zh. Obshch. Khim.*, 1979, **49**, 83 (*Engl. transl.* p. 68) (*uv, raman*)
Griffiths, W.R. *et al*, *Phosphorus Sulfur*, 1978, **4**, 341 (*ms*)
Sergienko, L.M. *et al*, *Zh. Obshch. Khim.*, 1980, **50**, 1958 (*Engl. transl.* p. 1578) (*uv*)
Zverev, V.V. *et al*, *Zh. Strukt. Khim.*, 1981, **22**, 22 (*Engl. transl.* p. 659) (*pe, struct*)
Ye, H. *et al*, *J. Magn. Reson.*, 1983, **51**, 313 (*cmr, P nmr, pmr*)

Phenylphosphonic difluoride, 9CI — P-00224

[657-39-6]

$PhP(O)F_2$

$C_6H_5F_2OP$ M 162.075

Liq. d_{20}^{20} 1.30. Bp 186-187°, Bp_{15} 78°. n_D^{20} 1.4680.

Schmutzler, R., *J. Chem. Soc.*, 1964, 4551 (*synth, P nmr*)
Szafraniec, L.L., *Org. Mass. Spectrom.*, 1974, **6**, 565 (*nmr*)
Daasch, L.W. *et al*, *Phosphorus*, 1975, **5**, 189 (*ms*)
Dorokhova, V.V. *et al*, *Zh. Obshch. Khim.*, 1979, **49**, 83 (*Engl. transl.* p.68) (*uv, raman*)

P-Phenylphosphonic dihydrazide, 9CI — P-00225

$PhP(O)(^1NH^2NH_2)_2$

$C_6H_{11}N_4OP$ M 186.153

Plates (EtOH). Insol. Et_2O, sol. H_2O with hydrol. Mp 131°.

$N^1,N^{1'}$-*Di-Me:* [54529-67-8]. *1,1′-Dimethyl*-P-*phenylphosphonic dihydrazide.*
$C_8H_{15}N_4OP$ M 214.206
Prisms. Mp 125-126°, 139-143°.

$N^2,N^{2'}$-*Di-Ph:* [54529-69-0]. P,*2,2′-Triphenylphosphonic dihydrazide.* Mp 188-190° dec.

Smith, W.C. *et al*, *J. Org. Chem.*, 1956, **21**, 113 (*synth, props*)
Cates, L.A. *et al*, *J. Pharm. Sci.*, 1974, **63**, 1736 (*deriv*)
Majoral, J.P. *et al*, *Tetrahedron*, 1976, **32**, 2633 (*deriv, props*)
Mathis, R. *et al*, *Spectrochim. Acta, Part A*, 1981, **37**, 677 (*deriv, ir*)

Phenylphosphonic diisocyanate, 9CI — P-00226

[1078-84-8]

$PhP(O)(NCO)_2$

$C_8H_5N_2O_3P$ M 208.113

Reactive liq. d_4^{20} 1.35. Mp ca. −5°. Bp_2 123°. n_D^{20} 1.5480. Hydrolysed to *P*-Phenylphosphonic diamide, P-00222 .

Popoff, I.C. *et al*, *J. Polym. Sci., Part B*, 1963, **1**, 247 (*synth, props*)
Bai, L.I. *et al*, *Zh. Obshch. Khim.*, 1964, **34**, 3609 (*Engl. transl.* p. 3656) (*synth*)
Utvary, K. *et al*, *Monatsh. Chem.*, 1966, **97**, 679 (*synth*)

Phenylphosphonisothiocyanatidothioic acid, 9CI P-00227

S(=)P(OH)(Ph)(NCS) ⇌ HS–P(=O)(Ph)(NCS)

$C_7H_6NOPS_2$ M 215.224

O-*Me ester:* [20145-77-1]. O-*Methyl phenylphosphonoisothiocyanatidothioate.*
$C_8H_8NOPS_2$ M 229.251
Liq. $Bp_{0.03}$ 105°. n_D^{20} 1.6225.

O-*Et ester:* [20145-78-2]. O-*Ethyl phenylphosphonisothiocyanatidothioate.*
$C_9H_{10}NOPS_2$ M 243.278
Liq. $Bp_{0.03}$ 108°. n_D^{20} 1.6097.

O-*Isopropyl ester:* O-*Isopropyl phenylphosphonisothiocyanatidothioate.*
$C_{10}H_{12}NOPS_2$ M 257.305
Liq. $Bp_{0.02}$ 110°. n_D^{20} 1.6225.

O-*Ph ester:* [20145-82-8]. O-*Phenyl phenylphosphonisothiocyanatidothioate.*
$C_{13}H_{10}NOPS_2$ M 291.322
Liq. $Bp_{0.02}$ 162°. n_D^{20} 1.6451.

U.S.P., 3 342 583, (*1967*); *CA*, **69**, 59371
Rahil, J. *et al*, *J. Am. Chem. Soc.*, 1981, **103**, 1723 (*synth*)

Phenylphosphonobromidodithioic acid, 9CI P-00228

PhPBr(S)SH

$C_6H_6BrPS_2$ M 253.111

Me ester: Methyl phenylphosphonobromidodithioate.
$C_7H_8BrPS_2$ M 267.138
Liq. $Bp_{0.01}$ 175°.

Fluck, E. *et al*, *Angew. Chem.*, 1967, **79**, 243 (*synth, P nmr*)

Phenylphosphonochloridic acid, 9CI P-00229

[62808-39-3]

PhPCl(O)OH

$C_6H_6ClO_2P$ M 176.539

Me ester: [41761-00-6]. *Methyl phenylphosphonochloridate.*
$C_7H_8ClO_2P$ M 190.566
Liq. which dec. on dist.

Et ester: [5284-12-8]. *Ethyl phenylphosphonochloridate.*
$C_8H_{10}ClO_2P$ M 204.593
Unstable liq.

Butyl ester: [18351-25-2]. *Butyl phenylphosphonochloridate.*
$C_{10}H_{14}ClO_2P$ M 232.646
Liq. $Bp_{0.75}$ 116-117.5°.

Ph ester: [61274-57-5]. *Phenyl phenylphosphonochloridate.*
$C_{12}H_{10}ClO_2P$ M 252.637
Liq. $Bp_{0.3}$ 152-155°. n_D^{31} 1.5718.
▷TA3700000.

Cyclohexyl ester: Cyclohexyl phenylphosphonochloridate.
$C_{12}H_{16}ClO_2P$ M 258.684
Liq. n_D^{25} 1.5328.

Marsi, K.L. *et al*, *J. Am. Chem. Soc.*, 1956, **78**, 3063 (*synth*)
Hafner, L.S. *et al*, *J. Med. Chem.*, 1970, **13**, 1025 (*synth*)
Michalski, J. *et al*, *J. Am. Chem. Soc.*, 1978, **100**, 5386 (*P nmr*)

Phenylphosphonochloridodithioic acid, 9CI P-00230

PhPCl(S)SH

$C_6H_6ClPS_2$ M 208.660

Et ester: [5120-49-0]. *Ethyl phenylphosphonochloridodithioate.*
$C_8H_{10}ClPS_2$ M 236.714
Liq. d_4^{20} 1.27-1.29. $Bp_{0.2}$ 110-111°. n_D^{20} 1.6354.

Ph ester: [21890-13-1]. *Phenyl phenylphosphonochloridodithioate.*
$C_{12}H_{10}ClPS_2$ M 284.758
Cryst. (CCl_4/pet. ether). Mp 83-84.5°.

Grapov, A.F. *et al*, *Zh. Obshch. Khim.*, 1968, **38**, 2658 (*Engl. transl.* p. 2572) (*phenyl ester*)
Krasil'nikova, E.A. *et al*, *Zh. Obshch. Khim.*, 1972, **42**, 2578 (*Engl. transl.* p. 2570) (*ethyl ester*)

Phenylphosphonochloridoselenous acid, 9CI, 8CI P-00231

PhP(Cl)SeH

C_6H_6ClPSe M 223.500

Me ester: [55776-68-6]. *Methyl phenylphosphonochloridoselenoite.*
C_7H_8ClPSe M 237.527
No phys. props. reported.

Anderson, J.W. *et al*, *Inorg. Nucl. Chem. Lett.*, 1975, **11**, 233 (*pmr*)

Phenylphosphonochloridothioic acid, 9CI P-00232

PhPCl(S)OH ⇌ PhPCl(O)SH

C_6H_6ClOPS M 192.600

O-*Me ester:* [20147-96-0]. O-*Methyl phenylphosphonochloridothioate.*
C_7H_8ClOPS M 206.626
Liq. $Bp_{0.005}$ 90°. n_D^{20} 1.6173.

S-*Me ester:* [13113-95-6]. S-*Methyl phenylphosphonochloridothioate.*
C_7H_8ClOPS M 206.626
Liq. d_4^{20} 1.32. $Bp_{0.28}$ 104.5-105.5°. n_D^{20} 1.5914.

O-*Et ester: see* O-*Ethyl phenylphosphonochloridothioate, E-00112*

S-*Et ester:* [13113-96-7]. S-*Ethyl phenylphosphonochloridothioate.*
$C_8H_{10}ClOPS$ M 220.653
Liq. d_4^{20} 1.28. $Bp_{0.1}$ 107-109°. n_D^{20} 1.5838.

S-*Propyl ester:* S-*Propyl phenylphosphonochloridothioate.*
$C_9H_{12}ClOPS$ M 234.680
Liq. d_4^{20} 1.24. $Bp_{0.25}$ 112-114°. n_D^{20} 1.5704.

O-*Isopropyl ester:* [13231-84-0]. O-*Isopropyl phenylphosphonochloridothioate.* O-(*1-Methylethyl*) *phenylphosphonochloridothioate.*
$C_9H_{12}ClOPS$ M 234.680
No phys. props. reported.

S-*Isopropyl ester:* [13114-00-6]. S-*Isopropyl phenylphosphonochloridothionate.* S-(*1-Methylethyl*) *phenylphosphonochloridothioate.*
$C_9H_{12}ClOPS$ M 234.680

Liq. d_4^{20} 1.23. $Bp_{0.5}$ 115-117°. n_D^{20} 1.5690.
O-*Ph ester:* [20148-06-5]. O-*Phenyl phenylphosphonochloridothioate.*
$C_{12}H_{10}ClOPS$ M 268.697
Liq. $Bp_{0.8}$ 143-145°. n_D^{23} 1.6199.
S-*Ph ester:* [20433-65-2]. S-*Phenyl phenylphosphonochloridothioate.*
$C_{12}H_{10}ClOPS$ M 268.697
Liq. d_4^{20} 1.32. n_D^{20} 1.6310.

Neimysheva, A.A. *et al, Zh. Obshch. Khim.*, 1966, **36**, 500 (*Engl. transl.* p. 520) (*S-alkyl esters, synth*)
Neimysheva, A.A. *et al, Zh. Obshch. Khim.*, 1967, **37**, 1822 (*Engl. transl.* p. 1736) (*hydrol*)
Shitov, L.N. *et al, Zh. Obshch. Khim.*, 1968, **38**, 2340 (*Engl. transl.* p. 2268) (*S-phenyl ester, synth*)
Newallis, P.E. *et al, J. Chem. Eng. Data*, 1970, **15**, 455 (*O-aryl esters*)

Phenylphosphonochloridothious acid, 9CI — P-00233

PhPCl(SH)

C_6H_6ClPS M 176.600
Et ester: [23588-02-5]. *Ethyl phenylphosphonochloridothioite.*
$C_8H_{10}ClPS$ M 204.654
Liq. d_4^{20} 1.21. $Bp_{1.5}$ 106-108°, $Bp_{0.07}$ 66-68°. n_D^{20} 1.6041 (1.6110).
Isopropyl ester: [26990-23-8]. *Isopropyl phenylphosphonochloridothioite.*
$C_9H_{12}ClPS$ M 218.681
Liq. d_4^{20} 1.16. $Bp_{0.07}$ 77-79°. n_D^{20} 1.5950.
Ph ester: [26990-26-1]. *Phenyl phenylphosphonochloridothioite.*
$C_{12}H_{10}ClPS$ M 252.698
Liq. d_4^{20} 1.25. $Bp_{0.07}$ 121-122°. n_D^{20} 1.6638.

Shitov, L.N. *et al, Zh. Obshch. Khim.*, 1969, **39**, 1251 (*Engl. transl.* p. 1220) (*synth, P nmr*)
Rizpolozhenskii, N.I. *et al, Izv. Akad. Nauk SSSR, Ser. Khim.*, 1970, 622 (*Engl. transl.* p. 571) (*synth*)

Phenylphosphonochloridous acid, 9CI — P-00234

PhPCl(OH)

C_6H_6ClOP M 160.540
Et ester: [40618-55-1]. *Ethyl phenylphosphonochloridite.*
$C_8H_{10}ClOP$ M 188.593
Liq. d_4^{20} 1.18. $Bp_{0.006}$ 55-65°. n_D^{20} 1.5685.

Steininger, E., *Chem. Ber.*, 1962, **95**, 2993 (*synth*)
Eliseenkova, R.M. *et al, Izv. Akad. Nauk SSSR, Ser. Khim.*, 1972, 2760 (*Engl. transl.* p. 2690) (*synth*)
Andreev, N.A. *et al, Zh. Obshch. Khim.*, 1980, **50**, 803 (*Engl. transl.* p. 641) (*synth, ir, nmr*)

Phenylphosphonodiselenous acid — P-00235

$PhP(SeH)_2$

$C_6H_7PSe_2$ M 268.015
Di-Me ester: [55776-58-4]. *Dimethyl phenylphosphonodiselenoite.*
$C_8H_{11}PSe_2$ M 296.069
Viscous yellow oil. Thermally stable.
Bis-trifluoromethyl ester: [69646-20-4]. *Bis(trifluoromethyl) phenylphosphonodiselenoite.*
$C_8H_5F_6PSe_2$ M 404.012
No phys. props. reported.

Anderson, J.W. *et al, Inorg. Nucl. Chem. Lett.*, 1975, **11**, 233 (*dimethyl ester, synth, ir, raman, pmr*)
Darmadi, A. *et al, Z. Anorg. Allg. Chem.*, 1979, **448**, 35 (*ester, synth, F and P nmr*)

Phenylphosphonodithioic acid, 9CI — P-00236

$PhP(O)(SH)_2 \rightleftharpoons PhP(S)(OH)(SH)$

$C_6H_7OPS_2$ M 190.214
O-*Et ester: see O-Ethyl phenylphosphonodithioate, E-00113*
S,S-*Di-Et ester:* [51805-04-0]. S,S-*Diethyl phenylphosphonodithioate.*
$C_{10}H_{15}OPS_2$ M 246.322
Liq. Bp_4 163-165°, $Bp_{0.1}$ 118°.
O-*Butyl ester:* [6230-97-3]. O-*Butyl phenylphosphonodithioate.*
$C_{10}H_{15}OPS_2$ M 246.322
Liq. d_4^{20} 1.16. $Bp_{0.0014}$ 86-88°. n_D^{20} 1.5683.
Chloride: see Phenylphosphonochloridodithioic acid, P-00230

Mizrakh, L.I. *et al, Zh. Obshch. Khim.*, 1966, **36**, 469 (*Engl. transl.* p. 487) (*butyl ester, synth*)
Yoshifuji, M. *et al, J. Chem. Soc., Perkin Trans. 1*, 1973, 2065 (*diethyl ester, synth, pmr, P nmr*)

Phenylphosphonodithioic acid anhydrosulfide — P-00237

Diphenylthiodiphosphonic acid, 9CI

Ph–P(=S)(OH)–S–P(=S)(OH)–Ph

$C_{12}H_{12}O_2P_2S_3$ M 346.353
Note 9CI name same as that of Phenylphosphonothioic acid monoanhydride, P-00245 .
O,*O-Di-Et ester:* [39177-91-8]. O-*Ethyl* S-*(ethoxyphenylphosphinothioyl) phenylphosphonodithioate.* O-*Ethyl phenylphosphonodithioic anhydrosulfide.*
$C_{16}H_{20}O_2P_2S_3$ M 402.460
Solid. Mp 113°.
O,O-*Diisopropyl ester:* [39177-93-0]. O-*Isopropyl* S-*(isopropoxyphenylphosphinothioyl) phenylphosphonodithioate.* O-*Isopropyl phenylphosphonodithioic anhydrosulfide.*
$C_{18}H_{24}O_2P_2S_3$ M 430.514
Solid. Mp 74-75°.
Difluoride:
$C_{12}H_{10}F_2P_2S_3$ M 350.335
Mixture of stereoisomers obtained only in impure form.

Almasi, L. *et al, Monatsh. Chem.*, 1972, **103**, 1027 (*esters*)
Harris, R.K. *et al, J. Chem. Soc., Dalton Trans.*, 1972, 1590 (*difluoride, pmr, F and P nmr*)

Phenylphosphonodithious acid, 9CI — P-00238

$PhP(SH)_2$

$C_6H_7PS_2$ M 174.215
Di-Et ester: [1486-37-9]. *Diethyl phenylphosphonodithioite.*
$C_{10}H_{15}PS_2$ M 230.322
Liq. $Bp_{0.8}$ 122°. n_D^{22} 1.6165.
Di-Ph ester: [38476-60-7]. *Diphenyl phenylphosphonodithioite.*
$C_{18}H_{15}PS_2$ M 326.410

Liq. $Bp_{0.1}$ 221-223°.

Gallagher, M.J. *et al*, *J. Chem. Soc. (C)*, 1966, 2176 (*diethyl ester, synth, pmr*)
Peake, S.C. *et al*, *J. Chem. Soc. (A)*, 1970, 1049 (*diphenyl ester, synth, P nmr*)
Razumov, A.I. *et al*, *Zh. Obshch. Khim.*, 1972, **42**, 1250 (*Engl. transl.* p. 1245) (*esters, synth, P nmr*)

Phenylphosphonofluoridic acid, 9CI — P-00239

[45778-98-1]

PhP(O)(F)OH

$C_6H_6FO_2P$ M 160.084
Solid. Mp 50°.
Anilinium salt: Solid. Mp 155-158°.
Me ester: [650-99-7]. *Methyl phenylphosphonofluoridate.*
$C_7H_8FO_2P$ M 174.111
Liq. d_4^{20} 1.25. Bp_4 85-86°, $Bp_{0.09}$ 52°. n_D^{20} 1.4887.
Et ester: [703-06-0]. *Ethyl phenylphosphonofluoridate.*
$C_8H_{10}FO_2P$ M 188.138
Liq. d_4^{20} 1.19. Bp_2 95°, $Bp_{0.15}$ 53-54°. n_D^{20} 1.4802.
Cyclohexyl ester: [1426-81-9]. *Cyclohexyl phenylphosphonofluoridate.*
$C_{12}H_{16}FO_2P$ M 242.229
Liq. $Bp_{0.05}$ 94°. n_D^{25} 1.5021.
Ph ester: Phenyl phenylphosphonofluoridate.
$C_{12}H_{10}FO_2P$ M 236.182
Liq. $Bp_{0.005}$ 105°. n_D^{22} 1.5439.

Yagupol'skii, L.M. *et al*, *Zh. Obshch. Khim.*, 1960, **30**, 1284 (*Engl. transl.* p. 1310) (*esters*)
Hafner, L.S. *et al*, *J. Med. Chem.*, 1970, **13**, 1025 (*ester*)
Reddy, G.S. *et al*, *Z. Naturforsch., B*, 1970, **25**, 1199 (*ester, synth, ir, pmr, F and P nmr*)
Bender, R. *et al*, *Phosphorus*, 1974, **4**, 183 (*synth, pmr, nmr*)
Fridland, S.V. *et al*, *Zh. Obshch. Khim.*, 1976, **46**, 2654 (*Engl. transl.* p. 2536) (*esters*)
Horner, L. *et al*, *Phosphorus Sulfur*, 1981, **11**, 157, 339 (*esters, ir, pmr*)

Phenylphosphonofluoridous acid, 10CI — P-00240

PhPF(OH)

C_6H_6FOP M 144.085

Jansen, A.F. *et al*, *Can. J. Chem.*, 1979, **57**, 1903 (*esters, F and P nmr*)

P-Phenylphosphonohydrazidic acid, 9CI — P-00241

PhP(O)(OH)NHNH$_2$

$C_6H_9N_2O_2P$ M 172.123
Me ester: [65111-20-8]. *Methyl* P-*phenylphosphonohydrazidate.*
$C_7H_{11}N_2O_2P$ M 186.150
Cryst. (C_6H_6). Mp 67-68°.

Felcht, U. *et al*, *Justus Liebigs Ann. Chem.*, 1977, 1309 (*synth, ir, pmr, props*)

Phenylphosphonoperoxoic acid, 9CI — P-00242

[25836-60-6]

PhP(O)(OH)OOH

$C_6H_7O_4P$ M 174.093
Cryst. (C_6H_6/pet. ether). Mp 88.5° dec.
Na salt: [25836-59-3]. Solid. Mp 237° dec.
OO-tert-*Butyl*, O-*Et ester:* [31459-97-9]. OO-tert-*Butyl* O-*ethyl phenylphosphonoperoxoate.*
$C_{12}H_{19}O_4P$ M 258.253
Liq. d_4^{20} 1.05. $Bp_{0.005}$ 20-25°. n_D^{20} 1.4852.

Cubbon, R.C. *et al*, *J. Chem. Soc. (C)*, 1970, 501 (*synth, ir, props*)
Maslennikov, V.P. *et al*, *Zh. Obshch. Khim.*, 1970, **40**, 1906 (*Engl. transl.* p. 1888) (*ester, synth*)
Maslennikov, V.P. *et al*, *Zh. Org. Khim.*, 1971, **7**, 686 (*Engl. transl.* p. 696) (*ester, synth, props*)

Phenylphosphonoselenoic acid, 9CI — P-00243

PhP(Se)(OH)OH ⇌ PhP(O)(SeH)OH

$C_6H_7O_2PSe$ M 221.054
O,O-*Di-Me ester:* [20180-12-5]. O,O-*Dimethyl phenylphosphonoselenoate.*
$C_8H_{11}O_2PSe$ M 249.107
Liq. $Bp_{0.8}$ 100-101°.
O,O-*Di-Et ester:* [39181-18-5]. O,O-*Diethyl phenylphosphonoselenoate.*
$C_{10}H_{15}O_2PSe$ M 277.161
Liq.
O,O-*Bis(trimethylsilyl) ester:* [66481-72-9]. O,O-*Bis(trimethylsilyl) phenylphosphonoselenoate.*
$C_{12}H_{23}O_2PSeSi_2$ M 365.417
Liq. d_4^{20} 1.17. $Bp_{0.06}$ 73°. n_D^{20} 1.5166.
Difluoride:
$C_6H_5F_2PSe$ M 225.036
Liq. $Bp_{0.01}$ 25°.
Dichloride: [39078-30-7].
$C_6H_5Cl_2PSe$ M 257.945
Liq. $Bp_{0.01}$ 60°.
Bis(dimethylamide): [23389-78-8]. N,N,N′,N′-*Tetramethyl*-P-*phenylphosphonoselenoic diamide.*
$C_{10}H_{17}N_2PSe$ M 275.191
No phys. props. reported.

Stec, W.J. *et al*, *Phosphorus*, 1972, **2**, 97 (*ester, P and Se nmr*)
McFarlane, W. *et al*, *J. Chem. Soc., Dalton Trans.*, 1973, 2162 (*pmr, P and Se nmr*)
Roesky, H.W. *et al*, *Z. Naturforsch., B*, 1973, **28**, 697 (*dichloride, difluoride, ir, pmr, P nmr*)
Stangeland, L.J. *et al*, *Acta Chem. Scand.*, 1973, **27**, 3919 (*ester*)
Nuretdinov, I.A. *et al*, *Izv. Akad. Nauk SSSR, Ser. Khim.*, 1975, 327 (*Engl. transl.* p. 263) (*dichloride, nqr*)
Varnavskaya-Samarina, O.A. *et al*, *Izv. Akad. Nauk SSSR, Ser. Khim.*, 1978, 363 (*Engl. transl.* p. 313) (*trimethylsilyl ester*)

Phenylphosphonothioic acid, 9CI — P-00244

[25331-57-1]

PhP(S)(OH)$_2$ ⇌ PhP(O)(OH)(SH)

$C_6H_7O_2PS$ M 174.154
O-*Me ester: see* O-*Methyl phenylphosphonothioate*, M-00243
S-*Me ester: see* S-*Methyl phenylphosphonothioate*, M-00244
O,O-*Di-Me ester:* [6840-11-5]. O,O-*Dimethyl phenylphosphonothioate.*
$C_8H_{11}O_2PS$ M 202.207

No phys. props. reported.

O-*Et ester: see O-Ethyl phenylphosphonothioate, E-00114*

S-*Et*, O-*Me ester:* [40618-53-9]. S-*Ethyl* O-*methyl phenylphosphonothioate.*

$C_9H_{13}O_2PS$ M 216.234

Liq. $Bp_{0.6}$ 117°. n_D^{25} 1.5520.

O,O-*Di-Et ester:* [6231-03-4]. O,O-*Diethyl phenylphosphonothioate.*

$C_{10}H_{15}O_2PS$ M 230.261

Liq. $Bp_{0.3}$ 94-96°. n_D^{20} 1.5365.

O,S-*Di-Et ester:* [57557-80-9]. O,S-*Diethyl phenylphosphonothioate.*

$C_{10}H_{15}O_2PS$ M 230.261

Liq. d_0^{25} 1.15. Bp_5 145-146°.

O-*Et* O-*4-nitrophenyl diester: see O-Ethyl O-4-nitrophenyl phenylphosphonothionate, E-00099*

O-*Et*-O-*2,4-dichlorophenyl diester: see S-Seven, S-00004*

O-*Et*-O-*4-cyanophenyl diester: see Cyanofenphos, C-00213*

O-*Et*, S-*Benzyl diester: see Inezin, I-00009*

O-*Et*-O-*8-quinolinyl diester: see Quintiofos, Q-00006*

O,O-*Di(2-propenyl) ester:* O,O-*Di-2-propenyl phenylphosphonothioate.* O,O-*Diallyl phenylphosphonothioate.*

$C_{12}H_{15}O_2PS$ M 254.283

Liq. d_{25}^{25} 1.12. Bp_1 120-129°. n_D^{25} 1.5508.

O,O-*Di-Ph ester:* [88239-51-4]. O,O-*Diphenyl phenylphosphonothioate.*

$C_{18}H_{15}O_2PS$ M 326.349

Liq. $Bp_{2.5}$ 223-225°.

O,O-*Bis(trimethylsilyl) ester:* [66481-71-8]. O,O-*Bis-(trimethylsilyl) phenylphosphonothioate.*

$C_{12}H_{23}O_2PSSi_2$ M 318.517

Liq. d_4^{20} 1.04. $Bp_{0.06}$ 76°. n_D^{20} 1.4991.

Difluoride: see Phenylphosphonothioic difluoride, P-00248

Dichloride: see Phenylphosphonothioic dichloride, P-00247

Dibromide: [6231-02-3].

$C_6H_5Br_2PS$ M 299.947

Liq. d_4^{20} 1.89. $Bp_{0.001}$ 88.5-90°. n_D^{20} 1.6968.

Diisothiocyanate:

$C_8H_5N_2PS_3$ M 256.295

Liq. $Bp_{0.08}$ 128-130°.

Diazide:

$C_6H_5N_6PS$ M 224.179

Liq. $Bp_{0.1}$ ca. 80°.

▷Shows no sensitivity in drop test, but dec. violently on dist.

Monoamide: see P-Phenylphosphonamidothioic acid, P-00219

Monoanilide: see N,P-Diphenylphosphonamidothioic acid, D-01104

Diamide: see P-Phenylphosphonothioic diamide, P-00246

Dihydrazide: see P-Phenylphosphonothioic dihydrazide, P-00249

Toy, A.D.F. *et al*, *J. Am. Chem. Soc.*, 1954, **76**, 2191 (*diallyl ester*)

Morrison, D.C., *J. Org. Chem.*, 1956, **21**, 705 (*diethyl ester*)

Green, B.S. *et al*, *Chem. Ind.* (*London*), 1960, 1306 (*isothiocyanate, synth, ir*)

Harvey, R.G. *et al*, *J. Am. Chem. Soc.*, 1963, **85**, 1618 (*ester, ir*)

Maier, L., *Helv. Chim. Acta*, 1964, **47**, 120 (*dibromide*)

Baldwin, R.A., *J. Org. Chem.*, 1965, **30**, 3866 (*azide, synth, ir, props*)

Misrakh, L.I. *et al*, *Zh. Obshch. Khim.*, 1966, **36**, 469 (*Engl. transl.* p. 487) (*diethyl ester, dibromide*)

Modro, T.A., *Can. J. Chem.*, 1977, **55**, 3681 (*diethyl ester, cmr*)

Varnavskaya-Samanina, O.A. *et al*, *Izv. Akad. Nauk SSSR, Ser. Khim.*, 1978, 363 (*Engl. transl.* p. 313) (*trimethylsilyl ester*)

Chrzanowski, R.L. *et al*, *J. Agric. Food Chem.*, 1981, **29**, 580 (*methyl esters, ms*)

Purnanand, R.H.D. *et al*, *Synthesis*, 1983, 731 (*diphenyl ester, synth, ir*)

Phenylphosphonothioic acid monoanhydride **P-00245**

Diphenylthiodiphosphonic acid, 9CI

```
     S       S
     ||      ||
Ph—P—O—P—Ph
     |       |
    HO      OH
```

$C_{12}H_{12}O_3P_2S_2$ M 330.293

Note 9CI name same as that of Phenylphosphonodithioic acid anhydrosulfide, P-00237 .

Dipiperidinium salt: Needles. Mp 188-189°. Presumably a mixt. of diastereoisomers.

Ecker, A. *et al*, *Monatsh. Chem.*, 1972, **103**, 736 (*synth, pmr*)

P-Phenylphosphonothioic diamide, 9CI **P-00246**

[3969-46-8]

$$PhP(S)(NH_2)_2$$

$C_6H_9N_2PS$ M 172.184

Cryst. (EtOH) or plates (Et_2O). Mp 38-40°, 51°. Rather unstable.

N,N′-*Di-Me:* [18994-99-5]. N,N′-*Dimethyl*-P-*phenylphosphonothioic diamide. Phenylphosphonothioic bis(methylamide).*

$C_8H_{13}N_2PS$ M 200.238

Cryst. (C_6H_6/pet. ether). Mp 69.5-70.5°.

N,N′-*Di-Et:* [6278-47-3]. N,N′-*Diethyl*-P-*phenylphosphonothioic diamide. Phenylphosphonothioic acid bis(ethylamide).*

$C_{10}H_{17}N_2PS$ M 228.291

Cryst. (Et_2O) or needles (EtOH). Mp 86-87°.

N,N′-*Diisopropyl:* [13789-70-3]. N,N′-*Diisopropyl*-P-*phenylphosphonothioic diamide.*

$C_{12}H_{21}N_2PS$ M 256.345

Needles (hexane). Mp 49-50°.

N,N′-*Di*-tert-*butyl:* [15917-00-7]. N,N′-*Di*-tert-*butyl*-P-*phenylphosphonothioic diamide.*

$C_{14}H_{25}N_2PS$ M 284.399

Cryst. (EtOH). Mp 106-106.5°.

N,N′-*Di-Ph:* [18995-01-2]. P-*Phenylphosphonothioic dianilide.*

$C_{18}H_{17}N_2PS$ M 324.379

Cryst. (MeOH). Mp 175-176°.

N,N′-*Dibenzyl:* N,N′-*Dibenzyl*-P-*phenylphosphonothioic diamide.*

$C_{20}H_{21}N_2PS$ M 352.433

Cryst. (CCl_4 or EtOH). Mp 80-81°, Mp 94°.

Smith, W.C. *et al*, *J. Org. Chem.*, 1957, **22**, 265 (*synth*)

Reist, E.J. *et al*, *J. Org. Chem.*, 1960, **25**, 666 (*synth, derivs*)

Trippett, S., *J. Chem. Soc.*, 1962, 4731 (*dimethyl, dibenzyl*)

Lane, A.P. *et al*, *J. Chem. Soc.* (*A*), 1967, 1492 (*derivs, ir, synth*)

Flint, C.D. *et al*, *J. Chem. Soc.* (*A*), 1971, 3513 (*diethyl, synth, props*)

Issleib, K. *et al*, *Z. Anorg. Allg. Chem.*, 1977, **428**, 16 (*synth, derivs*)

Phenylphosphonothioic dichloride, 9CI, 8CI P-00247

Phenylthiophosphonic dichloride

[3497-00-5]

PhP(S)Cl_2

$C_6H_5Cl_2PS$ M 211.045

Synthetic intermed. e.g. in synth. of phenylphosphonothioic derivs. and benzothiazoles. Liq. d_4^{20} 1.41. Bp 270°, Bp_4 110°, $Bp_{0.05}$ 72-75°. n_D^{20} 1.6227.

▷TB2200000.

Maier, L., *Helv. Chim. Acta*, 1964, **47**, 120 (*synth, P nmr*)
Patel, N.K. *et al*, *J. Org. Chem.*, 1967, **32**, 2999 (*synth*)
Vilesov, F.I. *et al*, *Z. Phys. Chem. (Leipzig)*, 1974, **255**, 661 (*pe*)
Jones, T.R.B. *et al*, *Org. Mass Spectrom.*, 1977, **12**, 317 (*ms*)
Modro, T.A., *Can. J. Chem.*, 1977, **55**, 3681 (*cmr*)
Dorokhova, V.V. *et al*, *Zh. Obshch. Khim.*, 1979, **49**, 83 (*Engl. transl.* p. 68) (*uv, raman*)
Yoshifuji, M. *et al*, *Bull. Chem. Soc. Jpn.*, 1982, **55**, 870 (*use*)
Lindner, E. *et al*, *J. Organomet. Chem.*, 1983, **255**, 245 (*complexes*)

Phenylphosphonothioic difluoride, 9CI P-00248

Phenylthiophosphonic difluoride

[657-40-9]

PhP(S)F_2

$C_6H_5F_2PS$ M 178.136

Liq. stable to heat and to cold H_2O. d_{20}^{18} 1.30. Bp 187-188°, Bp_3 47-49°. n_D^{18} 1.4650.

Tullock, C.W. *et al*, *J. Org. Chem.*, 1960, **25**, 2016 (*synth*)
Yagupol'skii, L.M. *et al*, *Zh. Obshch. Khim.*, 1960, **30**, 1284 (*Engl. transl.* p. 1310) (*synth, props*)
Reddy, G.S. *et al*, *Z. Naturforsch., B*, 1970, **25**, 1199 (*pmr, F and P nmr*)
Jones, T.R.B. *et al*, *Org. Mass. Spectrom.*, 1977, **12**, 317 (*ms*)
Dorokhova, V.V. *et al*, *Zh. Obshch. Khim.*, 1979, **49**, 83 (*Engl. transl.* p. 68) (*uv, raman*)

P-Phenylphosphonothioic dihydrazide, 9CI P-00249

[5395-21-1]

PhP(S)($NHNH_2)_2$

$C_6H_{11}N_4PS$ M 202.213

Cryst. (EtOH). Mp 115°.

$N^1,N^{1'}$-*Di-Me:* [54529-68-9]. $N^1,N^{1'}$-*Dimethyl*-P-*phosphonothioic dihydrazide.*
$C_8H_{15}N_4PS$ M 230.267
Solid. Mp 95-96°.

$N^2,N^{2'}$-*Di-Ph:* [20491-26-3]. $N^2,N^{2'}$,P-*Triphenylphosphonothioic dihydrazide.*
$C_{18}H_{19}N_4PS$ M 354.409
Cryst. (C_6H_6). Mp 199.5-200.5°.

Dibenzylidene: $N^2,N^{2'}$-*Dibenzylidene*-P-*phenylphosphonothioic dihydrazide.*
$C_{15}H_{17}N_4PS$ M 316.360
Cryst. (EtOH). Mp 109-110°.

Smith, W.C. *et al*, *J. Org. Chem.*, 1956, **21**, 113 (*synth, derivs, props*)
Scola, D.A. *et al*, *J. Chem. Eng. Data*, 1968, **13**, 571 (*diphenyl, synth, ir*)
Majoral, J.P. *et al*, *Tetrahedron*, 1976, **32**, 2633 (*dimethyl, synth, props, pmr, P nmr*)
Grapov, A.F. *et al*, *Zh. Obshch. Khim.*, 1977, **47**, 1704 (*Engl. transl.* p. 1560) (*props*)
Kornuta, P.P. *et al*, *Zh. Obshch. Khim.*, 1981, **51**, 2449 (*Engl. transl.* p. 2111) (*props*)
Mathis, R. *et al*, *Spectrochim. Acta, Part A*, 1981, **37**, 677 (*ir*)

Phenylphosphonotrithioic acid, 9CI P-00250

PhP(S)($SH)_2$

$C_6H_7PS_3$ M 206.275

Cyclic bis(anhydrosulfide): see 2,4-Diphenyl-1,3,2,4-dithiadiphosphetane 2,4-disulfide, D-01019

Di-Et ester: [34309-87-0]. *Diethyl phenylphosphonotrithioate.*
$C_{10}H_{15}PS_3$ M 262.382
Liq. d_4^{20} 1.21. Bp_2 164-168°, $Bp_{0.05}$ 109-117°. n_D^{20} 1.6355.

Dipropyl ester: [59568-74-0]. *Dipropyl phenylphosphonotrithioate.*
$C_{12}H_{19}PS_3$ M 290.436
Liq. d_4^{20} 1.14. Bp_2 163-168°. n_D^{20} 1.6210.

Bis(trimethylsilyl)ester: [63853-25-8]. *Bis(trimethylsilyl) phenylphosphonotrithioate.*
$C_{12}H_{23}PS_3Si_2$ M 350.638
Yellow viscous liq. Thermally unstable.

Ecker, A. *et al*, *Chem. Ber.*, 1973, **106**, 1453 (*diethyl ester, ms*)
Yoshifuji, M. *et al*, *J. Chem. Soc., Perkin Trans. 1*, 1973, 2069 (*diethyl ester, synth, pmr, P nmr*)
Murav'ev, I.V. *et al*, *Zh. Obshch. Khim.*, 1976, **46**, 789, 1262 (*Engl. transl.*, pp 787, 1241) (*synth*)
Roesky, H.W. *et al*, *Z. Anorg. Allg. Chem.*, 1977, **431**, 221 (*silyl ester, synth, ir, pmr, P nmr*)
Hahn, J. *et al*, *Z. Anorg. Allg. Chem.*, 1986, **543**, 7 (*silyl ester, synth, P nmr*)

Phenylphosphonous acid, 9CI P-00251

Benzenephosphonous acid

[121-70-0]

PhP($OH)_2$

$C_6H_7O_2P$ M 142.094

The free acid exists in the tautomeric phosphoryl form Phenylphosphinic acid, P-00210 .

Na salt: [4297-95-4]. *Sodium phenylphosphinite. Sodium phenylphosphonite.* Antioxidant in nylon etc. Cryst. Pptd. from aq. soln. with Me_2CO or THF.

Mono-Me ester: see Phenylphosphinic acid, P-00210

Di-Me ester: [2946-61-4]. *Dimethyl phenylphosphonite. Dimethyl benzenephosphonite.*
$C_8H_{11}O_2P$ M 170.147
Liq. d_4^{20} 1.08. Bp_{17} 98°. n_D^{25} 1.5261.

Mono-Et ester: see Phenylphosphinic acid, P-00210

Di-Et ester: see Diethyl phenylphosphonite, D-00311

Divinyl ester: Divinyl phenylphosphonite. Diethenyl phenylphosphonite.
$C_{10}H_{11}O_2P$ M 194.169
Liq. d_4^{20} 1.06. Bp_2 76-78°. n_D^{20} 1.5385.

Di-2-propenyl ester: [833-57-8]. *Di-2-propenyl phenylphosphonite. Diallyl phenylphosphonite.*
$C_{12}H_{15}O_2P$ M 222.223
Liq. d_4^{20} 1.04. $Bp_{0.5}$ 92-93°. n_D^{20} 1.5300.

Monoisopropyl ester: see Phenylphosphinic acid, P-00210
Diisopropyl ester: [36238-99-0]. *Diisopropyl phenylphosphonite. Bis(1-methylethyl) phenylphosphonite.*
$C_{12}H_{19}O_2P$ M 226.255
Liq. d^0_0 1.01. Bp_{10} 121-122°. n_D^{18} 1.5021.
Dibutyl ester: [3030-90-8]. *Dibutyl phenylphosphonite.*
$C_{14}H_{23}O_2P$ M 254.308
Liq. Bp_3 116-117°. n_D^{20} 1.4995.
Mono-Ph ester: see Phenylphosphinic acid, P-00210
Di-Ph ester: [13410-61-2]. *Diphenyl phenylphosphonite.*
$C_{18}H_{15}O_2P$ M 294.289
Liq. d^{20}_0 1.18. Bp_3 175-176°. n_D^{20} 1.6027.
Dibenzyl ester: [62292-07-3]. *Dibenzyl phenylphosphonite. Bis(phenylmethyl) phenylphosphonite.*
$C_{20}H_{19}O_2P$ M 322.343
Liq. $Bp_{0.1}$ 65-70°.
Difluoride: see Phenylphosphonous difluoride, P-00256
Dichloride: see Phenylphosphonous dichloride, P-00254
Dibromide: see Phenylphosphonous dibromide, P-00253
Diiodide: [20472-19-9]. *Diiodophenylphosphine.*
$C_6H_5I_2P$ M 361.888
Dark-red liq. Bp_1 142-143°.
Dicyanide: see Phenylphosphonous dicyanide, P-00255
Diisocyanate: [2736-49-4].
$C_8H_5N_2O_2P$ M 192.113
Liq. Bp_3 118-122°, $Bp_{0.5}$ 75-77°.
Diamide: see P-Phenylphosphonous diamide, P-00252
Diisothiocyanate: [71354-74-0].
$C_8H_5N_2PS_2$ M 224.235
Liq. $Bp_{0.02}$ 135°.

Anand, N. *et al*, *J. Chem. Soc.*, 1951, 1867 (*dibenzyl ester*)
Haven, A.C., *J. Am. Chem. Soc.*, 1956, **78**, 842 (*diisocyanate*)
Harwood, H.J. *et al*, *J. Am. Chem. Soc.*, 1960, **82**, 423 (*dimethyl ester, synth, props*)
Lutsenko, I.F. *et al*, *Dokl. Akad. Nauk SSSR, Ser. Sci. Khim.*, 1960, **132**, 612 (*Engl. transl.* p. 577) (*divinyl ester*)
Hoffmann, H. *et al*, *Chem. Ber.*, 1961, **94**, 186 (*diiodide*)
Nishizawa, Y., *Bull. Chem. Soc. Jpn.*, 1961, **34**, 1170 (*diisopropyl ester*)
Petrov, K.A. *et al*, *Zh. Obshch. Khim.*, 1962, **32**, 1974 (*Engl. transl.* p. 1954) (*dibutyl, diphenyl esters*)
Feshchenko, N.G. *et al*, *Zh. Obshch. Khim.*, 1969, **39**, 2184 (*Engl. transl.* p. 2133) (*diiodide*)
Verstuyft, A.W. *et al*, *Inorg. Chem.*, 1977, **16**, 2776 (*dibenzyl ester, nmr, pmr*)
Colle, K.S. *et al*, *J. Org. Chem.*, 1978, **43**, 571 (*dimethyl ester, synth, nmr*)
Pudovik, A.N. *et al*, *Zh. Obshch. Khim.*, 1979, **49**, 1425 (*Engl. transl.* p. 1248) (*diisothiocyanate*)
Vincent, E. *et al*, *J. Mol. Struct.*, 1980, **65**, 239 (*diiodide, nmr*)

P-Phenylphosphonous diamide **P-00252**

$PhP(NH_2)_2$

$C_6H_9N_2P$ M 140.124
N,N′-*Di-Et:* [774-49-2]. N,N′-*Diethyl*-P-*phenylphosphonous diamide. Phenylphosphonous bis(ethylamide).*
$C_{10}H_{17}N_2P$ M 196.231
Liq. Mp 7.5°. $Bp_{0.001}$ 53-56°. n_D^{22} 1.5486.
N,N,N′,N′-*Tetra-Et:* [1636-14-2]. N,N,N′,N′-*Tetraethyl*-P-*phenylphosphonous diamide. Phenylphosphonous bis(diethylamide).*
$C_{14}H_{25}N_2P$ M 252.339
Liq. Bp_2 113-116°.
N,N′-*Diisopropyl:* [716-85-8]. N,N′-*Diisopropyl*-P-*phenylphosphonous diamide. Phenylphosphonous di(isopropylamide).* N,N′-*Bis(1-methylethyl)*-P-*phenylphosphonous diamide.* Liq. or solid. Mp 13.5-14.5°. $Bp_{0.001}$ 52°. n_D^{22} 1.5289.
N,N′-*Di*-tert-*butyl:* [1516-96-8]. N,N′-*Di*-tert-*butyl*-P-*phenylphosphonous diamide. Phenylphosphonous di*(tert-*butylamide*). N,N′-*Bis(1,1-dimethylethyl)*-P-*phenylphosphonous diamide.* Liq. or solid. Mp 13.5-14.5°. $Bp_{0.001}$ 64-65°. n_D^{20} 1.5198.
N,N,N′,N′-*Tetra-Ph:* [26447-90-5]. *Pentaphenylphosphonous diamide.*
$C_{30}H_{25}N_2P$ M 444.515
Cryst. (EtOH or Et_2O). Mp 139-140°. V. sensitive to air and moisture. Even in dry, cold conditions, cryst. turn yellow and then bluish-green.
N,N,N′,N′-*Tetrabenzyl:* [63067-80-7]. N,N,N′,N′-*Tetrabenzyl*-P-*phenylphosphonous diamide.* P-*Phenyl*-N,N,N′,N′-*tetrakis(phenylmethyl)phosphonous diamide. Phenylphosphonous bis(dibenzylamide).*
$C_{34}H_{33}N_2P$ M 500.622
No phys. props. reported.

Lane, A.P. *et al*, *J. Chem. Soc. (A)*, 1967, 1492 (*dialkyl, synth, pmr*)
Hnoosh, M.H. *et al*, *Can. J. Chem.*, 1969, **47**, 4679 (*tetraphenyl, synth*)
Gray, G.A. *et al*, *J. Am. Chem. Soc.*, 1977, **99**, 3243 (*derivs, N nmr*)
Gray, G.A. *et al*, *Org. Magn. Reson.*, 1980, **14**, 8 (*derivs, cmr*)
Modro, T.A. *et al*, *Can. J. Chem.*, 1980, **55**, 3681 (*tetraethyl, cmr*)

Phenylphosphonous dibromide **P-00253**

Dibromophenylphosphine
[1073-47-8]

$PhPBr_2$

$C_6H_5Br_2P$ M 267.887
Fuming liq. d^{20}_4 1.88. Bp 259-261°, Bp_{14} 132-134°, Bp_5 117-119°. n_D^{25} 1.6534.

Finch, A. *et al*, *J. Chem. Soc. (B)*, 1966, 1162 (*synth, ir, P nmr*)
Bliznyuk, N.K. *et al*, *Zh. Obshch. Khim.*, 1967, **37**, 890 (*Engl. transl.* p. 840) (*synth*)
Vincent, E. *et al*, *J. Mol. Struct.*, 1980, **65**, 289 (*P nmr*)
Hinke, A. *et al*, *Phosphorus Sulfur*, 1983, **15**, 93 (*synth, P nmr*)

Phenylphosphonous dichloride, 9CI **P-00254**

Dichlorophenylphosphine
[644-97-3]

$PhPCl_2$

$C_6H_5Cl_2P$ M 178.985
Pungent, odourous liq. d^{20}_4 1.33. Bp_5 99-101°. n_D^{20} 1.5947.
▷TB2478000.

Bliznyuk, N.I. *et al*, *Zh. Obshch. Khim.*, 1967, **37**, 890 (*Engl. transl.* p. 840) (*synth*)
Angetelli, J.M. *et al*, *J. Am. Chem. Soc.*, 1969, **91**, 4500 (*ir*)
Rake, A.T. *et al*, *Org. Mass Spectrom.*, 1970, **3**, 237 (*ms*)
Betteridge, D. *et al*, *Anal. Chem.*, 1972, **44**, 2005 (*pe*)
Chernova, A.V. *et al*, *Izv. Akad. Nauk SSSR, Ser. Khim.*, 1972, 722 (*Engl. transl.* p. 693) (*uv, raman*)
Kabachnik, M.I. *et al*, *Aust. J. Chem.*, 1975, **28**, 755 (*ir*)
Muylle, E. *et al*, *Spectrochim. Acta, Part A*, 1975, **31**, 1039; 1976, **32**, 599 (*P nmr*)

Weinberg, K.G., *J. Org. Chem.*, 1975, **40**, 3586 (*synth*)
Parr, W.J.E., *J. Chem. Soc., Faraday Trans. 2*, 1978, **74**, 933 (*nmr, cmr, struct*)
Bodner, G.M. *et al, J. Organomet. Chem.*, 1983, **243**, 305 (*cmr*)

Phenylphosphonous dicyanide **P-00255**

Dicyanophenylphosphine

[2946-59-0]

PhP(CN)$_2$

$C_8H_5N_2P$ M 160.115

Solid. Mp 36-37°. Bp_{35} 160-162°, Bp_2 98-99°, $Bp_{0.1}$ 67-68°.

▷SY9100000.

Kirk, P.G. *et al, J. Chem. Soc. (A)*, 1969, 2190 (*synth*)
Jones, C.E. *et al, Inorg. Chem.*, 1971, **10**, 1536 (*synth, ms, P nmr, pmr*)
Uznanski, B. *et al, Synthesis*, 1975, 735 (*synth, ir, P nmr*)
Lazukina, L.A. *et al, Zh. Obshch. Khim.*, 1980, **50**, 985 (*Engl. transl.* p. 783) (*synth*)
Wilkie, C.A. *et al, Inorg. Chem.*, 1980, **19**, 1499 (*cmr, P nmr*)

Phenylphosphonous difluoride, 9CI **P-00256**

Difluorophenylphosphine

[657-97-6]

$PhPF_2$

$C_6H_5F_2P$ M 146.076

Liq. d_4^{20} 1.22. Bp_{20} 30-31°. n_D^{20} 1.4903. Forms Mo and Ni complexes. Disproportionates on keeping.

Schmutzler, R., *Chem. Ber.*, 1965, **98**, 552 (*synth, F and P nmr*)
Drozd, G.I. *et al, Zh. Obshch. Khim.*, 1967, **37**, 958, 1343 (*Engl. transl.* pp. 906, 1269) (*synth, F and P nmr*)
Brown, C. *et al, J. Chem. Soc. (C)*, 1970, 878 (*synth, props*)
Green, J.H.S. *et al, Bull. Soc. Chim. Belg.*, 1970, **79**, 567 (*ir, raman*)
Modro, T.A., *Can. J. Chem.*, 1977, **55**, 3681 (*cmr*)
Parr, W.J.E., *J. Chem. Soc., Faraday Trans. 2*, 1978, **74**, 933 (*pmr, cmr, struct*)
Marat, R.K. *et al, Inorg. Chem.*, 1980, **19**, 798 (*synth, props*)

Phenylphosphoramidic acid, 9CI, 8CI **P-00257**

Phosphoric anilide

[1445-36-9]

PhNHP(O)(OH)$_2$

$C_6H_8NO_3P$ M 173.108

Solid. Mp 267-271° (255-257°).

Monoanilinium salt: [36097-59-3]. Solid. Mp 265-267°.

Di-Me ester: see Dimethyl phenylphosphoramidate, D-00817

Di-Et ester: see Diethyl phenylphosphoramidate, D-00312

Dipropyl ester: [26245-77-2]. *Dipropyl phenylphosphoramidate.*
$C_{12}H_{20}NO_3P$ M 257.269
Solid. Mp 55°.

Diisopropyl ester: [1666-10-0]. *Diisopropyl phenylphosphoramidate. Bis(1-methylethyl) phenylphosphoramidate.*
$C_{12}H_{20}NO_3P$ M 257.269
Cryst. (MeOH aq.) or needles (H_2O or Et_2O/pet. ether). Mp 121-121.5°.

Dibutyl ester: [13024-84-5]. *Dibutyl phenylphosphoramidate.*
$C_{14}H_{24}NO_3P$ M 285.322
Liq. Bp_{14} 222°, $Bp_{0.01}$ 148°.

Bis(2-methylpropyl) ester: Bis(2-methylpropyl) phenylphosphoramidate. Diisobutyl phenylphosphoramidate.
$C_{14}H_{24}NO_3P$ M 285.322
Needles (MeOH). Mp 43.5-45°.

Bis(1-methylpropyl) ester: Bis(1-methylpropyl) phenylphosphoramidate. Di-sec-butyl phenylphosphoramidate.
$C_{14}H_{24}NO_3P$ M 285.322
Solid. Mp 42-44°. Bp_{14} 201°.

Dipentyl ester: see Dipentyl phosphate, D-00979

Bis(3-methylbutyl) ester: see Bis(3-methylbutyl) phosphate, B-00302

Dibenzyl ester: see Dibenzyl phosphate, D-00053

Di-Ph ester: see Diphenyl phenylphosphoramidate, D-01035

Di-2-naphthyl ester: see Di-2-naphthyl phosphate, D-00938

Difluoride: [701-61-1].
$C_6H_6F_2NOP$ M 177.090
Solid. Mp 48-49°. Bp_5 103-104°.

Dichloride: see Phenylphosphoramidic dichloride, P-00258

Diamide: Phenylphosphoric triamide.
$C_6H_{10}N_3OP$ M 171.138
Cryst. Sol. H_2O, EtOH, Me_2CO. Mp ca. 140°. Loses NH_3 at 170°.

McOmbie, H. *et al, J. Chem. Soc.*, 1945, 380, 921 (*esters*)
Cook, H.G. *et al, J. Chem. Soc.*, 1945, 873; 1949, 2921 (*synth, esters*)
Atherton, F.R. *et al, J. Chem. Soc.*, 1948, 1106 (*esters*)
Goldwhite, H. *et al, J. Chem. Soc.*, 1957, 2409 (*synth*)
Olah, G.A. *et al, J. Org. Chem.*, 1959, **24**, 1443 (*difluoride*)
Kobayashi, E. *et al, Bull. Chem. Soc. Jpn.*, 1973, **46**, 183 (*diamide, synth, ir*)
Parkhe, A.B. *et al, Indian J. Chem., Sect. B*, 1981, **20**, 79 (*synth*)
Calvo, K.C. *et al, J. Am. Chem. Soc.*, 1983, **105**, 2827 (*P nmr*)

Phenylphosphoramidic dichloride, 9CI **P-00258**

[6955-57-3]

PhNHP(O)Cl_2

$C_6H_6Cl_2NOP$ M 209.999

Silky needles (C_6H_6/pet. ether). Mp 93-94°.

Caven, R.M., *J. Chem. Soc.*, 1902, **81**, 1362 (*synth*)
Zhmurova, I.N. *et al, Zh. Obshch. Khim.*, 1962, **32**, 2576 (*Engl. transl.* p. 2540) (*synth*)
Cremlyn, R.J.W. *et al, J. Chem. Soc. (C)*, 1971, 300 (*synth, ir*)
Modro, T.A., *Phosphorus Sulfur*, 1979, **5**, 331 (*cmr*)

Phenylphosphoramidochloridic acid, 9CI **P-00259**

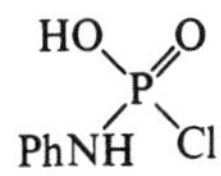

$C_6H_7ClNO_2P$ M 191.554

Derivs. may exist in chiral forms.

(±)-*form*

Et ester: Ethyl phenylphosphoramidochloridate.
$C_8H_{11}ClNO_2P$ M 219.607
Liq. Bp_{18} 98-100°.

2,2,2-Trichloroethyl ester: 2,2,2-Trichloroethyl phenylphosphoramidochloridate.
$C_8H_8Cl_4NO_2P$ M 322.942

Phosphorylating agent. Cryst. (pentane). Mp 83-83.5°.
2,2,2-Tribromoethyl ester: 2,2,2-Tribromoethyl phenylphosphoramidochloridate.
$C_8H_8Br_3ClNO_2P$ M 456.295
Phosphorylating agent for synth. of phospholipids and polysaccharide phosphates. Cryst. (Et_2O/pet. ether). Mp 110-113°.
Ph ester: see Phenyl phenylphosphoramidochloridate, P-00190
2-Chlorophenyl ester: 2-Chlorophenyl phenylphosphoramidochloridate.
$C_{12}H_{10}Cl_2NO_2P$ M 302.096
Phosphorylating agent. Solid. Mp 91-95°.
4-Chlorophenyl ester: 4-Chlorophenyl phenylphosphoramidochloridate.
$C_{12}H_{10}Cl_2NO_2P$ M 302.096
Phosphorylating agent. Solid. Mp 145-147°.
4-Nitrophenyl ester: 4-Nitrophenyl phenylphosphoramidochloridate.
$C_{12}H_{10}ClN_2O_4P$ M 312.649
Phosphorylating agent. Cryst. (C_6H_6). Mp 126-128°.

Cook, H.G. *et al*, *J. Chem. Soc.*, 1949, 2921 (*ethyl ester, synth*)
Zielinski, W.S. *et al*, *J. Chem. Soc., Chem. Commun.*, 1976, 772 (*synth, P nmr, use*)
Lammners, J.G. *et al*, *Recl. Trav. Chim. Pays-Bas*, 1979, **98**, 243 (*synth, ir, pmr, use*)
Ohtsuke, E. *et al*, *J. Am. Chem. Soc.*, 1979, **101**, 6409 (*synth, uv, use*)
Lesnikowsi, Z.J. *et al*, *Synthesis*, 1980, 397 (*use*)
Niewiarowski, W. *et al*, *J. Chem. Soc., Chem. Commun.*, 1980, 524 (*synth, P nmr, ms, use*)
Van Boekel, C.A.A., *Recl. Trav. Chim. Pays-Bas*, 1983, **102**, 526 (*synth, use*)

Phenylphosphoramidochloridothioic acid, 9CI, 8CI P-00260

PhNHPCl(S)OH ⇌ PhNHPCl(O)SH

$C_6H_7ClNOPS$ M 207.614

(±)-*form*

O-*Ph ester:* O-*Phenyl phenylphosphoramidochloridothioate.* O-*Phenyl chloro(phenylamido)thiophosphate.*
$C_{12}H_{11}ClNOPS$ M 283.712
Oil. d_4^{25} 1.31. n_D^{20} 1.6242.
O-*4-Nitrophenyl ester:* O-*(4-Nitrophenyl) phenylphosphoramidochloridothioate.* O-*(4-Nitrophenyl) chloro(phenylamido)thiophosphate.*
$C_{12}H_{10}ClN_2O_3PS$ M 328.709
Nucleoside phosphorylating agent. Cryst. (CCl_4). Mp 116-119°.

Blair, E.H. *et al*, *J. Org. Chem.*, 1960, **25**, 1620 (*synth*)
Lesiak, K. *et al*, *Pol. J. Chem.*, 1979, **53**, 2041 (*synth, P nmr, use*)

Phenylphosphoramidoselenoic acid, 9CI P-00261

Phosphoroselenoic acid anilide

PhNHP(Se)(OH)$_2$ ⇌ PhNHP(O)(SeH)(OH)

$C_6H_8NO_2PSe$ M 236.068

O,O-*Di-Et ester:* [57237-67-9]. O,O-*Diethyl phenylphosphoramidoselenoate.*
$C_{10}H_{16}NO_2PSe$ M 292.176
Liq. $Bp_{0.25}$ 122°.

Stec, W.J. *et al*, *J. Org. Chem.*, 1976, **41**, 227 (*synth, props, P nmr*)

Phenylphosphoramidothioic acid, 9CI, 8CI P-00262

Thiophosphoric acid monoanilide

PhNHP(S)(OH)$_2$ ⇌ PhNHP(O)(OH)(SH)

$C_6H_8NO_2PS$ M 189.168

O,O-*Di-Me ester:* [83436-57-1]. O,O-*Dimethyl phenylphosphoramidothioate, 9CI.* O,O-*Dimethyl thiophosphoric anilide.* O,O-*Dimethyl phenylamidothiophosphate.*
$C_8H_{12}NO_2PS$ M 217.222
Liq. d_4^{20} 1.23. Bp_3 131-132°. n_D^{20} 1.5724.
O,O-*Di-Et ester:* [3694-54-0]. O,O-*Diethyl phenylphosphoramidothioate, 9CI.* O,O-*Diethyl thiophosphoric anilide.* O,O-*Diethyl phenylamidothiophosphate.*
$C_{10}H_{16}NO_2PS$ M 245.276
Liq. d_4^{20} 1.15. Bp_2 134°. n_D^{20} 1.5482.
O,O-*Diisopropyl ester:* [64723-48-4]. O,O-*Bis(1-methylethyl) phenylphosphoramidothioate, 9CI.* O,O-*Diisopropyl thiophosphoric anilide.* O,O-*Diisopropyl phenylamidothiophosphate.*
$C_{12}H_{20}NO_2PS$ M 273.329
Solid. Mp 89-90°.
O,O-*Di-Ph ester:* [38945-95-6]. O,O-*Diphenyl phenylphosphoramidothioate, 9CI.* O,O-*Diphenyl thiophosphoric anilide.* O,O-*Diphenyl phenylamidothiophosphate.*
$C_{18}H_{16}NO_2PS$ M 341.364
Solid. Mp 92-92.5°.
Difluoride:
$C_6H_6F_2NPS$ M 193.151
Liq. Bp_1 86°. n_D^{20} 1.5414.
Dichloride: [38568-76-2].
$C_6H_6Cl_2NPS$ M 226.060
Liq. or cryst. ($CHCl_3$/pet. ether). Mp 181-183°. Bp_8 133-134°. n_D^{20} 1.6288.

Kabachnik, M.I. *et al*, *Dokl. Akad. Nauk SSSR, Ser. Sci. Khim.*, 1954, **96**, 991; *CA*, **49**, 8842 (*esters*)
Olah, G.A. *et al*, *Justus Liebigs Ann. Chem.*, 1959, **625**, 1959 (*dichloride*)
Olah, G.A. *et al*, *J. Org. Chem.*, 1959, **24**, 1443 (*difluoride*)
Keat, R., *J. Chem. Soc., Dalton Trans.*, 1972, 2189 (*dichloride, P nmr*)
Mathis, R. *et al*, *C.R. Hebd. Seances Acad. Sci., Ser. C*, 1975, **281**, 437; 1977, **284**, 767 (*esters, ir*)
Ayed, N. *et al*, *C.R. Hebd. Seances Acad. Sci., Ser. C*, 1977, **285**, 221 (*esters, ir*)
Khaskin, B.A. *et al*, *Zh. Obshch. Khim.*, 1977, **47**, 1912 (*Engl. transl.* p. 1748) (*esters*)
Al-Rawi, J.M.A. *et al*, *Org. Magn. Reson.*, 1983, **21**, 75 (*diethyl ester, cmr*)
Pasmitskene, G.I., *Izv. Akad. Nauk SSSR, Ser. Khim.*, 1983, 2373 (*Engl. transl.* p. 2138) (*dichloride*)

Phenylphosphoramidous acid, 9CI P-00263

PhNHP(OH)$_2$

$C_6H_8NO_2P$ M 157.108

Di-Et ester: [5156-98-9]. *Diethyl phenylphosphoramidite.*
$C_{10}H_{16}NO_2P$ M 213.216
Rgt. for converting carboxylic acids into their anilides. Liq. d_4^{20} 1.07. Bp_7 135°. n_D^{20} 1.5272.

Nifant'ev, É.E. *et al*, *Zh. Obshch. Khim.*, 1971, **41**, 2011 (*Engl. transl.* p. 2032) (*synth*)

Pudovik, M.A. *et al, Zh. Obshch. Khim.*, 1976, **46**, 773 (*Engl. transl.* p. 772) (*synth*)
Muratova, A.A. *et al, Zh. Obshch. Khim.*, 1976, **46**, 1729 (*Engl. transl.* p. 1680) (*synth, ir*)

Phenylphosphorimidic acid, 10CI, 9CI, 8CI **P-00264**

PhN=P(OH)$_3$

$C_6H_8NO_3P$ M 173.108

Free acid is tautomeric with Phenylphosphoramidic acid, P-00257 .

Tri-Me ester: [7077-62-5]. *Trimethyl phenylphosphorimidate.* P,P,P-*Trimethoxy*-N-*phenylphosphine imide.* P,P,P-*Trimethoxy*-N-*phenylphosphazene.*
$C_9H_{14}NO_3P$ M 215.188
Liq. d_4^{20} 1.17. $Bp_{0.0002}$ 44-45°. n_D^{20} 1.5232.

Tri-Et ester: [2397-47-9]. *Triethyl phenylphosphorimidate.* P,P,P-*Triethoxy*-N-*phenylphosphine imide.* P,P,P-*Triethoxy*-N-*phenylphosphazene.*
$C_{12}H_{20}NO_3P$ M 257.269
Liq. d_4^{20} 1.07. $Bp_{1.5}$ 118-119°. n_D^{20} 1.5022.

Triisopropyl ester: Triisopropyl phenylphosphorimidate. Tris(1-methylethyl) phenylphosphorimidate. P,P,P-*Triisopropoxy*-N-*phenylphosphine imide.*
$C_{15}H_{26}NO_3P$ M 299.349
Liq. d_4^{20} 1.01. $Bp_{0.0002}$ 49-50°. n_D^{20} 1.4880.

Tri-Ph ester: Triphenyl phenylphosphorimidate. P,P,P-*Triphenoxy*-N-*phenylphosphine imide.* P,P,P-*Triphenoxy*-N-*phenylphosphazene.*
$C_{24}H_{20}NO_3P$ M 401.401
Thick liq. d_4^{20} 1.21. $Bp_{0.05}$ 222-224°. n_D^{20} 1.6005. Monomeric.

Trichloride: [5290-43-7]. P,P,P-*Trichloro*-N-*phenylphosphine imide.* P,P,P-*Trichloro*-N-*phenylphosphazene.* P,P,P-*Trichloro*-N-*phenyliminophosphorane.*
$C_6H_{10}Cl_3NP$ M 233.484
Solid. Mp 180-182°. Dimeric in solid and in soln. Completely stable to H_2O.

Kabachnik, M.I. *et al, Izv. Akad. Nauk SSSR, Ser. Khim.*, 1956, 790 (*Engl. transl.* p. 809) (*esters, synth*)
Zhmurova, I.N. *et al, Zh. Obshch. Khim.*, 1960, **30**, 3044; 1962, **32**, 2576 (*Engl. transl.* pp. 3018, 2540) (*trichloride*)
Zhmurova, I.N. *et al, Zh. Obshch. Khim.*, 1961, **31**, 3741 (*Engl. transl.* p. 3495) (*triphenyl ester*)
Gilyarov, V.A. *et al, Zh. Obshch. Khim.*, 1966, **36**, 708 (*Engl. transl.* p. 722) (*trimethyl ester, synth, use*)
Wiegräbe, W. *et al, Chem. Ber.*, 1968, **101**, 1414 (*synth*)
Buchanan, G.W. *et al, Can. J. Chem.*, 1980, **58**, 2442 (*Trimethyl ester, N nmr*)
Buchanan, G.W. *et al, Org. Magn. Reson.*, 1980, **14**, 517 (*trimethyl ester, cmr*)
Mathis, R. *et al, Spectrochim. Acta, Part A*, 1982, **38**, 1181 (*ir*)

1-Phenylphosphorinane, 9CI **P-00265**

Cyclopentamethylenephenylphosphine. 1-Phenylphosphinane
[3302-83-8]

$C_{11}H_{15}P$ M 178.213
Liq. d_4^{20} 1.029. $Bp_{22\text{-}24}$ 154-155°, Bp_3 119°. n_D^{20} 1.5882.

B,MeI: [3302-94-1]. *1-Methyl-1-phenylphosphorinanium iodide, 8CI.*
$C_{12}H_{18}IP$ M 320.152
Cryst. (EtOH/Et_2O). Mp 176°.

B,PhBr: [59432-47-2]. *1,1-Diphenylphosphorinanium bromide.*
$C_{17}H_{20}BrP$ M 335.223
Mp 262-263°.

B,PhCH$_2$Br: [31082-05-0]. *1-Phenyl-1-phenylmethylphosphorinanium bromide. 1-Benzyl-1-phenylphosphorinanium bromide.*
$C_{18}H_{22}BrP$ M 349.250
No phys. props. reported.

Oxide: [4963-95-5].
$C_{11}H_{15}OP$ M 194.213
Cryst. (cyclohexane). Mp 134-135°, Mp 125-127°. $Bp_{0.2}$ 141°.

Sulfide: [4963-94-4].
$C_{11}H_{15}PS$ M 210.273
Cryst. (pet. ether). Mp 86°.

Issleib, K. *et al, Chem. Ber.*, 1961, **94**, 113 (*synth, deriv*)
Aksnes, G. *et al, Acta Chem. Scand.*, 1965, **19**, 931 (*synth, deriv*)
Davies, J.H. *et al, J. Chem. Soc. (C)*, 1966, 245 (*ir, synth, oxide, sulfide, ms, ir*)
Gray, G., *J. Org. Chem.*, 1972, **37**, 3458 (*oxide, cmr, P nmr*)
Quin, L.D. *et al, J. Am. Chem. Soc.*, 1975, **97**, 4349 (*P nmr, conformn*)
Gray, G., *J. Am. Chem. Soc.*, 1976, **98**, 2109 (*cmr, pmr, oxide, P nmr*)
Lambert, J.B. *et al, J. Am. Chem. Soc.*, 1976, **98**, 3778 (*cmr*)
Koch, C.W. *et al, Org. Mass Spectrom.*, 1977, **12**, 624 (*oxide, ms*)

1-Phenyl-2-phosphorinanol, 9CI **P-00266**

1-Phenyl-2-phosphinanol. 2-Hydroxy-1-phenylphosphorinane
[54552-87-3]

$C_{11}H_{15}OP$ M 194.213
Liq. Bp_1 137-140°.

Issleib, K., *Z. Anorg. Allg. Chem.*, 1974, **408**, 266 (*synth*)

1-Phenyl-2-phosphorinanone, 9CI **P-00267**

1-Phenyl-2-phosphinanone
[54552-95-3]

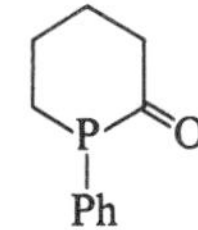

$C_{11}H_{13}OP$ M 192.197
Liq. Bp_3 143°. Forms an unstable methiodide.

Issleib, K. *et al, Z. Anorg. Allg. Chem.*, 1974, **408**, 266 (*synth*)

1-Phenyl-4-phosphorinanone, 9CI, 8CI **P-00268**

1-Phenyl-4-phosphinanone
[23855-87-0]

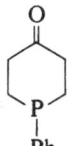

$C_{11}H_{13}OP$ M 192.197
Mp 43.5-44.0°. Bp_1 185-190°, $Bp_{0.02}$ 120-122°. n_D^{24} 1.6041.

Oxide: [38707-15-2].
$C_{11}H_{13}O_2P$ M 208.196
Cryst. Mp 166°.
Sulfide: [66685-93-6].
$C_{11}H_{13}OPS$ M 224.257
Cryst. Mp 144-145°.
B,MeI: 1-Methyl-4-oxo-1-phenyl-4-phosphorinanium iodide.
$C_{12}H_{16}IOP$ M 334.136
Solid. Mp 154-156°.

Welcher, R.P. *et al*, *J. Am. Chem. Soc.*, 1960, **82**, 4437 (*synth, derivs*)
McPhail, A.T. *et al*, *J. Am. Chem. Soc.*, 1971, **93**, 2574 (*cryst struct*)
Org. Synth., 1973, **53**, 98 (*synth*)
Venkataramu, S.D. *et al*, *Phosphorus Sulfur*, 1979, **7**, 133 (*cmr, oxide, sulfide, ir, pmr, cryst struct*)
Ramarajan, K. *et al*, *Phosphorus Sulfur*, 1981, **11**, 199 (*P nmr*)

Phenyl phosphorochloridic acid P-00269

PhOPCl(O)OH

$C_6H_6ClO_3P$ M 192.538
Derivs. may exist in chiral forms.
Me ester: Methyl phenyl phosphorochloridate.
$C_7H_8ClO_3P$ M 206.565
Liq. $Bp_{0.1}$ 90-91°.
Et ester: Ethyl phenyl phosphorochloridate.
$C_8H_{10}ClO_3P$ M 220.592
Liq. d_4^{20} 1.23. Bp_1 96-97°, $Bp_{0.01}$ 94-95°. n_D^{20} 1.4950.
Propyl ester: Phenyl propyl phosphorochloridate.
$C_9H_{12}ClO_3P$ M 234.619
Liq. $Bp_{0.01}$ 105-108°.
Isopropyl ester: Isopropyl phenyl phosphorochloridate.
$C_9H_{12}ClO_3P$ M 234.619
Liq. d_4^{20} 1.19. $Bp_{0.1}$ 80-82°. n_D^{20} 1.4685.
Butyl ester: Butyl phenyl phosphorochloridate.
$C_{10}H_{14}ClO_3P$ M 248.646
Liq. d_4^{20} 1.12. $Bp_{0.8}$ 102-105°, $Bp_{0.1}$ 120-122°. n_D^{20} 1.4615.

Navech, J. *et al*, *C.R. Hebd. Seances Acad. Sci.*, 1958, **246**, 2001.
Maklyaev, F.L. *et al*, *Zh. Obshch. Khim.*, 1962, **32**, 3421 (*Engl. transl.* p. 3357)

Phenyl phosphorochloridisocyanatidate, 9CI, 8CI P-00270

[21050-06-6]

PhOPCl(O)NCO

$C_7H_5ClNO_3P$ M 217.548

(±)-*form*
Reactive liq. with unpleasant odour. d_4^{20} 1.43. $Bp_{0.2}$ 105-108°. n_D^{20} 1.5330.

Derkach, G.I. *et al*, *Zh. Obshch. Khim.*, 1968, **38**, 1779 (*Engl. transl.* p. 1734) (*synth*)

Phenyl phosphorochloridofluoridate, 9CI, 8CI P-00271

Phenyl chlorofluorophosphate

PhOP(O)ClF

$C_6H_5ClFO_2P$ M 194.529

Liq. d_{20}^{20} 1.38. Bp 205°, Bp_5 66-70°. n_D^{20} 1.4793.

Olah, G.A. *et al*, *Can. J. Chem.*, 1962, **40**, 1917 (*synth*)
Börner, K.-B. *et al*, *Chem. Ber.*, 1963, **96**, 1328 (*synth, props*)

Phenyl phosphorodiamidate, 9CI, 8CI P-00272

[7450-69-3]

$PhOP(O)(NH_2)_2$

$C_6H_9N_2O_2P$ M 172.123
Converts lactams into cyclic aminoenamines. Cryst. (EtOH). Mp 185-190°.

Goehring, M. *et al*, *Chem. Ber.*, 1956, **89**, 1768.
Nielsen, M.L. *et al*, *J. Phys. Chem.*, 1964, **68**, 152 (*nmr*)
Marcus, P., *C. R. Hebd. Seances Acad. Sci.*, 1966, **262**, 1048 (*synth*)
Bullen, G.J. *et al*, *Acta Crystallogr., Sect. B*, 1973, **29**, 331 (*cryst struct*)
Stoss, P. *et al*, *Chem. Ber.*, 1976, **109**, 2097 (*use*)
Malhotra, R.C. *et al*, *Indian J. Chem., Sect. B*, 1978, **16**, 329 (*use*)
Fieser, M. *et al*, *Reagents for Organic Synthesis*, Wiley, 1967-84, **5**, 517.

N-Phenylphosphorodiamidic acid, 9CI, 8CI P-00273

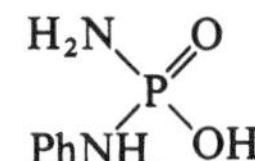

$C_6H_9N_2O_2P$ M 172.123
Plates. Spar. sol. cold H_2O. Mp 157-158°.
Et ester: Ethyl N*-phenylphosphorodiamidate.*
$C_8H_{13}N_2O_2P$ M 200.177
Prisms (H_2O). Mp 127°.

Caven, R.M., *J. Chem. Soc.*, 1902, **81**, 1362.

2-Phenylphosphorodiamidic hydrazide, 9CI P-00274

$PhNHNHP(O)(NH_2)_2$

$C_6H_{11}N_4OP$ M 186.153
Tetra-Me: N,N,N′,N′*-Tetramethyl-2-phenylphosphorodiamidic hydrazide.*
$C_{10}H_{19}N_4OP$ M 242.260
Cryst. (EtOH aq.). Mp 159-160°.
Tetra-Et: N,N,N′,N′*-Tetraethyl-2-phenylphosphorodiamidic hydrazide.*
$C_{14}H_{27}N_4OP$ M 298.367
Cryst. (EtOH). Mp 152°.

Ger. Pat., 1 042 581, (*1956*); *CA*, **50**, 4448
Bock, H. *et al*, *Chem. Ber.*, 1965, **98**, 2844.

Phenyl phosphorodiamidite, 9CI P-00275

$PhOP(NH_2)_2$

$C_6H_9N_2OP$ M 156.124
Yellowish amorph. solid. Mp 110-112° dec.

Becke-Goehring, M. *et al*, *Chem. Ber.*, 1958, **91**, 1188 (*synth, props*)

O-Phenyl phosphorodiamidothioate, 9CI P-00276

O*-Phenyl thiophosphoryl diamide.* O*-Phenyl diamidothiophosphate*

[3969-50-4]

$PhOP(S)(NH_2)_2$

$C_6H_9N_2OPS$ M 188.184

Cryst. (H_2O or $CHCl_3$). Sl. sol. EtOH, Et_2O, hot H_2O. Mp 119°.

Ephraim, F., *Ber.*, 1911, **44**, 3414 (*synth*)
Goehring, M. *et al*, *Chem. Ber.*, 1956, **89**, 1768 (*synth*)

Phenyl phosphorodibromidite, 9CI, 8CI — P-00277

Phenyl dibromophosphite

[70445-77-1]

$PhOPBr_2$

$C_6H_5Br_2OP$ M 283.887

Pungent liq. Bp_{11} 130-132°. n_D^{20} 1.6202.

Fluck, E. *et al*, *J. Am. Chem. Soc.*, 1959, **81**, 6363.
Frazer, M.J. *et al*, *Chem. Ind. (London)*, 1959, 728 (*synth*)
Gloede, J. *et al*, *J. Prakt. Chem.*, 1979, **321**, 1029.
Tseng, C.K., *J. Org. Chem.*, 1979, **44**, 2793.
Gloede, J. *et al*, *Z. Anorg. Allg. Chem.*, 1980, **471**, 147 (*P nmr*)

O-Phenyl phosphorodibromidothioate, 9CI, 8CI — P-00278

O-*Phenyl thiophosphoryl dibromide.* O-*Phenyl dibromothiophosphate*

$PhOP(S)Br_2$

C_6H_5BrOPS M 236.043

Liq. Bp_{11} 156-157°.

Strecker, W. *et al*, *Ber.*, 1916, **49**, 63 (*synth*)

Phenyl phosphorodichloridate, 9CI, 8CI — P-00279

Phenyl dichlorophosphate. Phenyl phosphoryl dichloride

[770-12-7]

$PhOP(O)Cl_2$

$C_6H_5Cl_2O_2P$ M 210.984

Acetylcholinesterase inhibitor. Activating reagent for the conversion of carboxylic acids into amides, azides, thio esters, etc. d_4^{20} 1.412. Bp 241-243°, Bp_5 99.5-101.5°. n_D^{25} 1.5230.

Freeman, H.F., *J. Am. Chem. Soc.*, 1938, **60**, 750 (*synth*)
Orloff, H.D. *et al*, *J. Am. Chem. Soc.*, 1958, **80**, 727 (*synth*)
Raevskii, O.A., *J. Mol. Struct.*, 1973, **19**, 275 (*ir, conformn*)
Owen, G.R. *et al*, *Synthesis*, 1974, 704 (*synth*)
Zverev, V.V. *et al*, *Izv. Akad. Nauk SSSR, Ser. Khim.*, 1975, 1051 (*Engl. transl.* p. 961) (*pe*)
Glukhikh, V.I. *et al*, *Dokl. Akad. Nauk SSSR, Ser. Sci. Khim.*, 1979, **248**, 142 (*Phys. Chem., Engl. transl.*, p. 744) (*cmr*)
Hsung-Jang Liu, *et al*, *Can. J. Chem.*, 1980, **58**, 2645 (*use*)
Hansen, J. *et al*, *Org. Magn. Reson.*, 1981, **15**, 29 (*P nmr, pmr*)
Lago, J.M. *et al*, *Synth. Commun.*, 1983, **13**, 289, 653 (*use*)

Phenyl phosphorodichloridite, 9CI, 8CI — P-00280

Phenyl dichlorophosphite

[3426-89-9]

$PhOPCl_2$

$C_6H_5Cl_2OP$ M 194.985

Polymerisation catalyst. Liq. d_4^{20} 1.35. Bp_{10} 90°. n_D^{20} 1.5588.

▷TD4440000.

Schwarz, R. *et al*, *Chem. Ber.*, 1957, **90**, 952 (*synth*)
Tolkmith, H., *J. Org. Chem.*, 1958, **23**, 1682 (*synth*)

Phenyl phosphorodichloridodithioate, 9CI, 8CI — P-00281

Phenyl dichlorodithiophosphate

[10258-61-4]

$PhSP(S)Cl_2$

$C_6H_5Cl_2PS_2$ M 243.105

Liq. Bp_{16} 168-170°.

Michaelis, A. *et al*, *Ber.*, 1907, **40**, 3419 (*synth*)

O-Phenyl phosphorodichloridothioate, 9CI, 8CI — P-00282

O-*Phenyl thiophosphoryl dichloride.* O-*Phenyl dichlorothiophosphate*

[18961-96-1]

$PhOP(S)Cl_2$

$C_6H_5Cl_2OPS$ M 227.045

Liq. d_4^{20} 1.41. Bp_{15} 88-89°, Bp_8 107-107.5°. n_D^{20} 1.5738.

Ketelaar, J.A. *et al*, *Recl. Trav. Chim. Pays-Bas*, 1958, **77**, 982 (*synth*)
Godovikov, N.N. *et al*, *Zh. Obshch. Khim.*, 1961, **31**, 1628 (*Engl. transl.* p. 1516) (*synth*)

S-Phenyl phosphorodichloridothioate, 9CI, 8CI — P-00283

S-*Phenyl dichlorothiophosphate*

[21186-90-3]

$PhSP(O)Cl_2$

$C_6H_5Cl_2OPS$ M 227.045

Low-melting solid. Mp 38°. Bp_1 95-97°. Hydrolysed in moist air.

Shitov, L.N. *et al*, *Zh. Obshch. Khim.*, 1968, **38**, 2340 (*Engl. transl.* p. 2268) (*synth*)
Aksenov, V.I. *et al*, *Zh. Obshch. Khim.*, 1971, **41**, 484 (*Engl. transl.* p. 478) (*synth, P nmr*)

Phenyl phosphorodichloridothioite, 9CI — P-00284

Phenyl thiodichlorophosphite. Phenyl dichlorothiophosphite

[14684-25-4]

$PhSPCl_2$

$C_6H_5Cl_2PS$ M 211.045

Liq. d_{15}^{15} 1.26. Bp_{10} 125°.

▷TD4410000.

Michaelis, A. *et al*, *Ber.*, 1907, **40**, 3419 (*synth*)

Phenyl phosphorodifluoridate, 9CI, 8CI — P-00285

Phenyl phosphoryl difluoride. Phenyl difluorophosphate

[1126-52-9]

$PhOP(O)F_2$

$C_6H_5F_2O_2P$ M 178.075

Liq. Bp_{80} 100°, Bp_{16} 64°. n_D^{25} 1.4358.

Reddy, G.S. *et al*, *Z. Naturforsch., B*, 1970, **25**, 1199 (*synth*)
Effenberger, F. *et al*, *Synthesis*, 1981, 70 (*synth, ir, P nmr*)

Phenyl phosphorodifluoridite, 9CI, 8CI **P-00286**

Phenyl difluorophosphite

[3965-01-3]

$PhOPF_2$

$C_6H_5F_2OP$ M 162.075

Ligand for Ni and Mo. d_{20}^{18} 1.25. Bp 126-127°, Bp_{60} 58°. n_D^{18} 1.4613.

Schmutzler, R., *Chem. Ber.*, 1963, **96**, 2435 (*synth, F nmr*)
Ivanova, Zh.M., *Zh. Obshch. Khim.*, 1964, **34**, 858 (*Engl. transl.* p. 852) (*synth*)
Binder, H. *et al*, *Z. Naturforsch., B*, 1972, **27**, 753 (*synth, P nmr*)
Micoud, M.-H. *et al*, *Bull. Soc. Chim. Fr.*, 1972, 3774 (*synth, P nmr, complexes*)

***O*-Phenyl phosphorodifluoridothioate,** 9CI, 8CI **P-00287**

O-*Phenyl difluorothiophosphate.* O-*Phenyl thiophosphoryl difluoride*

[658-35-5]

$PhOP(S)F_2$

$C_6H_5F_2OPS$ M 194.135

Liq. Bp_{11} 50°. n_D^{24} 1.460.

Olah, G. *et al*, *J. Org. Chem.*, 1960, **25**, 603 (*synth*)
Reddy, G.S. *et al*, *Z. Naturforsch., B*, 1970, **25**, 1199 (*synth, pmr, F and P nmr*)

***S*-Phenyl phosphorodifluoridothioate,** 9CI, 8CI **P-00288**

S-*Phenyl difluorothiophosphate*

[17620-69-8]

$PhSP(O)F_2$

$C_6H_5F_2OPS$ M 194.135

Low-melting solid. Mp 28-29°. Bp_{20} 88-89°, $Bp_{0.6}$ 62°.

Roesky, H.W., *Chem. Ber.*, 1968, **101**, 636 (*synth, ir, F and P nmr*)
Shitov, L.N. *et al*, *Zh. Obshch. Khim.*, 1968, **38**, 2268 (*Engl. transl.* p. 2340) (*synth*)

2-Phenylphosphorodifluoridothioic hydrazide, 8CI **P-00289**

2-(Difluorothiophosphoryl)phenylhydrazine

[14278-36-5]

$PhNHNHP(S)F_2$

$C_6H_7F_2N_2PS$ M 208.165

Solid. Mp 44-46°.

Horn, H.G. *et al*, *Chem. Ber.*, 1967, **100**, 2258 (*synth, ir, F and P nmr*)

Phenyl phosphorodihydrazidate, 9CI, 8CI **P-00290**

Phenyl phosphoric dihydrazide. Phenyl phosphoryl dihydrazide

[53426-77-0]

$PhOP(O)(NH^2NH_2)_2$

$C_6H_{11}N_4O_2P$ M 202.152

Cryst. (EtOH). Mp 111° (103°).

2,2′-Diisopropylidene:

$C_{12}H_{19}N_4O_2P$ M 282.281

Cryst. (Me_2CO). Mp 156.5°.

2,2′-Dibenzylidene:

$C_{20}H_{19}N_4O_2P$ M 378.369

Cryst. (EtOH aq.). Mp 174°.

Audrieth, L.F. *et al*, *J. Org. Chem.*, 1955, **20**, 1288 (*synth, derivs*)
Klement, R. *et al*, *Chem. Ber.*, 1960, **93**, 834 (*synth, derivs*)
Engelhardt, U. *et al*, *Z. Naturforsch., B*, 1979, **34**, 1107 (*synth, pmr, P nmr*)
Steppan, H. *et al*, *Justus Liebigs Ann. Chem.*, 1982, 2135 (*use*)

***O*-Phenyl phosphorodihydrazidothioate,** 9CI, 8CI **P-00291**

O-*Phenyl thiophosphoryl dihydrazide*

[55003-02-6]

$PhOP(S)(NHNH_2)_2$

$C_6H_{11}N_4OPS$ M 218.213

Forms Ni, Cd, and Mn complexes. Cryst. (MeOH). Mp 96°.

$N^2,N^{2'}$-*Bis(trimethylsilyl):*

$C_{12}H_{27}N_4OPSSi_2$ M 362.576

Solid. Mp 37-39°.

Steger, L. *et al*, *Z. Naturforsch., B*, 1975, **30**, 634 (*deriv, ir, raman*)
Engelhardt, U. *et al*, *Z. Naturforsch., B*, 1976, **31**, 1553 (*synth, pmr, ir, raman, complexes*)
Engelhardt, U., *Acta Crystallogr., Sect. B*, 1979, **35**, 3116 (*cryst struct*)

Phenyl phosphorodiisocyanatidate, 9CI, 8CI **P-00292**

Phenyl phosphoryl diisocyanate

[1844-12-8]

$PhOP(O)(NCO)_2$

$C_8H_5N_2O_4P$ M 224.112

Low-melting solid. Mp 28°. $Bp_{0.15}$ 87-88°. n_D^{20} 1.5188. Highly reactive to protic substances.

Shokol, V.A. *et al*, *Zh. Obshch. Khim.*, 1969, **39**, 1041 (*Engl. transl.* p. 1012) (*synth, ir*)

***O*-Phenyl phosphorodiisocyanatidothioate,** 9CI, 8CI **P-00293**

O-*Phenyl thiophosphoryl diisocyanate*

[20039-34-3]

$PhOP(S)(NCO)_2$

$C_8H_5N_2O_3PS$ M 240.173

Reactive liq. d_4^{20} 1.34. $Bp_{0.08}$ 70-71°. n_D^{20} 1.5500. Sensitive to moisture.

Shokol, V.A. *et al*, *Zh. Obshch. Khim.*, 1969, **39**, 1712; 1971, **41**, 2380 (*Engl. transl.* pp. 1678, 2407) (*synth, props*)
Borovikov, Yu.Ya. *et al*, *Zh. Obshch. Khim.*, 1977, **47**, 328 (*Engl. transl.* p. 304) (*conformn*)

Phenyl phosphorodi(isothiocyanatate), 9CI, 8CI **P-00294**

Phenylphosphoryl diisothiocyanate

[21049-95-6]

$PhOP(O)(NCS)_2$

$C_8H_5N_2O_2PS_2$ M 256.233
Moisture-sensitive liq. d_4^{20} 1.36. $Bp_{0.1}$ 110-111°. n_D^{20} 1.6165.

Derkach, G.I. *et al*, *Zh. Obshch. Khim.*, 1968, **38**, 1779 (*Engl. transl.* p. 1734) (*synth, props*)

Phenyl phosphorofluoridic acid P-00295

Phenyl phosphorofluoridate, 9CI

$PhOPF(O)OH$

$C_6H_6FO_3P$ M 176.084
Derivatives may exist in chiral forms.
NH_4 salt: [21758-85-0]. Cryst. (EtOH aq.). Mp 208-210°.
Cyclohexylammonium salt: Cryst. (MeCN). Mp 148-149°.
Amide: Phenyl phosphoramidofluoridate.
$C_6H_7FNO_2P$ M 175.099
Solid. Mp 86°.

Wittman, R., *Chem. Ber.*, 1963, **96**, 771 (*synth*)
Clark, V.M. *et al*, *J. Chem. Soc. (C)*, 1969, 233 (*synth*)
Roesky, H.W. *et al*, *Z. Anorg. Allg. Chem.*, 1970, **375**, 140 (*amide, synth, ms, ir, pmr, nmr*)

2-Phenylphosphorohydrazidothioic acid, 9CI, 8CI P-00296

$PhNHNHP(S)(OH)_2 \rightleftharpoons PhNHNHP(O)(OH)(SH)$

$C_6H_9N_2O_2PS$ M 204.183
O,O-*Di-Me ester:* O,O-*Dimethyl 2-phenylphosphorohydrazidothioate.* O,O-*Dimethyl phosphorothioic acid phenylhydrazide.*
$C_8H_{13}N_2O_2PS$ M 232.237
Solid. Mp 83-85°.
O,O-*Di-Et ester:* O,O-*Diethyl 2-phenylphosphorohydrazidothioate.* O,O-*Diethyl phosphorothioic acid phenylhydrazide.*
$C_{10}H_{17}N_2O_2PS$ M 260.290
Solid. Mp 68-69°.
O,O-*Diisopropyl ester:* O,O-*Diisopropyl 2-phenylphosphorohydrazidothioate.* O,O-*Diisopropyl phosphorothioic acid phenylhydrazide.*
$C_{12}H_{21}N_2O_2PS$ M 288.344
Solid. Mp 42-45°.

Mastin, T.W. *et al*, *J. Am. Chem. Soc.*, 1945, **67**, 1662 (*diethyl ester*)
Mel'nikov, N.N. *et al*, *Zh. Obshch. Khim.*, 1953, **25**, 828 (*Engl. transl.* p. 793) (*esters, synth*)

S-Phenyl phosphorothioate, 9CI, 8CI P-00297

S-*Phenyl dihydrogen phosphorothioate.* S-*Phenyl phosphoric acid.* S-*Phenyl thiophosphate*
[18852-83-0]

$PhSP(O)(OH)_2$

$C_6H_7O_3PS$ M 190.153
Isolated as Ba salt. Metab. and photodegradation product of Edifenphos, E-00002 .
Difluoride: see S-*Phenyl phosphorodifluoridothioate, P-00288*
Dichloride: see S-*Phenyl phosphorodichloridothioate, P-00283*

Murai, T. *et al*, *Agric. Biol. Chem.*, 1977, **41**, 71.
Ueyama, I. *et al*, *Agric. Biol. Chem.*, 1978, **42**, 885.
Sekine, M. *et al*, *Bull. Chem. Soc. Jpn.*, 1982, **55**, 239 (*synth, uv*)

(1-Phenyl-1,2-propadienyl)phosphonic acid, 9CI P-00298

1-Phenyl-1-phosphonoallene. 1-Phenylallenephosphonic acid

$H_2C{=}C{=}CPhP(O)(OH)_2$

$C_9H_9O_3P$ M 196.142
Di-Et ester: [3095-10-1]. *Diethyl (1-phenyl-1,2-propadienyl)phosphonate.*
$C_{13}H_{17}O_3P$ M 252.249
d_4^{20} 1.12. $Bp_{0.6}$ 157-159°. n_D^{20} 1.5290.

Pudovik, A.N. *et al*, *Zh. Obshch. Khim.*, 1965, **35**, 1210 (*Engl. transl.* p. 1214) (*synth*)
Pudovik, A.N. *et al*, *Zh. Obshch. Khim.*, 1973, **43**, 2329 (*Engl. transl.* p. 2317) (*props*)

Phenyl(1-propenyl)phosphinic acid, 9CI P-00299

$H_3CCH{=}CHP(O)(OH)Ph$

$C_9H_{11}O_2P$ M 182.158
Mp 65°.
Cyclohexylammonium salt: Cryst. (Me_2CO/MeOH). Mp >200° (variable, rapid heating).
Me ester: Methyl phenyl(1-propenyl)phosphinate.
$C_{10}H_{13}O_2P$ M 196.185
Liq. d_4^{20} 1.12. $Bp_{0.1}$ 98-99°. Mixt. of *E*- and *Z*-forms.

Davies, J.H. *et al*, *J. Chem. Soc.*, 1964, 3425 (*synth, ir*)
Gareev, R.D. *et al*, *Zh. Obshch. Khim.*, 1978, **48**, 226; 1979, **49**, 503 (*Engl. transl.* pp. 200, 442) (*esters, pmr, nmr*)

Phenyl(2-propenyl)phosphinic acid, 9CI P-00300

Allylphenylphosphinic acid, 8CI
[68896-35-5]

$H_2C{=}CHCH_2P(O)(OH)Ph$

$C_9H_{11}O_2P$ M 182.158
Viscous oil.
Cyclohexylammonium salt: Cryst. (Me_2CO/MeOH). Mp 192°.
Me ester: Methyl phenyl(2-propenyl)phosphinate. Methyl allylphenylphosphinate.
$C_{10}H_{13}O_2P$ M 196.185
Liq. d_4^{20} 1.22. $Bp_{0.5}$ 100-101°. n_D^{20} 1.5332. Known in opt. active forms.
Et ester: [59611-88-0]. *Ethyl phenyl(2-propenyl)phosphinate. Ethyl allylphenylphosphinate.*
$C_{11}H_{15}O_2P$ M 210.212
Liq. $Bp_{0.4}$ 114-115°.
2-Propenyl ester: [14655-46-0]. *2-Propenyl phenyl(2-propenyl)phosphinate. Allyl allylphenylphosphinate.*
$C_{12}H_{15}O_2P$ M 222.223
Liq. d_4^{20} 1.10. $Bp_{0.5}$ 129-130°. n_D^{20} 1.5250.

Davies, J.H. *et al*, *J. Chem. Soc.*, 1964, 3425 (*synth*)
Pudovik, A.N. *et al*, *Zh. Obshch. Khim.*, 1967, **37**, 700 (*Engl. transl.* p. 656) (*allyl ester*)
Weichmann, H. *et al*, *J. Prakt. Chem.*, 1976, **318**, 87 (*ethyl ester*)
Gareev, R.D. *et al*, *Zh. Obshch. Khim.*, 1979, **49**, 503 (*Engl. transl.* p. 442) (*methyl ester*)
Koizumi, T. *et al*, *Tetrahedron Lett.*, 1981, **22**, 571 (*methyl ester*)

(1-Phenyl-1-propenyl)phosphonic acid, 9CI **P-00301**

1-Phenyl-1-phosphonopropene

$(HO)_2P(O)C(Ph)=CH(CH_3)$ (*E*)-form

$C_9H_{11}O_3P$ M 198.158

(***E***)***-form*** [10562-82-0]

Cryst. ($CHCl_3$). Mp 157-161°.

Mono-Me ester:

$C_{10}H_{13}O_3P$ M 212.185

Cryst. (C_6H_6/hexane). Mp 148-149.5°.

(***Z***)***-form*** [10562-87-5]

Not obtained in pure form.

Mono-Me ester: Cryst. ($CHCl_3$/hexane). Mp 156-159°.

Satterthwait, A.C. *et al*, *J. Am. Chem. Soc.*, 1978, **100**, 3197 (*synth, pmr*)

3-Phenyl-2-propenylphosphonic acid, 9CI **P-00302**

1-Phenyl-3-phosphonopropene

[58922-31-9]

$PhCH{=}CHCH_2P(O)(OH)_2$

$C_9H_{12}O_3P$ M 199.166

Di-Et ester: [58922-31-9]. *Diethyl 3-phenyl-2-propenylphosphonate.*

$C_{13}H_{19}O_3P$ M 254.265

Liq. d_4^{20} 1.09. Bp_4 169-171°. n_D^{20} 1.5268.

Ionin, B.I. *et al*, *Zh. Obshch. Khim.*, 1963, **33**, 432 (*Engl. transl.* p. 426) (*synth, ir, uv*)
Blicke, F.F. *et al*, *J. Org. Chem.*, 1964, **29**, 2036 (*synth*)
Piechucki, C., *Synthesis*, 1976, 187 (*use*)
Nicolau, H.C. *et al*, *J. Am. Chem. Soc.*, 1982, **104**, 5555, 5557, 5558 (*use*)

Phenylpropylphosphinic acid, 9CI **P-00303**

[69387-00-4]

$H_3CCH_2CH_2-P(O)(OR)-Ph$ (*R*)$_P$-form of esters

$C_9H_{13}O_2P$ M 184.174

(***R***)$_P$***-form***

(−)-*Menthyl ester:* [16934-90-0].

$C_{19}H_{32}O_3P$ M 339.434

Solid. Mp 86°. $[\alpha]_D^{23\text{-}6}$ −14° (c, 1-3 in C_6H_6).

(***S***)$_P$***-form***

(−)-*Menthyl ester:* [16934-91-1]. Cryst. (hexane). Mp 40°. $[\alpha]_D^{23\text{-}6}$ −81° (c, 1-3 in C_6H_6).

(±)***-form***

Propyl ester: Propyl phenylpropylphosphinate.

$C_{12}H_{19}O_2P$ M 226.255

Liq. d_0^{16} 1.05. Bp_{14} 163°, $Bp_{0.01}$ 92°. $n_D^{22.5}$ 1.4979.

Korpiun, O. *et al*, *J. Am. Chem. Soc.*, 1968, **90**, 4842.
Appel, R. *et al*, *Chem. Ber.*, 1976, **109**, 805.

Phenyl propyl phosphonate, 9CI **P-00304**

Phenyl propyl phosphite

[16390-99-1]

$H_3CCH_2CH_2OPH(O)OPh \rightleftharpoons H_3CCH_2CH_2OP(OH)OPh$

$C_9H_{13}O_3P$ M 200.174

Tautomeric. Liq. d_4^{20} 1.13. $Bp_{0.5}$ 118-120°. n_D^{20} 1.5000.

Wolf, R. *et al*, *Bull. Soc. Chim. Fr.*, 1960, 124 (*synth, ir*)
Houalla, D. *et al*, *Bull. Soc. Chim. Fr.*, 1960, 129 (*ir*)
Mandel'baum, Ya.A. *et al*, *Zh. Obshch. Khim.*, 1972, **42**, 502 (*Engl. transl.* p. 500) (*synth*)

(3-Phenylpropyl)phosphonic acid, 9CI **P-00305**

3-Phenyl-1-propanephosphonic acid

[50577-84-9]

$PhCH_2CH_2CH_2P(O)(OH)_2$

$C_9H_{13}O_3P$ M 200.174

Cryst. (EtOAc/methylcyclohexane). Mp 123-125°.

Di-Me ester: [55340-00-6]. *Dimethyl (3-phenylpropyl)-phosphonate.*

$C_{11}H_{17}O_3P$ M 228.227

Liq. $Bp_{0.45}$ 120-121°.

Di-Et ester: [52221-91-7]. *Diethyl (3-phenylpropyl)-phosphonate.*

$C_{13}H_{21}O_3P$ M 256.281

Liq. $Bp_{0.1}$ 125-128°. n_D^{25} 1.4908.

Kagan, F. *et al*, *J. Am. Chem. Soc.*, 1959, **81**, 3026 (*synth, ester*)
Rosini, G. *et al*, *Synthesis*, 1975, 44 (*ester, synth, ir, pmr*)

Phenyl 8-quinolyl phosphate **P-00306**

Phenyl 8-quinolinyl phosphate. Phenyl 8-quinolyl hydrogen phosphate

[41255-51-0]

PhOP(O)O– (8-quinolyl)

$C_{15}H_{11}NO_3P$ M 284.231

Phosphorylating agent for synth. of nucleotides and pyrophosphates. Greenish-yellow oil.

Takaku, H. *et al*, *Chem. Pharm. Bull.*, 1973, **21**, 445 (*synth, ir, use*)
Takaku, H. *et al*, *Bull. Chem. Soc. Jpn.*, 1974, **47**, 779 (*use*)

4-Phenyl-4*H*-1,4-selenaphosphorin, 9CI **P-00307**

*4-Phenyl-4*H*-1,4-selenaphosphinin*

$C_{10}H_9PSe$ M 239.115

4-Oxide: [57044-92-5].

$C_{10}H_9OPSe$ M 255.114

Cryst. (toluene). Mp 155-157°.

Naaktgeboren, A. *et al*, *Recl. Trav. Chim. Pays-Bas*, 1975, **94**, 92 (*synth, pmr*)

2-Phenylspiro[1,3,2-benzothiazaphosphole-2(3*H*),2′-[1,3,2]dioxaphospholane], 10CI **P-00308**

2,3-Dihydro-2-phenylspiro[1,3,2-benzothiazaphosphole-2,2′-[1,3,2]dioxaphospholane, 9CI

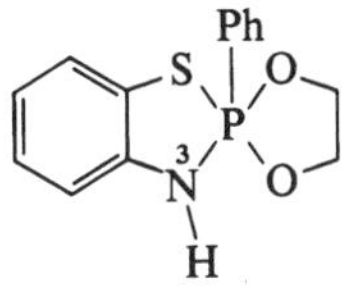

$C_{14}H_{14}NO_2PS$ M 291.304

3-(2,6-Dimethylphenyl): [58498-98-9].
$C_{22}H_{22}NO_2PS$ M 395.455
Cryst. (CH_2Cl_2/Et_2O). Mp 162-164.5°. $Bp_{0.05}$ 218-227°.

3-(2,4,6-Trimethylphenyl): [58498-96-7].
$C_{23}H_{24}NO_2PS$ M 409.482
Cryst. (CH_2Cl_2/Et_2O). Mp 165-167°. $Bp_{0.05}$ 220-228°.

Cadogan, J.I.G. *et al*, *J. Chem. Soc., Chem. Commun.*, 1975, 773 (*cryst struct*)
Cadogan, J.I.G. *et al*, *J. Chem. Soc., Perkin Trans. 1*, 1979, 1278 (*synth, pmr, P nmr*)
Dennis, L.W. *et al*, *J. Am. Chem. Soc.*, 1982, **104**, 230 (*P nmr*)

2-Phenyl-2,2′-spirobi[1,3,2-benzodioxaphosphole], 8CI **P-00309**

[21229-05-0]

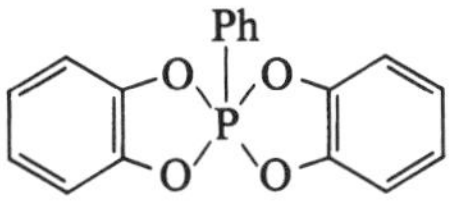

$C_{18}H_{13}O_4P$ M 324.272
Cryst. (toluene). Mp 182-185° (168-172°).

Wieber, M. *et al*, *Monatsh. Chem.*, 1970, **101**, 776 (*synth*)
Doak, G.O. *et al*, *J. Chem. Soc. (A)*, 1971, 1295 (*synth, P nmr*)
Brown, R.K. *et al*, *J. Am. Chem. Soc.*, 1977, **99**, 3326 (*cryst struct*)
Wunderlich, H., *Acta Crystallogr., Sect. B*, 1978, **34**, 342 (*cryst struct*)
Sau, A.C. *et al*, *J. Organomet. Chem.*, 1981, **217**, 157 (*pmr*)
Gloede, J., *Z. Anorg. Allg. Chem.*, 1983, **500**, 59 (*synth, P nmr*)
Poutasse, C.A. *et al*, *J. Am. Chem. Soc.*, 1984, **106**, 3814 (*props*)

[(Phenylsulfinyl)methyl]phosphonic acid, 9CI **P-00310**

$PhSOCH_2P(O)(OH)_2$

$C_7H_9O_4PS$ M 220.179
Esters are Wittig-Horner reagents for the synth. of phenyl vinyl sulfoxides.

(±)-***form***

Di-Me ester: Dimethyl [(phenylsulfinyl)methyl]-phosphonate.
$C_9H_{13}O_4PS$ M 248.233
Oil.

Di-Et ester: [65915-25-5]. *Diethyl [(phenylsulfinyl)-methyl]phosphonate.*
$C_{11}H_{17}O_4PS$ M 276.287
Oil. n_D^{20} 1.5305 (1.5236).

Mikolajczyk, M. *et al*, *Synthesis*, 1973, 669; 1975, 278 (*pmr, use*)
Drabowicz, J. *et al*, *Synthesis*, 1978, 758 (*P nmr*)
Mikolajczyk, M. *et al*, *J. Org. Chem.*, 1978, **43**, 2518 (*props*)
Hauske, J. *et al*, *J. Org. Chem.*, 1979, **44**, 2472 (*synth, use*)
De Jong, B.E. *et al*, *Recl. Trav. Chim. Pays-Bas*, 1981, **100**, 410 (*use*)

[(Phenylsulfonyl)methyl]phosphonic acid, 8CI **P-00311**

$PhSO_2CH_2P(O)(OH)_2$

$C_7H_9O_5PS$ M 236.179
Hygroscopic cryst. (EtOH aq.). Mp 261°.

Di-Et ester: [56069-39-7]. *Diethyl [(phenylsulfonyl)-methyl]phosphonate.*
$C_{11}H_{17}O_5PS$ M 292.286
Oil. $Bp_{0.25}$ 182-184°.

Kreutzkamp, N. *et al*, *Arch. Pharm. (Weinheim, Ger.)*, 1962, **295**, 773 (*synth*)
Mikolajczyk, M. *et al*, *Synthesis*, 1975, 278; 1976, 396 (*use*)
Ellingsen, P.O. *et al*, *Acta Chem. Scand., Ser. B*, 1979, **33**, 528 (*use*)
De Jong, B.E. *et al*, *Recl. Trav. Chim. Pays-Bas*, 1981, **100**, 410 (*use*)
Blumenkopf, T.A., *Synth. Commun.*, 1986, **16**, 139 (*ester, synth, ir, pmr, ms*)

Phenylsulfonylphosphoramidic acid, 9CI **P-00312**

[4737-12-6]

$PhSO_2NHP(O)(OH)_2$

$C_6H_8NO_5PS$ M 237.167
Prisms. Mp 147-148°.

Di-Me ester: [4140-56-0]. *Dimethyl phenylsulfonylphosphoramidate.*
$C_8H_{12}NO_5PS$ M 265.220
Cubes (H_2O). Mp 108-109°.

Di-Et ester: [1467-28-3]. *Diethyl phenylsulfonylphosphoramidate.*
$C_{10}H_{16}NO_5PS$ M 293.274
Cryst. (CCl_4), prisms (EtOH aq.). Sol. hot H_2O. Mp 111-112°.

Di-Ph ester: [16079-34-8]. *Diphenyl phenylsulfonylphosphoramidate.*
$C_{18}H_{16}NO_3PS$ M 357.363
Needles (CCl_4). Insol. H_2O. Mp 145-146°.

Dichloride: see Phenylsulfonylphosphoramidic dichloride, P-00313

Monobromide:
$C_6H_7BrNO_4PS$ M 300.063
Needles. Mp 147-150°.

Dibromide:
$C_6H_6Br_2NO_3PS$ M 362.960
Prisms (C_6H_6). Mp 122-124°.

Kirsanov, A.V. *et al*, *Zh. Obshch. Khim.*, 1954, **24**, 474, 882, 1980 (*Engl. transl.* pp. 483, 879, 1949) (*esters, synth*)
Kirsanov, A.V. *et al*, *Zh. Obshch. Khim.*, 1955, **25**, 571 (*Engl. transl.* p. 541) (*synth, bromides*)
Raetz, R., *J. Org. Chem.*, 1957, **22**, 372 (*diethyl ester, derivs*)

Phenylsulfonylphosphoramidic dichloride, 9CI **P-00313**

[16767-55-8]

$PhSO_2NHP(O)Cl_2$

$C_6H_6Cl_2NO_3PS$ M 274.058
Cryst. (CH_2Cl_2). Mp 129-130°. Forms metal derivs.

Anilinium salt: Needles ($CHCl_3$). Mp 107-110°.

Levchenko, E.S. *et al*, *Zh. Obshch. Khim.*, 1957, **27**, 3078 (*Engl. transl.* p. 3118) (*deriv*)
Haubold, W. *et al*, *Z. Anorg. Allg. Chem.*, 1967, **352**, 113 (*synth, P nmr*)

Phenylsulfonylphosphoramidothioic acid, 9CI, 8CI **P-00314**

$PhSO_2NHP(S)(OH)_2 \rightleftharpoons PhSO_2NHP(O)(OH)(SH)$

$C_6H_8NO_4PS$ M 221.167

O,O-Di-Ph ester: [70575-81-4]. O,O-*Diphenyl phenylsulfonylphosphoramidothioate.*
$C_{18}H_{16}NO_4PS$ M 373.362
Solid. Mp 106-107°. pK_a 2.69 (95% EtOH aq.).

Almasi, L. *et al*, *Rev. Roum. Chim.*, 1979, **24**, 3 (*synth, P nmr*)

Phenylsulfonylphosphorimidic acid, 9CI, 8CI **P-00315**

$PhSO_2N{=}PO(OH)_3$

$C_6H_8NO_5PS$ M 237.167

Tri-Me ester: [62461-25-0]. *Trimethyl phenysulfonylphosphorimidate.* P,P,P-*Trimethoxy*-N-*phenylsulfonylphosphazene.* P,P,P-*Trimethoxy*-N-*phenylsulfonylphosphine imide.*
$C_9H_{14}NO_5PS$ M 279.247
No phys. props. reported.

Tri-Et ester: Triethyl phenylsulfonylphosphorimidate. P,P,P-*Triethoxy*-N-*phenylsulfonylphosphazene.* P,P,P-*Triethoxy*-N-*phenylsulfonylphosphine imide.*
$C_{12}H_{20}NO_5PS$ M 321.327
Liq. d_4^{20} 1.23. $Bp_{0.3}$ 162-164°, $Bp_{0.0001}$ 116-117°. n_D^{20} 1.5060, n_D^{25} 1.4775.

Tri-Ph ester: Triphenyl phenylsulfonylphosphorimidate. P,P,P-*Triphenoxy*-N-*phenylsulfonylphosphazene.* P,P,P-*Triphenoxy*-N-*phenylsulfonylphosphine imide.*
$C_{24}H_{20}NO_5PS$ M 465.459
Rods (EtOH). Mp 85°.

Trichloride: *see Phenylsulfonylphosphorimidic trichloride, P-00316*

Tribromide: P,P,P-*Tribromo*-N-*phenylsulfonylphosphazene.* P,P,P-*Tribromo*-N-*phenylsulfonylphosphine imide.*
$C_6H_5Br_3NO_2PS$ M 425.857
Needles. Mp 111-112° (94-97°).

Goerdeler, J. *et al*, *Chem. Ber.*, 1961, **94**, 1067 (*esters, synth*)
Gilyarov, V.A. *et al*, *Zh. Obshch. Khim.*, 1966, **36**, 274 (*Engl. transl.* p. 285) (*triethyl ester*)
Lutskii, A.E. *et al*, *Zh. Obshch. Khim.*, 1967, **37**, 2034 (*Engl. transl.* p. 1930) (*triphenyl ester, uv*)
Goldstein, J.A., *J. Org. Chem.*, 1977, **42**, 2466 (*trimethyl ester, synth, pmr*)

Phenylsulfonylphosphorimidic trichloride, 9CI **P-00316**

P,P,P-*Trichloro*-N-*phenylsulfonylphosphine imide.* P,P,P-*Trichloro*-N-*phenylsulfonylphosphazene.* P,P,P-*Trichloro*-N-*phenylsulfonyliminophosphorane. Benzenesulfonylphosphorimidic trichloride*
[5666-55-7]

$PhSO_2N{=}PCl_3$

$C_6H_5Cl_3NO_2P$ M 260.444
Cryst. (CCl_4 or pet. ether). Mp 54-55°.

Kirsanov, A.V., *Zh. Obshch. Khim.*, 1952, **22**, 269 (*Engl. transl.* p. 329) (*synth*)
Levchenko, E.S. *et al*, *Zh. Obshch. Khim.*, 1959, **29**, 1813 (*Engl. transl.* p. 1784) (*synth*)
Glidewell, C., *Inorg. Chim. Acta*, 1976, **18**, 51 (*ms*)
Romanenko, E.A. *et al*, *Teor. Eksp. Khim.*, 1977, **13**, 70 (*Engl. transl.* p. 50) (*nqr*)

4-Phenyl-4*H*-1,4-telluraphosphorin, 9CI **P-00317**

*4-Phenyl-4*H*-1,4-telluraphosphinin*

$C_{10}H_9PTe$ M 287.755

4-Oxide: [57044-96-9].
$C_{10}H_9OPTe$ M 303.754
Cryst. (toluene). Mp 193-194°.

Naaktgeboren, A. *et al*, *Recl. Trav. Chim. Pays-Bas*, 1975, **94**, 92 (*synth, pmr*)

Phenyl tetraethylphosphorodiamidate, 9CI, 8CI **P-00318**

[4519-33-9]

$(Et_2N)_2P(O)OPh$

$C_{14}H_{25}N_2O_2P$ M 284.337

Used in conversion of acetanilides into 2-diethylaminoquinolines, and of oximes into amidines. Liq. $Bp_{2.5}$ 144-145°, $Bp_{0.015}$ 84-85°.

Pedersen, E.B. *et al*, *Synthesis*, 1977, 890 (*use*)
Jensen, H.G. *et al*, *Acta Chem. Scand., Ser. B*, 1979, **33**, 319 (*use*)
Modro, T.A. *et al*, *Phosphorus Sulfur*, 1979, **5**, 331 (*synth, cmr*)

1-Phenyl-1,2,3,4-tetrahydrobenzo[*h*]-phosphinoline, 9CI **P-00319**

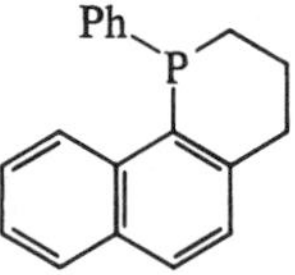

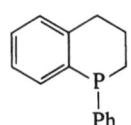

$C_{19}H_{17}P$ M 276.317

B,EtBr: 1-Ethyl-1,2,3,4-tetrahydro-1-phenylbenzo[h]*-phosphinolinium bromide.*
$C_{21}H_{22}BrP$ M 385.283
Inhibits growth of *B. subtilis* and has confirmed activity against P388 lymphocytic leukemia.

Holbrook, S.R. *et al*, *Phosphorus*, 1975, **6**, 15 (*cryst struct*)

1-Phenyl-1,2,3,4-tetrahydrophosphinoline, 9CI **P-00320**

[52221-68-8]

$C_{15}H_{15}P$ M 226.257

Air-sensitive solid. Mp 80-81°. $Bp_{0.001}$ 120° (oven).

1-Oxide: [52221-89-3].
$C_{15}H_{15}OP$ M 242.257

Sl. hygroscopic needles. Mp 107-108°. $Bp_{0.005}$ 170° (oven).

Rowley, L.E. *et al*, *Aust. J. Chem.*, 1974, **27**, 801 (*synth, ir, pmr, ms*)

7-Phenyl-1,3,4,6-tetrakis(trifluoromethyl)-7-aza-2,5-diphosphatetracyclo[4.1.0.0^{2,4}.0^{3,5}]-heptane, 9CI P-00321

[74930-73-7]

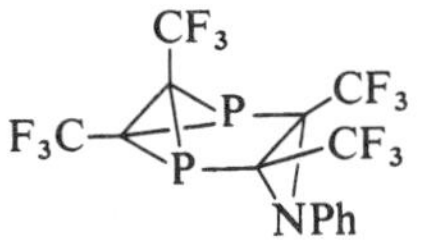

$C_{14}H_5F_{12}NP_2$ M 477.129

Pale-yellow oil.

Kobayashi, Y. *et al*, *J. Org. Chem.*, 1980, **45**, 4683 (*synth, ir, pmr, F nmr*)

2-Phenyl-4,4,5,5-tetramethyl-1,3,2-dioxaphospholane, 9CI P-00322

Tetramethylethylene phenylphosphonite. 4,4,5,5-Tetramethyl-2-phenyl-1,3-dioxa-2-phosphacyclopentane

[14812-63-6]

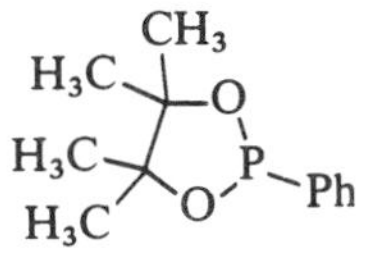

$C_{12}H_{17}O_2P$ M 224.239

Solid by subl. Mp 103-104°.

2-Oxide: [32084-79-0]. *Tetramethylethylene phenylphosphonate.*

$C_{12}H_{17}O_3P$ M 240.238

Cryst. (C_6H_6/hexane). Mp 108-110° (101°).

Fontal, B. *et al*, *Tetrahedron*, 1966, **22**, 3275 (*synth, pmr*)
Ishmaeva, E.A. *et al*, *Izv. Akad. Nauk SSSR, Ser. Khim.*, 1978, 2164 (*Engl. transl.* p. 1911) (*oxide*)
Baceiredo, A. *et al*, *Nouv. J. Chim.*, 1983, **7**, 255 (*synth, P nmr*)

3-Phenyl-2-thia-3-phosphabicyclo[2.2.2]-oct-5-ene, 9CI P-00323

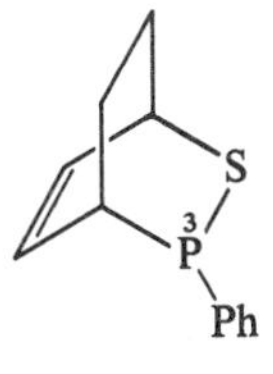

$C_{12}H_{13}PS$ M 220.268

3-Sulfide:

$C_{12}H_{13}PS_2$ M 252.328

Cryst. (Et_2O). Mp 92-93.5°. Mixt. of stereoisomers.

Nakayama, S. *et al*, *Bull. Chem. Soc. Jpn.*, 1975, **48**, 546 (*synth, ir, pmr*)

2-Phenyl-1,2-thiaphospholane, 10CI P-00324

[99632-61-8]

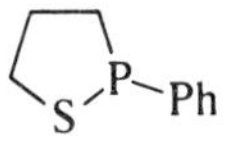

$C_9H_{11}PS$ M 182.220

Liq. $Bp_{0.5}$ 102-104°. With reactive alkylating agents at 80° undergoes reversible ring opening polymerisation *via* phosphonium salts.

2-Sulfide: [61157-03-7].

$C_9H_{11}PS_2$ M 214.280

Cryst. Mp 74°. Forms Fe complexes with $Fe_2(CO)_9$.

Mathey, F. *et al*, *J. Organomet. Chem.*, 1976, **117**, 377 (*sulfide, synth, pmr*)
Mathey, F. *et al*, *J. Chem. Soc., Chem. Commun.*, 1979, 417 (*complexes*)
Kobayashi, S.K. *et al*, *Bull. Chem. Soc. Jpn.*, 1985, **58**, 2153 (*synth, pmr, cmr, P nmr*)
Kobayashi, S.K. *et al*, *Macromolecules*, 1986, **19**, 462 (*props*)

2-Phenyl-1,2-thiaphosphorinane, 11CI P-00325

[99632-63-0]

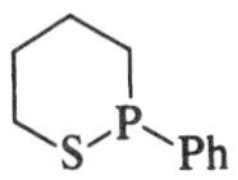

$C_{10}H_{13}PS$ M 196.246

Liq. $Bp_{0.3}$ 99-103°. With alkylating agents at 80°, undergoes reversible ring-opening *via* phosphonium intermediates.

Kobayashi, S. *et al*, *Bull. Chem. Soc. Jpn.*, 1985, **58**, 2153 (*synth, pmr, cmr, P nmr*)
Kobayashi, S. *et al*, *Macromolecules*, 1986, **19**, 462 (*props*)

[(Phenylthio)methyl]phosphonic acid, 8CI P-00326

$PhSCH_2P(O)(OH)_2$

$C_7H_9O_3PS$ M 204.180

Esters are Wittig-Horner reagents for the synth. of 1-phenylthio-1-alkenes, and thence aldehydes and ketones. These ester anions may be alkylated or acylated. Plates. Mp 242°.

Di-Me ester: [70369-42-5]. *Dimethyl [(phenylthio)-methyl]phosphonate.*

$C_9H_{13}O_3PS$ M 232.234

Liq. $Bp_{0.5}$ 140-143°.

Di-Et ester: [38066-16-9]. *Diethyl [(phenylthio)-methyl]phosphonate.*

$C_{11}H_{17}O_3PS$ M 260.287

Liq. $Bp_{0.4}$ 126-128°. n_D^{20} 1.5350.

S-*Oxide:* *see [(Phenylsulfinyl)methyl]phosphonic acid, P-00310*

S-*Dioxide:* *see [(Phenylsulfonyl)methyl]phosphonic acid, P-00311*

Kreutzkamp, N. *et al*, *Arch. Pharm. (Weinheim, Ger.)*, 1962, **295**, 773 (*synth*)
Green, M., *J. Chem. Soc.*, 1963, 1324 (*ester, synth, use*)
Mikolajczyk, M. *et al*, *Synthesis*, 1973, 669 (*ester, synth*)
Coutrot, P. *et al*, *Synthesis*, 1976, 107 (*ester, synth, use*)
Hauske, J. *et al*, *J. Org. Chem.*, 1979, **44**, 2472 (*ester, synth, pmr, use*)
Smith, J.C. *et al*, *J. Org. Chem.*, 1983, **48**, 1110 (*ester, synth, pmr, ir, ms*)

Phenyl(trichloromethyl)phosphinic acid, 9CI P-00327

[40103-72-8]

$Cl_3CPPh(O)OH$

$C_7H_6Cl_3O_2P$ M 259.456

Cryst. (EtOAc). Mp 162.5-163.5°. pK_a 6.90 (Me_2CO).

Me ester: Methyl phenyl(trichloromethyl)phosphinate.
$C_8H_8Cl_3O_2P$ M 273.483
Solid. Mp 108°.
Et ester: [54944-19-3]. *Ethyl phenyl(trichloromethyl)-phosphinate.*
$C_9H_{10}Cl_3O_2P$ M 287.510
Solid. Mp 79-80°. $Bp_{0.5}$ 131-132°.
Chloride: [52940-07-5].
$C_7H_5Cl_4OP$ M 277.902
Cryst. (CCl_4). Mp 87°.
Anhydride:
$C_{14}H_{10}Cl_6O_3P_2$ M 500.897
Cryst. (CCl_4). Mp 170-172°.

Kamai, G., *Zh. Obshch. Khim.*, 1948, **18**, 443; *CA*, **42**, 7723 (*esters*)
Yagupol'skii, L.M. *et al*, *Zh. Obshch. Khim.*, 1960, **30**, 1294 (*Engl. transl.* p. 1322) (*synth, anhydride, chloride*)
Kharrasova, F.M. *et al*, *Zh. Obshch. Khim.*, 1967, **37**, 2532 (*Engl. transl.* p. 2410) (*synth*)
Appel, R. *et al*, *Z. Anorg. Allg. Chem.*, 1975, **417**, 161 (*chloride, ir, P nmr*)

N-Phenyl-*P*-trichloromethylphosphonamidic acid P-00328

Cl_3C–P(=O)(NHPh)(OH)

$C_7H_7Cl_3NO_2P$ M 274.471
Isol. as the anilinium salt.
Anilinium salt: Solid. Mp 190-191°.
Me ester: Methyl N*-phenyl-*P*-trichloromethylphosphonamidate.*
$C_8H_9Cl_3NO_2P$ M 288.497
Cryst. (MeOH). Mp 146-147°.
Chloride: [59360-59-7].
$C_7H_6Cl_4NOP$ M 292.916
Prisms (CCl_4/hexane) or cryst. (Et_2O). Mp 119°.

Yakubovich, A.Ya. *et al*, *Zh. Obshch. Khim.*, 1954, **24**, 1465 (*Engl. transl.* p. 1455) (*chloride, synth*)
Kennard, K.C. *et al*, *J. Am. Chem. Soc.*, 1955, **77**, 1156 (*synth, derivs*)
Carallo, M. *et al*, *Phosphorus Sulfur*, 1978, **4**, 19 (*synth, ir, ms*)

Phenyl(trifluoromethyl)phosphinic acid, 9CI P-00329

[1513-46-8]

$F_3CPPh(O)OH$

$C_7H_6F_3O_2P$ M 210.092
Hygroscopic cryst. Mp 67° dec., 84-86°.
Anhydride:
$C_{14}H_{10}F_6O_3P_2$ M 402.169
Solid. Mp 264-268°. $Bp_{0.008}$ 154-170°.

Beg, M.A.A. *et al*, *Can. J. Chem.*, 1961, **39**, 564 (*synth*)
Sartori, P. *et al*, *Z. Anorg. Allg. Chem.*, 1972, **394**, 156; 1974, **404**, 161 (*synth, anhydride, ir, pmr, F nmr*)

[Phenyl(trimethylsilyl)methylene]-phosphinous chloride, 9CI P-00330

[74483-17-3]

$ClP{=}CPhSiMe_3$

$C_{10}H_{14}ClPSi$ M 228.733
Used in the synth. of Ph-subst. phosphorines. Yellow-green liq. $Bp_{0.001}$ 51°. May be stored at 0° for several weeks.

Appel, R. *et al*, *Angew. Chem., Int. Ed. Engl.*, 1980, **19**, 556 (*synth, P nmr, props*)
Burckett-St. Laurent, J.C.T.R. *et al*, *J. Mol. Spectrosc.*, 1982, **92**, 158 (*synth, props*)
Märkl, G. *et al*, *Angew. Chem., Int. Ed. Engl.*, 1982, **21**, 370 (*use*)

[Phenyl[(trimethylsilyl)oxy]methyl]-phosphonic acid, 9CI P-00331

[31675-43-1]

$Me_3SiOCHPhP(O)(OH)_2$

$C_{10}H_{17}O_4PSi$ M 260.301
Esters are α-hydroxybenzyl anion synthons which may be alkylated or acylated, and the products hydrolysed to functionalised, eg. hydroxy, ketones.

(±)-form

Di-Me ester: Dimethyl [phenyl[(trimethylsilyl)oxy]-methyl]phosphonate.
$C_{12}H_{21}O_4PSi$ M 288.355
Liq. d_4^{20} 1.10. Bp_8 154°, $Bp_{0.08}$ 101-106°. n_D^{20} 1.4862.
Di-Et ester: Diethyl [phenyl[(trimethylsilyl)oxy]-methyl]phosphonate.
$C_{14}H_{25}O_4PSi$ M 316.408
Liq. d_4^{20} 1.06. $Bp_{0.3}$ 105°. n_D^{20} 1.4767.

Nesterov, L.V. *et al*, *Zh. Obshch. Khim.*, 1971, **41**, 2449 (*Engl. transl.* p. 2474) (*synth*)
Evans, D.A. *et al*, *J. Am. Chem. Soc.*, 1978, **100**, 3467 (*synth, ir, pmr*)
Sekiguchi, A. *et al*, *Bull. Chem. Soc. Jpn.*, 1978, **51**, 337 (*synth, ir, pmr*)
Lebedev, E.P. *et al*, *Zh. Obshch. Khim.*, 1979, **49**, 1731 (*Engl. transl.* p. 1517) (*synth, ir, pmr*)
Koenigkramer, R.E. *et al*, *J. Org. Chem.*, 1980, **45**, 3994 (*synth, pmr, ms, use*)
Creary, X. *et al*, *J. Am. Chem. Soc.*, 1982, **104**, 4151 (*use*)
Sekine, M. *et al*, *Bull. Chem. Soc. Jpn.*, 1982, **55**, 218, 224 (*props, use*)

Phenyl(triphenylmethyl)phosphinic acid, 9CI P-00332

$Ph_3CP(O)(OH)Ph$

$C_{25}H_{21}O_2P$ M 384.413
Solid. Mp 277-279°.
Chloride: [21310-06-5].
$C_{25}H_{20}ClOP$ M 402.859
Cryst. (EtOAc). Mp 208-210°.

Brophy, J.J. *et al*, *J. Chem. Soc. (C)*, 1968, 2760.

1-Phenyl-2-(triphenylphosphoranylidene)-ethanone, 9CI P-00333

2-Triphenylphosphoranylideneacetophenone, 8CI.
Benzoylmethylidenetriphenylphosphorane
[859-65-4]

$Ph_3P{=}CHCOPh$

$C_{26}H_{21}OP$ M 380.425
Exists almost completely in the *cisoid*-(*Z*)-form. In 10% EtOH at pH 10, $t_{1/2}$ = ca. 22 hr. Wittig reagent. Also demonstrates hypoglycemic activity in rats. Cryst. (EtOAc, C_6H_6/Et_2O or C_6H_6/pet. ether). Mp 181-182°. pK_a 6.5 (50% EtOH).
▷AN0540000.

Ramirez, F. *et al*, *J. Org. Chem.*, 1957, **22**, 41 (*synth, props, uv*)
Bestmann, H.J., *Chem. Ber.*, 1962, **95**, 1513 (*synth, ir*)
Shevchuk, M.I. *et al*, *Zh. Obshch. Khim.*, 1972, **42**, 2630 (*Engl. transl.* p. 2621) (*ir, derivs*)
Wilson, I.W. *et al*, *J. Chem. Soc., Perkin Trans. 1*, 1972, 31 (*conformn, uv, pmr*)
Alexander, R.G. *et al*, *Org. Mass Spectrom.*, 1973, **7**, 963 (*ms*)
Blank, B. *et al*, *J. Med. Chem.*, 1975, **18**, 952 (*pharmacol*)
Senō, M. *et al*, *Bull. Chem. Soc. Jpn.*, 1975, **48**, 2001 (*pe, pmr, cmr, nmr*)
Albright, T.A. *et al*, *J. Am. Chem. Soc.*, 1976, **98**, 6249 (*cmr, nmr*)
Froeyen, P. *et al*, *Acta Chem. Scand.*, 1977, **31**, 256 (*nmr*)
Franza, G. *et al*, *J. Organomet. Chem.*, 1978, **157**, 299 (*nmr*)
Brittain, J.M. *et al*, *Tetrahedron*, 1979, **35**, 1139 (*nmr*)

1-Phenyl-1-(triphenylphosphoranylidene)-2-propanone P-00334

(α-Acetylbenzylidene)triphenylphosphorane
[76068-72-9]

$Ph_3P{=}CPhCOCH_3$

$C_{27}H_{23}OP$ M 394.452
A resonance-stabilized ylide. Cryst. (EtOH aq.). Mp 171°. Pyrolysis at 280°/10 mm gives $H_3CC{\equiv}CPh$. With $O_3 \rightarrow Ph_3PO + PhCOCOCH_3$.

Bestmann, H.J. *et al*, *Chem. Ber.*, 1962, **95**, 1513 (*synth*)
Gough, S.T.D. *et al*, *J. Chem. Soc.*, 1962, 2333 (*synth, props*)
Danion, D. *et al*, *Angew. Chem., Int. Ed. Engl.*, 1981, **20**, 113 (*use*)
Caminade, A.M. *et al*, *Phosphorus Sulfur*, 1983, **14**, 381 (*props*)

1-Phenyl-2-(triphenylphosphoranylidene)-1-propanone, 9CI P-00335

(1-Benzoylethylidene)triphenylphosphorane
[1450-07-3]

$Ph_3P{=}C(COPh)CH_3$

$C_{27}H_{23}OP$ M 394.452
Resonance-stabilised ylide. Cryst. (C_6H_6/Et_2O). Mp 170-172°. Pyrolysis at 200°→$MeC{\equiv}CPh$.

Bestmann, H.J. *et al*, *Chem. Ber.*, 1962, **95**, 1513 (*synth, ir*)
Wilson, I.F. *et al*, *J. Chem. Soc., Perkin Trans. 1*, 1972, 31 (*uv, pmr, struct*)
Blank, B. *et al*, *J. Med. Chem.*, 1975, **18**, 952 (*synth*)
Bestmann, H.J. *et al*, *Angew. Chem., Int. Ed. Engl.*, 1976, **15**, 298 (*props*)

1-Phenyl-3-(triphenylphosphoranylidene)-2-propanone, 9CI P-00336

2-Oxo-3-phenylpropylidenetriphenylphosphorane
[1174-61-4]

$Ph_3P{=}CHCOCH_2Ph$

$C_{27}H_{23}OP$ M 394.452
Stabilised ylide used in Wittig reaction. Cryst. (EtOAc/pet. ether). Mp 98-99°, 147-148°. Half-time ca. 15 hr. (10% EtOH aq., pH 10).

Bestmann, H.J. *et al*, *Chem. Ber.*, 1962, **95**, 1513 (*synth*)
Issleib, K. *et al*, *Justus Liebigs Ann. Chem.*, 1968, **713**, 12 (*props*)
Zeliger, H.I. *et al*, *Tetrahedron Lett.*, 1970, 3313 (*pmr*)
Bestmann, H.J. *et al*, *Chem. Ber.*, 1976, **109**, 1694 (*synth*)
Doleschall, G., *Synthesis*, 1981, 478 (*synth*)

Phenylvinylphosphinic acid P-00337

Ethenylphenylphosphinic acid, 9CI
[55743-26-5]

$H_2C{=}CHP(O)(OH)Ph$

$C_8H_9O_2P$ M 168.132
Cryst. (C_6H_6). Mp 89-90°, 120-121°. pK_a 2.26 (7% EtOH aq.), 3.53 (50% EtOH aq.), 4.29 (80% EtOH aq.).
Me ester: Methyl ethenylphenylphosphinate. Methyl phenylvinylphosphinate.
$C_9H_{11}O_2P$ M 182.158
d_4^{20} 1.19. Bp_1 93-95°. n_D^{20} 1.5336.
Et ester: [7766-52-1]. *Ethyl ethenylphenylphosphinate. Ethyl phenylvinylphosphinate.*
$C_{10}H_{13}O_2P$ M 196.185
Liq. d_4^{20} 1.11. $Bp_{0.5}$ 95-96°. n_D^{20} 1.5272.
Propyl ester: Propyl ethenylphenylphosphinate. Propyl phenylvinylphosphinate. Liq. Has been obt. in opt. active forms.
Ph ester: Phenyl ethenylphenylphosphinate. Phenyl phenylvinylphosphinate.
$C_{14}H_{13}O_2P$ M 244.229
d_4^{20} 1.12. Bp_1 130-133°. n_D^{20} 1.5391.
Chloride: [5290-57-3].
C_8H_8ClOP M 186.577
Liq. which rapidly hydrolyses in air. d_4^{20} 1.29. Bp_3 129-130°. n_D^{20} 1.5610.

Gefter, E.L. *et al*, *Zh. Obshch. Khim.*, 1961, **31**, 955 (*Engl. transl.* p. 883) (*synth, esters*)
Kabachnik, M.I. *et al*, *Zh. Obshch. Khim.*, 1963, **33**, 382 (*Engl. transl.* p. 375) (*synth, props*)
Gefter, E.L. *et al*, *Zh. Obshch. Khim.*, 1966, **36**, 79 (*Engl. transl.* p. 82) (*chloride*)
Neimysheva, A.A. *et al*, *Zh. Obshch. Khim.*, 1966, **36**, 1090 (*Engl. transl.* p. 1105) (*chloride*)
Gareev, R.D. *et al*, *Zh. Obshch. Khim.*, 1979, **49**, 503 (*Engl. transl.* p. 442) (*ethyl ester*)
Molinari, H. *et al*, *Synth. Commun.*, 1982, **12**, 749 (*ester*)

Phorate, BSI P-00338

O,O-*Diethyl* S-[*(ethylthio)methyl*] *phosphorodithioate, 9CI. Thimet*
[298-02-2]

$(EtO)_2P(S)SCH_2SEt$

$C_7H_{17}O_2PS_3$ M 260.364
Insecticide, acaricide. Mobile liq. d^{25} 1.167. $Bp_{0.8}$ 118-120°. n_D^{25} 1.5349. Metab. *via* the sulfoxide and sulfone.
▷V. highly toxic, TLV 0.05. TD9450000.
S-*Oxide:* [2588-03-6]. O,O-*Diethyl* S-[*(ethylsulfinyl)-methyl*] *phosphorodithioate.*
$C_7H_{17}O_3PS_2$ M 244.303
Metabolite of phorate. Liq. d_4^{25} 1.23. $Bp_{0.1}$ 150°. n_D^{30} 1.5365.
▷TD8750000.
S,S-*Dioxide:* [2588-04-7]. OO-*Diethyl* S-[*(ethylsulfonyl)methyl*] *phosphorodithioate.*
$C_7H_{17}O_4PS_2$ M 260.303
Metabolite of phorate. Liq. d_4^{25} 1.26. $Bp_{0.1}$ 150°. n_D^{30} 1.5251.
▷TD9100000.

U.S.P., 2 586 655, (*1952*); *CA*, **46**, 8144e
Bowman, J.S. *et al*, *J. Agric. Food Chem.*, 1957, **5**, 192 (*metab*)
Metcalf, R.L. *et al*, *J. Econ. Entomol.*, 1957, **50**, 338 (*metab*)
Keith, L.H. *et al*, *J. Assoc. Off. Anal. Chem.*, 1968, **51**, 1063 (*pmr*)

Holmstead, R.L. *et al*, *J. Assoc. Off. Anal. Chem.*, 1974, **57**, 1050 (*ms, metab*)
Szalontai, G., *J. Chromatogr.*, 1976, **124**, 9 (*hplc*)
Stan, H.-J. *et al*, *Fresenius' Z. Anal. Chem.*, 1977, **287**, 271; *Biomed. Mass Spectrom.*, 1982, **9**, 483 (*glc, ms*)
Ripley, B.-D. *et al*, *J. Assoc. Off. Anal. Chem.*, 1983, **66**, 1084 (*derivs, glc*)
Stan, H.-J. *et al*, *J. Chromatogr.*, 1983, **279**, 173 (*glc*)
Pesticide Manual, 6th Ed., 420; 7th Ed., 435.
The Agrochemicals Handbook, Royal Society of Chemistry, 1983, A323.
Sax, N.I., *Dangerous Properties of Industrial Materials*, 6th Ed., Van Nostrand-Reinhold, 1984, 907.

Phosalacine — P-00339

γ-(Hydroxymethylphosphinyl)-L-α-aminobutyryl-L-alanyl-L-leucine, 9CI. Phosphinothricylalanylleucine
[92567-89-0]

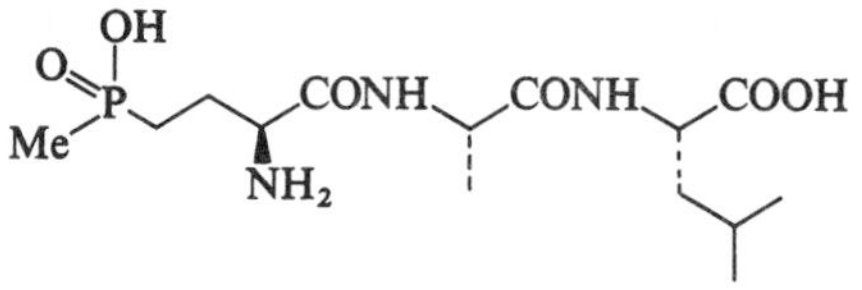

$C_{14}H_{28}N_3O_6P$ M 365.365

Oligopeptide antibiotic. Isol. from *Kitasatosporia phosalacinea.* Active against gram-positive and -negative bacteria and fungi on a chemically defined minimal medium. (Activity reversed by L-Glutamine). Herbicide. Inhibits glutamine synthetase. Amorph. powder. Mp >225° dec. $[\alpha]_D^{25}$ −38.8° (c, 0.65 in H_2O). pK_{a1} <3.0, pK_{a2} 4.3, pK_{a3} 8.3.

Omura, S. *et al*, *J. Antibiot.*, 1984, **37**, 829, 939 (*isol, uv, ir, pmr, cmr*)

Phosalone, BSI — P-00340

S-*[(6-Chloro-2-oxo-3(2*H*)-benzoxazolyl)methyl]* O,O-*diethyl phosphorodithioate, 9CI. Phosphorodithioic acid* O,O-*diethyl ester,* S-*ester with 6-chloro-3-(mercaptomethyl)-2-benzoxazolinone, 8CI.* S-*(6-Chloro-2-oxobenzoxazolin-3-yl)methyl diethyl phosphorodithioate. Zolone*
[2310-17-0]

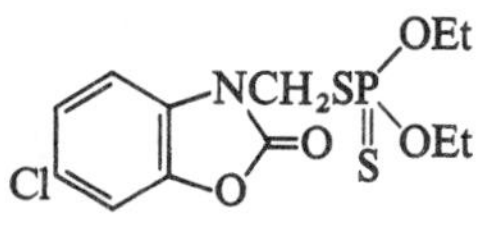

$C_{12}H_{15}ClNO_4PS_2$ M 367.802

Insecticide, acaricide, molluscicide. Insol. H_2O, hydrocarbons. Mp 47.5-48.0°.

▷TD5175000.

Colinese, D.L. *et al*, *Chem. Ind. (London)*, 1968, 1507 (*rev*)
Anderson, B.S., *Can. J. Spectrosc.*, 1974, **19**, 37 (*ir*)
Szalontai, G. *et al*, *J. Chromatogr.*, 1976, **124**, 9 (*hplc*)
Stan, H.-J. *et al*, *Fresenius' Z. Anal. Chem.*, 1977, **287**, 271; *Biomed. Mass Spectrom.*, 1982, **9**, 483 (*glc, ms*)
Ripley, B.D. *et al*, *J. Assoc. Offic. Anal. Chem.*, 1983, **66**, 1084 (*glc*)
Pesticide Manual, 6th Ed., 421.
The Agrochemicals Handbook, The Royal Society of Chemistry, 1983, A324.

Phosbutyl — P-00341

O-*Ethyl* S-*phenyl butylphosphoramidodithioate, 9CI*
[4205-52-1]

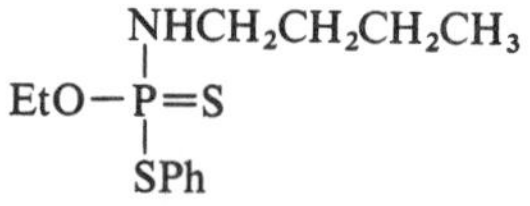

$C_{12}H_{20}NOPS_2$ M 289.390

▷TB4680000.

(±)-*form*

Fungicide. Oil. d_4^{20} 1.14. $Bp_{0.5}$ 127°. n_D^{20} 1.5680.

Mandel'baum, Ya.A. *et al*, *Zh. Obshch. Khim.*, 1967, **37**, 2540 (*Engl. transl.* p. 2417) (*synth*)
Abramova, G.L. *et al*, *CA*, 1974, **81**, 34462 (*use*)

Phosfolan, BSI — P-00342

Diethyl 1,3-dithiolan-2-ylidenephosphoramidate, 9CI. 2-(Diethoxyphosphinylimino)-1,3-dithiolan. Cyolane
[947-02-4]

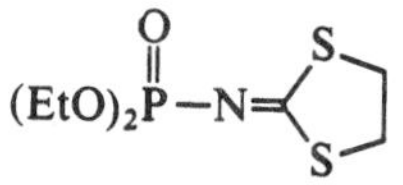

$C_7H_{14}NO_3PS_2$ M 255.286

Systemic insecticide. Colourless to yellow solid. Sol. H_2O, Me_2CO, C_6H_6, EtOH. Mp 39.5°. $Bp_{0.001}$ 100-105°. Stable at pH 2-9. n_D^{25} 1.5463.

▷Highly toxic. NJ6475000.

Belg. Pat., 618 155, (*1962*); *CA*, **59**, 10066f (*synth*)
Chia, C.Ku. *et al*, *ACS Symp. Ser.*, 1981, **158**, 97 (*metab*)
Abou-Donia, S.A. *et al*, *J. Chromatogr.*, 1982, **240**, 532 (*hplc*)
Fakhr, I.M.I. *et al*, *Isot. Rad. Res.*, 1982, **14**, 129 (*synth*)
Pesticide Manual, 6th Ed., 422; 7th Ed., 437.

Phosmet, BSI — P-00343

S-*[(1,3-Dihydro-1,3-dioxo-2*H*-isoindol-2-yl)methyl]* O,O-*dimethyl phosphorodithioate, 9CI.* O,O-*Dimethyl* S-*phthalimidomethyl phosphorodithioate. Imidan*
[732-11-6]

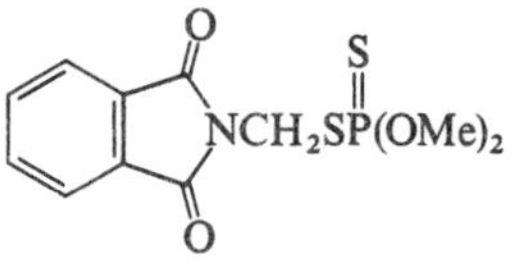

$C_{11}H_{12}NO_4PS_2$ M 317.314

Nonsystemic insecticide and acaricide. Solid. Mp 72-72.5°.

▷Fairly toxic. TE2275000.

Babad, H. *et al*, *Anal. Chim. Acta*, 1968, **41**, 259 (*pmr*)
Getz, M.E. *et al*, *J. Assoc. Off. Anal. Chem.*, 1968, **51**, 1101 (*tlc*)
Ross, R.T. *et al*, *Anal. Chim. Acta*, 1970, **52**, 139 (*P nmr*)
Stan, H.-J. *et al*, *Fresenius' Z. Anal. Chem.*, 1977, **287**, 271 (*glc, ms*)
Stan, H.-J. *et al*, *Biomed. Mass Spectrom.*, 1982, **9**, 483 (*ms*)
Stan, H.-J. *et al*, *J. Chromatogr.*, 1983, **279**, 173 (*glc*)
Agrochemicals Handbook, Royal Society of Chemistry, London, 1983, A325.
Pesticide Manual, 7th Ed., 438.

1-Phosphabicyclo[2.2.1]heptane, 9CI, 8CI **P-00344**

1-Phosphanorbornane

[29310-25-6]

$C_6H_{12}P$ M 115.135

p-*Oxide:* [40614-39-9].

$C_6H_{12}OP$ M 131.134

Hygroscopic prisms. Mp 207-210° (sealed tube).

Kenyon, G.L. *et al*, *J. Org. Chem.*, 1976, **41**, 2417 (*ms*)
Millbrath, D.S. *et al*, *J. Am. Chem. Soc.*, 1978, **100**, 3167 (*cryst struct, ir*)
Wetzel, R.B. *et al*, *J. Am. Chem. Soc.*, 1974, **96**, 5189 (*synth, ms, ir, pmr, cmr*)

1-Phosphabicyclo[3.3.1]nonane, 10CI **P-00345**

[61500-34-3]

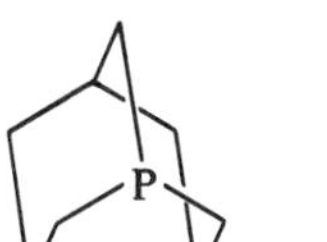

$C_8H_{15}P$ M 142.180

Cryst. Mp 169-173° (under Ar). Bp_{10} 80° subl. (bath).

B,MeI: 1-Methyl-1-phosphoniabicyclo[3.3.1]nonane iodide.

$C_9H_{18}IP$ M 284.119

Solid. Mp 427-429° (under Ar).

1-Oxide: [61500-41-2].

$C_8H_{15}OP$ M 158.180

Solid by subl. Mp 265-267°.

1-Sulfide: [61500-36-5].

$C_8H_{15}PS$ M 174.240

Solid. Mp 208-210°.

Krech, F. *et al*, *Z. Anorg. Allg. Chem.*, 1976, **425**, 209 (*synth, P nmr, pmr, ms, derivs*)

1-Phosphabicyclo[2.2.2]octane, 9CI **P-00346**

[280-35-3]

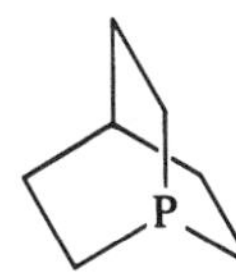

$C_7H_{13}P$ M 128.153

1-Oxide: [41809-52-3].

$C_7H_{13}OP$ M 144.153

Prisms. Mp 291-293° (sealed tube). Bp_1 80° subl.

Wetzel, R.B. *et al*, *J. Am. Chem. Soc.*, 1974, **96**, 5189 (*synth, ms, ir, cmr, pmr*)
Kenyon, G.L. *et al*, *J. Org. Chem.*, 1976, **41**, 2417 (*ms*)
Milbrath, D.S. *et al*, *J. Am. Chem. Soc.*, 1978, **100**, 3167 (*ir*)

Phosphamidon, BSI **P-00347**

2-Chloro-3-(diethylamino)-1-methyl-3-oxo-1-propenyl dimethyl phosphate, 9CI. *2-Chloro-2-diethylcarbamoyl-1-methylvinyl dimethyl phosphate. Dimecron*

[13171-21-6]

O
$(MeO)_2PO$ $CONEt_2$
C=C
H_3C Cl (*E*)-*form*

$C_{10}H_{19}ClNO_5P$ M 299.691

The technical prod. is a mixt. of 30% (*E*)- and 70% (*Z*)-isomer. Systemic insecticide. Pale-yellow liq. Sol. H_2O, most org. solvs. d_4^{25} 1.21. $Bp_{1.5}$ 162°, $Bp_{0.0001}$ 120°. n_D^{20} 1.4718.

▷Highly toxic. TC2800000.

(*E*)-*form* [297-99-4]

Isol. by glc.

▷More toxic than (*Z*)-form, and more powerful anti-cholinesterase

(*Z*)-*form* [23783-98-4]

Isol. by glc.

Anliker, R. *et al*, *Helv. Chim. Acta*, 1961, **44**, 1622 (*synth, ir, props*)
Getz, M.E. *et al*, *J. Assoc. Off. Anal. Chem.*, 1968, **51**, 1101 (*tlc*)
Keith, L.H. *et al*, *J. Assoc. Off. Anal. Chim.*, 1968, **51**, 1063 (*pmr*)
Ross, R.T. *et al*, *Anal. Chim. Acta*, 1970, **52**, 139 (*P nmr*)
Anliker, R. *et al*, *Residue Rev.*, 1971, **37** (*rev, bibl*)
Szalantai, G., *J. Chromatogr.*, 1976, **124**, 9 (*hlpc*)
Stan, H.-J. *et al*, *Fresenius' Z. Anal. Chem.*, 1977, **287**, 271 (*glc, ms*)
Stan, H.-J. *et al*, *Biomed. Mass Spectrom.*, 1982, **9**, 483 (*ms*)
Ripley, B.D. *et al*, *J. Assoc. Off. Anal. Chem.*, 1983, **66**, 1086 (*glc*)
Stan, H.-J. *et al*, *J. Chromatogr.*, 1983, **279**, 173 (*glc*)
Pesticide Manual, 6th Ed., 424; 7th Ed., 439.
The Agrochemicals Handbook, Royal Society of Chemistry, 1983, A326.
Sax, N.I., *Dangerous Properties of Industrial Materials*, 6th Ed., Van Nostrand-Reinhold, 1984, 491.

Phosphanthridene, 9CI, 8CI **P-00348**

9-Phosphaphenanthrene

[161-95-5]

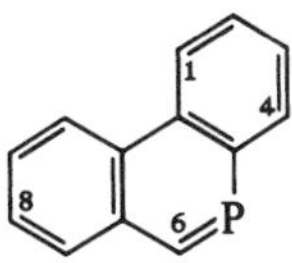

$C_{13}H_9P$ M 196.188

Known only in soln.

5,6-Dihydro: See 5,6-Dihydrophosphanthridene, D-00568

5,5-Dihydro, 5,5-di-Me: [82404-44-2].

$C_{15}H_{15}P$ M 226.257

A cyclic ylide. Needles (C_6H_6/pentane). Mp 160-162°.

DeKoe, P. *et al*, *Angew. Chem., Int. Ed. Engl.*, 1968, **7**, 465 (*synth*)
Costa, T. *et al*, *Chem. Ber.*, 1982, **115**, 1367 (*ylide, cmr, pmr, P nmr*)

5-Phospha(5-P^V)spiro[4.5]deca-1,3,5,7,9-pentaene, 9CI **P-00349**

[71466-49-4]

As yet unknown.

Böhm, M.C. *et al, J. Chem. Soc., Perkin Trans. 2*, 1979, 443 (*struct*)

5-Phospha(5-P^V)spiro[4.4]nonane, 9CI **P-00350**

[19549-57-6]

$C_8H_{17}P$ M 144.196

5-Me: [63702-97-6]. *5-Methyl-5-phospha(5-*P^V*)-spiro[4.4]nonane.*
$C_9H_{19}P$ M 158.223
Liq. with characteristic odour. Mp ca. −40°. Bp_1 46°.

5-Methoxy: [73410-82-9]. *5-Methoxy-5-phospha(5-*P^V*)spiro[4.4]nonane.*
$C_9H_{19}OP$ M 174.222
Liq. Bp_1 43°.

Schmidbaur, H. *et al, Z. Anorg. Allg. Chem.*, 1979, **458**, 249 (*synth, cmr, P nmr, pmr, ms*)

6-Phospha(6-P^V)spiro[5.5]undeca-1(6)-ene, 10CI **P-00351**

[75780-25-5]

A spirocyclic ylide. Cryst. (pentane). Mp 41°.

Schmidbaur, H. *et al, Z. Naturforsch., B*, 1980, **35**, 990 (*synth, cmr, P nmr, pmr*)
Schmidbaur, H. *et al, Chem. Ber.*, 1981, **114**, 3161 (*props*)

Phosphazomycin *A* **P-00352**

Struct. unknown

Isol. from *Streptomyces* sp. RK-803. Active against fungi and yeasts. Cryst. (EtOH aq.). Mp 155-158° dec. $[\alpha]_D^{20}$ +87.8° (c, 0.6 in MeOH). Mol. formula $C_{37-38}H_{56-60}NO_{12-13}P$.

Uramoto, M. *et al, J. Antibiot.*, 1985, **38**, 665 (*isol, uv, ir, pmr, cmr, props*)

Phosphemide **P-00353**

P,P-*Bis(1-aziridinyl)*-N-*2-pyrimidinylphosphinic amide, 9CI. Fosfemid*

[882-58-6]

$C_8H_{12}N_5OP$ M 225.189

Exhibits antitumour activity. Used experimentally and clinically. Cryst. (C_6H_6). Mp 128-129°.

▷SZ5835000.

Kropacheva, A.A. *et al, Zh. Obshch. Khim.*, 1961, **31**, 3601 (*Engl. transl.* p. 3357) (*synth*)
Noell, C.W. *et al, J. Med. Chem.*, 1967, **11**, 63 (*pharmacol, uv*)
Linberg, N.F. *et al, Khim. Geterosikl. Soedin*, 1975, 263 (*Engl. transl.* p. 226) (*ms, metab*)
Linberg, N.F. *et al, Khim.-Farm. Zh.*, 1976, **10**, 10 (*Engl. transl.* p. 563) (*metab, glc*)
Savin, Yu.I. *et al, Khim.-Farm. Zh.*, 1977, **11**, 3 (*metab*)

Phosphinecarbonitrile, 9CI **P-00354**

Phosphinous cyanide

[16777-05-2]

H_2PCN

CH_2NP M 59.007

Unstable, rearranges to Phosphinidynemethylamine, P-00356 .

Matveev, I.S., *Zh. Strukt. Khim.*, 1974, **15**, 145 (*Engl. transl.* p. 131); 1975, **16**, 1008 (*Engl. transl.* p. 926) (*struct*)

Phosphinetricarboxylic acid, 9CI **P-00355**

$P(COOH)_3$

$C_3H_3O_6P$ M 166.027

Esters are unstable towards H_2O_2/Me_2CO (giving H_3PO_2 and H_3PO_3), and to Br_2/EtOH (giving diethyl phosphonate). Aq. NaOH at 100° gives PH_3.

Tri-Et ester: [31081-90-0].
Triethoxycarbonylphosphine.
$C_9H_{15}O_6P$ M 250.188
Liq. Bp_1 130°. pK_a −10.9 (calc.). n_D^{20} 1.4678.

Tributyl ester: [31128-88-8].
Tributoxycarbonylphosphine.
$C_{15}H_{27}O_6P$ M 334.348
Liq. $Bp_{0.2}$ 147°. pK_a −10.4 (calc.). n_D^{20} 1.4646.

Tri-Ph ester: [31128-89-9].
Triphenoxycarbonylphosphine.
$C_{21}H_{15}O_6P$ M 394.320
Pale-yellow needles (EtOH). Spar. sol. CCl_4. Mp 125-126°.

Frank, A.W. *et al, J. Org. Chem.*, 1971, **36**, 3461 (*derivs, ir, pmr, nmr*)
Chervin, I.I. *et al, Izv. Akad. Nauk SSSR, Ser. Khim.*, 1981, 1769 (*Engl. transl.* p. 1438) (*triethyl ester, nmr*)

Phosphinidynemethylamine **P-00356**

1-Phosphinidynemethanamine, 9CI

[56764-36-4]

$H_2NC{\equiv}P$

CH_2NP M 59.007

Matveev, I.S., *Zh. Struct. Khim.*, 1974, **15**, 145 (*Engl. trans.* p. 131) (*synth, ir, nmr*)
Matveev, I.S., *Zh. Strukt. Khim.*, 1975, **16**, 1008 (*Engl. trans.* p. 926) (*struct, ir, pmr*)
Matveev, I.S., *CA*, 1980, **93**, 95347, 168341 (*props, derivs*)
Matveev, I.S., *CA*, 1981, **94**, 121641, 121642, 121643 (*use*)
Matveev, I.S. *et al, CA*, 1982, **97**, 163136 (*struct*)

3,3′,3″-Phosphinidynetrispropanoic acid, 10CI **P-00357**

Tris(3-carboxyethyl)phosphine

[5961-85-3]

$P(CH_2CH_2COOH)_3$

$C_9H_{15}O_6P$ M 250.188

Tri-Me ester: [29269-17-8].
$C_{12}H_{21}O_6P$ M 292.268
Oil. d_4^{20} 1.12. n_D^{20} 1.4868.
Tri-Me ester, oxide: [66250-56-4]. *3,3',3''-Phosphinylidynetrispropanoic acid trimethyl ester.*
$C_{12}H_{21}O_7P$ M 308.267
Yellow oil. Insol. CCl_4, hydrocarbons. d_4^{20} 1.25. n_D^{20} 1.4845.
Tri-Et ester: [67991-12-2].
$C_{15}H_{27}O_6P$ M 334.348
Light-yellow viscous oil. d_4^{20} 1.08. n_D^{20} 1.4803.
Tri-Et ester, oxide: [4115-99-5]. *3,3',3''-Phosphinyldynetrispropanoic acid triethyl ester.*
$C_{15}H_{27}O_7P$ M 350.348
Oil.

Yakovenko, T.V. *et al, Zh. Obshch. Khim.*, 1978, **48**, 1540 (*Engl. transl.* p. 1411) (*ethyl ester*)
Valetdinov, R.K. *et al, Zh. Obshch. Khim.*, 1978, **48**, 1726 (*Engl. transl.* p. 1577) (*methyl ester*)

3-Phosphinopropanenitrile, 9CI P-00358

2-Cyanoethylphosphine
[6783-71-7]

$H_2PCH_2CH_2CN$

C_3H_6NP M 87.061
Ligand for Fe, Ni, Pd, and Pt. Liq. Bp_9 54-59°. n_D^{25} 1.4831.

Rauhut, M.M. *et al, J. Am. Chem. Soc.*, 1959, **81**, 1103 (*synth*)
Maier, L., *Helv. Chim. Acta*, 1966, **49**, 1718 (*nmr*)

Phosphinothricin P-00359

2-Amino-4-(hydroxymethylphosphinyl)butanoic acid, 9CI. 2-Amino-4-(methylphosphino)butyric acid. Glufosinate, BSI
[51276-47-2]

COOH
H_2N–C–H O
$CH_2CH_2P(=O)(OH)$–Me (*S*)-*form*

$C_5H_{12}NO_4P$ M 181.128
Prob. exists as betaine.

(*S*)-*form* [35597-44-5]
L-form
N-Terminal residue of Antibiotic SF 1293, A-00141 . Shows herbicidal and acaricidal activity. Active glutamine synthetase inhibitor. Cryst. (EtOH). Diastereoisomers at P have been recognised but not characterised.

(±)-*form* [53369-07-6]
Cryst. (EtOH). Mp 241-242°. Forms a monohydrate.
B,HCl: Mp 195-198°.

Bayer, E. *et al, Helv. Chim. Acta*, 1972, **55**, 224 (*isol, pmr, ms, synth*)
Ogawa, Y. *et al, CA*, 1974, **81**, 37788m (*synth, ir, pmr*)
Gruszecka, E. *et al, Rocz. Chem.*, 1975, **49**, 2127; 1979, **53**, 937 (*synth*)
Gross, H. *et al, J. Prakt. Chem.*, 1976, **318**, 157 (*synth*)
Paulus, E.F. *et al, Z. Kristallogr., Kristallgeom., Kristallphys., Kristallchem.*, 1982, **160**, 39, 63 (*cryst struct*)
Seto, H. *et al, J. Antibiot.*, 1982, **35**, 1719 (*biosynth*)
Maier, L. *et al, Phosphorus*, 1983, **17**, 1, 21 (*synth, derivs, analogues*)
Minowa, N. *et al, Tetrahedron Lett.*, 1983, **24**, 2391; 1984, **25**, 1147 (*synth*)
Pesticide Manual, 7th Ed., 302.

[Phosphinylidynetris(methylene)]-trisphosphonic acid, 9CI P-00360

Tris(phosphonomethyl)phosphine oxide
[21852-02-8]

$O{=}P[CH_2P(O)(OH)_2]_3$

$C_3H_{12}O_{10}P_4$ M 332.017
Complexing agent. Viscous oil.
*Tris-*p*-toluidine salt:* Solid. Mp 203-204°.
Hexa-Me ester: [18788-36-8]. *Hexamethyl [phosphinylidynetris(methylene)]trisphosphonate.*
$C_9H_{24}O_{10}P_4$ M 416.178
Solid. Mp 169-171°.
Hexa-Et ester: [18788-35-7]. *Hexaethyl [phosphinylidynetris(methylene)]trisphosphonate.*
$C_{15}H_{36}O_{10}P_4$ M 500.338
Solid. Mp 168-170°.
Hexaisopropyl ester: Hexaisopropyl [phosphinylidynetris(methylene)]trisphosphonate.
$C_{21}H_{48}O_{10}P_4$ M 584.499
Solid. Mp 85-87°.

Medved, T.Ya. *et al, Izv. Akad. Nauk SSSR, Ser. Khim.*, 1968, 2062 (*Engl. transl.* p. 1953) (*synth, derivs*)
Maier, L. *et al, Helv. Chim. Acta*, 1969, **52**, 858 (*synth, pmr, P nmr, derivs*)
Anderegg, G., *Z. Naturforsch., B*, 1977, **32**, 547 (*complexes*)

Phosphirane, 9CI, 8CI P-00361

[6569-82-0]

C_2H_5P M 60.035
Mp −121.4° to −120.9°. Bp 36.5° (calcd.). Unstable to heat, dec. at 25°.
P-*Me:* [21658-91-3]. *1-Methylphosphirane.*
C_3H_7P M 74.062
Characterised spectroscopically.
P-*Ph: see 1-Phenylphosphirane, P-00214*

Wagner, R.I. *et al, J. Am. Chem. Soc.*, 1967, **89**, 1102 (*synth, ms, ir, pmr*)
Bowers, M.T. *et al, J. Am. Chem. Soc.*, 1969, **91**, 17 (*microwave struct*)
Chan, S. *et al, Tetrahedron*, 1969, **25**, 1097 (*deriv, ms, ir, P nmr, pmr*)
Mitchell, R.W. *et al, Spectrochim. Acta, Part A*, 1969, **25**, 819 (*ir, raman*)
Chan, S., *Spectrochim. Acta, Part A*, 1970, **26**, 249 (*ir*)
Albright, T., *Org. Magn. Reson.*, 1976, **8**, 489 (*cmr, P nmr, struct*)
Aue, D. *et al, J. Am. Chem. Soc.*, 1980, **102**, 5151 (*pe*)

N-Phosphoalanine P-00362

(1-Carboxyethyl)phosphoramidic acid. 2-Phosphoramidopropanoic acid. N-*Phosphonoalanine*

$H_3CCH(COOH)NHP(O)(OH)_2$

$C_3H_8NO_5P$ M 169.074

(±)-*form*
P,P-*Di-Me ester,* C-*amide: Dimethyl [1-(aminocarbonyl)ethyl]phosphoramidate.*
$C_5H_{13}N_2O_4P$ M 196.142
Solid. Mp 111-112°.
P,P-*Dibenzyl ester,* C-*amide: Dibenzyl [1-(aminocarbonyl)ethyl]phosphoramidate.*
$C_{17}H_{21}N_2O_4P$ M 348.338

Solid. Mp 97-99°.

P,P-*Dibenzyl ester,* P-*Me ester: Dibenzyl [1-(methoxycarbonyl)ethyl]phosphoramidate. Methyl 2-[(dibenzyloxyphosphinyl)amino]propanoate.*
$C_{18}H_{22}NO_5P$ M 363.349
Solid. Mp 40-41°.

C,P,P-*Tribenzyl ester: Dibenzyl [1-(benzyloxycarbonyl)ethyl]phosphoramidate. Benzyl 2-[(dibenzyloxyphosphinyl)amino]propanoate.*
$C_{24}H_{26}NO_5P$ M 439.447
Viscous oil.

Li, S.-O. *et al, J. Am. Chem. Soc.,* 1952, **74**, 5959; 1955, **77**, 1866 (*derivs*)

Phosphoarginine P-00363

N[5]-*[Imino(phosphonoamino)methyl]ornithine,* 9CI. N[5]-*(Phosphonoamidino)ornithine,* 8CI. N-*Phosphonoarginine*

COOH
H_2N–C–H OH
$(CH_2)_3NHCNHP{=}O$
NH OH

$C_6H_{15}N_4O_5P$ M 254.182
Mp 175-180°.

(*S*)-form [1189-11-3]
L-form
Constit. of crayfish muscle. Occurs in human brain. Energy store in invertebrates. Cryst. + $2H_2O$ (Me_2CO aq.). Mp 175-180°. Stored as Ba salt; rapidly dec. at −10° in free state.

P,P-*Dibenzyl ester,* C-*Me ester: Dibenzyl [[[(4-amino-4-methoxycarbonyl)butyl]amino](imino)methyl]-phosphoramidate.*
$C_{22}H_{32}N_4O_5P$ M 463.492
Solid. Mp 91-93°.

(±)-form
Cryst. (Me_2CO aq.). Mp 175-180°.

Si-Oh, L. *et al, J. Am. Chem. Soc.,* 1955, **77**, 1866 (*ester*)
Cramer, F. *et al, Chem. Ber.,* 1962, **95**, 1670 (*synth*)
Marcus, F. *et al, Biochem. J.,* 1964, **92**, 429 (*isol, synth*)
Poat, P.C. *et al, Biochem. Biophys. Acta,* 1980, **613**, 410 (*synth*)

N-Phosphocreatine P-00364

N-*[Imino(phosphonoamino)methyl]*-N-*methylglycine,* 9CI. *Creatine phosphate.* N[ω]-*Phosphonocreatine.* N-(N-*Phosphonoamidino)sarcosine*
[67-07-2]

$HN{=}C(N(Me)CH_2COOH)NHP(O)(OH)_2 \rightleftharpoons H_2N{-}C(N(Me)CH_2COOH){=}NP(O)(OH)_2$

$C_4H_{10}N_3O_5P$ M 211.114
Energy source in vertebrate muscle and nerve tissue. Isol. as Ca salt.

P,P-*Di-Ph ester:* N-*[Imino(diphenoxyphosphinylamino)methyl]*-N-*methylglycine.*
$C_{16}H_{18}N_3O_5P$ M 363.309
Cryst. (MeOH aq.). Mp 141-142°.

P,P-*Di-Ph ester,* C-*benzyl ester: Benzyl* N-*[Imino(diphenoxyphosphinylamino)methyl]*-N-*methylglycinate.*
$C_{23}H_{24}N_3O_5P$ M 453.433
Cryst. (MeOH aq.). Mp 103°.

Cramer, F. *et al, Chem. Ber.,* 1959, **92**, 392 (*synth*)
Berlet, H.H., *Anal. Biochem.,* 1974, **60**, 347 (*isol*)
Richard, R.E. *et al, J. Chem. Soc., Perkin Trans. 2,* 1974, 368 (*N nmr*)
Maudsley, A.A. *et al, J. Magn. Reson.,* 1983, **51**, 147 (*P nmr*)

N-Phosphocysteine P-00365

O COOH
$(HO)_2PNH$–C–H (*R*)-*form*
CH_2SH

$C_3H_8NO_5PS$ M 201.134

(*R*)-form
L-form

P,P-*Diisopropyl ester,* C-*Me ester: Methyl 2-[(diisopropoxyphosphinyl)amino]-3-mercaptopropanoate.*
$C_{10}H_{22}NO_5PS$ M 299.321
Solid. Mp 22° (softens).

(±)-form

P,P-*Dibenzyl ester,* C-*Me ester: Methyl 2-[(dibenzyloxyphosphinyl)amino]-3-mercaptopropanoate.*
$C_{18}H_{22}NO_5PS$ M 395.409
Solid. Mp 96-100°.

Plapinger, R.E. *et al, J. Am. Chem. Soc.,* 1953, **75**, 5757.
Li, S.-O. *et al, J. Am. Chem. Soc.,* 1955, **77**, 1866.

Phosphoenolpyruvic acid P-00366

2-(Phosphonooxy)-2-propenoic acid, 9CI. *2-Hydroxyacrylic acid dihydrogen phosphate,* 8CI
[138-08-9]

$H_2C{=}C(COOH)OPO_3H_2$

$C_3H_5O_6P$ M 168.043
Metab. intermed.

Monocyclohexylammonium salt: [10526-80-4]. Cryst. ($MeOH/Et_2O$). Mp 143-146° dec.

Tris(cyclohexylammonium) salt: [35556-70-8]. Solid. Mp 197-198° dec.

Et ester: [22065-56-1]. *Ethyl 2-(phosphonoxy)-2-propenoate.* Cryst. ($MeOH/Et_2O$) as biscyclohexylammonium salt. Mp 184-185°.

Amide, P,P-*dibenzyl ester: 2-[(Dibenzyloxyphosphinyl)-oxy]-2-propenoamide.*
$C_{17}H_{18}NO_5P$ M 347.307
Solid. Mp 48-50°.

Tris(trimethylsilyl) ester: Trimethylsilyl 2-[bis(trimethylsilyloxy)phosphinyl]oxy-2-propenoate.
$C_{12}H_{29}O_6PSi_3$ M 384.588
Liq. $Bp_{2.2}$ 112-116°.

Ferdman, D.L. *et al, Science,* 1940, **91**, 365 (*synth*)
Lardy, H.A. *et al, J. Biol. Chem.,* 1945, **159**, 343 (*biosynth*)
Biochem. Prep., 1966, **11**, 101 (*synth*)
Benkovic, S.J. *et al, Biochemistry,* 1968, **7**, 4090, 4097 (*props*)
Cohn, M. *et al, J. Am. Chem. Soc.,* 1970, **92**, 4095 (*pmr*)
Stubbs, J.A. *et al, Biochemistry,* 1972, **11**, 338 (*derivs*)
Harvey, D.J. *et al, J. Chromatogr.,* 1973, **76**, 51 (*gc*)
Watson, D.G. *et al, Acta Crystallogr., Sect. B,* 1973, **29**, 2358 (*cryst struct*)
Vogeli, U. *et al, Org. Magn. Reson.,* 1975, **7**, 617 (*ester, pmr, cmr*)
Davies, D.D., *Ann. Rev. Plant Physiol.,* 1979, **30**, 131 (*rev*)
Katti, S.K. *et al, Acta Crystallogr., Sect. B,* 1981, **37**, 834 (*cryst struct*)
Sekine, M. *et al, J. Chem. Soc., Perkin Trans. 1,* 1982, 2509 (*synth, pmr, derivs*)

Bartlett, P.A. *et al*, *J. Org. Chem.*, 1983, **48**, 3854 (*synth*)
Kluger, R. *et al*, *J. Am. Chem. Soc.*, 1984, **106**, 4017 (*derivs*)

N-Phosphoglutamic acid **P-00367**

2-[(Dihydroxyphosphinyl)amino]pentanedioic acid. 2-Phosphoramidopentanedioic acid. 2-Phosphoramidoglutaric acid. Phaseotoxin A. N*-Phosphonoglutamic acid*

$$\begin{array}{c} \overset{O}{\overset{\|}{(HO)_2PNH}} - \overset{COOH}{\overset{|}{C}} - H \\ \qquad\qquad | \\ \qquad\qquad CH_2CH_2COOH \end{array}$$

(*S*)*-form*

$C_5H_{10}NO_7P$ M 227.110

(S)-form [59360-03-1]

L-form

Active component of *Pseudomonas phaseolicata.*

P,P*-Di-Me ester,* C,C*-diamide: 2-[(Dimethoxyphosphinyl)amino]pentanedioic diamide.*
$C_7H_{16}N_3O_5P$ M 253.194
Solid. Mp 117-120° dec.

P,P*-Dibenzyl ester,* C,C*-di-Me ester: Dimethyl 2-[(dibenzyloxyphosphinyl)amino]pentanedioate.*
$C_{21}H_{27}NO_7P$ M 436.421
Viscous oil.

Tetrabenzyl ester: Dibenzyl 2-[(dibenzyloxyphosphinyl)-amino]pentanedioate.
$C_{33}H_{34}NO_7P$ M 587.608
Waxy solid. Mp 45-47°.

(±)-*form*

P,P*-Diphenyl ester,* C,C*-di-Et ester: Diethyl 2-[(diphenoxyphosphinyl)amino]pentanedioate.*
$C_{21}H_{26}NO_7P$ M 435.413
Solid. Mp 73.5-74°.

Sciarini, L.J. *et al*, *J. Am. Chem. Soc.*, 1949, **71**, 2940 (*synth*)
Li, S.-O. *et al*, *J. Am. Chem. Soc.*, 1955, **77**, 1866 (*synth*)
Patil, S.S. *et al*, *Biochem. Biophys. Res. Commun.*, 1976, **69**, 1019 (*isol, synth*)
Smith, A.G. *et al*, *Physiol. Plant Pathol.*, 1979, **15**, 269 (*synth, tlc, ir*)

N-Phosphoglycine **P-00368**

(Carboxymethyl)phosphoramidic acid. N*-Phosphonoglycine*

[5259-81-4]

$$(HO)_2P(O)NHCH_2COOH$$

$C_2H_6NO_5P$ M 155.047

Tri-Et ester: Triethyl phosphonoglycinate. Ethyl (diethoxyphosphinylamino)acetate.
$C_8H_{18}NO_5P$ M 239.208
Liq. $Bp_{0.3}$ 123-128°. n_D^{30} 1.4338.

Tripropyl ester: [69093-80-7]. *Tripropyl phosphonoglycinate. Propyl (dipropoxyphosphinylamino)acetate.*
$C_{11}H_{24}NO_5P$ M 281.288
Liq. d_4^{20} 1.09. $Bp_{0.04}$ 143-144°. n_D^{20} 1.4398.

Tribenzyl ester: Tribenzyl phosphonoglycinate. Benzyl (dibenzyloxyphosphinylamino)acetate.
$C_{23}H_{24}NO_5P$ M 425.420
Syrup which slowly cryst. Mp 143-144°.

Li, S.-O. *et al*, *J. Am. Chem. Soc.*, 1955, **77**, 1866 (*ester*)
Ives, T. *et al*, *J. Am. Chem. Soc.*, 1953, **75**, 5755 (*ester*)
Tomilets, V.A. *et al*, *Khim.-Farm. Zh.*, 1981, **15**, 34 (*Engl. transl.* p. 638) (*ester*)
Tsuhako, M. *et al*, *Chem. Pharm. Bull.*, 1982, **30**, 3882 (*synth*)

Phosphoglycocyamine **P-00369**

N$^{\omega}$*-Phosphonoglycocyamine.* N$^{\omega}$*-Phosphonoguanidinoacetic acid.* N*-(*N*-Phosphonoamidino)glycine.* N*-[Imino(phosphonoamino)methyl]glycine*

[5115-19-5]

$$HOOCCH_2NHC(NH_2){=}NP(O)(OH)_2$$

$C_3H_8N_3O_5P$ M 197.087

P,P*-Di-Ph ester:* N*-[Imino(diphenoxyphosphinylamino)methyl]glycine.*
$C_{15}H_{16}N_3O_5P$ M 349.282
Cryst. (EtOH aq.). Mp 144°.

P,P*-Di-Ph ester,* C*-Me ester: Methyl* N*-[imino(diphenoxyphosphinylamino)methyl]glycinate.*
$C_{16}H_{18}N_3O_5P$ M 363.309
Cryst. (C_6H_6/cyclohexane). Mp 125-126°.

P,P*-Di-Ph ester,* C*-Et ester: Ethyl* N*-[imino(diphenoxyphosphinylamino)methyl]glycinate.*
$C_{17}H_{20}N_3O_5P$ M 377.336
Cryst. (EtOH aq or C_6H_6/cyclohexane). Mp 122-123°.

P,P*-Di-Ph ester,* C*-Benzyl ester: Benzyl* N*-[imino(diphenoxyphosphinylamino)methyl]glycinate.*
$C_{22}H_{22}N_3O_5P$ M 439.407
Cryst. (C_6H_6/cyclohexane). Mp 103-104°.

Cramer, F. *et al*, *Chem. Ber.*, 1959, **92**, 392 (*synth*)

Phosphoguanidine **P-00370**

(Aminoiminomethyl)phosphoramidic acid. Guanidinephosphoric acid. N*-Amidinophosphoramidic acid.* N*-Phosphonoguanidine*

[3019-36-1]

$$HN{=}C(NH_2)NHP(O)(OH)_2$$

$CH_6N_3O_3P$ M 139.050
Cryst. (Me_2CO aq.). Mp 132°.

Di-Ph ester: Diphenyl (aminoiminomethyl)-phosphoramidate.
$C_{13}H_{14}N_3O_3P$ M 291.246
Cryst. (EtOH aq.). Mp 118°.

Dibenzyl ester: Dibenzyl (aminoiminomethyl)-phosphoramidate.
$C_{15}H_{18}N_3O_3P$ M 319.299
Cryst. (EtOH aq.). Mp 123°.

Cramer, F. *et al*, *Chem. Ber.*, 1958, **91**, 911 (*synth*)

Phospholane, 9CI **P-00371**

Tetramethylenephosphine. Phosphacyclopentane

[3466-00-0]

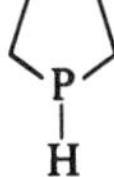

C_4H_9P M 88.089
Liq. Mp −88°. Bp 100-103°.

1-Chloro: see 1-Chlorophospholane, C-00167
1-Me: see 1-Methylphospholane, M-00280
1-Ph: see 1-Phenylphospholane, P-00216
1-Hydroxy, 1-oxide: see 1-Hydroxyphospholane 1-oxide, H-00185

Burg, A.B. *et al*, *J. Am. Chem. Soc.*, 1960, **82**, 2148 (*synth, ir*)
Sommer, K., *Z. Anorg. Allg. Chem.*, 1970, **379**, 56 (*synth*)
Weigert, F.J. *et al*, *Inorg. Chem.*, 1973, **12**, 313 (*cmr*)

1*H*-Phosphole, 8CI **P-00372**

Phosphacyclopentadiene

[288-01-7]

C_4H_5P M 84.057

Unknown. Rapidly forms dimer of 2*H*-form.

1-Me: see 2,5-Dihydro-3-methyl-1H-phosphole, D-00530

1-Ph: see 1-Phenyl-1H-phosphole, P-00217

Hase, H.L. *et al*, *Tetrahedron*, 1973, **29**, 469 (*struct*)

Palmer, M.H. *et al*, *J. Chem. Soc., Perkin Trans. 2*, 1974, 420; 1975, 974 (*struct*)

Hughes, A.N. *et al*, *J. Heterocycl. Chem.*, 1976, **13**, 1 (*bibl*)

Von Niessen, W. *et al*, *J. Am. Chem. Soc.*, 1976, **98**, 2066 (*struct*)

de Lauzon, G. *et al*, *J. Am. Chem. Soc.*, 1980, **102**, 994 (*derivs*)

Charrier, C. *et al*, *J. Am. Chem. Soc.*, 1983, **105**, 6871 (*P nmr, dimer*)

3,8-Phosphonanedione, 10CI **P-00373**

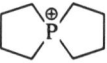

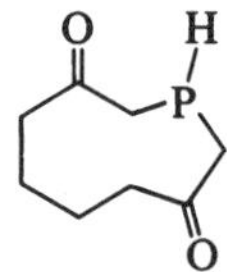

$C_9H_{13}O_3P$ M 200.174

1-Hydroxy, 1-oxide: [65114-89-8]. *1-Hydroxy-3,8-phosphonanedione 1-oxide.*

$C_9H_{13}O_4P$ M 216.173

Cryst. (MeCN). Mp 102-104°.

1-Me, 1-oxide: [65114-88-7]. *1-Methyl-3,8-phosphonanedione 1-oxide.*

$C_{10}H_{15}O_3P$ M 214.200

Cryst. (C_6H_6). Mp 130-132°.

1-Ph, 1-oxide: [80754-60-5]. *1-Phenyl-3,8-phosphonanedione 1-oxide.*

$C_{15}H_{17}O_3P$ M 276.271

Cryst. (EtOH). Mp 129-131°.

Quin, L.D. *et al*, *J. Am. Chem. Soc.*, 1982, **104**, 1893 (*synth, ir, cmr, P nmr, cryst struct*)

Quin, L.D. *et al*, *J. Org. Chem.*, 1982, **47**, 905.

5-Phosphoniaspiro[4.4]nonane, 9CI, 8CI **P-00374**

[176-50-1]

$C_8H_{16}P^{\oplus}$ M 143.188 (ion)

Chloride: [63702-96-5].

$C_8H_{16}ClP$ M 178.641

Needles (cyclohexane). Mp 231°.

Bromide: [20553-92-8].

$C_8H_{16}BrP$ M 223.092

Needles (cyclohexanone). Mp 236-237°.

Iodide: [20554-01-2].

$C_8H_{16}IP$ M 270.093

Needles (cyclohexanone). Mp 204-205°.

Tetrafluoroborate: [21162-48-1].

$C_8H_{16}BF_4IP$ M 356.896

Needles (cyclohexanone). Mp 136-137°.

Derkach, N.Ya. *et al*, *Zh. Obshch. Khim.*, 1968, **38**, 331 (*Engl. transl.* p. 332) (*synth*)

6-Phosphoniaspiro[5.5]undecane, 10CI **P-00375**

[181-07-7]

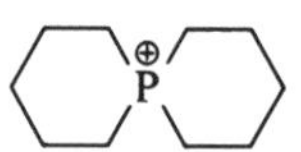

$C_{10}H_{20}P^{\oplus}$ M 171.242 (ion)

Chloride: [75780-28-8].

$C_{10}H_{20}ClP$ M 206.695

Cryst. ($CHCl_3/Et_2O/MeOH$). Mp 300° dec.

Schmidbaur, H. *et al*, *Z. Naturforsch., B*, 1980, **35**, 990 (*synth, cmr, P nmr, pmr*)

Phosphonoacetic acid **P-00376**

Dihydroxyphosphinylacetic acid, 9CI

[4408-78-0]

$(HO)_2P(O)CH_2COOH$

$C_2H_5O_5P$ M 140.032

Tribasic acid. Antiviral agent, used as Na salt. Rhombs. (AcOH). Insol. C_6H_6, $CHCl_3$, Et_2O. Mp 142-143°. pK_{a1} ~2.0, pK_{a2} 5.11, pK_{a3} 8.69.

Me ester: [40962-37-6]. *Methyl dihydroxyphosphinylacetate. Methyl phosphonoacetate.*

$C_3H_7O_5P$ M 154.059

No data available.

Et ester: [35752-46-6]. *Ethyl dihydroxyphosphinylacetate. Ethyl phosphonoacetate.*

$C_4H_9O_5P$ M 168.086

Viscous liq. Insol. C_6H_6, sl. sol. H_2O, EtOH, Et_2O.

Et ester, monoanilinium salt: Cryst. (Me_2CO). Mp 134-136°.

tert-*Butyl ester:*

$C_6H_{13}O_5P$ M 196.139

Cryst. (Me_2CO), as monoanilinium salt. Mp 115-117°.

Benzyl ester:

$C_9H_{11}O_5P$ M 230.157

Cryst. (Me_2CO), as monoanilinium salt. Mp 123-125°.

Tri-Me ester: see (Dimethoxyphosphinyl)acetic acid, D-00678

Tri-Et ester: see (Diethoxyphosphinyl)acetic acid, D-00241

Nylen, P., *Chem. Ber.*, 1924, **57**, 1023 (*synth*)

Maier, L. *et al*, *Phosphorus Sulfur*, 1978, **5**, 45 (*synth*)

Heubel, P.-H.C. *et al*, *J. Soln. Chem.*, 1979, **8**, 615 (*props, cmr, P nmr*)

Machida, Y. *et al*, *Synth. Commun.*, 1979, **9**, 97 (*ester, pmr*)

Fild, M. *et al*, *Chem. Ber.*, 1980, **113**, 142 (*synth, P nmr*)

Morita, T. *et al*, *Bull. Chem. Soc. Jpn.*, 1981, **54**, 267 (*esters, synth, pmr*)

Sekine, M. *et al*, *J. Org. Chem.*, 1981, **46**, 2097 (*ester*)

4-(Phosphonoamino)benzoic acid **P-00377**

N-*Phosphono-4-aminobenzoic acid. (4-Carboxyphenyl)-phosphoramidic acid*

$C_7H_8NO_5P$ M 217.118

Di-Ph ester: N-*(Diphenoxyphosphinyl)-4-aminobenzoic acid. Diphenyl (4-carboxyphenyl)phosphoramidate.*

$C_{19}H_{16}NO_5P$ M 369.313

Needles (H_2O). Mp 194-195°.

Cates, L.A. *et al*, *J. Pharm. Sci.*, 1964, **53**, 691 (*synth*)

2-Phosphonobenzoic acid, 9CI **P-00378**

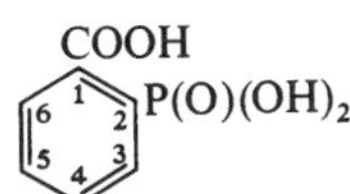

$C_7H_7O_5P$ M 202.103

Solid. Mp 175.5-179°.

Tri-Et ester: [28036-11-5]. *Ethyl [2-(diethoxyphosphinyl)]benzoate.*

$C_{13}H_{19}O_5P$ M 286.264

Liq. $Bp_{0.05}$ 132-134°. n_D^{20} 1.4953.

Nitrile: see (2-Cyanophenyl)phosphonic acid, C-00223

Freedman, L.D. *et al*, *J. Am. Chem. Soc.*, 1953, **75**, 1379 (*synth*)
Tavs, P., *Chem. Ber.*, 1970, **103**, 2428.
Hirao, T. *et al*, *Bull. Chem. Soc. Jpn.*, 1982, **55**, 909.

3-Phosphonobenzoic acid, 9CI **P-00379**

[14899-31-1]

$C_7H_7O_5P$ M 202.103

Needles (EtOH). Mp 245-246°. pK_{a1} 1.85 (H_2O), 3.13 (50% EtOH aq.), pK_{a2} 5.03 (H_2O).

Tri-Et ester: [26342-16-5]. *Ethyl [3-(diethoxyphosphinyl)]benzoate.*

$C_{13}H_{19}O_5P$ M 286.264

Liq. $Bp_{0.25}$ 149-151°. n_D^{20} 1.4936.

Trichloride: (3-Chloroformylphenyl)phosphonic dichloride.

$C_7H_4Cl_3O_2P$ M 257.440

Solid. Mp 61°.

Michaelis, A., *Justus Liebigs Ann. Chem.*, 1896, **293**, 261 (*synth, trichloride*)
Tavs, P., *Chem. Ber.*, 1970, **103**, 2428 (*ester*)
Ewen, G.D. *et al*, *J. Chem. Res. (S)*, 1983, 14 (*ester, synth, nmr*)

4-Phosphonobenzoic acid, 9CI **P-00380**

[618-21-3]

$C_7H_7O_5P$ M 202.103

Cryst. (H_2O). Mp 377-379°. pK_{a1} 1.8 (H_2O), 3.10 (50% EtOH aq.), pK_{a2} 6.89 (H_2O).

Et ester: Ethyl 4-phosphonobenzoate.

$C_9H_{11}O_5P$ M 230.157

Needles. Mp 78°.

Tri-Et ester: [17067-92-4]. *Ethyl [4-(diethoxyphosphinyl)]benzoate.*

$C_{13}H_{19}O_5P$ M 286.264

Liq. $Bp_{0.05}$ 147-148°. n_D^{20} 1.4946.

Trichloride: (4-Chlorocarbonyl)phenylphosphonic dichloride.

$C_7H_4Cl_3O_2P$ M 257.440

Plates (CCl_4/Et_2O). Mp 84.5-85.5°.

Carboxamide: 4-Phosphonobenzamide.

$C_7H_8NO_4P$ M 201.118

Solid. Mp >300°.

Trianilide: P-*(4-*N*-phenylcarbamoylphenyl)-*N,N′*-diphenylphosphonic diamide.*

$C_{25}H_{22}N_3O_2P$ M 427.441

Cryst. (EtOH). Mp 242°.

Michaelis, A., *Justus Liebigs Ann. Chem.*, 1896, **293**, 261 (*synth, derivs*)
Jaffé, H.H. *et al*, *J. Am. Chem. Soc.*, 1953, **75**, 2209 (*props*)
Tavs, P., *Chem. Ber.*, 1970, **103**, 2428 (*esters*)
Fields, E.K. *et al*, *Chem. Ind. (London)*, 1960, 999 (*synth*)
Kharrasova, F.M. *et al*, *Zh. Obshch. Khim.*, 1973, **43**, 2642 (*Engl. transl.* p. 2621) (*trichloride*)

Phosphonobutanedioic acid, 9CI **P-00381**

Phosphonosuccinic acid

[5768-48-9]

$HOOCCH_2CH(COOH)P(O)(OH)_2$

$C_4H_7O_7P$ M 198.069

(±)-*form*

Di-Et ester: Diethyl phosphonobutanedioate. Diethyl phosphonosuccinate. Diethyl (dihydroxyphosphinyl)-succinate.

$C_8H_{15}O_7P$ M 254.176

Liq. Bp 265-268°.

Pudovik, A.N. *et al*, *Zh. Obshch. Khim.*, 1959, **29**, 3338 (*Engl. transl.* p. 3301)

2-Phosphono-1,2,4-butanetricarboxylic acid, 9CI **P-00382**

[37971-96-1]

$(HO)_2P(=O)-C(COOH)(CH_2COOH)(CH_2CH_2COOH)$

$C_7H_{11}O_9P$ M 270.132

(±)-*form*

Complexing agent. Component of dentifrices.

Penta-Me ester: Trimethyl (2-dimethoxyphosphinyl)-1,2,4-butanetrioate. Pentamethyl 2-phosphono-1,2,4-butanetricarboxylate.

$C_{12}H_{21}O_9P$ M 340.266

Oil. Bp_1 175-183°.

Ger. Pat., 2 061 838, (*1972*); *CA*, **77**, 114565 (*synth*)
U.S.P., 4 348 381, (*1981*); *CA*, **97**, 188116 (*use*)

3-Phosphono-1,2,3-butanetricarboxylic acid, 9CI **P-00383**

[51757-41-6]

$HOOCCH_2-CH(COOH)-C(COOH)(CH_3)-P(=O)(OH)_2$

$C_7H_{11}O_9P$ M 270.132

Sterochem. not defined. Complexing agent. Hygroscopic cryst. + $1H_2O$. Mp 88-90°.

O^1,O^2,P,P-*Tetra-Et*, O^3-*Me ester:* O^1,O^2-*Diethyl* O^3-*methyl 3-(diethoxyphosphinyl)-1,2,3-butanetrioate.*

$C_{16}H_{29}O_9P$ M 396.373

Oil. $Bp_{0.05}$ 164-170°.

Penta-Me ester: Trimethyl 3-(dimethoxyphosphinyl)-1,2,3-butanetrioate.

$C_{12}H_{21}O_9P$ M 340.266

Oil. $Bp_{0.075-0.1}$ 180-190°.

Worms, K.-H. *et al*, *Z. Anorg. Allg. Chem.*, 1979, **457**, 219 (*synth*)

4-Phosphono-1,2,3-butanetricarboxylic acid, 9CI **P-00384**

[51395-43-8]

HOOCCH$_2$CH(COOH)CH(COOH)CH$_2$P(O)(OH)$_2$

$C_7H_{11}O_9P$ M 270.132
Sterochem. not defined. Complexing reagent. Solid. Mp 154-155°.

Worms, K.-H. *et al*, *Z. Anorg. Allg. Chem.*, 1979, **457**, 219 (*synth*)

3-Phosphonobutanoic acid, 9CI P-00385

[4422-66-6]

(HO)$_2$P(O)CH(CH$_3$)CH$_2$COOH

$C_4H_9O_5P$ M 168.086

(±)-*form*
Solid. Mp 145°.
Et ester: Ethyl 3-phosphonobutanoate.
$C_6H_{13}O_5P$ M 196.139
No phys. props. reported.
Et ester, monoanilinium salt: Solid. Mp 150-152°.

Sekine, M. *et al*, *Chem. Lett.*, 1977, 485 (*synth*)

4-Phosphonobutanoic acid, 9CI P-00386

(3-Carboxypropyl)phosphonic acid
[4378-43-2]

(HO)$_2$P(O)CH$_2$CH$_2$CH$_2$COOH

$C_4H_9O_5P$ M 168.086
Cryst. (H_2O or THF/$CHCl_3$). Mp 127-128.5°.

Nylen, P., *Ber.*, 1926, **59**, 1119 (*synth*)
Wasielewski, C. *et al*, *Synthesis*, 1981, 540 (*synth*)

Phosphonoformic acid P-00387

Dihydroxyphosphinecarboxylic acid oxide, 9CI
[4428-95-9]

(HO)$_2$P(O)COOH

CH_3O_5P M 126.005
Oral, topical, and nasal prepns., or salts, are useful for treating viral infections, e.g. influenza and herpes, in man. pK_{a1} 0.49 (P-OH), pK_{a2} 7.27 (P-OH), pK_{a3} 3.41 (COOH).
Tri-Na salt: [63585-09-1]. *Foscarnet sodium.* More effective than esters against herpes virus DNA polymerase. Cryst. + $6H_2O$.
▷SY8200000.
C-*Me ester:* [55920-68-8]. *Methyl (dihydroxyphosphinyl)formate.*
$C_2H_5O_5P$ M 140.032
Na salts used as plant growth regulators.
C-*Et ester:* [55920-71-3]. *Ethyl (dihydorxyphosphinyl)-formate.*
$C_3H_7O_5P$ M 154.059
Salts used as plant growth regulators.
C-*Et ester, dianilinium salt:* Cryst. (EtOH). Mp 143-145°.
C,P-*di-Et ester, anilinium salt:*
$C_{11}H_{18}NO_5P$ M 275.241
Solid. Mp 163° dec.
P,P-*di-Et ester: see Diethoxyphosphinecarboxylic acid oxide, D-00236*

Naqvi, R.R., *CA*, 1978, **89**, 107672 (*struct*)
Heubel, P-H.C. *et al*, *J. Solution Chem.*, 1979, **8**, 615 (*props, cmr, nmr*)
McKenna, C.E. *et al*, *J. Chem. Soc., Chem. Commun.*, 1979, 739 (*ester, props*)
Sekine, M. *et al*, *Bull. Chem. Soc. Jpn.*, 1982, **55**, 239 (*ester*)

N^5-Phosphonomethionine sulfoximine P-00388

2-Amino-4-(S-methyl-N-phosphonosulfonimidoyl)-butanoic acid, 9CI

COOH
H$_2$N►C◄H
CH$_2$CH$_2$S(=O)(Me)=N–P(O)(OH)–OH

$C_5H_{13}N_2O_6P$ M 228.141

(S,S)-form [21869-93-2]
Prod. by an unclassified *Streptomyces* sp. Present as a tripeptide with 2 Ala.

Preuss, D.L. *et al*, *J. Antibiot.*, 1973, **26**, 261.

(*N*-Phosphono)methionine-*S*-sulfoximinylalanylalanine P-00389

α-Amino-γ-(S-methyl-N-phosphonosulfonimidoyl)-butyrylalanylalanine, 9CI

(HO)$_2$P(O)N=S(O)(Me)CH$_2$CH$_2$CH(NH$_2$)CONHCH(CH$_3$)CONHCH(CH$_3$)COOH (L, L)

$C_{11}H_{23}N_4O_8P_5$ M 494.194
Peptide antibiotic.

all-L-form [41928-08-9]
From *Streptomyces* sp. X 13152. Deactivator for glutamine synthetase. Active against gram-positive bacteria. Amorph. Mp 292° dec.

Pruess, D.L. *et al*, *J. Antibiot.*, 1973, **26**, 261 (*isol*)
Diddens, H. *et al*, *J. Antibiot.*, 1979, **32**, 87 (*props*)

Phosphonooxyacetic acid, 9CI P-00390

Phosphoglycolic acid. Carboxymethyl phosphate
[13147-57-4]

HOOCCH$_2$OP(O)OH)$_2$

$C_2H_5O_6P$ M 156.032
Intermediate in photorespiration.

Andrews, T.J. *et al*, *Biochemistry*, 1973, **12**, 11 (*biosynth*)
Bassham, J.A. *et al*, *Plant Physiol.*, 1973, **52**, 407.
Takabe, T. *et al*, *Biochem. Biophys. Res. Commun.*, 1973, **53**, 1173 (*synth*)
Kent, S.S. *et al*, *J. Chromatogr.*, 1979, **177**, 372 (*hplc*)
Beutler, E. *et al*, *Anal. Biochem.*, 1980, **106**, 163 (*synth*)

2-(Phosphonooxy)benzoic acid, 9CI P-00391

Salicyclic acid dihydrogen phosphate, 8CI. Fosfosal
[6064-83-1]

(structure: benzene ring with HOOC and OP(O)(OH)–OH)

$C_7H_7O_6P$ M 218.102
Nonnarcotic analgesic with antiinflammatory and antipyretic props. Solid. Mp 168-170°.
▷VO2770000.
Mono-Na salt: Solid. Mp 172-175°.
Di-Na salt: Solid. Mp >350°.

Tri-Na salt: Solid. Mp >350°.
Tri-Me ester: Methyl 2-(dimethoxyphosphinyl)-oxybenzoate.
$C_{10}H_{13}O_6P$ M 260.183
Liq. $Bp_{0.05}$ 132°. n_D^{29} 1.502.
Tri-Et ester: [32565-98-3]. *Ethyl 2-(diethoxyphosphinyl)oxybenzoate. Salicyclic acid ethyl ester, diethyl phosphate, 8CI.*
$C_{13}H_{19}O_6P$ M 302.263
Liq. d_4^{20} 1.19. $Bp_{0.008}$ 126-128°. n_D^{20} 1.4832.

Bowers, G.N. *et al, Clin. Chem.*, 1967, **14**, 608 (*uv*)
Pashinkin, A.P. *et al, Izv. Akad. Nauk SSSR, Ser. Khim.*, 1971, 437 (*CA*, **75**, 48604) (*ester*)
Ger. Pat., 2 641 526, (*1978*); *CA*, **88**, 136317 (*synth, tox, use*)
Rafanell, J.G. *et al, Arzneim.-Forsch.*, 1980, **30**, 1091 (*pharmacol*)
Sanchez, M.S. *et al, Arzneim.-Forsch*, 1980, **30**, 1098 (*tox*)

2-Phosphonooxypropanoic acid, 9CI **P-00392**

Phospholactic acid. 1-Carboxyethyl phosphate
[18365-82-7]

$(HO)_2P(O)OCH(CH_3)COOH$

$C_3H_7O_6P$ M 170.058

(±)-*form* [32450-46-7]
Me ester: [54857-00-0].
$C_4H_9O_6P$ M 184.085
Cryst. (EtOH). Mp 177-180°.
Tri-Me ester: see under 2-[(Dimethoxyphosphinyl)oxy]-propanoic acid, D-00682
Tri-Et ester: see under 2-[(Diethoxyphosphinyl)oxy]-propanoic acid, D-00255

Nowak, T. *et al, Biochemistry*, 1972, **11**, 2813, 2819; 1973, **12**, 1690.
Damiani, A. *et al, J. Theor. Biol.*, 1975, **52**, 383 (*conformn*)
Breathnach, R. *et al, Biochemistry*, 1977, **16**, 3054 (*tribenyl ester, pmr, P nmr*)

α-Phosphonophenylacetic acid **P-00393**

Phosphonobenzeneacetic acid
[38654-93-2]

$(HO)_2P(O)CHPhCOOH$

$C_8H_9O_5P$ M 216.130

(±)-*form*
Weak inhibitor of DNA polymerases. Hygroscopic solid.
4-Methylphenylammonium salt: Solid. Mp 202°.
Tri-Et ester: see under α-(Diethoxyphosphinyl)-benzeneacetic acid, D-00242

Kreutzkamp, N. *et al, Arch. Pharm. (Weinheim, Ger.)*, 1962, **295**, 276 (*synth*)
Derse, D. *et al, J. Biol. Chem*, 1982, **257**, 10251 (*props*)
Ericksson, B. *et al, Biochem. Biophys. Acta*, 1982, **696**, 115 (*props*)

2-Phosphono-2-(2-phosphonoethyl)-butanedioic acid, 9CI **P-00394**

2,4-Diphosphono-1,2-butanedicarboxylic acid. 2-Phosphono-2-(2-phosphonoethyl)succinic acid
[65402-32-6]

```
         O
         ‖
         P(OH)2   O
         |        ‖
HOOCCH2CCH2CH2P(OH)2
         |
         COOH
```

$C_6H_{12}O_{10}P_2$ M 306.102

Complexing agent.
Hexa-Et ester: Diethyl 2-diethoxyphosphinyl-2-[2-(diethoxyphosphinyl)ethyl]butanedioate.
$C_{18}H_{36}O_{10}P_2$ M 474.424
Liq. $Bp_{0.15}$ 212-220°.

Worms, K.-H. *et al, Z. Anorg. Allg. Chem.*, 1979, **457**, 219 (*synth*)

2-Phosphonopropanoic acid, 9CI **P-00395**

[5962-41-4]

$(HO)_2P(O)CH(CH_3)COOH$

$C_3H_7O_5P$ M 154.059

(±)-*form*
Inhibitor of some DNA polymerases incl. that of herpes simplex virus. Syrup which slowly cryst. Mp 75-95°, 105-109°, 119-132°. pK_{a1} 1.8, pK_{a2} 5.15, pK_{a3} 8.54 (H_2O, 25°).

Heubel, P.H.C. *et al, J. Soln. Chem.*, 1979, **8**, 615 (*props*)
Eriksson, B. *et al, Biochem. Biophys. Acta*, 1982, **696**, 115 (*props*)

3-Phosphonopropanoic acid, 9CI **P-00396**

(2-Carboxyethyl)phosphonic acid
[5962-42-5]

$HOOCCH_2CH_2P(O)(OH)_2$

$C_3H_7O_5P$ M 154.059
Cryst. (AcOH/$CHCl_3$). Mp 167°, Mp 178°. pK_{a1} 2.26, pK_{a2} 4.63, pK_{a3} 7.75 (H_2O, 25°).
▷May be cytotoxic
Mono-p-*toluidinium salt:* Cryst. (EtOH). Mp 158°.
Et ester: [72563-46-3]. *Ethyl 3-phosphonopropanoate.*
$C_5H_{11}O_5P$ M 182.113
Plates (C_6H_6). Mp 64.5°.
Triethyl ester: see 3-(Diethoxyphosphinyl)propanoic acid, D-00259
P,P-*Dibutyl O-Me triester: Methyl 3-(dibutoxyphosphinyl)propanoate.*
$C_{12}H_{25}O_5P$ M 280.300
Liq. d_4^{20} 1.06. $Bp_{0.001}$ 78-90°. n_D^{20} 1.4383.

Nylén, P., *Ber.*, 1926, **59**, 1119 (*synth, deriv*)
Bochwic, B. *et al, Rocz. Chem.*, 1951, **25**, 338 (*synth, derivs*)
Ginsberg, V.A. *et al, Zh. Obshch. Khim.*, 1960, **30**, 3987 (*Engl. transl.* p. 3944) (*synth*)
Kreutzkamp, N. *et al, Arch. Pharm. (Weinheim, Ger.)*, 1962, **295**, 773 (*synth*)
Heubel, P.H.C. *et al, J. Soln. Chem.*, 1979, **8**, 615 (*props*)

4-(Phosphonoxy)benzoic acid, 9CI **P-00397**

[53497-47-5]

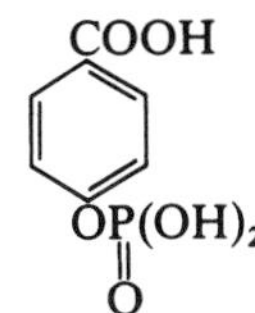

$C_7H_7O_6P$ M 218.102
Et ester: [75378-51-7]. *Ethyl 4-(phosphonoxy)benzoate.*
$C_9H_{11}O_6P$ M 246.156
Isol. as bis(cyclohexylammonium salt).
Et ester, bis(cyclohexylammonium) salt: Prisms (EtOH aq.). Mp 163-164°.

Blackburn, G.M. *et al, J. Chem. Soc., Perkin Trans. 1*, 1980, 1150 (*deriv, synth, pmr*)

3-Phosphonoxypropanenitrile, 9CI **P-00398**

Hydracrylonitrile dihydrogen phosphate (ester), 8CI. *2-Cyanoethyl dihydrogen phosphate. (2-Cyanoethyl) phosphoric acid. Mono(2-cyanoethyl) phosphate. 2-Cyanoethyl phosphate*

[2212-88-6]

$NCCH_2CH_2OP(O)(OH)_2$

$C_3H_6NO_4P$ M 151.058

Isol. as Ba salt trihydrate or organic amine salts. Phosphorylating agent.

Monocyclohexylammonium salt: [72236-74-9]. Solid. Mp 100-103°.

Bis(cyclohexylammonium) salt: Cryst. + $0.5H_2O$ (MeOH/EtOAc). Mp 165°.

Monoanilinium salt: Cryst. (EtOH). Mp 154-155°.

Cherbuliez, E. *et al*, *Helv. Chim. Acta*, 1956, **39**, 1455 (*synth*)
Tener, G.M. *et al*, *J. Am. Chem. Soc.*, 1961, **83**, 159 (*synth, use*)
Obata, T. *et al*, *J. Org. Chem.*, 1967, **32**, 1063 (*synth, deriv*)
Markowska, A. *et al*, *Bull. Acad. Polon. Sci., Ser. Sci. Chim.*, 1979, **27**, 115 (*synth*)
Fieser, M. *et al*, *Reagents for Organic Synthesis*, Wiley, 1967-84, **1**, 172.

3-Phosphonoxypropanoic acid, 9CI **P-00399**

Mono(2-carboxyethyl) phosphate

[61777-92-2]

$HOOCCH_2CH_2OP(O)(OH)_2$

$C_3H_7O_6P$ M 170.058

C-*Benzyl* P,P-*di-Ph ester: Benzyl 3-[(diphenoxyphosphinyl)oxy]propanoate.*
$C_{22}H_{21}O_6P$ M 412.378
No phys. props. described.

Breathnach, R. *et al*, *Biochemistry*, 1977, **16**, 3054 (*synth, props, derivs, pmr, P nmr*)

***N*-Phosphophenylalanine** **P-00400**

3-Phenyl-2-(phosphonoamino)propanoic acid. N-*Phosphonophenylalanine*

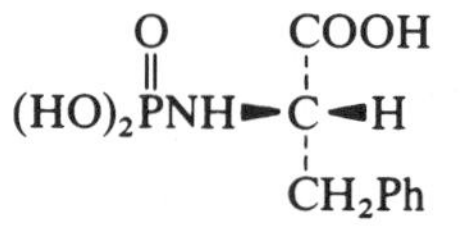

$(HO)_2P(O)NH-C(H)(COOH)-CH_2Ph$ (*S*)-*form*

$C_9H_{12}NO_5P$ M 245.171

(*S*)-*form*

L-form

P,P-*Di-Ph ester,* C-*benzyl ester: Benzyl 2-[(diphenoxyphosphinyl)amino]-3-phenylpropanoate.*
$C_{28}H_{26}NO_5P$ M 487.491
Cryst. (Et_2O). Mp 86°. $[\alpha]_D^{21}$ −5.2° (CCl_4).

(±)-*form*

Cryst. (MeOH/Et_2O). Mp 143-145°, Mp 163-164° dec.

P,P-*Di-Me ester,* C-*amide: 2-[(Dimethoxyphosphinyl)amino]-3-phenylpropanamide.*
$C_{11}H_{17}N_2O_4P$ M 272.240
Solid. Mp 148-149°.

P,P-*Di-Ph ester,* C-*Et ester: Ethyl 2-[(diphenoxyphosphinyl)amino]-3-phenylpropanoate.*
$C_{23}H_{24}NO_5P$ M 425.420
Solid. Mp 78-79°.

P,P-*Dibenzyl ester,* C-*Me ester: Methyl 2-[(dibenzyloxyphosphinyl)amino]-3-phenylpropanoate.*
$C_{24}H_{21}NO_6P$ M 450.407
Solid. Mp 82-83°.

P,P-*Diphenyl ester,* C-*benzyl ester:* Cryst. (Et_2O). Mp 90-91°.

Tribenzyl ester: Benzyl 2-[(dibenzyloxyphosphinyl)amino]-3-phenylpropanoate. Tribenzyl N-*phosphonophenylalaninate.*
$C_{30}H_{30}NO_5P$ M 515.544
Solid. Mp 67-69°.

Sciarini, L.J. *et al*, *J. Am. Chem. Soc.*, 1949, **71**, 2940 (*synth*)
Li, S.-O., *J. Am. Chem. Soc.*, 1952, **74**, 5959; 1955, **77**, 1866 (*synth*)
Clark, V.M. *et al*, *Tetrahedron, Suppl. 7*, 1966, 307 (*synth*)

Phosphophenylalanylarginine **P-00401**

FMP 1. Antibiotic FMP 1

[82064-34-4]

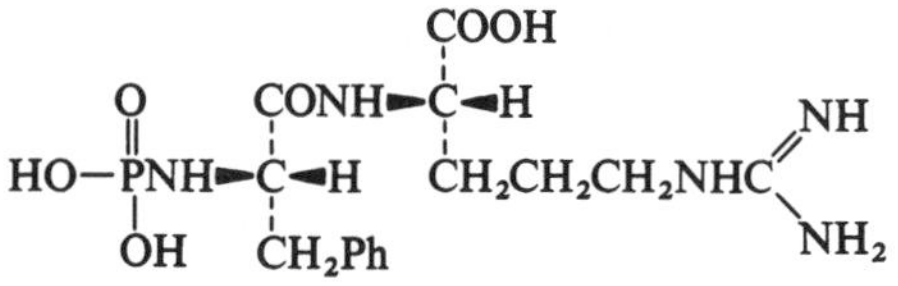

$HO-P(O)(OH)-NH-C(H)(CH_2Ph)-CONH-C(H)(COOH)-CH_2CH_2CH_2NHC(=NH)NH_2$

$C_{15}H_{24}N_5O_6P$ M 401.358

Peptide antibiotic.

L-L-form

Prod. by *Streptomyces rishiriensis*. Inhibits metalloproteinase. Powder. λ_{max} 247, 252, 258, 260 (sh), 264 268 nm (0.1*M* MeOH).

Murao, S. *et al*, *Agric. Biol. Chem.*, 1982, **46**, 855, 2697 (*isol, struct, nmr, props*)

Phosphoramidon **P-00402**

N-[N-[[(*6-Deoxy-α-L-mannopyranosyl)oxy*]-*hydroxyphinyl*]-*L-leucyl*]-*L-tryptophan*

[36357-77-4]

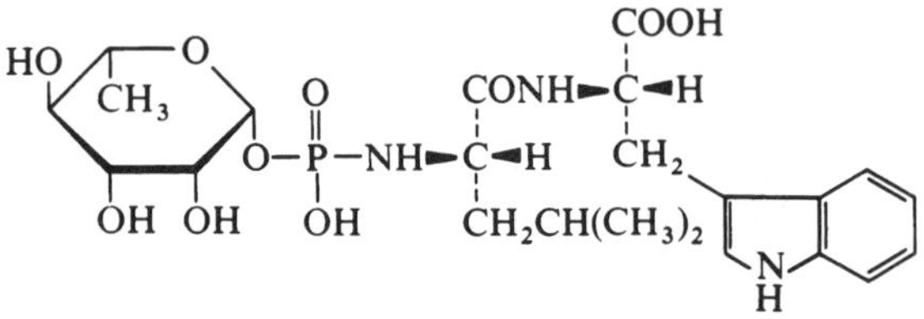

(structure: 6-deoxy-mannopyranosyl–O–P(O)(OH)–NH–CH($CH_2CH(CH_3)_2$)–CONH–CH(COOH)–CH_2–indol-3-yl)

$C_{23}H_{34}N_3O_{10}P$ M 543.509

Nucleotide antibiotic. Metabolite of *Streptomyces tanashiensis* and various other actinomycetes. Specific inhibitor of Thermolysin and the related enzyme. Inhibits the formation of spontaneous metastases. Related to Talopeptin.

Dicyclohexylammonium salt: Cryst. (2-propanol/isopropyl ether). Mp 130-148° dec. $[\alpha]_D^{21}$ −21° (c, 1.0 in H_2O).

Na salt: Mp 173-178° dec. $[\alpha]_D^{20}$ −33.6° (c, 1.0 in H_2O).

N-*Ac, Me ester:*
$C_{26}H_{38}N_3O_{11}P$ M 599.573
Mp 76-78°. $[\alpha]_D^{20}$ −10° ($CHCl_3$).

Umezawa, H., *Methods Enzymol.*, 1976, **45**, 693 (*rev*)
Weaver, L.H. *et al*, *J. Mol. Biol.*, 1977, **114**, 119 (*cryst struct*)
Giraldi, T. *et al*, *Anticancer Res.*, 1984, **4**, 221 (*props*)
Rose, M.E. *et al*, *Biomed. Mass Spectrom.*, 1984, **11**, 10 (*ms*)

Phosphorin, 9CI **P-00403**

Phosphabenzene. Phosphinine

[289-68-9]

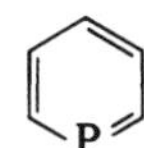

C_5H_5P M 96.068

Volatile liq. $Bp_{0.4}$ 30°. Air-sensitive but stable in absence of air.

Ashe, A.J., *J. Am. Chem. Soc.*, 1971, **93**, 3293 (*synth, ms, uv, ir*)
Batich, C. *et al, J. Am. Chem. Soc.*, 1973, **95**, 928 (*pe*)
Clarke, D.T. *et al, J. Chem. Soc., Faraday Trans. 2*, 1974, **70**, 1222 (*struct*)
Palmer, M.H. *et al, J. Chem. Soc., Perkin Trans. 2*, 1974, 420 (*struct*)
Wong, T.C. *et al, J. Chem. Phys.*, 1974, **61**, 2840 (*ed, microwave*)
von Niessen, W. *et al, Chem. Phys.*, 1975, **10**, 345 (*pe, struct*)
Ashe, A.J. *et al, J. Am. Chem. Soc.*, 1976, **98**, 5451 (*P nmr, pmr, cmr*)
Schäfer, W. *et al, J. Am. Chem. Soc.*, 1976, **98**, 4410 (*pe, struct*)
Ashe, A.J., *Acc. Chem. Res.*, 1978, **11**, 153 (*rev*)
Ashe, A.J. *et al, J. Mol. Struct.*, 1982, **78**, 169 (*synth, ir, raman*)
Dimroth, K., *Acc. Chem. Res.*, 1982, **15**, 58 (*derivs, rev*)

Phosphorinane, 9CI, 8CI **P-00404**

Phosphane. Phosphacyclohexane. Pentamethylenephosphine. Phosphinane

[4743-40-2]

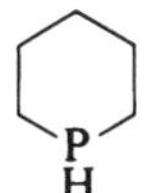

$C_5H_{11}P$ M 102.116

Mp 19°. Bp 118-121°.

B,MeI: Mp 220-225°.

Sommer, K., *Z. Anorg. Allg. Chem.*, 1970, **379**, 56 (*synth*)
Lambert, J.B. *et al*, Tet, 1971, **27**, 4245 (*synth, pmr, deriv, ms*)

Phosphorisocyanatidic dichloride, 9CI, 8CI **P-00405**

[870-30-4]

$OCNP(O)Cl_2$

CCl_2NO_2P M 159.896

Reactive liq with fishy odour. d_4^{15} 1.65. Bp 138-139°, Bp_5 19.7-20°. n_D^{15} 1.470.

Kirsanov, A.V., *Zh. Obshch. Khim.*, 1954, **24**, 1033 (*Engl. transl.* p. 1031) (*synth*)
Egorov, Yu.P. *et al, Zh. Strukt. Khim.*, 1973, **14**, 240 (*Engl. transl.* p. 216) (*ir*)
Naumov, V.A. *et al, Dokl. Akad. Nauk SSSR, Ser. Sci. Khim.*, 1973, **209**, 118 (*Engl. transl.* p. 188) (*ed, struct*)
Kisilenko, A.A. *et al, Zh. Prikl. Spektrosk.*, 1978, **28**, 480 (*Engl. transl.* p. 333) (*ir*)

Phosphorisocyanatidic difluoride, 9CI, 8CI **P-00406**

[1495-54-1]

$OCNP(O)F_2$

CF_2NO_2P M 126.987

Highly reactive liq. d^{25} 1.59. Bp 66-68.5°. n_D^{25} 1.3381.

Kuhn, S.I. *et al, Can. J. Chem.*, 1962, **40**, 1951 (*synth, ir*)
Glemser, O. *et al, Chem. Ber.*, 1967, **100**, 1082 (*synth, ir, ms, F nmr*)
Roesky, H.W., *Chem. Ber.*, 1967, **100**, 2147 (*synth*)

Phosphor(isothiocyanatidic) dichloride, 9CI **P-00407**

[1858-38-4]

$SCNP(O)Cl_2$

CCl_2NOPS M 175.957

V. reactive liq. Bp_1 40°.

Fluck, E. *et al, Z. Anorg. Allg. Chem.*, 1965, **338**, 58 (*P nmr*)
Schmitt, R. *et al, Z. Anorg. Allg. Chem.*, 1968, **358**, 38 (*synth, ir, raman*)

Phosphor(isothiocyanatidic) difluoride, 9CI, 8CI **P-00408**

[14526-13-7]

$SCNP(O)F_2$

CF_2NOPS M 143.048

Liq. Bp_{87} 35-36°.

Roesky, H.W., *Chem. Ber.*, 1967, **100**, 2142 (*synth, F and P nmr*)
Sprenger, G. *et al, J. Fluorine Chem.*, 1974, **4**, 201 (*synth*)

Phosphorocyanidous bromide iodide, 9CI **P-00409**

[60212-92-2]

BrIPCN

CBrINP M 263.800

Obt. only in soln.

Dillon, K.B. *et al, J. Inorg. Nucl. Chem.*, 1976, **38**, 1149 (*P nmr*)

Phosphorocyanidous dibromide **P-00410**

[60212-86-4]

Br_2PCN

CBr_2NP M 216.799

Obt. only in soln.

Dillon, K.B. *et al, J. Inorg. Nucl. Chem.*, 1976, **38**, 1149 (*P nmr*)

Phosphorodiisocyanatidic bromide, 9CI, 8CI **P-00411**

[17848-08-7]

$(OCN)_2P(O)Br$

$C_2BrN_2O_3P$ M 210.911

Highly reactive liq. d_4^{20} 1.93. Bp_{20} 83-84°. n_D^{20} 1.5060.

Derkach, G.I. *et al, Zh. Obshch. Khim.*, 1967, **37**, 2069 (*Engl. transl.* p. 1961) (*synth, ir*)

Phosphorodiisocyanatidic chloride, 9CI, 8CI **P-00412**

[1858-37-3]

$(OCN)_2P(O)Cl$

$C_2ClN_2O_3P$ M 166.460

Reactive liq. d_4^{20} 1.60. Bp_{12} 66-68°. n_D^{20} 1.4720.

Derkach, G.I. *et al*, *Zh. Obshch. Khim.*, 1967, **37**, 2069 (*Engl. transl.* p. 1961) (*synth, ir*)
Kisilenko, A.A. *et al*, *Zh. Prikl. Spektrosk.*, 1978, **28**, 480 (*Engl. transl.* p. 333) (*ir*)

Phosphorodiisocyanatidic fluoride, 9CI, 8CI — P-00413

[17848-06-5]

$(OCN)_2P(O)F$

$C_2FN_2O_3P$ M 150.006

Highly reactive liq. d_4^{20} 1.60. Mp 144-146°. n_D^{20} 1.4160.

Derkach, G. I. *et al*, *Zh. Obshch. Khim.*, 1967, **37**, 2069 (*Engl. transl.* p. 1961) (*synth, ir*)

Phosphorodi(isothiocyanatidic) fluoride, 9CI, 8CI — P-00414

[14526-14-8]

$(SCN)_2P(O)F$

$C_2FN_2OPS_2$ M 182.127

Highly reactive liq. $Bp_{3.5}$ 64-65°.

Roesky, H.W., *Z. Anorg. Allg. Chem.*, 1967, **353**, 265 (*ir*)
Roesky, H.W., *Chem. Ber.*, 1967, **100**, 2142 (*synth, F and P nmr*)

Phosphoryl isocyanate, 9CI — P-00415

[1858-24-8]

$(OCN)_3P{=}O$

$C_3N_3O_4P$ M 173.024

Liq. Mp 5°. Bp 193°, Bp_{10} 88°. Highly reactive to protic substances.

Miller, F.A. *et al*, *Spectrochem. Acta*, 1963, **18**, 1311 (*synth, ir, raman*)
Fluck, E., *Z. Naturforsch, B*, 1964, **19**, 869 (*P nmr*)
Borovikov, Yu.Ya. *et al*, *Zh. Obshch. Khim.*, 1975, **45**, 2377 (*Engl. transl.* p. 2335) (*struct*)

Phosphoryl isothiocyanate, 9CI — P-00416

[1858-25-9]

$(SCN)_3P{=}O$

$C_3N_3OPS_3$ M 221.206

Liq. $Bp_{0.5}$ 106-107°, $Bp_{0.05}$ 119°. Very reactive to protic substances.

Sowerby, D.B., *J. Inorg. Nucl. Chem.*, 1961, **22**, 205 (*synth, ir*)
Fluck, E. *et al*, *Z. Anorg. Allg. Chem.*, 1965, **338**, 58 (*synth, P nmr*)
Pudovik, A.N. *et al*, *Zh. Obshch. Khim.*, 1979, **49**, 1425 (*Engl. transl.* p. 1248) (*synth, ir, P nmr*)

Phosphoserine — P-00417

Serine dihydrogen phosphate, 9CI. *2-Amino-3-hydroxypropanoyl 3-phosphate*. O-*Phosphonoserine*

[7331-08-0]

COOH
H–C–NH_2 (*R*)-*form*
$CH_2OP(O)(OH)_2$

$C_3H_8NO_6P$ M 185.073

Found in casein and mammalian plasma. Occurs in brain.

(*R*)-form [73913-63-0]

D-form

Cryst. Mp 170-173°. $[\alpha]_D^{21}$ −15.6° (c, 3.2 in 2M HCl).

(*S*)-form [407-41-0]

L-form

Cryst. (EtOH aq.). Mp 175-176° dec. (167°). $[\alpha]_D^{21}$ +16.2° (c, 3 in 2M HCl), $[\alpha]_D^{23}$ +7.2° (c, 4 in 2M HCl).

(±)-form [17885-08-4]

Cryst. (H_2O or EtOH aq.). Mp 186°, Mp 153-156° (monohydrate).

Brucine salt: Solid. Mp 130° dec.

Mono-P-Ph ester: 2-Amino-3-[(hydroxyphenoxyphosphinyl)oxy]propanoic acid.
$C_9H_{12}NO_6P$ M 261.171
Cryst. (H_2O). Mp 163-165°.

P,P-*Di-Ph ester: 2-Amino-3-[(diphenoxyphosphinyl)oxy]propanoic acid.*
$C_{15}H_{16}NO_6P$ M 337.268
Solid. Mp 129-130° dec.

P,P-*Di-Ph ester*, C-*Et ester: Ethyl 2-amino-3-[(diphenoxyphosphinyl)oxy]propanoate.*
$C_{17}H_{20}NO_6P$ M 365.322
Mp 99-100° (as hydrochloride). Also forms a hydrobromide, Mp 67-68°.

Agren, G. *et al*, *Acta Chem. Scand.*, 1951, **5**, 324 (*isol*)
Plapinger, R.E. *et al*, *J. Am. Chem. Soc.*, 1953, **75**, 5757 (*derivs*)
Fölsch, G. *et al*, *Acta Chem. Scand.*, 1957, **11**, 1237 (*synth, ir*)
Putkey, E.F. *et al*, *Acta Crystallogr., Sect. B*, 1970, **26**, 782 (*cryst struct*)
Sundaralingam, M. *et al*, *Acta Crystallogr., Sect. B*, 1970, **26**, 790 (*cryst struct*)
Pogliani, L. *et al*, *Org. Magn. Reson.*, 1977, **10**, 26 (*pmr, conformn*)
Bolton, P.H., *J. Magn. Reson.*, 1981, **45**, 239 (*nmr*)
Vogel, H.J. *et al*, *Biochemistry*, 1982, **21**, 1126 (*P nmr*)
Mohan, M.S. *et al*, *Inorg. Chem.*, 1983, **22**, 714 (*complexes*)

N-Phosphoserine — P-00418

3-Hydroxy-2-phosphonoaminopropanoic acid. N-*Phosphonoserine*

$(HO)_2P(O)NHCH(COOH)CH_2OH$

$C_3H_8NO_6P$ M 185.073

(±)-form

Solid. Mp 166-167°.

P,P-*Diisopropyl ester*, C-*Me ester: Methyl 3-hydroxy-2-[(diisoproproxyphosphinyl)amino]propanoate.*
$C_{10}H_{22}NO_6P$ M 283.261
Waxy solid. Mp 48-50°.

P,P-*Dibenzyl ester*, C-*Me ester: Methyl 3-hydroxy-2-[(dibenzyloxyphosphinyl)amino]propanoate.*
$C_{18}H_{22}NO_6P$ M 379.349
Viscous syrup.

Plapinger, R.E. *et al*, *J. Am. Chem. Soc.*, 1953, **75**, 5757 (*synth*)
Li, S.-O. *et al*, *J. Am. Chem. Soc.*, 1955, **77**, 1866.

N^{δ}-Phosphosulfamylornithine — P-00419

8-Amino-1,1-dihydroxy-3-thia-2,4-diaza-1-phosphanonan-9-oic acid 1,3,3-trioxide, 9CI. *PSOrn*

[88286-55-9]

$HOOCCH(NH_2)(CH_2)_3NHSO_2NHP(O)(OH)_2$

$C_5H_{14}N_3O_7PS$ M 291.215

Isol. from *Pseudomonas syringae* pv. *phaseolicola* (*P. phaseolicola*). Inhibits ornithine carbamoyltransferase and is phytotoxic. Amphoteric.

Mitchell, R.E. *et al*, *Phytochemistry*, 1976, **15**, 1941 (*isol, props*)
Bublitz, F. *et al*, *Z. Allg. Mikrobiol.*, 1983, **23**, 485 (*isol, props*)

N-Phosphothreonine P-00420

3-Hydroxy-2-[(dihydroxyphosphinyl)amino]butanoic acid. 3-Hydroxy-2-phosphoramidobutanoic acid. N-*Phosphonothreonine*

$H_3CCH(OH)CH(COOH)NHP(O)(OH)_2$

$C_4H_{10}NO_6P$ M 199.100

(±)-form

Solid. Mp 184°. Mixt. of diastereoisomers.

P,P-*Diisopropyl ester,* C-*Me ester: Methyl 3-hydroxy-2-[(diisopropoxyphosphinyl)amino]butanoate.*
$C_{11}H_{24}NO_6P$ M 297.287
No phys. props. reported.

P,P-*Dibenzyl ester,* C-*Me ester: Methyl 3-hydroxy-2-[(dibenzyloxyphosphinyl)amino]butanoate.*
$C_{19}H_{24}NO_6P$ M 393.375
Waxy solid. Mp 52-54°.

Wagner Jauregg, T. *et al*, *J. Am. Chem. Soc.*, 1951, **73**, 5202.
Plapinger, R.E. *et al*, *J. Am. Chem. Soc.*, 1953, **75**, 5757 (*synth*)
Li, S.-O. *et al*, *J. Am. Chem. Soc.*, 1955, **77**, 1866.

Phosphotryptophan P-00421

*[[1-Amino-2-(1*H*-indol-3-yl]ethyl]phosphonic acid. Phosphonotryptophan*

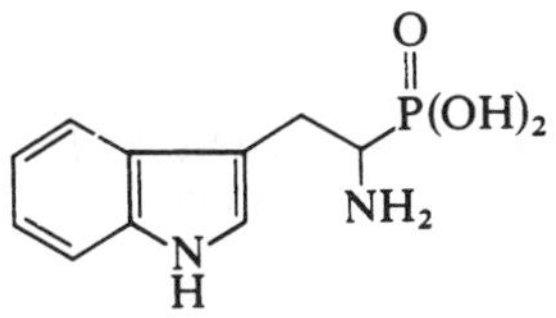

$C_{10}H_{13}N_2O_3P$ M 240.198

(±)-form

Solid. Mp 280-281° dec.

*Mono-Me ester: Methyl hydrogen [[1-amino-2-(1*H*-indol-3-yl)]ethyl]phosphonate.*
$C_{11}H_{15}N_2O_3P$ M 254.225
Solid (as *tert*-butylmethylamine salt). Mp 272-275°.

*Di-Me ester: Dimethyl [[1-amino-2-(1*H*-indol-3-yl)]-ethyl]phosphonate.*
$C_{12}H_{17}N_2O_3P$ M 268.252
Solid (as oxalate). Mp 172-173°.

Di-Me ester, N-*benzyloxycarbonyl:*
$C_{18}H_{19}N_2O_5P$ M 374.332
Solid. Mp 130-131°.

Chen, S.F. *et al*, *Tetrahedron Lett.*, 1983, **24**, 5461 (*synth, derivs, pmr*)

N-Phosphotyrosine P-00422

*[1-Carboxy-2-(4-hydroxyphenyl)ethyl]phosphoramidic acid. 2-*p*-Hydroxybenzyl-1-phosphoramidopropanoic acid.* N-*Phosphonotyrosine*

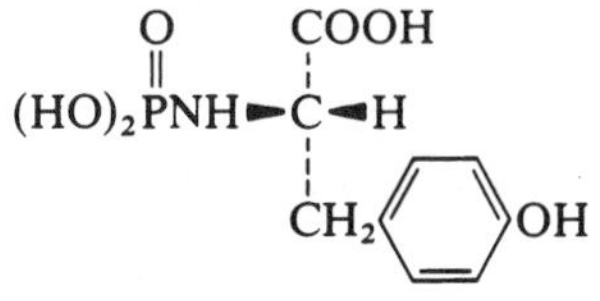

$C_9H_{12}NO_6P$ M 261.171

(*S*)-form

L-form

P,P-*Di-Ph ester,* C-*Et ester: Diphenyl [1-ethoxycarbonyl-2-(4-hydroxybenzyl)ethyl]phosphoramidate.*
$C_{23}H_{24}NO_6P$ M 441.419
Cryst. (Et_2O). Mp 93-94°.

P,P-*Bis-4-iodobenzyl ester,* C-*Et ester: Bis-4-iodobenzyl [1-ethoxycarbonyl-2-(4-hydroxybenzyl)ethyl]-phosphoramidate.*
$C_{25}H_{26}I_2NO_6P$ M 721.266
Cryst. (EtOH). Mp 143°.

P,P-*Dibenzyl,* C-*Et ester: Dibenzyl [1-ethoxycarbonyl-2-(4-hydroxybenzyl)ethyl]phosphoramidate.*
$C_{25}H_{28}NO_6P$ M 469.473
Solid. Mp 104-105°.

C,P,P-*Tribenzyl ester: Dibenzyl [1-benzyloxycarbonyl-2-(4-hydroxyphenyl)ethyl]phosphoramidate.*
$C_{30}H_{30}NO_6P$ M 531.544
Solid. Mp 54-55°.

Li, S.-O. *et al*, *J. Am. Chem. Soc.*, 1955, **77**, 1866 (*synth*)
Zervas, L. *et al*, *J. Am. Chem. Soc.*, 1955, **77**, 5351 (*synth*)

O-Phosphotyrosine P-00423

*Tyrosine-*O*-phosphate. Tyrosine* O*-dihydrogen phosphate.* O*-Phosphonotyrosine*

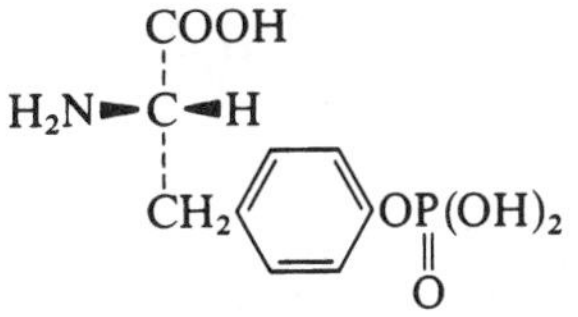

$C_9H_{12}NO_6P$ M 261.171

(*S*)-form [21820-51-9]

L-form

Product of phosphorylation of tyrosine in cellular proteins in malignant tumour viruses e.g. Rous sarcoma virus. Cryst. (EtOH aq. or H_2O). Mp 226-227°, Mp 253° dec. $[\alpha]_D^{28}$ −7.8° (c, 1 in 2M HCl aq.), $[\alpha]_D^{20}$ −5.5° (c, 1 in H_2O).

Gill, G.N. *et al*, *Nature (London)*, 1981, **293**, 305
Ross, A.H. *et al*, *Nature (London)*, 1981, **294**, 654
Patschinsky, T. *et al*, *Proc. Natl. Acad. Sci. USA*, 1982, **79**, 973.
Alewood, P.F. *et al*, *Synthesis*, 1983, 30 (*synth, ir, uv, cmr, pmr, P nmr*)
Chang, W.-C. *et al*, *Anal. Biochem.*, 1983, **132**, 342 (*tlc*)

Phoxim, BSI, ISO, BAN, INN P-00424

4-Ethoxy-7-phenyl-3,5-dioxa-8-aza-4-phosphaoct-6-ene-8-nitrile 4-sulfide, 9CI. *α-[[(Diethoxyphosphinothioyl)oxy]imino]-benzeneacetonitrile. Phenylglyoxylonitrile oxime* O-(O,O-*diethylphosphorothioate*), 8CI. O,O-*Diethyl α-cyanobenzylideneaminooxyphosphonothioate. Phoxime. Baythion*

[14816-18-3]

$(EtO)_2P(S)ON{=}C(CN)Ph$

$C_{12}H_{15}N_2O_3PS$ M 298.296

Wide spectrum veterinary insecticide and anthelmintic of short duration. Liq. d^{20} 1.176. Fp 5-6°. $Bp_{0.01}$ 102° dec. Tends to dec. on dist. Mod. cholinesterase inhibitor of low tox.

▷MD4740000.

Mason, W.A. *et al*, *J. Agric. Food Chem.*, 1973, **21**, 762 (*synth*)
Stan, H.-J. *et al*, *Fresenius' Z. Anal. Chem.*, 1977, **287**, 271 (*glc, ms*)
Stan, H.-J. *et al*, *Biomed. Mass Spectrom.*, 1982, **9**, 483 (*ms*)
The Agrochemicals Handbook, Royal Society of Chemistry, 1983, A328.
Pesticide Manual, 7th Ed., 440.

2-Phthalimidoethyl phosphorodichloridate **P-00425**

*2-(1,3-Dihydro-1,3-dioxo-2*H*-isoindol-2-yl)ethyl phosphorodichloridate, 9CI*

[52198-45-5]

$NCH_2CH_2OPCl_2$

$C_{10}H_8Cl_2NO_4P$ M 308.057

Reagent used in phospholipid synth. Powder. Mp 72-73°. Easily hydrolysed in moist air.

Hirt, R. *et al*, *Helv. Chim. Acta*, 1957, **40**, 1929 (*synth, use*)
Berchtold, R., *Chem. Phys. Lipids*, 1981, **28**, 55 (*use*)
Tsushina, S. *et al*, *Chem. Pharm. Bull.*, 1982, **30**, 3260 (*use*)

N-Phthalimidomethylphosphonic acid **P-00426**

*[(1,3-Dihydro-1,3-dioxo-2*H*-isoindol-2-yl)methyl]-phosphonic acid, 9CI.* N*-Phosphonomethylphthalimide*

[49594-18-5]

$NCH_2P(O)(OH)_2$

$C_9H_8NO_5P$ M 241.140

The acid and its esters are intermeds. in the synth. of aminomethylphosphonic acid. Cryst. (MeOH aq.). Mp 280-282°.

Di-Me ester: [28447-26-9]. *Dimethyl [(1,3-dihydro-1,3-dioxo-2*H*-isoindol-2-yl)methyl]phosphonate.*
$C_{11}H_{12}NO_5P$ M 269.193
Cryst. ($CHCl_3/CCl_4$). Mp 117.5-119°.

Mono-Et ester: [38416-66-9]. *Ethyl hydrogen [(1,3-dihydro-1,3-diazo-2*H*-isoindol-2-yl)methyl]-phosphonate.*
$C_{11}H_{12}NO_5P$ M 269.193
Cryst. (H_2O). Mp 202-204°.

Di-Et ester: [33512-26-4]. *Diethyl [(1,3-dihydro-1,3-dioxo-2*H*-isoindol-2-yl)methyl]phosphonate.*
$C_{13}H_{16}NO_5P$ M 297.247
Cryst. (hexane). Mp 66-67°.

Seyferth, D. *et al*, *J. Org. Chem.*, 1971, **36**, 1379 (*ester, synth, pmr, ir, use*)
Griffith, W.R. *et al*, *Phosphorus*, 1975, **5**, 273; *Phosphorus Sulfur*, 1978, **5**, 101 (*ms*)
Wasielewski, C. *et al*, *Rocz. Chem.*, 1976, **50**, 1795 (*synth, use*)
Hoffmann, M., *Pol. J. Chem.*, 1981, **55**, 1695 (*ester*)
Baraldi, P.G. *et al*, *Synthesis*, 1982, 653 (*ester, ir, pmr*)

Phytonadiol diphosphate **P-00427**

2-Methyl-3-phytyl-1,4-naphthalenedioldiphosphate

$OP(O)(OH)_2$, CH_3, $OP(O)(OH)_2$

$C_{31}H_{50}O_8P_2$ M 612.679

Di-Na salt: [5988-22-7]. *Phytonadiol sodium diphosphate*, INN. Antihaemorrhagic agent.

U.S.P., 3 065 255, (*1962*); *CA*, **58**, 6762d (*synth, pharmacol*)

Picofosforic acid, INN **P-00428**

4,4′-(2-Pyridinylmethylene)bisphenol bis(dihydrogen phosphate)

[36175-06-1]

$C_{18}H_{17}O_8P_2$ M 423.275

Laxative.

Bruzzese, T. *et al*, *Arzneim.-Forsch.*, 1972, **22**, 531 (*synth, pharmacol*)

(1-Piperidinylmethyl)phosphonic acid, 9CI **P-00429**

N*-(Phosphonomethyl)piperidine*

[4672-35-9]

$CH_2P(O)(OH)_2$

$C_6H_{14}NO_3P$ M 179.155

Cryst. (EtOH aq.). Mp 249-250°.

Di-Et ester: [4972-40-1]. *Diethyl (1-piperidinylmethyl)-phosphonate.*
$C_{10}H_{22}NO_3P$ M 235.262
Liq. Bp_3 124°. pK_a 6.54 (H_2O), 4.05 (EtOH), 25°.

Fields, E.K., *J. Am. Chem. Soc.*, 1952, **74**, 1528 (*ester*)
Bel'skii, V.E. *et al*, *Izv. Akad. Nauk SSSR, Ser. Khim.*, 1975, **24**, 1047, 1624, 2346 (*Engl. transl.* pp. 958, 1511, 2232) (*pmr, ir, nmr*)
Fredericks, P.M. *et al*, *Z. Naturforsch, C*, 1981, **36**, 242 (*synth*)

1-Piperidinylphosphonic acid, 9CI **P-00430**

Piperidinophosphonic acid, 8CI. *1-Piperidinephosphonic acid. 1-Phosphonopiperidine*

[4764-18-5]

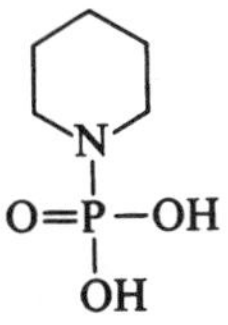

$C_5H_{12}NO_3P$ M 165.128

Di-Me ester: [597-24-0]. *Dimethyl 1-piperidinylphosphonate.*

$C_7H_{16}NO_3P$ M 193.182

Liq. Bp_3 85-90°. n_D^{20} 1.4513.

Di-Et ester: [4972-36-5]. *Diethyl 1-piperidinylphosphonate.*

$C_9H_{20}NO_3P$ M 221.236

Liq. $Bp_{0.3}$ 80°.

Di-Ph ester: [6214-09-1]. *Diphenyl 1-piperidinylphosphonate.*

$C_{17}H_{20}NO_3P$ M 317.324

Cryst. (pet. ether). Mp 75-76°.

Dichloride: [1498-56-2].

$C_5H_{10}Cl_2NOP$ M 202.020

Liq. d_0^{18} 1.32. Bp 257°, Bp_{11} 124°.

▷TA1840500.

Michaelis, A., *Justus Liebigs Ann. Chem.*, 1903, **326**, 129, 172 (*dichloride, esters*)
Stock, J.A. *et al*, *J. Chem. Soc.* (*C*), 1966, 637 (*ester*)
Buchanan, G.W. *et al*, *Can. J. Chem.*, 1979, **57**, 21 (*ester, cmr*)
Gray, G.A. *et al*, *J. Org. Chem.*, 1979, **44**, 1768 (*cmr, N nmr*)
Torii, S. *et al*, *Tetrahedron Lett.*, 1979, 4471 (*esters, synth*)

1-Piperidinylphosphonothioic acid, 9CI **P-00431**

Piperidinophosphonothioic acid, 8CI

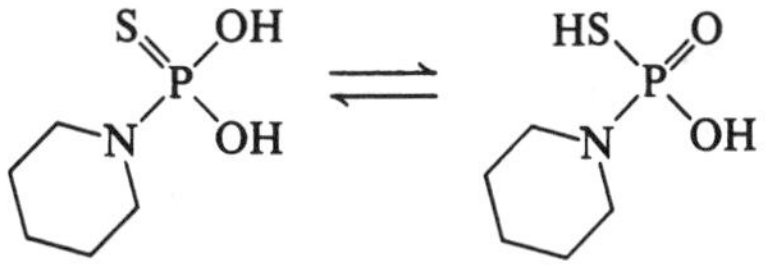

$C_5H_{12}NO_2PS$ M 181.189

Isolated as di-Na salt, dihydrate.

O,O-*Di-Me ester:* O,O-*Dimethyl 1-piperidinylphosphonothioate.*

$C_7H_{16}NO_2PS$ M 209.243

Liq. $Bp_{0.3}$ 80°.

Dichloride:

$C_5H_{10}Cl_2NPS$ M 218.080

Liq. Bp_{19} 145°. n_D^{20} 1.5587.

Christol, C. *et al*, *J. Chim. Phys.*, 1965, **62**, 246 (*dichloride, ir, raman*)
Cooks, R.G. *et al*, *J. Chem. Soc.* (*B*), 1968, 1327 (*ester*)
Karimova, P.M. *et al*, *Izv. Akad. Nauk SSSR, Ser. Khim.*, 1982, 2402 (*Engl. transl.* p. 2118) (*synth*)

1-Piperidinylphosphonous acid, 9CI **P-00432**

Piperidinophosphonous acid, 8CI. *1-Piperidinephosphonous acid*

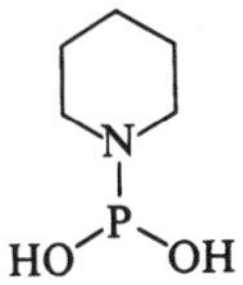

$C_5H_{12}NO_2P$ M 149.129

Prob. tautomeric.

Di-Et ester: [4972-39-8]. *Diethyl 1-piperidinylphosphonite.*

$C_9H_{20}NO_2P$ M 205.236

Liq. Bp_{10} 74°. n_D^{20} 1.4570.

Diisopropyl ester: Diisopropyl 1-piperidinylphosphonite.

$C_{11}H_{24}NO_2P$ M 233.290

Liq. Bp_{10} 82°.

Difluoride: [1073-52-5].

$C_5H_{10}F_2NP$ M 153.111

Liq. Bp_{30} 47°. n_D^{25} 1.4256.

Dichloride: [27325-49-1].

$C_5H_{10}Cl_2NP$ M 186.020

Liq. Bp_{18} 102°. n_D^{20} 1.5340.

Bis(dimethylamide): [87934-75-6]. N,N,N′,N′-*Tetramethyl*-P-*1-piperidinylphosphonous diamide.*

$C_9H_{22}N_3P$ M 203.267

Liq. Bp_{10} 95-96°. n_D^{20} 1.4890.

Dipiperidide: see Tri-1-piperidinophosphine, T-00704

Schmutzler, R., *Inorg. Chem.*, 1964, **3**, 415 (*difluoride*)
Barlow, C.G. *et al*, *J. Chem. Soc.* (*A*), 1966, 228 (*difluoride, synth, pmr, F and P nmr*)
Novikova, Z.S. *et al*, *Zh. Obshch. Khim.*, 1976, **46**, 2213 (*Engl. transl.* p. 2128) (*esters*)
Kukhar', V.P. *et al*, *Zh. Obshch. Khim.*, 1982, **52**, 562 (*Engl. transl.* p. 492) (*dichloride, synth, P nmr*)
Marchenko, A.P. *et al*, *Zh. Obshch. Khim.*, 1983, **53**, 1513 (*Engl. transl.* p. 1364) (*dimethylamide*)

Piperophos, BSI **P-00433**

S-[*2-(2-Methyl-1-piperidinyl)-2-oxoethyl*] O,O-*dipropyl phosphorothioate,* 9CI. S-*2-Methylpiperidinocarbonylmethyl* O,O-*dipropyl phosphorothioate*

[24151-93-7]

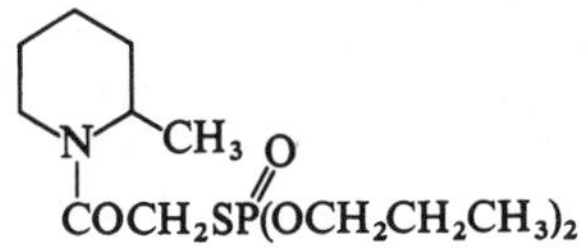

$C_{14}H_{28}NO_4PS$ M 337.413

Herbicide, selective against grasses. Liq.

▷TE3690000.

Pesticide Manual, 7th Ed., 445.

Pirimiphos-ethyl, BSI **P-00434**

O-(*2-Diethylamino-6-methyl-4-pyrimidinyl*) O,O-*diethyl phosphorothioate. Pirimicid. Fernex*

[23505-41-1]

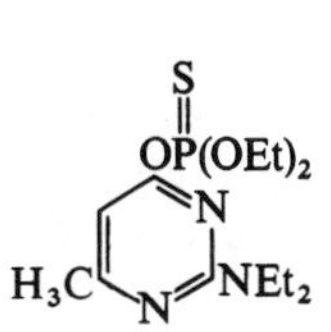

$C_{13}H_{24}N_3O_3PS$ M 333.385

Broad range insecticide and acaricide. Liq. d^{20} 1.14. n_D^{25} 1.520. Dec. >130°.

▷TF1610000.

Ripley, B.D. *et al*, *J. Assoc. Off. Anal. Chem.*, 1983, **66**, 1084 (*glc*)
Ding, X.D. *et al*, *J. Agric. Food Chem.*, 1984, **32**, 622 (*hplc*)
Pesticide Manual, 7th Ed., 448.

Pirimiphos-methyl, BAN, BSI — P-00435

O-[*2-(Diethylamino)-6-methyl-4-pyrimidinyl*] O,O-*dimethyl phosphorothioate*, 9CI, 8CI. *Actellic*

[29232-93-7]

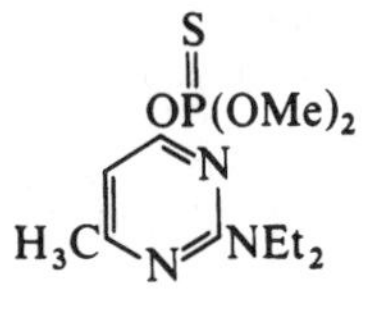

$C_{11}H_{20}N_3O_3PS$ M 305.331

Insecticide, acaricide. Straw-coloured oil. V. spar. sol. H_2O, misc. most org. solvs. d^{30} 1.157. n_D^{25} 1.527.

▷Less toxic than di-Et ester. TF1410000.

B.P., 1 019 227, (*1963*); *CA*, **64**, 14197b (*synth, use*)
Bagness, J.E. *et al*, *Analyst* (*London*), 1974, **99**, 225 (*glc*)
Ripley, B.D. *et al*, *J. Assoc. Off. Anal. Chem.*, 1983, **66**, 1084 (*glc*)
Stan, H.-J. *et al*, *J. Chromatogr.*, 1983, **279**, 173 (*glc*)
Bottomley, P. *et al*, *Analyst* (*London*), 1984, **109**, 85 (*hplc*)
Pesticide Manual, 6th Ed., 434; 7th Ed., 449.
The Agrochemicals Handbook, Royal Society of Chemistry, 1983, A335.

Platelet activating factor — P-00436

*1-O-Alkyl-2-acetyl-*sn*-glyceryl-3-phosphocholine. Antihypertensive polar renomedullary lipid. PAF. APRL*

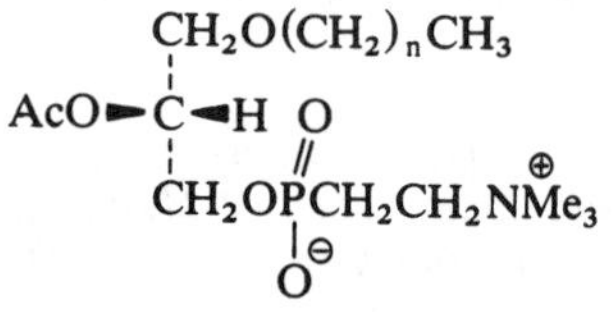

n of natural material not unequivocally detd., prob. a mixt. Isol. from renal medulla and basophils. Chemical mediator, exerts powerful antihypertensive action, causes aggregation of blood platelets.

Blank, M.L. *et al*, *Biochem. Biophys. Res. Commun.*, 1979, **90**, 1194.
Demopoulos, C.A. *et al*, *J. Biol. Chem*, 1979, **254**, 9355.

Plumbemycin *A* — P-00437

Alanylaspartyl-(2-amino-5-phosphoryl-3-pentenoic acid). N1409A. Antibiotic N1409A

[62896-18-8]

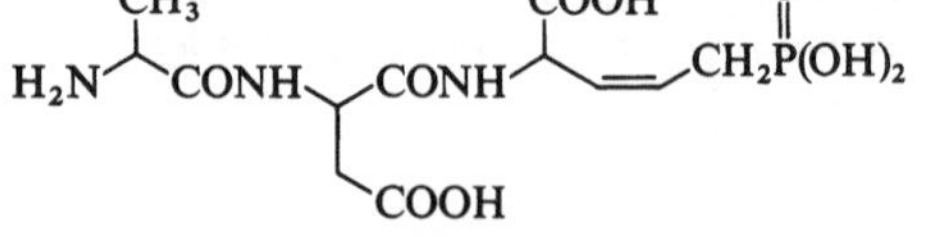

$C_{12}H_{20}N_3O_9P$ M 381.278

Peptide-type antibiotic, R = COOH. Isol. from *Streptomyces plumbeus*. Active against gram-positive bacteria. Powder. Mp 172-175° dec. $[\alpha]_D^{21}$ −10.93° (c, 1.09 in H_2O), +4.65° (c, 1.07 in 5.78M HCl).

▷RD2310000.

Carboxamide: [62896-17-7]. ***Plumbemycin* B**. *N 1409. Antibiotic N 1409.*

$C_{12}H_{21}N_4O_8P$ M 380.294

From *S. plumbeus*. Active against gram-positive bacteria. Needles. Mp 218-220° dec. $[\alpha]_D^{21}$ −6.57° (c, 1.06 in H_2O), +3.66° (c, 1.09 in 5.78M HCl). R = $-CONH_2$.

▷RD2300000.

Park, B.K. *et al*, *Agric. Biol. Chem.*, 1976, **40**, 1905; 1977, **41**, 161, 573 (*isol, struct, spectra*)
Japan. Pat., 78 87 314, (*1978*); *CA*, **89**, 195402a (*isol*)
Diddens, H. *et al*, *J. Antibiot.*, 1979, **32**, 87 (*props*)

Prasinomycin — P-00438

[12687-95-5]

$C_{62-74}H_{96-103}N_{5-6}O_{32-40}P$ M 1454.428-1747.618

A group of five closely related glycolipid antibiotics. Struct. unknown but believed to be complex phosphoric acid diesters. Prod. by a strain of *Streptomyces prasinus*. Active against gram-positive bacteria. Hydrolysis of mixt. gave lipid fractions identical to those from hydrolysis of Moenomycins. Strain also prod. Prasinomycin *D* [11021-57-1] and *E* [11021-58-2] (no data).

Prasinomycin A [11021-54-8]
Amorph. Mp 166-169° dec. $[\alpha]_D$ +0.8° (H_2O). λ_{max} 246 nm (0.1N HCl).

Prasinomycin B [11021-55-9]
Amorph. Mp 167-170° dec. $[\alpha]_D$ +2.8° (H_2O). λ_{max} 246 nm (0.1N HCl).

Prasinomycin C [11021-56-0]
Amorph. Mp 178-180° dec. $[\alpha]_D$ +4.4° (H_2O). λ_{max} 244 nm (0.1N HCl).

Weisenborn, F.L. *et al*, *Nature* (*London*), 1967, **213**, 1092 (*isol, uv, props*)
Slusarchyk, W.A. *et al*, *Tetrahedron Lett.*, 1969, 659 (*ir, ms, nmr*)
Van Heijenoort, Y. *et al*, *FEBS Lett.*, 1980, **110**, 241 (*props*)

Prenomycin — P-00439

Antibiotic 901A

[56748-95-9]

Struct. unknown

$C_{72}H_{125}N_6O_{42}P$ M 1777.768

Glycolipid antibiotic. Isol. from *Streptomyces ambofaciens*. Active against gram-positive and -negative bacteria. Animal growth promotor. Possibly related to Ensanchomycin. Similar to Moenomycin.

U.S.P., 3 891 753, (*1975*); *CA*, **83**, 162171x (*isol*)

Profenofos, BSI — P-00440

O-(*4-Bromo-2-chlorophenyl*) O-*ethyl* S-*propyl phosphorothioate*, 9CI. *Curacron. Selecron*

[41198-08-7]

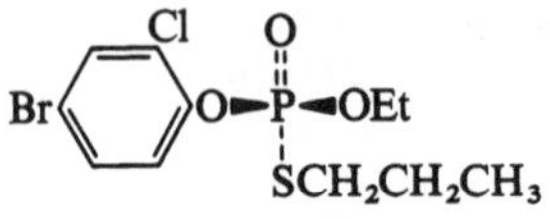

(R)-form

$C_{11}H_{15}BrClO_3PS$ M 373.628

▷TE6975000.

(R)-form

$[\alpha]_D^{20}$ +15.9° (c, 0.6 in $CHCl_3$). Tentative abs. config. assignment. Less toxic than (*S*)-form.

(S)-form [81123-19-5]
$[\alpha]_D^{20}$ −15.8° (c, 0.3 in $CHCl_3$). More toxic than (+)-form to insects and mammals.
(±)-form [92760-41-3]
Non systemic, broad range insecticide with contact and stomach action. Pale-yellow liq. V. spar. sol. H_2O. $Bp_{0.001}$ 110°.

Leader, H. *et al*, *J. Agric. Food Chem.*, 1982, **30**, 546 (*synth, use, resoln, pmr, P nmr*)
Glickman, A.H. *et al*, *Toxicol. Appl. Pharmacol.*, 1984, **73**, 16 (*tox*)
Hirashima, A. *et al*, *J. Agric. Food Chem.*, 1984, **32**, 1302 (*synth, pmr, P nmr*)
Wing, K.D. *et al*, *Pestic. Biochem. Physiol.*, 1984, **21**, 22 (*metab*)
Pesticide Manual, 6th Ed., 438; 7th Ed., 454.

1,2-Propadienylphosphonic acid, 9CI P-00441

Allenylphosphonic acid. Allenephosphonic acid
[34163-96-7]

$H_2C{=}C{=}CHP(O)(OH)_2$

$C_3H_5O_3P$ M 120.044
Di-Me ester: [18356-17-7]. *Dimethyl 1,2-propadienylphosphonate. Dimethyl allenylphosphonate.*
$C_5H_9O_3P$ M 148.098
d_4^{20} 1.17. $Bp_{0.4}$ 75-76°. n_D^{20} 1.4672.
Di-Et ester: [1609-72-9]. *Diethyl 1,2-propadienylphosphonate. Diethyl allenylphosphonate.*
$C_7H_{13}O_3P$ M 176.152
Liq. d_4^{20} 1.08. $Bp_{0.025}$ 60-62°. n_D^{20} 1.4615.
Diisopropyl ester: [3201-76-1]. *Bis(1-methylethyl) 1,2-propadienylphosphonate. Diisopropyl allenylphosphonate.*
$C_9H_{17}O_3P$ M 204.205
Liq. d_4^{20} 1.02. Bp_2 83-84°. n_D^{20} 1.4550.
*Di-*tert*-butyl ester:* [25383-48-6]. *Bis(1,1-dimethylethyl) 1,2-propadienylphosphonate. Di-*tert*-butyl allenylphosphonate.*
$C_{11}H_{21}O_3P$ M 232.259
Liq. $Bp_{0.1}$ 54-56°.
Difluoride: [17166-37-9].
$C_3H_3F_2OP$ M 124.027
Liq. d_4^{20} 1.30. Bp_{14} 36-36.5°. n_D^{20} 1.4053.
Dichloride: see Propadienylphosphonic dichloride, P-00443
Diamide: see P-Propadienylphosphonic diamide, P-00442

Pudovik, A.N. *et al*, *Zh. Obshch. Khim.*, 1965, **35**, 1210 (*Engl. transl.* p. 1214) (*esters*)
Ignat'ev, V.M. *et al*, *Zh. Obshch. Khim.*, 1967, **37**, 1898 (*Engl. transl.* p. 1807) (*difluoride, synth, ir*)
Timofeeva, T.M. *et al*, *Zh. Obshch. Khim.*, 1969, **39**, 2446 (*Engl. transl.* p. 2386) (*ester, synth, ir*)
Glamkowski, E.J. *et al*, *J. Org. Chem.*, 1970, **35**, 3510 (*ester, ir, synth, use*)
Khusainova, N.G. *et al*, *Zh. Obshch. Khim.*, 1982, **52**, 1040 (*Engl. transl.* p. 904) (*esters, props*)
Angelov, Kh. *et al*, *Zh. Obshch. Khim.*, 1982, **52**, 538 (*Engl. transl.* p. 472) (*esters, props*)

P-Propadienylphosphonic diamide, 9CI P-00442

P-*Allenylphosphonic diamide*

$H_2C{=}C{=}CHP(O)(NH_2)_2$

$C_3H_7N_2OP$ M 118.075
N,N,N′,N′-*Tetra-Me:* [3356-37-4]. N,N,N′,N′-*Tetramethyl-*P*-1,2-propadienylphosphonic diamide. Allenylphosphonic bis(dimethylamide).*
$C_7H_{15}N_2OP$ M 174.182
Liq. d_4^{20} 1.05. $Bp_{0.25}$ 80-83°. n_D^{20} 1.5021.

Ionin, B.I. *et al*, *Zh. Obshch. Khim.*, 1967, **37**, 1863 (*Engl. transl.* p. 1774) (*P nmr*)
Ignat'ev, V.M. *et al*, *Zh. Obshch. Khim.*, 1967, **37**, 1898 (*Engl. transl.* p. 1807) (*synth, ir*)
Khusainova, N.G. *et al*, *Zh. Org. Khim.*, 1978, **14**, 2555 (*Engl. transl.* p. 2353) (*props*)

Propadienylphosphonic dichloride, 9CI P-00443

Allenylphosphonic dichloride
[17166-36-8]

$H_2C{=}C{=}CHP(O)Cl_2$

$C_3H_3Cl_2OP$ M 156.936
Liq. d_4^{20} 1.41. Bp_3 73-74°. n_D^{20} 1.5232.

Ionin, B.I. *et al*, *Zh. Obshch. Khim.*, 1967, **37**, 1863 (*Engl. transl.* p. 1774) (*pmr*)
Ignat'ev, V.M. *et al*, *Zh. Obshch. Khim.*, 1967, **37**, 1898 (*Engl. transl.* p. 1807) (*synth, ir*)
Ivin, S.Z. *et al*, *Zh. Obshch. Khim.*, 1968, **38**, 2069 (*Engl. transl.* p. 2004) (*synth*)
Krudy, G.A. *et al*, *J. Org. Chem.*, 1978, **43**, 4656 (*cmr*)

1,3-Propanediphosphonic acid P-00444

1,3-Propanediylbis[phosphonic acid], *9CI*. *Trimethylenediphosphonic acid*, *8CI*
[4671-82-3]

$(HO)_2P(O)CH_2CH_2CH_2P(O)(OH)_2$

$C_3H_{10}O_6P_2$ M 204.056
Cryst. (H_2O). Mp 178°.
Tetra-Et ester: [22401-25-8]. *Tetraethyl 1,3-propanediylbisphosphonate. 1,3-Bis(diethoxyphosphinyl)-propane.*
$C_{11}H_{26}O_6P_2$ M 316.270
Liq. Bp_1 178-180°.
Bis-dichloride: [1499-31-6].
$C_3H_6Cl_4O_2P_2$ M 277.839
V. hygroscopic solid. Mp 59.5-61.5°. $Bp_{0.005}$ 100-110°.

Kosolapoff, G.M., *J. Am. Chem. Soc.*, 1944, **66**, 1511 (*synth, ester*)
Moedritzer, K. *et al*, *J. Inorg. Nucl. Chem.*, 1961, **22**, 297 (*synth, ester, ir, P nmr*)
Maier, L., *Helv. Chim. Acta*, 1965, **48**, 133 (*chloride, synth, ir*)
Kosolapoff, G.M. *et al*, *J. Chem. Soc. (C)*, 1966, 757 (*chloride*)
Sommer, K., *Z. Anorg. Allg. Chem.*, 1970, **376**, 37 (*synth, P nmr*)
Gebert, E. *et al*, *J. Phys. Chem.*, 1977, **81**, 471 (*cryst struct*)

2,2-Propanediphosphonic acid P-00445

(1-Methylethylidene)bisphosphonic acid, *9CI*. *Dimethylmethylenediphosphonic acid*
[6145-32-0]

$(HO)_2P(O)C(CH_3)_2P(O)(OH)_2$

$C_3H_{10}O_6P_2$ M 204.056
Disodium salt used in dentifrices. Cryst. Mp 228.5-229.5°.
Tetra-Et ester: [51346-85-1]. *Tetraethyl 2,2-propanediyldiphosphonate.*
$C_{11}H_{26}O_6P_2$ M 316.270
Liq. $Bp_{0.03}$ 86°. n_D^{25} 1.4375.
Tetraisopropyl ester: [10038-58-1]. *Tetraisopropyl 2,2-propanediphosphonate.*
$C_{15}H_{34}O_6P_2$ M 372.378

Oil. $Bp_{0.12}$ 102-105°, $Bp_{0.005}$ 65-69°. n_D^{20} 1.4332.

Hays, H.R. *et al, J. Org. Chem.*, 1966, **31**, 3391 (*derivs, P nmr*)
Quimby, O.T. *et al, J. Organomet. Chem.*, 1968, **13**, 199 (*synth, P nmr*)
Seyferth, D. *et al, J. Organomet. Chem.*, 1973, **59**, 237 (*ester, synth, pmr*)
Menge, M. *et al, Arch. Pharm.* (*Weinheim, Ger.*), 1981, **314**, 218 (*synth, pmr*)

1,3-Propanediphosphoramidothioic acid P-00446

1,3-Propanediylbisphosphoramidothioic acid, 9CI. Trimethylenediphosphoramidothioic acid

[34670-60-5]

$(HO)_2P(S)NH(CH_2)_3NHP(S)(OH)_2$

$C_3H_{12}N_2O_4P_2S_2$ M 266.206

Tautomeric.

O,O,O′,O′-Tetra Ph ester: [34670-60-5]. *O,O,O′,O′-Tetraphenyl 1,3-propanediphosphoramidothioate.*
$C_{27}H_{28}N_2O_4P_2S_2$ M 570.597
Oil.

Edmundson, R.S., *J. Chem. Soc.* (*C*), 1971, 3614 (*synth*)

1,3-Propanediylbis[phenylphosphinic acid], 9CI P-00447

[4851-72-3]

$$Ph-\underset{OH}{\overset{O}{\overset{\|}{\underset{|}{P}}}}CH_2CH_2CH_2\underset{OH}{\overset{O}{\overset{\|}{\underset{|}{P}}}}-Ph$$

$C_{15}H_{18}O_4P_2$ M 324.252

Cryst. (EtOH). Mp 156-157°.

Diisopropyl ester:
$C_{21}H_{30}O_4P_2$ M 408.413
Solid. Mp 86-87°.

Mastalerz, P., *Rocz. Chem.*, 1965, **39**, 1129 (*synth*)
Abramov, V.S. *et al, Zh. Obshch. Khim.*, 1968, **38**, 1794 (*Engl. transl.* p. 1748) (*ester*)
Sommer, K., *Z. Anorg. Allg. Chem.*, 1970, **376**, 31 (*nmr*)
Garst, M.E., *Synth. Commun.*, 1979, **9**, 261 (*synth*)

1,3-Propanediylbis[triphenylphosphonium](2+), 9CI P-00448

Trimethylenebis[triphenylphosphonium](2+). 1,3-Bis(triphenylphosphonio)propane(2+)

[35823-43-9]

$Ph_3P^{\oplus}CH_2CH_2CH_2PPh_3^{\oplus}$

$C_{39}H_{36}P_2^{\oplus\oplus}$ M 566.661 (ion)

Dibromide: [7333-67-7].
$C_{39}H_{36}Br_2P_2$ M 726.469
Cryst. ($EtOH/Et_2O$). Mp 335° dec. LiOEt or PhLi → bisylide.

▷Exhibits AChE activity in vertebrates and schistosomes

Bisylide: [63591-90-2]. *1,3-Propanediylidenebis[triphenylphosphorane].*
$C_{39}H_{34}P_2$ M 564.645
Wittig reagent used in synth. of annulenones and benzannelated annulenes. Red.

Horner, L. *et al, Chem. Ber.*, 1962, **95**, 581 (*bromide, synth*)
Cresp, T.M. *et al, J. Chem. Soc., Perkin Trans. 1*, 1974, 2145 (*use*)
Wood, G.W. *et al, J. Org. Chem.*, 1975, **40**, 636 (*bromide, ms*)
Finch, N. *et al, J. Org. Chem.*, 1979, **44**, 2804 (*use*)
Rabinowitz, M. *et al, Tetrahedron*, 1979, **35**, 667 (*use*)
Willcockson, W.S. *et al, Comp. Biochem. Physiol., C*, 1982, **72**, 101 (*tox*)

1,1,3-Propanetriphosphonic acid P-00449

1,1,3-Propanetriyltrisphosphonic acid, 9CI. 1,1,3-Triphosphonopropane

$(HO)_2P(O)CH_2CH_2CH[P(O)(OH)_2]_2$

$C_3H_{11}O_9P_3$ M 284.036

Hexa-Et ester: Hexaethyl 1,1,3-propanetriyltrisphosphonate.
$C_{15}H_{35}O_9P_3$ M 452.357
Liq. d_4^{20} 1.18. $Bp_{0.08}$ 183-185°. n_D^{20} 1.4540.

Pudovik, A.N. *et al, Zh. Obshch. Khim.*, 1970, **40**, 499 (*Engl. transl.* p. 462) (*synth*)

1,2,3-Propanetriphosphonic acid P-00450

1,2,3-Propanetriyltrisphosphonic acid, 9CI. 1,2,3-Triphosphonopropane

[25404-72-2]

$(HO)_2P(O)CH[CH_2P(O)(OH)_2]_2$

$C_3H_{11}O_9P_3$ M 284.036

Anticariogenic agent. Cryst.

Hexa-Et ester: [25091-07-0]. *Hexaethyl 1,2,3-propanetriyltrisphosphonate.*
$C_{15}H_{35}O_9P_3$ M 452.357
Liq. d_4^{20} 1.17. Bp_2 194°, $Bp_{0.1}$ 170°. n_D^{20} 1.4532.

Hexaisopropyl ester: Hexaisopropyl 1,2,3-propanetriyltrisphosphonate.
$C_{21}H_{47}O_9P_3$ M 536.518
d_4^{20} 1.12. $Bp_{0.7}$ 187-188°. n_D^{20} 1.4495.

Cilley, W.A. *et al, J. Am. Chem. Soc.*, 1970, **92**, 1685 (*synth, P nmr*)
Pudovik, A.N. *et al, Zh. Obshch. Khim.*, 1975, **45**, 2403 (*Engl. transl.* p. 2362) (*esters, synth*)
Sturtz, G. *et al, J. Chem. Res.* (*S*), 1978, 89 (*ester, ir, pmr*)

Propaphos, BSI P-00451

4-(Methylthio)phenyl dipropyl phosphate, 9CI, 8CI. Kayaphos

[7292-16-2]

$$4\text{-}MeS\text{-}C_6H_4\text{-}OP(O)(OCH_2CH_2CH_3)_2$$

$C_{13}H_{21}O_4PS$ M 304.340

Systemic insecticide and bactericide. Liq. $Bp_{0.85}$ 175-177°. Stable in neutral or acid media, but hydrol. in alkali. Rapidly degraded in aq. med. by uv.

▷Toxic

Fukuhara, T. *et al, CA*, 1979, **90**, 17322 (*pharmacol, tox*)
Pesticide Manual, 7th Ed., 463.

2-Propenylphosphinic acid, 9CI P-00452

Allylphosphinic acid, 8CI

[66899-05-6]

$H_2C{=}CHCH_2PH(O)OH$

$C_3H_7O_2P$ M 106.061

Tautomeric with 2-propenylphosphonous acid.

Butyl ester: Butyl 2-propenylphosphinate. Butyl allylphosphinate.
$C_7H_{15}O_2P$ M 162.168
Liq. d_4^{20} 0.99. $Bp_{1.5}$ 68-68.5°. n_D^{20} 1.4495.

Kabachnik, M.I. *et al*, *Zh. Obshch. Khim.*, 1962, **32**, 3351 (*Engl. transl.* p. 3288) (*ester*)
Nifant'ev, É.E. *et al*, *Zh. Obshch. Khim.*, 1982, **52**, 2459 (*Engl. transl.* p. 2173) (*synth*)

1-Propenylphosphonic acid, 9CI P-00453

1-Phosphonopropene
[4672-37-1]

$H_3CCH{=}CHP(O)(OH)_2$

$C_3H_7O_3P$ M 122.060

(E)-form

Monobenzylammonium salt: Solid. Mp 155-157°.
Di-Me ester: [25362-06-5]. *Dimethyl 1-propenylphosphonate.*
$C_5H_{11}O_3P$ M 150.114
Liq. d_4^{20} 1.11. Bp_{10} 95-96°. n_D^{20} 1.4465.
Di-Et ester: [18689-32-2]. *Diethyl 1-propenylphosphonate.*
$C_7H_{15}O_3P$ M 178.167
Liq. $Bp_{2.5}$ 78-80°. n_D^{20} 1.4325.
Diisopropyl ester: Diisopropyl 1-propenylphosphonate.
$C_9H_{19}O_3P$ M 206.221
Liq. Bp_3 76-77°.
Difluoride: [26076-14-2].
$C_3H_5F_2OP$ M 126.042
Liq. d_4^{20} 1.22. Bp_{135} 75°. n_D^{20} 1.3760.
Dichloride: [20408-30-4].
$C_3H_5Cl_2OP$ M 158.952
Liq. d_4^{20} 1.33. Bp_7 74-75°. n_D^{20} 1.4915.

(Z)-form

Hygroscopic cryst. Mp 55-57°.
Di-Me ester: [25921-18-0]. Liq. d_4^{20} 1.12. Bp_7 78-78.5°. n_D^{20} 1.4418.
Di-Et ester: [18689-36-6]. Liq. d_4^{20} 1.04. Bp_{13} 142°. n_D^{20} 1.4359.
*Di-*tert*-butyl ester:* [25383-05-5]. *Di-*tert*-butyl 1-propenylphosphonate.*
$C_9H_{19}O_3P$ M 206.221
Liq. $Bp_{0.1}$ 45-46°.
Difluoride: [25921-17-9]. d_4^{20} 1.31. Bp_{12} 26-27°. n_D^{20} 1.4062.
Dichloride: [25522-46-7]. Liq. d_4^{20} 1.38. Bp_7 67-68°. n_D^{20} 1.4922.

Pudovik, A.N. *et al*, *Zh. Obshch. Khim.*, 1961, **31**, 1693 (*Engl. transl.* p. 1510) (*ester*)
Petrov, K.A. *et al*, *Tetrahedron Lett.*, 1968, 15 (*ester*)
Timofeeva, T.N. *et al*, *Zh. Obshch. Khim.*, 1968, **38**, 1255 (*Engl. transl.* p. 1208) (*dichloride, synth, pmr*)
Timofeeva, T.N. *et al*, *Zh. Obshch. Khim.*, 1969, **39**, 2446 (*Engl. transl.* p. 2386) (*dichloride, difluoride, ester, synth, pmr, ir*)
Glamkowski, E.J. *et al*, *J. Org. Chem.*, 1970, **35**, 3510 (*synth, ester, ir*)
Slates, H.L. *et al*, *Chem. Ind.* (*London*), 1978, 430 (*synth*)
Koizumi, T. *et al*, *Synthesis*, 1982, 917 (*ester*)
Galvez-Ruano, E. *et al*, *J. Mol. Struct.*, 1986, **142**, 397 (*esters, ir, raman, pmr, cmr*)

2-Propenylphosphonic acid, 9CI P-00454

Allylphosphonic acid, 8CI
[6833-67-6]

$(HO)_2P(O)CH_2CH{=}CH_2$

$C_3H_7O_3P$ M 122.060

Deliquescent cryst. (after prolonged drying). Mp 47-50°.

Monocyclohexylammonium salt: Needles (EtOH aq.). Mp 236° dec.
Di-Me ester: [757-54-0]. *Dimethyl 2-propenylphosphonate. Dimethyl allylphosphonate.*
$C_5H_{11}O_3P$ M 150.114
Monomer for fire-resistant polymers. Liq. Bp_{12} 85-86°. n_D^{20} 1.4340.
Di-Et ester: [1067-87-4]. *Diethyl 2-propenylphosphonate. Diethyl allylphosphonate.*
$C_7H_{15}O_3P$ M 178.167
Fungicide. Liq. d_4^{20} 1.034. Bp_{16} 97-98°, Bp_2 78-79°. n_D^{20} 1.4313.
Diisopropyl ester: [1067-70-5]. *Diisopropyl allylphosphonate. Bis(1-methylethyl) 2-propenylphosphonate.* Liq. Bp_{19} 106-108°, $Bp_{0.25}$ 52°. n_D^{20} 1.4258.
Di(2-propenyl) ester: [3479-30-9]. *Di-2-propenyl 2-propenylphosphonate. Diallyl allylphosphonate.* Liq. d_4^{20} 1.005. $Bp_{0.5}$ 82°, Bp_{12} 124°. n_D^{20} 1.4620.
Dibutyl ester: [4762-64-5]. *Dibutyl 2-propenylphosphonate. Dibutyl allylphosphonate.* Synergist for insecticides. Liq. d_4^{20} 0.98. Bp_1 110-112°. n_D^{20} 1.4375.
Diamide: see P-2-Propenylphosphonic diamide, P-00455
Dichloride: see 2-Propenylphosphonic dichloride, P-00456

Kennedy, J. *et al*, *J. Appl. Chem.*, 1958, **8**, 459 (*diallyl ester*)
Bride, M.H. *et al*, *J. Appl. Chem.*, 1961, **11**, 352 (*esters*)
Davies, J.H. *et al*, *J. Chem. Soc.*, 1964, 3425 (*synth*)
Meisters, A. *et al*, *Aust. J. Chem.*, 1965, **18**, 163 (*esters*)
Ernst, L., *Org. Magn. Reson.*, 1977, **9**, 35 (*ester, nmr, cmr*)
Mango, L.A., *J. Polym. Sci., Polym. Chem. Ed.*, 1977, **15**, 513 (*ester, synth, ir, pmr*)
Nifant'ev, É.E. *et al*, *Zh. Obshch. Khim.*, 1982, **52**, 2459 (*Engl. transl.* p. 2173) (*esters*)

P-2-Propenylphosphonic diamide, 9CI P-00455

P-*Allylphosphonic diamide, 8CI*

$H_2C{=}CHCH_2P(O)(NH_2)_2$

$C_3H_9N_2OP$ M 120.091

N,N,N′,N′-*Tetra-Me:* N,N,N′,N′-*Tetramethyl-*P-*2-propenylphosphonic diamide. Allylphosphonic bis(dimethylamide).*
$C_7H_{17}N_2OP$ M 176.198
Oil which solidifies at r.t. $Bp_{0.03}$ 61°. Gives a mesomeric carbanion with butyllithium.
N,N′-*Di-Ph:* N,N′-*Diphenyl-*P-*2-propenylphosphonic diamide. Allylphosphonic dianilide.*
$C_{15}H_{17}N_2OP$ M 272.286
Solid. Mp 136-137°.

Kinnear, A.M. *et al*, *J. Chem. Soc.*, 1952, 3437 (*dianilide*)
Corey, E.J. *et al*, *J. Org. Chem.*, 1969, **34**, 3053 (*deriv, synth, ir, pmr, props, use*)

2-Propenylphosphonic dichloride, 9CI P-00456

Allylphosphonic dichloride, 8CI
[1498-47-1]

$H_2C{=}CHCH_2P(O)Cl_2$

$C_3H_5Cl_2OP$ M 158.952
Liq. d^{20} 1.34, 1.38. Bp_3 55°. n_D^{20} 1.4830.

Kinnear, A.M. *et al, J. Chem. Soc.*, 1952, 3437 (*synth*)
Christol, C. *et al, J. Chem. Phys.*, 1965, **62**, 246 (*ir, raman*)
Gruzdev, V.G. *et al, Zh. Obshch. Khim.*, 1967, **37**, 450 (*Engl. transl.* p. 419) (*synth*)
Zverev, V.V. *et al, Dokl. Akad. Nauk SSSR, Ser. Sci. Khim.*, 1979, **246**, 1368 (*Engl. transl.* p. 312) (*pe*)
Morita, T. *et al, Chem. Lett.*, 1980, 435 (*synth*)

2-Propenylphosphonodithioic acid, 9CI P-00457

Allylphosphonodithioic acid, 8CI

$H_2C{=}CHCH_2P(O)(SH)_2 \rightleftharpoons H_2C{=}CHCH_2P(S)(OH)(SH)$

$C_3H_7OPS_2$ M 154.181
S,S-Dialkyl esters are said to be nematocides.
S,S-*Di-Me ester:* [24838-92-4]. S,S-*Dimethyl 2-propenylphosphonodithioate.*
$C_5H_{11}OPS_2$ M 182.235
Liq. $Bp_{0.1}$ 84°. n_D^{24} 1.5696.
S,S-*Di-Et ester:* [24838-93-5]. S,S-*Diethyl 2-propenylphosphonodithioate.*
$C_7H_{15}OPS_2$ M 210.289
Liq. $Bp_{0.02}$ 92°. n_D^{20} 1.5462.
S,S-*Diisopropyl ester:* S,S-*Diisopropyl 2-propenylphosphonodithioate.*
$C_9H_{19}OPS_2$ M 238.342
Liq. $Bp_{0.03}$ 92°. n_D^{25} 1.5265.

U.S.P., 3 467 736, (*1969*); *CA*, **72**, 32037

(2-Propenyl)phosphonoselenoic acid P-00458

Allylphosphonoselenoic acid

$H_2C{=}CHCH_2P(Se)(OH)_2 \rightleftharpoons H_2C{=}CHCH_2P(O)(SeH)(OH)$

$C_3H_7O_2PSe$ M 185.021
O,O-*Di-Et ester:* [77659-30-4]. O,O-*Diethyl (2-propenyl)phosphonoselenoate.* O,O-*Diethyl allylphosphonoselenoate.*
$C_7H_{15}O_2PSe$ M 241.128
Liq. d_4^{20} 1.26. Bp_{10} 101-102°. n_D^{20} 1.5008.
Diisopropyl ester: Bis(1-methylethyl) (2-propenyl)-phosphonoselenoate, 9CI. Diisopropyl allylphosphonoselenoate, 8CI.
$C_9H_{19}O_2PSe$ M 269.182
Liq. d_4^{20} 1.18. Bp_7 101-102°. n_D^{20} 1.4958.

Razumov, A.I. *et al, Zh. Obshch. Khim.*, 1964, **34**, 1851 (*Engl. transl.* p. 1863) (*synth*)
Zverev, V.V. *et al, Zh. Obshch. Khim.*, 1981, **51**, 303 (*Engl. transl.* p. 242) (*pe*)

P-(2-Propenyl)phosphonoselenoic diamide P-00459

P-*Allylphosphonoselenoic diamide*

$H_2C{=}CHCH_2P(Se)(NH_2)_2$

$C_3H_9N_2PSe$ M 183.051
N,N,N′,N′-*Tetra-Me:* N,N,N′,N′-*Tetramethyl*-P-*(2-propenyl)phosphonoselenoic diamide.* P-*Allyl-N,N,N′,N′-tetramethylphosphonoselenoic diamide.*
$C_7H_{17}N_2PSe$ M 239.158
Liq. d_4^{20} 1.27. $Bp_{0.035}$ 87-88°. n_D^{20} 1.5541.
N,N,N′,N′-*Tetra-Et:* N,N,N′,N′-*Tetraethyl*-P-*(2-propenyl)phosphonoselenoic diamide.* P-*Allyl-N,N,N′,N′-tetramethylphosphonoselenoic diamide.*
$C_{11}H_{25}N_2PSe$ M 295.266
Liq. d_4^{20} 1.16. $Bp_{0.053}$ 118-119°. n_D^{20} 1.5337.

Razumov, A.I. *et al, Zh. Obshch. Khim.*, 1970, **40**, 1252 (*Engl. transl.* p. 1242) (*synth*)

(2-Propenyl)phosphonotelluroic acid P-00460

Allylphosphonotelluroic acid

$H_2C{=}CHCH_2P(=Te)(OH)_2 \rightleftharpoons H_2C{=}CHCH_2P(=O)(TeH)(OH)$

$C_3H_7O_2PTe$ M 233.661
Tautomeric.
Di-Et ester: Diethyl (2-propenyl)phosphonotelluroate.
$C_7H_{15}O_2PTe$ M 289.768
Liq. d_4^{20} 1.28. $Bp_{0.32}$ 65-66°. n_D^{20} 1.5070.

Razumov, A.I. *et al, Zh. Obshch. Khim.*, 1964, **34**, 1851 (*Engl. transl.* p. 1863) (*synth*)

2-Propenylphosphonothioic acid, 9CI P-00461

Allylphosphonothioic acid, 8CI

$H_2C{=}CHCH_2P(S)(OH)_2 \rightleftharpoons H_2C{=}CHCH_2P(O)(OH)(SH)$

$C_3H_7O_2PS$ M 138.121
O,O-*Di-Et ester:* [69695-70-1]. O,O-*Diethyl 2-propenylphosphonothioate.*
$C_7H_{15}O_2PS$ M 194.228
Liq. d_4^{20} 1.04. Bp_{11} 91-91.5°. n_D^{20} 1.4723.
O,O-*Diisopropyl ester:* O,O-*Bis(1-methylethyl) 2-propenylphosphonothioate.* O,O-*Diisopropyl allylphosphonothioate.*
$C_9H_{19}O_2PS$ M 222.282
Liq. Bp_7 89-90°. n_D^{20} 1.4642.
Dichloride: [1498-61-9].
$C_3H_5Cl_2PS$ M 175.012
Liq. d_4^{20} 1.31. Bp_{20} 80°, Bp_9 65-66°. n_D^{20} 1.5553.
Diamide: see P-2-Propenylphosphonothioic diamide, P-00462

Razumov, A.I. *et al, Zh. Obshch. Khim.*, 1964, **34**, 1851 (*Engl. transl.* p. 1863) (*dichloride, esters, synth*)
Christol, C. *et al, J. Chim. Phys.*, 1965, **62**, 246 (*dichloride, ir, raman*)
Gruzdev, V.G. *et al, Zh. Obshch. Khim.*, 1967, **37**, 450 (*Engl. transl.* p. 419) (*dichloride*)
Kaushik, M.P. *et al, J. Org. Chem.*, 1980, **45**, 2270 (*dichloride*)
Zverev, V.V. *et al, Zh. Obshch. Khim.*, 1981, **51**, 303 (*Engl. transl.* p. 242) (*ester, pe*)

P-2-Propenylphosphonothioic diamide, 9CI P-00462

Allylphosphonothioic diamide, 8CI. Allylthiophosphonic diamide

$H_2C{=}CHCH_2P(S)(NH_2)_2$

$C_3H_9N_2PS$ M 136.151
N,N,N′,N′-*Tetra-Me:* [31480-11-2]. N,N,N′,N′-*Tetramethyl*-P-*2-propenylphosphonothioic diamide. Allylphosphonothioic bis(dimethylamide).*
$C_7H_{17}N_2PS$ M 192.258
Liq. d_4^{20} 1.03. Bp_{14} 134-135°. n_D^{20} 1.5244.
N,N,N′,N′-*Tetra-Et:* N,N,N′,N′-*Tetraethyl*-P-*2-propenylphosphonothioic diamide.*
$C_{11}H_{25}N_2PS$ M 248.366
Liq. $Bp_{0.09}$ 109-111°. n_D^{20} 1.5139.
N,N′-*Di-Ph:* N,N′-*Diphenyl*-P-*2-propenylphosphonothioic diamide. Allylphosphonothioic dianilide.*
$C_{15}H_{17}N_2PS$ M 288.346

Pale-yellow cryst. (MeOH). Mp 97-98°.

Razumov, A.I. *et al*, *Zh. Obshch. Khim.*, 1964, **34**, 1851 (*Engl. transl.* p. 1863) (*dianilide*)
Razumov, A.I. *et al*, *Zh. Obshch. Khim.*, 1970, **40**, 1252 (*Engl. transl.* p. 1242) (*derivs, synth, ir*)

2-Propenylphosphonous acid, 9CI P-00463

Allylphosphonous acid

$H_2C{=}CHCH_2P(OH)_2 \rightleftharpoons H_2C{=}CHCH_2PH(O)OH$

$C_3H_7O_2P$ M 106.061

Free acid exists in the tautomeric struct 2-Propenylphosphinic acid, P-00452 .

Di-Et ester: Diethyl 2-propenylphosphonite. Diethyl allylphosphonite.
$C_7H_{15}O_2P$ M 162.168
Liq. Bp_{13} 53-55°. n_D^{20} 1.4428.

Diisopropyl ester: [31080-62-3]. *Diisopropyl allylphosphonite. Bis(1-methylethyl) 2-propenylphosphonite.*
$C_9H_{19}O_2P$ M 190.222
Liq. n_D^{20} 1.4375.

Monobutyl ester: see 2-Propenylphosphinic acid, P-00452

Dichloride: [31078-91-8]. *Allyldichlorophosphine.*
$C_3H_5Cl_2P$ M 142.952
Liq. d_4^{20} 1.23. Bp_{46} 51-52°. n_D^{20} 1.5112.

Bis(dimethylamide): [31080-60-1]. N,N,N′,N′-*Tetramethyl*-P-*2-propenylphosphonous diamide.*
$C_7H_{17}N_2P$ M 160.198
Liq. Bp_{12} 64°. n_D^{20} 1.4834.

Razumov, A.I. *et al*, *Zh. Obshch. Khim.*, 1964, **34**, 1851 (*Engl. transl.* p. 1863) (*dichloride, esters, synth*)
Razumov, A.I. *et al*, *Zh. Obshch. Khim.*, 1970, **40**, 1252 (*Engl. transl.* p. 1242) (*diamides, synth*)
Razumov, A.I. *et al*, *Zh. Obshch. Khim.*, 1970, **40**, 1704 (*Engl. transl.* p. 1691) (*derivs, P nmr*)

2-Propenylphosphoramidic acid, 9CI P-00464

Allylphosphoramidic acid

$H_2C{=}CHCH_2NHP(O)(OH)_2$

$C_3H_8NO_3P$ M 137.075

Di-Et ester: [7355-27-3]. *Diethyl 2-propenylphosphoramidate. Diethyl allylphosphoramidate.*
$C_7H_{16}NO_3P$ M 193.182
No phys. props. reported.

Di-Ph ester: Diphenyl 2-propenylphosphoramidate. Diphenyl allylphosphoramidate.
$C_{15}H_{16}NO_3P$ M 289.270
Needles (pet. ether). Sol. H_2O. Mp 54-54.5°.

Dichloride:
$C_3H_6Cl_2NOP$ M 173.966
Liq. d_4^{20} 1.35. $Bp_{0.0001}$ 50°. n_D^{20} 1.4840.

Petrov, K.A. *et al*, *Zh. Obshch. Khim.*, 1962, **32**, 915 (*Engl. transl.* p. 904) (*dichloride*)
Arni, P.C. *et al*, *J. Appl. Chem.*, 1964, **14**, 221 (*diphenyl ester, synth, ir*)
Zwierzak, A., *Synthesis*, 1982, 920 (*diethyl ester, synth, ir, pmr, P nmr*)

2-Propenyl phosphorodichloridite, 9CI, 8CI P-00465

Allyl phosphorodichloridite. Allyl dichlorophosphite

[41003-33-2]

$H_2C{=}CHCH_2OPCl_2$

$C_3H_5Cl_2OP$ M 158.952

Pungent liq. d_0^0 1.29, d_0^{18} 1.22. Bp_{742} 140°.

Podladtschikoff, M., *J. Russ. Phys. Chem. Soc.*, 1899, **31**, 30; *Chem. Zentralbl.*, 1899, I, 1067.

Propetamphos, BSI P-00466

1-Methylethyl 3-[[(ethylamino)-methoxyphosphinothioyl]oxy]-2-butenoate, 9CI. O-*2-Isopropoxycarbonyl-1-methylvinyl* O-*methyl ethylphosphoramidothioate. Safrotin*

[58995-37-2]

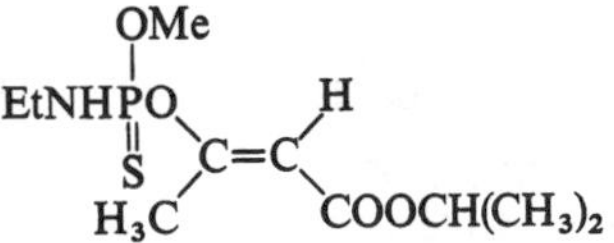

$C_{10}H_{20}NO_4PS$ M 281.306

▷GQ4750000.

(*E*)-*form* [31218-83-4]

Contact insecticide, used for household and public health purposes. Spar. sol. H_2O. d_4^{20} 1.13. $Bp_{0.005}$ 87-89°. n_D^{20} 1.495.

Ger. Pat., 2 739 310, (*1978*); *CA*, **89**, 42424
Pesticide Manual, 6th Ed, 449; 7th Ed., 466.
Agrochemicals Handbook, Royal Society of Chemistry, 1983, A345.

(1-Propylheptyl)phosphonic acid, 8CI P-00467

4-Decylphosphonic acid

$(HO)_2P(O)CH(CH_2CH_2CH_3)CH_2(CH_2)_4CH_3$

$C_{10}H_{23}O_3P$ M 222.264

(±)-*form*

Dichloride: [16474-70-7].
$C_{10}H_{21}Cl_2OP$ M 259.155
Liq. $Bp_{0.02}$ 50°. n_D^{25} 1.4650.

Geiseler, G. *et al*, *Ber. Bunsenges. Phys. Chem.*, 1967, **71**, 478 (*synth, ir, raman*)

O-Propyl methylphosphonochloridothioate, 9CI P-00468

[18005-37-3]

$$Cl{\blacktriangleright}P(=S)(OCH_2CH_2CH_3)(Me)$$

(*S*)-*form*

$C_4H_{10}ClOPS$ M 172.609

(*S*)-*form* [38315-82-1]

Oil. $[\alpha]_D$ +10.25° (neat).

(±)-*form*

Liq. Bp_2 54-55°. n_D^{25} 1.4886.

Hoffmann, F.W. *et al*, *J. Am. Chem. Soc.*, 1958, **80**, 3945 (*synth*)
Chupp, J.P. *et al*, *J. Org. Chem.*, 1962, **27**, 3832 (*synth*)
Mikolajczyk, M. *et al*, *Tetrahedron*, 1972, **28**, 3855 (*synth*)

O-Propyl methylphosphonodithioate P-00469

[1000-53-9]

Me(P=S)(SH)OCH$_2$CH$_2$CH$_3$

$C_4H_{11}OPS_2$ M 170.224

Liq. $Bp_{0.02}$ 50°. $n_D^{22.5}$ 1.5356.

Chupp, J.P. *et al*, *J. Org. Chem.*, 1962, **27**, 3832 (*synth*)

Propyl phosphinate, 9CI P-00470

Propyl hypophosphite

[18108-08-2]

$H_3CCH_2CH_2OP(O)H_2 \rightleftharpoons H_3CCH_2CH_2OP(OH)H$

$C_3H_9O_2P$ M 108.077

Liq. Rather unstable. Bp_{10} 79-87°.

Ivanov, B.E. *et al*, *Izv. Akad. Nauk SSSR, Ser. Khim.*, 1967, 1498 (*Engl. transl.* p. 1447); *CA*, **68**, 78359 (*synth*)

Karlstédt, N.B. *et al*, *Zh. Obshch. Khim.*, 1976, **46**, 2018 (*Engl. transl.* p. 1942) (*synth, P nmr*)

Propylphosphine, 9CI P-00471

[40200-59-7]

$H_2PCH_2CH_2CH_3$

C_3H_9P M 76.078

Foul-smelling liq., sensitive to air. Bp_{750} 54°.

Kreutzkampf, N., *Chem. Ber.*, 1954, **87**, 919 (*synth*)

Propylphosphinic acid, 9CI P-00472

[75779-78-1]

$H_3CCH_2CH_2PH(O)OH$

$C_3H_9O_2P$ M 108.077

Syrup. Rather insol. Et_2O, EtOH. d^{13} 1.14.

Guichard, F., *Ber.*, 1899, **32**, 1572.

Nifant'ev, É.E. *et al*, *Zh. Obshch. Khim.*, 1980, **50**, 1744 (*Engl. transl.* p. 1416)

Propylphosphonic acid, 9CI P-00473

1-Propanephosphonic acid

[4672-38-2]

$H_3CCH_2CH_2P(O)(OH)_2$

$C_3H_9O_3P$ M 124.076

Monoalkyl esters control plant growth and fruit ripening. Cryst. (C_6H_6). Mp 73°. pK_{a1} 2.49, pK_{a2} 8.18 (H_2O, 25°).

▷TA0420000.

Di-Me ester: [18755-43-6]. *Dimethyl propylphosphonate.*
$C_5H_{13}O_3P$ M 152.130
Bp_{14} 80-82°. n_D^{20} 1.4210.

Di-Et ester: [18812-51-6]. *Diethyl propylphosphonate.*
$C_7H_{17}O_3P$ M 180.183
Liq. d_4^{20} 1.01. Bp_9 88-89°. n_D^{20} 1.4172.

Dipropyl ester: [1789-95-3]. *Dipropyl propylphosphonate.*
$C_9H_{21}O_3P$ M 208.237
Liq. Bp_{18} 126°.

Dibutyl ester: Dibutyl propylphosphonate.
$C_{11}H_{25}O_3P$ M 236.290
Liq. d_4^{20} 0.96. Bp_{10} 133-137°. n_D^{20} 1.4272.

Dichloride: see Propylphosphonic dichloride, P-00475

Crofts, P.C. *et al*, *J. Am. Chem. Soc.*, 1953, **75**, 3379 (*synth*)

Ionin, B.I. *et al*, *Zh. Obshch. Khim.*, 1964, **34**, 2630 (*Engl. transl.* p. 2651) (*ester, synth, struct*)

Orlov, Yu.F. *et al*, *Zh. Obshch. Khim.*, 1965, **35**, 2046 (*Engl. trans.* p. 2037) (*ester, ir*)

Dietze, U., *J. Prakt. Chem.*, 1974, **316**, 293 (*ir*)

Jentzsch, R. *et al*, *J. Prakt. Chem.*, 1975, **317**, 721 (*ester*)

Ernst, L., *Org. Magn. Reson.*, 1977, **9**, 35 (*ester, cmr, P nmr*)

P-Propylphosphonic diamide, 9CI P-00474

$H_3CCH_2CH_2P(O)(NH_2)_2$

$C_3H_{11}N_2OP$ M 122.106

N,N,N′,N′-*Tetra-Me:* [14655-70-0]. N,N,N′,N′-*Tetramethyl*-P-*propylphosphonic diamide. Propylphosphonic bis(dimethylamide).*
$C_7H_{19}N_2OP$ M 178.214
d_4^{20} 0.98. Bp_{33} 149°, $Bp_{0.4}$ 70°. n_D^{22} 1.4553.

N,N,N′,N′-*Tetra-Et:* N,N,N′,N′-*Tetraethyl*-P-*propylphosphonic diamide. Propylphosphonic bis(diethylamide).*
$C_{11}H_{27}N_2OP$ M 234.321
Liq. Bp_{28} 189°. n_D^{20} 1.4580.

N,N,N′,N′-*Tetrapropyl: Pentapropylphosphonic diamide.*
$C_{15}H_{35}N_2OP$ M 290.428
Liq. $Bp_{1.2}$ 141°. n_D^{20} 1.4620.

Kosolapoff, G.M. *et al*, *J. Org. Chem.*, 1956, **21**, 413 (*synth*)

Normant, H. *et al*, *C.R. Hebd. Seances Acad. Sci., Ser. C*, 1967, **264**, 707 (*synth*)

Propylphosphonic dichloride, 9CI P-00475

[4708-04-7]

$H_3CCH_2CH_2P(O)Cl_2$

$C_3H_7Cl_2OP$ M 160.967

Liq. d_4^{25} 1.27. Bp_{15} 83-85°. n_D^{20} 1.4648.

Geiseler, G. *et al*, *Ber. Bunsenges. Phys. Chem.*, 1967, **71**, 478 (*ir, raman*)

Tsvetkov, E.N. *et al*, *Izv. Akad. Nauk SSSR, Ser. Khim.*, 1967, 2375 (*Engl. transl.* p. 2267) (*synth, nqr*)

Propylphosphonochloridic acid, 9CI P-00476

$H_3CCH_2CH_2PCl(O)OH$

$C_3H_8ClO_2P$ M 142.522

Cyclohexyl ester: Cyclohexyl propylphosphonochloridate.
$C_9H_{18}ClO_2P$ M 224.667
Liq. $Bp_{0.05}$ 79°. n_D^{25} 1.4688.

Hafner, L.S. *et al*, *J. Med. Chem.*, 1970, **13**, 1025.

Propylphosphonochloridothioic acid, 9CI P-00477

$H_3CCH_2CH_2P(=S)(OH)Cl \rightleftharpoons H_3CCH_2CH_2P(=O)(SH)Cl$

C_3H_8ClOPS M 158.582

S-*Me ester:* [19057-03-5]. S-*Methyl propylphosphonochloridothioate.*
$C_4H_{10}ClOPS$ M 172.609

Liq. d_4^{20} 1.23. Bp_2 84-86°. n_D^{20} 1.5083.
S-*Propyl ester:* [19057-04-6]. S-*Propyl propylphosphonochloridothioate.*
$C_6H_{14}ClOPS$ M 200.663
Liq. d_4^{20} 1.17. $Bp_{1.5}$ 96-97°. n_D^{20} 1.4975.
S-*Isopropyl ester:* S-*Isopropyl propylphosphonochloridothioate.*
$C_6H_{14}ClOPS$ M 200.663
Liq. d_4^{20} 1.14. Bp_5 105-106°. n_D^{20} 1.4930.

Neimysheva, A.A. *et al*, *Zh. Obshch. Khim.*, 1967, **37**, 1822 (*Engl. transl.* p. 1736) (*synth, props*)

Propylphosphonodithioic acid, 9CI — P-00478

$H_3CCH_2CH_2P(O)(SH)_2 \rightleftharpoons H_3CCH_2CH_3P(S)(OH)(SH)$

$C_3H_9OPS_2$ M 156.197
O-*Et ester:* [51926-09-1]. O-*Ethyl hydrogen propylphosphonodithioate.*
$C_5H_{13}OPS_2$ M 184.251
Liq. $Bp_{0.1}$ 46.5-48°.
S,S-*Dipropyl ester:* S,S-*Dipropyl propylphosphonodithioate.*
$C_9H_{21}OPS_2$ M 240.358
Liq. $Bp_{0.1}$ 89-91°. n_D^{25} 1.5176.
S,S-*Di-Ph ester:* [29703-21-7]. S,S-*Diphenyl propylphosphonodithioate.*
$C_{15}H_{17}OPS_2$ M 308.392
Cryst. (ligroin). Mp 66-67°.

Ger. Pat., 1 902 928, (*1970*); *CA*, **73**, 119693 (*diphenyl ester*)
U.S.P., 3 837 834, (*1974*); *CA*, **82**, 12288

Propylphosphonofluoridic acid, 9CI — P-00479

$H_3CCH_2CH_2PF(O)OH$

$C_3H_8FO_2P$ M 126.067
Cyclohexyl ester: [28364-21-8]. *Cyclohexyl propylphosphonofluoridate.*
$C_9H_{18}FO_2P$ M 208.212
Liq. $Bp_{0.03}$ 43°. n_D^{25} 1.4378.

Hafner, L.S. *et al*, *J. Med. Chem.*, 1970, **13**, 1025 (*synth*)

Propylphosphonothioic acid, 9CI — P-00480

$H_3CCH_2CH_2P(S)(OH)_2 \rightleftharpoons H_3CCH_2CH_2P(O)(OH)(SH)$

$C_3H_9O_2PS$ M 140.137
Na salt: Solid. Mp 197-199°.
O,O-*Di-Et ester:* O,O-*Diethyl propylphosphonothioate.*
$C_7H_7O_2PS$ M 186.165
Liq. d_4^{20} 1.02. Bp_2 63.5-65.5°. n_D^{20} 1.4596.
O,S-*Di-Et ester:* [2511-13-9]. O,S-*Diethyl propylphosphonothioate.* Liq. d_4^{20} 1.04. Bp_3 85-86.5°. n_D^{20} 1.4733.
Dichloride: [2524-01-8].
$C_3H_7Cl_2PS$ M 177.028
Pungent liq. d^{20} 1.30. Bp_{10} 66-67°. n_D^{20} 1.5360.
Dianilide: N,N'-*Diphenyl*-P-*propylphosphonothioic diamide.*
$C_{15}H_{19}N_2PS$ M 290.362
Solid. Mp 133°.

Kabachnik, M.I. *et al*, *Dokl. Akad. Nauk SSSR, Ser. Sci. Khim.*, 1956, **110**, 217 (*Engl. transl.* p. 549) (*dichloride, dianilide, synth*)
Kabachnik, M.I. *et al*, *Izv. Akad. Nauk SSSR, Ser. Khim.*, 1956, 193 (*Engl. transl.* p. 185) (*esters, synth*)
Kabachnik, M.I. *et al*, *Zh. Obshch. Khim.*, 1956, **26**, 2228 (*Engl. transl.* p. 2491) (*esters, synth, ir*)

P-Propylphosphonothioic diamide, 9CI — P-00481

$H_3CCH_2CH_2P(S)(NH_2)_2$

$C_3H_{11}N_2PS$ M 138.167
N,N'-*Di-Ph:* N,N'-*Diphenyl*-P-*propylphosphonothioic diamide. Propylphosphonothioic dianilide.*
$C_{15}H_{19}N_2PS$ M 290.362
Solid. Mp 133°.

Kabachnik, M.I. *et al*, *Dokl. Akad. Nauk SSSR, Ser. Sci. Khim.*, 1956, **110**, 217; *CA*, **51**, 4982.

Propylphosphonous acid — P-00482

$H_3CCH_2CH_2P(OH)_2$

$C_3H_9O_2P$ M 108.077
Free acid exists as the phosphoryl tautomer Propylphosphinic acid, P-00472 .
Di-Me ester: [6131-74-4]. *Dimethyl propylphosphonite.*
$C_5H_{13}O_2P$ M 136.130
Liq. Bp_9 30-31°.
Di-Et ester: [51825-26-4]. *Diethyl propylphosphonite.*
$C_7H_{17}O_2P$ M 164.184
Liq. Bp_{12} 52-53°. n_D^{20} 1.4275.
Diisopropyl ester: [31078-90-7]. *Diisopropyl propylphosphonite.*
$C_9H_{21}O_2P$ M 192.237
Liq. Bp_{15} 67°. n_D^{20} 1.4250.
Dichloride: see Propylphosphonous dichloride, P-00483
Bis(dipropylamide): [32596-69-3]. *Pentapropylphosphonous diamide.*
$C_{15}H_{35}N_2P$ M 274.429
Liq. Bp_{10} 140°. n_D^{20} 1.4695.

Razumov, A.I. *et al*, *Zh. Obshch. Khim.*, 1970, **40**, 1704 (*Engl. transl.* p. 1691) (*diisopropyl ester, synth, dimethylamide, P nmr*)
Zentil, M. *et al*, *Bull. Soc. Chim. Fr.*, 1971, 376 (*dipropylamide, synth*)
Efimova, V.D. *et al*, *Zh. Obshch. Khim.*, 1974, **44**, 55 (*Engl. transl.* p. 51) (*esters, synth*)
Quin, L.D. *et al*, *Org. Magn. Reson.*, 1974, **6**, 503 (*dimethyl ester, synth, cmr*)

Propylphosphonous dichloride — P-00483

Dichloropropylphosphine
[15573-31-3]

$H_3CCH_2CH_2PCl_2$

$C_3H_7Cl_2P$ M 144.968
Pungent Liq. d_4^{27} 1.17. Bp 134.5°, Bp_{90} 70°. n_D^{25} 1.4842.

Fox, R.B., *J. Am. Chem. Soc.*, 1950, **72**, 4147 (*synth*)
Goetz, H. *et al*, *Justus Liebigs Ann. Chem.*, 1967, **704**, 1 (*synth*)
Razumov, A.I. *et al*, *Zh. Obshch. Khim.*, 1970, **40**, 1704 (*Engl. transl.* p. 1691) (*P nmr*)
Quin, L.D. *et al*, *Org. Magn. Reson.*, 1974, **6**, 503 (*cmr*)

Propylphosphoramidic acid, 9CI — P-00484

$H_3CCH_2CH_2NHP(O)(OH)_2$

$C_3H_{10}NO_3P$ M 139.091

Di-Me ester: [20465-00-3]. *Dimethyl propylphosphoramidate.*

$C_5H_{14}NO_3P$ M 167.144

Liq. d_4^{20} 1.11. Bp_{11} 132-134°. n_D^{20} 1.4290.

Di-Et ester: [2672-32-4]. *Diethyl propylphosphoramidate.*

$C_7H_{18}NO_3P$ M 195.198

Liq. d_4^{20} 1.03. Bp_{10} 140°. n_D^{20} 1.4278.

Dipropyl ester: [17222-03-6]. *Dipropyl propylphosphoramidate.*

$C_9H_{22}NO_3P$ M 223.251

Liq. d_4^{20} 1.00. $Bp_{0.03}$ 96-98°. n_D^{20} 1.4310.

Di-Ph ester: [5756-03-6]. *Diphenyl propylphosphoramidate.*

$C_{15}H_{18}NO_3P$ M 291.286

Cryst. (pet. ether or cyclohexane). Mp 56-57°. Bp_8 208°.

Dichloride: [53931-67-2].

$C_3H_8Cl_2NOP$ M 175.982

Liq. Bp_{16} 146°.

Stock, J.A. *et al*, *J. Chem. Soc.* (*C*), 1966, 637 (*diphenyl ester*)
Cadogan, J.I.G. *et al*, *J. Chem. Soc.* (*C*), 1967, 1356 (*esters*)
Nikonorov, K.V. *et al*, *Izv. Akad. Nauk SSSR, Ser. Khim.*, 1968, 587 (*Engl. transl.* p. 567) (*esters, synth*)
Roth, H.J. *et al*, *Arch. Pharm.* (*Weinheim, Ger.*), 1981, **314**, 85 (*dichloride, synth, pmr*)
Zwierzak, A., *Synthesis*, 1982, 920 (*diethyl ester, synth, ir, pmr, P nmr*)

Propylphosphoramidothioic acid, 9CI, 8CI — P-00485

$H_3CCH_2CH_2NHP(S)(OH)_2 \rightleftharpoons H_3CCH_2CH_2NHP(O)(OH)(SH)$

$C_3H_{10}NO_2PS$ M 155.151

O,O-*Di-Et ester:* [36592-34-4]. O,O-*Diethyl propylphosphoramidothioate.*

$C_7H_{18}NO_2PS$ M 211.258

Yellow oil. d_0^{15} 1.01. Bp_{11} 98°.

Dichloride: [6141-80-6].

$C_3H_8Cl_2NPS$ M 192.043

Oil. Bp_{17} 121°.

Michaelis, A., *Justus Liebigs Ann. Chem.*, 1903, **326**, 129, 201.
Bock, H. *et al*, *Chem. Ber.*, 1966, **99**, 377 (*dichloride*)

Propylphosphoramidous acid, 9CI — P-00486

$H_3CCH_2CH_2NHP(OH)_2$

$C_3H_{10}NO_2P$ M 123.091

Di-Et ester: [24310-72-3]. *Diethyl propylphosphoramidite.*

$C_7H_{18}NO_2P$ M 179.198

Liq. Bp_{10} 76-80°. n_D^{20} 1.4350.

Dichloride:

$C_3H_8Cl_2NP$ M 159.983

Liq. d^{15} 1.23. Bp_{10} 97°.

Michaelis, A., *Justus Liebigs Ann. Chem.*, 1903, **326**, 129 (*dichloride*)
Tupchienko, S.K. *et al*, *Zh. Obshch. Khim.*, 1981, **51**, 1015 (*Engl. transl.* p. 847) (*ester*)

Propyl phosphorodichloridate, 9CI, 8CI — P-00487

Propyl phosphoryl dichloride. Propyl dichlorophosphate

[10173-43-0]

$H_3CCH_2CH_2OP(O)Cl_2$

$C_3H_7Cl_2O_2P$ M 176.967

Pungent liq. d_4^{20} 1.31. Bp_{10} 67°. n_D^{20} 1.4380.

▷Hydrolyses in moist air

Grunze, H., *Chem. Ber.*, 1959, **92**, 850 (*synth*)
Gazizov, M.B. *et al*, *Zh. Obshch. Khim.*, 1973, **43**, 2087 (*Engl. transl.* p. 2073) (*synth*)

Propyl phosphorodichloridite, 9CI, 8CI — P-00488

Propyl dichlorophosphite

[13040-68-1]

$H_3CCH_2CH_2OPCl_2$

$C_3H_7Cl_2OP$ M 160.967

Pungent liq. d_4^{20} 1.23. Bp 143-145°. n_D 1.466.

Kowalewsky, W.A., *J. Russ. Phys. Chem. Soc.*, 1897, **29**, 217; *Chem. Zentralbl.*, 1897, II, 333.

O-Propyl phosphorodichloridothioate, 9CI, 8CI — P-00489

O-*Propyl thiophosphoryl dichloride.* O-*Propyl dichlorothiophosphate*

[2523-96-8]

$H_3CCH_2CH_2OP(S)Cl_2$

$C_3H_7Cl_2OPS$ M 193.027

Liq. d_0^0 1.33. Bp_{20} 84°.

Pishchimuka, P., *Ber.*, 1908, **41**, 3854 (*synth*)
Shagidullin, R.R. *et al*, *Dokl. Akad. Nauk SSSR, Ser. Sci. Khim.*, 1975, **222**, 897 (*Engl. transl.* p. 564) (*uv*)
Shagidullin, R.R. *et al*, *Izv. Akad. Nauk SSSR, Ser. Khim.*, 1975, 1527 (*Engl. transl.* p. 1414) (*ir, raman, conformn*)
Baeuerlein, E. *et al*, *Phosphorus Sulfur*, 1978, **5**, 53 (*props*)

S-Propyl phosphorodichloridothioate, 9CI, 8CI — P-00490

S-*Propyl dichlorothiophosphate*

[18281-78-2]

$H_3CCH_2CH_2SP(O)Cl_2$

$C_3H_7Cl_2OPS$ M 193.027

Bp_{10} 105-110°.

Kobayashi, K. *et al*, *CA*, 1970, **72**, 100196 (*synth*)
Grosse, J. *et al*, *Angew. Chem., Int. Ed. Engl.*, 1982, **21**, 542 (*synth, P nmr*)

Propyl phosphorodichloridothioite, 9CI — P-00491

Propyl dichlorothiophosphite

[36696-24-9]

$H_3CCH_2CH_2SPCl_2$

$C_3H_7Cl_2PS$ M 177.028

Pungent, odorous liq. d_4^{20} 1.28. Bp_{10} 66-67°. n_D^{20} 1.5493. Slowly disproportionates to PCl_3 and Dipropyl phosphorochloridodithioite, D-01220 .

Stepashkina, L.V. *et al*, *Izv. Akad. Nauk SSSR, Ser. Khim.*, 1972, 380 (*Engl. transl.* p. 330) (*synth*)

Propyl phosphorodifluoridite, 9CI, 8CI — P-00492

Propyl difluorophosphite

[3964-95-2]

$H_3CCH_2CH_2OPF_2$

$C_3H_7F_2OP$ M 128.058
Ligand for Ni. Liq. Bp 44.5°. n_D^{20} 1.3400.

Schmutzler, R., *Chem. Ber.*, 1963, **96**, 2435 (*synth, F nmr*)

S-Propyl phosphorothioate, 9CI, 8CI **P-00493**
S-Propyl dihydrogen phosphorothioate. S-Propyl dihydrogen thiophosphate. S-Propyl thiophosphonic acid

$H_3CCH_2CH_2SP(O)(OH)_2$

$C_3H_9O_3PS$ M 156.136
Monodicyclohexylammonium salt: Solid. Mp 161-162°.
Dichloride: see S-Propyl phosphorodichloridothioate, P-00490

Zwierzak, A. *et al*, *Z. Naturforsch., B*, 1971, **26**, 386 (*synth, ir, pmr*)

1-Propynylphosphonic acid, 9CI **P-00494**
1-Phosphonopropyne
[5518-69-4]

$H_3CC{\equiv}CP(O)(OH)_2$

$C_3H_5O_3P$ M 120.044
p-*Anisidinium salt:* Tan plates + $0.5H_2O$ (EtOH). Mp 207°.
Di-Me ester: [5131-05-5]. *Dimethyl 1-propynylphosphonate.*
$C_5H_9O_3P$ M 148.098
Liq. d_4^{20} 1.16. Bp_1 90-92°. n_D^{20} 1.4500.
Di-Et ester: [1067-88-5]. *Diethyl 1-propynylphosphonate.*
$C_7H_{13}O_3P$ M 176.152
Liq. d_4^{20} 1.07. Bp_1 91-91.5°. n_D^{20} 1.4472.
Diisopropyl ester: [3201-79-4]. *Diisopropyl 1-propynylphosphonate.*
$C_9H_{17}O_3P$ M 204.205
Liq. d_4^{20} 1.02. $Bp_{0.3}$ 105-107°. n_D^{20} 1.4425.
Di-Ph ester: [3095-09-8]. *Diphenyl 1-propynylphosphonate.*
$C_{15}H_{13}O_3P$ M 272.240
Liq. d_4^{20} 1.19. $Bp_{0.45}$ 171-173°. n_D^{20} 1.5576.
Difluoride: [18026-42-1].
$C_3H_3F_2OP$ M 124.027
Liq. Bp_{12} 26°.
Dichloride: see 1-Propynylphosphonic dichloride, P-00496
Bis(N,N-*dimethylamide*): [16400-23-0]. N,N,N′,N′-*Tetramethyl*-P-*1-propynylphosphonic diamide.*
$C_7H_{15}N_2OP$ M 174.182
Liq. $Bp_{0.5}$ 95°.

Pudovik, A.N. *et al*, *Zh. Obshch. Khim.*, 1965, **35**, 1210 (*Engl. transl.* p. 1214) (*esters*)
Gordon, M. *et al*, *J. Org. Chem.*, 1966, **31**, 333 (*esters, synth, ir*)
Sturtz, G., *Bull. Soc. Chim. Fr.*, 1966, 1707; 1967, 1345 (*ester*)
Charrier, C. *et al*, *C.R. Hebd. Seances Acad. Sci., Ser. C*, 1967, **264**, 995 (*dimethylamide, pmr*)
Fluck, E. *et al*, *Z. Anorg. Allg. Chem.*, 1972, **393**, 126 (*difluoride, synth, F and P nmr*)
Lequan, R.-M. *et al*, *Org. Magn. Reson.*, 1975, **7**, 392 (*ester, pmr, cmr, P nmr*)
Blackburn, G.M. *et al*, *J. Chem. Soc., Perkin Trans. 1*, 1980, 1150 (*synth*)
Kirilov, M. *et al*, *Monatsh. Chem.*, 1980, **111**, 1351 (*ester*)

2-Propynylphosphonic acid, 9CI **P-00495**
Propargylphosphonic acid. 3-Phosphono-1-propyne

$HC{\equiv}CCH_2P(O)(OH)_2$

$C_3H_5O_3P$ M 120.044
Di-Et ester: [3201-73-8]. *Diethyl 2-propynylphosphonate.*
$C_7H_{17}O_3P$ M 180.183
Liq. $Bp_{0.8}$ 78-80°.
Dibutyl ester: [82756-86-3]. *Dibutyl 2-propynylphosphonate.*
$C_{11}H_{21}O_3P$ M 232.259
Employed in synth. of tritium-labelled Fosfonomycin.

U.S.P., 2 843 617, (*1956*); *CA*, **53**, 2090 (*ethyl ester*)
Mertel, H.E. *et al*, *J. Labelled Compd. Radiopharm.*, 1982, **19**, 405 (*dibutyl ester, use*)

1-Propynylphosphonic dichloride, 9CI **P-00496**
[4981-30-0]

$H_3CC{\equiv}CP(O)Cl_2$

$C_3H_3Cl_2OP$ M 156.936
Liq. d_4^{20} 1.40. $Bp_{1.5}$ 56-57°. n_D^{25} 1.4912.

Mashlyakovskii, L.N. *et al*, *Zh. Obshch. Khim.*, 1965, **35**, 1577 (*Engl. transl.* p. 1582) (*synth, ir, pmr*)
Tarasov, V.V. *et al*, *Zh. Obshch. Khim.*, 1968, **38**, 130 (*Engl. transl.* p. 129) (*ir*)
Fluck, E. *et al*, *Z. Anorg. Allg. Chem.*, 1972, **393**, 126 (*pmr, P nmr*)
Lequan, R.-M. *et al*, *Org. Magn. Reson.*, 1975, **7**, 392 (*pmr, P nmr, cmr*)

2-Propynyl phosphorodichloridate, 9CI **P-00497**
2-Propynyl dichlorophosphate
[53799-86-3]

$HC{\equiv}CCH_2OP(O)Cl_2$

$C_3H_3Cl_2O_2P$ M 172.935
Liq. $Bp_{0.01}$ 62°.

Sturtz, G. *et al*, *Synthesis*, 1974, 730 (*synth, ir, pmr*)

Prothiofos, BSI, ISO **P-00498**
O-(*2,4-Dichlorophenyl*) O-*ethyl* S-*propyl phosphorodithioate, 9CI. Tokuthion. Bideron*
[34643-46-4]

2,4-$Cl_2C_6H_3$–OP(S)(OEt)$S(CH_2)_2CH_3$

$C_{11}H_{15}Cl_2O_2PS_2$ M 345.238
Agricultural and public health insecticide. Liq. V. spar. sol. H_2O. d_4^{20} 1.3. $Bp_{0.1}$ 125-128°.
▷TD5680000.

Wada, Y. *et al*, *CA*, 1979, **91**, 187899 (*uv, ir, P nmr, ms*)
Ueyama, I, *et al*, *Pestic. Biochem. Physiol.*, 1980, **14**, 98 (*metab, ms*)
Pesticide Manual, 6th Ed., 455; 7th Ed., 474.
The Agrochemicals Handbook, Royal Society of Chemistry, 1983, A352.

Prothoate, BSI **P-00499**

O,O-*Diethyl* S-[*2-(1-methylethyl)amino-2-oxoethyl*] *phosphorodithioate*, 9CI. O,O-*Diethyl phosphorodithioate*, S-*ester with* N-*isopropyl-2-mercaptoacetamide*, 8CI. O,O-*Diethyl* S-(*isopropylcarbamoyl*)*methyl phosphorodithioate. Fac*

[2275-18-5]

$(EtO)_2P(S)SCH_2CONHCH(CH_3)_2$

$C_9H_{20}NO_3PS_2$ M 285.356

Systemic insecticide and acaricide. Solid. Spar. sol. H_2O. Mp 28.5°. $Bp_{0.1}$ 135°. Slowly hydrol. by acids.

▷Toxic, LD_{50} (mice) ca. 35 mg/Kg. TD8225000.

Chen, P.R.S. *et al, Pestic. Biochem. Physiol.*, 1971, **1**, 340 (*synth, metab, tox*)
Bazzi, B. *et al, Pestic. Sci.*, 1974, **5**, 511 (*glc*)
Pesticide Manual, 6th Ed., 456; 7th Ed., 475.

Proticin, 9CI **P-00500**

[12689-28-0]

Struct. unknown

$C_{31}H_{44}NaO_7P$ M 582.648

Triene antibiotic. Isol. from *Bacillus licheniformis*. Shows activity against gram-positive bacteria and mycobacteria. Amorph. solid. $[\alpha]_D^{22}$ −78° (c, 0.35 in EtOH). Contains phosphoric acid monoester function.

▷UL0352000.

Nesemann, G. *et al, Naturwissenschaften*, 1972, **59**, 81.
Präve, P. *et al, J. Antibiot.*, 1972, **25**, 1 (*isol, props*)
Vértesy, L. *et al, J. Antibiot.*, 1972, **25**, 4.

(1*H*-Pyrazol-3-yl)phosphonic acid, 9CI **P-00501**

3-Phosphonopyrazole

$C_3H_5N_2O_3P$ M 148.058

Di-Et ester: Diethyl (*1*H-*pyrazol-3-yl*)*phosphonate.*

$C_7H_{15}N_2O_3P$ M 206.181

Liq. $Bp_{0.5}$ 125-127°. n_D^{20} 1.4779.

Pudovik, A.N. *et al, Zh. Obshch. Khim.*, 1964, **34**, 3942 (*Engl. transl.* p. 4003); 1971, **41**, 1017 (*Engl. transl.* p. 1021) (*synth, ir*)

Pyrazophos, BSI **P-00502**

Ethyl 2-[(*diethoxyphosphinothioyl*)*oxy*]*-5-methylpyrazolo*[*1,5-*a]*pyrimidine-6-carboxylate*, 9CI. *Ethyl 2-hydroxy-5-methylpyrazolo*[*1,5-*c]*pyrimidine-6-carboxylate*, O-*ester with* O,O-*diethyl phosphorothioate*, 8CI. O-*6-Ethoxycarbonyl-5-methylpyrazolo*[*1,5-*a]*-pyrimidin-2-yl* O,O-*diethyl phosphorothioate. Afugan. Curamil*

[13457-18-6]

$C_{14}H_{20}N_3O_5PS$ M 373.363

Systemic agricultural fungicide. Solid. Mp 50-51° (38-40°). Hydrol. by acid and by alkali.

▷TF2035000.

U.S.P., 3 496 178, (*1965*); *CA*, **66**, 37946b (*synth*)
Stan, H.-J. *et al, Biomed. Mass Spectrom.*, 1982, **9**, 483 (*ms*)
Stan, H.-J. *et al, J. Chromatogr.*, 1983, **279**, 173 (*glc*)
Pesticide Manual, 6th Ed, 458; 7th Ed., 477.
The Agrochemicals Handbook, Royal Society of Chemistry, 1983, A356.

2-Pyridinephosphonic acid **P-00503**

2-Pyridinylphosphonic acid, 9CI. *2-Pyridylphosphonic acid*, 8CI. *2-Phosphonopyridine*

[26384-86-1]

$C_5H_6NO_3P$ M 159.081

Cryst. (EtOH aq.). Mp 217-224°, 224-227° (190-191°). pK_{a1} 4.13, pK_{a2} 7.71.

Di-Me ester: [61864-97-9]. *Dimethyl 2-pyridinylphosphonate.*

$C_7H_{10}NO_3P$ M 187.135

$Bp_{0.6}$ 115°.

Di-Et ester: [23081-78-9]. *Diethyl 2-pyridinylphosphonate.*

$C_9H_{14}NO_3P$ M 215.188

$Bp_{0.1}$ 134°, $Bp_{0.03}$ 105-107°. n_D^{20} 1.4940.

Dichloride:

$C_5H_4Cl_2NOP$ M 195.972

Liq. d_4^{20} 1.56. $Bp_{0.04}$ 82°. n_D^{20} 1.4682.

Bis(*diethylamide*)*:* N,N,N′,N′-*Tetraethyl*-P-*2-pyridinylphosphonic diamide.*

$C_{13}H_{24}N_3OP$ M 269.326

Cryst. (Et_2O). Mp 50-51°. $Bp_{0.03}$ 105°, $Bp_{0.002}$ 135-137°. n_D^{20} 1.5032.

Redmore, D., *J. Org. Chem.*, 1973, **38**, 1306 (*nmr, ms, uv*)
Eliseenkov, V.N. *et al, Khim. Geterotsikl. Soedin.*, 1974, 1354 (*Engl. transl.* p. 1182) (*derivs*)
Predivoditelev, D.A. *et al, Khim. Geterotsikl. Soedin.*, 1975, 377 (*Engl. transl.* p. 330) (*derivs, synth, ir, pmr*)
Loran, J.S. *et al, J. Chem. Soc., Perkin Trans. 2*, 1976, 1444 (*synth*)
Boduszek, B. *et al, Synthesis*, 1979, 452 (*synth, ir, pmr, ms*)
Redmore, D., *Phosphorus Sulfur*, 1979, **5**, 271 (*cmr, P nmr*)
Bulot, J.J. *et al, Phosphorus Sulfur*, 1984, **21**, 197 (*ester*)

3-Pyridinephosphonic acid **P-00504**

3-Pyridinylphosphonic acid, 9CI. *3-Pyridylphosphonic acid. 3-Phosphonopyridine*

[53340-11-7]

$C_5H_6NO_3P$ M 159.081

Needles (EtOH aq.). Mp 257-259°.

Di-Et ester: [53340-10-6].

$C_9H_{14}NO_3P$ M 215.188

$Bp_{0.04}$ 95-98° (oven). n_D^{22} 1.4871.

Collins, D.J. *et al, Aust. J. Chem.*, 1974, **27**, 1355 (*synth, uv, ir, pmr, ms, P nmr*)
Hirao, T. *et al, Bull. Chem. Soc. Jpn.*, 1982, **55**, 909 (*ester, synth, ir*)
Bulot, J.J. *et al, Phosphorus Sulfur*, 1984, **21**, 197 (*ester*)

4-Pyridinephosphonic acid **P-00505**

4-Pyridinylphosphonic acid, 9CI. *4-Pyridylphosphonic acid. 4-Phosphonopyridine*

[58816-01-6]

$C_5H_6NO_3P$ M 159.081
Cryst. + $0.5H_2O$ (EtOH aq.). Mp 270-275°, 318°. pK_{a1} 4.53, pK_{a2} 6.61.
Di-Me ester: [78133-46-7]. *Dimethyl 4-pyridinylphosphonate.*
$C_7H_{10}NO_3P$ M 187.135
Plates. Mp 139-141°.
Di-Et ester: [37175-34-1]. *Diethyl 4-pyridinylphosphonate.*
$C_9H_{14}NO_3P$ M 215.188
$Bp_{0.5}$ 101-103°.

Redmore, D., *J. Org. Chem.*, 1976, **41**, 2148 (*synth, P nmr, pmr*)
Boduszek, B. *et al*, *Synthesis*, 1979, 452 (*synth, ir, pmr, ms*)
Katritzky, A.R. *et al*, *J. Chem. Soc., Perkin Trans. 1*, 1981, 668 (*esters, ir, pmr*)

(2-Pyridinylmethyl)phosphonic acid, 9CI P-00506

(2-Pyridylmethyl)phosphonic acid. 2-(Phosphonomethyl)pyridine
[80241-45-8]

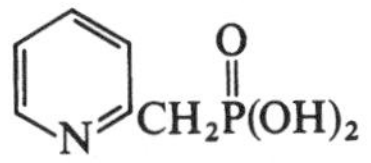

$C_6H_8NO_3P$ M 173.108
N-*Oxide:* [35469-53-5].
$C_6H_8NO_4P$ M 189.107
Cryst. (EtOH). Mp 224-228° dec.
Di-Et ester: [39996-87-7]. *Diethyl (2-pyridinylmethyl)-phosphonate.*
$C_{10}H_{16}NO_3P$ M 229.215
Oil. $Bp_{0.2}$ 112°. n_D^{25} 1.4947.

Bednarek, P. *et al*, *Bull. Acad. Polon. Sci., Ser. Sci. Chim.*, 1963, **11**, 507 (*synth*)
Bunel, E. *et al*, *J. Chem. Res. (S)*, 1981, 285 (*synth*)

(4-Pyridinylmethyl)phosphonic acid, 9CI P-00507

(4-Pyridylmethyl)phosphonic acid. 4-(Phosphonomethyl)pyridine
[80241-43-6]
$C_6H_8NO_3P$ M 173.108
Cryst. (EtOH aq.). Mp 273°.
Di-Et ester: [77047-42-8]. *Diethyl (4-pyridinylmethyl)-phosphonate.*
$C_{10}H_{16}NO_3P$ M 229.215
Oil. $Bp_{0.01}$ 95°. n_D^{20} 1.4943.
Di-Et ester, N-*oxide:* [35469-52-4].
$C_{10}H_{16}NO_4P$ M 245.214
Characterised spectroscopically.

Bednarek, P. *et al*, *Bull. Acad. Polon. Sci., Ser. Sci. Chim.*, 1963, **11**, 507 (*ester*)
Bodalski, R.M. *et al*, *Collect. Czech. Chem. Commun.*, 1971, **36**, 4079 (*oxide, ir, pmr*)
Bunel, E. *et al*, *J. Chem. Res. (S)*, 1981, 285 (*synth*)

Pyridoxal phosphate P-00508

3-Hydroxy-2-methyl-5-[(phosphonooxy)methyl]-4-pyridinecarboxaldehyde, 9CI. Pyridoxal 5-monophosphate. Codecarboxylase
[54-47-7]

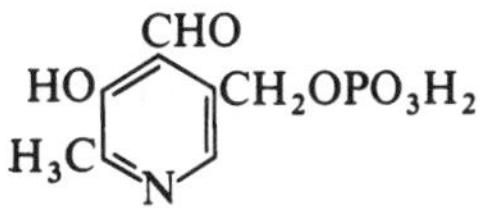

$C_8H_{10}NO_6P$ M 247.144
Coenzyme in racemisation, decarboxylation and transamination reactions of α-amino-acids. White powder. Bright yellow in alk. soln.
▷UV1207000.
Ca salt: Bright yellow powder. λ_{max} 228, 307.5, 390 nm (H_2O at pH 11).
Oxime:
$C_8H_{11}N_2O_6P$ M 262.158
Cryst. (H_2O contg. HCl). Mp 229-230° dec.
Acridine salt: Lemon-yellow cryst.

Gunsalus, I.C. *et al*, *J. Biol. Chem.*, 1945, **161**, 743.
Metzler, D.E. *et al*, *J. Am. Chem. Soc.*, 1954, **76**, 648.
Boyer, P.D. *et al*, *Enzymes*, 1960, Academic Press, N.Y., Vol. 2, Chap. 6 (*rev*)
Osbond, J.M., *Vitamin. Horm. (N.Y.)*, 1964, **22**, 367 (*synth, rev*)

Pyridoxamine 5′-phosphate P-00509

4-Aminomethyl-5-hydroxy-6-methyl-3-(dihydrogen phosphate)-3-pyridinemethanol, 9CI
[529-96-4]

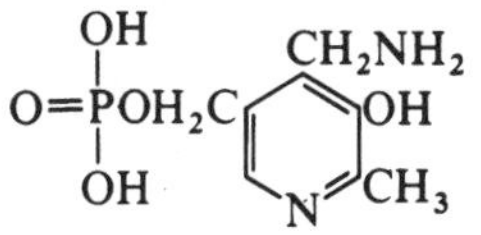

$C_8H_{13}N_2O_5P$ M 248.175
Rhombic plates + $2H_2O$ (EtOH aq.), loses $2H_2O$ over P_2O_5. Oxidn. → Pyridoxal phosphate, P-00508 .

Peterson, E.A. *et al*, *J. Am. Chem. Soc.*, 1952, **74**, 570 (*synth*)
Peterson, E.A. *et al*, *Biochem. Prep.*, 1953, **3**, 29 (*synth*)
Korytnyk, W. *et al*, *J. Am. Chem. Soc.*, 1963, **85**, 2813 (*nmr*)

2-Pyridylphosphoramidic acid, 8CI P-00510

(2-Pyridinyl)phosphoramidic acid, 9CI

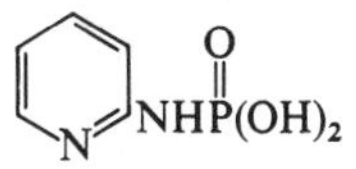

$C_5H_7N_2O_3P$ M 174.096
Di-Me ester: Dimethyl 2-pyridylphosphoramidate. 2-(Dimethoxyphosphinylamino)pyridine.
$C_7H_{11}N_2O_3P$ M 202.149
Prisms (C_6H_6/pet. ether). Readily sol. H_2O. Mp 108-109°.
Di-Et ester: Diethyl 2-pyridylphosphoramidate. 2-(Diethoxyphosphinylamino)pyridine.
$C_9H_{15}N_2O_3P$ M 230.203
Needles (H_2O). Mp 86-88°.
Diisopropyl ester: Diisopropyl 2-pyridylphosphoramidate. 2-(Diisopropoxyphosphinylamino)pyridine.
$C_{11}H_{19}N_2O_3P$ M 258.256
Needles (EtOH aq.). Mp 135-136°.
Di-Ph ester: [21915-78-6]. *Diphenyl 2-pyridylphosphoramidate. 2-(Diphenoxyphosphinylamino)pyridine.*
$C_{17}H_{15}N_2O_3P$ M 326.291

Solid. Mp 147°.
Dianilide: N,N′-*Diphenyl*-N″-*2-pyridylphosphoric triamide.*
$C_{17}H_{17}N_4OP$ M 324.321
Prisms (EtOH aq.). Mp 217-218°.

Arbuzov, B.A. *et al*, *Izv. Akad. Nauk SSSR, Ser. Khim.*, 1961, 2163 (*Engl. transl.* p. 2023) (*derivs, synth*)
Cates, L.A. *et al*, *J. Pharm. Sci.*, 1965, **54**, 331, 465 (*diethyl ester, synth*)
Omara, M.M. *et al*, *Rev. Roum. Chim.*, 1980, **25**, 253 (*diphenyl ester, synth, ir*)

3-Pyridylphosphoramidic acid, 8CI **P-00511**

(*3-Pyridinyl*)*phosphoramidic acid*, *9CI*
$C_5H_7N_2O_3P$ M 174.096
Di-Me ester: [14076-14-3]. *Dimethyl 3-pyridylphosphoramidate. 3-(Dimethoxyphosphinylamino)-pyridine.*
$C_7H_{11}N_2O_3P$ M 202.149
Thick liq. n_D^{20} 1.5180.
Di-Et ester: [14173-53-6]. *Diethyl 3-pyridylphosphoramidate. 3-(Diethoxyphosphinylamino)pyridine.*
$C_9H_{15}N_2O_3P$ M 230.203
Solid. Mp ca. 40°.
Diisopropyl ester: [14076-16-5]. *Diisopropyl 3-pyridylphosphoramidate. Bis(1-methylethyl) 3-pyridinylphosphoramidate. 3-(Diisopropoxyphosphinylamino)-pyridine.*
$C_{11}H_{19}N_2O_3P$ M 258.256
Solid. Sol. H_2O. Mp 109-109.5°.
Di-Ph ester: Diphenyl 3-pyridylphosphoramidate. 3-(Diphenoxyphosphinylamino)pyridine.
$C_{17}H_{15}N_2O_3P$ M 326.291
Cryst. (EtOH aq.). Mp 151-152°.
Dianilide: N,N′-*Diphenyl*-N″-*3-pyridylphosphoric triamide.*
$C_{17}H_{17}N_4OP$ M 324.321
Solid. Mp 240-241°.

Arbuzov, B.A. *et al*, *Zh. Org. Khim.*, 1966, **2**, 2190 (*Engl. transl.* p. 2148)

4-Pyridylphosphoramidic acid, 8CI **P-00512**

(*4-Pyridinyl*)*phosphoramidic acid*, *9CI*
$C_5H_7N_2O_3P$ M 174.096
Di-Ph ester: Diphenyl 4-pyridylphosphoramidate. 4-(Diphenoxyphosphinylamino)pyridine.
$C_{17}H_{15}N_2O_3P$ M 326.291
Solid. Mp 190-191°.

Dregval, G.F. *et al*, *CA*, 1969, **70**, 87504.

2-Pyridylphosphoramidothioic acid, 8CI **P-00513**

2-Pyridinylphosphoramidothioic acid, *9CI*

$C_5H_7N_2O_2PS$ M 190.156
OO-*Di-Ph ester:* [21915-79-7]. OO-*Diphenyl 2-pyridinylphosphoramidothioate. 2-[(Diphenoxyphosphinothioyl)amino]pyridine.*
$C_{17}H_{15}N_2O_2PS$ M 342.351
Solid. Mp 103-104°.

Dregval, G.F. *et al*, *CA*, 1969, **70**, 87504.

4-Pyridylphosphoramidothioic acid, 8CI **P-00514**

4-Pyridinylphosphoramidothioic acid, *9CI*
$C_5H_7N_2O_2PS$ M 190.156
O,O-*Di-Ph ester:* [21923-29-5]. O,O-*Diphenyl 4-pyridinylphosphoramidothioate. 4-[(Diphenoxyphosphinothioyl)amino]pyridine.*
$C_{17}H_{15}N_2O_2PS$ M 342.351
Solid. Mp 151-152°.

Dregval, G.F. *et al*, *CA*, 1969, **70**, 87504.

2-Pyrimidinephosphonic acid **P-00515**

(*2-Pyrimidinyl*)*phosphonic acid*, *9CI. 2-Phosphonopyrimidine*

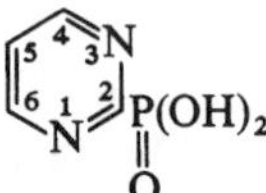

$C_4H_5N_2O_3P$ M 160.069
Yellowish needles. Mp 212-214°.
Diisopropyl ester: Diisopropyl (2-pyrimidinyl)-phosphonate.
$C_{10}H_{17}N_2O_3P$ M 244.230
Cryst. (Et_2O/pet. ether). Mp 59.5-61°. $Bp_{0.03}$ 125-126°. Unstable when impure.

Kosolapoff, G.M. *et al*, *J. Org. Chem.*, 1961, **26**, 1895 (*synth*)

4-Pyrimidinephosphonic acid **P-00516**

(*4-Pyrimidinyl*)*phosphonic acid*, *9CI. 4-Phosphonopyrimidine*
$C_4H_5N_2O_3P$ M 160.069
Very hygroscopic solid.
Diisopropyl ester: Diisopropyl (4-pyrimidinyl)-phosphonate.
$C_{10}H_{17}N_2O_3P$ M 244.230
Oil which solidifies. Mp 30.5-31°. $Bp_{0.015}$ 91-92°.

Kosolapoff, G.M. *et al*, *J. Org. Chem.*, 1961, **26**, 1895 (*synth*)

N″-2-Pyrimidinylphosphoric triamide, 9CI **P-00517**

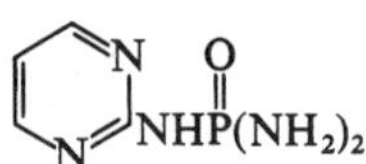

$C_4H_8N_5OP$ M 173.114
N,N′-*Bis(2-chloroethyl):* [17802-67-4].
$C_8H_{14}Cl_2N_5OP$ M 298.111
Metab. of Phosphemide. Solid. Mp 91-93°.
N′-*Et*-N-*(2-chloroethyl):* [62510-54-7].
$C_8H_{15}ClN_5OP$ M 263.666
Metab. of Phosphemide.

Savin, Yu.I. *et al*, *Khim. Farm. Zh.*, 1977, **11**, 3 (*Engl. transl.* p. 1303); *CA*, **88**, 44681 (*isol*)

1-Pyrrolidinylphosphonic acid, 9CI **P-00518**

1-Pyrrolidinephosphonic acid. 1-Phosphonopyrrolidine

[33876-55-0]

$C_4H_{10}NO_3P$ M 151.102

Di-Me ester: [24058-95-5]. *Dimethyl 1-pyrrolidinylphosphonate.*

$C_6H_{14}NO_3P$ M 179.155

Liq. Bp_3 90-93°, $Bp_{0.6}$ 56-57°. n_D^{20} 1.4498.

Buchanan, G.W. *et al*, *Can. J. Chem.*, 1979, **57**, 21 (*synth, cmr, nmr*)

Gray, G.A. *et al*, *J. Org. Chem.*, 1979, **44**, 1768 (*N nmr*)

Duangthai, S. *et al*, *Org. Magn. Reson.*, 1982, **20**, 33 (*nmr, struct*)

1-Pyrrolidinylphosphonochloridous acid, 11CI **P-00519**

C_4H_9ClNOP M 153.548

Me ester: [86030-44-6]. *Methyl 1-pyrrolidinylphosphonochloridite.*

$C_5H_{11}ClNOP$ M 167.575

Reagent for phosphitylation of nucleosides and synth. of polydeoxyribonucleotides. Liq. $Bp_{0.01}$ 28-30°.

McBride, L.J. *et al*, *Tetrahedron Lett.*, 1983, **24**, 245 (*synth, P nmr, use*)

Schwarz, M.W. *et al*, *Tetrahedron Lett.*, 1984, **25**, 5513 (*use*)

1-Pyrrolidinylphosphonous acid, 9CI, 8CI **P-00520**

1-Pyrrolidinephosphonous acid. 1-Pyrrolidylphosphonous acid

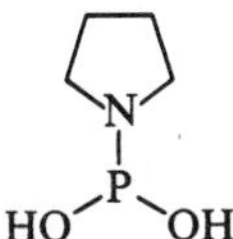

$C_4H_{10}NO_2P$ M 135.102

Di-Et ester: [24037-30-7]. *Diethyl 1-pyrrolidinylphosphonite.*

$C_8H_{18}NO_2P$ M 191.209

Liq. Bp_1 41-42.5°. n_D^{20} 1.4580. Easily oxidised.

Diisopropyl ester: Diisopropyl 1-pyrrolidinylphosphonite. Bis(1-methylethyl) 1-pyrrolidinylphosphonite.

$C_{10}H_{22}NO_2P$ M 219.263

Liq. Bp_{11} 85-87°. n_D^{20} 1.4483.

Difluoride:

$C_4H_8F_2P$ M 125.078

Liq. Bp_7 32-37°. At 50-60° dec. to PF_3.

Dichloride:

$C_4H_8Cl_2P$ M 157.987

Liq. Bp_{13} 92-93°. Easily hydrolysed.

Barlow, C.G. *et al*, *J. Chem. Soc. (A)*, 1966, 228 (*dichloride, difluroide, synth, F and P nmr, pmr, ir*)

Grechkin, N.P. *et al*, *Izv. Akad. Nauk SSSR, Ser. Khim.*, 1969, 1608 (*Engl. transl.* p. 1493) (*esters, synth*)

1*H*-Pyrrol-2-ylphosphonic acid **P-00521**

2-Pyrrolephosphonic acid. 2-Phosphonopyrrole

$C_4H_6NO_3P$ M 147.070

*Di-Et ester: Diethyl 1*H*-pyrrol-2-ylphosphonate.*

$C_8H_{14}NO_3P$ M 203.177

Liq. Bp_{14} 117-119°, $Bp_{0.06}$ 77°. Dec. by NaOH aq. to give *N*-ethylpyrrole.

N-*Me, Mono-Et ester:*

$C_7H_{12}NO_3P$ M 189.150

Cryst. (EtOH/C_6H_6) (as dicyclohexylamine salt). Mp 131-132° (dicyclohexylamine salt). When heated, an aq. soln. deposits a tar.

N-*Me, Di-Et ester:* [1012-43-7]. *Diethyl 1-methyl-1*H*-pyrrol-2-ylphosphonate.*

$C_9H_{16}NO_3P$ M 217.204

Bp_3 122.5-123°, $Bp_{0.3}$ 109-120°.

N-*Me, Di-Ph ester: Diphenyl 1-methyl-1*H*-pyrrol-2-ylphosphonate.*

$C_{17}H_{16}NO_3P$ M 313.292

Cryst. (hexane). Mp 73°.

Griffin, C.E. *et al*, *J. Org. Chem.*, 1965, **30**, 91 (*synth, uv, ir, props*)

Allen, D.W. *et al*, *J. Chem. Soc., Perkin Trans. 2*, 1977, 789 (*synth, pmr, P nmr*)

Q

Quinalphos, BSI **Q-00001**

O,O-Diethyl O-2-quinoxalinyl phosphorothioate, 9CI, 8CI. Bayrusil. Ekaluk

[13593-03-8]

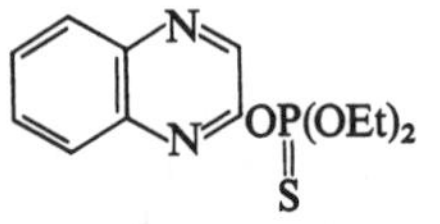

$C_{12}H_{15}N_2O_3PS$ M 298.296

Agricultural insecticide and acaricide with contact and stomach action. Cryst. Mp 31-32°. $Bp_{0.0003}$ 142° dec.

▷TF6125000.

B.P., 1 085 340, (*1966*); *CA*, **67**, 73620 (*synth*)
Pesticide Manual, 6th Ed., 462; 7th Ed., 481.

2-Quinolinephosphonic acid, 9CI **Q-00002**

[14646-14-1]

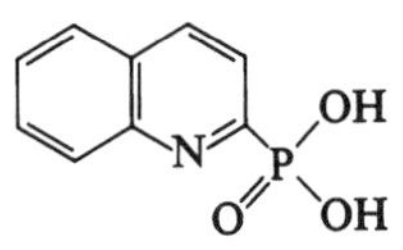

$C_9H_8NO_3P$ M 209.141

Solid. Mp 200°, 348-350°.

Sheinkman, A.K. *et al*, *Dokl. Akad. Nauk SSSR, Ser. Sci. Khim.*, 1971, **196**, 1377 (*Engl. transl.* p. 169) (*synth*)
U.S.P., 3 888 626, (*1973*); *CA*, **83**, 97569 (*synth, use*)

4-Quinolinephosphonic acid, 9CI **Q-00003**

[51835-37-1]

$C_9H_8NO_3P$ M 209.141

Di-Me ester: [78133-51-4]. *Dimethyl 4-quinolinephosphonate.*
$C_{11}H_{12}NO_3P$ M 237.194
Needles (EtOH). Mp 145-147°.

Katritzky, A.R. *et al*, *J. Chem. Soc., Perkin Trans. 1*, 1981, 668 (*synth, ir, pmr*)

[(2-Quinolinyl)methyl]phosphonic acid, 8CI **Q-00004**

2-(Phosphonomethyl)quinoline. (2-Quinolylmethyl)-phosphonic acid

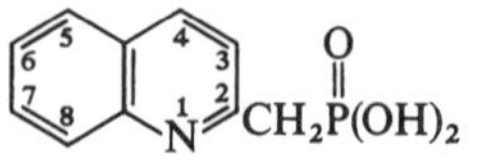

$C_{10}H_{10}NO_3P$ M 223.168

Mono-Et ester: Ethyl hydrogen[(2-quinolinyl)methyl]-phosphonate.
$C_{12}H_{14}NO_3P$ M 251.221
Solid. Mp 156-157°.

Di-Et ester: [42333-51-7]. *Diethyl [(2-quinolinyl)-methyl]phosphonate.*
$C_{14}H_{18}NO_3P$ M 279.275
Liq. $Bp_{0.05}$ 130-132°, $Bp_{0.04}$ 154-156°. n_D^{18} 1.5610.

Bednarek, P. *et al*, *Bull. Acad. Pol. Sci., Ser. Sci. Chim.*, 1963, **11**, 507 (*synth*)
Jagodić, V. *et al*, *J. Heterocycl. Chem.*, 1980, **17**, 685 (*synth, ir, pmr*)

[(8-Quinolinyl)methyl]phosphonic acid, 10CI **Q-00005**

8-(Phosphonomethyl)quinoline. (8-Quinolylmethyl)-phosphonic acid

[75355-39-4]

$C_{10}H_{10}NO_3P$ M 223.168

Cryst. (EtOH). Mp 185-187°.

Di-Et ester: [75355-36-1]. *Diethyl [(8-quinolinyl)-methyl]phosphonate.*
$C_{14}H_{18}NO_3P$ M 279.275
Liq. $Bp_{0.005}$ 142-144°.

Jagodić, V. *et al*, *J. Heterocycl. Chem.*, 1980, **17**, 685 (*synth, ir, pmr*)

Quintiofos **Q-00006**

O-Ethyl O′-quinolin-8-yl phenylphosphonothionate, 9CI. Bacdip. Oxinothiofos

[1776-83-6]

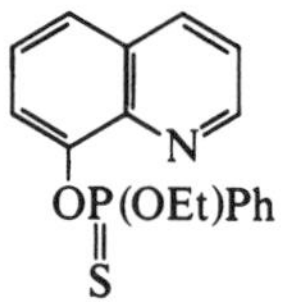

$C_{17}H_{16}NO_2PS$ M 329.353

Veterinary insecticide. Brown oil.

▷Quite toxic

Belg. Pat., 638 710, (*1965*); *U.S.P.*, 3 284 455, (*1966*); *CA*, **62**, 10462

S

Salithion **S-00001**

2-Methoxy-4H-1,3,2-benzodioxaphosphorin 2-sulfide, 9CI. *Cyclic* OO-*(methylene-*O*-phenylene)-phosphorothioate* O*-methyl ester*, 8CI

[3811-49-2]

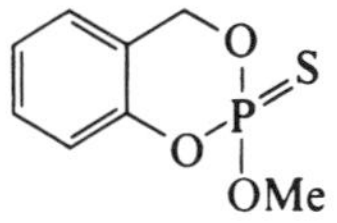

$C_8H_9O_3SP$ M 216.191

▷DF4375000.

***(R)*-form** [90293-16-6]

$[\alpha]_D$ +4.0° (c, 2.5 in MeCN).

***(S)*-form** [90293-10-0]

$[\alpha]_D$ −4.0° (c, 0.75 in MeCN).

(±)-form [92470-71-8]

Insecticide. Needles (MeOH). Mp 51-53°. Readily isom. on heating.

▷Toxic

Eto, M. *et al*, *Agric. Biol. Chem.*, 1963, **27**, 789 (*synth, ir, uv*)
Eto, M. *et al*, *Biochem. Toxicol. Insect.*, 1970, 93 (*ms*)
Eto, M. *et al*, *Anal. Method. Pest. Plant Growth Regul*, 1973, **7**, 431 (*anal*)
Iio, M. *et al*, *Agric. Biol. Chem.*, 1973, **37**, 115 (*props*)
Mihara, K. *et al*, *Agric. Biol. Chem.*, 1974, **38**, 1913 (*metab*)
Eto, M. *et al*, *Rev. Plant. Prot. Res.*, 1976, **9**, 1 (*rev*)
Hirashima, A. *et al*, *Agric. Biol. Chem.*, 1983, **47**, 2831 (*resoln, pmr, P nmr*)
Okamoto, Y. *et al*, *Bull. Chem. Soc. Jpn.*, 1984, **57**, 1681 (*resoln*)
Pesticide Manual, 6th Ed., 347, 7th Ed., 366.
Sax, N.I., *Dangerous Properties of Industrial Materials*, 6th Ed., Van Nostrand-Reinhold, 1984, 963.

Sarin **S-00002**

1-Methylethyl methylphosphonofluoridate, 9CI. *Isopropyl methylphosphonofluoridate*, 8CI

[107-44-8]

$(H_3C)_2CHOPF(O)Me$

$C_4H_{10}FO_2P$ M 140.094

Abs. config. not certain. Nerve gas which exhibits neurological effects.

▷Extremely toxic. TA8400000.

(+)-form [6171-94-4]

Liq. $[\alpha]^{25}_D$ +12.22° (neat). n_D^{25} 1.3812. Abs. config. probably (*R*).

(±)-form

d^{25} 1.09 (1.24). n_D^{25} 1.3810.

Bryant, P.J.R. *et al*, *J. Chem. Soc.*, 1960, 1553 (*synth*)
Quinchon, J. *et al*, *Bull. Soc. Chim. Fr.*, 1961, 1084, 1086 (*synth, rev, ir*)
Boter, H.L. *et al*, *Recl. Trav. Chim. Pays-Bas*, 1966, **85**, 147, 919 (*synth, props*)
Benschop, H.P. *et al*, *Recl. Trav. Chim. Pays-Bas*, 1968, **87**, 387, 957 (*config*)
Landau, M.A. *et al*, *Zh. Strukt. Khim.*, 1970, **11**, 513 (*Engl. transl.* p. 467) (*struct*)
Van den Berg, G.R. *et al*, *Recl. Trav. Chim. Pays-Bas*, 1972, **91**, 929 (*config*)
Margalit, Y. *et al*, *Phosphorus Sulfur*, 1977, **3**, 315 (*pmr, F nmr*)
Sass, S. *et al*, *Org. Mass Spectrom.*, 1979, **14**, 257 (*ms*)
Clement, J.G., *Fundam. Appl. Toxicol.*, 1981, **1**, 193 (*tox*)
Ellin, R.I. *et al*, *J. Environ. Sci., Health, Pt. B*, 1981, **16**, 713 (*props*)
Mager, P.P., *Toxicol. Lett.*, 1982, **11**, 67 (*tox*)

Selenocoenzyme *A* **S-00003**

$C_{21}H_{36}N_7O_{16}P_3Se$ M 814.434

Isol. in diselenide form as hexa-Li salt + $12H_2O$.

Günther, W.H.H. *et al*, *J. Am. Chem. Soc.*, 1965, **87**, 2708 (*synth*)

***S*-Seven** **S-00004**

O*-2,4-Dichlorophenyl* O*-ethyl phenylphosphonothioate*, 9CI

[3792-59-4]

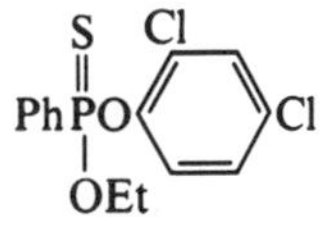

$C_{14}H_{13}Cl_2O_2PS$ M 347.195

Soil insecticide and fungicide. Light-brown oil (tech. prod.). Prac. insol. H_2O, sol. most org. solvs. d_4^{24} 1.312. Bp_5 206°. Bp_2 168-170°. n_D^{20} 1.5956.

▷TB1735000.

U.S.P., 3 318 764, (*1962*); *CA*, **67**, 31901g (*synth*)
Steurbaut, W. *et al*, *Bull. Soc. Chim. Belg.*, 1975, **84**, 791 (*synth, ir, pmr*)
Purnanard, R.K.D., *Synthesis*, 1983, 731 (*synth, ir, pmr*)
Pesticide Manual, 6th Ed., 173; 7th Ed., 179.

Soman **S-00005**

1,2,2-Trimethylpropyl methylphosphonofluoridate, 9CI. *Pinacolyl methylphosphonofluoridate*

[96-64-0]

$(H_3C)_3CCH(CH_3)OPF(O)Me$

$C_7H_{16}FO_2P$ M 182.174

Nerve gas. Liq. Consists of mixt. of diastereoisomers.

▷Extremely toxic, more so than Sarin. TA8750000.

Margalit, Y. *et al*, *Phosphorus Sulfur*, 1977, **3**, 315 (*pmr, F nmr*)
Sass, S. *et al*, *Org. Mass Spectrom.*, 1979, **14**, 257 (*ms*)
Benschop, H.P. *et al*, *Fundam. Appl. Toxicol.*, 1981, **1**, 177 (*stereochem, glc*)
Wolthuis, O.L. *et al*, *Fundam. Appl. Toxicol.*, 1981, **1**, 183.
Bošković, B. *et al*, *Fundam. Appl. Toxicol.*, 1981, **1**, 203.
Ellin, R.I. *et al*, *J. Environ. Sci. Health, Part B*, 1981, **16**, 713 (*props*)
Degenhardt, C.E.A.M. *et al*, *J. Am. Chem. Soc.*, 1986, **108**, 8290 (*resoln*)

Sparfosic acid, INN **S-00006**

N*-(Phosphonoacetyl)aspartic acid*, 9CI. *NSC 224131*

$HOOCCH_2CH(COOH)NHCOCH_2P(O)(OH)_2$

$C_6H_{10}NO_8P$ M 255.121

(S)-form [51321-79-0]

L-form

Antineoplastic agent, aspartate transcarboxylase inhibitor.

▷CI9469000.

Yoshida, T. *et al*, *J. Biol. Chem*, 1974, **249**, 6951 (*pharmacol*)

Spiro[1,3,2-benzodioxaphosphole-2,2′-[1,3,2]benzoxathiaphosphole], 10CI **S-00007**

[70117-03-2]

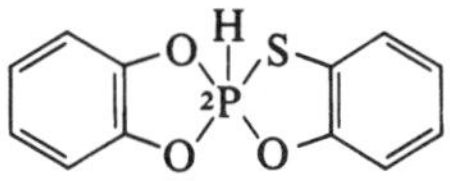

$C_{12}H_9O_3PS$ M 264.235

2-Phenoxy: [69774-92-1].

$C_{18}H_{13}O_4PS$ M 356.332

Cryst. (CH_2Cl_2/pet. ether or EtOAc/pet. ether). Mp 80-82°.

2-Dimethylamino: N,N-*Dimethylspiro[1,3,2-benzodioxaphosphole-2,2′-[1,3,2]benzoxathiaphosphol]-2-amine.*

$C_{14}H_{14}NO_3PS$ M 307.303

Solid. Mp 119-120°.

Singh, S. *et al*, *J. Chem. Soc., Perkin Trans. 1*, 1978, 1438 (*synth, P nmr*)

Spiro[1,3,2-benzodioxaphosphole-2,2′(3′*H*)-[1,3,2]benzoxazaphosphole], 9CI **S-00008**

[35722-10-2]

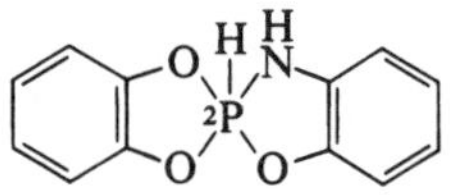

$C_{12}H_{10}NO_3P$ M 247.190

Trigonal bipyramidal struct. No tautom. with tricoordinate form. Cryst. (Et_2O). Mp 130-131° (120°).

2-Methoxy: [75080-75-0].

$C_{13}H_{12}NO_4P$ M 277.216

Solid. Mp 115-117°.

2-Ethoxy:

$C_{14}H_{14}NO_4P$ M 291.243

Solid. Mp 75-77°.

2-Phenoxy: [75080-74-9].

$C_{18}H_{14}NO_4P$ M 339.287

Cryst. (toluene). Mp 135-137°.

2-Anilino: [75080-73-8]. N-*Phenylspiro[1,3,2-benzodioxaphosphole-2,2′(3′*H*)-[1,3,2]-benzoxazaphosphol]-2-amine.*

$C_{12}H_{15}N_2O_3P$ M 266.236

Cryst. (toluene/pet. ether). Mp 137-139°.

2-Chloro: [75080-72-7].

$C_{12}H_9ClNO_3P$ M 281.635

Cryst. (C_6H_6/pet. ether). Dec. >100°.

Mathis, R. *et al*, *C.R. Hebd. Seances Acad. Sci., Ser. C*, 1972, **274**, 1156; 1975, **280**, 809 (*ir*)

Burgada, R. *et al*, *J. Organomet. Chem.*, 1974, **66**, 255 (*synth, P nmr, struct*)

Clark, T.E. *et al*, *Inorg. Chem.*, 1979, **18**, 1653 (*cryst struct*)

Kukhar', V.P. *et al*, *Zh. Obshch. Khim.*, 1980, **50**, 1017 (*Engl. transl.* p. 812) (*derivs*)

Spiro[1,3,2-benzodioxaphosphole-2,2′-[1,3,2]oxazaphospholidine], 9CI **S-00009**

[25336-93-0]

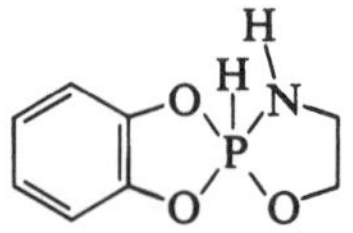

$C_8H_{10}NO_3P$ M 199.146

Contains 100% pentacoordinate tautomer at 20°, 83% at 100°. Cryst. Mp 62°.

Burgada, R. *et al*, *C.R. Hebd. Seances Acad. Sci., Ser. C*, 1972, **274**, 419 (*synth, ir, P nmr, pmr*)

Burgada, R. *et al*, *J. Organomet. Chem.*, 1974, **66**, 255 (*struct, deriv, P nmr*)

Spiro[1,3,2-benzodioxaphosphole-2,1′-[3]-phospholene], 9CI, 8CI **S-00010**

2′,5′-Dihydrospiro[1,3,2-benzodioxaphosphole-2,1′-[1H]phosphole], 10CI

[311-33-1]

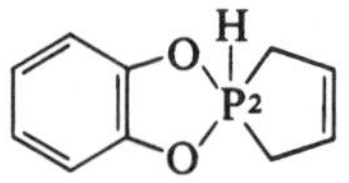

$C_{10}H_{11}O_2P$ M 194.169

2-Methoxy: [21940-99-8].

$C_{11}H_{13}O_3P$ M 224.196

Oil which cryst. slowly. d_4^{20} 1.25. Mp 112-114°. Bp_2 100-101°. n_D^{20} 1.5530.

2-Ph: [24901-17-5].

$C_{16}H_{15}O_2P$ M 270.267

Solid. Mp 127-129°.

2-Fluoro: [21940-92-1].

$C_{10}H_{10}FO_2P$ M 212.160

Solid. Mp 82-83°.

2-Bromo: [21940-87-4].

$C_{10}H_{10}BrO_2P$ M 273.066

Solid. Mp 67-68°.

Razumova, N.A. *et al*, *Zh. Obshch. Khim.*, 1969, **39**, 176, 2368, 2369 (*Engl. transl.* pp. 162, 2305, 2306) (*synth, ir, pmr*)

Spiro[1,3,2-benzoxathiaphosphole-2,2′-[1,3,2]dioxaphospholane], 9CI **S-00011**

[62974-08-7]

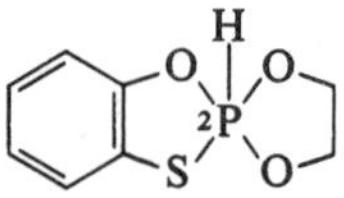

$C_8H_9O_3PS$ M 216.191

2-Ph: [71559-37-0].

$C_{14}H_{13}O_3PS$ M 292.289

Cryst. ($CHCl_3/Et_2O$). Mp 96-97°. Polymerises at 170°.

Cadogan, J.I.G. *et al*, *J. Chem. Soc., Chem. Commun.*, 1979, 191 (*synth, P nmr*)

Saegusa, T. *et al*, *Macromolecules*, 1980, **13**, 447 (*synth, pmr, P nmr, props*)

Spiro[1,3,2-benzoxazaphosphole-2(3*H*),2′-[1,3,2]dioxaphospholane], 9CI, 8CI S-00012

[16779-00-3]

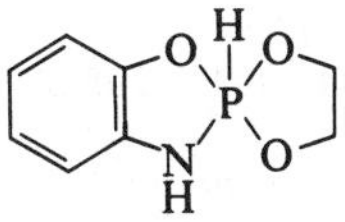

$C_8H_{10}NO_3P$ M 199.146

Exists completely as the pentacoordinate tautomer.

Sanchez, M. *et al, Bull. Soc. Chim. Fr.*, 1967, 3930 (*synth, ir, pmr, P nmr*)
Mathis, R. *et al, Spectrochim. Acta, Part A*, 1969, **25**, 1201 (*ir*)
Munoz, A. *et al, Bull. Soc. Chim. Fr., Part II*, 1977, 728 (*struct, deriv*)

Spiro[1,3,2-benzoxazaphosphole-2(3*H*),2′-[1,2]oxaphospholan]-5′-one, 10CI S-00013

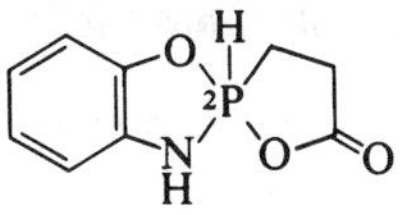

$C_9H_{10}NO_3P$ M 211.157

2-Et: [68694-30-4].
$C_{11}H_{14}NO_3P$ M 239.210
Solid. Mp 141-144°.

Terent'eva, S.A. *et al, Izv. Akad. Nauk SSSR, Ser. Khim.*, 1978, 2185; 1979, 1152 (*Engl. transl.* pp. 1931, 1078) (*synth, P nmr*)

2,2′-Spirobi[1,3,2-benzodioxaphosphole], 9CI, 8CI S-00014

[181-85-1]

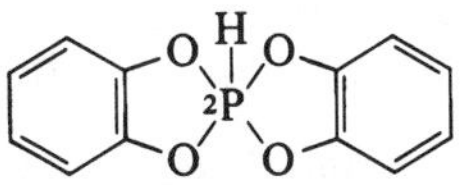

$C_{12}H_9O_4P$ M 248.174

Tautomeric. Completely in pentaco-ordinate form in CH_2Cl_2. At 100°, contains 24% monocyclic, tricoordinate form. Cryst. Mp 113-115° (sealed tube).

2-Fluoro: see 2-Fluoro-2,2′-spirobi[1,3,2-benzodioxaphosphole], F-00048
2-Chloro: see 2-Chloro-2,2′-spirobi[1,3,2-benzodioxaphosphole], C-00174
2-Phenoxy: see 2-Phenoxy-2,2′-spirobi[1,3,2-benzodioxaphosphole], P-00074
2-Me: see 2-Methyl-2,2′-spirobi[1,3,2-benzodioxaphosphole], M-00397
2-Ph: see 2-Phenyl-2,2′-spirobi[1,3,2-benzodioxaphosphole], P-00309
2-Trimethylsilyl:
$C_{15}H_{17}O_4PSi$ M 320.356
Yellow needles. Sol. THF, Et_2O, sl. sol. C_6H_6. Mp 142° dec.

Gloede, J. *et al, J. Prakt. Chem.*, 1972, **314**, 184 (*synth, pmr, P nmr*)
Bernard, D. *et al, J. Organomet. Chem.*, 1973, **47**, 113 (*struct*)
Munoz, A. *et al, Bull. Soc. Chim. Fr., Part II*, 1974, 2193 (*synth, P nmr, tautom*)
Starke, U. *et al, Phosphorus Sulfur*, 1986, **27**, 297 (*synth, pmr, cmr, P nmr*)

2,2′-Spirobi[1,3,2-benzodithiaphosphole], 9CI S-00015

[41029-31-6]

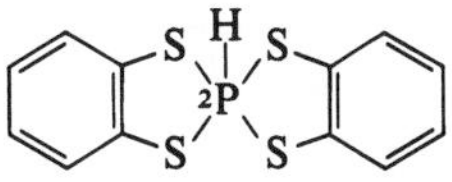

$C_{12}H_9PS_4$ M 312.417

2-Ph: [40585-73-7].
$C_{18}H_{13}PS_4$ M 388.514
Yellow needles (C_6H_6). Mp 133-135°.

Sau, A.C. *et al, J. Organomet. Chem.*, 1978, **156**, 253 (*synth*)
Day, R.O. *et al, J. Am. Chem. Soc.*, 1979, **101**, 3790 (*cryst struct*)

2,2′-Spirobi[1,3,2-benzodithiaphospholium](1+), 9CI S-00016

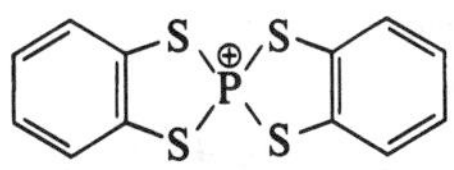

$C_{12}H_8PS_4^{\oplus}$ M 311.409 (ion)

Hexafluorophosphate: [41136-82-7].
$C_{12}H_8F_6P_2S_4$ M 456.373
Solid.

Eisenhut, M. *et al, J. Chem. Soc., Chem. Commun.*, 1973, 144 (*synth, cryst struct*)

1,1′(3*H*,3′*H*)-Spirobi[2,1-benzoxaphosphole], 10CI, 9CI S-00017

[57551-08-3]

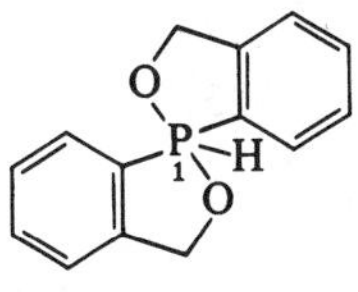

$C_{14}H_{13}O_2P$ M 244.229

1-Ph: [65715-52-8].
$C_{20}H_{17}O_2P$ M 320.327
Solid. Mp 121-122°.

Hellwinkel, D. *et al, Chem. Ber.*, 1978, **111**, 13 (*synth, ir, pmr, P nmr*)

1,1′(3*H*,3′*H*)-Spirobi[2,1-benzoxaphosphole]-3,3′-dione, 9CI S-00018

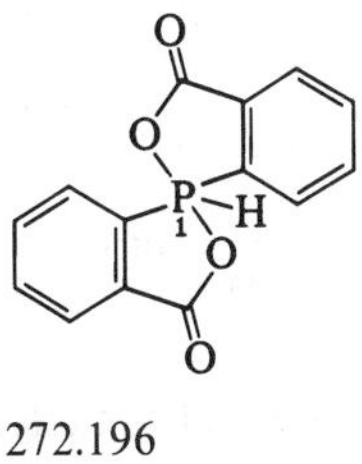

$C_{14}H_9O_4P$ M 272.196

1-Me: [57322-19-7].
$C_{15}H_{11}O_4P$ M 286.223
Solid. Mp 167°.
1-Ph: [57322-18-6].
$C_{20}H_{13}O_4P$ M 348.294
Cryst. (EtOH). Mp 232° (225°).

Petrov, K.A. *et al, Zh. Obshch. Khim.*, 1978, **48**, 91 (*Engl. transl.* p. 74) (*synth, ir, ms, pmr*)
Segall, Y. *et al, J. Am. Chem. Soc.*, 1978, **100**, 5130 (*synth, ir, ms, pmr, P nmr, cryst struct*)

2,2′-Spirobi[1,3,2-benzoxathiaphosphole], 9CI S-00019

[68129-82-8]

$C_{12}H_9O_2PS_2$ M 280.296

2-Ph: [67826-78-2].

$C_{18}H_{13}O_2PS_2$ M 356.393

Cryst. (hexane). Mp 116-117°.

Sau, A.C. *et al, J. Organomet. Chem.*, 1978, **156**, 253; 1981, **217**, 157 (*synth, pmr, P nmr*)

Day, R.O. *et al, J. Am. Chem. Soc.*, 1979, **101**, 3790 (*cryst struct*)

2,2(3H,3′H)-Spirobi[1,3,2-benzoxazaphosphole], 9CI, 8CI S-00020

[16841-91-1]

$C_{12}H_{11}N_2O_2P$ M 246.205

Plates (C_6H_6). Mp 160-162°.

2-Me: [57602-63-8].

$C_{13}H_{13}N_2O_2P$ M 260.232

Solid. Mp 160°.

2-Et: [39063-55-3].

$C_{14}H_{15}N_2O_2P$ M 274.258

Cryst. (hexane). Mp 121-122°. $Bp_{0.04}$ 160-170°.

2-Ph: [55983-44-3].

$C_{18}H_{15}N_2O_2P$ M 322.302

Solid. Mp 208-212°.

2-Phenoxy: [52961-94-1].

$C_{18}H_{15}N_2O_3P$ M 338.302

Cryst. (2-propanol). Mp 209-211°.

2-Fluoro:

$C_{12}H_{10}FN_2O_2P$ M 264.195

Solid. Mp 166° dec.

Sanchez, M. *et al, Bull. Soc. Chim. Fr.*, 1967, 3930 (*synth, ir, pmr, P nmr*)

Mathis, R. *et al, Spectrochim. Acta, Part A*, 1969, **25**, 1201 (*ir*)

Burgada, R. *et al, J. Organomet. Chem.*, 1974, **66**, 255 (*P nmr, struct*)

Malavaud, C. *et al, Tetrahedron Lett.*, 1975, 497, 3077 (*derivs, P nmr*)

Pudovik, M.A. *et al, Zh. Obshch. Khim.*, 1975, **45**, 266; 1977, **47**, 1230 (*Engl. transl.* pp. 252, 1133) (*synth, derivs*)

Meunier, P.F. *et al, Inorg. Chem.*, 1978, **17**, 3270 (*cryst struct*)

Reddy, C. *et al, Synthesis*, 1980, 1004 (*phenoxy, synth, ms, ir, pmr*)

Bartsch, R. *et al, Z. Anorg. Allg. Chem.*, 1986, **537**, 53 (*fluoro, synth, ms, F and P nmr*)

5,5′-Spirobi[5H-dibenzophosphole], 9CI, 8CI S-00021

[1443-62-5]

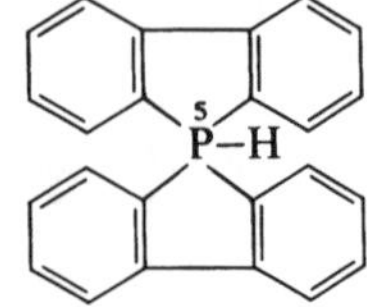

$C_{24}H_{17}P$ M 336.372

Some derivs. are resolvable. Needles. Mp 95° dec.

5-Me: [3151-21-1]. *Methylbisbiphenylenephosphorane.*

$C_{25}H_{19}P$ M 350.399

Needles (Me_2CO). Mp 217-218°.

5-Ph: [3572-91-6]. *Phenylbisbiphenylenephosphorane.*

$C_{30}H_{21}P$ M 412.470

Cryst. (C_6H_6). Mp 201.5-205.5°.

5-Amino: [30354-22-4]. *Aminobisbiphenylenephosphorane.*

$C_{24}H_{18}NP$ M 351.387

Cryst. (C_6H_6). Mp 210-215° dec.

Wittig, G. *et al, Chem. Ber.*, 1964, **97**, 741, 747 (*phenyl*)

Hellwinkel, D., *Chem. Ber.*, 1966, **99**, 3628 (*synth*)

Latsche, H.P., *Z. Naturforsch, B*, 1968, **23**, 139 (*P nmr*)

Hellwinkel, D. *et al, Chem. Ber.*, 1969, **102**, 527 (*synth, pmr, P nmr*)

Hellwinkel, D. *et al, Justus Liebigs Ann. Chem.*, 1970, **742**, 163 (*amino, synth, P nmr*)

Hellwinkel, D. *et al, Phosphorus*, 1973, **2**, 167 (*ms*)

Day, R.O. *et al, Inorg. Chem.*, 1980, **19**, 3616 (*cryst struct*)

5,5′-Spirobi[5H-dibenzophospholium](1+), 9CI, 8CI S-00022

Bis-2,2′-biphenylylenephosphonium(1+)

[25003-28-5]

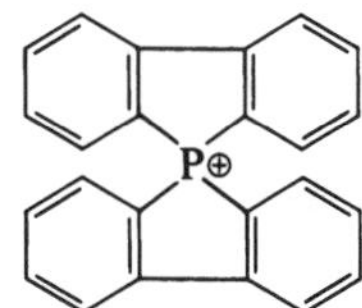

$C_{24}H_{16}P^{\oplus}$ M 335.364 (ion)

Bromide: [21955-86-2].

$C_{24}H_{16}BrP$ M 415.268

Needles ($EtOH/Et_2O$). Mp 398-405°.

Iodide: [3151-19-7].

$C_{24}H_{16}IP$ M 462.269

Characterised spectroscopically.

Hellwinkel, D. *et al, Chem. Ber.*, 1969, **102**, 548 (*bromide, synth, ir*)

Hellwinkel, D. *et al, Justus Liebigs Ann. Chem.*, 1970, **742**, 163 (*iodide, props*)

Shain, A.L., *J. Chem. Phys.*, 1972, **56**, 6201 (*iodide, esr*)

Hellwinkel, D. *et al, Phosphorus*, 1976, **6**, 151 (*uv, P nmr*)

10,10′(5H,5′H)-Spirobiphenophosphazinium(1+), 9CI, 8CI S-00023

[34392-13-7]

$C_{24}H_{18}N_2P^{\oplus}$ M 365.393 (ion)

Chloride: [34283-79-9].

$C_{24}H_{18}ClN_2P$ M 400.846

Cryst. (AcOH).

Jenkins, R.N. *et al, J. Chem. Soc., Dalton Trans.*, 1971, 1213 (*cryst struct*)

Jenkins, R.N. *et al, J. Org. Chem.*, 1975, **40**, 766 (*synth, ir*)

Spiro[1,3,2-dioxaphospholane-2,1′-[2,8,9]-trioxa[1]phosphatricyclo[3.3.1.1^{3,7}]-decane], 9CI **S-00024**

Spiro[1,3,2-dioxaphospholane-2,1′-[2,8,9]trioxa[1]-phosphaadamantane], 8CI

[34442-16-5]

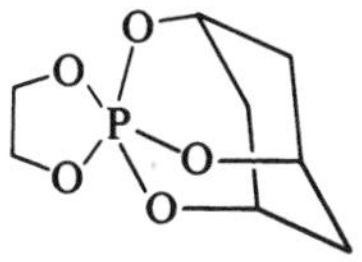

$C_8H_{13}O_5P$ M 220.161

Readily dec.

Navech, J. *et al*, *Tetrahedron Lett.*, 1980, **21**, 1449 (*synth, P nmr*)

Benhammon, M. *et al*, *Phosphorus Sulfur*, 1982, **14**, 105 (*synth, P nmr*)

Spiro[1,3,2-dioxaphosphorinane-2,2′-phenanthro[9,10-*d*][1,3,2]dioxaphosphole], 8CI **S-00025**

[19535-24-1]

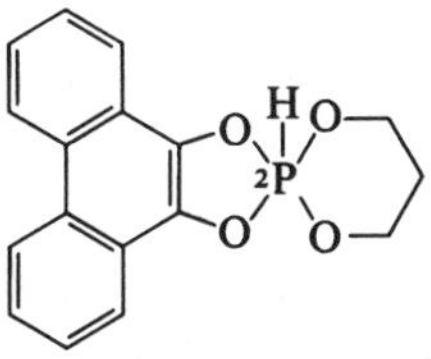

$C_{17}H_{15}O_4P$ M 314.277

2-Dimethylamino: [17454-11-4]. N,N-*Dimethyl-spiro[1,3,2-dioxaphosphorinane-2,2′-phenanthro[9,10-*d*][1,3,2]dioxaphosphol]-2-amine.*
$C_{19}H_{20}NO_4P$ M 357.345
Cryst. (C_6H_6). Mp 136-137°.

Ramirez, F. *et al*, *Tetrahedron*, 1968, **24**, 1785 (*synth, ir, pmr, P nmr*)

Sufosfamide, INN **S-00026**

*2-[[3-(2-Chloroethyl)tetrahydro-2*H*-1,3,2-oxazaphosphorin-2-yl]amino]ethanol methanesulfonate* P-*oxide, 9CI. Asta 5122. Cytimun*

[37753-10-9]

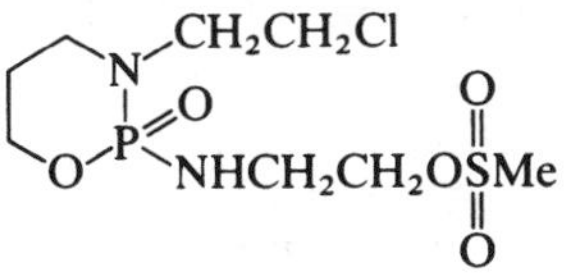

$C_8H_{18}ClN_2O_5PS$ M 320.727

Immunosuppressant, antineoplastic agent.

▷KK1580000.

Ger. Pat., 2 107 936, (*1972*); *CA*, **77**, 152238m (*synth, pharmacol*)

Brock, N. *et al*, *Arzneim.-Forsch.*, 1974, **24**, 1149 (*pharmacol*)

Sulprofos, BSI, ISO **S-00027**

O-*Ethyl* O-*[(4-methylthio)phenyl]* S-*propyl phosphorodithioate, 9CI. Bolstar. Helothion*

[35400-43-2]

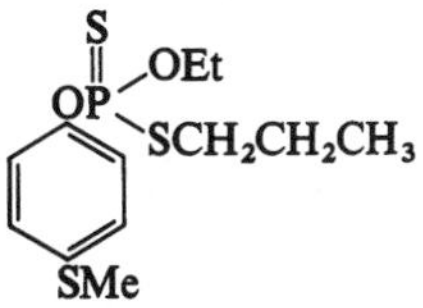

$C_{12}H_{19}O_2PS_3$ M 322.435

▷TE4165000.

(+)-*form*

$[\alpha]^{20}_{D}$ +14.0° (c, 0.6 in $CHCl_3$). More toxic to flies than (−)-form.

(−)-*form*

$[\alpha]^{20}_{D}$ −13.3° (c, 0.8 in $CHCl_3$).

(±)-*form*

Insecticide. Oil. d^{20}_{4} 1.20. $Bp_{0.1}$ 155-158°. n_D^{20} 1.5859.

B.P., 1 295 418, (*1970*); *CA*, **76**, 14087q (*synth*)

Agrochemicals Handbook, Royal Society of Chemistry, London, 1983, A374.

Hirashima, A. *et al*, *J. Agric. Food Chem.*, 1984, **32**, 1302 (*resoln, synth, pmr, P nmr, props*)

Pesticide Manual, 6th Ed., 489; 7th Ed., 501.

T

Tabun **T-00001**

Ethyl dimethylphosphoramidocyanidate, 9CI

[77-81-6]

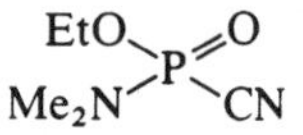

$C_5H_{11}N_2O_2P$ M 162.128

▷TB4550000.

(±)-**form**

Nerve gas. Liq. with cyanide 'almond' odour. Misc. H_2O. d_4^{20} 1.08. Mp −50°. Bp 240°, Bp_{10} 120°. Easily hydrolysed. Destroyed by alkali or bleaching powder.

▷Extremely toxic. Inhibits chymotrypsin, trypsin and acetylcholinesterase. Produces intense myosis. (+)-form more toxic than (−) or (±)-forms in mice

Ger. Pat., 767 511, (*1937*); *CA*, **49**, 14795 (*synth*)
Heath, D.F., *Organophosphorus Poisons*, Pergamon, London, 1961.
Sass, S. *et al*, *Org. Mass Spectrom.*, 1979, **14**, 257 (*ms*)
Gordon, J.J. *et al*, *Arch. Toxicol.*, 1983, **52**, 71 (*tox*)
Degenhardt, C.E.A.M. *et al*, *J. Am. Chem. Soc.*, 1986, **108**, 8290 (*resoln*)

Tagetitoxin **T-00002**

4-(Acetyloxy)-5-amino-2,3,7-trihydroxy-6-(phosphonooxy)-2,7-thiocanedicarboxylic acid, 9CI

[87913-21-1]

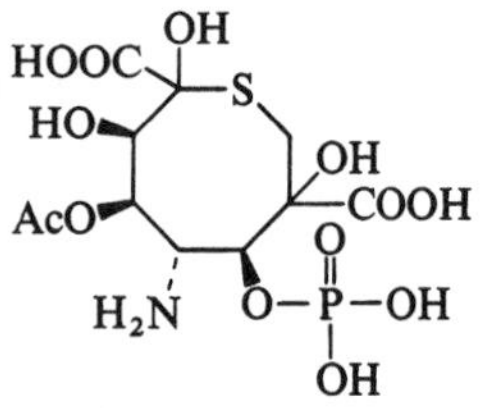

$C_{11}H_{18}NO_{13}PS$ M 435.296

Isol. from *Pseudomonas syringae* pv. *tagetis*. Phytotoxin, causing the apical chlorosis symptom. Glass.

Mitchell, R.E. *et al*, *Physiol. Plant Pathol.*, 1981, **18**, 157; *Phytochemistry*, 1983, **22**, 1425 (*isol, ir, pmr, cmr, props*)

Temephos, BSI **T-00003**

O,O′-(*Thiodi-4,1-phenylene*) *bis*(O,O-*dimethyl phosphorothioate*), 9CI. O,O,O′,O′-*Tetramethyl* O,O′-*thiodi*-p-*phenylene bis*(*phosphorothioate*). *Abate. Abathion*

[3383-96-8]

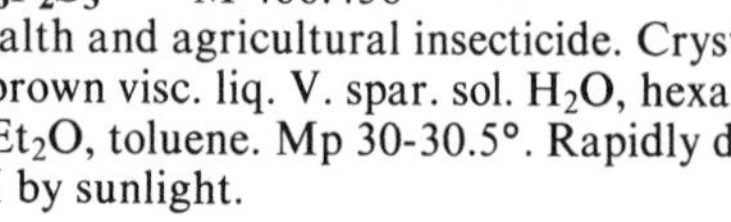

$C_{16}H_{20}O_6P_2S_3$ M 466.458

Public health and agricultural insecticide. Cryst. solid. Tech. brown visc. liq. V. spar. sol. H_2O, hexane, sol. CCl_4, Et_2O, toluene. Mp 30-30.5°. Rapidly decomp. in MeOH by sunlight.

▷TF6890000.

U.S.P., 3 317 636, (*1963*); *CA*, **63**, 11433e
Babad, H. *et al*, *Anal. Chim. Acta*, 1968, **41**, 259 (*pmr*)
Ross, R.T. *et al*, *Anal. Chem. Acta*, 1970, **52**, 139 (*P nmr*)
Stan, H.-J. *et al*, *Fresenius' Z. Anal. Chem.*, 1977, **287**, 271 (*glc, ms*)
Ripley, B.D. *et al*, *J. Assoc. Off. Anal. Chim.*, 1983, **66**, 1084 (*glc*)
Ding, X.D. *et al*, *J. Agric. Food Chem.*, 1984, **32**, 622 (*hplc*)
Opong-Mensah, K., *Residue Rev.*, 1984, **91**, 47 (*rev*)
Pesticide Manual, 6th Ed., 497; 7th Ed., 509.

Terbufos, BSI **T-00004**

S-[[(*1,1-Dimethylethyl*)*thio*]*methyl*] O,O-*diethyl phosphorodithioate*, 9CI. S-tert-*Butylthiomethyl* O,O-*diethyl phosphorodithioate. Counter*

[13071-79-9]

$(EtO)_2P(S)SCH_2SC(CH_3)_3$

$C_9H_{21}O_2PS_3$ M 288.417

Soil insecticide. Pale-yellow liq. V. spar. sol. H_2O, sol. most. org. solvs. d^{24} 1.105. Mp −29°. $Bp_{0.01}$ 69°. Dec. >120°.

▷Highly toxic. TD7200000.

S-*Oxide*: [10548-10-4]. S-[(tert-*Butylsulfinyl*)*methyl*] O,O-*diethyl phosphorodithioate.*
$C_9H_{21}O_3PS_3$ M 304.417
Metab. of terbufos.

S,S-*Dioxide*: [56070-16-7]. S-[(tert-*Butylsulfonyl*)-*methyl*] O,O-*diethyl phosphorodithioate.*
$C_9H_{21}O_4PS_3$ M 320.416
Metab. of terbufos.

Japan. Pat., 66 16 779, (*1963*); *CA*, **66**, 18510p
Ripley, B.D. *et al*, *J. Assoc. Off. Anal. Chem.*, 1983, **66**, 1084 (*glc*)
The Agrochemicals Handbook, Royal Society of Chemistry, 1983, A382.
Pesticide Manual, 6th Ed., 500; 7th Ed., 512.

1,1,4,4-Tetraamino-2,3-diphosphabutadiene **T-00005**

$(H_2N)_2C{=}P{-}P{=}C(NH_2)_2$

$C_2H_8N_4P_2$ M 150.060

N,N,N′,N′,N″,N″,N‴N‴-*Octa-Me*: *1,1,4,4-Tetrakis(dimethylamino)-2,3-diphosphabutadiene*, 11CI.
$C_{10}H_{24}N_4P_2$ M 262.274
Yellow liq. $Bp_{0.05}$ 78-81°.

N,N,N′,N′,N″,N″,N‴,N‴-*Octa-Et*: *1,1,4,4-Tetrakis(diethylamino)-2,3-diphosphabutadiene*, 11CI.
$C_{18}H_{30}N_4P_2$ M 364.409
Yellow liq. $Bp_{0.05}$ 140-143°.

Romanenko, V.D. *et al*, *Zh. Obshch. Khim.*, 1985, **55**, 2140 (*Engl. transl.* p. 1898) (*synth, pmr, P nmr*)

Tetrabenzyl diphosphate **T-00006**

Tetrakis(*phenylmethyl*) *diphosphate*, 9CI. *Tetrabenzyl pyrophosphate*, 8CI

[990-91-0]

$(PhCH_2O)_2P(O){-}O{-}P(O)(OCH_2Ph)_2$

$C_{28}H_{28}O_7P_2$ M 538.473

Useful phosphorylating agent for alcohols and amines. Needles (cyclohexane or C_6H_6/pet. ether). Mp 62°. Easily hydrolysed.

Atherton, F.R. *et al*, *J. Chem. Soc.*, 1947, 674 (*synth*)
Mason, H.S. *et al*, *J. Chem. Soc.*, 1951, 2267 (*synth*)
Corby, N.S. *et al*, *J. Chem. Soc.*, 1952, 1234 (*synth*)
Kenner, G.W. *et al*, *J. Chem. Soc.*, 1956, 1231 (*synth*)
Atkinson, R.E. *et al*, *J. Chem. Soc.* (*C*), 1967, 1356 (*synth*, *props*)

Tetrabenzyldiphosphine — T-00007

Tetrakis(phenylmethyl)diphosphine, 9CI

$(PhCH_2)_2PP(CH_2Ph)_2$

$C_{28}H_{28}P_2$ M 426.477

Dioxide:

$C_{28}H_{28}O_2P_2$ M 458.476

Used in catalysts for carbonylation of MeOH. Cryst. Mp 158-159°.

Disulfide: [26978-43-8].

$C_{28}H_{28}P_2S_2$ M 490.597

Cryst. (MeOH). Mp 145-150°.

Issleib, K. *et al*, *Chem. Ber.*, 1959, **92**, 704 (*disulfide*)
Quin, L.D. *et al*, *J. Org. Chem.*, 1966, **31**, 1206 (*dioxide*)

1,6,11,16-Tetrabenzyl-1,6,11,16-tetraphosphacycloeicosane — T-00008

1,6,11,16-Tetrakis(phenylmethyl)-1,6,11,16-tetraphosphacycloeicosane, 9CI

[57978-17-3]

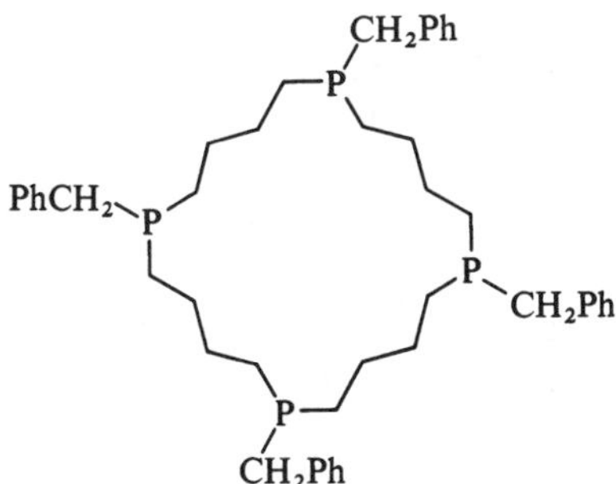

$C_{44}H_{60}P_4$ M 712.853

Solid. Mp 100-110°.

B,4PhCH₂Br: [57978-15-1]. *1,1,6,6,11,11,16,16-Octabenzyl-1,6,11,16-tetraphosphoniacycloeicosane tetrabromide.*

$C_{72}H_{90}Br_4P_4$ M 1399.014

Cryst. + $1H_2O$ (MeOH). Mp 370° dec.

Horner, L. *et al*, *Phosphorus Sulfur*, 1978, **5**, 171.

1,5,10,14-Tetrabenzyl-1,5,10,14-tetraphosphacyclooctadecane — T-00009

1,5,10,14-Tetrakis(phenylmethyl)-1,5,10,14-tetraphosphacyclooctadecane, 9CI

[69573-06-4]

$C_{42}H_{56}P_4$ M 684.799

Solid. Mp 145°.

B,4PhCH₂Br: 1,1,5,5,10,10,14,14-Octabenzyl-1,5,10,14-tetraphosphoniacyclooctadecane tetrabromide.

$C_{70}H_{84}Br_4P_4$ M 1368.945

Cryst. + $2H_2O$ (EtOH). Mp 360°.

1,5,10,14-Tetraoxide:

$C_{42}H_{56}O_4P_4$ M 748.797

Solid. Mp 230-250°.

Horner, L. *et al*, *Phosphorus Sulfur*, 1978, **5**, 171.

Tetrabromomethylphosphorane, 8CI — T-00010

$MePBr_4$

CH_3Br_4P M 365.624

Hygroscopic powder, dec. at 200°. Easily hydrolyzed.

Maier, L., *Helv. Chim. Acta*, 1963, **46**, 2667 (*synth*)

Tetrabromophenylphosphorane, 9CI — T-00011

[70430-44-3]

$PhPBr_4$

$C_6H_5Br_4P$ M 427.695

Possesses an ionic struct. Orange cryst. (chlorobenzene) which are v. readily hydrolysed by atm. moisture. Insol. C_6H_6, toluene. Mp 134-136°.

Br₂ adduct:

$C_6H_5Br_6P$ M 587.503

Red. Mp 110° subl.

Kovaleva, T.V. *et al*, *Zh. Obshch. Khim.*, 1979, **49**, 546 (*Engl. transl.* p. 476) (*synth*, *props*)
Dillon, K.B. *et al*, *Polyhedron*, 1982, **1**, 123 (*nmr*)

Tetra-*tert*-butylcyclotetraphosphine — T-00012

Tetrakis(1,1-dimethylethyl)tetraphosphetane, 9CI. *Tetra-*tert*-butylcyclotetraphosphine*

[5995-07-3]

$C_{16}H_{36}P_4$ M 352.355

Monoclinic cryst. V. sol. C_6H_6, THF, CCl_4, insol. EtOH. Mp 167-169°.

*B,MeI: Tetra-*tert*-butylmethyltetraphosphetanium iodide.*

$C_{17}H_{39}IP_4$ M 494.295

Solid. Sol. MeOH, insol. C_6H_6, Et_2O. Mp 183-187°.

Issleib, K. *et al, Chem. Ber.*, 1966, **99**, 1320 (*synth, deriv, props*)
Smith, L.R. *et al, J. Am. Chem. Soc.*, 1976, **98**, 3852 (*synth, ir, raman, ms, P nmr, pmr*)
Albrand, J.P. *et al, J. Am. Chem. Soc.*, 1978, **100**, 2600 (*P nmr, pmr*)
Cowley, A.H. *et al, J. Am. Chem. Soc.*, 1978, **100**, 3349 (*pe*)
Gleiter, R. *et al, Chem. Ber.*, 1981, **114**, 1004 (*pe*)
Weigand, W. *et al, Acta Crystallogr., Sect. B*, 1981, **37**, 1631 (*cryst struct*)

Tetra-*tert*-butyl-1,2,3,4-diphosphadiarsetane T-00013

1,2,3,4-Tetrakis(1,1-dimethylethyl)-1,2,3,4-diphosphadiarsetane

$$\begin{array}{ccc} (H_3C)_3C\diagdown & & \diagup C(CH_3)_3 \\ & As-P & \\ & |\quad\; | & \\ & As-P & \\ (H_3C)_3C\diagup & & \diagdown C(CH_3)_3 \end{array}$$

$C_{16}H_{36}As_2P_2$ M 440.251

Baudler, M. *et al, Angew. Chem., Int. Ed. Engl.*, 1979, **18**, 877 (*P nmr*)
Baudler, M. *et al, Z. Naturforsch., B*, 1981, **36**, 527 (*synth, ms, P nmr*)

Tetrabutyldiphosphine, 9CI T-00014

[13904-54-6]

$$(H_3CCH_2CH_2CH_2)_2PP(CH_2CH_2CH_2CH_3)_2$$

$C_{16}H_{36}P_2$ M 290.408

Liq. Bp_{18} 185°, Bp_2 150-152°. n_D^{20} 1.504. P—P bond cleaved by action of RI, I_2 or CCl_4 at 60°.

Monooxide: [27502-55-2].
$C_{16}H_{36}OP_2$ M 306.407
Liq. $Bp_{0.01}$ 120-125°.

Dioxide:
$C_{16}H_{36}O_2P_2$ M 322.407
Liq. $Bp_{0.01}$ 150-152°.

Monosulfide: [62969-14-6].
$C_{16}H_{36}P_2S$ M 322.468
Oil. Readily isomerises to PSP isomer.

Disulfide: [5958-53-2].
$C_{16}H_{36}P_2S_2$ M 354.528
Ligand for Co, Hg and Zn. Cryst. (C_6H_6, MeOH or EtOH). Mp 74-75°.

Niebergall, H. *et al, Chem. Ber.*, 1962, **95**, 64 (*synth*)
Voigt, D. *et al, Bull. Soc. Chim. Fr.*, 1968, 3561 (*synth*)
Bogolyubov, G.M. *et al, Zh. Obshch. Khim.*, 1969, **39**, 1808 (*Engl. transl.* p. 1772) (*ms*)
Issleib, K. *et al, J. Organomet. Chem.*, 1970, **22**, 375 (*monooxide*)
McQuillan, G.P. *et al, Spectrochim. Acta, Part A*, 1978, **34**, 33 (*disulfide, ir, raman*)
Keck, H. *et al, Org. Mass Spectrom.*, 1979, **14**, 149 (*disulfide, ms*)
Timokhin, B.V. *et al, Zh. Obshch. Khim.*, 1979, **49**, 1235 (*Engl. transl.* p. 1083) (*disulfide*)
Alagna, L. *et al, Z. Naturforsch., A*, 1981, **36**, 68 (*disulfide, pe*)
Foss, V.L. *et al, Zh. Obshch. Khim.*, 1982, **52**, 1054 (*Engl. transl.* p. 916)
Tuturjian, P.N. *et al, J. Magn. Reson.*, 1982, **49**, 155 (*nmr*)

Tetra-*tert*-butyldiphosphine T-00015

Tetrakis(1,1-dimethylethyl)diphosphine, 9CI

[5995-06-2]

$$[(H_3C)_3C]_2P-P[C(CH_3)_3]_2$$

$C_{16}H_{36}P_2$ M 290.408

Ligand for Co and Ni. Solid. Mp 48°. Bp_1 124-127°. P—P bond stable to PhLi, but cleaved by MeI.

Monooxide: [60714-30-9].
$C_{16}H_{36}OP_2$ M 306.407
Dec. at >100°.

Issleib, K. *et al, J. Organomet. Chem.*, 1968, **13**, 283 (*synth*)
Baudler, M. *et al, Chem. Ber.*, 1972, **105**, 3844 (*synth*)
Aime, S. *et al, J. Chem. Soc., Dalton Trans.*, 1976, 2144 (*pmr, cmr, nmr*)
Foss, V.L. *et al, Zh. Obshch. Khim.*, 1979, **49**, 1724 (*Engl. transl.*, p. 1510) (*monooxide*)
Lutsenko, I.F. *et al, Pure Appl. Chem.*, 1980, **52**, 917 (*rev*)
McFarlane, H.C.E. *et al, J. Chem. Soc., Dalton Trans.*, 1980, 240 (*pmr, nmr*)

Tetrabutyl hypodiphosphite, 9CI, 10CI T-00016

Tetrabutoxydiphosphine

[30884-93-6]

$$(H_3CCH_2CH_2CH_2O)_2P-P(OCH_2CH_2CH_2CH_3)_2$$

$C_{16}H_{36}O_4P_2$ M 354.406

Liq. Bp_3 143°, $Bp_{0.2}$ 110°. n_D^{20} 1.4450. P—P bond split by activated alkynes.

Proskurnina, M.V. *et al, Zh. Obshch. Khim.*, 1973, **43**, 66 (*Engl. transl.* p. 63) (*synth*)
Troy, S. *et al, Rev. Chim. Mineral.*, 1976, **13**, 589 (*synth, P nmr*)
Ponomarev, S.V. *et al, Zh. Obshch. Khim.*, 1978, **48**, 231 (*Engl. transl.* p. 207) (*synth*)
Proskurnina, M.V. *et al, Zh. Obshch. Khim.*, 1979, **49**, 1910 (*Engl. transl.* p. 1682) (*props*)

Tetrabutyl hypophosphate, 9CI T-00017

Tetrabutoxydiphosphine dioxide

[679-39-0]

$$(H_3CCH_2CH_2CH_2O)_2P(O)-P(O)(OCH_2CH_2CH_2CH_3)_2$$

$C_{16}H_{36}O_6P_2$ M 386.404

Liq. d_4^{20} 1.04. $Bp_{0.05}$ 120-122°. n_D^{20} 1.4434.

Michalski, J. *et al, Bull. Acad. Polon. Sci., Ser. Sci. Chim.*, 1965, **13**, 253 (*synth*)

N,N,N',N'-Tetrabutylphosphonic diamide, 9CI T-00018

Phosphonic bis(dibutylamide). N,N,N',N'-*Tetrabutylphosphorodiamidous acid*

[5843-29-8]

$$[(H_3CCH_2CH_2CH_2)_2N]_2P(O)H \rightleftharpoons [(H_3CCH_2CH_2CH_2)_2N]_2POH$$

$C_{16}H_{37}N_2OP$ M 304.455

Tautomeric. Liq. $Bp_{0.002}$ 105-107°. n_D^{20} 1.4575.

Zwierzak, A., *Bull. Acad. Polon. Sci., Ser. Sci. Chim.*, 1965, **13**, 609 (*synth, ir*)
Nofant'ev, E.E. *et al, Zh. Obshch. Khim.*, 1972, **42**, 1936 (*Engl. transl.* p. 1929) (*synth, props*)

Tetrabutylphosphonium(1+) T-00019

[15853-37-9]

$$P(CH_2CH_2CH_2CH_3)_4^{\oplus}$$

$C_{16}H_{36}P^{\oplus}$ M 259.434 (ion)

Chloride: [2304-30-5].
$C_{16}H_{36}ClP$ M 294.887
Reagent for perchlorate extraction. Corrosion inhibitor. Protects bone marrow progenitor cells against mechlorethamine cytotoxicity. Cryst. Mp 67°.
▷TA2419000.
Bromide: [3115-68-2].
$C_{16}H_{36}BrP$ M 339.338
Polymerisation catalyst. Phase-transfer catalyst. Cryst. (Me_2CO/Et_2O). Mp 112° (100-103°).
▷TA2417000.
Iodide: [3115-66-0].
$C_{16}H_{36}IP$ M 386.339
Phase-transfer catalyst. Cryst. (THF/Et_2O). Mp 98-99°.
▷TA2420000.
Acetate: [30345-49-4].
$C_{18}H_{39}O_2P$ M 318.479
Bactericide.

Speziale, A.J. *et al*, *J. Am. Chem. Soc.*, 1962, **84**, 1868 (*bromide*)
Kanai, K. *et al*, *Nippon Kagaku Zasshi*, 1965, **86**, 534; *CA*, **63**, 6586.
U.S.P., 3 341 580, (*1967*); *CA*, **68**, 13170 (*acetate*)
Swartz, W.E. *et al*, *Anal. Chem.*, 1971, **43**, 1066 (*nmr*)
Weigert, F.J. *et al*, *Inorg. Chem.*, 1973, **12**, 316 (*cmr*)
Tundo, P. *et al*, *Synthesis*, 1979, 952 (*iodide, use*)
Landini, D. *et al*, *J. Org. Chem.*, 1982, **47**, 2264 (*bromide, use*)
Nanjokaitis, S.A. *et al*, *Chem. Biol. Interact.*, 1982, **40**, 133; *CA*, **97**, 86779 (*chloride*)

Tetra-*tert*-butylphosphonium(1+) T-00020

Tetrakis(1,1-dimethylethyl)phosphonium(1+), 9CI

$[(H_3C)_3C]_4P^{\oplus}$

$C_{16}H_{36}P^{\oplus}$ M 259.434 (ion)
Tetrafluoroborate:
$C_{16}H_{36}BF_4P$ M 346.238
Solid.
Iodide:
$C_{16}H_{36}IP$ M 386.339
Cryst. (EtOH). Mp 258°.

Schmidbaur, H. *et al*, *Chem. Ber.*, 1980, **113**, 1612 (*ir, raman, cryst struct*)

Tetrabutylphosphorodiamidic acid, 9CI, 8CI T-00021

Phosphoric acid bis(dibutylamide)

$[(H_3CCH_2CH_2CH_2)_2N]_2P(O)OH$

$C_{16}H_{37}N_2O_2P$ M 320.454
Et ester: Ethyl tetrabutylphosphorodiamidate.
$C_{18}H_{41}N_2O_2P$ M 348.508
Liq. Insol. H_2O. $Bp_{0.5}$ 150-155°. n_D^{27} 1.4455.
Butyl ester: [40882-06-2]. *Butyl tetrabutylphosphorodiamidate.*
$C_{20}H_{45}N_2O_2P$ M 376.561
Extractant for rare earth metal nitrates. Liq. Bp_1 176°. n_D^{20} 1.4507.
Chloride: see Tetrabutylphosphorodiamidic chloride, T-00022

Loev, B. *et al*, *J. Org. Chem.*, 1957, **22**, 1186 (*ethyl ester*)
Coustures, Y. *et al*, *Bull. Soc. Chim. Fr. Part I*, 1973, 926 (*butyl ester, ir, P nmr*)

Tetrabutylphosphorodiamidic chloride, 9CI T-00022

Bis(dibutylamino)phosphoryl chloride.
Bis(dibutylamino)phosphinic chloride
[40881-96-7]

$[(H_3CCH_2CH_2CH_2)_2N]_2P(O)Cl$

$C_{16}H_{36}ClN_2OP$ M 338.900
Reactive liq. $Bp_{0.4}$ 144°. n_D^{20} 1.4660.

Coustures, Y. *et al*, *Bull. Soc. Chim. Fr. Part I*, 1973, 926 (*synth, ir, P nmr*)
Labarre, M-C. *et al*, *J. Chim. Phys. Phys. Chim. Biol.*, 1973, **70**, 534 (*ir, struct*)

Tetrabutylphosphorodiamidodithioic acid, 9CI, 8CI T-00023

Dithiophosphoric acid bis(dibutylamide)

$[(H_3CCH_2CH_2CH_2)_2N]_2P(S)SH$

$C_{16}H_{37}N_2PS_2$ M 352.575
Benzyl ester: Benzyl tetrabutylphosphorodiamidodithioate.
$C_{23}H_{43}N_2PS_2$ M 442.700
Yellow oil. n_D^{20} 1.5931.

Stuebe, C. *et al*, *J. Am. Chem. Soc.*, 1956, **78**, 976.

Tetrabutyl pyrophosphate, 8CI T-00024

Tetrabutyl diphosphate, 9CI. Butyl pyrophosphate
[1474-75-5]

$(H_3CCH_2CH_2CH_2O)_2P(O)$-
$OP(O)(OCH_2CH_2CH_2CH_3)_2$

$C_{16}H_{36}O_7P_2$ M 402.404
d_4^{25} 1.05. Bp_{11} 183-185°, $Bp_{0.01}$ 143-146°. n_D^{25} 1.4296.
▷Highly toxic. LD_{50} 14 mg/Kg (mice)

Toy, A.D.F., *J. Am. Chem. Soc.*, 1948, **70**, 3882 (*synth*)
Bliznyuk, N.K. *et al*, *Zh. Obshch. Khim.*, 1967, **37**, 1119 (*Engl. transl.* p. 1061) (*synth*)
Turpin, R. *et al*, *Bull Soc. Chim. Fr.*, 1971, 3878 (*synth*)
Appel, R. *et al*, *Z. Anorg. Allg. Chem.*, 1975, **414**, 236 (*synth*)

Tetrabutyl pyrophosphite, 8CI T-00025

Tetrabutyl diphosphite, 9CI
[54305-86-1]

$(H_3CCH_2CH_2CH_2O)_2P-O-P(OCH_2CH_2CH_2CH_3)_2$

$C_{16}H_{36}O_5P_2$ M 370.405
Liq. Bp_2 136-137°. n_D^{20} 1.4466.

Arbuzov, B.A. *et al*, *CA*, 1958, **52**, 241.

1,2,4,5-Tetra-*tert*-butyl-1,2,4,5-tetraphosphaspiro[2.2]pentane T-00026

1,2,4,5-Tetrakis(1,1-dimethylethyl)-1,2,4,5-tetraphosphaspiro[2.2]pentane, 9CI

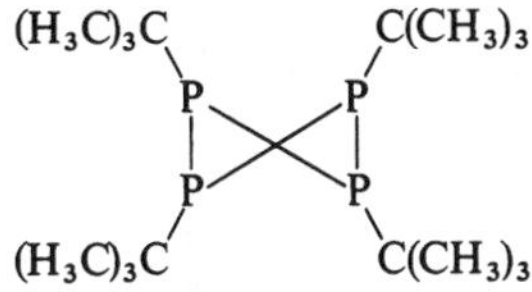

$C_{17}H_{36}P_4$ M 364.366
Stable cryst. Mp 141-143°.

Baudler, M. *et al*, *Angew. Chem., Int. Ed. Engl.*, 1983, **22**, 632 (*synth, struct*)

Tebbe, K.F. *et al*, *Z. Kristallogr., Kristallgeom., Kristallphys., Kristallchem.*, 1985, **172**, 89 (*cryst struct*)

1,2,4,5-Tetra-*tert*-butyl-1,2,4,5-tetraphosphorinane T-00027

*1,2,4,5-Tetrakis(1,1-dimethylethyl)-1,2,4,5-tetraphosphorinane. 1,2,4,5-Tetra-*tert*-butyl-1,2,4,5-tetraphosphinane. 1,2,4,5-Tetra-*tert*-butyl-3,6-dicarba-1,2,4,5-tetraphosphan. 1,2,4,5-Tetra-*tert*-butyl-1,2,4,5-tetraphosphacyclohexane*

[68969-74-4]

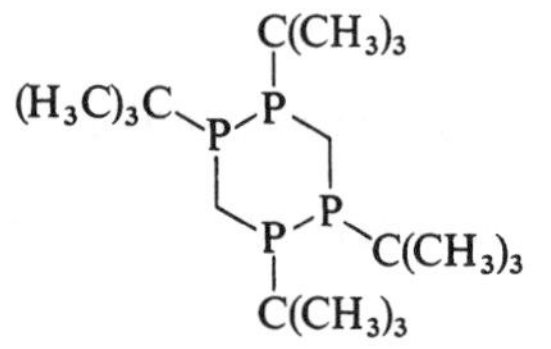

$C_{18}H_{40}P_4$ M 380.409

Solid. Mp 141-142°.

Baudler, M. *et al*, *Z. Naturforsch., B*, 1978, **33**, 1208 (*synth, ir, ms, P nmr, pmr*)

Prishchenko, A.A. *et al*, *Zh. Obshch. Khim.*, 1980, **50**, 689, 1881; *CA*, **93**, 168342, 239536 (*synth, pmr, P nmr*)

2,2,4,6-Tetrachloro-4,6-bis(dimethylamino)-2,2,4,4,6,6-hexahydro-1,3,5-triaza-2,4,6-triphosphorine, 9CI T-00028

2,2,4,6-Tetrachloro-4,6-bis(dimethylamino)-2,2,4,4,6,6-hexahydrocyclotriphosphazene

[2203-74-9]

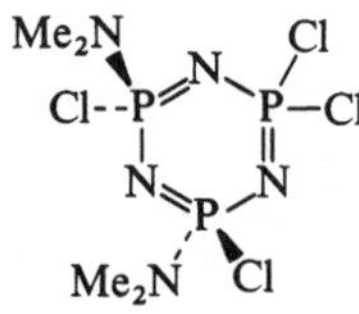

(*4RS,6RS*)-*form*

$C_4H_{12}Cl_4N_5P_3$ M 364.906

Cis- and *trans* forms equilibrate in soln. or in presence of $AlCl_3$.

(4RS, 6RS)-form [1081-87-4]

trans-*form*

Solid or cryst. (pet. ether). Mp 103°. Almost flat ring with sl. chair conformation.

(4RS, 6SR)-form [1018-07-1]

cis-*form*

Solid or cryst. (pet. ether). Mp 86°. Almost flat ring with sl. chair conformation.

Keat, R. *et al*, *J. Chem. Soc.*, 1965, 2215 (*synth, pmr, struct*)

Strahlberg, R. *et al*, *Spectrochim. Acta, Part A*, 1967, **23**, 2005 (*ir, raman*)

Keat, R. *et al*, *J. Chem. Soc. (A)*, 1968, 703 (*pmr*)

Goldschmidt, J.M.E. *et al*, *Inorg. Nucl. Chem. Lett.*, 1973, **9**, 161, 163 (*synth*)

Green, B. *et al*, *J. Chem. Soc., Dalton Trans.*, 1973, 1042 (*pe*)

Dagleish, W.H. *et al*, *J. Chem. Soc., Dalton Trans.*, 1975, 308 (*nqr*)

Keat, R. *et al*, *J. Chem. Soc., Dalton Trans.*, 1976, 1582 (*P nmr*)

Connelly, A. *et al*, *J. Magn. Reson.*, 1978, **30**, 439 (*nqr, struct*)

Ahmed, F.R. *et al*, *Acta Crystallogr., Sect. B*, 1980, **36**, 1456 (*cryst struct*)

Friedmann, N. *et al*, *J. Chem. Soc., Dalton Trans.*, 1981, 103 (*props*)

Thomas, B. *et al*, *Z. Anorg. Allg. Chem.*, 1982, **489**, 131 (*N and P nmr*)

2,2,4,6-Tetrachloro-2,2-dihydro-1,3,5,2-triazaphosphorine, 9CI T-00029

2,2,4,6-Tetrachloro-2,2-dihydro-1,3,5,2-triazaphosphinine

[26236-17-9]

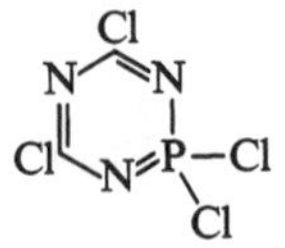

$C_2Cl_4N_3P$ M 238.828

Cryst. (CCl_4 or C_6H_6). Insol. pet. ether. Mp 141-143°.

Becke-Goehring, M. *et al*, *Z. Anorg. Allg. Chem.*, 1970, **372**, 233 (*synth, ir, nmr*)

Pinkert, W. *et al*, *Z. Anorg. Allg. Chem.*, 1977, **436**, 136 (*props*)

Kornuta, P.P. *et al*, *Zh. Obshch. Khim.*, 1978, **48**, 2218 (*Engl. transl.* p. 2015) (*synth*)

Romanenko, E.A. *et al*, *Teor. Eksp. Khim.*, 1979, **15**, 73; 1980, **16**, 308 (*Engl. transl.* p. 55, 246 (*uv, struct*)

Romanenko, E.A. *et al*, *Teor. Eksp. Khim.*, 1982, **18**, 710 (*Engl. transl.* p. 654) (*ir, raman*)

Romanenko, E.A. *et al*, *Teor. Eksp. Khim.*, 1983, **19**, 357 (*Engl. transl.* p. 330) (*nqr, struct*)

2,2,4,6-Tetrachloro-2,2,4,4,6,6-hexahydro-4,6-bis(methylamino)-1,3,5-triaza-2,4,6-triphosphorine, 9CI T-00030

2,2,4,6-Tetrachloro-2,2,4,4,6,6-hexahydro-4,6-bis(methylamino)cyclotriphosphazene

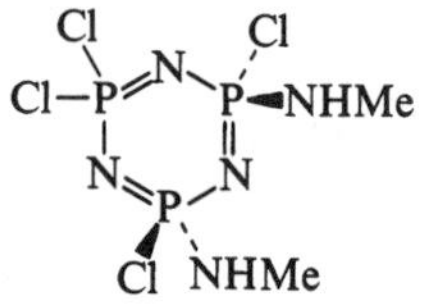

(*4RS,6RS*)-*form*

$C_2H_8Cl_4N_5P_3$ M 336.852

(4RS, 6RS)-form [16416-60-7]

trans-*form*

No phys. props. reported.

(4RS, 6SR)-form [16400-31-0]

cis-*form*

Cryst. (cyclohexane). Mp 145°.

Ray, S.K. *et al*, *J. Chem. Soc.*, 1961, 872 (*synth*)

Lehr, W., *Z. Anorg. Allg. Chem.*, 1967, **352**, 27 (*synth*)

Biran, Z. *et al*, *Synth, React. Inorg. Met. Org. Chem.*, 1978, **8**, 185 (*synth, glc*)

Thomas, B. *et al*, *Z. Anorg. Allg. Chem.*, 1982, **489**, 131 (*N and P nmr*)

Bamgboye, T.T. *et al*, *Spectrochim. Acta, Part A*, 1984, **40**, 329 (*pmr*)

2,2,4,4-Tetrachloro-2,2,4,4,6,6-hexahydro-6,6-diphenyl-1,3,5-triaza-2,4,6-triphosphorine, 9CI T-00031

2,2,4,4-Tetrachloro-2,2,4,4,6,6-hexahydro-6,6-diphenylcyclotriphosphazene

[2846-32-4]

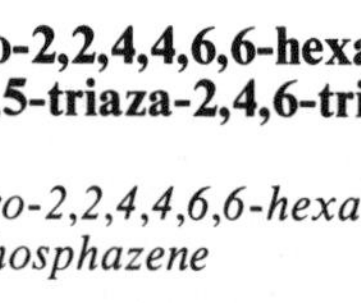

$C_{12}H_{10}Cl_4N_3P_3$ M 430.964

Cryst. (hexane or CH_2Cl_2/pet. ether). Mp 95°.

McBee, E.T. *et al*, *Inorg. Chem.*, 1965, **4**, 1672 (*synth*)

Desai, V.B. *et al*, *J. Chem. Soc. (A)*, 1969, 1977 (*synth, props*)
Wagner, A.J. *et al*, *J. Inorg. Nucl. Chem.*, 1971, **33**, 1307 (*uv*)
Keat, R. *et al*, *J. Chem. Soc., Dalton Trans.*, 1972, 1648 (*nqr*)
Keat, R. *et al*, *J. Chem. Soc., Dalton Trans.*, 1976, 1582 (*P nmr*)
Allen, C.W. *et al*, *J. Chem. Soc., Dalton Trans.*, 1978, 173 (*ms*)
Krishnamurthy, S.S. *et al*, *Org. Magn. Reson.*, 1981, **15**, 205 (*cmr*)

Tetrachloromethylphosphorane, 9CI, 8CI — T-00032

Trichloromethylphosphonium chloride

[2725-68-0]

$MePCl_4$ or $MePCl_3^{\oplus}Cl^{\ominus}$

CH_3Cl_4P M 187.820

Possesses the ionic struct. in solid state and ionizing solvents, but bipyramidal covalent struct. in non-ionizing solvents.

Baumgärtner, R. *et al*, *Z. Anorg. Allg. Chem.*, 1964, **333**, 171 (*struct*)
Beattie, I.R. *et al*, *J. Chem. Soc. (A)*, 1968, 1 (*ir, raman*)
Dillon, K.B. *et al*, *J. Chem. Soc., Dalton Trans.*, 1976, 1243 (*synth, nmr, nqr*)
Dillon, K.B. *et al*, *Polyhedron*, 1982, **1**, 123 (*nmr*)

Tetrachlorophenoxyphosphorane, 8CI — T-00033

[19579-04-5]

$PhOPCl_4$

$C_6H_5Cl_4OP$ M 265.891

Prepd. *in situ*.

Ramirez, F. *et al*, *Tetrahedron*, 1968, **24**, 5041 (*synth, props*)

Tetrachlorophenylphosphorane, 9CI, 8CI — T-00034

[4895-65-2]

$PhPCl_4$

$C_6H_5Cl_4P$ M 249.891

Nonionic in the solid state. Needles (CCl_4). Mp 75-76°.

Herring, D.L. *et al*, *Inorg. Chem.*, 1964, **3**, 428 (*synth*)
Whitehead, M.A. *et al*, *J. Chem. Soc. (A)*, 1971, 1738 (*nqr*)
Biryukov, I.P. *et al*, *Zh. Obshch. Khim.*, 1973, **43**, 1934 (*Engl. transl.* p. 1918) (*nqr*)
Dillon, K.B. *et al*, *J. Chem. Soc., Dalton Trans.*, 1976, 1243 (*struct*)
Timokhin, B.V. *et al*, *Zh. Obshch. Khim.*, 1978, **48**, 1421 (*Engl. transl.* p. 1304) (*nmr*)
Sergienko, L.M. *et al*, *Zh. Obshch. Khim.*, 1979, **49**, 317 (*Engl. transl.* p. 275) (*uv*)

(2,3,5,6-Tetrachloro-4-pyridinyl)phosphonic acid, 9CI — T-00035

2,3,5,6-Tetrachloro-4-phosphonopyridine

[23995-96-2]

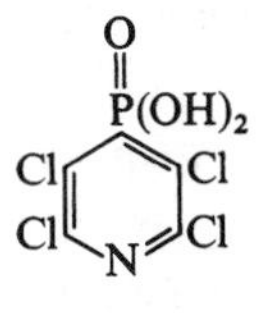

$C_5H_2Cl_4NO_3P$ M 296.861

Cryst. (acidified H_2O). Mp 226-228°.

Di-Me ester: [51066-52-5]. *Dimethyl (2,3,5,6-tetrachloro-4-pyridinyl)phosphonate.*
$C_7H_6Cl_4NO_3P$ M 324.915
Solid. Mp 109°.

Di-Et ester: [24138-46-3]. *Diethyl (2,3,5,6-tetrachloro-4-pyridinyl)phosphonate.*
$C_9H_{10}Cl_4NO_3P$ M 352.969
Possesses insecticidal props. Stimulates growth of oat seedlings. Oil which solidifies. Mp 45°, 51°. Bp_1 159-160°.

Di-Ph ester: Diphenyl (2,3,5,6-tetrachloro-4-pyridinyl)-phosphonate.
$C_{17}H_{10}Cl_4NO_3P$ M 449.057
Cryst. (2-propanol, aq.). Mp 117-119°.

Dianilide: N,N′-*Diphenyl* P-*(2,3,5,6-tetrachloro-4-pyridinyl)phosphonic diamide.*
$C_{17}H_{12}Cl_4N_3OP$ M 447.087
Cryst. (C_6H_6/pet. ether). Mp 227-229°.

Dichloride:
C_5Cl_6NOP M 333.753
Needles (pet. ether). Mp 107°. $Bp_{0.15}$ 150°.

Ivashchenko, Yu.N. *et al*, *Zh. Obshch. Khim.*, 1969, **39**, 1695 (*Engl. transl.* p. 1662) (*synth, derivs*)
Bratt, J. *et al*, *J. Chem. Soc., Perkin Trans. 1*, 1973, 1689 (*esters*)
Moshchitskii, S.D. *et al*, *Zh. Obshch. Khim.*, 1974, **44**, 2782 (*Engl. transl.* p. 2734) (*ester*)
Boenigk, W. *et al*, *Chem. Ber.*, 1983, **116**, 2418 (*esters, synth, ms, ir, P nmr*)

4,5,6,7-Tetrachloro-4′,4′,5′,5′-tetramethylspiro[1,3,2-benzodioxaphosphole-2,2′-[1,3,2]dioxaphospholane], 9CI — T-00036

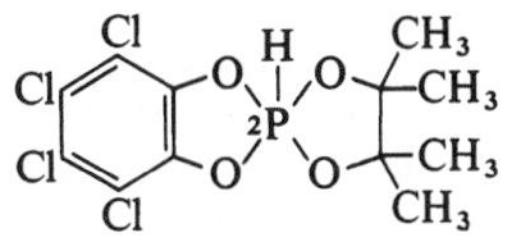

$C_{12}H_{13}Cl_4O_4P$ M 394.018

2-Methoxy: [54622-70-7].
$C_{13}H_{15}Cl_4O_5P$ M 424.044
Cryst. (pet. ether). Mp 116-118°.

2-Phenoxy: [54622-67-2].
$C_{18}H_{17}Cl_4O_5P$ M 486.115
Cryst. (pet. ether). Mp 159-160°.

2-Phenylthio: [54622-68-3].
$C_{18}H_{17}Cl_4O_4PS$ M 502.176
Cryst. (pet. ether). Mp 150.5-151.5°.

*2-Dimethylamino: 4,5,6,7-Tetrachloro-*N,N*-dimethylspiro[1,3,2-benzodioxaphosphole-2,2′-[1,3,2]-dioxaphospholan]-2-amine.*
$C_{14}H_{18}Cl_4NO_4P$ M 437.086
Cryst. (pet. ether). Mp 132-134°.

Bone, S. *et al*, *J. Chem. Soc., Perkin Trans. 1*, 1974, 2125 (*synth, P nmr*)

Tetrachloro(trichloromethyl)phosphorane, 9CI, 8CI — T-00037

[3582-10-3]

Cl_4PCCl_3

CCl_7P M 291.156

Thought to consist of unionized trigonal bipyramid with CCl_3 group axial. Dec. at 125-126°.

Quin, L.D. *et al*, *J. Org. Chem.*, 1958, **23**, 1693 (*synth*)

Kozlov, É.S. *et al*, *Zh. Obshch. Khim.*, 1972, **42**, 756 (*Engl. transl.* p. 748) (*struct*)
Dmitriev, V.I. *et al*, *Zh. Obshch. Khim.*, 1980, **50**, 2230 (*Engl. transl.* p. 1799) (*nmr, nqr*)
Sergienko, L.M. *et al*, *Zh. Obshch. Khim.*, 1981, **51**, 494; *CA*, **94**, 207998 (*uv*)
Kozlov, É.S. *et al*, *Zh. Obshch. Khim.*, 1982, **52**, 1077 (*Engl. transl.* p. 936) (*ir, raman*)

Tetrachlorvinphos, BSI — T-00038

2-Chloro-1-(2,4,5-trichlorophenyl)ethenyl dimethyl phosphate, 9CI. *2-Chloro-1-(2,4,5-trichlorophenyl)vinyl dimethyl phosphate*, 8CI. *2,4,5-Trichloro-α-(chloromethylene)benzyl dimethyl phosphate. Dietreen. Rabon. Gardona*

[961-11-5]

(*E*)-*form*

$C_{10}H_9Cl_4O_4P$ M 365.964

Coml. samples consist of ca. 25-30% (*E*)-form and 70-75% (*Z*)-form. Insecticide. Sol. $CHCl_3$, almost insol. H_2O. Unstable to base.

▷TB9050000.

(*E*)-*form* [22350-76-1]

Much lower biological activity than (*Z*)-form. LC_{50} for caterpillars 0.1%.

(*Z*)-*form* [22248-79-9]

Stirifos. Stiriphos

Mp 97-98°.

▷TB9100000.

Ger. Pats., 1 903 356, (*1969*) and 1 947 350, (*1970*); *CA*, **72**, 3564, 121166 (*synth*)
Bunton, W.B. *et al*, *J. Agric. Food Chem.*, 1972, **20**, 1180 (*synth*)
Beynon, K.I. *et al*, *Residue Rev.*, 1973, **47**, 55 (*rev*)
Szalontai, G., *J. Chromatogr.*, 1976, **124**, 9 (*hplc*)
Stan, H.-J. *et al*, *Fresenius' Z. Anal. Chem.*, 1977, **287**, 271 (*glc, ms*)
Stan, H.-J. *et al*, *Biomed. Mass Spectrom.*, 1982, **9**, 483 (*ms*)
Agrochemicals Handbook, Royal Society of Chemistry, London, 1983, A386.
Pesticide Manual, 6th Ed., 504; 7th Ed., 518.

Tetracyclohexyldiphosphine, 9CI — T-00039

[2359-99-1]

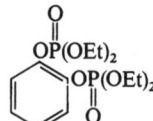

$C_{24}H_{44}P_2$ M 394.559

Ligand for Ni and Hg. Solid. Mp 169°. P—P bond cleaved by CCl_4 at 60°.

Monoxide: [66193-25-7].

$C_{24}H_{44}OP$ M 379.585

Mp 167-168°.

Dioxide:

$C_{24}H_{44}O_2P_2$ M 426.558

Cryst. (C_6H_6/pet. ether). Mp 205°.

Disulfide: [3676-98-0].

$C_{24}H_{44}P_2S_2$ M 458.679

Cryst. (CS_2). Mp 205°.

Issleib, K. *et al*, *Chem. Ber.*, 1959, **92**, 2681 (*synth, dioxide, disulfide*)
Issleib, K. *et al*, *Chem. Ber.*, 1963, **96**, 1544 (*synth*)
Issleib, K. *et al*, *Z. Kristallogr., Kristallgeom., Kristallphys., Kristallchem.*, 1964, **119**, 472 (*disulfide, cryst struct*)
Fluck, E. *et al*, *Chem. Ber.*, 1965, **98**, 2674 (*nmr, disulfide*)
Richter, R. *et al*, *Acta Crystallogr., Sect. B*, 1977, **33**, 1887 (*cryst struct*)
Foss, V.L. *et al*, *Zh. Obshch. Khim.*, 1979, **49**, 2418 (*Engl. transl.* p. 2134) (*monooxide*)

1,2,3,4-Tetracyclohexyl-1,2,3,4-tetraphosphorinane, 9CI — T-00040

*1,2,3,4-Tetracyclohexyl-1,2,3,4-tetraphosphinane. 1,2,3,4-Tetracyclohexyl-*cyclo*-5,6-dicarba-1,2,3,4-tetraphosphan*

[63923-74-0]

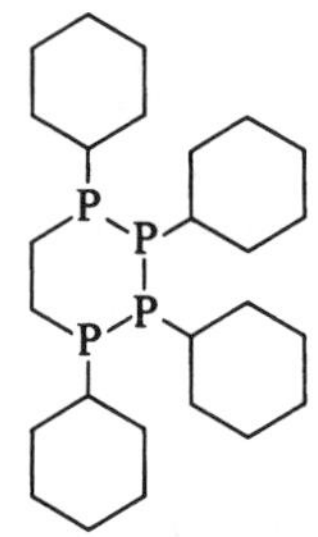

$C_{26}H_{48}P_4$ M 484.560

Solid. Mp 166-169°.

Issleib, K. *et al*, *Synth. React. Inorg. Metal-Org. Chem.*, 1977, 7, 253 (*synth*)

1,1,3,3-Tetraethoxy-2-ethyltriphosphine 1,3-dioxide, 10CI — T-00041

[73469-37-1]

$EtP[P(O)(OEt)_2]_2$

$C_{10}H_{25}O_6P$ M 272.278

Liq. d_4^{20} 1.15. $Bp_{0.02}$ 100°. n_D^{20} 1.4777.

Kabachnik, M.M. *et al*, *Zh. Obshch. Khim.*, 1982, **52**, 763 (*Engl. transl.* p. 662) (*synth*)

Tetraethyl 1,2-benzenediol bis(phosphate), 9CI — T-00042

1,2-Bis[(diethoxyphosphinyl)oxy]benzene

[37521-98-5]

$C_{14}H_{24}O_8P_2$ M 382.286

Oil. n_D^{20} 1.4710.

Hetnarski, B. *et al*, *J. Agric. Food Chem.*, 1972, **20**, 543 (*synth*)

Tetraethyldiphosphine, 9CI, 8CI — T-00043

[3040-63-9]

Et_2PPEt_2

$C_8H_{20}P_2$ M 178.194

Widely used ligand. Liq. Bp 220-221°, Bp_1 55-60°. P–P bond cleaved by CCl_4 or PhLi.

Monosulfide:
$C_8H_{20}P_2S$ M 210.254
Liq.
Disulfide: see Tetraethyldiphosphine disulfide, T-00044

Issleib, K. *et al, Chem. Ber.*, 1960, **93**, 1852 (*synth*)
Maier, L., *J. Inorg. Nucl. Chem.*, 1962, **24**, 275 (*monosulfide, nmr*)
Niebergall, H. *et al, Chem. Ber.*, 1962, **95**, 64 (*synth, derivs*)
Fluck, E. *et al, Z. Anorg. Allg. Chem.*, 1967, **354**, 113 (*synth, nmr*)
Bogolyubov, G.M. *et al, Zh. Obshch. Khim.*, 1969, **39**, 1808 (*Engl. transl.* p. 1772) (*ms*)
Aime, S. *et al, J. Chem. Soc., Dalton Trans.*, 1976, 2144 (*cmr, nmr, pmr*)
Troy, D. *et al, Bull. Soc. Chim. Fr., Part 1*, 1979, 241 (*uv*)
Albrand, J.P., *Org. Magn. Reson.*, 1983, **21**, 246 (*cmr, nmr, pmr*)

Tetraethyldiphosphine disulfide, 9CI, 8CI — T-00044

[3790-23-6]

$Et_2P(S)-P(S)Et_2$

$C_8H_{20}P_2S_2$ M 242.314
Forms complexes with Sn, Ti, Co, Cu, and V. Cryst. (Me_2CO aq.). Mp 77-78°.

Dutton, S.N. *et al, Acta Crystallogr.*, 1961, **14**, 178 (*cryst struct*)
Kuchen, W. *et al, Justus Liebigs Ann. Chem.*, 1962, **652**, 28 (*synth, props*)
Bogolyubov, G.M. *et al, Zh. Obshch. Khim.*, 1969, **39**, 1759 (*Engl. transl.* p. 1723) (*uv*)
Emoto, T. *et al, Bull. Chem. Soc. Jpn.*, 1973, **46**, 898 (*props*)
Aime, S. *et al, J. Chem. Soc., Dalton Trans.*, 1976, 2144 (*cmr, nmr*)
Harris, R.K. *et al, J. Chem. Soc., Faraday Trans. 2*, 1976, **72**, 2291 (*nmr*)
McQuillan, G.P. *et al, Spectrochim. Acta, Part A*, 1978, **34**, 33 (*ir, raman*)
Keck, H. *et al, Org. Mass Spectrom.*, 1979, **14**, 149 (*ms*)
Alagna, L. *et al, Z. Naturforsch., A*, 1981, **36**, 68 (*pe*)
Tutunjian, P.N. *et al, J. Chem. Phys.*, 1982, **76**, 1223 (*nmr*)

O,O,O,O-Tetraethyl dithiopyrophosphate — T-00045

Tetraethyl thiodiphosphate, 9CI. Sulfotep, BSI
[3689-24-5]

$(EtO)_2P(S)-O-P(S)(OEt)_2$

$C_8H_{20}O_5P_2S_2$ M 322.311
The CA name is ambiguous. Nonsystemic insecticide of brief persistence. Pale-yellow mobile liq. d_4^{25} 1.19. $Bp_{0.2}$ 110-113°. n_D^{25} 1.4753.
▷Toxic, LD_{50} (rats) 5 mg/Kg. Absorbed through skin. XN4375000.

Toy, A.D.F., *J. Am. Chem. Soc.*, 1951, **73**, 4670 (*synth*)
Schrader, G. *et al, Angew. Chem.*, 1958, **70**, 690 (*synth, tox*)
Harris, R.K. *et al, J. Chem. Soc. (A)*, 1967, 37 (*pmr, P nmr*)
Kimmerle, G. *et al, Arch. Toxicol.*, 1974, **33**, 1 (*tox*)
Stan, H.-J. *et al, Fresenius' Z. Anal. Chem.*, 1977, **287**, 271 (*glc, ms*)
Stan, H.-J. *et al, Biomed. Mass Spectrom.*, 1982, **9**, 483 (*ms*)
Stan, H.-J. *et al, J. Chromatogr.*, 1983, **279**, 173 (*glc*)
Pesticide Manual, 7th Ed., 498.
Sax, N.I., *Dangerous Properties of Industrial Materials*, 6th Ed., Van Nostrand-Reinhold, 1984, 1014.

unsym-O,O,O,O-Tetraethyl dithiopyrophosphate — T-00046

Tetraethyl thiodiphosphate, 9CI. O,O-*Diethyl* S-*(diethoxyphosphinyl) phosphorodithioate*
[15108-81-3]

$(EtO)_2P(O)—S—P(S)(OEt)_2$

$C_8H_{20}O_5P_2S_2$ M 322.311
Liq. d_4^{20} 1.22. $Bp_{0.1}$ 114-115°. n_D^{20} 1.4980. Isomerises to *O,O,O,O*-Tetraethyl dithiopyrophosphate, T-00045 when heated.

Almasi, L. *et al, Monatsh. Chem.*, 1968, **99**, 187 (*synth, props*)

O,O,O′,O′-Tetraethyl *S,S′*-ethanediyl bisphosphorodithioate, 9CI — T-00047

1,2-[Bis(diethoxyphosphinothioyl)thio]ethane

$(EtO)_2P(S)SCH_2CH_2SP(S)(OEt)_2$

$C_{10}H_{24}O_4P_2S_2$ M 334.365
Liq. d_4^{20} 1.22. $Bp_{0.25}$ 187°. n_D^{20} 1.5427.

Shvetsova-Shilovskaya, K.D. *et al, Zh. Obshch. Khim.*, 1959, **29**, 3593 (*Engl. transl.* p. 3554)

O,O,O′,O′-Tetraethyl hydrazodiphosphonothioate, 8CI — T-00048

N,N′-*Bis(diethoxyphosphinothioyl)hydrazine*
[35529-13-4]

$(EtO)_2P(S)-NHNH-P(S)(OEt_2)_2$

$C_8H_{22}N_2O_4P_2S_2$ M 336.340
Cryst. (EtOH). Mp 85-87°.

Tolkmith, H., *J. Am. Chem. Soc.*, 1962, **84**, 2097 (*synth*)
Lomakina, V.I. *et al, Zh. Obshch. Khim.*, 1971, **41**, 1204 (*Engl. transl.* p. 1215) (*synth*)

Tetraethyl hypodiphosphite, 9CI — T-00049

Tetraethoxydiphosphine
[31335-59-8]

$(EtO)_2P-P(OEt)_2$

$C_8H_{20}O_4P_2$ M 242.191
Liq. Bp_3 78°.

Proskurnina, M.V. *et al, Zh. Obshch. Khim.*, 1973, **43**, 66 (*Engl. transl.* p. 63) (*synth*)
Ponomarev, S.V. *et al, Zh. Obshch. Khim.*, 1978, **48**, 231 (*Engl. transl.* p. 201) (*synth, P nmr*)

Tetraethyl hypophosphate, 9CI — T-00050

Tetraethoxydiphosphine dioxide
[679-37-8]

$(EtO)_2P(O)-P(O)(OEt)_2$

$C_8H_{20}O_6P_2$ M 274.190
Liq. d_4^{20} 1.17. $Bp_{0.01}$ 88-89°. n_D^{20} 1.4394.

Baudler, M., *Z. Anorg. Allg. Chem.*, 1956, **288**, 171 (*synth, props, raman*)
Michalski, J. *et al, Bull. Acad. Pol. Sci., Ser. Sci. Chim.*, 1965, **13**, 253 (*synth*)
Harris, R.K. *et al, J. Chem. Soc. (A)*, 1967, 37 (*pmr, P nmr*)
Mowthorpe, D.J. *et al, Spectrochim. Acta, Part A*, 1967, **23**, 451 (*P nmr, pmr*)
Stec, W.J. *et al, Org. Mass Spectrom.*, 1975, **10**, 485 (*ms*)

O,O,O′,O′-Tetraethyl imidodiphosphate, 9CI — T-00051

O,O,O′,O′-*Tetraethyl iminodiphosphate, 8CI*
[2423-99-6]

$(EtO)_2P(O)NHP(O)(OEt)_2 \rightleftharpoons (EtO)_2P(OH){=}N{-}P(O)(OEt)_2$

$C_8H_{21}NO_6P_2$ M 289.205
Monoprotic acid largely in the iminol form. Liq. d_4^{20} 1.203. $Bp_{0.001}$ 79-80°. n_D^{20} 1.4415.

Cyclohexylammonium salt: Cryst. (hexane). Mp 77-78°.

N-*Trimethylsilyl: Tetraethyl trimethylsilylimidodiphosphate.*
$C_{11}H_{29}NO_6P_2Si$ M 361.386
Liq. $Bp_{0.001}$ 90-93°.

Kabachnik, M.I. *et al*, *Izv. Akad. Nauk SSSR, Ser. Khim.*, 1961, 819, 1022 (*Engl. transl.* pp. 745, 945) (*synth, ir, struct*)
Riesel, L. *et al*, *Z. Anorg. Allg. Chem.*, 1977, **430**, 227 (*synth, ir, silyl deriv*)
Gilyarov, V.A. *et al*, *Zh. Obshch. Khim.*, 1980, **50**, 44 (*Engl. transl.* p. 35) (*synth, deriv*)
Reynolds, M.A. *et al*, *J. Am. Chem. Soc.*, 1983, **105**, 6475 (*P nmr*)
Steinbach, J. *et al*, *Z. Anorg. Allg. Chem.*, 1984, **511**, 51 (*P nmr*)

sym-Tetraethyl monothiopyrophosphate T-00052

Tetraethyl thiodiphosphate, 9CI. O,O-*Diethyl* S-*(diethoxyphosphinyl) phosphorothioate.* OO-*Diethyl phosphorothioate anhydrosulfide*
[7342-94-1]

$(EtO)_2P(O){-}S{-}P(O)(OEt)_2$

$C_8H_{20}O_6P_2S$ M 306.250
Liq. n_D^{20} 1.4625. When heated, rapidly isomerises to *unsym*-Tetraethyl monothiopyrophosphate, T-00053 .

Michalski, J. *et al*, *Chem. Ber.*, 1969, **102**, 90 (*synth, ir*)
Michalski, J. *et al*, *J. Chem. Soc., Perkin Trans. 1*, 1974, 319 (*synth, ir, P nmr, props*)
Lebedev, E.P. *et al*, *Zh. Obshch. Khim.*, 1979, **49**, 1730 (*Engl. transl.* p. 1515) (*ir, P nmr*)

unsym-Tetraethyl monothiopyrophosphate T-00053

Tetraethyl thiodiphosphate, 9CI. O,O-*Diethyl* O-*(diethoxyphosphinyl) phosphorothioate*
[645-78-3]

$(EtO)_2P(O){-}O{-}P(S)(OEt)_2$

$C_8H_{20}O_6P_2S$ M 306.250
More stable than the isomer *sym*-Tetraethyl monothiopyrophosphate, T-00052 . Liq. $Bp_{0.05}$ 83-84°. n_D^{25} 1.4480.

▷LD_{50} (rat), 1 mg/kg. QC3550000.

Popov, E.M. *et al*, *Zh. Obshch. Khim.*, 1959, **29**, 1998 (*Engl. transl.* p. 1967) (*ir*)
Harris, R.K. *et al*, *J. Chem. Soc. (A)*, 1967, 37 (*pmr, nmr*)
Michalski, J. *et al*, *Chem. Ber.*, 1969, **102**, 90 (*synth, ir*)
Michalski, J. *et al*, *J. Chem. Soc. (C)*, 1970, 703; *J. Chem. Soc., Perkin Trans. 1*, 1974, 313 (*synth, P nmr*)
Mlotkowska, B. *et al*, *J. Prakt. Chem.*, 1978, **320**, 777 (*P nmr*)
Reynolds, M.A. *et al*, *J. Am. Chem. Soc.*, 1983, **105**, 6663 (*O and P nmr*)

N,N,N′,N′-Tetraethyl-*N″*-phenylphosphorodiamidimidic acid T-00054

$PhN{=}P(OH)(NEt_2)_2$

$C_{14}H_{26}N_3OP$ M 283.353
Free acid tautomeric with *N,N,N′,N′*-Tetraethyl-*N″*-phenylphosphoric triamide.

Ph ester: [77713-27-0]. *Phenyl* N,N,N′,N′-*tetraethyl*-N″-*phenylphosphorodiamidimidate.* P,P-*Bis(diethylamino)*-P-*phenoxy*-N-*phenylphosphazene.* P,P-*Bis(diethylamino)*-P-*phenoxy*-N-*phenylphosphine imide.*
$C_{20}H_{30}N_3OP$ M 359.450
Liq. d_4^{20} 1.07. n_D^{20} 1.5520.

Fluoride: [86601-02-7]. P,P-*Bis(diethylamino)*-P-*fluoro*-N-*phenylphosphazene.* P,P-*Bis(diethylamino)*-P-*fluoro*-N-*phenylphosphine imide.*
$C_{14}H_{25}FN_3P$ M 285.344
Liq. $Bp_{0.02}$ 94-95°. n_D^{20} 1.5445.

Chloride: [3185-83-9]. P-*Chloro*-P,P-*bis(diethylamino)*-N-*phenylphosphazene.* P-*Chloro*-P,P-*bis(diethylamino)*-N-*phenylphosphine imide.*
$C_{14}H_{25}ClN_3P$ M 301.798
No phys. props. reported.

Iodide: P,P-*Bis(diethylamino)*-P-*iodo*-N-*phenylphosphazene.* P,P-*Bis(diethylamino)*-P-*iodo*-N-*phenylphosphine imide.*
$C_{14}H_{25}IN_3P$ M 393.250
Viscous liq.

Gorbatenko, Zh.K. *et al*, *Zh. Obshch. Khim.*, 1981, **51**, 717 (*Engl. transl.* p. 583) (*ester, iodide, synth, ir, pmr, P nmr*)
Marchenko, A.P. *et al*, *Zh. Obshch. Khim.*, 1983, **53**, 698 (*Engl. transl.* p. 608) (*fluoride, F nmr, synth*)

N,N,N′,N′-Tetraethylphosphonic diamide, 9CI T-00055

Phosphonic bis(diethylamide). N,N,N′,N′-*Tetraethylphosphorodiamidous acid*
[3560-15-4]

$(Et_2N)_2P(O)H \rightleftharpoons (Et_2N)_2POH$

$C_4H_{21}N_2OP$ M 144.196
Tautomeric. Liq. $Bp_{0.03}$ 54-55°. n_D^{20} 1.4551.

Zwierzak, A., *Bull. Acad. Pol. Sci., Ser. Sci. Chim.*, 1965, **13**, 609 (*synth, ir*)
Wolf, R., *Spectrochim. Acta, Part A*, 1967, **23**, 1641 (*ir*)
Shilov, I.V. *et al*, *Zh. Prikl. Khim.*, 1971, **44**, 2581 (*Engl. transl.* p. 2660) (*synth, pmr*)
Foss, V.L. *et al*, *Zh. Obshch. Khim.*, 1980, **50**, 1236 (*Engl. transl.*, p. 1000) (*synth*)

Tetraethylphosphonium(1+) T-00056

[13983-95-4]

$PEt_4^{\oplus}$

$C_8H_{20}P^{\oplus}$ M 147.220 (ion)

Chloride: [7368-65-2].
$C_8H_{20}ClP$ M 182.673
Cryst. Mp 300° dec.

Bromide: [4317-07-1].
$C_8H_{20}BrP$ M 227.124
Needles ($EtOH/Et_2O$). Mp 320° dec.

Iodide: [4317-06-0].
$C_8H_{20}IP$ M 274.124
Cryst. ($EtOH/Et_2O$). Mp 270-278°, 294°, 306-308°.

▷TA2435000.

Horner, L. *et al*, *Chem. Ber.*, 1959, **92**, 2088 (*bromide, iodide*)
Grayson, M. *et al*, *J. Am. Chem. Soc.*, 1960, **82**, 3919 (*iodide*)
Schmidbaur, H. *et al*, *Chem. Ber.*, 1970, **103**, 3007 (*use*)
Koole, N.J. *et al*, *J. Magn. Reson.*, 1977, **25**, 375 (*nmr*)
Lipkowitz, K.B. *et al*, *Tetrahedron Lett.*, 1980, **21**, 1297 (*pmr*)

N,N,N',N'-Tetraethylphosphonothioic diamide, 9CI T-00057

Thiophosphonic bis(diethylamide)
[57039-78-8]

$(Et_2N)_2P(S)H$

$C_8H_{21}N_2PS$ M 208.301
Liq. d^{20} 1.00. Mp −37°. n_D^{20} 1.5108.

Falius, H. *et al*, *Z. Anorg. Allg. Chem.*, 1976, **420**, 65 (*synth, ir, raman, pmr, P nmr*)
Nifant'ev, Ê.E. *et al*, *Zh. Obshch. Khim.*, 1977, **47**, 299 (*Engl. transl.* p. 276) (*synth, nmr, pmr*)

N,N,N',N'-Tetraethylphosphoric triamide, 9CI T-00058

[38590-11-3]

$(Et_2N)_2P(O)NH_2$

$C_8H_{22}N_3OP$ M 207.255
N″-*Et:* [80920-38-3]. *Pentaethylphosphoric triamide.*
$C_{10}H_{26}N_3OP$ M 235.309
Liq. $Bp_{0.05}$ 94-95°. n_D^{20} 1.4625.
N″,N″-*Di-Et: see Hexaethylphosphorous triamide, H-00015*
N″-*Ph:* [3185-66-8]. *Tetraethylphosphorodiamidic anilide.*
$C_{14}H_{26}N_3OP$ M 283.353
Cryst. (cyclohexane). Mp 108-109°.
N″-*Trimethylsilyl:*
$C_{11}H_{30}N_3OPSi$ M 279.437
Solid. Mp 94°.

Modro, T.A., *Phosphorus Sulfur*, 1979, **5**, 331 (*phenyl deriv, cmr*)
Riesel, L. *et al*, *Z. Chem.*, 1980, **20**, 151 (*trimethylsilyl deriv, synth, pmr, nmr*)
Kovenya, V.A., *Zh. Obshch. Khim.*, 1981, **51**, 2678 (*Engl. transl.* p. 2310) (*derivs, synth*)
Miroshnichenko, V.V. *et al*, *Zh. Obshch. Khim.*, 1982, **52**, 2517 (*Engl. transl.* p. 2222) (*ethyl deriv, synth, P nmr*)

Tetraethylphosphorodiamidic acid, 9CI, 8CI T-00059

$(Et_2N)_2P(O)OH$

$C_8H_{21}N_2O_2P$ M 208.240
Me ester: Methyl tetraethylphosphorodiamidate.
$C_9H_{23}N_2O_2P$ M 222.267
Liq. Sol. H_2O. Bp_1 96°. n_D^{25} 1.4410.
Et ester: [3644-89-1]. *Ethyl tetraethylphosphorodiamidate.*
$C_{10}H_{25}N_2O_2P$ M 236.293
Liq. Sol. H_2O. d_4^{20} 0.98. Bp_9 119-120°. n_D^{20} 1.4452.
Heptyl ester: [28084-39-4]. *Heptyl tetraethylphosphorodiamidate.*
$C_{15}H_{36}N_2O_2P$ M 307.435
Extractant for metals of Groups IIIB, IVB.
Ph ester: see Phenyl tetraethylphosphorodiamidate, P-00318
Fluoride: [562-17-4].
$C_8H_{20}FN_2OP$ M 210.231
Liq. Bp_{22} 127-128°.
▷TD3850000.
Chloride: see Tetraethylphosphorodiamidic chloride, T-00060
Isothiocyanate: Tetraethyl phosphor(isothiocyanatidic) diamide.
$C_9H_{20}N_3OPS$ M 249.310
Liq. d_4^{20} 1.07. $Bp_{0.1}$ 82-84°. n_D^{20} 1.5030.
Isocyanate: [18025-89-3]. *Tetraethyl phosphor(isocyanatidic) diamide.*
$C_9H_{20}N_3O_2PS$ M 265.310
Reactive liq. d^{20} 1.08. Bp_1 112-114°. n_D^{20} 1.4629.
Azide: [59740-66-8].
$C_8H_{20}N_5OP$ M 233.253
Liq. $Bp_{0.8}$ 104°. n_D^{20} 1.4678.
Amide: see N,N,N',N'-Tetraethylphosphoric triamide, T-00058
Phenylhydrazide: see 2-Phenylphosphorodiamidic hydrazide, P-00274
Anhydride: see Octaethyldiphosphoramide, O-00001

Heap, R. *et al*, *J. Chem. Soc.*, 1948, 1313 (*fluoride*)
Loev, B. *et al*, *J. Org. Chem.*, 1957, **22**, 1186 (*esters, synth, props*)
Michalski, J. *et al*, *Rocz. Chem.*, 1957, **31**, 879; *CA*, **52**, 8037 (*isothiocyanate*)
Cheymol, J. *et al*, *C.R. Hebd. Seances Acad. Sci.*, 1959, **249**, 1240 (*methyl ester*)
Scott, F.L. *et al*, *J. Org. Chem.*, 1962, **27**, 4255 (*azide, synth, ir*)
Shtepanek, A.S. *et al*, *CA*, 1968, **68**, 114695 (*isocyanate*)
Abramov, V.S. *et al*, *Zh. Obshch. Khim.*, 1969, **39**, 1003, 2234 (*Engl. transl.* p. 974, 2180) (*ethyl ester*)
Landau, M.A. *et al*, *Zh. Strukt. Khim.*, 1970, **11**, 513 (*Engl. transl.* p. 467) (*fluoride, struct*)
Litvincheva, A.S. *et al*, *Izv. Akad. Nauk SSSR, Ser. Khim.*, 1970, 1935 (*Engl. transl.* p. 1823) (*heptyl ester, use*)

Tetraethylphosphorodiamidic chloride, 9CI, 8CI T-00060

Bis(diethylamino) phosphoryl chloride.
Bis(diethylamino)phosphinic chloride
[1794-24-7]

$(Et_2N)_2P(O)Cl$

$C_8H_{20}ClN_2OP$ M 226.686
Pungent liq. d_4^{20} 1.07. $Bp_{0.3}$ 101°, $Bp_{0.05}$ 72-73°. n_D^{20} 1.4671.

Scott, F.L. *et al*, *J. Org. Chem.*, 1962, **27**, 4255 (*synth*)
Coustures, Y. *et al*, *Bull. Soc. Chim. Fr.*, 1973, 926 (*ir, P nmr*)
Ko, E.C.F. *et al*, *Can. J. Chem.*, 1973, **51**, 597 (*synth, pmr, props*)
Labarre, M.-C. *et al*, *J. Chim. Phys.*, 1973, **70**, 534 (*ir, struct*)

Tetraethylphosphorodiamidoselenoic acid, 8CI T-00061

Phosphoroselenoic acid bis(diethylamide)

$(Et_2N)_2P(Se)OH \rightleftharpoons (Et_2N)_2P(O)SeH$

$C_8H_{21}N_2OPSe$ M 271.200
O-*Me ester:* [26452-49-3]. O-*Methyl tetraethylphosphorodiamidoselenoate.*
$C_9H_{23}N_2OPSe$ M 285.227
Liq. d_4^{20} 1.20. $Bp_{0.07}$ 83°. n_D^{20} 1.5055.
O-*Et ester:* [26452-50-6]. O-*Ethyl tetraethylphosphorodiamidoselenoate.*
$C_{10}H_{25}N_2OPSe$ M 299.254
Liq. d_4^{20} 1.16. $Bp_{0.05}$ 84-85°. n_D^{20} 1.4955.
Se-*Et ester:* [64673-73-0]. Se-*Ethyl tetraethylphosphorodiamidoselenoate.*
$C_{10}H_{25}N_2OPSe$ M 299.254
Liq. d_4^{20} 1.20. $Bp_{0.04}$ 94°. n_D^{20} 1.5039.
O-*Butyl ester:* [26452-52-8]. O-*Butyl tetraethylphosphorodiamidoselenoate.*
$C_{12}H_{29}N_2OPSe$ M 327.308

Liq. d_4^{20} 1.13. $Bp_{0.08}$ 98-99°. n_D^{20} 1.4964.
Se-*Butyl ester:* Se-*Butyl tetraethylphosphorodiamidoselenoate.*
$C_{12}H_{29}N_2OPSe$ M 327.308
Liq. d_4^{20} 1.15. $Bp_{0.08}$ 108.5°. n_D^{20} 1.5010.
O-*Ph ester:* [16604-66-3]. O-*Phenyl tetraethylphosphorodiamidoselenoate.*
$C_{14}H_{25}N_2OPSe$ M 347.298
Liq. d_4^{20} 1.22. $Bp_{0.0001}$ 145° (bath). n_D^{20} 1.5495.
Chloride: [25408-77-9].
$C_8H_{20}ClN_2PSe$ M 289.646
Liq. d_4^{20} 1.26. $Bp_{0.5}$ 97°. n_D^{20} 1.5290.

Byina, N.A. *et al, Izv. Akad. Nauk SSSR, Ser. Khim.*, 1967, 1606 (*Engl. transl.* p. 1545) (*phenyl ester, synth*)
Nuretdinova, I.A. *et al, Izv. Akad. Nauk SSSR, Ser. Khim.*, 1969, 1535 (*Engl. transl.* p. 1423) (*chloride*)
Nuretdinova, I.A. *et al, Zh. Obshch. Khim.*, 1969, **39**, 2265 (*Engl. transl.* p. 2209) (*O-esters*)
Nuretdinova, I.A. *et al, Izv. Akad. Nauk SSSR, Ser. Khim.*, 1971, 2095 (*Engl. transl.* p. 1989) (*O-esters*)
Shagidullin, R.R. *et al, Izv. Akad. Nauk SSSR, Ser. Khim.*, 1976, 184 (*Engl. transl.* p. 174) (*uv*)
Nikonorova, L.K. *et al, Zh. Obshch. Khim.*, 1977, **47**, 1740 (*Engl. transl.* p. 1591) (*Se*-esters)

Tetraethylphosphorodiamidotelluroic acid T-00062

$(Et_2N)_2P(Te)OH \rightleftharpoons (Et_2N)_2P(O)TeH$

$C_8H_{21}N_2OPTe$ M 319.840
O-*Et ester:* [51462-00-1]. O-*Ethyl tetraethylphosphorodiamidotelluroate.*
$C_{10}H_{25}N_2OPTe$ M 347.894
No phys. props. reported.

Nuretdinov, I.A. *et al, Izv. Akad. Nauk SSSR, Ser. Khim.*, 1973, 2827 (*Engl. transl.* p. 2765) (*nmr*)

Tetraethylphosphorodiamidothioic acid, 9CI, 8CI T-00063

$(Et_2N)_2P(S)OH \rightleftharpoons (Et_2N)_2P(O)SH$

$C_8H_{21}N_2OPS$ M 224.300
Diethylammonium salt: Solid. Mp 50-53°.
O-*Et ester:* [19946-96-4]. O-*Ethyl tetraethylphosphorodiamidothioate.*
$C_{10}H_{25}N_2OPS$ M 252.354
Liq. d_4^{20} 1.00. Bp_6 112-113°, $Bp_{0.5}$ 81.5°. n_D^{20} 1.4804 (1.5010). Variations may be due to presence of *S*-Et isomer.
S-*Et ester:* [24635-39-0]. S-*Ethyl tetraethylphosphorodiamidothioate.*
$C_{10}H_{25}N_2OPS$ M 252.354
Liq. d_4^{20} 1.02. Bp_1 102-104°, $Bp_{0.05}$ 76°. n_D^{20} 1.4900.
O-*Ph ester:* [16604-64-1]. O-*Phenyl tetraethylphosphorodiamidothioate.*
$C_{14}H_{25}N_2OPS$ M 300.398
Liq. d_4^{20} 1.07. $Bp_{0.0001}$ 95°. n_D^{20} 1.5330.
S-*Ph ester:* [18551-98-9]. S-*Phenyl tetraethylphosphorodiamidothioate.*
$C_{14}H_{25}N_2OPS$ M 300.398
Pale-yellow liq. $Bp_{0.1}$ 131°. n_D^{20} 1.5440.
Chloride: see Tetraethylphosphorodiamidothioic chloride, T-00065

Buina, N.A. *et al, Izv. Akad. Nauk SSSR, Ser. Khim.*, 1967, 1606 (*Engl. transl.* p. 1545) (*phenyl ester*)
Blindheim, U. *et al, Spectrochim. Acta, Part A*, 1969, **25**, 1105 (*esters, synth, ir*)
Nifant'ev, E.E. *et al, Zh. Obshch. Khim.*, 1973, **43**, 2658 (*Engl. transl.* p. 2636) (*esters, synth*)

Tetraethylphosphorodiamidothioic azide, 9CI T-00064

Bis(diethylamino) thiophosphoryl azide.
Bis(diethylamino)phosphinothioic azide
[59998-83-3]

$(Et_2N)_2P(S)N_3$

$C_8H_{20}N_5PS$ M 249.313
Liq. d_4^{20} 1.08. $Bp_{0.01}$ 84°. n_D^{20} 1.5058.

Zaslavskaya, N.N. *et al, Izv. Akad. Nauk SSSR, Ser. Khim.*, 1976, 931 (*Engl. transl.* p. 911) (*synth*)

Tetraethylphosphorodiamidothioic chloride, 9CI, 8CI T-00065

Bis(diethylamino)phosphinothioyl chloride
[4234-61-1]

$(Et_2N)_2P(S)Cl$

$C_8H_{20}ClN_2PS$ M 242.746
Liq. d_4^{20} 1.09. Bp_1 117-119°. n_D^{20} 1.5130.

Cowley, A.H. *et al, J. Am. Chem. Soc.*, 1965, **87**, 4454 (*pmr*)
Nuretdinova, O.N. *et al, Zh. Obshch. Khim.*, 1965, **35**, 1880 (*Engl. transl.* p. 1815) (*synth*)

Tetraethylphosphorodiamidous acid, 9CI T-00066

$(Et_2N)_2POH \rightleftharpoons (Et_2N)_2P(O)H$

$C_8H_{21}N_2OP$ M 192.240
Free acid exists as the phosphoryl tautomer *N,N,N',N'*-Tetraethylphosphonic diamide, T-00055 .
Et ester: [2632-88-4]. *Ethyl tetraethylphosphorodiamidite.*
$C_{10}H_{25}N_2OP$ M 220.294
In the presence of diethyl azodicarboxylate, converts carboxylic acids to their ethyl esters. Liq. Bp_{10} 105°.
Isopropyl ester: [3402-28-6]. *Isopropyl tetraethylphosphorodiamidite.*
$C_{11}H_{27}N_2OP$ M 234.321
Liq. $Bp_{0.5}$ 45-46°. n_D^{20} 1.4499.
Ph ester: [14684-28-7]. *Phenyl tetraethylphosphorodiamidite.*
$C_{14}H_{25}N_2OP$ M 268.338
Liq. $Bp_{0.5}$ 95°. n_D^{20} 1.5148.
Trimethylsilyl ester: [55274-09-4]. *Trimethylsilyl tetraethylphosphorodiamidite.*
$C_{11}H_{29}N_2OPSi$ M 264.422
Liq. Bp_{10} 86-90°. n_D^{20} 1.4499.
Chloride: see Tetraethylphosphorodiamidous chloride, T-00067
Bromide: [53764-97-9].
$C_8H_{20}BrN_2P$ M 255.137
Liq. Bp_{10} 135°. n_D^{20} 1.5075.
Iodide: [59612-01-0].
$C_8H_{20}IN_2P$ M 302.138
Orange liq. Bp_2 98°, $Bp_{0.07}$ 73-75°.

Fluck, E., *Z. Anorg. Allg. Chem.*, 1960, **307**, 38 (*phenyl ester, synth, ir*)
Houalla, D. *et al, Bull. Soc. Chim. Fr.*, 1965, 2368 (*esters, synth, P nmr*)
Mitsunobu, O. *et al, Bull. Chem. Soc. Jpn.*, 1971, **44**, 3427 (*ester, use*)

Novikova, Z.S. *et al, Zh. Obshch. Khim.*, 1974, **44**, 1857 (*Engl. transl.* p. 1825) (*bromide, synth, P nmr*)
Pudovik, A.N. *et al, Zh. Obshch. Khim.*, 1975, **45**, 248 (*Engl. transl.* p. 240) (*trimethylsilyl ester*)
Novikova, Z.S. *et al, Zh. Obshch. Khim.*, 1976, **46**, 2213 (*Engl. transl.* p. 2128) (*ethyl ester, synth, P nmr*)
Gorbatenko, Zh.K. *et al, Zh. Obshch. Khim.*, 1977, **47**, 1915 (*Engl. transl.* p. 1752) (*iodide*)
Troy, D. *et al, Bull. Soc. Chim. Fr., Part I*, 1979, 241 (*ethyl ester, uv*)
Romanenko, V.D. *et al, Synthesis*, 1980, 823 (*iodide, synth, P nmr*)

Tetraethylphosphorodiamidous chloride, 9CI — T-00067

[685-83-6]

$(Et_2N)_2PCl$

$C_8H_{20}ClN_2P$ M 210.686

Viscous liq. $Bp_{0.02}$ 60°. n_D^{20} 1.4900. Air-sensitive.

▷Reacts violently with H_2O

Eliseenkov, V.N. *et al, Zh. Obshch. Khim.*, 1970, **40**, 498 (*Engl. transl.* p. 461) (*synth*)
Barlos, K. *et al, Z. Naturforsch., B*, 1978, **33**, 515 (*N nmr*)
Van Lindthoudt, J.P. *et al, Spectrochim. Acta, Part A*, 1980, **36**, 315 (*pmr, cmr, P nmr*)
Butters, T. *et al, Chem. Ber.*, 1984, **117**, 990 (*synth*)

O,O,O′,O′-Tetraethyl *S,S′*-propanediyl bisphosphorodithioate, 9CI — T-00068

1,3-[Bis(diethoxyphosphinothioyl)thio]propane

$(EtO)_2P(S)-SCH_2CH_2CH_2S-P(S)(OEt)_2$

$C_{11}H_{26}O_4P_2S_4$ M 412.512

Liq. d_4^{20} 1.19. $Bp_{0.0001}$ 110°. n_D^{20} 1.5402.

Zemlyanskii, N.I. *et al, Zh. Obshch. Khim.*, 1970, **40**, 1713 (*Engl. transl.* p. 1701)

Tetraethyl pyrophosphate, 8CI — T-00069

Tetraethyl diphosphate, 9CI. TEPP, BSI

[107-49-3]

$(EtO)_2P(O)-O-P(O)(OEt)_2$

$C_8H_{20}O_7P_3$ M 321.163

Non-systemic insecticide. Misc. H_2O, most org. solvs., spar. sol. pet. ether. d_4^{24} 1.19. $Bp_{0.08}$ 104-110°. n_D^{25} 1.4180. Corrodes metals, hydrol. by H_2O. Dec. at 170° → ethylene.

▷V. highly toxic, TLV 0.05. UX7051000.

Toy, A.D.F., *J. Am. Chem. Soc.*, 1948, **70**, 3882 (*synth*)
Mowthorpe, D.J. *et al, Spectrochim. Acta, Part A*, 1967, **23**, 451 (*nmr, pmr*)
Turpin, R. *et al, Bull. Soc. Chim. Fr.*, 1971, 3878 (*synth*)
Petrov, K.A. *et al, Zh. Obshch. Khim.*, 1976, **46**, 1242 (*Engl. transl.* p. 1222) (*synth*)
Sosnovsky, G. *et al, Z. Naturforsch, B*, 1976, **31**, 820 (*synth, pmr*)
Mizraki, V. *et al, J. Org. Chem.*, 1982, **47**, 3533 (*ms*)
Pesticide Manual, 6th Ed., 498; 7th Ed., 510.
Sax, N.I., *Dangerous Properties of Industrial Materials*, 6th Ed., Van Nostrand-Reinhold, 1984, 1015.

Tetraethyl pyrophosphite, 8CI — T-00070

Tetraethyl diphosphite, 9CI

[21646-99-1]

$(EtO)_2POP(OEt)_2$

$C_8H_{20}O_5P_2$ M 258.191

Reagent for peptide synth. Ligand for Cu, Fe, Mo, Pt, Re, and Mn. Liq. d_0^0 1.057. Bp_1 87-89°. n_D^{10} 1.4377.

Ressler, C. *et al, J. Am. Chem. Soc.*, 1957, **79**, 4511 (*use*)
Samuel, D. *et al, J. Org. Chem.*, 1963, **28**, 1155 (*synth, ir*)
Stec, W.J. *et al, Org. Mass. Spectrom.*, 1975, **10**, 485 (*ms*)
Du Preez, A.L. *et al, J. Organomet. Chem.*, 1977, **141**, C10 (*complexes*)
Kabachnik, M.M. *et al, Zh. Obshch. Khim.*, 1979, **49**, 1446 (*Engl. transl.* p. 1264) (*synth*)
Fieser, M. *et al, Reagents for Organic Synthesis*, Wiley, 1967-84, **1**, 1138.

Tetraethyltetraphosphetane, 9CI — T-00071

Tetraethylcyclotetraphosphine

[3040-70-8]

cyclo-$(EtP)_4$ (four-membered P–P ring, each P bearing Et)

$C_8H_{20}P_4$ M 240.141

Liq. Bp_3 160°, $Bp_{0.05}$ 124-129°.

B,MeI: Tetraethylmethyltetraphosphetanium iodide.
$C_9H_{23}IP_4$ M 382.080
Sol. MeOH, Me_2CO, THF, insol. C_6H_6, Et_2O. Mp 84° dec.

Henderson, W.A. *et al, J. Am. Chem. Soc.*, 1963, **85**, 2462 (*synth, P nmr*)
Issleib, K. *et al, Z. Anorg. Allg. Chem.*, 1968, **360**, 77 (*synth, deriv, P nmr*)
Albrand, J.P. *et al, J. Am. Chem. Soc.*, 1978, **100**, 2600 (*synth, P nmr, pmr*)

Tetraethyl thiohypophosphate, 9CI — T-00072

Tetraethoxydiphosphine disulfide

[5935-39-7]

$(EtO)_2P(S)—P(S)(OEt)_2$

$C_8H_{20}O_4P_2S_2$ M 306.311

Possesses fungicidal props. Liq. d_4^{20} 1.16. Mp 0-1°. $Bp_{0.1}$ 109°. n_D^{20} 1.5080.

Almasi, L. *et al, Chem. Ber.*, 1963, **96**, 2024 (*synth, ir*)
Harris, R.K. *et al, J. Chem. Soc. (A)*, 1967, 37 (*pmr, P nmr*)
Proskurnina, M.V. *et al, Zh. Obshch. Khim.*, 1973, **43**, 66 (*Engl. transl.* p. 63) (*synth*)
Stec, W.J. *et al, Org. Mass. Spectrom.*, 1975, **10**, 485 (*ms*)

O,O,O′,O′-Tetraethyl thiohypophosphate, 9CI — T-00073

Tetraethoxydiphosphine 1-oxide 2-sulfide

[5935-34-2]

$(EtO)_2P(O)—P(S)(OEt)_2$

$C_8H_{20}O_5P_2S$ M 290.251

Liq. d_4^{20} 1.17. $Bp_{0.01}$ 99°. n_D^{20} 1.4724.

Michalski, J. *et al, Bull. Acad. Pol. Sci., Ser. Sci. Chim.*, 1965, **13**, 677 (*synth, struct*)
Harris, R.K. *et al, J. Chem. Soc. (A)*, 1967, 37 (*pmr, P nmr*)
Stec, W.J. *et al, J. Chem. Soc., Perkin Trans. 2*, 1972, 463 (*P nmr*)

Stec, W.J. *et al*, *Org. Mass Spectrom.*, 1975, **10**, 485 (*ms*)
Foss, V.L. *et al*, *Zh. Obshch. Khim.*, 1978, **48**, 1713 (*Engl. transl.* p. 1565) (*synth*)

_O,O,O',O'_-Tetraethyl thioimidodiphosphate, 9CI T-00074

O,O-*Diethyl (diethoxyphosphinothioyl)-phosphoramidate*. O,O-*Diethyl (diethoxyphosphinyl)-phosphoramidothioate*. O,O,O',O'-*Tetraethyl thioiminodiphosphate*

[7109-08-2]

$(EtO)_2P(O)-NH-P(S)(OEt)_2$

$C_8H_{21}NO_5PS$ M 274.291

Liq. d_4^{20} 1.21. $Bp_{0.0001}$ 78-80°. n_D^{20} 1.4728.

Gilyarov, V.A. *et al*, *Zh. Obshch. Khim.*, 1966, **36**, 274 (*Engl. transl.* p. 285)

***N,N,N',N'*-Tetraethyl-*P*-(trichloromethyl)-phosphonimidic diamide, 9CI** T-00075

P,P-*Bis(diethylamino)*-P-*(trichloromethyl)-phosphazene*. P,P-*Bis(diethylamino)*-P-*(trichloromethyl)phosphine imide*

[77339-54-9]

$(Et_2N)_2P(=NH)CCl_3$

$C_9H_{21}Cl_3N_3P$ M 308.618

Liq. $Bp_{0.02}$ 80-82°. pK_a 10.50 (H_2O). n_D^{20} 1.5127.

B, HCl: [77339-55-0]. *Aminobis(diethylamino)(trichloromethyl)-phosphonium chloride.*
$C_9H_{22}Cl_4N_3P$ M 345.079
Cryst. (C_6H_6/hexane). Mp 133-134°.

N''-*Chloro:* [77339-59-4]. N''-*Chloro*-N,N,N',N'-*tetraethyl*-P-*(trichloromethyl)phosphonimidic diamide.*
$C_9H_{20}Cl_4N_3P$ M 343.063
Unstable oil.

Kozlov, E.S. *et al*, *Zh. Obshch. Khim.*, 1980, **50**, 2672 (*Engl. transl.* p. 2156) (*synth, ir, derivs, P nmr*)
Kozlov, E.S. *et al*, *Zh. Obshch. Khim.*, 1982, **52**, 2239 (*Engl. transl.* p. 1992) (*ir, P nmr*)

_O,O,O',O'_-Tetraethyl trithiopyrophosphate T-00076

Tetraethyl thiodiphosphate, 10CI, 9CI. O,O-*Diethyl phosphorodithioate anhydrosulfide*

[4328-22-7]

$(EtO)_2P(S)-S-P(S)(OEt)_2$

$C_8H_{20}O_4P_2S_3$ M 338.371

Constit. of crude *O,O*-Diethyl phosphorodithioate, D-00384 . Cryst. (hexane or MeOH). Mp 50° (42.5-43°). $Bp_{0.01}$ 110°. n_D^{20} 1.5098.

▷LD_{50} (rats) 100 mg/kg

Schrader, G. *et al*, *Angew. Chem.*, 1958, **70**, 692 (*synth, tox*)
Mel'nikov, N.N. *et al*, *Zh. Obshch. Chem.*, 1960, **30**, 2319 (*Engl. transl.* p. 2300) (*synth*)
Almasi, L. *et al*, *Chem. Ber.*, 1964, **97**, 661 (*synth*)
Lippman, A.E. *et al*, *J. Org. Chem.*, 1966, **31**, 471 (*P nmr*)

3,3,4,4-Tetrafluoro-1,2-bis(diphenylphosphino)cyclobutene T-00077

(3,3,4,4-Tetrafluoro-1-cyclobutene-1,2-diyl)-bis[diphenylphosphine], 9CI

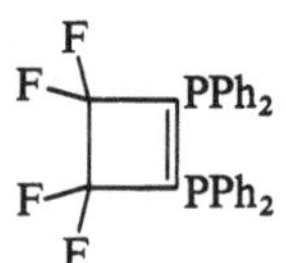

$C_{28}H_{20}F_4P_2$ M 494.407

Ligand for Fe. Cryst. (Me_2CO or EtOH). Mp 129.5-130.5°. Forms Fe complexes.

Cullen, W.R. *et al*, *Can. J. Chem.*, 1967, **45**, 683 (*synth, ir, pmr, nmr*)
Stockel, R.F. *et al*, *Can. J. Chem.*, 1969, **47**, 867 (*synth, ir, nmr*)

3,3,4,4-Tetrafluoro-1-cyclobutene-1,2-diphosphonic acid T-00078

(3,3,4,4-Tetrafluoro-1-cyclobutene-1,2-diyl)-bis(phosphonic acid), 9CI

$C_4H_4F_4O_6P_2$ M 286.013

Tetra-Me ester: [21669-30-7]. *Tetramethyl (3,3,4,4-tetrafluoro-1-cyclobutene-1,2-diyl)bisphosphonate.*
$C_8H_{12}F_4O_6P_2$ M 342.120
Liq. $Bp_{0.02}$ 98°. n_D^{25} 1.4159.

Tetra-Et ester: [4545-92-0]. *Tetraethyl (3,3,4,4-tetrafluoro-1-cyclobutene-1,2-diyl)bisphosphonate.*
$C_{12}H_{20}F_4O_6P_2$ M 398.228
Liq. Undistillable. n_D^{25} 1.4208.

Frank, A.W., *J. Org. Chem.*, 1965, **30**, 3663 (*synth, ir*)
Bauer, G. *et al*, *Z. Naturforsch., B*, 1979, **34**, 1252 (*synth, pmr, F and P nmr*)

2,2,4,6-Tetrafluoro-2,2,4,4,6,6-hexahydro-4,6-diphenyl-1,3,5-triaza-2,4,6-triphosphorine, 9CI T-00079

2,2,4,6-Tetrafluoro-2,2,4,4,6,6-hexahydro-4,6-diphenylcyclotriphosphazene

[73502-97-3]

(*4RS,6RS*)*-form*

$C_{12}H_{10}F_4N_3P_3$ M 365.146

(_4RS,6RS_)-form [22341-00-0]
trans-*form*
Solid. Mp 78-79°.

(_4RS,6SR_)-form [21079-47-0]
cis-*form*
Solid. Mp 74-75°.

Allen, C.W. *et al*, *Inorg. Chem.*, 1968, **7**, 2177 (*synth*)
Wagner, A.J. *et al*, *J. Inorg. Nucl. Chem.*, 1971, **33**, 1307 (*uv*)
Allen, C.W. *et al*, *J. Chem. Soc., Dalton Trans.*, 1974, 1685 (*ms*)

Kauffman, G.B. *et al*, *J. Chromatogr.*, 1976, **123**, 448 (*tlc*)
Allen, C.W., *J. Organomet. Chem.*, 1977, **125**, 215 (*cmr*)
Allen, C.W. *et al*, *Inorg. Chem.*, 1980, **19**, 1719 (*pe*)

Tetrafluoromethoxyphosphorane, 8CI — T-00080

[20107-85-1]

MeOPF_4

CH_3F_4OP M 138.001
Obt. only in impure form.

Brown, D.H. *et al*, *J. Chem. Soc.* (*A*), 1969, 872 (*synth, pmr, nmr, ms*)

Tetrafluoromethylphosphorane, 9CI, 8CI — T-00081

[420-64-4]

MePF_4

CH_3F_4P M 122.002
Gas or liq. Mp −50°. Bp_{756} 12.5°. V. sensitive to moisture.

Downs, A.J. *et al*, *Spectrochim. Acta, Part A*, 1967, **23**, 681 (*ir*)
Blazer, T.A. *et al*, *Z. Naturforsch., B*, 1969, **24**, 1081 (*ms*)
Inorg. Synth., 1971, **13**, 37 (*synth*)
Appel, R. *et al*, *Chem. Ber.*, 1974, **107**, 2169 (*synth, nmr*)
Eisenhut, M. *et al*, *J. Am. Chem. Soc.*, 1974, **96**, 5385 (*nmr, struct*)
Robert, D.U. *et al*, *Org. Magn. Reson.*, 1975, **7**, 291 (*pmr, nmr*)
Russegger, P. *et al*, *J. Chem. Phys.*, 1975, **62**, 1086 (*struct*)
Fastenakel, D. *et al*, *Mol. Phys.*, 1980, **40**, 361 (*struct*)

Tetrafluoromethylthiophosphorane, 9CI — T-00082

[22606-68-4]

MeSPF_4

CH_3F_4PS M 154.062
Volatile liq. at −45°. Dec. at 0°.

Brown, D.H. *et al*, *J. Chem. Soc.* (*A*), 1970, 914 (*synth, ms, ir, pmr*)
Peake, S.C. *et al*, *J. Chem. Soc.* (*A*), 1970, 1049 (*synth, nmr*)
Cavell, R.G. *et al*, *Inorg. Chem.*, 1979, **18**, 3400 (*synth, struct, nmr*)

Tetrafluorophenoxyphosphorane, 9CI — T-00083

[79303-16-5]

PhOPF_4

$C_6H_5F_4OP$ M 200.072
Possesses a bipyramidal covalent struct.

Ruppert, I., *Z. Anorg. Allg. Chem.*, 1981, **477**, 59 (*synth, nmr, cmr*)
Il'in, E.G. *et al*, *Dokl. Akad. Nauk SSSR, Ser. Sci. Khim.*, 1982, **266**, 123 (*Engl. transl.* p. 300) (*struct, nmr*)

Tetrafluorophenylphosphorane, 9CI — T-00084

[666-23-9]

PhPF_4

$C_6H_5F_4P$ M 184.073
Reacts with silyl ethers to prod. alkyl fluorides and with carboxylic acid anhydrides to form acyl fluorides. d_4^{15} 1.39. Bp 136°, Bp_{60} 58°. n_D^{20} 1.4245. Mod. stable to moisture, but fumes in atm.

Komikov, I.P. *et al*, *Zh. Obshch. Khim.*, 1962, **32**, 301 (*Engl. transl.* p. 295) (*synth*)
Inorg. Synth., 1967, **9**, 64 (*synth*)
Blazer, T.A. *et al*, *Z. Naturforsch., B*, 1969, **24**, 1081 (*ms*)
Robert, D.U. *et al*, *Tetrahedron*, 1973, **29**, 1877 (*use*)
Appel, R. *et al*, *Chem. Ber.*, 1974, **107**, 2169 (*synth, nmr*)
Robert, D.U. *et al*, *Org. Magn. Reson.*, 1975, **7**, 291 (*pmr, nmr*)
Dittebrandt, C. *et al*, *J. Mol. Struct.*, 1980, **63**, 227 (*ed*)
Robert, D. *et al*, *Inorg. Chem.*, 1982, **21**, 1805 (*pmr, nmr, props*)

Tetrafluorophenylthiophosphorane, 8CI — T-00085

PhSPF_4

$C_6H_5F_4PS$ M 216.133
Possesses an unionized trigonal bipyramidal struct. Mobile fuming liq. Bp_{40} 75-76°, Bp_{10} 58°. Dec. at r.t. to PF_5, F_3PO, and $(PhS)_2$.

Peake, S.C. *et al*, *J. Chem. Soc.* (*A*), 1970, 1049 (*synth, nmr*)
Norbury, A.H. *et al*, *Spectrochim. Acta, Part A*, 1971, **27**, 151 (*ir, raman, nmr*)

Tetrafluoro(trifluoromethyl)phosphorane, 9CI, 8CI — T-00086

[1184-81-2]

F_4PCF_3

CF_7P M 175.974
Molecule consists of unionized trigonal bipyramid. Gas. Mp −117°. Bp −35°.

Mahler, W., *Inorg. Chem.*, 1963, **2**, 230 (*synth, ir*)
Muetterties, E.L. *et al*, *Inorg. Chem.*, 1963, **2**, 613 (*ir, nmr*)
Nixon, J.F. *et al*, *Spectrochim. Acta*, 1964, **20**, 1835 (*nmr*)
Cohen, E.A. *et al*, *Inorg. Chem.*, 1968, **7**, 398 (*struct*)
Griffiths, J.E., *J. Chem. Phys.*, 1968, **49**, 1307 (*ir, raman*)
Cavell, R.G. *et al*, *J. Am. Chem. Soc.*, 1977, **99**, 7841 (*cmr*)
Oberhammer, H. *et al*, *Inorg. Chem.*, 1982, **21**, 275 (*ed*)

2,2,4,4-Tetrahydro-1*H*-2,4-benzodiphosphepin, 10CI — T-00087

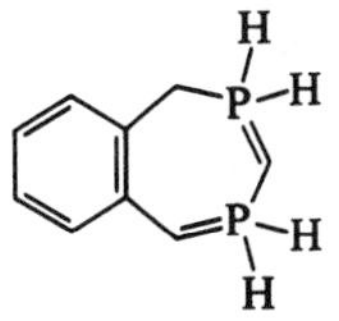

$C_9H_{12}P_2$ M 182.141
Double ylides exhibiting fluxionality.

2,2,4,4-Tetra-Me: [78532-93-1]. *2,2,4,4-Tetrahydro-2,2,4,4-tetramethyl-1*H*-2,4-benzodiphosphepin.*
$C_{13}H_{20}P_2$ M 238.249
Cryst. (toluene/pentane). Mp 105°. $Bp_{0.001}$ 100° subl.

2,2,4,4-Tetra-Ph: [78532-94-2]. *2,2,4,4-Tetrahydro-2,2,4,4-tetraphenyl-1*H*-2,4-benzodiphosphepin.*
$C_{33}H_{28}P_2$ M 486.532
Orange cryst. (toluene/pentane). Mp 122°.

Schmidbaur, H. *et al*, *Chem. Ber.*, 1981, **114**, 1428 (*synth, ms, complexes*)

2,2,2,3-Tetrahydro-1,2-benzoxaphosphole, 9CI T-00088

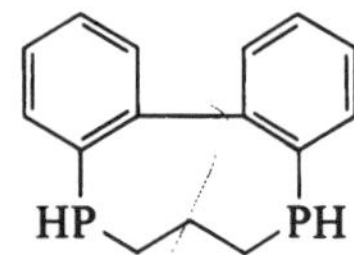

C_7H_9OP M 140.121

2,2,2-Triethoxy: [70385-78-3].
$C_{13}H_{21}O_4P$ M 272.280
Liq. $Bp_{0.001}$ 88-90°. n_D^{20} 1.4995.

Bel'skii, V.E. *et al*, *Zh. Obshch. Khim.*, 1979, **49**, 344 (*Engl. transl.* p. 298) (*synth, P nmr, props*)

6,7,8,9-Tetrahydro-5*H*-dibenzo[*f,h*][1,5]-diphosphonin, 8CI T-00089

5,9-Diphospha-1,2:3,4-dibenzocyclononadiene

$C_{15}H_{16}P_2$ M 258.239

5,5,9,9-Tetra-Et, dibromide: [5274-21-5]. *5,5,9,9-Tetraethyl-6,7,8,9-tetrahydro-5*H*-dibenzo*[f,h][*1,5*]*-diphosphoninium dibromide. 5,5,9,9-Tetraethyl-5,9-diphosphonia-1,2:3,4-dibenzocyclononadiene dibromide.*
$C_{23}H_{34}Br_2P_2$ M 532.277
Cryst. + $1.5H_2O$ (EtOH). Mp 335-336° dec.

Allen, D.W. *et al*, *J. Chem. Soc. (C)*, 1967, 1869.

4,5,9,10-Tetrahydro-4,9-dihydroxyphosphorino[2,3,4,5-*lmn*]phosphanthridine 4,9-dioxide, 9CI T-00090

[40964-81-6]

$C_{14}H_{12}O_4P_2$ M 306.194
Solid. Mp 392-395° dec.

4,9-Dimethoxy (*di-Me ester*):
$C_{16}H_{16}O_4P_2$ M 334.248
Solid (EtOAc). Mp 269-272°. A mixt. of *cis-* and *trans-*isomers.

Robinson, C.N. *et al*, *Tetrahedron Lett.*, 1972, 4977 (*synth, uv, ir, pmr*)

1,2,3,4-Tetrahydro-1,4-dimethyl-1,4-benzophospharsenine T-00091

1,2,3,4-Tetrahydro-1,4-dimethyl-1,4-benzophospharsinine

$C_{10}H_{14}AsP$ M 240.116

B,2MeBr: 1,1,4,4-Tetramethyl-1,2,3,4-tetrahydro-1,4-benzophospharseninium dibromide.
$C_{14}H_{20}AsBr_2P$ M 454.015
Cryst. + $2H_2O$ (MeOH).

Jones, E.R.H. *et al*, *J. Chem. Soc.*, 1955, 4472.

2,3,4,5-Tetrahydro-1,5-dimethyl-1*H*-1,5-benzophospharsepine T-00092

$C_{11}H_{16}AsP$ M 254.143

*B,2MeBr: 2,3,4,5-Tetrahydro-1,1,5,5-tetramethyl-1*H*-1,5-benzophospharsepinium dibromide.*
$C_{13}H_{22}AsBr_2P$ M 444.020
Cryst. + $1H_2O$ (MeOH/EtOH).

Jones, E.R.H. *et al*, *J. Chem. Soc.*, 1955, 4472.

Tetrahydro-1,7-dimethyl-1*H*,5*H*-[1,3,2]-diazaphospholo[1,2-*a*][1,3,2]-diazaphosphole, 10CI T-00093

2,8-Dimethyl-2,5,8-triaza-1-phosphabicyclo[*3.3.0*]*-octane*

[62051-26-7]

$C_6H_{14}N_3P$ M 159.170
Oil. $Bp_{0.1}$ 54-55°, $Bp_{0.01}$ 35-36°. Rapidly polymerises; depolymerises readily on dist.

Denney, D.B. *et al*, *J. Am. Chem. Soc.*, 1980, **102**, 7072 (*synth, pmr, cmr, P nmr*)

Tetrahydro-2,6-dimethyl-4*H*-1,3,6,2-dioxazaphosphocine, 9CI T-00094

$C_6H_{14}NO_2P$ M 163.156

2-Oxide: [52202-92-3].
$C_6H_{14}NO_3P$ M 179.155
Solid. Mp 53-54°. Bp_3 135°, Bp_1 88-90°.

2-Oxide; B,MeI: Solid. Mp 189-190°.

2-Sulfide: [56152-43-3].
$C_6H_{14}NO_2PS$ M 195.216
Solid. Mp 65-66°. Bp_1 82°.

2-Sulfide; B,MeI: Solid. Mp 166-168°.

Godovikov, N.N. *et al*, *Zh. Obshch. Khim.*, 1975, **45**, 728 (*Engl. transl.* p. 717) (*derivs, synth*)
Sharma, R.K. *et al*, *J. Chem. Res. (S)*, 1980, 12 (*synth, ir, pmr*)
Patsanovskii, I.I. *et al*, *Zh. Obshch. Khim.*, 1981, **51**, 980 (*Engl. transl.* p. 818) (*conformn*)
Vitkovskii, V.Yu., *et al*, *Zh. Obshch. Khim.*, 1981, **51**, 1769 (*Engl. transl.* p. 1514) (*oxide, ms*)

1,2,3,6-Tetrahydro-4,5-dimethyl-1,2-diphosphorine, 8CI T-00095

1,2,3,6-Tetrahydro-4,5-dimethyl-1,2-diphosphinine

$C_6H_{12}P_2$ M 146.108

1,2-Di-Me: [14410-06-1]. *1,2,3,6-Tetrahydro-1,2,4,5-tetramethyl-1,2-diphosphinine.*
$C_8H_{16}P_2$ M 174.162
Liq. Bp_8 100°.

1,2-Di-Me, 1,2-disulfide: [14410-00-5].
$C_8H_{16}P_2S_2$ M 238.282
Cryst. (EtOAc). Mp 193°.

1,2-Di-Ph: [14463-05-9]. *1,2,3,6-Tetrahydro-4,5-dimethyl-1,2-diphenyl-1,2-diphosphinine.*
$C_{18}H_{20}P_2$ M 298.304
Liq. $Bp_{0.005}$ 150°.

1,2-Di-Ph, 1,2-disulfide: [14463-04-8].
$C_{18}H_{20}P_2S_2$ M 362.424
Cryst. (EtOAc). Mp 178-179°.

Schmidt, U. *et al*, *Chem. Ber.*, 1968, **101**, 1381.

1,3,4,5-Tetrahydro-4,5-dimethyl-2,5-methano-2*H*-2-benzophosphepin, 9CI T-00096

3,4-Benzo-5,6-dimethyl-1-phosphabicyclo[3.2.1]oct-3-ene

[76232-78-5]

$C_{13}H_{18}P$ M 205.259

B,MeI: [76232-79-6]. *1,3,4,5-Tetrahydro-2,4,5-trimethyl-2,5-methano-2*H*-2-benzophosphepinium iodide.*
$C_{14}H_{21}IP$ M 347.198
Solid. Mp 208.5-210°.

Oxide: [76232-77-4].
$C_{13}H_{18}OP$ M 221.258
Cryst. (hexane). Mp 110-111.5°.

MacDiarmid, J.E. *et al*, *J. Org. Chem.*, 1981, **46**, 1451 (*derivs, pmr, cmr, P nmr*)
Quin, L.D. *et al*, *Tetrahedron Lett.*, 1982, **23**, 2529 (*synth, props, pmr, P nmr, cmr*)

Tetrahydro-2,6-dimethyl-[1,3,2]-oxazaphospholo[2,3-*b*][1,3,2]-oxazaphosphole, 10CI T-00097

$C_6H_{12}NO_2P$ M 161.140

Ligand for Mo and W. Liq. $Bp_{0.05}$ 65-70°. Mixt. of racemic and meso forms.

Houalla, D. *et al*, *Tetrahedron Lett.*, 1977, 3041 (*synth, P nmr*)
Houalla, D. *et al*, *Nouv. J. Chim.*, 1979, **3**, 507 (*pe*)
Grec, D. *et al*, *J. Am. Chem. Soc.*, 1980, **102**, 7133 (*complex, cryst struct*)
Bonningue, C. *et al*, *J. Chem. Soc., Perkin Trans. 2*, 1981, 19 (*stereochem*)

Tetrahydro-5,5-dimethyl-2-phenoxy-2*H*-1,3,2-oxazaphosphorine, 9CI T-00098

5,5-Dimethyl-2-phenoxy-1,3,2-oxazaphosphorinane

$C_{11}H_{16}NO_2P$ M 225.227

2-Oxide: [85289-24-3].
$C_{11}H_{16}NO_3P$ M 241.226
Cryst. (EtOAc/cyclohexane). Mp 130-130.5°.

2-Sulfide: [85289-25-4].
$C_{11}H_{16}NO_2PS$ M 257.287
Cryst. (cyclohexane). Mp 90.5-91°.

Edmundson, R.S., *Org. Mass Spectrom.*, 1982, **17**, 558; 1983, **18**, 150 (*synth, ir, ms, pe*)

3*a*,4,7,7*a*-Tetrahydro-1,8-dimethyl-4,7-phosphinidene-1*H*-phosphindole, 10CI T-00099

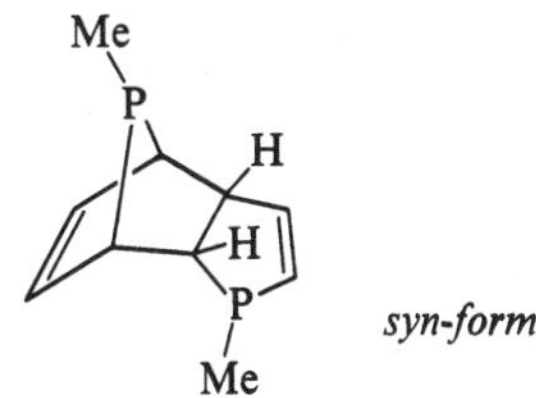

syn-form

$C_{10}H_{14}P_2$ M 196.168

The dimer of 1-Methyl-1*H*-phosphole, M-00282 . Liq. $Bp_{0.03}$ 75-78°.

B,2MeI: *3*a*,4,7,7*a*-Tetrahydro-1,1,8,8-tetramethyl-4,7-phosphinidene-1*H*-phosphindolium diiodide.*
$C_{12}H_{20}I_2P_2$ M 480.047
Solid. Mp 197-199° dec.

1,8-Dioxide:
$C_{10}H_{14}O_2P_2$ M 228.167
Solid. Mp 212.5-214°.

1,8-Disulfide:
$C_{10}H_{14}P_2S_2$ M 260.288
Cryst. (MeOH). Mp 198-199°.

Quin, L.D. *et al*, *J. Chem. Soc., Chem. Commun.*, 1980, 959 (*synth, deriv, P nmr*)
Quin, L.D. *et al*, *Org. Magn. Reson.*, 1982, **20**, 83 (*derivs, cmr*)

3*a*,4,7,7*a*-Tetrahydro-5,6-dimethyl-2,3*a*,7*a*,8-tetrakis(trifluoromethyl)-1,2,3-metheno-2*H*-benzodiphosphole, 10CI T-00100

8,9-Dimethyl-1,3,4,6-tetrakis(trifluoromethyl)-2,5-diphosphatetracyclo[4.4.0.0^{2,4}.0^{3,5}]dec-8-ene

[74930-70-4]

$C_{14}H_{10}F_{12}P_2$ M 468.161

Pale-yellow oil which cryst. slowly. Mp 55-56°.

Kobayashi, Y. *et al*, *J. Org. Chem.*, 1980, **45**, 4683 (*synth, ir, ms, pmr, F nmr*)

2,3,4,5-Tetrahydro-1,5-diphenyl-1*H*-1,5-benzodiphosphepin T-00101

$C_{21}H_{20}P_2$ M 334.337

cis-form [78269-37-1]
Cryst. Mp 144-145°.

Kyba, E.P. *et al*, *J. Am. Chem. Soc.*, 1981, **103**, 3868 (*synth, ms, P nmr, pmr*)

1,2,3,4-Tetrahydro-1,4-diphenyl-1,4-benzodiphosphorin, 9CI T-00102

(1*RS*,4*SR*)-*form*

$C_{20}H_{18}P_2$ M 320.310

(*1RS,4SR*)-*form* [38234-86-5]
cis-*form*
Needles (Me_2CO aq.). Forms Pt complex.
B,2MeI: 1,2,3,4-Tetrahydro-1,4-dimethyl-1,4-diphenyl-1,4-benzodiphosphorinium diiodide.
$C_{22}H_{24}I_2P_2$ M 604.188
Rhombs (MeOH). Mp 348-349° dec.
1,4-Disulfide: [38234-87-6].
$C_{20}H_{18}P_2S_2$ M 384.430
Needles (Me_2CO aq.). Mp 207-208°.

Mann, F.G. *et al*, *J. Chem. Soc., Perkin Trans. 1*, 1972, 2548 (*synth, ir, ms, pmr, complexes*)
Sanz, F. *et al*, *Phosphorus*, 1972, **2**, 135 (*cryst struct*)

1,2,3,6-Tetrahydro-1,2-diphenyl-1,2-diphosphorin T-00103
1,2,3,6-Tetrahydro-1,2-diphenyl-1,2-diphosphinine

(1*RS*,2*RS*)-*form*

$C_{16}H_{16}P_2$ M 270.250

(*1RS,2RS*)-*form*
(±)-trans-*form*
1,2-Disulfide: [81500-74-5].
$C_{16}H_{16}P_2S_2$ M 334.370
Cryst. (Me_2CO). Mp 198.2-199.5°.

(*1RS,2SR*)-*form*
cis-*form*
1,2-Disulfide: [81500-73-4]. Cryst. (Me_2CO). Mp 183-185° dec.

Kawashima, T. *et al*, *Heterocycles*, 1982, **17**, 341 (*synth, ir, ms, P nmr, pmr*)

2,2,2,3-Tetrahydro-3,5-diphenyl-1,3,4,2-oxadiazaphosphole, 9CI T-00104
2,2-Dihydro-3,5-diphenyl-1,3,4,2-oxadiazaphospholine

$C_{13}H_{13}N_2OP$ M 244.232
2,2,2-Trimethoxy: [18334-55-9].
$C_{16}H_{19}N_2O_4P$ M 334.311
Yellow cryst. (pet. ether). Mp 78-80°.
2,2,2-Triphenoxy: [21610-28-6].
$C_{31}H_{25}N_2O_4P$ M 520.523
Cryst. (pet. ether). Mp 116-117°.
2,2,2-Trichloro:
$C_{13}H_{10}Cl_3N_2OP$ M 347.568
Oil.

Arbuzov, B.A. *et al*, *Izv. Akad. Nauk SSSR, Ser. Khim.*, 1967, 1605; 1968, 2525 (*Engl. transl.* pp. 1543, 2391) (*synth, ir, P nmr*)
Schmidpeter, A. *et al*, *Chem. Ber.*, 1975, **108**, 820 (*trichloro, P nmr*)

3*a*,4,7,7*a*-Tetrahydro-1,8-diphenyl-4,7-phosphinidene-1*H*-phosphindole, 9CI, 8CI T-00105

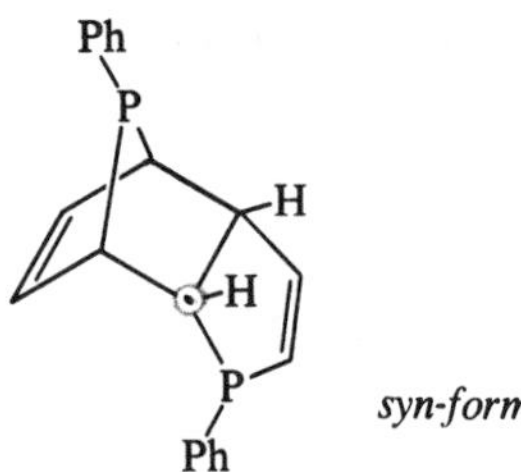

$C_{20}H_{18}P_2$ M 320.310
Dimer of 1-Phenyl-1*H*-phosphole, P-00217 . Solid. Mp 124-127°.
1,8-Dioxide: [20342-05-6].
$C_{20}H_{18}O_2P_2$ M 352.309
Cryst. (C_6H_6). Mp 234°.
1,8-Disulfide: [20701-99-9].
$C_{20}H_{18}P_2S_2$ M 384.430
Solid. Mp 183-184°.

Märkl, G. *et al*, *Tetrahedron Lett.*, 1968, 1755 (*derivs*)
Hughes, A.N. *et al*, *J. Heterocycl. Chem.*, 1979, **16**, 1417 (*dioxide, ir, ms, pmr*)
Quin, L.D. *et al*, *J. Chem. Soc., Chem. Commun.*, 1980, 959 (*synth, P nmr*)
Quin, L.D. *et al*, *Org. Magn. Reson.*, 1982, **20**, 83 (*dioxide, cmr*)

3*a*,4,7,7*a*-Tetrahydro-1,2,3,5,6,8-hexamethyl-4,7-phosphinidene-1*H*-phosphindole, 9CI, 8CI T-00106

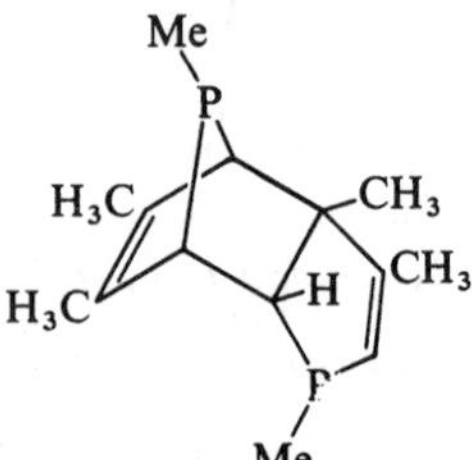

$C_{14}H_{22}P_2$ M 252.275

Solid. Mp 77-80°.

1,8-Dioxide:

$C_{14}H_{22}O_2P_2$ M 284.274

Solid. Mp 224-226°.

1,8-Disulfide:

$C_{14}H_{22}P_2S_2$ M 316.395

Solid. Mp 158-159°.

Quin, L.D. *et al, J. Chem. Soc., Chem. Commun.*, 1980, 959 (*synth, disulfide, P nmr*)

Quin, L.D. *et al, Org. Magn. Reson.*, 1982, **20**, 83 (*dioxide, cmr*)

5a,6,11,11a-Tetrahydro-1,2,3,6,11,12-hexaphenyl-6,11-phosphinidene-1*H*-naphth[2,3-*e*]isophosphindole, 9CI — T-00107

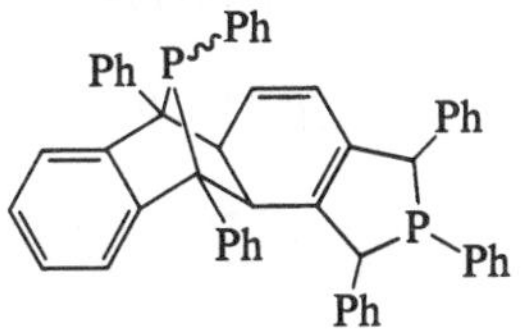

$C_{52}H_{40}P_2$ M 726.836

2,12-Dioxide: [42003-54-3].

$C_{52}H_{40}O_2P_2$ M 758.834

Yellow prisms (C_6H_6/pet. ether). Mp 237-239°.

Holland, J.M. *et al, J. Chem. Soc., Perkin Trans. 1*, 1973, 927 (*synth, ir, uv, pmr, props*)

Tetrahydro-2-hydroxy-2*H*-1,3,2-oxazaphosphorine 2-sulfide — T-00108

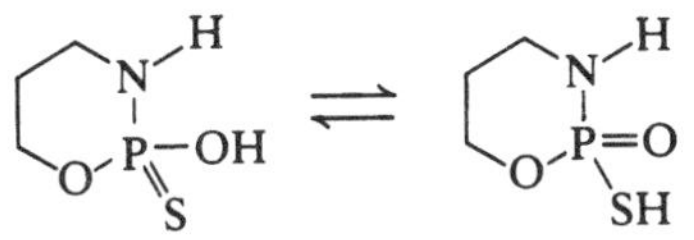

$C_3H_8NO_2PS$ M 153.135

(±)-*form*

Cryst. (MeOH). Mp 168-172°. Tautomeric.

Tetramethylammonium salt: Solid. Mp 218-221.5°.

Dicyclohexylammonium salt: Cryst. (Me_2CO/MeOH). Mp 186-191°.

2-Methoxy (O-*Me ester*): *see Tetrahydro-2-methoxy-2H-1,3,2-oxazaphosphorine, T-00111*

2-Methylthio (S-*Me ester*): *Tetrahydro-2-methylthio-2*H*-oxazaphosphorine 2-oxide.* Liq. $Bp_{0.4}$ 140° (bath). n_D^{24} 1.5360.

Mikolajczyk, M. *et al, Tetrahedron*, 1982, **38**, 2183 (*synth, derivs, pmr, cmr, P nmr*)

2,3,4,7-Tetrahydro-1*H*-isophosphindole 2-oxide, 10CI — T-00109

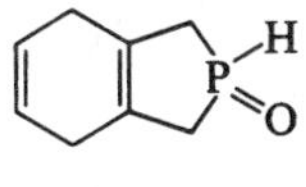

$C_8H_{11}OP$ M 154.148

2-Me: [70179-64-5].

$C_9H_{13}OP$ M 168.175

Hygroscopic solid. Mp 107-110°. $Bp_{0.02}$ 119-124°.

2-Ph: [70179-63-4].

$C_{14}H_{15}OP$ M 230.246

Cryst. (C_6H_6/ligroin). Mp 120-121°. $Bp_{0.02}$ 154-160°.

2-Hydroxy: [70179-65-6].

$C_8H_{11}O_2P$ M 170.147

Needles (EtOH). Mp 189-192°.

Middlemas, E.D. *et al, J. Org. Chem.*, 1979, **44**, 2587 (*synth, cmr, P nmr, pmr*)

1,2,3,4-Tetrahydroisophosphinoline, 8CI — T-00110

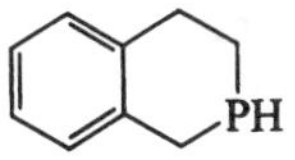

$C_9H_{11}P$ M 150.160

2-Et: 2-Ethyl-1,2,3,4-tetrahydroisophosphinoline.

$C_{11}H_{15}P$ M 178.213

Liq. Bp_{15} 129-130°.

2-Ph: 1,2,3,4-Tetrahydro-2-phenylisophosphinoline.

$C_{15}H_{15}P$ M 226.257

Oil. $Bp_{0.2}$ 130-160°.

Holliman, F.G. *et al, J. Chem. Soc.*, 1943, 547 (*phenyl*)

Beeby, M.H. *et al, J. Chem. Soc.*, 1951, 411 (*ethyl*)

Tetrahydro-2-methoxy-2*H*-1,3,2-oxazaphosphorine — T-00111

*2-Methoxy-1,3,2-oxazaphosphorinane. Tetrahydro-2-methoxy-2*H*-1,3,2-oxazaphosphinine*

[88142-41-0]

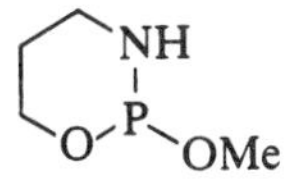

$C_4H_{10}NO_2P$ M 135.102

2-Sulfide: [84451-63-8].

$C_4H_{10}NO_2PS$ M 167.162

Liq. $Bp_{0.3}$ 140° (bath). n_D^{20} 1.5291.

Mikolajczyk, M. *et al, Tetrahedron*, 1982, **38**, 2183 (*synth, ms, cmr, pmr, P nmr*)

3,3a,4,5-Tetrahydro-3-methyl-2*H*-benzo[*e*]phosphindole 3-oxide, 9CI — T-00112

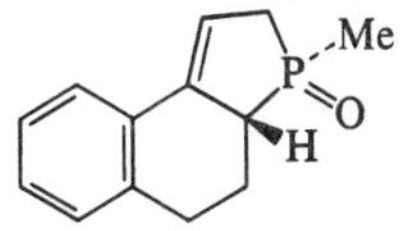

$C_{13}H_{15}OP$ M 218.235

(3RS,3aRS)-form [57065-70-0]

(±)-cis-*form*

Solid. Sol. H_2O. Mp 111-115°. $Bp_{0.05}$ 100° subl.

Symmes, C. *et al, J. Org. Chem.*, 1976, **41**, 238 (*synth, ir, pmr, nmr*)

1,2,3,4-Tetrahydro-4-methyl-2,2-diphenylbenz[*h*]isophosphinolinium(1+), 9CI — T-00113

[54293-26-4]

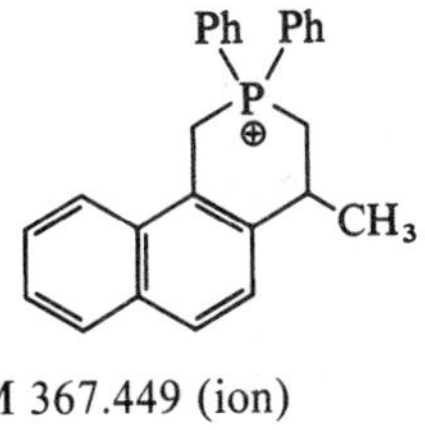

$C_{26}H_{24}P^{\oplus}$ M 367.449 (ion)

Hexafluorophosphate: [54293-27-5].
$C_{26}H_{24}F_6P_2$ M 512.414
Solid. Mp 219-220°.

Dilbeck, G.A. *et al, J. Org. Chem.*, 1975, **40**, 1150 (*synth, ir, pmr*)

2,3,4,5-Tetrahydro-5-methyl-2,2-diphenyl-1*H*-2-benzophosphepinium(1+), 9CI T-00114

[54229-99-1]

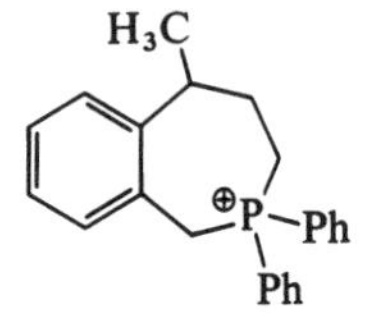

$C_{23}H_{24}P^{\oplus}$ M 331.416 (ion)
Hexafluorophosphate: [54230-00-1].
$C_{23}H_{24}F_6P_2$ M 476.381
Solid. Mp 214-216°.

Dilbeck, G.A. *et al, J. Org. Chem.*, 1975, **40**, 1150 (*synth, ir, pmr, P nmr*)

3,4,5,6-Tetrahydro-4-methyl-2*H*-1,5-methano-4,1-benzazaphosphocine, 9CI T-00115

[52427-87-9]

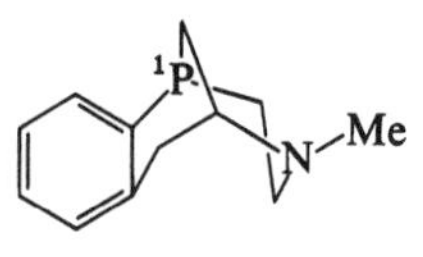

$C_{12}H_{16}NP$ M 205.239
Liq. with typical phosphine odour. $Bp_{0.005}$ 76°. Rel. insensitive to O_2.
1-Oxide: [52427-85-7].
$C_{12}H_{16}NOP$ M 221.238
Hygroscopic cryst. (C_6H_6/cyclohexane).
1-Oxide; B,HCl: [52427-86-8]. Hygroscopic cryst. (EtOH aq.). Mp >250°.
1-Sulfide: [52427-88-0].
$C_{12}H_{16}NPS$ M 237.299
Cryst. (C_6H_6/cyclohexane). Mp 137-138°.
1-Sulfide; B,HCl: [52427-89-1]. Cryst. (EtOH). Mp 252-253°.

Collins, D.J. *et al, Aust. J. Chem.*, 1974, **27**, 815 (*synth, derivs, ir, uv, pmr, ms*)

Tetrahydro-4-methyl-2*H*-1,3,4,2-oxadiazaphosphorine 2-oxide, 10CI, 9CI T-00116

*Tetrahydro-4-methyl-2*H*-1,3,4,2-oxadiazaphosphinine 2-oxide*

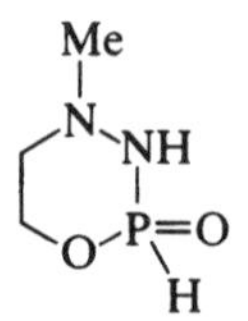

$C_3H_9N_2O_2P$ M 136.090
2-Phenoxy: [40535-35-1].
$C_9H_{13}N_2O_3P$ M 228.187
Cryst. (C_6H_6). Mp 118-119°.
2-Cyclohexylamino: [33063-22-8].
$C_9H_{20}N_3O_2P$ M 233.250
Needles (C_6H_6 or H_2O). Mp 154-155°.

Dorn, H. *et al, Z. Chem.*, 1971, **11**, 173 (*cyclohexylamino, synth*)
Cates, L.A. *et al, J. Heterocycl. Chem.*, 1973, **10**, 111 (*phenoxy, synth, ir, pmr*)
Arshinova, R. *et al, Org. Magn. Reson.*, 1975, **7**, 309 (*phenoxy, pmr, cmr*)

Tetrahydro-3-methyl-2,4,5,6-tetraphenyl-2*H*-1,3,5-oxazaphosphorine, 9CI T-00117

*Tetrahydro-3-methyl-2,4,5,6-tetraphenyl-2*H*-1,3,5-oxazaphosphinine*

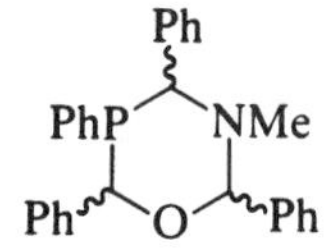

$C_{28}H_{26}NOP$ M 423.493
Cryst. (EtOH or EtOH/C_6H_6). Poorly sol. Et_2O, hexane. Mp 153-155°.

Oehme, H. *et al, Synth. React. Inorg. Met.-Org. Chem.*, 1974, **4**, 453 (*synth, pmr*)

2,3,4,5-Tetrahydro-6-methyl-2,4,4-triphenyl-1,2,3-diazaphosphorine, 10CI T-00118

2,3,4,5-Tetrahydro-6-methyl-2,4,4-triphenyl-1,2,3-diazaphosphinine

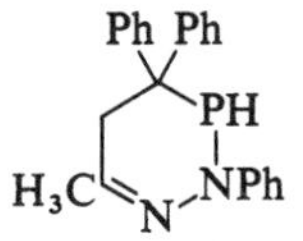

$C_{22}H_{21}N_2P$ M 344.395
3-Chloro: [70111-30-7].
$C_{22}H_{20}ClN_2P$ M 378.840
Cryst. (pet. ether). Mp 151-152.5°.
3-Methoxy: [70111-27-2].
$C_{23}H_{23}N_2OP$ M 374.421
No phys. props. reported.
3-Methoxy, 3-sulfide: [70111-33-0].
$C_{23}H_{23}N_2OPS$ M 406.481
Cryst. (pet. ether). Mp 141-142.5°.
3-Me, 3-oxide: [70111-32-9].
$C_{23}H_{23}N_2OP$ M 374.421
Cryst. (pet. ether). Mp 154-155.5°.

Arbuzov, B.A. *et al, Dokl. Akad. Nauk SSSR, Ser. Sci. Khim.*, 1979, **244**, 117 (*Engl. transl.* p. 59) (*synth, ir, nmr*)
Arbuzov, B.A. *et al, Dokl. Akad. Nauk SSSR, Ser. Sci. Khim.*, 1979, **247**, 1150 (*Engl. transl.* p. 376) (*methoxy, cryst struct*)
Arbuzov, B.A. *et al, Izv. Akad. Nauk SSSR, Ser. Khim.*, 1982, 2730 (*Engl. transl.* p. 2414) (*props, pmr, nmr*)

2,2,2,3-Tetrahydro-1,2-oxaphosphole, 9CI T-00119

2,2-Dihydro-1,2-oxaphosphol-4-ene, 8CI

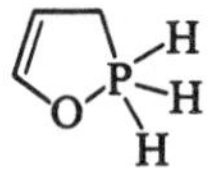

C_3H_7OP M 90.061
Known only as derivs.
2,2,2-Trimethoxy: [15096-31-8].
$C_6H_{13}O_4P$ M 180.140
Liq. d_4^{20} 1.17. $Bp_{0.01}$ 21-22°. n_D^{20} 1.4485.
2,2,2-Triethoxy: [15110-15-3].
$C_9H_{19}O_4P$ M 222.220
Liq. d_4^{20} 1.06. $Bp_{0.01}$ 48-50°. n_D^{20} 1.4352.

Arbuzov, B.A. *et al, Dokl. Akad. Nauk SSSR, Ser. Sci. Khim.*, 1967, **173**, 335 (*Engl. transl.* p. 231) (*synth, pmr, P nmr, ir, props*)

Arbuzov, B.A. *et al*, *Izv. Akad. Nauk SSSR, Ser. Khim.*, 1968, 2290 (*Engl. transl.* p. 2164)

Tetrahydro-2*H*,6*H*-[1.3.2]-oxazaphosphorino[2,3-*b*][1.3.2]-oxazaphosphorine, 9CI — T-00120

2,10-Dioxa-6-aza-1-phosphabicyclo[4.4.0]decane

[74378-80-6]

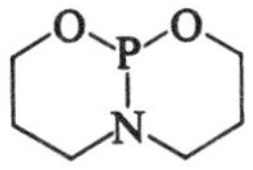

$C_6H_{12}NO_2P$ M 161.140

Liq. $Bp_{0.08}$ 56-60°.

Denney, D.B. *et al*, *J. Am. Chem. Soc.*, 1980, **102**, 5073 (*synth, pmr, cmr, P nmr*)

(Tetrahydro-2-oxo-2-furanyl)phosphonic acid, 9CI — T-00121

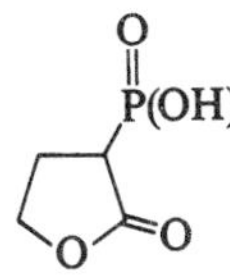

$C_4H_7O_5P$ M 166.070

(±)-***form***

Di-Et ester: Diethyl(tetrahydro-2-oxo-2-furanyl)-phosphonate.

$C_8H_{15}O_5P$ M 222.177

Wittig-Horner reagent. Liq. $Bp_{0.02}$ 115°.

Büchel, K.-H. *et al*, *Justus Liebigs Ann. Chem.*, 1965, **685**, 10 (*synth, ir, uv, derivs, props*)

Minami, T. *et al*, *J. Org. Chem.*, 1974, **39**, 3236; 1978, **43**, 2149 (*use*)

Hoye, T.R. *et al*, *J. Org. Chem.*, 1981, **46**, 1198 (*use*)

1,2,3,4-Tetrahydro-4-oxo-1-phenylphosphinoline, 8CI — T-00122

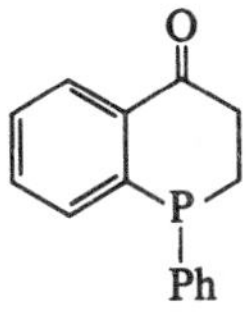

$C_{15}H_{13}OP$ M 240.241

Cryst. (EtOH aq.). Mp 46-47°. $Bp_{0.05}$ 143-145°.

1-Oxide:

$C_{15}H_{13}O_2P$ M 256.240

Cryst. (C_6H_6/cyclohexane). Mp 124-126°.

Gallagher, M.J. *et al*, *J. Chem. Soc.*, 1963, 4846 (*synth, derivs, uv*)

1,2,5,6-Tetrahydro-1,1,6,6,9-pentamethyl-4*H*-phospholo[3,2,1-*ij*]phosphinoline, 10CI — T-00123

[75558-19-9]

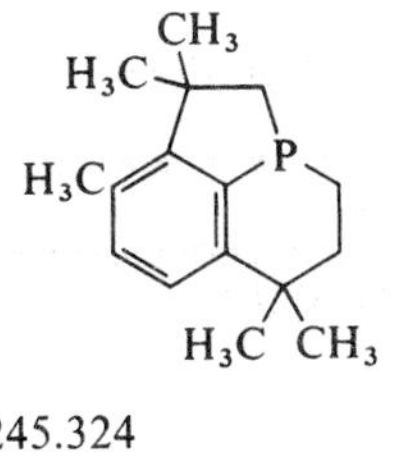

$C_{16}H_{22}P$ M 245.324

Liq. $Bp_{0.8}$ 133-134°.

B,MeI: [75558-27-9]. *1,2,5,6-Tetrahydro-1,1,3,6,6,9-hexamethyl-4H-phospholo[3,2,1-i,j]phosphinolinium iodide.*

$C_{17}H_{25}IP$ M 387.263

Solid. Mp 257-260° dec.

3-Oxide: [75558-26-8].

$C_{16}H_{22}OP$ M 261.323

Cryst. (hexane). Mp 160-166°.

Chen, C.H. *et al*, *J. Org. Chem.*, 1981, **46**, 361 (*synth, derivs, ms, uv, P nmr, pmr*)

2,3,6,7-Tetrahydro-1,1,7,7,9-pentamethyl-1*H*,5*H*-phosphorino[3,2,1-*ij*]-phosphinoline, 10CI — T-00124

[75558-20-2]

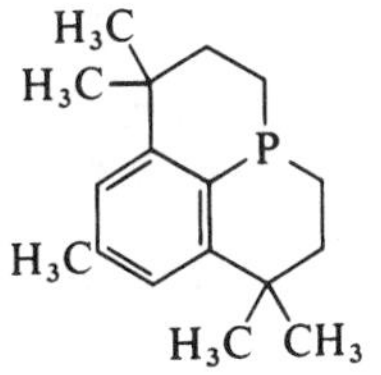

$C_{17}H_{25}P$ M 260.358

Liq. $Bp_{0.6}$ 130-132°.

4-Oxide: [75558-32-6].

$C_{17}H_{25}OP$ M 276.358

Cryst. (hexane). Mp 145-146°.

Chen, C.H. *et al*, *J. Org. Chem.*, 1981, **46**, 361 (*synth, oxide, ms, uv, P nmr, pmr*)

2,2,2,3-Tetrahydrophenanthro[9,10-*d*]-1,3,2-oxazaphosphole, 8CI — T-00125

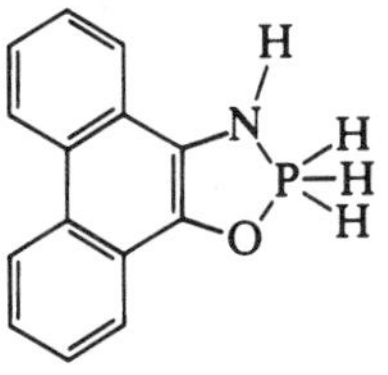

$C_{14}H_{12}NOP$ M 241.229

2,2,2-Trialkoxy derivs. have been characterised spectroscopically, but are unstable.

Sidky, M.M. *et al*, *Tetrahedron Lett.*, 1971, 2313 (*synth, ir, pmr*)

Tetrahydro-2-phenoxy-2*H*-1,3,2-oxazaphosphorine, 9CI — T-00126

2-Phenoxy-1,3,2-oxazaphosphorinane

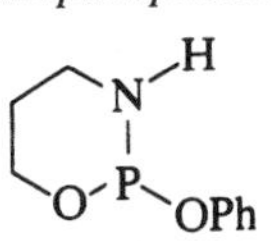

$C_9H_{12}NO_2P$ M 197.173

2-Oxide: [50742-52-4].

$C_9H_{12}NO_3P$ M 213.172

Oil or solid. Mp 60-61°.

2-Sulfide: [71093-77-1].

$C_9H_{12}NO_2PS$ M 229.233

Oil which cryst. slowly. Mp 38-39°. $Bp_{0.05}$ 162°. n_D^{20} 1.5840.

Durrieu, J. *et al*, *Org. Magn. Reson.*, 1973, **5**, 407 (*oxide, pmr*)

Karolek-Wojciechowska, J. *et al*, *J. Chem. Soc., Perkin Trans. 1*, 1979, 146 (*sulfide, synth, ir, pmr, props*)

Just, G. *et al*, *Can. J. Chem.*, 1983, **61**, 1730 (*oxide, synth, ir, ms, pmr, use*)

Tetrahydro-2-phenoxy-3-phenyl-2*H*-1,3,2-oxazaphosphorine, 9CI — T-00127

2-Phenoxy-3-phenyl-1,3,2-oxazaphosphorinane

$C_{15}H_{16}NO_2P$ M 273.271

2-Oxide:

$C_{15}H_{16}NO_3P$ M 289.270

Solid. Mp 52-53°.

Roca, C. *et al*, *Org. Magn. Reson.*, 1976, **8**, 407 (*synth, ir, pmr*)

1,2,3,4-Tetrahydro-1-phenylbenzo[*h*]phosphinoline, 9CI, 8CI — T-00128

1,2,3,4-Tetrahydro-4-phenyl-4-phosphaphenanthrene

[30597-71-8]

$C_{19}H_{17}P$ M 276.317

Solid. Mp 120.5-121.5°.

B, EtBr: [30541-74-3]. *1-Ethyl-1,2,3,4-tetrahydro-1-phenylbenzo*[h]*phosphinolinium bromide.*

$C_{19}H_{22}BrP$ M 361.261

Needles ($CHCl_3/Et_2O$). Mp 227.5-228.5°. Has been resolved.

1-Oxide: [30541-81-2].

$C_{19}H_{17}OP$ M 292.316

Cryst. (heptane). Mp 147.5-148.5°.

Berlin, K.D. *et al*, *J. Org. Chem.*, 1971, **36**, 2791 (*synth, pmr, oxide, ir*)

2,2,2,3-Tetrahydro-5-phenyl-3,3-bis(trifluoromethyl)-1,4,2-thiazaphosphole, 10CI — T-00129

$C_{10}H_8F_6NPS$ M 319.204

2,2,2-Trimethoxy: [66298-61-1].

$C_{13}H_{14}F_6NO_3PS$ M 409.283

Mp 62-63° dec.

2,2,2-Triphenoxy: [75374-08-2].

$C_{28}H_{20}F_6NO_3PS$ M 595.495

Solid. Mp 97-8°.

Burger, K. *et al*, *Synthesis*, 1978, 44 (*synth, pmr, F nmr*)

Burger, K. *et al*, *Chem. Ber.*, 1980, **113**, 2699 (*synth*)

5,6,7,12-Tetrahydro-12-phenyldibenz[*c,f*][1,5]azaphosphocine, 10CI — T-00130

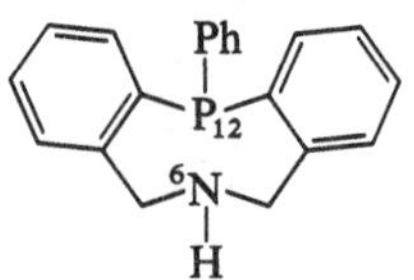

$C_{20}H_{18}NP$ M 303.343

12-Oxide: [68669-07-8].

$C_{20}H_{18}NOP$ M 319.342

Cryst. ($C_6H_6/CHCl_3$). Mp 193°.

6-Benzyl, 12-oxide: [68669-04-5].

$C_{27}H_{24}NOP$ M 409.466

Cryst. (C_6H_6). Mp 220-222°.

Petrov, K.A. *et al*, *Zh. Obshch. Khim.*, 1978, **48**, 2025 (*Engl. transl.* p. 1841) (*synth, ir, ms, pmr*)

Chauzov, V.A. *et al*, *Zh. Obshch. Khim.*, 1983, **53**, 364 (*synth*)

2,2,2,3-Tetrahydro-5-phenyl-1,3,4,2-oxadiazaphosphole — T-00131

$C_7H_9N_2OP$ M 168.135

2,2,2-Trifluoro: [76174-11-3].

$C_7H_6F_3N_2OP$ M 222.106

No phys. props. reported.

2,2,2-Trichloro: [55498-89-0].

$C_7H_6Cl_3N_2OP$ M 271.470

Cryst. (C_6H_6). Mp 145-148°. Previously formulated as Benzoylaminophosphorimidic trichloride.

Matyushecheva, G.I. *et al*, *Zh. Org. Khim.*, 1967, **3**, 2254 (*Engl. transl.* p. 2205) (*trichloride, synth*)

Schmidpeter, A. *et al*, *Chem. Ber.*, 1975, **108**, 820 (*trichloride, synth, ms, P nmr*)

Mathis, R. *et al*, *Spectrochim. Acta, Part A*, 1979, **35**, 745 (*trichloride, ir*)

Day, R.O. *et al*, *Inorg. Chem.*, 1981, **20**, 1229 (*trifluoride, F and P nmr*)

Tetrahydro-2-phenyl-2*H*-1,3,2-oxazaphosphorine, 9CI — T-00132

2-Phenyl-1,3,2-oxazaphosphorinane. 1,3-Propanolamine cyclic phenylphosphonamidite

[51992-59-7]

$C_9H_{12}NOP$ M 181.174

Liq. d_4^{20} 1.17. $Bp_{0.025}$ 86°. n_D^{20} 1.5850.

2-Sulfide: [50742-53-5].

$C_9H_{12}NOPS$ M 213.234

Solid. Mp 82-83°.

Durrieu, J. *et al*, *Org. Magn. Reson.*, 1973, **5**, 407 (*sulfide, synth, pmr*)

Pudovik, M.A. *et al*, *Zh. Obshch. Khim.*, 1974, **44**, 501 (*Engl. transl.* p. 482) (*synth*)

Shagidullin, R.R. *et al*, *Zh. Obshch. Khim.*, 1984, **54**, 1283 (*Engl. transl.* p. 1148) (*ir*)

Tetrahydro-3-phenyl-2-(phenylamino)-2*H*-1,3,2-oxazaphosphorine **T-00133**

N,3-Diphenyltetrahydro-2H-1,3,2-oxazaphosphorin-2-amine, 9CI. 2-Anilinotetrahydro-3-phenyl-2H-1,3,2-oxaphosphorine

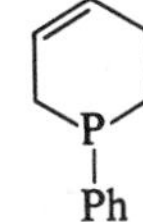

$C_{15}H_{17}N_2OP$ M 272.286

2-Oxide: [52463-56-6].
$C_{15}H_{17}N_2O_2P$ M 288.285
Cryst. (EtOH). Mp 183°.

Brault, J.F. *et al*, *J. Organomet. Chem.*, 1974, **66**, 71 (*synth*)
Roca, C. *et al*, *Org. Magn. Reson.*, 1976, **8**, 407 (*synth, ir*)

1,2,3,6-Tetrahydro-1-phenylphosphorin **T-00134**

1,2,3,6-Tetrahydro-1-phenylphosphinine. 3,4-Dehydro-1-phenylphosphinane
[23855-91-6]

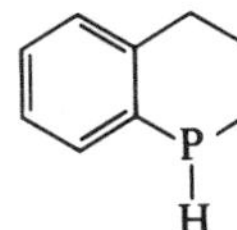

$C_{11}H_{13}P$ M 176.197
Liq. $Bp_{0.7}$ 86°, $Bp_{0.01}$ 65°. n_D^{20} 1.6084.

B,PhCH₂Br: [23855-92-7]. *1,2,3,6-Tetrahydro-1-phenyl-1-phenylmethylphosphininium bromide. 1-Benzyl-1,2,3,6-tetrahydro-1-phenylphosphorinium bromide.*
$C_{18}H_{20}PBr$ M 347.234
Solid. Mp 170° dec.

Mathey, F. *et al*, *C.R. Hebd. Seances Acad. Sci., Ser. C*, 1969, **269**, 158 (*synth, deriv, ir*)
Morris, D. *et al*, *Phosphorus*, 1972, **1**, 305 (*synth, ir, ms, pmr*)

1,2,3,4-Tetrahydrophosphinoline **T-00135**

$C_9H_{11}P$ M 150.160

1-Chloro: 1-Chloro-1,2,3,4-tetrahydrophosphinoline.
$C_9H_{10}ClP$ M 184.605
Fuming liq. $Bp_{0.001}$ 55-56°.

1-Et, 1-oxide: [52293-08-0]. *1-Ethyl-1,2,3,4-tetrahydrophosphinoline 1-oxide.*
$C_{11}H_{15}OP$ M 194.213
Hygroscopic oil.

1-Et, 1-sulfide: [52221-88-2].
$C_{11}H_{15}PS$ M 210.273
Hygroscopic oil. $Bp_{0.01}$ 150° (oven).

Rowley, L.E. *et al*, *Aust. J. Chem.*, 1974, **27**, 801 (*synth, ir, ms, pmr*)

[[(Tetrahydro-2*H*-pyran-2-yl)oxy]methyl]-phosphonic acid, 9CI **T-00136**

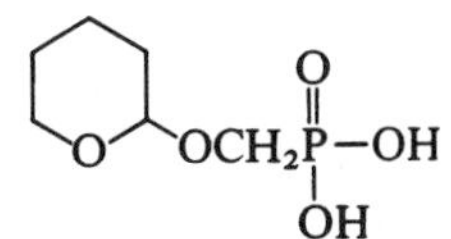

$C_6H_{13}O_5P$ M 196.139

(±)-*form*

Di-Et ester: Diethyl[[(tetrahydro-2H-pyran-2-yl)oxy]-methyl]phosphonate.
$C_{10}H_{21}O_5P$ M 252.247
Liq. $Bp_{0.1}$ 105° (oven).

Dibutyl ester: Dibutyl[[(tetrahydro-2H-pyran-2-yl)-oxy]methyl]phosphonate.
$C_{14}H_{29}O_5P$ M 308.354
Undergoes Wittig-Horner reaction with ketones to give vinyl tetrahydropyranyl ethers. Liq. $Bp_{0.1}$ 138-140°.

Kluge, A.K. *et al*, *J. Org. Chem.*, 1979, **44**, 4847 (*esters, synth, pmr, use*)
Cloudsdale, I.S. *et al*, *J. Org. Chem.*, 1982, **47**, 919 (*use*)

1,1′,3,3′-Tetrahydro-2,2′-spirobi[2*H*-isophosphindolium](1+), 9CI **T-00137**

Di-o-xylylenephosphonium(1+)

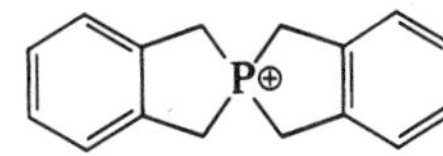

$C_{16}H_{16}P^{\oplus}$ M 239.276 (ion)

Bromide: [42092-11-5].
$C_{16}H_{16}BrP$ M 319.180
Cryst. (H_2O). Mp 338-345°.

Robinson, C.N. *et al*, *J. Heterocycl. Chem.*, 1973, **10**, 395 (*synth, pmr*)

2,2,7,7-Tetrahydro-3,4,5,6-tetrakis(methoxycarbonyl)-1*H*-1,2,7-azadiphosphepine **T-00138**

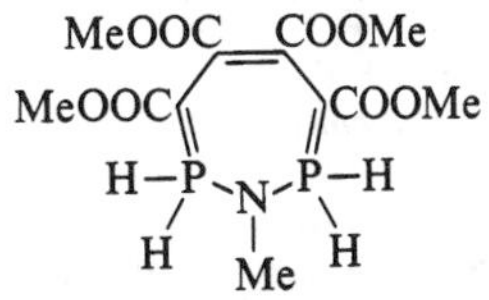

$C_{13}H_{19}NO_8P_2$ M 379.243
A cyclic bis-ylide.

2,2,7,7-Tetra-Me: [66881-05-8]. *2,2,7,7-Tetrahydro-3,4,5,6-tetrakis(methoxycarbonyl)-1,2,2,5,5-pentamethyl-1H-1,2,7-azadiphosphepine.*
$C_{17}H_{27}NO_8P_2$ M 435.350
Solid. Mp 120-130° dec.

2,2-Di-Me, 7,7-di-Ph: [66881-07-0].
$C_{27}H_{31}NO_8P_2$ M 559.491
Solid. Mp 125-130° dec.

2,2,7,7-Tetra-Ph: [66881-06-9].
$C_{37}H_{35}NO_8P_2$ M 683.633
Solid. Mp 175-178° dec.

Zeiss, W. *et al*, *Chem. Ber.*, 1978, **111**, 1655 (*synth, pmr, P nmr*)

2,3,4,5-Tetrahydro-2,2,4,4-tetramethyl-1*H*-2,4-benzodiphosphepinium(2+), 10CI T-00139

$C_{13}H_{22}P_2^{\oplus\oplus}$ M 240.264 (ion)

Dibromide: [78532-89-5].
$C_{13}H_{22}Br_2P_2$ M 400.072
Cryst. (MeOH/Et_2O). Mp 266°.

Schmidbaur, H. *et al*, *Chem. Ber.*, 1981, **114**, 1428 (*synth, ms*)

1,3*a*,3*b*,6-Tetrahydro-3,4,8,8-tetramethyl-1,6-diphenyl-8*H*-[1,3]diphospholo[1,5-*e*;3,4-*c'*]bis[1,2,3]diazaphosphole, 10CI T-00140

2,2,6,9-Tetramethyl-4,11-diphenyl-4,5,10,11-tetraaza-1,3-diphosphatricyclo[6.3.0.0^{3,7}]undeca-5,9-diene
[82685-97-0]

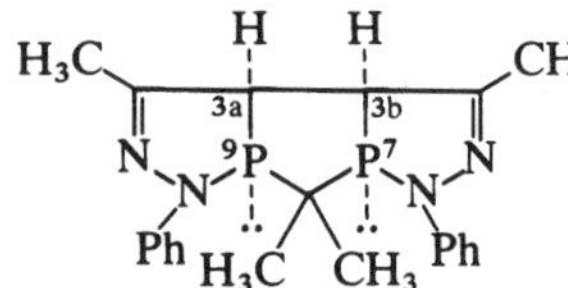

(3aα,3bα,7α,9α)-*form*

$C_{21}H_{24}N_4P_2$ M 394.395

(3aα,3bα,7α,9α)-form [94799-21-0]
cis-*form*
Cryst. (MeCN). Mp 150-152°.

(3aβ,3bα,7α,9α)-form [85272-73-7]
trans-*form*
Cryst. (EtOH or MeOH). Mp 260°. Bp 220° subl.

Disulfide:
$C_{21}H_{24}N_4P_2S_2$ M 458.515
Cryst. (C_6H_6). Mp 212-214°.

Litvinov, I.A. *et al*, *Izv. Akad. Nauk SSSR, Ser. Khim.*, 1982, 2718; 1984, 2023 (*synth, ir, pmr, P nmr, cryst struct*)
Il'yasov, A.V. *et al*, *Zh. Obshch. Khim.*, 1984, **54**, 1511 (*Engl. transl.* p. 1346) (*synth, pmr, cmr, P nmr*)
Arbusov, B.A. *et al*, *Zh. Obshch. Khim.*, 1985, **55**, 3 (*Engl. transl.* p. 1) (*synth, pmr, cmr, ir, disulfide, P nmr*)

3*a*,4,7,7*a*-Tetrahydro-2,3,5,6-tetramethyl-1,8-diphenyl-4,7-phosphinidene-1*H*-phosphindole, 9CI, 8CI T-00141

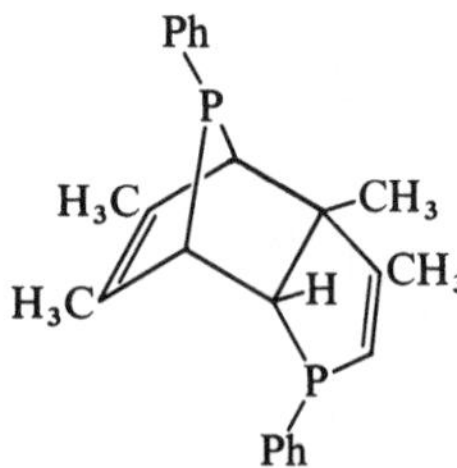

$C_{24}H_{26}P_2$ M 376.417

1,8-Dioxide: [37740-00-4].
$C_{24}H_{26}O_2P_2$ M 408.416
Solid. Mp 243-246°, 277°.

Mathey, F. *et al*, *Bull. Soc. Chim. Fr.*, 1970, 4433 (*dioxide, struct, pmr, P nmr*)
Mathey, F. *et al*, *Org. Magn. Reson.*, 1972, **4**, 171 (*dioxide, pmr, P nmr*)
Hughes, A.N. *et al*, *J. Heterocycl. Chem.*, 1979, **16**, 1417 (*dioxide, synth, ir, ms, pmr*)

1,1',3,3'-Tetrahydro-1,1',3,3'-tetramethyl-2,2'-spirobi[2*H*-1,3,2-benzodiazaphosphole], 9CI T-00142

$C_{16}H_{21}N_4P$ M 300.342

2-Me: [77490-41-6]. *1,1',3,3'-Tetrahydro-1,1',2,3,3'-pentamethyl-2,2'-spirobi[2*H*-1,3,2-benzodiazaphosphole]*.
$C_{17}H_{23}N_4P$ M 314.369
Cryst. (C_6H_6/Et_2O). Mp 168°. Air-sensitive.

Wieber, M. *et al*, *Z. Anorg. Allg. Chem.*, 1981, **477**, 108 (*synth, pmr, P nmr*)
Wunderlich, H. *et al*, *Acta Crystallogr., Sect. B*, 1981, **37**, 995 (*cryst struct*)

3,4,5,5-Tetrahydro-2,2,5,5-tetraphenyl-1*H*-1,2,5-azadiphospholium(1+) T-00143

$C_{26}H_{24}NP_2^{\oplus}$ M 412.430 (ion)

Perchlorate: [38452-89-0].
$C_{26}H_{24}ClNO_4P_2$ M 511.880
Solid. Mp 230-232° dec.

Appel, R. *et al*, *Chem. Ber.*, 1972, **105**, 2476 (*synth*)

2,3,4,5-Tetrahydro-2,2,4,4-tetraphenyl-1*H*-2,4-benzodiphosphonium(2+), 10CI T-00144

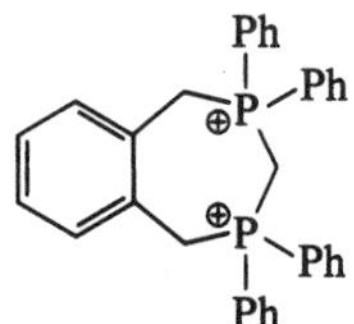

$C_{33}H_{30}P_2^{\oplus\oplus}$ M 488.548 (ion)

Dibromide: [78532-90-8].
$C_{33}H_{30}Br_2P_2$ M 648.356
Cryst. (MeOH/Et_2O). Mp 310°.

Schmidbaur, H. *et al*, *Chem. Ber.*, 1981, **114**, 1428 (*synth, ms*)

1,1,3,3-Tetrahydro-1,1,3,3-tetraphenyl-2,4-bis(triphenylphosphonio)-1,3-diphosphete, 9CI T-00145

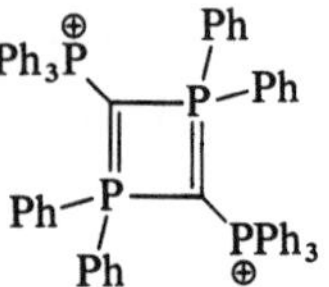

$C_{62}H_{50}P_2^{\oplus\oplus}$ M 857.025 (ion)

Dichloride: [62337-53-5].
$C_{62}H_{50}Cl_2P_2$ M 927.931
Cryst. (MeOH). Mp 385°. Forms adduct with $2Br_2$, Mp 365°.

Appel, R. *et al*, *Angew. Chem.*, *Int. Ed. Engl.*, 1977, **16**, 402 (*synth, ms, cmr, P nmr*)

2,2,4,4-Tetrahydro-2,2,4,4-tetraphenyl-1,5,2,4-diazadiphosphorine, 9CI **T-00146**

2,2,4,4-Tetrahydro-2,2,4,4-tetraphenyl-1,5,2,4-diazadiphosphinine

[38453-00-8]

$C_{26}H_{22}N_2P_2$ M 424.421

A cyclic bis-ylide.

B,HCl: Mp 258-261° dec.

Appel, R. *et al*, *Chem. Ber.*, 1972, **105**, 2476 (*synth, pmr*)

5,6,11,12-Tetrahydro-5,6,11,12-tetraphenyldibenzo[*c,g*][1,2,5,6]-tetraphosphocin, 9CI **T-00147**

[38234-91-2]

$C_{36}H_{28}P_4$ M 584.512

Needles (Me_2CO). Mp 194.5-196°. Fails to form a tetrasulfide.

5,11-(or 5,6)-Disulfide: [38240-96-9].
$C_{36}H_{28}P_4S_2$ M 648.632
Needles (Me_2CO). Mp 247-248°.

Mann, F.G. *et al*, *J. Chem. Soc.*, *Perkin Trans. 1*, 1972, 2548 (*synth, deriv, ir*)

1,1,3,3-Tetrahydro-1,1,3,3-tetraphenyl-1,3-diphosphacyclohepta-1,2-diene, 10CI **T-00148**

[74144-36-8]

$C_{29}H_{28}P_2$ M 438.488

Diylide. Yellow cryst. Sol. org. solvs. Mp 127° dec.

Schmidbaur, H. *et al*, *Angew. Chem.*, *Int. Ed. Engl.*, 1980, **19**, 555 (*synth, props, cmr, P nmr, pmr*)

1,1,3,3-Tetrahydro-1,1,3,3-tetraphenyl-1,3-diphosphacyclohexa-1,2-diene, 9CI **T-00149**

[74144-35-7]

$C_{28}H_{26}P_2$ M 424.461

Cryst. (pentane):thermally labile. Spar. sol. org. solvs. Mp 35° dec. Methylated at C-2 by MeI. Anhydrous HX(X = Cl, Br,I) yields 1,3-diphosphininium salts (see 42400-5).

Schmidbaur, H. *et al*, *Angew. Chem.*, *Int. Ed. Engl.*, 1980, **19**, 555 (*synth, props*)
Schmidbaur, H. *et al*, *Chem. Ber.*, 1981, **114**, 3063 (*synth, struct, ms, cmr, P nmr, pmr*)
Schubert, U. *et al*, *Chem. Ber.*, 1981, **114**, 3070 (*struct*)

1,1,3,3-Tetrahydro-1,1,3,3-tetraphenyl-1,3-diphosphacyclopenta-1,2-diene, 10CI **T-00150**

[74144-34-6]

$C_{27}H_{24}P_2$ M 410.434

Cyclic diylide. Poorly sol. org. solvs. Dec. within a few hr. at 20°.

Schmidbaur, H. *et al*, *Angew. Chem.*, *Int. Ed. Engl.*, 1980, **19**, 555 (*synth, props, P nmr, pmr*)

1,2,3,6-Tetrahydro-1,2,4,5-tetraphenyl-1,2-diphosphorine, 9CI **T-00151**

1,2,3,6-Tetrahydro-1,2,4,5-tetraphenyl-1,2-diphosphinine

$C_{28}H_{24}P_2$ M 422.445

1,2-Disulfide: [55781-95-8].
$C_{28}H_{24}P_2S_2$ M 486.565
Cryst. (C_6H_6). Mp 233-234°.

Nakayama, S. *et al*, *Bull. Chem. Soc. Jpn.*, 1975, **48**, 546 (*synth, ir, ms, pmr*)

1,2,3,4-Tetrahydro-1,1,4,4-tetraphenyl-1,4-diphosphorinium(2+) **T-00152**

1,2,3,4-Tetrahydro-1,1,4,4-tetraphenyl-1,4-diphosphininium(2+)

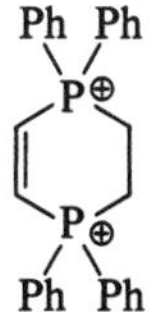

$C_{28}H_{26}P_2^{\oplus\oplus}$ M 424.461 (ion)

Dibromide: [13274-97-0].
$C_{28}H_{26}Br_2P_2$ M 584.269
Hygroscopic cryst. (Me_2CO/MeOH or EtOH/Et_2O). Mp 298-300° dec.

Aguiar, A.M. *et al*, *J. Am. Chem. Soc.*, 1966, **88**, 4090 (*synth, props, P nmr, pmr*)
Brophy, J.J. *et al*, *Aust. J. Chem.*, 1967, **20**, 503 (*synth*)
Brophy, J.J. *et al*, *Aust. J. Chem.*, 1969, **22**, 1399, 1405 (*props*)

4,5,5,5-Tetrahydro-3,5,5,5-tetraphenyl-1,2,5-oxazaphosphole, 9CI T-00153

5,5-Dihydro-3,5,5,5-tetraphenyl-Δ²-1,2,5-oxazaphospholine, 8CI

[14264-70-1]

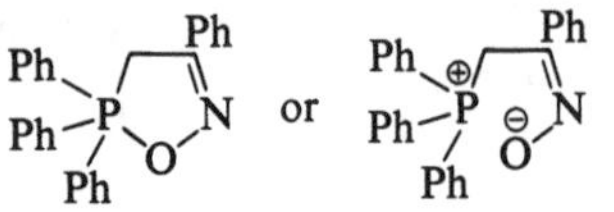

$C_{26}H_{22}NOP$ M 395.440

May consist of, or be in equilibrium with, the open-chain zwitterion. Needles (MeCN). Mp 131° dec,.

Masaki, M. *et al, J. Org. Chem.*, 1967, **32**, 3564 (*synth, ir, pmr*)
Guadiano, G. *et al, J. Org. Chem.*, 1968, **33**, 4431 (*synth, pmr*)

Tetrahydro-2*H*-1,3,2-thiazaphosphorine 2-oxide, 8CI T-00154

*Tetrahydro-2*H*-1,3,2-thiazaphosphinine 2-oxide*

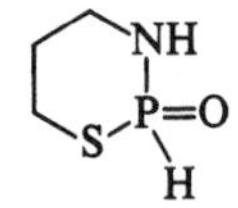

C_3H_8NOPS M 137.136

2-Ethoxy: [20043-26-9].
$C_5H_{12}NO_2PS$ M 181.189
Liq. $Bp_{0.02}$ 137-138°. n_D^{20} 1.5231.

2-Phenoxy: [20043-24-7].
$C_9H_{12}NO_2PS$ M 229.233
Solid. Mp 152°.

2-Me: [21852-18-6].
$C_4H_{10}NOPS$ M 151.163
Solid. Mp 86-88°. Dec. by H_2O.

2-Ph: [21852-19-7].
$C_9H_{12}NOPS$ M 213.234
Solid. Mp 109-110°.

Chabrier, P. *et al, C.R. Hebd. Seances Acad. Sci., Ser. C*, 1968, **267**, 1166 (*synth*)
Savignac, P. *et al, C.R. Hebd. Seances Acad. Sci., Ser. C*, 1968, **266**, 1791 (*synth*)

9,10,15,16-Tetrahydrotribenzo[*b,d,h*][1,6]-diphosphecin, 8CI T-00155

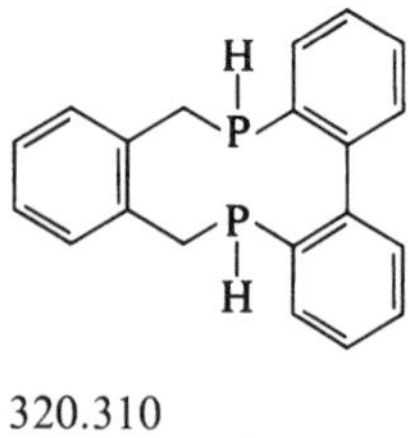

$C_{20}H_{18}P_2$ M 320.310

9,9,16,16-Tetra-Et, dibromide: [16523-79-8]. *9,9,16,16-Tetraethyl-9,10,15,16-tetrahydrotribenzo*[b,d,h][*1,6*]-*diphosphecinium dibromide.*
$C_{28}H_{36}Br_2P_2$ M 594.348
Solid. Spar. sol. org. solvs. Mp 360°.

Allen, D.W. *et al, J. Chem. Soc.* (*C*), 1967, 1869 (*synth*)

2,2,2,3-Tetrahydro-2,2,2-trimethoxy-3-methyl-4,5-diphenyl-1,3,2-oxazaphosphole, 9CI T-00156

2,2-Dihydro-2,2,2-trimethoxy-3-methyl-4,5-diphenyl-Δ⁴-1,3,2-oxazaphospholine, 8CI

[33452-31-2]

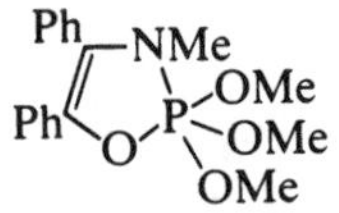

$C_{18}H_{22}NO_4P$ M 347.350

Bernard, D. *et al, C.R. Hebd. Seances Acad. Sci., Ser. C*, 1971, **272**, 2077 (*synth, nmr*)

2,2,2,3-Tetrahydro-2,2,2-trimethoxy-5-methyl-1,2-oxaphosphole, 9CI T-00157

2,2,2-Trimethoxy-5-methyl-Δ⁴-oxaphospholene

[26192-22-3]

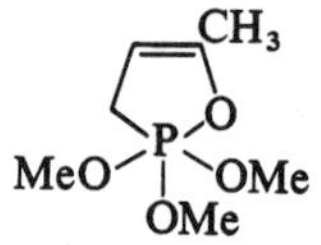

$C_7H_{15}O_4P$ M 194.167

Forms methyl esters and ethers from carboxylic acids and phenols in high yield at r.t. Bp_3 56-57°.

Gorenstein, D. *et al, J. Am. Chem. Soc.*, 1970, **92**, 634 (*synth, ir, pmr*)
Voncken, W.G. *et al, Recl. Trav. Chim. Pays-Bas*, 1974, **93**, 210 (*synth, use*)
Buano, G. *et al, J. Am. Chem. Soc.*, 1981, **103**, 4532 (*pmr, cmr, P nmr*)
Castelijno, M.M.C.F. *et al, J. Org. Chem.*, 1981, **46**, 47 (*synth, pmr, P nmr*)
Van Ool, P.J.J.M. *et al, Recl. Trav. Chim. Pays-Bas*, 1983, **102**, 215 (*struct, P nmr, cmr, pmr*)

2,2,2,3-Tetrahydro-2,2,2-trimethoxy-5-phenyl-3,3-bis(trifluoromethyl)-1,4,2-oxazaphosphole, 9CI T-00158

2,2-Dihydro-2,2,2-trimethoxy-5-phenyl-3,3-bis(trifluoromethyl)-Δ⁴-1,4,2-oxazaphospholine, 8CI

[33550-42-4]

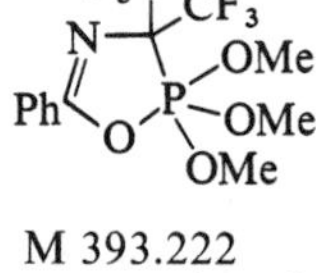

$C_{13}H_{14}F_6NO_4P$ M 393.222

Reagent used for the generation of phenyl-*N*-bis(trifluoromethyl) nitrile ylide which may be trapped by dipolarophiles to give heterocyclic compds. Solid.

Burger, K. *et al, Chem. Ber.*, 1971, **104**, 1826 (*synth, ir, ms, P nmr, pmr*)
Burger, K. *et al, Chem. Ber.*, 1972, **105**, 3814; 1973, **106**, 1, 3421; 1975, **108**, 2737 (*use*)
Burger, K. *et al, Synthesis*, 1975, 731 (*use*)
Burger, K. *et al, Z. Naturforsch., B*, 1980, **35**, 1426; 1981, **36**, 345; 1983, **38**, 769 (*use*)

2,2,2,3-Tetrahydro-2,2,2-trimethoxy-5-phenyl-3,3-bis(trifluoromethyl)-1-(2,4,6-trimethylphenyl)-1*H*-1,4,2-diazaphosphole, 10CI T-00159

[67565-40-6]

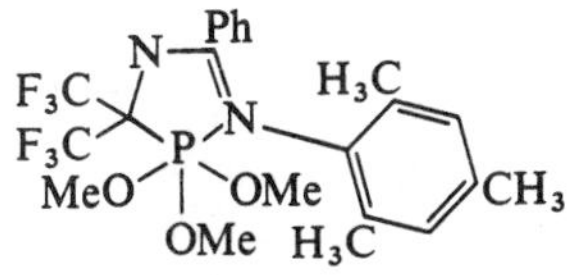

$C_{22}H_{25}F_6N_2O_3P$ M 510.415

Solid. Mp 112-113° dec.

Burger, K. *et al*, *Synthesis*, 1978, 526 (*synth*, *ir*)
Burger, K. *et al*, *Z. Naturforsch.*, *B*, 1980, **33**, 749 (*ir*, *nmr*)

Tetrahydro-1*H*,5*H*-[1,2,3]-triphospholo[2,1-*a*][1,2,3]triphosphole, 9CI T-00160

1,2,5,6-Tetraphosphabicyclo[3.3.0]octane

[69019-76-7]

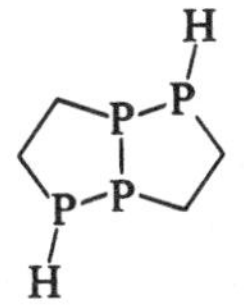

$C_4H_{10}P_4$ M 182.018

Cryst. (THF/pentane). Mp 86-88° dec. Forms di-Li salt.

Baudler, M. *et al*, *Chem. Ber.*, 1978, **111**, 3838 (*synth*, *ir*, *P nmr*)

2,4,6,8-Tetrahydroxy-1,5-dimethyl-3,7,9-trioxa-2,4,6,8-tetraphosphabicyclo[3.3.1]nonane T-00161

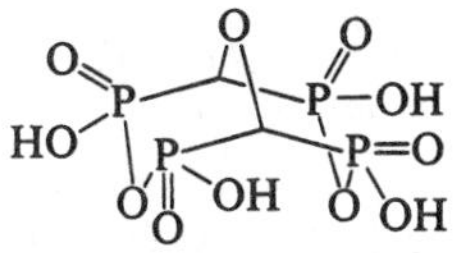

$C_2H_6O_{11}P_4$ M 329.958

Tetra-Na salt: [21923-73-9]. Solid.

Prentice, J.B. *et al*, *J. Am. Chem. Soc.*, 1972, **94**, 6119 (*synth*, *pmr*, *P nmr*)

1,1,3,7-Tetrahydroxy-7-methyl-2,4-dioxa-1,3-diphosphanonan-9-oic acid 1,3-dioxide, 10CI T-00162

5-Pyrophosphomevalonic acid

[1492-08-6]

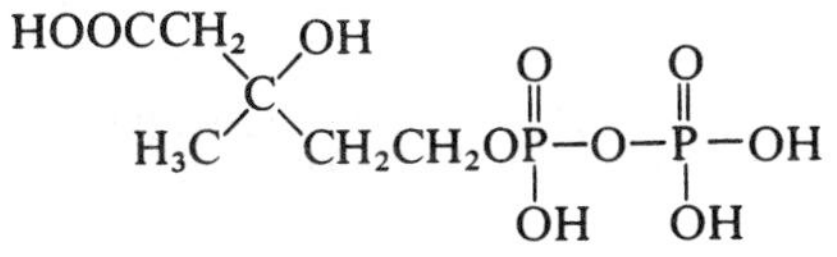

$C_6H_{14}O_{10}P_2$ M 308.118

Biosynthetic intermed.

Dugan, R.E. *et al*, *Anal. Biochem.*, 1968, **22**, 249.
Knotz, J. *et al*, *Plant Physiol.*, 1977, **60**, 81 (*biosynth*)
Linares, A. *et al*, *Biochem. Biophys. Res. Commun.*, 1977, **77**, 974 (*enzymic synth*)
Pérez, L.M. *et al*, *Phytochemistry*, 1983, **22**, 431.

Tetraiodomethylphosphorane, 9CI T-00163

[71640-53-4]

$MePI_4$

CH_3I_4P M 553.626

Cryst. Mp 156°.

Ginsburg, V.A. *et al*, *Zh. Obshch. Khim.*, 1958, **28**, 736 (*Engl. transl.* p. 716) (*synth*, *props*)
Mel'nichuk, E.A. *et al*, *Zh. Obshch. Khim.*, 1979, **49**, 1668 (*Engl. transl.* p. 1557)

Tetraiodophenoxyphosphorane, 9CI T-00164

[72975-01-0]

$PhOPI_4$

$C_6H_5I_4OP$ M 631.697

Blackish-green cryst. Insol. C_6H_6, $CHCl_3$, pet. ether, sol. alkyl iodides. Mp 74-78°.

Kostina, V.G. *et al*, *Zh. Obshch. Khim.*, 1979, **49**, 2452 (*Engl. transl.* p. 2165) (*synth*)

Tetraiodophenylphosphorane T-00165

[26852-27-7]

$PhPI_4$

$C_6H_5I_4P$ M 615.697

Red-brown cryst. (C_6H_6). Mp 90-91°. Subl. at 150-60°.

Feshchenko, N.G. *et al*, *Zh. Obshch. Khim.*, 1969, **39**, 2184 (*Engl. transl.* p. 2133)
Romanenko, V.D. *et al*, *Synthesis*, 1980, 823.

Tetraisopropyl diphosphate T-00166

Tetrakis(1-methylethyl) diphosphate, 9CI. *Tetraisopropyl pyrophosphate*

[5836-28-2]

$[(H_3C)_2CHO]_2P(O)-O-P(O)[OCH(CH_3)_2]_2$

$C_{12}H_{28}O_7P_2$ M 346.297

Liq. d_4^{25} 1.09. Mp 14-15°. $Bp_{0.2}$ 104°. n_D^{25} 1.4163.

▷Highly toxic. LD_{50} 13 mg/kg (mice). UX7100000.

Toy, A.D.F., *J. Am. Chem. Soc.*, 1948, **70**, 3882 (*synth*)
Sosnovsky, G. *et al*, *Z. Naturforsch.*, *B*, 1976, **31**, 820 (*synth*)

Tetraisopropyldiphosphine T-00167

Tetrakis(1-methylethyl)diphosphine, 9CI

[20491-55-8]

$[(H_3C)_2CH]_2P-P[CH(CH_3)_2]_2$

$C_{12}H_{28}P_2$ M 234.301

Liq. Bp_{10} 110°.

Monooxide: [66193-23-5].
$C_{12}H_{28}OP$ M 219.326
Bp_2 120°. n_D^{20} 1.5068.

Dioxide: [67949-88-6].
$C_{12}H_{28}O_2P_2$ M 266.300
No phys. props. reported.

Monosulfide: [62969-12-4].
$C_{12}H_{28}P_2S$ M 266.361
Oil.

Disulfide: [20491-56-9].
$C_{12}H_{28}P_2S_2$ M 298.421
Cryst. (MeOH). Mp 96°.

Issleib, K. *et al*, *J. Organomet. Chem.*, 1968, **13**, 283 (*synth*, *disulfide*)

Aime, S. *et al*, *J. Chem. Soc., Dalton Trans.*, 1976, 2144 (*pmr, cmr, nmr*)
Foss, V.L. *et al*, *Phosphorus Sulfur*, 1977, **3**, 299 (*monooxide*)
Foss, V.L. *et al*, *Zh. Obshch. Khim.*, 1977, **47**, 477 (*Engl. transl.* p. 437) (*monosulfide*)
Lutsenko, I.F. *et al*, *Pure Appl. Chem.*, 1980, **52**, 917 (*rev*)
Foss, V.L. *et al*, *Zh. Obshch. Khim.*, 1982, **52**, 1054, 1063 (*Engl. transl.* p. 916, 924) (*props*)
Albrand, J.P. *et al*, *Org. Magn. Reson.*, 1983, **21**, 246 (*pmr, cmr, nmr*)

sym-Tetraisopropyl dithiopyrophosphate **T-00168**

Tetrakis(1-methylethyl) thiodiphosphate, 9CI. *Tetraisopropyl thiodiphosphate*. O,O-*Diisopropyl phosphorothioate anhydride*

[61614-71-9]

$[(H_3C)_2CHO]_2P(S)—O—P(S)[OCH(CH_3)_2]_2$

$C_{12}H_{28}O_5P_2S_2$ M 378.418

Liq. d_4^{25} 1.09. $Bp_{0.004}$ 102-103°. n_D^{25} 1.4620.

▷XN4390000.

Toy, A.D.F., *J. Am. Chem. Soc.*, 1951, **73**, 4670 (*synth*)

Tetraisopropyl hypodiphosphite **T-00169**

Tetrakis(1-methylethyl) hypodiphosphite, 9CI. *Tetraisopropoxydiphosphine*

[40651-79-4]

$[(H_3C)_2CHO]_2P–P[OCH(CH_3)_2]_2$

$C_{12}H_{28}O_4P_2$ M 298.298

Liq. Bp_3 82°.

Proskurnina, M.V. *et al*, *Zh. Obshch. Khim.*, 1973, **43**, 66 (*Engl. transl.* p. 63) (*synth*)
Ponomarev, S.V. *et al*, *Zh. Obshch. Khim.*, 1978, **48**, 231 (*Engl. transl.* p. 207) (*synth, P nmr*)

Tetraisopropyl hypophosphate, 9CI **T-00170**

Tetrakis(1-methylethyl) hypophosphate, 9CI. *Tetraisopropoxydiphosphine dioxide*

[75340-94-2]

$[(H_3C)_2CHO]_2P(O)–P(O)[OCH(CH_3)_2]_2$

$C_{12}H_{28}O_6P_2$ M 330.297

Liq. d_4^{20} 1.07. $Bp_{0.001}$ 87-88°. n_D^{20} 1.4313.

Michalski, J. *et al*, *Bull. Acad. Polon. Sci., Ser. Sci. Chim.*, 1965, **13**, 253 (*synth*)

N,N,N',N'-Tetraisopropylphosphonic diamide, 10CI **T-00171**

N,N,N',N'-*Tetrakis(1-methylethyl)phosphonic diamide*. *Phosphonic tetraisopropyldiamide*. N,N,N',N'-*Tetraisopropylphosphorodiamidite*. *Phosphonic bis(diisopropylamide)*

[75292-68-1]

$[[(H_3C)_2CH]_2N]_2P(O)H \rightleftharpoons [[(H_3C)_2CH]_2N]_2POH$

$C_{12}H_{29}N_2OP$ M 248.348

Cryst. (hexane). Mp 62-63°.

Foss, V.L. *et al*, *Zh. Obshch. Khim.*, 1980, **50**, 1236 (*Engl. transl.* p. 1000) (*synth, P nmr*)

Tetraisopropylphosphorodiamidous acid **T-00172**

Tetrakis(1-methylethyl)phosphorodiamidous acid, 9CI

$[(H_3C)_2CH]_2NP(OH)N[CH(CH_3)_2]_2$

$C_{12}H_{29}N_2OP$ M 248.348

Me ester: [92611-10-4]. *Methyl tetraisopropylphosphorodiamidite*.
$C_{13}H_{21}N_2OP$ M 252.295
Reagent for oligoribonucleotide synthesis in liq. and solid phases. Liq. d^{26} 0.91. Bp_{30} 86-88°, $Bp_{0.45}$ 74-75°.

2-Cyanoethyl ester: [102691-36-1].
$C_{15}H_{32}N_3OP$ M 301.411
Reagent for oligoribonucleotide synthesis.

Barone, A.D. *et al*, *Nucleic Acids Res.*, 1984, **12**, 4051 (*synth, pmr, P nmr, ms, use*)
Lee, H.-J. *et al*, *Chem. Lett.*, 1984, 1229 (*synth, use*)
Moore, M.F. *et al*, *J. Org. Chem.*, 1985, **50**, 2019 (*synth, pmr, P nmr*)
Kierzek, R. *et al*, *Nucleic Acids Res.*, 1986, **14**, 4751 (*use*)
Nielson, J. *et al*, *J. Chem. Res. (S)*, 1986, 26 (*props*)
Tanaka, T. *et al*, *Nucleic Acids Res.*, 1986, **14**, 6265 (*use*)

O,O,O',O'-Tetraisopropyl thiohypophosphate **T-00173**

O,O,O',O'-*Tetrakis(1-methylethyl) thiohypophosphate*, 9CI. *Tetrapropoxydiphosphine disulfide*

[40651-74-9]

$[(H_3C)_2CHO]_2P(S)–P(S)[OCH(CH_3)_2]_2$

$C_{12}H_{28}O_4P_2S_2$ M 362.418

Possesses fungicidal props. Liq. which slowly cryst. d_4^{20} 1.07. Mp 43-44°. $Bp_{0.1}$ 113°. n_D^{20} 1.4865.

Almasi, L. *et al*, *Chem. Ber.*, 1963, **96**, 2024 (*synth*)
Proskurnina, M.V. *et al*, *Zh. Obshch. Khim.*, 1973, **43**, 66 (*Engl. transl.* p. 63) (*synth*)

O,O,O',O'-Tetraisopropyl trithiopyrophosphate **T-00174**

Tetrakis(1-methylethyl) thiodiphosphate, 9CI. O,O-*Diisopropyl phosphorodithioate anhydrosulfide*

[3253-29-0]

$[(H_3C)_2CHO]_2P(S)—S—P(S)[OCH(CH_3)_2]_2$

$C_{12}H_{28}O_4P_2S_3$ M 394.478

Liq. d_4^{20} 1.13. $Bp_{0.5}$ 140°. n_D^{20} 1.5113.

Schrader, G. *et al*, *Angew. Chem.*, 1958, **70**, 692 (*synth, tox*)
Almasi, L. *et al*, *Chem. Ber.*, 1964, **97**, 661 (*synth*)
Pudovik, A.N. *et al*, *Zh. Obshch. Khim.*, 1978, **48**, 2645 (*Engl. transl.* p. 2399) (*synth, P nmr*)

N,N',N'',N'''-Tetrakis(dihydroxyphosphinylmethyl)-1,4,7,10-tetraazacyclododecane **T-00175**

$(HO)_2P(O)CH_2$ $CH_2P(O)(OH)_2$
$(HO)_2P(O)CH_2$ $CH_2P(O)(OH)_2$

$C_{12}H_{32}N_4O_{12}P_4$ M 548.299

Forms complexes with Group II metals. Cryst. (Me_2CO aq.). Mp 250-255° dec. pK_{a1} <1, pK_{a2} <2, pK_{a3} 4.88, pK_{a4} 5.73, pK_{a5} 7.28, pK_{a6} 8.46, pK_{a7} 11.52, pK_{a8} 12.11.

B,MeI: Cryst. (Me_2CO). Mp >320°.
Biscyclohexylammonium salt: Solid. Mp 300° dec.

Kabachnik, M.I. *et al*, *Izv. Akad. Nauk SSSR, Ser. Khim.*, 1984, 844 (*Engl. transl.* p. 777) (*synth, derivs, pmr, P nmr, use*)

1,1,3,3-Tetrakis(dimethylamino)-1,3-diphosph(P^V)ete T-00176

[103678-18-8]

$C_{10}H_{26}N_4P_2$ M 264.290
Cryst. (pentane at −70°).

Svara, J. *et al*, *Z. Naturforsch., B*, 1985, **40**, 1258 (*synth, pmr, cmr, P nmr, ir, ms, cryst struct*)

Tetrakis(dimethylamino)diphosphine T-00177

Octamethylhypodiphosphorous tetraamide, 9CI

$(Me_2N)_2P{-}P(NMe_2)_2$

$C_8H_{24}N_4P_2$ M 238.252
Ligand for Rh. Needles. Mp 48°. $Bp_{0.01}$ 50°.

Nöth, H. *et al*, *Chem. Ber.*, 1961, **94**, 1505 (*synth, derivs*)
Foss, V.L. *et al*, *Zh. Obshch. Khim.*, 1980, **50**, 1236 (*Engl. transl.* p. 1000) (*monooxide, synth, nmr, props*)

Tetrakis(dimethylamino)iodophosphorane, 8CI T-00178

Iodo-N,N,N′,N′,N″,N″,N‴,N‴-octamethylphosphoranetetramine, 9CI

[21163-77-9]

$(Me_2N)_4PI$

$C_8H_{24}IN_4P$ M 334.183
Cryst. (Me_2CO). Mp 153° dec., 325°.

El Nigumi, Y.O. *et al*, *J. Inorg. Nucl. Chem.*, 1970, **32**, 3211 (*synth, ir*)
Haasemann, P. *et al*, *Z. Anorg. Allg. Chem.*, 1974, **408**, 293 (*synth, ir, raman*)
Koidan, G.N. *et al*, *Zh. Obshch. Khim.*, 1982, **52**, 2001 (*Engl. transl.* p. 1779) (*synth*)

Tetrakis[(diphenylphosphino)methyl]methane T-00179

[2,2-Bis[(diphenylphosphino)methyl]-1,3-propanediyl]-bis[diphenylphosphine]

[5032-78-0]

$C(CH_2PPh_2)_4$

$C_{53}H_{48}P_4$ M 808.857
Ligand for Co and Rh. Needles (EtOH). Mp 176-178°.

Tetraoxide: Tetrakis[(diphenylphosphinyl)methyl]-methane.
$C_{53}H_{48}O_4P_4$ M 872.855
Cryst. (Me_2CO/pet. ether). Mp 258-260°.

Tetrasulfide: Tetrakis[(diphenylphosphinothioyl)-methyl]methane.
$C_{53}H_{48}P_4S_4$ M 937.097
Cryst. (C_6H_6/pet. ether). Rather insol. Et_2O, EtOH, pet. ether. Mp 298-300°.

Tetraselenide: Tetrakis[(diphenylphosphinoselenoyl)-methyl]methane.
$C_{53}H_{48}P_4Se_4$ M 1124.697
Insol. Et_2O, MeCN, EtOH, pet. ether. Mp 337° dec.

Ellermann, J. *et al*, *Chem. Ber.*, 1966, **99**, 653 (*synth, ir*)
Ellermann, J. *et al*, *Chem. Ber.*, 1967, **100**, 2220 (*derivs*)

1,1,3,3-Tetrakis(diphenylphosphino)-1,2-propadiene T-00180

Tetrakis(diphenylphosphino)allene

$(Ph_2P)_2C{=}C{=}C(PPh_2)_2$

$C_{51}H_{10}P_4$ M 746.535
Cryst. (EtOH). Spar. sol. halogenated hydrocarbons. Mp 224° dec.

B,MeI:
$C_{52}H_{13}IP_4$ M 888.474
Solid. Mp 203°.

Tetraoxide:
$C_{51}H_{10}O_4P_4$ M 810.533
Solid. Mp 258-260°.

Schmidbaur, H. *et al*, *Angew. Chem., Int. Ed. Engl.*, 1986, **25**, 348 (*synth, derivs, ir, pmr, cmr, P nmr*)

Tetrakis(hydroxymethyl)phosphonium(1+) T-00181

$P(CH_2OH)_4^{\oplus}$

$C_4H_{12}O_4P^{\oplus}$ M 155.110 (ion)

Chloride: [124-64-1].
$C_4H_{12}ClO_4P$ M 190.563
Flameproofing agent. Hygroscopic solid. Sol. H_2O, MeOH, insol. Et_2O. Mp 151°.
▷TA2450000.

Acetate: [7580-37-2].
$C_6H_{15}O_6P$ M 214.155
Flameproofing agent.
▷TA2440000.

Hoffman, A., *J. Am. Chem. Soc.*, 1921, **43**, 1684 (*synth*)
Ger. Pat., 2 154 811, (*1970*); *CA*, **80**, 61008 (*use*)
Ellzey, S.E. *et al*, *Can. J. Chem.*, 1971, **49**, 3581 (*pmr*)
Valetdinov, R.K. *et al*, *CA*, 1971, **75**, 89084 (*synth*)

1,2,4,5-Tetrakis(methylene)-3-phosphoniaspiro[2.2]pentane, 9CI T-00182

[71466-47-2]

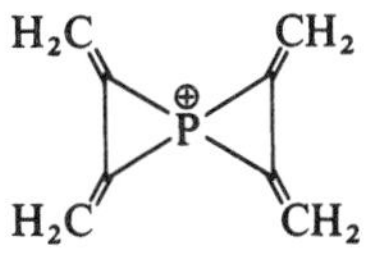

$C_8H_8P^{\oplus}$ M 135.125 (ion)
Currently unknown.

Böhm, M.C. *et al*, *J. Chem. Soc., Perkin Trans. 2*, 1979, 443 (*struct*)

Tetrakis(4-nitrophenyl) diphosphate, 9CI **T-00183**

Tetra(p-*nitrophenyl*) *pyrophosphate*

[52625-61-3]

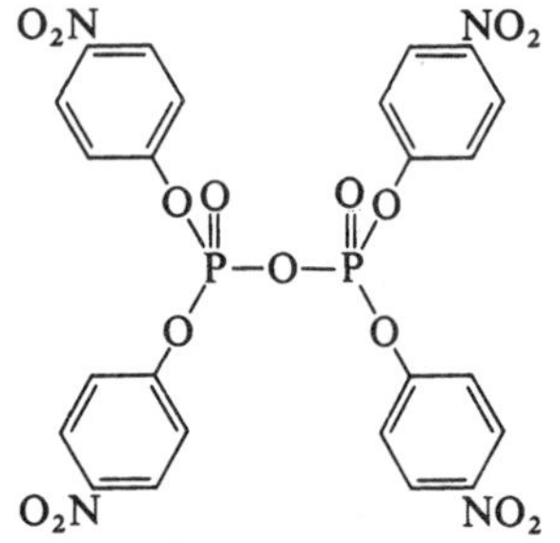

$C_{24}H_{16}N_4O_{15}P_2$ M 662.356

Phosphorylating agent. Cryst. (dioxan). Mp 126-127° (resolidifes and remelts at 146-148°).

Fr. Pat., 1 361 963, (*1964*); *CA*, **61**, 14586 (*synth*)
Fieser, M. *et al*, *Reagents for Organic Synthesis*, Wiley, 1967-84, **1**, 1148.

Tetrakis(pentafluorophenyl)diphosphine, 9CI **T-00184**

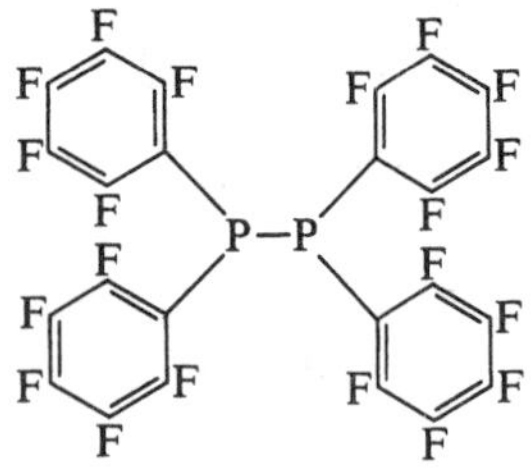

$C_{24}F_{20}P_2$ M 730.180

Ligand for Fe. Mp 165-170° dec. Cl_2 cleaves P—P bond.

Ang, H.G. *et al*, *Chem. Ind.* (*London*), 1966, 944 (*synth, ms, ir, complexes*)

Tetrakis(pentafluorophenyl)-tetraphosphetane, 8CI **T-00185**

Tetrakis(*pentafluorophenyl*)*cyclotetraphosphine*

[14655-88-0]

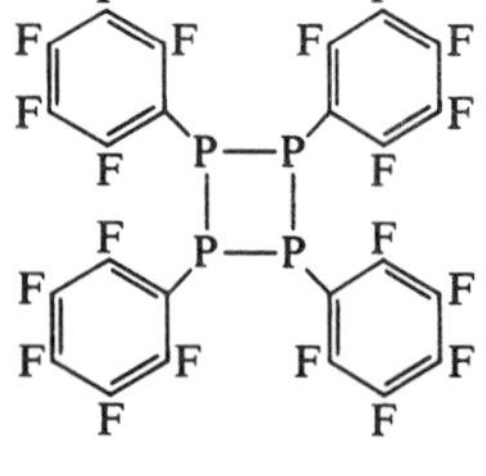

$C_{24}F_{20}P_4$ M 792.127

Solid. Mp 151°.

Fild, M. *et al*, *Naturwissenschaften*, 1967, **54**, 89 (*synth*)
Sanz, F. *et al*, *J. Chem. Soc.* (*A*), 1971, 1083 (*cryst struct*)
Baudler, M. *et al*, *Z. Naturforsch.*, *B*, 1976, **31**, 558 (*P nmr*)

1,3,4,6-Tetrakis(trifluoromethyl)-2,5-diphosphatricyclo[3.1.0.0^{2,6}]hex-3-ene **T-00186**

[65114-90-1]

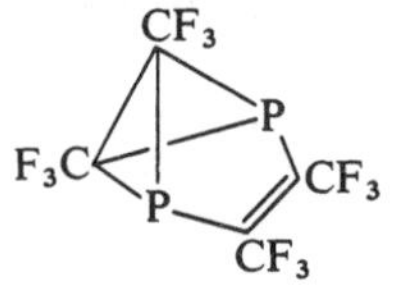

$C_8F_{12}P_2$ M 386.016

Oil which becomes solid at −75°.

Kobayashi, Y. *et al*, *J. Org. Chem.*, 1980, **45**, 4683 (*synth, props, P nmr, cmr, ms*)

2,2,3,3-Tetrakis(trifluoromethyl)-1,4,6,10-tetraoxa-5-phospha(5-P^V)spiro[4.5]-decane, 9CI **T-00187**

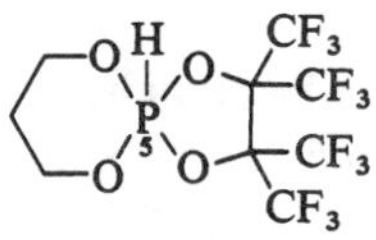

$C_9H_7F_{12}O_4P$ M 438.106

5-Phenoxy: [62013-25-6].
$C_{15}H_{11}F_{12}O_5P$ M 530.203
Solid. Mp 93-97°.

Bone, S.A. *et al*, *J. Chem. Soc.*, *Perkin Trans. 1*, 1977, 80 (*synth, ms, pmr, F and P nmr*)

Tetrakis(trifluoromethyl)tetraphosphetane, 9CI, 8CI **T-00188**

Tetrakis(*trifluoromethyl*)*cyclotetraphosphine*

[393-02-2]

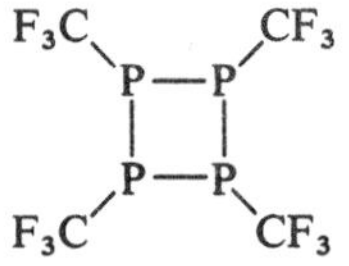

$C_4F_{12}P_4$ M 399.920

Solid. Sol. C_6H_6, Me_2CO, CCl_4, Et_2O, pet. ether. d 1.54 (liq. at Mp). Mp 66.3°. 10% aq. NaOH yields CHF_3.

▷Spont. flammable in air

Mahler, W. *et al*, *J. Am. Chem. Soc.*, 1958, **80**, 6161 (*synth, props, uv*)
Palenik, G.J. *et al*, *Acta Crystallogr.*, 1962, **15**, 564 (*cryst struct*)
Cowley, A.H. *et al*, *Inorg. Chem.*, 1966, **5**, 1459 (*ms*)
Cavell, R.G. *et al*, *Inorg. Chem.*, 1968, **7**, 690 (*ms, monosulfide*)
Baudler, M. *et al*, *Z. Naturforsch.*, *B*, 1976, **31**, 538 (*P nmr*)
Cowley, A.H. *et al*, *J. Am. Chem. Soc.*, 1978, **100**, 3349 (*pe, struct*)

2,3,7,8-Tetrakis(trifluoromethyl)-1,4,6,9-tetrathia-5-phospha(5-P^V)spiro[4.4]nona-2,7-diene, 9CI **T-00189**

$C_8HF_{12}PS_4$ M 484.290

5-Ph: [68823-56-3].
$C_{14}H_5F_{12}PS_4$ M 560.388
Yellow cryst. at −70°. Obt. only in impure form. Dec. on attempted purifn.

Burros, B.C. *et al*, *J. Am. Chem. Soc.*, 1978, **100**, 7300.

2,2,3,3-Tetrakis(trifluoromethyl)-1,4,6-trioxa-10-aza-5-phospha(5-P^V)-spiro[4.5]decane, 10CI T-00190

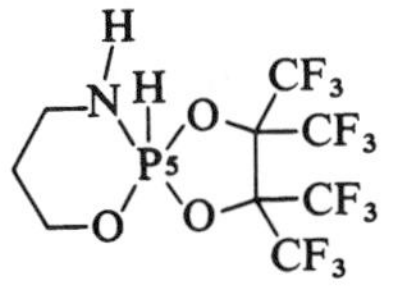

$C_9H_8F_{12}NO_3P$ M 437.122

5-(4-Bromophenoxy): [62608-26-8].
$C_{15}H_{11}BrF_{12}NO_3P$ M 592.115
Cryst.

Barlow, J.H. *et al*, *J. Chem. Soc., Chem. Commun.*, 1976, 1031 (*cryst struct*)

3,3,4,4-Tetramethyl-1,2-azaphospholidine 2-oxide, 8CI T-00191

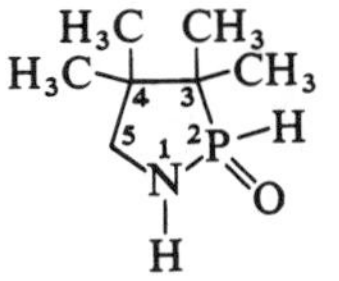

$C_7H_{16}NOP$ M 161.183

2-Methoxy: [32503-87-0]. *2-Methoxy-3,3,4,4-tetramethyl-1,2-azaphospholidine 2-oxide.*
$C_8H_{18}NO_2P$ M 191.209
Cryst. (pet. ether). Mp 104-107°.

Harger, M.J.P., *J. Chem. Soc., Perkin Trans. 1*, 1974, 2604 (*synth, ir, pmr*)

4,4,5,5-Tetramethyl-1,2-azaphospholidine 2-oxide T-00192

$C_7H_{16}NOP$ M 161.183

2-Methoxy: [32503-86-9]. *2-Methoxy-4,4,5,5-tetramethyl-1,2-azaphospholidine 2-oxide.*
$C_8H_{18}NO_2P$ M 191.209
Cryst. (pet. ether). Mp 65-66°.

Harger, M.J.P., *J. Chem. Soc., Perkin Trans. 1*, 1974, 2604 (*synth, ir, pmr*)

1,1,3,3-Tetramethyl-6,7-bis(trifluoromethyl)-5,8-dithia-4-phospha(4-P^V)-spiro[3.4]oct-6-ene, 10CI T-00193

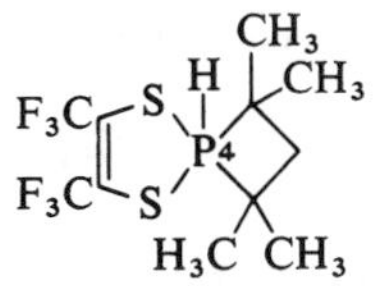

$C_{11}H_{15}F_6PS_2$ M 356.324

4-Ph: [38046-64-9].
$C_{17}H_{19}F_6PS_2$ M 432.421
Yellow prisms (CCl_4). Mp 109-112° (sealed tube). Dec. at 150°.

Burros, B.C. *et al*, *J. Am. Chem. Soc.*, 1978, **100**, 7300 (*synth, pmr, F nmr*)

1,2,3,4-Tetramethyl-1,3,2,4-diazadiphosphetidine, 9CI, 8CI T-00194

Dimethylphosphine imide dimer.
Tetramethylcyclodiphosphazane

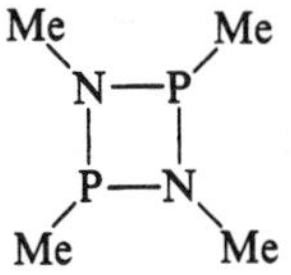

$C_4H_{12}N_2P_2$ M 150.100

2,4-Disulfide: [13789-71-4].
$C_4H_{12}N_2P_2S_2$ M 214.220
Cryst. (C_6H_6). Mp 129°, Mp 197-199°.

Mel'nikov, N.N. *et al*, *Zh. Obshch. Khim.*, 1967, **37**, 239 (*Engl. transl.* p. 222) (*disulfide, synth, ir, uv, raman*)
Granzhan, V.A. *et al*, *Zh. Obshch. Khim.*, 1969, **39**, 1501 (*Engl. transl.* p. 1470) (*disulfide, config*)
Wannagat, U. *et al*, *Z. Anorg. Allg. Chem.*, 1976, **420**, 119 (*disulfide, synth, ir, pmr, P nmr*)
Keck, H. *et al*, *Phosphorus Sulfur*, 1978, **4**, 173 (*ms*)

2,2,7,7-Tetramethyl-1,6-dioxa-4,9-diaza-5-phospha(5-P^V)spiro[4.4]nonane, 9CI T-00195

[51675-96-8]

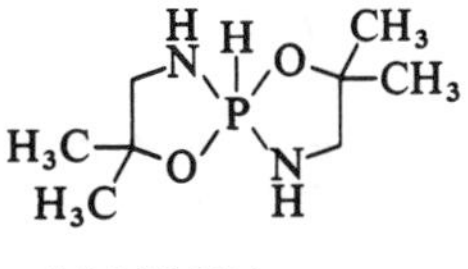

$C_8H_{19}N_2O_2P$ M 206.224

Exists completely as penta-coordinate tautomer in the range 20-150°. Cryst. (Et_2O). Mp 98°.

Burgada, R. *et al*, *J. Organomet. Chem.*, 1974, **66**, 255 (*synth, P nmr, struct*)

3,3,8,8-Tetramethyl-1,6-dioxa-4,9-diaza-5-phospha(P^V)spiro[4.4]nonane T-00196

[2318-47-0]

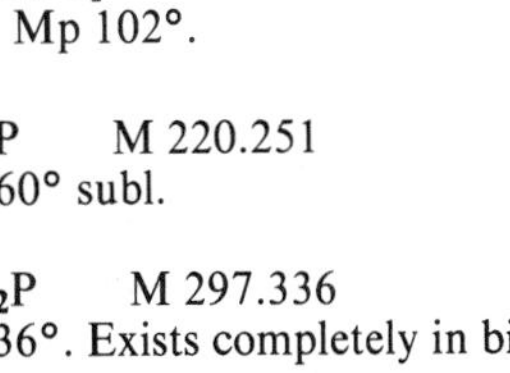

$C_8H_{19}N_2O_2P$ M 206.224

Exists completely as the pentacoordinate tautomer.

(±)-*form* [51572-21-5]
Cryst. (C_6H_6). Mp 102°.

5-Me:
$C_9H_{21}N_2O_2P$ M 220.251
Solid. $Bp_{0.1}$ 60° subl.

5-Anilino:
$C_{14}H_{24}N_3O_2P$ M 297.336
Solid. Mp 136°. Exists completely in bicyclic tautomeric form.

Sanchez, M. *et al*, *Bull. Soc. Chim. Fr.*, 1968, 773 (*synth*)
Burgada, R. *et al*, *J. Organomet. Chem.*, 1974, **66**, 255 (*P nmr, struct*)
Houalla, D. *et al*, *Org. Magn. Reson.*, 1974, **6**, 340 (*pmr*)
Malavaud, C. *et al*, *Tetrahedron Lett.*, 1975, 497 (*methyl, synth, P nmr*)
Sanchez, M. *et al*, *Nouv. J. Chim.*, 1976, **3**, 775 (*deriv, synth, pmr, P nmr*)

4,4,5,5-Tetramethyl-1,3,2-dioxaphospholane, 9CI T-00197

Tetramethyl-1,3-dioxa-2-phosphacyclopentane. Tetramethylethylene phosphonite

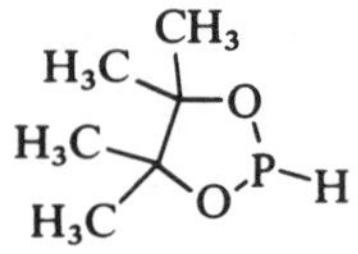

$C_6H_{13}O_2P$ M 148.141

2-Oxide: [16352-18-4]. *Tetramethylethylene phosphonate. Tetramethylethylene phosphite.*
$C_6H_{13}O_3P$ M 164.141
Cryst. (THF or C_6H_6). Mp 109°. pK_a 1.71 (10% EtOH aq.), pK_a 3.40 (90% EtOH aq.). Tautomeric.

2-Sulfide: [70372-09-7]. O,O-*Tetramethylethylene phosphonothioate.* O,O-*Tetramethylethylene thiophosphite.*
$C_6H_{13}O_2PS$ M 180.201
pK_a 4.76 (50% EtOH aq.), pK_a 8.50 (propanol). Tautomeric.

Zwierzak, A. *et al*, *Can. J. Chem.*, 1967, **45**, 2501 (*oxide, synth, ir*)
Sanchez, M. *et al*, *Bull. Soc. Chim. Fr.*, 1968, 773 (*oxide*)
Burgada, R., *Bull. Soc. Chim. Fr.*, 1972, 4161 (*oxide, P nmr*)
Ovchinnikov, V.V. *et al*, *Zh. Obshch. Khim.*, 1978, **48**, 2423; 1979, **49**, 1693 (*Engl. transl.* pp. 2199, 1482) (*P nmr, props*)

2,4,8,10-Tetramethyl-3,9-dioxo-2,4,8,10-tetraaza-6-phosphoniaspiro[5.5]-undecane(1+), 9CI T-00198

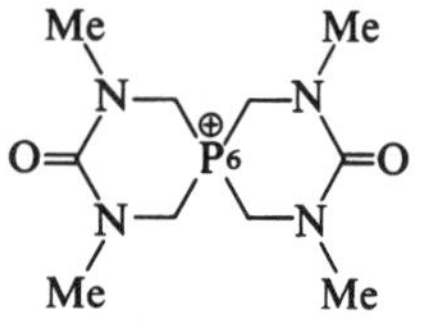

$C_{10}H_{20}N_4O_2P^{\oplus}$ M 259.267 (ion)

Chloride:
$C_{10}H_{20}ClN_4OP$ M 278.721
Cryst. (2-propanol). Mp 161-162°. Solidifies and remelts at 217°.

Sulfate:
$C_{20}H_{42}N_8O_8P_2S$ M 616.608
Solid. Mp 177° dec.

Frank, A.W., *Phosphorus Sulfur*, 1981, **10**, 147, 207 (*synth, ir, pmr, P nmr, props*)

3,3′,4,4′-Tetramethyl-1,1′-diphenyl-2,2′-bi-1*H*-phosphole T-00199

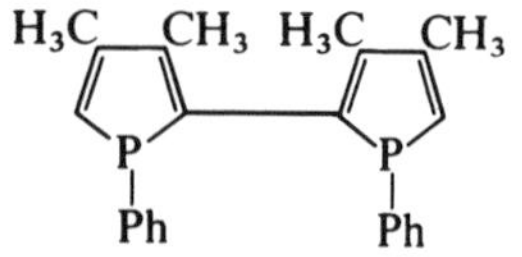

$C_{24}H_{24}P_2$ M 374.401
Ligand for Mo, Fe, Mn. Sl. yellow oil which cryst. on standing. Mp 108°.

1,1′-Disulfide:
$C_{24}H_{24}P_2S_2$ M 438.521
Pale-yellow solid. Mixt. of two isomers.

Mercier, F. *et al*, *J. Organomet. Chem.*, 1986, **316**, 271 (*synth, ms, pmr, cmr, P nmr*)

Tetramethyldiphosphine, 9CI T-00200

Tetramethylbiphosphine
[3676-91-3]

Me_2PPMe_2

$C_4H_{12}P_2$ M 122.086
Widely used ligand. Liq. Mp −2.15°. Bp 140-142°, Bp_{16} 38-40°. n_D^{20} 1.5252. Stable up to 300°.

▷Ignites in air

Monooxide: [66436-14-4].
$C_4H_{12}OP_2$ M 138.086
Solid. Mp 71-73°.

Dioxide: [69844-21-9].
$C_4H_{12}O_2P_2$ M 154.085
Solid. Mp 132.5-132.7°.

Monosulfide: [26978-38-1].
$C_4H_{12}P_2S$ M 154.146
Solid at r.t. $Bp_{0.5}$ 65-70°. V. air-sensitive.

Disulfide: see Tetramethyldiphosphine disulfide, T-00201

1-Oxide 2-sulfide: [64516-41-2].
$C_4H_{12}OP_2S$ M 170.146
Cryst. (C_6H_6). V. sol. CH_2Cl_2, $CHCl_3$, less sol. C_6H_6, MeCN. Mp 142° dec.

Diselenide: [73731-14-3].
$C_4H_{12}P_2Se_2$ M 280.006
Cryst. (air-stable). Mp 278°.

1-Selenide 2-sulfide: [73731-13-2].
$C_4H_{12}P_2SSe$ M 233.106
Cryst. Mp 199°.

Parshall, G.W., *J. Inorg. Nucl. Chem.*, 1960, **14**, 291 (*synth*)
Burg, A.B., *J. Am. Chem. Soc.*, 1961, **83**, 2226 (*synth, props*)
Griffith, J.E. *et al*, *J. Am. Chem. Soc.*, 1962, **84**, 3442 (*dioxide*)
Niebergall, H. *et al*, *Chem. Ber.*, 1962, **95**, 64 (*synth, props*)
Harris, R.K. *et al*, *Can. J. Chem.*, 1964, **42**, 2282 (*pmr, monosulfide*)
Durig, J.R. *et al*, *J. Mol. Struct.*, 1973, **17**, 426 (*raman*)
Inorg. Synth., 1974, **15**, 187 (*synth*)
Cowley, A.H. *et al*, *J. Am. Chem. Soc.*, 1974, **96**, 2648, 3666 (*pe*)
Wanczek, K.-P., *Z. Naturforsch, A*, 1976, **31**, 414 (*ms*)
Aime, S. *et al*, *J. Chem. Soc., Dalton Trans.*, 1976, 2144 (*pmr, cmr, nmr*)
Appel, R. *et al*, *Chem. Ber.*, 1977, **110**, 3201 (*oxide, sulfide, synth, props, ms, pmr, nmr, cmr*)
Volkhotz, M. *et al*, *Chem. Ber.*, 1978, **111**, 890 (*monooxide, pmr, nmr*)
Galasso, V., *J. Magn. Reson.*, 1979, **36**, 181 (*nmr*)
Schweig, A. *et al*, *J. Am. Chem. Soc.*, 1979, **101**, 80 (*pe*)
Troy, D. *et al*, *Bull. Soc. Chim. Fr., Part I*, 1979, 241 (*uv*)
McFarlane, H.C.E. *et al*, *J. Chem. Soc., Dalton Trans.*, 1980, 240 (*derivs, pmr, nmr*)
Albrand, J.P., *Org. Magn. Reson.*, 1983, **21**, 246 (*pmr, cmr, nmr*)
Bretherick, L., *Handbook of Reactive Chemical Hazards*, 2nd Ed., Butterworths, London and Boston, 1979, 513.

Tetramethyldiphosphine disulfide, 9CI, 8CI T-00201

[3676-97-9]

$Me_2P(S){-}P(S)Me_2$

$C_4H_{12}P_2S_2$ M 186.206
Ligand for metals of Groups IB, IIB, VIII and also Sn and W. Cryst. (2-ethoxyethanol or toluene/EtOH). Mp 233-235° (223-234°).

Cowley, A.H. *et al*, *Inorg. Chem.*, 1965, **4**, 1827; *Spectrochim. Acta*, 1966, **22**, 1431 (*synth, ir, raman, uv, ms*)
Pedone, C. *et al*, *J. Chem. Phys.*, 1967, **47**, 339 (*cryst struct*)
Inorg. Synth., 1974, **15**, 185 (*synth*)
Fluck, E. *et al*, *Z. Anorg. Allg. Chem.*, 1975, **412**, 47 (*pe*)

Harris, R.K. *et al*, *J. Chem. Soc., Faraday Trans. 2*, 1976, **72**, 2291 (*nmr*)
McQuillan, G.P. *et al*, *Spectrochim. Acta, Part A*, 1977, **33**, 233 (*ir, raman*)
Colquhoun, I.J. *et al*, *J. Magn. Reson.*, 1978, **31**, 63 (*nmr*)
Keck, H. *et al*, *Org. Mass Spectrom.*, 1979, **14**, 149 (*ms*)
Schröder, H.Fr. *et al*, *Z. Anorg. Allg. Chem.*, 1979, **451**, 158 (*ir, Raman*)
McFarlane, H.C.A. *et al*, *J. Chem. Soc., Dalton Trans.*, 1980, 240 (*pmr, nmr*)
Alagna, L. *et al*, *Z. Naturforsch., A*, 1981, **36**, 68 (*pe*)

2,4,8,10-Tetramethyl-3,9-dithia-6-phosphoniaspiro[5.5]undecane, 9CI **T-00202**

[55402-79-4]

H3C CH3 S P⊕ S H3C CH3

$C_{12}H_{24}PS_2^{\oplus}$ M 263.415 (ion)

Perchlorate: [54970-66-0].
$C_{12}H_{24}ClO_4PS_2$ M 362.866
Solid. Mp 234-238° dec.

Samaan, S., *Tetrahedron Lett.*, 1974, 3927 (*synth*)

***sym-O,O,O,O*-Tetramethyl dithiopyrophosphate** **T-00203**

O,O,O,O-*Tetramethyl thiodiphosphate*, *9CI*. O-(*Dimethoxyphosphinothioyl*) O,O-*dimethyl phosphorothioate*. O,O-*Dimethyl phosphorothioate anhydride*

[51120-35-5]

$(MeO)_2P(S)—O—P(S)(OMe)_2$

$C_4H_{12}O_5P_2S_2$ M 266.203

Note 9CI name identical with that for *unsym*-Tetramethyl monothiopyrophosphate, T-00207 . Impurity in tech. Malathion, M-00001 . Liq.

▷Very toxic. XN4500000.

Toia, R.F. *et al*, *J. Agric. Food Chem.*, 1980, **28**, 599 (*glc, tox*)
Greenhalgh, R. *et al*, *J. Agric. Food Chem.*, 1983, **31**, 710 (*P nmr*)

Tetramethyl hypophosphate, 9CI **T-00204**

Tetramethoxydiphosphine dioxide

[15103-99-8]

$(MeO)_2P(O)-P(O)(OMe)_2$

$C_4H_{12}O_6P_2$ M 218.083

Liq. d_4^{20} 1.33. $Bp_{0.0001}$ 85-87°. n_D^{20} 1.4421.

Baudler, M., *Z. Anorg. Allg. Chem.*, 1956, **288**, 171 (*synth*)
Mowthorpe, D.J. *et al*, *Spectrochim. Acta, Part A*, 1967, **23**, 451 (*pmr, P nmr*)
Suzuki, N. *et al*, *Bull. Chem. Soc. Jpn.*, 1980, **53**, 1421 (*synth, ir*)

***O,O,O′,O′*-Tetramethyl *S,S′*-methylene bisphosphorodithioate, 9CI** **T-00205**

Bis[(O,O-*dimethoxyphosphinothioyl*)*thio*]*methane*

[53718-38-0]

$(MeO)_2P(S)-SCH_2S-P(S)(OMe)_2$

$C_5H_{14}O_4P_2S_4$ M 328.351

Liq. d_4^{20} 1.36. n_D^{20} 1.5741.

Khaskin, B.A. *et al*, *Zh. Obshch. Khim.*, 1980, **50**, 1990 (*Engl. transl.* p. 1606)

***sym*-Tetramethyl monothiopyrophosphate** **T-00206**

Tetramethyl thiodiphosphate, *9CI*. O,O-*Dimethyl* S-(*dimethoxyphosphinyl*) *phosphorothioate*

[71861-22-8]

$(MeO)_2P(O)—S—P(O)(OMe)_2$

$C_4H_{12}O_6P_2S$ M 250.143

Liq. Easily isomerises to the unsym. compd., *unsym*-Tetramethyl monothiopyrophosphate, T-00207 .

Lebedev, E.P. *et al*, *Zh. Obshch. Khim.*, 1979, **49**, 1730 (*Engl. transl.* p. 1515) (*ir, P nmr*)

***unsym*-Tetramethyl monothiopyrophosphate** **T-00207**

O,O,O,O-*Tetramethyl thiodiphosphate*, *9CI*. O,O,O,O-*Tetramethyl monothiopyrophosphate*. O,O-*Dimethyl* O-(*dimethoxyphosphinyl*) *phosphorothioate*

[18764-12-0]

$(MeO)_2P(O)—O—P(S)(OMe)_2$

$C_4H_{12}O_6P_2S$ M 250.143

More stable than the thiol isomer, *sym*-Tetramethyl monothiopyrophosphate, T-00206 Note that the 9CI name is identical with that for *sym-O,O,O,O*-Tetramethyl dithiopyrophosphate, T-00203 . Liq. d_4^{25} 1.34. Bp_2 127-129°, $Bp_{0.01}$ 64°. n_D^{20} 1.4510.

▷LD_{50} (rat) 20 mg/Kg

Schrader, G. *et al*, *Angew. Chem.*, 1958, **70**, 690 (*synth, tox*)
Lebedev, E.P. *et al*, *Zh. Obshch. Khim.*, 1979, **49**, 1730 (*Engl. transl.* p. 1515) (*synth*)

3,3,5,5-Tetramethyl-1,2-oxaphospholidine 2-oxide **T-00208**

CH3 CH3 H3C H3C N P—H O H

$C_7H_{16}NOP$ M 161.183

2-Methoxy: [54882-43-8]. *2-Methoxy-3,3,5,5-tetramethyl-1,2-azaphospholidine 2-oxide.*
$C_8H_{18}NO_2P$ M 191.209
Cryst. (pet. ether). Mp 126-128°.

Harger, M.J.P., *J. Chem. Soc., Perkin Trans. 1*, 1974, 2604 (*synth, ir, pmr*)

2,2,5,5-Tetramethyl-1-phenyl-1-phospha-2,5-disilacyclopentane, 9CI **T-00209**

[63746-57-6]

Me Me Si P Ph Si Me Me

$C_{12}H_{21}PSi_2$ M 252.443

Liq. $Bp_{0.6}$ 92°.

Couret, C. *et al*, *J. Organomet. Chem.*, 1977, **132**, C5; 1982, **224**, 247 (*synth, pmr, P nmr*)
Andriamizaka, J.D. *et al*, *Phosphorus Sulfur*, 1982, **12**, 265 (*synth, pmr, P nmr*)

2,2,6,6-Tetramethyl-1-phenyl-4-phosphorinanone, 8CI T-00210

2,2,6,6-Tetramethyl-1-phenyl-4-phosphinanone
[13887-05-3]

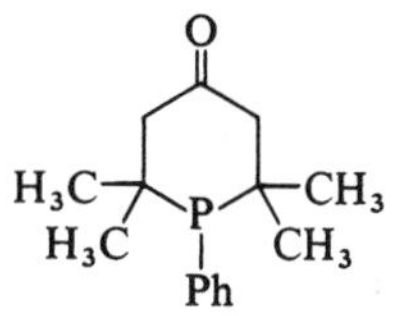

$C_{15}H_{21}OP$ M 248.304
Cryst. Mp 91-92°. $Bp_{0.5}$ 137-139°. pK_a 4.61.
B,MeI: 4-Oxo-1,2,2,6,6-pentamethyl-1-phenylphosphorinanium iodide.
$C_{16}H_{24}IOP$ M 390.243
Cryst. (MeCN). Mp 229-230°. $Bp_{0.3}$ 100° subl.
B,PhCH₂Br: 1-Benzyl-4-oxo-2,2,6,6-tetramethyl-1-phenylphosphorinanium bromide.
$C_{22}H_{28}BrOP$ M 419.340
Cryst. (MeCN). Mp 233-235°.
Oxide: [21230-89-7].
$C_{15}H_{21}O_2P$ M 264.303
Needles (xylene). Mp 207-208°.
Sulfide: [1216-38-2].
$C_{15}H_{21}OPS$ M 280.364
Cryst. (MeOH). Mp 138-139°.

Welcher, R.P. *et al*, *J. Org. Chem.*, 1962, **27**, 1824 (*synth, derivs*)
Asinger, F. *et al*, *Monatsh. Chem.*, 1968, **99**, 1695 (*synth*)
Rampal, J.B. *et al*, *J. Org. Chem.*, 1981, **46**, 1156 (*synth, derivs, ir, cmr, P nmr, pmr*)

2,7,10,15-Tetramethyl-17-phenyl-17*H*-tetrabenzo[*b,d,f,h*]phosphonin, 9CI T-00211

[59341-95-6]

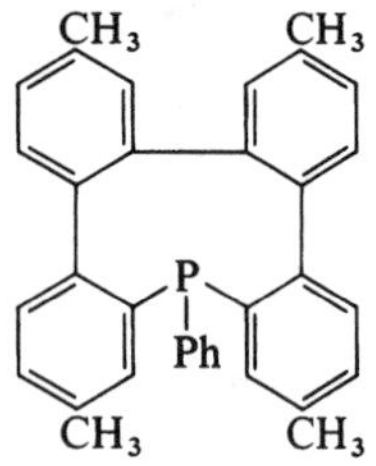

$C_{34}H_{29}P$ M 468.577
Cryst. (EtOH). Mp 180-181°.
*B, PhCH₂I: 17-Benzyl-2,7,10,15-tetramethyl-17-phenyl-17*H*-tetrabenzo*[b,d,f,h]*phosphoninium iodide.*
$C_{41}H_{36}IP$ M 686.614
Cryst. (EtOH). Mp 280-287°.

Hellwinkel, D. *et al*, *Chem. Ber.*, 1976, **109**, 1497 (*synth, pmr, P nmr, stereochem*)

1,1,3,3-Tetramethyl-1-phospha-3-silacyclohex-1-ene, 9CI T-00212

[57027-30-2]

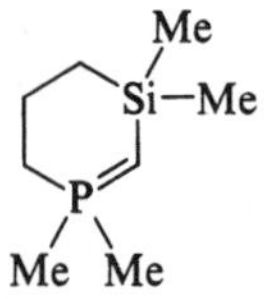

$C_8H_{19}PSi$ M 174.297
A cyclic ylide. Liq. Bp_4 69°.

Schmidbaur, H. *et al*, *Chem. Ber.*, 1975, **108**, 2842 (*synth, pmr, cmr, P nmr*)

Tetramethylphosphinous amide, 9CI T-00213

Dimethyl(dimethylamino)phosphine
[683-84-1]

Me_2PNMe_2

$C_4H_{12}NP$ M 105.119
Liq. Bp 98-99°.

Issleib, K. *et al*, *Chem. Ber.*, 1959, **92**, 2681 (*synth*)
Laurent, J.P. *et al*, *J. Inorg. Nucl. Chem.*, 1969, **31**, 1353 (*pmr, P nmr*)
Barlos, K. *et al*, *Z. Naturforsch., B*, 1978, **33**, 515 (*N nmr*)
Gouesnard, J.-P. *et al*, *Can. J. Chem.*, 1980, **58**, 1295 (*N and P nmr*)

***N,N,N′,N′*-Tetramethylphosphonic diamide, 9CI** T-00214

Phosphonic bis(dimethylamide). N,N,N′,N′-*Tetramethylphosphorodiamidous acid*
[5843-26-5]

$(Me_2N)_2P(O)H \rightleftharpoons (Me_2N)_2POH$

$C_4H_{13}N_2OP$ M 136.133
Tautomeric. Liq. $Bp_{1.3}$ 56°. $n_D^{24.5}$ 1.4515.

Zwierzak, A., *Bull. Acad. Pol. Sci., Ser. Sci. Chim.*, 1965, **13**, 609 (*synth, ir*)
Gallagher, M.J. *et al*, *J. Chem. Soc. (C)*, 1966, 2176 (*synth, ir, pmr*)
Wolf, R. *et al*, *Spectrochim. Acta, Part A*, 1967, **23**, 1641 (*ir*)
Corey, E.J. *et al*, *J. Org. Chem.*, 1969, **34**, 3053 (*synth, ir, pmr*)
Burgada, R. *et al*, *Bull. Soc. Chim. Fr.*, 1970, 192 (*synth, ir, pmr*)
Fluck, E. *et al*, *Z. Anorg. Allg. Chem.*, 1980, **461**, 187 (*P nmr*)

Tetramethylphosphonium(1+) T-00215

[32589-80-3]

$PMe_4^{\oplus}$

$C_4H_{12}P^{\oplus}$ M 91.113 (ion)
Chloride: [1941-19-1].
$C_4H_{12}ClP$ M 126.566
Hygroscopic cryst. (EtOH/Et_2O). Mp 345°, 396-397°, 406°.
Bromide: [4519-28-2].
$C_4H_{12}BrP$ M 171.017
Hygroscopic cryst. (MeOH/Et_2O). Mp 160-180°, >300°.
Iodide: [993-11-3].
$C_4H_{12}IP$ M 218.017
Cryst. (EtOH/Et_2O). Mp >360° (sealed tube).
▷TA2625000.

Doering, W. von E. *et al*, *J. Am. Chem. Soc.*, 1955, **77**, 521 (*iodide*)

Baumgärtner, R. *et al*, *Z. Anorg. Allg. Chem.*, 1964, **333**, 170 (*chloride, ir, raman*)
Maier, L., *Helv. Chim. Acta*, 1966, **49**, 2458 (*chloride, bromide*)
Marsi, K.L. *et al*, *J. Am. Chem. Soc.*, 1973, **95**, 200 (*bromide*)
Ohkubo, K., *Bull. Chem. Soc. Jpn.*, 1974, **47**, 557 (*struct*)
Ang, T.T. *et al*, *Can. J. Chem.*, 1976, **54**, 1985 (*pmr, cryst struct*)
Karsch, H.H., *Phosphorus Sulfur*, 1982, **12**, 217 (*pmr, nmr*)
Makovetskii, Yu.P. *et al*, *Zh. Obshch. Khim.*, 1982, **52**, 2235 (*Engl. transl.* p. 1989) (*polyiodides, uv*)

N,N,N',N'-Tetramethylphosphoric triamide, 9CI — T-00216

[3732-86-3]

$(Me_2N)_2P(O)NH_2$

$C_4H_{14}N_3OP$ M 151.148

N″*-Me: see Pentamethylphosphoric triamide, P-00026*

N″*-(2-Propenyl):* N,N,N′,N′*-Tetramethyl-*N″*-(2-propenyl)phosphoric triamide.* N″*-Allyl-*N,N,N′,N′*-tetramethylphosphoric triamide.*
$C_7H_{18}N_3OP$ M 191.212
Liq. $Bp_{0.05}$ 128°. n_D^{18} 1.4721.

N″,N″*-Di-Me: see Hexamethylphosphoric triamide, H-00075*

N″*-Benzyl:* [59950-71-9]. N″*-Benzyl-*N,N,N′,N′*-tetramethylphosphoric triamide.* N,N,N′,N′*-Tetramethyl-*N″*-(phenylmethyl)phosphoric triamide.*
$C_{11}H_{20}N_3OP$ M 241.272
Hygroscopic solid. Mp 93°.

N″*-Trimethylsilyl:*
$C_7H_{22}N_3OPSi$ M 223.330
Solid. Mp 128°.

Keat, R. *et al*, *J. Chem. Soc. (A)*, 1968, 703 (*pmr*)
Baldwin, M.A. *et al*, *Org. Mass Spectrom.*, 1977, **12**, 279 (*ms*)
Tomoi, M. *et al*, *Bull. Chem. Soc. Jpn.*, 1979, **52**, 1653 (*derivs, synth*)
Corbel, B. *et al*, *Can. J. Chem.*, 1980, **58**, 2183 (*derivs, synth*)
Riesel, L. *et al*, *Z. Chem.*, 1980, **20**, 151 (*trimethylsilyl deriv, synth, pmr, P nmr*)

Tetramethyl phosphor(isothiocyanatido)thioic diamide, 9CI — T-00217

Bis(dimethylamino)phosphinothioic isothiocyanate. Phosphor(isothiocyanatido)thioic bis(dimethylamide)

[54434-92-3]

$(Me_2N)_2P(S)NCS$

$C_5H_{12}N_3PS_2$ M 209.264
Reactive liq. d 1.17. $Bp_{0.3}$ 84°. n_D^{25} 1.5704. Sensitive to moisture.

Vetter, H.J., *Z. Naturforsch., B*, 1964, **19**, 168 (*synth, ir*)

Tetramethylphosphorodiamidic acid, 9CI, 8CI — T-00218

Phosphoric acid bis(dimethylamide)

[27972-73-2]

$(Me_2N)_2P(O)OH$

$C_4H_{13}N_2O_2P$ M 152.133

Tetramethylammonium salt: Solid. Mp 180-182°.

Me ester: see Methyl tetramethylphosphorodiamidate, M-00406

Et ester: [2404-65-1]. *Ethyl tetramethylphosphorodiamidate.*
$C_6H_{17}N_2O_2P$ M 180.186
Fireproofing agent. Liq. Bp_{20} 103-104°. n_D^{20} 1.4362.

Allyl ester: [50775-60-5]. *2-Propenyl tetramethylphosphorodiamidate. Allyl tetramethylphosphorodiamidate.*
$C_7H_{17}N_2O_2P$ M 192.197
Reagent for oxidn. of cyclic ketones to lactones. Liq. $Bp_{0.01}$ 57°.

Butyl ester: [52604-86-1]. *Butyl tetramethylphosphorodiamidate.*
$C_8H_{21}N_2O_2P$ M 208.240
Liq. Sol. H_2O. Bp_{15} 123-125°. n_D^{26} 1.4360.

Ph ester: [7393-13-7]. *Phenyl tetramethylphosphorodiamidate.*
$C_{10}H_{17}N_2O_2P$ M 228.230
Liq. Bp_9 154-155°. n_D^{25} 1.5037.

Fluoride: see Dimefox, D-00669

Chloride: see Tetramethylphosphorodiamidic chloride, T-00220

Bromide:
$C_4H_{12}BrN_2OP$ M 215.029
Liq. Bp_{12} 122°.

Azide: see Tetramethylphosphorodiamidic azide, T-00219

Isothiocyanate: [38500-29-7]. *Tetramethyl phosphor(isothiocyanatidic) diamide.*
$C_5H_{12}N_3OPS$ M 193.203
Liq. d_4^{20} 1.16. $Bp_{0.3}$ 83-85°. n_D^{20} 1.5318.

Amide: see N,N,N′,N′-Tetramethylphosphoric triamide, T-00216

Phenylhydrazide: see 2-Phenylphosphorodiamidic hydrazide, P-00274

Anhydride: see Octamethyldiphosphoramide, O-00029

Loev, B. *et al*, *J. Org. Chem.*, 1957, **22**, 1186 (*esters*)
Michalski, J. *et al*, *Rocz. Chem.*, 1957, **31**, 879; *CA*, **52**, 8037 (*isothiocyanate*)
Cheymol, J., *C.R. Hebd. Seances Acad. Sci.*, 1959, **249**, 1240 (*synth*)
Vetter, H-J., *Z. Naturforsch., B*, 1964, **19**, 72 (*bromide*)
Keat, R. *et al*, *J. Chem. Soc. (A)*, 1968, 703 (*phenyl ester, pmr*)
Perregaard, J. *et al*, *Recl. Trav. Chim. Pays-Bas*, 1974, **93**, 252 (*aryl esters*)
Sturtz, G. *et al*, *Synthesis*, 1980, 289 (*allyl ester, synth, use*)

Tetramethylphosphorodiamidic azide, 9CI — T-00219

Bis(dimethylamino)phosphoryl azide. Mazidox

[7219-78-5]

$(Me_2N)_2P(O)N_3$

$C_4H_6N_5OP$ M 171.098
Liq. Bp_2 93-94°. n_D^{21} 1.4673.

Vetter, H.-J., *Z. Naturforsch, B*, 1964, **19**, 168 (*synth*)
Buder, W. *et al*, *Z. Anorg. Allg. Chem.*, 1975, **418**, 72 (*synth, ir, raman*)

Tetramethylphosphorodiamidic chloride, 9CI — T-00220

Bis(dimethylamino) phosphoryl chloride. Bis(dimethylamino)phosphinic chloride

[1605-65-8]

$(Me_2N)_2P(O)Cl$

$C_4H_{12}ClN_2OP$ M 170.578
Liq. d_4^{20} 1.823. Bp_{30} 133-134°, Bp_{10} 110°, $Bp_{0.6}$ 79-82°. n_D^{25} 1.4661.

Nielsen, M.L. *et al*, *J. Phys. Chem.*, 1964, **68**, 152 (*P nmr*)
Cowley, A.H. *et al*, *J. Am. Chem. Soc.*, 1965, **87**, 4451 (*pmr*)
Keat, R. *et al*, *J. Chem. Soc.* (*A*), 1968, 703 (*pmr*)
Köttgen, D. *et al*, *Z. Anorg. Allg. Chem.*, 1971, **385**, 56 (*ir, raman*)
Osokin, D.Ya. *et al*, *Org. Magn. Reson.*, 1972, **4**, 831 (*nqr*)
Laskorin, B.N. *et al*, *Dokl. Akad. Nauk SSSR*, 1974, **215**, 595 (*Engl. transl.* p. 174) (*ir, P nmr, pmr*)
Zverev, V.V. *et al*, *Izv. Akad. Nauk SSSR, Ser. Khim.*, 1975, 1051 (*Engl. transl.* p. 961) (*pe*)
Poignant, S. *et al*, *Can. J. Chem.*, 1980, **58**, 946 (*pmr, P nmr*)
Traylor, P.S. *et al*, *J. Am. Chem. Soc.*, 1965, **87**, 553 (*props*)

Tetramethylphosphorodiamidic cyanide, 9CI — T-00221

Bis(dimethylamino)phosphoryl cyanide. Bis(dimethylamino)phosphinyl cyanide
[14445-60-4]

$(Me_2N)_2P(O)CN$

$C_5H_{12}N_3OP$ M 161.143
Liq. $Bp_{0.9}$ 91-94°. n_D^{20} 1.4518.

Larsson, L., *Acta Chem. Scand.*, 1952, **6**, 1470 (*synth, ir*)
Horn, H.-G. *et al*, *Z. Naturforsch., B*, 1966, **21**, 729 (*ir*)

Tetramethylphosphorodiamidodithioic acid, 9CI, 8CI — T-00222

Dithiophosphoric acid bis(dimethylamide)

$(Me_2N)_2P(S)SH$

$C_4H_{13}N_2PS_2$ M 184.254
Dimethylammonium salt: [77318-59-3]. Cryst. Mp 120-122°.
Me ester: [54975-91-6]. *Methyl tetramethylphosphorodiamidodithioate.*
$C_5H_{15}N_2PS_2$ M 198.281
Liq. d_4^{20} 1.13. Bp_1 90-93°. n_D^{20} 1.5495.
Et ester: [54975-93-8]. *Ethyl tetramethylphosphorodiamidodithioate.*
$C_6H_{17}N_2PS_2$ M 212.307
Liq. d_4^{20} 1.09. Bp_1 105-107°. n_D^{20} 1.5457.
Trimethylsilyl ester: [63853-21-4]. *Trimethylsilyl tetramethylphosphorodiamidodithioate.*
$C_7H_{21}N_3PS_2Si$ M 270.442
Liq. $Bp_{0.001}$ 82°.

Nifantev, É.E. *et al*, *Zh. Obshch. Khim.*, 1975, **45**, 295 (*Engl. transl.* p. 282) (*esters, pmr*)
Roesky, H.W. *et al*, *Z. Anorg. Allg. Chem.*, 1977, **431**, 221 (*trimethylsilyl ester, ir, ms, pmr, P nmr*)
Fluck, E. *et al*, *Z. Anorg. Allg. Chem.*, 1981, **473**, 51 (*synth, P nmr*)

Tetramethylphosphorodiamidoselenoic acid, 8CI — T-00223

Phosphoroselenoic acid bis(dimethylamide)

$(Me_2N)_2P(Se)OH \rightleftharpoons (Me_2N)_2P(O)SeH$

$C_4H_{13}N_2OPSe$ M 215.093
O-*Me ester:* [56595-16-5]. O-*Methyl tetramethylphosphorodiamidoselenoate.*
$C_5H_{15}N_2OPSe$ M 229.120
Solid or liq. Mp 18-19°. $Bp_{5.5}$ 89°.
Chloride: [25408-76-8].
$C_4H_{12}ClN_2PSe$ M 233.539
Liq. d_4^{20} 1.41. $Bp_{0.1}$ 69°. n_D^{20} 1.5508. Sensitive to air and moisture.

Nuretdinov, I.A. *et al*, *Izv. Akad. Nauk SSSR, Ser. Khim.*, 1969, 1535 (*Engl. transl.* p. 1423) (*chloride*)
Pohl, W. *et al*, *An. Quim.*, 1974, **70**, 1209; *CA*, **83**, 123600 (*methyl ester, synth, ir, raman*)

Tetramethylphosphorodiamidothioic azide, 9CI — T-00224

Bis(dimethylamino) thiophosphoryl azide. Bis(dimethylamino)phosphinothioic azide
[59998-82-2]

$(Me_2N)_2P(S)N_3$

$C_4H_6N_5PS$ M 187.159
Liq. d_4^{20} 1.16. $Bp_{1.5}$ 80°, $Bp_{0.01}$ 62°. n_D^{20} 1.5188.

Vetter, H.J., *Z. Naturforsch., B*, 1964, **19**, 168 (*synth, ir*)
Zaslavskaya, N.N. *et al*, *Izv. Akad. Nauk SSSR, Ser. Khim.*, 1976, 931 (*Engl. transl.* p. 911) (*synth, props*)

Tetramethylphosphorodiamidothioic chloride, 9CI — T-00225

Bis(dimethylamino)phosphinothioic chloride. Bis(dimethylamino) thiophosphoryl chloride
[3732-81-8]

$(Me_2N)_2P(S)Cl$

$C_4H_{12}ClN_2PS$ M 186.639
Mp 22°. Bp_{10} 104-105°, Bp_1 57-58°. n_D^{25} 1.5209.

Godovikov, N.N. *et al*, *Zh. Obshch. Khim.*, 1961, **31**, 1628 (*Engl. transl.* p. 1516) (*synth*)
Tolkmith, H., *J. Am. Chem. Soc.*, 1962, **84**, 2097 (*synth*)
Cowley, A.H. *et al*, *J. Am. Chem. Soc.*, 1965, **87**, 4454 (*pmr*)
Keat, R. *et al*, *J. Chem. Soc.* (*A*), 1968, 703 (*pmr*)
Köttgen, D. *et al*, *Z. Anorg. Allg. Chem.*, 1971, **385**, 56 (*ir, raman*)
Osokin, D.Ya. *et al*, *Org. Magn. Reson.*, 1972, **4**, 831 (*nqr*)
Vilesov, F.I. *et al*, *Z. Phys. Chem.* (*Leipzig*), 1974, **255**, 661 (*pe*)
Shagidullin, R.R. *et al*, *Dokl. Akad. Nauk SSSR, Ser. Sci. Khim.*, 1975, **222**, 897 (*Engl. transl.* p. 564) (*uv*)

Tetramethylphosphorodiamidothioic fluoride, 9CI, 8CI — T-00226

Bis(dimethylamino)phosphinothioyl fluoride
[918-47-8]

$(Me_2N)_2P(S)F$

$C_4H_{12}FN_2PS$ M 170.184
Bp_{11} 82°.
▷TD4375000.

Reddy, G.S. *et al*, *Z. Naturforsch., B*, 1970, **25**, 1199 (*P nmr*)
Köttgen, D. *et al*, *Z. Phys. Chem.* (*Frankfurt am Main*), 1974, **92**, 285 (*synth, ir, raman*)

Tetramethylphosphorodiamidous acid, 9CI — T-00227

$(Me_2N)_2POH \rightleftharpoons (Me_2N)_2P(O)H$

$C_4H_{13}N_2OP$ M 136.133
The free acid exists as the phosphoryl tautomer *N,N,N′,N′*-Tetramethylphosphonic diamide, T-00214
Me ester: [17166-16-4]. *Methyl tetramethylphosphorodiamidite.*
$C_5H_{15}N_2OP$ M 150.160
Liq. Bp_{24} 53-54°. n_D^{23} 1.4474.

Et ester: [3402-24-2]. *Ethyl tetramethylphosphorodiamidite.*
$C_6H_{17}N_2OP$ M 164.187
Liq. Bp_{10} 51°. n_D^{20} 1.4482.
Isopropyl ester: [36055-83-1]. *Isopropyl tetramethylphosphorodiamidite.*
$C_7H_{19}N_2OP$ M 178.214
Liq. Bp_{20} 61-62°.
tert-*Butyl ester:* [55666-82-5]. tert-*Butyl tetramethylphosphorodiamidite.*
$C_8H_{21}N_2OP$ M 192.240
Liq. Bp_{10} 54°. n_D^{19} 1.444.
Ph ester: [26546-75-8]. *Phenyl tetramethylphosphorodiamidite.*
$C_{10}H_{17}N_2OP$ M 212.231
Liq. $Bp_{0.1}$ 52-53°.
Benzyl ester: Benzyl tetramethylphosphorodiamidite.
$C_{11}H_{19}N_2OP$ M 226.258
Liq. $Bp_{0.01}$ 60-63°.
Fluoride: [1735-82-6].
$C_4H_{12}FN_2P$ M 138.124
Ligand for Co and Ni. Liq. Mp −89°. Bp 126°. Stable *in vacuo* for several days at r.t.
Chloride: see Tetramethylphosphorodiamidous chloride, T-00228
Bromide: [20502-36-7].
$C_4H_{12}BrN_2P$ M 199.030
Liq. d_4^{20} 1.37. Mp −13°. Bp 214° dec., Bp_2 45°. n_D^{20} 1.5522.
Cyanide:
$C_5H_6N_3P$ M 139.096
Ligand for Ag. Solid. Mp 21-23°. Bp_1 41°. n_D^{21} 1.4716.

Burgeda, R. *et al*, *Bull. Soc. Chim. Fr.*, 1963, 2154 (*tert*-butyl ester, synth, pmr, ir)
Nöth, H. *et al*, *Chem. Ber.*, 1963, **96**, 1109, 1816 (*bromide, cyanide*)
Houalla, D. *et al*, *Bull. Soc. Chim. Fr.*, 1965, 2368 (*ethyl ester, synth, P nmr*)
Mitsch, R.A., *J. Am. Chem. Soc.*, 1967, **89**, 6297 (*methyl ester*)
Bentrude, W.G., *J. Org. Chem.*, 1972, **37**, 462 (*methyl, phenyl esters, synth, pmr*)
Fleming, S. *et al*, *Inorg. Chem.*, 1972, **11**, 1 (*fluoride, synth, ir, pmr, complexes*)
Osokin, D.Ya. *et al*, *Org. Magn. Reson.*, 1972, **4**, 831 (*nqr*)
Cowley, A.H. *et al*, *J. Am. Chem. Soc.*, 1973, **95**, 6506 (*fluoride, pe*)
Hargis, J.H. *et al*, *J. Am. Chem. Soc.*, 1974, **96**, 5927 (*methyl ester, use*)
Barlos, K. *et al*, *Z. Naturforsch.*, *B*, 1978, **33**, 515 (*bromide, N nmr*)
Gouesnard, J.-P. *et al*, *Can. J. Chem.*, 1980, **58**, 1295 (*bromide, P nmr, pmr*)

Tetramethylphosphorodiamidous chloride, 9CI T-00228

Bis(dimethylamino)chlorophosphine. N,N,N′,N′-*Tetramethyldiaminophosphorochloridite*
[3348-44-5]

$$(Me_2N)_2PCl$$

$C_4H_{12}ClN_2P$ M 154.579
Liq. d_4^{20} 1.070. Mp −33°. Bp_{726} 184°, Bp_{10} 64°. n_D^{20} 1.495.
▷Inflames on standing in air. Reacts explosively with H_2O

Nöth, H. *et al*, *Chem. Ber.*, 1963, **96**, 1109 (*synth*)
v. Wazer, J.R. *et al*, *J. Am. Chem. Soc.*, 1964, **86**, 811 (*P nmr*)
Wolf, R. *et al*, *Bull. Soc. Chim. Fr.*, 1965, 2368 (*pmr, ir*)
Osokin, D.Ya. *et al*, *Org. Magn. Reson.*, 1972, **4**, 831 (*nqr*)
Whitesides, G. *et al*, *J. Am. Chem. Soc.*, 1974, **96**, 5398 (*synth*)
Lappert, M.F. *et al*, *J. Chem. Soc., Dalton Trans.*, 1975, 1207 (*pe*)
Barlos, K. *et al*, *Z. Naturforsch*, *B*, 1978, **33**, 515 (*N nmr*)
Bulloch, G. *et al*, *J. Chem. Soc., Dalton Trans.*, 1978, 764 (*cmr, pmr*)
Gouesnard, J-P. *et al*, *Can. J. Chem.*, 1980, **58**, 1295 (*N and P nmr*)

Tetramethyl pyrophosphate, 8CI T-00229

Tetramethyl diphosphate, 9CI
[690-49-3]

$$(MeO)_2P(O)OP(O)(OMe)_2$$

$C_4H_{12}O_7P_2$ M 234.082
Liq. d_4^{25} 1.36. $Bp_{0.5}$ 114-116°, $Bp_{0.05}$ 105°. n_D^{25} 1.4121.
▷Highly toxic, LD_{50} 2 mg/kg (mice). UX7175000.

Toy, A.D.F., *J. Am. Chem. Soc.*, 1948, **70**, 3882 (*synth, props*)
Cheymol, J. *et al*, *C. R. Hebd. Seances Acad. Sci.*, 1960, **251**, 1171 (*synth, props*)
Mowthorpe, D.J. *et al*, *Spectrochim. Acta, Part A*, 1967, **23**, 451 (*pmr, nmr*)
Schep, R.A. *et al*, *Inorg. Chem.*, 1973, **12**, 2711 (*synth*)

4′,4′,5′,5′-Tetramethylspiro[1,3,2-benzodioxaphosphole-2,2′-[1,3,2]-dioxaphospholane], 9CI T-00230

[33706-78-4]

$C_{12}H_{17}O_4P$ M 256.238
Parent compd. exists mainly (99%) as pentacoordinate tautomer at 100°.
2-Methoxy: [57301-51-6].
$C_{13}H_{19}O_5P$ M 286.264
Solid. Mp 80-81.5°.
2-Phenoxy: [54622-74-1].
$C_{18}H_{21}O_5P$ M 348.335
Cryst. (pet. ether). Mp 81-81.5°. Easily hydrolysed.
2-Phenylthio: [54622-76-3].
$C_{18}H_{21}O_4PS$ M 364.395
Cryst. (pet. ether). Mp 136-137°.
2-Phenylseleno: [81236-66-0].
$C_{18}H_{21}O_4PSe$ M 411.295
Cryst. (pet. ether). Mp 124°.
2-Dimethylamino: N,N,*4′,4′,5′,5′-Hexamethylspiro[1,3,2-benzodioxaphosphole-2,2′-[1,3,2]-dioxaphospholan]-2-amine.*
$C_{14}H_{22}NO_4P$ M 299.306
Solid. Mp 104.5-106.5°.

Bernard, D. *et al*, *J. Organomet. Chem.*, 1973, **47**, 113 (*synth, P nmr, struct*)
Bone, S.A. *et al*, *J. Chem. Soc., Perkin Trans. 1*, 1974, 2125; 1976, 156 (*derivs, synth, pmr, P nmr*)
Antczak, S. *et al*, *J. Chem. Soc., Perkin Trans. 1*, 1977, 278 (*derivs*)
Johnson, M.P. *et al*, *J. Chem. Soc., Perkin Trans. 1*, 1981, 3074 (*deriv, pmr, P nmr, struct*)

4′,4′,5′,5′-Tetramethylspiro[1,3,2-benzoxazaphosphole-2(3*H*),2′-[1,3,2]dioxaphospholane], 8CI T-00231

[51675-70-8]

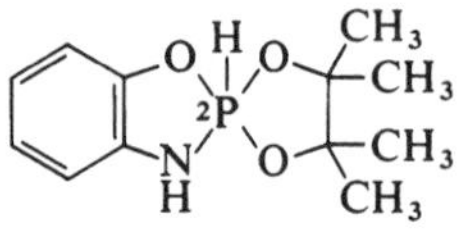

$C_{12}H_{18}NO_3P$ M 255.253

Parent compd. exists completely in pentaco-ordinate form.

2-Methoxy: [69775-02-6].

$C_{13}H_{20}NO_4P$ M 285.279

Solid. Mp 123-124°.

2-Phenoxy: [69775-03-7].

$C_{18}H_{22}NO_4P$ M 347.350

Solid. Mp 164-165°.

2-Dimethylamino: [69775-01-5]. *4,4,5,5,*N,N*-Hexamethylspiro[1,3,2-benzoxazaphosphole-2(3*H*),2′-[1,3,2]dioxaphospholan]-2-amine.*

$C_{14}H_{23}N_2O_3P$ M 298.321

Solid. Mp 144-145°.

Burgada, R. *et al*, *J. Organomet. Chem.*, 1974, **66**, 255 (*synth, struct*)

Singh, S. *et al*, *J. Chem. Soc., Perkin Trans. 1*, 1978, 1438 (*derivs, synth, P nmr*)

3,3,3′,3′-Tetramethyl-1,1′(3*H*,3′*H*)-spirobi[2,1-benzoxaphosphole], 10CI T-00232

[68823-65-4]

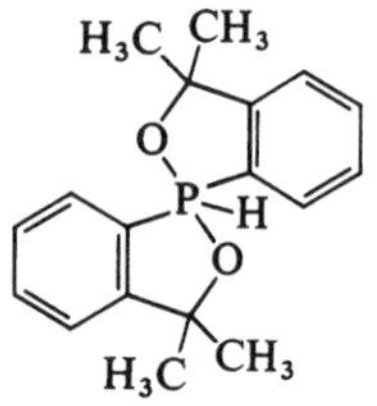

$C_{18}H_{21}O_2P$ M 300.336

Solid. Mp 99°. Forms Na and Li derivs.

1-Ph: [71433-47-1].

$C_{24}H_{25}O_2P$ M 376.434

Cryst. (MeOH). Mp 195°.

1-Benzyl:

$C_{25}H_{27}O_2P$ M 390.461

Cryst. (pentane). Mp 118°.

1-Chloro: [71401-87-1].

$C_{18}H_{20}ClO_2P$ M 334.782

Unstable. Possesses ionic, phosphonium struct.

Granoth, I. *et al*, *J. Am. Chem. Soc.*, 1979, **101**, 4618, 4623 (*synth, derivs, ir, ms, pmr, P nmr*)

Granoth, I. *et al*, *J. Chem. Soc., Perkin Trans. 1*, 1982, 735 (*phenyl, ms, cmr, pmr, P nmr*)

3,3,3′,3′-Tetramethyl-1,1′(3*H*,3′*H*)-spirobi[2,1-benzoxaphospholium](1+), 9CI T-00233

[68823-62-1]

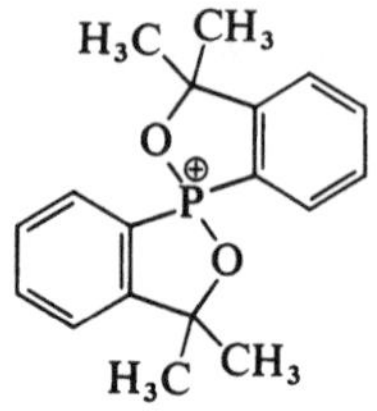

$C_{18}H_{20}O_2P^{\oplus}$ M 299.329 (ion)

Chloride:

$C_{18}H_{20}ClO_2P$ M 334.782

Thermally unstable.

Trifluoromethanesulfonate: [68823-63-2].

$C_{19}H_{20}F_3O_5PS$ M 448.393

Cryst. (THF/Et_2O). Mp 153°. Nonhygroscopic but easily hydrolysed.

Granoth, I. *et al*, *J. Am. Chem. Soc.*, 1979, **101**, 4618 (*synth, props, P nmr*)

1,3,5,7-Tetramethyl-1,3,5,7-tetraaza-4-phospha(4-*P*^V)spiro[3.3]heptane-2,6-dione, 9CI T-00234

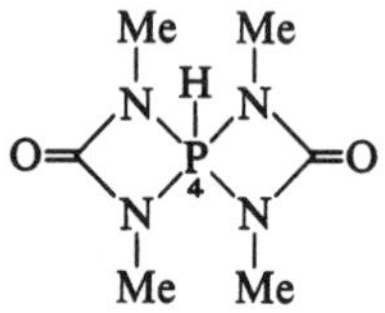

$C_6H_{13}N_4O_2P$ M 204.168

4-Chloro: [77507-70-1].

$C_6H_{12}ClN_4O_2P$ M 238.613

Cryst. (MeCN). Mp 115-117°.

4-Ph: [74411-89-5].

$C_{12}H_{17}N_4O_2P$ M 280.266

Solid. Mp 156-157° dec.

4-Azido: [89982-28-5].

$C_6H_{12}N_7O_2P$ M 245.180

Cryst. Mp 100-104°.

Roesky, H.W. *et al*, *Chem. Ber.*, 1980, **113**, 1847; 1981, **114**, 1554 (*phenyl, chloro, synth, ir, ms, pmr, P nmr*)

Baceiredo, A. *et al*, *J. Am. Chem. Soc.*, 1984, **106**, 7065 (*azido, synth, ir, cmr, pmr, P nmr*)

2,2,3,3-Tetramethyl-1,4,6,9-tetraoxa-5-phospha(5-*P*^V)spiro[4.4]nonane, 9CI, 8CI T-00235

[18389-64-5]

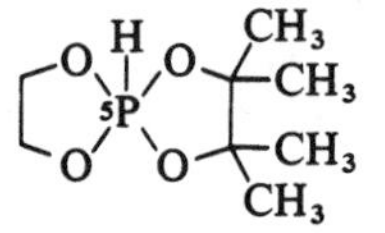

$C_8H_{17}O_4P$ M 208.194

Tautomeric with monocyclic form having tervalent phosphorus. Contains 81% pentaco-ordinate form at r.t. Liq. $Bp_{0.0001}$ 55°. n_D^{20} 1.455.

5-Phenoxy: [57301-53-8].

$C_{14}H_{21}O_5P$ M 300.291

Liq. $Bp_{0.2}$ 120°.

5-Dimethylamino: N,N*,2,2,3,3-Hexamethyl-1,4,6,9-tetraoxa-5-phospha(5-P^V)spiro[4.4]nonan-5-amine.*

$C_{10}H_{22}NO_4P$ M 251.262

Solid. Mp 37-39°.

Sanchez, M. *et al*, *Bull. Soc. Chim. Fr.*, 1968, 773 (*synth, P nmr*)

Burgada, R. *et al*, *Tetrahedron*, 1971, **27**, 5833 (*synth, P nmr*)

Bernard, D. *et al*, *J. Organomet. Chem.*, 1973, **47**, 113 (*P nmr, struct*)

Houalla, D. *et al*, *Org. Magn. Reson.*, 1974, **6**, 340 (*pmr*)

Bone, S.A. *et al*, *J. Chem. Soc., Perkin Trans. 1*, 1976, 156 (*derivs, synth, pmr, P nmr*)

3,3,9,9-Tetramethyl-1,5,7,11-tetraoxa-6-phospha(6-P^V)spiro[5.5]undecane T-00236

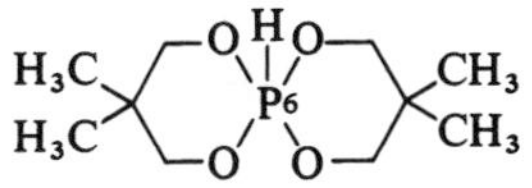

$C_{10}H_{21}O_4P$ M 236.247

6-Ethoxy: [34736-77-1].
$C_{12}H_{25}O_5P$ M 280.300
Liq. $Bp_{0.1}$ 86-90°.

6-Ph: [34736-78-2].
$C_{16}H_{25}O_4P$ M 312.345
Solid. Mp 69-72° (61.5-63°).

6-Benzyl:
$C_{17}H_{27}O_4P$ M 326.372
Solid. Mp 61.5-63°.

Denny, D.B. *et al*, *J. Am. Chem. Soc.*, 1971, **93**, 4004 (*derivs, synth, pmr, P nmr*)
Antczak, S. *et al*, *J. Chem. Soc., Perkin Trans. 1*, 1977, 278 (*phenyl, synth, pmr, P nmr*)
Bone, S.A. *et al*, *J. Chem. Soc., Perkin Trans. 1*, 1977, 437 (*derivs, struct*)

2,4,6,7-Tetramethyl-2,6,7-triaza-1-phosphabicyclo[2.2.2]octane, 9CI, 8CI T-00237

[14418-26-9]

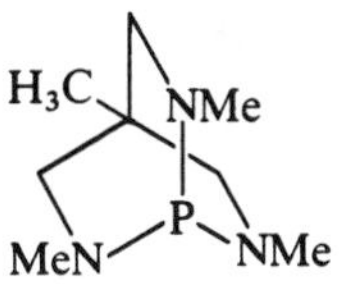

$C_8H_{18}N_3P$ M 187.224
Ligand for metals of Group VI B. Liq. Bp_{10} 82-86°.

1-Oxide: [15199-21-0].
$C_8H_{18}N_3OP$ M 203.223
Cryst. (EtOH). Mp 97-99°.

1-Sulfide: [15199-22-1].
$C_8H_{18}N_3PS$ M 219.284
Cryst. Mp 92.5-94.5°. Bp_{vac} 50° subl.

1-Selenide: [68378-99-4].
$C_8H_{18}N_3PSe$ M 266.184
Solid. Mp 118-120°. $Bp_{0.01}$ 60° subl.

Laube, B.L. *et al*, *Inorg. Chem.*, 1967, **6**, 173 (*synth, derivs, pmr*)
Mosbo, J.A. *et al*, *J. Magn. Reson.*, 1972, **8**, 243 (*pmr, complexes*)
Clardy, J.C. *et al*, *Phosphorus*, 1974, **4**, 133 (*synth, oxide, cryst struct*)
Cowley, A.H. *et al*, *Inorg. Chem.*, 1977, **16**, 854; 1982, **21**, 543 (*pe*)
Kroshefsky, R.D. *et al*, *Inorg. Chem.*, 1979, **18**, 469 (*selenide, synth, P nmr*)
Kroshefsky, R.D. *et al*, *Phosphorus Sulfur*, 1979, **6**, 391, 397 (*ir, cmr, derivs*)

2,2,3,3-Tetramethyl-1,4,6-trioxa-9-aza-5-phospha(5-P^V)spiro[4.4]nonane T-00238

[18389-65-6]

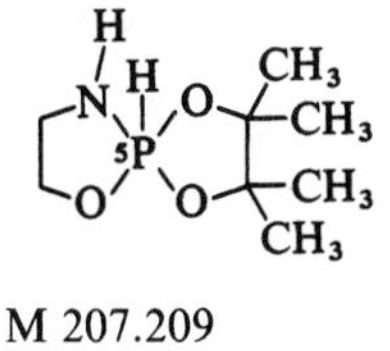

$C_8H_{18}NO_3P$ M 207.209
Tautomeric with tricoordinate form (9% at 25°, 50% at 100°). Liq. d_4^{20} 1.14. $Bp_{0.05}$ 80°. n_D^{20} 1.474.

5-Dimethylamino: [75386-25-3]. N,N,*2,2,3,3-Hexamethyl-1,4,6-trioxa-9-aza-5-phospha(5-*P^V*)-spiro[4.4]nonan-5-amine.*
$C_{10}H_{23}N_2O_3P$ M 250.277
Liq. $Bp_{0.08}$ 85-86°. n_D^{20} 1.4730.

Sanchez, M. *et al*, *Bull. Soc. Chim. Fr.*, 1968, 773 (*synth, ir*)
Burgada, R. *et al*, *J. Organomet. Chem.*, 1974, **66**, 255 (*P nmr, struct*)
Houalla, D. *et al*, *Org. Magn. Reson.*, 1974, **6**, 340 (*pmr*)
Gonçalves, H. *et al*, *Phosphorus Sulfur*, 1980, **8**, 147 (*derivs, pmr, P nmr, ir*)

3,3,5,5-Tetramethyl-1,2,4-triphenyl-1,2,4-triphospholane, 9CI T-00239

[66089-08-5]

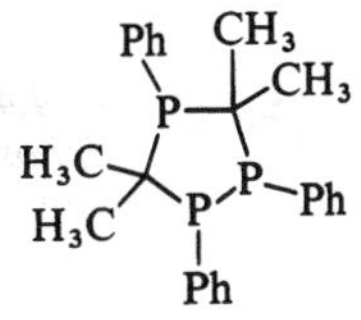

$C_{24}H_{27}P_3$ M 408.399

Baudler, M. *et al*, *Z. Naturforsch., B*, 1977, **32**, 1490 (*P nmr*)

***O,O,O',O'*-Tetramethyl trithiopyrophosphate** T-00240

Tetramethyl thiodiphosphate, 9CI. O,O-Dimethyl phosphorodithioate anhydrosulfide

[5930-73-4]

$(MeO)_2P(S)—S—P(S)(OMe)_2$

$C_4H_{12}O_4P_2S_3$ M 282.264
Constit. of crude *O,O*-Dimethyl phosphorodithioate, D-00898 . Low melting solid. Mp 34°. $Bp_{0.07}$ 108.5-111°.

Schrader, G. *et al*, *Angew. Chem.*, 1958, **70**, 692 (*synth, props*)
Mel'nikov, N.N. *et al*, *Zh. Obshch. Khim.*, 1960, **30**, 2319 (*Engl. transl.* p. 2300) (*synth*)
Lippman, A.E., *J. Org. Chem.*, 1966, **31**, 471 (*P nmr*)

Tetramyristoyldiphosphatidylglycerol T-00241

Cardiolipin

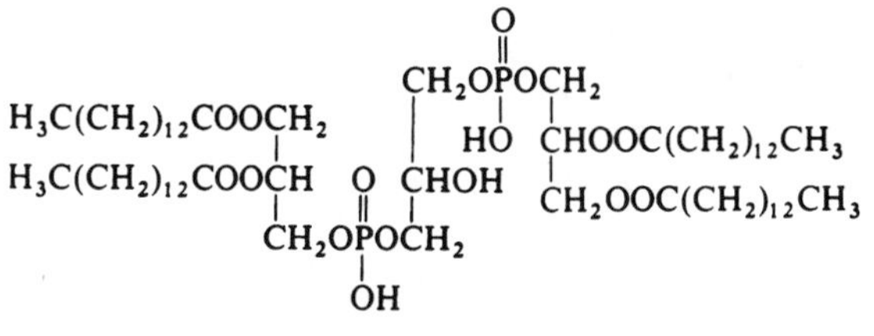

$C_{65}H_{126}O_{17}P_2$ M 1241.648

NH_4 salt: Dihydrate. Mp 181-182°. $[\alpha]_D^{24}$ +6.90° (c, 2.75 in $CHCl_3$).

Ramirez, F. *et al*, *Synthesis*, 1976, 769 (*synth*)

Tetra-9-octadecenoylphosphatidylglycerol **T-00242**

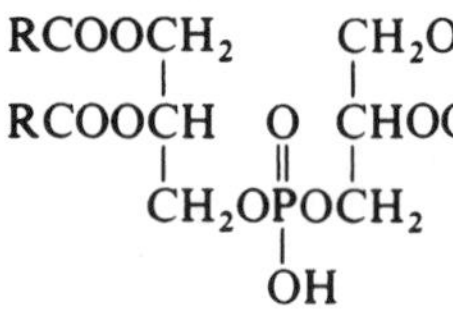

```
RCOOCH2     CH2OOCR
   |           |
RCOOCH  O   CHOOCR
   |    ||     |
   CH2OPOCH2
        |
        OH
```

R = $CH_3(CH_2)_7CH{=}CH(CH_2)_7$—

$C_{78}H_{143}O_{12}P$ M 1303.954

(all-Z)-form [17708-93-9]

Tetraoleoylphosphatidylglycerol. Olein 1,2-dihydrogenphosphate

$[\alpha]_D$ +6.1° (c, 10 in $CHCl_3$).

Baer, E. *et al*, *Arch. Biochem. Biophys.*, 1958, **78**, 294.
Baer, E., *Prog. Chem. Fats Other Lipids*, 1963, **6**, 33.

1,4,6,12-Tetraoxa-5-phospha(5-P^V)-spiro[4.7]dodecane, 10CI **T-00243**

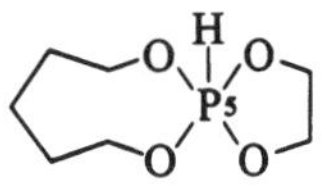

$C_7H_{15}O_4P$ M 194.167

5-Ph: [71559-34-7].

$C_{13}H_{19}O_4P$ M 270.264

First known spirophosphorane with 8-membered ring. Oil.

Cadogan, J.I.G. *et al*, *J. Chem. Soc., Chem. Commun.*, 1979, 191 (*synth, P nmr*)

1,4,6,9-Tetraoxa-5-phospha(5-P^V)-spiro[4.4]nonane, 9CI **T-00244**

[3646-10-4]

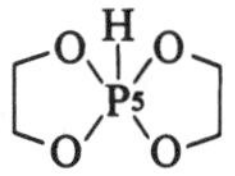

$C_4H_9O_4P$ M 152.086

Consists of 1:1 mixt. of tri- and pentacoordinate tautomers at 100°.

5-Ph:

$C_{10}H_{13}O_4P$ M 228.184

Solid. Mp 123°.

5-Methoxy: [61890-82-2].

$C_5H_{11}O_5P$ M 182.113

Cryst. $Bp_{0.1}$ 90-98°.

5-Ethoxy:

$C_6H_{13}O_5P$ M 196.139

Solid. Mp 36-38°.

Burgada, R. *et al*, *Tetrahedron*, 1971, **27**, 5833 (*synth, P nmr*)
Bernard, D. *et al*, *J. Organomet. Chem.*, 1973, **47**, 113 (*P nmr, struct*)
Houalla, D. *et al*, *Org. Magn. Reson.*, 1974, **6**, 340 (*pmr*)
Munoz, A. *et al*, *Bull. Soc. Chim. Fr. Part II*, 1974, 2193 (*struct*)
Malavaud, C. *et al*, *Tetrahedron Lett.*, 1975, 497 (*phenyl, P nmr*)
Laurenço, C. *et al*, *Tetrahedron*, 1976, **32**, 2253 (*methoxy, pmr, P nmr*)
Brierley, J. *et al*, *J. Chem. Soc., Perkin Trans. 1*, 1977, 273 (*derivs, pmr, P nmr*)
Antczak, S. *et al*, *J. Chem. Soc., Perkin Trans. 1*, 1977, 279 (*ethoxy, P nmr*)

1,4,6,11-Tetraoxa-5-phospha(5-P^V)-spiro[4.6]undecane, 10CI **T-00245**

[71990-10-8]

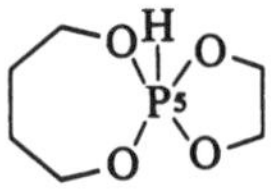

$C_6H_{13}O_4P$ M 180.140

5-Ph: [71559-33-6].

$C_{12}H_{17}O_4P$ M 256.238

First known spirophosphorane with seven membered ring. Solid. Mp 70-72°.

Cadogan, J.I.G. *et al*, *J. Chem. Soc., Chem. Commun.*, 1979, 191 (*synth, P nmr*)

Tetrapalmitoyldiphosphatidylglycerol **T-00246**

Cardiolipin

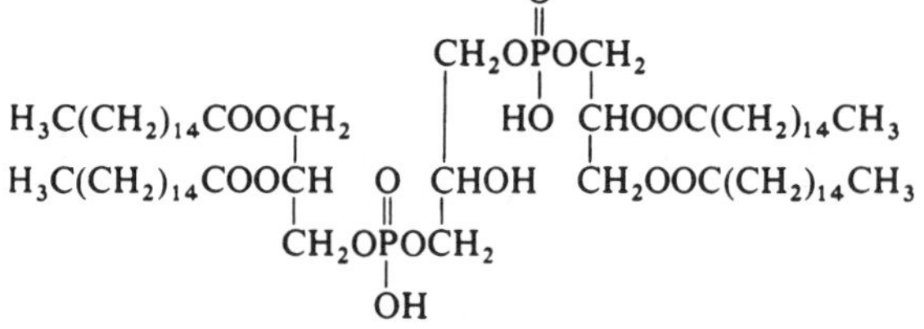

```
                              O
                              ||
                          CH2OPOCH2
                           |  |  |
H3C(CH2)14COOCH2           |  HO CHOOC(CH2)14CH3
           |               |     |
H3C(CH2)14COOCH  O  CHOH   CH2OOC(CH2)14CH3
           |     ||  |
           CH2OPOCH2
                |
                OH
```

$C_{73}H_{142}O_{17}P_2$ M 1353.862

NH_4 salt: Dihydrate. Mp 177-178°. $[\alpha]_D^{24}$ +6.35° (c, 2.75 in $CHCl_3$).

Ramirez, F. *et al*, *Synthesis*, 1976, 769 (*synth*)

Tetrapalmitoylphosphatidylglycerol **T-00247**

Tetrahexadecanoylphosphatidylglycerol. Bis(1,2-dihexadecanoylglycero)-3-phosphate

[84905-98-6]

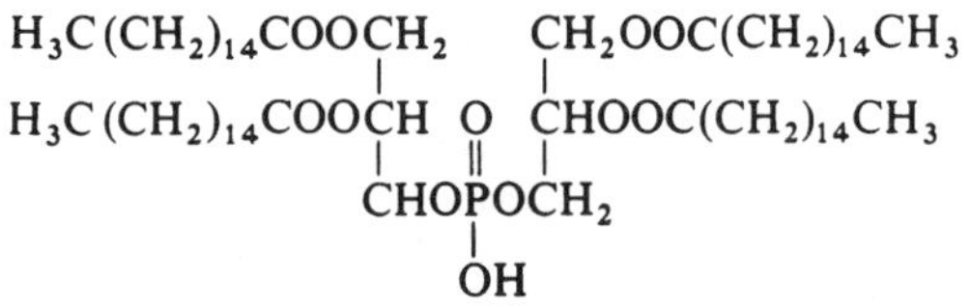

```
H3C(CH2)14COOCH2    CH2OOC(CH2)14CH3
           |          |
H3C(CH2)14COOCH  O  CHOOC(CH2)14CH3
           |     ||   |
           CHOPOCH2
                |
                OH
```

$C_{70}H_{135}O_{12}P$ M 1199.803

Cryst. (C_6H_6/Me_2CO). Mp 62-63°. $[\alpha]_D^{23}$ +6.7° (c, 4 in C_6H_6).

Baer, E., *J. Biol. Chem.*, 1952, **198**, 853.
Baer, E., *Prog. Chem. Fats Other Lipids*, 1963, **6**, 33.
Dang, Q.Q. *et al*, *Chem. Phys. Lipids*, 1983, **33**, 33 (*synth, nmr*)

2,2′,5,5′-Tetraphenyl-1*H*,1′*H*-biphosphole **T-00248**

[83085-52-3]

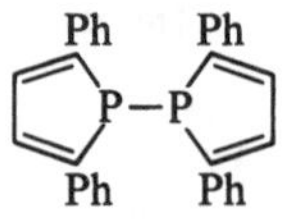

$C_{32}H_{24}P_2$ M 470.489

Orange cryst. (toluene/hexane). Mp 151°.

Charrier, C. *et al*, *J. Organomet. Chem.*, 1982, **231**, 361 (*synth, ms, pmr, P nmr*)

2,2′,6,6′-Tetraphenyl-4,4′-biphosphorine, 8CI T-00249

2,2′,6,6′-Tetraphenyl-4,4′-biphosphinine

[26092-87-5]

$C_{34}H_{24}P_2$ M 494.511

Solid. Mp 236-238° (sealed tube).

Märkl, G. *et al, Tetrahedron Lett.*, 1970, 645 (*synth, uv, pmr*)

3,5,8,10-Tetraphenyl-1*H*,5*H*-bisphospholo[1,2-*a*:1′,2′-*d*][1,4]diphosphorin, 9CI T-00250

[55153-57-6]

$C_{34}H_{26}P_2$ M 496.527

Cryst. ($CHCl_3$/EtOH). Mp 242-243°.

Märkl, G. *et al, Phosphorus*, 1974, **4**, 279 (*synth, uv, pmr, ms*)

2,3,5,6-Tetraphenyl-1,4-bis[(trimethylsilyl)-oxy]-2,3,5,6-tetraphosphabicyclo[2.2.0]-hexane, 10CI T-00251

[71838-20-5]

$C_{32}H_{38}O_2P_4Si_2$ M 634.717

Solid. Mp 148-150°.

Appel, R. *et al, Angew. Chem., Int. Ed. Engl.*, 1979, **18**, 469, 872 (*synth, nmr, cryst struct*)

(Tetraphenyl-2,4-cyclopentadien-1-ylidene)triphenylphosphorane, 9CI T-00252

1,2,3,4-Tetraphenyl-5-(triphenylphosphoranylidene)-1,3-cyclopentadiene

[15096-99-8]

$C_{47}H_{35}P$ M 630.767

Yellow cryst. ylide (DMF). Mp 306-308°. With HCl, gives 2:1 adduct Mp 232-3°.

Regitz, M. *et al, Tetrahedron*, 1967, **23**, 2701 (*synth, uv*)

1,2,3,4-Tetraphenyl-1,3,2,4-diazadiphosphetidine, 9CI T-00253

Diphenylphosphine imide dimer. Tetraphenylcyclodiphosphazane

$C_{24}H_{20}N_2P_2$ M 398.383

2,4-Dioxide:

$C_{24}H_{20}N_2O_2P_2$ M 430.382

Cryst. (C_6H_6). Mp 290°.

2,4-Disulfide: [13789-75-8].

$C_{24}H_{20}N_2P_2S_2$ M 462.503

Solid. Mp 220-221°.

trans-form

2,4-Disulfide: [40760-48-3]. Solid. Mp 265°.

Michaelis, A. *et al, Justus Liebigs Ann. Chem.*, 1915, **407**, 316 (*dioxide*)

Mel'nikov, N.N. *et al, Zh. Obshch. Khim.*, 1967, **37**, 239 (*Engl. transl.* p. 222) (*disulfide, synth, ir, uv, raman*)

Granzhan, V.A. *et al, Zh. Obshch. Khim.*, 1969, **39**, 1501 (*Engl. transl.* p. 1470) (*disulfide, config*)

Petersen, M.B. *et al, J. Chem. Soc., Dalton Trans.*, 1973, 106 (*disulfide, cryst struct*)

2,4,5,6-Tetraphenyl-1,3,5-dioxaphosphorinane, 9CI T-00254

2,4,5,6-Tetraphenyl-1,3,5-dioxaphosphinane. 2,4,5,6-Tetraphenyl-1,3-dioxa-5-phosphacyclohexane

[4722-60-5]

$C_{27}H_{23}O_2P$ M 410.451

Cryst. (MeOH or MeCN). Mp 205° (195-198°).

5-Oxide: [72564-40-0].

$C_{27}H_{23}O_3P$ M 426.451

Cryst. (MeCN). Mp 248-249°, Mp 275-278°.

5-Sulfide: [72564-41-1].

$C_{27}H_{23}O_2PS$ M 442.511

Cryst. (MeCN). Mp 223°.

Epstein, M. *et al, Tetrahedron*, 1962, **18**, 1231 (*synth, oxide*)

Arbuzov, B.A. *et al, Izv. Akad. Nauk SSSR, Ser. Khim.*, 1979, 2136 (*Engl. transl.* p. 1966) (*synth, derivs, P nmr, pmr, stereochem*)

4,7,13,16-Tetraphenyl-1,10-dioxa-4,7,13,16-tetraphosphacyclooctadecane, 10CI T-00255

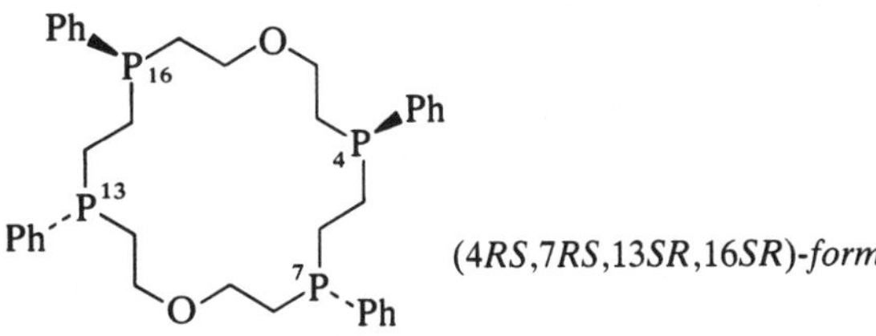

$C_{26}H_{44}O_2P_4$ M 512.527

Exists in three meso and two racemic modifications. Forms Ni and Co complexes.

(4RS,7RS,13SR,16SR)-form [74807-53-7]
Form least sol. in C_6H_6. Mp 202-203°.

(4RS,7RS,13RS,16SR)-form [74783-07-6]
Solid. Mp 112-114°. Three remaining forms have Mps 165-166°, 172-174°, 112-114°; stereochem. unassigned.

Crampoline, M. *et al*, *Inorg. Chem.*, 1982, **21**, 489 (*synth, nmr, complexes, cryst struct*)

Tetraphenyl diphosphate, 9CI T-00256

Tetraphenyl pyrophosphate, 8CI
[10448-49-4]

$(PhO)_2P(O)-O-P(O)(OPh)_2$

$C_{24}H_{20}O_7P_2$ M 482.365
Useful phosphorylating agent for alcohols and amines. Thick liq. $Bp_{0.1}$ 176-180°. n_D^{20} 1.5610.

Mason, H.S. *et al*, *J. Chem. Soc.*, 1951, 2267 (*synth*)
Corby, N.S. *et al*, *J. Chem. Soc.*, 1952, 1234 (*synth*)
Atkinson, R.E. *et al*, *J. Chem. Soc. (C)*, 1967, 1356 (*synth, props*)
Ramirez, F. *et al*, *J. Org. Chem.*, 1983, **48**, 847 (*P nmr*)

1,1,3,3-Tetraphenyl-1,3-diphosphepanium, 10CI T-00257

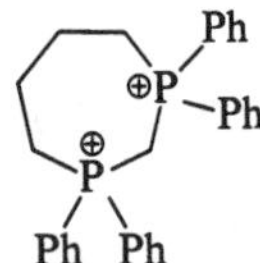

$C_{29}H_{30}P_2^{⊕⊕}$ M 440.504 (ion)
Dibromide: [74144-33-5].
$C_{29}H_{30}Br_2P_2$ M 600.312
Solid. Mp 232° dec.

Schmidbaur, H. *et al*, *Z. Naturforsch., B*, 1982, **37**, 677 (*synth, pmr, P nmr*)

Tetraphenyldiphosphine, 9CI T-00258

Tetraphenylbiphosphine
[1101-41-3]

Ph_2PPPh_2

$C_{24}H_{20}P_2$ M 370.370
Air-sensitive cryst. (C_6H_6/Et_2O). Mp 125-126°. Bp_1 258-260°. Cleaved by MeI, yields Ph_2P· radicals on uv irradn.

Mono-oxide: [2096-83-5].
$C_{24}H_{20}OP_2$ M 386.369
Cryst. (Me_2CO aq.), stable in air for short periods. Sl. sol. cold C_6H_6, toluene, EtOH, spar. sol. Et_2O. Mp 157-160°. Cleaved by Br_2 and by aromatic carboxylic acids.

Dioxide: see Tetraphenyldiphosphine dioxide, T-00259

Monosulfide: [26978-37-0].
$C_{24}H_{20}P_2S$ M 402.430
Cryst. (ligroin). Mp 138°.

Disulfide: see Tetraphenyldiphosphine disulfide, T-00260

Mono-oxide, monosulfide:
$C_{24}H_{20}OP_2S$ M 418.429
Needles (Me_2CO aq.). Mp 166-170°.

Kuchen, W. *et al*, *Chem. Ber.*, 1958, **91**, 2871; 1965, **98**, 480 (*synth*)
Niebergall, H. *et al*, *Chem. Ber.*, 1962, **95**, 64 (*synth*)
Issleib, K. *et al*, *Chem. Ber.*, 1962, **95**, 375 (*synth*)
Chatt, J. *et al*, *J. Chem. Soc.*, 1964, 1005 (*synth, complexes*)
McKechnie, J. *et al*, *J. Chem. Soc.*, 1965, 3500 (*props, derivs, ir*)
Davidson, R.S. *et al*, *J. Chem. Soc. (C)*, 1967, 1547 (*props*)
Highsmith, R.E. *et al*, *Inorg. Chem.*, 1968, **7**, 1740 (*synth*)
Matschiner, H. *et al*, *Z. Anorg. Allg. Chem.*, 1969, **371**, 256 (*monosulfide*)
Spanier, E.J. *et al*, *J. Am. Chem. Soc.*, 1970, **92**, 3348 (*synth, esr*)
Aime, S. *et al*, *J. Chem. Soc., Dalton Trans.*, 1976, 2144 (*nmr*)
Harris, R.K. *et al*, *J. Chem. Soc., Faraday Trans. 2*, 1976, **72**, 2291 (*nmr*)
Appel, R. *et al*, *Chem. Ber.*, 1977, **110**, 376 (*synth, nmr*)
Troy, D. *et al*, *Bull. Soc. Chim. Fr. Part II*, 1979, 241 (*uv*)
Alagna, L. *et al*, *Z. Naturforsch., A*, 1981, **36**, 68 (*pe*)

Tetraphenyldiphosphine dioxide, 9CI, 8CI T-00259

[1054-59-7]

$Ph_2P(O)-P(O)Ph_2$

$C_{24}H_{20}O_2P_2$ M 402.368
Cryst. (toluene or Me_2CO). Mp 167°, 179.5-181.5°. Easily oxidized by peroxides to diphenylphosphinic acid anhydride. Dec. rapidly when in contact with moisture.

Mallion, K.B. *et al*, *J. Chem. Soc.*, 1964, 6121 (*synth*)
McKechnie, J. *et al*, *J. Chem. Soc.*, 1965, 3500 (*ir*)
Quin, L.D. *et al*, *J. Org. Chem.*, 1966, **31**, 1206 (*synth*)
Spanier, E.J. *et al*, *J. Am. Chem. Soc.*, 1970, **92**, 3348 (*synth*)
Emoto, T. *et al*, *Bull. Chem. Soc. Jpn.*, 1973, **46**, 898 (*props*)
Alagna, L. *et al*, *Z. Naturforsch., A*, 1981, **36**, 68 (*pe*)
Blake, A.J. *et al*, *J. Mol. Struct.*, 1982, **78**, 265 (*ir, raman, cryst struct*)

Tetraphenyldiphosphine disulfide, 9CI, 8CI T-00260

[1054-60-0]

$Ph_2P(S)-P(S)Ph_2$

$C_{24}H_{20}P_2S_2$ M 434.490
Cryst. (EtOH). Mp 168-169°. Possesses the *anti*-conformation in the cryst.

Kuchen, W. *et al*, *Chem. Ber.*, 1958, **91**, 2871; 1965, **98**, 480 (*synth, ir, struct*)
Niebergall, H. *et al*, *Chem. Ber.*, 1962, **95**, 64 (*synth, props*)
Fluck, E. *et al*, *Chem. Ber.*, 1965, **98**, 2674 (*nmr*)
Cowley, A.H. *et al*, *Spectrochim. Acta*, 1966, **22**, 1431 (*ir*)
Spanier, E.J. *et al*, *J. Am. Chem. Soc.*, 1970, **92**, 3348 (*synth*)
Emoto, T. *et al*, *Bull. Chem. Soc. Jpn.*, 1973, **46**, 898 (*props*)
Keek, H. *et al*, *Org. Mass Spectrom.*, 1979, **14**, 149 (*ms*)
Alagna, L. *et al*, *Z. Naturforsch., A*, 1981, **36**, 68 (*pe*)
Blake, A.J. *et al*, *Acta Crystallogr., Sect. B*, 1981, **37**, 966 (*cryst struct*)
Blake, A.J. *et al*, *J. Mol. Struct.*, 1982, **78**, 265 (*ir, raman*)

1,1,5,5-Tetraphenyl-1,5-diphosphocanium(2+), 8CI T-00261

1,1,5,5-Tetraphenyl-1,5-diphosphoniacyclooctane

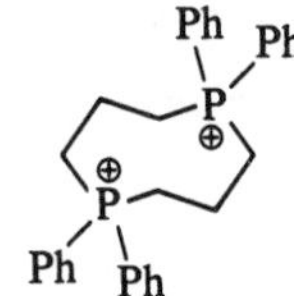

$C_{30}H_{32}P_2^{⊕⊕}$ M 454.530 (ion)

Dibromide:
$C_{30}H_{32}Br_2P_2$ M 614.338
Solid. Mp 43° dec.
Diiodide:
$C_{30}H_{32}I_2P_2$ M 708.339
Solid. Mp 204-205°.

Grim, S.O. *et al*, *Angew. Chem., Int. Ed. Engl.*, 1963, **2**, 486 (*diiodide*)
Issleib, K. *et al*, *J. Prakt. Chem.*, 1970, **312**, 578.

1,1,3,3-Tetraphenyl-1,3-diphosphorinanium(2+) T-00262

1,1,3,3-Tetraphenyl-1,3-diphosphinanium(2+)

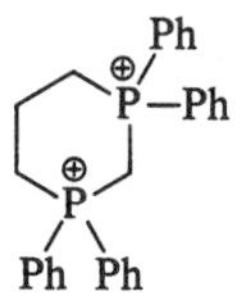

$C_{28}H_{28}P_2^{\oplus\oplus}$ M 426.477 (ion)
Dichloride:
$C_{28}H_{28}Cl_2P_2$ M 497.383
Cryst. (MeOH or EtOH). Mp 215°.
Dibromide: [74144-32-4].
$C_{28}H_{28}Br_2P_2$ M 586.285
Cryst. (MeOH or EtOH). Mp 260° dec.

Schmidbaur, H. *et al*, *Chem. Ber.*, 1981, **114**, 3063 (*synth*)

1,1,4,4-Tetraphenyl-1,4-diphosphorinanium(2+), 9CI T-00263

1,1,4,4-Tetraphenyl-1,4-diphosphoninanium(2+)
[13275-01-9]

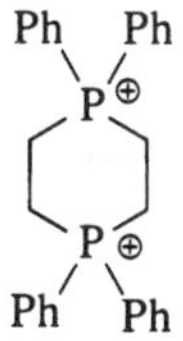

$C_{28}H_{28}P_2^{\oplus\oplus}$ M 426.477 (ion)
Dibromide: [2316-28-1].
$C_{28}H_{28}Br_2P_2$ M 586.285
Cryst. (MeOH/MeCN). Mp 324-325°. Aq. $OH^- \rightarrow$ C_6H_6.

Aguiar, A.M. *et al*, *J. Am. Chem. Soc.*, 1965, **87**, 671; 1966, **88**, 4090 (*synth, props, pmr, P-31 nmr*)

4,7,13,16-Tetraphenyl-1,10-dipropyl-1,10-diaza-4,7,13,16-tetraphosphacyclooctadecane, 9CI T-00264

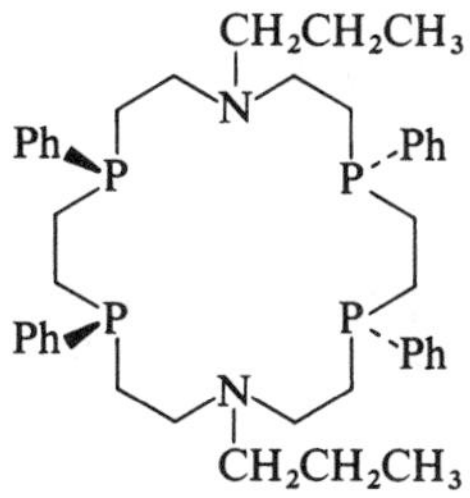

$C_{42}H_{58}N_2P_4$ M 714.829
(4*RS*,7*SR*,13*SR*,16*RS*)-form (illus.) characterised. Ligand for Co and Ni. Solid. Mp 136-138°.

Crampolini, M. *et al*, *Inorg. Chim. Acta*, 1983, **76**, L17 (*synth*)
Mangani, S. *et al*, *Inorg. Chim. Acta*, 1984, **85**, 65 (*complexes, cryst struct*)

Mealli, C. *et al*, *J. Chem. Soc., Dalton Trans.*, 1985, 479.

O,O,O',O'-Tetraphenyl imidodiphosphate, 9CI T-00265

O,O,O',O'*-Tetraphenyl iminodiphosphate*, *8CI*
[3848-53-1]

$(PhO)_2P(O)NHP(O)(OPh)_2$

$C_{24}H_{21}NO_6P_2$ M 481.381
Dimeric in C_6H_6. A monoprotic acid. Cryst. (Me_2CO, C_6H_6, or CCl_4). Mp 113°.

Riesel, L. *et al*, *Z. Anorg. Allg. Chem.*, 1977, **430**, 227 (*synth, P nmr*)
Richter, H. *et al*, *Z. Anorg. Allg. Chem.*, 1982, **491**, 266; 1983, **496**, 109 (*synth, ir, pmr, P nmr, derivs, cryst struct*)
Herrmann, E. *et al*, *Collect. Czech. Chem. Commun.*, 1984, **49**, 201 (*synth, props, complexes*)
Nöth, H. *et al*, *Z. Naturforsch., B*, 1984, **39**, 744 (*cryst struct*)

3,3,6,6-Tetraphenyl-1,3,6-oxadiphosphepanium(2+), 8CI T-00266

3,3,6,6-Tetraphenyl-1-oxa-3,6-diphosphoniacycloheptane

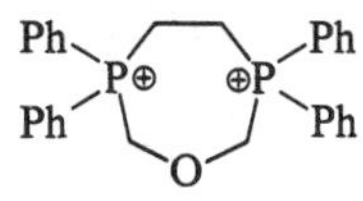

$C_{28}H_{28}OP_2^{\oplus\oplus}$ M 442.476 (ion)
Dibromide: [13119-11-4].
$C_{28}H_{28}Br_2OP_2$ M 602.284
Cryst. (MeCN/MeOH). Mp 292-293°.
Dipicrate: [13143-83-4].
$C_{40}H_{32}N_6O_{15}P_2$ M 898.672
Orange cryst. (MeCN). Mp 223-225° dec.

Aguiar, A.-M. *et al*, *J. Org. Chem.*, 1967, **32**, 2383 (*synth*)

3,3,5,5-Tetraphenyl-1,3,5-oxadiphosphorinanium(2+), 9CI, 8CI T-00267

3,3,5,5-Tetraphenyl-1-oxa-3,5-diphosphoniacyclohexane(2+). 3,3,5,5-Tetraphenyl-1,3,5-oxadiphosphinanium(2+)

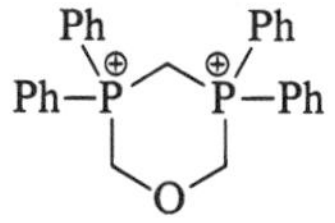

$C_{27}H_{26}OP_2^{\oplus\oplus}$ M 428.449 (ion)
Dichloride: [13119-12-5].
$C_{27}H_{26}Cl_2OP_2$ M 499.355
Deliquescent cryst. (MeCN/MeOH).
Dibromide: [13119-13-6].
$C_{27}H_{26}Br_2OP_2$ M 588.257
Cryst. (EtOAc/MeOH). Mp 235-241°.

Aguiar, A.M. *et al*, *J. Org. Chem.*, 1967, **32**, 2383 (*synth*)
Sammons, M.C. *et al*, *Anal. Chem.*, 1975, **47**, 1165 (*ms*)

2,3,3,4-Tetraphenyl-1,2-oxaphosphetane 2-oxide, 9CI T-00268

[52364-38-2]

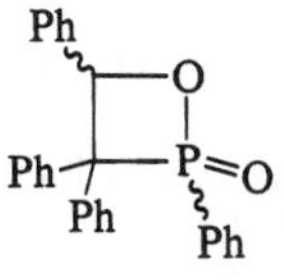

$C_{26}H_{21}O_2P$ M 396.424

Stereochem. not assigned. Cryst. (butanol). Mp 245-246°.

Regitz, M. *et al, Chem. Ber.*, 1980, **113**, 3303 (*synth*)

Tetraphenylphosphine imide, 9CI — T-00269

[2325-27-1]

$Ph_3P{=}NPh$

$C_{24}H_{20}NP$ M 353.402

Cryst. (Et_2O). Mp 131-132°. pK_a 7.66 (95% EtOH aq.), pK_a 9.78 (H_2O), pK_a 16.74 ($MeNO_2$).

Horner, L. *et al, Justus Liebigs Ann. Chem.*, 1959, **627**, 142 (*synth, uv, ir*)
Reddy, G.S. *et al, J. Org. Chem.*, 1963, **28**, 1822 (*P nmr*)
Schuster, P., *Monatsh. Chem.*, 1967, **98**, 1310 (*struct*)
Wiegräbe, W. *et al, Chem. Ber.*, 1968, **101**, 1414 (*synth, ir*)
Tökés, L. *et al, Org. Mass Spectrom.*, 1970, **4**, 59 (*ms*)

1,2,3,4-Tetraphenyl-5-phosphoniaspiro[4.5]deca-1,3-diene, 9CI — T-00270

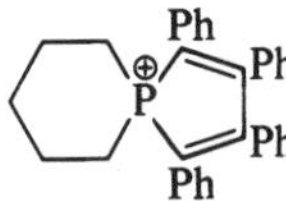

$C_{37}H_{30}P^{\oplus}$ M 505.618 (ion)

Bromide: [34667-31-7].
$C_{37}H_{30}BrP$ M 585.522
Yellow solid. Mp 357°.

Braye, E.H. *et al, Tetrahedron*, 1971, **27**, 5523 (*synth*)

N,N,N′,N′-Tetraphenylphosphonic diamide, 9CI — T-00271

Phosphonic bis(diphenylamide). N,N,N′,N′-*Tetraphenylphosphorodiamidite*

[58521-17-8]

$(Ph_2N)_2P(O)H \rightleftharpoons (Ph_2N)_2POH$

$C_{24}H_{21}N_2OP$ M 384.416

Cryst. (Et_2O at −30°). Sl. sol. C_6H_6, $CHCl_3$. Mp 57-58°.

Falius, H. *et al, Z. Anorg. Allg. Chem.*, 1976, **420**, 65 (*synth, pmr, P nmr*)

Tetraphenylphosphonium(1+) — T-00272

[18198-39-5]

$PPh_4^{\oplus}$

$C_{24}H_{20}P^{\oplus}$ M 339.396 (ion)

Chloride: [2001-45-8].
$C_{24}H_{20}ClP$ M 374.849
Cryst. (EtOH/Et_2O). Mp 265-267°, 272°.

Bromide: [2751-90-8].
$C_{24}H_{20}BrP$ M 419.300
Cryst. + $2H_2O$ (H_2O or MeOH/Et_2O). Mp 287-289°. Forms CBr_4 adduct, also a tribromide.

Iodide: [2065-67-0].
$C_{24}H_{20}IP$ M 466.300
Cryst. ($CHCl_3$/C_6H_6 or MeOH/Et_2O). Mp 337° (330-332°).

Bromide, CBr_4 adduct: Monoclinic needles (CH_2Cl_2/Et_2O). Mp 235°.

Tribromide: [3138-57-6].
$C_{24}H_{20}Br_3P$ M 579.108
Orange needles.

Methoxide:
$C_{25}H_{23}OP$ M 370.430
Cryst. (cyclohexane). Mp 162°. Stable in vacuo; instantly hydrolysed in air. At 170-180° → MeOH, C_6H_6, Ph_3P, and methoxybenzene.

Willard, H.H. *et al, J. Am. Chem. Soc.*, 1948, **70**, 737 (*chloride*)
Wittig, G. *et al, Justus Liebigs Ann. Chem.*, 1970, **732**, 97 (*bromide, iodide*)
Goetz, H. *et al, Phosphorus*, 1972, **1**, 217 (*iodide, nmr*)
Fluck, E. *et al, Z. Naturforsch., B*, 1974, **29**, 603 (*pe*)
Ohkuho, K. *et al, Bull. Chem. Soc. Jpn.*, 1974, **47**, 557 (*uv, struct*)
Albright, T.A. *et al, J. Am. Chem. Soc.*, 1975, **97**, 2946 (*cmr, nmr*)
Razavaev, G.A. *et al, J. Organomet. Chem.*, 1975, **99**, 93 (*methoxide*)
Effenberger, F. *et al, Chem. Ber.*, 1976, **109**, 306 (*props*)
Lindner, H.J. *et al, Chem. Ber.*, 1976, **109**, 314 (*cryst struct*)
Bogaard, M.P. *et al, Cryst. Struct. Commun.*, 1982, **11**, 175 (*perbromide, cryst struct*)
Cristau, H.J. *et al, J. Organomet. Chem.*, 1983, **241**, C1 (*props*)

N,N,N′,N′-Tetraphenylphosphonothioic diamide, 9CI — T-00273

Thiophosphonic acid bis(diphenylamide)

[58521-21-4]

$(Ph_2N)_2P(S)H$

$C_{24}H_{21}N_2PS$ M 400.477

Cryst. Sol. Et_2O, C_6H_6.

Falius, H. *et al, Z. Anorg. Allg. Chem.*, 1976, **420**, 65 (*synth, nmr, pmr, props*)

Tetraphenylphosphorodiamidous acid, 9CI — T-00274

$(Ph_2N)_2POH \rightleftharpoons (Ph_2N)_2P(O)H$

$C_{24}H_{21}N_2OP$ M 384.416

Free acid exists as the phosphoryl tautomer *N,N,N′,N′*-Tetraphenylphosphonic diamide, T-00271 .

Me ester: [58521-18-9]. *Methyl tetraphenylphosphorodiamidite.*
$C_{25}H_{23}N_2OP$ M 398.443
Cryst. (pet. ether). Dec. above Mp.

Ph ester: [58521-19-0]. *Phenyl tetraphenylphosphorodiamidite.*
$C_{30}H_{25}N_2OP$ M 460.514
Cryst. (Et_2O). Mp 27-28°.

Chloride: [58521-13-4].
$C_{24}H_{20}ClN_2P$ M 402.862
Cryst. Mp 44.5°.

Falius, H. *et al, Z. Anorg. Allg. Chem.*, 1976, **420**, 65 (*derivs, synth, P nmr, pmr*)

2,3,7,8-Tetraphenyl-1,4,6,9-tetraoxa-5-phospha(5-P^V)spiro[4.4]nona-2,7-diene, 9CI, 8CI — T-00275

$C_{28}H_{21}O_4P$ M 452.445

5-Ph: [18005-43-1].
$C_{34}H_{25}O_4P$ M 528.543

Prod. from trapping of PhP by benzil. Cryst. (butyronitrile/EtOH). Mp 216-218°.

Schmidt, U. *et al*, *Chem. Ber.*, 1968, **101**, 1381 (*synth*)

1,5,9,13-Tetraphenyl-1,5,9,13-tetraphosphacyclohexadecene, 9CI **T-00276**

[57978-09-3]

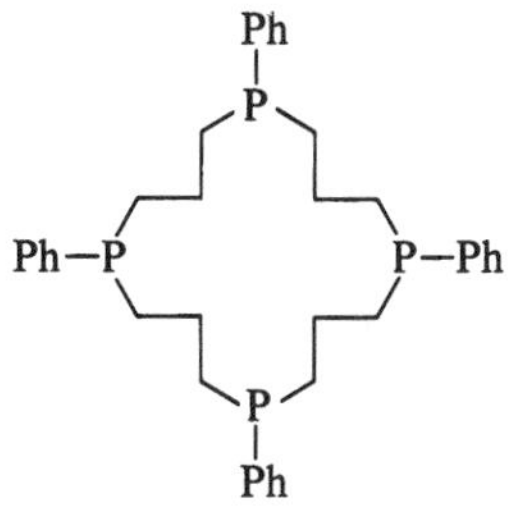

$C_{36}H_{44}P_4$ M 600.639

Solid. Mp 88-93°.

B,4PhCH₂Br: [57978-07-1]. *1,5,9,13-Tetrabenzyl-1,5,9,13-tetraphenyl-1,5,9,13-tetraphosphoniacyclohexadecane tetrabromide.*

$C_{64}H_{72}Br_4P_4$ M 1284.784

Mp 362°.

Tetraoxide: [57978-08-2].

$C_{36}H_{44}O_4P_4$ M 664.636

Solid. Mp 178-180°.

Horner, L. *et al*, *Phosphorus*, 1975, **6**, 63 (*synth, oxide*)
Horner, L. *et al*, *Phosphorus Sulfur*, 1978, **5**, 171 (*deriv*)

1,4,6,9-Tetraphenyl-1,4,6,9-tetraphospha-5-silaspiro[4.4]nonane, 10CI **T-00277**

[63746-61-2]

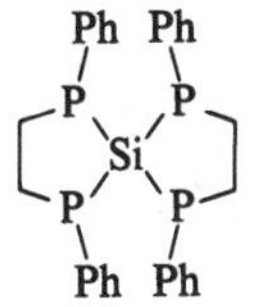

$C_{28}H_{28}P_4Si$ M 516.510

Solid. Mp 65-70°.

Couret, C. *et al*, *J. Organomet. Chem.*, 1977, **132**, C5 (*synth, pmr, P nmr*)

Tetraphenyltetraphosphetane, 9CI **T-00278**

Tetraphenylcyclotetraphosphine

[1104-52-5]

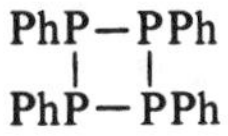

PhP—PPh
PhP—PPh

$C_{24}H_{20}P_4$ M 432.317

Light-yellow cryst. (MeCN). Mp 153-154°. Forms Ni, Cr, and Mo complexes.

Tetrasulfide: [1447-25-2].

$C_{24}H_{20}P_4S_4$ M 560.557

No phys. props. reported.

Ang, H.G. *et al*, *Aust. J. Chem.*, 1967, **20**, 1133 (*synth*)
Feshchenko, N.G. *et al*, *Zh. Obshch. Khim.*, 1972, **42**, 284 (*Engl. transl.* p. 273) (*synth*)
Baudler, M. *et al*, *Z. Naturforsch.*, *B*, 1976, **31**, 558 (*pmr*)

1,3,5,7-Tetraphenyl-1,3,5,7-tetraphosphocane **T-00279**

1,3,5,7-Tetraphenyl-1,3,5,7-tetraphosphacyclooctane

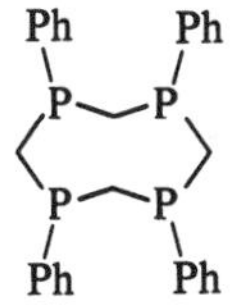

$C_{28}H_{28}P_4$ M 488.424

Cryst. (EtOH). Mp 125-127°.

Maier, L., *Helv. Chim. Acta*, 1965, **48**, 1034.

1,2,3,4-Tetraphenyltetraphospholane, 9CI **T-00280**

1,2,3,4-Tetraphenylcyclo-5-carba-1,2,3,4-tetraphosphane

[40425-94-3]

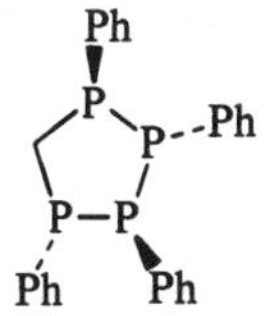

$C_{25}H_{22}P_4$ M 446.344

Needles (MeCN, EtOH, or EtOH/CS_2). Mp 141-143° (134-138°).

(1α,2β,3α,4β)-form

1-Sulfide: [64001-72-5].

$C_{25}H_{22}P_4S$ M 478.404

Solid. Mp 131-149°.

1,4-Disulfide: [63295-53-4].

$C_{25}H_{22}P_4S_2$ M 510.464

Cryst. (MeCN). Mp 151-153°.

Baudler, M. *et al*, *Angew. Chem., Int. Ed. Engl.*, 1971, **10**, 940 (*synth*)
Kaska, W.C. *et al*, *Helv. Chim. Acta*, 1974, **57**, 2550 (*synth*)
Baudler, M. *et al*, *Z. Anorg. Allg. Chem.*, 1977, **431**, 39 (*sulfides, ir, pmr, P nmr*)
Lex, M. *et al*, *Z. Anorg. Allg. Chem.*, 1977, **431**, 49 (*cryst struct*)

1,2,4,5-Tetraphenyl-1,2,4,5-tetraphosphorinane, 9CI **T-00281**

1,2,4,5-Tetraphenyl-1,2,4,5-tetraphosphinane. 1,2,4,5-Tetraphenyl-3,6-dicarba-1,2,4,5-tetraphosphan. 1,2,4,5-Tetraphenyl-1,2,4,5-tetraphosphacyclohexane

[50964-97-1]

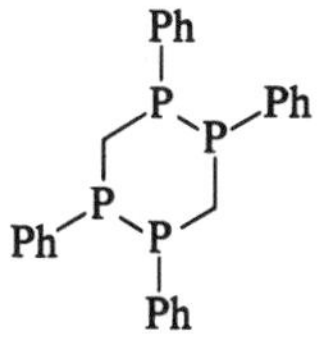

$C_{26}H_{24}P_4$ M 460.371

Platelets (MeCN). Mp 195-198°.

Baudler, M. *et al*, *Z. Naturforsch.*, *B*, 1973, **28**, 224 (*synth, ir, ms, nmr, pmr*)

Tetraphenylthiatetraphospholane, 9CI **T-00282**

[55658-70-3]

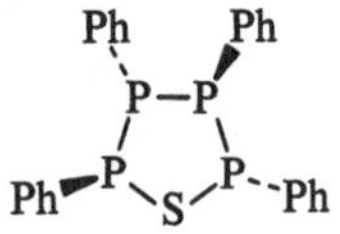

$C_{24}H_{20}P_4S$ M 464.377

(2α,3β,4α,5β)-form [50696-02-1]

Yellow needles (C_6H_6). Mp 154°.

Baudler, M. *et al*, *Z. Anorg. Allg. Chem.*, 1974,. **408**, 225 (*synth, ir, raman, P nmr, struct*)
Calhoun, H.P. *et al*, *J. Chem. Soc., Dalton Trans.*, 1974, 386 (*cryst struct*)
Hoffman, P.R. *et al*, *Inorg. Chem.*, 1975, **14**, 1997 (*pmr, P nmr*)
Issleib, K. *et al*, *Synth. React. Inorg. Met.-Org. Chem.*, 1977, **7**, 253 (*synth*)

***O,O,O′,O′*-Tetraphenyl thiohypophosphate, 9CI** **T-00283**

Tetraphenoxydiphosphine disulfide

[73608-45-4]

$(PhO)_2P(S){-}P(S)(OPh)_2$

$C_{24}H_{20}O_4P_2S_2$ M 498.487

Liq. $Bp_{0.05}$ 218-222°. n_D^{20} 1.6058.

Mazitova, F.N. *et al*, *Zh. Obshch. Khim.*, 1980, **50**, 815 (*Engl. transl.* p. 652) (*synth, P nmr*)

***O,O,O′,O′*-Tetraphenyl trithiopyrophosphate** **T-00284**

Tetraphenyl thiodiphosphate, 9CI. O,O-Diphenyl phosphorodithioate anhydrosulfide

[29516-95-8]

$(PhO)_2P(S){-}S{-}P(S)(OPh)_2$

$C_{24}H_{20}O_4P_2S_3$ M 530.547

Cryst. (C_6H_6/hexane). Mp 129-131°.

Mazitova, F.N. *et al*, *Zh. Obshch. Khim.*, 1980, **50**, 815 (*Engl. transl.* p. 652) (*synth, P nmr*)

Tetrapropyl diphosphate, 9CI **T-00285**

Tetrapropyl pyrophosphate

[3583-94-6]

$(H_3CCH_2CH_2O)_2P(O){-}O{-}P(O)(OCH_2CH_2CH_3)_2$

$C_{12}H_{28}O_7P_2$ M 346.297

Liq. d_4^{25} 1.10-1.12. $Bp_{0.01}$ 112-116°. n_D^{25} 1.4248.

▷Highly toxic. LD_{50} 10 mg/kg (mice). UX7200000.

Toy, A.D.F., *J. Am. Chem. Soc.*, 1948, **70**, 3882 (*synth*)
Turpin, R. *et al*, *Bull. Soc. Chim. Fr.*, 1971, 3878 (*synth*)

Tetrapropyldiphosphine, 10CI, 9CI **T-00286**

[20491-60-5]

$(H_3CCH_2CH_2)_2PP(CH_2CH_2CH_3)_2$

$C_{12}H_{28}P_2$ M 234.301

Liq. Bp_{20} 148°, Bp_5 112-113°. n_D^{20} 1.492. P—P bond cleaved by CCl_4 or PhLi.

Disulfide: [6830-45-1].

$C_{12}H_{28}P_2S_2$ M 298.421

Used for sepn. of Pt from Pd. Ligand for Cd, Co, Hg, Zn. Cryst. (MeOH or EtOH). Mp 147-148°. Desulfurized by Fe. P—P bond broken by I_2.

Niebergall, H. *et al*, *Chem. Ber.*, 1962, **95**, 64 (*synth*)
Voigt, D. *et al*, *Bull. Soc. Chim. Fr.*, 1968, 3561 (*synth*)
Appel, R. *et al*, *Chem. Ber.*, 1975, **108**, 1783 (*props*)
Aime, S. *et al*, *J. Chem. Soc., Dalton Trans.*, 1976, 2144 (*disulfide, cmr, nmr*)
McQuillan, G.P. *et al*, *Spectrochim. Acta, Part A*, 1978, **34**, 33 (*disulfide, ir, raman*)
Keck, H. *et al*, *Org. Mass Spectrom.*, 1979, **14**, 148 (*disulfide, ms*)
Alagna, L. *et al*, *Z. Naturforsch., A*, 1981, **36**, 68 (*disulfide, pe*)

Tetrapropyl hypodiphosphite, 9CI **T-00287**

Tetrapropoxydiphosphine

[40651-78-3]

$(H_3CCH_2CH_2O)_2P{-}P(OCH_2CH_2CH_3)_2$

$C_{12}H_{28}O_4P_2$ M 298.298

Liq. Bp_3 100-102°.

Arbuzov, A.E. *et al*, *Zh. Obshch. Khim.*, 1937, **7**, 1762; *CA*, **32**, 484.
Proskurnina, M.V. *et al*, *Zh. Obshch. Khim.*, 1973, **43**, 66 (*Engl. transl.* p. 63) (*synth*)
Ponomarev, S.V. *et al*, *Zh. Obshch. Khim.*, 1978, **48**, 231 (*Engl. transl.* p. 207) (*synth, P nmr*)

Tetrapropyl hypophosphate, 9CI **T-00288**

Tetrapropoxydiphosphine dioxide

[679-38-9]

$(H_3CCH_2CH_2O)_2P(O){-}P(O)(OCH_2CH_2CH_3)_2$

$C_{12}H_{28}O_6P_2$ M 330.297

Liq. d_4^{20} 1.09. $Bp_{0.03}$ 100-102°. n_D^{20} 1.4394.

Michalski, J. *et al*, *Bull. Acad. Polon. Sci., Ser. Sci. Chim.*, 1965, **13**, 253 (*synth*)

***N,N,N′,N′*-Tetrapropylphosphonic diamide, 8CI** **T-00289**

Phosphonic bis(dipropylamide). N,N,N′,N′-Tetrapropylphosphorodiamidite

[5843-28-7]

$[(H_3CCH_2CH_2)_2N]_2P(O)H \rightleftharpoons [(H_3CCH_2CH_2)_2N]_2POH$

$C_{12}H_{29}N_2OP$ M 248.348

Liq. $Bp_{0.001}$ 100-105° (bath). n_D^{20} 1.4586.

Zwierzak, A., *Bull. Acad. Polon. Sci., Ser. Sci. Chim.*, 1965, **13**, 609 (*synth, ir*)

Tetrapropylphosphonium(1+), 9CI, 8CI **T-00290**

$(H_3CCH_2CH_2)_4P^{\oplus}$

$C_{12}H_{28}P^{\oplus}$ M 203.327 (ion)

Chloride: [60931-61-5].

$C_{12}H_{28}ClP$ M 238.780

No phys. props. reported.

Bromide: [63462-98-6].

$C_{12}H_{28}BrP$ M 283.231

No phys. props. reported.

Iodide: [7259-36-1].

$C_{12}H_{28}IP$ M 330.231

Solid. Mp 267-269°.

Goetz, H. *et al*, *Phosphorus*, 1972, **1**, 217 (*synth, nmr*)
Dehmlow, E.V. *et al*, *Tetrahedron Lett.*, 1976, 1783 (*use*)
Landini, D. *et al*, *J. Chem. Soc., Chem. Commun.*, 1977, 112 (*use*)

Tetrapropylphosphorodiamidic acid, 9CI, 8CI — T-00291

Phosphoric acid bis(dipropylamide)

$[(H_3CCH_2CH_2)_2N]_2P(O)OH$

$C_{12}H_{29}N_2O_2P$ M 264.347

Me ester: Methyl tetrapropylphosphorodiamidate.
$C_{13}H_{31}N_2O_2P$ M 278.374
Liq. Bp_1 107°.

Et ester: Ethyl tetrapropylphosphorodiamidate.
$C_{14}H_{33}N_2O_2P$ M 292.401
Oil. Bp_{10} 164-166°.

Propyl ester: [40882-05-1]. *Propyl tetrapropylphosphorodiamidate.*
$C_{15}H_{35}N_2O_2P$ M 306.427
Liq. d_4^{20} 0.95. $Bp_{0.65}$ 133°. n_D^{20} 1.4471.

Chloride: [40881-95-6].
$C_{12}H_{28}ClN_2OP$ M 282.793
Liq. d_4^{20} 1.02. $Bp_{0.2}$ 118°. n_D^{20} 1.4660.

Michaelis, A., *Justus Liebigs Ann. Chem.*, 1903, **326**, 129 (*ethyl ester*)
Cheymol, J. *et al*, *C.R. Hebd. Seances Acad. Sci.*, 1959, **249**, 1240 (*methyl ester*)
Coustures, Y. *et al*, *Bull. Soc. Chim. Fr.*, 1973, 926 (*chloride, propyl ester, P nmr*)

Tetrapropyl pyrophosphite, 8CI — T-00292

Tetrapropyl diphosphite, 9CI

$(H_3CCH_2CH_2O)_2P{-}O{-}P(OCH_2CH_2CH_3)_2$

$C_{12}H_{28}O_5P_2$ M 314.298
Liq. d_0^0 1.07. Bp_6 147.5-149°. $n_D^{16.5}$ 1.4408.

Arbuzov, A.E. *et al*, *Zh. Obshch. Khim.*, 1937, **7**, 1762; *CA*, **32**, 484 (*synth*)

O,O,O′,O′-Tetrapropyl thiohypophosphate, 9CI — T-00293

Tetrapropoxydiphosphine disulfide. Tetrapropyl dithiohypophosphate
[51761-80-9]

$(H_3CCH_2CH_2O)_2P(S){-}P(S)(OCH_2CH_2CH_3)_2$

$C_{12}H_{28}O_4P_2S_2$ M 362.418
Possesses fungicidal props. Liq. d_4^{20} 1.09. $Bp_{0.07}$ 123°. n_D^{20} 1.4970.

Almasi, L. *et al*, *Chem. Ber.*, 1963, **96**, 2024 (*synth, struct*)

Tetrastearoyldiphosphatidylglycerol — T-00294

Cardiolipin

```
                             O
                             ‖
                       CH2OPOCH2
                        |    |
H3C(CH2)16COOCH2        |   HO  CHOOC(CH2)16CH3
              |         |        |
H3C(CH2)16COOCH    O   CHOH     CH2OOC(CH2)16CH3
              |    ‖    |
             CH2OPOCH2
                   |
                   OH
```

$C_{81}H_{158}O_{17}P_2$ M 1466.077
NH_4 salt: Dihydrate. Mp 182-183°. $[\alpha]_D^{24}$ +5.82° (c, 2.75 in $CHCl_3$).

Ramirez, F. *et al*, *Synthesis*, 1976, 769 (*synth*)

Tetrastearoylphosphatidylglycerol — T-00295

```
H3C(CH2)16COOCH2         CH2OOC(CH2)16CH3
              |          |
H3C(CH2)16COOCH    O    CHOOC(CH2)16CH3
              |    ‖    |
             CH2OPOCH2
                   |
                   OH
```

$C_{78}H_{151}O_{12}P$ M 1312.017
Mp 69.5-70.5°. $[\alpha]_D^{24}$ +6.2° (c, 4 in C_6H_6).

Baer, E., *J. Biol. Chem.*, 1952, **198**, 153.
Baer, E. *et al*, *Arch. Biochem. Biophys.*, 1958, **78**, 294.
Baer, E., *Prog. Chem. Fats Other Lipids*, 1963, **6**, 33.

Tetravinylphosphonium(1+) — T-00296

Tetraethenylphosphonium(1+)

$P(CH{=}CH_2)_4^{\oplus}$

$C_8H_{12}P^{\oplus}$ M 139.157 (ion)

Bromide:
$C_8H_{12}BrP$ M 219.061
Yellowish-red cryst. Mp 105-140° dec.

Maier, L. *et al*, *J. Am. Chem. Soc.*, 1957, **79**, 5884.

Thiamine diphosphate — T-00297

3-[(4-Amino-2-methyl-5-pyrimidyl)methyl]-4-methyl-5-(4,6,6-trihydroxy-3,5-dioxa-4,6-diphosphahex-1-yl)-thiazolium chloride P,P′*-dioxide, 9CI. Vitamin* B_1 *pyrophosphoric ester. Cocarboxylase*
[154-87-0]

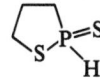

```
                                O     O
                                ‖     ‖
              H3C    CH2CH2OP—O—POH
     H2N                        |     |
     N      CH2N⊕  S           OH    OH
  H3C    N           Cl⊖
```

$C_{12}H_{19}ClN_4O_7P_2S$ M 460.765
Coenzyme of the yeast enzyme carboxylase. Catalyses the decarboxylation of α-ketoacids. Cryst. (EtOH contg. HCl). Mp 240°.
▷XI7552000.

Lohmann, K. *et al*, *Biochem. Z.*, 1937, **294**, 188.
Weijlard, J. *et al*, *J. Am. Chem. Soc.*, 1938, **60**, 2263.
Weijlard, J., *J. Am. Chem. Soc.*, 1941, **63**, 1160.
Karrer, P. *et al*, *Helv. Chim. Acta*, 1946, **29**, 711.
Ullrich, *et al*, *Vitam. Horm. (N.Y.)*, 1970, **28**, 365 (*rev, biochem*)
Florkin, M. *et al*, *Comprehensive Biochemistry*, 1971, Elsevier, Amsterdam, Vol. 11, 3 (*rev, biosynth*)

1,2-Thiaphospholane 2-sulfide, 9CI — T-00298

```
  /‾‾\
  \  P=S
   S   \
        H
```

$C_3H_7PS_2$ M 138.182

2-Me: [73627-87-9].
$C_4H_9PS_2$ M 152.209
Liq. Bp_2 140°.

2-Ph: [61157-03-7].
$C_9H_{11}PS_2$ M 214.280
Oil.

Mathey, F. *et al*, *J. Organomet. Chem.*, 1976, **117**, 377 (*phenyl, synth, pmr*)
Kleiner, H.-J. *et al*, *Justus Liebigs Ann. Chem.*, 1980, 324 (*methyl, synth, props*)

4H-1,4-Thiaphosphorin, 9CI **T-00299**
4H-1,4-Thiaphosphinin

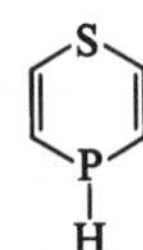

C_4H_5PS M 116.117
4-Chloro, 4-oxide: [82359-29-3].
C_4H_4ClOPS M 166.562
Solid. Mp 156-158°.
4-Ph: [55500-03-3].
$C_{10}H_9PS$ M 192.215
Oil.

Schoufs, M. *et al*, *Recl. Trav. Chim. Pays-Bas*, 1974, **93**, 241 (*phenyl, synth, pmr*)
Fridland, S.V. *et al*, *Zh. Obshch. Khim.*, 1982, **52**, 850 (*Engl. transl.* p. 738) (*chloro, synth, pmr*)

1,4-Thiaphosphorinane 1,1,4-trioxide, 9CI **T-00300**
1,4-Thiaphosphinane 1,1,4-trioxide

$C_4H_9O_3PS$ M 168.147
4-Me: [59466-63-6].
$C_5H_{11}O_3PS$ M 182.174
Solid. Bp >270° subl.
4-Ph: [59466-62-5].
$C_{10}H_{13}O_3PS$ M 244.245
Solid. Mp 240-242°.

Nagao, Y. *et al*, *Chem. Lett.*, 1976, 379 (*synth, pmr, nmr*)

2-Thia-1,3,5-triaza-7-phosphatricyclo[3.3.1.1^{3,7}]decane, 9CI **T-00301**
2-Thia-1,3,5-triaza-7-phosphaadamantane
[56299-23-1]

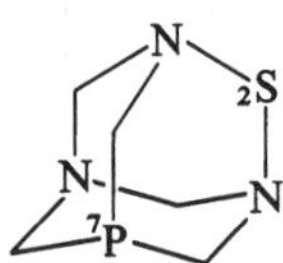

$C_5H_{10}N_3PS$ M 175.188
2,2-Dioxide: [55776-69-7].
$C_5H_{10}N_3O_2PS$ M 207.187
Cryst. (H_2O). Mp 274-275°.
2,2-Dioxide; B,MeI: [55776-71-1]. *7-Methyl-2-thia-1,3,5-triaza-7-phosphoniaadamantane iodide.*
$C_6H_{13}IN_3O_2PS$ M 349.126
Cryst. (MeOH/EtOAc). Mp 202-203°.
2,2,7-Trioxide: [55776-70-0].
$C_5H_{10}N_3O_3PS$ M 223.186
Cryst. (propanol). Mp 245-246°.

Daigle, D.J. *et al*, *J. Heterocycl. Chem.*, 1974, **11**, 1085 (*synth, derivs, ir, pmr*)
Delerno, J.R. *et al*, *J. Heterocycl. Chem.*, 1976, **13**, 757 (*cryst struct*)

1,3,2-Thiazaphospholidine 2-oxide **T-00302**

C_2H_6NOPS M 123.109
2-Ethoxy: see 2-Ethoxy-1,3,2-thiazaphospholidine, E-00053
2-Butoxy: [20043-27-0].
$C_6H_{14}NO_2PS$ M 195.216
Liq. $Bp_{0.02}$ 148-150°. n_D^{20} 1.5007.

Savignac, P. *et al*, *C.R. Hebd. Seances Acad. Sci., Ser. C*, 1968, **266**, 1791 (*synth*)

1,3,2-Thiazaphospholidine 2-sulfide **T-00303**

$C_2H_6NPS_2$ M 139.170
2-Methoxy: [13346-77-5].
$C_3H_8NOPS_2$ M 169.196
Liq. d_4^{20} 1.36. n_D^{20} 1.6000. Unstable to dist.
2-Ethoxy: see 2-Ethoxy-1,3,2-thiazaphospholidine, E-00053
2-Propoxy:
$C_5H_{12}NOPS_2$ M 197.250
Oil. $Bp_{0.03}$ 140-141°. n_D^{20} 1.5130.
2-Phenoxy: [13346-75-3].
$C_8H_{10}NOPS_2$ M 231.267
Oil. d_4^{20} 1.35. n_D^{20} 1.6800.
2-Me:
$C_3H_8NPS_2$ M 153.197
Oil. $Bp_{0.05}$ 134-135°. Hydrolyses v. rapidly.
2-Ph: [13346-73-1].
$C_8H_{10}NPS_2$ M 215.267
Oil. d_4^{20} 1.32. n_D^{20} 1.6807. Thermally unstable.

U.S.P., 3 285 999, (*1966*); *CA*, **66**, 28778 (*synth*)
Fr. Pat., 1 537 175, (*1968*); *CA*, **71**, 30585 (*synth*)

(2-Thienylmethyl)phosphonic acid, 9CI **T-00304**
2-Thenylphosphonic acid. 2-(Phosphonomethyl)-thiophene

$C_5H_7O_3PS$ M 178.142
Yellowish plates (H_2O). Mp 108-109°.
Di-Me ester: Dimethyl (2-thienylmethyl)phosphonate.
$C_7H_{11}O_3PS$ M 206.196
Oil. Bp_{10} 152-154°. n_D^{20} 1.5200.
Di-Et ester: [2026-42-8]. *Diethyl (2-thienylmethyl)-phosphonate.*
$C_9H_{15}O_3PS$ M 234.249
Employed in Wittig-Horner reactions to prep. thiophene-fused polycyclic compds. Oil. Bp_{31} 180-184°, $Bp_{2.2}$ 138-140°, $Bp_{0.1}$ 100-102°. n_D^{20} 1.5037.

Kosolapoff, G.M., *J. Am. Chem. Soc.*, 1947, **69**, 2248 (*synth*)
Bastian, J.M. *et al*, *Helv. Chim. Acta*, 1966, **49**, 214 (*ester, use*)
Kellog, R.M. *et al*, *J. Org. Chem.*, 1967, **32**, 3093 (*esters*)

Allen, D.W. *et al*, *J. Chem. Soc., Perkin Trans. 2*, 1977, 789 (*diethyl ester, synth, props, pmr, P nmr*)
Iwao, M. *et al*, *J. Heterocycl. Chem.*, 1980, **17**, 1259 (*use*)

(3-Thienylmethyl)phosphonic acid, 9CI **T-00305**
3-Thenylphosphonic acid. 3-(Phosphonomethyl)-thiophene

$C_5H_7O_3PS$ M 178.142

Di-Me ester: Dimethyl (3-thienylmethyl)phosphonate.
$C_7H_{11}O_3PS$ M 206.196
Oil. Bp_{24} 164-172°. n_D^{20} 1.5206.

Di-Et ester: [21382-79-6]. *Diethyl (3-thienylmethyl)-phosphonate.*
$C_9H_{15}O_3PS$ M 234.249
Employed in Wittig-Horner reactions. Oil. $Bp_{0.7}$ 123-133°, $Bp_{0.2}$ 90-95°.

Kellog, R.M. *et al*, *J. Org. Chem.*, 1967, **32**, 3093 (*dimethyl ester*)
Yom-Tov, B. *et al*, *J. Heterocycl. Chem.*, 1978, **15**, 285 (*use*)
Franke, A. *et al*, *Synthesis*, 1979, 712 (*diethyl ester*)
Iwao, M. *et al*, *J. Heterocycl. Chem.*, 1980, **17**, 1259 (*diethyl ester, synth, pmr, use*)

2-Thienylphosphinic acid, 9CI **T-00306**
[75263-36-4]

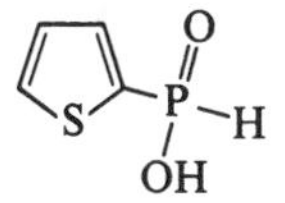

$C_4H_5O_2PS$ M 148.116
Tautomeric with 2-thienylphosphonous acid. Needles. Mp 70°.

Sachs, H., *Ber.*, 1892, **25**, 1514 (*synth, derivs*)

2-Thienylphosphonic acid, 9CI **T-00307**
2-Phosphonothiophene

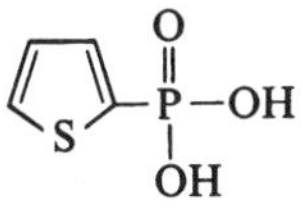

$C_4H_5O_3PS$ M 164.115

Di-Me ester: [13640-94-3]. *Dimethyl 2-thienylphosphonate.*
$C_6H_9O_3PS$ M 192.169
Liq. d_4^{20} 1.29. Bp_{12} 143-144°, $Bp_{0.3}$ 101-103°. n_D^{20} 1.5195.

Di-Et ester: [13640-95-4]. *Diethyl 2-thienylphosphonate.*
$C_8H_{13}O_3PS$ M 220.223
Liq. $Bp_{0.1}$ 103-104°. n_D^{20} 1.5001.

Di-Ph ester: [63818-46-2]. *Diphenyl 2-thienylphosphonate.*
$C_{16}H_{13}O_3PS$ M 316.311
Cryst. (hexane). Mp 75°.

Obrycki, R. *et al*, *J. Org. Chem.*, 1968, **33**, 632 (*dimethyl ester*)
Tavs, P., *Chem. Ber.*, 1970, **103**, 2428 (*diethyl ester*)
Allen, D.W. *et al*, *J. Chem. Soc., Perkin Trans. 2*, 1972, 63; 1977, 789 (*esters, synth, pmr, nmr, props*)
Khairullin, V.K. *et al*, *Zh. Obshch. Khim.*, 1974, **44**, 2120 (*Engl. transl.* p. 2083) (*dimethyl ester, dichloride*)

2-Thienylphosphonous acid, 9CI **T-00308**
2-Thiophenephosphonous acid

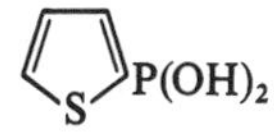

$C_4H_5O_2PS$ M 148.116
Free acid exists as the phosphoryl tautomer 2-Thienylphosphinic acid, T-00306 .

Di-Me ester: Dimethyl 2-thienylphosphonite.
$C_6H_9O_2PS$ M 176.170
Liq. d^{25} 1.19. $Bp_{0.9}$ 57°. n_D^{20} 1.5444.

Di-Et ester: [1725-19-5]. *Diethyl 2-thienylphosphonite.*
$C_8H_{13}O_2PS$ M 204.223
Liq. d^{25} 1.10. Bp_3 81-83°. n_D^{25} 1.5208.

Diisopropyl ester: Diisopropyl 2-thienylphosphonite. Bis(1-methylethyl) 2-thienylphosphonite.
$C_{10}H_{17}O_2PS$ M 232.277
Liq. d^{25} 1.06. $Bp_{0.5}$ 70°. n_D^{25} 1.5030.

Di-Ph ester: [34585-19-8]. *Diphenyl 2-thienylphosphonite.*
$C_{16}H_{13}O_2PS$ M 300.311
Oil. n_D^{20} 1.5310. Dec. on attempted distn.

Dichloride: [1726-72-3]. *Dichloro-2-thienylphosphine.*
$C_4H_3Cl_2PS$ M 185.007
Liq. Bp_{18} 108°. n_D^{14} 1.6281.

Bentov, M. *et al*, *J. Chem. Soc.*, 1964, 4750 (*dichloride, esters*)
Akhmedzade, D.A. *et al*, *Zh. Obshch. Khim.*, 1971, **41**, 1701 (*Engl. transl.* p. 1709) (*ester*)

2,2′-Thiobis[5,5-dimethyl-1,3,2-dioxaphosphorinane], 9CI **T-00309**
Bis(5,5-dimethyl-1,3,2-dioxaphosphorinan-2-yl) sulfide

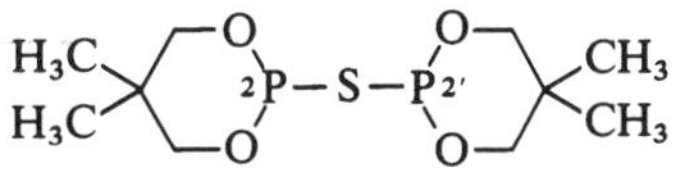

$C_{10}H_{20}O_4P_2S$ M 298.273

2,2′-Dioxide: see 2,2′-Thiobis[5,5-dimethyl-1,3,2-dioxaphosphorinane] 2,2′-dioxide, T-00310

2-Oxide, 2′-Sulfide: [4090-53-3]. *2-[(5,5-Dimethyl-1,3,2-dioxaphosphorinan-2-yl)thio]-5,5-dimethyl-1,3,2-dioxaphosphorinane* P*-oxide 2-sulfide.*
$C_{10}H_{20}O_5P_2S_2$ M 346.333
Cryst. (EtOAc). Mp 171-173°.

2,2′-Disulfide: [4090-54-4].
$C_{10}H_{20}O_4P_2S_3$ M 362.393
Cryst. ($CHCl_3$). Mp 223°.

Edmundson, R.S., *Tetrahedron*, 1965, **21**, 2379 (*synth, ir*)
Bartle, K.D. *et al*, *Tetrahedron*, 1967, **23**, 1701 (*deriv, pmr*)
Stec, W.J. *et al*, *J. Phys. Chem.*, 1971, **75**, 3975 (*deriv, pe, P nmr*)
Burkowska-Strzyzewska, M. *et al*, *Acta Crystallogr., Sect. B*, 1981, **37**, 724 (*deriv, cryst struct*)
Edmundson, R.S., *Phosphorus Sulfur*, 1981, **9**, 307 (*derivs, ms*)

2,2′-Thiobis[5,5-dimethyl-1,3,2-dioxaphosphorinane] 2,2′-dioxide, 9CI **T-00310**
[16956-55-1]

$C_{10}H_{20}O_6P_2S$ M 330.272
Cryst. ($C_6H_6/CHCl_3$). Mp 184-186°.

Edmundson, R.S., *Chem. Ind. (London)*, 1963, 784 (*synth, ir*)
Zwierzak, A. *et al*, *Tetrahedron*, 1969, **25**, 5177 (*synth*)
Stec, W.J. *et al*, *J. Phys. Chem.*, 1971, **75**, 3975 (*pe, P nmr*)

Bukowska-Strzyzewska, M. *et al*, *Acta Crystallogr., Sect. B*, 1976, **32**, 2605 (*cryst struct*)
Reimschussel, W. *et al*, *Org. Mass Spectrom.*, 1980, **15**, 302 (*ms*)
Van Nuffel, P. *et al*, *J. Mol. Struct.*, 1984, **125**, 1 (*struct*)
Paneth, P. *et al*, *J. Am. Chem. Soc.*, 1985, **107**, 1407 (*props*)

Thiobis[methanephosphonic acid] T-00311

[*Thiobis(methylene)*]*bis(phosphonic acid)*, 9CI. *Dimethyl sulfide-1,1'-diphosphonic acid*
[69404-53-1]

$$(HO)_2P(O)CH_2SCH_2P(O)(OH)_2$$

$C_2H_8O_6P_2S$ M 222.089
Cryst. + $1H_2O$.
Tetra-Et ester: [69404-52-0]. *Tetraethyl* [*thiobis(methylene)*]*bisphosphonate.*
$C_{10}H_{24}O_6P_2S$ M 334.304
Liq. d^{20} 1.19. $Bp_{0.2}$ 172-176°. n_D^{23} 1.4669.
Tetra-Et ester, S-*oxide:*
$C_{10}H_{24}O_7P_2S$ M 350.303
Liq. n_D^{23} 1.4696.
Tetra-Et ester, S,S-*dioxide:*
$C_{10}H_{24}O_8P_2S$ M 366.302
Cryst. (pet. ether). Mp 76-78°.
Tetraisopropyl ester: Tetraisopropyl [*thiobis(methylene)*]*bisphosphonate.*
$C_{14}H_{32}O_6P_2S$ M 390.411
Liq. d_4^{20} 1.10. Bp_2 164-165°. n_D^{20} 1.4548.

Petrov, K.A. *et al*, *Zh. Obshch. Khim.*, 1960, **30**, 1960 (*Engl. transl.* p. 1939) (*esters, synth*)
Maier, L. *et al*, *Phosphorus Sulfur*, 1978, **5**, 45 (*synth*)
Mikołajczyk, M. *et al*, *Phosphorus Sulfur*, 1981, **10**, 369 (*ester, oxides, synth, use, cmr, pmr, P nmr*)

[Thiobis[methylene]]-bis[triphenylphosphonium](2+), 9CI T-00312

$$Ph_3P^{\oplus}CH_2SCH_2P^{\oplus}Ph_3$$

$C_{38}H_{34}P_2S^{\oplus\oplus}$ M 584.694 (ion)
Dichloride: [55337-91-2].
$C_{38}H_{34}Cl_2P_2S$ M 655.600
Ylide formed *in situ* on treatment with butyllithium.
Dibromide: [4973-47-1].
$C_{38}H_{34}Br_2P_2S$ M 744.502
Solid. Mp 302-304°. LiOEt yields the bis-ylide.
Bis-ylide: [51593-74-9]. [*Thiobis*[*methylene*]]-*bis*[*triphenylphosphorane*].
$C_{38}H_{32}PS$ M 551.705
Used in synth. of heteroannulenes.

Dimroth, K. *et al*, *Chem. Ber.*, 1966, **99**, 642 (*bromide, use*)
Garrett, P.J. *et al*, *J. Am. Chem. Soc.*, 1972, **94**, 1022 (*ylide, use*)
Ogawa, H. *et al*, *Tetrahedron*, 1973, **29**, 809 (*use*)
Wife, R.L. *et al*, *J. Am. Chem. Soc.*, 1975, **97**, 640, 641 (*chloride, use*)
Ojima, J. *et al*, *Bull. Chem. Soc. Jpn.*, 1980, **53**, 1127 (*bromide, ylide, use*)

Thiometon, BSI T-00313

S-[*2-(Ethylthio)ethyl*] O,O-*dimethyl phosphorodithioate*, 9CI. S-*2-Ethylthioethyl* O,O-*dimethyl phosphorodithioate. Ekatin*
[640-15-3]

$$(MeO)_2P(S)SCH_2CH_2SEt$$

$C_6H_{15}O_2PS_3$ M 246.337
Systemic agricultural insecticide and acaricide. Oil. V. spar. sol. H_2O, sol. most org. solvs. d_4^{20} 1.209. $Bp_{0.2}$ 97°, $Bp_{0.1}$ 110°, $Bp_{0.01}$ 77°. n_D^{20} 1.5515; unstable in pure form; stable in non-polar solvents.
▷TE4375000.

S-*Oxide: see Oxydemeton-methyl, O-00098*

U.S.P., 3 041 367, (*1962*); *CA*, **57**, 16399 (*synth*)
Schrader, G., *Die Entwicklung neuer insektizider Phosphorsäure-Ester*, 1963, Verlag Chemie, Weinheim, 386.
U.S.P., 3 082 240, (*1963*); *CA*, **59**, 5077 (*synth*)
Krijgsman, W. *et al*, *J. Chromatogr.*, 1976, **117**, 201 (*glc*)
The Agrochemicals Handbook, Royal Society of Chemistry, 1983, A395.
Pesticide Manual, 6th Ed., 515; 7th Ed., 530.
Sax, N.I., *Dangerous Properties of Industrial Materials*, 6th Ed., Van Nostrand-Reinhold, 1984, 639.

Thionazin, BSI T-00314

O,O-*Diethyl* O-*(2-pyrazinyl) phosphorothioate*, 9CI, 8CI. *Nemafos. Zinophos. Zinofos*
[297-97-2]

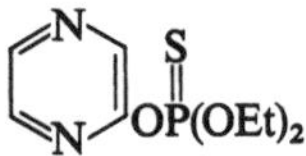

$C_8H_{13}N_2O_3PS$ M 248.236
Soil insecticide and nematocide. Oil. Sl. sol. H_2O, misc. most org. solvs. d^{25} 1.207. Mp −1.7°. $Bp_{0.0013}$ 80°. n_D^{25} 1.5080. Rel. stable to acids; rapidly hydrolysed by alkali.
▷TF5775000.

U.S.P., 2 918 468, (*1957*); *CA*, **54**, 9971b (*synth*)
Getz, M.E. *et al*, *J. Assoc. Off. Anal. Chem.*, 1968, **51**, 1101 (*tlc*)
Keith, L.H. *et al*, *J. Assoc. Off. Anal. Chem.*, 1968, **51**, 1063 (*pmr*)
Ross, R.T. *et al*, *Anal. Chim. Acta*, 1970, **52**, 139 (*P nmr*)
Stan, H.-J. *et al*, *Fresenius' Z. Anal. Chem.*, 1977, **287**, 271 (*glc, ms*)
Stan, H.-J. *et al*, *Biomed. Mass Spectrom.*, 1982, **9**, 483 (*ms*)
Ripley, B.D. *et al*, *J. Assoc. Off. Anal. Chem.*, 1983, **66**, 1084 (*glc*)
Pesticide Manual, 6th Ed., 516; 7th Ed., 531.

Thiophosphoryl isocyanate, 9CI T-00315

[17382-94-4]

$$(OCN)_3P{=}S$$

$C_3N_3O_3PS$ M 189.085
Viscous liq. d^{25} 1.54. Mp 9°. Bp 215°. n_D^{25} 1.5116. Rapidly hydrol. by H_2O.

Forbes, G.S. *et al*, *J. Am. Chem. Soc.*, 1943, **65**, 2271 (*synth, props*)
Borovikov, Yu.Ya. *et al*, *Zh. Obshch. Khim.*, 1975, **45**, 2377 (*Engl. transl.* p. 2335) (*struct*)

Thiophosphoryl isothiocyanate, 9CI T-00316

[1858-26-0]

$$(SCN)_3P{=}S$$

$C_3N_3PS_4$ M 237.267
Liq. $Bp_{0.1}$ 124°.

Sowerby, D.B., *J. Inorg. Nucl. Chem.*, 1961, **22**, 205 (*synth, ir*)
Fluck, E. *et al*, *Z. Anorg. Allg. Chem.*, 1965, **338**, 58 (*synth, P nmr*)
Pudovik, A.N. *et al*, *Zh. Obshch. Khim.*, 1979, **49**, 1425 (*Engl. transl.* p. 1248) (*synth*)

Thiotepa, BAN **T-00317**

1,1′,1″-Phosphinothioylidenetrisaziridine, 9CI. *Tris(1-aziridinyl)phosphine sulfide*, 8CI. N,N′,N″-*Triethylenethiophosphoramide. Phosphorothionic triethenamide. Tifosyl. Tespamin*

[52-24-4]

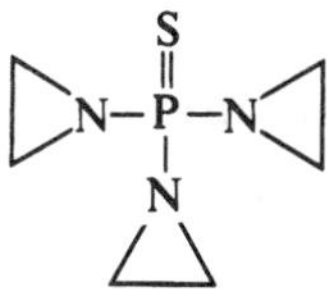

$C_6H_{12}N_3PS$ M 189.215

Antineoplastic drug and insect chemosterilant. Cryst. (pentane, Et_2O or C_6H_6/pet. ether). V. sol. EtOH, sol. C_6H_6, Et_2O, $CHCl_3$. Mp 52°.

▷SZ2975000.

Cheymol, J. *et al*, *Biol. Med. (Paris)*, 1967, **56**, 519 (*pharmacol*)
Fedotova, L.A. *et al*, *Khim. Geterotsikl. Soedin.*, 1969, 570(*Engl. transl.* p. 525); *CA*, **71**, 124078s (*prep*)
Bowman, M.C., *J. Chromatogr. Sci.*, 1975, **13**, 307 (*anal, glc*)
Zon, G. *et al*, *Biochem. Pharmacol.*, 1976, **25**, 989 (*props, nmr*)
Karstadt, M. *et al*, *CA*, 1983, **98**, 1469 (*epidemiol*)
Hisozumi, H., *et al*, *CA*, 1983, **98**, 191392 (*tox*)
Sax, N.I., *Dangerous Properties of Industrial Materials*, 6th Ed., Van Nostrand-Reinhold, 1984, 1071.

Threonine ethanolamine phosphate **T-00318**

Threonine 2-aminoethyl hydrogen phosphate, 9CI

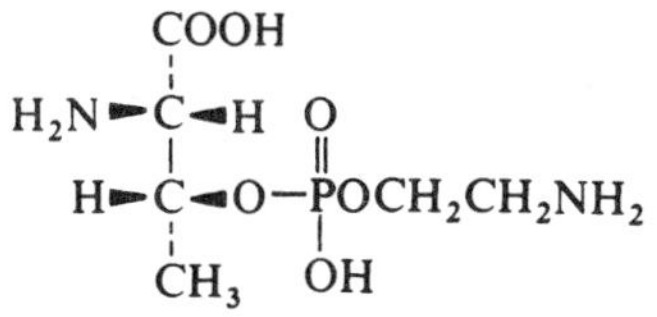

$C_6H_{15}N_2O_6P$ M 242.168

(S)-form [1935-19-9]

L-form

Constit. of fish, e.g. cod/haddock lenses. Mp 176° dec. $[\alpha]_D$ −37.3° (c, 0.94 in H_2O).

Rosenberg, H. *et al*, *Biochem. J.*, 1962, **84**, 536; *Comp. Biochem. Physiol.*, 1968, **27**, 695 (*isol, synth, ir*)
Porcellati, G. *et al*, *Comp. Biochem. Physiol.*, 1965, **14**, 413.
Allen, A.K. *et al*, *Comp. Biochem. Physiol.*, 1968, **27**, 695 (*biosynth*)
v. Heyningen, R. *et al*, *Exp. Eye Res.*, 1976, **23**, 29.

Thuringiensin **T-00319**

Bacillus thuringiensis Exotoxin. *β-Exotoxin*

[23526-02-5]

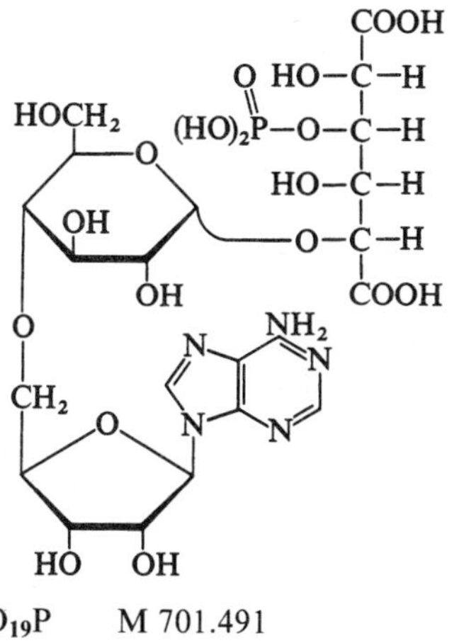

$C_{22}H_{32}N_5O_{19}P$ M 701.491

Nucleotide toxin. Isol. from *Bacillus thuringiensis* var. *gelechiae*. Specific inhibitor of DNA-dependent RNA polymerase. Insecticide. Cytotoxic. $[\alpha]_D^{25}$ +30.9° (c, 0.5 in H_2O). Thermostable.

Sebesta, K. *et al*, *Collect. Czech. Chem. Commun.*, 1969, **34**, 891.
Farkas, J. *et al*, *Collect. Czech. Chem. Commun.*, 1977, **42**, 909 (*uv, ir, pmr, cmr, ms, struct, bibl*)

Thymidine 5′-pyrophosphate **T-00320**

Thymidine 5′-(trihydrogen diphosphate), 9CI. *Thymidine diphosphate. TDP*

[491-97-4]

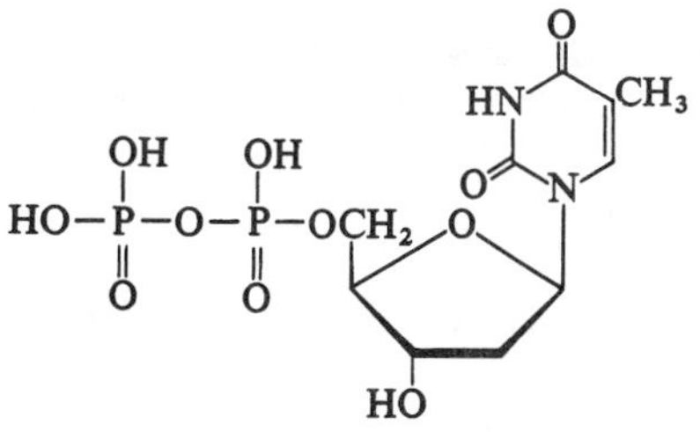

$C_{10}H_{16}N_2O_{11}P_2$ M 402.191

Griffin, B.E. *et al*, *J. Chem. Soc.*, 1958, 1389 (*synth*)
Michelson, A.M., *J. Chem. Soc.*, 1958, 1957 (*synth*)
Furusawa, K. *et al*, *J. Chem. Soc., Perkin Trans. 1*, 1976, 1711.
Labotka, R.J. *et al*, *J. Am. Chem. Soc.*, 1976, **98**, 3699 (*pmr*)

Toclofos-Methyl, BSI **T-00321**

O-*(2,6-Dichloro-4-methylphenyl)* O,O-*dimethyl phosphorothioate*, 9CI. O-*2,6-Dichloro*-p-*tolyl* O,O-*dimethyl phosphorothioate*

[57018-04-9]

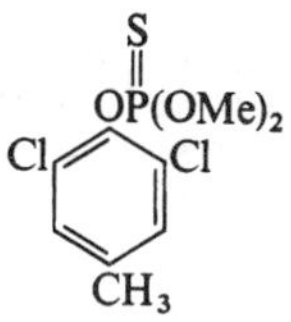

$C_9H_{11}Cl_2O_3PS$ M 301.124

Fungicide. Solid. Mp 78-80°. Stable to heat, light and moisture. Relatively non-toxic.

Ohtsuki, S. *et al*, *Jpn. Pestic. Inf.*, 1982, 21 (*rev*)
Pesticide Manual, 7th Ed., 536.

Toldimfos, BAN, INN **T-00322**

[4-(Dimethylamino)-2-methylphenyl]phosphinic acid, 9CI. *4-Dimethylamino*-o-*tolylphosphinic acid*

[57808-64-7]

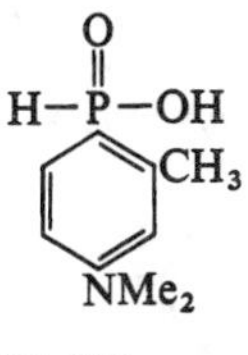

$C_9H_{14}NO_2P$ M 199.189

Phosphorus source in tonics, used in veterinary medicine. Used mainly as Na salt.

Na salt: [575-75-7]. *Tonophosphan. Novofosfan. Phinifos. Phosodyl. Tonofosfan.* Cryst. + $3H_2O$. Sl. sol. cold H_2O, hot EtOH.

Ger. Pat., 397 813; *Chem. Zentralbl.*, 1924, **2**, 1271 (*synth*)

Tolfamide, USAN, INN T-00323

N-(*Diaminophosphinyl*)-*2-methylbenzamide*, *9CI*. N-(*Diaminophosphinyl*)-o-*toluamide*. *EU-4584*

[70788-29-3]

O
CONHP(NH₂)₂ → $CONHP(O)(NH_2)_2$, CH_3

$C_8H_{12}N_3O_2P$ M 213.175

Urease inhibitor.

U.S.P., 4 182 881, (*1980*); *CA*, **92**, 146458b (*synth*)

Trehalose 6-(dihydrogen phosphate), 8CI T-00324

α-D-*Glucopyranosyl* α-D-*glucopyranoside 6-(dihydrogen phosphate)*, *9CI*

[4484-88-2]

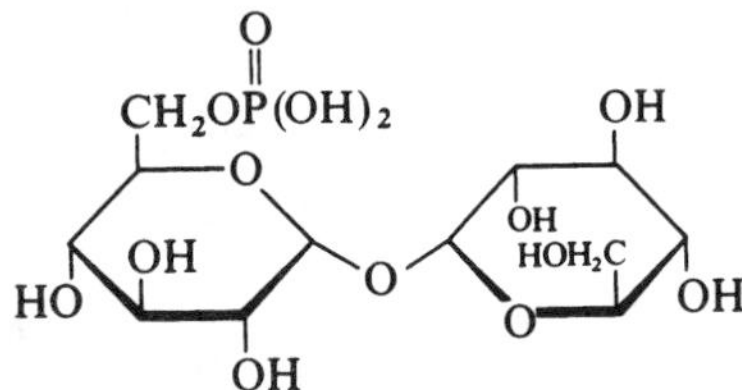

$C_{12}H_{23}O_{14}P$ M 422.279

Yeast metab. formed by *Streptomyces hygroscopicus* and *Mycobacterium smegmatis*. $[\alpha]_D$ +185° (H_2O).

Ba salt: $[\alpha]_D$ +132° (+99°) (H_2O).

Brucine salt: $[\alpha]_D$ +31° (H_2O).

Robison, R. *et al*, *Biochem. J.*, 1928, **22**, 1277 (*synth*)
Cabib, E. *et al*, *J. Biol. Chem.*, 1958, 259 (*biosynth*)
MacDonald, D.L. *et al*, *Biochim. Biophys. Acta*, 1964, **86**, 390 (*synth*)
Elbein, A.D. *et al*, *J. Bacteriol.*, 1968, **96**, 1623 (*enzymic synth*)
Lapp, D. *et al*, *J. Biol. Chem.*, 1971, **246**, 4567 (*enzymic synth*)
MacDonald, D.L., *The Carbohydrates*, Academic Press, 1972, 2nd Ed., **1A**, 253 (*rev*)

Triacetyl phosphate T-00325

Acetic acid, trianhydride with phosphoric acid, *9CI*

[56858-93-6]

$(AcO)_3P{=}O$

$C_6H_9O_7P$ M 224.107

V. hygroscopic plates (Et_2O). Mp 59-61°. Easily hydrolysed→Ac_2O and polyphosphoric acid.

Lynen, F., *Ber.*, 1940, **73**, 367 (*synth, props*)

Triamiphos, BSI T-00326

P-(*5-Amino-3-phenyl-1*H-*1,2,4-triazol-1-yl*)-N,N,N′,N′-*tetramethylphosphonic diamide*, *9CI*. *5-Amino-1-[bis(dimethylamino)phosphinyl]-3-phenyl-1*H-*1,2,4-triazole*. *Wepsyn 155*

[1031-47-6]

Ph, H_2N, N, N, N, $O{=}P(NMe_2)_2$

$C_{12}H_{19}N_6OP$ M 294.295

Fungicide with some systemic props. Solid (EtOH aq.). Sol. most org. solvs., v. spar. sol. H_2O. Mp 167-168°. Apparently superseded.

▷TA1400000.

Van den Bos, B.G. *et al*, *Recl. Trav. Chim. Pays-Bas*, 1960, **79**, 807, 1129 (*synth, struct, uv*)
Keith, L.H. *et al*, *J. Assoc. Off. Anal. Chem.*, 1968, **51**, 1063 (*pmr*)
Stan, H.-J. *et al*, *Biomed. Mass Spectrom.*, 1982, **9**, 483 (*ms*)
Stan, H.J. *et al*, *J. Chromatogr.*, 1983, **279**, 173 (*glc*)
Pesticide Manual, 6th Ed., 525.

1,3,5-Triaza-7-phosphatricyclo[3.3.1.1^{3,7}]decane, 9CI T-00327

1,3,5-Triaza-7-phosphaadamantane. *Monophosphaurotropine*

[53597-69-6]

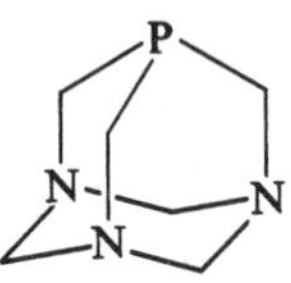

$C_6H_{12}N_3P$ M 157.155

Together with its *P*-derivs., quaternises on N. Cryst. (MeOH). Mp 266° dec.

7-Oxide: [53597-70-9].
$C_6H_{12}N_3OP$ M 173.154
Solid. Mp 264°.

B,MeI:
$C_7H_{15}IN_3P$ M 299.094
Needles (DMSO/Me_2CO). Mp 208-209° (196° dec.).

7-Oxide; B,MeI:
$C_7H_{15}IN_3OP$ M 315.093
Cryst. (MeOH/EtOAc). Mp 215-216°.

7-Sulfide: [56796-56-6].
$C_6H_{12}N_3PS$ M 189.215
Needles. Mp 270° dec.

7-Sulfide; B,MeI:
$C_7H_{15}IN_3PS$ M 331.154
Needles. Mp 217° dec.

Daigle, D.J. *et al*, *J. Heterocycl. Chem.*, 1974, **11**, 407; 1975, **12**, 579 (*synth, derivs, ir, pmr*)
Fluck, E. *et al*, *Chem.-Ztg.*, 1975, **99**, 246 (*synth, derivs, pmr, P nmr*)
Fluck, E. *et al*, *Z. Naturforsch., B*, 1977, **32**, 499 (*ir, raman, cryst struct*)
Jogun, K.H. *et al*, *Phosphorus Sulfur*, 1978, **4**, 199 (*oxide, sulfide, ir, raman, cryst struct*)
Benhammon, M. *et al*, *Phosphorus Sulfur*, 1982, **14**, 105 (*oxide, P nmr*)
Cowley, A.H. *et al*, *Inorg. Chem.*, 1982, **21**, 543 (*pe*)

Tri(1-azetidinyl)phosphine T-00328

1,1′,1″-Phosphinidynetrisazetidine, *9CI*

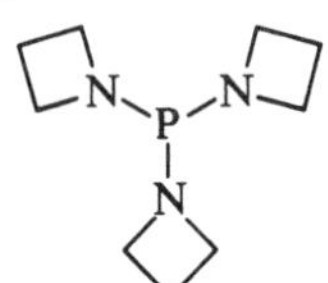

$C_9H_{18}N_3P$ M 199.235

Oxide: [3132-52-3]. *1,1′,1″-Phosphinylidynetrisazetidine*.
$C_9H_{18}N_3OP$ M 215.234
Cryst. (ligroin). Mp 82-87°.

Terry, P.H. *et al*, *J. Agric. Food Chem.*, 1973, **21**, 500.
Bollinger, J.C. *et al*, *Can. J. Chem.*, 1983, **61**, 328.

1,3,5-Triazine-2,4,6-triphosphonic acid **T-00329**

1,3,5-Triazine-2,4,6-triyltrisphosphonic acid, 9CI. *Triphosphono-*s*-triazine*

$C_3H_6N_3O_9P_3$ M 321.016

Hexa-Me ester: [903-22-0]. *Hexamethyl 1,3,5-triazine-2,4,6-triyltriphosphonate.*
$C_9H_{24}N_3O_9P_3$ M 411.225
Cryst. (C_6H_6). Mp 124-125°.

Hexa-Et ester: [912-40-3]. *Hexaethyl 1,3,5-triazine-2,4,6-triyltriphosphonate.*
$C_{15}H_{30}N_3O_9P_3$ M 489.338
Cryst. (C_6H_6/cyclohexane). Mp 82-84°, 95-96°.

Hexaisopropyl ester: Hexaisopropyl 1,3,5-triazine-2,4,6-triylphosphonate.
$C_{21}H_{42}N_3O_9P_3$ M 573.499
Cryst. (Et_2O/pet. ether). Mp 67-70°.

Hexa-Ph ester: Hexaphenyl 1,3,5-triazine-2,4,6-triylphosphonate.
$C_{39}H_{30}N_3O_9P_3$ M 777.602
Cryst. (C_6H_6/pet. ether). Mp 94-95°.

Morrison, D.C., *J. Org. Chem.*, 1957, **22**, 444 (*synth*)
Hewertson, W. *et al*, *J. Chem. Soc.*, 1963, 1670 (*synth*)

1,3,5-Triazine-2,4,6-triphosphoramidic acid **T-00330**

1,3,5-Triazine-2,4,6-triyltriphosphoramidic acid, 9CI

$C_3H_9N_6O_9P_3$ M 366.060

Hexa-Me ester: [18895-84-6]. *Hexamethyl 1,3,5-triazine-2,4,6-triyltriphosphoramidate.*
$C_9H_{21}N_6O_9P_3$ M 450.221
Solid. Mp 97-98°.

Hexa-Et ester: [18895-85-7]. *Hexaethyl 1,3,5-triazine-2,4,6-triyltriphosphoramidate.*
$C_{15}H_{33}N_6O_9P_3$ M 534.382
Solid. Mp 110-111°.

Bukovskii, M.I. *et al*, *CA*, 1968, **69**, 36083 (*synth*)

Tri-(1-aziridinyl)phosphine **T-00331**

1,1′,1″-Phosphinidynetrisaziridine, 9CI
[1194-53-2]

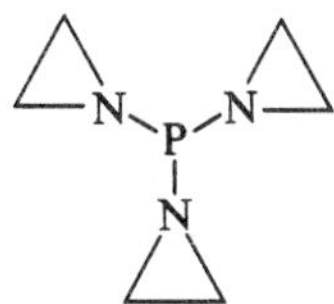

$C_6H_{12}N_3P$ M 157.155
Liq. d_4^{20} 1.10. Bp_{10} 82-83°. n_D^{20} 1.5279.

▷May explode when heated to 120°

Oxide: [545-55-1]. *1,1′,1″-Phosphinylidynetrisaziridine. Triethylenephosphoric triamide. Tepa.*
$C_6H_{12}N_3OP$ M 173.154
Insect sterilant, polymer crosslinking agent, fireproofer. Solid. Sol. H_2O. Mp 41°. $Bp_{0.3}$ 90-91°.

▷Exp. carcinogen, teratogen. Highly toxic, emits v. toxic fumes on heating. SZ1750000.

Sulfide: see Thiotepa, T-00317

Selenide: [68064-15-3]. *1,1′,1″-Phosphinoselenoylidynetrisaziridine.*
$C_6H_{12}N_3PSe$ M 236.115
Cryst. (C_6H_6/pet. ether). Mp 74-75°.

Bestian, K., *Justus Liebigs Ann. Chem.*, 1950, **566**, 210 (*oxide*)
Vilkov, L.V. *et al*, *Zh. Strukt. Khim.*, 1972, **13**, 7 (*Engl. transl.* p. 4) (*ed*)
White, D.W. *et al*, *J. Am. Chem. Soc.*, 1979, **101**, 4921 (*synth, derivs, cmr, P nmr, pmr*)
Bolinger, J.C. *et al*, *J. Mol. Struct.*, 1980, **69**, 273 (*oxide, struct*)
Konieczny, M. *et al*, *Z. Naturforsch., B*, 1981, **36**, 88 (*selenide, pharmacol*)

(1*H*-1,2,4-Triazol-1-yl)phosphonous acid, 9CI **T-00332**

*1*H*-1,2,4-Triazole-1-phosphonous acid*

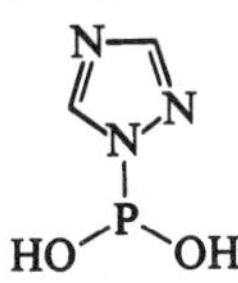

$C_2H_4N_3O_2P$ M 133.046

Di-Et ester: [58673-12-4]. *Diethyl (1*H*-1,2,4-triazol-1-yl)phosphonite. Diethyl phosphorotriazolidite.*
$C_6H_{12}N_3O_2P$ M 189.153
Used in peptide and deoxyoligonucleotide synth. Liq. $Bp_{0.01}$ 75-77°. n_D^{20} 1.4710.

Kricheldorf, H.R. *et al*, *Angew. Chem., Int. Ed. Engl.*, 1976, **15**, 305 (*synth, use*)
Kricheldorf, H.R. *et al*, *Makromol. Chem.*, 1977, **178**, 3141; 1979, **180**, 147 (*use*)
Kricheldorf, H.R. *et al*, *Justus Liebigs Ann. Chem.*, 1978, 1817 (*use*)
Matteucci, M.D. *et al*, *J. Am. Chem. Soc.*, 1981, **103**, 3185 (*use*)

Triazophos, BSI, ISO **T-00333**

O,O-*Diethyl* O-*(1-phenyl-1*H*-1,2,4-triazol-3-yl) phosphorothioate,* 9CI, 8CI. *Hostathion*
[24017-47-8]

$C_{12}H_{16}N_3O_3PS$ M 313.310
Insecticide and acaricide with some nematocidal props. Light-brown or yellowish oil. Sol. most org. solvs., v. spar. sol. H_2O. d_4^{20} 1.25 (1.43). Mp 5°. n_D^{20} 1.5500. Hydrolysed by acids and alkalis.

▷TF5635000.

South African Pat., 68/03 471, (*1967*); *CA*, **71**, 101861c
Bock, R. *et al*, *Pestic. Sci.*, 1976, **7**, 307 (*metab*)
Stan, H.-J. *et al*, *Biomed. Mass Spectrom.*, 1982, **9**, 483 (*ms*)
Stan, H.-J. *et al*, *J. Chromatogr.*, 1983, **279**, 173 (*glc*)
Pesticide Manual, 6th Ed., 526; 7th Ed., 542.
The Agrochemicals Handbook, Royal Society of Chemistry, 1983, A406.

Tri(2-benzothiazolyl)phosphine T-00334

2,2′,2″-Phosphinidynetrisbenzothiazole

[80679-24-9]

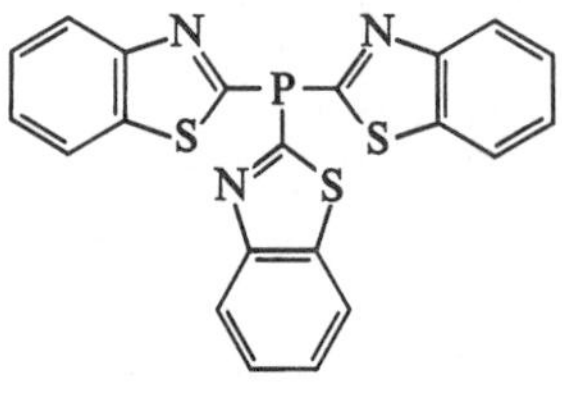

$C_{21}H_{12}N_3PS_3$ M 433.500

Needles (CH_2Cl_2/pentane). Mp 201-202°.

Moore, S.S. *et al, J. Org. Chem.*, 1982, **47**, 1489 (*synth, pmr, ir, nmr, ms*)

Tribenzoylphosphine, 10CI T-00335

[35696-22-1]

$P(COPh)_3$

$C_{21}H_{15}O_3P$ M 346.321

Benzoylating agent. Yellow cryst. (EtOH). Mp 149°. Stable to H_2O and dil. acid, dec. by dil. alkali.

Tyka, R. *et al, Bull. Acad. Pol. Sci., Ser. Sci. Chim.*, 1961, **9**, 577 (*synth, ir, uv*)
Kost, D. *et al, Tetrahedron Lett.*, 1979, 1983 (*cmr, nmr*)
Cogne, A. *et al, Org. Magn. Reson.*, 1980, **13**, 72 (*synth, cmr, nmr*)

Tribenzoyl phosphite T-00336

Benzoic acid(3:1)anhydride with phosphorous acid, 9CI

[54862-45-2]

$(PhCOO)_3P$

$C_{21}H_{15}O_6P$ M 394.320

Solid. Mp 93-95°. Unstable. Distn. gives benzoic anhydride. Phosphorylates alcohols.

Cade, J.A. *et al, J. Chem. Soc.*, 1954, 2030 (*synth, props*)
Petrov, K.A. *et al, Dokl. Akad. Nauk SSSR, Ser. Sci. Khim.*, 1968, **151**, 859; *CA*, **59**, 12627 (*synth, props*)

Tribenzyl phosphate, 8CI T-00337

Tris(phenylmethyl) phosphate, 9CI

[1707-92-2]

$(PhCH_2O)_3P{=}O$

$C_{21}H_{21}O_4P$ M 368.368

Prisms. Mp 65°.

Breusch, F.L. *et al, CA*, 1944, **38**, 1483 (*synth*)
Clark, V.M. *et al, J. Chem. Soc.*, 1949, 815; 1950, 2023, 2030 (*synth, props*)

Tribenzylphosphine, 8CI T-00338

Tris(phenylmethyl)phosphine, 9CI

[7650-89-7]

$P(CH_2Ph)_3$

$C_{21}H_{21}P$ M 304.371

Air-sensitive cryst. Mp 92-95°. $Bp_{0.2}$ 183-185°.

B,MeI: [40985-25-9]. *Tribenzylmethylphosphonium iodide. Methyltris(phenylmethyl)phosphonium iodide.*
$C_{22}H_{24}IP$ M 446.310
Cryst. (EtOH). Mp 170-171°, 183-184°.

B,PhCH₂Cl: Tetrabenzylphosphonium chloride. Tetrakis(phenylmethyl)phosphonium chloride.
$C_{28}H_{28}ClP$ M 430.956
Mp 225°.

Oxide: [4538-55-0].
$C_{21}H_{21}OP$ M 320.370
Ligand for Zn, Co, Ni, Zr. Cryst. (EtOH, Et_2O/pet. ether or Me_2CO). Mp 210-212°. Forms 2:1 complexes with HCl, HI, HNO_3, $HClO_4$.

▷SZ1500000.

Sulfide: [21187-15-5].
$C_{21}H_{21}PS$ M 336.431
Cryst. (toluene). Mp 265-267°, 282-283.5°.

Hinton, R.C. *et al, J. Chem. Soc.*, 1959, 2835 (*synth*)
Sander, M., *Chem. Ber.*, 1960, **93**, 1220 (*synth*)
Maier, L., *Chem. Ber.*, 1961, **94**, 3043 (*sulfide*)
Schindlbauer, H., *Monatsh. Chem.*, 1963, **94**, 99 (*uv, oxide*)
Crofts, P.C. *et al, J. Chem. Soc. (B)*, 1968, 1416 (*sulfide*)
Zhuravelva, L.P. *et al, Zh. Obshch. Khim.*, 1972, **42**, 526 (*Engl. transl.* p. 524) (*synth, oxide*)
Verstuyft, A.W. *et al, Inorg. Chem.*, 1977, **16**, 2776 (*pmr, cmr, nmr, derivs, complexes*)

***P,P,P*-Tribenzylphosphine imide** T-00339

P,P,P-*Tris(phenylmethyl)phosphine imide, 9CI.* P,P,P-*Tribenzylphosphine imine. Iminotribenzylphosphorane.* P,P,P-*Tribenzylphosphazene*

[51501-23-6]

$HN{=}P(CH_2Ph)_3$

$C_{21}H_{22}NP$ M 319.385

Completely cleaved by KNH_2/NH_3(l).

B,HCl: Mp 170°.

N-*Me:* [75794-09-1]. N-*Me.* P,P,P-*Tribenzyl-*N*-methylphosphine imide.*
$C_{22}H_{24}NP$ M 333.412
Solid. Mp 85°. Completely cleaved by KNH_2/NH_3(l).

N-*Me; B,HCl:* Solid. Mp 162-165°.

Ross, B. *et al, Chem. Ber.*, 1974, **107**, 2720 (*synth, props*)
Ross, B. *et al, Z. Anorg. Allg. Chem.*, 1980, **466**, 203 (*synth, ms, cmr, P nmr, pmr, deriv*)

Tribenzyl phosphite, 8CI T-00340

Tris(phenylmethyl) phosphite, 9CI

[15205-57-9]

$(PhCH_2O)_3P$

$C_{21}H_{21}O_3P$ M 352.369

Ligand for Pd. Needles (pet. ether). Mp 52°. $Bp_{0.05}$ 180-195°, $Bp_{0.02}$ 142-148°.

Landauer, S.R. *et al, J. Chem. Soc.*, 1953, 2224 (*synth, props*)
Ramirez, F. *et al, J. Am. Chem. Soc.*, 1968, **90**, 751 (*synth, ir, pmr, P nmr*)
Verstuyft, A.W. *et al, Inorg. Chem.*, 1977, **16**, 2776 (*pmr, P nmr, complexes*)
Campbell, M.M. *et al, Tetrahedron*, 1982, **38**, 2513 (*synth, ir, pmr*)

Tribenzyl phosphorotetrathioate, 8CI T-00341

Tris(phenylmethyl)phosphorotetrathioate, 9CI. Tribenzyl tetrathiophosphate

$(PhCH_2S)_3P{=}S$

$C_{21}H_{21}PS_4$ M 432.611
Liq., cryst. (EtOH) at −13°.

Rosnati, L., *Gazz. Chim. Ital.*, 1946, **76**, 272; *CA*, **42**, 876 (*synth*)

S,S,S-Tribenzyl phosphorotrithioate, 8CI **T-00342**
S,S,S-*Tris(phenylmethyl) phosphorotrithioate*, 9CI. S,S,S-*Tribenzyl trithiophosphate*
[14974-76-6]

$(PhCH_2S)_3P{=}O$

$C_{21}H_{21}OPS_3$ M 416.550
Solid. Mp 54°.

Dwek, R.A. *et al*, *J. Chem. Soc.* (*A*), 1970, 1173 (*P nmr*)
Shaw, R.A. *et al*, *Phosphorus*, 1971, **1**, 41, 191 (*synth, ir, pmr*)

Tribenzyl phosphorotrithioite, 8CI **T-00343**
Tris(phenylmethyl) phosphorotrithioite, 9CI. *Tribenzyl trithiophosphite*
[1656-65-1]

$(PhCH_2S)_3P$

$C_{21}H_{21}PS_3$ M 400.551
No phys. props. reported.

U.S.P., 2 542 370, (*1951*); *CA*, **45**, 5712 (*synth*)
Verstuyft, A.W. *et al*, *Inorg. Chem.*, 1977, **16**, 2776 (*P nmr, pmr, complexes*)

N,N′,N″-Tribenzylphosphorous triamide, 8CI **T-00344**
N,N′,N″-*Tris(phenylmethyl)phosphorous triamide*, 9CI. *Tri(benzylamino)phosphine. Tris(benzylamino)-phosphine*

$(PhCH_2NH)_3P$

$C_{21}H_{24}N_3P$ M 349.414
Oxide: [37624-66-1]. N,N′,N″-*Tribenzylphosphoric triamide*.
$C_{21}H_{24}N_3OP$ M 365.414
V. sol. $CHCl_3$, v. sl. sol. CCl_4. Mp 98-99°. pK_{a1} 10.18, pK_{a2} 7.02 ($MeNO_2$). Dec. at 250°.
Sulfide: [53820-05-6]. N,N′,N″-*Tribenzylphosphorothioic triamide*.
$C_{21}H_{24}N_3PS$ M 381.474
Cryst. (EtOH). Mp 127°.

Audrieth, L.F. *et al*, *J. Am. Chem. Soc.*, 1942, **64**, 1553 (*oxide*)
Buck, A.C. *et al*, *J. Am. Chem. Soc.*, 1948, **70**, 744 (*sulfide*)
Tolkmith, H., *J. Am. Chem. Soc.*, 1963, **85**, 3246 (*sulfide*)
Marchenko, A.P. *et al*, *Zh. Obshch. Khim.*, 1974, **44**, 67 (*Engl. transl.* p. 63) (*oxide, props*)
Healy, J.D. *et al*, *Phosphorus Sulfur*, 1978, **5**, 239 (*sulfide, synth, props*)

Tri-2-biphenylyl phosphate **T-00345**
Tris[(*1,1′-biphenyl*)-*2-yl*] *phosphate*, 9CI

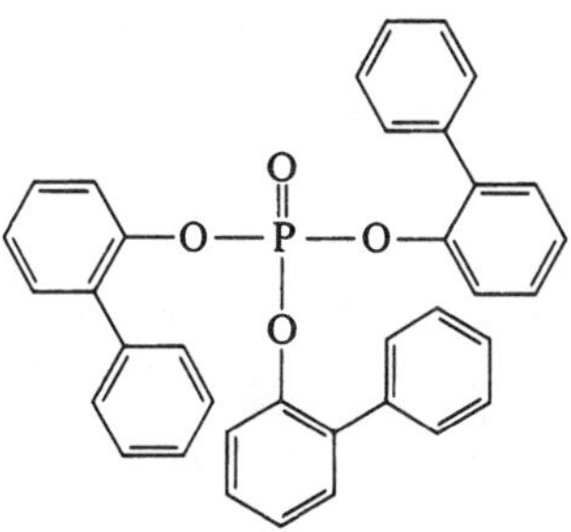

$C_{36}H_{27}O_4P$ M 554.581
Cryst. (EtOH). Mp 114°. Bp_{10} >300°.

U.S.P., 1 858 659, (*1932*)

Tri-3-biphenylyl phosphate **T-00346**
Tris[(*1,1′-biphenyl*)-*3-yl*] *phosphate*, 9CI
$C_{36}H_{27}O_4P$ M 554.581
Solid. Mp 137.5°.

U.S.P., 2 117 291, (*1938*)

Tri-4-biphenylyl phosphate **T-00347**
Tris[(*1,1′-biphenyl*)-*4-yl*] *phosphate*, 9CI
$C_{36}H_{27}O_4P$ M 554.581
Solid. Mp 84-86°. Bp_{10} 384°.

U.S.P., 2 117 290, (*1938*)

Tri-2-biphenylylphosphine **T-00348**
Tris[[*1,1′-biphenyl*]-*2-yl*]*phosphine*, 9CI
[13885-11-5]

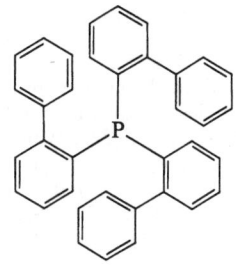

$C_{36}H_{27}P$ M 490.583
Solid, dec. at 384° (EtOH aq.). Mp 151-152°, 245°.
B,MeI: Methyl-2-biphenylylphosphonium iodide. Solid. Mp >250° dec.
Oxide:
$C_{36}H_{27}OP$ M 506.582
Cryst. (EtOH aq.). Mp 184-185°.

Worrall, D.E. *et al*, *J. Am. Chem. Soc.*, 1940, **62**, 2514 (*synth, oxide*)
Mitterhofer, F. *et al*, *Monatsh. Chem.*, 1967, **98**, 208 (*synth*)

Tri-3-biphenylylphosphine **T-00349**
Tris[[*1,1′-biphenyl*]-*3-yl*]*phosphine*, 9CI
$C_{36}H_{27}P$ M 490.583
Glass. Dec. at 372°.

Mitterhofer, F. *et al*, *Monatsh. Chem.*, 1967, **98**, 206.

Tri-4-biphenylylphosphine **T-00350**
Tris[[*1,1′-biphenyl*]-*4-yl*]*phosphine*, 9CI

[13885-05-7]
$C_{36}H_{27}P$ M 490.583
Needles (C_6H_6). Mp 173°. Dec. at 405°.
Oxide:
$C_{36}H_{27}OP$ M 506.582
Plates (EtOH). Mp 233-234°, 244-248°.
Sulfide:
$C_{36}H_{27}PS$ M 522.643
Cryst. ($CHCl_3$/MeOH). Mp 248-253° (241-242°).

Worrall, D.E., *J. Am. Chem. Soc.*, 1930, **52**, 2933 (*synth*)
Schiemenz, G.P., *Chem. Ber.*, 1966, **99**, 504 (*oxide*)
Mitterhofer, F. *et al*, *Monatsh. Chem.*, 1967, **98**, 206 (*synth*)

Tri([1,1-biphenyl]-2-yl) phosphite, 9CI **T-00351**

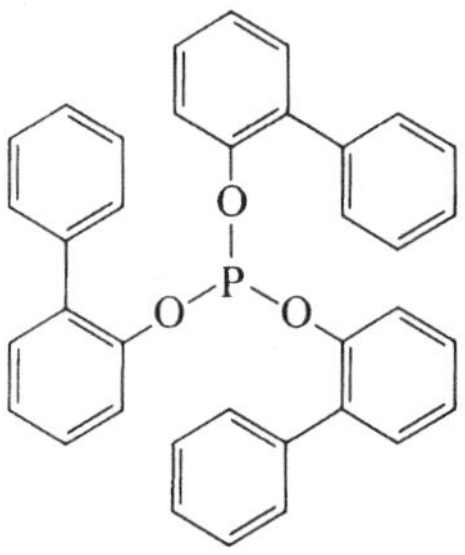

$C_{36}H_{27}O_3P$ M 538.581
Solid. Mp 95°. Bp_5 336-340°.

U.S.P., 2 220 845, (*1940*); *CA*, **35**, 1897 (*synth*)

2,2,2-Tribromo-2,2-dihydro-1,3,2-benzodioxaphosphole **T-00352**
[3712-44-5]

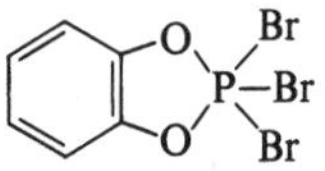

$C_6H_4Br_3O_2P$ M 378.782
Converts carboxylic acids into acyl bromides, alcohols into alkyl bromides and aldehydes into dibromo compds. Fuming solid. Generally prepd. *in situ*.

Gross, H. *et al*, *J. Prakt. Chem.*, 1965, **29**, 315 (*use*)
Gloede, J. *et al*, *Chem. Ber.*, 1967, **100**, 1770 (*use*)
Doerr, I.L. *et al*, *J. Org. Chem.*, 1973, **38**, 3878 (*use*)
Gloede, J. *et al*, *Z. Anorg. Allg. Chem.*, 1979, **458**, 108; 1980, **471**, 147 (*props, P nmr*)
Deng, R.M.K. *et al*, *J. Chem. Soc., Dalton Trans.*, 1984, 1917 (*complexes*)

2,2,2-Tribromo-2,2-dihydro-4,4,5,5-tetrakis(trifluoromethyl)-1,3,2-dioxaphospholane, 10CI **T-00353**
[70311-69-2]

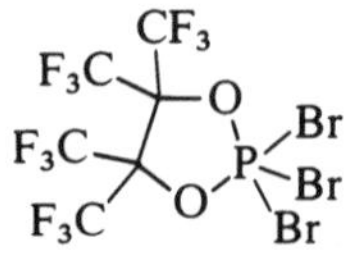

$C_6Br_3F_{12}O_2P$ M 602.731
Liq. $Bp_{0.01}$ 51°.

Roeschenthaler, G.-V. *et al*, *Z. Anorg. Allg. Chem.*, 1979, **450**, 79 (*synth, F and P nmr*)

Tribromodiphenoxyphosphorane **T-00354**

$(PhO)_2PBr_3$ or $(PhO)_3PBr^{\oplus}(PhO)_3P^{\ominus}Br_3$

$C_{12}H_{10}Br_3O_2P$ M 456.896
Probably possesses ionic struct. in MeCN.

Rydon, H.N. *et al*, *J. Chem. Soc.*, 1956, 3043 (*synth*)

Tribromodiphenylphosphorane, 9CI **T-00355**
[65200-23-9]

Ph_2PBr_3

$C_{12}H_{10}Br_3P$ M 424.897

Dillon, K.B. *et al*, *Polyhedron*, 1982, **1**, 123 (*synth, nmr*)

2,2,2-Tribromoethyl 4-*tert*-butylphenylphosphoramidic chloride **T-00356**
2,2,2-Tribromoethyl 4-(1,1-dimethylethyl)-phenylphosphoramidic chloride, 9CI
[71260-66-7]

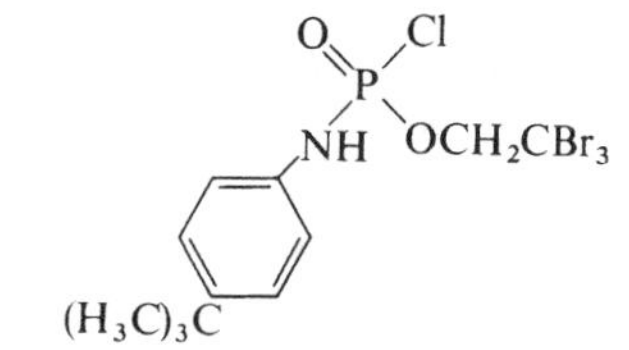

$C_{12}H_{16}Br_3ClNO_2P$ M 512.403
Phosphorylating agent used in phospholipid synth. Oil.

Lammers, J.G. *et al*, *Recl. Trav. Chim. Pays-Bas*, 1979, **98**, 243 (*synth, ir, use*)

2,2,2-Tribromoethyl 4-morpholinylphosphonochloridate, 9CI **T-00357**
[57575-15-2]

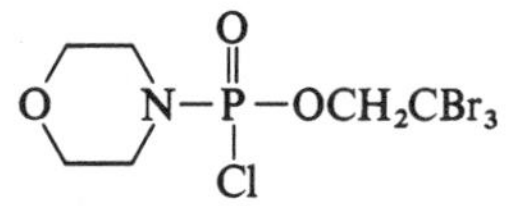

$C_6H_{10}Br_3ClNO_3P$ M 450.289
Phosphorylating reagent, specific to primary OH, used in the synth. of ribonucleoside di- and tri-phosphates. Cryst. (cyclohexane/pentane). Mp 79°.

Van Boom, J.H. *et al*, *Tetrahedron Lett.*, 1975, 2779 (*synth*)
Den Hartog, J.A.J. *et al*, *Recl. Trav. Chim. Pays-Bas*, 1979, **98**, 469 (*use*)
Den Hartog, J.A.J. *et al*, *J. Org. Chem.*, 1981, **46**, 2242 (*use*)

2,4,6-Tribromo-2,2,4,4,6,6-hexahydro-2,4,6-triphenyl-1,3,5-triaza-2,4,6-triphosphorine, 9CI **T-00358**
2,4,6-Tribromo-2,2,4,4,6,6-hexahydro-2,4,6-triphenylcyclotriphosphazene

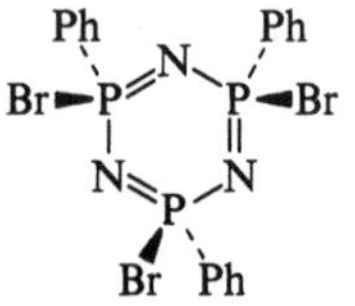

(2α,4α,6α)-*form*

$C_{18}H_{15}Br_3N_3P_3$ M 605.970
(2α,4α,6α)-form [19322-22-6]
cis-*form*
Cryst. (MeCN or heptane). Mp 152-153°, Mp 194-195° (dimorph.). Concd. solns. afford the *trans* form.
▷XX9440000.

(2α,4α,6β)-form [19329-39-6]
trans-*form*
Cryst. (heptane). Mp 162-163°.

Moeller, T. *et al*, *Inorg. Chem.*, 1963, **2**, 896 (*synth*)
Inorg. Synth., 1968, **11**, 201 (*synth*)
Kamminga, P.A. *et al*, *Cryst. Struct. Commun.*, 1979, **8**, 743 (*cryst struct*)

(2,4,6-Tribromophenyl)phosphoramidic acid, 9CI, 8CI — T-00359

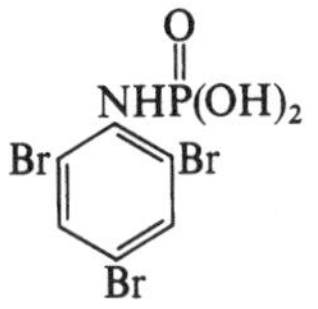

$C_6H_5Br_3NO_3P$ M 409.796
Di-Et ester: Diethyl (2,4,6-tribromophenyl)-phosphoramidate.
$C_{10}H_{13}Br_3NO_3P$ M 465.903
Needles (pet. ether). Mp 156.5-157.5°.
Di-Ph ester: [66392-89-0]. *Diphenyl (2,4,6-tribromophenyl)phosphoramidate.*
$C_{18}H_{13}Cl_3NO_3P$ M 428.638
Cryst. (EtOH). Mp 165-166°.
Dichloride:
$C_6H_3Br_3Cl_2NOP$ M 446.688
Needles. Mp 148°.

Michaelis, A., *Justus Liebigs Ann. Chem.*, 1903, **326**, 129 (*dichloride*)
Zhmurova, I.N. *et al*, *Zh. Obshch. Khim.*, 1961, **31**, 3741 (*Engl. transl.* p. 3495) (*diphenyl ester*)
Zwierzak, A. *et al*, *Tetrahedron*, 1973, **29**, 3899 (*diethyl ester, synth, ir, pmr, P nmr*)

Tri-3-butenylphosphine, 9CI — T-00360
[42585-47-7]

$P(CH_2CH_2CH{=}CH_2)_3$

$C_{12}H_{21}P$ M 196.272
Ligand for Pd, Rh, and Ir. Liq. Readily oxidized in air. $Bp_{1.0}$ 65-75°.

Clark, P.W. *et al*, *Can. J. Chem.*, 1974, **52**, 1714 (*synth, pmr, nmr*)
Clark, P.W. *et al*, *Inorg. Chem.*, 1979, **18**, 2067 (*cmr, complexes*)

Tri-2-butenyl phosphite, 9CI — T-00361
Tricrotyl phosphite

$(H_3CCH{=}CHCH_2O)_3P$

$C_{12}H_{21}O_3P$ M 244.270
Liq. Bp_1 98-99°. n_D^{20} 1.4680. Probably largely (*E,E,E*)-isomer.

Pudovik, A.N., *Zh. Obshch. Khim.*, 1957, **27**, 2755 (*Engl. transl.* p. 2794)

1,2,3-Tri-*tert*-butyldiazaphosphiridine — T-00362
1,2,3-Tris(1,1-dimethylethyl)diazaphosphiridine, 9CI

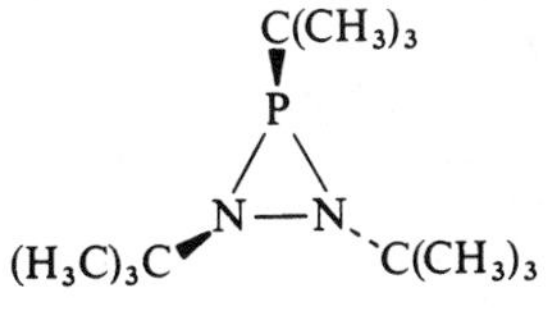

$C_{12}H_{27}N_2P$ M 230.332

(1α,2β,3α)-form
3-Oxide: [56908-15-7].
$C_{12}H_{27}N_2OP$ M 246.332
Cryst. (hexane at −75°). Mp 62-64°. $Bp_{0.000006}$ 65° subl. Ring cleaved by MeOH.

Quast, H. *et al*, *Synthesis*, 1976, 117 (*use*)
Quast, H. *et al*, *Justus Liebigs Ann. Chem.*, 1981, 967 (*synth, ms, pmr, P nmr*)

Tri-*tert*-butyldiphospharsirane — T-00363
Tris(1,1-dimethylethyl)diphospharsirane, 9CI
[77614-73-4]

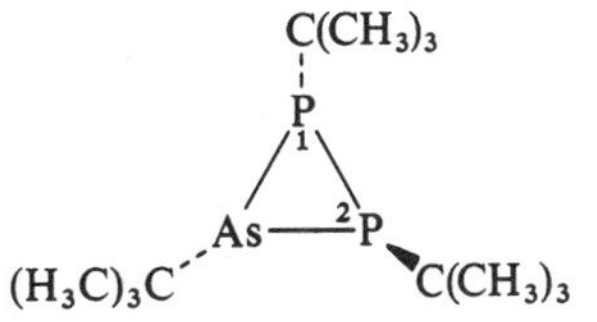

$C_{12}H_{27}AsP_2$ M 308.214
(1α,2β,3α)-form (shown) is more stable. Cryst. Mp 44°. Dec. at 150° giving cyclic polyphosphines. Easily oxid.

Baudler, M. *et al*, *Z. Naturforsch., B*, 1981, **36**, 527 (*synth, ir, raman, P nmr*)
Gleiter, R. *et al*, *Chem. Ber.*, 1981, **114**, 1004 (*pe*)

Tributyl(hexadecyl)phosphonium(1+), 9CI — T-00364
[66997-36-2]

$(H_3CCH_2CH_2CH_2)_3P^{\oplus}(CH_2)_{15}CH_3$

$C_{28}H_{60}P^{\oplus}$ M 427.756 (ion)
Salts can behave as phase transfer agents.
Chloride: [41272-12-2].
$C_{28}H_{60}ClP$ M 463.209
No phys. props. reported.
Bromide: [14937-45-2].
$C_{28}H_{60}BrP$ M 507.660
No phys. props. reported.
Iodide: [56772-64-6].
$C_{28}H_{60}IP$ M 554.660
No phys. props. reported.

Landini, D. *et al*, *Synthesis*, 1975, 430 (*use*)
Landini, D. *et al*, *J. Am. Chem. Soc.*, 1978, **100**, 2796 (*use*)
Fieser, M. *et al*, *Reagents for Organic Synthesis*, Wiley, 1967-84, **5**, 322; **6**, 271; **7**, 166 (*use*)

Tri-*tert*-butylmethylenephosphorane — T-00365
Tris(1,1-dimethylethyl)methylenephosphorane, 9CI
[64286-40-6]

$[(H_3C)_3C]_3P{=}CH_2$

$C_{13}H_{29}P$ M 216.346
Cubes. Mp 91-92°. Dec. at 200° liberating isobutene.

Schmidbaur, H. *et al*, *Z. Naturforsch, B*, 1977, **32**, 757 (*synth, pmr, cmr, nmr*)
Schmidbaur, H. *et al*, *Chem. Ber.*, 1980, **113**, 1480 (*props*)

Tributyl(methylphenylamino)-phosphonium(1+) — T-00366
Tributyl(methylbenzenaminato)phosphorus(1+), 10CI

$(H_3CCH_2CH_2CH_2)_3P^{\oplus}NMePh$

$C_{19}H_{35}NP^{\oplus}$ M 308.466 (ion)

Iodide: [67660-23-5]. *1,1,1-Tributyl-1-iodo-N-methyl-N-phenylphosphoranamine, 9CI.*
$C_{19}H_{35}IP$ M 421.364
Reagent for regio- and stereoselective γ-alkylations of allylic alcs. by organocuprate compds., and for the coupling of propynyl or enyne alcohols. Cryst. (EtOAc). Mp 120-120.5°.

Tanigawa, Y. *et al, J. Am. Chem. Soc.*, 1978, **100**, 4610 (*synth, use*)
Tanigawa, Y. *et al, J. Org. Chem.*, 1980, **45**, 4536 (*use*)
Goering, H.L. *et al, J. Org. Chem.*, 1981, **46**, 2144 (*use*)

1-(2,4,6-Tri-*tert*-butylphenyl)-4,4-diphenyl-1-phospha-1,2,3-butatriene T-00367

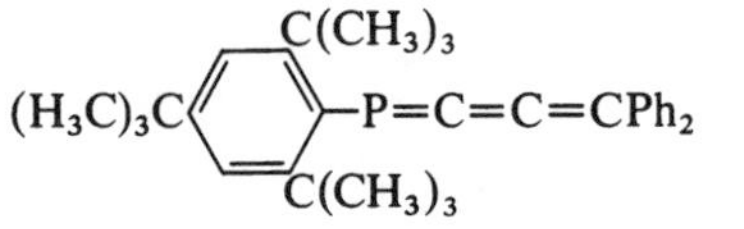

$(H_3C)_3C$—[ring with two $C(CH_3)_3$]—$P{=}C{=}C{=}CPh_2$

$C_{33}H_{39}P$ M 466.645
Yellow leaflets (EtOH). Mp 156-158°.

Märkl, G. *et al, Angew. Chem., Int. Ed. Engl.*, 1986, **25**, 1003 (*synth, ir, uv, ms, pmr, cmr, P nmr, cryst struct*)

(2,4,6-Tri-*tert*-butylphenyl)-methylidynephosphine T-00368

[2,4,6-Tris(1,1-dimethylethyl)phenyl]-methylidynephosphine

$(H_3C)_3C$—[ring with two $C(CH_3)_3$]—$C{\equiv}P$

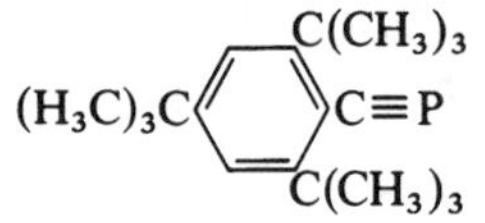

$C_{19}H_{29}P$ M 288.412
Isolable phosphine with (phosphorus) coordination number one. Cryst. ($MeNO_2$). Mp 127-128°.

Märkl, G. *et al, Tetrahedron Lett.*, 1986, **27**, 171 (*synth, ms, pmr, cmr, P nmr*)

(2,4,6-Tri-*tert*-butylphenyl)phosphine T-00369

[2,4,6-Tris(1,1-dimethylethyl)phenyl]phosphine, 11CI
[83115-12-2]

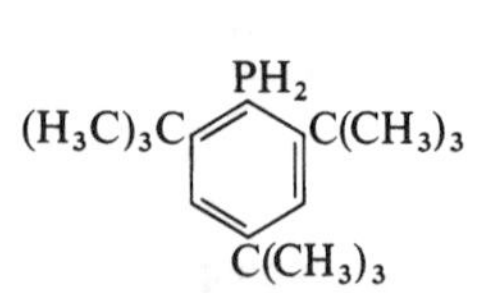

PH_2; $(H_3C)_3C$, $C(CH_3)_3$, $C(CH_3)_3$

$C_{18}H_{31}P$ M 278.417
Odourless cryst. (EtOH). Stable in air. Mp 150-152° (80°).

Oxide: [86539-32-4].
$C_{18}H_{31}OP$ M 294.416
Cryst. ($EtOH/Et_2O$). Mp 172-172.5°. Tautomeric with (2,4,6-Tri-*tert*-butylphenyl)phosphinic acid.

Sulfide: [89176-09-0].
$C_{18}H_{31}PS$ M 310.477
Cryst. (pentane). Mp 114-117°. Tautomeric with (2,4,6-Tri-*tert*-butylphenyl)phosphinothioic acid.

Cowley, A.H. *et al, J. Am. Chem. Soc.*, 1982, **104**, 5820 (*synth, pmr, P nmr*)
Issleib, K. *et al, Z. Anorg. Allg. Chem.*, 1982, **488**, 75 (*synth, pmr, P nmr*)
Zschunke, A. *et al, Z. Anorg. Allg. Chem.*, 1982, **495**, 115 (*cmr, P nmr*)
Yoshifuji, M. *et al, J. Am. Chem. Soc.*, 1983, **105**, 2495 (*synth, P nmr, props*)
Yoshifuji, M. *et al, Tetrahedron Lett.*, 1983, **24**, 4227 (*synth, derivs, ms, ir, cmr, pmr, P nmr*)
Oshikawa, T. *et al, Chem. Ind. (London)*, 1985, 126 (*synth, pmr*)

Tributyl phosphate, 9CI, 8CI T-00370

[126-73-8]

$(H_3CCH_2CH_2CH_2O)_3P{=}O$

$C_{12}H_{27}O_4P$ M 266.317
Used in extraction of lanthanide and actinide elements and in nuclear fuel reprocessing. Plasticiser for cellulose esters, lacquers, plastics and vinyl resins. Odourless liq. Mod. sol. H_2O, misc. org. solvs. d_4^{20} 0.98. Mp <−80°. Bp 289° dec., Bp_{50} 196°, Bp_9 148-150°. n_D^{20} 1.4246.

▷Mod. toxic. TC7700000.

Gerrard, W., *J. Chem. Soc.*, 1940, 1464 (*synth*)
Cox, J.R. *et al, J. Am. Chem. Soc.*, 1958, **80**, 5441 (*synth*)
Brown, C. *et al, Phosphorus Sulfur*, 1979, **6**, 481 (*synth*)
Cload, P.A. *et al, Org. Mass Spectrom.*, 1983, **18**, 57 (*ms*)
Suzuki, T. *et al, J. Agric. Food Chem.*, 1984, **32**, 603, 1278 (*metab*)
Science and Technology of Tributyl Phosphate, Ed. W.W. Schulz and J.D. Narratil, CRC Press, Florida, 1986 (*rev, bibl*)
Sax, N.I., *Dangerous Properties of Industrial Materials*, 6th Ed., Van Nostrand-Reinhold, 1984, 1041.

Tri-*tert*-butyl phosphate, 8CI T-00371

Tris(1,1-dimethylethyl) phosphate, 9CI
[20224-50-4]

$[(H_3C)_3CO]_3P{=}O$

$C_{12}H_{27}O_4P$ M 266.317
Cryst. (pet. ether). Mp 73-73.5°.

Cox, J.R., Jr. *et al, J. Org. Chem.*, 1969, **34**, 2600 (*synth*)
Chang, L.L. *et al, Phosphorus*, 1975, **6**, 69 (*synth*)

Tributylphosphine, 9CI, 8CI T-00372

[998-40-3]

$P(CH_2CH_2CH_2CH_3)_3$

$C_{12}H_{27}P$ M 202.319
Reagent for deoxygenation of epoxides and peroxides, and desulfurizations of thiiranes, and polysulfides. Ligand for metals of Groups IB, IIB, IVB, VB, VIB, VIII. Liq. with garlic-like odour. Misc. org. solvs. Fp 40°. Bp 240°, Bp_{20} 130°. pK_a 8.4. n_D^{20} 1.4635.

▷Pyrophoric

B,MeI: Tributylmethylphosphonium iodide.
$C_{13}H_{30}IP$ M 344.258
Solid. Mp 130-133°.
▷SZ1575000.

Oxide: see Tributylphosphine oxide, T-00376
▷*SZ2800000.*
Sulfide: see Tributylphosphine sulfide, T-00378
Selenide: see Tributylphosphine selenide, T-00377

Telluride: [2935-46-8].
$C_{12}H_{27}PTe$ M 329.919
Yellow cryst. (pet. ether). Mp 35-35.5°. Rather unstable to light and glass with liberation of Te.

Inorg. Synth., 1960, **6**, 87 (*synth*)
Cumper, C.W.R. *et al*, *J. Chem. Soc.*, 1964, 430 (*synth*)
Fritzsche, H. *et al*, *Chem. Ber.*, 1964, **97**, 1988; 1965, **98**, 171 (*synth*)
Zingaro, R.A., *J. Organomet. Chem.*, 1965, **4**, 320 (*telluride*)
Maier, L., *Helv. Chim. Acta*, 1966, **49**, 2458 (*synth*)
Fluck, E. *et al*, *Z. Naturforsch.*, *B*, 1967, **22**, 1095 (*nmr, deriv*)
Mironova, Z.N. *et al*, *Zh. Obshch. Khim.*, 1967, **37**, 2747 (*Engl. transl.* p. 2614) (*synth*)
Chremos, G.N. *et al*, *J. Organomet. Chem.*, 1970, **22**, 637 (*telluride, ir*)
Kostyanovskii, R.G. *et al*, *Org. Mass Spectrom.*, 1972, **6**, 1183 (*ms*)
Fluck, E. *et al*, *Z. Naturforsch.*, *B*, 1974, **29**, 603 (*pe*)
Quin, L.D. *et al*, *Org. Magn. Reson.*, 1974, **6**, 503 (*cmr*)
Goetz, H. *et al*, *Phosphorus*, 1978, **4**, 309 (*struct*)
Butler, I.S. *et al*, *Spectrochim. Acta, Part A*, 1979, **35**, 425 (*ir, raman*)
Sax, N.I., *Dangerous Properties of Industrial Materials*, 6th Ed., Van Nostrand-Reinhold, 1984, 1042.

Tri-*tert*-butylphosphine, 8CI — T-00373

Tris(1,1-dimethylethyl)phosphine, 9CI
[13716-12-6]

$$P[C(CH_3)_3]_3$$

$C_{12}H_{27}P$ M 202.319
Liq. or solid. Mp ca. 30°. Bp_{13} 102-103°.

B,MeBr: Tri-tert-*butylmethylphosphonium bromide.*
$C_{13}H_{30}BrP$ M 297.258
Mp 248° dec.

B,MeI: Tri-tert-*butylmethylphosphonium iodide.*
$C_{13}H_{30}IP$ M 344.258
Mp >360°.

B,EtBr: Tri-tert-*butylethylphosphonium bromide.*
$C_{14}H_{32}BrP$ M 311.285
Powder. Mp 241°.

B,(H₃C)₂CHBr: Tri-tert-*butylisopropylphosphonium bromide.*
$C_{15}H_{34}BrP$ M 325.311
Cryst. (EtOH). Mp 197°.

B,(H₃C)₃Cl: see Tetra-tert-butylphosphonium(1+), T-00020

Oxide: [6866-70-2].
$C_{12}H_{27}OP$ M 218.318
Cryst. by subl. V. sol. most org. solvs. Mp 77°. Stable to 250°. Hygroscopic.

Sulfide: [7441-04-5].
$C_{12}H_{27}PS$ M 234.379
Solid (pet. ether). Mp 220°.

Selenide:
$C_{12}H_{27}PSe$ M 281.279
Solid. Mp 215-217°.

Telluride:
$C_{12}H_{27}PTe$ M 329.919
Solid. Mp 125°.

Hoffmann, H. *et al*, *Chem. Ber.*, 1967, **100**, 692 (*synth, deriv*)
Bushweller, C.H. *et al*, *J. Am. Chem. Soc.*, 1973, **95**, 5949 (*pmr*)
Labarre, M.C. *et al*, *J. Mol. Struct.*, 1975, **26**, 17 (*struct*)
Lappert, M.F. *et al*, *J. Chem. Soc., Dalton Trans.*, 1975, 1207 (*pe*)
Müller, H. *et al*, *J. Organomet. Chem.*, 1977, **140**, C17 (*cmr*)
Schmidbaur, H. *et al*, *Z. Naturforsch.*, *B*, 1977, **32**, 757 (*derivs, pmr, nmr, ir, raman, cryst struct*)
Schmidbaur, H. *et al*, *Z. Naturforsch.*, *B*, 1978, **33**, 1556 (*oxide, pmr, cmr, nmr*)
van Linthoudt, J.P. *et al*, *Spectrochim. Acta, Part A*, 1980, **36**, 17 (*pmr, cmr, nmr*)
Schmidbaur, H. *et al*, *Chem. Ber.*, 1980, **113**, 1612 (*derivs*)
du Mont, W.W., *Z. Naturforsch.*, *B*, 1985, **40**, 1483 (*telluride, selenide, synth, conformn, P nmr, ir, ms*)
Rankin, D.W.H. *et al*, *J. Chem. Soc., Dalton Trans.*, 1985, 827 (*oxide, ed*)

P,P,P-Tributylphosphine imide, 9CI — T-00374

P,P,P-*Tributylphosphine imine. Iminotributylphosphorane.* P,P,P-*Tributylphosphazene*
[57831-27-3]

$$HN{=}P(CH_2CH_2CH_2CH_3)_3$$

$C_{12}H_{28}NP$ M 217.334
Oil. $Bp_{0.1}$ 104°.

N-*Trimethylsilyl:* P,P,P-*Tributyl*-N-*trimethylsilylphosphine imide.*
$C_{15}H_{36}NPSi$ M 289.515
Liq. Bp_{11} 149°, $Bp_{0.1}$ 94°.

Birkofer, L. *et al*, *Chem. Ber.*, 1964, **97**, 2100 (*synth, deriv*)
Finch, N. *et al*, *J. Org. Chem.*, 1980, **45**, 3416.

P,P,P-Tri-*tert*-butylphosphine imide — T-00375

P,P,P-*Tris(1,1-dimethylethyl)phosphine imide, 9CI. Iminotri*-tert-*butylphosphorane.* P,P,P-*Tri*-tert-*butylphosphazene*
[69277-58-3]

$$HN{=}P[C(CH_3)_3]_3$$

$C_{12}H_{28}NP$ M 217.334
Solid. Mp 109° (101°). $Bp_{0.1}$ 75° subl.

N-*Trimethylsilyl:* [53561-53-8]. P,P,P-*Tri*-tert-*butyl*-N-*trimethylsilylphosphine imide.*
$C_{15}H_{36}NPSi$ M 289.515
Solid. Mp 94-95°. $Bp_{0.1}$ 90-100° (oven).

Buchner, W. *et al*, *Z. Naturforsch.*, *B*, 1974, **29**, 328 (*silyl deriv, cmr, P nmr*)
Schmidbaur, H. *et al*, *Z. Naturforsch.*, *B*, 1978, **33**, 1556 (*synth, cmr, P nmr, pmr, ir*)
Wolfsberger, W., *Z. Naturforsch.*, *B*, 1978, **33**, 1452 (*synth, props, salts, deriv*)
Ross, B. *et al*, *Chem. Ber.*, 1979, **112**, 1756 (*synth, cmr, P nmr, pmr*)
Rankin, D.W.H. *et al*, *J. Chem. Soc., Dalton Trans.*, 1985, 827 (*ed*)

Tributylphosphine oxide, 9CI, 8CI — T-00376

[814-29-9]

$$(H_3CCH_2CH_2CH_2)_3P{=}O$$

$C_{12}H_{27}OP$ M 218.318
Useful metal extractant. Ligand for metals of Groups IIIA, IVA, IIIB, VB, VIB, and VIII, as well as rare earth metals. Hygroscopic needles. Mp 68-69°. Bp 300°, Bp_2 134-135°. pK_{a1} 8.54, pK_{a2} 8.75 ($MeNO_2$).
▷SZ1575000.

Cumper, C.W.N. *et al*, *J. Chem. Soc.*, 1964, 430 (*synth, props*)
Cuddy, B.D. *et al*, *Tetrahedron Lett.*, 1971, 4433 (*pmr*)
Quin, L.D. *et al*, *Org. Magn. Reson.*, 1974, **6**, 503 (*cmr*)
Albright, T.A. *et al*, *J. Org. Chem.*, 1975, **40**, 3437 (*cmr, nmr*)
Fluck, E. *et al*, *Z. Anorg. Allg. Chem.*, 1975, **412**, 47 (*pe*)
Pudovik, A.N. *et al*, *Izv. Akad. Nauk SSSR, Ser. Khim.*, 1979, 2644 (*Engl. transl.* p. 2461) (*synth, ir, nmr*)
Sax, N.I., *Dangerous Properties of Industrial Materials*, 6th Ed., Van Nostrand-Reinhold, 1984, 1042.

Tributylphosphine selenide, 9CI, 8CI — T-00377

[39181-26-5]

$(H_3CCH_2CH_2CH_2)_3P{=}Se$

$C_{12}H_{27}PSe$ M 281.279

Converts epoxides to alkenes. Reduces selenoxides. Pale-yellow liq. $Bp_{0.8}$ 150-151°. n_D^{27} 1.5150. Converted into the oxide by DMSO. Forms Hg and Cd complexes.

Zingaro, R.A. *et al*, *J. Chem. Eng. Data*, 1963, **8**, 226 (*synth*)
Zingaro, R.A., *Inorg. Chem.*, 1963, **2**, 192 (*ir*)
Schmidpeter, A. *et al*, *Z. Naturforsch., B*, 1969, **24**, 179 (*synth, nmr*)
Stec, W.J. *et al*, *Phosphorus*, 1972, **2**, 97 (*nmr*)
Shagidullin, R.R. *et al*, *Dokl. Akad. Nauk SSSR, Ser. Sci. Khim.*, 1973, **211**, 1363 (*Engl. transl.* p. 694) (*ir*)
Mathey, F. *et al*, *C.R. Hebd. Seances Acad. Sci., Ser. C*, 1975, **281**, 881 (*use*)
Shagidullin, R.R. *et al*, *Izv. Akad. Nauk SSSR, Ser. Khim.*, 1976, 184 (*Engl. transl.* p. 174) (*uv*)
Sakaki, K. *et al*, *Chem. Lett.*, 1977, 1003 (*use*)
Carr, S.W. *et al*, *Aust. J. Chem.*, 1981, **34**, 35 (*nmr*)

Tributylphosphine sulfide, 9CI, 8CI — T-00378

[3084-50-2]

$(H_3CCH_2CH_2CH_2)_3P{=}S$

$C_{12}H_{27}PS$ M 234.379

Extractant for metals. Converts oxiranes to thiiranes. Pale-yellow oil. d_4^{24} 1.03. $Bp_{0.1}$ 111°. n_D^{20} 1.5045.

▷SZ2800000.

Christen, P.J. *et al*, *Recl. Trav. Chim. Pays-Bas*, 1959, **78**, 161 (*synth, ir*)
Zingaro, R.A. *et al*, *J. Org. Chem.*, 1961, **26**, 5205 (*synth, ir*)
Inorg. Synth., 1967, **9**, 71 (*prep*)
Chan, T.H. *et al*, *J. Am. Chem. Soc.*, 1972, **94**, 2880 (*use*)
Quin, L.D. *et al*, *Org. Magn. Reson.*, 1974, **6**, 503 (*cmr*)
Albright, T.A. *et al*, *J. Org. Chem.*, 1975, **40**, 3437 (*cmr, nmr*)
Cattrall, R.W., *J. Inorg. Nucl. Chem.*, 1978, **40**, 687 (*complexes*)
Timokhin, B.V. *et al*, *Zh. Obshch. Khim.*, 1979, **49**, 1235 (*Engl. transl.* p. 1083) (*synth*)
Sax, N.I., *Dangerous Properties of Industrial Materials*, 6th Ed., Van Nostrand-Reinhold, 1984, 1042.

Tributyl phosphite, 9CI, 8CI — T-00379

[102-85-2]

$P(OCH_2CH_2CH_2CH_3)_3$

$C_{12}H_{27}O_3P$ M 250.317

Forms complexes containing metals of Groups IB, VIB, and VIII. Liq. with sickly odour. d_4^{20} 0.93. Bp_{12} 122°. n_D^{20} 1.4320. Hydrol. by H_2O.

▷TH0875000.

Gerrard, W., *J. Chem. Soc.*, 1940, 1464 (*synth*)
Mel'nikov, N.N. *et al*, *Zh. Obshch. Khim.*, 1958, **28**, 2473 (*Engl. transl.* p. 2507) (*synth*)
Moedritzer, K., *J. Inorg. Nucl. Chem.*, 1962, **22**, 19 (*P nmr*)
Occolowitz, J.L. *et al*, *Anal. Chem.*, 1963, **35**, 1179 (*ms*)
Turkevich, V.V. *et al*, *Zh. Obshch. Khim.*, 1977, **47**, 2388 (*Engl. transl.* p. 2181) (*ir*)
Brown, C. *et al*, *Phosphorus Sulfur*, 1979, **6**, 481 (*synth*)
Vasil'ev, V.V. *et al*, *Zh. Obshch. Khim.*, 1981, **51**, 2134 (*Engl. transl.* p. 1836) (*O nmr*)
Sax, N.I., *Dangerous Properties of Industrial Materials*, 6th Ed., Van Nostrand-Reinhold, 1984, 1042.

Tri-*tert*-butyl phosphite, 8CI — T-00380

Tris(1,1-dimethylethyl) phosphite, 9CI

[15205-62-6]

$P[OC(CH_3)_3]_3$

$C_{12}H_{27}O_3P$ M 250.317

Liquid freezing to feathery cryst. Mp 4-5°, 10°. Bp_3 67-69°. Rapidly oxidised to the phosphate in air. Dec. on dist. in the presence of minute traces of acid.

Mark, V. *et al*, *J. Org. Chem.*, 1964, **29**, 1006 (*synth, ir, P nmr, pmr*)
Cox, J.R. *et al*, *J. Org. Chem.*, 1969, **34**, 2600 (*synth, ir*)

1-(Tributylphosphoranylidene)-2-heptanone, 9CI — T-00381

[35563-52-1]

$(H_3CCH_2CH_2CH_2)_3P{=}CHCO(CH_2)_4CH_3$

$C_{19}H_{39}OP$ M 314.490

Widely used in synth. of prostaglandins. Liq. which darkens on exposure to air. $Bp_{0.01}$ 140-143°.

Katsube, J. *et al*, *Agric. Biol. Chem.*, 1971, **35**, 1768 (*use*)
Finch, N. *et al*, *J. Org. Chem.*, 1973, **38**, 4412 (*synth, use*)
Jones, D.N. *et al*, *J. Chem. Soc., Perkin Trans. 1*, 1982, 1333 (*use*)

2,4,6-Tri-*tert*-butylphosphorin — T-00382

2,4,6-Tris(1,1-dimethylethyl)phosphorin, 9CI. *2,4,6-Tris(1,1-dimethylethyl)phosphinine*

[17420-29-0]

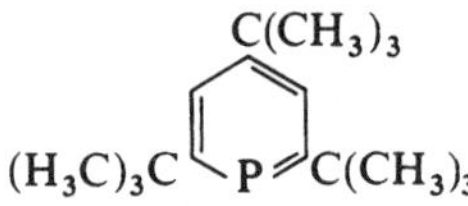

$C_{17}H_{29}P$ M 264.390

Cryst. (MeOH aq.). Mp 88°.

Dimroth, K. *et al*, *Angew. Chem., Int. Ed. Engl.*, 1968, **7**, 460 (*synth, pmr, P nmr*)
Oehling, H. *et al*, *Angew. Chem., Int. Ed. Engl.*, 1971, **10**, 656 (*pe*)
Schweig, A. *et al*, *Angew. Chem., Int. Ed. Engl.*, 1972, **11**, 631 (*pe*)
Bundgaard, T. *et al*, *Tetrahedron Lett.*, 1974, 3179 (*cmr*)

O,Se,Se-Tributyl phosphorodiselenoite — T-00383

[66498-95-1]

$(H_3CCH_2CH_2CH_2Se)_2PO(CH_2)_3CH_3$

$C_{12}H_{27}OPSe_2$ M 376.238

Liq. d_4^{20} 1.29. $Bp_{0.01}$ 93°. n_D^{20} 1.5430.

Kolodii, Ya.I. *et al*, *Zh. Obshch. Khim.*, 1978, **48**, 331 (*Engl. transl.* p. 296) (*synth*)

O,O,S-Tributyl phosphorodithioate, 9CI, 8CI — T-00384

O,O,S-*Tributyl dithiophosphate*

[17610-56-9]

$(H_3CCH_2CH_2CH_2O)_2P(S)S(CH_2)_3CH_3$

$C_{12}H_{27}O_2PS_2$ M 298.438

Extractant for Ga, In, and transition metals. Liq. d_4^{20} 1.01. Bp_5 160-163°. n_D^{20} 1.4831.

Petrov, K.A. *et al*, *Zh. Obshch. Khim.*, 1961, **31**, 1361 (*Engl. transl.* p. 1260)

O,O,O-Tributyl phosphoroselenoate, 9CI, 8CI **T-00385**

O,O,O-*Tributyl selenophosphate*

[7441-05-6]

$(H_3CCH_2CH_2CH_2O)_3P{=}Se$

$C_{12}H_{27}O_3PSe$ M 329.277

Liq. d_4^{20} 1.14. $Bp_{0.04}$ 83-84°. n_D^{20} 1.4663.

Nuretdinov, I.A. *et al*, *Izv. Akad. Nauk SSSR*, *Ser. Khim.*, 1968, 2831 (*Engl. transl.* p. 2685) (*synth, ir*)
Loginova, É.I. *et al*, *Teor. Eksp. Khim.*, 1974, **10**, 75 (*Engl. transl.* p. 47) (*nmr*)
Ando, F. *et al*, *Bull. Chem. Soc. Jpn.*, 1979, **52**, 807 (*pmr, ir*)

Tributyl phosphorotetrathioate, 9CI, 8CI **T-00386**

Tributyl tetrathiophosphate

[1642-47-3]

$(H_3CCH_2CH_2CH_2S)_3P{=}S$

$C_{12}H_{27}PS_4$ M 330.559

Liq. d_4^{20} 1.08. $Bp_{1.5}$ 168-170°. n_D^{20} 1.5675.

Menefee, A. *et al*, *J. Org. Chem.*, 1957, **22**, 792 (*ir*)
Maier, L. *et al*, *J. Am. Chem. Soc.*, 1962, **84**, 3054 (*synth*)
Blagoveshchenskii, V.S. *et al*, *Zh. Obshch. Khim.*, 1971, **41**, 1032 (*Engl. transl.* p. 1036) (*synth*)
Shaw, R.A. *et al*, *Phosphorus*, 1971, **1**, 41, 191 (*synth, ir, pmr*)
Khokhlov, P.S. *et al*, *Zh. Obshch. Khim.*, 1984, **54**, 2545 (*Engl. transl.* p. 2274) (*synth*)

O,O,O-Tributyl phosphorothioate, 9CI, 8CI **T-00387**

O,O,O-*Tributyl thiophosphate*

[78-47-7]

$(H_3CCH_2CH_2CH_2O)_3P{=}S$

$C_{12}H_{27}O_3PS$ M 282.377

Liq. Bp_{11} 156-157°. n_D^{20} 1.4497.

Michalski, J. *et al*, *Bull. Acad. Pol. Sci.*, *Ser. Sci. Chim.*, 1957, **5**, 917 (*synth*)
Fluck, E. *et al*, *Z. Anorg. Allg. Chem.*, 1967, **354**, 139 (*P nmr*)
Engel, R.R. *et al*, *J. Chem. Soc. (C)*, 1970, 523 (*props*)

O,O,S-Tributyl phosphorothioate, 9CI, 8CI **T-00388**

O,O,S-*Tributyl thiophosphate*

[26818-24-6]

$(H_3CCH_2CH_2CH_2O)_2P(O)S(CH_2)_3CH_3$

$C_{12}H_{27}O_3PS$ M 282.377

Metal extractant, similar in props. to Tributyl phosphate, T-00370 . Liq. d_4^{20} 1.00. Bp_8 160°. n_D^{20} 1.4582.

Petrov, K.A. *et al*, *Zh. Obshch. Khim.*, 1961, **31**, 1361 (*Engl. transl.* p. 1260) (*synth*)
Rozen, A.M. *et al*, *Zh. Obshch. Khim.*, 1982, **52**, 1232 (*Engl. transl.* p. 1083) (*props, use*)

O,Se,Se-Tributyl phosphorotriselenoate **T-00389**

[80821-21-2]

$(H_3CCH_2CH_2CH_2Se)_2P(Se)OCH_2CH_2CH_2CH_3$

$C_{12}H_{27}OPSe_3$ M 455.198

Liq. $Bp_{0.01}$ 110°. n_D^{20} 1.5620.

Mel'nik, Ya.I. *et al*, *Ukr. Khim. Zh.*, 1981, **47**, 1289; *CA*, **96**, 85674.

O,S,S-Tributyl phosphorotrithioate, 9CI, 8CI **T-00390**

O,S,S-*Tributyl trithiophosphate*

[13362-44-2]

$(H_3CCH_2CH_2CH_2S)_2P(S)O(CH_2)_3CH_3$

$C_{12}H_{27}OPS_3$ M 314.498

Odorous liq. d_4^{20} 1.04. Bp_1 153-154°. n_D^{20} 1.5275.

Murav'ev, I.V. *et al*, *Zh. Obshch. Khim.*, 1968, **38**, 133 (*Engl. transl.* p. 133) (*synth*)
Kotovich, B.P. *et al*, *Zh. Obshch. Khim.*, 1968, **38**, 1282 (*Engl. transl.* p. 1235) (*synth*)

S,S,S-Tri-*tert*-butyl phosphorotrithioate, 8CI **T-00391**

S,S,S-*Tris(1,1-dimethylethyl) phosphorotrithioate*, 9CI.
S,S,S-*Tri*-tert-*butyl trithiophosphate*

[78788-16-6]

$[(H_3C)_3CS]_3P{=}O$

$C_{12}H_{27}OPS_3$ M 314.498

Odourous liq. d^{29} 1.09. $Bp_{0.2}$ 90-91°. n_D^{25} 1.5222.

Ibrahim, E.H. *et al*, *Egypt. J. Chem.*, 1979, **22**, 403 (*synth, ir*)

S,S,S-Tributyl phosphorotrithioate, 9CI, 8CI **T-00392**

S,S,S-*Tributyl trithiophosphate. Butiphos. Morphos oxide. DEF*

[78-48-8]

$(H_3CCH_2CH_2CH_2S)_3P{=}O$

$C_{12}H_{27}OPS_3$ M 314.498

Insecticide; defoliant for cotton. Liq. d^{20} 1.06. Bp_1 167-170°. n_D^{20} 1.5345. Rel. stable to heat and to acids. Slowly hydrolysed by aq. alkali. Strongly phytotoxic; weak cholinesterase inhibitor.

▷TG5425000.

Keith, L.H. *et al*, *J. Assoc. Off. Anal. Chem.*, 1968, **51**, 1063 (*pmr*)
Shaw, R.A. *et al*, *Phosphorus*, 1971, **1**, 41; 1972, **1**, 191 (*synth, pmr*)
Kilgore, W. *et al*, *Residue Rev.*, 1984, **91**, 71 (*tox, metab*)
McElroy, R.D. *et al*, *J. Agric. Food Chem.*, 1984, **32**, 119 (*synth, tox, pharmacol*)
Pesticide Manual, 7th Ed., 543.

Tributyl phosphorotrithioite, 9CI **T-00393**

Tributyl trithiophosphite. Folex. Merphos

[150-50-5]

$(H_3CCH_2CH_2CH_2S)_3P$

$C_{12}H_{27}PS_3$ M 298.499

Defoliant. Pale-yellow liq. V. sol. most org. solvs., spar. sol. H_2O. d_4^{25} 1.04. Bp_{15} 174-178°, Bp_1 145-148°. n_D^{25} 1.5307.

▷Highly toxic. TG5600000.

Lippert, A. *et al*, *J. Am. Chem. Soc.*, 1938, **60**, 2370 (*synth, deriv, complexes*)

Keith, L.H. *et al*, *J. Assoc. Off. Anal. Chem.*, 1968, **51**, 1063 (*pmr*)
Bowman, M.C. *et al*, *J. Assoc. Off. Anal. Chem.*, 1970, **53**, 499 (*glc*)
Ross, R.T. *et al*, *Anal. Chim. Acta*, 1970, **52**, 139 (*nmr*)
Gore, R.C. *et al*, *J. Assoc. Off. Anal. Chem.*, 1971, **54**, 1040 (*ir, uv*)
Nifant'ev, E.É. *et al*, *Zh. Obshch. Khim.*, 1972, **42**, 1936 (*Engl. transl.* p. 1929) (*synth*)
Nicholas, M.L. *et al*, *J. Assoc. Off. Anal. Chem.*, 1976, **59**, 1071 (*raman*)
Pesticide Manual, 6th Ed., 527, 528; 7th Ed., 544.
Sax, N.I., *Dangerous Properties of Industrial Materials*, 6th Ed., Van Nostrand-Reinhold, 1984, 1042.

N,N′,N″-Tributylphosphorous triamide, 9CI — T-00394

Tris(butylamino)phosphine. Tri(butylamino)phosphine

$(H_3CCH_2CH_2CH_2NH)_3P$

$C_{12}H_{30}N_3P$ M 247.363

Oxide: [23344-69-6]. N,N′,N″-*Tributylphosphoric triamide.*
$C_{12}H_{30}N_3OP$ M 263.362
Has been used in extraction of uranium. pK_{a1} 12.16, pK_{a2} 7.00 ($MeNO_2$).

Sulfide: [53364-04-8]. N,N′,N″-*Tributylphosphorothioic triamide.*
$C_{12}H_{30}N_3PS$ M 279.423
Cryst. (pet. ether). Mp 54°.

Wise, G. *et al*, *J. Am. Chem. Soc.*, 1952, **74**, 529 (*sulfide*)
Healy, J.D. *et al*, *Phosphorus Sulfur*, 1978, **5**, 239 (*sulfide*)

N,N′,N″-Tri-*tert*-butylphosphorous triamide, 8CI — T-00395

N,N′,N″-*Tris(1,1-dimethylethyl)phosphorous triamide*, 9CI

$P[NHC(CH_3)_3]_3$

$C_{12}H_{30}N_3P$ M 247.363

Oxide: [1808-64-6]. N,N′,N″-*Tris(1,1-dimethylethyl)-phosphoric triamide. Tri(*tert*-butylamino)phosphine oxide.*
$C_{12}H_{30}N_3OP$ M 263.362
Cryst. (C_6H_6). Mp 245-246°. pK_{a1} 10.08, pK_{a2} 4.42 ($MeNO_2$).

Buchikhin, E.P. *et al*, *Zh. Obshch. Khim.*, 1974, **44**, 1354 (*Engl. transl.* p. 1330) (*props*)
Quast, H. *et al*, *Justus Liebigs Ann. Chem.*, 1981, 943 (*synth, pmr, ir, P nmr, ms*)

2,4,6-Tributyl-1,3,5,2,4,6-trioxatriphosphorinane, 9CI — T-00396

2,4,6-Tributyl-1,3,5,2,4,6-trioxatriphosphinane

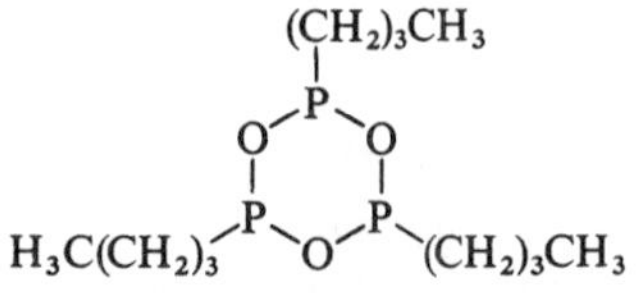

$C_{12}H_{27}O_3P_3$ M 312.265

2,4,6-Trisulfide: [58160-80-8]. *Butylphosphonothioic acid trimer trianhydride.*
$C_{12}H_{27}O_3P_3S_3$ M 408.445
Viscous liq. d_4^{20} 1.20. $Bp_{0.08}$ 150-152°. n_D^{20} 1.5438.

Grishina, O.N. *et al*, *Zh. Obshch. Khim.*, 1975, **45**, 2344 (*Engl. transl.* p. 2302) (*synth, P nmr*)

Tri-*tert*-butyltriphosphirane — T-00397

Tris(1,1-dimethylethyl)triphosphirane, 9CI. *Tri-*tert*-butylcyclotriphosphine*

[61695-12-3]

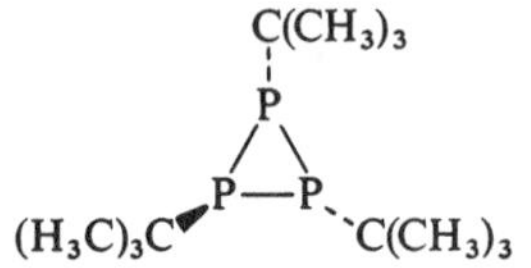

(1α,2α,3β)-*form*

$C_{12}H_{27}P_3$ M 264.267

Solid. $Bp_{0.2}$ 76-80°. Ring cleaved by the action of Cl_2, Br_2, or I_2.

(1α,2α,3β)-form [81739-37-9]
Subl. at r.t./0.001 mm.

Baudler, M. *et al*, *Z. Naturforsch., B*, 1976, **31**, 1305, 1311 (*synth, ir, P nmr*)
Baudler, M. *et al*, *Z. Naturforsch., B*, 1981, **36**, 266 (*synth*)
Baudler, M. *et al*, *Z. Anorg. Allg. Chem.*, 1981, **480**, 129 (*props*)
Gleiter, R. *et al*, *Chem. Ber.*, 1981, **114**, 1004 (*struct*)
Hahn, J. *et al*, *Z. Naturforsch., B*, 1982, **37**, 797 (*cmr, P nmr, pmr, cryst struct*)

(Trichloroacetyl)phosphorimidic trichloride, 9CI — T-00398

P,P,P-*Trichloro*-N-*(trichloroacetyl)phosphine imide.* P,P,P-*Trichloro*-N-*(trichloroacetyl)phosphazene.* P,P,P-*Trichloro*-N-*(trichloroacetyl)iminophosphorane*

[14335-48-9]

$Cl_3CCON{=}PCl_3$

C_2Cl_6NOP M 297.720

Dyadyusha, G.G. *et al*, *Zh. Strukt. Khim.*, 1972, **13**, 155 (*Engl. transl.* p. 142) (*ir*)
Kyuntsel', I.A. *et al*, *Zh. Obshch. Khim.*, 1975, **45**, 1989 (*Engl. transl.* p. 1954) (*nqr*)

Trichlorobis(trichloromethyl)phosphorane, 9CI, 8CI — T-00399

[21089-18-9]

$Cl_3P(CCl_3)_2$

C_2Cl_9P M 374.073

As solid and in soln. in benzene, molecule consists of unionized trigonal bipyramid with axial CCl_3 groups. Prisms (pet. ether). Mp 192-193° (sealed tube).

Frank, A.W., *Can. J. Chem.*, 1968, **46**, 3573 (*synth, ir*)
Kozlov, É.S. *et al*, *Zh. Obshch. Khim.*, 1969, **39**, 933 (*Engl. transl.* p. 902) (*synth*)
Kozlov, É.S. *et al*, *Zh. Obshch. Khim.*, 1972, **42**, 756 (*Engl. transl.* p. 748) (*ir, nqr, struct*)
Dmitriev, V.I. *et al*, *Zh. Obshch. Khim.*, 1980, **50**, 2230 (*Engl. transl.* p. 1799) (*nmr, nqr, struct*)
Sergienko, L.M. *et al*, *Zh. Obshch. Khim.*, 1981, **51**, 494; *CA*, **94**, 207998 (*uv*)
Kozlov, É.S. *et al*, *Zh. Obshch. Khim.*, 1982, **52**, 2513 (*Engl. transl.* p. 2219) (*ir, Raman*)

2,2,2-Trichloro-2,2-dihydro-1,3,2-benzodioxaphosphole, 9CI T-00400

[2007-97-8]

$C_6H_4Cl_3O_2P$ M 245.429

Converts carboxylic anhydrides into acid chlorides and trialkyl phosphites into dialkyl phosphorochloridites. Chlorinating agent. Pale-yellow solid. V. sol. pet. ether. Bp_{11} 132°, $Bp_{1.3}$ 104°.

Khwaja, T. *et al*, *J. Chem. Soc. (C)*, 1970, 2092 (*synth*)
Gloede, J. *et al*, *J. Prakt. Chem.*, 1974, **316**, 703 (*use*)
Dillon, K.B. *et al*, *J. Inorg. Nucl. Chem.*, 1976, **38**, 1439 (*synth, P nmr*)
Dillon, K.B. *et al*, *J. Chem. Soc., Dalton Trans.*, 1978, 1465 (*complexes*)
Gloede, J. *et al*, *Z. Chem.*, 1982, **22**, 126 (*rev*)
Kukhor', V.P. *et al*, *Zh. Obshch. Khim.*, 1982, **52**, 2227 (*Engl. transl.* p. 1982) (*props, use*)

2,2,2-Trichloro-2,2-dihydro-1,3-dimethyl-6,8-bis(trifluoromethyl)-1,3,5,7,9-pentaaza-2,4-diphospha(4-P^V)spiro[3.5]nona-4,6,8-triene, 10CI T-00401

4,4,4-Trichloro-1,3-dimethyl-4',6'-bis(trifluoromethyl)-spiro[1,3,2(P^V),4(P^V)-diazaphosphetidine-2,2'-[1,3,5,2(P^V)-triazaphosphorin]

[64595-13-7]

$C_6H_6Cl_3F_6N_5P$ M 399.470

Solid. Mp 106°. $Bp_{0.01}$ 40° subl.

Schöning, G. *et al*, *Chem. Ber.*, 1977, **110**, 3231 (*synth, ir, ms, pmr, F and P nmr*)

Trichlorodimethylphosphorane, 9CI T-00402

Dichlorodimethylphosphonium chloride

[2725-67-9]

Me_2PCl_3 or $Me_2PCl_2^{\oplus}Cl^{\ominus}$

$C_2H_6Cl_3P$ M 167.402

Shown to have ionic struct. Cryst. Mp 183° dec.

Baumgärtner, R. *et al*, *Z. Anorg. Allg. Chem.*, 1964, **333**, 171 (*synth, ir, raman, struct*)
Dillon, K.B. *et al*, *J. Chem. Soc., Dalton Trans.*, 1976, 1243 (*nmr, nqr*)
Dillon, K.B. *et al*, *Polyhedron*, 1982, **1**, 123 (*nmr*)

Trichlorodiphenoxyphosphorane, 9CI, 8CI T-00403

[15247-35-5]

$(PhO)_2PCl_3$

$C_{12}H_{10}Cl_3O_2P$ M 323.543

Viscous oil or pale-yellow solid.

Tikhonina, N.A. *et al*, *Izv. Akad. Nauk SSSR, Ser. Khim.*, 1976, 2624 (*Engl. transl.* p. 2442) (*synth, props*)

Trichlorodiphenylphosphorane, 9CI, 8CI T-00404

[1017-89-6]

Ph_2PCl_3

$C_{12}H_{10}Cl_3P$ M 291.544

Unionized in solid state. Needles (C_6H_6). Mp 199-201°.

Herring, D.L. *et al*, *Inorg. Chem.*, 1964, **3**, 428 (*synth*)
Whitehead, M.A. *et al*, *J. Chem. Soc. (A)*, 1971, 1738 (*nqr*)
Dillon, K.B. *et al*, *J. Chem. Soc., Dalton Trans.*, 1976, 1243 (*nmr, nqr, struct*)
Timokhin, B.V. *et al*, *Zh. Obshch. Khim.*, 1978, **48**, 1421 (*Engl. transl.* p. 1304) (*nmr*)
Sergienko, L.M. *et al*, *Zh. Obshch. Khim.*, 1979, **49**, 317 (*Engl. transl.* p. 275) (*uv*)
Kuchen, W. *et al*, *Chem. Ber.*, 1981, **114**, 3485 (*synth, pmr, nmr*)
Lindner, E. *et al*, *Chem. Ber.*, 1982, **115**, 2181.

2,2,2-Trichloroethyl 2-chlorophenyl phosphorochloridate, 9CI T-00405

[59819-52-2]

$C_8H_6Cl_5O_3P$ M 358.372

Phosphorylating agent used in nucleotide synth. Oil. $Bp_{0.15}$ 130-135°.

Van Boom, J.H. *et al*, *Tetrahedron Lett.*, 1976, 869 (*synth, use*)
Van Boom, J.H. *et al*, *Tetrahedron*, 1978, **34**, 1999 (*use*)
De Rooij, J.F.M. *et al*, *Recl. Trav. Chim. Pays-Bas*, 1979, **98**, 537 (*use*)
Den Hartog, J.A.J. *et al*, *J. Org. Chem.*, 1981, **46**, 2242 (*use*)
Flockerzi, D. *et al*, *Helv. Chim. Acta*, 1983, **66**, 2069 (*use*)

2,2,2-Trichloroethyl phosphorodichloridite, 9CI T-00406

2,2,2-Trichloroethyl dichlorophosphite

[60010-51-7]

$Cl_3CCH_2OPCl_2$

$C_2H_2Cl_5OP$ M 250.276

Useful reagent for synth. of internucleotide links. Pungent liq. d^{22} 1.70. $Bp_{0.1}$ 42°. n_D^{22} 1.5140. Easily hydrolysed.

Gerrard, W., *et al*, *J. Chem. Soc.*, 1954, 1148 (*synth*)
Ogilvie, K.K., *et al*, *J. Am. Chem. Soc.*, 1977, **99**, 7741 (*use*)
Melnick, B.P. *et al*, *J. Org. Chem.*, 1980, **45**, 2715 (*use*)
Ogilvie, K.K. *et al*, *Can. J. Chem.*, 1980, **58**, 2686 (*synth, P nmr, use*)
Imai, J. *et al*, *J. Org. Chem.*, 1981, **46**, 4015 (*synth*)

2,4,6-Trichloro-2,2,4,4,6,6-hexahydro-2,4,6-triphenyl-1,3,5-triaza-2,4,6-triphosphorine, 9CI T-00407

2,4,6-Trichloro-2,2,4,4,6,6-hexahydro-2,4,6-triphenylcyclotriphosphazene

[3006-66-4]

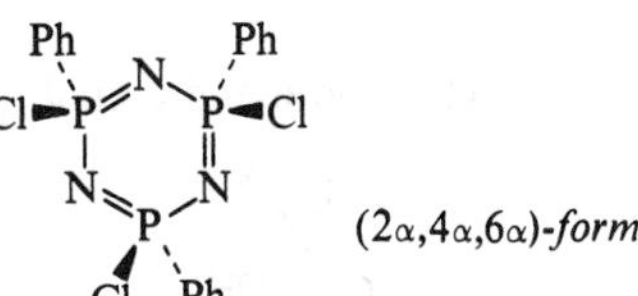

(2α,4α,6α)-*form*

$C_{18}H_{15}Cl_3N_3P_3$ M 472.617

(2α,4α,6α)-form [3606-85-7]
cis-*form*
Solid. Mp 191-192°. Forms a 1:1 solvate with C_6H_6.
(2α,4α,6β)-form [3587-00-6]
trans-*form*
Solid. Mp 158-159°.

Dagliesh, W.H. *et al*, *J. Chem. Soc., Dalton Trans.*, 1975, 309 (*nqr*)
Keat, R. *et al*, *J. Chem. Soc., Dalton Trans.*, 1976, 1582 (*P nmr*)
Allen, C.W. *et al*, *J. Chem. Soc., Dalton Trans.*, 1978, 173 (*ms*)
Krishnamurthy, S.S. *et al*, *Org. Magn. Reson.*, 1981, **15**, 205 (*cmr*)

(2,2,2-Trichloro-1-hydroxyethyl)-phosphonic acid, 9CI — T-00408

2,2,2-Trichloro-1-phosphonoethanol
[684-17-3]

$Cl_3CCH(OH)P(O)(OH)_2$

$C_2H_4Cl_3O_4P$ M 229.384

(±)-form
Di-Me ester: see Trichlorphon, T-00423
Di-Et ester: [993-86-2]. *Diethyl (2,2,2-trichloro-1-hydroxyethyl)phosphonate.*
$C_6H_{12}Cl_3O_4P$ M 285.491
Solid. Mp 57-58°.
▷SZ8570000.
Diisopropyl ester: [996-42-9]. *Diisopropyl (2,2,2-trichloro-1-hydroxyethyl)phosphonate.*
$C_8H_{16}Cl_3O_4P$ M 313.545
Solid. Mp 105-106.5°.

Barthel, W.F. *et al*, *J. Am. Chem. Soc.*, 1954, **76**, 4186 (*synth*)
Biryukov, I.P. *et al*, *Zh. Obshch. Khim.*, 1972, **42**, 1223 (*Engl. transl.* p. 1217) (*nqr*)
Nikonorov, K.V. *et al*, *Zh. Obshch. Khim.*, 1973, **43**, 1925 (*Engl. transl.* p. 1910) (*props*)
Gazizov, T.Kh. *et al*, *Zh. Obshch. Khim.*, 1976, **46**, 2383 (*Engl. transl.* p. 2281) (*P nmr*)

(4-Trichloromethylphenyl)phosphonic acid, 8CI — T-00409

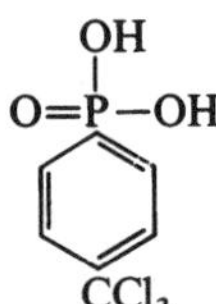

$C_7H_6Cl_3O_3P$ M 275.455
Dichloride:
$C_7H_4Cl_5OP$ M 312.347
Liq. d_{20}^{21} 1.59. $Bp_{0.05}$ 129-131°. n_D^{21} 1.5830.
Dianilide: N,N′-*Diphenyl*-P-[(*4-trichloromethyl*)-*phenyl*]*phosphonic diamide.*
$C_{19}H_{16}Cl_3N_2OP$ M 425.681
Needles (MeOH). Mp 194-195°.

Yagupol'skii, L.M. *et al*, *Zh. Obshch. Khim.*, 1960, **30**, 4026 (*Engl. transl.* p. 3986)

P-Trichloromethylphosphonamidic acid — T-00410

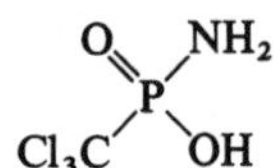

$CH_3Cl_3NO_2P$ M 198.373

Et ester: [42003-31-6]. *Ethyl* P-*trichloromethylphosphonamidate.*
$C_3H_7Cl_3NO_2P$ M 226.427
Cryst. (CCl_4). Mp 128-130°.
Ph ester: [42003-32-7]. *Phenyl* P-*trichloromethylphosphonamidate.*
$C_7H_7Cl_3NO_2P$ M 274.471
Cryst. (C_6H_6). Mp 147-148°.

Shokol, V.A. *et al*, *Zh. Obshch. Khim.*, 1973, **43**, 267 (*Engl. transl.* p. 266)

(Trichloromethyl)phosphonic acid, 9CI — T-00411

[5994-41-2]

$Cl_3CP(O)(OH)_2$

$CH_2Cl_3O_3P$ M 199.358
Cryst. + $2H_2O$. Mp 81-82°. pK_{a1} 1.93, pK_{a2} 3.31 (4.93) (H_2O, 25°).
Mono-anilinium salt: Solid. Mp 262-263°.
Di-Me ester: [29238-81-1]. *Dimethyl (trichloromethyl)-phosphonate.*
$C_3H_6Cl_3O_3P$ M 227.411
d_0^{15} 1.46. Bp_{12} 121-122°. n_D^{14} 1.4580.
Di-Et ester: [866-23-9]. *Diethyl (trichloromethyl)-phosphonate.*
$C_5H_{10}Cl_3O_3P$ M 255.465
Converts carboxylic acids into their ethyl esters. With Li → anion of diethyl ester of (Dichloromethyl)-phosphonic acid, D-00175 . Liq. d_0^{14} 1.37. Bp_{16} 129-130.5°. n_D^{14} 1.4585.
▷TA0988000.
Di-2-propenyl ester: Di-2-propenyl (trichloromethyl)-phosphonate. Diallyl (trichloromethyl)phosphonate.
$C_7H_{10}Cl_3O_3P$ M 279.487
Liq. d_0^{20} 1.50. Bp_{10} 136-138°. n_D^{20} 1.4552.
Diisopropyl ester: Diisopropyl (trichloromethyl)-phosphonate.
$C_7H_{14}Cl_3O_3P$ M 283.519
Liq. d_0^{22} 1.22. Bp_{12} 127-130°. n_D^{20} 1.4478.
Di-Ph ester: [23614-63-3]. *Diphenyl (trichloromethyl)-phosphonate.*
$C_{13}H_{10}Cl_3O_3P$ M 351.553
Cryst. (2-propanol). Mp 54°, 64-65°. $Bp_{0.1}$ 158°.
Difluoride: see (Trichloromethyl)phosphonic difluoride, T-00414
Dichloride: see (Trichloromethyl)phosphonic dichloride, T-00413
Diisocyanate: [20416-73-3].
$C_3Cl_3N_2O_3P$ M 249.377
Solid. Mp 82-84°.
Diisothiocyanate:
$C_3Cl_3N_2OPS_2$ M 281.499
Solid. Mp 36-37°. $Bp_{0.03}$ 110-111°.
Diamide: see P-(Trichloromethyl)phosphonic diamide, T-00412

Kamai, G. *et al*, *Zh. Obshch. Khim.*, 1946, **16**, 1521 (*esters*)
Bengelsdorf, I.S. *et al*, *J. Am. Chem. Soc.*, 1955, **77**, 2869 (*synth, esters*)
Kennard, K.C. *et al*, *J. Am. Chem. Soc.*, 1955, **77**, 1156 (*synth, derivs*)
Derkach, G.I. *et al*, *CA*, 1968, **69**, 67484 (*diisocyanate, diisothiocyanate*)
Derkach, G.I. *et al*, *Zh. Obshch. Khim.*, 1968, **38**, 1784 (*Engl. transl.* p. 1739) (*diisocyanate*)
Berlin, K.D. *et al*, *J. Chem. Eng. Data*, 1970, **15**, 579 (*aryl esters, P nmr*)

Bel'skii, V.E. *et al*, *Zh. Obshch. Khim.*, 1972, **42**, 2427 (*Engl. transl.* p. 2421) (*ester, P nmr*)
Seyferth, D. *et al*, *J. Organomet. Chem.*, 1973, **59**, 237 (*ester, props, use*)
Rosin, H. *et al*, *J. Org. Chem.*, 1975, **40**, 3298 (*ester, props*)
Van Der Veken, B.J. *et al*, *J. Mol. Struct.*, 1976, **32**, 393 (*ir, raman*)
Corallo, M. *et al*, *Phosphorus Sulfur*, 1978, **4**, 19 (*esters, ir, pmr, ms*)
Griffiths, W.R. *et al*, *Phosphorus Sulfur*, 1978, **5**, 101 (*esters, ms*)
Downie, I.M. *et al*, *Tetrahedron*, 1982, **38**, 1457 (*ester, use*)
Hall, C.R. *et al*, *J. Chem. Soc., Perkin Trans. 1*, 1984, 669.

P-(Trichloromethyl)phosphonic diamide, 9CI **T-00412**

[69098-18-6]

$Cl_3CP(O)(NH_2)_2$

$CH_4Cl_3N_2OP$ M 197.388

N,N′-*Di-Me:* (*Trichloromethyl*)*phosphonic bis*(*methylamide*). P-(*Trichloromethyl*)-N,N′-*dimethylphosphonic diamide.*
$C_3H_8Cl_3N_2OP$ M 225.442
Cryst. (H_2O). Mp 133-135°.

N,N,N′,N′-*Tetra-Et:* [77339-53-8]. P-(*Trichloromethyl*)-N,N,N′,N′-*tetraethylphosphonic diamide.* (*Trichloromethyl*)*phosphonic bis*(*diethylamide*).
$C_9H_{20}Cl_3N_2OP$ M 309.603
Liq. d_4^{20} 1.28. $Bp_{0.03}$ 84-86°. n_D^{20} 1.5020.

N,N′-*Di-Ph:* P-(*Trichloromethyl*)-N,N′-*diphenylphosphonic diamide.* (*Trichloromethyl*)*phosphonic dianilide.*
$C_{13}H_{12}Cl_3N_2OP$ M 349.583
Solid. Mp 172°.

N,N′-*Dicyclohexyl:* P-(*Trichloromethyl*)-N,N′-*dicyclohexylphosphonic diamide.* (*Trichloromethyl*)-*phosphonic bis*(*cyclohexylamide*).
$C_{13}H_{24}Cl_3N_2OP$ M 361.678
Cryst. (EtOH). Mp 191-192°.

Yakubovich, A.Ya. *et al*, *Zh. Obshch. Khim.*, 1954, **24**, 1465 (*dianilide*)
Kennard, K.C. *et al*, *J. Am. Chem. Soc.*, 1955, **77**, 1156 (*biscyclohexylamide*)
Arceneaux, R.L. *et al*, *J. Org. Chem.*, 1959, **24**, 1419 (*bismethylamide*)
Kozlov, É.S. *et al*, *Zh. Obshch. Khim.*, 1980, **50**, 2672 (*Engl. transl.* p. 2156) (*tetraethyl, ir, pmr, P nmr*)

(Trichloromethyl)phosphonic dichloride, 9CI **T-00413**

[21510-59-8]

$Cl_3CP(O)Cl_2$

CCl_5OP M 236.249
Solid. Mp 156°. Bp_1 95°.

Kinnear, A.M. *et al*, *J. Chem. Soc.*, 1952, 3437 (*synth*)
Ponomarenko, F.I. *et al*, *Zh. Obshch. Khim.*, 1969, **39**, 382 (*Engl. transl.* p. 359) (*synth*)
Schmidtberg, G. *et al*, *Org. Mass Spectrom.*, 1974, **9**, 844 (*ms*)
Griffiths, W.R. *et al*, *Phosphorus Sulfur*, 1978, **4**, 341 (*ms*)

(Trichloromethyl)phosphonic difluoride, 9CI **T-00414**

(*Trichloromethyl*)*phosphonyl difluoride*

[17919-24-3]

Cl_3CPOF_2

CCl_3F_2OP M 203.340
Solid. Mp 62-64°. Bp 96°, 104-105°. Fumes in air, rapidly attacked by moisture.

Shokol, V.A. *et al*, *Zh. Obshch. Khim.*, 1967, **37**, 2528 (*Engl. transl.* p. 2406) (*synth*)
Bender, R. *et al*, *Phosphorus*, 1974, **4**, 186 (*synth, nmr*)

(Trichloromethyl)phosphonothioic acid, 9CI **T-00415**

$Cl_3CP(S)(OH)_2 \rightleftharpoons Cl_3CP(S)(OH)(SH)$

$CH_2Cl_3O_2PS$ M 215.418

O,O-*Di-Et ester:* [17023-23-3]. O,O-*Diethyl* (*trichloromethyl*)*phosphonothioate.*
$C_5H_{10}Cl_3O_2PS$ M 271.526
Liq. d_4^{20} 1.33-1.36. Bp_{14} 127-128°, $Bp_{0.07}$ 62°. n_D^{20} 1.5035.

Dichloride: [17544-45-5].
CCl_5PS M 252.310
Cryst. (hexane). Mp 178°. Bp_{10} 95°. Sublimes.

Almasi, L. *et al*, *Chem. Ber.*, 1964, **97**, 623 (*ester, synth, ir*)
Atkinson, R.E. *et al*, *J. Chem. Soc. (C)*, 1967, 2542 (*ester, synth, ir*)
Frank, A.W., *Can. J. Chem.*, 1968, **46**, 3573 (*dichloride, synth, ir*)
Grishina, O.N. *et al*, *Zh. Obshch. Khim.*, 1975, **45**, 289 (*Engl. transl.*, p. 276) (*ester, dichloride, synth*)

(Trichloromethyl)phosphonous acid **T-00416**

$Cl_3CP(OH)_2$

$CH_2Cl_3O_2P$ M 183.358

Di-Et ester: [17051-95-5]. *Diethyl* (*trichloromethyl*)-*phosphonite.*
$C_5H_{10}Cl_3O_2P$ M 239.466
Liq. Bp_8 86-87°.

Difluoride: [1112-03-4].
CCl_3F_2P M 187.341
Liq. Mp 16°. Bp 73° (est.).

Dichloride: see Trichloromethylphosphonous dichloride, T-00417

Nixon, J.F., *Chem. Ind.* (*London*), 1963, 1555 (*difluoride, synth, F nmr*)
Nixon, J.F., *J. Chem. Soc.*, 1964, 2469 (*difluoride, ir*)
Nixon, J.F., *Spectrochim. Acta*, 1964, **20**, 1835 (*difluoride, P nmr*)
Atkinson, R.E. *et al*, *J. Chem. Soc. (C)*, 1967, 2542 (*ester, synth*)
Lappert, M.J. *et al*, *J. Chem. Soc., Dalton Trans.*, 1975, 1207 (*difluoride, pe*)

Trichloromethylphosphonous dichloride, 9CI **T-00417**

Dichloro(*trichloromethyl*)*phosphine*

[3582-11-4]

Cl_3CPCl_2

CCl_5P M 220.250
Solid. Mp 47°. Bp_{750} 171-172°, Bp_{23} 69-70°.

Quin, L.D. *et al*, *J. Org. Chem.*, 1958, **23**, 1693 (*synth, props*)
Ivin, S.Z. *et al*, *Zh. Obshch. Khim.*, 1966, **36**, 950 (*Engl. transl.* p. 966) (*synth*)
Inorg. Synth., 1970, **12**, 290 (*synth*)
McGlinchey, M.J. *et al*, *J. Chem. Soc. (A)*, 1970, 31 (*ms*)
Lappert, M. *et al*, *J. Chem. Soc., Dalton Trans.*, 1975, 1207 (*pe*)
Frenking, G. *et al*, *Phosphorus Sulfur*, 1980, **8**, 337 (*pe, struct*)
Prishchenko, A.A. *et al*, *Zh. Obshch. Khim.*, 1985, **55**, 340; *J. Gen. Chem. USSR* (*Engl. Transl.*), 299 (*cmr*)

(Trichloromethyl)phosphorimidic trichloride, 9CI T-00418

P,P,P-*Trichloro*-N-*(trichloromethyl)phosphine imide.*
P,P,P-*Trichloro*-N-*(trichloromethyl)phosphazene.*
P,P,P-*Trichloro*-N-*(trichloromethyl)iminophosphorane*
[10545-19-4]

$Cl_3CN{=}PCl_3$

CCl_6NP M 269.709
Oil. d_4^{20} 1.79. Bp_{12} 102-103°, $Bp_{0.01}$ 56°. n_D^{20} 1.5502. Monomeric in C_6H_6. Strong heat → PCl_5 + cyanuric chloride.

Kozlov, É.S. *et al, Zh. Obshch. Khim.*, 1966, **36**, 760 (*Engl. transl.* p. 774) (*synth*)
Fluck, E. *et al, Z. Anorg. Allg. Chem.*, 1968, **356**, 307 (*synth, P nmr*)
Tarasevich, A.S. *et al, Teor. Eksp. Khim.*, 1971, **7**, 828 (*Engl. transl.* p. 676) (*P nmr, struct*)
Romanenko, E.A. *et al, Teor. Eksp. Khim.*, 1977, **13**, 70 (*Engl. transl.* p. 50) (*nqr*)

Trichloronate, BSI T-00419

O-*Ethyl* O-*(2,4,5-trichlorophenyl) ethylphosphonothioate, 9CI. Agrisil. Agritox. Phytosol*
[327-98-0]

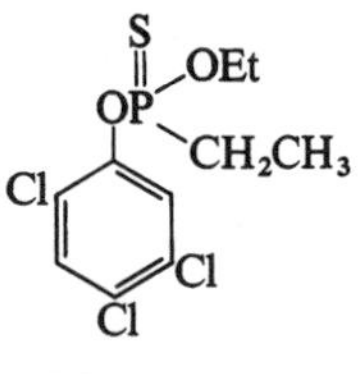

$C_{10}H_{12}Cl_3O_2PS$ M 333.596
Nonsystemic soil insecticide. Amber liq. d_4^{20} 1.365. $Bp_{0.01}$ 108°.
▷TB0700000.

B.P., 889 346, (*1959*); *CA*, **56**, 8748e
Menn, J.J. *et al, Residue Rev.*, 1974, **53**, 35 (*rev*)
Francis, B.M. *et al, J. Environ. Sci. Health, Part B*, 1980, **15**, 313 (*tox*)
Stan, H.J. *et al, J. Chromatogr.*, 1983, **279**, 173 (*glc*)
The Agrochemicals Handbook, Royal Society of Chemistry, 1983, 407 (*rev*)
Verschueren, K., *Handbook of Environmental Data*, Van Nostrand, 2nd Ed., 1983, 1136 (*rev*)
Pesticide Manual, 6th Ed., 530; 7th Ed., 546.

(2,4,6-Trichlorophenyl)phosphoramidic acid, 9CI, 8CI T-00420

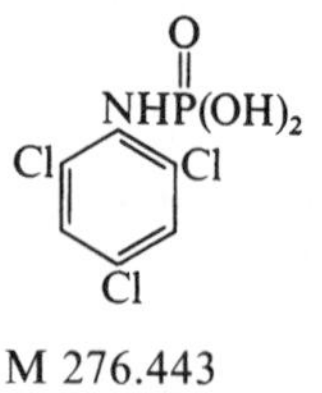

$C_6H_5Cl_3NO_3P$ M 276.443
Di-Ph ester: Diphenyl (2,4,6-trichlorophenyl)-phosphoramidate.
$C_{18}H_{13}Cl_3NO_3P$ M 428.638
Cryst. (EtOH). Mp 142-143°.
Dichloride:
$C_6H_3Cl_5NOP$ M 313.335
Cryst. (pet. ether). Mp 128°.

Michaelis, A., *Justus Liebigs Ann. Chem.*, 1903, **326**, 129 (*dichloride*)
Zhmurova, I.N. *et al, Zh. Obshch. Khim.*, 1961, **31**, 3741 (*Engl. transl.* p. 3495) (*ester*)

2,2,4-Trichloro-4,6,6-tris(dimethylamino)-2,2,4,4,6,6-hexahydro-1,3,5-triaza-2,4,6-triphosphorine, 9CI T-00421

2,2,4-Trichloro-4,6,6-tris(dimethylamino)-2,2,4,4,6,6-hexahydrocyclotriphosphazene
[957-07-3]

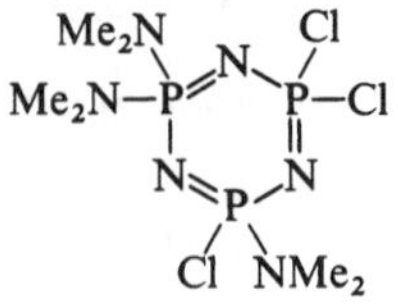

$C_6H_{18}Cl_3N_6P_3$ M 373.529
Cryst. (pentane). Mp 71°.

Keat, R. *et al, J. Chem. Soc.*, 1965, 2215 (*synth, struct, pmr*)
Strahlberg, R. *et al, Spectrochim. Acta, Part A*, 1967, **23**, 2005 (*ir, raman*)
Keat, R. *et al, J. Chem. Soc. (A)*, 1968, 703 (*pmr*)
Ahmed, F.R. *et al, Acta Crystallogr., Sect. B*, 1972, **28**, 513 (*cryst struct*)
Keat, R. *et al, J. Chem. Soc., Dalton Trans.*, 1972, 1648 (*nqr*)
Green, B. *et al, J. Chem. Soc., Dalton Trans.*, 1973, 1042 (*pe*)
Dagliesh, W.H. *et al, J. Chem. Soc., Dalton Trans.*, 1975, 309 (*nqr*)
Faucher, J.-P. *et al, J. Mol. Struct.*, 1975, **25**, 109 (*struct*)
Keat, R. *et al, J. Chem. Soc., Dalton Trans.*, 1976, 1582 (*P nmr*)

2,4,6-Trichloro-2,4,6-tris(dimethylamino)-2,2,4,4,6,6-hexahydro-1,3,5-triaza-2,4,6-triphosphorine T-00422

2,4,6-Trichloro-2,4,6-tris(dimethylamino)-2,2,4,4,6,6-hexahydrocyclotriphosphazene
[3721-13-9]

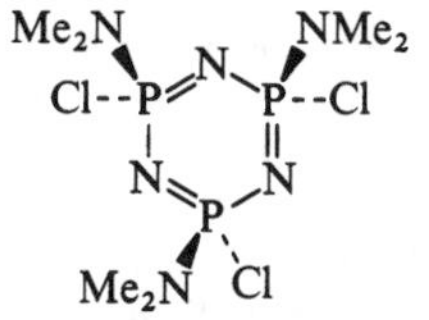

(2α,4α,6α)-*form*

$C_6H_{18}Cl_3N_6P_3$ M 373.529
Cis- and *trans* forms equilibrate in presence of $AlCl_3$.

(2α,4α,6α)-form [1216-69-9]
Cis-*form*
Solid. Mp 152°.

(2α,4α,6β)-form [957-10-8]
Trans-*form*
Cryst. (pet. ether). Mp 105°.

Keat, R. *et al, J. Chem. Soc.*, 1965, 2215 (*synth, struct, pmr*)
Strahlberg, R. *et al, Spectrochim. Acta, Part A*, 1967, **23**, 2005 (*ir, raman*)
Keat, R. *et al, J. Chem. Soc. (A)*, 1968, 703 (*pmr*)
Ahmed, F.R. *et al, Acta Crystallogr., Sect. B*, 1972, **28**, 3530; 1975, **31**, 1028 (*cryst struct*)
Green, B. *et al, J. Chem. Soc., Dalton Trans.*, 1973, 1042 (*pe*)
Dagleish, W.H. *et al, J. Chem. Soc., Dalton Trans.*, 1975, 309 (*nqr*)
Nabi, S.N. *et al, J. Chem. Soc., Dalton Trans.*, 1975, 588 (*synth*)
Keat, R. *et al, J. Chem. Soc., Dalton Trans.*, 1976, 1582 (*P nmr*)
Connelly, A. *et al, J. Magn. Reson.*, 1978, **30**, 439 (*nqr*)

Trichlorphon, BSI T-00423

Dimethyl (2,2,2-trichloro-1-hydroxyethyl)phosphonate, 9CI, 8CI. Metriphonate, BAN. Metrifonate, INN. Dipterex. Neguvon. Tugon. Dylox

[52-68-6]

$(MeO)_2P(O)CH(OH)CCl_3$

$C_4H_8Cl_3O_4P$ M 257.438

Agricultural insecticide and ectoparacticide. Cryst. Sol. C_6H_6, EtOH, mod. sol. H_2O, spar. sol. Et_2O, pet. ether. Mp 83-84°. Stable at room temp. but dec. by hot H_2O.

▷Toxic cholinesterase inhibitor. TA0700000.

Barthel, W.F. *et al*, *J. Am. Chem. Soc.*, 1954, **76**, 4186 (*synth*)
Lorenz, W. *et al*, *J. Am. Chem. Soc.*, 1955, **77**, 2554 (*synth, ir*)
Babad, H. *et al*, *Anal. Chim. Acta*, 1968, **41**, 259 (*pmr*)
Nikonorov, K.V. *et al*, *Izv. Akad. Nauk SSSR, Ser. Khim.*, 1976, **25**, 1398 (*Engl. transl.* p.1341) (*pmr*)
Stan, H.-H. *et al*, *Z. Anal. Chem.*, 1977, **287**, 271 (*glc, ms*)
Brienne, M.J. *et al*, *Nouv. J. Chim.*, 1978, **2**, 19 (*config*)
Klima, R. *et al*, *Cryst. Struct. Commun.*, 1981, **10**, 901 (*cryst struct*)
Sohr, H. *et al*, *Z. Naturforsch., B*, 1983, **38**, 819 (*polymorphism, props*)
Stobiecki, S. *et al*, *Przem. Chem.*, 1984, **63**, 22, 33; *CA*, **100**, 174954.
Pesticide Manual, 6th Ed., 531; 7th Ed., 547.

Triclofos, BAN T-00424

Mono(2,2,2-trichloroethyl) phosphate, 10CI. 2,2,2-Trichloroethanol dihydrogen phosphate, 9CI

$Cl_3CCH_2OP(O)(OH)_2$

$C_2H_4Cl_3O_4P$ M 229.384

Sedative. Cryst. (C_6H_6). Mp 120-121°.

Na salt: [7246-20-0]. *Triclofos sodium, USAN. Triclos. Tricloryl.*

▷KM4500000.

Monoanilinium salt: Solid. Mp 116-118°.

Monocyclohexylammonium salt: Solid. Mp 139-140°.

Dichloride: [18868-46-7]. *2,2,2-Trichloroethyl phosphorodichloridate. 2,2,2-Trichloroethyl phosphoryl dichloride.*

$C_2H_2Cl_5O_2P$ M 266.275

Precursor to reagents employed in nucleotide and phospholipid synth. Liq. Bp_{14} 116-118°.

Gerrard, W. *et al*, *J. Chem. Soc.*, 1954, 1148 (*dichloride*)
Hems, B.A. *et al*, *Br. Med. J.*, 1962, **1**, 1834 (*synth*)
Murakami, Y. *et al*, *J. Chem. Soc., Perkin Trans. 2*, 1973, 1235 (*ir, props*)
9wen, G.R. *et al*, *Synthesis*, 1974, 704 (*synth, dichloride, ir*)
Gajda, T. *et al*, *Synthesis*, 1977, 623 (*synth*)
Lammers, J.G. *et al*, *Recl. Trav. Chim. Pays-Bas*, 1977, **96**, 216 (*use*)
Markowska, A. *et al*, *Bull. Acad. Pol. Sci., Ser. Sci. Chim.*, 1979, **27**, 115 (*synth*)

Tricyclohexyl phosphate, 9CI T-00425

[2528-40-7]

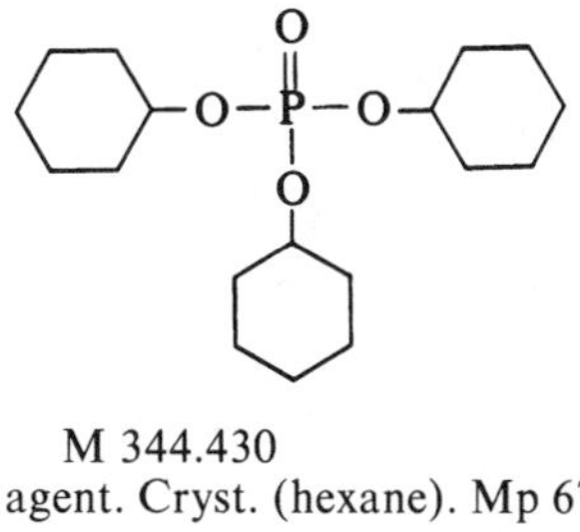

$C_{18}H_{33}O_4P$ M 344.430

Fireproofing agent. Cryst. (hexane). Mp 67-68°.

Klement, R. *et al*, *Chem. Ber.*, 1963, **96**, 1916 (*synth*)

Tricyclohexylphosphine, 9CI T-00426

[2622-14-2]

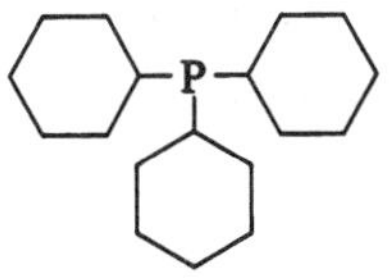

$C_{18}H_{33}P$ M 280.432

Ligand for metals of Groups of IB, IIB, VIB, VIIB and VIII. Air-sensitive cryst. Mp 76-78°. pK_a 9.70 (H_2O).

B,MeI: Tricyclohexylmethylphosphonium iodide.

$C_{19}H_{36}IP$ M 422.372

Solid. Mp 185-186°.

Oxide: [13689-19-5].

$C_{18}H_{33}OP$ M 296.432

Cryst. (EtOH). Mp 155-157°. pK_a 9.39 ($MeNO_2$).

▷SZ1600000.

Sulfide: [42201-98-9].

$C_{18}H_{33}PS$ M 312.492

Useful for extraction of lanthanides and actinides. Cryst. (EtOH or Me_2CO). Mp 185°.

Selenide: [52784-98-2].

$C_{18}H_{33}PSe$ M 359.392

Cryst. (EtOH or C_6H_6). Mp 192°.

Telluride:

$C_{18}H_{33}PTe$ M 408.032

Yellow cryst. (toluene/pet. ether). Mp 184-187°. Dec. by light and glass.

Issleib, K. *et al*, *Z. Anorg. Allg. Chem.*, 1954, **277**, 258 (*synth, oxide*)
Screttas, C. *et al*, *J. Org. Chem.*, 1962, **27**, 2573 (*synth, oxide, sulfide, selenide*)
Zingaro, R.A., *Inorg. Chem.*, 1963, **2**, 192 (*sulfide, selenide, ir*)
Zingaro, R.A. *et al*, *J. Organomet. Chem.*, 1965, **4**, 320 (*telluride, synth, props*)
Feshchenko, N.G. *et al*, *Zh. Obshch. Khim.*, 1968, **38**, 122; *CA*, **69**, 52210 (*synth*)
Olah, G.A. *et al*, *J. Org. Chem.*, 1969, **34**, 1832 (*nmr*)
Chremos, G.N. *et al*, *J. Organomet. Chem.*, 1970, **22**, 637 (*telluride, ir*)
Timokhin, B.V. *et al*, *Zh. Obshch. Khim.*, 1975, **45**, 2561 (*sulfide*)
Kerr, K.A. *et al*, *Can. J. Chem.*, 1977, **55**, 3081 (*sulfide, cryst struct*)
Dean, P.A.W., *Can. J. Chem.*, 1979, **57**, 754 (*selenide, nmr*)

P,P,P-Tricyclohexylphosphine imide, 9CI T-00427

P,P,P-*Tricyclohexylphosphine imine. Iminotricyclohexylphosphorane.* P,P,P-*Tricyclohexylphosphazene*

[54098-11-2]

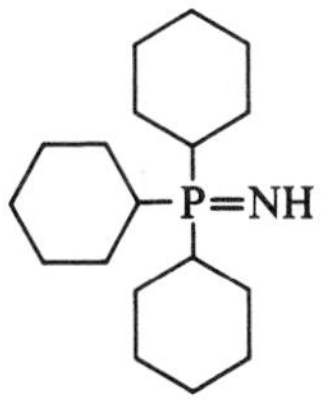

$C_{18}H_{34}NP$ M 295.447

Solid. Mp 149°.

B,HCl: Cryst. Sol. H_2O with little hydrol. Mp 256°.

N-*Me:* [75794-10-4]. P,P,P-*Tricyclohexyl*-N-*methylphosphine imide.*

$C_{19}H_{13}NP$ M 286.292

Hygroscopic cryst. Mp 75-77°.
N-*Me; B,HCl:* Hygroscopic cryst. Mp 190°.

Ross, B. *et al, Chem. Ber.*, 1974, **107**, 2720 (*synth, P nmr, pmr, derivs*)
Ross, B. *et al, Z. Anorg. Allg. Chem.*, 1980, **466**, 203 (*deriv, ms, cmr, P nmr, pmr*)

Tricyclohexyl phosphite, 9CI, 8CI — T-00428

[15205-58-0]

$C_{18}H_{33}O_3P$ M 328.431
Hygroscopic solid. Mp 73-74°. Bp_1 175-176°, Bp_2 141-146°. n_D^{20} 1.4796.

Saunders, B.C. *et al*, Tet, 1958, **4**, 169 (*synth*)
Ovchinnikova, N.K. *et al, Zh. Org. Khim.*, 1975, **11**, 1839 (*Engl. transl.* p. 1849) (*synth*)

N,N′,N″-Tricyclohexylphosphorous triamide, 9CI — T-00429

Tri(cyclohexylamino)phosphine. Tris(cyclohexylamino)phosphine

$C_{18}H_{36}N_3P$ M 325.476
Oxide: [31160-09-5]. N,N′,N″-*Tricyclohexylphosphoric triamide.*
$C_{18}H_{36}N_3OP$ M 341.476
Solid. Mp 245-246° dec. pK_{a1} 11.64, pK_{a2} 7.41 ($MeNO_2$).
Sulfide: [5332-68-3]. N,N′,N″-*Tricyclohexylphosphorothioic triamide.*
$C_{18}H_{36}N_3PS$ M 357.536
Cryst. (EtOH). Mp 143-144.5°.
▷TG4400000.

Audrieth, L. *et al, J. Am. Chem. Soc.*, 1942, **64**, 1553 (*oxide, sulfide, synth*)
Tolkmith, H., *J. Am. Chem. Soc.*, 1963, **85**, 3246 (*sulfide, pmr, P nmr*)
Marchenko, A.P. *et al, Zh. Obshch. Khim.*, 1974, **44**, 67 (*Engl. transl.* p. 63) (*oxide, props*)
Healy, J.D. *et al, Phosphorus Sulfur*, 1978, **5**, 239 (*sulfide, synth, props*)
Ibánez, W. *et al, Spectrochim. Acta, Part A*, 1982, **38**, 351 (*oxide, sulfide, ir*)

2,4,6-Tricyclohexyl-1,3,5,2,4,6-trioxatriphosphorinane, 9CI — T-00430

2,4,6-Tricyclohexyl-1,3,5,2,4,6-trioxatriphosphinane

$C_{18}H_{33}O_3P_3$ M 390.378
2,4,6-Trisulfide: [37862-86-5]. *Cyclohexylphosphonothioic acid trimer trianhydride.*
$C_{18}H_{33}O_3P_3S_3$ M 486.558
Cryst. (hexane or C_6H_6). Mp 159-161°, Mp 176-177°.

Grishina, O.N. *et al, Zh. Obshch. Khim.*, 1975, **45**, 2344 (*Engl. transl.* p. 2302) (*synth, P nmr*)
Maier, L., *Phosphorus*, 1975, **5**, 253 (*synth, P nmr, pmr, ir*)
Keck, H. *et al, Phosphorus Sulfur*, 1978, **4**, 173 (*ms*)

Tricyclopentylphosphine — T-00431

[7650-88-6]

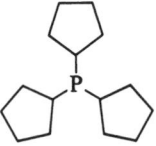

$C_{15}H_{27}P$ M 238.352
B,MeI: Methyltricyclopentylphosphonium iodide.
$C_{16}H_{30}IP$ M 380.291
Solid. Mp 293-294°.

Fluck, E. *et al, Z. Anorg. Allg. Chem.*, 1967, **354**, 139 (*P nmr, derivs*)
Fluck, E. *et al, Z. Naturforsch., B*, 1967, **22**, 1095 (*P nmr, derivs*)

Tricyclopropylphosphine, 9CI — T-00432

[13118-24-6]

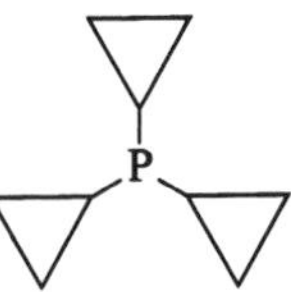

$C_9H_{15}P$ M 154.191
Ligand for Ni, Rh and W. Liq. Bp_{14} 68°, Bp_2 25°. pK_a 7.6 ($MeNO_2$).
B,MeI: Tricyclopropylmethylphosphonium iodide.
$C_{10}H_{18}IP$ M 296.130
Solid. Mp 273°.
Oxide:
$C_9H_{15}OP$ M 170.191
V. hygroscopic solid. Mp <40° subl.
Sulfide: [23270-16-8].
$C_9H_{15}PS$ M 186.251
Hygroscopic solid. Mp 53° (subl.).
Selenide: [23270-18-0].
$C_9H_{15}PSe$ M 233.151
Solid. Mp 44° subl.

Denney, D.B. *et al, J. Org. Chem.*, 1967, **32**, 2445 (*synth, ir, pmr, oxide*)

Cowley, A.H. *et al*, *J. Am. Chem. Soc.*, 1969, **91**, 2915 (*synth, ir, pmr, derivs*)
Fell, B. *et al*, *J. Mol. Catal.*, 1977, **2**, 211 (*use*)
Schmidbaur, H. *et al*, *Chem. Ber.*, 1981, **114**, 3385 (*pmr, nmr, derivs*)

Tri(decyl)phosphine **T-00433**
Tris(decyl)phosphine
[17621-07-7]

$P[(CH_2)_9CH_3]_3$

$C_{30}H_{63}P$ M 454.801
Cryst. Mp 48-50°. Bp_2 243-244°.
Oxide:
$C_{30}H_{63}OP$ M 470.801
Low-melting solid. Mp 40-42°. Bp_{3-4} 278-283°.
Selenide: [20612-75-3].
$C_{30}H_{63}PS$ M 486.861
Liq. $Bp_{0.03}$ 239-241°.

Blake, C.A. *et al*, *CA*, 1956, **50**, 15320 (*oxide*)
Feshchenko, N.G. *et al*, *Zh. Obshch. Khim.*, 1968, **38**, 122 (*Engl. transl.* p. 121) (*synth*)
Pavlenko, A.F. *et al*, *CA*, 1968, **69**, 29031 (*selenide*)

Tridecyl phosphite, 9CI, 8CI **T-00434**

$P[O(CH_2)_9CH_3]_3$

$C_{30}H_{63}O_3P$ M 502.800
Liq. Bp_3 254-255°. n_D^{20} 1.4557.

Petrov, K.A. *et al*, *Zh. Obshch. Khim.*, 1963, **33**, 899 (*Engl. transl.* p. 885) (*synth*)

Tridodecylphosphine, 9CI **T-00435**
[6411-24-1]

$P[(CH_2)_{11}CH_3]_3$

$C_{36}H_{75}P$ M 538.962
Co complexes useful in hydroformylation of C_{11-14} alkenes. Soft, waxy solid. Mp 44-45°. $Bp_{0.05}$ 255-275°.
Oxide: [13176-23-3].
$C_{36}H_{75}OP$ M 554.962
Cryst. (Me_2CO). Mp 50-52°, 60°.

Hays, H.R., *J. Org. Chem.*, 1966, **31**, 3817 (*synth, derivs*)
Borovikov, Yu.Yu. *et al*, *Zh. Obshch. Khim.*, 1970, **40**, 1957 (*Engl. transl.* p. 1942) (*sulfide, struct*)
Franks, S. *et al*, *J. Chem. Soc., Perkin Trans. 1*, 1979, 3029 (*synth, oxide, ir, pmr*)
Kohl, G. *et al*, *Chem. Tech.* (*Leipzig*), 1981, **33**, 629 (*use*)

2,2,2-Triethoxy-2,2-dihydro-1,3,2-dioxaphospholane, 9CI **T-00436**

OEt, OEt, OEt

$C_8H_{19}O_5P$ M 226.209
Liq. $Bp_{0.01}$ 81-82°.

Denney, D.B. *et al*, *J. Am. Chem. Soc.*, 1971, **93**, 4004 (*synth, pmr, P nmr*)

2,2,2-Triethoxy-2,2-dihydro-1,3,2-dioxaphosphorinane, 9CI **T-00437**
2,2,2-Triethoxy-2,2-dihydro-1,3,2-dioxaphosphinane
[23988-36-5]

OEt, OEt, OEt

$C_9H_{21}O_5P$ M 240.236
Rather unstable.

Denney, D.B. *et al*, *J. Am. Chem. Soc.*, 1969, **91**, 5821 (*synth, P nmr, props*)

Triethylamine-2,2′,2″-triphosphonic acid **T-00438**
(*Nitrilo-2,1-ethanediyl*)*trisphosphonic acid*, 9CI. *2,2′,2″-Triphosphonotriethylamine*
[34595-52-3]

$N[CH_2CH_2P(O)(OH)_2]_3$

$C_6H_{18}NO_9P_3$ M 341.131
Cryst. (MeOH aq.). Mp 226-228°. A pentabasic acid possessing poor chelating props. for polyvalent metal ions.
Hexa-Et ester: Hexaethyl (*nitrilotri-2,1-ethanediyl*)-*trisphosphonate.*
$C_{18}H_{32}NO_9P_3$ M 499.373
Oil. n_D^{20} 1.4699.

Maier, L., *Phosphorus*, 1971, **1**, 67 (*synth, deriv, pmr, P nmr*)

Triethyl (azidosulfonyl)phosphorimidate, 10CI **T-00439**
(*Azidosulfonylimino*)*triethoxyphosphorane.* N-*Azidosulfonyl*-P,P,P-*triethoxyphosphine imide.* N-*Azidosulfonyl*-P,P,P-*triethoxyphosphazene*
[72250-10-3]

$N_3SO_2N{=}P(OEt)_3$

$C_6H_{15}N_4O_5PS$ M 286.242
Liq.

Nojima, M. *et al*, *J. Chem. Soc., Perkin Trans. 1*, 1979, 1811 (*synth, ir, pmr, props*)

Triethyl phosphate, 9CI **T-00440**
[78-40-0]

$(EtO)_3P{=}O$

$C_6H_{15}O_4P$ M 182.156
Ethylating agent for nitrogen heterocyclic compds. Pleasant smelling liq. Misc. H_2O. d_4^{25} 1.06. Bp 215°, Bp_{50} 128°, Bp_5 75°. n_D^{20} 1.4053.
▷TC7900000.

Evans, D.P. *et al*, *J. Chem. Soc.*, 1930, 1310 (*synth*)
Cox, J.R. *et al*, *J. Am. Chem. Soc.*, 1958, **80**, 5441 (*synth*)
Mitsunobu, O. *et al*, *Bull. Chem. Soc. Jpn.*, 1961, **40**, 935 (*synth*)
Hata, T. *et al*, *Bull. Chem. Soc. Jpn.*, 1962, **35**, 1106 (*synth*)
Mukaiyama, T. *et al*, *J. Org. Chem.*, 1962, **27**, 1815; 1963, **28**, 481 (*synth*)
Crofts, P.C. *et al*, *J. Chem. Soc.*, 1963, 2559 (*synth*)
Strukov, O.G. *et al*, *Zh. Prikl. Spektrosk.*, 1967, **7**, 209 (*Engl. transl.* p. 150) (*ir*)
Betteridge, D. *et al*, *Anal. Chem.*, 1972, **44**, 2005 (*pe*)
Gerken, T.A. *et al*, *J. Magn. Reson.*, 1976, **24**, 155 (*pmr*)

Kawazoe, Y. *et al*, *Chem. Pharm. Bull.*, 1982, **30**, 2077 (*props, use*)
Cload, P.A. *et al*, *Org. Mass Spectrom.*, 1983, **18**, 57 (*ms*)

Triethylphosphine, 9CI, 8CI T-00441

Phosphorus triethyl
[554-70-1]

Et_3P

$C_6H_{15}P$ M 118.158
Liq. with vile, nauseating odour. Misc. EtOH, Et_2O, insol. H_2O. Bp_{745} 127°. pK_a 8.69, 8.86 (H_2O). n_D^{18} 1.4580.

▷Forms explosive product by reacn. with O_2

B,MeI: Triethylmethylphosphonium iodide.
$C_7H_{18}IP$ M 260.097
Solid. Mp 316-319°.
Oxide: see Triethylphosphine oxide, T-00443
Sulfide: see Triethylphosphine sulfide, T-00444
Selenide: [21522-01-0].
$C_6H_{15}PSe$ M 197.118
Cryst. (Et_2O or EtOH). Mp 111°.
Telluride:
$C_6H_{15}PTe$ M 245.758
Golden-yellow cryst. (toluene/pet. ether). Mp 76-78°.

Van Meerssche, M. *et al*, *Bull. Soc. Chim. Belg.*, 1960, **69**, 45; *CA*, **54**, 16982 (*selenide, cryst struct*)
Screttas, C. *et al*, *J. Org. Chem.*, 1962, **27**, 2573 (*synth*)
Cumper, C.W.N. *et al*, *J. Chem. Soc.*, 1964, 430 (*synth*)
Rabinowitz, R. *et al*, *J. Polym. Sci., Ser. A*, 1965, **3**, 2055 (*synth*)
Zingaro, R.A. *et al*, *J. Organomet. Chem.*, 1965, **4**, 320 (*telluride*)
Durig, J.R. *et al*, *J. Mol. Spectrosc.*, 1968, **28**, 444 (*selenide, ir, raman*)
Wolfsberger, W. *et al*, *Synth. React. Inorg. Met.-Org. Chem.*, 1974, **4**, 149 (*synth*)
Gerken, T.A. *et al*, *J. Magn. Reson.*, 1976, **24**, 155 (*cmr, nmr, pmr*)
Crocker, C. *et al*, *J. Chem. Soc., Dalton Trans.*, 1978, 388 (*ir, raman*)
Goetz, H. *et al*, *Phosphorus*, 1978, **4**, 309 (*struct*)
Buina, N.A. *et al*, *Zh. Obshch. Khim.*, 1979, **49**, 2386 (*Engl. transl.* p. 2105) (*selenide, pmr*)
Van Linthoudt, J.P. *et al*, *Spectrochim. Acta, Part A*, 1979, **35**, 1307; 1980, **36**, 17 (*cmr, nmr, pmr*)
Gonbeau, D. *et al*, *Inorg. Chem.*, 1981, **20**, 1966 (*pe*)
Bretherick, L., *Handbook of Reactive Chemical Hazards*, 2nd Ed., Butterworths, London and Boston, 1979, 609.
Sax, N.I., *Dangerous Properties of Industrial Materials*, 6th Ed., Van Nostrand-Reinhold, 1984, 1054.
Hazards in the Chemical Laboratory, (Bretherick, L., Ed.), 3rd Ed., Royal Society of Chemistry, London, 1981, 519.

P,P,P-Triethylphosphine imide, 9CI T-00442

Iminotriethylphosphorane. P,P,P-*Triethylphosphine imine.* P,P,P-*Triethylphosphazene*
[53098-45-6]

$Et_3P{=}NH$

$C_6H_{16}NP$ M 133.173
Liq. Bp_{11} 94°.
B,HCl: Hygroscopic solid. Mp 108°.
N-*Diphenylphosphino:* P,P,P-*Triethyl*-N-*(diphenylphosphino)phosphine imide.*
$C_{18}H_{25}NP_2$ M 317.350
Mp 39-40°. $Bp_{0.001}$ 142-144°.
N-*Trimethylsilyl:* P,P,P-*Triethyl*-N-*trimethylsilylphosphine imide.*
$C_9H_{24}NPSi$ M 205.355
Liq. Bp_{11} 89.5°.

Birkofer, L. *et al*, *Chem. Ber.*, 1964, **97**, 2100 (*synth, deriv*)
Ross, B. *et al*, *Chem. Ber.*, 1974, **107**, 2720 (*derivs*)
Wolfsberger, W., *Z. Naturforsch., B*, 1974, **29**, 35 (*deriv, ir, pmr, P nmr*)

Triethylphosphine oxide, 9CI, 8CI T-00443

[597-50-2]

$Et_3P{=}O$

$C_6H_{15}OP$ M 134.158
V. hygroscopic cryst. Mp 47°. Bp 238-240°, Bp_{18} 124.5-124.7°. pK_a 8.70 ($MeNO_2$).

Cumper, W.N. *et al*, *J. Chem. Soc.*, 1964, 430 (*synth, props*)
Cuddy, B.D. *et al*, *Tetrahedron Lett.*, 1971, 4433 (*pmr*)
Gray, G.A. *et al*, *J. Org. Chem.*, 1972, **37**, 3458 (*cmr*)
Raevskii, O.A. *et al*, *Izv. Akad. Nauk SSSR, Ser. Khim.*, 1973, 2491 (*Engl. transl.* p. 2432) (*struct*)
Skvortsov, N.K. *et al*, *Zh. Obshch. Khim.*, 1973, **43**, 981 (*Engl. transl.*, p. 976); 1976, **46**, 521 (*Engl. transl.* p. 518) (*pmr, nmr*)
Mayer, U. *et al*, *Monatsh. Chem.*, 1975, **106**, 1235 (*synth, nmr*)
Kenyon, G.L. *et al*, *J. Org. Chem.*, 1976, **41**, 2417 (*ms*)
Gerken, T.A. *et al*, *J. Magn. Reson.*, 1976, **24**, 155 (*cmr, nmr, pmr*)

Triethylphosphine sulfide, 9CI, 8CI T-00444

[597-51-3]

$Et_3P{=}S$

$C_6H_{15}PS$ M 150.218
Cryst. (pet. ether or EtOH). Mp 94.5-95.5°.

Christen, P.J. *et al*, *Recl. Trav. Chim. Pays-Bas*, 1959, **78**, 161 (*synth*)
Van Meerssche, M. *et al*, *Acta Crystallogr.*, 1959, **12**, 1053 (*cryst struct*)
Patel, N.K. *et al*, *J. Org. Chem.*, 1967, **32**, 2999 (*synth*)
Bogolyubov, G.M. *et al*, *Zh. Obshch. Khim.*, 1969, **39**, 1759 (*Engl. transl.* p. 1723) (*uv*)
Fridlyanskii, G.V. *et al*, *Zh. Obshch. Khim.*, 1971, **41**, 1707 (*Engl. transl.* p. 1714) (*ms*)
Lindner, E. *et al*, *Chem. Ber.*, 1972, **105**, 3261 (*synth, ir*)
Kincaid, J. *et al*, *Spectrochim. Acta, Part A*, 1974, **30**, 2091 (*ir, raman*)
Skvortsov, N.K. *et al*, *Zh. Obshch. Khim.*, 1976, **46**, 521 (*Engl. transl.* p. 518) (*pmr, nmr*)
Postle, S.R., *Phosphorus Sulfur*, 1977, **3**, 269 (*pmr, cmr*)

Triethyl phosphite, 9CI, 8CI T-00445

[122-52-1]

$(EtO)_3P$

$C_6H_{15}O_3P$ M 166.156
Widely used organophosphorus reagent. Deoxygenates epoxides to alkenes, peroxides to ethers, hydroperoxides to alcohols, desulfurizes the analogous sulfur compds. Condensing agent with Py/Br_2 or Py/I_2 for synth. of esters and amides. Deoxygenates *tert*-amine *N*-oxides. Deoxygenates aryl nitro groups *via* nitrenes giving syntheses of heterocyclic compds. including phenazines, phenoxazines, phenothiazines, indazoles, triazoles and also azepines. Converts aromatic dicarbonyl compds. to epoxides. Carbonyl compds. may be reductively dimerised, e.g. phthalic anhydride → biphthalyl. Widely used in Arbuzov reactions for synth. of phosphonic and phosphinic di-Et-esters, for use in Wadsworth-Emmons reactions. Liq. with characteristic, obnoxious, phosphite odour. d_4^{20} 0.963. Bp_{16} 57-58°. n_D^{25} 1.4105.

▷TH1130000.

Halman, M., *J. Chem. Soc.*, 1963, 2853 (*uv*)
Gerken, T.A. *et al*, *J. Magn. Reson.*, 1976, **24**, 155 (*pmr, cmr, P nmr*)
Labintsev, V.B. *et al*, *Zh. Org. Khim.*, 1977, **13**, 1141 (*Engl. transl.* p. 1050) (*ms*)
Cadogan, J.I.G., Ed., *Organophosphorus Reagents in Organic Synthesis*, Academic Press, 1979 (*uses*)
Chattopadhyay, S. *et al*, *J. Electron Spectrosc. Relat. Phenom.*, 1981, **24**, 27 (*pe*)
Vasil'ev, V.V. *et al*, *Zh. Obshch. Khim.*, 1981, **51**, 2134 (*Engl. transl.* p. 1836) (*O nmr*)
Cload, P.A. *et al*, *Org. Mass Spectrom.*, 1983, **18**, 57 (*ms*)

Triethylphosphonamidic acid, 9CI — T-00446

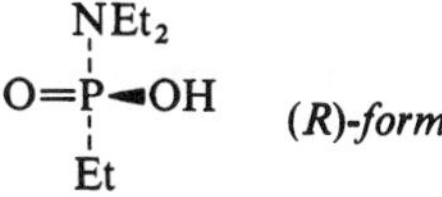

$C_6H_{16}NO_2P$ M 165.172

(R)-form

Et ester: [63842-89-7]. Liq. $Bp_{0.01}$ 37-38°. $[\alpha]_D$ +11.69°. n_D^{20} 1.4370.

(S)-form

Et ester: [63842-88-6]. Liq. $Bp_{0.1}$ 56-58°. $[\alpha]_D$ −17.34°. n_D^{20} 1.4365.

(±)-form

Me ester: Methyl triethylphosphonamidate.
$C_7H_{18}NO_2P$ M 179.198
Liq. d_4^{20} 1.00. Bp_4 87-87.5°. n_D^{20} 1.4381.

Et ester: [63842-91-1]. *Ethyl triethylphosphonamidate.*
$C_8H_{20}NO_2P$ M 193.225
Liq. $Bp_{0.02}$ 40-41°. n_D^{20} 1.4370 (1.4315).

Isopropyl ester: Isopropyl triethylphosphonamidate.
$C_9H_{22}NO_2P$ M 207.252
Liq. $Bp_{0.5}$ 87-88.5°. n_D^{20} 1.4381.
▷Markedly neurotoxic

Benzyl ester: Benzyl triethylphosphonamidate.
$C_{13}H_{22}NO_3P$ M 271.295
Liq. d_4^{20} 1.06. Bp_2 162-163°. n_D^{20} 1.4975.

Fluoride:
$C_6H_{15}FNOP$ M 167.163
Liq. d_4^{20} 1.00. Bp_{15} 103°. n_D^{20} 1.4130.

Chloride:
$C_6H_{15}ClNOP$ M 183.617
Liq. d_4^{20} 1.10. $Bp_{3.5}$ 101.5°. n_D^{20} 1.4643.

Razumov, A.I. *et al*, *Zh. Obshch. Khim.*, 1957, **27**, 2389 (*Engl. transl.* p. 2450) (*esters*)
Razumov, A.I. *et al*, *Zh. Obshch. Khim.*, 1958, **28**, 194 (*Engl. transl.* p. 194) (*chloride, fluoride*)
Michalski, J. *et al*, *Tetrahedron Lett.*, 1968, 3565 (*chloride, methyl ester, synth*)
Pudovik, A.N. *et al*, *Zh. Obshch. Khim.*, 1969, **39**, 1890 (*Engl. transl.* p. 1851) (*ester, synth*)
Berina, N.A. *et al*, *Izv. Akad. Nauk SSSR, Ser. Khim.*, 1977, 1437 (*Engl. transl.* p. 1324) (*ethyl ester, config*)
Mager, P.P., *Toxicol. Lett.*, 1982, **11**, 67 (*isopropyl ester, tox*)

Triethylphosphonamidodithioic acid, 9CI — T-00447

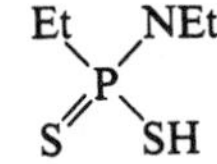

$C_6H_{16}NPS_2$ M 197.293

Et ester: [18032-85-4]. *Ethyl triethylphosphonamidodithioate.*
$C_8H_{20}NPS_2$ M 225.346
Liq. d_4^{20} 1.06. $Bp_{0.025}$ 73-74°. n_D^{20} 1.5432.

Isopropyl ester: Isopropyl triethylphosphonamidodithioate.
$C_9H_{22}NPS_2$ M 239.373
Liq. d_4^{20} 1.03. $Bp_{0.03}$ 72-74°. n_D^{20} 1.5300.

Akamsin, V.D. *et al*, *Izv. Akad. Nauk SSSR, Ser. Khim.*, 1967, 1983 (*Engl. transl.* p. 1900)

Triethylphosphonamidothioic acid, 9CI — T-00448

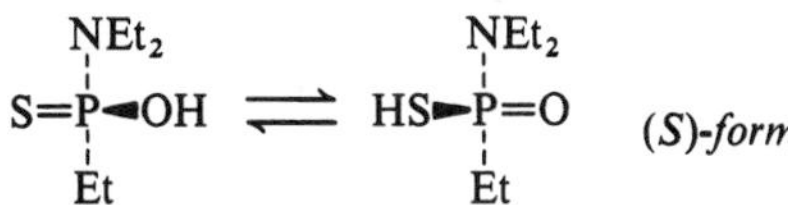

$C_6H_{16}NOPS$ M 181.232

(S)-form

O-*Et ester:* [20698-87-7]. O-*Ethyl triethylphosphonamidothioate.*
$C_8H_{20}NOPS$ M 209.286
$[\alpha]_D$ +55.15°.

(±)-form

O-*Me ester:* O-*Methyl triethylphosphonamidothioate.*
$C_7H_{18}NOPS$ M 195.259
Liq. d_4^{20} 1.03. Bp_{10} 112-113°. n_D^{20} 1.4882.

O-*Et ester:* [24680-56-6]. Liq. d_4^{20} 1.00. Bp_9 115-116°. n_D^{20} 1.4828.

S-*Et ester:* S-*Ethyl triethylphosphonamidothioate.*
$C_8H_{20}NOPS$ M 209.286
Liq. d_0^{20} 1.03. Bp_2 106-108°. n_D^{20} 1.4892. Has been obt. in optically active forms.

O-*Isopropyl ester:* O-*Isopropyl triethylphosphonamidothioate.*
$C_9H_{22}NOPS$ M 223.313
Liq. Bp_9 118.5-119-5°. n_D^{20} 1.4781.

Chloride:
$C_6H_{15}ClNPS$ M 199.678
Liq. d_4^{20} 1.11. Bp_1 94-95°. n_D^{20} 1.5205.

Arbuzov, B.A. *et al*, *Izv. Akad. Nauk SSSR, Ser. Khim.*, 1955, 1021 (*Engl. transl.* p. 935) (*chloride, esters, synth*)
Razumov, A.I. *et al*, *CA*, 1958, **52**, 237 (*ester, synth*)
Mikolajczyk, M. *et al*, *Tetrahedron*, 1972, **28**, 4357 (*ester, cd, uv, config*)
Michalski, J. *et al*, *Tetrahedron Lett.*, 1968, 3565 (*ester, synth, props*)

N,N,N'-Triethylphosphoramidimidic dichloride, 9CI — T-00449

[51992-60-0]

$EtN{=}PCl_2NEt_2$

$C_6H_{15}Cl_2N_2P$ M 217.078
Monomeric liq. d_4^{20} 1.15. Bp_{12} 91-92°. n_D^{25} 1.4880. In absence of solvent, slowly polymerises (30-40% in 8-10 days).

Pinchuk, A.M. *et al*, *Zh. Obshch. Khim.*, 1974, **44**, 705 (*Engl. transl.* p. 673) (*synth, ir, props*)
Kovenya, V.A. *et al*, *Zh. Obshch. Khim.*, 1978, **48**, 2686 (*Engl. transl.* p. 2436) (*synth*)

O,S,S-Triethyl phosphorodithioate, 9CI — T-00450

[2404-78-6]

$(EtS)_2P(O)OEt$

$C_6H_{15}O_2PS_2$ M 214.277
No phys. props. reported.
▷Toxic

Verschoyle, R.D. *et al*, *Arch. Toxicol.*, 1982, **51**, 221 (*tox*)

_O,O,S_-Triethyl phosphorodithioate, 9CI, 8CI **T-00451**
O,O,S-*Triethyl dithiophosphate*
[2524-09-6]

$(EtO)_2P(=S)SEt$

$C_6H_{15}O_2PS_2$ M 214.277
Bp_{16} 121-123°, Bp_2 74-77°. n_D^{20} 1.5010.

McIvor, R.A. *et al*, *Can. J. Chem.*, 1958, **36**, 820 (*ir*)
Lippman, A.E., *J. Org. Chem.*, 1966, **31**, 471 (*synth, P nmr*)
Ishmaeva, E.A. *et al*, *Zh. Obshch. Khim.*, 1974, **44**, 2625 (*Engl. transl.* p. 2582) (*synth, conformn*)
Meyer, H.J. *et al*, *Bull. Soc. Chim. Belg.*, 1978, **87**, 517 (*ms*)
Zimin, M.G. *et al*, *Zh. Obshch. Khim.*, 1978, **48**, 1020 (*Engl. transl.* p. 930) (*synth, P nmr*)

Triethyl phosphoro(dithioperoxoate), 9CI **T-00452**
O,O-*Diethyl* S-(*ethylthio*) *phosphorothioate*
[50704-15-9]

$(EtO)_2P(O)SSEt$

$C_6H_{15}O_3PS_2$ M 230.276
Liq. d_4^{20} 1.12. $Bp_{3.5}$ 82-85°. n_D^{20} 1.4721.

Kabachnik, M.I. *et al*, *Zh. Obshch. Khim.*, 1955, **25**, 1924 (*Engl. transl.* p. 1867)

Triethyl phosphoro(dithioperoxo)thioate, 9CI **T-00453**
[4403-52-5]

$(EtO)_2P(S)-S-S-Et$

$C_6H_{15}O_2PS_3$ M 246.337
Liq. d_4^{20} 1.18. Bp_3 106-107°. n_D^{20} 1.5431.

Almasi, L. *et al*, *Chem. Ber.*, 1962, **95**, 1582 (*synth*)

_O,O,O_-Triethyl phosphoroselenoate, 9CI, 8CI **T-00454**
Triethyl selenophosphate
[2651-89-0]

$(EtO)_3P{=}Se$

$C_6H_{15}O_3PSe$ M 245.116
Liq. d_4^{20} 1.304. Bp_{20} 117°, Bp_9 85°. n_D^{20} 1.4730.

Nuretdinov, I.A. *et al*, *Izv. Akad. Nauk SSSR, Ser. Khim.*, 1968, 2831 (*Engl. transl.* p. 2685) (*synth, ir*)
Kainosho, M., *J. Phys. Chem.*, 1970, **74**, 2853 (*pmr*)
Stec, W.J. *et al*, *Phosphorus*, 1972, **2**, 97 (*P nmr*)
Shagidullin, R.R. *et al*, *Izv. Akad. Nauk SSSR, Ser. Khim.*, 1976, 184 (*Engl. transl.* p. 174) (*uv*)
Glidewell, C. *et al*, *J. Chem. Soc., Dalton Trans.*, 1977, 527 (*ir, P nmr*)
Turkevich, V.V. *et al*, *Zh. Obshch. Khim.*, 1977, **47**, 2388 (*Engl. transl.* p. 2181) (*ir*)
Ando, F. *et al*, *Bull. Chem. Soc. Jpn.*, 1979, **52**, 807 (*ir, pmr*)
Zverev, V.V. *et al*, *Zh. Obshch. Khim.*, 1981, **51**, 303 (*Engl. transl.* p. 242) (*pe*)

_O,O,Se_-Triethyl phosphoroselenoate, 9CI, 8CI **T-00455**
O,O,Se-*Triethyl selenophosphate*
[39181-35-6]

$(EtO)_2P(O)SeEt$

$C_6H_{15}O_3PSe$ M 245.116
Has insecticidal props. Liq. d_0^0 1.359. Bp_2 102-105°.

B.P., 691 267, (*1954*); *CA*, **48**, 7047 (*synth, props*)
Stec, W.J. *et al*, *Phosphorus*, 1972, **2**, 97 (*P nmr*)
Bel'skii, I.A. *et al*, *Zh. Obshch. Khim.*, 1973, **43**, 1255 (*Engl. transl.* p. 1244) (*props*)

Triethyl phosphorotetrathioate, 9CI, 8CI **T-00456**
Triethyl tetrathiophosphate
[1642-43-9]

$(EtS)_3P{=}S$

$C_6H_{15}PS_4$ M 246.398
Liq. d_4^{20} 1.19. $Bp_{1.5}$ 124-127°. n_D^{20} 1.6180.

Scott, C.B. *et al*, *J. Org. Chem.*, 1957, **22**, 789 (*synth*)
Menefee, A. *et al*, *J. Org. Chem.*, 1957, **22**, 792 (*ir*)
Blagoveshchenski, V.S. *et al*, *Zh. Obshch. Khim.*, 1971, **41**, 1032 (*Engl. transl.* p. 1036) (*synth*)
Shaw, R.A. *et al*, *Phosphorus*, 1971, **1**, 41, 191 (*synth, ir, pmr*)
Khokhlov, P.S. *et al*, *Zh. Obshch. Khim.*, 1984, **54**, 2545 (*Engl. transl.* p. 2274) (*synth*)

_O,O,S_-Triethyl phosphorothioate, 9CI, 8CI **T-00457**
O,O,S-*Triethyl thiophosphate*
[1186-09-0]

$(EtO)_2P(O)SEt$

$C_6H_{15}O_3PS$ M 198.216
Liq. d_4^{20} 1.11. Bp_3 83-85°. n_D^{20} 1.4590.
▷High oral toxicity. TG3700000.

McIvor, R.A. *et al*, *Can. J. Chem.*, 1958, **36**, 820 (*ir*)
Mastryukova, T.A. *et al*, *Zh. Obshch. Khim.*, 1964, **34**, 94 (*Engl. transl.* p. 92) (*synth*)
Zimin, M.G. *et al*, *Zh. Obshch. Khim.*, 1972, **49**, 2189 (*Engl. transl.* p. 1922) (*synth, P nmr*)
Mallipudi, N.M. *et al*, *J. Agric. Food Chem.*, 1979, **27**, 463 (*tox*)
Verschoyle, R.D. *et al*, *Arch. Toxicol.*, 1982, **51**, 221 (*tox*)

_O,O,O_-Triethyl phosphorothioate, 9CI **T-00458**
O,O,O-*Triethyl thiophosphate*
[126-68-1]

$(EtO)_3P{=}S$

$C_6H_{15}O_3PS$ M 198.216
Liq. d_0^{20} 1.076. Bp_8 88-89°. n_D^{20} 1.4488. Isomerises to *O,O,S*-Triethyl phosphorothioate, T-00457 when heated.
▷Highly toxic. TG3675000.

McIvor, R.A. *et al*, *Can. J. Chem.*, 1950, **36**, 820 (*ir*)
Cadogan, J.I.G. *et al*, *J. Chem. Soc.*, 1961, 3067, 5524 (*synth*)
Zemlyanskii, N.I. *et al*, *Zh. Obshch. Khim.*, 1961, **30**, 4056 (*Engl. transl.* p. 4018) (*raman*)
Williamson, M.P. *et al*, *J. Phys. Chem.*, 1968, **72**, 4043 (*pmr*)
Zwierzak, A., *Tetrahedron*, 1969, **25**, 5177 (*synth*)

Nifant'ev, É.E. *et al*, *Zh. Obshch. Khim.*, 1975, **45**, 295 (*Engl. transl.* p. 282) (*synth*)
Gurley, T.W. *et al*, *Anal. Chem.*, 1976, **48**, 1137 (*P nmr*)
Sax, N.I., *Dangerous Properties of Industrial Materials*, 6th Ed., Van Nostrand-Reinhold, 1984, 1054.

O,S,S-Triethyl phosphorotrithioate, 9CI, 8CI T-00459

O,S,S-*Triethyl trithiophosphate*

[3347-30-6]

$(EtS)_2P(S)OEt$

$C_6H_{15}OPS_3$ M 230.338

Odorous liq. d_4^{20} 1.15. Bp_1 98-100°. n_D^{20} 1.5570.

▷TG5500000.

McIvor, R.A. *et al*, *Can. J. Chem.*, 1956, **34**, 1611 (*ir*)
Murav'ev, I.V. *et al*, *Zh. Obshch. Khim.*, 1968, **38**, 133 (*Engl. transl.* p. 133) (*synth*)

S,S,S-Triethyl phosphorotrithioate, 9CI, 8CI T-00460

S,S,S-*Triethyl trithiophosphate*

[1486-39-1]

$(EtS)_3P{=}O$

$C_6H_{15}OPS_3$ M 230.338

Liq. d_4^{20} 1.17. Bp_2 110°. n_D^{20} 1.5700.

▷Toxic (less so than 48299-7)

McIvor, R.A. *et al*, *Can. J. Chem.*, 1956, **34**, 1611, 1819; 1958, **36**, 820 (*synth, ir*)
Shaw, R.A. *et al*, *Phosphorus*, 1971, **1**, 41, 191 (*synth, ir, pmr*)
Sinyashin, O.G. *et al*, *Zh. Obshch. Khim.*, 1984, **54**, 540 (*Engl. transl.* p. 475) (*synth, P nmr*)

Triethyl phosphorotrithioite, 9CI T-00461

Triethyl trithiophosphite

[688-62-0]

$(EtS)_3P$

$C_6H_{15}PS_3$ M 214.338

Ligand for Hg and Au. Liq. d_4^{25} 1.16. Mp −32°. Bp_{18} 140-143°.

▷TG5700000.

B,MeI: Tri(ethylthio)methylphosphonium iodide.
$C_7H_{18}IPS_3$ M 356.277
Solid. Mp 191°.

B,EtI: Ethyltri(ethylthio)phosphonium iodide.
$C_8H_{20}IPS_3$ M 370.304
Solid. Mp 125-127°.

Lippert, A. *et al*, *J. Am. Chem. Soc.*, 1938, **60**, 2370 (*synth, derivs*)
Chernova, A.V. *et al*, *Zh. Obshch. Khim.*, 1979, **49**, 2002 (*Engl. transl.* p. 1760) (*uv, struct*)
Kostin, V.P. *et al*, *Zh. Obshch. Khim.*, 1982, **52**, 498, 2698 (*Engl. transl.* pp. 437, 2379) (*synth, P nmr*)
Nifant'ev, É.E. *et al*, *Zh. Obshch. Khim.*, 1983, **53**, 2695 (*Engl. transl.* p. 2429) (*synth, P nmr*)
Zverev, V.V. *et al*, *Zh. Obshch. Khim.*, 1983, **53**, 1968 (*Engl. transl.* p. 1775) (*pe*)

N,N′,N″-Triethylphosphorous triamide, 9CI T-00462

Tris(ethylamino)phosphine. Tri(ethylamino)phosphine

$(EtNH)_3P$

$C_6H_{18}N_3P$ M 163.202

Oxide: [19622-51-6]. N,N′,N″-*Triethylphosphoric triamide. Tri(ethylamino)phosphine oxide.*
$C_6H_{18}N_3OP$ M 179.201
Solid. Mp 58°. $Bp_{0.5}$ 134°.

Sulfide: [22965-10-2]. N,N′,N″-*Triethylphosphorothioic triamide. Tri(ethylamino)phosphine sulfide.*
$C_6H_{18}N_3PS$ M 195.262
Cryst. (cyclohexane). Mp 66-67°.

Tolkmith, H., *J. Am. Chem. Soc.*, 1963, **85**, 3246 (*sulfide*)
Healy, J.D. *et al*, *Phosphorus Sulfur*, 1978, **5**, 239 (*sulfide, synth, props*)
Yvernault, T. *et al*, *C.R. Hebd. Seances Acad. Sci., Ser. C*, 1978, 519 (*oxide*)

2,4,6-Triethyl-1,3,5,2,4,6-trioxatriphosphorinane, 9CI T-00463

2,4,6-Triethyl-1,3,5,2,4,6-trioxatriphosphinane

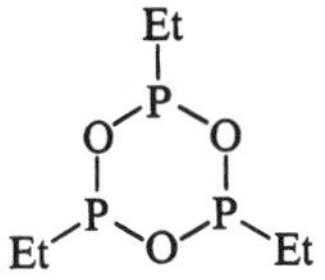

$C_6H_{15}O_3P_3$ M 228.104

2,4,6-Trisulfide: [17620-78-9]. *Ethylphosphonothioic acid trimer trianhydride.*
$C_6H_{15}O_3P_3S_3$ M 312.273
Cryst. (C_6H_6/pet. ether). Mp 48°.

Roesky, H.W. *et al*, *Chem. Ber.*, 1968, **101**, 630 (*synth, ir, pmr, ms*)

Triferrocenylphosphine T-00464

1,1″,1‴-Phosphinidynetrisferrocene, 9CI

[1292-82-6]

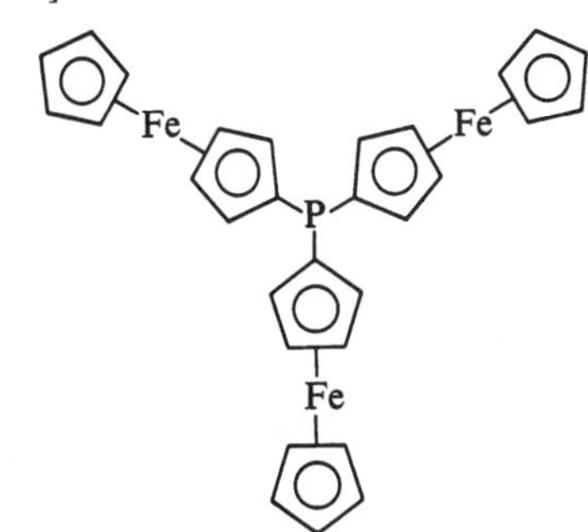

$C_{30}H_{27}Fe_3P$ M 586.058

Yellow cryst. (C_6H_6/pet. ether). Mp 271-273°.

Oxide: [1277-92-5]. *1′,1″,1‴-Phosphinylidynetrisferrocene, 9CI.*
$C_{30}H_{27}Fe_3OP$ M 602.057
Infusible light-yellow powder (C_6H_6/pet. ether). Forms hydrates readily. Diss. in H_2SO_4 with liberation of ionic Fe. Forms reversible benzene complex.

Sulfide: 1′,1″,1‴-Phosphinothioylidynetrisferrocene, 9CI.
$C_{30}H_{27}Fe_3PS$ M 618.118
Yellow needles (C_6H_6/heptane). Mp 291-293° dec.

Selenide: 1′,1″,1‴-Phosphinoselenoylidynetrisferrocene.
$C_{30}H_{27}Fe_3PSe$ M 665.018
Yellow needles (C_6H_6/heptane). Mp 297-299° dec.

Sollott, G.P. *et al*, *J. Org. Chem.*, 1962, **27**, 4034; *J. Organomet. Chem.*, 1965, **4**, 491 (*synth*)
Neuse, E.W., *J. Organomet. Chem.*, 1967, **7**, 349 (*oxide, uv*)
Nesmeyanov, A.N. *et al*, *Zh. Strukt. Khim.*, 1973, **14**, 49 (*cmr*)
Nesmeyanov, A.N. *et al*, *Izv. Akad. Nauk SSSR, Ser. Khim.*, 1975, 706; 1977, 2263 (*Engl. transl.* pp. 635, 2098) (*oxide, pmr*)

Trifluorobis(trifluoromethyl)phosphorane, 9CI — T-00465

[1184-82-3]

$F_3P(CF_3)_2$

C_2F_9P M 225.981

Molecule consists of unionized trigonal bipyramid with CF_3 groups axial. Gas. Mp −74°. Bp −5°.

Mahler, W., *Inorg. Chem.*, 1963, **2**, 230 (*synth, ir*)
Muetterties, E.L. *et al*, *Inorg. Chem.*, 1963, **2**, 613 (*ir, nmr*)
Nixon, J.F. *et al*, *Spectrochim. Acta*, 1964, **20**, 1835 (*nmr*)
Cavell, R.G. *et al*, *J. Am. Chem. Soc.*, 1977, **99**, 7841 (*cmr*)
Oberhammer, H. *et al*, *Inorg. Chem.*, 1982, **21**, 275 (*ed*)

1,1,1-Trifluoro-1,1-dihydrophospholane — T-00466

[646-44-6]

$C_4H_8F_3P$ M 144.076

Liq. Bp_{90} 61-62°.

Schmutzler, R., *Inorg. Chem.*, 1964, **3**, 421 (*synth, ir, ms*)
Reddy, G.S. *et al*, *Z. Naturforsch., B*, 1970, **25**, 1199 (*F and P nmr*)

2,2,2-Trifluoro-2,2-dihydro-4,4,5,5-tetrakis(trifluoromethyl)-1,3,2-dioxaphospholane — T-00467

[54799-74-5]

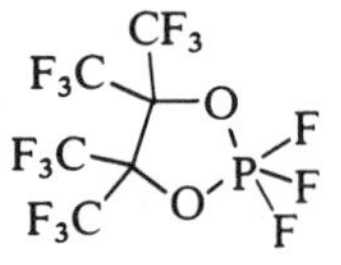

$C_6F_{15}O_2P$ M 420.015

Liq. Mp −11°. Bp 55°.

Gibson, J.A. *et al*, *J. Chem. Soc., Dalton Trans.*, 1975, 918; 1976, 1440 (*synth, ir, cmr, F and P nmr*)
Roeschenthaler, G.-V. *et al*, *Chem. Ber.*, 1977, **110**, 611 (*props*)

Trifluorodimethoxyphosphorane, 8CI — T-00468

Tetramethoxyphosphonium hexafluorophosphate

$(MeO)_2PF_3$ or $(MeO)_4P^{\oplus}PF_6^{\ominus}$

$C_2H_6F_3O_2P$ M 150.037

Thought to have the ionic struct. in the solid state. Hygroscopic cryst. Mp 108°.

Kolditz, L. *et al*, *Z. Anorg. Allg. Chem.*, 1968, **360**, 259 (*synth, pmr, nmr*)
Gontar, A.F. *et al*, *Izv. Akad. Nauk SSSR, Ser. Khim.*, 1981, 2632 (*Engl. transl.* p. 2188) (*props*)

Trifluorodimethylphosphorane, 9CI, 8CI — T-00469

[811-79-0]

Me_2PF_3

$C_2H_6F_3P$ M 118.038

Trigonal bipyramidal struct. Liq. d_4^{20} 1.22. Bp_{745} 60-61°. n_D^{20} 1.3230. Easily hydrolysed. A weak electrolyte in MeCN.

Komikov, I.P. *et al*, *Zh. Obshch. Khim.*, 1962, **32**, 301 (*Engl. transl.* p. 295) (*synth*)
Bartell, L.S. *et al*, *Inorg. Chem.*, 1965, **4**, 1777 (*ed*)
Downs, A.J. *et al*, *Spectrochim. Acta, Part A*, 1967, **23**, 681 (*ir, raman*)
Inorg. Synth., 1967, **9**, 67 (*synth*)
Blazer, T.A. *et al*, *Z. Naturforsch., B*, 1969, **24**, 1081 (*ms*)
Appel, R. *et al*, *Chem. Ber.*, 1974, **107**, 2169 (*synth, nmr*)
Robert, D.U. *et al*, *Org. Magn. Reson.*, 1975, **7**, 291 (*pmr, nmr*)
Deiters, J.A. *et al*, *J. Am. Chem. Soc.*, 1977, **99**, 5461 (*struct*)

Trifluorodiphenoxyphosphorane, 9CI, 8CI — T-00470

Tetraphenoxyphosphonium hexafluorophosphate

[20758-93-4]

$(PhO)_2PF_3$ or $(PhO)_4P^{\oplus}PF_6^{\ominus}$

$C_{12}H_{10}F_3O_2P$ M 274.179

Has pentavalent P in solid state, ionic struct. in MeCN soln. Solid. Mp 98°.

Kolditz, L. *et al*, *Z. Anorg. Allg. Chem.*, 1968, **360**, 259 (*synth, nmr, struct*)
Ruppert, I., *Z. Anorg. Allg. Chem.*, 1981, **477**, 59 (*synth, cmr, nmr*)
Il'in, E.G. *et al*, *Dokl. Akad. Nauk SSSR, Ser. Sci. Khim.*, 1982, **266**, 123 (*Engl. transl.* p. 300) (*nmr*)

Trifluorodiphenylphosphorane, 9CI, 8CI — T-00471

[1138-99-4]

Ph_2PF_3

$C_{12}H_{10}F_3P$ M 242.180

Converts primary and sec. alcohols into alkyl fluorides. Bp_2 106-107°, $Bp_{0.4}$ 92-93°. n_D^{23} 1.5425.

Schmutzler, R., *Inorg. Chem.*, 1964, **3**, 410 (*synth*)
Inorg. Synth., 1967, **9**, 63 (*synth*)
Kobayashi, Y. *et al*, *Chem. Pharm. Bull.*, 1968, **16**, 1784 (*synth, use*)
Blazer, T.A. *et al*, *Z. Naturforsch., B*, 1969, **24**, 1081 (*ms*)
Moreland, G.G. *et al*, *J. Am. Chem. Soc.*, 1973, **95**, 255 (*nmr, struct*)
Robert, D.U. *et al*, *Org. Magn. Reson.*, 1975, **7**, 291 (*pmr, nmr*)

Trifluoromethanethiol phosphorodichloridite — T-00472

Dichloro(trifluoromethylthio)phosphine

[1495-32-5]

F_3CSPCl_2

CCl_2F_3PS M 202.946

Mp 98°.

Emeléus, H.J. *et al*, *J. Chem. Soc.*, 1960, 1108 (*synth, props, ir*)
Haas, A. *et al*, *Z. Anorg. Allg. Chem.*, 1980, **468**, 68 (*synth, nmr, ms*)

(4-Trifluoromethylphenyl)phosphonic acid, 8CI **T-00473**

OH
|
O=P—OH
CF_3

$C_7H_6F_3O_3P$ M 226.092
Cryst. (H_2O or C_6H_6/Et_2O). Sol. Et_2O, Me_2CO; insol. C_6H_6, $CHCl_3$, pet. ether. Mp 177-179°.
Difluoride:
$C_7H_4F_5OP$ M 230.074
Liq. Bp 186-187°.
Dichloride:
$C_7H_4Cl_2F_3OP$ M 262.983
Liq. d^{20}_{20} 1.53. n_D^{23} 1.4940.

Yagupol'skii, L.M. *et al*, *Zh. Obshch. Khim.*, 1960, **30**, 4026 (*Engl. transl.* p. 3986) (*synth, derivs*)

(Trifluoromethyl)phosphine, 9CI **T-00474**
[420-52-0]

F_3CPH_2

CH_2F_3P M 101.996
Forms complexes with group VIB and VIII metals. Gas. Bp −26.5°. Dec. at 1000° to give HF and difluoromethylenephosphine.

Cavell, R.G. *et al*, *J. Chem. Soc. (A)*, 1967, 1308 (*synth*)
Cavell, R.G. *et al*, *Inorg. Chem.*, 1968, **7**, 101 (*ms*)
Brown, J.D. *et al*, *J. Chem. Soc., Dalton Trans.*, 1973, 1691 (*ir, raman*)
Buerger, A. *et al*, *Spectrochim. Acta, Part A*, 1973, **29**, 943 (*ir, raman*)
Ansari, S. *et al*, *Z. Naturforsch., B*, 1975, **30**, 651 (*synth*)
Cowley, A.H. *et al*, *J. Am. Chem. Soc.*, 1975, **97**, 3653 (*uv, pe*)
Whangbo, M.H. *et al*, *Inorg. Chem.*, 1982, **21**, 1720 (*struct*)

(Trifluoromethyl)phosphinic acid, 9CI **T-00475**

$F_3CPH(O)OH$

$CH_2F_3O_2P$ M 133.995
Dimeric in liq. or vapour at lower temps. Liq. Bp 217° (calc.). pK_a 1.01. Yields CHF_3 at pH >11.5 or in aq. soln. at 100°.
Isopropyl ester: [71545-04-5]. *Isopropyl (trifluoromethyl)phosphinate.*
$C_4H_8F_3O_2P$ M 176.075
Liq. Bp_{12} 43-45°.

Bennett, F.W. *et al*, *J. Chem. Soc.*, 1954, 3598, 3896 (*synth, props*)
Emeléus, H.J. *et al*, *J. Chem. Soc.*, 1955, 563 (*synth, props, ir*)
Burg, A.B. *et al*, *J. Am. Chem. Soc.*, 1961, **83**, 4333 (*synth, props, ir*)
Maslennikov, I.G. *et al*, *Zh. Obshch. Khim.*, 1979, **49**, 1498 (*Engl. transl.* p. 1307) (*ester, P nmr*)

(Trifluoromethyl)phosphinous bromide, 9CI **T-00476**
[58310-40-0]

F_3CPHBr

$CHBrF_3P$ M 180.892
Liq.

Dobbie, R.C. *et al*, *J. Chem. Soc., Dalton Trans.*, 1975, 2368 (*ir, F nmr*)

(Trifluoromethyl)phosphinous chloride, 9CI **T-00477**
[58310-41-1]

F_3CPHCl

$CHClF_3P$ M 136.441

Dobbie, R.C. *et al*, *J. Chem. Soc., Dalton Trans.*, 1975, 2368 (*F nmr, ir*)

(Trifluoromethyl)phosphonic acid, 9CI **T-00478**
[374-09-4]

$F_3CP(O)(OH)_2$

$CH_2F_3O_3P$ M 149.994
Hygroscopic cryst. Mp 81-82°. pK_{a1} 1.46 (H_2O), pK_{a2} 2.43 (H_2O).
Di-Et ester: [2708-87-4]. *Diethyl (trifluoromethyl)-phosphonate.*
$C_5H_{10}F_3O_3P$ M 206.101
Liq. Bp_8 65-67.5°.
Difluoride: [19162-94-8].
CF_5OP M 153.976
Gas. Mp −79°. Bp −20°.
Dichloride: see (Trifluoromethyl)phosphonic dichloride, T-00479

Bennett, F.W. *et al*, *J. Chem. Soc.*, 1954, 3598 (*synth, props, salts*)
Griffiths, J.E., *Spectrochim. Acta, Part A*, 1968, **24**, 115 (*difluoride, ir, F nmr*)
Burton, D.J. *et al*, *Synthesis*, 1979, 615 (*ester, ir, cmr, P nmr, pmr*)
Mahmood, T. *et al*, *Inorg. Chem.*, 1986, **25**, 3128 (*pmr, F and P nmr*)
Mahmood, T. *et al*, *Synth. Commun.*, 1987, **17**, 71 (*ester*)

(Trifluoromethyl)phosphonic dichloride, 9CI **T-00479**
[51965-64-1]

F_3CPOCl_2

CCl_2F_3OP M 186.885
Liq. Mp −25.3°. Bp 70.2°.

Griffiths, J.E., *Spectrochim. Acta, Part A*, 1968, **24**, 303 (*synth, F nmr, ir*)
Dagnac, P. *et al*, *J. Chem. Soc., Dalton Trans.*, 1979, 155 (*nqr, cmr*)

(Trifluoromethyl)phosphonodithious acid, 9CI **T-00480**

$F_3CP(SH)_2$

$CH_2F_3PS_2$ M 166.116
Di-Me ester: [1698-62-0]. *Dimethyl (trifluoromethyl)-phosphonodithioite.*
$C_3H_6F_3PS_2$ M 194.169
Liq. Mp −65°. Bp 168° (est.).

Burg, A.B. *et al*, *J. Am. Chem. Soc.*, 1965, **87**, 2113 (*synth, ir*)
Cowley, A.H. *et al*, *J. Am. Chem. Soc.*, 1969, **91**, 6609 (*synth, pmr, F nmr*)
Burg, A.B. *et al*, *Inorg. Nucl. Chem. Lett.*, 1977, **13**, 199 (*cmr, F and P nmr*)

(Trifluoromethyl)phosphonothioic acid **T-00481**

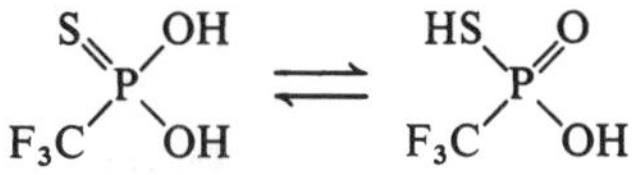

$CH_2F_3O_2PS$ M 166.055

O,O-*Di-Et ester:* [89262-65-7]. O,O-*Diethyl (trifluoromethyl)phosphonothioate.*
$C_5H_{10}F_3O_2PS$ M 222.162
Liq. Bp_{15} 59-61°.

O,S-*Di-Et ester:* [89262-66-8]. O,S-*Diethyl (trifluoromethyl)phosphonothioate.*
$C_5H_{10}F_3O_2PS$ M 222.162
Liq. Bp_{13} 55-58°.

Dichloride: [18799-78-5].
CCl_2F_3PS M 202.946
Volatile liq. Mp −28.7° to −28.5°. Bp 85.7°.

Bis(dimethylamide): [18894-82-1]. P-*Trifluoromethyl-N,N,N′,N′-tetramethylphosphonothioic diamide.*
$C_5H_{12}F_3N_2PS$ M 220.192
Low-melting solid. Mp 10.3°. Bp 211°.

Gosling, K. *et al*, *J. Am. Chem. Soc.*, 1968, **90**, 2011 (*chloride, amide, synth, ir*)
Poulin, D.D. *et al*, *Inorg. Chem.*, 1974, **13**, 2324 (*amide, synth, ir, pmr*)
Maslennikov, I.G. *et al*, *Zh. Obshch. Khim.*, 1983, **53**, 2681 (*Engl. transl.* p. 2417) (*esters, synth, ir, P nmr*)

(Trifluoromethyl)phosphonous acid, 9CI **T-00482**

$F_3CP(OH)_2$

$CH_2F_3O_2P$ M 133.995

Free acid probably exists as the phosphoryl tautomer, (Trifluoromethyl)phosphinic acid, T-00475 . Liq. Bp 217° (est.).

Di-Me ester: [684-56-0]. *Dimethyl (trifluoromethyl)-phosphonite.*
$C_3H_6F_3O_2P$ M 162.048
Liq. Bp 89°.

Di-Et ester: [64149-08-2]. *Diethyl (trifluoromethyl)-phosphonite.*
$C_5H_{10}F_3O_2P$ M 190.102
Liq. Bp_{10} 23-25°.

Monoisopropyl ester: see (Trifluoromethyl)phosphinic acid, T-00475

Diisopropyl ester: [71544-98-4]. *Diisopropyl (trifluoromethyl)phosphonite. Bis(1-methylethyl) (trifluoromethyl)phosphonite.*
$C_7H_{14}F_3O_2P$ M 218.155
Liq. Bp_{40} 62-63°.

Difluoride: see Trifluoromethylphosphonous difluoride, T-00484

Dichloride: see (Trifluoromethyl)phosphonous dichloride, T-00483

Dibromide: [2712-92-7]. *Dibromo(trifluoromethyl)-phosphine.*
CBr_2F_3P M 259.788
Liq. Bp 86.7° (est.).

Diiodide: [421-59-0]. *(Trifluoromethyl)-diiodophosphine.*
CF_3I_2P M 353.789
Liq. Bp_{29} 69°. n_D^{20} 1.630.

Dicyanide: [58310-46-6].
$C_3F_3N_2P$ M 152.015
No phys. props. reported.

Bennett, F.W. *et al*, *J. Chem. Soc.*, 1953, 1565 (*diiodide, synth, uv, ir*)
Burg, A.B. *et al*, *J. Am. Chem. Soc.*, 1960, **82**, 3514 (*dibromide, synth, ir*)
Burg, A.B. *et al*, *J. Am. Chem. Soc.*, 1961, **83**, 4333 (*synth, dimethyl ester, ir*)
Cavell, R.G. *et al*, *Inorg. Chem.*, 1968, **7**, 101 (*dihalides, ms*)
Burger, H. *et al*, *Spectrochim. Acta, Part A*, 1975, **31**, 1955 (*dihalides, ir, raman*)
Elbel, S. *et al*, *Z. Naturforsch., B*, 1976, **31**, 1472 (*dicyanide, pe*)
Burg, A.B., *Inorg. Nucl. Chem. Lett.*, 1977, **13**, 199 (*dihalides, cmr, F and P nmr*)
Maslennikov, I.G. *et al*, *Zh. Obshch. Khim.*, 1979, **49**, 1498 (*Engl. transl.* p. 1307) (*esters, synth, F and P nmr*)
Prokof'eva, G.N. *et al*, *Zh. Obshch. Khim.*, 1982, **52**, 934 (*Engl. transl.* p. 815) (*diisopropyl ester, synth, pmr, P nmr*)

(Trifluoromethyl)phosphonous dichloride, 9CI **T-00483**

Dichloro(trifluoromethyl)phosphine
[421-58-9]

F_3CPCl_2

CCl_2F_3P M 170.886
Liq. d^{20} 1.55. Mp −129.6°. Bp 37°.

Bennett, F.W. *et al*, *J. Chem. Soc.*, 1953, 1565 (*synth, ir*)
Nixon, J.F. *et al*, *J. Chem. Soc.*, 1964, 5983 (*synth*)
Nixon, J.F., *J. Chem. Soc.*, 1965, 777 (*F nmr*)
Cavell, R.G. *et al*, *Inorg. Chem.*, 1968, **7**, 101 (*ms*)
Brown, J.D. *et al*, *J. Chem. Soc., Dalton Trans.*, 1973, 1691 (*ir, raman*)
Bürger, H. *et al*, *Spectrochim. Acta, Part A*, 1975, **31**, 1955 (*ir, raman*)
Cowley, A.H. *et al*, *J. Am. Chem. Soc.*, 1975, **97**, 3653 (*pe*)
Burg, A.B., *Inorg. Nucl. Chem. Lett.*, 1977, **13**, 199 (*cmr, F and P nmr*)

Trifluoromethylphosphonous difluoride, 9CI **T-00484**

Difluoro(trifluoromethyl)phosphine
[1112-04-5]

F_3CPF_2

CF_5P M 137.977
Gas. Bp −43° to −44°.

Kulakova, V.N. *et al*, *Zh. Obshch. Khim.*, 1959, **29**, 3927 (*Engl. transl.* p. 3916) (*synth*)
Cavell, R.G. *et al*, *Inorg. Chem.*, 1968, **7**, 101 (*ms*)
Ang, H.G. *et al*, *J. Chem. Soc. (A)*, 1969, 702 (*synth*)
Brown, J.D. *et al*, *J. Chem. Soc., Dalton Trans.*, 1973, 1691 (*ir, raman*)
Bürger, H. *et al*, *Spectrochim. Acta, Part A*, 1975, **31**, 1955 (*ir, raman*)
Burg, A.B., *Inorg. Nucl. Chem. Lett.*, 1977, **13**, 199 (*cmr, F and P nmr*)

Trifosfamide **T-00485**

N,N,*3-Tris(2-chloroethyl)tetrahydro-2*H*-1,3,2-oxazaphosphorin-2-amine 2-oxide. 2-Bis(2-chloroethyl)-amino-3-chloroethyltetrahydro-2*H*-1,3,2-oxazaphosphorine 2-oxide. Trofosfamide*
[22089-22-1]

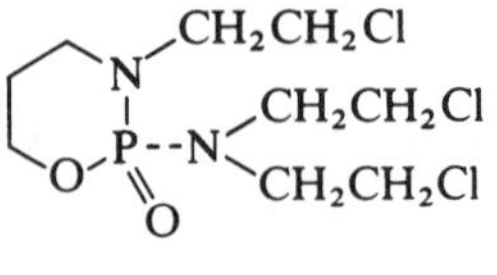

(*R*)-*form*

$C_9H_{18}Cl_3N_2O_2P$ M 323.586
Carcinostatic agent.
▷RP5880000.

(R)-form [72282-84-9]
Oil. $[\alpha]_D^{25}$ −28.6° (c, 2 in MeOH).
(S)-form [72282-85-0]
Oil. $[\alpha]_D^{25}$ +28.6° (MeOH).
(±)-form
Cryst. (Et_2O). Mp 49-51°.

Schulten, H.-R. *et al, Biomed. Mass Spectrom.*, 1974, **1**, 223 (*ms*)
Brassfield, H.A. *et al, Cryst. Struct. Commun.*, 1976, **5**, 417 (*cryst struct*)
Perales, A. *et al, Acta Crystallogr., Sect. B*, 1977, **33**, 1939 (*cryst struct*)
Okruszek, A. *et al, Phosphorus Sulfur*, 1979, **7**, 235 (*synth, pmr, P nmr, complex*)
Pankiewicz, K. *et al, J. Am. Chem. Soc.*, 1979, **101**, 7712 (*resoln, ms, P nmr*)
Wroblewski, A.E. *et al, Inorg. Chem.*, 1980, **19**, 3713 (*resoln, cmr, P nmr*)
Smith, H.W. *et al, Acta Crystallogr., Sect. B*, 1981, **37**, 857 (*cryst struct*)
Buess, M.L. *et al, Org. Magn. Reson.*, 1984, **22**, 67 (*nqr*)

Tri-2-furanylphosphine, 9CI — T-00486

Tri-2-furylphosphine
[5518-52-5]

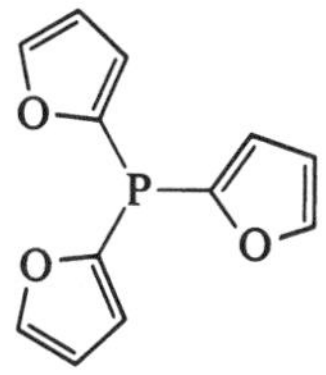

$C_{12}H_9O_3P$ M 232.175
Cryst. (C_6H_6/hexane). Mp 63°. Bp_4 136°.
B,MeI: Methyltri-2-furanylphosphonium iodide.
$C_{13}H_{12}IO_3P$ M 374.114
Mp 104-105°.
Oxide: [1021-20-1].
$C_{12}H_9O_4P$ M 248.174
Cryst. (C_6H_6/hexane). Mp 119°.
Sulfide: [5518-70-7].
$C_{12}H_9O_3PS$ M 264.235
Cryst. (C_6H_6/hexane). Mp 151°.
Selenide: [5593-72-6].
$C_{12}H_9O_3PSe$ M 311.135
Cryst. (C_6H_6/hexane). Mp 174°.

Niwa, E. *et al, Chem. Ber.*, 1966, **99**, 712 (*synth, uv, oxide, sulfide, selenide*)
Jakobsen, H.J. *et al, J. Mol. Spectrosc.*, 1970, **33**, 474 (*pmr*)
Allen, D.W. *et al, J. Chem. Soc., Dalton Trans.*, 1982, 51 (*nmr, complexes*)
Allen, D.W. *et al, J. Chem. Soc., Dalton Trans.*, 1985, 2505 (*P and Se nmr, cryst struct*)

Tri-3-furanylphosphine, 10CI, 9CI — T-00487

[33829-64-0]
$C_{12}H_9O_3P$ M 232.175
Oil. Mp ~20°. $Bp_{0.2}$ 93-95°.
B,MeI: [32488-66-7]. *Methyl(tri-3-furanyl)-phosphonium iodide.*
$C_{13}H_{12}IO_3P$ M 374.114
Solid. Mp 195-197°.
Oxide: [32488-63-4].
$C_{12}H_9O_4P$ M 248.174
Cryst. (MeOAc/hexane). Mp 87°, Mp 111-112°.
Sulfide:
$C_{12}H_9O_3PS$ M 264.235
Solid. Mp 82-83°.
Selenide: [32488-65-6].
$C_{12}H_9O_3PSe$ M 311.135
Cryst. (EtOH). Mp 107-108°.

Jakobsen, H.J., *J. Mol. Spectrosc.*, 1971, **38**, 243 (*synth, derivs, pmr*)
Allen, D.W. *et al, J. Chem. Soc., Dalton Trans.*, 1982, 51 (*synth, selenide, nmr*)

Triheptyl phosphate, 9CI — T-00488

[4621-50-5]

$[H_3C(CH_2)_6O]_3P{=}O$

$C_{21}H_{45}O_4P$ M 392.558
Extractant for transplutonium elements. Liq. $Bp_{0.0005}$ 172°.

Henning, H.G. *et al, J. Prakt. Chem.*, 1966, **33**, 188 (*synth*)
Chudinov, É.G. *et al, Radiokhimiya*, 1978, **20**, 14 (*Engl. transl.* p. 9) (*use*)

Triheptylphosphine, 9CI — T-00489

[17621-04-4]

$P[(CH_2)_6CH_3]_3$

$C_{21}H_{45}P$ M 328.560
Ligand for Ni. Liq. or solid. Mp ca. 20°. Bp_{50} 260°, Bp_2 181-183°.
Oxide: [17262-51-0].
$C_{21}H_{45}OP$ M 344.560
Extractant for Pt, U, Np, and Pu. Inhibits erythrocyte and choline esterases. Phytotoxic. pK_a 8.78 ($MeNO_2$).
▷Toxic, LD_{50} (mice) 150-180 mg/Kg
Sulfide: [21722-49-6].
$C_{21}H_{45}PS$ M 360.620
Possesses lubricating props.
Selenide: [20612-72-0].
$C_{21}H_{45}PSe$ M 407.520
Possesses lubricating props. Separates Pd from Pt, Ir, and Rh. $Bp_{0.03}$ 220-222°.

Jackson, I.K. *et al, J. Chem. Soc.*, 1931, 2109 (*synth, derivs*)
Pavlenko, A.F. *et al, CA*, 1968, **69**, 29031 (*selenide*)
Pavlenko, A.F. *et al, CA*, 1970, **73**, 122123 (*sulfide*)
Buchiklin, E.P. *et al, Zh. Obshch. Khim.*, 1974, **44**, 1354 (*Engl. transl.* p. 1330) (*oxide*)
Volovenko, S.I. *et al, Zh. Obshch. Khim.*, 1976, **46**, 2613 (*Engl. transl.* p. 2491) (*sulfide, ir*)
Yakshin, V.V. *et al, Dokl. Akad. Nauk SSSR, Ser. Sci. Khim.*, 1979, **247**, 128 (*Engl. transl.*, p. 344) (*oxide*)

Triheptyl phosphite, 9CI, 8CI — T-00490

$P[O(CH_2)_6CH_3]_3$

$C_{21}H_{45}O_3P$ M 376.558
Liq. $Bp_{0.7}$ 164-165°. n_D^{20} 1.4465.

Petrov, K.A. *et al, Zh. Obshch. Khim.*, 1963, **33**, 899 (*Engl. transl.* p. 885) (*synth*)

Trihexyl phosphate, 9CI — T-00491

[2528-39-4]

$[H_3C(CH_2)_5O]_3P{=}O$

$C_{18}H_{39}O_4P$ M 350.477
Extractant for Ru and U. Liq. Bp_2 187-188°. n_D^{20} 1.4340.

Arbuzov, B.A. *et al, Izv. Akad. Nauk SSSR, Ser. Khim.*, 1952, 865 (*Engl. transl.* p. 773) (*synth*)

Sax, N.I., *Dangerous Properties of Industrial Materials*, 6th Ed., Van Nostrand-Reinhold, 1984, 1056.

Trihexylphosphine, 9CI **T-00492**

[4168-73-4]

$P[(CH_2)_5CH_3]_3$

$C_{18}H_{39}P$ M 286.480

Ligand for Co, Ni, Fe, Al and Ir. Liq. or solid. Mp ca. 20°. Bp_{50} 227°, Bp_5 163.5-165°. pK_a 9.70.

Oxide: [3084-48-8].

$C_{18}H_{39}OP$ M 302.479

Extractant for Co, Pu, U, Nb, and Zr. Cryst. pK_a 8.78 ($MeNO_2$).

▷SZ1650000.

Sulfide: [18803-12-8].

$C_{18}H_{39}PS$ M 318.540

Possesses lubricating props.

Selenide: [20612-71-9].

$C_{18}H_{39}PSe$ M 365.440

Possesses lubricating props. Phytotoxic. $Bp_{0.04}$ 197-199°.

Jackson, I.K. *et al*, *J. Chem. Soc.*, 1931, 2109 (*synth, derivs*)
Feshchenko, N.G. *et al*, *Zh. Obshch. Khim.*, 1968, **38**, 122 (*Engl. transl.* p. 121) (*synth*)
Borovikov, Yu.Ya. *et al*, *Zh. Obshch. Khim.*, 1974, **44**, 1024 (*Engl. transl.* p.987) (*oxide, props, ir*)
Yakshin, V.V. *et al*, *Dokl. Akad. Nauk SSSR, Ser. Sci. Khim.*, 1979, **247**, 128 (*Engl. transl.*, p. 344) (*oxide*)
Bodner, G.M. *et al*, *J. Organomet. Chem.*, 1982, **226**, 85 (*cmr*)

Trihexyl phosphite, 9CI, 8CI **T-00493**

[6095-42-7]

$P[O(CH_2)_5CH_3]_3$

$C_{18}H_{39}O_3P$ M 334.478

Liq. d_0^{20} 0.90. Bp_3 166-168°, Bp_1 138-139°. n_D^{20} 1.4426.

Petrov, K.A. *et al*, *Zh. Obshch. Khim.*, 1961, **31**, 2377 (*Engl. transl.* p. 2214) (*synth*)
Ovchinnikova, N.K. *et al*, *Zh. Org. Khim.*, 1975, **11**, 1839 (*Engl. transl.* p. 1849) (*synth*)

2,4,6-Trihydroxy-1,2,4,6-oxatriphosphorinane 2,4,6-trioxide, 9CI **T-00494**

2,4,6-Trihydroxy-1,2,4,6-oxatriphosphinane 2,4,6-trioxide

[36792-53-7]

$C_2H_7O_7P_3$ M 235.994

Dimethylene analogue of trimeric metaphosphate. A tribasic acid. Unstable in acid soln.

Tricyclohexylammonium salt: [30798-15-8]. Cryst. + $1H_2O$. Mp 220-230° dec.

Trowbridge, D.B. *et al*, *J. Am. Chem. Soc.*, 1972, **94**, 3816 (*synth, derivs, props*)

Tri-1-imidazolylphosphine **T-00495**

1,1′,1″-Phosphinidynetris(1H-imidazole), 9CI. Tris(1-imidazolyl)phosphine

[73946-92-6]

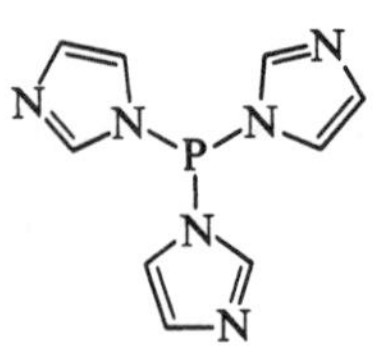

$C_9H_9N_6P$ M 232.184

Phosphorylating agent in oligonucleotide synth. Hygroscopic solid. Mp 154°.

Oxide: [16062-77-4]. *1,1′,1″-Phosphinylidynetris(1H-imidazole).*

$C_9H_9N_6OP$ M 248.183

Phosphorylating agent for oligonucleotide synth. Dehydrates aldoximes to cyanides. Cryst. (Et_2O). Mp 139°.

Cramer, F. *et al*, *Chem. Ber.*, 1961, **94**, 1612 (*oxide*)
Birkofer, L. *et al*, *Angew. Chem., Int. Ed. Engl.*, 1962, **1**, 351 (*synth*)
Sergeeva, N.F. *et al*, *Dokl. Akad. Nauk SSSR, Ser. Sci. Khim.*, 1977, **234**, 607 (*Engl. transl.* p. 280) (*oxide, use*)
Konieczny, M. *et al*, *Z. Naturforsch., B*, 1978, **33**, 1033 (*oxide, use*)
Shimidzu, T. *et al*, *Nucleic Acids Symp. Ser.*, 1980, **8**, 81; 1981, **9**, 183; 1982, **11**, 89 (*use*)
Shimidzu, T. *et al*, *J. Chem. Soc., Perkin Trans. 1*, 1981, 2294 (*use*)

Tri(2-imidazolyl)phosphine **T-00496**

2,2′,2″-Phosphinidynetrisimidazole

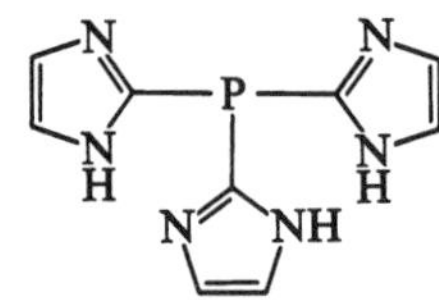

$C_9H_9N_6P$ M 232.184

N,N′,N″-*Tri-Me:* [80679-27-2]. *2,2′,2″-Phosphinidynetris[1-methylimidazole].*

$C_{12}H_{15}N_6P$ M 274.264

Cryst. Mp 203-205°.

Moore, S.S. *et al*, *J. Org. Chem.*, 1982, **47**, 1489 (*synth, pmr, nmr, ms, ir*)

Triiododiphenylphosphorane, 9CI **T-00497**

[59357-76-5]

Ph_2PI_3

$C_{12}H_{10}I_3P$ M 565.898

Hygroscopic brown solid. Insol. C_6H_6. Mp 132-135°. Hydrolysed to diphenylphosphinic acid.

Hoffmann, H. *et al*, *Chem. Ber.*, 1960, **93**, 861 (*props*)
Feshchenko, N.G. *et al*, *Zh. Obshch. Khim.*, 1976, **46**, 252 (*Engl. transl.* p. 248) (*synth*)

2,4,6-Triisopropyl-5-mercapto-1,3,5-dioxaphosphorinane 5-sulfide T-00498

5-Mercapto-2,4,6-tris(1-methylethyl)-1,3,5-dioxaphosphorinane 5-sulfide

S CH(CH₃)₂
HS–P O
(H₃C)₂CH O CH(CH₃)₂

$C_{12}H_{25}O_2PS_2$ M 296.422
Solid. Mp 49-52°.
Na salt: Cryst. (C_6H_6/Me_2CO). Mp 286-287° dec.
NH_4 salt: [6200-26-6]. Cryst. (C_6H_6/Me_2CO). Mp ca. 185° dec.

Rauhut, M.M. *et al*, *J. Org. Chem.*, 1961, **26**, 5133 (*synth*)

Triisopropyl phosphate, 8CI T-00499

Tris(1-methylethyl)phosphate, 9CI
[513-02-0]

$[(H_3C)_2CHO]_3PO$

$C_9H_{21}O_4P$ M 224.236
Catalyst for polymerisation. Liq. Sol. H_2O. d_4^{20} 0.99. Bp 218-220°, Bp_5 83.5°. n_D^{20} 1.4057.

Vogel, A.I. *et al*, *J. Chem. Soc.*, 1943, 16 (*synth*)
Cox, J.R. *et al*, *J. Am. Chem. Soc.*, 1958, **80**, 5441 (*synth*)
Fluck, E. *et al*, *Z. Anorg. Allg. Chem.*, 1967, **354**, 139 (*nmr*)
Ando, F. *et al*, *Bull. Chem. Soc. Jpn.*, 1979, **52**, 807 (*synth, ir*)
Cload, P.A. *et al*, *Org. Mass. Spectrom.*, 1983, **18**, 57 (*ms*)

Triisopropylphosphine, 8CI T-00500

Tris(1-methylethyl)phosphine, 9CI
[6476-36-4]

$P[CH(CH_3)_2]_3$

$C_9H_{21}P$ M 160.239
Air-sensitive liq. Mp −72°. Bp 176-178°, Bp_{22} 81°.
B,MeI: Triisopropylmethylphosphonium iodide.
$C_{10}H_{24}IP$ M 302.178
Solid. Mp >360°.
Oxide: [17513-58-5].
$C_9H_{21}OP$ M 176.238
Liq. solidifying at r.t. Bp_{12} 126-130°, $Bp_{0.8}$ 110-114°.
Sulfide: [17513-59-6].
$C_9H_{21}PS$ M 192.299
Solid. Mp 35°.

Davies, W.C., *J. Chem. Soc.*, 1933, 1043 (*synth*)
Fluck, E. *et al*, *Z. Naturforsch.*, *B*, 1967, **22**, 1095 (*nmr, derivs*)
Cowley, A.H. *et al*, *J. Am. Chem. Soc.*, 1969, **91**, 2915 (*synth, ir, pmr*)
Kincaid, J. *et al*, *Spectrochim. Acta, Part A*, 1974, **30**, 2091 (*oxide, sulfide, synth, ir, pmr, complexes*)
Petrushanskaya, N.V. *et al*, *Zh. Org. Khim.*, 1974, **10**, 1402 (*Engl. transl.* p. 1414) (*use, complexes*)
Chervin, I.I. *et al*, *Izv. Akad. Nauk SSSR, Ser. Khim.*, 1981, **30**, 1769 (*Engl. transl.* p. 1438) (*nmr*)
Bretherick, L., *Handbook of Reactive Chemical Hazards*, 2nd Ed., Butterworths, London and Boston, 1979, 682.
Hazards in the Chemical Laboratory, (Bretherick, L., Ed.), 3rd Ed., Royal Society of Chemistry, London, 1981, 520.

***P,P,P*-Triisopropylphosphine imide** T-00501

P,P,P-*Tris(1-methylethyl)phosphine imide*, 9CI. P,P,P-*Triisopropylphosphine imine. Iminotriisopropylphosphorane*
[63387-15-5]

$HN{=}P[CH(CH_3)_2]_3$

$C_9H_{22}NP$ M 175.253
Very hygroscopic liq. $Bp_{0.3}$ 68°.
B,HCl: Hygroscopic solid. Mp 138-139°.
N-*Me:* P,P,P-*Triisopropyl*-N-*methylphosphine imide.*
$C_{10}H_{24}NP$ M 189.280
Hygroscopic liq. $Bp_{0.05}$ 67°.
N-*Me; B,HCl:* Hygroscopic cryst. Mp 243-245°.
N-*Trimethylsilyl:* P,P,P-*Triisopropyl*-N-*trimethylsilylphosphine imide.*
$C_{12}H_{30}NPSi$ M 247.435
Liq. $Bp_{0.3}$ 53-54°.

Buchner, W. *et al*, *Z. Naturforsch.*, *B*, 1974, **29**, 328 (*silyl deriv, cmr, P nmr*)
Ross, B. *et al*, *Chem. Ber.*, 1979, **112**, 1756 (*synth, derivs, ms, cmr, P nmr, pmr*)

Triisopropyl phosphite T-00502

Tris(1-methylethyl) phosphite, 9CI
[116-17-6]

$P[OCH(CH_3)_2]_3$

$C_9H_{21}O_3P$ M 208.237
Mild dechlorinating agent and synthetic reagent. Ligand for metals of groups IB, IIB, VIB, VIIB, and VIII. Liq. Bp_{11} 63-64°. n_D^{20} 1.4101.
▷TH2800000.

Arbuzov, A.E., *Ber.*, 1905, **38**, 1171 (*synth*)
Ford-Moore, A. *et al*, *J. Chem. Soc.*, 1947, 1465 (*synth*)
Goodell, L.J. *et al*, *Can. J. Chem.*, 1969, **47**, 2461 (*pmr*)
Engel, R. *et al*, *J. Chem. Soc., Perkin Trans. 1*, 1972, 1233 (*pmr*)
Verstuyft, A.W. *et al*, *Inorg. Chem.*, 1976, **15**, 732 (*cmr, complexes*)
Labintsev, V.B. *et al*, *Zh. Org. Khim.*, 1977, **13**, 1141 (*Engl. transl.* p. 1050) (*ms*)
Vasil'ev, V.V. *et al*, *Zh. Obshch. Khim.*, 1981, **51**, 2134 (*Engl. transl.* p. 1836) (*O nmr*)
Fieser, M. *et al*, *Reagents for Organic Synthesis*, Wiley, 1967-84, **1**, 1229.

***O,O,S*-Triisopropyl phosphorodithioate,** 8CI T-00503

O,O,S-*Tris(1-methylethyl) phosphorodithioate*, 9CI. O,O,S-*Triisopropyl dithiophosphate*
[20442-31-3]

$[(H_3C)_2CHO]_2P(S)SCH(CH_3)_2$

$C_9H_{21}O_2PS_2$ M 256.357
Liq. d_4^{20} 1.03-1.04. Bp_3 92-93°, $Bp_{0.07}$ 66-67°. n_D^{20} 1.4815.

Zimin, M.G. *et al*, *Zh. Obshch. Khim.*, 1978, **48**, 1020 (*Engl. transl.* p. 930) (*synth, P nmr*)
Pudovik, A.N. *et al*, *Izv. Akad. Nauk SSSR, Ser. Khim.*, 1979, 861 (*Engl. transl.* p. 805) (*synth*)

***O,O,O*-Triisopropyl phosphoroselenoate,** 8CI T-00504

O,O,O-*Tris(1-methylethyl) phosphoroselenoate*, 9CI. O,O,O-*Triisopropyl selenophosphate*
[14540-49-9]

$[(H_3C)_2CHO]_3P{=}Se$

$C_9H_{21}O_3PSe$ M 287.197

Liq. d_4^{20} 1.16. $Bp_{0.2}$ 63°. n_D^{20} 1.4557.

Nuretdinov, I.A., *Izv. Akad. Nauk SSSR, Ser. Khim.*, 1968, 2831 (*Engl. transl.* p. 2685) (*synth, ir*)
Stec, W.J. *et al*, *Phosphorus*, 1972, **2**, 97 (*P nmr*)
Ando, F. *et al*, *Bull. Chem. Soc. Jpn.*, 1979, **52**, 807 (*ir, pmr*)

Triisopropyl phosphorotetrathioate, 8CI — T-00505

Tris(1-methylethyl) phosphorotetrathioate, 9CI. Triisopropyl tetrathiophosphate

[2386-41-6]

$[(H_3C)_2CHS]_3P{=}S$

$C_9H_{21}PS_4$ M 288.479

Odorous liq. d_4^{20} 1.11. Bp_1 125-126°. n_D^{20} 1.5790.

Scott, C.B. *et al*, *J. Org. Chem.*, 1957, **22**, 789 (*synth*)
Menefee, A. *et al*, *J. Org. Chem.*, 1957, **22**, 792 (*ir*)
Murav'ev, I.V. *et al*, *Zh. Obshch. Khim.*, 1975, **45**, 1746 (*Engl. transl.* p. 1711) (*synth*)

O,O,O-Triisopropyl phosphorothioate, 8CI — T-00506

O,O,O-*Tris(1-methylethyl) phosphorothioate, 9CI.*
O,O,O-*Triisopropyl thiophosphate*

[2464-03-1]

$[(H_3C)_2CHO]_3P{=}S$

$C_9H_{27}O_3PS$ M 246.344

Photodecomp. product of *S*-Benzyl *O,O*-diisopropyl phosphorothioate, B-00045 . Liq.

Fluck, E. *et al*, *Z. Anorg. Allg. Chem.*, 1967, **354**, 139 (*P nmr*)
Murai, T. *et al*, *Agric. Biol. Chem.*, 1977, **41**, 803 (*ms, glc*)
Ando, F. *et al*, *Bull. Chem. Soc. Jpn.*, 1979, **52**, 807; 1981, **54**, 3495.

O,S,S-Triisopropyl phosphorotrithioate, 8CI — T-00507

O,S,S-*Tris(1-methylethyl) phosphorotrithioate, 9CI.*
O,S,S-*Triisopropyl trithiophosphate*

$[(H_3C)_2CHS]_2P(S)OCH(CH_3)_2$

$C_9H_{21}OPS_3$ M 272.418

Odorous liq. d_4^{20} 1.07. Bp_1 101-103°. n_D^{20} 1.5314.

Murav'ev, I.V. *et al*, *Zh. Obshch. Khim.*, 1968, **38**, 133 (*Engl. transl.* p. 133) (*synth*)
Kotovich, B.P. *et al*, *Zh. Obshch. Khim.*, 1968, **38**, 1282 (*Engl. transl.* p. 1235) (*synth*)

S,S,S-Triisopropyl phosphorotrithioate, 8CI — T-00508

S,S,S-*Tris(1-methylethyl) phosphorotrithioate, 9CI.*
S,S,S-*Triisopropyl trithiophosphate*

[78788-15-5]

$[(H_3C)_2CHS]_3P{=}O$

$C_9H_{21}OPS_3$ M 272.418

Liq.

McElroy, R.D. *et al*, *J. Agric. Food Chem.*, 1984, **32**, 119 (*synth, chromatog*)

N,N',N''-Triisopropylphosphorous triamide — T-00509

N,N',N''-*Tris(1-methylethyl)phosphorous triamide. Tri(isopropylamino)phosphine. Tris(isopropylamino)-phosphine*

$P[NHCH(CH_3)_2]_3$

$C_9H_{24}N_3P$ M 205.282

Stabilized as a Mo complex.

Oxide: N,N',N''-*Triisopropylphosphoric triamide.*
$C_9H_{24}N_3OP$ M 221.282
Solid. Mp 108-110°.

Sulfide: [69695-90-5]. N,N',N''-*Triisopropylphosphorothioic triamide.*
$C_9H_{24}N_3PS$ M 237.342
Solid. Mp 93°.

Sosnovsky, G. *et al*, *Z. Naturforsch., B*, 1976, **31**, 1526 (*oxide*)
Healy, J.D. *et al*, *Phosphorus Sulfur*, 1978, **5**, 239 (*sulfide*)
Maisch, H. *Z. Naturforsch., B*, 1979, **34**, 784 (*complex*)

5,5,6-Trimethyl-4,7-dioxa-3-phospha(3-P^V)spiro[2.4]heptane, 9CI — T-00510

[69084-18-0]

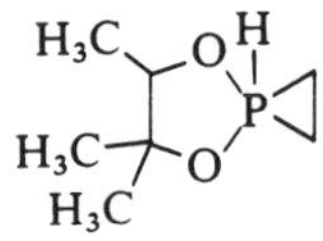

$C_7H_{15}O_2P$ M 162.168

Mixt. of stereoisomers. Stable at −80°, dec. at r.t. → ethylene.

Campbell, B.C. *et al*, *J. Chem. Soc., Chem. Commun.*, 1978, 854 (*synth*)

2,4,5-Trimethyl-1,3,2-dioxaphospholane, 9CI — T-00511

2,4,5-Trimethyl-1,3-dioxa-2-phosphacyclopentane. 2,3-Butylene methylphosphonite

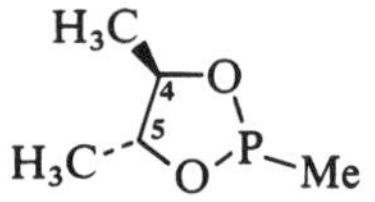

(4*R*,5*R*)-*form*

$C_5H_{11}O_2P$ M 134.114

(4*R*,5*R*)-*form* [80093-92-1]
Liq. Bp 133-135°. $[\alpha]_D^{20}$ +30.3° (c, 15 in THF).

(4*RS*,5*RS*)-*form*

2-Oxide: [80125-96-8]. *2,3-Butylene methylphosphonate.*
$C_5H_{11}O_3P$ M 150.114
Liq. $Bp_{0.001}$ 58-60°.

Richter, W.J., *Phosphorus Sulfur*, 1981, **10**, 395 (*synth, pmr, P nmr*)

2,5,5-Trimethyl-1,3,2-dioxaphosphorinane, 9CI — T-00512

2,5,5-Trimethyl-1,3,2-dioxaphosphinane. 2,5,5-Trimethyl-1,3-dioxa-2-phosphacyclohexane. Neopentylene methylphosphonite

[36315-13-6]

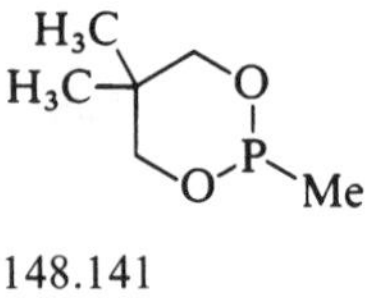

$C_6H_{13}O_2P$ M 148.141

Mobile liq. with phosphine-like odour. Bp_{55} 92-98°. Undergoes slow ring-opening oligomerization.

2-Oxide: [873-97-2]. *Neopentylene methylphosphonate.*
$C_6H_{13}O_3P$ M 164.141
Cryst. (EtOAc/pet. ether). Mp 119-121°.

2-Sulfide: [53121-78-1]. O,O-*Neopentylene methylphosphonothioate.*
$C_6H_{13}O_2PS$ M 180.201
Cryst. (C_6H_6). Mp 60°.

2-Selenide: O,O-*Neopentylene methylphosphonoselenoate.*
$C_6H_{18}O_2PSe$ M 232.141
Solid. Mp 84-86°.

Bartle, K.D. *et al, Tetrahedron*, 1967, **23**, 1701 (*oxide, pmr*)
White, D.W., *Phosphorus*, 1971, **1**, 33 (*synth, props, pmr, P nmr*)
Albrand, J.P. *et al, J. Am. Chem. Soc.*, 1974, **96**, 4584 (*synth, sulfide, pmr, P nmr*)
Francis, G.W. *et al, Acta Chem. Scand., Ser. B*, 1976, **30**, 31 (*oxide, ms*)
Dale, A.J., *Acta Chem. Scand., Ser. B*, 1976, **30**, 255 (*pmr*)
Grand, A. *et al, Acta Crystallogr., Sect. B*, 1976, **34**, 199 (*sulfide, cryst struct*)
Shagidullin, R.R. *et al, Izv. Akad. Nauk SSSR, Ser. Khim.*, 1981, 1156 (*Engl. transl.* p. 916) (*selenide, raman, P nmr*)

4,4,6-Trimethyl-1,3,2-dioxaphosphorinane T-00513

(2*RS*,4*RS*)-*form*

$C_6H_{13}O_2P$ M 148.141
Liq. Bp_1 26-28°. Mixt. of diastereoisomers.

2-Sulfide: [40651-48-7].
$C_6H_{13}O_2PS$ M 180.201
Oil which gradually cryst. $Bp_{0.007}$ 72-73° (bath). n_D^{20} 1.5095. Mixt. of diastereoisomers.

(2RS,4RS)-form [69220-15-1]
(±)-cis-*form*

2-Oxide: [34882-99-0].
$C_6H_{13}O_3P$ M 164.141
Cryst. (Et_2O). Mp 66-67°. Bp_1 128-129°.

(2RS,4SR)-form [69220-16-2]
(±)-trans-*form*

2-Oxide: [34882-98-9]. Cryst. (Et_2O). Mp 24-25°. Bp_1 99-100°.

Nifant'ev, É.E. *et al, Zh. Obshch. Khim.*, 1971, **41**, 2368 (*Engl. transl.* p. 2394) (*oxide, synth, pmr, P nmr*)
Borisenko, A.A. *et al, J. Chem. Soc., Chem. Commun.*, 1972, 406 (*oxide, cmr*)
Predvoditelev, D.A. *et al, Zh. Obshch. Khim.*, 1973, **43**, 73 (*Engl. transl.* p. 70) (*sulfide, ir, pmr, P nmr*)
Stec, W.J., *Z. Naturforsch., B*, 1974, **29**, 109 (*oxide, P nmr*)
Nifant'ev, É.E. *et al, Tetrahedron*, 1981, **37**, 3183 (*synth, pmr, P nmr*)

3,7,11-Trimethyl-1,6,10-dodecatrien-3-yl dihydrogen phosphate T-00514

Nerolidyl phosphate. Nerolidyl dihydrogen phosphate

$C_{15}H_{27}O_4P$ M 302.350

(S)-form

Di-NH_4 salt: Solid. Spar. sol. H_2O, MeOH. Mp 133-136° dec.

Cramer, F. *et al, Justus Liebigs Ann. Chem.*, 1962, **654**, 180 (*synth, ir*)

3,7,11-Trimethyl-2,6,10-dodecatrien-1-yl dihydrogen phosphate, 8CI T-00515

Farnesyl phosphate. Farnesyl dihydrogen phosphate
[15416-86-1]

(*E*,*E*)-*form*

$C_{15}H_{27}O_4P$ M 302.350

(E,E)-form [63023-44-9]
In Et_2O, dec. within 1 week → hydrocarbons incl. farnesenes and biosabolenes.

Monocyclohexylammonium salt: Cryst. (H_2O). Sol. MeOH, EtOH, warm Me_2CO, C_6H_6, spar. sol. cold H_2O. Mp 164-165° dec.

Cramer, F. *et al, Justus Liebigs Ann. Chem.*, 1962, **654**, 180 (*synth, salt, ir*)
Dugan, R.E. *et al, Anal. Biochem.*, 1968, **22**, 249 (*isol*)
Miller, J.A. *et al, J. Chem. Soc. (C)*, 1968, 1837 (*derivs*)
Larkin, J.P. *et al, J. Chem. Soc., Perkin Trans. 1*, 1976, 2524 (*props*)

Trimethylgermylphosphine, 10CI, 9CI T-00516

Trimethylphosphinogermane
[20519-92-0]

Me_3GePH_2

$C_3H_{11}GeP$ M 150.684
Liq. Mp −97.2°. Bp 95.8°. Readily oxidised.

Norman, D.A., *Inorg. Chem.*, 1970, **9**, 870 (*synth, ir*)
Dahl, A.R. *et al, Inorg. Chem.*, 1975, **14**, 1093, 2562.
Ansari, S. *et al, Z. Naturforsch., B*, 1975, **30**, 523.

2,4,10-Trimethyl-1,3,4,5,7,10-hexaaza-3-phosphatricyclo[3.3.1.1^{3,7}]decane, 9CI T-00517

2,4,10-Trimethyl-1,2,4,5,7,10-hexaaza-3-phosphaadamantane
[75543-58-7]

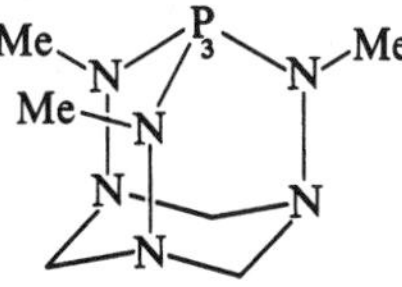

$C_6H_{15}N_6P$ M 202.198
Cryst. (C_6H_6/hexane). Mp 43-44°.

3-Oxide: [56634-34-5].
$C_6H_{15}N_6OP$ M 218.198
Solid. Mp 155-156° (150-151°).

3-Sulfide: [56634-35-6].
$C_6H_{15}N_6PS$ M 234.258
Solid. Mp 157°.

Majoral, J.P. *et al, Tetrahedron*, 1976, **32**, 2633 (*synth, derivs, pmr, P nmr*)
Jaud, J. *et al, Z. Kristallogr., Kristallgeom., Kristallphys., Kristallchem.*, 1982, **160**, 67 (*synth, P nmr, cryst struct*)

Trimethylmethylenephosphorane, 9CI T-00518

Trimethylphosphoranylidenemethane

[14580-91-7]

$Me_3P{=}CH_2$

$C_4H_{11}P$ M 90.105

Liq. or solid. Mp 13-14°. Bp_{750} 118-120°.

Schmidbaur, H. *et al*, *Chem. Ber.*, 1968, **101**, 595, 604 (*synth, pmr, props*)

Sawodny, W., *Z. Anorg. Allg. Chem.*, 1969, **368**, 284 (*ir, raman*)

Schmidbaur, H. *et al*, *Chem. Ber.*, 1975, **108**, 2649; 1977, **110**, 677 (*cmr, nmr*)

Ostoja Starzewski, K.A. *et al*, *J. Am. Chem. Soc.*, 1976, **98**, 8486 (*pe*)

Ostoja Starzewski, K.A. *et al*, *Phosphorus*, 1976, **6**, 177 (*pmr, cmr, nmr*)

Inorg. Synth., 1978, **18**, 135 (*synth*)

3,5,5-Trimethyl-1,2-oxaphospholan-3-ol 2-oxide, 9CI T-00519

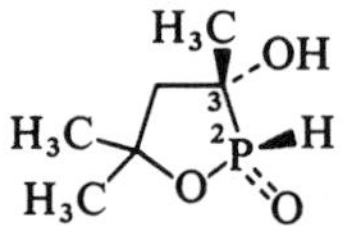

(2*RS*,3*RS*)-*form*

$C_6H_{13}O_3P$ M 164.141

(*2RS,3RS*)-*form*

(±)-cis-*form*

2-Methoxy: [86703-69-7]. *2-Methoxy-3,5,5-trimethyl-1,2-oxaphospholan-3-ol 2-oxide.*

$C_7H_{15}O_3P$ M 178.167

Solid. Mp 127.5-128°.

2-Ph: [39770-56-4]. *3,5,5-Trimethyl-2-phenyl-1,2-oxaphospholan-3-ol 2-oxide.*

$C_{12}H_{17}O_3P$ M 240.238

Solid. Mp 135-136°.

2-Dimethylamino: [66830-78-2]. *3,5,5-Trimethyl-2-dimethylamino-1,2-oxaphospholan-3-ol 2-oxide.*

$C_8H_{18}NO_3P$ M 207.209

Solid. Mp 140°.

2-Diethylamino: [39770-55-3]. *2-Diethylamino-3,5,5-trimethyl-1,2-oxaphospholan-3-ol 2-oxide.*

$C_{10}H_{22}NO_3P$ M 235.262

Solid. Mp 125-126°.

(*2RS,3SR*)-*form*

(±)-trans-*form*

2-Methoxy: [86703-68-6]. Solid. Mp 63.5-64°.

2-Ethoxy: [83627-25-2]. *2-Ethoxy-3,5,5-trimethyl-1,2-oxaphospholan-3-ol 2-oxide.*

$C_8H_{17}O_3P$ M 192.194

Solid. Mp 100-101°.

2-Phenoxy:

$C_{12}H_{17}O_4P$ M 256.238

Solid. Mp 126-127°.

2-Ph: [39770-56-4]. *3,5,5-Trimethyl-2-phenyl-1,2-oxaphospholan-3-ol 2-oxide.*

$C_{12}H_{17}O_3P$ M 240.238

Solid. Mp 164-165°.

2-Dimethylamino: [41157-86-2]. Solid. Mp 140-141°.

2-Diethylamino: [39770-51-9]. Solid. Mp 118-119°.

Shagidullin, R.R. *et al*, *Izv. Akad. Nauk SSSR, Ser. Khim.*, 1972, 1604, 2585 (*Engl. transl.* pp. 1545, 2510); 1978, 667 (*Engl. transl.* p. 575) (*synth, ir, pmr, config*)

Pilgram, K.H. *et al*, *Z. Naturforsch., B*, 1983, **38**, 1122 (*ethoxy, methoxy, pmr, cmr, nmr, cryst struct, config*)

Shagidullin, R.R. *et al*, *Zh. Obshch. Khim.*, 1983, **53**, 1035 (*Engl. transl.* p. 915) (*ir*)

Wroblewski, A.E. *et al*, *Tetrahedron*, 1983, **39**, 1809 (*methoxy, pmr, nmr, cmr*)

4,4,6-Trimethyl-2-phenoxy-1,3,2-dioxaphosphorinane, 9CI T-00520

4,4,6-Trimethyl-2-phenoxy-1,3-dioxa-2-phosphacyclohexane

[1083-44-9]

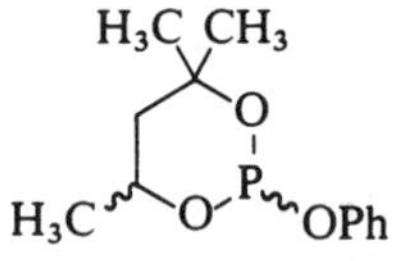

$C_{12}H_{17}O_3P$ M 240.238

Liq. $Bp_{0.05}$ 75°. Mixt. of *cis-* and *trans-*isomers.

Majoral, J.P. *et al*, *Spectrochim. Acta, Part A*, 1972, **28**, 2247 (*derivs, ir*)

Maria, P.C. *et al*, *Tetrahedron Lett.*, 1972, 1485 (*derivs, P nmr*)

El Khatib, F. *et al*, *Phosphorus Sulfur*, 1984, **20**, 55 (*synth, P nmr, props*)

2,4,6-Trimethyl-5-phenyl-1,3,5-dioxaphosphorinane, 9CI T-00521

2,4,6-Trimethyl-5-phenyl-1,3,5-dioxaphosphinane

[68100-16-3]

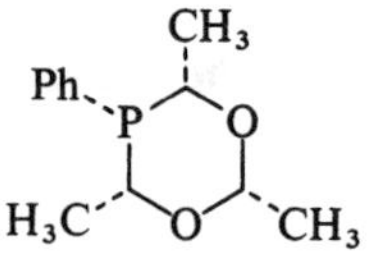

(2α,4α,5α,6α)-*form*

$C_{12}H_{17}O_2P$ M 224.239

Liq. d_4^{20} 1.11. $Bp_{0.1}$ 87-88°. n_D^{20} 1.5450 (1.5545). A mixture of stereoisomers.

(*2α,4α,5α,6α*)-*form* [71485-27-3]

5-Oxide: [71485-34-2].

$C_{12}H_{17}O_3P$ M 240.238

Solid. Mp 89°. $Bp_{0.1}$ 144°.

5-Sulfide: [68128-37-0].

$C_{12}H_{17}O_2PS$ M 256.299

Solid. Mp 88°.

5-Selenide: [81439-66-9].

$C_{12}H_{17}O_2PSe$ M 303.199

Solid. Mp 86°.

(*2α,4α,5β,6α*)-*form* [71485-28-4]

5-Oxide: [71444-56-9]. Cryst. (CCl_4). Mp 184°.

5-Sulfide: [68100-17-4]. Solid. Mp 165°.

5-Selenide: [81495-87-6]. Cryst. (MeOH). Mp 162°.

Arbuzov, B.A. *et al*, *Izv. Akad. Nauk SSSR, Ser. Khim.*, 1979, 2239 (*Engl. transl.* p. 2061) (*synth, derivs, pmr, nmr*)

Arbuzov, B.A. *et al*, *Dokl. Akad. Nauk SSSR, Ser. Sci. Khim.*, 1979, **246**, 326 (*Engl. transl.* p. 231) (*nmr, stereochem*)

Kadyrov, R.A. *et al*, *Izv. Akad. Nauk SSSR, Ser. Khim.*, 1982, 594 (*Engl. transl.* p. 525) (*uv*)

Arbuzov, B.A. *et al*, *Izv. Akad. Nauk SSSR, Ser. Khim.*, 1982, 127 (*Engl. transl.* p. 119) (*selenides, synth, nmr, stereochem*)

2,2,5-Trimethyl-3-phenyl-1,3-oxaphospholane, 9CI T-00522

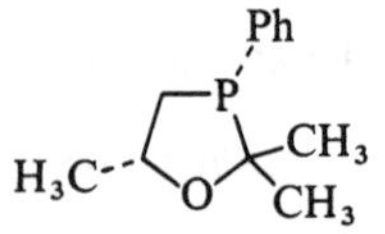

(3*RS*,5*RS*)-*form*

$C_{12}H_{17}OP$ M 208.239

(3RS,5RS)-form [61009-60-7]
(±)-cis-*form*
Liq. Bp_3 97°. Obt. only in impure form.
B,PhCH₂Br: [61009-63-0]. *3-Benzyl-2,2,5-trimethyl-3-phenyl-1,3-oxaphospholanium bromide.*
$C_{19}H_{24}BrOP$ M 379.276
Cryst. + 1 EtOH (EtOAc/EtOH). Mp 202°.
(3RS,5SR)-form [61009-58-3]
(±)-trans-*form*
Liq. $Bp_{3.5}$ 95°. Obt. stereoisomerically pure.
B,PhCH₂Br: [61009-62-9]. Cryst. (EtOH/EtOAc). Mp 208-209°.
3-Oxide: [61009-61-8].
$C_{12}H_{17}O_2P$ M 224.239
Solid. Mp 78-80°. $Bp_{0.07}$ 115°.

Marsi, K.L *et al*, *J. Org. Chem.*, 1977, **42**, 778 (*synth, derivs, pmr*)

2,3,4-Trimethyl-5-phenyl-1,3,2-oxazaphospholidine, 9CI T-00523

2,3,4-Trimethyl-5-phenyl-1,3,2-oxazaphospholane

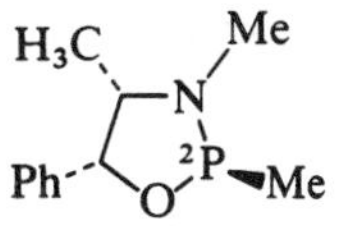

(*2R,4S,5R*)-*form*

$C_{11}H_{16}NOP$ M 209.227
Deriv. used in synth. of chiral methylphosphonic and methylphosphonothioic (selenoic) esters.
(2R,4S,5R)-form
2-Oxide: $[\alpha]_D$ −65°.
2-Sulfide: Cryst. (pet. ether). Mp 88-90°. $[\alpha]_D$ −25° ($CHCl_3$). Has 2*S*-config.
2-Selenide: Cryst. (diisopropyl ether). Mp 100-101°. $[\alpha]_D$ −8.6° ($CHCl_3$). Has 2*S*-config.
(2S,4S,5R)-form
2-Oxide:
$C_{11}H_{16}NO_2P$ M 225.227
$[\alpha]_D$ −81° ($CHCl_3$).
2-Sulfide:
$C_{11}H_{16}NOPS$ M 241.287
Cryst. (pet. ether). Mp 73-74°. Has 2*R*-config.
2-Selenide:
$C_{11}H_{16}NOPSe$ M 288.187
Syrup. $[\alpha]_D$ −97.5° ($CHCl_3$). Has 2*R*-config.

Cooper, D.B. *et al*, *J. Chem. Soc., Perkin Trans. 1*, 1977, 1969 (*synth, use*)
Hall, C.R. *et al*, *Pol. J. Chem.*, 1980, **54**, 489 (*props, use*)
Hall, C.R. *et al*, *J. Chem. Soc., Perkin Trans. 1*, 1981, 2368, 2746 (*props, use*)

(2,4,5-Trimethylphenyl)phosphinic acid, 9CI T-00524

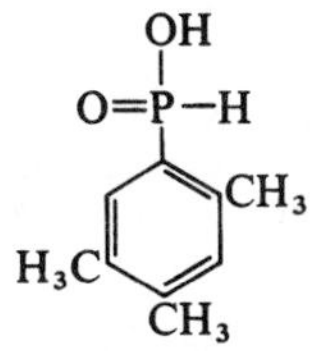

$C_9H_{13}O_2P$ M 184.174
Tautomeric with (2,4,5-trimethylphenyl)phosphonous acid, but free acid probably exists in phosphinic acid form. Solid. Spar. sol. EtOH, mod. sol. Et_2O. Mp 128°.
Phenylhydrazinium salt: Solid. Mp 180°.

Michaelis, A., *Justus Liebigs Ann. Chem.*, 1897, **294**, 1 (*synth, derivs*)

(2,4,6-Trimethylphenyl)phosphinic acid, 9CI T-00525

Mesitylphosphinic acid

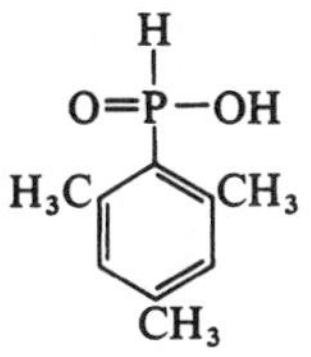

$C_9H_{13}O_2P$ M 184.174
Solid. Mp 146°.

Brazier, J-F. *et al*, *C.R. Hebd. Seances Acad. Sci., Ser. C*, 1966, **262**, 1393.

(2,4,5-Trimethylphenyl)phosphonic acid T-00526

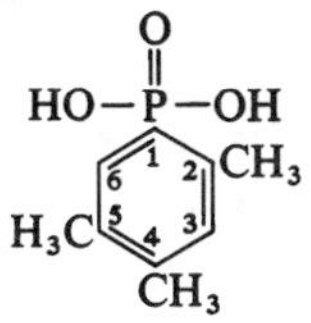

$C_9H_{13}O_3P$ M 200.174
Solid. Mp 212°.
Di-Ph ester: Diphenyl (2,4,5-trimethylphenyl)-phosphonate.
$C_{21}H_{21}O_3P$ M 352.369
Solid. Mp 62-65°.
Dichloride:
$C_9H_{11}Cl_2OP$ M 237.065
Solid. Mp 63°. Bp 307-308°.
Dianilide: P-(*2,4,5-Trimethylphenyl*)-N,N′-*diphenyl-phosphonic diamide.*
$C_{21}H_{23}N_2OP$ M 350.399
Needles (EtOH). Mp 197°.

Michaelis, A., *Justus Liebigs Ann. Chem.*, 1896, **294**, 1 (*synth, derivs, props*)

(2,4,6-Trimethylphenyl)phosphonic acid T-00527

Mesitylphosphonic acid
$C_9H_{13}O_3P$ M 200.174
Bp_{18} 158-161°, $Bp_{0.05}$ 111-112°.
Di-Et ester: [28036-01-3]. *Diethyl(2,4,6-trimethylphenyl)phosphonate. Diethyl mesitylphosphonate.*
$C_{13}H_{21}O_3P$ M 256.281
Liq.

Tavs, P., *Chem. Ber.*, 1970, **103**, 2428 (*synth*)
Ernst, L., *Org. Magn. Reson.*, 1977, **9**, 35 (*synth, cmr*)
Osuka, A. *et al*, *Synthesis*, 1983, 69 (*synth*)

Trimethyl phosphate, 9CI T-00528

[512-56-1]

$(MeO)_3P{=}O$

$C_3H_9O_4P$ M 140.075
Solvent for aromatic halogenations and nitrations. Methylating agent for nitrogen heterocyclic compds. Pleasant smelling liq. Sol. H_2O, organic solvents. d_4^{20} 1.21. Bp 196°, Bp_{50} 110°, Bp_5 62°. n_D^{20} 1.3963.

▷Shows delayed toxicity. TC8225000.

Evans, D.P. *et al*, *J. Chem. Soc.*, 1930, 1310 (*synth*)
Yamauchi, K. *et al*, *J. Chem. Soc., Perkin Trans. 1*, 1973, 391 (*props*, *use*)
Pearson, D.E. *et al*, *Synthesis*, 1976, 621 (*use*)
Cowley, A.H. *et al*, *Inorg. Chim. Acta*, 1977, **25**, L151 (*pe*)
Burkhardt, W.D. *et al*, *Z. Anorg. Allg. Chem.*, 1978, **442**, 19 (*ir*, *raman*)
Lowe, G. *et al*, *J. Chem. Soc., Chem. Commun.*, 1979, 733 (*O and P nmr*)
Wolff, R. *et al*, *Z. Phys. Chem.* (*Leipzig*), 1980, **261**, 726 (*struct*)
Kawazoe, Y. *et al*, *Chem. Pharm. Bull.*, 1982, **30**, 2077 (*props*, *use*)
Cload, P.A. *et al*, *Org. Mass Spectrom.*, 1983, **18**, 57 (*ms*)
Cohen, J.S. *et al*, *J. Magn. Reson.*, 1984, **59**, 181 (*pmr*)
Cowley, A.H. *et al*, *Inorg. Chem.*, 1984, **23**, 3378 (*pe*)
Yarkov, A.V. *et al*, *Zh. Obshch. Khim.*, 1985, **55**, 50 (*J. Gen. Chem. USSR* p. 42) (*ir*, *raman*, *conformn*)

2,3,5-Trimethyl-3-phosphatricyclo[4.2.1.0^{2,5}]nonane T-00529

$C_{11}H_{19}P$ M 182.245

Oxide: [76287-39-3].
$C_{11}H_{19}OP$ M 198.244
Triclinic cryst. (Et_2O/pentane). Mp 67-69°. Has ($1\alpha,2\alpha,3\beta,5\alpha,6\alpha$)-config. as shown.

Vilkas, E. *et al*, *J. Chem. Soc., Perkin Trans. 1*, 1980, 2136 (*synth*, *ms*, *pmr*, *cryst struct*)

Trimethylphosphine, 9CI, 8CI T-00530

[594-09-2]

Me_3P

C_3H_9P M 76.078
Insol. H_2O. Mp −85°. Bp 37.8°. pK_a 8.65 (H_2O).

▷Emits highly toxic fumes on contact with acids. May ignite in air

B,MeI: see Tetramethylphosphonium(1+), T-00215
Oxide: see Trimethylphosphine oxide, T-00532
Sulfide: see Trimethylphosphine sulfide, T-00534
Selenide: see Trimethylphosphine selenide, T-00533
Telluride: [76773-16-5].
C_3H_9PTe M 203.678
Yellow cryst. (toluene at −50°). Rapidly oxidized in air → Te.

Screttas, C. *et al*, *J. Org. Chem.*, 1962, **27**, 2573 (*synth*)
Maier, L., *Helv. Chim. Acta*, 1966, **49**, 2458 (*synth*, *derivs*)
Peterson, D.J. *et al*, *J. Inorg. Nucl. Chem.*, 1966, **28**, 53 (*synth*)
Inorg. Synth., 1967, **9**, 59 (*synth*)
Brodie, A.M. *et al*, *J. Chem. Soc.* (*A*), 1969, 2927 (*telluride*)
Guest, M.F. *et al*, *J. Chem. Soc., Faraday Trans. 2*, 1972, **68**, 867 (*struct*)
Kostyanovskii, R.G. *et al*, *Org. Mass Spectrom.*, 1972, **6**, 1183 (*ms*)
Wolfsberger, W. *et al*, *Synth. React. Inorg. Met.-Org. Chem.*, 1974, **4**, 149 (*synth*)
Lappert, M.F. *et al*, *J. Chem. Soc., Dalton Trans.*, 1975, 1207 (*pe*)
Albright, T.A., *Org. Magn. Reson.*, 1976, **8**, 489 (*cmr*, *nmr*, *pmr*, *struct*)
Inorg. Synth., 1976, **16**, 153 (*synth*)
Rajhantalab, H. *et al*, *Spectrochim. Acta, Part A*, 1976, **32**, 519 (*raman*)
McKean, D.C. *et al*, *J. Mol. Struct.*, 1978, **49**, 275 (*ir*)
Stein, J. *et al*, *J. Am. Chem. Soc.*, 1981, **103**, 2192 (*derivs*)
Watari, F., *Inorg. Chem.*, 1981, **20**, 1776 (*telluride*, *ir*, *raman*)
Bretherick, L., *Handbook of Reactive Chemical Hazards*, 2nd Ed., Butterworths, London and Boston, 1979, 453.
Sax, N.I., *Dangerous Properties of Industrial Materials*, 6th Ed., Van Nostrand-Reinhold, 1984, 1062.
Hazards in the Chemical Laboratory, (Bretherick, L., Ed.), 3rd Ed., Royal Society of Chemistry, London, 1981, 524.

P,P,P-Trimethylphosphine imide, 9CI T-00531

Iminotrimethylphosphorane. P,P,P-*Trimethylphosphazene*

[15107-02-5]

$Me_3P{=}NH$

$C_3H_{10}NP$ M 91.092
Solid. Mp 59-60°. Bp_1 70°. Yields Li deriv. when treated with butyllithium.

▷K deriv. is pyrophoric when dry

B,HCl: Hygroscopic solid. Mp 121°.
N-*Me:* [42437-75-2]. *Tetramethylphosphine imide. Tetramethylphosphazene.*
$C_4H_{12}NP$ M 105.119
Liq.
N-tert-*butyl:* [71328-66-0]. N-tert-*Butyl*-P,P,P-*trimethylphosphine imide.*
$C_7H_{18}NP$ M 147.200
Liq.
N-*Ph:* P,P,P-*Trimethyl*-N-*phenylphosphine imide.*
$C_9H_{14}NP$ M 167.190
Solid.
N-(*Trimethylsilyl*)*:* P,P,P-*Trimethyl*-N-*trimethylsilylphosphine imide.*
$C_6H_{18}NPSi$ M 163.274
Liq. Bp_{11} 57°.

Schmidbaur, H. *et al*, *Chem. Ber.*, 1968, **101**, 1271 (*synth*, *ir*, *pmr*, *deriv*)
Bragin, J. *et al*, *J. Phys. Chem.*, 1973, **77**, 1506 (*deriv*, *ir*, *raman*)
Schmidbaur, H. *et al*, *Chem. Ber.*, 1973, **106**, 1251 (*cmr*, *P nmr*)
Ross, B. *et al*, *Chem. Ber.*, 1974, **107**, 2720 (*synth*, *derivs*)
Avanzino, S.C. *et al*, *Inorg. Chem.*, 1975, **14**, 1595 (*synth*, *pe*)
Ostoja Starzewski, K.A. *et al*, *Inorg. Chem.*, 1979, **18**, 3307 (*derivs*, *pe*, *struct*, *props*)

Trimethylphosphine oxide, 9CI, 8CI T-00532

[676-96-0]

$Me_3P{=}O$

C_3H_9OP M 92.077
Ligand for metals of Group IIB, and many from Groups IIIA, IVA, IB, and VIII, and additionally U, Np, and Pu. Solid. Mp 140°. pK_a 8.70 ($MeNO_2$).

Fenton, G.W. *et al*, *J. Chem. Soc.*, 1929, 2342 (*synth*)
Wang, H.K., *Acta Chem. Scand.*, 1965, **19**, 879 (*ed*)
Choplin, F. *et al*, *J. Mol. Struct.*, 1972, **11**, 381 (*struct*)
Seel, F. *et al*, *Chem. Ber.*, 1972, **105**, 406 (*synth*)
Schmidbauer, H. *et al*, *Chem. Ber.*, 1973, **106**, 1251 (*cmr*, *nmr*, *pmr*)
Bula, M.J. *et al*, *Can. J. Chem.*, 1975, **53**, 326 (*adducts*)
Perry, W.B. *et al*, *J. Am. Chem. Soc.*, 1975, **97**, 4899 (*pe*)
Wilkins, C.J. *et al*, *J. Am. Chem. Soc.*, 1975, **97**, 6352 (*ed*)
Elbel, S. *et al*, *J. Chem. Soc., Dalton Trans.*, 1976, 1757 (*pe*)
Rojhantalab, H. *et al*, *Spectrochim. Acta, Part A*, 1976, **32**, 519 (*raman*)

Trimethylphosphine selenide **T-00533**

[20819-54-9]

Me_3PSe

C_3H_9PSe M 155.038

Needles (EtOH). Mp 140.5-141°.

▷SZ2450000.

Brodie, A.M. *et al, J. Chem. Soc. (A)*, 1969, 2927 (*synth*)
McFarlane, W. *et al, J. Chem. Soc., Dalton Trans.*, 1973, 2162 (*nmr*)
Albrand, J.P. *et al, Chem. Phys. Lett.*, 1977, **48**, 524 (*cmr, nmr, struct*)
Jacob, E.J. *et al, J. Am. Chem. Soc.*, 1977, **99**, 5656 (*ed*)
Montana, A.J. *et al, J. Am. Chem. Soc.*, 1977, **99**, 4290 (*nmr*)
Cogne, A. *et al, J. Am. Chem. Soc.*, 1980, **102**, 2238 (*cmr, nmr*)

Trimethylphosphine sulfide, 9CI, 8CI **T-00534**

[2404-55-9]

$Me_3P{=}S$

C_3H_9PS M 108.138

Forms complexes with salts of Group IIIB and Group VIII metals. Cryst. (toluene, cyclohexane or EtOH). Mp 155-156°.

Screttas, C. *et al, J. Org. Chem.*, 1962, **27**, 2573 (*synth*)
Harwood, H.J. *et al, J. Org. Chem.*, 1963, **28**, 3430 (*synth*)
Crofts, P.C. *et al, J. Chem. Soc. (B)*, 1968, 1416 (*synth*)
Pantzer, R. *et al, Z. Anorg. Allg. Chem.*, 1975, **416**, 297 (*ir, raman*)
Wilkins, C.J. *et al, J. Am. Chem. Soc.*, 1975, **97**, 6352 (*struct*)
Elbel, S. *et al, J. Chem. Soc., Dalton Trans.*, 1976, 1757 (*pe*)
Rojhantalab, H. *et al, Spectrochim. Acta, Part A*, 1976, **32**, 519 (*raman*)
Albrand, J.P. *et al, Chem. Phys. Lett.*, 1977, **48**, 524 (*cmr, nmr, struct*)

Trimethyl phosphite, 9CI, 8CI **T-00535**

[121-45-9]

$P(OMe)_3$

$C_3H_9O_3P$ M 124.076

Reagent for dehydrohalogenations, debrominations and deoxygenations, used in Corey-Winter synth. Uses similar to those of Triethyl phosphite, T-00445 . Air-sensitive liq. with powerful sickly odour. d_4^{20} 1.052. Mp −78°. Bp 111-112°, Bp_{23} 22°. n_D^{20} 1.4080. Forms complexes with cuprous halides.

▷TLV 2-6. Flammable. Powerful obnoxious odour. Induces headache. TH1400000.

Meyrick, C.I. *et al, J. Chem. Soc.*, 1950, 225 (*ir, raman*)
Halman, M. *et al, J. Chem. Soc.*, 1963, 2853 (*uv*)
Goodell, L.J. *et al, Can. J. Chem.*, 1969, **47**, 2461 (*pmr*)
Verstuyft, A.N. *et al, Inorg. Chem.*, 1976, **15**, 732 (*cmr*)
Cowley, A.H. *et al, Inorg. Chem.*, 1977, **16**, 854; 1984, **23**, 3378 (*pe*)
Griend, L.J.V. *et al, J. Am. Chem. Soc.*, 1977, **99**, 2459 (*props, struct*)
Gray, G.A. *et al, J. Am. Chem. Soc.*, 1977, **99**, 3243 (*nmr*)
Labintsev, V.B. *et al, Zh. Org. Khim.*, 1977, **13**, 1141 (*Engl. transl.* p. 1050) (*ms*)
Organophosphorus Reagents in Organic Synthesis, Ed. J.I.G. Cadogan, Academic Press, 1979 (*uses*)
Chattopadhyay, S. *et al, J. Electron. Spectrosc. Relat. Phenom.*, 1981, **24**, 27 (*pe*)
Vasil'ev, V.V. *et al, Zh. Obshch. Khim.*, 1981, **51**, 2134 (*Engl. transl.* p. 1836) (*O nmr*)
Worley, S.D. *et al, J. Electron. Spectrosc. Relat. Phenom.*, 1982, **25**, 135 (*struct*)
Bart, J.C.J. *et al, Phosphorus Sulfur*, 1983, **17**, 205 (*conformn, complexes*)
Cload, P.A. *et al, Org. Mass Spectrom.*, 1983, **18**, 57 (*ms*)
Sodhi, R.N. *et al, J. Electron. Spectrosc. Relat. Phenom.*, 1983, **32**, 283 (*struct, pe*)
Holtzclaw, J.R. *et al, Org. Mass Spectrom.*, 1985, **20**, 90 (*ms, struct*)
Bretherick, L., *Handbook of Reactive Chemical Hazards*, 2nd Ed., Butterworths, London and Boston, 1979, 452.
Sax, N.I., *Dangerous Properties of Industrial Materials*, 6th Ed., Van Nostrand-Reinhold, 1984, 1062.
Hazards in the Chemical Laboratory, (Bretherick, L., Ed.), 3rd Ed., Royal Society of Chemistry, London, 1981, 524.

1,3,4-Trimethyl-1*H*-phosphole, 8CI **T-00536**

[37739-99-4]

H_3C CH_3 P Me

$C_7H_{11}P$ M 126.138

Ligand for Cr, Ru, Pt and Pd. Liq. Bp_{500} 135-136°, Bp_{15} 60°. Dimerises during oxidation.

B,MeI: [37737-13-6]. *1,1,3,4-Tetramethyl-1*H*-phospholium iodide.*

$C_8H_{14}P^{\oplus}$ M 141.172 (ion)

Monomeric solid; cryst. ($MeOH/Et_2O$). Mp 181-183°.

Sulfide:

$C_7H_{11}PS$ M 158.198

Monomeric solid. Mp 100-101°.

Mathey, F. *et al, Org. Magn. Reson.*, 1972, **4**, 171 (*synth, derivs, pmr, P nmr*)
Quin, L.D. *et al, Org. Magn. Reson.*, 1973, **5**, 161 (*cmr*)
Quin, L.D. *et al, J. Org. Chem.*, 1973, **38**, 1858, 1954 (*synth, derivs, pmr, P nmr*)
Gray, G.A. *et al, Org. Magn. Reson.*, 1980, **14**, 14 (*cmr*)
MacDougall, J.J. *et al, Inorg. Chem.*, 1980, **19**, 709 (*pmr, cmr, P nmr, complexes*)

***N,N,P*-Trimethylphosphonamidic acid, 9CI** **T-00537**

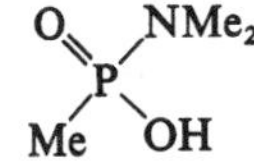

$C_3H_{10}NO_2P$ M 123.091

Me ester: [7351-34-0]. *Methyl trimethylphosphonamidate.*

$C_4H_{12}NO_2P$ M 137.118

Liq. d_4^{20} 1.02. Bp_{32} 118-120°. n_D^{20} 1.4382.

Et ester: Ethyl trimethylphosphonamidate.

$C_5H_{14}NO_2P$ M 151.145

Liq. $Bp_{1.0}$ 55°. n_D^{24} 1.4303.

Isopropyl ester: Isopropyl trimethylphosphonamidate.

$C_6H_{16}NO_2P$ M 165.172

Liq. $Bp_{0.04}$ 43°. n_D^{22} 1.4300.

Coe, D.G. *et al, J. Org. Chem.*, 1959, **24**, 1018 (*synth*)
Goldwhite, H. *et al, J. Am. Chem. Soc.*, 1966, **88**, 3572 (*synth, pmr*)
Goldwhite, H. *et al, Spectrochim. Acta, Part A*, 1970, **26**, 1403 (*ir*)
Nifant'ev, E.E. *et al, Zh. Obshch. Khim.*, 1970, **40**, 1492 (*Engl. transl.* p. 1478) (*synth*)

Trimethylphosphonamidic fluoride, 9CI **T-00538**

[661-60-9]

$MePF(O)NMe_2$

C_3H_9FNOP M 125.082

Liq.

▷V. toxic

Landau, M.A. *et al*, *Zh. Strukt. Khim.*, 1970, **11**, 513 (*Engl. transl.* p. 467) (*struct, P nmr*)
Ryzhikov, B.D. *et al*, *Zh. Strukt. Khim.*, 1975, **16**, 754 (*Engl. transl.* p. 700) (*F nmr*)

Trimethylphosphonamidodithioic acid, 9CI **T-00539**

$MeP(=S)(NMe_2)SH$

$C_3H_{10}NPS_2$ M 155.212

Cryst. (H_2O). Mp 89°.

Ph ester: Phenyl trimethylphosphonoamidodithioate.
$C_9H_{14}NPS_2$ M 231.310
Oil. $Bp_{0.01}$ 102°.
▷Toxic

Ger. Pat. 1 139 119, (*1962*); *CA*, **58**, 12601 (*ester*)
Seel, F. *et al*, *Chem. Ber.*, 1980, **113**, 1837 (*synth, P nmr, complexes*)

Trimethylphosphonamidothioic acid, 9CI **T-00540**

$MeP(=S)(NMe_2)OH \rightleftharpoons MeP(=O)(NMe_2)SH$

$C_3H_{10}NOPS$ M 139.152

O-*Isopropyl ester:* O-*Isopropyl trimethylphosphonamidothioate.*
$C_6H_{16}NOPS$ M 181.232
Liq. Bp_2 67-73°.

S-*Me ester:* [67242-36-8]. S-*Methyl trimethylphosphonamidothioate.*
$C_4H_{12}NOPS$ M 153.179
Solid. Mp 46-48°.

Fluoride: [661-61-0].
C_3H_9FNPS M 141.143
Oil. Bp_2 50°.
▷Extremely toxic

Chloride: [3450-34-8].
C_3H_9ClNPS M 157.598
Liq. d^{20} 1.25. Bp_{15} 103-105°.

Bromide:
C_3H_9BrNPS M 202.049
Solid. Mp 40°. Bp_{10} 120-125°.

Ger. Pat. 1 099 532, (*1959*); *CA*, **56**, 3515 (*fluoride, synth*)
Godovikov, N.N. *et al*, *Zh. Obshch. Khim.*, 1961, **31**, 1628 (*Engl. transl.* p. 1516) (*chloride*)
Maier, L., *Helv. Chim. Acta*, 1964, **47**, 1448 (*bromide, ester, synth*)
Reddy, G.S. *et al*, *Z. Naturforsch., B*, 1970, **25**, 1199 (*fluoride, F and P nmr*)
Wustner, D.A. *et al*, *J. Agric. Food Chem.*, 1978, **26**, 1104 (*methyl ester, synth, tox*)

***O,S,S*-Trimethyl phosphorodithioate,** 9CI **T-00541**

O,S,S-*Trimethyl dithiophosphate*

[22608-53-3]

$(MeS)_2P(O)OMe$

$C_3H_9O_2PS_2$ M 172.197

Liq. d_4^{20} 1.25. $Bp_{0.05}$ 60-62°. n_D^{20} 1.5340.

▷High oral tox. TE4780000.

Itskova, A.L. *et al*, *Zh. Obshch. Khim.*, 1968, **38**, 2471 (*Engl. transl.* p. 2471) (*synth, ir*)
Goubeau, J. *et al*, *Spectrochim. Acta, Part A*, 1971, **27**, 1703 (*ir*)
Wafa, O.A. *et al*, *Z. Anorg. Allg. Chem.*, 1971, **380**, 128 (*synth, ir, raman*)
Santaro, E. *et al*, *Org. Mass. Spectrom.*, 1973, **7**, 589 (*ms*)
Benschop, H.P. *et al*, *J. Chem. Soc., Chem. Commun.*, 1980, 370 (*props*)
Verschoyle, R.D. *et al*, *Arch. Toxicol.*, 1982, **51**, 221 (*tox*)
Greenhalgh, R. *et al*, *J. Agric. Food Chem.*, 1983, **31**, 170 (*P nmr, tox*)

***O,O,S*-Trimethyl phosphorodithioate,** 9CI, 8CI **T-00542**

Trimethyl dithiophosphate

[2953-29-9]

$(MeO)_2P(=S)SMe$

$C_3H_9O_2PS_2$ M 172.197

d_4^{20} 1.23. Bp_8 86°, $Bp_{0.2}$ 51-52°. n_D^{20} 1.5200.

▷TE4750000.

Norman, G.R. *et al*, *J. Am. Chem. Soc.*, 1952, **74**, 161 (*synth*)
McIvor, R.A. *et al*, *Can. J. Chem.*, 1958, **36**, 820 (*ir*)
Mel'nikov, N.N. *et al*, *Zh. Obshch. Khim.*, 1960, **30**, 200 (*Engl. transl.* p. 213) (*synth*)
Lippman, A.E., *J. Org. Chem.*, 1966, **31**, 471 (*synth, P nmr*)
Santaro, E., *Org. Mass Spectrom.*, 1973, **7**, 589 (*ms*)
Paasivirta, J. *et al*, *Org. Magn. Reson.*, 1977, **9**, 708 (*cmr, pmr, P nmr*)
Verschoyle, R.D. *et al*, *Arch. Toxicol.*, 1982, **51**, 221 (*tox*)

***O,O,O*-Trimethyl phosphoroselenoate,** 9CI, 8CI **T-00543**

O,O,O-*Trimethyl selenophosphate*

[152-19-2]

$(MeO)_3P{=}Se$

$C_3H_9O_3PSe$ M 203.036

Liq. d_4^{20} 1.51. Mp −41° to −40°. Bp_{11} 73-74°. n_D^{20} 1.4487.

Nuretdinov, I.A. *et al*, *Izv. Akad. Nauk SSSR, Ser. Khim.*, 1968, 2831 (*Engl. transl.* p. 2685) (*synth, ir*)
Kainosho, M., *J. Phys. Chem.*, 1970, **74**, 2853 (*pmr*)
Marsault-Hérail, F., *J. Chim. Phys. Phys. Chim. Biol.*, 1971, **68**, 274 (*ir, raman, conformn*)
McFarlane, W. *et al*, *J. Chem. Soc., Dalton Trans.*, 1973, 2162 (*P,Se nmr*)
Pohl, W. *et al*, *An. Quim.*, 1974, **70**, 1209 (*synth, ir, raman*)
Glidewell, C. *et al*, *J. Chem. Soc., Dalton Trans.*, 1977, 527 (*P nmr, ir*)

Trimethyl phosphorotetrathioate, 9CI, 8CI **T-00544**

Trimethyl tetrathiophosphate

[2381-38-1]

$(MeS)_3P{=}S$

$C_3H_9PS_4$ M 204.318

Liq. or solid. Mp 29-30°. Bp_2 110-112°, $Bp_{0.1}$ 124-125°.

Scott, C.B. *et al*, *J. Org. Chem.*, 1957, **22**, 789 (*synth*)
Shaw, R.A. *et al*, *Phosphorus*, 1971, **1**, 41, 191 (*synth, ir, pmr*)
Pantzer, R. *et al*, *Z. Anorg. Allg. Chem.*, 1973, **395**, 262 (*ir, raman*)
Murav'ev, I.V. *et al*, *Zh. Obshch. Khim.*, 1975, **45**, 1746 (*Engl. transl.* p. 1711) (*synth*)
Khokhlov, P.S. *et al*, *Zh. Obshch. Khim.*, 1984, **54**, 2545 (*Engl. transl.* p. 2274) (*synth*)

O,O,O-Trimethyl phosphorothioate, 9CI, 8CI — T-00545

Trimethyl thiophosphate
[152-18-1]

$(MeO)_3PS$

$C_3H_9O_3PS$ M 156.136
Liq. d_0^{20} 1.219. Bp_3 75°. n_D^{20} 1.4545. Isomerises on heating to *O,O,S*-Trimethyl phosphorothioate, T-00546 Forms adducts with Hg, Fe, Pt halides. Rel. nontoxic.

▷TG3850000.

Pishchimuka, P., *J. Prakt. Chem.*, 1911, **84**, 746.
Durig, J.R. *et al*, *J. Mol. Struct.*, 1969, **3**, 179 (*ir, raman*)
Kainosho, M., *J. Phys. Chem.*, 1970, **74**, 2853 (*pmr*)
Shagidullin, R.R. *et al*, *Dokl. Akad. Nauk SSSR, Ser. Sci. Khim.*, 1975, **222**, 897 (*Engl. transl.* p. 564) (*uv*)
Cowley, A.H. *et al*, *Inorg. Chim. Acta*, 1977, **25**, L151 (*pe*)
Toia, R.F. *et al*, *J. Agric. Food Chem.*, 1980, **28**, 599 (*glc, tox*)
Hammond, P.S. *et al*, *Pestic. Biochem. Physiol.*, 1982, **18**, 90 (*tox*)
Loewus, D.I. *et al*, *J. Am. Chem. Soc.*, 1983, **105**, 3287 (*synth, pmr, P nmr*)

O,O,S-Trimethyl phosphorothioate, 9CI, 8CI — T-00546

O,O,S-*Trimethyl thiophosphate*
[152-20-5]

$(MeO)_2P(O)SMe$

$C_3H_9O_3PS$ M 156.136
Liq. d_4^{20} 1.25. Bp_{11} 98-101°. n_D^{20} 1.4654.

▷Very toxic, with delayed action. TG3880000.

Hilgetag, G. *et al*, *J. Prakt. Chem.*, 1959, **8**, 121; 1960, **12**, 1 (*synth*)
Belskii, V.E. *et al*, *Zh. Obshch. Khim.*, 1969, **39**, 181 (*Engl. transl.* p. 167) (*synth, props*)
Santoro, E., *Org. Mass Spectrom.*, 1973, **7**, 589 (*ms*)
Stec, W.J. *et al*, *J. Org. Chem.*, 1976, **41**, 1291 (*props, P nmr*)
Mallipudi, N.M. *et al*, *J. Agric. Food Chem.*, 1979, **27**, 463 (*tox*)
Hammond, P.S. *et al*, *Pestic. Biochem. Physiol.*, 1982, **18**, 77, 90 (*tox*)
Thompson, C.M. *et al*, *J. Org. Chem.*, 1984, **49**, 1696 (*cmr*)

Trimethyl phosphorotriselenoite, 9CI, 8CI — T-00547

Trimethyl triselenophosphite
[55776-59-5]

$(MeSe)_3P$

$C_3H_9PSe_3$ M 312.958
Viscous yellow oil with repulsive odour. $Bp_{0.1}$ 45-50°.

Anderson, J.W. *et al*, *Inorg. Nucl. Chem. Lett.*, 1975, **11**, 233 (*synth, pmr*)
Maier, L., *Helv. Chim. Acta*, 1976, **59**, 252 (*synth, pmr, P nmr*)

O,S,S-Trimethyl phosphorotrithioate, 9CI, 8CI — T-00548

O,S,S-*Trimethyl trithiophosphate*
[3347-28-2]

$(MeS)_2P(S)OMe$

$C_3H_9OPS_3$ M 188.257
Odorous liq. d_4^{20} 1.28. Bp_1 76-77°. n_D^{20} 1.5953.

McIver, R.A. *et al*, *Can. J. Chem.*, 1956, **34**, 1611 (*ir*)
Murav'ev, I.V. *et al*, *Zh. Obshch. Khim.*, 1968, **38**, 133 (*Engl. transl.* p. 133) (*synth*)
Kotovich, B.P. *et al*, *Zh. Obshch. Khim.*, 1968, **38**, 1282 (*Engl. transl.* p. 1235) (*synth*)

S,S,S-Trimethyl phosphorotrithioate, 9CI, 8CI — T-00549

S,S,S-*Trimethyl trithiophosphate*
[681-71-0]

$(MeS)_3P{=}O$

$C_3H_9OPS_3$ M 188.257
Liq. d_4^{20} 1.27. n_D^{20} 1.6000.

▷High oral toxicity. TG5510000.

Gupalo, A.P. *et al*, *Zh. Obshch. Khim.*, 1968, **38**, 2550 (*Engl. transl.* p. 2464) (*synth*)
Shaw, R.A. *et al*, *Phosphorus*, 1971, **1**, 41, 191 (*synth, ir, pmr*)
Pantzer, R. *et al*, *Z. Anorg. Allg. Chem.*, 1973, **395**, 262 (*ir, raman*)
Cowley, A.H. *et al*, *Inorg. Chem.*, 1984, **23**, 3378 (*pe*)

Trimethyl phosphorotrithioite, 9CI — T-00550

Trimethyl trithiophosphite
[816-80-8]

$(MeS)_3P$

$C_3H_9PS_3$ M 172.258
Liq. Bp_2 72°.

▷TG5740000.

Fritzowsky, N. *et al*, *Z. Anorg. Allg. Chem.*, 1971, **386**, 67 (*synth, ir, raman*)
Razumov, A.I. *et al*, *Zh. Obshch. Khim.*, 1972, **42**, 1250 (*Engl. transl.* p. 1245) (*P nmr*)
Chernova, A.V. *et al*, *Zh. Obshch. Khim.*, 1979, **49**, 2002 (*Engl. transl.* p. 1760) (*uv, struct*)
Ofitserov, E.N. *et al*, *Zh. Obshch. Khim.*, 1983, **53**, 511 (*Engl. transl.* p. 443) (*ir, raman, struct*)
Zverev, V.V. *et al*, *Zh. Obshch. Khim.*, 1983, **53**, 1968 (*Engl. transl.* p. 1775) (*pe*)

N,N',N''-Trimethylphosphorous triamide, 9CI — T-00551

Tris(methylamino)phosphine. Tri(methylamino)-phosphine

$(MeNH)_3P$

$C_3H_{12}N_3P$ M 121.122
Stabilised as Mo complex.

Oxide: [6326-72-3]. N,N',N''-*Trimethylphosphoric triamide. Tri(methylamino)phosphine oxide.*
$C_3H_{12}N_3OP$ M 137.121
Used as an insect sterilant, and in textile fireproofing. Hygroscopic, cryst. powder.

Sulfide: [6141-78-2]. N,N',N''-*Trimethylphosphorothioic triamide. Tri(methylamino)phosphine sulfide.*
$C_3H_{12}N_3PS$ M 153.182

Solid. Mp 108°.

Tolkmith, H., *J. Am. Chem. Soc.*, 1963, **85**, 3246 (*sulfide, P nmr, pmr*)
Bock, H. *et al*, *Chem. Ber.*, 1966, **99**, 377 (*sulfide*)
Fluck, E. *et al*, *Z. Anorg. Allg. Chem.*, 1968, **359**, 102 (*sulfide, synth, P nmr*)
Healy, J.D. *et al*, *Phosphorus Sulfur*, 1978, **5**, 239 (*sulfide, synth, props*)
Maisch, H., *Z. Naturforsch., B*, 1979, **34**, 784 (*complex*)
Fluck, E. *et al*, *Z. Anorg. Allg. Chem.*, 1980, **461**, 187 (*oxide, props, P nmr, pmr*)

[[2-(Trimethylsilyl)ethoxy]methyl]-triphenylphosphonium(1+) T-00552

$Ph_3P^{\oplus}CH_2OCH_2CH_2SiMe_3$

$C_{24}H_{30}OPSi^{\oplus}$ M 393.560 (ion)
Reagent for homologenation of aldehydes RCH=O → RCH_2CH=O.
Chloride:
$C_{24}H_{30}ClOPSi$ M 429.013
Affords the ylide with NaH/DMSO. Cryst. (CH_2Cl_2/EtOAc). Mp 140-142°.
Ylide: [[*2-(Trimethylsilyl)ethoxy*]*methylene*]*-triphenylphosphorane.*
$C_{24}H_{29}OPSi$ M 392.552
Reactive ylide used in Wittig reactions.

Schönauer, K., *Tetrahedron Lett.*, 1983, **24**, 573 (*synth, pmr, ylide, use*)

[2-(Trimethylsilyl)ethyl]-triphenylphosphonium(1+), 9CI T-00553

$Ph_3P^{\oplus}CH_2CH_2SiMe_3$

$C_{23}H_{28}PSi^{\oplus}$ M 363.533 (ion)
Iodide: [63922-84-9].
$C_{23}H_{28}IPSi$ M 490.438
Treatment with butyllithium gives the ylide. Cryst. (H_2O). Mp 163-164.5°.
Ylide: [63922-69-0]. [*2-(Trimethylsilyl)ethylidene*]*-triphenylphosphorane.*
$C_{23}H_{27}PSi$ M 362.526
Reactive Wittig reagent prepd. *in situ*. Especially reactive towards electrophiles.

Seyferth, D. *et al*, *J. Org. Chem.*, 1977, **42**, 3104 (*synth, ylide*)
Fleming, I. *et al*, *Synthesis*, 1979, 446 (*ylide, use*)
Seyferth, D. *et al*, *J. Organomet. Chem.*, 1979, **181**, 293 (*synth, ir, pmr*)
Seyferth, D. *et al*, *J. Organomet. Chem.*, 1981, **205**, 301 (*ylide, use*)

2-(Trimethylsilyloxy)-1,3,2-dioxaphospholane, 9CI T-00554

Ethylene trimethylsilyl phosphite
[58068-62-5]

(ring: O–P(OSiMe₃)–O, five-membered)

$C_5H_{13}O_3PSi$ M 180.215
Liq. d_4^{20} 1.04. Bp_{10} 55°. n_D^{20} 1.4316.
2-Oxide: [73075-73-7]. *Ethylene trimethylsilyl phosphate.*
$C_5H_{13}O_4PSi$ M 196.215
Liq. Bp_1 132°.

Hay, R.S. *et al*, *J. Chem. Soc., Perkin Trans. 2*, 1978, 770 (*synth*)
Pudovik, M.A. *et al*, *Zh. Obshch. Khim.*, 1978, **48**, 2648 (*Engl. transl.* p. 2402) (*synth*)
Pudovik, A.N. *et al*, *Izv. Akad. Nauk SSSR, Ser. Khim.*, 1979, 2644 (*Engl. transl.* p. 2461) (*oxide, ir, P nmr*)

[1-[(Trimethylsilyl)oxy]ethyl]phosphonic acid, 9CI T-00555

$Me_3SiOCH(CH_3)P(O)(OH)_2$

$C_5H_{15}O_4PSi$ M 198.230
Esters may be alkylated and the products hydrol. under basic condition to give Me ketones.
(±)-*form*
Di-Me ester: Dimethyl [*1-*[(*trimethylsilyl*)*oxy*]*ethyl*]*-phosphonate.*
$C_7H_{19}O_4PSi$ M 226.284
Liq. d_4^{20} 1.04. Bp_{10} 98-100°. n_D^{20} 1.4281.
Di-Et ester: Diethyl [*1-*[(*trimethylsilyl*)*oxy*]*ethyl*]*-phosphonate.*
$C_9H_{23}O_4PSi$ M 254.338
Bp_8 109-110°, $Bp_{0.4}$ 62-63°. n_D^{20} 1.4210.

Nesterov, L.V. *et al*, *Zh. Obshch. Khim.*, 1971, **41**, 2449 (*Engl. transl.* p. 2474) (*synth, pmr*)
Lebedev, E.P. *et al*, *Zh. Obshch. Khim.*, 1979, **49**, 1731 (*Engl. transl.* p. 1517) (*synth, ir*)
Sekine, M. *et al*, *Bull. Chem. Soc. Jpn.*, 1982, **55**, 224 (*synth, pmr, use*)

[(Trimethylsilyloxy)methyl]phosphonic acid, 9CI T-00556

$Me_3SiOCH_2P(O)(OH)_2$

$C_4H_{13}O_4PSi$ M 184.204
Di-Et ester: [4631-50-9]. *Diethyl*[(*trimethylsilyloxy*)*-methyl*]*phosphonate.*
$C_8H_{21}O_4PSi$ M 240.311
Forms a carbanion which may be alkylated; on hydrol. with alkali, the products yield aldehydes. Liq. $Bp_{0.6}$ 72-75°.

Sekine, M. *et al*, *Tetrahedron Lett.*, 1979, 4475; *Bull. Chem. Soc. Jpn.*, 1982, **55**, 224 (*synth, pmr, use*)

(Trimethylsilyl)phosphorimidic acid, 9CI T-00557

$Me_3SiN{=}P(OH)_3$

$C_3H_{12}NO_3PSi$ M 169.192
Free acid tautomeric with trimethylsilylphosphoramidic acid.
Tri-Me ester: [58942-11-3]. *Trimethyl* (*trimethylsilyl*)*-phosphorimidate.* P,P,P-*Trimethoxy*-N-*trimethylsilylphosphazene.* P,P,P-*Trimethoxy*-N-*trimethylsilylphosphine imide.*
$C_6H_{18}NO_3PSi$ M 211.272
Liq. Bp_{14} 68-70°.
Tri-Et ester: [39071-56-2]. *Triethyl* (*trimethylsilyl*)*-phosphorimidate.* P,P,P-*Triethoxy*-N-*trimethylsilylphosphazene.* P,P,P-*Triethoxy*-N-*trimethylsilylphosphine imide.*
$C_9H_{24}NO_3PSi$ M 253.353
Liq. Bp_{18} 95-97°. n_D^{20} 1.4160.
Triisopropyl ester: Triisopropyl (*trimethylsilyl*)*-phosphorimidate.* P,P,P-*Triisopropoxy*-N-*trimethylsilylphosphazene.* P,P,P-*Triisopropoxy*-N-*trimethylsilylphosphine imide.*
$C_{12}H_{30}NO_3PSi$ M 295.433

Liq. $Bp_{0.2}$ 61.5-64°. n_D^{20} 1.4163.
Trichloride: [40678-60-2]. P,P,P-*Trichloro*-N-*trimethylsilylphosphazene*. P,P,P-*Trichloro*-N-*trimethylsilylphosphine imide*.
$C_3H_9Cl_3NPSi$ M 224.529
Liq. Bp_{10} 38°. Dec. at 100° → Me_3SiCl.

Niecke, E. *et al*, *Inorg. Nucl. Chem. Lett.*, 1973, **9**, 127 (*trichloride, synth, ir, pmr, P nmr, props*)
Schlak, O. *et al*, *Z. Anorg. Allg. Chem.*, 1976, **419**, 275 (*trimethyl ester, synth, ir, pmr, P and Si nmr*)
Flindt, E.-P. *et al*, *Z. Anorg. Allg. Chem.*, 1977, **430**, 155 (*esters, synth, ir, ms, pmr, P and Si nmr*)

Trimethylsilyl phosphorodichloridate, 9CI — T-00558

[18026-83-0]

$Me_3SiOP(O)Cl_2$

$C_3H_9Cl_2O_2PSi$ M 207.068
Liq. d^{25} 1.21. Mp −20.5°. Bp_{29} 85-7°, $Bp_{0.1}$ 40°. n_D^{20} 1.4290.

Fertig, J. *et al*, *J. Chem. Soc.*, 1957, 1488 (*synth, props*)
Schmidt, M. *et al*, *Chem. Ber.*, 1960, **93**, 872 (*synth, props*)
Yamaguchi, K. *et al*, *Chem. Lett.*, 1979, 1057 (*synth, use*)

Trimethylsilyl phosphorodifluoridate, 9CI, 8CI — T-00559

[4414-25-9]

$Me_3SiOP(O)F_2$

$C_3H_9F_2O_2PSi$ M 174.159
Liq. Bp 118°, Bp_{36} 36°. n_D^{20} 1.3495.

Kreshkov, A.P. *et al*, *Zh. Obshch. Khim.*, 1966, **36**, 525 (*Engl. transl.* p. 544) (*synth, ir*)
Roesky, H.W., *Chem. Ber.*, 1967, **100**, 2147 (*synth, ir, pmr*)
Cavell, R.G. *et al*, *Inorg. Chem.*, 1972, **11**, 2573 (*synth*)
Häbich, D. *et al*, *Synthesis*, 1978, 755 (*synth*)

1-Trimethylsilyl-2-trimethylsilyloxy-2-(2,4,6-tri-*tert*-butylphenyl)-methylenephosphine — T-00560

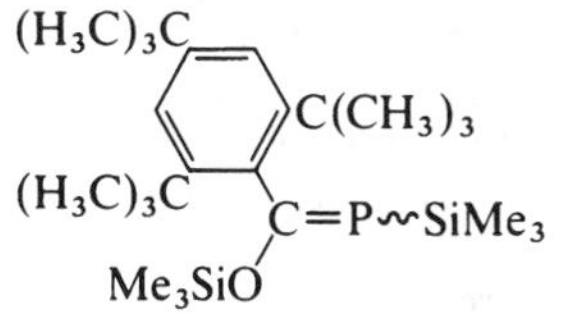

$C_{25}H_{47}OPSi_2$ M 450.790
Cryst. (MeCN). Mp 108-111°.

Märkl, G. *et al*, *Tetrahedron Lett.*, 1986, **27**, 171 (*synth, ms, cmr, P nmr*)

[2-(Trimethylstannyl)ethyl]-triphenylphosphonium, 10CI, 9CI — T-00561

$Ph_3P^{\oplus}CH_2CH_2SnMe_3$

$C_{23}H_{28}PSn^{\oplus}$ M 454.138 (ion)
Source of ylide which can be used as a vinylating agent in Wittig reactions.
Iodide: [63922-83-8].
$C_{23}H_{28}IPSn$ M 581.042
Forms the ylide when treated with MeLi, $LiNCH(CH_3)_2$, or better, $LiN(SiMe_3)_2$. Cryst. (EtOH/EtOAc). Mp 122.5-123.5° dec.
Ylide: [57241-88-0]. [2-(*Trimethylstannyl*)*ethylidene*]-*triphenylphosphorane*. *Triphenyl*[2-(*trimethylstannyl*)*ethylidene*]*phosphorane*, 9CI.
$C_{23}H_{27}PSn$ M 453.130
Cranberry-red ylide prepd. *in situ*.

Hannon, S.J. *et al*, *J. Chem. Soc., Chem. Commun.*, 1975, 630 (*ylide, synth, use*)
Seyferth, D. *et al*, *J. Org. Chem.*, 1977, **42**, 3104; *J. Organomet. Chem.*, **137**, C17 (*iodide, synth, use*)
Seyferth, D. *et al*, *J. Organomet. Chem.*, 1979, **179**, 25 (*ylide, use*)
Fieser, M. *et al*, *Reagents for Organic Synthesis*, Wiley, 1967-84, **6**, 640.

1,2,4-Trimethyl-1,2,4,3,5-triazadiphospholidine, 9CI — T-00562

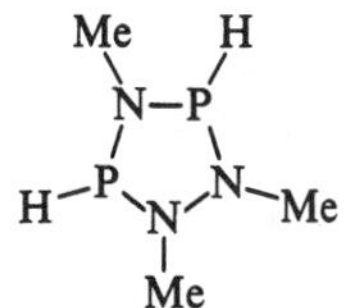

$C_3H_{11}N_3P_2$ M 151.088
3,5-Dichloro: [59725-15-4].
$C_3H_9Cl_2N_3P_2$ M 219.978
Solid. Mp 86°.
3,5-Dimethoxy: [59725-17-6].
$C_5H_{15}N_3O_2P_2$ M 211.140
Liq. $Bp_{0.001}$ 60°.
3,5-Bis(methylthio): [59725-18-7].
$C_5H_{15}N_3P_2S_2$ M 243.261
Oil.
3,5-Di-Me: [59725-16-5]. *1,2,3,4,5-Pentamethyl-1,2,4,3,5-triazadiphospholidine.*
$C_5H_{15}N_3P_2$ M 179.141
Oil, unstable at r.t., stable at −30°.

Nöth, H. *et al*, *Chem. Ber.*, 1976, **109**, 1942 (*synth, pmr, P nmr, ms*)
Ullmann, R. *et al*, *Chem. Ber.*, 1976, **109**, 2581 (*dichloro, cryst struct*)
Bulloch, G. *et al*, *J. Chem. Soc., Dalton Trans.*, 1978, 764 (*dichloro, pmr, cmr*)

2,8,9-Trimethyl-2,8,9-triaza-1-phosphatricyclo[3.3.1.1^3,7]decane, 9CI — T-00563

2,8,9-Trimethyl-2,8,9-triaza-1-phosphaadamantane
[50636-19-6]

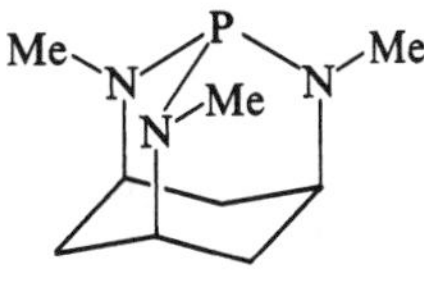

$C_9H_{18}N_3P$ M 199.235
Cryst. $Bp_{0.2}$ 50° subl.
Oxide: [50636-18-5].
$C_9H_{18}N_3OP$ M 215.234
Cryst. (pet. ether). Mp 185-188°. $Bp_{0.2}$ 140° subl.
Sulfide: [50636-20-9].
$C_9H_{18}N_3PS$ M 231.295
Cryst. (pentane). Mp 157°. $Bp_{0.2}$ 120-130° subl.

Stetter, H. *et al*, *Chem. Ber.*, 1973, **106**, 2523 (*synth, ir, pmr*)

Trimethyl(trimethylgermyl)-methylenephosphorane, 9CI — T-00564

[21458-56-0]

$Me_3P{=}CHGeMe_3$

$C_7H_{19}GeP$ M 206.791
Liq. Mp −33° to −32°. Bp_{14} 77-80°.

Schmidbaur, H. *et al*, *Chem. Ber.*, 1968, **101**, 3545 (*synth, pmr*)
Schmidbaur, H. *et al*, *Chem. Ber.*, 1977, **110**, 677 (*cmr, nmr*)

Trimethyl[(trimethylphosphoranylidene)-methyl]phosphonium(1+), 9CI **T-00565**

$Me_3P{=}CHPMe_3{}^{\oplus}$

$C_7H_{19}P_2{}^{\oplus}$ M 165.175 (ion)
Chloride: [57432-16-3].
$C_7H_{19}ClP$ M 169.654
Solid. Mp 168°.
Bromide: [81487-72-1].
$C_7H_{19}BrP$ M 214.105
No phys. props. reported.
Iodide: [20477-07-0].
$C_7H_{19}IP$ M 261.105
Solid. Mp 84-87°.

Schmidbaur, H. *et al*, *Chem. Ber.*, 1977, **110**, 3501 (*chloride, pmr, nmr*)
Karsch, H.H., *Z. Naturforsch., B*, 1979, **34**, 1178 (*iodide*)
Karsch, H.H., *Phosphorus Sulfur*, 1982, **12**, 217 (*bromide, pmr, nmr*)

Trimethyl(trimethylsilylmethylene)-phosphorane, 9CI **T-00566**
[3272-86-4]

$Me_3P{=}CHSiMe_3$

$C_7H_{19}PSi$ M 162.286
Liq. Mp −36°. Bp_{11} 66°. Desilylated on methanolysis.

Schmidbaur, H. *et al*, *Chem. Ber.*, 1967, **100**, 1032 (*synth, ir, pmr*)
Ostoja Starzewski, K.A. *et al*, *J. Organomet. Chem.*, 1974, **65**, 311 (*pe*)
Schmidbaur, H. *et al*, *Chem. Ber.*, 1975, **108**, 2649; 1977, **110**, 677 (*cmr, nmr*)
Inorg. Synth., 1978, **18**, 135 (*synth*)

2,4,6-Trimethyl-1,3,5,2,4,6-trioxatriphosphorinane, 9CI **T-00567**
2,4,6-Trimethyl-1,3,5,2,4,6-trioxatriphosphinane

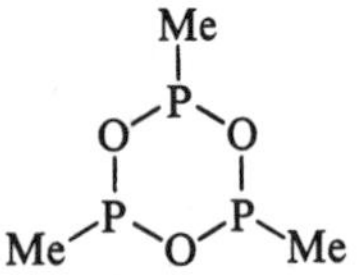

$C_3H_9O_3P_3$ M 186.024
2,4,6-Trisulfide: [17726-41-9]. *Methylphosphonothioic acid trimer trianhydride.*
$C_3H_9O_3P_3S_3$ M 282.204
Needles (C_6H_6) or plates (hexane). Mp 108°, Mp 118-123°.

Roesky, H.W. *et al*, *Chem. Ber.*, 1968, **101**, 630 (*synth, ir, ms*)
Maier, L., *Phosphorus*, 1975, **5**, 253 (*trisulfide, synth, pmr, P nmr, ir*)
Keck, H. *et al*, *Phosphorus Sulfur*, 1978, **4**, 173 (*ms*)

1,2,4-Trimethyl-1,2,4-triphospholane, 9CI **T-00568**
[66093-69-4]

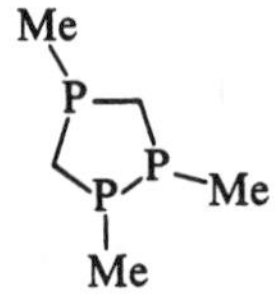

$C_5H_{13}P_3$ M 166.079
Liq. Bp_{12} 111°.

Baudler, M. *et al*, *Z. Anorg. Allg. Chem.*, 1977, **435**, 21 (*synth*)

Trimorpholinophosphine **T-00569**
4,4′,4″-Phosphinidynetrismorpholine, 9CI.
Tris(morpholino)phosphine
[5815-61-2]

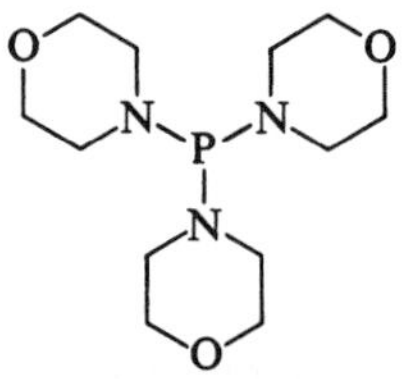

$C_{12}H_{24}N_3O_3P$ M 289.314
Cryst. (C_6H_6 or MeCN). Mp 157°.
B,MeI: Methyltris(morpholino)phosphonium iodide.
$C_{13}H_{27}IN_3O_3P$ M 431.253
Solid. Mp 310° dec.
Oxide: [4441-12-7]. *4,4′,4″-Phosphinylidynetrismorpholine.*
$C_{12}H_{24}N_3O_4P$ M 305.313
Solid. V. sol. H_2O, $CHCl_3$, sl. sol. CCl_4. Mp 191-192°. Dec. at 250°.
Sulfide: [14129-98-7]. *4,4′,4″-Phosphinothioylidynetrismorpholine.*
$C_{12}H_{24}N_3O_3PS$ M 321.374
Insect sterilant. Solid. Mp 145.5-146°.
▷SZ2815000.
Selenide: [51639-96-4]. *4,4′,4″-Phosphinoselenoylidynetrismorpholine.*
$C_{12}H_{24}N_3O_3PSe$ M 368.274
Solid. Mp 154-155°.
Telluride: [75404-54-5]. *4,4′,4″-Phosphinotelluroylidynetrismorpholine.*
$C_{12}H_{24}N_3O_3PTe$ M 416.914
Cream-coloured cryst. (MeOH or hexane). Mp 130-132° dec. Dec. in benzene or MeCN at r.t.

Audrieth, L.F. *et al*, *J. Am. Chem. Soc.*, 1942, **64**, 1553 (*oxide, sulfide, props*)
Thorstenson, T. *et al*, *Acta Chem. Scand., Ser. A*, 1976, **30**, 781 (*deriv*)
Rømming, C. *et al*, *Acta Chem. Scand., Ser. A*, 1978, **32**, 689 (*cryst struct*)
Konieczny, M. *et al*, *Z. Naturforsch., B*, 1978, **33**, 1040 (*selenide*)
Rømming, C. *et al*, *Acta Chem. Scand., Ser. A*, 1979, **33**, 187 (*selenide, cryst struct*)
Rømming, C. *et al*, *Acta Chem. Scand., Ser. A*, 1980, **34**, 333 (*uv, ir, telluride, cryst struct*)
Bergesen, K. *et al*, *Acta Chem. Scand., Ser. A*, 1981, **35**, 147 (*cmr*)
Gonbeau, D. *et al*, *Inorg. Chem.*, 1981, **20**, 1966 (*pe*)
Rømming, G. *et al*, *Acta Chem. Scand., Ser. A*, 1982, **36**, 665 (*oxide, cryst struct*)

Tri-1-naphthyl phosphate, 8CI **T-00570**
Tri-1-naphthalenyl phosphate, 9CI

$C_{30}H_{21}O_4P$ M 476.467
Cryst. (EtOH). Mp 148-149°.

Anschütz, L. *et al, Justus Liebigs Ann. Chem.*, 1939, **542**, 14 (*synth*)
Krishnakumar, V.K. *et al, Synthesis*, 1983, 558; *Synth. Commun.*, 1984, **14**, 189 (*synth, ir, pmr*)

Tri-2-naphthyl phosphate, 8CI **T-00571**
Tri-2-naphthalenyl phosphate, 9CI
$C_{30}H_{21}O_4P$ M 476.467
Cryst. (EtOH). Mp 111°.

Anschütz, L. *et al, Justus Liebigs Ann. Chem.*, 1939, **542**, 14 (*synth*)
Krishnakumar, V.K. *et al, Synthesis*, 1983, 558; *Synth. Commun.*, 1984, **14**, 189 (*synth, ir, pmr*)

Tri-1-naphthylphosphine **T-00572**
Tri-1-naphthalenylphosphine, 9CI
[3411-48-1]

$C_{30}H_{21}P$ M 412.470
Ligand for Ni. Cryst. (C_6H_6/Et_2O, EtOH or dioxan). Mp 263-265°, 278-280°. Forms complexes with $CHCl_3$ and CCl_4.
Oxide:
$C_{30}H_{21}OP$ M 428.469
Solid. Mp 341-343°. Forms a monohydrate.
Sulfide:
$C_{30}H_{21}PS$ M 444.530
Cryst. (EtOH or butanol). Mp 255.5-256.5°.

Anschütz, L. *et al, Justus Liebigs Ann. Chem.*, 1939, **542**, 14 (*oxide*)
Mikhailov, B.M. *et al, Zh. Obshch. Khim.*, 1952, **22**, 792; *CA*, **47**, 5388 (*synth, oxide*)
Tefteller, W. *et al, J. Chem. Eng. Data*, 1965, **10**, 301 (*synth, sulfide*)
Ziegler, C.B. *et al, J. Org. Chem.*, 1978, **43**, 2941 (*synth, nmr, props*)

Tri-2-naphthylphosphine **T-00573**
Tri-2-naphthalenylphosphine, 9CI
$C_{30}H_{21}P$ M 412.470
Oxide:
$C_{30}H_{21}OP$ M 428.469
Solid. Mp 248-249°.

Burger, A. *et al, J. Org. Chem.*, 1951, **16**, 1250.

Tri-1-naphthyl phosphite, 8CI **T-00574**
Tri-1-naphthalenyl phosphite, 9CI

$C_{30}H_{21}O_3P$ M 460.468
Hygroscopic cryst. (xylene). Sl. sol. Et_2O. Mp 91°. $Bp_{0.04}$ 250°.

Anschutz, L. *et al, Justus Liebigs Ann. Chem.*, 1939, **542**, 14 (*synth*)
Iselin, B.M. *et al, Helv. Chim. Acta*, 1957, **40**, 373 (*synth, use*)

Tri-2-naphthyl phosphite, 8CI **T-00575**
Tri-2-naphthalenyl phosphite, 9CI
$C_{30}H_{21}O_3P$ M 460.468
Solid or oil. Mp 97°. Bp_1 192-195°. n_D^{20} 1.4523.

Anschutz, L. *et al, Justus Liebigs Ann. Chem.*, 1939, **542**, 14 (*synth*)
Iselin, B.M. *et al, Helv. Chim. Acta*, 1957, **40**, 373 (*synth*)
Petrov, K.A. *et al, Zh. Obshch. Khim.*, 1963, **33**, 899 (*Engl. transl.* p. 885) (*synth*)

***N,N′,N″*-Tri-1-naphthylphosphorous triamide,** 8CI **T-00576**
N,N′,N″-*Tri-1-naphthalenylphosphorous triamide, 9CI. Tri(1-naphthylamino)phosphine*

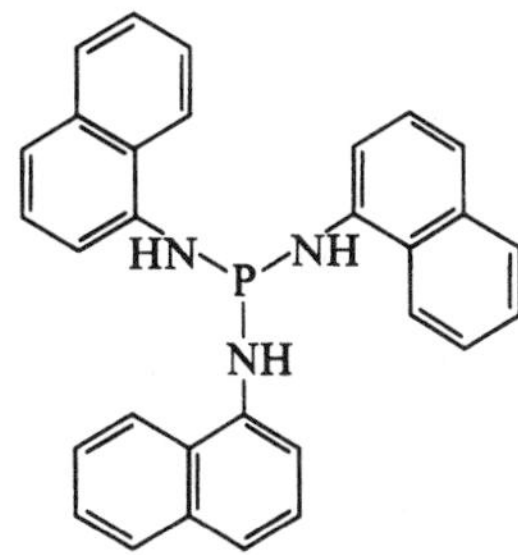

$C_{30}H_{24}N_3P$ M 457.513
Oxide: N,N′,N″-*Tri-1-naphthalenylphosphoric triamide.* N,N′,N″-*Tri-1-naphthylphosphoric triamide.*
$C_{30}H_{24}N_3OP$ M 473.513
Solid. Mp 216°.

Rudert, P., *Ber.*, 1893, **26**, 565.

***N,N′,N″*-Tri-2-naphthylphosphorous triamide,** 8CI **T-00577**
N,N′,N″-*Tri-2-naphthalenylphosphorous triamide, 9CI*
$C_{30}H_{24}N_3P$ M 457.513
Oxide: N,N′,N″-*Tri-2-naphthylphosphoric triamide.* N,N,N″-*Tri-2-naphthalenylphosphoric triamide.*
$C_{30}H_{24}N_3OP$ M 473.513
Cryst. (EtOH). Mp 168-170°.

Buck, A.C. *et al, J. Am. Chem. Soc.*, 1948, **70**, 744.

Trinonylphosphine, 9CI T-00578

[17621-06-6]

$P[(CH_2)_8CH_3]_3$

$C_{27}H_{57}P$ M 412.721

Mp 45-46°. Bp_2 225-227°, $Bp_{0.01}$ 160-162°.

Oxide: [17262-53-2].

$C_{27}H_{57}OP$ M 428.720

Solid. Mp 35-36°. Bp_4 235°. pK_a 8.78 ($MeNO_2$).

Sulfide: [21722-52-1].

$C_{27}H_{57}PS$ M 444.781

Possesses lubricating props.

Selenide: [20612-74-2].

$C_{27}H_{57}PSe$ M 491.681

Possesses lubricating props. Liq. $Bp_{0.03}$ 232-234°.

Zakharkin, L.I. *et al*, *Izv. Akad. Nauk SSSR, Ser. Khim.*, 1962, 2002 (*Engl. transl.* p. 1913) (*oxide*)
Feshchenko, N.G. *et al*, *Zh. Obshch. Khim.*, 1968, **38**, 122 (*Engl. transl.* p. 121) (*synth*)
Pavlenko, A.F. *et al*, *CA*, 1968, **69**, 29031 (*selenide*)
Pavlenko, A.F. *et al*, *CA*, 1970, **73**, 122123 (*sulfide*)
Yakshin, V.V. *et al*, *Dokl. Akad. Nauk SSSR, Ser. Sci. Khim.*, 1979, **247**, 128 (*Engl. transl.* p. 344) (*oxide*)

Trinonyl phosphite, 9CI, 8CI T-00579

[2549-63-5]

$P[O(CH_2)_8CH_3]_3$

$C_{27}H_{57}O_3P$ M 460.719

Liq. Bp_4 232-234°. n_D^{20} 1.4507.

Petrov, K.A. *et al*, *Zh. Obshch. Khim.*, 1963, **33**, 899 (*Engl. transl.* p. 885) (*synth*)
Ovchinnikova, N.K. *et al*, *Zh. Org. Khim.*, 1975, **11**, 1839 (*Engl. transl.* p. 1849) (*synth*)

Trioctadecyl phosphite, 9CI T-00580

Tristearyl phosphite, 8CI

[2082-80-6]

$[H_3C(CH_2)_{17}]_3P$

$C_{54}H_{101}O_3P$ M 829.364

Antioxidant and heat stabiliser for PVC. Low-melting solid. Mp 40°.

U.S.P., 3 047 608, (*1962*); *CA*, **57**, 16400 (*synth, use*)

Trioctyl phosphate, 9CI T-00581

[1806-54-8]

$[H_3C(CH_2)_7O]_3P{=}O$

$C_{24}H_{51}O_4P$ M 434.638

Extractant for U. Liq. Bp_2 225-227°. n_D^{20} 1.4410.

Arbuzov, B.A. *et al*, *Izv. Akad. Nauk SSSR, Ser. Khim.*, 1952, 865 (*Engl. transl.* p. 773) (*synth*)
Bliznyuk, N.K. *et al*, *Zh. Obshch. Khim.*, 1964, **34**, 1169 (*Engl. transl.* p. 1160) (*synth*)

Trioctylphosphine, 9CI T-00582

[4731-53-7]

$P[(CH_2)_7CH_3]_3$

$C_{24}H_{51}P$ M 370.641

Metal extractant. Exhibits synergistic activity with other extractants. Extracts mineral acids (HCl, HBr, HNO_3, H_2SO_4, $HClO_4$) from aq. soln. Used in catalysts for prep. of polymers, and for the dimerization of alkenes. Mp 30°, Mp 48°. Bp_{50} 291°, Bp_1 194-195° (234°). Bp_{50} 291°. n_D^{25} 1.4666.

B,MeI: Methyltrioctylphosphonium iodide.

$C_{25}H_{54}IP$ M 512.580

Liq.

Oxide:

$C_{24}H_{51}OP$ M 386.640

Extracts carboxylic acids from aq. solns. Widely used as a metal extractant, particularly for rare earth metals, and in the chromatography of these. Employed in catalysts for the cyclotrimerization of butadiene and for the cross-linking of epoxy resins. Solid. Mp 49-50°. Bp_3 225-228°. pK_a 8.77, 9.39 ($MeNO_2$).

Sulfide: [2551-53-3].

$C_{24}H_{51}PS$ M 402.701

Selective metal extractant. Possesses bacteriacidal and fungicidal activity. Liq. Bp_1 226-228°.

Horner, L. *et al*, *Chem. Ber.*, 1959, **92**, 2088 (*synth*)
Pass, F. *et al*, *Monatsh. Chem.*, 1959, **90**, 792 (*synth*)
Rauhut, M.M. *et al*, *J. Org. Chem.*, 1961, **26**, 5138 (*synth*)
Zakharkin, L.I. *et al*, *Izv. Akad. Nauk SSSR, Ser. Khim.*, 1962, 2002 (*Engl. transl.* p. 1913) (*oxide, synth*)
Elliott, D.E. *et al*, *Anal. Chim. Acta*, 1965, **33**, 237 (*sulfide, use*)
Burdett, J.L. *et al*, *Can. J. Chem.*, 1966, **44**, 111 (*oxide, ir, nmr, use*)
Venkateswarlu, K.S. *et al*, *Indian J. Chem.*, 1972, **10**, 748 (*use*)
Saraiya, V.N., *Indian J. Chem.*, 1973, **11**, 490 (*use*)
Wang, S-M., *CA*, 1974, **80**, 148891 (*use*)
Bodner, G.M. *et al*, *J. Organomet. Chem.*, 1982, **226**, 85 (*cmr*)

Trioctyl phosphite, 9CI, 8CI T-00583

[3028-88-4]

$P[O(CH_2)_7CH_3]_3$

$C_{24}H_{51}O_3P$ M 418.639

Employed in catalysts for the dimerisation of propylene. Liq. Bp_1 200-202°. n_D^{20} 1.4490.

▷TH1450000.

Ovchinnikova, N.K. *et al*, *Zh. Org. Khim.*, 1975, **11**, 1839 (*Engl. transl.* p. 1849) (*synth*)

2,8,9-Trioxa-5-aza-1-phosphabicyclo[3.3.3]undecane, 9CI T-00584

[36712-66-0]

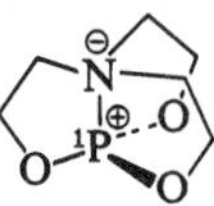

$C_6H_{12}NO_3P$ M 177.139

Not isolable. Forms both azonia and phosphonia salts.

1-Oxide: [10022-55-6].

$C_6H_{12}NO_4P$ M 193.139

Cryst. ($CHCl_3$ or CH_2Cl_2/hexane at low temp.). Mp 208-212°. $Bp_{0.1}$ 120° subl.

1-Sulfide: [60028-25-3].

$C_6H_{12}NO_3PS$ M 209.199

Cryst. (CH_2Cl_2/hexane, $MeCN/Et_2O$ or toluene at low temp.). Mp 218-220° dec. $Bp_{0.1}$ 120-130° subl. Forms complex with HBF_4.

1-Selenide:

$C_6H_{12}NO_3PSe$ M 256.099

Cryst. (CH_2Cl_2/hexane at low temp.). Mp 208-210° dec.

Casida, J.E. *et al*, *Toxicol. Appl. Pharmacol.*, 1976, **36**, 261 (*sulfide, tox*)
Clardy, J.C. *et al*, *J. Am. Chem. Soc.*, 1976, **98**, 623; 1977, **99**, 631 (*sulfide, cryst struct, ir*)
Milbrath, D.S. *et al*, *J. Am. Chem. Soc.*, 1977, **99**, 6607 (*synth, derivs, pmr, cmr, P nmr, ir*)

2,6,7-Trioxa-1,4-diphosphabicyclo[2.2.2]-octane, 9CI — T-00585

[4579-03-7]

$C_3H_6O_3P_2$ M 152.026

Ligand for Cr, Mo and W. Mp 75-76°. $Bp_{0.1}$ 50° subl.

1-Oxide:
$C_3H_6O_3P_2$ M 152.026
Cryst. (Me_2CO/hexane).

4-Oxide: [41097-27-2].
$C_3H_6O_3P_2$ M 152.026
Cryst. (C_6H_6/hexane).

▷PC3503500.

1,4-Dioxide: [4726-80-1].
$C_3H_6O_5P_2$ M 184.025
Needles. Mp 210-213°.

▷Mod. toxic. PC3503000.

1-Sulfide: [4646-06-4].
$C_3H_6O_3P_2S$ M 184.086
Cryst. (Me_2CO or MeCN). Mp 235-237°.

4-Sulfide:
$C_3H_6O_3P_2S$ M 184.086
Cryst. (MeCN) with foul odour.

1,4-Disulfide: [41097-29-4].
$C_3H_6O_3P_2S_2$ M 216.146
Cryst. (MeCN).

Coskran, K.J. *et al*, *Inorg. Chem.*, 1965, **4**, 1655 (*synth, ir, pmr, derivs*)
Boros, E.J. *et al*, *J. Am. Chem. Soc.*, 1966, **88**, 1140 (*pmr, P nmr*)
Allison, D.A. *et al*, *Phosphorus*, 1973, **2**, 257 (*derivs, ir, pmr, P nmr, complexes*)
Clardy, J.C. *et al*, *Phosphorus*, 1975, **5**, 85 (*sulfide, cryst struct*)
Cowley, A.H. *et al*, *Inorg. Chem.*, 1977, **16**, 854 (*pe*)
Kozlov, É.S. *et al*, *Zh. Obshch. Khim.*, 1980, **50**, 1499 (*Engl. transl.* p. 1210) (*synth*)

2,6,7-Trioxa-4-phospha-1-arsabicyclo[2.2.2]octane — T-00586

[24647-30-1]

$C_3H_6AsO_3P$ M 195.974

Ligand for Ni. Hygroscopic cryst. Mp 91-92°.

4-Oxide:
$C_3H_6AsO_4P$ M 211.973
Needles. Dec. at 175°.

Rathke, J.W. *et al*, *J. Org. Chem.*, 1970, **35**, 2310 (*synth, oxide, pmr, ir*)

Bertrand, R.D. *et al*, *Phosphorus*, 1973, **3**, 1 (*P nmr, complexes*)

2,6,7-Trioxa-1-phosphabicyclo[2.2.1]-heptane, 9CI, 8CI — T-00587

Glyceryl phosphite. Glycerol phosphite. Glycerol bicyclic phosphite

[279-53-8]

$C_3H_5O_3P$ M 120.044

Ligand for Ag. Liq. or cryst. (pentane, −78°). Bp_4 40°. Hydrol. → polymer.

1-Oxide: [13849-00-8].
$C_3H_5O_4P$ M 136.044
Solid (CH_2Cl_2/Et_2O). Dec. on heating. Easily hydrolysed.

Bertrand, R. *et al*, *J. Magn. Reson.*, 1970, **3**, 494 (*pmr, P nmr*)
Denney, D.B. *et al*, *Phosphorus*, 1973, **2**, 245 (*synth, oxide, pmr, props*)
Shaidullin, S.A. *et al*, *Zh. Strukt. Khim.*, 1978, **19**, 942 (*Engl. transl.* p. 809) (*ed, struct*)
Gehrmann, T. *et al*, *Makromol. Chem.*, 1981, **182**, 3069 (*props*)
Socol, S.M. *et al*, *Inorg. Chem.*, 1984, **23**, 3487 (*complex*)

2,6,7-Trioxa-1-phosphabicyclo[2.2.2]-octane, 9CI — T-00588

Bicyclic phosphite

[280-45-5]

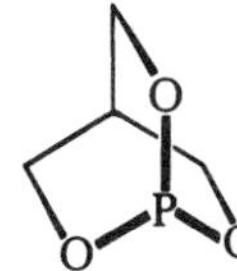

$C_4H_7O_3P$ M 134.071

Solid. Mp 126-127°. Sublimes.

1-Oxide: [873-13-2]. *Bicyclic phosphate.*
$C_4H_7O_4P$ M 150.071
Solid. Mp 245-247°. $Bp_{0.01}$ 100° subl.

Boros, E.J. *et al*, *J. Am. Chem. Soc.*, 1966, **88**, 1140 (*synth, deriv, pmr*)
Slafer, W.D. *et al*, *J. Mol. Struct.*, 1974, **50**, 435 (*microwave, struct*)
Voorhees, K.J. *et al*, *Org. Mass. Spectrom.*, 1979, **14**, 459 (*ms*)
Cowley, A.H. *et al*, *Inorg. Chem.*, 1984, **23**, 3378 (*pe*)

2,7,8-Trioxa-1-phosphabicyclo[3.2.1]-octane, 9CI — T-00589

[13232-18-3]

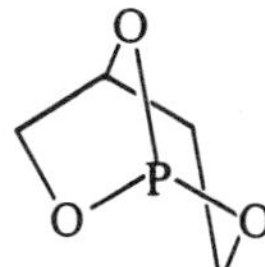

$C_4H_7O_3P$ M 134.071

Liq. with characteristic phosphite odour. Bp_{35} 83°, $Bp_{0.6}$ 52°. n_D^{21} 1.4753. Easily hydrolysed.

Kainosho, M. *et al*, *Tetrahedron*, 1969, **25**, 4071 (*synth, ir, pmr*)
Edmundson, R.S. *et al*, *J. Chem. Soc., Perkin Trans. 1*, 1971, 3179 (*synth, ir, props*)

2,6,7-Trioxa-1-phosphabicyclo[2.2.2]-octane-4-methanol, 9CI **T-00590**

4-Hydroxymethyl-2,6,7-trioxa-1-phosphabicyclo[2.2.2]octane. Hydroxymethyl bicyclic phosphite

[873-93-8]

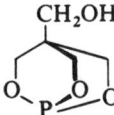

$C_5H_9O_4P$ M 164.097

Cryst. Mp 62.4-64.4° (58-59°). Bp_2 152-153°, $Bp_{0.5}$ 120°, $Bp_{0.1}$ 110° subl.

Oxide: [5301-78-0].
$C_5H_9O_5P$ M 180.097
No phys. props. reported.

Oxide, Me ether:
$C_6H_{11}O_5P$ M 194.124
Solid. Mp 136.5-137°.

Sulfide: [768-87-6].
$C_5H_9O_5PS$ M 212.157
Needles (H_2O or xylene). Mp 161°.

Wadsworth, W.S. *et al, J. Am. Chem. Soc.*, 1962, **84**, 610 (*synth, props*)
Rätz, R. *et al, J. Org. Chem.*, 1965, **30**, 438 (*sulfide, ir*)
Nifant'ev, É.E. *et al, Zh. Obshch. Khim.*, 1970, **40**, 2196 (*Engl. transl.* p. 2182) (*synth*)
Bellet, E.M. *et al, Science*, 1973, **182**, 1135 (*tox*)
Voorhees, K.J. *et al, Org. Mass Spectrom.*, 1979, **14**, 459 (*ms*)
Ozoe, Y. *et al, Agric. Biol. Chem.*, 1982, **46**, 411 (*derivs*)

2,8,9-Trioxa-1-phosphatricyclo[3.3.1.1^{3,7}]-decane, 10CI, 9CI **T-00591**

2,8,9-Trioxa-1-phosphaadamantane, 8CI

[281-33-4]

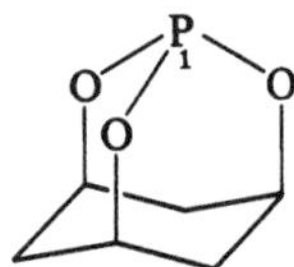

$C_6H_9O_3P$ M 160.109

Solid. Mp 207-208°.

1-Oxide: [875-12-7].
$C_6H_9O_4P$ M 176.108
Cryst. (butanol or dioxan). Insol. C_6H_6, Et_2O, Me_2CO, $CHCl_3$. Mp 267-268°.
▷GV7660000.

1-Sulfide: [58299-37-9].
$C_6H_9O_3PS$ M 192.169
Needles (MeOH). Insol. Et_2O. Mp 250-251°.

1-Selenide: [85000-72-2].
$C_6H_9O_3PSe$ M 239.069
Solid. Mp 262-264°.

Stetter, H. *et al, Chem. Ber.*, 1952, **85**, 451 (*synth, derivs*)
Verkade, J.G. *et al, Inorg. Chem.*, 1962, **1**, 948 (*pmr, P nmr, derivs*)
Boros, E.J. *et al, J. Am. Chem. Soc.*, 1966, **88**, 1140 (*oxide, pmr*)
Casida, J.E. *et al, Toxicol. Appl. Pharmacol.*, 1976, **36**, 261 (*tox*)
Sugawara, T. *et al, Bull. Chem. Soc. Jpn.*, 1979, **52**, 3391 (*O nmr*)
Lee, T.H. *et al, J. Am. Chem. Soc.*, 1980, **102**, 2631 (*pe, struct*)
Benhammou, M. *et al, Phosphorus Sulfur*, 1982, **14**, 105 (*synth, derivs, P nmr*)
Gordon, J.J. *et al, Arch. Toxicol.*, 1983, **52**, 71 (*tox*)
Cowley, A.H. *et al, Inorg. Chem.*, 1984, **23**, 3378 (*pe*)

1,3,6,2-Trioxaphosphocane, 9CI **T-00592**

1,3,6-Trioxa-2-phosphacyclooctane

[6573-35-9]

$C_4H_9O_3P$ M 136.087

2-Chloro: [71256-07-0].
$C_4H_8ClO_3P$ M 170.532
Characterised spectroscopically.

2-Methoxy: [71256-08-1].
$C_5H_{11}O_4P$ M 166.113
Characterised spectroscopically.

2-Phenoxy, 2-oxide: [68755-16-8].
$C_{10}H_{13}O_5P$ M 244.183
Viscous oil.

2-Me: [66295-37-2].
$C_5H_{11}O_3P$ M 150.114
Characterised spectroscopically.

2-Me, 2-sulfide: [66295-38-3].
$C_5H_{11}O_3PS$ M 182.174
Solid.

Dutasta, J.-P. *et al, J. Am. Chem. Soc.*, 1978, **100**, 1925 (*methyl, pmr, stereochem*)
Penney, C.L. *et al, Can. J. Chem.*, 1978, **56**, 2396 (*phenoxy, oxide, synth, pmr*)
Dutasta, J.-P. *et al, Tetrahedron Lett.*, 1979, 933 (*chloro, methoxy, pmr, cmr, P nmr, stereochem*)
Grand, A., *Cryst. Struct. Commun.*, 1982, **11**, 569 (*methyl, sulfide, cryst struct*)

Tripentyl phosphate, 9CI **T-00593**

Triamyl phosphate

[2528-38-3]

$[H_3C(CH_2)_4O]_3P{=}O$

$C_{15}H_{33}O_4P$ M 308.397

Liq. Bp_{50} 225°, Bp_5 167°. n_D^{20} 1.4319.

Evans, E.P. *et al, J. Chem. Soc.*, 1930, 1310 (*synth*)
Noller, C.R. *et al, J. Am. Chem. Soc.*, 1933, **55**, 424 (*synth*)
Bliznyuk, N.K. *et al, Zh. Obshch. Khim.*, 1964, **34**, 1169 (*Engl. transl.* p. 1160) (*synth*)

Tripentylphosphine, 9CI **T-00594**

Triamylphosphine

[10496-10-3]

$P[(CH_2)_4CH_3]_3$

$C_{15}H_{33}P$ M 244.399

Mp ca. 29°. Bp_{20} 165-166°, Bp_3 121-122°. n_D^{20} 1.4642.

Oxide: [3084-47-7].
$C_{15}H_{33}OP$ M 260.399
Solid. Mp 59°, 84°. $Bp_{0.4}$ 136°.
▷SZ1675000.

Sulfide: [17643-99-1].
$C_{15}H_{33}PS$ M 276.459
Liq. $Bp_{0.9}$ 165-167°. n_D^{29} 1.4902.

Selenide:
$C_{15}H_{33}PSe$ M 323.359
Pale-yellow viscous liq. $Bp_{0.75}$ 158°. n_D^{25} 1.5055.

Telluride:
$C_{15}H_{33}PTe$ M 371.999
Yellow compd. unstable to light and to glass surfaces, depositing Te.

Zingaro, R.A. *et al*, *J. Chem. Eng. Data*, 1963, **8**, 226 (*synth, derivs*)
Cumper, C.W.N. *et al*, *J. Chem. Soc.*, 1964, 430 (*synth, oxide*)
Mironova, Z.N. *et al*, *Zh. Obshch. Khim.*, 1967, **37**, 2747 (*Engl. transl.* p. 2614) (*synth*)
Chremos, G.N., *J. Organomet. Chem.*, 1970, **22**, 637 (*telluride*)
Mironova, Z.N. *et al*, *Zh. Obshch. Khim.*, 1974, **44**, 1217 (*Engl. transl.* p. 1195) (*oxide*)
Shmidt, F.K. *et al*, *Zh. Org. Khim.*, 1974, **10**, 155 (*Engl. transl.* p. 156) (*complex, use*)
Vdovenko, S.I. *et al*, *Zh. Obshch. Khim.*, 1976, **46**, 2613 (*Engl. transl.* p. 2491) (*sulfide, ir*)
Bodner, G.M. *et al*, *J. Organomet. Chem.*, 1982, **226**, 85 (*cmr*)

Tripentyl phosphite, 9CI, 8CI **T-00595**

Triamyl phosphite
[1990-22-3]

$P[O(CH_2)_4CH_3]_3$

$C_{15}H_{33}O_3P$ M 292.398
Liq. Bp_8 126-127°, Bp_7 151-153°. n_D^{20} 1.4303 (1.4382).

Petrov, K.A. *et al*, *Zh. Obshch. Khim.*, 1961, **31**, 2377 (*Engl. transl.* p. 2214) (*synth*)
Ovchinnikova, N.K. *et al*, *Zh. Org. Khim.*, 1975, **11**, 1839 (*Engl. transl.* p. 1849) (*synth*)

1,2,3-Triphenyl-1,3-azaphospholidin-5-one, 9CI **T-00596**

[51531-26-1]

$C_{21}H_{18}NOP$ M 331.353
Solid. Mp 150-152°.
B,MeI: 3-Methyl-1,2,3-triphenyl-5-oxo-1,3-azaphospholidinium iodide.
$C_{22}H_{21}INOP$ M 473.292
Solid. Mp 150° dec.
3-Oxide: [52111-99-6].
$C_{21}H_{18}NO_2P$ M 347.352
Solid. Mp 195-197°.
3-Sulfide: [52112-01-3].
$C_{21}H_{18}NO_2PS$ M 379.412
Solid. Mp 122-126°.

Oehme, H. *et al*, *Phosphorus*, 1973, **3**, 159 (*synth, derivs*)
Zschunke, A. *et al*, *Org. Magn. Reson.*, 1974, **6**, 568 (*conformn*)

Triphenyl[bis(trifluoromethyl)methylene]phosphorane **T-00597**

Triphenyl[2,2,2-trifluoro-1-(trifluoromethyl)ethylidene]phosphorane, 9CI. *1,1,1,3,3,3-Hexafluoro-2-(triphenylphosphoranylidene)propane*
[73607-10-0]

$Ph_3P{=}C(CF_3)_2$

$C_{21}H_{15}F_6P$ M 412.314
Wittig reagent, prepd. in situ, for synth. of terminal gem-bis(trifluoromethyl)alkenes.

Burton, D.J. *et al*, *Tetrahedron Lett.*, 1979, 3397 (*synth, use*)

4,6,7-Triphenyl-2,3-bis(trifluoromethyl)-1-phosphabicyclo[2.2.2]octa-2,5,7-triene **T-00598**

[20477-35-4]

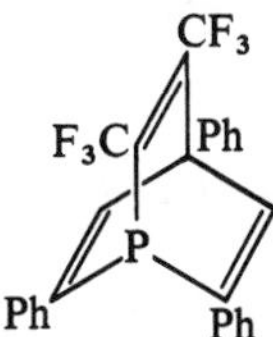

$C_{27}H_{17}F_6P$ M 486.395
Cryst. Mp 189°.

Märkl, G. *et al*, *Angew. Chem., Int. Ed. Engl.*, 1968, **7**, 733 (*synth, uv, ir, P nmr, pmr*)

6,8,10-Triphenyl-1,4-diaza-5-phospha(5-P^V)spiro[4.5]deca-5,7,9-triene, 9CI **T-00599**

[36231-70-6]

$C_{25}H_{23}N_2P$ M 382.444
Solid. Mp 194-196°.
1,4-Di-Me: [36231-69-3].
$C_{27}H_{27}N_2P$ M 410.497
Solid. Mp 106-107°.

Hettche, A. *et al*, *Tetrahedron Lett.*, 1972, 829 (*synth, uv, pmr*)

2,4,5-Triphenyl-2*H*-1,3,2-diazaphosphole, 8CI **T-00600**

$C_{20}H_{15}N_2P$ M 314.326
2-Oxide: [18054-52-9].
$C_{20}H_{15}N_2OP$ M 330.325
Yellow prisms (toluene). Mp 210-213°.

Tuchtenhagen, G. *et al*, *Justus Liebigs Ann. Chem.*, 1968, **711**, 174 (*synth*)

3,4,5-Triphenyl-4*H*-1,2,4-diazaphosphole **T-00601**

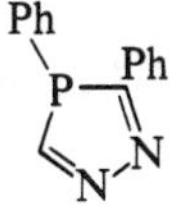

$C_{20}H_{15}N_2P$ M 314.326
Yellow cryst. (Me_2CO). Sol. C_6H_6, THF, dioxan, hot EtOH. Mp 195-196°. Dec. by hot H_2O.

Issleib, K. *et al*, *Chem. Ber.*, 1966, **99**, 1316 (*synth, props*)

2,4,5-Triphenyl-1,3,2-dioxaphosphole **T-00602**

$C_{20}H_{15}O_2P$ M 318.311

2-Sulfide: [34250-29-8].
$C_{20}H_{15}O_2PS$ M 350.371
Solid. Mp 126.5-127°.

Nakayama, S. *et al*, *J. Chem. Soc., Chem. Commun.*, 1971, 1186 (*synth, ms, nmr*)

1,2,3-Triphenyl-1,3,2-diphospharsolane, 9CI **T-00603**

[60600-13-7]

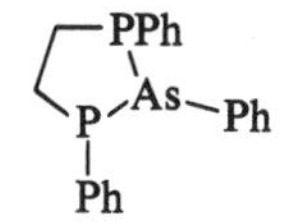

$C_{20}H_{19}AsP_2$ M 396.239
Solid. Mp 77°.

Issleib, K. *et al*, *Synth. React. Inorg. Metal-Org. Chem.*, 1976, **6**, 179 (*synth*)

6,8,10-Triphenyl-1,4-dithia-5-phospha(5-*P*^{V})spiro[4.5]deca-5,7,9-triene, 10CI **T-00604**

2′,4′,6′-Triphenylspiro[1,3,2-dithiaphosph(P^V)*olan-2,1′-phosph*(P^V)*orin*]
[62496-79-1]

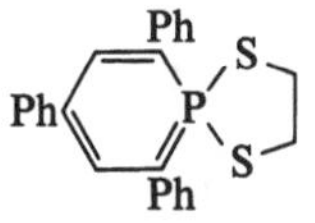

$C_{25}H_{21}PS_2$ M 416.535
Solid. Mp 165-170°.

Kanter, H. *et al*, *Chem. Ber.*, 1977, **110**, 395 (*synth, uv, ms, pmr*)

7,9,11-Triphenyl-1,5-dithia-6-phospha(6-*P*^{V})spiro[5.5]undeca-6,8,10-triene, 10CI **T-00605**

2′,4′,6′-Triphenylspiro[1,3,2-dithiaphosph(P^V)*orinan-2,1′-phosph*(P^V)*orin*]
[62496-80-4]

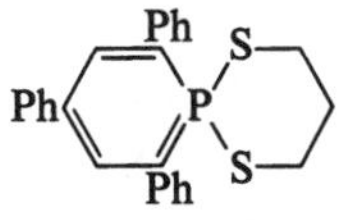

$C_{26}H_{23}PS_2$ M 430.561
Solid. Mp 150-152°.

Kanter, H. *et al*, *Chem. Ber.*, 1977, **110**, 395 (*synth, pmr, uv, ms, P nmr*)

2,4,5-Triphenyl-1,3,2,4,5-dithiatriphospholane, 9CI **T-00606**

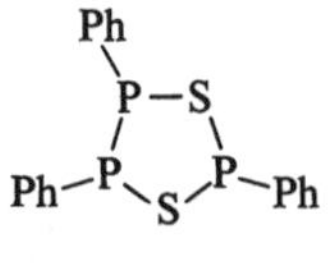

$C_{18}H_{15}P_3S_2$ M 388.358
4-Sulfide: [55499-33-7].
$C_{18}H_{15}P_3S_3$ M 420.418
Cryst. (C_6H_6). Mp 150.5°.

Baudler, M. *et al*, *Z. Anorg. Allg. Chem.*, 1975, **413**, 239 (*synth, ir, raman, P nmr, ms, bibl*)
Andreev, N.A. *et al*, *Zh. Obshch. Khim.*, 1982, **52**, 1530 (*Engl. transl.* p. 1352) (*synth, ir, P nmr*)

2,3,10-Triphenyl-4*H*-1,4-ethenophosphinoline, 9CI **T-00607**

2,3-Benzo-5,6,7-triphenyl-1-phosphabicyclo[2.2.2]octa-2,5,7-triene
[55523-21-2]

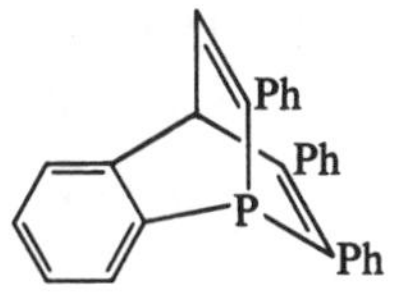

$C_{29}H_{21}P$ M 400.459
Prisms. Mp 148-150°.

Märkl, G. *et al*, *Tetrahedron Lett.*, 1974, 4369 (*synth, uv, ir, ms, pmr*)

Triphenylmethylphosphonic acid **T-00608**

Tritylphosphonic acid
[4759-28-8]

$Ph_3CP(O)(OH)_2$

$C_{19}H_{17}O_3P$ M 324.315
pK_{a1} 4.15 pK_{a2} 7.12 (50% EtOH aq.).
Di-Me ester: Dimethyl triphenylmethylphosphonate.
$C_{21}H_{21}O_3P$ M 352.369
Solid. Mp 157-158°.
Di-Et ester: Diethyl triphenylmethylphosphonate.
$C_{23}H_{25}O_3P$ M 380.422
Cryst. (Me_2CO or Et_2O/pet. ether). Mp 121-122°.
Dichloride: [3736-70-7]. *Boyd's chloride.*
$C_{19}H_{15}Cl_2OP$ M 361.207
Cryst. (Et_2O/C_6H_6 or $Et_2O/CHCl_3$). Mp 188-189°.

Hatt, H.H., *J. Chem. Soc.*, 1929, 2412; 1933, 776 (*dichloride, esters, synth*)
Dimroth, K. *et al*, *Chem. Ber.*, 1960, **93**, 1649 (*esters*)
Halmann, M. *et al*, *J. Chem. Soc.*, 1961, 3542 (*dichloride, ir, struct*)

6,8,10-Triphenyl-1-oxa-4-aza-5-phosphaspiro(5-*P*^{V})spiro[4.5]deca-5,7,9-triene, 9CI **T-00609**

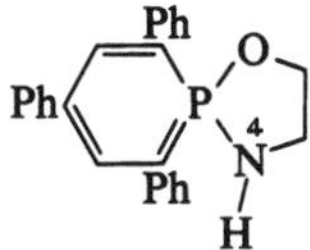

$C_{25}H_{22}NOP$ M 383.429
N-*Me:* [36231-71-7].
$C_{26}H_{24}NOP$ M 397.455
Laser dye. Solid. Mp 149-151°.

Hettche, A. *et al*, *Tetrahedron Lett.*, 1972, 829 (*synth, uv, pmr*)
Basting, D. *et al*, *Appl. Phys.*, 1974, **3**, 81 (*use*)

2,4,6-Triphenyl-4*H*-1,4-oxaphosphorin, 9CI **T-00610**

*2,4,6-Triphenyl-4-phospha-4*H*-pyran*

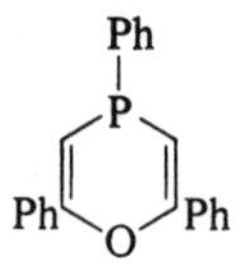

$C_{22}H_{17}OP$ M 328.349

B,PhBr: [21680-84-2]. *2,4,4,6-Tetraphenyl-4H-1,4-oxaphosphorinium bromide. 2,4,4,6-Tetraphenyl-4H-4-phosphoniapyran bromide.*
$C_{28}H_{22}BrOP$ M 485.359
Cryst. (AcOH). Mp 316-317°.
B,PhClO₄: [21680-93-3].
$C_{28}H_{22}ClO_5P$ M 504.906
Cryst. (AcOH). Mp 278-279°.
4-Oxide: [31398-97-7].
$C_{22}H_{17}O_2P$ M 344.349
Cryst. (EtOAc/pentane or cyclohexane). Mp 145°, Mp 176°.

Simalty, M. *et al*, *Bull. Soc. Chim. Fr.*, 1968, 4938 (*salts, synth, uv*)
Maumy, M., *Bull. Soc. Chim. Fr.*, 1972, 1600 (*oxide, synth, uv, ir, pmr*)
Simalty, M. *et al*, *Tetrahedron*, 1972, **28**, 3343 (*oxide, synth, pmr*)
Guilhem, J., *Cryst. Struct. Commun.*, 1974, **3**, 227 (*perchlorate, cryst struct*)

2,4,5-Triphenyl-1,3,4-oxathiaphospholane, 9CI — T-00611

[72564-42-2]

Ph–P₄–S ring with Ph–C–O–C–Ph

$C_{20}H_{17}OPS$ M 336.387
4-Sulfide: [72564-42-2].
$C_{20}H_{17}OPS_2$ M 368.447
Cryst. (MeOH). Mp 153°.

Arbuzov, B.A. *et al*, *Izv. Akad. Nauk SSSR, Ser. Khim.*, 1979, 2136 (*Engl. transl.* p. 1966) (*synth, pmr, P nmr*)

Triphenyl(1-phenylethyl)phosphonium(1+), 9CI — T-00612

$Ph_3P^{\oplus}CH(CH_3)Ph$

$C_{26}H_{24}P^{\oplus}$ M 367.449 (ion)
Bromide: [30537-09-8].
$C_{26}H_{24}BrP$ M 447.353
Solid.
Iodide: [72807-24-0].
$C_{26}H_{24}IP$ M 494.354
Solid. Mp 224-228°.
Ylide: [58594-19-7]. *Triphenyl(1-phenylethylidene)-phosphorane.*
$C_{26}H_{23}P$ M 366.441
Reactive Wittig reagent. Prepd. *in situ.*

Schweizer, E.E. *et al*, *J. Org. Chem.*, 1973, **38**, 1583.
Bestmann, H.J. *et al*, *Angew. Chem.*, 1976, **88**, 297 (*ylide, props*)
Casey, C.P. *et al*, *J. Am. Chem. Soc.*, 1977, **99**, 2533 (*synth, nmr, complex*)
Schaumann, E. *et al*, *Justus Liebigs Ann. Chem.*, 1979, 1702 (*iodide*)
Wheaton, G.A. *et al*, *J. Org. Chem.*, 1983, **48**, 917 (*ylide, use*)

Triphenyl(2-phenylethyl)phosphonium(1+) — T-00613

Phenethyltriphenylphosphonium(1+)

$Ph_3P^{\oplus}CH_2CH_2Ph$

$C_{26}H_{24}P^{\oplus}$ M 367.449 (ion)
Bromide: [53213-26-6].
$C_{26}H_{24}BrP$ M 447.353
Hygroscopic cryst. Mp 78-83°. Treatment with butyllithium or NaH yields the ylide.
Ylide: [53213-08-4]. *Triphenyl(2-phenylethylidene)-phosphorane.*
$C_{26}H_{23}P$ M 366.441
A Wittig reagent prepd. *in situ.*

Hunter, D.H. *et al*, *Can. J. Chem.*, 1977, **55**, 1229 (*synth, pmr, ir, ylide*)
Bestmann, H.J. *et al*, *Angew. Chem., Int. Ed. Engl.*, 1982, **21**, 545 (*ylide*)
Hine, J. *et al*, *J. Org. Chem.*, 1982, **47**, 4758 (*synth, ylide*)

Triphenyl(2-phenylethynyl)-phosphonium(1+), 9CI — T-00614

$Ph_3P^{\oplus}C{\equiv}CPh$

$C_{26}H_{20}P^{\oplus}$ M 363.418 (ion)
Used in synth. of P-contg. heterocyclic compds.
Chloride:
$C_{26}H_{20}ClP$ M 398.871
Solid. Mp 180°.
Bromide: [34387-64-9].
$C_{26}H_{20}BrP$ M 443.322
Solid. Mp 209-211°.

Dickstein, J.I. *et al*, *J. Org. Chem.*, 1972, **37**, 2168 (*bromide, complexes, ir*)
Tanaka, Y. *et al*, *Tetrahedron*, 1973, **29**, 3271 (*use*)
Bestmann, H.J. *et al*, *Angew. Chem., Int. Ed. Engl.*, 1977, **16**, 45 (*chloride, ir, nmr*)
Schweizer, E.E. *et al*, *J. Org. Chem.*, 1978, **43**, 2972; 1982, **47**, 1652 (*use*)
Monita, N. *et al*, *J. Chem. Soc., Perkin Trans. 1*, 1979, 2103 (*props*)

Triphenyl(3-phenyl-2-propenylidene)-phosphorane, 9CI — T-00615

Cinnamylidenetriphenylphosphorane
[41429-69-0]

$Ph_3P{=}CHCH{=}CHPh$

$C_{27}H_{23}P$ M 378.452
(*E*)-*form* [56374-74-4]
Isol. at −75°.

Misumi, S. *et al*, *Bull. Chem. Soc. Jpn.*, 1963, **36**, 39 (*synth, use*)
Hug, R. *et al*, *Helv. Chim. Acta*, 1972, **55**, 1828 (*synth, use*)
Bogdanović, B. *et al*, *Justus Liebigs Ann. Chem.*, 1975, 692 (*isol, pmr*)
Capuano, L. *et al*, *Chem. Ber.*, 1982, **115**, 3904 (*use*)
Wheaton, G.A. *et al*, *J. Org. Chem.*, 1983, **48**, 917 (*props, use*)

Triphenyl(3-phenyl-2-propenyl)-phosphonium(1+), 9CI — T-00616

Cinnamyltriphenylphosphonium(1+)

$Ph_3P^{\oplus}CH_2CH{=}CHPh$

$C_{27}H_{24}P^{\oplus}$ M 379.460 (ion)
(*E*)-*form*
Chloride: [69052-20-6].
$C_{27}H_{24}ClP$ M 414.913
Source of Triphenyl(3-phenyl-2-propenylidene)-phosphorane, T-00615 . Cryst. (Me_2CO). Mp 253-255°.
Bromide: [38633-40-8].
$C_{27}H_{24}BrP$ M 459.364
Source of Triphenyl(3-phenyl-2-propenylidene)-phosphorane, T-00615 . Pale-yellow solid ($EtOAc/CH_2Cl_2$). Mp 254°. NaH → ylide.

Ylide: see Triphenyl(3-phenyl-2-propenylidene)-phosphorane, T-00615

Friedrich, K. *et al*, *Chem. Ber.*, 1959, **92**, 2756 (*synth*)
Schweizer, E.E. *et al*, *J. Org. Chem.*, 1971, **36**, 4033 (*bromide, ir, pmr*)
Doll, M.H. *et al*, *J. Med. Chem.*, 1976, **19**, 1079 (*chloride*)
Courtot, P. *et al*, *Bull. Soc. Chim. Fr. Part 2*, 1977, 749 (*bromide, use*)

Triphenyl(3-phenylpropyl)-phosphonium(1+), 9CI — T-00617

[47562-49-2]

$Ph_3P^{\oplus}CH_2CH_2CH_2Ph$

$C_{27}H_{26}P^{\oplus}$ M 381.476 (ion)

Bromide: [7484-37-9].
$C_{27}H_{26}BrP$ M 461.380
Source of ylide. Needles. Mp 205°. Butyllithium → ylide.

Ylide: [39110-26-4]. *Triphenyl(3-phenylpropylidene)-phosphorane.*
$C_{27}H_{25}P$ M 380.468
Widely used in synth. of prostaglandins through the Wittig reaction.

Bestmann, H.J. *et al*, *Chem. Ber.*, 1970, **103**, 685 (*synth*)
Eberle, M.K. *et al*, *J. Org. Chem.*, 1982, **47**, 2210 (*use*)
Snider, B.N. *et al*, *J. Org. Chem.*, 1983, **48**, 1471 (*use*)

Triphenyl[1-(phenylthio)cyclopropyl]-phosphonium(1+), 9CI — T-00618

PhS, $PPh_3^{\oplus}$ (cyclopropane)

$C_{27}H_{24}PS^{\oplus}$ M 411.520 (ion)

Reagent for pentannulation *via* Wittig reaction.

Bromide: [58992-24-8].
$C_{27}H_{24}BrPS$ M 491.424
No phys. props. reported.

Tetrafluoroborate: [58992-26-0].
$C_{27}H_{24}BF_4PS$ M 498.324
Solid. Mp 201-202.5°.

Marino, J.P. *et al*, *J. Org. Chem.*, 1981, **46**, 1828 (*synth, use*)

Triphenyl[1-(phenylthio)ethenyl]-phosphonium(1+), 9CI — T-00619

$H_2C{=}C(SPh)PPh_3^{\oplus}$

$C_{26}H_{22}PS^{\oplus}$ M 397.494 (ion)

Salts used in the synth. of cyclopentanones.

Chloride: [69442-51-9].
$C_{26}H_{22}ClPS$ M 432.947
No phys. props. reported.

Iodide: [69442-52-0].
$C_{26}H_{22}IPS$ M 524.398
Yellow cryst. (MeCN/EtOAc). Mp 145-146°.

Hewson, A.T., *Tetrahedron Lett.*, 1978, 3267 (*use*)
Cameron, A.G. *et al*, *J. Chem. Soc., Perkin Trans. 1*, 1983, 2979 (*iodide, pmr, use*)

Triphenyl[(phenylthio)methylene]-phosphorane — T-00620

[34884-33-8]

$Ph_3P{=}CHSPh$

$C_{25}H_{21}PS$ M 384.475

Used in conversion of aldehydes and ketones into phenylthio(vinyl) ethers, and also in synth. of 2-deoxyaldoses, e.g. 2-deoxyglucose from arabinose. Deep-red ylide.

Mukaiyama, T. *et al*, *Tetrahedron Lett.*, 1968, 3787 (*use*)
Bestmann, H.J. *et al*, *Tetrahedron Lett.*, 1969, 3665 (*use*)
LaRochelle, R.W. *et al*, *J. Am. Chem. Soc.*, 1971, **93**, 6077 (*synth, use*)

Triphenyl(phenylthiomethyl)-phosphonium(1+), 9CI — T-00621

$Ph_3P^{\oplus}CH_2SPh$

$C_{25}H_{22}PS^{\oplus}$ M 385.483 (ion)

Salts provide Triphenyl[(phenylthio)methylene]-phosphorane, T-00620 used in prepn. of thiovinyl ethers, and in the prep. of deoxyaldopentoses.

Chloride: [13884-92-9].
$C_{25}H_{22}ClPS$ M 420.936
Cryst. ($MeNO_2$). Mp 154-156°, 224-226°, 232-234°.

Ylide: see Triphenyl[(phenylthio)methylene]-phosphorane, T-00620

Raasch, M.S., *J. Org. Chem.*, 1972, **37**, 1347 (*chloride, use*)
Bestmann, H.J. *et al*, *Justus Liebigs Ann. Chem.*, 1974, 2085 (*chloride, use*)
Hauske, J.R. *et al*, *J. Org. Chem.*, 1979, **44**, 2472 (*chloride, use*)

5,6,9-Triphenyl-5-phospha(5-P^V)-spiro[4.4]nona-1,3,6,8-tetraene-1,2,3,4-tetracarboxylic acid — T-00622

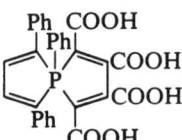

Ph, COOH, Ph, P, COOH, COOH, Ph, COOH

$C_{30}H_{21}O_8P$ M 540.465

Tetra-Me ester: Tetramethyl 1,2,5-triphenyl-5-phospha(5-P^V)spiro[4.4]nona-1,3,6,8-tetraene-1,2,3,4-tetracarboxylate.
$C_{34}H_{29}O_8P$ M 596.572
Solid. Mp 179-181°.

Hughes, A.N. *et al*, *Tetrahedron*, 1968, **24**, 3437 (*synth, ir*)

Triphenyl phosphate — T-00623

[115-86-6]

$(PhO)_3P{=}O$

$C_{18}H_{15}O_4P$ M 326.288

Plasticizer and stabilizer for polyesters. Fire-retardant. Synergist for malathion. Prisms (EtOH or EtOH/pet. ether). Mp 49-51°. Bp_{11} 245°.

▷Mod. toxic, TLV 3. TC8400000.

Cox, J.R. *et al*, *J. Am. Chem. Soc.*, 1958, **80**, 5441 (*synth*)
Glonak, T. *et al*, *J. Phys. Chem.*, 1976, **80**, 639 (*P nmr*)
Kodolov, V.I. *et al*, *Izv. Akad. Nauk SSSR, Ser. Khim.*, 1977, 165 (*Engl. transl.* p. 142) (*pe, pmr, cryst struct*)
Cheng, C.P. *et al*, *J. Am. Chem. Soc.*, 1980, **102**, 6418 (*O nmr*)
Cload, P.A. *et al*, *Org. Mass Spectrom.*, 1983, **18**, 57 (*ms*)
Krishnakumar, V.K. *et al*, *Synthesis*, 1983, 558 (*synth, ir, pmr*)
Sax, N.I., *Dangerous Properties of Industrial Materials*, 6th Ed., Van Nostrand-Reinhold, 1984, 1068.

Hazards in the Chemical Laboratory, (Bretherick, L., Ed.), 3rd Ed., Royal Society of Chemistry, London, 1981, 525.

Triphenylphosphine **T-00624**

[603-35-0]

Ph_3P

$C_{18}H_{15}P$ M 262.290

Nucleophile with many synthetic uses; deoxygenating and desulfurising reagent, reagent for alkene synth. by double extrusion. Converted by alkyl halides into quaternary triphenylphosphonium salts used to prepare ylides for Wittig reactions. Reagent for the deoxygenation of nitroso and nitro compds. (often with the formn. of *N*-containing heterocyclic compds.), oxiranes and peroxides, and for the desulfurization of thiiranes and di- and polysulfides. Reagent for the dehalogenation of 1,2-dihalogenoalkanes. Used in conjunction with CCl_4 or CBr_4 to convert alcohols into alkyl chlorides or bromides, and carboxamides into nitriles. Used in conjunction with diethyl azodicarboxylate in phosphorylations. Plates or prisms (Et_2O). Mp 80°. Bp >360° (under N_2). Triboluminescent.

▷Mod. toxic. SZ3500000.

B,HF: Liq.
B,2HCl: Solid. Mp 60.5-61°.
B,HBr: [6399-81-1]. Mp 191-196°.
B,HI: [6396-08-3]. Solid. Mp 226-228°.
Oxide: see Triphenylphosphine oxide, T-00627
Sulfide: see Triphenylphosphine sulfide, T-00629
Selenide: see Triphenylphosphine selenide, T-00628
Telluride:
$C_{18}H_{15}PTe$ M 389.890
Rather unstable solid. Mp 83-85° dec.

Williams, D.H. *et al*, *J. Am. Chem. Soc.*, 1968, **90**, 966 (*ms*)
Austad, T. *et al*, *Acta Chem. Scand.*, 1973, **27**, 1939 (*telluride*)
Spalding, T.R., *Org. Mass Spectrom.*, 1976, **11**, 1019 (*ms*)
Starzewskii, K.A.O. *et al*, *Phosphorus*, 1976, **6**, 177; *J. Am. Chem. Soc.*, 1976, **98**, 8486 (*cmr, pe*)
Clark, D.A. *et al*, *Synthesis*, 1977, 628 (*deriv, synth, use*)
Grim, S.O. *et al*, *J. Org. Chem.*, 1977, **42**, 1236 (*nmr*)
Masaki, M. *et al*, *Chem. Lett.*, 1977, 151 (*synth*)
Olah, G.A. *et al*, *J. Org. Chem.*, 1977, **42**, 2190 (*synth*)
Hoste, S. *et al*, *J. Electron Spectrosc. Relat. Phenom.*, 1979, **17**, 191 (*pe*)
Jakobsen, H.J. *et al*, *J. Magn. Reson.*, 1979, **33**, 477 (*cmr*)
Ratovskii, G.V. *et al*, *Zh. Obshch. Khim.*, 1979, **49**, 548 (*Engl. transl.* p. 479) (*uv, raman*)
Marshall, G. *et al*, *Org. Mass Spectrom.*, 1981, **16**, 272 (*ms*)
Mitsunoba, O., *Synthesis*, 1981, 1 (*rev, use*)
Allman, T. *et al*, *Can. J. Chem.*, 1982, **60**, 716 (*pmr, cmr, props*)
Fieser, M. *et al*, *Reagents for Organic Synthesis*, Wiley, 1967-84, **10**, 447 (*use*)
Sax, N.I., *Dangerous Properties of Industrial Materials*, 6th Ed., Van Nostrand-Reinhold, 1984, 1068.

N,1,1-Triphenylphosphinecarbonamide, 10CI **T-00625**

N-*Phenyl-diphenylphosphinoformamide*

[7067-72-3]

$Ph_2PCONHPh$

$C_{19}H_{16}NOP$ M 305.315

Solid. Mp 138-139°. Forms Ir, Mn, Fe, Mo, and Rh complexes.

Oxide: [7067-73-4].
$C_{19}H_{16}NO_2P$ M 321.315
Solid. Mp 159-161°.
Sulfide: [10494-05-0].
$C_{19}H_{16}NOPS$ M 337.375
Solid. Mp 127-129°.

Thewissen, D.H.M.W. *et al*, *Recl. Trav. Chim. Pays-Bas*, 1980, **99**, 344 (*ir, pmr, nmr*)
Kunze, U. *et al*, *J. Organomet. Chem.*, 1981, **215**, 187 (*synth, ir*)
Kunze, U. *et al*, *Z. Naturforsch., B*, 1982, **37**, 560 (*synth, derivs*)

P,P,P-Triphenylphosphine imide, 9CI **T-00626**

P,P,P-*Triphenylphosphazene*

[2240-47-3]

$Ph_3P{=}NH$

$C_{18}H_{16}NP$ M 277.305

Cryst. (cyclohexane). Mp 128°. pK_a 13.37 (H_2O), 20.65 ($MeNO_2$).

B,HCl: Cryst. ($MeOH/Et_2O$). V. sol. alcs, H_2O, MeCN. Mp 236°.
N-*Chloro:* N-*Chloro*-P,P,P-*triphenylphosphine imide.*
$C_{18}H_{15}ClNP$ M 311.750
Solid. Mp 179-180°.
N-*Bromo:* N-*Bromo*-P,P,P-*triphenylphosphine imide.*
$C_{18}H_{15}BrNP$ M 356.201
Solid. Mp 170-172°.
N-*Cyano:* N-*Cyano*-P,P,P-*triphenylphosphine imide.*
$C_{19}H_{15}N_2P$ M 302.315
Solid. Mp 194-196°.
N-*Me:* [17986-01-5]. N-*Methyl*-P,P,P-*triphenylphosphine imide.*
$C_{19}H_{18}NP$ M 291.332
Cryst. (ethylcyclohexane). Mp 67°.
N-*Me; B,HBr:* Solid. Mp 197-199°.
N-*Et:* N-*Ethyl*-P,P,P-*triphenylphosphine imide.*
$C_{20}H_{20}NP$ M 305.358
Cryst. (ethylcyclohexane). Mp 96°.
N-*Et; B,HBr:* Solid. Mp 246-247°.
N-*Propyl:* P,P,P-*Triphenyl*-N-*propylphosphine imide.*
$C_{21}H_{22}NP$ M 319.385
Cryst. (ethylcyclohexane). Mp 112-114°.
N-*Propyl; B,HBr:* Solid. Mp 190-192°.
N-*Isopropyl:* N-*Isopropyl*-P,P,P-*triphenylphosphine imide.*
$C_{21}H_{22}NP$ M 319.385
Cryst. (ethylcyclohexane). Mp 126-127°.
N-*Isopropyl; B,HBr:* Solid. Mp 242-243°.
N-tert-*Butyl:* [13989-64-5]. N-tert-*Butyl*-P,P,P-*triphenylphosphine imide.*
$C_{22}H_{24}NP$ M 333.412
Cryst. (ethylcyclohexane). Mp 146-148°.
N-tert-*Butyl; B,HBr:* Solid. Mp 223°.
N-*Ph: see Tetraphenylphosphine imide, T-00269*
N-*1-Naphthyl:* N-*1-Naphthyl*-P,P,P-*triphenylphosphine imide.*
$C_{28}H_{22}NP$ M 403.462
Solid. Mp 141-143°.
N-*2-Naphthyl:* N-*2-Naphthyl*-P,P,P-*triphenylphosphine imide.*
$C_{28}H_{22}NP$ M 403.462
Cryst. (EtOAc).
N-*Ac:* N-*Acetyl*-P,P,P-*triphenylphosphine imide.*
$C_{20}H_{18}NOP$ M 319.342

Prisms (C_6H_6/pet. ether). Mp 163-165°.

N-Benzoyl: *N-Benzoyl-P,P,P-triphenylphosphine imide.*
$C_{25}H_{20}NOP$ M 381.413
Cryst. ($CHCl_3$), needles (EtOH). Mp 196-197°.

N-Phenylsulfonyl: *P,P,P-Triphenyl-N-phenylsulfonylphosphine imide.*
$C_{24}H_{20}NO_2PS$ M 417.461
Cryst. (dioxan aq.). Mp 156-158°.

*N-Diphenylphosphinyl: see N-Diphenylphosphino-P,P,P-*triphenylphosphine imide, D-01088

N-Trimethylsilyl: N-Trimethylsilyl-P,P,P-*triphenylphosphine imide.*
$C_{21}H_{24}NPSi$ M 349.487
Solid. Mp 75-76°. $Bp_{0.5}$ 162-164°.

N-Trimethylgermyl: *N-Trimethylgermyl-P,P,P-triphenylphosphine imide.*
$C_{21}H_{24}GeNP$ M 393.991
Solid. Mp 78-80°. $Bp_{0.1}$ 164°.

N-Trimethylstannyl: *N-Trimethylstannyl-P,P,P-triphenylphosphine imide.*
$C_{21}H_{24}NPSn$ M 440.091
Solid. Mp 85-86°. $Bp_{0.5}$ 172-176°.

N-(Hydroxycarbonyl): see (Triphenylphosphoranylidene)carbamic acid, T-00648

Horner, L. *et al, Justus Liebigs Ann. Chem.*, 1956, **627**, 142 (*derivs*)
Appel, R. *et al, Chem. Ber.*, 1960, **93**, 405; 1965, **98**, 1355 (*synth*)
Appel, R. *et al, Z. Anorg. Allg. Chem.*, 1961, **311**, 290 (*chloro, bromo derivs*)
Derkach, G.I. *et al, Zh. Obshch. Khim.*, 1962, **32**, 1874 (*Engl. transl.* p. 1853) (*acetyl*)
Birkofer, L. *et al, Chem. Ber.*, 1963, **96**, 2750 (*trimethylsilyl, use*)
Zimmer, H. *et al, J. Org. Chem.*, 1963, **28**, 483 (*derivs, uv, pmr*)
Levchenko, E.S. *et al, Zh. Obshch. Khim.*, 1965, **35**, 2080 (*Engl. transl.* p. 2069) (*deriv*)
Schmidbaur, H. *et al, Chem. Ber.*, 1967, **100**, 1120 (*derivs, ir, pmr*)
Wiegräbe, W. *et al, Chem. Ber.*, 1968, **101**, 1414 (*derivs, ir*)
Ross, B. *et al, Chem. Ber.*, 1974, **107**, 2720 (*synth*)
Ostoja Starzewski, K.A. *et al, Phosphorus*, 1976, **6**, 177 (*cmr*)

Triphenylphosphine oxide, 9CI, 8CI — T-00627

[791-28-6]

$Ph_3P{=}O$

$C_{18}H_{15}OP$ M 278.290
Complexes with metals of Groups IIIA, IVA, IIIB, IVB, VB, VIB, VIIB, and VIII, and lanthanides. Cryst. (EtOH). Mp 152-153°. pK_a 2.9 ($MeNO_2$); 6.11 ($MeNO_2$). Triboluminescent. Forms complexes with SO_3, Br_2, I_2, HBr, HCl, Cl_3CCOOH, PhCOOH, and many metal halides.

Gilman, H. *et al, Recl. Trav. Chim. Pays-Bas*, 1929, **48**, 328 (*synth*)
Williams, D.H. *et al, J. Am. Chem. Soc.*, 1968, **90**, 966 (*ms*)
Miličev, S., *Spectrochim. Acta, Part A*, 1974, **30**, 255 (*ir, raman*)
Albright, T.A. *et al, J. Org. Chem.*, 1975, **40**, 3437 (*cmr, nmr*)
Ostoja Starzewski, K.A. *et al, Phosphorus*, 1976, **6**, 177 (*cmr*)
Hoste, S. *et al, J. Electron Spectrosc. Rel. Phenom.*, 1979, **17**, 191 (*pe*)
Ratovskii, G.V. *et al, Zh. Obshch. Khim.*, 1979, **49**, 548 (*Engl. transl.* p. 479) (*uv*)
Yakshin, V.V. *et al, Radiokhimiyia*, 1980, **22**, 517 (*Engl. transl.* p. 410) (*props*)
Marshall, G. *et al, Org. Mass Spectrom.*, 1981, **16**, 272 (*ms*)
Shifrina, R.R. *et al, Zh. Prikl. Spektrosk.*, 1981, **34**, 111 (*Engl. transl.* p. 81) (*uv*)
Bemi, L. *et al, J. Am. Chem. Soc.*, 1982, **104**, 438 (*nmr*)
Matrosov, E.I. *et al, Phosphorus Sulfur*, 1982, **13**, 69 (*props*)

Triphenylphosphine selenide, 9CI, 8CI — T-00628

[3878-44-2]

$Ph_3P{=}Se$

$C_{18}H_{15}PSe$ M 341.250
Cryst. (Me_2CO or EtOH). Mp 187-188°.

Screttas, C. *et al, J. Org. Chem.*, 1962, **27**, 2573 (*synth*)
Nicpon, P. *et al, Inorg. Chem.*, 1966, **5**, 1297 (*synth*)
McFarlane, W. *et al, J. Chem. Soc., Dalton Trans.*, 1973, 2162 (*pmr, nmr*)
De Ketelaere, R.F. *et al, J. Mol. Struct.*, 1974, **23**, 233 (*ir, raman*)
Wuyts, L.F. *et al, J. Organomet. Chem.*, 1977, **129**, 163 (*cmr*)
Codding, P.W. *et al, Acta Crystallogr., Sect. B*, 1979, **35**, 1261 (*cryst struct*)
Dean, P.A.W., *Can. J. Chem.*, 1979, **57**, 754 (*nmr*)
Robert, J.B. *et al, J. Magn. Reson.*, 1980, **38**, 357 (*nmr*)
Carr, S.W. *et al, Aust. J. Chem.*, 1981, **34**, 35 (*synth, nmr, ir*)

Triphenylphosphine sulfide, 9CI — T-00629

[3878-45-3]

$Ph_3P{=}S$

$C_{18}H_{15}PS$ M 294.350
Prod. of desulfurizations by Ph_3P. Reacts with oxiranes to produce thiiranes. Forms complexes with Sn, Pd, Pt and metals of Groups IB, 11B and VIIB. Cryst. (EtOH). Mp 162-164°. Not protonated in strong acids e.g. 100% H_2SO_4.

▷SZ2820000.

Markovskii, L.N. *et al, Zh. Org. Khim.*, 1972, **8**, 1822 (*Engl. transl.* p. 1869) (*synth*)
Grutsch, P.A., *Inorg. Chem.*, 1973, **12**, 1431 (*pe, rev*)
Albright, T.A. *et al, J. Org. Chem.*, 1975, **40**, 3437 (*cmr, nmr*)
Cauguis, G. *et al, Org. Mass Spectrom.*, 1975, **10**, 770 (*ms*)
Glidewell, C. *et al, J. Organomet. Chem.*, 1976, **116**, 199 (*synth, ms*)
Olah, G.A. *et al, J. Org. Chem.*, 1977, **42**, 2190 (*synth*)
Dillon, K.B. *et al, J. Chem. Soc., Dalton Trans.*, 1982, 465 (*props*)
Pierrard, J.C. *et al, J. Chem. Res. (S)*, 1982, 52 (*complexes*)

N,P,P-Triphenylphosphinic amide, 9CI — T-00630

[6190-28-9]

$Ph_2P(O)NHPh$

$C_{18}H_{16}NOP$ M 293.304
Needles (MeCN or EtOH). Mp 242-244°.

Hunt, B.B. *et al, J. Chem. Soc.*, 1957, 2413 (*synth*)
Tyssee, D.A. *et al, J. Am. Chem. Soc.*, 1973, **95**, 8066 (*synth, ir, pmr, props, derivs*)
Appel, R. *et al, Chem. Ber.*, 1975, **108**, 2349 (*synth, P nmr*)
Modro, T.A., *Phosphorus Sulfur*, 1979, **5**, 331 (*cmr*)
Bradamante, S., *J. Org. Chem.*, 1980, **45**, 114 (*cmr, pmr*)

N,P,P-Triphenylphosphinimidic acid — T-00631

N,P,P-Triphenylphosphinic amide

$Ph_2P(OH){=}NPh \rightleftharpoons Ph_2P(O)NHPh$

$C_{18}H_{16}NOP$ M 293.304
Acid exists only as tautomeric amide (see 45558-1).

Et ester: [17985-95-4]. *Ethyl* N,P,P-*triphenylphosphinimidate.* P-*Ethoxy*-N,P,P-*triphenylphosphazene.* P-*Ethoxy*-N,P,P-*triphenylphosphine imide.*
$C_{20}H_{20}NOP$ M 321.358

Cryst. (C_6H_6/pet. ether). Mp 57-58°, 66-68°. pK_a 15.35 ($MeNO_2$).

Chloride: [5290-45-9]. P-*Chloro*-N,P,P-*triphenylphosphazene.* P-*Chloro*-N,P,P-*triphenylphosphine imide.*
$C_{18}H_{15}ClNP$ M 311.750
Solid. Mp 96-97°. Turns red in sunlight.

Anilide: [17985-98-7]. N,N′,P,P-*Tetraphenylphosphinimidic amide.*
$C_{24}H_{21}N_2P$ M 368.417
Cryst. (Et_2O). Mp 178-179°.

▷SZ5950000.

Bock, H. *et al, Chem. Ber.*, 1966, **99**, 1068 (*chloride*)
Wiegräbe, W. *et al, Chem. Ber.*, 1968, **101**, 1414 (*ester, ir, anilide*)
Haubold, W. *et al, Z. Anorg. Allg. Chem.*, 1970, **372**, 273 (*chloride, P nmr, props*)
Genkina, G.K. *et al, Zh. Obshch. Khim.*, 1971, **41**, 80 (*Engl. transl.* p. 76) (*ester, props*)

1,2,3-Triphenyl-1*H*-phosphirene, 9CI T-00632

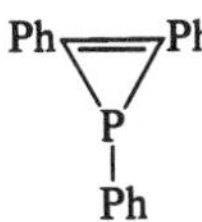

$C_{20}H_{15}P$ M 286.312
Cryst. (hexane). Mp 73°. Stable at r.t. In toluene, the half-life is >17 hr. Forms a stable sulfide, and stable Fe and W complexes.

Oxide: [29954-85-6]. Yellow oil. Struct. questioned by Quast who suggested that the correct structure of the oxide is diphenyl(1,2-diphenylethenyl)phosphine oxide. Later attempts to prepare the oxide failed.

Koos, E.W. *et al, J. Chem. Soc., Chem. Commun.*, 1972, 1085 (*oxide, ms, pmr, ir, props*)
Quast, H. *et al, J. Chem. Soc., Chem. Commun.*, 1979, 390 (*struct, oxide*)
Marinetti, A. *et al, J. Chem. Soc., Chem. Commun.*, 1984, 45 (*synth, P nmr, cmr, ms, cryst struct, sulfide*)

Triphenyl phosphite, 9CI, 8CI T-00633

[101-02-0]

$(PhO)_3P$

$C_{18}H_{15}O_3P$ M 310.288
Reagent for peptide synth., particularly in combination with imidazole. Used in conjunction with other reagents e.g. halogens, MeI, for synth. of alkyl and aryl halides. Solid or liq. d_{25}^{25} 1.18. Mp 21-23°. Bp_1 183-184°. n_D^{25} 1.5890. Easily hydrolysed. Forms stable quarternary salts.

▷Skin sensitizer. TH1575000.

Walsh, E.N., *J. Am. Chem. Soc.*, 1959, **81**, 3023 (*synth*)
Bochkarev, V.N. *et al, Zh. Obshch. Khim.*, 1972, **42**, 2348 (*Engl. transl.* p. 2345) (*ms*)
Mitin, Yu.V. *et al, Zh. Obshch. Khim.*, 1973, **43**, 203 (*Engl. transl.* p. 199) (*use*)
Hudson, H.R. *et al, J. Chem. Soc., Perkin Trans. 1*, 1974, 982 (*props, adducts*)
Yamazaki, N. *et al, Tetrahedron*, 1974, **30**, 1326 (*use*)
Bodner, G.M. *et al, J. Organomet. Chem.*, 1975, **101**, 63 (*cmr, complexes*)
Cadogan, J.I.G. *et al, Organophosphorus Reagents in Organic Synthesis*, Academic Press, 1979 (*rev, use*)
Barlett, P.D. *et al, J. Am. Chem. Soc.*, 1983, **105**, 1984 (*derivs*)
Cload, P.A., *et al, Org. Mass. Spectrom.*, 1983, **18**, 57 (*ms*)

1,2,5-Triphenyl-1*H*-phosphole, 8CI T-00634

1,2,5-Triphenylphosphacyclopentadiene
[1162-70-5]

$C_{22}H_{17}P$ M 312.350
Yellow needles ($CHCl_3$). Exhibits blue fluor. in EtOH soln., and a reversible thermochromic effect (yellow → red) in CCl_4. Photolysis → dimer.

*B,MeI: 1-Methyl-1,2,5-triphenyl-1*H*-phospholium iodide.*
$C_{23}H_{20}IP$ M 454.289
Cryst. + $1H_2O$ (MeOH/pet. ether). Mp 218.5-222° dec.

Oxide:
$C_{22}H_{17}OP$ M 328.349
Cryst. (EtOH/EtOAc). Mp 237-239°.

Sulfide:
$C_{22}H_{17}PS$ M 344.410
Yellow cryst. (butanone). Mp 215-216.5°.

Selenide: [2302-70-7].
$C_{22}H_{17}PSe$ M 391.310
Yellow needles (C_6H_6). Mp 205.5-206.5°.

Campbell, I.G.M. *et al, J. Chem. Soc.*, 1965, 2184 (*synth, derivs, uv*)
Märkl, G. *et al, Angew. Chem., Int. Ed. Engl.*, 1967, **6**, 86 (*synth*)
Barton, T.J. *et al, Tetrahedron Lett.*, 1969, 5037 (*photolysis*)
Osbirn, W.P. *et al, J. Chem. Soc., Chem. Commun.*, 1971, 1062 (*cryst struct*)
Farnham, W.B. *et al, J. Chem. Soc., Chem. Commun.*, 1972, 469 (*aromaticity*)
Allen, D.W. *et al, J. Chem. Res.(S)*, 1981, 220 (*selenide, P nmr*)
De Lauzon, G. *et al, Tetrahedron Lett.*, 1982, **23**, 511 (*props*)
Hocking, M.B. *et al, Can. J. Chem.*, 1982, **60**, 138 (*uv, thermochromism*)
Quin, L.D. *et al, J. Org. Chem.*, 1982, **47**, 905 (*P nmr, oxide*)

10-(Triphenylphosphoniomethyl)benzo[*g*]chrysene(1+) T-00635

(Benzo[g]*chrysen-10-ylmethyl)-triphenylphosphonium(1+)*, 9CI

$C_{41}H_{30}P^{\oplus}$ M 553.662 (ion)

Bromide: [53156-59-5].
$C_{41}H_{30}BrP$ M 633.566
Source of ylide. Solid.

Ylide: (Benzo[g]*chrysen-10-ylmethylene)-triphenylphosphorane.*
$C_{41}H_{29}P$ M 552.654
Used in Wittig reactions to prepare helicenes.

Tinnemans, A.H.A. *et al, J. Am. Chem. Soc.*, 1974, **96**, 4617 (*use*)
Laarhoven, W.H. *et al, Tetrahedron*, 1976, **32**, 2445 (*use*)

2-(Triphenylphosphoniomethyl)benzo[*c*]-phenanthrene(1+) **T-00636**

(Benzo[c]*phenanthren-2-ylmethyl)-triphenylphosphonium(1+)*, *9CI*

$C_{37}H_{28}P^{\oplus}$ M 503.602 (ion)

Bromide: [35160-98-6].

$C_{37}H_{28}BrP$ M 583.506

Source of ylide. Solid. Mp 314-315° dec., 320-321°.

Ylide: (Benzo[e]*phenanthren-2-ylmethylene)-triphenylphosphorane.*

$C_{37}H_{27}P$ M 502.594

Used in Wittig reactions to give helicenes.

Bernstein, W.J. *et al*, *J. Am. Chem. Soc.*, 1972, **94**, 494 (*synth, use*)
Lightner, D.A. *et al*, *J. Am. Chem. Soc.*, 1972, **94**, 3492 (*synth, use*)
Martin, R.H. *et al*, *Tetrahedron*, 1972, **28**, 1749 (*use*)
Tinnemans, A.H.A. *et al*, *J. Chem. Soc., Perkin Trans. 2*, 1976, 1104 (*use*)
Nakazaki, M. *et al*, *J. Org. Chem.*, 1981, **46**, 1985 (*use*)

3-(Triphenylphosphoniomethyl)-phenanthrene(1+) **T-00637**

(3-Phenanthrenylmethyl)triphenylphosphonium(1+), *9CI*

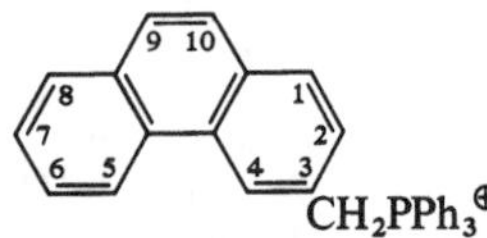

$C_{33}H_{26}P^{\oplus}$ M 453.542 (ion)

Used in Wittig reactions leading to helicenes.

Bromide: [33895-27-1].

$C_{33}H_{26}BrP$ M 533.446

Source of ylide. Cubes (H_2O). Mp 273-276°, 295-298°.

Ylide: (3-Phenanthrenylmethylene)-triphenylphosphorane.

$C_{33}H_{25}P$ M 452.534

Akiyama, S. *et al*, *Bull. Chem. Soc. Jpn.*, 1971, **44**, 2231 (*synth, use*)
Martin, R.H. *et al*, *Tetrahedron*, 1972, **28**, 1749; 1975, **31**, 2135 (*use*)
Tinnemans, A.H.A. *et al*, *J. Chem. Soc., Perkin Trans. 2*, 1976, 1104 (*use*)

9-(Triphenylphosphoniomethyl)-phenanthrene(1+) **T-00638**

(9-Phenenthrenylmethyl)triphenylphosphonium(1+), *9CI*

$C_{33}H_{26}P^{\oplus}$ M 453.542 (ion)

Used in Wittig reactions leading to stilbenes.

Chloride: [79926-86-6].

$C_{33}H_{26}ClP$ M 488.995

Source of ylide. Solid. Mp 315° dec.

Bromide: [33895-28-2].

$C_{33}H_{26}BrP$ M 533.446

Source of ylide. Solid. Mp 286-291°.

Ylide: (9-Phenanthrenylmethylene)-triphenylphosphorane.

$C_{33}H_{25}P$ M 452.534

Tinnemans, A.H.A. *et al*, *J. Am. Chem. Soc.*, 1974, **96**, 4617 (*bromide, use*)
Listvan, V.N. *et al*, *Zh. Org. Khim.*, 1981, **17**, 1711 (*Engl. transl.* p. 1528) (*chloride, use*)

Triphenylphosphoranylideneacetaldehyde, 9CI, 8CI **T-00639**

Formylmethylidenetriphenylphosphorane. 2-Oxoethylidenetriphenylphosphorane. Formylmethylenetriphenylphosphorane

[2136-75-6]

$Ph_3P{=}CHCHO$

$C_{20}H_{17}OP$ M 304.327

Used in Wittig synth. of α,β-unsaturated aldehydes. Cryst. (Me_2CO). Mp 187-188° dec. Alkylation→enol ether. Forms complexes of Fe and W.

Trippett, S. *et al*, *J. Chem. Soc.*, 1961, 1266 (*synth*)
Dale, A.J. *et al*, *Acta Chem. Scand.*, 1970, **24**, 2681 (*ir, pmr*)
Snyder, J.P. *et al*, *Tetrahedron Lett.*, 1970, 3317 (*conformn, pmr, nmr*)
Bestmann, H.J. *et al*, *Angew. Chem., Int. Ed. Engl.*, 1979, **18**, 687 (*use*)
Brittain, J.M. *et al*, *Tetrahedron*, 1979, **35**, 1139 (*nmr*)
Olstein, R. *et al*, *Aust. J. Chem.*, 1979, **32**, 681 (*use*)

(Triphenylphosphoranylidene)acetic acid, 9CI **T-00640**

[15677-02-8]

$Ph_3P{=}CHCOOH$

$C_{20}H_{17}O_2P$ M 320.327

Wittig reagent.

Me ester: see Methyl triphenylphosphoranylideneacetate, M-00433

Et ester: see Ethyl triphenylphosphoranylideneacetate, E-00198

tert-*Butyl ester:*

$C_{24}H_{25}O_2P$ M 376.434

Cryst. (EtOH/pet. ether). Mp 154-155° (147°).

Benzyl ester:

$C_{27}H_{23}O_2P$ M 410.451

Cryst. (EtOH/pet. ether). Mp 120°.

Knorr, H. *et al*, *Justus Liebigs Ann. Chem.*, 1977, 545 (*esters*)
Montgomery, J.A. *et al*, *J. Org. Chem.*, 1981, **46**, 594 (*use*)
Cook, M.P. *et al*, *J. Org. Chem.*, 1982, **47**, 4955 (*ester*)

Triphenylphosphoranylideneacetonitrile, 9CI, 8CI **T-00641**

Cyanomethylidenetriphenylphosphorane. Triphenylphosphoranylidenemethanenitrile

[16640-68-9]

$Ph_3P{=}CHCN$

$C_{20}H_{16}NP$ M 301.327

Wittig reagent. Cryst. (C_6H_6, pet. ether, EtOH, or C_6H_6/cyclohexane). Mp 196-197°.

Trippett, S., *J. Chem. Soc.*, 1959, 3874 (*synth, uv*)
Schiemenz, G.P. *et al*, *Chem. Ber.*, 1961, **94**, 578 (*synth, props*)
Speziale, A.J. *et al*, *J. Am. Chem. Soc.*, 1963, **85**, 2790 (*props, nmr*)
Bestmann, H.J. *et al*, *Justus Liebigs Ann. Chem.*, 1974, 1688 (*synth, ir*)
Corsaro, A. *et al*, *J. Chem. Soc., Perkin Trans. 1*, 1977, 2154 (*prop, use*)
Kukhar, V.P. *et al*, *Zh. Obshch. Khim.*, 1979, **49**, 1025 (*Engl. transl.* p. 889) (*synth, ir, nmr*)

Wätjen, F. *et al*, *Tetrahedron Lett.*, 1982, **23**, 4741 (*props*)

[[(Triphenylphosphoranylidene)amino]-sulfonyl]phosphorimidic acid, 9CI T-00642

[(Triphenylphosphoranylidene)sulfamoyl]-phosphorimidic acid

$Ph_3P{=}NSO_2N{=}P(OH)_3$

$C_{18}H_{18}N_2O_5P_2S$ M 436.358

Free acid tautomeric with [[(triphenylphosphoranylidene)amino]sulfonyl]-phosphoramidic acid.

Tri-Me ester: [56843-02-8]. *Trimethyl [[(triphenylphosphoranylidene)amino]sulfonyl]-phosphorimidate.*
$C_{21}H_{24}N_2O_5P_2S$ M 478.439
Cryst. (C_6H_6/cyclohexane). Mp 137-138°.

Tri-Et ester: [56843-03-9]. *Triethyl [[(triphenylphosphoranylidene)amino]sulfonyl]-phosphorimidate.*
$C_{24}H_{30}N_2O_5P_2S$ M 520.519
Cryst. (C_6H_6/cyclohexane). Mp 91-92°.

Tri-Ph ester: [41309-18-6]. *Triphenyl [[(triphenylphosphoranylidene)amino]sulfonyl]-phosphorimidate.*
$C_{36}H_{30}N_2O_5P_2S$ M 664.651
Prisms (Et_2O/dichloroethane or EtOH). Mp 169.5-172.5°.

Shtepanek, A.S. *et al*, *Zh. Obshch. Khim.*, 1973, **43**, 25 (*Engl. transl.* p. 21) (*triphenyl ester*)
Arrington, D.E., *J. Chem. Soc., Dalton Trans.*, 1975, 1221 (*esters, synth*)

α-(Triphenylphosphoranylidene)-benzeneacetaldehyde, 9CI T-00643

(α-Formylbenzylidene)triphenylphosphorane. Phenyl(triphenylphosphoranylidene)acetaldehyde

[33078-07-8]

$Ph_3P{=}CPhCHO$

$C_{26}H_{21}OP$ M 380.425

Resonance-stabilised ylide. Cryst. (EtOAc). Mp 157-158°.

Le Corre, M., *Tetrahedron Lett.*, 1974, 1037 (*use*)
Devlin, C.J. *et al*, *J. Chem. Soc., Perkin Trans. 1*, 1974, 453 (*synth*)
Brittain, J.M. *et al*, *Tetrahedron*, 1979, **35**, 1139 (*nmr*)

α-(Triphenylphosphoranylidene)-benzeneacetic acid, 9CI T-00644

Phenyl(triphenylphosphoranylidene)acetic acid

$Ph_3P{=}CPhCOOH$

$C_{26}H_{21}O_2P$ M 396.424

Me ester: [1106-06-5]. *Methyl α-(triphenylphosphoranylidene)benzeneacetate, 9CI. Methyl phenyl(triphenylphosphoranylidene)acetate. [α-(Methoxycarbonyl)benzylidene]-triphenylphosphorane.*
$C_{27}H_{23}O_2P$ M 410.451
Stabilised ylide for Wittig reacns. Mp 155°. With aq. alkali gives $PhCH_2COOH$. Photooxygenation gives PhCOCOOMe.

Bestmann, H.J. *et al*, *Justus Liebigs Ann. Chem.*, 1964, **674**, 11 (*synth, props*)
Cooks, R.G. *et al*, *Tetrahedron*, 1968, **24**, 3289 (*ms*)
Zeliger, H.I. *et al*, *Tetrahedron Lett.*, 1969, 2199 (*pmr*)
Jefford, C.W. *et al*, *Tetrahedron Lett.*, 1977, 4531 (*props*)

α-(Triphenylphosphoranylidene)-benzenepropanoic acid T-00645

α-(Triphenylphosphoranylidene)hydrocinnamic acid, 8CI

$Ph_3P{=}C(COOH)CH_2Ph$

$C_{27}H_{23}O_2P$ M 410.451

Me ester: [26480-92-2]. *Methyl α-(triphenylphosphoranylidene)hydrocinnamate, 8CI.*
$C_{28}H_{25}O_2P$ M 424.478
Resonance-stabilised ylide used in Wittig reacns. Cryst. Mp 187-188°.

Et ester: Ethyl α-(triphenylphosphoranylidene)-hydrocinnamate.
$C_{29}H_{27}O_2P$ M 438.505
Cryst. Mp 143-144°.

Bestmann, H.J. *et al*, *Chem. Ber.*, 1962, **95**, 2921 (*synth, ir*)
Shevchuk, M.I. *et al*, *Zh. Obshch. Khim.*, 1970, **40**, 57 (*Engl. transl.* p. 54) (*synth, uv, use*)

(Triphenylphosphoranylidene)butanedioic acid T-00646

(Triphenylphosphoranylidene)succinic acid

[14438-23-4]

$Ph_3P{=}C(^1COOH)CH_2{}^4COOH$

$C_{22}H_{19}O_4P$ M 378.363

All esters are useful synthetic intermeds. for Wittig reactions.

1-Me ester: [57367-54-1]. *α-Methoxycarbonyltriphenyl-phosphoranylidenepropanoic acid.*
$C_{23}H_{21}O_4P$ M 392.390
Cryst. (C_6H_6/EtOAc). Mp 142-144°.

1-Et ester: [68434-72-2]. *α-Ethoxycarbonyltriphenyl-phosphoranylidenepropanoic acid.*
$C_{24}H_{23}O_4P$ M 406.417
No phys. props. reported.

Di-Me ester: [1104-78-5]. *Dimethyl (triphenylphosphoranylidene)butanedioate. Dimethyl (triphenylphosphoranylidene)succinate.*
$C_{24}H_{23}O_4P$ M 406.417
Cryst. (EtOAc). Mp 157-158°.

Di-Et ester: [23360-63-6]. *Diethyl (triphenylphosphoranylidene)butanedioate. Diethyl (triphenylphosphoranylidene)succinate.*
$C_{26}H_{27}O_4P$ M 434.471
Cryst. (cyclohexane). Mp 104-106°.

*Di-*tert*-butyl ester:* [72649-11-7]. *Di-*tert*-butyl (triphenylphosphoranylidene)butanedioate. Di-*tert*-butyl (triphenylphosphoranylidene)succinate.*
$C_{30}H_{35}O_4P$ M 490.578
No phys. props. reported.

*1-Me, 4-*tert*-butyl ester:* [57367-56-3]. tert*-Butyl α-methoxycarbonyltriphenylphosphoranylidenepropanoate.*
$C_{27}H_{29}O_4P$ M 448.497
Cryst. Mp 140-142°.

1-Me ester, betaine: Cryst. (C_6H_6/EtOAc). Mp 142-144°.

1-Et ester, betaine: Cryst. (EtOAc or C_6H_6/pet. ether). Mp 126-127°.

Anhydride: see Dihydro(triphenylphosphoranylidene)-2,5-furandione, D-00612

Hoffman, H., *Chem. Ber.*, 1961, **94**, 1331 (*synth*)
Bestmann, H.J. *et al*, *Chem. Ber.*, 1962, **95**, 2921 (*synth, ir*)
Hudson, R.F. *et al*, *Helv. Chim. Acta*, 1963, **46**, 2178 (*synth*)
Cameron, A.F. *et al*, *J. Chem. Soc., Perkin Trans. 2*, 1975, 1030 (*synth, cryst struct, ir, pmr, cmr*)

McMurray, J.E. *et al*, *Tetrahedron Lett.*, 1977, 2869 (*synth, use*)
Flitsch, W. *et al*, *Chem. Ber.*, 1979, **112**, 3577 (*synth*)
Cooke, M.P., *Tetrahedron Lett.*, 1981, **22**, 381 (*synth, use*)
Ramage, R. *et al*, *Tetrahedron, Suppl.*, 1981, 157 (*synth, use*)
Sargent, M.V. *et al*, *J. Chem. Soc., Perkin Trans. 1*, 1982, 1605, 2373 (*synth, use*)

4-(Triphenylphosphoranylidene)-2-butenoic acid, 9CI — T-00647

(3-Carboxy-2-propenylidene)triphenylphosphorane

$Ph_3P{=}CHCH{=}CHCOOH$

$C_{22}H_{19}O_2P$ M 346.365
Esters are Wittig reagents used in prostaglandin synth. and in synth. of biotin from D-arabinose.

***(E)*-form**

Me ester: [65866-00-4]. *Methyl 4-(triphenylphosphoranylidene)-2-butanoate.*
$C_{23}H_{21}O_2P$ M 360.391
Orange-yellow leaflets. Mp 175-179° (sinters at 168°).

Et ester:
$C_{24}H_{23}O_2P$ M 374.418
No phys. props. reported.

Buchta, E. *et al*, *Chem. Ber.*, 1959, **92**, 3111 (*ester, synth*)
Darby, N. *et al*, *J. Org. Chem.*, 1977, **42**, 1960 (*ester, synth, use*)
Bredenkamp, M. *et al*, *Tetrahedron Lett.*, 1980, **21**, 4199.
Vogel, F.G.M. *et al*, *Justus Liebigs Ann. Chem.*, 1980, 1972 (*use*)
Clemo, N.G. *et al*, *J. Chem. Soc., Perkin Trans. 1*, 1981, 1448 (*ester*)

(Triphenylphosphoranylidene)carbamic acid, 9CI — T-00648

Triphenylphosphazocarbonic acid

$Ph_3P{=}NCOOH$

$C_{19}H_{16}NO_2P$ M 321.315
Esters are Wittig-type reagents. Pyrolysis of alkyl esters yields alkyl isocyanates.

Me ester: [40438-23-1]. *Methyl (triphenylphosphoranylidene)carbamate.*
$C_{20}H_{18}NO_2P$ M 335.341
Cryst. (EtOAc/pet. ether). Mp 134-136°.

Et ester: [17437-51-3]. *Ethyl (triphenylphosphoranylidene)carbamate.*
$C_{21}H_{20}NO_2P$ M 349.368
Prisms (Et_2O at low temp., or ligroin). Mp 136-138°.

Propyl ester: Propyl (triphenylphosphoranylidene)-carbamate.
$C_{22}H_{22}NO_2P$ M 363.395
Cryst. (Et_2O/pet. ether). Mp 89-90°.

Isopropyl ester: Isopropyl (triphenylphosphoranylidene)carbamate.
$C_{22}H_{22}NO_2P$ M 363.395
Solid. Mp 134-135°.

Shokol, V.A. *et al*, *Zh. Obshch. Khim.*, 1969, **39**, 874 (*Engl. transl.* p. 839) (*P nmr*)
Kricheldorf, H.R., *Synthesis*, 1972, 695 (*synth, ir*)
Seyferth, D. *et al*, *J. Org. Chem.*, 1974, **39**, 2336 (*use*)
Whitfield, G.H. *et al*, *J. Org. Chem.*, 1974, **39**, 2148 (*synth, ir, pmr*)
Niclas, H.J. *et al*, *Tetrahedron*, 1978, **34**, 703 (*synth, props*)
Yim, A.S. *et al*, *Can. J. Chem.*, 1978, **56**, 289 (*props*)
Tamura, Y. *et al*, *J. Org. Chem.*, 1981, **46**, 1732 (*synth*)
Bittner, S. *et al*, *J. Org. Chem.*, 1985, **50**, 1712 (*synth, ir, pmr*)

4-Triphenylphosphoranylidene-2,5-cyclohexadien-1-one — T-00649

$Ph_3P{=}C_6H_4{=}O$

$C_{24}H_{19}OP$ M 354.387
Cryst. + $2H_2O$ (H_2O). Mp 310°. With MeI → (4-Methoxyphenyl)diphenylphosphine, M-00058 .

Horner, L. *et al*, *Chem. Ber.*, 1958, **91**, 52.

(Triphenylphosphoranylidene)ethenethione, 9CI — T-00650

(Triphenylphosphoranylidene)thioketene
[17507-47-0]

$Ph_3P{=}C{=}C{=}S$

$C_{20}H_{15}PS$ M 318.372
Cryst. (C_6H_6). Mp 224-226°.

Daly, J.J., *J. Chem. Soc. (A)*, 1967, 1913 (*cryst struct*)
Lumbroso, H. *et al*, *J. Organomet. Chem.*, 1978, **161**, 347 (*struct*)
Albright, T.A. *et al*, *Z. Naturforsch, B*, 1980, **35**, 343 (*struct*)
Bestmann, H.J. *et al*, *Chem. Ber.*, 1980, **113**, 274, 3369 (*synth, ir, cmr, nmr*)
Bestmann, H.J. *et al*, *Chem. Ber.*, 1980, **113**, 912 (*props*)
Bestmann, H.J. *et al*, *Houben-Weyl Method. Org. Chem.*, Band E1, 1982, 752 (*rev*)
Bestmann, H.J. *et al*, *Chem. Ber.*, 1985, **118**, 1709 (*use*)

(Triphenylphosphoranylidene)ethenone, 9CI — T-00651

Triphenylphosphoranylideneketene
[15596-07-3]

$Ph_3P{=}C{=}C{=}O$

$C_{20}H_{15}OP$ M 302.312
Employed in synth. of steroids and heterocyclic compds. Needles (toluene). Mp 171-172°.

Daly, J.J. *et al*, *J. Chem. Soc. (A)*, 1966, 1703 (*cryst struct*)
Lumbroso, H. *et al*, *J. Organomet. Chem.*, 1978, **161**, 347 (*struct*)
Albright, T.A. *et al*, *Z. Naturforsch., B*, 1980, **35**, 343 (*struct*)
Bestmann, H.J. *et al*, *Chem. Ber.*, 1980, **113**, 274 (*synth, nmr, ir*)
Bestmann, H.J. *et al*, *Chem. Ber.*, 1980, **113**, 912 (*props*)
Nickisch, K. *et al*, *Chem. Ber.*, 1980, **113**, 2038, 3086 (*use*)
Bestmann, H.J. *et al*, *Houben-Weyls Method. Chem. Org.*, Band E1, 1982, 752 (*rev*)
Bestmann, H.J. *et al*, *Chem. Ber.*, 1985, **118**, 1709 (*use*)

N-[(Triphenylphosphoranylidene)-ethylidene]aniline — T-00652

N-[*(Triphenylphosphoranylidene)ethenylidene*]-*benzenamine*, 9CI. *Triphenyl(phenyliminovinylidene)-phosphorane*
[21385-80-8]

$Ph_3P{=}C{=}C{=}NPh$

$C_{26}H_{20}NP$ M 377.424
Wittig reagent employed in synth. of heterocyclic compds. and bicycloalkenones. Cryst. (EtOAc). Mp 151-152°. When melted, forms a dimer. Mp 200°.

Bestmann, H.J. *et al*, *Angew. Chem., Int. Ed. Engl.*, 1974, **13**, 273, 473; 1977, **16**, 349 (*synth, rev*)

Burzhoff, H. *et al*, *Chem. Ber.*, 1977, **110**, 3168 (*cryst struct*)
Albright, T.A. *et al*, *Z. Naturforsch.*, *B*, 1980, **35**, 343 (*struct*)
Bestmann, H.J. *et al*, *Chem. Ber.*, 1980, **113**, 3369 (*synth, ir, cmr, nmr*)
Bestmann, H.J. *et al*, *Chem. Ber.*, 1980, **113**, 3937 (*dimer, cryst struct*)
Bestmann, H.J. *et al*, *Tetrahedron Lett.*, 1982, **23**, 3543 (*use*)
Houben-Weyls Method. Org. Chem., Band E1, 1982, 759 (*rev*)
Bestmann, H.J. *et al*, *Chem. Ber.*, 1985, **118**, 1709 (*use*)

[(Triphenylphosphoranylidene)hydrazono]-acetic acid, 9CI **T-00653**

[(Carboxymethylene)hydrazono]triphenylphosphorane, 8CI

$Ph_3P{=}NN{=}CHCOOH$

$C_{20}H_{19}N_2O_2P$ M 350.356

Et ester: [22610-15-9]. *Ethyl [(triphenylphosphoranylidene)hydrazono]acetate. [(Ethoxycarbonylmethylene)hydrazono]-triphenylphosphorane.*
$C_{22}H_{21}N_2O_2P$ M 376.394
Cryst. (C_6H_6/Et_2O). Mp 113-114°.

Ph ester: Phenyl [(triphenylphosphoranylidene)-hydrazono]acetate. [(Phenoxycarbonylmethylene)-hydrazono]triphenylphosphorane.
$C_{26}H_{23}N_2O_2P$ M 426.454
Solid. Mp 129° dec.

Staudinger, H. *et al*, *Helv. Chim. Acta*, 1919, **2**, 619 (*synth*)
Zeeh, B. *et al*, *Org. Mass Spectrom.*, 1968, **1**, 791 (*ms*)
Albright, T.A. *et al*, *J. Org. Chem.*, 1976, **41**, 2716; 1977, **42**, 3691 (*cmr, P nmr, props, use*)
Bestmann, H.J. *et al*, *J. Organomet. Chem.*, 1980, **192**, 177 (*ir, pmr, cmr, P nmr*)

3-(Triphenylphosphoranylidene)-2,4-pentanedione, 9CI **T-00654**

(Diacetylmethylene)triphenylphosphorane

[1474-32-4]

$Ph_3P{=}C(COCH_3)_2$

$C_{23}H_{21}O_2P$ M 360.391

Stabilised ylide. Solid. Mp 167-169°. Pyrolysis → $H_3CC{\equiv}CCOCH_3$.

Chopard, P.A. *et al*, *J. Org. Chem.*, 1965, **30**, 1015 (*synth, ir*)
Zeliger, H.I. *et al*, *Tetrahedron Lett.*, 1970, 3313 (*pmr*)
Alexander, R.G. *et al*, *Org. Mass Spectrom.*, 1973, **7**, 963 (*ms*)
Cooke, M.P. *et al*, *J. Am. Chem. Soc.*, 1973, **95**, 7891 (*dianion*)
Brittain, J.M. *et al*, *Tetrahedron*, 1979, **35**, 1139 (*P nmr*)

5-(Triphenylphosphoranylidene)pentanoic acid, 9CI **T-00655**

[39968-97-3]

$Ph_3P{=}CH(CH_2)_3COOH$

$C_{23}H_{23}O_2P$ M 362.407

Wittig reagent prepd. *in situ*, and widely used in prostaglandin synth.

Et ester: [63129-92-0].
$C_{25}H_{27}O_2P$ M 390.461
Reactive Wittig reagent. Prepd. *in situ*.

House, H.O. *et al*, *J. Org. Chem.*, 1963, **28**, 90 (*ester, props*)
Corey, E.J. *et al*, *J. Am. Chem. Soc.*, 1969, **91**, 5675; 1970, **92**, 397 (*use*)
Bestmann, H.J. *et al*, *Justus Liebigs Ann. Chem.*, 1981, 1705 (*ester, use*)
Maryanoff, B.E. *et al*, *Tetrahedron Lett.*, 1981, **22**, 4185 (*props*)
Johnson, F. *et al*, *J. Am. Chem. Soc.*, 1982, **104**, 2190 (*use*)

(Triphenylphosphoranylidene)-phosphoramidic dichloride, 9CI **T-00656**

Triphenylphosphazophosphoryl dichloride. Triphenylphosphinimine dichlorophosphate

[19085-97-3]

$Ph_3P{=}NP(O)Cl_2$

$C_{18}H_{15}Cl_2NOP_2$ M 394.176

Cryst. (MeCN). Mp 186°.

Appel, R. *et al*, *Z. Anorg. Allg. Chem.*, 1963, **320**, 3 (*synth*)
Dagleish, W.H. *et al*, *J. Chem. Soc., Dalton Trans.*, 1977, 1505 (*nqr*)
Cameron, A.F. *et al*, *Acta Crystallogr., Sect. B*, 1979, **35**, 1373 (*cryst struct*)
Fluck, E. *et al*, *Z. Anorg. Allg. Chem.*, 1979, **458**, 103 (*synth*)
Zasorina, V.A. *et al*, *Zh. Obshch. Khim.*, 1982, **52**, 1081 (*Engl. transl.* p. 941) (*synth*)

2-(Triphenylphosphoranylidene)propanal, 9CI **T-00657**

2-(Triphenylphosphoranylidene)propionaldehyde, 8CI. (1-Formylethylidene)triphenylphosphorane

[24720-64-7]

$Ph_3P{=}C(CH_3)CHO$

$C_{21}H_{19}OP$ M 318.354

Stabilised ylide used in polyene synth. Cryst. (C_6H_6/pet. ether). Mp 220-222°. Forms complex contg. Fe. Exists largely in transoid form in chloroform.

Trippett, S. *et al*, *J. Chem. Soc.*, 1961, 1266 (*synth*)
Dale, A.J. *et al*, *Acta. Chem. Scand.*, 1970, **24**, 3772 (*ir, pmr*)
Wilson, I.F. *et al*, *J. Chem. Soc., Perkin Trans. 1*, 1972, 31 (*uv*)
Demole, E. *et al*, *Helv. Chim. Acta*, 1973, **56**, 2053 (*use*)
Le Corre, M., *Tetrahedron Lett.*, 1974, 1037 (*use*)

(Triphenylphosphoranylidene)propanedioic acid, 9CI **T-00658**

(Triphenylphosphoranylidene)malonic acid

$Ph_3P{=}C(COOH)_2$

$C_{21}H_{17}O_4P$ M 364.337

Di-Me ester: [19491-23-7]. *Dimethyl (triphenylphosphoranylidene)propanedioate, 9CI. Di(methoxycarbonyl)methylenetriphenylphosphorane.*
$C_{23}H_{21}O_4P$ M 392.390
Stabilized ylide. Cream prisms (EtOH) or yellow cryst. (cyclohexane). Mp 142-143°, 180-182°.

Di-Et ester: [7509-48-0]. *Diethyl (triphenylphosphoranylidene)propanedioate, 9CI. Di(ethoxycarbonyl)methylenetriphenylphosphorane.*
$C_{25}H_{25}O_4P$ M 420.444
Stabilized ylide. Yellow cryst. (cyclohexane, EtOAc/pet. ether or CH_2Cl_2/EtOAc). Mp 106-107°. pK_{a1} 2.78 (50% EtOH), pK_{a2} 12.3 ($MeNO_2$) (conj. phosphonium halide).

Horner, L. *et al*, *Justus Liebigs Ann. Chem.*, 1956, **627**, 142 (*synth, ir*)
Horner, L. *et al*, *Chem. Ber.*, 1958, **91**, 437 (*synth, uv*)
Cooks, R.G. *et al*, *Tetrahedron*, 1968, **24**, 3289 (*ms*)
Mastryukova, T.A. *et al*, *Phosphorus*, 1972, **1**, 159 (*props*)
Alexander, R.G. *et al*, *Org. Mass Spectrom.*, 1973, **7**, 963 (*ms*)
Wulfmann, D.S. *et al*, *J. Chem. Soc., Dalton Trans.*, 1975, 522 (*synth, ir, pmr*)

Bestmann, H.J. *et al*, *J. Chem. Res. (S)*, 1979, 313 (*synth*)
Gompper, R. *et al*, *Justus Liebigs Ann. Chem.*, 1979, 1388 (*use*)
Malenko, D.M. *et al*, *Zh. Obshch. Khim.*, 1979, **49**, 308 (*Engl. transl.* p. 267) (*synth*)

2-(Triphenylphosphoranylidene)-propanenitrile, 9CI **T-00659**

(1-Cyanoethylidene)triphenylphosphorane

[43055-47-6]

$Ph_3P{=}C(CH_3)CN$

$C_{21}H_{18}NP$ M 315.354

Reagent for Wittig reactions. Yellow cryst. (toluene). Mp 172-173°.

Bestmann, H.J. *et al*, *Justus Liebigs Ann. Chem.*, 1974, 1688 (*synth, use*)
Plieninger, H. *et al*, *Justus Liebigs Ann. Chem.*, 1976, 1475 (*use*)

2-(Triphenylphosphoranylidene)propanoic acid **T-00660**

$Ph_3P{=}C(CH_3)COOH$

$C_{21}H_{19}O_2P$ M 334.354

Me ester: see Methyl (2-triphenylphosphoranylidene)-propanoate, M-00434

Et ester: see Ethyl 2-triphenylphosphoranylidenepropanoate, E-00199

tert-*Butyl ester:* [56904-86-0]. tert-*Butyl 2-(triphenylphosphoranylidene)propanoate.*
$C_{25}H_{28}O_2P$ M 391.469
Solid. Mp 168-169°.

Ph ester: [41343-62-8]. *Phenyl 2-(triphenylphosphoranylidene)propanote.*
$C_{27}H_{23}O_2P$ M 410.451
No phys. props. reported.

Benzyl ester: [68613-50-3]. *Benzyl 2-(triphenylphosphoranylidene)propanoate.*
$C_{28}H_{25}O_2P$ M 424.478
Oil.

Le Corre, M., *C.R. Hebd. Seances Acad. Sci., Ser. C*, 1973, **276**, 936 (*synth*)
Stolter, P.L. *et al*, *Tetrahedron Lett.*, 1975, 1679 (*synth*)
Gossauer, A. *et al*, *Justus Liebigs Ann. Chem.*, 1977, 664 (*synth, pmr, use*)

3-(Triphenylphosphoranylidene)propanoic acid, 9CI **T-00661**

[63129-91-9]

$Ph_3P{=}CHCH_2COOH$

$C_{21}H_{19}O_2P$ M 334.354

The acid, and its esters, are widely used in prostaglandin synth. Reactive ylide, prepd. *in situ.*

Me ester: [40955-14-4].
$C_{22}H_{21}O_2P$ M 348.380
No phys. props. reported.

Et ester: [54356-04-6].
$C_{23}H_{23}O_2P$ M 362.407
No phys. props. reported.

Corey, H.S. *et al*, *J. Am. Chem. Soc.*, 1964, **86**, 1884 (*synth, use*)
Oda, R. *et al*, *Tetrahedron Lett.*, 1964, 1653 (*ester*)
Ingham, C.F. *et al*, *Aust. J. Chem.*, 1974, **27**, 1491 (*ester, use*)
Hashimoto, S. *et al*, *Tetrahedron Lett.*, 1980, **21**, 2857 (*ester*)

1-Triphenylphosphoranylidene-2-propanone, 9CI **T-00662**

Acetonylidenetriphenylphosphorane, 8CI. Acetylmethylidenetriphenylphosphorane. Triphenylphosphoranylideneacetone

[1439-36-7]

$Ph_3P{=}CHCOCH_3$

$C_{21}H_{19}OP$ M 318.354

In 10% EtOH at pH 10, $t_{1/2}$ = 5.3 hr. Wittig reagent. Cryst. (EtOAc or MeOH aq.). Mp 205-206°.

▷UC3900000.

Ramirez, F. *et al*, *J. Org. Chem.*, 1957, **22**, 41 (*synth, props, uv, iv*)
Bestmann, H.J. *et al*, *Chem. Ber.*, 1962, **95**, 1513 (*synth, props*)
Dale, A.J. *et al*, *Acta Chem. Scand.*, 1970, **24**, 3772 (*ir, struct*)
Gara, A.P. *et al*, *Aust. J. Chem.*, 1970, **23**, 307 (*ms*)
Wilson, I.F. *et al*, *J. Chem. Soc., Perkin Trans. 1*, 1972, 31 (*pmr, struct*)
Alexander, R.G. *et al*, *Org. Mass Spectrom.*, 1973, **7**, 963 (*ms*)
Cooke, M.P., *J. Org. Chem.*, 1973, **38**, 4082 (*deriv, use*)
Albright, T.A. *et al*, *J. Am. Chem. Soc.*, 1976, **98**, 6249 (*cmr, nmr*)
Chamberlin, K.S. *et al*, *Synth. Commun.*, 1978, **8**, 579 (*props, use*)
Doleschall, G. *et al*, *Synthesis*, 1981, 478 (*synth*)

3-(Triphenylphosphoranylidene)-2,5-pyrrolidenedione, 9CI **T-00663**

Triphenylphosphoranylidenesuccinimide

[28118-79-8]

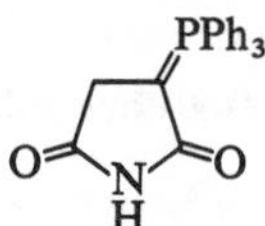

$C_{22}H_{18}NO_2P$ M 359.363

Promoter for polymerizations. Cryst. (Me_2CO). Mp 220°.

N-*Ph:*
$C_{28}H_{24}NO_2P$ M 437.477
Cryst. (Me_2CO). Mp 176.5-178.5°.

Hedaya, E. *et al*, *Tetrahedron*, 1968, **24**, 2241 (*synth, ir, pmr*)

2,2′,2″-Triphenylphosphoric trihydrazide **T-00664**

$(PhNHNH)_3P{=}O$

$C_{18}H_{21}N_6OP$ M 368.377

Cryst. (EtOH or dioxan). Mp 188°.

Audrieth, L.F. *et al*, *J. Am. Chem. Soc.*, 1942, **64**, 1553 (*synth*)
Tolkmith, H., *J. Am. Chem. Soc.*, 1962, **84**, 2097 (*synth*)
Cremlyn, R.J.W. *et al*, *J. Chem. Soc. (C)*, 1971, 300 (*synth*)

2,4,6-Triphenylphosphorin, 9CI **T-00665**

2,4,6-Triphenylphosphabenzene. 2,4,6-Triphenylphosphinine

[13497-36-4]

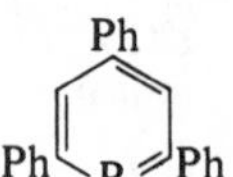

$C_{23}H_{17}P$ M 324.361

Pale-yellowish needles (EtOH). Mp 172-173°.

Märkl, G. *et al*, *Angew. Chem., Int. Ed. Engl.*, 1966, **5**, 846; 1967, **6**, 458, 944 (*synth, uv, pmr, P nmr*)
Onken, H. *et al*, *Naturwissenschaften*, 1967, **54**, 560 (*cryst struct*)
Deberitz, J. *et al*, *Chem. Ber.*, 1970, **103**, 2541 (*ir, pmr, complex*)
Nettche, A. *et al*, *Chem. Ber.*, 1973, **106**, 1001 (*props*)
Bundgaard, T. *et al*, *Tetrahedron Lett.*, 1974, 3179 (*cmr*)
Schafer, W. *et al*, *J. Am. Chem. Soc.*, 1976, **98**, 4410 (*pe, struct*)

1,2,6-Triphenyl-4-phosphorinanone, 9CI **T-00666**
1,2,6-Triphenyl-4-phosphinanone

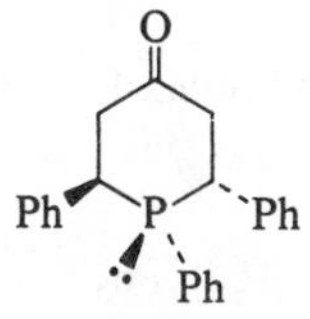

O, Ph, P, Ph, Ph — (1α,2α,6β)-*form*

$C_{23}H_{19}OP$ M 342.376
(1α,2α,6β)-*form* [76189-76-9]
Needles (MeCN). Mp 171-172°.
Oxide: [76156-74-6].
$C_{23}H_{19}O_2P$ M 358.376
Cryst. (EtOH). Mp 253-254°.
Sulfide: [76156-75-7].
$C_{23}H_{19}OPS$ M 374.436
Cryst. (C_6H_6/EtOH). Mp 240-242°.
(1α,2β,6β)-*form* [76189-77-0]
Cryst. (MeCN). Mp 181-182°.
Oxide: Solid. Mp 286-287°.
Sulfide: Solid. Mp 205-206°.

Welcher, R.P. *et al*, *J. Org. Chem.*, 1962, **27**, 1824 (*synth, derivs, props*)
Märkl, G. *et al*, *Tetrahedron Lett.*, 1970, 645 (*ir, pmr*)
Rampal, J.B. *et al*, *J. Am. Chem. Soc.*, 1981, **103**, 7602 (*synth, derivs, ir, cmr, P nmr, pmr, uv, cd*)
Rampal, J.B. *et al*, *J. Org. Chem.*, 1981, **46**, 1156 (*synth, derivs, cryst struct*)

***O,O,O*-Triphenyl phosphoroselenoate,** 9CI **T-00667**
O,O,O-*Triphenyl selenophosphate*
[7248-72-8]

$(PhO)_3P{=}Se$

$C_{18}H_{15}O_3PSe$ M 389.248
Needles (MeOH). Sol. alcohols, less sol. Et_2O, $CHCl_3$, C_6H_6. Mp 73-74°.

Strecker, W. *et al*, *Ber.*, 1916, **49**, 63 (*synth*)
Morgan, W.E. *et al*, *Inorg. Chem.*, 1971, **10**, 926 (*P nmr, pe*)

Triphenyl phosphorotetraselenoate **T-00668**
Triphenyl tetraselenophosphate

$(PhSe)_3P{=}Se$

$C_{18}H_{15}PSe_4$ M 578.130
Yellow cryst. (Me_2CO). Sol. C_6H_6, Et_2O. Mp 70-72°.

Zemlyanskii, N.I. *et al*, *Zh. Obshch. Khim.*, 1977, **47**, 62 (*Engl. transl.* p. 55)

Triphenyl phosphorotetrathioate, 9CI, 8CI **T-00669**
Triphenyl tetrathiophosphate
[3820-71-1]

$(PhS)_3P{=}S$

$C_{18}H_{15}PS_4$ M 390.530
Cryst. (EtOH). Mp 86-88°.

Maier, L. *et al*, *J. Am. Chem. Soc.*, 1962, **84**, 3054 (*synth, P nmr*)
Khokhlov, P.S. *et al*, *Zh. Obshch. Khim.*, 1984, **54**, 2545 (*Engl. transl.* p. 2274) (*synth*)

***O,O,O*-Triphenyl phosphorothioate** **T-00670**
Triphenyl thiophosphate
[597-82-0]

$(PhO)_3PS$

$C_{18}H_{15}O_3PS$ M 342.348
Cryst. (MeOH). Mp 53°. Bp_1 148-150°, $Bp_{0.1}$ 178-180°.

Fluck, E. *et al*, *Z. Anorg. Allg. Chem.*, 1967, **354**, 139 (*P nmr*)
Michalski, J. *et al*, *J. Chem. Soc. (C)*, 1970, 703 (*synth*)
Morgan, W.E. *et al*, *Inorg. Chem.*, 1971, **10**, 926 (*pe*)
Mazitova, F.N. *et al*, *Zh. Obshch. Khim.*, 1980, **50**, 815 (*Engl. transl.* p. 652) (*synth, P nmr*)

***O,O,S*-Triphenyl phosphorothioate,** 9CI, 8CI **T-00671**
O,O,S-*Triphenyl thiophosphate*
[70562-38-8]

$(PhO)_2P(O)SPh$

$C_{18}H_{15}O_3PS$ M 342.348
Liq. $Bp_{0.02}$ 106-110°.

Torii, S. *et al*, *J. Org. Chem.*, 1979, **44**, 2938 (*synth, ir, pmr*)

***Se,Se,Se*-Triphenyl phosphorotriselenoate,** 9CI **T-00672**
[20459-30-7]

$(PhSe)_3P{=}O$

$C_{18}H_{15}OPSe_3$ M 515.170
Yellowish cryst. (pet. ether or C_6H_6/hexane). Mp 110° dec.

Petragnani, N. *et al*, *Chem. Ber.*, 1968, **101**, 3070 (*synth*)
Maier, L., *Helv. Chim. Acta*, 1976, **59**, 252 (*synth*)
Zemlyanskii, N.I. *et al*, *Zh. Obshch. Khim.*, 1977, **47**, 62 (*Engl. transl.* p. 55) (*synth*)

Triphenyl phosphorotriselenoite, 9CI, 8CI **T-00673**
Triphenyl triselenophosphite
[58558-73-9]

$(PhSe)_3P$

$C_{18}H_{15}PSe_3$ M 499.170
Yellow cryst. (Me_2CO). Spar. sol. CCl_4, CS_2, cyclohexane. Mp 70-72°, Mp 91-96°, Mp 150° dec.

Drake, J.E. *et al*, *J. Chem. Soc., Dalton Trans.*, 1976, 1730 (*synth, raman*)
Maier, L., *Helv. Chim. Acta*, 1976, **59**, 252 (*synth, pmr, props*)
Zemlyanskii, N.I., *Zh. Obshch. Khim.*, 1977, **47**, 62 (*Engl. transl.* p. 55) (*synth*)

Se,Se,Se-Triphenyl phosphorotriselenothioate T-00674

[58558-74-0]

$(PhSe)_3P{=}S$

$C_{18}H_{15}PSSe_3$ M 531.230

Yellow cryst. (EtOH at −20°). Mp 55-58°.

Maier, L., *Helv. Chim. Acta*, 1976, **59**, 252 (*synth*)

S,S,S-Triphenyl phosphorotrithioate, 9CI, 8CI T-00675

S,S,S-*Triphenyl trithiophosphate*

[597-83-1]

$(PhS)_3P{=}O$

$C_{18}H_{15}OPS_3$ M 374.470

Cryst. (Et_2O or cyclohexane). Mp 115-116°.

Michaelis, A. *et al*, *Ber.*, 1907, **40**, 3419 (*synth*)
Buckler, S.A. *et al*, *J. Org. Chem.*, 1962, **27**, 794 (*synth*)
Ibrahim, E.H. *et al*, *Egypt. J. Chem.*, 1979, **22**, 403 (*synth, ir*)

Triphenyl phosphorotrithioite, 9CI T-00676

Triphenyl trithiophosphite

[1095-04-1]

$(PhS)_3P$

$C_{18}H_{15}PS_3$ M 358.470

Cryst. (Et_2O, EtOH or cyclohexane). Mp 76-77°. Does not form an adduct with MeI.

Michaelis, A. *et al*, *Ber.*, 1907, **40**, 3419 (*synth, props*)
Razumov, A.I. *et al*, *Zh. Obshch. Khim.*, 1972, **42**, 1250 (*Engl. transl.* p. 1245) (*P nmr*)
Shaw, R.A. *et al*, *Phosphorus*, 1972, **1**, 191.

N,N′,N″-Triphenylphosphorous triamide, 9CI T-00677

Phosphorous trianilide. Tris(phenylamino)phosphine. Trianilinophosphine

[15159-51-0]

$(PhNH)_3P$

$C_{18}H_{18}N_3P$ M 307.334

Prisms (Et_2O). Mp 95-96°. Stable in the solid phase, but in soln. liberates $PhNH_2$. When heated, forms 1,3-diphenyl-2,4-bis(phenylamino)-1,3,2,4-diazadiphosphetidine.

Oxide: [5326-10-3]. N,N′,N″-*Triphenylphosphoric triamide. Phosphoric trianilide. Tris(phenylamino)-phosphine oxide.*

$C_{18}H_{18}N_3OP$ M 323.333

Insect sterilant. Prisms (EtOH aq.). Insol. CCl_4, sl. sol. $CHCl_3$. Mp 213-215°. pK_a 5.13 ($MeNO_2$). Sl. dec. at 250°.

Sulfide: [4743-38-8]. N,N′,N″-*Triphenylphosphorothioic triamide. Phosphorothioic trianilide. Tris(phenylamino)phosphine sulfide.*

$C_{18}H_{19}N_3PS$ M 340.402

Cryst. (EtOH). Mp 153-154°.

Buck, A.C. *et al*, *J. Am. Chem. Soc.*, 1948, **70**, 2398 (*oxide, sulfide*)
Trishin, Yu.G. *et al*, *Zh. Org. Khim.*, 1975, **11**, 1752 (*Engl. transl.* p. 1750) (*synth, oxide, sulfide*)
Ibanez, W. *et al*, *Spectrochim. Acta, Part A*, 1982, **38**, 351 (*derivs, ir*)
Tarassoli, A. *et al*, *Inorg. Chem.*, 1982, **21**, 2684 (*synth, props, ir, pmr, nmr, cryst struct*)
Thompson, M.L. *et al*, *Inorg. Chem.*, 1982, **21**, 1287 (*props*)

Triphenylpropadienylidenephosphorane T-00678

Triphenylpropargylidenephosphorane. Triphenylphosphoranylideneallene

$Ph_3P{=}C{=}C{=}CH_2$

$C_{21}H_{17}P$ M 300.339

Prepd. from Triphenyl-2-propynylphosphonium(1+), T-00681 . Used in synth. of enynes.

Corey, E.J. *et al*, *Tetrahedron Lett.*, 1973, 1495 (*synth*)

Triphenyl-2-propenylphosphonium(1+), 9CI T-00679

Allyltriphenylphosphonium(1+), 8CI

[15912-76-2]

$H_2C{=}CHCH_2P^{\oplus}Ph_3$

$C_{21}H_{20}P^{\oplus}$ M 303.363 (ion)

PhLi or $NaNH_2$→ ylide.

Chloride: [18480-23-4].

$C_{21}H_{20}ClP$ M 338.816

Cryst. Mp 234° (229-231°).

Bromide: [1560-54-9].

$C_{21}H_{20}BrP$ M 383.267

Cryst. (CH_2Cl_2/AcOH). Mp 222-225°.

▷TA1843000.

Ylide: [15935-94-1]. *Triphenyl-2-propenylidenephosphorane.*

$C_{21}H_{19}P$ M 302.355

Used in Wittig reactions. Red.

Keough, P. *et al*, *J. Org. Chem.*, 1964, **29**, 631 (*synth, props*)
Wood, G.W., *J. Org. Chem.*, 1975, **40**, 636 (*ms*)
Albright, T.A. *et al*, *J. Am. Chem. Soc.*, 1976, **98**, 6249 (*cmr, nmr*)
Nesmeyanov, N.A. *et al*, *J. Organomet. Chem.*, 1977, **129**, 41 (*tautom, pmr, nmr*)
Padwa, A. *et al*, *J. Org. Chem.*, 1980, **45**, 4555 (*use*)
Schlosser, M. *et al*, *Chimia*, 1982, **36**, 396 (*ylide*)

Triphenylpropylphosphonium(1+), 9CI T-00680

Propyltriphenylphosphonium(1+)

[15912-75-1]

$Ph_3P^{\oplus}CH_2CH_2CH_3$

$C_{21}H_{22}P^{\oplus}$ M 305.379 (ion)

Treatment with $NaNH_2$ or butyllithium gives the ylide.

Chloride: [16721-43-0].

$C_{21}H_{22}ClP$ M 340.832

Solid. Mp 221-223° dec.

Bromide: [6228-47-3].

$C_{21}H_{22}BrP$ M 385.283

Cryst. (CH_2Cl_2/AcOH or CH_2Cl_2/EtOAc). Mp 238-240°.

Iodide: [14350-50-6].

$C_{21}H_{22}IP$ M 432.283

Solid. Mp 203-204° dec.

Ylide: [16666-78-7].

$C_{21}H_{21}P$ M 304.371

Wittig reagent.

Bergelson, L.D. *et al*, *Tetrahedron*, 1967, **23**, 2709 (*ylide, props*)
Grim, S.O. *et al*, *J. Chem. Soc., Chem. Commun.*, 1967, 1191 (*nmr*)

Schlosser, M. *et al*, *Justus Liebigs Ann. Chem.*, 1967, **708**, 1 (*synth*)
Senyavina, L.B. *et al*, *Zh. Obshch. Khim.*, 1967, **37**, 499 (*Engl. transl.* p. 469) (*synth, ir, pmr*)
Grim, S.O. *et al*, *J. Org. Chem.*, 1968, **33**, 2993 (*ylide, uv*)
Wood, W.G. *et al*, *J. Org. Chem.*, 1975, **40**, 636 (*ms*)
Doleschall, G., *Synthesis*, 1981, 478 (*iodide*)
Le Bigot, Y. *et al*, *Synth. Commun.*, 1982, **12**, 107 (*ylide, use*)
Schlosser, M. *et al*, *Chimia*, 1982, **36**, 396 (*ylide*)

Triphenyl-2-propynylphosphonium(1+), 9CI T-00681

Propargyltriphenylphosphonium

$Ph_3P^{\oplus}CH_2C{\equiv}CH$

$C_{21}H_{18}P^{\oplus}$ M 301.347 (ion)

Reagent used in synth. of heterocyclic systems, incl. 2-Me-quinolines, and for prep. of enzymes. In neutral or basic soln. salts isomerize to triphenyl(1,2-propadienyl)phosphonium. With NH_3 → ylide. Source of Triphenylpropadienylidenephosphorane, T-00678 .

Bromide: [2091-46-5].
$C_{21}H_{18}BrP$ M 381.251
Cryst. (EtOH or 2-propanol). Mp 179°.

Ylide: [72184-68-0]. *Triphenyl(propynylidene)-phosphorane.*
$C_{21}H_{17}P$ M 300.339
Used in Wittig reactions to prepare conj. enzymes.

Appleyard, G.D. *et al*, *J. Chem. Soc.* (*C*), 1969, 1904 (*synth, ir*)
Albright, T.A. *et al*, *J. Am. Chem. Soc.*, 1975, **97**, 2946 (*nmr*)
Schweizer, E.E. *et al*, *J. Org. Chem.*, 1977, **42**, 200 (*synth, ir, nmr, use*)
Schweizer, E.E. *et al*, *J. Org. Chem.*, 1982, **47**, 1652 (*use*)

Triphenyl(tetradecyl)phosphonium(1+), 9CI, 8CI T-00682

$Ph_3P^{\oplus}(CH_2)_{13}CH_3$

$C_{32}H_{44}P^{\oplus}$ M 459.673 (ion)

Bromide: [25791-20-2].
$C_{32}H_{44}BrP$ M 539.577
Ylide source with RLi. Cryst. (Me_2CO/Et_2O). Mp 94-96°.

Ylide: Triphenyltetradecylidenephosphorane.
$C_{32}H_{43}P$ M 458.665
Used in Wittig reactions leading to fatty acid esters.

Reist, E.J. *et al*, *J. Org. Chem.*, 1970, **35**, 3521 (*use, synth*)
Chasin, D.G. *et al*, *Chem. Phys. Lipids*, 1971, **6**, 8 (*synth, use*)

Triphenyl[[3-(tetrahydro-2*H*-pyran-2-yl)-oxy]propyl]phosphonium(1+) T-00683

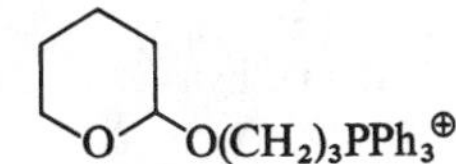

$C_{26}H_{30}O_2P^{\oplus}$ M 405.496 (ion)

Reagent for conversion of RCH=O → RCH=CHCH$_2$CH$_2$OH.

Bromide: [70665-02-0].
$C_{26}H_{30}BrO_2P$ M 485.400
Amorph. solid. Mp 192-195°.

Iodide: [52103-13-6].
$C_{26}H_{30}IO_2P$ M 532.400
Solid. Mp 162-164°.

Ylide: [71436-82-3]. *Triphenyl[3-[(tetrahydro-2*H*-pyran-2-yl)oxy]propylidene]phosphorane.*
$C_{26}H_{29}O_2P$ M 404.488
Prepd. *in situ*. Wittig reagent containing protected OH gp.

Hodgson, G.L. *et al*, *J. Chem. Soc., Perkin Trans. 1*, 1973, 2113 (*iodide, ylide*)
Schow, S.R. *et al*, *J. Org. Chem.*, 1979, **44**, 3760 (*bromide, ylide, use*)
Heath, R.R. *et al*, *J. Org. Chem.*, 1980, **45**, 2910 (*iodide, ylide*)
Iwamoto, M. *et al*, *Agric. Biol. Chem.*, 1983, **47**, 117 (*ylide, use*)

Triphenyl(tetrahydro-2*H*-pyran-4-yl)-phosphonium(1+), 9CI T-00684

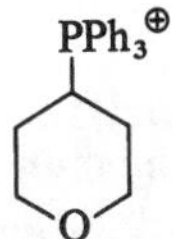

$C_{23}H_{24}OP^{\oplus}$ M 347.416 (ion)

Bromide: [22836-03-9].
$C_{23}H_{24}BrOP$ M 427.320
Source of ylide. Cryst. (H_2O). Mp 262-264°.

*Ylide: 4-Triphenylphosphoranylidene(tetrahydro-2*H*-pyran).*
$C_{23}H_{23}OP$ M 346.408

Bestmann, H.J. *et al*, *Chem. Ber.*, 1976, **109**, 1694 (*synth, use*)

Triphenyl(tetrahydro-2*H*-thiopyran-4-yl)-phosphonium(1+), 8CI T-00685

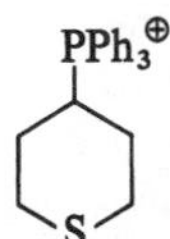

$C_{23}H_{24}PS^{\oplus}$ M 363.476 (ion)

Rgt. for Wittig reactions. With butyllithium → ylide.

Bromide: [22836-04-0].
$C_{23}H_{24}BrPS$ M 443.380
Source of ylide. Cryst. (H_2O). Mp 293-296°.

*Ylide: 4-Triphenylphosphoranylidene(tetrahydro-2*H*-thiopyran).*
$C_{23}H_{23}PS$ M 362.468

Bestmann, H.J. *et al*, *Chem. Ber.*, 1969, **102**, 1802 (*synth, use*)

2,4,6-Triphenyl-4*H*-1,4-thiaphosphorin, 9CI T-00686

*2,4,6-Triphenyl-4*H*-1,4-thiaphosphinin*

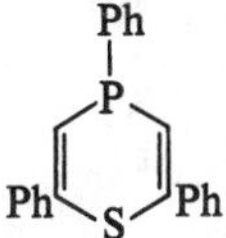

$C_{22}H_{17}PS$ M 344.410

4-Oxide: [50694-63-8].
$C_{22}H_{17}OPS$ M 360.409
Solid. Mp 182-183°.

1,1,4-Trioxide: [50694-65-0].
$C_{22}H_{17}O_3PS$ M 392.408
Solid. Mp 206-209°.

Chattha, M.S. *et al*, *Phosphorus*, 1973, **3**, 65 (*oxides, synth, ir, pmr*)

2,3,4-Triphenyl-1,3,2,4-thiazadiphospheti-dine, 8CI T-00687

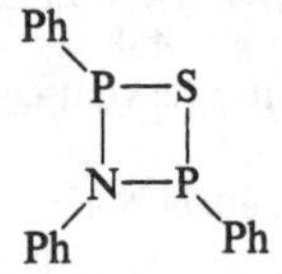

$C_{18}H_{15}NP_2S$ M 339.331

2,4-Disulfide: [15435-17-3].

$C_{18}H_{15}NP_2S_3$ M 403.451

Cryst. (C_6H_6). Mp 183°. Probably possesses the 2,4-*trans* structure.

Fluck, E. *et al*, *Z. Anorg. Allgem. Chem.*, 1967, **354**, 113 (*synth, P nmr*)

Triphenyl[(2,6,6-trimethyl-1,3-cyclohexa-dien-1-yl)methyl]phosphonium(1+), 8CI T-00688

2,6,6-Trimethyl-1-(triphenylphosphoniomethyl)-1,3-cyclohexadiene(1+)

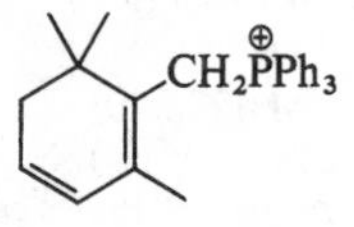

$C_{28}H_{30}P^{\oplus}$ M 397.519 (ion)

Bromide: [23069-03-6].

$C_{28}H_{30}BrP$ M 477.423

Source of ylide. Cryst. (Me_2CO/Et_2O). Mp 124-125°.

Ylide: Triphenyl[(2,6,6-trimethyl-1,3-cyclohexadien-1-yl)methylene]phosphorane.

$C_{28}H_{29}P$ M 396.511

Used in Wittig reactions.

Surmatis, J.D. *et al*, *J. Org. Chem.*, 1970, **35**, 1053.

Triphenyl[(2,6,6-trimethyl-1-cyclohexen-1-yl)methyl]phosphonium(1+), 9CI T-00689

β-Cyclogeranyltriphenylphosphonium(1+)

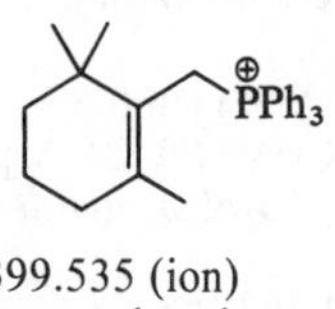

$C_{28}H_{32}P^{\oplus}$ M 399.535 (ion)

Employed in Wittig reactions in synth. of naturally occurring polyenes.

Chloride: [73410-08-9].

$C_{28}H_{32}ClP$ M 434.988

No phys. props. reported.

Bromide: [56013-01-5].

$C_{28}H_{32}BrP$ M 479.439

No phys. props. reported.

Gedye, R.N. *et al*, *Can. J. Chem.*, 1975, **53**, 1943 (*use*)
Ramamurthy, V. *et al*, *Tetrahedron*, 1975, **31**, 193 (*use*)
Motto, M.G. *et al*, *J. Am. Chem. Soc.*, 1980, **102**, 7947 (*synth, use*)

Triphenyl(3,7,11-trimethyldodecyl)-phosphonium(1+), 9CI T-00690

(Hexahydrofarnesyl)triphenylphosphonium(1+)

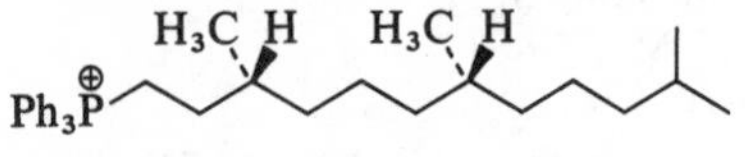

(3*R*,7*R*)-*form*

$C_{33}H_{46}P^{\oplus}$ M 473.700 (ion)

BuLi on salts yields the ylide.

(3R,7R)-form

Bromide: [60919-78-0].

$C_{33}H_{46}BrP$ M 553.604

Hygroscopic cryst. (Me_2CO/Et_2O). Mp 92-93.5°.

Ylide: Triphenyl(3,7,11-trimethyldodecylidene)-phosphorane.

$C_{33}H_{45}P$ M 472.692

Used in polyene synth.

(3RS,7RS)-form

Bromide: Cryst. (Me_2CO/Et_2O). Mp 81-83°.

Mayer, H. *et al*, *Helv. Chim. Acta*, 1963, **46**, 650 (*synth, use*)
Cohen, N. *et al*, *J. Am. Chem. Soc.*, 1979, **101**, 6710 (*use*)

Triphenyl[4-(trimethylsilyl)-3-butynyl]-phosphonium(1+), 9CI T-00691

$Ph_3P^{\oplus}CH_2CH_2C{\equiv}CSiMe_3$

$C_{25}H_{28}PSi^{\oplus}$ M 387.555 (ion)

Used in synth. of polyacetylenic compds.

Iodide: [41345-58-8].

$C_{25}H_{28}IPSi$ M 514.460

Source of ylide. Cryst. (EtOAc/THF). Mp 135.5-136.5°. NaH → ylide.

Ylide: [4-(Trimethylsilyl)-3-butyn-1-ylidene]-triphenylphosphorane.

$C_{25}H_{27}PSi$ M 386.548

Used in Wittig reactions.

Fallis, A.G. *et al*, *J. Chem. Soc., Perkin Trans. 1*, 1973, 743 (*synth, use*)

Triphenyl[(trimethylsilyl)methylene]-phosphorane, 9CI T-00692

[(Trimethylsilyl)methylene]triphenylphosphorene

[3739-97-7]

$Ph_3P{=}CHSiMe_3$

$C_{22}H_{25}PSi$ M 348.499

Wittig reagent; stable ylide. Solid. Mp 76-77°. Bp_1 150-153°.

Schmidbaur, H. *et al*, *Chem. Ber.*, 1967, **100**, 1032 (*ir, pmr, synth*)
Schmidbaur, H. *et al*, *Angew. Chem., Int. Ed. Engl.*, 1973, **12**, 321.
Fieser, M. *et al*, *Reagents for Organic Synthesis*, Wiley, 1967-84, 1975, **5**, 723.
Ostoja, S. *et al*, *J. Am. Chem. Soc.*, 1976, **98**, 8486 (*pe*)
Bestmann, H.J. *et al*, *Angew. Chem., Int. Ed. Engl.*, 1982, **21**, 542 (*use*)

Triphenyl[(trimethylsilyl)methyl]-phosphonium(1+), 10CI T-00693

$Ph_3P^{\oplus}CH_2SiMe_3$

$C_{22}H_{26}PSi^{\oplus}$ M 349.507 (ion)

Salts are source of [(trimethylsilyl)methylene]-triphenylphosphorane.

Chloride:

$C_{22}H_{26}ClPSi$ M 384.960

Solid. Mp 170-180° dec.

Iodide: [3739-98-8].

$C_{22}H_{26}IPSi$ M 476.411

Cryst. (EtOH/EtOAc). Mp 168-169°.

Tetrafluoroborate:

$C_{22}H_{26}BF_4PSi$ M 436.310

Solid. Mp 193-197°.

Schmidbaur, H. *et al*, *Chem. Ber.*, 1967, **100**, 1039 (*synth*)
Reith, B.A. *et al*, *J. Org. Chem.*, 1974, **39**, 2728 (*iodide, ir, ylide*)
Singh, G. *et al*, *J. Org. Chem.*, 1979, **44**, 1057 (*iodide, pmr, cmr, nmr*)

2,4,6-Triphenyl-1,3,5,2,4,6-trioxatriphosphorinane, 9CI T-00694

2,4,6-Triphenyl-1,3,5,2,4,6-trioxatriphosphinane

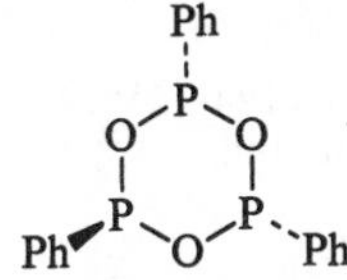

(2α,4α,6β)-*form*

$C_{18}H_{15}O_3P_3$ M 372.236

2,4,6-Trioxide: [57156-84-0]. *Phenylphosphonic acid trimer trianhydride.*
$C_{18}H_{15}O_6P_3$ M 420.234
Solid. Mp 209-212°.

(2α,4α,6β)-form

2,4,6-Trisulfide: [51371-50-7]. *Phenylphosphonothioic acid trimer trianhvdride.*
$C_{18}H_{15}O_3P_3S_3$ M 468.416
Solid. Mp 139-143°.

Cherbuliez, E. *et al*, *Helv. Chim. Acta*, 1961, **44**, 1812 (*trioxide, synth*)
Daly, J.J. *et al*, *Helv. Chim. Acta*, 1972, **53**, 1991 (*trisulfide, P nmr, pmr, ir, cryst struct*)
Ecker, A. *et al*, *Monatsh. Chem.*, 1972, **103**, 736 (*trisulfide, synth, ir, pmr*)
Daly, J.J. *et al*, *J. Chem. Soc., Dalton Trans.*, 1973, 2032 (*trisulfide, cryst struct*)
Grishina, O.N. *et al*, *Zh. Obshch. Khim.*, 1975, **45**, 2344 (*Engl. transl.* p. 2303) (*synth, P nmr*)
Maier, L., *Phosphorus*, 1975, **5**, 253 (*trisulfide, synth, ir, pmr, P nmr, cryst struct*)
Keck, H. *et al*, *Phosphorus Sulfur*, 1978, **4**, 173 (*trisulfide, ms*)

Triphenyl(triphenylgermyl)phosphinimine T-00695

1,1,1-Triphenyl-N-triphenylphosphoranylidenegermanamine, 10CI

[68669-95-4]

$Ph_3GeN{=}PPh_3$

$C_{36}H_{30}GeNP$ M 580.203
Air-stable cryst. Mp 97-98°.

Glidewell, C., *J. Organomet. Chem.*, 1978, **159**, 23 (*struct*)
Bajpai, K. *et al*, *Synth. React. Inorg. Metal-Org. Chem.*, 1982, **12**, 47 (*synth, ir*)

Triphenyl[[2-[2-(triphenylphosphonio)ethyl]phenyl]methyl]phosphonium(2+), 9CI T-00696

1-[2-(Triphenylphosphonio)ethyl]-2-(triphenylphosphoniomethyl)benzene(2+)

$C_6H_4(CH_2CH_2PPh_3^{\oplus})(CH_2PPh_3^{\oplus})$

$C_{45}H_{40}P_2^{\oplus\oplus}$ M 642.759 (ion)

Dibromide: [10038-37-6].
$C_{45}H_{40}Br_2P_2$ M 802.567
Used in prepn. of benzannelated annulenes. Mp 148°. LiOEt affords bis-ylide *in situ.*

Diperiodate:
$C_{45}H_{40}I_2O_8P_2$ M 1024.563
Solid. Mp 145-147°. Gives indene with $NaNH_2/Fe(NO_3)_3$.

Bisylide: [*2-[2-(Triphenylphosphoranylidenemethyl)-phenyl]ethylidenetriphenylphosphorane.*
$C_{45}H_{38}P_2$ M 640.743

Bestmann, H.J. *et al*, *Chem. Ber.*, 1969, **102**, 2259 (*periodate, synth, use*)
Rabinowitz, M. *et al*, *Tetrahedron*, 1979, **35**, 667 (*bromide, synth, use*)

Triphenyl[(triphenylphosphoranylidene)methyl]phosphonium(1+) T-00697

$Ph_3P{=}CHPPh_3^{\oplus}$

$C_{37}H_{31}P_2^{\oplus}$ M 537.599 (ion)
Source of Bis(triphenylphosphoranylidene)methane, B-00459 .

Bromide:
$C_{37}H_{31}BrP_2$ M 617.503
Cryst. (CH_2Cl_2/hexane). Mp 272-274°.

Chloride: [58513-98-7].
$C_{37}H_{31}ClP_2$ M 573.052
Cryst. (CH_2Cl_2/Et_2O). Mp 263-265°, Mp 274-276°.

Iodide:
$C_{37}H_{31}IP_2$ M 664.504
Photochromic (becoming yellow).

Tetraphenylborate:
$C_{61}H_{51}BP_2$ M 856.831
Photochromic (becoming orange).

Ramirez, F. *et al*, *J. Am. Chem. Soc.*, 1962, **84**, 1745 (*bromide, synth*)
Matthews, C.N. *et al*, *J. Am. Chem. Soc.*, 1962, **84**, 4349 (*photochromism*)
Driscoll, J.S. *et al*, *J. Org. Chem.*, 1964, **29**, 2423 (*chloride, pmr, nmr*)
Grisley, D.W. *et al*, *Tetrahedron*, 1965, **21**, 5 (*bromide, synth*)
Dillon, K.B. *et al*, *Z. Anorg. Allg. Chem.*, 1982, **488**, 7 (*chloride, cryst struct*)

Triphenyltriphosphirane, 9CI T-00698

Triphenylcyclotriphosphine

[40633-23-6]

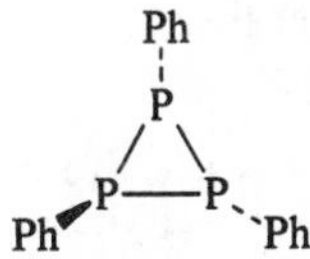

$C_{18}H_{15}P_3$ M 324.238

(1α,2α,3β)-form [64599-80-4]

This isomer predominates in soln. Concn. decreases continuously with the simultaneous formn. of oligomeric polyphosphines.

Baudler, M. *et al*, *Z. Anorg. Allg. Chem.*, 1977, **432**, 67 (*synth, P nmr*)

1,2,3-Triphenyl-1,2,3-triphospholane, 9CI T-00699

Cyclo-4,5-carba-1,2,3-triphenyl-1,2,3-triphosphine

[40012-64-4]

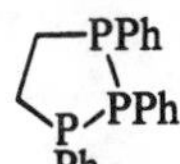

Cryst. (MeCN). Mp 54-56°.

Baudler, M. *et al*, *Z. Naturforsch., B*, 1972, **27**, 1007 (*synth, ir, P nmr, pmr, ms*)
Issleib, K. *et al*, *Z. Anorg. Allg. Chem.*, 1974, **406**, 178 (*synth*)

Triphenyl(2,5-undecadienylidene)-phosphorane, 9CI — T-00700

$Ph_3P{=}CHCH{=}CHCH_2CH{=}CH(CH_2)_4CH_3$

$C_{29}H_{33}P$ M 412.553

(Z,Z)-form [73958-02-8]
Generated from phosphonium mesylate by $LiN[CH(CH_3)_2]_2$ in hexamethylphosphoramide. Used in synth. of leukotrienes.

Corey, E.J. *et al*, *J. Am. Chem. Soc.*, 1980, **102**, 1436 (*synth, use*)
Baker, S.R. *et al*, *Tetrahedron Lett.*, 1980, **21**, 4123 (*synth, use*)

Triphenylundecylphosphonium(1+), 9CI — T-00701

$Ph_3P^{\oplus}(CH_2)_{10}CH_3$

$C_{29}H_{38}P^{\oplus}$ M 417.593 (ion)

Bromide: [60669-22-9].
$C_{29}H_{38}BrP$ M 497.497
Source of ylide, obt. with RLi. Cryst. (C_6H_6/hexane). Mp 94-95° (sealed tube).

Ylide: Triphenylundecylidenephosphorane.
$C_{29}H_{38}P$ M 417.593
Used in Wittig reactions for synth. of fatty acid esters, incl. insect pheromones. Red-orange.

Starratt, A.N., *Chem. Phys. Lipids*, 1976, **16**, 215 (*synth, use*)
Hoshino, C. *et al*, *Agric. Biol. Chem.*, 1980, **44**, 3007 (*use*)

Triphenylvinylphosphonium(1+) — T-00702

Ethenyltriphenylphosphonium(1+), 9CI. Vinyltriphenylphosphonium(1+). Schweizer's reagent
[38066-33-0]

$Ph_3P^{\oplus}CH{=}CH_2$

$C_{20}H_{18}P^{\oplus}$ M 289.336 (ion)
Used in heterocycle synth.

Bromide: [5044-52-0].
$C_{20}H_{18}BrP$ M 369.240
Reagent for the stereoselective synth. of (*E*)-allylamines and of (*E*)-acrylic acids from aldehydes. Employed in steroid synth., synth. of thiophenes, and in Diels-Alder reactions. Cryst. (EtOAc). Mp 186-190°.
▷Produces an allergic sneezing reaction

Schweizer, E.E. *et al*, *J. Org. Chem.*, 1964, **29**, 1746 (*synth, pmr*)
Swan, J.M. *et al*, *Aust. J. Chem.*, 1971, **24**, 777 (*synth, props*)
Wood, G.W. *et al*, *J. Org. Chem.*, 1975, **40**, 636 (*ms*)
Albright, T.A. *et al*, *J. Org. Chem.*, 1975, **40**, 3437 (*cmr, nmr*)
Kampmeier, J.A. *et al*, *J. Am. Chem. Soc.*, 1981, **103**, 1478 (*pmr, ir*)
Meyers, A.I. *et al*, *J. Org. Chem.*, 1981, **46**, 3119 (*synth, pmr, use*)
Posner, G.A. *et al*, *Tetrahedron*, 1981, **37**, 3921 (*use*)
Fieser, M. *et al*, *Reagents for Organic Synthesis*, Wiley, 1967-84, **6**, 666.

Triphosphetan-4-one — T-00703

O═C—PH
| |
HP—PH

CH_3OP_3 M 123.955

*1,3-Di-*tert*-butyl, 2-Me:*
$C_{10}H_{21}OP_3$ M 250.197
Liq. $Bp_{0.1}$ 59-60°.

*1,2,3-Tri-*tert*-butyl:*
$C_{13}H_{27}OP_3$ M 292.277
Cryst. (MeCN/toluene). Mp 124-125°. Uv irradiation gives tri-*tert*-butyltriphosphirane.

*1,3-Di-*tert*-butyl, 2-Ph:*
$C_{15}H_{23}OP_3$ M 312.267
Solid. Mp 128-130°.

Appel, R. *et al*, *Chem. Ber.*, 1983, **116**, 2371 (*synth, props, ir, pmr, P nmr*)

Tri-1-piperidinophosphine — T-00704

1,1',1''-Phosphinidynetrispiperidine, 9CI. Tris(piperidino)phosphine
[13954-38-6]

$C_{15}H_{30}N_3P$ M 283.396
Cryst. (Et_2O at −75°). Mp 37°. $Bp_{0.2}$ 120-130°.
▷SZ3600000.

B,MeI: Methyltris(piperidino)phosphonium iodide.
$C_{16}H_{33}IN_3P$ M 425.335
Cryst. (MeCN or MeCN/Et_2O). Mp 254°.

Oxide: [4441-17-2]. *1,1',1''-Phosphinylidynetrispiperidine. Tris(piperidino)phosphine oxide.*
$C_{15}H_{30}N_3OP$ M 299.395
Solid. Mp 75-76°. pK_{a1} 10.94, pK_{a2} 7.61 ($MeNO_2$).
▷SZ1678000.

Sulfide: [6789-42-4]. *1,1',1''-Phosphinothioylidynetrispiperidine. Tris(piperidino)phosphine sulfide.*
$C_{15}H_{30}N_3PS$ M 315.456
Solid. Mp 121-122°.

Selenide: [68541-88-8]. *1,1',1''-Phosphinoselenoylidynetrispiperidine. Tris(piperidino)phosphine selenide.*
$C_{15}H_{30}N_3PSe$ M 362.356
No phys. props. reported.
▷SZ2460000.

Telluride: [75404-55-6]. *1,1',1''-Phosphinotelluroylidynetrispiperidine. Tris(piperidino)phosphine telluride.*
$C_{15}H_{30}N_3PTe$ M 410.996
Solid. Mp 94-95° dec.

Audrieth, L.F. *et al*, *J. Am. Chem. Soc.*, 1942, **64**, 1553 (*sulfide*)
Burgada, R., *Ann. Chim. (Paris)*, 1963, **8**, 374 (*deriv*)
Thorstenson, T. *et al*, *Acta Chem. Scand., Ser. A*, 1976, **30**, 781 (*synth, deriv*)
Rømming, C. *et al*, *Acta Chem. Scand., Ser. A*, 1978, **32**, 689 (*cryst struct*)
Rømming, C. *et al*, *Acta Chem. Scand., Ser. A*, 1979, **33**, 187 (*selenide, cryst struct*)

Rømming, C. *et al*, *Acta Chem. Scand., Ser. A*, 1980, **34**, 333 (*selenide, telluride, ir, uv*)
Bergesen, K. *et al*, *Acta Chem. Scand., Ser. A*, 1981, **35**, 147 (*cmr*)
Gonbeau, D. *et al*, *Inorg. Chem.*, 1981, **20**, 1966 (*pe*)

Tri-2-propenyl phosphate, 9CI — T-00705

Triallyl phosphate, 8CI
[1623-19-4]

$$(H_2C{=}CHCH_2O)_3P{=}O$$

$C_9H_{15}O_4P$ M 218.189
Employed in dental compositions. *N*-Allylating agent for nitrogen heterocyclic compds. Liq. $Bp_{0.5}$ 78°. n_D^{20} 1.449.
▷TC8575000.

Kennedy, J. *et al*, *J. Appl. Chem.*, 1958, **8**, 459 (*synth, props*)
Tanabe, T. *et al*, *Bull. Chem. Soc. Jpn.*, 1979, **52**, 259 (*use*)

Tri-1-propenylphosphine, 9CI — T-00706

$$P(CH{=}CHCH_3)_3$$

$C_9H_{15}P$ M 154.191
(E,E,E)-form
Liq., stable to uv or heat. Bp_{10} 69-70°. n_D^{20} 1.5160.
B,MeI: Methyltri-1-propenylphosphonium iodide.
$C_{10}H_{18}IP$ M 296.130
Solid. Mp 48°.
(Z,Z,Z)-form
Liq. Bp_{20} 76-77°. n_D^{20} 1.5172. Heat or uv → (*all-E*)-form.
B,MeI: Solid. Mp 104-105°.

Borisov, A.E. *et al*, *Izv. Akad. Nauk SSSR, Ser. Khim.*, 1962, 1258 (*Engl. transl.* p. 1182) (*synth, ir, deriv*)

Tri-2-propenylphosphine, 9CI — T-00707

Triallylphosphine, 8CI
[16523-89-0]

$$P(CH_2CH{=}CH_2)_3$$

$C_9H_{15}P$ M 154.191
Air-sensitive liq. Bp_{13} 69°.
B,MeI: [17586-54-0]. *Methyltri-2-propenylphosphonium iodide. Triallylmethylphosphonium iodide.*
$C_{10}H_{18}IP$ M 296.130
Cryst. ($EtOH/Et_2O$). Mp 115-117°.
Oxide: [2946-60-3].
$C_9H_{15}OP$ M 170.191
Liq. Mp 15-17°. $Bp_{0.1}$ 78°.

Jones, W.J. *et al*, *J. Chem. Soc.*, 1947, 1446 (*synth*)
Kennedy, J. *et al*, *J. Appl. Chem.*, 1958, **8**, 459 (*oxide*)
Fluck, E. *et al*, *Z. Naturforsch., B*, 1967, **22**, 1095 (*nmr*)
Davidson, G. *et al*, *Spectrochim. Acta, Part A*, 1979, **35**, 83 (*ir, raman*)

Tri-2-propenyl phosphite, 9CI — T-00708

Triallyl phosphite
[102-84-1]

$$(H_2C{=}CHCH_2O)_3P$$

$C_9H_{15}O_3P$ M 202.189
Employed in synth. of fire retardants. Liq. d_0^0 1.05. Bp_9 85-86°. n_D 1.4596.
▷Mod. toxicity

Kamai, G., *CA*, 1957, **51**, 6503 (*synth*)
Clark, P.W. *et al*, *Can. J. Chem.*, 1974, **52**, 1714 (*pmr, P nmr*)
Verstuyft, A.W. *et al*, *Inorg. Chem.*, 1976, **15**, 732 (*cmr*)

N,N′,N″-Tri-2-propenylphosphorous triamide, 9CI — T-00709

N,N′,N″-*Triallylphosphorous triamide, 8CI. Tris(allylamino)phosphine. Tri(allylamino)phosphine*
[41999-14-8]

$$(H_2C{=}CHCH_2NH)_3P$$

$C_9H_{18}N_3P$ M 199.235
Liq. d_4^{20} 1.10. $Bp_{0.0001}$ 90-100°. n_D^{20} 1.5560.
Oxide: [35542-86-0].
$C_9H_{18}N_3OP$ M 215.234
Used in creaseproofing and fireproofing processes. Liq. d_4^{20} 1.07. $Bp_{0.0001}$ 110-115°. pK_{a1} 9.52, pK_{a2} 5.91 ($MeNO_2$). n_D^{20} 1.5028.
Sulfide: [53820-06-7].
$C_9H_{18}N_3PS$ M 231.295
Liq. d_4^{20} 1.09. $Bp_{0.0001}$ 80-90°. n_D^{20} 1.5461.

Petrov, K.A. *et al*, *Zh. Obshch. Khim.*, 1962, **32**, 915 (*Engl. transl.* p. 904) (*synth*)
Pepperman, A.B. *et al*, *J. Fire Retard. Chem.*, 1976, **3**, 276; *CA*, **86**, 107915 (*oxide*)
Yakshin, V.V. *et al*, *Dokl. Akad. Nauk SSSR, Ser. Sci. Khim.*, 1979, **247**, 128 (*Engl. transl.* p. 344) (*oxide, props*)

Tripropyl phosphate, 9CI, 8CI — T-00710

[513-08-6]

$$(H_3CCH_2CH_2O)_3P{=}O$$

$C_9H_{21}O_4P$ M 224.236
Fireproofing agent. Liq. Sol. H_2O. d_4^{20} 1.01. Bp 252°, Bp_{50} 161°, Bp_{10} 121°. n_D^{20} 1.4165.

Evans, D.P. *et al*, *J. Chem. Soc.*, 1930, 1310 (*synth*)
Vogel, A.I. *et al*, *J. Chem. Soc.*, 1943, 16 (*synth*)
Sax, N.I., *Dangerous Properties of Industrial Materials*, 6th Ed., Van Nostrand-Reinhold, 1984, 1070.

Tripropylphosphine, 9CI — T-00711

[2234-97-1]

$$P(CH_2CH_2CH_3)_3$$

$C_9H_{21}P$ M 160.239
Forms Pt and Pd complexes. Air-sensitive liq. Mp −83°. Bp 187.5°, Bp_{20} 81.8-81.9°. pK_a 8.64. n_D^{20} 1.458, n_D^{27} 1.5071.
▷Burns in CCl_4
B,MeI: Methyltripropylphosphonium iodide.
$C_{10}H_{24}IP$ M 302.178
Cryst. (EtOH). Mp 212-215°.
Oxide: [1496-94-2].
$C_9H_{21}OP$ M 176.238
Mp 39°. Bp 260-265°, Bp_7 120.3-120.5°. pK_a 8.77 ($MeNO_2$).
Sulfide: [13639-72-0].
$C_9H_{21}PS$ M 192.299
Bp_1 108-110°. n_D^{27} 1.5071.
Selenide:
$C_9H_{21}PSe$ M 239.199
Solid. Mp 32°. $Bp_{0.95}$ 116°.
Telluride:
$C_9H_{21}PTe$ M 287.839
Yellow cryst. (pet. ether). Mp 45-46°.

Davies, W.C. *et al*, *J. Chem. Soc.*, 1929, 1262 (*deriv*)
Rothstein, E. *et al*, *J. Chem. Soc.*, 1953, 3994 (*synth*)
Screttas, C. *et al*, *J. Org. Chem.*, 1962, **27**, 2573 (*synth, props*)
Zingaro, R.A. *et al*, *J. Chem. Eng. Data*, 1963, **8**, 226 (*selenide*)
Cumper, C.W.N. *et al*, *J. Chem. Soc.*, 1964, 430 (*oxide*)
Zingaro, R.A. *et al*, *J. Organomet. Chem.*, 1965, **4**, 320 (*telluride*)
Patel, N.K. *et al*, *J. Org. Chem.*, 1967, **32**, 2999 (*sulfide*)
Chremos, G.N. *et al*, *J. Organomet. Chem.*, 1970, **22**, 637 (*telluride, ir*)
Allen, E.A. *et al*, *Spectrochim. Acta, Part A*, 1974, **30**, 1219 (*ir, complexes*)
Wolfsberger, W. *et al*, *Synth. React. Inorg. Met.-Org. Chem.*, 1974, **4**, 149 (*synth*)
Bodner, G.M. *et al*, *J. Organomet. Chem.*, 1982, **226**, 85 (*cmr*)

Tripropyl phosphite, 9CI, 8CI — T-00712

[923-99-9]

$P(OCH_2CH_2CH_3)_3$

$C_9H_{21}O_3P$ M 208.237

Polymerisation catalyst. Liq. Bp_{10} 83°. n_D^{20} 1.4290. Ligand for Cu, Cr, and Pd.

▷TH1750000.

McCombie, H. *et al*, *J. Chem. Soc.*, 1945, 380 (*synth*)
Mel'nikov, N.N. *et al*, *Zh. Obshch. Khim.*, 1958, **28**, 2473 (*Engl. transl.* p. 2507) (*synth*)
Engel, R. *et al*, *J. Chem. Soc. (C)*, 1971, 1761 (*pmr*)
Fluck, E. *et al*, *Z. Naturforsch. B*, 1974, **29**, 603 (*pe*)
Labintsev, V.B. *et al*, *Zh. Org. Khim.*, 1976, **12**, 1597; 1977, **13**, 1141 (*Engl transl.* pp. 1573, 1050) (*ms*)

O,O,O-Tripropyl phosphoroselenoate, 9CI, 8CI — T-00713

O,O,O-*Tripropyl selenophosphate*

[7322-77-2]

$(H_3CCH_2CH_2O)_3P{=}Se$

$C_9H_{21}O_3PSe$ M 287.197

Liq. d_4^{20} 1.20. $Bp_{0.09}$ 67-68°. n_D^{20} 1.4680.

Nuretdinov, I.A. *et al*, *Izv. Akad. Nauk SSSR, Ser. Khim.*, 1968, 2831 (*Engl. transl.* p. 2685) (*synth, ir*)
Stec, W.J. *et al*, *Phosphorus*, 1972, **2**, 97 (*P nmr*)

Tripropyl phosphorotetrathioate, 9CI, 8CI — T-00714

Tripropyl tetrathiophosphate

[1642-48-4]

$(H_3CCH_2CH_2S)_3P{=}S$

$C_9H_{21}PS_4$ M 288.479

Odorous liq. d_4^{20} 1.115. Bp_1 130-132°. n_D^{20} 1.5854.

Menefee, A. *et al*, *J. Org. Chem.*, 1957, **22**, 792 (*ir*)
Maier, L. *et al*, *J. Am. Chem. Soc.*, 1962, **84**, 3054 (*synth*)
Voigt, D. *et al*, *C.R. Hebd. Seances Acad. Sci.*, 1965, **260**, 2210 (*synth*)
Murav'ev, I.V. *et al*, *Zh. Obshch. Khim.*, 1975, **45**, 1746 (*Engl. transl.* p. 1711) (*synth*)
Khokhlov, P.S. *et al*, *Zh. Obshch. Khim.*, 1984, **54**, 2545 (*Engl. transl.* p. 2274) (*synth*)

O,O,O-Tripropyl phosphorothioate, 9CI, 8CI — T-00715

O,O,O-*Tripropyl thiophosphate*

[2272-08-4]

$(H_3CCH_2CH_2O)_3P{=}S$

$C_9H_{21}O_3PS$ M 240.297

Liq. d_4^{20} 1.04. Bp_{20} 130-131°, Bp_8 95-97°. n_D^{20} 1.4510.

Aksnes, D.W. *et al*, *Acta Chem. Scand.*, 1973, **27**, 3277 (*pmr*)
Nifant'ev, E.Ê. *et al*, *Zh. Obshch. Khim.*, 1975, **45**, 295 (*Engl. transl.* p. 282) (*synth*)
Mazitova, F.N. *et al*, *Zh. Obshch. Khim.*, 1980, **50**, 1718 (*Engl. transl.* p. 1393)

O,O,S-Tripropyl phosphorothioate, 9CI, 8CI — T-00716

O,O,S-*Tripropyl thiophosphate*

[64008-12-4]

$(H_3CCH_2CH_2O)_2P(O)SCH_2CH_2CH_3$

$C_9H_{21}O_3PS$ M 240.297

Liq. d_0^0 1.05. Bp_{20} 156°.

Pishchimuka, P., *J. Russ. Phys. Chem. Soc.*, 1912, **44**, 1406; *CA*, **7**, 987.

S,S,S-Tripropyl phosphorotrithioate, 9CI, 8CI — T-00717

S,S,S-*Tripropyl trithiophosphate*

[1642-44-0]

$(H_3CCH_2CH_2S)_3P{=}O$

$C_9H_{21}OPS_3$ M 272.418

Liq. d_4^{20} 1.10. $Bp_{0.9}$ 131.5°. n_D^{20} 1.5497.

Voigt, D. *et al*, *C.R. Hebd. Seances Acad. Sci.*, 1965, **260**, 2210 (*synth*)
Ibrahim, E.H. *et al*, *Egypt. J. Chem.*, 1979, **22**, 403 (*synth*)
Sinyashin, O.G. *et al*, *Zh. Obshch. Khim.*, 1984, **54**, 540 (*Engl. transl.* p. 475) (*synth, P nmr*)

Tripropyl phosphorotrithioite, 9CI — T-00718

Tripropyl trithiophosphite

[869-56-7]

$(H_3CCH_2CH_2S)_3P$

$C_9H_{21}PS_3$ M 256.419

Ligand for Hg and Au. Liq. Bp_1 112-114°. n_D^{20} 1.5353.

B,MeI: Methyltri(propylthio)phosphonium iodide.
$C_{10}H_{24}IPS_3$ M 398.358
Solid. Mp 191°.

Lippert, A. *et al*, *J. Am. Chem. Soc.*, 1938, **60**, 2370 (*synth, deriv, complexes*)
Kostin, V.P. *et al*, *Zh. Obshch. Khim.*, 1982, **52**, 2698 (*Engl. transl.* p. 2379) (*synth, P nmr*)
Nifant'ev, É.E. *et al*, *Zh. Obshch. Khim.*, 1983, **53**, 2695 (*Engl. transl.* p. 2429) (*synth, P nmr*)

N,N′,N″-Tripropylphosphorous triamide, 9CI — T-00719

Tris(propylamino)phosphine. Tri(propylamino)-phosphine

$(H_3CCH_2CH_2NH)_3P$

$C_9H_{24}N_3P$ M 205.282

Oxide: [19622-52-7]. N,N′,N″-*Tripropylphosphoric triamide.*
$C_9H_{24}N_3OP$ M 221.282
pK_{a1} 12.11, pK_{a2} 7.09 ($MeNO_2$).

Sulfide: [5395-86-8]. N,N′,N″-*Tripropylphosphorothioic triamide.*
$C_9H_{24}N_3PS$ M 237.342

Needles (pet. ether). Mp 75°.

Buck, A. *et al, J. Am. Chem. Soc.*, 1948, **70**, 744 (*sulfide*)
Bock, H. *et al, Chem. Ber.*, 1966, **99**, 377 (*sulfide*)
Laskorin, B.N. *et al, Dokl. Akad. Nauk SSSR, Ser. Sci. Khim.*, 1976, **227**, 884 (*Engl. transl.* p. 274) (*oxide, ir*)
Healy, J.D. *et al, Phosphorus Sulfur*, 1978, **5**, 239 (*sulfide, synth, props*)
Yakshin, V.V. *et al, Dokl. Akad. Nauk SSSR, Ser. Sci. Khim.*, 1979, **247**, 128 (*Engl. transl.* p. 344) (*oxide, props*)

2,4,6-Tripropyl-1,3,5,2,4,6-trioxatriphosphorinane, 9CI — T-00720

2,4,6-Tripropyl-1,3,5,2,4,6-trioxatriphosphinane

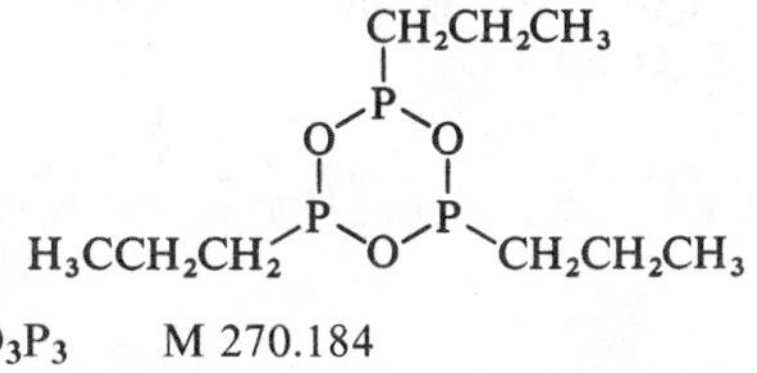

$C_9H_{21}O_3P_3$ M 270.184

2,4,6-Trioxide: [68957-94-8]. *Propylphosphonic acid trimer trianhydride.*
$C_9H_{21}O_6P_3$ M 318.183
Acts as a peptide-coupling agent. Syrup. Sol. CH_2Cl_2, THF, DMF. $Bp_{0.3}$ 200°. Stable at r.t.

2,4,6-Trisulfide: Propylphosphonothioic acid trimer trianhydride.
$C_9H_{21}O_3P_3S_3$ M 366.364
Oil.

Maier, L., *Phosphorus*, 1975, **5**, 253 (*trisulfide, synth, pmr, P nmr, ir*)
Wissmann, H., *Angew. Chem., Int. Ed. Engl.*, 1980, **19**, 133 (*trioxide, synth, use*)

Tri-1-propynylphosphine, 9CI — T-00721

[33909-73-8]

$(H_3CC{\equiv}C)_3P$

C_9H_9P M 148.144

Oxide: [27258-71-5].
C_9H_9OP M 164.143
No phys. props. reported.

Sulfide:
C_9H_9PS M 180.204
Solid. Mp 198-199°.

Bogolyubov, G.M. *et al, Zh. Obshch. Khim.*, 1965, **35**, 704, 1566 (*Engl. transl.* p. 705, 1570) (*sulfide, uv*)
Mootz, D. *et al, Z. Kristallogr., Kristallgeom., Kristallphys., Kristallchem.*, 1969, **130**, 239; *CA*, **72**, 71872 (*oxide, cryst struct*)
Cullen, W.R. *et al, J. Fluorine Chem.*, 1971, **1**, 227 (*uv*)
Rosenberg, D. *et al, Tetrahedron*, 1971, **27**, 3893 (*oxide, cmr, nmr*)
Sacher, R.E. *et al, Spectrochim. Acta, Part A*, 1972, **28**, 1361 (*ir, raman*)
Lequar, R.-M. *et al, Org. Magn. Reson.*, 1975, **7**, 392 (*pmr, cmr, nmr*)

Tri(1*H*-pyrazol-1-yl)phosphine — T-00722

1,1′,1″-Phosphinidynetris(1H-pyrazole), 9CI. *Tris(1H-pyrazol-1-yl)phosphine*

[54877-55-3]

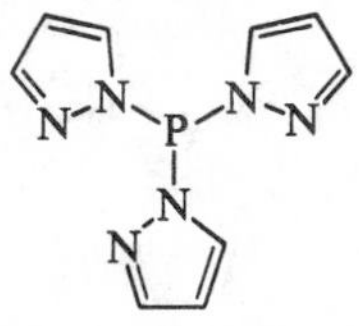

$C_9H_9N_6P$ M 232.184

Hygroscopic solid. Mp 104°.

Oxide: [61324-13-8]. *1,1′,1″-Phosphinylidynetris(1H-pyrazole).*
$C_9H_9N_6OP$ M 248.183
Solid by subl. Mp 76-78°. $Bp_{0.3}$ 160° subl.

Sulfide: [61324-15-0]. *1,1′,1″-Phosphinothioyldiynetris(1H-pyrazole).*
$C_9H_9N_6PS$ M 264.244
Mp 149-151°. $Bp_{0.3}$ 130° subl.

Fischer, S. *et al, Can. J. Chem.*, 1974, **52**, 3981 (*synth, pmr*)
Cobbledick, R.E. *et al, Acta Crystallogr., Sect. B*, 1975, **31**, 2731 (*cryst struct*)
Fischer, S. *et al, Can. J. Chem.*, 1976, **54**, 2710 (*synth, derivs, props*)

Tri-2-pyridylphosphine — T-00723

2,2′,2″-Phosphinidynetrispyridine, 9CI. *Tri-2-pyridinylphosphine*

[26437-48-9]

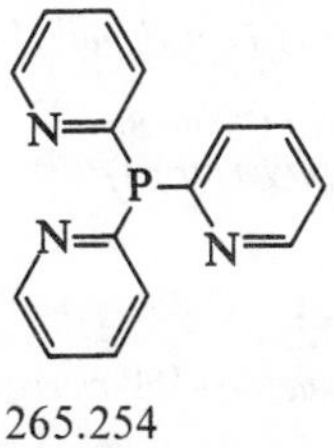

$C_{15}H_{12}N_3P$ M 265.254

Ligand for Cu, Zn, Co, Ni, Fe, Mn, and Ru. Cryst. (MeOH). Mp 115°. $Bp_{0.15}$ 210°.

B,3HCl: Solid. Mp 207.5-209.5°.

B,MeI: Solid. Mp 143.5-144.5° dec.

P-*Oxide:*
$C_{15}H_{12}N_3OP$ M 281.253
Ligand for Cu, Zn, Co, Ni, Fe, Mn and Ru. Cryst. (EtOH). Mp 209°.

Sulfide:
$C_{15}H_{12}N_3PS$ M 297.314
Cryst. (EtOH). Mp 161°.

Mann, F.G. *et al, J. Org. Chem.*, 1948, **13**, 502 (*synth, derivs*)
Griffin, G.E. *et al, J. Chem. Soc. (B)*, 1970, 477 (*derivs, pmr*)
Boggess, R.K. *et al, J. Coord. Chem.*, 1975, **4**, 217 (*complexes*)
Kurtev, K. *et al, J. Chem. Soc., Dalton Trans.*, 1980, 55 (*synth, complexes*)
Schmidbaur, H. *et al, Z. Naturforsch., B*, 1980, **35**, 1329 (*synth, deriv*)

Tri-1-pyrrolidylphosphine **T-00724**
1,1′,1″-Phosphinidynetrispyrrolidine, 9CI.
Tris(pyrrolidino)phosphine
[5666-12-6]

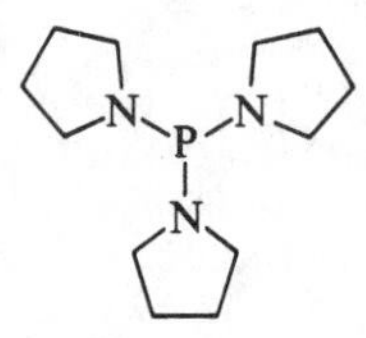

$C_{12}H_{24}N_3P$ M 241.315
Liq. d_4^{20} 1.05. $Bp_{0.1}$ 103-104°. n_D^{19} 1.5268.
Oxide: [6415-07-2]. *1,1′,1″-Phosphinylidynetrispyrrolidine. Tris(pyrrolidino)phosphine oxide.*
$C_{12}H_{24}N_3OP$ M 257.315
Highly polar aprotic solvent with electron-donating power greater than any other known solv. d_4^{25} 1.15. Mp 14°. $Bp_{0.5}$ 143°. n_D^{25} 1.5115.
Selenide: [75404-57-8]. *1,1′,1″-Phosphinoselenoylidynetrispyrrolidine. Tris(pyrrolidino)phosphine selenide.*
$C_{12}H_{24}N_3PSe$ M 320.275
Cryst. (hexane). Mp 51°.
Telluride: [75404-56-7]. *1,1′,1″-Phosphinotelluroylidynetrispyrrolidine. Tris(pyrrolidino)phosphine telluride.*
$C_{12}H_{24}N_3PTe$ M 368.915
Cream-coloured cryst. (hexane). Mp 84-85°.

Burgada, R., *Ann. Chim. (Paris)*, 1963, **8**, 374 (*synth*)
Mathis, R. *et al*, *Spectrochim. Acta, Part A*, 1974, **30**, 357 (*ir*)
Yvernault, T. *et al*, *C.R. Hebd. Seances Acad. Sci., Ser. C*, 1978, **287**, 519 (*oxide, synth, props*)
Rømming, C. *et al*, *Acta Chem. Scand., Ser. A*, 1980, **34**, 333 (*selenide, telluride, ir, uv*)
Bergesen, K. *et al*, *Acta Chem. Scand., Ser. A*, 1981, **35**, 147 (*cmr*)
Rømming, C. *et al*, *Acta Chem. Scand., Ser. A*, 1984, **38**, 349 (*selenide, telluride, cryst struct*)

Tri-1-pyrrolylphosphine **T-00725**
*1,1′,1″-phosphinidynetris-1*H*-pyrrole*, 10CI
[60259-30-5]

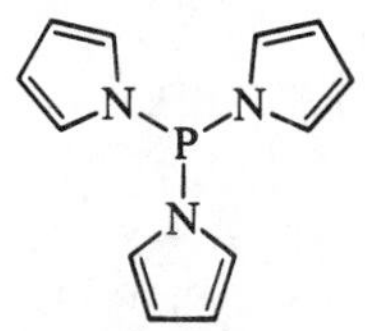

$C_{12}H_{12}N_3P$ M 229.221
Cryst. (MeOH). d_4^{20} 1.2357. Mp 43-44°. $Bp_{0.04}$ 85°. n_D^{20} 1.5920.
Oxide: [60090-34-8]. *1,1′,1″-Phosphinylidynetrispyrrole.*
$C_{12}H_{12}N_3OP$ M 245.220
Solid. Mp 81°.

Fischer, S. *et al*, *Can. J. Chem.*, 1976, **54**, 2706 (*synth, pmr, P nmr, oxide*)
Marschner, F. *et al*, *Phosphorus Sulfur*, 1976, **6**, 135 (*synth, oxide*)
Atwood, J.L. *et al*, *Inorg. Chem.*, 1982, **21**, 1354 (*cryst struct*)
Gurevich, P.A. *et al*, *Zh. Obshch. Khim.*, 1983, **53**, 238 (*Engl. transl.* p. 209) (*synth, ir, cmr, P nmr, pmr*)

Tri-2-pyrrolylphosphine, 8CI **T-00726**
2,2′,2″-Phosphinidynetrispyrrole, 9CI

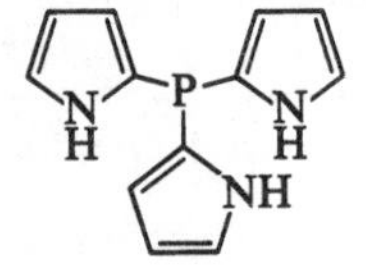

$C_{12}H_{12}N_3P$ M 229.221
Oxide: [953-60-6]. *2,2′,2″-Phosphinylidynetrispyrrole.*
$C_{12}H_{12}N_3OP$ M 245.220
Cryst. (EtOH aq.). Mp 136-137°.
Tri-N-Me, oxide: 2,2′,2″-Phosphinylidynetris(1-methylpyrrole).
$C_{15}H_{18}N_3OP$ M 287.300
Cryst. (Et_2O/pet. ether). Mp 136-137.5°.

Griffin, C.E. *et al*, *J. Org. Chem.*, 1965, **30**, 97 (*synth, uv, ir*)
Kemp, R.H. *et al*, *J. Chem. Soc. (B)*, 1969, 527 (*deriv, pmr*)
Penkovskii, V.V. *et al*, *Phosphorus*, 1974, **3**, 247 (*struct*)

Tris(4-aminophenyl) phosphate, 9CI **T-00727**

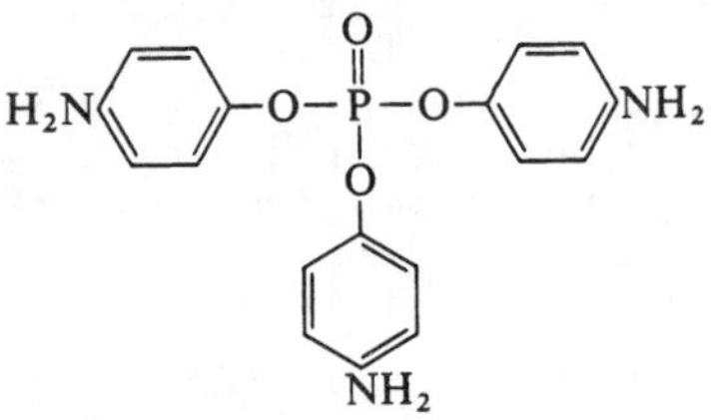

$C_{18}H_{18}N_3O_4P$ M 371.332
Solid. Mp 153-155°.

B.P., 1 027 059, (*1960*); *CA*, **65**, 2172

Tris(2-aminophenyl) phosphite, 9CI **T-00728**
Tri(2-aminophenyl) phosphite

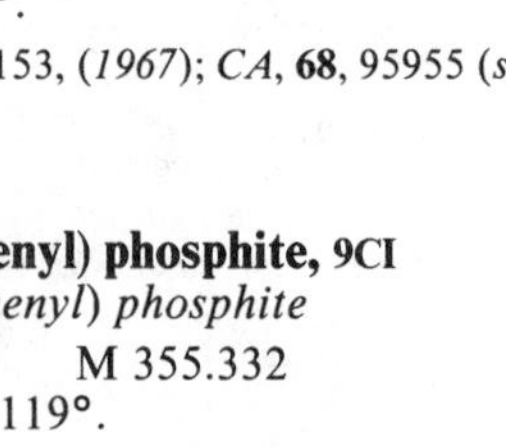

$C_{18}H_{18}N_3O_3P$ M 355.332
Cryst. Mp 145°.

Ger. Pat., 1 257 153, (*1967*); *CA*, **68**, 95955 (*synth*)

Tris(3-aminophenyl) phosphite, 9CI **T-00729**
Tri(3-aminophenyl) phosphite
$C_{18}H_{18}N_3O_3P$ M 355.332
Solid. Mp 115-119°.

Ger. Pat., 1 257 153, (*1967*); *CA*, **68**, 95955 (*synth*)

Tris[1,2-benzenediolato(2−)-*O*,*O*′] phosphate(1−), 9CI T-00730

Tris[pyrocatecholato(2−)] phosphate(1−), 8CI

[47383-70-0]

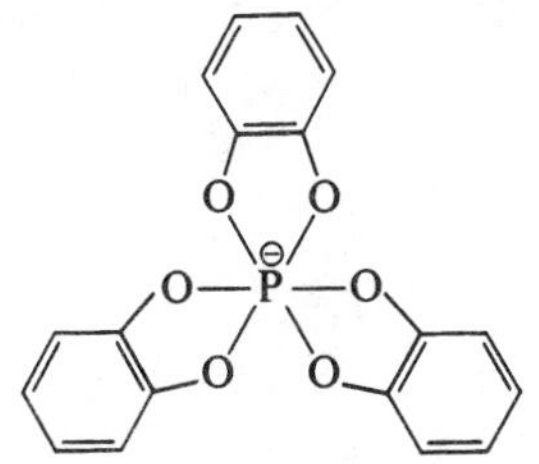

$C_{18}H_{12}O_6P^{\ominus}$ M 355.263 (ion)

Dimethylammonium salt: [38641-40-6]. Solid. Mp 246°.

Tetramethylammonium salt: [28304-87-2]. Needles (Me_2CO). Dec. >350°.

Triethylammonium salt: [39043-29-3]. Cryst. (DMSO). Mp 305-310°.

4-Methylmorpholinium salt: [90501-97-6]. Cryst. (DMF). Mp 193-194°.

Hellwinkel, D. *et al, Chem. Ber.*, 1970, **103**, 1056 (*salts*)
Lopez, L. *et al, C.R. Hebd. Seances Acad. Sci., Ser. C*, 1972, **275**, 295 (*synth, P nmr*)
Allcock, H.R. *et al, J. Am. Chem. Soc.*, 1973, **95**, 3154 (*cryst struct*)
Gallagher, M. *et al, J. Chem. Soc., Chem. Commun.*, 1976, 321 (*synth, P nmr*)
Kukhar', V.P. *et al, Zh. Obshch. Khim.*, 1979, **49**, 1671 (*Engl. transl.* p. 1461) (*synth, P nmr*)
Osman, F.H. *et al, Phosphorus Sulfur*, 1982, **14**, 1 (*synth, pmr, P nmr*)
Kostin, V.P. *et al, Zh. Obshch. Khim.*, 1984, **54**, 218 (*Engl. transl.* p. 193) (*synth, P nmr*)

Tris[[1,1′-biphenyl]-2,2′-diyl]phosphate(1−), 9CI T-00731

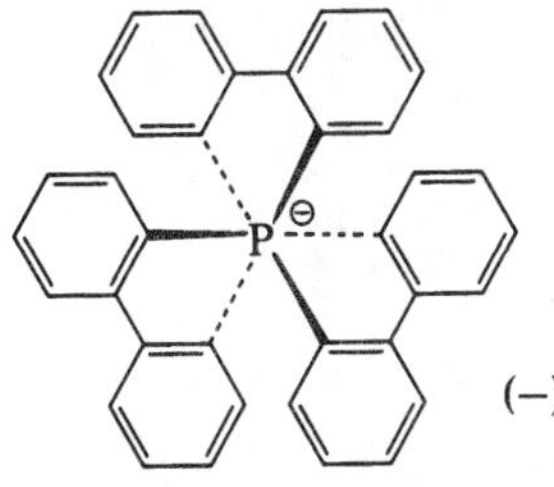

(−)-*form*

$C_{36}H_{24}P^{\ominus}$ M 487.559 (ion)

(+)-***form***

K salt:

$C_{36}H_{24}KP$ M 526.658

Needles (THF or Me_2CO/Et_2O). $[\alpha]_D^{24}$ +1930° (c, 0.6 in Me_2CO).

19-Methylbrucinium salt: [19066-48-9].

$C_{60}H_{53}N_2O_4P$ M 897.063

Cryst. + 0.5MeOH. $[\alpha]_D^{24}$ +962° (c, 0.86 in CH_2Cl_2).

5,5′-Spirobi[5H-dibenzophospholium] salt: [16871-93-5].

$C_{60}H_{40}P_2$ M 822.924

$[\alpha]_D^{24}$ +1198° (c, 0.8 in DMF).

(−)-***form***

K salt: [19107-17-6]. Needles (THF or Me_2CO/Et_2O). $[\alpha]_D^{24}$ −1930° (c, 0.5 in Me_2CO).

19-Methylbrucinium salt: [19681-77-7]. Cryst. + 0.5Me_2CO. Mp 230° dec. $[\alpha]_D^{24}$ −1250° (c, 0.7-1.0 in CH_2Cl_2).

5,5′-Spirobi[5H-dibenzophospholium] salt: $[\alpha]_D^{24}$ −1186° (c, 0.8 in DMF).

(±)-***form***

Li salt:

$C_{36}H_{24}LiP$ M 494.500

Cryst. + 4.5THF (THF). Dec. at 247-255°.

Na salt:

$C_{36}H_{24}NaP$ M 510.549

Cryst. + 2THF (THF). Dec. at 266-268°.

K salt: Needles (THF or Me_2CO/Et_2O).

5,5′-Spirobi[5H-dibenzophospholium] salt: [19049-13-9]. Cryst. (DMF/EtOH). Mp 255-256° dec.

Hellwinkel, D., *Chem. Ber.*, 1965, **98**, 576; 1966, **99**, 3668 (*synth, resoln, deriv, props, P nmr*)
Hellwinkel, D. *et al, J. Chem. Soc. (B)*, 1970, 640 (*uv, cd, abs config*)
Rothuis, R. *et al, Recl. Trav. Chim. Pays-Bas*, 1972, **91**, 836 (*deriv, props*)

Tris[(4-bromophenyl)methyl] phosphate, 9CI T-00732

Tris(p-bromobenzyl) phosphate, 8CI

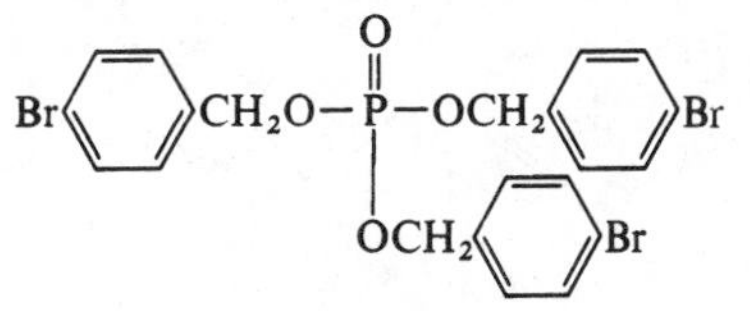

$C_{21}H_{18}Br_3O_4P$ M 605.057

Cryst. (EtOH); needles (EtOAc). Mp 132-133°.

Baddiley, J. *et al, J. Chem. Soc.*, 1949, 815 (*synth*)
Dilaris, I. *et al, J. Org. Chem.*, 1969, **30**, 686 (*synth*)

Tris[(4-bromophenyl)methyl]phosphine, 9CI T-00733

Tris(p-bromobenzyl)phosphine

[42844-49-5]

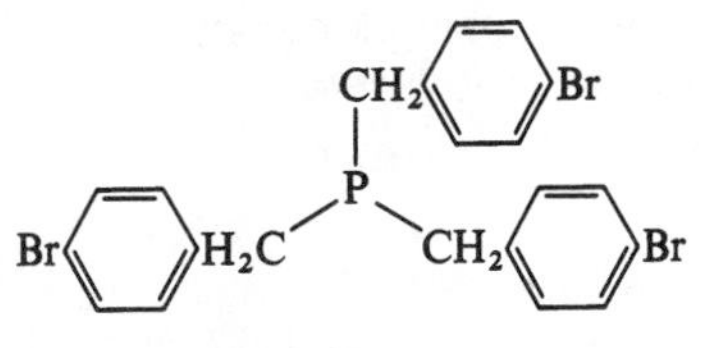

$C_{21}H_{18}Br_3P$ M 541.059

Oxide:

$C_{21}H_{18}Br_3OP$ M 557.058

Cryst. (CCl_4). Insol. Et_2O, pet. ether. Mp 182-183°.

Zhuravleva, L.P. *et al, Zh. Obshch. Khim.*, 1968, **38**, 342 (*Engl. transl.* p. 341)

Tris(3-bromophenyl)phosphine, 9CI T-00734

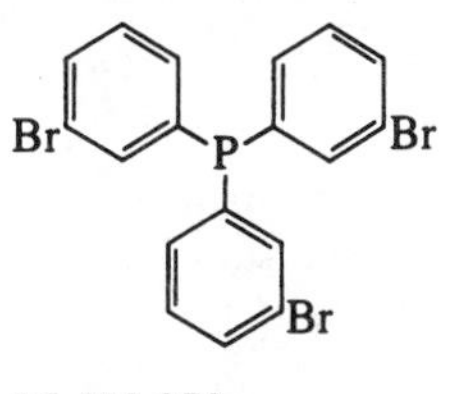

$C_{18}H_{12}Br_3P$ M 498.979

Oxide: [38019-09-9].

$C_{18}H_{12}Br_3OP$ M 514.978

No phys. props. reported.

Shvets, A.A. *et al, Zh. Obshch. Khim.*, 1972, **42**, 829 (*Engl. transl.* p. 821) (*oxide, ir*)

Tris(4-bromophenyl)phosphine, 9CI **T-00735**

[29949-81-3]

$C_{18}H_{12}Br_3P$ M 498.979

Solid. Mp 163-164°.

Oxide: [900-99-2].

$C_{18}H_{12}Br_3OP$ M 514.978

Solid. Mp 179-180°.

Sulfide: [54560-62-2].

$C_{18}H_{12}Br_3PS$ M 531.039

Solid. Mp 182-183°.

Benassi, R. *et al*, *J. Chem. Soc., Perkin Trans. 2*, 1974, 1338 (*synth, pmr, nmr*)
Amarski, E.G. *et al*, *Zh. Obshch. Khim.*, 1975, **45**, 898 (*Engl. transl.* p. 881) (*oxide, ir*)
Shapiro, B.L. *et al*, *J. Phys. Chem. Ref. Data*, 1977, **6**, 919 (*nmr*)

Tris(4-*tert*-butylphenyl) phosphite, 8CI **T-00736**

Tris[4-(1,1-dimethylethyl)phenyl] phosphite, 9CI

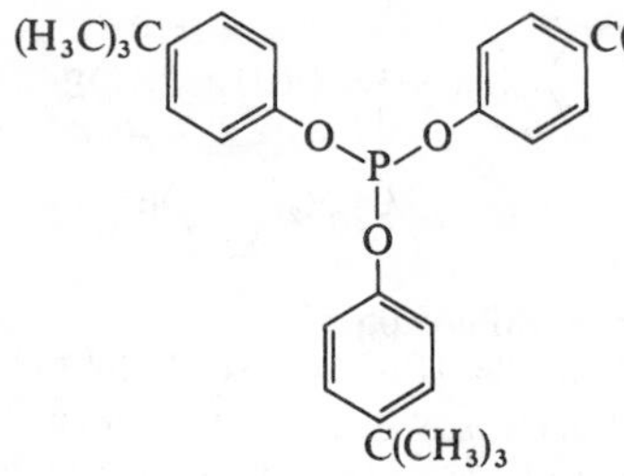

$C_{30}H_{39}O_3P$ M 478.610

Solid. Mp 75°. Bp_1 253-254°.

Walsh, E.N., *J. Am. Chem. Soc.*, 1959, **81**, 3023 (*synth*)

Tris(*tert*-butylphosphino)phosphine **T-00737**

Tris[(1,1-dimethylethyl)phosphino]phosphine

$P[PHC(CH_3)_3]_3$

$C_{12}H_{30}P_4$ M 298.264

Cryst. (pentane at −100°). As prepd., contains two diastereomeric forms in the ratio 1:3.

Baudler, M. *et al*, *Z. Naturforsch., B*, 1983, **38**, 537 (*synth, nmr, raman*)

Tris(2-carbamoylethyl)phosphine **T-00738**

3,3′,3″-Phosphinidynetris[propanamide], 9CI

[4343-60-6]

$P(CH_2CH_2CONH_2)_3$

$C_9H_{18}N_3O_3P$ M 247.233

Waxy solid. Mp 53-60°.

Oxide: [4116-00-1]. *3,3′,3″-Phosphinylidynetris[propanamide]*.

$C_9H_{18}N_3O_4P$ M 263.233

Used in treatment of cotton fabrics. Solid (MeOH or DMF). Mp 161-163°, 208-210°.

Rauhut, M.M. *et al*, *J. Org. Chem.*, 1963, **28**, 478 (*oxide*)
Morris, C.E. *et al*, *CA*, 1971, **75**, 50319 (*oxide, use*)
Yakovenko, T.V. *et al*, *Zh. Obshch. Khim.*, 1976, **46**, 278 (*Engl. transl.* p. 275) (*synth, derivs, props*)

Tris(3-carboxyphenyl)phosphine **T-00739**

3,3′,3″-Phosphinidynetris[benzoic acid], 9CI

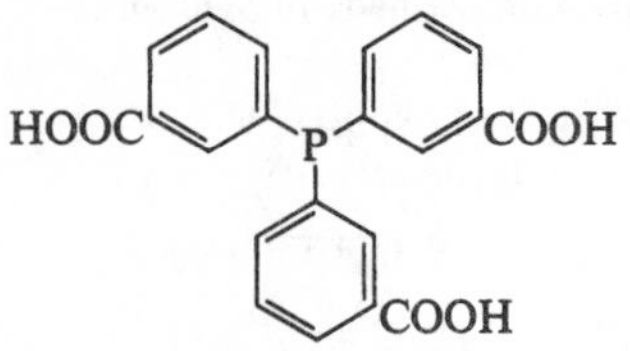

$C_{21}H_{15}O_6P$ M 394.320

Solid (MeOH aq.). Mp 276-279°.

Tri-Me ester:

$C_{24}H_{21}O_6P$ M 436.400

Cryst. (MeOH). Mp 115-116°.

Oxide: 3,3′,3″-Phosphinylidynetris[benzoic acid].

$C_{21}H_{15}O_7P$ M 410.319

Cryst. (MeOH aq.). Mp 335-337° (321-335°).

Oxide, tri-Me ester:

$C_{24}H_{21}O_7P$ M 452.399

Cryst. (MeOH). Mp 139-140°.

Schiemenz, G.P. *et al*, *Chem. Ber.*, 1969, **102**, 1883 (*derivs, ir*)

Tris(4-carboxyphenyl)phosphine **T-00740**

4,4′,4″-Phosphinidynetris[benzoic acid], 9CI

[22836-27-7]

$C_{21}H_{15}O_6P$ M 394.320

Cryst. (MeOH aq.). Mp 270-274°.

Tri-Me ester: [66417-54-7].

$C_{24}H_{21}O_6P$ M 436.400

No phys. props. reported.

Oxide: [807-19-2]. *4,4′,4″-Phosphinylidynetris[benzoic acid]*.

$C_{21}H_{15}O_7P$ M 410.319

Solid, cryst. (MeOH aq.). Mp 323-330°, 347-357°.

Oxide, tri-Me ester: [809-44-9].

$C_{24}H_{21}O_7P$ M 452.399

Cryst. (C_6H_6/pet. ether). Mp 123-125°.

Sulfide: 4,4′,4″-Phosphinothioylidynetris[benzoic acid].

$C_{21}H_{15}O_6PS$ M 426.380

Cryst. (MeOH aq.). Mp 321-323°.

Sulfide, tri-Me ester: [22836-36-8].

$C_{24}H_{21}O_6PS$ M 468.460

Cryst. (MeOH). Mp 178-179°.

Morgan, P.W. *et al*, *J. Am. Chem. Soc.*, 1952, **74**, 4526 (*oxide, synth*)
Schiemenz, G.P. *et al*, *Chem. Ber.*, 1969, **102**, 1883 (*synth, derivs*)
Ratovskii, G.V. *et al*, *Zh. Obshch. Khim.*, 1979, **49**, 548 (*Engl. transl.* p. 479) (*ir*)
Goncharova, L.V. *et al*, *Zh. Obshch. Khim.*, 1980, **50**, 321 (*Engl. transl.* p. 258) (*ir*)
Rowley, A.G. *et al*, *Chem. Ind. (London)*, 1981, 365 (*esters*)

Tris(2-chloroethyl) phosphate, 9CI, 8CI **T-00741**

Fyrol CEF

$(ClCH_2CH_2O)_3P{=}O$

$C_6H_{12}Cl_3O_4P$ M 285.491

Flame retardent for textiles. Liq. d_4^{20} 1.42. Bp_{25} 214°, Bp_1 146°. n_D^{20} 1.4731.

Jones, W.J. *et al*, *J. Chem. Soc.*, 1946, 824 (*synth*)
Frank, A.W. *et al*, *J. Org. Chem.*, 1966, **31**, 872 (*synth*)
Pudovik, A.N. *et al*, *Zh. Obshch. Khim.*, 1966, **36**, 1454 (*Engl. transl.* p. 1461) (*synth*)

Fluck, E. *et al*, *Z. Anorg. Allg. Chem.*, 1967, **354**, 139 (*nmr*)
Sax, N.I., *Dangerous Properties of Industrial Materials*, 6th Ed., Van Nostrand-Reinhold, 1984, 1071.

Tris(2-chloroethyl) phosphite, 9CI, 8CI **T-00742**
Tri(2-chloroethyl) phosphite

$(ClCH_2CH_2O)_3P$

$C_6H_{12}Cl_3O_3P$ M 269.492
Ligand for Cr, Mo. Intermed. for synth. of vinylphosphonic acid derivs. Oil. n_D^{20} 1.4877. Isom. on heating → bis(2-chloroethyl) 2-chloroethylphosphonate.

Ger. Pat., 1 244 757, (*1967*); *CA*, **68**, 21521 (*synth*)

***O,O,O*-Tris(2-chloroethyl) phosphorothioate, 9CI, 8CI** **T-00743**
O,O,O-*Tris(2-chloroethyl) thiophosphate*
[10235-09-3]

$(ClCH_2CH_2O)_3P{=}S$

$C_6H_{12}Cl_3O_3PS$ M 301.552
Flameproofer. Liq. d_4^{20} 1.48. Bp_9 142-150°. n_D^{20} 1.5650.
▷TG4025000.

Galashina, M.L. *et al*, *Zh. Obshch. Khim.*, 1953, **23**, 433 (*Engl. transl.* p. 441) (*synth*)
Fluck, E. *et al*, *Z. Anorg. Allg. Chem.*, 1967, **354**, 139 (*P nmr*)

Tris[(4-chlorophenyl)methyl]phosphine, 9CI **T-00744**
*Tris(*p*-chlorobenzyl)phosphine*

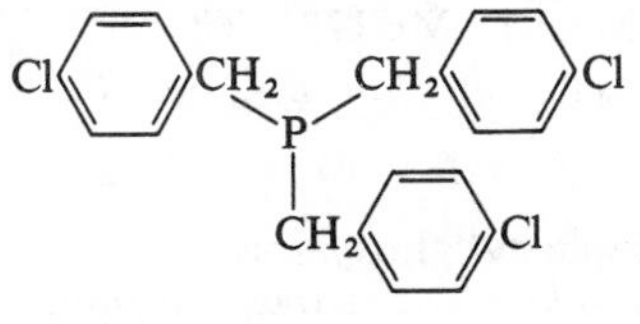

$C_{21}H_{18}Cl_3P$ M 407.706
Oxide: [4145-68-0].
$C_{21}H_{18}Cl_3OP$ M 423.705
Used in flame retardant prepns. Mp 179-180°.

U.S.P., 3 213 057, (*1965*); *CA*, **63**, 18153 (*synth*)
U.S.P., 3 316 293, (*1967*); *CA*, **67**, 43921 (*synth*)

Tris(2-chlorophenyl) phosphate, 9CI **T-00745**
[631-44-7]

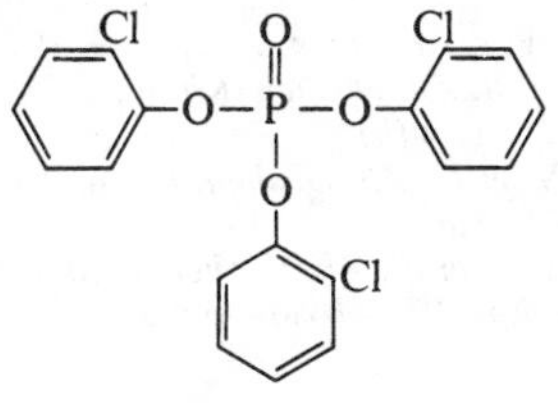

$C_{18}H_{12}Cl_3O_4P$ M 429.623
Low-melting solid. d^{40} 1.41. Mp 37°. $Bp_{17.5}$ 309°.

U.S.P., 2 033 916, (*1936*)

Tris(4-chlorophenyl) phosphate, 9CI **T-00746**
[3871-31-6]

$C_{18}H_{12}Cl_3O_4P$ M 429.623
Cryst. (EtOH). Mp 116-117°.

Krishnakumar, V.K. *et al*, *Synthesis*, 1983, 558; *Synth. Commun.*, 1984, **14**, 189 (*synth, ir, pmr*)

Tris(2-chlorophenyl)phosphine, 9CI **T-00747**
[6962-87-4]

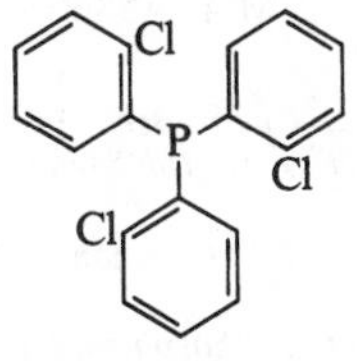

$C_{18}H_{12}Cl_3P$ M 365.626
Cryst. (EtOH). Mp 66-67°.
Oxide: [61102-88-3].
$C_{18}H_{12}Cl_3OP$ M 381.625
Cryst. + $0.5H_2O$ (2-ethoxyethanol). Mp 226-236°.

Mann, F.G. *et al*, *J. Chem. Soc.*, 1937, 527 (*synth*)
Shvets, A.A. *et al*, *Zh. Obshch. Khim.*, 1976, **46**, 1701 (*Engl. transl.* p. 1654) (*oxide, ir*)
Grim, S.O. *et al*, *Phosphorus Sulfur*, 1977, **3**, 191 (*nmr*)

Tris(3-chlorophenyl)phosphine, 9CI **T-00748**
[29949-85-7]
$C_{18}H_{12}Cl_3P$ M 365.626
Cryst. (EtOH). Mp 67°.
Oxide: [54300-36-6].
$C_{18}H_{12}Cl_3OP$ M 381.625
Solid. Mp 135°.
Sulfide: [54300-45-7].
$C_{18}H_{12}ClPS$ M 326.780
Solid. Mp 104°.
Selenide: [54300-56-0].
$C_{18}H_{12}Cl_3PSe$ M 444.586
Solid. Mp 126° (66°).

De Ketelaere, R.F. *et al*, *J. Mol. Struct.*, 1974, **23**, 233 (*derivs, ir, raman*)
De Ketelaere, R.F. *et al*, *J. Organomet. Chem.*, 1974, **73**, 251 (*ir, raman*)
De Ketelaere, R.F. *et al*, *Phosphorus*, 1974, **5**, 43 (*derivs, ms*)
Hoste, S. *et al*, *J. Electron Spectrosc. Relat. Phenom.*, 1974, **5**, 227; 1979, **17**, 191 (*pe*)
Van der Kelen, G.P. *et al*, *J. Mol. Struct.*, 1974, **23**, 329 (*nqr*)
Shvets, A.A. *et al*, *Zh. Obshch. Khim.*, 1976, **46**, 1701 (*Engl. transl.* p. 1654); 1981, **51**, 642 (*Engl. transl.*, p. 512) (*derivs, ir*)
Grim, S.O. *et al*, *Phosphorus Sulfur*, 1977, **3**, 191 (*synth, nmr*)
Goncharova, L.V. *et al*, *Zh. Obshch. Khim.*, 1980, **50**, 321 (*Engl. transl.* p. 258); 1981, **51**, 1450 (*Engl. transl.* p. 1230) (*derivs, ir*)

Tris(4-chlorophenyl)phosphine, 9CI **T-00749**
[1159-54-2]
$C_{18}H_{12}Cl_3P$ M 365.626
Important ligand for metals of Groups IB, IIB, VB, VIB, and VIII. Cryst. (EtOH). Mp 66-67°, 103-104°. $Bp_{0.05}$ 192-200°. pK_{a1} 2.86, 2.73, pK_{a2} 1.03 ($MeNO_2$).
B,MeBr: Tris(4-chlorophenyl)methylphosphonium bromide.
$C_{19}H_{15}BrCl_3P$ M 460.564
Solid. Mp 230-231°.
B,MeI: Tris(4-chlorophenyl)methylphosphonium iodide.
$C_{19}H_{15}Cl_3IP$ M 507.565
Cryst. (H_2O or MeOH aq.). Mp 252-262°.

Oxide: [4576-56-1].
$C_{18}H_{12}Cl_3OP$ M 381.625
Cryst. (EtOH). Mp 175°.
▷SZ1780000.
Sulfide: [5032-62-2].
$C_{18}H_{12}Cl_3PS$ M 397.686
Solid or cryst. (MeOH). Mp 139°, 152-153°.
Selenide: [41398-45-2].
$C_{18}H_{12}Cl_3PSe$ M 444.586
Solid. Mp 146°.

Schiemenz, G.P. *et al, Chem. Ber.*, 1966, **99**, 504 (*synth, derivs*)
De Ketelaere, R.F. *et al, J. Mol. Struct.*, 1974, **23**, 233 (*derivs, ir, raman*)
De Ketelaere, R.F. *et al, J. Organomet. Chem.*, 1974, **73**, 251 (*ir, raman*)
De Ketelaere, R.F. *et al, Phosphorus*, 1974, **5**, 43 (*derivs, ms*)
Van der Kelen, G.P. *et al, J. Mol. Struct.*, 1974, **23**, 329 (*nqr*)
De Ketelaere, R.F. *et al, J. Mol. Struct.*, 1975, **27**, 25, 363 (*derivs, nmr*)
Weiner, M.A. *et al, J. Org. Chem.*, 1975, **40**, 1292 (*pe*)
Claeys, E.G. *et al, J. Mol. Struct.*, 1977, **40**, 97 (*derivs, struct*)
Casper, J.M. *et al, Spectrochim. Acta, Part A*, 1978, **34**, 1 (*ir, raman*)
Ratovskii, G.V. *et al, Zh. Obshch. Khim.*, 1978, **48**, 1520 (*Engl. transl.*, p. 1394) (*uv*)
Hoste, S. *et al, J. Electron Spectrosc. Relat. Phenom.*, 1979, **17**, 191 (*oxide, pe*)
Marshall, G. *et al, Org. Mass Spectrom.*, 1981, **16**, 272 (*ms*)
Allman, T. *et al, Can. J. Chem.*, 1982, **60**, 716 (*pmr, cmr*)

Tris(2-chlorophenyl) phosphite, 9CI — T-00750

Tri(2-chlorophenyl) phosphite
[24460-31-9]

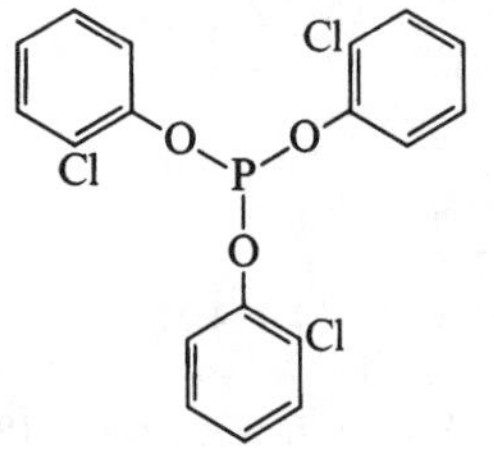

$C_{18}H_{12}Cl_3O_3P$ M 413.624
Ligand for Fe. Oil. Bp_3 230°. n_D^{25} 1.6041.

Walsh, E.N., *J. Am. Chem. Soc.*, 1959, **81**, 3023 (*synth*)

Tris(4-chlorophenyl) phosphite, 9CI — T-00751

Tri(4-chlorophenyl) phosphite
[5679-61-8]
$C_{18}H_{12}Cl_3O_3P$ M 413.624
Ligand for Cu and metals of Groups VIB and VIII. Cryst. (C_6H_6/hexane). Mp 48-50°. $Bp_{1.5}$ 207°.

Walsh, E.N., *J. Am. Chem. Soc.*, 1959, **81**, 3023 (*synth*)

Tris(2-cyanoethyl)phosphine — T-00752

3,3′,3″-Phosphinidynetris[propanenitrile], 9CI
[4023-53-4]

$P(CH_2CH_2CN)_3$

$C_9H_{12}N_3P$ M 193.188
Cryst. Mp 97-98°. pK_a 1.37 (H_2O).
B,MeI: Tris(2-cyanoethyl)methylphosphonium iodide.
$C_{10}H_{15}IN_3P$ M 335.127
Cryst. (MeCN). Mp 238-239°.
Oxide: [1439-41-4]. *3,3′,3″-Phosphinylidynetris[propanenitrile]*.
$C_9H_{12}N_3OP$ M 209.187
Needles (2-propanol aq.). Mp 173°.
Sulfide: [6783-73-9]. *3,3′,3″-Phosphinothioylidynetris[propanenitrile]*.
$C_9H_{12}N_3PS$ M 225.248
Cryst. (H_2O). Mp 141-142°.
▷SZ2980000.
Selenide: [71309-04-1]. *3,3′,3″-Phosphinoselenoylidynetris[propanenitrile]*.
$C_9H_{12}N_3PSe$ M 272.148
No phys. props. reported.

Rauhut, M.M. *et al, J. Am. Chem. Soc.*, 1959, **81**, 1103 (*synth, derivs*)
Valetdinov, R.K. *et al, Zh. Obshch. Khim.*, 1969, **39**, 1744 (*Engl. transl.* p. 1708) (*synth, oxide*)
Valetdinov, R.K. *et al, Zh. Obshch. Khim.*, 1974, **44**, 284 (*Engl. transl.* p. 269) (*props*)
Yakovenko, T.V. *et al, Zh. Obshch. Khim.*, 1976, **46**, 278 (*Engl. transl.* p. 275) (*props*)
Blake, A.J. *et al, Acta Crystallogr., Sect. B*, 1981, **37**, 997, 1959 (*cryst struct, oxide, sulfide, selenide*)
Cotton, F.A. *et al, Inorg. Chem.*, 1981, **20**, 1869 (*cryst struct*)
Foxman, B.M. *et al, Acta Crystallogr., Sect. B*, 1982, **38**, 1622 (*cryst struct, oxide*)

Tris(3-cyanophenyl)phosphine — T-00753

3,3′,3″-Phosphinidynetris[benzonitrile], 9CI
$C_{21}H_{12}N_3P$ M 337.320
Cryst. (MeOH). Mp 154-155°.
Oxide: 3,3′,3″-Phosphinylidynetris[benzonitrile].
$C_{21}H_{12}N_3OP$ M 353.319
Cryst. (C_6H_6/ligroin). Mp 198-200°.
Sulfide: 3,3′,3″-Phosphinothioylidynetris[benzonitrile].
$C_{21}H_{12}N_3PS$ M 369.380
Cryst. (MeOH). Mp 175-176°.

Schiemenz, G.P. *et al, Chem. Ber.*, 1969, **102**, 1883 (*synth, ir, derivs*)

Tris(4-cyanophenyl)phosphine — T-00754

4,4′,4″-Phosphinidynetris[benzonitrile], 9CI
[22836-30-2]
$C_{21}H_{12}N_3P$ M 337.320
Cryst. (EtOH). Mp 186-189°.
Oxide: [22836-23-3]. *4,4′,4″-Phosphinylidynetris[benzonitrile]*.
$C_{21}H_{12}N_3OP$ M 353.319
Cryst. (C_6H_6/ligroin). Mp 215-219°.
Sulfide: 4,4′,4″-Phosphinothioylidynetris[benzonitrile].
$C_{21}H_{12}N_3PS$ M 369.380
Cryst. (MeOH). Mp 255-260°.

Schiemenz, G.P. *et al, Chem. Ber.*, 1969, **102**, 1883 (*ir, synth*)
Shvets, A.A. *et al, Zh. Obshch. Khim.*, 1978, **48**, 2185 (*Engl. transl.* p. 479) (*synth*)
Ratovskii, G.V. *et al, Zh. Obshch. Khim.*, 1979, **49**, 548 (*Engl. transl.* p. 479) (*uv*)
Goncharova, L.V. *et al, Zh. Obshch. Khim.*, 1980, **50**, 321 (*Engl. transl.* p. 258) (*oxide, ir*)

Tris(4-cyanophenyl) phosphite T-00755

Tri(4-cyanophenyl) phosphite

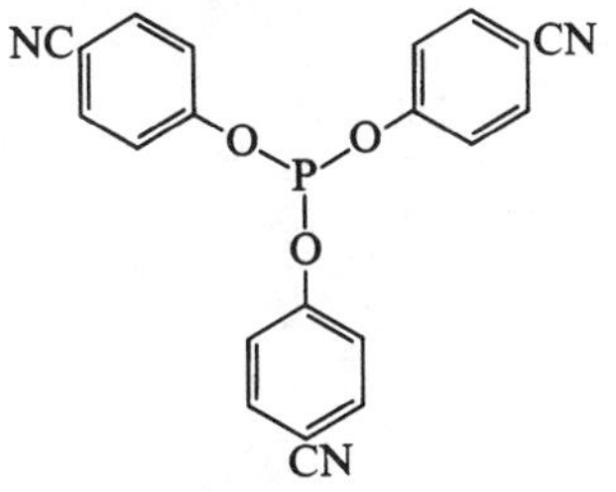

$C_{21}H_{12}N_3O_3P$ M 385.318

Reagent for peptide synth. Cryst. (pet. ether). Mp 129-130°.

Iselin, B.M. *et al*, *Helv. Chim. Acta*, 1957, **40**, 373.

2,4,6-Tris(diethylamino)-1,3,5,2,4,6-trioxatriphosphorinane 2,4,6-trioxide T-00756

N,N,N′,N′,N″,N″*-Hexaethyl-1,3,5,2,4,6-trioxatriphosphorinane-2,4,6-triamine 2,4,6-trioxide*, 9CI

[26227-69-0]

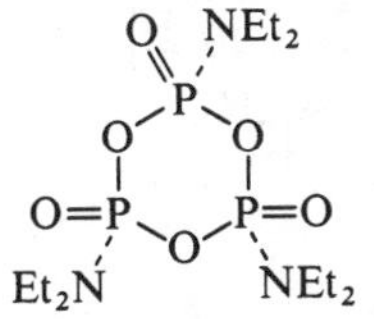

(2α,4α,6α)*-form*

$C_{12}H_{30}N_3O_6$ M 312.386

(2α,4α,6α)-form [63814-77-7]

Monoclinic cryst. Exists in the chair form.

(2α,4α,4β)-form [63814-78-8]

Cryst. Exists in distorted chair form.

Andrianov, V.I. *et al*, *Zh. Strukt. Khim.*, 1969, **10**, 865 (*Engl. transl.* p. 751) (*cryst struct*)

Cherepinskii-Malov, V.D. *et al*, *Zh. Strukt. Khim.*, 1971, **12**, 126 (*Engl. transl.* p. 105) (*cryst struct*)

***N,N′,N″*-Tris(dihydroxyphosphinylmethyl)-1,4,7-triazacyclononane** T-00757

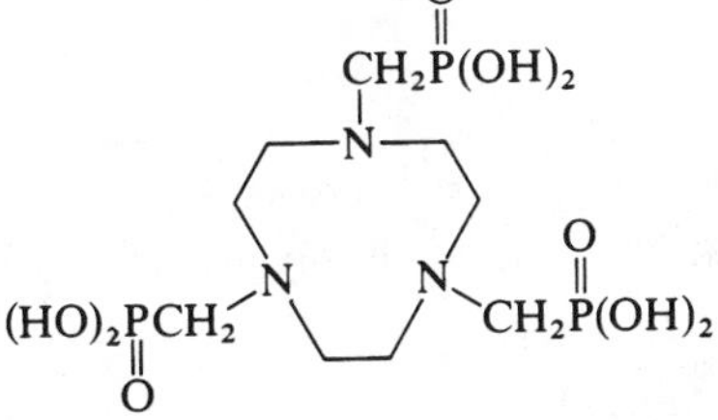

$C_9H_{24}N_3O_9P_3$ M 411.225

Forms complexes with metals of Groups IIA and IIB. Cryst. (H_2O). Mp 263-265° dec. pK_{a1} <2, pK_{a2} 2.53, pK_{a3} 5.38, pK_{a4} 7.09, pK_{a5} 8.65, pK_{a6} 11.79.

Kabachnik, M.I. *et al*, *Izv. Akad. Nauk SSSR, Ser. Khim.*, 1984, 835 (*Engl. transl.* p. 769) (*synth, ir, props, use*)

Tris(2,4-dimethoxyphenyl)phosphine, 9CI T-00758

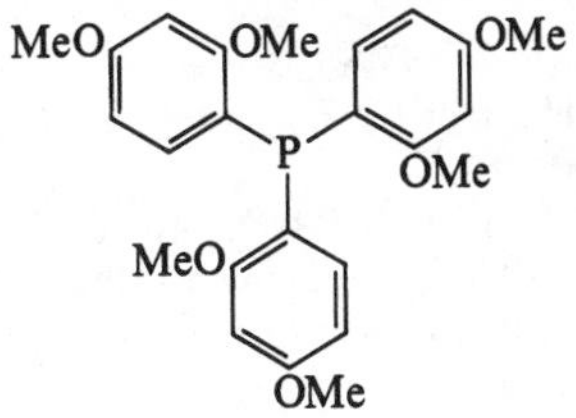

$C_{24}H_{27}O_6P$ M 442.447

Needles (EtOH). Mp 187-188°.

B,MeI: Tris(2,4-dimethoxyphenyl)methylphosphonium iodide.

$C_{25}H_{30}IO_6P$ M 584.387

Needles (EtOH). Mp 192-194°.

Protopopov, I.S. *et al*, *Zh. Obshch. Khim.*, 1964, **34**, 1446 (*Engl. transl.* p. 1451)

Tris(3,5-dimethoxyphenyl)phosphine, 9CI T-00759

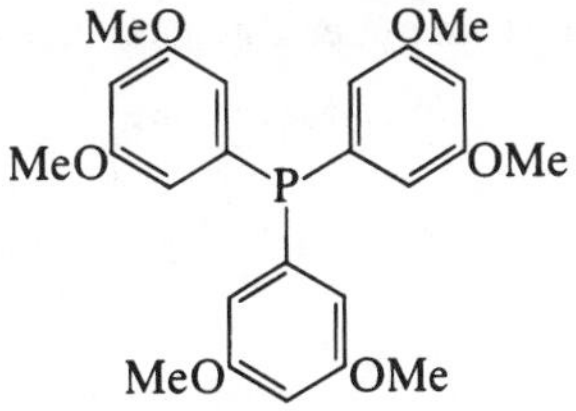

$C_{24}H_{27}O_6P$ M 442.447

Plates (EtOH). Mp 131.5-132.5°.

Oxide:

$C_{24}H_{27}O_7P$ M 458.447

Needles (pet. ether). Mp 139-141°.

Lamza, L., *J. Prakt. Chem.*, 1964, **25**, 294.

Tris(dimethylamino)difluorophosphorane, 8CI T-00760

Difluoro-N,N,N′,N′,N″,N″*-hexamethylphosphoranetriamine*, 9CI

[7549-83-9]

$(Me_2N)_3PF_2$

$C_6H_{18}F_2N_3P$ M 201.199

Liq. d^{25} 1.06. Bp_3 35°. n_D^{25} 1.4190.

Ramirez, F. *et al*, *Tetrahedron Lett.*, 1966, 3651 (*synth, ms, ir, pmr, nmr*)

Cowley, A.H. *et al*, *J. Am. Chem. Soc.*, 1973, **95**, 6506 (*pe*)

Oberhammer, H. *et al*, *J. Chem. Soc., Dalton Trans.*, 1976, 1454 (*synth, ed, nmr, struct, cndo*)

Cowley, A.H. *et al*, *Inorg. Chem.*, 1977, **16**, 854 (*pe*)

Tris[2-(dimethylamino)phenyl] phosphate, 9CI T-00761

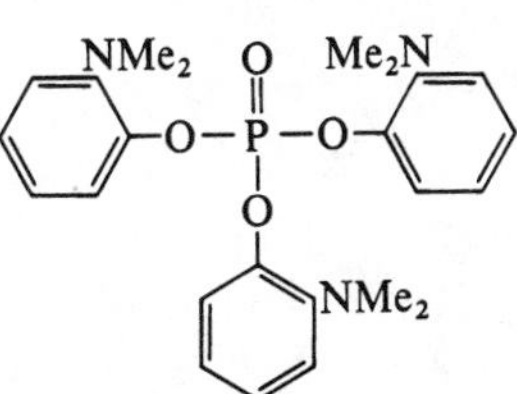

$C_{24}H_{30}N_3O_4P$ M 455.492

Pale-yellow oil. n_D^{25} 1.5792.

U.S.P., 2 759 961, (*1956*); *CA*, **51**, 482

Tris[3-(dimethylamino)phenyl] phosphate, 9CI **T-00762**

$C_{24}H_{30}N_3O_4P$ M 455.492
Cryst. (EtOH aq. or C_6H_6/pet. ether). Mp 48-49°. $Bp_{0.001}$ 200°.
B,3MeI: Solid; cryst. + $1H_2O$. Mp 207-212° (anhyd.), Mp 193-194.5° dec. (monohydrate).

U.S.P., 2 759 961, (*1956*); *CA*, **51**, 482

Tris[4-(dimethylamino)phenyl] phosphate, 9CI **T-00763**

$C_{24}H_{30}N_3O_4P$ M 455.492
Needles (2-propanol or C_6H_6/pet. ether). Mp 92.5-93°.
B,3MeI: Solid + $2H_2O$. Mp 160-161° dec.

U.S.P., 2 759 961, (*1956*); *CA*, **51**, 482

Tris(2-dimethylaminophenyl)phosphine, 9CI **T-00764**

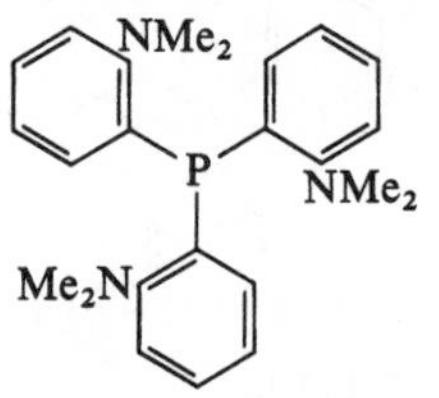

$C_{24}H_{30}N_3P$ M 391.495
Solid. Mp 106°.

Horner, L. *et al*, *Phosphorus Sulfur*, 1983, **15**, 165 (*synth*)

Tris(4-dimethylaminophenyl)phosphine, 9CI **T-00765**

$C_{24}H_{30}N_3P$ M 391.495
Cryst. (C_6H_6 or EtOH aq.). Solid dil. HCl. Mp 254°, 282-287°. $AgClO_4$ yields a green radical.
Oxide:
$C_{24}H_{30}N_3OP$ M 407.494
Needles (EtOH). Insol. Et_2O, EtOAc. Mp 262°, 275-280°, 290°, 305-306°. Forms a monohydrate Mp 321°.
Sulfide:
$C_{24}H_{30}N_3PS$ M 423.555
Cryst. (chlorobenzene). Mp 258°.

Koenigs, E. *et al*, *Justus Liebigs Ann. Chem.*, 1934, **509**, 138 (*oxide*)
Schiemenz, G.P., *Tetrahedron Lett.*, 1964, 2729 (*oxide, uv*)
Schiemenz, G.P., *Chem. Ber.*, 1965, **98**, 65 (*synth, derivs, ir*)
Goetz, H. *et al*, *Justus Liebigs Ann. Chem.*, 1968, **715**, 1 (*synth*)
Grindley, B.T. *et al*, *Aust. J. Chem.*, 1975, **28**, 327 (*derivs, ir*)
Werner, M.A. *et al*, *J. Org. Chem.*, 1975, **40**, 1292 (*pe*)
Shvets, A.A. *et al*, *Zh. Obshch. Khim.*, 1976, **46**, 1701 (*Engl. transl.* p. 1654) (*ir*)

Tris(4-dimethylaminophenyl) phosphite, 9CI **T-00766**
Tri(4-dimethylaminophenyl) phosphite

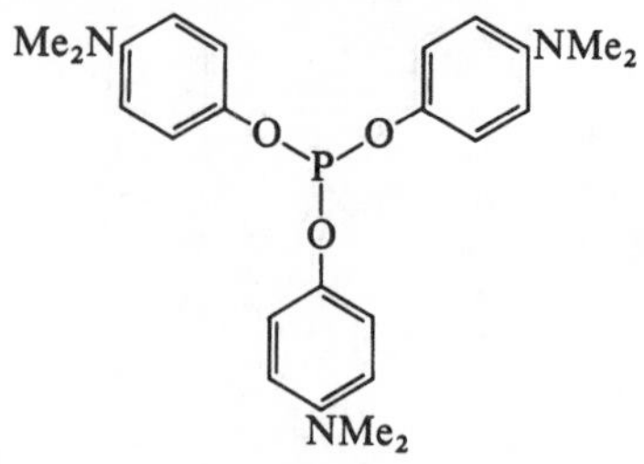

$C_{24}H_{30}N_3O_3P$ M 439.493
Solid. Mp 71°.

Iselin, B.M. *et al*, *Helv. Chim. Acta*, 1957, **40**, 373 (*synth*)

Tris(2,5-dimethylphenyl) phosphate, 9CI **T-00767**

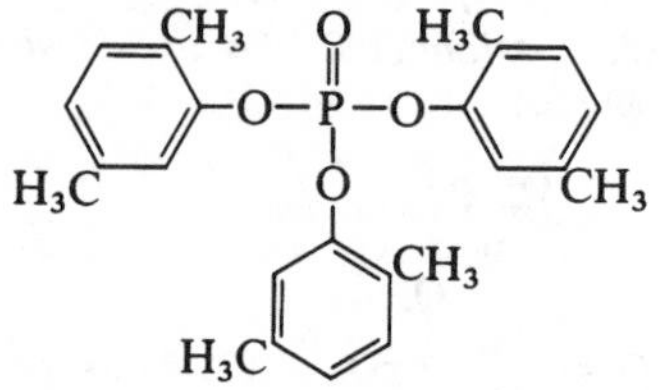

$C_{24}H_{27}O_4P$ M 410.449
Cryst. Mp 77°.

Breusch, F.L. *et al*, *CA*, 1944, **38**, 1483 (*synth*)

Tris(2,4-dimethylphenyl)phosphine, 9CI **T-00768**
[49676-42-8]

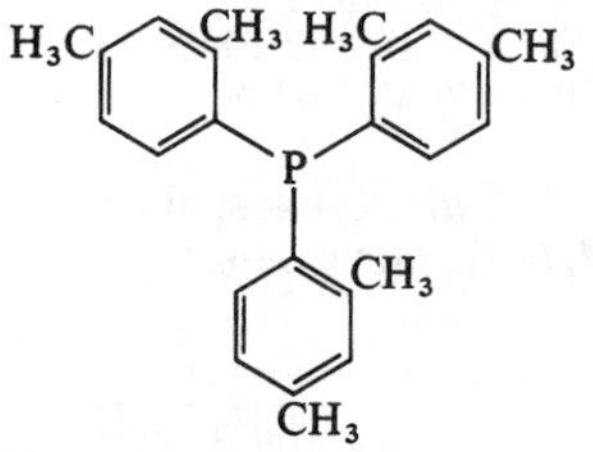

$C_{24}H_{27}P$ M 346.451
Cryst. (EtOH). Mp 158.5-159.5°. pK_a 9.66 ($MeNO_2$).
B,MeI: Methyltris(2,4-dimethylphenyl)phosphonium iodide.
$C_{25}H_{30}IP$ M 488.390
Solid. Mp 230.5°.
Oxide: [52944-44-0].
$C_{24}H_{27}OP$ M 362.450
Cryst. (pet. ether). Mp 157.5-158.5°.
Sulfide:
$C_{24}H_{27}PS$ M 378.511
Solid. Mp 167°.

Michaelis, A., *Justus Liebigs Ann. Chem.*, 1901, **315**, 43 (*synth, derivs*)
Bokanov, A.I. *et al*, *Zh. Obshch. Khim.*, 1974, **44**, 760 (*Engl. transl.* p. 732) (*synth, ir, oxide, uv*)
Shifrina, R.R. *et al*, *Zh. Prikl. Spektrosk.*, 1981, **34**, 111 (*Engl. transl.* p. 84) (*oxide, ir*)

Tris(2,5-dimethylphenyl)phosphine, 9CI **T-00769**

$C_{24}H_{27}P$ M 346.451

Ligand for Co and Ni. Needles. Mp 155°.
B,MeI: Methyltris(2,5-dimethylphenyl)phosphonium iodide.
$C_{25}H_{30}IP$ M 488.390
Mp 169°.
Oxide:
$C_{24}H_{27}OP$ M 362.450
Solid. Mp 173°.
Sulfide:
$C_{24}H_{27}PS$ M 378.511
Solid. Mp 170°.

Michaelis, A., *Justus Liebigs Ann. Chem.*, 1901, **315**, 43 (*synth, derivs*)

Tris(2,6-dimethylphenyl)phosphine, 9CI T-00770

[50341-15-6]

$C_{24}H_{27}P$ M 346.451
Cryst. (EtOH). Mp 146.5-147.5°. pK_a 11.65 ($MeNO_2$).
Oxide: [52944-83-9].
$C_{24}H_{27}OP$ M 362.450
Cryst. (heptane). Mp 188-189°.

Bokanov, A.I. *et al*, *Zh. Obshch. Khim.*, 1974, **44**, 760 (*Engl. transl.* p. 732) (*synth, uv, ir, oxide*)
Sobolev, A.N. *et al*, *Zh. Strukt. Khim.*, 1976, **17**, 103 (*Engl. transl.* p. 83) (*cryst struct*)
Spalding, T.R., *Org. Mass Spectrom.*, 1976, **11**, 1019 (*ms*)
Negrebetskii, V.V. *et al*, *Zh. Obshch. Khim.*, 1978, **48**, 1308 (*Engl. transl.* p. 1196) (*cmr*)
Shifrina, R.R. *et al*, *Zh. Prikl. Spektrosk.*, 1981, **34**, 111 (*Engl. transl.* p. 84) (*oxide, ir*)

Tris(dimethylphosphino)methane T-00771

Methylidynetris[dimethylphosphine], *10CI*
[70355-41-8]

$(Me_2P)_3CH$

$C_7H_{19}P_3$ M 196.148
Ligand for Ru, Ag. Cryst. by subl. Mp 45-47°.
Trisulfide: [81431-68-7].
Tris(dimethylphosphinothioyl)methane.
$C_7H_{19}P_3S_3$ M 292.328
Needles (toluene). Mp 240-243°.

Karsch, H.H. *et al*, *Angew. Chem., Int. Ed. Engl.*, 1979, **18**, 484 (*struct, synth*)
Karsch, H.H. *et al*, *Z. Naturforsch., B*, 1979, **34**, 1171 (*synth, ms, pmr, cmr, nmr*)
Karsch, H.H. *et al*, *Chem. Ber.*, 1982, **115**, 818 (*sulfide*)

Tris(2,2-dimethylpropyl) phosphite, 9CI T-00772

Trineopentyl phosphite. Tri(2,2-dimethylpropyl) phosphite

$[(H_3C)_3CCH_2O]_3P$

$C_{15}H_{33}O_3P$ M 292.398
Solid. Mp 55-57°. $Bp_{0.15}$ 80°.

Kosolapoff, G., *Organophosphorus Compounds*, 1st Ed., 1950, Wiley.
Bellamy, L.J. *et al*, *J. Chem. Soc.*, 1952, 475 (*ir*)
Verstuyft, A.W. *et al*, *Inorg. Chem.*, 1976, **15**, 732 (*cmr*)

Tris[(2-diphenylphosphino)ethyl]amine T-00773

2-(Diphenylphosphino)-N,N-bis[2-(diphenylphosphino)ethyl]ethanamine, *10CI*. *2,2′,2″-Tris(diphenylphosphino)triethylamine*
[15114-55-3]

$N(CH_2CH_2PPh_2)_3$

$C_{42}H_{42}NP_3$ M 653.722
Ligand for Co, Fe, Ni and Pd. Syrup.

Sacconi, L., *Proc. 6th Conf. Coord. Chem.*, 1976, **6**, 225; *CA*, **90**, 72287 (*rev*)

Tris[2-(diphenylphosphino)ethyl]phosphine, 9CI T-00774

Tetraphos-2
[23582-03-8]

$P(CH_2CH_2PPh_2)_3$

$C_{42}H_{42}P_4$ M 670.689
Ligand for V, Cr, Mn, Cu and metals of Group VIII. Air-stable cryst. Mp 129°.
Tetraoxide: [23725-53-3].
$C_{42}H_{42}O_4P_4$ M 734.686
Cryst. (xylene/MeOH). Insol. Et_2O, C_6H_6. Mp 204-206°, 216-217° dec.
Tetrasulfide:
$C_{42}H_{42}P_4S_4$ M 798.929
Cryst. (CH_2Cl_2/hexane). Insol. Et_2O, MeOH. Mp 207-208° dec.

King, R.B. *et al*, *J. Am. Chem. Soc.*, 1971, **93**, 4153 (*synth, ir, pmr, nmr*)
Maier, L, *Phosphorus*, 1972, **1**, 245 (*oxide, pmr, nmr*)
King, R.B. *et al*, *Phosphorus*, 1974, **3**, 209 (*pmr, nmr, derivs*)
King, R.B. *et al*, *Inorg. Chem.*, 1975, **14**, 1550 (*nmr, complexes*)
King, R.B., *Adv. Chem. Ser.*, 1982, **196**, 313 (*rev*)

Tris(diphenylphosphino)methane T-00775

Methylidynetris[diphenylphosphine], *10CI*
[28926-65-0]

$(Ph_2P)_3CH$

$C_{37}H_{31}P_3$ M 568.573
"Tripod" ligand for Cr, Ir, Ni, Fe, Pd, Rh, Ru and W. Cryst. (EtOH). Mp 175-176°.
Dioxide:
$C_{37}H_{31}O_2P_3$ M 600.572
Solid. Insol. Et_2O, pet. ether. Mp 214-218°.
Trioxide: [89915-89-9].
$C_{37}H_{31}O_3P_3$ M 616.571
Solid. Mp 216°.
Monosulfide:
$C_{37}H_{31}P_3S$ M 600.633
Cryst. (EtOH). Mp 224-227°.
Disulfide:
$C_{37}H_{31}P_3S_2$ M 632.693
Solid. Mp 199-200°.
Trisulfide: [28926-66-1].
$C_{37}H_{31}P_3S_3$ M 664.753
Cryst. (Me_2CO or toluene/Et_2O). Mp 220-223°, Mp 230°. Forms stabilized carbanion.
Dioxide, monomethiodide:
$C_{38}H_{34}IO_3P_2$ M 727.537
Solid. Mp 300°.
Dioxide, monosulfide: [89915-88-8].
$C_{37}H_{31}O_2P_3S$ M 632.632
Solid. Mp 201°.
Oxide, disulfide: [102615-39-4].
$C_{37}H_{31}OP_3S_2$ M 648.693
Solid. Mp 215°.

Issleib, K. *et al*, *J. Prakt. Chem.*, 1970, **312**, 456 (*synth, deriv*)
Grim, S.O. *et al*, *Phosphorus Sulfur*, 1980, **9**, 123 (*derivs*)

Grim, S.O. *et al*, *Z. Naturforsch.*, *B*, 1980, **35**, 832 (*synth, pmr, nmr, cmr, derivs*)
Colquhoun, I.J. *et al*, *J. Chem. Soc., Dalton Trans.*, 1981, 1645 (*synth, sulfides, cryst struct*)
Colquhoun, I.J. *et al*, *J. Magn. Reson.*, 1981, **42**, 186 (*pmr, cmr, nmr*)
Bahsoun, A.A. *et al*, *Organometallics*, 1982, **1**, 1114 (*synth, pmr, nmr, ir, ms, complexes*)
Grim, S.O., *Angew. Chem., Int. Ed. Engl.*, 1983, **22**, 254 (*cryst struct*)
Grim, S.O. *et al*, *Inorg. Chem.*, 1986, **25**, 2699 (*derivs, synth, ir, raman, cmr, pmr, P nmr*)

1,1,1-Tris(diphenylphosphinomethyl)ethane T-00776

[2-[(Diphenylphosphino)methyl]-2-methyl-1,3-propanediyl]bis[diphenylphosphine], 9CI
[22031-12-5]

$(Ph_2PCH_2)_3CCH_3$

$C_{41}H_{39}P_3$ M 624.680
Air-sensitive cryst. (EtOH or ligroin). Mp 100-101°. Forms complexes with Group VIII metals.
B,3MeI: 1,1,1-Tris(methyldiphenylphosphoniomethyl)-ethane.
$C_{44}H_{48}I_3P_3$ M 1050.498
Cryst. ($MeNO_2$). Mp 310-311°.

Hewertson, W. *et al*, *J. Chem. Soc.*, 1962, 1490 (*synth, derivs*)
Sacconi, L., *Proc. 6th Conf. Coord. Chem.*, 1976, **6**, 225 (*complexes, rev*)
Sanger, A.R., *J. Mol. Catal.*, 1978, **3**, 221 (*use*)
Davies, S.G. *et al*, *J. Chem. Soc., Chem. Commun.*, 1981, 341 (*complexes*)
Mealli, C. *et al*, *Acta Crystallogr., Sect. B*, 1982, **38**, 1040 (*cryst struct*)

Tris(2-diphenylphosphinophenyl)phosphine, 9CI T-00777

[53218-00-1]

$C_{54}H_{42}P_4$ M 814.821
Ligand for metals of Groups VIB and VIII. Prisms + 1DMF (DMF). Mp 221-223°.

Hartley, J.G. *et al*, *J. Chem. Soc.*, 1963, 3930 (*synth, complexes*)
Mynatt, R.J. *et al*, *J. Coord. Chem.*, 1973, **3**, 145 (*complexes*)

Tris(2-ethoxyethyl) phosphate, 9CI T-00778

$(EtOCH_2CH_2O)_3P{=}O$

$C_{12}H_{27}O_7P$ M 314.315
Liq. d^{20} 1.08. Bp_{20} 225°. n_D^{20} 1.437.

U.S.P., 1 944 530, (*1934*)

Tris(2-ethylhexyl) phosphate T-00779

[78-42-2]

$[H_3CCH_2CH_2CH_2CH(CH_2CH_3)CH_2O]_3P{=}O$

$C_{24}H_{51}O_4P$ M 434.638
Liq. d_4^{20} 0.99. Bp_5 210-220°. n_D^{20} 1.4452. Doubtless a mixt. of diastereoisomers.
▷MP0770000.

Smeykel, K. *et al*, *J. Prakt. Chem.*, 1963, **22**, 186 (*synth*)
Frank, A.W. *et al*, *J. Org. Chem.*, 1966, **31**, 872 (*synth*)
Ger. Pat., 2 944 778, (*1980*); *CA*, **93**, 167882 (*synth*)
Rozen, A.M. *et al*, *Dokl. Akad. Nauk SSSR, Ser. Sci. Khim.*, 1984, **274**, 1139 (*Engl. transl.* p. 177) (*props*)

Tris(2-ethylhexyl) phosphite T-00780

Tri(2-ethylhexyl) phosphite, 9CI
[301-13-3]

$[H_3C(CH_2)_3CH(CH_2CH_3)CH_2O]_3P$

$C_8H_{51}O_3P$ M 226.463
A light stabiliser for PVC. Liq. Bp_1 155-156°. n_D^{20} 1.4476.
▷May be skin-irritant. Rel. nontoxic by oral admin.

Bellamy, L.J. *et al*, *J. Chem. Soc.*, 1952, 475 (*ir*)
Petrov, K.A. *et al*, *Zh. Obshch. Khim.*, 1961, **31**, 2377 (*Engl. transl.* p. 2214) (*synth*)

Tris(2-fluorophenyl)phosphine, 9CI T-00781

[84350-73-2]

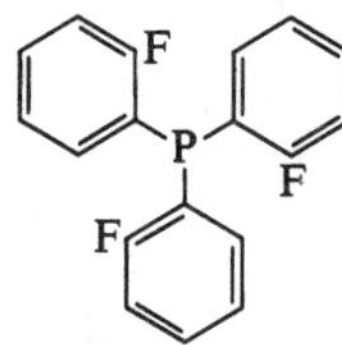

$C_{18}H_{12}F_3P$ M 316.262
Cryst. (EtOH). Mp 147°.
Oxide:
$C_{18}H_{12}F_3OP$ M 332.261
Cryst. (toluene). Mp 157°.

Stegman, H.B. *et al*, *Phosphorus Sulfur*, 1982, **13**, 331 (*synth, cmr, nmr, ms, oxide*)

Tris(3-fluorophenyl)phosphine, 9CI T-00782

[23039-94-3]

$C_{18}H_{12}F_3P$ M 316.262
Ligand for metals of Groups VIIB and VIII. Cryst. (MeOH). Mp 64°. $Bp_{0.1}$ 150°.
B,MeI: Methyltris(3-fluorophenyl)phosphonium iodide.
$C_{19}H_{15}F_3IP$ M 458.201
Cryst. (EtOH). Mp 216°.
Oxide: [54300-34-4].
$C_{18}H_{12}F_3OP$ M 332.261
Cryst. (C_6H_6/pet. ether), needles (hexane). Mp 104°.
Sulfide: [54300-42-4].
$C_{18}H_{12}F_3PS$ M 348.322
Solid. Mp 68°.
Selenide: [54300-50-4].
$C_{18}H_{12}F_3PSe$ M 395.222
Solid. Mp 109°.

Schindlbauer, H. *et al*, *Chem. Ber.*, 1969, **102**, 2914 (*synth, oxide, methiodide*)
De Ketelaere, R.F. *et al*, *J. Mol. Struct.*, 1974, **23**, 233 (*derivs, ir, raman*)
De Ketelaere, R.F. *et al*, *J. Organomet. Chem.*, 1974, **73**, 251 (*ir, raman*)
De Ketelaere, R.F. *et al*, *Phosphorus*, 1974, **5**, 43 (*derivs, ms*)

De Ketelaere, R.F. *et al*, *J. Mol. Struct.*, 1975, **27**, 25 (*derivs, pmr*)
Weiner, M.A. *et al*, *J. Org. Chem.*, 1975, **40**, 1292 (*pe*)
Grim, S.O. *et al*, *J. Org. Chem.*, 1977, **42**, 1236; *Phosphorus Sulfur*, 1977, **3**, 191 (*pmr, nmr*)
Hoste, S. *et al*, *J. Electron. Spectrosc. Relat. Phenom.*, 1979, **17**, 191 (*derivs, pe*)

Tris(4-fluorophenyl)phosphine, 9CI **T-00783**

[18437-78-0]

$C_{18}H_{12}F_3P$ M 316.262

Ligand for metals of Groups IB, VIB, and VIII, as well as for Mn, V, and Hg. Cryst. Mp 77-79°. $Bp_{0.1}$ 160°.

B,MeI: Methyltris(4-fluorophenyl)phosphonium iodide.
$C_{19}H_{15}F_3IP$ M 458.201
Cryst. (EtOH), needles ($EtOH/Et_2O$). Mp 306-307°, 318-320°.

Oxide:
$C_{18}H_{12}F_3OP$ M 332.261
Cryst. (C_6H_6 or MeOH). Mp 121-123°.

Sulfide: [18437-80-4].
$C_{18}H_{12}F_3PS$ M 348.322
Solid. Mp 139-141°.

Selenide: [54300-47-9].
$C_{18}H_{12}F_3PSe$ M 395.222
Solid. Mp 169°.

Schindlbauer, H., *Chem. Ber.*, 1967, **100**, 3432 (*synth, derivs*)
Johnson, A.W. *et al*, *J. Am. Chem. Soc.*, 1968, **90**, 5232 (*synth, derivs*)
De Ketelaere, R.F. *et al*, *J. Mol. Struct.*, 1974, **23**, 233 (*derivs, ir, raman*)
De Ketelaere, R.F. *et al*, *J. Organomet. Chem.*, 1974, **73**, 251 (*ir, raman*)
De Ketelaere, R.F. *et al*, *Phosphorus*, 1974, **5**, 43 (*derivs, ms*)
De Ketelaere, R.F. *et al*, *J. Mol. Struct.*, 1975, **27**, 25 (*derivs, pmr*)
Weiner, M.A. *et al*, *J. Org. Chem.*, 1975, **40**, 1292 (*pe*)
Grim, S.O. *et al*, *Phosphorus Sulfur*, 1977, **3**, 191; *J. Org. Chem.*, 1977, **42**, 1236 (*pmr, nmr*)
Hoste, S. *et al*, *J. Electron. Spectrosc. Relat. Phenom.*, 1979, **17**, 191 (*derivs, pe*)

Tris(hydroxymethyl)phosphine **T-00784**

Phosphinidynetrismethanol

[2767-80-8]

$P(CH_2OH)_3$

$C_3H_9O_3P$ M 124.076

Viscous liq. or hygroscopic needles (butanol). Mp 55°. $Bp_{2.5}$ 111-114° dec. Isomerises in AcOH or HCl at 180-90°.

B,MeI: Methyltris(hydroxymethyl)phosphonium iodide.
$C_4H_{12}IO_3P$ M 266.015
Oil. Insol. Et_2O, Me_2CO.

B,MeBPh₄: Methyltris(hydroxymethyl)phosphonium tetraphenylborate. Solid (Me_2CO/C_6H_6). Mp 170-171° dec.

Tri-Ac: [18788-02-8]. *Tris(acetoxymethyl)phosphine. Phosphinidynetris(methyl acetate).*
$C_9H_{15}O_6P$ M 250.188
Forms water-sol. Rh, Pd, and Pt complexes. In AcOH at 180-90°, isomerizes into $MeP(O)(CH_2OAc)_2$.

Oxide: [1067-12-5]. *Phosphinylidynetrismethanol.*
$C_3H_9O_4P$ M 140.075
Cryst. (EtOH or MeOH). Sol. H_2O, insol. Et_2O, C_6H_6. Mp 44-45°, 50-52°, 69°, 70°. Complexes with Cd, Co, Cr, Cu, Fe, Mn, Ni, and Zn.

Oxide; tri-Ac: [4851-97-2].
$C_9H_{15}O_7P$ M 266.187
Cryst. (Me_2CO/pet. ether). Mp 66-68°. Bp_4 210-211°.

Oxide, tribenzoyl: Tris(benzoyloxymethyl)phosphine oxide. Phosphinylidynetris(methyl benzoate).
$C_{24}H_{21}O_6P$ M 436.400
Solid. Mp 110°.

Grayson, M., *J. Am. Chem. Soc.*, 1963, **85**, 79 (*synth, derivs*)
Grinshtein, E.I. *et al*, *Zh. Obshch. Khim.*, 1966, **36**, 302 (*Engl. transl.* p. 311) (*synth*)
Valetdinov, R.K. *et al*, *Zh. Obshch. Khim.*, 1967, **37**, 2269 (*Engl. transl.* p. 2154) (*synth, props*)
Ellzey, S.E. *et al*, *J. Org. Chem.*, 1972, **37**, 3453 (*synth, oxide*)
Chatt, J. *et al*, *J. Chem. Soc., Dalton Trans.*, 1973, 2021 (*pmr, deriv, complexes*)
Mironova, Z.N. *et al*, *Zh. Obshch. Khim.*, 1974, **44**, 1217 (*Engl. transl.* p. 1195) (*oxide, derivs*)
Tsvetkov, E.N. *et al*, *Izv. Akad. Nauk SSSR, Ser. Khim.*, 1979, 1859 (*Engl. transl.* p. 1722) (*deriv*)

Tris(2-hydroxyphenyl) phosphate **T-00785**

1,2-Benzenediol phosphate(3:1), 9CI

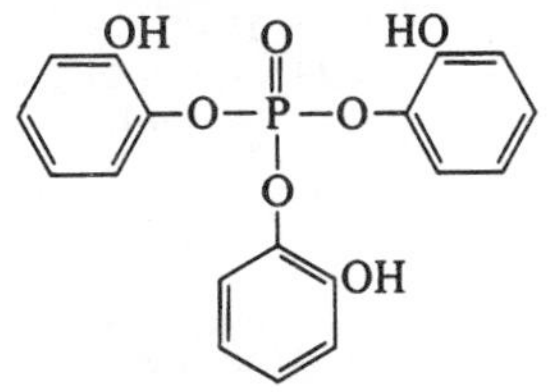

$C_{18}H_{15}O_7P$ M 374.286

Tri-Me ether: Tris(2-methoxyphenyl)phosphate, 9CI.
$C_{21}H_{27}O_7P$ M 422.414
Solid. Mp 90-91°. Bp_3 278-280°.

Breusch, F.L. *et al*, *CA*, 1944, **38**, 1483 (*synth*)
Kucherov, V.F., *Zh. Obshch. Khim.*, 1949, **19**, 126; *CA*, **43**, 6178 (*synth*)

Tris(3-hydroxyphenyl) phosphate **T-00786**

1,3-Benzenediol phosphate (3:1), 9CI

$C_{18}H_{15}O_7P$ M 374.286

Cryst. + $1H_2O$. Mp 75°.

Secretant, M., *Bull. Soc. Chim. Fr.*, 1896, **15**, 361.

Tris(4-hydroxyphenyl) phosphate **T-00787**

1,4-Benzenediol phosphate(3:1), 9CI

[3957-64-0]

$C_{18}H_{15}O_7P$ M 374.286

Tri-Me ether: Tris(4-methoxyphenyl) phosphate, 9CI.
$C_{21}H_{21}O_7P$ M 416.366
Liq. $Bp_{0.2}$ 255°.

Perregaard, J. *et al*, *Recl. Trav. Chim. Pays-Bas*, 1974, **93**, 252 (*synth, pmr*)

Tris(2-hydroxyphenyl)phosphine **T-00788**

2,2',2''-Phosphinidynetrisphenol, 9CI

[77013-89-9]

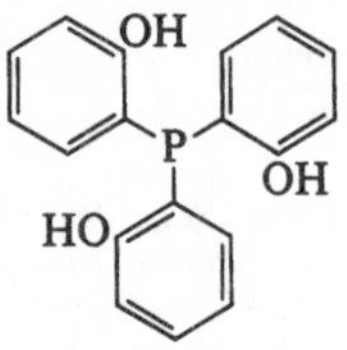

$C_{18}H_{15}O_3P$ M 310.288

Cryst. + $1H_2O$. Mp 182-183°.

B,MeI: Tris(2-hydroxyphenyl)methylphosphonium iodide.

$C_{19}H_{18}IO_3P$ M 452.228

Solid. Mp 297-299° dec.

Oxide: 3,3',3''-Phosphinylidynetrisphenol.

$C_{18}H_{15}O_4P$ M 326.288

Cryst. (EtOH). Mp 214.5-216°.

Oxide, tri-O-Ac: Tris(2-acetyloxyphenyl)phosphine oxide.

$C_{24}H_{21}O_7P$ M 452.399

Solid. Mp 197-199°.

Kennedy, J. *et al*, *J. Chem. Soc.*, 1956, 4670 (*oxide*)
Neunhoeffer, O. *et al*, *Chem. Ber.*, 1961, **94**, 2514 (*synth*)
Tzschach, A. *et al*, *Z. Chem.*, 1980, **20**, 341 (*synth, nmr*)

Tris(3-hydroxyphenyl)phosphine **T-00789**

3,3',3''-Phosphinidynetrisphenol, 9CI

$C_{18}H_{15}O_3P$ M 310.288

Oxide: [42405-96-9]. *3,3',3''-Phosphinylidynetrisphenol.*

$C_{18}H_{15}O_4P$ M 326.288

Lustrous cryst. Mp 270-272°.

Lamza, L., *J. Prakt. Chem.*, 1964, **25**, 294.

Tris(4-hydroxyphenyl)phosphine **T-00790**

4,4',4''-Phosphinidynetrisphenol, 9CI

[26707-09-5]

$C_{18}H_{15}O_3P$ M 310.288

Cryst. + $1H_2O$ in 2 forms. Mp 134-137°, 188-189° (dimorph.).

B,HBr: Cryst. (MeOH/HBr). Mp 227-228° dec.

Tri-Ac:

$C_{24}H_{21}O_6P$ M 436.400

Cryst. (EtOH). Mp 155-156°.

Tri-Ac; B,MeI:

$C_{25}H_{24}IO_6P$ M 578.339

Mp 87-89°. Dec. at 105°.

Oxide: [797-71-7]. *4,4',4''-Phosphinylidynetrisphenol.*

$C_{18}H_{15}O_4P$ M 326.288

Needles (H_2O or MeOH). Mp 278-280°.

Senear, A.E. *et al*, *J. Org. Chem.*, 1960, **25**, 2001 (*oxide*)
Neunhoeffer, O. *et al*, *Chem. Ber.*, 1961, **94**, 2514 (*synth, derivs*)

Tris(2-methoxyphenyl)phosphine, 9CI **T-00791**

Tri-o-anisylphosphine

[4731-65-1]

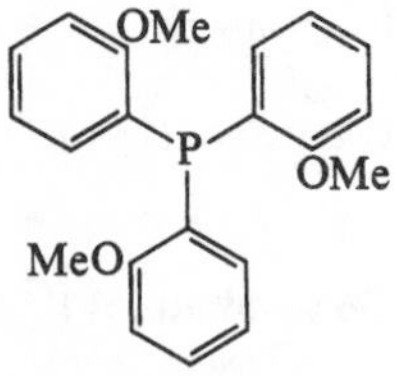

$C_{21}H_{21}O_3P$ M 352.369

Ligand for metals of Groups IB, IIB, and VIII. Cryst. (EtOH or C_6H_6). Mp 204°.

B,MeI: Tris(2-methoxyphenyl)methylphosphonium iodide. Tri-o-anisylmethylphosphonium iodide.

$C_{22}H_{24}IO_3P$ M 494.308

Solid. Mp 212-213°.

Oxide: [47467-89-0].

$C_{21}H_{21}O_4P$ M 368.368

Cryst. (H_2O). Mp 215-217°. Subl. at 175°.

Neunhoeffer, O. *et al*, *Chem. Ber.*, 1961, **94**, 2514 (*synth*)
Fritzsche, H. *et al*, *Chem. Ber.*, 1964, **97**, 1988 (*synth*)
Weiner, M.A. *et al*, *J. Org. Chem.*, 1975, **40**, 1292 (*pe*)
Shvets, A.A. *et al*, *Zh. Obshch. Khim.*, 1976, **46**, 1701 (*Engl. transl.* p. 1654) (*oxide, ir*)
Grim. S.O. *et al*, *J. Org. Chem.*, 1977, **42**, 1236 (*derivs, nmr*)
Grim, S.O. *et al*, *Phosphorus Sulfur*, 1977, **3**, 191 (*nmr*)

Tris(3-methoxyphenyl)phosphine, 9CI **T-00792**

Tri-m-anisylphosphine

[29949-84-6]

$C_{21}H_{21}O_3P$ M 352.369

Ligand for Cr, Mo, W, Ir, and Rh. Solid (EtOH). Mp 114-115°.

B,MeI: Tris(3-methoxyphenyl)methylphosphonium iodide. Tri-m-anisylmethylphosphonium iodide.

$C_{22}H_{24}IO_3P$ M 494.308

Solid. Mp 160-161°.

Oxide: [40331-46-2].

$C_{21}H_{21}O_4P$ M 368.368

Cryst. (EtOH or CCl_4). Mp 151-152°.

Mann, F.G. *et al*, *J. Chem. Soc.*, 1937, 527 (*synth, oxide*)
Lamza, L. *et al*, *J. Prakt. Chem.*, 1964, **25**, 294 (*synth, derivs*)
Grindley, B.T. *et al*, *Aust. J. Chem.*, 1975, **28**, 327 (*ir*)
Weiner, M.A. *et al*, *J. Org. Chem.*, 1975, **40**, 1292 (*pe*)
Grim, S.O. *et al*, *Phosphorus Sulfur*, 1977, **3**, 191 (*nmr*)

Tris(4-methoxyphenyl)phosphine, 9CI **T-00793**

[855-38-9]

$C_{21}H_{21}O_3P$ M 352.369

Ligand for metals of Groups IB, IIB, VIB, VIIB, and VIII. Cryst. (EtOH). Mp 131-132°. pK_a 3.15 (80% EtOH aq.), 4.57 ($MeNO_2$).

B,MeBr: Methyltris(4-methoxyphenyl)phosphonium bromide.

$C_{22}H_{24}BrO_3P$ M 447.308

Solid. Mp 198-210°.

B,MeI: Methyltris(4-methoxyphenyl)phosphonium iodide.

$C_{22}H_{24}IO_3P$ M 494.308

Cryst. (H_2O). Mp 220-224°.

Oxide: [803-17-8].

$C_{21}H_{21}O_4P$ M 368.368

Cryst. (C_6H_6). Mp 145-148°.

Sulfide: [14180-55-3].

$C_{21}H_{21}O_3PS$ M 384.429

Cryst. (MeOH). Mp 117-120°.

Neunhöffer, O. *et al*, *Chem. Ber.*, 1961, **94**, 2514 (*synth*)
Schiemenz, G.P., *Justus Liebigs Ann. Chem.*, 1971, **752**, 30 (*synth, derivs, uv*)
Yakutina, O.A. *et al*, *Zh. Obshch. Khim.*, 1972, **42**, 1733 (*Engl. transl.* p. 1722) (*uv, raman*)
Benassi, R. *et al*, *J. Chem. Soc., Perkin Trans. 2*, 1974, 1338 (*pmr, derivs*)
Grindley, B.T. *et al*, *Aust. J. Chem.*, 1975, **28**, 327 (*oxide, sulfide, ir*)
Weiner, M.A. *et al*, *J. Org. Chem.*, 1975, **40**, 1292 (*pe*)
Grim, S.O. *et al*, *J. Org. Chem.*, 1977, **42**, 1236; *Phosphorus Sulfur*, 1977, **3**, 191 (*nmr*)
Casper, J.M. *et al*, *Spectrochim. Acta, Part A*, 1978, **34**, 1 (*oxide, ir, raman*)
Marshall, G. *et al*, *Org. Mass Spectrom.*, 1981, **16**, 272 (*ms, oxide*)
Bemi, L. *et al*, *J. Am. Chem. Soc.*, 1982, **104**, 438 (*nmr*)
Allman, T. *et al*, *Can. J. Chem.*, 1982, **60**, 716 (*props, pmr, cmr*)

Tris(2-methoxyphenyl) phosphite, 9CI — T-00794

Tri-o-anisyl phosphite

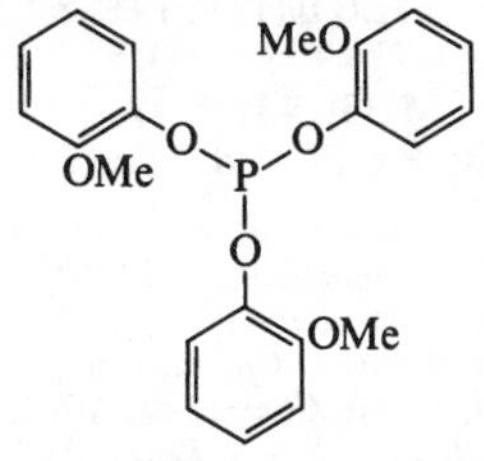

$C_{21}H_{21}O_6P$ M 400.367
Oil which cryst. slowly. Mp 59°. Bp_{13} 275-280°.

Dupuis, P., *C.R. Hebd. Seances Acad. Sci.*, 1910, **150**, 622 (*synth*)

Tris(4-methoxyphenyl) phosphorotritelluroite, 9CI — T-00795

Tris(4-methoxyphenyl) tritellurophosphite
[58558-75-1]

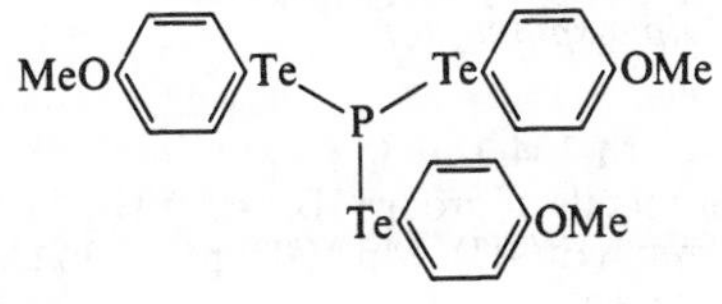

$C_{21}H_{21}O_3PTe_3$ M 735.169
Only example of its type.

Maier, L., *Helv. Chim. Acta*, 1976, **59**, 252 (*synth, pmr*)

Tris(3-methylbutyl) phosphate, 9CI — T-00796

Triisopentyl phosphate, 8CI
[919-62-0]

$[(H_3C)_2CHCH_2CH_2O]_3P{=}O$

$C_{15}H_{33}O_4P$ M 308.397
Liq. Bp_3 143°.

U.S.P., 2 008 478, (*1935*)

Tris(1-methylheptyl) phosphite, 9CI — T-00797

Tri(2-octyl) phosphite. Tri(1-methylheptyl) phosphite

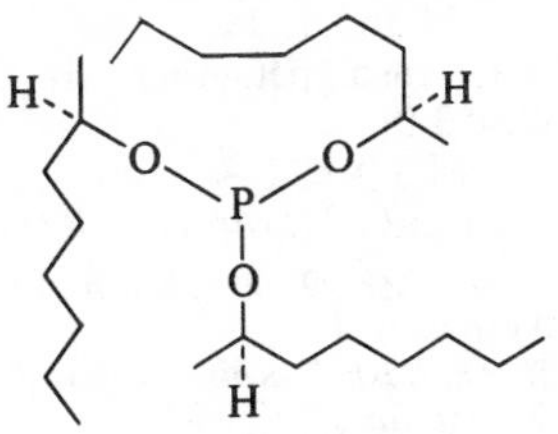

$C_{24}H_{51}O_3P$ M 418.639
(R,R,R)-form
Liq. Bp_2 162-164°. $[\alpha]^{16}_{D}$ −0.8° (neat). n_D^{22} 1.4449.

Gerrard, W., *J. Chem. Soc.*, 1944, 85 (*synth, props*)

Tris(6-methylheptyl) phosphite — T-00798

Triisooctyl phosphite, 8CI. Tri(6-methylheptyl) phosphite
[25103-12-2]

$[(H_3C)_2CH(CH_2)_5O]_3P$

$C_{24}H_{51}O_3P$ M 418.639
d^{29} 0.90. $Bp_{0.3}$ 161-164°. n_D^{25} 1.4520.
▷TH1150000.

Russ. Pat., 3 056 824, (*1962*); *CA*, **58**, 5575 (*synth*)

Tris(8-methylnonyl) phosphite — T-00799

Triisodecyl phosphite. Tri(8-methylnonyl) phosphite
[25448-25-3]

$[(H_3C)_2CH(CH_2)_7O]_3P$

$C_{30}H_{63}O_3P$ M 502.800
Antioxidant and additive to oil for high pressure use. Oil. n_D^{25} 1.5180.

U.S.P., 3 047 608, (*1962*); *CA*, **57**, 16400 (*synth*)

Tris[(2-methylphenyl)methyl]phosphine, 10CI, 9CI — T-00800

Tris(o-methylbenzyl)phosphine, 8CI

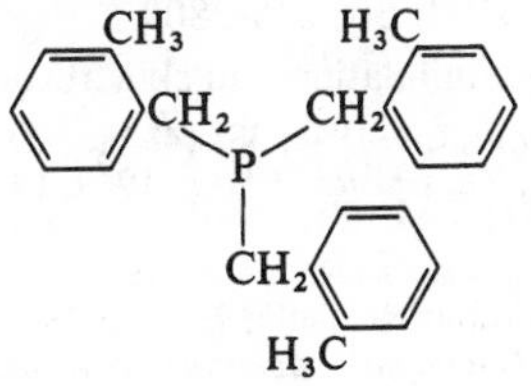

$C_{24}H_{27}P$ M 346.451
Oxide: [18945-69-2].
$C_{24}H_{27}OP$ M 362.450
Needles (CCl_4). Insol. pet. ether. Mp 143-144°.
Sulfide: [31675-49-7].
$C_{24}H_{27}PS$ M 378.511
Needles (C_6H_6 or EtOH). Mp 150-152°.

Zhuravleva, L.P. *et al*, *Zh. Obshch. Khim.*, 1968, **38**, 342 (*Engl. transl.* p. 341) (*sulfide*)
Z'ola, M.I. *et al*, *Zh. Obshch. Khim.*, 1970, **40**, 1937 (*Engl. transl.* p. 1922) (*sulfide*)

Tris[(3-methylphenyl)methyl]phosphine, 9CI — T-00801

Tris(m-methylbenzyl)phosphine, 8CI

$C_{24}H_{27}P$ M 346.451
Oxide: [18945-68-1].
$C_{24}H_{27}OP$ M 362.450
Needles (CCl_4). Insol. pet. ether. Mp 188-189°.
Sulfide: [31675-50-0].
$C_{24}H_{27}PS$ M 378.511
Cryst. (C_6H_6 or EtOH). Mp 186-187°.

Zhuravleva, L.P. *et al*, *Zh. Obshch. Khim.*, 1968, **38**, 342 (*Engl. transl.* p. 341) (*oxide*)
Z'ola, M.I. *et al*, *Zh. Obshch. Khim.*, 1970, **40**, 1937 (*Engl. transl.* p. 1922) (*sulfide*)

Tris[(4-methylphenyl)methyl]phosphine, 9CI **T-00802**
Tris(p-*methylbenzyl*)*phosphine, 8CI*
$C_{24}H_{27}P$ M 346.451
Oxide: [15049-10-2].
$C_{24}H_{27}OP$ M 362.450
Needles (CCl_4). Insol. pet. ether. Mp 168-169°.
Sulfide: [31675-51-1].
$C_{24}H_{27}PS$ M 378.511
Cryst. (C_6H_6 or EtOH). Mp 173-174°.

Zhuravleva, L.P. *et al*, *Zh. Obshch. Khim.*, 1968, **38**, 342 (*Engl. transl.* p. 341) (*oxide*)
Z'ola, M.I. *et al*, *Zh. Obshch. Khim.*, 1970, **40**, 1937 (*Engl. transl.* p. 1922) (*sulfide*)

Tris(2-methylphenyl) phosphate, 9CI **T-00803**
*Tri-*o*-tolyl phosphate. Tri-*o*-cresyl phosphate. TOCP. Phosflex 179-C*

CH3 O O–P=O O CH3 CH3

$C_{21}H_{21}O_4P$ M 368.368
Flame retardant in vinyl and cellulosic plastics. Solid. Mp 90-91°. Bp 410°, Bp_3 275-280°.
▷Highly toxic by inhalation and skin absorption, TLV 0.1

Breusch, F.L. *et al*, *CA*, 1944, **38**, 1483 (*synth*)
Kucherov, V.F., *Zh. Obshch. Khim.*, 1949, **19**, 126; *CA*, **43**, 6178.
Sax, N.I., *Dangerous Properties of Industrial Materials*, 6th Ed., Van Nostrand-Reinhold, 1984, 1048.
Hazards in the Chemical Laboratory, (Bretherick, L., Ed.), 3rd Ed., Royal Society of Chemistry, London, 1981, 526.

Tris(3-methylphenyl) phosphate, 9CI **T-00804**
*Tri-*m*-tolyl phosphate. Tri-*m*-cresyl phosphate*
$C_{21}H_{21}O_4P$ M 368.368
Low-melting solid. Mp 25-26°. Bp_4 258-263°.

Breusch, F.L. *et al*, *CA*, 1944, **38**, 1483 (*synth*)

Tris(4-methylphenyl) phosphate, 9CI **T-00805**
*Tri-*p*-tolyl phosphate. Tri-*p*-cresyl phosphate*
[78-32-0]
$C_{21}H_{21}O_4P$ M 368.368
Cryst. (EtOH). Mp 77.5-78°.

Breusch, F.L. *et al*, *CA*, 1944, **38**, 1483 (*synth*)
Krishnakumar, V.K. *et al*, *Synthesis*, 1983, 558; *Synth. Commun.*, 1984, **14**, 189 (*synth, ir, pmr*)

Tris(2-methylphenyl)phosphine, 9CI **T-00806**
*Tri-*o*-tolylphosphine, 8CI*
[6163-58-2]

H3C P CH3 CH3

$C_{21}H_{21}P$ M 304.371
Ligand for metals of Groups IB, VIB and VIII. Air-sensitive cryst. (EtOH). Mp 125-128°. pK_a 3.08, 8.42 ($MeNO_2$).
Oxide: [6163-63-9].
$C_{21}H_{21}OP$ M 320.370
Cryst. + ½H_2O (EtOH). Mp 153°.
Sulfide: [6163-61-7].
$C_{21}H_{21}PS$ M 336.431
Forms Cd, Hg, Pd, Pt, and Ta complexes. Cryst. Mp 163°.

Mann, F.G. *et al*, *J. Chem. Soc.*, 1937, 527 (*synth, oxide*)
Fritzche, H. *et al*, *Chem. Ber.*, 1964, **97**, 1988 (*synth*)
Schindlbauer, H., *Monatsh. Chem.*, 1965, **96**, 1793 (*uv*)
Pinnell, R.P. *et al*, *J. Am. Chem. Soc.*, 1973, **95**, 977 (*nmr*)
Bokanov, A.I. *et al*, *Zh. Obshch. Khim.*, 1974, **44**, 760 (*Engl. transl.* p. 732) (*props*)
Cameron, T.S. *et al*, *J. Chem. Soc., Perkin Trans. 2*, 1975, 1737 (*oxide, sulfide, selenide, cryst struct*)
Weiner, M.A. *et al*, *J. Org. Chem.*, 1975, **40**, 1292 (*pe*)
Shvets, A.A. *et al*, *Zh. Obshch. Khim.*, 1976, **46**, 1701 (*Engl. transl.* p. 1654) (*oxide, ir*)
Spalding, T.R. *et al*, *Org. Mass Spectrom.*, 1976, **11**, 1019 (*ms*)
Grim, S.O. *et al*, *J. Org. Chem.*, 1977, **42**, 1236 (*nmr, derivs*)
Dean, P.A.W., *Can. J. Chem.*, 1979, **57**, 754 (*selenide, nmr*)
Allman, T. *et al*, *Can. J. Chem.*, 1982, **60**, 716 (*props, pmr, cmr*)

Tris(3-methylphenyl)phosphine, 9CI **T-00807**
*Tri-*m*-tolylphosphine, 8CI*
[6224-63-1]
$C_{21}H_{21}P$ M 304.371
Ligand for metals of groups IB, IIB, VIB, VIIB, and VIII. Cryst. (EtOH). Mp 100°. pK_a 3.30 ($MeNO_2$).
Oxide: [6151-88-8].
$C_{21}H_{21}OP$ M 320.370
Cryst. (pet. ether). Mp 111°.
Sulfide: [6163-62-8].
$C_{21}H_{12}PS$ M 327.360
Cryst. Mp 156°.

Mann, F.G. *et al*, *J. Chem. Soc.*, 1937, 527 (*synth*)
Schindlbauer, H., *Monatsh. Chem.*, 1965, **96**, 1793 (*uv*)
Pinnell, R.P. *et al*, *J. Am. Chem. Soc.*, 1973, **95**, 977 (*nmr, selenide*)
Grindley, T.B. *et al*, *Aust. J. Chem.*, 1975, **28**, 327 (*ir, oxide*)
Weiner, M.A. *et al*, *J. Org. Chem.*, 1975, **40**, 1292 (*pe*)
Spalding, T.R., *Org. Mass Spectrom.*, 1976, **11**, 1019 (*ms*)
Grim, S.O. *et al*, *J. Org. Chem.*, 1977, **42**, 1236; *Phosphorus Sulfur*, 1977, **3**, 191 (*nmr*)
Cameron, T.S. *et al*, *Acta Crystallogr., Sect. B*, 1978, **34**, 1639 (*cryst struct, phosphine, oxide, sulfide*)
Allman, T. *et al*, *Can. J. Chem.*, 1982, **60**, 716 (*props, pmr, cmr*)

Tris(4-methylphenyl)phosphine, 9CI **T-00808**
*Tri-*p*-tolylphosphine, 8CI*

[1038-95-5]

$C_{21}H_{21}P$ M 304.371

Ligand for metals of Groups IB, IIB, VIB, VIIB and VIII. Rhombs (Me_2CO), silky needles or prisms (EtOH aq.). Mp 146°. pK_a 3.84 ($MeNO_2$). Air-sensitive.

▷SZ3880000.

B,MeBr: Methyltris(4-methylphenyl)phosphonium bromide.

$C_{22}H_{24}BrP$ M 399.309

Cryst. Mp 219-220°.

B,MeI: Methyltris(4-methylphenyl)phosphonium iodide.

$C_{22}H_{24}IP$ M 446.310

Cryst. (EtOH). Mp 191-192°.

Oxide: [797-70-6].

$C_{21}H_{21}OP$ M 320.370

Complexes with Fe, Ni, Pd, Ta, and Sn. Cryst. (C_6H_6). Mp 145-146°.

▷SZ2000000.

Sulfide: [6224-65-3].

$C_{21}H_{21}PS$ M 336.431

Complexes with Cd, Fe, Hg, Ni, Pd, Pt, and Ta. Cryst. (pet. ether). Mp 185-186°. $Bp_{0.3}$ 220-250°.

Selenide: [10089-43-7].

$C_{21}H_{21}PSe$ M 383.331

Needles. Mp 193°.

Morgan, P.W. *et al*, *J. Am. Chem. Soc.*, 1952, **74**, 4526 (*synth*)
Balian, V. *et al*, *J. Org. Chem.*, 1961, **25**, 1833 (*derivs*)
Maier, L., *Helv. Chim. Acta*, 1964, **47**, 2137 (*derivs*)
Mallion, K.B. *et al*, *J. Chem. Soc.*, 1964, 5716 (*synth, deriv*)
Cameron, T.S. *et al*, *Phosphorus*, 1973, **3**, 71 (*cryst struct, deriv*)
Pinnell, R.P. *et al*, *J. Am. Chem. Soc.*, 1973, **95**, 977 (*nmr, selenide*)
Bokanov, A.I. *et al*, *Zh. Obshch. Khim.*, 1974, **44**, 760 (*Engl. transl.* p. 732) (*props*)
Grindley, T.B. *et al*, *Aust. J. Chem.*, 1975, **28**, 327 (*ir, oxide, sulfide*)
Weiner, M.A. *et al*, *J. Org. Chem.*, 1975, **40**, 1292 (*pe*)
Spalding, T.R. *et al*, *Org. Mass Spectrom.*, 1976, **11**, 1019 (*ms*)
Grim, S.O. *et al*, *J. Org. Chem.*, 1977, **42**, 1232; *Phosphorus Sulfur*, 1977, **3**, 191 (*nmr*)
Casper, J.M. *et al*, *Spectrochim. Acta, Part A*, 1978, **34**, 1 (*oxide, ir, raman*)
Ratovskii, G.V. *et al*, *Zh. Obshch. Khim.*, 1978, **48**, 1520 (*Engl. transl.* p. 1394) (*uv*)
Dean, P.A.W., *Can. J. Chem.*, 1979, **57**, 754 (*selenide, nmr*)
Marshall, G. *et al*, *Org. Mass Spectrom.*, 1981, **16**, 272 (*ms, oxide*)
Allmann, T. *et al*, *Can. J. Chem.*, 1982, **60**, 716 (*props, pmr, cmr*)
Bemi, L. *et al*, *J. Am. Chem. Soc.*, 1982, **104**, 438 (*nmr*)
Sabolev, A.N. *et al*, *Zh. Strukt. Khim.*, 1983, **24**, 123 (*Engl. transl.* p. 434) (*cryst struct, uv*)

Tris(2-methylphenyl) phosphite, 9CI — T-00809

Tri-o-tolyl phosphite, 8CI. Tri-o-cresyl phosphite

[2622-08-4]

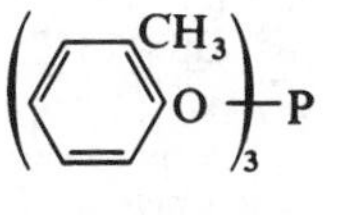

$C_{21}H_{21}O_3P$ M 352.369

Ligand for Mo and metals of group VIII. Oil. d^{20} 1.138. Bp_1 193-194°. n_D^{25} 1.5760.

▷TH1050000.

Walsh, E.N., *J. Am. Chem. Soc.*, 1959, **81**, 3023.

Tris(3-methylphenyl) phosphite, 9CI — T-00810

Tri-m-cresyl phosphite. Tri-m-tolyl phosphite

[620-38-2]

$C_{21}H_{21}O_3P$ M 352.369

Oil. d_{25}^{25} 1.12. Bp_1 188°. n_D^{25} 1.5734.

▷TH1000000.

Walsh, E.N., *J. Am. Chem. Soc.*, 1959, **81**, 3023 (*synth*)

Tris(4-methylphenyl) phosphite, 9CI — T-00811

Tri-p-cresyl phosphite. Tri-p-tolyl phosphite

[620-42-8]

$C_{21}H_{21}O_3P$ M 352.369

Ligand for Ir, Fe, Mn, Ni, and Pt. Oil. d_{25}^{25} 1.11. Bp_1 194°. n_D^{25} 1.5734.

▷TH1100000.

Strecker, W. *et al*, *Ber.*, 1916, **49**, 63 (*synth*)
Walsh, E.N., *J. Am. Chem. Soc.*, 1959, **81**, 3023 (*synth*)

O,O,O-Tris(2-methylphenyl) phosphorothioate, 9CI — T-00812

O,O,O-Tri-o-tolyl phosphorothioate, 8CI. O,O,O-Tri-o-tolyl thiophosphate

[631-45-8]

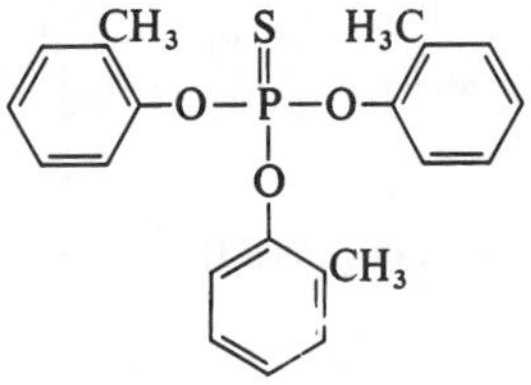

$C_{21}H_{21}O_3PS$ M 384.429

Solid. Mp 45-46°. Bp_1 260-265°, $Bp_{0.02}$ 169-170°.

Gottleib, H.B., *J. Am. Chem. Soc.*, 1932, **54**, 748 (*synth*)
Muller, N. *et al*, *J. Am. Chem. Soc.*, 1956, **78**, 3557 (*P nmr*)
Mazitova, F.N. *et al*, *Zh. Obshch. Khim.*, 1980, **50**, 815 (*Engl. transl.* p. 652) (*synth, P nmr*)

O,O,O-Tris(3-methylphenyl) phosphorothioate, 9CI — T-00813

O,O,O-Tri-m-tolyl phosphorothioate, 8CI. O,O,O-Tri-m-tolyl thiophosphate

$C_{21}H_{21}O_3PS$ M 384.429

Cryst. (EtOH). Mp 40-41°. Bp_{12} 270-272°.

Broeker, W., *J. Prakt. Chem.*, 1928, **118**, 287; *CA*, **22**, 1964.

O,O,O-Tris(4-methylphenyl) phosphorothioate, 9CI — T-00814

O,O,O-Tri-p-tolyl phosphorothioate, 8CI. O,O,O-Tri-p-tolyl thiophosphate

[597-84-2]

$C_{21}H_{21}O_3PS$ M 384.429

Cryst. (EtOH). Mp 93-94°. $Bp_{0.6}$ 234-235°.

Strecker, W. *et al*, *Ber.*, 1916, **49**, 63 (*synth*)
Dwek, R.A. *et al*, *J. Chem. Soc. (A)*, 1970, 1173 (*P nmr*)
Mazitova, F.N. *et al*, *Zh. Obshch. Khim.*, 1980, **50**, 815 (*Engl. transl.* p. 652) (*synth, P nmr*)

S,S,S-Tris(2-methylphenyl) phosphorotrithioate, 9CI **T-00815**

S,S,S-*Tri*-o-*tolyl phosphorotrithioate, 8CI.* S,S,S-*Tri*-o-*tolyl trithiophosphate*

[35029-38-0]

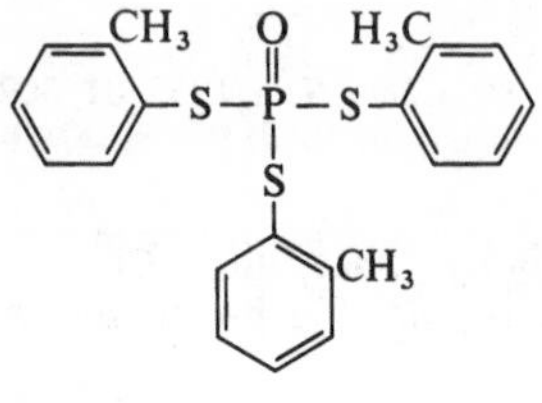

$C_{21}H_{21}OPS_3$ M 416.550

Cryst. (EtOH). Mp 62°.

Shaw, R.A. *et al*, *Phosphorus*, 1971, **1**, 41 (*synth, ir*)

S,S,S-Tris(4-methylphenyl) phosphorotrithioate, 9CI **T-00816**

S,S,S-*Tri*-p-*tolyl phosphorotrithioate, 8CI.* S,S,S-*Tri*-p-*tolyl trithiophosphate*

[13799-87-6]

$C_{21}H_{21}OPS_3$ M 416.550

Cryst. (EtOH). Mp 120°, Mp 138-140°.

Buckler, S.A. *et al*, *J. Org. Chem.*, 1962, **27**, 794 (*synth*)
Dwek, R.A. *et al*, *J. Chem. Soc.* (*A*), 1970, 1173 (*P nmr*)
Shaw, R.A. *et al*, *Phosphorus*, 1971, **1**, 41 (*synth, ir*)

N,N′,N″-Tris(2-methylphenyl)phosphorous triamide, 9CI **T-00817**

N,N′,N″-*Tri*-o-*tolylphosphorous triamide, 8CI. Tris*(o-*tolylamino*)*phosphine. Tri*(o-*tolylamino*)*phosphine*

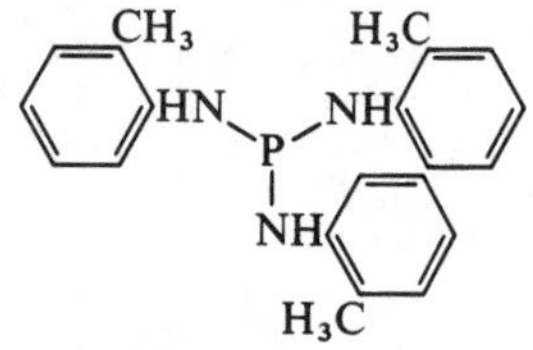

$C_{21}H_{24}N_3P$ M 349.414

Oxide: [31160-11-9]. N,N′,N″-*Tris*(*2-methylphenyl*)-*phosphoric triamide.*

$C_{21}H_{24}N_3OP$ M 365.414

Needles or plates. Sl. sol. EtOH, AcOH, $CHCl_3$, Me_2CO, insol. CCl_4. Mp 229-230° dec. pK_a 5.79 ($MeNO_2$).

Sulfide: [57858-64-7]. N,N′,N″-*Tris*(*2-methylphenyl*)-*phosphorothioic triamide.*

$C_{21}H_{24}N_3PS$ M 381.474

Cryst. (AcOH). Mp 134-135°.

Rudert, P., *Ber.*, 1893, **26**, 565 (*derivs*)
Autenrieth, W. *et al*, *Ber.*, 1900, **33**, 2099 (*oxide*)
Audrieth, L.F. *et al*, *J. Am. Chem. Soc.*, 1942, **64**, 1553 (*oxide*)

N,N′,N″-Tris(3-methylphenyl)phosphorous triamide, 9CI **T-00818**

N,N′,N″-*Tri*-m-*tolylphosphorous triamide. Tris*(m-*tolylamino*)*phosphine. Tri*(m-*tolylamino*)*phosphine*

$C_{21}H_{24}N_3P$ M 349.414

Oxide: [31160-12-0]. N,N′,N″-*Tris*(*3-methylphenyl*)-*phosphoric triamide.*

$C_{21}H_{24}N_3OP$ M 365.414

pK_a 5.23 ($MeNO_2$).

Yakshin, V.V. *et al*, *Dokl. Akad. Nauk SSSR, Ser. Sci. Khim.*, 1979, **247**, 128 (*Engl. transl.* p. 344) (*props*)

N,N′,N″-Tris(4-methylphenyl)phosphorous triamide, 9CI **T-00819**

N,N′,N″-*Tri*-p-*tolylphosphorous triamide. Tris*(p-*tolylamino*)*phosphine. Tri*(p-*tolylamino*)*phosphine*

$C_{21}H_{24}N_3P$ M 349.414

Oxide: [31160-10-8]. N,N′,N″-*Tris*(*4-methylphenyl*)-*phosphoric triamide.*

$C_{21}H_{24}N_3OP$ M 365.414

Cryst. (EtOH). Sl. sol. Me_2CO, EtOH, AcOH, $CHCl_3$, insol. CCl_4. Mp 198-199°, 250° dec. pK_a 5.85 ($MeNO_2$).

Sulfide: [57858-63-6]. N,N′,N″-*Tris*(*4-methylphenyl*)-*phosphorothioic triamide.*

$C_{21}H_{24}N_3PS$ M 381.474

Needles (EtOH). Mp 164°, 186°.

Rudert, P., *Ber.*, 1893, **26**, 565 (*derivs*)
Authenrieth, W. *et al*, *Ber.*, 1900, **33**, 2099, 2112 (*sulfide*)
Audrieth, L.F. *et al*, *J. Am. Chem. Soc.*, 1942, **64**, 1553 (*oxide*)
Cameron, T.S. *et al*, *Z. Naturforsch., B*, 1976, **31**, 1295 (*oxide, cryst struct*)

Tris(2-methyl-2-propenyl) phosphate, 9CI **T-00820**

$$[H_2C{=}C(CH_3)CH_2O]_3P{=}O$$

$C_{12}H_{21}O_4P$ M 260.269

Liq. Bp_5 134.5-140°, $Bp_{0.1}$ 90-92°. n_D^{25} 1.4454.

Kennedy, J. *et al*, *J. Appl. Chem.*, 1958, **8**, 459 (*synth, props*)

Tris(1-methylpropyl) phosphate, 9CI **T-00821**

Tri-sec-*butyl phosphate, 8CI*

[2528-45-2]

$$[H_3CCH_2CH(CH_3)O]_3P{=}O$$

$C_{12}H_{27}O_4P$ M 266.317

Liq. Bp_{8-12} 119-129°.

Noller, C.R. *et al*, *J. Am. Chem. Soc.*, 1933, **55**, 424 (*synth*)
Laskorin, B.N. *et al*, *Radiokhimiya*, 1984, **26**, 161 (*Engl. transl.* p. 147) (*ir, use*)

Tris(2-methylpropyl) phosphate, 9CI **T-00822**

Triisobutyl phosphate, 8CI

[126-71-6]

$$[H_3CCH(CH_3)CH_2O]_3PO$$

$C_{12}H_{27}O_4P$ M 266.317

Catalyst for polymerisation. Extractant for transuranic metals. Liq. Bp 264°, Bp_{10} 138°. n_D^{20} 1.4193.

Evans, D.P. *et al*, *J. Chem. Soc.*, 1930, 1310 (*synth*)
Frank, A.W. *et al*, *J. Org. Chem.*, 1966, **31**, 872 (*synth*)
Donaldson, B. *et al*, *Can. J. Chem.*, 1972, **50**, 2111 (*pmr*)
Laskorin, B.N. *et al*, *Radiokhimiya*, 1984, **26**, 161 (*Engl. transl.* p. 147) (*ir, use*)

Tris(1-methylpropyl)phosphine, 9CI **T-00823**

Tri-sec-*butylphosphine*

[17586-49-1]

P[CH(CH$_3$)CH$_2$CH$_3$]$_3$

$C_{12}H_{27}P$ M 202.319
Liq. Bp$_{11}$ 108°. n_D^{25} 1.4028.
B,MeI: Methyltris(1-methylpropyl)phosphonium iodide.
$C_{13}H_{30}IP$ M 344.258
Cryst. (H_2O). Mp 151-152°.
Oxide: [2959-65-1]. Liq. Bp$_{0.3}$ 105-111°. pK_a 9.49 ($MeNO_2$). n_D^{20} 1.4598.

Davies, W.C., *J. Chem. Soc.*, 1933, 1043 (*synth*)
Siddall, T.H. *et al*, *J. Chem. Eng. Data*, 1965, **10**, 303 (*oxide*)
Grim, S.O. *et al*, *J. Chem. Eng. Data*, 1970, **15**, 497 (*nmr*)

Tris(2-methylpropyl)phosphine, 9CI — T-00824

Triisobutylphosphine
[4125-25-1]

P[CH$_2$CH(CH$_3$)$_2$]$_3$

$C_{12}H_{27}P$ M 202.319
Ligand for Cr, Ni, Pd, and V. Liq. Bp 215°, Bp$_{0.1}$ 51-60°. pK_a 7.97. n_D^{25} 1.4530.
B,MeI: Methyltris(2-methylpropyl)phosphonium iodide.
$C_{13}H_{30}IP$ M 344.258
Solid. Mp 287-288°.
Oxide: [7682-87-3].
$C_{12}H_{27}OP$ M 218.318
Mp 89°, 123-125°. Bp$_4$ 119-120°.
Sulfide: [3982-87-4].
$C_{12}H_{27}PS$ M 234.379
Mp 59-60°.

Davies, W.C. *et al*, *J. Chem. Soc.*, 1929, 1262 (*synth*)
Malatesta, L., *Gazz. Chim. Ital.*, 1947, **77**, 509; *CA*, **42**, 5411 (*sulfide*)
Petrov, K.A. *et al*, *Zh. Obshch. Khim.*, 1960, **30**, 2995 (*Engl. transl.* p. 2967) (*oxide*)
Quin, L.D. *et al*, *Org. Magn. Reson.*, 1974, **6**, 503 (*cmr, nmr, sulfide*)
Wolfsberger, W. *et al*, *Synth. React. Inorg. Met.-Org. Chem.*, 1974, **4**, 149 (*synth*)

Tris(1-methylpropyl) phosphite, 9CI — T-00825

*Tri-*sec-*butyl phosphite. Tri(1-methylpropyl) phosphite*
[7504-61-2]

$C_{12}H_{27}O_3P$ M 250.317
(R,R,R)-form
Liq. Bp$_{10}$ 101°. $[\alpha]_D^{23}$ +1.68° (neat). n_D^{20} 1.4286.

Goodwin, D.G. *et al*, *J. Chem. Soc. (B)*, 1968, 1333 (*synth, P nmr*)

Tris(2-methylpropyl) phosphite, 9CI — T-00826

Triisobutyl phosphite, 8CI
[1606-96-8]

P[OCH$_2$CH(CH$_3$)$_2$]$_3$

$C_{12}H_{27}O_3P$ M 250.317
Liq. Bp 234-235°, Bp$_{14}$ 112°. n_D^{20} 1.4330.

Mel'nikov, N.N. *et al*, *Zh. Obshch. Khim.*, 1958, **28**, 2473; *CA*, **53**, 3032 (*synth*)
Ovchinnikova, N.K. *et al*, *Zh. Org. Khim.*, 1975, **11**, 1839 (*Engl. transl.* p. 1849) (*synth*)

O,O,O-Tris(2-methylpropyl) phosphoroselenoate, 9CI — T-00827

O,O,O-*Triisobutyl phosphoroselenoate.* O,O,O-*Triisobutyl selenophosphate*
[22230-88-2]

[(H$_3$C)$_2$CHCH$_2$O]$_3$P=Se

$C_{12}H_{27}O_3PSe$ M 329.277
Liq. d_4^{20} 1.12. Bp$_{0.01}$ 77-78°. n_D^{20} 1.4620.

Nuretdinov, I.A. *et al*, *Izv. Akad. Nauk SSSR, Ser. Khim.*, 1968, 2831 (*Engl. transl.* p. 2685) (*synth, ir*)

O,O,S-Tris(2-methylpropyl) phosphorothioate, 9CI — T-00828

O,O,S-*Triisobutyl phosphorothioate, 8CI.* O,O,S-*Triisobutyl thiophosphate*

[(H$_3$C)$_2$CHCH$_2$O]$_2$P(O)SCH$_2$CH(CH$_3$)$_2$

$C_{12}H_{27}O_3PS$ M 282.377
Liq. d_0^0 0.99. Bp$_{20}$ 155°.

Pishchimuka, P., *J. Russ. Phys. Chem. Soc.*, 1912, **44**, 1406; *CA*, **7**, 987.

N,N',N''-Tris(1-methylpropyl)phosphorous triamide — T-00829

*Tri(*sec-*butylamino)phosphine. Tris(*sec-*butylamino)phosphine*

P[NHCH(CH$_3$)CH$_2$CH$_3$]$_3$

$C_{12}H_{30}N_3P$ M 247.363
Oxide: N,N',N''-*Tris(1-methylpropyl)phosphoric triamide.*
$C_{12}H_{30}N_3OP$ M 263.362
Oil. pK_{a1} 11.91, pK_{a2} 5.47 ($MeNO_2$).
Sulfide: [69695-91-6]. N,N',N''-*Tris(1-methylpropyl)phosphorothioic triamide.*
$C_{12}H_{30}N_3PS$ M 279.423
Solid. Mp 94°.

Sosnovsky, G. *et al*, *Z. Naturforsch., B*, 1976, **31**, 1526 (*oxide*)
Healy, J.D. *et al*, *Phosphorus Sulfur*, 1978, **5**, 239 (*sulfide*)

N,N',N''-Tris(2-methylpropyl)phosphorous triamide, 9CI — T-00830

Tris(isobutylamino)phosphine. Tri(isobutylamino)phosphine. Triisobutylphosphorous triamide

[(H$_3$C)$_2$CHCH$_2$NH]$_3$P

$C_{12}H_{30}N_3P$ M 247.363
Oxide: [6141-77-1]. N,N',N''-*Tris(2-methylpropyl)phosphoric triamide. Tris(isobutylamino)phosphine oxide.*
$C_{12}H_{30}N_3OP$ M 263.362
Needles (pet. ether). Mp 57°. pK_{a1} 13.28, pK_{a2} 4.66 ($MeNO_2$).
Sulfide: [6141-79-3]. N,N',N''-*Tris(2-methylpropyl)phosphorothioic triamide. Tris(isobutylamino)phosphine sulfide.*
$C_{12}H_{30}N_3PS$ M 279.423

Needles (pet. ether). Mp 78°.

Bock, H. *et al*, *Chem. Ber.*, 1966, **99**, 377 (*derivs*)
Buchikhin, E.P. *et al*, *Zh. Obshch. Khim.*, 1974, **44**, 1354 (*Engl. transl.* p. 1330) (*oxide, props*)
Healy, J.D. *et al*, *Phosphorus Sulfur*, 1978, **5**, 239 (*sulfide, synth, props*)

Tris[2-(methylthio)phenyl]phosphine, 9CI — T-00831

[17617-66-2]

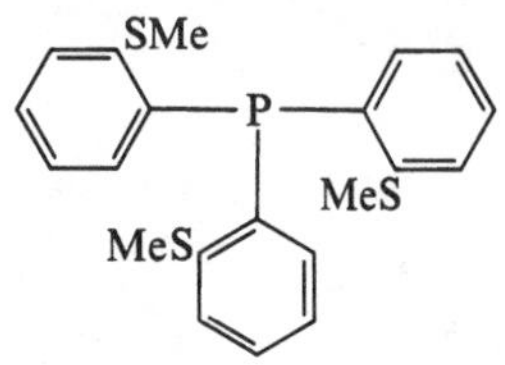

$C_{21}H_{21}S_3P$ M 400.551
Ligand. Cryst. (butanol).

Inorg. Synth., 1976, **16**, 168 (*synth*)

Tris[4-(methylthio)phenyl]phosphine, 9CI — T-00832

[29949-80-2]

$C_{21}H_{21}S_3P$ M 400.551
Ligand for Cr, Mo and W.
▷SZ3800000.

Grim, S.O. *et al*, *Phosphorus Sulfur*, 1977, **3**, 191 (*nmr*)

3,3,3-Tris-(4-morpholinophosphoranylidene)-1-(2,4,6-trinitrophenyl)triazene — T-00833

4,4′,4″-[[3-(2,4,6-Trinitrophenyl)-2-triazenylidene]-phosphoranylidene]trismorpholine, 11CI

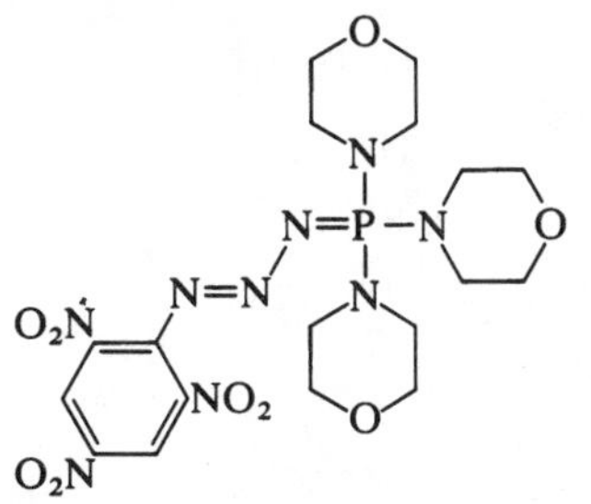

$C_{18}H_{26}N_9O_9P$ M 543.432
(*E*)-*form* [94721-88-7]
Unusual in being a PNNN structure stable to heat. Orange prisms (MeCN). Mp 155-156° dec.

Chernega, A.N. *et al*, *Zh. Obshch. Khim.*, 1984, **54**, 1979 (*Engl. transl.* p. 1766) (*cryst struct*)
Ponomarchuk, M.P. *et al*, *Zh. Obshch. Khim.*, 1984, **54**, 2468 (*Engl. transl.* p. 2204) (*synth, props*)

Tris[(4-nitrophenyl)methyl] phosphate, 9CI — T-00834

Tris(p-nitrobenzyl) phosphate, 8CI
[66777-93-3]

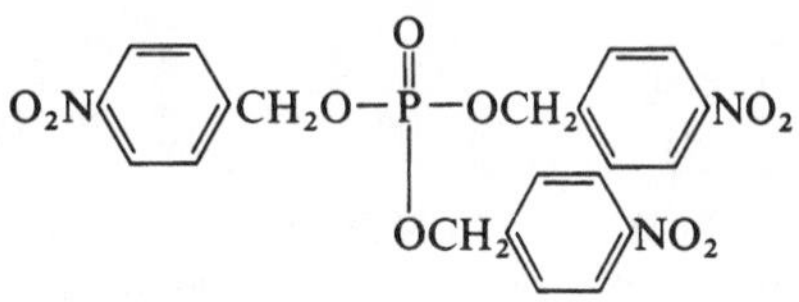

$C_{21}H_{18}N_3O_{10}P$ M 503.361

Pale-yellow plates (EtOH). Mp 127-128°.
Hydrogenolysis → *p*-toluidine and H_3PO_4.

Baddiley, J. *et al*, *J. Chem. Soc.*, 1949, 815 (*synth, props*)

Tris[(4-nitrophenyl)methyl]phosphine, 10CI — T-00835

Tri(p-nitrobenzyl)phosphine, 8CI

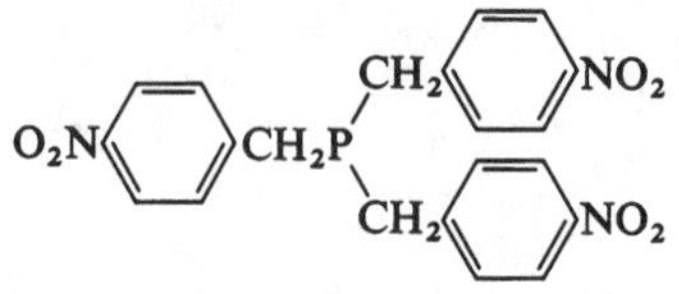

$C_{21}H_{18}N_3O_6P$ M 439.363
Oxide: [67265-07-4].
$C_{21}H_{18}N_3O_7P$ M 455.363
Almost colourless needles (AcOH aq.). Sl. sol. $CHCl_3$, Me_2CO, C_6H_6, EtOH, insol. pet. ether.

Challenger, F. *et al*, *J. Chem. Soc.*, 1929, 2610 (*oxide*)
Raevskii, O.A. *et al*, *Zh. Obshch. Khim.*, 1978, **48**, 1053 (*Engl. transl.* p. 959) (*ir*)

Tris(2-nitrophenyl) phosphate, 9CI — T-00836

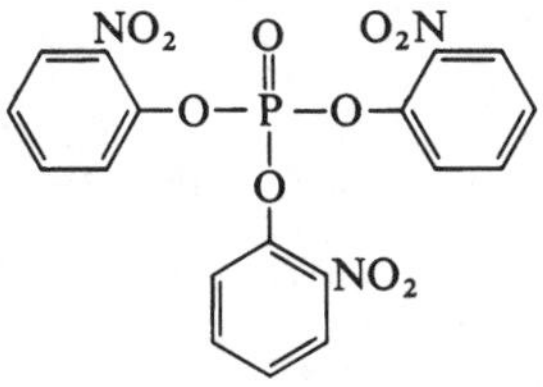

$C_{18}H_{12}N_3O_{10}P$ M 461.281
Solid. Mp 127-129°.

Kirsanov, A.V. *et al*, *Zh. Obshch. Khim.*, 1956, **26**, 2642 (*Engl. transl.* p. 2947)

Tris(4-nitrophenyl) phosphate, 9CI — T-00837

[3871-20-3]

$C_{18}H_{12}N_3O_{10}P$ M 461.281
Converts alkyl and activated aryl and heteroaryl halides into their *p*-nitrophenyl ethers. Peptide forming reagt. Cryst. (EtOAc or Me_2CO). Mp 156°.

Moffatt, J.G. *et al*, *J. Am. Chem. Soc.*, 1957, **79**, 3741 (*synth*)
Bel'skii, V.E. *et al*, *Izv. Akad. Nauk SSSR, Ser. Khim.*, 1979, 1633 (*Engl. transl.* p. 1510) (*P nmr*)
Mukaiyama, T. *et al*, *Chem. Lett.*, 1979, 1305 (*use*)
Effenberger, F. *et al*, *Synthesis*, 1981, 70 (*synth*)
Ohta, A. *et al*, *Synthesis*, 1982, 828 (*props, use*)

Tris(3-nitrophenyl)phosphine, 9CI T-00838

[31638-73-0]

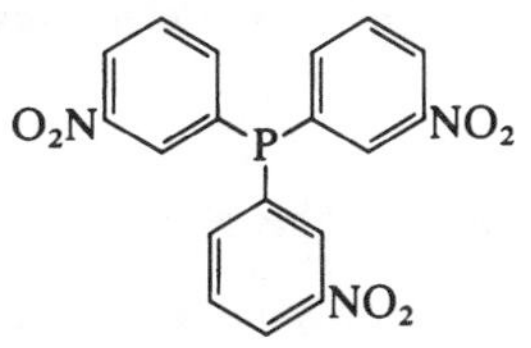

$C_{18}H_{12}N_3O_6P$ M 397.283

Cryst. (AcOH/EtOH). Mp 213°.

Oxide: [31638-89-8].

$C_{18}H_{12}N_3O_7P$ M 413.282

Forms complexes with Sn and Group VB metals. Pale-yellow cryst. (EtOH aq. or AcOH). Mp 232-235°, 242°.

Sulfide:

$C_{18}H_{12}N_3O_6PS$ M 429.343

Yellow plates (AcOH). Mp 140-141°.

Challenger, F. *et al*, *J. Chem. Soc.*, 1924, **125**, 2675 (*oxide, synth*)
Schiemenz, G.P. *et al*, *Chem. Ber.*, 1971, **104**, 1219 (*synth, derivs*)
Shvets, A.A. *et al*, *Zh. Obshch. Khim.*, 1976, **46**, 1701 (*Engl. transl.* p. 1654) (*ir*)

Tris(2-nitrophenyl) phosphite, 9CI T-00839

Tri(2-nitrophenyl) phosphite

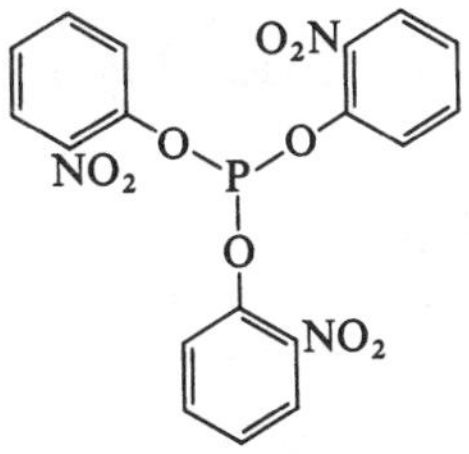

$C_{18}H_{12}N_3O_9P$ M 445.281

Solid. Mp 126°.

Kamai, G. *et al*, *CA*, 1956, **50**, 6346 (*synth*)

Tris(4-nitrophenyl) phosphite, 9CI T-00840

Tri(4-nitrophenyl) phosphite

[23485-35-0]

$C_{18}H_{12}N_3O_9P$ M 445.281

Reagent for peptide synth. Cryst. (AcOH). Mp 170-171°.

Strecker, W. *et al*, *Ber.*, 1916, **49**, 93 (*synth*)
Iselin, B.M. *et al*, *Helv. Chim. Acta*, 1957, **40**, 373 (*use*)

Tris(pentachlorophenyl) phosphate, 9CI T-00841

$$(C_6Cl_5O)_3P{=}O$$

$C_{18}Cl_{15}O_4P$ M 842.964

Fungicide. Cryst. + $4H_2O$. Dec. at 322-326°.

U.S.P., 2 993 934, (*1961*); *CA*, **57**, 12383

Tris(pentafluorophenyl) phosphate, 9CI T-00842

$$(C_6F_5O)_3P{=}O$$

$C_{18}F_{15}O_4P$ M 596.145

Viscous oil. Bp_1 164-168°.

U.S.P., 3 341 631, (*1967*); *CA*, **68**, 95520

Tris(2-phenylethenyl)phosphine, 10CI, 9CI T-00843

Tristyrylphosphine, 8CI

[24082-55-1]

$$P(CH{=}CHPh)_3$$

$C_{24}H_{21}P$ M 340.404

Cryst. ($EtOH/C_6H_6$). Mp 116.5-118.5°. pK_a 6.62 ($MeNO_2$) (conj. acid).

Oxide: [20435-15-8].

$C_{24}H_{21}OP$ M 356.403

Glistening needles (EtOH). Mp 236-237°.

Sulfide: [4319-13-5].

$C_{24}H_{21}PS$ M 372.464

Prisms ($EtOH/C_6H_6$). Mp 227.5-228°.

Bogolyubov, G.M. *et al*, *Zh. Obshch. Khim.*, 1965, **35**, 1566 (*Engl. transl.* p. 1570); 1966, **36**, 724 (*Engl. transl.* p. 737) (*sulfide, uv, struct*)
Fedorova, G.K. *et al*, *Zh. Obshch. Khim.*, 1967, **37**, 2686 (*Engl. transl.* p. 2557) (*synth, sulfide*)
Federova, G.K. *et al*, *Zh. Obshch. Khim.*, 1969, **39**, 1227 (*Engl. transl.* p. 1197) (*prep, sulfide*)
Chernova, A.V. *et al*, *Izv. Akad. Nauk SSSR, Ser. Khim.*, 1972, 722 (*Engl. transl.* p. 693) (*derivs, raman, ir*)

Tris(2-phenylethyl)phosphine, 9CI T-00844

Triphenethylphosphine, 8CI

$$P(CH_2CH_2Ph)_3$$

$C_{24}H_{27}P$ M 346.451

Liq. $Bp_{0.35}$ 218-224°. pK_a 6.60 (H_2O). n_D^{25} 1.5950.

Oxide: [29701-83-5].

$C_{24}H_{27}OP$ M 362.450

Cryst. (C_6H_6/heptane). Mp 156-157°, Mp 174-175°. pK_a 7.77 ($MeNO_2$) (conj. acid).

Rauhut, M.M. *et al*, *J. Org. Chem.*, 1961, **26**, 5138 (*synth*)
Petrov, K.A. *et al*, *Zh. Obshch. Khim.*, 1970, **40**, 2192 (*Engl. transl.* p. 2179) (*oxide*)

Tris(2-phenylethyl) phosphite, 9CI, 8CI T-00845

Tri(2-phenylethyl) phosphite

[59924-20-8]

$$(PhCH_2CH_2O)_3P$$

$C_{24}H_{27}O_3P$ M 394.449

Liq. d_0^{20} 1.11. $Bp_{0.05}$ 162-171°. n_D^{20} 1.5550.

Gerrard, W. *et al*, *J. Chem. Soc.*, 1953, 2069 (*synth, props*)

Tris(phenylethynyl)phosphine, 9CI T-00846

[4547-77-7]

$$P(C{\equiv}CPh)_3$$

$C_{24}H_{15}P$ M 334.356

Ligand for Cu and Ni. Cryst. (EtOH). Mp 92°.

Oxide: [31398-96-6].

$C_{24}H_{15}OP$ M 350.356

Solid. Mp 125.5-126°.

Sulfide: [2025-12-9].

$C_{24}H_{15}PS$ M 366.416

Cryst. (EtOH). Mp 138.5-139°. Two cryst. modifications exist.

Selenide: [27258-72-6].

$C_{24}H_{15}PSe$ M 413.316

Two cryst. modifications exist.

Bogolyubov, G.M. *et al*, *Zh. Obshch. Khim.*, 1965, **35**, 704, 1566 (*Engl. transl.* p. 705, 1570) (*sulfide, ir, pmr, uv*)
Mootz, D. *et al*, *Z. Kristallogr., Kristallgeom., Kristallphys., Kristallchem.*, 1969, **130**, 239 (*oxide, sulfide, selenide, cryst struct*)
Chekunina, L.I. *et al*, *Zh. Obshch. Khim.*, 1972, **42**, 995 (*Engl. transl.* p. 985) (*ir, uv, nmr*)
Fluck, E. *et al*, *Pure Appl. Chem.*, 1975, **44**, 373 (*pe*)
Timokhin, B.V. *et al*, *Zh. Obshch. Khim.*, 1976, **46**, 490 (*Engl. transl.* p. 488) (*synth*)
Skolimowski, J. *et al*, *Synthesis*, 1979, 109 (*oxide, use*)

2,4,6-Tris(2,4,6-tri-*tert*-butylphenoxy)-1,3,5,2,4,6-trioxatriphosphorine T-00847

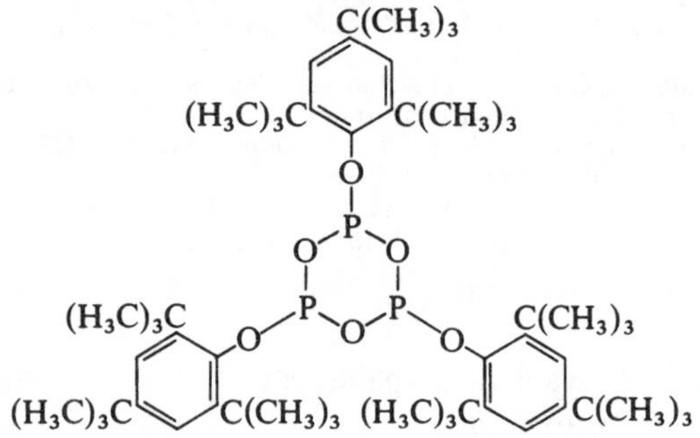

$C(CH_3)_3$ $(H_3C)_3C$ $C(CH_3)_3$ O P O O P O P O O $(H_3C)_3C$ $C(CH_3)_3$ $(H_3C)_3C$ $C(CH_3)_3$ $(H_3C)_3C$ $C(CH_3)_3$

$C_{54}H_{87}O_6P_3$ M 925.199
Unusual in that phosphorus is tervalent. Solid. Mp 212-220°. Stable to air and to moisture.

Chasar, D.W. *et al*, *J. Am. Chem. Soc.*, 1986, **108**, 5956 (*synth, ir, pmr, cryst struct, P nmr*)

Tris(2,2,2-trichloroethyl) phosphate, 9CI T-00848

$(Cl_3CCH_2O)_3P{=}O$

$C_6H_6Cl_9O_4P$ M 492.162
Solid. Mp 70-72°.

Markovskii, L.N. *et al*, *Zh. Obshch. Khim.*, 1980, **50**, 807 (*Engl. transl.* p. 644) (*synth, P nmr*)

Tris(triethylsilyl) phosphate T-00849

Triethylsilanol 3:1 phosphate, 9CI
[14579-57-8]

$(Et_3SiO)_3P{=}O$

$C_{18}H_{45}O_4PSi$ M 384.610
Liq. Bp_{11} 200.5°, Bp_1 166.5°. n_D^{20} 1.4455.

Voronkov, M.G. *et al*, *Zh. Obshch. Khim.*, 1955, **25**, 469; 1957, **27**, 1483 (*Engl. transl.* pp. 437, 1557) (*synth*)
Kolodyazhnyi, Yu.V. *et al*, *Zh. Obshch. Khim.*, 1975, **45**, 749 (*Engl. transl.* p. 738) (*struct*)

Tris(triethylsilyl) phosphite, 8CI T-00850

$(Et_3SiO)_3P$

$C_{18}H_{45}O_3PSi_3$ M 424.782
Liq. Bp_1 145-146°. n_D^{20} 1.4550.

Orlov, N.F. *et al*, *Zh. Obshch. Khim.*, 1969, **39**, 222 (*Engl. transl.* p. 211) (*synth*)

Tris(2,2,2-trifluoroethyl) phosphate, 9CI T-00851

$(F_3CCH_2O)_3P{=}O$

$C_6H_6F_9O_4P$ M 344.070
Liq. d^{20} 1.59. Mp −22°. Bp 186-189°. n_D^{20} 1.3198.

Chugunov, V.S. *et al*, *Zh. Obshch. Khim.*, 1968, **38**, 416 (*Engl. transl.* p. 412)

Tris[2-(trifluoromethyl)phenyl]phosphine, 9CI T-00852

[25688-42-0]

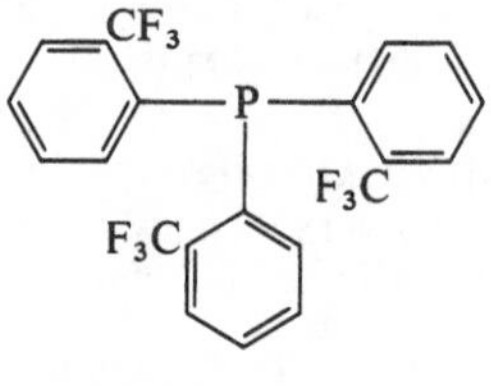

CF_3 P F_3C F_3C

$C_{21}H_{12}F_9P$ M 466.285
Cryst. (EtOH or C_6H_6/pet. ether). Mp 164-165°, 174-175°. Not v. air-sensitive.
Oxide: [74860-85-8].
$C_{21}H_{12}F_9OP$ M 482.285
Cryst. (Me_2CO/hexane). Mp 247-248°.

Miller, G.R. *et al*, *J. Chem. Phys.*, 1969, **51**, 3185 (*synth, nmr*)
Weiner, M.A. *et al*, *J. Org. Chem.*, 1975, **40**, 1292 (*pe*)
Grim, S.O. *et al*, *Phosphorus Sulfur*, 1977, **3**, 191 (*nmr*)
Eapen, K.C. *et al*, *J. Fluorine Chem.*, 1980, **15**, 239 (*synth, oxide, ir*)

Tris[3-(trifluoromethyl)phenyl]phosphine, 9CI T-00853

[25688-46-4]
$C_{21}H_{12}F_9P$ M 466.285
Ligand for Hg and metals of Group VIB. Liq. $Bp_{0.2}$ 130-132°. Not v. air-sensitive.
Oxide: [74038-21-4].
$C_{21}H_{12}F_9OP$ M 482.285
Cryst. (hexane). Mp 103-104°.
▷SZ2050000.
Selenide: [81358-38-5].
$C_{21}H_{12}F_9PSe$ M 545.245
Cryst. (hexane/EtOH). Mp 105°.

Grim, S.O. *et al*, *Phosphorus Sulfur*, 1977, **3**, 191 (*synth, nmr*)
Eapen, K.C. *et al*, *J. Fluorine Chem.*, 1980, **15**, 239 (*synth, oxide*)
Allen, D.W. *et al*, *J. Chem. Soc., Dalton Trans.*, 1982, 51 (*pmr, selenide*)
Allen, D.W. *et al*, *J. Chem. Soc., Dalton Trans.*, 1984, 483 (*selenide, cryst struct*)

Tris[4-(trifluoromethyl)phenyl]phosphine, 9CI T-00854

[13406-29-6]
$C_{21}H_{12}F_9P$ M 466.285
Ligand for metals of Groups VIB, VIIB and VIII. Cryst. (EtOH). Mp 68-70°, 78-80°. Air-sensitive.
Oxide: [13406-27-4].
$C_{21}H_{12}F_9OP$ M 482.285
Solid (hexane), needles (EtOH). Mp 181-183°, 191-193°.

Zhmurova, I.N. *et al*, *Zh. Obshch. Khim.*, 1966, **36**, 1248 (*Engl. transl.* p. 1265) (*synth*)
Weiner, M.A. *et al*, *J. Org. Chem.*, 1975, **40**, 1292 (*pe*)
Grim, S.O. *et al*, *Phosphorus Sulfur*, 1977, **3**, 191 (*synth, nmr*)
Casper, J.M. *et al*, *Spectrochim. Acta, Part A*, 1978, **34**, 1 (*oxide, ir*)

Eapen, K.C. *et al*, *J. Fluorine Chem.*, 1980, **15**, 239 (*synth, oxide, ir, nmr*)

Tris(trifluoromethyl)phosphine, 9CI T-00855

[432-04-2]

$(F_3C)_3P$

C_3F_9P M 237.992

Liq. or gas. Bp 173° (est.). Forms a glass at N_2(l) temp. Stable to 200° and in boiling H_2O. No reaction with S_8.

Oxide: see Tris(trifluoromethyl)phosphine oxide, T-00856

Sulfide: [2025-08-3].

C_3F_9PS M 270.052

Volatile solid. Mp 6.5-7°. Bp 47° (est.).

Bennett, F.W. *et al*, *J. Chem. Soc.*, 1953, 1565 (*synth, ir, uv, props*)
Cavell, R.G. *et al*, *J. Chem. Soc.*, 1964, 5896 (*sulfide, synth, ir, F nmr, props*)
Hawthorne, J.D. *et al*, *J. Organomet. Chem.*, 1968, **12**, 407 (*ms*)
Cowley, A.H. *et al*, *J. Am. Chem. Soc.*, 1975, **97**, 3653 (*pe*)
Marsden, C.J. *et al*, *Inorg. Chem.*, 1976, **15**, 2713 (*ed, raman*)
Apel, J. *et al*, *Z. Anorg. Allg. Chem.*, 1979, **453**, 28 (*cmr, F and P nmr*)
Burg, A.B., *Inorg. Chem.*, 1985, **24**, 3342 (*F and P nmr*)

Tris(trifluoromethyl)phosphine oxide, 9CI T-00856

[423-01-8]

$(F_3C)_3P{=}O$

C_3F_9OP M 253.992

Liq. Mp −89°. Bp 32°.

Burg, A.B. *et al*, *J. Am. Chem. Soc.*, 1965, **87**, 238 (*synth, ir*)
Dagnac, P. *et al*, *J. Chem. Soc., Dalton Trans.*, 1979, 155 (*cmr*)
Marsden, C.J., *Inorg. Chem.*, 1984, **23**, 1703 (*synth, ed*)
Burg, A.B., *Inorg. Chem.*, 1985, **24**, 3342 (*F and P nmr*)

Tris(trifluoromethyl) phosphorotriselenoite, 9CI, 8CI T-00857

Tris(trifluoromethyl) triselenophosphite

[70058-90-1]

$(F_3CSe)_3P$

$C_3F_9PSe_3$ M 474.872

Liq. Mp −55°. Dec. at ca. 20°.

Darmadi, A. *et al*, *Z. Anorg. Allg. Chem.*, 1979, **448**, 35 (*synth, ir, F nmr, props*)
Gombler, W., *Z. Naturforsch., B*, 1981, **36**, 535 (*F nmr, cmr*)

Tris(2,4-6-trimethoxyphenyl)phosphine, 9CI T-00858

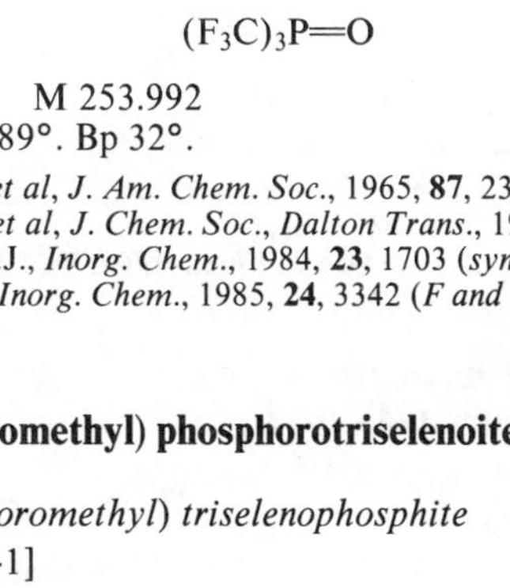

$C_{27}H_{33}O_9P$ M 532.526

Solid. Insol. Et_2O. Mp 146-147°.

B,MeI: Tris(2,4,6-trimethoxyphenyl)-methylphosphonium iodide.

$C_{28}H_{36}IO_9P$ M 674.465

Cryst. (H_2O). Mp 175-176°.

Selenide:

$C_{27}H_{33}O_9PSe$ M 611.486

Cryst. (EtOH). Mp 199-200°.

Protopopov, I.S. *et al*, *Zh. Obshch. Khim.*, 1963, **33**, 3050 (*Engl. transl.* p. 2975)
Allen, D.W. *et al*, *J. Chem. Soc., Dalton Trans.*, 1985, 2505 (*selenide*)

Tris(trimethylgermyl)phosphate T-00859

Hydroxytrimethylgermane phosphate

$(Me_3GeO)_3PO$

$C_9H_{27}Ge_3PO_4$ M 448.054

Solid. Mp 37°. Bp_1 102-103°.

Schmidt, M. *et al*, *Chem. Ber.*, 1962, **95**, 1434.

Tris(trimethylgermyl)phosphine, 10CI, 9CI T-00860

[13904-36-4]

$(Me_3Ge)_3P$

$C_9H_{27}Ge_3P$ M 384.056

Clear liq. $Bp_{0.1}$ 62-63°. Readily oxidised.

Schumann, I. *et al*, *Z. Naturforsch.*, 1966, **21**, 1105 (*synth, ir, pmr*)
Schumann, H. *et al*, *Chem. Ber.*, 1975, **106**, 1630.
Roesch, L. *et al*, *Z. Anorg. Allg. Chem.*, 1976, **426**, 99.
Schumann, H. *et al*, *Z. Naturforsch., B*, 1977, **32**, 513 (*nmr*)

Tris(2,4,6-trimethylphenyl)phosphine, 9CI T-00861

Trimesitylphosphine

[23897-15-6]

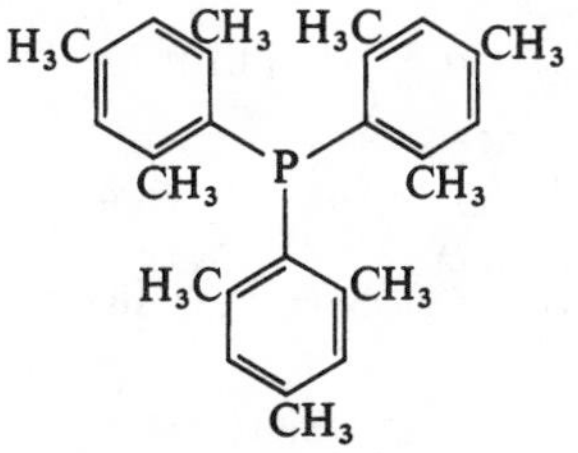

$C_{27}H_{33}P$ M 388.531

Ligand for Pt and Pd. Needles (EtOH or $CHCl_3$/EtOH). Mp 191.5-193°. pK_a 13.1 ($MeNO_2$). Triboluminescent. Readily metallated.

B,MeI: Methyltris(2,4,6-trimethylphenyl)phosphonium iodide.

$C_{28}H_{36}IP$ M 530.471

Cryst. (EtOH/Et_2O). Mp 269°, 315-317°.

B,MeBF$_4$: Solid. Mp 320-323°.

Oxide: [23897-17-8].

$C_{27}H_{33}OP$ M 404.531

No phys. props. reported.

Sulfide: [57368-24-8].

$C_{27}H_{33}PS$ M 420.591

Cryst. (EtOH). Mp 221-222°.

Halpern, E.J. *et al*, *J. Am. Chem. Soc.*, 1967, **89**, 5224 (*synth, deriv*)
Bokanov, A.I. *et al*, *Zh. Obshch. Khim.*, 1974, **44**, 760 (*Engl. transl.* p. 732) (*props, uv*)
Blount, J.F. *et al*, *Tetrahedron Lett.*, 1975, 913 (*cryst struct*)

Spepanov, B.I. *et al*, *Zh. Obshch. Khim.*, 1975, **45**, 2096 (*Engl. trans.* p. 2059) (*sulfide, pmr*)
Negrebetskii, V.V. *et al*, *Zh. Strukt. Khim.*, 1978, **19**, 628 (*Engl. transl.* p. 545) (*cmr, nmr, derivs*)
Ratovskii, G.V. *et al*, *Zh. Obshch. Khim.*, 1978, **48**, 1520 (*Engl. transl.* p. 1394) (*uv*)
Bellamy, A.J. *et al*, *J. Chem. Soc., Perkin Trans. 2*, 1981, 1093, 1098 (*derivs*)
Shifrina, R.R. *et al*, *Zh. Prikl. Spektrosk.*, 1981, **34**, 111 (*Engl. transl.* p. 84) (*oxide, ir*)

2,4,6-Tris(2,4,6-trimethylphenyl)-1,3,5,2,4,6-trithiatriphosphorinane T-00862

2,4,6-Tris(2,4,6-trimethylphenyl)-1,3,5,2,4,6-trithiatriphosphinane. 2,4,6-Trimesityl-1,3,5,2,4,6-trithiatriphosphorinane

[85505-00-6]

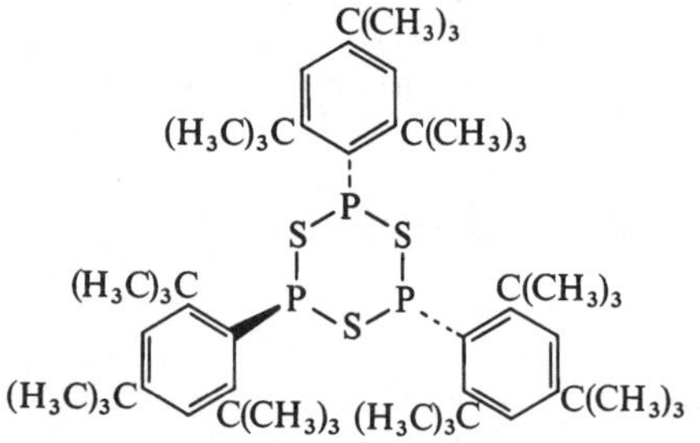

C(CH3)3; (H3C)3C; C(CH3)3; P; S; S; (H3C)3C; P; S; P; C(CH3)3; (H3C)3C; C(CH3)3; (H3C)3C; C(CH3)3

$C_{54}H_{87}P_3S_3$ M 925.383

Unusual in the presence of tervalent P in the $(PS)_3$ ring.

(2α,4α,6β)-form [83692-00-6]

Cryst. Mp 170-1°.

Cetinkaya, B. *et al*, *J. Chem. Soc., Chem. Commun.*, 1982, 691 (*cryst struct*)

Tris(trimethylplumbyl)phosphine, 8CI T-00863

[13355-82-3]

$(Me_3Pb)_3P$

$C_9H_{27}PPb_3$ M 787.886

White solid (C_6H_6). Mp 48-49° (46-47°). Air-sensitive.

Schumann, H. *et al*, *Inorg. Nucl. Chem. Lett.*, 1966, **2**, 311.
Schumann, H. *et al*, *Chem. Ber.*, 1969, **102**, 2900.

Tris(trimethylsilyl) phosphate T-00864

Trimethylsilanol 3:1 phosphate, 9CI

[10497-05-9]

$(Me_3SiO)_3P{=}O$

$C_9H_{27}O_4PSi$ M 258.369

Liq. Bp 232°, Bp_5 93-94°. n_D^{20} 1.4100.

▷TC9700000.

B,MeI: Methoxytris(trimethylsilyloxy)phosphonium iodide.

$C_{10}H_{30}O_4PSi$ M 273.404

Solid. Sol. CH_2Cl_2. Mp 95-96°.

B,Me₃SiI: Tetrakis(trimethylsilyloxy)phosphonium iodide.

$C_{12}H_{36}IO_4PSi_4$ M 514.634

Solid. Mp 151°.

Voronkov, M.G. *et al*, *Zh. Obshch. Khim.*, 1972, **42**, 2030 (*Engl. transl.* p. 2027) (*synth*)
Lebedev, E.P. *et al*, *Zh. Obshch. Khim.*, 1974, **44**, 787, 1769 (*Engl. transl.* pp. 759, 1736) (*synth*)
Schmidbaur, H. *et al*, *Chem. Ber.*, 1974, **107**, 1731 (*synth, derivs, ir, pmr*)
Kolodyazknyi, Yu.V. *et al*, *Zh. Obshch. Khim.*, 1975, **45**, 749 (*Engl. transl.* p. 738) (*struct*)

Tris(trimethylsilyl)phosphine, 9CI T-00865

[15573-38-3]

$P(SiMe_3)_3$

$C_9H_{27}PSi_3$ M 250.543

Versatile reagent for the prepn. of alkyl and acyl phosphines, ligand for B, Al, Mn, Co, Ni and metals of Group VIB. Air-sensitive liq. Mp 24°. Bp 242-243°, Bp_{16} 102-105°. n_D^{25} 1.5027.

▷Flammable in air

Parshall, G.W. *et al*, *J. Am. Chem. Soc.*, 1959, **81**, 6273 (*synth*)
Bürger, H., *Spectrochim. Acta, Part A*, 1970, **26**, 671 (*ir, raman*)
Schumann, H. *et al*, *Chem. Ber.*, 1974, **107**, 854 (*synth, props*)
Becker, G. *et al*, *Chem. Ber.*, 1975, **108**, 2484 (*synth*)
Müller, H. *et al*, *J. Organomet. Chem.*, 1977, **140**, C17 (*cmr*)
Schäfer, H., *Z. Anorg. Allg. Chem.*, 1979, **459**, 157 (*pmr, complexes*)
van Lindthardt, J.P. *et al*, *Spectrochim. Acta, Part A*, 1980, **36**, 17 (*synth, pmr, cmr, nmr*)
Becker, G. *et al*, *Z. Anorg. Allg. Chem.*, 1981, **480**, 21 (*use, props*)
Fritz, G. *et al*, *Z. Anorg. Allg. Chem.*, 1981, **481**, 185; 1982, **487**, 44 (*props*)
Sax, N.I., *Dangerous Properties of Industrial Materials*, 6th Ed., Van Nostrand-Reinhold, 1984, 1072.

Tris(trimethylsilyl) phosphite, 8CI T-00866

$(Me_3SiO)_3P$

$C_9H_{27}O_3PSi_3$ M 298.541

Liq. Bp_{25} 129°, Bp_1 58-60°. n_D^{20} 1.4145.

Voronkov, M.G. *et al*, *Zh. Obshch. Khim.*, 1965, **35**, 106 (*Engl. transl.* p. 105) (*synth, raman*)
Orlov, N.F. *et al*, *Zh. Obshch. Khim.*, 1969, **39**, 222 (*Engl. transl.* p. 211) (*synth*)

NNN'-Tris(trimethylsilyl)-phosphonamidimidic acid T-00867

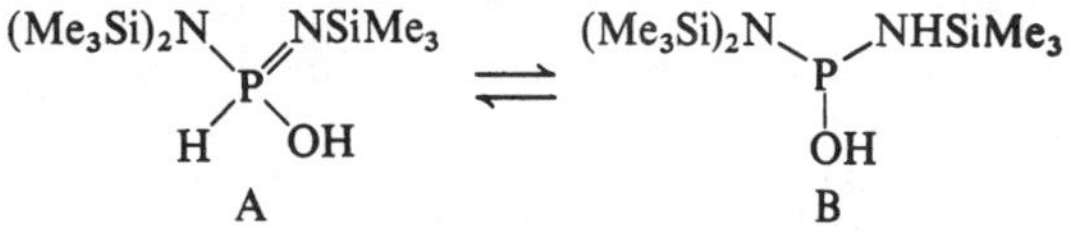

$(Me_3Si)_2N$, $NSiMe_3$, P, H, OH (A) ⇌ $(Me_3Si)_2N$, $NHSiMe_3$, P, OH (B)

$C_9H_{29}N_2OPSi_3$ M 296.571

Me ester: [50732-24-6]. *Methyl* N,N,N'-*tris(trimethylsilyl)phosphonamidimidate.* P-*Methoxy*-P-[*bis(trimethylsilyl)amino*]-N-*trimethylsilylphosphazene.* P-*Methoxy*-N-*trimethylsilyl*-P-[*bis(trimethylsilyl)amino*]*phosphine imide.*

$C_{10}H_{31}N_2OPSi_3$ M 310.598

Liq. $Bp_{0.02}$ 51-52°.

Isopropyl ester: Isopropyl N,N,N'-*tris(trimethylsilyl)-phosphonamidimidate.* P-*Isopropoxy*-N-*trimethylsilyl*-P-[*bis(trimethylsilyl)amino*]*phosphazene.* P-*Isopropoxy*-N-*trimethylsilyl*-P-[*bis(trimethylsilyl)-amino*]*phosphine imide.*

$C_{12}H_{35}N_2OPSi_3$ M 338.652

Liq. $Bp_{0.02}$ 74-76°.

tert-*Butyl ester:* [79629-50-8]. tert-*Butyl* N,N,N'-*tris(trimethylsilyl)phosphonamidimidate.* P-tert-*Butoxy*-N-*trimethylsilyl*-P-[*bis(trimethylsilyl)amino*]-*phosphazene.* P-tert-*Butoxy*-N-*trimethylsilyl*-P-[*bis(trimethylsilyl)amino*]*phosphine imide.*

$C_{13}H_{37}N_2OPSi_3$ M 352.678

Liq. $Bp_{0.02}$ 62-64°. Admixed with 15-20% of the ester of form B.

Ph ester: [79629-54-2]. *Phenyl* N,N,N'-*tris(trimethylsilyl)phosphonamidimidate.* P-*Phenoxy*-N-*trimethylsilyl*-P-[*bis(trimethylsilyl)amino*]-*phosphazene.* P-*Phenoxy*-N-*trimethylsilyl*-P-[*bis(trimethylsilyl)amino*]*phosphine imide.*
$C_{15}H_{33}N_2OPSi_3$ M 372.669
Liq. Dec. on attempted dist. Admixed with 40-50% of the ester of form B.

Trimethylsilyl ester: [89596-83-8]. *Trimethylsilyl* N,N,N'-*tris(trimethylsilyl)phosphonamidimidate.* N-*Trimethylsilyl*-P-[*bis(trimethylsilyl)amino*]-P-*trimethylsilyloxyphosphazene.* N-*Trimethylsilyl*-P-[*bis(trimethylsilyl)amino*]-P-*trimethylsilyloxyphosphine imide.*
$C_{12}H_{37}N_2OPSi_4$ M 368.753
Liq. $Bp_{0.02}$ 49-51°.

Niecke, E. *et al*, *Angew. Chem.*, 1973, **85**, 586 (*methyl ester, synth, ir, P nmr*)
Romanenko, V.D. *et al*, *Zh. Obshch. Khim.*, 1981, **51**, 1726 (*Engl. transl.* p. 1475) (*esters, synth, ir, pmr, P nmr, cmr*)
O'Neal, H.R. *et al*, *Inorg. Chem.*, 1984, **23**, 1372 (*silyl ester, synth, pmr, cmr, P nmr*)

N,N,N'-Tris(trimethylsilyl)-phosphoramidimidic acid, 9CI — T-00868

$$(Me_3Si)_2NP(OH)_2{=}NSiMe_3$$

$C_9H_{29}N_2O_2PSi_3$ M 312.571
Free acid tautomeric with *N,N,N'*-tris(trimethylsilyl)-phosphorodiamidic acid.

Di-Me ester: [39230-44-9]. *Dimethyl* N,N,N'-*tris(trimethylsilyl)phosphoramidimidate.* P,P-*Dimethoxy*-P-[*bis(trimethylsilyl)amino*]-N-*trimethylsilylphosphazene.*
$C_{11}H_{33}N_2O_2PSi_3$ M 340.624
Liq. Mp −19° to −18°. $Bp_{0.1}$ 55-57°. Exhibits silylotropy when heated.

Difluoride: [58972-02-4]. P,P-*Difluoro*-N-*trimethylsilyl*-P-[*bis(trimethylsilyl)amino*]*phosphazene.* P,P-*Difluoro*-N-*trimethylsilyl*-P-[*bis(trimethylsilyl)amino*]-*phosphine imide.*
$C_9H_{27}F_2N_2PSi_3$ M 316.553
Liq. Bp_{10} 74-75°, Bp_1 55°.

Dichloride: [58972-01-3]. P,P-*Dichloro*-N-*trimethylsilyl*-P-[*bis(trimethylsilyl)amino*]*phosphazene.* P,P-*Dichloro*-N-*trimethylsilyl*-P-[*bis(trimethylsilyl)*-*amino*]*phosphine imide.*
$C_9H_{27}Cl_2N_2PSi_3$ M 349.462
Liq. $Bp_{0.1}$ 84-85°.

Dibromide: [58972-03-5]. P,P-*Dibromo*-N-*trimethylsilyl*-P-[*bis(trimethylsilyl)amino*]*phosphazene.* P,P-*Dibromo*-N-*trimethylsilyl*-P-[*bis(trimethylsilyl)*-*amino*]*phosphine imide.*
$C_9H_{27}Br_2N_2PSi_3$ M 438.364
Viscous liq.

Diiodide: [58972-04-6]. P,P-*Diiodo*-N-*trimethylsilyl*-P-[*bis(trimethylsilyl)amino*]*phosphazene.* P,P-*Diiodo*-N-*trimethylsilyl*-P-[*bis(trimethylsilyl)amino*]-*phosphine imide.*
$C_9H_{27}I_2N_2PSi_3$ M 532.365
Viscous liq.

Scherer, O.J. *et al*, *Z. Naturforsch., B*, 1972, **27**, 1429 (*ester, synth, pmr*)
Niecke, E. *et al*, *Chem. Ber.*, 1976, **109**, 415 (*dihalides, ir, pmr, P and Si nmr*)
Wisian-Neilson, P. *et al*, *Inorg. Chem.*, 1977, **16**, 1460 (*difluoride, synth, ir, ms, pmr, F nmr*)

Tris(trimethylsilyl) phosphorotetrathioate, 10CI — T-00869

Tris(trimethylsilyl) tetrathiophosphate
[63853-23-6]

$$(Me_3SiS)_3P{=}S$$

$C_9H_{27}PS_4Si_3$ M 378.783
Light-golden cryst. Mp 69°. Sublimes.

Roesky, H.W. *et al*, *Z. Anorg. Allg. Chem.*, 1977, **431**, 221 (*synth, ir, ms, pmr, P nmr*)
Horn, H.G. *et al*, *Chem. Ztg.*, 1985, **109**, 77 (*synth, ms, pmr, Si and P nmr*)

Tris(triphenylgermyl)phosphine, 8CI — T-00870

[13371-32-9]

$$(Ph_3Ge)_3P$$

$C_{54}H_{45}Ge_3P$ M 942.693
White solid. Mp 128°. Oxidises and hydrolyses readily.

Engelhardt, G. *et al*, *Z. Naturforsch., B*, 1967, **22**, 382 (*raman, nmr*)
Schumann, H. *et al*, *Chem. Ber.*, 1969, **102**, 2900 (*synth, ir*)

Tris[(triphenylphosphoranylidene)methyl]-phosphine, 9CI, 8CI — T-00871

[24505-49-5]

$$(Ph_3P{=}CH)_3P$$

$C_{57}H_{48}P_4$ M 856.901
Dark-red solid. Mp 104-106° (under Ar). With HCl, forms a 1:3 adduct of the struct. $[(Ph_3PCH_2)_3P]^{\oplus}3Cl^{\ominus}$ Mp 110-112°. The oxide and sulfide behave similarly.

Oxide: [24442-28-2].
$C_{57}H_{48}OP_4$ M 872.901
Mp 69-71° (under Ar). Forms 'trihydrochloride' Mp 184-6° (under Ar).

Sulfide: [24442-30-6].
$C_{57}H_{48}P_4S$ M 888.961
Mp 90-92° (under Ar). Forms 'trihydrochloride' Mp 188-91° (under Ar).

Issleib, K. *et al*, *J. Prakt. Chem.*, 1969, **311**, 857 (*synth, derivs*)

Tris(triphenylphosphoranylidene)-phosphorous triamide, 9CI — T-00872

[56726-81-9]

$$(Ph_3P{=}N)_3P$$

$C_{54}H_{45}N_3P_4$ M 859.865
Prisms (CH_2Cl_2/Et_2O). Mp 198-200°. Complexes with Me_3SiCl.

Oxide: [3101-74-4]. *Tris(triphenylphosphoranylidene)*-*phosphoric triamide.*
$C_{54}H_{45}N_3OP_4$ M 875.864
Cryst. (dioxan). Mp 183-185°.

Shtepanek, A.S. *et al*, *Zh. Obshch. Khim.*, 1975, **45**, 1012 (*Engl. transl.* p. 999)

Tri(2-thiazolyl)phosphine T-00873

2,2′,2″-Phosphinidynetristhiazole

[80679-23-8]

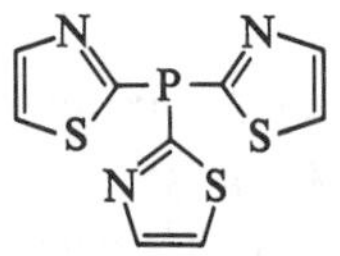

$C_9H_6N_3PS_3$ M 283.320

Cryst. (MeOH/hexane). Mp 97-99°.

Moore, S.S. *et al, J. Org. Chem.*, 1982, **47**, 1489 (*synth, pmr, nmr, ms, ir*)

Tri-2-thienylphosphine, 9CI T-00874

[24171-89-9]

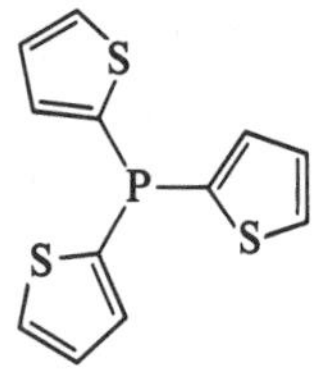

$C_{12}H_9PS_3$ M 280.357

Low-melting solid. Mp 35°. Bp_2 205°.

B,MeI: Methyltri-2-thienylphosphonium iodide.

$C_{13}H_{12}IPS_3$ M 422.296

Solid. Mp 185-186°.

Oxide: [1021-21-2].

$C_{12}H_9OPS_3$ M 296.356

Cryst. (MeOH/hexane). Mp 130°.

Sulfide:

$C_{12}H_9PS_4$ M 312.417

Needles (EtOH). Mp 138°.

Selenide: [26910-74-7].

$C_{12}H_9PS_3Se$ M 359.317

Solid. Mp 150-151°.

Griffith, C.E. *et al, J. Org. Chem.*, 1965, **30**, 97 (*oxide, uv*)
Jakobsen, H.J. *et al, J. Mol. Spectrosc.*, 1969, **31**, 230; 1970, **33**, 474 (*pmr, deriv*)
Jakobsen, H.J. *et al, Mol. Phys.*, 1972, **23**, 197 (*cmr, nmr*)
Horner, L. *et al, Phosphorus*, 1976, **6**, 147 (*synth*)
Allen, D.W. *et al, J. Chem. Soc., Dalton Trans.*, 1982, 51 (*complexes*)
Moore, S.S. *et al, J. Org. Chem.*, 1982, **47**, 1489 (*complexes*)

Tri-3-thienylphosphine, 10CI, 9CI T-00875

[23415-53-4]

$C_{12}H_9PS_3$ M 280.357

Cryst. (C_6H_6/pet. ether). Mp 69-70°. $Bp_{0.2}$ 150-156°.

B,MeI: Methyltri-3-thienylphosphonium iodide.

$C_{13}H_{12}IPS_3$ M 422.296

Solid. Mp 178-179°.

Oxide:

$C_{12}H_9OPS_3$ M 296.356

Solid. Mp 186-187°.

Sulfide:

$C_{12}H_9PS_4$ M 312.417

Solid. Mp 158-159°.

Selenide: [26944-07-0].

$C_{12}H_9PS_3Se$ M 359.317

Solid. Mp 171-172°.

Jakobsen, H.J. *et al, Acta Chem. Scand.*, 1969, **23**, 1070 (*synth, pmr*)
Jakobsen, H.J. *et al, J. Mol. Spectrosc.*, 1970, **33**, 474 (*derivs, pmr, cmr, nmr*)
Jakobsen, H.J. *et al, Mol. Phys.*, 1972, **23**, 197 (*cmr, nmr*)
Hazell, A.C. *et al, Acta Crystallogr., Sect. B*, 1977, **33**, 1105 (*cryst struct*)
Allen, D.W. *et al, J. Chem. Soc., Dalton Trans.*, 1982, 51 (*complexes*)

Tri(1,2,4-triazol-1-yl)phosphine T-00876

1,1′,1″-Phosphinidyetris(1H-1,2,4-triazole), 9CI. *Tris(1,2,4-triazol-1-yl)phosphine*

[72741-17-4]

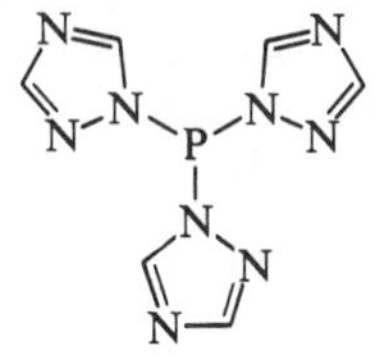

$C_6H_6N_9P$ M 235.147

Solid. Mp 120-125° dec.

Oxide: [72741-18-5]. *1,1′,1″-Phosphinylidynetris(1H-1,2,4-triazole).*

$C_6H_6N_9OP$ M 251.147

Prepd. *in situ* as a phosphorylating agent. Solid. Mp 114-116°.

Walter, W. *et al, Justus Liebigs Ann. Chem.*, 1979, 1756 (*synth, pmr, oxide*)
Kraszewski, A. *et al, Tetrahedron Lett.*, 1980, **21**, 2935 (*oxide, use*)

Trivinylphosphine, 8CI T-00877

Triethenylphosphine, 9CI

[3746-01-8]

$P(CH{=}CH_2)_3$

C_6H_9P M 112.111

Ligand for Cr and Ni. Bp 117-119°, Bp_{50} 44°.

B,MeI: [25743-68-4]. *Triethenylmethylphosphonium iodide. Methyltrivinylphosphonium iodide.*

$C_7H_{13}IP$ M 255.058

Crease-proofing and fire-proofing textile agent. Cryst. (EtOH/Et_2O). Mp 198°.

Oxide: [13699-67-7].

C_6H_9OP M 128.110

Feathery cryst. (Et_2O). Mp 98-99°. $Bp_{0.1}$ 120°.

Sulfide: [42495-86-3].

C_6H_9PS M 144.171

$Bp_{0.35}$ 85°.

Manatt, S.L. *et al, J. Am. Chem. Soc.*, 1963, **85**, 2664 (*pmr*)
Weiner, M.A. *et al, J. Org. Chem.*, 1967, **32**, 3707 (*synth, oxide, uv*)
Quin, L.D. *et al, J. Am. Chem. Soc.*, 1969, **91**, 3308 (*nmr*)
King, R.B. *et al, J. Am. Chem. Soc.*, 1973, **95**, 5083; 1975, **97**, 53 (*sulfide, pmr*)
Collins, D.J. *et al, Aust. J. Chem.*, 1974, **27**, 841 (*oxide, synth, uv, ir, pmr, ms*)
Davidson, G. *et al, Spectrochim. Acta, Part A*, 1978, **34**, 949 (*ir, raman*)
Weiner, M.A. *et al, Inorg. Chem.*, 1978, **17**, 1084 (*pe*)

Popov, E.M. *et al*, *Zh. Obshch. Khim.*, 1962, **32**, 3255 (*Engl. transl.* p. 3199) (*uv, raman*)
Timofeeva, T.N. *et al*, *Zh. Obshch. Khim.*, 1970, **40**, 1169 (*Engl. transl.* p. 1159) (*pmr*)
Naumov, V.A. *et al*, *Zh. Strukt. Khim.*, 1972, **17**, 304 (*Engl. transl.* p. 262) (*ed*)
Bel'skii, V.E. *et al*, *Zh. Obshch. Khim.*, 1974, **44**, 2657 (*Engl. transl.* p. 2612) (*P nmr*)
Zverev, V.V. *et al*, *Zh. Obshch. Khim.*, 1979, **49**, 1737 (*Engl. transl.* p. 1522) (*pe, struct*)
Dolhaine, H. *et al*, *J. Magn. Reson.*, 1980, **41**, 1 (*pmr*)
Morita, T. *et al*, *Chem. Lett.*, 1980, 435 (*synth*)

Vinylphosphonochloridothioic acid V-00006

Ethenylphosphonochloridothioic acid, 9CI

C_2H_4ClOPS M 142.540

O-*Et ester:* O-*Ethyl ethenylphosphonochloridothioate.*
C_4H_8ClOPS M 170.593
Liq. d_4^{20} 1.20. Bp_{12} 71-73°. n_D^{20} 1.5117.

Kabachnik, M.I. *et al*, *Izv. Akad. Nauk SSSR, Ser. Khim.*, 1961, 604 (*Engl. transl.* p. 556)

Vinylphosphonothioic acid, 8CI V-00007

Ethenylphosphonothioic acid, 9CI

$H_2C{=}CHP(S)(OH)_2 \rightleftharpoons H_2C{=}CHP(O)(OH)(SH)$

$C_2H_5O_2PS$ M 124.094
Viscous material.

Monoanilinium salt: Cryst. (EtOH). Mp 105-106° dec.
O,O-*Di-Me ester:* [50687-78-0]. O,O-*Dimethyl vinylphosphonothioate.*
$C_4H_9O_2PS$ M 152.148
Liq. d_4^{20} 1.14. Bp_{20} 82-83°. n_D^{20} 1.4912.
O,O-*Di-Et ester:* O,O-*Diethyl vinylphosphonothioate.*
$C_6H_{13}O_2PS$ M 180.201
Liq. d_4^{20} 1.06. Bp_{10} 75-76°, $Bp_{0.07}$ 43°. n_D^{20} 1.4788.
Difluoride:
$C_2H_3F_2PS$ M 128.076
Liq. Mp 71-73°.
Dichloride: [15849-99-7].
$C_2H_3Cl_2PS$ M 160.985
Liq. d_4^{20} 1.40. Bp_{12} 54-55°. n_D^{20} 1.5623.
Dianilide: P-*Ethenyl*-N,N'-*diphenylphosphosphonothioic diamide.*
$C_{14}H_{15}N_2PS$ M 274.320
Cryst. (EtOH). Mp 112-113°.

Kabachnik, M.I. *et al*, *Izv. Akad. Nauk SSSR, Ser. Khim.*, 1961, 604 (*Engl. transl.* p. 556) (*synth, derivs*)
Dorokhova, V.V. *et al*, *Zh. Obshch. Khim.*, 1973, **43**, 2172 (*Engl. transl.* p. 2164) (*dichloride, struct*)
Remizov, A.B. *et al*, *Zh. Obshch. Khim.*, 1974, **44**, 1863 (*Engl. transl.* p. 1831) (*dichloride, ir, conformn*)
Althoff, W. *et al*, *Chem. Ber.*, 1978, **111**, 1845 (*difluoride, ester, synth, cmr, nmr*)
Dolhaine, H. *et al*, *J. Magn. Reson.*, 1980, **41**, 1 (*dichloride, pmr*)

O-Vinyl phosphorodichloridothioate, 8CI V-00008

O-*Ethenyl phosphorodichloridothioate*, 9CI. O-*Vinyl dichlorothiophosphate.* O-*Vinyl thiophosphoryl dichloride*

[17973-39-6]

$H_2C{=}CHOP(S)Cl_2$

$C_2H_3Cl_2OPS$ M 176.985
Liq. d^{20} 1.43. Bp_{30} 53-55°. n_D^{20} 1.5180.

Gololobov, Yu.G. *et al*, *CA*, 1968, **68**, 114710.

Vitamin B_{12} V-00009

Cyanocobalamin, BAN. α-(*5,6-Dimethylbenzimidazoyl*)-*cobamide cyanide. Numerous proprietary synonyms*

[68-19-9]

R = CN, R' = $CONH_2$

$C_{63}H_{88}CoN_{14}O_{14}P$ M 1355.381

Antipernicious anaemia factor isol. from liver extracts; now obt. comly. from fermentation liquors of *Streptomyces griseus* and other microorganisms, e.g. *Propionibacterium shermanii.* Red needles. Sol. H_2O, alcohols, AcOH. Mp >300°, darkens at 210-220°. $[\alpha]_{656}^{23}$ −59°. λ_{max} 279, 306, 322, 361, 520, 550 nm. Isotopic variants contg. ^{57}Co ($t_{1/2}$ 270d), ^{58}Co ($t_{1/2}$ 71.3d) and ^{60}Co ($t_{1/2}$ 5.26y) are prepd. by fermentation in the presence of radiocobalt and are used for diagnostic purposes.

▷GG3750000.

13-Epimer: [32627-64-8]. *Cyanoneocobalamin. Neovitamin B_{12}.* Formed by acid-catalysed epimerisation. Red cryst. (Me_2CO aq.). CD; λ_{max} 330 nm, Δε + 8.4 (0.1M KCN). Dec. at ca. 200° without melting.

Bonnett, R., *Chem. Rev.*, 1963, **63**, 573.
Nockolds, C.K. *et al*, *Nature (London)*, 1967, **214**, 129 (*cryst struct*)
Pratt, J.M., *Inorg. Chem. of Vit. B_{12}*, 1972, Academic Press (*rev*)
Stöckli-Evans, H. *et al*, *J. Chem. Soc., Perkin Trans. 2*, 1972, 605 (*cryst struct*)
Bonnett, R., *Philos. Trans. R. Soc. London*, 1976, **273**, 295 (*epimer*)
Rajoria, D.S. *et al*, *J. Inorg. Nucl. Chem.*, 1977, **39**, 1291 (*ir*)
Schrauzer, G.N., *Angew. Chem.*, 1977, **89**, 239.
Florent, J., *Microbial Technology*, Peppler, H.J. *et al*, Eds., Academic Press, 1979, 2nd Ed., **1**, 497 (*rev, manuf*)
Toraya, T., *Vitamins*, 1977, **51**, 87.
Anton, D.L. *et al*, *J. Am. Chem. Soc.*, 1980, **102**, 2215 (*cmr*)
Battersby, A.R. *et al*, *Nature (London)*, 1980, **285**, 17 (*biosynth*)

Vitamin B_{12a} V-00010

α-(5,6-Dimethylbenzimidazolyl)aquocobamide. Aquocobalamin

[13422-52-1]

As Vitamin B_{12}, V-00009 with

$$R = H_2O, R' = CONH_2$$

$C_{62}H_{92}CoN_{13}O_{15}P$ M 1349.394

Red cryst. (Me_2CO aq.).

Bonnett, R., *Chem. Rev.*, 1963, **63**, 573.

X

(9*H*-Xanthen-9-yl)phosphonic acid, 9CI **X-00001**

9-Phosphonoxanthene

[36952-24-6]

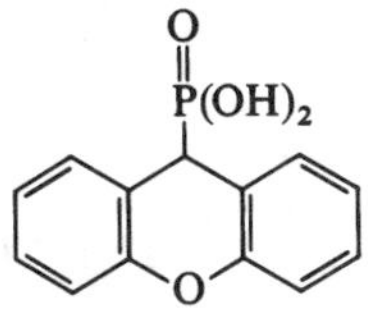

$C_{13}H_{11}O_4P$ M 262.201

Cryst. (C_6H_6). Mp 148-149°.

Di-Me ester: [14110-88-4]. *Dimethyl (9H-xanthen-9-yl)phosphonate.*

$C_{15}H_{15}O_4P$ M 290.255

Solid. Mp 132-133°.

Octahydro:

$C_{13}H_{19}O_4P$ M 270.264

Solid. Mp 187°.

Mustafa, A. *et al*, *Justus Liebigs Ann. Chem.*, 1966, **698**, 109 (*ester*)

Krivun, S.V. *et al*, *Zh. Obshch. Khim.*, 1973, **43**, 91 (*Engl. transl.* p. 87) (*synth*)

Ishikawa, K. *et al*, *Bull. Chem. Soc. Jpn.*, 1978, **51**, 2684 (*synth, props, ir, pmr*)

3′-Xanthylic acid, 9CI, 8CI **X-00002**

9β-D-Ribofuranosylxanthine 3-(dihydrogen phosphate)

[21089-32-7]

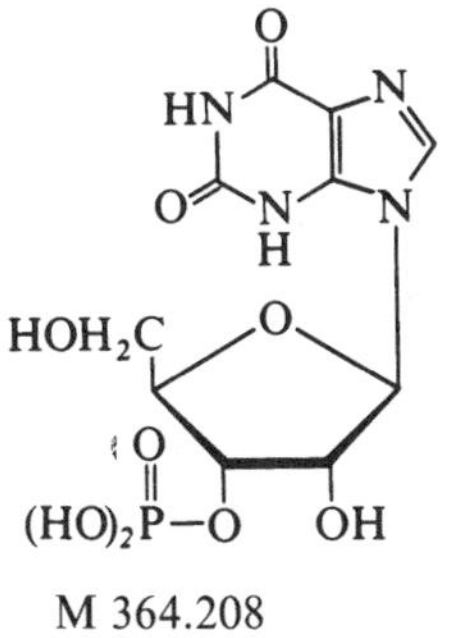

$C_{10}H_{13}N_4O_9P$ M 364.208

$[\alpha]_D^{20}$ −61.66° (c, 5.0 in NaOH).

Brucine salt: Mp 200°.

Levene, P.A. *et al*, *J. Biol. Chem.*, 1932, **95**, 757.

Holy, A., *Collect. Czech. Chem. Commun.*, 1968, **33**, 2259.

Xylose 3-dihydrogen phosphate **X-00003**

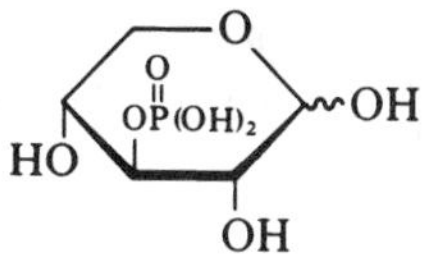

Pyranose-*form*

$C_5H_{11}O_8P$ M 230.111

D-form

Ba salt: $[\alpha]_D^{22}$ +1.27° (c, 5.13 in H_2O).

Moffatt, J.G. *et al*, *J. Am. Chem. Soc.*, 1956, **78**, 883 (*synth*)

MacDonald, D.L., *The Carbohydrates*, Academic Press, 1972, 2nd Ed., **1A**, 253 (*rev*)

Xylose 5-dihydrogen phosphate **X-00004**

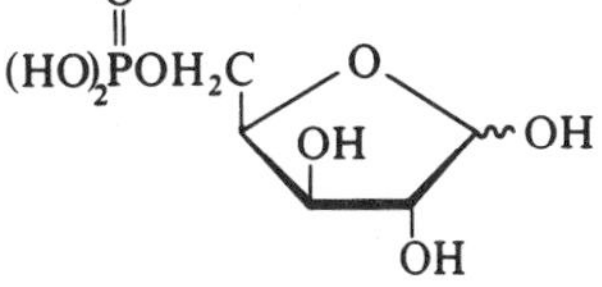

$C_5H_{11}O_8P$ M 230.111

D-form

$[\alpha]_D^{20}$ +25° (c, 2.0 in H_2O).

Ba salt: $[\alpha]_D$ +8° (H_2O).

Na salt: $[\alpha]_D^{20}$ +10° → +4.4° (c, 2.0 in H_2O).

Dibrucine salt: Mp 150° dec. $[\alpha]_D^{20}$ −37.8° (c, 2.02 in $CHCl_3$).

Levene, P.A. *et al*, *J. Biol. Chem.*, 1933, **102**, 347 (*synth*)

Barnwell, J.L. *et al*, *Chem. Ind.* (*London*), 1955, 173 (*synth*)

MacDonald, D.L., *The Carbohydrates*, Academic Press, 1972, 2nd Ed., **1A**, 253 (*rev*)

Z

Zilantel, USAN **Z-00001**

1,2-Ethanediyl bis(phenylmethyl)carbonimidodithioate, *9CI*. *Ethylene dibenzyl* P,P,P′,P′*-tetraethylphosphonodithioimidocarbonate*, *8CI*

[22012-72-2]

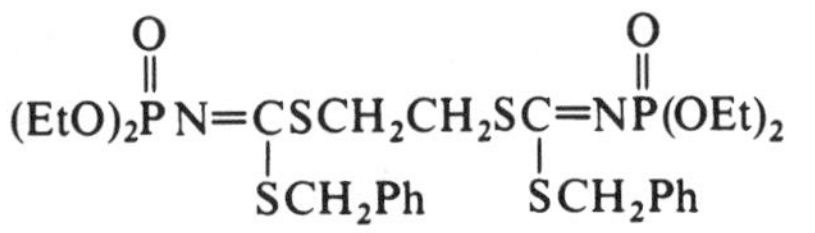

$$(EtO)_2\overset{O}{\overset{\|}{P}}N{=}\underset{SCH_2Ph}{\underset{|}{C}}SCH_2CH_2S\underset{SCH_2Ph}{\underset{|}{C}}{=}N\overset{O}{\overset{\|}{P}}(OEt)_2$$

$C_{26}H_{38}N_2O_6P_2S_4$ M 664.784

Anthelmintic. Cryst. (H_2O). Mp 67.5-68.5°.

South African Pat., 68 01 720, (*1968*); *CA*, **70**, 86572t (*synth, pharmacol*)

U.S.P., 3 691 283, (*1972*); *CA*, **77**, 156345 (*synth, ir*)

Zytron **Z-00002**

O-(*2,4-Dichlorophenyl*) O-*methyl* (*1-methylethyl*)-*phosphoramidothioate*, *9CI*. O-(*2,4-Dichlorophenyl*) O-*methyl isopropylphosphoramidothioate*

[299-85-4]

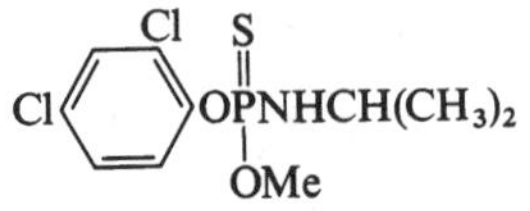

$$2,4\text{-}Cl_2C_6H_3O\overset{S}{\overset{\|}{P}}(OMe)NHCH(CH_3)_2$$

$C_{10}H_{14}Cl_2NO_2PS$ M 314.166

▷TB5075000.

(+)-***form*** [21248-23-7]

Oil. $[\alpha]_D$ +28.2° (c, 0.32 in C_6H_6).

(−)-***form*** [21248-24-8]

Oil. $[\alpha]_D$ −27.9° (c, 0.40 in C_6H_6).

(±)-***form*** [25137-74-0]

Herbicide. Solid. Mp 49-51°.

Getz, M.E. *et al*, *J. Assoc. Off. Anal. Chem.*, 1968, **51**, 1101 (*tlc*)

Keith, L.H. *et al*, *J. Assoc. Off. Anal. Chem.*, 1968, **51**, 1063 (*pmr*)

Seiber, J.N. *et al*, *Tetrahedron*, 1969, **25**, 381 (*synth*)

Ripley, B.D. *et al*, *J. Assoc. Off. Anal. Chem.*, 1983, **66**, 1084 (*glc*)

Name Index

Name Index

▷Abar, *see* L-00001
▷Abate, *see* T-00003
▷Abathion, *see* T-00003
▷Acconem, *see* F-00065
(5,6-Acenaphthenylenedimethylene)-bis[triphenylphosphonium](2+), A-00001
1-Acenaphthenylidenetriphenylphosphorane, A-00002
▷Acephate, A-00003
2-Acetamido-2-deoxy-α-D-glucose 1-(dihydrogen phosphate), *in* A-00067
2-Acetamido-2-deoxy-β-D-glucose 1-(dihydrogen phosphate), *in* A-00067
(Acetamidomethyl)phosphonic acid, *see* A-00012
▷Acethion, *see* A-00004
Acetic acid anhydride with dibenzyl hydrogen phosphate, *in* A-00025
Acetic acid anhydride with diethyl hydrogen phosphate, *in* A-00025
Acetic acid anhydride with diethyl phosphite, *see* A-00013
Acetic acid anhydride with dimethyl hydrogen phosphate, *in* A-00025
Acetic acid monoanhydride with methyl dihydrogen phosphate, *in* A-00025
Acetic acid, trianhydride with phosphoric acid, *see* T-00325
Acetic diphenylphosphinic anhydride, *in* D-01039
Acetoin diol cyclic pyrophosphate, *in* O-00088
Acetoinenediol cyclic pyrophosphate, *in* O-00088
Acetone-1,3-bis(triphenylphosphonium)(2+), *see* O-00073
Acetone 1,3-diphosphonic acid, *see* O-00072
Acetonyldiphenylphosphine, *see* D-01075
▷Acetonylidenetriphenylphosphorane, *see* T-00662
Acetonylphosphonic acid, *see* O-00075
P-Acetonylphosphonic diamide, *see* O-00077
Acetonylphosphonothioic acid, *see* O-00079
Acetonyltriphenylphosphonium(1+), *see* O-00080
▷Acetophos, A-00004
(α-Acetoxybenzyl)phosphonic acid, *see* A-00020
1-Acetoxy-1,1-ethanediphosphonic acid, A-00005
(2-Acetoxyethyl)phosphonic acid, A-00006
[(2-Acetoxymethyl)benzyl]-triphenylphosphonium(1+), *see* A-00019
(Acetoxymethyl)phosphonic acid, A-00007
(1-Acetoxypropyl)phosphonic acid, A-00008
(2-Acetoxypropyl)phosphonic acid, A-00009
3-Acetoxypropylphosphonic acid, A-00010
(1-Acetoxy-2,2,2-trichloroethyl)phosphonic acid, A-00011
[(Acetylamino)methyl]phosphonic acid, A-00012
(3-Acetylaminophenyl)-methyldiphenylphosphonium iodide, *in* D-01048
(α-Acetylbenzylidene)triphenylphosphorane, *see* P-00334
Acetyl dibenzyl phosphate, *in* A-00025
Acetyl diethyl phosphate, *in* A-00025
O-Acetyl diethylphosphinothioate, *in* D-00330
Acetyl diethyl phosphite, A-00013
P-Acetyl-*N*,*N*-diethylphosphonamidic acid, A-00014
Acetyl dimethyl phosphate, *in* A-00025
Acetyldimethylphosphine, A-00015
Acetyldiphenylphosphine, A-00016
N-Acetyl-*P*,*P*-diphenylphosphinic amide, *in* D-01040
S-Acetyl diphenylphosphinodithioate, *in* D-01065
N-Acetyl-*P*,*P*-diphenylphosphinothioic amide, *in* D-01083
Acetylenephosphonic acid, *see* E-00203
Acetylethylphosphinic acid, A-00017
[3-(Acetylhydroxyamino)-2-hydroxypropyl]-phosphonic acid, *see* A-00138
▷Acetylmethylidenetriphenylphosphorane, *see* T-00662
Acetylmethylphosphinic acid, A-00018
(1-Acetyl-2-oxopropyl)-triphenylphosphonium(1+), *see* D-00022
4-(Acetyloxy)-5-amino-2,3,7-trihydroxy-6-(phosphonooxy)-2,7-thiocanedicarboxylic acid, *see* T-00002
[1-(Acetyloxy)ethylidene]bisphosphonic acid, *see* A-00005
[2-(Acetyloxy)ethyl]phosphonic acid, *see* A-00006
[[2-[(Acetyloxy)methyl]phenyl]methyl]-triphenylphosphonium(1+), A-00019
[(Acetyloxy)methyl]phosphonic acid, *see* A-00007
[(Acetyloxy)phenylmethyl]phosphonic acid, A-00020
[1-(Acetyloxy)propyl]phosphonic acid, *see* A-00008
[2-(Acetyloxy)propyl]phosphonic acid, *see* A-00009
[(3-Acetyloxy)propyl]phosphonic acid, *see* A-00010
[1-(Acetyloxy)-2,2,2-trichloroethyl]-phosphonic acid, *see* A-00011
O-(2-Acetylphenyl) dichlorothiophosphate, *see* A-00022
O-(3-Acetylphenyl) dichlorothiophosphate, *see* A-00023
O-(4-Acetylphenyl) dichlorothiophosphate, *see* A-00024
Acetylphenylphosphinic acid, A-00021
O-(2-Acetylphenyl) phosphorodichloridothioate, A-00022
O-(3-Acetylphenyl) phosphorodichloridothioate, A-00023
O-(4-Acetylphenyl) phosphorodichloridothioate, A-00024
O-(2-Acetylphenyl) thiophosphoryl dichloride, *see* A-00022
O-(3-Acetylphenyl) thiophosphoryl dichloride, *see* A-00023
O-(4-Acetylphenyl) thiophosphoryl dichloride, *see* A-00024
Acetyl phosphate, A-00025
Acetylphosphonic acid, A-00026
P-Acetylphosphonic diamide, A-00027
Acetylphosphonothioic acid, A-00028
Acetylphosphoramidic acid, A-00029
P-Acetyl-*N*,*N*,*N'*,*N'*-tetraethylphosphonic diamide, *in* A-00027
Acetylthiophosphonic acid, *see* A-00028
N-Acetyl-*P*,*P*,*P*-triphenylphosphine imide, *in* T-00626
▷Acrasin, *see* C-00233
9-Acridanphosphonic acid, *see* D-00438
9-Acridinephosphonic acid, A-00030
9-Acridinylphosphonic acid, *see* A-00030
Acridophosphine, A-00031
▷Actellic, *see* P-00435
▷*N*-(1-Adamantyl)-*P*,*P*-bis(aziridinyl)-phosphinic amide, A-00032
1-Adamantylmethylidenephosphine, A-00033
1-Adamantylphosphonic acid, A-00034
2-Adamantylphosphonic acid, A-00035
1-Adamantylphosphoramidic acid, A-00036
2-Adamantylphosphoramidic acid, A-00037
▷Adenosidine 5'-(trihydrogen diphosphate) (5'→5') ester with 3-(aminocarbonyl)-1-β-D-ribofuranosylpyridinium hydroxide inner salt, *see* C-00202
Adenosine 2',5'-bis(phosphate), *see* A-00039
Adenosine 3',5'-bis(phosphate), *see* A-00040
▷Adenosine cyclic 3',5'-(hydrogen phosphate), *see* C-00233
Adenosine 2'-(dihydrogen phosphate), *see* A-00045
Adenosine 3'-(dihydrogen phosphate), *see* A-00046
▷Adenosine diphosphate, A-00038
Adenosine 2',5'-diphosphate, A-00039
Adenosine 3',5'-diphosphate, A-00040
Adenosine diphosphate ribose, A-00041
Adenosine 5'-diphosphoglucose, A-00042
▷Adenosine 5'-pyrophosphate, *see* A-00038
▷Adenosine 5'-(tetrahydrogen triphosphate), *see* A-00043
▷Adenosine 5'-(trihydrogen diphosphate), *see* A-00038
Adenosine 5'-(trihydrogen diphosphate) 2'-(dihydrogen phosphate) (5'→5') ester with 3-(aminocarbonyl)-1-β-D-ribofuranosylpyridinium hydroxide inner salt, *see* C-00203
Adenosine 5'-(trihydrogen diphosphate) 3'-(dihydrogen phosphate) 5'-[3-hydroxy-4-[[3-[(2-mercaptoethyl)amino]-3-oxopropyl]-amino]-2,2-dimethyl-4-oxobutyl] ester, *see* C-00200
Adenosine 5'-(trihydrogen diphosphate) 5'→5-ester with D-ribose, *see* A-00041
Adenosine 5'-(trihydrogen diphosphate) mono-α-D-glucopyranosyl ester, *see* A-00042
▷Adenosine triphosphate, A-00043
Adenosine 5'-uridine 5'-phosphate, A-00044
Adenylic acid *a*, *see* A-00045
Adenylic acid *b*, *see* A-00046
2'-Adenylic acid, A-00045
3'-Adenylic acid, A-00046
2'-Adenylic acid 5'-(dihydrogen phosphate), *see* A-00039
3'-Adenylic acid 5'-(dihydrogen phosphate), *see* A-00040
Adenylosuccinic acid, A-00047
Adetphos, *in* A-00043
▷ADP, *see* A-00038
ADPG, *see* A-00042
ADPR, *see* A-00041
▷Aflix, *see* F-00055
▷Afugan, *see* P-00502
▷Agrisil, *see* T-00419
▷Agritox, *see* T-00419
Agrocin 84, A-00048
AICAR, *see* A-00130
Alafosfalin, A-00049
Alafosfin, *see* A-00049
[2-(Alanylamino)ethyl]phosphonic acid, *see* A-00049
N-Alanyl-(2-aminoethyl)phosphonic acid, *see* A-00049
▷Alanylaspartyl-(2-amino-5-phosphoryl-3-pentenoic acid), *see* P-00437
Alcophosphamide, *in* B-00126
Aldophosphamide, A-00050
Aliette, *in* M-00473
1-*O*-Alkyl-2-acetyl-*sn*-glyceryl-3-phosphocholine, *see* P-00436
Allenephosphonic acid, *see* P-00441
P-Allenyl-*N*,*N*-diethylphosphonamidic acid, *see* D-00399
Allenyldiphenylphosphine, *see* D-01142
Allenylphosphonic acid, *see* P-00441
Allenylphosphonic bis(dimethylamide), *in* P-00442

P-Allenylphosphonic diamide, *see* P-00442
Allenylphosphonic dichloride, *see* P-00443
Allyl allylphenylphosphinate, *in* P-00300
Allylbenzylmethylphenylphosphonium bromide, *in* M-00264
Allyl diallylphosphinate, *in* D-01193
Allyldichlorophosphine, *in* P-00463
Allyl dichlorophosphite, *see* P-00465
▷Allyl dihydrogen phosphate, *see* M-00509
Allyldiphenylphosphine, *see* D-01144
Allyl diphenylphosphinite, *in* D-01089
Allyl diphenylphosphinodithioate, *in* D-01065
Allylmethylphenylphosphine, *see* M-00264
Allylmethylphenylpropylphosphonium bromide, *in* M-00265
Allylphenylphosphinic acid, *see* P-00300
▷Allyl phosphate, *see* M-00509
Allylphosphinic acid, *see* P-00452
Allylphosphonic acid, *see* P-00454
Allylphosphonic bis(dimethylamide), *in* P-00455
P-Allylphosphonic diamide, *see* P-00455
Allylphosphonic dianilide, *in* P-00455
Allylphosphonic dichloride, *see* P-00456
Allylphosphonodithioic acid, *see* P-00457
Allylphosphonoselenoic acid, *see* P-00458
P-Allylphosphonoselenoic diamide, *see* P-00459
Allylphosphonotelluroic acid, *see* P-00460
Allylphosphonothioic acid, *see* P-00461
Allylphosphonothioic bis(dimethylamide), *in* P-00462
Allylphosphonothioic diamide, *see* P-00462
Allylphosphonothioic dianilide, *in* P-00462
Allylphosphonous acid, *see* P-00463
Allylphosphoramidic acid, *see* P-00464
Allyl phosphorodichloridite, *see* P-00465
P-Allyl-*N*,*N*,*N'*,*N'*-tetramethylphosphonoselenoic diamide, *in* P-00459
P-Allyl-*N*,*N*,*N'*,*N'*-tetramethylphosphonoselenoic diamide, *in* P-00459
N''-Allyl-*N*,*N*,*N'*,*N'*-tetramethylphosphoric triamide, *in* T-00216
Allyl tetramethylphosphorodiamidate, *in* T-00218
Allylthiophosphonic diamide, *see* P-00462
Allyltriphenylphosphonium(1+), *see* T-00679
AM 630, *see* F-00058
AM 630B, *see* F-00059
N-Amidinophosphoramidic acid, *see* P-00370
▷Amidithion, A-00051
▷Amifostine, A-00052
(Aminobenzylidene)bisphosphonic acid, A-00053
(*α*-Aminobenzyl)phenylphosphinic acid, *see* A-00102
(*α*-Aminobenzyl)phosphonic acid, *see* A-00105
o-Aminobenzylphosphonic acid, *see* A-00106
p-Aminobenzylphosphonic acid, *see* A-00107
Aminobisbiphenylenephosphorane, *in* S-00021
Aminobis(diethylamino)(trichloromethyl)-phosphonium chloride, *in* T-00075
▷5-Amino-1-[bis(dimethylamino)phosphinyl]-3-phenyl-1*H*-1,2,4-triazole, *see* T-00326
(1-Aminobutyl)phosphonic acid, A-00054
(2-Aminobutyl)phosphonic acid, A-00055
(3-Aminobutyl)phosphonic acid, A-00056
(4-Aminobutyl)phosphonic acid, A-00057
[(Aminocarbonyl)amino]methylphosphonic acid, A-00058
P-Aminocarbonyl-*N*,*N*-diethylphosphonamidic acid, A-00059
[(Aminocarbonyl)methylene]-triphenylphosphorane, *in* A-00061
(Aminocarbonylmethyl)phosphonic acid, A-00060
[(Aminocarbonyl)methyl]-triphenylphosphonium(1+), A-00061
Aminocarbonylphosphonic acid, *see* C-00003
Aminocarbonylphosphoramidic acid, A-00062
P-Aminocarbonyl-*N*,*N*,*N'*,*N'*-tetraethylphosphonic diamide, *in* C-00003
2-Amino-3-(2-chloroethyl)-1,3,2-oxazaphosphorinane 2-oxide, *see* A-00063
2-Amino-3-(2-chloroethyl)tetrahydro-4*H*-1,3,2-oxazaphosphorine 2-oxide, A-00063
(10-Aminodecyl)phosphonic acid, A-00064
2-Amino-2-deoxy-*α*-D-galactopyranosyl phosphate, *in* A-00065
2-Amino-2-deoxygalactose 1-(dihydrogen phosphate), A-00065
2-Amino-2-deoxyglucitol, A-00066
2-Amino-2-deoxyglucose 1-(dihydrogen phosphate), A-00067
2-Amino-2-deoxyglucose 3-(dihydrogen phosphate), A-00068
2-Amino-2-deoxyglucose 6-(dihydrogen phosphate), A-00069
2-Amino-1,3-di-*tert*-butylazaphosphiridine, A-00070
Aminodi-*tert*-butylphosphine, *see* D-00124
3-Amino-2,3-di-*tert*-butyl-1,2,4,3-thiadiazaphosphetidine, A-00071
1-Amino-4-[(diethoxyphosphinothioyl)oxy]-benzene, *see* A-00099
1-Amino-4-[(diethoxyphosphinyl)oxy]benzene, *see* A-00098
Aminodiethylphosphine, *see* D-00334
2-Amino-2,3-dihydro-5-methyl-3-phenyl-3*H*-1,3,4,2-thiadiazaphosphole 2-sulfide, A-00072
8-Amino-1,1-dihydroxy-3-thia-2,4-diaza-1-phosphanonan-9-oic acid 1,3,3-trioxide, *see* P-00419
2-Amino-3-[(diphenoxyphosphinyl)oxy]-propanoic acid, *in* P-00417
1-Amino-1,1-ethanediphosphonic acid, A-00073
(2-Aminoethyl)butylphenylphosphine, A-00074
2-Aminoethyl dihydrogen phosphate, A-00075
(2-Aminoethyl)diphenylphosphine, *see* D-01066
(1-Aminoethylidene)bisphosphonic acid, *see* A-00073
(1-Aminoethylidene)diphosphonic acid, *see* A-00073
(1-Aminoethyl)methylphosphinic acid, A-00076
(1-Aminoethyl)phosphinic acid, A-00077
(1-Aminoethyl)phosphonic acid, A-00078
2-Aminoethylphosphonic acid, A-00079
(2-Aminoethyl)phosphoramidic acid, A-00080
Aminofenitrothion, *see* A-00090
2-Amino-2,2,4,4,6,6-hexahydro-2-methyl-4,4,6,6-tetraphenylcyclotriphosphazene, *see* A-00081
2-Amino-2,2,4,4,6,6-hexahydro-2-methyl-4,4,6,6-tetraphenyl-1,3,5-triaza-2,4,6-triphosphorine, A-00081
2-Amino-2,2,4,4,6,6-hexahydro-2,4,4,6,6-pentaphenoxycyclotriphosphazene, *see* A-00082
2-Amino-2,2,4,4,6,6-hexahydro-2,4,4,6,6-pentaphenoxy-1,3,5-triaza-2,4,6-triphosphorine, A-00082
(1-Amino-2-hydroxyethyl)phosphonic acid, A-00083
2-Amino-4-(hydroxymethylphosphinyl)butanoic acid, *see* P-00359
2-Amino-3-[(hydroxyphenoxyphosphinyl)oxy]-propanoic acid, *in* P-00417
Aminohydroxyphosphinylacetic acid methyl ester, *see* M-00028
2-Amino-3-hydroxypropanoyl 3-phosphate, *see* P-00417
(Aminoiminomethyl)phosphoramidic acid, *see* P-00370
[[1-Amino-2-(1*H*-indol-3-yl]ethyl]phosphonic acid, *see* P-00421
(1-Aminoisobutyl)phosphonic acid, *see* A-00094
(*α*-Aminoisopropyl)phosphonic acid, *see* A-00087
(1-Amino-2-mercaptoethyl)phosphonic acid, A-00084
Aminomethanediphosphonic acid, *see* A-00086
(Aminomethyl)diphenylphosphine oxide, A-00085
(Aminomethylene)bisphosphonic acid, A-00086
(1-Amino-1-methylethyl)phosphonic acid, A-00087
(2-Amino-1-methylethyl)phosphonic acid, A-00088
4-Aminomethyl-5-hydroxy-6-methyl-3-(dihydrogen phosphate)-3-pyridinemethanol, *see* P-00509
(Aminomethyl)methylphosphinic acid, A-00089
O-(4-Amino-3-methylphenyl) *O*,*O*-dimethyl phosphorothioate, A-00090
2-Amino-4-(methylphosphino)butyric acid, *see* P-00359
(Aminomethyl)phosphonic acid, A-00091
2-Amino-4-(*S*-methyl-*N*-phosphonosulfonimidoyl)butanoic acid, *see* P-00388
α-Amino-*γ*-(*S*-methyl-*N*-phosphonosulfonimidoyl)-butyrylalanylalanine, *see* P-00389
(1-Amino-2-methylpropyl)phosphinic acid, A-00092
(1-Amino-1-methylpropyl)phosphonic acid, A-00093
(1-Amino-2-methylpropyl)phosphonic acid, A-00094
▷*S*-[2-[[(4-Amino-2-methyl-5-pyrimidinyl)-methyl]formylamino]-1-[2-(phosphonoxy)ethyl]-1-propenyl]-benzenecarbothioate, *see* B-00002
▷3-[(4-Amino-2-methyl-5-pyrimidyl)methyl]-4-methyl-5-(4,6,6-trihydroxy-3,5-dioxa-4,6-diphosphahex-1-yl)thiazolium chloride *P*,*P'*-dioxide, *see* T-00297
(1-Amino-3-methylthiopropyl)phosphonic acid, A-00095
(2-Amino-2-oxoethyl)phosphonic acid, *see* A-00060
[1-[(2-Amino-1-oxopropyl)amino]ethyl]-phosphonic acid, *see* A-00049
1-Amino-2,2,3,4,4-Pentamethylphosphetane 1-oxide, *see* P-00021
(1-Aminopentyl)phosphonic acid, A-00096
(5-Aminopentyl)phosphonic acid, A-00097
4-Aminophenyl diethyl phosphate, A-00098
O-(4-Aminophenyl) *O*,*O*-diethyl phosphorothioate, A-00099
P-(4-Aminophenyl)-*N*,*N'*-dimethylphosphonic diamide, *in* A-00112
(2-Aminophenyl)diphenylphosphine, *see* D-01047
(3-Aminophenyl)diphenylphosphine, *see* D-01048
(4-Aminophenyl)diphenylphosphine, *see* D-01049
P-(4-Aminophenyl)-*N*,*N'*-diphenylphosphonic diamide, *in* A-00112
(1-Amino-1-phenylethyl)phosphonic acid, A-00100
(1-Amino-2-phenylethyl)phosphonic acid, A-00101
(3-Aminophenyl)methyldiphenylphosphonium iodide, *in* D-01048
(4-Aminophenyl)methyldiphenylphosphonium iodide, *in* D-01049
(Aminophenylmethyl)phenylphosphinic acid, A-00102
(2-Aminophenyl)methylphosphinic acid, A-00103
(4-Aminophenyl)methylphosphinic acid, A-00104
(Aminophenylmethyl)phosphonic acid, A-00105
[(2-Aminophenyl)methyl]phosphonic acid, A-00106
[(4-Aminophenyl)methyl]phosphonic acid, A-00107
(4-Aminophenyl)phosphinic acid, A-00108
▷(2-Aminophenyl)phosphonic acid, A-00109
▷(3-Aminophenyl)phosphonic acid, A-00110
▷(4-Aminophenyl)phosphonic acid, A-00111
P-(4-Aminophenyl)phosphonic diamide, A-00112
(2-Aminophenyl)phosphoramidic acid, A-00113
(3-Aminophenyl)phosphoramidic acid, A-00114
(4-Aminophenyl)phosphoramidic acid, A-00115
▷*P*-(5-Amino-3-phenyl-1*H*-1,2,4-triazol-1-yl)-*N*,*N*,*N'*,*N'*-tetramethylphosphonic diamide, *see* T-00326
3-Amino-3-phosphonobutanamide, *in* A-00118
2-Amino-3-phosphonobutanoic acid, A-00116
2-Amino-4-phosphonobutanoic acid, A-00117
3-Amino-3-phosphonobutanoic acid, A-00118
4-Amino-4-phosphonobutanoic acid, A-00119
2-Amino-3-phosphonooxybutanoic acid, A-00120
(3-Amino-3-phosphono)propanoamide, *in* A-00122
2-Amino-3-phosphonopropanoic acid, A-00121
3-Amino-3-phosphonopropanoic acid, A-00122

2-Amino-4-phosphonopyrimidine, *see* A-00128
5-Amino-1-(5-*O*-phosphonoribofuranosyl)-1*H*-imidazole-4-carboxamide, *see* A-00130
N-[[5-Amino-1-(5-*O*-phosphono-*β*-D-ribofuranosyl)-1*H*-imidazol-4-yl]-carbonyl]aspartic acid, *see* A-00129
2-Amino-5-phospho-3-pentenoic acid, A-00123
▷2-[(3-Aminopropyl)amino]ethanethiol dihydrogenphosphate, *see* E-00021
▷2-[(3-Aminopropyl)amino]ethanethiol dihydrogen phosphate (ester), *see* A-00052
▷*S*-[2-[(3-Aminopropyl)amino]ethyl]dihydrogen phosphorothioate, *see* A-00052
(1-Aminopropylidene)bisphosphonic acid, A-00124
(1-Aminopropyl)phosphonic acid, A-00125
(2-Aminopropyl)phosphonic acid, A-00126
(3-Aminopropyl)phosphonic acid, A-00127
(2-Amino-4-pyrimidinyl)phosphonic acid, A-00128
N-(5-Amino-1-*β*-D-ribofuranosylimidazole-4-carbonyl)-L-aspartic acid 5′-phosphate, A-00129
5-Amino-1-ribofuranosylimidazole-4-carboxamide 5′-(dihydrogen phosphate), A-00130
N-[(5-Amino-1-*β*-D-ribofuranosylimidazol-4-yl)carbonyl]-5′-(dihydrogen phosphate)-L-aspartic acid, *see* A-00129
N^5-[Amino(sulfoamino)phosphinyl]-L-ornithine, *see* O-00038
Aminotris(dimethylamino)phosphonium bromide, *in* H-00077
Aminotris(dimethylamino)phosphonium chloride, *in* H-00077
Amiprophos *M*, A-00131
▷Amiprophos-Methyl, A-00132
▷Amiton, A-00133
Amyl *sec*-butylphenylphosphinate, *in* M-00372
Amyl dichlorophosphite, *see* P-00054
O-Amyl dichlorothiophosphate, *see* P-00055
Amyl dihydrogen phosphate, *see* M-00504
Amyl hypophosphite, *see* P-00048
Amylphosphine, *see* P-00049
Amylphosphonic acid, *see* P-00050
tert-Amylphosphonic acid, *see* D-00911
Amylphosphonodithioic acid, *see* P-00051
Amylphosphonous acid, *see* P-00052
Amylphosphoramidic acid, *see* P-00053
Amyl phosphoric acid, *see* M-00504
Amyl phosphorodichloridite, *see* P-00054
Amyltriphenylphosphonium(1+), *see* P-00056
Anhydride of 1,1-dimethylethyl carbonate and diethyl phosphate, *see* B-00531
Anilino-*tert*-butylchlorophosphine phenylimide, *see* B-00559
5-Anilino-5,6-dihydrodibenzo[*b*,*d*]phosphorin 5-oxide, *in* D-00504
5-Anilino-5,6-dihydrophosphanthridene 5-oxide, *in* D-00504
2-Anilino-5,5-dimethyl-1,3-dioxa-2-phosphacyclohexane, *see* D-00792
2-Anilino-5,5-dimethyl-1,3,2-dioxaphosphorinane, *see* D-00792
2-Anilino-1,3-dioxa-2-phosphacyclopentane, *see* P-00083
2-Anilino-1,3,2-dioxaphospholane, *see* P-00083
Anilinodiphosphoric acid, *see* P-00168
2-Anilino-4-methyl-1,3-dioxa-2-phosphacyclohexane, *see* M-00193
2-Anilino-4-methyl-1,3,2-dioxaphosphorinane, *see* M-00193
2-Anilino-5-methyl-3-phenyl-Δ^4-1,3,4,2-oxadiazaphospholine, *see* D-00527
(Anilinomethyl)phosphonic acid, *see* P-00084
(3-Anilinopropyl)phosphonic acid, *see* P-00085
2-Anilinotetrahydro-3-phenyl-2*H*-1,3,2-oxaphosphorine, *see* T-00133
Anilofos, A-00134
m-Anisyldiphenylphosphine, *see* M-00057
o-Anisyldiphenylphosphine, *see* M-00056
p-Anisyldiphenylphosphine, *see* M-00058
p-Anisylmethyl-*α*-naphthylphenylphosphonium bromide, *in* M-00066
p-Anisylmethyl-*α*-naphthylphenylphosphonium chloride, *in* M-00066
o-Anisylmethylphenylphosphine, *see* M-00059
o-Anisyl-*β*-naphthylphenylphosphine, *see* M-00065
p-Anisyl-*α*-naphthylphenylphosphine, *see* M-00066
p-Anisylphosphine, *see* M-00067
o-Anisylphosphinic acid, *see* M-00068
p-Anisylphosphinic acid, *see* M-00069
m-Anisylphosphonic acid, *see* M-00071
o-Anisylphosphonic acid, *see* M-00070
p-Anisylphosphonic acid, *see* M-00072
p-Anisylphosphonic dianilide, *in* M-00072
o-Anisyl phosphoric acid, *see* M-00480
▷Anthio, *see* F-00055
[1,8-Anthracenediylbis[methylene]]-bis[triphenylphosphonium](2+), *see* B-00443
[1,8-Anthracenediylbis[methylene]]-bis[triphenylphosphorane], *in* B-00443
▷Antibiotic 833A, *in* M-00185
Antibiotic 901A, *see* P-00439
Antibiotic AM 630, *see* F-00058
Antibiotic AM 630B, *see* F-00059
Antibiotic BMG 59-R2, A-00135
Antibiotic CI 920, A-00136
Antibiotic CL 1565*A*, *see* A-00136
Antibiotic CL 1565*B*, *in* A-00136
Antibiotic CL 1565*C*, *in* A-00136
Antibiotic CL 1565*T*, *in* A-00136
Antibiotic FMP 1, *see* P-00401
Antibiotic FR 32863, A-00137
Antibiotic FR 33289, A-00138
Antibiotic FR 900136, *see* A-00137
Antibiotic I5*B*, A-00139
Antibiotic I5*B*1, *in* A-00139
Antibiotic I5*B*2, *in* A-00139
Antibiotic K 4, *in* A-00139
Antibiotic K 26, A-00140
▷Antibiotic MK 955, *in* M-00185
▷Antibiotic N 1409, *in* P-00437
▷Antibiotic N1409A, *see* P-00437
Antibiotic PD 110161, *see* A-00136
Antibiotic PD 113270, *in* A-00136
Antibiotic PD 113271, *in* A-00136
▷Antibiotic SF 1293, A-00141
Antihypertensive polar renomedullary lipid, *see* P-00436
▷Aphidan, A-00142
▷Apholate, *see* H-00060
APRL, *see* P-00436
Aquocobalamin, *see* V-00010
Arachidonoylstearoylphosphatidylethanolamine, *in* G-00065
▷Arphos, *see* D-00993
L-Aspartic acid 4-monoanhydride with phosphoric acid, *see* A-00143
β-Aspartyl phosphate, A-00143
▷Aspon, A-00144
▷Asta 5122, *see* S-00026
▷Asuntol, *see* C-00205
▷ATP, *see* A-00043
▷Avinar, *see* U-00001
Axerophytyltriphenylphosphonium(1+), *see* D-00929
▷Azadithion, *see* M-00003
Azaphosphatriptycene, *see* B-00015
AzaUMP, *see* A-00145
6-Azauridine 5′-phosphate, A-00145
6-Azauridylic acid, *see* A-00145
Azemethephos, A-00146
1-Azetidinephosphonic acid, A-00147
1-Azetidinylphosphonic acid, *see* A-00147
2-Azido-1,3,2-benzodioxaphosphole, A-00148
1-Azido-2,2,3,4,4-pentamethylphosphetane 1-oxide, *in* H-00168
(Azidosulfonylimino)triethoxyphosphorane, *see* T-00439
N-Azidosulfonyl-*P*,*P*,*P*-triethoxyphosphazene, *see* T-00439
N-Azidosulfonyl-*P*,*P*,*P*-triethoxyphosphine imide, *see* T-00439
Azidotris(dimethylamino)phosphorus(1+), A-00149
Azidotris(*N*-methylmethanaminato)-phosphorus(1+), *see* A-00149
Azinobis(triphenylphosphorane), *see* B-00458
▷Azinphos-ethyl, A-00150
▷Azinphos-Methyl, A-00151
1-Aziridinephosphonic acid, *see* A-00153
1-Aziridinylmethylphosphinic acid, *see* A-00152
1-Aziridinyl-*P*-methylphosphonamidic acid, A-00152
1-Aziridinylphosphonic acid, A-00153
▷1-Aziridinylphosphonic bis(dimethylamide), *in* A-00153
1-Aziridinylphosphonous acid, A-00154
1-Aziridinylphosphonous bis(diethylamide), *in* A-00155
1-Aziridinylphosphonous bis(dimethylamide), *in* A-00155
P-1-Aziridinylphosphonous diamide, A-00155
P-1-Aziridinyl-*N*,*N*,*N*′,*N*′-tetraethylphosphonous amide, *in* A-00155
▷*P*-1-Aziridinyl-*N*,*N*,*N*′,*N*′-tetramethylphosphonic diamide, *in* A-00153
P-1-Aziridinyl-*N*,*N*,*N*′,*N*′-tetramethylphosphonous diamide, *in* A-00155
Azodrin, *see* M-00459
(1-Azulenylmethylene)triphenylphosphorane, *in* A-00156
(1-Azulenylmethyl)triphenylphosphonium(1+), A-00156
▷Bacdip, *see* Q-00006
Bacillus thuringiensis *Exotoxin*, *see* T-00319
▷Basudin, *see* D-00027
Batestan, *see* B-00003
▷Bayrusil, *see* Q-00001
▷Baytex, *see* F-00007
▷Baythion, *see* P-00424
Benfosformin, B-00001
▷Benfotiamine, B-00002
Benoxafos, B-00003
▷Bensulide, B-00004
[*α*-(Benzamido)benzyl]phosphonic acid, *in* A-00105
[1-(Benzamido)ethyl]phosphonic acid, *in* A-00078
(Benzamidophenylmethyl)phosphonic acid, *in* A-00105
1*H*-1,3-Benzazaphosphole, B-00005
Benzenecarbothioic (*O*,*O*-diethylphosphorodithioic) anhydrosulfide, *see* D-00270
1,3-Benzenediol bis(dihydrogen phosphate), *see* P-00147
1,4-Benzenediol bis(dihydrogen phosphate), *see* P-00148
1,2-Benzenediol phosphate(3:1), *see* T-00785
1,3-Benzenediol phosphate (3:1), *see* T-00786
1,4-Benzenediol phosphate(3:1), *see* T-00787
1,2-Benzenediphosphinic acid, *see* B-00009
1,4-Benzenediphosphinic acid, *see* B-00010
1,2-Benzenediphosphonic acid, B-00006
1,3-Benzenediphosphonic acid, B-00007
1,4-Benzenediphosphonic acid, B-00008
1,2-Benzenediphosphonous acid, B-00009
1,4-Benzenediphosphonous acid, B-00010
1,3-Benzenediphosphoramidic acid, B-00011
1,4-Benzenediphosphoramidic acid, B-00012
▷Benzenephosphonic acid, *see* P-00221
Benzenephosphonous acid, *see* P-00251
▷*S*-2-Benzenesulfonamidoethyl *O*,*O*-diisopropyl dithiophosphate, *see* B-00004
N-Benzenesulfonyl-*P*,*P*-diphenylphosphinimidic acid, B-00013
N-Benzenesulfonyl-*P*-phenylphosphonamidic acid, *see* P-00191
Benzenesulfonylphosphorimidic trichloride, *see* P-00316
10*H*-5,10[1′,2′]Benzenoacridophosphine, B-00014
5,10[1′,2′]-Benzenophenophosphazine, B-00015
5,10-*o*-Benzenophosphazine, *see* B-00015
▷Benzhydroxamatophosphorus(III) chloride, *see* C-00126
Benzhydrylphenylphosphinic acid, *see* D-01023
5-Benzimidazolephosphonic acid, B-00016
(Benzimidazol-5-yl)phosphonic acid, *see* B-00016
(Benzo[*g*]chrysen-10-ylmethylene)-triphenylphosphorane, *in* T-00635
(Benzo[*g*]chrysen-10-ylmethyl)-triphenylphosphonium(1+), *see* T-00635

Benzodepa, B-00017
3,4-Benzo-5,6-dimethyl-1-phosphabicyclo[3.2.1]oct-3-ene, *see* T-00096
1,3,2-Benzodioxaphosphole, B-00018
(1,3-Benzodioxol-5-yl)phosphonic acid, B-00019
(1,3-Benzodithiol-2-ylidene)-triphenylphosphorane, *in* B-00021
1,3-Benzodithiol-2-ylphosphonic acid, B-00020
1,3-Benzodithiol-2-yltriphenylphosphonium(1+), B-00021
Benzoic acid, anhydride with diethyl phosphite, *see* B-00028
Benzoic acid, anhydride with *O,O*-diethylphosphorothioate, *see* B-00029
Benzoic acid(3:1)anhydride with phosphorous acid, *see* T-00336
Benzoic acid-phosphoric acid anhydride, *see* M-00442
Benzoic-*O,O*-diethyl phosphorothioic anhydride, *see* B-00029
Benzoic diphenylphosphinic anhydride, *in* D-01039
Benzoic-phosphoric (2:1) anhydride, *see* D-00042
2,3-Benzo-1,4,5,6,7-pentaphenyl-7-phosphabicyclo[2.2.1]hept-5-ene 7-oxide, *in* D-00538
[Benzo[*c*]phenanthrene-2,11-diylbis[methylene]]-bis[diphenylphosphine], *see* B-00246
(Benzo[*e*]phenanthren-2-ylmethylene)-triphenylphosphorane, *in* T-00636
(Benzo[*c*]phenanthren-2-ylmethyl)-triphenylphosphonium(1+), *see* T-00636
Benzo[*e*]phenazaphosphinic acid, *in* D-00440
7*H*-Benzo[*c*]phenothiaphosphine, B-00022
12*H*-Benzo[*c*]phenothiaphosphine, B-00023
5,10-[1′,2′]Benzophosphenanthrene, B-00024
1*H*-Benzo[*b*]phosphindole, *see* D-00040
Benzo[*c*]phosphinine, *see* I-00027
Benzo[*c*]phosphorin, *see* I-00027
▷Benzotef, *in* B-00092
1,3-Benzothiaphosphole, B-00025
2,3-Benzo-5,6,7-triphenyl-1-phosphabicyclo[2.2.2]octa-2,5,7-triene, *see* T-00607
1,3,2-Benzoxathiaphosphole, B-00026
6-Benzoxazolephosphonic acid, B-00027
(1,3-Benzoxazol-6-yl)phosphonic acid, *see* B-00027
(1-Benzoylaminobutyl)phosphonic acid, *in* A-00054
[1-(Benzoylamino)ethyl]phosphonic acid, *in* A-00078
2-(Benzoylamino)ethylphosphonic acid, *in* A-00079
(1-Benzoylamino-1-methylethyl)phosphonic acid, *in* A-00087
(*N*-Benzoylaminomethyl)phosphonic acid, *in* A-00091
(1-Benzoylamino-2-methylpropyl)phosphonic acid, *in* A-00094
(1-Benzoylaminopentyl)phosphonic acid, *in* A-00096
[(Benzoylamino)phenylmethyl]phosphonic acid, *in* A-00105
2-Benzoylamino-4-phosphonobutanoic acid, *in* A-00117
Benzoylaminophosphorimidic trichloride, *in* T-00131
[1-(*N*-Benzoylamino)propyl]phosphonic acid, *in* A-00125
▷*N*-Benzoyl-*P,P*-bis(1-aziridinyl)phosphinic amide, *in* B-00092
(Benzoyldiazomethyl)diphenylphosphine, *see* D-00029
N-Benzoyl-*P,P*-dichloro-*P*-phenylphosphazene, *see* B-00033
N-Benzoyl-*P,P*-dichloro-*P*-phenylphosphine imide, *see* B-00033
Benzoyl dichlorophosphate, *in* M-00442
N-Benzoyl-*P,P*-diethylphosphinothioic amide, *in* D-00331
Benzoyl diethyl phosphite, B-00028
O-Benzoyl *O,O*-diethylphosphorothioate, B-00029
O-Benzoyl *O,O*-diethyl thiophosphate, *see* B-00029
Benzoyl difluorophosphate, *in* M-00442
N-Benzoyl-*P,P*-dimethylphosphinothioic amide, *in* D-00840
Benzoyldiphenylphosphine, B-00030
N-Benzoyl-*P,P*-diphenylphosphinic amide, *in* D-01040
S-Benzoyl diphenylphosphinodithioate, *in* D-01065
N-Benzoyl-*P,P*-diphenylphosphinothioic amide, *in* D-01083
(1-Benzoylethylidene)triphenylphosphorane, *see* P-00335
[(1-Benzoyl-2-methoxycarbonyl)ethylidene]-triphenylphosphorane, *in* O-00082
▷Benzoylmethylidenetriphenylphosphorane, *see* P-00333
N-Benzoyl-*P*-methylphosphonamidic acid, B-00031
N-Benzoyl-*P*-methylphosphonamidic chloride, *in* B-00031
(Benzoylmethyl)triphenylphosphonium(1+), *see* O-00067
Benzoylphenylphosphinic acid, B-00032
N-Benzoyl-*P*-phenylphosphonimidic dichloride, B-00033
Benzoylphosphonic acid, B-00034
Benzoylphosphonic bis(diethylamide), *in* B-00035
P-Benzoylphosphonic diamide, B-00035
Benzoylphosphoramidic acid, B-00036
Benzoylphosphoramidic dianilide, *in* D-01114
Benzoylphosphoramidic dichloride, B-00037
Benzoylphosphoramidothioic acid, B-00038
Benzoylphosphorimidic acid, B-00039
Benzoylphosphorimidic trichloride, B-00040
Benzoyl phosphorodichloridate, *in* M-00442
Benzoyl phosphorodifluoridate, *in* M-00442
P-Benzoyl-*N,N,N′,N′*-tetraethylphosphonic diamide, *in* B-00035
N-Benzoyl-*P,P,P*-trichlorophosphazene, *see* B-00040
N-Benzoyl-*P,P,P*-trichlorophosphine imide, *see* B-00040
N-Benzoyl-*P,P,P*-triethoxyphosphazene, *in* B-00039
N-Benzoyl-*P,P,P*-triethoxyphosphine imide, *in* B-00039
N-Benzoyl-*P,P,P*-triisopropoxyphosphazene, *in* B-00039
N-Benzoyl-*P,P,P*-trimethoxyphosphazene, *in* B-00039
N-Benzoyl-*P,P,P*-trimethoxyphosphine imide, *in* B-00039
N-Benzoyl-*P,P,P*-triphenoxyphosphazene, *in* B-00039
N-Benzoyl-*P,P,P*-triphenoxyphosphine imide, *in* B-00039
N-Benzoyl-*P,P,P*-triphenylphosphine imide, *in* T-00626
Benzoyl(triphenylphosphoranylidene)acetic acid, *see* O-00083
Benzoyl(triphenylphosphoranylidene)-acetonitrile, *in* O-00083
3-Benzoyl-3-(triphenylphosphoranylidene)-propanoic acid, *see* O-00082
[(Benzylamidino)amidino]phosphoramidic acid, *see* B-00001
Benzylamine-*α,α*-diphosphonic acid, *see* A-00053
Benzyl 2-amino-4,6-*O*-benzylidene-2-deoxy-*α*-D-glucopyranoside 3-(dihydrogen phosphate), *in* A-00068
(*α*-Benzylaminobenzyl)phenylphosphinic acid, *in* A-00102
[*α*-(Benzylamino)benzyl]phosphonic acid, *see* P-00188
[1-(Benzylamino)butyl]phosphonic acid, *in* A-00054
[(1-Benzylamino)ethyl]phosphonic acid, *in* A-00078
(1-Benzylamino-1-methylethyl)phosphonic acid, *in* A-00087
[1-(Benzylamino)pentyl]phosphonic acid, *in* A-00096
[1-(Benzylamino)propyl]phosphonic acid, *in* A-00125
5-Benzyl-10*H*-5,10[1′,2′]-benzenoacridophosphonium bromide, *in* B-00014
Benzyl bis(1-aziridinyl)phosphinate, *in* B-00091
Benzyl [bis(1-aziridinyl)phosphinyl]-carbamate, *see* B-00017
N-Benzyl-*P,P*-bis(chloromethyl)phosphinic amide, *in* B-00128
Benzyl [bis(trimethylsilyloxy)phosphinyl]-acetate, *in* B-00434
Benzyl-4-bromophenyldiphenylphosphonium bromide, *in* B-00481
Benzyl-2-bromophenyldiphenylphosphonium chloride, *in* B-00480
▷*S*-Benzyl *O*-butyl *S*-ethyl dithiophosphate, *see* C-00204
▷*S*-Benzyl *O*-butyl *S*-ethyl phosphorodithioate, *see* C-00204
Benzyl-*tert*-butylmethylphenylphosphonium bromide, *in* B-00565
1-Benzyl-4-*tert*-butyl-1-phenylphosphorinanium bromide, *in* B-00586
3-Benzyl-2-chloro-1,3,2-oxazaphosphorinane, *see* B-00041
Benzyl(2-chlorophenyl)diphenylphosphonium chloride, *in* C-00127
3-Benzyl-2-chlorotetrahydro-2*H*-1,3,2-oxazaphosphorine, B-00041
Benzyl *N*-cyclohexyl-*P*-ethylphosphonamidate, *in* C-00246
Benzyl *N*-cyclohexyl-*P*-phenylphosphonamidate, *in* C-00251
Benzyl (dibenzyloxyphosphinylamino)acetate, *in* P-00368
Benzyl 2-[(dibenzyloxyphosphinyl)amino]-3-phenylpropanoate, *in* P-00400
Benzyl 2-[(dibenzyloxyphosphinyl)amino]-propanoate, *in* P-00362
Benzyl dibenzylphosphinodithioate, *in* D-00056
Benzyldibromophosphine, *in* B-00068
Benzyl (dibutoxyphosphinyl)acetate, *in* D-00073
O-Benzyl di-*tert*-butylphosphinothioate, *in* D-00118
1-Benzyl-4,5-dichloro-1-phenyl-1*H*-phosphepinium bromide, *in* D-00565
Benzyl dichlorophosphate, *in* M-00443
Benzyldichlorophosphine, *in* B-00068
Benzyl dichlorophosphite, *see* B-00070
Benzyl dicyclohexylphosphinodithioate, *in* D-00212
Benzyl 4-(diethoxyphosphinyl)butanoate, *in* D-00247
Benzyl diethylphosphinate, *in* D-00322
▷*S*-Benzyl *O,O*-diethyl phosphorothioate, *see* K-00001
Benzyldiferrocenylphenylphosphonium chloride, *in* D-00410
Benzyldifluorophosphine, *in* B-00068
1-Benzyl-2,5-dihydro-3,4-dimethyl-1-phenyl-1*H*-phospholium bromide, *in* D-00468
Benzyl dihydrogen phosphate, *see* M-00443
Benzyl dihydrogen phosphite, *see* M-00444
S-Benzyl dihydrogen thiophosphate, *see* B-00071
1-Benzyl-2,5-dihydro-3-methyl-1-phenyl-1*H*-phospholium bromide, *in* D-00528
5-Benzyl-5,5-dihydro-5-phenylacridophosphine, B-00042
5-Benzyl-5,10-dihydro-5-phenylacridophosphonium bromide, *in* D-00547
5-Benzyl-5,6-dihydro-6-phenylphosphanthridine, B-00043
1-Benzyl-2,5-dihydro-1*H*-phosphole, *in* D-00572
1-Benzyl-2,5-dihydro-1,3,4-trimethyl-1*H*-phospholium bromide, *in* D-00600
O-Benzyl *O,O*-diisopropyl phosphorothioate, B-00044
▷*S*-Benzyl *O,O*-diisopropyl phosphorothioate, B-00045
1-Benzyl-3,4-dimethyl-1-phenyl-1*H*-phospholium bromide, *in* D-00807

Benzyldimethylphenylphosphonium iodide, *in* B-00051
N-Benzyl-*P*,*P*-dimethylphosphinothioic amide, *in* D-00840
Benzyl 2-[(diphenoxyphosphinyl)amino]-3-phenylpropanoate, *in* P-00400
Benzyl *N*-(diphenoxyphosphinyl)carbamate, *in* D-00989
Benzyl 3-[(diphenoxyphosphinyl)oxy]-propanoate, *in* P-00399
Benzyldiphenyl(4-phenyl-1,3-butadienyl)-phosphonium bromide, *in* D-01031
Benzyl diphenylphosphinate, *in* D-01039
Benzyldiphenylphosphine, B-00046
N-Benzyl-*P*,*P*-diphenylphosphinic amide, *in* D-01040
P-Benzyl-*N*,*P*-diphenylphosphinic amide, *in* B-00056
Benzyl diphenylphosphinodithioate, *in* D-01065
O-Benzyl diphenylphosphinothioate, *in* D-01082
S-Benzyl diphenylphosphinothioate, *in* D-01082
Benzyl diphenylphosphinothioite, *in* D-01085
Benzyldiphenyl-2-propenylphosphonium bromide, *in* D-01144
1-Benzyl-2-ethoxyhexahydro-1,2-azaphosphorine 2-oxide, *in* B-00048
Benzylethylmethylphenylphosphonium iodide, *in* B-00047
Benzylethylmethylphenylphosphonium iodide, *in* E-00084
Benzylethylphenylphosphine, B-00047
S-Benzyl *O*-ethyl phenylphosphonothioate, *see* I-00009
Benzyl ethylphosphonobromidothioite, *in* E-00134
Benzyl ethylphosphonochloridate, *in* E-00135
Benzyl(ferrocenyl)diphenylphosphonium chloride, *in* D-01068
1-Benzylhexahydro-2-hydroxy-1,2-azaphosphinine 2-oxide, *see* B-00048
1-Benzylhexahydro-2-hydroxy-1,2-azaphosphorine 2-oxide, B-00048
1-Benzyl-1,2,2,3,4,4-hexamethylphosphetanium bromide, *in* H-00074
Benzyl hyophosphite, *see* B-00057
1-(Benzylideneamino)ethylphosphonic acid, B-00049
Benzylidenemethylphenylpropylphosphorane, *in* B-00052
Benzylidenetriphenylphosphorane, B-00050
Benzyl *N*-[imino(diphenoxyphosphinylamino)-methyl]glycinate, *in* P-00369
Benzyl *N*-[Imino(diphenoxyphosphinylamino)-methyl]-*N*-methylglycinate, *in* P-00364
Benzyl(2-methoxyphenyl)diphenylphosphonium bromide, *in* M-00056
Benzyl(3-methoxyphenyl)diphenylphosphonium chloride, *in* M-00057
Benzyl(4-methoxyphenyl)diphenylphosphonium chloride, *in* M-00058
Benzylmethyldiphenylphosphonium iodide, *in* B-00046
Benzyl methyl methylphosphonate, *in* M-00168
Benzylmethyl-2-naphthylphenylphosphonium bromide, *in* M-00177
Benzylmethylphenylphosphine, B-00051
N-Benzyl-*P*-methyl-*P*-phenylphosphinothioic amide, *in* M-00226
1-Benzyl-3-methyl-1-phenylphospholanium bromide, *in* M-00229
1-Benzyl-3-methyl-1-phenylphospholanium iodide, *in* M-00229
Benzyl methyl phenylphosphonate, *in* M-00488
Benzylmethylphenylpropylphosphonium(1+), B-00052
Benzylmethylphosphinic acid, B-00053
1-Benzyl-1-methylphospholanium bromide, *in* M-00280
Benzylmethylpropylphosphine, B-00054
Benzyl(2-methylthiophenyl)-diphenylphosphonium chloride, *in* M-00417
Benzylneomenthyldiphenylphosphonium bromide, *in* M-00164
6-Benzyl-2-oxa-6-phosphatricyclo[3.3.1.1^{3,7}]decane, B-00055
1-Benzyl-4-oxo-2,2,6,6-tetramethyl-1-phenylphosphorinanium bromide, *in* T-00210
[2-(Benzyloxy)phenyl]phosphonic acid, *in* H-00179
1-Benzyl-2,2,3,4,4-pentamethyl-1-phenylphosphetanium bromide, *in* P-00020
17-Benzyl-2,7,10,15,17-pentamethyl-17*H*-tetrabenzo[*b*,*d*,*f*,*h*]phosphoninium iodide, *in* P-00028
Benzyl 1,1,2,3,3-pentamethyltrimethylenephosphinate, *in* H-00168
N-Benzyl-*P*,*P*-(1,1,2,3,3-pentamethyltrimethylene)phosphinic amide, *in* P-00021
1-Benzyl-3-phenyl-1,3-azaphosphetidine, *in* P-00087
1-Benzyl-1-phenyl-1*H*-benzo[*b*]phosphindolium chloride, *in* P-00110
5-Benzyl-5-phenyl-5*H*-dibenzophospholium chloride, *in* P-00110
5-Benzyl-5-phenyl-(5-P^V)-dibenzo[*b*,*e*]-phosphorin, *see* B-00042
Benzyl phenyl phenylphosphonate, *in* M-00471
9-Benzyl-9-phenyl-9-phosphaanthracene, *see* B-00042
1-Benzyl-1-phenylphosphepanium bromide, *in* P-00205
Benzylphenylphosphinic acid, B-00056
1-Benzyl-1-phenylphosphocanium bromide, *in* P-00215
1-Benzyl-1-phenylphospholanium bromide, *in* P-00216
Benzyl *P*-phenylphosphonamidate, *in* P-00218
8-Benzyl-8-phenyl-8-phosphoniabicyclo[3.2.1]octan-3-one chloride, *in* P-00199
1-Benzyl-1-phenylphosphorinanium bromide, *in* P-00265
6-Benzyl-6-phospha-2-oxaadamantane, *see* B-00055
Benzyl phosphinate, B-00057
Benzylphosphine, B-00058
Benzylphosphinic acid, B-00059
Benzyl phosphite, *see* M-00444
1-Benzyl-3-phospholene, *in* D-00572
Benzylphosphonic acid, B-00060
Benzylphosphonic bis(diethylamide), *in* B-00061
Benzylphosphonic bis(dimethylamide), *in* B-00061
P-Benzylphosphonic diamide, B-00061
Benzylphosphonic dianilide, *in* B-00061
Benzylphosphonic dichloride, B-00062
Benzylphosphonochloridic acid, B-00063
Benzylphosphonochloridothioic acid, B-00064
Benzylphosphonodithioic acid, B-00065
Benzylphosphonofluoridic acid, B-00066
Benzylphosphonothioic acid, B-00067
Benzylphosphonous acid, B-00068
Benzylphosphoramidic acid, B-00069
Benzyl phosphoric acid, *see* M-00443
Benzyl phosphorodichloridate, *in* M-00443
Benzyl phosphorodichloridite, B-00070
S-Benzyl phosphorothioate, B-00071
S-Benzyl *O*,*O*-propylene dithiophosphate, *in* M-00016
S-Benzyl *O*,*O*-propylene phosphorodithioate, *in* M-00016
Benzyl propyl phenylphosphonate, *in* M-00511
Benzyl styrylphosphonate, *in* P-00155
Benzyl tetrabutylphosphorodiamidodithioate, *in* T-00023
1-Benzyl-1,2,3,6-tetrahydro-1-phenylphosphorinium bromide, *in* T-00134
17-Benzyl-2,7,10,15-tetramethyl-17-phenyl-17*H*-tetrabenzo[*b*,*d*,*f*,*h*]-phosphoninium iodide, *in* T-00211
N″-Benzyl-*N*,*N*,*N*′,*N*′-tetramethylphosphoric triamide, *in* T-00216
Benzyl tetramethylphosphorodiamidite, *in* T-00227
[(Benzylthio)methyl]phosphonic acid, B-00072
S-Benzyl thiophosphate, *see* B-00071
Benzyl triethylphosphonamidate, *in* T-00446
3-Benzyl-2,2,5-trimethyl-3-phenyl-1,3-oxaphospholanium bromide, *in* T-00522
Benzyltriphenylphosphonium(1+), B-00073
Benzyl 2-(triphenylphosphoranylidene)-propanoate, *in* T-00660
▷Betasan, *see* B-00004
▷Bialaphos, *see* A-00141
2,2′-Bi(1,3,2-benzodithiaphosphole), B-00074
Bicyclic phosphate, *in* T-00588
Bicyclic phosphite, *see* T-00588
Bicyclo[2.2.1]hept-5-ene-2,3-diylbis[diphenylphosphine], *see* B-00220
Bicyclo[2.2.1]hept-5-ene-2,3-diylbis(methylene)bis[diphenylphosphine], B-00075
Bicyclo[4.2.0]octa-1,3,5-trien-7,8-diylidenebis[triphenylphosphorane], *see* B-00460
Bicyclo[4.2.0]octa-1,3,5-trien-7,8-ylenebis[triphenylphosphonium](2+), *see* D-00447
▷Bideron, *see* P-00498
▷Bidrin, *see* D-00204
Binap, *see* B-00221
[1,1′-Binaphthalene]-2,2′-diylbis[diphenylphosphine], *see* B-00221
N,*N*′-[1,1′-Binaphthalene]-2,2′-diylbis[*P*,*P*′-diphenylphosphinous amide], *see* B-00217
[[1,1′-Binaphthalene]-2,2′-diylbis[methylene]]-bis[diphenylphosphine], *see* B-00248
[[1,1′-Binaphthalene]-2,2′-diylbis[methylene]]-bis[triphenylphosphonium](2+), *see* B-00444
1,1′-Binaphthyl 2,2′-cyclic phosphoric acid, *see* H-00131
O,*O*-1,1′-Binaphthylene hydrogen dithiophosphate, *see* M-00012
O,*O*-1,1′-Binaphthylene phosphorodithioate, *see* M-00012
2,2′-(1,1′-Binaphthyl) methyl phosphate, *in* H-00131
▷Biotamin, *see* B-00002
[1,1′-Biphenyl]-2,2′-diylbis(diethylmethylphosphonium)-diiodide, *in* B-00179
[1,1′-Biphenyl]-2,2′-diylbis(diethylphosphine), *see* B-00179
[1,1′-Biphenyl]-3,3′-diylbis(diphenylphosphine), *see* B-00222
[[1,1′-Biphenyl]-4,4′-diylbis(methylene)]-bis(phosphonic acid), *see* B-00391
[[1,1′-Biphenyl]-2,2′-diylbis[methylene]]-bis[triphenylphosphonium](2+), *see* B-00445
[[1,1′-Biphenyl]-3,3′-diylbis[methylene]]-bis[triphenylphosphonium](2+), *see* B-00446
[[1,1′-Biphenyl]-4,4′-diylbis[methylene]]-bis[triphenylphosphonium](2+), *see* B-00447
[[1,1′-Biphenyl]-2,2′-diylbis[methylene]]-bis[triphenylphosphorane], *in* B-00445
[[1,1′-Biphenyl]-3,3′-diylbis[methylene]]-bis[triphenylphosphorane], *in* B-00446
[[1,1′-Biphenyl]-4,4′-diylbis[methylene]]-bis[triphenylphosphorane], *in* B-00447
2,2′-Biphenyldiyl methyl phosphate, *in* M-00032
2,2′-Biphenyldiyl methyl phosphite, *see* M-00032
O,*O*-(2,2′-Biphenylene) cyclic phosphorothioate, *see* M-00007
(1,8-Biphenylenediyldimethylidyne)-bis[triphenylphosphorane], B-00076
O,*O*-(2,2′-Biphenylene) hydrogen thiophosphate, *see* M-00007
2,2′-Biphenylene phenylphosphonite, *see* P-00107
2,2′-Biphenylene phosphite, *in* D-00038
2,2′-Biphenylene phosphonate, *in* D-00038
4,5-(2′,2″-Biphenyleno)-2,2-dihydro-1,3,2-dioxaphosphole, *see* D-00541
4-Biphenylphosphonic acid, B-00077
O,*O*-2,2′-Biphenylyl cyclic dithiophosphate, *see* M-00008
[1,1′-Biphenyl]-2-yl dichlorophosphate, *in* M-00445

(1,1′-Biphenyl)-3-yl dichlorophosphate, *see* B-00082
(1,1′-Biphenyl)-4-yl dichlorophosphate, *in* M-00446
O-2-Biphenylyl dichlorothiophosphate, *see* B-00083
[1,1′-Biphenyl]-2-yl dihydrogen phosphate, *see* M-00445
(1,1′-Biphenyl)-4-yl dihydrogen phosphate, *see* M-00446
[1,1′-Biphenyl]-4-yldiphenylphosphine, B-00078
[1,1′-Biphenyl]-2-yl *N,N′*-diphenylphosphorodiamidate, *in* M-00445
(2,2′-Biphenylylenedimethylene)-bis[triphenylphosphonium](2+), *see* B-00445
(3,3′-Biphenylylenedimethylene)-bis[triphenylphosphonium](2+), *see* B-00446
(4,4′-Biphenylylenedimethylene)-bis[triphenylphosphonium](2+), *see* B-00447
O,O-2,2′-Biphenylyl hydrogen phosphorodithioate, *see* M-00008
P-[(1,1′-Biphenyl)-2-ylmethyl]-*N,N′*-diphenylphosphonic diamide, *in* B-00079
[1,1′-Biphenyl]-4-ylmethyl-1-naphthalenylphenylphosphonium bromide, *in* B-00081
[(1,1′-Biphenyl)-2-ylmethyl]phosphonic acid, B-00079
[(1,1′-Biphenyl)-4-ylmethyl]phosphonic acid, B-00080
[1,1′-Biphenyl]-4-yl-1-naphthalenylphenylphosphine, B-00081
[1,1′-Biphenyl]-4-ylphosphonic acid, *see* B-00077
[1,1′-Biphenyl]-2-yl phosphoric acid, *see* M-00445
O,O-(2,2′-Biphenylyl) phosphorochloridate, *in* C-00032
O,O′-(2,2′-Biphenylyl) phosphorochloridite, *see* C-00032
O,O-(2,2′-Biphenylyl) phosphorochloridoselenoate, *in* C-00032
O,O-(2,2′-Biphenylyl) phosphorochloridothioate, *in* C-00032
[1,1′-Biphenyl]-2-yl phosphorodichloridate, *in* M-00445
(1,1′-Biphenyl)-3-yl phosphorodichloridate, B-00082
(1,1′-Biphenyl)-4-yl phosphorodichloridate, *in* M-00446
O-[(1,1′-Biphenyl)-2-yl] phosphorodichloridothioate, B-00083
1,1′-Biphosphinane, *see* B-00085
1,1′-Biphospholane, B-00084
1,1′-Biphosphorinane, B-00085
Bis(acetoxymethyl)phenylphosphine, *in* B-00281
Bis(2-aminoethyl)phenylphosphine, B-00086
Bis(aminomethyl)phosphinic acid, B-00087
▷Bis(3-aminophenyl)phosphinic acid, B-00088
Bis(4-aminophenyl)phosphinic acid, B-00089
Bis(3-aminopropyl)phenylphosphine, B-00090
Bisazir, *in* B-00094
P,P-Bis(1-aziridinyl)-*N,N*-diethylphosphinic amide, *in* B-00092
P,P-Bis(1-aziridinyl)-*N,N*-diethylphosphinous amide, *in* B-00096
▷*P,P*-Bis(1-aziridinyl)-*N*-[2-(dimethylamino)-7-methyl-7*H*-purin-6-yl]phosphonic amide, *see* F-00053
▷*P,P*-Bis(1-aziridinyl)-*N,N*-dimethylphosphinic amide, *in* B-00092
P,P-Bis(1-aziridinyl)-*N,N*-dimethylphosphinous amide, *in* B-00096
▷*P,P*-Bis(1-aziridinyl)-*N*-ethylphosphinic amide, *in* B-00092
P,P-Bis(1-aziridinyl)-*N*-ethylphosphinothioic amide, *in* B-00094
▷*P,P*-Bis(1-aziridinyl)-*N*-methylphosphinic amide, *in* B-00092
P,P-Bis(1-aziridinyl)-*N*-methylphosphinothioic amide, *in* B-00094
P,P-Bis(1-aziridinyl)-*N*-2-naphthylphosphinic amide, *in* B-00092
P,P-Bis(1-aziridinyl)-*N*-phenylphosphinic amide, *in* B-00092
P,P-Bis(1-aziridinyl)-*N*-phenylphosphinothioic amide, *in* B-00094
Bis(1-aziridinyl)phosphinic acid, B-00091
▷*P,P*-Bis(1-aziridinyl)phosphinic amide, B-00092
Bis(1-aziridinyl)phosphinothioic acid, B-00093
▷*P,P*-Bis(1-aziridinyl)phosphinothioic amide, B-00094
Bis(1-aziridinyl)phosphinous acid, B-00095
P,P-Bis(1-aziridinyl)phosphinous amide, B-00096
Bis(1-aziridinyl)phosphinous diethylamide, *in* B-00096
Bis(1-aziridinyl)phosphinous dimethylamide, *in* B-00096
▷*P,P*-Bis(1-aziridinyl)-*N*-2-pyrimidinylphosphinic amide, *see* P-00353
▷*P,P*-Bis(1-aziridinyl)-*N*-tricyclo[3.3.1.1^{3,7}]dec-1-ylphosphinic amide, *see* A-00032
P,P-Bis(1*H*-benzimidazol-1-yl)-*N*-phenylphosphinic amide, *in* B-00097
Bis(1*H*-benzimidazol-1-yl)phosphinic acid, B-00097
Bis(1*H*-benzimidazol-1-yl)phosphinic anilide, *in* B-00097
O,O-Bis(1-benzotriazolyl) 2-chlorophenyl phosphate, B-00098
O,O-Bis(1-benzotriazolyl) morpholinyl phosphate, B-00099
O,O-Bis(1-benzotriazolyl) 2,2,2-tribromoethyl phosphate, B-00100
Bis(benzoylthiomethyl)phosphinic acid, *in* B-00292
Bis(*N*-benzylaminomethyl)phosphinic acid, *in* B-00087
Bis[(benzyloxy)phosphinyl]acetic acid, *see* B-00389
Bis(benzylsulfonylmethyl)phosphinic acid, *in* B-00101
Bis(benzylthiomethyl)phosphinic acid, B-00101
Bis[(1,1′-biphenyl)-2-yl] chlorophosphate, *see* B-00102
Bis-2,2′-biphenylylenephosphonium(1+), *see* S-00022
Bis[(1,1′-biphenyl)-2-yl] phosphorochloridate, B-00102
▷1,4-Bis[bis(1-aziridinyl)phosphinyl]-piperazine, *see* D-01189
Bis[bis(dimethylamino)phosphinothioyl]-methylamine, *see* B-00191
Bis[[bis(2-diphenylphosphino)ethyl]-phosphino]ethane, *see* B-00240
1,2-Bis[(bis-2-methylphenyl)phosphino]-ethane, B-00103
1,2-Bis[(bis-4-methylphenyl)phosphino]-ethane, B-00104
[Bis[bis(trimethylsilyl)methylene]]-chlorophosphorane, B-00105
Bis[bis(trimethylsilyl)methylene]-9*H*-fluoren-9-ylphosphorane, B-00106
Bis[bis(trimethylsilyl)methylene]-phenylphosphorane, B-00107
Bis-*p*-bromobenzyl hydrogen phosphate, *see* B-00109
Bis-*p*-bromobenzyl phosphate, *see* B-00109
Bis-*p*-bromobenzyl phosphoric acid, *see* B-00109
Bis(2-bromoethyl) (2-bromoethyl)phosphonate, *in* B-00472
Bis(bromomethyl)phosphinic acid, B-00108
Bis(4-bromophenyl) hydrogen phosphate, *see* B-00111
Bis[(4-bromophenyl)methyl] phenylphosphoramidate, *in* B-00109
Bis[(4-bromophenyl)methyl] phosphate, B-00109
Bis[(4-bromophenyl)methyl] phosphite, *see* B-00110
Bis[(4-bromophenyl)methyl] phosphonate, B-00110
Bis[(4-bromophenyl)methyl] phosphoramidate, *in* B-00109
Bis(4-bromophenyl) phosphate, B-00111
Bis(2-bromophenyl)phosphinic acid, B-00112
Bis(3-bromophenyl)phosphinic acid, B-00113
Bis(4-bromophenyl)phosphinic acid, B-00114
Bis(4-bromophenyl) phosphoric acid, *see* B-00111
Bis(*N*-*tert*-butylaminomethyl)phosphinic acid, *in* B-00087
Bis(1,3-butylene) pyrophosphite, *see* O-00096
Bis(2-*tert*-butylphenyl) chlorophosphate, *in* B-00115
Bis(2-*tert*-butylphenyl) hydrogen phosphate, *see* B-00115
Bis(2-*tert*-butylphenyl) phosphate, B-00115
Bis(4-*tert*-butylphenyl) phosphite, *see* B-00116
Bis(4-*tert*-butylphenyl) phosphonate, B-00116
Bis(2-*tert*-butylphenyl) phosphoric acid, *see* B-00115
Bis(2-*tert*-butylphenyl) phosphorochloridate, *in* B-00115
Bis(2-carboxyphenyl)phosphinic acid, B-00117
Bis(4-carboxyphenyl)phosphinic acid, B-00118
Bis(4-chlorocarbonylphenyl)phosphinic chloride, *in* B-00118
▷2-Bis(2-chloroethyl)amino-3-chloroethyltetrahydro-2*H*-1,3,2-oxazaphosphorine 2-oxide, *see* T-00485
2-Bis(2-chloroethyl)amino-5,6-dihydro-2*H*-1,3,2-oxazaphosphorine 2-oxide, *see* I-00004
2-[Bis(2-chloroethyl)amino]-1,3,2-oxazaphosphorinan-4-ol 2-oxide, *see* H-00115
▷2-[Bis(2-chloroethyl)amino]-1,3,2-oxazaphosphorinan-4-one 2-oxide, *see* O-00058
2-[Bis(2-chloroethyl)amino]tetrahydro-4-hydroperoxy-2*H*-1,3,2-oxazaphosphorine 2-oxide, *see* H-00106
2-[Bis(2-chloroethyl)amino]tetrahydro-2*H*-1,3,2-oxazaphosphorin-4-ol 2-oxide, *see* H-00115
▷2-[Bis(2-chloroethyl)amino]tetrahydro-2*H*-1,3,2-oxazaphosphorin-4-one 2-oxide, *see* O-00058
Bis(2-chloroethyl) (2-chloroethyl)-phosphonate, *in* C-00074
O,O-Bis(2-chloroethyl) chlorothiophosphate, *see* B-00124
N,N-Bis(2-chloroethyl)-5,6-dihydro-2*H*-1,3,2-oxazaphosphorin-2-amine 2-oxide, *see* I-00004
N,N-Bis(2-chloroethyl)-*N′,N″*-diphenylphosphoric triamide, *in* B-00121
Bis(2-chloroethyl) hydrogen phosphate, *see* B-00119
Bis(2-chloroethyl) phosphate, B-00119
▷Bis(2-chloroethyl)phosphite, *see* B-00120
▷Bis(2-chloroethyl) phosphonate, B-00120
Bis(2-chloroethyl) phosphoramidate, *in* B-00119
Bis(2-chloroethyl)phosphoramidic acid, B-00121
▷Bis(2-chloroethyl)phosphoramidic dichloride, B-00122
▷Bis(2-chloroethyl)phosphoramidothioic dichloride, B-00123
Bis(2-chloroethyl) phosphoric acid, *see* B-00119
▷Bis(2-chloroethyl) phosphorochloridate, *in* B-00119
O,O-Bis(2-chloroethyl) phosphorochloridothioate, B-00124
N,N′-Bis(2-chloroethyl)phosphorodiamidic acid, B-00125
N,N-Bis(2-chloroethyl)phosphorodiamidic acid, B-00126
Bis(2-chloroethyl) phosphorofluoridate, *in* B-00119
▷Bis(2-chloroethyl) phosphoryl chloride, *in* B-00119
Bis(2-chloroethyl) phosphoryl fluoride, *in* B-00119
▷*N,N*-Bis(2-chloroethyl)tetrahydro-2*H*-1,3,2-oxazaphosphorin-2-amine 2-oxide, *see* C-00283
▷*N*,3-Bis(2-chloroethyl)tetrahydro-2*H*-1,3,2-oxazaphosphorin-2-amine 2-oxide, *see* I-00001

O,O-Bis(2-chloroethyl) thiophosphoryl chloride, *see* B-00124
P,P-Bis(chloromethyl)-*N,N*-dimethylphosphinic amide, *in* B-00128
P,P-Bis(chloromethyl)-*N*-methylphosphinic amide, *in* B-00128
P,P-Bis(chloromethyl)-*N*-phenylphosphinic amide, *in* B-00128
Bis(chloromethyl)phosphinic acid, B-00127
P,P-Bis(chloromethyl)phosphinic amide, B-00128
Bis(chloromethyl)phosphinothioic acid, B-00129
Bis(2-chlorophenyl) chlorophosphate, *in* B-00131
Bis(3-chlorophenyl) chlorophosphate, *in* B-00132
Bis(4-chlorophenyl) chlorophosphate, *in* B-00133
P,P-Bis(3-chlorophenyl)-*N,N*-diethylphosphinous amide, *in* B-00141
P,P-Bis(4-chlorophenyl)-*N,N*-diethylphosphinous amide, *in* B-00142
O,O-Bis(2-chlorophenyl) dithiophosphoric acid, *see* B-00150
O,O-Bis(4-chlorophenyl) dithiophosphoric acid, *see* B-00151
O,O-Bis(2-chlorophenyl) hydrogen dithiophosphate, *see* B-00150
O,O-Bis(4-chlorophenyl) hydrogen dithiophosphate, *see* B-00151
Bis(2-chlorophenyl) hydrogen phosphate, *see* B-00131
Bis(3-chlorophenyl) hydrogen phosphate, *see* B-00132
Bis(4-chlorophenyl) hydrogen phosphate, *see* B-00133
▷*O,O*-Bis(4-chlorophenyl) (1-iminoethyl)-phosphoramidothioate, *in* I-00005
Bis[(4-chlorophenyl)methyl] phosphite, *see* B-00130
Bis[(4-chlorophenyl)methyl] phosphonate, B-00130
Bis(2-chlorophenyl) phosphate, B-00131
Bis(3-chlorophenyl) phosphate, B-00132
Bis(4-chlorophenyl) phosphate, B-00133
Bis(2-chlorophenyl)phosphine, B-00134
Bis(3-chlorophenyl)phosphine, B-00135
Bis(4-chlorophenyl)phosphine, B-00136
Bis(2-chlorophenyl)phosphinic acid, B-00137
Bis(3-chlorophenyl)phosphinic acid, B-00138
▷Bis(4-chlorophenyl)phosphinic acid, B-00139
Bis(2-chlorophenyl)phosphinous acid, B-00140
Bis(3-chlorophenyl)phosphinous acid, B-00141
Bis(4-chlorophenyl)phosphinous acid, B-00142
Bis(2-chlorophenyl) phosphonate, B-00143
Bis(4-chlorophenyl) phosphonate, B-00144
Bis(2-chlorophenyl) phosphoric acid, *see* B-00131
Bis(3-chlorophenyl) phosphoric acid, *see* B-00132
Bis(4-chlorophenyl) phosphoric acid, *see* B-00133
Bis(2-chlorophenyl) phosphorochloridate, *in* B-00131
Bis(3-chlorophenyl) phosphorochloridate, *in* B-00132
Bis(4-chlorophenyl) phosphorochloridate, *in* B-00133
Bis(2-chlorophenyl) phosphorochloridite, B-00145
Bis(3-chlorophenyl) phosphorochloridite, B-00146
Bis(4-chlorophenyl) phosphorochloridite, B-00147
O,O-Bis(3-chlorophenyl) phosphorochloridothioate, B-00148
O,O-Bis(4-chlorophenyl) phosphorochloridothioate, B-00149
O,O-Bis(2-chlorophenyl) phosphorodithioate, B-00150
O,O-Bis(4-chlorophenyl) phosphorodithioate, B-00151
Bis(2-chlorophenyl) phosphoryl chloride, *in* B-00131
Bis(3-chlorophenyl) phosphoryl chloride, *in* B-00132
Bis(4-chlorophenyl) phosphoryl chloride, *in* B-00133
Bis(2-cyanoethyl)methylphenylphosphonium iodide, *in* B-00152
Bis(2-cyanoethyl)phenylphosphine, B-00152
Bis(2-cyanoethyl)phosphine, B-00153
Bis(cyclenphosphorane), B-00154
2,4-Bis(cyclohexyl)-1,3,2,4-dithiadiphosphetane 2,4-disulfide, *see* D-00207
Bis(dibenzyloxyphosphinothioyl) disulfide, B-00155
Bis(2,5-dibromophenyl)phosphinic acid, B-00156
1,2-Bis(dibutoxyphosphinyl)ethane, *in* E-00006
Bis(dibutylamino)phosphinic chloride, *see* T-00022
Bis(dibutylamino)phosphoryl chloride, *see* T-00022
1,4-Bis(dibutylphosphino)butane, B-00157
1,2-Bis(dibutylphosphino)ethane, B-00158
1,2-Bis(dibutylphosphino)ethylene, B-00159
Bis(dibutylphosphino)methane, B-00160
1,5-Bis(dibutylphosphino)pentane, B-00161
1,4-Bis(dibutylphosphinyl)butane, *in* B-00157
1,2-Bis(dibutylphosphinyl)ethane, *in* B-00158
1,2-Bis(dibutylphosphinyl)ethylene, *in* B-00159
Bis(dibutylphosphinyl)methane, *in* B-00160
1,5-Bis(dibutylphosphinyl)pentane, *in* B-00161
Bis(2,4-dichlorophenyl) ethyl phosphate, B-00162
Bis(2,3-dichlorophenyl)phosphinic acid, B-00163
Bis(2,5-dichlorophenyl)phosphinic acid, B-00164
Bis(3,5-dichlorophenyl)phosphinic acid, B-00165
1,2-Bis(dichlorophosphinyl)ethane, *in* E-00006
Bis(dichlorophosphinyl)methane, *see* M-00145
1,4-Bis(dichlorophosphinyloxy)benzene, *in* P-00148
1,4-Bis(dicyclohexylphosphino)butane, B-00166
1,2-Bis(dicyclohexylphosphino)ethane, B-00167
4,5-Bis[(dicyclohexylphosphino)methyl]-2,2-dimethyl-1,3-dioxolane, B-00168
1,3-Bis(dicyclohexylphosphino)propane, B-00169
1,4-Bis(dicyclohexylphosphinothioyl)butane, *in* B-00166
1,3-Bis(dicyclohexylphosphinothioyl)propane, *in* B-00169
1,2-Bis(dicyclophosphinothioyl)ethane, *in* B-00167
1,2-Bis(dicyclophosphinyl)ethane, *in* B-00167
Bis(diethoxyphosphinothioyl) disulfide, B-00170
N,N′-Bis(diethoxyphosphinothioyl)hydrazine, *see* T-00048
1,2-[Bis(diethoxyphosphinothioyl)thio]-ethane, *see* T-00047
▷Bis[*S*-(diethoxyphosphinothioyl)thio]methane, *see* E-00022
1,3-[Bis(diethoxyphosphinothioyl)thio]-propane, *see* T-00068
Bis(diethoxyphosphinothioyl) trisulfide, B-00171
1,3-Bis(diethoxyphosphinyl)benzene, *in* B-00007
1,4-Bis(diethoxyphosphinyl)benzene, *in* B-00008
1,10-Bis(diethoxyphosphinyl)decane, *in* D-00008
▷Bis(diethoxyphosphinyl) disulfide, B-00172
1,2-Bis(diethoxyphosphinyl)ethane, *in* E-00006
2,3-Bis(diethoxyphosphinylmethyl)-naphthalene, *in* N-00005
2,7-Bis(diethoxyphosphinylmethyl)-naphthalene, *in* N-00007
1,5-Bis(diethoxyphosphinyl)naphthalene, *in* N-00002
1,6-Bis(diethoxyphosphinyl)naphthalene, *in* N-00003
1,9-Bis(diethoxyphosphinyl)nonane, *in* N-00069
1,8-Bis(diethoxyphosphinyl)octane, *in* O-00035
1,2-Bis[(diethoxyphosphinyl)oxy]benzene, *see* T-00042
1,1-Bis(diethoxyphosphinyl)pentane, *in* P-00031
1,5-Bis(diethoxyphosphinyl)pentane, *in* P-00032
1,3-Bis(diethoxyphosphinyl)propane, *in* P-00444
2,3-Bis(diethoxyphosphinyl)propanoic acid, B-00173
P,P-Bis(diethylamino)-*N*-ethyl-*P*-fluorophosphazene, *in* P-00012
P,P-Bis(diethylamino)-*N*-ethyl-*P*-fluorophosphine imide, *in* P-00012
P,P-Bis(diethylamino)-*P*-fluoro-*N*-phenylphosphazene, *in* T-00054
P,P-Bis(diethylamino)-*P*-fluoro-*N*-phenylphosphine imide, *in* T-00054
P,P-Bis(diethylamino)-*P*-iodo-*N*-phenylphosphazene, *in* T-00054
P,P-Bis(diethylamino)-*P*-iodo-*N*-phenylphosphine imide, *in* T-00054
P,P-Bis(diethylamino)-*P*-phenoxy-*N*-phenylphosphazene, *in* T-00054
P,P-Bis(diethylamino)-*P*-phenoxy-*N*-phenylphosphine imide, *in* T-00054
Bis(diethylamino)phosphinic chloride, *see* T-00060
Bis(diethylamino)phosphinothioic azide, *see* T-00064
Bis(diethylamino)phosphinothioyl chloride, *see* T-00065
N″-[Bis(diethylamino)phosphinothioyl]-*N,N,N′,N′*-tetraethylphosphorodiamidoimidic acid, B-00174
[Bis(diethylamino)phosphinyl]acetic acid, B-00175
[Bis(diethylamino)phosphinyl]phosphorimidic acid, B-00176
N″-[Bis(diethylamino)phosphinyl]-*N,N,N′,N′*-tetraethylphosphorodiamidoimidic acid, B-00177
N-[Bis(diethylamino)phosphinyl]-*P,P,P*-trimethoxyphosphazene, *in* B-00176
N-[Bis(diethylamino)phosphinyl]-*P,P,P*-trimethoxyphosphine imide, *in* B-00176
Bis(diethylamino) phosphoryl chloride, *see* T-00060
Bis(diethylamino) thiophosphoryl azide, *see* T-00064
P,P-Bis(diethylamino)-*P*-(trichloromethyl)-phosphazene, *see* T-00075
P,P-Bis(diethylamino)-*P*-(trichloromethyl)-phosphine imide, *see* T-00075
Bis(diethylphosphino)acetylene, B-00178
2,2′-Bis(diethylphosphino)biphenyl, B-00179
1,2-Bis(diethylphosphino)ethane, B-00180
1,2-Bis(diethylphosphino)ethylene, B-00181
1,2-Bis(diethylphosphinothioyl)ethane, *in* B-00180
2,2-Bis(diethylphosphinyl)biphenyl, *in* B-00179
1,2-Bis(diethylphosphinyl)ethane, *in* B-00180
2,6-Bis(diethylphosphinylmethyl)naphthalene, *in* N-00006
Bis(difluorophosphinyl)methane, *see* M-00146
1,4-Bis(difluorophosphinyloxy)benzene, *in* P-00148
Bis(1,2-dihexadecanoylglycero)-3-phosphate, *see* T-00247
1,2-Bis(dihydroxyphosphinyl)benzene, *see* B-00006
1,3-Bis(dihydroxyphosphinyl)benzene, *see* B-00007
1,4-Bis(dihydroxyphosphinyl)benzene, *see* B-00008
Bis[(dihydroxyphosphinyl)methyl]phosphinic acid, B-00182
1,2-Bis(diisopropoxyphosphinylamino)ethane, *in* E-00009
Bis(diisopropoxyphosphinyl) disulfide, B-00183
1,1-Bis(diisopropoxyphosphinyl)pentane, *in* P-00031
Bis(diisopropyoxyphosphinothioyl) disulfide, B-00184

▷Bis(dimethoxyphosphinothioyl) disulfide, B-00185
Bis[(*O,O*-dimethoxyphosphinothioyl)thio]-methane, *see* T-00205
1,2-Bis(dimethoxyphosphinyl)benzene, *in* B-00006
1,3-Bis(dimethoxyphosphinyl)benzene, *in* B-00007
1,4-Bis(dimethoxyphosphinyl)benzene, *in* B-00008
Bis(dimethoxyphosphinyl) disulfide, B-00186
1,2-Bis(dimethoxyphosphinyl)ethane, *in* E-00006
1,8-Bis(dimethoxyphosphinyl)naphthalene, *in* N-00004
▷Bis(dimethylamido)phosphoryl fluoride, *see* D-00669
▷Bis(dimethylamino)chlorophosphine, *see* T-00228
P,P'-Bis(dimethylamino)-*N,N'*-di-*tert*-butylcyclodiphosphazane, *see* D-00074
P,P-Bis(dimethylamino)-*P*-methylphosphazene, *see* P-00023
P,P-Bis(dimethylamino)-*P*-methylphosphine imide, *see* P-00023
2,4-Bis(dimethylamino)-3-methyl-1,3,2,4-thiazadiphosphetidine, B-00187
O,O-Bis(4-dimethylaminophenyl) *O*-isopropyl phosphorothioate, *in* B-00190
O,O-Bis(3-dimethylaminophenyl) phenylphosphoramidothioate, *in* B-00189
Bis(4-dimethylaminophenyl)phosphinic acid, B-00188
O,O-Bis(4-dimethylaminophenyl) phosphorobromidothioate, *in* B-00190
O,O-Bis(4-dimethylaminophenyl) phosphorochloridothioate, *in* B-00190
O,O-Bis(4-dimethylaminophenyl) phosphorofluoridothioate, *in* B-00190
O,O-Bis(3-dimethylaminophenyl) phosphorothioate, B-00189
O,O-Bis(4-dimethylaminophenyl) phosphorothioate, B-00190
O,O-Bis(4-dimethylaminophenyl) thiophosphoric acid, *see* B-00190
Bis(dimethylamino)phosphinic chloride, *see* T-00220
▷Bis(dimethylamino)phosphinic fluoride, *see* D-00669
Bis[(dimethylamino)phosphino]acetylene, *see* E-00200
Bis(dimethylamino)phosphinothioic azide, *see* T-00224
Bis(dimethylamino)phosphinothioic chloride, *see* T-00225
Bis(dimethylamino)phosphinothioic isothiocyanate, *see* T-00217
▷Bis(dimethylamino)phosphinothioyl fluoride, *see* T-00226
N''-Bis(dimethylamino)phosphinothioyl-*N,N,N',N',N''*-pentamethylphosphoric triamide, *see* N-00032
Bis(dimethylaminophosphinothioyl)-pentamethylthiophosphoric triamide, B-00191
[Bis(dimethylamino)phosphinothioyl]-phosphorimidic acid, B-00192
[Bis(dimethylamino)phosphinyl]acetic acid, B-00193
O-[Bis(dimethylamino)phosphinyl] bis(dimethylamino)phosphinothioate, *see* O-00031
Bis(dimethylamino)phosphinyl cyanide, *see* T-00221
[Bis(dimethylamino)phosphinyl]-phosphorimidic acid, B-00194
O-[Bis(dimethylamino)phosphinyl] tetramethylphosphorodiamidothioate, *see* O-00031
N-[Bis(dimethylamino)phosphinyl]-*P,P,P*-triphenoxyphosphazene, *in* B-00194
N-[Bis(dimethylamino)phosphinyl]-*P,P,P*-triphenoxyphosphine imide, *in* B-00194
Bis(dimethylamino)phosphoryl azide, *see* T-00219
Bis(dimethylamino) phosphoryl chloride, *see* T-00220
Bis(dimethylamino)phosphoryl cyanide, *see* T-00221
Bis[3-(dimethylamino)propyl]phenylphosphine, B-00195
Bis(dimethylamino) thiophosphoryl azide, *see* T-00224
Bis(dimethylamino) thiophosphoryl chloride, *see* T-00225
Bis(dimethylamino)trifluorophosphorane, B-00196
2,2'-Bis(5,5-dimethyl-1,3-dioxa-2-phosphacyclohexane), *see* B-00197
2,2'-Bis(5,5-dimethyl-1,3,2-dioxaphosphorinane), B-00197
2,2'-Bis(5,5-dimethyl-1,3,2-dioxaphosphorinane) 2,2'-dioxide, *in* B-00197
Bis(5,5-dimethyl-1,3,2-dioxaphosphorinan-2-yl) disulfide, *see* D-01251
Bis(5,5-dimethyl-1,3,2-dioxaphosphorinan-2-yl) oxide, *see* O-00089
Bis(5,5-dimethyl-1,3,2-dioxaphosphorinan-2-yl) sulfide, *see* T-00309
1,3-Bis(1,1-dimethylethyl)-2-azaphosphiridinamine, *see* A-00070
N,P-Bis(1,1-dimethylethyl)-carboimidophosphene, *see* D-00077
3,5-Bis-(1,1-dimethylethyl)-2,3-dihydro-1,3,2-oxaphosphole, *see* D-00083
2,6-Bis(1,1-dimethylethyl)-1,4-dihydro-4-phenyl-1,4-azaphosphorine, *see* D-00084
1,3-Bis(1,1-dimethylethyl)-2,4-dimethoxy-1,3,2,4-diazadiphosphetidine, *see* D-00085
1,3-Bis(1,1-dimethylethyl)-2,4-dimethyl-1,3,2,4-diazadiphosphetidine, *see* D-00086
4,8-Bis(1,1-dimethylethyl)-2,10-dimethyl-12*H*-dibenzo[*d,g*][1,3,2]dioxaphosphocin, *see* D-00088
1,2-Bis(1,1-dimethylethyl)-1,2-dimethyldiphosphine, *see* D-00089
O,O-Bis(1,1-dimethylethyl) dimethylphosphoramidothioate, *in* D-00871
1-[Bis(1,1-dimethylethyl)-diphosphinylidene]methanediamine, *see* D-00090
1,2-Bis(1,1-dimethylethyl)diphosphirane, *see* D-00091
O,O-Bis(1,1-dimethylethyl) ethylphosphoramidothioate, *in* E-00159
[[3,5-Bis(1,1-dimethylethyl)-4-hydroxyphenyl]methyl]phosphonic acid, *see* D-00094
[Bis[(1,1-dimethylethyl)imino]][bis(1-methylethyl)amino]phosphorane, *see* D-00095
2,3-Bis(1,1-dimethylethyl)-1-(1-methylethyl)phosphadiarsirane, *see* D-00096
3,4-Bis(1,1-dimethylethyl)-2-(4-methylphenyl)-1,3,4,2-thiadiazaphospholidin-5-one 2-sulfide, *see* D-00097
4,4'-Bis(1,1-dimethylethyl)-1,1',2,2',5,5',6,6'-octahydro-1,1'-diphenyl-2,2'-biphosphinine, *see* D-00100
4,4'-Bis(1,1-dimethylethyl)-1,1',2,2',5,5',6,6'-octahydro-1,1'-diphenyl-2,2'-biphosphorin, *see* D-00100
Bis(1,1-dimethylethyl) (phenylmethyl)-phosphoramidate, *in* B-00069
Bis[2-(1,1-dimethylethyl)phenyl] phosphate, *see* B-00115
Bis(1,1-dimethylethyl)phenylphosphine, *see* D-00103
Bis[4-(1,1-dimethylethyl)phenyl] phosphite, *see* B-00116
Bis[4-(1,1-dimethylethyl)phenyl] phosphonate, *see* B-00116
N,N'-Bis(1,1-dimethylethyl)-*P*-phenylphosphonous diamide, *in* P-00252
Bis(1,1-dimethylethyl) phosphate, *see* D-00105
Bis(1,1-dimethylethyl)phosphine, *see* D-00108
Bis(1,1-dimethylethyl)phosphinic acid, *see* D-00110
Bis(1,1-dimethylethyl)phosphinodithioic acid, *see* D-00113
P,P-Bis(1,1-dimethylethyl)phosphinoselenoic amide, *see* D-00115
Bis(1,1-dimethylethyl)phosphinotellurous acid, *see* D-00116
Bis(1,1-dimethylethyl)phosphinothioic acid, *see* D-00118
Bis(1,1-dimethylethyl)phosphinothioic amide, *see* D-00119
Bis(1,1-dimethylethyl)phosphinothious acid, *see* D-00120
Bis(1,1-dimethylethyl)phosphinous acid, *see* D-00122
P,P-Bis(1,1-dimethylethyl)phosphinous amide, *see* D-00124
Bis(1,1-dimethylethyl)phosphinous chloride, *see* D-00125
Bis(1,1-dimethylethyl) phosphonate, *see* D-00128
Bis(1,1-dimethylethyl)phosphoramidous acid, *see* D-00137
Bis(1,1-dimethylethyl) phosphorobromidate, *see* D-00138
Bis(1,1-dimethylethyl) phosphorochloridate, *see* D-00141
Bis(1,1-dimethylethyl) phosphorochloridite, *see* D-00143
O,O-Bis(1,1-dimethylethyl) phosphorothioate, *see* D-00156
Bis(1,1-dimethylethyl) 1,2-propadienylphosphonate, *in* P-00441
4,6-Bis(1,1-dimethylethyl)spiro[1,3,2-benzodioxaphosphole-2,1'-[2,8,9]-trioxa[1]phosphatricyclo[3.3.1.$1^{3,7}$]-decane], *see* D-00157
1,3-Bis(1,1-dimethylethyl)-*N,N,N',N'*-tetramethyl-1,3,2,4-diazadiphosphetidine-2,4-diamine, *see* D-00074
2,9-Bis(1,1-dimethylethyl)-1,3,8,10-tetraoxa-2,9-diphosphacyclotetradecane, *see* D-00158
2,8-Bis(1,1-dimethylethyl)-1,3,7,9-tetrathia-2,8-diphosphacyclododecane, *see* D-00159
2,4-Bis(1,1-dimethylethyl)-1,2,4,3-thiadiazaphosphetidin-3-amine, *see* A-00071
Bis(5,5-dimethyl-2-oxo-1,3,2-dioxaphosphorinan-2-yl) disulfide, *in* D-01251
Bis(2,5-dimethylphenyl)phosphinic acid, B-00198
1,2-Bis(dimethylphosphino)benzene, *see* P-00123
1,3-Bis(dimethylphosphino)benzene, *see* P-00124
1,4-Bis(dimethylphosphino)benzene, *see* P-00125
1,4-Bis(dimethylphosphino)butane, B-00199
1,2-Bis(dimethylphosphino)ethane, B-00200
1,6-Bis(dimethylphosphino)hexane, B-00201
Bis(dimethylphosphino)methane, B-00202
[Bis(dimethylphosphino)methylene]-trimethylphosphorane, B-00203
Bis[(dimethylphosphino)methyl]-methylphosphine, B-00204
1,5-Bis(dimethylphosphino)pentane, B-00205
1,3-Bis(dimethylphosphino)propane, B-00206
1,4-Bis(dimethylphosphinoselenoyl)benzene, *in* P-00125
1,3-Bis(dimethylphosphinothioyl)benzene, *in* P-00124
1,4-Bis(dimethylphosphinothioyl)benzene, *in* P-00125
1,4-Bis(dimethylphosphinothioyl)butane, *in* B-00199
1,2-Bis(dimethylphosphinothioyl)ethane, *in* B-00200
1,3-Bis(dimethylphosphinothioyl)propane, *in* B-00206
1,3-Bis(dimethylphosphinyl)benzene, *in* P-00124
1,4-Bis(dimethylphosphinyl)benzene, *in* P-00125
1,4-Bis(dimethylphosphinyl)butane, *in* B-00199
1,2-Bis(dimethylphosphinyl)ethane, *in* B-00200
1,6-Bis(dimethylphosphinyl)hexane, *in* B-00201

Bis(dimethylphosphinyl)methane, *in* B-00202
Bis[(dimethylphosphinyl)methyl]-methylphosphine, *in* B-00204
1,5-Bis(dimethylphosphinyl)pentane, *in* B-00205
1,3-Bis(dimethylphosphinyl)propane, *in* B-00206
Bis(1,1-dimethyl) phosphoramidate, *in* D-00105
O,O-Bis(1,1-dimethyl) phosphorodithioate, *see* D-00151
Bis(2,2-dimethylpropoxy)-triphenylphosphorane, B-00207
Bis(2,2-dimethylpropyl) phosphite, *see* B-00208
Bis(2,2-dimethylpropyl) phosphonate, B-00208
Bis(2,2-dimethylpropyl) phosphorochloridite, B-00209
Bis(5,5-dimethyl-2-selenoxo-1,3,2-dioxaphosphorinan-2-yl) disulfide, *in* D-01251
Bis(5,5-dimethyl-2-thioxo-1,3,2-dioxaphosphorinan-2-yl) disulfide, *in* D-01251
Bis(dimethylvinylene) pyrophosphate, *in* O-00088
Bis(2,4-dinitrophenyl) hydrogen phosphate, *see* B-00210
Bis(2,4-dinitrophenyl) phosphate, B-00210
Bis(2,4-dinitrophenyl) phosphoric acid, *see* B-00210
Bis(1,3-dioxa-2-phosphacyclohexan-2-yl) ether, *see* O-00092
Bis(1,3,2-dioxaphosphorinan-2-yl) oxide, *see* O-00092
1,2-Bis[(diphenoxyphosphinothioyl)amino]-ethane, *in* E-00010
Bis[(diphenoxyphosphino)thioyl] disulfide, B-00211
1,2-Bis(diphenoxyphosphinylamino)ethane, *in* E-00009
1,2-Bis(diphenoxyphosphinyl)ethane, *in* E-00006
▷Bis[2-(diphenylarsino)ethyl]phenylphosphine, B-00212
[Bis(diphenylarsino)methylene]-triphenylphosphorane, B-00213
Bis[2-(diphenylarsino)phenyl]-phenylphosphine, B-00214
2,2′-Bis(diphenylphosphinamido)-6,6′-dimethylbiphenyl, B-00215
Bisdiphenylphosphinic peroxide, *see* D-00977
1,3-Bis(diphenylphosphiniomethyl)benzene(2+), *see* P-00133
Bis(diphenylphosphino)acetylene, B-00216
Bis(diphenylphosphino)amine, *see* D-01063
2,2′-Bis[(diphenylphosphino)amino]-1,1′-binaphthyl, B-00217
2,3-Bis[(diphenylphosphino)amino]butane, B-00218
1,2-Bis[(*N*-diphenylphosphino)amino]-cyclohexane, B-00219
1,2-Bis(diphenylphosphino)benzene, *see* P-00126
1,3-Bis(diphenylphosphino)benzene, *see* P-00127
1,4-Bis(diphenylphosphino)benzene, *see* P-00128
2,3-Bis(diphenylphosphino)bicyclo[2.2.1]-heptane, *in* B-00220
5,6-Bis(diphenylphosphino)bicyclo[2.2.1]-hept-2-ene, B-00220
2,2′-Bis(diphenylphosphino)-1,1′-binaphthyl, B-00221
3,3′-Bis(diphenylphosphino)biphenyl, B-00222
1,4-Bis(diphenylphosphino)-1,3-butadiyne, B-00223
1,3-Bis(diphenylphosphino)butane, B-00224
1,4-Bis(diphenylphosphino)butane, B-00225
2,3-Bis(diphenylphosphino)butane, B-00226
1,4-Bis(diphenylphosphino)-2-butyne, B-00227
3,4-Bis(diphenylphosphino)-3-cyclobutene-1,2-dione, B-00228
1,1-Bis(diphenylphosphino)cyclopropane, B-00229
1,2-Bis(diphenylphosphino)cyclopropane, B-00230
2,2′-Bis(diphenylphosphino)diethylamine, B-00231
Bis(diphenylphosphino)diphenylgermane, B-00232
N,N-Bis(diphenylphosphino)-*P,P*-diphenylphosphinous amide, B-00233
▷1,2-Bis(diphenylphosphino)ethane, B-00234
1,2-Bis(diphenylphosphino)-1,2-ethanedione, B-00235
Bis[(2-Diphenylphosphino)ethyl]amine, *see* B-00231
N,N-Bis(diphenylphosphino)ethylamine, *see* D-01067
1,1-Bis(diphenylphosphino)ethylene, B-00236
1,2-Bis(diphenylphosphino)ethylene, B-00237
Bis[(2-diphenylphosphino)ethyl]ether, *see* O-00093
N,N-Bis[(2-diphenylphosphino)ethyl]-ethylamine, B-00238
Bis[2-(diphenylphosphino)ethyl]-phenylphosphine, B-00239
4,7-Bis[2-(diphenylphosphino)ethyl]-1,1,10,10-tetraphenyl-1,4,7,10-tetraphosphadecane, B-00240
1,2-Bis(diphenylphosphino)ethyne, *see* B-00216
1,1′-Bis(diphenylphosphino)ferrocene, B-00241
1,7-Bis(diphenylphosphino)heptane, B-00242
1,6-Bis(diphenylphosphino)hexane, B-00243
5,6-Bis(diphenylphosphino)-8-isopropyl-2-methylbicyclo[2.2.2]oct-2-ene, *see* P-00059
Bis(diphenylphosphino)methane, B-00244
Bis(diphenylphosphino)methanone, B-00245
2,11-Bis[(diphenylphosphino)methyl]-benzo[*c*]phenanthrene, B-00246
5,6-Bis(diphenylphosphino)-2-methylbicyclo[2.2.1]hept-2-ene, B-00247
2,2′-Bis[(diphenylphosphino)methyl]-1,1′-binaphthyl, B-00248
1,2-Bis(diphenylphosphinomethyl)cyclobutane, B-00249
1,2-Bis(diphenylphosphinomethyl)cyclohexane, B-00250
4,5-Bis[(diphenylphosphino)methyl]-2,2-dimethyl-1,3-dioxolane, B-00251
[Bis(diphenylphosphino)methylene]-triphenylphosphorane, B-00252
2,3-Bis(diphenylphosphinomethyl)-5-norbornene, *see* B-00075
Bis[(diphenylphosphino)methyl]-phenylphosphine, B-00253
[2,2-Bis[(diphenylphosphino)methyl]-1,3-propanediyl]bis[diphenylphosphine], *see* T-00179
1,4-Bis(diphenylphosphino)naphthalene, B-00254
1,5-Bis(diphenylphosphino)naphthalene, B-00255
2,6-Bis(diphenylphosphino)naphthalene, B-00256
2,7-Bis(diphenylphosphino)naphthalene, B-00257
5,6-Bis(diphenylphosphino)norbornene, *see* B-00220
1,8-Bis(diphenylphosphino)octane, B-00258
1,5-Bis(diphenylphosphino)pentane, B-00259
2,4-Bis(diphenylphosphino)pentane, B-00260
9,10-Bis(diphenylphosphino)phenanthrene, B-00261
1,2-Bis(diphenylphosphino)-1-phenylethane, *see* P-00150
Bis[2-(diphenylphosphino)phenyl]-phenylphosphine, B-00262
Bis[4-(diphenylphosphino)phenyl]-phenylphosphine, B-00263
1,2-Bis(diphenylphosphino)propane, B-00264
1,3-Bis(diphenylphosphino)propane, B-00265
Bis[3-(diphenylphosphino)propyl]-phenylphosphine, B-00266
3,4-Bis(diphenylphosphino)pyrrolidine, B-00267
1,4-Bis(diphenylphosphinoselenoyl)butane, *in* B-00225
1,2-Bis(diphenylphosphinoselenoyl)ethane, *in* B-00234
1,1-Bis(diphenylphosphinoselenoyl)ethylene, *in* B-00236
1,6-Bis(diphenylphosphinoselenoyl)hexane, *in* B-00243
Bis(diphenylphosphinoselenoyl)methane, *in* B-00244
1,5-Bis(diphenylphosphinoselenoyl)pentane, *in* B-00259
Bis[2-(diphenylphosphino)-3,4,5,6-tetrafluorophenyl]phenylphosphine, B-00268
1,4-Bis(diphenylphosphinothioyl)butane, *in* B-00225
1,4-Bis(diphenylphosphinothioyl)-2-butyne, *in* B-00227
1,2-Bis(diphenylphosphinothioyl)ethane, *in* B-00234
1,1-Bis(diphenylphosphinothioyl)ethylene, *in* B-00236
1,2-Bis(diphenylphosphinothioyl)ethyne, *in* B-00216
1,1′-Bis(diphenylphosphinothioyl)ferrocene, *in* B-00241
1,6-Bis(diphenylphosphinothioyl)hexane, *in* B-00243
Bis(diphenylphosphinothioyl)methane, *in* B-00244
Bis[(diphenylphosphinothioyl)methyl]-phenylphosphine sulfide, *in* B-00253
1,3-Bis(diphenylphosphinothioyl)propane, *in* B-00265
1,4-Bis(diphenylphosphinyl)benzene, *in* P-00128
5,6-Bis(diphenylphosphinyl)bicyclo[2.2.1]-hept-2-ene, *in* B-00220
1,4-Bis(diphenylphosphinyl)butane, *in* B-00225
1,4-Bis(diphenylphosphinyl)-2-butyne, *in* B-00227
1,1-Bis(diphenylphosphinyl)cyclopropane, *in* B-00229
1,2-Bis(diphenylphosphinyl)cyclopropane, *in* B-00230
1,2-Bis(diphenylphosphinyl)ethane, *in* B-00234
1,1-Bis(diphenylphosphinyl)ethylene, *in* B-00236
Bis[(2-diphenylphosphinyl)ethyl]ether, *in* O-00093
1,2-Bis(diphenylphosphinyl)ethyne, *in* B-00216
1,1′-Bis(diphenylphosphinyl)ferrocene, *in* B-00241
1,7-Bis(diphenylphosphinyl)heptane, *in* B-00242
1,2-Bis(diphenylphosphinyl)-3,3,4,4,5,5-hexafluorocyclopentene, *in* H-00016
1,6-Bis(diphenylphosphinyl)hexane, *in* B-00243
Bis(diphenylphosphinyl) ketone, *see* B-00245
Bis(diphenylphosphinyl)methane, *in* B-00244
Bis[(diphenylphosphinyl)methyl]-phenylphosphine oxide, *in* B-00253
1,4-Bis(diphenylphosphinyl)naphthalene, *in* B-00254
1,5-Bis(diphenylphosphinyl)naphthalene, *in* B-00255
2,6-Bis(diphenylphosphinyl)naphthalene, *in* B-00256
2,7-Bis(diphenylphosphinyl)naphthalene, *in* B-00257
1,8-Bis(diphenylphosphinyl)octane, *in* B-00258
1,5-Bis(diphenylphosphinyl)pentane, *in* B-00259
Bisdiphenylphosphinyl peroxide, *see* D-00977
1,3-Bis(diphenylphosphinyl)propane, *in* B-00265
1,8-Bis(diphenylphosphonio)-2,7-dimethyl-2,4,6-octatriene(2+), *see* D-00778
Bis(*O,O*-diphenyl phosphoroselenoyl) diselenide, B-00269
[Bis(diphenylstibino)methylene]-triphenylphosphorane, B-00270
1,2-Bis[(dipropoxyphosphinothioyl)amino]-ethane, *in* E-00010
1,2-Bis(dipropoxyphosphinylamino)ethane, *in* E-00009
P,P-Bis(dipropylamino)-*P*-fluoro-*N*-propylphosphazene, *in* P-00043
P,P-Bis(dipropylamino)-*P*-fluoro-*N*-propylphosphine imide, *in* P-00043

1,2-Bis(di-*o*-tolylphosphino)ethane, *see* B-00103
1,2-Bis(di-*p*-tolylphosphino)ethane, *see* B-00104
Bis(ethoxycarbonyl)phenylphosphine, *in* P-00209
O,O-Bis(2-ethoxyethyl) chlorothiophosphate, *see* B-00271
O,O-Bis(2-ethoxyethyl) phosphorochloridothioate, B-00271
O,O-Bis(2-ethoxyethyl) thiophosphoryl chloride, *see* B-00271
Bis(ethoxymethyl)phosphinic acid, B-00272
1,2-Bis(ethoxyphenylphosphinyl)ethane, *in* E-00015
▷Bis(ethyleneimido)phosphorylurethane, *see* U-00001
▷*O,O*-Bis(2-ethylhexyl) dithiophosphate, *see* B-00274
▷*O,O*-Bis(2-ethylhexyl) hydrogen phosphorodithioate, *see* B-00274
▷Bis(2-ethylhexyl) phosphite, *see* B-00273
▷Bis(2-ethylhexyl) phosphonate, B-00273
▷*O,O*-Bis(2-ethylhexyl) phosphorodithioate, B-00274
▷*O,O*-Bis(2-ethylhexyl) phosphorodithioic acid, *see* B-00274
2,4-Bis(4-fluorophenyl)-1,3,2,4-dioxaphosphetane 2,4-dioxide, B-00275
Bis(3-fluorophenyl)phosphine, B-00276
Bis(4-fluorophenyl)phosphine, B-00277
Bis(2-fluorophenyl)phosphinic acid, B-00278
Bis(3-fluorophenyl)phosphinic acid, B-00279
Bis(4-fluorophenyl)phosphinic acid, B-00280
Bis(hydroxymethyl)phenylphosphine, B-00281
Bis(hydroxymethyl)phosphinic acid, B-00282
Bis(2-hydroxyphenyl)phosphinic acid, B-00283
Bis(3-hydroxyphenyl)phosphinic acid, B-00284
Bis(4-hydroxyphenyl)phosphinic acid, B-00285
1,2-Bis(hydroxyphenylphosphinyl)ethane, *see* E-00015
Bis(3-hydroxypropyl)methylphenylphosphonium iodide, *in* B-00286
Bis(3-hydroxypropyl)phenylphosphine, B-00286
Bis(1-hydroxy-2,2,2-trichloroethyl)-phosphinic acid, B-00287
Bis-4-iodobenzyl [1-ethoxycarbonyl-2-(4-hydroxybenzyl)ethyl]phosphoramidate, *in* P-00422
Bis(iodomethyl)phosphinic acid, B-00288
Bis(2-iodophenyl)phosphinic acid, B-00289
Bis(3-iodophenyl)phosphinic acid, B-00290
Bis(4-iodophenyl)phosphinic acid, B-00291
Bis(mercaptomethyl)phosphinic acid, B-00292
Bis(methoxycarbonyl)phenylphosphine, *in* P-00209
Bis(4-methoxycarbonylphenyl)phosphinic acid, *in* B-00118
▷1,3-Bis(methoxycarbonyl)-1-propen-2-yl dimethyl phosphate, *see* B-00461
Bis(2-methoxyphenyl) chlorophosphate, *in* B-00294
Bis(4-methoxyphenyl) chlorophosphate, *in* B-00295
2,4-Bis(4-methoxyphenyl)-1,3,2,4-dithiadiphosphetane 2,4-disulfide, B-00293
Bis(2-methoxyphenyl) hydrogen phosphate, *see* B-00294
Bis(4-methoxyphenyl) hydrogen phosphate, *see* B-00295
1,2-Bis[(2-methoxyphenyl)phenylphosphino]-ethane, *see* E-00011
Bis(2-methoxyphenyl) phosphate, B-00294
Bis(4-methoxyphenyl) phosphate, B-00295
Bis(2-methoxyphenyl)phosphine, B-00296
Bis(3-methoxyphenyl)phosphine, B-00297
Bis(4-methoxyphenyl)phosphine, B-00298
Bis(2-methoxyphenyl)phosphinic acid, B-00299
Bis(3-methoxyphenyl)phosphinic acid, B-00300
Bis(4-methoxyphenyl)phosphinic acid, B-00301
Bis(4-methoxyphenyl) phosphoramidate, *in* B-00295
Bis(2-methoxyphenyl) phosphoric acid, *see* B-00294
Bis(4-methoxyphenyl) phosphoric acid, *see* B-00295
Bis(2-methoxyphenyl) phosphorochloridate, *in* B-00294
Bis(4-methoxyphenyl) phosphorochloridate, *in* B-00295
Bis(4-methoxyphenyl)phosphoryl amide, *in* B-00295
Bis(2-methoxyphenyl)phosphoryl chloride, *in* B-00294
Bis(4-methoxyphenyl) phosphoryl chloride, *in* B-00295
Bis(3-methylbutyl) hydrogen phosphate, *see* B-00302
Bis(3-methylbutyl) phenylphosphoramidate, *in* B-00302
Bis(3-methylbutyl) phosphate, B-00302
Bis(3-methylbutyl) phosphite, *see* B-00303
Bis(3-methylbutyl) phosphonate, B-00303
Bis(3-methylbutyl) phosphoramidate, *in* B-00302
Bis(3-methylbutyl) phosphoric acid, *see* B-00302
Bis(3-methylbutyl) phosphorochloridate, *in* B-00302
O,O-Bis(3-methylbutyl) phosphorodithioate, B-00304
Bis(3-methylbutyl) phosphorofluoridate, *in* B-00302
O,O-Bis(3-methylbutyl) phosphorothioate, B-00305
Bis(4-methyl-1,3,2-dioxaphosphorinan-2-yl) ether, *see* O-00096
Bis(4-methyl-1,3,2-dioxaphosphorinan-2-yl) oxide, *see* O-00096
1,1-Bis(methyldiphenylphosphonio)ethene diiodide, *in* B-00236
Bis[2-(methyldiphenylphosphonio)ethyl]-methylphenylphosphonium triiodide, *in* B-00239
1,1′-Bis(methyldiphenylphosphonio)ferrocene diiodide, *in* B-00241
1,3-Bis(1-methylethoxy)-1,3-diphosphetane 1,3-dioxide, *in* D-00620
[[Bis(1-methylethoxy)phosphinothioyl]-thio]acetic acid, *see* D-00632
[Bis(1-methylethoxy)phosphinyl]acetamide, *in* D-00633
[Bis(1-methylethoxy)phosphinyl]acetic acid, *see* D-00633
▷*S*-[2-[Bis(1-methylethyl)amino]ethyl] *O*-ethyl methylphosphonothioate, *see* E-00062
Bis(1-methylethyl) (2-amino-2-oxoethyl)-phosphonate, *in* A-00060
O,O-Bis(1-methylethyl) benzoylphosphoramidothioate, *in* B-00038
Bis(1-methylethyl) bis(2-methylpropyl)-phosphoramidate, *in* B-00355
Bis(1-methylethyl) 1-butenylphosphonate, *in* B-00524
Bis(1-methylethyl) butylphosphonite, *in* B-00615
▷Bis(1-methylethyl) chlorothiophosphonate, *see* D-00634
▷*O,O*-Bis(1-methylethyl) chlorothiophosphonothioate, *see* D-00635
Bis(1-methylethyl) diethylphosphoramidate, *in* D-00343
Bis(1-methylethyl) [(dimethoxyphosphinothioyl)thio]-butanedioate, *in* D-00677
Bis(1-methylethyl) (1,1-dimethylethyl)-phosphonite, *in* B-00616
Bis(1-methylethyl) dimethylphosphoramidate, *in* D-00855
Bis(1-methylethyl) dodecylphosphonite, *in* D-01265
O,O-Bis(1-methylethyl) ethylphosphoramidothioate, *in* E-00159
O,O-Bis(1-methylethyl) hydrogen phosphorodiselenoate, *see* D-00662
Bis(1-methylethyl) (2-hydroxyethyl)-phosphoramidate, *in* H-00148
Bis(1-methylethyl) (hydroxyphenylmethyl)-phosphonate, *in* H-00174
2,4-Bis(1-methylethylidene)-1,3,6,9-tetraoxa-5-phospha(5-P^V)spiro[4.4]-nonane, B-00306
Bis(1-methylethyl) (1-methylethyl)-phosphonite, *in* I-00056
N,N-Bis(1-methylethyl)-*P*-methylphosphonamidothioic acid, *see* D-00637
O,O-Bis(1-methylethyl)-methylphosphonothioate, *in* M-00313
Bis(1-methylethyl) (2-oxopropyl)phosphonate, *in* O-00075
Bis(1-methylethyl) [(phenylamino)methyl]-phosphonate, *in* P-00084
Bis(1-methylethyl) (phenylmethyl)-phosphoramidate, *in* B-00069
O,O-Bis(1-methylethyl) *O*-phenylmethyl phosphorothioate, *see* B-00044
▷*O,O*-Bis(1-methylethyl) *S*-phenylmethyl phosphorothioate, *see* B-00045
Bis(1-methylethyl)phenylphosphine, *see* D-00638
N,N′-Bis(1-methylethyl)-*P*-phenylphosphonic diamide, *in* P-00222
Bis(1-methylethyl) phenylphosphonite, *in* P-00251
N,N′-Bis(1-methylethyl)-*P*-phenylphosphonous diamide, *in* P-00252
Bis(1-methylethyl) phenylphosphoramidate, *in* P-00257
O,O-Bis(1-methylethyl) phenylphosphoramidothioate, *in* P-00262
▷*O,O*-Bis(1-methylethyl) *S*-[2-[(phenylsulfonyl)amino]ethyl] phosphorodithioate, *see* B-00004
Bis(1-methylethyl) phosphate, *see* D-00639
Bis(1-methylethyl)phosphine, *see* D-00640
Bis(1-methylethyl)phosphinic acid, *see* D-00641
P,P-Bis(1-methylethyl)phosphinic amide, *see* D-00642
Bis(1-methylethyl)phosphinodithioic acid, *see* D-00643
Bis(1-methylethyl)phosphinoselenoic acid, *see* D-00644
Bis(1-methylethyl)phosphinotellurous acid, *see* D-00645
Bis(1-methylethyl)phosphinothioic acid, *see* D-00646
Bis(1-methylethyl)phosphinous acid, *see* D-00648
P,P-Bis(1-methylethyl)phosphinous amide, *see* D-00649
Bis(1-methylethyl)phosphinous chloride, *see* D-00650
▷Bis(1-methylethyl) phosphite, *see* D-00651
▷Bis(1-methylethyl) phosphonate, *see* D-00651
O,O-Bis(1-methylethyl) phosphonothioate, *see* D-00652
Bis(1-methylethyl) phosphoramidate, *see* D-00653
Bis(1-methylethyl)phosphoramidochloridous acid, *see* D-00654
Bis(1-methylethyl)phosphoramidoselenoic acid, *see* D-00655
Bis(1-methylethyl)phosphoramidous acid, *see* D-00656
Bis(1-methylethyl) phosphorochloridate, *see* D-00657
Bis(1-methylethyl) phosphorochloridite, *see* D-00658
Bis(1-methylethyl) phosphorochloridodithioite, *see* D-00660
▷*O,O*-Bis(1-methylethyl) phosphorochloridothioate, *see* D-00659
N,N′-Bis(1-methylethyl)phosphorodiamidic acid, *see* D-00661
O,O-Bis(1-methylethyl) phosphorodiselenoate, *see* D-00662
O,O-Bis(1-methylethyl) phosphorodiselenoic acid, *see* D-00662
O,O-Bis(1-methylethyl) phosphorodithioate, *see* D-00663
S,S-Bis(1-methylethyl) phosphorodithioate, *see* D-00664
▷Bis(1-methylethyl) phosphorofluoridate, *see* D-00665
O,O-Bis(1-methylethyl) phosphoroselenoate, *see* D-00666
O,O-Bis(1-methylethyl) phosphoroselenoic acid, *see* D-00666
O,O-Bis(1-methylethyl) phosphorothioate, *see* D-00667
Bis(1-methylethyl) 1,2-propadienylphosphonate, *in* P-00441
Bis(1-methylethyl) 2-propenylphosphonate, *in* P-00454

Bis(1-methylethyl) 2-propenylphosphonite, *in* P-00463
Bis(1-methylethyl) (2-propenyl)-phosphonoselenoate, *in* P-00458
O,O-Bis(1-methylethyl) 2-propenylphosphonothioate, *in* P-00461
Bis(1-methylethyl) 3-pyridinylphosphoramidate, *in* P-00511
Bis(1-methylethyl) 1-pyrrolidinylphosphonite, *in* P-00520
Bis(1-methylethyl) 2-thienylphosphonite, *in* T-00308
Bis(1-methylethyl) (tricyclo[3.3.1.1^{3,7}]-dec-1-yl)phosphonate, *in* A-00034
Bis(1-methylethyl) (trifluoromethyl)-phosphonite, *in* T-00482
Bis(1-methylheptyl) phosphite, *see* B-00307
Bis(1-methylheptyl) phosphonate, B-00307
Bis(1-methylheptyl) phosphorochloridite, B-00308
Bis(4-methylphenyl) hydrogen phosphate, *see* B-00314
Bis(2-methylphenyl)phenylphosphine, B-00309
Bis(3-methylphenyl)phenylphosphine, B-00310
Bis(4-methylphenyl)phenylphosphine, B-00311
P,P-Bis(2-methylphenyl)-*N*-phenylphosphinic amide, *in* B-00318
P,P-Bis(4-methylphenyl)-*N*-phenylphosphinic amide, *in* B-00320
▷Bis(2-methylphenyl) phosphate, B-00312
Bis(3-methylphenyl) phosphate, B-00313
Bis(4-methylphenyl) phosphate, B-00314
▷Bis(2-methylphenyl)phosphine, B-00315
▷Bis(3-methylphenyl)phosphine, B-00316
Bis(4-methylphenyl)phosphine, B-00317
Bis(2-methylphenyl)phosphinic acid, B-00318
Bis(3-methylphenyl)phosphinic acid, B-00319
Bis(4-methylphenyl)phosphinic acid, B-00320
1,2-Bis(methylphenylphosphino)ethane, *see* E-00012
1,3-Bis(methylphenylphosphino)propane, B-00321
Bis(4-methylphenyl)phosphinothioic acid, B-00322
1,3-Bis(methylphenylphosphinothioyl)propane, *in* B-00321
Bis(2-methylphenyl)phosphinous acid, B-00323
Bis(3-methylphenyl)phosphinous acid, B-00324
Bis(4-methylphenyl)phosphinous acid, B-00325
Bis(2-methylphenyl) phosphite, *see* B-00326
Bis(3-methylphenyl) phosphite, *see* B-00327
Bis(4-methylphenyl) phosphite, *see* B-00328
Bis(2-methylphenyl) phosphonate, B-00326
Bis(3-methylphenyl) phosphonate, B-00327
Bis(4-methylphenyl) phosphonate, B-00328
Bis(2-methylphenyl) phosphorochloridate, *in* B-00312
Bis(3-methylphenyl) phosphorochloridate, *in* B-00313
Bis(4-methylphenyl) phosphorochloridate, B-00329
Bis(2-methylphenyl) phosphorochloridite, B-00330
Bis(3-methylphenyl) phosphorochloridite, B-00331
Bis(4-methylphenyl) phosphorochloridite, B-00332
O,O-Bis(2-methylphenyl) phosphorochloridothioate, B-00333
O,O-Bis(3-methylphenyl) phosphorochloridothioate, B-00334
O,O-Bis(4-methylphenyl) phosphorochloridothioate, B-00335
O,O-Bis(4-methylphenyl) phosphorodiselenoate, B-00336
O,O-Bis(4-methylphenyl) phosphorodiselenoic acid, *see* B-00336
O,O-Bis(2-methylphenyl) phosphorodithioate, B-00337
▷*O,O*-Bis(3-methylphenyl) phosphorodithioate, B-00338
O,O-Bis(4-methylphenyl) phosphorodithioate, B-00339
Bis(4-methylphenyl) phosphoryl chloride, *see* B-00329
Bis(2-methylpropyl) bis(2-methylpropyl)-phosphoramidate, *in* B-00355
2,4-Bis(1-methylpropyl)-1,3,2,4-dithiodiphosphetane 2,4-disulfide, *in* M-00382
Bis(2-methylpropyl) hydrogen phosphate, *see* B-00342
Bis(2-methylpropyl) (2-methylpropyl)-phosphonate, *in* M-00378
Bis(2-methylpropyl) (2-methylpropyl)-phosphonite, *in* M-00383
O,O-Bis(2-methylpropyl) (2-methylpropyl)-phosphonothioate, *in* M-00381
Bis(1-methylpropyl)phenylphosphine, B-00340
Bis(2-methylpropyl) phenylphosphonate, *in* P-00221
Bis(1-methylpropyl) phenylphosphoramidate, *in* P-00257
Bis(2-methylpropyl) phenylphosphoramidate, *in* P-00257
Bis(1-methylpropyl) phosphate, B-00341
Bis(2-methylpropyl) phosphate, B-00342
Bis(1-methylpropyl)phosphine, B-00343
Bis(2-methylpropyl)phosphine, B-00344
Bis(1-methylpropyl)phosphinic acid, B-00345
Bis(2-methylpropyl)phosphinic acid, B-00346
P,P-Bis(2-methylpropyl)phosphinic amide, *in* B-00346
Bis(1-methylpropyl)phosphinodithioic acid, B-00347
Bis(2-methylpropyl)phosphinodithioic acid, B-00348
Bis(1-methylpropyl)phosphinoselenoic acid, B-00349
Bis(1-methylpropyl)phosphinothioic acid, B-00350
Bis(2-methylpropyl)phosphinothioic acid, B-00351
Bis(2-methylpropyl)phosphinothious acid, B-00352
Bis(1-methylpropyl) phosphite, *see* B-00353
Bis(2-methylpropyl) phosphite, *see* B-00354
Bis(1-methylpropyl) phosphonate, B-00353
Bis(2-methylpropyl) phosphonate, B-00354
Bis(2-methylpropyl)phosphoramidic acid, B-00355
Bis(1-methylpropyl) phosphorochloridate, *in* B-00341
Bis(2-methylpropyl) phosphorochloridate, *in* B-00342
O,O-Bis(2-methylpropyl) phosphorochloridoselenoate, B-00356
O,O-Bis(2-methylpropyl) phosphorochloridothioate, B-00357
N,N'-Bis(2-methylpropyl)-phosphorodiamidodiselenoic acid, B-00358
O,O-Bis(1-methylpropyl) phosphorodithioate, B-00359
O,O-Bis(2-methylpropyl) phosphorodithioate, B-00360
S,S-Bis(2-methylpropyl) phosphorodithioate, B-00361
Bis(1-methylpropyl) phosphorofluoridate, *in* B-00341
Bis(2-methylpropyl) phosphorofluoridate, B-00362
O,O-Bis(2-methylpropyl) phosphorothioate, B-00363
[Bis(methylthio)methyl]phosphonic acid, B-00364
Bis(4-methyl-2-thioxo-1,3,2-dioxaphosphorinan-2-yl) oxide, B-00365
Bis(4-nitrophenyl) chlorophosphate, *see* B-00376
Bis(4-nitrophenyl)(dichloromethyl)-phosphonate, *in* D-00175
Bis(2-nitrophenyl) hydrogen phosphate, *see* B-00369
Bis(3-nitrophenyl) hydrogen phosphate, *see* B-00370
Bis(4-nitrophenyl) hydrogen phosphate, *see* B-00371
Bis[(2-nitrophenyl)methyl] phosphate, B-00366
Bis[(4-nitrophenyl)methyl] phosphate, B-00367
Bis[(4-nitrophenyl)methyl] phosphite, *see* B-00368
Bis[(4-nitrophenyl)methyl]phosphonate, B-00368
Bis[(2-nitrophenyl)methyl] phosphorochloridate, *in* B-00366
Bis[(4-nitrophenyl)methyl] phosphorochloridate, *in* B-00367
Bis(2-nitrophenyl) phosphate, B-00369
Bis(3-nitrophenyl) phosphate, B-00370
Bis(4-nitrophenyl) phosphate, B-00371
Bis(4-nitrophenyl)phosphine, B-00372
▷Bis(3-nitrophenyl)phosphinic acid, B-00373
Bis(4-nitrophenyl)phosphinic acid, B-00374
Bis(4-nitrophenyl) phosphite, *see* B-00375
Bis(4-nitrophenyl) phosphonate, B-00375
Bis(2-nitrophenyl) phosphoric acid, *see* B-00369
Bis(3-nitrophenyl) phosphoric acid, *see* B-00370
Bis(4-nitrophenyl) phosphoric acid, *see* B-00371
Bis(4-nitrophenyl) phosphorochloridate, B-00376
O,O-Bis(4-nitrophenyl) phosphorochloridothioate, B-00377
Bis(4-nitrophenyl) phosphoryl chloride, *see* B-00376
Bis(2-oxo-3-oxazolidinyl) diphenyl phosphoric anhydride, *in* B-00378
Bis(2-oxo-3-oxazolidinyl)phosphinic acid, B-00378
Bis(pentachlorophenyl) chlorophosphate, *in* B-00379
Bis(pentachlorophenyl) hydrogen phosphate, *see* B-00379
Bis(pentachlorophenyl) phosphate, B-00379
Bis(pentachlorophenyl) phosphoric acid, *see* B-00379
Bis(pentachlorophenyl) phosphorochloridate, *in* B-00379
Bis(pentachlorophenyl)phosphoryl chloride, *in* B-00379
Bis(pentafluorophenyl) chlorophosphate, *in* B-00381
Bis(pentafluorophenyl) hydrogen phosphate, *see* B-00381
Bis(pentafluorophenyl)phenylphosphine, B-00380
Bis(pentafluorophenyl) phosphate, B-00381
Bis(pentafluorophenyl)phosphinic acid, B-00382
Bis(pentafluorophenyl)phosphinothioic acid, B-00383
Bis(pentafluorophenyl) phosphoric acid, *see* B-00381
Bis(pentafluorophenyl) phosphorochloridate, *in* B-00381
Bis(pentafluorophenyl) phosphoryl chloride, *in* B-00381
Bis(phenoxycarbonyl)phenylphosphine, *in* P-00209
Bis(phenoxymethyl)phosphinic acid, B-00384
Bis(4-phenoxyphenyl)phosphinic acid, *in* B-00285
Bis(4-phenyl-1,3-butadienyl)phosphinic acid, B-00385
Bis(2-phenylethenyl)phosphinic acid, B-00386
Bis(2-phenylethyl)(2-phenylethyl)-phosphonate, *in* P-00165
Bis(2-phenylethyl)phosphine, B-00387
Bis(phenylethynyl)phosphinic acid, B-00388
P,P-Bis(phenylethynyl)phosphinic anilide, *in* B-00388
[Bis(phenylmethoxy)phosphinyl]acetic acid, B-00389
Bis(phenylmethyl) diethylphosphoramidate, *in* D-00343
1,6-Bis(phenylmethyl)-1,6-diphosphacycloundecane, *see* D-00046
1,4-Bis(phenylmethyl)-1,4-diphosphocane, *see* D-00048
1,5-Bis(phenylmethyl)-1,5-diphosphocane, *see* D-00049
1,4-Bis(phenylmethyl)-1,4-diphosphonane, *see* D-00050
Bis(phenylmethyl) ethylphosphonodithioite, *in* E-00141
Bis(phenylmethyl) ethyphosphonite, *in* E-00153
Bis(phenylmethyl)(4-methylphenyl)-phosphoramidate, *in* M-00253
Bis(phenylmethyl) methylphosphonodithioite, *in* M-00302
Bis(phenylmethyl) phenylphosphonite, *in* P-00251

Bis(phenylmethyl) phenylphosphoramidate, *in* D-00053
Bis(phenylmethyl) phosphate, *see* D-00053
Bis(phenylmethyl)phosphine, *see* D-00054
Bis(phenylmethyl)phosphinic acid, *see* D-00055
Bis(phenylmethyl)phosphinodithioic acid, *see* D-00056
Bis(phenylmethyl)phosphinothioic acid, *see* D-00057
Bis(phenylmethyl)phosphinous acid, *see* D-00058
▷Bis(phenylmethyl) phosphonate, *see* D-00059
Bis(phenylmethyl) phosphoramidate, *in* D-00053
▷Bis(phenylmethyl) phosphorochloridate, *see* D-00060
N,N'-Bis(phenylmethyl)phosphorodiamidic acid, *see* D-00061
N,N'-Bis(phenylmethyl)-phosphorodiamidothioic acid, *see* D-00062
O,O-Bis(phenylmethyl) phosphorodithioate, *see* D-00063
Bis(phenylmethylsulfonylmethyl)phosphinic acid, *in* B-00101
Bis(phenylmethylthiomethyl)phosphinic acid, *see* B-00101
1,2-Bis(phenylphosphino)ethane, B-00390
Bis(2-phenylvinyl)phosphinic acid, *see* B-00386
Bis(phosphino)dimethylgermanium, *see* D-00750
1,2-Bis(phosphonomethyl)benzene, *see* P-00129
1,3-Bis(phosphonomethyl)benzene, *see* P-00130
1,4-Bis(phosphonomethyl)benzene, *see* P-00131
1,4-Bis(phosphonomethyl)biphenyl, B-00391
▷*N,N*-Bis(phosphonomethyl)glycine, *see* G-00098
1,5-Bis(phosphonomethyl)naphthalene, *see* N-00002
1,6-Bis(phosphonomethyl)naphthalene, *see* N-00003
1,8-Bis(phosphonomethyl)naphthalene, *see* N-00004
2,3-Bis(phosphonomethyl)naphthalene, *see* N-00005
2,6-Bis(phosphonomethyl)naphthalene, *see* N-00006
2,7-Bis(phosphonomethyl)naphthalene, *see* N-00007
Bis[tartrato(4−)]phosphorane, *see* O-00036
▷Bis-*N,N,N',N'*-tetramethylphosphorodiamidic anhydride, *see* O-00029
1,2-Bis[1-(2,2,6,6-tetramethylpiperidino)]-diphosphene, B-00392
Bis(2-thienyl)phosphinic acid, *see* D-01250
1,4-Bis(2,4,6-tri-*tert*-butylphenyl)-1,4-bis(trimethylsilyloxy)-2,3-diphosphabutadiene, B-00393
P,P'-Bis(2,4,6-tri-*tert*-butylphenyl)-1,3-diphosphaallene, B-00394
Bis(2,4,6-tri-*tert*-butylphenyl)diphosphene, B-00395
2,3-Bis(2,4,6-tri-*tert*-butylphenyl)-1,2,3-selenadiphosphirane, B-00396
2,3-Bis(2,4,6-tri-*tert*-butylphenyl)-thiadiphosphirane, B-00397
1,4-Bis(2,4,6-tri-*tert*-butylphenyl)-3-trimethylsilyloxy-1,2,4-triphospha-1,3-butadiene, B-00398
Bis(2,2,2-trichloro-1,1-dimethylethyl) phosphorochloridate, B-00399
Bis(2,2,2-trichloroethyl) chlorophosphite, *see* B-00401
Bis(2,2,2-trichloroethyl) hydrogen phosphate, *see* B-00400
Bis(2,2,2-trichloroethyl) phenylphosphoramidate, *in* B-00400
Bis(2,2,2-trichloroethyl) phosphate, B-00400
Bis(2,2,2-trichloroethyl) phosphoric acid, *see* B-00400
Bis(2,2,2-trichloroethyl) phosphorobromidate, *in* B-00400
Bis(2,2,2-trichloroethyl) phosphorochloridate, *in* B-00400
Bis(2,2,2-trichloroethyl) phosphorochloridite, B-00401
Bis(2,2,2-trichloroethyl) phosphoryl bromide, *in* B-00400
Bis(2,2,2-trichloroethyl) phosphoryl chloride, *in* B-00400
Bis(trichloromethyl)phosphinic acid, B-00402
P,P-Bis(trichloromethyl)phosphinic amide, *in* B-00402
PP-Bis(trichloromethyl)phosphinimidic chloride, B-00403
[Bis(trichloromethyl)phosphinyl]-phosphorimidic trichloride, B-00404
Bis(2-trichloromethyl-2-propyl) chlorophosphate, *see* B-00399
Bis(2,3,6-trichlorophenyl)phosphinic acid, B-00405
Bis(2,4,5-trichlorophenyl)phosphinic acid, B-00406
Bis(triethylsilyl) [(acetyloxy)methyl]-phosphonate, *in* A-00007
Bis(triethylsilyl) (hydroxymethyl)-phosphonate, *in* H-00162
Bis(triethylsilyl) phosphite, *see* B-00407
Bis(triethylsilyl) phosphonate, B-00407
Bis(2,2,2-trifluoroethoxy)-triphenylphosphorane, B-00408
Bis(2,2,2-trifluoroethyl) phosphite, *see* B-00409
Bis(2,2,2-trifluoroethyl) phosphonate, B-00409
1,2-Bis(trifluoromethyl)diphosphine, B-00410
Bis(trifluoromethyl) phenylphosphonodiselenoite, *in* P-00235
Bis(trifluoromethyl)phosphinic acid, B-00411
Bis(trifluoromethyl)phosphinodithioic acid, B-00412
Bis(trifluoromethyl)phosphinoselenous acid, B-00413
Bis(trifluoromethyl)phosphinotellurous acid, B-00414
Bis(trifluoromethyl)phosphinothioic acid, B-00415
P,P-Bis(trifluoromethyl)phosphinothioic amide, B-00416
Bis(trifluoromethyl)phosphinothious acid, B-00417
Bis(trifluoromethyl)phosphinous acid, B-00418
▷*P,P*-Bis(trifluoromethyl)phosphinous amide, B-00419
Bis(trifluoromethyl)phosphinous anilide, *in* B-00419
▷Bis(trifluoromethyl)phosphinous chloride, B-00420
4′,5′-Bis(trifluoromethyl)spiro[1,3,2-benzodioxaphosphole-2,2′-[1,3,2]-dioxaphosphole], B-00421
4′,5′-Bis(trifluoromethyl)spiro[1,3,2-benzodioxaphosphole-2,2′-[1,3,2]-dithiaphosphole], B-00422
4,5-Bis(trifluoromethyl)spiro[1,3,2-dithiaphosphole-2,1′-[2,6,7]trioxa[1]-phosphabicyclo[2.2.1]heptane, B-00423
4,5-Bis(trifluoromethyl)spiro[1,3,2-dithiaphosphole-2,1′-[2,8,9]trioxa[1]-phosphatricyclo[3.3.1.1^{5,7}]decane], B-00424
3,4-Bis(trifluoromethyl)-1,2,5-thiadiphosphole, B-00425
Bis(2,4,6-trimethylphenyl)phosphine, B-00426
Bis(2,4,5-trimethylphenyl)phosphinic acid, B-00427
Bis(2,4,6-trimethylphenyl)phosphinic acid, B-00428
P,P-Bis(2,4,6-trimethylphenyl)phosphinic amide, *in* B-00428
Bis(trimethylphosphoranylidene)methane, B-00429
Bis(trimethylsilyl) (acetamidomethyl)-phosphonate, *in* A-00012
Bis(trimethylsilyl) acetonylphosphonate, *in* O-00075
Bis(trimethylsilyl) (2-acetyloxyethyl)-phosphonate, *in* A-00006
Bis(trimethylsilyl) acetylphosphonate, *in* A-00026
2-[Bis(trimethylsilyl)amino]-2,2-dihydro-3,3-bis(trifluoromethyl)-2-[(trimethylsilyl)imino]oxaphosphirane, B-00430
Bis(trimethylsilyl)(4-aminophenyl)-phosphonate, *in* A-00111
Bis(trimethylsilyl) benzoylphosphonate, *in* B-00034
Bis(trimethylsilyl) benzylphosphonate, *in* B-00060
Bis(trimethylsilyl)(4-bromophenyl)-phosphonate, *in* B-00492
Bis(trimethylsilyl) (2-chloroethyl)-phosphonate, *in* C-00074
Bis(trimethylsilyl)(4-chlorophenyl)-phosphonate, *in* C-00147
Bis(trimethylsilyl) (cyanomethyl)-phosphonate, *in* C-00215
Bis(trimethylsilyl)[3,5-di-*tert*-butyl-4-hydroxyphenyl)methyl]phosphonate, *in* D-00094
Bis(trimethylsilyl) (dimethylaminocarbonyl)phosphonate, *in* D-00691
Bis(trimethylsilyl) (*N,N*-dimethylcarbamoyl)phosphonate, *in* D-00691
Bis(trimethylsilyl) (3,3-dimethyl-2-oxobutyl)phosphonate, *in* D-00783
Bis(trimethylsilyl) ethylphosphonate, *in* E-00129
Bis(trimethylsilyl)(2-fluorophenyl)-phosphonate, *in* F-00036
Bis(trimethylsilyl)(4-fluorophenyl)-phosphonate, *in* F-00038
Bis(trimethylsilyl) 2-furanylmethylphosphonate, *in* F-00072
Bis(trimethylsilyl) 2-furanylphosphonate, *in* F-00073
O,O-Bis(trimethylsilyl) hexylphosphonothioate, *in* H-00100
O,O-Bis(trimethylsilyl) hydrogen phosphorothioate, *see* B-00439
Bis(trimethylsilyl) (2-hydroxyethyl)-phosphonate, *in* H-00146
Bis(trimethylsilyl)(2-methoxyphenyl)-phosphonate, *in* M-00070
2-[Bis(trimethylsilyl)methylene]-1,1-bis(1,1-dimethylethyl)diphosphine, *see* B-00431
2-[Bis(trimethylsilyl)methylene]-1,1-di-*tert*-butyldiphosphine, B-00431
[Bis(trimethylsilyl)methylene]-triphenylphosphorane, *in* B-00432
Bis(trimethylsilyl)(3-methylphenyl)-phosphonate, *in* M-00236
Bis(trimethylsilyl)(4-methylphenyl)-phosphonate, *in* M-00237
[Bis(trimethylsilyl)methyl]-triphenylphosphonium(1+), B-00432
Bis(trimethylsilyl) (1-oxoethyl)phosphonate, *in* A-00026
Bis(trimethylsilyl) (1-oxopropyl)-phosphonate, *in* O-00074
Bis(trimethylsilyl) (2-oxopropyl)-phosphonate, *in* O-00075
Bis(trimethylsilyloxy)phosphinecarboxylic acid oxide, B-00433
[Bis(trimethylsilyloxy)phosphinyl]acetic acid, B-00434
[Bis(trimethylsilyloxy)]phosphinylformic acid, *see* B-00433
O,O-Bis(trimethylsilyl) phenylphosphonoselenoate, *in* P-00243
O,O-Bis(trimethylsilyl) phenylphosphonothioate, *in* P-00244
Bis(trimethylsilyl) phenylphosphonotrithioate, *in* P-00250
Bis(trimethylsilyl) phosphite, *see* B-00435
Bis(trimethylsilyl) phosphonate, B-00435
▷Bis(trimethylsilyl) phosphonite, B-00436
Bis(trimethylsilyl) phosphonoformate, *see* B-00433
O,O-Bis(trimethylsilyl) phosphonothioate, B-00437
Bis(trimethylsilyl)phosphoramidous acid, B-00438
O,O-Bis(trimethylsilyl) phosphoroselenoite, *see* B-00440
O,O-Bis(trimethylsilyl) phosphorotelluroite, *see* B-00441
O,O-Bis(trimethylsilyl) phosphorothioate, B-00439
O,O-Bis(trimethylsilyl) phosphorothioic acid, *see* B-00439
Bis(trimethylsilyl) propanoylphosphonate, *in* O-00074

Bis(trimethylsilyl) selenophosphonate, B-00440
Bis(trimethylsilyl) tellurophosphonate, B-00441
O,O-Bis(trimethylsilyl) thiophosphate, see B-00439
O,O-Bis(trimethylsilyl) thiophosphite, see B-00437
Bis(triphenylphosphine)iminium(1+), B-00442
1,4-Bis(triphenylphosphonio)butane(2+), see B-00517
1,2-Bis(triphenylphosphonio)ethane, see E-00017
1,2-Bis(triphenylphosphonio)ethylene(2+), see E-00019
N,N′-Bis(triphenylphosphonio)hydrazine dichloride, in B-00458
Bis(triphenylphosphonio)methane, see M-00147
5,6-Bis(triphenylphosphoniomethyl)-acenaphene(2+), see A-00001
1,8-Bis(triphenylphosphoniomethyl)-anthracene(2+), B-00443
1,2-Bis(triphenylphosphoniomethyl)benzene, see P-00135
1,4-Bis(triphenylphosphoniomethyl)benzene, see P-00136
2,2′-Bis(triphenylphosphoniomethyl)-bibenzyl(2+), see M-00141
2,2′-Bis[(triphenylphosphonio)methyl]-1,1-binaphthyl(2+), B-00444
2,2′-Bis[(triphenylphosphonio)methyl]-biphenyl(2+), B-00445
3,3′-Bis[(triphenylphosphonio)methyl]-biphenyl(2+), B-00446
4,4′-Bis[(triphenylphosphonio)methyl]-biphenyl(2+), B-00447
2,5-Bis[(triphenylphosphonio)methyl]-furan(2+), B-00448
3,4-Bis[(triphenylphosphonio)methyl]-furan(2+), B-00449
1,4-Bis(triphenylphosphoniomethyl)-naphthalene(2+), B-00450
1,5-Bis(triphenylphosphoniomethyl)-naphthalene(2+), B-00451
1,8-Bis(triphenylphosphoniomethyl)-naphthalene(2+), B-00452
2,3-Bis(triphenylphosphoniomethyl)-naphthalene(2+), B-00453
2,6-Bis(triphenylphosphoniomethyl)-naphthalene(2+), B-00454
2,7-Bis(triphenylphosphoniomethyl)-naphthalene(2+), B-00455
2,5-Bis[(triphenylphosphonio)methyl]-thiophene(2+), B-00456
3,4-Bis[(triphenylphosphonio)methyl]-thiophene(1+), B-00457
▷1,5-Bis(triphenylphosphonio)pentane(2+), see P-00034
1,3-Bis(triphenylphosphonio)propane(2+), see P-00448
1,3-Bis(triphenylphosphonio)-2-propanone(2+), see O-00073
Bis(triphenylphosphoranylidene)hydrazine, B-00458
Bis(triphenylphosphoranylidene)methane, B-00459
1,8-Bis(triphenylphosphoranylidenemethyl)-biphenylene, see B-00076
1,3-Bis(triphenylphosphoranylidene)-2-propanone, in O-00073
1,2-Bis(triphosphoranylidene)-benzocyclobutene, B-00460
1,2-Bis(triphosphoranylidene)-bicyclo[4.2.0]octa-1,3,5-triene, see B-00460
Bis[2,4,6-tris(1,1-dimethylethyl)phenyl]-diphosphene, see B-00395
2,3-Bis[2,4,6-tris(1,1-dimethylethyl)-phenyl]selenadiphosphirane, see B-00396
BMG 59-R2, see A-00135
▷Bolstar, see S-00027
▷Bomyl, B-00461
Boyd's chloride, in T-00608
BPPM, see B-00532
▷Bromchlophos, see N-00001
▷Bromfenvinfos, B-00462
▷Bromfenvinfos-methyl, B-00463
(3-Bromoacetonyl)triphenylphosphonium, see B-00478
2-Bromo-1,3,2-benzodioxaphosphole, B-00464
(α-Bromobenzyl)phosphonic acid, see B-00482
o-Bromobenzylphosphonic acid, see B-00483
p-Bromobenzylphosphonic acid, see B-00484
Bromobis(diethylamino)(ethylamino)-phosphonium bromide, in P-00012
P-Bromo-P,P-bis(diethylamino)-N-ethylphosphazene, in P-00012
P-Bromo-P,P-bis(diethylamino)-N-ethylphosphine imide, in P-00012
Bromobis(dipropylamino)(propylamino)-phosphonium bromide, in P-00043
P-Bromo-P,P-bis(dipropylamino)-N-propylphosphazene, in P-00043
P-Bromo-P,P-bis(dipropylamino)-N-propylphosphine imide, in P-00043
Bromobis(phenylmethyl)phosphine, in D-00058
Bromobis(trifluoromethyl)phosphine, in B-00418
2-Bromo-5-bromomethyl-5-methyl-1,3-dioxa-2-phosphacyclohexane, see B-00465
2-Bromo-5-bromomethyl-5-methyl-1,3,2-dioxaphosphinane, see B-00465
2-Bromo-5-bromomethyl-5-methyl-1,3,2-dioxaphosphorinane, B-00465
Bromo(butylamino)bis(dibutylamino)-phosphonium bromide, in P-00004
P-Bromo-N-butyl-P,P-bis(dibutylamino)-phosphazene, in P-00004
P-Bromo-N-butyl-P,P-bis(dibutylamino)-phosphine imide, in P-00004
(4-Bromobutyl)phosphonic acid, B-00466
▷O-(4-Bromo-2-chlorophenyl) O-ethyl S-propyl phosphorothioate, see P-00440
Bromodibutylphosphine, in D-00121
Bromodi-tert-butylphosphine, in D-00122
▷O-(4-Bromo-2,5-dichlorophenyl) O,O-diethyl phosphorothioate, see B-00499
▷O-(4-Bromo-2,5-dichlorophenyl) O,O-dimethyl phosphorothioate, see B-00498
▷2-Bromo-1-(2,4-dichlorophenyl)ethenyl diethyl phosphate, see B-00462
▷2-Bromo-1-(2,4-dichlorophenyl)ethenyl dimethyl phosphate, see B-00463
(4-Bromo-2,5-dichlorophenyl)methyl phenylphosphonate, in M-00488
▷O-4-Bromo-2,5-dichlorophenyl O-methyl phenylphosphonothioate, see L-00001
▷2-Bromo-1-(2,4-dichlorophenyl)vinyl diethyl phosphate, see B-00462
▷2-Bromo-1-(2,4-dichlorophenyl)vinyl dimethyl phosphate, see B-00463
Bromodicyclohexylphosphine, in D-00214
Bromo(diethoxyphosphinyl)acetic acid, B-00467
Bromodiethylphosphine, in D-00333
(Bromodifluoromethyl)-triphenylphosphonium(1+), B-00468
1-Bromo-2,5-dihydro-3,4-dimethyl-1H-phosphole, B-00469
1-Bromo-2,5-dihydro-3,4-dimethyl-1H-phosphole 1-oxide, in D-00493
1-Bromo-2,5-dihydro-1H-phosphole, in D-00572
Bromodiisopropylphosphine, in D-00648
2-Bromo-5,5-dimethyl-1,3-dioxa-2-phosphacyclohexane, see B-00471
2-Bromo-4,5-dimethyl-1,3-dioxa-2-phosphacyclopentane, see B-00470
2-Bromo-5,5-dimethyl-1,3,2-dioxaphosphinane, see B-00471
2-Bromo-4,5-dimethyl-1,3,2-dioxaphospholane, B-00470
2-Bromo-5,5-dimethyl-1,3,2-dioxaphosphorinane, B-00471
Bromodimethylphosphine, in D-00844
1-Bromo-3,4-dimethyl-3-phospholene, see B-00469
Bromodiphenylphosphine, in D-01089
Bromodipropylphosphine, in D-01209
2-Bromoethyl dihydrogen phosphate, see M-00447
P-(2-Bromoethyl)-N,N′-diphenylphosphonic diamide, in B-00472
Bromoethylmethylphosphine, in E-00087
Bromoethylphenylphosphine, in E-00108
(2-Bromoethyl)phosphonic acid, B-00472
(2-Bromoethyl)phosphonic dianilide, in B-00472
2-Bromoethyl phosphoric acid, see M-00447
2-Bromoethyl phosphorodichloridate, in M-00447
Bromofonofos, in E-00070
2-Bromo-4-methyl-1,3-dioxa-2-phosphacyclohexane, see B-00474
2-Bromo-4-methyl-1,3,2-dioxaphosphinane, see B-00474
2-Bromo-4-methyl-1,3,2-dioxaphospholane, B-00473
2-Bromo-4-methyl-1,3,2-dioxaphosphorinane, B-00474
(Bromomethylene)triphenylphosphorane, in B-00477
Bromomethylphenylphosphine, in M-00227
(Bromomethyl)phosphonic acid, B-00475
(Bromomethyl)phosphonous dibromide, B-00476
1-Bromo-N-methyl-N,1,1,1-tetraphenylphosphoranamine, in M-00194
(Bromomethyl)triphenylphosphonium(1+), B-00477
(3-Bromo-2-oxopropyl)-triphenylphosphonium(1+), B-00478
Bromopentabutyltriaminophosphonium bromide, in P-00004
Bromopentaethyltriaminophosphonium bromide, in P-00012
4-Bromophenyldichlorophosphine, in B-00493
O-(4-Bromophenyl) dichlorothiophosphate, see B-00497
4-Bromophenyl dihydrogen phosphate, see M-00448
2-(4-Bromophenyl)-8,8-dihydro-4,6,8,8-tetraphenyl[1,2]oxaphospholo[2,3-b]-[1,2,5]dioxaphosphorin, B-00479
4-Bromophenyldiiodophosphine, in B-00493
(2-Bromophenyl)diphenylphosphine, B-00480
(4-Bromophenyl)diphenylphosphine, B-00481
S-4-Bromophenyl O-ethyl ethylphosphonodithioate, in E-00070
(Bromophenylmethyl)phosphonic acid, B-00482
[(2-Bromophenyl)methyl]phosphonic acid, B-00483
[(4-Bromophenyl)methyl]phosphonic acid, B-00484
4-(4-Bromophenyl)-2′,2′,3′,4′,4′-pentamethyl-3,3,6,6-tetrakis(trifluoromethyl)spiro[2,7-dioxa-1-phosphabicyclo[3.2.0]hept-3-ene-1,1′-phosphetane], B-00485
(2-Bromophenyl)phosphine, B-00486
(4-Bromophenyl)phosphine, B-00487
(3-Bromophenyl)phosphinic acid, B-00488
(4-Bromophenyl)phosphinic acid, B-00489
(2-Bromophenyl)phosphonic acid, B-00490
(3-Bromophenyl)phosphonic acid, B-00491
▷(4-Bromophenyl)phosphonic acid, B-00492
(4-Bromophenyl)phosphonous acid, B-00493
(2-Bromophenyl)phosphoramidic acid, B-00494
(3-Bromophenyl)phosphoramidic acid, B-00495
(4-Bromophenyl)phosphoramidic acid, B-00496
4-Bromophenyl phosphoric acid, see M-00448
O-(4-Bromophenyl) phosphorodichloridothioate, B-00497
O-(4-Bromophenyl) thiophosphoryl dichloride, see B-00497
▷Bromophos, B-00498
▷Bromophos-Et, B-00499
(1-Bromopropyl)phosphonic acid, B-00500
(3-Bromopropyl)phosphonic acid, B-00501
▷(3-Bromopropyl)triphenylphosphonium(1+), B-00502
Bromotetraphenoxyphosphorane, B-00503
Bromotrimethylphosphonium bromide, see D-00069
N-Bromo-P,P,P-triphenylphosphine imide, in T-00626
Bromotriphenylphosphonium bromide, see D-00071
Bromo(triphenylphosphoranylidene)acetic acid, B-00504
1-Bromo-1-(triphenylphosphoranylidene)-2-butanone, B-00505
1-Bromo-3-(triphenylphosphoranylidene)-2-propanone, B-00506
▷Bucladesine, in C-00233
1,3-Butadiene-1,4-diphosphonic acid, B-00507
1,3-Butadiene-1,4-diylbisphosphonic acid, see B-00507

P-(1,3-Butadienyl)-N,N-diethylphosphonamidic acid, B-00508
(1,2-Butadienyl)phosphonic acid, B-00509
(1,3-Butadienyl)phosphonic acid, B-00510
(1,3-Butadien-1-yl)phosphonothioic acid, B-00511
P-(1,3-Butadien-1-yl)-N,N,N′,N-tetraethylphosphonothioic diamide, in B-00511
(1,3-Butadienyl)triphenylphosphonium(1+), B-00512
1,3-Butadiyne-1,4-diylbis[diphenylphosphine], see B-00223
Butafosfan, see B-00542
▷Butafox, in D-00146
▷Butamifos, B-00513
1,4-Butanediphosphonic acid, B-00514
1,4-Butanediphosphonous acid, B-00515
P,P′-1,4-Butanediylbis[bis N,N-diethylphosphonous diamide], in B-00515
1,4-Butanediylbis[dibutylphosphine], see B-00157
1,4-Butanediylbis[dicyclohexylphosphine], see B-00166
1,4-Butanediylbis[dimethylphosphine], see B-00199
1,4-Butanediylbis[diphenylphosphine], see B-00225
1,4-Butanediylbis[phenylphosphinodithioic acid], B-00516
1,4-Butanediylbisphosphine, see D-01174
1,4-Butanediylbis(phosphonic acid), see B-00514
1,4-Butanediylbis(phosphonous acid), see B-00515
1,4-Butanediylbis[triphenylphosphonium](2+), B-00517
1,4-Butanediylidenebis[triphenylphosphorane], in B-00517
S,S′-1,3-Butanediyl O-methyl phosphorodithioite, in M-00138
S,S′-1,3-Butanediyl phosphonodithioite, see M-00138
S,S′-1,3-Butanediyl phosphorochloridodithioite, in M-00138
1,2,3,4-Butanetetraphosphonic acid, B-00518
1,2,3,4-Butanetetrayltetraphosphonic acid, see B-00518
2-Butene-1,4-diphosphonic acid, B-00519
2-Butene-1,4-diylbisphosphonic acid, see B-00519
3-Butenyl N,N′-bis(2-chloroethyl)-phosphorodiamidate, in B-00125
3-Butenyl N,N-bis(2-chloroethyl)-phosphorodiamidate, in B-00126
3-Butenyldiphenylphosphine, B-00521
(2-Buten-1-yl)diphenylphosphine, B-00520
2-Butenylidenetriphenylphosphorane, B-00522
(2-Butenyl)methylphosphinic acid, B-00523
1-Butenylphosphonic acid, B-00524
2-Butenylphosphonic acid, B-00525
3-Buten-1-ylphosphonic acid, B-00526
2-Butenylphosphonothioic acid, B-00527
2-Butenylphostonic acid, see D-00501
(2-Butenyl)triphenylphosphonium(1+), B-00528
3-Buten-1-ynylphosphonic acid, B-00529
▷Butiphos, see T-00392
▷Butonate, B-00530
P-Butoxy-P,P-bis(diethylamino)-N-[bis(diethylamino)phosphinothioyl]-phosphazene, in B-00174
P-Butoxy-P,P-bis(diethylamino)-N-[bis(diethylamino)phosphinyl]-phosphazene, in B-00177
tert-Butoxybis(trifluoromethyl)phosphine, in B-00418
tert-Butoxycarbonyl diethyl phosphate, B-00531
N-tert-Butoxycarbonyl-4-diphenylphosphino-2-diphenyl-phosphinomethylpyrrolidine, B-00532
2-tert-Butoxy-5,5-dimethyl-1,3-dioxa-2-phosphacyclohexane, see B-00534
2-tert-Butoxy-4,5-dimethyl-1,3-dioxa-2-phosphacyclopentane, see B-00533
2-tert-Butoxy-4,5-dimethyl-1,3,2-dioxaphospholane, B-00533
2-tert-Butoxy-5,5-dimethyl-1,3,2-dioxaphosphorinane, B-00534
2-tert-Butoxy-1,3,2-dioxaphospholane, B-00535
O-(2-Butoxyethyl) dichlorothiophosphate, in H-00149
O-(2-Butoxyethyl) phosphorodichloridothioate, in H-00149
O-(2-Butoxyethyl) thiophosphoryl dichloride, in H-00149
P-(Butoxymethyl)-N,N-diethylphosphonamidic acid, B-00536
2-tert-Butoxy-4-methyl-1,3-dioxa-2-phosphacyclohexane, see B-00537
2-tert-Butoxy-4-methyl-1,3,2-dioxaphosphorinane, B-00537
P-tert-Butoxy-N-trimethylsilyl-P-[bis(trimethylsilyl)amino]phosphazene, in T-00867
P-tert-Butoxy-N-trimethylsilyl-P-[bis(trimethylsilyl)amino]phosphine imide, in T-00867
Butyl allylphosphinate, in P-00452
(Butylamino)chlorobis(dibutylamino)-phosphonium chloride, in P-00004
2-tert-Butylamino-5,5-dimethyl-1,3,2-dioxaphosphorinane, B-00538
2-(tert-Butylamino)-1,3,2-dioxaphospholane, B-00539
2-[(Butylamino)-1-hexenyl]phosphonic acid, B-00540
2-tert-Butylamino-4-methyl-1,3,2-dioxaphosphorinane, B-00541
[1-(Butylamino)-1-methylethyl]phosphinic acid, B-00542
1-tert-Butylaza-2,4-diphosphacyclobutane, see B-00543
1-tert-Butyl-1,2,4-azadiphosphetidine, B-00543
2-tert-Butyl-1,3-benzoxaphosphole, B-00544
▷tert-Butyl bicyclic phosphate, in B-00640
▷tert-Butyl bicyclic phosphite, see B-00640
▷tert-Butyl bicyclic thiophosphate, in B-00640
Butyl bis(1-aziridinyl)phosphinate, in B-00091
Butyl bis(4-chlorophenyl)phosphinate, in B-00139
Butyl bis(4-chlorophenyl)phosphinite, in B-00142
N-Butyl-P,P-bis(dibutylamino)-P-fluorophosphazene, in P-00004
N-Butyl-P,P-bis(dibutylamino)-P-fluorophosphine imide, in P-00004
Butyl N″-[bis(diethylamino)-phosphinothioyl]-N,N,N′,N′-tetraethylphosphorodiamidoimidate, in B-00174
Butyl N″-[bis(diethylamino)phosphinyl]-N,N,N′,N′-tetraethylphosphorodiamidoimidate, in B-00177
Butyl bis(iodomethyl)phosphinate, in B-00288
Butyl bis(2-methylphenyl)phosphinite, in B-00323
O-Butyl bis(2-methylpropyl)phosphinothioate, in B-00351
Butyl bis(2-methylpropyl)phosphinothioite, in B-00352
Butyl bis(phenylmethyl)phosphinite, in D-00058
tert-Butyl bis(trifluoromethyl)phosphinite, in B-00418
tert-Butyl bis(trifluoromethyl)-phosphinothioite, in B-00417
tert-Butyl [bis(trimethylsilyloxy)-phosphinyl]acetate, in B-00434
tert-Butylbis(trimethylsilyl)phosphine, B-00545
P-tert-Butyl-N,N′-bis(trimethylsilyl)-phosphonous diamide, in B-00617
Butyl P-(butoxymethyl)-N,N-diethylphosphonamidate, in B-00536
tert-Butyl 1,3-butylene phosphate, in B-00537
tert-Butyl 1,3-butylene phosphite, see B-00537
tert-Butyl 2,3-butylene phosphite, see B-00533
O-Butyl OO-tert-butyl ethylphosphonoperoxoate, in E-00144
Butyl butylmethylphosphinate, in B-00566
O-Butyl OO-tert-butyl methylphosphonoperoxoate, in M-00310
Butyl-tert-butylphosphinic acid, B-00546
Butyl tert-butylphosphonochloridite, in D-00130
Butyl butylphosphonochoridite, in B-00605
tert-Butylcarbonate-diethyl phosphate anhydride, see B-00531
N-Butyl-P-chloro-P,P-bis(dibutylamino)-phosphazene, in P-00004
N-Butyl-P-chloro-P,P-bis(dibutylamino)-phosphine imide, in P-00004
5-tert-Butyl-2-chloro-1,3,2-dioxaphosphorinane, B-00547
N-sec-Butyl-P-chloromethyl-phosphoramidothioic acid, see C-00101
▷(4-tert-Butyl-2-chlorophenyl) methyl methylphosphoramidate, see C-00208
P-tert-Butyl-P-chloro-N-phenyl-P-phenylaminophosphine imide, see B-00559
(4-tert-Butyl-2-chlorophenyl) 2,2,2-tribromoethyl phosphorochloridate, B-00548
5-tert-Butyl-2-chlorotetrahydro-3-phenyl-1,3,2-oxazaphosphorinane, B-00549
5-tert-Butyl-2-chlorotetrahydro-3-phenyl-2H-1,3,2-oxazaphosphorine, B-00549
tert-Butyl cyano(triphenylphosphoranylidene)-acetate, in C-00231
(4-tert-Butylcyclohexyl)dichlorophosphine, in B-00550
4-tert-Butylcyclohexylphosphonous acid, B-00550
Butyl cyclopentylmethylphosphinate, in C-00279
Butyl dibenzylphosphinite, in D-00058
Butyl dibromophosphite, see B-00627
Butyl dibutylphosphinate, in D-00109
Butyl dibutylphosphinite, in D-00121
Butyl N,N-dibutylphosphonamidate, in D-00126
Butyl dibutylphosphoramidochloridite, in D-00135
Butyl N,N′-dibutylphosphorodiamidate, in D-00146
▷Butyl dichlorophosphate, see B-00628
Butyldichlorophosphine, see B-00618
tert-Butyldichlorophosphine, see B-00619
Butyl dichlorophosphite, see B-00629
O-Butyl dichlorothiophosphate, see B-00630
S-Butyl dichlorothiophosphate, see B-00631
Butyl dichlorothiophosphite, see B-00632
Butyl dicyclohexylphosphinodithioate, in D-00212
P-tert-Butyl-N,N-diethyl-P-methylphosphinic amide, in B-00567
Se-Butyl diethylphosphinoselenoate, in D-00327
S-Butyl diethylphosphinothioate, in D-00330
N-tert-Butyl-P,P-diethylphosphinous amide, in D-00334
Butyl N,N-diethylphosphonamidate, in D-00337
OO-tert-Butyl O,O-diethyl phosphoroperoxoate, B-00551
Butyl N,N-diethyl-P-propadienylphosphonamidate, in D-00399
Butyl N,N-diethyl-P-1-propynylphosphonamidate, in D-00403
Butyldifluorophosphine, in B-00615
▷tert-Butyldifluorophosphine, B-00620
Butyl difluorophosphite, see B-00633
N-Butyl-P,P-dihexylphosphinic amide, in D-00434
7-Butyl-2,3-dihydro-1,6-diphenyl-1H-1-benzophosphonin-2,3,4,5-tetracarboxylic acid, B-00552
Butyl dihydrogen phosphate, see M-00451
sec-Butyl dihydrogen phosphate, see M-00492
tert-Butyl dihydrogen phosphate, see M-00452
S-Butyl dihydrogen thiophosphate, see B-00634
1-tert-Butyl-1,4-dihydro-4-phenyl-1,4-phospharsenin, B-00553
Butyldiiodophosphine, in B-00615
tert-Butyldiiodophosphine, in B-00616
Butyl diisobutylphosphinodithioate, in B-00348

O-Butyl diisobutylphosphinothioate, in B-00351
Butyl diisobutylphosphinothioite, in B-00352
2-tert-Butyl-1,3-diisopropoxytriphosphetane 1,3-dioxide, in D-00627
Butyl diisopropylphosphinite, in D-00648
OO-tert-Butyl O,O-diisopropyl phosphoroperoxoate, B-00554
5-tert-Butyl-2-dimethylamino-1,3,2-dioxaphosphorinane, B-00555
5-tert-Butyl-2-dimethylamino-1,3,2-oxazaphosphorinane, see B-00635
5-tert-Butyl-2-dimethylamino-3-phenyl-1,3,2-oxazaphosphorinane, see B-00636
Butyl(1,1-dimethylethyl)phosphinic acid, see B-00546
tert-Butyldimethylphenylphosphonium iodide, in B-00565
Butyl dimethylphosphinite, in D-00844
Butyl dimethylphosphinodithioate, in D-00834
tert-Butyl N,N′-dimethyl-N,N′-trimethylene phosphorodiamidite, in H-00022
tert-Butyl dimethylvinylene phosphate, in H-00123
Butyl dioctylphosphinate, in D-00958
N-Butyl-P,P-dioctylphosphinic amide, in D-00958
tert-Butyl diphenylphosphinate, in D-01039
Butyldiphenylphosphine, B-00557
sec-Butyldiphenylphosphine, see M-00369
tert-Butyldiphenylphosphine, B-00558
P-tert-Butyl-N,P-diphenylphosphinic amide, in B-00580
Butyl diphenylphosphinite, in D-01089
tert-Butyl diphenylphosphinodithioate, in D-01065
N-tert-Butyl-P,P-diphenylphosphinothioic amide, in D-01083
Butyl diphenylphosphinothioite, in D-01085
P-tert-Butyl-N,N′-diphenylphosphonamidimidic chloride, B-00559
Butyl dipropylphosphinite, in D-01209
S-Butyl dipropylphosphinothioate, in D-01208
2-tert-Butyl-1,3,2-dithiaphosphorinane, in D-00159
S-Butyl divinylphosphinothioate, in D-01259
1,3-Butylene bromophosphate, in B-00474
1,3-Butylene bromophosphite, see B-00474
2,3-Butylene bromophosphite, see B-00470
O,O-1,3-Butylene bromothiophosphate, in B-00474
O,O-2,3-Butylene bromothiophosphate, in B-00470
1,3-Butylene tert-butylphosphoramidate, in B-00541
1,3-Butylene tert-butylphosphoramidite, see B-00541
O,O-1,3-Butylene tert-butylphosphoramidoselenoate, in B-00541
1,3-Butylene chlorophosphate, see C-00094
2,3-Butylene chlorophosphate, in C-00056
1,3-Butylene chlorophosphite, see C-00093
2,3-Butylene chlorophosphite, see C-00056
O,O-1,3-Butylene chlorothiophosphate, see C-00095
O,O-2,3-Butylene chlorothiophosphate, in C-00056
1,3-Butylene cyclic phosphoric acid, see H-00156
2,3-Butylene cyclic phosphoric acid, see H-00121
O,O-2,3-Butylene cyclic phosphorothioic acid, see H-00122
S,S-1,3-Butylene diethylphosphoramidodithioate, in M-00138
N,N′-1,3-Butylene N,N′-dimethylphosphonic diamide, in H-00057
1,3-Butylene dimethylphosphoramidate, see M-00118
2,3-Butylene dimethylphosphoramidite, see D-00725
O,O-1,3-Butylene dimethylphosphoramidoselenoate, in M-00117
O,O-1,3-Butylene dimethylphosphoramidothioate, in M-00117
O,O-2,3-Butylene dimethylphosphoramidothioate, in D-00725
N,N′-1,3-Butylene-N,N′-dimethylphosphorodiamidous chloride, in H-00057
O,O-1,3-Butylene dithiophosphoric acid, see M-00017
O,O-2,3-Butylene dithiophosphoric acid, see M-00009
2,3-Butylene ethyl phosphate, in E-00031
1,3-Butylene ethyl phosphite, see E-00041
2,3-Butylene ethyl phosphite, see E-00031
S,S-1,3-Butylene O-ethyl phosphorodithioite, see E-00042
O,O-Butylene O-ethyl phosphorothioate, in E-00031
S,S-1,3-Butylene O-ethyl phosphorotrithioate, in E-00042
O,O-Butylene O-ethyl thiophosphate, in E-00031
O,O-1,3-Butylene O-ethyl thiophosphate, in E-00041
▷O,O-2,3-Butylene S-ethyl thiophosphite, see E-00189
1,3-Butylene fluorophosphite, see F-00021
2,3-Butylene fluorophosphite, see F-00015
O,O-1,3-Butylene fluoroselenophosphate, in F-00021
O,O-1,3-Butylene fluorothiophosphate, in F-00021
O,O-2,3-Butylene fluorothiophosphate, in F-00015
O,O-1,3-Butylene hydrogen dithiophosphate, see M-00017
O,O-2,3-Butylene hydrogen dithiophosphate, see M-00009
1,3-Butylene hydrogen phosphate, see H-00156
2,3-Butylene hydrogen phosphate, see H-00121
2,3-Butylene hydrogen phosphite, in D-00733
O,O-1,3-Butylene hydrogen phosphoroselenoate, see H-00157
O,O-1,3-Butylene hydrogen phosphorothioate, see H-00158
O,O-2,3-Butylene hydrogen thiophosphate, see H-00122
O,O-2,3-Butylene hydrogen thiophosphite, in D-00733
O,O-1,3-Butylene S-methyl dithiophosphate, in M-00174
1,3-Butylene methyl phosphate, see M-00045
2,3-Butylene methyl phosphate, in M-00033
1,3-Butylene methyl phosphite, see M-00044
2,3-Butylene methyl phosphite, see M-00033
1,3-Butylene methylphosphonate, in D-00734
2,3-Butylene methylphosphonate, in T-00511
1,3-Butylene methylphosphonite, see D-00734
2,3-Butylene methylphosphonite, see T-00511
O,O-1,3-Butylene methylphosphonothioate, in D-00734
S,S′-1,3-Butylene O-methyl phosphorodithioite, in M-00138
O,O-1,3-Butylene O-methyl phosphoroselenoate, see M-00046
O,O-1,3-Butylene S-methyl phosphoroselenothioate, in M-00174
O,O-1,3-Butylene O-methyl phosphorothioate, see M-00047
O,O-2,3-Butylene O-methyl phosphorothioate, in M-00033
O,O-2,3-Butylene S-methyl phosphorothioate, in D-00765
O,O-1,3-Butylene S-methyl phosphorothioite, see M-00174
O,O-2,3-Butylene S-methyl phosphorothioite, see D-00765
O,O-1,3-Butylene O-methyl selenophosphate, see M-00046
O,O-1,3-Butylene O-methyl thiophosphate, see M-00047
O,O-1,3-Butylene S-methyl thiophosphate, in M-00174
O,O-2,3-Butylene O-methyl thiophosphate, in M-00033
O,O-2,3-Butylene S-methyl thiophosphate, in D-00765
O,O-2,3-Butylene S-methyl thiophosphite, see D-00765
O,O-1,3-Butylene methylthiophosphonate, in D-00734
O,O-2,3-Butylene S-phenyl dithiophosphate, in D-00824
1,3-Butylene phenyl phosphate, in M-00191
1,3-Butylene phenyl phosphite, see M-00191
2,3-Butylene phenyl phosphite, see D-00789
1,3-Butylene phenylphosphonate, in M-00199
2,3-Butylene phenylphosphonate, in D-00794
1,3-Butylene phenylphosphonite, see M-00199
2,3-Butylene phenylphosphonite, see D-00794
1,3-Butylene phenylphosphoramidate, in M-00193
1,3-Butylene phenylphosphoramidite, see M-00193
O,O-1,3-Butylene phenylphosphoramidoselenoate, in M-00193
O,O-1,3-Butylene phenylphosphoramidothioate, in M-00193
O,O-2,3-Butylene S-phenyl phosphorodithioate, in D-00824
O,O-2,3-Butylene S-phenyl phosphorothioite, see D-00824
O,O-1,3-Butylene O-phenyl thiophosphate, in M-00191
O,O-2,3-Butylene S-phenyl thiophosphite, see D-00824
1,3-Butylene phosphite, see M-00130
2,3-Butylene phosphite, in D-00733
1,3-Butylene phosphonate, see M-00130
2,3-Butylene phosphonate, in D-00733
1,3-Butylene phosphonite, see M-00129
2,3-Butylene phosphonite, see D-00733
S,S′-1,3-Butylene phosphonodithioite, see M-00138
O,O-1,3-Butylene phosphonothioate, see M-00131
O,O-2,3-Butylene phosphonothioate, in D-00733
2,3-Butylene phosphorobromidite, see B-00470
O,O-2,3-Butylene phosphorobromidothioate, in B-00470
2,3-Butylene phosphorochloridate, in C-00056
2,3-Butylene phosphorochloridite, see C-00056
S,S′-1,3-Butylene phosphorochloridodithioite, in M-00138
O,O-Butylene phosphorochloridoselenoate, in C-00093
O,O-2,3-Butylene phosphorodithioate, see M-00009
2,3-Butylene phosphorofluoridite, see F-00015
O,O-1,3-Butylene phosphorofluoridoselenoate, in F-00021
O,O-1,3-Butylene phosphorofluoridothioate, in F-00021
O,O-2,3-Butylene phosphorofluoridothioate, in F-00015
O,O-1,3-Butylene phosphoroselenoic acid, see H-00157
O,O-1,3-Butylene phosphorothioic acid, see H-00158
1,3-Butylene phosphoryl bromide, in B-00474
1,3-Butylene phosphoryl chloride, see C-00094
2,3-Butylene phosphoryl chloride, in C-00056
O,O-1,3-Butylene thiophosphonate, see M-00131
O,O-2,3-Butylene thiophosphonate, in D-00733
OO-1,3-Butylene thiophosphoryl bromide, in B-00474
O,O-2,3-Butylene thiophosphoryl bromide, in B-00470
O,O-1,3-Butylene thiophosphoryl chloride, see C-00095
O,O-2,3-Butylene thiophosphoryl chloride, in C-00056
2,3-Butylene trimethylsilyl phosphite, see D-00930
O,O-2,3-Butylene O-trimethylsilyl phosphorothioate, in D-00930
O,O-2,3-Butylene O-trimethylsilyl thiophosphate, in D-00930
Butyl ethenylphosphinate, in V-00002
Butylethyldiphenylphosphonium iodide, in B-00557
tert-Butyl ethylene phosphate, in B-00535
tert-Butyl ethylene phosphite, see B-00535
OO-tert-Butyl O-ethyl ethylphosphonoperoxoate, in E-00144

OO-tert-Butyl *O*-ethyl methylphosphonoperoxoate, *in* M-00310
▷*S*-Butyl *O*-ethyl methylphosphonothioate, *in* E-00093
▷*O*-Butyl *S*-ethyl-*S*-(phenylmethyl) phosphorodithioate, *see* C-00204
Butyl ethylphenylphosphinothioite, *in* E-00107
OO-tert-Butyl *O*-ethyl phenylphosphonoperoxoate, *in* P-00242
tert-Butyl ethylphosphinate, *in* E-00124
tert-Butylethylphosphinic acid, B-00560
tert-Butyl ethyl phosphite, *see* B-00561
tert-Butyl ethyl phosphonate, B-00561
Butyl ethylphosphonobromidothioite, *in* E-00134
tert-Butyl ethylphosphonochloridodithioate, *in* E-00136
Butyl ethylphosphonochloridothioite, *in* E-00138
Butyl ethynylphosphinate, *in* E-00202
tert-Butyl hydrogen bis(2-chloroethyl)-phosphoramidate, *in* B-00121
Butyl hydrogen ethylphosphonate, *in* E-00129
Butyl hypophosphite, *see* B-00587
Butylidenetriphenylphosphorane, *in* B-00641
sec-Butylidenetriphenylphosphorane, *in* M-00386
OO-tert-Butyl *O*-isopropyl ethylphosphonoperoxoate, *in* E-00144
OO-tert-Butyl *O*-isopropyl methylphosphonoperoxoate, *in* M-00310
Butyl isopropylphosphinate, *in* I-00049
tert-Butylisopropylphosphinic acid, B-00562
tert-Butyl α-methoxycarbonyltriphenyl-phosphoranylidenepropanoate, *in* T-00646
5-*tert*-Butyl-2-methoxy-1,3-dioxa-2-phosphacyclohexane, *see* B-00563
5-*tert*-Butyl-2-methoxy-1,3,2-dioxaphosphorinane, B-00563
5-*tert*-Butyl-2-methyl-1,3-dioxa-2-phosphacyclohexane, *see* B-00564
5-*tert*-Butyl-2-methyl-1,3,2-dioxaphosphorinane, B-00564
tert-Butylmethyldiphenylphosphonium iodide, *in* B-00558
OO-tert-Butyl *O*-methyl ethylphosphonoperoxoate, *in* E-00144
O-Butyl *O*-methyl hydrogen thiophosphate, *see* B-00574
(*tert*-Butylmethylidyne)phosphine, *see* D-00910
OO-tert-Butyl *O*-methyl methylphosphonoperoxoate, *in* M-00310
S-Butyl *O*-methyl methylphosphonothioate, *in* M-00171
tert-Butyl methylphenylphosphinate, *in* M-00219
tert-Butylmethylphenylphosphine, B-00565
Butyl methylphenylphosphinodithioate, *in* M-00224
N-Butyl-*P*-methyl-*P*-phenylphosphinothioic amide, *in* M-00226
4-*tert*-Butyl-1-methyl-1-phenylphosphorinanium bromide, *in* B-00586
Butylmethylphosphinic acid, B-00566
tert-Butylmethylphosphinic acid, B-00567
P-tert-Butyl-*P*-methylphosphinic amide, *in* B-00567
tert-Butylmethylphosphinodithioic acid, B-00568
tert-Butylmethylphosphinodithioic acid anhydrosulfide, B-00569
O-Butyl methylphosphinothioate, *in* M-00279
tert-Butylmethylphosphinothioic acid, B-00570
tert-Butylmethylphosphinothioic anhydride, B-00571
N-tert-Butyl-*P*-methylphosphonamidic acid, B-00572
S-tert-Butyl methylphosphonochloridodithioate, *in* M-00297
▷*O*-Butyl methylphosphonochloridothioate, B-00573
S-Butyl methylphosphonochloridothioate, *in* M-00298
Butyl methylphosphonochloridothioite, *in* M-00299
O-Butyl methylphosphonofluoridothioate, *in* M-00305
Butyl methylphosphonofluoridothioite, *in* M-00306
O-Butyl *O*-methyl phosphorodithioate, B-00574
O-Butyl *O*-methyl phosphorothioic acid, *see* B-00574
Butyl methyl(trifluoromethyl)phosphinate, *in* M-00420
O-Butyl *O*-1-naphthalenyl phosphorothioate, B-00575
O-Butyl *O*-1-naphthyl hydrogen thiophosphate, *see* B-00575
tert-Butyl neopentylene phosphate, *in* B-00534
tert-Butyl neopentylene phosphite, *see* B-00534
tert-Butyloxaphosphine, B-00576
2-*tert*-Butylphenyl dihydrogen phosphate, *see* M-00449
4-*tert*-Butylphenyl dihydrogen phosphate, *see* M-00450
5-*tert*-Butyl-2-phenyl-1,3-dioxa-2-phosphacyclohexane, *see* B-00577
5-*tert*-Butyl-2-phenyl-1,3,2-dioxaphosphorinane, B-00577
2-*tert*-Butylphenyl phosphate, *see* M-00449
Butylphenylphosphinic acid, B-00578
sec-Butylphenylphosphinic acid, *see* M-00372
tert-Butylphenylphosphinic acid, B-00579
P-tert-Butyl-*P*-phenylphosphinic amide, B-00580
tert-Butylphenylphosphinodithioic acid, B-00581
2-(Butylphenylphosphino)ethanamine, *see* A-00074
tert-Butylphenylphosphinoselenoic acid, B-00582
tert-Butylphenylphosphinothioic acid, B-00583
tert-Butylphenylphosphinothioselenoic acid, B-00584
N-(*tert*-Butylphenylphosphinyl)imidazole, *in* B-00579
Butyl phenyl phosphite, *see* B-00585
Butyl phenyl phosphonate, B-00585
Butyl phenylphosphonochloridate, *in* P-00229
O-Butyl phenylphosphonodithioate, *in* P-00236
4-*tert*-Butyl-1-phenylphosphorinane, B-00586
Butyl phenyl phosphorochloridate, *in* P-00269
2-*tert*-Butylphenyl phosphorodichloridate, *in* M-00449
4-*tert*-Butylphenyl phosphorodichloridate, *in* M-00450
Butyl phosphinate, B-00587
Butylphosphine, B-00588
tert-Butylphosphine, B-00589
Butylphosphinic acid, B-00590
tert-Butylphosphinic acid, B-00591
O-Butyl phosphinothioate, B-00592
Butylphosphinothioic acid, B-00593
Butylphosphonic acid, B-00594
sec-Butylphosphonic acid, *see* M-00377
tert-Butylphosphonic acid, B-00595
Butylphosphonic acid monoanhydride, *see* D-00092
Butylphosphonic bis(diethylamide), *in* B-00596
Butylphosphonic bis(dimethylamide), *in* B-00596
P-Butylphosphonic diamide, B-00596
P-tert-Butylphosphonic diamide, B-00597
sec-Butylphosphonic dianilide, *in* M-00377
tert-Butylphosphonic dianilide, *in* B-00597
Butylphosphonic *N,N′*-dibenzyldiamide, *in* B-00596
tert-Butylphosphonic *N,N′*-di-*tert*-butyldiamide, *in* B-00597
Butylphosphonic dichloride, B-00598
sec-Butylphosphonic dichloride, *see* M-00379
tert-Butylphosphonic dichloride, B-00599
tert-Butylphosphonic difluoride, B-00600
tert-Butylphosphonic *N,N′*-dimethyldiamide, *in* B-00597
Butylphosphonisocyanatidic acid, B-00601
tert-Butylphosphon(isothiocyanatidic) acid, B-00602
Butylphosphonochloridic acid, B-00603
Butylphosphonochloridothioic acid, B-00604
Butylphosphonochloridous acid, B-00605
Butylphosphonodithioic acid, B-00606
sec-Butylphosphonodithioic acid, *see* M-00380
Butylphosphonodithious acid, B-00607
Butylphosphonofluoridic acid, B-00608
Butylphosphonoselenoic acid, B-00609
tert-Butylphosphonoselenoic acid, B-00610
Butylphosphonothioic acid, B-00611
tert-Butylphosphonothioic acid, B-00612
Butylphosphonothioic acid trimer trianhydride, *in* T-00396
tert-Butylphosphonothioic dichloride, B-00613
Butylphosphonotrithioic acid, B-00614
sec-Butylphosphonotrithioic acid, *see* M-00382
Butylphosphonous acid, B-00615
tert-Butylphosphonous acid, B-00616
tert-Butylphosphonous bis(methylamide), *in* B-00617
tert-Butylphosphonous diamide, B-00617
Butylphosphonous dichloride, B-00618
tert-Butylphosphonous dichloride, B-00619
▷*tert*-Butylphosphonous difluoride, B-00620
Butylphosphoramidic acid, B-00621
tert-Butylphosphoramidic acid, B-00622
tert-Butylphosphoramidothioic acid, B-00623
tert-Butylphosphoramidous acid, B-00624
Butyl phosphoric acid, *see* M-00451
sec-Butyl phosphoric acid, *see* M-00492
tert-Butyl phosphoric acid, *see* M-00452
tert-Butylphosphorimidic acid, B-00625
tert-Butylphosphorimidic trichloride, B-00626
Butyl phosphorodibromidite, B-00627
▷Butyl phosphorodichloridate, B-00628
Butyl phosphorodichloridite, B-00629
O-Butyl phosphorodichloridothioate, B-00630
S-Butyl phosphorodichloridothioate, B-00631
Butyl phosphorodichloridothioite, B-00632
Butyl phosphorodifluoridite, B-00633
S-Butyl phosphorothioate, B-00634
▷Butyl phosphoryl dichloride, *see* B-00628
Butylphostonic acid, *see* H-00167
Butylphostonic chloride, *in* H-00167
Butyl 2-propenylphosphinate, *in* P-00452
▷Butyl pyrophosphate, *see* T-00024
S-[(*tert*-Butylsulfinyl)methyl] *O,O*-diethyl phosphorodithioate, *in* T-00004
S-[(*tert*-Butylsulfonyl)methyl] *O,O*-diethyl phosphorodithioate, *in* T-00004
Butyl tetrabutylphosphorodiamidate, *in* T-00021
P-Butyl-*N,N,N′,N′*-tetraethylphosphonic diamide, *in* B-00596
O-Butyl tetraethylphosphorodiamidoselenoate, *in* T-00061
Se-Butyl tetraethylphosphorodiamidoselenoate, *in* T-00061
5-*tert*-Butyltetrahydro-2-dimethylamino-2*H*-1,3,2-oxazaphosphorine, B-00635
5-*tert*-Butyltetrahydro-2-dimethylamino-3-phenyl-2*H*-1,3,2-oxazaphosphorine, B-00636
5-*tert*-Butyl-2,2,4,5-tetrahydro-2,2,4-triisopropoxy-1,2,4-oxadiphosphole, B-00637
4-*tert*-Butyl-1,1,3,3-tetrakis(dimethylamino)-1λ^5,3λ^5,5λ^3-triphosphabenzene, *see* B-00638
4-*tert*-Butyl-1,1,3,3-tetrakis(dimethylamino)-1,3,5-triphosphorin, B-00638
P-Butyl-*N,N,N′,N′*-tetramethylphosphonic diamide, *in* B-00596
Butyl tetramethylphosphorodiamidate, *in* T-00218
tert-Butyl tetramethylphosphorodiamidite, *in* T-00227
▷*S-tert*-Butylthiomethyl *O,O*-diethyl phosphorodithioate, *see* T-00004
tert-Butylthiophosphonic dichloride, *see* B-00613
S-Butyl thiophosphoric acid, *see* B-00634
O-Butyl thiophosphoryl dichloride, *see* B-00630
N-tert-Butyl-*P*-(2,4,6-tri-*tert*-butylphenyl)metaphosphonamidate, *in* B-00639

N-tert-Butyl-*P*-(2,4,6-tri-*tert*-butylphenyl)metaphosphonamidoselenoate, *in* B-00639
N-tert-Butyl-*P*-(2,4,6-tri-*tert*-butylphenyl)metaphosphonamidothioate, *in* B-00639
N-tert-Butyl-*P*-(2,4,6-tri-*tert*-butylphenyl)metaphosphonimidate, B-00639
N-tert-Butyl-*P*,*P*,*P*-trichloroiminophosphorane, *see* B-00626
N-tert-Butyl-*P*,*P*,*P*-trichlorophosphazene, *see* B-00626
N-tert-Butyl-*P*,*P*,*P*-trichlorophosphine imide, *see* B-00626
N-tert-Butyl-*P*,*P*,*P*-trimethoxyiminophosphorane, *in* B-00625
N-tert-Butyl-*P*,*P*,*P*-trimethoxyphosphazene, *in* B-00625
2-*tert*-Butyltrimethylene dimethylphosphoramidate, *in* B-00555
2-*tert*-Butyltrimethylene dimethylphosphoroamidite, *see* B-00555
2-*tert*-Butyltrimethylene phenylphosphonate, *in* B-00577
2-*tert*-Butyltrimethylene phenylphosphonite, *see* B-00577
N-tert-Butyl-*P*,*P*,*P*-trimethylphosphine imide, *in* T-00531
1-*tert*-Butyl-1,3,4-trimethyl-1*H*-phospholium iodide, *in* E-00004
▷4-*tert*-Butyl-2,6,7-trioxa-1-phosphabicyclo[2.2.2]octane, B-00640
N-tert-Butyl-*P*,*P*,*P*-triphenylphosphine imide, *in* T-00626
Butyltriphenylphosphonium(1+), B-00641
sec-Butyltriphenylphosphonium(1+), *see* M-00386
tert-Butyl 2-(triphenylphosphoranylidene)-propanoate, *in* T-00660
tert-Butyl *N*,*N*,*N'*-tris(trimethylsilyl)-phosphonamidimidate, *in* T-00867
2-*tert*-Butyl-1,3,6,2-trithiaphosphacyclooctane, *see* B-00642
2-*tert*-Butyl-1,3,6,2-trithiaphosphocane, B-00642
Butyl vinylphosphinate, *in* V-00002
2-Butynediyl-1,4-bis[diphenylphosphine], *see* B-00227
1-Butynylphosphonic acid, B-00643
2-Butynylphosphonic acid, B-00644
Calyculin *A*, C-00001
▷cAMP, *see* C-00233
Candiolin, *in* F-00067
P-Carbamoyl-*N*,*N*-diethylphosphonamidic acid, *see* A-00059
▷Carbamoyl dihydrogen phosphate, C-00002
(Carbamoylmethylene)triphenylphosphorane, *in* A-00061
(Carbamoylmethyl)phosphonic acid, *see* A-00060
(Carbamoylmethyl)triphenylphosphonium(1+), *see* A-00061
▷Carbamoyl phosphate, *see* C-00002
Carbamoylphosphonic acid, C-00003
Carbamoylphosphoramidic acid, *see* A-00062
▷Carbamoyl phosphoric monoanhydride, *see* C-00002
▷Carbamyl phosphate, *see* C-00002
Carbethoxycyclopropyltriphenyl-phosphonium(1+), *see* E-00025
(1-Carbethoxyethylidene)-triphenylphosphorane, *see* E-00199
Carbethoxymethylidenetriphenylphosphorane, *see* E-00198
Carbethoxymethyltriphenylphosphonium, *see* E-00047
(2-Carbethoxyphenyl)phosphoramidic acid, C-00004
(3-Carbethoxyphenyl)phosphoramidic acid, C-00005
(4-Carbethoxyphenyl)phosphoramidic acid, C-00006
▷Carbicron, *see* D-00204
▷Carbofenothion, *see* C-00008
Carbomethoxymethylidenetriphenyl-phosphorane, *see* M-00433
Carbomethoxymethyltriphenyl-phosphonium(1+), *see* M-00055
Carbonic acid anhydride with diethyl phosphate *tert*-butyl ester, *see* B-00531
Carbonylbis[diphenylphosphine], *see* B-00245
Carbonylbis[phosphorimidic trichloride], C-00007
▷Carbophenothion, C-00008
(4-Carboxybutyl)triphenylphosphonium(1+), C-00009
(2-Carboxyethyl)methylphosphinic acid, *see* H-00161
(2-Carboxyethyl)methylphosphinic cyclic anhydride, *in* O-00053
(2-Carboxyethyl)phenylphosphinic acid, *see* H-00178
(2-Carboxyethyl)phenylphosphinic cyclic anhydride, *in* O-00053
1-Carboxyethyl phosphate, *see* P-00392
(2-Carboxyethyl)phosphinic cyclic anhydride, *see* O-00053
▷(2-Carboxyethyl)phosphonic acid, *see* P-00396
(1-Carboxyethyl)phosphoramidic acid, *see* P-00362
(1-Carboxyethyl)triphenylphosphonium(1+), C-00010
(2-Carboxyethyl)triphenylphosphonium(1+), C-00011
(7-Carboxyheptylidene)triphenylphosphorane, *in* C-00012
(7-Carboxyheptyl)triphenylphosphonium(1+), C-00012
(6-Carboxyhexyl)triphenylphosphonium(1+), C-00013
[1-Carboxy-2-(4-hydroxyphenyl)ethyl]-phosphoramidic acid, *see* P-00422
3-Carboxy-1-hydroxyphosphetane 1-oxide, *see* H-00184
(Carboxymethyl)diphenylphosphine, *see* D-01045
[(Carboxymethylene)hydrazono]-triphenylphosphorane, *see* T-00653
Carboxymethyl phosphate, *see* P-00390
(Carboxymethyl)phosphoramidic acid, *see* P-00368
(Carboxymethyl)triphenylphosphonium(1+), C-00014
(Carboxymethyl)triphenylphosphonium ethyl ester, *see* E-00047
(Carboxymethyl)triphenylphosphonium methyl ester, *see* M-00055
(5-Carboxypentyl)triphenylphosphonium(1+), C-00015
(2-Carboxyphenyl)diphenylphosphine, *see* D-01053
(3-Carboxyphenyl)diphenylphosphine, *see* D-01054
(4-Carboxyphenyl)diphenylphosphine, *see* D-01055
(2-Carboxyphenyl)methyldiphenylphosphonium iodide, *in* D-01053
(4-Carboxyphenyl)phosphoramidic acid, *see* P-00377
[1-Carboxy-2-(4-phosphonoxyphenyl)ethyl]-phosphoramidic acid, *see* D-01188
(2-Carboxy-2-propenylidene)-triphenylphosphorane, *in* C-00016
(3-Carboxy-2-propenylidene)-triphenylphosphorane, *see* T-00647
(2-Carboxy-2-propenyl)-triphenylphosphonium(1+), C-00016
(3-Carboxypropyl)phosphonic acid, *see* P-00386
(3-Carboxypropyl)triphenylphosphonium(1+), C-00017
Cardiolipin, *see* T-00241
Cardiolipin, *see* T-00246
Cardiolipin, *see* T-00294
Castro reagent, *see* H-00111
Catechol cyclic phosphate, *see* H-00108
Catechol cyclic phosphoric acid anhydride, *in* O-00087
▷8088 C.B., *see* B-00002
▷CDP-Choline, *see* C-00291
CDP-ethanolamine, *see* C-00292
CDP-glycerol, *see* C-00293
Celathion, *see* C-00197
▷Cereb, *see* C-00291
cGMP, *see* C-00234
Chairphos, *see* B-00224
Chiraphos, *see* B-00226
▷Chloracetophos, *in* A-00011
▷Chlorfenvinphos, C-00018
▷Chlormephos, C-00019
(Chloroacetyl)phosphonic acid, C-00020
N-(4-Chlorobenzenesulfonyl)-*P*-ethylphosphonimidothioic acid, *see* C-00165
2-Chloro-1,3,2-benzodioxaphosphole, C-00021
2-Chloro-4*H*-1,3,2-benzodioxaphosphorinan-4-one, C-00022
2-Chloro-1,3,2-benzodithiaphosphole, C-00023
1-Chloro-1*H*-benzo[*b*]phosphindole, *see* C-00034
(*m*-Chlorobenzylidene)triphenylphosphorane, *in* C-00136
(*o*-Chlorobenzylidene)triphenylphosphorane, *in* C-00135
(*p*-Chlorobenzylidene)triphenylphosphorane, *in* C-00137
m-Chlorobenzylphosphonic acid, *see* C-00133
o-Chlorobenzylphosphonic acid, *see* C-00132
p-Chlorobenzylphosphonic acid, *see* C-00134
(*m*-Chlorobenzyl)triphenylphosphonium(1+), *see* C-00136
(*p*-Chlorobenzyl)triphenylphosphonium(1+), *see* C-00137
o-Chlorobenzyltriphenylphosphonium(1+), *see* C-00135
▷7-Chlorobicyclo[3.2.0]hepta-2,6-dien-6-yl dimethyl phosphate, *see* H-00004
Chlorobis(2-chlorophenyl)phosphine, *in* B-00140
Chlorobis(3-chlorophenyl)phosphine, *in* B-00141
Chlorobis(4-chlorophenyl)phosphine, *in* B-00142
Chlorobis(diethylamino)(ethylamino)-phosphonium chloride, *in* P-00012
P-Chloro-*P*,*P*-bis(diethylamino)-*N*-ethylphosphazene, *in* P-00012
P-Chloro-*P*,*P*-bis(diethylamino)-*N*-ethylphosphine imide, *in* P-00012
P-Chloro-*P*,*P*-bis(diethylamino)-*N*-phenylphosphazene, *in* T-00054
P-Chloro-*P*,*P*-bis(diethylamino)-*N*-phenylphosphine imide, *in* T-00054
3-Chloro-2,4-bis(1,1-dimethylethyl)-1,2,4,3-thiadiazaphosphetidine, *see* D-00079
P-Chloro-*P*,*P*-bis(dipropylamino)-*N*-propylphosphazene, *in* P-00043
P-Chloro-*P*,*P*-bis(dipropylamino)-*N*-propylphosphine imide, *in* P-00043
2-Chloro-4,5-bis(methoxycarbonyl)-1,3,2-dioxaphospholane, *in* C-00062
Chlorobis(3-methylphenyl)phosphine, *in* B-00324
Chlorobis(4-methylphenyl)phosphine, *in* B-00325
Chlorobis(phenylmethyl)phosphine, *in* D-00058
P-Chloro-*P*,*P*-bis(trichloromethyl)-phosphazene, *see* B-00403
P-Chloro-*P*,*P*-bis(trichloromethyl)phosphine imide, *see* B-00403
N-Chloro-*P*,*P*-bis(trichloromethyl)-phosphinimidic chloride, *in* B-00403
▷Chlorobis(trifluoromethyl)phosphine, *see* B-00420
P-(4-Chlorobutyl)-*N*,*N'*-diphenylphosphonic diamide, *in* C-00026
P-(4-Chlorobutyl)-*N*-phenylphosphonamidic acid, C-00024
(4-Chlorobutyl)phosphonic acid, C-00025
P-(4-Chlorobutyl)phosphonic diamide, C-00026
(4-Chlorobutyl)phosphonic dianilide, *in* C-00026
(4-Chlorobutyl)phosphonic *N*,*N'*-dibenzylamide, *in* C-00026
(4-Chlorocarbonyl)phenylphosphonic dichloride, *in* P-00380
2-(Chlorocarbonyl)phenyl phosphorodichloridate, C-00027
3-(Chlorocarbonyl)phenyl phosphorodichloridate, C-00028
4-(Chlorocarbonyl)phenyl phosphorodichloridate, C-00029
2-Chloro-3-(2-chloroethyl)-1,3,2-oxazaphosphorinane, *see* C-00030

2-Chloro-3-(2-chloroethyl)tetrahydro-2*H*-1,3,2-oxazaphosphinine, *see* C-00030
2-Chloro-3-(2-chloroethyl)tetrahydro-2*H*-1,3,2-oxazaphosphorine, C-00030
2-Chloro-5-chloromethyl-5-methyl-1,3-dioxa-2-phosphacyclohexane, *see* C-00031
2-Chloro-5-chloromethyl-5-methyl-1,3,2-dioxaphosphinane, *see* C-00031
2-Chloro-5-chloromethyl-5-methyl-1,3,2-dioxaphosphorinane, C-00031
6-Chlorodibenzo[*d,f*][1,3,2]dioxaphosphepin, C-00032
6-Chloro-6*H*-dibenzo[*c,e*][1,2]oxaphosphorin, C-00033
5-Chloro-5*H*-dibenzo[*b,f*]phosphepin 5-oxide, *in* H-00117
5-Chloro-5*H*-dibenzophosphole, C-00034
Chlorodibenzylphosphine, *in* D-00058
▷*S*-Chloro *O,O*-dibutyl phosphorothioate, *see* D-00080
▷2-Chloro-1-(2,4-dichlorophenyl)ethenyl diethyl phosphate, *see* C-00018
▷2-Chloro-1-(2,4-dichlorophenyl)ethenyl dimethyl phosphate, *see* C-00035
▷2-Chloro-1-(2,4-dichlorophenyl)vinyl diethyl phosphate, *see* C-00018
▷2-Chloro-1-(2,4-dichlorophenyl)vinyl dimethyl phosphate, C-00035
Chlorodicyclohexylphosphine, *in* D-00214
2-Chloro-α-[(diethoxyphosphinothioyloxy)-imino]benzeneacetonitrile, *see* C-00193
Chloro(diethoxyphosphinyl)acetic acid, C-00036
▷2-Chloro-3-(diethylamino)-1-methyl-3-oxo-1-propenyl dimethyl phosphate, *see* P-00347
▷2-Chloro-2-diethylcarbamoyl-1-methylvinyl dimethyl phosphate, *see* P-00347
Chlorodiethylphosphine, *see* D-00335
2-Chloro-2-diethylphosphonoacetic acid, *see* C-00036
▷*S*-Chloro *O,O*-diethyl phosphorodithioate, *see* D-00275
▷*S*-Chloro *O,O*-diethyl phosphorothioate, *see* D-00274
5-Chloro-5,10-dihydroacridophosphine, C-00037
3-Chloro-1,5-dihydro-2,4,3-benzodioxaphosphepin, C-00038
5-Chloro-10,11-dihydro-5*H*-dibenzo[*b,f*]-phosphepin 5-oxide, *in* D-00492
5-Chloro-5,10-dihydrodibenzo[*b,e*]-phosphinine, *see* C-00037
5-Chloro-5,10-dihydrodibenzo[*b,e*]phosphorin, *see* C-00037
5-Chloro-5,6-dihydrodibenzo[*b,d*]phosphorin 5-oxide, *in* D-00504
2-Chloro-2,3-dihydro-3,5-dimethyl-1,3,4,2-oxadiazaphosphole, C-00039
2-Chloro-2,3-dihydro-4,5-dimethyl-1,2-oxaphosphole, C-00040
1-Chloro-2,5-dihydro-3,4-dimethyl-1*H*-phosphole, C-00041
1-Chloro-2,5-dihydro-3,4-dimethyl-1*H*-phosphole 1-oxide, *in* D-00493
▷*S*-[2-Chloro-1-(1,3-dihydro-1,3-dioxo-2*H*-isoindol-2-yl)ethyl] *O,O*-diethyl phosphorodithioate, *see* D-00023
2-Chloro-2,3-dihydro-3,5-diphenyl-1,3,4,2-oxadiazaphosphole, C-00042
5-Chloro-5,10-dihydro-10-methylacridophosphine, *in* D-00515
2-Chloro-2,3-dihydro-3-methyl-1,3,2-benzothiazaphosphole, C-00043
2-Chloro-2,3-dihydro-5-methyl-3-phenyl-1,3,4,2-oxadiazaphosphole, C-00044
1-Chloro-2,3-dihydro-4-methyl-1*H*-phosphole, C-00045
1-Chloro-2,5-dihydro-2-methyl-1*H*-phosphole, C-00046
1-Chloro-2,5-dihydro-3-methyl-1*H*-phosphole, C-00047
4-Chloro-1,4-dihydro-2*H*-naphth[2,1-*c*]-[1,2]oxaphosphorin, C-00048
S-6-Chloro-2,3-dihydro-2-oxooxazolo[4,5-*b*]pyridin-3-ylmethyl *O,O*-dimethyl phosphorothioate, *see* A-00146
10-Chloro-5,10-dihydrophenophosphazine, C-00049
5-Chloro-5,10-dihydro-10-phenylacridophosphine, *in* D-00548
2-Chloro-2,3-dihydro-3-phenyl-4*H*-1,3,2-benzoxazaphosphorin-4-one, C-00050
5-Chloro-5,6-dihydrophosphanthridine 5-oxide, *in* D-00504
1-Chloro-2,3-dihydro-1*H*-phosphole, C-00051
1-Chloro-2,5-dihydro-1*H*-phosphole, C-00052
2-Chloro-3,6-dihydro-2*H*-1,2-thiaphosphorin 2-sulfide, C-00053
2-Chloro-2,3-dihydro-3,3,5-trimethyl-1,2-oxaphosphole, C-00054
Chlorodiisopropylphosphine, *see* D-00650
▷*S*-Chloro *O,O*-diisopropyl phosphorodithioate, *see* D-00635
▷*S*-Chloro *O,O*-diisopropyl phosphorothioate, *see* D-00634
2-Chloro-1,3-dimethyl-1,3,2-diazaphospholidine, C-00055
2-Chloro-5,5-dimethyl-1,3-dioxa-2-oxo-2-phosphacyclohexane, *see* C-00059
2-Chloro-5,5-dimethyl-1,3-dioxa-2-phosphacyclohexane, *see* C-00058
2-Chloro-4,5-dimethyl-1,3-dioxa-2-phosphacyclopentane, *see* C-00056
2-Chloro-5,5-dimethyl-1,3,2-dioxaphosphinane, *see* C-00058
2-Chloro-5,5-dimethyl-1,3,2-dioxaphosphinane 2-oxide, *see* C-00059
2-Chloro-5,5-dimethyl-1,3,2-dioxaphosphinane 2-sulfide, *see* C-00060
2-Chloro-4,5-dimethyl-1,3,2-dioxaphospholane, C-00056
2-Chloro-4,5-dimethyl-1,3,2-dioxaphosphole, C-00057
2-Chloro-5,5-dimethyl-1,3,2-dioxaphosphorinane, C-00058
2-Chloro-5,5-dimethyl-1,3,2-dioxaphosphorinane 2-oxide, C-00059
2-Chloro-5,5-dimethyl-1,3,2-dioxaphosphorinane 2-sulfide, C-00060
2-Chloro-5,5-dimethyl-1,3-dioxa-2-thioxo-2-phosphacyclohexane, *see* C-00060
2-Chloro-5-(1,1-dimethylethyl)-1,3,2-dioxaphosphorinane, *see* B-00547
▷2-Chloro-4-(1,1-dimethylethyl)phenyl methyl methylphosphoramidate, *see* C-00208
2-Chloro-4-(1,1-dimethylethyl)phenyl 2,2,2-tribromoethyl phosphorochloridate, *see* B-00548
2-Chloro-5-(1,1-dimethylethyl)tetrahydro-3-phenyl-2*H*-1,3,2-oxazaphosphorine, *see* B-00549
2-Chloro-5,5-dimethyl-4-(1-methylethyl)-1,3,2-dioxaphosphorinane, *see* C-00090
2-Chloro-3,5-dimethyl-Δ^4-1,3,4,2-oxadiazaphospholine, *see* C-00039
2-Chloro-4,5-dimethyl-1,2-oxaphosphol-4-ene, *see* C-00040
2-Chloro-5,5-dimethyl-4-phenyl-1,3,2-dioxaphosphorinane 2-oxide, *in* H-00129
2-Chloro-3,4-dimethyl-5-phenyl-1,3,2-oxazaphospholane, *see* C-00061
2-Chloro-3,4-dimethyl-5-phenyl-1,3,2-oxazaphospholidine, C-00061
▷Chlorodimethylphosphine, *see* D-00846
7-Chloro-1-[(dimethylphosphinyl)methyl]-1,3-dihydro-5-phenyl-2*H*-1,4-benzodiazepin-2-one, *see* F-00057
1-Chloro-3,4-dimethyl-3-phospholene, *see* C-00041
▷*S*-Chloro-*O,O*-dimethyl phosphorothioate, *see* D-00716
2-Chloro-1,3-dioxa-2-oxo-2-phosphacyclohexane, *see* C-00067
2-Chloro-1,3-dioxa-2-oxo-2-phosphacyclopentane, *see* C-00064
▷2-Chloro-1,3-dioxa-2-phosphacyclohexane, *see* C-00066
2-Chloro-1,3-dioxa-2-phosphacyclotridecane, *in* D-00966
2-Chloro-1,3-dioxa-2-phosphadecalin, *see* C-00085
▷2-Chloro-1,3,2-dioxaphosphinane, *see* C-00066
2-Chloro-1,3,2-dioxaphosphinane 2-oxide, *see* C-00067
2-Chloro-1,3,2-dioxaphosphinane 2-sulfide, *see* C-00068
2-Chloro-1,3,2-dioxaphospholan-4,5-dicarboxylic acid, C-00062
2-Chloro-1,3,2-dioxaphospholane, C-00063
2-Chloro-1,3,2-dioxaphospholane 2-oxide, C-00064
2-Chloro-1,3,2-dioxaphospholane 2-sulfide, C-00065
2-Chloro-1,3,2-dioxaphosphonane, *in* D-00971
▷2-Chloro-1,3,2-dioxaphosphorinane, C-00066
2-Chloro-1,3,2-dioxaphosphorinane 2-oxide, C-00067
2-Chloro-1,3,2-dioxaphosphorinane 2-sulfide, C-00068
2-Chloro-1,3-dioxa-2-thioxo-2-phosphacyclohexane, *see* C-00068
2-Chloro-1,3-dioxa-2-thioxo-2-phosphacyclopentane, *see* C-00065
2-Chloro-4,5-diphenyl-1,3,2-dioxaphospholane, C-00069
2-Chloro-4,5-diphenyl-1,3,2-dioxaphosphole, C-00070
2-Chloro-3,5-diphenyl-Δ^4-1,3,4,2-oxadiazaphospholine, *see* C-00042
4-Chloro-2,6-diphenyl-4*H*-1,4-oxaphosphorin 4-oxide, *in* H-00140
▷Chlorodiphenylphosphine, *see* D-01091
4-Chloro-2,6-diphenyl-4*H*-1,4-thiaphosphorin 4-oxide, *in* H-00143
Chlorodipropylphosphine, *in* D-01209
▷*S*-Chloro *O,O*-dipropyl phosphorodithioate, *see* D-01201
2-Chloro-1,3,2-dithiaphosphinane, *see* C-00199
2-Chloro-1,3,2-dithiaphospholane, C-00071
Chlorodi-*o*-tolylphosphine, *in* B-00323
2-[(2-Chloroethyl)amino]-1,3,2-oxazaphosphorinane 2-oxide, *see* C-00072
2-[(2-Chloroethyl)amino]tetrahydro-2*H*-1,3,2-oxazaphosphorine 2-oxide, C-00072
▷2-Chloroethyl-*N,N*-bis(2-chloroethyl)-*N'*-(3-hydroxypropyl)phosphorodiamidate, *see* D-00015
▷3-(2-Chloroethyl)-2-[(2-chloroethyl)-amino]tetrahydro-2*H*-1,3,2-oxazaphosphorine 2-oxide, *see* I-00001
1-Chloroethyl dichlorophosphate, *see* C-00079
O-(2-Chloroethyl) dichlorothiophosphate, *see* C-00080
S-(2-Chloroethyl) dichlorothiophosphate, *see* C-00081
2-Chloroethyl dihydrogen phosphate, *see* M-00454
Chloroethylmethylphosphine, *in* E-00087
Chloroethylphenylphosphine, *in* E-00108
(1-Chloroethyl)phosphonic acid, C-00073
▷(2-Chloroethyl)phosphonic acid, C-00074
(2-Chloroethyl)phosphonic bis(diethylamide), *in* C-00075
(2-Chloroethyl)phosphonic bis(dimethylamide), *in* C-00075
P-2-Chloroethylphosphonic diamide, C-00075
(2-Chloroethyl)phosphonic dichloride, C-00076
(2-Chloroethyl)phosphonochloridic acid, C-00077
(2-Chloroethyl)phosphonothioic acid, C-00078
2-Chloroethyl phosphoric acid, *see* M-00454
1-Chloroethyl phosphorodichloridate, C-00079
2-Chloroethyl phosphorodichloridate, *in* M-00454
O-(2-Chloroethyl) phosphorodichloridothioate, C-00080
S-(2-Chloroethyl) phosphorodichloridothioate, C-00081
1-Chloroethyl phosphoryl dichloride, *see* C-00079
2-Chloroethyl phosphoryl dichloride, *in* M-00454
P-(2-Chloroethyl)-*N,N,N',N'*-tetraethylphosphonic diamide, *in* C-00075
1-(2-Chloroethyl)tetrahydro-1*H*,5*H*-[1.3.2]-diazaphospholo[2,1-*b*][1.3.2]-oxazaphosphorine 9-oxide, C-00082
3-(2-Chloroethyl)tetrahydro-4*H*-1,3,2-oxazaphosphorin-2-amine 2-oxide, *see* A-00063
N-(2-Chloroethyl)tetrahydro-4*H*-1,3,2-oxazaphosphorin-2-amine 2-oxide, *see* C-00072

▷2-[[3-(2-Chloroethyl)tetrahydro-2*H*-1,3,2-oxazaphosphorin-2-yl]amino]ethanol methanesulfonate *P*-oxide, *see* S-00026
P-(2-Chloroethyl)-*N*,*N*,*N*′,*N*′-tetramethylphosphonic diamide, *in* C-00075
O-(2-Chloroethyl) thiophosphoryl dichloride, *see* C-00080
(2-Chloroethyl)triphenylphosphonium(1+), C-00083
[2-(Chloroformyl)ethyl]methylphosphinic chloride, *in* H-00161
2-[(Chloroformyl)oxy]-ethyltriphenylphosphonium(1+), C-00084
2-(Chloroformyl)phenyl dichlorophosphate, *see* C-00027
3-(Chloroformyl)phenyl dichlorophosphate, *see* C-00028
4-(Chloroformyl)phenyl dichlorophosphate, *see* C-00029
(3-Chloroformylphenyl)phosphonic dichloride, *in* P-00379
3-(Chloroformyl)phenyl phosphorodichloridate, *see* C-00028
4-(Chloroformyl)phenyl phosphorodichloridate, *see* C-00029
2-(Chloroformyl)phenyl phosphoryl dichloride, *see* C-00027
3-(Chloroformyl)phenyl phosphoryl dichloride, *see* C-00028
4-(Chloroformyl)phenyl phosphoryl dichloride, *see* C-00029
2-Chlorohexahydro-4*H*-1,3,2-benzodioxaphosphorin, C-00085
2-Chloro-2,2,4,4,6,6-hexahydro-2-methyl-4,4,6,6-tetraphenylcyclotriphosphazene, *see* C-00086
2-Chloro-2,2,4,4,6,6-hexahydro-2-methyl-4,4,6,6-tetraphenyl-1,3,5-triaza-2,4,6-triphosphorine, C-00086
2-Chloro-2,2,4,4,6,6-hexahydro-2,4,4,6,6-pentaphenoxycyclotriphosphazene, *see* C-00087
2-Chloro-2,2,4,4,6,6-hexahydro-2,4,4,6,6-pentaphenoxy-1,3,5-triaza-2,4,6-triphosphorine, C-00087
2-Chloro-2,2,4,4,6,6-hexahydro-2,4,4,6,6-pentaphenylcyclotriphosphazene, *see* C-00088
2-Chloro-2,2,4,4,6,6-hexahydro-2,4,4,6,6-pentaphenyl-1,3,5-triaza-2,4,6-triphosphorine, C-00088
2-Chlorohexahydropyrano[3,2-*d*]-1,3,2-dioxaphosphorin, C-00089
▷3-Chloro-7-hydroxy-4-methylcoumarin *O*-ester with *O*,*O*-diethyl phosphorothioate, *see* C-00205
S-4-Chloro-*N*-isopropylcarbaniloylmethyl *O*,*O*-dimethyl phosphorodithioate, *see* A-00134
2-Chloro-4-isopropyl-5,5-dimethyl-1,3-dioxa-2-phosphacyclohexane, *see* C-00090
2-Chloro-4-isopropyl-5,5-dimethyl-1,3,2-dioxaphosphorinane, C-00090
▷*O*-(5-Chloro-1-isopropyl-1*H*-1,2,4-triazol-3-yl) *O*,*O*-diethyl phosphorothioate, *see* I-00021
2-Chloro-3-(α-methylbenzyl)-1,3,2-oxazaphosphorinane 2-oxide, *see* C-00181
▷*P*-Chloromethyl-*O*-(2-chloro-4-methylphenyl)-*N*-(1-methylpropyl) phosphonamidothioate, *in* C-00101
P-Chloromethyl-*N*-(dichlorophosphinyl)-phosphonochloridimidic acid, C-00091
▷Chloromethyl *O*,*O*-diethyl dithiophosphate, *see* C-00019
▷*S*-Chloromethyl *O*,*O*-diethyl phosphorodithioate, *see* C-00019
▷(Chloromethyl)difluorophosphine, *in* C-00115
2-Chloro-4-methyl-1,3-dioxa-2-phosphacyclohexane, *see* C-00093
2-Chloro-4-methyl-1,3-dioxa-2-phosphacyclopentane, *see* C-00092
2-Chloro-4-methyl-1,3,2-dioxaphosphinane, *see* C-00093
2-Chloro-4-methyl-1,3,2-dioxaphosphinane 2-oxide, *see* C-00094
2-Chloro-4-methyl-1,3,2-dioxaphosphinane 2-sulfide, *see* C-00095
2-Chloro-4-methyl-1,3,2-dioxaphospholane, C-00092
2-Chloro-4-methyl-1,3,2-dioxaphosphorinane, C-00093
2-Chloro-4-methyl-1,3,2-dioxaphosphorinane 2-oxide, C-00094
2-Chloro-4-methyl-1,3,2-dioxaphosphorinane 2-sulfide, C-00095
P-Chloromethyl-*N*,*N*′-diphenylphosphonic diamide, *in* C-00107
P-Chloromethyl-*N*,*N*′-diphenylphosphonothioic diamide, *in* C-00114
2-Chloro-4-methyl-1,3,2-dithiaphospholane, *in* M-00136
▷*O*-[5-Chloro-1-(1-methylethyl)-1*H*-1,2,4-triazol-3-yl] *O*,*O*-diethyl phosphorothioate, *see* I-00021
5-Chloromethyl-2-fluoro-5-methyl-1,3-dioxa-2-phosphacyclohexane, *see* C-00096
5-Chloromethyl-2-fluoro-5-methyl-1,3,2-dioxaphosphinane, *see* C-00096
5-Chloromethyl-2-fluoro-5-methyl-1,3,2-dioxaphosphorinane, C-00096
5-Chloromethyl-2-hydroxy-5-methyl-1,3-dioxa-2-phosphacyclohexane, *see* C-00097
5-Chloromethyl-2-hydroxy-5-methyl-1,3,2-dioxaphosphinane, *see* C-00097
5-Chloromethyl-2-hydroxy-5-methyl-1,3,2-dioxaphosphorinane 2-oxide, C-00097
5-Chloromethyl-2-methoxy-5-methyl-1,3-dioxa-2-phosphacyclohexane, *see* C-00098
5-Chloromethyl-2-methoxy-5-methyl-1,3,2-dioxaphosphinane, *see* C-00098
5-Chloromethyl-2-methoxy-5-methyl-1,3,2-dioxaphosphorinane, C-00098
5-Chloromethyl-5-methyl-2-phenoxy-1,3-dioxa-2-phosphacyclohexane, *see* C-00099
5-Chloromethyl-5-methyl-2-phenoxy-1,3,2-dioxaphosphinane, *see* C-00099
5-Chloromethyl-5-methyl-2-phenoxy-1,3,2-dioxaphosphorinane, C-00099
(Chloromethyl)methylphosphinic acid, C-00100
P-Chloromethyl-*N*-(1-methylpropyl)-phosphonamidothioic acid, C-00101
(2-Chloromethyl-4-nitrophenyl) dichlorophosphate, *in* M-00455
(2-Chloromethyl-4-nitrophenyl) dihydrogen phosphate, *see* M-00455
(2-Chloromethyl-4-nitrophenyl) phosphoric acid, *see* M-00455
(2-Chloromethyl-4-nitrophenyl) phosphorodichloridate, *in* M-00455
(2-Chloromethyl-4-nitrophenyl) phosphoryl dichloride, *in* M-00455
2-Chloro-3-methyl-1-oxa-3-aza-2-phosphacyclohexane, *see* C-00178
2-Chloro-3-methyl-1,3,2-oxazaphospholane, *see* C-00102
2-Chloro-3-methyl-1,3,2-oxazaphospholidine, C-00102
2-Chloro-3-methyl-1,3,2-oxazaphosphorinane, *see* C-00178
▷*O*-(3-Chloro-4-methyl-2-oxo-2*H*-1-benzoypran-7-yl) *O*,*O*-diethyl phosphorothioate, *see* C-00205
2-Chloro-5-methyl-3-phenyl-Δ^4-1,3,4,2-oxadiazaphospholine, *see* C-00044
Chloromethylphenylphosphine, *see* M-00228
(Chloromethyl)phenylphosphinic acid, C-00103
(Chloromethyl)phosphine, C-00104
3-(Chloromethylphosphinyl)propanoyl chloride, *in* H-00161
1-Chloro-2-methyl-3-phospholene, *see* C-00046
1-Chloro-3-methyl-2-phospholene, *see* C-00045
1-Chloro-3-methyl-3-phospholene, *see* C-00047
P-Chloromethylphosphonamidic acid, C-00105
(Chloromethyl)phosphonic acid, C-00106
(Chloromethyl)phosphonic bis(diethylamide), *in* C-00107
(Chloromethyl)phosphonic bis(dimethylamide), *in* C-00107
P-Chloromethylphosphonic diamide, C-00107
Chloromethylphosphonic dianilide, *in* C-00107
(Chloromethyl)phosphonic dichloride, C-00108
(Chloromethyl)phosphonic difluoride, C-00109
(Chloromethyl)phosphonic diisocyanate, C-00110
(Chloromethyl)phosphonochloridic acid, C-00111
(Chloromethyl)phosphonofluoridic acid, C-00112
(Chloromethyl)phosphonothioic acid, C-00113
P-(Chloromethyl)phosphonothioic diamide, C-00114
(Chloromethyl)phosphonothioic dianilide, *in* C-00114
(Chloromethyl)phosphonous acid, C-00115
(Chloromethyl)phosphonous dichloride, C-00116
P-Chloromethyl-*N*,*N*,*N*′,*N*′-tetraethylphosphonic diamide, *in* C-00107
P-Chloromethyl-*N*,*N*,*N*′,*N*′-tetramethylphosphonic diamide, *in* C-00107
(Chloromethyl)triphenylphosphonium(1+), C-00117
2-Chloronaphtho[1,2-*d*]-1,3,2-dioxaphosphole, C-00118
2-Chloronaphtho[2,3-*d*]-1,3,2-dioxaphosphole, C-00119
2-Chloronaphtho[1,8-*de*]-1,3,2-dioxaphosphorin, C-00120
▷*O*-(3-Chloro-4-nitrophenyl) *O*,*O*-dimethyl phosphorothioate, *see* C-00196
▷3-Chloro-4-nitrophenyl dimethyl thiophosphate, *see* C-00196
2-Chloro-1,2-oxaphosphorinane 2-oxide, *in* H-00167
2-Chloro-1,3,2-oxathiaphospholane, *in* O-00056
2-Chloro-1,3,2-oxazaphosphinane, *see* C-00180
2-Chloro-1,3,2-oxazaphosphorinane, *see* C-00180
▷*S*-(6-Chloro-2-oxobenzoxazolin-3-yl)methyl diethyl phosphorodithioate, *see* P-00340
▷*S*-[(6-Chloro-2-oxo-3(2*H*)-benzoxazolyl)-methyl] *O*,*O*-diethyl phosphorodithioate, *see* P-00340
(2-Chloro-3-oxobutyl)methylphosphinic acid, C-00121
S-[(6-Chloro-2-oxooxazolo[4,5-*b*]pyridin-3(2*H*)-yl]methyl *O*,*O*-dimethyl phosphorothioate, *see* A-00146
[(1-Chloro-2-oxo-2-phenyl)ethylidene]-triphenylphosphorane, C-00122
(3-Chloro-2-oxopropyl)-triphenylphosphonium(1+), C-00123
Chloropentaethyltriaminophosphonium chloride, *in* P-00012
1-Chloro-2,2,3,4,4-pentamethylphosphetane, C-00124
1-Chloro-2,2,3,4,4-pentamethylphosphetane 1-oxide, *in* C-00124
1-Chloro-2,2,3,4,4-pentamethylphosphetane 1-sulfide, *in* C-00124
1,1′-[(2-Chlorophenoxy)phosphinylidene]-bis(oxy)bis(1*H*-benzotriazole), *see* B-00098
4-Chlorophenyl 5-chloro-8-quinolinyl hydrogen phosphate, *see* M-00456
4-Chlorophenyl 5-chloro-8-quinolinyl phosphorochloridate, *in* M-00456
4-Chlorophenyl 2-cyanoethyl phosphorochloridate, C-00125
4-Chlorophenyl 2-cyanoethyl phosphoryl chloride, *see* C-00125
2-Chlorophenyl dichlorophosphate, *see* C-00161
4-Chlorophenyl dichlorophosphate, *see* C-00162
O-(4-Chlorophenyl) dichlorothiophosphate, *see* C-00163
S-(4-Chlorophenyl) dichlorothiophosphate, *see* C-00164
4-Chlorophenyl difluorophosphate, *in* M-00458
(4-Chlorophenyl)difluorophosphine, *in* C-00153
2-Chlorophenyl dihydrogen phosphate, *see* M-00457
4-Chlorophenyl dihydrogen phosphate, *see* M-00458
(4-Chlorophenyl)diiodophosphine, *in* C-00153
▷2-Chloro-5-phenyl-1,3,4,2-dioxazaphosphole, C-00126
(2-Chlorophenyl)diphenylphosphine, C-00127
(3-Chlorophenyl)diphenylphosphine, C-00128
(4-Chlorophenyl)diphenylphosphine, C-00129

4-Chlorophenyl *N,N′*-diphenylphosphorodiamidate, *in* M-00458
7-(2-Chlorophenyl)-4-ethoxy-3,5-dioxa-6-aza-4-phosphaoct-6-ene-8-nitrile 4-sulfide, *see* C-00193
o-Chlorophenylglyoxylonitrile oxime *O,O*-diethyl phosphorothioate, *see* C-00193
2-Chlorophenyl (4-methoxyphenyl)-phosphoramidochloridate, C-00130
(2-Chlorophenylmethylene)-triphenylphosphorane, *in* C-00135
(3-Chlorophenylmethylene)-triphenylphosphorane, *in* C-00136
(4-Chlorophenylmethylene)-triphenylphosphorane, *in* C-00137
S-[2-[4-Chlorophenyl(1-methylethyl)amino]-2-oxoethyl] *O,O*-dimethyl phosphorodithioate, *see* A-00134
(4-Chlorophenyl)methylphosphinic acid, C-00131
[(2-Chlorophenyl)methyl]phosphonic acid, C-00132
[(3-Chlorophenyl)methyl]phosphonic acid, C-00133
[(4-Chlorophenyl)methyl]phosphonic acid, C-00134
[(2-Chlorophenyl)methyl]-triphenylphosphonium(1+), C-00135
[(3-Chlorophenyl)methyl]-triphenylphosphonium(1+), C-00136
[(4-Chlorophenyl)methyl]-triphenylphosphonium(1+), C-00137
2-Chloro-3-phenyl-1,3,2-oxazaphospholane, *see* C-00138
2-Chloro-3-phenyl-1,3,2-oxazaphospholidine, C-00138
2-Chlorophenyl phenylphosphoramidochloridate, *in* P-00259
4-Chlorophenyl phenylphosphoramidochloridate, *in* P-00259
(2-Chlorophenyl)phosphine, C-00139
(3-Chlorophenyl)phosphine, C-00140
(4-Chlorophenyl)phosphine, C-00141
(2-Chlorophenyl)phosphinic acid, C-00142
(3-Chlorophenyl)phosphinic acid, C-00143
(4-Chlorophenyl)phosphinic acid, C-00144
3-(Chlorophenylphosphinyl)propanoyl chloride, *in* H-00178
▷(2-Chlorophenyl)phosphonic acid, C-00145
▷(3-Chlorophenyl)phosphonic acid, C-00146
▷(4-Chlorophenyl)phosphonic acid, C-00147
(4-Chlorophenyl)phosphonic dichloride, C-00148
(4-Chlorophenyl)phosphonic difluoride, C-00149
2-Chlorophenyl phosphono-(4-anisido) chloridate, *see* C-00130
(4-Chlorophenyl)phosphonothioic acid, C-00150
(2-Chlorophenyl)phosphonous acid, C-00151
(3-Chlorophenyl)phosphonous acid, C-00152
(4-Chlorophenyl)phosphonous acid, C-00153
(4-Chlorophenyl)phosphonous dichloride, C-00154
(2-Chlorophenyl)phosphoramidic acid, C-00155
(3-Chlorophenyl)phosphoramidic acid, C-00156
(4-Chlorophenyl)phosphoramidic acid, C-00157
(2-Chlorophenyl)phosphoramidothioic acid, C-00158
(3-Chlorophenyl)phosphoramidothioic acid, C-00159
(4-Chlorophenyl)phosphoramidothioic acid, C-00160
2-Chlorophenyl phosphoric acid, *see* M-00457
4-Chlorophenyl phosphoric acid, *see* M-00458
2-Chlorophenyl phosphorodichloridate, C-00161
4-Chlorophenyl phosphorodichloridate, C-00162
O-(4-Chlorophenyl) phosphorodichloridothioate, C-00163
S-(4-Chlorophenyl) phosphorodichloridothioate, C-00164
4-Chlorophenyl phosphorodifluoridate, *in* M-00458
2-Chlorophenyl phosphoryl dichloride, *see* C-00161
4-Chlorophenyl phosphoryl dichloride, *see* C-00162
4-Chlorophenyl phosphoryl difluoride, *in* M-00458
N-[(4-Chlorophenyl)sulfonyl]-*P*-ethoxy-*P*-ethyl-*P*-methylthiophosphine imide, *in* C-00165
N-[(4-Chlorophenyl)sulfonyl]-*P*-ethylphosphonimidothioic acid, C-00165
N-[(4-Chlorophenyl)sulfonyl]-*P*-phenylphosphonimidothioic acid, C-00166
P-2-Chlorophenyl-*N,N,N′,N′*-tetraethylphosphonous diamide, *in* C-00151
P-3-Chlorophenyl-*N,N,N′,N′*-tetraethylphosphonous diamide, *in* C-00152
P-4-Chlorophenyl-*N,N,N′,N′*-tetramethylphosphonic diamide, *in* C-00147
P-4-Chlorophenyl-*N,N,N′,N′*-tetramethylphosphonothioic diamide, *in* C-00150
P-3-Chlorophenyl-*N,N,N′,N′*-tetramethylphosphonous diamide, *in* C-00152
P-4-Chlorophenyl-*N,N,N′,N′*-tetramethylphosphonous diamide, *in* C-00153
▷*S*-[[(4-Chlorophenyl)thio]methyl] *O,O*-diethyl phosphorodithioate, *see* C-00008
▷*S*-4-Chlorophenylthiomethyl *O,O*-diethyl phosphorodithioate, *see* C-00008
▷*S*-[[(4-Chlorophenyl)thio]methyl] *O,O*-dimethyl phosphorodithioate, *see* M-00437
O-(4-Chlorophenyl) thiophosphoryl dichloride, *see* C-00163
1-Chlorophosphacyclohexane, *see* C-00168
1-Chlorophosphinane, *see* C-00168
1-Chlorophospholane, C-00167
1-Chloro-2-phospholene, *see* C-00051
1-Chloro-3-phospholene, *see* C-00052
1-Chloro-1-phosphonoethane, *see* C-00073
1-Chlorophosphorinane, C-00168
▷*S*-2-Chloro-1-phthalimidoethyl *O,O*-diethyl phosphorodithioate, *see* D-00023
(3-Chloropropyl)diphenylphosphine, C-00169
P-(3-Chloropropyl)-*N,N′*-diphenylphosphonic diamide, *in* C-00173
P-(3-Chloropropyl)-*N,N′*-di-2-propenylphosphonic diamide, *in* C-00173
(3-Chloropropyl)methyldiphenylphosphonium tetraphenylborate, *in* C-00169
(1-Chloropropyl)phosphonic acid, C-00170
(2-Chloropropyl)phosphonic acid, C-00171
(3-Chloropropyl)phosphonic acid, C-00172
P-(3-Chloropropyl)phosphonic diamide, C-00173
(3-Chloropropyl)phosphonic dianilide, *in* C-00173
P-(3-Chloropropyl)-*N,N,N′,N′*-tetraethylphosphonic diamide, *in* C-00173
2-Chloro-2,2′-spirobi[1,3,2-benzodioxaphosphole], C-00174
2-(Chlorosulfonyl)phenyl phosphorodichloridate, C-00175
4-(Chlorosulfonyl)phenyl phosphorodichloridate, C-00176
N″-Chloro-*N,N,N′,N′*-tetraethyl-*P*-(trichloromethyl)phosphonimidic diamide, *in* T-00075
2-Chloro-1,2,3,4-tetrahydro-6-methyl-3-(4-methylphenyl)-1,3,2-benzodiazaphosphorine 2-oxide, C-00177
2-Chlorotetrahydro-3-methyl-2*H*-1,3,2-oxazaphosphorine, C-00178
4-Chloro-1,2,3,4-tetrahydronaphth[2,1-*c*]-[1,2]azaphosphorine, C-00179
2-Chlorotetrahydro-2*H*-1,3,2-oxazaphosphorine, C-00180
2-Chlorotetrahydro-3-(1-phenylethyl)-2*H*-1,3,2-oxazaphosphorine 2-oxide, C-00181
2-Chlorotetrahydro-3-phenylmethyl-2*H*-1,3,2-oxazaphosphorine, *see* B-00041
1-Chloro-1,2,3,4-tetrahydrophosphinoline, *in* T-00135
6-Chloro-5,6,7,12-tetrahydro-2,5,7,10-tetramethyldibenzo[*d,g*][1,3,2]-diazaphosphocine, C-00182
2-Chlorotetrahydro-3,5,5-trimethyl-2*H*-1,3,2-oxazaphosphorine, C-00183
2-Chloro-4,4,5,5-tetrakis(trifluoromethyl)-1,3,2-dioxaphospholane, C-00184
2-Chloro-4,4,5,5-tetramethyl-1,3-dioxa-2-phosphacyclopentane, *see* C-00185
2-Chloro-4,4,5,5-tetramethyl-1,3,2-dioxaphospholane, C-00185
2-Chloro-5,6-tetramethylene-1,3,2 dioxaphosphorinane, *see* C-00085
Chlorotetraphenoxyphosphorane, C-00186
7-Chloro-6-thia-7-phosphabicyclo[3.2.1]-octane, C-00187
3-Chloro-1,3-thiaphosphetane, C-00188
▷2-Chloro-1-(2,4,5-trichlorophenyl)ethenyl dimethyl phosphate, *see* T-00038
▷2-Chloro-1-(2,4,5-trichlorophenyl)vinyl dimethyl phosphate, *see* T-00038
2-Chloro-4,4,6-trimethyl-1,3-dioxa-2-phosphacyclohexane, *see* C-00189
2-Chloro-4,4,6-trimethyl-1,3,2-dioxaphosphorinane, C-00189
2-Chloro-3,5,5-trimethyl-1-oxa-3-aza-2-phosphacyclohexane, *see* C-00183
2-Chloro-3,3,5-trimethyl-1,2-oxaphosphol-4-ene, *see* C-00054
2-Chloro-3,5,5-trimethyl-1,3,2-oxazaphosphorinane, *see* C-00183
Chlorotrimethylphosphonium chloride, *see* D-00198
3-Chloro-2,4,7-trioxa-3-phosphabicyclo[4.4.0]decane, *see* C-00089
P-Chloro-*N,P,P*-triphenylphosphazene, *in* T-00631
N-Chloro-*P,P,P*-triphenylphosphine imide, *in* T-00626
P-Chloro-*N,P,P*-triphenylphosphine imide, *in* T-00631
Chloro(triphenylphosphoranylidene)acetic acid, C-00190
2-Chloro-2-(triphenylphosphoranylidene)-acetophenone, *see* C-00122
1-Chloro-3-(triphenylphosphoranylidene)-2-propanone, C-00191
N-Chlorotriphenylphosphoranyl-*P,P,P*-triphenylphosphine imide, *in* B-00442
Chlorphonium, C-00192
Chlorphoxim, C-00193
▷Chlorpyrifos, C-00194
▷Chlorpyrifos Methyl, C-00195
▷Chlorthion, C-00196
Chlorthiophos, C-00197
▷Choline hydroxide 5′ ester with cytidine 5′-(trihydrogen pyrophosphate), inner salt, *see* C-00291
Choline hydroxide hydrogen phosphate inner salt, *see* C-00198
Choline *O*-phosphate, C-00198
Chondrosamine 1-phosphate, *see* A-00065
2-Choro-1,3,2-dithiaphosphorinane, C-00199
CI 920, *see* A-00136
▷Ciafos, *see* C-00226
CIBA 18244, *see* D-00171
▷Cidial, *see* P-00078
Ciliatine, *see* A-00079
Cinnamylidenetriphenylphosphorane, *see* T-00615
Cinnamyltriphenylphosphonium(1+), *see* T-00616
Ciodrin, *see* C-00207
▷Citicoline, *see* C-00291
CL 1565*A*, *see* A-00136
CL 1565*B*, *in* A-00136
CL 1565*T*, *in* A-00136
Clodronic acid, *see* D-00172
▷Co I, *see* C-00202
CoA, *see* C-00200
▷Cocarboxylase, *see* T-00297
▷Codecarboxylase, *see* P-00508
▷Codehydrase I, *see* C-00202
Codehydrase II, *see* C-00203
▷Codehydrogenase I, *see* C-00202

Codehydrogenase II, *see* C-00203
Coenzyme *A*, C-00200
▷Coenzyme I, C-00202
Coenzyme II, C-00203
Coenzyme F_{420}, C-00201
Coforta, *see* B-00542
▷Colite, *see* C-00291
▷Conen, C-00204
▷Co-ral, *see* C-00205
▷Coumafos, *see* C-00205
▷Coumaphos, C-00205
▷Counter, *see* T-00004
▷Cozymase, *see* C-00202
Creatine phosphate, *see* P-00364
▷Creatinolfosfate, C-00206
▷Cremart, *see* B-00513
m-Cresyl dichlorophosphate, *in* M-00486
p-Cresyl dichlorophosphate, *see* M-00257
m-Cresyl phosphate, *see* M-00486
p-Cresyl phosphate, *see* M-00487
Crotoxyphos, C-00207
▷Crufomate, C-00208
▷Curacron, *see* P-00440
▷Curamil, *see* P-00502
(α-Cyanobenzyl)diphenylphosphine oxide, *see* D-01095
(α-Cyanobenzyl)phosphonic acid, *see* C-00219
o-Cyanobenzylphosphonic acid, *see* C-00220
p-Cyanobenzylphosphonic acid, *see* C-00221
▷Cyanobis(trifluoromethyl)phosphine, *in* B-00418
▷Cyanocobalamin, *see* V-00009
Cyanodiethylphosphine, *in* D-00333
4-Cyano-2,2-dihydro-3,5-diphenyl-3*H*-1,2-azaphosphole, *in* D-00461
Cyanodiisopropylphosphine, *in* D-00648
2-Cyano-5,5-dimethyl-1,3,2-dioxaphosphorinane, C-00209
▷Cyanodimethylphosphine, *see* D-00847
Cyanodipropylphosphine, *in* D-01209
2-Cyanoethyl dihydrogen phosphate, *see* P-00398
2-Cyanoethyl diisopropylphosphoramidochloridite, *in* D-00654
2-Cyanoethyl dimethylphosphoramidochloridite, *in* D-00862
(2-Cyanoethyl)diphenylphosphine, *see* D-01073
(1-Cyanoethylidene)triphenylphosphorane, *see* T-00659
(2-Cyanoethylidene)triphenylphosphorane, *in* C-00212
Cyanoethylphenylphosphine, *in* E-00108
2-Cyanoethyl phosphate, *see* P-00398
2-Cyanoethylphosphine, *see* P-00358
(1-Cyanoethyl)phosphonic acid, C-00210
(2-Cyanoethyl)phosphonic acid, C-00211
(2-Cyanoethyl) phosphoric acid, *see* P-00398
(2-Cyanoethyl)triphenylphosphonium(1+), C-00212
▷Cyanofenphos, C-00213
(Cyanomethyl)diphenylphosphine, *see* D-01046
Cyanomethylenebis(triphenylphosphonium)(2+), C-00214
[(Cyanomethylene)triphenylphosphoranyl]-triphenylphosphonium, *in* C-00214
▷*S*-[2-[(1-Cyano-1-methylethyl)amino]-2-oxoethyl] *O*,*O*-diethyl phosphorothioate, *see* C-00232
▷*S*-[*N*-(1-Cyano-1-methylethyl)-carbamoylmethyl] *O*,*O*-diethylphosphorothioate, *see* C-00232
Cyanomethylidenetriphenylphosphorane, *see* T-00641
(Cyanomethyl)phosphonic acid, C-00215
P-Cyanomethyl-*N*,*N*,*N*′,*N*′-tetramethylphosphonic diamide, C-00216
(Cyanomethyl)triphenylphosphonium(1+), C-00217
Cyanoneocobalamin, *in* V-00009
3-Cyano-2-oxo-3-(triphenylphosphoranylidene)propanoic acid, C-00218
α-Cyanophenacylidenetriphenylphosphorane, *in* O-00083
2-Cyanophenyl dihydrogen phosphate, *see* M-00460
▷*O*-(4-Cyanophenyl) *O*,*O*-dimethyl phosphorothioate, *see* C-00226
(2-Cyanophenyl)diphenylphosphine, *see* D-01056
(3-Cyanophenyl)diphenylphosphine, *see* D-01057
(4-Cyanophenyl)diphenylphosphine, *see* D-01058
4-Cyanophenyl ethyl phenylphosphonate, *in* M-00471
▷*O*-4-Cyanophenyl *O*-ethyl phenylphosphonothioate, *see* C-00213
(Cyanophenylmethyl)phosphonic acid, C-00219
[(2-Cyanophenyl)methyl]phosphonic acid, C-00220
[(4-Cyanophenyl)methyl]phosphonic acid, C-00221
(4-Cyanophenyl)phosphinic acid, C-00222
(2-Cyanophenyl)phosphonic acid, C-00223
(3-Cyanophenyl)phosphonic acid, C-00224
(4-Cyanophenyl)phosphonic acid, C-00225
2-Cyanophenyl phosphoric acid, *see* M-00460
▷Cyanophos, C-00226
Cyanophosphoramidic dichloride, C-00227
Cyanophosphorimidic acid, C-00228
(1-Cyanopropyl)phosphonic acid, C-00229
(3-Cyanopropyl)phosphonic acid, C-00230
N-Cyano-*P*,*P*,*P*-trimethoxyphosphazene, *in* C-00228
N-Cyano-*P*,*P*,*P*-trimethoxyphosphine imide, *in* C-00228
N-Cyano-*P*,*P*,*P*-triphenoxyphosphazene, *in* C-00228
N-Cyano-*P*,*P*,*P*-triphenoxyphosphine imide, *in* C-00228
N-Cyano-*P*,*P*,*P*-triphenylphosphine imide, *in* T-00626
Cyano(triphenylphosphoranylidene)acetic acid, C-00231
Cyano(triphenylphosphoranylidene)pyruvic acid, *see* C-00218
▷Cyanox, *see* C-00226
▷Cyanthoate, C-00232
▷Cyclic AMP, C-00233
Cyclic anhydrosulfide from cyclohexylphosphonothioic acid, *see* D-00207
Cyclic anhydrosulfide from methylphosphonotrithioic acid, *see* D-00752
Cyclic dianhydrosulfide from phenylphosphonotrithoic acid, *see* D-01019
Cyclic GMP, C-00234
Cyclic 2′,3′-(hydrogen phosphate)cytidine, *see* C-00294
▷Cyclic *OO*-(methylene-*O*-phenylene)-phosphorothioate *O*-methyl ester, *see* S-00001
[1,2-Cyclobutanediylbis[methylene]]-bis[diphenylphosphine], *see* B-00249
Cyclobutylidenetriphenylphosphorane, *in* C-00235
Cyclobutyltriphenylphosphonium(1+), C-00235
Cyclo-4,5-carba-1,2,3-triphenyl-1,2,3-triphosphine, *see* T-00699
β-Cyclogeranyltriphenylphosphonium(1+), *see* T-00689
(1,3,6-Cycloheptatrien-1-ylmethylene)-triphenylphosphorane, C-00236
(2,4,6-Cycloheptatrien-1-ylmethyl)-triphenylphosphonium(1+), C-00237
Cycloheptylidenetriphenylphosphorane, *in* C-00238
Cycloheptyltriphenylphosphonium(1+), C-00238
Cyclohexanecarbonylphosphonic acid, C-00239
1,2-Cyclohexanediphosphonic acid, C-00240
N,*N*′-1,2-Cyclohexanediylbis[*P*,*P*-diphenylphosphinous amide], *see* B-00219
[1,2-Cyclohexanediylbis[methylene]]-bis[diphenylphosphine], *see* B-00250
1,2-Cyclohexanediylbisphosphonic acid, *see* C-00240
Cyclohexanephosphonic acid, *see* C-00256
1-Cyclohexenephosphonic acid, C-00241
3-Cyclohexenephosphonic acid, C-00242
4-Cyclohexenephosphonic acid, C-00243
(3-Cyclohexen-1-ylhydroxymethyl)phosphinic acid, *see* F-00061
1-Cyclohexen-1-ylphosphonic acid, *see* C-00241
2-Cyclohexen-1-ylphosphonic acid, *see* C-00242
3-Cyclohexen-1-ylphosphonic acid, *see* C-00243
Cyclohexyl benzylphosphonochloridate, *in* B-00063
Cyclohexyl benzylphosphonofluoridate, *in* B-00066
1-Cyclohexyl-1,2-bis(diphenylphosphino)-ethane, C-00244
N-Cyclohexyl-*P*,*P*-bis(trichloromethyl)-phosphinic amide, *in* B-00402
Cyclohexyl butylphosphonochloridate, *in* B-00603
Cyclohexyl butylphosphonofluoridate, *in* B-00608
Cyclohexyl *N*-cyclohexyl-*P*-ethylphosphonamidate, *in* C-00246
Cyclohexyl dichlorophosphite, *see* C-00272
Cyclohexyl dicyclohexylphosphinate, *in* D-00211
Cyclohexyl dicyclohexylphosphinodithioate, *in* D-00212
P-Cyclohexyl-*N*,*N*-diethylcyclohexylphosphinic amide, *in* C-00253
P-Cyclohexyl-*N*,*N*-diethylphosphonamidothioic chloride, *in* C-00255
Cyclohexyl dihydrogen phosphate, *see* M-00461
Cyclohexyldiiodophosphine, *in* C-00267
4-Cyclohexyl-1,3-dimethyl-1,2,4-diazaphospholidine, *in* D-00718
P-Cyclohexyl-*N*,*N*-dimethylphosphonamidothioic chloride, *in* C-00255
Cyclohexyl dimethylphosphoramidocyanidate, *in* D-00863
Cyclohexyl di-4-morpholinylphosphinate, *in* D-00935
Cyclohexyldiphenylphosphine, C-00245
N-Cyclohexyl-*P*,*P*-diphenylphosphinic amide, *in* D-01040
P-Cyclohexyl-*N*,*N*′-diphenylphosphonic diamide, *in* C-00257
P-Cyclohexyl-*N*,*N*′-diphenylphosphonothioic diamide, *in* C-00265
(1-Cyclohexyl-1,2-ethanediyl)-bis[diphenylphosphine], *see* C-00244
O-Cyclohexyl *S*-ethyl cyclohexylphosphonodithioate, *in* C-00261
N-Cyclohexyl-*P*-ethylphosphonamidic acid, C-00246
Cyclohexyl hydrogen methylphosphonate, *in* M-00288
Cyclohexylidenetriphenylphosphorane, *in* C-00273
P-Cyclohexyl-*P*-methyl-*N*-phenylphosphinothioic amide, *in* C-00248
Cyclohexylmethylphosphinic acid, C-00247
Cyclohexylmethylphosphinothioic acid, C-00248
Cyclohexyl *P*-methylphosphonohydrazidate, *in* M-00308
Cyclohexylphenylphosphine, C-00249
Cyclohexylphenylphosphinic acid, C-00250
N-Cyclohexyl-*P*-phenylphosphonamidic acid, C-00251
Cyclohexyl phenylphosphonochloridate, *in* P-00229
Cyclohexyl phenylphosphonofluoridate, *in* P-00239
▷Cyclohexylphosphine, C-00252
Cyclohexylphosphinic acid, C-00253
P-Cyclohexylphosphonamidic fluoride, C-00254
P-Cyclohexylphosphonamidothioic chloride, C-00255
Cyclohexylphosphonic acid, C-00256
Cyclohexylphosphonic acid monoanhydride, *see* D-00206
Cyclohexylphosphonic bis(dimethylamide), *in* C-00257
P-Cyclohexylphosphonic diamide, C-00257
Cyclohexylphosphonic dianilide, *in* C-00257
Cyclohexylphosphonic dichloride, C-00258
Cyclohexylphosphonochloridic acid, C-00259
Cyclohexylphosphonochloridothioic acid, C-00260
Cyclohexylphosphonodithioic acid, C-00261

Cyclohexylphosphonodithious acid, C-00262
Cyclohexylphosphonoselenoic acid, C-00263
Cyclohexylphosphonothioic acid, C-00264
Cyclohexylphosphonothioic acid trimer trianhydride, *in* T-00430
Cyclohexylphosphonothioic bis(diethylamide), *in* C-00265
P-Cyclohexylphosphonothioic diamide, C-00265
Cyclohexylphosphonothioic dianilide, *in* C-00265
Cyclohexylphosphonothioic dichloride, C-00266
Cyclohexylphosphonous acid, C-00267
Cyclohexylphosphonous dichloride, C-00268
Cyclohexylphosphoramidic acid, C-00269
Cyclohexylphosphoramidochloridothioic acid, C-00270
Cyclohexylphosphoramidothioic acid, C-00271
Cyclohexyl phosphoric acid, *see* M-00461
Cyclohexyl phosphorodichloridite, C-00272
Cyclohexyl propylphosphonochloridate, *in* P-00476
Cyclohexyl propylphosphonofluoridate, *in* P-00479
P-Cyclohexyl-*N*,*N*,*N'*,*N'*-tetraethylphosphonothioic diamide, *in* C-00265
P-Cyclohexyl-*N*,*N*,*N'*,*N'*-tetraethylphosphonous diamide, *in* C-00267
P-Cyclohexyl-*N*,*N*,*N'*,*N'*-tetramethylphosphonic diamide, *in* C-00257
Cyclohexylthiophosphonic dichloride, *see* C-00266
Cyclohexyltriphenylphosphonium(1+), C-00273
2,4-Cyclopentadien-1-ylidenetriphenylphosphorane, C-00274
Cyclopentadienylphosphonous acid, C-00275
Cyclopentamethylenephenylphosphine, *see* P-00265
Cyclopentanephosphonic acid, C-00276
Cyclopentatrimethylene phosphoric acid, *see* H-00034
1-Cyclopentenephosphonic acid, C-00277
3-Cyclopentenephosphonic acid, C-00278
1-Cyclopenten-1-ylphosphonic acid, *see* C-00277
2-Cyclopenten-1-ylphosphonic acid, *see* C-00278
Cyclopentylidenetriphenylphosphorane, *in* C-00282
Cyclopentylmethylphosphinic acid, C-00279
Cyclopentylphosphonic acid, *see* C-00276
Cyclopentylphosphonothioic acid, C-00280
Cyclopentylphosphonous acid, C-00281
Cyclopentyltriphenylphosphonium(1+), C-00282
▷Cyclophosphamide, C-00283
1,2-Cyclopropanediylbis[diphenylphosphine], *see* B-00230
Cyclopropylidenebis[diphenylphosphine], *see* B-00229
Cyclopropylidenetriphenylphosphorane, C-00284
[α-Cyclopropyl(methoxycarbonyl)methyl]-triphenylphosphonium(1+), C-00285
(Cyclopropylmethylene)triphenylphosphorane, *in* C-00286
(Cyclopropylmethyl)triphenyl-phosphonium(1+), C-00286
Cyclopropyltriphenylphosphonium(1+), C-00287
Cycphos, *see* C-00244
CyDIOP, *see* B-00168
▷Cyflee, *see* D-00920
▷Cygon, *see* D-00671
▷Cyolane, *see* P-00342
▷Cyscholin, *see* C-00291
Cytidine 2'-(dihydrogen phosphate), C-00288
Cytidine 3'-(dihydrogen phosphate), C-00289
▷Cytidine 5'-(dihydrogen phosphate), C-00290
▷Cytidine diphosphate choline, C-00291
Cytidine diphosphate ethanolamine, C-00292
Cytidine diphosphate glycerol, C-00293
Cytidine 2',3'-phosphate, C-00294
▷Cytidine 5'-(trihydrogen diphosphate) hydroxide mono[2-(trimethylammonio)-ethyl ester, *see* C-00291
Cytidine 5'-(trihydrogen diphosphate) mono(2-aminoethyl) ester, *see* C-00292
Cytidine 5'-(trihydrogen diphosphate) mono(2,3-dihydroxypropyl) ester(*S*), *see* C-00293
Cytidine 5'-(trihydrogen pyrophosphate) mono(2-aminoethyl) ester, *see* C-00292
Cytidylic acid *a*, *see* C-00288
Cytidylic acid *b*, *see* C-00289
2'-Cytidylic acid, *see* C-00288
3'-Cytidylic acid, *see* C-00289
▷5'-Cytidylic acid, *see* C-00290
▷Cytimun, *see* S-00026
Cytosylic acid *a*, *see* C-00288
Cytosylic acid *b*, *see* C-00289
▷Cytoxan, *see* C-00283
▷Cytrolane, *see* M-00004
▷Dasanit, *see* F-00006
DBDECMP, *in* D-00268
▷D.D.V.P., *see* D-00203
7,8,10,11,13,14,16,17,19,24-Decahydro-5*H*-dibenzo[*o*,*r*][1,4,7,10,13,17]-pentaoxaphosphacycloeicosin, D-00001
Decahydro-1,5-dihydroxy-1*H*-1,5-benzodiphosphepin 1,5-dioxide, D-00002
Decahydro-1,4-dihydroxy-1,4-benzodiphosphorin 1,4-dioxide, D-00003
5,6,7,8,9,14,15,16,17,18-Decahydro-9,14-dimethyl-5,18-diphenyldibenzo[*b*,*i*]-[1,4,8,11]diphosphadiarsacyclotetradecin, D-00004
1,3,4,5,5,6,7,8,9,9*a*-Decahydro-5-methylene-2*H*-phosphinolizine, D-00005
5,6,7,8,9,14,15,16,17,18-Decahydro-5,9,14,18-tetraphenyldibenzo[*b*,*i*]-[1,4,8,11]tetraphosphacyclotetradecin, D-00006
Decahydro-1,3,5-trihydroxy-1*H*-1,3,5-benzotriphosphepin 1,3,5-trioxide, D-00007
Decamethylenebisphosphine, *see* D-01175
Decamethylenediphosphonic acid, *see* D-00008
Decamethylene phosphorochloridite, *in* D-00966
Decamethylene phosphorofluoridite, *in* D-00966
1,10-Decanediphosphonic acid, D-00008
1,10-Decanediylbisphosphine, *see* D-01175
1,10-Decanediylbisphosphonic acid, *see* D-00008
Decyl dihydrogen phosphate, *see* M-00462
Decylidenetriphenylphosphorane, *in* D-00014
Decylmethylphosphinic acid, D-00009
Decylphenylphosphinic acid, D-00010
Decylphosphine, D-00011
▷Decylphosphonic acid, D-00012
2-Decylphosphonic acid, *see* M-00181
4-Decylphosphonic acid, *see* P-00467
Decylphosphoramidic acid, D-00013
Decyl phosphoric acid, *see* M-00462
Decyltriphenylphosphonium(1+), D-00014
DEDECMP, *in* D-00268
▷DEF, *see* T-00392
▷Defosfamide, D-00015
3,4-Dehydro-1-phenylphosphinane, *see* T-00134
4,5-Dehydro-1,2,3-triphospholane, *see* D-00615
▷Delnav, *see* D-00974
Demanyl phosphate, D-00016
Demephion, *in* D-00017
Demephion-*O*, D-00017
▷Demephion-*S*, D-00018
▷Demeton, *in* D-00019
▷Demeton-*O*, D-00019
Demeton-*S*, *see* D-00294
▷Demeton-*S*-methyl, D-00020
▷Demeton-*S*-methyl sulfone, D-00021
N-[*N*-[[(6-Deoxy-α-L-mannopyranosyl)oxy]-hydroxyphinyl]-L-leucyl]-L-tryptophan, *see* P-00402
▷Dermaphos, *see* F-00003
▷Desbromoleptophos, *in* M-00243
Desmethyl salioxon, *see* H-00109
(Diacetylmethylene)triphenylphosphorane, *see* T-00654
(Diacetylmethyl)triphenylphosphonium(1+), D-00022
▷Dialifor, *see* D-00023
▷Dialifos, D-00023
Diallyl allylphosphonate, *in* P-00454
Diallyl chloromethylphosphonate, *in* C-00106
Diallyl (3-chloropropyl)phosphonate, *in* C-00172
N,*N'*-Diallyl-*P*-(3-chloropropyl)phosphonic diamide, *in* C-00173
O,*O*-Diallyl chlorothiophosphate, *see* D-01196
O,*O*-Diallyl dithiophosphoric acid, *see* D-01197
O,*O*-Diallyl hydrogen dithiophosphate, *see* D-01197
O,*O*-Diallyl hydrogen phosphorothioate, *see* D-01198
O,*O*-Diallyl hydrogen thiophosphate, *see* D-01198
Diallyl methylphosphonate, *in* M-00288
Diallyl phenyl phosphate, *see* P-00118
Diallylphenylphosphine, *see* P-00119
Diallyl phenylphosphonate, *in* P-00221
Diallyl phenylphosphonite, *in* P-00251
O,*O*-Diallyl phenylphosphonothioate, *in* P-00244
Diallyl phosphate, *see* D-01192
Diallylphosphinic acid, *see* D-01193
▷Diallyl phosphite, *see* D-01194
▷Diallyl phosphonate, *see* D-01194
Diallylphosphoramidic acid, *see* D-01195
Diallyl phosphorochloridate, *in* D-01192
O,*O*-Diallyl phosphorochloridothioate, *see* D-01196
O,*O*-Diallyl phosphorodithioate, *see* D-01197
Diallyl phosphorofluoridate, *in* D-01192
O,*O*-Diallyl phosphorothioate, *see* D-01198
O,*O*-Diallyl thiophosphoric acid, *see* D-01198
O,*O*-Diallyl thiophosphoryl chloride, *see* D-01196
Diallyl (trichloromethyl)phosphonate, *in* T-00411
▷Diamidafos, D-00024
2,2-Diamino-2,2,4,4,6,6-hexahydro-4,4,6,6-tetraphenylcyclotriphosphazene, *see* D-00025
2,2-Diamino-2,2,4,4,6,6-hexahydro-4,4,6,6-tetraphenyl-1,3,5-triaza-2,4,6-triphosphorine, D-00025
2-Diaminomethylene-1,1-di-*tert*-butyldiphosphine, *see* D-00090
▷*N*-(Diaminophosphinyl)-4-fluorobenzamide, *see* F-00051
N-(Diaminophosphinyl)-2-methylbenzamide, *see* T-00323
N-(Diaminophosphinyl)-*o*-toluamide, *see* T-00323
▷*S*-[(4,6-Diamino-*sym*-triazin-2-yl)methyl] *O*,*O'*-dimethyl dithiophosphate, *see* M-00003
▷*S*-[(4,6-Diamino-1,3,5-triazin-2-yl)methyl] *O*,*O*-dimethyl phosphorodithioate, *see* M-00003
Diamyl amylphosphonate, *in* P-00050
Diamyl amylphosphonite, *in* P-00052
Diamyl chlorophosphite, *see* D-00983
O,*O*-Diamyl chlorothiophosphate, *see* D-00984
O,*O*-Diamyl diselenophosphate, *see* D-00985
O,*O*-Diamyl dithiophosphoric acid, *see* D-00986
▷Diamyl fluorophosphate, *in* D-00979
Diamyl hydrogen phosphate, *see* D-00979
Diamylphenylphosphine, *see* D-00978
Diamyl phenylphosphoramidate, *in* D-00979
Diamyl phosphate, *see* D-00979
Diamylphosphine, *see* D-00980
Diamylphosphinic acid, *see* D-00981
Diamyl phosphite, *see* D-00982
Diamyl phosphonate, *see* D-00982
Diamyl phosphoric acid, *see* D-00979
Diamyl phosphorochloridate, *in* D-00979
Diamyl phosphorochloridite, *see* D-00983
O,*O*-Diamyl phosphorodiselenoate, *see* D-00985
O,*O*-Diamyl phosphorodithioate, *see* D-00986
▷Diamyl phosphorofluoridate, *in* D-00979
Diamyl phosphoryl chloride, *in* D-00979
▷Diamyl phosphoryl fluoride, *in* D-00979
O,*O*-Diamyl thiophosphoryl chloride, *see* D-00984
1,4:3,6-Dianhydro-2,5-deoxy-2,5-bis(diphenylphosphino)iditol, D-00026
▷Dianilido phosphorochloridate, *see* D-01126
▷Dianilido phosphoryl fluoride, *see* D-01127

▷Dianilinochlorophosphine oxide, *see* D-01126
Di-*m*-anisylphosphine, *see* B-00297
Di-*o*-anisylphosphine, *see* B-00296
Di-*p*-anisylphosphine, *see* B-00298
Di-*m*-anisylphosphinic acid, *see* B-00300
Di-*o*-anisylphosphinic acid, *see* B-00299
Di-*p*-anisylphosphinic acid, *see* B-00301
Diarachidonoylphosphatidylethanolamine, *in* G-00018
Diarachidonylphosphatidylcholine, *in* G-00017
▷Diarphos, *see* B-00212
▷Diazinon, D-00027
▷Diazitol, *see* D-00027
α-Diazobenzylphosphonic acid, *see* D-00037
(Diazocyclohexylmethyl)phosphonic acid, D-00028
2-Diazo-2-(diphenylphosphino)-1-phenylethanone, D-00029
2-Diazo-2-(diphenylphosphinothioyl)-1-phenylethanone, *in* D-00029
Diazo(diphenylphosphinyl)acetic acid, D-00030
2-Diazo-2-(diphenylphosphinyl)-1-phenylethanone, *in* D-00029
1-Diazo-1-(diphenylphosphinyl)-2-propanone, D-00031
(1-Diazoethyl)phenylphosphinic acid, D-00032
(1-Diazoethyl)phosphonic acid, D-00033
(Diazomethyl)diphenylphosphine oxide, D-00034
▷Diazomethylphosphonic acid, D-00035
(Diazophenylmethyl)diphenylphosphine oxide, D-00036
(Diazophenylmethyl)phosphonic acid, D-00037
Dibenzo[*d,f*][1,3,2]dioxaphosphepin, D-00038
2,3:22,23-Dibenzo-5,8,11,14,17,20-hexaoxa-1-phospha-2,22-cyclotricosadiene, *see* D-01261
Dibenzo(methylphosphinidene)-20-crown-7, *see* O-00019
6*H*-Dibenzo[*c,e*][1,2]oxaphosphorin, D-00039
2,3:19,20-Dibenzo-5,8,11,14,17-pentaoxa-1-phospha-2,19-cycloeicosadiene, *see* D-00001
Dibenzo[*b,e*]phosphinine, *see* A-00031
5*H*-Dibenzophosphole, D-00040
9*H*-Dibenzo[2,3:4,5]phospholo[1,2-*f*]-phosphanthridene, D-00041
Dibenzo[*b,e*]phosphorin, *see* A-00031
1,2,7,8-Dibenzo-9,12,15-trioxa-3,6-diphospha-3,6-diphenyl-15-crown-5, *see* O-00015
1,2-Dibenzoyl-1,2-bis(triphenylphosphoranylidene)ethane, *see* D-00997
(1,2-Dibenzoylethylidene)-triphenylphosphorane, *see* D-01162
(Dibenzoylmethylene)triphenylphosphorane, *see* D-01166
Dibenzoyl phosphate, D-00042
Dibenzoyl(triphenylphosphoranylidene)-methane, *see* D-01166
Dibenzyl 1-adamantylphosphoramidate, *in* A-00036
Dibenzyl amidophosphate, *in* D-00053
Dibenzyl [1-(aminocarbonyl)ethyl]-phosphoramidate, *in* P-00362
Dibenzyl (aminoiminomethyl)phosphoramidate, *in* P-00370
Dibenzyl [[[(4-amino-4-methoxycarbonyl)-butyl]amino](imino)methyl]-phosphoramidate, *in* P-00363
Dibenzyl benzoylphosphoramidate, *in* B-00036
Dibenzyl [1-(benzyloxycarbonyl)ethyl]-phosphoramidate, *in* P-00362
Dibenzyl [1-benzyloxycarbonyl-2-(4-hydroxyphenyl)ethyl]phosphoramidate, *in* P-00422
Dibenzyl benzylphosphoramidate, *in* B-00069
Dibenzylbromophosphine, *in* D-00058
Dibenzyl [1-carboxy-2-[4-(dibenzyloxyphosphinyloxy)phenyl]-ethyl]phosphoramidate, *in* D-01188
▷Dibenzyl chlorophosphate, *see* D-00060
▷Dibenzyl chlorophosphonate, *see* D-00060
Dibenzyl cyclohexylphosphoramidate, *in* C-00269
Dibenzyl 2-[(dibenzyloxyphosphinyl)amino]-pentanedioate, *in* P-00367
P,P-Dibenzyl-*N,N*-diethylphosphinous amide, *in* D-00058
Dibenzyl diethylphosphoramidate, *in* D-00343
1,2-Dibenzyl-1,2-dimethyldiphosphine, D-00043
P,P-Dibenzyl *P′,P′*-diphenyl diphosphate, D-00044
1,6-Dibenzyl-1,6-diphenyl-1,6-diphosphecanium dibromide, *in* D-01171
1,2-Dibenzyl-1,2-diphenyldiphosphine, D-00045
N,N-Dibenzyl-*P,P*-diphenylphosphinic amide, *in* D-01040
P,P-Dibenzyl *P′,P′*-diphenyl pyrophosphate, *see* D-00044
1,4-Dibenzyl-1,4-diphosphacycloheptane, *see* D-00047
1,4-Dibenzyl-1,4-diphosphacyclononane, *see* D-00050
1,5-Dibenzyl-1,5-diphosphacyclononane, *see* D-00051
1,4-Dibenzyl-1,4-diphosphacyclooctane, *see* D-00048
1,5-Dibenzyl-1,5-diphosphacyclooctane, *see* D-00049
1,6-Dibenzyl-1,6-diphosphacycloundecane, D-00046
1,6-Dibenzyl-1,6-diphosphecane, *in* D-01171
1,4-Dibenzyl-1,4-diphosphepane, D-00047
1,4-Dibenzyl-1,4-diphosphocane, D-00048
1,5-Dibenzyl-1,5-diphosphocane, D-00049
1,4-Dibenzyl-1,4-diphosphonane, D-00050
1,5-Dibenzyl-1,5-diphosphonane, D-00051
O,O-Dibenzyl dithiophosphoric acid, *see* D-00063
Dibenzyl [1-ethoxycarbonyl-2-(4-hydroxybenzyl)ethyl]phosphoramidate, *in* P-00422
Dibenzyl ethylphosphonate, *in* E-00129
Dibenzyl ethylphosphonite, *in* E-00153
Dibenzyl ethylphosphonodithioite, *in* E-00141
N,N′-Dibenzyl-*P*-ethylphosphonothioic diamide, *in* E-00149
O,O-Dibenzyl hydrogen dithiophosphate, *see* D-00063
Dibenzyl hydrogen phosphate, *see* D-00053
$N^2,N^{2'}$-Dibenzylidene-*P*-phenylphosphonothioic dihydrazide, *in* P-00249
Dibenzyl [1-(methoxycarbonyl)ethyl]-phosphoramidate, *in* P-00362
Dibenzyl methylphenylphosphoramidate, *in* M-00250
Dibenzyl (4-methylphenyl)phosphoramidate, *in* M-00253
Dibenzyl methylphosphonodithioite, *in* M-00302
Dibenzyl 2-naphthylphosphoramidate, *in* N-00029
[Di(benzyloxy)phosphinyl]acetic acid, *see* B-00389
2-[(Dibenzyloxyphosphinyl)oxy]-2-propenoamide, *in* P-00366
▷Dibenzylphenylphosphine, D-00052
P,P-Dibenzyl-*N*-phenylphosphinic amide, *in* D-00055
Dibenzyl phenylphosphonite, *in* P-00251
N,N′-Dibenzyl-*P*-phenylphosphonothioic diamide, *in* P-00246
Dibenzyl phenylphosphoramidate, *in* D-00053
Dibenzyl phosphate, D-00053
Dibenzylphosphine, D-00054
Dibenzylphosphinic acid, D-00055
Dibenzylphosphinic amide, *in* D-00055
P,P-Dibenzylphosphinic anilide, *in* D-00055
Dibenzylphosphinodithioic acid, D-00056
Dibenzylphosphinothioic acid, D-00057
Dibenzylphosphinous acid, D-00058
▷Dibenzyl phosphite, *see* D-00059
▷Dibenzyl phosphonate, D-00059
6,6-Dibenzyl-6-phosphonia-2-oxaadamantane chloride, *in* B-00055
Dibenzylphosphonium iodide, *in* D-00054
Dibenzyl phosphoramidate, *in* D-00053
Dibenzyl phosphoric acid, *see* D-00053
▷Dibenzyl phosphorochloridate, D-00060
N,N′-Dibenzylphosphorodiamidic acid, D-00061
N,N′-Dibenzylphosphorodiamidothioic acid, D-00062
O,O-Dibenzyl phosphorodithioate, D-00063
▷Dibrom, *see* N-00001
Di-*p*-bromobenzyl phenylphosphoramidate, *in* B-00109
Di-*p*-bromobenzyl phosphite, *see* B-00110
Di-*p*-bromobenzyl phosphonate, *see* B-00110
Di-*p*-bromobenzyl phosphoric amide, *in* B-00109
Dibromo(bromomethyl)phosphine, *see* B-00476
Dibromo-4-bromophenylphosphine, *in* B-00493
Dibromobutylphosphine, *in* B-00615
Dibromo-*tert*-butylphosphine, *in* B-00616
Dibromo(chloromethyl)phosphine, *in* C-00115
Dibromo(4-chlorophenyl)phosphine, *in* C-00153
Dibromocyclohexylphosphine, *in* C-00267
▷1,2-Dibromo-2,2-dichloroethyl dimethyl phosphate, *see* N-00001
Dibromo(dichloromethyl)phosphine, *in* D-00181
1,1-Dibromo-1,1-dihydro-2,4,6-triphenylphosphorin, *in* D-00613
Dibromoethylphosphine, *in* E-00153
Dibromo(4-fluorophenyl)phosphine, *in* F-00042
Dibromoisopropylphosphine, *in* I-00056
(Dibromomethylene)triphenylphosphorane, *in* D-00065
Dibromomethylphosphine, *see* M-00321
(Dibromomethyl)phosphonic acid, D-00064
P-(Dibromomethyl)-*N,N,N′,N′*-tetramethylphosphonic diamide, *in* D-00064
(Dibromomethyl)triphenylphosphonium(1+), D-00065
Dibromo-1-naphthylphosphine, *in* N-00025
Dibromooctylphosphine, *in* O-00044
N-(2,4-Dibromophenyl)-*N′,N″*-diphenylphosphoric triamide, *in* D-00067
Dibromophenylphosphine, *see* P-00253
(2,5-Dibromophenyl)phosphonic acid, D-00066
(2,4-Dibromophenyl)phosphoramidic acid, D-00067
(1,2-Dibromo-1-phenylpropyl)phosphonic acid, D-00068
Dibromo-*m*-tolylphosphine, *in* M-00247
Dibromo-*o*-tolylphosphine, *in* M-00246
Dibromo-*p*-tolylphosphine, *in* M-00248
Dibromo(trifluoromethyl)phosphine, *in* T-00482
Dibromotrimethylphosphorane, D-00069
P,P-Dibromo-*N*-trimethylsilyl-*P*-[bis(trimethylsilyl)amino]phosphazene, *in* T-00868
P,P-Dibromo-*N*-trimethylsilyl-*P*-[bis(trimethylsilyl)amino]phosphine imide, *in* T-00868
Dibromotriphenoxyphosphorane, D-00070
Dibromotriphenylphosphorane, D-00071
Di-3-butenylphenylphosphine, D-00072
(Dibutoxyphosphinyl)acetamide, *in* D-00073
(Dibutoxyphosphinyl)acetic acid, D-00073
Dibutoxyphosphinyl fluoride, *see* D-00153
1-(Dibutoxyphosphinyl)phosphonate, *in* D-00012
O,O-Dibutyl acetylphosphonothioate, *in* A-00028
Di-*tert*-butyl allenylphosphonate, *in* P-00441
Dibutyl allylphosphonate, *in* P-00454
S,S-Dibutyl amidodithiophosphate, *in* D-00152
▷Dibutyl amidophosphate, *see* D-00133
Di-*tert*-butyl amidophosphate, *in* D-00105
Dibutyl [(2-aminophenyl)methyl]phosphonate, *in* A-00106
▷Dibutyl benzylphosphonate, *in* B-00060
Di-*tert*-butyl benzylphosphoramidate, *in* B-00069
1,3-Di-*tert*-butyl-2,4-bis(dimethylamino)-1,3,2,4-diazadiphosphetidine, D-00074
Di-*tert*-butyl bromophosphate, *see* D-00138
Dibutyl bromophosphite, *see* D-00139
4,4-Dibutyl-1-*tert*-butyl-1,4-dihydro-1,4-phosphastannin, D-00075
▷Dibutyl butylphosphonate, D-00076
Dibutyl *sec*-butylphosphonate, *in* M-00377
Dibutyl butylphosphonite, *in* B-00615
S,S-Dibutyl butylphosphonodithioate, *in* B-00606
Dibutyl butylphosphonodithioite, *in* B-00607
O,O-Dibutyl butylphosphonothioate, *in* B-00611

Dibutyl butylphosphonotrithioate, *in* B-00614
Dibutyl butylphosphoramidate, *in* B-00621
Dibutyl *tert*-butylphosphoramidate, *in* B-00622
N,P-Di-*tert*-butylcarboimidophosphene, D-00077
Dibutyl (4-chlorobutyl)phosphonate, *in* C-00025
3,5-Di-*tert*-butyl-2-chloro-2,3-dihydro-1,3,2-oxazaphosphole, D-00078
S,S-Dibutyl chlorodithiophosphate, *see* D-00144
Dibutyl (chloromethyl)phosphonite, *in* C-00115
Dibutyl chlorophosphate, *see* D-00140
Di-*sec*-butyl chlorophosphate, *in* B-00341
Di-*tert*-butyl chlorophosphate, *see* D-00141
Dibutylchlorophosphine, *in* D-00121
Di-*tert*-butylchlorophosphine, *see* D-00125
Dibutyl chlorophosphite, *see* D-00142
Di-*tert*-butyl chlorophosphite, *see* D-00143
O,O-Dibutyl chloroselenophosphate, *in* D-00154
2,4-Di-*tert*-butyl-3-chloro-1,2,4,3-thiadiazaphosphetidine, D-00079
O,O-Dibutyl chlorothiophosphate, *see* D-00145
▷Dibutyl chlorothiophosphonate, D-00080
Dibutyl 4-cyanobenzylphosphonate, *in* C-00221
Dibutyl [(4-cyanophenyl)methyl]phosphonate, *in* C-00221
Dibutylcyanophosphine, *in* D-00121
O,O-Dibutyl cyclohexylphosphonothioate, *in* C-00264
Di-*tert*-butyl cyclohexylphosphoramidate, *in* C-00269
Dibutyl 2,4-cyclopentadienylphosphonite, *in* C-00275
Dibutyl cyclopentylphosphonite, *in* C-00281
O,O-Dibutyl cyclopentylphosphonothioate, *in* C-00280
Dibutyl decylphosphonate, *in* D-00012
Dibutyl decylphosphoramidate, *in* D-00013
Dibutyl dibutylphosphoramidate, *in* D-00134
Dibutyl dibutylphosphoramidite, *in* D-00136
Dibutyl (2-diethylamino-2-oxoethyl)-phosphonate, *in* D-00268
Dibutyl (diethylcarbamoylmethyl)phosphonate, *in* D-00268
P,P-Dibutyl-*N,N*-diethylphosphinic amide, *in* D-00111
P,P-Dibutyl-*N,N*-diethylphosphinoselenoic amide, *in* D-00114
P,P-Dibutyl-*N,N*-diethylphosphinous amide, *in* D-00123
Dibutyl diethylphosphoramidate, *in* D-00343
O,O-Dibutyl diethylphosphoramidothioate, *in* D-00356
1,3-Dibutyl-2,3-dihydro-4,5-dimethyl-1*H*-1,3,2-diazaphosphole 2-oxide, D-00081
2,6-Di-*tert*-butyl-1,4-dihydro-4,4-diphenyl-1,4-azaphosphorinium bromide, *in* D-00084
Di-*tert*-butyl dihydrogen phosphate, *see* D-00105
3,5-Di-*tert*-butyl-2,5-dihydro-1,2-oxaphosphole 2-oxide, D-00082
3,5-Di-*tert*-butyl-2,3-dihydro-1,3,2-oxazaphosphole, D-00083
2,6-Di-*tert*-butyl-1,4-dihydro-4-phenyl-1,4-azaphosphorine, D-00084
N,N'-Di-*tert*-butyl-*P,P'*-dimethoxycyclodiphosphazane, *see* D-00085
1,3-Di-*tert*-butyl-2,4-dimethoxy-1,3,2,4-diazadiphosphetidine, D-00085
N,N'-Di-*tert*-butyl-*P,P'*-dimethylcyclodiphosphazane, *see* D-00086
1,3-Di-*tert*-butyl-2,4-dimethyl-1,3,2,4-diazadiphosphetidine, D-00086
1,3-Di-*tert*-butyl-4,4-dimethyl-1,3,2-diazaphosphasiletidine, D-00087
4,8-Di-*tert*-butyl-2,10-dimethyl-12*H*-dibenzo[*d,g*][1,3,2]dioxaphosphocin, D-00088
1,2-Di-*tert*-butyl-1,2-dimethyldiphosphine, D-00089
P,P-Dibutyl-*N,N*-dimethylphosphinic amide, *in* D-00111
P,P-Dibutyl-*N,N*-dimethylphosphinous amide, *in* D-00123
Di-*tert*-butyl dimethylphosphoramidite, *in* D-00875
O,O-Di-*tert*-butyl dimethylphosphoramidothioate, *in* D-00871
1-(Di-*tert*-butyldiphosphinylidene)-methanediamine, D-00090
1,2-Di-*tert*-butyldiphosphirane, D-00091
Dibutyldiphosphonic acid, D-00092
O,O-Dibutyl diselenophosphate, *see* D-00148
2,4-Dibutyl-1,3,2,4-dithiaphosphetane 2,4-disulfide, *in* B-00614
S,S-Dibutyl dithiophosphate, *see* D-00152
O,O-Dibutyl dithiophosphoric acid, *see* D-00150
O,O-Di-*sec*-butyl dithiophosphoric acid, *see* B-00359
O,O-Di-*tert*-butyl dithiophosphoric acid, *see* D-00151
S,S-Dibutyl dithiophosphoric acid, *see* D-00152
P,P-Dibutyl-*N*-ethylphosphinic amide, *in* D-00111
P,P-Dibutyl-*N*-ethylphosphinous amide, *in* D-00123
Dibutyl ethylphosphonate, *in* E-00129
Dibutyl ethylphosphonodithioite, *in* E-00141
Dibutyl ethylphosphonotrithioate, *in* E-00152
Dibutyl ethylphosphoramidate, *in* E-00156
O,O-Di-*tert*-butyl ethylphosphoramidothioate, *in* E-00159
Se,Se-Dibutyl *O*-ethyl phosphorodiselenothioate, *in* D-00149
Dibutyl ethynylphosphonite, *in* E-00204
Dibutyl fluorophosphate, *see* D-00153
Di-*sec*-butyl fluorophosphate, *in* B-00341
▷Di-*tert*-butylfluorophosphine, *in* D-00122
O,O-Dibutyl heptylphosphonothioate, *in* H-00007
2,10-Di-*tert*-butyl-1,3,6,9,11,14-hexathia-2,10-diphosphacyclohexadecane, D-00093
Se,Se-Dibutyl hydrogen diselenothiophosphate, *see* D-00149
O,O-Dibutyl hydrogen dithiophosphate, *see* D-00150
O,O-Di-*sec*-butyl hydrogen dithiophosphate, *see* B-00359
O,O-Di-*tert*-butyl hydrogen dithiophosphate, *see* D-00151
S,S-Dibutyl hydrogen dithiophosphate, *see* D-00152
▷Dibutyl hydrogen phosphate, *see* D-00104
Di-*sec*-butyl hydrogen phosphate, *see* B-00341
O,O-Dibutyl hydrogen phosphorodiselenoate, *see* D-00148
O,O-Dibutyl hydrogen phosphoroselenoate, *see* D-00154
O,O-Dibutyl hydrogen phosphorothioate, *see* D-00155
O,O-Di-*tert*-butyl hydrogen phosphorothioate, *see* D-00156
O,O-Dibutyl hydrogen thiophosphate, *see* D-00155
O,O-Di-*tert*-butyl hydrogen thiophosphate, *see* D-00156
(3,5-Di-*tert*-butyl-4-hydroxybenzyl)-phosphonic acid, D-00094
(Di-*tert*-butylimino)(diisopropylamino)-phosphorane, D-00095
Dibutyliodophosphine, *in* D-00121
Di-*tert*-butyliodophosphine, *in* D-00122
Dibutyl isobutylphosphoramidate, *in* M-00384
2,3-Di-*tert*-butyl-1-isopropylphosphadiarsirane, D-00096
Dibutyl isopropylphosphonite, *in* I-00056
Dibutylmethyl[(dibutylphosphinyl)ethenyl]-phosphonium iodide, *in* B-00159
Dibutyl (1-methylethyl)phosphonite, *in* I-00056
Dibutylmethylphenylphosphonium iodide, *in* D-00102
3,4-Di-*tert*-butyl-2-(4-methylphenyl)-1,3,4,2-thiadiazaphospholidin-5-one 2-sulfide, D-00097
Dibutylmethylphosphine, D-00098
Di-*tert*-butylmethylphosphine, D-00099
P,P-Di-*tert*-butyl-*N*-methylphosphinothioic amide, *in* D-00119
P,P-Di-*tert*-butyl-*N*-methylphosphinous amide, *in* D-00124
Dibutyl methylphosphonate, *in* M-00288
Di-*tert*-butyl methylphosphonate, *in* M-00288
N,N'-Di-*tert*-butyl-*P*-methylphosphonic diamide, *in* M-00289
Di-*tert*-butylmethylphosphonium iodide, *in* D-00103
O,O-Dibutyl methylphosphonoselenoate, *in* M-00311
▷*O,S*-Dibutyl methylphosphonothioate, *in* M-00313
N,N'-Di-*tert*-butyl-*P*-methylphosphonous diamide, *in* M-00320
Dibutyl methylphosphoramidate, *in* M-00326
Dibutyl (1-methylpropyl)phosphonate, *in* M-00377
Dibutyl (2-methylpropyl)phosphoramidate, *in* M-00384
Dibutyl 4-morpholinylphosphonite, *in* M-00529
Dibutyl (nitromethyl)phosphonate, *in* N-00037
Dibutyl [(2-nitrophenyl)methyl]phosphonate, *in* N-00044
4,4'-Di-*tert*-butyl-1,1',2,2',5,5',6,6'-octahydro-1,1'-diphenyl-2,2'-biphosphorin, D-00100
Dibutyl octylphosphonate, *in* O-00042
3,5-Di-*tert*-butyl-1,2-oxaphosphol-3-ene, *see* D-00082
3,5-Di-*tert*-butyl-Δ^4-1,3,2-oxazaphospholine, *see* D-00083
Dibutyl (2-oxo-2-phenylethyl)phosphonate, *in* O-00066
2,7-Di-*tert*-butyl-1-phenylphosphepin, D-00101
▷Dibutylphenylphosphine, D-00102
Di-*sec*-butylphenylphosphine, *see* B-00340
Di-*tert*-butylphenylphosphine, D-00103
P,P-Di-*tert*-butyl-*N*-phenylphosphinoselenoic amide, *in* D-00115
P,P-Di-*tert*-butyl-*N*-phenylphosphinothioic amide, *in* D-00119
P,P-Dibutyl-*N*-phenylphosphinothioic amide, *in* D-00117
P,P-Di-*tert*-butyl-*N*-phenylphosphinous amide, *in* D-00124
Dibutyl phenylphosphonate, *in* P-00221
N,N'-Di-*tert*-butyl-*P*-phenylphosphonic diamide, *in* P-00222
Dibutyl phenylphosphonite, *in* P-00251
N,N'-Di-*tert*-butyl-*P*-phenylphosphonothioic diamide, *in* P-00246
N,N'-Di-*tert*-butyl-*P*-phenylphosphonous diamide, *in* P-00252
Dibutyl phenylphosphoramidate, *in* P-00257
Di-*sec*-butyl phenylphosphoramidate, *in* P-00257
▷Dibutyl phosphate, D-00104
Di-*sec*-butyl phosphate, *see* B-00341
Di-*tert*-butyl phosphate, D-00105
Di-*tert*-butylphosphazene, D-00106
Dibutylphosphine, D-00107
Di-*sec*-butylphosphine, *see* B-00343
Di-*tert*-butylphosphine, D-00108
▷Dibutylphosphinic acid, D-00109
Di-*sec*-butylphosphinic acid, *see* B-00345
Di-*tert*-butylphosphinic acid, D-00110
P,P-Dibutylphosphinic amide, D-00111
Dibutylphosphinodithioic acid, D-00112
Di-*sec*-butylphosphinodithioic acid, *see* B-00347
Di-*tert*-butylphosphinodithioic acid, D-00113
Di-*tert*-butylphosphinodithioic anhydrosulfide, *in* D-00113
Di-*sec*-butylphosphinoithioic acid, *see* B-00350
Dibutylphosphinoselenoic acid, D-00114
Di-*sec*-butylphosphinoselenoic acid, *see* B-00349
P,P-Di-*tert*-butylphosphinoselenoic amide, D-00115
Di-*tert*-butylphosphinoselenoic anilide, *in* D-00115
Di-*tert*-butylphosphinotellurous acid, D-00116
Dibutylphosphinothioic acid, D-00117
Di-*tert*-butylphosphinothioic acid, D-00118
Di-*tert*-butylphosphinothioic amide, D-00119

Dibutylphosphinothioic anilide, *in* D-00117
Di-*tert*-butylphosphinothioic anilide, *in* D-00119
Di-*tert*-butylphosphinothious acid, D-00120
Dibutylphosphinous acid, D-00121
Di-*tert*-butylphosphinous acid, D-00122
P,P-Dibutylphosphinous amide, D-00123
P,P-Di-*tert*-butylphosphinous amide, D-00124
Di-*tert*-butylphosphinous anilide, *in* D-00124
Di-*tert*-butylphosphinous chloride, D-00125
Dibutylphosphinous diethylamide, *in* D-00123
Dibutylphosphinous dimethylamide, *in* D-00123
Dibutylphosphinous ethylamide, *in* D-00123
▷Dibutyl phosphite, *see* D-00127
Di-*sec*-butyl phosphite, *see* B-00353
Di-*tert*-butyl phosphite, *see* D-00128
N,N-Dibutylphosphonamidic acid, D-00126
▷Dibutyl phosphonate, D-00127
Di-*sec*-butyl phosphonate, *see* B-00353
Di-*tert*-butyl phosphonate, D-00128
Dibutyl phosphonite, D-00129
Di-*tert*-butylphosphonochloridous acid, D-00130
2,6-Di-*tert*-butyl-4-(phosphonomethyl)phenol, *see* D-00094
O,O-Dibutyl phosphonoselenoate, D-00131
O,O-Dibutyl phosphonothioate, D-00132
▷Dibutyl phosphoramidate, D-00133
Di-*tert*-butyl phosphoramidate, *in* D-00105
▷Dibutyl phosphoramide, *see* D-00133
Dibutylphosphoramidic acid, D-00134
Dibutylphosphoramidochloridous acid, D-00135
S,S-Dibutyl phosphoramidodithioate, *in* D-00152
Dibutylphosphoramidous acid, D-00136
Di-*tert*-butylphosphoramidous acid, D-00137
N,N-Dibutylphosphoramidous acid, *see* D-00126
▷Dibutyl phosphoric acid, *see* D-00104
Di-*sec*-butyl phosphoric acid, *see* B-00341
Di-*tert*-butyl phosphoric acid, *see* D-00105
Di-*tert*-butyl phosphorobromidate, D-00138
Dibutyl phosphorobromidite, D-00139
Dibutyl phosphorochloridate, D-00140
Di-*sec*-butyl phosphorochloridate, *in* B-00341
Di-*tert*-butyl phosphorochloridate, D-00141
Dibutyl phosphorochloridite, D-00142
Di-*tert*-butyl phosphorochloridite, D-00143
S,S-Dibutyl phosphorochloridodithioate, D-00144
O,O-Dibutyl phosphorochloridoselenoate, *in* D-00154
O,O-Dibutyl phosphorochloridothioate, D-00145
N,N'-Dibutylphosphorodiamidic acid, D-00146
N,N'-Dibutylphosphorodiamidodithioic acid, D-00147
O,O-Dibutyl phosphorodiselenoate, D-00148
O,O-Dibutyl phosphorodiselenoic acid, *see* D-00148
Se,Se-Dibutyl phosphorodiselenothioate, D-00149
Se,Se-Dibutyl phosphorodiselenothioic acid, *see* D-00149
O,O-Dibutyl phosphorodithioate, D-00150
O,O-Di-*sec*-butyl phosphorodithioate, *see* B-00359
O,O-Di-*tert*-butyl phosphorodithioate, D-00151
S,S-Dibutyl phosphorodithioate, D-00152
Dibutyl phosphorofluoridate, D-00153
Di-*sec*-butyl phosphorofluoridate, *in* B-00341
O,O-Dibutyl phosphoroselenoate, D-00154
O,O-Dibutyl phosphoroselenoic acid, *see* D-00154
O,O-Dibutyl phosphoroselenoite, *see* D-00131
O,O-Dibutyl phosphorothioate, D-00155
O,O-Di-*tert*-butyl phosphorothioate, D-00156
O,O-Dibutylphosphorothioic acid, *see* D-00155
O,O-Di-*tert*-butyl phosphorothioic acid, *see* D-00156
▷Dibutyl phosphoryl amide, *see* D-00133
Di-*tert*-butyl phosphoryl bromide, *see* D-00138
Dibutyl phosphoryl chloride, *see* D-00140
Di-*sec*-butyl phosphoryl chloride, *in* B-00341
Di-*tert*-butyl phosphoryl chloride, *see* D-00141
Dibutyl phosphoryl fluoride, *see* D-00153
Di-*sec*-butyl phosphoryl fluoride, *in* B-00341
Di-*tert*-butyl 1-propenylphosphonate, *in* P-00453
Dibutyl 2-propenylphosphonate, *in* P-00454
Dibutyl propylphosphonate, *in* P-00473
Se,Se-Dibutyl *O*-propyl phosphorodiselenothioate, *in* D-00149
Dibutyl 2-propynylphosphonate, *in* P-00495
Dibutylpyrophosphonic acid, *see* D-00092
O,O-Dibutyl selenophosphoryl chloride, *in* D-00154
4,6-Di-*tert*-butylspiro[1,3,2-benzodioxaphosphole-2,1'-[2,8,9]-trioxa[1]phosphaadamantane], D-00157
Dibutyl[[(tetrahydro-2*H*-pyran-2-yl)oxy]-methyl]phosphonate, *in* T-00136
2,9-Di-*tert*-butyl-1,3,8,10-tetraoxa-2,9-diphosphacyclotetradecane, D-00158
2,8-Di-*tert*-butyl-1,3,7,9-tetrathia-2,8-diphosphacyclododecane, D-00159
O,O-Dibutyl thiophosphite, *see* D-00132
O,O-Dibutyl thiophosphoric acid, *see* D-00155
O,O-Dibutyl thiophosphoryl chloride, *see* D-00145
3,4-Di-*tert*-butyl-2-*p*-tolyl-1,3,4,2-thiadiazaphospholidin-5-one 2-sulfide, *see* D-00097
1,3-Di-*tert*-butyl-2,4,4-trimethyl-1,3-diaza-2-phospha-4-silacyclobutane, *in* D-00087
P,P-Di-*tert*-butyl-*N*-trimethylsilylphosphinous amide, *in* D-00124
Di-*tert*-butyl (triphenylphosphoranylidene)-butanedioate, *in* T-00646
Di-*tert*-butyl (triphenylphosphoranylidene)-succinate, *in* T-00646
▷Dibutyrylcyclic AMP, *in* C-00233
▷*S*-(1,2-Dicarbethoxyethyl) *O,O*-dimethyl dithiophosphate, *see* M-00001
(1,2-Dicarboxyethyl)triphenyl-phosphonium(1+), D-00160
Dicetyl phosphate, *see* D-00428
Dicetyl phosphoric acid, *see* D-00428
▷Dichlofenthion, D-00161
S-[(5,7-Dichlorobenzoxazol-2-yl)methyl] *O,O*-diethylphosphorodithioate, *see* B-00003
Di-*p*-chlorobenzyl phosphite, *see* B-00130
Di-*p*-chlorobenzyl phosphonate, *see* B-00130
2,4-Dichloro-2,4-bis(dimethylamino)-2,2,4,4,6,6-hexahydro-6,6-diphenylcyclotriphosphazene, *see* D-00162
2,4-Dichloro-2,4-bis(dimethylamino)-2,2,4,4,6,6-hexahydro-6,6-diphenyl-1,3,5-triaza-2,4,6-triphosphorine, D-00162
2,4-Dichloro-1,3-bis(1,1-dimethylethyl)-1,3,2,4-diazadiphosphetidine, *see* D-00163
Dichloro(2-chlorophenyl)phosphine, *in* C-00151
Dichloro(3-chlorophenyl)phosphine, *in* C-00152
Dichloro-4-chlorophenylphosphine, *see* C-00154
Dichlorocyclohexylphosphine, *see* C-00268
P,P'-Dichloro-*N,N'*-di-*tert*-butylcyclodiphosphazane, *see* D-00163
2,4-Dichloro-1,3-di-*tert*-butyl-1,3,2,4-diazadiphosphetidine, D-00163
Dichloro(dichloromethyl)phosphine, *in* D-00181
P,P-Dichloro-*N*-dichlorophosphinyl-*P*-ethoxyphosphazene, *see* E-00058
P,P-Dichloro-*N*-dichlorophosphinyl-*P*-ethoxyphosphine imide, *see* E-00058
P,P-Dichloro-*P*-diethylamino-*N*-phenylphosphazene, *in* D-00313
P,P-Dichloro-*P*-diethylamino-*N*-phenylphosphine imide, *in* D-00313
2,2-Dichloro-2,2-dihydro-3-phenyl-Δ^1-1,3,2-diazaphosphetin-4-one, *in* D-00557
5,10-Dichloro-5,10-dihydrophosphanthrene, D-00164
2,2-Dichloro-2,2-dihydro-4-trichloromethyl-1,3,2-diazaphosphete, D-00165
1,1-Dichloro-1,1-dihydro-2,4,6-triphenylphosphorin, *in* D-00613
N,N-Dichloro-*P,P*-diisopropylphosphinic amide, *in* D-00642
▷Dichloro(dimethylamino)phosphine, *see* D-00876
2,4-Dichloro-1,3-dimethyl-1,3,2,4-diazadiphosphetidine, D-00166
Dichlorodimethylphosphonium chloride, *see* T-00402
▷2,5-Dichloro-3,4-dimethyl-1,3,4,2,5-thiadiazadiphospholidine, D-00167
Dichlorododecylphosphine, *in* D-01265
▷*O*-(2,2-Dichloroethenyl) *O,O*-dimethyl phosphate, *see* D-00203
2,3-Dichloroethenyl methyl octyl phosphate, *see* D-00202
2,2-Dichloroethenyl phosphorodichloridate, D-00168
▷2-[Di(2-chloroethyl)amino]-1-oxa-3-aza-2-phosphacyclohexane 2-oxide, *see* C-00283
▷Dichloroethylphosphine, *see* E-00155
N,N-Di(2-chloroethyl)tetrahydro-4-hydroperoxy-2*H*-1,3,2-oxazaphosphorin-2-amine 2-oxide, *see* H-00106
Dichloro(2-fluorophenyl)phosphine, *in* F-00040
Dichloro(3-fluorophenyl)phosphine, *in* F-00041
Dichloro(4-fluorophenyl)phosphine, *in* F-00042
Dichloroheptylphosphine, *in* H-00008
2,2-Dichloro-2,2,4,4,6,6-hexahydro-4,4,6,6-tetraphenylcyclotriphosphazene, *see* D-00169
2,2-Dichloro-2,2,4,4,6,6-hexahydro-4,4,6,6-tetraphenyl-1,3,5-triaza-2,4,6-triphosphorine, D-00169
4,6-Dichloro-1,3,5,7,9,10-hexamethyl-1,3,5,7,9,10-hexaaza-4,6-diphospha(4,6-P^V)dispiro[3.1.3.1]-decane-2,8-dione, D-00170
Dichlorohexylphosphine, *in* H-00101
▷*O*-(2,5-Dichloro-4-iodophenyl) *O,O*-dimethyl phosphorothioate, *see* I-00011
O-(2,5-Dichloro-4-iodophenyl) *O*-ethyl ethylphosphonothioate, D-00171
Dichloroisobutylphosphine, *in* M-00383
Dichloroisopropylphosphine, *see* I-00057
Dichloromethanediphosphonic acid, *see* D-00172
(Dichloromethyl)difluorophosphine, *in* D-00181
P-Dichloromethyl-*N,N'*-diphenylphosphonic diamide, *in* D-00176
Dichloromethylenebisphosphonic acid, D-00172
1-Dichloromethylene-*N,N,N',N',N'',N''*-hexamethylphosphoranetriamine, *see* D-00173
(Dichloromethylene)triphenylphosphorane, *in* D-00182
Dichloromethylenetris(dimethylamino)-phosphorane, D-00173
O-(2,6-Dichloro-4-methylphenyl) *O,O*-dimethyl phosphorothioate, *see* T-00321
▷Dichloromethylphosphine, *see* M-00322
P-Dichloromethylphosphonamidic acid, D-00174
(Dichloromethyl)phosphonic acid, D-00175
P-(Dichloromethyl)phosphonic bis(diethylamide), *in* D-00176
P-(Dichloromethyl)phosphonic bis(dimethylamide), *in* D-00176
P-(Dichloromethyl)phosphonic diamide, D-00176
(Dichloromethyl)phosphonic dianilide, *in* D-00176
(Dichloromethyl)phosphonic dichloride, D-00177
(Dichloromethyl)phosphonic difluoride, D-00178
(Dichloromethyl)phosphonic diisocyanate, D-00179
(Dichloromethyl)phosphonothioic acid, D-00180
(Dichloromethyl)phosphonous acid, D-00181
P-Dichloromethyl-*N,N,N',N'*-tetraethylphosphonic diamide, *in* D-00176

P-Dichloromethyl-*N*,*N*,*N'*,*N'*-tetramethylphosphonic diamide, *in* D-00176
O-[2,5-Dichloro-4-(methylthio)phenyl] *O*,*O*-diethyl phosphorothiate, *see* C-00197
(Dichloromethyl)triphenylphosphonium(1+), D-00182
Dichloro-1-naphthylphosphine, *in* N-00025
Dichloro-2-naphthylphosphine, *in* N-00026
Dichlorooctylphosphine, *in* O-00044
Dichloropentylphosphine, *in* P-00052
P,*P*-Dichloro-*P*-phenoxy-*N*-(dichlorophosphinothioyl)-iminophosphorane, *see* P-00112
P,*P*-Dichloro-*P*-phenoxy-*N*-(dichlorophosphinothioyl)phosphazene, *see* P-00112
P,*P*-Dichloro-*P*-phenoxy-*N*-(dichlorophosphinothioyl)phosphine imide, *see* P-00112
Di-*m*-chlorophenyl chlorophosphite, *see* B-00146
Di-*o*-chlorophenyl chlorophosphite, *see* B-00145
Di-*p*-chlorophenyl chlorophosphite, *see* B-00147
O,*O*-Di-*m*-chlorophenyl chlorothiophosphate, *see* B-00148
O,*O*-Di-*p*-chlorophenyl chlorothiophosphate, *see* B-00149
▷*O*-2,4-Dichlorophenyl *O*,*O*-diethyl phosphorothioate, *see* D-00161
2,4-Dichlorophenyl dihydrogen phosphate, *see* M-00463
2,4-Dichlorophenyl *N*,*N'*-diphenylphosphorodiamidate, *in* M-00463
▷*O*-2,4-Dichlorophenyl *O*-ethyl phenylphosphonothioate, *see* S-00004
▷*O*-(2,4-Dichlorophenyl) *O*-ethyl *S*-propyl phosphorodithioate, *see* P-00498
▷*O*-(2,4-Dichlorophenyl) *O*-ethyl *S*-propyl phosphorothioate, *see* E-00005
▷*O*-(2,4-Dichlorophenyl) *O*-methyl isopropylphosphoramidothioate, *see* Z-00002
O-2,4-Dichlorophenyl *S*-methyl isopropylphosphoramidothioate, D-00183
▷*O*-(2,4-Dichlorophenyl) *O*-methyl (1-methylethyl)phosphoramidothioate, *see* Z-00002
O-2,4-Dichlorophenyl *S*-methyl (1-methylethyl)phosphoramidothioate, *see* D-00183
▷*O*-2,5-Dichlorophenyl *O*-methyl phenylphosphonothioate, *in* M-00243
2,4-Dichlorophenyl 4-morpholinylphosphonochloridate, *in* M-00524
▷Dichlorophenylphosphine, *see* P-00254
Di-*o*-chlorophenyl phosphite, *see* B-00143
Di-*p*-chlorophenyl phosphite, *see* B-00144
Di-*o*-chlorophenyl phosphonate, *see* B-00143
Di-*p*-chlorophenyl phosphonate, *see* B-00144
(2,3-Dichlorophenyl)phosphonic acid, D-00184
(2,5-Dichlorophenyl)phosphonic acid, D-00185
(3,4-Dichlorophenyl)phosphonic acid, D-00186
(3,5-Dichlorophenyl)phosphonic acid, D-00187
(2,4-Dichlorophenyl)phosphoramidic acid, D-00188
2,4-Dichlorophenyl phosphoric acid, *see* M-00463
2,4-Dichlorophenyl phosphorodichloridate, D-00189
▷*S*-[[(2,5-Dichlorophenyl)thio]methyl] *O*,*O*-diethyl phosphorodithioate, *see* P-00062
O,*O*-Di-*m*-chlorophenyl thiophosphoryl chloride, *see* B-00148
O,*O*-Di-*p*-chlorophenyl thiophosphoryl chloride, *see* B-00149
Dichloro-β-phenylvinylphosphine, *see* P-00162
▷Dichlorophos, *see* D-00203
(Dichlorophosphinyl)acetic acid, D-00190
1-(Dichlorophosphinyl)butane, *see* B-00598
Dichlorophosphinylcarbamic acid, D-00191
3-(Dichlorophosphinyl)propanoic acid, D-00192
Dichloropropylphosphine, *see* P-00483
3,3-Dichloro-1,1,2,2-tetrahydro-1,1,1,2,2,2-hexaphenyldiphosphirane, D-00193
2,4-Dichloro-2,4,6,6-tetrakis(dimethylamino)-2,2,4,4,6,6-hexahydrocyclotriphosphazene, *see* D-00194
2,4-Dichloro-2,4,6,6-tetrakis(dimethylamino)-2,2,4,4,6,6-hexahydro-1,3,5-triaza-2,4,6-triphosphorine, D-00194
7,7-Dichloro-2,3,5,6-tetrakis(trifluoromethyl)-1,4-diphosphabicyclo[2.2.1]hepta-2,5-diene, D-00195
3,9-Dichloro-2,4,8,10-tetraoxa-3,9-diphosphaspiro[5.5]undecane, D-00196
Dichloro-2-thienylphosphine, *in* T-00308
O-2,6-Dichloro-*p*-tolyl *O*,*O*-dimethyl phosphorothioate, *see* T-00321
Dichloro-*m*-tolylphosphine, *in* M-00247
Dichloro-*o*-tolylphosphine, *in* M-00246
Dichloro-*p*-tolylphosphine, *see* M-00249
Dichloro(trichloromethyl)phosphine, *see* T-00417
2,2-Dichloro-1,1,1-triethyldiphosphinium 2-oxide(1+), D-00197
Dichloro(trifluoromethyl)phosphine, *see* T-00483
Dichloro(trifluoromethylthio)phosphine, *see* T-00472
Dichlorotrimethylphosphorane, D-00198
P,*P*-Dichloro-*N*-trimethylsilyl-*P*-[bis(trimethylsilyl)amino]phosphazene, *in* T-00868
P,*P*-Dichloro-*N*-trimethylsilyl-*P*-[bis(trimethylsilyl)amino]phosphine imide, *in* T-00868
Dichlorotriphenoxyphosphorane, D-00199
Dichlorotriphenylphosphorane, D-00200
Dichlorotris(trichloromethyl)phosphorane, D-00201
▷2,2-Dichlorovinyl dimethyl phosphate, *see* D-00203
2,2-Dichlorovinyl methyl octyl phosphate, D-00202
2,2-Dichlorovinyl phosphorodichloridate, *see* D-00168
▷Dichlorvos, D-00203
▷Dicrotophos, D-00204
(Dicyanomethylene)triphenylphosphorane, D-00205
Dicyanomethylphosphine, *see* M-00323
▷Dicyanophenylphosphine, *see* P-00255
Dicyclohexyl chlorophosphate, *in* D-00209
Dicyclohexyl (2-diethylamino-2-oxoethyl)-phosphonate, *in* D-00268
Dicyclohexyl (diethylcarbamoylmethyl)-phosphonate, *in* D-00268
Dicyclohexyldiphosphonic acid, D-00206
2,4-Dicyclohexyl-1,3,2,4-dithiadiphosphetane 2,4-disulfide, D-00207
N,*N'*-Dicyclohexyl-*P*-ethylphosphonic diamide, *in* E-00130
▷Dicyclohexyl fluorophosphate, *in* D-00209
O,*O*-Dicyclohexyl hydrogen diselenophosphate, *see* D-00218
Dicyclohexyl hydrogen phosphate, *see* D-00209
Dicyclohexyliodophosphine, *in* D-00214
Dicyclohexylphenylphosphine, D-00208
N,*N'*-Dicyclohexyl-*P*-phenylphosphonic diamide, *in* P-00222
Dicyclohexyl phosphate, D-00209
Dicyclohexylphosphine, D-00210
Dicyclohexylphosphinic acid, D-00211
Dicyclohexylphosphinodithioic acid, D-00212
Dicyclohexylphosphinothioic acid, D-00213
Dicyclohexylphosphinous acid, D-00214
Dicyclohexyl phosphite, *see* D-00215
Dicyclohexyl phosphonate, D-00215
Dicyclohexyl phosphoric acid, *see* D-00209
Dicyclohexyl phosphorochloridate, *in* D-00209
N,*N'*-Dicyclohexylphosphorodiamidic acid, D-00216
N,*N'*-Dicyclohexylphosphorodiamidodithioic acid, D-00217
O,*O*-Dicyclohexyl phosphorodiselenoate, D-00218
O,*O*-Dicyclohexylphosphorodiselenoic acid, *see* D-00218
▷Dicyclohexyl phosphorofluoridate, *in* D-00209
Dicyclohexyl phosphoryl chloride, *in* D-00209
▷Dicyclohexyl phosphoryl fluoride, *in* D-00209
Dicyclohexylpyrophosphonic acid, *see* D-00206
Didecyl chlorophosphate, *in* D-00220
O,*O*-Didecyl chlorothiophosphate, *see* D-00223
O,*O*-Didecyl dithiophosphate, *see* D-00224
O,*O*-Didecyl dithiophosphoric acid, *see* D-00224
Didecyl hydrogen phosphate, *see* D-00220
O,*O*-Didecyl hydrogen phosphorodithioate, *see* D-00224
Didecylphenylphosphine, D-00219
Didecyl phosphate, D-00220
Didecylphosphinic acid, D-00221
Didecyl phosphite, *see* D-00222
Didecyl phosphonate, D-00222
Didecyl phosphoric acid, *see* D-00220
Didecyl phosphorochloridate, *in* D-00220
O,*O*-Didecyl phosphorochloridothioate, D-00223
O,*O*-Didecyl phosphorodithioate, D-00224
Didecyl phosphoryl chloride, *in* D-00220
O,*O*-Didecyl thiophosphoryl chloride, *see* D-00223
Didodecyl hydrogen phosphate, *see* D-00225
Didodecyl phosphate, D-00225
Didodecylphosphinic acid, D-00226
Didodecyl phosphite, *see* D-00227
Didodecyl phosphonate, D-00227
Didodecyl phosphoric acid, *see* D-00225
Dielaidylphosphatidylethanolamine, *in* G-00043
P,*P*-Diethenyl-*N*,*N*-diethylphosphinic amide, *in* D-01258
Diethenyl hydrogen phosphate, *see* D-01257
Diethenylphenylphosphine, *see* P-00122
Diethenyl phenylphosphonite, *in* P-00251
Diethenyl phosphate, *see* D-01257
Diethenylphosphinic acid, *see* D-01258
Diethenylphosphinothioic acid, *see* D-01259
Diethenyl phosphorobromidate, *in* D-01257
Diethenyl phosphorochloridate, *in* D-01257
Di(ethoxycarbonyl)-methylenetriphenylphosphorane, *in* T-00658
4,5-Diethoxy-2*H*-1,3,2-diazaphosphole, D-00228
P,*P*-Diethoxy-*P*-diethylamino-*N*-phenylphosphazene, *in* D-00313
P,*P*-Diethoxy-*P*-diethylamino-*N*-phenylphosphine imide, *in* D-00313
1,1-Diethoxy-2,2-diethyldiphosphine 1-oxide, D-00229
4,5-Diethoxy-1,3-dimethyl-2-phenyl-2*H*-1,3,2-diazaphosphole, D-00230
2,5-Diethoxy-1,3,2,5-dioxadiphosphinane, *see* D-00231
2,5-Diethoxy-1,3,2,5-dioxadiphosphorinane, D-00231
2,5-Diethoxy-1,4,2,5-dioxaphosphorinane 2,5-dioxide, *in* D-00618
(Diethoxyethenylidene)triphenylphosphorane, D-00232
(2,2-Diethoxyethylidene)-triphenylphosphorane, D-00233
(2,2-Diethoxyethyl)phosphonic acid, D-00234
4,5-Diethoxy-2-methyl-2*H*-1,3,2-diazaphosphole, D-00235
1,8-Diethoxyoctahydro-4,7-phosphinidene-1*H*-phosphindole 1,8-dioxide, *in* D-00625
Diethoxyphosphinecarboxylic acid oxide, D-00236
(Diethoxyphosphinothioyl)acetic acid, D-00237
4-[(Diethoxyphosphinothioyl)oxy]aniline, *see* A-00099
▷α-[[(Diethoxyphosphinothioyl)oxy]imino]-benzeneacetonitrile, *see* P-00424
(Diethoxyphosphinothioyl)phosphorimidic acid, D-00238
[(Diethoxyphosphinothioyl)thio]acetic acid, D-00239
[(Diethoxyphosphinothioyl)thio]butanedioic acid, D-00240
[(Diethoxyphosphinothioyl)thio]succinic acid, *see* D-00240

▷N-(Diethoxyphosphinothioyl)-P,P,P-triethoxyphosphine imide, *in* D-00238
N-(Diethoxyphosphinothioyl)-P,P,P-trimethoxyphosphazene, *in* D-00238
N-(Diethoxyphosphinothioyl)-P,P,P-trimethoxyphosphine imide, *see* D-00238
N-(Diethoxyphosphinothioyl)-P,P,P-trimethoxyphosphine imide, *in* D-00238
N-(Diethoxyphosphinothioyl)-P,P,P-triphenoxyphosphazene, *in* D-00238
N-(Diethoxyphosphinothioyl)-P,P,P-triphenoxyphosphine imide, *in* D-00238
N-(Diethoxyphosphinyl)acetamide, *in* A-00029
(Diethoxyphosphinyl)acetic acid, D-00241
2-(Diethoxyphosphinyl)acrylic acid, *see* D-00260
3-(Diethoxyphosphinyl)acrylic acid, *see* D-00261
2-(Diethoxyphosphinylamino)pyridine, *in* P-00510
3-(Diethoxyphosphinylamino)pyridine, *in* P-00511
N-(Diethoxyphosphinyl)aniline, *see* D-00312
α-(Diethoxyphosphinyl)benzeneacetic acid, D-00242
2-(Diethoxyphosphinyl)benzoic acid, D-00243
4-(Diethoxyphosphinyl)benzoic acid, D-00244
(Diethoxyphosphinyl)butanedioic acid, D-00245
2-(Diethoxyphosphinyl)butanoic acid, D-00246
4-(Diethoxyphosphinyl)butanoic acid, D-00247
4-(Diethoxyphosphinyl)-2-butenoic acid, D-00248
N-(Diethoxyphosphinyl)carbamic acid, D-00249
4-Diethoxyphosphinylcrotonic acid, *see* D-00248
Diethoxyphosphinylcyclohexane, *in* C-00256
1-(Diethoxyphosphinyl)decane, *in* D-00012
2-(Diethoxyphosphinyl)-1,3-dithian, *in* D-01240
2-(Diethoxyphosphinyl)ethanal, *in* O-00059
N-(Diethoxyphosphinyl)ethylenediamine, *in* A-00080
▷Diethoxyphosphinyl fluoride, *see* D-00387
(Diethoxyphosphinyl)fluoroacetic acid, D-00250
(Diethoxyphosphinyl)formamide, *in* C-00003
(Diethoxyphosphinyl)hydrazine, *see* D-00391
N-Diethoxyphosphinylimidazole, *in* I-00003
▷2-(Diethoxyphosphinylimino)-1,3-dithiolan, *see* P-00342
O-(Diethoxyphosphinyl)lactic acid, *see* D-00255
4-(Diethoxyphosphinyl)-3-methyl-2-butenoic acid, D-00251
2-Diethoxyphosphinyl-3-methylpyridine, *in* M-00389
2-Diethoxyphosphinyl-4-methylpyridine, *in* M-00391
2-Diethoxyphosphinyl-6-methylpyridine, *in* M-00392
(Diethoxyphosphinyl)oxoacetic acid, D-00252
▷2-(Diethoxyphosphinyl)-3-oxobutanoic acid, D-00253
4-(Diethoxyphosphinyl)-3-oxobutanoic acid, D-00254
4-[(Diethoxyphosphinyl)oxy]aniline, *see* A-00098
▷2-[(Diethoxyphosphinyl)oxy]-1H-benz[de]-isoquinoline-1,3(2H)-dione, *see* N-00008
2-[(Diethoxyphosphinyl)oxy]propanoic acid, D-00255
2-[(Diethoxyphosphinyl)oxy]-2-propenoic acid, D-00256
α-(Diethoxyphosphinyl)phenylacetic acid, *see* D-00242
Diethoxyphosphinyl phenyl disulfide, *see* D-00314
(Diethoxyphosphinyl)phosphorimidic acid, D-00257
3-(Diethoxyphosphinyl)propanenitrile, *in* D-00259
2-(Diethoxyphosphinyl)propanoic acid, D-00258
3-(Diethoxyphosphinyl)propanoic acid, D-00259
1-(Diethoxyphosphinyl)-2-propanol, *in* H-00191
3-Diethoxyphosphinyl-1-propanol acetate, *in* A-00010
2-(Diethoxyphosphinyl)propenoic acid, D-00260
3-(Diethoxyphosphinyl)propenoic acid, D-00261
(Diethoxyphosphinyl)succinic acid, *see* D-00245
Diethoxyphosphinylthiocholine, *see* E-00001
2-[(Diethoxyphosphinyl)thio]-N,N,N-trimethylethanaminium, *see* E-00001
N-Diethoxyphosphinyl-P,P,P-triethoxyphosphazene, *in* D-00257
N-Diethoxyphosphinyl-P,P,P-triethoxyphosphine imide, *in* D-00257
N-(Diethoxyphosphinyl)-P,P,P-trimethoxyphosphazene, *in* D-00257
N-(Diethoxyphosphinyl)-P,P,P-trimethoxyphosphine imide, *in* D-00257
N-Diethoxyphosphinyl-P,P,P-trisisopropoxyphosphine imide, *in* D-00257
N-Diethoxyphosphinyl-P,P,P-tris(1-methylethoxy)phosphazene, *in* D-00257
4,4-Diethoxy-2,4,4,5-tetrahydro-2-oxo-5-(4-nitrophenyl)-1,3,4-oxazaphosphole, D-00262
1,8-Diethoxy-3a,4,7,7a-tetrahydro-4,7-phosphinidene-1H-phosphinidole 1,8-dioxide, *in* D-00625
Diethoxytriphenylphosphorane, D-00263
1,1-Diethoxy-2-(triphenylphosphoranylidene)ethane, *see* D-00233
1,1-Diethoxy-2-(triphenylphosphoranylidene)ethylene, *see* D-00232
(Diethoxyvinylidene)triphenylphosphorane, *see* D-00232
Diethyl (acetamidomethyl)phosphonate, *in* A-00012
Diethyl acetonylphosphonate, *in* O-00075
O,O-Diethyl acetonylphosphonothioate, *in* O-00079
▷Diethyl 1-acetoxy-2,2,2-trichloroethylphosphonate, *in* A-00011
Diethyl (2-acetyloxyethyl)phosphonate, *in* A-00006
Diethyl [(acetyloxy)methyl]phosphonate, *in* A-00007
Diethyl [(acetyloxy)phenylmethyl]-phosphonate, *in* A-00020
Diethyl [1-(acetyloxy)propyl]phosphonate, *in* A-00008
Diethyl [2-(acetyloxy)propyl]phosphonate, *in* A-00009
Diethyl (3-acetyloxypropyl)phosphonate, *in* A-00010
Diethyl acetylphosphonate, *in* A-00026
O,O-Diethyl acetylphosphonothioate, *in* A-00028
Diethyl acetylphosphoramidate, *in* A-00029
Diethyl 9-acridinylphosphonate, *in* A-00030
Diethyl 1-adamantylphosphonate, *in* A-00034
Diethyl 1-adamantylphosphoramidate, *in* A-00036
Diethyl allenylphosphonate, *in* P-00441
Diethyl allylphosphonate, *in* P-00454
Diethyl allylphosphonite, *in* P-00463
O,O-Diethyl allylphosphonoselenoate, *in* P-00458
Diethyl allylphosphoramidate, *in* P-00464
O,S-Diethyl amidodithiophosphate, *see* D-00349
S,S-Diethyl amidodithiophosphate, *see* D-00350
Diethyl amidophosphate, *see* D-00342
O,O-Diethyl amidothiophosphate, *see* D-00354
O,S-Diethyl amidothiophosphate, *see* D-00355
▷Diethyl p-aminobenzylphosphonate, *in* A-00107
Diethyl (1-aminobutyl)phosphonate, *in* A-00054
Diethyl (2-aminobutyl)phosphonate, *in* A-00055
Diethyl (3-aminobutyl)phosphonate, *in* A-00056
Diethyl (4-aminobutyl)phosphonate, *in* A-00057
Diethyl [(aminocarbonyl)amino]-methylphosphonate, *in* A-00058
Diethyl aminocarbonylphosphonate, *in* C-00003
(Diethylaminocarbonyl)phosphonic acid, D-00264
Diethyl aminocarbonylphosphoramidate, *in* A-00062
(Diethylamino)dichlorophosphine, *see* D-00361
1-Diethylamino-2,5-dihydro-2-methyl-1H-phosphole 1-oxide, *in* D-00498
4-Diethylamino-2,6-dimethyl-1-phenyl-1,4(1H)-phosphaborin, *see* D-00285
2-Diethylamino-1,3,2-dioxaphospholan-4,5-dicarboxylic acid, D-00265
▷S-[2-(Diethylamino)ethyl] O,O-diethyl phosphorothioate, *see* A-00133
Diethyl (1-aminoethyl)phosphonate, *in* A-00078
Diethyl (2-aminoethyl)phosphoramidate, *in* A-00080
Diethyl (1-amino-1-methylethyl)phosphonate, *in* A-00087
Diethyl (2-amino-1-methylethyl)phosphonate, *in* A-00088
Diethyl (aminomethyl)phosphonate, *in* A-00091
Diethyl (1-amino-1-methylpropyl)phosphonate, *in* A-00093
Diethyl (1-amino-2-methylpropyl)phosphonate, *in* A-00094
▷O-(2-Diethylamino-6-methyl-4-pyrimidinyl) O,O-diethyl phosphorothioate, *see* P-00434
▷O-[2-(Diethylamino)-6-methyl-4-pyrimidinyl] O,O-dimethyl phosphorothioate, *see* P-00435
2-Diethylaminonaphtho[1,2-d]-1,3,2-dioxaphosphole, D-00266
2-Diethylaminonaphtho[1,8-de]-1,3,2-dioxaphosphorin, D-00267
Diethyl (2-amino-2-oxoethyl)phosphonate, *in* A-00060
(2-Diethylamino-2-oxoethyl)phosphonic acid, D-00268
▷O,O-Diethyl S-(2-amino-2-oxoethyl) phosphorodithioate, *in* D-00239
Diethyl (1-aminopentyl)phosphonate, *in* A-00096
Diethyl (5-aminopentyl)phosphonate, *in* A-00097
Diethyl (1-amino-1-phenylethyl)phosphonate, *in* A-00100
Diethyl (aminophenylmethyl)phosphonate, *in* A-00105
▷Diethyl [(4-aminophenyl)methyl]phosphonte, *in* A-00107
2-Diethylamino-3-phenyl-1,3,2-oxazaphospholane, *see* D-00269
2-Diethylamino-3-phenyl-1,3,2-oxazaphospholidine, D-00269
Diethyl (2-aminophenyl)phosphonate, *in* A-00109
Diethyl (4-aminophenyl)phosphonate, *in* A-00111
Diethylaminophosphonic dichloride, *see* D-00344
(Diethylamino)phosphonothioic dichloride, *see* D-00357
(Diethylamino)phosphonothioic difluoride, *see* D-00358
Diethylaminophosphorous dichloride, *see* D-00361
Diethylamino phosphoryl dichloride, *see* D-00344
Diethyl (1-aminopropyl)phosphonate, *in* A-00125
Diethyl (2-aminopropyl)phosphonate, *in* A-00126
Diethyl (3-aminopropyl)phosphonate, *in* A-00127
2-Diethylaminotetrahydro-3-phenyl-1,3,2-oxazaphosphole, *see* D-00269
2-Diethylamino-3,5,5-trimethyl-1,2-oxaphospholan-3-ol 2-oxide, *in* T-00519
▷O,O-Diethyl azidothiophosphate, *see* D-00363
Diethyl 1-aziridinylphosphonate, *in* A-00153
Diethyl 1-aziridinylphosphonite, *in* A-00154
▷Diethyl benzenephosphonate, *see* D-00310
Diethyl[[4-(2-benzothiazolyl)phenyl]-methyl]phosphonate, *see* F-00064

Diethyl (1-*N*-benzoylamino-2-methylpropyl)-phosphonate, *in* A-00094
Diethyl (1-benzoylamino-2-phenylethyl)-phosphonate, *in* A-00101
Diethyl [1-(*N*-benzoylamino)propyl]-phosphonate, *in* A-00125
Diethyl benzoylphosphonate, *in* B-00034
Diethyl benzoylphosphoramidate, *in* B-00036
O,O-Diethyl benzoylphosphoramidothioate, *in* B-00038
O,O-Diethyl *S*-benzoyl phosphorodithioate, D-00270
Diethyl [(benzylamino)methyl]phosphonate, *in* A-00091
Diethyl [(benzylideneamino)methyl]-phosphonate, D-00271
Diethyl 1-benzylideneaminophosphonate, *in* B-00049
Diethyl [(benzylimino)methyl]phosphonate, *see* D-00271
Diethyl (benzyloxycarbonylmethyl)-phosphonate, *in* D-00241
▷Diethyl benzylphosphonate, D-00272
Diethyl benzylphosphonite, *in* B-00068
S,S-Diethyl benzylphosphonodithioate, *in* B-00065
O,O-Diethyl benzylphosphonothioate, *in* B-00067
O,S-Diethyl benzylphosphonothioate, *in* B-00067
Diethyl benzylphosphoramidate, *in* B-00069
Diethyl [(benzylthio)methyl]phosphonate, *in* B-00072
Diethyl [(1,1′-biphenyl)-2-ylmethyl]-phosphonate, *in* B-00079
Diethyl[(1,1′-biphenyl)-4-ylmethyl]-phosphonate, *in* B-00080
Diethyl 4-biphenylylphosphonate, *in* B-00077
N,N-Diethyl-*P,P*-bis(chloromethyl)-phosphinothioic amide, *in* B-00129
Diethyl 2,5-bis(diethoxyphosphinyl)-hexanedioate, *in* D-01184
Diethyl [bis(diethoxyphosphinyl)methyl]-butanedioate, *in* D-01185
Diethyl 2,4-bis(diethoxyphosphinyl)-pentanedioate, *in* D-01186
N,N-Diethyl-*P,P*-bis(iodomethyl)phosphinic amide, *in* B-00288
Diethyl [bis(methylthio)methyl]phosphonate, *in* B-00364
P,P-Diethyl-*N,N*-bis(trimethylsilyl)-phosphinous amide, *in* D-00334
Diethyl bis(trimethylsilyl)phosphoramidite, *in* B-00438
Diethyl (α-bromobenzyl)phosphonate, *in* B-00482
Diethyl *o*-bromobenzylphosphonate, *in* B-00483
Diethyl (4-bromobutyl)phosphonate, *in* B-00466
Diethyl (2-bromoethyl)phosphonate, *in* B-00472
Diethyl (bromomethyl)phosphonate, *in* B-00475
Diethyl (bromophenylmethyl)phosphonate, *in* B-00482
Diethyl [(2-bromophenyl)methyl]phosphonate, *in* B-00483
Diethyl [(4-bromophenyl)methyl]phosphonate, *in* B-00484
Diethyl (2-bromophenyl)phosphonate, *in* B-00490
Diethyl (3-bromophenyl)phosphonate, *in* B-00491
Diethyl (4-bromophenyl)phosphonate, *in* B-00492
Diethyl (4-bromophenyl)phosphonite, *in* B-00493
Diethyl (4-bromophenyl)phosphoramidate, *in* B-00496
Diethyl bromophosphate, *see* D-00371
Diethyl bromophosphite, *see* D-00372
Diethyl bromophosphonate, *see* D-00371
Diethyl (1-bromopropyl)phosphonate, *in* B-00500
Diethyl (3-bromopropyl)phosphonate, *in* B-00501
Diethyl (1,2-butadienyl)phosphonate, *in* B-00509
Diethyl (1,3-butadienyl)phosphonate, *in* B-00510
O,O-Diethyl (1,3-butadien-1-yl)-phosphonothioate, *in* B-00511
Diethyl 1,4-butanediylbis[phenyl-phosphinodithioate], *in* B-00516
Diethyl 1-butenylphosphonate, *in* B-00524
Diethyl 2-butenylphosphonate, *in* B-00525
Diethyl 3-buten-1-ylphosphonate, *in* B-00526
O,O-Diethyl 2-butenylphosphonothioate, *in* B-00527
Diethyl 3-buten-1-ynylphosphonate, *in* B-00529
Diethyl *tert*-butoxycyanomethylphosphonate, D-00273
Diethyl [2-(butylamino)-1-hexenyl]-phosphonate, *in* B-00540
Diethyl butylphosphonate, *in* B-00594
Diethyl *sec*-butylphosphonate, *in* M-00377
Diethyl *tert*-butylphosphonate, *in* B-00595
Diethyl butylphosphonite, *in* B-00615
Diethyl *tert*-butylphosphonite, *in* B-00616
O,O-Diethyl butylphosphonodithioate, *in* B-00611
Diethyl butylphosphonodithioite, *in* B-00607
O,O-Diethyl butylphosphonoselenoate, *in* B-00609
O,S-Diethyl butylphosphonothioate, *in* B-00611
Diethyl butylphosphoramidate, *in* B-00621
Diethyl *tert*-butylphosphoramidate, *in* B-00622
Diethyl *tert*-butylphosphoramidite, *in* B-00624
O,O-Diethyl *tert*-butylphosphoramidothioate, *in* B-00623
Diethyl 1-butynylphosphonate, *in* B-00643
Diethyl 2-butynylphosphonate, *in* B-00644
Diethyl (carbamoylmethyl)phosphonate, *in* A-00060
(Diethylcarbamoylmethyl)phosphonic acid, *see* D-00268
[(*N,N*-Diethylcarbamoyl)methyl]phosphonic bis(diethylamide), *in* B-00175
▷*O,O*-Diethyl *S*-(carbamoylmethyl) phosphorodithioate, *in* D-00239
(Diethylcarbamoyl)phosphonic acid, *see* D-00264
Diethyl carboxychloromethylphosphonate, *see* C-00036
Diethyl (carboxymethyl)phosphonate, *see* D-00241
Diethyl carboxyphosphoramidate, *see* D-00249
Diethyl(chloroacetyl)phosphonate, *in* C-00020
Diethyl *m*-chlorobenzylphosphonate, *in* C-00133
Diethyl (4-chlorobutyl)phosphonate, *in* C-00025
O,O-Diethyl 2-chloro-α-cyanobenzylideneaminooxyphosphonothioate, *see* C-00193
▷*S,S*-Diethyl chlorodithiophosphate, *see* D-00375
Diethyl [chloro(ethoxycarbonyl)methyl]-phosphonate, *in* C-00036
Diethyl (1-chloroethyl)phosphonate, *in* C-00073
▷Diethyl (2-chloroethyl)phosphonate, *in* C-00074
Diethyl chloromethylphosphonate, *in* C-00106
Diethyl (chloromethyl)phosphonite, *in* C-00115
Diethyl [(2-chlorophenyl)methyl]phosphonate, *in* C-00132
Diethyl [(3-chlorophenyl)methyl]phosphonate, *in* C-00133
Diethyl [(4-chlorophenyl)methyl]phosphonate, *in* C-00134
Diethyl (3-chlorophenyl)phosphinate, *in* C-00152
Diethyl (2-chlorophenyl)phosphonate, *in* C-00145
Diethyl (3-chlorophenyl)phosphonate, *in* C-00146
Diethyl (4-chlorophenyl)phosphonate, *in* C-00147
Diethyl (4-chlorophenyl)phosphonite, *in* C-00153
O,O-Diethyl (4-chlorophenyl)-phosphonothioate, *in* C-00150
Diethyl (4-chlorophenyl)phosphoramidate, *in* C-00157
O,O-Diethyl (2-chlorophenyl)-phosphoramidothioate, *in* C-00158
O,O-Diethyl (3-chlorophenyl)-phosphoramidothioate, *in* C-00159
O,O-Diethyl (4-chlorophenyl)-phosphoramidothioate, *in* C-00160
▷Diethyl chlorophosphate, *see* D-00373
▷Diethyl chlorophosphite, *see* D-00374
▷Diethyl chlorophosphonate, *see* D-00373
Diethyl (1-chloropropyl)phosphonate, *in* C-00170
Diethyl (2-chloropropyl)phosphonate, *in* C-00171
Diethyl (3-chloropropyl)phosphonate, *in* C-00172
O,O-Diethyl chloroselenophosphate, *see* D-00377
▷*O,O*-Diethyl chlorothiophosphate, *see* D-00378
O,S-Diethyl chlorothiophosphate, *see* D-00379
▷Diethyl chlorothiophosphonate, D-00274
▷*O,O*-Diethyl chlorothiophosphonothioate, D-00275
▷Diethyl chlorotrithiophosphate, *see* D-00380
▷*O,O*-Diethyl α-cyanobenzylideneaminooxyphosphonothioate, *see* P-00424
Diethyl (α-cyanobenzyl)phosphonate, *in* C-00219
Diethyl 4-cyanobenzylphosphonate, *in* C-00221
Diethyl cyano(1,1-dimethylethyl)-methylphosphonate, *see* D-00273
Diethyl (1-cyanoethyl)phosphonate, *in* C-00210
▷Diethyl (2-cyanoethyl)phosphonate, *in* C-00211
Diethyl (cyanomethyl)phosphonate, *in* C-00215
Diethyl (cyanophenylmethyl)phosphonate, *in* C-00219
Diethyl [(4-cyanophenyl)methyl]phosphonate, *in* C-00221
Diethyl (2-cyanophenyl)phosphonate, *in* C-00223
Diethyl (3-cyanophenyl)phosphonate, *in* C-00224
Diethyl (4-cyanophenyl)phosphonate, *in* C-00225
Diethyl cyanophosphonate, *see* D-00381
Diethyl (1-cyanopropyl)phosphonate, *in* C-00229
Diethyl (3-cyanopropyl)phosphonate, *in* C-00230
▷*PP*-Diethyl cyclic propylene phosphonodithioimidocarbonate, *see* M-00004
Diethyl 1-cyclohexen-1-ylphosphonate, *in* C-00241
Diethyl 2-cyclohexen-1-ylphosphonate, *in* C-00242
Diethyl 3-cyclohexen-1-ylphosphonate, *in* C-00243
Diethyl 2-(cyclohexylamino)ethenyl phosphonate, *see* D-00278
Diethyl 2-(cyclohexylamino)-2-methylethenyl phosphonate, D-00276
Diethyl [2-(cyclohexylamino)-2-phenylethenyl]phosphonate, D-00277
Diethyl 2-(cyclohexylamino)vinylphosphonate, D-00278
Diethyl (cyclohexylcarbonyl)phosphonate, *in* C-00239
Diethyl(2-cyclohexylimino-2-phenylethyl)-phosphonate, *see* D-00277
Diethyl [2-(cyclohexylimino)propyl]-phosphonate, *see* D-00276
Diethyl cyclohexylphosphonate, *in* C-00256
Diethyl cyclohexylphosphonite, *in* C-00267
O,S-Diethyl cyclohexylphosphonodithioate, *in* C-00261
S,S-Diethyl cyclohexylphosphonodithioate, *in* C-00261
O,O-Diethyl cyclohexylphosphonothioate, *in* C-00264
Diethyl cyclohexylphosphoramidate, *in* C-00269
O,O-Diethyl cyclohexylphosphoramidothioate, *in* C-00271
O,S-Diethyl cyclohexylphosphoramidothioate, *in* C-00271

Diethyl 1-cyclopenten-1-ylphosphonate, *in* C-00277
Diethyl 2-cyclopenten-1-ylphosphonate, *in* C-00278
Diethyl decylphosphonate, *in* D-00012
Diethyl diallylphosphoramidate, *in* D-01195
Diethyl (1-diazoethyl)phosphonate, *in* D-00033
Diethyl (diazomethyl)phosphonate, *in* D-00035
Diethyl (dibromomethyl)phosphonate, *in* D-00064
Diethyl (2,5-dibromophenyl)phosphonate, *in* D-00066
Diethyl (2,4-dibromophenyl)phosphoramidate, *in* D-00067
Diethyl dibromophosphoramidate, D-00279
Diethyl dibutyldiphosphonate, *in* D-00092
Diethyl [(3,5-di-*tert*-butyl-4-hydroxyphenyl)methyl]phosphonate, *in* D-00094
▷Diethyl dibutylphosphoramidate, *in* D-00134
Diethyl dibutylphosphoramidite, *in* D-00136
Diethyl (dichloromethyl)phosphonate, *in* D-00175
Diethyl (2,5-dichlorophenyl)phosphonate, *in* D-00185
Diethyl (3,4-dichlorophenyl)phosphonate, *in* D-00186
Diethyl (2,4-dichlorophenyl)phosphoramidate, *in* D-00188
▷Diethyl dichlorophosphoramidate, D-00280
Diethyl (2,2-diethoxyethyl)phosphonate, *in* D-00234
O,O-Diethyl (diethoxyphosphinothioyl)-phosphoramidate, *see* T-00074
▷Diethyl [(diethoxyphosphinothioyl)thio]-butanedioate, *in* D-00240
Diethyl (diethoxyphosphinyl)butanedioate, *in* D-00245
Diethyl 2-diethoxyphosphinyl-2-[2-(diethoxyphosphinyl)ethyl]butanedioate, *in* P-00394
N,N-Diethyl(diethoxyphosphinyl)-fluoroacetamide, *in* D-00250
O,O-Diethyl (diethoxyphosphinyl)-methylphosphoramidothioate, *in* M-00413
O,O-Diethyl (diethoxyphosphinyl)-phosphoramidothioate, *see* T-00074
O,O-Diethyl *S*-(diethoxyphosphinyl) phosphorodithioate, *see* T-00046
▷*O,O*-Diethyl *O*-(diethoxyphosphinyl) phosphorothioate, *see* T-00053
O,O-Diethyl *S*-(diethoxyphosphinyl) phosphorothioate, *see* T-00052
Diethyl diethylamidophosphate, *see* D-00281
▷*O,O*-Diethyl *S*-2-diethylaminoethyl thiophosphate, *see* A-00133
Diethyl (2-diethylamino-2-oxoethyl)-phosphonate, *in* D-00268
Diethyl (diethylcarbamoylmethyl)phosphonate, *in* D-00268
Diethyl diethyldiphosphonate, *in* D-00290
Diethyl *N,N*-diethyl-*N'*-phenylphosphoramidimidate, *in* D-00313
Diethyl diethylphosphoramidate, D-00281
Diethyl diethylphosphoramidite, *in* D-00360
O,O-Diethyl diethylphosphoramidoselenoate, *in* D-00353
O,O-Diethyl diethylphosphoramidothioate, *in* D-00356
Diethyl diethylpyrophosphonate, *in* D-00290
O,O'-Diethyl diethylthiodiphosphonate, *in* D-00405
Diethyl (difluoromethyl)phosphonate, *in* D-00415
Diethyl (9,10-dihydro-9-acridinyl)-phosphonate, *in* D-00438
Diethyl (3,4-dihydro-1*H*-2-benzopyran-1-yl)phosphonate, *in* D-00443
5,7-Diethyl-6,7-dihydro-5*H*-dibenzo[*d,f*]-[1,3]diphosphepin, D-00282
Diethyl [(1,3-dihydro-1,3-dioxo-2*H*-isoindol-2-yl)methyl]phosphonate, *in* P-00426
▷*O,O*-Diethyl (1,3-dihydro-1,3-dioxo-2*H*-isoindol-2-yl)phosphonothioate, *see* D-01237
Diethyl (9,10-dihydro-5-methyl-9-acridinyl)phosphonate, *in* D-00438
N,N-Diethyl-2,3-dihydro-5-methyl-3-phenyl-1,3,4,2-thiadiazaphosphole-2(3*H*)-amine 2-sulfide, *in* A-00072
Diethyl [(5,6-dihydro-4,4,6-trimethyl-4*H*-1,3-oxazin-2-yl]phosphonate, *in* D-00598
Diethyl (dihydroxyphosphinyl)succinate, *in* P-00381
Diethyl (2,3-dihydroxypropyl)phosphonate, *in* D-00624
N,N-Diethyl-*P,P*-diisopropylphosphinous amide, *in* D-00649
O,O-Diethyl diisopropylphosphoramidoselenoate, *in* D-00655
Diethyl (dimethylamido)phosphate, *see* D-00286
Diethyl 2-dimethylaminoethyl phosphate, D-00283
O,O-Diethyl *S*-(2-dimethylaminoethyl) phosphorothioate, D-00284
O,O-Diethyl *S*-(2-dimethylaminoethyl) thiophosphate, *see* D-00284
Diethyl (2-dimethylamino-2-oxoethyl)-phosphonate, *in* D-00703
Diethyl (4-dimethylaminophenyl)phosphonate, *in* D-00706
Diethyl (dimethylcarbamoylmethyl)-phosphonate, *in* D-00703
Diethyl dimethyldiphosphonate, *in* D-00751
Diethyl (1,1-dimethylethyl)phosphonate, *in* B-00595
Diethyl (3,3-dimethyl-2-oxobutyl)-phosphonate, *in* D-00783
N,N-Diethyl-2,6-dimethyl-1-phenyl-1,4-phosphaborin-4(1*H*)-amine, D-00285
Diethyl (2,3-dimethylphenyl)phosphonate, *in* D-00811
Diethyl (2,4-dimethylphenyl)phosphonate, *in* D-00812
Diethyl (2,5-dimethylphenyl)phosphonate, *in* D-00813
Diethyl(3,4-dimethylphenyl)phosphonate, *in* D-00815
Diethyl (3,5-dimethylphenyl)phosphonate, *in* D-00816
Diethyl (2,4-dimethylphenyl)phosphoramidate, *in* D-00818
N,N-Diethyl-*P,P*-dimethylphosphinic amide, *in* D-00830
P,P-Diethyl-*N,N*-dimethylphosphinoselenoic amide, *in* D-00328
P,P-Diethyl-*N,N*-dimethylphosphinothioic amide, *in* D-00331
P,P-Diethyl-*N,N*-dimethylphosphinous amide, *in* D-00334
Diethyl dimethylphosphoramidate, D-00286
O,O-Diethyl dimethylphosphoramidoselenoate, *in* D-00869
O,O-Diethyl dimethylphosphoramidothioate, *in* D-00871
Diethyl dimethylpyrophosphonate, *in* D-00751
Diethyl (2,4-dinitrophenyl)phosphonate, *in* D-00946
Diethyl 2-[(diphenoxyphosphinyl)amino]-pentanedioate, *in* P-00367
N,N-Diethyl-*P,P*-diphenylphosphinic amide, *in* D-01040
P,P-Diethyl-*N,N'*-diphenylphosphinimidic amide, *in* D-00306
N,N-Diethyl-*P,P*-diphenylphosphinous amide, *in* D-01090
Diethyl diphenylphosphoramidate, *in* D-01108
N,N-Diethyl-2,2'-diphenylphosphoramidic dihydrazide, *in* D-01109
N,N-Diethyl-*N',N''*-diphenylphosphoric triamide, *in* D-00364
N,N-Diethyl-*N',N''*-diphenylphosphorothioic triamide, *in* D-00398
2,3-Diethyl-2,3-diphosphabicyclo[2.2.1]-hept-5-ene, D-00287
2,3-Diethyl-2,3-diphosphabicyclo[2.2.2]-octane, D-00288
P,P'-Diethyl diphosphate, D-00289
Diethyldiphosphonic acid, D-00290
P,P'-Diethyldiphosphonic diamide, D-00291
N,N-Diethyl-*P,P*-di-2-propenylphosphinic amide, *in* D-01193
Diethyl di-2-propenylphosphoramidate, *in* D-01195
Diethyl dipropyldiphosphonate, *in* D-01202
N,N-Diethyl-*P,P*-dipropylphosphinic amide, *in* D-01206
Diethyl dipropylphosphoramidate, *in* D-01214
N,N-Diethyl-*P,P*-di-1-pyrrolidinylphosphinic amide, *in* D-01233
O,O-Diethyl diselenophosphate, *see* D-00383
2,4-Diethyl-1,3,2,4-dithiadiphosphetane 2,4-disulfide, *in* E-00152
Diethyl 1,3-dithian-2-ylphosphonate, *in* D-01240
N,N-Diethyl-1,3,2-dithiaphospholane-2-amine, *in* D-01243
N,N-Diethyl-1,3,2-dithiaphospholane-2-amine 2-sulfide, *in* D-01245
▷Diethyl 1,3-dithietan-2-ylidenephosphoramidate, *see* F-00065
Diethyl dithiochlorophosphite, *see* D-00376
▷Diethyl 1,3-dithiolan-2-ylidenephosphoramidate, *see* P-00342
▷*O,O*-Diethyl dithiophosphate, *see* D-00384
S,S-Diethyl dithiophosphate, *see* D-00386
O,S-Diethyl dithiophosphoric acid, *see* D-00385
▷*O,O*-Diethyl dithiophosphoric acid, *see* D-00384
S,S-Diethyl dithiophosphoric acid, *see* D-00386
Diethyl 1,2-ethanediylbis(phenylphosphinate), *in* E-00015
Diethyl 1,2-ethenediylbis(phenylphosphinate), *in* E-00018
Diethyl ethenylphosphonate, *in* V-00003
▷Diethyl (ethoxycarbonylmethyl)phosphonate, *in* D-00241
Diethyl (2-ethoxyethyl)phosphonate, *in* H-00146
N,N-Diethyl *S,S*-ethylene phosphoramidodithioite, *in* D-01243
Diethyl [(ethylimino)methylethenyl]-phosphonate, *in* E-00081
▷*O,O*-Diethyl *S*-ethylmercaptoethyl dithiophosphate, *see* D-01236
Diethyl ethylphenylphosphoramidate, *in* E-00116
Diethyl ethylphosphonate, D-00292
Diethyl ethylphosphonite, *in* E-00153
O,S-Diethyl ethylphosphonodithioate, *in* E-00070
Diethyl ethylphosphonodithioite, *in* E-00141
O,Se-Diethyl ethylphosphonoselenoate, *in* E-00145
O,O-Diethyl ethylphosphonothioate, *in* E-00148
O,S-Diethyl ethylphosphonothioate, *in* E-00074
Diethyl ethylphosphonotrithioate, *in* E-00152
Diethyl ethylphosphoramidate, D-00293
O,O-Diethyl ethylphosphoramidothioate, *in* E-00159
▷*O,O*-Diethyl *S*-[(ethylsulfinyl)methyl] phosphorodithioate, *in* P-00338
▷*OO*-Diethyl *S*-[(ethylsulfonyl)methyl] phosphorodithioate, *in* P-00338
▷*O,O*-Diethyl *S*-(2-ethylthioethyl) phosphorodithioate, *see* D-01236
▷*O,O*-Diethyl *O*-[2-(ethylthio)ethyl] phosphorothioate, *see* D-00019
O,O-Diethyl *S*-[2-(ethylthio)ethyl] phosphorothioate, D-00294
Diethyl [(ethylthio)methyl]phosphonate, *in* E-00192
▷*O,O*-Diethyl *S*-[(ethylthio)methyl] phosphorodithioate, *see* P-00338
O,O-Diethyl *S*-(ethylthio) phosphorothioate, *see* T-00452
Diethyl ethynylphosphonate, *in* E-00203
Diethyl ethynylphosphonite, *in* E-00204
Diethyl (9*H*-fluoren-9-yl)phosphonate, *in* F-00011
O,S-Diethyl fluorodithiophosphate, *see* D-00388
S,S-Diethyl fluorodithiophosphate, *in* D-00386
Diethyl [(3-fluorophenyl)methyl]phosphonate, *in* F-00030
Diethyl [(4-fluorophenyl)methyl]phosphonate, *in* F-00031

Diethyl [(2-fluorophenyl)methyl]phosphonic acid, *in* F-00029
Diethyl (2-fluorophenyl)phosphonate, *in* F-00036
Diethyl (3-fluorophenyl)phosphonate, *in* F-00037
Diethyl (4-fluorophenyl)phosphonate, *in* F-00038
Diethyl (3-fluorophenyl)phosphoramidate, *in* F-00043
Diethyl (4-fluorophenyl)phosphoramidate, *in* F-00044
O,O-Diethyl (3-fluorophenyl)-phosphoramidothioate, *in* F-00045
O,O-Diethyl (4-fluorophenyl)-phosphoramidothioate, *in* F-00046
▷Diethyl fluorophosphate, *see* D-00387
O,O-Diethyl fluorothiophosphate, *see* D-00389
Diethyl fluorotrithiophosphate, *see* D-00390
Diethyl (2-furanylhydroxymethyl)phosphonate, *in* F-00071
Diethyl 2-furanylmethylphosphonate, *in* F-00072
Diethyl 2-furanylphosphonate, *in* F-00073
Diethyl 3-furanylphosphonate, *in* F-00074
Diethyl heptylphosphonate, *in* H-00006
O,O-Diethyl heptylphosphonothioate, *in* H-00007
▷Diethyl hexylphosphonate, *in* H-00097
O,O'-Diethyl hydrazobismethyl-phosphinothioate, *in* H-00105
S,S-Diethyl hydrogen dithiophosphate, *see* D-00386
Diethyl hydrogen phosphate, *see* D-00320
O,O-Diethyl hydrogen phosphorodiselenoate, *see* D-00383
O,S-Diethyl hydrogen phosphorodithioate, *see* D-00385
O,O-Diethyl hydrogen phosphoroselenoate, *see* D-00393
O,O-Diethyl hydrogen phosphoroselenothioate, *see* D-00394
O,O-Diethyl hydrogen phosphorotelluroate, *see* D-00395
O,S-Diethyl hydrogen phosphorothioate, *see* D-00397
O,O-Diethyl hydrogen selenophosphate, *see* D-00393
O,O-Diethyl hydrogen thiophosphate, *see* D-00396
Diethyl (α-hydroxybenzyl)phosphonate, *in* H-00174
Diethyl *o*-hydroxybenzylphosphonate, *in* H-00175
Diethyl *p*-hydroxybenzylphosphonate, *in* H-00176
Diethyl (1-hydroxybutyl)phosphonate, *in* H-00114
Diethyl (1-hydroxyethyl)phosphonate, *in* H-00145
Diethyl (2-hydroxyethyl)phosphonate, *in* H-00146
Diethyl (2-hydroxyethyl)phosphoramidate, *in* H-00148
Diethyl (α-hydroxy-2-furanylmethyl)-phosphonate, *in* F-00071
Diethyl (1-hydroxy-2-methylethyl)-phosphonate, *in* H-00159
Diethyl (hydroxymethyl)phosphonate, *in* H-00162
Diethyl (hydroxyphenylmethyl)phosphonate, *in* H-00174
Diethyl [(2-hydroxyphenyl)methyl]-phosphonate, *in* H-00175
Diethyl [(4-hydroxyphenyl)methyl]-phosphonate, *in* H-00176
Diethyl (2-hydroxyphenyl)phosphonate, *in* H-00179
Diethyl (4-hydroxyphenyl)phosphonate, *in* H-00181
Diethyl (1-hydroxypropyl)phosphonate, *in* H-00190
Diethyl (2-hydroxypropyl)phosphonate, *in* H-00191
Diethyl (3-hydroxypropyl)phosphonate, *in* H-00192
Diethyl 1*H*-imidazol-1-ylphosphonate, *in* I-00003
O,O-Diethyl (1-iminoethyl)-phosphoramidothioate, *in* I-00005
O,S-Diethyl (1-iminoethyl)-phosphoramidothioate, *in* I-00005
Diethyl 1*H*-inden-2-ylphosphonate, *in* I-00007
Diethyl 1*H*-1-indenylphosphonite, *in* I-00006
Diethyl (iodomethyl)phosphonate, *in* I-00013
Diethyl [(2-iodophenyl)methyl]phosphonate, *in* I-00015
Diethyliodophosphine, *in* D-00333
Diethyl iodophosphite, *see* D-00365
Diethyl isobutylphosphonate, *in* M-00378
Diethyl isobutylphosphoramidate, *in* M-00384
Diethyl (isocyanatomethyl)phosphonate, *in* I-00022
Diethyl (isocyanomethyl)phosphonate, *in* I-00024
▷*O,O*-Diethyl *S*-(isopropylcarbamoyl)methyl phosphorodithioate, *see* P-00499
Diethyl isopropylphosphonate, *in* I-00050
O,O-Diethyl isopropylphosphonothioate, *in* I-00054
Diethyl isopropylphosphoramidate, *in* I-00058
O,O-Diethyl isopropylphosphoramidothioate, *in* I-00059
Diethyl(2-mercaptoethyl)phosphine, *see* D-00326
Diethyl (mercaptomethyl)phosphonate, *in* M-00018
Diethyl mesitylphosphonate, *in* T-00527
Diethyl (methoxycarbonylmethyl)phosphonate, *in* D-00241
Diethyl[(2-methoxyethoxy)methyl]phosphonate, *in* M-00040
O,O-Diethyl (2-methoxyethyl)-phosphoramidothioate, *in* M-00041
N,N-Diethyl-*P*-(methoxymethyl)-phosphonamidic acid, D-00295
Diethyl [(3-methoxyphenyl)methyl]-phosphonate, *in* M-00060
Diethyl [(4-methoxyphenyl)methyl]-phosphonate, *in* M-00061
Diethyl (2-methoxyphenyl)phosphonate, *in* M-00070
Diethyl (3-methoxyphenyl)phosphonate, *in* M-00071
Diethyl (4-methoxyphenyl)phosphonate, *in* M-00072
Diethyl (4-methoxyphenyl)phosphoramidate, *in* M-00076
O,O-Diethyl (2-methoxyphenyl)-phosphoramidothioate, *in* M-00077
O,O-Diethyl (3-methoxyphenyl)-phosphoramidothioate, *in* M-00078
O,O-Diethyl (4-methoxyphenyl)-phosphoramidothioate, *in* M-00079
Diethyl (methylamido)phosphate, *see* D-00299
Diethyl (methylaminocarbonyl)phosphonate, *in* M-00088
Diethyl (2-methyl-1,3-butadienyl)-phosphonate, *in* M-00093
Diethyl (3-methyl-1,2-butadienyl)-phosphonate, *in* M-00094
Diethyl (3-methyl-1,3-butadienyl)-phosphonate, *in* M-00095
Diethyl (1-methyl-3-butenyl)phosphonate, *in* M-00099
Diethyl(3-methyl-1-butenyl)phosphonate, *in* M-00101
Diethyl (3-methyl-2-butenyl)phosphonate, *in* M-00102
Diethyl (3-methyl-3-butenyl)phosphonate, *in* M-00103
Diethyl (1-methylbutyl)phosphonate, *in* M-00105
Diethyl (3-methylbutyl)phosphonate, *in* M-00106
O[1],*O*[2]-Diethyl *O*[3]-methyl 3-(diethoxyphosphinyl)-1,2,3-butanetrioate, *in* P-00383
N,N-Diethyl-4-methyl-1,3,2-dithiaphosphorinan-2-amine, *in* M-00138
▷Diethyl (4-methyl-1,3-dithiolan-2-ylidene)phosphoramidate, *see* M-00004
Diethyl methylenebis[methylphosphinate], *in* M-00140
Diethyl (1-methylethenyl)phosphonate, *in* M-00149
▷*O,O*-Diethyl *S*-[2-(1-methylethyl)amino-2-oxoethyl] phosphorodithioate, *see* P-00499
▷*O,O*-Diethyl *O*-[6-methyl-2-(1-methylethyl)-4-pyrimidinyl] phosphorothioate, *see* D-00027
Diethyl (3-methyl-2-oxiranyl)phosphonate, *in* M-00185
Diethyl [(2-methylphenyl)methyl]phosphonate, *in* M-00206
Diethyl [(3-methylphenyl)methyl]phosphonate, *in* M-00207
Diethyl [(4-methylphenyl)methyl]phosphonate, *in* M-00208
N,N-Diethyl-*P*-methyl-*P*-phenylphosphinous amide, *in* M-00227
Diethyl (2-methylphenyl)phosphonate, *in* M-00235
Diethyl (3-methylphenyl)phosphonate, *in* M-00236
Diethyl (4-methylphenyl)phosphonate, *in* M-00237
Diethyl (4-methylphenyl)phosphonite, *in* M-00248
Diethyl methylphenylphosphoramidate, *in* M-00250
Diethyl (2-methylphenyl)phosphoramidate, *in* M-00251
Diethyl (3-methylphenyl)phosphoramidate, *in* M-00252
Diethyl (4-methylphenyl)phosphoramidate, *in* M-00253
O,O-Diethyl (2-methylphenyl)-phosphoramidothioate, *in* M-00254
O,O-Diethyl (3-methylphenyl)-phosphoramidothioate, *in* M-00255
O,O-Diethyl (4-methylphenyl)-phosphoramidothioate, *in* M-00256
P,P-Diethyl-*N*-methylphosphinic amide, *in* D-00323
P,P-Diethyl-*N*-methylphosphinoselenoic amide, *in* D-00328
P,P-Diethyl-*N*-methylphosphinous amide, *in* D-00334
N,N-Diethyl-*P*-methylphosphonamidic acid, D-00296
N,N-Diethyl-*P*-methylphosphonamidothioic acid, D-00297
Diethyl methylphosphonate, D-00298
N,N-Diethyl-*P*-methylphosphonazidic amide, *in* M-00287
N,N-Diethyl-*P*-methylphosphonisocyanatidic amide, *in* M-00293
▷Diethyl methylphosphonite, *in* M-00319
O,S-Diethyl methylphosphonodithioate, *in* M-00301
S,S-Diethyl methylphosphonodithioate, *in* M-00301
Diethyl methylphosphonodithioite, *in* M-00302
Diethyl methyl phosphonoformate, *in* D-00236
O,O-Diethyl methylphosphonoselenoate, *in* M-00311
O,O-Diethyl methylphosphonothioate, *in* M-00313
O,S-Diethyl methylphosphonothioate, *in* E-00093
Diethyl methylphosphonotrithioate, *in* M-00318
Diethyl methylphosphoramidate, D-00299
Diethyl methylphosphoramidite, *in* M-00335
O,O-Diethyl methylphosphoramidoselenoate, *in* M-00332
O,O-Diethyl methylphosphoramidothioate, *in* M-00333
O,O-Diethyl *SS*-methyl phosphoro(dithioperoxo)thioate, D-00300
O,O-Diethyl *Se*-methyl phosphoroselenoite, D-00301
O,O-Diethyl (2-methyl-1-propenyl)-phosphonothioate, *in* M-00366
Diethyl (1-methylpropyl)phosphonate, *in* M-00377
Diethyl (2-methylpropyl)phosphonate, *in* M-00378
Diethyl (2-methylpropyl)phosphoramidate, *in* M-00384
Diethyl (3-methyl-2-pyridinyl)phosphonate, *in* M-00389

Diethyl (4-methyl-2-pyridinyl)phosphonate, *in* M-00391
Diethyl (6-methyl-2-pyridinyl)phosphonate, *in* M-00392
Diethyl 1-methyl-1*H*-pyrrol-2-ylphosphonate, *in* P-00521
Diethyl [(methylsulfinyl)methyl]phosphonate, *in* M-00398
▷*O,O*-Diethyl *O*-[4-(methylsulfinyl)phenyl] phosphorothioate, *see* F-00006
Diethyl [(methylsulfonyl)methyl]phosphonate, *in* M-00400
▷*O,O*-Diethyl *O*-[4-(methylsulfonyl)phenyl] phosphorothioate, *in* D-00302
Diethyl methylsulfonylphosphoramidate, *in* M-00401
Diethyl [1-(methylthio)ethyl]phosphonate, *in* M-00412
Diethyl [(methylthio)methyl]phosphonate, *in* M-00415
Diethyl (4-methylthiophenyl)phosphonate, *in* M-00418
▷*O,O*-Diethyl *O*-[4-(methylthio)phenyl] phosphorothioate, D-00302
O,O-Diethyl *S*-(methylthio) phosphorodithioate, *see* D-00300
Diethyl (4-morpholinylmethyl)phosphonate, *in* M-00521
Diethyl 4-morpholinylphosphonate, *in* M-00523
Diethyl 4-morpholinylphosphonite, *in* M-00529
Diethyl [1,5-naphthalenediylbis[methylene]]-bisphosphonate, *in* N-00002
Diethyl [1,6-naphthalenediylbis[methylene]]-bisphosphonate, *in* N-00003
N,N-Diethylnaphtho[1,8-*de*]-1,3,2-dioxaphosphorin-2-amine, *see* D-00267
▷Diethyl 1-naphthylmethylphosphonate, *in* N-00013
Diethyl 2-naphthylmethylphosphonate, *in* N-00014
Diethyl 2-naphthylphosphonate, *in* N-00022
Diethyl 1-naphthylphosphonite, *in* N-00025
Diethyl 1-naphthylphosphoramidate, *in* N-00028
Diethyl 2-naphthylphosphoramidate, *in* N-00029
Diethyl (1-nitroethyl)phosphonate, *in* N-00035
Diethyl (nitromethyl)phosphonate, *in* N-00037
Diethyl[(2-nitrophenyl)methyl]phosphonate, *in* N-00044
Diethyl [(4-nitrophenyl)methyl]phosphonate, *in* N-00045
▷Diethyl 2-nitrophenyl phosphate, D-00303
▷Diethyl 3-nitrophenyl phosphate, D-00304
▷*O,O*-Diethyl *O*-(4-nitrophenyl) phosphate, *see* P-00001
Diethyl (2-nitrophenyl)phosphonate, *in* N-00055
Diethyl (4-nitrophenyl)phosphonate, *in* N-00057
Diethyl (3-nitrophenyl)phosphoramidate, *in* N-00059
Diethyl (4-nitrophenyl)phosphoramidate, *in* N-00060
▷*O,O*-Diethyl *O*-(4-nitrophenyl) phosphorothioate, *see* P-00002
▷Diethyl 4-nitrophenyl thiophosphate, *see* P-00002
Diethyl (1-nitroso-1-methylethyl)-phosphonate, *in* N-00066
Diethyl octylphosphonate, *in* O-00042
Diethyl octylphosphonite, *in* O-00044
Diethyl octylphosphoramidate, *in* O-00045
▷*O,O*-Diethyl *S*-[(4-oxo-1,2,3-benzotriazin-3(4*H*)-yl)methyl] phosphorodithioate, *see* A-00150
Diethyl (1-oxoethyl)phosphonate, *in* A-00026
Diethyl (2-oxoethyl)phosphonate, *in* O-00059
Diethyl (2-oxoheptyl)phosphonate, *in* O-00061
Diethyl (2-oxononyl)phosphonate, *in* O-00063
Diethyl (2-oxooctyl)phosphonate, *in* O-00064
Diethyl (2-oxo-4-phenylbutyl)phosphonate, *in* O-00065
Diethyl (2-oxo-2-phenylethyl)phosphonate, *in* O-00066
Diethyl (1-oxopropyl)phosphonate, *in* O-00074
Diethyl (2-oxopropyl)phosphonate, *in* O-00075
Diethyl (3-oxopropyl)phosphonate, *in* O-00076
O,O-Diethyl (2-oxopropyl)phosphonothioate, *in* O-00079
O,O-Diethyl (1-oxopropyl)phosphonothioate, *in* O-00078
Diethyl [2-oxo-2-(triphenylphosphoranylidene)propyl]-phosphonate, *in* O-00086
Diethyl 1,3-pentadienylphosphonate, *in* P-00009
Diethyl 1,3-pentadiyn-1-ylphosphonate, *in* P-00010
Diethyl (pentafluorophenyl)phosphonate, *in* P-00015
Diethyl (3-penten-1-ynyl)phosphonate, *in* P-00045
Diethyl (4-penten-2-yn-1-yl)phosphonate, *in* P-00046
Diethyl pentylphosphonate, *in* P-00050
Diethyl 2-pentylphosphonate, *in* M-00105
Diethyl pentylphosphoramidate, *in* P-00053
Diethyl 6-phenanthradinylphosphonate, *in* P-00060
O,O-Diethyl phenylamidothiophosphate, *in* P-00262
Diethyl (phenylamino)-carbonylphosphoramidate, *in* P-00081
O,O-Diethyl [(phenylamino)carbonyl]-phosphoramidothioate, *in* P-00082
O,S-Diethyl [(phenylamino)carbonyl]-phosphoramidothioate, *in* P-00082
Diethyl [(phenylamino)methyl]phosphonate, *in* P-00084
Diethyl 1,2-phenylenebis(phenylphosphinate), *in* P-00139
Diethyl (1-phenylethenyl)phosphonate, *in* P-00154
Diethyl (2-phenylethenyl)phosphonate, *in* P-00155
Diethyl (2-phenylethenyl)phosphonite, *in* P-00161
S,S-Diethyl (2-phenylethenyl)-phosphonodithioate, *in* P-00159
O,O-Diethyl (2-phenylethenyl)-phosphonothioate, *in* P-00160
▷Diethyl 1-phenylethylphosphonate, *in* P-00164
Diethyl 2-phenylethylphosphonate, *in* P-00165
Diethyl phenylethynylphosphonate, *in* P-00166
O,O-Diethyl (phenylethynyl)phosphonothioate, *in* P-00167
▷*O,O*-Diethyl *O*-(5-phenyl-3-isoxazolyl) phosphorothioate, *see* I-00068
Diethyl [[(phenylmethylene)amino]methyl]-phosphonate, *see* D-00271
▷Diethyl (phenylmethyl)phosphonate, *see* D-00272
▷*O,O*-Diethyl *S*-(phenylmethyl) phosphorothioate, *see* K-00001
Diethyl [[(phenylmethyl)thio]methyl]-phosphonate, *in* B-00072
N,N-Diethyl-3-phenyl-1,3,2-oxazaphospholidin-2-amine, *see* D-00269
Diethylphenylphosphine, D-00305
N,N-Diethyl-*P*-phenylphosphinic amide, *in* P-00210
P,P-Diethyl-*N*-phenylphosphinic amide, *see* D-00306
P,P-Diethyl-*N*-phenylphosphinimidic acid, D-00306
P,P-Diethyl-*N*-phenylphosphinoselenoic amide, *in* D-00328
P,P-Diethyl-*N*-phenylphosphinothioic amide, *in* D-00331
P,P-Diethyl-*N*-phenylphosphinous amide, *in* D-00334
N,N-Diethyl-*P*-phenylphosphonamidic acid, D-00307
N,N-Diethyl-*P*-phenylphosphonamidodithioic acid, D-00308
N,N-Diethyl-*P*-phenylphosphonamidothioic acid, D-00309
▷Diethyl phenylphosphonate, D-00310
N,N'-Diethyl-*P*-phenylphosphonic diamide, *in* P-00222
Diethyl phenylphosphonite, D-00311
S,S-Diethyl phenylphosphonodithioate, *in* P-00236
Diethyl phenylphosphonodithioite, *in* P-00238
O,O-Diethyl phenylphosphonoselenoate, *in* P-00243
O,O-Diethyl phenylphosphonothioate, *in* P-00244
O,S-Diethyl phenylphosphonothioate, *in* P-00244
N,N'-Diethyl-*P*-phenylphosphonothioic diamide, *in* P-00246
Diethyl phenylphosphonotrithioate, *in* P-00250
N,N'-Diethyl-*P*-phenylphosphonous diamide, *in* P-00252
Diethyl phenylphosphoramidate, D-00312
Diethyl phenylphosphoramidite, *in* P-00263
N,N-Diethyl-*N'*-phenylphosphoramidoimidic acid, D-00313
O,O-Diethyl phenylphosphoramidoselenoate, *in* P-00261
O,O-Diethyl phenylphosphoramidothioate, *in* P-00262
O,O-Diethyl *SS*-phenylphosphoro(dithioperoxoate), D-00314
O,O-Diethyl *SS*-phenyl phosphoro(dithioperoxo)thioate, D-00315
O,O-Diethyl 2-phenyl-phosphorohydrazidothioate, *in* P-00296
O,O-Diethyl *Se*-phenyl phosphoroselenoite, D-00316
O,O-Diethyl *S*-phenyl phosphorothioate, D-00317
Diethyl (1-phenyl-1,2-propadienyl)-phosphonate, *in* P-00298
Diethyl 3-phenyl-2-propenylphosphonate, *in* P-00302
Diethyl (3-phenylpropyl)phosphonate, *in* P-00305
Diethyl [(phenylsulfinyl)methyl]phosphonate, *in* P-00310
Diethyl [(phenylsulfonyl)methyl]phosphonate, *in* P-00311
Diethyl phenylsulfonylphosphoramidate, *in* P-00312
Diethyl [(phenylthio)methyl]phosphonate, *in* P-00326
O,O-Diethyl *S*-[(phenylthio)methyl] phosphorodithioate, D-00318
O,O-Diethyl *S*-phenyl thiophosphate, *see* D-00317
O,O-Diethyl *S*-(phenylthio) phosphorodithioate, *see* D-00315
O,O-Diethyl *S*-(phenylthio) phosphorothioate, *see* D-00314
▷*O,O*-Diethyl *O*-(1-phenyl-1*H*-1,2,4-triazol-3-yl) phosphorothioate, *see* T-00333
Diethyl [phenyl[(trimethylsilyl)oxy]-methyl]phosphonate, *in* P-00331
Diethyl[3-phenyl-1-[(trimethylsilyl)oxy]-2-propenyl]phosphonate, D-00319
Diethyl phosphate, D-00320
▷Diethylphosphine, D-00321
Diethylphosphinic acid, D-00322
P,P-Diethylphosphinic amide, D-00323
Diethylphosphinic diethylphosphinothioic anhydride, *in* D-00322
Diethylphosphinodiselenoic acid, D-00324
Diethylphosphinodithioic acid, D-00325
2-(Diethylphosphino)ethanethiol, D-00326
Diethylphosphinoselenoic acid, D-00327
P,P-Diethylphosphinoselenoic amide, D-00328
Diethylphosphinoselenothioic acid, D-00329
Diethylphosphinothioic acid, D-00330
P,P-Diethylphosphinothioic amide, D-00331
Diethylphosphinothious acid, D-00332
N-(Diethylphosphinothioyl)-*P,P*-diethylphosphinothioic amide, *in* D-00331
Diethylphosphinous acid, D-00333
P,P-Diethylphosphinous amide, D-00334
Diethylphosphinous anilide, *in* D-00334
Diethylphosphinous chloride, D-00335
N-(Diethylphosphinyl)-*N'*-phenylurea, *in* D-00323
▷Diethyl phosphite, *see* D-00338
N,P-Diethylphosphonamidic acid, D-00336
N,N-Diethylphosphonamidic acid, D-00337
▷Diethyl phosphonate, D-00338
Diethyl phosphonite, D-00339
Diethyl phosphonobutanedioate, *in* P-00381

4-Diethylphosphonocrotonic acid, *see* D-00248
NN-Diethylphosphonoformamide, *see* D-00264
O,O-Diethylphosphonoformic acid, *see* D-00236
O,O-Diethyl phosphonoselenoate, D-00340
Diethyl phosphonosuccinate, *in* P-00381
O,O-Diethyl phosphonothioate, D-00341
Diethyl phosphoramidate, D-00342
Diethyl phosphoramide, *see* D-00342
Diethylphosphoramidic acid, D-00343
Diethylphosphoramidic dianilide, *in* D-00364
Diethylphosphoramidic dichloride, D-00344
Diethylphosphoramidic difluoride, D-00345
Diethylphosphoramidochloridic acid, D-00346
Diethylphosphoramidochloridoselenoic acid, D-00347
Diethylphosphoramidochloridothioic acid, D-00348
O,S-Diethyl phosphoramidodithioate, D-00349
S,S-Diethyl phosphoramidodithioate, D-00350
Diethylphosphoramidofluoridic acid, D-00351
Diethylphosphoramidofluoridodithioic acid, D-00352
Diethylphosphoramidoselenoic acid, D-00353
O,O-Diethyl phosphoramidothioate, D-00354
O,S-Diethyl phosphoramidothioate, D-00355
Diethylphosphoramidothioic acid, D-00356
Diethylphosphoramidothioic dichloride, D-00357
Diethylphosphoramidothioic difluoride, D-00358
Diethylphosphoramidotrithioic acid, D-00359
Diethylphosphoramidous acid, D-00360
Diethylphosphoramidous dichloride, D-00361
Diethyl phosphoranilidate, *see* D-00312
▷Diethyl phosphorazidate, D-00362
▷*O,O*-Diethyl phosphorazidothioate, D-00363
Diethyl phosphoric acid, *see* D-00320
Diethyl phosphoric acid hydrazide, *see* D-00391
Diethyl phosphoric anilide, *see* D-00312
N,N-Diethylphosphoric triamide, D-00364
Diethyl phosphoriodidite, D-00365
Diethyl phosphorisocyanatidate, D-00366
O,O-Diethyl phosphorisocyanatidothioate, D-00367
O,S-Diethyl phosphorisocyanatidothioate, D-00368
Diethyl phosphor(isothiocyanatidate), D-00369
▷*O,O*-Diethyl phosphor(isothiocyanatido)-thioate, D-00370
Diethyl phosphorobromidate, D-00371
Diethyl phosphorobromidite, D-00372
▷Diethyl phosphorochloridate, D-00373
▷Diethyl phosphorochloridite, D-00374
▷*S,S*-Diethyl phosphorochloridodithioate, D-00375
Diethyl phosphorochloridodithioite, D-00376
O,O-Diethyl phosphorochloridoselenoate, D-00377
▷*O,O*-Diethyl phosphorochloridothioate, D-00378
O,S-Diethyl phosphorochloridothioate, D-00379
▷Diethyl phosphorochloridotrithioate, D-00380
Diethyl phosphorocyanidate, D-00381
N,N'-Diethylphosphorodiamidic acid, D-00382
O,O-Diethyl phosphorodiselenoate, D-00383
O,O-Diethyl phosphorodiselenoic acid, *see* D-00383
▷*O,O*-Diethyl phosphorodithioate, D-00384
O,S-Diethyl phosphorodithioate, D-00385
S,S-Diethyl phosphorodithioate, D-00386
▷*O,O*-Diethyl phosphorodithioate anhydrosulfide, *see* T-00076
▷*O,O*-Diethyl phosphorodithioate, *S*-ester with *N*-isopropyl-2-mercaptoacetamide, *see* P-00499
▷Diethyl phosphorofluoridate, D-00387
O,S-Diethyl phosphorofluoridodithioate, D-00388
S,S-Diethyl phosphorofluoridodithioate, *in* D-00386
O,O-Diethyl phosphorofluoridothioate, D-00389
Diethyl phosphorofluoridotrithioate, D-00390
Diethyl phosphorohydrazidate, D-00391
O,O-Diethyl phosphorohydrazidothioate, D-00392
O,O-Diethyl phosphoroselenoate, D-00393
O,O-Diethyl phosphoroselenoic acid, *see* D-00393
O,O-Diethyl phosphoroselenoite, *see* D-00340
O,O-Diethyl phosphoroselenothioate, D-00394
O,O-Diethyl phosphoroselenothioic acid, *see* D-00394
O,O-Diethyl phosphorotelluroate, D-00395
O,O-Diethyl phosphorotelluroic acid, *see* D-00395
O,O-Diethyl phosphorothioate, D-00396
O,S-Diethyl phosphorothioate, D-00397
OO-Diethyl phosphorothioate anhydrosulfide, *see* T-00052
▷*O,O*-Diethyl phosphorothioate, *S*-ester with *N*-(1-cyano-1-methylethyl-2-mercaptoacetamide), *see* C-00232
O,S-Diethyl phosphorothioic acid, *see* D-00397
O,O-Diethyl phosphorothioic acid phenylhydrazide, *in* P-00296
N,N-Diethylphosphorothioic triamide, *in* D-00356
N,N-Diethylphosphorothioic triamide, D-00398
Diethyl phosphorotriazolidite, *in* T-00332
Diethyl phosphoryl amide, *see* D-00342
▷Diethyl phosphoryl azide, *see* D-00362
Diethyl phosphoryl bromide, *see* D-00371
▷Diethyl phosphoryl chloride, *see* D-00373
Diethyl phosphoryl cyanide, *see* D-00381
▷Diethyl phosphoryl fluoride, *see* D-00387
N-Diethylphosphorylimidazole, *in* I-00003
Diethyl phosphoryl isocyanate, *see* D-00366
O,S-Diethyl phosphoryl isocyanate, *see* D-00368
Diethyl phosphoryl isothiocyanate, *see* D-00369
2-Diethylphosphoryl-3-methylpyridine, *in* M-00389
2-Diethylphosphoryl-4-methylpyridine, *in* M-00391
2-Diethylphosphoryl-6-methylpyridine, *in* M-00392
▷*O,O*-Diethyl phthalimidophosphonothioate, *see* D-01237
Diethyl (1-piperidinylmethyl)phosphonate, *in* P-00429
Diethyl 1-piperidinylphosphonate, *in* P-00430
Diethyl 1-piperidinylphosphonite, *in* P-00432
N,N-Diethyl-*P*-propadienylphosphonamidic acid, D-00399
Diethyl 1,2-propadienylphosphonate, *in* P-00441
Diethyl propanoyl phosphate, D-00400
Diethyl propanoylphosphonate, *in* O-00074
Diethyl 1-propenylphosphonate, *in* P-00453
Diethyl 2-propenylphosphonate, *in* P-00454
Diethyl 2-propenylphosphonite, *in* P-00463
S,S-Diethyl 2-propenylphosphonodithioate, *in* P-00457
O,O-Diethyl (2-propenyl)phosphonoselenoate, *in* P-00458
Diethyl (2-propenyl)phosphonotelluroate, *in* P-00460
O,O-Diethyl 2-propenylphosphonothioate, *in* P-00461
Diethyl 2-propenylphosphoramidate, *in* P-00464
N,N-Diethyl-*P*-propylphosphonamidic acid, D-00401
Diethyl propylphosphonate, *in* P-00473
Diethyl propylphosphonite, *in* P-00482
O,O-Diethyl propylphosphonothioate, *in* P-00480
O,S-Diethyl propylphosphonothioate, *in* P-00480
Diethyl propylphosphoramidate, *in* P-00484
Diethyl propylphosphoramidite, *in* P-00486
O,O-Diethyl propylphosphoramidothioate, *in* P-00485
O,O-Diethyl *SS*-propyl phosphoro(dithioperoxo)thioate, D-00402
O,O-Diethyl *S*-(propylthio) phosphorodithioate, *see* D-00402
N,N-Diethyl-*P*-1-propynylphosphonamidic acid, D-00403
Diethyl 1-propynylphosphonate, *in* P-00494
Diethyl 2-propynylphosphonate, *in* P-00495
▷*O,O*-Diethyl *O*-(2-pyrazinyl) phosphorothioate, *see* T-00314
Diethyl (1*H*-pyrazol-3-yl)phosphonate, *in* P-00501
Diethyl (2-pyridinylmethyl)phosphonate, *in* P-00506
Diethyl (4-pyridinylmethyl)phosphonate, *in* P-00507
Diethyl 2-pyridinylphosphonate, *in* P-00503
Diethyl 4-pyridinylphosphonate, *in* P-00505
Diethyl 2-pyridylphosphoramidate, *in* P-00510
Diethyl 3-pyridylphosphoramidate, *in* P-00511
P,P'-Diethyl pyrophosphate, *see* D-00289
Diethylpyrophosphonic acid, *see* D-00290
Diethyl 1-pyrrolidinylphosphonite, *in* P-00520
Diethyl 1*H*-pyrrol-2-ylphosphonate, *in* P-00521
Diethyl [(2-quinolinyl)methyl]phosphonate, *in* Q-00004
Diethyl [(8-quinolinyl)methyl]phosphonate, *in* Q-00005
▷*O,O*-Diethyl *O*-2-quinoxalinyl phosphorothioate, *see* Q-00001
O,O-Diethyl selenophosphoryl chloride, *see* D-00377
O,O-Diethyl selenophosphoryl methylamide, *in* M-00332
O,O-Diethyl selenothiophosphate, *see* D-00394
Diethyl styrylphosphonate, *in* P-00155
O,O-Diethyl tellurophosphate, *see* D-00395
Diethyl (2,3,5,6-tetrachloro-4-pyridinyl)-phosphonate, *in* T-00035
5,8-Diethyl-5,6,7,8-tetrahydrobibenzo[*e,g*][1,4]diphosphocin, D-00404
Diethyl(tetrahydro-2-oxo-2-furanyl)-phosphonate, *in* T-00121
Diethyl[[(tetrahydro-2*H*-pyran-2-yl)oxy]-methyl]phosphonate, *in* T-00136
▷*P,P'*-Diethyl-*N,N,N',N'*-tetramethyldiphosphonic diamide, *in* D-00291
Diethyl (2-thienylmethyl)phosphonate, *in* T-00304
Diethyl (3-thienylmethyl)phosphonate, *in* T-00305
Diethyl 2-thienylphosphonate, *in* T-00307
Diethyl 2-thienylphosphonite, *in* T-00308
Diethylthiodiphosphonic acid, D-00405
O,O-Diethyl thiophosphate, *see* D-00396
O,O-Diethyl thiophosphite, *see* D-00341
O,O-Diethyl thiophosphoric acid, *see* D-00396
O,O-Diethyl thiophosphoric anilide, *in* P-00262
O,O-Diethyl thiophosphoric hydrazide, *see* D-00392
O,O-Diethyl thiophosphoryl amide, *see* D-00354
▷*O,O*-Diethyl thiophosphoryl azide, *see* D-00363
▷Diethyl thiophosphoryl chloride, *see* D-00378
O,O-Diethyl thiophosphoryl fluoride, *see* D-00389
O,O-Diethyl thiophosphoryl hydrazide, *see* D-00392
O,O-Diethyl thiophosphoryl isocyanate, *see* D-00367
▷*O,O*-Diethyl thiophosphoryl isothiocyanate, *see* D-00370
Diethyl *m*-tolylphosphonate, *in* M-00236
Diethyl *o*-tolylphosphonate, *in* M-00235
Diethyl *p*-tolylphosphonate, *in* M-00237
Diethyl (1*H*-1,2,4-triazol-1-yl)phosphonite, *in* T-00332
Diethyl (2,4,6-tribromophenyl)-phosphoramidate, *in* T-00359
▷Diethyl (2,2,2-trichloro-1-hydroxyethyl)-phosphonate, *in* T-00408
▷Diethyl (trichloromethyl)phosphonate, *in* T-00411
Diethyl (trichloromethyl)phosphonite, *in* T-00416
O,O-Diethyl (trichloromethyl)-phosphonothioate, *in* T-00415
▷*O,O*-Diethyl *O*-(3,5,6-trichloro-2-pyridinyl) phosphorothioate, *see* C-00194
Diethyl (tricyclo[3.3.1.1^{3,7}]dec-1-yl)-phosphonate, *in* A-00034

Diethyl trifluoroacetyl phosphate, D-00406
Diethyl (trifluoromethyl)phosphonate, *in* T-00478
Diethyl (trifluoromethyl)phosphonite, *in* T-00482
O,O-Diethyl (trifluoromethyl)-phosphonothioate, *in* T-00481
O,S-Diethyl (trifluoromethyl)-phosphonothioate, *in* T-00481
Diethyl(2,4,6-trimethylphenyl)phosphonate, *in* T-00527
Diethyl [(trimethylsilyloxy)carbonyl]-phosphonate, *in* D-00241
Diethyl [1-[(trimethylsilyl)oxy]ethyl]-phosphonate, *in* T-00555
Diethyl[(trimethylsilyloxy)methyl]-phosphonate, *in* T-00556
Diethyl (trimethylsilylphenyl)-phosphoramidate, *in* D-00312
Diethyl trimethylsilyl phosphate, D-00407
P,P-Diethyl-*N*-trimethylsilylphosphinothioic amide, *in* D-00331
Diethyl trimethylsilyl phosphite, D-00408
Diethyl triphenylmethylphosphonate, *in* T-00608
Diethyl (triphenylphosphoranylidene)-butanedioate, *in* T-00646
Diethyl (triphenylphosphoranylidene)-propanedioate, *in* T-00658
Diethyl (triphenylphosphoranylidene)-succinate, *in* T-00646
Diethyl vinylphosphonate, *in* V-00003
O,O-Diethyl vinylphosphonothioate, *in* V-00007
P,P-Diethynyl-*N,N*-dimethylphosphinic amide, *in* D-00409
Diethynylphosphinic acid, D-00409
▷Dietreen, *see* T-00038
Diferrocenylmethylphenylphosphonium iodide, *in* D-00410
Diferrocenylphenyl(phenylmethyl)phosphonium chloride, *in* D-00410
Diferrocenylphenylphosphine, D-00410
Difficidin, D-00411
Difficol, *in* D-00411
2,2-Difluoro-2,2-dihydro-1,3-dimethyl-2-phenyl-1,3,2-diazaphosphetidin-4-one, *in* D-00459
2,2-Difluoro-2,2-dihydro-1,2,3-trimethyl-1,3,2-diazaphosphetidin-4-one, *in* D-00459
1,1-Difluoro-1,1-dihydro-2,4,6-triphenylphosphorin, *in* D-00613
Difluoro(dimethylamino)phosphine, *see* D-00877
Difluoro(3-fluorophenyl)phosphine, *in* F-00041
Difluoro(4-fluorophenyl)phosphine, *in* F-00042
2,2-Difluoro-2,2,4,4,6,6-hexahydro-4,4,6,6-tetraphenylcyclotriphosphazene, *see* D-00412
2,2-Difluoro-2,2,4,4,6,6-hexahydro-4,4,6,6-tetraphenyl-1,3,5-triaza-2,4,6-triphosphorine, D-00412
Difluoro-*N,N,N′,N′,N″,N″*-hexamethylphosphoranetriamine, *see* T-00760
2,5-Difluoro-1-methyl-1,2,5-azadiphospholidine, *in* M-00089
(Difluoromethylene)(trifluoromethyl)-phosphine, D-00413
(Difluoromethylene)triphenylphosphorane, D-00414
Difluoromethylphosphine, *see* M-00324
(Difluoromethyl)phosphonic acid, D-00415
Difluorophenylphosphine, *see* P-00256
2-(Difluorothiophosphoryl)phenylhydrazine, *see* P-00289
Difluoro-*p*-tolylphosphine, *in* M-00248
Difluoro(trifluoromethyl)phosphine, *see* T-00484
Difluorotrimethoxyphosphorane, D-00416
Difluorotrimethylphosphorane, D-00417
P,P-Difluoro-*N*-trimethylsilyl-*P*-[bis(trimethylsilyl)amino]phosphazene, *in* T-00868
P,P-Difluoro-*N*-trimethylsilyl-*P*-[bis(trimethylsilyl)amino]phosphine imide, *in* T-00868
Difluorotriphenoxyphosphorane, D-00418
Difluorotriphenylphosphorane, D-00419
Difluorotris(trifluoromethyl)phosphorane, D-00420
▷Diflupyl, *see* D-00665
Di-2-furanylphosphinic acid, D-00421
Di-2-furanylphosphinothioic acid, D-00422
Di-2-furylphosphinic acid, *see* D-00421
Di-2-furylphosphinothioic acid, *see* D-00422
Diguanosine 5′,5‴-(P^1,P^4-tetraphosphate), *see* D-00423
P^1,P^4-Diguanosine 5′-tetraphosphate, D-00423
Diheptadecanoylphosphatidylcholine, *in* G-00010
Diheptyl chlorophosphate, *in* D-00424
O,O-Diheptyl dithiophosphoric acid, *see* D-00427
Diheptyl heptylphosphonate, *in* H-00006
Diheptyl heptylphosphonite, *in* H-00008
O,O-Diheptyl hydrogen dithiophosphate, *see* D-00427
Diheptyl hydrogen phosphate, *see* D-00424
Diheptyl phosphate, D-00424
Diheptylphosphinic acid, D-00425
Diheptyl phosphite, *see* D-00426
Diheptyl phosphonate, D-00426
Diheptyl phosphoric acid, *see* D-00424
Diheptyl phosphorochloridate, *in* D-00424
O,O-Diheptyl phosphorodithioate, D-00427
Diheptyl phosphoryl chloride, *in* D-00424
1,2-Dihexadecanoyl-*sn*-glycerol 3-phosphate, *in* G-00011
1,2-Dihexadecanoylglycero-3-phosphocholine, *see* G-00012
1,2-Dihexadecanoylglycero-3-phosphoethanolamine, *see* G-00013
Dihexadecyl hydrogen phosphate, *see* D-00428
Dihexadecyl phosphate, D-00428
Dihexadecylphosphinic acid, D-00429
▷Dihexadecyl phosphite, *see* D-00430
▷Dihexadecyl phosphonate, D-00430
Dihexadecyl phosphoric acid, *see* D-00428
Dihexyl amidophosphate, *in* D-00432
Dihexyl chlorophosphate, *in* D-00432
▷Dihexyl chlorophosphite, *see* D-00436
Dihexyl (diethylcarbamoyl)phosphonate, *in* D-00264
O,O-Dihexyl dithiophosphoric acid, *see* D-00437
O,O-Dihexyl hydrogen dithiophosphate, *see* D-00437
Dihexyl hydrogen phosphate, *see* D-00432
Dihexylmethylphenylphosphonium iodide, *in* D-00431
Dihexylphenylphosphine, D-00431
Dihexyl phosphate, D-00432
Dihexylphosphine, D-00433
▷Dihexylphosphinic acid, D-00434
Dihexyl phosphite, *see* D-00435
Dihexyl phosphonate, D-00435
Dihexyl phosphoramidate, *in* D-00432
Dihexyl phosphoric acid, *see* D-00432
Dihexyl phosphorochloridate, *in* D-00432
▷Dihexyl phosphorochloridite, D-00436
O,O-Dihexyl phosphorodithioate, D-00437
Dihexyl phosphoryl amide, *in* D-00432
Dihexyl phosphoryl chloride, *in* D-00432
(9,10-Dihydro-9-acridinyl)phosphonic acid, D-00438
9,10-Dihydro-9,10-*o*-benzeno-9-phosphaanthracene, *see* B-00014
2,2-Dihydro-1,3,2-benzodioxaphosphole, D-00439
7,12-Dihydrobenzo[*e*]phenophosphazine, D-00440
4,5-Dihydro-3*H*-benzo[*e*]phosphindole, D-00441
2,3-Dihydro-1*H*-benzo[*e*]phosphindole 3-oxide, D-00442
(3,4-Dihydro-1*H*-2-benzopyran-1-yl)-phosphonic acid, D-00443
2,2-Dihydro-1,3,2-benzoxazaphosphole, D-00444
2,5-Dihydro-3,5-bis(1,1-dimethylethyl)-1,2-oxaphosphole 2-oxide, *see* D-00082
2,2-Dihydro-3,4-bis(methoxycarbonyl)-2,2-dimethyl-5,5-diphenyl-5*H*-1,2-azaphosphole, D-00445
1,1-Dihydro-1,1-bis(methylthio)-2,4,6-triphenylphosphorin, *in* D-00613
1,1-Dihydro-1,1-bis(phenylthio)-2,4,6-triphenylphosphorin, *in* D-00613
2,2-Dihydro-3,3-bis(trifluoromethyl)-1-[[2,2,2-trifluoro-1-(trifluoromethyl)-ethylidene]amino]azaphosphiridine, D-00446
1,2-Dihydro-1,2-bis(triphenylphosphonio)-benzocyclobutene(2+), D-00447
10,11-Dihydro-1*H*-dibenz[*b,e*][1,4]-azaphosphepine 5-oxide, D-00448
5,10-Dihydro-11*H*-dibenz[*b,e*][1,4]-azaphosphepin-11-one 5-oxide, D-00449
6,7-Dihydro-5*H*-dibenzo[*d,f*][1,3,2]-diazaphosphepine, D-00450
10,11-Dihydro-5*H*-dibenzo[*b,f*]phosphepin, D-00451
5,6-Dihydrodibenzo[*b,d*]phosphorin, *see* D-00568
1,3-Dihydro-1,3-dihydroxy-1,2,3-benziodoxaphosphole 3-oxide, *in* D-00488
9,10-Dihydro-9,10-dihydroxy-9,10-diphosphaanthracene 9,10-dioxide, *see* D-00452
5,10-Dihydro-5,10-dihydroxyphosphanthrene 5,10-dioxide, D-00452
1,1-Dihydro-1,1-dimethoxy-2,4,6-triphenylphosphorin, *in* D-00613
1,5-Dihydro-3-dimethylamino-2,4,3-benzodioxaphosphepin, D-00453
4,5-Dihydro-2-dimethylamino-1,3,2-benzodioxaphosphepin, D-00454
2,3-Dihydro-2-dimethylamino-5-methyl-3-phenyl-1,3,4,2-oxadiazaphosphole, D-00455
2,5-Dihydro-1-dimethylamino-3-methyl-1*H*-phosphole, *in* D-00530
2,3-Dihydro-1,3-dimethyl-1*H*-1,3,2-benzodiazaphosphole, D-00456
2,3-Dihydro-1,3-dimethyl-1,3,2-benzodiazaphosphorin-4(1*H*)-one, D-00457
2,3-Dihydro-2,7-dimethyl-1,4,2-benzodithiaphosphorin, D-00458
2,3-Dihydro-1,1-dimethyl-1*H*-benzo[*g*]-phosphindolinium iodide, *in* D-00518
2,2-Dihydro-1,3-dimethyl-1,3,2-diazaphosphetidin-4-one, D-00459
10,11-Dihydro-5,5-dimethyl-5*H*-dibenzo[*b,f*]phosphepinium iodide, *in* D-00524
2,3-Dihydro-4,5-dimethyl-2,7-diphenyl-1*H*-1,2-azaphosphepine, D-00460
2,2-Dihydro-2,2-dimethyl-3,5-diphenyl-3*H*-1,2-azaphosphole-4-carboxylic acid, D-00461
2,3-Dihydro-1,3-dimethyl-1,3-diphenyl-1*H*-1,3-benzodiphospholium dibromide, *in* D-00476
2,3-Dihydro-4,5-dimethyl-2,7-diphenyl-1,2-oxaphosphepin 2-oxide, D-00462
9,10-Dihydro-9,10-dimethyl-9,10-diphenylphosphaanthranium diiodide, *in* D-00482
2,3-Dihydro-4,5-dimethyl-2,2-diphenyl-1,2-thiaphosphepin, D-00463
2,5-Dihydro-5,5-dimethyl-1,2-oxaphosphole 2-oxide, D-00464
2,2-Dihydro-1,3-dimethyl-4-oxo-1,3,2-diazaphosphetidine, *see* D-00459
5,5-Dihydro-4,4-dimethyl-2,3,5,5,5-pentaphenyl-1,2,5-oxazaphospholidine, D-00465
5,10-Dihydro-5,10-dimethylphenophosphazine, D-00466
2,5-Dihydro-3,4-dimethyl-1-phenoxy-1*H*-phosphole 1-oxide, *in* D-00493
2,3-Dihydro-1,3-dimethyl-2-phenyl-1*H*-benzodiazaphosphole, D-00467
2,5-Dihydro-3,4-dimethyl-1-phenyl-1*H*-phosphole, D-00468
5,6-Dihydro-5,5-dimethylphosphanthridinium chloride, *in* D-00568
9,10-Dihydro-9,9-dimethyl-9-phosphoniaphenanthrene bromide, *in* D-00568
1,3-Dihydro-1,3-dimethylspiro[2*H*-1,3,2-benzodiazaphosphole-2,2′[1,3,2]-benzodioxaphosphole], D-00469
4′,5′-Dihydro-3′,5′-dimethylspiro[1,3,2-benzoxazaphosph(P^V)ole-2,2′-[1,3,2]-oxazaphospholidin]-4′-one, D-00470

2,3-Dihydro-3,5-dimethyl-1,3,4,2-thiadiazaphosphole 2-sulfide, D-00471
4,7-Dihydro-1,3,2-dioxaphosphepin, D-00472
4,7-Dihydro-1,3,2-dioxaphosphepin 2-sulfide, D-00473
2-(1,3-Dihydro-1,3-dioxo-2*H*-isoindol-2-yl)ethyl phosphorodichloridate, *see* P-00425
[2-(1,3-Dihydro-1,3-dioxo-2*H*-isoindol-2-yl)ethyl]triphenylphosphonium(1+), D-00474
▷*S*-[(1,3-Dihydro-1,3-dioxo-2*H*-isoindol-2-yl)methyl] *O*,*O*-dimethyl phosphorodithioate, *see* P-00343
[(1,3-Dihydro-1,3-dioxo-2*H*-isoindol-2-yl)-methyl]phosphonic acid, *see* P-00426
(1,3-Dihydro-1,3-dioxo-2*H*-isoindol-2-yl)-phosphonothioic acid, D-00475
2,2-Dihydro-2,2-diphenoxy-3-phenyl-Δ^1-1,3,2-diazaphosphetin-4-one, *in* D-00557
1,1-Dihydro-1,1-diphenoxy-2,4,6-triphenylphosphorin, *in* D-00613
5,10-Dihydro-5,10-diphenylacridophosphine 5-oxide, *in* D-00548
2,3-Dihydro-1,3-diphenyl-1*H*-1,3-benzodiphosphole, D-00476
1,4-Dihydro-1,4-diphenyl-1,4-benzodiphosphorin, D-00477
2,3-Dihydro-2,3-diphenyl-1,3,2-benzoxazaphosphole, D-00478
9,10-Dihydro-9,10-diphenyl-9,10-diphosphaanthracene, *see* D-00482
2,3-Dihydro-2,2-diphenyl-1*H*-isophosphindolium bromide, *in* D-00562
2,3-Dihydro-2,2-diphenyl-1*H*-isophosphindolium perchlorate, *in* D-00562
2,2-Dihydro-3,5-diphenyl-1,3,4,2-oxadiazaphospholine, *see* T-00104
8,8-Dihydro-8,8-diphenyl-2*H*,6*H*-[1,2]-oxaphospholo[4,3,2-*hi*][2,1]-benzoxaphosphole, D-00479
2,2-Dihydro-4,5-diphenyl-1,3,2-oxathiaphosphole, D-00480
10,10-Dihydro-2,3-diphenyl-[1,3,2]-oxazaphospholo[2,3-*b*][1,3,2]-benzoxazaphosphole, D-00481
5,10-Dihydro-5,10-diphenylphosphanthrene, D-00482
5,6-Dihydro-5,5-diphenylphosphanthridinium bromide, *in* D-00568
1,1-Dihydro-1,1-diphenyl-3*H*-phosphindole, D-00483
1,1-Dihydro-1,1-diphenylphosphinine, *see* D-00484
9,10-Dihydro-9,9-diphenyl-9-phosphoniaphenanthrene bromide, *in* D-00568
1,1-Dihydro-1,1-diphenylphosphorin, D-00484
2,2-Dihydro-4-ethoxycarbonyl-2,2-dimethyl-3,5-diphenyl-3*H*-1,2-azaphosphole, *in* D-00461
2,3-Dihydro-1-ethoxy-4-methyl-1*H*-phosphole 1-oxide, *in* D-00496
5,10-Dihydro-10-ethoxyphenophosphazine 10-sulfide, *in* D-00502
2,3-Dihydro-2-ethoxy-1,3,5-triphenyl-1*H*-1,2-azaphosphole 2-oxide, *in* D-00508
Dihydrogen octyl phosphate, *see* M-00501
Dihydrogen phenyl phosphate, *see* M-00507
Dihydrogen tribromoethyl phosphate, *see* M-00516
1,1-Dihydro-1,1,3,3,5,5-hexamethyl-1-phospha-3,5-disilacyclohex-1-ene, D-00485
1,1-Dihydro-1,1,3,3,4,4-hexamethyl-1-phospha-3,4-disilacyclopent-1-ene, D-00486
5,10-Dihydro-5-hydroxyacridophosphine 5-oxide, D-00487
1,3-Dihydro-1-hydroxy-1,2,3-benziodoxaphosphole 3-oxide, D-00488
4,5-Dihydro-4-hydroxybenzo[*lmn*]-phosphanthridine 4-oxide, D-00489
1,3-Dihydro-1-hydroxy-2,1-benzoxaphosphole 1-oxide, D-00490
6,7-Dihydro-6-hydroxy-5*H*-dibenzo[*c*,*e*]-phosphepin 6-oxide, D-00491
10,11-Dihydro-5-hydroxy-5*H*-dibenzo[*b*,*f*]-phosphepin 5-oxide, D-00492
5,10-Dihydro-5-hydroxydibenzo[*b*,*e*]-phosphinine 5-oxide, *see* D-00487
5,6-Dihydro-5-hydroxydibenzo[*b*,*d*]phosphorin 5-oxide, *see* D-00504
5,10-Dihydro-5-hydroxydibenzo[*b*,*e*]-phosphorin 5-oxide, *see* D-00487
▷2,5-Dihydro-1-hydroxy-3,4-dimethyl-1*H*-phosphole 1-oxide, D-00493
2,3-Dihydro-2-hydroxy-1*H*-isophosphindole 2-oxide, D-00494
1,2-Dihydro-2-hydroxyisophosphinoline 2-oxide, D-00495
5,10-Dihydro-5-hydroxy-10-methylacridophosphine 5-oxide, *in* D-00515
1,3-Dihydro-1-hydroxy-3-methyl-1,2,3-benzodoxaphosphole 3-oxide, *in* D-00488
▷2,3-Dihydro-1-hydroxy-4-methyl-1*H*-phosphole 1-oxide, D-00496
▷2,3-Dihydro-1-hydroxy-5-methyl-1*H*-phosphole 1-oxide, D-00497
2,5-Dihydro-1-hydroxy-2-methyl-1*H*-phosphole-1-oxide, D-00498
2,5-Dihydro-1-hydroxy-3-methyl-1*H*-phosphole 1-oxide, D-00499
3,6-Dihydro-2-hydroxy-2*H*-1,2-oxaphosphinin 2-oxide, *see* D-00501
2,5-Dihydro-2-hydroxy-1,2-oxaphosphole 2-oxide, D-00500
3,6-Dihydro-2-hydroxy-2*H*-1,2-oxaphosphorin 2-oxide, D-00501
▷5,10-Dihydro-10-hydroxyphenophosphazine 10-oxide, *see* P-00061
5,10-Dihydro-10-hydroxyphenophosphazine 10-sulfide, D-00502
5,10-Dihydro-5-hydroxy-10-phenylacridophosphine 5-oxide, *in* D-00548
1,3-Dihydro-1-hydroxy-3-phenyl-1,2,3-benziodoxaphosphole 3-oxide, *in* D-00488
1,2-Dihydro-1-hydroxy-1-phenyl-4*H*-3,1,2-benzothiazaphosphorin-4-one 1-sulfide, D-00503
5,6-Dihydro-5-hydroxyphosphanthridine 5-oxide, D-00504
4,5-Dihydro-4-hydroxy-4-phosphapyrene 4-oxide, *see* D-00489
1,2-Dihydro-1-hydroxyphosphinoline 1-oxide, D-00505
2,3-Dihydro-1-hydroxy-1*H*-phosphole 1-oxide, D-00506
2,5-Dihydro-1-hydroxy-1*H*-phosphole 1-oxide, D-00507
2,3-Dihydro-2-hydroxy-1,3,5-triphenyl-1*H*-1,2-azaphosphole 2-oxide, D-00508
2,3-Dihydro-1*H*-isophosphindole, D-00509
1,2-Dihydroisophosphinoline, D-00510
5,10-Dihydro-10-isopropoxyphenophosphazine 10-sulfide, *in* D-00502
2,3-Dihydro-2-mercapto-1,3,2-benzoxazaphosphole 2-sulfide, D-00511
2,5-Dihydro-1-mercaptophosphole 1-sulfide, D-00512
5,10-Dihydro-5-methoxyacridophosphine 5-oxide, *in* D-00487
2,3-Dihydro-2-methoxy-1,2-benzoxaphosphole 2-oxide, *in* H-00119
10,11-Dihydro-5-methoxy-5*H*-dibenzo[*b*,*f*]-phosphepin 5-oxide, *in* D-00492
2,5-Dihydro-1-methoxy-3,4-dimethyl-1*H*-phosphole 1-oxide, *in* D-00493
2,3-Dihydro-2-methoxy-1*H*-isophosphindole 2-oxide, *in* D-00494
2,3-Dihydro-2-methoxy-5-methyl-3-phenyl-1,3,4,2-oxadiazaphosphole, D-00513
2,5-Dihydro-1-methoxy-3-methyl-1*H*-phosphole, *in* D-00530
2,3-Dihydro-1-methoxy-4-methyl-1*H*-phosphole 1-oxide, *in* D-00496
▷2,5-Dihydro-1-methoxy-3-methyl-1*H*-phosphole 1-oxide, *in* D-00499
2,3-Dihydro-2-methoxy-5-methyl-1,2-thiaphosphole, D-00514
5,10-Dihydro-10-methoxyphenophosphazine 10-oxide, *in* P-00061
5,10-Dihydro-10-methoxyphenophosphazine 10-sulfide, *in* D-00502
2,3-Dihydro-1-methoxy-1*H*-phosphindole 1-oxide, *in* H-00120
1,2-Dihydro-1-methoxyphosphinoline 1-oxide, *in* D-00505
▷2,3-Dihydro-1-methoxy-1*H*-phosphole 1-oxide, *in* D-00506
▷2,5-Dihydro-1-methoxy-1*H*-phosphole 1-oxide, *in* D-00507
5,10-Dihydro-10-methylacridophosphine, D-00515
1,5-Dihydro-3-methyl-2,4,3-benzodioxaphosphepin, D-00516
2,3-Dihydro-2-methyl-1,4,2-benzodioxaphosphorin, D-00517
2,3-Dihydro-1-methyl-1*H*-benzo[*g*]-phosphindole, D-00518
4,5-Dihydro-3-methyl-3*H*-benzo[*e*]-phosphindole, *in* D-00441
2,3-Dihydro-2-methyl-1,3,2-benzothiazaphosphole, D-00519
2,3-Dihydro-2-methyl-1*H*-4,1,2-benzothiazaphosphorine, D-00520
3,4-Dihydro-2-methyl-2*H*-1,4,2-benzothiazaphosphorine, D-00521
3,4-Dihydro-2-methyl-2*H*-1,4,2-benzoxazaphosphinine, *see* D-00522
3,4-Dihydro-2-methyl-2*H*-1,4,2-benzoxazaphosphorine, D-00522
1,2-Dihydro-1-methyl-4*H*-3,1,2-benzoxazaphosphorin-4-one, D-00523
10,11-Dihydro-5-methyl-5*H*-dibenzo[*b*,*f*]-phosphepin, D-00524
5,10-Dihydro-5-methyl-5,10-diphenylacridophosphonium iodide, *in* D-00548
2,3-Dihydro-5-methyl-*N*,3-diphenyl-1,3,4,2-oxadiazaphosphol-2-amine, *see* D-00527
2,3-Dihydro-5-methyl-*N*,3-diphenyl-1,3,4,2-thiadiazaphosphole-2(3*H*)amine 2-sulfide, *in* A-00072
2,5-Dihydro-1-(1-methylethoxy)-1*H*-phosphole 1-oxide, *in* D-00507
2,3-Dihydro-5-methyl-1,2-oxaphosphole 2-oxide, D-00525
2,3-Dihydro-4-methyl-1-phenoxy-1*H*-phosphole 1-oxide, *in* D-00496
2,5-Dihydro-2-methyl-1-phenoxy-1*H*-phosphole 1-oxide, *in* D-00498
2,5-Dihydro-3-methyl-2-phenoxy-1*H*-phosphole 1-oxide, *in* D-00499
10,11-Dihydro-5-methyl-5-phenyl-5*H*-dibenzo[*b*,*f*]phosphepinium iodide, *in* D-00559
7,12-Dihydro-12-methyl-12-phenyl-5*H*-dibenz[*c*,*f*][1,5]oxaphosphocinium iodide, *in* D-00560
2,3-Dihydro-2-methyl-2-phenyl-1*H*-isophosphindolium iodide, *in* D-00562
5,10-Dihydro-5-methyl-10-phenylphenophosphazine, D-00526
2,3-Dihydro-5-methyl-3-phenyl-2-phenylamino-1,3,4,2-oxadiazaphosphole, D-00527
4,5-Dihydro-1-methyl-1-phenyl-1*H*-phosphepinium perchlorate, *in* D-00565
2,5-Dihydro-3-methyl-1-phenyl-1*H*-phosphole, D-00528
2,3-Dihydro-1-methyl-1-phenyl-1*H*-phospholium iodide, *in* D-00566
2,3-Dihydro-5-methyl-3-phenyl-1,3,4,2-thiadiazaphosphole-2(3*H*)-amine 2-sulfide, *see* A-00072
2,3-Dihydro-5-methyl-3-phenyl-1,3,4,2-thiadiazaphosphole 2-sulfide, D-00529
5,6-Dihydro-5-methylphosphanthridene 5-oxide, *in* D-00568
2,5-Dihydro-3-methyl-1*H*-phosphole, D-00530
2,2-Dihydro-4-methyl-2,2,2,3-tetraphenyl-1,2-oxaphosphetane, D-00531
1,2-Dihydro-1-methyl-1,2,4,6-tetraphenylphosphorinium iodide, *in* D-00584
5,10-Dihydro-10-methylthiophenophosphazine 10-oxide, *in* D-00502
2,3-Dihydro-1*H*-naphtho[1,8-*de*]-1,3,2-diazaphosphorine, D-00532
2,2-Dihydro-1,2-oxaphosphetane, D-00533

2,2-Dihydro-1,2-oxaphospholane, D-00534
2,2-Dihydro-1,2-oxaphosphol-4-ene, *see* T-00119
▷*S*-(3,4-Dihydro-4-oxobenzo[*d*][1,2,3]-triazin-3-ylmethyl) *O,O*-diethyl phosphorodithioate, *see* A-00150
▷*S*-(3,4-Dihydro-4-oxobenzo[*d*][1,2,3]-triazin-3-ylmethyl) *O,O*-dimethyl phosphorodithioate, *see* A-00151
(1,3-Dihydro-3-oxo-1-isobenzofuranyl)-phosphonic acid, D-00535
9,10-Dihydro-10-oxo-9-phenyl-9-phosphoniaanthracene chloride, *in* P-00080
▷2,3-Dihydro-1,2,3,4,5-pentakis(trifluoromethyl)-1*H*-1,2,3-triphosphole, D-00536
2,2-Dihydro-2,2,2,4,5-pentaphenyl-1,3,2-dioxaphospholane, D-00537
1,4-Dihydro-1,2,3,4,9-pentaphenyl-1,4-phosphinidenenaphthalene, D-00538
2,3-Dihydro-1,2,3,4,5-pentaphenyl-1*H*-1,2,3-triphosphole, D-00539
[(9,10-Dihydro-9-phenanthrenenyl)-methylene]triphenylphosphorane, *in* D-00540
(9,10-Dihydro-9-phenanthreneyl)-triphenylphosphonium(1+), D-00540
2,2-Dihydrophenanthro[9,10-*d*]-1,3,2-dioxaphosphole, D-00541
(9,10-Dihydro-9-phenanthryl)-triphenylphosphonium(1+), *see* D-00540
5,10-Dihydrophenophosphazine, D-00542
2,3-Dihydro-2-phenoxy-1*H*-1,3,2-benzodiazaphosphole, D-00543
1,5-Dihydro-2-phenoxy-2,4,3-benzodioxaphosphepin, D-00544
2,3-Dihydro-2-phenoxy-1,3,2-benzothiazaphosphole, D-00545
1,3-Dihydro-1-phenoxy-2,1-benzoxaphosphole, D-00546
2,3-Dihydro-1-phenoxy-1*H*-phosphole 1-oxide, *in* D-00506
2,5-Dihydro-1-phenoxy-1*H*-phosphole 1-oxide, *in* D-00507
5,10-Dihydro-5-phenylacridophosphine, D-00547
5,10-Dihydro-10-phenylacridophosphine, D-00548
4,5-Dihydro-3-phenyl-3*H*-1,3-azaphosphole, D-00549
1,5-Dihydro-3-phenyl-2,4,3-benzodioxaphosphepin, D-00550
2,3-Dihydro-1-phenyl-1*H*-benzo[*g*]-phosphindole, D-00551
4,5-Dihydro-3-phenyl-3*H*-benzo[*e*]-phosphindole, *in* D-00441
1,2-Dihydro-1-phenyl-4*H*-3,1,2-benzothiazaphosphorin-4-thione, D-00552
1,3-Dihydro-1-phenyl-2,1-benzoxaphosphole, D-00553
3,4-Dihydro-3-phenyl-1*H*-2,3-benzoxaphosphorin, D-00554
4,5-Dihydro-2-phenyl-1,3,2-benzoxathiaphosphepin, D-00555
2,3-Dihydro-3-phenyl-4*H*-1,3,2-benzoxazaphosphorin-4-one, D-00556
2,2-Dihydro-3-phenyl-Δ^1-1,3,2-diazaphosphetin-4-one, D-00557
2,3-Dihydro-2-phenyl-1*H*-dibenz[*e,g*]-isophosphindole, D-00558
10,11-Dihydro-5-phenyl-5*H*-dibenzo[*b,f*]-phosphepin, D-00559
7,12-Dihydro-12-phenyl-5*H*-dibenz[*c,f*]-[1,5]oxaphosphocin, D-00560
4,7-Dihydro-2-phenyl-1,3,2-dioxaphosphepin, D-00561
2,3-Dihydro-2-phenyl-1*H*-isophosphindole, D-00562
5,10-Dihydro-10-phenylphenophosphazine, D-00563
5,5-Dihydro-5-phenyl-5-(phenylmethyl)-acridophosphine, *see* B-00042
5,5-Dihydro-5-phenyl-5-(phenylmethyl)-dibenzo[*b,e*]phosphinine, *see* B-00042
5,6-Dihydro-6-phenyl-5-(phenylmethyl)-phosphanthridine, *see* B-00043
9,10-Dihydro-9-phenyl-9-phosphaanthracene, *see* D-00547
9,10-Dihydro-9-phenyl-9-phosphaanthracen-10-one, *see* P-00080
5,6-Dihydro-5-phenylphosphanthridene 5-oxide, *in* D-00568
2,7-Dihydro-1-phenyl-1*H*-phosphepin, D-00564
4,5-Dihydro-1-phenyl-1*H*-phosphepin, D-00565
2,3-Dihydro-1-phenyl-1*H*-phosphole, D-00566
2,5-Dihydro-1-phenyl-1*H*-phosphole, D-00567
2,3-Dihydro-2-phenylspiro[1,3,2-benzothiazaphosphole-2,2′-[1,3,2]-dioxaphospholane, *see* P-00308
9,9-Dihydro-9-phosphahomocubane, *see* D-00569
5,6-Dihydrophosphanthridene, D-00568
9,9-Dihydro-9-phosphapentacyclo[4.3.0.$0^{2,5}$.$0^{3,8}$.$0^{4,7}$]nonane, D-00569
9,10-Dihydro-9-phosphaphenanthrene, *see* D-00568
1,1-Dihydrophosphinane, *see* D-00573
2,3-Dihydro-1*H*-phosphindole 1-oxide, D-00570
2,3-Dihydro-1*H*-phosphole, D-00571
2,5-Dihydro-1*H*-phosphole, D-00572
9,10-Dihydro-9-phosphonoacridine, *see* D-00438
3,4-Dihydro-1-phosphono-1*H*-2-benzopyran, *see* D-00443
1,1-Dihydrophosphorinane, D-00573
1,3-Dihydrospiro[2*H*-1,3,2-benzodiazaphosphole-2,2′-[1,3,2]-benzodioxaphosphole], D-00574
1,3-Dihydrospiro[2*H*-1,3,2-benzodiazaphosphole-2,2′-[1,3,2]-dioxaphosholane], D-00575
2′,5′-Dihydrospiro[1,3,2-benzodioxaphosphole-2,1′-[1*H*]phosphole], *see* S-00010
1,4-Dihydrospiro[2*H*-3,1,2-benzoxazaphosphorine-2,2′-[1,3,2]-dioxaphospholane], D-00576
6,7-Dihydro-5,5,7,7-tetramethyl-5*H*-dibenzo[*d,f*][1,3]diphosphepinium(2+), D-00577
2,5-Dihydro-1,1,3,4-tetramethyl-1*H*-phospholium bromide, *in* D-00600
2,2-Dihydro-4,4,5,5-tetramethyl-2,2,2-triphenyl-1,3,2-dioxaphospholane, D-00578
3,4-Dihydro-2,3,4,5-tetraphenyl-2*H*-1,2,3-diazaphosphole, D-00579
3,4-Dihydro-2,3,4,5-tetraphenyl-2*H*-1,3,2-diazaphosphole, D-00580
1,2-Dihydro-1,2,3,4-tetraphenyl-1,2-diphosphete, D-00581
1,4-Dihydro-1,1,4,4-tetraphenyl-1,4-diphosphininium(2+), *see* D-00582
1,4-Dihydro-1,1,4,4-tetraphenyl-1,4-diphosphorinium(2+), D-00582
2,5-Dihydro-2,3,4,5-tetraphenyl-1,2,5-oxadiphosphole 2,5-dioxide, D-00583
5,5-Dihydro-3,5,5,5-tetraphenyl-Δ^2-1,2,5-oxazaphospholine, *see* T-00153
1,2-Dihydro-1,2,4,6-tetraphenylphosphinine, *see* D-00584
1,2-Dihydro-1,2,4,6-tetraphenylphosphorin, D-00584
2,5-Dihydro-2,3,4,5-tetraphenyl-1,2,5-thiadiphosphole, D-00585
3,6-Dihydro-2*H*-1,2-thiaphosphorin 2-sulfide, D-00586
N-(4,5-Dihydro-2-thiazolyl)-*P*-methylphosphonamidic acid, D-00587
N-(4,5-Dihydro-2-thiazolyl)-*P*-methylphosphonamidothioic acid, D-00588
2,2-Dihydro-2,2,2-trichloro-4,4,5,5-tetrakis(trifluoromethyl)-1,3,2-dioxaphospholane, D-00589
2,2-Dihydro-2,2,2-triethoxy-4-[(4-methylphenyl)sulfonyl]-1,4,2-oxazaphospholidine, D-00590
2,2-Dihydro-2,2,2-trimethoxy-3,5-bis(pentafluorophenyl)-1,4,2-dioxaphospholane, D-00591
2,2-Dihydro-2,2,2-trimethoxy-4,5-bis(trifluoromethyl)-1,3,2-dithiaphosphole, D-00592
2,2-Dihydro-2,2,2-trimethoxy-4,5-dimethyl-1,3,2-dioxaphosphole, D-00593
2,2-Dihydro-2,2,2-trimethoxy-3-methyl-4,5-diphenyl-Δ^4-1,3,2-oxazaphospholine, *see* T-00156
2,2-Dihydro-2,2,2-trimethoxynaphtho[1,2-*d*]-1,3,2-dioxaphosphole, D-00594
2,2-Dihydro-2,2,2-trimethoxy-5-phenyl-3,3-bis(trifluoromethyl)-Δ^4-1,4,2-oxazaphospholine, *see* T-00158
2,2-Dihydro-2,2,2-trimethoxy-4,4,5,5-tetrakis(trifluoromethyl)-1,3,2-dioxaphospholane, D-00595
2,2-Dihydro-2,2,2-trimethyl-1,2-oxaphosphinane, *see* D-00597
▷2,3-Dihydro-3,3,5-trimethyl-1,2-oxaphosphole 2-oxide, D-00596
2,2-Dihydro-2,2,2-trimethyl-1,2-oxaphosphorinane, D-00597
[(5,6-Dihydro-4,4,6-trimethyl-4*H*-1,3-oxazin-2-yl)methylene]-triphenylphosphorane, *in* D-00599
[(5,6-Dihydro-4,4,6-trimethyl-4*H*-1,3-oxazin-2-yl)methyl]phosphonic acid, D-00598
[(5,6-Dihydro-4,4,6-trimethyl-4*H*-1,3-oxazin-2-yl)methyl]-triphenylphosphonium(1+), D-00599
2,3-Dihydro-*N*,*N*,5-trimethyl-3-phenyl-1,3,4,2-oxadiazaphosphol-2-amine, *see* D-00455
2,5-Dihydro-1,3,4-trimethyl-1-phenyl-1*H*-phospholium bromide, *in* D-00468
2,3-Dihydro-*N*,*N*,5-trimethyl-3-phenyl-1,3,4,2-thiadiazaphosphole-2(3*H*)-amine-2-sulfide, *in* A-00072
2,5-Dihydro-1,3,4-trimethyl-1*H*-phosphole, D-00600
1,3-Dihydro-1,2,3-trimethylspiro[2*H*-1,3,2-benzodiazaphosphole-2,2′-[1,3,2]benzodioxaphosphole], *in* D-00469
2,2-Dihydro-2,2,2-trimethyl-4,4,5,5-tetrakis(trifluoromethyl)-1,3,2-dioxaphospholane, D-00601
2,3-Dihydro-1,2,3-trimethyl-1*H*-1,2,3-triphosphole, *in* D-00615
2,2-Dihydro-2,2,2-triphenoxy-1,3,2-benzodioxaphosphole, D-00602
4,4-Dihydro-4,4,4-triphenoxytrioxaphosphetane, D-00603
1,4-Dihydro-2,4,6-triphenyl-1,4-azaphosphinine, *see* D-00604
1,4-Dihydro-2,4,6-triphenyl-1,4-azaphosphorine, D-00604
2,3-Dihydro-1,2,3-triphenyl-1*H*-1,3,2-benzodiphospharsole, D-00605
2,3-Dihydro-1,2,3-triphenyl-1*H*-benzotriphosphole, D-00606
2,2-Dihydro-2,2,2-triphenyl-2*H*-1,2-benzoxaphosphorin, D-00607
2,2-Dihydro-2,2,2-triphenyl-4,4-bis(trifluoromethyl)-3-(triphenylphosphoranylidene)-1,2-oxaphosphetane, D-00608
2,2-Dihydro-2,2,3-triphenyl-Δ^1-1,3,2-diazaphosphetin-4-one, *in* D-00557
2,3-Dihydro-2,4,9-triphenyl-1*H*-1,3,2-diazaphosphonine 2-oxide, D-00609
5,5-Dihydro-5,5,5-triphenyl-5*H*-dibenzophosphole, D-00610
2,2-Dihydro-2,2,2-triphenyl-1,3,2-dioxaphospholane, D-00611
1,1-Dihydro-2,4,6-triphenylphosphinin, *see* D-00613
Dihydro(triphenylphosphoranylidene)-2,5-furandione, D-00612
1,1-Dihydro-2,4,6-triphenylphosphorin, D-00613
2,2-Dihydro-2,2,2-triphenyl-4,4,5,5-tetrakis(trifluoromethyl)-1,3,2-dioxaphospholane, D-00614
2,3-Dihydro-1,2,3-triphenyl-1*H*-1,2,3-triphosphole, *in* D-00615
2,3-Dihydro-1*H*-1,2,3-triphosphole, D-00615
1,3-Dihydroxy-1*H*-1,2,4,3-benziodadioxaphosphorin 3-oxide, D-00616
2,5-Dihydroxy-3,6-dimethyl-3,6-bis(dihydroxyphosphinyl)-1,4,2,5-dioxadiphosphorinane 2,5-dioxide, D-00617
[2,5-Dihydroxy-3,6-dimethyl-2,5-dioxo-1,4,2,5-dioxadiphosphorinan-3,6-diyl]-bisphosphonic acid, *see* D-00617

2,5-Dihydroxy-1,4,2,5-dioxadiphosphinane 2,5-dioxide, *see* D-00618
2,5-Dihydroxy-1,4,2,5-dioxadiphosphorinane 2,5-dioxide, D-00618
1,5-Dihydroxy-1,5-diphosphabicyclo[5.4.0]-undecane 1,5-dioxide, *see* D-00002
1,3-Dihydroxy-1,3-diphosphepane 1,3-dioxide, D-00619
1,3-Dihydroxy-1,3-diphosphetane 1,3-dioxide, D-00620
1,3-Dihydroxy-1,3-diphosphinane 1,3-dioxide, *see* D-00622
1,3-Dihydroxy-1,3-diphospholane 1,3-dioxide, D-00621
1,3-Dihydroxy-1,3-diphosphorinane 1,3-dioxide, D-00622
2,5-Dihydroxy-3-(1-methylpropyl)-1,3,2,5-oxazadiphospholidine 2,5-dioxide, D-00623
Dihydroxyphosphinecarboxylic acid oxide, *see* P-00387
Dihydroxyphosphinylacetic acid, *see* P-00376
2-[(Dihydroxyphosphinyl)amino]pentanedioic acid, *see* P-00367
Dihydroxyphosphinylformamide, *see* C-00003
2-[(Dihydroxyphosphinyl)oxy]-*N,N,N*-trimethylethanaminium hydroxide, inner salt, *see* C-00198
(2,3-Dihydroxypropyl)phosphonic acid, D-00624
1,8-Dihydroxy-3*a*,4,7,7*a*-tetrahydro-4,7-phosphinidene-1*H*-phosphindole 1,8-dioxide, D-00625
3,9-Dihydroxy-2,4,8,10-Tetraoxa-3,9-diphosphaspiro[5.5]undecane 3,9-dioxide, D-00626
1,3-Dihydroxytriphosphetane 1,3-dioxide, D-00627
1,2-Diicosa-5,8,11,14-tetraenoylglycero-3-phosphocholine, *see* G-00017
1,2-Diicosa-5,8,11,14-tetraenoylglycero-3-phosphoethanolamine, *see* G-00018
Di-1*H*-imidazol-1-ylphosphinic acid, D-00628
Diiodomethylphosphine, *see* M-00325
Diiodophenylphosphine, *in* P-00251
Diiodo-*p*-tolylphosphine, *in* M-00248
Diiodotrimethylphosphorane, D-00629
P,P-Diiodo-*N*-trimethylsilyl-*P*-[bis(trimethylsilyl)amino]phosphazene, *in* T-00868
P,P-Diiodo-*N*-trimethylsilyl-*P*-[bis(trimethylsilyl)amino]phosphine imide, *in* T-00868
Diiodotriphenoxyphosphorane, D-00630
▷Diiodotriphenylphosphorane, D-00631
Diisoamyl phosphate, *see* B-00302
Diisoamyl phosphoranilide, *in* B-00302
Diisobutoxyphosphinyl fluoride, *see* B-00362
O,O-Diisobutyl chloroselenophosphate, *see* B-00356
O,O-Diisobutyl dichlorothiophosphate, *see* B-00357
Diisobutyl diisobutylphosphoramidate, *in* B-00355
S,S-Diisobutyl dithiophosphate, *see* B-00361
O,O-Diisobutyl dithiophosphoric acid, *see* B-00360
S,S-Diisobutyl dithiophosphoric acid, *see* B-00361
Diisobutyl fluorophosphate, *see* B-00362
O,O-Diisobutyl hydrogen dithiophosphate, *see* B-00360
S,S-Diisobutyl hydrogen dithiophosphate, *see* B-00361
O,O-Diisobutyl hydrogen phosphorothioate, *see* B-00363
Diisobutyl isobutylphosphonate, *in* M-00378
Diisobutyl isobutylphosphonite, *in* M-00383
O,O-Diisobutyl isobutylphosphonothioate, *in* M-00381
Diisobutyl phenylphosphonate, *in* P-00221
Diisobutyl phenylphosphoramidate, *in* P-00257
Diisobutyl phosphate, *see* B-00342
Diisobutylphosphine, *see* B-00344
Diisobutylphosphinic acid, *see* B-00346
Diisobutylphosphinic amide, *in* B-00346
Diisobutylphosphinodithioic acid, *see* B-00348
Diisobutylphosphinothioic acid, *see* B-00351
Diisobutylphosphinothious acid, *see* B-00352
Diisobutyl phosphite, *see* B-00354
Diisobutyl phosphonate, *see* B-00354
Diisobutylphosphoramidic acid, *see* B-00355
Diisobutyl phosphoric acid, *see* B-00342
O,O-Diisobutyl phosphorochloridoselenoate, *see* B-00356
O,O-Diisobutyl phosphorochloridothioate, *see* B-00357
N,N'-Diisobutylphosphorodiamidodiselenoic acid, *see* B-00358
O,O-Diisobutyl phosphorodithioate, *see* B-00360
Diisobutyl phosphorofluoridate, *see* B-00362
O,O-Diisobutyl phosphorothioate, *see* B-00363
O,O-Diisobutyl phosphorothioic acid, *see* B-00363
Diisobutyl phosphoryl chloride, *in* B-00342
Diisobutyl phosphoryl fluoride, *see* B-00362
O,O-Diisobutyl selenophosphoryl chloride, *see* B-00356
O,O-Diisobutyl thiophosphoric acid, *see* B-00363
O,O-Diisobutyl thiophosphoryl chloride, *see* B-00357
Diisododecanoylphosphatidylcholine, *in* G-00025
Diisoheptadecanoylphosphatidylcholine, *in* G-00021
Diisohexadecanoylphosphatidylcholine, *in* G-00022
1,2-Diisooctadecanoylphosphatidylcholine, *in* G-00020
1,2-Diisopentadecanoylphosphatidylcholine, *in* G-00023
O,O-Diisopentyl dithiophosphoric acid, *see* B-00304
O,O-Diisopentyl hydrogen dithiophosphate, *see* B-00304
O,O-Diisopentyl hydrogen phosphorothioate, *see* B-00305
O,O-Diisopentyl hydrogen thiophosphate, *see* B-00305
Diisopentyl phenylphosphoramide, *in* B-00302
Diisopentyl phosphate, *see* B-00302
Diisopentyl phosphite, *see* B-00303
Diisopentyl phosphonate, *see* B-00303
Diisopentyl phosphoramide, *in* B-00302
O,O-Diisopentyl phosphorodithioate, *see* B-00304
O,O-Diisopentyl phosphorothioate, *see* B-00305
Diisopentyl phosphoryl chloride, *in* B-00302
Diisopentyl phosphoryl fluoridate, *in* B-00302
1,3-Diisopropoxy-1,3-diphosphepane 1,3-dioxide, *in* D-00619
1,3-Diisopropoxy-1,3-diphospholane 1,3-dioxide, *in* D-00621
1,3-Diisopropoxy-1,3-diphosphorinane 1,3-dioxide, *in* D-00622
1,3-Diisopropoxy-2-isopropyltriphosphetane 1,3-dioxide, *in* D-00627
(Diisopropoxyphosphinothioyl)thioacetic acid, D-00632
(Diisopropoxyphosphinyl)acetamide, *in* D-00633
(Diisopropoxyphosphinyl)acetic acid, D-00633
2-(Diisopropoxyphosphinylamino)pyridine, *in* P-00510
3-(Diisopropoxyphosphinylamino)pyridine, *in* P-00511
Diisopropoxyphosphinylcyclohexane, *in* C-00256
(Diisopropoxyphosphinyl)formamide, *in* C-00003
N-Diisopropoxyphosphinylimidazole, *in* I-00003
1-(Diisopropoxyphosphinyl)-2-propanol, *in* H-00191
Diisopropyl (acetamidomethyl)phosphonate, *in* A-00012
Diisopropyl acetonylphosphonate, *in* O-00075
Diisopropyl 1-acetoxy-2,2,2-trichloroethylphosphonate, *in* A-00011
Diisopropyl acetylphosphonate, *in* A-00026
Diisopropyl 1-adamantylphosphonate, *in* A-00034
Diisopropyl allenylphosphonate, *in* P-00441
Diisopropyl allylphosphonate, *in* P-00454
Diisopropyl allylphosphonite, *in* P-00463
Diisopropyl allylphosphonoselenoate, *in* P-00458
O,O-Diisopropyl allylphosphonothioate, *in* P-00461
Diisopropyl amidophosphate, *see* D-00653
Diisopropyl aminocarbonylphosphonate, *in* C-00003
Diisopropyl aminocarbonylphosphoramidate, *in* A-00062
Diisopropyl 2-aminoethyl phosphate, *in* A-00075
Diisopropyl (3-aminopropyl)phosphonate, *in* A-00127
Diisopropyl (2-amino-4-pyrimidinyl)-phosphonate, *in* A-00128
Diisopropyl 1-aziridinylphosphonite, *in* A-00154
Diisopropyl benzoylphosphonate, *in* B-00034
Diisopropyl benzoylphosphoramidate, *in* B-00036
O,O-Diisopropyl benzoylphosphoramidothioate, *in* B-00038
Diisopropyl benzylphosphonite, *in* B-00068
Diisopropyl benzylphosphoramidate, *in* B-00069
▷Diisopropyl bromophosphate, *in* D-00639
Diisopropyl 1-butenylphosphonate, *in* B-00524
Diisopropyl [2-(butylamino)-1-hexenyl)]-phosphonate, *in* B-00540
P,P-Diisopropyl-*N*-*tert*-butylphosphinous amide, *in* D-00649
Diisopropyl butylphosphonite, *in* B-00615
Diisopropyl *tert*-butylphosphonite, *in* B-00616
Diisopropyl (carbamoylmethyl)phosphonate, *in* A-00060
Diisopropyl(chloroacetyl)phosphonate, *in* C-00020
Diisopropyl chloromethylphosphonate, *in* C-00106
Diisopropyl (4-chlorophenyl)phosphonite, *in* C-00153
Diisopropyl chlorophosphate, *see* D-00657
Diisopropyl chlorophosphite, *see* D-00658
▷*O,O*-Diisopropyl chlorothiophosphate, *see* D-00659
▷Diisopropyl chlorothiophosphonate, D-00634
▷*O,O*-Diisopropyl chlorothiophosphonothioate, D-00635
Diisopropyl (2-cyanoethyl)phosphonate, *in* C-00211
Diisopropyl 2-cyclohexen-1-ylphosphonate, *in* C-00242
Diisopropyl (cyclohexylcarbonyl)phosphonate, *in* C-00239
Diisopropyl cyclohexylphosphonate, *in* C-00256
Diisopropyl cyclohexylphosphoramidate, *in* C-00269
Diisopropyl 1-cyclopenten-1-ylphosphonate, *in* C-00277
Diisopropyl 2-cyclopenten-1-ylphosphonate, *in* C-00278
Diisopropyl α-diazobenzylphosphonate, *in* D-00037
Diisopropyl (diazophenylmethyl)phosphonate, *in* D-00037
Diisopropyl dibutylphosphoramidate, *in* D-00134
Diisopropyl (dichloromethyl)phosphonite, *in* D-00181
Diisopropyl [(diethoxyphosphinothioyl)-thio]butanedioate, *in* D-00240
Diisopropyl (diethylcarbamoylmethyl)-phosphonate, *in* D-00268
Diisopropyl (diethylcarbamoyl)phosphonate, *in* D-00264
Diisopropyl diethylphosphoramidate, *in* D-00343
O,O-Diisopropyl diethylphosphoramidoselenoate, *in* D-00353
O,O-Diisopropyl diethylphosphoramidothioate, *in* D-00356
O,O'-Diisopropyl diethylthiodiphosphonate, *in* D-00405

O,O-Diisopropyl (1,3-dihydro-1,3-dioxo-2*H*-isoindol-2-yl)phosphonothioate, *in* D-00475
Diisopropyl (9,10-dihydro-5-methyl-9-acridinyl)phosphonate, *in* D-00438
Diisopropyl (1,3-dihydro-3-oxo-1-isobenzofuranyl)phosphonate, *in* D-00535
Diisopropyl diisobutylphosphoramidate, *in* B-00355
Diisopropyl [(dimethoxyphosphinothioyl)-thio]succinate, *in* D-00677
P,P-Diisopropyl-*N,N*-dimethylphosphinic amide, *in* D-00642
P,P-Diisopropyl-*N,N*-dimethylphosphinous amide, *in* D-00649
Diisopropyl dimethylphosphoramidate, *in* D-00855
O,O-Diisopropyl dimethylphosphoramidothioate, *in* D-00871
Diisopropyl (2,4-dinitrophenyl)phosphonate, *in* D-00946
N,N-Diisopropyl-*P,P*-diphenylphosphinous amide, *in* D-01090
O,O-Diisopropyl diselenophosphate, *see* D-00662
Diisopropyl dithiochlorophosphite, *see* D-00660
O,O-Diisopropyl dithiophosphate, *see* D-00663
S,S-Diisopropyldithiophosphate, *see* D-00664
S,S-Diisopropyl dithiophosphoric acid, *see* D-00664
Diisopropyl dodecylphosphonite, *in* D-01265
Diisopropyl 1,2-ethanediylbis(phenylphosphinate), *in* E-00015
Diisopropyl ethylphosphonite, *in* E-00153
O,O-Diisopropyl ethylphosphonothioate, *in* E-00148
O,O-Diisopropyl ethylphosphoramidothioate, *in* E-00159
Diisopropyl ethynylphosphonate, *in* E-00203
Diisopropyl fluoromethylphosphonate, *in* F-00023
Diisopropyl (3-fluorophenyl)phosphoramidate, *in* F-00043
Diisopropyl (4-fluorophenyl)phosphoramidate, *in* F-00044
O,O-Diisopropyl (4-fluorophenyl)-phosphoroamidothioate, *in* F-00046
▷Diisopropyl fluorophosphate, *see* D-00665
▷Diisopropyl fluorophosphonate, *see* D-00665
Diisopropyl heptylphosphonate, *in* H-00006
Diisopropyl hexylphosphonate, *in* H-00097
Diisopropyl hexylphosphonite, *in* H-00101
O,O-Diisopropyl hydrogen dithiophosphate, *see* D-00663
S,S-Diisopropyl hydrogen dithiophosphate, *see* D-00664
Diisopropyl hydrogen phosphate, *see* D-00639
O,O-Diisopropyl hydrogen phosphorodiselenoate, *see* D-00662
O,O-Diisopropyl hydrogen phosphoroselenoate, *see* D-00666
O,O-Diisopropyl hydrogen thiophosphate, *see* D-00667
Diisopropyl (α-hydroxybenzyl)phosphonate, *in* H-00174
Diisopropyl (2-hydroxyethyl)phosphoramidate, *in* H-00148
Diisopropyl α-hydroxy-2-furanylmethylphosphonate, *in* F-00071
Diisopropyl (2-hydroxypropyl)phosphonate, *in* H-00191
1,2:3,4-Di-*O*-isopropylidene-α-D-galactopyranose 6-dihydrogen phosphate, *in* G-00003
2,4-Diisopropylidene-1,3,6,9-tetraoxa-5-phospha(5-P^V)spiro[4.4]nonane, *see* B-00306
Diisopropyl 1*H*-imidazol-1-ylphosphonate, *in* I-00003
Diisopropyl (isocyanatomethyl)phosphonate, *in* I-00022
Diisopropyl isopropylphosphonite, *in* I-00056
Diisopropyl isopropylphosphoramidate, *in* I-00058
Diisopropyl methylenebis[methylphosphinate], *in* M-00140
Diisopropyl methylenebisphosphinate, *in* M-00143
Diisopropylmethylphenylphosphonium iodide, *in* D-00638
P,P-Diisopropyl-*N*-methylphosphinothioic amide, *in* D-00646
P,P-Diisopropyl-*N*-methylphosphinous amide, *in* D-00649
N,N-Diisopropyl-*P*-methylphosphonamidic acid, D-00636
N,N-Diisopropyl-*P*-methylphosphonamidothioic acid, D-00637
▷Diisopropyl methylphosphonate, *in* M-00288
Diisopropyl methylphosphonite, *in* M-00319
O,O-Diisopropyl methylphosphonothioate, *in* M-00313
Diisopropyl methylphosphoramidite, *in* M-00335
O,O-Diisopropyl methylphosphoramidoselenoate, *in* M-00332
Diisopropyl (3-methyl-4-pyridinyl)-phosphonate, *in* M-00390
Diisopropyl 1-naphthylphosphonite, *in* N-00025
Diisopropyl (2-nitrophenyl)phosphonate, *in* N-00055
Diisopropyl (3-nitrophenyl)phosphonate, *in* N-00056
Diisopropyl (1-nitroso-1-methylethyl)-phosphonate, *in* N-00066
Diisopropyl nonylphosphonate, *in* N-00076
Diisopropyl octylphosphonite, *in* O-00044
Diisopropyl pentylphosphonite, *in* P-00052
O,O-Diisopropyl phenylamidothiophosphate, *in* P-00262
Diisopropyl (phenylamino)-carbonylphosphoramidate, *in* P-00081
Diisopropyl [(phenylamino)methyl]-phosphonate, *in* P-00084
Diisopropyl (2-phenylethenyl)phosphonite, *in* P-00161
Diisopropylphenylphosphine, D-00638
P,P-Diisopropyl-*N*-phenylphosphinous amide, *in* D-00649
Diisopropyl phenylphosphonate, *in* P-00221
N,N'-Diisopropyl-*P*-phenylphosphonic diamide, *in* P-00222
Diisopropyl phenylphosphonite, *in* P-00251
N,N'-Diisopropyl-*P*-phenylphosphonothioic diamide, *in* P-00246
N,N'-Diisopropyl-*P*-phenylphosphonous diamide, *in* P-00252
Diisopropyl phenylphosphoramidate, *in* P-00257
O,O-Diisopropyl 2-phenylphosphorohydrazidothioate, *in* P-00296
▷*O,O*-Diisopropyl *S*-2-phenylsulfonylaminoethyl phosphorodithioate, *see* B-00004
Diisopropyl phosphate, D-00639
Diisopropylphosphine, D-00640
Diisopropylphosphinic acid, D-00641
P,P-Diisopropylphosphinic amide, D-00642
Diisopropylphosphinodithioic acid, D-00643
Diisopropylphosphinoselenoic acid, D-00644
Diisopropylphosphinotellurous acid, D-00645
Diisopropylphosphinothioic acid, D-00646
Diisopropylphosphinothious acid, D-00647
Diisopropylphosphinous acid, D-00648
P,P-Diisopropylphosphinous amide, D-00649
Diisopropylphosphinous anilide, *in* D-00649
Diisopropylphosphinous chloride, D-00650
▷Diisopropyl phosphite, *see* D-00651
▷Diisopropyl phosphonate, D-00651
O,O-Diisopropyl phosphonothioate, D-00652
Diisopropyl phosphoramidate, D-00653
Diisopropyl phosphoramide, *see* D-00653
Diisopropylphosphoramidochloridous acid, D-00654
Diisopropylphosphoramidoselenoic acid, D-00655
Diisopropylphosphoramidous acid, D-00656
Diisopropyl phosphoric acid, *see* D-00639
▷Diisopropyl phosphorobromidate, *in* D-00639
Diisopropyl phosphorochloridate, D-00657
Diisopropyl phosphorochloridite, D-00658
▷*O,O*-Diisopropyl phosphorochloridothioate, D-00659
Diisopropyl phosphorochloridothioite, D-00660
N,N'-Diisopropylphosphorodiamidic acid, D-00661
O,O-Diisopropyl phosphorodiselenoate, D-00662
O,O-Diisopropyl phosphorodiselenoic acid, *see* D-00662
O,O-Diisopropyl phosphorodithioate, D-00663
S,S-Diisopropyl phosphorodithioate, D-00664
O,O-Diisopropyl phosphorodithioate anhydrosulfide, *see* T-00174
▷Diisopropyl phosphorofluoridate, D-00665
O,O-Diisopropyl phosphoroselenoate, D-00666
O,O-Diisopropyl phosphoroselenoic acid, *see* D-00666
O,O-Diisopropyl phosphorothioate, D-00667
▷*O,O*-Diisopropyl phosphorothioate anhydride, *see* T-00168
O,O-Diisopropyl phosphorothioic acid phenylhydrazide, *in* P-00296
▷Diisopropyl phosphoryl bromide, *in* D-00639
Diisopropyl phosphoryl chloride, *see* D-00657
N-Diisopropylphosphorylimidazole, *in* I-00003
Diisopropyl 1-piperidinylphosphonite, *in* P-00432
Diisopropyl 1-propenylphosphonate, *in* P-00453
S,S-Diisopropyl 2-propenylphosphonodithioate, *in* P-00457
Diisopropyl propylphosphonite, *in* P-00482
Diisopropyl 1-propynylphosphonate, *in* P-00494
Diisopropyl 2-pyridylphosphoramidate, *in* P-00510
Diisopropyl 3-pyridylphosphoramidate, *in* P-00511
Diisopropyl (2-pyrimidinyl)phosphonate, *in* P-00515
Diisopropyl (4-pyrimidinyl)phosphonate, *in* P-00516
Diisopropyl 1-pyrrolidinylphosphonite, *in* P-00520
O,O-Diisopropyl selenophosphoryl methylamide, *in* M-00332
P,P'-Diisopropyl-*N,N,N',N'*-tetramethyldiphosphonic diamide, D-00668
Diisopropyl 2-thienylphosphonite, *in* T-00308
O,O-Diisopropyl thiophosphite, *see* D-00652
O,O-Diisopropyl thiophosphoric acid, *see* D-00667
O,O-Diisopropyl thiophosphoric anilide, *in* P-00262
▷*O,O*-Diisopropyl thiophosphoryl chloride, *see* D-00659
Diisopropyl *p*-tolylphosphonite, *in* M-00248
Diisopropyl (2,2,2-trichloro-1-hydroxyethyl)phosphonate, *in* T-00408
Diisopropyl (trichloromethyl)phosphonate, *in* T-00411
Diisopropyl (trifluoromethyl)phosphonite, *in* T-00482
Diisopropyl vinylphosphonate, *in* V-00003
1,2-Diisotetradecanoylphosphatidylcholine, *in* G-00024
1,2-Diisotridecanoylphosphatidylcholine, *in* G-00019
Dilauroylphosphatidylcholine, *in* G-00008
Dilauroylphosphatidylethanolamine, *see* G-00009
Dilauryl phosphate, *see* D-00225
1,2-Dilinolenoyl-*sn*-glycerol 3-phosphate, *in* G-00039
Dilinolenoylphosphatidylcholine, *in* G-00040
Dilinolenoylphosphatidylethanolamine, *in* G-00041
Dilinoleoylphosphatidylcholine, *in* G-00028
Dilinoleoylphosphatidylethanolamine, *in* G-00029
Dilinoleoylphosphatidylserine, *in* G-00031
Dilinoloylphosphatidylinositol, *see* G-00030
▷Dimatif, *see* B-00094
▷Dimecron, *see* P-00347
▷Dimefox, D-00669
3,9-Dimercapto-2,4,8,10-tetraoxa-3,9-diphosphaspiro[5.5]undecane 3,9-disulfide, D-00670
Dimesitylphosphine, *see* B-00426

Dimesitylphosphinic acid, *see* B-00428
▷Dimethoate, D-00671
P,P-Dimethoxy-*P*-[bis(trimethylsilyl)-amino]-*N*-trimethylsilylphosphazene, *in* T-00868
Di(methoxycarbonyl)-methylenetriphenylphosphorane, *in* T-00658
P,P-Dimethoxy-*N*-methyl-*P*-phenylphosphine imide, *in* M-00240
P,P-Dimethoxy-*P*-phenyl-*N*-phenylsulfonylphosphazene, *in* P-00192
P,P-Dimethoxy-*P*-phenyl-*N*-phenylsulfonylphosphine imide, *in* P-00192
Dimethoxyphosphinecarboxylic acid oxide, D-00672
▷*O*-(Dimethoxyphosphinothioyl) *O,O*-dimethyl phosphorothioate, *see* T-00203
[(Dimethoxyphosphino)thioyl]fluoride, *see* D-00902
3-[(Dimethoxyphosphinothioyl)oxy]-2-methyl-2-propenoic acid, D-00673
▷1-[(Dimethoxyphosphinothioyl)oxy]-4-nitrobenzene, *see* P-00003
(Dimethoxyphosphinothioyl)phosphorimidic acid, D-00674
▷[(Dimethoxyphosphinothioyl)thio]acetic acid, D-00675
α-[(Dimethoxyphosphinothioyl)thio]-benzeneacetic acid, D-00676
[(Dimethoxyphosphinothioyl)thio]butanedioic acid, D-00677
▷[(Dimethoxyphosphinothioyl)thio]butanedioic acid diethyl ester, *see* M-00001
[(Dimethoxyphosphinothioyl)thio]succinic acid, *see* D-00677
N-(Dimethoxyphosphinothioyl)-*P,P,P*-triphenoxyphosphazene, *in* D-00674
N-(Dimethoxyphosphinothioyl)-*P,P,P*-triphenoxyphosphine imide, *in* D-00674
N-(Dimethoxyphosphinyl)acetamide, *in* A-00029
(Dimethoxyphosphinyl)acetic acid, D-00678
3-(Dimethoxyphosphinyl)acrylic acid, *see* D-00686
2-[(Dimethoxyphosphinyl)amino]pentanedioic diamide, *in* P-00367
2-[(Dimethoxyphosphinyl)amino]-3-phenylpropanamide, *in* P-00400
2-(Dimethoxyphosphinylamino)pyridine, *in* P-00510
3-(Dimethoxyphosphinylamino)pyridine, *in* P-00511
N-(Dimethoxyphosphinyl)aniline, *see* D-00817
Dimethoxyphosphinyl bromide, *see* D-00886
2-(Dimethoxyphosphinyl)butanoic acid, D-00679
N-(Dimethoxyphosphinyl)carbamic acid, D-00680
Dimethoxyphosphinyl chloride, *see* D-00887
Dimethoxyphosphinyl cyanide, *see* D-00894
Dimethoxyphosphinylcyclohexane, *in* C-00256
1-(Dimethoxyphosphinyl)decane, *in* D-00012
2-(Dimethoxyphosphinyl)ethanal, *in* O-00059
▷Dimethoxyphosphinyl fluoride, *see* D-00901
(Dimethoxyphosphinyl)formamide, *in* C-00003
Dimethoxyphosphinylformic acid, *see* D-00672
Dimethoxyphosphinylhydrazine, *see* D-00904
N-Dimethoxyphosphinylimidazole, *in* I-00003
O-(Dimethoxyphosphinyl)lactic acid, *see* D-00682
3-[(Dimethoxyphosphinyl)oxy]-2-butenoic acid, D-00681
3-[(Dimethoxyphosphinyl)oxy]crotonic acid, *see* D-00681
2-[(Dimethoxyphosphinyl)oxy]propanoic acid, D-00682
(Dimethoxyphosphinyl)phosphorimidic acid, D-00683
2-(Dimethoxyphosphinyl)propanoic acid, D-00684
3-(Dimethoxyphosphinyl)propanoic acid, D-00685
1-(Dimethoxyphosphinyl)-2-propanol, *in* H-00191
3-Dimethoxyphosphinyl-1-propanol acetate, *in* A-00010
3-(Dimethoxyphosphinyl)propenoic acid, D-00686
(Dimethoxyphosphinyl)thioacetic acid, D-00687
▷1-[(Dimethoxyphosphinyl)thio]-4-nitrobenzene, *see* D-00773
Dimethyl (acetamidomethyl)phosphonate, *in* A-00012
Dimethyl acetonylphosphonate, *in* O-00075
▷Dimethyl 1-acetoxy-2,2,2-trichloroethylphosphonate, *in* A-00011
Dimethyl [3-(acetylhydroxyimino)-2-hydroxypropyl]phosphonate, *in* A-00138
Dimethyl (2-acetyloxyethyl)phosphonate, *in* A-00006
Dimethyl [(acetyloxy)methyl]phosphonate, *in* A-00007
Dimethyl [(acetyloxy)phenylmethyl]-phosphonate, *in* A-00020
Dimethyl (3-acetyloxypropyl)phosphonate, *in* A-00010
Dimethyl acetylphosphonate, *in* A-00026
Dimethyl acetylphosphoramidate, *in* A-00029
▷*O,S*-Dimethyl acetylphosphoramidothioate, *see* A-00003
Dimethyl 1-adamantylphosphonate, *in* A-00034
Dimethyl 2-adamantylphosphonate, *in* A-00035
Dimethyl 1-adamantylphosphoramidate, *in* A-00036
Dimethyl allenylphosphonate, *in* P-00441
Dimethyl allylphosphonate, *in* P-00454
(γ,γ-Dimethylallyl)phosphonic acid, *see* M-00102
γ,γ-Dimethylallyl pyrophosphate, *see* M-00482
O,S-Dimethyl amidodithiophosphate, *see* D-00865
S,S-Dimethyl amidodithiophosphate, *see* D-00866
O,O-Dimethyl amidothiophosphate, *see* D-00870
▷*O,S*-Dimethyl amidothiophosphate, *see* M-00021
Dimethylamine-1,1′-diphosphonic acid, D-00688
2-Dimethylamino-1,3,2-benzodioxaphosphole, D-00689
Dimethyl *p*-aminobenzylphosphonate, *in* A-00107
(4-Dimethylaminobutylidene)-triphenylphosphorane, *in* D-00690
[4-(Dimethylamino)butyl]-triphenylphosphonium(1+), D-00690
Dimethyl [1-(aminocarbonyl)ethyl]-phosphoramidate, *in* P-00362
Dimethyl aminocarbonylphosphonate, *in* C-00003
(Dimethylaminocarbonyl)phosphonic acid, D-00691
Dimethyl aminocarbonylphosphoramidate, *in* A-00062
4-Dimethylamino-2,2-dihydro-2,2-diphenyl-1,3,5,2-triazaphosphorine, D-00692
5-Dimethylamino-2,4-diisopropylidene-1,3,6,9-tetraoxa-5-phospha(5-P^V)-spiro[4.4]nonane, *in* B-00306
2-Dimethylamino-5,5-dimethyl-1,3,2-dioxaphosphorinane, D-00693
2-Dimethylamino-3,4-dimethyl-5-phenyl-1,3,2-oxazaphospholane, *see* D-00694
2-Dimethylamino-3,4-dimethyl-5-phenyl-1,3,2-oxazaphospholidine, D-00694
2-Dimethylamino-1,3-dioxa-2-phosphacyclopentane, *see* D-00695
2-Dimethylamino-1,3,2-dioxaphosphinane, *see* D-00696
2-Dimethylamino-1,3,2-dioxaphospholane, D-00695
2-Dimethylamino-1,3,2-dioxaphosphorinane, D-00696
(Dimethylamino)diphenylphosphine, *see* D-00747
1-[1-(Dimethylamino)ethyl]-2-(diphenylphosphino)ferrocene, D-00697
(2-Dimethylaminoethylidene)-triphenylphosphorane, *in* D-00698
2-Dimethylaminoethylphosphonic acid, *in* A-00079
(2-Dimethylaminoethyl)-triphenylphosphonium(1+), D-00698
Dimethyl [[1-amino-2-(1*H*-indol-3-yl)]-ethyl]phosphonate, *in* P-00421
2-Dimethylamino-4-isopropyl-5,5-dimethyl-1,3,2-dioxaphosphorinane, D-00699
(Dimethylaminomethyl)diphenylphosphine, D-00700
[(Dimethylamino)methylene]bisphosphonic acid, D-00701
Dimethyl (1-amino-1-methylethyl)phosphonate, *in* A-00087
▷3-(Dimethylamino)-1-methyl-3-oxo-1-propenyl dimethyl phosphate, *see* D-00204
2-Dimethylamino-5-methyl-3-phenyl-Δ^4-1,3,4,2-oxadiazaphospholine, *see* D-00455
[4-(Dimethylamino)-2-methylphenyl]-phosphinic acid, *see* T-00322
2-Dimethylamino-1,3,2-oxathiaphospholane, *in* O-00056
2-Dimethylamino-1,3,2-oxazaphospholidine, D-00702
(2-Dimethylamino-2-oxoethyl)phosphonic acid, D-00703
O,O-Dimethyl *S*-(2-amino-2-oxoethyl) phosphorodithioate, *in* D-00675
2-Dimethylamino-4-phenyl-1,3,2-dioxaphosphorinane, D-00704
(2-Dimethylaminophenyl)diphenylphosphine, *see* D-01059
(3-Dimethylaminophenyl)diphenylphosphine, *see* D-01060
(4-Dimethylaminophenyl)diphenylphosphine, *see* D-01061
(4-Dimethylaminophenyl)methylphosphinic acid, D-00705
Dimethyl [(2-aminophenyl)methyl]phosphonate, *in* A-00106
Dimethyl [(4-aminophenyl)methyl]phosphonate, *in* A-00107
Dimethyl (2-aminophenyl)phosphonate, *in* A-00109
Dimethyl (4-aminophenyl)phosphonate, *in* A-00111
(4-Dimethylaminophenyl)phosphonic acid, D-00706
Dimethyl (4-aminophenyl)phosphoramidate, *in* A-00115
▷(Dimethylamino)phosphonic dichloride, *see* D-00856
Dimethylaminophosphonic dicyanide, *see* D-00857
(Dimethylamino)phosphonothioic difluoride, *see* D-00873
(Dimethylamino)phosphonothioyl dichloride, *see* D-00872
▷Dimethylamino phosphoryl dichloride, *see* D-00856
Dimethylamino phosphoryl difluoride, *see* D-00858
[3-(Dimethylamino)propylidene]-triphenylphosphorane, *in* D-00707
[3-(Dimethylamino)propyl]-triphenylphosphonium(1+), D-00707
▷*O*-[4-(Dimethylamino)sulfonyl]phenyl *O,O*-dimethyl phosphorothioate, *see* F-00001
▷*O,O*-Dimethyl *O*-(4-aminosulfonylphenyl) phosphorothioate, *see* D-00920
(Dimethylamino)tetrafluorophosphorane, D-00708
(Dimethylamino)tetraiodophosphorane, D-00709
2-Dimethylamino-4,4,5,5-tetramethyl-1,3-dioxa-2-phosphacyclopentane, *see* D-00710
2-Dimethylamino-4,4,5,5-tetramethyl-1,3,2-dioxaphospholane, D-00710
4-Dimethylamino-*o*-tolylphosphinic acid, *see* T-00322
S,S-Dimethyl anilidodithiophosphate, *in* D-00866
1-(Dimethylarsino)-2-(dimethylphosphino)-benzene, *see* D-00711
[2-(Dimethylarsino)phenyl]dimethylphosphine, D-00711
[2-(Dimethylarsino)phenyl]-trimethylphosphonium iodide, *in* D-00711
Dimethyl 1-azetidinylphosphonate, *in* A-00147
O,O-Dimethyl azidothiophosphate, *see* D-00879
Dimethyl 1-aziridinylphosphonate, *in* A-00153
Dimethyl benzenephosphonate, *see* D-00810
Dimethyl benzenephosphonite, *in* P-00251
α-(5,6-Dimethylbenzimidazolyl)aquocobamide, *see* V-00010
▷α-(5,6-Dimethylbenzimidazoyl)cobamide cyanide, *see* V-00009

1,3-Dimethyl-1*H*-1,3,2-benzodiazaphospholine, *see* D-00456
N,N-Dimethyl-1,3,2-benzodioxaphosphol-2-amine, *see* D-00689
2,5-Dimethyl-1,3,2-benzodithiaphosphole, D-00712
1,3-Dimethylbenzo[*c*]phosphinine, *see* D-00758
1,3-Dimethylbenzo[*c*]phosphorin, *see* D-00758
Dimethyl benzoylphosphonate, *in* B-00034
Dimethyl benzoylphosphoramidate, *in* B-00036
Dimethyl benzylphosphonate, *in* B-00060
Dimethyl benzylphosphoramidate, *in* B-00069
N,N'-(6,6'-Dimethyl[1,1'-biphenyl]-2,2'-diyl)bis[*P,P*-diphenylphosphinous amide], *see* B-00215
N,N'-Dimethyl-*N,N'*-bis[[bis(dimethylamino)phosphinyl]-amino]ethane, D-00713
Dimethyl 1,2-bis(diethoxyphosphinyl)-succinate, *in* D-01182
Dimethyl 1,2-bis(dimethoxyphosphinyl)-butanedioate, *in* D-01182
N,P-Dimethyl-*P,P*-bis(dimethylamino)-phosphazene, *in* P-00023
4,4-Dimethyl-1,3-bis(1,1-dimethylethyl)-1,3-diaza-2-phospha-4-silacyclobutane, *see* D-00087
N,N-Dimethyl-2,4-bis(1-methylethylidene)-1,3,6,9-tetraoxa-5-phospha(5-P^V)-spiro[4.4]nonan-5-amine, *in* B-00306
Dimethyl [bis(methylthio)methyl]phosphonate, *in* B-00364
1,2-Dimethyl-1,2-bis(phenylmethyl)-diphosphine, *see* D-00043
1,4-Dimethyl-7,9-bis(trifluoromethyl)-1,4,6,8,10-pentaaza-5-phospha(5-P^V)-spiro[4.5]deca-5,7,9-triene, D-00714
N,N-Dimethyl-*P,P*-bis(trifluoromethyl)-phosphinic amide, *in* B-00411
N,N-Dimethyl-*P,P*-bis(trifluoromethyl)-phosphinous amide, *in* B-00419
1,3-Dimethyl-4',6'-bis(trifluoromethyl)-spiro[1,3,2(P^V)diazaphospholidine-2,2'-[1,3,5,2(P^V)triazaphosphorine], *see* D-00714
Dimethylbis(trimethylphosphonimethyl)-phosphonium triiodide, *in* B-00204
P,P-Dimethyl-*N,N*-bis(trimethylsilyl)-phosphinothioic amide, *in* D-00840
P,P-Dimethyl-*N,N*-bis(trimethylsilyl)-phosphinous amide, *in* D-00845
Dimethyl (2-bromophenyl)phosphonate, *in* B-00490
Dimethyl (3-bromophenyl)phosphonate, *in* B-00491
Dimethyl (4-bromophenyl)phosphonate, *in* B-00492
Dimethyl (4-bromophenyl)phosphonite, *in* B-00493
Dimethyl bromophosphate, *see* D-00886
Dimethyl (1,3-butadienyl)phosphonate, *in* B-00510
O,O-Dimethyl (1,3-butadien-1-yl)-phosphonothioate, *in* B-00511
Dimethyl 1,4-butanediylbis[phenyl-phosphinodithioate], *in* B-00516
Dimethyl 3-buten-1-ynylphosphonate, *in* B-00529
Dimethyl 4-*tert*-butylcyclohexylphosphonite, *in* B-00550
Dimethyl butylphosphonate, *in* B-00594
Dimethyl *sec*-butylphosphonate, *in* M-00377
Dimethyl butylphosphonite, *in* B-00615
Dimethyl *tert*-butylphosphonite, *in* B-00616
O,S-Dimethyl *tert*-butylphosphonothioate, *in* M-00107
Dimethyl butylphosphoramidate, *in* B-00621
Dimethyl *tert*-butylphosphoramidate, *in* B-00622
(Dimethylcarbamoylmethyl)phosphonic acid, *see* D-00703
(Dimethylcarbamoyl)phosphonic acid, *see* D-00691
Dimethyl (carboxymethyl)phosphonate, *see* D-00678
Dimethyl carboxyphosphoramidate, *see* D-00680
Dimethyl 2-chloro-1,3,2-dioxaphospholan-4,5-dicarboxylate, *in* C-00062
Dimethyl chlorodiselenophosphite, *see* D-00890
S,S-Dimethyl chlorodithiophosphate, *see* D-00888
Dimethyl (2-chloroethyl)phosphonate, *in* C-00074
Dimethyl chloromethylphosphonate, *in* C-00106
Dimethyl [(2-chlorophenyl)methyl]-phosphonate, *in* C-00132
Dimethyl (2-chlorophenyl)phosphonate, *in* C-00145
Dimethyl (3-chlorophenyl)phosphonate, *in* C-00146
Dimethyl (4-chlorophenyl)phosphonate, *in* C-00147
Dimethyl (4-chlorophenyl)phosphonite, *in* C-00153
Dimethyl (4-chlorophenyl)phosphoramidate, *in* C-00157
O,O-Dimethyl (4-chlorophenyl)-phosphoramidothioate, *in* C-00160
Dimethyl chlorophosphate, *see* D-00887
▷Dimethyl chlorophosphite, D-00715
Dimethyl (3-chloropropyl)phosphonate, *in* C-00172
▷*OO*-Dimethyl chlorothiophosphate, *see* D-00891
O,S-Dimethyl chlorothiophosphate, *see* D-00892
▷Dimethyl chlorothiophosphonate, D-00716
Dimethyl chlorotrithiophosphate, *see* D-00893
Dimethyl (2-cyanoethyl)phosphonate, *in* C-00211
Dimethyl [(2-cyanophenyl)methyl]phosphonate, *in* C-00220
Dimethyl (4-cyanophenyl)phosphonate, *in* C-00225
▷*O,O*-Dimethyl *O*-4-cyanophenyl thiophosphate, *see* C-00226
Dimethyl 2-cyclohexen-1-ylphosphonate, *in* C-00242
Dimethyl (cyclohexylcarbonyl)phosphonate, *in* C-00239
Dimethyl cyclohexylphosphonate, *in* C-00256
Dimethyl cyclohexylphosphonite, *in* C-00267
Dimethyl cyclohexylphosphoramidate, *in* C-00269
O,O-Dimethyl cyclohexylphosphoramidothioate, *in* C-00271
Dimethyl 1-cyclopenten-1-ylphosphonate, *in* C-00277
Dimethyl cyclopentylphosphonate, *in* C-00276
Dimethyl decylphosphonate, *in* D-00012
▷*O,O'*-Dimethyl *S*-[(4,6-diamino-1,3,5-triazin-2-yl)methyl] phosphorodithioate, *see* M-00003
2,10-Dimethyl-2,10-diarsa-6-phenyl-6-phosphabicyclo[9.4.0]pentadeca-11(1),-12,14-triene, *see* O-00007
1,2-Dimethyl-1,2-diaza-4-phosphacyclopentane, *see* D-00718
1,3-Dimethyl-1,3,2-diazaphosphetidin-4-one, D-00717
1,5-Dimethyl-1*H*-1,2,3-diazaphosphole, *in* M-00114
1,2-Dimethyl-1,2,4-diazaphospholidine, D-00718
1,3-Dimethyl-1,3,2-diazaphospholidine, D-00719
1,3-Dimethyl-1,3,2-diazaphospholidine 2-oxide, D-00720
1,3-Dimethyl-1,3,2-diazaphospholidin-4-one 2-oxide, D-00721
1,4-Dimethyl-1,4,2-diazaphospholidin-5-one 2-oxide, D-00722
Dimethyl α-diazobenzylphosphonate, *in* D-00037
Dimethyl (diazocyclohexylmethyl)phosphonate, *in* D-00028
Dimethyl (1-diazoethyl)phosphonate, *in* D-00033
Dimethyl (diazomethyl)phosphonate, *in* D-00035
Dimethyl (diazophenylmethyl)phosphonate, *in* D-00037
5,5-Dimethyl-5*H*-dibenzophospholium iodide, *in* M-00115
Dimethyl 2-[(dibenzyloxyphosphinyl)amino]-pentanedioate, *in* P-00367
Dimethyl (dibromomethyl)phosphonate, *in* D-00064
Dimethyl (1,2-dibromo-1-phenylpropyl)-phosphonate, *in* D-00068
Dimethyl dibutylphosphoramidate, *in* D-00134
Dimethyl (dichloromethyl)phosphonate, *in* D-00175
Dimethyl dichlorophosphoramidate, D-00723
▷*O,O*-Dimethyl *O*-(2,2-dichlorovinyl) phosphate, *see* D-00203
Dimethyl [(diethoxyphosphinothioyl)thio]-butanedioate, *in* D-00240
Dimethyl (diethoxyphosphinyl)butanedioate, *in* D-00245
Dimethyl (diethoxyphosphinyl)succinate, *in* D-00245
Dimethyl diethyldiphosphonate, *in* D-00290
Dimethyl diethylphosphoramidate, *in* D-00343
O,O-Dimethyl diethylphosphoramidoselenoate, *in* D-00353
O,O-Dimethyl diethylphosphoramidothioate, *in* D-00356
Dimethyl (9,10-dihydro-9-acridinyl)-phosphonate, *in* D-00438
N,N-Dimethyl-1,5-dihydro-2,4,3-benzodioxaphosphepin-3-amine, *see* D-00453
N,N-Dimethyl-4,5-dihydro-1,3,2-benzodioxaphosphepin-2-amine, *see* D-00454
Dimethyl [(1,3-dihydro-1,3-dioxo-2*H*-isoindol-2-yl)methyl]phosphonate, *in* P-00426
O,O-Dimethyl (1,3-dihydro-1,3-dioxo-2*H*-isoindol-2-yl)phosphonothioate, *in* D-00475
Dimethyl (9,10-dihydro-5-methyl-9-acridinyl)phosphonate, *in* D-00438
Dimethyl (1,3-dihydro-3-oxo-1-isobenzofuranyl)phosphonate, *in* D-00535
Dimethyl [(5,6-dihydro-4,4,6-trimethyl-4*H*-1,3-oxazin-2-yl]phosphonate, *in* D-00598
▷Dimethyl [(dimethoxyphosphinothioyl)thio]-butanedioate, *in* D-00677
O,O-Dimethyl (dimethoxyphosphinyl)-methylphosphoramidothioate, *in* M-00413
▷Dimethyl 3-[(dimethoxyphosphinyl)oxy]-2-pentenedioate, *see* B-00461
▷*O,O*-Dimethyl *O*-(dimethoxyphosphinyl) phosphorothioate, *see* T-00207
O,O-Dimethyl *S*-(dimethoxyphosphinyl) phosphorothioate, *see* T-00206
Dimethyl (dimethylamido)phosphate, *see* D-00727
1,3-Dimethyl-2-dimethylamino-1,3,2-diazaphospholidine, D-00724
4,5-Dimethyl-2-dimethylamino-1,3-dioxa-2-phosphacyclohexane, *see* D-00725
4,5-Dimethyl-2-dimethylamino-1,3,2-dioxaphospholane, D-00725
5,5-Dimethyl-2-dimethylamino-1,3,2-oxazaphosphorinane, *see* D-00726
Dimethyl (2-dimethylamino-2-oxoethyl)-phosphonate, *in* D-00703
Dimethyl(dimethylamino)phosphine, *see* T-00213
5,5-Dimethyl-2-dimethylaminotetrahydro-2*H*-1,3,2-oxazaphosphorine, D-00726
Dimethyl (dimethylcarbamoylmethyl)-phosphonate, *in* D-00703
▷Dimethyl 2-dimethylcarbamoyl-1-methylvinyl phosphate, *see* D-00204
Dimethyl dimethyldiphosphonate, *in* D-00751
5,5-Dimethyl-2-(1,1-dimethylethoxy)-1,3,2-dioxaphosphinane, *see* B-00534
4,5-Dimethyl-2-(1,1-dimethylethoxy)-1,3,2-dioxaphospholane, *see* B-00533
5,5-Dimethyl-2-(1,1-dimethylethoxy)-1,3,2-dioxaphosphorinane, *see* B-00534
N,N-Dimethyl-5-(1,1-dimethylethyl)-1,3,2-dioxaphosphorin-2-amine, *see* B-00555
5,5-Dimethyl-*N*-(1,1-dimethylethyl)-1,3,2-dioxaphosphorinan-2-amine, *see* B-00538
N,N'-Dimethyl-*P*-(1,1-dimethylethyl)-phosphonic diamide, *in* B-00597
N,N'-Dimethyl-*P*-(1,1-dimethylethyl)-phosphonous diamide, *in* B-00617

O,O-Dimethyl *SS*-(1,1-dimethylethyl) phosphoro(dithioperoxo)thioate, *see* B-00556
N,N-Dimethyl-5-(1,1-dimethylethyl)-tetrahydro-2*H*-1,3,2-oxazaphosphorin-2-amine, *see* B-00635
Dimethyl (3,3-dimethyl-2-oxoheptyl)-phosphonate, *in* D-00785
Dimethyl (3,3-dimethyl-2-oxo-4-phenylbutyl)phosphonate, *in* D-00786
Dimethyl (2,6-dimethylphenyl)phosphonate, *in* D-00814
P,P-Dimethyl-*N*-(dimethylphosphinothioyl)-phosphinothioic amide, *in* D-00840
P,P-Dimethyl-*N*-(dimethylphosphinyl)-phosphinic amide, *in* D-00830
Dimethyl dimethylphosphoramidate, D-00727
Dimethyl dimethylphosphoramidite, *in* D-00875
O,O-Dimethyl dimethylphosphoramidoselenoate, *in* D-00869
O,O-Dimethyl dimethylphosphoramidothioate, *in* D-00871
Dimethyl (2,2-dimethylpropyl)phosphonate, *in* D-00912
Dimethyl dimethylpyrophosphonate, *in* D-00751
Dimethyl (2,4-dinitrophenyl)phosphonate, *in* D-00946
6,9-Dimethyl-1,3-dioxa-6,9-diaza-2-phosphacycloundecane, D-00728
4,9-Dimethyl-1,6-dioxa-4,9-diaza-5-phospha(5-P^V)spiro[4.4]nonane, D-00729
2,6-Dimethyl-1,3-dioxa-2,6-diphosphacyclooctane, *see* D-00730
2,6-Dimethyl-1,3,2,6-dioxadiphosphocane, D-00730
2,4-Dimethyl-1,3-dioxa-2-phosphacyclohexane, *see* D-00734
5,5-Dimethyl-1,3-dioxa-2-phosphacyclohexane, *see* D-00735
2,4-Dimethyl-1,3-dioxa-2-phosphacyclopentane, *see* D-00732
4,5-Dimethyl-1,3-dioxa-2-phosphacyclopentane, *see* D-00733
2,2-Dimethyl-1,3-dioxa-5-phospha-2-silacyclohexane, D-00731
N,N-Dimethyl-1,3,2-dioxaphosphinan-2-amine, *see* D-00696
2,4-Dimethyl-1,3,2-dioxaphosphinane, *see* D-00734
5,5-Dimethyl-1,3,2-dioxaphosphinane, *see* D-00735
N,N-Dimethyl-1,3,2-dioxaphospholan-2-amine, *see* D-00695
2,4-Dimethyl-1,3,2-dioxaphospholane, D-00732
4,5-Dimethyl-1,3,2-dioxaphospholane, D-00733
1-(4,5-Dimethyl-1,3,2-dioxaphosphol-2-yl)-1*H*-imidazole *P*-oxide, *see* D-00753
N,N-Dimethyl-1,3,2-dioxaphosphorinan-2-amine, *see* D-00696
2,4-Dimethyl-1,3,2-dioxaphosphorinane, D-00734
5,5-Dimethyl-1,3,2-dioxaphosphorinane, D-00735
5,5-Dimethyl-1,3,2-dioxaphosphorinane-2-carbonitrile, *see* C-00209
2-[(5,5-Dimethyl-1,3,2-dioxaphosphorinan-2-yl)oxy]-5,5-dimethyl-1,3,2-dioxaphosphorinane *P*-oxide 2-sulfide, *see* O-00090
2-[(5,5-Dimethyl-1,3,2-dioxaphosphorinan-2-yl)thio]-5,5-dimethyl-1,3,2-dioxaphosphorinane *P*-oxide 2-sulfide, *in* T-00309
[(2,2-Dimethyl-1,3-dioxolane-4,5-diyl)-bis[methylene]]bis[dicyclohexylphosphine], *see* B-00168
[(2,2-Dimethyl-1,3-dioxolane-4,5-diyl)-bis[methylene]]bis[diphenylphosphine], *see* B-00251
4,5-Dimethyl-2,7-diphenyl-1-aza-2-phospha-4,5-cycloheptadiene, *see* D-00460
1,2-Dimethyl-4,5-diphenyl-1*H*-1,3-azaphosphole, D-00736
2,17-Dimethyl-6,13-diphenyl-2,17-diarsa-6,13-diphosphatricyclo[16.4.0.$0^{7,12}$]-docosa-7(12),8,10,18(1),19,21-hexaene, *see* D-00004
2,4-Dimethyl-9,13-diphenyl-1,5-diaza-9,13-diphosphacyclohexadeca-1,4-diene, D-00737
12,14-Dimethyl-4,8-diphenyl-1,11-diaza-4,8-diphosphacyclotetradeca-11,14-diene, D-00738
1,4-Dimethyl-3,5-diphenyl-1,4,2-diazaphospholidine 2-oxide, D-00739
3,8-Dimethyl-2,7-diphenyl-1,6-dioxa-4,9-diaza-5-phospha(5-P^V)spiro[4.4]nonane, D-00740
6,9-Dimethyl-2,3-diphenyl-1,4-dioxa-6,9-diaza-5-phospha(5-P^V)spiro[4.4]non-2-ene, D-00741
1,3-Dimethyl-6,7-diphenyl-5,8-dioxa-1,3-diaza-4-phospha(4-P^V)spiro[3.4]oct-6-en-2-one, D-00742
9,9-Dimethyl-2,8-diphenyl-3,7-dioxa-1-phosphabicyclo[3.3.1]nonane, D-00743
2,2-Dimethyl-1,3-diphenyl-1,3-diphosphinane, *see* D-00745
1,4-Dimethyl-1,4-diphenyl-1,4-diphosphinanium dibromide, *in* D-01017
1,2-Dimethyl-1,2-diphenyldiphosphine, D-00744
1,3-Dimethyl-1,3-diphenyl-1,3-diphospholanium dibromide, *in* D-01013
2,2-Dimethyl-1,3-diphenyl-1,3-diphosphorinane, D-00745
1,4-Dimethyl-1,4-diphenyl-1,4-diphosphorinanium dibromide, *in* D-01017
2,2-Dimethyl-5,6-diphenyl-1-oxa-5-phospha-2-silacyclohexane, D-00746
2,6-Dimethyl-4,4-diphenyl-1,4-oxaphosphinanium bromide, *in* D-00797
2,6-Dimethyl-4,4-diphenyl-1-oxa-4-phosphoniacyclohexane bromide, *in* D-00797
2,6-Dimethyl-4,4-diphenyl-1,4-oxaphosphorinanium bromide, *in* D-00797
N,N-Dimethyl-*P,P*-diphenylphosphinic amide, *in* D-01040
N,N-Dimethyl-*P,P*-diphenylphosphinoselenoic amide, *in* D-01079
N,N-Dimethyl-*P,P*-diphenylphosphinothioic amide, *in* D-01083
N,N-Dimethyl-*P,P*-diphenylphosphinous amide, D-00747
P,P-Dimethyl-*N,N*-diphenylphosphinous amide, *in* D-00845
Dimethyldiphenylphosphonium iodide, *in* M-00135
N,N-Dimethyl-2,2′-diphenylphosphoramidic dihydrazide, *in* D-01109
N,N-Dimethyl-*N′,N″*-diphenylphosphorothioic triamide, *in* D-00909
1,3-Dimethyl-5,7-diphenyl-1,3,5,7-tetraaza-4-phospha(4-P^V)spiro[3.3]-heptane-2,6-dione, D-00748
8,8-Dimethyl-2,3-diphenyl-1,4,6,10-tetraoxa-5-phospha(5-P^V)spiro[4.5]-dec-2-ene, D-00749
N,N-Dimethyl-2,3-diphenyl-1,4,6,9-tetraoxa-5-phospha(5-P^V)spiro[4.4]-non-2-en-5-amine, *in* D-01150
2,9-Dimethyl-2,9-diphosphadecane, *see* B-00201
2,6-Dimethyl-2,6-diphosphaheptane, *see* B-00206
2,5-Dimethyl-2,5-diphosphahexane, *see* B-00200
2,8-Dimethyl-2,8-diphosphanonane, *see* B-00205
2,6-Dimethyl-2,7-diphosphaoctane, *see* B-00199
Dimethyldiphosphinogermane, D-00750
1,4-Dimethyl-1,4-diphosphoniabicyclo[2.2.2]octane diiodide, *in* D-01169
Dimethyldiphosphonic acid, D-00751
2,4-Dimethyl-2,4-diphosphosphapentane, *see* B-00202
N,N-Dimethyl-*P,P*-di-2-propenylphosphinic amide, *in* D-01193
Dimethyl dipropylphosphoramidate, *in* D-01214
N,N-Dimethyl-*P,P*-dipropynylphosphinic amide, *in* D-01231
N,N-Dimethyl-*P,P*-di-1-pyrrolidinylphosphinic amide, *in* D-01233
O,O-Dimethyl diselenophosphate, *see* D-00897
2,4-Dimethyl-1,3,2,4-dithiadiphosphetane 2,4-disulfide, D-00752
N,N-Dimethyl-1,3,2-dithiaphosphetane-2-amine, *in* D-01243
N,N-Dimethyl-1,3,2-dithiaphospholane-2-amine 2-oxide, *in* D-01244
Dimethyl dithiochlorophosphite, *see* D-00889
O,S-Dimethyl dithiophosphate, *see* D-00899
▷*O,O*-Dimethyl dithiophosphate, *see* D-00898
O,S-Dimethyl dithiophosphoric acid, *see* D-00899
▷*O,O*-Dimethyl dithiophosphoric acid, *see* D-00898
S,S-Dimethyl dithiophosphoric acid, *see* D-00900
(1,2-Dimethyl-1,2-ethanediyl)-bis[diphenylphosphine], *see* B-00226
N,N′-(1,2-Dimethyl-1,2-ethanediyl)-bis[*P,P*-diphenylphosphinous amide], *see* B-00218
Dimethyl 1,2-ethenediylbis(phenylphosphinate), *in* E-00018
N-(1,2-Dimethylethenylenedioxyphosphoryl)-imidazole, D-00753
Dimethyl ethenylphosphonate, *in* V-00003
Dimethyl ether bis(triphenylphosphonium)(2+), *see* O-00097
Dimethyl ether 1,1′-diphosphonic acid, *see* O-00095
2-(1,1-Dimethylethoxy)-1,3,2-dioxaphospholane, *see* B-00535
1-(1,1-Dimethylethyl)-1,2,4-azadiphosphetidine, *see* B-00543
2-(1,1-Dimethylethyl)-1,3-benzoxaphosphole, *see* B-00544
(1,1-Dimethylethyl)bis(trimethylsilyl)-phosphine, *see* B-00545
[4-(1,1-Dimethylethyl)cyclohexyl]-phosphonous acid, *see* B-00550
1-(1,1-Dimethylethyl)-1,4-dihydro-4-phenyl-1,4-phospharsenin, *see* B-00553
1-(1,1-Dimethylethyl)-3,4-dimethyl-1*H*-phosphole, *see* E-00004
N-(1,1-Dimethylethyl)-1,3,2-dioxaphospholan-2-amine, *see* B-00539
(1,1-Dimethylethyl)diphenylphosphine, *see* B-00558
1,1-Dimethylethyl 4-(diphenylphosphino)-2-[(diphenylphosphino)methyl]-1-pyrrolidinecarboxylate, *see* B-00532
P-(1,1-Dimethylethyl)-*N,N′*-diphenylphosphonamidimidic chloride, *see* B-00559
P-(1,1-Dimethylethyl)-*N,N′*-diphenylphosphonic diamide, *in* B-00597
N,N′-Dimethyl-*N,N′*-ethylenephosphonic diamide, *see* D-00720
(1,1-Dimethylethyl)(1-methylethyl)-phosphinic acid, *see* B-00562
(1,1-Dimethylethyl)methylphosphinic acid, *see* B-00567
P-(1,1-Dimethylethyl)-*P*-methylphosphinic amide, *in* B-00567
(1,1-Dimethylethyl)methylphosphinodithioic acid, *see* B-00568
(1,1-Dimethylethyl)methylphosphinodithioic acid anhydrosulfide, *see* B-00569
(1,1-Dimethylethyl)methylphosphinothioic acid, *see* B-00570
(1,1-Dimethylethyl)methylphosphinothioic anhydride, *see* B-00571
(1,1-Dimethylethyl)oxaphosphine, *see* B-00576
5-(1,1-Dimethylethyl)-2-phenyl-1,3,2-dioxaphosphorinane, *see* B-00577
4-(1,1-Dimethylethyl)-1-phenyl-1-phenylmethylethylphosphorinanium bromide, *in* B-00586
4-(1,1-Dimethylethyl)-1-phenylphosphinane, *see* B-00586
(1,1-Dimethylethyl)phenylphosphinic acid, *see* B-00579
P-(1,1-Dimethylethyl)-*P*-phenylphosphinic amide, *see* B-00580
(1,1-Dimethylethyl)phenylphosphinodithioic acid, *see* B-00581
(1,1-Dimethylethyl)phenylphosphinoselenoic acid, *see* B-00582
(1,1-Dimethylethyl)phenylphosphinothioic acid, *see* B-00583
(1,1-Dimethylethyl)-phenylphosphinothioselenoic acid, *see* B-00584

Dimethylethylphenylphosphonium iodide, *in* E-00084
4-(1,1-Dimethylethyl)-1-phenylphosphorinane, *see* B-00586
(1,1-Dimethylethyl)phosphine, *see* B-00589
(1,1-Dimethylethyl)phosphinic acid, *see* B-00591
N-[(1,1-Dimethylethyl)phosphinidene]-2-methyl-2-propanamine, *see* D-00106
Dimethyl ethylphosphonate, D-00754
(1,1-Dimethylethyl)phosphonic acid, *see* B-00595
P-(1,1-Dimethylethyl)phosphonic diamide, *see* B-00597
(1,1-Dimethylethyl)phosphonic dichloride, *see* B-00599
(1,1-Dimethylethyl)phosphonic difluoride, *see* B-00600
(1,1-Dimethylethyl)-phosphon(isothiocyanatidic) acid, *see* B-00602
▷Dimethyl ethylphosphonite, *in* E-00153
(1,1-Dimethylethyl)phosphonochloridous acid, *see* D-00130
(1,1-Dimethylethyl)phosphonoselenoic acid, *see* B-00610
O,O-Dimethyl ethylphosphonothioate, *in* E-00148
(1,1-Dimethylethyl)phosphonothioic acid, *see* B-00612
(1,1-Dimethylethyl)phosphonothioic dichloride, *see* B-00613
(1,1-Dimethylethyl)phosphonous acid, *see* B-00616
(1,1-Dimethylethyl)phosphonous diamide, *see* B-00617
(1,1-Dimethylethyl)phosphonous dichloride, *see* B-00619
▷(1,1-Dimethylethyl)phosphonous difluoride, *see* B-00620
Dimethyl ethylphosphoramidate, *in* E-00156
(1,1-Dimethylethyl)phosphoramidic acid, *see* B-00622
(1,1-Dimethylethyl)phosphoramidothioic acid, *see* B-00623
(1,1-Dimethylethyl)phosphoramidous acid, *see* B-00624
(1,1-Dimethylethyl)phosphorimidic acid, *see* B-00625
(1,1-Dimethylethyl)phosphorimidic trichloride, *see* B-00626
5-(1,1-Dimethylethyl)tetrahydro-*N,N*-dimethyl-3-phenyl-2*H*-1,3,2-oxazaphosphorin-2-amine, *see* B-00636
5-(1,1-Dimethylethyl)-2,2,4,5-tetrahydro-2,2,4-tris(1-methylethoxy)-1,2,4-oxadiphosphole 4-oxide, *see* B-00637
▷*O,O*-Dimethyl 5-[2-(ethylthio)ethyl] phosphorothioate, *see* D-00020
▷*S*-[[(1,1-Dimethylethyl)thio]methyl] *O,O*-diethyl phosphorodithioate, *see* T-00004
O,O-Dimethyl *S*-[(ethylthio)methyl] phosphorodithioate, D-00755
▷4-(1,1-Dimethylethyl)-2,6,7-trioxa-1-phosphabicyclo[2.2.2]octane, *see* B-00640
N-(1,1-Dimethylethyl)-*P*-[2,4,6-tris(1,1-dimethylethyl)phenyl]phosphazene, *see* B-00639
2-(1,1-Dimethylethyl)-1,3,6,2-trithiaphosphocane, *see* B-00642
OO-Dimethyl ethylurethane phosphate, *in* D-00680
S,S-Dimethyl fluorodithiophosphate, *in* D-00900
Dimethyl (3-fluorophenyl)phosphonate, *in* F-00037
Dimethyl (4-fluorophenyl)phosphonate, *in* F-00038
Dimethyl (3-fluorophenyl)phosphonite, *in* F-00041
Dimethyl (4-fluorophenyl)phosphonite, *in* F-00042
Dimethyl (3-fluorophenyl)phosphoramidate, *in* F-00043
Dimethyl (4-fluorophenyl)phosphoramidate, *in* F-00044
▷Dimethyl fluorophosphate, *see* D-00901
O,O-Dimethyl fluorothiophosphate, *see* D-00902
Dimethyl fluorotrithiophosphate, *see* D-00903
Dimethyl (2-furanylhydroxymethyl)-phosphonate, *in* F-00071
Dimethyl 2-furanylphosphonate, *in* F-00073
Dimethyl 3-furanylphosphonate, *in* F-00074
(Dimethylgermylene)bisphosphine, *see* D-00750
Dimethylgermylphosphine, D-00756
Dimethyl heptylphosphonate, *in* H-00006
2,10-Dimethyl-1,3,6,9,11,14-hexaoxa-2,10-diphosphacyclohexadecane, D-00757
Dimethyl hexylphosphonate, *in* H-00097
Dimethyl hexylphosphoramidate, *in* H-00102
O,O-Dimethyl hydrogen diselenophosphate, *see* D-00897
S,S-Dimethyl hydrogen dithiophosphate, *see* D-00900
Dimethyl hydrogen phosphate, *see* D-00826
O,S-Dimethyl hydrogen phosphorodithioate, *see* D-00899
O,O-Dimethyl hydrogen phosphoroselenoate, *see* D-00906
O,S-Dimethyl hydrogen thiophosphate, *see* D-00908
O,O-Dimethyl hydrogen thiophosphate, *see* D-00907
Dimethyl (α-hydroxybenzyl)phosphonate, *in* H-00174
Dimethyl *o*-hydroxybenzylphosphonate, *in* H-00175
Dimethyl *p*-hydroxybenzylphosphonate, *in* H-00176
Dimethyl (1-hydroxybutyl)phosphonate, *in* H-00114
Dimethyl (1-hydroxyethyl)phosphonate, *in* H-00145
Dimethyl (2-hydroxyethyl)phosphonate, *in* H-00146
Dimethyl (α-hydroxy-2-furanylmethyl)-phosphonate, *in* F-00071
Dimethyl (hydroxymethyl)phosphonate, *in* H-00162
Dimethyl (hydroxyphenylmethyl)phosphonate, *in* H-00174
Dimethyl [(2-hydroxyphenyl)methyl]-phosphonate, *in* H-00175
Dimethyl [(4-hydroxyphenyl)methyl]-phosphonate, *in* H-00176
Dimethyl (2-hydroxyphenyl)phosphonate, *in* H-00179
Dimethyl (3-hydroxyphenyl)phosphonate, *in* H-00180
Dimethyl (4-hydroxyphenyl)phosphonate, *in* H-00181
Dimethyl (1-hydroxypropyl)phosphonate, *in* H-00190
Dimethyl (2-hydroxypropyl)phosphonate, *in* H-00191
Dimethyl (3-hydroxypropyl)phosphonate, *in* H-00192
4,5-Dimethyl-2-(1′-imidazolyl)-2-oxo-1,3,2-dioxaphosphole, *see* D-00753
Dimethyl 1*H*-imidazol-1-ylphosphonate, *in* I-00003
O,S-Dimethyl (1-iminoethyl)-phosphoramidothioate, *in* I-00005
Dimethyl (isocyanatomethyl)phosphonate, *in* I-00022
1,3-Dimethylisophosphinoline, D-00758
Dimethyl isopropylphosphonate, *in* I-00050
Dimethyl isopropylphosphoramidate, *in* I-00058
▷*O,O*-Dimethyl *S*-[(2-isopropylthio)ethyl] phosphorodithioate, *see* I-00067
▷*O,O*-Dimethyl *S*-(4-mercaptomethyl-2-methoxy-Δ^2-1,3,4-thiadiazolin-5-one) phosphorodithioate, *see* M-00025
Dimethyl [1-(methoxycarbonyl)ethyl]-phosphonate, *see* D-00684
O,O-Dimethyl (2-methoxyethyl)-phosphoramidothioate, *in* M-00041
▷*O,O*-Dimethyl *S*-[(5-methoxy-2-oxo-1,3,4-thiadiazol-3(2*H*)-yl)-methyl] phosphorodithioate, *see* M-00025
Dimethyl [(4-methoxyphenyl)methyl]-phosphonate, *in* M-00061
Dimethyl (2-methoxyphenyl)phosphonate, *in* M-00070
Dimethyl (3-methoxyphenyl)phosphonate, *in* M-00071
Dimethyl (4-methoxyphenyl)phosphonate, *in* M-00072
Dimethyl (2-methoxyphenyl)phosphoramidate, *in* M-00074
Dimethyl (3-methoxyphenyl)phosphoramidate, *in* M-00075
Dimethyl (4-methoxyphenyl)phosphoramidate, *in* M-00076
Dimethyl (methylaminocarbonyl)phosphonate, *in* M-00088
▷*O,O*-Dimethyl *S*-(2-methylamino-2-oxoethyl) phosphorodithioate, *see* D-00671
▷*O,O*-Dimethyl *S*-[2-(methylamino)-2-oxoethyl] phosphorothioate, *see* O-00048
Dimethyl (2-methyl-1,3-butadienyl)-phosphonate, *in* M-00093
Dimethyl (3-methyl-1,2-butadienyl)-phosphonate, *in* M-00094
Dimethyl (3-methyl-1,3-butadienyl)-phosphonate, *in* M-00095
N,N′-Dimethyl-*P*-(2-methyl-1,3-butadienyl)-phosphonic diamide, *in* M-00096
O,O-Dimethyl (2-methyl-1,3-butadienyl)-phosphonothioate, *in* M-00097
Dimethyl (3-methyl-2-butenyl)phosphonate, *in* M-00102
▷*O,O*-Dimethyl *S*-2-(1-methylcarbamoylethylthio)ethyl phosphorothioate, *see* V-00001
▷*O,O*-Dimethyl *S*-(methylcarbamoylmethyl) phosphorodithioate, *see* D-00671
▷*O,O*-Dimethyl *S*-methylcarbamoylmethyl phosphorothioate, *see* O-00048
Dimethyl methylenebis(methylphosphinate), *in* M-00140
Dimethylmethylenediphosphonic acid, *see* P-00445
Dimethyl (1-methylethenyl)phosphonate, *in* M-00149
5,5-Dimethyl-4-(1-methylethyl)-1,3,2-dioxaphosphorinane, *see* I-00028
2,6-Dimethyl-5-(1-methylethyl)-6-phosphabicyclo[3.1.1]hept-2-ene 6-oxide, *see* I-00029
O,S-Dimethyl *O*-(1-methylethyl) phosphorothioate, *see* I-00030
▷*O,O*-Dimethyl *S*-[2-[(1-methylethyl)thio]-ethyl] phosphorodithioate, *see* I-00067
▷*O,O*-Dimethyl *S*-[2-[[1-methyl-2-(methylamino)-2-oxoethyl]thio]ethyl] phosphorothioate, *see* V-00001
Dimethyl 1-methyl-3-(methylamino)-3-oxo-1-propenyl phosphate, *see* M-00459
Dimethyl 1-methyl-2-methylcarbamoylvinyl phosphate, *see* M-00459
O,O-Dimethyl *O*-[(3-methyl-4-methylsulfinyl)phenyl] phosphorothioate, *in* F-00007
O,O-Dimethyl *O*-[(3-methyl-4-methylsulfonyl)phenyl]phosphorothioate, *in* F-00007
▷*O,O*-Dimethyl *O*-[(3-methyl-4-methylthio)-phenyl] phosphorothioate, *see* F-00007
▷Dimethyl 3-methyl-4-nitrophenyl phosphate, *see* F-00004
▷*O,O*-Dimethyl *O*-(3-methyl-4-nitrophenyl) phosphorothioate, *see* F-00005
▷*O,O*-Dimethyl *O*-(3-methyl-4-nitrophenyl) thiophosphate, *see* F-00005
Dimethyl (3-methyloxiranyl)phosphonate, *in* M-00185
Dimethyl (3-methyl-2-oxoheptyl)phosphonate, *in* M-00187
Dimethyl 1-methyl-2-(1-phenylethoxycarbonyl)vinyl phosphate, *see* C-00207
Dimethyl (2-methylphenyl)phosphate, *in* M-00235
Dimethyl(2-methylphenyl)phosphine, D-00759
Dimethyl(3-methylphenyl)phosphine, D-00760
Dimethyl(4-methylphenyl)phosphine, D-00761
Dimethyl (3-methylphenyl)phosphonate, *in* M-00236
Dimethyl (4-methylphenyl)phosphonate, *in* M-00237
Dimethyl *N*-methyl-*P*-phenylphosphonimidate, *in* M-00240

Dimethyl (4-methylphenyl)phosphonite, *in* M-00248
Dimethyl methylphenylphosphoramidate, *in* M-00250
Dimethyl (2-methylphenyl)phosphoramidate, *in* M-00251
Dimethyl (3-methylphenyl)phosphoramidate, *in* M-00252
Dimethyl (4-methylphenyl)phosphoramidate, *in* M-00253
Dimethyl [[(4-methylphenyl)sulfonyl]-methyl]phosphonate, *in* M-00266
▷Dimethyl methylphosphonate, D-00762
Dimethyl methylphosphonite, *in* M-00319
O,S-Dimethyl methylphosphonodithioate, *in* M-00301
S,S-Dimethyl methylphosphonodithioate, *in* M-00301
Dimethyl methylphosphonodithioite, *in* M-00302
O,O-Dimethyl methylphosphonoselenoate, *in* M-00311
O,O-Dimethyl methylphosphonothioate, *in* M-00313
Dimethyl methylphosphonotrithioate, *in* M-00318
Dimethyl methylphosphoramidate, *in* M-00326
Dimethyl methylphosphoramidite, *in* M-00335
O,O-Dimethyl methylphosphoramidothioate, *in* M-00333
N,N-Dimethyl-*P*-(1-methylpropyl)-phosphonamidic chloride, D-00763
Dimethyl (1-methylpropyl)phosphonate, *in* M-00377
Dimethyl 2-methyl-4-pyridinephosphonate, *in* M-00388
Dimethyl (3-methyl-4-pyridinyl)phosphonate, *in* M-00390
5,5-Dimethyl-2-methylseleno-1,3-dioxa-2-phosphacyclohexane, *see* D-00764
5,5-Dimethyl-2-methylseleno-1,3,2-dioxaphosphinane, *see* D-00764
5,5-Dimethyl-2-methylseleno-1,3,2-dioxaphosphorinane, D-00764
Dimethyl [(methylsulfinyl)methyl]-phosphonate, *in* M-00398
Dimethyl [(methylsulfonyl)methyl]-phosphonate, *in* M-00400
Dimethyl methylsulfonylphosphoramidate, *in* M-00401
5,5-Dimethyl-2-methylthio-1,3-dioxa-2-phosphacyclohexane, *see* D-00766
4,5-Dimethyl-2-methylthio-1,3-dioxa-2-phosphacyclopentane, *see* D-00765
5,5-Dimethyl-2-methylthio-1,3,2-dioxaphosphinane, *see* D-00766
4,5-Dimethyl-2-methylthio-1,3,2-dioxaphospholane, D-00765
5,5-Dimethyl-2-methylthio-1,3,2-dioxaphosphorinane, D-00766
O,O-Dimethyl *O*-[2-(methylthio)ethyl] phosphorothioate, *see* D-00017
▷*O,O*-Dimethyl *S*-[(2-methylthio)ethyl] phosphorothioate, *see* D-00018
Dimethyl [(methylthio)methyl]phosphonate, *in* M-00415
▷Dimethyl 4-(methylthio)phenyl phosphate, D-00767
▷*O,O*-Dimethyl-*O*-(4-methylthio-*m*-tolyl) phosphorothioate, *see* F-00007
OO-Dimethyl methylurethane phosphate, *in* D-00680
▷Dimethyl 4-morpholinylphosphonate, *in* M-00523
O,O-Dimethyl 4-morpholinylphosphonothioate, *in* M-00527
Dimethyl[1,8-naphthalenediylbis[methylene]]-bisphosphonate, *in* N-00004
Dimethyl 1-naphthylphosphonate, *in* N-00021
Dimethyl 1-naphthylphosphonite, *in* N-00025
Dimethyl neopentylphosphonate, *in* D-00912
Dimethyl (1-nitroethyl)phosphonate, *in* N-00035
Dimethyl (2-nitroethyl)phosphonate, *in* N-00036
Dimethyl (nitromethyl)phosphonate, *in* N-00037
Dimethyl [(2-nitrophenyl)methyl]phosphonate, *in* N-00044
▷Dimethyl 2-nitrophenyl phosphate, D-00768
▷Dimethyl 3-nitrophenyl phosphate, D-00769
▷Dimethyl 4-nitrophenyl phosphate, D-00770
Dimethyl (2-nitrophenyl)phosphonate, *in* N-00055
Dimethyl (3-nitrophenyl)phosphonate, *in* N-00056
Dimethyl (3-nitrophenyl)phosphoramidate, *in* N-00059
Dimethyl (4-nitrophenyl)phosphoramidate, *in* N-00060
O,O-Dimethyl *Se,Se*-(2-nitrophenyl) phosphoro(diselenoperoxoate), D-00771
O,O-Dimethyl-*S,Se*-2-nitrophenylphosphoro(selenothioperoxo)-thioate, D-00772
▷*O,O*-Dimethyl *S*-(4-nitrophenyl) phosphorothioate, D-00773
▷*O,S*-Dimethyl *O*-4-nitrophenyl phosphorothioate, D-00774
▷*OO*-Dimethyl *O*-(4-nitrophenyl) phosphorothioate, *see* P-00003
▷*O,S*-Dimethyl *O*-*p*-nitrophenyl thiophosphate, *see* D-00774
▷Dimethyl 4-nitro-*m*-tolyl phosphate, *see* F-00004
▷*O,O*-Dimethyl *O*-4-nitro-*m*-tolyl phosphorothioate, *see* F-00005
Dimethyl nonylphosphonate, *in* N-00076
3,7-Dimethyl-2,6-octadien-1-ol pyrophosphate, *see* M-00465
3,7-Dimethyl-2,6-octadien-1-ol trihydrogen pyrophosphate, *see* M-00465
3,7-Dimethyl-1,6-octadien-3-yl dihydrogen phosphate, D-00775
3,7-Dimethyl-2,6-octadien-1-yl dihydrogen phosphate, D-00776
(3,7-Dimethyl-2,6-octadienylidene)-triphenylphosphorane, *in* D-00777
(3,7-Dimethyl-2,6-octadienyl)-triphenylphosphonium(1+), D-00777
(2,7-Dimethyl-2,4,6-octatrien-1,8-diyl)-bis[triphenylphosphonium](2+), D-00778
(2,7-Dimethyl-2,4,6-octatrien-1,8-diylidene)bis[triphenylphosphorane], *in* D-00778
Dimethyl octylphosphonate, *in* O-00042
Dimethyl octylphosphoramidate, *in* O-00045
2,7-Dimethyl-1,2,7-oxadiphosphepane 2,7-dioxide, D-00779
2,6-Dimethyl-1,2,6-oxadiphosphorinane 2,6-dioxide, D-00780
4,5-Dimethyl-2-oxa-3-phenyl-1-phosphabicyclo[2.2.1]hept-5-ene, D-00781
2,6-Dimethyl-1-oxa-4-phenyl-1-phosphacyclohexane, *see* D-00797
5,5-Dimethyl-1,2-oxaphosphol-3-ene 2-oxide, *see* D-00464
N,N-Dimethyl-1,3,2-oxaphospholidin-2-amine, *see* D-00702
N,N-Dimethyl-1,3,2-oxathiaphospholan-2-amine, *in* O-00056
▷*O,O*-Dimethyl *S*-[(4-oxo-1,2,3-benzotriazin-3(4*H*)-yl)methyl] phosphorodithioate, *see* A-00151
(1,1-Dimethyl-3-oxobutyl)ethylphosphinic acid, D-00782
(3,3-Dimethyl-2-oxobutylidene)-triphenylphosphorane, *see* D-00934
(3,3-Dimethyl-2-oxobutyl)phosphonic acid, D-00783
(3,3-Dimethyl-2-oxobutyl)-triphenylphosphonium(1+), D-00784
1,3-Dimethyl-4-oxo-1,3,2-diazaphosphetidine, *see* D-00717
Dimethyl (1-oxoethyl)phosphonate, *in* A-00026
Dimethyl (2-oxoethyl)phosphonate, *in* O-00059
Dimethyl (2-oxoheptyl)phosphonate, *in* O-00061
(3,3-Dimethyl-2-oxoheptyl)phosphonic acid, D-00785
Dimethyl (2-oxononyl)phosphonate, *in* O-00063
Dimethyl (2-oxooctyl)phosphonate, *in* O-00064
Dimethyl (2-oxo-4-phenylbutyl)phosphonate, *in* O-00065
(3,3-Dimethyl-2-oxo-4-phenylbutyl)-phosphonic acid, D-00786
Dimethyl (2-oxo-2-phenylethyl)phosphonate, *in* O-00066
Dimethyl(2-oxo-3-phenylpropyl)phosphonate, *in* O-00068
Dimethyl (1-oxopropyl)phosphonate, *in* O-00074
Dimethyl (2-oxopropyl)phosphonate, *in* O-00075
(3,3-Dimethyl-2-oxopropyl)-triphenylphosphonium(1+), D-00787
Dimethyl 1,3-pentadienylphosphonate, *in* P-00009
Dimethyl (pentafluorophenyl)phosphonate, *in* P-00015
N,N-Dimethyl-*P,P*-(1,1,2,3,3-pentamethyltrimethylene)phosphinic amide, *in* P-00021
Dimethyl pentylphosphonate, *in* P-00050
1,3-Dimethyl-2-phenoxy-1,3,2-diazaphospholidine, D-00788
1,3-Dimethyl-2-phenoxy-1,3-diazaphospholidin-4-one 2-oxide, *in* D-00721
5,5-Dimethyl-2-phenoxy-1,3-dioxa-2-phosphacyclohexane, *see* D-00790
4,5-Dimethyl-2-phenoxy-1,3-dioxa-2-phosphacyclopentane, *see* D-00789
5,5-Dimethyl-2-phenoxy-1,3,2-dioxaphosphinane, *see* D-00790
4,5-Dimethyl-2-phenoxy-1,3,2-dioxaphospholane, D-00789
5,5-Dimethyl-2-phenoxy-1,3,2-dioxaphosphorinane, D-00790
5,5-Dimethyl-2-phenoxy-1,3,2-oxazaphosphorinane, *see* T-00098
3,4-Dimethyl-2-phenoxy-5-phenyl-1,3,2-oxazaphospholane, *see* D-00791
3,4-Dimethyl-2-phenoxy-5-phenyl-1,3,2-oxazaphospholidine, D-00791
6,9-Dimethyl-5-phenoxy-2,2,3,3-tetrakis(trifluoromethyl)-1,4-dioxa-6,9-diaza-5-phospha(5-P^V)spiro[4.4]-nonane, *in* D-00922
O,O-Dimethyl phenylamidothiophosphate, *in* P-00262
Dimethyl (phenylamino)-carbonylphosphoramidate, *in* P-00081
O,S-Dimethyl [(phenylamino)carbonyl]-phosphoramidothioate, *in* P-00082
5,5-Dimethyl-2-phenylamino-1,3,2-dioxaphosphorinane, D-00792
1,3-Dimethyl-2-phenyl-1*H*-benzodiazaphospholine, *see* D-00467
1,2-Dimethyl-4-phenyl-1,2,4-diazaphospholidine, *in* D-00718
1,3-Dimethyl-2-phenyl-1,3,2-diazaphospholidine, D-00793
1,3-Dimethyl-2-phenyl-1,3-diazaphospholidin-4-one 2-oxide, *in* D-00721
3,5-Dimethylphenyl dichlorophosphate, *see* D-00822
2,6-Dimethylphenyl dihydrogen phosphate, *see* M-00466
5,5-Dimethyl-2-phenyl-1,3-dioxa-2-phosphacyclohexane, *see* D-00795
4,5-Dimethyl-2-phenyl-1,3-dioxa-2-phosphacyclopentane, *see* D-00794
5,5-Dimethyl-2-phenyl-1,3,2-dioxaphosphinane, *see* D-00795
4,5-Dimethyl-2-phenyl-1,3,2-dioxaphospholane, D-00794
5,5-Dimethyl-*N*-phenyl-1,3,2-dioxaphosphorinan-2-amine, *see* D-00792
N,N-Dimethyl-4-phenyl-1,3,2-dioxaphosphorinan-2-amine, *see* D-00704
5,5-Dimethyl-2-phenyl-1,3,2-dioxaphosphorinane, D-00795
Dimethyl 1,4-phenylenebis[methylphosphinate], *in* P-00137
Dimethyl 1,4-phenylenebis[methyl-phosphinodithioate], *in* P-00138
Dimethyl 1,2-phenylenebis(phenylphosphinate), *in* P-00139
Dimethyl (1-phenylethenyl)phosphonate, *in* P-00154
Dimethyl (2-phenylethenyl)phosphonite, *in* P-00161
Dimethyl 1-phenylethylphosphonate, *in* P-00164

Dimethyl 2-phenylethylphosphonate, *in* P-00165
Dimethyl phenylethynylphosphonate, *in* P-00166
O,O-Dimethyl (phenylethynyl)phosphonothioate, *in* P-00167
1,4-Dimethyl-2-phenyl-7-oxa-2-phosphabicyclo[2.2.1]heptane, D-00796
2,6-Dimethyl-4-phenyl-1,4-oxaphosphinane, *see* D-00797
2,6-Dimethyl-4-phenyl-1,4-oxaphosphorinane, D-00797
2,8-Dimethyl-10-phenyl-10*H*-phenothiaphosphine, D-00798
2,8-Dimethyl-10-phenyl-10*H*-phenoxaphosphine, D-00799
Dimethyl *P*-phenyl-*N*-phenylsulfonylphosphonimidate, *in* P-00192
2,6-Dimethyl-4-phenylphosphabenzene, *see* D-00819
1,4-Dimethyl-2-phenyl-2-phosphabicyclo[2.2.1]heptane 2-oxide, *in* D-00825
5,6-Dimethyl-7-phenyl-7-phosphabicyclo[2.2.1]hept-5-ene-2,3-dicarboxylic acid, D-00800
4,4-Dimethyl-1-phenyl-1-phospha-4-silacyclohexa-2,5-diene, D-00801
2,2-Dimethyl-1-phenyl-1-phospha-2-silacyclopentane, D-00802
Dimethylphenylphosphine, D-00803
(2,4-Dimethylphenyl)phosphinic acid, D-00804
(2,5-Dimethylphenyl)phosphinic acid, D-00805
(3,4-Dimethylphenyl)phosphinic acid, D-00806
P,P-Dimethyl-*N*-phenylphosphinoselenoic amide, *in* D-00836
P,P-Dimethyl-*N*-phenylphosphinothioic amide, *in* D-00840
1,3-Dimethyl-1-phenylphospholanium bromide, *in* M-00229
3,4-Dimethyl-1-phenyl-1*H*-phosphole, D-00807
3,4-Dimethyl-1-phenyl-3-phospholene, *see* D-00468
N,N-Dimethyl-*P*-phenylphosphonamidic acid, D-00808
N,N-Dimethyl-*P*-phenylphosphonamidic azide, D-00809
Dimethyl phenylphosphonate, D-00810
(2,3-Dimethylphenyl)phosphonic acid, D-00811
(2,4-Dimethylphenyl)phosphonic acid, D-00812
(2,5-Dimethylphenyl)phosphonic acid, D-00813
(2,6-Dimethylphenyl)phosphonic acid, D-00814
(3,4-Dimethylphenyl)phosphonic acid, D-00815
(3,5-Dimethylphenyl)phosphonic acid, D-00816
1,1′-Dimethyl-*P*-phenylphosphonic dihydrazide, *in* P-00225
Dimethyl phenylphosphonite, *in* P-00251
Dimethyl phenylphosphonodiselenoite, *in* P-00235
O,O-Dimethyl phenylphosphonoselenoate, *in* P-00243
O,O-Dimethyl phenylphosphonothioate, *in* P-00244
O,S-Dimethyl phenylphosphonothioate, *in* M-00243
N,N′-Dimethyl-*P*-phenylphosphonothioic diamide, *in* P-00246
N,P-Dimethyl-*N′*-phenylphosphonothioic diamide, *in* M-00245
Dimethyl phenylphosphoramidate, D-00817
(2,4-Dimethylphenyl)phosphoramidic acid, D-00818
S,S-Dimethyl phenylphosphoramidodithioate, *in* D-00866
O,O-Dimethyl phenylphosphoramidothioate, *in* P-00262
2,6-Dimethyl-4-phenylphosphorin, D-00819
2,5-Dimethyl-1-phenylphosphorinan-4-ol, D-00820
2,5-Dimethyl-1-phenylphosphorinan-4-one, D-00821
2,6-Dimethylphenyl phosphorodichloridate, *in* M-00466
3,5-Dimethylphenyl phosphorodichloridate, D-00822
O,O-Dimethyl 2-phenylphosphorohydrazidothioate, *in* P-00296
Dimethyl (3-phenylpropyl)phosphonate, *in* P-00305
Dimethylphenylpropylphosphonium iodide, *in* M-00265
5,6-Dimethyl-*N*-phenylspiro[1,3,2-benzodioxaphosphole-2,1′(3′*H*)-[2,1]-benzoxaphosphol]-2-amine, *in* D-00915
3′,4′-Dimethyl-5′-phenylspiro[1,3,2-benzoxazaphosphole-2(3*H*),2′-[1,3,2]-oxazaphospholidine], D-00823
3′,4′-Dimethyl-5′-phenylspiro[1,3,2-benzoxazaphospholine-2,2′-[1,3,2]-oxazaphospholidine], *see* D-00823
Dimethyl [(phenylsulfinyl)methyl]-phosphonate, *in* P-00310
Dimethyl phenylsulfonylphosphoramidate, *in* P-00312
4,5-Dimethyl-2-phenylthio-1,3-dioxa-2-phosphacyclopentane, *see* D-00824
4,5-Dimethyl-2-phenylthio-1,3,2-dioxaphospholane, D-00824
Dimethyl [(phenylthio)methyl]phosphonate, *in* P-00326
6,9-Dimethyl-5-phenylthio-2,2,3,3-tetrakis(trifluoromethyl)-1,4-dioxa-6,9-diaza-5-phospha(5-P^V)spiro[4.4]-nonane, *in* D-00922
Dimethyl [phenyl[(trimethylsilyl)oxy]-methyl]phosphonate, *in* P-00331
2,6-Dimethylphosphabenzene, *see* D-00881
1,4-Dimethyl-2-phosphabicyclo[2.2.1]heptane 2-oxide, D-00825
1,3-Dimethylphosphacyclohexane, *see* D-00882
1,4-Dimethylphosphacyclohexane, *see* D-00883
1,3-Dimethyl-2-phosphanaphthalene, *see* D-00758
Dimethyl phosphate, D-00826
Dimethyl phosphate, ester with 3-hydroxy-*N*-methylcrotonamide, *see* M-00459
1,3-Dimethylphosphinane, *see* D-00882
1,4-Dimethylphosphinane, *see* D-00883
1,1-Dimethylphosphinanium bromide, *in* M-00339
▷Dimethylphosphine, D-00827
Dimethylphosphine imide dimer, *see* T-00194
Dimethylphosphine oxide, D-00828
Dimethylphosphinic acid, D-00829
P,P-Dimethylphosphinic amide, D-00830
Dimethylphosphinic chloride, D-00831
Dimethylphosphinic fluoride, D-00832
P,P-Dimethylphosphinimidic amide, D-00833
2,6-Dimethylphosphinine, *see* D-00881
Dimethylphosphinodithioic acid, D-00834
Dimethylphosphinodithioic anhydrosulfide, *in* D-00834
1-(Dimethylphosphino)ethanone, *see* A-00015
Dimethyl(phosphino)germane, *see* D-00756
o-(Dimethylphosphinophenyl)dimethylarsine, *see* D-00711
Dimethylphosphinoselenoic acid, D-00835
P,P-Dimethylphosphinoselenoic amide, D-00836
Dimethylphosphinoselenoic anilide, *in* D-00836
Dimethylphosphinoselenothioic acid, D-00837
Dimethylphosphinoselenous acid, D-00838
Dimethylphosphinothioic acid, D-00839
P,P-Dimethylphosphinothioic amide, D-00840
P,P-Dimethylphosphinothioic anilide, *in* D-00840
Dimethylphosphinothioic bromide, D-00841
Dimethylphosphinothioic chloride, D-00842
Dimethylphosphinothious acid, D-00843
m-(Dimethylphosphino)toluene, *see* D-00760
o-(Dimethylphosphino)toluene, *see* D-00759
p-(Dimethylphosphino)toluene, *see* D-00761
▷(Dimethylphosphino)trimethylsilane, *see* D-00932
Dimethylphosphinous acid, D-00844
P,P-Dimethylphosphinous amide, D-00845
▷Dimethylphosphinous chloride, D-00846
▷Dimethylphosphinous cyanide, D-00847
▷Dimethylphosphinous fluoride, D-00848
1-(Dimethylphosphinyl)ethanone, *in* A-00015
N-(Dimethylphosphinyl)-*N′*-phenylurea, *in* D-00830
▷Dimethyl phosphite, *see* D-00851
N,P-Dimethylphosphonamidic acid, D-00849
N,P-Dimethylphosphonamidothioic acid, D-00850
▷*N,P*-Dimethylphosphonamidothioic fluoride, *in* D-00850
1,1-Dimethylphosphonanium iodide, *in* M-00285
▷Dimethyl phosphonate, D-00851
1,5-Dimethyl-1-phosphonia-5-silabicyclo[3.3.1]nonane iodide, *in* M-00274
Dimethyl phosphonite, D-00852
Dimethyl phosphonoformate, *see* D-00672
O,O-Dimethyl phosphonothioate, D-00853
$N^1,N^{1'}$-Dimethyl-*P*-phosphonothioic dihydrazide, *in* P-00249
O,O-Dimethyl phosphoramidate, D-00854
O,O-Dimethyl phosphoramide, *see* D-00854
Dimethylphosphoramidic acid, D-00855
▷Dimethylphosphoramidic dichloride, D-00856
Dimethylphosphoramidic dicyanide, D-00857
Dimethylphosphoramidic difluoride, D-00858
Dimethyl phosphoramidite, D-00859
Dimethylphosphoramidochloridic acid, D-00860
Dimethylphosphoramidochloridothioic acid, D-00861
Dimethylphosphoramidochloridous acid, D-00862
Dimethylphosphoramidocyanidic acid, D-00863
Dimethylphosphoramidocyanidothioic acid, D-00864
O,S-Dimethyl phosphoramidodithioate, D-00865
S,S-Dimethyl phosphoramidodithioate, D-00866
Dimethylphosphoramidofluoridodithioic acid, D-00867
Dimethylphosphoramidofluoridothioic acid, D-00868
Dimethylphosphoramidoselenoic acid, D-00869
O,O-Dimethyl phosphoramidothioate, D-00870
▷*O,S*-Dimethyl phosphoramidothioate, *see* M-00021
Dimethylphosphoramidothioic acid, D-00871
Dimethylphosphoramidothioic dichloride, D-00872
Dimethylphosphoramidothioic difluoride, D-00873
Dimethylphosphoramidotrithioic acid, D-00874
Dimethylphosphoramidous acid, D-00875
▷Dimethylphosphoramidous dichloride, D-00876
Dimethylphosphoramidous difluoride, D-00877
▷Dimethyl phosphorazidate, D-00878
O,O-Dimethyl phosphorazidothioate, D-00879
Dimethyl phosphoric acid, *see* D-00826
O,O-Dimethyl phosphoric amide, *see* D-00854
Dimethyl phosphoric anilide, *see* D-00817
N,N-Dimethylphosphoric triamide, D-00880
2,6-Dimethylphosphorin, D-00881
1,3-Dimethylphosphorinane, D-00882
1,4-Dimethylphosphorinane, D-00883
1,1-Dimethylphosphorinanium bromide, *in* M-00339
O,O-Dimethyl phosphorisocyanatidothioate, D-00884
O,S-Dimethyl phosphorisocyanatidothioate, D-00885
Dimethyl phosphorobromidate, D-00886
Dimethyl phosphorochloridate, D-00887
S,S-Dimethyl phosphorochloridodithioate, D-00888
Dimethyl phosphorochloridodithioite, D-00889
Dimethyl phosphorochloridoselenoite, D-00890
▷*O,O*-Dimethyl phosphorochloridothioate, D-00891
O,S-Dimethyl phosphorochloridothioate, D-00892
Dimethyl phosphorochloridotrithioate, D-00893
Dimethyl phosphorocyanidate, D-00894
N,N′-Dimethylphosphorodiamidic acid, D-00895
N,N′-Dimethylphosphorodiamidodithioic acid, D-00896
O,O-Dimethyl phosphorodiselenoate, D-00897
O,O-Dimethyl phosphorodiselenoic acid, *see* D-00897
▷*O,O*-Dimethyl phosphorodithioate, D-00898
O,S-Dimethyl phosphorodithioate, D-00899
S,S-Dimethyl phosphorodithioate, D-00900
O,O-Dimethyl phosphorodithioate anhydrosulfide, *see* T-00240
▷*O,O*-Dimethyl phosphorodithioate *S*-ester with 2-mercapto-*N*-methylacetamide, *see* D-00671

▷Dimethyl phosphorofluoridate, D-00901
S,S-Dimethyl phosphorofluoridodithioate, *in* D-00900
O,O-Dimethyl phosphorofluoridothioate, D-00902
Dimethyl phosphorofluoridotrithioate, D-00903
Dimethyl phosphorohydrazidate, D-00904
O,O-Dimethyl phosphorohydrazidothioate, D-00905
O,O-Dimethyl phosphoroselenoate, D-00906
O,O-Dimethyl phosphoroselenoic acid, *see* D-00906
O,O-Dimethyl phosphorothioate, D-00907
O,S-Dimethyl phosphorothioate, D-00908
▷*O,O*-Dimethyl phosphorothioate anhydride, *see* T-00203
▷*O,O*-Dimethyl phosphorothioate, ester with *p*-hydroxybenzene sulfonamide, *see* D-00920
▷*O,O*-Dimethyl phosphorothioate, *O*-ester with *p*-hydroxy-*N,N*-dimethylbenzenesulfonamide, *see* F-00001
▷*O,O*-Dimethyl phosphorothioate, *S*-ester with 2-mercapto-*N*-methylacetamide, *see* O-00048
O,S-Dimethyl phosphorothioic acid, *see* D-00908
O,O-Dimethyl phosphorothioic acid, *see* D-00907
O,O-Dimethyl phosphorothioic acid phenylhydrazide, *in* P-00296
NN-Dimethylphosphorothioic triamide, *in* D-00871
N,N-Dimethylphosphorothioic triamide, D-00909
O,O-Dimethyl phosphoryl amide, *see* D-00854
▷Dimethyl phosphoryl azide, *see* D-00878
Dimethyl phosphoryl bromide, *see* D-00886
Dimethyl phosphoryl chloride, *see* D-00887
Dimethyl phosphoryl cyanide, *see* D-00894
▷Dimethyl phosphoryl fluoride, *see* D-00901
N-Dimethylphosphorylimidazole, *in* I-00003
O,O-Dimethyl phosphoryl isocyanate, *see* D-00884
O,S-Dimethyl phosphoryl isocyanate, *see* D-00885
▷*O,O*-Dimethyl *S*-phthalimidomethyl phosphorodithioate, *see* P-00343
Dimethyl 1-piperidinylphosphonate, *in* P-00430
O,O-Dimethyl 1-piperidinylphosphonothioate, *in* P-00431
Dimethyl 1,2-propadienylphosphonate, *in* P-00441
2,2-Dimethyl-1,3-propanediol cyclic *P,P,P′,P′*-hypophosphate, *in* B-00197
(1,3-Dimethyl-1,3-propanediyl)-bis[diphenylphosphine], *see* B-00260
Dimethyl propanoylphosphonate, *in* O-00074
Dimethyl 1-propenylphosphonate, *in* P-00453
Dimethyl 2-propenylphosphonate, *in* P-00454
S,S-Dimethyl 2-propenylphosphonodithioate, *in* P-00457
(2,2-Dimethylpropylidene)-triphenylphosphorane, *in* D-00914
(2,2-Dimethylpropylidyne)phosphine, D-00910
Dimethyl propylphosphonate, *in* P-00473
(1,1-Dimethylpropyl)phosphonic acid, D-00911
(2,2-Dimethylpropyl)phosphonic acid, D-00912
Dimethyl propylphosphonite, *in* P-00482
Dimethyl propylphosphoramidate, *in* P-00484
2,2-Dimethylpropyl phosphorodichloridite, D-00913
(2,2-Dimethylpropyl)triphenyl-phosphonium(1+), D-00914
Dimethyl 1-propynylphosphonate, *in* P-00494
Dimethyl 2-pyridinylphosphonate, *in* P-00503
Dimethyl 4-pyridinylphosphonate, *in* P-00505
Dimethyl 2-pyridylphosphoramidate, *in* P-00510
Dimethyl 3-pyridylphosphoramidate, *in* P-00511
Dimethylpyrophosphonic acid, *see* D-00751
Dimethyl 1-pyrrolidinylphosphonate, *in* P-00518
Dimethyl 4-quinolinephosphonate, *in* Q-00003
5,6-Dimethylspiro[1,3,2-benzodioxaphosphole-2,1′(3′*H*)-[2,1]-benzoxaphosphole, D-00915
N,N-Dimethylspiro[1,3,2-benzodioxaphosphole-2,2′-[1,3,2]-benzoxathiaphosphol]-2-amine, *in* S-00007
2,3′-Dimethylspiro[1,3,2-benzodioxaphosphole-2,2′(3′*H*)-[1,3,2]-benzoxazaphosphole], *in* M-00395
3,3′-Dimethyl-2,2′-(3*H*,3′*H*)-spirobi[1,3,2-benzothiazaphosphole], D-00916
3,3-Dimethyl-2,2(3*H*,3′*H*)-spirobi[1,3,2-benzoxazaphosphole], D-00917
1,3-Dimethylspiro[1,3,2-diazaphospholidine-2,2′-phenanthro[9,10-*d*][1,3,2]dioxaphosphole], D-00918
4,4-Dimethylspiro[1,3,2-dioxaphospholane-2,1′-[2,6,7]trioxa[1]-phosphabicyclo[2.2.1]heptane], D-00919
N,N-Dimethylspiro[1,3,2-dioxaphosphorinane-2,2′-phenanthro[9,10-*d*][1,3,2]-dioxaphosphol]-2-amine, *in* S-00025
▷*O,O*-Dimethyl *O*-4-sulfamoylphenyl phosphorothioate, D-00920
Dimethyl sulfide-1,1′-diphosphonic acid, *see* T-00311
15,16-Dimethyl-1,4,8,11-tetraaza-15,16-diphosphatricyclo[9.3.1.1^{4,8}]hexadecane, D-00921
Dimethyl (2,3,5,6-tetrachloro-4-pyridinyl)phosphonate, *in* T-00035
6,9-Dimethyl-2,2,3,3-tetrakis(trifluoromethyl)-1,4-dioxa-6,9-diaza-5-phospha(5-P^V)spiro[4.4]-nonane, D-00922
6,10-Dimethyl-2,2,3,3-tetrakis(trifluoromethyl)-1,4-dioxa-6,10-diaza-5-phospha(5-P^V)spiro[4.5]-decane, D-00923
8,9-Dimethyl-1,3,4,6-tetrakis(trifluoromethyl)-2,5-diphosphatetracyclo[4.4.0.0^{2,4}.0^{3,5}]-dec-8-ene, *see* T-00100
▷*P,P′*-Dimethyl-*N,N,N′,N′*-tetramethyldiphosphonic diamide, *in* D-00751
2,3-Dimethyl-1,4,6,9-tetraoxa-5-phospha(5-P^V)spiro[4.4]non-2-ene, D-00924
Dimethyl (2-thienylmethyl)phosphonate, *in* T-00304
Dimethyl (3-thienylmethyl)phosphonate, *in* T-00305
Dimethyl 2-thienylphosphonate, *in* T-00307
Dimethyl 2-thienylphosphonite, *in* T-00308
Dimethylthiodiphosphonic acid, D-00925
Dimethylthiodiphosphonic acid difluoride, *in* M-00304
O,S-Dimethyl thiophosphate, *see* D-00908
O,O-Dimethyl thiophosphite, *see* D-00853
O,O-Dimethyl thiophosphoric acid, *see* D-00907
O,O-Dimethyl thiophosphoric anilide, *in* P-00262
O,O-Dimethyl thiophosphoryl amide, *see* D-00870
O,O-Dimethyl thiophosphoryl azide, *see* D-00879
▷*OO*-Dimethyl thiophosphoryl chloride, *see* D-00891
O,O-Dimethyl thiophosphoryl fluoride, *see* D-00902
O,O-Dimethyl thiophosphoryl hydrazide, *see* D-00905
Dimethyl-*m*-tolylphosphine, *see* D-00760
Dimethyl-*o*-tolylphosphine, *see* D-00759
Dimethyl-*p*-tolylphosphine, *see* D-00761
Dimethyl *m*-tolylphosphonate, *in* M-00236
Dimethyl *o*-tolylphosphonate, *in* M-00235
Dimethyl *p*-tolylphosphonate, *in* M-00237
2,10-Dimethyl-2,6,10-triaza-1-phosphabicyclo[4.4.0]decane, *see* O-00005
2,8-Dimethyl-2,5,8-triaza-1-phosphabicyclo[3.3.0]octane, *see* T-00093
1,5-Dimethyl-1*H*-1,2,4,3-triazaphosphole, D-00926
Dimethyl 2-(tributylphosphoranylidene)-1,3-dithiol-4,5-dicarboxylate, D-00927
▷Dimethyl (2,2,2-trichloro-1-hydroxyethyl)-phosphonate, *see* T-00423
▷Dimethyl (2,2,2-trichloro-1-hydroxyethyl)-phosphonate butyrate, *see* B-00530
Dimethyl (trichloromethyl)phosphonate, *in* T-00411
▷*O,O*-Dimethyl *O*-(2,4,5-trichlorophenyl) phosphorothioate, *see* F-00003
▷Dimethyl 3,5,6-trichloro-2-pyridinyl phosphate, *see* F-00063
Dimethyl (tricyclo[3.3.1.1^{3,7}]dec-1-yl)-phosphonate, *in* A-00034
Dimethyl (tricyclo[3.3.1.1^{3,7}]dec-2-yl)-phosphonate, *in* A-00035
N,N-Dimethyl-*P*-trifluoromethylphosphonamidothioic fluoride, D-00928
Dimethyl (trifluoromethyl)phosphonite, *in* T-00482
Dimethyl (trifluoromethyl)-phosphonodithioite, *in* T-00480
[3,7-Dimethyl-9-(2,6,6-trimethyl-1-cyclohexen-1-yl)-2,4,6,8-nonatetraenyl]triphenylphosphonium(1+), D-00929
N,N′-Dimethyl-*N,N′*-trimethylene-*P*-methylphosphonous diamide, *in* H-00022
N,N′-Dimethyl-*N,N′*-trimethylene-*P*-phenylphosphonic diamide, *in* H-00023
N,N′-Dimethyl-*N,N′*-trimethylene-*P*-phenylphosphonous diamide, *in* H-00022
N,N′-Dimethyl-*N,N′*-trimethylenephosphonic diamide, *see* H-00023
N″,N‴-Dimethyl-*N,N′*-trimethylenephosphoric triamide, *in* H-00020
N,N′-Dimethyl-*N,N′*-trimethylene phosphorodiamidous chloride, *in* H-00022
4,5-Dimethyl-2-trimethylsilyloxy-1,3-dioxa-2-phosphacyclopentane, *see* D-00930
4,5-Dimethyl-2-trimethylsilyloxy-1,3,2-dioxaphospholane, D-00930
Dimethyl [1-[(trimethylsilyl)oxy]ethyl]-phosphonate, *in* T-00555
Dimethyl trimethylsilyl phosphate, D-00931
▷Dimethyl(trimethylsilyl)phosphine, D-00932
P,P-Dimethyl-*N*-trimethylsilylphosphinothioic amide, *in* D-00840
8,8-Dimethyl-1,6,10-trioxa-5-phospha(5-P^V)spiro[4.5]dec-2-ene, D-00933
Dimethyl triphenylmethylphosphonate, *in* T-00608
Dimethyl (triphenylphosphoranylidene)-butanedioate, *in* T-00646
3,3-Dimethyl-1-triphenylphosphoranylidene-2-butanone, D-00934
Dimethyl (triphenylphosphoranylidene)-propanedioate, *in* T-00658
Dimethyl (triphenylphosphoranylidene)-succinate, *in* T-00646
P,P-Dimethyl-*N,N,N′*-tris(trimethylsilyl)-phosphinimidic amide, *in* D-00833
Dimethyl *N,N,N′*-tris(trimethylsilyl)-phosphoramidimidate, *in* T-00868
(2,2-Dimethylvinyl)diphenylphosphine, *see* M-00362
Dimethylvinylene chlorophosphate, *in* C-00057
Dimethylvinylene chlorophosphite, *see* C-00057
Dimethylvinylene hydrogen phosphate, *see* H-00123
Dimethylvinylene phenyl phosphate, *in* H-00123
Dimethylvinylene phosphoric acid, *see* H-00123
Dimethylvinylene trimethylsilyl phosphate, *in* H-00123
Dimethyl vinylphosphonate, *in* V-00003
O,O-Dimethyl vinylphosphonothioate, *in* V-00007
(2,2-Dimethylvinyl) phosphoric acid, *see* M-00491
Dimethyl (9*H*-xanthen-9-yl)phosphonate, *in* X-00001
Dimorpholinophosphinic acid, *see* D-00935
Di-4-morpholinyldiphosphonic acid, *in* M-00523

Di-4-morpholinylphosphinic acid, D-00935
Di-4-morpholinylphosphinous acid, D-00936
▷Dimpylate, *see* D-00027
1,2-Dimyristoyl-*sn*-glycero-3-phospho-D-serine, *in* G-00051
1,2-Dimyristoyl-*sn*-glycero-3-phospho-DL-serine, *in* G-00051
1,2-Dimyristoyl-*sn*-glycero-3-phospho-L-serine, *in* G-00051
Dimyristoyl-α-glycerylphosphoryl-2-amino-1-propanol, *see* G-00050
Dimyristoylphosphatidic acid, *see* G-00047
Dimyristoylphosphatidylcholine, *see* G-00048
Dimyristoylphosphatidylethanolamine, *see* G-00049
Dimyristoylphosphatidylserine, *see* G-00051
Di-1-naphthalenyl phosphate, *see* D-00937
Di-2-naphthalenyl phosphate, *see* D-00938
Di-1-naphthalenylphosphine, *see* D-00939
Di-2-naphthalenylphosphine, *see* D-00940
Di-1-naphthalenylphosphinic acid, *see* D-00941
Di-1-naphthalenyl phosphorochloridite, *see* D-00942
Di-2-naphthalenyl phosphorochloridite, *see* D-00943
O,O-Di-1-naphthalenyl phosphorodithioate, *see* D-00944
O,O-Di-2-naphthalenyl phosphorodithioate, *see* D-00945
O,O-Di-1-naphthyl dithiophosphoric acid, *see* D-00944
O,O-Di-2-naphthyl dithiophosphoric acid, *see* D-00945
O,O-Di-1-naphthyl hydrogen dithiophosphate, *see* D-00944
O,O-Di-2-naphthyl hydrogen dithiophosphate, *see* D-00945
Di-1-naphthyl hydrogen phosphate, *see* D-00937
Di-2-naphthyl hydrogen phosphate, *see* D-00938
P,P-Di-1-naphthyl-*N*-phenylphosphinic amide, *in* D-00941
Di-1-naphthyl phosphate, D-00937
Di-2-naphthyl phosphate, D-00938
Di-1-naphthylphosphine, D-00939
Di-2-naphthylphosphine, D-00940
Di-1-naphthylphosphinic acid, D-00941
Di-α-naphthylphosphinic anilide, *in* D-00941
Di-2-naphthyl phosphoramidate, *in* D-00938
Di-1-naphthyl phosphoric acid, *see* D-00937
Di-2-naphthyl phosphoric acid, *see* D-00938
Di-1-naphthyl phosphorochloridate, *in* D-00937
Di-1-naphthyl phosphorochloridite, D-00942
Di-2-naphthyl phosphorochloridite, D-00943
O,O-Di-1-naphthyl phosphorodithioate, D-00944
O,O-Di-2-naphthyl phosphorodithioate, D-00945
Dineopentyl chlorophosphite, *see* B-00209
Dineopentyl phosphite, *see* B-00208
Dineopentyl phosphonate, *see* B-00208
Di-*o*-nitrobenzyl hydrogen phosphate, *see* B-00366
Di-*p*-nitrobenzyl hydrogen phosphate, *see* B-00367
Di-*o*-nitrobenzyl phosphate, *see* B-00366
Di-*p*-nitrobenzyl phosphate, *see* B-00367
Di-*p*-nitrobenzyl phosphite, *see* B-00368
Di-*p*-nitrobenzyl phosphonate, *see* B-00368
Di-*p*-nitrobenzyl phosphoric acid, *see* B-00367
Di-*o*-nitrobenzyl phosphorochloridate, *in* B-00366
Di-*o*-nitrobenzyl phosphoryl chloride, *in* B-00366
Di-*p*-nitrobenzyl phosphoryl chloride, *in* B-00367
O,O-Di-*p*-nitrophenyl chlorothiophosphate, *see* B-00377
2,4-Dinitrophenyl dichlorophosphate, *in* M-00467
2,4-Dinitrophenyl dihydrogen phosphate, *see* M-00467
Di-*p*-nitrophenyl phosphite, *see* B-00375
Di(4-nitrophenyl) phosphonate, *see* B-00375
(2,4-Dinitrophenyl)phosphonic acid, D-00946
(2,4-Dinitrophenyl)phosphoramidic acid, D-00947
2,4-Dinitrophenyl phosphoric acid, *see* M-00467
O,O-Di-*p*-nitrophenyl phosphorochloridothioate, *see* B-00377
2,4-Dinitrophenyl phosphorodichloridate, *in* M-00467
2,4-Dinitrophenyl phosphoryl dichloride, *in* M-00467
O,O-Di-*p*-nitrophenyl thiophosphoryl chloride, *see* B-00377
Dinonadecanoylphosphatidylcholine, *in* G-00026
Dinonyl chlorophosphate, *in* D-00948
O,O-Dinonyl dithiophosphate, *see* D-00951
O,O-Dinonyl dithiophosphoric acid, *see* D-00951
Dinonyl hydrogen phosphate, *see* D-00948
O,O-Dinonyl hydrogen phosphorodithioate, *see* D-00951
Dinonyl phosphate, D-00948
Dinonylphosphinic acid, D-00949
Dinonyl phosphite, *see* D-00950
Dinonyl phosphonate, D-00950
Dinonyl phosphoric acid, *see* D-00948
Dinonyl phosphorochloridate, *in* D-00948
O,O-Dinonyl phosphorodithioate, D-00951
Dinonyl phosphoryl chloride, *in* D-00948
1,2-Dioctadeca-9,12-dienoylglycero-3-phospohocholine, *see* G-00028
1,2-Dioctadecanoyl-*sn*-glycerol 3-phosphate, *in* G-00032
1,2-Dioctadecanoylglycero-3-phosphocholine, *see* G-00034
Dioctadecanoyl-α-glycerylphosphoryl-2-amino-1-propanol, *see* G-00033
1,2-Di-9-octadecenoylglycerol-3-phosphocholine, *see* G-00042
1,2-Di-9-octadecenoylphosphatidylethanolamine, *see* G-00043
Dioctadecyl hydrogen phosphate, *see* D-00952
Dioctadecyl phosphate, D-00952
Dioctadecylphosphinic acid, D-00953
Dioctadecyl phosphite, *see* D-00954
Dioctadecyl phosphonate, D-00954
Dioctadecyl phosphoric acid, *see* D-00952
1,2-Di-9,12-octodecadienoylphosphatidylcholine, *see* G-00029
Dioctyl dichlorophosphate, *in* D-00956
Dioctyl (2-dimethylamino-2-oxoethyl)-phosphonate, *in* D-00703
Dioctyl (dimethylcarbamoylmethyl)-phosphonate, *in* D-00703
O,O-Dioctyl dithiophosphate, *see* D-00960
O,O-Dioctyl dithiophosphoric acid, *see* D-00960
Dioctyl hydrogen phosphate, *see* D-00956
O,O-Dioctyl hydrogen phosphorodithioate, *see* D-00960
Dioctylphenylphosphine, D-00955
▷Dioctyl phenylphosphonate, *in* P-00221
Dioctyl phosphate, D-00956
Dioctylphosphine, D-00957
Dioctylphosphinic acid, D-00958
Dioctyl phosphite, *see* D-00959
Di(2-octyl) phosphite, *see* B-00307
Dioctyl phosphonate, D-00959
Di(2-octyl) phosphonate, *see* B-00307
Dioctyl phosphoric acid, *see* D-00956
Dioctyl phosphorochloridate, *in* D-00956
O,O-Dioctyl phosphorodithioate, D-00960
Dioctyl phosphoryl chloride, *in* D-00956
1,2-Dioleoyl-*sn*-glycerol-3-phospho-1′-*sn*-glycerol, *in* G-00044
1,2-Dioleoyl-*sn*-glycero-3-phospho-D-serine, *in* G-00045
1,2-Dioleoyl-*sn*-glycero-3-phospho-DL-serine, *in* G-00045
1,2-Dioleoyl-*sn*-glycero-3-phospho-L-serine, *in* G-00045
Dioleoylphosphatidylcholine, *in* G-00042
Dioleoylphosphatidylethanolamine, *in* G-00043
Diop, *see* B-00251
2,10-Dioxa-6-aza-1-phosphabicyclo[4.4.0]-decane, *see* T-00120
2,8-Dioxa-5-aza-1-phosphabicyclo[3.3.0]-octane, D-00961
1,4-Dioxa-6-aza-5-phospha(5-P^{V})spiro[4.4]-nonan-7-one, D-00962
1,6-Dioxa-4,9-diaza-5-phospha(5-P^{V})-spiro[4.4]nonane, D-00963
1,4-Dioxa-6,9-dithia-5-phospha(5-P^{V})-spiro[4.4]nonane, D-00964
▷1,4-Dioxan-2,3-diyl *S,S*-di(*O,O*-diethyl phosphorodithioate), *see* D-00974
▷2,3-*p*-Dioxanedithiol *S,S*-bis(*O,O*-diethyl phosphorodithioate), *see* D-00974
▷*S,S*′-(1,4-Dioxane-2,3-diyl) *O,O,O*′*,O*′-tetraethyl di(phosphorothioate), *see* D-00974
[2-(1,3-Dioxan-2-yl)ethylidene]-triphenylphosphorane, *in* D-00965
[2-(1,3-Dioxan-2-yl)ethyl]-triphenylphosphonium(1+), D-00965
1,3-Dioxa-2-phosphacycloheptane, *see* D-00968
1,3-Dioxa-2-phosphacyclohexane, *see* D-00972
1,3-Dioxa-2-phosphacyclononane, *see* D-00971
1,3-Dioxa-2-phosphacyclopentane, *see* D-00970
1,3-Dioxa-2-phosphacyclotridecane, D-00966
1,4-Dioxa-5-phospha(5-P^{V})spiro[4.4]non-7-ene, D-00967
1,3,2-Dioxaphosphepane, D-00968
1,3,2-Dioxaphosphepane 2-sulfide, D-00969
1,3,2-Dioxaphosphinane, *see* D-00972
1,4,2-Dioxaphosphinane 2-oxide, *see* D-00973
1,3,2-Dioxaphospholane, D-00970
1,3,2-Dioxaphosphonane, D-00971
1,3,2-Dioxaphosphorinane, D-00972
1,4,2-Dioxaphosphorinane 2-oxide, D-00973
▷Dioxathion, D-00974
(1,4-Dioxo-1,4-butanediyl)bisphosphonic acid, D-00975
1,4-Dioxo-1,4-dihydroxy-1,4-diphosphabicyclo[4.4.0]decane, *see* D-00003
(1,2-Dioxo-1,2-ethanediyl)-bis[diphenylphosphine], *see* B-00235
[(1,3-Dioxolan-2-yl)methylene]-triphenylphosphorane, *in* D-00976
[(1,3-Dioxolan-2-yl)methyl]-triphenylphosphonium(1+), D-00976
(2,4-Dioxo-3-pentyl)triphenylphosphonium(1+), *see* D-00022
4,4′-Dioxybis[*N,N*-bis(2-chloroethyl)-amino]tetrahydro-2*H*-1,3,2-oxazaphosphorine 2,2′-dioxide, *see* P-00057
Dioxybis[diphenylphosphine oxide], D-00977
1,2-Dipalmitoyl-*sn*-glycero-L-serine, *in* G-00016
Dipalmitoylphosphatidic acid, *see* G-00011
Dipalmitoylphosphatidylcholine, *see* G-00012
Dipalmitoylphosphatidylethanolamine, *see* G-00013
Dipalmitoylphosphatidylglycerol, *see* G-00014
Dipalmitoylphosphatidylinositol, *see* G-00015
Dipalmitoylphosphatidylserine, *see* G-00016
DiPAMP, *see* E-00011
Dipentadecanoylphosphatidylcholine, *in* G-00046
▷Dipentyl fluorophosphate, *in* D-00979
O,O-Dipentyl hydrogen dithiophosphate, *see* D-00986
Dipentyl hydrogen phosphate, *see* D-00979
O,O-Dipentyl hydrogen phosphorodiselenoate, *see* D-00985
Dipentyl pentylphosphonate, *in* P-00050
Dipentyl pentylphosphonite, *in* P-00052
Dipentylphenylphosphine, D-00978
Dipentyl phenylphosphoramidate, *in* D-00979
Dipentyl phosphate, D-00979
Dipentylphosphine, D-00980
Dipentylphosphinic acid, D-00981
Dipentyl phosphite, *see* D-00982
Dipentyl phosphonate, D-00982
Dipentyl phosphoric acid, *see* D-00979
Dipentyl phosphorochloridate, *in* D-00979
Dipentyl phosphorochloridite, D-00983
O,O-Dipentyl phosphorochloridothioate, D-00984
O,O-Dipentyl phosphorodiselenoate, D-00985
O,O-Dipentyl phosphorodiselenoic acid, *see* D-00985
O,O-Dipentyl phosphorodithioate, D-00986
▷Dipentyl phosphorofluoridate, *in* D-00979

▷Dipentyl phosphoryl chloride, *in* D-00979
Dipentyl phosphoryl chloride, *in* D-00979
P,P-Diphenoxy-*P*-phenyl-*N*-phenylsulfonylphosphazene, *in* P-00192
P,P-Diphenoxy-*P*-phenyl-*N*-phenylsulfonylphosphine imide, *in* P-00192
2-[(Diphenoxyphosphinothioyl)amino]pyridine, *in* P-00513
4-[(Diphenoxyphosphinothioyl)amino]pyridine, *in* P-00514
Diphenoxyphosphinothioyl fluoride, *see* D-01134
(Diphenoxyphosphinothioyl)phosphorimidic acid, D-00987
N-(Diphenoxyphosphinothioyl)-*P,P,P*-tris(phenylamino)phosphazene, *in* D-00987
N-(Diphenoxyphosphinothioyl)-*P,P,P*-tris(phenylamino)phosphine imide, *in* D-00987
(Diphenoxyphosphinyl)acetic acid, D-00988
N-(Diphenoxyphosphinyl)-4-aminobenzoic acid, *in* P-00377
N-Diphenoxyphosphinyl-4-aminobenzoic acid ethyl ester, *in* C-00006
2-(Diphenoxyphosphinylamino)pyridine, *in* P-00510
3-(Diphenoxyphosphinylamino)pyridine, *in* P-00511
4-(Diphenoxyphosphinylamino)pyridine, *in* P-00512
N-(Diphenoxyphosphinyl)aniline, *see* D-01035
N-(Diphenoxyphosphinyl)carbamic acid, D-00989
Diphenoxyphosphinylcyclohexane, *in* C-00256
1-(Diphenoxyphosphinyl)decane, *in* D-00012
Diphenoxyphosphinyl fluoride, *see* D-01133
(Diphenoxyphosphinyl)formamide, *in* C-00003
(Diphenoxyphosphinyl)hydrazine, *see* D-01136
N-Diphenoxyphosphinylimidazole, *in* I-00003
[(Diphenoxyphosphinyl)methyl]-triphenylphosphonium(1+), D-00990
1-[(Diphenoxyphosphinyl)oxy]-2,5-pyrrolidinedione, *see* D-01148
3,9-Diphenoxy-2,4,8,10-tetraoxa-3,9-diphosphaspiro[5.5]undecane, D-00991
Diphenoxytriphenylphosphorane, D-00992
Diphenyl acetonylphosphonate, *in* O-00075
Diphenyl acetylphosphonate, *in* A-00026
Diphenyl 1-adamantylphosphoramidate, *in* A-00036
Diphenyl allylphosphoramidate, *in* P-00464
Diphenyl amidophosphate, *see* D-01107
O,O-Diphenyl amidoselenophosphate, *in* D-01139
Diphenyl aminocarbonylphosphonate, *in* C-00003
Diphenyl aminocarbonylphosphoramidate, *in* A-00062
Diphenyl (1-aminoethyl)phosphonate, *in* A-00078
Diphenyl (aminoiminomethyl)phosphoramidate, *in* P-00370
Diphenyl (1-amino-2-methylpropyl)-phosphonate, *in* A-00094
Diphenyl (aminophenylmethyl)phosphonate, *in* A-00105
Diphenyl (4-aminophenyl)phosphonate, *in* A-00111
Diphenyl (3-aminophenyl)phosphoramidate, *in* A-00114
▷[2-(Diphenylarsino)ethyl]diphenylphosphine, D-00993
[2-(Diphenylarsino)phenyl]diphenylphosphine, D-00994
2,10-Diphenyl-6-aza-2,10-diphosphabicyclo[9.4.0]pentadeca-11(1),12,14-triene, *see* O-00009
2,6-Diphenyl-1-aza-4-phosphabenzene, *see* D-00996
2,3-Diphenyl-1,3-azaphosphacyclopentane, *see* D-00995
1,3-Diphenyl-1,3-azaphosphetidine, *in* P-00087
1,2-Diphenyl-1,2-azaphospholidine, *in* P-00089
2,3-Diphenyl-1,3-azaphospholidine, D-00995
2,6-Diphenyl-1,4-azaphosphorine, D-00996
O,O-Diphenyl azidothiophosphate, *see* D-01113
Diphenyl 1-aziridinylphosphonite, *in* A-00154
1,1-Diphenyl-1*H*-benzo[*b*]phosphindolium iodide, *in* P-00110
2,3-Diphenyl-1,3,2-benzoxazaphospholine, *see* D-00478
Diphenyl benzoylphosphoramidate, *in* B-00036
O,O-Diphenyl benzoylphosphoramidothioate, *in* B-00038
Diphenyl benzylphosphonate, *in* B-00060
Diphenyl benzylphosphoramidate, *in* B-00069
1,3-Diphenyl-2,2:4,4-bi(*o*-phenylenedioxy)-1,3,2λ^5,4λ^5-diazadiphosphetidine, *see* D-01018
Diphenyl bis(2-chloroethyl)phosphoramidate, *in* B-00121
Diphenyl bis(2-methylpropyl)phosphoramidate, *in* B-00355
1,2-Diphenyl-1,2-bis(phenylmethyl)-diphosphine, *see* D-00045
P,P-Diphenyl *P′,P′*-bis(phenymethyl) diphosphate, *see* D-00044
P,P-Diphenyl-*N,N′*-bis(trimethylsilyl)-phosphinimidic amide, *in* D-01043
1,4-Diphenyl-2,3-bis(triphenylphosphoranylidene)-1,4-butanedione, D-00997
P,P-Diphenyl-*N,N′*-bis(triphenylsilyl)-phosphinimidic amide, *in* D-01043
Diphenyl (4-bromophenyl)phosphonite, *in* B-00493
Diphenyl (2-bromophenyl)phosphoramidate, *in* B-00494
Diphenyl (3-bromophenyl)phosphoramidate, *in* B-00495
Diphenyl (4-bromophenyl)phosphoramidate, *in* B-00496
Diphenyl bromophosphate, *in* D-01036
Diphenyl bromophosphite, *see* D-01118
O,O-Diphenyl bromoselenophosphate, *in* D-01139
O,O-Diphenyl bromothiophosphate, *see* D-01119
Diphenyl (1,3-butadienyl)phosphonate, *in* B-00510
Diphenyl butylphosphonate, *in* B-00594
Diphenyl *tert*-butylphosphonate, *in* B-00595
Diphenyl butylphosphonite, *in* B-00615
Diphenyl *tert*-butylphosphonite, *in* B-00616
S,S-Diphenyl butylphosphonodithioate, *in* B-00606
Diphenyl butylphosphonodithioite, *in* B-00607
Diphenyl butylphosphoramidate, *in* B-00621
Diphenyl *tert*-butylphosphoramidate, *in* B-00622
Diphenyl (carboxymethyl)phosphonate, *see* D-00988
Diphenyl (4-carboxyphenyl)phosphoramidate, *in* P-00377
Diphenyl (2-chloroethyl)phosphonate, *in* C-00074
Diphenyl chloromethylphosphonate, *in* C-00106
Diphenyl (2-chlorophenyl)phosphoramidate, *in* C-00155
Diphenyl (3-chlorophenyl)phosphoramidate, *in* C-00156
Diphenyl (4-chlorophenyl)phosphoramidate, *in* C-00157
Diphenyl chlorophosphate, *see* D-01120
Diphenyl chlorophosphite, *see* D-01121
Diphenyl (3-chloropropyl)phosphonate, *in* C-00172
O,O-Diphenyl chloroselenophosphate, *see* D-01122
O,O-Diphenyl chlorothiophosphate, *see* D-01123
Diphenyl (1-cyanoethyl)phosphonate, *in* C-00210
Diphenyl cyclohexylphosphonate, *in* C-00256
Diphenyl cyclohexylphosphonite, *in* C-00267
Diphenyl cyclohexylphosphonodithioite, *in* C-00262
Diphenyl cyclohexylphosphoramidate, *in* C-00269
Diphenyl cyclooctylphosphoramidate, D-00998
(2,4-Diphenyl-2,4-cyclopentadien-1-ylidene)triphenylphosphorane, D-00999
Diphenyl decylphosphonate, *in* D-00012
Diphenyl diallylphosphoramidate, *in* D-01195
3,5-Diphenyl-1*H*-1,2,4-diazaphosphole, D-01000
5,10-Diphenyldibenzo[1,4]phospharsenin, D-01001
5,5-Diphenyl-5*H*-dibenzophospholium iodide, *in* P-00110
Diphenyl (2,4-dibromophenyl)phosphoramidate, *in* D-00067
Diphenyl (dichloromethyl)phosphonate, *in* D-00175
Diphenyl (2,4-dichlorophenyl)-phosphoramidate, *in* D-00188
Diphenyl dichlorophosphoramidate, D-01002
Diphenyl (2,2-diethoxyethyl)phosphonate, *in* D-00234
Diphenyl diethylphosphoramidate, *in* D-00343
Diphenyl diethylphosphoramidite, *in* D-00360
O,O-Diphenyl diethylphosphoramidothioate, *in* D-00356
Diphenyl diethylphosphoramidotrithioate, *in* D-00359
Diphenyl diisopropylphosphoramidite, *in* D-00656
Diphenyl (1,1-dimethylethyl)phosphonate, *in* B-00595
Diphenyl (2,4-dimethylphenyl)-phosphoramidate, *in* D-00818
Diphenyl dimethylphosphoramidate, *in* D-00855
Diphenyl dimethylphosphoramidite, *in* D-00875
Diphenyl dimethylphosphoramidotrithioate, *in* D-00874
Diphenyl (2,2-dimethylpropyl)phosphonate, *in* D-00912
Diphenyl (2,4-dinitrophenyl)phosphoramidate, *in* D-00947
2,5-Diphenyl-1,3,5,2-dioxaphosphaborinane, D-01003
O,O-Diphenyl (diphenoxyphosphinothioyl)-phosphoramidothioate, D-01004
O,O-Diphenyl (diphenoxyphosphinyl)-phosphoramidothioate, D-01005
Diphenyl diphenyldiphosphonate, *in* D-01014
Diphenyl[(diphenylphosphinoselenoyl)-methyl]phosphine, *in* B-00244
P,P-Diphenyl-*N*-(diphenylphosphinothioyl)-phosphinimidic acid, D-01006
P,P-Diphenyl-*N*-(diphenylphosphinothioyl)-phosphonic amide, *see* D-01006
Diphenyl[(2-diphenylphosphinyl)ethyl]-phosphine, *in* B-00234
P,P-Diphenyl-*N*-diphenylphosphinylphosphinic amide, *see* D-01007
P,P-Diphenyl-*N*-(diphenylphosphinyl)-phosphinimidic acid, D-01007
Diphenyl diphenylphosphoramidate, *in* D-01108
Diphenyl diphenylphosphoramidite, *in* D-01111
2,10-Diphenyl-2,10-diphosphabicyclo[9.4.0]pentadeca-11(1),12,14-triene, *see* O-00010
1,12-Diphenyl-1,12-diphosphacyclodocosane, D-01008
1,3-Diphenyl-1,3-diphosphacyclohexane, *see* D-01016
1,10-Diphenyl-1,10-diphosphacyclooctadecane, D-01009
1,12-Diphenyl-1,12-diphosphacyclotetracosane, D-01010
6,13-Diphenyl-6,13-diphospha-2,17-dithiatricyclo[16.4.0.$0^{7,12}$]docosa-7(12),8,10,1(18),19,21-hexaene, *see* O-00013
6,17-Diphenyl-6,17-diphospha-2,13-dithiatricyclo[16.4.0.$0^{7,12}$]docosa-7(12),8,10,1(18),19,21-hexaene, *see* O-00014
13,17-Diphenyl-13,17-diphospha-2,6-dithiatricyclo[16.4.0.$0^{7,12}$]docosa-7(12),8,10,1(18),19,21-hexaene, *see* H-00026
P,P′-Diphenyl diphosphate, D-01011
2,10-Diphenyl-2,10-diphospha-6-thiabicyclo[9.4.0]pentadeca-11(1),-12,14-triene, *see* O-00011
1,2-Diphenyl-1,2-diphosphinane, *see* D-01015
1,3-Diphenyl-1,3-diphosphinane, *see* D-01016
1,4-Diphenyl-1,4-diphosphinane, *see* D-01017
1,2-Diphenyldiphosphine, D-01012
1,3-Diphenyl-1,3-diphospholane, D-01013

1,2-Diphenyl-1,2-diphospholane 1,2-disulfide, *in* D-01181
Diphenyldiphosphonic acid, D-01014
1,2-Diphenyl-1,2-diphosphorinane, D-01015
1,3-Diphenyl-1,3-diphosphorinane, D-01016
1,4-Diphenyl-1,4-diphosphorinane, D-01017
Diphenyl di-2-propenylphosphoramidate, *in* D-01195
Diphenyl dipropylphosphoramidate, *in* D-01214
Diphenyl dipropylphosphoramidite, *in* D-01215
O,O-Diphenyl diselenophosphate, *see* D-01130
Diphenyldiselenophosphinic acid, *see* D-01064
1′,3′-Diphenyldispiro[1,3,2-benzodioxaphosphole-2,2′-[1,3,2,4]-diazaphosphetidine-4′,2″-[1,3,2]-benzodioxaphosphole], D-01018
2,4-Diphenyl-1,3,2,4-dithiadiphosphetane 2,4-disulfide, D-01019
O,O-Diphenyl dithiophosphate, *see* D-01131
O,O-Diphenyl dithiophosphoric acid, *see* D-01131
S,S-Diphenyl dithiophosphoric acid, *see* D-01132
N,N-Diphenyl-*P,P*-1,2-ethanediylbis[phenylphosphinic amide], *in* E-00015
Diphenyl [1-ethoxycarbonyl-2-(4-hydroxybenzyl)ethyl]phosphoramidate, *in* P-00422
Diphenyl [4-(ethoxycarbonyl)phenyl]-phosphoramidate, *in* C-00006
Diphenylethylenedioxyphosphoric acid, *see* H-00139
Diphenyl ethylphosphonate, *in* E-00129
Diphenyl ethylphosphonite, *in* E-00153
Diphenyl ethylphosphonodithioite, *in* E-00141
O,O-Diphenyl ethylphosphonothioate, *in* E-00148
Diphenyl ethylphosphonotrithioate, *in* E-00152
Diphenyl ethylphosphoramidate, *in* E-00156
O,O-Diphenyl ethylphosphoramidothioate, *in* E-00159
O,O-Diphenyl fluorodithiophosphate, *see* D-01134
Diphenyl fluorophosphate, *see* D-01133
Diphenyl fluorotrithiophosphate, *see* D-01135
Diphenyl 2-furanylphosphonate, *in* F-00073
Diphenyl hexylphosphoramidate, *in* H-00102
O,O-Diphenyl hydrogen diselenophosphate, *see* D-01130
S,S-Diphenyl hydrogen dithiophosphate, *see* D-01132
Diphenyl hydrogen phosphate, *see* D-01036
O,O-Diphenyl hydrogen phosphorodithioate, *see* D-01131
O,O-Diphenyl hydrogen phosphoroselenoate, *see* D-01139
O,O-Diphenyl hydrogen phosphoroselenothioate, *see* D-01140
O,O-Diphenyl hydrogen phosphorothioate, *see* D-01141
O,O-Diphenyl hydrogen thiophosphate, *see* D-01141
Diphenyl (1-hydroxy-1-methylethyl)-phosphonate, *in* H-00159
Diphenyl hypodiphosphonous acid, D-01020
Diphenyl 1*H*-imidazol-1-ylphosphonate, *in* I-00003
Diphenyl (iodomethyl)phosphonate, *in* I-00013
Diphenyl iodophosphite, *see* D-01115
Diphenyl isobutylphosphonate, *in* M-00378
Diphenyl isobutylphosphoramidate, *in* M-00384
Diphenyl isopropylphosphonate, *in* I-00050
Diphenyl isopropylphosphonite, *in* I-00056
Diphenyl isopropylphosphoramidate, *in* I-00058
O,O-Diphenyl isopropylphosphoramidothioate, *in* I-00059
Diphenyl (4-methoxyphenyl)phosphonate, *in* M-00072
Diphenyl (4-methoxyphenyl)phosphoramidate, *in* M-00076
O,O-Diphenyl (4-methoxyphenyl)-phosphoramidothioate, *in* M-00079
[1-(Diphenylmethylamino)ethyl]phosphinic acid, *in* A-00077
Diphenyl (2-methyl-1,3-butadienyl)-phosphonate, *in* M-00093
(Diphenylmethylene)mesitylphosphine, *see* D-01021
(Diphenylmethylene)(2,4,6-trimethylphenyl)phosphine, D-01021
(Diphenylmethylene)triphenylphosphorane, D-01022
Diphenyl (1-methylethenyl)phosphonate, *in* M-00149
Diphenyl (3-methyl-2-oxiranyl)phosphonate, *in* M-00185
(Diphenylmethyl)phenylphosphinic acid, D-01023
Diphenyl (4-methylphenyl)phosphonate, *in* M-00237
Diphenyl (4-methylphenyl)phosphonite, *in* M-00248
Diphenyl methylphenylphosphoramidate, *in* M-00250
Diphenyl (2-methylphenyl)phosphoramidate, *in* M-00251
Diphenyl (3-methylphenyl)phosphoramidate, *in* M-00252
Diphenyl (4-methylphenyl)phosphoramidate, *in* M-00253
O,O-Diphenyl (2-methylphenyl)-phosphoramidothioate, *in* M-00254
O,O-Diphenyl (3-methylphenyl)-phosphoramidothioate, *in* M-00255
O,O-Diphenyl (4-methylphenyl)-phosphoramidothioate, *in* M-00256
Diphenyl methylphosphonate, D-01024
Diphenyl methylphosphonite, *in* M-00319
S,S-Diphenyl methylphosphonodithioate, *in* M-00301
Diphenyl methylphosphonodithioite, *in* M-00302
O,O-Diphenyl methylphosphonothioate, *in* M-00313
O,S-Diphenyl methylphosphonothioate, *in* M-00313
Diphenyl methylphosphonotrithioate, *in* M-00318
Diphenyl methylphosphoramidate, *in* M-00326
Diphenyl (2-methylpropyl)phosphonate, *in* M-00378
Diphenyl (2-methylpropyl)phosphoramidate, *in* M-00384
Diphenyl 1-methyl-1*H*-pyrrol-2-ylphosphonate, *in* P-00521
(Diphenylmethyl)triphenylphosphonium(1+), D-01025
▷Diphenyl 4-morpholinylphosphonate, *in* M-00523
Diphenyl 4-morpholinylphosphonite, *in* M-00529
O,O-Diphenyl 4-morpholinylphosphonothioate, *in* M-00527
Diphenyl 1-naphthylphosphonite, *in* N-00025
Diphenyl neopentylphosphonate, *in* D-00912
Diphenyl (2-nitroethyl)phosphonate, *in* N-00036
Diphenyl (2-nitrophenyl)phosphoramidate, *in* N-00058
Diphenyl (3-nitrophenyl)phosphoramidate, *in* N-00059
Diphenyl (4-nitrophenyl)phosphoramidate, *in* N-00060
Diphenyl octylphosphonate, *in* O-00042
2,10-Diphenyl-6-oxa-2,10-diphosphabicyclo[9.4.0]pentadeca-11(1),12,14-triene, *see* O-00012
2,3-Diphenyl-1-oxa-3-phosphacyclohexane, *see* D-01028
2,3-Diphenyl-1,3-oxaphosphepane, D-01026
2,3-Diphenyl-1,3-oxaphosphinane, *see* D-01028
2,3-Diphenyl-1,3-oxaphospholane, D-01027
2,6-Diphenyl-4*H*-1,4-oxaphosphorin-4-amine 4-oxide, *in* H-00140
2,3-Diphenyl-1,3-oxaphosphorinane, D-01028
2,3-Diphenyl-1,3,2-oxazaphospholane, *see* D-01029
2,3-Diphenyl-1,3,2-oxazaphospholidine, D-01029
Diphenyl (1-oxoethyl)phosphonate, *in* A-00026
(1,3-Diphenyl-3-oxopropyl)phenylphosphinic acid, D-01030
Diphenyl (2-oxopropyl)phosphonate, *in* O-00075
10,10-Diphenyl-10*H*-phenoxaphosphonium iodide, *in* P-00183
O,O-Diphenyl phenylamidothiophosphate, *in* P-00262
Diphenyl (phenylamino)-carbonylphosphoramidate, *in* P-00081
O,O-Diphenyl [(phenylamino)carbonyl]-phosphoramidothioate, *in* P-00082
Diphenyl(4-phenyl-1,3-butadienyl)phosphine, D-01031
N,P-Diphenyl-*P*-(4-phenyl-1,3-butadienyl)-phosphinic amide, *in* P-00184
Diphenyl(4-phenyl-1,3-butadiynyl)phosphine, D-01032
Diphenyl(2-phenylethenyl)phosphine, D-01033
Diphenyl (1-phenylethenyl)phosphonate, *in* P-00154
S,S-Diphenyl (2-phenylethenyl)-phosphonodithioate, *in* P-00159
Diphenyl(2-phenylethyl)phosphine, D-01034
Diphenyl(phenylmethyl)phosphine, *see* B-00046
N,N′-Diphenyl-*P*-(phenylmethyl)phosphonic diamide, *in* B-00061
Diphenyl *P*-phenyl-*N*-phenylsulfonylphosphonimidate, *in* P-00192
Diphenyl phenylphosphonate, *in* P-00221
Diphenyl phenylphosphonite, *in* P-00251
Diphenyl phenylphosphonodithioite, *in* P-00238
O,O-Diphenyl phenylphosphonothioate, *in* P-00244
Diphenyl phenylphosphoramidate, D-01035
O,O-Diphenyl phenylphosphoramidothioate, *in* P-00262
P,P-Diphenyl-*N*-phenylsulfonylphosphinic amide, *in* D-01040
P,P-Diphenyl-*N*-phenylsulfonylphosphinimidic acid, *see* B-00013
Diphenyl phenylsulfonylphosphoramidate, *in* P-00312
O,O-Diphenyl phenyl-sulfonylphosphoramidothioate, *in* P-00314
Diphenyl phosphate, D-01036
1,1-Diphenylphosphepanium bromide, *in* P-00205
▷Diphenylphosphine, D-01037
Diphenylphosphine imide dimer, *see* T-00253
Diphenylphosphine oxide, D-01038
▷Diphenylphosphinic acid, D-01039
P,P-Diphenylphosphinic amide, D-01040
Diphenylphosphinic chloride, D-01041
Diphenylphosphinic fluoride, D-01042
P,P-Diphenylphosphinimidic amide, D-01043
P,P-Diphenylphosphinimidic fluoride, D-01044
2-(Diphenylphosphino)acetamide, *in* D-01045
3-(Diphenylphosphino)acetanilide, *in* D-01048
(Diphenylphosphino)acetic acid, D-01045
(Diphenylphosphino)acetone, *see* D-01075
(Diphenylphosphino)acetonitrile, D-01046
Diphenylphosphinoacetylene, *see* E-00201
Diphenylphosphinoallene, *see* D-01142
2-(Diphenylphosphino)aniline, D-01047
3-(Diphenylphosphino)aniline, D-01048
4-(Diphenylphosphino)aniline, D-01049
m-(Diphenylphosphino)anisole, *see* M-00057
o-(Diphenylphosphino)anisole, *see* M-00056
p-(Diphenylphosphino)anisole, *see* M-00058
2-(Diphenylphosphino)benzaldehyde, D-01050
3-(Diphenylphosphino)benzaldehyde, D-01051
4-(Diphenylphosphino)benzaldehyde, D-01052
2-Diphenylphosphinobenzenamine, *see* D-01047
3-Diphenylphosphinobenzenamine, *see* D-01048
4-Diphenylphosphinobenzenamine, *see* D-01049
2-(Diphenylphosphino)benzoic acid, D-01053
3-(Diphenylphosphino)benzoic acid, D-01054
4-(Diphenylphosphino)benzoic acid, D-01055
2-(Diphenylphosphino)benzonitrile, D-01056
3-(Diphenylphosphino)benzonitrile, D-01057
4-(Diphenylphosphino)benzonitrile, D-01058
4-(Diphenylphosphino)biphenyl, *see* B-00078
2-(Diphenylphosphino)-*N,N*-bis[2-(diphenylphosphino)ethyl]ethanamine, *see* T-00773
o-(Diphenylphosphino)bromobenzene, *see* B-00480
p-(Diphenylphosphino)bromobenzene, *see* B-00481

1-Diphenylphosphino-2-butene, *see* B-00520
4-Diphenylphosphino-1-butene, *see* B-00521
Diphenylphosphinocyclohexane, *see* C-00245
[1-(Diphenylphosphino)cyclopropyl]-triphenylphosphonium chloride, *in* B-00229
2-(Diphenylphosphino)-*N*,*N*-dimethylaniline, D-01059
3-(Diphenylphosphino)-*N*,*N*-dimethylaniline, D-01060
4-Diphenylphosphino-*N*,*N*-dimethylaniline, D-01061
2-Diphenylphosphino-*N*,*N*-dimethyl-benzenamine, *see* D-01059
3-Diphenylphosphino-*N*,*N*-dimethyl-benzenamine, *see* D-01060
4-Diphenylphosphino-*N*,*N*-dimethyl-benzenamine, *see* D-01061
1-Diphenylphosphino-*N*,*N*-dimethyl-methanamine, *see* D-00700
▷1-Diphenylphosphino-2-diphenylarsinoethane, *see* D-00993
2-(Diphenylphosphino)-*N*-[2-(diphenylphosphino)ethyl]ethanamine, *see* B-00231
2-(Diphenylphosphino)-*N*-[2-(diphenylphosphino)ethyl]-*N*-ethylethanamine, *see* B-00238
4-(Diphenylphosphino)-2-(diphenylphosphinomethyl)pyrrolidine, D-01062
N-Diphenylphosphino-*P*,*P*-diphenylphosphinous amide, D-01063
Diphenylphosphinodiselenoic acid, D-01064
Diphenylphosphinodithioic acid, D-01065
2-(Diphenylphosphino)ethanamine, *see* D-01066
(Diphenylphosphino)ethanenitrile, *in* D-01046
1-(Diphenylphosphino)ethanone, *see* A-00016
2-(Diphenylphosphino)ethylamine, D-01066
N-(Diphenylphosphino)-*N*-ethyl-*P*,*P*-diphenylphinous amide, D-01067
Diphenylphosphinoethyne, *see* E-00201
(Diphenylphosphino)ferrocene, D-01068
α-[[2-(Diphenylphosphino)ferrocenyl]-ethyl]dimethylamine, *see* D-00697
1-Diphenylphosphinohexane, *see* H-00093
(2-Diphenylphosphino)-1-isopropyl-4-methylcyclohexane, *see* M-00164
3-(Diphenylphosphino)-*p*-menthane, *see* M-00164
(Diphenylphosphino)methanol, D-01069
m-(Diphenylphosphino)methoxybenzene, *see* M-00057
o-(Diphenylphosphino)methoxybenzene, *see* M-00056
[2-[(Diphenylphosphino)methyl]-2-methyl-1,3-propanediyl]bis[diphenylphosphine], *see* T-00776
1-(Diphenylphosphino)-2-methylpropene, *see* M-00362
1-(Diphenylphosphino)naphthalene, *see* N-00011
2-(Diphenylphosphino)naphthalene, *see* N-00012
1-Diphenylphosphinooctane, *see* O-00039
2-[(Diphenylphosphino)oxyimino]propane, *in* D-01097
2-(Diphenylphosphino)phenol, D-01070
3-(Diphenylphosphino)phenol, D-01071
4-(Diphenylphosphino)phenol, D-01072
1-Diphenylphosphino-4-phenyl-1,3-butadiene, *see* D-01031
1-Diphenylphosphino-4-phenyl-1,3-butadiyne, *see* D-01032
2-[(Diphenylphosphino)phenyl]diphenylarsine, *see* D-00994
3-(Diphenylphosphino)propanenitrile, D-01073
3-(Diphenylphosphino)-1-propanol, D-01074
1-(Diphenylphosphino)-2-propanone, D-01075
1-Diphenylphosphinopropene, *see* D-01143
1-(Diphenylphosphino)propyne, *see* D-01146
2-(Diphenylphosphino)pyridine, D-01076
4-(Diphenylphosphino)pyridine, D-01077
Diphenylphosphinoselenoic acid, D-01078
P,*P*-Diphenylphosphinoselenoic amide, D-01079
Diphenylphosphinoselenothioic acid, D-01080
Diphenylphosphinoselenous acid, D-01081
o-(Diphenylphosphino)thioanisole, *see* M-00417
Diphenylphosphinothioic acid, D-01082
P,*P*-Diphenylphosphinothioic amide, D-01083
Diphenylphosphinothioic anhydride, *in* D-01082
Diphenylphosphinothioic chloride, D-01084
Diphenylphosphinothious acid, D-01085
3-(Diphenylphosphinothioyl)acetanilide, *in* D-01048
(Diphenylphosphinothioyl)acetic acid, D-01086
(Diphenylphosphinothioyl)acetonitrile, *in* D-01046
3-(Diphenylphosphinothioyl)aniline, *in* D-01048
4-(Diphenylphosphinothioyl)aniline, *in* D-01049
p-(Diphenylphosphinothioyl)anisole, *in* M-00058
2-(Diphenylphosphinothioyl)benzaldehyde, *in* D-01050
3-(Diphenylphosphinothioyl)benzaldehyde, *in* D-01051
3-(Diphenylphosphinothioyl)benzoic acid, *in* D-01054
p-(Diphenylphosphinothioyl)benzoic acid, *in* D-01055
3-(Diphenylphosphinothioyl)benzonitrile, *in* D-01057
4-(Diphenylphosphinothioyl)benzonitrile, *in* D-01058
Diphenylphosphinothioylcyclohexane, *in* C-00245
3-Diphenylphosphinothioyl-*N*,*N*-dimethylaniline, *in* D-01060
4-Diphenylphosphinothioyl-*N*,*N*-dimethylaniline, *in* D-01061
N-Diphenylphosphinothioyl-*P*,*P*-diphenylphosphinothioic amide, D-01087
1-(Diphenylphosphinothioyl)hexane, *in* H-00093
1-(Diphenylphosphinothioyl)octane, *in* O-00039
2-(Diphenylphosphinothioyl)phenol, *in* D-01070
4-(Diphenylphosphinothioyl)phenol, *in* D-01072
1-(Diphenylphosphinothioyl)piperidine, *in* D-01082
3-(Diphenylphosphinothioyl)propanenitrile, *in* D-01073
2-Diphenylphosphinothioylpropionitrile, *in* D-01073
1-(Diphenylphosphinothioyl)propyne, *in* D-01146
2-(Diphenylphosphinothioyl)pyridine, *in* D-01076
N-Diphenylphosphinothioyl-*P*,*P*,*P*-triphenylphosphazene, *in* D-01165
N-Diphenylphosphinothioyl-*P*,*P*,*P*-triphenylphosphine imide, *in* D-01088
N-Diphenylphosphinothioyl-*P*,*P*,*P*-triphenylphosphine imide, *in* D-01165
Diphenylphosphinotrimethylsilane, *see* D-01156
(Diphenylphosphino)triphenylgermane, *see* D-01161
N-Diphenylphosphino-*P*,*P*,*P*-triphenylphosphazene, *see* D-01165
N-Diphenylphosphino-*P*,*P*,*P*-triphenylphosphine imide, D-01088
N-Diphenylphosphino-*P*,*P*,*P*-triphenylphosphine imide, *see* D-01165
(Diphenylphosphino)triphenyltin, *see* D-01167
Diphenylphosphinous acid, D-01089
P,*P*-Diphenylphosphinous amide, D-01090
▷Diphenylphosphinous chloride, D-01091
Diphenylphosphinous diethylamide, *in* D-01090
Diphenylphosphinous diisopropylamide, *in* D-01090
Diphenylphosphinous fluoride, D-01092
(Diphenylphosphinyl)acetic acid, D-01093
Diphenylphosphinylacetic acid hydrazide, D-01094
Diphenylphosphinylacetone, *in* D-01075
(Diphenylphosphinyl)acetonitrile, *in* D-01046
Diphenylphosphinylacetylene, *in* E-00201
2-(Diphenylphosphinyl)aniline, *in* D-01047
3-(Diphenylphosphinyl)aniline, *in* D-01048
4-(Diphenylphosphinyl)aniline, *in* D-01049
m-(Diphenylphosphinyl)anisole, *in* M-00057
p-(Diphenylphosphinyl)anisole, *in* M-00058
2-(Diphenylphosphinyl)benzaldehyde, *in* D-01050
3-(Diphenylphosphinyl)benzaldehyde, *in* D-01051
4-(Diphenylphosphinyl)benzaldehyde, *in* D-01052
α-(Diphenylphosphinyl)benzeneacetonitrile, D-01095
3-(Diphenylphosphinyl)benzoic acid, *in* D-01054
4-(Diphenylphosphinyl)benzoic acid, *in* D-01055
3-(Diphenylphosphinyl)benzonitrile, *in* D-01057
4-(Diphenylphosphinyl)benzonitrile, *in* D-01058
α-(Diphenylphosphinyl)benzyl cyanide, *see* D-01095
4-(Diphenylphosphinyl)biphenyl, *in* B-00078
Diphenylphosphinylcyclohexane, *in* C-00245
(Diphenylphosphinyl)diazomethane, *see* D-00034
2-Diphenylphosphinyl-*N*,*N*-dimethylaniline, *in* D-01059
3-Diphenylphosphinyl-*N*,*N*-dimethylaniline, *in* D-01060
4-Diphenylphosphinyl-*N*,*N*-dimethylaniline, *in* D-01061
(Diphenylphosphinyl)dimethyl ether, *in* D-01069
N-(Diphenylphosphinyl)-*P*,*P*-diphenylphosphinic amide, D-01096
1-(Diphenylphosphinyl)ethanone, *in* A-00016
Diphenylphosphinylethyne, *in* E-00201
Diphenylphosphinylferrocene, *in* D-01068
1-(Diphenylphosphinyl)hexane, *in* H-00093
O-(Diphenylphosphinyl)hydroxylamine, D-01097
(Diphenylphosphinyl)isocyanoacetic acid, D-01098
1-(Diphenylphosphinyl)methanamine, *see* A-00085
1-(Diphenylphosphinyl)methylamine, *see* A-00085
(Diphenylphosphinylmethylene)-triphenylphosphorane, D-01099
N-[(Diphenylphosphinyl)methyl]-*N*-methylaniline, D-01100
N-[(Diphenylphosphinyl)methyl]-*N*-methylbenzenamine, *see* D-01100
4-[(Diphenylphosphinyl)methyl]morpholine, *see* M-00519
1-Diphenylphosphinyl-2-methylpropene, *in* M-00362
[(Diphenylphosphinyl)methyl]-triphenylphosphonium(1+), D-01101
1-(Diphenylphosphinyl)naphthalene, *in* N-00011
1-(Diphenylphosphinyl)octane, *in* O-00039
4-(Diphenylphosphinyl)-3-oxobutanoic acid, D-01102
2-(Diphenylphosphinyl)phenol, *in* D-01070
4-(Diphenylphosphinyl)phenol, *in* D-01072
(α-Diphenylphosphinyl)phenyldiazomethane, *see* D-00036
Diphenylphosphinylpropadiene, *in* D-01142
3-(Diphenylphosphinyl)propanenitrile, *in* D-01073
1-(Diphenylphosphinyl)-2-propanone, *in* D-01075
2-Diphenylphosphinylpropionitrile, *in* D-01073
1-(Diphenylphosphinyl)propyne, *in* D-01146
2-(Diphenylphosphinyl)pyridine, *in* D-01076
4-(Diphenylphosphinyl)pyridine, *in* D-01077
N-Diphenylphosphinyl-*P*,*P*,*P*-triphenylphosphazene, *in* D-01165
N-Diphenylphosphinyl-*P*,*P*,*P*-triphenylphosphine imide, *in* D-01088
N-Diphenylphosphinyl-*P*,*P*,*P*-triphenylphosphine imide, *in* D-01165
Diphenyl phosphite, *see* D-01105
1,1-Diphenylphosphocanium bromide, *in* P-00215
1,1-Diphenylphospholanium bromide, *in* P-00216

N,P-Diphenylphosphonamidic acid, D-01103
N,P-Diphenylphosphonamidothioic acid, D-01104
Diphenyl phosphonate, D-01105
9,9-Diphenyl-9-phosphoniabicyclo[4.2.1]-nonane bromide, *in* P-00194
3,3-Diphenyl-3-phosphoniabicyclo[3.2.1]-octane bromide, *in* P-00198
9,9-Diphenyl-9-phosphoniahomocubane bromide, *in* P-00201
9,9-Diphenyl-9-phosphoniatricyclo[4.2.1.0^{2,5}]nonane bromide, *in* P-00204
N,P-Diphenylphosphonic diamide, *in* P-00222
O,O-Diphenyl phosphonothioate, D-01106
Diphenyl phosphoramidate, D-01107
Diphenyl phosphoramide, *see* D-01107
Diphenylphosphoramidic acid, D-01108
2,2′-Diphenylphosphoramidic dihydrazide, D-01109
Diphenyl phosphoramidite, D-01110
O,O-Diphenyl phosphoramidoselenoate, *in* D-01139
Diphenylphosphoramidous acid, D-01111
Diphenyl phosphoranilidate, *see* D-01035
Diphenyl phosphorazidate, D-01112
O,O-Diphenyl phosphorazidothioate, D-01113
Diphenyl phosphoric acid, *see* D-01036
N,N′-Diphenylphosphoric triamide, D-01114
1,1-Diphenylphosphorinanium bromide, *in* P-00265
Diphenyl phosphoriodidite, D-01115
Diphenyl phosphorisocyanatidate, D-01116
O,O-Diphenyl phosphorisocyanatidothioate, D-01117
Diphenyl phosphorobromidate, *in* D-01036
Diphenyl phosphorobromidite, D-01118
O,O-Diphenyl phosphorobromidoselenoate, *in* D-01139
O,O-Diphenyl phosphorobromidothioate, D-01119
Diphenyl phosphorochloridate, D-01120
Diphenyl phosphorochloridite, D-01121
O,O-Diphenyl phosphorochloridoselenoate, D-01122
O,O-Diphenyl phosphorochloridothioate, D-01123
Diphenyl phosphorocyanidate, D-01124
N,N′-Diphenylphosphorodiamidic acid, D-01125
▷*N,N′*-Diphenylphosphorodiamidic chloride, D-01126
▷*N,N′*-Diphenylphosphorodiamidic fluoride, D-01127
N,N′-Diphenylphosphorodiamidodithioic acid, D-01128
N,N′-Diphenylphosphorodiamidothioic acid, D-01129
O,O-Diphenyl phosphorodiselenoate, D-01130
O,O-Diphenyl phosphorodiselenoate, *see* D-01130
O,O-Diphenyl phosphorodithioate, D-01131
S,S-Diphenyl phosphorodithioate, D-01132
O,O-Diphenyl phosphorodithioate anhydrosulfide, *see* T-00284
Diphenyl phosphorofluoridate, D-01133
S,S-Diphenyl phosphorofluoridodithioate, *in* D-01132
O,O-Diphenyl phosphorofluoridothioate, D-01134
Diphenyl phosphorofluoridotrithioate, D-01135
Diphenyl phosphorohydrazidate, D-01136
O,O-Diphenyl phosphorohydrazidothioate, D-01137
Diphenyl phosphoroisothiocyanatidate, D-01138
O,O-Diphenyl phosphoroselenoate, D-01139
O,O-Diphenyl phosphoroselenoic acid, *see* D-01139
O,O-Diphenyl phosphoroselenothioate, D-01140
O,O-Diphenyl phosphoroselenothioic acid, *see* D-01140
O,O-Diphenyl phosphorothioate, D-01141
Diphenyl phosphoryl amide, *see* D-01107
Diphenyl phosphoryl azide, *see* D-01112
Diphenylphosphoryl bromide, *in* D-01036
Diphenyl phosphoryl chloride, *see* D-01120
Diphenyl phosphoryl cyanide, *see* D-01124
Diphenyl phosphoryl fluoride, *see* D-01133
Diphenyl phosphoryl hydrazine, *see* D-01136
N-Diphenylphosphorylimidazole, *in* I-00003
Diphenyl phosphoryl isocyanate, *see* D-01116
Diphenyl 1-piperidinylphosphonate, *in* P-00430
Diphenyl-1,2-propadienylphosphine, D-01142
Diphenyl-1-propenylphosphine, D-01143
Diphenyl-2-propenylphosphine, D-01144
N,N′-Diphenyl-*P*-2-propenylphosphonic diamide, *in* P-00455
N,N′-Diphenyl-*P*-2-propenylphosphonothioic diamide, *in* P-00462
Diphenyl 2-propenylphosphoramidate, *in* P-00464
Diphenylpropylphosphine, D-01145
P,P-Diphenyl-*N*-propylphosphinic amide, *in* D-01040
N,N′-Diphenyl-*P*-propylphosphonic diamide, *in* I-00051
S,S-Diphenyl propylphosphonodithioate, *in* P-00478
N,N′-Diphenyl-*P*-propylphosphonothioic diamide, *in* P-00480
N,N′-Diphenyl-*P*-propylphosphonothioic diamide, *in* P-00481
Diphenyl propylphosphoramidate, *in* P-00484
Diphenyl(1-propynyl)phosphine, D-01146
Diphenyl 1-propynylphosphonate, *in* P-00494
OO-Diphenyl 2-pyridinylphosphoramidothioate, *in* P-00513
O,O-Diphenyl 4-pyridinylphosphoramidothioate, *in* P-00514
Diphenyl-2-pyridylphosphine, *see* D-01076
Diphenyl-4-pyridylphosphine, *see* D-01077
Diphenyl 2-pyridylphosphoramidate, *in* P-00510
Diphenyl 3-pyridylphosphoramidate, *in* P-00511
Diphenyl 4-pyridylphosphoramidate, *in* P-00512
N,N′-Diphenyl-*N″*-2-pyridylphosphoric triamide, *in* P-00510
N,N′-Diphenyl-*N″*-3-pyridylphosphoric triamide, *in* P-00511
P,P′-Diphenyl pyrophosphate, *see* D-01011
Diphenylpyrophosphonic acid, *see* D-01014
O,O-Diphenyl selenophosphoryl bromide, *in* D-01139
O,O-Diphenyl selenophosphoryl chloride, *see* D-01122
N-Diphenylselenoyl-*P,P,P*-triphenylphosphazene, *in* D-01165
N-Diphenylselenoyl-*P,P,P*-triphenylphosphine imide, *in* D-01165
2,5′-Diphenylspiro[4*H*-1,3,2-benzodioxaphosphorin-2,2′-[1,3,2]-dioxaphospholane]-4,4′-dione, D-01147
Diphenylstyrylphosphine, *see* D-01033
Diphenyl succinimido phosphate, D-01148
N-Diphenyltelluroyl-*P,P,P*-triphenylphosphazene, *in* D-01165
N-Diphenyltelluroyl-*P,P,P*-triphenylphosphine imide, *in* D-01165
Diphenyl (2,3,5,6-tetrachloro-4-pyridinyl)phosphonate, *in* T-00035
N,N′-Diphenyl *P*-(2,3,5,6-tetrachloro-4-pyridinyl)phosphonic diamide, *in* T-00035
N,3-Diphenyltetrahydro-2*H*-1,3,2-oxazaphosphorin-2-amine, *see* T-00133
6,12-Diphenyl-1,4,8,11-tetraoxa-6,12-diaza-5,7-diphospha(5,7-P^V)-dispiro[4.1.4.1]dodecane, D-01149
2,3-Diphenyl-1,4,6,9-tetraoxa-5-phospha(5-P^V)spiro[4.4]non-2-ene, D-01150
2,3-Diphenyl-1,3-thiaphosphinane, *see* D-01152
2,3-Diphenyl-1,3-thiaphospholane, D-01151
2,3-Diphenyl-1,3-thiaphosphorinane, D-01152
Diphenyl 2-thienylphosphonate, *in* T-00307
Diphenyl 2-thienylphosphonite, *in* T-00308
Diphenylthiodiphosphonic acid, *see* P-00237
Diphenylthiodiphosphonic acid, *see* P-00245
Diphenylthiohypophosphonic acid, D-01153
O,O-Diphenyl thiophosphite, *see* D-01106
O,O-Diphenyl thiophosphoric acid, *see* D-01141
O,O-Diphenyl thiophosphoric anilide, *in* P-00262
O,O-Diphenyl thiophosphoric hydrazide, *see* D-01137
O,O-Diphenyl thiophosphoryl azide, *see* D-01113
O,O-Diphenyl thiophosphoryl bromide, *see* D-01119
O,O-Diphenyl thiophosphoryl chloride, *see* D-01123
O,O-Diphenyl thiophosphoryl fluoride, *see* D-01134
O,O-Diphenyl thiophosphoryl hydrazide, *see* D-01137
O,O-Diphenyl thiophosphoryl isocyanate, *see* D-01117
Diphenyl-*m*-tolylphosphine, *see* M-00201
Diphenyl-*o*-tolylphosphine, *see* M-00200
Diphenyl-*p*-tolylphosphine, *see* M-00202
Diphenyl *p*-tolylphosphonate, *in* M-00237
Diphenyl (2,4,6-tribromophenyl)-phosphoramidate, *in* T-00359
N,N′-Diphenyl-*P*-[(4-trichloromethyl)-phenyl]phosphonic diamide, *in* T-00409
Diphenyl (trichloromethyl)phosphonate, *in* T-00411
Diphenyl (2,4,6-trichlorophenyl)-phosphoramidate, *in* T-00420
Diphenyl(2-trifluoromethylphenyl)phosphine, D-01154
Diphenyl(3-trifluoromethylphenyl)phosphine, D-01155
Diphenyl (2,4,5-trimethylphenyl)phosphonate, *in* T-00526
Diphenyl(trimethylsilyl)phosphine, D-01156
P,P-Diphenyl-*N*-trimethylsilylphosphinic amide, *see* D-01157
P,P-Diphenyl-*N*-trimethylsilylphosphinimidic acid, D-01157
P,P-Diphenyl-*N*-trimethylsilylphosphinimidic fluoride, *in* D-01157
P,P-Diphenyl-*N*-trimethylsilylphosphinothioic amide, *in* D-01083
2,3-Diphenyl-1,6,10-trioxa-4-aza-5-phospha(5-P^V)spiro[4.5]dec-2-ene, D-01158
2,3-Diphenyl-1,4,6-trioxa-9-aza-5-phospha(5-P^V)spiro[4.4]non-2-ene, D-01159
P,P-Diphenyl-*N*-(triphenylarsoranylidene)-phosphinic amide, *in* D-01160
P,P-Diphenyl-*N*-(triphenylarsoranylidene)-phosphinous amide, D-01160
Diphenyl(triphenylgermyl)phosphine, D-01161
1,4-Diphenyl-2-(triphenylphosphoranylidene)-1,4-butanedione, D-01162
1,4-Diphenyl-5-(triphenylphosphoranylidene)-1,3-cyclopentadiene, *see* D-00999
Diphenyl [(triphenylphosphoranylidene)-methyl]phosphonate, D-01163
P,P-Diphenyl-*N*-(triphenylphosphoranylidene)phosphinic amide, *in* D-01165
P,P-Diphenyl-*N*-(triphenylphosphoranylidene)-phosphinimidic amide, D-01164
P,P-Diphenyl-*N*-(triphenylphosphoranylidene)-phosphinoselenoic amide, *in* D-01165
P,P-Diphenyl-*N*-(triphenylphosphoranylidene)-phosphinotelluroic amide, *in* D-01165
P,P-Diphenyl-*N*-(triphenylphosphoranylidene)-phosphinothioic amide, *in* D-01165
P,P-Diphenyl-*N*-(triphenylphosphoranylidene)phosphinous amide, D-01165
1,3-Diphenyl-2-(triphenylphosphoranylidene)-1,3-propanedione, D-01166
Diphenyl(triphenylstannyl)phosphine, D-01167
Diphenylvinylphosphine, D-01168
▷Diphos, *see* B-00234
Diphos, *see* P-00123

1,4-Diphosphabicyclo[2.2.2]octane, D-01169
1,5-Diphosphabicyclo[3.3.0]octane, D-01170
1,6-Diphosphacyclodecane, *see* D-01171
5,10-Diphospha-1,2:3,4-dibenzocyclodecadiene, *see* H-00021
5,9-Diphospha-1,2:3,4-dibenzocyclononadiene, *see* T-00089
1,6-Diphosphatriptycene, *see* B-00024
1,6-Diphosphecane, D-01171
1,1′-(1,2-Diphosphenediyl)bis(2,2,6,6-tetramethylpiperidine), *see* B-00392
1,2-Diphosphinobenzene, D-01172
1,2-Diphosphinobenzene, *see* D-01172
1,4-Diphosphinobenzene, D-01173
1,4-Diphosphinobutane, D-01174
1,10-Diphosphinodecane, D-01175
▷1,2-Diphosphinoethane, D-01176
1,6-Diphosphinohexane, D-01177
Diphosphinomethane, D-01178
1,5-Diphosphinopentane, D-01179
1,3-Diphosphinopropane, D-01180
1,2-Diphospholane, D-01181
2,5-Diphosphonoadipic acid, *see* D-01184
1,2-Diphosphonobenzene, *see* B-00006
1,3-Diphosphonobenzene, *see* B-00007
1,4-Diphosphonobenzene, *see* B-00008
1,4-Diphosphono-1,3-butadiene, *see* B-00507
1,4-Diphosphonobutane, *see* B-00514
2,4-Diphosphono-1,2-butanedicarboxylic acid, *see* P-00394
2,3-Diphosphonobutanedioic acid, D-01182
4,4-Diphosphonobutanoic acid, D-01183
1,1-Diphosphono-1-butanol, *see* H-00113
1,4-Diphosphono-2-butene, *see* B-00519
1,2-Diphosphonocyclohexane, *see* C-00240
1,10-Diphosphonodecane, *see* D-00008
1,1′-Diphosphonodimethylamine, *see* D-00688
1,2-Diphosphonoethane, *see* E-00006
1,1-Diphosphonoethene, *see* E-00020
2,4-Diphosphonoglutaric acid, *see* D-01186
1,6-Diphosphonohexane, *see* H-00085
2,5-Diphosphonohexanedioic acid, D-01184
1,1-Diphosphono-1-hexanol, *see* H-00153
(Diphosphonomethyl)butanedioic acid, D-01185
4-(Diphosphonomethyl)morpholine, *see* M-00520
(Diphosphonomethyl)succinic acid, *see* D-01185
1,5-Diphosphonopentane, *see* P-00032
2,4-Diphosphonopentanedioic acid, D-01186
3,3-Diphosphonopropanoic acid, D-01187
1,1-Diphosphono-1-propanol, *see* H-00189
1,3-Diphosphono-1-propanone, *see* O-00071
2,3-Diphosphonosuccinic acid, *see* D-01182
▷Diphosphopyridine nucleotide, *see* C-00202
N,O-Diphosphotyrosine, D-01188
▷Dipin, D-01189
Dipiperidinophosphinodithioic acid, *see* D-01191
Di-1-piperidinylphosphinic acid, D-01190
Di-1-piperidinylphosphinic anilide, *in* D-01190
Di-1-piperidinylphosphinodithioic acid, D-01191
Di-1-piperidylphosphinic acid, *see* D-01190
Di-2-propenyl chloromethylphosphonate, *in* C-00106
Di-2-propenyl (3-chloropropyl)phosphonate, *in* C-00172
Di-2-propenyl hydrogen phosphate, *see* D-01192
Di-2-propenyl (3-methyl-2-oxiranyl)-phosphonate, *in* M-00185
Di-2-propenyl methylphosphonate, *in* M-00288
Di-2-propenyl phenylphosphonate, *in* P-00221
Di-2-propenyl phenylphosphonite, *in* P-00251
O,O-Di-2-propenyl phenylphosphonothioate, *in* P-00244
Di-2-propenyl phosphate, D-01192
Di-2-propenylphosphinic acid, D-01193
▷Di-2-propenyl phosphite, *see* D-01194
▷Di-2-propenyl phosphonate, D-01194
Di-2-propenylphosphoramidic acid, D-01195
Di-2-propenyl phosphorochloridate, *in* D-01192
O,O-Di-2-propenyl phosphorochloridothioate, D-01196
O,O-Di-2-propenyl phosphorodithioate, D-01197
Di-2-propenyl phosphorofluoridate, *in* D-01192
O,O-Di-2-propenyl phosphorothioate, D-01198
Di-2-propenyl 2-propenylphosphonate, *in* P-00454
Di-2-propenyl (trichloromethyl)phosphonate, *in* T-00411
Dipropoxyphosphinothioyl chloride, *see* D-01221
(Dipropoxyphosphinyl)acetamide, *in* D-01199
(Dipropoxyphosphinyl)acetic acid, D-01199
Dipropoxyphosphinyl bromide, *in* D-01203
▷Dipropoxyphosphinyl fluoride, *see* D-01228
S,S-Dipropyl amidodithiophosphate, *in* D-01227
Dipropyl amidophosphate, *see* D-01213
O,O-Dipropyl azidothiophosphate, *see* D-01216
S,S-Dipropyl benzylphosphonodithioate, *in* B-00065
Dipropyl bromophosphate, *in* D-01203
S,S-Dipropyl chlorodithiophosphate, *see* D-01219
Dipropyl chlorophosphate, *see* D-01217
Dipropyl chlorophosphite, *see* D-01218
O,O-Dipropyl chlorothiophosphate, *see* D-01221
▷Dipropyl chlorothiophosphonate, D-01200
▷*O,O*-Dipropyl chlorothiophosphonothioate, D-01201
Dipropyl chlorotrithiophosphate, *see* D-01222
Dipropyl (α-cyanobenzyl)phosphonate, *in* C-00219
Dipropyl (cyanophenylmethyl)phosphonate, *in* C-00219
Dipropyldiphosphonic acid, D-01202
Dipropyl dipropylphosphoramidate, *in* D-01214
O,O-Dipropyl diselenophosphate, *see* D-01224
Dipropyl dithiochlorophosphite, *see* D-01220
O,O-Dipropyl dithiophosphate, *see* D-01225
O,S-Dipropyl dithiophosphoric acid, *see* D-01226
S,S-Dipropyl dithiophosphoric acid, *see* D-01227
Dipropyl ethylphosphoramidate, *in* E-00156
▷Dipropyl fluorophosphate, *see* D-01228
O,O-Dipropyl hydrogen dithiophosphate, *see* D-01225
O,S-Dipropyl hydrogen dithiophosphate, *see* D-01226
S,S-Dipropyl hydrogen dithiophosphate, *see* D-01227
Dipropyl hydrogen phosphate, *see* D-01203
O,O-Dipropyl hydrogen phosphorodiselenoate, *see* D-01224
O,O-Dipropyl hydrogen phosphoroselenoate, *see* D-01229
O,O-Dipropyl hydrogen phosphorothioate, *see* D-01230
O,O-Dipropyl hydrogen thiophosphate, *see* D-01230
Dipropyl (3-methyl-2-oxiranyl)phosphonate, *in* M-00185
Dipropyl methylphosphonate, *in* M-00288
Dipropyl methylphosphoramidate, *in* M-00326
Dipropyl phenylphosphonate, *in* P-00221
Dipropyl phenylphosphonotrithioate, *in* P-00250
Dipropyl phenylphosphoramidate, *in* P-00257
O,O-Dipropyl phenylphosphoramidothioate, *in* D-01230
Dipropyl phosphate, D-01203
Dipropylphosphine, D-01204
Dipropylphosphinic acid, D-01205
P,P-Dipropylphosphinic amide, D-01206
Dipropylphosphinodithioic acid, D-01207
Dipropylphosphinothioic acid, D-01208
Dipropylphosphinous acid, D-01209
Dipropyl phosphite, *see* D-01210
Dipropyl phosphonate, D-01210
O,O-Dipropyl phosphonoselenoate, D-01211
O,O-Dipropyl phosphonoselenoate, *in* E-00145
O,O-Dipropyl phosphonothioate, D-01212
Dipropyl phosphoramidate, D-01213
Dipropylphosphoramidic acid, D-01214
S,S-Dipropyl phosphoramidodithioate, *in* D-01227
Dipropylphosphoramidous acid, D-01215
O,O-Dipropyl phosphorazidothioate, D-01216
Dipropyl phosphoric acid, *see* D-01203
Dipropyl phosphorobromidate, *in* D-01203
Dipropyl phosphorochloridate, D-01217
Dipropyl phosphorochloridite, D-01218
S,S-Dipropyl phosphorochloridodithioate, D-01219
Dipropyl phosphorochloridodithioite, D-01220
O,O-Dipropyl phosphorochloridoselenoate, *in* D-01229
O,O-Dipropyl phosphorochloridothioate, D-01221
Dipropyl phosphorochloridotrithioate, D-01222
N,N′-Dipropylphosphorodiamidic acid, D-01223
O,O-Dipropyl phosphorodiselenoate, D-01224
O,O-Dipropyl phosphorodiselenoic acid, *see* D-01224
O,O-Dipropyl phosphorodithioate, D-01225
O,S-Dipropyl phosphorodithioate, D-01226
S,S-Dipropyl phosphorodithioate, D-01227
Dipropyl phosphorodithioic acid, *see* D-01225
▷Dipropyl phosphorofluoridate, D-01228
O,O-Dipropyl phosphoroselenoate, D-01229
O,O-Dipropyl phosphoroselenoic acid, *see* D-01229
O,O-Dipropyl phosphoroselenoite, *see* D-01211
O,O-Dipropyl phosphorothioate, D-01230
▷*O,O*-Dipropyl phosphorothioate anhydride, *see* A-00144
Dipropyl phosphoryl amide, *see* D-01213
Dipropyl phosphoryl bromide, *in* D-01203
Dipropyl phosphoryl chloride, *see* D-01217
▷Dipropyl phosphoryl fluoride, *see* D-01228
Dipropyl propylphosphonate, *in* P-00473
S,S-Dipropyl propylphosphonodithioate, *in* P-00478
Dipropyl propylphosphoramidate, *in* P-00484
Dipropylpyrophosphonic acid, *see* D-01202
O,O-Dipropyl thiophosphite, *see* D-01212
O,O-Dipropyl thiophosphoric acid, *see* D-01230
O,O-Dipropyl thiophosphoryl azide, *see* D-01216
O,O-Dipropyl thiophosphoryl chloride, *see* D-01221
P,P-Dipropyl-*N*-trimethylsilylphosphinothioic amide, *in* D-01208
▷*O,O*-Di-2-propynyl dithiophosphoric acid, *see* D-01232
▷*O,O*-Di-2-propynyl hydrogen dithiophosphate, *see* D-01232
Dipropynylphosphinic acid, D-01231
▷*O,O*-Di-2-propynyl phosphorodithioate, D-01232
▷Dipterex, *see* T-00423
P,P-Di-1-pyrrolidinylphosphinic acid, D-01233
2,2′-Diselenobis[5,5-dimethyl-1,3,2-dioxaphosphorinane], D-01234
2,2′-Diselenobis[5,5-dimethyl-2-thioxo-1,3,2-dioxaphosphorinane], *in* D-01234
Diselenoxo(2,4,6-tri-*tert*-butylphenyl)-phosphorane, D-01235
Diselenoxo[2,4,6-tris(1,1-dimethylethyl)-phenyl]phosphorane, *see* D-01235
1,2-Distearoyl-*sn*-glycero-3-phospo-L-serine, *in* G-00038
Distearoyl-α-glycerylphosphoryl-2-amino-1-propanol, *see* G-00033
Distearoylphosphatidic acid, *see* G-00032
Distearoylphosphatidylcholine, *see* G-00034
Distearoylphosphatidylethanolamine, *see* G-00037
Distearoylphosphatidylglycerol, *see* G-00035
Distearoylphosphatidylinositol, *see* G-00036
Distearoylphosphatidylserine, *see* G-00038
Distearyl phosphate, *see* D-00952
Distearyl phosphite, *see* D-00954
Distearyl phosphonate, *see* D-00954
Distyrylphosphinic acid, *see* B-00386
▷Disulfoton, D-01236
▷Disyston, *see* D-01236
▷Ditalimfos, D-01237
1,2-Ditetradecanoyl-*sn*-glycerol 3-phosphate, *in* G-00047
1,2-Ditetradecanoylglycero-3-phosphocholine, *see* G-00048

1,2-Ditetradecanoylglycero-3-phosphoethanolamine, *see* G-00049
Ditetradecanoyl-α-glycerylphosphoryl-2-amino-1-propanol, *see* G-00050
Ditetradecylphosphinic acid, D-01238
Ditetradecyl phosphite, *see* D-01239
Ditetradecyl phosphonate, D-01239
1,3-Dithiane-2-phosphonic acid, D-01240
1,3-Dithian-2-ylidenetriphenylphosphorane, *in* D-01242
(1,3-Dithian-2-ylmethylene)-triphenylphosphorane, *in* D-01241
(1,3-Dithian-2-ylmethyl)-triphenylphosphonium(1+), D-01241
1,3-Dithian-2-yltriphenylphosphonium(1+), D-01242
1,3,2-Dithiaphosphinane 2-oxide, *see* D-01248
1,3,2-Dithiaphosphinane 2-sulfide, *see* D-01249
1,3,2-Dithiaphospholane, D-01243
1,3,2-Dithiaphospholane 2-oxide, D-01244
1,3,2-Dithiaphospholane 2-sulfide, D-01245
1,3,2-Dithiaphosphole, D-01246
1,3,2-Dithiaphosphorinane, D-01247
1,3,2-Dithiaphosphorinane 2-oxide, D-01248
1,3,2-Dithiaphosphorinane 2-sulfide, D-01249
Di-2-thienylphosphinic acid, D-01250
2,2′-Dithiobis[5,5-dimethyl-1,3,2-dioxaphosphorinane], D-01251
▷Dithiodemeton, *see* D-01236
(1,3-Dithiolan-2-ylmethylene)-triphenylphosphorane, *in* D-01252
(1,3-Dithiolan-2-ylmethyl)-triphenylphosphonium(1+), D-01252
1,3-Dithiolan-2-yltriphenylphosphonium(1+), D-01253
1,3-Dithiol-2-ylidenetriphenylphosphorane, *in* D-01253
Dithiophosphoric acid bis(butylamide), *see* D-00147
Dithiophosphoric acid bis(cyclohexylamide), *see* D-00217
Dithiophosphoric acid bis(dibutylamide), *see* T-00023
Dithiophosphoric acid bis(dimethylamide), *see* T-00222
Dithiophosphoric acid dianilide, *see* D-01128
Dithiophosphoric acid methylamide, *see* M-00330
▷Dithiosystox, *see* D-01236
Dithioxo(2,4,6-tri-*tert*-butylphenyl)-phosphorane, D-01254
Di-*m*-tolyl chlorophosphate, *in* B-00313
Di-*o*-tolyl chlorophosphate, *in* B-00312
Di-*p*-tolyl chlorophosphate, *see* B-00329
Di-*m*-tolyl chlorophosphite, *see* B-00331
Di-*o*-tolyl chlorophosphite, *see* B-00330
Di-*p*-tolyl chlorophosphite, *see* B-00332
O,O-Di-*m*-tolyl chlorothiophosphate, *see* B-00334
O,O-Di-*o*-tolyl chlorothiophosphate, *see* B-00333
O,O-Di-*p*-tolyl chlorothiophosphate, *see* B-00335
O,O-Di-*p*-tolyl dithiophosphate, *see* B-00339
Di-*p*-tolyl dithiophosphoric acid, *see* B-00339
▷*O,O*-Di-*m*-tolyl dithiophosphoric acid, *see* B-00338
O,O-Di-*o*-tolyl dithiophosphoric acid, *see* B-00337
▷*O,O*-Di-*m*-tolyl hydrogen dithiophosphate, *see* B-00338
O,O-Di-*o*-tolyl hydrogen dithiophosphate, *see* B-00337
Di-*m*-tolyl hydrogen phosphate, *see* B-00313
▷Di-*o*-tolyl hydrogen phosphate, *see* B-00312
Di-*p*-tolyl hydrogen phosphate, *see* B-00314
O,O-Di-*p*-tolyl hydrogen phosphorodiselenoate, *see* B-00336
O,O-Di-*p*-tolyl hydrogen phosphorodithioate, *see* B-00339
Di-*m*-tolyl phosphate, *see* B-00313
▷Di-*o*-tolyl phosphate, *see* B-00312
Di-*p*-tolyl phosphate, *see* B-00314
▷Di-*m*-tolylphosphine, *see* B-00316
▷Di-*o*-tolylphosphine, *see* B-00315
Di-*p*-tolylphosphine, *see* B-00317
Di-*m*-tolylphosphinic acid, *see* B-00319
Di-*o*-tolylphosphinic acid, *see* B-00318
Di-*p*-tolylphosphinic acid, *see* B-00320
Di-*o*-tolylphosphinic anilide, *in* B-00318
Di-*p*-tolylphosphinic anilide, *in* B-00320
Di-*p*-tolylphosphinothioic acid, *see* B-00322
Di-*m*-tolylphosphinous acid, *see* B-00324
Di-*o*-tolylphosphinous acid, *see* B-00323
Di-*p*-tolylphosphinous acid, *see* B-00325
Di-*m*-tolyl phosphite, *see* B-00327
Di-*o*-tolyl phosphite, *see* B-00326
Di-*p*-tolyl phosphite, *see* B-00328
Di-*m*-tolyl phosphonate, *see* B-00327
Di-*o*-tolyl phosphonate, *see* B-00326
Di-*p*-tolyl phosphonate, *see* B-00328
Di-*m*-tolyl phosphoric acid, *see* B-00313
▷Di-*o*-tolyl phosphoric acid, *see* B-00312
Di-*p*-tolyl phosphoric acid, *see* B-00314
Di-*m*-tolyl phosphorochloridite, *see* B-00331
Di-*o*-tolyl phosphorochloridite, *see* B-00330
Di-*p*-tolyl phosphorochloridite, *see* B-00332
O,O-Di-*m*-tolyl phosphorochloridothioate, *see* B-00334
O,O-Di-*o*-tolyl phosphorochloridothioate, *see* B-00333
O,O-Di-*p*-tolyl phosphorochloridothioate, *see* B-00335
O,O-Di-*p*-tolyl phosphorodiselenoic acid, *see* B-00336
▷*O,O*-Di-*m*-tolyl phosphorodithioate, *see* B-00338
O,O-Di-*o*-tolyl phosphorodithioate, *see* B-00337
O,O-Di-*p*-tolyl phosphorodithioate, *see* B-00339
Di-*p*-tolyl phosphoryl chloride, *see* B-00329
O,O-Di-*m*-tolyl thiophosphoryl chloride, *see* B-00334
O,O-Di-*o*-tolyl thiophosphoryl chloride, *see* B-00333
O,O-Di-*p*-tolyl thiophosphoryl chloride, *see* B-00335
Ditridecanoylphosphatidylcholine, *in* G-00052
Diundecyl hydrogen phosphate, *see* D-01255
Diundecyl phosphate, D-01255
Diundecylphosphinic acid, D-01256
Diundecyl phosphoric acid, *see* D-01255
Divinyl bromophosphate, *in* D-01257
Divinyl chlorophosphate, *in* D-01257
Divinyl hydrogen phosphate, *see* D-01257
Divinyl phenylphosphonite, *in* P-00251
Divinyl phosphate, D-01257
Divinylphosphinic acid, D-01258
Divinylphosphinothioic acid, D-01259
Divinyl phosphoric acid, *see* D-01257
Divinyl phosphorobromidate, *in* D-01257
Divinyl phosphorochloridate, *in* D-01257
Di-*o*-xylylenephosphonium(1+), *see* T-00137
DMPE, *see* B-00200
2,2,4,4,6,6,8,8,9,9,10,10-Dodecaethyl-1,3,5,7-tetraphospha-2,4,6,8,9,10-hexagermatricyclo[3.3.1.1^{3,7}]decane, D-01260
7,8,10,11,13,14,16,17,19,20,22,27-Dodecahydro-5*H*-dibenzo[*r,u*]-[1,4,7,10,13,16,20]-hexaoxaphosphacyclotricosin, D-01261
2,3,4,5,6,7,8,9,10,11,12,13-Dodecahydro-2,5,9,12-tetraphenyl-1*H*-2,5,9,12-benzotetraphosphacyclopentadecin, D-01262
2,2,4,4,6,6,8,8,9,9,10,10-Dodecamethyl-1,3,5,7-tetraphospha-2,4,6,8,9,10-hexagermatricyclo[3.3.1.1^{3,7}]decane, D-01263
Dodecanoic acid 1-[[[(2-aminoethoxy)-hydroxyphosphinyl]oxy]methyl]-1,2-ethanediyl ester, *see* G-00009
Dodecyl dihydrogen phosphate, *see* M-00468
Dodecyldimethylphosphine, D-01264
Dodecylidenetriphenylphosphorane, *in* D-01266
Dodecylphosphonous acid, D-01265
Dodecyl phosphoric acid, *see* M-00468
Dodecyltrimethylphosphonium iodide, *in* D-01264
Dodecyltriphenylphosphonium(1+), D-01266
▷Dotan, *see* C-00019
▷Dowco 132, *see* C-00208
▷Dowco 217, *see* F-00063
DPDECMP, *in* D-00268
▷DPN, *see* C-00202
DPPA, *see* D-01112
Dualar, *see* B-00017
▷Dursban, *see* C-00194
▷Dyflos, *see* D-00665
▷Dyfonate, *see* F-00052
▷Dylox, *see* T-00423
Ecothiopate, E-00001
▷Ectoral, *see* F-00003
▷Edifenphos, E-00002
▷Ekaluk, *see* Q-00001
▷Ekamet, *see* E-00205
▷Ekatin, *see* T-00313
▷Ektafos, *see* D-00204
Elaidoylstearoylphosphatidylethanolamine, *in* G-00092
▷Elsan, *see* P-00078
▷Endoxane, *see* C-00283
▷Enolofos, *see* C-00018
Ensanchomycin, E-00003
▷Enzopride, *see* C-00202
▷EPN, *see* E-00099
▷EPN-Methyl, *in* M-00243
(1,2-Epoxypropyl)phosphonic acid, *see* M-00185
1-*tert*-Butyl-3,4-dimethyl-1*H*-phosphole, E-00004
▷Estox, *see* O-00099
▷Ethafos, E-00005
1,2-Ethanediphosphonic acid, E-00006
1,2-Ethanediphosphonothioic acid, E-00007
1,2-Ethanediphosphonous acid, E-00008
1,2-Ethanediphosphoramidic acid, E-00009
1,2-Ethanediphosphoramidothioic acid, E-00010
1,2-Ethanediylbis[bis(2-methylphenyl)-phosphine], *see* B-00103
1,2-Ethanediylbis[bis(4-methylphenyl)-phosphine], *see* B-00104
1,2-Ethanediylbis[dibutylphosphine], *see* B-00158
1,2-Ethanediylbis[dicyclohexylphosphine], *see* B-00167
1,2-Ethanediylbis[diethylphosphine], *see* B-00180
1,2-Ethanediylbis[dimethylphenylphosphonium] dibromide, *in* E-00012
1,2-Ethanediylbis[dimethylphosphine], *see* B-00200
▷1,2-Ethanediylbis[diphenylphosphine], *see* B-00234
1,2-Ethanediylbis[diphenylphosphine oxide], *in* B-00234
1,2-Ethanediylbis[[2-(diphenylphosphino)-ethyl]phosphine], *see* B-00240
1,2-Ethanediylbis(di-*o*-tolylphosphine), *see* B-00103
1,2-Ethanediylbis[di-*p*-tolylphosphine], *see* B-00104
1,2-Ethanediylbis[(2-methoxyphenyl)-phenylphosphine], E-00011
1,2-Ethanediylbis[methyldiphenylphosphonium] diiodide, *in* B-00234
1,2-Ethanediylbis[methylphenylphosphine], E-00012
1,2-Ethanediylbis[methylphosphinic acid], E-00013
[1,2-Ethanediylbis[nitrilobis[methylene]]]-tetrakisphosphonic acid, E-00014
1,2-Ethanediyl bis(phenylmethyl)-carbonimidodithioate, *see* Z-00001
1,2-Ethanediylbis[phenylphosphine], *see* B-00390
1,2-Ethanediylbis[phenylphosphinic acid], E-00015
1,2-Ethanediylbis[phenylphosphinodithioic acid], E-00016
▷1,2-Ethanediylbisphosphine, *see* D-01176
1,2-Ethanediylbisphosphonic acid, *see* E-00006
1,2-Ethanediylbis(phosphonic acid), *see* E-00008
1,2-Ethanediylbis[phosphonic dichloride], *in* E-00006
Ethanediylbis(phosphonothioic acid), *see* E-00007

1,2-Ethanediylbis(phosphonous acid), *see* E-00008
1,2-Ethanediylbis(phosphoramidic acid), *see* E-00009
1,2-Ethanediylbisphosphoramidothioic acid, *see* E-00010
P,P'-1,2-Ethanediylbis[*N,N,N',N'*-tetraethylphosphonous diamide], *in* E-00008
1,2-Ethanediylbis[triphenylphosphonium](2+), E-00017
N,N'-1,2-Ethanediyl-*N,N'*-dimethyl-*P*-phenylphosphonic diamide, *in* D-00793
N,N'-1,2-Ethanediyl-*N,N'*-dimethyl-*P*-phenylphosphonous diamide, *see* D-00793
Ethanolamine *O*-phosphate, *see* A-00075
Ethene-1,1-diphosphonic acid, *see* E-00020
1,2-Ethenediylbis[(dibutylmethyl)-phosphonium] diiodide, *in* B-00159
1,2-Ethenediylbis[dibutylphosphine], *see* B-00159
1,2-Ethenediylbis[diethylphosphine], *see* B-00181
1,2-Ethenediylbis[diphenylphosphine], *see* B-00237
1,2-Ethenediylbis(phenylphosphinic acid), E-00018
1,2-Ethenediylbis[triphenylphosphonium](2+), E-00019
▷Ethenyl dihydrogen phoshate, *see* M-00518
Ethenyldiphenylphosphine, *see* D-01168
P-Ethenyl-*N,N'*-diphenylphosphosphonothioic diamide, *in* V-00007
Ethenylidenebis(diphenylphosphine), *see* B-00236
Ethenylidenebisphosphonic acid, E-00020
Ethenylmethyldiphenylphosphonium iodide, *in* D-01168
Ethenylmethylphosphinic acid, *see* M-00438
Ethenylphenylphosphinic acid, *see* P-00337
Ethenylphosphinic acid, *see* V-00002
Ethenylphosphonic acid, *see* V-00003
P-Ethenylphosphonic diamide, *see* V-00004
Ethenylphosphonic dichloride, *see* V-00005
Ethenylphosphonochloridothioic acid, *see* V-00006
Ethenylphosphonothioic acid, *see* V-00007
Ethenyl phosphorodichloridate, *in* M-00518
O-Ethenyl phosphorodichloridothioate, *see* V-00008
P-Ethenyl-*N,N,N',N'*-tetraethylphosphonic diamide, *in* V-00004
P-Ethenyl-*N,N,N',N'*-tetramethylphosphonic diamide, *in* V-00004
Ethenyltriphenylphosphonium(1+), *see* T-00702
▷Ethephon, *see* C-00074
▷Ethiofos, *see* A-00052
▷Ethiofos, E-00021
▷Ethion, E-00022
▷Ethoprophos, E-00023
2-Ethoxyallylidenetriphenylphosphorane, *see* E-00049
2-Ethoxy-1,3,2-benzodioxaphosphole, E-00024
1-Ethoxy-1*H*-benzo[*b*]phosphindole 1-oxide, *in* H-00118
2-Ethoxy-1,3,2-benzoxazaphospholine, *see* E-00027
P-Ethoxy-*P,P*-bis(diethylamino)-*N*-[bis(diethylamino)phosphinothioyl]-phosphazene, *in* B-00174
P-Ethoxy-*P,P*-bis(diethylamino)-*N*-[bis(diethylamino)phosphinyl]-phosphazene, *in* B-00177
▷*S*-α-Ethoxycarbonylbenzyl *O,O*-dimethyl phosphorodithioate, *see* P-00078
1-(Ethoxycarbonyl)-cyclopropyltriphenylphosphonium(1+), E-00025
Ethoxycarbonyl diethyl phosphate, *see* E-00055
3-Ethoxycarbonyl-1-hydroxyphosphetane 1-oxide, *in* H-00184
▷*S*-(*N*-Ethoxycarbonyl-*N*-methylcarbamoylmethyl) *O,O*-diethyl phosphorodithioate, *see* M-00002
[(Ethoxycarbonylmethylene)hydrazono]-triphenylphosphorane, *in* T-00653
(Ethoxycarbonylmethyl)phosphonic bis(diethylamide), *in* B-00175
▷*O*-6-Ethoxycarbonyl-5-methylpyrazolo[1,5-*a*]pyrimidin-2-yl *O,O*-diethyl phosphorothioate, *see* P-00502
(α-Ethoxycarbonylphenacylidene)-triphenylphosphorane, *in* O-00083
2-(Ethoxycarbonylphenyl)phosphoramidic acid, *see* C-00004
3-(Ethoxycarbonylphenyl)phosphoramidic acid, *see* C-00005
(4-Ethoxycarbonylphenyl)phosphoramidic acid, *see* C-00006
(1-Ethoxycarbonyl)triphenylphosphonium bromide, *in* C-00010
α-Ethoxycarbonyltriphenyl-phosphoranylidenepropanoic acid, *in* T-00646
5-Ethoxy-5*H*-dibenzophosphole 5-oxide, *in* H-00118
2-Ethoxy-2,3-dihydro-1*H*-1,3,2-benzodiazaphosphole, E-00026
2-Ethoxy-2,3-dihydro-1,2-benzoxaphosphole 2-oxide, *in* H-00119
2-Ethoxy-2,3-dihydro-1,3,2-benzoxazaphosphole, E-00027
1-Ethoxy-2,5-dihydro-3,4-dimethyl-1*H*-phosphole 1-oxide, *in* D-00493
3-Ethoxy-2,3-dihydro-2-methyl-5-phenyl-1*H*-1,2,4,3-triazaphosphole, E-00028
1-Ethoxy-2,5-dihydro-3-methyl-1*H*-phosphole, *in* D-00530
1-Ethoxy-2,3-dihydro-5-methyl-1*H*-phosphole 1-oxide, *in* D-00497
1-Ethoxy-2,5-dihydro-2-methyl-1*H*-phosphole 1-oxide, *in* D-00498
1-Ethoxy-2,5-dihydro-3-methyl-1*H*-phosphole 1-oxide, *in* D-00499
1-Ethoxy-2,3-dihydro-1*H*-phosphole 1-oxide, *in* D-00506
▷1-Ethoxy-2,5-dihydro-1*H*-phosphole 1-oxide, *in* D-00507
4-Ethoxy-1,4-dihydro-1,3,5-triphenyl-1,2,4-diazaphosphorine 4-oxide, E-00029
2-Ethoxy-1,3-dimethyl-1,3,2-diazaphospholidine, E-00030
2-Ethoxy-5,5-dimethyl-1,3-dioxa-2-phosphacyclohexane, *see* E-00032
2-Ethoxy-4,5-dimethyl-1,3-dioxa-2-phosphacyclopentane, *see* E-00031
2-Ethoxy-5,5-dimethyl-1,3,2-dioxaphosphinane, *see* E-00032
2-Ethoxy-4,5-dimethyl-1,3,2-dioxaphospholane, E-00031
2-Ethoxy-5,5-dimethyl-1,3,2-dioxaphosphorinane, E-00032
2-Ethoxy-1,3-dioxa-2-phosphacyclohexane, *see* E-00037
2-Ethoxy-1,3,2-dioxaphosphepane 2-oxide, *in* H-00132
2-Ethoxy-1,3,2-dioxaphosphinane, *see* E-00037
2-Ethoxy-1,3,2-dioxaphospholan-4,5-dicarboxylic acid, E-00033
▷2-Ethoxy-1,3,2-dioxaphospholane, E-00034
2-Ethoxy-1,3,2-dioxaphospholane 2-oxide, E-00035
2-Ethoxy-1,3,2-dioxaphospholane 2-sulfide, E-00036
2-Ethoxy-1,3,2-dioxaphosphorinane, E-00037
P-Ethoxy-*P,P*-diphenyl-*N*-(diphenylphosphinyl)phosphazene, *in* D-01007
P-Ethoxy-*P,P*-diphenyl-*N*-(diphenylphosphinyl)phosphine imide, *in* D-01007
4-Ethoxy-2,6-diphenyl-4*H*-1,4-oxaphosphorin 4-oxide, *in* H-00140
P-Ethoxy-*P,P*-diphenyl-*N*-phenylsulfonylphosphazene, *in* B-00013
P-Ethoxy-*P,P*-diphenyl-*N*-phenylsulfonylphosphine imide, *in* B-00013
1-Ethoxy-3,4-diphenyl-1*H*-phosphole 1-oxide, *in* H-00142
4-Ethoxy-2,6-diphenyl-4*H*-1,4-thiaphosphorin 4-oxide, *in* H-00143
2-Ethoxy-1,3,2-dithiaphospholane, E-00038
2-Ethoxy-1,3,2-dithiaphosphorinane, *in* D-01247
(2-Ethoxyethenyl)triphenylphosphonium(1+), E-00039
O-(2-Ethoxyethyl) dichlorothiophosphate, *in* H-00149
O-(2-Ethoxyethyl) phosphorodichloridothioate, *in* H-00149
▷*O*-(6-Ethoxy-2-ethyl-4-pyrimidinyl) *O,O*-dimethyl phosphorothioate, *see* E-00205
O-(2-Ethoxyethyl) thiophosphoryl dichloride, *in* H-00149
2-[(Ethoxyhydroxyphosphinyl)oxy]-*N,N,N*-trimethylethanaminium hydroxide, inner salt, *in* C-00198
2-Ethoxy-4-methyl-1,3-dioxa-2-phosphacyclohexane, *see* E-00041
2-Ethoxy-4-methyl-1,3-dioxa-2-phosphacyclopentane, *see* E-00040
2-Ethoxy-4-methyl-1,3,2-dioxaphosphinane, *see* E-00041
2-Ethoxy-4-methyl-1,3,2-dioxaphospholane, E-00040
2-Ethoxy-4-methyl-1,3,2-dioxaphosphorinane, E-00041
2-Ethoxy-4-methyl-1,3,2-dithiaphosphinane, *see* E-00042
2-Ethoxy-4-methyl-1,3,2-dithiaphosphorinane, E-00042
O-(Ethoxymethylphosphinothioyl) *O'*-ethyl methylphosphonothioate, *in* D-00925
2-Ethoxy-3-methyl-1,3,2-thiazaphospholidine, E-00043
2-Ethoxynaphtho[1,2-*d*]-1,3,2-dioxaphosphole, E-00044
2-Ethoxynaphtho[1,8-*de*]-1,3,2-dioxaphosphorin, E-00045
2-Ethoxynaphtho[1,8-*de*]-1,3,2-dioxaphosphorin, E-00046
2-Ethoxy-1,2-oxaphosphepane 2-oxide, *in* H-00165
2-Ethoxy-1,2-oxaphospholane 2-oxide, *in* H-00166
2-Ethoxy-1,2-oxaphosphorinane 2-oxide, *in* H-00167
2-Ethoxy-1,3,2-oxazaphosphorinane, *see* E-00051
(2-Ethoxy-2-oxoethyl)triphenylphosphonium, E-00047
1-Ethoxy-2,2,3,4,4-pentamethylphosphetane 1-oxide, *in* H-00168
1-Ethoxy-2,2,3,4,4-pentamethylphosphetane 1-sulfide, *in* H-00169
▷4-Ethoxy-7-phenyl-3,5-dioxa-8-aza-4-phosphaoct-6-ene-8-nitrile 4-sulfide, *see* P-00424
2-Ethoxy-1-(2-phenylethyl)-1,2-azaphosphetidine 2-oxide, *in* H-00107
2-Ethoxy-3-phenyl-1,3,2-oxazaphospholane, *see* E-00048
2-Ethoxy-3-phenyl-1,3,2-oxazaphospholidine, E-00048
▷Ethoxyphos, *see* A-00004
2-Ethoxyphospholane 1-oxide, *in* H-00185
1-Ethoxyphosphorinane 1-oxide, *in* H-00188
(2-Ethoxy-2-propenylidene)-triphenylphosphorane, E-00049
(2-Ethoxypropenyl)triphenylphosphonium(1+), E-00050
2-Ethoxytetrahydro-2*H*-1,3,2-oxazaphosphinine, *see* E-00051
2-Ethoxytetrahydro-2*H*-1,3,2-oxazaphosphorine, E-00051
2-Ethoxytetrahydro-3-phenyl-1,3,2-oxazophosphole, *see* E-00048
1-Ethoxy-1,2,3,4-tetrahydrophosphinoline 1-oxide, *in* H-00195
4-Ethoxy-1,4,5,6-tetrahydro-1,3,5-triphenyl-1,2,4-diazaphosphorine 4-oxide, *in* E-00029
2-Ethoxy-4,4,5,5-tetramethyl-1,3-dioxa-2-phosphacyclopentane, *see* E-00052
2-Ethoxy-4,4,5,5-tetramethyl-1,3,2-dioxaphospholane, E-00052
1-Ethoxy-2,3,4,5-tetraphenyl-1*H*-phosphole 1-oxide, *in* H-00200
2-Ethoxy-1,3,2-thiazaphospholane, *see* E-00053
2-Ethoxy-1,3,2-thiazaphospholidine, E-00053
2-Ethoxy-3,5,5-trimethyl-1,2-oxaphospholan-3-ol 2-oxide, *in* T-00519

P-Ethoxy-*N,P,P*-triphenylphosphazene, *in* T-00631
P-Ethoxy-*N,P,P*-triphenylphosphine imide, *in* T-00631
(2-Ethoxyvinyl)triphenylphosphonium(1+), *see* E-00039
▷Ethrel, *see* C-00074
Ethyl *P*-acetyl-*N,N*-diethylphosphonamidate, *in* A-00014
Ethyl acetylethylphosphinate, *in* A-00017
Ethyl acetylphenylphosphinate, *in* A-00021
Ethyl allylphenylphosphinate, *in* P-00300
O-Ethyl amidofluorothiophosphate, *see* E-00158
Ethylamine-1,1-diphosphonic acid, *see* A-00073
Ethylaminobis(dichlorophosphine), *see* E-00080
Ethyl *P*-aminocarbonyl-*N,N*-diethylphosphonamidate, *in* A-00059
Ethyl 2-amino-3-[(diphenoxyphosphinyl)-oxy]butanoate, *in* A-00120
Ethyl 2-amino-3-[(diphenoxyphosphinyl)-oxy]propanoate, *in* P-00417
Ethyl (aminophenylmethyl)phenylphosphinate, *in* A-00102
Ethyl 1-aziridinylmethylphosphinate, *in* A-00152
Ethyl 1-aziridinyl-*P*-methylphosphonamidate, *in* A-00152
Ethyl benzhydrylphenylphosphinate, *in* D-01023
Ethyl benzoylphenylphosphinate, *in* B-00032
Ethyl benzylmethylphosphinate, *in* B-00053
Ethyl 2-(benzyloxycarbonyl)amino-3-[(diphenoxyphosphinoyl)oxy]butanoate, *in* A-00120
Ethyl benzylphenylphosphinate, *in* B-00056
Ethyl benzylphosphonochloridate, *in* B-00063
▷Ethyl bicyclic phosphate, *in* E-00196
▷Ethyl bicyclic phosphite, *see* E-00196
Ethyl bicyclic phosphite ozonide, *see* E-00187
Ethyl bicyclic selenophosphate, *in* E-00196
▷Ethyl bicyclic thiophosphate, *in* E-00196
▷Ethyl bis(1-aziridinyl)phosphinate, *in* B-00091
Ethyl bis(1-aziridinyl)phosphinite, *in* B-00095
O-Ethyl bis(1-aziridinyl)phosphinothioate, *in* B-00093
S-Ethyl bis(1-aziridinyl)phosphinothioate, *in* B-00093
▷Ethyl [bis(1-aziridinyl)phosphinyl]-carbamate, *see* U-00001
Ethyl bis(2-bromophenyl)phosphinate, *in* B-00112
Ethyl bis(chloromethyl)phosphinate, *in* B-00127
O-Ethyl bis(chloromethyl)phosphinothioate, *in* B-00129
S-Ethyl bis(chloromethyl)phosphinothioate, *in* B-00129
Ethyl bis(2-chlorophenyl)phosphinate, *in* B-00140
Ethyl bis(3-chlorophenyl)phosphinate, *in* B-00138
Ethyl bis(3-chlorophenyl)phosphinite, *in* B-00141
Ethyl bis(4-chlorophenyl)phosphinite, *in* B-00142
Ethyl bis[(diethoxyphosphinyl)methyl]-phosphinate, *in* B-00182
Ethyl 2,3-bis(diethoxyphosphinyl)propanoate, *in* B-00173
Ethyl 3,3-bis(diethoxyphosphinyl)propanoate, *in* D-01187
Ethyl *N″*-[bis(diethylamino)-phosphinothioyl]-*N,N,N′,N′*-tetraethylphosphorodiamidoimidate, *in* B-00174
Ethyl [bis(diethylamino)phosphinyl]acetate, *in* B-00175
Ethyl *N″*-[bis(diethylamino)phosphinyl]-*N,N,N′,N′*-tetraethylphosphorodiamidoimidate, *in* B-00177
Ethyl bis(4-dimethylaminophenyl)phosphinate, *in* B-00188
Ethyl [bis(dimethylamino)phosphinyl]acetate, *in* B-00193
Ethyl [bis(2,2-dimethyl-1-aziridinyl)-phosphinyl]carbamate, *see* M-00439
Ethyl bis(1,1-dimethylethyl)phosphinite, *in* D-00122
Ethyl bis(2-ethoxycarbonylphenyl)-phosphinate, *in* B-00117
Ethyl bis(4-ethoxycarbonylphenyl)-phosphinate, *in* B-00118
Ethyl bis(ethoxymethyl)phosphinate, *in* B-00272
Ethyl bis(hydroxymethyl)phosphinate, *in* B-00282
Ethyl bis(2-hydroxyphenyl)phosphinate, *in* B-00283
Ethyl bis(1-hydroxy-2,2,2-trichloroethyl)-phosphinate, *in* B-00287
Ethyl bis(iodomethyl)phosphinate, *in* B-00288
Ethyl bis(2-methylphenyl)phosphinite, *in* B-00323
Ethyl bis(4-methylphenyl)phosphinite, *in* B-00325
▷*O*-Ethyl *O,O*-bis(4-nitrophenyl) phosphorothioate, E-00054
O-Ethyl bis(pentafluorophenyl)-phosphinothioate, *in* B-00383
Ethyl bis(phenoxymethyl)phosphinate, *in* B-00384
Ethyl bis(phenylethynyl)phosphinate, *in* B-00388
Ethyl bis(phenylmethyl)phosphinite, *in* D-00058
Ethyl bis(2-thienyl)phosphinate, *in* D-01250
O-Ethyl bis(trifluoromethyl)-phosphinothioate, *in* B-00415
S-Ethyl bis(trifluoromethyl)-phosphinothioate, *in* B-00415
Ethyl bis(trimethylsilyloxy)-phosphinecarboxylate oxide, *in* B-00433
Ethyl [bis(trimethylsilyloxy)phosphinyl]-acetate, *in* B-00434
Ethyl bromo(diethoxyphosphinyl)acetate, *in* B-00467
Ethyl bromofluorophosphate, *see* E-00162
Ethyl (4-bromophenyl)phosphinate, *in* B-00489
Ethyl bromo(triphenylphosphoranylidene)-acetate, *in* B-00504
Ethyl *P*-(1,3-butadienyl)-*N,N*-diethylphosphonamidate, *in* B-00508
Ethyl (2-butenyl)methylphosphinate, *in* B-00523
Ethyl *P*-(butoxymethyl)-*N,N*-diethylphosphonamidate, *in* B-00536
Ethyl *N*-*tert*-butyl-*P*-methylphosphonamidate, *in* B-00572
Ethyl butylphosphinate, *in* B-00590
Ethyl *tert*-butylphosphinate, *in* B-00591
Ethyl butylphosphonisocyanatidate, *in* B-00601
Ethyl butylphosphonochloridite, *in* B-00605
O-Ethyl butylphosphonodithioate, *in* B-00606
Ethyl butylphostonate, *in* H-00167
Ethyl carbamoylphosphonate, *see* M-00469
Ethyl carbonate-diethyl phosphate anhydride, E-00055
Ethyl chloro(diethoxyphosphinyl)acetate, *in* C-00036
O-Ethyl chlorodimethylamidothiophosphate, *in* D-00861
Ethyl (2-chloroethyl)phosphonochloridate, *in* C-00077
Ethyl chlorofluorophosphate, *see* E-00164
Ethyl (chloromethyl)phenylphosphinate, *in* C-00103
Ethyl *P*-chloromethylphosphonamidate, *in* C-00105
▷*O*-Ethyl *S*-4-chlorophenyl ethylphosphonodithioate, *in* E-00070
Ethyl (4-chlorophenyl)phosphinate, *in* C-00144
Ethyl (chlorosulfonyl)-phosphorodichlorimidate, E-00056
Ethyl chloro(triphenylphosphoranylidene)-acetate, *in* C-00190
Ethyl cyano(diethoxyphosphinyl)acetate, E-00057
O-Ethyl cyano(dimethylamido)thiophosphate, *in* D-00864
Ethyl cyano(triphenylphosphoranylidene)-pyruvate, *in* C-00218
Ethyl cyclohexylphenylphosphinate, *in* C-00250
Ethyl cyclohexylphosphinate, *in* C-00253
Ethyl cyclohexylphosphonochloridate, *in* C-00259
Ethyl decylphenylphosphinate, *in* D-00010
Ethyl diallylphosphinate, *in* D-01193
O-Ethyl diamidothiophosphate, *in* E-00182
Ethyl diazo(diphenylphosphinyl)acetate, *in* D-00030
Ethyl (1-diazoethyl)phenylphosphinate, *in* D-00032
Ethyl dibenzylphosphinate, *in* D-00055
Ethyl dibenzylphosphinite, *in* D-00058
Ethyl dibromophosphite, *see* E-00166
Ethyl (dibutoxyphosphinyl)acetate, *in* D-00073
Ethyl dibutylphosphinate, *in* D-00109
Ethyl di-*tert*-butylphosphinate, *in* D-00110
Ethyl dibutylphosphinite, *in* D-00121
Ethyl di-*tert*-butylphosphinite, *in* D-00122
Ethyl dibutylphosphinodithioate, *in* D-00112
O-Ethyl dibutylphosphinothioate, *in* D-00117
Ethyl *N,N*-dibutylphosphonamidate, *in* D-00126
Ethyl dibutylphosphoramidochloridite, *in* D-00135
Ethyl *N,N′*-dibutylphosphorodiamidate, *in* D-00146
Ethyl dichlorodithiophosphate, *see* E-00169
Ethyl *P*-dichloromethylphosphonamidate, *in* D-00174
▷Ethyl dichlorophosphate, *see* E-00167
Ethyl dichlorophosphinylacetate, *in* D-00190
Ethyl dichlorophosphinylcarbamate, *in* D-00191
Ethyl (dichlorophosphinyl)-phosphorodichloroimidate, E-00058
Ethyl 3-dichlorophosphinylpropanoate, *in* D-00192
▷Ethyl dichlorophosphite, *see* E-00168
O-Ethyl dichlorothiophosphate, *see* E-00170
S-Ethyl dichlorothiophosphate, *see* E-00171
Ethyl dichlorothiophosphite, *see* E-00172
Ethyl dicyclohexylphosphinate, *in* D-00211
Ethyl dicyclohexylphosphinite, *in* D-00214
Ethyl *N,N′*-dicyclohexylphosphorodiamidate, *in* D-00216
Ethyl diethenylphosphinate, *in* D-01258
Ethyl (diethoxyphosphinothioyl)acetate, *in* D-00237
▷Ethyl 2-[(diethoxyphosphinothioyl)oxy]-5-methylpyrazolo[1,5-*a*]pyrimidine-6-carboxylate, *see* P-00502
▷Ethyl [(diethoxyphosphinothioyl)thio]-acetate, *see* A-00004
▷Ethyl [[(diethoxyphosphinothioyl)thio]-acetyl] methylcarbamate, *see* M-00002
Ethyl (diethoxyphosphinylamino)acetate, *in* P-00368
Ethyl 2-[(diethoxyphosphinyl)amino]benzoate, *in* C-00004
Ethyl 3-[(diethoxyphosphinyl)amino]benzoate, *in* C-00005
Ethyl α-(diethoxyphosphinyl)benzeneacetate, *in* D-00242
Ethyl [2-(diethoxyphosphinyl)]benzoate, *in* P-00378
Ethyl [3-(diethoxyphosphinyl)]benzoate, *in* P-00379
Ethyl [4-(diethoxyphosphinyl)]benzoate, *in* P-00380
Ethyl 2-(diethoxyphosphinyl)butanoate, *in* D-00246
Ethyl 4-(diethoxyphosphinyl)butanoate, *in* D-00247
Ethyl 4-(diethoxyphosphinyl)-2-butenoate, *in* D-00248
Ethyl *N*-(diethoxyphosphinyl)carbamate, *in* D-00249
Ethyl (diethoxyphosphinyl)fluoroacetate, *in* D-00250
Ethyl (diethoxyphosphinyl)oxoacetate, *in* D-00252
▷Ethyl 2-(diethoxyphosphinyl)-3-oxobutanoate, *in* D-00253
Ethyl 4-(diethoxyphosphinyl)-3-oxobutanoate, *in* D-00254

Ethyl 2-(diethoxyphosphinyl)oxybenzoate, *in* P-00391
Ethyl 2-[(diethoxyphosphinyl)oxy]propanoate, *in* D-00255
Ethyl 2-[(diethoxyphosphinyl)oxy]-2-propenoate, *in* D-00256
Ethyl 2-(diethoxyphosphinyl)propanoate, E-00059
Ethyl 3-(diethoxyphosphinyl)propanoate, *in* D-00259
Ethyl 2-(diethoxyphosphinyl)propenoate, *in* D-00260
Ethyl 3-(diethoxyphosphinyl)propenoate, *in* D-00261
Ethyl *N,N*-diethyl-*P*-(methoxymethyl)-phosphonamidate, *in* D-00295
Ethyl *N,N*-diethyl-*P*-methylphosphonamidate, *in* D-00296
Ethyl *P,P*-diethyl-*N*-phenylphosphinimidate, *in* D-00306
Ethyl *N,N*-diethyl-*P*-phenylphosphonamidodithioate, *in* D-00308
S-Ethyl *N,N*-diethyl-*P*-phenylphosphonamidothioate, *in* D-00309
Ethyl diethylphosphinate, *in* D-00322
▷Ethyl diethylphosphinite, *in* D-00333
Ethyl diethylphosphinodiselenoate, *in* D-00324
Ethyl diethylphosphinodithioate, *in* D-00325
O-Ethyl diethylphosphinoselenoate, *in* D-00327
S-Ethyl diethylphosphinoselenothioate, *in* D-00329
Se-Ethyl diethylphosphinoselenothioate, *in* D-00329
O-Ethyl diethylphosphinothioate, *in* D-00330
S-Ethyl diethylphosphinothioate, *in* D-00330
Ethyl diethylphosphinothioite, *in* D-00332
Ethyl *N*-(diethylphosphinyl)carbamate, *in* D-00323
Ethyl *N,N*-diethylphosphonamidate, *in* D-00337
Ethyl *N,P*-diethylphosphonamidate, *in* D-00336
Ethyl diethylphosphoramidochloridate, *in* D-00346
Ethyl diethylphosphoramidochloridite, E-00060
O-Ethyl diethylphosphoramidochloridoselenoate, *in* D-00347
O-Ethyl diethylphosphoramidochloridothioate, *in* D-00348
Ethyl diethylphosphoramidofluoridate, *in* D-00351
Ethyl diethylphosphoramidofluoridodithioate, *in* D-00352
Ethyl *N,N'*-diethylphosphorodiamidate, *in* D-00382
Ethyl *N,N*-diethyl-*P*-propadienylphosphonamidate, *in* D-00399
Ethyl *N,N*-diethyl-*P*-1-propynylphosphonamidate, *in* D-00403
▷Ethyl difluorophosphate, *see* E-00173
Ethyldifluorophosphine, *in* E-00153
Ethyl difluorophosphite, *see* E-00174
O-Ethyl difluorothiophosphate, *see* E-00175
S-Ethyl difluorothiophosphate, *see* E-00176
Ethyl di-2-furanylphosphinate, *in* D-00421
O-Ethyl di-2-furanylphosphinothioate, *in* D-00422
Ethyl dihexylphosphinate, *in* D-00434
Ethyl (dihydorxyphosphinyl)formate, *in* P-00387
Ethyl dihydrogen phosphate, *see* M-00472
S-Ethyl dihydrogen thiophosphate, *see* E-00183
1-Ethyl-2,5-dihydro-3-methyl-1-phenyl-1*H*-phospholium iodide, *in* D-00528
4'-Ethyl-1,3-dihydro-5-phenyl-3,3-bis(trifluoromethyl)spiro[2*H*-1,4,2-diazaphosphole-2,1'-[2,6,7]trioxa[1]-phosphabicyclo[2.2.2]octane], E-00061
Ethyl *N*-(4,5-dihydro-2-thiazolyl)-*P*-methylphosphonamidate, *in* D-00587
O-Ethyl *N*-(4,5-dihydro-2-thiazolyl)-*P*-methylphosphonamidothioate, *in* D-00588
Ethyl dihydroxyphosphinylacetate, *in* P-00376
Ethyldiiodophosphine, *in* E-00153
Ethyl (diisopropoxyphosphinothioyl)-thioacetate, *in* D-00632
Ethyl (diisopropoxyphosphinyl)acetate, *in* D-00633
▷*O*-Ethyl *S*-[2-(diisopropylamino)ethyl] methylphosphonothioate, E-00062
Ethyl *N,N*-diisopropyl-*P*-methylphosphonamidate, *in* D-00636
Ethyl diisopropylphosphinate, *in* D-00641
Ethyl diisopropylphosphinite, *in* D-00648
Ethyl diisopropylphosphinodithioate, *in* D-00643
Ethyl [(dimethoxyphosphinothioyl)thio]-acetate, *in* D-00675
▷Ethyl α-[(dimethoxyphosphinothioyl)thio]-benzeneacetate, *see* P-00078
1-Ethyl [(dimethoxyphosphinothioyl)thio]-butanedioate, *in* D-00677
2-Ethyl [(dimethoxyphosphinothioyl)thio]-butanedioate, *in* D-00677
Ethyl (dimethoxyphosphinyl)acetate, *in* D-00678
Ethyl *N*-(dimethoxyphosphinyl)carbamate, *in* D-00680
Ethyl dimethoxyphosphinylformate, *in* D-00672
Ethyl 3-[(dimethoxyphosphinyl)oxy]crotonate, *in* D-00681
Ethyl 2-[(dimethoxyphosphinyl)oxy]-propanoate, *in* D-00682
Ethyl 2-(dimethoxyphosphinyl)propanoate, *in* D-00684
▷Ethyl (dimethoxyphosphinyl)thioacetate, *in* D-00687
Ethyl (dimethylamido)fluorodithiophosphate, *in* D-00352
Ethyl dimethylethylene phosphite, *see* E-00031
Ethyl(1,1-dimethylethyl)phosphinic acid, *see* B-00560
Ethyl (1,1-dimethylethyl) phosphite, *see* B-00561
Ethyl (1,1-dimethylethyl) phosphonate, *see* B-00561
Ethyl (1,1-dimethyl-3-oxobutyl)-ethylphosphinate, *in* D-00782
Ethyl (2,5-dimethylphenyl)phosphinate, *in* D-00805
P-Ethyl-*N,N*-dimethyl-*P*-phenylphosphinothioic amide, *in* E-00105
P-Ethyl-*N,N*-dimethyl-*P*-phenylphosphinothioic amide, *in* E-00105
Ethyl dimethylphosphinate, *in* D-00829
Ethyl dimethylphosphinodithioate, *in* D-00834
Ethyl *N,P*-dimethylphosphonamidate, *in* D-00849
P-Ethyl-*N,N*-dimethylphosphonamidodithioic acid, E-00063
O-Ethyl *N,P*-dimethylphosphonamidothioate, *in* D-00850
Ethyl dimethylphosphoramido chloridate, *in* D-00860
Ethyl dimethylphosphoramidochloridite, *in* D-00862
O-Ethyl dimethylphosphoramidochloridothioate, *in* D-00861
▷Ethyl dimethylphosphoramidocyanidate, *see* T-00001
O-Ethyl dimethylphosphoramidocyanidothioate, *in* D-00864
▷Ethyl dimethylphosphoramidofluoridate, E-00064
Ethyl dimethylphosphoramidofluoridodithioate, *in* D-00867
O-Ethyl dimethylphosphoramidofluoridothioate, *in* D-00868
Ethyl *N,N'*-dimethylphosphorodiamidate, *in* D-00895
O-Ethyl *O,S*-dimethyl phosphorothioate, E-00065
O-Ethyl *O,S*-dimethyl thiophosphate, *see* E-00065
Ethyl *N,N'*-dimethyl-*N,N'*-trimethylene phosphorodiamidite, *in* H-00022
Ethyl dimethylvinylene phosphate, *in* H-00123
Ethyl di-4-morpholinylphosphinate, *in* D-00935
Ethyl di-4-morpholinylphosphinite, *in* D-00936
Ethyl di-1-naphthylphosphinate, *in* D-00941
Ethyl 2-[(diphenoxyphosphinyl)amino]-benzoate, *in* C-00004
Ethyl 4-[(diphenoxyphosphinyl)amino]-benzoate, *in* C-00006
Ethyl 2-[(diphenoxyphosphinyl)amino]-3-phenylpropanoate, *in* P-00400
Ethyl *N*-(diphenoxyphosphinyl)carbamate, *in* D-00989
1-Ethyl-2,3-diphenyl-1,3-azaphosphacyclopentane, *in* D-00995
Ethyl *P,P*-diphenyl-*N*-(diphenylphosphinyl)-phosphinimidate, *in* D-01007
Ethyl(diphenylmethyl)phenylphosphinate, *in* D-01023
Ethyl *P,P*-diphenyl-*N*-phenylsulfonylphosphinimidate, *in* B-00013
Ethyl diphenylphosphinate, *in* D-01039
Ethyldiphenylphosphine, E-00066
P-Ethyl-*N,P*-diphenylphosphinic amide, *in* E-00102
Ethyl diphenylphosphinite, *in* D-01089
Ethyl (diphenylphosphino)acetate, *in* D-01045
Ethyl diphenylphosphinodithioate, *in* D-01065
O-Ethyl diphenylphosphinoselenoate, *in* D-01078
S-Ethyl diphenylphosphinoselenothioate, *in* D-01080
O-Ethyl diphenylphosphinothioate, *in* D-01082
S-Ethyl diphenylphosphinothioate, *in* D-01082
Ethyl diphenylphosphinothioite, *in* D-01085
Ethyl (diphenylphosphinothioyl)acetate, *in* D-01086
Ethyl 4-(diphenylphosphinyl)-3-oxobutanoate, *in* D-01102
P-Ethyl-*N,N*-diphenylphosphonamidic acid, E-00067
P-Ethyl-*N,N'*-diphenylphosphonic diamide, *in* E-00130
P-Ethyl-*N,N'*-diphenylphosphonoselenoic diamide, *in* E-00146
P-Ethyl-*N,N'*-diphenylphosphonothioic diamide, *in* E-00149
Ethyl *N,N'*-diphenylphosphorodiamidate, *in* D-01125
▷*O*-Ethyl *S,S*-diphenyl phosphorodithioate, *see* E-00002
Ethyl di-1-piperidinylphosphinate, *in* D-01190
Ethyl di-2-propenylphosphinate, *in* D-01193
Ethyl (dipropoxyphosphinyl)acetate, *in* D-01199
Ethyl dipropylphosphinate, *in* D-01205
Ethyl *N,N'*-dipropylphosphorodiamidate, *in* D-01223
▷*O*-Ethyl *S,S*-dipropyl phosphorodithioate, *see* E-00023
Ethyl divinylphosphinate, *in* D-01258
O-Ethyl divinylphosphinothioate, *in* D-01259
Ethylenebis[bis[2-(diphenylphosphino)-ethyl]phosphine], *see* B-00240
Ethylenebis[dibutylphosphine], *see* B-00158
Ethylenebis(methylphosphinic acid), *see* E-00013
Ethylenebis[phenylphosphine], *see* B-00390
Ethylenebis(phenylphosphinic acid), *see* E-00018
▷Ethylenebisphosphine, *see* D-01176
1,2-Ethylenebis[triphenylphosphonium](2+), *see* E-00017
Ethylene *tert*-butylphosphoramidate, *in* B-00539
Ethylene *tert*-butylphosphoramidite, *see* B-00539
Ethylene chlorodithiophosphite, *see* C-00071
Ethylene chlorophosphate, *see* C-00064
Ethylene chlorophosphite, *see* C-00063
O,O-Ethylene chlorothiophosphate, *see* C-00065
Ethylene dibenzyl *P,P,P',P'*-tetraethylphosphonodithioimidocarbonate, *see* Z-00001
S,S-Ethylene diethylphosphoramidodithioite, *in* D-01243
S,S-Ethylene diethylphosphoramidodithioite, *in* D-01245
Ethylene dimethylphosphoramidate, *in* D-00695
Ethylene dimethylphosphoramidite, *see* D-00695
S,S-Ethylene *N,N*-dimethyl phosphoramidodithioate, *in* D-01244

Ethylene dimethylphosphoramidoselenoate, *in* D-00695
Ethylene dimethylphosphoramidothioate, *in* D-00695
Ethylenedioxyphosphinic acid, *see* H-00136
Ethylenedioxytriphenylphosphorane, *see* D-00611
Ethylene-1,1-diphosphonic acid, *see* E-00020
1,2-Ethylenediphosphonic acid, *see* E-00006
1,2-Ethylenediphosphonous acid, *see* E-00008
Ethylene fluorophosphite, *see* F-00017
Ethylene glycol cyclic *P,P:P′,P′*-pyrophosphate, *in* O-00091
▷Ethylene glycol cyclic *P,P:P′,P′*-pyrophosphite, *see* O-00091
▷Ethylene glycol hydrogen phosphite, *see* H-00133
Ethylene hydrogen phosphate, *see* H-00134
O,O-Ethylene hydrogen thiophosphate, *see* H-00135
Ethylene isobutyl phosphite, *see* M-00368
Ethylene isopropyl phosphite, *see* M-00152
O,S-Ethylene isopropylphosphonodithioate, *in* I-00042
O,S-Ethylene isopropylphosphonothioate, *see* I-00042
O,S-Ethylene isopropylphosphonothioate, *in* I-00042
Ethylene methyl phosphate, *see* M-00037
▷Ethylene methyl phosphite, *see* M-00036
Ethylene methylphosphonate, *in* M-00126
Ethylene methylphosphonite, *see* M-00126
S,S-Ethylene methylphosphonodithioate, *in* M-00139
S,S-Ethylene methylphosphonodithioite, *see* M-00139
S,S-Ethylene methylphosphonoselenodithioate, *in* M-00139
Ethylene methylphosphonothioate, *in* M-00126
Ethylene methylphosphonotrithioate, *in* M-00139
S,S-Ethylene *O*-methyl phosphorodithioate, *in* D-01244
O,S-Ethylene *S*-methyl phosphorodithioite, *in* O-00056
S,S-Ethylene-*O*-methylphosphorodithioite, *in* D-01243
O,O-Ethylene *O*-methyl phosphoroselenoate, *in* M-00036
O,S-Ethylene *O*-methyl phosphorothioite, *in* O-00056
Ethylene methyl phosphorotrithioite, *in* D-01243
Ethylene phenyl phosphate, *in* P-00067
Ethylene phenyl phosphite, *see* P-00067
Ethylene phenyl phosphite ozonide, *see* P-00181
Ethylene phenylphosphonate, *in* P-00115
Ethylene phenylphosphonite, *see* P-00115
S,S-Ethylene phenylphosphonodithioate, *in* P-00121
S,S-Ethylene phenylphosphonodithioite, *see* P-00121
O,O-Ethylene phenylphosphonothioate, *in* P-00115
O,S-Ethylene phenylphosphonothioite, *in* O-00056
Ethylene phenylphosphonotrithioate, *in* P-00121
Ethylene phenylphosphoramidate, *in* P-00083
Ethylene phenylphosphoramidite, *see* P-00083
Ethylene phenylphosphoramidothioate, *in* P-00083
O,O-Ethylene *O*-phenyl phosphorothioate, *in* P-00067
O,S-Ethylene *O*-phenyl phosphorothioate, *in* P-00072
O,S-Ethylene *O*-phenyl phosphorothioite, *see* P-00072
S,S-Ethylene *O*-phenyl phosphorotrithioate, *in* D-01245
O,O-Ethylene *O*-phenyl thiophosphate, *in* P-00067
▷Ethylene phosphite, *in* D-00970
▷Ethylene phosphonate, *in* D-00970
Ethylene phosphonite, *see* D-00970
O,O-Ethylene phosphonothioate, *in* D-00970
O,S-Ethylene phosphonothioite, *see* O-00056
Ethylene phosphoric acid, *see* H-00134
Ethylene phosphorochloridate, *see* C-00064
Ethylene phosphorochloridite, *see* C-00063
S,S-Ethylene phosphorochloridodithioate, *in* C-00071
Ethylene phosphorochloridodithioite, *see* C-00071
O,S-Ethylene phosphorochloridothioate, *in* O-00056
O,O-Ethylene phosphorochloridothioate, *see* C-00065
Ethylene phosphorochloridotrithioate, *in* C-00071
Ethylene phosphorofluoridite, *see* F-00017
Ethylene phosphorofluoridodithioite, *in* D-01243
O,O-Ethylene phosphorothioate, *see* H-00135
Ethylene pyrophosphate, *in* O-00091
▷Ethylene pyrophosphite, *see* O-00091
O,O-Ethylene thiophosphite, *in* D-00970
O,O-Ethylene thiophosphoric acid, *see* H-00135
Ethylene trimethylsilyl phosphate, *in* T-00554
Ethylene trimethylsilyl phosphite, *see* T-00554
Ethyl *N,N′*-1,2-ethanediyl-*N,N′*-dimethylphosphorodiamidate, *in* E-00030
Ethyl *N,N′*-1,2-ethanediyl-*N,N′*-dimethylphosphorodiamidite, *see* E-00030
Ethyl ethenylphenylphosphinate, *in* P-00337
O-Ethyl ethenylphosphonochloridothioate, *in* V-00006
Ethyl [2-(ethoxycarbonyl)ethyl]-methylphosphinate, *in* H-00161
Ethyl [(ethoxycarbonyl)-(ethylthiothiocarbonyl)methylene]-triphenylphosphorane, *in* E-00194
▷Ethyl 6-ethoxy-2-methyl-7-oxa-5-thia-2-aza-9-phosphanonanoate 6-sulfide, *see* M-00002
Ethyl 3-(ethoxymethylphosphinyl)propanoate, *in* H-00161
O-Ethyl *S*-(ethoxyphenylphosphinothioyl) phenylphosphonodithioate, *in* P-00237
Ethyl 3-(ethoxyphenylphosphinyl)propanoate, *in* H-00178
Ethyl *P*-ethyl-*N,N*-diphenylphosphonamidate, *in* E-00067
S-Ethyl *O,O*-ethylene dithiophosphate, *in* E-00191
O-Ethyl ethylene dithiophosphite, *see* E-00038
Ethyl ethylene phosphate, *see* E-00035
▷Ethyl ethylene phosphite, *see* E-00034
S-Ethyl *O,O*-ethylene phosphorodithioate, *in* E-00191
O-Ethyl ethylene phosphorodithioite, *see* E-00038
O-Ethyl *O,O*-ethylene phosphorothioate, *see* E-00036
▷*S*-Ethyl ethylene phosphorothioite, *see* E-00191
Ethyl ethylene thiophosphate, *see* E-00036
▷*S*-Ethyl ethylene thiophosphite, *see* E-00191
Ethyl ethylmethylphosphinate, *in* E-00085
Ethyl ethylmethylphosphinite, *in* E-00087
Ethyl ethylmethylphosphinodithioate, *in* E-00086
Ethyl *N*-ethyl-*P*-methylphosphonamidate, *in* E-00088
Ethyl ethylphenylphosphinate, *in* E-00101
Ethyl ethylphenylphosphinite, *in* E-00108
Ethyl ethylphenylphosphinodithioate, *in* E-00103
O-Ethyl ethylphenylphosphinoselenoate, *in* E-00104
S-Ethyl ethylphenylphosphinothioate, *in* E-00105
Ethyl ethylphenylphosphinothioite, *in* E-00107
S-Ethyl ethylphenylphosphinothioselenoate, *in* E-00106
Ethyl *P*-ethyl-*N*-phenylphosphonamidate, *in* E-00110
Ethyl ethylphosphinate, E-00068
O-Ethyl ethylphosphinothioate, *in* E-00125
Ethyl *P*-ethylphosphonamidate, *in* E-00126
S-Ethyl *P*-ethylphosphonamidothioate, *in* E-00128
Ethyl ethylphosphonisocyanatidate, *in* E-00133
Ethyl ethylphosphonobromidothioite, *in* E-00134
Ethyl ethylphosphonochloridate, *in* E-00135
Ethyl ethylphosphonochloridite, *in* E-00139
Ethyl ethylphosphonochloridodithioate, *in* E-00136
O-Ethyl ethylphosphonochloridothioate, E-00069
S-Ethyl ethylphosphonochloridothioate, *in* E-00137
Ethyl ethylphosphonochloridothioite, *in* E-00138
O-Ethyl ethylphosphonodithioate, E-00070
O-Ethyl ethylphosphonodithioic anhydrosulfide, *in* D-00405
Ethyl ethylphosphonofluoridate, *in* E-00142
O-Ethyl ethylphosphonofluoridothioate, E-00071
O-Ethyl ethylphosphonoselenoate, E-00072
O-Ethyl ethylphosphonoselenothioate, E-00073
O-Ethyl ethylphosphonothioate, E-00074
Ethyl ethylpropylphosphinodithioate, *in* E-00185
Ethyl 3-(ethylthio)-3-thioxo-2-(triphenylphosphoranylidene)propanoate, *in* E-00194
Ethyl fluoro(dimethylamido)dithiophosphate, *in* D-00867
Ethyl fluoro(dimethylamido)dithiophosphate, *in* D-00867
O-Ethyl fluorodimethylamidothiophosphate, *in* D-00868
5-Ethyl-6,6*a*,7,8,9,9*a*-hexahydro-5-phenyl-1*H*-benzo[*h*]cyclopenta[*c*]phosphinolinium perchlorate, *in* H-00043
2-Ethylhexyl dihydrogen phosphate, *see* M-00470
▷2-Ethylhexyl diphenyl phosphate, E-00075
2-Ethylhexyl phosphoric acid, *see* M-00470
Ethyl hydrogen (α-aminobenzyl)phosphonate, *in* A-00105
Ethyl hydrogen (aminocarbonyl)phosphonate, *see* M-00469
Ethyl hydrogen 2-aminoethyl phosphate, *in* A-00075
Ethyl hydrogen (1-amino-1-methylethyl)-phosphonate, *in* A-00087
Ethyl hydrogen (1-amino-1-methylpropyl)-phosphonate, *in* A-00093
Ethyl hydrogen (2-aminophenyl)phosphonate, *in* A-00109
Ethyl hydrogen (4-aminophenyl)phosphonate, *in* A-00111
Ethyl hydrogen benzylphosphonate, *in* B-00060
O-Ethyl hydrogen benzylphosphonothioate, *in* B-00067
Ethyl hydrogen bis(2-chloroethyl)-phosphoramidate, *in* B-00121
Ethyl hydrogen *tert*-butylphosphonate, *in* B-00595
Ethyl hydrogen carbamoylphosphonate, *see* M-00469
Ethyl hydrogen chloromethylphosphonate, *in* C-00106
Ethyl hydrogen cyclohexylphosphonate, *in* C-00256
O-Ethyl hydrogen cyclohexylphosphonodithioate, *in* C-00261
O-Ethyl hydrogen cyclohexylphosphonothioate, *in* C-00264
Ethyl hydrogen [(1,3-dihydro-1,3-diazo-2*H*-isoindol-2-yl)methyl]phosphonate, *in* P-00426
Ethyl hydrogen ethylphosphonate, *in* E-00129
O-Ethyl hydrogen ethylphosphonodithioate, *see* E-00070
O-Ethyl hydrogen ethylphosphonoselenoate, *see* E-00072
O-Ethyl hydrogen hexylphosphonodithioate, *in* H-00099
Ethyl hydrogen (4-hydroxyphenyl)phosphonate, *in* H-00181
O-Ethyl hydrogen isopropylphosphonothioate, *see* E-00082

Ethyl hydrogen methylphosphonate, *see* E-00089
O-Ethyl hydrogen methylphosphonoselenoate, *see* E-00092
O-Ethyl hydrogen methylphosphonothioate, *see* E-00093
O-Ethyl hydrogen (1-methylpropyl)-phosphonodithioate, *in* M-00380
Ethyl hydrogen (2-nitrophenyl)phosphonate, *in* N-00055
Ethyl hydrogen (4-nitrophenyl)phosphonate, *in* N-00057
O-Ethyl hydrogen pentylphosphonodithioate, *in* P-00051
Ethyl hydrogen phenylphosphonate, *see* M-00471
O-Ethyl hydrogen phosphorothioate, *see* E-00182
O-Ethyl hydrogen phosphorotriselenoate, *see* E-00184
O-Ethyl hydrogen propylphosphonodithioate, *in* P-00478
O-Ethyl hydrogen *S*-propyl phosphorodithioate, *see* E-00186
Ethyl hydrogen[(2-quinolinyl)methyl]-phosphonate, *in* Q-00004
Ethyl (α-hydroxybenzyl)phenylphosphinate, *in* H-00173
▷Ethyl 2-hydroxy-5-methylpyrazolo[1,5-*c*]-pyrimidine-6-carboxylate, *O*-ester with *O,O*-diethyl phosphorothioate, *see* P-00502
▷Ethyl hypophosphite, *see* E-00122
Ethylidenetriphenylphosphorane, E-00076
Ethylidynephosphine, E-00077
Ethylimidobisdiphenylphosphine, *see* D-01067
Ethylimidobisphosphoryl fluoride, *in* E-00078
Ethylimidodiphosphoric acid, E-00078
Ethylimidodiphosphoric acid, *see* E-00078
Ethylimidodi(phosphorous acid), E-00079
Ethylimidodiphosphorous tetrachloride, E-00080
Ethyl *N*-[imino(diphenoxyphosphinylamino)-methyl]glycinate, *in* P-00369
Ethyliminodiphosphoric acid, *see* E-00078
Ethyliminodi(phosphorous acid), *see* E-00079
Ethyliminodiphosphorous tetrachloride, *see* E-00080
[(Ethylimino)methylethenyl]phosphonic acid, E-00081
Ethyl (iodomethyl)phenylphosphinate, *in* I-00012
Ethyl isobutylphosphinate, *in* M-00376
▷*O*-Ethyl *O*-2-isopropoxycarbonylphenyl isopropylphosphoramidothioate, *see* I-00025
Ethyl *N*-isopropyl-*P*-methylphosphonamidate, *in* I-00036
O-Ethyl *N*-isopropyl-*P*-methylphosphonamidothioate, *in* I-00037
S-Ethyl *O*-isopropyl methylphosphonothioate, *in* I-00040
S-Ethyl *O*-isopropyl methylphosphonothioate, *in* I-00041
S-Ethyl isopropylphosphonochloridothioate, *in* I-00053
O-Ethyl isopropylphosphonothioate, E-00082
Ethyl [(methoxycarbonyl)methyl]-phosphonamidate, *in* M-00028
Ethyl (methoxymethyl)phenylphosphinate, *in* M-00052
Ethyl (4-methoxyphenyl)phosphinate, *in* M-00069
1-Ethyl-2-methyl-1,2,3-azaphospharsolidine, E-00083
O-Ethyl *S*-methyl *N*-[(4-chlorophenyl)-sulfonyl]-*P*-ethylphosphonamidothioate, *in* C-00165
O-Ethyl *S*-methyl *N*-[(4-chlorophenyl)-sulfonyl]-*P*-phenylphosphonimidothioate, *in* C-00166
O-Ethyl *Se*-methyl chloroselenophosphite, *see* E-00094
Ethylmethyldiphenylphosphonium iodide, *in* E-00066
O-Ethyl *S*-methyl dithiophosphate, *see* E-00096
S-Ethyl *O*-methyl dithiophosphate, *see* E-00097
O-Ethyl *O*-methyl dithiophosphoric acid, *see* E-00095
O-Ethyl *S*-methyl dithiophosphoric acid, *see* E-00096
S-Ethyl *O*-methyl dithiophosphoric acid, *see* E-00097
O-Ethyl *Se*-methyl ethylphosphonoselenoate, *in* E-00072
O-Ethyl *O*-methyl ethylphosphonothioate, *in* M-00155
O-Ethyl *S*-methyl ethylphosphonothioate, *in* E-00074
O-Ethyl (1-methylethyl)phosphonothioate, *see* E-00082
O-Ethyl *O*-methyl hydrogen dithiophosphate, *see* E-00095
S-Ethyl *O*-methyl hydrogen dithiophosphate, *see* E-00097
O-Ethyl *S*-methyl hydrogen phosphorodithioate, *see* E-00096
O-Ethyl *O*-methyl hydrogen phosphorothioate, *see* E-00098
O-Ethyl *O*-methyl isopropylphosphonothioate, *in* M-00166
O-Ethyl *Se*-methyl methylphosphonoselenoate, *in* E-00092
▷Ethyl 3-methyl-4-(methylthio)phenyl 1-methylethylphosphoramidate, *see* F-00002
O-Ethyl *O*-(4-methyl-2-nitrophenyl) isopropylphosphoramidothioate, *see* A-00131
S-Ethyl *O*-(4-methyl-2-nitrophenyl) isopropylphosphoramidothioate, *in* M-00180
O-Ethyl *O*-(4-methyl-2-nitrophenyl) (1-methylethyl)phosphoramidothioate, *see* A-00131
▷*O*-Ethyl *O*-(5-methyl-2-nitrophenyl) 1-methylpropylphosphoramidothioate, *see* B-00513
Ethyl *P*-(4-methylphenyl)-*N*-phenylphosphonamidate, *in* M-00214
Ethyl methylphenylphosphinate, *in* M-00219
Ethylmethylphenylphosphine, E-00084
O-Ethyl methylphenylphosphinothioate, *in* M-00225
S-Ethyl methylphenylphosphinothioate, *in* M-00225
Ethyl *P*-methyl-*N*-phenylphosphonamidate, *in* M-00231
S-Ethyl *P*-methyl-*N*-phenylphosphonamidothioate, *in* M-00233
O-Ethyl *O*-methyl phenylphosphonothioate, *in* M-00243
▷*O*-Ethyl *S*-methyl phenylphosphonothioate, *in* M-00244
S-Ethyl *O*-methyl phenylphosphonothioate, *in* P-00244
Ethyl methylphosphinate, *in* M-00278
Ethylmethylphosphinic acid, E-00085
Ethylmethylphosphinodithioic acid, E-00086
O-Ethyl methylphosphinothioate, *in* M-00279
Ethylmethylphosphinous acid, E-00087
Ethyl *P*-methylphosphonamidate, *in* M-00283
N-Ethyl-*P*-methylphosphonamidic acid, E-00088
O-Ethyl *P*-methylphosphonamidothioate, *in* M-00284
S-Ethyl *P*-methylphosphonamidothioate, *in* M-00284
Ethyl methylphosphonate, E-00089
Ethyl methylphosphonazidate, *in* M-00286
Ethyl methylphosphonisocyanatidate, *in* M-00292
Ethyl methylphosphonisothiocyanatidate, *in* M-00294
O-Ethyl methylphosphonobromidothioate, *in* M-00295
Ethyl methylphosphonochloridate, *in* M-00296
O-Ethyl methylphosphonochloridothioate, E-00090
S-Ethyl methylphosphonochloridothioate, *in* M-00298
Ethyl methylphosphonochloridothioite, *in* M-00299
O-Ethyl methylphosphonodithioate, E-00091
Ethyl methylphosphonofluoridate, *in* M-00303
Ethyl methylphosphonofluoridite, *in* M-00307
O-Ethyl methylphosphonofluoridothioate, *in* M-00305
S-Ethyl methylphosphonofluoridothioate, *in* M-00305
Ethyl *P*-methylphosphonohydrazidate, *in* M-00308
O-Ethyl *P*-methylphosphonohydrazidothioate, *in* M-00309
O-Ethyl methylphosphonoselenoate, E-00092
O-Ethyl methylphosphonothioate, E-00093
O-Ethyl methylphosphonothioic monoanhydride, *in* D-00925
S-Ethyl methylphosphonothioic monoanhydride, *in* D-00925
O-Ethyl *Se*-methyl phosphorochloridoselenoite, E-00094
O-Ethyl *O*-methyl phosphorodithioate, E-00095
O-Ethyl *S*-methyl phosphorodithioate, E-00096
S-Ethyl *O*-methyl phosphorodithioate, E-00097
O-Ethyl *O*-methyl phosphorothioate, E-00098
Ethyl (2-methylpropyl)phosphinate, *in* M-00376
▷*O*-Ethyl *O*-[(4-methylthio)phenyl] *S*-propyl phosphorodithioate, *see* S-00027
O-Ethyl *O*-methyl thiophosphoric acid, *see* E-00098
▷Ethyl 4-methylthio-*m*-tolyl isopropylphosphoramidate, *see* F-00002
Ethyl methyl(trichloromethyl)phosphinate, *in* M-00419
Ethyl 4-morpholinylphosphinate, *in* M-00522
Ethyl 4-morpholinylphosphonochloridite, *in* M-00526
Ethyl 1,2-naphthylene phosphate, *in* E-00044
Ethyl 1,8-naphthylene phosphate, *in* E-00045
Ethyl 1,8-naphthylene phosphate, *in* E-00046
Ethyl 1,2-naphthylene phosphite, *see* E-00044
Ethyl 1,8-naphthylene phosphite, *see* E-00045
Ethyl 1,8-naphthylene phosphite, *see* E-00046
O-Ethyl *O,O*-1,2-naphthylene thiophosphate, *in* E-00044
O-Ethyl *O,O*-1,8-naphthylene thiophosphate, *in* E-00045
O-Ethyl *O,O*-1,8-naphthylene thiophosphate, *in* E-00046
Ethyl 2-naphthylphenylphosphinate, *in* N-00018
Ethyl neopentylene phosphate, *in* E-00032
Ethyl neopentylene phosphite, *see* E-00032
S-Ethyl *O,O*-neopentylene phosphorodithioate, *in* E-00190
O-Ethyl *O,O*-neopentylene phosphorothioate, *in* E-00032
S-Ethyl *O,O*-neopentylene phosphorothioate, *in* E-00190
S-Ethyl *O,O*-neopentylene phosphorothioite, *see* E-00190
▷*O*-Ethyl *O*-*p*-nitrophenyl benzenethiophosphonate, *see* E-00099
Ethyl (2-nitrophenyl)methylphosphinate, *in* N-00042
▷Ethyl 4-nitrophenyl phenylphosphonate, *in* M-00471
▷*O*-Ethyl *O*-4-nitrophenyl phenylphosphonothionate, E-00099
▷*O*-Ethyl *O*-(6-nitro-*m*-tolyl) *sec*-butylphosphoramidothioate, *see* B-00513
Ethyl nonylphenylphosphinate, *in* N-00074
N‴-Ethyl-*N*,*N*,*N*′,*N*′,*N*″,*N*″,*N*‴,*N*‴-octamethylimidodiphosphorous tetraamide, *in* E-00079
Ethyl octylphenylphosphinate, *in* O-00040
Ethyl β-oxo-α-(triphenylphosphoranylidene)-benzenepropanoate, *in* O-00083
Ethyl pentamethylenephosphinate, *in* H-00188
Ethyl 1,1,2,3,3-pentamethyltrimethylenephosphinate, *in* H-00168
O-Ethyl 1,1,2,3,3-pentamethyltrimethylenephosphinothioate, *in* H-00169
S-Ethyl 1,1,2,3,4-pentamethyltrimethylenephosphinothioate, *in* H-00169
1-Ethyl-2-phenyl-1,3-azaphospholidine, *in* P-00090
1-Ethyl-3-phenyl-1,3-azaphospholidine, *in* P-00091

Ethyl phenyl 1-aziridinylphosphonite, *in* A-00154
▷*O*-Ethyl *S*-phenyl butylphosphoramidodithioate, *see* P-00341
O-Ethyl *O*-phenyl chlorothiophosphate, *see* E-00117
O-Ethyl *S*-phenyl chlorothiophosphate, *see* E-00118
S-Ethyl *O*-phenyl chlorothiophosphate, *see* E-00119
S-Ethyl *O*-phenyl cyclohexylphosphonodithioate, *in* C-00261
O-Ethyl *S*-phenyl dithiophosphate, *see* E-00121
O-Ethyl *O*-phenyl dithiophosphoric acid, *see* E-00120
O-Ethyl *S*-phenyl dithiophosphoric acid, *see* E-00121
Ethyl *o*-phenylene phosphate, *in* E-00024
Ethyl *o*-phenylene phosphite, *see* E-00024
Ethyl *N*,*N*′-*o*-phenylenephosphorodiamidite, *see* E-00026
O-Ethyl *N*,*N*′-*o*-phenylenephosphorodiamidothioate, *in* E-00026
O-Ethyl *O*,*O*-*o*-phenylene thiophosphate, *in* E-00024
Ethyl (2-phenylethenyl)phosphinate, *in* P-00153
▷*O*-Ethyl *S*-phenyl ethylphosphonodithioate, *see* F-00052
O-Ethyl *S*-phenyl ethylphosphonothioate, *in* E-00074
O-Ethyl *O*-phenyl hydrogen dithiophosphate, *see* E-00120
O-Ethyl *S*-(phenylmethyl) phenylphosphonothioate, *see* I-00009
Ethyl phenyl methylphosphonate, E-00100
O-Ethyl-*S*-phenyl methylphosphoramidodithioate, *in* M-00330
Ethyl phenyl (1-methylpropyl)-phosphonotrithioate, *in* M-00382
Ethyl phenyl(2-phenylethenyl)phosphinate, *in* P-00186
Ethylphenyl(phenylmethyl)phosphine, *see* B-00047
Ethyl phenyl phenylphosphonate, *in* M-00471
Ethyl *P*-phenyl-*N*-phenylsulfonylphosphonamidate, *in* P-00191
▷Ethyl phenylphosphinate, *in* P-00210
Ethylphenylphosphinic acid, E-00101
P-Ethyl-*P*-phenylphosphinic amide, E-00102
Ethylphenylphosphinodithioic acid, E-00103
Ethylphenylphosphinoselenoic acid, E-00104
O-Ethyl phenylphosphinothioate, *in* P-00213
Ethylphenylphosphinothioic acid, E-00105
Ethylphenylphosphinothioselenoic acid, E-00106
Ethylphenylphosphinothious acid, E-00107
Ethylphenylphosphinous acid, E-00108
Ethyl phenyl phosphite, *see* E-00111
Ethyl *P*-phenylphosphonamidate, *in* P-00218
N-Ethyl-*P*-phenylphosphonamidic acid, E-00109
P-Ethyl-*N*-phenylphosphonamidic acid, E-00110
Ethyl phenyl phosphonate, E-00111
O-Ethyl phenylphosphonisothiocyanatidothioate, *in* P-00227
Ethyl phenylphosphonochloridate, *in* P-00229
Ethyl phenylphosphonochloridite, *in* P-00234
Ethyl phenylphosphonochloridodithioate, *in* P-00230
▷*O*-Ethyl phenylphosphonochloridothioate, E-00112
S-Ethyl phenylphosphonochloridothioate, *in* P-00232
Ethyl phenylphosphonochloridothioite, *in* P-00233
O-Ethyl phenylphosphonodithioate, E-00113
O-Ethyl phenylphosphonodithioic anhydrosulfide, *in* P-00237
Ethyl phenylphosphonofluoridate, *in* P-00239
O-Ethyl phenylphosphonothioate, E-00114
Ethyl phenyl phosphoramidate, E-00115
Ethylphenylphosphoramidic acid, E-00116
Ethyl phenylphosphoramidochloridate, *in* P-00259
Ethyl phenyl phosphorochloridate, *in* P-00269
O-Ethyl *O*-phenyl phosphorochloridothioate, E-00117
O-Ethyl *S*-phenyl phosphorochloridothioate, E-00118
S-Ethyl *O*-phenyl phosphorochloridothioate, E-00119
Ethyl *N*-phenylphosphorodiamidate, *in* P-00273
O-Ethyl *O*-phenyl phosphorodithioate, E-00120
O-Ethyl *S*-phenyl phosphorodithioate, E-00121
Ethyl phenyl(2-propenyl)phosphinate, *in* P-00300
Ethyl phenylstyrylphosphinate, *in* P-00186
O-Ethyl *O*-phenyl thiophosphoryl chloride, *see* E-00117
Ethyl phenyl(trichloromethyl)phosphinate, *in* P-00327
Ethyl phenylvinylphosphinate, *in* P-00337
▷Ethyl phosphinate, E-00122
▷Ethylphosphine, E-00123
Ethylphosphinic acid, E-00124
Ethylphosphinothioic acid, E-00125
P-Ethylphosphonamidic acid, E-00126
P-Ethylphosphonamidic anhydride, *see* D-00291
P-Ethylphosphonamidic fluoride, *in* E-00126
P-Ethylphosphonamidodithioic acid, E-00127
P-Ethylphosphonamidothioic acid, E-00128
Ethylphosphonic acid, E-00129
Ethylphosphonic acid monoanhydride, *see* D-00290
Ethylphosphonic bis(dimethylamide), *in* E-00130
P-Ethylphosphonic diamide, E-00130
Ethylphosphonic dianilide, *in* E-00130
▷Ethylphosphonic dichloride, E-00131
Ethylphosphonic difluoride, E-00132
Ethylphosphonisocyanatidic acid, E-00133
Ethyl phosphonoacetate, *in* P-00376
Ethyl *N*-phosphono-2-aminobenzoate, *see* C-00004
Ethyl *N*-phosphono-3-aminobenzoate, *see* C-00005
Ethyl *N*-phosphono-4-aminobenzoate, *see* C-00006
Ethyl 4-phosphonobenzoate, *in* P-00380
Ethylphosphonobromidothious acid, E-00134
Ethyl 3-phosphonobutanoate, *in* P-00385
Ethylphosphonochloridic acid, E-00135
Ethylphosphonochloridodithioic acid, E-00136
Ethylphosphonochloridothioic acid, E-00137
Ethylphosphonochloridothious acid, E-00138
Ethylphosphonochloridous acid, E-00139
Ethylphosphonodithioic acid, E-00140
Ethylphosphonodithioic acid anhydrosulfide, *see* D-00405
Ethylphosphonodithious acid, E-00141
Ethylphosphonofluoridic acid, E-00142
Ethylphosphonofluoridodithioic acid, E-00143
Ethylphosphonoperoxoic acid, E-00144
Ethyl 3-phosphonopropanoate, *in* P-00396
Ethylphosphonoselenoic acid, E-00145
Ethylphosphonoselenoic bis(diethylamide), *in* E-00146
Ethylphosphonoselenoic bis(dimethylamide), *in* E-00146
P-Ethylphosphonoselenoic diamide, E-00146
Ethylphosphonoselenoic dianilide, *in* E-00146
Ethylphosphonoselenoic dichloride, E-00147
Ethylphosphonothioic acid, E-00148
Ethylphosphonothioic acid trimer trianhydride, *in* T-00463
Ethylphosphonothioic bis(dimethylamide), *in* E-00149
P-Ethylphosphonothioic diamide, E-00149
Ethylphosphonothioic dianilide, *in* E-00149
Ethylphosphonothioic dichloride, E-00150
Ethylphosphonothioic difluoride, E-00151
Ethylphosphonotrithioic acid, E-00152
Ethylphosphonous acid, E-00153
Ethylphosphonous bis(diethylamide), *in* E-00154
Ethylphosphonous diamide, E-00154
▷Ethylphosphonous dichloride, E-00155
Ethyl 4-(phosphonoxy)benzoate, *in* P-00397
Ethyl 2-(phosphonoxy)-2-propenoate, *in* P-00366
Ethylphosphoramidic acid, E-00156
Ethylphosphoramidochloridothioic acid, E-00157
O-Ethyl phosphoramidofluoridothioate, E-00158
Ethylphosphoramidothioic acid, E-00159
Ethylphosphoramidous acid, E-00160
Ethyl phosphoric acid, *see* M-00472
Ethyl phosphoric dianilide, *in* D-01125
Ethylphosphorimidic acid, E-00161
Ethyl phosphorobromidofluoridate, E-00162
Ethyl phosphorochloridisocyanatidate, E-00163
Ethyl phosphorochloridofluoridate, E-00164
Ethyl phosphorodiamidite, E-00165
O-Ethyl phosphorodiamidothioate, *in* E-00182
Ethyl phosphorodibromidite, E-00166
▷Ethyl phosphorodichloridate, E-00167
▷Ethyl phosphorodichloridite, E-00168
Ethyl phosphorodichloridodithioate, E-00169
O-Ethyl phosphorodichloridothioate, E-00170
S-Ethyl phosphorodichloridothioate, E-00171
Ethyl phosphorodichloridothioite, E-00172
▷Ethyl phosphorodifluoridate, E-00173
Ethyl phosphorodifluoridite, E-00174
O-Ethyl phosphorodifluoridothioate, E-00175
S-Ethyl phosphorodifluoridothioate, E-00176
Ethyl phosphorodiisocyanatidate, E-00177
O-Ethyl phosphorodiisocyanatidothioate, E-00178
Ethyl phosphorofluoridate, E-00179
Ethyl phosphorofluoridisocyanatidate, E-00180
O-Ethyl phosphorofluoridoisocyanatidothioate, E-00181
O-Ethyl phosphorothioate, E-00182
S-Ethyl phosphorothioate, E-00183
S-Ethyl phosphorothioate dianilide, *in* D-01129
O-Ethyl phosphorothioic acid, *see* E-00182
O-Ethyl phosphorotriselenoate, E-00184
O-Ethyl phosphorotriselenoic acid, *see* E-00184
▷Ethyl phosphoryl dichloride, *see* E-00167
▷Ethyl phosphoryl difluoride, *see* E-00173
Ethylphosphoryl diisocyanate, *see* E-00177
Ethyl pinacolyl phosphite, *see* E-00052
O-Ethyl *S*-propyl dithiophosphate, *see* E-00186
Ethyl propylene phosphate, *in* E-00040
Ethyl propylene phosphite, *see* E-00040
O-Ethyl *O*,*O*-propylene phosphorothioate, *in* E-00040
O-Ethyl *O*,*O*-propylene thiophosphate, *in* E-00040
O-Ethyl *S*-propyl methylphosphoramidodithioate, *in* M-00330
Ethylpropylphosphinodithioic acid, E-00185
O-Ethyl *S*-propyl phosphorodithioate, E-00186
Ethyl propylphostonate, *in* H-00166
▷*O*-Ethyl *O*′-quinolin-8-yl phenylphosphonothionate, *see* Q-00006
O-Ethyl saligenin cyclic thiophosphate, *in* H-00110
S-Ethyl saligenin cyclic thiophosphate, *in* H-00110
4-Ethylspiro[2,8,9-trioxa-1-phosphatricyclo[3.3.1.$1^{3,7}$]decane-1,4′-trioxaphosphetane], E-00187
Ethyl styrylphosphinate, *in* P-00153
▷*S*-[2-(Ethylsulfinyl)ethyl] *O*,*O*-dimethyl phosphorothioate, *see* O-00098
▷*S*-2-Ethylsulfinylethyl *O*,*O*-dimethyl phosphorothioate, *see* O-00098
▷*S*-[(Ethylsulfinyl)methyl] *O*,*O*-bis(1-methylethyl) phosphorodithioate, *see* A-00142
▷*S*-[2-(Ethylsulfinyl)-1-methylethyl] *O*,*O*-dimethyl phosphorothioate, *see* O-00099
▷*S*-[2-(Ethylsulfonyl)ethyl] *O*,*O*-dimethyl phosphorothioate, *see* D-00021
▷*S*-Ethylsulphinylmethyl *O*,*O*-diisopropyl phosphorodithioate, *see* A-00142
Ethyl tetrabutylphosphorodiamidate, *in* T-00021
Ethyl tetraethylphosphorodiamidate, *in* T-00059
Ethyl tetraethylphosphorodiamidite, *in* T-00066
O-Ethyl tetraethylphosphorodiamidoselenoate, *in* T-00061

Se-Ethyl tetraethylphosphorodiamidoselenoate, *in* T-00061
O-Ethyl tetraethylphosphorodiamidotelluroate, *in* T-00062
O-Ethyl tetraethylphosphorodiamidothioate, *in* T-00063
S-Ethyl tetraethylphosphorodiamidothioate, *in* T-00063
2-Ethyl-1,2,3,4-tetrahydroisophosphinoline, *in* T-00110
2-Ethyl-2,3,4,5-tetrahydro-2-methyl-1*H*-1,3-benzazaphosphepine, E-00188
1-Ethyl-1,2,3,4-tetrahydro-1-phenylbenzo[*h*]phosphinolinium bromide, *in* P-00319
1-Ethyl-1,2,3,4-tetrahydro-1-phenylbenzo[*h*]phosphinolinium bromide, *in* T-00128
1-Ethyl-1,2,3,4-tetrahydrophosphinoline 1-oxide, *in* T-00135
P-Ethyl-*N*,*N*,*N'*,*N'*-tetrakis(trimethylsilyl)phosphonous diamide, *in* E-00154
Ethyl tetramethylene phosphate, *in* H-00132
Ethyl tetramethylenephosphinate, *in* H-00185
Ethyl tetramethylethylene phosphite, *see* E-00052
S-Ethyl *O*,*O*-tetramethylethylene phosphorodithioate, *in* E-00193
O-Ethyl *O*,*O*-tetramethylethylene phosphorothioate, *in* E-00052
S-Ethyl *O*,*O*-tetramethylethylene phosphorothioite, *see* E-00193
P-Ethyl-*N*,*N*,*N'*,*N'*-tetramethylphosphonic diamide, *in* E-00130
P-Ethyl-*N*,*N*,*N'*,*N'*-tetramethylphosphonoselenoic diamide, *in* E-00146
P-Ethyl-*N*,*N*,*N'*,*N'*-tetramethylphosphonothioic diamide, *in* E-00149
Ethyl tetramethylphosphorodiamidate, *in* T-00218
Ethyl tetramethylphosphorodiamidite, *in* T-00227
Ethyl tetramethylphosphorodiamidodithioate, *in* T-00222
Ethyl tetrapropylphosphorodiamidate, *in* T-00291
1-Ethylthio-1,5-dihydro-3-methyl-1*H*-phosphole, *in* D-00530
10-Ethylthio-5,10-dihydrophenophosphazine 10-oxide, *in* D-00502
2-Ethylthio-5,5-dimethyl-1,3-dioxa-2-phosphacyclohexane, *see* E-00190
▷2-Ethylthio-4,5-dimethyl-1,3-dioxa-2-phosphacyclopentane, *see* E-00189
▷2-Ethylthio-4,5-dimethyl-1,3,2-dioxaphospholane, E-00189
2-Ethylthio-5,5-dimethyl-1,3,2-dioxaphosphorinane, E-00190
Ethyl thiodimethylenephosphinate, *in* H-00201
▷2-Ethylthio-1,3,2-dioxaphospholane, E-00191
▷*S*-[2-(Ethylthio)ethyl] *O*,*O*-dimethyl phosphorodithioate, *see* T-00313
▷*S*-2-Ethylthioethyl *O*,*O*-dimethyl phosphorodithioate, *see* T-00313
O-[(Ethylthio)methylphosphinyl] *S*-ethyl methylphosphonothioate, *in* D-00925
[(Ethylthio)methyl]phosphonic acid, E-00192
1-Ethylthio-2,2,3,4,4-pentamethylphosphetane 1-oxide, *in* H-00169
Ethylthiophosphonic diamide, *see* E-00149
Ethylthiophosphonic dichloride, *see* E-00150
Ethylthiophosphonic difluoride, *see* E-00151
O-Ethyl thiophosphoric acid, *see* E-00182
S-Ethyl thiophosphoric acid, *see* E-00183
O-Ethyl thiophosphoryl diamide, *in* E-00182
O-Ethyl thiophosphoryl difluoride, *see* E-00175
O-Ethyl thiophosphoryl diisocyanate, *see* E-00178
2-Ethylthio-4,4,5,5-tetramethyl-1,3,2-dioxaphospholane, E-00193
3-(Ethylthio)-3-thioxo-2-(triphenylphosphoranylidene)propanoic acid, E-00194
Ethyl *P*-trichloromethylphosphonamidate, *in* T-00410
▷*O*-Ethyl *O*-(2,4,5-trichlorophenyl) ethylphosphonothioate, *see* T-00419
Ethyl triethylphosphonamidate, *in* T-00446
Ethyl triethylphosphonamidodithioate, *in* T-00447
O-Ethyl triethylphosphonamidothioate, *in* T-00448
S-Ethyl triethylphosphonamidothioate, *in* T-00448
Ethyltri(ethylthio)phosphonium iodide, *in* T-00461
N-Ethyl-*P*,*P*,*P*-trimethoxyphosphazene, *in* E-00161
N-Ethyl-*P*,*P*,*P*-trimethoxyphosphine imide, *in* E-00161
Ethyl trimethylene phosphate, *in* E-00037
Ethyl trimethylene phosphite, *see* E-00037
O-Ethyl *S*,*S*-trimethylene phosphorodithioite, *in* D-01247
O-Ethyl *O*,*O*-trimethylene phosphorothioate, *in* E-00037
O-Ethyl *O*,*O*-trimethylene thiophosphate, *in* E-00037
Ethyl trimethylphosphonamidate, *in* T-00537
O-Ethyl *O*-trimethylsilyl ethylphosphonoselenoate, *in* E-00072
4-Ethyl-2,6,7-trimethyl-2,6,7-triaza-1-phosphabicyclo[2.2.2]octane, E-00195
▷4-Ethyl-2,6,7-trioxa-1-phosphabicyclo[2.2.2]octane, E-00196
N-Ethyl-*P*,*P*,*P*-triphenoxyphosphazene, *in* E-00161
N-Ethyl-*P*,*P*,*P*-triphenoxyphosphine imide, *in* E-00161
N-Ethyl-*P*,*P*,*P*-triphenylphosphine imide, *in* T-00626
Ethyl *N*,*P*,*P*-triphenylphosphinimidate, *in* T-00631
▷Ethyltriphenylphosphonium(1+), E-00197
Ethyl triphenylphosphoranylideneacetate, E-00198
Ethyl (triphenylphosphoranylidene)carbamate, *in* T-00648
Ethyl [(triphenylphosphoranylidene)-hydrazono]acetate, *in* T-00653
Ethyl α-(triphenylphosphoranylidene)-hydrocinnamate, *in* T-00645
Ethyl 2-triphenylphosphoranylidenepropanoate, E-00199
O-Ethyl triselenophosphate, *see* E-00184
1,2-Ethynediylbis[diethylphosphine], *see* B-00178
1,2-Ethynediylbis[diphenylphosphine], *see* B-00216
P,*P'*-1,2-Ethynediylbis[*N*,*N*,*N'*,*N'*-tetramethylphosphonous diamide], E-00200
Ethynyldiphenylphosphine, E-00201
Ethynylphosphinic acid, E-00202
Ethynylphosphonic acid, E-00203
Ethynylphosphonous acid, E-00204
P-Ethynyl-*N*,*N*,*N'*,*N'*-tetramethylphosphonic diamide, *in* E-00203
▷Eticol, *see* P-00001
Etidronic acid, *see* H-00144
Etkafos, *in* C-00003
▷ETPB, *see* E-00196
▷Etrimfos, E-00205
EU-4584, *see* T-00323
β-Exotoxin, *see* T-00319
F_{420}, *see* C-00201
▷Fac, *see* P-00499
Factor F_{420}, *see* C-00201
▷Famphur, F-00001
Farnesyl dihydrogen phosphate, *see* T-00515
Farnesyl phosphate, *see* T-00515
Farnesyl pyrophosphate, *in* M-00517
▷Fenamiphos, F-00002
▷Fenchlorphos, F-00003
▷Fenitrooxon, F-00004
▷Fenitrothion, F-00005
▷Fensulfothion, F-00006
▷Fenthion, F-00007
Fenthion sulfone, *in* F-00007
Fenthion sulfoxide, *in* F-00007
▷Fernex, *see* P-00434
Ferrocene-1,1'-bis(methyldiphenylphosphonium) diiodide, *in* B-00241
1,1'-Ferrocenediylbis(diphenylphosphine), *see* B-00241
(1,1'-Ferrocenediyl)phenylphosphine, *see* P-00211
Ferrocenyldiphenyl(phenylmethyl)phosphonium chloride, *in* D-01068
Ferrocenyldiphenylphosphine, *see* D-01068
Ferrocenylmethyldiphenylphosphonium iodide, *in* D-01068
(Ferrocenylmethylene)triphenylphosphorane, *in* F-00008
(Ferrocenylmethyl)triphenylphosphonium(1+), F-00008
Flavomycin, *in* M-00441
▷Florel, *see* C-00074
▷Floropryl, *see* D-00665
(9*H*-Fluoren-9-ylideneethenylidene)-triphenylphosphorane, F-00009
9*H*-Fluoren-9-ylidenetriphenylphosphorane, F-00010
2-(9-Fluorenylidene)-1-(triphenylphosphoranylidene)ethene, *see* F-00009
(9*H*-Fluoren-9-yl)phosphonic acid, F-00011
9*H*-Fluoren-9-yltriphenylphosphonium(1+), F-00012
2-Fluoro-1,3,2-benzodioxaphosphole, F-00013
(*p*-Fluorobenzylidene)triphenylphosphorane, *in* F-00032
m-Fluorobenzylphosphonic acid, *see* F-00030
o-Fluorobenzylphosphonic acid, *see* F-00029
p-Fluorobenzylphosphonic acid, *see* F-00031
(*p*-Fluorobenzyl)triphenylphosphonium(1+), *see* F-00032
Fluorobis(trifluoromethyl)phosphine, *in* B-00418
1-Fluoro-2,5-dihydro-1*H*-phosphole 1-oxide, *in* D-00507
2-Fluoro-1,3-dimethyl-1,3,2-diazaphospholidine, F-00014
2-Fluoro-5,5-dimethyl-1,3-dioxa-2-phosphacyclohexane, *see* F-00016
2-Fluoro-4,5-dimethyl-1,3-dioxa-2-phosphacyclopentane, *see* F-00015
2-Fluoro-5,5-dimethyl-1,3,2-dioxaphosphinane, *see* F-00016
2-Fluoro-4,5-dimethyl-1,3,2-dioxaphospholane, F-00015
2-Fluoro-5,5-dimethyl-1,3,2-dioxaphosphorinane, F-00016
▷Fluorodimethylphosphine, *see* D-00848
2-Fluoro-1,3-dioxa-2-phosphacyclohexane, *see* F-00018
2-Fluoro-1,3-dioxa-2-phosphacyclopentane, *see* F-00017
2-Fluoro-1,3-dioxa-2-phosphacyclotridecane, *in* D-00966
2-Fluoro-1,3,2-dioxaphosphinane, *see* F-00018
2-Fluoro-1,3,2-dioxaphospholane, F-00017
2-Fluoro-1,3,2-dioxaphosphonane, *in* D-00971
2-Fluoro-1,3,2-dioxaphosphorinane, F-00018
P-Fluoro-*P*,*P*-diphenylphosphazene, *see* D-01044
Fluorodiphenylphosphine, *see* D-01092
P-Fluoro-*P*,*P*-diphenylphosphine imide, *see* D-01044
P-Fluoro-*P*,*P*-diphenyl-*N*-trimethylsilylphosphazene, *in* D-01157
P-Fluoro-*P*,*P*-diphenyl-*N*-trimethylsilylphosphine imide, *in* D-01157
2-Fluoro-1,3,2-dithiaphospholane, *in* D-01243
O-(2-Fluoroethyl) dichlorothiophosphate, *see* F-00019
O-(2-Fluoroethyl) phosphorodichloridothioate, F-00019
O-(2-Fluoroethyl) thiophosphoryl dichloride, *see* F-00019
2-Fluoro-4-methyl-1,3-dioxa-2-phosphacyclohexane, *see* F-00021
2-Fluoro-4-methyl-1,3-dioxa-2-phosphacyclopentane, *see* F-00020
2-Fluoro-4-methyl-1,3,2-dioxaphosphinane, *see* F-00021
2-Fluoro-4-methyl-1,3,2-dioxaphospholane, F-00020
2-Fluoro-4-methyl-1,3,2-dioxaphosphorinane, F-00021

(Fluoromethylene)triphenylphosphorane, *in* F-00025
(Fluoromethyl)phenylphosphinic acid, F-00022
(Fluoromethyl)phosphonic acid, F-00023
(Fluoromethyl)phosphonochloridic acid, F-00024
(Fluoromethyl)triphenylphosponium(1+), F-00025
1-Fluoro-2,2,3,4,4-pentamethylphosphetane 1-oxide, *in* H-00168
(4-Fluorophenyl)diiodophosphine, *in* F-00042
(2-Fluorophenyl)diphenylphosphine, F-00026
(3-Fluorophenyl)diphenylphosphine, F-00027
(4-Fluorophenyl)diphenylphosphine, F-00028
N′-(4-Fluorophenyl)-*P*,*P*-diphenyl-*N*-(triphenylphosphoranylidene)-phosphinimidic amide, *in* D-01164
[(4-Fluorophenyl)methylene]-triphenylphosphorane, *in* F-00032
[(2-Fluorophenyl)methyl]phosphonic acid, F-00029
[(3-Fluorophenyl)methyl]phosphonic acid, F-00030
[(4-Fluorophenyl)methyl]phosphonic acid, F-00031
[(4-Fluorophenyl)methyl]-triphenylphosphonium(1+), F-00032
(3-Fluorophenyl)phosphine, F-00033
(4-Fluorophenyl)phosphine, F-00034
(3-Fluorophenyl)phosphinic acid, F-00035
(2-Fluorophenyl)phosphonic acid, F-00036
(3-Fluorophenyl)phosphonic acid, F-00037
(4-Fluorophenyl)phosphonic acid, F-00038
(4-Fluorophenyl)phosphonic dichloride, F-00039
(2-Fluorophenyl)phosphonous acid, F-00040
(3-Fluorophenyl)phosphonous acid, F-00041
(4-Fluorophenyl)phosphonous acid, F-00042
(3-Fluorophenyl)phosphoramidic acid, F-00043
(4-Fluorophenyl)phosphoramidic acid, F-00044
(3-Fluorophenyl)phosphoramidothioic acid, F-00045
(4-Fluorophenyl)phosphoramidothioic acid, F-00046
P-4-Fluorophenyl-*N*,*N*,*N*′,*N*′-tetraethylphosphonous diamide, *in* F-00042
P-3-Fluorophenyl-*N*,*N*,*N*′,*N*′-tetramethylphosphonous diamide, *in* F-00041
1-Fluorophosphacyclohexane, *see* F-00047
1-Fluorophosphinane, *see* F-00047
1-Fluorophosphorinane, F-00047
2-Fluoro-2,2′-spirobi[1,3,2-benzodioxaphosphole], F-00048
Fluorotetramethylphosphorane, F-00049
3-Fluoro-1,3-thiaphosphetane, F-00050
▷Fluostigmine, *see* D-00665
▷Flurfamide, *see* F-00051
▷Flurofamide, F-00051
FMP 1, *see* P-00401
▷Folex, *see* T-00393
▷Folimat, *see* O-00048
▷Folithion, *see* F-00005
▷Fonofos, F-00052
▷Fonofos oxon, *in* E-00074
▷Fopurine, F-00053
Formaldehyde (triphenylphosphoranylidene)-hydrazone, F-00054
▷Formothion, F-00055
(α-Formylbenzylidene)triphenylphosphorane, *see* T-00643
(α-Formylbenzyl)triphenylphosphonium(1+), F-00056
(1-Formylethylidene)triphenylphosphorane, *see* T-00657
(2-Formylethyl)phosphonic acid, *see* O-00076
[3-(Formylhydroxyamino)-1-propenyl]-phosphonic acid, *see* A-00137
[3-(Formylhydroxyamino)propyl]phosphonic acid, *see* F-00062
▷*S*-[2-(Formylmethylamino)-2-oxoethyl] *O*,*O*-dimethyl phosphorodithioate, *see* F-00055
▷*S*-(*N*-Formyl-*N*-methylcarbamoylmethyl) *O*,*O*-dimethyl phosphorodithioate, *see* F-00055
Formylmethylenetriphenylphosphorane, *see* T-00639
Formylmethylidenetriphenylphosphorane, *see* T-00639
(Formylmethyl)phosphonic acid, *see* O-00059
(Formylmethyl)triphenylphosphonium(1+), *see* O-00060
[α-Formyl(phenylmethyl)]-triphenylphosphonium(1+), *see* F-00056
▷Fosamine ammonium, *in* M-00469
Fosazepam, F-00057
▷Foscarnet sodium, *in* P-00387
Fosenazide, *see* D-01094
Fosetyl, *in* M-00473
Fosfabenzid, *see* D-01094
Fosfazinomycin *A*, F-00058
Fosfazinomycin *B*, F-00059
▷Fosfemid, *see* P-00353
Fosfocreatinine, F-00060
▷Fosfomycin, *in* M-00185
Fosfonochlorin, *see* C-00020
▷Fosfosal, *see* P-00391
Fosmenic acid, F-00061
Fosmidomycin, F-00062
▷Fospirate, F-00063
Fostedil, F-00064
▷Fosthietan, F-00065
▷Fostion MM, *see* D-00671
▷Fotrin, F-00066
FR 32863, *see* A-00137
FR 33289, *see* A-00138
FR 900136, *see* A-00137
Fructose 1,6-bis(dihydrogen phosphate), F-00067
Fructose 1-dihydrogen phosphate, F-00068
Fructose 2-dihydrogen phosphate, F-00069
Fructose 6-dihydrogen phosphate, F-00070
1,6-Fructose diphosphoric acid, *see* F-00067
Fructose-6-phosphoric acid, *see* F-00070
Fructosyl phosphate, *see* F-00068
[2,5-Furandiylbis[methylene]]-bis[triphenylphosphonium](2+), *see* B-00448
[3,4-Furandiylbis[methylene]]-bis[triphenylphosphonium](2+), *see* B-00449
[2,5-Furandiylbis[methylene]]-bis[triphenylphosphorane], *in* B-00448
[3,4-Furandiylbis[methylene]]-bis[triphenylphosphorane], *in* B-00449
(2-Furanylhydroxymethyl)phosphonic acid, F-00071
(2-Furanylmethyl)phosphonic acid, F-00072
2-Furanylphosphonic acid, F-00073
3-Furanylphosphonic acid, F-00074
2-Furylphosphonic acid, *see* F-00073
3-Furylphosphonic acid, *see* F-00074
Fyrol CEF, *see* T-00741
Galactosamine 1-phosphate, *see* A-00065
Galactose 1-dihydrogen phosphate, G-00001
Galactose 3-dihydrogen phosphate, G-00002
Galactose 6-dihydrogen phosphate, G-00003
▷Gammaphos, *see* A-00052
▷Gammaphos, *see* E-00021
▷Gardona, *see* T-00038
GDP, *see* G-00099
▷Geofos, *see* F-00065
Geranyl dihydrogen phosphate, *in* D-00776
Geranylgeranyl pyrophosphate, *in* M-00515
Geranyl phosphate, *in* D-00776
Geranyl pyrophosphate, *in* M-00465
Geranyltriphenylphosphonium(1+), *in* D-00777
Gidifen, *see* D-01094
α-D-Glucopyranosyl α-D-glucopyranoside 6-(dihydrogen phosphate), *see* T-00324
Glucosamine 6-(dihydrogen phosphate), *see* A-00069
Glucosaminitol, *see* A-00066
Glucosaminol, *see* A-00066
Glucose 2-dihydrogen phosphate, G-00004
Glucose 3-dihydrogen phosphate, G-00005
Glucose 4-dihydrogen phosphate, G-00006
Glucose 6-dihydrogen phosphate, G-00007
Glufosinate, *see* P-00359
Glyceraldehyde 3-dihydrogen phosphate, *see* H-00187
Glyceraldehyde-3-phosphate, *see* H-00187
Glycerol bicyclic phosphite, *see* T-00587
Glycerol 1,2-didodecanoate 3-phosphocholine, G-00008
Glycerol 1,2-didodecanoate 3-phosphoethanolamine, G-00009
Glycerol 1,2-diheptadecanoate 3-phosphocholine, G-00010
Glycerol 1,2-dihexadecanoate 3-phosphate, G-00011
Glycerol 1,2-dihexadecanoate 3-phosphocholine, G-00012
Glycerol 1,2-dihexadecanoate 3-phosphoethanolamine, G-00013
Glycerol 1,2-dihexadecanoate 3-phospho-1′-glycerol, G-00014
Glycerol 1,2-dihexadecanoate 3-phosphoinositol, G-00015
Glycerol 1,2-dihexadecanoate 3-phosphoserine, G-00016
Glycerol 1,2-di-5,8,11,14-icosatetraenoate 3-phosphocholine, G-00017
Glycerol 1,2-di-5,8,11,14-icosatetraenoate 3-phosphoethanolamine, G-00018
Glycerol 1,2-di-11-methyldodecanoate 3-phosphocholine, G-00019
Glycerol 1,2-di-16-methylheptadecanoate 3-phosphocholine, G-00020
Glycerol 1,2-di-15-methylhexadecanoate 3-phosphocholine, G-00021
Glycerol 1,2-di-14-methylpentadecanoate 3-phosphocholine, G-00022
Glycerol 1,2-di-13-methyltetradecanoate 3-phosphocholine, G-00023
Glycerol 1,2-di-12-methyltridecanoate 3-phosphocholine, G-00024
Glycerol 1,2-di-10-methylundecanoate 3-phosphocholine, G-00025
Glycerol 1,2-dinonadecanoate 3-phosphocholine, G-00026
Glycerol 1,2-di-9,12-octadecadienoate 3-phosphate, G-00027
Glycerol 1,2-di-9,12-octadecadienoate 3-phosphocholine, G-00028
Glycerol 1,2-di-9,12-octadecadienoate 3-phosphoethanolamine, G-00029
Glycerol 1,2-di-9,12-octadecadienoate 3-phosphoinositol, G-00030
Glycerol 1,2-di-9,12-octadecadienoate 3-phosphoserine, G-00031
Glycerol 1,2-dioctadecanoate 3-phosphate, G-00032
Glycerol 1,2-dioctadecanoate 3-phospho-2-amino-1-propanol, G-00033
Glycerol 1,2-dioctadecanoate 3-phosphocholine, G-00034
Glycerol 1,2-dioctadecanoate 3-phospho-1′-glycerol, G-00035
Glycerol 1,2-dioctadecanoate 3-phosphoinositol, G-00036
Glycerol 1,2-dioctadecanoate 3-phosphonoethanolamine, G-00037
Glycerol 1,2-dioctadecanoate 3-phosphoserine, G-00038
Glycerol 1,2-di-9,12,15-octadecatrienoate 3-phosphate, G-00039
Glycerol 1,2-di-9,12,15-octadecatrienoate 3-phosphocholine, G-00040
Glycerol 1,2-di-9,12,15-octadecatrienoate 3-phosphoethanolamine, G-00041
Glycerol 1,2-di-9-octadecenoate 3-phosphocholine, G-00042
Glycerol 1,2-di-9-octadecenoate 3-phosphoethanolamine, G-00043
Glycerol 1,2-di-9-octadecenoate 3-phospho-1′-glycerol, G-00044
Glycerol 1,2-di-9-octadecenoate 3-phosphoserine, G-00045
Glycerol 1,2-dipentadecanoate 3-phosphocholine, G-00046
Glycerol 1,2-ditetradecanoate 3-phosphate, G-00047
Glycerol 1,2-ditetradecanoate 3-phosphocholine, G-00048
Glycerol 1,2-ditetradecanoate 3-phosphoethanolamine, G-00049
Glycerol 1,2-ditetradecanoate 3-phosphono-2-amino-1-propanol, G-00050
Glycerol 1,2-ditetradecanoate 3-phosphoserine, G-00051
Glycerol 1,2-ditridecanoate 3-phosphocholine, G-00052

Glycerol 1-dodecanoate 2-octadecanoate 3-phosphocholine, G-00053
Glycerol 1-dodecanoate 2-9-octadecenoate 3-phosphoinositol, G-00054
Glycerol 1-hexadecanoate 2-5,8,11,14-icosatetraenoate 3-phosphocholine, G-00055
Glycerol 1-hexadecanoate 2-9,12-octadecadienoate 3-phosphocholine, G-00056
Glycerol 1-hexadecanoate 2-octadecanoate 3-phosphocholine, G-00057
Glycerol 1-hexadecanoate 2-9,12,15-octadecatrienoate 3-phosphocholine, G-00058
Glycerol 1-hexadecanoate 2-9,12,15-octadecatrienoate 3-phosphoethanolamine, G-00059
Glycerol 1-hexadecanoate 2-9-octadecenoate 3-phosphate, G-00060
Glycerol 1-hexadecanoate 2-9-octadecenoate 3-phosphocholine, G-00061
Glycerol 1-hexadecanoate 2-9-octadecenoate 3-phosphoethanolamine, G-00062
Glycerol-1-hexadecanoate 2-9-octadecenoate 3-phosphoinositol, G-00063
Glycerol 1-hexadecanoate 2-tetradecanoate 3-phosphocholine, G-00064
Glycerol 1-5,8,11,14-icosatetraenoate 2-octadecanoate 3-phosphoethanolamine, G-00065
Glycerol 1-monophosphate, G-00066
Glycerol 2-monophosphate, G-00067
Glycerol 1-9,12-octadecadienoate 2-hexadecanoate 3-phosphocholine, G-00068
Glycerol 1-9,12-octadecadienoate 2-hexadecanoate 3-phosphoethanolamine, G-00069
Glycerol 1-9,12-octadecadienoate 2-octadecanoate 3-phosphocholine, G-00070
Glycerol 1-9,12-octadecadienoate 2-octadecanoate 3-phosphoethanolamine, G-00071
Glycerol 1-9,12-octadecadienoate 2-9-octadecenoate 3-phosphocholine, G-00072
Glycerol 1-9,12-octadecadienoate 2,9-octadecenoate 3-phosphoinositol, G-00073
Glycerol 1-octadecanoate 2-dodecanoate 3-phosphocholine, G-00074
Glycerol 1-octadecanoate 2-hexadecanoate 3-phosphocholine, G-00075
Glycerol 1-octadecanoate 2-5,8,11,14-icosatetraenoate 3-phosphocholine, G-00076
Glycerol 1-octadecanoate 2-5,8,11,14-icosatetraenoate 3-phosphoethanolamine, G-00077
Glycerol 1-octadecanoate 2-9,12-octadecadienoate 3-phosphocholine, G-00078
Glycerol 1-octadecanoate 2-9,12-octadecadienoate 3-phosphoethanolamine, G-00079
Glycerol 1-octadecanoate 2-9,12,15-octadecatrienoate 3-phosphocholine, G-00080
Glycerol 1-octadecanoate 2-9,12,15-octadecatrienoate 3-phosphoethanolamine, G-00081
Glycerol 1-octadecanoate 2-9-octadecenoate 3-phosphocholine, G-00082
Glycerol 1-octadecanoate 2-9-octadecenoate 3-phosphoethanolamine, G-00083
Glycerol 1-octadecanoate 2-9-octadecenoate 3-phosphoinositol, G-00084
Glycerol 1-9,12,15-octadecatrienoate 2-hexadecanoate 3-phosphoethanolamine, G-00085
Glycerol 1-9-octadecenoate 2-hexadecanoate 3-phosphate, G-00086
Glycerol 1-9-octadecenoate 2-hexadecanoate 3-phosphocholine, G-00087
Glycerol 1-9-octadecenoate 2-hexadecanoate 3-phospho-1′-glycerol, G-00088
Glycerol 1-9-octadecenoate 2-9,12-octadecadienoate 3-phosphoethanolamine, G-00089
Glycerol 1-9-octadecenoate 2-9,12-octadecadienoate 3-phosphoserine, G-00090
Glycerol 1-9-octadecenoate 2-octadecanoate 3-phosphocholine, G-00091
Glycerol 1-9-octadecenoate 2-octadecanoate 3-phosphoethanolamine, G-00092
Glycerol 1-9-octadecenoate 2-octadecanoate 3-phosphoserine, G-00093
Glycerol 1-9-octadecenoate 2-tetradecanoate 3-phosphate, G-00094
Glycerol 1-9-octadecenoate 2-tetradecanoate 3-phosphoglycerophosphate, G-00095
Glycerol phosphite, *see* T-00587
Glycerol 1-tetradecanoate 2-hexadecanoate 3-phosphocholine, G-00096
α-Glycerophosphoric acid, *see* G-00066
β-Glycerophosphoric acid, *see* G-00067
Glyceryl phosphite, *see* T-00587
▷Glyphosate, G-00097
▷Glyphosine, G-00098
▷GMP, *see* G-00103
GP_4G, *see* D-00423
▷GTP, *see* G-00101
Guanidinephosphoric acid, *see* P-00370
O^3-(2-Guanidinoethoxyphosphinicoserine), *see* L-00002
Guanosine cyclic 3′,5′-(hydrogen phosphate), *see* C-00234
Guanosine 3′-(dihydrogen phosphate), *see* G-00102
▷Guanosine 5′-(dihydrogen phosphate), *see* G-00103
Guanosine 5′-diphosphate, G-00099
Guanosine diphosphate mannose, G-00100
Guanosine 5′-(pentahydrogen tetraphosphate) 5′,5′-ester of guanosine, *see* D-00423
▷Guanosine 5′-(tetrahydrogen triphosphate), *see* G-00101
Guanosine 5′-(trihydrogen diphosphate), *see* G-00099
Guanosine 5′-(trihydrogen diphosphate)-mono-α-D-mannopyranosyl ester, *see* G-00100
Guanosine 5′-(trihydrogen pyrophosphate), *see* G-00099
Guanosine 5′-(trihydrogen pyrophosphate)-mono-α-D-mannopyranosyl ester, *see* G-00100
▷Guanosine 5′-triphosphate, G-00101
Guanylic acid *b*, *see* G-00102
3′-Guanylic acid, G-00102
▷5′-Guanylic acid, G-00103
▷Gusathion *A*, *see* A-00150
Harden-Young ester, *see* F-00067
HDBPP, *in* O-00066
▷HEDSPA, *in* H-00144
▷Helothion, *see* S-00027
Heptamethylenediphosphonic acid, *see* H-00002
N,N,2,2,3,4,4-Heptamethylphosphetanamine 1-oxide, *in* P-00021
1,1,2,2,3,4,4-Heptamethylphosphetanium bromide, *in* H-00074
1,1,2,2,3,4,4-Heptamethylphosphetanium iodide, *in* H-00074
N,N,N′,N′,N″,N″,N‴-Heptamethylphosphorimidic triamide, H-00001
Heptamethyltetraaminophosphonium iodide, *in* H-00001
1,7-Heptanediphosphonic acid, H-00002
1,7-Heptanediylbis[diphenylphosphine], *see* B-00242
1,7-Heptanediylbis(phosphonic acid), *see* H-00002
1,7-Heptanediylbis(phosphonic dichloride), *in* H-00002
1-Heptanephosphonic acid, *see* H-00006
2,4,6,7,7,8,8-Heptaphenyl-1-phosphabicyclo[4.2.0]octa-2,4-diene, H-00003
▷Heptenophos, H-00004
Heptyl dichlorophosphite, *see* H-00010
Heptyl diheptylphosphinate, *in* D-00425
Heptyl dihydrogen phosphate, *see* M-00474
Heptylidenetriphenylphosphorane, *in* H-00011
Heptylphosphine, H-00005
Heptylphosphonic acid, H-00006
Heptylphosphonic bis(dimethylamide), *in* H-00006
Heptylphosphonothioic acid, H-00007
Heptylphosphonous acid, H-00008
Heptylphosphoramidic acid, H-00009
Heptyl phosphoric acid, *see* M-00474
Heptyl phosphorodichloridite, H-00010
Heptyl tetraethylphosphorodiamidate, *in* T-00059
P-Heptyl-*N,N,N′,N′*-tetramethylphosphonic diamide, *in* H-00006
Heptyltriphenylphosphonium(1+), H-00011
Hexabenzylphosphorous triamide, H-00012
Hexabutylphosphoric triamide, *in* H-00013
Hexabutylphosphorothioic triamide, *in* H-00013
Hexabutylphosphorous triamide, H-00013
Hexadecanoic acid 1-[[[(2-aminoethoxy)-hydroxyphosphinyl]oxy]methyl]-1,2-ethanediyl ester, *see* G-00013
Hexadecanoic acid 1-[[[(2,3-dihydroxypropoxy)hydroxyphosphinyl]-oxy]methyl]-1,2-ethanediyl ester, *see* G-00014
Hexadecanoic acid 1-[(phosphonoxy)methyl]-1,2-ethanediyl ester, *see* G-00011
1-Hexadecanoyl-2-icosa-5,8,11,14-tetraenoylglycero-3-phosphocholine, *see* G-00055
1-Hexadecanoyl-2-octadeca-9,12-dienoylglycero-3-phosphocholine, *see* G-00056
1-Hexadecanoyl-2-octadecanoylglycero-3-phosphocholine, *see* G-00057
1-Hexadecanoyl-2-octadec-9-enoylglycero-3-phosphocholine, *see* G-00061
1-Hexadecanoyl-2-octadec-9-enoylglycero-3-phosphoethanolamine, *see* G-00062
1-Hexadecanoyl-2-tetradecanoylglycero-3-phosphocholine, *see* G-00064
Hexadecyl dihydrogen phosphate, *see* M-00475
Hexadecyl phosphoric acid, *see* M-00475
Hexadecyltriphenylphosphonium(1+), H-00014
▷Hexaethyldiphosphonic diamide, *in* D-00291
Hexaethyl (nitrilotri-2,1-ethanediyl)-trisphosphonate, *in* T-00438
Hexaethyl [nitrilotris(methylene)]-trisphosphonate, *in* N-00033
Hexaethyl [phosphinylidynetris(methylene)]-trisphosphonate, *in* P-00360
Hexaethylphosphoramide, *in* H-00015
Hexaethylphosphoric triamide, *in* H-00015
Hexaethylphosphoroselenoic triamide, *in* H-00015
Hexaethylphosphorothioic triamide, *in* H-00015
Hexaethylphosphorous triamide, H-00015
Hexaethyl 1,1,3-propanetriyltrisphosphonate, *in* P-00449
Hexaethyl 1,2,3-propanetriyltrisphosphonate, *in* P-00450
Hexaethyl 1,3,5-triazine-2,4,6-triyltriphosphonate, *in* T-00329
Hexaethyl 1,3,5-triazine-2,4,6-triyltriphosphoramidate, *in* T-00330
N,N,N′,N′,N″,N″-Hexaethyl-1,3,5,2,4,6-trioxatriphosphorinane-2,4,6-triamine 2,4,6-trioxide, *see* T-00756
3,3,4,4,5,5-Hexafluoro-1,2-bis(diphenylphosphino)cyclopentene, H-00016
(Hexafluorocyclobutylidene)-triphenylphosphorane, H-00017
3,3,4,4,5,5-Hexafluoro-1,2-cyclopentenediphosphonic acid, H-00018
(3,3,4,4,5,5-Hexafluoro-1-cyclopentene-1,2-diyl)bis[diphenylphosphine], *see* H-00016
(3,3,4,4,5,5-Hexafluoro-1-cyclopentene-1,2-diyl)bisphosphonic acid, *see* H-00018
3,3,4,4,5,5-Hexafluoro-1,2-diphosphonocyclopentene, *see* H-00018
1,1,2,2,3,3-Hexafluoro-4-(triphenylphosphoranylidene)cyclobutane, *see* H-00017

1,1,1,3,3,3-Hexafluoro-2-(triphenylphosphoranylidene)propane, *see* T-00597
Hexahydro-1,5-bis(1-phenylethyl)-1*H*-1,5,3-diazaphosphepine, H-00019
Hexahydro-8*b*,8*b*′-bi-8*bH*-2*a*,4*a*,6*a*,8*a*-tetraaza-8*b*-phosphapentaleno[1,6-*cd*]-pentalene, *see* B-00154
Hexahydro-1,3,2-diazaphosphinine 2-oxide, *see* H-00020
Hexahydro-1,3,2-diazaphosphorine 2-oxide, H-00020
5,6,7,8,9,10-Hexahydrodibenzo[*b*,*d*][1,6]-diphosphecin, H-00021
Hexahydro-1,3-dimethyl-1,3,2-diazaphosphinine, *see* H-00022
Hexahydro-1,3-dimethyl-1,3,2-diazaphosphorine, H-00022
Hexahydro-1,3-dimethyl-1,3,2-diazaphosphorine 2-oxide, H-00023
Hexahydro-2,4-dimethyl-3-phenoxy-1,2,4,5,3-tetrazaphosphinine, *see* H-00024
Hexahydro-2,4-dimethyl-3-phenoxy-1,2,4,5,3-tetrazaphosphorine, H-00024
2,2,4,4,6,6-Hexahydro-2,2-dimethyl-4,4,6,6-tetraphenylcyclotriphosphazene, *see* H-00025
2,2,4,4,6,6-Hexahydro-2,2-dimethyl-4,4,6,6-tetraphenyl-1,3,5-triaza-2,4,6-triphosphorine, H-00025
Hexahydro-1,3-diphenyl-1,3-azaphosphorine, *in* H-00041
Hexahydro-1,4-diphenyl-1,4-azaphosphorine, *in* H-00042
4*b*,5,6,10*b*,11,12-Hexahydro-5,5-diphenylbenzo[*c*]phosphanthridinium hexafluorophosphate, *in* H-00045
2,3,3*a*,4,5,9*b*-Hexahydro-4,4-diphenyl-1*H*-cyclopent[*c*]isophosphinolinium bromide, *in* H-00048
2,3,3*a*,4,5,9*b*-Hexahydro-4,4-diphenyl-1*H*-cyclopent[*c*]isophosphinolinium hexafluorophosphate, *in* H-00048
2,3,3*a*,4,5,9*b*-Hexahydro-4,4-diphenyl-1*H*-cyclopent[*c*]isophosphinolinum hexafluorophosphate, *see* H-00048
7,8,15,16,17,18-Hexahydro-14,18-diphenyl-6*H*,18*H*-1,6*H*-dibenzo[*b*,*i*][1,11,4,8]-dithiadiphosphacyclotetradecin, H-00026
(Hexahydrofarnesyl)triphenylphosphonium(1+), *see* T-00690
2,2,4,4,6,6-Hexahydro-2,2,4,4,6,6-hexakis(methylamino)cyclotriphosphazene, *see* H-00027
2,2,4,4,6,6-Hexahydro-2,2,4,4,6,6-hexakis(methylamino)-1,3,5-triaza-2,4,6-triphosphorine, H-00027
1,1*a*,2,3,6*a*,6*b*-Hexahydro-3,4,5,6*a*,6*b*,7-hexakis(trifluoromethyl)-1,2,3,6-ethanediylidene-6*H*-3*a*,6-diphosphacyclobut[*c*,*d*]*indene*, *see* H-00065
2,2,4,4,6,6-Hexahydro-2,2,4,4,6,6-hexamethylcyclotriphosphazene, *see* H-00028
2,2,4,4,6,6-Hexahydro-2,2,4,4,6,6-hexamethyl-1,3,5-triaza-2,4,6-triphosphorine, H-00028
Hexahydro-2,2,4,4,6,6-hexamethyl-1,3,5-triphenyl-1,3,5,2,4,6-triphosphatrigermanin, H-00029
2,2,4,4,6,6-Hexahydro-2,2,4,4,6,6-hexaphenoxycyclotriphosphazene, *see* H-00030
2,2,4,4,6,6-Hexahydro-2,2,4,4,6,6-hexaphenoxy-1,3,5-triaza-2,4,6-triphosphorine, H-00030
1,3*a*,3*b*,4,6*a*,6*b*-Hexahydro-1,2,3*b*,4,5,6*b*-hexaphenylcyclobuta[1,2-*b*:3,4-*b*′]-diphosphole, H-00031
2,2,4,4,6,6-Hexahydro-2,2,4,4,6,6-hexaphenylcyclotriphosphazene, *see* H-00032
2,2,4,4,6,6-Hexahydro-2,2,4,4,6,6-hexaphenyl-1,3,5-triaza-2,4,6-triphosphorine, H-00032
2,2,4,4,6,6-Hexahydro-2,2,4,4,6,6-hexapropoxycyclotriphosphazene, *see* H-00033
2,2,4,4,6,6-Hexahydro-2,2,4,4,6,6-hexapropoxy-1,3,5-triaza-2,4,6-triphosphorine, H-00033
Hexahydro-2-hydroxycyclopenta[*d*]-1,3,2-dioxaphosphorin 2-oxide, H-00034
Hexahydro-2-hydroxy-1-(phenylmethyl)-1,2-azaphosphorine 2-oxide, *see* B-00048
2,3,4,5,6,7-Hexahydro-1*H*-isophosphindole 2-oxide, H-00035
Hexahydro-1-methyl-4-phenyl-1,4-azaphosphinine, *see* H-00036
Hexahydro-1-methyl-4-phenyl-1,4-azaphosphorine, H-00036
Hexahydro-1,2,2,3,4,4,5,6,6-nonaphenyl-*s*-triphosphatrigermanin, *see* N-00070
Hexahydro-2*H*-[1,2]oxaphosphorino[2,3-*b*]-[1,2]oxaphosphorin, H-00037
Hexahydro-2-phenoxy-1*H*-1,3,2-diazaphosphepine, H-00038
Hexahydro-2-phenoxy-1,3,2-diazaphosphinine, *see* H-00039
Hexahydro-2-phenoxy-1,3,2-diazaphosphorine, H-00039
Hexahydro-3-phenyl-1*H*-1,3-azaphosphepine, H-00040
Hexahydro-3-phenyl-1,3-azaphosphinine, *see* H-00041
Hexahydro-4-phenyl-1,4-azaphosphinine, *see* H-00042
Hexahydro-3-phenyl-1,3-azaphosphorine, H-00041
Hexahydro-4-phenyl-1,4-azaphosphorine, H-00042
6,6*a*,7,8,9,9*a*-Hexahydro-5-phenyl-1*H*-benzo[*h*]cyclopenta[*c*]phosphinoline, H-00043
3,4,5,6,7,8-Hexahydro-5-phenyl-2*H*-1,9,5-benzodithiaphosphacycloundecin, H-00044
4*b*,5,6,10*b*,11,12-Hexahydro-5-phenylbenzo[*c*]phosphanthridine, H-00045
1,2,3,4,7,8-Hexahydro-1-phenyl-5,6-bis[(trimethylsilyl)oxy]phosphocin, H-00046
2,3,4,7,8,9-Hexahydro-1-phenyl-5,6-bis[(trimethylsilyl)oxy]-1*H*-phosphonine, H-00047
2,3,3*a*,4,5,9*b*-Hexahydro-4-phenyl-1*H*-cyclopent[*c*]isophosphinoline, H-00048
2,3,5,6,8,8-Hexahydro-8-phenyl-[1,3,2]-oxazaphospholo[2,3-*b*][1,3,2]-oxazaphosphole, H-00049
1,4,4*a*,9,9*a*,10-Hexahydro-11-phenyl-9,10-phosphinideneanthracene, H-00050
Hexahydro-3-phenyl-1,2,4,3-triazaphosphinine, *see* H-00051
Hexahydro-3-phenyl-1,2,4,3-triazaphosphorine, H-00051
Hexahydrospiro[1,3,2-benzoxazaphosphole-2(3*H*),1′-phospholane], H-00052
1,2,3,4,5,6-Hexahydro-2,2,5,5-tetraphenyl-2,5-benzodiphosphocinium(2+), H-00053
2,2,4,4,6,6-Hexahydro-4,4,6,6-tetraphenyl-1,3,5-triaza-2,4,6-triphosphorine-2,2-diamine, *see* D-00025
Hexahydro-3*H*,6*H*-2*a*,5*a*,8*a*-triaza-8*b*-phosphaacenaphthylene, H-00054
Hexahydro-2*a*,4*a*,6*a*-triaza-6*b*-phosphacyclopenta[*cd*]pentalene, H-00055
2,2,2,3,4,5-Hexahydro-2,2,2-trimethoxy-3-phenylnaphth[2,1-*d*]-1,2-oxaphosphole, H-00056
Hexahydro-6,7,8-trimethoxypyrano[2,3-*e*]-1,3,2-dioxaphosphorin 2-oxide, *see* M-00119
Hexahydro-6,7,8-trimethoxypyrano[2,3-*e*]-1,3,2-dioxaphosphorin 2-sulfide, *see* M-00120
Hexahydro-6,7,8-trimethoxypyrano[3,2-*d*]-[1,3,2]oxathiaphosphorine 2-oxide, *see* M-00121
Hexahydro-6,7,8-trimethoxypyrano[2,3-*e*]-1,3,2-oxathiaphosphorin 2-oxide, *see* M-00122
Hexahydro-6,7,8-trimethoxypyrano[3,2-*d*]-[1,3,2]oxathiaphosphorin 2-sulfide, *see* M-00123
Hexahydro-1,3,4-trimethyl-1,3,2-diazaphosphinine, *see* H-00057
Hexahydro-1,3,4-trimethyl-1,3,2-diazaphosphorine, H-00057
Hexahydro-1,3,5-triphenyl-1,3,5-diazaphosphorine, H-00058
2,2,2,3,3*a*,11*b*-Hexahydro-2,2,2-triphenylphenanthro[9,10-*d*]-1,3,2-oxazaphosphole, H-00059
Hexaisobutylphosphoric triamide, *in* H-00063
Hexaisobutylphosphorous triamide, *see* H-00063
Hexaisopropyl [phosphinylidynetris(methylene)]trisphosphonate, *in* P-00360
Hexaisopropyl 1,2,3-propanetriyltrisphosphonate, *in* P-00450
Hexaisopropyl 1,3,5-triazine-2,4,6-triylphosphonate, *in* T-00329
▷2,2,4,4,6,6-Hexakis(1-aziridinyl)-2,2,4,4,6,6-hexahydro-1,3,5-triaza-2,4,6-triphosphorine, H-00060
adamanta-Hexakis(diethylgermyl)-tetraphosphane, *see* D-01260
▷2,2,4,4,6,6-Hexakis(dimethylamino)-2,2,4,4,6,6-hexahydrocyclophosphazene, *see* H-00061
▷2,2,4,4,6,6-Hexakis(dimethylamino)-2,2,4,4,6,6-hexahydro-1,3,5-triaza-2,4,6-triphosphorine, H-00061
Hexakis(dimethylamino)-*μ*-oxodiphosphorus(2+), H-00062
adamanta-Hexakis(dimethylgermyl)-tetraphosphane, *see* D-01263
Hexakis(*N*-methylmethanaminato)-*μ*-oxodiphosphorus(2+), *see* H-00062
Hexakis(2-methylpropyl)phosphoric triamide, *in* H-00063
Hexakis(2-methylpropyl)phosphorous triamide, H-00063
Hexakis(phenylmethyl)phosphorous triamide, *see* H-00012
2,3,5,6,7,8-Hexakis(trifluoromethyl)-1,4-diphosphabarrelene, *see* H-00064
2,3,5,6,7,8-Hexakis(trifluoromethyl)-1,4-diphosphabicyclo[2.2.2]octa-2,5,7-triene, H-00064
2,5,7,10,11,12-Hexakis(trifluoromethyl)-1,6-diphosphahexacyclo[4.4.2.$0^{2,5}$,$0^{3,9}$,$0^{4,8}$,$0^{7,10}$]-dodec-11-ene, H-00065
4,5,9,10,11,12-Hexakis(trifluoromethyl)-1,8-diphosphatetracyclo[6.2.2.$0^{2,7}$.$0^{3,6}$]-dodecane, H-00066
2,7,9,10,11,12-Hexakis(trifluoromethyl)-1,8-diphosphatetracyclo[6.2.2.$0^{2,7}$.$0^{3,6}$]-dodeca-4,9,11-triene, H-00067
N,*N*,*N*′,*N*′,6,7-Hexamethyl-1,3-bis(trimethylsilyl)-5,8-dioxa-1,3-diaza-2,4-diphospha(4-P^V)spiro[3.4]-oct-6-ene-2,4-diamine, H-00068
Hexamethyl-1,3,2-dioxaphospholan-2-amine, *see* D-00710
N,*N*,*N*′,*N*′,*N*″,*N*″-Hexamethyl-*N*‴-diphenylphosphinophosphorimidic triamide, H-00069
▷Hexamethyldiphosphonamide, *in* D-00751
Hexamethyl 2,3-diphosphonobutanedioate, *in* D-01182
Hexamethylenebis[dimethylphosphine], *see* B-00201
Hexamethylenebis[diphenylphosphine], *see* B-00243
Hexamethylenebisphosphine, *see* D-01177
Hexamethylene chlorophosphite, *in* D-00971
Hexamethylene cyclic phosphoric acid, *see* H-00137
Hexamethylenediphosphonic acid, *see* H-00085
Hexamethylenediphosphonous acid, *see* H-00086
Hexamethylene fluorophosphite, *in* D-00971
Hexamethylene hydrogen phosphate, *see* H-00137
Hexamethylene phenyl phosphate, *in* H-00137
2,3,5,6,7,8-Hexamethyl-2,3,5,6,7,8-hexaaza-1,4-diphosphabicyclo[2.2.2]-octane, H-00070
2,4,6,8,9,10-Hexamethyl-2,4,6,8,9,10-hexaaza-1,3,5,7-tetraphosphaadamantane, *see* H-00071
2,4,6,8,9,10-Hexamethyl-2,4,6,8,9,10-hexaaza-1,3,5,7-tetraphosphatricyclo[3.3.1.$1^{3,7}$]decane, H-00071

2,3,4,6,7,8-Hexamethyl-2,3,4,6,7,8-hexaphosphabicyclo[3.3.0]octane, *see* O-00017
N,N,N′,N′,N″,N″-Hexamethyl-1-methylenephosphoranetriamide, H-00072
1,2,2,3,4,4-Hexamethyl-1-phenylphosphetanium bromide, *in* P-00020
N,N,N′,N′,N″,N″-Hexamethyl-*N‴*-phenylphosphorimidic triamide, H-00073
N,2,2,3,4,4-Hexamethylphosphetanamine 1-oxide, *in* P-00021
1,2,2,3,4,4-Hexamethylphosphetane, H-00074
Hexamethyl [phosphinylidynetris(methylene)]-trisphosphonate, *in* P-00360
1,1,3,3,5,5-Hexamethyl-1-phosphonia-3,5-disilacyclohexane chloride, *in* D-00485
Hexamethylphosphonimidic diamide, *in* P-00023
▷Hexamethylphosphoramide, *see* H-00075
▷Hexamethylphosphoric triamide, H-00075
2,2,2′,2′,2″,2″-Hexamethylphosphoric trihydrazide, H-00076
N,N,N′,N′,N″,N″-Hexamethylphosphorimidic triamide, H-00077
Hexamethylphosphoroselenoic triamide, H-00078
▷Hexamethylphosphorothioic triamide, H-00079
▷Hexamethylphosphorous triamide, H-00080
▷Hexamethylpyrophosphonamide, *in* D-00751
N,N,4,4′,5′,5′-Hexamethylspiro[1,3,2-benzodioxaphosphole-2,2′-[1,3,2]-dioxaphospholan]-2-amine, *in* T-00230
4,4,5,5,*N,N*-Hexamethylspiro[1,3,2-benzoxazaphosphole-2(3*H*),2′-[1,3,2]-dioxaphospholan]-2-amine, *in* T-00231
N,N,N′,N′,N″,N″-Hexamethyltetraaminophosphonium bromide, *in* H-00077
N,N,N′,N′,N″,N″-Hexamethyltetraaminophosphonium chloride, *in* H-00077
2,7,10,15,17,17-Hexamethyl-17*H*-tetrabenzo[*b,d,f,h*]phosphoninium iodide, *in* P-00028
2,5,5,8,11,11-Hexamethyl-1,3,7,9-tetraoxa-2,8-diphosphacyclododecane, H-00081
N,N,2,2,3,3-Hexamethyl-1,4,6,9-tetraoxa-5-phospha(5-P^V)spiro[4.4]nonan-5-amine, *in* T-00235
Hexamethyl 1,3,5-triazine-2,4,6-triyltriphosphonate, *in* T-00329
Hexamethyl 1,3,5-triazine-2,4,6-triyltriphosphoramidate, *in* T-00330
N,N,N′,N′,N″,N″-Hexamethyl-*N‴*-trimethylsilylphosphorimidic triamide, H-00082
N,N,2,2,3,3-Hexamethyl-1,4,6-trioxa-9-aza-5-phospha(5-P^V)spiro[4.4]nonan-5-amine, *in* T-00238
2,2,3,3,7,7-Hexamethyl-1,4,6-trioxa-9-aza-5-phospha(5-P^V)spiro[4.4]nonane, H-00083
2,2,3,3,8,8-Hexamethyl-1,4,6-trioxa-9-aza-5-phospha(5P^V)spiro[4.4]nonane, H-00084
1,1,3,3,5,5-Hexamethyl-2,4,6-triphenyl-1,3,5-trigerma-2,4,6-triphosphane, *see* H-00029
1,6-Hexanediphosphonic acid, H-00085
1,6-Hexanediphosphonous acid, H-00086
P,P′-1,6-Hexanediylbis[bis *N,N*-diethylphosphonous diamide], *in* H-00086
1,6-Hexanediylbis[dimethylphosphine], *see* B-00201
1,6-Hexanediylbis[diphenylphosphine], *see* B-00243
1,6-Hexanediylbisphosphine, *see* D-01177
1,6-Hexanediylbisphosphonic acid, *see* H-00085
1,6-Hexanediylbis(phosphonous acid), *see* H-00086
1-Hexanephosphonic acid, *see* H-00097
1-Hexanephosphonothioic acid, *see* H-00100
Hexaphenylcyclohexaphosphine, *see* H-00089
Hexaphenyl cyclophosphonitrilate, *see* H-00030
Hexaphenyl-1,3-diphospha-2,4-digermacyclobutane, H-00087
1,2,2,3,4,4-Hexaphenyl-1,3,2,4-diphosphdigermetane, *see* H-00087
1,2,2,3,4,4-Hexaphenyl-1,3-diphosphetane, H-00088
Hexaphenylhexaphosphinane, *see* H-00089
Hexaphenylhexaphosphorinane, H-00089
Hexaphenylphosphoric triamide, *in* H-00090
Hexaphenylphosphorous triamide, H-00090
1,1,4,7,10,10-Hexaphenyl-1,4,7,10-tetraphosphadecane, H-00091
Hexaphenyl 1,3,5-triazine-2,4,6-triylphosphonate, *in* T-00329
Hexaphenyltriethylenetetraphosphine, *see* H-00091
Hexapropylphosphoric triamide, *in* H-00092
Hexapropylphosphorothioic triamide, *in* H-00092
Hexapropylphosphorous triamide, H-00092
Hexose diphosphate, *see* F-00067
Hexose monophosphate, *see* F-00070
Hexyl dichlorophosphite, *see* H-00103
Hexyl dihexylphosphinate, *in* D-00434
Hexyl dihydrogen phosphate, *see* M-00476
Hexyldiphenylphosphine, H-00093
Hexylidenetriphenylphosphorane, *in* H-00104
Hexylmethylphosphinic acid, H-00094
Hexyl phosphinate, H-00095
Hexylphosphine, H-00096
Hexylphosphonic acid, H-00097
Hexylphosphonic dichloride, H-00098
Hexylphosphonodithioic acid, H-00099
Hexylphosphonothioic acid, H-00100
Hexylphosphonous acid, H-00101
Hexylphosphoramidic acid, H-00102
Hexyl phosphoric acid, *see* M-00476
Hexyl phosphorodichloridite, H-00103
▷Hexyltriphenylphosphonium(1+), H-00104
▷Hinosan, *see* E-00002
HMDP, *see* H-00154
▷HMPA, *see* H-00075
▷HMPT, *see* H-00075
HMPT Dehydrodimer, *see* D-00713
HOE 2910, *see* B-00003
Homocubylphosphorane, *see* D-00569
Homocubyltrimethylphosphorane, *in* D-00569
Homocubyltriphenylphosphorane, *in* D-00569
▷Hostaflot, *in* D-00384
▷Hostaquick, *see* H-00004
▷Hostathion, *see* T-00333
Hydracrylonitrile dihydrogen phosphate (ester), *see* P-00398
Hydrazobis[methylphosphinothioic acid], H-00105
Hydrogen diphenyl phosphate, *see* D-01036
Hydrogen 4-nitrophenyl (dichloromethyl)-phosphonate, *in* D-00175
Hydrogen propyl methylphosphonate, *in* M-00288
4-Hydroperoxycyclophosphamide, H-00106
2-Hydroxyacrylic acid anhydride with diethyl phosphate, *see* D-00256
2-Hydroxyacrylic acid dihydrogen phosphate, *see* P-00366
2-Hydroxy-1,2-azaphosphetidine 2-oxide, H-00107
2-Hydroxy-1,3,2-benzodioxaphosphole 2-oxide, H-00108
2-Hydroxy-4*H*-1,3,2-benzodioxaphosphorin 2-oxide, H-00109
2-Hydroxy-4*H*-1,3,2-benzodioxaphosphorin 2-sulfide, H-00110
1-Hydroxy-1*H*-benzo[*b*]phosphindole 1-oxide, *see* H-00118
(1-Hydroxy-1*H*-benzotriazolato-*O*)-tris(dimethylamino)phosphorus(1+), H-00111
(1-Hydroxy-1*H*-benzotriazolato-*O*)tris(*N*-methylmethanaminato)phosphorus(1+), *see* H-00111
2-Hydroxy-2*H*-1,2-benzoxaphosphorin 2-oxide, H-00112
(*o*-Hydroxybenzylidene)triphenylphosphorane, *in* H-00177
(α-Hydroxybenzyl)phenylphosphinic acid, *see* H-00173
(α-Hydroxybenzyl)phosphonic acid, *see* H-00174
o-Hydroxybenzylphosphonic acid, *see* H-00175
p-Hydroxybenzylphosphonic acid, *see* H-00176
2-*p*-Hydroxybenzyl-1-phosphoramidopropanoic acid, *see* P-00422
(*o*-Hydroxybenzyl)triphenylphosphonium(1+), *see* H-00177
1-Hydroxy-1,1-butanediphosphonic acid, H-00113
▷3-Hydroxy-2-butenoic acid, methyl ester, dimethyl phosphate, *see* M-00440
(1-Hydroxybutylidene)bisphosphonic acid, *see* H-00113
(1-Hydroxybutyl)phosphonic acid, H-00114
3-Hydroxycrotonic acid α-methylbenzyl ester dimethylphosphate, *see* C-00207
4-Hydroxycyclophosphamide, H-00115
2-Hydroxydibenzo[*d,f*][1,3,2]dioxaphosphepin 2-oxide, H-00116
2-Hydroxydibenzo[*d,f*][1,3,2]dioxaphosphepin 2-sulfide, *see* M-00007
5-Hydroxy-5*H*-dibenzo[*b,f*]phosphepin 5-oxide, H-00117
5-Hydroxy-5*H*-dibenzophosphole 5-oxide, H-00118
2-Hydroxy-2,3-dihydro-1,2-benzoxaphosphole 2-oxide, H-00119
1-Hydroxy-2,3-dihydro-1*H*-phosphindole 1-oxide, H-00120
3-Hydroxy-2-[(dihydroxyphosphinyl)amino]-butanoic acid, *see* P-00420
2-Hydroxy-5,5-dimethyl-1,3-dioxa-2-oxo-2-phosphacyclohexane, *see* H-00124
2-Hydroxy-4,5-dimethyl-1,3-dioxa-2-oxo-2-phosphacyclopentane, *see* H-00121
2-Hydroxy-5,5-dimethyl-1,3,2-dioxaphosphinane 2-oxide, *see* H-00124
2-Hydroxy-5,5-dimethyl-1,3,2-dioxaphosphinane 2-selenide, *see* H-00125
2-Hydroxy-4,5-dimethyl-1,3,2-dioxaphospholane 2-oxide, H-00121
2-Hydroxy-4,5-dimethyl-1,3,2-dioxaphospholane 2-sulfide, H-00122
2-Hydroxy-4,5-dimethyl-1,3,2-dioxaphosphole 2-oxide, H-00123
2-Hydroxy-5,5-dimethyl-1,3,2-dioxaphosphorinane 2-oxide, H-00124
2-Hydroxy-5,5-dimethyl-1,3,2-dioxaphosphorinane 2-selenide, H-00125
2-Hydroxy-5,5-dimethyl-1,3-dioxa-2-seleno-2-phosphacyclohexane, *see* H-00125
3-Hydroxy-2,2-dimethyl-1,3-oxaphosphetane 3-oxide, H-00126
10-Hydroxy-2,8-dimethyl-10*H*-phenothiaphosphine 10-oxide, H-00127
10-Hydroxy-2,8-dimethyl-10*H*-phenoxaphosphine 10-oxide, H-00128
2-Hydroxy-5,5-dimethyl-4-phenyl-1,3,2-dioxaphoshorinane 2-oxide, H-00129
4-Hydroxy-2,5-dimethyl-1-phenylphosphacyclohexane, *see* D-00820
▷1-Hydroxy-3,4-dimethyl-3-phospholene 1-oxide, *see* D-00493
1-Hydroxy-3,4-dimethyl-1*H*-phosphole 1-oxide, H-00130
4-Hydroxydinaphtho[2,1-*d*:1′,2′-*f*][1,3,2]-dioxaphosphepin 4-oxide, H-00131
2-Hydroxy-1,3-dioxa-2-oxo-2-phosphacyclohexane, *see* H-00138
2-Hydroxy-1,3-dioxa-2-oxo-2-phosphacyclopentene, *see* H-00136
2-Hydroxy-2,4-dioxa-3-phosphabicyclo[4.3.0]nonane 3-oxide, *see* H-00034
2-Hydroxy-1,3-dioxa-2-phosphadecalin 2-oxide, *see* H-00152
2-Hydroxy-1,3,2-dioxaphosphepane 2-oxide, H-00132
2-Hydroxy-1,3,2-dioxaphosphinane 2-oxide, *see* H-00138
▷2-Hydroxy-1,3,2-dioxaphospholane, *in* D-00970
▷2-Hydroxy-1,3,2-dioxaphospholane, H-00133
2-Hydroxy-1,3,2-dioxaphospholane 2-oxide, H-00134
2-Hydroxy-1,3,2-dioxaphospholane 2-sulfide, H-00135
2-Hydroxy-1,3,2-dioxaphosphole 2-oxide, H-00136
2-Hydroxy-1,3,2-dioxaphosphonane 2-oxide, H-00137

2-Hydroxy-1,3,2-dioxaphosphorinane 2-oxide, H-00138
2-Hydroxy-4,5-diphenyl-1,3,2-dioxaphosphole 2-oxide, H-00139
4-Hydroxy-2,6-diphenyl-1,4-oxaphosphacyclohexa-2,5-diene 4-oxide, *see* H-00140
4-Hydroxy-2,6-diphenyl-4*H*-1,4-oxaphosphorin 4-oxide, H-00140
N-Hydroxy-*P*,*P*-diphenylphosphinic amide, H-00141
1-Hydroxy-3,4-diphenyl-1*H*-phosphole 1-oxide, H-00142
4-Hydroxy-2,6-diphenyl-1,4-thiaphospha-2,5-cyclohexadiene 2-oxide, *see* H-00143
4-Hydroxy-2,6-diphenyl-4*H*-1,4-thiaphosphorine 2-oxide, H-00143
Hydroxyethane-1,1-diphosphonic acid, *see* H-00144
O-(2-Hydroxyethyl) dichlorothiophosphate, *see* H-00149
(1-Hydroxyethylidene)bisphosphonic acid, H-00144
▷1-(2-Hydroxyethyl)-1-methylguanidine dihydrogen phosphate, *see* C-00206
(1-Hydroxyethyl)phosphonic acid, H-00145
(2-Hydroxyethyl)phosphonic acid, H-00146
(2-Hydroxyethyl)phosphonothioic acid, H-00147
(2-Hydroxyethyl)phosphoramidic acid, H-00148
O-(2-Hydroxyethyl) phosphorodichloridothioate, H-00149
(2-Hydroxyethyl)trimethylphosphonium(1+), H-00150
(2-Hydroxyethyl)triphenylphosphonium(1+), H-00151
▷3-Hydroxyglutaconic acid dimethyl ester dimethyl phosphate, *see* B-00461
2-Hydroxyhexahydro-4*H*-1,3,2-benzodioxaphosphorin 2-oxide, H-00152
1-Hydroxy-1,1-hexanediphosphonic acid, H-00153
1-Hydroxyhexylidenebisphosphonic acid, *see* H-00153
2-[(Hydroxyisopropoxyphosphinyl)oxy]-*N*,*N*,*N*-trimethylethanaminium hydroxide, inner salt, *in* C-00198
3-Hydroxy-2-methacrylic acid *O*-ester with *O*,*O*-dimethyl phosphorothioate, *see* D-00673
Hydroxymethanediphosphonic acid, H-00154
1-Hydroxy-3-methoxy-1*H*-1,2,4,3-benziodadioxaphosphorine 3-oxide, *in* D-00616
2-[(Hydroxymethoxyphosphinyl)oxy]-*N*,*N*,*N*-trimethylethanaminium hydroxide, inner salt, *in* C-00198
Hydroxymethyl bicyclic phosphite, *see* T-00590
2-Hydroxy-4-methyl-1,3,2-dioxaphosphinane 2-oxide, *see* H-00156
2-Hydroxy-4-methyl-1,3,2-dioxaphospholane, *in* M-00127
2-Hydroxy-4-methyl-1,3,2-dioxaphospholane 2-oxide, H-00155
2-Hydroxy-4-methyl-1,3,2-dioxaphospholane 2-sulfide, *see* M-00015
2-Hydroxy-4-methyl-1,3,2-dioxaphosphorinane 2-oxide, H-00156
2-Hydroxy-4-methyl-1,3,2-dioxaphosphorinane 2-selenide, H-00157
2-Hydroxy-4-methyl-1,3,2-dioxaphosphorinane 2-sulfide, H-00158
Hydroxymethyldiphenylphosphine, *see* D-01069
(Hydroxymethylene)bis(phosphonic acid), *see* H-00154
(1-Hydroxy-1-methylethyl)phosphonic acid, H-00159
(Hydroxymethyl)phenylphosphinic acid, H-00160
▷γ-(Hydroxymethylphosphinyl)-L-α-aminobutyryl-L-alanyl-L-alanine, *see* A-00141
γ-(Hydroxymethylphosphinyl)-L-α-aminobutyryl-L-alanyl-L-leucine, *see* P-00339
3-(Hydroxymethylphosphinyl)propanoic acid, H-00161
1-Hydroxy-2-methyl-3-phospholene 1-oxide, *see* D-00498
▷1-Hydroxy-3-methyl-2-phospholene 1-oxide, *see* D-00496
1-Hydroxy-3-methyl-3-phospholene 1-oxide, *see* D-00499
(Hydroxymethyl)phosphonic acid, H-00162
▷3-Hydroxy-2-methyl-5-[(phosphonooxy)-methyl]-4-pyridinecarboxaldehyde, *see* P-00508
4-Hydroxymethyl-2,6,7-trioxa-1-phosphabicyclo[2.2.2]octane, *see* T-00590
Hydroxymethyltriphenylphosphonium(1+), H-00163
▷*N*-Hydroxynaphthalimide diethyl phosphate, *see* N-00008
2-Hydroxynaphtho[2,3-*d*]-1,3,2-dioxaphosphole 2-oxide, H-00164
2-Hydroxy-1-oxa-2-oxo-2-phosphaindane, *see* H-00119
2-Hydroxy-1,2-oxaphosphepane 2-oxide, H-00165
2-Hydroxy-1,2-oxaphosphinane 2-oxide, *see* H-00167
2-Hydroxy-1,2-oxaphospholane 2-oxide, H-00166
2-Hydroxy-1,2-oxaphosphol-3-ene 2-oxide, *see* D-00500
2-Hydroxy-1,2-oxaphosphorinane 2-oxide, H-00167
1-Hydroxy-1-oxo-1-phospha-2,3:6,7-dibenzocyclohepta-2,6-diene, *see* D-00492
1-Hydroxy-1-oxo-1-phospho-2,3:6,7-dibenzocyclohepta-2,4,6-triene, *see* H-00117
1-Hydroxy-2,2,3,4,4-pentamethylphosphetane 1-oxide, H-00168
1-Hydroxy-2,2,3,4,4-pentamethylphosphetane 1-sulfide, H-00169
10-Hydroxy-10*H*-phenoxaphosphine 10-oxide, H-00170
2-[(Hydroxyphenoxyphosphinyl)oxy]-*N*,*N*,*N*-trimethylethanaminium hydroxide, inner salt, *in* C-00198
2-Hydroxy-4-phenyl-1,3,2-dioxaphosphorinane 2-oxide, H-00171
(2-Hydroxyphenyl)diphenylphosphine, *see* D-01070
(3-Hydroxyphenyl)diphenylphosphine, *see* D-01071
(4-Hydroxyphenyl)diphenylphosphine, *see* D-01072
(Hydroxyphenylmethyl)bisphosphonic acid, H-00172
(2-Hydroxyphenyl)methyldiphenylphosphonium iodide, *in* D-01070
(Hydroxyphenylmethyl)phenylphosphinic acid, H-00173
(Hydroxyphenylmethyl)phosphonic acid, H-00174
[(2-Hydroxyphenyl)methyl]phosphonic acid, H-00175
[(4-Hydroxyphenyl)methyl]phosphonic acid, H-00176
[(2-Hydroxyphenyl)methyl]-triphenylphosphonium(1+), H-00177
[(2-Hydroxyphenyl)methyl]-triphenylphosphorane, *in* H-00177
3-(Hydroxyphenylphosphinyl)propanoic acid, H-00178
(2-Hydroxyphenyl)phosphonic acid, H-00179
▷(3-Hydroxyphenyl)phosphonic acid, H-00180
▷(4-Hydroxyphenyl)phosphonic acid, H-00181
2-Hydroxy-1-phenylphosphorinane, *see* P-00266
(2-Hydroxyphenyl)triphenylphosphonium(1+), H-00182
(4-Hydroxyphenyl)triphenylphosphonium(1+), H-00183
1-Hydroxy-3-phosphetanecarboxylic acid 1-oxide, H-00184
1-Hydroxyphosphinane 1-oxide, *see* H-00188
[(Hydroxyphosphinylidene)bis(methylene)]-bis(phosphonic acid), *see* B-00182
1-Hydroxyphospholane 1-oxide, H-00185
1-Hydroxy-2-phospholene 1-oxide, *see* D-00506
1-Hydroxy-3-phospholene 1-oxide, *see* D-00507
1-Hydroxy-1*H*-phosphole 1-oxide, H-00186
1-Hydroxy-3,8-phosphonanedione 1-oxide, *in* P-00373
3-Hydroxy-2-phosphonoaminopropanoic acid, *see* P-00418
2-Hydroxy-3-(phosphonoxy)propanal, H-00187
3-Hydroxy-2-phosphoramidobutanoic acid, *see* P-00420
1-Hydroxyphosphorinane 1-oxide, H-00188
1-Hydroxy-1,1-propanediphosphonic acid, H-00189
(3-Hydroxypropyl)diphenylphosphine, *see* D-01074
(1-Hydroxypropylidene)bisphosphonic acid, *see* H-00189
(1-Hydroxypropyl)phosphonic acid, H-00190
(2-Hydroxypropyl)phosphonic acid, H-00191
(3-Hydroxypropyl)phosphonic acid, H-00192
(3-Hydroxypropyl)triphenylphosphonium(1+), H-00193
1-Hydroxy-1,1′(3*H*,3′*H*)-spirobi[2,1-benzoxaphosphole]-3,3′-dione, H-00194
1-Hydroxy-1,2,3,4-tetrahydrophosphinoline 1-oxide, H-00195
2-Hydroxy-4,4,5,5-tetramethyl-1,3,2-dioxaphospholane 2-oxide, H-00196
2-Hydroxy-4,4,5,5-tetramethyl-1,3,2-dioxaphospholane 2-sulfide, H-00197
2-Hydroxy-5,6-tetramethylene-1,3,2-dioxaphosphorinane 2-oxide, *see* H-00152
4-Hydroxy-2,2,6,6-tetramethyl-1-oxa-4-phospha-2,6-disilacyclohexane 4-oxide, H-00198
1-Hydroxy-3,3,3′,3′-tetramethyl-1,1′(3*H*,3′*H*)-spirobi[2,1-benzoxaphosphole], H-00199
1-Hydroxy-2,3,4,5-tetraphenyl-1*H*-phosphole 1-oxide, H-00200
3-Hydroxy-1,3-thiaphosphetane 3-oxide, H-00201
[5-(4-Hydroxy-2,6,6-trimethyl-1-cyclohexen-1-yl)-3-methyl-2,4-pentadienyl]triphenylphosphonium(1+), H-00202
1-Hydroxy-2,4,6-trimethyl-1,3,5-dioxaphosphorinane 5-oxide, H-00203
2-Hydroxy-5,6-trimethylene-1,3,2-dioxaphosphorin 2-oxide, *see* H-00034
Hydroxytrimethylgermane phosphate, *see* T-00859
5-Hydroxy-2,4,6-triphenyl-1,3,5-dioxaphosphorinane 5-oxide, H-00204
Hypophosphoric acid cyclic bis(2,2-dimethyltrimethylene) ester, *in* B-00197
I5*B*1, *in* A-00139
I5*B*2, *in* A-00139
2-Icosa-1-octadecanoyl-5,8,11,14-tetraenoylglycero-3-phosphocholine, *see* G-00076
5,8,11,14-Icosatetraenoic acid 1-[[[(2-aminoethoxy)-hydroxyphosphinyl]oxy]methyl]-2-[(1-oxooctadecyl)oxy]ethyl ester, *see* G-00077
5,8,11,14-Icosatetraenoic acid 3-[[(2-aminoethoxy)hydroxyphosphinyl]-oxy]-2-[(1-oxooctadecyl)oxy]propyl ester, *see* G-00065
1-Icosa-5,8,11,14-tetraenoyl-2-octadecanoylglycero-3-phosphoethanolamine, *see* G-00065
▷Ifosfamide, I-00001
Imcarbofos, I-00002
▷Imidan, *see* P-00343
1-Imidazolephosphonic acid, *see* I-00003
1*H*-Imidazol-1-ylphosphonic acid, I-00003
P-Imidazol-1-yl-*N*,*N*,*N*′,*N*′-tetramethylphosphonic diamide, *in* I-00003
Iminobis(methanesphosphonic acid), *see* D-00688
Iminobis(methylene)bisphosphonic acid, *see* D-00688
Iminocyclophosphamide, I-00004
(Iminodimethylene)diphosphonic acid, *see* D-00688
N-[Imino(diphenoxyphosphinylamino)methyl]-glycine, *in* P-00369
N-[Imino(diphenoxyphosphinylamino)methyl]-*N*-methylglycine, *in* P-00364

(1-Iminoethyl)phosphoramidothioic acid, I-00005
[Imino[[imino[(phenylmethyl)amino]methyl]-amino]methyl]phosphoramidic acid, *see* B-00001
N-[Imino(phosphonoamino)methyl]glycine, *see* P-00369
N-[Imino(phosphonoamino)methyl]-*N*-methylglycine, *see* P-00364
N^5-[Imino(phosphonoamino)methyl]ornithine, *see* P-00363
Iminotribenzylphosphorane, *see* T-00339
Iminotributylphosphorane, *see* T-00374
Iminotri-*tert*-butylphosphorane, *see* T-00375
Iminotricyclohexylphosphorane, *see* T-00427
Iminotriethylphosphorane, *see* T-00442
Iminotriisopropylphosphorane, *see* T-00501
▷Iminotrimethylphosphorane, *see* T-00531
1*H*-Indene-1-phosphonous acid, I-00006
1-Indenylidenetriphenylphosphorane, *in* I-00008
1*H*-Inden-2-ylphosphonic acid, I-00007
1*H*-1-Indenylphosphonous acid, *see* I-00006
1-Indenyltriphenylphosphonium(1+), I-00008
Inezin, I-00009
myo-Inositol mono[[2,3-bis(1-oxohexadecyl)oxy]propyl] 1-(dihydrogen phosphate) ester, *in* G-00015
myo-Inositol 1-[2-[(1-oxo-9-octadecenyl)-oxy]-3-[(1-oxododecyl)oxy]propyl hydrogen phosphate, *see* G-00054
o-Iodobenzylphosphonic acid, *see* I-00015
Iodobis(trifluoromethyl)phosphine, *in* B-00418
Iododiisopropylphosphine, *in* D-00648
2-Iodo-5,5-dimethyl-1,3-dioxa-2-phosphacyclohexane, *see* I-00010
2-Iodo-5,5-dimethyl-1,3,2-dioxaphosphinane, *see* I-00010
2-Iodo-5,5-dimethyl-1,3,2-dioxaphosphorinane, I-00010
Iododimethylphosphine, *in* D-00844
Iododiphenylphosphine, *in* D-01089
Iododipropylphosphine, *in* D-01209
▷Iodofenphos, I-00011
(Iodomethyl)phenylphosphinic acid, I-00012
(Iodomethyl)phosphonic acid, I-00013
1-Iodo-*N*-methyl-*N*,1,1,1-tetraphenylphosphoranamine, *in* M-00194
Iodo-*N*,*N*,*N'*,*N'*,*N''*,*N''*,*N'''*,*N'''*-octamethylphosphoranetetramine, *see* T-00178
[(1-Iodo-2-oxo-2-phenyl)ethylidene]-triphenylphosphorane, I-00014
O-(4-Iodophenyl) dichlorothiophosphate, *see* I-00019
[(2-Iodophenyl)methyl]phosphonic acid, I-00015
(2-Iodophenyl)phosphonic acid, I-00016
(3-Iodophenyl)phosphonic acid, I-00017
(4-Iodophenyl)phosphonic acid, I-00018
O-(4-Iodophenyl) phosphorodichloridothioate, I-00019
O-(4-Iodophenyl) thiophosphoryl dichloride, *see* I-00019
(2-Iodosophenyl)methylphosphinic acid, *in* D-00488
(2-Iodosophenyl)phenylphosphinic acid, *in* D-00488
2-Iodosophenylphosphonic acid, *in* D-00488
(4-Iodosophenyl)phosphonic acid, *in* I-00018
2-Iodosophenylphosphoric acid, *see* D-00616
Iodotetraphenoxyphosphorane, I-00020
Iodotrimethylphosphonium iodide, *see* D-00629
Iodotriphenylphosphonium triiodide, *in* D-00631
2-Iodo-2-(triphenylphosphoranylidene)-acetophenone, *see* I-00014
(*β*-Ionylideneethyl)triphenylphosphonium(1+), *see* M-00421
β-Ionylidenetriphenylphosphonium(1+), *see* M-00422
β-Ionylidenetriphenylphosphorane, *in* M-00422
▷IPSP, *see* A-00142
Irganox, *in* D-00094
Irganox 1222A, *in* D-00094
▷Isazofos, I-00021
Isoamyl dichlorophosphite, *see* M-00108
Isoamyl phosphate, *see* M-00484
Isoamyl phosphoric acid, *see* M-00484
Isoamyltriphenylphosphonium(1+), *see* M-00111
Isobutenyl dihydrogen phosphate, *see* M-00491
Isobutenyldiphenylphosphine, *see* M-00362
Isobutenylphosphonothioic acid, *see* M-00366
Isobutenyl phosphoric acid, *see* M-00491
2-Isobutoxy-1,3,2-dioxaphospholan, *see* M-00368
Isobutyl butylmethylphosphinate, *in* B-00566
O-Isobutyl cyclohexylphosphonochloridothioate, *in* C-00260
Isobutyl decylmethylphosphinate, *in* D-00009
Isobutyl dichlorophosphite, *see* M-00385
Isobutyl *N*,*P*-diethylphosphonamidate, *in* D-00336
Isobutyl dihydrogen phosphate, *see* M-00493
Isobutyl diisobutylphosphinate, *in* B-00346
Isobutyldiphenylphosphine, *see* M-00370
N-Isobutyl-*N'*,*N''*-diphenylphosphoric triamide, *in* M-00384
Isobutyl ethylmethylphosphinate, *in* E-00085
Isobutyl ethylphosphonochloridate, *in* E-00135
Isobutyl hexylmethylphosphinate, *in* H-00094
Isobutylidenetriphenylphosphorane, *see* M-00371
Isobutyl isobutylmethylphosphinate, *in* M-00172
Isobutyl isobutylphenylphosphinate, *in* M-00373
Isobutylmethylphosphinic acid, *see* M-00172
Isobutyl *P*-methylphosphonohydrazidate, *in* M-00308
Isobutyl methylpropylphosphinate, *in* M-00375
Isobutylphenylphosphinic acid, *see* M-00373
Isobutylphosphine, *see* M-00374
Isobutylphosphinic acid, *see* M-00376
Isobutylphosphonic acid, *see* M-00378
Isobutylphosphonothioic acid, *see* M-00381
Isobutylphosphonous acid, *see* M-00383
Isobutylphosphoramidic acid, *see* M-00384
Isobutyl phosphoric acid, *see* M-00493
Isobutyltriphenylphosphonium(1+), *see* M-00387
(Isocyanatomethyl)phosphonic acid, I-00022
2-Isocyano-5,5-dimethyl-1,3,2-dioxaphosphorinane, I-00023
(Isocyanomethyl)phosphonic acid, I-00024
▷Isoendoxan, *see* I-00001
▷Isofenphos, I-00025
▷Isoflurophate, *see* D-00665
Isopentenylphosphonic acid, *see* M-00102
Δ^3-Isopentenyl pyrophosphate, *see* M-00483
3-Isopentenyl trihydrogen pyrophosphate, *see* M-00483
Isopentenyltriphenylphosphonium(1+), *see* M-00104
O-Isopentyl dichlorothiophosphate, *see* M-00109
S-Isopentyl dichlorothiophosphate, *see* M-00110
Isopentyl dihydrogen phosphate, *see* M-00484
Isopentylidenetriphenylphosphorane, *in* M-00111
Isopentylphosphonic acid, *see* M-00106
Isopentyl phosphoric acid, *see* M-00484
Isopentyl phosphorodichloridite, *see* M-00108
O-Isopentyl phosphorodichloridothioate, *see* M-00109
S-Isopentyl phosphorodichloridothioate, *see* M-00110
O-Isopentyl thiophosphoryl dichloride, *see* M-00109
Isopentyltriphenylphosphonium(1+), *see* M-00111
▷Isophos-3, *in* C-00101
▷Isophosphamide, *see* I-00001
1*H*-Isophosphindole-1,3(2*H*)-dione, I-00026
Isophosphindoline, *see* D-00509
Isophosphinoline, I-00027
Isophosphoramide mustard, *see* B-00125
Isopropenyl dichlorophosphate, *see* M-00151
Isopropenylphosphonic acid, *see* M-00149
Isopropenylphosphonic dichloride, *see* M-00150
Isopropenyl phosphorodichloridate, *see* M-00151
Isopropenyl phosphoryl dichloride, *see* M-00151
▷*O*-2-Isopropoxycarbonyl-1-methylvinyl *O*-methyl ethylphosphoramidothioate, *see* P-00466
2-Isopropoxy-1,3,2-dioxaphospholane, *see* M-00152
P-Isopropoxy-*N*-trimethylsilyl-*P*-[bis(trimethylsilyl)amino]phosphazene, *in* T-00867
P-Isopropoxy-*N*-trimethylsilyl-*P*-[bis(trimethylsilyl)amino]phosphine imide, *in* T-00867
Isopropyl benzylphenylphosphinate, *in* B-00056
▷Isopropyl bis(1-aziridinyl)phosphinate, *in* B-00091
Isopropyl bis(1-aziridinyl)phosphinite, *in* B-00095
Isopropyl bis(chloromethyl)phosphinate, *in* B-00127
O-Isopropyl bis(trifluoromethyl)-phosphinothioate, *in* B-00415
S-Isopropyl bis(trifluoromethyl)-phosphinothioate, *in* B-00415
Isopropyl (2-butenyl)methylphosphinate, *in* B-00523
Isopropyl butylphosphinate, *in* B-00590
O-Isopropyl butylphosphonodithioate, *in* B-00606
Isopropyl cyclohexylphosphinate, *in* C-00253
O-Isopropyl dibutylphosphinothioate, *in* D-00117
Isopropyl dichlorophosphate, *see* I-00060
Isopropyl dichlorophosphinylcarbamate, *in* D-00191
Isopropyl dichlorophosphite, *see* I-00061
O-Isopropyl dichlorothiophosphate, *see* I-00062
S-Isopropyl dichlorothiophosphate, *in* I-00064
Isopropyl dichlorothiophosphite, *see* I-00063
Isopropyl *N*,*N*-diethyl-*P*-methylphosphonamidate, *in* D-00296
Isopropyl *N*,*N*-diethyl-*P*-phenylphosphonamidodithioate, *in* D-00308
S-Isopropyl *N*,*N*-diethyl-*P*-phenylphosphonamidothioate, *in* D-00309
Isopropyl diethylphosphinate, *in* D-00322
▷Isopropyl diethylphosphinite, *in* D-00333
Isopropyl diethylphosphinodithioate, *in* D-00325
O-Isopropyl diethylphosphinoselenoate, *in* D-00327
O-Isopropyl diethylphosphinothioate, *in* D-00330
S-Isopropyl diethylphosphinothioate, *in* D-00330
Isopropyl diethylphosphinothioite, *in* D-00332
Isopropyl diethylphosphoramidochloridate, *in* D-00346
Isopropyl diethylphosphoramidofluoridate, *in* D-00351
▷Isopropyl dihydrogen phosphate, *see* M-00477
S-Isopropyl dihydrogen thiophosphate, *see* I-00064
Isopropyl (diisopropoxyphosphinyl)acetate, *in* D-00633
Isopropyl diisopropylphosphinate, *in* D-00641
4-Isopropyl-5,5-dimethyl-1,3,2-dioxaphosphorinane, I-00028
Isopropyl (1,1-dimethyl-3-oxobutyl)-ethylphosphinate, *in* D-00782
Isopropyl (2,5-dimethylphenyl)phosphinate, *in* D-00805
P-Isopropyl-*N*,*N*-dimethyl-*P*-phenylphosphinothioic amide, *in* I-00046
5-Isopropyl-2,6-dimethyl-6-phosphabicyclo[3.1.1]hept-2-ene 6-oxide, I-00029
Isopropyl *N*,*P*-dimethylphosphonamidate, *in* D-00849
P-Isopropyl-*N*,*N*-dimethylphosphonamidic anhydride, *see* D-00668
Isopropyl dimethylphosphoramidocyanidate, *in* D-00863
O-Isopropyl *O*,*S*-dimethyl phosphorothioate, I-00030

O-Isopropyl O,S-dimethyl thiophosphate, *see* I-00030
Isopropyl N-(diphenoxyphosphinyl)carbamate, *in* D-00989
Isopropyl diphenylphosphinate, *in* D-01039
Isopropyldiphenylphosphine, I-00031
N-Isopropyl-P,P-diphenylphosphinic amide, *in* D-01040
P-Isopropyl-N,P-diphenylphosphinic amide, *in* I-00044
Isopropyl diphenylphosphinodithioate, *in* D-01065
Isopropyl N,N′-diphenylphosphorodiamidate, *in* D-01125
Isopropyl ethylmethylphosphinodithioate, *in* E-00103
Isopropyl N-ethyl-P-methylphosphonamidate, *in* E-00088
Isopropyl ethylphenylphosphinate, *in* E-00101
Isopropyl ethylphenylphosphinothioite, *in* E-00107
Isopropyl ethylphosphinate, *in* E-00124
O-Isopropyl ethylphosphinothioate, *in* E-00125
Isopropyl ethylphosphonochloridate, *in* E-00135
Isopropyl ethylphosphonochloridodithioate, *in* E-00136
S-Isopropyl ethylphosphonochloridothioate, *in* E-00137
Isopropyl ethylphosphonochloridothioite, *in* E-00138
O-Isopropyl ethylphosphonodithioic anhydrosulfide, *in* D-00405
▷Isopropyl ethylphosphonofluoridate, *in* E-00142
Isopropyl (fluoromethyl)phosphonochloridate, *in* F-00024
Isopropyl hydrogen cyclohexylphosphonate, *in* C-00256
O-Isopropyl hydrogen ethylphosphonoselenoate, *in* E-00145
Isopropyl hydrogen methylphosphonate, *see* I-00038
O-Isopropyl hydrogen methylphosphonodithioate, *see* I-00040
O-Isopropyl hydrogen methylphosphonothioate, *see* I-00041
O-Isopropyl hydrogen (1-methylpropyl)-phosphonodithioate, *in* M-00380
O-Isopropyl hydrogen pentylphosphonodithioate, *in* P-00051
2,3-O-Isopropylidene-2,3-dihydroxy-1,4-bis(diphenylphosphino)butane, *see* B-00251
7-Isopropylidene-4,8,8-trimethyl-1,6-dioxa-4-aza-5-phospha(5-P^V)spiro[4.4]-nonan-9-one, I-00032
Isopropylidenetriphenylphosphorane, *see* M-00153
O-Isopropyl S-(isopropoxyphenyl-phosphinothioyl) phenylphosphonodithioate, *in* P-00237
Isopropyl N-isopropyl-P-methylphosphonamidate, *in* I-00036
Isopropyl isopropylphenylphosphinate, *in* I-00043
4-Isopropyl-2-methoxy-5,5-dimethyl-1,3,2-dioxaphosphorinane, I-00033
Isopropylmethyldiphenylphosphonium iodide, *in* I-00031
Isopropyl methyl methylphosphonate, *in* M-00168
O-Isopropyl O-methyl methylphosphonothioate, *in* M-00171
O-Isopropyl S-methyl methylphosphonothioate, *in* I-00041
Isopropyl methyl-4-morpholinylphosphinate, *in* M-00175
Isopropyl P-methyl-4-morpholinylphosphonamidate, *in* M-00175
S-Isopropyl methylphenylhosphinothioate, *in* M-00225
Isopropyl methylphenylphosphinate, *in* M-00219
Isopropyl (4-methylphenyl)phosphinate, *in* M-00222
O-Isopropyl methylphenylphosphinothioate, *in* M-00225
Isopropyl P-methyl-N-phenylphosphonamidate, *in* M-00231
Isopropyl methylphosphinate, I-00034
Isopropylmethylphosphinic acid, I-00035
O-Isopropyl methylphosphinothioate, *in* M-00279
N-Isopropyl-P-methylphosphonamidic acid, I-00036
N-Isopropyl-P-methylphosphonamidothioic acid, I-00037
Isopropyl methylphosphonate, I-00038
Isopropyl methylphosphonisothiocyanatidate, *in* M-00294
O-Isopropyl methylphosphonobromidothioate, *in* M-00295
Isopropyl methylphosphonochloridate, *in* M-00296
O-Isopropyl methylphosphonochloridothioate, I-00039
S-Isopropyl methylphosphonochloridothioate, *in* M-00298
O-Isopropyl methylphosphonodithioate, I-00040
▷Isopropyl methylphosphonofluoridate, *see* S-00002
Isopropyl methylphosphonofluoridite, *in* M-00307
O-Isopropyl methylphosphonofluoridothioate, *in* M-00305
Isopropyl P-methylphosphonohydrazidate, *in* M-00308
O-Isopropyl methylphosphonothioate, I-00041
Isopropyl methyl-1-piperidinylphosphinate, *in* M-00361
Isopropyl P-methyl-1-piperidinylphosphonamidate, *in* M-00361
▷2-Isopropyl-6-methyl-4-pyrimidyl phosphorothioic acid O,O′-diethyl ester, *see* D-00027
Isopropyl methyl(trichloromethyl)-phosphinate, *in* M-00419
2-Isopropyl 1,3,2-oxathiaphospholane, I-00042
2-Isopropyl-1,3,2-oxathiaphospholane 2-oxide, *in* I-00042
2-Isopropyl-1,3,2-oxathiaphospholane 2-sulfide, *in* I-00042
Isopropyl 1,1,2,3,3-pentamethyltrimethylenephosphinate, *in* H-00168
Isopropyl phenyl methylphosphonate, *in* I-00038
O-Isopropyl S-phenyl methylphosphonothioate, *in* I-00041
O-Isopropyl S-phenyl methylphosphoramidodithioate, *in* M-00330
Isopropyl phenylphosphinate, *in* P-00210
Isopropylphenylphosphinic acid, I-00043
P-Isopropyl-P-phenylphosphinic amide, I-00044
Isopropylphenylphosphinodithioic acid, I-00045
Isopropylphenylphosphinothioic acid, I-00046
Isopropyl phenyl phosphite, *see* I-00047
Isopropyl phenyl phosphonate, I-00047
O-Isopropyl phenyl-phosphonisothiocyanatidothioate, *in* P-00227
O-Isopropyl phenylphosphonochloridothioate, *in* P-00232
Isopropyl phenylphosphonochloridothioite, *in* P-00233
S-Isopropyl phenylphosphonochloridothionate, *in* P-00232
O-Isopropyl phenylphosphonodithioic anhydrosulfide, *in* P-00237
Isopropyl phenyl phosphorochloridate, *in* P-00269
Isopropylphosphine, I-00048
Isopropylphosphinic acid, I-00049
Isopropylphosphonic acid, I-00050
Isopropylphosphonic bis(dimethylamide), *in* I-00051
P-Isopropylphosphonic diamide, I-00051
Isopropylphosphonic dianilide, *in* I-00051
Isopropylphosphonic dichloride, I-00052
Isopropylphosphonochloridothioic acid, I-00053
Isopropylphosphonothioic acid, I-00054
Isopropylphosphonothioic dichloride, I-00055
Isopropylphosphonous acid, I-00056
Isopropylphosphonous dichloride, I-00057
Isopropylphosphoramidic acid, I-00058
Isopropylphosphoramidothioic acid, I-00059
▷Isopropyl phosphoric acid, *see* M-00477
Isopropyl phosphoric dianilide, *in* D-01125
Isopropyl phosphorodichloridate, I-00060
Isopropyl phosphorodichloridite, I-00061
O-Isopropyl phosphorodichloridothioate, I-00062
S-Isopropyl phosphorodichloridothioate, *in* I-00064
Isopropyl phosphorodichloridothioite, I-00063
S-Isopropyl phosphorothioate, I-00064
Isopropyl phosphoryl dichloride, *see* I-00060
S-Isopropyl propylphosphonochloridothioate, *in* P-00477
Isopropyl styrylphosphinate, *in* P-00153
Isopropyl tetraethylphosphorodiamidite, *in* T-00066
P-Isopropyl-N,N,N′,N′-tetramethylphosphonic diamide, *in* I-00051
Isopropyl tetramethylphosphorodiamidite, *in* T-00227
Isopropylthiophosphonic dichloride, *see* I-00055
S-Isopropyl thiophosphoric acid, *see* I-00064
O-Isopropyl thiophosphoryl dichloride, *see* I-00062
▷Isopropyl triethylphosphonamidate, *in* T-00446
Isopropyl triethylphosphonamidodithioate, *in* T-00447
O-Isopropyl triethylphosphonamidothioate, *in* T-00448
Isopropyl (trifluoromethyl)phosphinate, *in* T-00475
Isopropyl trimethylphosphonamidate, *in* T-00537
O-Isopropyl trimethylphosphonamidothioate, *in* T-00540
N-Isopropyl-P,P,P-triphenylphosphine imide, *in* T-00626
Isopropyltriphenylphosphonium(1+), I-00065
Isopropyl (triphenylphosphoranylidene)-carbamate, *in* T-00648
Isopropyl N,N,N′-tris(trimethylsilyl)-phosphonamidimidate, *in* T-00867
2-Isoselenocyanato-5,5-dimethyl-1,3,2-dioxaphosphorinane, I-00066
▷Isothioate, I-00067
▷Isoxathion, I-00068
JAV 852, *in* B-00001
K 4, *in* A-00139
K 26, *see* A-00140
▷Karphos, *see* I-00068
▷Kayaphos, *see* P-00451
KB-944, *see* F-00064
▷4-Ketocyclophosphamide, *see* O-00058
▷Kilval, *see* V-00001
▷Kitazin, K-00001
▷Kitazin L, *see* B-00045
▷Kitazin P, *see* B-00045
▷Korlan, *see* F-00003
▷Krenite, *in* M-00469
Lauroylstearoylphosphatidylcholine, *in* G-00053
Lawesson's reagent, *see* B-00293
▷Lebaycid, *see* F-00007
Le BOP reagent, *see* H-00111
▷Leptophos, L-00001
Leptophos-Ethyl, *in* E-00114
Leptophos Oxon, *in* M-00488
Linaloyl dihydrogen phosphate, *see* D-00775
Linaloyl phosphate, *see* D-00775
Linoleoyloleoylphosphatdiylcholine, *in* G-00072
Linoleoylpalmitoylphosphatidylcholine, *in* G-00068
Linoleoylpalmitoylphosphatidylethanolamine, *in* G-00069
Linoleoylpalmitoylphosphatidylethanolamine, *in* G-00085
Linoleoylstearoylphosphatidylethanolamine, *in* G-00071
Lombricine, L-00002

LS 74 783, *in* M-00473
▷Malathion, M-00001
▷Maretin, *see* N-00008
Mazidox, *see* T-00219
MDPP, *in* M-00164
N-Me, *in* T-00339
▷Mecarbam, M-00002
Medronic acid, *see* M-00022
▷Menazon, M-00003
(−)-Menthyl (2-aminophenyl)-methylphosphinate, *in* A-00103
Menthyl *tert*-butylphenylphosphinate, *in* B-00579
Menthyl cyclohexylmethylphosphinate, *in* C-00247
Menthyldiphenylphosphine, *in* M-00164
Menthyl methylphenylphosphinate, *in* M-00219
Menthyl phenylphosphinate, *in* P-00210
▷Mephosfolan, M-00004
2-Mercapto-1,3,2-benzodioxaphosphole 2-sulfide, M-00005
2-Mercapto-4*H*-1,3,2-benzodioxaphosphorin 2-sulfide, M-00006
2-Mercaptodibenzo[*d,f*][1,3,2]-dioxaphosphepin 2-oxide, M-00007
6-Mercaptodibenzo[*d,f*][1,3,2]-dioxaphosphepin 6-sulfide, M-00008
2-Mercapto-5,5-dimethyl-1,3-dioxa-2-oxo-2-phosphacyclohexane, *see* M-00010
2-Mercapto-5,5-dimethyl-1,3,2-dioxaphosphinane 2-sulfide, *see* M-00011
2-Mercapto-4,5-dimethyl-1,3,2-dioxaphospholane 2-oxide, *see* H-00122
2-Mercapto-4,5-dimethyl-1,3,2-dioxaphospholane 2-sulfide, M-00009
2-Mercapto-5,5-dimethyl-1,3,2-dioxaphosphorinane 2-oxide, M-00010
2-Mercapto-5,5-dimethyl-1,3,2-dioxaphosphorinane 2-sulfide, M-00011
2-Mercapto-5,5-dimethyl-2-thioxo-1,3-dioxa-2-phosphacyclohexane, *see* M-00011
4-Mercaptodinaphtho[2,1-*d*:1′,2′-*f*][1,3,2]-dioxaphosphepin 4-sulfide, M-00012
2-Mercapto-1,3,2-dioxaphosphinane 2-sulfide, *see* M-00013
2-Mercapto-1,3,2-dioxaphospholane, *in* D-00970
2-Mercapto-1,3,2-dioxaphosphorinane 2-sulfide, M-00013
2-Mercapto-1,3,2-dithiaphosphorinane 2-sulfide, M-00014
(2-Mercaptoethyl)trimethylammonium ester with diethyl thiophosphoric acid, *see* E-00001
▷Mercaptofos, *see* F-00007
2-Mercapto-4-methyl-1,3,2-dioxaphosphinane 2-sulfide, *see* M-00017
2-Mercapto-4-methyl-1,3,2-dioxaphospholane 2-oxide, M-00015
2-Mercapto-4-methyl-1,3,2-dioxaphospholane 2-sulfide, M-00016
2-Mercapto-4-methyl-1,3,2-dioxaphosphorinane 2-sulfide, M-00017
2-Mercapto-4-methyl-1,3-dioxa-2-thioxo-2-phosphacyclohexane, *see* M-00017
2-Mercapto-4-methyl-1,3,2-dioxophospholane, *in* M-00127
(Mercaptomethyl)phosphonic acid, M-00018
2-Mercapto-4,4,5,5-tetramethyl-1,3,2-dioxaphospholane 2-oxide, *see* H-00197
2-Mercapto-4,4,5,5-tetramethyl-1,3,2-dioxaphospholane 2-sulfide, M-00019
2-Mercapto-1,3,5,2-triazaphosphorine-4,6(1*H*,5*H*)-dione 2-sulfide, M-00020
5-Mercapto-2,4,6-tris(1-methylethyl)-1,3,5-dioxaphosphorinane 5-sulfide, *see* T-00498
▷Merphos, *see* T-00393
Mesitylphosphinic acid, *see* T-00525
Mesitylphosphonic acid, *see* T-00527
▷Metasystox-*R*, *see* O-00098
▷Metasystox *S*, *see* O-00099
Methacrifos, *in* D-00673
Methafos, *in* C-00003
▷Methamidophos, M-00021
Methanediphosphinic acid, *see* M-00143
Methanediphosphonic acid, M-00022
Methanediphosphonic tetrachloride, *see* M-00145
Methanediphosphonic tetrachloride, *see* M-00146
Methanediphosphonothioic acid, M-00023
Methanediphosphonous acid, M-00024
▷Methanephosphonic acid, *see* M-00288
Methanetetraylbis(trimethylphosphorane), *see* B-00429
Methanetetraylbis[triphenylphosphorane], *see* B-00459
Methanetetraylbis[2,4,6-tris(1,1-dimethylethyl)phenyl]phosphine, *see* B-00394
▷Methidathion, M-00025
3-Methoxyallylidenetriphenylphosphorane, *in* M-00084
2-Methoxy-1,3,2-benzodioxaphosphole, M-00026
▷2-Methoxy-4*H*-1,3,2-benzodioxaphosphorin 2-sulfide, *see* S-00001
1-Methoxy-1*H*-benzo[*b*]phosphindole 1-oxide, *in* H-00118
(2-Methoxybenzylidene)triphenylphosphorane, *in* M-00062
(3-Methoxybenzylidene)triphenylphosphorane, *in* M-00063
4-Methoxybenzylidenetriphenylphosphorane, *in* M-00064
m-Methoxybenzylphosphonic acid, *see* M-00060
p-Methoxybenzylphosphonic acid, *see* M-00061
(*m*-Methoxybenzyl)triphenylphosphonium(1+), *see* M-00063
(*o*-Methoxybenzyl)triphenylphosphonium(1+), *see* M-00062
(*p*-Methoxybenzyl)triphenylphosphonium(1+), *see* M-00064
Methoxybis(trifluoromethyl)phosphine, *in* B-00418
P-Methoxy-*P*-[bis(trimethylsilyl)amino]-*N*-trimethylsilylphosphazene, *in* T-00867
[α-(Methoxycarbonyl)benzylidene]-triphenylphosphorane, *in* T-00644
(*m*-Methoxycarbonylbenzylidene)-triphenylphosphorane, *in* M-00030
(*o*-Methoxycarbonylbenzylidene)-triphenylphosphorane, *in* M-00029
(*p*-Methoxycarbonylbenzylidene)-triphenylphosphorane, *in* M-00031
(*m*-Methoxycarbonylbenzyl)-triphenylphosphonium(1+), *see* M-00030
(*o*-Methoxycarbonylbenzyl)-triphenylphosphonium(1+), *see* M-00029
(*p*-Methoxycarbonylbenzyl)-triphenylphosphonium(1+), *see* M-00031
Methoxycarbonyl dihydrogen phosphate, *see* M-00479
(1-Methoxycarbonylethyl)-triphenylphosphonium(1+), M-00027
(6-Methoxycarbonyl-3-methyl-2,4-hexadienyl)triphenylphosphonium(1+), *see* M-00050
[(Methoxycarbonyl)methyl]phosphonamidic acid, M-00028
▷2-Methoxycarbonyl-1-methylvinyl dimethyl phosphate, *see* M-00440
[[2-(Methoxycarbonyl)phenyl]methylene]-triphenylphosphorane, *in* M-00029
[[4-(Methoxycarbonyl)phenyl]methylene]-triphenylphosphorane, *in* M-00031
[[3-(3-Methoxycarbonyl)phenyl]methylene]-triphenylphosphorane, *in* M-00030
[[2-(Methoxycarbonyl)phenyl]methyl]-triphenylphosphonium(1+), M-00029
[[3-(Methoxycarbonyl)phenyl]methyl]-triphenylphosphonium(1+), M-00030
[[4-(Methoxycarbonyl)phenyl]methyl]-triphenylphosphonium(1+), M-00031
α-Methoxycarbonyltriphenyl-phosphoranylidenepropanoic acid, *in* T-00646
6-Methoxydibenzo[*d,f*][1,3,2]dioxaphosphepin, M-00032
5-Methoxy-5*H*-dibenzo[*b,f*]phosphepin 5-oxide, *in* H-00117
5-Methoxy-5*H*-dibenzophosphole 5-oxide, *in* H-00118
▷Methoxydichlorophosphine, *see* M-00346
Methoxydichlorophosphine oxide, *see* M-00345
5-Methoxy-5,6-dihydrodibenzo[*b,d*]phosphorin 5-oxide, *in* D-00504
5-Methoxy-5,6-dihydrophosphanthridene 5-oxide, *in* D-00504
2-Methoxy-5,5-dimethyl-1,3-dioxa-2-phosphacyclohexane, *see* M-00034
2-Methoxy-4,5-dimethyl-1,3-dioxa-2-phosphacyclopentane, *see* M-00033
2-Methoxy-5,5-dimethyl-1,3,2-dioxaphosphinane, *see* M-00034
2-Methoxy-4,5-dimethyl-1,3,2-dioxaphospholane, M-00033
2-Methoxy-5,5-dimethyl-1,3,2-dioxaphosphorinane, M-00034
2-Methoxy-5,5-dimethyl-4-(1-methylethyl)-1,3,2-dioxaphosphorinane, *see* I-00033
2-Methoxy-3,4-dimethyl-5-phenyl-1,3,2-oxazaphospholane, *see* M-00035
2-Methoxy-3,4-dimethyl-5-phenyl-1,3,2-oxazaphospholidine, M-00035
1-Methoxy-3,4-dimethyl-1*H*-phosphole 1-oxide, *in* H-00130
4-Methoxydinaphtho[2,1-*d*,1′,2′-*f*][1,3,2]-dioxaphosphepin 4-oxide, *in* H-00131
2-Methoxy-1,3-dioxa-2-oxo-2-phosphacyclopentane, *see* M-00037
▷2-Methoxy-1,3-dioxa-2-phosphacyclohexane, *see* M-00038
▷2-Methoxy-1,3-dioxa-2-phosphacyclopentane, *see* M-00036
▷2-Methoxy-1,3,2-dioxaphosphinane, *see* M-00038
▷2-Methoxy-1,3,2-dioxaphospholane, M-00036
2-Methoxy-1,3,2-dioxaphospholane 2-oxide, M-00037
▷2-Methoxy-1,3,2-dioxaphosphorinane, M-00038
1-Methoxy-1,3-dioxo-2,4-dimethyl-1-phospha-2,4-diazacyclopentane, *in* D-00722
2-Methoxy-4,5-diphenyl-1,3-dioxa-2-phosphacyclopentane, *see* M-00039
2-Methoxy-4,5-diphenyl-1,3,2-dioxaphospholane, M-00039
P-(*O*-Methoxy)-*P,P*-diphenyl-*N*-(diphenylphosphinothioyl)phosphazene, *in* D-01006
P-(*O*-Methoxy)-*P,P*-diphenyl-*N*-(diphenylphosphinothioyl)phosphine imide, *in* D-01006
P-Methoxy-*P,P*-diphenyl-*N*-phenylsulfonylphosphazene, *in* B-00013
P-Methoxy-*P,P*-diphenyl-*N*-phenylsulfonylphosphine imide, *in* B-00013
4-Methoxy-2,6-diphenyl-4*H*-1,4-thiaphosphorin 4-oxide, *in* H-00143
2-Methoxy-1,3,2-dithiaphospholane, *in* D-01243
2-Methoxy-1,3,2-dithiaphospholane 2-oxide, *in* D-01244
2-Methoxy-1,3,2-dithiaphosphorinane, *in* D-01247
[(2-Methoxyethoxy)methyl]phosphonic acid, M-00040
▷*S*-[2-[(2-Methoxyethyl)amino]-2-oxoethyl] *O,O*-dimethyl phosphorothioate, *see* A-00051
O-(2-Methoxyethyl) dichlorothiophosphate, *in* H-00149
(2-Methoxyethyl)phosphoramidothioic acid, M-00041
O-(2-Methoxyethyl) phosphorodichloridothioate, *in* H-00149
7-Methoxy-2,3,5,6,7,8-hexakis(trifluoromethyl)-1,4-diphosphabicyclo[2.2.2]octa-2,5-diene, M-00042
1-Methoxy-3-methoxycarbonylphosphetane 1-oxide, *in* H-00184
2-Methoxy-4-methyl-1,3-dioxa-2-oxo-2-phosphacyclohexane, *see* M-00045
2-Methoxy-4-methyl-1,3-dioxa-2-phosphacyclohexane, *see* M-00044
2-Methoxy-4-methyl-1,3-dioxa-2-phosphacyclopentane, *see* M-00043
2-Methoxy-4-methyl-1,3,2-dioxaphosphinane, *see* M-00044
2-Methoxy-4-methyl-1,3,2-dioxaphosphinane 2-oxide, *see* M-00045
2-Methoxy-4-methyl-1,3,2-dioxaphosphinane 2-selenide, *see* M-00046
2-Methoxy-4-methyl-1,3,2-dioxaphosphinane 2-sulfide, *see* M-00047

2-Methoxy-4-methyl-1,3,2-dioxaphospholane, M-00043
2-Methoxy-4-methyl-1,3,2-dioxaphosphorinane, M-00044
2-Methoxy-4-methyl-1,3,2-dioxaphosphorinane 2-oxide, M-00045
2-Methoxy-4-methyl-1,3,2-dioxaphosphorinane 2-selenide, M-00046
2-Methoxy-4-methyl-1,3,2-dioxaphosphorinane 2-sulfide, M-00047
2-Methoxy-4-methyl-1,3-dioxa-2-selenoxo-2-phosphacyclohexane, *see* M-00046
2-Methoxy-4-methyl-1,3-dioxa-2-thioxo-2-phosphacyclohexane, *see* M-00047
Methoxymethyldiphenylphosphine, *in* D-01069
2-Methoxy-4-methyl-1,3,2-dithiaphospholane, *in* M-00136
Methoxymethylenetriphenylphosphorane, *in* M-00053
3-Methoxy-15-methyl-18-nor-15-phosphaestra-1,3,5(10),6,8,13-hexaene, M-00048
2-Methoxy-4-methyl-1,2-oxaphospholane, M-00049
(6-Methoxy-3-methyl-6-oxo-2,4-hexadienyl)-triphenylphosphonium(1+), M-00050
2-Methoxy-5-methyl-3-phenyl-1,3,4,2-oxadiazaphospholine, *see* D-00513
3-Methoxy-17-methyl-15-phenyl-15-phosphaestra-1,3,5(10),8,16-pentaene, M-00051
(Methoxymethyl)phenylphosphinic acid, M-00052
2-Methoxy-5-methyl-1,2-thiaphosphol-4-ene, *see* D-00514
(Methoxymethyl)triphenylphosphonium(1+), M-00053
3-Methoxy-18-nor-17-phosphaestra-1,3,5(10),9(11)-tetraen-15-one, M-00054
2-Methoxy-1,3,2-oxazaphosphorinane, *see* T-00111
(2-Methoxy-2-oxoethyl)-triphenylphosphonium(1+), M-00055
1-Methoxy-1,2,3,4,4-pentamethylphosphetane 1-oxide, *in* H-00168
2-Methoxyphenyl dichlorophosphate, *in* M-00480
3-Methoxyphenyl dichlorophosphate, *see* M-00080
4-Methoxyphenyl dichlorophosphate, *see* M-00081
O-(4-Methoxyphenyl) dichlorothiophosphate, *see* M-00082
4-Methoxyphenyl difluorophosphate, *in* M-00481
2-Methoxyphenyl dihydrogen phosphate, *see* M-00480
4-Methoxyphenyl dihydrogen phosphate, *see* M-00481
(2-Methoxyphenyl)diphenylphosphine, M-00056
(3-Methoxyphenyl)diphenylphosphine, M-00057
(4-Methoxyphenyl)diphenylphosphine, M-00058
P-(4-Methoxyphenyl)-*N*,*N'*-diphenylphosphonic diamide, *in* M-00072
4-Methoxyphenyl *N*,*N'*-diphenylphosphorodiamidate, *in* M-00481
p-Methoxyphenyldithioxophosphorane dimer, *see* B-00293
[(2-Methoxy-1,4-phenylene)-bis(iminocarbonothioyl)]-bis(phosphoramidic acid) tetraethyl ester, *see* I-00002
(2-Methoxyphenyl)methyldiphenylphosphonium iodide, *in* M-00056
(4-Methoxyphenyl)methyldiphenylphosphonium iodide, *in* M-00058
[(2-Methoxyphenyl)methylene]-triphenylphosphorane, *in* M-00062
[(3-Methoxyphenyl)methylene]-triphenylphosphorane, *in* M-00063
[(4-Methoxyphenyl)methylene]-triphenylphosphorane, *in* M-00064
(4-Methoxyphenyl)methyl-1-naphthalenylphenylphosphonium bromide, *in* M-00066
(4-Methoxyphenyl)methyl-1-naphthalenylphenylphosphonium chloride, *in* M-00066
(2-Methoxyphenyl)methylphenylphosphine, M-00059
[(3-Methoxyphenyl)methyl]phosphonic acid, M-00060
[(4-Methoxyphenyl)methyl]phosphonic acid, M-00061
[(2-Methoxyphenyl)methyl]-triphenylphosphonium(1+), M-00062
[(3-Methoxyphenyl)methyl]-triphenylphosphonium(1+), M-00063
[(4-Methoxyphenyl)methyl]-triphenylphosphonium(1+), M-00064
(2-Methoxyphenyl)-2-naphthalenylphenylphosphine, M-00065
(4-Methoxyphenyl)-1-naphthalenylphenylphosphine, M-00066
(4-Methoxyphenyl)phosphine, M-00067
(2-Methoxyphenyl)phosphinic acid, M-00068
(4-Methoxyphenyl)phosphinic acid, M-00069
(2-Methoxyphenyl)phosphonic acid, M-00070
(3-Methoxyphenyl)phosphonic acid, M-00071
(4-Methoxyphenyl)phosphonic acid, M-00072
(4-Methoxyphenyl)phosphonic dichloride, M-00073
(2-Methoxyphenyl)phosphoramidic acid, M-00074
(3-Methoxyphenyl)phosphoramidic acid, M-00075
(4-Methoxyphenyl)phosphoramidic acid, M-00076
(2-Methoxyphenyl)phosphoramidothioic acid, M-00077
(3-Methoxyphenyl)phosphoramidothioic acid, M-00078
(4-Methoxyphenyl)phosphoramidothioic acid, M-00079
2-Methoxyphenyl phosphoric acid, *see* M-00480
4-Methoxyphenyl phosphoric acid, *see* M-00481
2-Methoxyphenyl phosphorodichloridate, *in* M-00480
3-Methoxyphenyl phosphorodichloridate, M-00080
4-Methoxyphenyl phosphorodichloridate, M-00081
O-(4-Methoxyphenyl) phosphorodichloridothioate, M-00082
4-Methoxyphenyl phosphorodifluoridate, *in* M-00481
2-Methoxyphenyl phosphoryl dichloride, *in* M-00480
3-Methoxyphenyl phosphoryl dichloride, *see* M-00080
4-Methoxyphenyl phosphoryl dichloride, *see* M-00081
4-Methoxyphenyl phosphoryl difluoride, *in* M-00481
2-(4-Methoxyphenyl)-1,3,2-thiazaphosphetidine 2-sulfide, M-00083
p-Methoxyphenylthionophosphine sulfide dimer, *see* B-00293
O-(4-Methoxyphenyl) thiophosphoryl dichloride, *see* M-00082
5-Methoxy-5-phospha(5-P^V)spiro[4.4]nonane, *in* P-00350
2-Methoxyphospholane 1-oxide, *in* H-00185
(3-Methoxy-2-propenylidene)-triphenylphosphorane, *in* M-00084
(3-Methoxy-2-propenyl)-triphenylphosphonium(1+), M-00084
2-Methoxy-3,3,4,4-tetramethyl-1,2-azaphospholidine 2-oxide, *in* T-00191
2-Methoxy-3,3,5,5-tetramethyl-1,2-azaphospholidine 2-oxide, *in* T-00208
2-Methoxy-4,4,5,5-tetramethyl-1,2-azaphospholidine 2-oxide, *in* T-00192
2-Methoxy-4,4,5,5-tetramethyl-1,3,2-dioxaphospholane, M-00085
1-Methoxy-2,3,4,5-tetraphenyl-1*H*-phosphole 1-oxide, *in* H-00200
2-Methoxy-3,5,5-trimethyl-1,2-oxaphospholan-3-ol 2-oxide, *in* T-00519
P-Methoxy-*N*-trimethylsilyl-*P*-[bis(trimethylsilyl)amino]phosphine imide, *in* T-00867
Methoxytris(trimethylsilyloxy)phosphonium iodide, *in* T-00864
Methylacetophos, M-00086
Methyl *P*-acetyl-*N*,*N*-diethylphosphonamidate, *in* A-00014
Methyl acetylmethylphosphinate, *in* A-00018
Methyl acetylphenylphosphinate, *in* A-00021
10-Methylacridophosphine, M-00087
Methyl allylphenylphosphinate, *in* P-00300
O-Methyl amidofluorothiophosphate, *see* M-00331
Methylamine-1,1-diphosphonic acid, *see* A-00086
Methylaminobis(dichlorophosphine), *see* M-00158
Methylaminobis(difluorophosphine), *see* M-00159
(Methylaminocarbonyl)phosphonic acid, M-00088
Methyl [3-amino-3-(dimethoxyphosphinyl)]-propanoate, *in* A-00122
Methyl (aminoethoxyphosphinyl)acetate, *in* M-00028
2-Methylaminoethylphosphonic acid, *in* A-00079
Methyl aminohydroxyphosphinylacetate, *see* M-00028
Methyl 3-amino-3-phosphonobutanoate, *in* A-00118
Methyl 4-amino-4-phosphonobutanoate, *in* A-00119
Methyl 3-amino-3-phosphonopropanoate, *in* A-00122
(2-Methylaminopropyl)phosphonic acid, *in* A-00126
1-Methyl-1-aza-2,5-diphosphacyclopentane, *see* M-00089
1-Methyl-1,2,5-azadiphospholidine, M-00089
(*N*-Methylbenzenaminato)-triphenylphosphorus(1+), *see* M-00194
5-Methyl-10*H*-5,10[1',2']-benzenoacridophosphonium iodide, *in* B-00014
2-Methyl-1,3,2-benzodioxaphosphole, M-00090
10-Methyl-5,10[1',2']-benzophenophosphazinium iodide, *in* B-00015
1-Methyl-1*H*-benzo[*b*]phosphindole, *see* M-00115
3-Methylbenzo[*c*]phosphinine, *see* M-00163
3-Methylbenzo[*c*]phosphorin, *see* M-00163
Methyl *N*-benzoyl-*P*-methylphosphonamidate, *in* B-00031
Methyl benzoylphenylphosphinate, *in* B-00032
Methyl 3-benzoyl-3-(triphenylphosphoranylidene)propionate, *in* O-00082
α-Methylbenzyl 3-(dimethoxyphosphinyloxy)-isocrotonate, *see* C-00207
m-Methylbenzylidenetriphenylphosphorane, *in* M-00210
o-Methylbenzylidenetriphenylphosphorane, *in* M-00209
p-Methylbenzylidenetriphenylphosphorane, *in* M-00211
Methyl benzylmethylphosphinate, *in* B-00053
Methyl benzylphenylphosphinate, *in* B-00056
m-Methylbenzylphosphonic acid, *see* M-00207
o-Methylbenzylphosphonic acid, *see* M-00206
p-Methylbenzylphosphonic acid, *see* M-00208
(2-Methylbenzyl)triphenylphosphonium(1+), *see* M-00209
(3-Methylbenzyl)triphenylphosphonium(1+), *see* M-00210
(4-Methylbenzyl)triphenylphosphonium(1+), *see* M-00211
▷Methyl bicyclic phosphate, *in* M-00424
▷Methyl bicyclic phosphite, *see* M-00424
Methyl bicyclic selenophosphate, *in* M-00424
▷Methyl bicyclic thiophosphate, *in* M-00424
(5-Methylbicyclo[2.2.1]hept-5-ene-2,3-diyl)bis[diphenylphosphine], *see* B-00247
Methyl-2-biphenylylphosphonium iodide, *in* T-00348
▷Methyl bis(1-aziridinyl)phosphinate, *in* B-00091
Methyl bis(1-aziridinyl)phosphinite, *in* B-00095
O-Methyl bis(1-aziridinyl)phosphinothioate, *in* B-00093
Methylbisbiphenylenephosphorane, *in* S-00021
Methyl *N*,*N*-bis(2-chloroethyl)-phosphorodiamidate, *in* B-00126

Methyl bis(chloromethyl)phosphinate, *in* B-00127
Methyl bis(4-chlorophenyl)phosphinate, *in* B-00139
Methyl bis(4-chlorophenyl)phosphinite, *in* B-00142
Methyl 4,4-bis(diethoxyphosphinyl)butanoate, *in* D-01183
Methyl 2,3-bis(diethoxyphosphinyl)-propanoate, *in* B-00173
Methyl 3,3-bis(dimethoxyphosphinyl)-propanoate, *in* D-01187
Methylbis(dimethylamino)iminophosphorane, *see* P-00023
N-Methyl-*P*,*P*-bis(dimethylamino)-*P*-phenylphosphazene, *see* N-00079
Methyl bis(4-dimethylaminophenyl)-phosphinate, *in* B-00188
N-Methyl-*P*,*P*-bis(dimethylamino)-*P*-phenylphosphine imide, *see* N-00079
Methylbis(1,1-dimethylethyl)phosphine, *see* D-00099
Methyl bis(1,1-dimethylethyl)phosphinite, *in* D-00122
N-Methyl-*P*,*P*-bis(1,1-dimethylethyl)-phosphinous amide, *in* D-00124
Methyl 2,3-bis-*O*-diphenylphosphino-4,6-*O*-benzylideneglucopyranoside, M-00091
Methyl bis(hydroxymethyl)phosphinate, *in* B-00282
Methyl bis(1-hydroxy-2,2,2-trichloroethyl)phosphinate, *in* B-00287
Methyl bis(3-methoxyphenyl)phosphinate, *in* B-00300
Methyl bis(4-methoxyphenyl)phosphinate, *in* B-00301
P-Methyl-*N*,*N*-bis(1-methylethyl)-phosphonamidic acid, *see* D-00636
Methylbis(2-methylphenyl)phenylphosphonium bromide, *in* B-00309
Methyl bis(2-methylphenyl)phosphinate, *in* B-00318
Methyl bis(4-methylphenyl)phosphinate, *in* B-00320
Methyl bis(4-methylphenyl)phosphinite, *in* B-00325
O-Methyl bis(pentafluorophenyl)-phosphinothioate, *in* B-00383
N-Methyl-*P*,*P*-bis(trichloromethyl)phosphinic amide, *in* B-00402
N-Methyl-*P*,*P*-bis(trichloromethyl)-phosphinimidic chloride, *in* B-00403
Methyl bis(trifluoromethyl)phosphinate, *in* B-00411
Methyl bis(trifluoromethyl)phosphinite, *in* B-00418
Methyl bis(trifluoromethyl)-phosphinodithioate, *in* B-00412
Methyl bis(trifluoromethyl)-phosphinoselenoite, *in* B-00413
Methyl bis(trifluoromethyl)-phosphinotelluroite, *in* B-00414
O-Methyl bis(trifluoromethyl)-phosphinothioate, *in* B-00415
S-Methyl bis(trifluoromethyl)-phosphinothioate, *in* B-00415
Methyl bis(trifluoromethyl)phosphinothioite, *in* B-00417
N-Methyl-*P*,*P*-bis(trifluoromethyl)-phosphinous amide, *in* B-00419
4′-Methyl-4,5-bis(trifluoromethyl)-spiro[1,3,2-dithiaphosphole-2,1′-[2,6,7]trioxa[1]phosphabicyclo[2.2.2]-octane], M-00092
Methyl bis(trimethylsilyloxy)-phosphinecarboxylate oxide, *in* B-00433
Methyl [bis(trimethylsilyloxy)phosphinyl]-acetate, *in* B-00434
Methyl bromo(diethoxyphosphinyl)acetate, *in* B-00467
(2-Methyl-2-bromomethyl)trimethylene bromophosphite, *see* B-00465
(2-Methyl-2-bromomethyl)trimethylene phosphoryl bromide, *in* B-00465
Methyl 3-bromophenylphosphinate, *in* B-00488
Methyl bromo(triphenylphosphoranylidene)-acetate, *in* B-00504
P-(2-Methyl-1,3-butadienyl)-*N*,*N*′-bis(1-methylethyl)phosphonic diamide, *in* M-00096
(2-Methyl-1,3-butadienyl)phosphonic acid, M-00093
(3-Methyl-1,2-butadienyl)phosphonic acid, M-00094
(3-Methyl-1,3-butadienyl)phosphonic acid, M-00095
(2-Methyl-1,3-butadienyl)phosphonic bis(diethylamide), *in* M-00096
(2-Methyl-1,3-butadienyl)phosphonic bis(dimethylamide), *in* M-00096
(2-Methyl-1,3-butadienyl)phosphonic bis(isopropylamide), *in* M-00096
(2-Methyl-1,3-butadienyl)phosphonic bis(methylamide), *in* M-00096
(2-Methyl-1,3-butadienyl)phosphonic diamide, M-00096
(2-Methyl-1,3-butadienyl)phosphonothioic acid, M-00097
3-Methyl-2-butenylidenetriphenylphosphorane, M-00098
(1-Methyl-3-butenyl)phosphonic acid, M-00099
(2-Methyl-1-butenyl)phosphonic acid, M-00100
(3-Methyl-1-butenyl)phosphonic acid, M-00101
(3-Methyl-2-butenyl)phosphonic acid, M-00102
(3-Methyl-3-butenyl)phosphonic acid, M-00103
3-Methyl-2-butenyl pyrophosphate, *see* M-00482
(3-Methyl-2-butenyl)triphenyl-phosphonium(1+), M-00104
O-(3-Methylbutyl) dichlorothiophosphate, *see* M-00109
S-(3-Methylbutyl) dichlorothiophosphate, *see* M-00110
Methyl *N*,*N*′-1,3-butylene-*N*,*N*′-dimethylphosphorodiamidite, *in* H-00057
3-Methylbutylidenetriphenylphosphorane, *in* M-00111
Methyl *tert*-butylmethylphosphinate, *in* B-00567
O-Methyl *tert*-butylmethylphosphinothioate, *in* B-00570
Methyl *N*-*tert*-butyl-*P*-methylphosphonamidate, *in* B-00572
Methyl *tert*-butylphenylphosphinate, *in* B-00579
Se-Methyl *tert*-butylphenylphosphinoselenoate, *in* B-00582
O-Methyl *tert*-butylphenylphosphinothioate, *in* B-00583
S-Methyl *tert*-butylphenylphosphinothioate, *in* B-00583
Methyl butylphosphinate, *in* B-00590
Methyl *tert*-butylphosphinate, *in* B-00591
(1-Methylbutyl)phosphonic acid, M-00105
(3-Methylbutyl)phosphonic acid, M-00106
(2-Methyl-2-butyl)phosphonic acid, *see* D-00911
Methyl *tert*-butylphosphon(isocyanatidate), *in* B-00602
O-Methyl butylphosphonodithioate, *in* B-00606
O-Methyl *tert*-butylphosphonothioate, M-00107
3-Methylbutyl phosphorodichloridite, M-00108
O-(3-Methylbutyl) phosphorodichloridothioate, M-00109
S-(3-Methylbutyl) phosphorodichloridothioate, M-00110
(3-Methylbutyl)triphenylphosphonium(1+), M-00111
N-Methylcarbamoylphosphonic acid, *see* M-00088
Methyl carbonate-phosphoric acid anhydride, *see* M-00479
O-Methyl chlorodimethylamidothiophosphate, *in* D-00861
Methyl chlorofluorophosphate, *see* M-00341
Methyl (chloromethyl)methylphosphinate, *in* C-00100
Methyl (chloromethyl)phenylphosphinate, *in* C-00103
Methyl (2-chloro-3-oxobutyl)-methylphosphinate, *in* C-00121
Methyl (4-chlorophenyl)methylphosphinate, *in* C-00131
Methyl (3-chlorophenyl)phosphinate, *in* C-00143
Methyl chloro(triphenylphosphoranylidene)-acetate, *in* C-00190
Methylcoppertris(triphenylphosphine), *see* M-00435
Methyl cyano(triphenylphosphoranylidene)-acetate, *in* C-00231
Methyl cyano(triphenylphosphoranylidene)-pyruvate, *in* C-00218
Methyl cyclohexylmethylphosphinate, *in* C-00247
Methyl cyclohexylphenylphosphinate, *in* C-00250
Methyl *N*-cyclohexyl-*P*-phenyl-phosphonamidate, *in* C-00251
Methyl cyclohexylphosphinate, *in* C-00253
Methyl cyclohexylphosphonochloridate, *in* C-00259
O-Methyl cyclohexylphosphoramidochloridothioate, *in* C-00270
Methyl 6-deoxy-2,3-di-*O*-methyl-6-methylaminoglucopyranoside 4,6-cyclic phosphonamide, M-00112
Methyl 6-deoxy-2,3-di-*O*-methyl-6-methylaminoglucopyranoside 4,6-cyclic thiophosphonamide, M-00113
Methyl diallylphosphinate, *in* D-01193
O-Methyl diamidothiophosphate, *in* M-00359
S-Methyl diamidothiophosphate, *see* M-00343
5-Methyl-1*H*-1,2,3-diazaphosphole, M-00114
Methyl (1-diazoethyl)phenylphosphinate, *in* D-00032
10-Methyldibenzo[*be*]phosphinine, *see* M-00087
5-Methyl-5*H*-dibenzophosphole, M-00115
10-Methyldibenzo[*b*,*e*]phosphorin, *see* M-00087
Methyl 2-[(dibenzyloxyphosphinyl)amino]-3-mercaptopropanoate, *in* P-00365
Methyl 2-[(dibenzyloxyphosphinyl)amino]-3-phenylpropanoate, *in* P-00400
Methyl 2-[(dibenzyloxyphosphinyl)amino]-propanoate, *in* P-00362
Methyl dibenzylphosphinate, *in* D-00055
Methyl dibromophosphite, *see* M-00344
Methyl 3-(dibutoxyphosphinyl)propanoate, *in* P-00396
Methyl dibutylphosphinate, *in* D-00109
Methyl di-*tert*-butylphosphinate, *in* D-00110
Methyl di-*tert*-butylphosphinite, *in* D-00122
Methyl dibutylphosphinodithioate, *in* D-00112
O-Methyl di-*tert*-butylphosphinothioate, *in* D-00118
S-Methyl dibutylphosphinothioate, *in* D-00117
S-Methyl di-*tert*-butylphosphinothioate, *in* D-00118
Methyl di-*tert*-butylphosphinothioite, *in* D-00120
Methyl dichlorodithiophosphate, *see* M-00347
Methyl *P*-dichloromethylphosphonamidate, *in* D-00174
Methyl dichlorophosphate, *see* M-00345
Methyl dichlorophosphinylacetate, *in* D-00190
Methyl dichlorophosphinylcarbamate, *in* D-00191
Methyl 3-dichlorophosphinylpropanoate, *in* D-00192
▷Methyl dichlorophosphite, *see* M-00346
Methyl dichloroselenophosphite, *see* M-00348
O-Methyl dichlorothiophosphate, *see* M-00349
S-Methyl dichlorothiophosphate, *see* M-00351
Methyl dichlorothiophosphite, *see* M-00350
Methyl dicyclohexylphosphinate, *in* D-00211
Methyl dicyclohexylphosphinodithioate, *in* D-00212
Methyl diethenylphosphinate, *in* D-01258
Methyl [(diethoxyphosphinothioyl)thio]-acetate, *see* M-00086
Methyl 2-(diethoxyphosphinyl)benzoate, *in* D-00243
Methyl 4-(diethoxyphosphinyl)benzoate, *in* D-00244
Methyl 4-(diethoxyphosphinyl)butanoate, *in* D-00247
Methyl 4-(diethoxyphosphinyl)-2-butenoate, *in* D-00248
Methyl *N*-(diethoxyphosphinyl)carbamate, *in* D-00249
Methyl (diethoxyphosphinyl)fluoroacetate, *in* D-00250
Methyl (diethoxyphosphinyl)oxoacetate, *in* D-00252
Methyl 2-[(diethoxyphosphinyl)oxy]-propanoate, *in* D-00255

Methyl 2-[(diethoxyphosphinyl)oxy]-2-propenoate, *in* D-00256
Methyl 2-(diethoxyphosphinyl)propanoate, *in* D-00258
Methyl 3-(diethoxyphosphinyl)propanoate, *in* D-00259
Methyl 2-(diethoxyphosphinyl)propenoate, *in* D-00260
Methyl 3-(diethoxyphosphinyl)propenoate, *in* D-00261
Methyl *N*,*N*-diethyl-*P*-methylphosphonamidate, *in* D-00296
Methyl *N*,*N*-diethyl-*P*-phenylphosphonamidate, *in* D-00307
O-Methyl *N*,*N*-diethyl-*P*-phenylphosphonamidothioate, *in* D-00309
S-Methyl *N*,*N*-diethyl-*P*-phenylphosphonamidothioate, *in* D-00309
Methyl diethylphosphinate, *in* D-00322
Methyl diethylphosphinodithioate, *in* D-00325
O-Methyl diethylphosphinoselenoate, *in* D-00327
O-Methyl diethylphosphinothioate, *in* D-00330
S-Methyl diethylphosphinothioate, *in* D-00330
Methyl 3-(diethylphosphinyl)acrylate, *in* D-00261
Methyl *N*,*N*-diethylphosphonamidate, *in* D-00337
Methyl diethylphosphoramidofluoridodithioate, *in* D-00352
Methyl *N*,*N'*-diethylphosphorodiamidate, *in* D-00382
▷Methyl difluorophosphate, *see* M-00352
Methyl difluorophosphite, *see* M-00353
O-Methyl difluorothiophosphate, *see* M-00354
S-Methyl difluorothiophosphate, *see* M-00355
▷Methyl dihydrogen phosphate, *see* M-00489
O-Methyl dihydrogen thiophosphate, *see* M-00359
S-Methyl dihydrogen thiophosphate, *see* M-00360
Methyl dihydroxyphosphinylacetate, *in* P-00376
Methyl (dihydroxyphosphinyl)formate, *in* P-00387
Methyl 2-[(diisopropoxyphosphinyl)amino]-3-mercaptopropanoate, *in* P-00365
3-Methyl-4-(diisopropoxyphosphinyl)pyridine, *in* M-00390
Methyl diisopropylphosphinate, *in* D-00641
Methyl diisopropylphosphinodithioate, *in* D-00643
O-Methyl diisopropylphosphinothioate, *in* D-00646
S-Methyl diisopropylphosphinothioate, *in* D-00646
P-Methyl-*N*,*N'*-diisopropylphosphonothioic diamide, *in* M-00314
Methyl diisopropylphosphoramidochloridite, *in* D-00654
3-Methyl-4-(diisopropylphosphoryl)pyridine, *in* M-00390
Methyl 3-[(dimethoxyphosphinothioyl)oxy]-2-methyl-2-propenoate, *in* D-00673
Methyl [(dimethoxyphosphinothioyl)thio]-acetate, *in* D-00675
Methyl (dimethoxyphosphinyl)acetate, *in* D-00678
Methyl 2-(dimethoxyphosphinyl)butanoate, *in* D-00679
Methyl *N*-(dimethoxyphosphinyl)carbamate, *in* D-00680
Methyl 2-(dimethoxyphosphinyl)oxybenzoate, *in* P-00391
▷Methyl 3-[(dimethoxyphosphinyl)oxy]-2-butenoate, *see* M-00440
Methyl 2-[(dimethoxyphosphinyl)oxy]-propanoate, *in* D-00682
Methyl 2-(dimethoxyphosphinyl)propanoate, *in* D-00684
Methyl 3-(dimethoxyphosphinyl)propanoate, *in* D-00685
Methyl 3-(dimethoxyphosphinyl)propenoate, *in* D-00686
3-Methyl-4-dimethoxyphosphinylpyridine, *in* M-00390
Methyl (dimethylamido)fluorodithiophosphate, *in* D-00352
Methyl [3-dimethylamino-3-(diethoxyphosphinyl)]propanoate, *in* A-00122
4-Methyl-2-dimethylamino-1,3-dioxa-2-oxo-2-phosphacyclohexane, *see* M-00118
4-Methyl-2-dimethylamino-1,3-dioxa-2-phosphacyclohexane, *see* M-00117
4-Methyl-2-dimethylamino-1,3-dioxa-2-phosphacyclopentane, *see* M-00116
4-Methyl-2-dimethylamino-1,3,2-dioxaphospholane, M-00116
4-Methyl-2-dimethylamino-1,3,2-dioxaphosphorinane, M-00117
4-Methyl-2-dimethylamino-1,3,2-dioxaphosphorinane 2-oxide, M-00118
Methyl (4-dimethylaminophenyl)-methylphosphinate, *in* D-00705
4-Methyl-2-(1,1-dimethylethoxy)-1,3,2-dioxaphosphorinane, *see* B-00537
4-Methyl-*N*-(1,1-dimethylethyl)-1,3,2-dioxaphosphorinan-2-amine, *see* B-00541
Methyl(1,1-dimethylethyl)phenylphosphine, *see* B-00565
P-Methyl-*N*-(1,1-dimethylethyl)-phosphonamidic acid, *see* B-00572
O-Methyl (1,1-dimethylethyl)-phosphonothioate, *see* M-00107
Methyl 2,3-di-*O*-methylglucopyranoside 4,6-cyclic phosphonate, M-00119
Methyl 2,3-di-*O*-methylglucopyranoside 4,6-cyclic phosphonothioate, M-00120
Methyl (1,1-dimethyl-3-oxobutyl)-ethylphosphinate, *in* D-00782
Methyl (2,5-dimethylphenyl)phosphinate, *in* D-00805
Methyl (3,4-dimethylphenyl)phosphinate, *in* D-00806
Methyl *N*,*N*-dimethyl-*P*-phenyl-phosphonamidate, *in* D-00808
Methyl dimethylphosphinate, *in* D-00829
Methyl dimethylphosphinite, *in* D-00844
Methyl dimethylphosphinodithioate, *in* D-00834
O-Methyl dimethylphosphinoselenoate, *in* D-00835
Methyl dimethylphosphinoselenoite, *in* D-00838
Se-Methyl dimethylphosphinoselenothioate, *in* D-00837
O-Methyl dimethylphosphinothioate, *in* D-00839
S-Methyl dimethylphosphinothioate, *in* D-00839
Methyl dimethylphosphinothioite, *in* D-00843
Methyl *N*-(dimethylphosphinyl)carbamate, *in* D-00830
Methyl *N*,*P*-dimethylphosphonamidate, *in* D-00849
S-Methyl *N*,*P*-dimethylphosphonamidothioate, *in* D-00850
Methyl dimethylphosphoramidochloridate, *in* D-00860
Methyl dimethylphosphoramidochloridite, *in* D-00862
O-Methyl dimethyl-phosphoramidochloridothioate, *in* D-00861
Methyl dimethylphosphoramidocyanidate, *in* D-00863
Methyl dimethyl-phosphoramidofluoridodithioate, *in* D-00867
3-Methyl-4-dimethylphosphorylpyridine, *in* M-00390
Methyl 2,3-di-*O*-methyl-4-thioglucopyranoside 4,6-cyclic phosphonate, M-00121
Methyl 2,3-di-*O*-methyl-6-thioglucopyranoside 4,6-cyclic phosphonate, M-00122
Methyl 2,3-di-*O*-methyl-4-thioglucopyranoside 4,6-cyclic phosphonothioate, M-00123
Methyl *N*,*N'*-dimethyl-*N*,*N'*-trimethylene phosphorodiamidite, *in* H-00022
Methyl dimethylvinylene phosphate, *in* H-00123
Methyl di-4-morpholinylphosphinite, *in* D-00936
Methyldi-1-naphthylphosphonium iodide, *in* D-00939
Methyldioctylphenylphosphonium iodide, *in* D-00955
4-Methyl-2,6-dioxa-7-aza-1-phosphabicyclo[2.2.2]octane 1-oxide, M-00124
2-Methyl-1,3-dioxa-2-phosphacyclohexane, *see* M-00128
4-Methyl-1,3-dioxa-2-phosphacyclohexane, *see* M-00129
2-Methyl-1,3-dioxa-2-phosphacyclopentane, *see* M-00126
2-Methyl-1,3,2-dioxaphosphepane, M-00125
2-Methyl-1,3,2-dioxaphosphinane, *see* M-00128
4-Methyl-1,3,2-dioxaphosphinane, *see* M-00129
4-Methyl-1,3,2-dioxaphosphinane 2-oxide, *see* M-00130
4-Methyl-1,3,2-dioxaphosphinane 2-sulfide, *see* M-00131
2-Methyl-1,3,2-dioxaphospholane, M-00126
4-Methyl-1,3,2-dioxaphospholane, M-00127
2-Methyl-1,3,2-dioxaphosphorinane, M-00128
4-Methyl-1,3,2-dioxaphosphorinane, M-00129
4-Methyl-1,3,2-dioxaphosphorinane 2-oxide, M-00130
4-Methyl-1,3,2-dioxaphosphorinane 2-sulfide, M-00131
4-Methyl-1,3-dioxa-2-thioxo-2-phosphacyclohexane, *see* M-00131
[3-(2-Methyl-1,3-dioxolan-2-yl)-propylidene]triphenylphosphorane, *in* M-00132
[3-(2-Methyl-1,3-dioxolan-2-yl)propyl]-triphenylphosphonium, M-00132
Methyldipentylphenylphosphonium iodide, *in* D-00978
Methyl *N*-(diphenoxyphosphinyl)carbamate, *in* D-00989
1-Methyl-2,5-diphenyl-1,2,5-azadiphospholidine, *in* M-00089
5-Methyl-5,10-diphenyldibenzo[1,4]-phosphoniaarsenin, *in* D-01001
O-Methyl *P*,*P*-diphenyl-*N*-(diphenylphosphinothioyl)-phosphinimidate, *in* D-01006
5-Methyl-2,3-diphenyl-1,3-oxaphospholane, M-00133
3-Methyl-2,3-diphenyl-1,3-oxaphospholanium iodide, *in* D-01027
3-Methyl-2,3-diphenyl-1,3-oxaphosphorinanium iodide, *in* D-01028
Methyl diphenyl(phenylethenyl)phosphonium iodide, *in* D-01033
Methyldiphenyl(phenylmethyl)phosphonium iodide, *in* B-00046
Methyl *P*,*P*-diphenyl-*N*-phenylsulfonylphosphinimidate, *in* B-00013
6-Methyl-1,5-diphenyl-6-phosphabicyclo[3.2.1]octane, M-00134
Methyl diphenylphosphinate, *in* D-01039
Methyldiphenylphosphine, M-00135
N-Methyl-*P*,*P*-diphenylphosphinic amide, *in* D-01040
P-Methyl-*N*,*P*-diphenylphosphinic amide, *in* M-00223
N-Methyl-*P*,*P*-diphenylphosphinimidic fluoride, *in* D-01044
Methyl diphenylphosphinite, *in* D-01089
Methyl 3-(diphenylphosphino)benzoate, *in* D-01054
Methyl 4-(diphenylphosphino)benzoate, *in* D-01055
Methyl diphenylphosphinodithioate, *in* D-01065
O-Methyl diphenylphosphinoselenoate, *in* D-01078
Methyl diphenylphosphinoselenoite, *in* D-01081
O-Methyl diphenylphosphinothioate, *in* D-01082
S-Methyl diphenylphosphinothioate, *in* D-01082
N-Methyl-*P*,*P*-diphenylphosphinothioic amide, *in* D-01083
Methyl diphenylphosphinothioite, *in* D-01085
Methyl (4-diphenylphosphinothioyl)benzoate, *in* D-01055
Methyl (diphenylphosphinyl)acetate, *in* D-01093

Methyl 4-(diphenylphosphinyl)benzoate, *in* D-01055
Methyl *N,P*-diphenylphosphonamidate, *in* D-01103
P-Methyl-*N,N'*-diphenylphosphonic diamide, *in* M-00289
P-Methyl-*N,N'*-diphenylphosphonothioic diamide, *in* M-00314
Methyl *N,N'*-diphenylphosphorodiamidate, *in* D-01125
S-Methyl *N,N'*-diphenylphosphorodiamidothioate, *in* D-01129
Methyldiphenyl-2-propenylphosphonium iodide, *in* D-01144
Methyldiphenyl-(1-propynyl)phosphonium iodide, *in* D-01146
Methyldiphenyl-4-pyridinylphosphonium bromide, *in* D-01077
Methyldiphenyl(3-trifluoromethylphenyl)-phosphonium iodide, *in* D-01155
Methyldiphenylvinylphosphonium iodide, *in* D-01168
Methyl di-1-piperidinylphosphinate, *in* D-01190
Methyl di-2-propenylphosphinate, *in* D-01193
O-Methyl dipropylphosphinothioate, *in* D-01208
S-Methyl dipropylphosphinothioate, *in* D-01208
2-Methyl-1,3-dithia-2-phosphacyclopentane, *see* M-00139
4-Methyl-1,3,2-dithiaphosphinane, *see* M-00138
4-Methyl-1,3,2-dithiaphospholane, M-00136
2-Methyl-1,3,2-dithiaphosphorinane, M-00137
4-Methyl-1,3,2-dithiaphosphorinane, M-00138
2-Methyl-1,3,2-dithiophospholane, M-00139
Methyl di-*o*-tolylphosphinate, *in* B-00318
Methyl di-*p*-tolylphosphinate, *in* B-00320
Methyl divinylphosphinate, *in* D-01258
Methylenebis[dibutylphosphine], *see* B-00160
Methylenebis[dimethylphosphine], *see* B-00202
Methylenebis[diphenylphosphine], *see* B-00244
Methylenebis[methylphosphinic acid], M-00140
[Methylenebis(2,1-phenylenemethylene)]-bis[triphenylphosphonium](2+), M-00141
[Methylenebis(2,1-phenylenemethylene)]-bis[triphenylphosphorane], *in* M-00141
Methylenebis[phenylphosphinic acid], M-00142
Methylenebisphosphine, *see* D-01178
Methylenebis[phosphinic acid], M-00143
Methylenebisphosphonic acid, *see* M-00022
P,P'-Methylenebis(phosphonic diamide), M-00144
Methylenebis(phosphonic dichloride), M-00145
Methylenebisphosphonic difluoride, M-00146
Methylenebisphosphonothioic acid, *see* M-00023
Methylenebisphosphonous acid, *see* M-00024
Methylenebis[*N,N,N',N'*-tetraethylphosphonous diamide], *in* M-00024
P,P'-Methylenebis[(*N,N,N',N'*-tetramethyl)-phosphonodiamidothioate], *in* M-00023
Methylenebis[triphenylphosphonium](2+), M-00147
3,4-Methylenedioxybenzenephosphonic acid, *see* B-00019
Methylenediphosphonic acid, *see* M-00022
Methylenediphosphonous acid, *see* M-00024
(Methylenehydrazono)triphenylphosphorane, *see* F-00054
▷*S,S'*-Methylene-*O,O,O',O'*-tetraethyl di(phosphorodithioate), *see* E-00022
Methylenetriphenylphosphorane, M-00148
(1-Methyl-1,2-ethanediyl)-bis[diphenylphosphine], *see* B-00264
Methyl *N,N'*-ethanediyl-*N,N'*-dimethylphosphorodiamidite, *in* D-00719
Methyl ethenylmethylphosphinate, *in* M-00438
Methyl ethenylphenylphosphinate, *in* P-00337
(1-Methylethenyl)phosphonic acid, M-00149
(1-Methylethenyl)phosphonic dichloride, M-00150
1-Methylethenyl phosphorodichloridate, M-00151
2-(1-Methylethoxy)-1,3,2-dioxaphospholane, M-00152
1-Methylethoxy-2,2,3,4,4-pentamethylphosphetane 1-oxide, *in* H-00168
1-Methylethyl bis(1-aziridinyl)phosphinite, *in* B-00095
1-Methylethyl diethylphosphinothioite, *in* D-00332
1-Methylethyl diethylphosphoramidochloridate, *in* D-00346
1-Methylethyl diethylphosphoramidofluoridate, *in* D-00351
(1-Methylethyl)diphenylphosphine, *see* I-00031
▷1-Methylethyl 2-[[ethoxy[(1-methylethyl)-amino]phosphinothioyl]oxy]benzoate, *see* I-00025
▷1-Methylethyl 3-[[(ethylamino)-methoxyphosphinothioyl]oxy]-2-butenoate, *see* P-00466
O-1-Methylethyl ethylphosphonoselenoate, *in* E-00145
(1-Methylethylidene)bisphosphonic acid, *see* P-00445
(1-Methylethylidene)triphenylphosphorane, M-00153
1-Methylethyl methylphosphinate, *see* I-00034
Methyl ethylmethylphosphinite, *in* E-00087
Methyl *N*-ethyl-*P*-methylphosphonamidate, *in* E-00088
1-Methylethyl methylphosphonate, *see* I-00038
O-(1-Methylethyl) methylphosphonochloridothioate, *see* I-00039
O-(1-Methylethyl) methylphosphonodithioate, *see* I-00040
▷1-Methylethyl methylphosphonofluoridate, *see* S-00002
O-(1-Methylethyl) methylphosphonothioate, *see* I-00041
1-Methylethyl phenyl methylphosphonate, *in* I-00038
Methyl ethylphenylphosphinate, *in* E-00101
1-Methylethyl phenylphosphinate, *in* P-00210
(1-Methylethyl)phenylphosphinic acid, *see* I-00043
P-(1-Methylethyl)-*P*-phenylphosphinic amide, *see* I-00044
Methyl ethylphenylphosphinite, *in* E-00108
(1-Methylethyl)phenylphosphinodithioic acid, *see* I-00045
O-Methyl ethylphenylphosphinoselenoate, *in* E-00104
O-Methyl ethylphenylphosphinothioate, *in* E-00105
S-Methyl ethylphenylphosphinothioate, *in* E-00105
(1-Methylethyl)phenylphosphinothioic acid, *see* I-00046
(1-Methylethyl) phenyl phosphite, *see* I-00047
Methyl *N*-ethyl-*P*-phenylphosphonamidate, *in* E-00109
Methyl *P*-ethyl-*N*-phenylphosphonamidate, *in* E-00110
(1-Methylethyl) phenyl phosphonate, *see* I-00047
O-(1-Methylethyl) phenylphosphonochloridothioate, *in* P-00232
S-(1-Methylethyl) phenylphosphonochloridothioate, *in* P-00232
Methyl ethylphosphinate, *in* E-00124
(1-Methylethyl)phosphine, *see* I-00048
(1-Methylethyl)phosphinic acid, *see* I-00049
Methyl *P*-ethylphosphonamidodithioate, *in* E-00127
S-Methyl *P*-ethylphosphonamidothioate, *in* E-00128
(1-Methylethyl)phosphonic acid, *see* I-00050
P-(1-Methylethyl)phosphonic diamide, *see* I-00051
(1-Methylethyl)phosphonic dichloride, *see* I-00052
O-Methyl ethylphosphonochloridothioate, M-00154
S-Methyl ethylphosphonochloridothioate, *in* E-00137
(1-Methylethyl)phosphonochloridothioic acid, *see* I-00053
Methyl ethylphosphonofluoridate, *in* E-00142
O-Methyl ethylphosphonothioate, M-00155
(1-Methylethyl)phosphonothioic acid, *see* I-00054
(1-Methylethyl)phosphonothioic dichloride, *see* I-00055
(1-Methylethyl)phosphonous acid, *see* I-00056
(1-Methylethyl)phosphonous dichloride, *see* I-00057
(1-Methylethyl)phosphoramidic acid, *see* I-00058
O-Methyl ethylphosphoramidochloridothioate, *in* E-00157
(1-Methylethyl)phosphoramidothioic acid, *see* I-00059
1-Methylethyl phosphorodichloridate, *see* I-00060
1-Methylethyl phosphorodichloridite, *see* I-00061
O-(1-Methylethyl) phosphorodichloridothioate, *see* I-00062
1-Methylethyl phosphorodichloridothioite, *see* I-00063
S-(1-Methylethyl) phosphorothioate, *see* I-00064
(1-Methylethyl)triphenylphosphonium(1+), *see* I-00065
Methyl fluoro(dimethylamido)dithiophosphate, *in* D-00867
Methylglycerol bicyclic phosphate, *in* M-00423
Methylglycerol bicyclic selenophosphate, *in* M-00423
Methylglycerol phosphite, *see* M-00423
Methyl hydrogen acetyl phosphate, *in* A-00025
Methyl hydrogen [[1-amino-2-(1*H*-indol-3-yl)]ethyl]phosphonate, *in* P-00421
Methyl hydrogen (2-aminophenyl)phosphonate, *in* A-00109
Methyl hydrogen (4-aminophenyl)phosphonate, *in* A-00111
Methyl hydrogen benzylphosphonate, *in* B-00060
Methyl hydrogen chloromethylphosphonate, *in* C-00106
Methyl hydrogen (1,2-dibromo-1-phenylpropyl)phosphonate, *in* D-00068
O-Methyl hydrogen hexylphosphonodithioate, *in* H-00099
Methyl hydrogen (α-hydroxybenzyl)-phosphonate, *in* H-00174
Methyl hydrogen [[(4-methylphenyl)-sulfinyl]methyl]phosphonate, *in* M-00266
Methyl hydrogen methylphosphonate, *see* M-00168
O-Methyl hydrogen methylphosphonothioate, *see* M-00171
Methyl hydrogen (2-nitrophenyl)phosphonate, *in* N-00055
Methyl hydrogen (4-nitrophenyl)phosphonate, *in* N-00057
O-Methyl hydrogen pentylphosphonodithioate, *in* P-00051
Methyl hydrogen (2-phenylethenyl)-phosphonate, *in* P-00155
Methyl hydrogen phenylphosphonate, *see* M-00488
▷Methyl hydrogen phosphonate, *see* M-00490
Methyl hydrogen phosphoromorpholidothioate, *in* M-00527
Methyl 3-hydroxy-2-[(dibenzyloxyphosphinyl)amino]butanoate, *in* P-00420
Methyl 3-hydroxy-2-[(dibenzyloxyphosphinyl)amino]-propanoate, *in* P-00418
Methyl 3-hydroxy-2-[(diisopropoxyphosphinyl)amino]-butanoate, *in* P-00420
Methyl 3-hydroxy-2-[(diisoproproxyphosphinyl)amino]-propanoate, *in* P-00418
Methyl (hydroxymethyl)phenylphosphinate, *in* H-00160
▷2-Methyl-1-hydroxy-2-phospholene 1-oxide, *see* D-00497

Methyl hypophosphite, *see* M-00276
Methylidynetris[dimethylphosphine], *see* T-00771
Methylidynetris[diphenylphosphine], *see* T-00775
Methylimidobisphosphoric acid, *see* M-00156
Methylimido bis(phosphoryl dimethylamide), *see* N-00068
Methylimidobisphosphoryl fluoride, *in* M-00156
Methylimidodiphosphoric acid, M-00156
Methylimidodiphosphorothioic acid, *see* M-00161
Methylimidodi(phosphorous acid), M-00157
Methylimidodiphosphorous tetrachloride, M-00158
Methylimidodiphosphorous tetrafluoride, M-00159
Methylimidodiphosphoryl chloride, M-00160
Methylimidothiodiphosphoric acid, M-00161
Methyl *N*-[imino(diphenoxyphosphinylamino)-methyl]glycinate, *in* P-00369
Methyliminodiphosphoric acid, *see* M-00156
Methyliminodi(phosphorous acid), *see* M-00157
Methyliminodiphosphorous tetrachloride, *see* M-00158
Methyliminodiphosphorous tetrafluoride, *see* M-00159
Methyliminodiphosphoryl tetrachloride, *see* M-00160
Methyliminothiodiphosphoric acid, *see* M-00161
Methylimino trimethyl phosphate, *in* M-00336
1-Methylisophosphinoline, M-00162
3-Methylisophosphinoline, M-00163
(5-Methyl-2-isopropylcyclohexyl)-diphenylphosphine, M-00164
Methyl *N*-isopropyl-*P*-methylphosphonamidate, *in* I-00036
Methyl isopropylphenylphosphinate, *in* I-00043
S-Methyl isopropylphenylphosphinothioate, *in* I-00046
P-Methyl-*N*-isopropyl-*N'*-phenylphosphonothioic diamide, *in* M-00245
O-Methyl isopropylphosphonochloridothioate, M-00165
O-Methyl isopropylphosphonothioate, M-00166
▷4-Methylmercapto-3-methylphenyl dimethyl thiophosphate, *see* F-00007
Methyl [2-(methoxycarbonyl)ethyl]-methylphosphinate, *in* H-00161
Methyl 3-(methoxymethylphosphinyl)-propanoate, *in* H-00161
Methyl(2-methoxyphenyl)phenylphosphine, *see* M-00059
Methyl 3-(methoxyphenylphosphinyl)-propanoate, *in* H-00178
Methyl *P*-methyl-*N*,*N*-dimethylphosphonamidodithioate, *in* E-00063
6-Methyl-4-methylene-4*H*-1,3,2-dioxaphosphinine, *see* M-00167
6-Methyl-4-methylene-4*H*-1,3,2-dioxaphosphorin, M-00167
[5-Methyl-7-(1-methylethyl)bicyclo[2.2.2]-oct-5-ene-2,3-diyl]-bis[diphenylphosphine], *see* P-00059
[5-Methyl-2-(1-methylethyl)cyclohexyl]-diphenylphosphine, *see* M-00164
Methyl 1-methylethyl methylphosphonate, *in* M-00168
Methyl(1-methylethyl)phosphinic acid, *see* I-00035
P-Methyl-*N*-(1-methylethyl)phosphonamidic acid, *see* I-00036
P-Methyl-*N*-(1-methylethyl)-phosphonamidothioic acid, *see* I-00037
O-Methyl (1-methylethyl)-phosphonochloridothioate, *see* M-00165
Methyl methyl-1-naphthylphosphinate, *in* M-00178
▷*O*-Methyl *O*-(4-methyl-2-nitrophenyl) isopropylphosphoramidothioate, *see* A-00132
S-Methyl *O*-(4-methyl-2-nitrophenyl) isopropylphosphoramidothioate, *in* M-00180
▷*O*-Methyl *O*-(4-methyl-2-nitrophenyl) (1-methylethyl)phosphoramidothioate, *see* A-00132
Methyl (4-methylphenyl)methylphosphinate, *in* M-00204
Methyl *P*-(4-methylphenyl)-*N*-phenylphosphonamidate, *in* M-00214
Methyl methylphenylphosphinate, *in* M-00219
Methyl (4-methylphenyl)phosphinate, *in* M-00222
Methyl methylphenylphosphinite, *in* M-00227
O-Methyl methylphenylphosphinothioate, *in* M-00225
S-Methyl methylphenylphosphinothioate, *in* M-00225
Methyl *N*-methyl-*P*-phenylphosphonamidate, *in* M-00230
Methyl *P*-methyl-*N*-phenylphosphonamidate, *in* M-00231
Methyl *P*-methyl-*N*-phenylphosphonamidodithioate, *in* M-00232
Methyl methylphosphinate, *in* M-00278
O-Methyl *P*-methylphosphonamidothioate, *in* M-00284
S-Methyl *P*-methylphosphonamidothioate, *in* M-00284
Methyl methylphosphonate, M-00168
Methyl methylphosphonisothiocyanatidate, *in* M-00294
Methyl methylphosphonochloridate, *in* M-00296
Methyl methylphosphonochloridite, *in* M-00300
▷Methyl methylphosphonochloridodithioate, *in* M-00297
O-Methyl methylphosphonochloridothioate, M-00169
S-Methyl methylphosphonochloridothioate, *in* M-00298
Methyl methylphosphonochloridothioite, *in* M-00299
O-Methyl methylphosphonodithioate, M-00170
Methyl methylphosphonofluoridate, *in* M-00303
Methyl methylphosphonofluoridite, *in* M-00307
O-Methyl methylphosphonofluoridothioate, *in* M-00305
O-Methyl methylphosphonothioate, M-00171
Methyl(1-methylpropyl)diphenylphosphonium iodide, *in* M-00369
Methyl(2-methylpropyl)phosphinic acid, M-00172
4-Methyl-2-methylthio-1,3-dioxa-2-phosphacyclohexane, *see* M-00174
4-Methyl-2-methylthio-1,3-dioxa-2-phosphacyclopentane, *see* M-00173
4-Methyl-2-methylthio-1,3,2-dioxaphosphinane, *see* M-00174
4-Methyl-2-methylthio-1,3,2-dioxaphospholane, M-00173
4-Methyl-2-methylthio-1,3,2-dioxaphosphorinane, M-00174
Methyl methyl-*p*-tolylphosphinate, *in* M-00204
Methyl methyl(trifluoromethyl)phosphinate, *in* M-00420
Methyl methylvinylphosphinate, *in* M-00438
Methyl-4-morpholinylphosphinic acid, *see* M-00175
P-Methyl-4-morpholinylphosphonamidic acid, M-00175
Methyl 4-morpholinylphosphonochloridite, *in* M-00526
Methyl-2-naphthalenylphenyl(phenylmethyl)-phosphonium bromide, *in* M-00177
Methyl-2-naphthalenylphenylphosphine, *see* M-00177
Methyl(1-naphthalenyl)phosphinic acid, *see* M-00178
O-Methyl *O*-1-naphthalenyl phosphorothioate, M-00176
O-Methyl *O*-1-naphthyl hydrogen phosphorothioate, *see* M-00176
Methyl 1-naphthylphenylphosphinate, *in* N-00017
Methyl 2-naphthylphenylphosphinate, *in* N-00018
Methyl-2-naphthylphenylphosphine, M-00177
Methyl-1-naphthylphosphinic acid, M-00178
O-Methyl naphthylphosphonothioate, *in* N-00023
O-Methyl *O*-1-naphthyl phosphorothioic acid, *see* M-00176
Methyl neopentylene phosphate, *in* M-00034
Methyl neopentylene phosphite, *see* M-00034
Se-Methyl *O*,*O*-neopentylene phosphorodiselenoate, *in* D-00764
S-Methyl *O*,*O*-neopentylene phosphorodithioate, *in* D-00766
Se-Methyl *O*,*O*-neopentylene phosphoroselenoate, *in* D-00764
Se-Methyl *O*,*O*-neopentylene phosphoroselenoite, *see* D-00764
Se-Methyl *O*,*O*-neopentylene phosphoroselenothioate, *in* D-00764
O-Methyl *O*,*O*-neopentylene phosphorothioate, *in* M-00034
S-Methyl *O*,*O*-neopentylene phosphorothioate, *in* D-00766
S-Methyl *O*,*O*-neopentylene phosphorothioite, *see* D-00766
O-Methyl *O*-4-nitrophenyl cyclohexylphosphoramidothioate, M-00179
Methyl(3-nitrophenyl)diphenylphosphonium iodide, *in* N-00039
Methyl(4-nitrophenyl)diphenylphosphonium iodide, *in* N-00040
O-(4-Methyl-2-nitrophenyl) isopropylphosphoramidothioate, M-00180
O-(4-Methyl-2-nitrophenyl) (1-methylethyl)phosphoramidothioate, *see* M-00180
▷*O*-Methyl *O*-4-nitrophenyl phenylphosphonothioate, *in* M-00243
Methyl nonylphenylphosphinate, *in* N-00074
(1-Methylnonyl)phosphonic acid, M-00181
Methyl norphos, *see* B-00247
Methyloldiphenylphosphine, *see* D-01069
2-Methyl-1,2-oxaphospholane, M-00182
5-Methyl-1,2-oxaphospholane 2-oxide, M-00183
2-Methyl-1,2-oxaphospholan-5-one 2-oxide, *in* O-00053
5-Methyl-1,2-oxaphosphol-4-ene 2-oxide, *see* D-00525
5-Methyl-1,3,2-oxathiaphosphole 2-oxide, M-00184
(3-Methyloxiranyl)phosphonic acid, M-00185
(3-Methyl-2-oxo-2(5*H*)-furanyl)-triphenylphosphonium(1+), M-00186
(3-Methyl-2-oxoheptyl)phosphonic acid, M-00187
(1-Methyl-4-oxo-2-imidazolidinylene)-phosphoramidic acid, *see* F-00060
(1-Methyl-2-oxo-2-phenylethyl)-triphenylphosphonium(1+), M-00188
3-Methyl-2-oxo-3-phenyl-1,3-oxaphosphorinanium iodide, *in* P-00177
1-Methyl-4-oxo-1-phenyl-4-phosphorinanium iodide, *in* P-00268
2-(1-Methyl-2-oxopropylidene)-phosphorohydrazidothioate oxime, M-00189
Methyl γ-oxo-β-(triphenylphosphoranylidene)-benzenebutanoate, *in* O-00082
Methyl 3-oxo-2-(triphenylphosphoranylidene)butanoate, *in* O-00084
Methyl 4-oxo-5-(triphenylphosphoranylidene)pentanoate, *in* O-00085
▷Methyl paraoxon, *see* D-00770
▷Methyl parathion, *see* P-00003
Methylpentamethylenephosphine, *see* M-00339
Methyl 1,1,2,3,3-pentamethyltrimethylenephosphinate, *in* H-00168
N-Methyl-*P*,*P*-(1,1,2,3,3-pentamethyltrimethylene)phosphinic amide, *in* P-00021
S-Methyl 1,1,2,3,3-pentamethyltrimethylenephosphinothioate, *in* H-00169
2-Methyl-2,4-pentanediyl chlorophosphite, *see* C-00189
2-Methyl-2,4-pentanediyl chlorothiophosphate, *in* C-00189
9-Methyl-1,2,3,4,9-pentaphenyl-1*H*-tribenzo[*b*,*d*,*f*]phosphepinium iodide, *in* P-00042
Methyl phenazaphosphinate, *in* P-00061
4-Methyl-2-phenoxy-1,3-dioxa-2-phosphacyclohexane, *see* M-00191

4-Methyl-2-phenoxy-1,3-dioxa-2-phosphacyclopentane, *see* M-00190
4-Methyl-2-phenoxy-1,3,2-dioxaphosphinane, *see* M-00191
4-Methyl-2-phenoxy-1,3,2-dioxaphospholane, M-00190
4-Methyl-2-phenoxy-1,3,2-dioxaphosphorinane, M-00191
3-Methyl-2-phenoxy-1,3,2-oxazaphospholane, *see* M-00192
3-Methyl-2-phenoxy-1,3,2-oxazaphospholidine, M-00192
3-Methyl-2-phenoxy-1,3,2-oxazaphosphorinane, *see* M-00404
4-Methyl-2-phenylamino-1,3,2-dioxaphosphorinane, M-00193
(Methylphenylamino)triphenylphosphonium(1+), M-00194
Methyl phenyl 1-aziridinylphosphonite, *in* A-00154
5-Methyl-2-phenyl-1,3,2-benzodithiaphosphole, M-00195
1-Methyl-1-phenyl-1*H*-benzo[*b*]phosphindolium iodide, *in* P-00110
3-Methyl-2-phenyl-1,3-diaza-2-phosphabicyclo[3.3.1]nonane, M-00196
5-Methyl-2-phenyl-2*H*-1,2,3-diazaphosphole, M-00197
5-Methyl-5-phenyl-5*H*-dibenzophospholium iodide, *in* P-00110
7-Methyl-7-phenyl-7*H*-dibenzo[*d,f*]-phosphoninium iodide, *in* P-00111
4-Methylphenyl difluorophosphate, *in* M-00487
2,4-Methyl-2-phenyl-1,3-dioxa-2-phosphacyclohexane, *see* M-00199
4-Methyl-2-phenyl-1,3-dioxa-2-phosphacyclopentane, *see* M-00198
4-Methyl-2-phenyl-1,3,2-dioxaphosphinane, *see* M-00199
4-Methyl-2-phenyl-1,3,2-dioxaphospholane, M-00198
4-Methyl-*N*-phenyl-1,3,2-dioxaphosphorinan-2-amine, *see* M-00193
4-Methyl-2-phenyl-1,3,2-dioxaphosphorinane, M-00199
(2-Methylphenyl)diphenylphosphine, M-00200
(3-Methylphenyl)diphenylphosphine, M-00201
(4-Methylphenyl)diphenylphosphine, M-00202
P-(2-Methylphenyl)-*N,N'*-diphenylphosphonic diamide, *in* M-00235
P-(4-Methylphenyl)-*N,N'*-diphenylphosphonic diamide, *in* M-00238
N-(4-Methylphenyl)-*N',N''*-diphenylphosphoric triamide, *in* M-00253
4-Methyl-2-phenyl-1,3,2-dithiaphospholane, *in* M-00136
O-Methyl *O*-phenyl dithiophosphoric acid, *see* M-00263
Methyl *o*-phenylene phosphate, *in* M-00026
Methyl *o*-phenylene phosphite, *see* M-00026
Methyl (2-phenylethenyl)phosphinate, *in* P-00153
O-Methyl *S*-phenyl ethylphosphonodithioate, *in* E-00140
O-Methyl *O*-phenyl hydrogen dithiophosphate, *see* M-00263
[(2-Methylphenyl)methylene]-triphenylphosphorane, *in* M-00209
[(3-Methylphenyl)methylene]-triphenylphosphorane, *in* M-00210
[(4-Methylphenyl)methylene]-triphenylphosphorane, *in* M-00211
Methyl phenylmethyl methylphosphonate, *in* M-00168
Methyl(phenylmethyl)phosphinic acid, *see* B-00053
(2-Methylphenyl)methylphosphinic acid, M-00203
(4-Methylphenyl)methylphosphinic acid, M-00204
Methyl phenyl methylphosphonate, M-00205
[(2-Methylphenyl)methyl]phosphonic acid, M-00206
[(3-Methylphenyl)methyl]phosphonic acid, M-00207
[(4-Methylphenyl)methyl]phosphonic acid, M-00208
O-Methyl *S*-phenyl methylphosphonothioate, *in* M-00171
O-Methyl-*S*-phenyl methylphosphoramidodithioate, *in* M-00330
Methyl(phenylmethyl)propylphosphine, *see* B-00054
[(2-Methylphenyl)methyl]-triphenylphosphonium(1+), M-00209
[(3-Methylphenyl)methyl]-triphenylphosphonium(1+), M-00210
[(4-Methylphenyl)methyl]-triphenylphosphonium(1+), M-00211
4-Methyl-4-phenyl-1,4-oxaphosphinanium iodide, *in* P-00176
4-Methyl-4-phenyl-1-oxa-4-phosphoniacyclohexane iodide, *in* P-00176
4-Methyl-4-phenyl-1,4-oxaphosphorinanium iodide, *in* P-00176
2-Methyl-3-phenyl-1,3,2-oxazaphospholane, *see* M-00212
3-Methyl-2-phenyl-1,3,2-oxazaphospholane, *see* M-00213
2-Methyl-3-phenyl-1,3,2-oxazaphospholidine, M-00212
3-Methyl-2-phenyl-1,3,2-oxazaphospholidine, M-00213
10-Methyl-10-phenyl-10*H*-phenothiaphosphinium 5,5-dioxide iodide, *in* P-00182
10-Methyl-10-phenyl-10*H*-phenothiaphosphinium iodide, *in* P-00182
10-Methyl-10-phenyl-10*H*-phenoxaphosphinium iodide, *in* P-00183
Methyl phenyl(2-phenylethenyl)phosphinate, *in* P-00186
Methylphenyl(phenylmethylene)-propylphosphorane, *in* B-00052
Methylphenyl(phenylmethyl)phosphine, *see* B-00051
Methylphenyl(phenylmethyl)-2-propenylphosphonium bromide, *in* M-00264
Methylphenyl(phenylmethyl)-propylphosphonium(1+), *see* B-00052
P-(4-Methylphenyl)-*N*-phenylphosphonamidic acid, M-00214
Methyl phenyl phenylphosphonate, *in* M-00488
Methyl *P*-phenyl-*N*-phenylsulfonylphosphonamidate, *in* P-00191
Methyl phenylphosphinate, *in* P-00210
Methylphenylphosphine, M-00215
(2-Methylphenyl)phosphine, M-00216
(3-Methylphenyl)phosphine, M-00217
(4-Methylphenyl)phosphine, M-00218
Methylphenylphosphinic acid, M-00219
(2-Methylphenyl)phosphinic acid, M-00220
(3-Methylphenyl)phosphinic acid, M-00221
▷(4-Methylphenyl)phosphinic acid, M-00222
P-Methyl-*P*-phenylphosphinic amide, M-00223
Methylphenylphosphinodithioic acid, M-00224
Methylphenylphosphinothioic acid, M-00225
Methylphenylphosphinothioic amide, M-00226
Methylphenylphosphinous acid, M-00227
Methylphenylphosphinous chloride, M-00228
Methyl phenyl phosphite, *see* M-00234
3-Methyl-1-phenylphospholane, M-00229
1-Methyl-1-phenylphospholanium iodide, *in* P-00216
Methyl *P*-phenylphosphonamidate, *in* P-00218
N-Methyl-*P*-phenylphosphonamidic acid, M-00230
P-Methyl-*N*-phenylphosphonamidic acid, M-00231
P-Methyl-*N*-phenylphosphonamidodithioic acid, M-00232
P-Methyl-*N*-phenylphosphonamidothioic acid, M-00233
Methyl phenyl phosphonate, M-00234
Methyl phenylphosphonazidate, *in* P-00220
P-Methyl-*N*-phenylphosphonazidic amide, *in* M-00287
9-Methyl-9-phenyl-9-phosphoniabicyclo[4.2.1]nonane iodide, *in* P-00194
9-Methyl-9-phenyl-9-phosphoniabicyclo[4.2.1]nona-2,4,7-triene iodide, *in* P-00196
8-Methyl-8-phenyl-8-phosphoniabicyclo[3.2.1]octan-3-one iodide, *in* P-00199
9-Methyl-9-phenyl-9-phosphoniahomocubane iodide, *in* P-00201
(2-Methylphenyl)phosphonic acid, M-00235
(3-Methylphenyl)phosphonic acid, M-00236
▷(4-Methylphenyl)phosphonic acid, M-00237
P-(4-Methylphenyl)phosphonic diamide, M-00238
(4-Methylphenyl)phosphonic dichloride, M-00239
N-Methyl-*P*-phenylphosphonimidic acid, M-00240
Methyl phenylphosphonobromidodithioate, *in* P-00228
Methyl phenylphosphonochloridate, *in* P-00229
Methyl phenylphosphonochloridoselenoite, *in* P-00231
O-Methyl phenylphosphonochloridothioate, *in* P-00232
S-Methyl phenylphosphonochloridothioate, *in* P-00232
Methyl phenylphosphonofluoridate, *in* P-00239
Methyl *P*-phenylphosphonohydrazidate, *in* P-00241
P-Methyl-N^2-phenylphosphonohydrazidic acid, M-00241
P-Methyl-N^2-phenylphosphonohydrazidothioic acid, M-00242
O-Methyl phenyl-phosphonoisothiocyanatidothioate, *in* P-00227
O-Methyl phenylphosphonothioate, M-00243
S-Methyl phenylphosphonothioate, M-00244
P-Methyl-*N*-phenylphosphonothioic diamide, M-00245
(2-Methylphenyl)phosphonous acid, M-00246
(3-Methylphenyl)phosphonous acid, M-00247
(4-Methylphenyl)phosphonous acid, M-00248
(4-Methylphenyl)phosphonous dichloride, M-00249
Methylphenylphosphoramidic acid, M-00250
(2-Methylphenyl)phosphoramidic acid, M-00251
(3-Methylphenyl)phosphoramidic acid, M-00252
(4-Methylphenyl)phosphoramidic acid, M-00253
(2-Methylphenyl)phosphoramidothioic acid, M-00254
(3-Methylphenyl)phosphoramidothioic acid, M-00255
(4-Methylphenyl)phosphoramidothioic acid, M-00256
2-Methylphenyl phosphoric acid, *see* M-00485
4-Methylphenyl phosphoric acid, *see* M-00487
1-Methyl-1-phenylphosphorinanium iodide, *in* P-00265
Methyl phenyl phosphorochloridate, *in* P-00269
2-Methylphenyl phosphorodichloridate, *in* M-00485
3-Methylphenyl phosphorodichloridate, *in* M-00486
4-Methylphenyl phosphorodichloridate, M-00257
2-Methylphenyl phosphorodichloridite, M-00258
3-Methylphenyl phosphorodichloridite, M-00259
4-Methylphenyl phosphorodichloridite, M-00260
O-(2-Methylphenyl) phosphorodichloridothioate, M-00261
O-(4-Methylphenyl) phosphorodichloridothioate, M-00262
2-Methylphenyl phosphorodifluoridate, *in* M-00485
4-Methylphenyl phosphorodifluoridate, *in* M-00487
O-Methyl *O*-phenyl phosphorodithioate, M-00263
4-Methylphenyl phosphoryl difluoride, *in* M-00487
Methyl phenyl(1-propenyl)phosphinate, *in* P-00299
Methyl phenyl(2-propenyl)phosphinate, *in* P-00300
Methylphenyl-2-propenylphosphine, M-00264
Methylphenyl(2-propenyl)propylphosphonium bromide, *in* M-00265
Methylphenylpropylphosphine, M-00265
[[(4-Methylphenyl)sulfinyl]methyl]-phosphonic acid, M-00266

[[(4-Methylphenyl)sulfonyl]methyl]-triphenylphosphonium(1+), M-00267
[(4-Methylphenyl)sulfonyl]phosphorimidic acid, M-00268
1-Methyl-5-phenyl-1*H*-1,2,4,3-triazaphosphole, M-00269
2-Methyl-5-phenyl-2*H*-1,2,4,3-triazaphosphole, M-00270
Methyl phenyl(trichloromethyl)phosphinate, *in* P-00327
Methyl *N*-phenyl-*P*-trichloromethylphosphonamidate, *in* P-00328
Methyl phenyl(triphenylphosphoranylidene)-acetate, *in* T-00644
4-Methylphenyl triphenylphosphoranylidenemethyl sulfone, M-00271
Methyl phenylvinylphosphinate, *in* P-00337
10-Methyl-9-phosphaanthracene, *see* M-00087
2-Methylphosphabenzene, *see* M-00337
8-Methyl-8-phosphabicyclo[3.2.1]octane, M-00272
8-Methyl-8-phosphabicyclo[3.2.1]oct-6-ene 8-oxide, M-00273
1-Methylphosphacyclohexane, *see* M-00339
1-Methylphosphacyclononane, *see* M-00285
3-Methyl-1-phospha-3-cyclopentene, *see* D-00530
1-Methyl-1-phospha-2,3:6,7-dibenzocyclohepta-2,6-diene, *see* D-00524
9-Methyl-9-phosphafluorene, *see* M-00115
1-Methyl-2-phosphanaphthalene, *see* M-00162
3-Methyl-2-phosphanaphthalene, *see* M-00163
5-Methyl-1-phospha-5-silabicyclo[3.3.1]-nonane, M-00274
5-Methyl-5-phospha(5-P^{V})spiro[4.4]nonane, *in* P-00350
4-Methyl-4-phosphatetracyclo[3.3.0.$0^{2,8}$.$0^{3,6}$]-octane, M-00275
1-Methylphosphinane, *see* M-00339
1-Methylphosphinanium iodide, *in* M-00339
1-Methyl-4-phosphinanone, *see* M-00340
Methyl phosphinate, M-00276
▷Methylphosphine, M-00277
Methylphosphinic acid, M-00278
2-Methylphosphinine, *see* M-00337
4-Methylphosphinine, *see* M-00338
Methylphosphinothioic acid, M-00279
1-Methylphosphirane, *in* P-00361
▷Methyl phosphite, *see* M-00490
1-Methylphospholane, M-00280
3-Methylphospholane, M-00281
1-Methyl-1*H*-phosphole, M-00282
3-Methyl-3-phospholene, *see* D-00530
P-Methylphosphonamidic acid, M-00283
P-Methylphosphonamidic fluoride, *in* M-00283
P-Methylphosphonamidothioic acid, M-00284
1-Methylphosphonane, M-00285
1-Methyl-3,8-phosphonanedione 1-oxide, *in* P-00373
Methylphosphonazidic acid, M-00286
P-Methylphosphonazidic amide, M-00287
P-Methylphosphonazidic anilide, *in* M-00287
1-Methyl-1-phosphoniabicyclo[3.3.1]nonane iodide, *in* P-00345
▷Methylphosphonic acid, M-00288
Methylphosphonic acid monoanhydride, *see* D-00751
Methylphosphonic bis(*tert*-butylamide), *in* M-00289
Methylphosphonic bis(diethylamide), *in* M-00289
Methylphosphonic bis(dimethylamide), *in* M-00289
Methylphosphonic bis(methylamide), *in* M-00289
P-Methylphosphonic diamide, M-00289
Methylphosphonic dianilide, *in* M-00289
▷Methylphosphonic dichloride, M-00290
▷Methylphosphonic difluoride, M-00291
Methylphosphonisocyanatidic acid, M-00292
P-Methylphosphonisocyanatidic amide, M-00293
Methylphosphonisothiocyanatidic acid, M-00294
Methyl phosphonoacetate, *in* P-00376
Methylphosphonobromidothioic acid, M-00295
3-Methyl-1-phosphono-1,3-butadiene, *see* M-00095
2-Methyl-1-phosphono-1-butene, *see* M-00100
2-Methyl-4-phosphono-1-butene, *see* M-00103
Methylphosphonochloridic acid, M-00296
Methylphosphonochloridodithioic acid, M-00297
Methylphosphonochloridothioic acid, M-00298
Methylphosphonochloridothious acid, M-00299
Methylphosphonochloridous acid, M-00300
Methylphosphonodithioic acid, M-00301
Methylphosphonodithious acid, M-00302
▷Methylphosphonofluoridic acid, M-00303
Methylphosphonofluoridodithioic acid, M-00304
Methylphosphonofluoridothioic acid, M-00305
Methylphosphonofluoridothious acid, M-00306
Methylphosphonofluoridous acid, M-00307
P-Methylphosphonohydrazidic acid, M-00308
P-Methylphosphonohydrazidothioic acid, M-00309
▷*N*-Methyl-*N*-[2-(phosphonooxy)ethyl]guanidine, *see* C-00206
Methylphosphonoperoxoic acid, M-00310
2-Methyl-4-phosphonopyridine, *see* M-00388
2-Methyl-6-phosphonopyridine, *see* M-00392
3-Methyl-2-phosphonopyridine, *see* M-00389
3-Methyl-4-phosphonopyridine, *see* M-00390
4-Methyl-2-phosphonopyridine, *see* M-00391
Methylphosphonoselenoic acid, M-00311
Methylphosphonoselenoic dichloride, M-00312
Methylphosphonothioic acid, M-00313
Methylphosphonothioic acid trimer trianhydride, *in* T-00567
Methylphosphonothioic bis(diethylamide), *in* M-00314
Methylphosphonothioic bis(dimethylamide), *in* M-00314
P-Methylphosphonothioic diamide, M-00314
Methylphosphonothioic dianilide, *in* M-00314
Methylphosphonothioic dibromide, M-00315
▷Methylphosphonothioic dichloride, M-00316
Methylphosphonothioic difluoride, M-00317
Methylphosphonothioic monoanhydride, *see* D-00925
Methylphosphonotrithioic acid, M-00318
Methylphosphonous acid, M-00319
Methylphosphonous bis(diethylamide), *in* M-00320
Methylphosphonous bis(dimethylamide), *in* M-00320
Methylphosphonous diamide, M-00320
Methylphosphonous dibromide, M-00321
▷Methylphosphonous dichloride, M-00322
Methylphosphonous dicyanide, M-00323
Methylphosphonous difluoride, M-00324
Methylphosphonous diiodide, M-00325
Methylphosphoramidic acid, M-00326
Methylphosphoramidic dichloride, M-00327
Methylphosphoramidic difluoride, M-00328
Methylphosphoramidochloridothioic acid, M-00329
Methylphosphoramidodithioic acid, M-00330
O-Methyl phosphoramidofluoridothioate, M-00331
Methylphosphoramidoselenoic acid, M-00332
Methylphosphoramidothioic acid, M-00333
Methylphosphoramidothioic dichloride, M-00334
Methylphosphoramidous acid, M-00335
▷Methyl phosphoric acid, *see* M-00489
Methyl phosphoric dianilide, *in* D-01125
Methylphosphorimidic acid, M-00336
2-Methylphosphorin, M-00337
4-Methylphosphorin, M-00338
1-Methylphosphorinane, M-00339
1-Methylphosphorinanium iodide, *in* M-00339
1-Methyl-4-phosphorinanone, M-00340
Methyl phosphorochloridofluoridate, M-00341
Methyl phosphorodiamidite, M-00342
O-Methyl phosphorodiamidothioate, *in* M-00359
S-Methyl phosphorodiamidothioate, M-00343
Methyl phosphorodibromidite, M-00344
Methyl phosphorodichloridate, M-00345
▷Methyl phosphorodichloridite, M-00346
Methyl phosphorodichloridodithioate, M-00347
Methyl phosphorodichloridoselenoite, M-00348
O-Methyl phosphorodichloridothioate, M-00349
Methyl phosphorodichloridothioite, M-00350
S-Methyl phosphorodichlorodothioate, M-00351
▷Methyl phosphorodifluoridate, M-00352
Methyl phosphorodifluoridite, M-00353
O-Methyl phosphorodifluoridothioate, M-00354
S-Methyl phosphorodifluoridothioate, M-00355
Methyl phosphorodiisocyanatidate, M-00356
O-Methyl phosphorodiisocyanatidothioate, M-00357
O-Methyl phosphorofluoridoisothiocyanatidothioate, M-00358
O-Methyl phosphorothioate, M-00359
S-Methyl phosphorothioate, M-00360
▷Methyl phosphoryl difluoride, *see* M-00352
Methyl phosphoryl diisocyanate, *see* M-00356
2-Methyl-3-phytyl-1,4-naphthalenedioldiphosphate, *see* P-00427
Methyl pinacolyl phosphite, *see* M-00085
▷*S*-2-Methylpiperidinocarbonylmethyl *O,O*-dipropyl phosphorothioate, *see* P-00433
▷*S*-[2-(2-Methyl-1-piperidinyl)-2-oxoethyl] *O,O*-dipropyl phosphorothioate, *see* P-00433
Methyl-1-piperidinylphosphinic acid, *see* M-00361
P-Methyl-1-piperidinylphosphonamidic acid, M-00361
(1-Methyl-1,3-propanediyl)-bis[diphenylphosphine], *see* B-00224
(1-Methyl-1,3-propanediyl)-bis[methyldiphenylphosphonium] hexafluorophosphate, *in* B-00224
2-Methyl-1-propenyl dihydrogen phosphate, *see* M-00491
(2-Methyl-1-propenyl)diphenylphosphine, M-00362
(2-Methyl-2-propenylidene)-triphenylphosphorane, M-00363
Methyl(2-propenyl)phosphinic acid, M-00364
(2-Methyl-1-propenyl)phosphonic dichloride, M-00365
(2-Methyl-1-propenyl)phosphonothioic acid, M-00366
(2-Methyl-2-propenyl)-triphenylphosphonium(1+), M-00367
2-(2-Methylpropoxy)-1,3,2-dioxaphospholane, M-00368
2-Methylpropyl bis(2-methylpropyl)-phosphinate, *in* B-00346
O-(2-Methylpropyl) cyclohexylphosphonochloridothioate, *in* C-00260
2-Methylpropyl decylmethylphosphinate, *in* D-00009
2-Methylpropyl *N,P*-diethylphosphonamidate, *in* D-00336
(1-Methylpropyl)diphenylphosphine, M-00369
(2-Methylpropyl)diphenylphosphine, M-00370
P-(1-Methylpropyl)-*N,N'*-diphenylphosphonic diamide, *in* M-00377
N-2-Methylpropyl-*N',N''*-diphenylphosphoric triamide, *in* M-00384
S-Methyl *O,O*-propylene dithiophosphate, *in* M-00173
Methyl propylene phosphate, *in* M-00043
Methyl propylene phosphite, *see* M-00043
S-Methyl *O,O*-propylene phosphorodithioate, *in* M-00173
O-Methyl *O,O*-propylene phosphorothioate, *in* M-00043
S-Methyl *O,O*-propylene phosphorothioate, *in* M-00173
S-Methyl *O,O*-propylene phosphorothioite, *see* M-00173
O-Methyl *O,O*-propylene thiophosphate, *in* M-00043
S-Methyl *O,O*-propylene thiophosphate, *in* M-00173
S-Methyl *O,O*-propylene thiophosphite, *see* M-00173
2-Methylpropyl ethylmethylphosphinate, *in* E-00085
2-Methylpropyl ethylphosphonochloridate, *in* E-00135
2-Methylpropyl hexylmethylphosphinate, *in* H-00094
(1-Methylpropylidene)triphenylphosphorane, *in* M-00386
(2-Methylpropylidene)triphenylphosphorane, M-00371

2-Methylpropyl methyl(2-methylpropyl)-phosphinate, *in* M-00172
2-Methylpropyl *P*-methylphosphonohydrazidate, *in* M-00308
2-Methylpropyl (2-methylpropyl)-phenylphosphinate, *in* M-00373
2-Methylpropyl methylpropylphosphinate, *in* M-00375
(1-Methylpropyl)phenylphosphinic acid, M-00372
(2-Methylpropyl)phenylphosphinic acid, M-00373
(2-Methylpropyl)phosphine, M-00374
Methylpropylphosphinic acid, M-00375
(2-Methylpropyl)phosphinic acid, M-00376
(1-Methylpropyl)phosphonic acid, M-00377
(2-Methylpropyl)phosphonic acid, M-00378
(1-Methylpropyl)phosphonic dichloride, M-00379
S-Methyl propylphosphonochloridothioate, *in* P-00477
(1-Methylpropyl)phosphonodithioic acid, M-00380
(2-Methylpropyl)phosphonothioic acid, M-00381
(1-Methylpropyl)phosphonotrithioic acid, M-00382
(2-Methylpropyl)phosphonous acid, M-00383
(2-Methylpropyl)phosphoramidic acid, M-00384
2-Methylpropyl phosphorodichloridite, M-00385
(1-Methylpropyl)triphenylphosphonium(1+), M-00386
(2-Methylpropyl)triphenylphosphonium(1+), M-00387
(2-Methyl-4-pyridinyl)phosphonic acid, M-00388
(3-Methyl-2-pyridinyl)phosphonic acid, M-00389
(3-Methyl-4-pyridinyl)phosphonic acid, M-00390
(4-Methyl-2-pyridinyl)phosphonic acid, M-00391
(6-Methyl-2-pyridinyl)phosphonic acid, M-00392
Methyl 1-pyrrolidinylphosphonochloridite, *in* P-00519
[2-(Methylseleno)phenyl]diphenylphosphine, M-00393
3′-Methylspiro[1,3,2-benzodioxaphosphole-2,2(3′*H*)-[1,3,2]benzothiazaphosphole], M-00394
3′-Methylspiro[1,3,2-benzodioxaphosphole-2,2′(3′*H*)-[1,3,2]benzoxazaphosphole], M-00395
5′-Methylspiro[1,3,2-benzodioxaphosphole-2,2′(3*H*)-[1,2]oxaphosphole], M-00396
2-Methyl-2,2′-spirobi[1,3,2-benzodioxaphosphole], M-00397
Methyl styrylphosphinate, *in* P-00153
[(Methylsulfinyl)methyl]phosphonic acid, M-00398
[(Methylsulfonyl)methylene]-triphenylphosphorane, M-00399
[(Methylsulfonyl)methyl]phosphonic acid, M-00400
Methylsulfonylphosphoramidic acid, M-00401
(Methylsulfonyl)phosphorimidic acid, M-00402
N-Methylsulfonyl-*P,P,P*-triphenoxyphosphazene, *in* M-00402
N-Methylsulfonyl-*P,P,P*-triphenoxyphosphine imide, *in* M-00402
Methyl tetraethylphosphorodiamidate, *in* T-00059
O-Methyl tetraethylphosphorodiamidoselenoate, *in* T-00061
17-Methyl-6,7,9,10-tetrahydrodibenzo[*d,m*]-[1,3,6,9,12,2]-pentaoxaphosphacyclotetradecin, M-00403
3-Methyltetrahydro-2-phenoxy-2*H*-1,3,2-oxazaphosphorine, M-00404
Methyl tetraisopropylphosphorodiamidite, *in* T-00172
4′-Methyl-4,4,5,5-tetrakis(trifluoromethyl)spiro[1,3,2-dioxaphospholane-2,1′-[2,6,7]-trioxa[1]phosphabicyclo[2.2.2]octane], M-00405
P-Methyl-*N,N,N′,N′*-tetrakis(trimethylsilyl)phosphonous diamide, *in* M-00320
Methyl tetramethylenephosphinate, *in* H-00185
Methyltetramethylenephosphine, *see* M-00280
Methyl tetramethylene phosphite, *in* D-00968
Methyl tetramethylethylene phosphate, *in* M-00085
Methyl tetramethylethylene phosphite, *see* M-00085
Methyl tetramethylphosphorodiamidate, M-00406
Methyl tetramethylphosphorodiamidite, *in* T-00227
Methyl tetramethylphosphorodiamidodithioate, *in* T-00222
O-Methyl tetramethyl-phosphorodiamidoselenoate, *in* T-00223
Methyl tetraphenylphosphorodiamidite, *in* T-00274
Methyl tetrapropylphosphorodiamidate, *in* T-00291
5-Methyl-1,2-thiaphospholane 2-sulfide, M-00407
7-Methyl-2-thia-1,3,5-triaza-7-phosphoniaadamantane iodide, *in* T-00301
3-Methyl-1,3,2-thiazaphospholidine, M-00408
P-Methyl-*N*-2-thiazolylphosphonamidic acid, M-00409
2-Methylthioadenosine 5′-(dihydrogen phosphate), M-00410
2-Methylthio-AMP, *see* M-00410
2-Methylthio-4*H*-1,3,2-benzodioxaphosphorin, M-00411
Methyl thiodimethylenephosphinate, *in* H-00201
2-Methylthio-1,3,2-dioxaphosphorinane 2-sulfide, *in* M-00013
2-(Methylthio)-1,3,2-dithiaphospholane, *in* D-01243
[1-(Methylthio)ethyl]phosphonic acid, M-00412
Methylthioimidodiphosphoric acid, *see* M-00161
Methylthioimidodiphosphoric acid, M-00413
(Methylthio)imidodiphosphoryl chloride, M-00414
(Methylthio)iminodiphosphoryl tetrachloride, *see* M-00414
[(Methylthio)methylene]triphenylphosphorane, *in* M-00416
[(Methylthio)methyl]phosphonic acid, M-00415
[(Methylthio)methyl]triphenyl-phosphonium(1+), M-00416
1-Methylthio-2,2,3,4,4-pentamethylphosphetane 1-oxide, *in* H-00169
[2-(Methylthio)phenyl]diphenylphosphine, M-00417
▷4-(Methylthio)phenyl dipropyl phosphate, *see* P-00451
(4-Methylthiophenyl)phosphonic acid, M-00418
Methylthiophosphonic dibromide, *see* M-00315
▷Methylthiophosphonic dichloride, *see* M-00316
Methylthiophosphonic difluoride, *see* M-00317
O-Methyl thiophosphoric acid, *see* M-00359
S-Methyl thiophosphoric acid, *see* M-00360
O-Methyl thiophosphoryl diamide, *in* M-00359
O-Methyl thiophosphoryl dichloride, *see* M-00349
O-Methyl thiophosphoryl difluoride, *see* M-00354
O-Methyl thiophosphoryl diisocyanate, *see* M-00357
Methyl-*o*-tolylphosphinic acid, *see* M-00203
Methyl-*p*-tolylphosphinic acid, *see* M-00204
Methyl(trichloromethyl)phosphinic acid, M-00419
Methyltricyclopentylphosphonium iodide, *in* T-00431
Methyl triethylphosphonamidate, *in* T-00446
O-Methyl triethylphosphonamidothioate, *in* T-00448
Methyl(trifluoromethyl)phosphinic acid, M-00420
Methyltri-2-furanylphosphonium iodide, *in* T-00486
Methyl(tri-3-furanyl)phosphonium iodide, *in* T-00487
[3-Methyl-5-(2,6,6-trimethyl-1-cyclohexen-1-yl)-2,4-pentadienylidene]triphenylphosphorane, *in* M-00421
[3-Methyl-5-(2,6,6-trimethyl-1-cyclohexen-1-yl)-2,4-pentadienyl]-triphenylphosphonium(1+), M-00421
[1-Methyl-3-(2,6,6-trimethyl-1-cyclohexen-1-yl)-2-propenyl]-triphenylphosphonium(1+), M-00422
Methyl trimethylene phosphate, *in* M-00038
▷Methyl trimethylene phosphite, *see* M-00038
P-Methyl-*N,N′*-trimethylenephosphonic diamide, *in* H-00020
O-Methyl *S,S*-trimethylene phosphonotrithioate, *in* D-01249
Methyl *N,N′*-trimethylenephosphorodiamidate, *in* H-00020
O-Methyl *S,S*-trimethylene phosphorodithioite, *in* D-01247
O-Methyl *O,O*-trimethylene phosphoroselenoate, *in* M-00038
Methyl trimethylene phosphorotetrathioate, *in* M-00014
O-Methyl *O,O*-trimethylene phosphorothioate, *in* M-00038
Methyl trimethylphosphonamidate, *in* T-00537
S-Methyl trimethylphosphonamidothioate, *in* T-00540
O-Methyl *O*-trimethylsilyl *tert*-butylphosphonothioate, *in* M-00107
Methyltrioctylphosphonium iodide, *in* T-00582
4-Methyl-2,6,7-trioxa-1-phosphabicyclo[2.2.1]heptane, M-00423
▷4-Methyl-2,6,7-trioxa-1-phosphabicyclo[2.2.2]octane, M-00424
4-Methyl-3,5,8-trioxa-1-phosphabicyclo[2.2.2]octane, M-00425
1-Methyl-2,6,7-trioxa-4-phospha-1-silabicyclo[2.2.2]octane, M-00426
7-Methyl-1,4,6-trioxa-5-phospha(5-P^V)-spiro[4.4]nonan-8-one, M-00427
9-Methyl-1,4,6-trioxa-5-phospha(5-P^V)-spiro[4.4]non-7-ene, M-00428
Methyltriphenoxyphosphonium(1+), M-00429
Methyltriphenoxyphosphorus(1+), *see* M-00429
6-Methyl-2,3,4-triphenyl-2*H*-1,5,2-dioxaphosphinine 2-oxide, *see* M-00430
6-Methyl-2,3,4-triphenyl-2*H*-1,5,2-dioxaphosphorin 2-oxide, M-00430
4-Methyl-4-triphenylmethyl-2,6,7-trioxa-1-phosphoniabicyclo[2.2.2]octane perchlorate, *in* M-00424
4-Methyl-4-triphenylmethyl-2,6,7-trioxa-1-phosphoniabicyclo[2.2.2]octane tetrafluoroborate, *in* M-00424
3-Methyl-1,2,3-triphenyl-5-oxo-1,3-azaphospholidinium iodide, *in* T-00596
6-Methyl-3,5,7-triphenyl-6-phosphatricyclo[3.3.1.$0^{2,7}$]non-3-ene, M-00431
N-Methyl-*P,P,P*-triphenylphosphine imide, *in* T-00626
1-Methyl-1,2,5-triphenyl-1*H*-phospholium iodide, *in* T-00634
Methyltriphenylphosphonium(1+), M-00432
Methyl triphenylphosphoranylideneacetate, M-00433
Methyl α-(triphenylphosphoranylidene)-benzeneacetate, *in* T-00644
Methyl 4-(triphenylphosphoranylidene)-2-butanoate, *in* T-00647
Methyl (triphenylphosphoranylidene)-carbamate, *in* T-00648
Methyl α-(triphenylphosphoranylidene)-cyclopropaneacetate, *in* C-00285
3-Methyl-5-(triphenylphosphoranylidene)-2(5*H*)-furanone, *in* M-00186
Methyl α-(triphenylphosphoranylidene)-hydrocinnamate, *in* T-00645
Methyl (2-triphenylphosphoranylidene)-propanoate, M-00434
Methyltri-1-propenylphosphonium iodide, *in* T-00706
Methyltri-2-propenylphosphonium iodide, *in* T-00707
Methyltripropylphosphonium iodide, *in* T-00711

Methyltri(propylthio)phosphonium iodide, *in* T-00718
2-Methyl-*N*-[[2,4,6-tris(1,1-dimethylethyl)phenyl]phosphinidene]-2-propanamine, *see* B-00639
2-Methyl-*N*-[[2,4,6-tris(1,1,-dimethylethyl)phenyl]-phosphinoselenoylidene]-2-propanamine, *in* B-00639
2-Methyl-*N*-[[2,4,6-tris(1,1-dimethylethyl)phenyl]-phosphinothioylidene]-2-propanamine, *in* B-00639
2-Methyl-*N*-[[2,4,6-tris(1,1-dimethylethyl)phenyl]phosphinylidene]-2-propanamine, *in* B-00639
Methyltris(2,4-dimethylphenyl)phosphonium iodide, *in* T-00768
Methyltris(2,5-dimethylphenyl)phosphonium iodide, *in* T-00769
Methyltris(3-fluorophenyl)phosphonium iodide, *in* T-00782
Methyltris(4-fluorophenyl)phosphonium iodide, *in* T-00783
Methyltris(hydroxymethyl)phosphonium iodide, *in* T-00784
Methyltris(hydroxymethyl)phosphonium tetraphenylborate, *in* T-00784
Methyltris(4-methoxyphenyl)phosphonium bromide, *in* T-00793
Methyltris(4-methoxyphenyl)phosphonium iodide, *in* T-00793
Methyltris(4-methylphenyl)phosphonium bromide, *in* T-00808
Methyltris(4-methylphenyl)phosphonium iodide, *in* T-00808
Methyltris(1-methylpropyl)phosphonium iodide, *in* T-00823
Methyltris(2-methylpropyl)phosphonium iodide, *in* T-00824
Methyltris(morpholino)phosphonium iodide, *in* T-00569
Methyltris(phenylmethyl)phosphonium iodide, *in* T-00338
Methyltris(piperidino)phosphonium iodide, *in* T-00704
Methyltris(2,4,6-trimethylphenyl)-phosphonium iodide, *in* T-00861
Methyl *N,N,N'*-tris(trimethylsilyl)-phosphonamidimidate, *in* T-00867
Methyltris(triphenylphosphine)copper, M-00435
4-Methyl-2,6,7-trithia-1-phosphabicyclo[2.2.2]octane, M-00436
Methyltri-2-thienylphosphonium iodide, *in* T-00874
Methyltri-3-thienylphosphonium iodide, *in* T-00875
▷Methyl trithion, M-00437
Methyltrivinylphosphonium iodide, *in* T-00877
1-Methylvinyl dichlorophosphate, *see* M-00151
Methyl vinylene phosphate, *in* H-00136
Methylvinylphosphinic acid, M-00438
(1-Methylvinyl)phosphonic acid, *see* M-00149
1-Methylvinyl phosphorodichloridate, *see* M-00151
1-Methylvinyl phosphoryl dichloride, *see* M-00151
▷Metrifonate, *see* T-00423
▷Metriphonate, *see* T-00423
Meturedepa, M-00439
▷Mevinphos, M-00440
▷Mintacol, *see* P-00001
▷Miotisal *A*, *see* P-00001
▷Mipafox, *in* D-00661
▷Mitoxana, *see* I-00001
▷MK 955, *in* M-00185
▷Mocap, *see* E-00023
Moenomycin, M-00441
Moenomycin *A*, *in* M-00441
Moenomycin *C*, *in* M-00441
Moenomycins *D-H*, *in* M-00441
Moenomycins B_1 and B_2, *in* M-00441
Monitor, *see* D-00865
▷Monoallyl phosphate, *see* M-00509
Monoallyl phosphite, *see* M-00510
Monoallyl phosphonate, *see* M-00510
Monoamyl phosphate, *see* M-00504
Mono-*o*-anisyl phosphate, *see* M-00480
Monobenzoyl phosphate, M-00442
Monobenzyl phosphate, M-00443
Monobenzyl phosphite, *see* M-00444
Monobenzyl phosphonate, M-00444
Mono(2-biphenylyl) phosphate, M-00445
Mono-(1,1'-biphenyl)-4-yl phosphate, M-00446
Mono(2-bromoethyl) phosphate, M-00447
Mono(4-bromophenyl) phosphate, M-00448
Monobutyl ethylphosphonate, *in* E-00129
Mono-2-*tert*-butylphenyl phosphate, M-00449
Mono-4-*tert*-butylphenyl phosphate, M-00450
Monobutyl phosphate, M-00451
Mono-*tert*-butyl phosphate, M-00452
▷Monobutyl phosphite, *see* M-00453
▷Monobutyl phosphonate, M-00453
Mono(2-carboxyethyl) phosphate, *see* P-00399
Mono(2-chloroethyl) phosphate, M-00454
Mono(2-chloromethyl-4-nitrophenyl) phosphate, M-00455
Mono(4-chlorophenyl) mono(5-chloro-8-quinolinyl) phosphate, M-00456
Mono(2-chlorophenyl) phosphate, M-00457
Mono(4-chlorophenyl) phosphate, M-00458
Mono-*o*-cresyl phosphate, *see* M-00485
Monocrotophos, M-00459
Mono(2-cyanoethyl) phosphate, *see* P-00398
Mono(2-cyanophenyl) phosphate, M-00460
Monocyclohexyl phosphate, M-00461
Monodecyl phosphate, M-00462
Mono(2,4-dichlorophenyl) phosphate, M-00463
Mono-2-dimethylaminoethanol dihydrogen phosphate, *see* D-00016
Mono[2-(dimethylamino)-4-nitrophenyl] phosphate, M-00464
Mono 2-(1,1-dimethylethyl)phenyl phosphate, *see* M-00449
Mono-4-(1,1-dimethylethyl)phenyl phosphate, *see* M-00450
Mono(1,1-dimethylethyl) phosphate, *see* M-00452
Mono(3,7-dimethyl-2,6-octadienyl) diphosphate, M-00465
Mono-2,6-dimethylphenyl phosphate, M-00466
Mono(2,4-dinitrophenyl) phosphate, M-00467
Monododecyl phosphate, M-00468
▷Monoethenyl phosphate, *see* M-00518
Monoethyl (aminocarbonyl)phosphonate, M-00469
Monoethyl benzylthiophosphonate, *in* B-00067
Monoethyl butylphosphonate, *in* B-00594
Monoethyl chloromethylphosphonate, *in* C-00106
Monoethyl cyclohexylphosphonate, *in* C-00256
Monoethyl (1,1-dimethylethyl)phosphonate, *in* B-00595
Monoethyl ethylphosphonate, *in* E-00129
O-Monoethyl ethylphosphonoselenoate, *see* E-00072
Mono(2-ethylhexyl) phosphate, M-00470
Monoethyl methylphosphonate, *see* E-00089
Monoethyl (2-phenylethenyl)phosphonate, *in* P-00155
Monoethyl phenylphosphonate, M-00471
Monoethyl phosphate, M-00472
Monoethyl phosphite, *see* M-00473
Monoethyl phosphonate, M-00473
Monoethyl styrylphosphonate, *in* P-00155
Monoheptyl phosphate, M-00474
Monohexadecyl phosphate, M-00475
Monohexyl phosphate, M-00476
Monoisobutenyl phosphate, *see* M-00491
Monoisobutyl phosphate, *see* M-00493
Monoisopentyl phosphate, *see* M-00484
Monoisopropyl cyclohexylphosphonate, *in* C-00256
▷Monoisopropyl phosphate, M-00477
Monoisopropyl phosphite, *see* M-00478
Monoisopropyl phosphonate, M-00478
Monomethoxycarbonyl phosphate, M-00479
Mono(2-methoxyphenyl) phosphate, M-00480
Mono(4-methoxyphenyl) phosphate, M-00481
Mono(3-methyl-2-butenyl) diphosphate, M-00482
Mono(3-methyl-3-butenyl) diphosphate, M-00483
Mono(3-methylbutyl) phosphate, M-00484
▷Mono(1-methylethyl) phosphate, *see* M-00477
Mono(1-methylethyl) phosphonate, *see* M-00478
Monomethyl methylphosphonate, *see* M-00168
Monomethyl 4-morpholinylphosphonothioate, *in* M-00527
Mono(2-methylphenyl) phosphate, M-00485
Mono(3-methylphenyl) phosphate, M-00486
Mono(4-methylphenyl) phosphate, M-00487
Monomethyl phenylphosphonate, M-00488
▷Monomethyl phosphate, M-00489
▷Monomethyl phosphite, *see* M-00490
▷Monomethyl phosphonate, M-00490
Mono(2-methyl-1-propenyl) phosphate, M-00491
Mono(1-methylpropyl) phosphate, M-00492
Mono(2-methylpropyl) phosphate, M-00493
Monomethyl styrylphosphonate, *in* P-00155
Mono-1-naphthalenyl phosphate, *see* M-00494
Mono-2-naphthalenyl phosphate, *see* M-00495
Mono-1-naphthyl phosphate, M-00494
Mono-2-naphthyl phosphate, M-00495
Mono-1-naphthyl phosphite, *see* M-00496
Mono-2-naphthyl phosphite, *see* M-00497
Mono-1-naphthyl phosphonate, M-00496
Mono-2-naphthyl phosphonate, M-00497
Mono(2-nitrophenyl) phosphate, M-00498
Mono(3-nitrophenyl) phosphate, M-00499
Mono(4-nitrophenyl) phosphate, M-00500
Monooctyl phosphate, M-00501
Mono(pentachlorophenyl) phosphate, M-00502
Mono(pentafluorophenyl) phosphate, M-00503
Monopentyl phosphate, M-00504
Mono(2-phenoxyphenyl) phosphate, M-00505
Monophenyl cyclohexylphosphonate, *in* C-00256
Monophenyl (dichloromethyl)phosphonate, *in* D-00175
Mono(2-phenylethyl) phosphate, M-00506
Mono(phenylmethyl) (2-phenylethenyl)-phosphonate, *in* P-00155
Mono(phenylmethyl) phosphate, *see* M-00443
Monophenyl methylphosphonate, *see* P-00171
Monophenyl phosphate, M-00507
Monophenyl phosphite, *see* M-00508
Monophenyl phosphonate, M-00508
Monophosphaurotropine, *see* T-00327
▷Mono-2-propenyl phosphate, M-00509
Mono-2-propenyl phosphite, *see* M-00510
Mono-2-propenyl phosphite, *see* M-00510
Mono-2-propenyl phosphonate, M-00510
Monopropyl phenylphosphonate, M-00511
Monopropyl phosphate, M-00512
Monopropyl phosphite, *see* M-00513
Monopropyl phosphonate, M-00513
Monotetradecyl phosphate, M-00514
Mono(3,7,11,15-tetramethyl-2,6,10,14-hexadecatetraenyl) diphosphate, M-00515
Mono-*m*-tolyl phosphate, *see* M-00486
Mono-*o*-tolyl phosphate, *see* M-00485
Mono-*p*-tolyl phosphate, *see* M-00487
Mono(2,2,2-tribromoethyl) phosphate, M-00516
Mono(2,2,2-trichloroethyl) phosphate, *see* T-00424
Mono(3,7,11-trimethyl-2,6,10-dodecatrienyl) diphosphate, M-00517
▷Monovinyl phosphate, M-00518
▷Montrel, *see* C-00208
4-Morpholinephosphinic acid, *see* M-00522
4-Morpholinephosphonic acid, *see* M-00523
4-Morpholinephosphonous acid, *see* M-00529
4-Morpholinomethanediphosphonic acid, *see* M-00520
(Morpholinomethyl)diphenylphosphine oxide, M-00519
4-Morpholinophosphinic acid, *see* M-00522
Morpholinophosphonic acid, *see* M-00523
Morpholinophosphonothioic acid, *see* M-00527
Morpholinophosphonous acid, *see* M-00529
P-Morpholinyl-*N,N'*-diphenylphosphonothioic diamide, *in* M-00528
(4-Morpholinylmethylene)bisphosphonic acid, M-00520
(4-Morpholinylmethyl)phosphonic acid, M-00521
4-Morpholinylphosphinic acid, M-00522
1,1'-(4-Morpholinylphosphinylidene)-bis[oxy]bis[1-*H*-benzotriazole], *see* B-00099
4-Morpholinylphosphonic acid, M-00523
4-Morpholinylphosphonic diamide, *in* M-00523
4-Morpholinylphosphonochloridic acid, M-00524

4-Morpholinylphosphonochloridothioic acid, M-00525
4-Morpholinylphosphonochloridous acid, M-00526
4-Morpholinylphosphonothioic acid, M-00527
P-4-Morpholinylphosphonothioic diamide, M-00528
P-Morpholinylphosphonothioic dianilide, *in* M-00528
4-Morpholinylphosphonous acid, M-00529
▷Morphos oxide, *see* T-00392
▷Murfotox, *see* M-00002
▷Muscatox, *see* C-00205
▷Myco 63, *see* H-00060
▷Myko 63, *see* H-00060
Myristoylpalmitoylphosphatidylcholine, *see* G-00096
N-1409, *in* A-00123
▷N 1409, *in* P-00437
▷N1409A, *see* P-00437
▷NAD, *see* C-00202
▷Nadide, *see* C-00202
NADP, *see* C-00203
▷Naftalophos, *see* N-00008
▷Naled, N-00001
▷Nankor, *see* F-00003
Naphos[1,1], *see* B-00248
1,4-Naphthalenediylbis[diphenylphosphine], *see* B-00254
1,5-Naphthalenediylbis[diphenylphosphine], *see* B-00255
2,6-Naphthalenediylbis[diphenylphosphine], *see* B-00256
2,7-Naphthalenediylbis[diphenylphosphine], *see* B-00257
[1,5-Naphthalenediylbis[methylene]]-bis[phosphonic acid], N-00002
[1,6-Naphthalenediylbis[methylene]]-bis[phosphonic acid], N-00003
[1,8-Naphthalenediylbis[methylene]]-bis[phosphonic acid], N-00004
[2,3-Naphthalenediylbis[methylene]]-bis[phosphonic acid], N-00005
[2,6-Naphthalenediylbis[methylene]]-bis[phosphonic acid], N-00006
[2,7-Naphthalenediylbis[methylene]]-bis[phosphonic acid], N-00007
1,4-[Naphthalenediylbis[methylene]]-bis(triphenylphosphonium)(2+), *see* B-00450
[1,5-Naphthalenediylbis[methylene]]-bis[triphenylphosphonium](2+), *see* B-00451
[1,8-Naphthalenediylbis[methylene]]-bis[triphenylphosphonium](2+), *see* B-00452
[2,3-Naphthalenediylbis[methylene]]-bis[triphenylphosphonium](2+), *see* B-00453
[2,6-Naphthalenediylbis[methylene]]-bis[triphenylphosphonium](2+), *see* B-00454
[2,7-Naphthalenediylbis[methylene]]-bis[triphenylphosphonium](2+), *see* B-00455
[1,5-Naphthalenediylbis[methylene]]-bis[triphenylphosphorane], *in* B-00451
[1,8-Naphthalenediylbis[methylene]]-bis[triphenylphosphorane], *in* B-00452
[(2,3-Naphthalenediylbis[methylene]]-bis[triphenylphosphorane], *in* B-00453
[2,7-Naphthalenediylbis[methylene]]-bis[triphenylphosphorane], *in* B-00455
1-Naphthalenyldiphenylphosphine, *see* N-00011
2-Naphthalenyldiphenylphosphine, *see* N-00012
1-Naphthalenyl hydrogen phosphonate, *see* M-00496
2-Naphthalenyl hydrogen phosphonate, *see* M-00497
1-Naphthalenylmethylenetriphenylphosphorane, *in* N-00015
2-Naphthalenylmethylenetriphenylphosphorane, *in* N-00016
1-Naphthalenylmethylphosphonic acid, *see* N-00013
2-Naphthalenylmethylphosphonic acid, *see* N-00014
(1-Naphthalenylmethyl)-triphenylphosphonium(1+), *see* N-00015
(2-Naphthalenylmethyl)-triphenylphosphonium(1+), *see* N-00016
(1-Naphthalenyl)phenylphosphinic acid, *see* N-00017
(2-Naphthalenyl)phenylphosphinic acid, *see* N-00018
1-Naphthalenylphosphinic acid, *see* N-00019
2-Naphthalenylphosphinic acid, *see* N-00020
1-Naphthalenylphosphonic acid, *see* N-00021
2-Naphthalenylphosphonic acid, *see* N-00022
1-Naphthalenylphosphonothioic acid, *see* N-00023
2-Naphthalenylphosphonothioic acid, *see* N-00024
1-Naphthalenylphosphonous acid, *see* N-00025
2-Naphthalenylphosphonous acid, *see* N-00026
1-Naphthalenylphosphoramidic acid, *see* N-00028
2-Naphthalenylphosphoramidic acid, *see* N-00029
O-2-Naphthalenyl phosphorodiamidothioate, *see* N-00027
1-Naphthalenyl phosphorodichloridite, *see* N-00030
2-Naphthalenyl phosphorodichloridite, *see* N-00031
▷Naphthalophos, N-00008
Naphtho[1,2-*d*]-1,3,2-dioxaphosphole, N-00009
Naphtho[1,8-*de*]-1,3,2-dioxaphosphorin, N-00010
1-Naphthyl benzylphosphonochloridothioate, *in* B-00064
2-Naphthyl benzylphosphonochloridothioate, *in* B-00064
2,3-Naphthyl cyclic phosphoric acid, *see* H-00164
O-2-Naphthyl diamidothiophosphate, *see* N-00027
1-Naphthyl dichlorophosphate, *in* M-00494
2-Naphthyl dichlorophosphate, *in* M-00495
1-Naphthyl dichlorophosphite, *see* N-00030
2-Naphthyl dichlorophosphite, *see* N-00031
1-Naphthyl difluorophosphate, *in* M-00494
1-Naphthyl dihydrogen phosphate, *see* M-00494
2-Naphthyl dihydrogen phosphate, *see* M-00495
1-Naphthyl diphenylphosphinate, *in* D-01039
1-Naphthyldiphenylphosphine, N-00011
2-Naphthyldiphenylphosphine, N-00012
N-1-Naphthyl-*P*,*P*-diphenylphosphinic amide, *in* D-01040
N-2-Naphthyl-*P*,*P*-diphenylphosphinic amide, *in* D-01040
2,3-Naphthylene chlorophosphate, *in* C-00119
1,2-Naphthylene chlorophosphite, *see* C-00118
1,8-Naphthylene chlorophosphite, *see* C-00120
2,3-Naphthylene chlorophosphite, *see* C-00119
2,3-Naphthylene chlorothiophosphate, *in* C-00119
1,2-Naphthylene diethylphosphoramidate, *in* D-00266
1,8-Naphthylene diethylphosphoramidate, *in* D-00267
1,2-Naphthylene diethylphosphoramidite, *see* D-00266
1,8-Naphthylene diethylphosphoramidite, *see* D-00267
O,*O*-1,2-Naphthylene diethylphosphoramidothioate, *in* D-00266
O,*O*-1,8-Naphthylene diethylphosphoramidothioate, *in* D-00267
1,8-Naphthylene phenyl phosphate, *in* P-00070
1,8-Naphthylene phenyl phosphite, *see* P-00070
1,2-Naphthylene phosphonate, *in* N-00009
1,8-Naphthylene phosphonate, *in* N-00010
1,2-Naphthylene phosphonite, *see* N-00009
1,8-Naphthylene phosphonite, *see* N-00010
2,3-Naphthylene phosphoric acid, *see* H-00164
2,3-Naphthylene phosphorochloridate, *in* C-00119
1,2-Naphthylene phosphorochloridite, *see* C-00118
1,8-Naphthylene phosphorochloridite, *see* C-00120
2,3-Naphthylene phosphorochloridite, *see* C-00119
2,3-Naphthylene phosphorochloridothioate, *in* C-00119
(1-Naphthylmethylene)triphenylphosphorane, *in* N-00015
(2-Naphthylmethylene)triphenylphosphorane, *in* N-00016
(1-Naphthylmethyl)phosphonic acid, N-00013
(2-Naphthylmethyl)phosphonic acid, N-00014
(1-Naphthylmethyl)triphenylphosphonium(1+), N-00015
(2-Naphthylmethyl)triphenylphosphonium(1+), N-00016
1-Naphthylphenylphosphinic acid, N-00017
2-Naphthylphenylphosphinic acid, N-00018
1-Naphthylphosphinic acid, N-00019
2-Naphthylphosphinic acid, N-00020
1-Naphthylphosphonic acid, N-00021
2-Naphthylphosphonic acid, N-00022
1-Naphthylphosphonothioic acid, N-00023
2-Naphthylphosphonothioic acid, N-00024
1-Naphthylphosphonous acid, N-00025
2-Naphthylphosphonous acid, N-00026
O-2-Naphthyl phosphoradiamidothioate, N-00027
1-Naphthylphosphoramidic acid, N-00028
2-Naphthylphosphoramidic acid, N-00029
1-Naphthyl phosphoric acid, *see* M-00494
2-Naphthylphosphoric acid, *see* M-00495
1-Naphthyl phosphorodichloridate, *in* M-00494
2-Naphthyl phosphorodichloridate, *in* M-00495
1-Naphthyl phosphorodichloridite, N-00030
2-Naphthyl phosphorodichloridite, N-00031
1-Naphthyl phosphorodifluoridate, *in* M-00494
O-2-Naphthyl tetramethyl-phosphorodiamidothioate, *in* N-00027
N-1-Naphthyl-*P*,*P*,*P*-triphenylphosphine imide, *in* T-00626
N-2-Naphthyl-*P*,*P*,*P*-triphenylphosphine imide, *in* T-00626
N″-Bis(dimethylamino)phosphinyl-*N*,*N*,*N*′,*N*′,*N*″-pentamethyl-phosphorotriamidothioate, N-00032
▷Negasunt, *see* C-00205
▷Neguvon, *see* T-00423
▷Nemacur, *see* F-00002
▷Nemafos, *see* T-00314
▷Nem-a-tak, *see* F-00065
▷Neocidol, *see* D-00027
Neomenthyldiphenylphosphine, *in* M-00164
Neopentyl dichlorophosphite, *see* D-00913
Neopentylene *tert*-butylphosphoramidate, *in* B-00538
Neopentylene *tert*-butylphosphoramidite, *see* B-00538
Neopentylene *tert*-butylphosphoramidothioate, *in* B-00538
Neopentylene cyclic phosphoric acid, *see* H-00124
O,*O*-Neopentylene cyclic phosphoroselenoic acid, *see* H-00125
O,*O*-Neopentylene cyclic phosphorothioic acid, *see* M-00010
Neopentylene dimethylphosphoramidate, *in* D-00693
Neopentylene dimethylphosphoramidite, *see* D-00693
Neopentylene dimethylphosphoramidoselenoate, *in* D-00693
Neopentylene dimethylphosphoramidothioate, *in* D-00693
O,*O*-Neopentylene hydrogen dithiophosphate, *see* M-00011
O,*O*-Neopentylene hydrogen phosphoroselenoate, *see* H-00125
O,*O*-Neopentylene hydrogen thiophosphate, *see* M-00010
Neopentylene methylphosphonate, *in* T-00512
Neopentylene methylphosphonite, *see* T-00512
O,*O*-Neopentylene methylphosphonoselenoate, *in* T-00512
O,*O*-Neopentylene methylphosphonothioate, *in* T-00512
Neopentylene phenyl phosphate, *in* D-00790
Neopentylene phenyl phosphite, *see* D-00790
Neopentylene phenylphosphonate, *in* D-00795
Neopentylene phenylphosphonite, *see* D-00795
O,*O*-Neopentylene phenylphosphonothioate, *in* D-00795
Neopentylene phenylphosphoramidate, *in* D-00792

Neopentylene phenylphosphoramidite, *see* D-00792
Neopentylene phenylphosphoramidoselenoate, *in* D-00792
Neopentylene phenylphosphoramidothioate, *in* D-00792
O,O-Neopentylene *O*-phenyl phosphoroselenoate, *in* D-00790
O,O-Neopentylene *O*-phenyl phosphorothioate, *in* D-00790
Neopentylene phosphite, *in* D-00735
Neopentylene phosphonate, *in* D-00735
Neopentylene phosphonite, *see* D-00735
O,O-Neopentylene phosphonothioate, *in* D-00735
Neopentylene phosphorobromidate, *in* B-00471
Neopentylene phosphorobromidite, *see* B-00471
Neopentylene phosphorobromidothioate, *in* B-00471
Neopentylene phosphorochloridate, *see* C-00059
Neopentylene phosphorochloridite, *see* C-00058
Neopentylene phosphorochloridoselenoate, *in* C-00058
Neopentylene phosphorochloridothioate, *see* C-00060
O,O-Neopentylene phosphorodithioate, *see* M-00011
Neopentylene phosphorofluoridate, *in* F-00016
Neopentylene phosphorofluoridite, *see* F-00016
Neopentylene phosphoroiodidite, *see* I-00010
Neopentylene phosphoroiodidothioate, *in* I-00010
O,O-Neopentylene thiophosphite, *in* D-00735
Neopentylphosphonic acid, *see* D-00912
Neopentyl phosphorodichloridite, *see* D-00913
Neovitamin B_{12}, *in* V-00009
Nerolidyl dihydrogen phosphate, *see* T-00514
Nerolidyl phosphate, *see* T-00514
Neryl dihydrogen phosphate, *in* D-00776
Neryl phosphate, *in* D-00776
Neryl pyrophosphate, *in* M-00465
Neryltriphenylphosphonium(1+), *in* D-00777
Neuberg ester, *see* F-00070
▷Neurostop, *see* B-00002
▷Nexion, *see* B-00498
▷Niagara 1240, *see* E-00022
▷Nialate, *see* E-00022
Nicotinamide adenine dinucleotide phosphate, *see* C-00203
▷Nictinamide adenine dinucleotide, *see* C-00202
μ-Nitridobis(triphenylphosphorus)(1+), *see* B-00442
(Nitrilo-2,1-ethanediyl)trisphosphonic acid, *see* T-00438
▷Nitrilotris(methylenephosphonic acid), *see* N-00033
▷[Nitrilotris(methylene)]trisphosphonic acid, N-00033
[(4-Nitrobenzoyl)methyl]-triphenylphosphonium(1+), *see* N-00051
m-Nitrobenzylidenetriphenylphosphorane, *in* N-00048
o-Nitrobenzylidenetriphenylphosphorane, *in* N-00047
p-Nitrobenzylidenetriphenylphosphorane, *see* N-00041
m-Nitrobenzylphosphonic acid, *see* N-00045
o-Nitrobenzylphosphonic acid, *see* N-00044
p-Nitrobenzylphosphonic acid, *see* N-00046
(*m*-Nitrobenzyl)triphenylphosphonium(1+), *see* N-00048
(*o*-Nitrobenzyl)triphenylphosphonium(1+), *see* N-00047
(*p*-Nitrobenzyl)triphenylphosphonium(1+), *see* N-00049
(2-Nitro-2,4-cyclopentadienylidene)-triphenylphosphorane, N-00034
(1-Nitroethyl)phosphonic acid, N-00035
(2-Nitroethyl)phosphonic acid, N-00036
(Nitromethyl)phosphonic acid, N-00037
(Nitromethyl)triphenylphosphonium(1+), N-00038
O-(4-Nitrophenyl) chloro(phenylamido)-thiophosphate, *in* P-00260
2-Nitrophenyl dichlorophosphate, *in* M-00498
3-Nitrophenyl dichlorophosphate, *in* M-00499
4-Nitrophenyl dichlorophosphate, *see* N-00064
O-(*p*-Nitrophenyl) dichlorothiophosphate, *see* N-00065
2-Nitrophenyl difluorophosphate, *in* M-00498
4-Nitrophenyl difluorophosphate, *in* M-00500
2-Nitrophenyl dihydrogen phosphate, *see* M-00498
3-Nitrophenyl dihydrogen phosphate, *see* M-00499
4-Nitrophenyl dihydrogen phosphate, *see* M-00500
(3-Nitrophenyl)diphenylphosphine, N-00039
(4-Nitrophenyl)diphenylphosphine, N-00040
P-(3-Nitrophenyl)-*N,N'*-diphenylphosphonic diamide, *in* N-00056
P-(4-Nitrophenyl)-*N,N'*-diphenylphosphonic diamide, *in* N-00057
N-(3-Nitrophenyl)-*N',N''*-diphenylphosphoric triamide, *in* N-00059
N-(4-Nitrophenyl)-*N',N''*-diphenylphosphoric triamide, *in* N-00060
3-Nitrophenyl *N,N'*-diphenylphosphorodiamidate, *in* M-00499
4-Nitrophenyl *N,N'*-diphenylphosphorodiamidate, *in* M-00500
N'-(4-Nitrophenyl)-*P,P*-diphenyl-*N*-(triphenylphosphoranylidene)-phosphinimidic amide, *in* D-01164
[(2-Nitrophenyl)methylene]-triphenylphosphorane, *in* N-00047
[(3-Nitrophenyl)methylene]-triphenylphosphorane, *in* N-00048
[(4-Nitrophenyl)methylene]-triphenylphosphorane, N-00041
4-Nitrophenyl methylphenylphosphinate, *in* M-00219
(2-Nitrophenyl)methylphosphinic acid, N-00042
(4-Nitrophenyl)methylphosphinic acid, N-00043
[(2-Nitrophenyl)methyl]phosphonic acid, N-00044
[(3-Nitrophenyl)methyl]phosphonic acid, N-00045
[(4-Nitrophenyl)methyl]phosphonic acid, N-00046
[(2-Nitrophenyl)methyl]-triphenylphosphonium(1+), N-00047
[(3-Nitrophenyl)methyl]-triphenylphosphonium(1+), N-00048
[(4-Nitrophenyl)methyl]-triphenylphosphonium(1+), N-00049
4-Nitrophenyl 4-morpholinylphosphonochloridate, *in* M-00524
(4-Nitrophenyl-2-oxoethylidene)-triphenylphosphorane, N-00050
[2-(4-Nitrophenyl)-2-oxoethyl]-triphenylphosphonium(1+), N-00051
4-Nitrophenyl phenyl chlorophosphate, *in* N-00052
4-Nitrophenyl phenyl hydrogen phosphate, *see* N-00052
4-Nitrophenyl phenyl phosphate, N-00052
4-Nitrophenyl phenylphosphoramidochloridate, *in* P-00259
O-(4-Nitrophenyl) phenylphosphoramidochloridothioate, *in* P-00260
4-Nitrophenyl phenyl phosphoric acid, *see* N-00052
4-Nitrophenyl phenyl phosphorochloridate, *in* N-00052
(2-Nitrophenyl)phosphinic acid, N-00053
(4-Nitrophenyl)phosphinic acid, N-00054
(2-Nitrophenyl)phosphonic acid, N-00055
(3-Nitrophenyl)phosphonic acid, N-00056
▷(4-Nitrophenyl)phosphonic acid, N-00057
(2-Nitrophenyl)phosphoramidic acid, N-00058
(3-Nitrophenyl)phosphoramidic acid, N-00059
(4-Nitrophenyl)phosphoramidic acid, N-00060
2-Nitrophenyl phosphoric acid, *see* M-00498
3-Nitrophenyl phosphoric acid, *see* M-00499
4-Nitrophenyl phosphoric acid, *see* M-00500
(2-Nitrophenyl)phosphorimidic acid, N-00061
(3-Nitrophenyl)phosphorimidic acid, N-00062
(4-Nitrophenyl)phosphorimidic acid, N-00063
2-Nitrophenyl phosphorodichloridate, *in* M-00498
3-Nitrophenyl phosphorodichloridate, *in* M-00499
4-Nitrophenyl phosphorodichloridate, N-00064
O-(4-Nitrophenyl) phosphorodichloridothioate, N-00065
2-Nitrophenyl phosphorodifluoridate, *in* M-00498
4-Nitrophenyl phosphorodifluoridate, *in* M-00500
2-Nitrophenyl phosphoryl dichloride, *in* M-00498
3-Nitrophenyl phosphoryl dichloride, *in* M-00499
4-Nitrophenyl phosphoryl dichloride, *see* N-00064
2-Nitrophenyl phosphoryl difluoride, *in* M-00498
4-Nitrophenyl phosphoryl difluoride, *in* M-00500
N-(2-Nitrophenyl)-*P*,P,P-triphenoxyphosphine imide, *in* N-00061
O-(*p*-Nitrophenyl) thiophosphoryl dichloride, *see* N-00065
N-(2-Nitrophenyl)-*P,P,P*-triphenoxyphosphazene, *in* N-00061
N-(4-Nitrophenyl)-*P,P,P*-triphenoxyphosphazene, *in* N-00063
N-(4-Nitrophenyl-*P,P,P*-triphenoxyphosphine imide, *in* N-00063
▷Nitrophos, *see* F-00005
(1-Nitroso-1-methylethyl)phosphonic acid, N-00066
1-Nitro-5-(triphenylphosphoranylidene)-1,3-cyclopentadiene, *see* N-00034
NMDPP, *in* M-00164
Nonamethylenediphosphonic acid, *see* N-00069
2,5,5,8,11,11,14,17,17-Nonamethyl-1,3,7,9,13,15-hexaoxa-2,8,14-triphosphacyclooctadecane, N-00067
Nonamethyl imidodiphosphoramide, N-00068
Nonamethylimidodiphosphorous tetraamide, *in* M-00157
Nonamethyl iminodiphosphoramide, *see* N-00068
Nonamethyl thioimidodiphosphoramide, *see* B-00191
Nonamethyl thioimidodiphosphoramide, *see* N-00032
1,9-Nonanediphosphonic acid, N-00069
1,9-Nonanediylbisphosphonic acid, *see* N-00069
Nonaphenyl-1,3,5-triphospha-2,4,6-trigermacyclohexane, N-00070
2-Nonenylidenetriphenylphosphorane, N-00071
3-Nonenylidenetriphenylphosphorane, *in* N-00072
4-Nonenylidenetriphenylphosphorane, *in* N-00073
3-Nonenyltriphenylphosphonium(1+), N-00072
4-Nonenyltriphenylphosphonium(1+), N-00073
Nonylidenetriphenylphosphorane, *in* N-00078
Nonylphenylphosphinic acid, N-00074
Nonylphosphine, N-00075
Nonylphosphonic acid, N-00076
Nonylphosphoramidic acid, N-00077
Nonyltriphenylphosphonium(1+), N-00078
Norphos, *see* B-00220
Novofosfan, *in* T-00322
N,N,N',N',N''-Pentamethyl-*P*-phenylphosphonimidic diamide, N-00079
▷NRPPRA, *see* C-00202
NSC 224131, *see* S-00006
▷NSC 296961, *see* A-00052
▷NSC-296961, *see* E-00021
▷Nucidol, *see* D-00027
Nuvacron, *see* M-00459
▷Nuvanol *N*, *see* I-00011
1,1,6,6,11,11,16,16-Octabenzyl-1,6,11,16-tetraphosphoniacycloeicosane tetrabromide, *in* T-00008
1,1,5,5,10,10,14,14-Octabenzyl-1,5,10,14-tetraphosphoniacyclooctadecane tetrabromide, *in* T-00009
9,12-Octadecadienoic acid 1-[[[(2-aminoethoxy)hydroxyphosphinyl]oxy]-methyl]-1,2-ethanediyl ester, *see* G-00029

9,12-Octadecadienoic acid 1-[[[(2-aminoethoxy)hydroxyphosphinyl]oxy]-methyl]-2-[(1-oxooctadecyl)oxy]ethyl ester, *see* G-00079
1-Octadeca-9,12-dienoyl-2-hexadecanoylglycero-3-phosphocholine, *see* G-00068
1-Octadeca-9,12-dienoyl-2-hexadecanoylglycero-3-phosphoethanolamine, *see* G-00069
1,9,12-Octadecadienoyl-2-octadecanoylglycero-3-phosphocholine, *see* G-00070
1-Octadeca-9,12-dienoyl-2-octadec-9-enoylglycero-3-phosphocholine, *see* G-00072
Octadecanoic acid 1-[[[(2-aminoethoxy)-hydroxyphosphinyl]oxy]methyl]-1,2-ethanediyl ester, *see* G-00037
Octadecanoic acid 1-[[[(2,3-dihydroxypropoxy)hydroxyphosphinyl]-oxy]methyl]-1,2-ethanediyl ester, *see* G-00035
Octadecanoic acid 1-[(phosphonoxy)methyl]-1,2-ethanediyl ester, *see* G-00032
1-Octadecanoyl-2-icosa-5,8,11,14-tetraenoylglycero-3-phosphoethanolamine, *see* G-00077
1-Octadecanoyl-2,9,12-octadecadienoylglycero-3-phosphocholine, *see* G-00078
1-Octadecanoyl-2-octadeca-9,12-dienoylglycero-3-phosphoethanolamine, *see* G-00079
1-Octadecanoyl-2,9-octadecenoylglycero-3-phosphocholine, *see* G-00082
2-Octadecanoyl-1,9-octadecenoylglycero-3-phosphocholine, *see* G-00091
1-Octadecanoyl-2-octadec-9-enoylglycero-3-phosphoethanolamine, *see* G-00083
9,12,15-Octadecatrienoic acid 1-[[[(2-aminoethoxy)-hydroxyphosphinyl]oxy]methyl]-1,2-ethanediyl ester, *see* G-00041
9,12,15-Octadecatrienoic acid 1-[[[(2-aminoethoxy)-hydroxyphosphinyl]oxy]methyl]-2-[(1-oxooctadecyl)oxy]ethyl ester, *see* G-00081
9,12,15-Octadecatrienoic acid 1-[(phosphonoxy)methyl]-1,2-ethanediyl ester, *see* G-00039
1-Octadec-9,12-dienoyl-2-octadecanoylglycero-3-phosphocholine, *see* G-00071
Octadecenoic acid 1-[[[(2-aminoethoxy)-hydroxyphosphinyl]oxy]methyl]-2-[(1-oxooctadecyl)oxy]ethyl ester, *see* G-00083
9-Octadecenoic acid 1-[[[(2-aminoethoxy)-hydroxyphosphinyl]oxy]methyl]-1,2-ethanediyl ester, *see* G-00043
9-Octadecenoic acid 1-[[[(2-aminoethoxy)-hydroxyphosphinyl]oxy]methyl]-2-[(1-oxohexadecyloxy)ethyl ester, *see* G-00062
9-Octadecenoic acid 3-[[[(2-Aminoethoxy)-hydroxyphosphinyl]oxy]methyl]-2-[(1-oxooctadecyl)oxy]propyl ester, *see* G-00092
9-Octadecenoic acid 1-[[(1-oxohexadecyl)-oxy]methyl]-2-(phosphonoxy)ethyl ester, *see* G-00086
1-Octadec-9-enoyl-2-octadeca-9,12-dienoylglycero-3-phosphoethanolamine, *see* G-00089
1-Octadec-9-enoyl-2-octadecanoylglycero-3-phosphoethanolamine, *see* G-00092
1-Octadec-9-enyl-2-hexadecanoylglycero-3-phosphocholine, *see* G-00087
1-Octadecyl-2-hexadecanoylglycerol-3-phosphocholine, *see* G-00075
Octaethyl 1,2,3,4-butanetetrayltetraphosphonate, *in* B-00518
Octaethyldiphosphoramide, O-00001
3,3,4,4,5,5,6,6-Octafluoro-1,2-bis(diphenylphosphino)cyclohexene, O-00002
3,3,4,4,5,5,6,6-Octafluoro-1,2-cyclohexenediphosphonic acid, O-00003
(3,3,4,4,5,5,6,6-Octafluoro-1-cyclohexene-1,2-diyl)-bis[diphenylphosphine], *see* O-00002
(3,3,4,4,5,5,6,6-Octafluoro-1-cyclohexene-1,2-diyl)bisphosphonic acid, *see* O-00003
Octahydro-1*H*-cyclopropa[*c*]phosphindole 3-oxide, O-00004
Octahydro-1,8-dihydroxy-4,7-phosphinidene-1*H*-phosphindole 1,8-dioxide, *in* D-00625
Octahydro-1,9-dimethyl-[1,3,2]-diazaphosphorino[1,2-*a*][1,3,2]-diazaphosphorine, O-00005
Octahydro-3,8-dimethyl-2,7-diphenyl-1,6,3,8,2,7-dioxadiazadiphosphecine, O-00006
2,3,4,5,6,7,8,9-Octahydro-1,9-dimethyl-5-phenyl-1*H*-5,1,9-benzophosphadiarsacycloundecin, O-00007
13,14,16,17,31,32,34,35-Octahydro-6,24-dimethyltetrabenzo[*d*,*m*,*r*,*a*′]-[1,3,6,9,12,15,20,23,26,2,16]-decaoxadiphosphacyclooctacosin, O-00008
Octahydro-3,7-dimethyl-1,3,5,7-tetraphenyl-1,5,3,7-diazadiphosphocinium iodide, *in* O-00026
2,3,4,5,6,7,8,9-Octahydro-1,9-diphenyl-1*H*-5,1,9-benzazadiphosphacycloundecine, O-00009
2,3,4,5,6,7,8,9-Octahydro-1,9-diphenyl-1*H*-1,9-benzodiphosphacycloundecin, O-00010
1,2,3,4,6,7,8,9-Octahydro-1,9-diphenyl-5,9-benzothiadiphosphacycloundecin, O-00011
1,2,3,4,6,7,8,9-Octahydro-1,9-diphenyl-5,1,9-benzoxadiphosphacycloundecin, O-00012
6,7,8,9,14,15,16,17-Octahydro-9,14-diphenyldibenzo[*b*,*i*][1,4,8,11]-dithiadiphosphacyclotetradecin, O-00013
6,7,8,9,15,16,17,18-Octahydro-9,18-diphenyldibenzo[*b*,*i*][1,8,4,11]-dithiadiphosphacyclotetradecin, O-00014
6,7,9,10,16,17,18,19-Octahydro-16,19-diphenyldibenzo[*h*,*n*][1,4,7,10,13]-trioxadiphosphacyclopentadecin, O-00015
Octahydro-2,4-diphenyl-1,3,5-metheno-1*H*-2,4-diphosphacyclobuta[*c*,*d*]pentalene, O-00016
Octahydro-1,2,3,4,5,6-hexamethyl-[1,2,3]-triphospholo[4,5-*d*]-1,2,3-triphosphole, O-00017
8,9,10,11,12,13,14,15-Octahydro-4-hydroxydinaphtho[2,1-*d*:1′,2′-*f*]-[1,3,2]dioxaphosphepin 4-oxide, O-00018
Octahydro-3-methoxy-1*H*-cyclopropa[*c*]-phosphindole 3-oxide, *in* O-00004
1,2,3*a*,3*b*,4,5,11,11*a*-Octahydro-7-methoxy-3*H*-naphtho[2,1-*c*]phosphindol-3-one, *see* M-00054
Octahydro-3-methyl-1*H*-cyclopropa[*c*]-phosphindole 3-oxide, *in* O-00004
6,7,11,12,19,20,22,23-Octahydro-9-methyldibenzo[*g*,*p*]-[1,3,6,9,12,15,18,21]-heptaoxaphosphacycloeicosin, O-00019
Octahydro-3-methylpyrano[2,3-*e*]-1,3,2-oxazaphosphorine 2-oxide, *see* M-00112
Octahydro-3-methylpyrano[2,3-*e*]-1,3,2-oxazaphosphorine 2-sulfide, *see* M-00113
Octahydro-2-(4-nitrophenoxy)-2*H*-1,3,2-benzoxazaphosphorine, O-00020
1,2,3,4,5,8,9,10-Octahydro-1-phenyl-6,7-bis[(trimethylsilyl)oxy]phosphecine, O-00021
Octahydro-3-phenyl-1*H*-cyclopropa[*c*]-phosphindole 3-oxide, *in* O-00004
Octahydro-3*H*,8*H*,10*bH*-2*a*,5*a*,7*a*,10*a*-tetraaza-10*b*-phosphaindeno[7,1-*cd*]-indene, O-00022
Octahydro-7*H*,9*bH*-2*a*,4*a*,6*a*,9*a*-tetraaza-9*b*-phosphapentaleno[1,6-*cd*]indene, O-00023
Octahydro-5*H*,8*H*,10*bH*-2*a*,4*a*,7*a*,10*a*-tetraaza-10*b*-phosphapentaleno[1,6-*de*]-naphthalene, O-00024
Octahydro-8*bH*-2*a*,4*a*,6*a*,8*a*-tetraaza-8*b*-phosphapentaleno[1,6-*cd*]pentalene, O-00025
Octahydro-1,3,5,7-tetraphenyl-1,5,3,7-diazadiphosphocine, O-00026
2,3,4,5,6,7,8,9-Octahydro-1,5,9-triphenyl-1*H*-1,9,5-benzodiphosphaarsacycloundecin, O-00027
2,3,4,5,6,7,8,9-Octahydro-1,5,9-triphenyl-1*H*-1,5,9-benzotriphosphacycloundecin, O-00028
▷Octamethyldiphosphoramide, O-00029
Octamethylenediphosphonic acid, *see* O-00035
Octamethylhypodiphosphorous tetraamide, *see* T-00177
sym-Octamethylmonothiopyrophosphoramide, O-00030
unsym-Octamethylmonothiopyrophosphoramide, O-00031
sym-Octamethylmonothiopyrophosphoric tetraamide, *see* O-00030
unsym-Octamethylmonothiopyrophosphoric tetraamide, *see* O-00031
N,*N*,*N*′,*N*′,*N*″,*N*″,*N*‴,*N*‴-Octamethyl-*P*,*P*′-1,2-phenylenebis[phosphonous diamide], *in* B-00009
▷Octamethylpyrophosphoramide, *see* O-00029
2,2,3,3,7,7,8,8-Octamethyl-1,4,6,9-tetraoxa-5-phospha(5-P^V)spiro[4.4]-nonane, O-00032
1,2,7,8,9,10,15,16-Octamethyl-3,6,11,14-tetraphenyltetrakisphospholo[1,2-*b*:2′,1′-*d*:1″,2″-*f*:2‴,1‴-*h*][1,2,5,6]-tetraphosphocin, O-00033
Octamethyl thiodiphosphoramide, *see* O-00030
Octamethyl thiodiphosphoramide, *see* O-00031
Octamethyl thiodiphosphoramide, O-00034
1,8-Octanediphosphonic acid, O-00035
1,8-Octanediylbis[diphenylphosphine], *see* B-00258
1,8-Octanediylbisphosphonic acid, *see* O-00035
1-Octanephosphonic acid, *see* O-00042
2,7,9,14,15,16,17,18-Octaoxa-1,8-diphospha(1,8-P^V)-pentacyclo[10.2.1.1^{1,4}.1^{5,8}.1^{8,11}]-octadecane-3,6,10,13-tetrone, O-00036
1,2,3,4,9,10,11,12-Octaphenyl-5,8-diphosphoniadispiro[4.2.4.2]deca-1,3,9,11-tetraene(2+), O-00037
Octicidin, O-00038
▷Octicizer, *see* E-00075
O-Octyl cyclohexylphosphonochloridothioate, *in* C-00260
Octyl dichlorophosphite, *see* O-00046
Octyl dihydrogen phosphate, *see* M-00501
Octyldiphenylphosphine, O-00039
Octylidenetriphenylphosphorane, *in* O-00047
Octyl (4-methylphenyl)phosphinate, *in* M-00222
Octylphenylphosphinic acid, O-00040
Octylphosphine, O-00041
Octylphosphonic acid, O-00042
Octylphosphonic dichloride, O-00043
Octylphosphonous acid, O-00044
Octylphosphoramidic acid, O-00045
Octyl phosphoric acid, *see* M-00501
Octyl phosphorodichloridite, O-00046
Octyltriphenylphosphonium(1+), O-00047
▷Oftanol, *see* I-00025
Olein 1,2-dihydrogenphosphate, *in* T-00242
Olein 2-myristo-1,2,3-dihydroxypropyl hydrogen phosphate 3-dihydrogen phosphate, *in* G-00095
Olein 2-palmito-1-(2,3-dihydroxypropyl)-hydrogen phosphate, *in* G-00088
Oleyllinoleoylphosphatidylethanolamine, *in* G-00089
Oleyllinoleoylphosphatidylserine, *in* G-00090
1-Oleoyl-2-myristoylglycerol-3-phospho-*rac*-1′-glycerol-3′-phosphate, *in* G-00095
1-Oleoyl-2-palmitoyl-*sn*-glycerol 3-phosphate, *in* G-00094
Oleoylpalmitoylphosphatidylcholine, *in* G-00087
Oleoylpalmitoylphosphatidylserine, *in* G-00093

Oleoylstearoylphosphatidylcholine, *in* G-00091
Oleoylstearoylphosphatidylethanolamine, *in* G-00092
▷Omethoate, O-00048
▷OMPA, *see* O-00029
O,*O*-*o*-Phenylene dithiophosphoric acid, *see* M-00005
O,*O*-*o*-Phenylene hydrogen dithiophosphate, *see* M-00005
O,*O*-*o*-Phenylene phosphorothioic acid, *see* M-00005
O,*O*-*o*-Phenylene selenophosphonate, *in* M-00090
▷Orthene, *see* A-00003
▷Ortho 12420, *see* A-00003
O-*o*-Tolyl dichlorothiophosphate, *see* M-00261
O-*o*-Tolyl thiophosphoryl dichloride, *see* M-00261
Oxaloylbis[diphenylphosphine], *see* B-00235
1-Oxa-4-phenylphosphacyclohexane, *see* P-00176
1-Oxa-2-phosphacyclohexane, *see* O-00054
1-Oxa-4-phosphacyclohexane, *see* O-00055
1-Oxa-5-phospha(5-P^V)spiro[4.5]decane, O-00049
6-Oxa-5-phospha(5-P^V)spiro[4.5]decane, O-00050
1-Oxa-5-phospha(5-P^V)spiro[4.4]nonane, O-00051
1-Oxa-6-phospha(6-P^V)spiro[5.5]undecane, O-00052
1,2-Oxaphosphinane, *see* O-00054
1,4-Oxaphosphinane, *see* O-00055
1,2-Oxaphospholan-5-one 2-oxide, O-00053
1,2-Oxaphosphorinane, O-00054
1,4-Oxaphosphorinane, O-00055
1,3,2-Oxathiaphosphinane 2-oxide, *see* O-00057
1,3,2-Oxathiaphospholane, O-00056
1,3,2-Oxathiaphosphorinane 2-oxide, O-00057
Oxidronic acid, *see* H-00154
▷Oxinothiofos, *see* Q-00006
▷4-Oxocyclophosphamide, O-00058
2-Oxoethylidenetriphenylphosphorane, *see* T-00639
(1-Oxoethyl)phosphonic acid, *see* A-00026
(2-Oxoethyl)phosphonic acid, O-00059
(2-Oxoethyl)triphenylphosphonium(1+), O-00060
(2-Oxoheptyl)phosphonic acid, O-00061
(2-Oxoheptyl)triphenylphosphonium(1+), O-00062
2-Oxononylphosphonic acid, O-00063
(2-Oxooctyl)phosphonic acid, O-00064
4-Oxo-1,2,2,6,6-pentamethyl-1-phenylphosphorinanium iodide, *in* T-00210
(2-Oxo-4-phenylbutyl)phosphonic acid, O-00065
(2-Oxo-2-phenylethyl)phosphonic acid, O-00066
(2-Oxo-2-phenylethyl)-triphenylphosphonium(1+), O-00067
2-Oxo-3-phenylpropylidenetriphenyl-phosphorane, *see* P-00336
(2-Oxo-3-phenylpropyl)phosphonic acid, O-00068
(2-Oxo-1-phenylpropyl)-triphenylphosphonium(1+), O-00069
(3-Oxo-3-phenylpropyl)-triphenylphosphonium(1+), O-00070
1-Oxo-1,3-propanediphosphonic acid, O-00071
2-Oxo-1,3-propanediphosphonic acid, O-00072
(1-Oxo-1,3-propanediyl)bisphosphonic acid, *see* O-00071
(2-Oxo-1,3-propanediyl)bisphosphonic acid, *see* O-00072
(2-Oxo-1,3-propanediyl)-bis(triphenylphosphonium)(2+), O-00073
3-Oxopropyl *N*,*N*-bis(2-chloroethyl)-phosphorodiamidate, *see* A-00050
(1-Oxopropyl)phosphonic acid, O-00074
(2-Oxopropyl)phosphonic acid, O-00075
(3-Oxopropyl)phosphonic acid, O-00076
P-(2-Oxopropyl)phosphonic diamide, O-00077
(1-Oxopropyl)phosphonothioic acid, O-00078
(2-Oxopropyl)phosphonothioic acid, O-00079
(2-Oxopropyl)triphenylphosphonium(1+), O-00080
5-Oxo-1,1,3,3-tetraphenyl-1,3-diphosphinanium(2+), *see* O-00081
5-Oxo-1,1,3,3-tetraphenyl-1,3-diphosphorinanium(2+), O-00081
γ-Oxo-β-(triphenylphosphoranylidene)-benzenebutanoic acid, O-00082
β-Oxo-α-(triphenylphosphoranylidene)-benzenepropanoic acid, O-00083
3-Oxo-2-(triphenylphosphoranylidene)-butanoic acid, O-00084
4-Oxo-5-(triphenylphosphoranylidene)-pentanoic acid, O-00085
[2-Oxo-2-(triphenylphosphoranylidene)-propyl]phosphonic acid, O-00086
2,2′-Oxybis[1,3,2-benzodioxaphosphole], O-00087
5,5′-Oxybis[5*H*-benzophosphindole] 5,5′-dioxide, *in* H-00118
2,2′-Oxybis[4,5-dimethyl-1,3,2-dioxaphosphole], O-00088
2,2′-Oxybis[5,5-dimethyl-1,3,2-dioxaphosphorinane], O-00089
2,2′-Oxybis[5,5-dimethyl-1,3,2-dioxaphosphorinane] 2-oxide 2′-sulfide, O-00090
2,2′-Oxybis[1,3,2-dioxaphosphinane], *see* O-00092
▷2,2′-Oxybis[1,3,2-dioxaphospholane], O-00091
2,2′-Oxybis[1,3,2-dioxaphosphorinane], O-00092
Oxybis(2,6-diphenyl-4*H*-1,4-oxaphosphorin)-4,4′-dioxide, *in* H-00140
Oxybis[2-(diphenylphosphino)ethane], O-00093
Oxybis[2-(diphenylphosphinyl)ethane], *in* O-00093
Oxybis(1,2′-ethanediyl)bis(phosphonic acid), *see* O-00094
2,2′-Oxybis[ethanephosphonic acid], O-00094
Oxybis[methanephosphonic acid], O-00095
2,2′-Oxybis[4-methyl-1,3,2-dioxaphosphorinane], O-00096
[Oxybis(methylene)]bisphosphonic acid, *see* O-00095
[Oxybis[methylene]]-bis[triphenylphosphonium](2+), O-00097
▷Oxydemeton-methyl, O-00098
▷Oxydeprofos, O-00099
Oxy(di-2,1-ethanediyl)bis[diphenylphosphine], *see* O-00093
Oxydifficidin, *in* D-00411
Oxydifficol, *in* D-00411
PAF, *see* P-00436
Palmitoylarachidonoylphosphatidylcholine, *in* G-00055
Palmitoyllinolenoylphosphatidylcholine, *in* G-00058
Palmitoyllinoleoylphosphatidylcholine, *in* G-00056
Palmitoyllinoleoylphosphatidylethanolamine, *in* G-00059
Palmitoylmyristoylphosphatidylcholine, *see* G-00064
1-Palmitoyl-2-oleoyl-*sn*-glycerol 3-phosphate, *in* G-00060
1-D-1-*O*-(1′-Palmitoyl-2′-oleoyl-*sn*-glycero-3′-phosphate)myoinositol, *in* G-00063
1-L-1-*O*-(1′-Palmitoyl-2′-oleoyl-*sn*-glycero-3′-phosphate)myoinositol, *in* G-00063
DL-1-*O*-(1′-Palmitoyl-2′-oleoyl-*sn*-glycero-3′-phosphate)myoinositol, *in* G-00063
Palmitoyloleoylphosphatidylcholine, *in* G-00061
Palmitoyloleoylphosphatidylethanolamine, *in* G-00062
Palmitoylstearoylphosphatidylcholine, *see* G-00057
PAMP, *see* M-00059
Panclar, *see* D-00016
▷Papthion, *see* P-00078
▷Paraoxon, P-00001
▷Parathion, P-00002
▷Parathion methyl, P-00003
PB-1, *see* D-00998
PD 110161, *see* A-00136
PD 113270, *in* A-00136
PD 113271, *in* A-00136
PDMEA, *see* D-00016
Pentabutylchlorotriaminophosphonium chloride, *in* P-00004
Pentabutylphosphonous diamide, *in* B-00615
Pentabutylphosphoric triamide, *see* P-00004
Pentabutylphosphorodiamidimidic acid, P-00004
(Pentachloroethyl)phosphorimidic trichloride, P-00005
Pentachlorophenyl dichlorophosphate, *see* P-00006
O-(Pentachlorophenyl) dichlorothiophosphate, *see* P-00007
S-(Pentachlorophenyl) dichlorothiophosphate, *see* P-00008
Pentachlorophenyl dihydrogen phosphate, *see* M-00502
Pentachlorophenyl phosphoric acid, *see* M-00502
Pentachlorophenyl phosphorodichloridate, P-00006
O-(Pentachlorophenyl) phosphorodichloridothioate, P-00007
S-(Pentachlorophenyl) phosphorodichloridothioate, P-00008
Pentachlorophenyl phosphoryl dichloride, *see* P-00006
1,3-Pentadienylphosphonic acid, P-00009
1,3-Pentadiyn-1-ylphosphonic acid, P-00010
Pentaerythritol bis(cyclic hydrogen dithiophosphate), *see* D-00670
Pentaerythritol bis(cyclic phosphorodithioate), *see* D-00670
Pentaerythritol diphenyl phosphite, *see* D-00991
Pentaethoxyphosphorane, P-00011
Pentaethyl 2,3-diphosphonopropanoate, *in* B-00173
▷Pentaethyl imidodiphosphate, *in* E-00078
▷Pentaethyl iminodiphosphate, *in* E-00078
Pentaethylphosphonic diamide, *in* E-00130
Pentaethylphosphonoselenoic diamide, *in* E-00146
Pentaethylphosphonothioic diamide, *in* E-00149
Pentaethylphosphonous diamide, *in* E-00154
Pentaethylphosphoric triamide, *in* T-00058
Pentaethylphosphorodiamidimidic acid, P-00012
(Pentafluoroethyl)phosphorimidic trichloride, P-00013
Pentafluorophenyl dichlorophosphate, *in* M-00503
Pentafluorophenyl dichlorophosphate, *see* P-00017
Pentafluorophenyl dihydrogen phosphate, *see* M-00503
(Pentafluorophenyl)diphenylphosphine, P-00014
Pentafluorophenylphosphonic acid, P-00015
(Pentafluorophenyl)phosphoramidic acid, *see* P-00016
Pentafluorophenyl phosphoric acid, *see* M-00503
(Pentafluorophenyl)phosphorimidic acid, P-00016
Pentafluorophenyl phosphorodichloridate, *in* M-00503
Pentafluorophenyl phosphorodichloridate, P-00017
Pentafluorophenyl phosphoryl dichloride, *in* M-00503
Pentafluorophenyl phosphoryl dichloride, *see* P-00017
▷2,2,4,4,6-Pentakis(1-aziridinyl)-2,2,4,4,6,6-hexahydro-6-(4-morpholinyl)-1,3,5-triaza-2,4,6-triphosphorine, *see* F-00066
▷1,2,3,4,5-Pentakis(trifluoromethyl)-1,2,3-phosphol-4-ene, *see* D-00536
Pentamethoxyphosphorane, P-00018
Pentamethylcyclopentaphosphine, *see* P-00019
Pentamethylenebis[dibutylphosphine], *see* B-00161
Pentamethylenebis[dimethylphosphine], *see* B-00205
Pentamethylenebis(diphenylphosphine), *see* B-00259

Pentamethylenebisphosphine, *see* D-01179
▷Pentamethylenebis[triphenylphosphonium](2+), *see* P-00034
Pentamethylenediphosphonic acid, *see* P-00032
Pentamethylenediphosphonous acid, *see* P-00033
Pentamethylenephosphine, *see* P-00404
Pentamethylenephosphinic acid, *see* H-00188
Pentamethylpentaphospholane, P-00019
2,2,3,4,4-Pentamethyl-1-phenylphosphetane, P-00020
2,2,3,4,4-Pentamethylphosphetanamine 1-oxide, P-00021
2,2,3,4,4-Pentamethylphosphetane, P-00022
Pentamethylphosphinimidic amide, *in* D-00833
Pentamethylphosphonic diamide, *in* M-00289
N,N,N',N',P-Pentamethylphosphonimidic diamide, P-00023
Pentamethyl 2-phosphono-1,2,4-butanetricarboxylate, *in* P-00382
Pentamethylphosphonoselenoic diamide, P-00024
Pentamethylphosphonotelluroic diamide, P-00025
Pentamethylphosphonothioic diamide, *in* M-00314
Pentamethylphosphonous diamide, *in* M-00320
▷Pentamethylphosphoric triamide, P-00026
2',2',3',4',4'-Pentamethylspiro[1,3,2-benzodioxaphosphole-2,1'-phosphetane], P-00027
2,7,10,15,17-Pentamethyl-17*H*-tetrabenzo[*b,d,f,h*]phosphonin, P-00028
1,1,2,3,3-Pentamethyl-6,6,7,7-tetrakis(trifluoromethyl)-5,8-dioxa-4-phospha(4-P^V)spiro[3.4]octane, P-00029
1,1,2,3,3-Pentamethyl-6,6,8,8-tetrakis(trifluoromethyl)-5,7-dioxa-4-phospha(4-P^V)spiro[3.4]octane, P-00030
N,N,N',N',3-Pentamethyl-1,3,2,4-thiazadiphosphetidin-2,4-diamine, *see* B-00187
1,2,3,4,5-Pentamethyl-1,2,4,3,5-triazadiphospholidine, *in* T-00562
1,1,2,3,3-Pentamethyltrimethylenephosphinic acid, *see* H-00168
1,1,2,3,3-Pentamethyltrimethylenephosphinic azide, *in* H-00168
1,1,2,3,3-Pentamethyltrimethylenephosphinic chloride, *in* C-00124
1,1,2,3,3-Pentamethyltrimethylenephosphinic fluoride, *in* H-00168
1,1,2,3,3-Pentamethyltrimethylenephosphinothioic acid, *see* H-00169
1,1,2,3,3-Pentamethyltrimethylenephosphinothioic chloride, *in* C-00124
1,1-Pentanediphosphonic acid, P-00031
1,5-Pentanediphosphonic acid, P-00032
1,5-Pentanediphosphonous acid, P-00033
P,P'-1,5-Pentanediylbis[bis-*N,N*-diethylphosphonous diamide], *in* P-00033
1,5-Pentanediylbis[dibutylphosphine], *see* B-00161
1,5-Pentanediylbis[dimethylphosphine], *see* B-00205
1,5-Pentanediylbis[diphenylphosphine], *see* B-00259
1,5-Pentanediylbisphosphine, *see* D-01179
1,5-Pentanediylbisphosphonic acid, *see* P-00032
1,5-Pentanediylbis(phosphonous acid), *see* P-00033
▷1,5-Pentanediylbis[triphenylphosphonium](2+), P-00034
1,5-Pentanediylidenebis[triphenylphosphorane], *in* P-00034
Pentaphenoxyphosphorane, P-00035
Pentaphenylcyclopentaphosphine, *see* P-00037
▷1,1,4,7,7-Pentaphenyl-1,7-diarsa-4-phosphaheptane, *see* B-00212
2,2,3,4,6-Pentaphenyl-2*H*-1,5,2-dioxaphosphinin 2-oxide, *see* P-00036
2,2,3,4,6-Pentaphenyl-2*H*-1,5,2-dioxaphosphorin 2-oxide, P-00036
Pentaphenylpentaphospholane, P-00037
Pentaphenylphosphacyclopentadiene, *see* P-00038
2,3,4,5,6-Pentaphenylphosphinine, *see* P-00040
Pentaphenyl-1*H*-phosphole, P-00038
1,2,3,4,5-Pentaphenyl-1,2,3-phosphol-4-ene, *see* D-00539
Pentaphenylphosphonous diamide, *in* P-00252
Pentaphenylphosphorane, P-00039
2,3,4,5,6-Pentaphenylphosphorin, P-00040
3,3,3,6,6-Pentaphenyl-3,3,3,6-tetrahydro-4,5-bis(methoxycarbonyl)-1,2,3-diazaphosphorine, P-00041
3,3,3,6,6-Pentaphenyl-3,3,3,6-tetrahydro-4,5-bis(methoxycarbonyl)-1,2,3-diazaphosphinine, *see* P-00041
1,2,3,4,9-Pentaphenyl-1*H*-tribenzo[*b,d,f*]-phosphepin, P-00042
1,1,4,7,7-Pentaphenyl-1,4,7-triphosphaheptane, *see* B-00239
1,1,5,9,9-Pentaphenyl-1,5,9-triphosphanonane, *see* B-00266
1,1,3,5,5-Pentaphenyl-1,3,5-triphosphapentane, *see* B-00253
Pentapropylphosphonic diamide, *in* P-00474
Pentapropylphosphonous diamide, *in* P-00482
Pentapropylphosphoric triamide, *see* P-00043
Pentapropylphosphorodiamidimidic acid, P-00043
4-Pentene-2-phosphonic acid, *see* M-00099
4-Pentenylidenetriphenylphosphorane, *in* P-00044
4-Pentenyltriphenylphosphonium(1+), P-00044
2-Penten-4-ynylidenetriphenylphosphorane, *in* P-00047
(3-Penten-1-ynyl)phosphonic acid, P-00045
(4-Penten-2-yn-1-yl)phosphonic acid, P-00046
2-Penten-4-ynyltriphenylphosphonium(1+), P-00047
O-Pentyl dichlorothiophosphate, *see* P-00055
Pentyl dihydrogen phosphate, *see* M-00504
Pentyl dipentylphosphinate, *in* D-00981
N-Pentyl-*N',N''*-diphenylphosphoric triamide, *in* P-00053
Pentylidenebisphosphonic acid, *see* P-00031
Pentylidenetriphenylphosphorane, *in* P-00056
Pentyl (1-methylpropyl)phenylphosphinate, *in* M-00372
Pentyl phosphinate, P-00048
Pentylphosphine, P-00049
Pentylphosphonic acid, P-00050
2-Pentylphosphonic acid, *see* M-00105
Pentylphosphonodithioic acid, P-00051
Pentylphosphonous acid, P-00052
Pentylphosphoramidic acid, P-00053
Pentyl phosphoric acid, *see* M-00504
Pentyl phosphorodichloridite, P-00054
O-Pentyl phosphorodichloridothioate, P-00055
O-Pentyl thiophosphoryl dichloride, *see* P-00055
Pentyltriphenylphosphonium(1+), P-00056
Peroxycyclophosphamide, P-00057
▷Pestan, *see* M-00002
▷Pestox III, *see* O-00029
▷Pestox XIV, *see* D-00669
▷Phaseolotoxin, P-00058
Phaseotoxin *A*, *see* P-00367
Phellanphos, P-00059
Phenacylphosphonic acid, *see* O-00066
Phenacyltriphenylphosphonium, *see* O-00067
▷Phenamiphos, *see* F-00002
9,10-Phenanthrenediylbis[diphenylphosphine], *see* B-00261
(3-Phenanthrenylmethylene)-triphenylphosphorane, *in* T-00637
(9-Phenanthrenylmethylene)-triphenylphosphorane, *in* T-00638
(3-Phenanthrenylmethyl)-triphenylphosphonium(1+), *see* T-00637
6-Phenanthridinephosphonic acid, P-00060
6-Phenanthridinylphosphonic acid, *see* P-00060
▷Phenazaphosphinic acid, P-00061
(9-Phenenthrenylmethyl)-triphenylphosphonium(1+), *see* T-00638
Phenethyl dichlorophosphate, *in* M-00506
Phenethyl dihydrogen phosphate, *see* M-00506
Phenethyldiphenylphosphine, *see* D-01034
2-Phenethylphosphine, *see* P-00163
Phenethylphosphonic acid, *see* P-00165
Phenethyl phosphoric acid, *see* M-00506
Phenethyl phosphorodichloridate, *in* M-00506
Phenethyltriphenylphosphonium(1+), *see* T-00613
Phenidet, *in* B-00092
▷Phenkapton, P-00062
Phenoxaphosphinic acid, *see* H-00170
2-Phenoxybenzo[*d*][1,3,2]-dioxaphosphacyclohept-4-ene, *see* P-00066
2-Phenoxy-4*H*-1,3,2-benzodioxaphosphinin, *see* P-00064
2-Phenoxy-1,3,2-benzodioxaphosphole, P-00063
2-Phenoxy-4*H*-1,3,2-benzodioxaphosphorin, P-00064
[(Phenoxycarbonylmethylene)hydrazono]-triphenylphosphorane, *in* T-00653
6-Phenoxy-6*H*-dibenzo[*c,e*][1,2]oxaphosphorin, P-00065
2-Phenoxy-1,5-dihydro-2,4,3-benzodioxaphosphepin, P-00066
5-Phenoxy-5,6-dihydrodibenzo[*b,d*]phosphorin 5-oxide, *in* D-00504
5-Phenoxy-5,6-dihydrophosphanthridene 5-oxide, *in* D-00504
2-Phenoxy-1,3-dioxa-2-phosphacyclohexane, *see* P-00068
2-Phenoxy-1,3-dioxa-2-phosphacyclopentane, *see* P-00067
2-Phenoxy-1,3,2-dioxaphosphepane 2-oxide, *in* H-00132
2-Phenoxy-1,3,2-dioxaphosphinane, *see* P-00068
2-Phenoxy-1,3,2-dioxaphospholane, P-00067
2-Phenoxy-1,3,2-dioxaphosphonane 2-oxide, *in* H-00137
2-Phenoxy-1,3,2-dioxaphosphorinane, P-00068
2-Phenoxy-4,5-diphenyl-1,3,2-dioxaphospholane, P-00069
P-Phenoxy-*P,P*-diphenyl-*N*-phenylsulfonylphosphazene, *in* B-00013
P-Phenoxy-*P,P*-diphenyl-*N*-phenylsulfonylphosphine imide, *in* B-00013
1-Phenoxy-3,4-diphenyl-1*H*-phosphole 1-oxide, *in* H-00142
2-Phenoxy-1,3,2-dithiaphospholane 2-sulfide, *in* D-01245
O-(Phenoxymethylphosphinothioyl) *O'*-phenyl methylphosphonothioate, *in* D-00925
2-Phenoxynaphtho[1,8-*de*]-1,3,2-dioxaphosphorine, P-00070
1-Phenoxy-2-oxa-1-phosphaindane, *see* D-00546
2-Phenoxy-1,2-oxaphosphorinane 2-oxide, *in* H-00167
2-Phenoxy-1,3,2-oxaselenaphosphinane 2-oxide, *see* P-00071
2-Phenoxy-1,3,2-oxaselenaphosphorinane 2-oxide, P-00071
2-Phenoxy-1,3,2-oxathiaphospholane, P-00072
2-Phenoxy-1,3,2-oxazaphosphorinane, *see* T-00126
1-Phenoxy-2,2,3,4,4-pentamethylphosphetane 1-oxide, *in* H-00168
2-Phenoxyphenyl dichlorophosphate, *in* M-00505
2-Phenoxyphenyl dihydrogen phosphate, *see* M-00505
2-Phenoxy-3-phenyl-1,3,2-oxazaphospholane, *see* P-00073
2-Phenoxy-3-phenyl-1,3,2-oxazaphospholidine, P-00073
2-Phenoxy-3-phenyl-1,3,2-oxazaphosphorinane, *see* T-00127
2-Phenoxyphenyl phosphoric acid, *see* M-00505
2-Phenoxyphenyl phosphorodichloridate, *in* M-00505
2-Phenoxyphenyl phosphoryl dichloride, *in* M-00505
2-Phenoxy-2,2'-spirobi[1,3,2-benzodioxaphosphole], P-00074
2-Phenoxy-4,4,5,5-tetramethyl-1,3,2-dioxaphospholane, P-00075
1-Phenoxy-2,3,4,5-tetraphenyl-1*H*-phosphole 1-oxide, *in* H-00200
Phenoxytetraphenylphosphorane, P-00076
P-Phenoxy-*N*-trimethylsilyl-*P*-[bis(trimethylsilyl)amino]phosphazene, *in* T-00867
P-Phenoxy-*N*-trimethylsilyl-*P*-[bis(trimethylsilyl)amino]phosphine imide, *in* T-00867

Phenoxytriphenylphosphorus(1+), P-00077
Phenphos, *see* P-00150
▷Phenthoate, P-00078
10-Phenylacridophosphine, P-00079
5-Phenyl-10(5*H*)-acridophosphinone, P-00080
1-Phenylallenephosphonic acid, *see* P-00298
(Phenylamino)carbonylphosphoramidic acid, P-00081
[(Phenylamino)carbonyl]phosphoramidothioic acid, P-00082
1-Phenylamino-1,3,2-dioxaphospholane, P-00083
[(Phenylamino)methyl]phosphonic acid, P-00084
[3-(Phenylamino)propyl]phosphonic acid, P-00085
7-Phenyl-6-aza-7-phosphabicyclo[3.2.1]-octane, P-00086
2-Phenyl-1-aza-2-phosphacyclopentane, *see* P-00089
3-Phenyl-1-aza-3-phosphacyclopentane, *see* P-00091
3-Phenyl-1-aza-3-phosphacyclopentane-2-thione, *see* P-00093
3-Phenyl-1,3-azaphosphetidine, P-00087
2-Phenyl-1*H*-1,3-azaphosphole, P-00088
2-Phenyl-1,2-azaphospholidine, P-00089
2-Phenyl-1,3-azaphospholidine, P-00090
3-Phenyl-1,3-azaphospholidine, P-00091
1-Phenyl-1,2-azaphospholidin-5-one, P-00092
3-Phenyl-1,3-azaphospholidin-2-thione, P-00093
3-Phenyl-Δ^1-1,3-azaphospholine, *see* D-00549
2-Phenyl-1,3,2-benzodioxaphosphole, P-00094
2-Phenyl-4*H*-1,3,2-benzodioxaphosphorin, P-00095
3-Phenyl-3*H*-benzophosphepin, P-00096
1-Phenyl-1*H*-benzo[*b*]phosphindole, *see* P-00110
2-Phenylbenzo[*b*]phosphinine, *see* P-00212
2-Phenylbenzo[*b*]phosphorin, *see* P-00212
2-Phenyl-4*H*-1,3,2-benzoxathiaphosphorin, P-00097
Phenyl bis(1-aziridinyl)phosphinate, *in* B-00091
Phenyl bis(1-aziridinyl)phosphinite, *in* B-00095
Phenyl bis(1*H*-benzimidazol-1-yl)phosphinate, *in* B-00097
Phenylbisbiphenylenephosphorane, *in* S-00021
Phenyl bis(4-bromophenyl)phosphinate, *in* B-00114
Phenyl *N*,*N*-bis(2-chloroethyl)-*N*′-phenylphosphorodiamidate, *in* B-00121
Phenyl *N*,*N*-bis(2-chloroethyl)-phosphorodiamidate, *in* B-00126
Phenyl bis(chloromethyl)phosphinate, *in* B-00127
O-Phenyl bis(chloromethyl)phosphinothioate, *in* B-00129
N-Phenyl-*P*,*P*-bis(chloromethyl)-phosphinothioic amide, *in* B-00129
Phenyl bis(1,1-dimethylethyl)phosphinite, *in* D-00122
Phenyl bis(iodomethyl)phosphinate, *in* B-00288
Phenyl bis(2-methylpropyl)phosphinothioite, *in* B-00352
O-Phenyl bis(pentafluorophenyl)-phosphinothioate, *in* B-00383
Phenyl bis(4-phenyl-1,3-butadienyl)-phosphinate, *in* B-00385
N-Phenyl-*P*,*P*-bis(4-phenyl-1,3-butadienyl)-phosphinic amide, *in* B-00385
Phenylbis[2-(phenylethynyl)phenyl]phosphine, P-00098
Phenyl bis(phenylethynyl)phosphinate, *in* B-00388
Phenylbis(phenylethynyl)phosphine, P-00099
N-Phenyl-*P*,*P*-bis(phenylethynyl)phosphinic amide, *in* B-00388
▷Phenylbis(phenylmethyl)phosphine, *see* D-00052
Phenyl bis(phenylmethyl)phosphinite, *in* D-00058
O-Phenyl *N*,*N*′-bis(phenylmethyl)-phosphorodiamidothioate, *in* D-00062
Phenyl bis(1-pyrrolidinyl)phosphinate, *in* D-01233
N-Phenyl-*P*,*P*-bis(trichloromethyl)-phosphinimidic chloride, *in* B-00403
N-Phenyl-*P*,*P*-bis(trifluoromethyl)-phosphinous amide, *in* B-00419
N-Phenyl-*P*,*P*-bis(2,4,6-trimethylphenyl)-phosphinic amide, *in* B-00428
1-Phenyl-6,7-bis[(trimethylsilyl)oxy]-phosphacycloundec-6-ene, P-00100
Phenyl [bis(trimethylsilyloxy)phosphinyl]-acetate, *in* B-00434
Phenylbis(triphenylgermyl)phosphine, P-00101
P-Phenyl-*N*,*N*′-bis(triphenyl-phosphoranylidene)phosphonic diamide, *in* P-00102
P-Phenyl-*N*,*N*′-bis(triphenyl-phosphoranylidene)phosphonothioic diamide, *in* P-00102
P-Phenyl-*N*,*N*′-bis(triphenyl-phosphoranylidene)phosphonous diamide, P-00102
Phenylbis[3-(triphenylphosphoranylidene)-propenyl]phosphine, P-00103
Phenyl *N*,*N*′-1,3-butylene-*N*,*N*′-dimethylphosphorodiamidite, *in* H-00057
Phenyl *tert*-butylmethylphosphinate, *in* B-00567
Phenyl *tert*-butylphenylphosphinate, *in* B-00579
Phenyl butylphosphinate, *in* B-00590
Phenyl *tert*-butylphosphinate, *in* B-00591
Phenyl butylphostonate, *in* H-00167
P-(4-*N*-phenylcarbamoylphenyl)-*N*,*N*′-diphenylphosphonic diamide, *in* P-00380
Phenylcarbamoylphosphoramidic acid, *see* P-00081
Phenylcarbamoylphosphoramidothioic acid, *see* P-00082
O-Phenyl chlorodimethylamidothiophosphate, *in* D-00861
Phenyl (2-chloroethyl)phosphonochloridate, *in* C-00077
Phenyl chlorofluorophosphate, *see* P-00271
O-Phenyl chloro(methylamido)thiophosphate, *in* M-00329
Phenyl *P*-chloromethyl-*N*-(dichlorophosphinyl)-phosphonochloridimidate, *in* C-00091
Phenyl *P*-chloromethylphosphonamidate, *in* C-00105
Phenyl (chloromethyl)phosphonochloridate, *in* C-00111
O-Phenyl chloro(phenylamido)thiophosphate, *in* P-00260
Phenyl cyclohexylphosphoramidochloridate, P-00104
O-Phenyl diamidothiophosphate, *see* P-00276
9-Phenyl-1,5-diaza-9-phosphaadamantane, *see* P-00106
9-Phenyl-1,5-diaza-9-phosphabicyclo[3.3.1]nonane, P-00105
9-Phenyl-1,5-diaza-9-phosphatricyclo[3.3.1.$1^{3,7}$]decane, P-00106
6-Phenyldibenzo[*d*,*f*][1,3,2]dioxaphosphepin, P-00107
5-Phenyl-5*H*-dibenzo[*b*,*f*]phosphepin, P-00108
6-Phenyl-5*H*-dibenzo[*c*,*e*]phosphepin-5,7(6*H*)-dione, P-00109
10-Phenyldibenzo[*b*,*e*]phosphinine, *see* P-00079
5-Phenyl-5*H*-dibenzophosphole, P-00110
7-Phenyl-7*H*-dibenzo[*d*,*f*]phosphonin, P-00111
10-Phenyldibenzo[*b*,*e*]phosphorin, *see* P-00079
Phenyl dibenzylphosphinite, *in* D-00058
Phenyl *N*,*N*′-dibenzylphosphorodiamidate, *in* D-00061
O-Phenyl-*N*,*N*′-dibenzylphosphorodiamidothioate, *in* D-00062
Phenyl dibromophosphite, *see* P-00277
O-Phenyl dibromothiophosphate, *see* P-00278
Phenyl di-*tert*-butylphosphinate, *in* D-00110
Phenyl di-*tert*-butylphosphinite, *in* D-00122
O-Phenyl di-*tert*-butylphosphinothioate, *in* D-00118
Phenyl *N*,*N*′-dibutylphosphorodiamidate, *in* D-00146
Phenyl dichlorodithiophosphate, *see* P-00281
Phenyl dichlorophosphate, *see* P-00279
Phenyl (dichlorophosphinothioyl)-phosphorodichlorimidate, P-00112
▷Phenyl dichlorophosphite, *see* P-00280
O-Phenyl dichlorothiophosphate, *see* P-00282
S-Phenyl dichlorothiophosphate, *see* P-00283
▷Phenyl dichlorothiophosphite, *see* P-00284
Phenyl dicyclohexylphosphinate, *in* D-00211
S-Phenyl dicyclohexylphosphinothioate, *in* D-00213
Phenyl *N*,*N*′-dicyclohexylphosphorodiamidate, *in* D-00216
Phenyl diethenylphosphinate, *in* D-01258
Phenyl 2-(diethoxyphosphinyl)butanoate, *in* D-00246
Phenyl (2-diethoxyphosphinyl)propanoate, *in* D-00258
Phenyl *N*,*N*-diethyl-*P*-methylphosphonamidate, *in* D-00296
O-Phenyl *N*,*N*-diethyl-*P*-methylphosphonamidothioate, *in* D-00297
S-Phenyl *N*,*N*-diethyl-*P*-methylphosphonamidothioate, *in* D-00297
Phenyl *N*,*N*-diethyl-*P*-phenylphosphonamidate, *in* D-00307
S-Phenyl *N*,*N*-diethyl-*P*-phenylphosphonamidothioate, *in* D-00309
Phenyl diethylphosphinate, *in* D-00322
Phenyl diethylphosphinodithioate, *in* D-00325
O-Phenyl diethylphosphinoselenoate, *in* D-00327
O-Phenyl diethylphosphinothioate, *in* D-00330
Phenyl diethylphosphinothioite, *in* D-00332
Phenyl diethylphosphoramidochloridate, *in* D-00346
O-Phenyl diethylphosphoramidochloridoselenoate, *in* D-00347
O-Phenyl diethylphosphoramidochloridothioate, *in* D-00348
Phenyl diethylphosphoramidofluoridodithioate, *in* D-00352
Phenyl *N*,*N*′-diethylphosphorodiamidate, *in* D-00382
Phenyl difluorophosphate, *see* P-00285
Phenyl difluorophosphite, *see* P-00286
O-Phenyl difluorothiophosphate, *see* P-00287
S-Phenyl difluorothiophosphate, *see* P-00288
Phenyl dihydrogen phosphate, *see* M-00507
S-Phenyl dihydrogen phosphorothioate, *see* P-00297
Phenyl *N*-(4,5-dihydro-2-thiazolyl)-*P*-methylphosphonamidate, *in* D-00587
O-Phenyl *N*-(4,5-dihydro-2-thiazolyl)-*P*-methylphosphonamidothioate, *in* D-00588
Phenyl di-1*H*-imidazol-1-ylphosphinate, *in* D-00628
Phenyl diisobutylphosphinothioite, *in* B-00352
O-Phenyl diisopropylphosphinothioate, *in* D-00646
Phenyl *N*,*N*′-diisopropylphosphorodiamidate, *in* D-00661
Phenyl 3-[(dimethoxyphosphinothioyl)oxy]-2-methyl-2-propenoate, *in* D-00673
Phenyl (dimethoxyphosphinyl)acetate, *in* D-00678
Phenyl (dimethylamido)fluorodithiophosphate, *in* D-00352
Phenyl dimethylphosphinate, *in* D-00829
Phenyl dimethylphosphinite, *in* D-00844
Phenyl dimethylphosphinodithioate, *in* D-00834
Phenyl dimethylphosphinoselenoite, *in* D-00838
O-Phenyl dimethylphosphinothioate, *in* D-00839
Phenyl *N*,*P*-dimethylphosphonamidate, *in* D-00849
Phenyl dimethylphosphoramidochloridate, *in* D-00860
Phenyl dimethylphosphoramidochloridite, *in* D-00862
O-Phenyl dimethyl-phosphoramidochloridothioate, *in* D-00861
Phenyl dimethyl-phosphoramidofluoridodithioate, *in* D-00867
O-Phenyl dimethyl-phosphoramidofluoridothioate, *in* D-00868
▷Phenyl *N*,*N*′-dimethylphosphorodiamidate, *see* D-00024
▷Phenyl di-4-morpholinylphosphinate, *in* D-00935

Phenyl di-4-morpholinylphosphinite, *in* D-00936
4-Phenyl-3,5-dioxa-1-phosphabicyclo[2.2.1]heptane, P-00113
2-Phenyl-1,3-dioxa-2-phosphacyclopentane, *see* P-00115
2-Phenyl-1,3,2-dioxaphosphepane, P-00114
2-Phenyl-1,3,2-dioxaphosphinane, *see* P-00116
N-Phenyl-1,3,2-dioxaphospholan-2-amine, *see* P-00083
2-Phenyl-1,3,2-dioxaphospholane, P-00115
2-Phenyl-1,3,2-dioxaphosphorinane, P-00116
2-Phenyl-1,3,6,2-dioxathiaphosphocane, P-00117
Phenyl(diphenoxyphosphinyl)acetate, *in* D-00988
Phenyl *N*-(diphenoxyphosphinyl)carbamate, *in* D-00989
5-Phenyl-5,10-diphenyldibenzo[1,4]-phosphoniaarsenin, *in* D-01001
Phenyl *P,P*-diphenyl-*N*-phenylsulfonylphosphinimidate, *in* B-00013
▷Phenyl diphenylphosphinate, *in* D-01039
Phenyl diphenylphosphinite, *in* D-01089
Phenyl diphenylphosphinodithioate, *in* D-01065
N-Phenyl-diphenylphosphinoformamide, *see* T-00625
O-Phenyl diphenylphosphinoselenoate, *in* D-01078
Se-Phenyl diphenylphosphinoselenoate, *in* D-01078
Phenyl diphenylphosphinoselenoite, *in* D-01081
O-Phenyl diphenylphosphinothioate, *in* D-01082
S-Phenyl diphenylphosphinothioate, *in* D-01082
Phenyl diphenylphosphinothioite, *in* D-01085
▷Phenyl *N,P*-diphenylphosphonamidate, *in* D-01103
O-Phenyl *N,P*-diphenylphosphonamidothioate, *in* D-01104
Phenyl *N,N'*-diphenylphosphorodiamidate, *in* D-01125
O-Phenyl *N,N'*-diphenyl-phosphorodiamidothioate, *in* D-01129
Phenyl di-1-piperidinylphosphinate, *in* D-01190
N-Phenyl-*P,P*-di-1-piperidinylphosphinic amide, *in* D-01190
Phenyl di-2-propenyl phosphate, P-00118
Phenyldi-2-propenylphosphine, P-00119
Phenyldipropylphosphine, P-00120
Phenyl *N,N'*-dipropylphosphorodiamidate, *in* D-01223
2-Phenyl-1,3-dithia-2-phosphacyclopentane, *see* P-00121
2-Phenyl-1,3,2-dithiaphospholane, P-00121
2-Phenyl-1,3,2-dithiaphosphorinane, *in* D-01247
Phenyldi-*m*-tolylphosphine, *see* B-00310
Phenyldi-*o*-tolylphosphine, *see* B-00309
Phenyldi-*p*-tolylphosphine, *see* B-00311
N-Phenyl di-*o*-tolylphosphinic amide, *in* B-00318
N-Phenyl di-*p*-tolylphosphinic amide, *in* B-00320
Phenyl divinylphosphinate, *in* D-01258
Phenyldivinylphosphine, P-00122
1,3-Phenylenebis(dichlorophosphite), *see* P-00141
1,4-Phenylenebis(dichlorophosphite), *see* P-00142
1,2-Phenylenebis[dimethylphosphine], P-00123
1,3-Phenylenebis[dimethylphosphine], P-00124
1,4-Phenylenebis[dimethylphosphine], P-00125
1,2-Phenylenebis[diphenylphosphine], P-00126
1,3-Phenylenebis[diphenylphosphine], P-00127
1,4-Phenylenebis[diphenylphosphine], P-00128
1,3-Phenylenebis(methyldiphenylphosphonium)-diiodide, *in* P-00127
[1,2-Phenylenebis(methylene)]bisphosphonic acid, P-00129
[1,3-Phenylenebis(methylene)]bisphosphonic acid, P-00130
[1,4-Phenylenebis(methylene)]bisphosphonic acid, P-00131
1,2-[Phenylenebis[methylene]]-bis[triphenylphosphonium](2+), P-00132
[1,3-Phenylenebis[methylene]]-bis[triphenylphosphonium](2+), P-00133
1,4-[Phenylenebis[methylene]]-bis[triphenylphosphonium](2+), P-00134
[1,2-Phenylenebis[methylene]]-bis[triphenylphosphorane], P-00135
[1,3-Phenylenebis[methylene]]-bis[triphenylphosphorane], *in* P-00133
[1,4-Phenylenebis(methylene)]-bis[triphenylphosphorane], P-00136
1,4-Phenylenebis[methylphosphinic acid], P-00137
1,4-Phenylenebis[methylphosphinodithioic acid], P-00138
1,2-Phenylenebis[phenylphosphinic acid], P-00139
1,4-Phenylenebis[phenylphosphinic acid], P-00140
1,2-Phenylenebisphosphine, *see* D-01172
1,4-Phenylenebisphosphine, *see* D-01173
1,2-Phenylenebisphosphonic acid, *see* B-00006
1,3-Phenylenebisphosphonic acid, *see* B-00007
1,4-Phenylenebisphosphonic acid, *see* B-00008
1,2-Phenylenebis(phosphonous acid), *see* B-00009
1,4-Phenylenebis(phosphonous acid), *see* B-00010
1,3-Phenylenebis(phosphoramidic acid), *see* B-00011
1,4-Phenylenebis(phosphoramidic acid), *see* B-00012
1,3-Phenylene bisphosphorodichloridate, *in* P-00147
1,4-Phenylenebis(phosphorodichloridate), *in* P-00148
1,3-Phenylenebis[phosphorodichloridite], P-00141
1,4-Phenylenebis[phosphorodichloridite], P-00142
1,4-Phenylenebis(phosphorodifluoridate), *in* P-00148
N,N'-1,2-Phenylenebis[*P,P,P*-triphenylphosphine imide], P-00143
N,N'-1,3-Phenylenebis[*P,P,P*-triphenylphosphine imide], P-00144
N,N'-1,4-Phenylenebis[*P,P,P*-triphenylphosphine imide], P-00145
P,P'-1,4-Phenylenebis[*N,P,P*-triphenylphosphine imide], P-00146
o-Phenylene bromophosphite, *see* B-00464
o-Phenylene chlorophosphite, *see* C-00021
(*o*-Phenylenedimethyl)-bis[triphenylphosphonium](2+), *see* P-00132
(*p*-Phenylenedimethylene)-bis[triphenylphosphonium](2+), *see* P-00134
(*m*-Phenylenedimethylidyne)-bis[triphenylphosphorane], *in* P-00133
o-Phenylene dimethylphosphoramidate, *in* D-00689
o-Phenylene dimethylphosphoramidite, *see* D-00689
o-Phenylenedioxy hydrogen phosphate, *see* H-00108
o-Phenylenedioxy phosphate, *see* H-00108
o-Phenylenedioxyphosphinyl azide, *see* A-00148
o-Phenylenediphosphonous acid, *see* B-00009
p-Phenylenediphosphonous acid, *see* B-00010
1,3-Phenylene diphosphoric acid, P-00147
1,4-Phenylenediphosphoric acid, P-00148
(1,2-Phenylenedi-2-propene-3,1-diyl)-bis[triphenylphosphonium](2+), P-00149
(1,2-Phenylenedi-2-propene-3,1-diylidene)-bis[triphenylphosphorane], *in* P-00149
(1,2-Phenylenediyldimethylidyne)-bis[triphenylphosphorane], *see* P-00135
(1,4-Phenylenediyldimethylidyne)-bis(triphenylphosphorane), *see* P-00136
o-Phenylene fluorophosphite, *see* F-00013
o-Phenylene methylphosphonate, *in* M-00090
o-Phenylene methylphosphonite, *see* M-00090
o-Phenylene phenylphosphonate, *in* P-00094
o-Phenylene phenylphosphonite, *see* P-00094
o-Phenylene phosphite, *in* B-00018
o-Phenylene phosphonate, *in* B-00018
o-Phenylene phosphonite, *see* B-00018
o-Phenylene phosphorobromidate, *in* B-00464
o-Phenylene phosphorobromidite, *see* B-00464
o-Phenylene phosphorochloridite, *see* C-00021
o-Phenylene phosphorofluoridite, *see* F-00013
o-Phenylene phosphoryl azide, *in* A-00148
O,O-*o*-Phenylene thiophosphonate, *in* M-00090
(1-Phenyl-1,2-ethanediyl)-bis[diphenylphosphine], P-00150
Phenyl *N,N'*-1,2-ethanediyl-*N,N'*-dimethylphosphorodiamidite, *see* D-00788
12-Phenyl-10*H*-5,10-ethenoacridophosphine, P-00151
Phenyl ethenylphenylphosphinate, *in* P-00337
(2-Phenylethenyl)phosphine, P-00152
(2-Phenylethenyl)phosphinic acid, P-00153
(1-Phenylethenyl)phosphonic acid, P-00154
2-Phenylethenylphosphonic acid, P-00155
(2-Phenylethenyl)phosphonic bis(dibutylamide), *in* P-00156
(2-Phenylethenyl)phosphonic bis(diethylamide), *in* P-00156
(2-Phenylethenyl)phosphonic bis(dimethylamide), *in* P-00156
(2-Phenylethenyl)phosphonic bis(dipropylamide), *in* P-00156
P-(2-Phenylethenyl)phosphonic diamide, P-00156
(2-Phenylethenyl)phosphonic dichloride, P-00157
(2-Phenylethenyl)phosphonic difluoride, P-00158
(2-Phenylethenyl)phosphonodithioic acid, P-00159
(2-Phenylethenyl)phosphonothioic acid, P-00160
(2-Phenylethenyl)phosphonous acid, P-00161
(2-Phenylethenyl)phosphonous dichloride, P-00162
P-(2-Phenylethenyl)-*N,N,N',N'*-tetrapropylphosphonic diamide, *in* P-00156
2-Phenylethyl dichlorophosphate, *in* M-00506
2-Phenylethyl dihydrogen phosphate, *see* M-00506
1-Phenylethyl 3-[(dimethoxyphosphinyl)-oxy]-2-butenoate, *see* C-00207
Phenyl *P*-ethyl-*N,N*-dimethylphosphonamidodithioate, *in* E-00063
Phenyl *N,N'*-ethylene-*N,N'*-dimethylphosphorodiamidite, *see* D-00788
Phenyl *N*-ethyl-*P*-methylphosphonamidate, *in* E-00088
Phenyl ethylphenylphosphinite, *in* E-00108
Phenyl ethylphosphinate, *in* E-00124
(2-Phenylethyl)phosphine, P-00163
1-Phenylethylphosphonic acid, P-00164
(2-Phenylethyl)phosphonic acid, P-00165
O-Phenyl ethylphosphonochloridothioate, *in* E-00137
O-Phenyl ethylphosphonothioic monoanhydride, *in* D-00925
O-Phenyl ethylphosphoramidochloridothioate, *in* E-00157
2-Phenylethylphosphoric acid, *see* M-00506
2-Phenylethyl phosphorodichloridate, *in* M-00506
Phenylethynylphosphonic acid, P-00166
(Phenylethynyl)phosphonothioic acid, P-00167
O-Phenyl fluorodimethylamidothiophosphate, *in* D-00868
▷Phenylglyoxylonitrile oxime *O*-(*O,O*-diethylphosphorothioate), *see* P-00424
Phenyl hydrogen 2-aminoethyl phosphate, *in* A-00075
Phenyl hydrogen (2-aminoethyl)-phosphoramidate, *in* A-00080
Phenyl hydrogen (4-aminophenyl)phosphonate, *in* A-00111
Phenyl hydrogen (2-aminophenyl)-phosphoramidate, *in* A-00113
Phenyl hydrogen bis(2-chloroethyl)-phosphoramidate, *in* B-00121
Phenyl hydrogen cyclohexylphosphonate, *in* C-00256

Phenyl hydrogen methylphosphonate, *see* P-00171
Phenyl hydrogen (4-nitrophenyl)phosphonate, *in* N-00057
Phenyl hydrogen phenylphosphonate, *in* P-00221
Phenyl hydrogen phenylphosphoramidate, *see* P-00189
Phenylimidodiphosphoric acid, P-00168
Phenyliminodiphosphoric acid, *see* P-00168
2-Phenyl-2*H*-isophosphindole 2-oxide, P-00169
2-Phenylisophosphindoline, *see* D-00562
Phenyl *N*-isopropyl-*P*-methylphosphonamidate, *in* I-00036
Phenyl [(methoxycarbonyl)methyl]-phosphonamidate, *in* M-00028
1-Phenylmethoxy-2,2,3,4,4-pentamethylphosphetane 1-oxide, *in* H-00168
[2-(Phenylmethoxy)phenyl]phosphonic acid, *in* H-00179
(1-Phenylmethylamino-1-methylethyl)-phosphonic acid, *in* A-00087
[(Phenylmethylamino)phenylmethyl]-phenylphosphinic acid, *in* A-00102
Phenylmethyl [bis(1-aziridinyl)-phosphinyl]carbamate, *see* B-00017
Phenylmethyl diphenylphosphinothioite, *in* D-01085
[1-[(Phenylmethylene)amino]ethyl]phosphonic acid, *see* B-00049
Phenylmethyl ethylphosphonobromidothioite, *in* E-00134
(Phenylmethylidyne)phosphine, P-00170
6-(Phenylmethyl)-2-oxa-6-phosphatricyclo[3.3.1.1^{3,7}]decane, *see* B-00055
Phenyl *P*-(4-methylphenyl)-*N*-phenylphosphonamidate, *in* M-00214
Phenyl methylphenylphosphinate, *in* M-00219
O-Phenyl methylphenylphosphinothioate, *in* M-00225
S-Phenyl methylphenylphosphinothioate, *in* M-00225
Phenyl *P*-methyl-*N*-phenylphosphonamidate, *in* M-00231
Phenyl *P*-methyl-N^2-phenylphosphonohydrazidate, *in* M-00241
O-Phenyl *P*-methyl-N^2-phenylphosphonohydrazidothioate, *in* M-00242
1-Phenyl-3-methyl-1-phospha-3-cyclopentene, *see* D-00528
Phenylmethyl phosphinate, *see* B-00057
Phenyl methylphosphinate, *in* M-00278
(Phenylmethyl)phosphine, *see* B-00058
Phenylmethylphosphinic acid, *see* B-00059
Phenyl *P*-methylphosphonamidate, *in* M-00283
Phenyl methylphosphonate, P-00171
Phenyl methylphosphonazidate, *in* M-00286
(Phenylmethyl)phosphonic acid, *see* B-00060
P-(Phenylmethyl)phosphonic diamide, *see* B-00061
Phenylmethylphosphonic dichloride, *see* B-00062
Phenyl methylphosphonisocyanatidate, *in* M-00292
Phenyl methylphosphonisothiocyanatidate, *in* M-00294
Phenyl methylphosphonochloridate, *in* M-00296
(Phenylmethyl)phosphonochloridic acid, *see* B-00063
Phenyl methylphosphonochloridodithioate, *in* M-00297
O-Phenyl methylphosphonochloridothioate, *in* M-00298
S-Phenyl methylphosphonochloridothioate, *in* M-00298
Phenylmethylphosphonochloridothioic acid, *see* B-00064
Phenyl methylphosphonochloridothioite, *in* M-00299
O-Phenyl methylphosphonodithioate, P-00172
Phenylmethylphosphonodithioic acid, *see* B-00065
(Phenylmethyl)phosphonofluoridic acid, *see* B-00066
Phenyl methylphosphonofluoridite, *in* M-00307
O-Phenyl *P*-methylphosphonohydrazidothioate, *in* M-00309
Phenylmethylphosphonothioic acid, *see* B-00067
(Phenylmethyl)phosphonous acid, *see* B-00068
(Phenylmethyl)phosphoramidic acid, *see* B-00069
O-Phenyl methylphosphoramidochloridothioate, *in* M-00329
Phenylmethyl phosphorodichloridate, *in* M-00443
Phenylmethyl phosphorodichloridite, *see* B-00070
S-(Phenylmethyl) phosphorothioate, *see* B-00071
Phenylmethyl propyl phenylphosphonate, *in* M-00511
Phenyl *P*-methyl-*N*-2-thiazolylphosphonamidate, *in* M-00409
[[(Phenylmethyl)thio]methyl]phosphonic acid, *see* B-00072
Phenyl 4-morpholinylphosphonochloridate, *in* M-00524
O-Phenyl 4-morpholinylphosphonochloridothioate, *in* M-00525
2-Phenyl-1-oxa-4-aza-5-phospha(5-P^V)-spiro[4.4]nonane, P-00173
2-Phenyl-1-oxa-2-phosphacyclopentane, *see* P-00174
2-Phenyl-1-oxa-2-phospha-3-thiacycloheptane, *see* P-00178
4-Phenyl-1,4-oxaphosphinane, *see* P-00176
3-Phenyl-1,3-oxaphosphinan-2-one, *see* P-00177
2-Phenyl-1,2-oxaphospholane, P-00174
2-Phenyl-1,2-oxaphospholan-2-one 2-oxide, *in* O-00053
4-Phenyl-4*H*-1,4-oxaphosphorin, P-00175
4-Phenyl-1,4-oxaphosphorinane, P-00176
3-Phenyl-1,3-oxaphosphorinan-2-one, P-00177
2-Phenyl-1,3,2-oxathiaphosphepane, P-00178
3-Phenyl-1,3,2-oxazaphospholane, *see* P-00180
2-Phenyl-1,3,2-oxazaphospholidine, P-00179
3-Phenyl-1,3,2-oxazaphospholidine, P-00180
2-Phenyl-1,3,2-oxazaphosphorinane, *see* T-00132
N-Phenyl-2,2,3,4,4-pentamethylphosphetanamine 1-oxide, *in* P-00021
Phenyl 1,1,2,3,3-pentamethyltrimethylenephosphinate, *in* H-00168
N-Phenyl-*P*,*P*-(1,1,2,3,3-pentamethyltrimethylene)phosphinic amide, *in* P-00021
4-Phenyl-1,2,3,5,8-pentaoxa-4-phospha(4-P^V)spiro[3.4]octane, P-00181
10-Phenyl-10*H*-phenothiaphosphine, P-00182
10-Phenyl-10*H*-phenoxaphosphine, P-00183
Phenyl(4-phenyl-1,3-butadienyl)phosphinic acid, P-00184
Phenyl(phenylbutadiynyl)phosphinic acid, P-00185
Phenyl *o*-phenylene phosphate, *in* P-00063
Phenyl *o*-phenylene phosphite, *see* P-00063
Phenyl *N*,*N*′-*o*-phenylenephosphorodiamidate, *in* D-00543
Phenyl *N*,*N*′-*o*-phenylenephosphorodiamidite, *see* D-00543
O-Phenyl *N*,*N*′-*o*-phenylenephosphorodiamidothioate, *in* D-00543
O-Phenyl *o*-phenylene thiophosphate, *in* P-00063
Phenyl(2-phenylethenyl)phosphinic acid, P-00186
Phenyl(phenylethynyl)phosphinic acid, P-00187
[Phenyl[(phenylmethyl)amino]methyl]-phosphonic acid, P-00188
3-Phenyl-1-phenylmethyl-1,3-azaphosphetidine, *in* P-00087
Phenyl(phenylmethyl)phosphinic acid, *see* B-00056
N-[Phenyl(phenylmethyl)phosphinyl]-morpholine, *in* B-00056
1-Phenyl-1-phenylmethylphosphorinanium bromide, *in* P-00265
Phenyl phenyl(4-phenyl-1,3-butadienyl)-phosphinate, *in* P-00184
Phenyl *P*-phenyl-*N*-phenylsulfonylphosphonamidate, *in* P-00191
Phenyl phenylphosphinate, *in* P-00210
Phenyl *P*-phenylphosphonamidate, *in* P-00218
O-Phenyl phenyl-phosphonisothiocyanatidothioate, *in* P-00227
▷Phenyl phenylphosphonochloridate, *in* P-00229
Phenyl phenylphosphonochloridodithioate, *in* P-00230
O-Phenyl phenylphosphonochloridothioate, *in* P-00232
S-Phenyl phenylphosphonochloridothioate, *in* P-00232
Phenyl phenylphosphonochloridothioite, *in* P-00233
Phenyl phenylphosphonofluoridate, *in* P-00239
Phenyl phenylphosphoramidic acid, P-00189
Phenyl phenylphosphoramidochloridate, P-00190
O-Phenyl phenylphosphoramidochloridothioate, *in* P-00260
P-Phenyl-*N*-phenylsulfonylphosphonamidic acid, P-00191
P-Phenyl-*N*-phenylsulfonylphosphonimidic acid, P-00192
Phenyl phenylvinylphosphinate, *in* P-00337
Phenyl(2-phenylvinyl)phosphinic acid, *see* P-00186
10-Phenyl-9-phosphaanthracene, *see* P-00079
9-Phenyl-9-phosphabicyclo[3.3.1]nonane, P-00193
9-Phenyl-9-phosphabicyclo[4.2.1]nonane, P-00194
9-Phenyl-9-phosphabicyclo[3.3.1]nonan-3-one, P-00195
9-Phenyl-9-phosphabicyclo[4.2.1]nona-2,4,7-triene, P-00196
9-Phenyl-9-phosphabicyclo[6.1.0]none-2,4,6-triene, P-00197
3-Phenyl-3-phosphabicyclo[3.2.1]octane, P-00198
8-Phenyl-8-phosphabicyclo[3.2.1]octan-3-one, P-00199
8-Phenyl-8-phosphabicyclo[3.2.1]oct-6-ene, P-00200
3-Phenyl-3-phosphabicyclo[3.2.1]oct-6-ene 3-oxide, *in* P-00198
1-Phenylphosphacyclooctane, *see* P-00215
1-Phenyl-1-phospha-2,3:6,7-dibenzocyclohepta-2,6-diene, *see* D-00559
1-Phenyl-1-phospha-2,3:6,7-dibenzocyclohepta-2,4,6-triene, *see* P-00108
6-Phenyl-6-phospha-2,10-dithiabicyclo[9.4.0]pentadeca-11(1),-12,14-triene, *see* H-00044
1-Phenyl-1-phospha[1]ferrocenophane, *see* P-00211
9-Phenyl-9-phosphafluorene, *see* P-00110
9-Phenyl-9-phosphahomocubane, *see* P-00201
2-Phenyl-1-phosphanaphthalene, *see* P-00212
9-Phenyl-9-phosphapentacyclo[4.3.0.0^{2,5}.0^{3,8}.0^{4,7}]nonane, P-00201
4-Phenyl-4-phospha-4*H*-pyran, *see* P-00175
9-Phenyl-9-phosphatricyclo[3.3.1.0^{2,8}]-nona-3,6-diene, P-00202
9-Phenyl-9-phosphatricyclo[4.2.1.0^{2,5}]-nona-3,7-diene, P-00203
9-Phenyl-9-phosphatricyclo[4.2.1.0^{2,5}]-nonane, P-00204
1-Phenylphosphepane, P-00205
1-Phenyl-1*H*-phosphepin, P-00206
1-Phenylphosphinane, *see* P-00265
1-Phenyl-2-phosphinanol, *see* P-00266
1-Phenyl-2-phosphinanone, *see* P-00267
1-Phenyl-4-phosphinanone, *see* P-00268
1-Phenyl-1*H*-phosphindole, P-00207
▷Phenylphosphine, P-00208
Phenylphosphinedicarboxylic acid, P-00209
Phenylphosphinic acid, P-00210
3,3′-(Phenylphosphinidene)bis[*N*,*N*-dimethyl-1-propanamine], *see* B-00195
2,2′-(Phenylphosphinidene)bis-1-ethanamine, *see* B-00086
1,1″-(Phenylphosphinidene)bisferrocene, *see* D-00410

(Phenylphosphinidene)bismethanol, *see* B-00281
Phenylphosphinidenebis[methyl acetate], *in* B-00281
3,3′-(Phenylphosphinidene)bis-1-propanamine, *see* B-00090
3,3′-(Phenylphosphinidene)bispropanenitrile, *see* B-00152
3,3′-(Phenylphosphinidene)bis[1-propanol], *see* B-00286
N,N′-(Phenylphosphinidene)bis[*P,P,P*-triphenylphosphazene], *see* P-00102
N,N′-(Phenylphosphinidene)bis[*P,P,P*-triphenylphosphine imide], *see* P-00102
1,1′-(Phenylphosphinidene)diylferrocene, *see* P-00211
1,1′-(Phenylphosphinidene)ferrocene, P-00211
2-Phenylphosphinoline, P-00212
Phenylphosphinothioic acid, P-00213
(Phenylphosphinothioylidene)bismethanol, *in* B-00281
3,3′-(Phenylphosphinothioylidene)bis-1-propanamine, *in* B-00090
3,3′-(Phenylphosphinothioylidene)bis[1-propanol], *in* B-00286
3,3′-(Phenylphosphinothioylidene)bis[1-propyl acetate], *in* B-00286
N,N′-(Phenylphosphinothioylidene)-bis[*P,P,P*-triphenylphosphine imide], *in* P-00102
(Phenylphosphinyldene)bis[methyl acetate], *in* B-00281
1,1″-(Phenylphosphinylidene)bisferrocene, *in* D-00410
(Phenylphosphinylidene)bismethanol, *in* B-00281
3,3′-(Phenylphosphinylidene)bis[1-propanol], *in* B-00286
3,3′-(Phenylphosphinylidene)bis[1-propyl acetate], *in* B-00286
N,N′-(Phenylphosphinylidene)bis[*P,P,P*-triphenylphosphine imide], *in* P-00102
1-Phenylphosphirane, P-00214
1-Phenylphosphocane, P-00215
1-Phenylphospholane, P-00216
1-Phenyl-1*H*-phosphole, P-00217
1-Phenyl-2-phospholene, *see* D-00566
1-Phenyl-3-phospholene, *see* D-00567
P-Phenylphosphonamidic acid, P-00218
P-Phenylphosphonamidothioic acid, P-00219
1-Phenyl-3,8-phosphonanedione 1-oxide, *in* P-00373
Phenylphosphonazidic acid, P-00220
▷Phenylphosphonic acid, P-00221
Phenylphosphonic acid monoanhydride, *see* D-01014
Phenylphosphonic acid trimer trianhydride, *in* T-00694
Phenylphosphonic bis(cyclohexylamide), *in* P-00222
▷*P*-Phenylphosphonic diamide, P-00222
▷Phenylphosphonic dianilide, *in* P-00222
Phenylphosphonic dichloride, P-00223
Phenylphosphonic di(ethylamide), *in* P-00222
Phenylphosphonic difluoride, P-00224
P-Phenylphosphonic dihydrazide, P-00225
Phenylphosphonic diisocyanate, P-00226
Phenylphosphonisothiocyanatidothioic acid, P-00227
1-Phenyl-2-phosphonoacetylene, *see* P-00166
1-Phenyl-1-phosphonoallene, *see* P-00298
3-Phenyl-2-(phosphonoamino)propanoic acid, *see* P-00400
Phenylphosphonobromidodithioic acid, P-00228
Phenylphosphonochloridic acid, P-00229
Phenylphosphonochloridodithioic acid, P-00230
Phenylphosphonochloridoselenous acid, P-00231
Phenylphosphonochloridothioic acid, P-00232
Phenylphosphonochloridothious acid, P-00233
Phenylphosphonochloridous acid, P-00234
Phenylphosphonodiselenous acid, P-00235
Phenylphosphonodithioic acid, P-00236
Phenylphosphonodithioic acid anhydrosulfide, P-00237
Phenylphosphonodithious acid, P-00238
1-Phenyl-1-phosphonoethylene, *see* P-00154
1-Phenyl-2-phosphonoethylene, *see* P-00155
Phenylphosphonofluoridic acid, P-00239
Phenylphosphonofluoridous acid, P-00240
P-Phenylphosphonohydrazidic acid, P-00241
Phenylphosphonoperoxoic acid, P-00242
1-Phenyl-1-phosphonopropene, *see* P-00301
1-Phenyl-3-phosphonopropene, *see* P-00302
Phenylphosphonoselenoic acid, P-00243
Phenylphosphonothioic acid, P-00244
Phenylphosphonothioic acid bis(ethylamide), *in* P-00246
Phenylphosphonothioic acid monoanhydride, P-00245
Phenylphosphonothioic acid trimer trianhydride, *in* T-00694
Phenylphosphonothioic bis(methylamide), *in* P-00246
P-Phenylphosphonothioic diamide, P-00246
P-Phenylphosphonothioic dianilide, *in* P-00246
▷Phenylphosphonothioic dichloride, P-00247
Phenylphosphonothioic difluoride, P-00248
P-Phenylphosphonothioic dihydrazide, P-00249
Phenylphosphonotrithioic acid, P-00250
Phenylphosphonous acid, P-00251
Phenylphosphonous bis(dibenzylamide), *in* P-00252
Phenylphosphonous bis(diethylamide), *in* P-00252
Phenylphosphonous bis(ethylamide), *in* P-00252
P-Phenylphosphonous diamide, P-00252
Phenylphosphonous dibromide, P-00253
Phenylphosphonous di(*tert*-butylamide), *in* P-00252
▷Phenylphosphonous dichloride, P-00254
▷Phenylphosphonous dicyanide, P-00255
Phenylphosphonous difluoride, P-00256
Phenylphosphonous di(isopropylamide), *in* P-00252
Phenylphosphoramidic acid, P-00257
Phenylphosphoramidic dichloride, P-00258
Phenylphosphoramidochloridic acid, P-00259
Phenylphosphoramidochloridothioic acid, P-00260
Phenyl phosphoramidofluoridate, *in* P-00295
Phenylphosphoramidoselenoic acid, P-00261
Phenylphosphoramidothioic acid, P-00262
Phenylphosphoramidous acid, P-00263
Phenyl phosphoric acid, *see* M-00507
S-Phenyl phosphoric acid, *see* P-00297
Phenyl phosphoric dianilide, *in* D-01125
Phenyl phosphoric dihydrazide, *see* P-00290
Phenyl phosphoric monoanilide, *see* P-00189
Phenylphosphoric triamide, *in* P-00257
Phenylphosphorimidic acid, P-00264
1-Phenylphosphorinane, P-00265
1-Phenyl-2-phosphorinanol, P-00266
1-Phenyl-2-phosphorinanone, P-00267
1-Phenyl-4-phosphorinanone, P-00268
Phenyl phosphorochloridic acid, P-00269
Phenyl phosphorochloridisocyanatidate, P-00270
Phenyl phosphorochloridofluoridate, P-00271
Phenyl phosphorodiamidate, P-00272
N-Phenylphosphorodiamidic acid, P-00273
2-Phenylphosphorodiamidic hydrazide, P-00274
Phenyl phosphorodiamidite, P-00275
O-Phenyl phosphorodiamidothioate, P-00276
Phenyl phosphorodibromidite, P-00277
O-Phenyl phosphorodibromidothioate, P-00278
Phenyl phosphorodichloridate, P-00279
▷Phenyl phosphorodichloridite, P-00280
Phenyl phosphorodichloridodithioate, P-00281
O-Phenyl phosphorodichloridothioate, P-00282
S-Phenyl phosphorodichloridothioate, P-00283
▷Phenyl phosphorodichloridothioite, P-00284
Phenyl phosphorodifluoridate, P-00285
Phenyl phosphorodifluoridite, P-00286
O-Phenyl phosphorodifluoridothioate, P-00287
S-Phenyl phosphorodifluoridothioate, P-00288
2-Phenylphosphorodifluoridothioic hydrazide, P-00289
Phenyl phosphorodihydrazidate, P-00290
O-Phenyl phosphorodihydrazidothioate, P-00291
Phenyl phosphorodiisocyanatidate, P-00292
O-Phenyl phosphorodiisocyanatidothioate, P-00293
Phenyl phosphorodi(isothiocyanatate), P-00294
Phenyl phosphorofluoridate, *see* P-00295
Phenyl phosphorofluoridic acid, P-00295
2-Phenylphosphorohydrazidothioic acid, P-00296
S-Phenyl phosphorothioate, P-00297
O-Phenyl phosphorothioate dianilide, *in* D-01129
Phenyl phosphoryl dichloride, *see* P-00279
Phenyl phosphoryl difluoride, *see* P-00285
Phenyl phosphoryl dihydrazide, *see* P-00290
Phenyl phosphoryl diisocyanate, *see* P-00292
Phenylphosphoryl diisothiocyanate, *see* P-00294
(1-Phenyl-1,2-propadienyl)phosphonic acid, P-00298
3-Phenyl-1-propanephosphonic acid, *see* P-00305
Phenyl(1-propenyl)phosphinic acid, P-00299
Phenyl(2-propenyl)phosphinic acid, P-00300
(1-Phenyl-1-propenyl)phosphonic acid, P-00301
3-Phenyl-2-propenylphosphonic acid, P-00302
Phenyl propylene phosphate, *in* M-00190
Phenyl propylene phosphite, *see* M-00190
Phenyl propyl phenylphosphonate, *in* M-00511
Phenylpropylphosphinic acid, P-00303
Phenyl propyl phosphite, *see* P-00304
Phenyl propyl phosphonate, P-00304
(3-Phenylpropyl)phosphonic acid, P-00305
Phenyl propyl phosphorochloridate, *in* P-00269
Phenyl 8-quinolinyl phosphate, *see* P-00306
Phenyl 8-quinolyl hydrogen phosphate, *see* P-00306
Phenyl 8-quinolyl phosphate, P-00306
▷Phenyl saligenin cyclic phosphate, *in* P-00064
Phenyl saligenin cyclic phosphite, *see* P-00064
O-Phenyl saligenin cyclic thiophosphate, *in* P-00064
S-Phenyl saligenin cyclic thiophosphate, *in* H-00110
4-Phenyl-4*H*-1,4-selenaphosphinin, *see* P-00307
4-Phenyl-4*H*-1,4-selenaphosphorin, P-00307
N-Phenylspiro[1,3,2-benzodioxaphosphole-2,2′(3′*H*)-[1,3,2]benzoxazaphosphol]-2-amine, *in* S-00008
2-Phenylspiro[1,3,2-benzothiazaphosphole-2(3*H*),2′-[1,3,2]dioxaphospholane], P-00308
2-Phenyl-2,2′-spirobi[1,3,2-benzodioxaphosphole], P-00309
Phenylstyrylphosphinic acid, *see* P-00186
[(Phenylsulfinyl)methyl]phosphonic acid, P-00310
[(Phenylsulfonyl)methyl]phosphonic acid, P-00311
Phenylsulfonylphosphoramidic acid, P-00312
Phenylsulfonylphosphoramidic dichloride, P-00313
Phenylsulfonylphosphoramidothioic acid, P-00314
Phenylsulfonylphosphorimidic acid, P-00315
Phenylsulfonylphosphorimidic trichloride, P-00316
4-Phenyl-4*H*-1,4-telluraphosphinin, *see* P-00317
4-Phenyl-4*H*-1,4-telluraphosphorin, P-00317
Phenyl *N,N,N′,N′*-tetraethyl-*N″*-phenylphosphorodiamidimidate, *in* T-00054
Phenyl tetraethylphosphorodiamidate, P-00318
Phenyl tetraethylphosphorodiamidite, *in* T-00066
O-Phenyl tetraethylphosphorodiamidoselenoate, *in* T-00061
O-Phenyl tetraethylphosphorodiamidothioate, *in* T-00063
S-Phenyl tetraethylphosphorodiamidothioate, *in* T-00063
1-Phenyl-1,2,3,4-tetrahydrobenzo[*h*]-phosphinoline, P-00319
1-Phenyl-1,2,3,4-tetrahydrophosphinoline, P-00320
P-Phenyl-*N,N,N′,N′*-tetrakis(phenylmethyl)-phosphonous diamide, *in* P-00252

7-Phenyl-1,3,4,6-tetrakis(trifluoromethyl)-7-aza-2,5-diphosphatetracyclo[4.1.0.$0^{2,4}$.$0^{3,5}$]-heptane, P-00321
2-Phenyl-4,4,5,5-tetramethyl-1,3,2-dioxaphospholane, P-00322
Phenyl tetramethylene phosphate, *in* H-00132
Phenyltetramethylenephosphine, *see* P-00216
Phenyl tetramethylphosphorodiamidate, *in* T-00218
Phenyl tetramethylphosphorodiamidite, *in* T-00227
Phenyl tetraphenylphosphorodiamidite, *in* T-00274
3-Phenyl-2-thia-3-phosphabicyclo[2.2.2]-oct-5-ene, P-00323
2-Phenyl-1,2-thiaphospholane, P-00324
2-Phenyl-1,2-thiaphosphorinane, P-00325
▷Phenyl thiodichlorophosphite, *see* P-00284
Phenyl thiodimethylenephosphinate, *in* H-00201
[(Phenylthio)methyl]phosphonic acid, P-00326
S-Phenyl thiophosphate, *see* P-00297
▷Phenylthiophosphonic dichloride, *see* P-00247
Phenylthiophosphonic difluoride, *see* P-00248
O-Phenyl thiophosphoryl diamide, *see* P-00276
O-Phenyl thiophosphoryl dibromide, *see* P-00278
O-Phenyl thiophosphoryl dichloride, *see* P-00282
O-Phenyl thiophosphoryl difluoride, *see* P-00287
O-Phenyl thiophosphoryl dihydrazide, *see* P-00291
O-Phenyl thiophosphoryl diisocyanate, *see* P-00293
N-Phenyl-*P*-*p*-tolylphosphonamidic acid, *see* M-00214
Phenyl(trichloromethyl)phosphinic acid, P-00327
Phenyl *P*-trichloromethylphosphonamidate, *in* T-00410
N-Phenyl-*P*-trichloromethylphosphonamidic acid, P-00328
Phenyl(trifluoromethyl)phosphinic acid, P-00329
Phenyl *N*,*N*′-trimethylenephosphorodiamidate, *in* H-00039
O-Phenyl *N*,*N*′-trimethyl-enephosphorodiamidothioate, *in* H-00039
▷Phenyl trimethylphosphonoamidodithioate, *in* T-00539
[Phenyl(trimethylsilyl)methylene]-phosphinous chloride, P-00330
[Phenyl[(trimethylsilyl)oxy]methyl]-phosphonic acid, P-00331
Phenyl(triphenylmethyl)phosphinic acid, P-00332
Phenyl(triphenylphosphoranylidene)-acetaldehyde, *see* T-00643
Phenyl(triphenylphosphoranylidene)acetic acid, *see* T-00644
▷1-Phenyl-2-(triphenylphosphoranylidene)-ethanone, P-00333
Phenyl [(triphenylphosphoranylidene)-hydrazono]acetate, *in* T-00653
1-Phenyl-1-(triphenylphosphoranylidene)-2-propanone, P-00334
1-Phenyl-2-(triphenylphosphoranylidene)-1-propanone, P-00335
1-Phenyl-3-(triphenylphosphoranylidene)-2-propanone, P-00336
Phenyl 2-(triphenylphosphoranylidene)-propanote, *in* T-00660
Phenyl *N*,*N*,*N*′-tris(trimethylsilyl)-phosphonamidimidate, *in* T-00867
N-Phenylurea-*N*′-phosphonic acid, *see* P-00081
(2-Phenylvinyl)phosphine, *see* P-00152
Phenylvinylphosphinic acid, P-00337
1-Phenylvinylphosphonic acid, *see* P-00154
2-Phenylvinylphosphonic acid, *see* P-00155
P-(2-Phenylvinyl)phosphonic diamide, *see* P-00156
(2-Phenylvinyl)phosphonothioic acid, *see* P-00160
β-Phenylvinylphosphonous acid, *see* P-00161
β-Phenylvinylphosphonous dichloride, *see* P-00162
Phinifos, *in* T-00322
▷Phorate, P-00338
▷Phorpurine, *see* F-00053
▷Phosacetim, *in* I-00005
Phosalacine, P-00339
▷Phosalone, P-00340
▷Phosbutyl, P-00341
▷Phosdrin, *see* M-00440
Phosethyl AL, *in* M-00473
▷Phosflex 179-C, *see* T-00803
▷Phosfolan, P-00342
▷Phosmet, P-00343
Phosodyl, *in* T-00322
▷Phosphaaniline, *see* P-00208
9-Phosphaanthracene, *see* A-00031
Phosphabenzene, *see* P-00403
1-Phosphabicyclo[2.2.1]heptane, P-00344
1-Phosphabicyclo[3.3.1]nonane, P-00345
1-Phosphabicyclo[2.2.2]octane, P-00346
Phosphacholine, *see* H-00150
▷Phosphacol, *see* P-00001
Phosphacyclohexane, *see* P-00404
Phosphacyclopentadiene, *see* P-00372
Phosphacyclopentane, *see* P-00371
1-Phospha-2,3:6,7-dibenzocyclohepta-2,6-diene, *see* D-00451
1-Phospha-2,10-dioxabicyclo[4.4.0]decane, *see* H-00037
9-Phosphafluorene, *see* D-00040
17-Phospha-9*H*-fluoreno[9,9*a*-*b*]phenanthrene, *see* D-00041
Phosphafluorinic acid, *see* H-00118
▷Phosphamidon, P-00347
2-Phosphanaphthalene, *see* I-00027
Phosphane, *see* P-00404
▷Phosphanilic acid, *see* A-00111
Phosphanilic diamide, *see* A-00112
Phosphanilic dianilide, *in* A-00112
Phosphanilic dimethylamide, *in* A-00112
1-Phosphanorbornane, *see* P-00344
Phosphanthridene, P-00348
9-Phosphaphenanthrene, *see* P-00348
1-Phosphapropyne, *see* E-00077
5-Phospha(5-P^V)spiro[4.5]deca-1,3,5,7,9-pentaene, P-00349
5-Phospha(5-P^V)spiro[4.4]nonane, P-00350
6-Phospha(6-P^V)spiro[5.5]undeca-1(6)-ene, P-00351
10-Phospha-1,4,7-triazatricyclo[5.2.1.$0^{4,10}$]decane, *see* H-00055
2-Phosphatricyclo[4.4.0.$0^{4,6}$]decane 2-oxide, *see* O-00004
Phosphatriptycene, *see* B-00014
Phosphazomycin *A*, P-00352
▷Phosphemide, P-00353
Phosphiden, *see* B-00162
Phosphinane, *see* P-00404
Phosphinecarbonitrile, P-00354
Phosphinetricarboxylic acid, P-00355
2,2′-Phosphinicobis[benzoic acid], *see* B-00117
4,4′-Phosphinicobis[benzoic acid], *see* B-00118
3,3′-Phosphinidenebis(propanenitrile), *see* B-00153
1,1′,1″-Phosphinidyetris(1*H*-1,2,4-triazole), *see* T-00876
1-Phosphinidynemethanamine, *see* P-00356
Phosphinidynemethylamine, P-00356
1,1′,1″-Phosphinidynetrisazetidine, *see* T-00328
▷1,1′,1″-Phosphinidynetrisaziridine, *see* T-00331
3,3′,3″-Phosphinidynetris[benzoic acid], *see* T-00739
4,4′,4″-Phosphinidynetris[benzoic acid], *see* T-00740
3,3′,3″-Phosphinidynetris[benzonitrile], *see* T-00753
4,4′,4″-Phosphinidynetris[benzonitrile], *see* T-00754
2,2′,2″-Phosphinidynetrisbenzothiazole, *see* T-00334
1,1″,1‴-Phosphinidynetrisferrocene, *see* T-00464
2,2′,2″-Phosphinidynetrisimidazole, *see* T-00496
1,1′,1″-Phosphinidynetris(1*H*-imidazole), *see* T-00495
Phosphinidynetrismethanol, *see* T-00784
Phosphinidynetris(methyl acetate), *in* T-00784
2,2′,2″-Phosphinidynetris[1-methylimidazole], *in* T-00496
4,4′,4″-Phosphinidynetrismorpholine, *see* T-00569
2,2′,2″-Phosphinidynetrisphenol, *see* T-00788
3,3′,3″-Phosphinidynetrisphenol, *see* T-00789
4,4′,4″-Phosphinidynetrisphenol, *see* T-00790
▷1,1′,1″-Phosphinidynetrispiperidine, *see* T-00704
3,3′,3″-Phosphinidynetris[propanamide], *see* T-00738
3,3′,3″-Phosphinidynetris[propanenitrile], *see* T-00752
3,3′,3″-Phosphinidynetrispropanoic acid, P-00357
1,1′,1″-Phosphinidynetris(1*H*-pyrazole), *see* T-00722
2,2′,2″-Phosphinidynetrispyridine, *see* T-00723
2,2′,2″-Phosphinidynetrispyrrole, *see* T-00726
1,1′,1″-phosphinidynetris-1*H*-pyrrole, *see* T-00725
1,1′,1″-Phosphinidynetrispyrrolidine, *see* T-00724
2,2′,2″-Phosphinidynetristhiazole, *see* T-00873
Phosphinine, *see* P-00403
3-Phosphinopropanenitrile, P-00358
1,1′,1″-Phosphinoselenoylidynetrisaziridine, *in* T-00331
1′,1″,1‴-Phosphinoselenoylidynetrisferrocene, *in* T-00464
4,4′,4″-Phosphinoselenoylidynetrismorpholine, *in* T-00569
▷1,1′,1″-Phosphinoselenoylidynetrispiperidine, *in* T-00704
3,3′,3″-Phosphinoselenoylidynetris[propanenitrile], *in* T-00752
1,1′,1″-Phosphinoselenoylidynetrispyrrolidine, *in* T-00724
4,4′,4″-Phosphinotelluroylidynetrismorpholine, *in* T-00569
1,1′,1″-Phosphinotelluroylidynetrispiperidine, *in* T-00704
1,1′,1″-Phosphinotelluroylidynetrispyrrolidine, *in* T-00724
1,1′,1″-Phosphinothioyldiynetris(1*H*-pyrazole), *in* T-00722
▷1,1′,1″-Phosphinothioylidenetrisaziridine, *see* T-00317
4,4′,4″-Phosphinothioylidynetris[benzoic acid], *in* T-00740
3,3′,3″-Phosphinothioylidynetris[benzonitrile], *in* T-00753
4,4′,4″-Phosphinothioylidynetris[benzonitrile], *in* T-00754
1′,1″,1‴-Phosphinothioylidynetrisferrocene, *in* T-00464
▷4,4′,4″-Phosphinothioylidynetrismorpholine, *in* T-00569
1,1′,1″-Phosphinothioylidynetrispiperidine, *in* T-00704
▷3,3′,3″-Phosphinothioylidynetris[propanenitrile], *in* T-00752
Phosphinothricin, P-00359
▷Phosphinothricylalanylalanine, *see* A-00141
Phosphinothricylalanylleucine, *see* P-00339
Phosphinous cyanide, *see* P-00354
3,3′,3″-Phosphinyldynetrispropanoic acid triethyl ester, *in* P-00357
1,1′,1″-Phosphinylidynetrisazetidine, *in* T-00328
▷1,1′,1″-Phosphinylidynetrisaziridine, *in* T-00331
3,3′,3″-Phosphinylidynetris[benzoic acid], *in* T-00739
4,4′,4″-Phosphinylidynetris[benzoic acid], *in* T-00740
3,3′,3″-Phosphinylidynetris[benzonitrile], *in* T-00753
4,4′,4″-Phosphinylidynetris[benzonitrile], *in* T-00754

1′,1″,1‴-Phosphinylidynetrisferrocene, in T-00464
1,1′,1″-Phosphinylidynetris(1H-imidazole), in T-00495
Phosphinylidynetrismethanol, in T-00784
Phosphinylidynetris(methyl benzoate), in T-00784
[Phosphinylidynetris(methylene)]-trisphosphonic acid, P-00360
2,2′,2″-Phosphinylidynetris(1-methylpyrrole), in T-00726
4,4′,4″-Phosphinylidynetrismorpholine, in T-00569
3,3′,3″-Phosphinylidynetrisphenol, in T-00788
3,3′,3″-Phosphinylidynetrisphenol, in T-00789
4,4′,4″-Phosphinylidynetrisphenol, in T-00790
▷1,1′,1″-Phosphinylidynetrispiperidine, in T-00704
3,3′,3″-Phosphinylidynetris[propanamide], in T-00738
3,3′,3″-Phosphinylidynetris[propanenitrile], in T-00752
3,3′,3″-Phosphinylidynetrispropanoic acid trimethyl ester, in P-00357
1,1′,1″-Phosphinylidynetris(1H-pyrazole), in T-00722
1,1′,1″-Phosphinylidynetrispyrrole, in T-00725
2,2′,2″-Phosphinylidynetrispyrrole, in T-00726
1,1′,1″-Phosphinylidynetrispyrrolidine, in T-00724
1,1′,1″-Phosphinylidynetris(1H-1,2,4-triazole), in T-00876
Phosphirane, P-00361
N-Phosphoalanine, P-00362
Phosphoarginine, P-00363
Phosphobenzene A, see P-00037
Phosphobenzene B, see H-00089
N-Phosphocreatine, P-00364
N-Phosphocysteine, P-00365
Phosphoenolpyruvic acid, P-00366
O-Phosphoethanolamine, see A-00075
N-Phosphoglutamic acid, P-00367
3-Phosphoglyceraldehyde, see H-00187
N-Phosphoglycine, P-00368
Phosphoglycocyamine, P-00369
Phosphoglycolic acid, see P-00390
Phosphoguanidine, P-00370
Phospholactic acid, see P-00392
Phospholane, P-00371
1H-Phosphole, P-00372
2-Phospholene, see D-00571
3-Phospholene, see D-00572
Phosphon, see C-00192
3,8-Phosphonanedione, P-00373
5-Phosphoniaspiro[4.4]nonane, P-00374
6-Phosphoniaspiro[5.5]undecane, P-00375
Phosphonic bis(dibutylamide), see T-00018
Phosphonic bis(diethylamide), see T-00055
Phosphonic bis(diisopropylamide), see T-00171
Phosphonic bis(dimethylamide), see T-00214
Phosphonic bis(diphenylamide), see T-00271
Phosphonic bis(dipropylamide), see T-00289
Phosphonic tetraisopropyldiamide, see T-00171
Phosphonoacetaldehyde, see O-00059
Phosphonoacetic acid, P-00376
Phosphonoacetone, see O-00075
N-(Phosphonoacetyl)aspartic acid, see S-00006
Phosphonoacetylene, see E-00203
9-Phosphonoacridine, see A-00030
3-Phosphonoalanine, see A-00121
N-Phosphonoalanine, see P-00362
N-(N-Phosphonoamidino)glycine, see P-00369
N^5-(Phosphonoamidino)ornithine, see P-00363
N-(N-Phosphonoamidino)sarcosine, see P-00364
4-(Phosphonoamino)benzoic acid, P-00377
N-Phosphono-4-aminobenzoic acid, see P-00377
N-Phosphono-2-aminobenzoic acid ethyl ester, see C-00004
N-Phosphono-3-aminobenzoic acid ethyl ester, see C-00005
N-Phosphono-4-aminobenzoic acid ethyl ester, see C-00006
▷m-Phosphonoaniline, see A-00110
▷o-Phosphonoaniline, see A-00109
▷p-Phosphonoaniline, see A-00111
N-Phosphonoarginine, see P-00363
1-Phosphonoazetidine, see A-00147
4-Phosphonobenzamide, in P-00380
Phosphonobenzeneacetic acid, see P-00393
5-Phosphonobenzimidazole, see B-00016
5-Phosphono-1,3-benzodioxole, see B-00019
2-Phosphonobenzoic acid, P-00378
3-Phosphonobenzoic acid, P-00379
4-Phosphonobenzoic acid, P-00380
6-Phosphonobenzoxazole, see B-00027
1-Phosphono-1,2-butadiene, see B-00509
1-Phosphono-1,3-butadiene, see B-00510
Phosphonobutanedioic acid, P-00381
2-Phosphono-1,2,4-butanetricarboxylic acid, P-00382
3-Phosphono-1,2,3-butanetricarboxylic acid, P-00383
4-Phosphono-1,2,3-butanetricarboxylic acid, P-00384
3-Phosphonobutanoic acid, P-00385
4-Phosphonobutanoic acid, P-00386
1-Phosphono-1-butanol, see H-00114
1-Phosphono-1-butene, see B-00524
1-Phosphono-2-butene, see B-00525
4-Phosphono-1-butene, see B-00526
1-Phosphono-1-butyne, see B-00643
1-Phosphono-2-butyne, see B-00644
N^{ω}-Phosphonocreatine, see P-00364
Phosphonocyclohexane, see C-00256
1-Phosphonocyclohexene, see C-00241
3-Phosphonocyclohexene, see C-00242
4-Phosphonocyclohexene, see C-00243
1-Phosphonocyclopentene, see C-00277
3-Phosphonocyclopentene, see C-00278
Phosphonocysteine, see A-00084
2-Phosphonoethanal, see O-00059
1-Phosphonoethanol, see H-00145
2-Phosphonoethanol, see H-00146
Phosphonoformamide, see C-00003
Phosphonoformic acid, P-00387
2-Phosphonofuran, see F-00073
3-Phosphonofuran, see F-00074
N-Phosphonoglutamic acid, see P-00367
N-Phosphonoglycine, see P-00368
N^{ω}-Phosphonoglycocyamine, see P-00369
N-Phosphonoguanidine, see P-00370
N^{ω}-Phosphonoguanidinoacetic acid, see P-00369
1-Phosphonoheptane, see H-00006
1-Phosphono-2-heptanone, see O-00061
1-Phosphonohexane, see H-00097
1-Phosphonoimidazole, see I-00003
Phosphonomethanol, see H-00162
Phosphonomethionine, see A-00095
N^5-Phosphonomethionine sulfoximine, P-00388
(N-Phosphono)methionine-S-sulfoximinylalanylalanine, P-00389
2-(Phosphonomethyl)aniline, see A-00106
4-(Phosphonomethyl)aniline, see A-00107
2-(Phosphonomethyl)benzonitrile, see C-00220
4-(Phosphonomethyl)benzonitrile, see C-00221
2-(Phosphonomethyl)biphenyl, see B-00079
4-(Phosphonomethyl)biphenyl, see B-00080
1-Phosphono-3-methylbutane, see M-00106
2-(Phosphonomethyl)furan, see F-00072
▷N-(Phosphonomethyl)glycine, see G-00097
1-(Phosphonomethyl)naphthalene, see N-00013
2-(Phosphonomethyl)naphthalene, see N-00014
N-Phosphonomethylphthalimide, see P-00426
N-(Phosphonomethyl)piperidine, see P-00429
2-(Phosphonomethyl)pyridine, see P-00506
4-(Phosphonomethyl)pyridine, see P-00507
2-(Phosphonomethyl)quinoline, see Q-00004
8-(Phosphonomethyl)quinoline, see Q-00005
2-(Phosphonomethyl)thiophene, see T-00304
3-(Phosphonomethyl)thiophene, see T-00305
4-Phosphonomorpholine, see M-00523
▷Phosphonomycin, in M-00185
1-Phosphono-2-nonanone, see O-00063
1-Phosphono-2-octanone, see O-00064
Phosphonooxyacetic acid, P-00390
▷2-(Phosphonooxy)benzoic acid, P-00391
2-Phosphonooxypropanoic acid, P-00392
2-(Phosphonooxy)-2-propenoic acid, see P-00366
2-[5-O-(Phosphonooxy)-β-D-ribofuranosyl]-1,2,4-triazine-3,5(2H,4H)-dione, see A-00145
1-Phosphono-1,3-pentadiene, see P-00009
1-Phosphono-1,3-pentadiyne, see P-00010
4-Phosphono-1-pentene, see M-00099
6-Phosphonophenanthridine, see P-00060
α-Phosphonophenylacetic acid, P-00393
N-Phosphonophenylalanine, see P-00400
2-Phosphono-2-(2-phosphonoethyl)butanedioic acid, P-00394
2-Phosphono-2-(2-phosphonoethyl)succinic acid, see P-00394
3-Phosphonophthalide, see D-00535
1-Phosphonopiperidine, see P-00430
3-Phosphonopropanal, see O-00076
2-Phosphonopropane, see I-00050
3-Phosphono-1,2-propanediol, see D-00624
2-Phosphonopropanoic acid, P-00395
▷3-Phosphonopropanoic acid, P-00396
3-Phosphonopropanol, see H-00192
1-Phosphono-1-propanol, see H-00190
1-Phosphono-2-propanol, see H-00191
2-Phosphono-2-propanol, see H-00159
1-Phosphonopropene, see P-00453
2-Phosphonopropene, see M-00149
1-Phosphonopropyne, see P-00494
3-Phosphono-1-propyne, see P-00495
3-Phosphonopyrazole, see P-00501
2-Phosphonopyridine, see P-00503
3-Phosphonopyridine, see P-00504
4-Phosphonopyridine, see P-00505
2-Phosphonopyrimidine, see P-00515
4-Phosphonopyrimidine, see P-00516
2-Phosphonopyrrole, see P-00521
1-Phosphonopyrrolidine, see P-00518
N-[9-(5-O-Phosphono-β-D-ribofuranosyl)-9H-purin-6-yl]-L-aspartic acid, see A-00047
Phosphonoserine, see A-00083
N-Phosphonoserine, see P-00418
O-Phosphonoserine, see P-00417
Phosphonosuccinic acid, see P-00381
2-Phosphonothiophene, see T-00307
N-Phosphonothreonine, see P-00420
Phosphonotryptophan, see P-00421
N-Phosphonotyrosine, see P-00422
O-Phosphonotyrosine, see P-00423
Phosphonourea, see A-00062
9-Phosphonoxanthene, see X-00001
4-(Phosphonoxy)benzoic acid, P-00397
2-p-Phosphonoxybenzyl-1-phosphoramidopropanoic acid, see D-01188
3-Phosphonoxypropanenitrile, P-00398
3-Phosphonoxypropanoic acid, P-00399
N-Phosphophenylalanine, P-00400
Phosphophenylalanylarginine, P-00401
Phosphoramide mustard, see B-00126
2-Phosphoramidoglutaric acid, see P-00367
Phosphoramidon, P-00402
2-Phosphoramidopentanedioic acid, see P-00367
2-Phosphoramidopropanoic acid, see P-00362
Phosphoric acid bis(dibutylamide), see T-00021
Phosphoric acid bis(dimethylamide), see T-00218
Phosphoric acid bis(dipropylamide), see T-00291
Phosphoric acid bismethylamide, see D-00895
Phosphoric acid dianilide, see D-01125
Phosphoric acid diethylamide, see D-00343
Phosphoric acid dimethylamide, see D-00855
Phosphoric acid methylamide, see M-00326
Phosphoric acid mono[2-(dimethylamino)-ethyl] ester, see D-00016
Phosphoric anilide, see P-00257
Phosphoric trianilide, in T-00677
Phosphorin, P-00403
Phosphorinane, P-00404
Phosphorisocyanatidic dichloride, P-00405
Phosphorisocyanatidic difluoride, P-00406
Phosphor(isothiocyanatidic) dichloride, P-00407
Phosphor(isothiocyanatidic) difluoride, P-00408

Phosphor(isothiocyanatido)thioic bis(dimethylamide), *see* T-00217
Phosphorocyanidous bromide iodide, P-00409
Phosphorocyanidous dibromide, P-00410
Phosphorodiisocyanatidic bromide, P-00411
Phosphorodiisocyanatidic chloride, P-00412
Phosphorodiisocyanatidic fluoride, P-00413
Phosphorodi(isothiocyanatidic) fluoride, P-00414
▷Phosphorodithioic acid *O,O*-diethyl ester, *S*-ester with 6-chloro-3-(mercaptomethyl)-2-benzoxazolinone, *see* P-00340
Phosphoroselenoic acid anilide, *see* P-00261
Phosphoroselenoic acid bis(diethylamide), *see* T-00061
Phosphoroselenoic acid bis(dimethylamide), *see* T-00223
Phosphoroselenoic acid diisopropylamide, *see* D-00655
Phosphoroselenoic acid methylamide, *see* M-00332
Phosphorothioic trianilide, *in* T-00677
▷Phosphorothionic triethenamide, *see* T-00317
▷Phosphorous acid cyclic ethylene ester, *see* H-00133
Phosphorous trianilide, *see* T-00677
▷Phosphorus triethyl, *see* T-00441
Phosphoryldimethylaminoethanol, *see* D-00016
Phosphoryldimethylcholamine, *see* D-00016
Phosphoryldimethylethanolamine, *see* D-00016
Phosphoryl isocyanate, P-00415
Phosphoryl isothiocyanate, P-00416
Phosphoserine, P-00417
N-Phosphoserine, P-00418
N^{δ}-Phosphosulfamylornithine, P-00419
N-Phosphothreonine, P-00420
O-Phosphothreonine, *see* A-00120
Phosphotryptophan, P-00421
N-Phosphotyrosine, P-00422
O-Phosphotyrosine, P-00423
▷Phosvel, *see* L-00001
▷Phoxim, P-00424
▷Phoxime, *see* P-00424
Phthalide-3-phosphonic acid, *see* D-00535
Phthalimide *N*-phosphonothioic acid, *see* D-00475
2-Phthalimidoethyl phosphorodichloridate, P-00425
N-Phthalimidomethylphosphonic acid, P-00426
Phthalimidophosphonothioic acid, *see* D-00475
Phytonadiol diphosphate, P-00427
Phytonadiol sodium diphosphate, *in* P-00427
▷Phytosol, *see* T-00419
Picofosforic acid, P-00428
Pinacol cyclic dithiophosphoric acid, *see* M-00019
Pinacol cyclic phosphoric acid, *see* H-00196
▷Pinacolyl methylphosphonofluoridate, *see* S-00005
1-Piperidinephosphonic acid, *see* P-00430
1-Piperidinephosphonous acid, *see* P-00432
Piperidinophosphonic acid, *see* P-00430
Piperidinophosphonothioic acid, *see* P-00431
Piperidinophosphonous acid, *see* P-00432
(1-Piperidinylmethyl)phosphonic acid, P-00429
1-Piperidinylphosphonic acid, P-00430
1-Piperidinylphosphonothioic acid, P-00431
1-Piperidinylphosphonous acid, P-00432
▷Piperophos, P-00433
▷Pirimicid, *see* P-00434
▷Pirimiphos-ethyl, P-00434
▷Pirimiphos-methyl, P-00435
Platelet activating factor, P-00436
▷Plumbemycin *A*, P-00437
▷Plumbemycin B, *in* P-00437
$(S)_P$-*O*-Menthyl ethylphenylphosphinothioate, *in* E-00105
▷Polaris, *see* G-00098
▷PP 175, *see* M-00003
PPFA, *see* D-00697
P,P,P-Trichloro-*N*-(2-nitrophenyl)-iminophosphorane, *in* N-00061
Prasinomycin, P-00438
Prasinomycin *A*, *in* P-00438
Prasinomycin *B*, *in* P-00438
Prasinomycin *C*, *in* P-00438
▷Prefar, *see* B-00004
Prenomycin, P-00439
▷Procytox, *see* C-00283
▷Profenofos, P-00440
1,2-Propadienylphosphonic acid, P-00441
P-Propadienylphosphonic diamide, P-00442
Propadienylphosphonic dichloride, P-00443
1,3-Propanediphosphonic acid, P-00444
2,2-Propanediphosphonic acid, P-00445
1,3-Propanediphosphoramidothioic acid, P-00446
1,3-Propanediylbis[dicyclohexylphosphine], *see* B-00169
1,3-Propanediylbis[dimethylphosphine], *see* B-00206
1,3-Propanediylbis[diphenylphosphine], *see* B-00265
1,3-Propanediylbis[methylphenylphosphine], *see* B-00321
1,3-Propanediylbis[phenylphosphinic acid], P-00447
1,3-Propanediylbisphosphine, *see* D-01180
1,3-Propanediylbis[phosphonic acid], *see* P-00444
1,3-Propanediylbisphosphoramidothioic acid, *see* P-00446
1,3-Propanediylbis[triphenylphosphonium](2+), P-00448
S,S'-1,3-Propanediyl *N,N*-diethylphosphoramidodithioate, *in* D-01248
1,3-Propanediylidenebis[triphenylphosphorane], *in* P-00448
S,S'-Propanediyl methylphosphonodithioate, *in* M-00137
S,S'-Propanediyl methylphosphonodithioite, *see* M-00137
S,S'-Propanediyl methylphosphonoselenodithioate, *in* M-00137
1,3-Propanediyl methylphosphonotrithioate, *in* M-00137
S,S'-1,3-Propanediyl phenylphosphonotrithioate, *in* D-01249
S,S'-1,3-Propanediyl phosphonodithioate, *see* D-01248
S,S'-1,3-Propanediyl phosphonotrithioate, *see* D-01249
▷1-Propanephosphonic acid, *see* P-00473
1,1,3-Propanetriphosphonic acid, P-00449
1,2,3-Propanetriphosphonic acid, P-00450
1,1,3-Propanetriyltrisphosphonic acid, *see* P-00449
1,2,3-Propanetriyltrisphosphonic acid, *see* P-00450
Propanoic acid anhydride with diethyl phosphate, *see* D-00400
Propanoic-diethyl phosphoric anhydride, *see* D-00400
1,3-Propanolamine cyclic phenylphosphonamidite, *see* T-00132
Propanoylphosphonic acid, *see* O-00074
Propanoylphosphonothioic acid, *see* O-00078
▷Propaphos, P-00451
Propargylphosphonic acid, *see* P-00495
Propargyltriphenylphosphonium, *see* T-00681
2-Propenyl diphenylphosphinite, *in* D-01089
2-Propenyl diphenylphosphinodithioate, *in* D-01065
2-Propenyl di-2-propenylphosphinate, *in* D-01193
2-Propenyl phenyl(2-propenyl)phosphinate, *in* P-00300
2-Propenylphosphinic acid, P-00452
(2-Propenyl)phosphinic dichloride, *see* M-00150
1-Propenylphosphonic acid, P-00453
2-Propenylphosphonic acid, P-00454
P-2-Propenylphosphonic diamide, P-00455
2-Propenylphosphonic dichloride, P-00456
2-Propenylphosphonodithioic acid, P-00457
(2-Propenyl)phosphonoselenoic acid, P-00458
P-(2-Propenyl)phosphonoselenoic diamide, P-00459
(2-Propenyl)phosphonotelluroic acid, P-00460
2-Propenylphosphonothioic acid, P-00461
P-2-Propenylphosphonothioic diamide, P-00462
2-Propenylphosphonous acid, P-00463
2-Propenylphosphoramidic acid, P-00464
2-Propenyl phosphorodichloridite, P-00465
2-Propenyl tetramethylphosphorodiamidate, *in* T-00218
▷Propetamphos, P-00466
Prophos, *see* B-00264
▷Prophos, *see* E-00023
Propionylphosphonothioic acid, *see* O-00078
Propionylthiophosphonic acid, *see* O-00078
Propyl acetylethylphosphinate, *in* A-00017
Propylamine-1,1-diphosphonic acid, *see* A-00124
O-Propyl bis(1-aziridinyl)phosphinothioate, *in* B-00093
O-Propyl butylphosphinothioate, *in* B-00593
O-Propyl butylphosphonochloridothioate, *in* B-00604
▷Propyl dichlorophosphate, *see* P-00487
Propyl dichlorophosphite, *see* P-00488
O-Propyl dichlorothiophosphate, *see* P-00489
S-Propyl dichlorothiophosphate, *see* P-00490
Propyl dichlorothiophosphite, *see* P-00491
Propyl *N,N*-diethyl-*P*-(methoxymethyl)-phosphonamidate, *in* D-00295
Propyl diethylphosphinodiselenoate, *in* D-00324
Se-Propyl diethylphosphinoselenothioate, *in* D-00329
Propyl difluorophosphite, *see* P-00492
Propyl dihydrogen phosphate, *see* M-00512
S-Propyl dihydrogen phosphorothioate, *see* P-00493
S-Propyl dihydrogen thiophosphate, *see* P-00493
Propyldiphenylphosphine, *see* D-01145
Propyl (dipropoxyphosphinylamino)acetate, *in* P-00368
O-Propyl dipropylphosphinothioate, *in* D-01208
Propylene bromophosphite, *see* B-00473
Propylene chlorophosphate, *in* C-00092
Propylene chlorophosphite, *see* C-00092
Propylene chlorothiophosphate, *in* C-00092
Propylene cyclic phosphate, *see* H-00155
Propylene cyclic thiophosphoric acid, *see* M-00015
Propylene dimethylphosphoramidate, *in* M-00116
Propylene dimethylphosphoramidite, *see* M-00116
S,S-1,2-Propylene *N,N*-dimethylphosphoramidodithioite, *in* M-00136
O,O-Propylene dimethyl-phosphoramidoselenoate, *in* M-00116
O,O-Propylene dimethylphosphoramidothioate, *in* M-00116
O,O-Propylene dithiophosphate, *see* M-00016
Propylene fluorophosphite, *see* F-00020
Propylene fluorothiophosphate, *in* F-00020
O,O-Propylene hydrogen dithiophosphate, *see* M-00016
1,2-Propylene hydrogen phosphate, *see* H-00155
O,O-1,2-Propylene hydrogen thiophosphate, *see* M-00015
Propylene methylphosphonate, *in* D-00732
Propylene methylphosphonite, *see* D-00732
S,S-1,2-Propylene *O*-methyl phosphorodithioite, *in* M-00136
Propylene phenylphosphonite, *see* M-00198
1,2-Propylene phenylphosphonodithioite, *in* M-00136
Propylene phosphite, *in* M-00127
Propylene phosphonate, *in* M-00127
Propylene phosphonite, *see* M-00127
Propylene phosphorobromidite, *see* B-00473
Propylene phosphorochloridate, *in* C-00092
Propylene phosphorochloridite, *see* C-00092
1,2-Propylene phosphorochloridodithioite, *in* M-00136
Propylene phosphorochloridothioate, *in* C-00092
1,2-Propylene phosphorofluoridite, *see* F-00020
1,2-Propylene phosphorofluoridothioate, *in* F-00020
Propylene phosphoryl chloride, *in* C-00092

O,O-Propylene thiophosphite, *in* M-00127
O,O-Propylene thiophosphonate, *in* M-00127
Propylene thiophosphoryl chloride, *in* C-00092
Propyl ethenylphenylphosphinate, *in* P-00337
S-Propyl ethylphenylphosphinothioate, *in* E-00105
O-Propyl ethylphosphonochloridothioate, *in* E-00137
S-Propyl ethylphosphonochloridothioate, *in* E-00137
(1-Propylheptyl)phosphonic acid, P-00467
O-Propyl hydrogen methylphosphonoselenoate, *in* M-00311
Propyl hydrogen phenylphosphonate, *see* M-00511
O-Propyl hydrogen phosphonoselenoate, *in* E-00145
Propyl hypophosphite, *see* P-00470
O-Propyl isopropylphosphonochloridothioate, *in* I-00053
Propyl (4-methylphenyl)phenylphosphinate, *in* M-00204
S-Propyl *P*-methyl-*N*-phenylphosphonamidothioate, *in* M-00233
O-Propyl methylphosphinothioate, *in* M-00279
O-Propyl methylphosphonochloridothioate, P-00468
S-Propyl methylphosphonochloridothioate, *in* M-00298
O-Propyl methylphosphonodithioate, P-00469
O-Propyl methylphosphonoselenoate, *in* M-00311
O-Propyl phenylphosphinothioate, *in* P-00213
S-Propyl phenylphosphonochloridothioate, *in* P-00232
Propyl phenylpropylphosphinate, *in* P-00303
Propyl phenylvinylphosphinate, *in* P-00337
Propyl phosphinate, P-00470
Propylphosphine, P-00471
Propylphosphinic acid, P-00472
▷Propylphosphonic acid, P-00473
Propylphosphonic acid monoanhydride, *see* D-01202
Propylphosphonic acid trimer trianhydride, *in* T-00720
Propylphosphonic bis(diethylamide), *in* P-00474
Propylphosphonic bis(dimethylamide), *in* P-00474
P-Propylphosphonic diamide, P-00474
Propylphosphonic dichloride, P-00475
Propylphosphonochloridic acid, P-00476
Propylphosphonochloridothioic acid, P-00477
Propylphosphonodithioic acid, P-00478
Propylphosphonofluoridic acid, P-00479
Propylphosphonothioic acid, P-00480
Propylphosphonothioic acid trimer trianhydride, *in* T-00720
P-Propylphosphonothioic diamide, P-00481
Propylphosphonothioic dianilide, *in* P-00481
Propylphosphonous acid, P-00482
Propylphosphonous dichloride, P-00483
Propylphosphoramidic acid, P-00484
Propylphosphoramidothioic acid, P-00485
Propylphosphoramidous acid, P-00486
Propyl phosphoric acid, *see* M-00512
▷Propyl phosphorodichloridate, P-00487
Propyl phosphorodichloridite, P-00488
O-Propyl phosphorodichloridothioate, P-00489
S-Propyl phosphorodichloridothioate, P-00490
Propyl phosphorodichloridothioite, P-00491
Propyl phosphorodifluoridite, P-00492
S-Propyl phosphorothioate, P-00493
▷Propyl phosphoryl dichloride, *see* P-00487
Propylphostonic acid, *see* H-00166
S-Propyl propylphosphonochloridothioate, *in* P-00477
Propyl tetrapropylphosphorodiamidate, *in* T-00291
S-Propyl thiophosphonic acid, *see* P-00493
O-Propyl thiophosphoryl dichloride, *see* P-00489
Propyltriphenylphosphonium(1+), *see* T-00680
Propyl (triphenylphosphoranylidene)-carbamate, *in* T-00648
2-Propynyl dichlorophosphate, *see* P-00497
1-Propynylphosphonic acid, P-00494
2-Propynylphosphonic acid, P-00495
1-Propynylphosphonic dichloride, P-00496
2-Propynyl phosphorodichloridate, P-00497
▷Prothiofos, P-00498
▷Prothoate, P-00499
▷Proticin, P-00500
PSOrn, *see* P-00419
O-p-Tolyl dichlorothiophosphate, *see* M-00262
P-p-Tolylphosphonic diamide, *see* M-00238
O-p-Tolyl thiophosphoryl dichloride, *see* M-00262
(1*H*-Pyrazol-3-yl)phosphonic acid, P-00501
▷Pyrazophos, P-00502
2-Pyridinephosphonic acid, P-00503
3-Pyridinephosphonic acid, P-00504
4-Pyridinephosphonic acid, P-00505
4,4′-(2-Pyridinylmethylene)bisphenol bis(dihydrogen phosphate), *see* P-00428
(2-Pyridinylmethyl)phosphonic acid, P-00506
(4-Pyridinylmethyl)phosphonic acid, P-00507
2-Pyridinylphosphonic acid, *see* P-00503
3-Pyridinylphosphonic acid, *see* P-00504
4-Pyridinylphosphonic acid, *see* P-00505
(2-Pyridinyl)phosphoramidic acid, *see* P-00510
(3-Pyridinyl)phosphoramidic acid, *see* P-00511
(4-Pyridinyl)phosphoramidic acid, *see* P-00512
2-Pyridinylphosphoramidothioic acid, *see* P-00513
4-Pyridinylphosphoramidothioic acid, *see* P-00514
▷Pyridoxal 5-monophosphate, *see* P-00508
▷Pyridoxal phosphate, P-00508
Pyridoxamine 5′-phosphate, P-00509
(2-Pyridylmethyl)phosphonic acid, *see* P-00506
(4-Pyridylmethyl)phosphonic acid, *see* P-00507
2-Pyridylphosphonic acid, *see* P-00503
3-Pyridylphosphonic acid, *see* P-00504
4-Pyridylphosphonic acid, *see* P-00505
2-Pyridylphosphoramidic acid, P-00510
3-Pyridylphosphoramidic acid, P-00511
4-Pyridylphosphoramidic acid, P-00512
2-Pyridylphosphoramidothioic acid, P-00513
4-Pyridylphosphoramidothioic acid, P-00514
2-Pyrimidinephosphonic acid, P-00515
4-Pyrimidinephosphonic acid, P-00516
(2-Pyrimidinyl)phosphonic acid, *see* P-00515
(4-Pyrimidinyl)phosphonic acid, *see* P-00516
N″-2-Pyrimidinylphosphoric triamide, P-00517
5-Pyrophosphomevalonic acid, *see* T-00162
2-Pyrrolephosphonic acid, *see* P-00521
3,4-Pyrrolidinediylbis[diphenylphosphine], *see* B-00267
1-Pyrrolidinephosphonic acid, *see* P-00518
1-Pyrrolidinephosphonous acid, *see* P-00520
1-Pyrrolidinylphosphonic acid, P-00518
1-Pyrrolidinylphosphonochloridous acid, P-00519
1-Pyrrolidinylphosphonous acid, P-00520
1-Pyrrolidylphosphonous acid, *see* P-00520
1*H*-Pyrrol-2-ylphosphonic acid, P-00521
▷Quinalphos, Q-00001
2-Quinolinephosphonic acid, Q-00002
4-Quinolinephosphonic acid, Q-00003
[(2-Quinolinyl)methyl]phosphonic acid, Q-00004
[(8-Quinolinyl)methyl]phosphonic acid, Q-00005
(2-Quinolylmethyl)phosphonic acid, *see* Q-00004
(8-Quinolylmethyl)phosphonic acid, *see* Q-00005
▷Quintiofos, Q-00006
▷Rabon, *see* T-00038
▷Ragadan, *see* H-00004
▷Rametin, *see* N-00008
▷Resitox, *see* C-00205
Resorcinol bis(phosphoric acid), *see* P-00147
Resorcinol diphosphate, *see* P-00147
2-*β*-D-Ribofuranosyl-5′-(dihydrogen phosphate)-*as*-triazine-3,5(2*H*,4*H*)-dione, *see* A-00145
N-(9-*β*-D-Ribofuranosyl-9*H*-purin-6-yl)-L-aspartic acid 5′-(dihydrogen phosphate), *see* A-00047
9*β*-D-Ribofuranosylxanthine 3-(dihydrogen phosphate), *see* X-00002
Ro 03-7008, *see* A-00049
Robison ester, *in* G-00007
▷Rogor, *see* D-00671
▷Ronnel, *see* F-00003
▷Roundup, *in* G-00097
▷Roxion, *see* D-00671
▷Ruelene, *see* C-00208
▷Safrotin, *see* P-00466
▷Salicyclic acid dihydrogen phosphate, *see* P-00391
Salicyclic acid ethyl ester, diethyl phosphate, *in* P-00391
Saligenin cyclic thiophosphoric acid, *see* H-00110
Salioxon, *in* H-00109
▷Salithion, S-00001
Sandoflam, *in* O-00089
▷Saphicol, *see* M-00003
▷Sarin, S-00002
▷Schradan, *see* O-00029
Schweizer's reagent, *see* T-00702
P-sec-Butyl-*N,N*-dimethylphosphamidic chloride, *see* D-00763
▷Selecron, *see* P-00440
Selenocoenzyme *A*, S-00003
Selenophosphoric acid diethylamide, *see* D-00353
Selenophosphoric acid dimethylamide, *see* D-00869
Selenoxo[2,4,6-tris(1,1-dimethylethyl)-phenyl]phosphine selenide, *see* D-01235
▷Sendoxan, *see* C-00283
Serine 2-[(aminoiminomethyl)amino]ethyl hydrogen phosphate (ester), *see* L-00002
Serine 2,3-bis[(1-oxohexadecyl)oxy]propyl hydrogen phosphate ester, *see* G-00016
Serine 2,3-bis[(1-oxooctadecyl)oxy]propyl hydrogen phosphate ester, *see* G-00038
Serine dihydrogen phosphate, *see* P-00417
Serine 2-[(1-oxooctadecenyl)oxy]-3-[(1-oxooctadecyl)oxy]propyl ester, dihydrogen phosphate ester, *see* G-00093
▷*S*-Seven, S-00004
▷SF 1293, *see* A-00141
Skewphos, *see* B-00260
Sodium phenylphosphinite, *in* P-00251
Sodium phenylphosphonite, *in* P-00251
▷Solugliacit, *see* P-00001
▷Soman, S-00005
Sparfosic acid, S-00006
▷Spectacide, *see* D-00027
Spiro[1,3,2-benzodioxaphosphole-2,2′-[1,3,2]benzoxathiaphosphole], S-00007
Spiro[1,3,2-benzodioxaphosphole-2,2′(3′*H*)-[1,3,2]benzoxazaphosphole], S-00008
Spiro[1,3,2-benzodioxaphosphole-2,2′-[1,3,2]oxazaphospholidine], S-00009
Spiro[1,3,2-benzodioxaphosphole-2,1′-[3]-phospholene], S-00010
Spiro[1,3,2-benzoxathiaphosphole-2,2′-[1,3,2]dioxaphospholane], S-00011
Spiro[1,3,2-benzoxazaphosphole-2(3*H*),2′-[1,3,2]dioxaphospholane], S-00012
Spiro[1,3,2-benzoxazaphosphole-2(3*H*),2′-[1,2]oxaphospholan]-5′-one, S-00013
2,2′-Spirobi[1,3,2-benzodioxaphosphole], S-00014
2,2′-Spirobi[1,3,2-benzodithiaphosphole], S-00015
2,2′-Spirobi[1,3,2-benzodithiaphospholium](1+), S-00016
1,1′(3*H*,3′*H*)-Spirobi[2,1-benzoxaphosphole], S-00017
1,1′(3*H*,3′*H*)-Spirobi[2,1-benzoxaphosphole]-3,3′-dione, S-00018
2,2′-Spirobi[1,3,2-benzoxathiaphosphole], S-00019
2,2(3*H*,3′*H*)-Spirobi[1,3,2-benzoxazaphosphole], S-00020
5,5′-Spirobi[5*H*-dibenzophosphole], S-00021
5,5′-Spirobi[5*H*-dibenzophospholium](1+), S-00022
10,10′(5*H*,5′*H*)-Spirobiphenophosphazinium(1+), S-00023
Spiro[1,3,2-dioxaphospholane-2,1′-[2,8,9]-trioxa[1]phosphaadamantane], *see* S-00024

Spiro[1,3,2-dioxaphospholane-2,1′-[2,8,9]-trioxa[1]phosphatricyclo[3.3.1.1^{3,7}]-decane], S-00024
Spiro[1,3,2-dioxaphosphorinane-2,2′-phenanthro[9,10-*d*][1,3,2]dioxaphosphole], S-00025
Stearoylarachidonoylphosphatidylcholine, *in* G-00076
Stearoylarachidonylphosphatidylethanolamine, *in* G-00077
Stearoylelaidoylphosphatidylethanolamine, *in* G-00083
Stearoyllauroylphosphatidylcholine, *in* G-00074
Stearoyllinolenoylphosphatidylcholine, *in* G-00080
Stearoyllinolenoylphosphatidylethanolamine, *in* G-00081
Stearoyllinoleoylphosphatidylcholine, *in* G-00078
Stearoyllinoleoylphosphatidylethanolamine, *in* G-00079
Stearoyloleoylphosphatidylcholine, *in* G-00082
Stearoyloleoylphosphatidylcholine, *in* G-00083
Stearoylpalmitoylphosphatidylcholine, *see* G-00075
▷Stirifos, *in* T-00038
▷Stiriphos, *in* T-00038
Styrylphosphine, *see* P-00152
Styrylphosphinic acid, *see* P-00153
α-Styrylphosphonic acid, *see* P-00154
β-Styrylphosphonic acid, *see* P-00155
P-Styrylphosphonic diamide, *see* P-00156
Styrylphosphonic dichloride, *see* P-00157
β-Styrylphosphonic difluoride, *see* P-00158
Styrylphosphonodithioic acid, *see* P-00159
β-Styrylphosphonothioic acid, *see* P-00160
Styrylphosphonous acid, *see* P-00161
Styrylphosphonous dichloride, *see* P-00162
▷Sufosfamide, S-00026
▷Sulfotep, *see* T-00045
▷Sulprofos, S-00027
▷Sumioxon, *see* F-00004
▷Sumithion, *see* F-00005
▷Supracide, *see* M-00025
▷Surecide, *see* C-00213
▷Systox, *in* D-00019
▷Sytam, *see* O-00029
▷Tabun, T-00001
Tagetitoxin, T-00002
▷Tamaron, *see* M-00021
TDP, *see* T-00320
▷Temephos, T-00003
▷Tepa, *in* T-00331
▷TEPP, *see* T-00069
▷Terbufos, T-00004
▷Terracur *P*, *see* F-00006
O-*tert*-Butyl *tert*-butylmethylphosphinothioate, *in* B-00570
N-*tert*-Butyl-*P*,*P*-diisopropylphosphinic amide, *in* D-00642
N-*tert*-Butyl-*P*,*P*-dimethylphosphinic amide, *in* D-00830
SS-*tert*-Butyl *O*,*O*-dimethyl phosphoro(dithioperoxo)thioate, B-00556
N-*tert*-Butyl-*P*,*P*-diphenylphosphinic amide, *in* D-01040
P-*tert*-Butyl-*P*-phenylphosphinic anilide, *in* B-00580
S-*tert*-Butylthio *O*,*O*-dimethyl phosphorodithioate, *see* B-00556
▷Tespamin, *see* T-00317
▷Tetraallyl methylenediphosphonate, *in* M-00022
1,1,4,4-Tetraamino-2,3-diphosphabutadiene, T-00005
Tetrabenzyl diphosphate, T-00006
1,1,6,6-Tetrabenzyl-1,6-diphosphecanium dibromide, *in* D-01171
1,1,4,4-Tetrabenzyl-1,4-diphosphepanium dibromide, *in* D-00047
Tetrabenzyldiphosphine, T-00007
1,1,4,4-Tetrabenzyl-1,4-diphosphocanium dibromide, *in* D-00048
1,1,5,5-Tetrabenzyl-1,5-diphosphocanium dibromide, *in* D-00049
1,1,4,4-Tetrabenzyl-1,4-diphosphonanium dibromide, *in* D-00050
1,1,5,5-Tetrabenzyl-1,5-diphosphonanium dibromide, *in* D-00051
1,1,6,6-Tetrabenzyl-1,6-diphosphoniacycloundecane dibromide, *in* D-00046
N,*N*,*N*′,*N*′-Tetrabenzyl-*P*-phenylphosphonous diamide, *in* P-00252
Tetrabenzylphosphonium chloride, *in* T-00338
Tetrabenzyl pyrophosphate, *see* T-00006
1,5,9,13-Tetrabenzyl-1,5,9,13-tetraphenyl-1,5,9,13-tetraphosphoniacyclohexadecane tetrabromide, *in* T-00276
1,6,11,16-Tetrabenzyl-1,6,11,16-tetraphosphacycloeicosane, T-00008
1,5,10,14-Tetrabenzyl-1,5,10,14-tetraphosphacyclooctadecane, T-00009
Tetrabenzyl thioperoxydiphosphate, *see* B-00155
Tetrabromomethylphosphorane, T-00010
Tetrabromophenylphosphorane, T-00011
Tetrabutoxydiphosphine, *see* T-00016
Tetrabutoxydiphosphine dioxide, *see* T-00017
Tetrabutyl (aminophenylmethyl)-bisphosphonate, *in* A-00053
Tetra-*tert*-butylcyclotetraphosphine, T-00012
Tetra-*tert*-butylcyclotetraphosphine, *see* T-00012
1,2,4,5-Tetra-*tert*-butyl-3,6-dicarba-1,2,4,5-tetraphosphan, *see* T-00027
Tetra-*tert*-butyl-1,2,3,4-diphosphadiarsetane, T-00013
▷Tetrabutyl diphosphate, *see* T-00024
Tetrabutyldiphosphine, T-00014
Tetra-*tert*-butyldiphosphine, T-00015
Tetrabutyl diphosphite, *see* T-00025
Tetrabutyl 1,2-ethanediylbisphosphonate, *in* E-00006
Tetrabutyl hypodiphosphite, T-00016
Tetrabutyl hypophosphate, T-00017
Tetra-*tert*-butylmethyltetraphosphetanium iodide, *in* T-00012
Tetrabutyl [oxybis(methylene)]-bisphosphonate, *in* O-00095
N,*N*,*N*′,*N*′-Tetrabutyl-*P*-(2-phenylethenyl)-phosphonic diamide, *in* P-00156
N,*N*,*N*′,*N*′-Tetrabutylphosphonic diamide, T-00018
Tetrabutylphosphonium(1+), T-00019
Tetra-*tert*-butylphosphonium(1+), T-00020
Tetrabutylphosphorodiamidic acid, T-00021
Tetrabutylphosphorodiamidic chloride, T-00022
Tetrabutylphosphorodiamidodithioic acid, T-00023
N,*N*,*N*′,*N*′-Tetrabutylphosphorodiamidous acid, *see* T-00018
▷Tetrabutyl pyrophosphate, T-00024
Tetrabutyl pyrophosphite, T-00025
1,2,4,5-Tetra-*tert*-butyl-1,2,4,5-tetraphosphacyclohexane, *see* T-00027
1,2,4,5-Tetra-*tert*-butyl-1,2,4,5-tetraphosphaspiro[2.2]pentane, T-00026
1,2,4,5-Tetra-*tert*-butyl-1,2,4,5-tetraphosphinane, *see* T-00027
1,2,4,5-Tetra-*tert*-butyl-1,2,4,5-tetraphosphorinane, T-00027
2,2,4,6-Tetrachloro-4,6-bis(dimethylamino)-2,2,4,4,6,6-hexahydrocyclotriphosphazene, *see* T-00028
2,2,4,6-Tetrachloro-4,6-bis(dimethylamino)-2,2,4,4,6,6-hexahydro-1,3,5-triaza-2,4,6-triphosphorine, T-00028
2,2,4,6-Tetrachloro-2,2-dihydro-1,3,5,2-triazaphosphinine, *see* T-00029
2,2,4,6-Tetrachloro-2,2-dihydro-1,3,5,2-triazaphosphorine, T-00029
4,5,6,7-Tetrachloro-*N*,*N*-dimethylspiro[1,3,2-benzodioxaphosphole-2,2′-[1,3,2]-dioxaphospholan]-2-amine, *in* T-00036
2,2,4,6-Tetrachloro-2,2,4,4,6,6-hexahydro-4,6-bis(methylamino)-cyclotriphosphazene, *see* T-00030
2,2,4,6-Tetrachloro-2,2,4,4,6,6-hexahydro-4,6-bis(methylamino)-1,3,5-triaza-2,4,6-triphosphorine, T-00030
2,2,4,4-Tetrachloro-2,2,4,4,6,6-hexahydro-6,6-diphenylcyclotriphosphazene, *see* T-00031
2,2,4,4-Tetrachloro-2,2,4,4,6,6-hexahydro-6,6-diphenyl-1,3,5-triaza-2,4,6-triphosphorine, T-00031
Tetrachloromethylphosphorane, T-00032
Tetrachlorophenoxyphosphorane, T-00033
Tetrachlorophenylphosphorane, T-00034
2,3,5,6-Tetrachloro-4-phosphonopyridine, *see* T-00035
(2,3,5,6-Tetrachloro-4-pyridinyl)phosphonic acid, T-00035
4,5,6,7-Tetrachloro-4′,4′,5′,5′-tetramethylspiro[1,3,2-benzodioxaphosphole-2,2′-[1,3,2]-dioxaphospholane], T-00036
Tetrachloro(trichloromethyl)phosphorane, T-00037
▷Tetrachlorvinphos, T-00038
1,2,3,4-Tetracyclohexyl-*cyclo*-5,6-dicarba-1,2,3,4-tetraphosphan, *see* T-00040
Tetracyclohexyldiphosphine, T-00039
1,2,3,4-Tetracyclohexyl-1,2,3,4-tetraphosphinane, *see* T-00040
1,2,3,4-Tetracyclohexyl-1,2,3,4-tetraphosphorinane, T-00040
Tetradecanoic acid 1-[[[(2-aminoethoxy)-hydroxyphosphinyl]oxy]methyl]-1,2-ethanediyl ester, *see* G-00049
Tetradecanoic acid 1-[(phosphonoxy)-methyl]-1,2-ethanediyl ester, *see* G-00047
1-Tetradecanoyl-2-hexadecanoylglycero-3-phosphocholine, *see* G-00096
Tetradecyl dihydrogen phosphate, *see* M-00514
Tetradecyl phosphoric acid, *see* M-00514
Tetraethenylphosphonium(1+), *see* T-00296
Tetraethoxydiphosphine, *see* T-00049
Tetraethoxydiphosphine dioxide, *see* T-00050
Tetraethoxydiphosphine disulfide, *see* T-00072
Tetraethoxydiphosphine 1-oxide 2-sulfide, *see* T-00073
1,1,3,3-Tetraethoxy-2-ethyltriphosphine 1,3-dioxide, T-00041
Tetraethyl (1-aminoethylidene)-bisphosphonate, *in* A-00073
Tetraethyl (aminophenylmethyl)-bisphosphonate, *in* A-00053
Tetraethyl 1,2-benzenediol bis(phosphate), T-00042
Tetraethyl [[1,1′-biphenyl]-4,4′-diylbis(methylene)]bisphosphonate, *in* B-00391
Tetraethyl 1,3-butadiene-1,4-diylbisphosphonate, *in* B-00507
Tetraethyl 1,4-butanediylbisphosphonate, *in* B-00514
Tetraethyl 1,4-butanediylbisphosphonite, *in* B-00515
Tetraethyl (2-butene-1,4-diyl)-bisphosphonate, *in* B-00519
Tetraethyl carbamoylphosphonate, *in* D-00264
Tetraethyl (1-cyano-1,2-ethanediyl)-bisphosphonate, *in* B-00173
Tetraethyl 1,2-cyclohexanediylbisphosphonate, *in* C-00240
Tetraethylcyclotetraphosphine, *see* T-00071
Tetraethyl 1,10-decanediylbisphosphonate, *in* D-00008
Tetraethyl dichloromethylenebisphosphonate, *in* D-00172
5,5,7,7-Tetraethyl-6,7-dihydro-5*H*-dibenzo[*d*,*f*][1,3]diphosphepinium dibromide, *in* D-00282
Tetraethyl [(dimethylamino)methylene]-bisphosphonate, *in* D-00701
▷Tetraethyl diphosphate, *see* T-00069
Tetraethyldiphosphine, T-00043
Tetraethyldiphosphine disulfide, T-00044
Tetraethyl diphosphite, *see* T-00070
5,5,10,10-Tetraethyl-5,10-diphosphonia-1,2:3,4-dibenzocyclodecadiene dibromide, *in* H-00021

5,5,9,9-Tetraethyl-5,9-diphosphonia-1,2:3,4-dibenzocyclononadiene dibromide, *in* T-00089
5,5,8,8-Tetraethyl-5,8-diphosphonia-1,2:3,4-dibenzocyclooctadiene dibromide, *in* D-00404
▷*O,O,O,O*-Tetraethyl dithiopyrophosphate, T-00045
unsym-*O,O,O,O*-Tetraethyl dithiopyrophosphate, T-00046
Tetraethyl 1,2-ethanediylbisphosphonate, *in* E-00006
O,O,O′,O′-Tetraethyl *S,S′*-ethanediyl bisphosphorodithioate, T-00047
Tetraethyl ethenylidenebisphosphonate, *in* E-00020
Tetraethyl [(ethoxyphosphinylidene)-bis(methylene)]bisphosphonate, *in* B-00182
▷Tetraethyl ethylimidodiphosphate, *in* E-00078
Tetraethyl ethylimidodiphosphite, *in* E-00079
Tetraethyl 1,7-heptanediylbisphosphonate, *in* H-00002
N,N,N′,N′-Tetraethyl-*P*-heptylphosphonous diamide, *in* H-00008
Tetraethyl (3,3,4,4,5,5-hexafluoro-1-cyclopentene-1,2-diyl)bisphosphonate, *in* H-00018
5,5,10,10-Tetraethyl-5,6,7,8,9,10-hexahydrodibenzo[*b,d*][1,6]-diphosphecinium dibromide, *in* H-00021
Tetraethyl hexamethylenediphosphonite, *in* H-00086
Tetraethyl 1,6-hexanediylbisphosphonate, *in* H-00085
Tetraethyl 1,6-hexanediylbisphosphonite, *in* H-00086
O,O,O′,O′-Tetraethyl hydrazodiphosphonothioate, T-00048
Tetraethyl (1-hydroxyethylidene)-bisphosphonate, *in* H-00144
Tetraethyl hypodiphosphite, T-00049
Tetraethyl hypophosphate, T-00050
O,O,O′,O′-Tetraethyl imidodiphosphate, T-00051
▷Tetraethyl iminobis(methylene)-bisphosphonate, *in* D-00688
O,O,O′,O′-Tetraethyl iminodiphosphate, *see* T-00051
N,N,N′,N′-Tetraethyl-*P*-(2-methyl-1,3-butadienyl)phosphonic diamide, *in* M-00096
N,N,N′,N′-Tetraethyl-*P*-(2-methyl-1,3-butadienyl)phosphonothioic diamide, *in* M-00097
▷Tetraethyl methylenebisphosphonate, *in* M-00022
Tetraethyl methylenebisphosphonite, *in* M-00024
O,O,O,O-Tetraethyl methylenebisphosphonothioate, *in* M-00023
▷*O,O,O′,O′*-Tetraethyl *S,S′*-methylene bis(phosphorodithioate), *see* E-00022
Tetraethyl methylimidodiphosphate, *in* M-00156
Tetraethyl methylimidodiphosphite, *in* M-00157
O,O,O,O-Tetraethyl methylimidodiphosphorothioate, *in* M-00161
Tetraethyl methyliminodiphosphate, *in* M-00156
N,N,N′,N′-Tetraethyl-*P*-(3-methylphenyl)-phosphonous diamide, *in* M-00247
N,N,N′,N′-Tetraethyl-*P*-(2-methylphenyl)-phosphonous diamide, *in* M-00246
N,N,N′,N′-Tetraethyl-*P*-(4-methylphenyl)-phosphonous diamide, *in* M-00248
N,N,N′,N′-Tetraethyl-*P*-methylphosphonic diamide, *in* M-00289
N,N,N′,N′-Tetraethyl-*P*-methylphosphonothioic diamide, *in* M-00314
N,N,N′,N′-Tetraethyl-*P*-methylphosphonous diamide, *in* M-00320
Tetraethylmethyltetraphosphetanium iodide, *in* T-00071
O,O,O,O-Tetraethyl methylthioimidodiphosphate, *in* M-00161
sym-Tetraethyl monothiopyrophosphate, T-00052
▷*unsym*-Tetraethyl monothiopyrophosphate, T-00053
Tetraethyl (4-morpholinylmethylene)-bisphosphonate, *in* M-00520
Tetraethyl [2,3-naphthalenediylbis[methylene]]-bisphosphonate, *in* N-00005
Tetraethyl [2,6-naphthalenediylbis[methylene]]-bisphosphonate, *in* N-00006
Tetraethyl [2,7-naphthalenediylbis[methylene]]-bisphosphonate, *in* N-00007
Tetraethyl 1,9-nonanediylbisphosphonate, *in* N-00069
Tetraethyl (3,3,4,4,5,5,6,6,-octafluoro-1-cyclohexene-1,2-diyl)bisphosphonate, *in* O-00003
Tetraethyl 1,8-octanediylbisphosphonate, *in* O-00035
Tetraethyl (1-oxo-1,3-propanediyl)-bisphosphonate, *in* O-00071
Tetraethyl (2-oxo-1,3-propanediyl)-bisphosphonate, *in* O-00072
N,N,N′,N′-Tetraethyl-*P*-(2-oxopropyl)-phosphonic diamide, *in* O-00077
Tetraethyl oxybis(1,2′-ethanediyl)-bisphosphonate, *in* O-00094
Tetraethyl [oxybis(methylene)]-bisphosphonate, *in* O-00095
Tetraethyl pentamethylenediphosphonite, *in* P-00033
Tetraethyl 1,5-pentanediylbisphosphonate, *in* P-00032
Tetraethyl 1,5-pentanediylbisphosphonite, *in* P-00033
Tetraethyl pentylidenebisphosphonate, *in* P-00031
Tetraethyl [1,2-phenylenebis(methylene)]-bisphosphonate, *in* P-00129
Tetraethyl [1,2-phenylenebis(methylene)]-bisphosphonate, *in* P-00130
Tetraethyl [1,4-phenylenebis(methylene)]-bisphosphonate, *in* P-00131
Tetraethyl 1,3-phenylenebisphosphonate, *in* B-00007
Tetraethyl 1,4-phenylenebisphosphonate, *in* B-00008
Tetraethyl 1,2-phenylenebisphosphonite, *in* B-00009
Tetraethyl 1,4-phenylenebisphosphonite, *in* B-00010
Tetraethyl 1,4-phenylenebisphosphoramidate, *in* B-00012
N,N,N′,N′-Tetraethyl-*P*-(2-phenylethenyl)-phosphonic diamide, *in* P-00156
N,N,N′,N′-Tetraethyl-*P*-(2-phenylethenyl)-phosphonous diamide, *in* P-00161
Tetraethyl phenylimidodiphosphate, *in* P-00168
Tetraethyl phenyliminodiphosphate, *in* P-00168
N,N,N′,N′-Tetraethyl-*P*-(phenylmethyl)-phosphonic diamide, *in* B-00061
N,N,N′,N′-Tetraethyl-*P*-phenylphosphonous diamide, *in* P-00252
N,N,N′,N′-Tetraethyl-2-phenylphosphorodiamidic hydrazide, *in* P-00274
N,N,N′,N′-Tetraethyl-*N″*-phenylphosphorodiamidimidic acid, T-00054
Tetraethylphosphinic amide, *in* D-00323
Tetraethylphosphinoselenoic amide, *in* D-00328
Tetraethylphosphinothioic amide, *in* D-00331
Tetraethylphosphinous amide, *in* D-00334
N,N,N′,N′-Tetraethylphosphonic diamide, T-00055
Tetraethylphosphonium(1+), T-00056
Tetraethyl phosphonobutanedioate, *in* D-00245
Tetraethyl phosphonosuccinate, *in* D-00245
N,N,N′,N′-Tetraethylphosphonothioic diamide, T-00057
N,N,N′,N′-Tetraethylphosphoric triamide, T-00058
Tetraethyl phosphor(isocyanatidic) diamide, *in* T-00059
Tetraethyl phosphor(isothiocyanatidic) diamide, *in* T-00059
Tetraethylphosphorodiamidic acid, T-00059
Tetraethylphosphorodiamidic acid anhydride, *see* O-00001
Tetraethylphosphorodiamidic anilide, *in* T-00058
Tetraethylphosphorodiamidic chloride, T-00060
Tetraethylphosphorodiamidoselenoic acid, T-00061
Tetraethylphosphorodiamidotelluroic acid, T-00062
Tetraethylphosphorodiamidothioic acid, T-00063
Tetraethylphosphorodiamidothioic azide, T-00064
Tetraethylphosphorodiamidothioic chloride, T-00065
Tetraethylphosphorodiamidous acid, T-00066
N,N,N′,N′-Tetraethylphosphorodiamidous acid, *see* T-00055
▷Tetraethylphosphorodiamidous chloride, T-00067
Tetraethyl 1,3-propanediylbisphosphonate, *in* P-00444
O,O,O′,O′-Tetraethyl *S,S′*-propanediyl bisphosphorodithioate, T-00068
Tetraethyl 2,2-propanediyldiphosphonate, *in* P-00445
N,N,N′,N′-Tetraethyl-*P*-(2-propenyl)-phosphonoselenoic diamide, *in* P-00459
N,N,N′,N′-Tetraethyl-*P*-2-propenylphosphonothioic diamide, *in* P-00462
N,N,N′,N′-Tetraethyl-*P*-propylphosphonic diamide, *in* P-00474
N,N,N′,N′-Tetraethyl-*P*-2-pyridinylphosphonic diamide, *in* P-00503
▷Tetraethyl pyrophosphate, T-00069
Tetraethyl pyrophosphite, T-00070
Tetraethyl (3,3,4,4-tetrafluoro-1-cyclobutene-1,2-diyl)bisphosphonate, *in* T-00078
5,5,8,8-Tetraethyl-5,6,7,8-tetrahydrodibenzo[*e,g*][1,4]-diphosphocinium dibromide, *in* D-00404
5,5,9,9-Tetraethyl-6,7,8,9-tetrahydro-5*H*-dibenzo[*f,h*][1,5]diphosphoninium dibromide, *in* T-00089
9,9,16,16-Tetraethyl-9,10,15,16-tetrahydrotribenzo[*b,d,h*][1,6]-diphosphecinium dibromide, *in* T-00155
Tetraethyl tetramethylenediphosphonite, *in* B-00515
Tetraethyltetraphosphetane, T-00071
Tetraethyl [thiobis(methylene)]-bisphosphonate, *in* T-00311
▷Tetraethyl thiodiphosphate, *see* T-00045
Tetraethyl thiodiphosphate, *see* T-00046
Tetraethyl thiodiphosphate, *see* T-00052
▷Tetraethyl thiodiphosphate, *see* T-00053
▷Tetraethyl thiodiphosphate, *see* T-00076
Tetraethyl thiohypophosphate, T-00072
O,O,O′,O′-Tetraethyl thiohypophosphate, T-00073
O,O,O′,O′-Tetraethyl thioimidodiphosphate, T-00074
O,O,O′,O′-Tetraethyl thioiminodiphosphate, *see* T-00074
Tetraethyl thioperoxydiphosphate, *see* B-00170
▷Tetraethyl thioperoxydiphosphate, *see* B-00172
N,N,N′,N′-Tetraethyl-*P*-(trichloromethyl)-phosphonimidic diamide, T-00075
Tetraethyl trimethylsilylimidodiphosphate, *in* T-00051
▷*O,O,O′,O′*-Tetraethyl trithiopyrophosphate, T-00076
3,3,4,4-Tetrafluoro-1,2-bis(diphenylphosphino)cyclobutene, T-00077
3,3,4,4-Tetrafluoro-1-cyclobutene-1,2-diphosphonic acid, T-00078
(3,3,4,4-Tetrafluoro-1-cyclobutene-1,2-diyl)bis[diphenylphosphine], *see* T-00077

(3,3,4,4-Tetrafluoro-1-cyclobutene-1,2-diyl)bis(phosphonic acid), *see* T-00078
Tetrafluorodimethylaminophosphorane, *see* D-00708
Tetrafluoro-*N*,*N*-dimethylphosphoranamine, *see* D-00708
2,2,4,6-Tetrafluoro-2,2,4,4,6,6-hexahydro-4,6-diphenylcyclotriphosphazene, *see* T-00079
2,2,4,6-Tetrafluoro-2,2,4,4,6,6-hexahydro-4,6-diphenyl-1,3,5-triaza-2,4,6-triphosphorine, T-00079
Tetrafluoromethoxyphosphorane, T-00080
Tetrafluoromethylphosphorane, T-00081
Tetrafluoromethylthiophosphorane, T-00082
Tetrafluorophenoxyphosphorane, T-00083
Tetrafluorophenylphosphorane, T-00084
Tetrafluorophenylthiophosphorane, T-00085
Tetrafluoro(trifluoromethyl)phosphorane, T-00086
Tetrahexadecanoylphosphatidylglycerol, *see* T-00247
2,2,4,4-Tetrahydro-1*H*-2,4-benzodiphosphepin, T-00087
2,2,2,3-Tetrahydro-1,2-benzoxaphosphole, T-00088
6,7,8,9-Tetrahydro-5*H*-dibenzo[*f*,*h*][1,5]-diphosphonin, T-00089
4,5,9,10-Tetrahydro-4,9-dihydroxyphosphorino[2,3,4,5-*lmn*]-phosphanthridine 4,9-dioxide, T-00090
1,2,3,4-Tetrahydro-1,4-dimethyl-1,4-benzophospharsenine, T-00091
2,3,4,5-Tetrahydro-1,5-dimethyl-1*H*-1,5-benzophospharsepine, T-00092
1,2,3,4-Tetrahydro-1,4-dimethyl-1,4-benzophospharsinine, *see* T-00091
Tetrahydro-1,7-dimethyl-1*H*,5*H*-[1,3,2]-diazaphospholo[1,2-*a*][1,3,2]-diazaphosphole, T-00093
Tetrahydro-2,6-dimethyl-4*H*-1,3,6,2-dioxazaphosphocine, T-00094
1,2,3,4-Tetrahydro-1,4-dimethyl-1,4-diphenyl-1,4-benzodiphosphorinium diiodide, *in* T-00102
1,2,3,6-Tetrahydro-4,5-dimethyl-1,2-diphenyl-1,2-diphosphinine, *in* T-00095
1,2,3,6-Tetrahydro-4,5-dimethyl-1,2-diphosphinine, *see* T-00095
1,2,3,6-Tetrahydro-4,5-dimethyl-1,2-diphosphorine, T-00095
1,3,4,5-Tetrahydro-4,5-dimethyl-2,5-methano-2*H*-2-benzophosphepin, T-00096
Tetrahydro-2,6-dimethyl-[1,3,2]-oxazaphospholo[2,3-*b*][1,3,2]-oxazaphosphole, T-00097
Tetrahydro-5,5-dimethyl-2-phenoxy-2*H*-1,3,2-oxazaphosphorine, T-00098
3*a*,4,7,7*a*-Tetrahydro-1,8-dimethyl-4,7-phosphinidene-1*H*-phosphindole, T-00099
3*a*,4,7,7*a*-Tetrahydro-5,6-dimethyl-2,3*a*,7*a*,8-tetrakis(trifluoromethyl)-1,2,3-metheno-2*H*-benzodiphosphole, T-00100
2,3,4,5-Tetrahydro-1,5-diphenyl-1*H*-1,5-benzodiphosphepin, T-00101
1,2,3,4-Tetrahydro-1,4-diphenyl-1,4-benzodiphosphorin, T-00102
1,2,3,6-Tetrahydro-1,2-diphenyl-1,2-diphosphinine, *see* T-00103
1,2,3,6-Tetrahydro-1,2-diphenyl-1,2-diphosphorin, T-00103
2,2,2,3-Tetrahydro-3,5-diphenyl-1,3,4,2-oxadiazaphosphole, T-00104
3*a*,4,7,7*a*-Tetrahydro-1,8-diphenyl-4,7-phosphinidene-1*H*-phosphindole, T-00105
Tetrahydro-1*H*,5*H*-[1,2]diphospholo[1,2-*a*]-[1,2]diphosphole, *see* D-01170
3*a*,4,7,7*a*-Tetrahydro-1,2,3,5,6,8-hexamethyl-4,7-phosphinidene-1*H*-phosphindole, T-00106
1,2,5,6-Tetrahydro-1,1,3,6,6,9-hexamethyl-4*H*-phospholo[3,2,1-*i*,*j*]-phosphinolinium iodide, *in* T-00123
5*a*,6,11,11*a*-Tetrahydro-1,2,3,6,11,12-hexaphenyl-6,11-phosphinidene-1*H*-naphth[2,3-*e*]isophosphindole, T-00107
Tetrahydro-2-hydroxy-2*H*-1,3,2-oxazaphosphorine 2-sulfide, T-00108
2,3,4,7-Tetrahydro-1*H*-isophosphindole 2-oxide, T-00109
1,2,3,4-Tetrahydroisophosphinoline, T-00110
2,3,10,11-Tetrahydro-7-methoxy-3-methyl-1*H*-naphtho[1,2-*g*]phosphindole, *see* M-00048
Tetrahydro-2-methoxy-2*H*-1,3,2-oxazaphosphinine, *see* T-00111
Tetrahydro-2-methoxy-2*H*-1,3,2-oxazaphosphorine, T-00111
1,2,3,4-Tetrahydro-1-methoxyphosphinoline 1-oxide, *in* H-00195
3,3*a*,4,5-Tetrahydro-3-methyl-2*H*-benzo[*e*]-phosphindole 3-oxide, T-00112
1,2,3,4-Tetrahydro-4-methyl-2,2-diphenylbenz[*h*]isophosphinolinium(1+), T-00113
2,3,4,5-Tetrahydro-5-methyl-2,2-diphenyl-1*H*-2-benzophosphepinium(1+), T-00114
Tetrahydro-4-methyl-1*H*,5*H*-[1,2]-diphospholo[1,2-*a*][1,2]diphospholium iodide, *in* D-01170
3,4,5,6-Tetrahydro-4-methyl-2*H*-1,5-methano-4,1-benzazaphosphocine, T-00115
Tetrahydro-4-methyl-2*H*-1,3,4,2-oxadiazaphosphinine 2-oxide, *see* T-00116
Tetrahydro-4-methyl-2*H*-1,3,4,2-oxadiazaphosphorine 2-oxide, T-00116
Tetrahydro-3-methyl-2,4,5,6-tetraphenyl-2*H*-1,3,5-oxazaphosphinine, *see* T-00117
Tetrahydro-3-methyl-2,4,5,6-tetraphenyl-2*H*-1,3,5-oxazaphosphorine, T-00117
Tetrahydro-2-methylthio-2*H*-oxazaphosphorine 2-oxide, *in* T-00108
2,3,4,5-Tetrahydro-6-methyl-2,4,4-triphenyl-1,2,3-diazaphosphinine, *see* T-00118
2,3,4,5-Tetrahydro-6-methyl-2,4,4-triphenyl-1,2,3-diazaphosphorine, T-00118
2,2,2,3-Tetrahydro-1,2-oxaphosphole, T-00119
Tetrahydro-2*H*,6*H*-[1.3.2]-oxazaphosphorino[2,3-*b*][1.3.2]-oxazaphosphorine, T-00120
(Tetrahydro-2-oxo-2-furanyl)phosphonic acid, T-00121
1,2,3,4-Tetrahydro-4-oxo-1-phenylphosphinoline, T-00122
1,2,5,6-Tetrahydro-1,1,6,6,9-pentamethyl-4*H*-phospholo[3,2,1-*ij*]phosphinoline, T-00123
2,3,6,7-Tetrahydro-1,1,7,7,9-pentamethyl-1*H*,5*H*-phosphorino[3,2,1-*ij*]-phosphinoline, T-00124
1,1′,3,3′-Tetrahydro-1,1′,2,3,3′-pentamethyl-2,2′-spirobi[2*H*-1,3,2-benzodiazaphosphole], *in* T-00142
2,2,2,3-Tetrahydrophenanthro[9,10-*d*]-1,3,2-oxazaphosphole, T-00125
Tetrahydro-2-phenoxy-2*H*-1,3,2-oxazaphosphorine, T-00126
Tetrahydro-2-phenoxy-3-phenyl-2*H*-1,3,2-oxazaphosphorine, T-00127
1,2,3,4-Tetrahydro-1-phenylbenzo[*h*]-phosphinoline, T-00128
2,2,2,3-Tetrahydro-5-phenyl-3,3-bis(trifluoromethyl)-1,4,2-thiazaphosphole, T-00129
5,6,7,12-Tetrahydro-12-phenyldibenz[*c*,*f*]-[1,5]azaphosphocine, T-00130
1,2,3,4-Tetrahydro-2-phenylisophosphinoline, *in* T-00110
2,2,2,3-Tetrahydro-5-phenyl-1,3,4,2-oxadiazaphosphole, T-00131
Tetrahydro-2-phenyl-2*H*-1,3,2-oxazaphosphorine, T-00132
Tetrahydro-3-phenyl-2-(phenylamino)-2*H*-1,3,2-oxazaphosphorine, T-00133
1,2,3,6-Tetrahydro-1-phenyl-1-phenylmethylphosphininium bromide, *in* T-00134
1,2,3,4-Tetrahydro-4-phenyl-4-phosphaphenanthrene, *see* T-00128
1,2,3,6-Tetrahydro-1-phenylphosphinine, *see* T-00134
1,2,3,6-Tetrahydro-1-phenylphosphorin, T-00134
1,2,3,4-Tetrahydrophosphinoline, T-00135
[[(Tetrahydro-2*H*-pyran-2-yl)oxy]methyl]-phosphonic acid, T-00136
1,1′,3,3′-Tetrahydro-2,2′-spirobi[2*H*-isophosphindolium](1+), T-00137
2,2,7,7-Tetrahydro-3,4,5,6-tetrakis(methoxycarbonyl)-1*H*-1,2,7-azadiphosphepine, T-00138
2,2,7,7-Tetrahydro-3,4,5,6-tetrakis(methoxycarbonyl)-1,2,2,5,5-pentamethyl-1*H*-1,2,7-azadiphosphepine, *in* T-00138
2,2,4,4-Tetrahydro-2,2,4,4-tetramethyl-1*H*-2,4-benzodiphosphepin, *in* T-00087
2,3,4,5-Tetrahydro-2,2,4,4-tetramethyl-1*H*-2,4-benzodiphosphepinium(2+), T-00139
2,3,4,5-Tetrahydro-1,1,5,5-tetramethyl-1*H*-1,5-benzophospharsepinium dibromide, *in* T-00092
1,3*a*,3*b*,6-Tetrahydro-3,4,8,8-tetramethyl-1,6-diphenyl-8*H*-[1,3]diphospholo[1,5-*e*;3,4-*c*′]bis[1,2,3]diazaphosphole, T-00140
3*a*,4,7,7*a*-Tetrahydro-2,3,5,6-tetramethyl-1,8-diphenyl-4,7-phosphinidene-1*H*-phosphindole, T-00141
1,2,3,6-Tetrahydro-1,2,4,5-tetramethyl-1,2-diphosphinine, *in* T-00095
Tetrahydro-5,6-tetramethylene-2-(4-nitrophenoxy)-2*H*-1,3,2-oxazaphosphorine, *see* O-00020
Tetrahydro-*N*,*N*,5,5-tetramethyl-2*H*-1,3,2-oxazaphosphorin-2-amine, *see* D-00726
3*a*,4,7,7*a*-Tetrahydro-1,1,8,8-tetramethyl-4,7-phosphinidene-1*H*-phosphindolium diiodide, *in* T-00099
1,1′,3,3′-Tetrahydro-1,1′,3,3′-tetramethyl-2,2′-spirobi[2*H*-1,3,2-benzodiazaphosphole], T-00142
3,4,5,5-Tetrahydro-2,2,5,5-tetraphenyl-1*H*-1,2,5-azadiphospholium(1+), T-00143
1,2,5,6-Tetrahydro-2,2,5,5-tetraphenyl-2,5-benzodiphocinium dibromide, *in* H-00053
2,2,4,4-Tetrahydro-2,2,4,4-tetraphenyl-1*H*-2,4-benzodiphosphepin, *in* T-00087
2,3,4,5-Tetrahydro-2,2,4,4-tetraphenyl-1*H*-2,4-benzodiphosphonium(2+), T-00144
1,1,3,3-Tetrahydro-1,1,3,3-tetraphenyl-2,4-bis(triphenylphosphonio)-1,3-diphosphete, T-00145
2,2,4,4-Tetrahydro-2,2,4,4-tetraphenyl-1,5,2,4-diazadiphosphinine, *see* T-00146
2,2,4,4-Tetrahydro-2,2,4,4-tetraphenyl-1,5,2,4-diazadiphosphorine, T-00146
5,6,11,12-Tetrahydro-5,6,11,12-tetraphenyldibenzo[*c*,*g*][1,2,5,6]-tetraphosphocin, T-00147
1,1,3,3-Tetrahydro-1,1,3,3-tetraphenyl-1,3-diphosphacyclohepta-1,2-diene, T-00148
1,1,3,3-Tetrahydro-1,1,3,3-tetraphenyl-1,3-diphosphacyclohexa-1,2-diene, T-00149
1,1,3,3-Tetrahydro-1,1,3,3-tetraphenyl-1,3-diphosphacyclopenta-1,2-diene, T-00150
1,2,3,6-Tetrahydro-1,2,4,5-tetraphenyl-1,2-diphosphinine, *see* T-00151
1,2,3,4-Tetrahydro-1,1,4,4-tetraphenyl-1,4-diphosphininium(2+), *see* T-00152
1,2,3,6-Tetrahydro-1,2,4,5-tetraphenyl-1,2-diphosphorine, T-00151
1,2,3,4-Tetrahydro-1,1,4,4-tetraphenyl-1,4-diphosphorinium(2+), T-00152
4,5,5,5-Tetrahydro-3,5,5,5-tetraphenyl-1,2,5-oxazaphosphole, T-00153
Tetrahydro-2*H*-1,3,2-thiazaphosphinine 2-oxide, *see* T-00154
Tetrahydro-2*H*-1,3,2-thiazaphosphorine 2-oxide, T-00154
9,10,15,16-Tetrahydrotribenzo[*b*,*d*,*h*][1,6]-diphosphecin, T-00155
2,2,2,3-Tetrahydro-2,2,2-trimethoxy-3-methyl-4,5-diphenyl-1,3,2-oxazaphosphole, T-00156
2,2,2,3-Tetrahydro-2,2,2-trimethoxy-5-methyl-1,2-oxaphosphole, T-00157
2,2,2,3-Tetrahydro-2,2,2-trimethoxy-5-phenyl-3,3-bis(trifluoromethyl)-1,4,2-oxazaphosphole, T-00158

2,2,2,3-Tetrahydro-2,2,2-trimethoxy-5-phenyl-3,3-bis(trifluoromethyl)-1-(2,4,6-trimethylphenyl)-1*H*-1,4,2-diazaphosphole, T-00159
1,3,4,5-Tetrahydro-2,4,5-trimethyl-2,5-methano-2*H*-2-benzophosphepinium iodide, *in* T-00096
Tetrahydro-1*H*,5*H*-[1,2,3]triphospholo[2,1-*a*][1,2,3]triphosphole, T-00160
2,4,6,8-Tetrahydroxy-1,5-dimethyl-3,7,9-trioxa-2,4,6,8-tetraphosphabicyclo[3.3.1]-nonane, T-00161
1,1,3,7-Tetrahydroxy-7-methyl-2,4-dioxa-1,3-diphosphanonan-9-oic acid 1,3-dioxide, T-00162
Tetraiisopropyl thioperoxydiphosphate, *see* B-00184
Tetraiododimethylaminophosphorane, *see* D-00709
Tetraiodo-*N*,*N*-dimethylphosphoranamine, *see* D-00709
Tetraiodomethylphosphorane, T-00163
Tetraiodophenoxyphosphorane, T-00164
Tetraiodophenylphosphorane, T-00165
Tetraisopropoxydiphosphine, *see* T-00169
Tetraisopropoxydiphosphine dioxide, *see* T-00170
Tetraisopropyl 1,4-butanediylbisphosphonite, *in* B-00515
Tetraisopropyl dichloromethylenebisphosphonate, *in* D-00172
▷Tetraisopropyl diphosphate, T-00166
Tetraisopropyldiphosphine, T-00167
▷*sym*-Tetraisopropyl dithiopyrophosphate, T-00168
Tetraisopropyl 1,2-ethanediylbisphosphoramidate, *in* E-00009
Tetraisopropyl ethenylidenebisphosphonate, *in* E-00020
Tetraisopropyl ethyliminodiphosphate, *in* E-00078
Tetraisopropyl hexamethylenebisphosphonite, *in* H-00086
Tetraisopropyl 1,6-hexanediylbisphosphonite, *in* H-00086
Tetraisopropyl hypodiphosphite, T-00169
Tetraisopropyl hypophosphate, T-00170
Tetraisopropyl methylenediphosphonate, *in* M-00022
Tetraisopropyl methyliminodiphosphate, *in* M-00156
Tetraisopropyl oxybis(1,2′-ethanediyl)-bisphosphonate, *in* O-00094
Tetraisopropyl pentylidenebisphosphonate, *in* P-00031
N,*N*,*N*′,*N*′-Tetraisopropylphosphonic diamide, T-00171
N,*N*,*N*′,*N*′-Tetraisopropylphosphorodiamidite, *see* T-00171
Tetraisopropylphosphorodiamidous acid, T-00172
Tetraisopropyl 2,2-propanediphosphonate, *in* P-00445
▷Tetraisopropyl pyrophosphate, *see* T-00166
Tetraisopropyl tetramethylenediphosphonite, *in* B-00515
Tetraisopropyl [thiobis(methylene)]-bisphosphonate, *in* T-00311
▷Tetraisopropyl thiodiphosphate, *see* T-00168
O,*O*,*O*′,*O*′-Tetraisopropyl thiohypophosphate, T-00173
O,*O*,*O*′,*O*′-Tetraisopropyl trithiopyrophosphate, T-00174
1,1,4,4-Tetrakis(diethylamino)-2,3-diphosphabutadiene, *in* T-00005
N,*N*′,*N*″,*N*‴-Tetrakis(dihydroxyphosphinylmethyl)-1,4,7,10-tetraazacyclododecane, T-00175
1,1,4,4-Tetrakis(dimethylamino)-2,3-diphosphabutadiene, *in* T-00005
1,1,3,3-Tetrakis(dimethylamino)-1,3-diphosph(P^V)ete, T-00176
Tetrakis(dimethylamino)diphosphine, T-00177
Tetrakis(dimethylamino)iodophosphorane, T-00178
N,*N*,1,3-Tetrakis(1,1-dimethylethyl)-4,4-dimethyl-1,3-diaza-2-phosphacyclobutane-2-amine, *in* D-00087
1,2,3,4-Tetrakis(1,1-dimethylethyl)-1,2,3,4-diphosphadiarsetane, *see* T-00013
Tetrakis(1,1-dimethylethyl)diphosphine, *see* T-00015
Tetrakis(1,1-dimethylethyl)phosphonium(1+), *see* T-00020
1,2,4,5-Tetrakis(1,1-dimethylethyl)-1,2,4,5-tetraphosphaspiro[2.2]pentane, *see* T-00026
Tetrakis(1,1-dimethylethyl)tetraphosphetane, *see* T-00012
1,2,4,5-Tetrakis(1,1-dimethylethyl)-1,2,4,5-tetraphosphorinane, *see* T-00027
Tetrakis(diphenylphosphino)allene, *see* T-00180
Tetrakis[(diphenylphosphino)methyl]methane, T-00179
1,1,3,3-Tetrakis(diphenylphosphino)-1,2-propadiene, T-00180
Tetrakis[(diphenylphosphinoselenoyl)-methyl]methane, *in* T-00179
Tetrakis[(diphenylphosphinothioyl)methyl]-methane, *in* T-00179
Tetrakis[(diphenylphosphinyl)methyl]methane, *in* T-00179
Tetrakis(hydroxymethyl)phosphonium(1+), T-00181
1,2,4,5-Tetrakis(methylene)-3-phosphoniaspiro[2.2]pentane, T-00182
▷Tetrakis(1-methylethyl) diphosphate, *see* T-00166
Tetrakis(1-methylethyl)diphosphine, *see* T-00167
Tetrakis(1-methylethyl) ethylimidodiphosphate, *in* E-00078
Tetrakis(1-methylethyl) hypodiphosphite, *see* T-00169
Tetrakis(1-methylethyl) hypophosphate, *see* T-00170
Tetrakis(1-methylethyl) methylenebisphosphonate, *in* M-00022
Tetrakis(1-methylethyl) methylimidodiphosphate, *in* M-00156
N,*N*,*N*′,*N*′-Tetrakis(1-methylethyl)phosphonic diamide, *see* T-00171
Tetrakis(1-methylethyl)phosphorodiamidous acid, *see* T-00172
▷Tetrakis(1-methylethyl) thiodiphosphate, *see* T-00168
Tetrakis(1-methylethyl) thiodiphosphate, *see* T-00174
O,*O*,*O*′,*O*′-Tetrakis(1-methylethyl) thiohypophosphate, *see* T-00173
Tetrakis(1-methylethyl) thioperoxydiphosphate, *see* B-00183
Tetrakis(1-methylethyl) thioperoxydiphosphate, *see* B-00184
Tetrakis-*p*-nitrobenzyl pyrophosphate, *in* B-00367
Tetrakis(4-nitrophenyl) diphosphate, T-00183
Tetrakis[(4-nitrophenyl)methyl] diphosphate, *in* B-00367
Tetrakis(pentafluorophenyl)-cyclotetraphosphine, *see* T-00185
Tetrakis(pentafluorophenyl)diphosphine, T-00184
Tetrakis(pentafluorophenyl)tetraphosphetane, T-00185
Tetrakis(phenylmethyl) diphosphate, *see* T-00006
Tetrakis(phenylmethyl)diphosphine, *see* T-00007
Tetrakis(phenylmethyl)phosphonium chloride, *in* T-00338
1,6,11,16-Tetrakis(phenylmethyl)-1,6,11,16-tetraphosphacycloeicosane, *see* T-00008
1,5,10,14-Tetrakis(phenylmethyl)-1,5,10,14-tetraphosphacyclooctadecane, *see* T-00009
Tetrakis(phenylmethyl) thioperoxydiphosphate, *see* B-00155
N,*N*,*N*′,*N*′-Tetrakis(phosphonomethyl)-1,2-ethanediamine, *see* E-00014
▷Tetrakis(trifluoromethyl)-cyclotetraphosphine, *see* T-00188
1,3,4,6-Tetrakis(trifluoromethyl)-2,5-diphosphatricyclo[3.1.0.0^{2,6}]hex-3-ene, T-00186
Tetrakis(trifluoromethyl)ethylene chlorophosphate, *in* C-00184
Tetrakis(trifluoromethyl)ethylene chlorophosphite, *see* C-00184
Tetrakis(trifluoromethyl)ethylene phosphorochloridate, *in* C-00184
Tetrakis(trifluoromethyl)ethylene phosphorochloridite, *see* C-00184
2,2,3,3-Tetrakis(trifluoromethyl)-1,4,6,10-tetraoxa-5-phospha(5-P^V)-spiro[4.5]decane, T-00187
▷Tetrakis(trifluoromethyl)tetraphosphetane, T-00188
2,3,7,8-Tetrakis(trifluoromethyl)-1,4,6,9-tetrathia-5-phospha(5-P^V)-spiro[4.4]nona-2,7-diene, T-00189
2,2,3,3-Tetrakis(trifluoromethyl)-1,4,6-trioxa-10-aza-5-phospha(5-P^V)-spiro[4.5]decane, T-00190
Tetrakis(trimethylsilyloxy)phosphonium iodide, *in* T-00864
▷Tetram, *see* A-00133
Tetramethoxydiphosphine dioxide, *see* T-00204
Tetramethoxyphosphonium hexafluorophosphate, *see* T-00468
Tetramethyl [1-(acetoxy)ethylidene]-bisphosphonate, *in* A-00005
Tetramethyl (aminomethylene)bisphosphonate, *in* A-00086
3,3,4,4-Tetramethyl-1,2-azaphospholidine 2-oxide, T-00191
4,4,5,5-Tetramethyl-1,2-azaphospholidine 2-oxide, T-00192
▷Tetramethylbiphosphine, *see* T-00200
N,*N*,4,4-Tetramethyl-1,3-bis(1,1-dimethylethyl)-1,3-diaza-2-phosphacyclobutane-2-amine, *in* D-00087
5,5,5′,5′-Tetramethyl-2,2′-bis-1,3,2-dioxaphosphorinane, *see* B-00197
5,5,5′,5′-Tetramethyl-2,2′-bis-1,3,2-dioxaphosphorinane 2,2′-dioxide, *in* B-00197
N,*N*,*N*′,*N*′-Tetramethyl-*P*,*P*′-bis(1-methylethyl)diphosphonic diamide, *see* D-00668
1,1,3,3-Tetramethyl-6,7-bis(trifluoromethyl)-5,8-dithia-4-phospha(4-P^V)spiro[3.4]oct-6-ene, T-00193
Tetramethyl 1,4-butanediylbisphosphonite, *in* B-00515
Tetramethyl (2-butene-1,4-diyl)-bisphosphonate, *in* B-00519
Tetramethylcyclodiphosphazane, *see* T-00194
▷*N*,*N*,*N*′,*N*′-Tetramethyl-diaminophosphorochloridite, *see* T-00228
1,2,3,4-Tetramethyl-1,3,2,4-diazadiphosphetidine, T-00194
N,*N*,1,3-Tetramethyl-1,3,2-diazaphospholidin-2-amine, *see* D-00724
Tetramethyl [(dimethylamino)methylene]-bisphosphonate, *in* D-00701
2,2,7,7-Tetramethyl-1,6-dioxa-4,9-diaza-5-phospha(5-P^V)spiro[4.4]nonane, T-00195
3,3,8,8-Tetramethyl-1,6-dioxa-4,9-diaza-5-phospha(P^V)spiro[4.4]nonane, T-00196
4,4,5,5-Tetramethyl-1,3-dioxa-2-phenoxy-2-phosphacyclopentane, *see* P-00075
Tetramethyl-1,3-dioxa-2-phosphacyclopentane, *see* T-00197
N,*N*,5,5-Tetramethyl-1,3,2-dioxaphosphinin-2-amine, *see* D-00693
N,*N*,4,5-Tetramethyl-1,3,2-dioxaphospholan-2-amine, *see* D-00725
4,4,5,5-Tetramethyl-1,3,2-dioxaphospholane, T-00197
N,*N*,5,5-Tetramethyl-1,3,2-dioxaphosphorin-2-amine, *see* D-00693
Tetramethyl (1,4-dioxo-1,4-butanediyl)-bisphosphonate, *in* D-00975
2,4,8,10-Tetramethyl-3,9-dioxo-2,4,8,10-tetraaza-6-phosphoniaspiro[5.5]-undecane(1+), T-00198
3,3′,4,4′-Tetramethyl-1,1′-diphenyl-2,2′-bi-1*H*-phosphole, T-00199
N,*N*,6,9-Tetramethyl-2,3-diphenyl-1,4-dioxa-6,9-diaza-5-phospha(5-P^V)-spiro[4.4]non-2-en-5-amine, *in* D-00741

N,N,N′,N′-Tetramethyl-1′,3′-diphenyldispiro[1,3,2-benzodioxophosphole-2,2′-[1,3,2,4]-diazaphosphetidine-4′,2″-[1,3,2]-benzodioxaphosphol]-2,2′-diamine, *in* D-01018
2,2,6,9-Tetramethyl-4,11-diphenyl-4,5,10,11-tetraaza-1,3-diphosphatricyclo[6.3.0.0^{3,7}]undeca-5,9-diene, *see* T-00140
▷Tetramethyl diphosphate, *see* T-00229
5,5,7,7-Tetramethyl[*d,f*][1,λ^{5}3]-diphosphepinium bromide, *in* D-00577
▷Tetramethyldiphosphine, T-00200
Tetramethyldiphosphine disulfide, T-00201
2,4,8,10-Tetramethyl-3,9-dithia-6-phosphoniaspiro[5.5]undecane, T-00202
▷*sym-O,O,O,O*-Tetramethyl dithiopyrophosphate, T-00203
Tetramethylenebis[dibutylphosphine], *see* B-00157
Tetramethylenebis[dicyclohexylphosphine], *see* B-00166
Tetramethylenebis[dimethylphosphine], *see* B-00199
Tetramethylenebisphosphine, *see* D-01174
Tetramethylenebis[triphenylphosphonium](2+), *see* B-00517
Tetramethylene *tert*-butylphosphonite, *in* D-00968
Tetramethylenediphosphonic acid, *see* B-00514
P,P-Tetramethylenediphosphonous acid, *see* B-00515
Tetramethylene hydrogen phosphate, *see* H-00132
Tetramethylene methylphosphonate, *in* M-00125
Tetramethylene methylphosphonite, *see* M-00125
O,O-Tetramethylene methylphosphonothioate, *in* M-00125
5,6-Tetramethylene-2-(4-nitrophenoxy)-1,3,2-oxazaphosphorinane, *see* O-00020
Tetramethylene phenyl phosphite, *in* D-00968
Tetramethylene phenylphosphonate, *in* P-00114
Tetramethylene phenylphosphonite, *see* P-00114
O,O-Tetramethylene phenylphosphonothioate, *in* P-00114
O,O-Tetramethylene *O*-phenyl phosphorothioate, *in* D-00969
Tetramethylenephosphine, *see* P-00371
Tetramethylenephosphinic acid, *see* H-00185
Tetramethylenephosphinic chloride, *in* C-00167
Tetramethylenephosphinothioic chloride, *in* C-00167
O,O′-Tetramethylene phosphonothioate, *see* D-00969
Tetramethylene phosphoric acid, *see* H-00132
Tetramethylene phosphorochloridite, *in* D-00968
Tetramethylene phosphorofluoridite, *in* D-00968
Tetramethyl 1,2-ethanediylbisphosphonate, *in* E-00006
Tetramethyl ethenylidenebisphosphonate, *in* E-00020
Tetramethylethylene chlorophosphate, *in* C-00185
Tetramethylethylene chlorophosphite, *see* C-00185
Tetramethylethylene chlorothiophosphate, *in* C-00185
4,4,5,5-Tetramethylethylene dimethylphosphoramidate, *in* D-00710
4,4,5,5-Tetramethylethylene dimethylphosphoramidite, *see* D-00710
4,4,5,5-Tetramethylethylene dimethylphosphoramidothioate, *in* D-00710
O,O-Tetramethylethylene dithiophosphate, *see* M-00019
O,O-Tetramethylethylene hydrogen phosphorodithioate, *see* M-00019
O,O-Tetramethylethylene hydrogen phosphorothioate, *see* H-00197
4,4,5,5-Tetramethylethylene phenyl phosphite, *see* P-00075
Tetramethylethylene phenylphosphonate, *in* P-00322
Tetramethylethylene phenylphosphonite, *see* P-00322
Tetramethylethylene phosphite, *in* T-00197
Tetramethylethylene phosphonate, *in* T-00197
Tetramethylethylene phosphonite, *see* T-00197
O,O-Tetramethylethylene phosphonothioate, *in* T-00197
Tetramethylethylene phosphoryl chloride, *in* C-00185
O,O-Tetramethylethylene thiophosphate, *see* H-00197
O,O-Tetramethylethylene thiophosphite, *in* T-00197
Tetramethylethylene thiophosphoryl chloride, *in* C-00185
3,7,11,15-Tetramethyl-2,6,10,14-hexadecatetren-1-ol trihydrogen pyrophosphate, *see* M-00515
Tetramethyl (3,3,4,4,5,5-hexafluoro-1-cyclopentene-1,2-diyl)bisphosphonate, *in* H-00018
Tetramethyl hexamethylenediphosphonite, *in* H-00086
Tetramethyl 1,6-hexanediylbisphosphonite, *in* H-00086
Tetramethyl (1-hydroxyethylidene)-bisphosphonate, *in* H-00144
Tetramethyl (hydroxyphenylmethyl)-bisphosphonate, *in* H-00172
Tetramethyl hypophosphate, T-00204
N,N,N′,N′-Tetramethyl-*P*-(2-methyl-1,3-butadienyl)phosphonic diamide, *in* M-00096
N,N,N′,N′-Tetramethyl-*P*-(2-methyl-1,3-butadienyl)phosphonothioic diamide, *in* M-00097
Tetramethyl methylenebisphosphonate, *in* M-00022
Tetramethyl methylenebisphosphonite, *in* M-00024
O,O,O′,O′-Tetramethyl *S,S′*-methylene bisphosphorodithioate, T-00205
N,N,5,5-Tetramethyl-4-(1-methylethyl)-1,3,2-dioxaphosphorinan-2-amine, *see* D-00699
Tetramethyl methylimidodiphosphate, *in* M-00156
Tetramethyl methylimidodiphosphite, *in* M-00157
O,O,O,O-Tetramethyl methylimidodiphosphorothioate, *in* M-00161
Tetramethyl methyliminodiphosphate, *in* M-00156
O,O,O,O-Tetramethyl methylthioimidodiphosphate, *in* M-00161
▷*O,O,O,O*-Tetramethyl monothiopyrophosphate, *see* T-00207
sym-Tetramethyl monothiopyrophosphate, T-00206
▷*unsym*-Tetramethyl monothiopyrophosphate, T-00207
N,N,N′,N′-Tetramethyl-*P*-4-morpholinylphosphonothioic diamide, *in* M-00528
N,N,N′,N′-Tetramethyl-*P*-4-morpholinylphosphonous diamide, *in* M-00529
3,3,5,5-Tetramethyl-1,2-oxaphospholidine 2-oxide, T-00208
N,N,N′,N′-Tetramethyl-*P*-(2-oxo-2-phenylethyl)phosphonic diamide, *in* O-00066
Tetramethyl (2-oxo-1,3-propanediyl)-bisphosphonate, *in* O-00072
4,4,5,5-Tetramethyl-2-phenyl-1,3-dioxa-2-phosphacyclopentane, *see* P-00322
Tetramethyl [1,4-phenylenebis(methylene)]-bisphosphonate, *in* P-00131
Tetramethyl 1,2-phenylenebisphosphonate, *in* B-00006
Tetramethyl 1,3-phenylenebisphosphonate, *in* B-00007
Tetramethyl 1,4-phenylenebisphosphonate, *in* B-00008
Tetramethyl 1,4-phenylenebisphosphonite, *in* B-00010
N,N,N′,N′-Tetramethyl-*P*-(2-phenylethenyl)-phosphonic diamide, *in* P-00156
N,N,N′,N-Tetramethyl-*P*-(phenylmethyl)-phosphonic diamide, *in* B-00061
N,N,N′,N′-Tetramethyl-*N″*-(phenylmethyl)-phosphoric triamide, *in* T-00216
N,N,3,4-Tetramethyl-5-phenyl-1,3,2-oxazaphospholidin-2-amine, *see* D-00694
2,2,5,5-Tetramethyl-1-phenyl-1-phospha-2,5-disilacyclopentane, T-00209
4,4,5,5-Tetramethyl phenyl phosphate, *in* P-00075
2,2,6,6-Tetramethyl-1-phenyl-4-phosphinanone, *see* T-00210
N,N,N′,N′-Tetramethyl-*P*-phenylphosphonoselenoic diamide, *in* P-00243
2,2,6,6-Tetramethyl-1-phenyl-4-phosphorinanone, T-00210
N,N,N′,N′-Tetramethyl-2-phenylphosphorodiamidic hydrazide, *in* P-00274
2,7,10,15-Tetramethyl-17-phenyl-17*H*-tetrabenzo[*b,d,f,h*]phosphonin, T-00211
1,1,3,3-Tetramethyl-1-phospha-3-silacyclohex-1-ene, T-00212
Tetramethylphosphazene, *in* T-00531
Tetramethylphosphine imide, *in* T-00531
Tetramethylphosphinic amide, *in* D-00830
N,N,P,P-Tetramethylphosphinimidic amide, *in* D-00833
N,N′,P,P-Tetramethylphosphinimidic amide, *in* D-00833
Tetramethylphosphinoselenoic amide, *in* D-00836
Tetramethylphosphinothioic amide, *in* D-00840
Tetramethylphosphinous amide, T-00213
1,1,3,4-Tetramethyl-1*H*-phospholium iodide, *in* T-00536
N,N,N′,N′-Tetramethylphosphonic diamide, T-00214
Tetramethylphosphonium(1+), T-00215
N,N,N′,N′-Tetramethylphosphoric triamide, T-00216
N,N,N′,N″-Tetramethylphosphoric triamide, *in* D-00880
Tetramethyl phosphor(isothiocyanatidic) diamide, *in* T-00218
Tetramethyl phosphor(isothiocyanatido)-thioic diamide, T-00217
Tetramethylphosphorodiamidic acid, T-00218
Tetramethylphosphorodiamidic azide, T-00219
Tetramethylphosphorodiamidic chloride, T-00220
Tetramethylphosphorodiamidic cyanide, T-00221
▷Tetramethylphosphorodiamidic fluoride, *see* D-00669
Tetramethylphosphorodiamidodithioic acid, T-00222
Tetramethylphosphorodiamidoselenoic acid, T-00223
Tetramethyl phosphorodiamidothioic anhydride, *see* O-00034
Tetramethylphosphorodiamidothioic azide, T-00224
Tetramethylphosphorodiamidothioic chloride, T-00225
▷Tetramethylphosphorodiamidothioic fluoride, T-00226
Tetramethylphosphorodiamidous acid, T-00227
N,N,N′,N′-Tetramethylphosphorodiamidous acid, *see* T-00214
▷Tetramethylphosphorodiamidous chloride, T-00228
N,N,N′,N′-Tetramethyl-*P*-1-piperidinylphosphonous diamide, *in* P-00432
N,N,N′,N′-Tetramethyl-*P*-1,2-propadienylphosphonic diamide, *in* P-00442
N,N,N′,N′-Tetramethyl-*P*-2-propenylphosphonic diamide, *in* P-00455
N,N,N′,N′-Tetramethyl-*P*-(2-propenyl)-phosphonoselenoic diamide, *in* P-00459
N,N,N′,N′-Tetramethyl-*P*-2-propenylphosphonothioic diamide, *in* P-00462
N,N,N′,N′-Tetramethyl-*P*-2-propenylphosphonous diamide, *in* P-00463

N,N,N′,N′-Tetramethyl-*N″*-(2-propenyl)-phosphoric triamide, *in* T-00216
N,N,N′,N′-Tetramethyl-*P*-propylphosphonic diamide, *in* P-00474
N,N,N′,N′-Tetramethyl-*P*-1-propynylphosphonic diamide, *in* P-00494
▷Tetramethyl pyrophosphate, T-00229
N,N,5,6-Tetramethylspiro[1,3,2-benzodioxaphosphole-2,1′(3′*H*)-[2,1]-benzoxaphosphol]-2-amine, *in* D-00915
4′,4′,5′,5′-Tetramethylspiro[1,3,2-benzodioxaphosphole-2,2′-[1,3,2]-dioxaphospholane], T-00230
4′,4′,5′,5′-Tetramethylspiro[1,3,2-benzoxazaphosphole-2(3*H*),2′-[1,3,2]-dioxaphospholane], T-00231
3,3,3′,3′-Tetramethyl-1,1′(3*H*,3′*H*)-spirobi[2,1-benzoxaphosphole], T-00232
3,3,3′,3′-Tetramethyl-1,1′(3*H*,3′*H*)-spirobi[2,1-benzoxaphospholium](1+), T-00233
N,N,1,3-Tetramethylspiro[1,3,2-diazaphospholidine-2,2′-phenanthro[9,10-*d*][1,3,2]-dioxaphosphol]-2-amine, *in* D-00918
1,3,5,7-Tetramethyl-1,3,5,7-tetraaza-4-phospha(4-P^{V})spiro[3.3]heptane-2,6-dione, T-00234
Tetramethyl (3,3,4,4-tetrafluoro-1-cyclobutene-1,2-diyl)bisphosphonate, *in* T-00078
1,1,4,4-Tetramethyl-1,2,3,4-tetrahydro-1,4-benzophospharseninium dibromide, *in* T-00091
6,9,*N,N*-Tetramethyl-2,2,3,3-tetrakis(trifluoromethyl)-1,4-dioxa-6,9-diaza-5-phospha(P^{V})spiro[4.4]-nonan-5-amine, *in* D-00922
Tetramethyl tetramethylenediphosphonite, *in* B-00515
2,2,3,3-Tetramethyl-1,4,6,9-tetraoxa-5-phospha(5-P^{V})spiro[4.4]nonane, T-00235
N,N,2,3-Tetramethyl-1,4,6,9-tetraoxa-5-phospha(5-P^{V})spiro[4.4]non-2-en-5-amine, *in* D-00924
3,3,9,9-Tetramethyl-1,5,7,11-tetraoxa-6-phospha(6-P^{V})spiro[5.5]undecane, T-00236
▷*O,O,O′,O′*-Tetramethyl *O,O′*-thiodi-*p*-phenylene bis(phosphorothioate), *see* T-00003
Tetramethyl thiodiphosphate, *see* T-00206
Tetramethyl thiodiphosphate, *see* T-00240
▷*O,O,O,O*-Tetramethyl thiodiphosphate, *see* T-00203
▷*O,O,O,O*-Tetramethyl thiodiphosphate, *see* T-00207
▷Tetramethyl thioperoxydiphosphate, *see* B-00185
Tetramethyl thioperoxydiphosphate, *see* B-00186
2,4,6,7-Tetramethyl-2,6,7-triaza-1-phosphabicyclo[2.2.2]octane, T-00237
N,N′,N″,N″-Tetramethyl-*N,N′*-trimethylenephosphoric triamide, *in* H-00023
N,N,P,P-Tetramethyl-*N′*-trimethylsilylphosphinimidic amide, *in* D-00833
2,2,3,3-Tetramethyl-1,4,6-trioxa-9-aza-5-phospha(5-P^{V})spiro[4.4]nonane, T-00238
Tetramethyl 1,2,5-triphenyl-5-phospha(5-P^{V})spiro[4.4]nona-1,3,6,8-tetraene-1,2,3,4-tetracarboxylate, *in* T-00622
3,3,5,5-Tetramethyl-1,2,4-triphenyl-1,2,4-triphospholane, T-00239
O,O,O′,O′-Tetramethyl trithiopyrophosphate, T-00240
Tetramyristoyldiphosphatidylglycerol, T-00241
Tetra(*p*-nitrophenyl) pyrophosphate, *see* T-00183
Tetra-9-octadecenoylphosphatidylglycerol, T-00242
Tetraoleoylphosphatidylglycerol, *in* T-00242
1,4,6,12-Tetraoxa-5-phospha(5-P^{V})-spiro[4.7]dodecane, T-00243
1,4,6,9-Tetraoxa-5-phospha(5-P^{V})-spiro[4.4]nonane, T-00244
1,4,6,11-Tetraoxa-5-phospha(5-P^{V})-spiro[4.6]undecane, T-00245
Tetrapalmitoyldiphosphatidylglycerol, T-00246
Tetrapalmitoylphosphatidylglycerol, T-00247
Tetraphenoxydiphosphine disulfide, *see* T-00283
Tetraphenoxyphosphonium hexafluorophosphate, *see* T-00470
Tetraphenoxyphosphonium iodide, *see* I-00020
1,1,7,7-Tetraphenyl-4-aza-1,7-diphosphaheptane, *see* B-00231
Tetraphenylbiphosphine, *see* T-00258
2,2′,6,6′-Tetraphenyl-4,4′-biphosphinine, *see* T-00249
2,2′,5,5′-Tetraphenyl-1*H*,1′*H*-biphosphole, T-00248
2,2′,6,6′-Tetraphenyl-4,4′-biphosphorine, T-00249
3,5,8,10-Tetraphenyl-1*H*,5*H*-bisphospholo[1,2-*a*:1′,2′-*d*][1,4]-diphosphorin, T-00250
2,3,5,6-Tetraphenyl-1,4-bis[(trimethylsilyl)oxy]-2,3,5,6-tetraphosphabicyclo[2.2.0]hexane, T-00251
1,2,3,4-Tetraphenylcyclo-5-carba-1,2,3,4-tetraphosphane, *see* T-00280
Tetraphenylcyclodiphosphazane, *see* T-00253
(Tetraphenyl-2,4-cyclopentadien-1-ylidene)triphenylphosphorane, T-00252
Tetraphenylcyclotetraphosphine, *see* T-00278
1,2,3,4-Tetraphenyl-1,3,2,4-diazadiphosphetidine, T-00253
1,2,4,5-Tetraphenyl-Δ^{3}-1,3,2-diazaphospholine, *see* D-00580
1,2,4,5-Tetraphenyl-3,6-dicarba-1,2,4,5-tetraphosphan, *see* T-00281
2,4,5,6-Tetraphenyl-1,3-dioxa-5-phosphacyclohexane, *see* T-00254
2,4,5,6-Tetraphenyl-1,3,5-dioxaphosphinane, *see* T-00254
2,4,5,6-Tetraphenyl-1,3,5-dioxaphosphorinane, T-00254
4,7,13,16-Tetraphenyl-1,10-dioxa-4,7,13,16-tetraphosphacyclooctadecane, T-00255
Tetraphenyl diphosphate, T-00256
Tetraphenyldiphosphazane, *see* D-01063
1,1,3,3-Tetraphenyl-1,3-diphosphepanium, T-00257
1,1,3,3-Tetraphenyl-1,3-diphosphinanium(2+), *see* T-00262
1,1,4,4-Tetraphenyl-1,4-diphosphinanium dibromide, *in* D-01017
Tetraphenyldiphosphine, T-00258
Tetraphenyldiphosphine dioxide, T-00259
Tetraphenyldiphosphine disulfide, T-00260
1,1,5,5-Tetraphenyl-1,5-diphosphocanium(2+), T-00261
1,1,3,3-Tetraphenyl-1,3-diphospholanium dibromide, *in* D-01013
Tetraphenyl-1,4-diphosphonia-6,7-benzocycloocta-2,6-diene, *in* H-00053
Tetraphenyl-1,4-diphosphonia-6,7-benzocyclooct-6-ene, *in* H-00053
1,1,5,5-Tetraphenyl-1,5-diphosphoniacyclooctane, *see* T-00261
1,1,4,4-Tetraphenyl-1,4-diphosphoninanium(2+), *see* T-00263
1,1,3,3-Tetraphenyl-1,3-diphosphorinanium(2+), T-00262
1,1,4,4-Tetraphenyl-1,4-diphosphorinanium(2+), T-00263
1,1,4,4-Tetraphenyl-1,4-diphosphorinanium dibromide, *in* D-01017
4,7,13,16-Tetraphenyl-1,10-dipropyl-1,10-diaza-4,7,13,16-tetraphosphacyclooctadecane, T-00264
Tetraphenyldithioimidophosphinic acid, *see* D-01087
Tetraphenyl 1,2-ethanediylbisphosphonate, *in* E-00006
Tetraphenyl 1,2-ethanediylbisphosphonite, *in* E-00008
Tetraphenyl 1,2-ethanediylbisphosphoramidate, *in* E-00009
O,O,O,O-Tetraphenyl 1,2-ethanediylbisphosphoramidothioate, *in* E-00010
Tetraphenyl ethylimidodiphosphite, *in* E-00079
O,O,O′,O′-Tetraphenyl imidodiphosphate, T-00265
O,O,O′,O′-Tetraphenyl iminodiphosphate, *see* T-00265
Tetraphenyl methylenebisphosphonate, *in* M-00022
Tetraphenyl methylenebisphosphonite, *in* M-00024
Tetraphenyl methylimidodiphosphate, *in* M-00156
Tetraphenyl methylimidodiphosphite, *in* M-00157
Tetraphenyl methyliminodiphosphate, *in* M-00156
3,3,6,6-Tetraphenyl-1,3,6-oxadiphosphepanium(2+), T-00266
3,3,5,5-Tetraphenyl-1,3,5-oxadiphosphinanium(2+), *see* T-00267
3,3,6,6-Tetraphenyl-1-oxa-3,6-diphosphoniacycloheptane, *see* T-00266
3,3,5,5-Tetraphenyl-1-oxa-3,5-diphosphoniacyclohexane(2+), *see* T-00267
3,3,5,5-Tetraphenyl-1,3,5-oxadiphosphorinanium(2+), T-00267
2,3,3,4-Tetraphenyl-1,2-oxaphosphetane 2-oxide, T-00268
2,4,4,6-Tetraphenyl-4*H*-1,4-oxaphosphorinium bromide, *in* T-00610
Tetraphenyl pentamethylenediphosphonite, *in* P-00033
Tetraphenyl 1,5-pentanediylbisphosphonite, *in* P-00033
Tetraphenylphenoxyphosphorane, *see* P-00076
Tetraphenyl [1,4-phenylenebis(methylene)]-bisphosphonate, *in* P-00131
Tetraphenyl 1,3-phenylenebisphosphoramidate, *in* B-00011
Tetraphenyl 1,4-phenylenebisphosphoramidate, *in* B-00012
Tetraphenylphosphine imide, T-00269
Tetraphenylphosphinic amide, *in* D-01040
▷*N,N′,P,P*-Tetraphenylphosphinimidic amide, *in* T-00631
2,4,4,6-Tetraphenyl-4*H*-4-phosphoniapyran bromide, *in* T-00610
1,2,3,4-Tetraphenyl-5-phosphoniaspiro[4.5]deca-1,3-diene, T-00270
N,N,N′,N′-Tetraphenylphosphonic diamide, T-00271
Tetraphenylphosphonium(1+), T-00272
N,N,N′,N′-Tetraphenylphosphonothioic diamide, T-00273
N,N′,N″,N″-Tetraphenylphosphoric triamide, *in* D-01108
N,N,N′,N′-Tetraphenylphosphorodiamidite, *see* T-00271
Tetraphenylphosphorodiamidous acid, T-00274
O,O,O′,O′-Tetraphenyl 1,3-propanediphosphoramidothioate, *in* P-00446
Tetraphenyl pyrophosphate, *see* T-00256
O,O,O′,O′-Tetraphenyl selenoperoxydiphosphate, *see* B-00269
2,3,7,8-Tetraphenyl-1,4,6,9-tetraoxa-5-phospha(5-P^{V})spiro[4.4]nona-2,7-diene, T-00275
1,5,9,13-Tetraphenyl-1,5,9,13-tetraphosphacyclohexadecene, T-00276
1,2,4,5-Tetraphenyl-1,2,4,5-tetraphosphacyclohexane, *see* T-00281
1,3,5,7-Tetraphenyl-1,3,5,7-tetraphosphacyclooctane, *see* T-00279
1,4,6,9-Tetraphenyl-1,4,6,9-tetraphospha-5-silaspiro[4.4]nonane, T-00277
2,6,13,17-Tetraphenyl-2,6,13,17-tetraphosphatricyclo[16.4.0.$0^{7,12}$]-docosa-7(12),8,10,1(18),19,21-hexaene, *see* D-00006
Tetraphenyltetraphosphetane, T-00278
1,2,4,5-Tetraphenyl-1,2,4,5-tetraphosphinane, *see* T-00281
1,3,5,7-Tetraphenyl-1,3,5,7-tetraphosphocane, T-00279
1,2,3,4-Tetraphenyltetraphospholane, T-00280
1,2,4,5-Tetraphenyl-1,2,4,5-tetraphosphorinane, T-00281

Tetraphenylthiatetraphospholane, T-00282
Tetraphenyl thiodiphosphate, *see* T-00284
O,O,O′,O′-Tetraphenyl thiohypophosphate, T-00283
O,O,O′,O′-Tetraphenyl thioimidodiphosphate, *see* D-01004
O,O,O′,O′-Tetraphenyl thioimidodiphosphate, *see* D-01005
Tetraphenyl thioperoxydiphosphate, *see* B-00211
1,2,3,4-Tetraphenyl-5-(triphenylphosphoranylidene)-1,3-cyclopentadiene, *see* T-00252
O,O,O′,O′-Tetraphenyl trithiopyrophosphate, T-00284
Tetraphos, *see* H-00091
Tetraphos-2, *see* T-00774
1,2,5,6-Tetraphosphabicyclo[3.3.0]octane, *see* T-00160
1,2,3,4-Tetraphosphonobutane, *see* B-00518
closo-Tetraphosphorus hexakis(methylimide), *see* H-00071
Tetrapropoxydiphosphine, *see* T-00287
Tetrapropoxydiphosphine dioxide, *see* T-00288
Tetrapropoxydiphosphine disulfide, *see* T-00173
Tetrapropoxydiphosphine disulfide, *see* T-00293
▷Tetrapropyl diphosphate, T-00285
Tetrapropyldiphosphine, T-00286
Tetrapropyl diphosphite, *see* T-00292
Tetrapropyl dithiohypophosphate, *see* T-00293
▷*O,O,O′,O′*-Tetrapropyl dithiopyrophosphate, *see* A-00144
Tetrapropyl 1,2-ethanediylbisphosphoramidate, *in* E-00009
O,O,O,O-Tetrapropyl 1,2-ethanediylbisphosphoramidothioate, *in* E-00010
Tetrapropyl hypodiphosphite, T-00287
Tetrapropyl hypophosphate, T-00288
Tetrapropyl [1,4-phenylenebis(methylene)]-bisphosphonate, *in* P-00131
N,N,N′,N′-Tetrapropylphosphonic diamide, T-00289
Tetrapropylphosphonium(1+), T-00290
Tetrapropylphosphorodiamidic acid, T-00291
N,N,N′,N′-Tetrapropylphosphorodiamidite, *see* T-00289
▷Tetrapropyl pyrophosphate, *see* T-00285
Tetrapropyl pyrophosphite, T-00292
▷Tetrapropyl thiodiphosphate, *see* A-00144
O,O,O′,O′-Tetrapropyl thiohypophosphate, T-00293
Tetrastearoyldiphosphatidylglycerol, T-00294
Tetrastearoylphosphatidylglycerol, T-00295
Tetravinylphosphonium(1+), T-00296
Thalassemine, *in* L-00002
2-Thenylphosphonic acid, *see* T-00304
3-Thenylphosphonic acid, *see* T-00305
▷Thiamine diphosphate, T-00297
1,4-Thiaphosphinane 1,1,4-trioxide, *see* T-00300
4*H*-1,4-Thiaphosphinin, *see* T-00299
1,2-Thiaphospholane 2-sulfide, T-00298
4*H*-1,4-Thiaphosphorin, T-00299
1,4-Thiaphosphorinane 1,1,4-trioxide, T-00300
2-Thia-1,3,5-triaza-7-phosphaadamantane, *see* T-00301
2-Thia-1,3,5-triaza-7-phosphatricyclo[3.3.1.1^{3,7}]decane, T-00301
1,3,2-Thiazaphospholidine 2-oxide, T-00302
1,3,2-Thiazaphospholidine 2-sulfide, T-00303
(2-Thienylmethyl)phosphonic acid, T-00304
(3-Thienylmethyl)phosphonic acid, T-00305
2-Thienylphosphinic acid, T-00306
2-Thienylphosphonic acid, T-00307
2-Thienylphosphonous acid, T-00308
▷Thimet, *see* P-00338
2,2′-Thiobis[5,5-dimethyl-1,3,2-dioxaphosphorinane], T-00309
2,2′-Thiobis[5,5-dimethyl-1,3,2-dioxaphosphorinane] 2,2′-dioxide, T-00310
Thiobis[methanephosphonic acid], T-00311
[Thiobis(methylene)]bis(phosphonic acid), *see* T-00311
[Thiobis[methylene]]-bis[triphenylphosphonium](2+), T-00312
[Thiobis[methylene]]-bis[triphenylphosphorane], *in* T-00312
Thiodimethylenephosphinic acid, *see* H-00201
Thiodimethylenephosphinous chloride, *see* C-00188
Thiodimethylenephosphinous fluoride, *see* F-00050
▷*O,O′*-(Thiodi-4,1-phenylene) bis(*O,O*-dimethyl phosphorothioate), *see* T-00003
▷Thiohypochlorous acid anhydrosulfide with bis(1-methylethyl) phosphorothioate, *see* D-00634
▷Thiohypochlorous acid, anhydrosulfide with *O,O*-bis(1-methylethyl) dithioate, *see* D-00635
▷Thiohypochlorous acid anhydrosulfide with dibutyl phosphorothioate, *see* D-00080
▷Thiohypochlorous acid anhydrosulfide with diethyl phosphorothioate, *see* D-00274
▷Thiohypochlorous acid anhydrosulfide with *O,O*-diethyl phosphorodithioate, *see* D-00275
▷Thiohypochlorous acid anhydrosulfide with *O,O*-diethyl phosphorothioate, *see* D-00716
▷Thiohypochlorous acid anhydrosulfide with dipropyl phosphorothioate, *see* D-01200
▷Thiohypochlorous acid anhydrosulfide with *O,O*-dipropyl phosphorodithioate, *see* D-01201
Thiohypophosphoric acid cyclic *P,P,P′,P′*-bis(2,2-dimethyltrimethylene) ester, *in* B-00197
Thiohypophosphoric acid cyclic *P,P,P′,P′*-bis(2,2-dimethyltrimethylene) ester, *in* B-00197
▷Thiometon, T-00313
▷Thionazin, T-00314
[2,5-Thiophenediylbis[methylene]]-bis[triphenylphosphonium](2+), *see* B-00456
[3,4-Thiophenediylbis[methylene]]-bis[triphenylphosphonium](1+), *see* B-00457
[2,5-Thiophenediylbis[methylene]]-bis[triphenylphosphorane], *in* B-00456
[3,4-Thiophenediylbis[methylene]]-bis[triphenylphosphorane], *in* B-00457
2-Thiophenephosphonous acid, *see* T-00308
Thiophosphonic acid bis(diphenylamide), *see* T-00273
Thiophosphonic bis(diethylamide), *see* T-00057
Thiophosphoric acid dianilide, *see* D-01129
Thiophosphoric acid dibenzylamide, *see* D-00062
Thiophosphoric acid *O*-ethyl ester-fluoride-amide, *see* E-00158
Thiophosphoric acid *O*-methyl ester-fluoride-amide, *see* M-00331
Thiophosphoric acid monoanilide, *see* P-00262
Thiophosphoric diethylamide, *see* D-00356
Thiophosphoric dimethylamide, *see* D-00871
Thiophosphoryl isocyanate, T-00315
Thiophosphoryl isothiocyanate, T-00316
▷Thiotepa, T-00317
Thioxo[2,4,6-tris(1,1-dimethylethyl)-phenyl]phosphine sulfide, *see* D-01254
Threonine 2-aminoethyl hydrogen phosphate, *see* T-00318
Threonine dihydrogen phosphate, *see* A-00120
Threonine ethanolamine phosphate, T-00318
Threonine phosphate, *see* A-00120
Thuringiensin, T-00319
Thymidine diphosphate, *see* T-00320
Thymidine 5′-pyrophosphate, T-00320
Thymidine 5′-(trihydrogen diphosphate), *see* T-00320
▷Tifosyl, *see* T-00317
Tinox, *in* D-00017
Toclofos-Methyl, T-00321
▷TOCP, *see* T-00803
▷Tokuthion, *see* P-00498
Toldimfos, T-00322
Tolfamide, T-00323
p-Toluenesulfonylphosphorimidic acid, *see* M-00268
N-p-Toluenesulfonyl-*P,P,P*-triphenoxyphosphazene, *in* M-00268
N-p-Toluenesulfonyl-*P,P,P*-triphenoxyphosphine imide, *in* M-00268
m-Tolyl dichlorophosphate, *in* M-00486
o-Tolyl dichlorophosphate, *in* M-00485
m-Tolyl dichlorophosphite, *see* M-00259
o-Tolyl dichlorophosphite, *see* M-00258
p-Tolyl dichlorophosphite, *see* M-00260
o-Tolyl difluorophosphate, *in* M-00485
p-Tolyl difluorophosphate, *in* M-00487
m-Tolyl dihydrogen phosphate, *see* M-00486
o-Tolyl dihydrogen phosphate, *see* M-00485
p-Tolyl dihydrogen phosphate, *see* M-00487
m-Tolylphosphine, *see* M-00217
o-Tolylphosphine, *see* M-00216
p-Tolylphosphine, *see* M-00218
m-Tolylphosphinic acid, *see* M-00221
o-Tolylphosphinic acid, *see* M-00220
▷*p*-Tolylphosphinic acid, *see* M-00222
o-Tolyl phosphodifluoridate, *in* M-00485
m-Tolylphosphonic acid, *see* M-00236
o-Tolylphosphonic acid, *see* M-00235
▷*p*-Tolylphosphonic acid, *see* M-00237
P-Tolylphosphonic dianilide, *in* M-00238
p-Tolylphosphonic dichloride, *see* M-00239
m-Tolylphosphonous acid, *see* M-00247
o-Tolylphosphonous acid, *see* M-00246
p-Tolylphosphonous acid, *see* M-00248
m-Tolylphosphoramidic acid, *see* M-00252
o-Tolylphosphoramidic acid, *see* M-00251
p-Tolylphosphoramidic acid, *see* M-00253
m-Tolylphosphoramidothioic acid, *see* M-00255
o-Tolylphosphoramidothioic acid, *see* M-00254
p-Tolylphosphoramidothioic acid, *see* M-00256
m-Tolyl phosphorodichloridate, *in* M-00486
o-Tolyl phosphorodichloridate, *in* M-00485
p-Tolyl phosphorodichloridate, *see* M-00257
m-Tolyl phosphorodichloridite, *see* M-00259
o-Tolyl phosphorodichloridite, *see* M-00258
p-Tolyl phosphorodichloridite, *see* M-00260
O-o-Tolyl phosphorodichloridothioate, *see* M-00261
O-p-Tolyl phosphorodichloridothioate, *see* M-00262
p-Tolyl phosphorodifluoridate, *in* M-00487
p-Tolyl phosphoryl dichloride, *see* M-00257
p-Tolyl phosphoryl difluoride, *in* M-00487
[(*p*-Tolylsulfinyl)methyl]phosphonic acid, *see* M-00266
Tonofosfan, *in* T-00322
Tonophosphan, *in* T-00322
▷Torak, *see* D-00023
▷Torelle, *see* F-00063
Tosylphosphorimidic acid, *see* M-00268
TPN, *see* C-00203
Trehalose 6-(dihydrogen phosphate), T-00324
Triacetyl phosphate, T-00325
Tri(allylamino)phosphine, *see* T-00709
Triallylmethylphosphonium iodide, *in* T-00707
▷Triallyl phosphate, *see* T-00705
Triallylphosphine, *see* T-00707
▷Triallyl phosphite, *see* T-00708
N,N′,N″-Triallylphosphorous triamide, *see* T-00709
Triaminobromopentapropylphosphonium bromide, *in* P-00043
Tri(2-aminophenyl) phosphite, *see* T-00728
Tri(3-aminophenyl) phosphite, *see* T-00729
▷Triamiphos, T-00326
Triamyl phosphate, *see* T-00593
Triamylphosphine, *see* T-00594
Triamyl phosphite, *see* T-00595
P,P,P-Trianilino-*N*-(diphenoxyphosphinothioyl)phosphine imide, *in* D-00987
Trianilinophosphine, *see* T-00677
Tri-*m*-anisylmethylphosphonium iodide, *in* T-00792
Tri-*o*-anisylmethylphosphonium iodide, *in* T-00791
Tri-*m*-anisylphosphine, *see* T-00792
Tri-*o*-anisylphosphine, *see* T-00791
Tri-*o*-anisyl phosphite, *see* T-00794
1,3,5-Triaza-7-phosphaadamantane, *see* T-00327

1,3,5-Triaza-7-phosphatricyclo[3.3.1.1^{3,7}]decane, T-00327
Tri(1-azetidinyl)phosphine, T-00328
1,3,5-Triazine-2,4,6-triphosphonic acid, T-00329
1,3,5-Triazine-2,4,6-triphosphoramidic acid, T-00330
1,3,5-Triazine-2,4,6-triyltriphosphoramidic acid, *see* T-00330
1,3,5-Triazine-2,4,6-triyltrisphosphonic acid, *see* T-00329
▷Tri-(1-aziridinyl)phosphine, T-00331
1*H*-1,2,4-Triazole-1-phosphonous acid, *see* T-00332
(1*H*-1,2,4-Triazol-1-yl)phosphonous acid, T-00332
▷Triazophos, T-00333
Tri(2-benzothiazolyl)phosphine, T-00334
Tribenzoylphosphine, T-00335
Tribenzoyl phosphite, T-00336
Tri(benzylamino)phosphine, *see* T-00344
P,P,P-Tribenzyl-*N*-methylphosphine imide, *in* T-00339
Tribenzylmethylphosphonium iodide, *in* T-00338
Tribenzyl phosphate, T-00337
P,P,P-Tribenzylphosphazene, *see* T-00339
Tribenzylphosphine, T-00338
P,P,P-Tribenzylphosphine imide, T-00339
P,P,P-Tribenzylphosphine imine, *see* T-00339
Tribenzyl phosphite, T-00340
Tribenzyl phosphonoglycinate, *in* P-00368
Tribenzyl *N*-phosphonophenylalaninate, *in* P-00400
N,N′,N″-Tribenzylphosphoric triamide, *in* T-00344
Tribenzyl phosphorotetrathioate, T-00341
N,N′,N″-Tribenzylphosphorothioic triamide, *in* T-00344
S,S,S-Tribenzyl phosphorotrithioate, T-00342
Tribenzyl phosphorotrithioite, T-00343
N,N′,N″-Tribenzylphosphorous triamide, T-00344
Tribenzyl tetrathiophosphate, *see* T-00341
S,S,S-Tribenzyl trithiophosphate, *see* T-00342
Tribenzyl trithiophosphite, *see* T-00343
Tri-2-biphenylyl phosphate, T-00345
Tri-3-biphenylyl phosphate, T-00346
Tri-4-biphenylyl phosphate, T-00347
Tri-2-biphenylylphosphine, T-00348
Tri-3-biphenylylphosphine, T-00349
Tri-4-biphenylylphosphine, T-00350
Tri([1,1-biphenyl]-2-yl) phosphite, T-00351
P,P,P-Tribromo-*N*-*tert*-butyliminophosphorane, *in* B-00625
P,P,P-Tribromo-*N*-*tert*-butylphosphazene, *in* B-00625
P,P,P-Tribromo-*N*-*tert*-butylphosphine imide, *in* B-00625
2,2,2-Tribromo-2,2-dihydro-1,3,2-benzodioxaphosphole, T-00352
2,2,2-Tribromo-2,2-dihydro-4,4,5,5-tetrakis(trifluoromethyl)-1,3,2-dioxaphospholane, T-00353
Tribromodiphenoxyphosphorane, T-00354
Tribromodiphenylphosphorane, T-00355
1,1′-[(2,2,2-Tribromoethoxy)-phosphinylidene]bis[oxy]bis(1*H*-benzotriazole)], *see* B-00100
2,2,2-Tribromoethyl 4-*tert*-butylphenylphosphoramidic chloride, T-00356
2,2,2-Tribromoethyl dichlorophosphate, *in* M-00516
2,2,2-Tribromoethyl 4-(1,1-dimethylethyl)-phenylphosphoramidic chloride, *see* T-00356
2,2,2-Tribromoethyl 4-morpholinylphosphonochloridate, T-00357
2,2,2-Tribromoethyl phenylphosphoramidochloridate, *in* P-00259
Tribromoethyl phosphoric acid, *see* M-00516
2,2,2-Tribromoethyl phosphorodichloridate, *in* M-00516
2,2,2-Tribromoethyl phosphoryl dichloride, *in* M-00516
2,4,6-Tribromo-2,2,4,4,6,6-hexahydro-2,4,6-triphenylcyclotriphosphazene, *see* T-00358
2,4,6-Tribromo-2,2,4,4,6,6-hexahydro-2,4,6-triphenyl-1,3,5-triaza-2,4,6-triphosphorine, T-00358
(2,4,6-Tribromophenyl)phosphoramidic acid, T-00359
P,P,P-Tribromo-*N*-phenylsulfonylphosphazene, *in* P-00315
P,P,P-Tribromo-*N*-phenylsulfonylphosphine imide, *in* P-00315
P,P,P-Tribromo-*N*-*p*-toluenesulfonylphosphazene, *in* M-00268
P,P,P-Tribromo-*N*-*p*-toluenesulfonylphosphine imide, *in* M-00268
Tri-3-butenylphosphine, T-00360
Tri-2-butenyl phosphite, T-00361
P,P,P-Tributoxy-*N*-[bis(dimethylamino)-phosphinothioyl]phosphazene, *in* B-00192
P,P,P-Tributoxy-*N*-[bis(dimethylamino)-phosphinothioyl]phosphine imide, *in* B-00192
Tributoxycarbonylphosphine, *in* P-00355
P,P,P-Tributoxy-*N*-(dimethoxyphosphinothioyl)phosphazene, *in* D-00674
P,P,P-Tributoxy-*N*-(dimethoxyphosphinothioyl)phosphine imide, *in* D-00674
Tri(butylamino)phosphine, *see* T-00394
Tri(*sec*-butylamino)phosphine, *see* T-00829
Tri(*tert*-butylamino)phosphine oxide, *in* T-00395
Tributyl [bis(dimethylamino)-phosphinothioyl]phosphorimidate, *in* B-00192
Tri-*tert*-butylcyclotriphosphine, *see* T-00397
1,2,3-Tri-*tert*-butyldiazaphosphiridine, T-00362
Tributyl(2,4-dichlorobenzyl)phosphonium, *see* C-00192
Tributyl[(2,4-dichlorophenyl)methyl]-phosphonium, *see* C-00192
Tributyl (dimethoxyphosphinothioyl)-phosphorimidate, *in* D-00674
Tri-*tert*-butyldiphospharsirane, T-00363
O,O,S-Tributyl dithiophosphate, *see* T-00384
Tri-*tert*-butylethylphosphonium bromide, *in* T-00373
Tributyl(hexadecyl)phosphonium(1+), T-00364
1,1,1-Tributyl-1-iodo-*N*-methyl-*N*-phenylphosphoranamine, *in* T-00366
Tri-*tert*-butylisopropylphosphonium bromide, *in* T-00373
Tributyl(methylbenzenaminato)-phosphorus(1+), *see* T-00366
Tri-*tert*-butylmethylenephosphorane, T-00365
Tributyl(methylphenylamino)-phosphonium(1+), T-00366
Tri-*tert*-butylmethylphosphonium bromide, *in* T-00373
▷Tributylmethylphosphonium iodide, *in* T-00372
Tri-*tert*-butylmethylphosphonium iodide, *in* T-00373
1-(2,4,6-Tri-*tert*-butylphenyl)-4,4-diphenyl-1-phospha-1,2,3-butatriene, T-00367
2,4,6-Tri-*tert*-butylphenylmetadiselenophosphonate, *see* D-01235
2,4,6-Tri-*tert*-butylphenylmetadithiophosphonate, *see* D-01254
(2,4,6-Tri-*tert*-butylphenyl)-methylidynephosphine, T-00368
(2,4,6-Tri-*tert*-butylphenyl)phosphine, T-00369
(2,4,6-Tri-*tert*-butylphenyl)phosphinic acid, *in* T-00369
(2,4,6-Tri-*tert*-butylphenyl)phosphinothioic acid, *in* T-00369
▷Tributyl phosphate, T-00370
Tri-*sec*-butyl phosphate, *see* T-00821
Tri-*tert*-butyl phosphate, T-00371
P,P,P-Tributylphosphazene, *see* T-00374
P,P,P-Tri-*tert*-butylphosphazene, *see* T-00375
▷Tributylphosphine, T-00372
Tri-*sec*-butylphosphine, *see* T-00823
Tri-*tert*-butylphosphine, T-00373
P,P,P-Tributylphosphine imide, T-00374
P,P,P-Tri-*tert*-butylphosphine imide, T-00375
P,P,P-Tributylphosphine imine, *see* T-00374
▷Tributylphosphine oxide, T-00376
Tributylphosphine selenide, T-00377
▷Tributylphosphine sulfide, T-00378
▷Tributyl phosphite, T-00379
Tri-*sec*-butyl phosphite, *see* T-00825
Tri-*tert*-butyl phosphite, T-00380
Tributyl phosphonoacetate, *in* D-00073
1-(Tributylphosphoranylidene)-2-heptanone, T-00381
N,N′,N″-Tributylphosphoric triamide, *in* T-00394
2,4,6-Tri-*tert*-butylphosphorin, T-00382
O,Se,Se-Tributyl phosphorodiselenoite, T-00383
O,Se,Se-Tributyl phosphorodiselenothioate, *in* D-00149
O,O,S-Tributyl phosphorodithioate, T-00384
O,O,O-Tributyl phosphoroselenoate, T-00385
Tributyl phosphorotetrathioate, T-00386
O,O,O-Tributyl phosphorothioate, T-00387
O,O,S-Tributyl phosphorothioate, T-00388
N,N′,N″-Tributylphosphorothioic triamide, *in* T-00394
O,Se,Se-Tributyl phosphorotriselenoate, T-00389
O,S,S-Tributyl phosphorotrithioate, T-00390
S,S,S-Tri-*tert*-butyl phosphorotrithioate, T-00391
▷*S,S,S*-Tributyl phosphorotrithioate, T-00392
▷Tributyl phosphorotrithioite, T-00393
N,N′,N″-Tributylphosphorous triamide, T-00394
N,N′,N″-Tri-*tert*-butylphosphorous triamide, T-00395
O,O,O-Tributyl selenophosphate, *see* T-00385
Tributyl tetrathiophosphate, *see* T-00386
O,O,O-Tributyl thiophosphate, *see* T-00387
O,O,S-Tributyl thiophosphate, *see* T-00388
P,P,P-Tributyl-*N*-trimethylsilylphosphine imide, *in* T-00374
P,P,P-Tri-*tert*-butyl-*N*-trimethylsilylphosphine imide, *in* T-00375
2,4,6-Tributyl-1,3,5,2,4,6-trioxatriphosphinane, *see* T-00396
2,4,6-Tributyl-1,3,5,2,4,6-trioxatriphosphorinane, T-00396
Tri-*tert*-butyltriphosphirane, T-00397
O,S,S-Tributyl trithiophosphate, *see* T-00390
▷*S,S,S*-Tributyl trithiophosphate, *see* T-00392
S,S,S-Tri-*tert*-butyl trithiophosphate, *see* T-00391
▷Tributyl trithiophosphite, *see* T-00393
(Trichloroacetyl)phosphorimidic trichloride, T-00398
P,P,P-Trichloro-*N*-[bis(trichloromethyl)-phosphinyl]iminophosphorane, *see* B-00404
P,P,P-Trichloro-*N*-[bis(trichloromethyl)-phosphinyl]phosphazene, *see* B-00404
P,P,P-Trichloro-*N*-[bis(trichloromethyl)-phosphinyl]phosphine imide, *see* B-00404
Trichlorobis(trichloromethyl)phosphorane, T-00399
▷2,4,5-Trichloro-α-(chloromethylene)benzyl dimethyl phosphate, *see* T-00038
2,2,2-Trichloro-2,2-dihydro-1,3,2-benzodioxaphosphole, T-00400
2,2,2-Trichloro-2,2-dihydro-1,3-dimethyl-6,8-bis(trifluoromethyl)-1,3,5,7,9-pentaaza-2,4-diphospha(4-P^V)-spiro[3.5]nona-4,6,8-triene, T-00401
2,2,2-Trichloro-2,2-dihydro-5-phenyl-1,3,4,2-dioxazaphosphole, *in* C-00126
▷[2,2,2-Trichloro-1-(dimethoxyphosphinyl)-ethyl] butanoate, *see* B-00530
4,4,4-Trichloro-1,3-dimethyl-4′,6′-bis(trifluoromethyl)spiro[1,3,2(P^V),-4(P^V)-diazaphosphetidine-2,2′-[1,3,5,2(P^V)-triazaphosphorin], *see* T-00401
Trichlorodimethylphosphorane, T-00402
P,P,P-Trichloro-*N*-(diphenoxyphosphinothioyl)phosphazene, *in* D-00987

P,P,P-Trichloro-N-(diphenoxyphosphinylthioyl)phosphine imide, in D-00987
Trichlorodiphenoxyphosphorane, T-00403
Trichlorodiphenylphosphorane, T-00404
2,2,2-Trichloroethanol dihydrogen phosphate, see T-00424
2,2,2-Trichloroethyl 2-chlorophenyl phosphorochloridate, T-00405
2,2,2-Trichloroethyl dichlorophosphite, see T-00406
2,2,2-Trichloroethyl phenylphosphoramidochloridate, in P-00259
Tri(2-chloroethyl) phosphite, see T-00742
2,2,2-Trichloroethyl phosphorodichloridate, in T-00424
2,2,2-Trichloroethyl phosphorodichloridite, T-00406
2,2,2-Trichloroethyl phosphoryl dichloride, in T-00424
2,4,6-Trichloro-2,2,4,4,6,6-hexahydro-2,4,6-triphenylcyclotriphosphazene, see T-00407
2,4,6-Trichloro-2,2,4,4,6,6-hexahydro-2,4,6-triphenyl-1,3,5-triaza-2,4,6-triphosphorine, T-00407
(2,2,2-Trichloro-1-hydroxyethyl)phosphonic acid, T-00408
P-(Trichloromethyl)-N,N'-dicyclohexylphosphonic diamide, in T-00412
P-(Trichloromethyl)-N,N'-dimethylphosphonic diamide, in T-00412
P-(Trichloromethyl)-N,N'-diphenylphosphonic diamide, in T-00412
(4-Trichloromethylphenyl)phosphonic acid, T-00409
P,P,P-Trichloro-N-methylphosphazene, in M-00336
P,P,P-Trichloro-N-methylphosphine imide, in M-00336
P-Trichloromethylphosphonamidic acid, T-00410
(Trichloromethyl)phosphonic acid, T-00411
(Trichloromethyl)phosphonic bis(cyclohexylamide), in T-00412
(Trichloromethyl)phosphonic bis(diethylamide), in T-00412
(Trichloromethyl)phosphonic bis(methylamide), in T-00412
P-(Trichloromethyl)phosphonic diamide, T-00412
(Trichloromethyl)phosphonic dianilide, in T-00412
(Trichloromethyl)phosphonic dichloride, T-00413
(Trichloromethyl)phosphonic difluoride, T-00414
Trichloromethylphosphonium chloride, see T-00032
(Trichloromethyl)phosphonothioic acid, T-00415
(Trichloromethyl)phosphonous acid, T-00416
Trichloromethylphosphonous dichloride, T-00417
(Trichloromethyl)phosphonyl difluoride, see T-00414
(Trichloromethyl)phosphorimidic trichloride, T-00418
P,P,P-Trichloro-N-methylsulfonylphosphazene, in M-00402
P-(Trichloromethyl)-N,N,N',N'-tetraethylphosphonic diamide, in T-00412
▷Trichloronate, T-00419
P,P,P-Trichloro-N-(3-nitrophenyl)-iminophosphorane, in N-00062
P,P,P-Trichloro-N-(2-nitrophenyl)-phosphazene, in N-00061
P,P,P-Trichloro-N-(3-nitrophenyl)-phosphazene, in N-00062
P,P,P-Trichloro-N-(4-nitrophenyl)-phosphazene, in N-00063
P,P,P-Trichloro-N-(2-nitrophenyl)phosphine imide, in N-00061
P,P,P-Trichloro-N-(3-nitrophenyl)phosphine imide, in N-00062
P,P,P-Trichloro-N-(4-nitrophenyl)phosphine imide, in N-00063
P,P,P-Trichloro-N-(pentachloroethyl)-iminophosphorane, see P-00005
P,P,P-Trichloro-N-(pentachloroethyl)-phosphazene, see P-00005
P,P,P-Trichloro-N-(pentachloroethyl)-phosphine imide, see P-00005
P,P,P-Trichloro-N-(pentafluoroethyl)-iminophosphorane, see P-00013
P,P,P-Trichloro-N-(pentafluoroethyl)-phosphazene, see P-00013
P,P,P-Trichloro-N-(pentafluoroethyl)-phosphine imide, see P-00013
P,P,P-Trichloro-N-(pentafluorophenyl)-phosphazene, in P-00016
P,P,P-Trichloro-N-(pentafluorophenyl)-phosphine imide, in P-00016
P,P,P-Trichloro-N-phenyliminophosphorane, in P-00264
P,P,P-Trichloro-N-phenylphosphazene, in P-00264
P,P,P-Trichloro-N-phenylphosphine imide, in P-00264
Tri(2-chlorophenyl) phosphite, see T-00750
Tri(4-chlorophenyl) phosphite, see T-00751
(2,4,6-Trichlorophenyl)phosphoramidic acid, T-00420
P,P,P-Trichloro-N-phenylsulfonyliminophosphorane, see P-00316
P,P,P-Trichloro-N-phenylsulfonylphosphazene, see P-00316
P,P,P-Trichloro-N-phenylsulfonylphosphine imide, see P-00316
2,2,2-Trichloro-1-phosphonoethanol, see T-00408
▷O-(3,5,6-Trichloro-2-pyridinyl) O,O-dimethyl phosphorothioate, see C-00195
P,P,P-Trichloro-N-p-toluenesulfonylphosphazene, in M-00268
P,P,P-Trichloro-N-p-toluenesulfonylphosphine imide, in M-00268
P,P,P-Trichloro-N-(trichloroacetyl)-iminophosphorane, see T-00398
P,P,P-Trichloro-N-(trichloroacetyl)-phosphazene, see T-00398
P,P,P-Trichloro-N-(trichloroacetyl)-phosphine imide, see T-00398
P,P,P-Trichloro-N-(trichloromethyl)-iminophosphorane, see T-00418
P,P,P-Trichloro-N-(trichloromethyl)-phosphazene, see T-00418
P,P,P-Trichloro-N-(trichloromethyl)-phosphine imide, see T-00418
P,P,P-Trichloro-N-trimethylsilylphosphazene, in T-00557
P,P,P-Trichloro-N-trimethylsilylphosphine imide, in T-00557
2,2,4-Trichloro-4,6,6-tris(dimethylamino)-2,2,4,4,6,6-hexahydrocyclotriphosphazene, see T-00421
2,4,6-Trichloro-2,4,6-tris(dimethylamino)-2,2,4,4,6,6-hexahydrocyclotriphosphazene, see T-00422
2,2,4-Trichloro-4,6,6-tris(dimethylamino)-2,2,4,4,6,6-hexahydro-1,3,5-triaza-2,4,6-triphosphorine, T-00421
2,4,6-Trichloro-2,4,6-tris(dimethylamino)-2,2,4,4,6,6-hexahydro-1,3,5-triaza-2,4,6-triphosphorine, T-00422
▷Trichlorphon, T-00423
Triclofos, T-00424
▷Triclofos sodium, in T-00424
▷Tricloryl, in T-00424
▷Triclos, in T-00424
Tri-m-cresyl phosphate, see T-00804
▷Tri-o-cresyl phosphate, see T-00803
Tri-p-cresyl phosphate, see T-00805
▷Tri-m-cresyl phosphite, see T-00810
▷Tri-o-cresyl phosphite, see T-00809
▷Tri-p-cresyl phosphite, see T-00811
Tricrotyl phosphite, see T-00361
Tri(4-cyanophenyl) phosphite, see T-00755
(Tricyclo[3.3.1.1^{3,7}]dec-1-yl)phosphonic acid, see A-00034
(Tricyclo[3.3.1.1^{3,7}]dec-2-yl)phosphonic acid, see A-00035
Tricyclo[3.3.1.1.^{3,7}]dec-1-ylphosphoramidic acid, see A-00036
Tri(cyclohexylamino)phosphine, see T-00429
P,P,P-Tricyclohexyl-N-methylphosphine imide, in T-00427
Tricyclohexylmethylphosphonium iodide, in T-00426
Tricyclohexyl phosphate, T-00425
P,P,P-Tricyclohexylphosphazene, see T-00427
Tricyclohexylphosphine, T-00426
P,P,P-Tricyclohexylphosphine imide, T-00427
P,P,P-Tricyclohexylphosphine imine, see T-00427
Tricyclohexyl phosphite, T-00428
N,N',N''-Tricyclohexylphosphoric triamide, in T-00429
▷N,N',N''-Tricyclohexylphosphorothioic triamide, in T-00429
N,N',N''-Tricyclohexylphosphorous triamide, T-00429
2,4,6-Tricyclohexyl-1,3,5,2,4,6-trioxatriphosphinane, see T-00430
2,4,6-Tricyclohexyl-1,3,5,2,4,6-trioxatriphosphorinane, T-00430
Tricyclopentylphosphine, T-00431
Tricyclopropylmethylphosphonium iodide, in T-00432
Tricyclopropylphosphine, T-00432
Tri(decyl)phosphine, T-00433
Tridecyl phosphite, T-00434
Tri(4-dimethylaminophenyl) phosphite, see T-00766
Tri(2,2-dimethylpropyl) phosphite, see T-00772
Tridodecylphosphine, T-00435
Triethenylmethylphosphonium iodide, in T-00877
Triethenylphosphine, see T-00877
P,P,P-Triethoxy-N-[bis(diethylamino)-phosphinyl]phosphazene, in B-00176
P,P,P-Triethoxy-N-[bis(diethylamino)-phosphinyl]phosphine imide, in B-00176
P,P,P-Triethoxy-N-[bis(dimethylamino)-phosphinothioyl]phosphazene, in B-00192
P,P,P-Triethoxy-N-[bis(dimethylamino)-phosphinothioyl]phosphine imide, in B-00192
P,P,P-Triethoxy-N-[bis(dimethylamino)-phosphinyl]phosphazene, in B-00194
P,P,P-Triethoxy-N-[bis(dimethylamino)-phosphinyl]phosphine imide, in B-00194
Triethoxycarbonylphosphine, in P-00355
▷P,P,P-Triethoxy-N-(diethoxyphosphinothioyl)phosphazene, in D-00238
2,2,2-Triethoxy-2,2-dihydro-1,3,2-dioxaphosphinane, see T-00437
2,2,2-Triethoxy-2,2-dihydro-1,3,2-dioxaphospholane, T-00436
2,2,2-Triethoxy-2,2-dihydro-1,3,2-dioxaphosphorinane, T-00437
P,P,P-Triethoxy-N-(dimethoxyphosphinothioyl)phosphazene, in D-00674
P,P,P-Triethoxy-N-(dimethoxyphosphinothioyl)phosphine imide, in D-00674
P,P,P-Triethoxy-N-(dimethoxyphosphinyl)-phosphazene, in D-00683
P,P,P-Triethoxy-N-(dimethoxyphosphinyl)-phosphine imide, in D-00683
P,P,P-Triethoxy-N-(diphenoxyphosphinothioyl)phosphazene, in D-00987
P,P,P-Triethoxy-N-(diphenoxyphosphinothioyl)phosphine imide, in D-00987
P,P,P-Triethoxy-N-ethylphosphazene, in E-00161
P,P,P-Triethoxy-N-ethylphosphine imide, in E-00161
P,P,P-Triethoxy-N-methylphosphine imide, in M-00336
P,P,P-Triethoxy-N-methylsulfonylphosphazene, in M-00402
P,P,P-Triethoxy-N-methylsulfonylphosphine imide, in M-00402
P,P,P-Triethoxy-N-(2-nitrophenyl)-phosphazene, in N-00061

P,P,P-Triethoxy-*N*-(4-nitrophenyl)-phosphazene, *in* N-00063
P,P,P-Triethoxy-*N*-(2-nitrophenyl)phosphine imide, *in* N-00061
P,P,P-Triethoxy-*N*-(4-nitrophenyl)phosphine imide, *in* N-00063
P,P,P-Triethoxy-*N*-pentafluorophenylphosphazine, *in* P-00016
P,P,P-Triethoxy-*N*-(pentafluorophenyl)-phosphine imide, *in* P-00016
P,P,P-Triethoxy-*N*-phenylphosphazene, *in* P-00264
P,P,P-Triethoxy-*N*-phenylphosphine imide, *in* P-00264
P,P,P-Triethoxy-*N*-phenylsulfonylphosphazene, *in* P-00315
P,P,P-Triethoxy-*N*-phenylsulfonylphosphine imide, *in* P-00315
P,P,P-Triethoxy-*N*-*p*-toluenesulfonylphosphazene, *in* M-00268
P,P,P-Triethoxy-*N*-*p*-toluenesulfonylphosphine imide, *in* M-00268
P,P,P-Triethoxy-*N*-trimethylsilylphosphazene, *in* T-00557
P,P,P-Triethoxy-*N*-trimethylsilylphosphine imide, *in* T-00557
Triethylamine-2,2′,2″-triphosphonic acid, T-00438
Tri(ethylamino)phosphine, *see* T-00462
Tri(ethylamino)phosphine oxide, *in* T-00462
Tri(ethylamino)phosphine sulfide, *in* T-00462
Triethyl (azidosulfonyl)phosphorimidate, T-00439
Triethyl benzoylphosphorimidate, *in* B-00039
Triethyl [bis(diethylamino)phosphinyl]-phosphorimidate, *in* B-00176
Triethyl [bis(dimethylamino)-phosphinothioyl]phosphorimidate, *in* B-00192
Triethyl [bis(dimethylamino)phosphinyl]-phosphorimidate, *in* B-00194
Triethyl cyanophosphonoacetate, *see* E-00057
▷Triethyl (diethoxyphosphinothioyl)-phosphorimidate, *in* D-00238
Triethyl (diethoxyphosphinyl)-phosphorimidate, *in* D-00257
Triethyl (dimethoxyphosphinothioyl)-phosphorimidate, *in* D-00674
Triethyl (dimethoxyphosphinyl)-phosphorimidate, *in* D-00683
Triethyl (diphenoxyphosphinothioyl)-phosphorimidate, *in* D-00987
P,P,P-Triethyl-*N*-(diphenylphosphino)-phosphine imide, *in* T-00442
O,O,S-Triethyl dithiophosphate, *see* T-00451
▷Triethylenephosphoric triamide, *in* T-00331
▷*N,N′,N″*-Triethylenethiophosphoramide, *see* T-00317
Triethyl ethylphosphorimidate, *in* E-00161
▷Tri(2-ethylhexyl) phosphite, *see* T-00780
Triethyl methylimido phosphate, *in* M-00336
Triethyl [(4-methylphenyl)sulfonyl]-phosphorimidate, *in* M-00268
Triethylmethylphosphonium iodide, *in* T-00441
Triethyl methylphosphorimidate, *in* M-00336
Triethyl methylsulfonylphosphorimidate, *in* M-00402
Triethyl (2-nitrophenyl)phosphorimidate, *in* N-00061
Triethyl (4-nitrophenyl)phosphorimidate, *in* N-00063
▷Triethyl 3-oxo-2-phosphonobutanoate, *in* D-00253
Triethyl 3-oxo-4-phosphonobutanoate, *in* D-00254
Triethyl pentafluorophenylphosphorimidate, *in* P-00016
N,N,P-Triethyl-*P*-phenylphosphinous amide, *in* E-00108
Triethyl phenylphosphorimidate, *in* P-00264
Triethyl phenylsulfonylphosphorimidate, *in* P-00315
▷Triethyl phosphate, T-00440
P,P,P-Triethylphosphazene, *see* T-00442
▷Triethylphosphine, T-00441
P,P,P-Triethylphosphine imide, T-00442
P,P,P-Triethylphosphine imine, *see* T-00442
Triethylphosphine oxide, T-00443
Triethylphosphine sulfide, T-00444
N,N,P-Triethylphosphinic amide, *in* E-00124
N,P,P-Triethylphosphinic amide, *in* D-00323
▷Triethyl phosphite, T-00445
Triethyl phosphoenolpyruvate, *in* D-00256
Triethyl phospholactate, *in* D-00255
Triethylphosphonamidic acid, T-00446
Triethylphosphonamidodithioic acid, T-00447
Triethylphosphonamidothioic acid, T-00448
Triethylphosphoniophosphoryl chloride, *see* D-00197
▷Triethyl phosphonoacetate, *in* D-00241
Triethyl 2-phosphonoacrylate, *in* D-00260
Triethyl 3-phosphonoacrylate, *in* D-00261
Triethyl *N*-phosphono-2-aminobenzoate, *in* C-00004
Triethyl *N*-phosphonoanthranilate, *in* C-00004
Triethyl 3-(*N*-phosphono)benzoate, *in* C-00005
Triethyl 2-phosphonobutanoate, *in* D-00246
Triethyl 4-phosphonobutanoate, *in* D-00247
Triethyl 4-phosphonocrotonate, *in* D-00248
Triethyl 2-phosphono-2-cyanoacetate, *see* E-00057
Triethyl phosphonoformate, *in* D-00236
Triethyl phosphonoglycinate, *in* P-00368
Triethyl *O*-phosphonolactate, *in* D-00255
Triethyl 2-phosphono-2-phenylacetate, *in* D-00242
Triethyl 3-phosphonopropanoate, *in* D-00259
Triethyl 2-phosphonopropionate, *see* E-00059
Triethyl 2-phosphonoxy-2-propenoate, *in* D-00256
N,N,N′-Triethylphosphoramidimidic dichloride, T-00449
Triethyl phosphoramidofluoride, *in* D-00351
N,N′,N″-Triethylphosphoric triamide, *in* T-00462
▷*O,S,S*-Triethyl phosphorodithioate, T-00450
O,O,S-Triethyl phosphorodithioate, T-00451
Triethyl phosphoro(dithioperoxoate), T-00452
Triethyl phosphoro(dithioperoxo)thioate, T-00453
O,O,O-Triethyl phosphoroselenoate, T-00454
O,O,Se-Triethyl phosphoroselenoate, T-00455
Triethyl phosphorotetrathionate, T-00456
▷*O,O,S*-Triethyl phosphorothioate, T-00457
▷*O,O,O*-Triethyl phosphorothioate, T-00458
N,N′,N″-Triethylphosphorothioic triamide, *in* T-00462
▷*O,S,S*-Triethyl phosphorotrithioate, T-00459
▷*S,S,S*-Triethyl phosphorotrithioate, T-00460
▷Triethyl phosphorotrithioite, T-00461
N,N′,N″-Triethylphosphorous triamide, T-00462
Triethyl selenophosphate, *see* T-00454
O,O,Se-Triethyl selenophosphate, *see* T-00455
Triethylsilanol 3:1 phosphate, *see* T-00849
Triethyl tetrathiophosphate, *see* T-00456
Tri(ethylthio)methylphosphonium iodide, *in* T-00461
▷*O,O,O*-Triethyl thiophosphate, *see* T-00458
▷*O,O,S*-Triethyl thiophosphate, *see* T-00457
P,P,P-Triethyl-*N*-trimethylsilylphosphine imide, *in* T-00442
Triethyl (trimethylsilyl)phosphorimidate, *in* T-00557
2,4,6-Triethyl-1,3,5,2,4,6-trioxatriphosphinane, *see* T-00463
2,4,6-Triethyl-1,3,5,2,4,6-trioxatriphosphorinane, T-00463
Triethyl [[(triphenylphosphoranylidene)-amino]sulfonyl]phosphorimidate, *in* T-00642
▷*O,S,S*-Triethyl trithiophosphate, *see* T-00459
▷*S,S,S*-Triethyl trithiophosphate, *see* T-00460
▷Triethyl trithiophosphite, *see* T-00461
Triferrocenylphosphine, T-00464
Trifluoroacetic acid anhydride with diethyl phosphate, *see* D-00406
Trifluoroacetic-diethyl phosphoric anhydride, *see* D-00406
Trifluorobis(trifluoromethyl)phosphorane, T-00465
1,1,1-Trifluoro-1,1-dihydrophosphinane, *in* D-00573
1,1,1-Trifluoro-1,1-dihydrophospholane, T-00466
1,1,1-Trifluoro-1,1-dihydrophosphorinane, *in* D-00573
2,2,2-Trifluoro-2,2-dihydro-4,4,5,5-tetrakis(trifluoromethyl)-1,3,2-dioxaphospholane, T-00467
Trifluorodimethoxyphosphorane, T-00468
Trifluorodimethylphosphorane, T-00469
Trifluorodiphenoxyphosphorane, T-00470
Trifluorodiphenylphosphorane, T-00471
Trifluoromethanethiol phosphorodichloridite, T-00472
Trifluoromethyl bis(trifluoromethyl)-phosphinoselenoite, *in* B-00413
Trifluoromethyl bis(trifluoromethyl)-phosphinothioite, *in* B-00417
(Trifluoromethyl)diiodophosphine, *in* T-00482
Trifluoromethyl dimethylphosphinoselenoite, *in* D-00838
Trifluoromethyl dimethylphosphinothioite, *in* D-00843
(4-Trifluoromethylphenyl)phosphonic acid, T-00473
P,P,P-Trifluoro-*N*-methylphosphazene, *in* M-00336
(Trifluoromethyl)phosphine, T-00474
P,P,P-Trifluoro-*N*-methylphosphine imide, *in* M-00336
(Trifluoromethyl)phosphinic acid, T-00475
(Trifluoromethyl)phosphinous bromide, T-00476
(Trifluoromethyl)phosphinous chloride, T-00477
(Trifluoromethyl)phosphonic acid, T-00478
(Trifluoromethyl)phosphonic dichloride, T-00479
(Trifluoromethyl)phosphonodithious acid, T-00480
(Trifluoromethyl)phosphonothioic acid, T-00481
(Trifluoromethyl)phosphonous acid, T-00482
(Trifluoromethyl)phosphonous dichloride, T-00483
Trifluoromethylphosphonous difluoride, T-00484
P-Trifluoromethyl-*N,N,N′,N′*-tetramethylphosphonothioic diamide, *in* T-00481
Trifluoro-*N,N,N′,N′*-tetramethylphosphoranediamine, *see* B-00196
▷Trifosfamide, T-00485
Tri-2-furanylphosphine, T-00486
Tri-3-furanylphosphine, T-00487
Tri-2-furylphosphine, *see* T-00486
Triheptyl phosphate, T-00488
Triheptylphosphine, T-00489
Triheptyl phosphite, T-00490
Trihexyl phosphate, T-00491
Trihexylphosphine, T-00492
Trihexyl phosphite, T-00493
2,4,6-Trihydroxy-1,2,4,6-oxatriphosphinane 2,4,6-trioxide, *see* T-00494
2,4,6-Trihydroxy-1,2,4,6-oxatriphosphorinane 2,4,6-trioxide, T-00494
Tri-1-imidazolylphosphine, T-00495
Tri(2-imidazolyl)phosphine, T-00496
Triiododiphenylphosphorane, T-00497
P,P,P-Triisobutoxy-*N*-(dimethoxyphosphinyl)phosphazene, *in* D-00683
P,P,P-Triisobutoxy-*N*-(dimethoxyphosphinyl)phosphine imide, *in* D-00683
Tri(isobutylamino)phosphine, *see* T-00830
Triisobutyl (dimethoxyphosphinyl)-phosphorimidate, *in* D-00683
Triisobutyl phosphate, *see* T-00822
Triisobutylphosphine, *see* T-00824
Triisobutyl phosphite, *see* T-00826
O,O,O-Triisobutyl phosphoroselenoate, *see* T-00827
O,O,S-Triisobutyl phosphorothioate, *see* T-00828
Triisobutylphosphorous triamide, *see* T-00830
O,O,O-Triisobutyl selenophosphate, *see* T-00827
O,O,S-Triisobutyl thiophosphate, *see* T-00828

Triisodecyl phosphite, *see* T-00799
▷Triisooctyl phosphite, *see* T-00798
Triisopentyl phosphate, *see* T-00796
P,P,P-Triisopropoxy-*N*-phenylphosphine imide, *in* P-00264
P,P,P-Triisopropoxy-*N*-trimethylsilylphosphazene, *in* T-00557
P,P,P-Triisopropoxy-*N*-trimethylsilylphosphine imide, *in* T-00557
Tri(isopropylamino)phosphine, *see* T-00509
Triisopropyl benzoylphosphorimidate, *in* B-00039
Triisopropyl (diethoxyphosphinyl)-phosphorimidate, *in* D-00257
O,O,S-Triisopropyl dithiophosphate, *see* T-00503
2,4,6-Triisopropyl-5-mercapto-1,3,5-dioxaphosphorinane 5-sulfide, T-00498
P,P,P-Triisopropyl-*N*-methylphosphine imide, *in* T-00501
Triisopropylmethylphosphonium iodide, *in* T-00500
Triisopropyl phenylphosphorimidate, *in* P-00264
Triisopropyl phosphate, T-00499
Triisopropylphosphine, T-00500
P,P,P-Triisopropylphosphine imide, T-00501
P,P,P-Triisopropylphosphine imine, *see* T-00501
▷Triisopropyl phosphite, T-00502
N,N′,N″-Triisopropylphosphoric triamide, *in* T-00509
O,O,S-Triisopropyl phosphorodithioate, T-00503
O,O,O-Triisopropyl phosphoroselenoate, T-00504
Triisopropyl phosphorotetrathioate, T-00505
O,O,O-Triisopropyl phosphorothioate, T-00506
N,N′,N″-Triisopropylphosphorothioic triamide, *in* T-00509
O,S,S-Triisopropyl phosphorotrithioate, T-00507
S,S,S-Triisopropyl phosphorotrithioate, T-00508
N,N′,N″-Triisopropylphosphorous triamide, T-00509
O,O,O-Triisopropyl selenophosphate, *see* T-00504
Triisopropyl tetrathiophosphate, *see* T-00505
O,O,O-Triisopropyl thiophosphate, *see* T-00506
P,P,P-Triisopropyl-*N*-trimethylsilylphosphine imide, *in* T-00501
Triisopropyl (trimethylsilyl)-phosphorimidate, *in* T-00557
O,S,S-Triisopropyl trithiophosphate, *see* T-00507
S,S,S-Triisopropyl trithiophosphate, *see* T-00508
▷Tri-Me, *see* M-00437
Trimesitylphosphine, *see* T-00861
2,4,6-Trimesityl-1,3,5,2,4,6-trithiatriphosphorinane, *see* T-00862
P,P,P-Trimethoxy-*N*-(dimethoxyphosphinyl)-phosphazene, *in* D-00683
P,P,P-Trimethoxy-*N*-(dimethoxyphosphinyl)-phosphine imide, *in* D-00683
Trimethoxy(dimethylethylenedioxy)-phosphorane, *see* D-00593
8,9,10-Trimethoxy-2,7-dioxa-4-thia-3-phosphabicyclo[4.4.0]decane 3-oxide, *see* M-00122
8,9,10-Trimethoxy-4,7-dioxa-2-thia-3-phosphabicyclo[4.4.0]decane 3-oxide, *see* M-00121
8,9,10-Trimethoxy-4,7-dioxa-2-thia-3-phosphabicyclo[4.4.0]decane 3-sulfide, *see* M-00123
P,P,P-Trimethoxy-*N*-(diphenoxyphosphinothioyl)phosphazene, *in* D-00987
P,P,P-Trimethoxy-*N*-(diphenoxyphosphinothioyl)phosphine imide, *in* D-00987
8,9,10-Trimethoxy-4-methyl-2,7,4-dioxaza-3-phosphabicyclo[4.4.0]decane 3-oxide, *see* M-00112
8,9,10-Trimethoxy-4-methyl-2,7,4-dioxaza-3-phosphabicyclo[4.4.0]decane 3-sulfide, *see* M-00113
2,2,2-Trimethoxy-5-methyl-Δ^4-oxaphospholene, *see* T-00157
P,P,P-Trimethoxy-*N*-methylphosphine imide, *in* M-00336
P,P,P-Trimethoxy-*N*-methylsulfonylphosphazene, *in* M-00402
P,P,P-Trimethoxy-*N*-methylsulfonylphosphine imide, *in* M-00402
P,P,P-Trimethoxy-*N*-(4-nitrophenyl)-phosphazene, *in* N-00063
P,P,P-Trimethoxy-*N*-(4-nitrophenyl)phosphine imide, *in* N-00063
P,P,P-Trimethoxy-*N*-phenylphosphazene, *in* P-00264
P,P,P-Trimethoxy-*N*-phenylphosphine imide, *in* P-00264
P,P,P-Trimethoxy-*N*-phenylsulfonylphosphazene, *in* P-00315
P,P,P-Trimethoxy-*N*-phenylsulfonylphosphine imide, *in* P-00315
P,P,P-Trimethoxy-*N*-*p*-toluenesulfonylphosphazene, *in* M-00268
P,P,P-Trimethoxy-*N*-*p*-toluenesulfonylphosphine imide, *in* M-00268
P,P,P-Trimethoxy-*N*-trimethylsilylphosphazene, *in* T-00557
P,P,P-Trimethoxy-*N*-trimethylsilylphosphine imide, *in* T-00557
8,9,10-Trimethoxy-2,4,7-trioxa-3-phosphabicyclo[4.4.0]decane 3-oxide, *see* M-00119
8,9,10-Trimethoxy-2,4,7-trioxa-3-phosphabicyclo[4.4.0]decane 3-sulfide, *see* M-00120
Trimethylamine-1,1-diphosphonic acid, *see* D-00701
▷*sym*-Trimethylaminetriphosphonic acid, *see* N-00033
Tri(methylamino)phosphine, *see* T-00551
Tri(methylamino)phosphine oxide, *in* T-00551
Tri(methylamino)phosphine sulfide, *in* T-00551
Trimethyl benzoylphosphorimidate, *in* B-00039
Trimethyl [bis(diethylamino)phosphinyl]-phosphorimidate, *in* B-00176
Trimethyl *tert*-butylphosphorimidate, *in* B-00625
Trimethyl cyanophosphorimidate, *in* C-00228
4-[(2,6,6-Trimethyl-1-cyclohexen-1-yl)-3-buten-2-yl]triphenylphosphonium(1+), *see* M-00422
1,2,3-Trimethyl-1,3-diazaphospholidin-4-one 2-oxide, *in* D-00721
Trimethyl (diethoxyphosphinothioyl)-phosphorimidate, *in* D-00238
Trimethyl (diethoxyphosphinyl)-phosphorimidate, *in* D-00257
Trimethyl (2-dimethoxyphosphinyl)-1,2,4-butanetrioate, *in* P-00382
Trimethyl 3-(dimethoxyphosphinyl)-1,2,3-butanetrioate, *in* P-00383
Trimethyl (dimethoxyphosphinyl)-phosphorimidate, *in* D-00683
3,5,5-Trimethyl-2-dimethylamino-1,2-oxaphospholan-3-ol 2-oxide, *in* T-00519
Trimethyl(2-dimethylphosphinophenyl)-phosphonium iodide, *in* P-00123
O,O-Trimethyl dimethylphosphoramidothioate, *in* D-00696
2,5,5-Trimethyl-1,3-dioxa-2-phosphacyclohexane, *see* T-00512
2,4,5-Trimethyl-1,3-dioxa-2-phosphacyclopentane, *see* T-00511
5,5,6-Trimethyl-4,7-dioxa-3-phospha(3-P^V)-spiro[2.4]heptane, T-00510
N,N,4-Trimethyl-1,3,2-dioxaphosphinan-2-amine, *see* M-00117
2,5,5-Trimethyl-1,3,2-dioxaphosphinane, *see* T-00512
N,N,4-Trimethyl-1,3,2-dioxaphospholan-2-amine, *see* M-00116
2,4,5-Trimethyl-1,3,2-dioxaphospholane, T-00511
N,N,4-Trimethyl-1,3,2-dioxaphosphorinan-2-amine, *see* M-00117
N,N,4-Trimethyl-1,3,2-dioxaphosphorinan-2-amine 2-oxide, *see* M-00118
2,5,5-Trimethyl-1,3,2-dioxaphosphorinane, T-00512
4,4,6-Trimethyl-1,3,2-dioxaphosphorinane, T-00513
Trimethyl (diphenoxyphosphinothioyl)-phosphorimidate, *in* D-00987
1,9,9-Trimethyl-2,8-diphenyl-3,7-dioxa-1-phosphoniabicyclo[3.3.1]nonane, *in* D-00743
N,N,4-Trimethyl-1,3,2-dithiaphospholan-2-amine, *in* M-00136
▷Trimethyl dithiophosphate, *see* T-00542
▷*O,S,S*-Trimethyl dithiophosphate, *see* T-00541
3,7,11-Trimethyl-2,6,10-dodecatrien-1-ol trihydrogen phosphate, *see* M-00517
3,7,11-Trimethyl-1,6,10-dodecatrien-3-yl dihydrogen phosphate, T-00514
3,7,11-Trimethyl-2,6,10-dodecatrien-1-yl dihydrogen phosphate, T-00515
3,7,11-Trimethyl-2,6,10-dodecatrienyl pyrophosphate, *see* M-00517
Trimethylenebis[dicyclohexylphosphine], *see* B-00169
Trimethylenebis[dimethylphosphine], *see* B-00206
Trimethylenebis[methylphenylphosphine], *see* B-00321
Trimethylenebisphosphine, *see* D-01180
Trimethylenebis[triphenylphosphonium](2+), *see* P-00448
▷Trimethylene chlorophosphite, *see* C-00066
Trimethylene dimethylphosphoramidite, *see* D-00696
Trimethylenediphosphonic acid, *see* P-00444
Trimethylenediphosphoramidothioic acid, *see* P-00446
O,O-Trimethylene dithiophosphoric acid, *see* M-00013
▷Trimethylene fluorophosphate, *in* F-00018
Trimethylene fluorophosphite, *see* F-00018
O,O-Trimethylene hydrogen dithiophosphate, *see* M-00013
Trimethylene hydrogen phosphate, *see* H-00138
Trimethylene methylphosphonate, *in* M-00128
Trimethylene methylphosphonite, *see* M-00128
O,O-Trimethylene methylphosphonoselenoate, *in* M-00128
O,O-Trimethylene methylphosphonothioate, *in* M-00128
Trimethylene methylphosphonotrithioate, *in* M-00137
O,O-Trimethylene *S*-methyl phosphorodithioate, *in* M-00013
Trimethylene phenyl phosphate, *in* P-00068
Trimethylene phenyl phosphite, *see* P-00068
Trimethylene phenylphosphonate, *in* P-00116
N,N′-Trimethylene-*P*-phenylphosphonic diamide, *in* H-00020
Trimethylene phenylphosphonite, *see* P-00116
Trimethylene phenylphosphonodithioite, *in* D-01247
O,O-Trimethylene phenylphosphonothioate, *in* P-00116
Trimethylene phenylphosphonotrithioate, *in* D-01249
Trimethylene phosphite, *in* D-00972
Trimethylene phosphonate, *in* D-00972
N,N′-Trimethylenephosphonic diamide, *see* H-00020
Trimethylene phosphonite, *see* D-00972
S,S′-Trimethylene phosphonodithioate, *see* D-01248
O,O-Trimethylene phosphonothioate, *in* D-00972
Trimethylene phosphonotrithioate, *see* D-01249
Trimethylene phosphoric acid, *see* H-00138
Trimethylene phosphorochloridate, *see* C-00067
▷Trimethylene phosphorochloridite, *see* C-00066
Trimethylene phosphorochloridodithioite, *see* C-00199
O,O-Trimethylene phosphorochloridoselenoate, *in* C-00066
O,O-Trimethylene phosphorochloridothioate, *see* C-00068

O,O-Trimethylene phosphorodithioate, *see* M-00013
▷Trimethylene phosphorofluoridate, *in* F-00018
Trimethylene phosphorofluoridite, *see* F-00018
Trimethylene phosphorotetrathioate, *see* M-00014
O,O-Trimethylene phosphorothioite, *in* D-00972
Trimethylene phosphoryl chloride, *see* C-00067
O,O-Trimethylene thiophosphite, *in* D-00972
O,O-Trimethylene thiophosphoryl chloride, *see* C-00068
Trimethyl ethylphosphorimidate, *in* E-00161
Trimethylgermylphosphine, T-00516
N-Trimethylgermyl-*P,P,P*-triphenylphosphine imide, *in* T-00626
Tri(1-methylheptyl) phosphite, *see* T-00797
▷Tri(6-methylheptyl) phosphite, *see* T-00798
2,4,10-Trimethyl-1,2,4,5,7,10-hexaaza-3-phosphaadamantane, *see* T-00517
2,4,10-Trimethyl-1,3,4,5,7,10-hexaaza-3-phosphatricyclo[3.3.1.1^{3,7}]decane, T-00517
Trimethylmethylenephosphorane, T-00518
4,8,8-Trimethyl-7-(1-methylethylidene)-1,6-dioxa-4-aza-5-phospha(5-P^V)-spiro[4.4]nonan-9-one, *see* I-00032
Trimethyl [(4-methylphenyl)sulfonyl]-phosphorimidate, *in* M-00268
Trimethyl methylphosphorimidate, *in* M-00336
Trimethyl methylsulfonylphosphorimidate, *in* M-00402
Trimethyl (4-nitrophenyl)phosphorimidate, *in* N-00063
Tri(8-methylnonyl) phosphite, *see* T-00799
▷Trimethylolethane phosphate, *in* M-00424
▷Trimethylolethane phosphite, *see* M-00424
▷Trimethylolethane thiophosphate, *in* M-00424
▷Trimethylolpropane phosphate, *in* E-00196
▷Trimethylolpropane phosphite, *see* E-00196
Trimethylolpropane selenophosphate, *in* E-00196
▷Trimethylolpropane thiophosphate, *in* E-00196
Trimethylol selenophosphate, *in* M-00424
3,5,5-Trimethyl-1,2-oxaphospholan-3-ol 2-oxide, T-00519
▷3,3,5-Trimethyl-1,2-oxaphosphol-4-ene 2-oxide, *see* D-00596
1,2,5-Trimethyl-4-oxo-1-phenylphosphorinanium iodide, *in* D-00821
4,4,6-Trimethyl-2-phenoxy-1,3-dioxa-2-phosphacyclohexane, *see* T-00520
4,4,6-Trimethyl-2-phenoxy-1,3,2-dioxaphosphorinane, T-00520
2,4,6-Trimethyl-5-phenyl-1,3,5-dioxaphosphinane, *see* T-00521
2,4,6-Trimethyl-5-phenyl-1,3,5-dioxaphosphorinane, T-00521
2,5,5-Trimethyl-2-phenyl-1,3,2-dioxaphosphorinanium iodide, *in* D-00795
P-(2,4,5-Trimethylphenyl)-*N,N′*-diphenylphosphonic diamide, *in* T-00526
2,2,5-Trimethyl-3-phenyl-1,3-oxaphospholane, T-00522
3,5,5-Trimethyl-2-phenyl-1,2-oxaphospholan-3-ol 2-oxide, *in* T-00519
3,5,5-Trimethyl-2-phenyl-1,2-oxaphospholan-3-ol 2-oxide, *in* T-00519
2,3,4-Trimethyl-5-phenyl-1,3,2-oxazaphospholane, *see* T-00523
2,3,4-Trimethyl-5-phenyl-1,3,2-oxazaphospholidine, T-00523
2,8,10-Trimethyl-10-phenyl-10*H*-phenoxaphosphinium iodide, *in* D-00799
P,P,P-Trimethyl-*N*-phenylphosphine imide, *in* T-00531
(2,4,5-Trimethylphenyl)phosphinic acid, T-00524
(2,4,6-Trimethylphenyl)phosphinic acid, T-00525
N,N,P-Trimethyl-*P*-phenylphosphinothioic amide, *in* M-00226
1,3,4-Trimethyl-1-phenyl-1*H*-phospholium iodide, *in* D-00807
(2,4,5-Trimethylphenyl)phosphonic acid, T-00526
(2,4,6-Trimethylphenyl)phosphonic acid, T-00527
Trimethylphenylphosphonium bromide, *in* D-00803
Trimethyl phenylphosphorimidate, *in* P-00264
Trimethyl phenysulfonylphosphorimidate, *in* P-00315
1,2,4-Trimethyl-2-phosphabicyclo[2.2.1]-heptane 2-oxide, *in* D-00825
▷Trimethyl phosphate, T-00528
2,3,5-Trimethyl-3-phosphatricyclo[4.2.1.0^{2,5}]nonane, T-00529
▷*P,P,P*-Trimethylphosphazene, *see* T-00531
1,1,3-Trimethylphosphinanium iodide, *in* D-00882
1,1,4-Trimethylphosphinanium iodide, *in* D-00883
▷Trimethylphosphine, T-00530
Trimethylphosphine dibromide, *see* D-00069
Trimethylphosphine dichloride, *see* D-00198
Trimethylphosphine difluoride, *see* D-00417
Trimethylphosphine diiodide, *see* D-00629
▷*P,P,P*-Trimethylphosphine imide, T-00531
Trimethylphosphine oxide, T-00532
▷Trimethylphosphine selenide, T-00533
Trimethylphosphine sulfide, T-00534
Trimethylphosphinogermane, *see* T-00516
N,P,P-Trimethylphosphinoselenoic amide, *in* D-00836
▷Trimethyl phosphite, T-00535
Trimethyl phosphite difluoride, *see* D-00416
Trimethyl phospholactate, *in* D-00682
1,3,4-Trimethyl-1*H*-phosphole, T-00536
1,3,4-Trimethyl-3-phospholene, *see* D-00600
N,N,P-Trimethylphosphonamidic acid, T-00537
▷Trimethylphosphonamidic fluoride, T-00538
Trimethylphosphonamidodithioic acid, T-00539
Trimethylphosphonamidothioic acid, T-00540
N,N′,P-Trimethylphosphonic diamide, *in* M-00289
2-(Trimethylphosphonio)ethanol, *see* H-00150
N,N,P-Trimethylphosphonisocyanatidic amide, *in* M-00293
Trimethyl phosphonoacetate, *in* D-00678
Trimethyl 3-phosphonoacrylate, *in* D-00686
Trimethyl 2-phosphonobutanoate, *in* D-00679
Trimethyl phosphonocarbamate, *in* D-00680
Trimethyl phosphonoformate, *in* D-00672
Trimethyl *O*-phosphonolactate, *in* D-00682
Trimethyl 3-phosphonopropanoate, *in* D-00685
Trimethyl 2-phosphonopropionate, *in* D-00684
1,1,4-Trimethylphosphoranium iodide, *in* D-00883
Trimethylphosphoranylidenemethane, *see* T-00518
N,N′,N″-Trimethylphosphoric triamide, *in* T-00551
1,1,3-Trimethylphosphorinanium iodide, *in* D-00882
▷*O,S,S*-Trimethyl phosphorodithioate, T-00541
▷*O,O,S*-Trimethyl phosphorodithioate, T-00542
O,O,O-Trimethyl phosphoroselenoate, T-00543
Trimethyl phosphorotetrathioate, T-00544
▷*O,O,O*-Trimethyl phosphorothioate, T-00545
▷*O,O,S*-Trimethyl phosphorothioate, T-00546
N,N′,N″-Trimethylphosphorothioic triamide, *in* T-00551
Trimethyl phosphorotriselenoite, T-00547
O,S,S-Trimethyl phosphorotrithioate, T-00548
▷*S,S,S*-Trimethyl phosphorotrithioate, T-00549
▷Trimethyl phosphorotrithioite, T-00550
N,N′,N″-Trimethylphosphorous triamide, T-00551
▷1,2,2-Trimethylpropyl methylphosphonofluoridate, *see* S-00005
Tri(1-methylpropyl) phosphite, *see* T-00825
O,O,O-Trimethyl selenophosphate, *see* T-00543
▷Trimethylsilanol 3:1 phosphate, *see* T-00864
Trimethylsilyl bis(chloromethyl)phosphinate, *in* B-00127
Trimethylsilyl bis(1,1-dimethylethyl)-phosphinite, *in* D-00122
Trimethylsilyl bis(1-methylethyl)-phosphinite, *in* D-00648
O-Trimethylsilyl bis(trifluoromethyl)-phosphinothioate, *in* B-00415
Trimethylsilyl bis[*N*-(trimethylsilyl)-aminomethyl]phosphinate, *in* B-00087
N-Trimethylsilyl-*P*-[bis(trimethylsilyl)-amino]-*P*-trimethylsilyloxyphosphazene, *in* T-00867
N-Trimethylsilyl-*P*-[bis(trimethylsilyl)-amino]-*P*-trimethylsilyloxyphosphine imide, *in* T-00867
Trimethylsilyl bis(trimethylsilyloxymethyl)phosphinate, *in* B-00282
Trimethylsilyl 2-[bis(trimethylsilyloxy)-phosphinyl]oxy-2-propenoate, *in* P-00366
[4-(Trimethylsilyl)-3-butyn-1-ylidene]-triphenylphosphorane, *in* T-00691
Trimethylsilyl cyclohexylphosphinate, *in* C-00253
Trimethylsilyl di-*tert*-butylphosphinate, *in* D-00110
Trimethylsilyl dibutylphosphinite, *in* D-00121
Trimethylsilyl di-*tert*-butylphosphinite, *in* D-00122
O-Trimethylsilyl dibutylphosphinothioate, *in* D-00117
O-Trimethylsilyl di-*tert*-butylphosphinothioate, *in* D-00118
Trimethylsilyl dicyclohexylphosphinite, *in* D-00214
Trimethylsilyl 2-(diethoxyphosphinyl)-propanoate, *in* D-00258
Trimethylsilyl diethylphosphinate, *in* D-00322
Trimethylsilyl diethylphosphinodithioate, *in* D-00325
O-Trimethylsilyl diethylphosphinothioate, *in* D-00330
Trimethylsilyl diisopropylphosphinite, *in* D-00648
Trimethylsilyl dimethylphosphinate, *in* D-00829
Trimethylsilyl dimethylphosphinodithioate, *in* D-00834
Trimethylsilyl diphenylphosphinodithioate, *in* D-01065
O-Trimethylsilyl diphenylphosphinothioate, *in* D-01082
Trimethylsilyl *P,P*-diphenyl-*N*-trimethylsilylphosphinimidate, *in* D-01157
Trimethylsilyl dipropylphosphinate, *in* D-01205
[[2-(Trimethylsilyl)ethoxy]methylene]-triphenylphosphorane, *in* T-00552
[[2-(Trimethylsilyl)ethoxy]methyl]-triphenylphosphonium(1+), T-00552
[2-(Trimethylsilyl)ethylidene]-triphenylphosphorane, *in* T-00553
[2-(Trimethylsilyl)ethyl]-triphenylphosphonium(1+), T-00553
[(Trimethylsilyl)methylene]-triphenylphosphorene, *see* T-00692
2-(Trimethylsilyloxy)-1,3,2-dioxaphospholane, T-00554
[1-[(Trimethylsilyl)oxy]ethyl]phosphonic acid, T-00555
[(Trimethylsilyloxy)methyl]phosphonic acid, T-00556
Trimethylsilyl phenylphosphinate, *in* P-00210
(Trimethylsilyl)phosphorimidic acid, T-00557
Trimethylsilyl phosphorodichloridate, T-00558
Trimethylsilyl phosphorodifluoridate, T-00559
Trimethylsilyl tetraethylphosphorodiamidite, *in* T-00066
Trimethylsilyl tetramethylphosphorodiamidodithioate, *in* T-00222
1-Trimethylsilyl-2-trimethylsilyloxy-2-(2,4,6-tri-*tert*-butylphenyl)-methylenephosphine, T-00560
N-Trimethylsilyl-*P,P,P*-triphenylphosphine imide, *in* T-00626
Trimethylsilyl *N,N,N′*-tris(trimethylsilyl)phosphonamidimidate, *in* T-00867
3′,*N,N*-Trimethylspiro[1,3,2-benzodioxaphosphole-2,2′(3′*H*)-[1,3,2]-benzothiazaphosphol]-2-amine, *in* M-00394

2,3,3′-Trimethyl-2,2′(3*H*,3′*H*)-spirobi[1,3,2-benzoxazaphosphole], *in* D-00917
[2-(Trimethylstannyl)ethylidene]-triphenylphosphorane, *in* T-00561
[2-(Trimethylstannyl)ethyl]-triphenylphosphonium, T-00561
N-Trimethylstannyl-*P*,*P*,*P*-triphenylphosphine imide, *in* T-00626
Trimethyl tetrathiophosphate, *see* T-00544
▷Trimethyl thiophosphate, *see* T-00545
▷*O*,*O*,*S*-Trimethyl thiophosphate, *see* T-00546
1,2,4-Trimethyl-1,2,4,3,5-triazadiphospholidine, T-00562
2,8,9-Trimethyl-2,8,9-triaza-1-phosphaadamantane, *see* T-00563
2,8,9-Trimethyl-2,8,9-triaza-1-phosphatricyclo[3.3.1.1^{3,7}]decane, T-00563
1,3,3-Trimethyltrimethylene chlorophosphite, *see* C-00189
1,3,3-Trimethyltrimethylene chlorothiophosphate, *in* C-00189
Trimethyl(trimethylgermyl)-methylenephosphorane, T-00564
Trimethyl[(trimethylphosphoranylidene)-methyl]phosphonium(1+), T-00565
Trimethyl(trimethylsilylmethylene)-phosphorane, T-00566
P,*P*,*P*-Trimethyl-*N*-trimethylsilylphosphine imide, *in* T-00531
Trimethyl (trimethylsilyl)phosphorimidate, *in* T-00557
2,4,6-Trimethyl-1,3,5,2,4,6-trioxatriphosphinane, *see* T-00567
2,4,6-Trimethyl-1,3,5,2,4,6-trioxatriphosphorinane, T-00567
2,6,6-Trimethyl-1-(triphenylphosphoniomethyl)-1,3-cyclohexadiene(1+), *see* T-00688
Trimethyl [[(triphenylphosphoranylidene)-amino]sulfonyl]phosphorimidate, *in* T-00642
2,4,6-Trimethyl-2,4,6-triphosphaheptane, *see* B-00204
1,2,4-Trimethyl-1,2,4-triphospholane, T-00568
Trimethyl triselenophosphite, *see* T-00547
O,*S*,*S*-Trimethyl trithiophosphate, *see* T-00548
▷*S*,*S*,*S*-Trimethyl trithiophosphate, *see* T-00549
▷Trimethyl trithiophosphite, *see* T-00550
Trimorpholinophosphine, T-00569
Tri-1-naphthalenyl phosphate, *see* T-00570
Tri-2-naphthalenyl phosphate, *see* T-00571
Tri-1-naphthalenylphosphine, *see* T-00572
Tri-2-naphthalenylphosphine, *see* T-00573
Tri-1-naphthalenyl phosphite, *see* T-00574
Tri-2-naphthalenyl phosphite, *see* T-00575
N,*N*′,*N*″-Tri-1-naphthalenylphosphoric triamide, *in* T-00576
N,*N*,*N*″-Tri-2-naphthalenylphosphoric triamide, *in* T-00577
N,*N*′,*N*″-Tri-2-naphthalenylphosphorous triamide, *see* T-00577
N,*N*′,*N*″-Tri-1-naphthalenylphosphorous triamide, *see* T-00576
Tri(1-naphthylamino)phosphine, *see* T-00576
Tri-1-naphthyl phosphate, T-00570
Tri-2-naphthyl phosphate, T-00571
Tri-1-naphthylphosphine, T-00572
Tri-2-naphthylphosphine, T-00573
Tri-1-naphthyl phosphite, T-00574
Tri-2-naphthyl phosphite, T-00575
N,*N*′,*N*″-Tri-1-naphthylphosphoric triamide, *in* T-00576
N,*N*′,*N*″-Tri-2-naphthylphosphoric triamide, *in* T-00577
N,*N*′,*N*″-Tri-1-naphthylphosphorous triamide, T-00576
N,*N*′,*N*″-Tri-2-naphthylphosphorous triamide, T-00577
Trineopentyl phosphite, *see* T-00772
Tri(*p*-nitrobenzyl)phosphine, *see* T-00835
Tri(2-nitrophenyl) phosphite, *see* T-00839
Tri(4-nitrophenyl) phosphite, *see* T-00840
4,4′,4″-[[3-(2,4,6-Trinitrophenyl)-2-triazenylidene]phosphoranylidene]-trismorpholine, *see* T-00833
Trinonylphosphine, T-00578
Trinonyl phosphite, T-00579
Trinosin, *in* A-00043
Trioctadecyl phosphite, T-00580
Trioctyl phosphate, T-00581
Trioctylphosphine, T-00582
▷Trioctyl phosphite, T-00583
Tri(2-octyl) phosphite, *see* T-00797
2,8,9-Trioxa-5-aza-1-phosphabicyclo[3.3.3]undecane, T-00584
2,6,7-Trioxa-1,4-diphosphabicyclo[2.2.2]-octane, T-00585
2,8,9-Trioxa-1-phosphaadamantane, *see* T-00591
2,6,7-Trioxa-4-phospha-1-arsabicyclo[2.2.2]octane, T-00586
2,6,7-Trioxa-1-phosphabicyclo[2.2.1]heptane, T-00587
2,6,7-Trioxa-1-phosphabicyclo[2.2.2]octane, T-00588
2,7,8-Trioxa-1-phosphabicyclo[3.2.1]octane, T-00589
2,6,7-Trioxa-1-phosphabicyclo[2.2.2]-octane-4-methanol, T-00590
1,3,6-Trioxa-2-phosphacyclooctane, *see* T-00592
2,8,9-Trioxa-1-phosphatricyclo[3.3.1.1^{3,7}]decane, T-00591
1,3,6,2-Trioxaphosphocane, T-00592
Tripentyl phosphate, T-00593
Tripentylphosphine, T-00594
Tripentyl phosphite, T-00595
Triphenethylphosphine, *see* T-00844
Triphenoxycarbonylphosphine, *in* P-00355
P,*P*,*P*-Triphenoxy-*N*-(diphenoxyphosphinothioyl)phosphazene, *in* D-00987
P,*P*,*P*-Triphenoxy-*N*-(diphenoxyphosphinothioyl)phosphine imide, *in* D-00987
P,*P*,*P*-Triphenoxy-*N*-(3-nitrophenyl)-phosphazene, *in* N-00062
P,*P*,*P*-Triphenoxy-*N*-(3-nitrophenyl)phosphine imide, *in* N-00062
P,*P*,*P*-Triphenoxy-*N*-phenylphosphazene, *in* P-00264
P,*P*,*P*-Triphenoxy-*N*-phenylphosphine imide, *in* P-00264
P,*P*,*P*-Triphenoxy-*N*-phenylsulfonylphosphazene, *in* P-00315
P,*P*,*P*-Triphenoxy-*N*-phenylsulfonylphosphine imide, *in* P-00315
2,6,10-Triphenyl-6-arsa-2,10-diphosphabicyclo[9.4.0]pentadeca-1(11),12,14-triene, *see* O-00027
1,2,3-Triphenyl-1,3-azaphosphacyclopentane, *in* D-00995
1,2,3-Triphenyl-1,3-azaphospholidin-5-one, T-00596
Triphenyl[(benzoylphenacyl)methylene]-phosphorane, *see* D-01162
Triphenyl benzoylphosphorimidate, *in* B-00039
Triphenylbiphenylenephosphorane, *see* D-00610
Triphenyl [bis(dimethylamino)phosphinyl]-phosphorimidate, *in* B-00194
Triphenyl[bis(trifluoromethyl)methylene]-phosphorane, T-00597
4,6,7-Triphenyl-2,3-bis(trifluoromethyl)-1-phosphabicyclo[2.2.2]octa-2,5,7-triene, T-00598
Triphenyl cyanophosphorimidate, *in* C-00228
Triphenylcyclotriphosphine, *see* T-00698
6,8,10-Triphenyl-1,4-diaza-5-phospha(5-P^V)spiro[4.5]deca-5,7,9-triene, T-00599
2,4,5-Triphenyl-2*H*-1,3,2-diazaphosphole, T-00600
3,4,5-Triphenyl-4*H*-1,2,4-diazaphosphole, T-00601
Triphenyl (diethoxyphosphinothioyl)-phosphorimidate, *in* D-00238
Triphenyl (dimethoxyphosphinothioyl)-phosphorimidate, *in* D-00674
2,4,5-Triphenyl-1,3,2-dioxaphosphole, T-00602
Triphenyl (diphenoxyphosphinothioyl)-phosphorimidate, *in* D-00987
N,*N*′,*N*″-Triphenyl (diphenoxyphosphinothioyl)-phosphorotriamidoimidate, *in* D-00987
P,*P*,*P*-Triphenyl-*N*-diphenylphosphinoiminophosphorane, *see* D-01165
1,2,3-Triphenyl-1,3,2-diphospharsolane, T-00603
6,8,10-Triphenyl-1,4-dithia-5-phospha(5-P^V)spiro[4.5]deca-5,7,9-triene, T-00604
7,9,11-Triphenyl-1,5-dithia-6-phospha(6-P^V)spiro[5.5]undeca-6,8,10-triene, T-00605
2,4,5-Triphenyl-1,3,2,4,5-dithiatriphospholane, T-00606
2,3,10-Triphenyl-4*H*-1,4-ethenophosphinoline, T-00607
Tri(2-phenylethyl) phosphite, *see* T-00845
Triphenyl ethylphosphorimidate, *in* E-00161
Triphenyl [(4-methylphenyl)sulfonyl]-phosphorimidate, *in* M-00268
Triphenylmethylphosphonic acid, T-00608
Triphenyl methylsulfonylphosphorimidate, *in* M-00402
Triphenyl (2-nitrophenyl)phosphorimidate, *in* N-00061
Triphenyl (3-nitrophenyl)phosphorimidate, *in* N-00062
Triphenyl (4-nitrophenyl)phosphorimidate, *in* N-00063
6,8,10-Triphenyl-1-oxa-4-aza-5-phosphaspiro(5-P^V)spiro[4.5]deca-5,7,9-triene, T-00609
2,4,6-Triphenyl-4*H*-1,4-oxaphosphorin, T-00610
2,4,5-Triphenyl-1,3,4-oxathiaphospholane, T-00611
Triphenyl(1-phenylethylidene)phosphorane, *in* T-00612
Triphenyl(2-phenylethylidene)phosphorane, *in* T-00613
Triphenyl(1-phenylethyl)phosphonium(1+), T-00612
Triphenyl(2-phenylethyl)phosphonium(1+), T-00613
Triphenyl(2-phenylethynyl)phosphonium(1+), T-00614
Triphenyl(phenyliminovinylidene)phosphorane, *see* T-00652
Triphenyl(phenylmethylene)phosphorane, *see* B-00050
Triphenyl(phenylmethyl)phosphonium(1+), *see* B-00073
Triphenyl phenylphosphorimidate, *in* P-00264
Triphenyl(3-phenyl-2-propenylidene)-phosphorane, T-00615
Triphenyl(3-phenyl-2-propenyl)-phosphonium(1+), T-00616
Triphenyl(3-phenylpropylidene)phosphorane, *in* T-00617
Triphenyl(3-phenylpropyl)phosphonium(1+), T-00617
P,*P*,*P*-Triphenyl-*N*-phenylsulfonylphosphine imide, *in* T-00626
Triphenyl phenylsulfonylphosphorimidate, *in* P-00315
Triphenyl[1-(phenylthio)cyclopropyl]-phosphonium(1+), T-00618
Triphenyl[1-(phenylthio)ethenyl]-phosphonium(1+), T-00619
Triphenyl[(phenylthio)methylene]phosphorane, T-00620
Triphenyl(phenylthiomethyl)phosphonium(1+), T-00621
2,4,6-Triphenylphosphabenzene, *see* T-00665
1,2,5-Triphenylphosphacyclopentadiene, *see* T-00634
2,4,6-Triphenyl-4-phospha-4*H*-pyran, *see* T-00610
5,6,9-Triphenyl-5-phospha(5-P^V)-spiro[4.4]nona-1,3,6,8-tetraene-1,2,3,4-tetracarboxylic acid, T-00622
▷Triphenyl phosphate, T-00623
P,*P*,*P*-Triphenylphosphazene, *see* T-00626
Triphenylphosphazocarbonic acid, *see* T-00648
Triphenylphosphazophosphoryl dichloride, *see* T-00656
1,2,6-Triphenyl-4-phosphinanone, *see* T-00666
▷Triphenylphosphine, T-00624
N,1,1-Triphenylphosphinecarbonamide, T-00625
Triphenylphosphine dibromide, *see* D-00071

Triphenylphosphine dichloride, *see* D-00200
Triphenylphosphine difluoride, *see* D-00419
P,P,P-Triphenylphosphine imide, T-00626
Triphenylphosphine oxide, T-00627
Triphenylphosphine selenide, T-00628
▷Triphenylphosphine sulfide, T-00629
N,P,P-Triphenylphosphinic amide, T-00630
N,P,P-Triphenylphosphinic amide, *see* T-00631
▷Triphenylphosphinic diiodide, *see* D-00631
N,P,P-Triphenylphosphinimidic acid, T-00631
Triphenylphosphinimine dichlorophosphate, *see* T-00656
2,4,6-Triphenylphosphinine, *see* T-00665
N,P,P-Triphenylphosphinoselenoic amide, *in* D-01079
N,P,P-Triphenylphosphinothioic amide, *in* D-01083
1,2,3-Triphenyl-1*H*-phosphirene, T-00632
▷Triphenyl phosphite, T-00633
Triphenyl phosphite dibromide, *see* D-00070
Triphenyl phosphite dichloride, *see* D-00199
Triphenyl phosphite difluoride, *see* D-00418
Triphenyl phosphite diiodide, *see* D-00630
Triphenylphosphite methiodide, *in* M-00429
Triphenylphosphite ozonide, *see* D-00603
1,2,5-Triphenyl-1*H*-phosphole, T-00634
▷*N,N′,P*-Triphenylphosphonic diamide, *in* P-00222
P,2,2′-Triphenylphosphonic dihydrazide, *in* P-00225
(Triphenylphosphonio)acetic acid, *see* C-00014
4-(Triphenylphosphonio)butanoic acid, *see* C-00009
2-(Triphenylphosphonio)ethanol(1+), *see* H-00151
1-[2-(Triphenylphosphonio)ethyl]-2-(triphenylphosphoniomethyl)benzene(2+), *see* T-00696
9-(Triphenylphosphonio)fluorene(1+), *see* F-00012
7-(Triphenylphosphonio)heptanoic acid, *see* C-00012
6-(Triphenylphosphonio)hexanoic acid, *see* C-00013
6-(Triphenylphosphonio)hexanoic acid, *see* C-00015
(Triphenylphosphonio)methanol, *see* H-00163
10-(Triphenylphosphoniomethyl)benzo[*g*]chrysene(1+), T-00635
2-(Triphenylphosphoniomethyl)benzo[*c*]phenanthrene(1+), T-00636
3-(Triphenylphosphoniomethyl)phenanthrene(1+), T-00637
9-(Triphenylphosphoniomethyl)phenanthrene(1+), T-00638
2-(Triphenylphosphoniomethyl)-2-propenoic acid(1+), *see* C-00016
2-(Triphenylphosphonio)propanoic acid, *see* C-00010
3-(Triphenylphosphonio)propanoic acid, *see* C-00011
3-(Triphenylphosphonio)-1-propanol, *see* H-00193
Triphenylphosphonium bis(diphenylstibino)methylide, *see* B-00270
Triphenyl phosphonoacetate, *in* D-00988
Triphenylphosphonoacetone(1+), *see* O-00080
N^2,$N^{2'}$,*P*-Triphenylphosphonothioic dihydrazide, *in* P-00249
1-Triphenylphosphoranylideneacenaphthene, *see* A-00002
Triphenylphosphoranylideneacetaldehyde, T-00639
(Triphenylphosphoranylidene)acetic acid, T-00640
▷Triphenylphosphoranylideneacetone, *see* T-00662
Triphenylphosphoranylideneacetonitrile, T-00641
▷2-Triphenylphosphoranylideneacetophenone, *see* P-00333
Triphenylphosphoranylideneallene, *see* T-00678
[[(Triphenylphosphoranylidene)amino]sulfonyl]phosphorimidic acid, T-00642
α-(Triphenylphosphoranylidene)benzeneacetaldehyde, T-00643
α-(Triphenylphosphoranylidene)benzeneacetic acid, T-00644
α-(Triphenylphosphoranylidene)benzenepropanoic acid, T-00645
2-Triphenylphosphoranylidene-1,3-benzodithiole, *in* B-00021
(Triphenylphosphoranylidene)butanedioic acid, T-00646
4-(Triphenylphosphoranylidene)butanoic acid, *in* C-00017
4-(Triphenylphosphoranylidene)-2-butenoic acid, T-00647
(Triphenylphosphoranylidene)carbamic acid, T-00648
4-Triphenylphosphoranylidene-2,5-cyclohexadien-1-one, T-00649
5-Triphenylphosphoranylidene-1,3-cyclopentadiene, *see* C-00274
2-(Triphenylphosphoranylidene)-1,1-dithiomalonic acid, *see* E-00194
(Triphenylphosphoranylidene)ethenethione, T-00650
(Triphenylphosphoranylidene)ethenone, T-00651
N-[(Triphenylphosphoranylidene)ethenylidene]benzenamine, *see* T-00652
N-[(Triphenylphosphoranylidene)ethylidene]aniline, T-00652
9-(Triphenylphosphoranylidene)fluorene, *see* F-00010
1-(Triphenylphosphoranylidene)-2-heptanone, *in* O-00062
6-(Triphenylphosphoranylidene)hexanoic acid, *in* C-00015
[(Triphenylphosphoranylidene)hydrazono]acetic acid, T-00653
α-(Triphenylphosphoranylidene)hydrocinnamic acid, *see* T-00645
Triphenylphosphoranylideneketene, *see* T-00651
5-(Triphenylphosphoranylidene)levulinic acid, *see* O-00085
(Triphenylphosphoranylidene)malonic acid, *see* T-00658
Triphenylphosphoranylidenemethanenitrile, *see* T-00641
[2-[2-(Triphenylphosphoranylidenemethyl)phenyl]ethylidenetriphenylphosphorane, *in* T-00696
3-(Triphenylphosphoranylidene)-2,4-pentanedione, T-00654
5-(Triphenylphosphoranylidene)pentanoic acid, T-00655
(Triphenylphosphoranylidene)phosphoramidic dichloride, T-00656
2-(Triphenylphosphoranylidene)propanal, T-00657
2-(Triphenylphosphoranylidene)-1,3-propanedinitrile, *see* D-00205
(Triphenylphosphoranylidene)propanedioic acid, T-00658
2-(Triphenylphosphoranylidene)propanenitrile, T-00659
3-Triphenylphosphoranylidenepropanenitrile, *in* C-00212
2-(Triphenylphosphoranylidene)propanoic acid, T-00660
3-(Triphenylphosphoranylidene)propanoic acid, T-00661
▷1-Triphenylphosphoranylidene-2-propanone, T-00662
2-(Triphenylphosphoranylidene)propionaldehyde, *see* T-00657
3-(Triphenylphosphoranylidene)-2,5-pyrrolidenedione, T-00663
(Triphenylphosphoranylidene)succinic acid, *see* T-00646
Triphenylphosphoranylidenesuccinic anhydride, *see* D-00612
Triphenylphosphoranylidenesuccinimide, *see* T-00663
[(Triphenylphosphoranylidene)sulfamoyl]phosphorimidic acid, *see* T-00642
4-Triphenylphosphoranylidene(tetrahydro-2*H*-pyran), *in* T-00684
4-Triphenylphosphoranylidene(tetrahydro-2*H*-thiopyran), *in* T-00685
(Triphenylphosphoranylidene)thioketene, *see* T-00650
N,N′,N″-Triphenylphosphoric triamide, *in* T-00677
2,2′,2″-Triphenylphosphoric trihydrazide, T-00664
2,4,6-Triphenylphosphorin, T-00665
1,2,6-Triphenyl-4-phosphorinanone, T-00666
(Triphenylphosphoroanylidene)malononitrile, *see* D-00205
O,O,O-Triphenyl phosphoroselenoate, T-00667
Triphenyl phosphorotetraselenoate, T-00668
Triphenyl phosphorotetrathioate, T-00669
O,O,O-Triphenyl phosphorothioate, T-00670
O,O,S-Triphenyl phosphorothioate, T-00671
N,N′,N″-Triphenylphosphorothioic triamide, *in* T-00677
Se,Se,Se-Triphenyl phosphorotriselenoate, T-00672
Triphenyl phosphorotriselenoite, T-00673
Se,Se,Se-Triphenyl phosphorotriselenothioate, T-00674
S,S,S-Triphenyl phosphorotrithioate, T-00675
Triphenyl phosphorotrithioite, T-00676
N,N′,N″-Triphenylphosphorous triamide, T-00677
Triphenyl(2-phthalimidoethyl)phosphonium(1+), *see* D-00474
Triphenylpropadienylidenephosphorane, T-00678
Triphenylpropargylidenephosphorane, *see* T-00678
Triphenyl-2-propenylidenephosphorane, *in* T-00679
Triphenyl-2-propenylphosphonium(1+), T-00679
P,P,P-Triphenyl-*N*-propylphosphine imide, *in* T-00626
Triphenylpropylphosphonium(1+), T-00680
Triphenyl(propynylidene)phosphorane, *in* T-00681
Triphenyl-2-propynylphosphonium(1+), T-00681
Triphenyl(retinyl)phosphonium(1+), *see* D-00929
O,O,O-Triphenyl selenophosphate, *see* T-00667
2′,4′,6′-Triphenylspiro[1,3,2-dithiaphosph(P^{V})olan-2,1′-phosph(P^{V})orin], *see* T-00604
2′,4′,6′-Triphenylspiro[1,3,2-dithiaphosph(P^{V})orinan-2,1′-phosph(P^{V})orin], *see* T-00605
Triphenylstannyldiphenylphosphine, *see* D-01167
Triphenyltetradecylidenephosphorane, *in* T-00682
Triphenyl(tetradecyl)phosphonium(1+), T-00682
Triphenyl[3-[(tetrahydro-2*H*-pyran-2-yl)oxy]propylidene]phosphorane, *in* T-00683
Triphenyl[[3-(tetrahydro-2*H*-pyran-2-yl)oxy]propyl]phosphonium(1+), T-00683
Triphenyl(tetrahydro-2*H*-pyran-4-yl)phosphonium(1+), T-00684
Triphenyl(tetrahydro-2*H*-thiopyran-4-yl)phosphonium(1+), T-00685
Triphenyl tetraselenophosphate, *see* T-00668
Triphenyl tetrathiophosphate, *see* T-00669
2,4,6-Triphenyl-4*H*-1,4-thiaphosphinin, *see* T-00686
2,4,6-Triphenyl-4*H*-1,4-thiaphosphorin, T-00686
2,3,4-Triphenyl-1,3,2,4-thiazadiphosphetidine, T-00687
Triphenyl thiophosphate, *see* T-00670
O,O,S-Triphenyl thiophosphate, *see* T-00671
Triphenyl(tosylmethyl)phosphonium(1+), *see* M-00267
Triphenyl[2,2,2-trifluoro-1-(trifluoromethyl)ethylidene]phosphorane, *see* T-00597
Triphenyl[(2,6,6-trimethyl-1,3-cyclohexadien-1-yl)methylene]phosphorane, *in* T-00688
Triphenyl[(2,6,6-trimethyl-1,3-cyclohexadien-1-yl)methyl]phosphonium(1+), T-00688
Triphenyl[(2,6,6-trimethyl-1-cyclohexen-1-yl)methyl]phosphonium(1+), T-00689
Triphenyl(3,7,11-trimethyldodecylidene)phosphorane, *in* T-00690
Triphenyl(3,7,11-trimethyldodecyl)phosphonium(1+), T-00690

Triphenyl[4-(trimethylsilyl)-3-butynyl]-phosphonium(1+), T-00691
Triphenyl[(trimethylsilyl)methylene]-phosphorane, T-00692
Triphenyl[(trimethylsilyl)methyl]-phosphonium(1+), T-00693
Triphenyl[2-(trimethylstannyl)ethylidene]-phosphorane, *in* T-00561
2,4,6-Triphenyl-1,3,5,2,4,6-trioxatriphosphinane, *see* T-00694
2,4,6-Triphenyl-1,3,5,2,4,6-trioxatriphosphorinane, T-00694
Triphenyl(triphenylgermyl)phosphinimine, T-00695
Triphenyl[[2-[2-(triphenylphosphonio)-ethyl]phenyl]methyl]phosphonium(2+), T-00696
Triphenyl [[(triphenylphosphoranylidene)-amino]sulfonyl]phosphorimidate, *in* T-00642
1,1,1-Triphenyl-*N*-triphenylphosphoranylidenegermanamine, *see* T-00695
Triphenyl[(triphenylphosphoranylidene)-methyl]phosphonium(1+), T-00697
N′,*P*,*P*-Triphenyl-*N*-(triphenylphosphoranylidene)-phosphinimidic amide, *in* D-01164
2,6,10-Triphenyl-2,6,10-triphosphabicyclo[9.4.0]pentadeca-1(11),12,14-triene, *see* O-00028
1,2,3-Triphenyl-1,2,3-triphosphaindane, *see* D-00606
Triphenyltriphosphirane, T-00698
1,2,3-Triphenyl-1,2,3-triphospholane, T-00699
Triphenyl triselenophosphite, *see* T-00673
S,*S*,*S*-Triphenyl trithiophosphate, *see* T-00675
Triphenyl trithiophosphite, *see* T-00676
Triphenyl(2,5-undecadienylidene)phosphorane, T-00700
Triphenylundecylidenephosphorane, *in* T-00701
Triphenylundecylphosphonium(1+), T-00701
Triphenylvinylphosphonium(1+), T-00702
Triphenyl-*m*-xylylphosphonium(1+), *see* M-00210
Triphenyl-*o*-xylylphosphonium(1+), *see* M-00209
Triphenyl-*p*-xylylphosphonium(1+), *see* M-00211
Triphos, *see* B-00239
Triphosphetan-4-one, T-00703
1,2,3-Triphosphol-4-ene, *see* D-00615
1,1,3-Triphosphonopropane, *see* P-00449
1,2,3-Triphosphonopropane, *see* P-00450
Triphosphono-*s*-triazine, *see* T-00329
2,2′,2″-Triphosphonotriethylamine, *see* T-00438
Triphosphopyridine nucleotide, *see* C-00203
▷Tri-1-piperidinophosphine, T-00704
▷Tri-2-propenyl phosphate, T-00705
Tri-1-propenylphosphine, T-00706
Tri-2-propenylphosphine, T-00707
▷Tri-2-propenyl phosphite, T-00708
N,*N′*,*N″*-Tri-2-propenylphosphorous triamide, T-00709
Tri(propylamino)phosphine, *see* T-00719
Tripropyl phosphate, T-00710
▷Tripropylphosphine, T-00711
▷Tripropyl phosphite, T-00712
Tripropyl phosphonoglycinate, *in* P-00368
N,*N′*,*N″*-Tripropylphosphoric triamide, *in* T-00719
O,*O*,*O*-Tripropyl phosphoroselenoate, T-00713
Tripropyl phosphorotetrathioate, T-00714
O,*O*,*O*-Tripropyl phosphorothioate, T-00715
O,*O*,*S*-Tripropyl phosphorothioate, T-00716
N,*N′*,*N″*-Tripropylphosphorothioic triamide, *in* T-00719
S,*S*,*S*-Tripropyl phosphorotrithioate, T-00717
Tripropyl phosphorotrithioite, T-00718
N,*N′*,*N″*-Tripropylphosphorous triamide, T-00719
O,*O*,*O*-Tripropyl selenophosphate, *see* T-00713
Tripropyl tetrathiophosphate, *see* T-00714
O,*O*,*O*-Tripropyl thiophosphate, *see* T-00715
O,*O*,*S*-Tripropyl thiophosphate, *see* T-00716
2,4,6-Tripropyl-1,3,5,2,4,6-trioxatriphosphinane, *see* T-00720
2,4,6-Tripropyl-1,3,5,2,4,6-trioxatriphosphorinane, T-00720
S,*S*,*S*-Tripropyl trithiophosphate, *see* T-00717
Tripropyl trithiophosphite, *see* T-00718
Tri-1-propynylphosphine, T-00721
Tri(1*H*-pyrazol-1-yl)phosphine, T-00722
Tri-2-pyridinylphosphine, *see* T-00723
Tri-2-pyridylphosphine, T-00723
Tri-1-pyrrolidylphosphine, T-00724
Tri-1-pyrrolylphosphine, T-00725
Tri-2-pyrrolylphosphine, T-00726
Tris(acetoxymethyl)phosphine, *in* T-00784
Tris(2-acetyloxyphenyl)phosphine oxide, *in* T-00788
Tris(allylamino)phosphine, *see* T-00709
Tris(4-aminophenyl) phosphate, T-00727
Tris(2-aminophenyl) phosphite, T-00728
Tris(3-aminophenyl) phosphite, T-00729
▷Tris(1-aziridinyl)phosphine sulfide, *see* T-00317
Tris[1,2-benzenediolato(2−)-*O*,*O′*] phosphate(1−), T-00730
Tris(benzoyloxymethyl)phosphine oxide, *in* T-00784
Tris(benzylamino)phosphine, *see* T-00344
Tris[[1,1′-biphenyl]-2,2′-diyl]phosphate(1−), T-00731
Tris[(1,1′-biphenyl)-2-yl] phosphate, *see* T-00345
Tris[(1,1′-biphenyl)-3-yl] phosphate, *see* T-00346
Tris[(1,1′-biphenyl)-4-yl] phosphate, *see* T-00347
Tris[[1,1′-biphenyl]-2-yl]phosphine, *see* T-00348
Tris[[1,1′-biphenyl]-3-yl]phosphine, *see* T-00349
Tris[[1,1′-biphenyl]-4-yl]phosphine, *see* T-00350
Tris(*p*-bromobenzyl) phosphate, *see* T-00732
Tris(*p*-bromobenzyl)phosphine, *see* T-00733
Tris[(4-bromophenyl)methyl] phosphate, T-00732
Tris[(4-bromophenyl)methyl]phosphine, T-00733
Tris(3-bromophenyl)phosphine, T-00734
Tris(4-bromophenyl)phosphine, T-00735
Tris(butylamino)phosphine, *see* T-00394
Tris(*sec*-butylamino)phosphine, *see* T-00829
Tris(4-*tert*-butylphenyl) phosphite, T-00736
Tris(*tert*-butylphosphino)phosphine, T-00737
Tris(2-carbamoylethyl)phosphine, T-00738
Tris(3-carboxyethyl)phosphine, *see* P-00357
Tris(3-carboxyphenyl)phosphine, T-00739
Tris(4-carboxyphenyl)phosphine, T-00740
Tris(*p*-chlorobenzyl)phosphine, *see* T-00744
Tris(2-chloroethyl) phosphate, T-00741
Tris(2-chloroethyl) phosphite, T-00742
▷*O*,*O*,*O*-Tris(2-chloroethyl) phosphorothioate, T-00743
▷*N*,*N*,3-Tris(2-chloroethyl)tetrahydro-2*H*-1,3,2-oxazaphosphorin-2-amine 2-oxide, *see* T-00485
▷*O*,*O*,*O*-Tris(2-chloroethyl) thiophosphate, *see* T-00743
Tris[(4-chlorophenyl)methyl]phosphine, T-00744
Tris(4-chlorophenyl)methylphosphonium bromide, *in* T-00749
Tris(4-chlorophenyl)methylphosphonium iodide, *in* T-00749
Tris(2-chlorophenyl) phosphate, T-00745
Tris(4-chlorophenyl) phosphate, T-00746
Tris(2-chlorophenyl)phosphine, T-00747
Tris(3-chlorophenyl)phosphine, T-00748
Tris(4-chlorophenyl)phosphine, T-00749
Tris(2-chlorophenyl) phosphite, T-00750
Tris(4-chlorophenyl) phosphite, T-00751
Tris(2-cyanoethyl)methylphosphonium iodide, *in* T-00752
Tris(2-cyanoethyl)phosphine, T-00752
Tris(3-cyanophenyl)phosphine, T-00753
Tris(4-cyanophenyl)phosphine, T-00754
Tris(4-cyanophenyl) phosphite, T-00755
Tris(cyclohexylamino)phosphine, *see* T-00429
Tris(decyl)phosphine, *see* T-00433
Tris(dibenzylamino)phosphine, *see* H-00012
Tris(*N*,*N*-dibutylamino)phosphine, *see* H-00013
Tris(*N*,*N*-dibutylamino)phosphine oxide, *in* H-00013
Tris(*N*,*N*-dibutylamino)phosphine sulfide, *in* H-00013
Tris(diethylamino)phosphine, *see* H-00015
2,4,6-Tris(diethylamino)-1,3,5,2,4,6-trioxatriphosphorinane 2,4,6-trioxide, T-00756
N,*N′*,*N″*-Tris(dihydroxyphosphinylmethyl)-1,4,7-triazacyclononane, T-00757
Tris(diisobutylamino)phosphine, *see* H-00063
Tris(2,4-dimethoxyphenyl)methylphosphonium iodide, *in* T-00758
Tris(2,4-dimethoxyphenyl)phosphine, T-00758
Tris(3,5-dimethoxyphenyl)phosphine, T-00759
Tris(dimethylamino)difluorophosphorane, T-00760
P,*P*,*P*-Tris(dimethylamino)-*N*-(diphenylphosphino)phosphazene, *see* H-00069
P,*P*,*P*-Tris(dimethylamino)-*N*-(diphenylphosphino)phosphine imide, *see* H-00069
P,*P*,*P*-Tris(dimethylamino)iminophosphorane, *see* H-00077
Tris(dimethylamino)(methylamino)-phosphonium iodide, *in* H-00001
Tris(dimethylamino)methylenephosphorane, *see* H-00072
P,*P*,*P*-Tris(dimethylamino)-*N*-methyliminophosphorane, *see* H-00001
P,*P*,*P*-Tris(dimethylamino)-*N*-methylphosphazene, *see* H-00001
P,*P*,*P*-Tris(dimethylamino)-*N*-methylphosphine imide, *see* H-00001
P,*P*,*P*-Tris(dimethylamino)-*N*-phenyliminophosphorane, *see* H-00073
Tris[2-(dimethylamino)phenyl] phosphate, T-00761
Tris[3-(dimethylamino)phenyl] phosphate, T-00762
Tris[4-(dimethylamino)phenyl] phosphate, T-00763
P,*P*,*P*-Tris(dimethylamino)-*N*-phenylphosphazene, *see* H-00073
Tris(2-dimethylaminophenyl)phosphine, T-00764
Tris(4-dimethylaminophenyl)phosphine, T-00765
P,*P*,*P*-Tris(dimethylamino)-*N*-phenylphosphine imide, *see* H-00073
Tris(4-dimethylaminophenyl) phosphite, T-00766
P,*P*,*P*-Tris(dimethylamino)phosphazene, *see* H-00077
▷Tris(dimethylamino)phosphine, *see* H-00080
P,*P*,*P*-Tris(dimethylamino)phosphine imide, *see* H-00077
▷Tris(dimethylamino)phosphine sulfide, *see* H-00079
P,*P*,*P*-Tris(dimethylamino)-*N*-(trimethylsilyl)phosphazene, *see* H-00082
P,*P*,*P*-Tris(dimethylamino)-*N*-(trimethylsilyl)phosphine imide, *see* H-00082
1,2,3-Tris(1,1-dimethylethyl)-diazaphosphiridine, *see* T-00362
Tris(1,1-dimethylethyl)diphospharsirane, *see* T-00363
Tris(1,1-dimethylethyl)methylenephosphorane, *see* T-00365
[2,4,6-Tris(1,1-dimethylethyl)phenyl]-methylidynephosphine, *see* T-00368
[2,4,6-Tris(1,1-dimethylethyl)phenyl]-phosphine, *see* T-00369
Tris[4-(1,1-dimethylethyl)phenyl] phosphite, *see* T-00736
Tris(1,1-dimethylethyl) phosphate, *see* T-00371
Tris(1,1-dimethylethyl)phosphine, *see* T-00373
P,*P*,*P*-Tris(1,1-dimethylethyl)phosphine imide, *see* T-00375
2,4,6-Tris(1,1-dimethylethyl)phosphinine, *see* T-00382

Tris[(1,1-dimethylethyl)phosphino]phosphine, *see* T-00737
Tris(1,1-dimethylethyl) phosphite, *see* T-00380
N,N′,P-Tris(1,1-dimethylethyl)phosphonic diamide, *in* B-00597
N,N′,N″-Tris(1,1-dimethylethyl)phosphoric triamide, *in* T-00395
2,4,6-Tris(1,1-dimethylethyl)phosphorin, *see* T-00382
S,S,S-Tris(1,1-dimethylethyl) phosphorotrithioate, *see* T-00391
N,N′,N″-Tris(1,1-dimethylethyl)phosphorous triamide, *see* T-00395
Tris(1,1-dimethylethyl)triphosphirane, *see* T-00397
Tris(2,5-dimethylphenyl) phosphate, T-00767
Tris(2,4-dimethylphenyl)phosphine, T-00768
Tris(2,5-dimethylphenyl)phosphine, T-00769
Tris(2,6-dimethylphenyl)phosphine, T-00770
Tris(dimethylphosphino)methane, T-00771
Tris(dimethylphosphinothioyl)methane, *in* T-00771
Tris(2,2-dimethylpropyl) phosphite, T-00772
Tris(*N,N*-diphenylamino)phosphine, *see* H-00090
Tris(diphenylphosphino)amine, *see* B-00233
Tris[(2-diphenylphosphino)ethyl]amine, T-00773
Tris[2-(diphenylphosphino)ethyl]phosphine, T-00774
Tris(diphenylphosphino)methane, T-00775
1,1,1-Tris(diphenylphosphinomethyl)ethane, T-00776
Tris(2-diphenylphosphinophenyl)phosphine, T-00777
2,2′,2″-Tris(diphenylphosphino)-triethylamine, *see* T-00773
Tris(*N,N*-dipropylamino)phosphine, *see* H-00092
Tris(*N,N*-dipropylamino)phosphine oxide, *in* H-00092
Tris(*N,N*-dipropylamino)phosphine sulfide, *in* H-00092
Tris(2-ethoxyethyl) phosphate, T-00778
Tris(ethylamino)phosphine, *see* T-00462
▷Tris(2-ethylhexyl) phosphate, T-00779
▷Tris(2-ethylhexyl) phosphite, T-00780
Tris(2-fluorophenyl)phosphine, T-00781
Tris(3-fluorophenyl)phosphine, T-00782
Tris(4-fluorophenyl)phosphine, T-00783
Tris(hydroxymethyl)phosphine, T-00784
Tris(2-hydroxyphenyl)methylphosphonium iodide, *in* T-00788
Tris(2-hydroxyphenyl) phosphate, T-00785
Tris(3-hydroxyphenyl) phosphate, T-00786
Tris(4-hydroxyphenyl) phosphate, T-00787
Tris(2-hydroxyphenyl)phosphine, T-00788
Tris(3-hydroxyphenyl)phosphine, T-00789
Tris(4-hydroxyphenyl)phosphine, T-00790
Tris(1-imidazolyl)phosphine, *see* T-00495
Tris(isobutylamino)phosphine, *see* T-00830
Tris(isobutylamino)phosphine oxide, *in* T-00830
Tris(isobutylamino)phosphine sulfide, *in* T-00830
Tris(isopropylamino)phosphine, *see* T-00509
Tris(2-methoxyphenyl)methylphosphonium iodide, *in* T-00791
Tris(3-methoxyphenyl)methylphosphonium iodide, *in* T-00792
Tris(2-methoxyphenyl)phosphate, *in* T-00785
Tris(4-methoxyphenyl) phosphate, *in* T-00787
Tris(2-methoxyphenyl)phosphine, T-00791
Tris(3-methoxyphenyl)phosphine, T-00792
Tris(4-methoxyphenyl)phosphine, T-00793
Tris(2-methoxyphenyl) phosphite, T-00794
Tris(4-methoxyphenyl) phosphorotritelluroite, T-00795
Tris(4-methoxyphenyl) tritellurophosphite, *see* T-00795
Tris(methylamino)phosphine, *see* T-00551
Tris(*m*-methylbenzyl)phosphine, *see* T-00801
Tris(*o*-methylbenzyl)phosphine, *see* T-00800
Tris(*p*-methylbenzyl)phosphine, *see* T-00802
Tris(3-methylbutyl) phosphate, T-00796
1,1,1-Tris(methyldiphenylphosphoniomethyl)-ethane, *in* T-00776
Tris(1-methylethyl) benzoylphosphorimidate, *in* B-00039
Tris(1-methylethyl) phenylphosphorimidate, *in* P-00264
Tris(1-methylethyl)phosphate, *see* T-00499
Tris(1-methylethyl)phosphine, *see* T-00500
P,P,P-Tris(1-methylethyl)phosphine imide, *see* T-00501
▷Tris(1-methylethyl) phosphite, *see* T-00502
O,O,S-Tris(1-methylethyl) phosphorodithioate, *see* T-00503
O,O,O-Tris(1-methylethyl) phosphoroselenoate, *see* T-00504
Tris(1-methylethyl) phosphorotetrathioate, *see* T-00505
O,O,O-Tris(1-methylethyl) phosphorothioate, *see* T-00506
O,S,S-Tris(1-methylethyl) phosphorotrithioate, *see* T-00507
S,S,S-Tris(1-methylethyl) phosphorotrithioate, *see* T-00508
N,N′,N″-Tris(1-methylethyl)phosphorous triamide, *see* T-00509
Tris(1-methylheptyl) phosphite, T-00797
▷Tris(6-methylheptyl) phosphite, T-00798
Tris(8-methylnonyl) phosphite, T-00799
Tris[(2-methylphenyl)methyl]phosphine, T-00800
Tris[(3-methylphenyl)methyl]phosphine, T-00801
Tris[(4-methylphenyl)methyl]phosphine, T-00802
▷Tris(2-methylphenyl) phosphate, T-00803
Tris(3-methylphenyl) phosphate, T-00804
Tris(4-methylphenyl) phosphate, T-00805
Tris(2-methylphenyl)phosphine, T-00806
Tris(3-methylphenyl)phosphine, T-00807
▷Tris(4-methylphenyl)phosphine, T-00808
▷Tris(2-methylphenyl) phosphite, T-00809
▷Tris(3-methylphenyl) phosphite, T-00810
▷Tris(4-methylphenyl) phosphite, T-00811
N,N′,N″-Tris(2-methylphenyl)phosphoric triamide, *in* T-00817
N,N′,N″-Tris(3-methylphenyl)phosphoric triamide, *in* T-00818
N,N′,N″-Tris(4-methylphenyl)phosphoric triamide, *in* T-00819
O,O,O-Tris(2-methylphenyl) phosphorothioate, T-00812
O,O,O-Tris(3-methylphenyl) phosphorothioate, T-00813
O,O,O-Tris(4-methylphenyl) phosphorothioate, T-00814
N,N′,N″-Tris(2-methylphenyl)phosphorothioic triamide, *in* T-00817
N,N′,N″-Tris(4-methylphenyl)phosphorothioic triamide, *in* T-00819
S,S,S-Tris(2-methylphenyl) phosphorotrithioate, T-00815
S,S,S-Tris(4-methylphenyl) phosphorotrithioate, T-00816
N,N′,N″-Tris(2-methylphenyl)phosphorous triamide, T-00817
N,N′,N″-Tris(3-methylphenyl)phosphorous triamide, T-00818
N,N′,N″-Tris(4-methylphenyl)phosphorous triamide, T-00819
Tris(2-methyl-2-propenyl) phosphate, T-00820
Tris(2-methylpropyl) (dimethoxyphosphinyl)phosphorimidate, *in* D-00683
Tris(1-methylpropyl) phosphate, T-00821
Tris(2-methylpropyl) phosphate, T-00822
Tris(1-methylpropyl)phosphine, T-00823
Tris(2-methylpropyl)phosphine, T-00824
Tris(1-methylpropyl) phosphite, T-00825
Tris(2-methylpropyl) phosphite, T-00826
N,N′,N″-Tris(1-methylpropyl)phosphoric triamide, *in* T-00829
N,N′,N″-Tris(2-methylpropyl)phosphoric triamide, *in* T-00830
O,O,O-Tris(2-methylpropyl) phosphoroselenoate, T-00827
O,O,S-Tris(2-methylpropyl) phosphorothioate, T-00828
N,N′,N″-Tris(1-methylpropyl)phosphorothioic triamide, *in* T-00829
N,N′,N″-Tris(2-methylpropyl)phosphorothioic triamide, *in* T-00830
N,N′,N″-Tris(1-methylpropyl)phosphorous triamide, T-00829
N,N′,N″-Tris(2-methylpropyl)phosphorous triamide, T-00830
Tris[2-(methylthio)phenyl]phosphine, T-00831
▷Tris[4-(methylthio)phenyl]phosphine, T-00832
Tris(morpholino)phosphine, *see* T-00569
3,3,3-Tris-(4-morpholinophosphoranylidene)-1-(2,4,6-trinitrophenyl)triazene, T-00833
Tris(*p*-nitrobenzyl) phosphate, *see* T-00834
Tris[(4-nitrophenyl)methyl] phosphate, T-00834
Tris[(4-nitrophenyl)methyl]phosphine, T-00835
Tris(2-nitrophenyl) phosphate, T-00836
Tris(4-nitrophenyl) phosphate, T-00837
Tris(3-nitrophenyl)phosphine, T-00838
Tris(2-nitrophenyl) phosphite, T-00839
Tris(4-nitrophenyl) phosphite, T-00840
Tris(pentachlorophenyl) phosphate, T-00841
Tris(pentafluorophenyl) phosphate, T-00842
Tris(phenylamino)phosphine, *see* T-00677
Tris(phenylamino)phosphine oxide, *in* T-00677
Tris(phenylamino)phosphine sulfide, *in* T-00677
Tris(2-phenylethenyl)phosphine, T-00843
Tris(2-phenylethyl)phosphine, T-00844
Tris(2-phenylethyl) phosphite, T-00845
Tris(phenylethynyl)phosphine, T-00846
Tris(phenylmethyl) phosphate, *see* T-00337
Tris(phenylmethyl)phosphine, *see* T-00338
P,P,P-Tris(phenylmethyl)phosphine imide, *see* T-00339
Tris(phenylmethyl) phosphite, *see* T-00340
Tris(phenylmethyl)phosphorotetrathioate, *see* T-00341
S,S,S-Tris(phenylmethyl) phosphorotrithioate, *see* T-00342
Tris(phenylmethyl) phosphorotrithioite, *see* T-00343
N,N′,N″-Tris(phenylmethyl)phosphorous triamide, *see* T-00344
Tris(phosphonomethyl)phosphine oxide, *see* P-00360
▷Tris(piperidino)phosphine, *see* T-00704
▷Tris(piperidino)phosphine oxide, *in* T-00704
▷Tris(piperidino)phosphine selenide, *in* T-00704
Tris(piperidino)phosphine sulfide, *in* T-00704
Tris(piperidino)phosphine telluride, *in* T-00704
Tris(propylamino)phosphine, *see* T-00719
Tris(1*H*-pyrazol-1-yl)phosphine, *see* T-00722
Tris[pyrocatecholato(2−)] phosphate(1−), *see* T-00730
Tris(pyrrolidino)phosphine, *see* T-00724
Tris(pyrrolidino)phosphine oxide, *in* T-00724
Tris(pyrrolidino)phosphine selenide, *in* T-00724
Tris(pyrrolidino)phosphine telluride, *in* T-00724
Tristearyl phosphite, *see* T-00580
Tris(*m*-tolylamino)phosphine, *see* T-00818
Tris(*o*-tolylamino)phosphine, *see* T-00817
Tris(*p*-tolylamino)phosphine, *see* T-00819
Tris(1,2,4-triazol-1-yl)phosphine, *see* T-00876
2,4,6-Tris(2,4,6-tri-*tert*-butylphenoxy)-1,3,5,2,4,6-trioxatriphosphorine, T-00847
Tris(2,2,2-trichloroethyl) phosphate, T-00848
Tris(triethylsilyl) phosphate, T-00849
Tris(triethylsilyl) phosphite, T-00850
Tris(2,2,2-trifluoroethyl) phosphate, T-00851
Tris[2-(trifluoromethyl)phenyl]phosphine, T-00852
Tris[3-(trifluoromethyl)phenyl]phosphine, T-00853
Tris[4-(trifluoromethyl)phenyl]phosphine, T-00854
Tris(trifluoromethyl)phosphine, T-00855
Tris(trifluoromethyl)phosphine oxide, T-00856
Tris(trifluoromethyl) phosphorotriselenoite, T-00857

Tris(trifluoromethyl) triselenophosphite, *see* T-00857
Tris(2,4,6-trimethoxyphenyl)-methylphosphonium iodide, *in* T-00858
Tris(2,4-6-trimethoxyphenyl)phosphine, T-00858
Tris(trimethylgermyl)phosphate, T-00859
Tris(trimethylgermyl)phosphine, T-00860
Tris(2,4,6-trimethylphenyl)phosphine, T-00861
2,4,6-Tris(2,4,6-trimethylphenyl)-1,3,5,2,4,6-trithiatriphosphinane, *see* T-00862
2,4,6-Tris(2,4,6-trimethylphenyl)-1,3,5,2,4,6-trithiatriphosphorinane, T-00862
Tris(trimethylplumbyl)phosphine, T-00863
▷Tris(trimethylsilyl) phosphate, T-00864
▷Tris(trimethylsilyl)phosphine, T-00865
Tris(trimethylsilyl) phosphite, T-00866
NNN'-Tris(trimethylsilyl)phosphonamidimidic acid, T-00867
N,N,N'-Tris(trimethylsilyl)-phosphoramidimidic acid, T-00868
Tris(trimethylsilyl) phosphorotetrathioate, T-00869
Tris(trimethylsilyl) tetrathiophosphate, *see* T-00869
Tris(triphenylgermyl)phosphine, T-00870
Tris[(triphenylphosphoranylidene)methyl]-phosphine, T-00871
Tris(triphenylphosphoranylidene)phosphoric triamide, *in* T-00872
Tris(triphenylphosphoranylidene)phosphorous triamide, T-00872
Tristyrylphosphine, *see* T-00843
Tri(2-thiazolyl)phosphine, T-00873
Tri-2-thienylphosphine, T-00874
Tri-3-thienylphosphine, T-00875
▷Trithion, *see* C-00008
Trithiophosphoric acid diethylamide, *see* D-00359
Tri(*m*-tolylamino)phosphine, *see* T-00818
Tri(*o*-tolylamino)phosphine, *see* T-00817
Tri(*p*-tolylamino)phosphine, *see* T-00819
Tri-*m*-tolyl phosphate, *see* T-00804
▷Tri-*o*-tolyl phosphate, *see* T-00803
Tri-*p*-tolyl phosphate, *see* T-00805
Tri-*m*-tolylphosphine, *see* T-00807
Tri-*o*-tolylphosphine, *see* T-00806
▷Tri-*p*-tolylphosphine, *see* T-00808
▷Tri-*m*-tolyl phosphite, *see* T-00810
▷Tri-*o*-tolyl phosphite, *see* T-00809
▷Tri-*p*-tolyl phosphite, *see* T-00811
O,O,O-Tri-*m*-tolyl phosphorothioate, *see* T-00813
O,O,O-Tri-*o*-tolyl phosphorothioate, *see* T-00812
O,O,O-Tri-*p*-tolyl phosphorothioate, *see* T-00814
S,S,S-Tri-*o*-tolyl phosphorotrithioate, *see* T-00815
S,S,S-Tri-*p*-tolyl phosphorotrithioate, *see* T-00816
N,N',N''-Tri-*m*-tolylphosphorous triamide, *see* T-00818
N,N',N''-Tri-*o*-tolylphosphorous triamide, *see* T-00817
N,N',N''-Tri-*p*-tolylphosphorous triamide, *see* T-00819
O,O,O-Tri-*m*-tolyl thiophosphate, *see* T-00813
O,O,O-Tri-*o*-tolyl thiophosphate, *see* T-00812
O,O,O-Tri-*p*-tolyl thiophosphate, *see* T-00814
S,S,S-Tri-*o*-tolyl trithiophosphate, *see* T-00815
S,S,S-Tri-*p*-tolyl trithiophosphate, *see* T-00816
Tri(1,2,4-triazol-1-yl)phosphine, T-00876
Tritylphosphonic acid, *see* T-00608
Trivinylphosphine, T-00877
▷Trofosfamide, *see* T-00485
▷Trolene, *see* F-00003
▷Tugon, *see* T-00423
Turloc, *see* M-00439
Tyrosine *O*-dihydrogen phosphate, *see* P-00423
Tyrosine-*O*-phosphate, *see* P-00423
UDPG, *see* U-00002
▷Ultracide, *see* M-00025
▷Uredepa, U-00001
Ureidomethylphosphonic acid, *see* A-00058
Ureidophosphonic acid, *see* A-00062
▷Urethimine, *see* U-00001
Uridine diphosphate glucose, U-00002
Uridine 5'-(trihydrogen diphosphate)mono-α-D-glucopyranosyl ester, *see* U-00002
Uridylyl-(5',5')-adenosine, *see* A-00044
▷Vamidothion, V-00001
▷Vapona, *see* D-00203
▷Varbex, *see* F-00001
Vincofos, *see* D-00202
Vingard, *see* D-00202
Vinyl dichlorophosphate, *in* M-00518
O-Vinyl dichlorothiophosphate, *see* V-00008
▷Vinyl dihydrogen phosphate, *see* M-00518
1,2-Vinylenebis[diphenylphosphine], *see* B-00237
Vinylenephosphoric acid, *see* H-00136
Vinylidenebis(diphenylphosphine), *see* B-00236
Vinylidenebisphosphonic acid, *see* E-00020
▷Vinyl phosphate, *see* M-00518
Vinylphosphinic acid, V-00002
Vinylphosphonic acid, V-00003
Vinylphosphonic bis(diethylamide), *in* V-00004
Vinylphosphonic bis(dimethylamide), *in* V-00004
P-Vinylphosphonic diamide, V-00004
Vinylphosphonic dichloride, V-00005
Vinylphosphonochloridothioic acid, V-00006
Vinylphosphonothioic acid, V-00007
▷Vinyl phosphoric acid, *see* M-00518
Vinyl phosphorodichloridate, *in* M-00518
O-Vinyl phosphorodichloridothioate, V-00008
O-Vinyl thiophosphoryl dichloride, *see* V-00008
Vinyltriphenylphosphonium(1+), *see* T-00702
▷Vitamin B_{12}, V-00009
Vitamin B_{12a}, V-00010
▷Vitamin B_1 pyrophosphoric ester, *see* T-00297
▷Vitanervil, *see* B-00002
▷VX agent, *see* E-00062
▷Wepsyn 155, *see* T-00326
▷WR 2721, *see* A-00052
▷WR-2721, *see* E-00021
(9*H*-Xanthen-9-yl)phosphonic acid, X-00001
3'-Xanthylic acid, X-00002
m-Xenyl dichlorophosphate, *see* B-00082
Xylose 3-dihydrogen phosphate, X-00003
Xylose 5-dihydrogen phosphate, X-00004
2,6-Xylyl dichlorophosphate, *in* M-00466
2-Xylylenebis[triphenylphosphonium](2+), *see* P-00132
4-Xylylenebis[triphenylphosphonium](2+), *see* P-00134
m-Xylylenediphosphonic acid, *see* P-00130
o-Xylylenediphosphonic acid, *see* P-00129
p-Xylylenediphosphonic acid, *see* P-00131
2,6-Xylyl phosphate, *see* M-00466
3,4-Xylylphosphinic acid, *see* D-00806
p-Xylylphosphinic acid, *see* D-00805
3,5-Xylyl phosphorodichloridate, *see* D-00822
Yeast adenylic acid, *see* A-00046
Zilantel, Z-00001
▷Zinofos, *see* T-00314
▷Zinophos, *see* T-00314
▷Zolone, *see* P-00340
▷Zytron, Z-00002

Molecular Formula Index

Molecular Formula Index

CBrINP
Phosphorocyanidous bromide iodide, P-00409

CBr_2F_3P
Dibromo(trifluoromethyl)phosphine, *in* T-00482

CBr_2NP
Phosphorocyanidous dibromide, P-00410

CCl_2F_3OP
(Trifluoromethyl)phosphonic dichloride, T-00479

CCl_2F_3P
(Trifluoromethyl)phosphonous dichloride, T-00483

CCl_2F_3PS
Trifluoromethanethiol phosphorodichloridite, T-00472
(Trifluoromethyl)phosphonothioic acid; Dichloride, *in* T-00481

CCl_2NOPS
Phosphor(isothiocyanatidic) dichloride, P-00407

CCl_2NO_2P
Phosphorisocyanatidic dichloride, P-00405

CCl_3F_2OP
(Trichloromethyl)phosphonic difluoride, T-00414

CCl_3F_2P
(Trichloromethyl)phosphonous acid; Difluoride, *in* T-00416

CCl_5OP
(Trichloromethyl)phosphonic dichloride, T-00413

CCl_5P
Trichloromethylphosphonous dichloride, T-00417

CCl_5PS
(Trichloromethyl)phosphonothioic acid; Dichloride, *in* T-00415

CCl_6NP
(Trichloromethyl)phosphorimidic trichloride, T-00418

$CCl_6N_2OP_2$
Carbonylbis[phosphorimidic trichloride], C-00007

CCl_7P
Tetrachloro(trichloromethyl)phosphorane, T-00037

CF_2NOPS
Phosphor(isothiocyanatidic) difluoride, P-00408

CF_2NO_2P
Phosphorisocyanatidic difluoride, P-00406

CF_3I_2P
(Trifluoromethyl)diiodophosphine, *in* T-00482

CF_5OP
(Trifluoromethyl)phosphonic acid; Difluoride, *in* T-00478

CF_5P
Trifluoromethylphosphonous difluoride, T-00484

CF_7P
Tetrafluoro(trifluoromethyl)phosphorane, T-00086

$CHBrF_3P$
(Trifluoromethyl)phosphinous bromide, T-00476

$CHBr_2Cl_2P$
Dibromo(dichloromethyl)phosphine, *in* D-00181

$CHBr_2F_2OP$
(Dibromomethyl)phosphonic acid; Difluoride, *in* D-00064

$CHBr_4OP$
(Dibromomethyl)phosphonic acid; Dibromide, *in* D-00064

$CHClF_3P$
(Trifluoromethyl)phosphinous chloride, T-00477

$CHCl_2F_2OP$
(Dichloromethyl)phosphonic difluoride, D-00178

$CHCl_2F_2P$
(Dichloromethyl)difluorophosphine, *in* D-00181

$CHCl_2F_2PS$
(Dichloromethyl)phosphonothioic acid; Difluoride, *in* D-00180

$CHCl_2N_2OP$
Cyanophosphoramidic dichloride, C-00227

$CHCl_4OP$
(Dichloromethyl)phosphonic dichloride, D-00177

$CHCl_4P$
Dichloro(dichloromethyl)phosphine, *in* D-00181

$CHCl_4PS$
(Dichloromethyl)phosphonothioic acid; Dichloride, *in* D-00180

CH_2BrCl_2OP
(Bromomethyl)phosphonic acid; Dichloride, *in* B-00475

CH_2Br_2ClP
Dibromo(chloromethyl)phosphine, *in* C-00115

CH_2Br_3OP
(Bromomethyl)phosphonic acid; Dibromide, *in* B-00475

CH_2Br_3P
(Bromomethyl)phosphonous dibromide, B-00476

$CH_2Br_4P_2$
Methanediphosphonous acid; Bis(dibromide), *in* M-00024

$CH_2Br_4P_2S_2$
Methanediphosphonothioic acid; Tetrabromide, *in* M-00023

CH_2ClF_2OP
(Chloromethyl)phosphonic difluoride, C-00109

CH_2ClF_2P
▷(Chloromethyl)difluorophosphine, *in* C-00115

CH_2ClF_2PS
(Chloromethyl)phosphonothioic acid; Difluoride, *in* C-00113

CH_2Cl_2FOP
(Fluoromethyl)phosphonic acid; Dichloride, *in* F-00023

$CH_2Cl_2NO_3P$
Dichlorophosphinylcarbamic acid, D-00191

CH_2Cl_3OP
(Chloromethyl)phosphonic dichloride, C-00108

$CH_2Cl_3O_2P$
(Trichloromethyl)phosphonous acid, T-00416

$CH_2Cl_3O_2PS$
(Trichloromethyl)phosphonothioic acid, T-00415

$CH_2Cl_3O_3P$
(Trichloromethyl)phosphonic acid, T-00411

CH_2Cl_3P
(Chloromethyl)phosphonous dichloride, C-00116

CH_2Cl_3PS
(Chloromethyl)phosphonothioic acid; Dichloride, *in* C-00113

$CH_2Cl_4O_2P_2$
Methylenebis(phosphonic dichloride), M-00145

$CH_2Cl_4P_2$
Methanediphosphonous acid; Bis(dichloride), *in* M-00024

$CH_2Cl_4P_2S_2$
Methanediphosphonothioic acid; Tetrachloride, *in* M-00023

$CH_2F_3O_2P$
(Trifluoromethyl)phosphinic acid, T-00475
(Trifluoromethyl)phosphonous acid, T-00482

$CH_2F_3O_2PS$
(Trifluoromethyl)phosphonothioic acid, T-00481

$CH_2F_3O_3P$
(Trifluoromethyl)phosphonic acid, T-00478

CH_2F_3P
(Trifluoromethyl)phosphine, T-00474

$CH_2F_3PS_2$
(Trifluoromethyl)phosphonodithious acid, T-00480

$CH_2F_4O_2P_2$
Methylenebisphosphonic difluoride, M-00146

$CH_2F_4P_2$
Methanediphosphonous acid; Bis(difluoride), *in* M-00024

$CH_2F_4P_2S_2$
Methanediphosphonothioic acid; Tetrafluoride, *in* M-00023

CH_2NP
Phosphinecarbonitrile, P-00354
Phosphinidynemethylamine, P-00356

CH_3Br_2OP
Methylphosphonic acid; Dibromide, *in* M-00288
Methyl phosphorodibromidite, M-00344

$CH_3Br_2O_3P$
(Dibromomethyl)phosphonic acid, D-00064

CH_3Br_2P
Methylphosphonous dibromide, M-00321

CH_3Br_2PS
Methylphosphonothioic dibromide, M-00315

CH_3Br_2PSe
Methylphosphonoselenoic acid; Dibromide, *in* M-00311

CH_3Br_4P
Tetrabromomethylphosphorane, T-00010

CH_3ClFO_2P
(Chloromethyl)phosphonofluoridic acid, C-00112
(Fluoromethyl)phosphonochloridic acid, F-00024
Methyl phosphorochloridofluoridate, M-00341

CH_3ClN_3OP
Methylphosphonazidic acid; Chloride, *in* M-00286

$CH_3Cl_2NO_2PS$
P,P,P-Trichloro-*N*-methylsulfonylphosphazene, *in* M-00402

$CH_3Cl_2N_2O_2P$
Aminocarbonylphosphoramidic acid; Dichloride, *in* A-00062

CH_3Cl_2OP
▷Methylphosphonic dichloride, M-00290
▷Methyl phosphorodichloridite, M-00346

CH_3Cl_2OPS
O-Methyl phosphorodichloridothioate, M-00349
S-Methyl phosphorodichlorodothioate, M-00351

$CH_3Cl_2O_2P$
(Chloromethyl)phosphonochloridic acid, C-00111
(Dichloromethyl)phosphonous acid, D-00181
Methyl phosphorodichloridate, M-00345

$CH_3Cl_2O_2PS$
(Dichloromethyl)phosphonothioic acid, D-00180

$CH_3Cl_2O_3P$
(Dichloromethyl)phosphonic acid, D-00175

CH_3Cl_2P
▷Methylphosphonous dichloride, M-00322

CH_3Cl_2PS
▷Methylphosphonothioic dichloride, M-00316
Methyl phosphorodichloridothioite, M-00350

$CH_3Cl_2PS_2$
Methyl phosphorodichloridodithioate, M-00347

CH_3Cl_2PSe
Methylphosphonoselenoic dichloride, M-00312
Methyl phosphorodichloridoselenoite, M-00348

$CH_3Cl_3NO_2P$
P-Trichloromethylphosphonamidic acid, T-00410

CH_3Cl_3P
P,P,P-Trichloro-*N*-methylphosphazene, *in* M-00336

$CH_3Cl_4NO_2P_2$
P-Chloromethyl-*N*-(dichlorophosphinyl)-phosphonochloridimidic acid, C-00091
Methylimidodiphosphoryl chloride, M-00160

$CH_3Cl_4NP_2$
Methylimidodiphosphorous tetrachloride, M-00158

$CH_3Cl_4NP_2S_2$
(Methylthio)imidodiphosphoryl chloride, M-00414

CH_3Cl_4P
Tetrachloromethylphosphorane, T-00032

CH_3F_2OP
▷Methylphosphonic difluoride, M-00291
Methyl phosphorodifluoridite, M-00353

CH_3F_2OPS
O-Methyl phosphorodifluoridothioate, M-00354
S-Methyl phosphorodifluoridothioate, M-00355

$CH_3F_2O_2P$
▷Methyl phosphorodifluoridate, M-00352

$CH_3F_2O_3P$
(Difluoromethyl)phosphonic acid, D-00415

CH_3F_2P
Methylphosphonous difluoride, M-00324

CH_3F_2PS
Methylphosphonothioic difluoride, M-00317

CH_3F_2PSe
Methylphosphonoselenoic acid; Difluoride, *in* M-00311

CH_3F_3P
P,P,P-Trifluoro-*N*-methylphosphazene, *in* M-00336

$CH_3F_4NO_2P_2$
Methylimidobisphosphoryl fluoride, *in* M-00156

$CH_3F_4NP_2$
Methylimidodiphosphorous tetrafluoride, M-00159

CH_3F_4OP
Tetrafluoromethoxyphosphorane, T-00080

CH_3F_4P
Tetrafluoromethylphosphorane, T-00081

CH_3F_4PS
Tetrafluoromethylthiophosphorane, T-00082

CH_3I_2P
Methylphosphonous diiodide, M-00325

$CH_3I_4NP_2$
Methylimidodi(phosphorous acid); Tetraiodide, *in* M-00157

CH_3I_4P
Tetraiodomethylphosphorane, T-00163

$CH_3N_2O_3P$
Cyanophosphorimidic acid, C-00228
▷Diazomethylphosphonic acid, D-00035

CH_3N_6OP
Methylphosphonic acid; Diazide, *in* M-00288

CH_3OP_3
Triphosphetan-4-one, T-00703

CH_3O_5P
Phosphonoformic acid, P-00387

CH_4BrOPS
Methylphosphonobromidothioic acid, M-00295

CH_4BrO_3P
(Bromomethyl)phosphonic acid, B-00475

CH_4ClOP
Methylphosphonochloridic acid, M-00296
Methylphosphonochloridous acid, M-00300

CH_4ClOPS
Methylphosphonochloridothioic acid, M-00298

CH_4ClO_2P
(Chloromethyl)phosphonous acid, C-00115

CH_4ClO_2PS
(Chloromethyl)phosphonothioic acid, C-00113

CH_4ClO_3P
(Chloromethyl)phosphonic acid, C-00106

CH_4ClP
(Chloromethyl)phosphine, C-00104

CH_4ClPS
Methylphosphonochloridothious acid, M-00299

CH_4ClPS_2
Methylphosphonochloridodithioic acid, M-00297

CH_4Cl_2NOP
Methylphosphoramidic dichloride, M-00327

$CH_4Cl_2NO_2P$
P-Dichloromethylphosphonamidic acid, D-00174

$CH_4Cl_2NO_3PS$
Methylsulfonylphosphoramidic acid; Dichloride, *in* M-00401

CH_4Cl_2NPS
Methylphosphoramidothioic dichloride, M-00334

$CH_4Cl_2O_6P_2$
Dichloromethylenebisphosphonic acid, D-00172

$CH_4Cl_3N_2OP$
P-(Trichloromethyl)phosphonic diamide, T-00412

CH_4FOP
Methylphosphonofluoridous acid, M-00307

CH_4FOPS
Methylphosphonofluoridothioic acid, M-00305

CH_4FO_2P
▷Methylphosphonofluoridic acid, M-00303

CH_4FO_3P
(Fluoromethyl)phosphonic acid, F-00023

CH_4FPS
Methylphosphonofluoridothious acid, M-00306

CH_4FPS_2
Methylphosphonofluoridodithioic acid, M-00304

CH_4F_2NOP
Methylphosphoramidic difluoride, M-00328

CH_4F_2NP
Methylphosphoramidous acid; Difluoride, *in* M-00335

CH_4IO_3P
(Iodomethyl)phosphonic acid, I-00013

CH_4NO_4P
Carbamoylphosphonic acid, C-00003

CH_4NO_5P
▷Carbamoyl dihydrogen phosphate, C-00002
(Nitromethyl)phosphonic acid, N-00037

$CH_4N_3O_2P$
Methylphosphonazidic acid, M-00286

$CH_5ClNOPS$
Methylphosphoramidochloridothioic acid, M-00329

CH_5ClNO_2P
P-Chloromethylphosphonamidic acid, C-00105

CH_5ClN_2OP
P-(Dichloromethyl)phosphonic diamide, D-00176

CH_5FNOP
P-Methylphosphonamidic fluoride, *in* M-00283

CH_5FNOPS
O-Methyl phosphoramidofluoridothioate, M-00331

CH_5FNPS
P-Methylphosphonamidothioic acid; Fluoride, *in* M-00284

$CH_5N_2O_4P$
Aminocarbonylphosphoramidic acid, A-00062

CH_5N_4OP
P-Methylphosphonazidic amide, M-00287

CH_5OP
Methyl phosphinate, M-00276

CH_5OPS
Methylphosphinothioic acid, M-00279

CH_5OPS_2
Methylphosphonodithioic acid, M-00301

CH_5O_2P
Methylphosphinic acid, M-00278
Methylphosphonous acid, M-00319

CH_5O_2PS
Methylphosphonothioic acid, M-00313

CH_5O_2PSe
Methylphosphonoselenoic acid, M-00311

CH_5O_3P
▷Methylphosphonic acid, M-00288
▷Monomethyl phosphonate, M-00490

CH_5O_3PS
(Mercaptomethyl)phosphonic acid, M-00018
O-Methyl phosphorothioate, M-00359
S-Methyl phosphorothioate, M-00360

CH_5O_4P
(Hydroxymethyl)phosphonic acid, H-00162
Methylphosphonoperoxoic acid, M-00310
▷Monomethyl phosphate, M-00489

$CH_5O_4P_3$
1,3-Dihydroxytriphosphetane 1,3-dioxide, D-00627

CH_5O_6P
Monomethoxycarbonyl phosphate, M-00479

CH_5P
▷Methylphosphine, M-00277

CH_5PS_2
Methylphosphonodithious acid, M-00302

CH_5PS_3
Methylphosphonotrithioic acid, M-00318

CH_6ClN_2OP
P-Chloromethylphosphonic diamide, C-00107

CH_6ClN_2PS
P-(Chloromethyl)phosphonothioic diamide, C-00114

CH_6NOPS
P-Methylphosphonamidothioic acid, M-00284

CH_6NOPS_2
Methylphosphoramidodithioic acid, M-00330

CH_6NO_2P
P-Methylphosphonamidic acid, M-00283
Methylphosphoramidous acid, M-00335

CH_6NO_2PS
Methylphosphoramidothioic acid, M-00333

CH_6NO_2PSe
Methylphosphoramidoselenoic acid, M-00332

CH_6NO_3P
(Aminomethyl)phosphonic acid, A-00091
Methylphosphoramidic acid, M-00326
Methylphosphorimidic acid, M-00336

CH_6NO_5PS
Methylsulfonylphosphoramidic acid, M-00401
(Methylsulfonyl)phosphorimidic acid, M-00402

$CH_6N_3O_3P$
Phosphoguanidine, P-00370

$CH_6O_4P_2$
Methanediphosphonous acid, M-00024
Methylenebis[phosphinic acid], M-00143

$CH_6O_4P_2S_2$
Methanediphosphonothioic acid, M-00023

$CH_6O_6P_2$
Methanediphosphonic acid, M-00022

$CH_6O_7P_2$
Hydroxymethanediphosphonic acid, H-00154

CH_6P_2
Diphosphinomethane, D-01178

$CH_7NO_4P_2$
Methylimidodi(phosphorous acid), M-00157

$CH_7NO_4P_2S_2$
Methylimidothiodiphosphoric acid, M-00161

$CH_7NO_5P_2S$
Methylthioimidodiphosphoric acid, M-00413

$CH_7NO_6P_2$
(Aminomethylene)bisphosphonic acid, A-00086
Methylimidodiphosphoric acid, M-00156

CH_7N_2OP
P-Methylphosphonic diamide, M-00289
Methyl phosphorodiamidite, M-00342

CH_7N_2OPS
P-Methylphosphonohydrazidothioic acid, M-00309
O-Methyl phosphorodiamidothioate, *in* M-00359
S-Methyl phosphorodiamidothioate, M-00343

$CH_7N_2O_2P$
P-Methylphosphonohydrazidic acid, M-00308

CH_7N_2P
Methylphosphonous diamide, M-00320

CH_7N_2PS
P-Methylphosphonothioic diamide, M-00314

$CH_{10}N_4O_2P_2$
P,P'-Methylenebis(phosphonic diamide), M-00144

C_2BrF_6OP
Bis(trifluoromethyl)phosphinic acid; Bromide, *in* B-00411

C_2BrF_6P
Bromobis(trifluoromethyl)phosphine, *in* B-00418

C_2BrF_6PS
Bis(trifluoromethyl)phosphinothioic acid; Bromide, *in* B-00415

$C_2BrN_2O_3P$
Phosphorodiisocyanatidic bromide, P-00411

C_2ClF_6OP
Bis(trifluoromethyl)phosphinic acid; Chloride, *in* B-00411

C_2ClF_6P
▷Bis(trifluoromethyl)phosphinous chloride, B-00420

C_2ClF_6PS
Bis(trifluoromethyl)phosphinothioic acid; Chloride, *in* B-00415

$C_2ClN_2O_3P$
Phosphorodiisocyanatidic chloride, P-00412

$C_2Cl_3F_5NP$
(Pentafluoroethyl)phosphorimidic trichloride, P-00013

$C_2Cl_4N_3P$
2,2,4,6-Tetrachloro-2,2-dihydro-1,3,5,2-triazaphosphorine, T-00029

$C_2Cl_5N_2P$
2,2-Dichloro-2,2-dihydro-4-trichloromethyl-1,3,2-diazaphosphete, D-00165

C_2Cl_6NOP
(Trichloroacetyl)phosphorimidic trichloride, T-00398

C_2Cl_7OP
Bis(trichloromethyl)phosphinic acid; Chloride, *in* B-00402

C_2Cl_8NP
N-Chloro-*P,P*-bis(trichloromethyl)-phosphinimidic chloride, *in* B-00403
(Pentachloroethyl)phosphorimidic trichloride, P-00005

$C_2Cl_9NOP_2$
[Bis(trichloromethyl)phosphinyl]-phosphorimidic trichloride, B-00404

C_2Cl_9P
Trichlorobis(trichloromethyl)phosphorane, T-00399

$C_2FN_2OPS_2$
Phosphorodi(isothiocyanatidic) fluoride, P-00414

$C_2FN_2O_3P$
Phosphorodiisocyanatidic fluoride, P-00413

C_2F_5P
(Difluoromethylene)(trifluoromethyl)-phosphine, D-00413

C_2F_6IP
Iodobis(trifluoromethyl)phosphine, *in* B-00418

C_2F_6IPS
Bis(trifluoromethyl)phosphinothioic acid; Iodide, *in* B-00415

C_2F_7OP
Bis(trifluoromethyl)phosphinic acid; Fluoride, *in* B-00411

C_2F_7P
Fluorobis(trifluoromethyl)phosphine, *in* B-00418

C_2F_7PS
Bis(trifluoromethyl)phosphinothioic acid; Fluoride, *in* B-00415

C_2F_9P
Trifluorobis(trifluoromethyl)phosphorane, T-00465

$C_2HCl_3O_2P$
2,2-Dichloroethenyl phosphorodichloridate, D-00168

$C_2HCl_6O_2P$
Bis(trichloromethyl)phosphinic acid, B-00402

C_2HCl_7NP
PP-Bis(trichloromethyl)phosphinimidic chloride, B-00403

C_2HF_6OP
Bis(trifluoromethyl)phosphinous acid, B-00418

C_2HF_6OPS
Bis(trifluoromethyl)phosphinothioic acid, B-00415

$C_2HF_6O_2P$
Bis(trifluoromethyl)phosphinic acid, B-00411

C_2HF_6PS
Bis(trifluoromethyl)phosphinothious acid, B-00417

$C_2HF_6PS_2$
Bis(trifluoromethyl)phosphinodithioic acid, B-00412

C_2HF_6PSe
Bis(trifluoromethyl)phosphinoselenous acid, B-00413

C_2HF_6PTe
Bis(trifluoromethyl)phosphinotellurous acid, B-00414

$C_2H_2Br_3Cl_2O_2P$
2,2,2-Tribromoethyl phosphorodichloridate, *in* M-00516

$C_2H_2ClPS_2$
1,3,2-Dithiaphosphole; 2-Chloro, *in* D-01246

$C_2H_2Cl_2NO_2P$
(Isocyanatomethyl)phosphonic acid; Dichloride, *in* I-00022

$C_2H_2Cl_3O_2P$
(Dichlorophosphinyl)acetic acid; Chloride, *in* D-00190

$C_2H_2Cl_5OP$
2,2,2-Trichloroethyl phosphorodichloridite, T-00406

$C_2H_2Cl_5O_2P$
2,2,2-Trichloroethyl phosphorodichloridate, *in* T-00424

$C_2H_2Cl_6NOP$
P,P-Bis(trichloromethyl)phosphinic amide, *in* B-00402

$C_2H_2F_2NO_2P$
(Isocyanatomethyl)phosphonic acid; Difluoride, *in* I-00022

$C_2H_2F_6NP$
▷*P,P*-Bis(trifluoromethyl)phosphinous amide, B-00419

$C_2H_2F_6NPS$
P,P-Bis(trifluoromethyl)phosphinothioic amide, B-00416

$C_2H_2F_6P_2$
1,2-Bis(trifluoromethyl)diphosphine, B-00410

$C_2H_2O_4P$
Acetylphosphonic acid, A-00026

$C_2H_3Cl_2OP$
Vinylphosphonic dichloride, V-00005

$C_2H_3Cl_2OPS$
O-Vinyl phosphorodichloridothioate, V-00008

$C_2H_3Cl_2O_2P$
Ethenyl phosphorodichloridate, *in* M-00518

$C_2H_3Cl_2O_3P$
(Dichlorophosphinyl)acetic acid, D-00190

$C_2H_3Cl_2PS$
Vinylphosphonothioic acid; Dichloride, *in* V-00007

$C_2H_3Cl_4OP$
Methyl(trichloromethyl)phosphinic acid; Chloride, *in* M-00419

$C_2H_3FNOPS_2$
O-Methyl phosphorofluoridoisothiocyanatidothioate, M-00358

$C_2H_3F_2OP$
Vinylphosphonic acid; Difluoride, *in* V-00003

$C_2H_3F_2PS$
Vinylphosphonothioic acid; Difluoride, *in* V-00007

$C_2H_3N_4O_2P$
Methylphosphonisocyanatidic acid; Azide, *in* M-00292

$C_2H_3O_2P$
Ethynylphosphinic acid, E-00202
Ethynylphosphonous acid, E-00204

$C_2H_3O_3P$
Ethynylphosphonic acid, E-00203

$C_2H_3O_4P$
2-Hydroxy-1,3,2-dioxaphosphole 2-oxide, H-00136

C_2H_3P
Ethylidynephosphine, E-00077

$C_2H_3PS_2$
1,3,2-Dithiaphosphole, D-01246

$C_2H_4BrCl_2OP$
(2-Bromoethyl)phosphonic acid; Dichloride, *in* B-00472

$C_2H_4BrCl_2O_2P$
2-Bromoethyl phosphorodichloridate, *in* M-00447

$C_2H_4Br_3OP$
Bis(bromomethyl)phosphinic acid; Bromide, *in* B-00108
(2-Bromoethyl)phosphonic acid; Dibromide, *in* B-00472

$C_2H_4Br_3O_4P$
Mono(2,2,2-tribromoethyl) phosphate, M-00516

$C_2H_4Br_4P_2$
1,2-Ethanediphosphonous acid; Bis(dibromide), *in* E-00008

$C_2H_4Br_4P_2S_2$
1,2-Ethanediphosphonothioic acid; Tetrabromide, *in* E-00007

C_2H_4ClOPS
2-Chloro-1,3,2-oxathiaphospholane, *in* O-00056
3-Chloro-1,3-thiaphosphetane; 3-Oxide, *in* C-00188
Vinylphosphonochloridothioic acid, V-00006

$C_2H_4ClOPS_2$
S,S-Ethylene phosphorochloridodithioate, *in* C-00071

$C_2H_4ClO_2P$
2-Chloro-1,3,2-dioxaphospholane, C-00063

$C_2H_4ClO_2PS$
2-Chloro-1,3,2-dioxaphospholane 2-sulfide, C-00065

$C_2H_4ClO_3P$
2-Chloro-1,3,2-dioxaphospholane 2-oxide, C-00064

$C_2H_4ClO_4P$
(Chloroacetyl)phosphonic acid, C-00020

C_2H_4ClPS
3-Chloro-1,3-thiaphosphetane, C-00188
3-Chloro-1,3-thiaphosphetane; 3-Sulfide, *in* C-00188

$C_2H_4ClPS_2$
2-Chloro-1,3,2-dithiaphospholane, C-00071

$C_2H_4ClPS_3$
Ethylene phosphorochloridotrithioate, *in* C-00071

$C_2H_4Cl_2FOPS$
O-(2-Fluoroethyl) phosphorodichloridothioate, F-00019

$C_2H_4Cl_2NO_3P$
Methyl dichlorophosphinylcarbamate, *in* D-00191

$C_2H_4Cl_3OP$
Bis(chloromethyl)phosphinic acid; Chloride, *in* B-00127
(2-Chloroethyl)phosphonic dichloride, C-00076

$C_2H_4Cl_3OPS$
O-(2-Chloroethyl) phosphorodichloridothioate, C-00080
S-(2-Chloroethyl) phosphorodichloridothioate, C-00081

$C_2H_4Cl_3O_2P$
1-Chloroethyl phosphorodichloridate, C-00079
2-Chloroethyl phosphorodichloridate, *in* M-00454
Methyl(trichloromethyl)phosphinic acid, M-00419

$C_2H_4Cl_3O_4P$
(2,2,2-Trichloro-1-hydroxyethyl)phosphonic acid, T-00408
Triclofos, T-00424

$C_2H_4Cl_3PS$
Bis(chloromethyl)phosphinothioic acid; Chloride, *in* B-00129
(2-Chloroethyl)phosphonothioic acid; Dichloride, *in* C-00078

$C_2H_4Cl_4O_2P_2$
1,2-Ethanediylbis[phosphonic dichloride], *in* E-00006

$C_2H_4Cl_4P_2$
1,2-Ethanediphosphonous acid; Bis(dichloride), *in* E-00008

$C_2H_4Cl_4P_2S_2$
1,2-Ethanediphosphonothioic acid; Tetrachloride, *in* E-00007

$C_2H_4FO_2P$
2-Fluoro-1,3,2-dioxaphospholane, F-00017

C_2H_4FPS
3-Fluoro-1,3-thiaphosphetane, F-00050

$C_2H_4FPS_2$
2-Fluoro-1,3,2-dithiaphospholane, *in* D-01243
3-Fluoro-1,3-thiaphosphetane; 3-Sulfide, *in* F-00050

$C_2H_4F_2ClPS$
(2-Chloroethyl)phosphonothioic acid; Difluoride, *in* C-00078

$C_2H_4F_3O_2P$
Methyl(trifluoromethyl)phosphinic acid, M-00420

$C_2H_4NO_2PS$
Methylphosphonisothiocyanatidic acid, M-00294

$C_2H_4NO_3P$
(Cyanomethyl)phosphonic acid, C-00215
(Isocyanomethyl)phosphonic acid, I-00024
Methylphosphonisocyanatidic acid, M-00292

$C_2H_4NO_4P$
(Isocyanatomethyl)phosphonic acid, I-00022

$C_2H_4N_3O_2P$
(1*H*-1,2,4-Triazol-1-yl)phosphonous acid, T-00332

$C_2H_4N_3O_2PS_2$
2-Mercapto-1,3,5,2-triazaphosphorine-4,6(1*H*,5*H*)-dione 2-sulfide, M-00020

$C_2H_5BrFO_2P$
Ethyl phosphorobromidofluoridate, E-00162

$C_2H_5Br_2OP$
Ethyl phosphorodibromidite, E-00166

$C_2H_5Br_2O_2P$
Bis(bromomethyl)phosphinic acid, B-00108

$C_2H_5Br_2P$
Dibromoethylphosphine, *in* E-00153

$C_2H_5Br_2PS$
Ethylphosphonothioic acid; Dibromide, *in* E-00148

$C_2H_5ClFO_2P$
Ethyl phosphorochloridofluoridate, E-00164

$C_2H_5Cl_2OP$
(Chloromethyl)methylphosphinic acid; Chloride, *in* C-00100
▷Ethylphosphonic dichloride, E-00131
▷Ethyl phosphorodichloridite, E-00168

$C_2H_5Cl_2OPS$
Bis(chloromethyl)phosphinothioic acid, B-00129
O-Ethyl phosphorodichloridothioate, E-00170
S-Ethyl phosphorodichloridothioate, E-00171

$C_2H_5Cl_2O_2P$
Bis(chloromethyl)phosphinic acid, B-00127
(2-Chloroethyl)phosphonochloridic acid, C-00077
▷Ethyl phosphorodichloridate, E-00167

$C_2H_5Cl_2O_2PS$
O-(2-Hydroxyethyl) phosphorodichloridothioate, H-00149

$C_2H_5Cl_2P$
▷Ethylphosphonous dichloride, E-00155

$C_2H_5Cl_2PS$
Ethylphosphonothioic dichloride, E-00150
Ethyl phosphorodichloridothioite, E-00172

$C_2H_5Cl_2PS_2$
Ethyl phosphorodichloridodithioate, E-00169

$C_2H_5Cl_2PSe$
Ethylphosphonoselenoic dichloride, E-00147

$C_2H_5Cl_3NO_3PS$
Ethyl (chlorosulfonyl)phosphorodichlorimidate, E-00056

$C_2H_5Cl_4NO_2P_2$
Ethyl (dichlorophosphinyl)-phosphorodichloroimidate, E-00058

$C_2H_5Cl_4NP_2$
Ethylimidodiphosphorous tetrachloride, E-00080

$C_2H_5F_2OP$
Ethylphosphonic difluoride, E-00132
Ethyl phosphorodifluoridite, E-00174

$C_2H_5F_2OPS$
O-Ethyl phosphorodifluoridothioate, E-00175
S-Ethyl phosphorodifluoridothioate, E-00176

$C_2H_5F_2O_2P$
▷Ethyl phosphorodifluoridate, E-00173

$C_2H_5F_2P$
Ethyldifluorophosphine, *in* E-00153

$C_2H_5F_2PS$
Ethylphosphonothioic difluoride, E-00151

$C_2H_5F_4NO_2P_2$
Ethylimidobisphosphoryl fluoride, *in* E-00078

$C_2H_5I_2O_2P$
Bis(iodomethyl)phosphinic acid, B-00288

$C_2H_5I_2P$
Ethyldiiodophosphine, *in* E-00153

$C_2H_5N_2O_2P$
P-Methylphosphonisocyanatidic amide, M-00293

$C_2H_5N_2O_3P$
(1-Diazoethyl)phosphonic acid, D-00033

C_2H_5OPS
1,3,2-Oxathiaphospholane, O-00056

$C_2H_5OPS_2$
1,3,2-Dithiaphospholane 2-oxide, D-01244

$C_2H_5O_2P$
1,3,2-Dioxaphospholane, D-00970
Vinylphosphinic acid, V-00002

$C_2H_5O_2PS$
3-Hydroxy-1,3-thiaphosphetane 3-oxide, H-00201
2-Mercapto-1,3,2-dioxaphospholane, *in* D-00970
Vinylphosphonothioic acid, V-00007

$C_2H_5O_3P$
▷2-Hydroxy-1,3,2-dioxaphospholane, *in* D-00970
▷2-Hydroxy-1,3,2-dioxaphospholane, H-00133
Vinylphosphonic acid, V-00003

$C_2H_5O_3PS$
Acetylphosphonothioic acid, A-00028
2-Hydroxy-1,3,2-dioxaphospholane 2-sulfide, H-00135

$C_2H_5O_4P$
2-Hydroxy-1,3,2-dioxaphospholane 2-oxide, H-00134
▷Monovinyl phosphate, M-00518
(2-Oxoethyl)phosphonic acid, O-00059

$C_2H_5O_5P$
Acetyl phosphate, A-00025
Methyl (dihydroxyphosphinyl)formate, *in* P-00387
Phosphonoacetic acid, P-00376

$C_2H_5O_6P$
Phosphonooxyacetic acid, P-00390

C_2H_5P
Phosphirane, P-00361

$C_2H_5PS_2$
1,3,2-Dithiaphospholane, D-01243

$C_2H_5PS_3$
1,3,2-Dithiaphospholane 2-sulfide, D-01245

$C_2H_5P_3$
2,3-Dihydro-1*H*-1,2,3-triphosphole, D-00615

$C_2H_6BrO_3P$
(2-Bromoethyl)phosphonic acid, B-00472
Dimethyl phosphorobromidate, D-00886

$C_2H_6BrO_4P$
Mono(2-bromoethyl) phosphate, M-00447

C_2H_6BrP
Bromodimethylphosphine, *in* D-00844

C_2H_6BrPS
Dimethylphosphinothioic bromide, D-00841
Ethylphosphonobromidothious acid, E-00134

C_2H_6BrPSe
Dimethylphosphinoselenoic acid; Bromide, *in* D-00835

$C_2H_6Br_2NP$
Dimethylphosphoramidous acid; Dibromide, *in* D-00875

C_2H_6ClOP
Dimethylphosphinic chloride, D-00831
Ethylphosphonochloridous acid, E-00139
Methyl methylphosphonochloridate, *in* M-00296
Methyl methylphosphonochloridite, *in* M-00300

C_2H_6ClOPS
Ethylphosphonochloridothioic acid, E-00137
O-Methyl methylphosphonochloridothioate, M-00169
S-Methyl methylphosphonochloridothioate, *in* M-00298

$C_2H_6ClOPS_2$
S,S-Dimethyl phosphorochloridodithioate, D-00888

$C_2H_6ClO_2P$
(Chloromethyl)methylphosphinic acid, C-00100
▷Dimethyl chlorophosphite, D-00715
Ethylphosphonochloridic acid, E-00135

$C_2H_6ClO_2PS$
(2-Chloroethyl)phosphonothioic acid, C-00078
▷*O,O*-Dimethyl phosphorochloridothioate, D-00891
O,S-Dimethyl phosphorochloridothioate, D-00892

$C_2H_6ClO_3P$
(1-Chloroethyl)phosphonic acid, C-00073
▷(2-Chloroethyl)phosphonic acid, C-00074
Dimethyl phosphorochloridate, D-00887
Methyl hydrogen chloromethylphosphonate, *in* C-00106

$C_2H_6ClO_3PS$
▷Dimethyl chlorothiophosphonate, D-00716

$C_2H_6ClO_4P$
Mono(2-chloroethyl) phosphate, M-00454

C_2H_6ClP
▷Dimethylphosphinous chloride, D-00846

C_2H_6ClPS
Dimethylphosphinothioic chloride, D-00842
Ethylphosphonochloridothious acid, E-00138
Methyl methylphosphonochloridothioite, *in* M-00299

$C_2H_6ClPS_2$
Dimethyl phosphorochloridodithioite, D-00889
Ethylphosphonochloridodithioic acid, E-00136
▷Methyl methylphosphonochloridodithioate, *in* M-00297

$C_2H_6ClPS_3$
Dimethyl phosphorochloridotrithioate, D-00893

C_2H_6ClPSe
Dimethylphosphinoselenoic acid; Chloride, *in* D-00835

$C_2H_6ClPSe_2$
Dimethyl phosphorochloridoselenoite, D-00890

$C_2H_6Cl_2NOP$
P,P-Bis(chloromethyl)phosphinic amide, B-00128
▷Dimethylphosphoramidic dichloride, D-00856
Ethylphosphoramidic acid; Dichloride, *in* E-00156

$C_2H_6Cl_2NO_2P$
Methyl *P*-dichloromethylphosphonamidate, *in* D-00174

$C_2H_6Cl_2NO_3P$
Dimethyl dichlorophosphoramidate, D-00723

$C_2H_6Cl_2NP$
▷Dimethylphosphoramidous dichloride, D-00876
Ethylphosphoramidous acid; Dichloride, *in* E-00160

$C_2H_6Cl_2NPS$
Dimethylphosphoramidothioic dichloride, D-00872
Ethylphosphoramidothioic acid; Dichloride, *in* E-00159

$C_2H_6Cl_2N_2O_2P_2$
2,4-Dichloro-1,3-dimethyl-1,3,2,4-diazadiphosphetidine; 2,4-Dioxide, *in* D-00166

$C_2H_6Cl_2N_2P_2$
2,4-Dichloro-1,3-dimethyl-1,3,2,4-diazadiphosphetidine, D-00166

$C_2H_6Cl_2N_2P_2S$
▷2,5-Dichloro-3,4-dimethyl-1,3,4,2,5-thiadiazadiphospholidine, D-00167

$C_2H_6Cl_2N_2P_2S_2$
2,4-Dichloro-1,3-dimethyl-1,3,2,4-diazadiphosphetidine; 2,4-Disulfide, *in* D-00166

$C_2H_6Cl_3P$
Trichlorodimethylphosphorane, T-00402

C_2H_6FOP
Dimethylphosphinic fluoride, D-00832
Methyl methylphosphonofluoridite, *in* M-00307

C_2H_6FOPS
O-Methyl methylphosphonofluoridothioate, *in* M-00305

$C_2H_6FOPS_2$
S,S-Dimethyl phosphorofluoridodithioate, *in* D-00900

$C_2H_6FO_2P$
Ethylphosphonofluoridic acid, E-00142
Methyl methylphosphonofluoridate, *in* M-00303

$C_2H_6FO_2PS$
O,O-Dimethyl phosphorofluoridothioate, D-00902

$C_2H_6FO_3P$
▷Dimethyl phosphorofluoridate, D-00901
Ethyl phosphorofluoridate, E-00179

C_2H_6FP
▷Dimethylphosphinous fluoride, D-00848

C_2H_6FPS
Dimethylphosphinothioic acid; Fluoride, *in* D-00839

$C_2H_6FPS_2$
Ethylphosphonofluoridodithioic acid, E-00143

$C_2H_6FPS_3$
Dimethyl phosphorofluoridotrithioate, D-00903

$C_2H_6F_2NOP$
Dimethylphosphoramidic difluoride, D-00858
Ethylphosphoramidic acid; Difluoride, *in* E-00156

$C_2H_6F_2NP$
Dimethylphosphoramidous difluoride, D-00877
Ethylphosphoramidous acid; Difluoride, *in* E-00160

$C_2H_6F_2NPS$
Dimethylphosphoramidothioic difluoride, D-00873

$C_2H_6F_2NPSe$
Dimethylphosphoramidoselenoic acid; Difluoride, *in* D-00869

$C_2H_6F_2P_2S_3$
Dimethylthiodiphosphonic acid; Difluoride, *in* D-00925
Dimethylthiodiphosphonic acid difluoride, *in* M-00304

$C_2H_6F_3O_2P$
Trifluorodimethoxyphosphorane, T-00468

$C_2H_6F_3P$
Trifluorodimethylphosphorane, T-00469

$C_2H_6F_4NP$
(Dimethylamino)tetrafluorophosphorane, D-00708

C_2H_6IP
Iododimethylphosphine, *in* D-00844

C_2H_6IPS
Dimethylphosphinothioic acid; Iodide, *in* D-00839

$C_2H_6I_2NP$
Dimethylphosphoramidous acid; Diiodide, *in* D-00875

$C_2H_6I_4NP$
(Dimethylamino)tetraiodophosphorane, D-00709

C_2H_6NOPS
1,3,2-Thiazaphospholidine 2-oxide, T-00302

$C_2H_6NO_2P$
1-Aziridinylphosphonous acid, A-00154
2-Hydroxy-1,2-azaphosphetidine 2-oxide, H-00107

$C_2H_6NO_3P$
1-Aziridinylphosphonic acid, A-00153

$C_2H_6NO_4P$
Acetylphosphoramidic acid, A-00029
(Aminocarbonylmethyl)phosphonic acid, A-00060
(Methylaminocarbonyl)phosphonic acid, M-00088

$C_2H_6NO_5P$
(1-Nitroethyl)phosphonic acid, N-00035
(2-Nitroethyl)phosphonic acid, N-00036
N-Phosphoglycine, P-00368

$C_2H_6NPS_2$
1,3,2-Thiazaphospholidine 2-sulfide, T-00303

$C_2H_6N_3OP$
Dimethylphosphinic acid; Azide, *in* D-00829

$C_2H_6N_3O_2PS$
O,O-Dimethyl phosphorazidothioate, D-00879

$C_2H_6N_3O_3P$
▷Dimethyl phosphorazidate, D-00878

$C_2H_6N_3PS$
Dimethylphosphinothioic acid; Azide, *in* D-00839

$C_2H_6N_3PSe$
Dimethylphosphinoselenoic acid; Azide, *in* D-00835

$C_2H_6O_4P_2$
1,3-Dihydroxy-1,3-diphosphetane 1,3-dioxide, D-00620

$C_2H_6O_6P_2$
2,5-Dihydroxy-1,4,2,5-dioxadiphosphorinane 2,5-dioxide, D-00618
Ethenylidenebisphosphonic acid, E-00020

$C_2H_6O_{11}P_4$
2,4,6,8-Tetrahydroxy-1,5-dimethyl-3,7,9-trioxa-2,4,6,8-tetraphosphabicyclo[3.3.1]nonane, T-00161

C_2H_7ClNOP
Dimethylphosphoramidochloridous acid, D-00862

$C_2H_7ClNOPS$
Dimethylphosphoramidochloridothioic acid, D-00861
Ethylphosphoramidochloridothioic acid, E-00157

$C_2H_7ClNO_2P$
Dimethylphosphoramidochloridic acid, D-00860

C_2H_7FNOP
N,P-Dimethylphosphonamidic acid; Fluoride, *in* D-00849
P-Ethylphosphonamidic fluoride, *in* E-00126

C_2H_7FNOPS
Dimethylphosphoramidofluoridothioic acid, D-00868
O-Ethyl phosphoramidofluoridothioate, E-00158

C_2H_7FNPS
▷*N,P*-Dimethylphosphonamidothioic fluoride, *in* D-00850
P-Ethylphosphonamidothioic acid; Fluoride, *in* E-00128

$C_2H_7FNPS_2$
Dimethylphosphoramidofluoridodithioic acid, D-00867

$C_2H_7N_2OP$
P-Vinylphosphonic diamide, V-00004

$C_2H_7N_2O_2P$
P-Acetylphosphonic diamide, A-00027

$C_2H_7N_2O_2PS$
(1-Iminoethyl)phosphoramidothioic acid, I-00005

$C_2H_7N_2O_4P$
[(Aminocarbonyl)amino]methylphosphonic acid, A-00058

C_2H_7OP
2,2-Dihydro-1,2-oxaphosphetane, D-00533
Dimethylphosphine oxide, D-00828
Dimethylphosphinous acid, D-00844

C_2H_7OPS
Dimethylphosphinothioic acid, D-00839
Ethylphosphinothioic acid, E-00125

$C_2H_7OPS_2$
Ethylphosphonodithioic acid, E-00140
O-Methyl methylphosphonodithioate, M-00170

C_2H_7OPSe
Dimethylphosphinoselenoic acid, D-00835

$C_2H_7OPSe_3$
O-Ethyl phosphorotriselenoate, E-00184

$C_2H_7O_2P$
Dimethylphosphinic acid, D-00829
Dimethyl phosphonite, D-00852
▷Ethyl phosphinate, E-00122
Ethylphosphinic acid, E-00124
Ethylphosphonous acid, E-00153
Methyl methylphosphinate, *in* M-00278

$C_2H_7O_2PS$
O,O-Dimethyl phosphonothioate, D-00853
Ethylphosphonothioic acid, E-00148
O-Methyl methylphosphonothioate, M-00171

$C_2H_7O_2PS_2$
Bis(mercaptomethyl)phosphinic acid, B-00292
▷*O,O*-Dimethyl phosphorodithioate, D-00898
O,S-Dimethyl phosphorodithioate, D-00899
S,S-Dimethyl phosphorodithioate, D-00900

$C_2H_7O_2PSe$
Ethylphosphonoselenoic acid, E-00145

$C_2H_7O_2PSe_2$
O,O-Dimethyl phosphorodiselenoate, D-00897

$C_2H_7O_3P$
▷Dimethyl phosphonate, D-00851
Ethylphosphonic acid, E-00129
Methyl methylphosphonate, M-00168
Monoethyl phosphonate, M-00473

$C_2H_7O_3PS$
O,O-Dimethyl phosphorothioate, D-00907
O,S-Dimethyl phosphorothioate, D-00908
O-Ethyl phosphorothioate, E-00182
S-Ethyl phosphorothioate, E-00183
(2-Hydroxyethyl)phosphonothioic acid, H-00147
[(Methylthio)methyl]phosphonic acid, M-00415

$C_2H_7O_3PSe$
O,O-Dimethyl phosphoroselenoate, D-00906

$C_2H_7O_4P$
Bis(hydroxymethyl)phosphinic acid, B-00282
Dimethyl phosphate, D-00826
Ethylphosphonoperoxoic acid, E-00144
(1-Hydroxyethyl)phosphonic acid, H-00145
(2-Hydroxyethyl)phosphonic acid, H-00146
Monoethyl phosphate, M-00472

$C_2H_7O_4PS$
[(Methylsulfinyl)methyl]phosphonic acid, M-00398

$C_2H_7O_5PS$
[(Methylsulfonyl)methyl]phosphonic acid, M-00400

$C_2H_7O_7P_3$
2,4,6-Trihydroxy-1,2,4,6-oxatriphosphorinane 2,4,6-trioxide, T-00494

C_2H_7P
▷Dimethylphosphine, D-00827
▷Ethylphosphine, E-00123

C_2H_7PS
Dimethylphosphine; Sulfide, *in* D-00827
Dimethylphosphinothious acid, D-00843

C_2H_7PSSe
Dimethylphosphinoselenothioic acid, D-00837

$C_2H_7PS_2$
Dimethylphosphinodithioic acid, D-00834
Ethylphosphonodithious acid, E-00141

$C_2H_7PS_3$
Ethylphosphonotrithioic acid, E-00152

C_2H_7PSe
Dimethylphosphinoselenous acid, D-00838

$C_2H_8ClN_2OP$
P-2-Chloroethylphosphonic diamide, C-00075

$C_2H_8ClN_2PS$
(2-Chloroethyl)phosphonothioic acid; Diamide, *in* C-00078

$C_2H_8Cl_4N_5P_3$
2,2,4,6-Tetrachloro-2,2,4,4,6,6-hexahydro-4,6-bis(methylamino)-1,3,5-triaza-2,4,6-triphosphorine, T-00030

C_2H_8NOP
P,P-Dimethylphosphinic amide, D-00830

C_2H_8NOPS
N,P-Dimethylphosphonamidothioic acid, D-00850
P-Ethylphosphonamidothioic acid, E-00128
O-Methyl *P*-methylphosphonamidothioate, *in* M-00284
S-Methyl *P*-methylphosphonamidothioate, *in* M-00284

$C_2H_8NOPS_2$
O,S-Dimethyl phosphoramidodithioate, D-00865
S,S-Dimethyl phosphoramidodithioate, D-00866

$C_2H_8NO_2P$
(1-Aminoethyl)phosphinic acid, A-00077
(Aminomethyl)methylphosphinic acid, A-00089
N,P-Dimethylphosphonamidic acid, D-00849
Dimethyl phosphoramidite, D-00859
Dimethylphosphoramidous acid, D-00875
P-Ethylphosphonamidic acid, E-00126
Ethylphosphoramidous acid, E-00160

$C_2H_8NO_2PS$
O,O-Dimethyl phosphoramidothioate, D-00870
Dimethylphosphoramidothioic acid, D-00871
Ethylphosphoramidothioic acid, E-00159
▷Methamidophos, M-00021

$C_2H_8NO_2PSe$
Dimethylphosphoramidoselenoic acid, D-00869

$C_2H_8NO_3P$
(1-Aminoethyl)phosphonic acid, A-00078
2-Aminoethylphosphonic acid, A-00079
O,O-Dimethyl phosphoramidate, D-00854
Dimethylphosphoramidic acid, D-00855

Ethylphosphoramidic acid, E-00156
Ethylphosphorimidic acid, E-00161

$C_2H_8NO_3PS$
(1-Amino-2-mercaptoethyl)phosphonic acid, A-00084

$C_2H_8NO_4P$
2-Aminoethyl dihydrogen phosphate, A-00075
(1-Amino-2-hydroxyethyl)phosphonic acid, A-00083
(2-Hydroxyethyl)phosphoramidic acid, H-00148

C_2H_8NP
P,P-Dimethylphosphinous amide, D-00845

C_2H_8NPS
P,P-Dimethylphosphinothioic amide, D-00840

$C_2H_8NPS_2$
P-Ethylphosphonamidodithioic acid, E-00127

$C_2H_8NPS_3$
Dimethylphosphoramidotrithioic acid, D-00874

C_2H_8NPSe
P,P-Dimethylphosphinoselenoic amide, D-00836

$C_2H_8N_3P$
P-1-Aziridinylphosphonous diamide, A-00155

$C_2H_8N_4P_2$
1,1,4,4-Tetraamino-2,3-diphosphabutadiene, T-00005

$C_2H_8O_3PS_2$
Dimethylthiodiphosphonic acid, D-00925

$C_2H_8O_4P_2$
1,2-Ethanediphosphonous acid, E-00008

$C_2H_8O_4P_2S_2$
1,2-Ethanediphosphonothioic acid, E-00007

$C_2H_8O_5P_2$
Dimethyldiphosphonic acid, D-00751

$C_2H_8O_6P_2$
1,2-Ethanediphosphonic acid, E-00006

$C_2H_8O_6P_2S$
Thiobis[methanephosphonic acid], T-00311

$C_2H_8O_7P_2$
(1-Hydroxyethylidene)bisphosphonic acid, H-00144
Oxybis[methanephosphonic acid], O-00095

$C_2H_8P_2$
▷1,2-Diphosphinoethane, D-01176

C_2H_9GeP
Dimethylgermylphosphine, D-00756

$C_2H_9NO_4P$
Ethylimidodi(phosphorous acid), E-00079

$C_2H_9NO_5P_2S$
(Dimethoxyphosphinothioyl)phosphorimidic acid, D-00674

$C_2H_9NO_6P_2$
1-Amino-1,1-ethanediphosphonic acid, A-00073
(Dimethoxyphosphinyl)phosphorimidic acid, D-00683
Dimethylamine-1,1′-diphosphonic acid, D-00688
Ethylimidodiphosphoric acid, E-00078

$C_2H_9N_2OP$
Dimethylphosphinic acid; Hydrazide, *in* D-00829
P-Ethylphosphonic diamide, E-00130
Ethyl phosphorodiamidite, E-00165

$C_2H_9N_2OPS$
O-Ethyl phosphorodiamidothioate, *in* E-00182

$C_2H_9N_2O_2P$
Bis(aminomethyl)phosphinic acid, B-00087
N,N′-Dimethylphosphorodiamidic acid, D-00895

$C_2H_9N_2O_2PS$
O,O-Dimethyl phosphorohydrazidothioate, D-00905

$C_2H_9N_2O_3P$
(2-Aminoethyl)phosphoramidic acid, A-00080
Dimethyl phosphorohydrazidate, D-00904

$C_2H_9N_2P$
P,P-Dimethylphosphinimidic amide, D-00833
Ethylphosphonous diamide, E-00154

$C_2H_9N_2PS$
P-Ethylphosphonothioic diamide, E-00149

$C_2H_9N_2PS_2$
N,N′-Dimethylphosphorodiamidodithioic acid, D-00896

$C_2H_9N_2PSe$
P-Ethylphosphonoselenoic diamide, E-00146

$C_2H_9O_8P_3$
Bis[(dihydroxyphosphinyl)methyl]phosphinic acid, B-00182

$C_2H_{10}GeP_2$
Dimethyldiphosphinogermane, D-00750

$C_2H_{10}N_2O_2P_2S_2$
Hydrazobis[methylphosphinothioic acid], H-00105

$C_2H_{10}N_2O_4P_2S_2$
1,2-Ethanediphosphoramidothioic acid, E-00010

$C_2H_{10}N_2O_6P_2$
1,2-Ethanediphosphoramidic acid, E-00009

$C_2H_{10}N_3OP$
N,N-Dimethylphosphoric triamide, D-00880

$C_2H_{10}N_3PS$
NN-Dimethylphosphorothioic triamide, *in* D-00871
N,N-Dimethylphosphorothioic triamide, D-00909

$C_3Cl_3N_2OPS_2$
(Trichloromethyl)phosphonic acid; Diisothiocyanate, *in* T-00411

$C_3Cl_3N_2O_3P$
(Trichloromethyl)phosphonic acid; Diisocyanate, *in* T-00411

$C_3Cl_{11}P$
Dichlorotris(trichloromethyl)phosphorane, D-00201

$C_3F_3N_2P$
(Trifluoromethyl)phosphonous acid; Dicyanide, *in* T-00482

C_3F_6NOP
Bis(trifluoromethyl)phosphinous acid; Isocyanate, *in* B-00418

C_3F_6NP
▷Cyanobis(trifluoromethyl)phosphine, *in* B-00418

C_3F_6NPS
Bis(trifluoromethyl)phosphinous acid; Isothiocyanate, *in* B-00418

C_3F_9OP
Tris(trifluoromethyl)phosphine oxide, T-00856

C_3F_9P
Tris(trifluoromethyl)phosphine, T-00855

C_3F_9PS
Trifluoromethyl bis(trifluoromethyl)-phosphinothioite, *in* B-00417
Tris(trifluoromethyl)phosphine; Sulfide, *in* T-00855

C_3F_9PSe
Trifluoromethyl bis(trifluoromethyl)-phosphinoselenoite, *in* B-00413

$C_3F_9PSe_3$
Tris(trifluoromethyl) phosphorotriselenoite, T-00857

$C_3F_{11}P$
Difluorotris(trifluoromethyl)phosphorane, D-00420

$C_3HCl_2N_2O_3P$
(Dichloromethyl)phosphonic diisocyanate, D-00179

$C_3H_2ClN_2O_3P$
(Chloromethyl)phosphonic diisocyanate, C-00110

$C_3H_3Cl_2OP$
Propadienylphosphonic dichloride, P-00443
1-Propynylphosphonic dichloride, P-00496

$C_3H_3Cl_2O_2P$
2-Propynyl phosphorodichloridate, P-00497

$C_3H_3Cl_7NP$
N-Methyl-*P,P*-bis(trichloromethyl)-phosphinimidic chloride, *in* B-00403

$C_3H_3F_2OP$
1,2-Propadienylphosphonic acid; Difluoride, *in* P-00441
1-Propynylphosphonic acid; Difluoride, *in* P-00494

$C_3H_3F_6OP$
Methyl bis(trifluoromethyl)phosphinite, *in* B-00418

$C_3H_3F_6OPS$
O-Methyl bis(trifluoromethyl)phosphinothioate, *in* B-00415
S-Methyl bis(trifluoromethyl)phosphinothioate, *in* B-00415

$C_3H_3F_6O_2P$
Methyl bis(trifluoromethyl)phosphinate, *in* B-00411

$C_3H_3F_6PS$
Methyl bis(trifluoromethyl)phosphinothioite, *in* B-00417

$C_3H_3F_6PS_2$
Methyl bis(trifluoromethyl)phosphinodithioate, *in* B-00412

$C_3H_3F_6PSe$
Methyl bis(trifluoromethyl)phosphinoselenoite, *in* B-00413

$C_3H_3F_6PTe$
Methyl bis(trifluoromethyl)phosphinotelluroite, *in* B-00414

$C_3H_3N_2OP$
Methylphosphonic acid; Dicyanide, *in* M-00288

$C_3H_3N_2OPS_2$
Methylphosphonic acid; Diisothiocyanate, *in* M-00288

$C_3H_3N_2O_2P$
Methylphosphonous acid; Diisocyanate, *in* M-00319

$C_3H_3N_2O_3P$
Methylphosphonic acid; Diisocyanate, *in* M-00288

$C_3H_3N_2O_3PS$
O-Methyl phosphorodiisocyanatidothioate, M-00357

$C_3H_3N_2O_4P$
Methyl phosphorodiisocyanatidate, M-00356

$C_3H_3N_2P$
Methylphosphonous dicyanide, M-00323

$C_3H_3N_2PS_2$
Methylphosphonous acid; Diisothiocyanate, *in* M-00319

$C_3H_3O_6P$
Phosphinetricarboxylic acid, P-00355

$C_3H_4Cl_6NOP$
N-Methyl-*P,P*-bis(trichloromethyl)phosphinic amide, *in* B-00402

$C_3H_4F_6NP$
N-Methyl-*P,P*-bis(trifluoromethyl)phosphinous amide, *in* B-00419

$C_3H_5ClNO_3P$
Ethyl phosphorochloridisocyanatidate, E-00163

$C_3H_5Cl_2OP$
(1-Methylethenyl)phosphonic dichloride, M-00150
1-Propenylphosphonic acid; Dichloride, *in* P-00453
2-Propenylphosphonic dichloride, P-00456
2-Propenyl phosphorodichloridite, P-00465

$C_3H_5Cl_2O_2P$
1-Methylethenyl phosphorodichloridate, M-00151
(2-Oxopropyl)phosphonic acid; Dichloride, *in* O-00075

$C_3H_5Cl_2O_3P$
3-(Dichlorophosphinyl)propanoic acid, D-00192
Methyl dichlorophosphinylacetate, *in* D-00190

$C_3H_5Cl_2P$
Allyldichlorophosphine, *in* P-00463

$C_3H_5Cl_2PS$
2-Propenylphosphonothioic acid; Dichloride, *in* P-00461

$C_3H_5FNO_2PS$
O-Ethyl phosphorofluoridoisocyanatidothioate, E-00181

$C_3H_5FNO_3P$
Ethyl phosphorofluoridisocyanatidate, E-00180

$C_3H_5F_2OP$
1-Propenylphosphonic acid; Difluoride, *in* P-00453

$C_3H_5N_2O_3P$
1*H*-Imidazol-1-ylphosphonic acid, I-00003
(1*H*-Pyrazol-3-yl)phosphonic acid, P-00501

$C_3H_5N_2P$
5-Methyl-1*H*-1,2,3-diazaphosphole, M-00114

$C_3H_5OPS_2$
1,3,2-Dithiaphosphole; 2-Methoxy, *in* D-01246

$C_3H_5O_2PS$
5-Methyl-1,3,2-oxathiaphosphole 2-oxide, M-00184

$C_3H_5O_3P$
2,5-Dihydro-2-hydroxy-1,2-oxaphosphole 2-oxide, D-00500
1,2-Oxaphospholan-5-one 2-oxide, O-00053
1,2-Propadienylphosphonic acid, P-00441
1-Propynylphosphonic acid, P-00494
2-Propynylphosphonic acid, P-00495
2,6,7-Trioxa-1-phosphabicyclo[2.2.1]heptane, T-00587

$C_3H_5O_4P$
Methyl vinylene phosphate, *in* H-00136
2,6,7-Trioxa-1-phosphabicyclo[2.2.1]heptane; 1-Oxide, *in* T-00587

$C_3H_5O_6P$
Phosphoenolpyruvic acid, P-00366

$C_3H_6AsO_3P$
2,6,7-Trioxa-4-phospha-1-arsabicyclo[2.2.2]-octane, T-00586

$C_3H_6AsO_4P$
2,6,7-Trioxa-4-phospha-1-arsabicyclo[2.2.2]-octane; 4-Oxide, *in* T-00586

$C_3H_6BrCl_2OP$
▷(3-Bromopropyl)phosphonic acid; Dichloride, *in* B-00501

$C_3H_6BrO_2P$
2-Bromo-4-methyl-1,3,2-dioxaphospholane, B-00473

$C_3H_6ClN_2OP$
2-Chloro-2,3-dihydro-3,5-dimethyl-1,3,4,2-oxadiazaphosphole, C-00039

$C_3H_6ClN_2PS_2$
2,3-Dihydro-3,5-dimethyl-1,3,4,2-thiadiazaphosphole 2-sulfide; 2-Chloro, *in* D-00471

C_3H_6ClOP
Methylvinylphosphinic acid; Chloride, *in* M-00438

$C_3H_6ClOPS_2$
2-Choro-1,3,2-dithiaphosphorinane; 2-Oxide, *in* C-00199

$C_3H_6ClO_2P$
▷2-Chloro-1,3,2-dioxaphosphorinane, C-00066
2-Chloro-4-methyl-1,3,2-dioxaphospholane, C-00092

$C_3H_6ClO_2PS$
2-Chloro-1,3,2-dioxaphosphorinane 2-sulfide, C-00068
Propylene phosphorochloridothioate, *in* C-00092

$C_3H_6ClO_2PSe$
O,O-Trimethylene phosphorochloridoselenoate, *in* C-00066

$C_3H_6ClO_3P$
2-Chloro-1,3,2-dioxaphosphorinane 2-oxide, C-00067
2-Chloro-4-methyl-1,3,2-dioxaphospholane; 2-Oxide, *in* C-00092
Propylene phosphorochloridate, *in* C-00092

$C_3H_6ClPS_2$
2-Chloro-4-methyl-1,3,2-dithiaphospholane, *in* M-00136
2-Choro-1,3,2-dithiaphosphorinane, C-00199

$C_3H_6ClPS_3$
2-Choro-1,3,2-dithiaphosphorinane; 2-Sulfide, *in* C-00199

$C_3H_6Cl_2NOP$
2-Propenylphosphoramidic acid; Dichloride, *in* P-00464

$C_3H_6Cl_2NO_3P$
Ethyl dichlorophosphinylcarbamate, *in* D-00191

$C_3H_6Cl_3N_2OP$
2,2-Dihydro-1,3-dimethyl-1,3,2-diazaphosphetidin-4-one; 2,2,2-Trichloro, *in* D-00459

$C_3H_6Cl_3OP$
(2-Chloropropyl)phosphonic acid; Dichloride, *in* C-00171
(3-Chloropropyl)phosphonic acid; Dichloride, *in* C-00172

$C_3H_6Cl_3O_3P$
Dimethyl (trichloromethyl)phosphonate, *in* T-00411

$C_3H_6Cl_4O_2P_2$
1,3-Propanediphosphonic acid; Bis-dichloride, *in* P-00444

$C_3H_6FO_2P$
2-Fluoro-1,3,2-dioxaphosphorinane, F-00018
2-Fluoro-4-methyl-1,3,2-dioxaphospholane, F-00020

$C_3H_6FO_2PS$
Propylene fluorothiophosphate, *in* F-00020

$C_3H_6FO_3P$
▷Trimethylene fluorophosphate, *in* F-00018

$C_3H_6F_3N_2OP$
2,2-Dihydro-1,3-dimethyl-1,3,2-diazaphosphetidin-4-one; 2,2,2-Trifluoro, *in* D-00459

$C_3H_6F_3O_2P$
Dimethyl (trifluoromethyl)phosphonite, *in* T-00482
Methyl methyl(trifluoromethyl)phosphinate, *in* M-00420

$C_3H_6F_3PS$
Trifluoromethyl dimethylphosphinothioite, *in* D-00843

$C_3H_6F_3PS_2$
Dimethyl (trifluoromethyl)phosphonodithioite, *in* T-00480

$C_3H_6F_3PSe$
Trifluoromethyl dimethylphosphinoselenoite, *in* D-00838

$C_3H_6F_4NPS$
N,N-Dimethyl-*P*-trifluoromethyl-phosphonamidothioic fluoride, D-00928

C_3H_6NOP
Dimethylphosphinic acid; Cyanide, *in* D-00829

C_3H_6NOPS
Dimethylphosphinic acid; Isothiocyanate, *in* D-00829

$C_3H_6NO_2P$
Dimethylphosphinic acid; Isocyanate, *in* D-00829

$C_3H_6NO_2PS$
Methyl methylphosphonisothiocyanatidate, *in* M-00294

$C_3H_6NO_3P$
(1-Cyanoethyl)phosphonic acid, C-00210
(2-Cyanoethyl)phosphonic acid, C-00211
Dimethyl phosphorocyanidate, D-00894
Ethylphosphonisocyanatidic acid, E-00133

$C_3H_6NO_3PS$
O,O-Dimethyl phosphorisocyanatidothioate, D-00884
O,S-Dimethyl phosphorisocyanatidothioate, D-00885

$C_3H_6NO_4P$
3-Phosphonoxypropanenitrile, P-00398

C_3H_6NP
▷Dimethylphosphinous cyanide, D-00847
3-Phosphinopropanenitrile, P-00358

$C_3H_6N_3O_9P_3$
1,3,5-Triazine-2,4,6-triphosphonic acid, T-00329

$C_3H_6O_3P_2$
2,6,7-Trioxa-1,4-diphosphabicyclo[2.2.2]octane, T-00585
2,6,7-Trioxa-1,4-diphosphabicyclo[2.2.2]octane; 1-Oxide, *in* T-00585
▷2,6,7-Trioxa-1,4-diphosphabicyclo[2.2.2]-octane; 4-Oxide, *in* T-00585

$C_3H_6O_3P_2S$
2,6,7-Trioxa-1,4-diphosphabicyclo[2.2.2]octane; 1-Sulfide, *in* T-00585
2,6,7-Trioxa-1,4-diphosphabicyclo[2.2.2]octane; 4-Sulfide, *in* T-00585

$C_3H_6O_3P_2S_2$
2,6,7-Trioxa-1,4-diphosphabicyclo[2.2.2]octane; 1,4-Disulfide, *in* T-00585

$C_3H_6O_5P_2$
▷2,6,7-Trioxa-1,4-diphosphabicyclo[2.2.2]-octane; 1,4-Dioxide, *in* T-00585

$C_3H_7Br_2O_3P$
Dimethyl (dibromomethyl)phosphonate, *in* D-00064

$C_3H_7Br_2P$
Dibromoisopropylphosphine, *in* I-00056

C_3H_7ClNOP
2-Chloro-3-methyl-1,3,2-oxazaphospholidine, C-00102
2-Chlorotetrahydro-2*H*-1,3,2-oxazaphosphorine, C-00180

$C_3H_7ClNOPS$
2-Chlorotetrahydro-2*H*-1,3,2-oxazaphosphorine; 2-Sulfide, *in* C-00180
3-Methyl-1,3,2-thiazaphospholidine; 2-Chloro, 2-oxide, *in* M-00408

$C_3H_7ClNO_2P$
2-Chlorotetrahydro-2*H*-1,3,2-oxazaphosphorine; 2-Oxide, *in* C-00180

$C_3H_7ClNPS_2$
3-Methyl-1,3,2-thiazaphospholidine; 2-Chloro, 2-sulfide, *in* M-00408

$C_3H_7Cl_2OP$
Isopropylphosphonic dichloride, I-00052

Isopropyl phosphorodichloriditе, I-00061
Propylphosphonic dichloride, P-00475
Propyl phosphorodichloridite, P-00488

$C_3H_7Cl_2OPS$

O-Isopropyl phosphorodichloridothioate, I-00062
S-Isopropyl phosphorodichloridothioate, *in* I-00064
O-Propyl phosphorodichloridothioate, P-00489
S-Propyl phosphorodichloridothioate, P-00490

$C_3H_7Cl_2O_2P$

Isopropyl phosphorodichloridate, I-00060
Methyl bis(chloromethyl)phosphinate, *in* B-00127
▷Propyl phosphorodichloridate, P-00487

$C_3H_7Cl_2O_2PS$

O-(2-Methoxyethyl) phosphorodichloridothioate, *in* H-00149

$C_3H_7Cl_2O_3P$

Dimethyl (dichloromethyl)phosphonate, *in* D-00175

$C_3H_7Cl_2P$

Isopropylphosphonous dichloride, I-00057
Propylphosphonous dichloride, P-00483

$C_3H_7Cl_2PS$

Isopropylphosphonothioic dichloride, I-00055
Isopropyl phosphorodichloridothioite, I-00063
Propylphosphonothioic acid; Dichloride, *in* P-00480
Propyl phosphorodichloridothioite, P-00491

$C_3H_7Cl_3NO_2P$

Ethyl *P*-trichloromethylphosphonamidate, *in* T-00410

$C_3H_7F_2NP_2$

2,5-Difluoro-1-methyl-1,2,5-azadiphospholidine, *in* M-00089

$C_3H_7F_2OP$

Isopropylphosphonic acid; Difluoride, *in* I-00050
Propyl phosphorodifluoridite, P-00492

$C_3H_7N_2OP$

1,3-Dimethyl-1,3,2-diazaphosphetidin-4-one, D-00717
P-Propadienylphosphonic diamide, P-00442

$C_3H_7N_2OPS$

Dimethylphosphoramidocyanidothioic acid, D-00864

$C_3H_7N_2O_2P$

Dimethylphosphoramidocyanidic acid, D-00863

$C_3H_7N_2O_3P$

Dimethyl (diazomethyl)phosphonate, *in* D-00035

$C_3H_7N_2PS$

2,3-Dihydro-3,5-dimethyl-1,3,4,2-thiadiazaphosphole 2-sulfide, D-00471

$C_3H_7N_3P$

1,5-Dimethyl-1*H*-1,2,4,3-triazaphosphole, D-00926

C_3H_7OP

2,2,2,3-Tetrahydro-1,2-oxaphosphole, T-00119

$C_3H_7OPS_2$

1,3,2-Dithiaphosphorinane 2-oxide, D-01248
S,S-Ethylene methylphosphonodithioate, *in* M-00139
O,S-Ethylene *S*-methyl phosphorodithioite, *in* O-00056
2-Methoxy-1,3,2-dithiaphospholane, *in* D-01243
2-Propenylphosphonodithioic acid, P-00457

$C_3H_7O_2P$

1,3,2-Dioxaphosphorinane, D-00972
2-Methyl-1,3,2-dioxaphospholane, M-00126
4-Methyl-1,3,2-dioxaphospholane, M-00127
Methylvinylphosphinic acid, M-00438
2-Propenylphosphinic acid, P-00452
2-Propenylphosphonous acid, P-00463

$C_3H_7O_2PS$

Ethylene methylphosphonothioate, *in* M-00126
O,S-Ethylene *O*-methyl phosphorothioite, *in* O-00056
2-Mercapto-4-methyl-1,3,2-dioxophospholane, *in* M-00127
Methyl thiodimethylenephosphinate, *in* H-00201
1,3,2-Oxathiaphosphorinane 2-oxide, O-00057
2-Propenylphosphonothioic acid, P-00461
O,O-Trimethylene phosphonothioate, *in* D-00972

$C_3H_7O_2PS_2$

2-Mercapto-1,3,2-dioxaphosphorinane 2-sulfide, M-00013
2-Mercapto-4-methyl-1,3,2-dioxaphospholane 2-sulfide, M-00016
2-Methoxy-1,3,2-dithiaphospholane 2-oxide, *in* D-01244

$C_3H_7O_2PSe$

(2-Propenyl)phosphonoselenoic acid, P-00458

$C_3H_7O_2PTe$

(2-Propenyl)phosphonotelluroic acid, P-00460

$C_3H_7O_3P$

Acetylmethylphosphinic acid, A-00018
1,4,2-Dioxaphosphorinane 2-oxide, D-00973
Ethylene methylphosphonate, *in* M-00126
2-Hydroxy-4-methyl-1,3,2-dioxaphospholane, *in* M-00127
2-Hydroxy-1,2-oxaphospholane 2-oxide, H-00166
▷2-Methoxy-1,3,2-dioxaphospholane, M-00036
(1-Methylethenyl)phosphonic acid, M-00149
Mono-2-propenyl phosphonate, M-00510
1-Propenylphosphonic acid, P-00453
2-Propenylphosphonic acid, P-00454
Trimethylene phosphonate, *in* D-00972

$C_3H_7O_3PS$

2-Mercapto-4-methyl-1,3,2-dioxaphospholane 2-oxide, M-00015
(1-Oxopropyl)phosphonothioic acid, O-00078

$C_3H_7O_3PS_2$

(2-Oxopropyl)phosphonothioic acid, O-00079

$C_3H_7O_3PSe$

O,O-Ethylene *O*-methyl phosphoroselenoate, *in* M-00036

$C_3H_7O_4P$

2-Hydroxy-1,3,2-dioxaphosphorinane 2-oxide, H-00138
2-Hydroxy-4-methyl-1,3,2-dioxaphospholane 2-oxide, H-00155
2-Methoxy-1,3,2-dioxaphospholane 2-oxide, M-00037
(3-Methyloxiranyl)phosphonic acid, M-00185
▷Mono-2-propenyl phosphate, M-00509
(1-Oxopropyl)phosphonic acid, O-00074
(2-Oxopropyl)phosphonic acid, O-00075
(3-Oxopropyl)phosphonic acid, O-00076

$C_3H_7O_5P$

Acetic acid monoanhydride with methyl dihydrogen phosphate, *in* A-00025
(Acetoxymethyl)phosphonic acid, A-00007
Dimethoxyphosphinecarboxylic acid oxide, D-00672
Ethyl (dihydorxyphosphinyl)formate, *in* P-00387
Methyl dihydroxyphosphinylacetate, *in* P-00376
2-Phosphonopropanoic acid, P-00395
▷3-Phosphonopropanoic acid, P-00396

$C_3H_7O_6P$

2-Hydroxy-3-(phosphonoxy)propanal, H-00187
2-Phosphonooxypropanoic acid, P-00392
3-Phosphonoxypropanoic acid, P-00399

C_3H_7P

1-Methylphosphirane, *in* P-00361

$C_3H_7PS_2$

1,3,2-Dithiaphosphorinane, D-01247
4-Methyl-1,3,2-dithiaphospholane, M-00136
2-Methyl-1,3,2-dithiophospholane, M-00139
1,2-Thiaphospholane 2-sulfide, T-00298

$C_3H_7PS_2Se$

S,S-Ethylene methylphosphonoselenodithioate, *in* M-00139

$C_3H_7PS_3$

1,3,2-Dithiaphosphorinane 2-sulfide, D-01249
Ethylene methylphosphonotrithioate, *in* M-00139
2-(Methylthio)-1,3,2-dithiaphospholane, *in* D-01243

$C_3H_7PS_4$

2-Mercapto-1,3,2-dithiaphosphorinane 2-sulfide, M-00014

C_3H_8BrOPS

O-Ethyl methylphosphonobromidothioate, *in* M-00295

$C_3H_8BrO_3P$

(1-Bromopropyl)phosphonic acid, B-00500
(3-Bromopropyl)phosphonic acid, B-00501

C_3H_8BrP

Bromoethylmethylphosphine, *in* E-00087

C_3H_8ClOP

Ethylmethylphosphinic acid; Chloride, *in* E-00085
Ethyl methylphosphonochloridate, *in* M-00296

C_3H_8ClOPS

O-Ethyl methylphosphonochloridothioate, E-00090
S-Ethyl methylphosphonochloridothioate, *in* M-00298
Isopropylphosphonochloridothioic acid, I-00053
O-Methyl ethylphosphonochloridothioate, M-00154
S-Methyl ethylphosphonochloridothioate, *in* E-00137
Propylphosphonochloridothioic acid, P-00477

$C_3H_8ClOPSe$

O-Ethyl *Se*-methyl phosphorochloridoselenoite, E-00094

$C_3H_8ClO_2P$

Methyl (chloromethyl)methylphosphinate, *in* C-00100
Propylphosphonochloridic acid, P-00476

$C_3H_8ClO_3P$

(1-Chloropropyl)phosphonic acid, C-00170
(2-Chloropropyl)phosphonic acid, C-00171
(3-Chloropropyl)phosphonic acid, C-00172
Dimethyl chloromethylphosphonate, *in* C-00106
Monoethyl chloromethylphosphonate, *in* C-00106

C_3H_8ClP

Chloroethylmethylphosphine, *in* E-00087

C_3H_8ClPS

Ethyl methylphosphonochloridothioite, *in* M-00299

$C_3H_8Cl_2NOP$

P,P-Bis(chloromethyl)-*N*-methylphosphinic amide, *in* B-00128
Isopropylphosphoramidic acid; Dichloride, *in* I-00058
Propylphosphoramidic acid; Dichloride, *in* P-00484

$C_3H_8Cl_2NO_2P$
Ethyl *P*-dichloromethylphosphonamidate, *in* D-00174

$C_3H_8Cl_2NP$
Propylphosphoramidous acid; Dichloride, *in* P-00486

$C_3H_8Cl_2NPS$
Propylphosphoramidothioic acid; Dichloride, *in* P-00485

$C_3H_8Cl_3N_2OP$
(Trichloromethyl)phosphonic bis(methylamide), *in* T-00412

C_3H_8FOP
Ethyl methylphosphonofluoridite, *in* M-00307

C_3H_8FOPS
O-Ethyl methylphosphonofluoridothioate, *in* M-00305
S-Ethyl methylphosphonofluoridothioate, *in* M-00305

$C_3H_8FO_2P$
Ethyl methylphosphonofluoridate, *in* M-00303
Methyl ethylphosphonofluoridate, *in* E-00142
Propylphosphonofluoridic acid, P-00479

$C_3H_8F_2NOP$
Isopropylphosphoramidic acid; Difluoride, *in* I-00058

C_3H_8NOPS
Tetrahydro-2*H*-1,3,2-thiazaphosphorine 2-oxide, T-00154

$C_3H_8NOPS_2$
1,3,2-Thiazaphospholidine 2-sulfide; 2-Methoxy, *in* T-00303

$C_3H_8NO_2P$
1-Aziridinyl-*P*-methylphosphonamidic acid, A-00152

$C_3H_8NO_2PS$
Tetrahydro-2-hydroxy-2*H*-1,3,2-oxazaphosphorine 2-sulfide, T-00108

$C_3H_8NO_3P$
1-Azetidinephosphonic acid, A-00147
2-Propenylphosphoramidic acid, P-00464

$C_3H_8NO_4P$
[(Acetylamino)methyl]phosphonic acid, A-00012
Dimethyl aminocarbonylphosphonate, *in* C-00003
(Dimethylaminocarbonyl)phosphonic acid, D-00691
[(Methoxycarbonyl)methyl]phosphonamidic acid, M-00028
Monoethyl (aminocarbonyl)phosphonate, M-00469
(1-Nitroso-1-methylethyl)phosphonic acid, N-00066

$C_3H_8NO_5P$
2-Amino-3-phosphonopropanoic acid, A-00121
3-Amino-3-phosphonopropanoic acid, A-00122
N-(Dimethoxyphosphinyl)carbamic acid, D-00680
Dimethyl (nitromethyl)phosphonate, *in* N-00037
▷Glyphosate, G-00097
N-Phosphoalanine, P-00362

$C_3H_8NO_5PS$
N-Phosphocysteine, P-00365

$C_3H_8NO_6P$
Phosphoserine, P-00417
N-Phosphoserine, P-00418

C_3H_8NPS
3-Methyl-1,3,2-thiazaphospholidine, M-00408

$C_3H_8NPS_2$
1,3,2-Thiazaphospholidine 2-sulfide; 2-Me, *in* T-00303

$C_3H_8N_3O_2P$
Ethyl methylphosphonazidate, *in* M-00286

$C_3H_8N_3O_5P$
Phosphoglycocyamine, P-00369

$C_3H_8O_4P_2$
1,3-Dihydroxy-1,3-diphospholane 1,3-dioxide, D-00621

$C_3H_8O_7P_2$
1-Oxo-1,3-propanediphosphonic acid, O-00071
2-Oxo-1,3-propanediphosphonic acid, O-00072

$C_3H_8O_8P_2$
3,3-Diphosphonopropanoic acid, D-01187

$C_3H_8P_2$
1,2-Diphospholane, D-01181

C_3H_9BrNPS
Trimethylphosphonamidothioic acid; Bromide, *in* T-00540

$C_3H_9Br_2P$
Dibromotrimethylphosphorane, D-00069

C_3H_9ClNOP
Methyl dimethylphosphoramidochloridite, *in* D-00862

$C_3H_9ClNOPS$
O-Methyl dimethylphosphoramidochloridothioate, *in* D-00861
O-Methyl ethylphosphoramidochloridothioate, *in* E-00157

$C_3H_9ClNO_2P$
Ethyl *P*-chloromethylphosphonamidate, *in* C-00105
Methyl dimethylphosphoramidochloridate, *in* D-00860

C_3H_9ClNPS
Trimethylphosphonamidothioic acid; Chloride, *in* T-00540

$C_3H_9Cl_2N_3P_2$
1,2,4-Trimethyl-1,2,4,3,5-triazadiphospholidine; 3,5-Dichloro, *in* T-00562

$C_3H_9Cl_2O_2PSi$
Trimethylsilyl phosphorodichloridate, T-00558

$C_3H_9Cl_2P$
Dichlorotrimethylphosphorane, D-00198

$C_3H_9Cl_3NPSi$
P,P,P-Trichloro-*N*-trimethylsilylphosphazene, *in* T-00557

C_3H_9FNOP
▷Trimethylphosphonamidic fluoride, T-00538

C_3H_9FNPS
▷Trimethylphosphonamidothioic acid; Fluoride, *in* T-00540

$C_3H_9FNPS_2$
Methyl dimethylphosphoramidofluoridodithioate, *in* D-00867

$C_3H_9F_2O_2PSi$
Trimethylsilyl phosphorodifluoridate, T-00559

$C_3H_9F_2O_3P$
Difluorotrimethoxyphosphorane, D-00416

$C_3H_9F_2P$
Difluorotrimethylphosphorane, D-00417

$C_3H_9I_2P$
Diiodotrimethylphosphorane, D-00629

$C_3H_9NP_2$
1-Methyl-1,2,5-azadiphospholidine, M-00089

$C_3H_9N_2OP$
2,2-Dihydro-1,3-dimethyl-1,3,2-diazaphosphetidin-4-one, D-00459
Hexahydro-1,3,2-diazaphosphorine 2-oxide, H-00020
P-2-Propenylphosphonic diamide, P-00455

$C_3H_9N_2O_2P$
P-(2-Oxopropyl)phosphonic diamide, O-00077
Tetrahydro-4-methyl-2*H*-1,3,4,2-oxadiazaphosphorine 2-oxide, T-00116

$C_3H_9N_2O_4P$
3-Amino-3-phosphonopropanoic acid; Amide, *in* A-00122
Dimethyl aminocarbonylphosphoramidate, *in* A-00062

$C_3H_9N_2PS$
P-2-Propenylphosphonothioic diamide, P-00462

$C_3H_9N_2PSe$
P-(2-Propenyl)phosphonoselenoic diamide, P-00459

$C_3H_9N_6O_9P_3$
1,3,5-Triazine-2,4,6-triphosphoramidic acid, T-00330

C_3H_9OP
2,2-Dihydro-1,2-oxaphospholane, D-00534
Ethylmethylphosphinous acid, E-00087
Methyl dimethylphosphinite, *in* D-00844
Trimethylphosphine oxide, T-00532

C_3H_9OPS
O-Ethyl methylphosphinothioate, *in* M-00279
O-Methyl dimethylphosphinothioate, *in* D-00839
S-Methyl dimethylphosphinothioate, *in* D-00839

$C_3H_9OPS_2$
O,S-Dimethyl methylphosphonodithioate, *in* M-00301
S,S-Dimethyl methylphosphonodithioate, *in* M-00301
O-Ethyl methylphosphonodithioate, E-00091
Propylphosphonodithioic acid, P-00478

$C_3H_9OPS_3$
O,S,S-Trimethyl phosphorotrithioate, T-00548
▷*S,S,S*-Trimethyl phosphorotrithioate, T-00549

$C_3H_9O_2P$
Dimethyl methylphosphonite, *in* M-00319
Ethyl methylphosphinate, *in* M-00278
Ethylmethylphosphinic acid, E-00085
Isopropylphosphinic acid, I-00049
Isopropylphosphonous acid, I-00056
Methyl dimethylphosphinate, *in* D-00829
Methyl ethylphosphinate, *in* E-00124
Propyl phosphinate, P-00470
Propylphosphinic acid, P-00472
Propylphosphonous acid, P-00482

$C_3H_9O_2PS$
O,O-Dimethyl methylphosphonothioate, *in* M-00313
O-Ethyl methylphosphonothioate, E-00093
Isopropylphosphonothioic acid, I-00054
O-Methyl ethylphosphonothioate, M-00155
Propylphosphonothioic acid, P-00480

$C_3H_9O_2PS_2$
O-Ethyl *O*-methyl phosphorodithioate, E-00095
O-Ethyl *S*-methyl phosphorodithioate, E-00096
S-Ethyl *O*-methyl phosphorodithioate, E-00097
▷*O,S,S*-Trimethyl phosphorodithioate, T-00541
▷*O,O,S*-Trimethyl phosphorodithioate, T-00542

$C_3H_9O_2PSe$
O,O-Dimethyl methylphosphonoselenoate, *in* M-00311
O-Ethyl methylphosphonoselenoate, E-00092

$C_3H_9O_3P$
▷Dimethyl methylphosphonate, D-00762
Ethyl methylphosphonate, E-00089
Isopropylphosphonic acid, I-00050
Monoisopropyl phosphonate, M-00478

Monopropyl phosphonate, M-00513
▷Propylphosphonic acid, P-00473
▷Trimethyl phosphite, T-00535
Tris(hydroxymethyl)phosphine, T-00784

$C_3H_9O_3PS$

O-Ethyl *O*-methyl phosphorothioate, E-00098
[(Ethylthio)methyl]phosphonic acid, E-00192
S-Isopropyl phosphorothioate, I-00064
[1-(Methylthio)ethyl]phosphonic acid, M-00412
S-Propyl phosphorothioate, P-00493
▷*O,O,O*-Trimethyl phosphorothioate, T-00545
▷*O,O,S*-Trimethyl phosphorothioate, T-00546

$C_3H_9O_3PS_2$

[Bis(methylthio)methyl]phosphonic acid, B-00364

$C_3H_9O_3PSe$

O,O,O-Trimethyl phosphoroselenoate, T-00543

$C_3H_9O_3P_3$

2,4,6-Trimethyl-1,3,5,2,4,6-trioxatriphosphorinane, T-00567

$C_3H_9O_3P_3S_3$

Methylphosphonothioic acid trimer trianhydride, *in* T-00567

$C_3H_9O_4P$

Dimethyl (hydroxymethyl)phosphonate, *in* H-00162
(1-Hydroxy-1-methylethyl)phosphonic acid, H-00159
(1-Hydroxypropyl)phosphonic acid, H-00190
(2-Hydroxypropyl)phosphonic acid, H-00191
(3-Hydroxypropyl)phosphonic acid, H-00192
Methyl bis(hydroxymethyl)phosphinate, *in* B-00282
▷Monoisopropyl phosphate, M-00477
Monopropyl phosphate, M-00512
Phosphinylidynetrismethanol, *in* T-00784
▷Trimethyl phosphate, T-00528

$C_3H_9O_5P$

(2,3-Dihydroxypropyl)phosphonic acid, D-00624

$C_3H_9O_6P$

Glycerol 1-monophosphate, G-00066
Glycerol 2-monophosphate, G-00067

C_3H_9P

Isopropylphosphine, I-00048
Propylphosphine, P-00471
▷Trimethylphosphine, T-00530

C_3H_9PS

Methyl dimethylphosphinothioite, *in* D-00843
Trimethylphosphine sulfide, T-00534

C_3H_9PSSe

Se-Methyl dimethylphosphinoselenothioate, *in* D-00837

$C_3H_9PS_2$

Dimethyl methylphosphonodithioite, *in* M-00302
Ethylmethylphosphinodithioic acid, E-00086
Methyl dimethylphosphinodithioate, *in* D-00834

$C_3H_9PS_3$

Dimethyl methylphosphonotrithioate, *in* M-00318
▷Trimethyl phosphorotrithioite, T-00550

$C_3H_9PS_4$

Trimethyl phosphorotetrathioate, T-00544

C_3H_9PSe

Methyl dimethylphosphinoselenoite, *in* D-00838
▷Trimethylphosphine selenide, T-00533

$C_3H_9PSe_3$

Trimethyl phosphorotriselenoite, T-00547

C_3H_9PTe

Trimethylphosphine; Telluride, *in* T-00530

$C_3H_{10}ClN_2OP$

P-(3-Chloropropyl)phosphonic diamide, C-00173

$C_3H_{10}NOPS$

O-Ethyl *P*-methylphosphonamidothioate, *in* M-00284
S-Ethyl *P*-methylphosphonamidothioate, *in* M-00284
S-Methyl *N,P*-dimethylphosphonamidothioate, *in* D-00850
S-Methyl *P*-ethylphosphonamidothioate, *in* E-00128
Trimethylphosphonamidothioic acid, T-00540

$C_3H_{10}NO_2P$

(1-Aminoethyl)methylphosphinic acid, A-00076
Dimethyl methylphosphoramidite, *in* M-00335
Ethyl *P*-methylphosphonamidate, *in* M-00283
N-Ethyl-*P*-methylphosphonamidic acid, E-00088
Methyl *N,P*-dimethylphosphonamidate, *in* D-00849
Propylphosphoramidous acid, P-00486
N,N,P-Trimethylphosphonamidic acid, T-00537

$C_3H_{10}NO_2PS$

O,O-Dimethyl methylphosphoramidothioate, *in* M-00333
Isopropylphosphoramidothioic acid, I-00059
Propylphosphoramidothioic acid, P-00485

$C_3H_{10}NO_3P$

(1-Amino-1-methylethyl)phosphonic acid, A-00087
(2-Amino-1-methylethyl)phosphonic acid, A-00088
(1-Aminopropyl)phosphonic acid, A-00125
(2-Aminopropyl)phosphonic acid, A-00126
(3-Aminopropyl)phosphonic acid, A-00127
Dimethyl methylphosphoramidate, *in* M-00326
Isopropylphosphoramidic acid, I-00058
2-Methylaminoethylphosphonic acid, *in* A-00079
Propylphosphoramidic acid, P-00484

$C_3H_{10}NO_3PS$

(2-Methoxyethyl)phosphoramidothioic acid, M-00041

$C_3H_{10}NO_5PS$

Dimethyl methylsulfonylphosphoramidate, *in* M-00401

$C_3H_{10}NP$

▷*P,P,P*-Trimethylphosphine imide, T-00531

$C_3H_{10}NPS_2$

Methyl *P*-ethylphosphonamidodithioate, *in* E-00127
Trimethylphosphonamidodithioic acid, T-00539

$C_3H_{10}NPSe$

N,P,P-Trimethylphosphinoselenoic amide, *in* D-00836

$C_3H_{10}O_4P_2$

Methylenebis[methylphosphinic acid], M-00140

$C_3H_{10}O_6P_2$

1,3-Propanediphosphonic acid, P-00444
2,2-Propanediphosphonic acid, P-00445

$C_3H_{10}O_7P_2$

1-Hydroxy-1,1-propanediphosphonic acid, H-00189

$C_3H_{10}P_2$

1,3-Diphosphinopropane, D-01180

$C_3H_{11}GeP$

Trimethylgermylphosphine, T-00516

$C_3H_{11}NO_6P_2$

(1-Aminopropylidene)bisphosphonic acid, A-00124
[(Dimethylamino)methylene]bisphosphonic acid, D-00701

$C_3H_{11}N_2OP$

P-Isopropylphosphonic diamide, I-00051
P-Propylphosphonic diamide, P-00474
N,N′,P-Trimethylphosphonic diamide, *in* M-00289

$C_3H_{11}N_2OPS$

O-Ethyl *P*-methylphosphonohydrazidothioate, *in* M-00309

$C_3H_{11}N_2O_2P$

Ethyl *P*-methylphosphonohydrazidate, *in* M-00308

$C_3H_{11}N_2PS$

P-Propylphosphonothioic diamide, P-00481

$C_3H_{11}N_3P_2$

1,2,4-Trimethyl-1,2,4,3,5-triazadiphospholidine, T-00562

$C_3H_{11}O_9P_3$

1,1,3-Propanetriphosphonic acid, P-00449
1,2,3-Propanetriphosphonic acid, P-00450

$C_3H_{12}NO_3PSi$

(Trimethylsilyl)phosphorimidic acid, T-00557

$C_3H_{12}NO_9P_3$

▷[Nitrilotris(methylene)]trisphosphonic acid, N-00033

$C_3H_{12}N_2O_4P_2S_2$

1,3-Propanediphosphoramidothioic acid, P-00446

$C_3H_{12}N_3OP$

N,N′,N″-Trimethylphosphoric triamide, *in* T-00551

$C_3H_{12}N_3P$

N,N′,N″-Trimethylphosphorous triamide, T-00551

$C_3H_{12}N_3PS$

N,N′,N″-Trimethylphosphorothioic triamide, *in* T-00551

$C_3H_{12}O_{10}P_4$

[Phosphinylidynetris(methylene)]-trisphosphonic acid, P-00360

$C_3N_3OPS_3$

Phosphoryl isothiocyanate, P-00416

$C_3N_3O_3PS$

Thiophosphoryl isocyanate, T-00315

$C_3N_3O_4P$

Phosphoryl isocyanate, P-00415

$C_3N_3PS_4$

Thiophosphoryl isothiocyanate, T-00316

$C_4F_6P_2S$

3,4-Bis(trifluoromethyl)-1,2,5-thiadiphosphole, B-00425

$C_4F_{12}OP_2$

Bis(trifluoromethyl)phosphinous acid; Anhydride, *in* B-00418

$C_4F_{12}O_3P_2$

Bis(trifluoromethyl)phosphinic acid; Anhydride, *in* B-00411

$C_4F_{12}P_2S$

Bis(trifluoromethyl)phosphinothious acid; Anhydrosulfide, *in* B-00417

$C_4F_{12}P_2S_2$

Bis(trifluoromethyl)phosphinothioic acid; Anhydride, *in* B-00415

$C_4F_{12}P_2S_3$

Bis(trifluoromethyl)phosphinodithioic acid; Anhydrosulfide, *in* B-00412

$C_4F_{12}P_2Se$

Bis(trifluoromethyl)phosphinoselenous acid; Anhydroselenide, *in* B-00413

$C_4F_{12}P_4$
▷Tetrakis(trifluoromethyl)tetraphosphetane, T-00188

$C_4H_3Cl_2OP$
3-Buten-1-ynylphosphonic acid; Dichloride, *in* B-00529

$C_4H_3Cl_2O_2P$
3-Furanylphosphonic acid; Dichloride, *in* F-00074

$C_4H_3Cl_2PS$
Dichloro-2-thienylphosphine, *in* T-00308

$C_4H_3N_2O_3P$
Vinylphosphonic acid; Diisocyanate, *in* V-00003

$C_4H_3O_2P$
Diethynylphosphinic acid, D-00409

$C_4H_3O_3P$
3-Buten-1-ynylphosphonic acid, B-00529

$C_4H_4BrCl_6O_3P$
Bis(2,2,2-trichloroethyl) phosphorobromidate, *in* B-00400

C_4H_4ClOPS
4*H*-1,4-Thiaphosphorin; 4-Chloro, 4-oxide, *in* T-00299

$C_4H_4ClO_6P$
2-Chloro-1,3,2-dioxaphospholan-4,5-dicarboxylic acid, C-00062

$C_4H_4Cl_7O_2P$
Bis(2,2,2-trichloroethyl) phosphorochloridite, B-00401

$C_4H_4F_4O_6P_2$
3,3,4,4-Tetrafluoro-1-cyclobutene-1,2-diphosphonic acid, T-00078

$C_4H_5Cl_2OP$
(1,3-Butadienyl)phosphonic acid; Dichloride, *in* B-00510

$C_4H_5Cl_6O_4P$
Bis(1-hydroxy-2,2,2-trichloroethyl)phosphinic acid, B-00287
Bis(2,2,2-trichloroethyl) phosphate, B-00400

$C_4H_5F_6OPS$
O-Ethyl bis(trifluoromethyl)phosphinothioate, *in* B-00415
S-Ethyl bis(trifluoromethyl)phosphinothioate, *in* B-00415

$C_4H_5F_6O_3P$
Bis(2,2,2-trifluoroethyl) phosphonate, B-00409

$C_4H_5N_2O_3P$
Ethylphosphonic acid; Diisocyanate, *in* E-00129
2-Pyrimidinephosphonic acid, P-00515
4-Pyrimidinephosphonic acid, P-00516

$C_4H_5N_2O_3PS$
O-Ethyl phosphorodiisocyanatidothioate, E-00178

$C_4H_5N_2O_4P$
Ethyl phosphorodiisocyanatidate, E-00177

$C_4H_5O_2P$
1-Hydroxy-1*H*-phosphole 1-oxide, H-00186

$C_4H_5O_2PS$
2-Thienylphosphinic acid, T-00306
2-Thienylphosphonous acid, T-00308

$C_4H_5O_3PS$
2-Thienylphosphonic acid, T-00307

$C_4H_5O_4P$
2-Furanylphosphonic acid, F-00073
3-Furanylphosphonic acid, F-00074

C_4H_5P
1*H*-Phosphole, P-00372

C_4H_5PS
4*H*-1,4-Thiaphosphorin, T-00299

$C_4H_6BrO_3P$
Diethenyl phosphorobromidate, *in* D-01257

C_4H_6BrP
1-Bromo-2,5-dihydro-1*H*-phosphole, *in* D-00572

C_4H_6ClOP
1-Chloro-2,3-dihydro-1*H*-phosphole; 1-Oxide, *in* C-00051
1-Chloro-2,5-dihydro-1*H*-phosphole; 1-Oxide, *in* C-00052
Divinylphosphinic acid; Chloride, *in* D-01258

$C_4H_6ClO_2P$
2-Chloro-4,5-dimethyl-1,3,2-dioxaphosphole, C-00057
2,3-Dihydro-5-methyl-1,2-oxaphosphole 2-oxide; 2-Chloro, *in* D-00525

$C_4H_6ClO_2PS$
4,7-Dihydro-1,3,2-dioxaphosphepin 2-sulfide; 2-Chloro, *in* D-00473

$C_4H_6ClO_3P$
Dimethylvinylene chlorophosphate, *in* C-00057
Divinyl phosphorochloridate, *in* D-01257

C_4H_6ClP
1-Chloro-2,3-dihydro-1*H*-phosphole, C-00051
1-Chloro-2,5-dihydro-1*H*-phosphole, C-00052

C_4H_6ClPS
1-Chloro-2,3-dihydro-1*H*-phosphole; 1-Sulfide, *in* C-00051
1-Chloro-2,5-dihydro-1*H*-phosphole; 1-Sulfide, *in* C-00052
Divinylphosphinothioic acid; Chloride, *in* D-01259

$C_4H_6ClPS_2$
2-Chloro-3,6-dihydro-2*H*-1,2-thiaphosphorin 2-sulfide, C-00053

$C_4H_6Cl_3O_5P$
(1-Acetoxy-2,2,2-trichloroethyl)phosphonic acid, A-00011

C_4H_6FOP
1-Fluoro-2,5-dihydro-1*H*-phosphole 1-oxide, *in* D-00507

$C_4H_6F_2OP$
(1,3-Butadienyl)phosphonic acid; Difluoride, *in* B-00510

$C_4H_6F_6NP$
N,*N*-Dimethyl-*P*,*P*-bis(trifluoromethyl)-phosphinous amide, *in* B-00419

$C_4H_6NO_3P$
1*H*-Pyrrol-2-ylphosphonic acid, P-00521

$C_4H_6N_3OP$
Dimethylphosphoramidic dicyanide, D-00857

$C_4H_6N_3O_3P$
(2-Amino-4-pyrimidinyl)phosphonic acid, A-00128

$C_4H_6N_3P$
Dimethylphosphoramidous acid; Dicyanide, *in* D-00875

$C_4H_6N_5OP$
Tetramethylphosphorodiamidic azide, T-00219

$C_4H_6N_5PS$
Tetramethylphosphorodiamidothioic azide, T-00224

$C_4H_6O_4P$
Dimethyl acetylphosphonate, *in* A-00026

$C_4H_7Br_2Cl_2O_4P$
▷Naled, N-00001

$C_4H_7Cl_2OP$
(2-Methyl-1-propenyl)phosphonic dichloride, M-00365

$C_4H_7Cl_2O_2P$
3-(Chloromethylphosphinyl)propanoyl chloride, *in* H-00161

$C_4H_7Cl_2O_3P$
Ethyl dichlorophosphinylacetate, *in* D-00190
Methyl 3-dichlorophosphinylpropanoate, *in* D-00192

$C_4H_7Cl_2O_4P$
▷Dichlorvos, D-00203

$C_4H_7N_2O_2PS$
P-Methyl-*N*-2-thiazolylphosphonamidic acid, M-00409

$C_4H_7N_2P$
1,5-Dimethyl-1*H*-1,2,3-diazaphosphole, *in* M-00114

C_4H_7OPS
Divinylphosphinothioic acid, D-01259

$C_4H_7O_2P$
4,7-Dihydro-1,3,2-dioxaphosphepin, D-00472
2,3-Dihydro-1-hydroxy-1*H*-phosphole 1-oxide, D-00506
2,5-Dihydro-1-hydroxy-1*H*-phosphole 1-oxide, D-00507
2,3-Dihydro-5-methyl-1,2-oxaphosphole 2-oxide, D-00525
Divinylphosphinic acid, D-01258

$C_4H_7O_2PS$
(1,3-Butadien-1-yl)phosphonothioic acid, B-00511
4,7-Dihydro-1,3,2-dioxaphosphepin 2-sulfide, D-00473

$C_4H_7O_3P$
(1,2-Butadienyl)phosphonic acid, B-00509
(1,3-Butadienyl)phosphonic acid, B-00510
1-Butynylphosphonic acid, B-00643
2-Butynylphosphonic acid, B-00644
3,6-Dihydro-2-hydroxy-2*H*-1,2-oxaphosphorin 2-oxide, D-00501
2-Methyl-1,2-oxaphospholan-5-one 2-oxide, *in* O-00053
4-Methyl-2,6,7-trioxa-1-phosphabicyclo[2.2.1]-heptane, M-00423
2,6,7-Trioxa-1-phosphabicyclo[2.2.2]octane, T-00588
2,7,8-Trioxa-1-phosphabicyclo[3.2.1]octane, T-00589

$C_4H_7O_3PSe$
Methylglycerol bicyclic selenophosphate, *in* M-00423

$C_4H_7O_4P$
Bicyclic phosphate, *in* T-00588
Divinyl phosphate, D-01257
2-Hydroxy-4,5-dimethyl-1,3,2-dioxaphosphole 2-oxide, H-00123
1-Hydroxy-3-phosphetanecarboxylic acid 1-oxide, H-00184
Methylglycerol bicyclic phosphate, *in* M-00423

$C_4H_7O_5P$
(Tetrahydro-2-oxo-2-furanyl)phosphonic acid, T-00121

$C_4H_7O_7P$
Phosphonobutanedioic acid, P-00381

C_4H_7P
2,3-Dihydro-1*H*-phosphole, D-00571
2,5-Dihydro-1*H*-phosphole, D-00572

$C_4H_7PS_2$
2,5-Dihydro-1-mercaptophosphole 1-sulfide, D-00512
3,6-Dihydro-2*H*-1,2-thiaphosphorin 2-sulfide, D-00586

$C_4H_8BrN_2O_2P$
1,4-Dimethyl-1,4,2-diazaphospholidin-5-one 2-oxide; 2-Bromo, *in* D-00722

$C_4H_8BrO_2P$
2-Bromo-4,5-dimethyl-1,3,2-dioxaphospholane, B-00470
2-Bromo-4-methyl-1,3,2-dioxaphosphorinane, B-00474

$C_4H_8BrO_2PS$
O,*O*-2,3-Butylene bromothiophosphate, *in* B-00470
OO-1,3-Butylene thiophosphoryl bromide, *in* B-00474

$C_4H_8BrO_3P$
1,3-Butylene phosphoryl bromide, *in* B-00474

$C_4H_8Br_2NOP$
4-Morpholinylphosphonous acid; Dibromide, *in* M-00529

$C_4H_8Br_4O_3P_2$
Bis(bromomethyl)phosphinic acid; Anhydride, *in* B-00108

$C_4H_8Br_4P_2$
1,4-Butanediphosphonous acid; Bis(dibromide), *in* B-00515

C_4H_8ClOP
Methyl(2-propenyl)phosphinic acid; Chloride, *in* M-00364
Tetramethylenephosphinic chloride, *in* C-00167

C_4H_8ClOPS
O-Ethyl ethenylphosphonochloridothioate, *in* V-00006

$C_4H_8ClO_2P$
2-Chloro-4,5-dimethyl-1,3,2-dioxaphospholane, C-00056
2-Chloro-4-methyl-1,3,2-dioxaphosphorinane, C-00093
2-Chloro-1,2-oxaphosphorinane 2-oxide, *in* H-00167
Tetramethylene phosphorochloridite, *in* D-00968

$C_4H_8ClO_2PS$
O,O-2,3-Butylene chlorothiophosphate, *in* C-00056
2-Chloro-4-methyl-1,3,2-dioxaphosphorinane 2-sulfide, C-00095

$C_4H_8ClO_2PSe$
O,O-Butylene phosphorochloridoselenoate, *in* C-00093

$C_4H_8ClO_3P$
2-Chloro-4-methyl-1,3,2-dioxaphosphorinane 2-oxide, C-00094
1,3,6,2-Trioxaphosphocane; 2-Chloro, *in* T-00592

C_4H_8ClP
1-Chlorophospholane, C-00167

C_4H_8ClPS
Tetramethylenephosphinothioic chloride, *in* C-00167

$C_4H_8ClPS_2$
S,S'-1,3-Butylene phosphorochloridodithioite, *in* M-00138
5-Methyl-1,2-thiaphospholane 2-sulfide; 2-Chloro, *in* M-00407

$C_4H_8Cl_2FO_3P$
Bis(2-chloroethyl) phosphorofluoridate, *in* B-00119

$C_4H_8Cl_2NOP$
4-Morpholinylphosphonous acid; Dichloride, *in* M-00529

$C_4H_8Cl_2NOPS$
4-Morpholinylphosphonothioic acid; Dichloride, *in* M-00527

$C_4H_8Cl_2NO_2P$
4-Morpholinylphosphonic acid; Dichloride, *in* M-00523

$C_4H_8Cl_2NO_3P$
Isopropyl dichlorophosphinylcarbamate, *in* D-00191

$C_4H_8Cl_2P$
1-Pyrrolidinylphosphonous acid; Dichloride, *in* P-00520

$C_4H_8Cl_3NOP$
▷Bis(2-chloroethyl)phosphoramidic dichloride, B-00122

$C_4H_8Cl_3OP$
(4-Chlorobutyl)phosphonic acid; Dichloride, *in* C-00025

$C_4H_8Cl_3O_2P$
Ethyl methyl(trichloromethyl)phosphinate, *in* M-00419

$C_4H_8Cl_3O_2PS$
O,O-Bis(2-chloroethyl) phosphorochloridothioate, B-00124

$C_4H_8Cl_3O_3P$
▷Bis(2-chloroethyl) phosphorochloridate, *in* B-00119

$C_4H_8Cl_3O_4P$
▷Trichlorphon, T-00423

$C_4H_8Cl_4NPS$
▷Bis(2-chloroethyl)phosphoramidothioic dichloride, B-00123

$C_4H_8Cl_4O_2P_2$
1,4-Butanediphosphonic acid; Bisdichloride, *in* B-00514

$C_4H_8Cl_4O_3P_2$
Bis(chloromethyl)phosphinic acid; Anhydride, *in* B-00127

$C_4H_8Cl_4P_2$
1,4-Butanediphosphonous acid; Bis(dichloride), *in* B-00515

$C_4H_8FO_2P$
2-Fluoro-4,5-dimethyl-1,3,2-dioxaphospholane, F-00015
2-Fluoro-4-methyl-1,3,2-dioxaphosphorinane, F-00021
Tetramethylene phosphorofluoridite, *in* D-00968

$C_4H_8FO_2PS$
O,O-1,3-Butylene fluorothiophosphate, *in* F-00021
O,O-2,3-Butylene fluorothiophosphate, *in* F-00015

$C_4H_8FO_2PSe$
O,O-1,3-Butylene fluoroselenophosphate, *in* F-00021

$C_4H_8F_2P$
1-Pyrrolidinylphosphonous acid; Difluoride, *in* P-00520

$C_4H_8F_3O_2P$
Isopropyl (trifluoromethyl)phosphinate, *in* T-00475

$C_4H_8F_3P$
1,1,1-Trifluoro-1,1-dihydrophospholane, T-00466

$C_4H_8NO_2P$
2,8-Dioxa-5-aza-1-phosphabicyclo[3.3.0]octane, D-00961

$C_4H_8NO_2PS$
Ethyl methylphosphonisothiocyanatidate, *in* M-00294

$C_4H_8NO_3P$
(1-Cyanopropyl)phosphonic acid, C-00229
(3-Cyanopropyl)phosphonic acid, C-00230
Ethyl methylphosphonisocyanatidate, *in* M-00292

$C_4H_8NO_4P$
Dimethyl (isocyanatomethyl)phosphonate, *in* I-00022
4-Morpholinylphosphonic acid, M-00523

$C_4H_8NO_5P$
Antibiotic FR 32863, A-00137

$C_4H_8NO_7P$
β-Aspartyl phosphate, A-00143

$C_4H_8N_3O_4P$
Fosfocreatinine, F-00060

$C_4H_8N_5OP$
N''-2-Pyrimidinylphosphoric triamide, P-00517

$C_4H_8O_5P_2$
▷2,2'-Oxybis[1,3,2-dioxaphospholane], O-00091

$C_4H_8O_6P_2$
1,3-Butadiene-1,4-diphosphonic acid, B-00507

$C_4H_8O_7P_2$
Ethylene glycol cyclic *P,P:P',P'*-pyrophosphate, *in* O-00091

$C_4H_8O_8P_2$
(1,4-Dioxo-1,4-butanediyl)bisphosphonic acid, D-00975

$C_4H_8O_{10}P_2$
2,3-Diphosphonobutanedioic acid, D-01182

$C_4H_9Br_2OP$
Butyl phosphorodibromidite, B-00627

$C_4H_9Br_2P$
Dibromobutylphosphine, *in* B-00615
Dibromo-*tert*-butylphosphine, *in* B-00616

$C_4H_9Br_2PS$
tert-Butylphosphonothioic acid; Dibromide, *in* B-00612

$C_4H_9Br_3NP$
P,P,P-Tribromo-*N*-*tert*-butylphosphazene, *in* B-00625

$C_4H_9ClFO_2P$
Isopropyl (fluoromethyl)phosphonochloridate, *in* F-00024

C_4H_9ClNOP
2-Chlorotetrahydro-3-methyl-2*H*-1,3,2-oxazaphosphorine, C-00178
1-Pyrrolidinylphosphonochloridous acid, P-00519

$C_4H_9ClNO_2P$
4-Morpholinylphosphonochloridous acid, M-00526

$C_4H_9ClNO_2PS$
4-Morpholinylphosphonochloridothioic acid, M-00525

$C_4H_9ClNO_3P$
4-Morpholinylphosphonochloridic acid, M-00524

$C_4H_9Cl_2OP$
Butylphosphonic dichloride, B-00598
tert-Butylphosphonic dichloride, B-00599
Butyl phosphorodichloridite, B-00629
(2-Methylpropyl)phosphonic acid; Dichloride, *in* M-00378
(1-Methylpropyl)phosphonic dichloride, M-00379
2-Methylpropyl phosphorodichloridite, M-00385

$C_4H_9Cl_2OPS$
O-Butyl phosphorodichloridothioate, B-00630
S-Butyl phosphorodichloridothioate, B-00631
O-Ethyl bis(chloromethyl)phosphinothioate, *in* B-00129
S-Ethyl bis(chloromethyl)phosphinothioate, *in* B-00129

$C_4H_9Cl_2O_2P$
▷Butyl phosphorodichloridate, B-00628
Ethyl bis(chloromethyl)phosphinate, *in* B-00127
Ethyl (2-chloroethyl)phosphonochloridate, *in* C-00077

$C_4H_9Cl_2O_2PS$
O-(2-Ethoxyethyl) phosphorodichloridothioate, *in* H-00149

$C_4H_9Cl_2O_3P$
▷Bis(2-chloroethyl) phosphonate, B-00120

$C_4H_9Cl_2O_4P$
Bis(2-chloroethyl) phosphate, B-00119

$C_4H_9Cl_2P$
Butylphosphonous dichloride, B-00618
tert-Butylphosphonous dichloride, B-00619
Dichloroisobutylphosphine, *in* M-00383

$C_4H_9Cl_2PS$
Butylphosphonothioic acid; Dichloride, *in* B-00611
tert-Butylphosphonothioic dichloride, B-00613
Butyl phosphorodichloridothioite, B-00632

(2-Methylpropyl)phosphonothioic acid; Dichloride, *in* M-00381

$C_4H_9Cl_2PSe$

Butylphosphonoselenoic acid; Dichloride, *in* B-00609

$C_4H_9Cl_3NP$

tert-Butylphosphorimidic trichloride, B-00626

$C_4H_9F_2N_2OP$

2,2-Difluoro-2,2-dihydro-1,2,3-trimethyl-1,3,2-diazaphosphetidin-4-one, *in* D-00459

$C_4H_9F_2OP$

Butylphosphonic acid; Difluoride, *in* B-00594
tert-Butylphosphonic difluoride, B-00600
Butyl phosphorodifluoridite, B-00633

$C_4H_9F_2P$

Butyldifluorophosphine, *in* B-00615
▷*tert*-Butylphosphonous difluoride, B-00620

$C_4H_9I_2O_2P$

Ethyl bis(iodomethyl)phosphinate, *in* B-00288

$C_4H_9I_2P$

Butyldiiodophosphine, *in* B-00615
tert-Butyldiiodophosphine, *in* B-00616

$C_4H_9N_2OP$

Bis(1-aziridinyl)phosphinous acid, B-00095

$C_4H_9N_2OPS$

Bis(1-aziridinyl)phosphinothioic acid, B-00093

$C_4H_9N_2OPS_2$

N-(4,5-Dihydro-2-thiazolyl)-*P*-methylphosphonamidothioic acid, D-00588

$C_4H_9N_2O_2P$

Bis(1-aziridinyl)phosphinic acid, B-00091
1,3-Dimethyl-1,3,2-diazaphosphetidin-4-one; 2-Methoxy, *in* D-00717
1,3-Dimethyl-1,3,2-diazaphospholidin-4-one 2-oxide, D-00721
1,4-Dimethyl-1,4,2-diazaphospholidin-5-one 2-oxide, D-00722
Methyl dimethylphosphoramidocyanidate, *in* D-00863
N,N,P-Trimethylphosphonisocyanatidic amide, *in* M-00293

$C_4H_9N_2O_2PS$

N-(4,5-Dihydro-2-thiazolyl)-*P*-methylphosphonamidic acid, D-00587

$C_4H_9N_2O_3P$

Dimethyl (1-diazoethyl)phosphonate, *in* D-00033
Trimethyl cyanophosphorimidate, *in* C-00228

C_4H_9OP

Acetyldimethylphosphine, A-00015
tert-Butyloxaphosphine, B-00576
2-Methyl-1,2-oxaphospholane, M-00182
1,2-Oxaphosphorinane, O-00054
1,4-Oxaphosphorinane, O-00055

C_4H_9OPS

2-Methyl-1,2-oxaphospholane; 2-Sulfide, *in* M-00182

$C_4H_9OPS_2$

2-Ethoxy-1,3,2-dithiaphospholane, E-00038
2-Methoxy-1,3,2-dithiaphosphorinane, *in* D-01247
2-Methoxy-4-methyl-1,3,2-dithiaphospholane, *in* M-00136
S,S'-Propanediyl methylphosphonodithioate, *in* M-00137

$C_4H_9OPS_3$

O-Methyl *S,S*-trimethylene phosphonotrithioate, *in* D-01249

$C_4H_9O_2P$

2,4-Dimethyl-1,3,2-dioxaphospholane, D-00732
4,5-Dimethyl-1,3,2-dioxaphospholane, D-00733
1-(Dimethylphosphinyl)ethanone, *in* A-00015
1,3,2-Dioxaphosphepane, D-00968
2-Methyl-1,3,2-dioxaphosphorinane, M-00128
4-Methyl-1,3,2-dioxaphosphorinane, M-00129
Methyl methylvinylphosphinate, *in* M-00438
2-Methyl-1,2-oxaphospholane; 2-Oxide, *in* M-00182
5-Methyl-1,2-oxaphospholane 2-oxide, M-00183
Methyl(2-propenyl)phosphinic acid, M-00364

$C_4H_9O_2PS$

2-Butenylphosphonothioic acid, B-00527
O,O-2,3-Butylene phosphonothioate, *in* D-00733
4,5-Dimethyl-1,3,2-dioxaphospholane; 2-Sulfide, *in* D-00733
O,O-Dimethyl vinylphosphonothioate, *in* V-00007
1,3,2-Dioxaphosphepane 2-sulfide, D-00969
Ethyl thiodimethylenephosphinate, *in* H-00201
▷2-Ethylthio-1,3,2-dioxaphospholane, E-00191
4-Methyl-1,3,2-dioxaphosphorinane 2-sulfide, M-00131
4-Methyl-2-methylthio-1,3,2-dioxaphospholane, M-00173
(2-Methyl-1-propenyl)phosphonothioic acid, M-00366
O,O-Trimethylene methylphosphonothioate, *in* M-00128

$C_4H_9O_2PS_2$

1,4-Dioxa-6,9-dithia-5-phospha(5-P^V)-spiro[4.4]nonane, D-00964
S-Ethyl *O,O*-ethylene dithiophosphate, *in* E-00191
2-Mercapto-4,5-dimethyl-1,3,2-dioxaphospholane 2-sulfide, M-00009
2-Mercapto-4-methyl-1,3,2-dioxaphosphorinane 2-sulfide, M-00017
S-Methyl *O,O*-propylene phosphorodithioate, *in* M-00173
2-Methylthio-1,3,2-dioxaphosphorinane 2-sulfide, *in* M-00013

$C_4H_9O_2PSe$

O,O-Trimethylene methylphosphonoselenoate, *in* M-00128

$C_4H_9O_3P$

Acetylethylphosphinic acid, A-00017
1-Butenylphosphonic acid, B-00524
2-Butenylphosphonic acid, B-00525
3-Buten-1-ylphosphonic acid, B-00526
2,3-Butylene phosphonate, *in* D-00733
4,5-Dimethyl-1,3,2-dioxaphospholane; 2-Oxide, *in* D-00733
Dimethyl ethenylphosphonate, *in* V-00003
▷2-Ethoxy-1,3,2-dioxaphospholane, E-00034
3-Hydroxy-2,2-dimethyl-1,3-oxaphosphetane 3-oxide, H-00126
2-Hydroxy-1,2-oxaphosphorinane 2-oxide, H-00167
▷2-Methoxy-1,3,2-dioxaphosphorinane, M-00038
2-Methoxy-4-methyl-1,3,2-dioxaphospholane, M-00043
Methyl acetylmethylphosphinate, *in* A-00018
4-Methyl-1,3,2-dioxaphosphorinane 2-oxide, M-00130
Propylene methylphosphonate, *in* D-00732
Trimethylene methylphosphonate, *in* M-00128
1,3,6,2-Trioxaphosphocane, T-00592

$C_4H_9O_3PS$

2-Ethoxy-1,3,2-dioxaphospholane 2-sulfide, E-00036
2-Hydroxy-4,5-dimethyl-1,3,2-dioxaphospholane 2-sulfide, H-00122
2-Hydroxy-4-methyl-1,3,2-dioxaphosphorinane 2-sulfide, H-00158
S-Methyl *O,O*-propylene phosphorothioate, *in* M-00173
O-Methyl *O,O*-propylene thiophosphate, *in* M-00043
O-Methyl *O,O*-trimethylene phosphorothioate, *in* M-00038
1,4-Thiaphosphorinane 1,1,4-trioxide, T-00300

$C_4H_9O_3PS_2$

1,3-Dithiane-2-phosphonic acid, D-01240

$C_4H_9O_3PSe$

2-Hydroxy-4-methyl-1,3,2-dioxaphosphorinane 2-selenide, H-00157
O-Methyl *O,O*-trimethylene phosphoroselenoate, *in* M-00038

$C_4H_9O_3PSi$

1-Methyl-2,6,7-trioxa-4-phospha-1-silabicyclo[2.2.2]octane, M-00426

$C_4H_9O_4P$

Dimethyl (2-oxoethyl)phosphonate, *in* O-00059
2-Hydroxy-4,5-dimethyl-1,3,2-dioxaphospholane 2-oxide, H-00121
2-Hydroxy-1,3,2-dioxaphosphepane 2-oxide, H-00132
2-Hydroxy-4-methyl-1,3,2-dioxaphosphorinane 2-oxide, H-00156
3-(Hydroxymethylphosphinyl)propanoic acid, H-00161
Methyl propylene phosphate, *in* M-00043
Methyl trimethylene phosphate, *in* M-00038
Mono(2-methyl-1-propenyl) phosphate, M-00491
1,4,6,9-Tetraoxa-5-phospha(5-P^V)spiro[4.4]-nonane, T-00244

$C_4H_9O_4PS_2$

▷[(Dimethoxyphosphinothioyl)thio]acetic acid, D-00675

$C_4H_9O_5P$

Acetic acid anhydride with dimethyl hydrogen phosphate, *in* A-00025
(2-Acetoxyethyl)phosphonic acid, A-00006
(Dimethoxyphosphinyl)acetic acid, D-00678
Ethyl dihydroxyphosphinylacetate, *in* P-00376
3-Phosphonobutanoic acid, P-00385
4-Phosphonobutanoic acid, P-00386
Trimethyl phosphonoformate, *in* D-00672

$C_4H_9O_5PS$

(Dimethoxyphosphinyl)thioacetic acid, D-00687

$C_4H_9O_6P$

2-Phosphonooxypropanoic acid; Me ester, *in* P-00392

C_4H_9P

Phospholane, P-00371

$C_4H_9PS_2$

2-Methyl-1,3,2-dithiaphosphorinane, M-00137
4-Methyl-1,3,2-dithiaphosphorinane, M-00138
5-Methyl-1,2-thiaphospholane 2-sulfide, M-00407
1,2-Thiaphospholane 2-sulfide; 2-Me, *in* T-00298

$C_4H_9PS_2Se$

S,S'-Propanediyl methylphosphonoselenodithioate, *in* M-00137

$C_4H_9PS_3$

Trimethylene methylphosphonotrithioate, *in* M-00137

$C_4H_9PS_4$

Methyl trimethylene phosphorotetrathioate, *in* M-00014

$C_4H_{10}BrOPS$

O-Isopropyl methylphosphonobromidothioate, *in* M-00295

$C_4H_{10}BrO_2P$

Diethyl phosphorobromidite, D-00372

$C_4H_{10}BrO_3P$

(4-Bromobutyl)phosphonic acid, B-00466

Diethyl phosphorobromidate, D-00371

$C_4H_{10}BrP$

Bromodiethylphosphine, *in* D-00333

$C_4H_{10}BrPS$

Diethylphosphinothioic acid; Bromide, *in* D-00330
Ethyl ethylphosphonobromidothioite, *in* E-00134

$C_4H_{10}BrPSe$

Diethylphosphinoselenoic acid; Bromide, *in* D-00327

$C_4H_{10}Br_2NO_3P$

Diethyl dibromophosphoramidate, D-00279

$C_4H_{10}Br_2NP$

Diethylphosphoramidous acid; Dibromide, *in* D-00360

$C_4H_{10}ClN_2OP$

2-Chloro-1,3-dimethyl-1,3,2-diazaphospholidine; 2-Oxide, *in* C-00055

$C_4H_{10}ClN_2P$

2-Chloro-1,3-dimethyl-1,3,2-diazaphospholidine, C-00055

$C_4H_{10}ClN_2PS$

2-Chloro-1,3-dimethyl-1,3,2-diazaphospholidine; 2-Sulfide, *in* C-00055

$C_4H_{10}ClOP$

Butylphosphonochloridous acid, B-00605
Di-*tert*-butylphosphonochloridous acid, D-00130
Diethylphosphinic acid; Chloride, *in* D-00322
Ethyl ethylphosphonochloridite, *in* E-00139
Isopropylmethylphosphinic acid; Chloride, *in* I-00035
Isopropyl methylphosphonochloridate, *in* M-00296
Methylpropylphosphinic acid; Chloride, *in* M-00375

$C_4H_{10}ClOPS$

Butylphosphonochloridothioic acid, B-00604
O-Ethyl ethylphosphonochloridothioate, E-00069
S-Ethyl ethylphosphonochloridothioate, *in* E-00137
O-Isopropyl methylphosphonochloridothioate, I-00039
S-Isopropyl methylphosphonochloridothioate, *in* M-00298
O-Methyl isopropylphosphonochloridothioate, M-00165
S-Methyl propylphosphonochloridothioate, *in* P-00477
O-Propyl methylphosphonochloridothioate, P-00468
S-Propyl methylphosphonochloridothioate, *in* M-00298

$C_4H_{10}ClOPS_2$

▷*S,S*-Diethyl phosphorochloridodithioate, D-00375

$C_4H_{10}ClO_2P$

Butylphosphonochloridic acid, B-00603
▷Diethyl phosphorochloridite, D-00374
Ethyl ethylphosphonochloridate, *in* E-00135

$C_4H_{10}ClO_2PS$

▷*O,O*-Diethyl phosphorochloridothioate, D-00378
O,S-Diethyl phosphorochloridothioate, D-00379

$C_4H_{10}ClO_2PS_2$

▷*O,O*-Diethyl chlorothiophosphonothioate, D-00275

$C_4H_{10}ClO_2PSe$

O,O-Diethyl phosphorochloridoselenoate, D-00377

$C_4H_{10}ClO_3P$

(4-Chlorobutyl)phosphonic acid, C-00025
▷Diethyl phosphorochloridate, D-00373

$C_4H_{10}ClO_3PS$

▷Diethyl chlorothiophosphonate, D-00274

$C_4H_{10}ClP$

Diethylphosphinous chloride, D-00335

$C_4H_{10}ClPS$

Diethylphosphinothioic acid; Chloride, *in* D-00330
Ethyl ethylphosphonochloridothioite, *in* E-00138

$C_4H_{10}ClPS_2$

Diethyl phosphorochloridodithioite, D-00376
Ethyl ethylphosphonochloridodithioate, *in* E-00136

$C_4H_{10}ClPS_3$

▷Diethyl phosphorochloridotrithioate, D-00380

$C_4H_{10}ClPSe$

Diethylphosphinoselenoic acid; Chloride, *in* D-00327

$C_4H_{10}Cl_2NOP$

P,P-Bis(chloromethyl)-*N,N*-dimethylphosphinic amide, *in* B-00128
Butylphosphoramidic acid; Dichloride, *in* B-00621
tert-Butylphosphoramidic acid; Dichloride, *in* B-00622
Diethylphosphoramidic dichloride, D-00344
(2-Methylpropyl)phosphoramidic acid; Dichloride, *in* M-00384

$C_4H_{10}Cl_2NO_3P$

Bis(2-chloroethyl) phosphoramidate, *in* B-00119
Bis(2-chloroethyl)phosphoramidic acid, B-00121
▷Diethyl dichlorophosphoramidate, D-00280

$C_4H_{10}Cl_2NP$

tert-Butylphosphoramidous acid; Dichloride, *in* B-00624
Diethylphosphoramidous dichloride, D-00361

$C_4H_{10}Cl_2NPS$

Diethylphosphoramidothioic dichloride, D-00357

$C_4H_{10}Cl_2O_3P_2$

(Chloromethyl)methylphosphinic acid; Anhydride, *in* C-00100

$C_4H_{10}FN_2OP$

2-Fluoro-1,3-dimethyl-1,3,2-diazaphospholidine; 2-Oxide, *in* F-00014

$C_4H_{10}FN_2P$

2-Fluoro-1,3-dimethyl-1,3,2-diazaphospholidine, F-00014

$C_4H_{10}FN_2PS$

2-Fluoro-1,3-dimethyl-1,3,2-diazaphospholidine; 2-Sulfide, *in* F-00014

$C_4H_{10}FOP$

Diethylphosphinic acid; Fluoride, *in* D-00322
Isopropyl methylphosphonofluoridite, *in* M-00307

$C_4H_{10}FOPS$

O-Ethyl ethylphosphonofluoridothioate, E-00071
O-Isopropyl methylphosphonofluoridothioate, *in* M-00305

$C_4H_{10}FOPS_2$

O,S-Diethyl phosphorofluoridodithioate, D-00388
S,S-Diethyl phosphorofluoridodithioate, *in* D-00386

$C_4H_{10}FO_2P$

Butylphosphonofluoridic acid, B-00608
Ethyl ethylphosphonofluoridate, *in* E-00142
▷Sarin, S-00002

$C_4H_{10}FO_2PS$

O,O-Diethyl phosphorofluoridothioate, D-00389

$C_4H_{10}FO_3P$

▷Diethyl phosphorofluoridate, D-00387

$C_4H_{10}FPS$

Diethylphosphinothioic acid; Fluoride, *in* D-00330

$C_4H_{10}FPS_3$

Diethyl phosphorofluoridotrithioate, D-00390

$C_4H_{10}F_2NOP$

Butylphosphoramidic acid; Difluoride, *in* B-00621
tert-Butylphosphoramidic acid; Difluoride, *in* B-00622
Diethylphosphoramidic difluoride, D-00345

$C_4H_{10}F_2NO_2PSe$

Diethylphosphoramidoselenoic acid; Difluoride, *in* D-00353

$C_4H_{10}F_2NP$

tert-Butylphosphoramidous acid; Difluoride, *in* B-00624
Diethylphosphoramidous acid; Difluoride, *in* D-00360

$C_4H_{10}F_2NPS$

tert-Butylphosphoramidothioic acid; Difluoride, *in* B-00623
Diethylphosphoramidothioic difluoride, D-00358

$C_4H_{10}F_2P_2S_3$

Diethylthiodiphosphonic acid; Difluoride, *in* D-00405

$C_4H_{10}IO_2P$

Diethyl phosphoriodidite, D-00365

$C_4H_{10}IP$

Diethyliodophosphine, *in* D-00333

$C_4H_{10}IPS$

Diethylphosphinothioic acid; Iodide, *in* D-00330

$C_4H_{10}I_2NOP$

N,N-Diethyl-*P,P*-bis(iodomethyl)phosphinic amide, *in* B-00288

$C_4H_{10}I_2NP$

Diethylphosphoramidous acid; Diiodide, *in* D-00360

$C_4H_{10}NOPS$

N,N-Dimethyl-1,3,2-oxathiaphospholan-2-amine, *in* O-00056
2-Ethoxy-1,3,2-thiazaphospholidine, E-00053
Tetrahydro-2*H*-1,3,2-thiazaphosphorine 2-oxide; 2-Me, *in* T-00154

$C_4H_{10}NOPS_2$

2-Ethoxy-1,3,2-thiazaphospholidine; 2-Sulfide, *in* E-00053
S,S-Ethylene *N,N*-dimethyl phosphoramidodithioate, *in* D-01244
3-Methyl-1,3,2-thiazaphospholidine; 2-Methoxy, 2-sulfide, *in* M-00408

$C_4H_{10}NO_2P$

2-Dimethylamino-1,3,2-dioxaphospholane, D-00695
1-Pyrrolidinylphosphonous acid, P-00520
Tetrahydro-2-methoxy-2*H*-1,3,2-oxazaphosphorine, T-00111

$C_4H_{10}NO_2PS$

2-Ethoxy-1,3,2-thiazaphospholidine; 2-Oxide, *in* E-00053
Ethylene dimethylphosphoramidothioate, *in* D-00695
3-Methyl-1,3,2-thiazaphospholidine; 2-Methoxy, 2-oxide, *in* M-00408
Tetrahydro-2-methoxy-2*H*-1,3,2-oxazaphosphorine; 2-Sulfide, *in* T-00111

$C_4H_{10}NO_2PS_3$

Ethylene dimethylphosphoramidoselenoate, *in* D-00695

$C_4H_{10}NO_3P$

Dimethyl 1-aziridinylphosphonate, *in* A-00153
Ethylene dimethylphosphoramidate, *in* D-00695
Methyl *N*-(dimethylphosphinyl)carbamate, *in* D-00830
4-Morpholinylphosphinic acid, M-00522
4-Morpholinylphosphonous acid, M-00529
1-Pyrrolidinylphosphonic acid, P-00518

$C_4H_{10}NO_3PS$

▷Acephate, A-00003
4-Morpholinylphosphonothioic acid, M-00527

$C_4H_{10}NO_3PS_2$

O,O-Dimethyl *S*-(2-amino-2-oxoethyl) phosphorodithioate, *in* D-00675

$C_4H_{10}NO_4P$

Dimethyl acetylphosphoramidate, *in* A-00029
(2-Dimethylamino-2-oxoethyl)phosphonic acid, D-00703
Dimethyl (methylaminocarbonyl)phosphonate, *in* M-00088

$C_4H_{10}NO_5P$

2-Amino-3-phosphonobutanoic acid, A-00116
2-Amino-4-phosphonobutanoic acid, A-00117
3-Amino-3-phosphonobutanoic acid, A-00118
4-Amino-4-phosphonobutanoic acid, A-00119
Dimethyl (1-nitroethyl)phosphonate, *in* N-00035
Dimethyl (2-nitroethyl)phosphonate, *in* N-00036
Fosmidomycin, F-00062
Methyl 3-amino-3-phosphonopropanoate, *in* A-00122
Methyl *N*-(dimethoxyphosphinyl)carbamate, *in* D-00680

$C_4H_{10}NO_6P$

2-Amino-3-phosphonooxybutanoic acid, A-00120
N-Phosphothreonine, P-00420

$C_4H_{10}NPS_2$

S,S-Ethylene diethylphosphoramidodithioite, *in* D-01243

$C_4H_{10}N_3OP$

▷*P,P*-Bis(1-aziridinyl)phosphinic amide, B-00092
Diethylphosphinic acid; Azide, *in* D-00322

$C_4H_{10}N_3O_2PS$

▷*O,O*-Diethyl phosphorazidothioate, D-00363

$C_4H_{10}N_3O_3P$

▷Diethyl phosphorazidate, D-00362

$C_4H_{10}N_3O_5P$

N-Phosphocreatine, P-00364

$C_4H_{10}N_3P$

P,P-Bis(1-aziridinyl)phosphinous amide, B-00096

$C_4H_{10}N_3PS$

▷*P,P*-Bis(1-aziridinyl)phosphinothioic amide, B-00094
Diethylphosphinothioic acid; Azide, *in* D-00330

$C_4H_{10}N_3PSe$

Diethylphosphinoselenoic acid; Azide, *in* D-00327

$C_4H_{10}O_4P_2$

1,3-Dihydroxy-1,3-diphosphorinane 1,3-dioxide, D-00622

$C_4H_{10}O_6P_2$

2-Butene-1,4-diphosphonic acid, B-00519

$C_4H_{10}O_8P_2$

1-Acetoxy-1,1-ethanediphosphonic acid, A-00005
4,4-Diphosphonobutanoic acid, D-01183

$C_4H_{10}P_2S_4$

2,4-Diethyl-1,3,2,4-dithiadiphosphetane 2,4-disulfide, *in* E-00152

$C_4H_{10}P_4$

Tetrahydro-1*H*,5*H*-[1,2,3]triphospholo[2,1-*a*][1,2,3]triphosphole, T-00160

$C_4H_{11}ClNOP$

Ethyl dimethylphosphoramidochloridite, *in* D-00862

$C_4H_{11}ClNOPS$

Diethylphosphoramidochloridothioic acid, D-00348
O-Ethyl dimethylphosphoramidochloridothioate, *in* D-00861

$C_4H_{11}ClNOPSe$

Diethylphosphoramidochloridoselenoic acid, D-00347

$C_4H_{11}ClNO_2P$

(4-Chlorobutyl)phosphonic acid; Monoamide, *in* C-00025
Diethylphosphoramidochloridic acid, D-00346
Ethyl dimethylphosphoramido chloridate, *in* D-00860

$C_4H_{11}Cl_2N_2O_2P$

N,N'-Bis(2-chloroethyl)phosphorodiamidic acid, B-00125
N,N-Bis(2-chloroethyl)phosphorodiamidic acid, B-00126

$C_4H_{11}FNOPS$

O-Ethyl dimethylphosphoramidofluoridothioate, *in* D-00868

$C_4H_{11}FNO_2P$

Diethylphosphoramidofluoridic acid, D-00351
▷Ethyl dimethylphosphoramidofluoridate, E-00064

$C_4H_{11}FNPS_2$

Diethylphosphoramidofluoridodithioic acid, D-00352
Ethyl dimethylphosphoramidofluoridodithioate, *in* D-00867

$C_4H_{11}NO_8P_2$

▷Glyphosine, G-00098

$C_4H_{11}NPSe$

P,P-Diethylphosphinoselenoic amide, D-00328

$C_4H_{11}N_2OP$

2-Dimethylamino-1,3,2-oxazaphospholidine, D-00702
1,3-Dimethyl-1,3,2-diazaphospholidine 2-oxide, D-00720
P-Methyl-*N,N'*-trimethylenephosphonic diamide, *in* H-00020

$C_4H_{11}N_2O_2P$

2-Dimethylamino-1,3,2-oxazaphospholidine; 2-Oxide, *in* D-00702
1,6-Dioxa-4,9-diaza-5-phospha(5-P^V)spiro[4.4]-nonane, D-00963
Methyl *N,N'*-trimethylenephosphorodiamidate, *in* H-00020

$C_4H_{11}N_2O_2PS$

O,S-Dimethyl (1-iminoethyl)-phosphoramidothioate, *in* I-00005

$C_4H_{11}N_2O_4P$

3-Amino-3-phosphonobutanamide, *in* A-00118

$C_4H_{11}N_2P$

1,2-Dimethyl-1,2,4-diazaphospholidine, D-00718
1,3-Dimethyl-1,3,2-diazaphospholidine, D-00719

$C_4H_{11}OP$

Diethylphosphine; Oxide, *in* D-00321
Diethylphosphinous acid, D-00333
Methyl ethylmethylphosphinite, *in* E-00087

$C_4H_{11}OPS$

O-Butyl phosphinothioate, B-00592
Butylphosphinothioic acid, B-00593
Diethylphosphinothioic acid, D-00330
O-Ethyl ethylphosphinothioate, *in* E-00125
O-Isopropyl methylphosphinothioate, *in* M-00279
O-Propyl methylphosphinothioate, *in* M-00279

$C_4H_{11}OPSSe$

O-Ethyl ethylphosphonoselenothioate, E-00073

$C_4H_{11}OPS_2$

Butylphosphonodithioic acid, B-00606
O-Ethyl ethylphosphonodithioate, E-00070
O-Isopropyl methylphosphonodithioate, I-00040
(1-Methylpropyl)phosphonodithioic acid, M-00380
O-Propyl methylphosphonodithioate, P-00469

$C_4H_{11}OPSe$

Diethylphosphinoselenoic acid, D-00327

$C_4H_{11}O_2P$

Butyl phosphinate, B-00587
Butylphosphinic acid, B-00590
tert-Butylphosphinic acid, B-00591
Butylphosphonous acid, B-00615
tert-Butylphosphonous acid, B-00616
Diethylphosphinic acid, D-00322
Diethyl phosphonite, D-00339
▷Dimethyl ethylphosphonite, *in* E-00153
Ethyl dimethylphosphinate, *in* D-00829
Ethyl ethylphosphinate, E-00068
Isopropyl methylphosphinate, I-00034
Isopropylmethylphosphinic acid, I-00035
Methylpropylphosphinic acid, M-00375
(2-Methylpropyl)phosphinic acid, M-00376
(2-Methylpropyl)phosphonous acid, M-00383

$C_4H_{11}O_2PS$

Butylphosphonothioic acid, B-00611
tert-Butylphosphonothioic acid, B-00612
O,O-Diethyl phosphonothioate, D-00341
O,O-Dimethyl ethylphosphonothioate, *in* E-00148
O-Ethyl ethylphosphonothioate, E-00074
O-Isopropyl methylphosphonothioate, I-00041
O-Methyl isopropylphosphonothioate, M-00166
(2-Methylpropyl)phosphonothioic acid, M-00381

$C_4H_{11}O_2PSSe$

O,O-Diethyl phosphoroselenothioate, D-00394

$C_4H_{11}O_2PS_2$

▷*O,O*-Diethyl phosphorodithioate, D-00384
O,S-Diethyl phosphorodithioate, D-00385
S,S-Diethyl phosphorodithioate, D-00386

$C_4H_{11}O_2PSe$

Butylphosphonoselenoic acid, B-00609
tert-Butylphosphonoselenoic acid, B-00610
O,O-Diethyl phosphonoselenoate, D-00340
O-Ethyl ethylphosphonoselenoate, E-00072
O-Ethyl methylphosphonoselenoate; *Se*-Me ester, *in* E-00092
O-Propyl methylphosphonoselenoate, *in* M-00311

$C_4H_{11}O_2PSe_2$

O,O-Diethyl phosphorodiselenoate, D-00383

$C_4H_{11}O_2PSi$

2,2-Dimethyl-1,3-dioxa-5-phospha-2-silacyclohexane, D-00731

$C_4H_{11}O_3P$

Butylphosphonic acid, B-00594
tert-Butylphosphonic acid, B-00595
▷Diethyl phosphonate, D-00338
Dimethyl ethylphosphonate, D-00754
Ethyl hydrogen ethylphosphonate, *in* E-00129
Hydrogen propyl methylphosphonate, *in* M-00288
Isopropyl methylphosphonate, I-00038
(1-Methylpropyl)phosphonic acid, M-00377
(2-Methylpropyl)phosphonic acid, M-00378
▷Monobutyl phosphonate, M-00453

$C_4H_{11}O_3PS$

S-Butyl phosphorothioate, B-00634

O,O-Diethyl phosphorothioate, D-00396
O,S-Diethyl phosphorothioate, D-00397
Dimethyl [(methylthio)methyl]phosphonate, *in* M-00415
O-Ethyl *O,S*-dimethyl phosphorothioate, E-00065

$C_4H_{11}O_3PSe$
O,O-Diethyl phosphoroselenoate, D-00393

$C_4H_{11}O_3PTe$
O,O-Diethyl phosphorotelluroate, D-00395

$C_4H_{11}O_4P$
Diethyl phosphate, D-00320
Dimethyl (1-hydroxyethyl)phosphonate, *in* H-00145
Dimethyl (2-hydroxyethyl)phosphonate, *in* H-00146
Ethyl bis(hydroxymethyl)phosphinate, *in* B-00282
(1-Hydroxybutyl)phosphonic acid, H-00114
Monobutyl phosphate, M-00451
Mono-*tert*-butyl phosphate, M-00452
Mono(1-methylpropyl) phosphate, M-00492
Mono(2-methylpropyl) phosphate, M-00493

$C_4H_{11}O_4PS$
Dimethyl [(methylsulfinyl)methyl]phosphonate, *in* M-00398

$C_4H_{11}O_5PS$
Dimethyl [(methylsulfonyl)methyl]-phosphonate, *in* M-00400

$C_4H_{11}P$
Butylphosphine, B-00588
tert-Butylphosphine, B-00589
▷Diethylphosphine, D-00321
(2-Methylpropyl)phosphine, M-00374
Trimethylmethylenephosphorane, T-00518

$C_4H_{11}PS$
Diethylphosphine; Sulfide, *in* D-00321
Diethylphosphinothious acid, D-00332

$C_4H_{11}PSSe$
Diethylphosphinoselenothioic acid, D-00329

$C_4H_{11}PS_2$
Butylphosphonodithious acid, B-00607
Diethylphosphinodithioic acid, D-00325
Ethyl dimethylphosphinodithioate, *in* D-00834

$C_4H_{11}PS_3$
Butylphosphonotrithioic acid, B-00614
(1-Methylpropyl)phosphonotrithioic acid, M-00382

$C_4H_{11}PSe_2$
Diethylphosphinodiselenoic acid, D-00324

$C_4H_{12}BrN_2OP$
Tetramethylphosphorodiamidic acid; Bromide, *in* T-00218

$C_4H_{12}BrN_2P$
Tetramethylphosphorodiamidous acid; Bromide, *in* T-00227

$C_4H_{12}BrP$
Tetramethylphosphonium(1+); Bromide, *in* T-00215

$C_4H_{12}ClN_2OP$
P-(4-Chlorobutyl)phosphonic diamide, C-00026
N,N'-Diethylphosphorodiamidic acid; Chloride, *in* D-00382
Tetramethylphosphorodiamidic chloride, T-00220

$C_4H_{12}ClN_2P$
▷Tetramethylphosphorodiamidous chloride, T-00228

$C_4H_{12}ClN_2PS$
Tetramethylphosphorodiamidothioic chloride, T-00225

$C_4H_{12}ClN_2PSe$
Tetramethylphosphorodiamidoselenoic acid; Chloride, *in* T-00223

$C_4H_{12}ClO_4P$
▷Tetrakis(hydroxymethyl)phosphonium(1+); Chloride, *in* T-00181

$C_4H_{12}ClP$
Tetramethylphosphonium(1+); Chloride, *in* T-00215

$C_4H_{12}Cl_4N_5P_3$
2,2,4,6-Tetrachloro-4,6-bis(dimethylamino)-2,2,4,4,6,6-hexahydro-1,3,5-triaza-2,4,6-triphosphorine, T-00028

$C_4H_{12}FN_2OP$
▷*N,N'*-Diethylphosphorodiamidic acid; Fluoride, *in* D-00382
▷Dimefox, D-00669

$C_4H_{12}FN_2P$
Tetramethylphosphorodiamidous acid; Fluoride, *in* T-00227

$C_4H_{12}FN_2PS$
▷Tetramethylphosphorodiamidothioic fluoride, T-00226

$C_4H_{12}FP$
Fluorotetramethylphosphorane, F-00049

$C_4H_{12}F_3PN_2$
Bis(dimethylamino)trifluorophosphorane, B-00196

$C_4H_{12}IO_3P$
Methyltris(hydroxymethyl)phosphonium iodide, *in* T-00784

$C_4H_{12}IP$
▷Tetramethylphosphonium(1+); Iodide, *in* T-00215

$C_4H_{12}NOP$
P,P-Diethylphosphinic amide, D-00323
Tetramethylphosphinic amide, *in* D-00830

$C_4H_{12}NOPS$
O-Ethyl *N,P*-dimethylphosphonamidothioate, *in* D-00850
S-Ethyl *P*-ethylphosphonamidothioate, *in* E-00128
N-Isopropyl-*P*-methylphosphonamidothioic acid, I-00037
S-Methyl trimethylphosphonamidothioate, *in* T-00540

$C_4H_{12}NOPS_2$
O,S-Diethyl phosphoramidodithioate, D-00349
S,S-Diethyl phosphoramidodithioate, D-00350

$C_4H_{12}NO_2P$
(1-Amino-2-methylpropyl)phosphinic acid, A-00092
tert-Butylphosphoramidous acid, B-00624
N,P-Diethylphosphonamidic acid, D-00336
N,N-Diethylphosphonamidic acid, D-00337
Diethylphosphoramidous acid, D-00360
Dimethyl dimethylphosphoramidite, *in* D-00875
Ethyl *N,P*-dimethylphosphonamidate, *in* D-00849
Ethyl *P*-ethylphosphonamidate, *in* E-00126
N-Isopropyl-*P*-methylphosphonamidic acid, I-00036
Methyl *N*-ethyl-*P*-methylphosphonamidate, *in* E-00088
Methyl trimethylphosphonamidate, *in* T-00537

$C_4H_{12}NO_2PS$
tert-Butylphosphoramidothioic acid, B-00623
O,O-Diethyl phosphoramidothioate, D-00354
O,S-Diethyl phosphoramidothioate, D-00355
Diethylphosphoramidothioic acid, D-00356
O,O-Dimethyl dimethylphosphoramidothioate, *in* D-00871

$C_4H_{12}NO_2PSe$
Diethylphosphoramidoselenoic acid, D-00353
O,O-Dimethyl dimethyl-phosphoramidoselenoate, *in* D-00869

$C_4H_{12}NO_3P$
(1-Aminobutyl)phosphonic acid, A-00054
(2-Aminobutyl)phosphonic acid, A-00055
(3-Aminobutyl)phosphonic acid, A-00056
(4-Aminobutyl)phosphonic acid, A-00057
(1-Amino-1-methylpropyl)phosphonic acid, A-00093
(1-Amino-2-methylpropyl)phosphonic acid, A-00094
Butylphosphoramidic acid, B-00621
tert-Butylphosphoramidic acid, B-00622
tert-Butylphosphorimidic acid, B-00625
Diethyl phosphoramidate, D-00342
Diethylphosphoramidic acid, D-00343
2-Dimethylaminoethylphosphonic acid, *in* A-00079
Dimethyl dimethylphosphoramidate, D-00727
Dimethyl ethylphosphoramidate, *in* E-00156
(2-Methylaminopropyl)phosphonic acid, *in* A-00126
(2-Methylpropyl)phosphoramidic acid, M-00384
Trimethyl methylphosphorimidate, *in* M-00336

$C_4H_{12}NO_3PS$
(1-Amino-3-methylthiopropyl)phosphonic acid, A-00095

$C_4H_{12}NO_4P$
Demanyl phosphate, D-00016
Ethyl hydrogen 2-aminoethyl phosphate, *in* A-00075

$C_4H_{12}NO_4PS$
(1-Amino-3-methylthiopropyl)phosphonic acid; *S*-Oxide, *in* A-00095

$C_4H_{12}NO_5PS$
(1-Amino-3-methylthiopropyl)phosphonic acid; *S,S*-Dioxide, *in* A-00095
Trimethyl methylsulfonylphosphorimidate, *in* M-00402

$C_4H_{12}NP$
P,P-Diethylphosphinous amide, D-00334
Tetramethylphosphine imide, *in* T-00531
Tetramethylphosphinous amide, T-00213

$C_4H_{12}NPS$
P,P-Diethylphosphinothioic amide, D-00331
Tetramethylphosphinothioic amide, *in* D-00840

$C_4H_{12}NPS_2$
P-Ethyl-*N,N*-dimethylphosphonamidodithioic acid, E-00063

$C_4H_{12}NPS_3$
Diethylphosphoramidotrithioic acid, D-00359

$C_4H_{12}NPSe$
Tetramethylphosphinoselenoic amide, *in* D-00836

$C_4H_{12}N_2P_2$
1,2,3,4-Tetramethyl-1,3,2,4-diazadiphosphetidine, T-00194

$C_4H_{12}N_2P_2S_2$
1,2,3,4-Tetramethyl-1,3,2,4-diazadiphosphetidine; 2,4-Disulfide, *in* T-00194

$C_4H_{12}N_3OPS$
P-4-Morpholinylphosphonothioic diamide, M-00528

$C_4H_{12}N_3O_2P$
4-Morpholinylphosphonic diamide, *in* M-00523

$C_4H_{12}N_3O_4P$
▷Creatinolfosfate, C-00206

$C_4H_{12}OP_2$
Tetramethyldiphosphine; Monooxide, *in* T-00200

$C_4H_{12}OP_2S$
Tetramethyldiphosphine; 1-Oxide 2-sulfide, *in* T-00200

$C_4H_{12}OP_2S_2$
Dimethylphosphinothioic acid; Anhydride, *in* D-00839

$C_4H_{12}O_2P_2$
Tetramethyldiphosphine; Dioxide, *in* T-00200

$C_4H_{12}O_2P_2S_3$
Diethylthiodiphosphonic acid, D-00405

$C_4H_{12}O_3P_2$
Dimethylphosphinic acid; Anhydride, *in* D-00829

$C_4H_{12}O_4P^{\oplus}$
Tetrakis(hydroxymethyl)phosphonium(1+), T-00181

$C_4H_{12}O_4P_2$
1,4-Butanediphosphonous acid, B-00515
1,2-Ethanediylbis[methylphosphinic acid], E-00013

$C_4H_{12}O_4P_2S_3$
O,O,O′,O′-Tetramethyl trithiopyrophosphate, T-00240

$C_4H_{12}O_4P_2S_4$
▷Bis(dimethoxyphosphinothioyl) disulfide, B-00185

$C_4H_{12}O_5P_2$
Diethyldiphosphonic acid, D-00290
Dimethyl dimethyldiphosphonate, *in* D-00751

$C_4H_{12}O_5P_2S_2$
▷*sym-O,O,O,O*-Tetramethyl dithiopyrophosphate, T-00203

$C_4H_{12}O_6P_2$
1,4-Butanediphosphonic acid, B-00514
Tetramethyl hypophosphate, T-00204

$C_4H_{12}O_6P_2S$
sym-Tetramethyl monothiopyrophosphate, T-00206
▷*unsym*-Tetramethyl monothiopyrophosphate, T-00207

$C_4H_{12}O_6P_2S_2$
Bis(dimethoxyphosphinyl) disulfide, B-00186

$C_4H_{12}O_7P_2$
P,P′-Diethyl diphosphate, D-00289
1-Hydroxy-1,1-butanediphosphonic acid, H-00113
2,2′-Oxybis[ethanephosphonic acid], O-00094
▷Tetramethyl pyrophosphate, T-00229

$C_4H_{12}O_{12}P_4$
2,5-Dihydroxy-3,6-dimethyl-3,6-bis(dihydroxyphosphinyl)-1,4,2,5-dioxadiphosphorinane 2,5-dioxide, D-00617

$C_4H_{12}P^{\oplus}$
Tetramethylphosphonium(1+), T-00215

$C_4H_{12}P_2$
1,4-Diphosphinobutane, D-01174
▷Tetramethyldiphosphine, T-00200

$C_4H_{12}P_2S$
Tetramethyldiphosphine; Monosulfide, *in* T-00200

$C_4H_{12}P_2SSe$
Tetramethyldiphosphine; 1-Selenide 2-sulfide, *in* T-00200

$C_4H_{12}P_2S_2$
Tetramethyldiphosphine disulfide, T-00201

$C_4H_{12}P_2S_3$
Dimethylphosphinodithioic anhydrosulfide, *in* D-00834

$C_4H_{12}P_2Se_2$
Tetramethyldiphosphine; Diselenide, *in* T-00200

$C_4H_{13}NO_2P_2$
P,P-Dimethyl-*N*-(dimethylphosphinyl)-phosphinic amide, *in* D-00830

$C_4H_{13}NO_5P_2S$
(Diethoxyphosphinothioyl)phosphorimidic acid, D-00238

$C_4H_{13}NO_6P_2$
(Diethoxyphosphinyl)phosphorimidic acid, D-00257

$C_4H_{13}NP_2S_2$
P,P-Dimethyl-*N*-(dimethylphosphinothioyl)-phosphinothioic amide, *in* D-00840

$C_4H_{13}N_2OP$
P-Butylphosphonic diamide, B-00596
P-tert-Butylphosphonic diamide, B-00597
N,N,N′,N′-Tetramethylphosphonic diamide, T-00214
Tetramethylphosphorodiamidous acid, T-00227

$C_4H_{13}N_2OPSe$
Tetramethylphosphorodiamidoselenoic acid, T-00223

$C_4H_{13}N_2O_2P$
N,N′-Diethylphosphorodiamidic acid, D-00382
Ethyl *N,N′*-dimethylphosphorodiamidate, *in* D-00895
Isopropyl *P*-methylphosphonohydrazidate, *in* M-00308
Tetramethylphosphorodiamidic acid, T-00218

$C_4H_{13}N_2O_2PS$
O,O-Diethyl phosphorohydrazidothioate, D-00392

$C_4H_{13}N_2O_3P$
Diethyl phosphorohydrazidate, D-00391

$C_4H_{13}N_2P$
tert-Butylphosphonous diamide, B-00617
N,N,P,P-Tetramethylphosphinimidic amide, *in* D-00833
N,N′,P,P-Tetramethylphosphinimidic amide, *in* D-00833

$C_4H_{13}N_2PS_2$
Tetramethylphosphorodiamidodithioic acid, T-00222

$C_4H_{13}O_4PSi$
[(Trimethylsilyloxy)methyl]phosphonic acid, T-00556

$C_4H_{14}N_2O_3P_2$
P,P′-Diethyldiphosphonic diamide, D-00291

$C_4H_{14}N_3OP$
N,N-Diethylphosphoric triamide, D-00364
N,N,N′,N′-Tetramethylphosphoric triamide, T-00216
N,N,N′,N″-Tetramethylphosphoric triamide, *in* D-00880

$C_4H_{14}N_3PS$
N,N-Diethylphosphorothioic triamide, *in* D-00356
N,N-Diethylphosphorothioic triamide, D-00398

$C_4H_{14}O_{12}P_4$
1,2,3,4-Butanetetraphosphonic acid, B-00518

$C_4H_{15}N_3O_3P_2S$
[Bis(dimethylamino)phosphinothioyl]-phosphorimidic acid, B-00192

$C_4H_{15}N_3O_4P_2$
[Bis(dimethylamino)phosphinyl]phosphorimidic acid, B-00194

$C_4H_{21}N_2OP$
N,N,N′,N′-Tetraethylphosphonic diamide, T-00055

$C_5Cl_4F_6O_2P_2$
3,3,4,4,5,5-Hexafluoro-1,2-cyclopentenediphosphonic acid; Bis(dichloride), *in* H-00018

C_5Cl_6NOP
(2,3,5,6-Tetrachloro-4-pyridinyl)phosphonic acid; Dichloride, *in* T-00035

$C_5H_2Cl_4NO_3P$
(2,3,5,6-Tetrachloro-4-pyridinyl)phosphonic acid, T-00035

$C_5H_4Cl_2NOP$
2-Pyridinephosphonic acid; Dichloride, *in* P-00503

$C_5H_4F_6O_6P_2$
3,3,4,4,5,5-Hexafluoro-1,2-cyclopentenediphosphonic acid, H-00018

$C_5H_5Br_2P$
Cyclopentadienylphosphonous acid; Bromide, *in* C-00275

$C_5H_5Cl_2P$
Cyclopentadienylphosphonous acid; Dichloride, *in* C-00275

$C_5H_5O_3P$
1,3-Pentadiyn-1-ylphosphonic acid, P-00010

C_5H_5P
Phosphorin, P-00403

$C_5H_6NO_3P$
2-Pyridinephosphonic acid, P-00503
3-Pyridinephosphonic acid, P-00504
4-Pyridinephosphonic acid, P-00505

$C_5H_6N_3P$
Tetramethylphosphorodiamidous acid; Cyanide, *in* T-00227

C_5H_7ClOP
1-Chloro-2,3-dihydro-4-methyl-1*H*-phosphole; 1-Oxide, *in* C-00045
1-Chloro-2,5-dihydro-2-methyl-1*H*-phosphole; 1-Oxide, *in* C-00046
1-Chloro-2,5-dihydro-3-methyl-1*H*-phosphole; 1-Oxide, *in* C-00047

C_5H_7ClP
1-Chloro-2,3-dihydro-4-methyl-1*H*-phosphole, C-00045
1-Chloro-2,5-dihydro-2-methyl-1*H*-phosphole, C-00046
1-Chloro-2,5-dihydro-3-methyl-1*H*-phosphole, C-00047

C_5H_7ClPS
1-Chloro-2,5-dihydro-2-methyl-1*H*-phosphole; 1-Sulfide, *in* C-00046
1-Chloro-2,5-dihydro-3-methyl-1*H*-phosphole; 1-Sulfide, *in* C-00047

$C_5H_7Cl_2OP$
(2-Methyl-1,3-butadienyl)phosphonic acid; Dichloride, *in* M-00093
(3-Methyl-1,2-butadienyl)phosphonic acid; Dichloride, *in* M-00094
(3-Methyl-1,3-butadienyl)phosphonic acid; Dichloride, *in* M-00095
1,3-Pentadienylphosphonic acid; Dichloride, *in* P-00009

$C_5H_7Cl_2PS$
(2-Methyl-1,3-butadienyl)phosphonothioic acid; Dichloride, *in* M-00097

$C_5H_7Cl_6O_4P$
Methyl bis(1-hydroxy-2,2,2-trichloroethyl)-phosphinate, *in* B-00287

$C_5H_7F_2OP$
(3-Methyl-1,2-butadienyl)phosphonic acid; Difluoride, *in* M-00094
(3-Methyl-1,3-butadienyl)phosphonic acid; Difluoride, *in* M-00095
1,3-Pentadienylphosphonic acid; Difluoride, *in* P-00009

$C_5H_7F_6OPS$
O-Isopropyl bis(trifluoromethyl)-phosphinothioate, *in* B-00415
S-Isopropyl bis(trifluoromethyl)-phosphinothioate, *in* B-00415

$C_5H_7N_2O_2PS$
2-Pyridylphosphoramidothioic acid, P-00513
4-Pyridylphosphoramidothioic acid, P-00514

$C_5H_7N_2O_3P$
Isopropylphosphonic acid; Diisocyanate, *in* I-00050
2-Pyridylphosphoramidic acid, P-00510

3-Pyridylphosphoramidic acid, P-00511
4-Pyridylphosphoramidic acid, P-00512

$C_5H_7O_2P$
Cyclopentadienylphosphonous acid, C-00275
6-Methyl-4-methylene-4*H*-1,3,2-dioxaphosphorin, M-00167

$C_5H_7O_3P$
(3-Penten-1-ynyl)phosphonic acid, P-00045
(4-Penten-2-yn-1-yl)phosphonic acid, P-00046

$C_5H_7O_3PS$
(2-Thienylmethyl)phosphonic acid, T-00304
(3-Thienylmethyl)phosphonic acid, T-00305

$C_5H_7O_4P$
(2-Furanylmethyl)phosphonic acid, F-00072

$C_5H_7O_5P$
(2-Furanylhydroxymethyl)phosphonic acid, F-00071

C_5H_7P
1-Methyl-1*H*-phosphole, M-00282

C_5H_8ClOP
2-Chloro-2,3-dihydro-4,5-dimethyl-1,2-oxaphosphole, C-00040

$C_5H_8ClO_2P$
2-Chloro-2,3-dihydro-4,5-dimethyl-1,2-oxaphosphole; 2-Oxide, *in* C-00040
2,5-Dihydro-5,5-dimethyl-1,2-oxaphosphole 2-oxide; 2-Chloro, *in* D-00464

$C_5H_8Cl_2O_4P_2$
3,9-Dichloro-2,4,8,10-tetraoxa-3,9-diphosphaspiro[5.5]undecane, D-00196

$C_5H_8Cl_2O_4P_2S_2$
3,9-Dichloro-2,4,8,10-tetraoxa-3,9-diphosphaspiro[5.5]undecane; 3,9-Disulfide, *in* D-00196

$C_5H_8Cl_2O_6P_2$
3,9-Dichloro-2,4,8,10-tetraoxa-3,9-diphosphaspiro[5.5]undecane; 3,9-Dioxide, *in* D-00196

$C_5H_9Br_2O_2P$
2-Bromo-5-bromomethyl-5-methyl-1,3,2-dioxaphosphorinane, B-00465

$C_5H_9Br_2O_3P$
(2-Methyl-2-bromomethyl)trimethylene phosphoryl bromide, *in* B-00465

$C_5H_9ClFO_2P$
5-Chloromethyl-2-fluoro-5-methyl-1,3,2-dioxaphosphorinane, C-00096

$C_5H_9ClFO_2PS$
5-Chloromethyl-2-fluoro-5-methyl-1,3,2-dioxaphosphorinane; 2-Sulfide, *in* C-00096

$C_5H_9ClFO_3P$
5-Chloromethyl-2-fluoro-5-methyl-1,3,2-dioxaphosphorinane; 2-Oxide, *in* C-00096

$C_5H_9Cl_2OP$
Cyclopentanephosphonic acid; Dichloride, *in* C-00276
(2-Methyl-1-butenyl)phosphonic acid; Dichloride, *in* M-00100
(3-Methyl-1-butenyl)phosphonic acid; Dichloride, *in* M-00101

$C_5H_9Cl_2O_2P$
2-Chloro-5-chloromethyl-5-methyl-1,3,2-dioxaphosphorinane, C-00031

$C_5H_9Cl_2O_2PS$
2-Chloro-5-chloromethyl-5-methyl-1,3,2-dioxaphosphorinane; 2-Sulfide, *in* C-00031

$C_5H_9Cl_2O_3P$
2-Chloro-5-chloromethyl-5-methyl-1,3,2-dioxaphosphorinane; 2-Oxide, *in* C-00031
Ethyl 3-dichlorophosphinylpropanoate, *in* D-00192

$C_5H_9Cl_2PS$
Cyclopentylphosphonothioic acid; Dichloride, *in* C-00280

$C_5H_9F_6OPSSi$
O-Trimethylsilyl bis(trifluoromethyl)-phosphinothioate, *in* B-00415

$C_5H_9N_2O_3P$
Dimethyl 1*H*-imidazol-1-ylphosphonate, *in* I-00003

C_5H_9OPS
2,3-Dihydro-2-methoxy-5-methyl-1,2-thiaphosphole, D-00514

$C_5H_9OPS_3$
4-Methyl-2,6,7-trithia-1-phosphabicyclo[2.2.2]-octane; 1-Oxide, *in* M-00436

$C_5H_9O_2P$
2,5-Dihydro-5,5-dimethyl-1,2-oxaphosphole 2-oxide, D-00464
▷2,3-Dihydro-1-hydroxy-4-methyl-1*H*-phosphole 1-oxide, D-00496
▷2,3-Dihydro-1-hydroxy-5-methyl-1*H*-phosphole 1-oxide, D-00497
2,5-Dihydro-1-hydroxy-2-methyl-1*H*-phosphole-1-oxide, D-00498
2,5-Dihydro-1-hydroxy-3-methyl-1*H*-phosphole 1-oxide, D-00499
▷2,3-Dihydro-1-methoxy-1*H*-phosphole 1-oxide, *in* D-00506
▷2,5-Dihydro-1-methoxy-1*H*-phosphole 1-oxide, *in* D-00507
Methyl diethenylphosphinate, *in* D-01258

$C_5H_9O_2PS$
4,7-Dihydro-1,3,2-dioxaphosphepin 2-sulfide; 2-Me, *in* D-00473
2,3-Dihydro-2-methoxy-5-methyl-1,2-thiaphosphole; 2-Oxide, *in* D-00514
(2-Methyl-1,3-butadienyl)phosphonothioic acid, M-00097

$C_5H_9O_3P$
1-Cyclopentenephosphonic acid, C-00277
3-Cyclopentenephosphonic acid, C-00278
2,5-Dihydro-5,5-dimethyl-1,2-oxaphosphole 2-oxide; 2-Hydroxy, *in* D-00464
3,6-Dihydro-2-hydroxy-2*H*-1,2-oxaphosphorin 2-oxide; 2-Methoxy (Me ester), *in* D-00501
2,3-Dihydro-5-methyl-1,2-oxaphosphole 2-oxide; 2-Methoxy, *in* D-00525
Dimethyl 1,2-propadienylphosphonate, *in* P-00441
Dimethyl 1-propynylphosphonate, *in* P-00494
(2-Methyl-1,3-butadienyl)phosphonic acid, M-00093
(3-Methyl-1,2-butadienyl)phosphonic acid, M-00094
(3-Methyl-1,3-butadienyl)phosphonic acid, M-00095
▷4-Methyl-2,6,7-trioxa-1-phosphabicyclo[2.2.2]-octane, M-00424
4-Methyl-3,5,8-trioxa-1-phosphabicyclo[2.2.2]-octane, M-00425
1,3-Pentadienylphosphonic acid, P-00009

$C_5H_9O_3PS$
5-Methyl-1,3,2-oxathiaphosphole 2-oxide; 2-Ethoxy, *in* M-00184
▷Trimethylolethane thiophosphate, *in* M-00424

$C_5H_9O_3PSe$
Trimethylol selenophosphate, *in* M-00424

$C_5H_9O_4P$
Methyl dimethylvinylene phosphate, *in* H-00123
▷Trimethylolethane phosphate, *in* M-00424
2,6,7-Trioxa-1-phosphabicyclo[2.2.2]octane-4-methanol, T-00590

$C_5H_9O_5P$
3-(Dimethoxyphosphinyl)propenoic acid, D-00686
2,6,7-Trioxa-1-phosphabicyclo[2.2.2]octane-4-methanol; Oxide, *in* T-00590

$C_5H_9O_5PS$
2,6,7-Trioxa-1-phosphabicyclo[2.2.2]octane-4-methanol; Sulfide, *in* T-00590

C_5H_9P
2,5-Dihydro-3-methyl-1*H*-phosphole, D-00530
(2,2-Dimethylpropylidyne)phosphine, D-00910

$C_5H_9PS_3$
4-Methyl-2,6,7-trithia-1-phosphabicyclo[2.2.2]-octane, M-00436

$C_5H_9PS_4$
▷4-Methyl-2,6,7-trithia-1-phosphabicyclo[2.2.2]-octane; 1-Sulfide, *in* M-00436

$C_5H_{10}BrO_2P$
2-Bromo-5,5-dimethyl-1,3,2-dioxaphosphorinane, B-00471

$C_5H_{10}BrO_2PS$
Neopentylene phosphorobromidothioate, *in* B-00471

$C_5H_{10}BrO_3P$
Neopentylene phosphorobromidate, *in* B-00471

$C_5H_{10}Br_4P_2$
1,5-Pentanediphosphonous acid; Bis(dibromide), *in* P-00033

$C_5H_{10}ClN_2OP$
2-Cyanoethyl dimethyl-phosphoramidochloridite, *in* D-00862

$C_5H_{10}ClOP$
1-Chlorophosphorinane; Oxide, *in* C-00168

$C_5H_{10}ClO_2P$
2-Chloro-5,5-dimethyl-1,3,2-dioxaphosphorinane, C-00058

$C_5H_{10}ClO_2PS$
2-Chloro-5,5-dimethyl-1,3,2-dioxaphosphorinane 2-sulfide, C-00060

$C_5H_{10}ClO_2PSe$
Neopentylene phosphorochloridoselenoate, *in* C-00058

$C_5H_{10}ClO_3P$
2-Chloro-5,5-dimethyl-1,3,2-dioxaphosphorinane 2-oxide, C-00059
(2-Chloro-3-oxobutyl)methylphosphinic acid, C-00121

$C_5H_{10}ClO_4P$
5-Chloromethyl-2-hydroxy-5-methyl-1,3,2-dioxaphosphorinane 2-oxide, C-00097

$C_5H_{10}ClP$
1-Chlorophosphorinane, C-00168

$C_5H_{10}ClPS$
1-Chlorophosphorinane; Sulfide, *in* C-00168

$C_5H_{10}ClPS_2$
Isopropyl ethylphosphonochloridodithioate, *in* E-00136

$C_5H_{10}Cl_2NOP$
2-Chloro-3-(2-chloroethyl)tetrahydro-2*H*-1,3,2-oxazaphosphorine, C-00030
▷1-Piperidinylphosphonic acid; Dichloride, *in* P-00430

$C_5H_{10}Cl_2NO_2P$
2-Chloro-3-(2-chloroethyl)tetrahydro-2*H*-1,3,2-oxazaphosphorine; 2-Oxide, *in* C-00030

$C_5H_{10}Cl_2NP$
1-Piperidinylphosphonous acid; Dichloride, *in* P-00432

$C_5H_{10}Cl_2NPS$
1-Piperidinylphosphonothioic acid; Dichloride, *in* P-00431

$C_5H_{10}Cl_3O_2P$
Diethyl (trichloromethyl)phosphonite, *in* T-00416
Isopropyl methyl(trichloromethyl)phosphinate, *in* M-00419

$C_5H_{10}Cl_3O_2PS$
O,O-Diethyl (trichloromethyl)-phosphonothioate, *in* T-00415

$C_5H_{10}Cl_3O_3P$
▷Diethyl (trichloromethyl)phosphonate, *in* T-00411

$C_5H_{10}Cl_4O_2P_2$
1,1-Pentanediphosphonic acid; Bis(dichloride), *in* P-00031

1,5-Pentanediphosphonic acid; Bis(dichloride), *in* P-00032

$C_5H_{10}Cl_4P_2$

1,5-Pentanediphosphonous acid; Bis(dichloride), *in* P-00033

$C_5H_{10}FOP$

1-Fluorophosphorinane; Oxide, *in* F-00047

$C_5H_{10}FO_2P$

2-Fluoro-5,5-dimethyl-1,3,2-dioxaphosphorinane, F-00016
▷Isopropyl ethylphosphonofluoridate, *in* E-00142

$C_5H_{10}FO_3P$

Neopentylene phosphorofluoridate, *in* F-00016

$C_5H_{10}FP$

1-Fluorophosphorinane, F-00047

$C_5H_{10}FPS$

1-Fluorophosphorinane; Sulfide, *in* F-00047

$C_5H_{10}F_2NP$

1-Piperidinylphosphonous acid; Difluoride, *in* P-00432

$C_5H_{10}F_3O_2P$

Diethyl (trifluoromethyl)phosphonite, *in* T-00482

$C_5H_{10}F_3O_2PS$

O,O-Diethyl (trifluoromethyl)-phosphonothioate, *in* T-00481
O,S-Diethyl (trifluoromethyl)phosphonothioate, *in* T-00481

$C_5H_{10}F_3O_3P$

Diethyl (trifluoromethyl)phosphonate, *in* T-00478

$C_5H_{10}F_3P$

1,1,1-Trifluoro-1,1-dihydrophosphorinane, *in* D-00573

$C_5H_{10}IO_2P$

2-Iodo-5,5-dimethyl-1,3,2-dioxaphosphorinane, I-00010

$C_5H_{10}IO_2PS$

Neopentylene phosphoroiodidothioate, *in* I-00010

$C_5H_{10}NOPS$

Diethylphosphinic acid; Isothiocyanate, *in* D-00322

$C_5H_{10}NO_2P$

Diethylphosphinic acid; Isocyanate, *in* D-00322

$C_5H_{10}NO_2PS$

tert-Butylphosphon(isothiocyanatidic) acid, B-00602
Isopropyl methylphosphonisothiocyanatidate, *in* M-00294

$C_5H_{10}NO_2PS_2$

▷*O,O*-Diethyl phosphor(isothiocyanatido)thioate, D-00370

$C_5H_{10}NO_3P$

Butylphosphonisocyanatidic acid, B-00601
Diethyl phosphorocyanidate, D-00381
Dimethyl (2-cyanoethyl)phosphonate, *in* C-00211
1,4-Dioxa-6-aza-5-phospha(5-P^V)spiro[4.4]-nonan-7-one, D-00962
Ethyl ethylphosphonisocyanatidate, *in* E-00133
[(Ethylimino)methylethenyl]phosphonic acid, E-00081
4-Methyl-2,6-dioxa-7-aza-1-phosphabicyclo[2.2.2]octane 1-oxide, M-00124

$C_5H_{10}NO_3PS$

O,O-Diethyl phosphorisocyanatidothioate, D-00367
O,S-Diethyl phosphorisocyanatidothioate, D-00368
Diethyl phosphor(isothiocyanatidate), D-00369

$C_5H_{10}NO_4P$

Diethyl phosphorisocyanatidate, D-00366

$C_5H_{10}NO_5P$

2-Amino-5-phospho-3-pentenoic acid, A-00123

$C_5H_{10}NO_7P$

N-Phosphoglutamic acid, P-00367

$C_5H_{10}NP$

Cyanodiethylphosphine, *in* D-00333

$C_5H_{10}N_3O_2PS$

2-Thia-1,3,5-triaza-7-phosphatricyclo[3.3.1.1^{3,7}]decane; 2,2-Dioxide, *in* T-00301

$C_5H_{10}N_3O_3PS$

2-Thia-1,3,5-triaza-7-phosphatricyclo[3.3.1.1^{3,7}]decane; 2,2,7-Trioxide, *in* T-00301

$C_5H_{10}N_3PS$

2-Thia-1,3,5-triaza-7-phosphatricyclo[3.3.1.1^{3,7}]decane, T-00301

$C_5H_{10}O_4P_2S_4$

3,9-Dimercapto-2,4,8,10-tetraoxa-3,9-diphosphaspiro[5.5]undecane 3,9-disulfide, D-00670

$C_5H_{10}O_8P_2$

3,9-Dihydroxy-2,4,8,10-Tetraoxa-3,9-diphosphaspiro[5.5]undecane 3,9-dioxide, D-00626

$C_5H_{10}O_{10}P_2$

(Diphosphonomethyl)butanedioic acid, D-01185
2,4-Diphosphonopentanedioic acid, D-01186

$C_5H_{10}P$

3-Methylphospholane, M-00281

$C_5H_{11}Br_2O_3P$

Diethyl (dibromomethyl)phosphonate, *in* D-00064

$C_5H_{11}ClNOP$

Methyl 1-pyrrolidinylphosphonochloridite, *in* P-00519

$C_5H_{11}ClNO_2P$

Methyl 4-morpholinylphosphonochloridite, *in* M-00526

$C_5H_{11}Cl_2NO_2P_2$

1-*tert*-Butyl-1,2,4-azadiphosphetidine; 2,4-Dichloro, 2,4-dioxide, *in* B-00543

$C_5H_{11}Cl_2OP$

(2,2-Dimethylpropyl)phosphonic acid; Dichloride, *in* D-00912
2,2-Dimethylpropyl phosphorodichloridite, D-00913
(1-Methylbutyl)phosphonic acid; Dichloride, *in* M-00105
(3-Methylbutyl)phosphonic acid; Dichloride, *in* M-00106
3-Methylbutyl phosphorodichloridite, M-00108
Pentylphosphonic acid; Dichloride, *in* P-00050
Pentyl phosphorodichloridite, P-00054

$C_5H_{11}Cl_2OPS$

O-(3-Methylbutyl) phosphorodichloridothioate, M-00109
S-(3-Methylbutyl) phosphorodichloridothioate, M-00110
O-Pentyl phosphorodichloridothioate, P-00055

$C_5H_{11}Cl_2O_2P$

Isopropyl bis(chloromethyl)phosphinate, *in* B-00127

$C_5H_{11}Cl_2O_3P$

Diethyl (dichloromethyl)phosphonate, *in* D-00175

$C_5H_{11}Cl_2P$

Dichloropentylphosphine, *in* P-00052

$C_5H_{11}F_2O_3P$

Diethyl (difluoromethyl)phosphonate, *in* D-00415

$C_5H_{11}N_2OP$

Methyl bis(1-aziridinyl)phosphinite, *in* B-00095
(2-Methyl-1,3-butadienyl)phosphonic diamide, M-00096

$C_5H_{11}N_2OPS$

O-Ethyl dimethylphosphoramidocyanidothioate, *in* D-00864
O-Methyl bis(1-aziridinyl)phosphinothioate, *in* B-00093

$C_5H_{11}N_2OPS_2$

2,3-Dihydro-3,5-dimethyl-1,3,4,2-thiadiazaphosphole 2-sulfide; 2-Ethoxy, *in* D-00471

$C_5H_{11}N_2O_2P$

▷Methyl bis(1-aziridinyl)phosphinate, *in* B-00091
▷Tabun, T-00001
1,2,3-Trimethyl-1,3-diazaphospholidin-4-one 2-oxide, *in* D-00721

$C_5H_{11}N_2O_3P$

Diethyl (diazomethyl)phosphonate, *in* D-00035
1-Methoxy-1,3-dioxo-2,4-dimethyl-1-phospha-2,4-diazacyclopentane, *in* D-00722

$C_5H_{11}OP$

2-Methoxy-4-methyl-1,2-oxaphospholane, M-00049
1-Methylphospholane; Oxide, *in* M-00280

$C_5H_{11}OPS$

2-Isopropyl 1,3,2-oxathiaphospholane, I-00042

$C_5H_{11}OPS_2$

S,S'-1,3-Butylene *O*-methyl phosphorodithioite, *in* M-00138
S,S-Dimethyl 2-propenylphosphonodithioate, *in* P-00457
1,3,2-Dithiaphospholane 2-oxide; 2-Isopropyl, *in* D-01244
2-Ethoxy-1,3,2-dithiaphosphorinane, *in* D-01247
2-Isopropyl-1,3,2-oxathiaphospholane 2-sulfide, *in* I-00042

$C_5H_{11}O_2P$

(2-Butenyl)methylphosphinic acid, B-00523
Cyclopentylphosphonous acid, C-00281
2,4-Dimethyl-1,3,2-dioxaphosphorinane, D-00734
5,5-Dimethyl-1,3,2-dioxaphosphorinane, D-00735
1-Hydroxyphosphorinane 1-oxide, H-00188
2-Methoxy-4-methyl-1,2-oxaphospholane; 2-Oxide, *in* M-00049
2-Methyl-1,3,2-dioxaphosphepane, M-00125
5-Methyl-1,2-oxaphospholane 2-oxide; 2-Methoxy, *in* M-00183
Methyl tetramethylenephosphinate, *in* H-00185
2,4,5-Trimethyl-1,3,2-dioxaphospholane, T-00511

$C_5H_{11}O_2PS$

O,O-1,3-Butylene methylphosphonothioate, *in* D-00734
Cyclopentylphosphonothioic acid, C-00280
4,5-Dimethyl-2-methylthio-1,3,2-dioxaphospholane, D-00765
2-Isopropyl-1,3,2-oxathiaphospholane 2-oxide, *in* I-00042
4-Methyl-2-methylthio-1,3,2-dioxaphosphorinane, M-00174
O,O-Neopentylene phosphonothioate, *in* D-00735
O,O-Tetramethylene methylphosphonothioate, *in* M-00125

$C_5H_{11}O_2PSSe$

O,O-1,3-Butylene *S*-methyl phosphoroselenothioate, *in* M-00174

$C_5H_{11}O_2PS_2$

O,O-1,3-Butylene *S*-methyl dithiophosphate, *in* M-00174
2-Mercapto-5,5-dimethyl-1,3,2-dioxaphosphorinane 2-sulfide, M-00011

$C_5H_{11}O_3P$

1,3-Butylene methylphosphonate, *in* D-00734
2,3-Butylene methylphosphonate, *in* T-00511
Cyclopentanephosphonic acid, C-00276
Dimethyl (1-methylethenyl)phosphonate, *in* M-00149
Dimethyl 1-propenylphosphonate, *in* P-00453
Dimethyl 2-propenylphosphonate, *in* P-00454
1,4,2-Dioxaphosphorinane 2-oxide; 2-Et, *in* D-00973
2-Ethoxy-1,3,2-dioxaphosphorinane, E-00037
2-Ethoxy-4-methyl-1,3,2-dioxaphospholane, E-00040
2-Ethoxy-1,2-oxaphospholane 2-oxide, *in* H-00166
2-Hydroxy-1,2-oxaphosphepane 2-oxide, H-00165
2-Methoxy-4,5-dimethyl-1,3,2-dioxaphospholane, M-00033
2-Methoxy-4-methyl-1,3,2-dioxaphosphorinane, M-00044
(1-Methyl-3-butenyl)phosphonic acid, M-00099
(2-Methyl-1-butenyl)phosphonic acid, M-00100
(3-Methyl-1-butenyl)phosphonic acid, M-00101
(3-Methyl-2-butenyl)phosphonic acid, M-00102
(3-Methyl-3-butenyl)phosphonic acid, M-00103
2-(1-Methylethoxy)-1,3,2-dioxaphospholane, M-00152
Methyl tetramethylene phosphite, *in* D-00968
Neopentylene phosphonate, *in* D-00735
Tetramethylene methylphosphonate, *in* M-00125
1,3,6,2-Trioxaphosphocane; 2-Me, *in* T-00592

$C_5H_{11}O_3PS$

O,O-1,3-Butylene *S*-methyl thiophosphate, *in* M-00174
O,O-2,3-Butylene *O*-methyl thiophosphate, *in* M-00033
O,O-2,3-Butylene *S*-methyl thiophosphate, *in* D-00765
O-Ethyl *O,O*-propylene thiophosphate, *in* E-00040
O-Ethyl *O,O*-trimethylene thiophosphate, *in* E-00037
2-Mercapto-5,5-dimethyl-1,3,2-dioxaphosphorinane 2-oxide, M-00010
2-Methoxy-4-methyl-1,3,2-dioxaphosphorinane 2-sulfide, M-00047
1,4-Thiaphosphorinane 1,1,4-trioxide; 4-Me, *in* T-00300
1,3,6,2-Trioxaphosphocane; 2-Me, 2-sulfide, *in* T-00592

$C_5H_{11}O_3PS_2$

1,4-Dioxa-6,9-dithia-5-phospha(5-P^V)-spiro[4.4]nonane; 5-Methoxy, *in* D-00964

$C_5H_{11}O_3PSe$

2-Hydroxy-5,5-dimethyl-1,3,2-dioxaphosphorinane 2-selenide, H-00125
2-Methoxy-4-methyl-1,3,2-dioxaphosphorinane 2-selenide, M-00046

$C_5H_{11}O_4P$

2,3-Butylene methyl phosphate, *in* M-00033
Dimethyl (3-methyloxiranyl)phosphonate, *in* M-00185
Dimethyl (1-oxopropyl)phosphonate, *in* O-00074
Dimethyl (2-oxopropyl)phosphonate, *in* O-00075
1,4,2-Dioxaphosphorinane 2-oxide; 2-Ethoxy, *in* D-00973
Ethyl propylene phosphate, *in* E-00040
Ethyl trimethylene phosphate, *in* E-00037
2-Hydroxy-5,5-dimethyl-1,3,2-dioxaphosphorinane 2-oxide, H-00124
2-Methoxy-4-methyl-1,3,2-dioxaphosphorinane 2-oxide, M-00045
1,3,6,2-Trioxaphosphocane; 2-Methoxy, *in* T-00592

$C_5H_{11}O_4PS_2$

Methyl [(dimethoxyphosphinothioyl)thio]-acetate, *in* D-00675

$C_5H_{11}O_5P$

(1-Acetoxypropyl)phosphonic acid, A-00008
(2-Acetoxypropyl)phosphonic acid, A-00009
3-Acetoxypropylphosphonic acid, A-00010
Diethoxyphosphinecarboxylic acid oxide, D-00236
2-(Dimethoxyphosphinyl)propanoic acid, D-00684
3-(Dimethoxyphosphinyl)propanoic acid, D-00685
Dimethyl [(acetyloxy)methyl]phosphonate, *in* A-00007
Ethyl dimethoxyphosphinylformate, *in* D-00672
Ethyl 3-phosphonopropanoate, *in* P-00396
Methyl (dimethoxyphosphinyl)acetate, *in* D-00678
1,4,6,9-Tetraoxa-5-phospha(5-P^V)spiro[4.4]-nonane; 5-Methoxy, *in* T-00244

$C_5H_{11}O_6P$

2-[(Dimethoxyphosphinyl)oxy]propanoic acid, D-00682

$C_5H_{11}O_8P$

Xylose 3-dihydrogen phosphate, X-00003
Xylose 5-dihydrogen phosphate, X-00004

$C_5H_{11}P$

1-Methylphospholane, M-00280
Phosphorinane, P-00404

$C_5H_{11}PS$

1-Methylphospholane; Sulfide, *in* M-00280

$C_5H_{11}PS_3$

1,3,2-Dithiaphospholane 2-sulfide; 2-Isopropyl, *in* D-01245

$C_5H_{11}P_3$

2,3-Dihydro-1,2,3-trimethyl-1*H*-1,2,3-triphosphole, *in* D-00615

$C_5H_{12}BrO_3P$

Diethyl (bromomethyl)phosphonate, *in* B-00475

$C_5H_{12}BrPS$

tert-Butylmethylphosphinothioic acid; Bromide, *in* B-00570

$C_5H_{12}ClN_2O_2P$

2-Amino-3-(2-chloroethyl)tetrahydro-4*H*-1,3,2-oxazaphosphorine 2-oxide, A-00063
2-[(2-Chloroethyl)amino]tetrahydro-2*H*-1,3,2-oxazaphosphorine 2-oxide, C-00072

$C_5H_{12}ClN_2P$

N,N'-Dimethyl-*N,N'*-trimethylene phosphorodiamidous chloride, *in* H-00022

$C_5H_{12}ClOP$

Butylmethylphosphinic acid; Chloride, *in* B-00566
tert-Butylmethylphosphinic acid; Chloride, *in* B-00567

$C_5H_{12}ClOPS$

▷*O*-Butyl methylphosphonochloridothioate, B-00573
S-Butyl methylphosphonochloridothioate, *in* M-00298
S-Ethyl isopropylphosphonochloridothioate, *in* I-00053
S-Isopropyl ethylphosphonochloridothioate, *in* E-00137
O-Methyl *tert*-butylphosphonothioate; Chloride, *in* M-00107
O-Propyl ethylphosphonochloridothioate, *in* E-00137
S-Propyl ethylphosphonochloridothioate, *in* E-00137

$C_5H_{12}ClO_2P$

Diethyl (chloromethyl)phosphonite, *in* C-00115
Isopropyl ethylphosphonochloridate, *in* E-00135

$C_5H_{12}ClO_2PS_2$

▷Chlormephos, C-00019

$C_5H_{12}ClO_3P$

Diethyl chloromethylphosphonate, *in* C-00106
Dimethyl (3-chloropropyl)phosphonate, *in* C-00172

$C_5H_{12}ClPS$

tert-Butylmethylphosphinothioic acid; Chloride, *in* B-00570
Butyl methylphosphonochloridothioite, *in* M-00299
Isopropyl ethylphosphonochloridothioite, *in* E-00138

$C_5H_{12}ClPS_2$

S-tert-Butyl methylphosphonochloridodithioate, *in* M-00297

$C_5H_{12}Cl_2NOP$

Pentylphosphoramidic acid; Dichloride, *in* P-00053

$C_5H_{12}FOPS$

O-Butyl methylphosphonofluoridothioate, *in* M-00305

$C_5H_{12}FPS$

Butyl methylphosphonofluoridothioite, *in* M-00306

$C_5H_{12}F_3N_2PS$

P-Trifluoromethyl-*N,N,N',N'*-tetramethyl-phosphonothioic diamide, *in* T-00481

$C_5H_{12}IO_3P$

Diethyl (iodomethyl)phosphonate, *in* I-00013

$C_5H_{12}NOPS$

2-Ethoxy-3-methyl-1,3,2-thiazaphospholidine, E-00043

$C_5H_{12}NOPS_2$

2-Ethoxy-3-methyl-1,3,2-thiazaphospholidine; 2-Sulfide, *in* E-00043
3-Methyl-1,3,2-thiazaphospholidine; 2-Ethoxy, 2-sulfide, *in* M-00408
1,3,2-Thiazaphospholidine 2-sulfide; 2-Propoxy, *in* T-00303

$C_5H_{12}NO_2P$

2-Dimethylamino-1,3,2-dioxaphosphorinane, D-00696
2-Ethoxytetrahydro-2*H*-1,3,2-oxazaphosphorine, E-00051
Ethyl 1-aziridinyl-*P*-methylphosphonamidate, *in* A-00152
4-Methyl-2-dimethylamino-1,3,2-dioxaphospholane, M-00116
1-Piperidinylphosphonous acid, P-00432

$C_5H_{12}NO_2PS$

2-Ethoxy-3-methyl-1,3,2-thiazaphospholidine; 2-Oxide, *in* E-00043
2-Ethoxytetrahydro-2*H*-1,3,2-oxazaphosphorine; 2-Sulfide, *in* E-00051
3-Methyl-1,3,2-thiazaphospholidine; 2-Ethoxy, 2-oxide, *in* M-00408
1-Piperidinylphosphonothioic acid, P-00431
O,O-Propylene dimethylphosphoramidothioate, *in* M-00116
Tetrahydro-2*H*-1,3,2-thiazaphosphorine 2-oxide; 2-Ethoxy, *in* T-00154
O,O-Trimethyl dimethylphosphoramidothioate, *in* D-00696

$C_5H_{12}NO_2PSe$

O,O-Propylene dimethyl-phosphoramidoselenoate, *in* M-00116

$C_5H_{12}NO_3P$

Dimethyl 1-azetidinylphosphonate, *in* A-00147
P-Methyl-4-morpholinylphosphonamidic acid, M-00175
1-Piperidinylphosphonic acid, P-00430
Propylene dimethylphosphoramidate, *in* M-00116

$C_5H_{12}NO_3PS$
Monomethyl 4-morpholinylphosphonothioate, *in* M-00527

$C_5H_{12}NO_3PS_2$
▷Dimethoate, D-00671

$C_5H_{12}NO_4P$
Diethyl aminocarbonylphosphonate, *in* C-00003
(Diethylaminocarbonyl)phosphonic acid, D-00264
Dimethyl (acetamidomethyl)phosphonate, *in* A-00012
Ethyl [(methoxycarbonyl)methyl]-phosphonamidate, *in* M-00028
(4-Morpholinylmethyl)phosphonic acid, M-00521
Phosphinothricin, P-00359

$C_5H_{12}NO_4PS$
▷Omethoate, O-00048

$C_5H_{12}NO_5P$
N-(Diethoxyphosphinyl)carbamic acid, D-00249
Diethyl (nitromethyl)phosphonate, *in* N-00037
Ethyl *N*-(dimethoxyphosphinyl)carbamate, *in* D-00680
Methyl 3-amino-3-phosphonobutanoate, *in* A-00118
Methyl 4-amino-4-phosphonobutanoate, *in* A-00119

$C_5H_{12}NO_6P$
Antibiotic FR 33289, A-00138

$C_5H_{12}NPS_2$
S,S-1,2-Propylene *N,N*-dimethyl-phosphoramidodithioite, *in* M-00136

$C_5H_{12}N_2O_6P_2$
3,9-Dihydroxy-2,4,8,10-Tetraoxa-3,9-diphosphaspiro[5.5]undecane 3,9-dioxide; 3,9-Diamide, *in* D-00626

$C_5H_{12}N_3OP$
▷*P,P*-Bis(1-aziridinyl)-*N*-methylphosphinic amide, *in* B-00092
tert-Butylmethylphosphinic acid; Azide, *in* B-00567
1,3-Dimethyl-1,3,2-diazaphosphetidin-4-one; 2-Dimethylamino, *in* D-00717
Tetramethylphosphorodiamidic cyanide, T-00221

$C_5H_{12}N_3OPS$
Tetramethyl phosphor(isothiocyanatidic) diamide, *in* T-00218

$C_5H_{12}N_3PS$
P,P-Bis(1-aziridinyl)-*N*-methylphosphinothioic amide, *in* B-00094

$C_5H_{12}N_3PS_2$
Tetramethyl phosphor(isothiocyanatido)thioic diamide, T-00217

$C_5H_{12}O_3P_2$
2,6-Dimethyl-1,2,6-oxadiphosphorinane 2,6-dioxide, D-00780

$C_5H_{12}O_4P_2$
1,3-Dihydroxy-1,3-diphosphepane 1,3-dioxide, D-00619

$C_5H_{12}O_7P_2$
Mono(3-methyl-2-butenyl) diphosphate, M-00482
Mono(3-methyl-3-butenyl) diphosphate, M-00483

$C_5H_{13}AsNP$
1-Ethyl-2-methyl-1,2,3-azaphospharsolidine, E-00083

$C_5H_{13}Br_2N_2OP$
P-(Dibromomethyl)-*N,N,N',N'*-tetramethyl-phosphonic diamide, *in* D-00064

$C_5H_{13}ClNOP$
N-tert-Butyl-*P*-methylphosphonamidic acid; Chloride, *in* B-00572
N,N-Diethyl-*P*-methylphosphonamidic acid; Chloride, *in* D-00296

$C_5H_{13}ClNOPS$
P-Chloromethyl-*N*-(1-methylpropyl)-phosphonamidothioic acid, C-00101

$C_5H_{13}ClNPS$
N,N-Diethyl-*P*-methylphosphonamidothioic acid; Chloride, *in* D-00297

$C_5H_{13}Cl_2N_2OP$
P-(Dichloromethyl)phosphonic bis(dimethyl-amide), *in* D-00176

$C_5H_{13}Cl_2N_2O_2P$
Methyl *N,N*-bis(2-chloroethyl)-phosphorodiamidate, *in* B-00126

$C_5H_{13}Cl_2O_2PSi$
Trimethylsilyl bis(chloromethyl)phosphinate, *in* B-00127

$C_5H_{13}FNPS$
N,N-Diethyl-*P*-methylphosphonamidothioic acid; Fluoride, *in* D-00297

$C_5H_{13}FNPS_2$
Methyl diethylphosphoramidofluoridodithioate, *in* D-00352

$C_5H_{13}NO_5P$
2,5-Dihydroxy-3-(1-methylpropyl)-1,3,2,5-oxazadiphospholidine 2,5-dioxide, D-00623

$C_5H_{13}NO_7P_2$
(4-Morpholinylmethylene)bisphosphonic acid, M-00520

$C_5H_{13}NPSe$
P,P-Diethyl-*N*-methylphosphinoselenoic amide, *in* D-00328

$C_5H_{13}NP_2$
1-*tert*-Butyl-1,2,4-azadiphosphetidine, B-00543

$C_5H_{13}N_2OP$
Hexahydro-1,3-dimethyl-1,3,2-diazaphosphorine 2-oxide, H-00023
Methyl *N,N'*-ethanediyl-*N,N'*-dimethyl-phosphorodiamidite, *in* D-00719

$C_5H_{13}N_2O_2P$
N,N-Diethyl-*P*-methylphosphonisocyanatidic amide, *in* M-00293

$C_5H_{13}N_2O_3P$
P-Aminocarbonyl-*N,N*-diethylphosphonamidic acid, A-00059
Dimethyl phosphorohydrazidate; N^2-Isopropylidene, *in* D-00904

$C_5H_{13}N_2O_4P$
Alafosfalin, A-00049
Diethyl aminocarbonylphosphoramidate, *in* A-00062
Dimethyl [1-(aminocarbonyl)ethyl]-phosphoramidate, *in* P-00362

$C_5H_{13}N_2O_6P$
N^5-Phosphonomethionine sulfoximine, P-00388

$C_5H_{13}N_2P$
1,3-Dimethyl-1,3,2-diazaphospholidine; 2-Me, *in* D-00719
Hexahydro-1,3-dimethyl-1,3,2-diazaphosphorine, H-00022

$C_5H_{13}N_4OP$
N,N-Diethyl-*P*-methylphosphonazidic amide, *in* M-00287

$C_5H_{13}OP$
Ethyl ethylmethylphosphinite, *in* E-00087

$C_5H_{13}OPS$
O-Butyl methylphosphinothioate, *in* M-00279
tert-Butylmethylphosphinothioic acid, B-00570
O-Isopropyl ethylphosphinothioate, *in* E-00125
O-Methyl diethylphosphinothioate, *in* D-00330
S-Methyl diethylphosphinothioate, *in* D-00330

$C_5H_{13}OPS_2$
O,S-Diethyl methylphosphonodithioate, *in* M-00301
S,S-Diethyl methylphosphonodithioate, *in* M-00301
O-Ethyl hydrogen propylphosphonodithioate, *in* P-00478
O-Ethyl *S*-propyl phosphorodithioate, E-00186
O-Methyl butylphosphonodithioate, *in* B-00606
Pentylphosphonodithioic acid, P-00051

$C_5H_{13}OPSe$
O-Methyl diethylphosphinoselenoate, *in* D-00327

$C_5H_{13}O_2P$
Butylmethylphosphinic acid, B-00566
tert-Butylmethylphosphinic acid, B-00567
tert-Butylphosphinic acid; Me ester, *in* B-00591
▷Diethyl methylphosphonite, *in* M-00319
Dimethyl propylphosphonite, *in* P-00482
Ethyl ethylmethylphosphinate, *in* E-00085
Isopropyl ethylphosphinate, *in* E-00124
Methyl butylphosphinate, *in* B-00590
Methyl diethylphosphinate, *in* D-00322
Methyl(2-methylpropyl)phosphinic acid, M-00172
Pentyl phosphinate, P-00048
Pentylphosphonous acid, P-00052

$C_5H_{13}O_2PS$
O,O-Diethyl methylphosphonothioate, *in* M-00313
O,S-Diethyl methylphosphonothioate, *in* E-00093
O-Ethyl isopropylphosphonothioate, E-00082
O-Ethyl *O*-methyl ethylphosphonothioate, *in* M-00155
O-Ethyl *S*-methyl ethylphosphonothioate, *in* E-00074
O-Isopropyl *O*-methyl methylphosphonothioate, *in* M-00171
O-Isopropyl *S*-methyl methylphosphonothioate, *in* I-00041
O-Methyl *tert*-butylphosphonothioate, M-00107

$C_5H_{13}O_2PSSi$
2,2-Dimethyl-1,3-dioxa-5-phospha-2-silacyclohexane; 5-Me, 5-sulfide, *in* D-00731

$C_5H_{13}O_2PS_2$
O-Butyl *O*-methyl phosphorodithioate, B-00574

$C_5H_{13}O_2PS_3$
O,O-Diethyl *SS*-methyl phosphoro(dithioperoxo)thioate, D-00300
O,O-Dimethyl *S*-[(ethylthio)methyl] phosphorodithioate, D-00755

$C_5H_{13}O_2PSe$
tert-Butylphosphonoselenoic acid; *O*-Me ester, *in* B-00610
O,O-Diethyl methylphosphonoselenoate, *in* M-00311
O,O-Diethyl *Se*-methyl phosphoroselenoite, D-00301
O-Ethyl *Se*-methyl ethylphosphonoselenoate, *in* E-00072
O-Isopropyl hydrogen ethylphosphonoselenoate, *in* E-00145
O-Propyl hydrogen phosphonoselenoate, *in* E-00145

$C_5H_{13}O_2PSi$
2,2-Dimethyl-1,3-dioxa-5-phospha-2-silacyclohexane; 5-Me, *in* D-00731

$C_5H_{13}O_3P$
Diethyl methylphosphonate, D-00298
Dimethyl isopropylphosphonate, *in* I-00050
Dimethyl propylphosphonate, *in* P-00473
(1,1-Dimethylpropyl)phosphonic acid, D-00911
(2,2-Dimethylpropyl)phosphonic acid, D-00912
Isopropyl methyl methylphosphonate, *in* M-00168
(1-Methylbutyl)phosphonic acid, M-00105
(3-Methylbutyl)phosphonic acid, M-00106

Pentylphosphonic acid, P-00050

$C_5H_{13}O_3PS$

Diethyl (mercaptomethyl)phosphonate, *in* M-00018
O-Isopropyl *O*,*S*-dimethyl phosphorothioate, I-00030

$C_5H_{13}O_3PS_2$

Demephion-*O*, D-00017
▷Demephion-*S*, D-00018
Dimethyl [bis(methylthio)methyl]phosphonate, *in* B-00364

$C_5H_{13}O_3PSi$

2-(Trimethylsilyloxy)-1,3,2-dioxaphospholane, T-00554

$C_5H_{13}O_4P$

Diethyl (hydroxymethyl)phosphonate, *in* H-00162
Dimethyl (1-hydroxypropyl)phosphonate, *in* H-00190
Dimethyl (2-hydroxypropyl)phosphonate, *in* H-00191
Dimethyl (3-hydroxypropyl)phosphonate, *in* H-00192
Mono(3-methylbutyl) phosphate, M-00484
Monopentyl phosphate, M-00504

$C_5H_{13}O_4PSi$

Ethylene trimethylsilyl phosphate, *in* T-00554

$C_5H_{13}O_6P$

Glycerol 1-monophosphate; Di-Me ether, Na salt, *in* G-00066

$C_5H_{13}P$

1,1-Dihydrophosphorinane, D-00573
Pentylphosphine, P-00049

$C_5H_{13}PS_2$

tert-Butylmethylphosphinodithioic acid, B-00568
Ethyl ethylmethylphosphinodithioate, *in* E-00086
Ethylpropylphosphinodithioic acid, E-00185
Methyl diethylphosphinodithioate, *in* D-00325

$C_5H_{13}PS_3$

Diethyl methylphosphonotrithioate, *in* M-00318

$C_5H_{13}P_3$

1,2,4-Trimethyl-1,2,4-triphospholane, T-00568

$C_5H_{14}ClN_2OP$

P-Chloromethyl-*N*,*N*,*N'*,*N'*-tetramethylphosphonic diamide, *in* C-00107

$C_5H_{14}ClOP$

(2-Hydroxyethyl)trimethylphosphonium(1+); Chloride, *in* H-00150

$C_5H_{14}IOP$

(2-Hydroxyethyl)trimethylphosphonium(1+); Iodide, *in* H-00150

$C_5H_{14}NOP$

P,*P*-Diethyl-*N*-methylphosphinic amide, *in* D-00323
P-(1,1-Dimethylethyl)-*P*-methylphosphinic amide, *in* B-00567

$C_5H_{14}NOPS$

N,*N*-Diethyl-*P*-methylphosphonamidothioic acid, D-00297

$C_5H_{14}NO_2P$

N-*tert*-Butyl-*P*-methylphosphonamidic acid, B-00572
N,*N*-Diethyl-*P*-methylphosphonamidic acid, D-00296
Diethyl methylphosphoramidite, *in* M-00335
Ethyl *N*-ethyl-*P*-methylphosphonamidate, *in* E-00088
Ethyl trimethylphosphonamidate, *in* T-00537
Isopropyl *N*,*P*-dimethylphosphonamidate, *in* D-00849
Methyl *N*,*N*-diethylphosphonamidate, *in* D-00337
Methyl *N*-isopropyl-*P*-methylphosphonamidate, *in* I-00036

$C_5H_{14}NO_2PS$

O,*O*-Diethyl methylphosphoramidothioate, *in* M-00333

$C_5H_{14}NO_2PSe$

O,*O*-Diethyl methylphosphoramidoselenoate, *in* M-00332

$C_5H_{14}NO_3P$

(1-Aminopentyl)phosphonic acid, A-00096
(5-Aminopentyl)phosphonic acid, A-00097
Diethyl (aminomethyl)phosphonate, *in* A-00091
Dimethyl (1-amino-1-methylethyl)phosphonate, *in* A-00087
Dimethyl isopropylphosphoramidate, *in* I-00058
Dimethyl propylphosphoramidate, *in* P-00484
Ethyl hydrogen (1-amino-1-methylethyl)-phosphonate, *in* A-00087
Pentylphosphoramidic acid, P-00053
Trimethyl ethylphosphorimidate, *in* E-00161

$C_5H_{14}NO_3PS$

O,*O*-Dimethyl (2-methoxyethyl)-phosphoramidothioate, *in* M-00041

$C_5H_{14}NO_4P$

Choline *O*-phosphate, C-00198

$C_5H_{14}NO_5PS$

Diethyl methylsulfonylphosphoramidate, *in* M-00401

$C_5H_{14}NP$

P,*P*-Diethyl-*N*-methylphosphinous amide, *in* D-00334

$C_5H_{14}NPS_2$

Methyl *P*-methyl-*N*,*N*-dimethylphosphonamidodithioate, *in* E-00063

$C_5H_{14}N_3OP$

N'',*N''*-Dimethyl-*N*,*N'*-trimethylenephosphoric triamide, *in* H-00020

$C_5H_{14}N_3O_7PS$

N^{δ}-Phosphosulfamylornithine, P-00419

$C_5H_{14}OP^{\oplus}$

(2-Hydroxyethyl)trimethylphosphonium(1+), H-00150

$C_5H_{14}O_2P_2$

Bis(dimethylphosphinyl)methane, *in* B-00202

$C_5H_{14}O_4P_2$

Dimethyl methylenebis(methylphosphinate), *in* M-00140
1,5-Pentanediphosphonous acid, P-00033
Tetramethyl methylenebisphosphonite, *in* M-00024

$C_5H_{14}O_4P_2S_4$

O,*O*,*O'*,*O'*-Tetramethyl *S*,*S'*-methylene bisphosphorodithioate, T-00205

$C_5H_{14}O_6P_2$

1,1-Pentanediphosphonic acid, P-00031
1,5-Pentanediphosphonic acid, P-00032
Tetramethyl methylenebisphosphonate, *in* M-00022

$C_5H_{14}P_2$

Bis(dimethylphosphino)methane, B-00202
1,5-Diphosphinopentane, D-01179

$C_5H_{15}INO_3PS$

(1-Amino-3-methylthiopropyl)phosphonic acid; B,MeI, *in* A-00095

$C_5H_{15}NO_4PS_2$

O,*O*,*O*,*O*-Tetramethyl methylthioimidodiphosphate, *in* M-00161

$C_5H_{15}NO_4P_2$

Tetramethyl methylimidodiphosphite, *in* M-00157

$C_5H_{15}NO_5P_2S$

O,*O*-Dimethyl (dimethoxyphosphinyl)-methylphosphoramidothioate, *in* M-00413

$C_5H_{15}NO_6P_2$

Tetramethyl (aminomethylene)bisphosphonate, *in* A-00086
Tetramethyl methylimidodiphosphate, *in* M-00156
Trimethyl (dimethoxyphosphinyl)-phosphorimidate, *in* D-00683

$C_5H_{15}N_2OP$

Methyl tetramethylphosphorodiamidite, *in* T-00227
Pentamethylphosphonic diamide, *in* M-00289

$C_5H_{15}N_2OPSe$

O-Methyl tetramethyl-phosphorodiamidoselenoate, *in* T-00223

$C_5H_{15}N_2O_2P$

Methyl *N*,*N'*-diethylphosphorodiamidate, *in* D-00382
2-Methylpropyl *P*-methyl-phosphonohydrazidate, *in* M-00308
Methyl tetramethylphosphorodiamidate, M-00406

$C_5H_{15}N_2O_3PS$

▷Amifostine, A-00052
▷Ethiofos, E-00021

$C_5H_{15}N_2P$

Pentamethylphosphinimidic amide, *in* D-00833
Pentamethylphosphonous diamide, *in* M-00320

$C_5H_{15}N_2PS$

Pentamethylphosphonothioic diamide, *in* M-00314

$C_5H_{15}N_2PS_2$

Methyl tetramethylphosphorodiamidodithioate, *in* T-00222

$C_5H_{15}N_2PSe$

Pentamethylphosphonoselenoic diamide, P-00024

$C_5H_{15}N_2PTe$

Pentamethylphosphonotelluroic diamide, P-00025

$C_5H_{15}N_3O_2P_2$

1,2,4-Trimethyl-1,2,4,3,5-triazadiphospholidine; 3,5-Dimethoxy, *in* T-00562

$C_5H_{15}N_3P_2$

1,2,3,4,5-Pentamethyl-1,2,4,3,5-triazadiphospholidine, *in* T-00562

$C_5H_{15}N_3P_2S$

2,4-Bis(dimethylamino)-3-methyl-1,3,2,4-thiazadiphosphetidine, B-00187

$C_5H_{15}N_3P_2S_2$

1,2,4-Trimethyl-1,2,4,3,5-triazadiphospholidine; 3,5-Bis(methylthio), *in* T-00562

$C_5H_{15}N_3P_2S_3$

2,4-Bis(dimethylamino)-3-methyl-1,3,2,4-thiazadiphosphetidine; 2,4-Disulfide, *in* B-00187

$C_5H_{15}N_4O_6PS$

Octicidin, O-00038

$C_5H_{15}OPSSi$

Dimethylphosphinothioic acid; *O*-Trimethylsilyl ester, *in* D-00839

$C_5H_{15}O_2PSi$

Trimethylsilyl dimethylphosphinate, *in* D-00829

$C_5H_{15}O_3P_3S_3$

Ethylphosphonothioic acid trimer trianhydride, *in* T-00463

$C_5H_{15}O_4PSi$

Dimethyl trimethylsilyl phosphate, D-00931
[1-[(Trimethylsilyl)oxy]ethyl]phosphonic acid, T-00555

$C_5H_{15}O_5P$

Pentamethoxyphosphorane, P-00018

$C_5H_{15}PS_2Si$

Trimethylsilyl dimethylphosphinodithioate, *in* D-00834

$C_5H_{15}PSi$
▷Dimethyl(trimethylsilyl)phosphine, D-00932

$C_5H_{15}P_5$
Pentamethylpentaphospholane, P-00019

$C_5H_{16}NPSSi$
P,P-Dimethyl-*N*-trimethylsilylphosphinothioic amide, *in* D-00840

$C_5H_{16}N_3OP$
▷Pentamethylphosphoric triamide, P-00026

$C_5H_{16}N_3P$
N,N,N',N',P-Pentamethylphosphonimidic diamide, P-00023

$C_5H_{23}O_3PS$
O,O-Bis(3-methylbutyl) phosphorothioate, B-00305

$C_6Br_3F_{12}O_2P$
2,2,2-Tribromo-2,2-dihydro-4,4,5,5-tetrakis(trifluoromethyl)-1,3,2-dioxaphospholane, T-00353

$C_6ClF_{12}O_2P$
2-Chloro-4,4,5,5-tetrakis(trifluoromethyl)-1,3,2-dioxaphospholane, C-00184

$C_6ClF_{12}O_3P$
Tetrakis(trifluoromethyl)ethylene chlorophosphate, *in* C-00184

$C_6Cl_2F_5O_2P$
Pentafluorophenyl phosphorodichloridate, *in* M-00503
Pentafluorophenyl phosphorodichloridate, P-00017

$C_6Cl_3F_5NP$
P,P,P-Trichloro-*N*-(pentafluorophenyl)-phosphazene, *in* P-00016

$C_6Cl_3F_{12}O_2P$
2,2-Dihydro-2,2,2-trichloro-4,4,5,5-tetrakis(trifluoromethyl)-1,3,2-dioxaphospholane, D-00589

C_6Cl_7OPS
O-(Pentachlorophenyl) phosphorodichloridothioate, P-00007
S-(Pentachlorophenyl) phosphorodichloridothioate, P-00008

$C_6Cl_7O_2P$
Pentachlorophenyl phosphorodichloridate, P-00006

C_6F_7OP
Pentafluorophenylphosphonic acid; Difluoride, *in* P-00015

$C_6F_{15}O_2P$
2,2,2-Trifluoro-2,2-dihydro-4,4,5,5-tetrakis(trifluoromethyl)-1,3,2-dioxaphospholane, T-00467

$C_6H_2Cl_5O_4P$
Mono(pentachlorophenyl) phosphate, M-00502

$C_6H_2F_5O_3P$
Pentafluorophenylphosphonic acid, P-00015

$C_6H_2F_5O_4P$
Mono(pentafluorophenyl) phosphate, M-00503

$C_6H_3Br_2Cl_2OP$
(2,5-Dibromophenyl)phosphonic acid; Dichloride, *in* D-00066

$C_6H_3Br_3Cl_2NOP$
(2,4,6-Tribromophenyl)phosphoramidic acid; Dichloride, *in* T-00359

$C_6H_3Cl_2N_2O_6P$
2,4-Dinitrophenyl phosphorodichloridate, *in* M-00467

$C_6H_3Cl_4O_2P$
2,4-Dichlorophenyl phosphorodichloridate, D-00189

$C_6H_3Cl_5NOP$
(2,4,6-Trichlorophenyl)phosphoramidic acid; Dichloride, *in* T-00420

$C_6H_3F_5NO_3P$
(Pentafluorophenyl)phosphorimidic acid, P-00016

$C_6H_3F_{12}N_2P$
2,2-Dihydro-3,3-bis(trifluoromethyl)-1-[[2,2,2-trifluoro-1-(trifluoromethyl)ethylidene]-amino]azaphosphiridine, D-00446

$C_6H_4BrCl_2OP$
(2-Bromophenyl)phosphonic acid; Dichloride, *in* B-00490
(3-Bromophenyl)phosphonic acid; Dichloride, *in* B-00491
(4-Bromophenyl)phosphonic acid; Dichloride, *in* B-00492

$C_6H_4BrCl_2OPS$
O-(4-Bromophenyl) phosphorodichloridothioate, B-00497

$C_6H_4BrCl_2P$
4-Bromophenyldichlorophosphine, *in* B-00493

$C_6H_4BrI_2P$
4-Bromophenyldiiodophosphine, *in* B-00493

C_6H_4BrOPS
1,3,2-Benzoxathiaphosphole; 2-Bromo, *in* B-00026

$C_6H_4BrO_2P$
2-Bromo-1,3,2-benzodioxaphosphole, B-00464

$C_6H_4BrO_3P$
o-Phenylene phosphorobromidate, *in* B-00464

$C_6H_4Br_2ClP$
Dibromo(4-chlorophenyl)phosphine, *in* C-00153

$C_6H_4Br_2Cl_2NOP$
(2,4-Dibromophenyl)phosphoramidic acid; Dichloride, *in* D-00067

$C_6H_4Br_2FP$
Dibromo(4-fluorophenyl)phosphine, *in* F-00042

$C_6H_4Br_3O_2P$
2,2,2-Tribromo-2,2-dihydro-1,3,2-benzodioxaphosphole, T-00352

$C_6H_4Br_3P$
Dibromo-4-bromophenylphosphine, *in* B-00493

$C_6H_4ClF_2OP$
(4-Chlorophenyl)phosphonic difluoride, C-00149

$C_6H_4ClF_2O_2P$
4-Chlorophenyl phosphorodifluoridate, *in* M-00458

$C_6H_4ClF_2P$
(4-Chlorophenyl)difluorophosphine, *in* C-00153

$C_6H_4ClI_2P$
(4-Chlorophenyl)diiodophosphine, *in* C-00153

C_6H_4ClOPS
1,3,2-Benzoxathiaphosphole; 2-Chloro, *in* B-00026

$C_6H_4ClO_2P$
2-Chloro-1,3,2-benzodioxaphosphole, C-00021

$C_6H_4ClPS_2$
2-Chloro-1,3,2-benzodithiaphosphole, C-00023

$C_6H_4Cl_2FOP$
(3-Fluorophenyl)phosphonic acid; Dichloride, *in* F-00037
(4-Fluorophenyl)phosphonic dichloride, F-00039

$C_6H_4Cl_2FP$
Dichloro(2-fluorophenyl)phosphine, *in* F-00040
Dichloro(3-fluorophenyl)phosphine, *in* F-00041
Dichloro(4-fluorophenyl)phosphine, *in* F-00042

$C_6H_4Cl_2IOPS$
O-(4-Iodophenyl) phosphorodichloridothioate, I-00019

$C_6H_4Cl_2NOP$
2,2-Dihydro-1,3,2-benzoxazaphosphole; 2,2-Dichloro, *in* D-00444
(6-Methyl-2-pyridinyl)phosphonic acid; Dichloride, *in* M-00392

$C_6H_4Cl_2NO_3P$
(3-Nitrophenyl)phosphonic acid; Dichloride, *in* N-00056
(4-Nitrophenyl)phosphonic acid; Dichloride, *in* N-00057

$C_6H_4Cl_2NO_3PS$
O-(4-Nitrophenyl) phosphorodichloridothioate, N-00065

$C_6H_4Cl_2NO_4P$
2-Nitrophenyl phosphorodichloridate, *in* M-00498
3-Nitrophenyl phosphorodichloridate, *in* M-00499
4-Nitrophenyl phosphorodichloridate, N-00064

$C_6H_4Cl_3N_2O_2P$
P,P,P-Trichloro-*N*-(2-nitrophenyl)phosphazene, *in* N-00061

$C_6H_4Cl_3N_2O_5P$
P,P,P-Trichloro-*N*-(3-nitrophenyl)phosphazene, *in* N-00062
P,P,P-Trichloro-*N*-(4-nitrophenyl)phosphine imide, *in* N-00063

$C_6H_4Cl_3OP$
(2-Chlorophenyl)phosphonic acid; Dichloride, *in* C-00145
(3-Chlorophenyl)phosphonic acid; Dichloride, *in* C-00146
(4-Chlorophenyl)phosphonic dichloride, C-00148

$C_6H_4Cl_3OPS$
O-(4-Chlorophenyl) phosphorodichloridothioate, C-00163
S-(4-Chlorophenyl) phosphorodichloridothioate, C-00164

$C_6H_4Cl_3O_2P$
2-Chlorophenyl phosphorodichloridate, C-00161
4-Chlorophenyl phosphorodichloridate, C-00162
2,2,2-Trichloro-2,2-dihydro-1,3,2-benzodioxaphosphole, T-00400

$C_6H_4Cl_3O_4PS$
2-(Chlorosulfonyl)phenyl phosphorodichloridate, C-00175
4-(Chlorosulfonyl)phenyl phosphorodichloridate, C-00176

$C_6H_4Cl_3P$
(4-Chlorophenyl)phosphonous dichloride, C-00154
Dichloro(2-chlorophenyl)phosphine, *in* C-00151
Dichloro(3-chlorophenyl)phosphine, *in* C-00152

$C_6H_4Cl_3PS$
(4-Chlorophenyl)phosphonothioic acid; Dichloride, *in* C-00150

$C_6H_4Cl_4NOP$
(2,4-Dichlorophenyl)phosphoramidic acid; Dichloride, *in* D-00188

$C_6H_4Cl_4O_2P_2$
1,3-Phenylenebis[phosphorodichloridite], P-00141
1,4-Phenylenebis[phosphorodichloridite], P-00142

$C_6H_4Cl_4O_4P_2$
1,3-Phenylene bisphosphorodichloridate, *in* P-00147
1,4-Phenylenebis(phosphorodichloridate), *in* P-00148

$C_6H_4Cl_4P_2$
1,2-Benzenediphosphonous acid; Bis(dichloride), *in* B-00009
1,4-Benzenediphosphonous acid; Bis(dichloride), *in* B-00010

$C_6H_4FI_2P$
(4-Fluorophenyl)diiodophosphine, *in* F-00042

$C_6H_4FO_2P$
2-Fluoro-1,3,2-benzodioxaphosphole, F-00013

$C_6H_4F_2NO_4P$
2-Nitrophenyl phosphorodifluoridate, *in* M-00498
4-Nitrophenyl phosphorodifluoridate, *in* M-00500

$C_6H_4F_3P$
Difluoro(3-fluorophenyl)phosphine, *in* F-00041
Difluoro(4-fluorophenyl)phosphine, *in* F-00042

$C_6H_4F_4O_4P_2$
1,4-Phenylenebis(phosphorodifluoridate), *in* P-00148

$C_6H_4F_8O_6P_2$
3,3,4,4,5,5,6,6-Octafluoro-1,2-cyclohexenediphosphonic acid, O-00003

$C_6H_4I_4P_2$
1,4-Benzenediphosphonous acid; Bis(diiodide), *in* B-00010

$C_6H_4N_3O_2P$
2-Azido-1,3,2-benzodioxaphosphole, A-00148

$C_6H_4N_3O_3P$
o-Phenylene phosphoryl azide, *in* A-00148

$C_6H_5BrCl_2NOP$
(3-Bromophenyl)phosphoramidic acid; Dichloride, *in* B-00495
(4-Bromophenyl)phosphoramidic acid; Dichloride, *in* B-00496

C_6H_5BrOPS
O-Phenyl phosphorodibromidothioate, P-00278

$C_6H_5Br_2OP$
Phenylphosphonic acid; Dibromide, *in* P-00221
Phenyl phosphorodibromidite, P-00277

$C_6H_5Br_2O_3P$
(2,5-Dibromophenyl)phosphonic acid, D-00066

$C_6H_5Br_2P$
Phenylphosphonous dibromide, P-00253

$C_6H_5Br_2PS$
Phenylphosphonothioic acid; Dibromide, *in* P-00244

$C_6H_5Br_3NO_2PS$
P,P,P-Tribromo-*N*-phenylsulfonylphosphazene, *in* P-00315

$C_6H_5Br_3NO_3P$
(2,4,6-Tribromophenyl)phosphoramidic acid, T-00359

$C_6H_5Br_4P$
Tetrabromophenylphosphorane, T-00011

$C_6H_5Br_6P$
Tetrabromophenylphosphorane; Br_2 adduct, *in* T-00011

$C_6H_5ClFO_2P$
Phenyl phosphorochloridofluoridate, P-00271

$C_6H_5Cl_2N_2O_3P$
(3-Nitrophenyl)phosphoramidic acid; Dichloride, *in* N-00059
(4-Nitrophenyl)phosphoramidic acid; Dichloride, *in* N-00060

$C_6H_5Cl_2OP$
Phenylphosphonic dichloride, P-00223
▷Phenyl phosphorodichloridite, P-00280

$C_6H_5Cl_2OPS$
O-Phenyl phosphorodichloridothioate, P-00282
S-Phenyl phosphorodichloridothioate, P-00283

$C_6H_5Cl_2O_2P$
Phenyl phosphorodichloridate, P-00279

$C_6H_5Cl_2O_3P$
(2,3-Dichlorophenyl)phosphonic acid, D-00184
(2,5-Dichlorophenyl)phosphonic acid, D-00185
(3,4-Dichlorophenyl)phosphonic acid, D-00186
(3,5-Dichlorophenyl)phosphonic acid, D-00187

$C_6H_5Cl_2O_4P$
Mono(2,4-dichlorophenyl) phosphate, M-00463

$C_6H_5Cl_2P$
▷Phenylphosphonous dichloride, P-00254

$C_6H_5Cl_2PS$
▷Phenylphosphonothioic dichloride, P-00247
▷Phenyl phosphorodichloridothioite, P-00284

$C_6H_5Cl_2PS_2$
Phenyl phosphorodichloridodithioate, P-00281

$C_6H_5Cl_2PSe$
Phenylphosphonoselenoic acid; Dichloride, *in* P-00243

$C_6H_5Cl_3NOP$
(4-Chlorophenyl)phosphoramidic acid; Dichloride, *in* C-00157

$C_6H_5Cl_3NO_2P$
Phenylsulfonylphosphorimidic trichloride, P-00316

$C_6H_5Cl_3NO_3P$
(2,4,6-Trichlorophenyl)phosphoramidic acid, T-00420

$C_6H_5Cl_4NOPS$
Phenyl (dichlorophosphinothioyl)-phosphorodichlorimidate, P-00112

$C_6H_5Cl_4OP$
Tetrachlorophenoxyphosphorane, T-00033

$C_6H_5Cl_4P$
Tetrachlorophenylphosphorane, T-00034

$C_6H_5F_2OP$
Phenylphosphonic difluoride, P-00224
Phenyl phosphorodifluoridite, P-00286

$C_6H_5F_2OPS$
O-Phenyl phosphorodifluoridothioate, P-00287
S-Phenyl phosphorodifluoridothioate, P-00288

$C_6H_5F_2O_2P$
Phenyl phosphorodifluoridate, P-00285

$C_6H_5F_2P$
Phenylphosphonous difluoride, P-00256

$C_6H_5F_2PS$
Phenylphosphonothioic difluoride, P-00248

$C_6H_5F_2PSe$
Phenylphosphonoselenoic acid; Difluoride, *in* P-00243

$C_6H_5F_4OP$
Tetrafluorophenoxyphosphorane, T-00083

$C_6H_5F_4P$
Tetrafluorophenylphosphorane, T-00084

$C_6H_5F_4PS$
Tetrafluorophenylthiophosphorane, T-00085

$C_6H_5I_2P$
Diiodophenylphosphine, *in* P-00251

$C_6H_5I_4OP$
Tetraiodophenoxyphosphorane, T-00164

$C_6H_5I_4P$
Tetraiodophenylphosphorane, T-00165

$C_6H_5N_2O_7P$
(2,4-Dinitrophenyl)phosphonic acid, D-00946

$C_6H_5N_2O_8P$
Mono(2,4-dinitrophenyl) phosphate, M-00467

$C_6H_5N_6OP$
▷Phenylphosphonic acid; Diazide, *in* P-00221

$C_6H_5N_6PS$
▷Phenylphosphonothioic acid; Diazide, *in* P-00244

C_6H_5OPS
1,3,2-Benzoxathiaphosphole, B-00026

$C_6H_5O_2P$
1,3,2-Benzodioxaphosphole, B-00018

$C_6H_5O_2PS_2$
2-Mercapto-1,3,2-benzodioxaphosphole 2-sulfide, M-00005

$C_6H_5O_3P$
o-Phenylene phosphite, *in* B-00018

$C_6H_5O_4P$
2-Hydroxy-1,3,2-benzodioxaphosphole 2-oxide, H-00108

$C_6H_6BrO_2P$
(3-Bromophenyl)phosphinic acid, B-00488
(4-Bromophenyl)phosphinic acid, B-00489
(4-Bromophenyl)phosphonous acid, B-00493

$C_6H_6BrO_3P$
(2-Bromophenyl)phosphonic acid, B-00490
(3-Bromophenyl)phosphonic acid, B-00491
▷(4-Bromophenyl)phosphonic acid, B-00492

$C_6H_6BrO_4P$
Mono(4-bromophenyl) phosphate, M-00448

C_6H_6BrP
(2-Bromophenyl)phosphine, B-00486
(4-Bromophenyl)phosphine, B-00487

$C_6H_6BrPS_2$
Phenylphosphonobromidodithioic acid, P-00228

$C_6H_6Br_2NO_3P$
(2,4-Dibromophenyl)phosphoramidic acid, D-00067

$C_6H_6Br_2NO_3PS$
Phenylsulfonylphosphoramidic acid; Dibromide, *in* P-00312

$C_6H_6ClN_4OP$
Di-1*H*-imidazol-1-ylphosphinic acid; Chloride, *in* D-00628

C_6H_6ClOP
Phenylphosphonochloridous acid, P-00234

C_6H_6ClOPS
Phenylphosphonochloridothioic acid, P-00232

$C_6H_6ClO_2P$
(2-Chlorophenyl)phosphinic acid, C-00142
(3-Chlorophenyl)phosphinic acid, C-00143
(4-Chlorophenyl)phosphinic acid, C-00144
(2-Chlorophenyl)phosphonous acid, C-00151
(3-Chlorophenyl)phosphonous acid, C-00152
(4-Chlorophenyl)phosphonous acid, C-00153
Phenylphosphonochloridic acid, P-00229

$C_6H_6ClO_2PS$
(4-Chlorophenyl)phosphonothioic acid, C-00150

$C_6H_6ClO_3P$
▷(2-Chlorophenyl)phosphonic acid, C-00145
▷(3-Chlorophenyl)phosphonic acid, C-00146
▷(4-Chlorophenyl)phosphonic acid, C-00147
Phenyl phosphorochloridic acid, P-00269

$C_6H_6ClO_4P$
Mono(2-chlorophenyl) phosphate, M-00457

Mono(4-chlorophenyl) phosphate, M-00458

C_6H_6ClP
(2-Chlorophenyl)phosphine, C-00139
(3-Chlorophenyl)phosphine, C-00140
(4-Chlorophenyl)phosphine, C-00141

C_6H_6ClPS
Phenylphosphonochloridothious acid, P-00233

$C_6H_6ClPS_2$
Phenylphosphonochloridodithioic acid, P-00230

C_6H_6ClPSe
Phenylphosphonochloridoselenous acid, P-00231

$C_6H_6Cl_2NOP$
Phenylphosphoramidic dichloride, P-00258

$C_6H_6Cl_2NO_3P$
(2,4-Dichlorophenyl)phosphoramidic acid, D-00188

$C_6H_6Cl_2NO_3PS$
Phenylsulfonylphosphoramidic dichloride, P-00313

$C_6H_6Cl_2NP$
N-Benzoyl-*P*-phenylphosphonimidic dichloride, B-00033

$C_6H_6Cl_2NPS$
Phenylphosphoramidothioic acid; Dichloride, *in* P-00262

$C_6H_6Cl_3F_6N_5P$
2,2,2-Trichloro-2,2-dihydro-1,3-dimethyl-6,8-bis(trifluoromethyl)-1,3,5,7,9-pentaaza-2,4-diphospha(4-P^V)spiro[3.5]nona-4,6,8-triene, T-00401

$C_6H_6Cl_9O_4P$
Tris(2,2,2-trichloroethyl) phosphate, T-00848

C_6H_6FOP
Phenylphosphonofluoridous acid, P-00240

$C_6H_6FO_2P$
(3-Fluorophenyl)phosphinic acid, F-00035
(2-Fluorophenyl)phosphonous acid, F-00040
(3-Fluorophenyl)phosphonous acid, F-00041
(4-Fluorophenyl)phosphonous acid, F-00042
Phenylphosphonofluoridic acid, P-00239

$C_6H_6FO_3P$
(2-Fluorophenyl)phosphonic acid, F-00036
(3-Fluorophenyl)phosphonic acid, F-00037
(4-Fluorophenyl)phosphonic acid, F-00038
Phenyl phosphorofluoridic acid, P-00295

C_6H_6FP
(3-Fluorophenyl)phosphine, F-00033
(4-Fluorophenyl)phosphine, F-00034

$C_6H_6F_2NOP$
Phenylphosphoramidic acid; Difluoride, *in* P-00257

$C_6H_6F_2NPS$
Phenylphosphoramidothioic acid; Difluoride, *in* P-00262

$C_6H_6F_9O_4P$
Tris(2,2,2-trifluoroethyl) phosphate, T-00851

$C_6H_6IO_3P$
1,3-Dihydro-1-hydroxy-1,2,3-benziodoxaphosphole 3-oxide, D-00488
(2-Iodophenyl)phosphonic acid, I-00016
(3-Iodophenyl)phosphonic acid, I-00017
(4-Iodophenyl)phosphonic acid, I-00018

$C_6H_6IO_4P$
1,3-Dihydro-1,3-dihydroxy-1,2,3-benziodoxaphosphole 3-oxide, *in* D-00488
(4-Iodosophenyl)phosphonic acid, *in* I-00018

$C_6H_6IO_5P$
1,3-Dihydroxy-1*H*-1,2,4,3-benziodadioxaphosphorin 3-oxide, D-00616

C_6H_6NOP
2,2-Dihydro-1,3,2-benzoxazaphosphole, D-00444

$C_6H_6NOPS_2$
2,3-Dihydro-2-mercapto-1,3,2-benzoxazaphosphole 2-sulfide, D-00511

$C_6H_6NO_4P$
(2-Nitrophenyl)phosphinic acid, N-00053
(4-Nitrophenyl)phosphinic acid, N-00054

$C_6H_6NO_5P$
(2-Nitrophenyl)phosphonic acid, N-00055
(3-Nitrophenyl)phosphonic acid, N-00056
▷(4-Nitrophenyl)phosphonic acid, N-00057

$C_6H_6NO_6P$
Mono(2-nitrophenyl) phosphate, M-00498
Mono(3-nitrophenyl) phosphate, M-00499
Mono(4-nitrophenyl) phosphate, M-00500

$C_6H_6N_3O_2P$
Phenylphosphonazidic acid, P-00220

$C_6H_6N_3O_7P$
(2,4-Dinitrophenyl)phosphoramidic acid, D-00947

$C_6H_6N_9OP$
1,1′,1″-Phosphinylidynetris(1*H*-1,2,4-triazole), *in* T-00876

$C_6H_6N_9P$
Tri(1,2,4-triazol-1-yl)phosphine, T-00876

$C_6H_7BrNO_3P$
(2-Bromophenyl)phosphoramidic acid, B-00494
(3-Bromophenyl)phosphoramidic acid, B-00495
(4-Bromophenyl)phosphoramidic acid, B-00496

$C_6H_7BrNO_4PS$
Phenylsulfonylphosphoramidic acid; Monobromide, *in* P-00312

$C_6H_7ClNOPS$
Phenylphosphoramidochloridothioic acid, P-00260

$C_6H_7ClNO_2P$
Phenylphosphoramidochloridic acid, P-00259

$C_6H_7ClNO_2PS$
(2-Chlorophenyl)phosphoramidothioic acid, C-00158
(3-Chlorophenyl)phosphoramidothioic acid, C-00159
(4-Chlorophenyl)phosphoramidothioic acid, C-00160

$C_6H_7ClNO_3P$
(2-Chlorophenyl)phosphoramidic acid, C-00155
(3-Chlorophenyl)phosphoramidic acid, C-00156
(4-Chlorophenyl)phosphoramidic acid, C-00157

C_6H_7FNOP
P-Phenylphosphonamidic acid; Fluoride, *in* P-00218

$C_6H_7FNO_2P$
Phenyl phosphoramidofluoridate, *in* P-00295

$C_6H_7FNO_2PS$
(3-Fluorophenyl)phosphoramidothioic acid, F-00045
(4-Fluorophenyl)phosphoramidothioic acid, F-00046

$C_6H_7FNO_3P$
(3-Fluorophenyl)phosphoramidic acid, F-00043
(4-Fluorophenyl)phosphoramidic acid, F-00044

C_6H_7FNPS
P-Phenylphosphonamidothioic acid; Fluoride, *in* P-00219

$C_6H_7F_2N_2PS$
2-Phenylphosphorodifluoridothioic hydrazide, P-00289

C_6H_7NOP
P,*P*-Diethynyl-*N*,*N*-dimethylphosphinic amide, *in* D-00409

$C_6H_7N_2O_5P$
(2-Nitrophenyl)phosphoramidic acid, N-00058
(3-Nitrophenyl)phosphoramidic acid, N-00059
(4-Nitrophenyl)phosphoramidic acid, N-00060
(2-Nitrophenyl)phosphorimidic acid, N-00061
(3-Nitrophenyl)phosphorimidic acid, N-00062
(4-Nitrophenyl)phosphorimidic acid, N-00063

$C_6H_7N_4O_2P$
Di-1*H*-imidazol-1-ylphosphinic acid, D-00628

C_6H_7OPS
Phenylphosphinothioic acid, P-00213

$C_6H_7OPS_2$
Phenylphosphonodithioic acid, P-00236

$C_6H_7O_2P$
2,2-Dihydro-1,3,2-benzodioxaphosphole, D-00439
Dipropynylphosphinic acid, D-01231
Phenylphosphinic acid, P-00210
Phenylphosphonous acid, P-00251

$C_6H_7O_2PS$
Phenylphosphonothioic acid, P-00244

$C_6H_7O_2PS_2$
▷*O*,*O*-Di-2-propynyl phosphorodithioate, D-01232

$C_6H_7O_2PSe$
Phenylphosphonoselenoic acid, P-00243

$C_6H_7O_3P$
Monophenyl phosphonate, M-00508
▷Phenylphosphonic acid, P-00221

$C_6H_7O_3PS$
S-Phenyl phosphorothioate, P-00297

$C_6H_7O_4P$
(2-Hydroxyphenyl)phosphonic acid, H-00179
▷(3-Hydroxyphenyl)phosphonic acid, H-00180
▷(4-Hydroxyphenyl)phosphonic acid, H-00181
Monophenyl phosphate, M-00507
Phenylphosphonoperoxoic acid, P-00242

C_6H_7P
2-Methylphosphorin, M-00337
4-Methylphosphorin, M-00338
▷Phenylphosphine, P-00208

$C_6H_7PS_2$
Phenylphosphonodithious acid, P-00238

$C_6H_7PS_3$
Phenylphosphonotrithioic acid, P-00250

$C_6H_7PSe_2$
Phenylphosphonodiselenous acid, P-00235

$C_6H_8ClN_2OP$
Bis(2-oxo-3-oxazolidinyl)phosphinic acid; Chloride, *in* B-00378

$C_6H_8ClO_6P$
Dimethyl 2-chloro-1,3,2-dioxaphospholan-4,5-dicarboxylate, *in* C-00062

C_6H_8NOPS
P-Phenylphosphonamidothioic acid, P-00219

$C_6H_8NO_2P$
(4-Aminophenyl)phosphinic acid, A-00108
P-Phenylphosphonamidic acid, P-00218
Phenylphosphoramidous acid, P-00263

$C_6H_8NO_2PS$
Phenylphosphoramidothioic acid, P-00262

$C_6H_8NO_2PSe$
Phenylphosphoramidoselenoic acid, P-00261

$C_6H_8NO_3P$
▷(2-Aminophenyl)phosphonic acid, A-00109
▷(3-Aminophenyl)phosphonic acid, A-00110
▷(4-Aminophenyl)phosphonic acid, A-00111
(2-Methyl-4-pyridinyl)phosphonic acid, M-00388

(3-Methyl-2-pyridinyl)phosphonic acid, M-00389
(3-Methyl-4-pyridinyl)phosphonic acid, M-00390
(4-Methyl-2-pyridinyl)phosphonic acid, M-00391
(6-Methyl-2-pyridinyl)phosphonic acid, M-00392
Phenylphosphoramidic acid, P-00257
Phenylphosphorimidic acid, P-00264
(2-Pyridinylmethyl)phosphonic acid, P-00506
(4-Pyridinylmethyl)phosphonic acid, P-00507

$C_6H_8NO_4P$

(2-Pyridinylmethyl)phosphonic acid; *N*-Oxide, *in* P-00506

$C_6H_8NO_4PS$

Phenylsulfonylphosphoramidothioic acid, P-00314

$C_6H_8NO_5PS$

Phenylsulfonylphosphoramidic acid, P-00312
Phenylsulfonylphosphorimidic acid, P-00315

$C_6H_8N_5OP$

Bis(2-oxo-3-oxazolidinyl)phosphinic acid; Azide, *in* B-00378

$C_6H_8O_4P_2$

1,2-Benzenediphosphonous acid, B-00009
1,4-Benzenediphosphonous acid, B-00010

$C_6H_8O_6P_2$

1,2-Benzenediphosphonic acid, B-00006
1,3-Benzenediphosphonic acid, B-00007
1,4-Benzenediphosphonic acid, B-00008

$C_6H_8O_8P_2$

1,3-Phenylene diphosphoric acid, P-00147
1,4-Phenylenediphosphoric acid, P-00148

$C_6H_8P_2$

1,2-Diphosphinobenzene, D-01172
1,4-Diphosphinobenzene, D-01173

$C_6H_9Cl_2OP$

1-Cyclohexenephosphonic acid; Dichloride, *in* C-00241
3-Cyclohexenephosphonic acid; Dichloride, *in* C-00242

$C_6H_9Cl_6O_4P$

Ethyl bis(1-hydroxy-2,2,2-trichloroethyl)-phosphinate, *in* B-00287

$C_6H_9F_6OP$

tert-Butyl bis(trifluoromethyl)phosphinite, *in* B-00418

$C_6H_9F_6PS$

tert-Butyl bis(trifluoromethyl)phosphinothioite, *in* B-00417

$C_6H_9NO_6P_2$

Phenylimidodiphosphoric acid, P-00168

$C_6H_9NO_8P$

2-Amino-2-deoxyglucitol; 3-Phosphate, *in* A-00066

$C_6H_9N_2OP$

Bis(2-cyanoethyl)phosphine; Oxide, *in* B-00153
▷*P*-Phenylphosphonic diamide, P-00222
Phenyl phosphorodiamidite, P-00275

$C_6H_9N_2OPS$

O-Phenyl phosphorodiamidothioate, P-00276

$C_6H_9N_2O_2P$

P-Phenylphosphonohydrazidic acid, P-00241
Phenyl phosphorodiamidate, P-00272
N-Phenylphosphorodiamidic acid, P-00273

$C_6H_9N_2O_2PS$

2-Phenylphosphorohydrazidothioic acid, P-00296

$C_6H_9N_2O_3P$

(2-Aminophenyl)phosphoramidic acid, A-00113
(3-Aminophenyl)phosphoramidic acid, A-00114
(4-Aminophenyl)phosphoramidic acid, A-00115

$C_6H_9N_2O_6P$

Bis(2-oxo-3-oxazolidinyl)phosphinic acid, B-00378

$C_6H_9N_2P$

Bis(2-cyanoethyl)phosphine, B-00153
P-Phenylphosphonous diamide, P-00252

$C_6H_9N_2PS$

P-Phenylphosphonothioic diamide, P-00246

C_6H_9OP

Trivinylphosphine; Oxide, *in* T-00877

$C_6H_9O_2P$

1-Hydroxy-3,4-dimethyl-1*H*-phosphole 1-oxide, H-00130
1-Hydroxy-1*H*-phosphole 1-oxide; Et ester, *in* H-00186

$C_6H_9O_2PS$

Dimethyl 2-thienylphosphonite, *in* T-00308

$C_6H_9O_3P$

Dimethyl 3-buten-1-ynylphosphonate, *in* B-00529
6-Methyl-4-methylene-4*H*-1,3,2-dioxaphosphorin; 2-Me, 2-oxide, *in* M-00167
2,8,9-Trioxa-1-phosphatricyclo[3.3.1.1^{3,7}]-decane, T-00591

$C_6H_9O_3PS$

Dimethyl 2-thienylphosphonate, *in* T-00307
2,8,9-Trioxa-1-phosphatricyclo[3.3.1.1^{3,7}]-decane; 1-Sulfide, *in* T-00591

$C_6H_9O_3PSe$

2,8,9-Trioxa-1-phosphatricyclo[3.3.1.1^{3,7}]-decane; 1-Selenide, *in* T-00591

$C_6H_9O_4P$

Dimethyl 2-furanylphosphonate, *in* F-00073
Dimethyl 3-furanylphosphonate, *in* F-00074
▷2,8,9-Trioxa-1-phosphatricyclo[3.3.1.1^{3,7}]-decane; 1-Oxide, *in* T-00591

$C_6H_9O_7P$

2-Ethoxy-1,3,2-dioxaphospholan-4,5-dicarboxylic acid, E-00033
Triacetyl phosphate, T-00325

C_6H_9P

Trivinylphosphine, T-00877

C_6H_9PS

Trivinylphosphine; Sulfide, *in* T-00877

$C_6H_{10}BrOP$

1-Bromo-2,5-dihydro-3,4-dimethyl-1*H*-phosphole; 1-Oxide, *in* B-00469
1-Bromo-2,5-dihydro-3,4-dimethyl-1*H*-phosphole 1-oxide, *in* D-00493

$C_6H_{10}BrP$

1-Bromo-2,5-dihydro-3,4-dimethyl-1*H*-phosphole, B-00469

$C_6H_{10}BrPS$

1-Bromo-2,5-dihydro-3,4-dimethyl-1*H*-phosphole; 1-Sulfide, *in* B-00469

$C_6H_{10}Br_3ClNO_3P$

2,2,2-Tribromoethyl 4-morpholinylphosphonochloridate, T-00357

$C_6H_{10}ClN_2O_2P$

4,5-Diethoxy-2*H*-1,3,2-diazaphosphole; 2-Chloro, *in* D-00228

$C_6H_{10}ClOP$

1-Chloro-2,5-dihydro-3,4-dimethyl-1*H*-phosphole; 1-Oxide, *in* C-00041
1-Chloro-2,5-dihydro-3,4-dimethyl-1*H*-phosphole 1-oxide, *in* D-00493
2-Chloro-2,3-dihydro-3,3,5-trimethyl-1,2-oxaphosphole, C-00054
Di-2-propenylphosphinic acid; Chloride, *in* D-01193

$C_6H_{10}ClO_2P$

2-Chloro-2,3-dihydro-3,3,5-trimethyl-1,2-oxaphosphole; 2-Oxide, *in* C-00054

$C_6H_{10}ClO_2PS$

O,O-Di-2-propenyl phosphorochloridothioate, D-01196

$C_6H_{10}ClO_3P$

2-Chlorohexahydropyrano[3,2-*d*]-1,3,2-dioxaphosphorin, C-00089
Dimethyl (2-chloroethyl)phosphonate, *in* C-00074
Di-2-propenyl phosphorochloridate, *in* D-01192
Hexahydro-2-hydroxycyclopenta[*d*]-1,3,2-dioxaphosphorin 2-oxide; Chloride, *in* H-00034

$C_6H_{10}ClO_3PS$

2-Chlorohexahydropyrano[3,2-*d*]-1,3,2-dioxaphosphorin; 2-Sulfide, *in* C-00089

$C_6H_{10}ClO_4P$

7-Methyl-1,4,6-trioxa-5-phospha(5-P^V)-spiro[4.4]nonan-8-one; 5-Chloro, *in* M-00427

$C_6H_{10}ClP$

1-Chloro-2,5-dihydro-3,4-dimethyl-1*H*-phosphole, C-00041

$C_6H_{10}ClPS$

1-Chloro-2,5-dihydro-3,4-dimethyl-1*H*-phosphole; 1-Sulfide, *in* C-00041
7-Chloro-6-thia-7-phosphabicyclo[3.2.1]octane, C-00187

$C_6H_{10}ClPS_2$

7-Chloro-6-thia-7-phosphabicyclo[3.2.1]octane; 7-Sulfide, *in* C-00187

$C_6H_{10}Cl_3NP$

P,P,P-Trichloro-*N*-phenylphosphine imide, *in* P-00264

$C_6H_{10}Cl_3O_5P$

▷Dimethyl 1-acetoxy-2,2,2-trichloroethylphosphonate, *in* A-00011

$C_6H_{10}FO_3P$

Di-2-propenyl phosphorofluoridate, *in* D-01192

$C_6H_{10}F_3O_5P$

Diethyl trifluoroacetyl phosphate, D-00406

$C_6H_{10}NO_2P$

2-Cyano-5,5-dimethyl-1,3,2-dioxaphosphorinane, C-00209
2-Isocyano-5,5-dimethyl-1,3,2-dioxaphosphorinane, I-00023

$C_6H_{10}NO_2PSe$

2-Cyano-5,5-dimethyl-1,3,2-dioxaphosphorinane; 2-Selenide, *in* C-00209
2-Isoselenocyanato-5,5-dimethyl-1,3,2-dioxaphosphorinane, I-00066

$C_6H_{10}NO_3P$

2-Cyano-5,5-dimethyl-1,3,2-dioxaphosphorinane; 2-Oxide, *in* C-00209
2-Isocyano-5,5-dimethyl-1,3,2-dioxaphosphorinane; 2-Oxide, *in* I-00023

$C_6H_{10}NO_3PSe$

2-Isoselenocyanato-5,5-dimethyl-1,3,2-dioxaphosphorinane; 2-Oxide, *in* I-00066

$C_6H_{10}NO_5$

2-Amino-2-deoxyglucitol, A-00066

$C_6H_{10}NO_8P$

Sparfosic acid, S-00006

$C_6H_{10}N_2O_6P_2$

1,3-Benzenediphosphoramidic acid, B-00011
1,4-Benzenediphosphoramidic acid, B-00012

$C_6H_{10}N_3OP$

P-(4-Aminophenyl)phosphonic diamide, A-00112
Phenylphosphoric triamide, *in* P-00257

$C_6H_{11}Br_2P$

Dibromocyclohexylphosphine, *in* C-00267

$C_6H_{11}Cl_2OP$

Cyclohexylphosphonic dichloride, C-00258
Cyclohexyl phosphorodichloridite, C-00272

$C_6H_{11}Cl_2P$

Cyclohexylphosphonous dichloride, C-00268

$C_6H_{11}Cl_2PS$

Cyclohexylphosphonothioic dichloride, C-00266

$C_6H_{11}Cl_2PSe$
Cyclohexylphosphonoselenoic acid; Dichloride, *in* C-00263

$C_6H_{11}F_2OP$
Cyclohexylphosphonic acid; Difluoride, *in* C-00256

$C_6H_{11}F_2PSe$
Cyclohexylphosphonoselenoic acid; Difluoride, *in* C-00263

$C_6H_{11}I_2P$
Cyclohexyldiiodophosphine, *in* C-00267

$C_6H_{11}N_2O_2P$
4,5-Diethoxy-2*H*-1,3,2-diazaphosphole, D-00228

$C_6H_{11}N_2O_4PS_3$
▷Methidathion, M-00025

$C_6H_{11}N_4OP$
P-Phenylphosphonic dihydrazide, P-00225
2-Phenylphosphorodiamidic hydrazide, P-00274

$C_6H_{11}N_4OPS$
O-Phenyl phosphorodihydrazidothioate, P-00291

$C_6H_{11}N_4O_2P$
Phenyl phosphorodihydrazidate, P-00290

$C_6H_{11}N_4PS$
P-Phenylphosphonothioic dihydrazide, P-00249

$C_6H_{11}OP$
2,5-Dihydro-1-methoxy-3-methyl-1*H*-phosphole, *in* D-00530
1-Methyl-4-phosphorinanone, M-00340

$C_6H_{11}OPS$
O-Ethyl divinylphosphinothioate, *in* D-01259

$C_6H_{11}O_2P$
Butyl ethynylphosphinate, *in* E-00202
Diethyl ethynylphosphonite, *in* E-00204
▷2,5-Dihydro-1-hydroxy-3,4-dimethyl-1*H*-phosphole 1-oxide, D-00493
2,3-Dihydro-1-methoxy-4-methyl-1*H*-phosphole 1-oxide, *in* D-00496
▷2,5-Dihydro-1-methoxy-3-methyl-1*H*-phosphole 1-oxide, *in* D-00499
▷2,3-Dihydro-3,3,5-trimethyl-1,2-oxaphosphole 2-oxide, D-00596
1,4-Dioxa-5-phospha(5-P^V)spiro[4.4]non-7-ene, D-00967
Di-2-propenylphosphinic acid, D-01193
1-Ethoxy-2,3-dihydro-1*H*-phosphole 1-oxide, *in* D-00506
▷1-Ethoxy-2,5-dihydro-1*H*-phosphole 1-oxide, *in* D-00507
Ethyl diethenylphosphinate, *in* D-01258
1-Methyl-4-phosphorinanone; Oxide, *in* M-00340

$C_6H_{11}O_2PS$
O,O-Dimethyl (1,3-butadien-1-yl)-phosphonothioate, *in* B-00511

$C_6H_{11}O_2PS_2$
O,O-Di-2-propenyl phosphorodithioate, D-01197

$C_6H_{11}O_3P$
1-Cyclohexenephosphonic acid, C-00241
3-Cyclohexenephosphonic acid, C-00242
4-Cyclohexenephosphonic acid, C-00243
Diethyl ethynylphosphonate, *in* E-00203
2,5-Dihydro-5,5-dimethyl-1,2-oxaphosphole 2-oxide; 2-Methoxy, *in* D-00464
Dimethyl (1,3-butadienyl)phosphonate, *in* B-00510
▷Di-2-propenyl phosphonate, D-01194
▷4-Ethyl-2,6,7-trioxa-1-phosphabicyclo[2.2.2]-octane, E-00196
9-Methyl-1,4,6-trioxa-5-phospha(5-P^V)-spiro[4.4]non-7-ene, M-00428

$C_6H_{11}O_3PS$
O,O-Di-2-propenyl phosphorothioate, D-01198
▷Ethyl bicyclic thiophosphate, *in* E-00196

$C_6H_{11}O_3PSe$
Ethyl bicyclic selenophosphate, *in* E-00196

$C_6H_{11}O_4P$
2,3-Dimethyl-1,4,6,9-tetraoxa-5-phospha(5-P^V)spiro[4.4]non-2-ene, D-00924
Di-2-propenyl phosphate, D-01192
3-Ethoxycarbonyl-1-hydroxyphosphetane 1-oxide, *in* H-00184
▷Ethyl bicyclic phosphate, *in* E-00196
Ethyl dimethylvinylene phosphate, *in* H-00123
Hexahydro-2-hydroxycyclopenta[*d*]-1,3,2-dioxaphosphorin 2-oxide, H-00034
1-Methoxy-3-methoxycarbonylphosphetane 1-oxide, *in* H-00184
7-Methyl-1,4,6-trioxa-5-phospha(5-P^V)-spiro[4.4]nonan-8-one, M-00427

$C_6H_{11}O_5P$
Methyl 3-(dimethoxyphosphinyl)propenoate, *in* D-00686
7-Methyl-1,4,6-trioxa-5-phospha(5-P^V)-spiro[4.4]nonan-8-one; 5-Hydroxy, *in* M-00427
2,6,7-Trioxa-1-phosphabicyclo[2.2.2]octane-4-methanol; Oxide, Me ether, *in* T-00590

$C_6H_{11}O_5PS$
3-[(Dimethoxyphosphinothioyl)oxy]-2-methyl-2-propenoic acid, D-00673

$C_6H_{11}O_6P$
(Diethoxyphosphinyl)oxoacetic acid, D-00252
3-[(Dimethoxyphosphinyl)oxy]-2-butenoic acid, D-00681
4-Ethylspiro[2,8,9-trioxa-1-phosphatricyclo[3.3.1.1^{3,7}]decane-1,4′-trioxaphosphetane], E-00187

$C_6H_{11}O_6PS_2$
[(Dimethoxyphosphinothioyl)thio]butanedioic acid, D-00677

$C_6H_{11}PS_2$
2,5-Dihydro-1-mercaptophosphole 1-sulfide; 2-Ethylthio (Et ester), *in* D-00512
3,6-Dihydro-2*H*-1,2-thiaphosphorin 2-sulfide; 2-Et, *in* D-00586

$C_6H_{12}BrO_5P$
Bromo(diethoxyphosphinyl)acetic acid, B-00467

$C_6H_{12}Br_3O_3P$
Bis(2-bromoethyl) (2-bromoethyl)phosphonate, *in* B-00472

$C_6H_{12}Br_4P_2$
1,6-Hexanediphosphonous acid; Bis(dibromide), *in* H-00086

$C_6H_{12}ClN_4O_2P$
1,3,5,7-Tetramethyl-1,3,5,7-tetraaza-4-phospha(4-P^V)spiro[3.3]heptane-2,6-dione; 4-Chloro, *in* T-00234

$C_6H_{12}ClOP$
Cyclopentylmethylphosphinic acid; Chloride, *in* C-00279

$C_6H_{12}ClOPS$
Cyclohexylphosphonochloridothioic acid, C-00260

$C_6H_{12}ClO_2P$
2-Chloro-1,3,2-dioxaphosphonane, *in* D-00971
2-Chloro-4,4,5,5-tetramethyl-1,3,2-dioxaphospholane, C-00185
2-Chloro-4,4,6-trimethyl-1,3,2-dioxaphosphorinane, C-00189
Cyclohexylphosphonochloridic acid, C-00259

$C_6H_{12}ClO_2PS$
Tetramethylethylene chlorothiophosphate, *in* C-00185
1,3,3-Trimethyltrimethylene chlorothiophosphate, *in* C-00189

$C_6H_{12}ClO_3P$
5-Chloromethyl-2-methoxy-5-methyl-1,3,2-dioxaphosphorinane, C-00098
Methyl (2-chloro-3-oxobutyl)-methylphosphinate, *in* C-00121

$C_6H_{12}ClO_3PS$
[(Diethoxyphosphinothioyl)thio]acetic acid; Chloride, *in* D-00239

$C_6H_{12}ClO_4P$
5-Chloromethyl-2-methoxy-5-methyl-1,3,2-dioxaphosphorinane; 2-Oxide, *in* C-00098
Diethyl(chloroacetyl)phosphonate, *in* C-00020

$C_6H_{12}ClO_5P$
Chloro(diethoxyphosphinyl)acetic acid, C-00036

$C_6H_{12}Cl_2NOP$
Cyclohexylphosphoramidic acid; Dichloride, *in* C-00269

$C_6H_{12}Cl_3O_3P$
Bis(2-chloroethyl) (2-chloroethyl)phosphonate, *in* C-00074
Tris(2-chloroethyl) phosphite, T-00742

$C_6H_{12}Cl_3O_3PS$
▷*O,O,O*-Tris(2-chloroethyl) phosphorothioate, T-00743

$C_6H_{12}Cl_3O_4P$
▷Diethyl (2,2,2-trichloro-1-hydroxyethyl)-phosphonate, *in* T-00408
Tris(2-chloroethyl) phosphate, T-00741

$C_6H_{12}Cl_4O_2P_2$
1,6-Hexanediphosphonic acid; Bisdichloride, *in* H-00085

$C_6H_{12}Cl_4P_2$
1,6-Hexanediphosphonous acid; Bis(dichloride), *in* H-00086

$C_6H_{12}FO_2P$
2-Fluoro-1,3,2-dioxaphosphonane, *in* D-00971

$C_6H_{12}FO_3PS$
[(Diethoxyphosphinothioyl)thio]acetic acid; Fluoride, *in* D-00239

$C_6H_{12}FO_5P$
(Diethoxyphosphinyl)fluoroacetic acid, D-00250

$C_6H_{12}F_2NPS$
Cyclohexylphosphoramidothioic acid; Difluoride, *in* C-00271

$C_6H_{12}F_3O_2P$
Butyl methyl(trifluoromethyl)phosphinate, *in* M-00420

$C_6H_{12}NOP$
3,5-Di-*tert*-butyl-2,3-dihydro-1,3,2-oxazaphosphole, D-00083
1-Methyl-4-phosphorinanone; Oxime, *in* M-00340

$C_6H_{12}NO_2P$
4,7-Dihydro-1,3,2-dioxaphosphepin; 2-Dimethylamino, *in* D-00472
1-Methyl-4-phosphorinanone; Oxide, oxime, *in* M-00340
Tetrahydro-2,6-dimethyl-[1,3,2]-oxazaphospholo[2,3-*b*][1,3,2]-oxazaphosphole, T-00097
Tetrahydro-2*H*,6*H*-[1.3.2]-oxazaphosphorino[2,3-*b*][1.3.2]-oxazaphosphorine, T-00120

$C_6H_{12}NO_2PS$
Methyl *tert*-butylphosphon(isocyanatidate), *in* B-00602

$C_6H_{12}NO_3P$
Diethyl (cyanomethyl)phosphonate, *in* C-00215
Diethyl (isocyanomethyl)phosphonate, *in* I-00024
Di-2-propenylphosphoramidic acid, D-01195
2,8,9-Trioxa-5-aza-1-phosphabicyclo[3.3.3]-undecane, T-00584

$C_6H_{12}NO_3PS$
2,8,9-Trioxa-5-aza-1-phosphabicyclo[3.3.3]-undecane; 1-Sulfide, *in* T-00584

$C_6H_{12}NO_3PS_2$
▷Fosthietan, F-00065

$C_6H_{12}NO_3PSe$
2,8,9-Trioxa-5-aza-1-phosphabicyclo[3.3.3]-undecane; 1-Selenide, *in* T-00584

$C_6H_{12}NO_4P$
Diethyl (isocyanatomethyl)phosphonate, *in* I-00022
2,8,9-Trioxa-5-aza-1-phosphabicyclo[3.3.3]-undecane; 1-Oxide, *in* T-00584

$C_6H_{12}NO_4PS_2$
▷Formothion, F-00055

$C_6H_{12}N_3OP$
Hexahydro-2*a*,4*a*,6*a*-triaza-6*b*-phosphacyclopenta[*cd*]pentalene; 6*b*-Oxide, *in* H-00055
▷1,1′,1″-Phosphinylidynetrisaziridine, *in* T-00331
1,3,5-Triaza-7-phosphatricyclo[3.3.1.1^{3,7}]-decane; 7-Oxide, *in* T-00327

$C_6H_{12}N_3O_2P$
Diethyl (1*H*-1,2,4-triazol-1-yl)phosphonite, *in* T-00332

$C_6H_{12}N_3P$
Hexahydro-2*a*,4*a*,6*a*-triaza-6*b*-phosphacyclopenta[*cd*]pentalene, H-00055
1,3,5-Triaza-7-phosphatricyclo[3.3.1.1^{3,7}]-decane, T-00327
▷Tri-(1-aziridinyl)phosphine, T-00331

$C_6H_{12}N_3PS$
Hexahydro-2*a*,4*a*,6*a*-triaza-6*b*-phosphacyclopenta[*cd*]pentalene; 6*b*-Sulfide, *in* H-00055
▷Thiotepa, T-00317
1,3,5-Triaza-7-phosphatricyclo[3.3.1.1^{3,7}]-decane; 7-Sulfide, *in* T-00327

$C_6H_{12}N_3PSe$
Hexahydro-2*a*,4*a*,6*a*-triaza-6*b*-phosphacyclopenta[*cd*]pentalene; 6*b*-Selenide, *in* H-00055
1,1′,1″-Phosphinoselenoylidynetrisaziridine, *in* T-00331

$C_6H_{12}N_5O_2PS_2$
▷Menazon, M-00003

$C_6H_{12}N_7O_2P$
1,3,5,7-Tetramethyl-1,3,5,7-tetraaza-4-phospha(4-P^V)spiro[3.3]heptane-2,6-dione; 4-Azido, *in* T-00234

$C_6H_{12}OP$
1-Phosphabicyclo[2.2.1]heptane; *p*-Oxide, *in* P-00344

$C_6H_{12}O_2P_2$
1,4-Diphosphabicyclo[2.2.2]octane; Dioxide, *in* D-01169

$C_6H_{12}O_5P_2$
2,2′-Oxybis[1,3,2-dioxaphosphorinane], O-00092

$C_6H_{12}O_7P_2$
2,2′-Oxybis[1,3,2-dioxaphosphorinane]; 2,2′-Dioxide, *in* O-00092

$C_6H_{12}O_{10}P_2$
2,5-Diphosphonohexanedioic acid, D-01184
2-Phosphono-2-(2-phosphonoethyl)butanedioic acid, P-00394

$C_6H_{12}P$
1-Phosphabicyclo[2.2.1]heptane, P-00344

$C_6H_{12}P_2$
1,4-Diphosphabicyclo[2.2.2]octane, D-01169
1,5-Diphosphabicyclo[3.3.0]octane, D-01170
1,2,3,6-Tetrahydro-4,5-dimethyl-1,2-diphosphorine, T-00095

$C_6H_{12}P_2S$
1,5-Diphosphabicyclo[3.3.0]octane; Monosulfide, *in* D-01170

$C_6H_{12}P_2S_2$
1,2-Bis(dimethylphosphinothioyl)ethane, *in* B-00200
1,4-Diphosphabicyclo[2.2.2]octane; Disulfide, *in* D-01169
1,5-Diphosphabicyclo[3.3.0]octane; Disulfide, *in* D-01170

$C_6H_{13}ClNOP$
2-Chlorotetrahydro-3,5,5-trimethyl-2*H*-1,3,2-oxazaphosphorine, C-00183

$C_6H_{13}ClNOPS$
2-Chlorotetrahydro-3,5,5-trimethyl-2*H*-1,3,2-oxazaphosphorine; 2-Sulfide, *in* C-00183
Cyclohexylphosphoramidochloridothioic acid, C-00270

$C_6H_{13}ClNO_2P$
2-Chlorotetrahydro-3,5,5-trimethyl-2*H*-1,3,2-oxazaphosphorine; 2-Oxide, *in* C-00183
Ethyl 4-morpholinylphosphonochloridite, *in* M-00526

$C_6H_{13}ClNPS$
P-Cyclohexylphosphonamidothioic chloride, C-00255

$C_6H_{13}Cl_2OP$
Hexylphosphonic dichloride, H-00098
Hexyl phosphorodichloridite, H-00103

$C_6H_{13}Cl_2P$
Dichlorohexylphosphine, *in* H-00101

$C_6H_{13}Cl_2PS$
Hexylphosphonothioic acid; Dichloride, *in* H-00100

$C_6H_{13}FNOP$
P-Cyclohexylphosphonamidic fluoride, C-00254

$C_6H_{13}F_2OP$
Hexylphosphonic acid; Difluoride, *in* H-00097

$C_6H_{13}IN_3O_2PS$
7-Methyl-2-thia-1,3,5-triaza-7-phosphoniaadamantane iodide, *in* T-00301

$C_6H_{13}I_2O_2P$
Butyl bis(iodomethyl)phosphinate, *in* B-00288

$C_6H_{13}NO_2P$
2-(*tert*-Butylamino)-1,3,2-dioxaphospholane, B-00539

$C_6H_{13}NO_3P$
Ethylene *tert*-butylphosphoramidate, *in* B-00539

$C_6H_{13}N_2OP$
Ethyl bis(1-aziridinyl)phosphinite, *in* B-00095
P-Ethynyl-*N,N,N′,N′*-tetramethylphosphonic diamide, *in* E-00203

$C_6H_{13}N_2OPS$
O-Ethyl bis(1-aziridinyl)phosphinothioate, *in* B-00093
S-Ethyl bis(1-aziridinyl)phosphinothioate, *in* B-00093

$C_6H_{13}N_2OPS_2$
O-Ethyl *N*-(4,5-dihydro-2-thiazolyl)-*P*-methylphosphonamidothioate, *in* D-00588

$C_6H_{13}N_2O_2P$
▷Ethyl bis(1-aziridinyl)phosphinate, *in* B-00091
Isopropyl dimethylphosphoramidocyanidate, *in* D-00863

$C_6H_{13}N_2O_2PS$
Ethyl *N*-(4,5-dihydro-2-thiazolyl)-*P*-methylphosphonamidate, *in* D-00587

$C_6H_{13}N_2O_3P$
Diethyl (1-diazoethyl)phosphonate, *in* D-00033

$C_6H_{13}N_4O_2P$
1,3,5,7-Tetramethyl-1,3,5,7-tetraaza-4-phospha(4-P^V)spiro[3.3]heptane-2,6-dione, T-00234

$C_6H_{13}OP$
1-Methylphosphorinane; Oxide, *in* M-00339

$C_6H_{13}OPS_2$
Cyclohexylphosphonodithioic acid, C-00261
2-Ethoxy-4-methyl-1,3,2-dithiaphosphorinane, E-00042

$C_6H_{13}OPS_3$
S,S-1,3-Butylene *O*-ethyl phosphorotrithioate, *in* E-00042

$C_6H_{13}O_2P$
Butyl vinylphosphinate, *in* V-00002
Cyclohexylphosphinic acid, C-00253
Cyclohexylphosphonous acid, C-00267
Cyclopentylmethylphosphinic acid, C-00279
1,3,2-Dioxaphosphonane, D-00971
Ethyl tetramethylenephosphinate, *in* H-00185
Methyl 3-(methoxymethylphosphinyl)-propanoate, *in* H-00161
4,4,5,5-Tetramethyl-1,3,2-dioxaphospholane, T-00197
2,5,5-Trimethyl-1,3,2-dioxaphosphorinane, T-00512
4,4,6-Trimethyl-1,3,2-dioxaphosphorinane, T-00513

$C_6H_{13}O_2PS$
O-Acetyl diethylphosphinothioate, *in* D-00330
Cyclohexylphosphonothioic acid, C-00264
O,O-Diethyl vinylphosphonothioate, *in* V-00007
5,5-Dimethyl-2-methylthio-1,3,2-dioxaphosphorinane, D-00766
▷2-Ethylthio-4,5-dimethyl-1,3,2-dioxaphospholane, E-00189
O,O-Neopentylene methylphosphonothioate, *in* T-00512
O,O-Tetramethylethylene phosphonothioate, *in* T-00197
4,4,6-Trimethyl-1,3,2-dioxaphosphorinane; 2-Sulfide, *in* T-00513

$C_6H_{13}O_2PSSe$
Se-Methyl *O,O*-neopentylene phosphoroselenothioate, *in* D-00764

$C_6H_{13}O_2PS_2$
2-Ethylthio-4,5-dimethyl-1,3,2-dioxaphospholane; 2-Sulfide, *in* E-00189
2-Mercapto-4,4,5,5-tetramethyl-1,3,2-dioxaphospholane 2-sulfide, M-00019
S-Methyl *O,O*-neopentylene phosphorodithioate, *in* D-00766

$C_6H_{13}O_2PSe$
Cyclohexylphosphonoselenoic acid, C-00263
5,5-Dimethyl-2-methylseleno-1,3,2-dioxaphosphorinane, D-00764

$C_6H_{13}O_2PSe_2$
Se-Methyl *O,O*-neopentylene phosphorodiselenoate, *in* D-00764

$C_6H_{13}O_3P$
2-*tert*-Butoxy-1,3,2-dioxaphospholane, B-00535
Cyclohexylphosphonic acid, C-00256
Diethyl 2-butenylphosphonate, *in* B-00525
Diethyl ethenylphosphonate, *in* V-00003
2-Ethoxy-4,5-dimethyl-1,3,2-dioxaphospholane, E-00031
2-Ethoxy-4-methyl-1,3,2-dioxaphosphorinane, E-00041
2-Ethoxy-1,2-oxaphosphorinane 2-oxide, *in* H-00167
Ethyl acetylethylphosphinate, *in* A-00017
2-Methoxy-5,5-dimethyl-1,3,2-dioxaphosphorinane, M-00034
2-(2-Methylpropoxy)-1,3,2-dioxaphospholane, M-00368
Neopentylene methylphosphonate, *in* T-00512
Tetramethylethylene phosphonate, *in* T-00197
4,4,6-Trimethyl-1,3,2-dioxaphosphorinane; 2-Oxide, *in* T-00513
3,5,5-Trimethyl-1,2-oxaphospholan-3-ol 2-oxide, T-00519

$C_6H_{13}O_3PS$
O,O-Butylene *O*-ethyl thiophosphate, *in* E-00031
O,O-1,3-Butylene *O*-ethyl thiophosphate, *in* E-00041

O,O-Diethyl acetylphosphonothioate, *in* A-00028
2-Hydroxy-4,4,5,5-tetramethyl-1,3,2-dioxaphospholane 2-sulfide, H-00197
O-Methyl *O,O*-neopentylene phosphorothioate, *in* M-00034
S-Methyl *O,O*-neopentylene phosphorothioate, *in* D-00766
1,3,2-Oxathiaphosphorinane 2-oxide; 2-Isopropoxy, *in* O-00057

$C_6H_{13}O_3PSe$

Se-Methyl *O,O*-neopentylene phosphoroselenoate, *in* D-00764

$C_6H_{13}O_4P$

Acetyl diethyl phosphite, A-00013
2,3-Butylene ethyl phosphate, *in* E-00031
tert-Butyl ethylene phosphate, *in* B-00535
Diethyl acetylphosphonate, *in* A-00026
Diethyl (2-oxoethyl)phosphonate, *in* O-00059
(3,3-Dimethyl-2-oxobutyl)phosphonic acid, D-00783
2-Ethoxy-1,3,2-dioxaphosphepane 2-oxide, *in* H-00132
2-Hydroxy-1,3,2-dioxaphosphonane 2-oxide, H-00137
2-Hydroxy-4,4,5,5-tetramethyl-1,3,2-dioxaphospholane 2-oxide, H-00196
1-Hydroxy-2,4,6-trimethyl-1,3,5-dioxaphosphorinane 5-oxide, H-00203
Methyl neopentylene phosphate, *in* M-00034
Monocyclohexyl phosphate, M-00461
2,2,2,3-Tetrahydro-1,2-oxaphosphole; 2,2,2-Trimethoxy, *in* T-00119
1,4,6,11-Tetraoxa-5-phospha(5-P^V)spiro[4.6]-undecane, T-00245

$C_6H_{13}O_4PS$

(Diethoxyphosphinothioyl)acetic acid, D-00237

$C_6H_{13}O_4PS_2$

[(Diethoxyphosphinothioyl)thio]acetic acid, D-00239
Ethyl [(dimethoxyphosphinothioyl)thio]acetate, *in* D-00675

$C_6H_{13}O_5P$

Acetic acid anhydride with diethyl hydrogen phosphate, *in* A-00025
(Diethoxyphosphinyl)acetic acid, D-00241
Diethyl methyl phosphonoformate, *in* D-00236
2-(Dimethoxyphosphinyl)butanoic acid, D-00679
Dimethyl (2-acetyloxyethyl)phosphonate, *in* A-00006
Ethyl (dimethoxyphosphinyl)acetate, *in* D-00678
Ethyl 3-phosphonobutanoate, *in* P-00385
Methyl 2-(dimethoxyphosphinyl)propanoate, *in* D-00684
Methyl 3-(dimethoxyphosphinyl)propanoate, *in* D-00685
Phosphonoacetic acid; *tert*-Butyl ester, *in* P-00376
[[(Tetrahydro-2*H*-pyran-2-yl)oxy]methyl]-phosphonic acid, T-00136
1,4,6,9-Tetraoxa-5-phospha(5-P^V)spiro[4.4]-nonane; 5-Ethoxy, *in* T-00244

$C_6H_{13}O_5PS$

▷Ethyl (dimethoxyphosphinyl)thioacetate, *in* D-00687

$C_6H_{13}O_6P$

Methyl 2-[(dimethoxyphosphinyl)oxy]-propanoate, *in* D-00682

$C_6H_{13}O_9P$

Fructose 1-dihydrogen phosphate, F-00068
Fructose 2-dihydrogen phosphate, F-00069
Fructose 6-dihydrogen phosphate, F-00070
Galactose 1-dihydrogen phosphate, G-00001
Galactose 3-dihydrogen phosphate, G-00002
Galactose 6-dihydrogen phosphate, G-00003
Glucose 2-dihydrogen phosphate, G-00004
Glucose 3-dihydrogen phosphate, G-00005
Glucose 4-dihydrogen phosphate, G-00006
Glucose 6-dihydrogen phosphate, G-00007

$C_6H_{13}P$

▷Cyclohexylphosphine, C-00252
1-Methylphosphorinane, M-00339

$C_6H_{13}PS$

1-Methylphosphorinane; Sulfide, *in* M-00339

$C_6H_{13}PS_2$

Cyclohexylphosphonodithious acid, C-00262
5-Methyl-1,2-thiaphospholane 2-sulfide; 2-Et, *in* M-00407

$C_6H_{14}BrO_3P$

Diethyl (2-bromoethyl)phosphonate, *in* B-00472
▷Diisopropyl phosphorobromidate, *in* D-00639
Dipropyl phosphorobromidate, *in* D-01203

$C_6H_{14}BrP$

Bromodiisopropylphosphine, *in* D-00648
Bromodipropylphosphine, *in* D-01209

$C_6H_{14}BrPS$

Butyl ethylphosphonobromidothioite, *in* E-00134
Dipropylphosphinothioic acid; Bromide, *in* D-01208

$C_6H_{14}ClN_2P$

N,N'-1,3-Butylene-*N,N'*-dimethyl-phosphorodiamidous chloride, *in* H-00057

$C_6H_{14}ClOP$

tert-Butylethylphosphinic acid; Chloride, *in* B-00560
Diisopropylphosphinic acid; Chloride, *in* D-00641
Dipropylphosphinic acid; Chloride, *in* D-01205
Ethyl butylphosphonochloridite, *in* B-00605

$C_6H_{14}ClOPS$

S-Isopropyl propylphosphonochloridothioate, *in* P-00477
O-Propyl isopropylphosphonochloridothioate, *in* I-00053
S-Propyl propylphosphonochloridothioate, *in* P-00477

$C_6H_{14}ClOPS_2$

S,S-Dipropyl phosphorochloridodithioate, D-01219

$C_6H_{14}ClO_2P$

Diisopropyl phosphorochloridite, D-00658
Dipropyl phosphorochloridite, D-01218
2-Methylpropyl ethylphosphonochloridate, *in* E-00135

$C_6H_{14}ClO_2PS$

▷*O,O*-Diisopropyl phosphorochloridothioate, D-00659
O,O-Dipropyl phosphorochloridothioate, D-01221

$C_6H_{14}ClO_2PS_2$

▷*O,O*-Diisopropyl chlorothiophosphonothioate, D-00635
▷*O,O*-Dipropyl chlorothiophosphonothioate, D-01201

$C_6H_{14}ClO_2PSe$

O,O-Dipropyl phosphorochloridoselenoate, *in* D-01229

$C_6H_{14}ClO_3P$

Diethyl (1-chloroethyl)phosphonate, *in* C-00073
▷Diethyl (2-chloroethyl)phosphonate, *in* C-00074
Diisopropyl phosphorochloridate, D-00657
Dipropyl phosphorochloridate, D-01217

$C_6H_{14}ClO_3PS$

▷Diisopropyl chlorothiophosphonate, D-00634
▷Dipropyl chlorothiophosphonate, D-01200

$C_6H_{14}ClP$

Chlorodipropylphosphine, *in* D-01209
Diisopropylphosphinous chloride, D-00650

$C_6H_{14}ClPS$

Butyl ethylphosphonochloridothioite, *in* E-00138
Diisopropylphosphinothioic acid; Chloride, *in* D-00646
Dipropylphosphinothioic acid; Chloride, *in* D-01208

$C_6H_{14}ClPS_2$

tert-Butyl ethylphosphonochloridodithioate, *in* E-00136
Diisopropyl phosphorochloridothioite, D-00660
Dipropyl phosphorochloridodithioite, D-01220

$C_6H_{14}ClPS_3$

Dipropyl phosphorochloridotrithioate, D-01222

$C_6H_{14}Cl_2NOP$

N,N-Dichloro-*P,P*-diisopropylphosphinic amide, *in* D-00642
Dipropylphosphoramidic acid; Dichloride, *in* D-01214
Hexylphosphoramidic acid; Dichloride, *in* H-00102

$C_6H_{14}Cl_2NO_3P$

Ethyl hydrogen bis(2-chloroethyl)-phosphoramidate, *in* B-00121

$C_6H_{14}Cl_2NP$

Diisopropylphosphoramidous acid; Dichloride, *in* D-00656
Dipropylphosphoramidous acid; Dichloride, *in* D-01215

$C_6H_{14}Cl_2NPS$

N,N-Diethyl-*P,P*-bis(chloromethyl)-phosphinothioic amide, *in* B-00129

$C_6H_{14}FOP$

Diisopropylphosphinic acid; Fluoride, *in* D-00641
Dipropylphosphinic acid; Fluoride, *in* D-01205

$C_6H_{14}FO_3P$

▷Diisopropyl phosphorofluoridate, D-00665
▷Dipropyl phosphorofluoridate, D-01228

$C_6H_{14}FPS$

Dipropylphosphinothioic acid; Fluoride, *in* D-01208

$C_6H_{14}F_2NOP$

Dipropylphosphoramidic acid; Difluoride, *in* D-01214

$C_6H_{14}F_2NP$

Diisopropylphosphoramidous acid; Difluoride, *in* D-00656

$C_6H_{14}IP$

Iododiisopropylphosphine, *in* D-00648
Iododipropylphosphine, *in* D-01209
1-Methylphosphorinanium iodide, *in* M-00339

$C_6H_{14}IPS$

Dipropylphosphinothioic acid; Iodide, *in* D-01208

$C_6H_{14}I_2NP$

Dipropylphosphoramidous acid; Diiodide, *in* D-01215

$C_6H_{14}NO_2P$

Diethyl 1-aziridinylphosphonite, *in* A-00154
4,5-Dimethyl-2-dimethylamino-1,3,2-dioxaphospholane, D-00725
4-Methyl-2-dimethylamino-1,3,2-dioxaphosphorinane, M-00117
P-Methyl-1-piperidinylphosphonamidic acid, M-00361
Tetrahydro-2,6-dimethyl-4*H*-1,3,6,2-dioxazaphosphocine, T-00094

$C_6H_{14}NO_2PS$

O,O-1,3-Butylene dimethyl-phosphoramidothioate, *in* M-00117
O,O-2,3-Butylene dimethyl-phosphoramidothioate, *in* D-00725

Cyclohexylphosphoramidothioic acid, C-00271
Tetrahydro-2,6-dimethyl-4*H*-1,3,6,2-dioxazaphosphocine; 2-Sulfide, *in* T-00094
1,3,2-Thiazaphospholidine 2-oxide; 2-Butoxy, *in* T-00302

$C_6H_{14}NO_2PSe$

O,*O*-1,3-Butylene dimethylphosphoramidoselenoate, *in* M-00117

$C_6H_{14}NO_3P$

P-Acetyl-*N*,*N*-diethylphosphonamidic acid, A-00014
Cyclohexylphosphoramidic acid, C-00269
Diethyl 1-aziridinylphosphonate, *in* A-00153
Dimethyl 1-pyrrolidinylphosphonate, *in* P-00518
Ethyl 4-morpholinylphosphinate, *in* M-00522
4-Methyl-2-dimethylamino-1,3,2-dioxaphosphorinane 2-oxide, M-00118
(1-Piperidinylmethyl)phosphonic acid, P-00429
Tetrahydro-2,6-dimethyl-4*H*-1,3,6,2-dioxazaphosphocine; 2-Oxide, *in* T-00094

$C_6H_{14}NO_3PS$

O,*O*-Dimethyl 4-morpholinylphosphonothioate, *in* M-00527

$C_6H_{14}NO_3PS_2$

▷*O*,*O*-Diethyl *S*-(2-amino-2-oxoethyl) phosphorodithioate, *in* D-00239

$C_6H_{14}NO_4P$

Diethyl acetylphosphoramidate, *in* A-00029
Diethyl (2-amino-2-oxoethyl)phosphonate, *in* A-00060
(2-Diethylamino-2-oxoethyl)phosphonic acid, D-00268
Diethyl (methylaminocarbonyl)phosphonate, *in* M-00088
Dimethyl (dimethylcarbamoylmethyl)-phosphonate, *in* D-00703
▷Dimethyl 4-morpholinylphosphonate, *in* M-00523

$C_6H_{14}NO_5P$

Diethyl (1-nitroethyl)phosphonate, *in* N-00035
Methyl [3-amino-3-(dimethoxyphosphinyl)]-propanoate, *in* A-00122
Methyl *N*-(diethoxyphosphinyl)carbamate, *in* D-00249

$C_6H_{14}NO_8P$

2-Amino-2-deoxygalactose 1-(dihydrogen phosphate), A-00065
2-Amino-2-deoxyglucose 1-(dihydrogen phosphate), A-00067
2-Amino-2-deoxyglucose 3-(dihydrogen phosphate), A-00068
2-Amino-2-deoxyglucose 6-(dihydrogen phosphate), A-00069

$C_6H_{14}NPS_2$

N,*N*-Diethyl *S*,*S*-ethylene phosphoramidodithioite, *in* D-01243

$C_6H_{14}NPS_3$

S,*S*-Ethylene diethylphosphoramidodithioite, *in* D-01245

$C_6H_{14}N_3OP$

▷*P*,*P*-Bis(1-aziridinyl)-*N*,*N*-dimethylphosphinic amide, *in* B-00092
▷*P*,*P*-Bis(1-aziridinyl)-*N*-ethylphosphinic amide, *in* B-00092
P-Cyanomethyl-*N*,*N*,*N'*,*N'*-tetramethylphosphonic diamide, C-00216
Diisopropylphosphinic acid; Azide, *in* D-00641

$C_6H_{14}N_3O_2PS$

O,*O*-Dipropyl phosphorazidothioate, D-01216

$C_6H_{14}N_3P$

P,*P*-Bis(1-aziridinyl)-*N*,*N*-dimethylphosphinous amide, *in* B-00096
Tetrahydro-1,7-dimethyl-1*H*,5*H*-[1,3,2]-diazaphospholo[1,2-*a*][1,3,2]diazaphosphole, T-00093

$C_6H_{14}N_3PS$

P,*P*-Bis(1-aziridinyl)-*N*-ethylphosphinothioic amide, *in* B-00094

$C_6H_{14}O_2P_2$

2,6-Dimethyl-1,3,2,6-dioxadiphosphocane, D-00730

$C_6H_{14}O_2P_2S_2$

2,6-Dimethyl-1,3,2,6-dioxadiphosphocane; 2,6-Disulfide, *in* D-00730

$C_6H_{14}O_2P_2Se_2$

2,6-Dimethyl-1,3,2,6-dioxadiphosphocane; 2,6-Diselenide, *in* D-00730

$C_6H_{14}O_3P_2$

2,7-Dimethyl-1,2,7-oxadiphosphepane 2,7-dioxide, D-00779

$C_6H_{14}O_4P$

2,5-Diethoxy-1,3,2,5-dioxadiphosphorinane, D-00231

$C_6H_{14}O_5P$

2,5-Diethoxy-1,3,2,5-dioxadiphosphorinane; 5-Oxide, *in* D-00231

$C_6H_{14}O_5PS$

2,5-Diethoxy-1,3,2,5-dioxadiphosphorinane; 5-Oxide, 2-sulfide, *in* D-00231

$C_6H_{14}O_6P_2$

1,2-Cyclohexanediphosphonic acid, C-00240
2,5-Diethoxy-1,4,2,5-dioxaphosphorinane 2,5-dioxide, *in* D-00618
Tetramethyl ethenylidenebisphosphonate, *in* E-00020

$C_6H_{14}O_{10}P_2$

1,1,3,7-Tetrahydroxy-7-methyl-2,4-dioxa-1,3-diphosphanonan-9-oic acid 1,3-dioxide, T-00162

$C_6H_{14}O_{12}P_2$

Fructose 1,6-bis(dihydrogen phosphate), F-00067

$C_6H_{15}ClNOP$

Diisopropylphosphoramidochloridous acid, D-00654
N,*N*-Dimethyl-*P*-(1-methylpropyl)-phosphonamidic chloride, D-00763
Ethyl diethylphosphoramidochloridite, E-00060
Triethylphosphonamidic acid; Chloride, *in* T-00446

$C_6H_{15}ClNOPS$

O-Ethyl diethylphosphoramidochloridothioate, *in* D-00348

$C_6H_{15}ClNOPSe$

O-Ethyl diethylphosphoramidochloridoselenoate, *in* D-00347

$C_6H_{15}ClNO_2P$

Ethyl diethylphosphoramidochloridate, *in* D-00346

$C_6H_{15}ClNPS$

Triethylphosphonamidothioic acid; Chloride, *in* T-00448

$C_6H_{15}Cl_2N_2P$

N,*N*,*N'*-Triethylphosphoramidimidic dichloride, T-00449

$C_6H_{15}Cl_2OP_2^{\oplus}$

2,2-Dichloro-1,1,1-triethyldiphosphinium 2-oxide(1+), D-00197

$C_6H_{15}Cl_3OP_2$

2,2-Dichloro-1,1,1-triethyldiphosphinium 2-oxide(1+); Chloride, *in* D-00197

$C_6H_{15}FNOP$

Triethylphosphonamidic acid; Fluoride, *in* T-00446

$C_6H_{15}FNO_2P$

Ethyl diethylphosphoramidofluoridate, *in* D-00351

$C_6H_{15}FNPS_2$

Ethyl diethylphosphoramidofluoridodithioate, *in* D-00352

$C_6H_{15}NPSe$

P,*P*-Diethyl-*N*,*N*-dimethylphosphinoselenoic amide, *in* D-00328

$C_6H_{15}N_2OP$

N,*N'*-1,3-Butylene *N*,*N'*-dimethylphosphonic diamide, *in* H-00057
P-Cyclohexylphosphonic diamide, C-00257
P-Ethenyl-*N*,*N*,*N'*,*N'*-tetramethylphosphonic diamide, *in* V-00004
2-Ethoxy-1,3-dimethyl-1,3,2-diazaphospholidine, E-00030
Methyl *N*,*N'*-dimethyl-*N*,*N'*-trimethylene phosphorodiamidite, *in* H-00022

$C_6H_{15}N_2O_2P$

4,9-Dimethyl-1,6-dioxa-4,9-diaza-5-phospha(5-P^{v})spiro[4.4]nonane, D-00729
Ethyl *N*,*N'*-1,2-ethanediyl-*N*,*N'*-dimethylphosphorodiamidate, *in* E-00030

$C_6H_{15}N_2O_2PS$

O,*O*-Diethyl (1-iminoethyl)-phosphoramidothioate, *in* I-00005
O,*S*-Diethyl (1-iminoethyl)-phosphoramidothioate, *in* I-00005

$C_6H_{15}N_2O_3P$

[Bis(dimethylamino)phosphinyl]acetic acid, B-00193

$C_6H_{15}N_2O_4P$

Diethyl [(aminocarbonyl)amino]-methylphosphonate, *in* A-00058

$C_6H_{15}N_2O_6P$

Threonine ethanolamine phosphate, T-00318

$C_6H_{15}N_2P$

Hexahydro-1,3,4-trimethyl-1,3,2-diazaphosphorine, H-00057

$C_6H_{15}N_2PS$

P-Cyclohexylphosphonothioic diamide, C-00265

$C_6H_{15}N_4O_5P$

Phosphoarginine, P-00363

$C_6H_{15}N_4O_5PS$

Triethyl (azidosulfonyl)phosphorimidate, T-00439

$C_6H_{15}N_4O_6P$

Lombricine, L-00002

$C_6H_{15}N_6OP$

2,4,10-Trimethyl-1,3,4,5,7,10-hexaaza-3-phosphatricyclo[3.3.1.1^{3,7}]decane; 3-Oxide, *in* T-00517

$C_6H_{15}N_6P$

2,4,10-Trimethyl-1,3,4,5,7,10-hexaaza-3-phosphatricyclo[3.3.1.1^{3,7}]decane, T-00517

$C_6H_{15}N_6PS$

2,4,10-Trimethyl-1,3,4,5,7,10-hexaaza-3-phosphatricyclo[3.3.1.1^{3,7}]decane; 3-Sulfide, *in* T-00517

$C_6H_{15}OP$

Butyl dimethylphosphinite, *in* D-00844
2,2-Dihydro-1,2-oxaphospholane; 2,2,2-Tri-Me, *in* D-00534
Diisopropylphosphine; Oxide, *in* D-00640
Diisopropylphosphinous acid, D-00648
Dipropylphosphine; Oxide, *in* D-01204
Dipropylphosphinous acid, D-01209
▷Ethyl diethylphosphinite, *in* D-00333
Triethylphosphine oxide, T-00443

$C_6H_{15}OPS$

Diisopropylphosphinothioic acid, D-00646
Dipropylphosphinothioic acid, D-01208
O-Ethyl diethylphosphinothioate, *in* D-00330
S-Ethyl diethylphosphinothioate, *in* D-00330
O-Methyl *tert*-butylmethylphosphinothioate, *in* B-00570

$C_6H_{15}OPS_2$

O,*S*-Diethyl ethylphosphonodithioate, *in* E-00070
O-Ethyl butylphosphonodithioate, *in* B-00606

O-Ethyl hydrogen (1-methylpropyl)-phosphonodithioate, *in* M-00380
S-Ethyl *O*-isopropyl methylphosphonothioate, *in* I-00040
▷Ethylphosphonodithioic acid; *S,S*-Di-Et ester, *in* E-00140
Hexylphosphonodithioic acid, H-00099
O-Methyl hydrogen pentylphosphonodithioate, *in* P-00051

$C_6H_{15}OPS_3$

▷*O,S,S*-Triethyl phosphorotrithioate, T-00459
▷*S,S,S*-Triethyl phosphorotrithioate, T-00460

$C_6H_{15}OPSe$

Diisopropylphosphinoselenoic acid, D-00644
O-Ethyl diethylphosphinoselenoate, *in* D-00327

$C_6H_{15}O_2P$

tert-Butyl ethylphosphinate, *in* E-00124
tert-Butylethylphosphinic acid, B-00560
Diethyl ethylphosphonite, *in* E-00153
Diisopropylphosphinic acid, D-00641
Dimethyl butylphosphonite, *in* B-00615
Dimethyl *tert*-butylphosphonite, *in* B-00616
Dipropylphosphinic acid, D-01205
Ethyl butylphosphinate, *in* B-00590
Ethyl *tert*-butylphosphinate, *in* B-00591
Ethyl diethylphosphinate, *in* D-00322
Ethyl (2-methylpropyl)phosphinate, *in* M-00376
Hexyl phosphinate, H-00095
Hexylphosphonous acid, H-00101
Methyl *tert*-butylmethylphosphinate, *in* B-00567

$C_6H_{15}O_2PS$

S-Butyl *O*-methyl methylphosphonothioate, *in* M-00171
Butylphosphonothioic acid; Mono-*O*-Et ester, *in* B-00611
O,O-Diethyl ethylphosphonothioate, *in* E-00148
O,S-Diethyl ethylphosphonothioate, *in* E-00074
O,O-Diisopropyl phosphonothioate, D-00652
O,O-Dipropyl phosphonothioate, D-01212
S-Ethyl *O*-isopropyl methylphosphonothioate, *in* I-00041
O-Ethyl *O*-methyl isopropylphosphonothioate, *in* M-00166
Hexylphosphonothioic acid, H-00100
O-Methyl *tert*-butylphosphonothioate; *S*-Me ester, *in* M-00107

$C_6H_{15}O_2PS_2$

O,O-Diisopropyl phosphorodithioate, D-00663
S,S-Diisopropyl phosphorodithioate, D-00664
O,O-Dipropyl phosphorodithioate, D-01225
O,S-Dipropyl phosphorodithioate, D-01226
S,S-Dipropyl phosphorodithioate, D-01227
▷*O,S,S*-Triethyl phosphorodithioate, T-00450
O,O,S-Triethyl phosphorodithioate, T-00451

$C_6H_{15}O_2PS_3$

SS-*tert*-Butyl *O,O*-dimethyl phosphoro(dithioperoxo)thioate, B-00556
▷Thiometon, T-00313
Triethyl phosphoro(dithioperoxo)thioate, T-00453

$C_6H_{15}O_2PSe$

O,Se-Diethyl ethylphosphonoselenoate, *in* E-00145
O,O-Dipropyl phosphonoselenoate, D-01211
Ethylphosphonoselenoic acid; *O,O*-Di-Et ester, *in* E-00145

$C_6H_{15}O_2PSe_2$

O,O-Diisopropyl phosphorodiselenoate, D-00662
O,O-Dipropyl phosphorodiselenoate, D-01224

$C_6H_{15}O_3P$

tert-Butyl ethyl phosphonate, B-00561
Diethyl ethylphosphonate, D-00292
Dimethyl butylphosphonate, *in* B-00594
Dimethyl (1-methylpropyl)phosphonate, *in* M-00377
Dipropyl phosphonate, D-01210
Hexylphosphonic acid, H-00097
Monobutyl ethylphosphonate, *in* E-00129
Monoethyl butylphosphonate, *in* B-00594
Monoethyl (1,1-dimethylethyl)phosphonate, *in* B-00595
▷Triethyl phosphite, T-00445

$C_6H_{15}O_3PS$

O,O-Diisopropyl phosphorothioate, D-00667
O,O-Dipropyl phosphorothioate, D-01230
▷*O,O,S*-Triethyl phosphorothioate, T-00457
▷*O,O,O*-Triethyl phosphorothioate, T-00458

$C_6H_{15}O_3PS_2$

▷Demeton-*S*-methyl, D-00020
Triethyl phosphoro(dithioperoxoate), T-00452

$C_6H_{15}O_3PSe$

O,O-Diisopropyl phosphoroselenoate, D-00666
O,O-Dipropyl phosphoroselenoate, D-01229
O,O,O-Triethyl phosphoroselenoate, T-00454
O,O,Se-Triethyl phosphoroselenoate, T-00455

$C_6H_{15}O_3P_3$

2,4,6-Triethyl-1,3,5,2,4,6-trioxatriphosphorinane, T-00463

$C_6H_{15}O_4P$

Bis(ethoxymethyl)phosphinic acid, B-00272
OO-*tert*-Butyl *O*-methyl methyl-phosphonoperoxoate, *in* M-00310
Diethyl (1-hydroxyethyl)phosphonate, *in* H-00145
Diethyl (2-hydroxyethyl)phosphonate, *in* H-00146
Diisopropyl phosphate, D-00639
▷Diisopropyl phosphonate, D-00651
Dipropyl phosphate, D-01203
Monohexyl phosphate, M-00476
▷Triethyl phosphate, T-00440

$C_6H_{15}O_4PS$

Diethyl [(methylsulfinyl)methyl]phosphonate, *in* M-00398

$C_6H_{15}O_4PS_2$

▷Oxydemeton-methyl, O-00098

$C_6H_{15}O_5P$

[(2-Methoxyethoxy)methyl]phosphonic acid, M-00040

$C_6H_{15}O_5PS$

Diethyl [(methylsulfonyl)methyl]phosphonate, *in* M-00400

$C_6H_{15}O_5PS_2$

▷Demeton-*S*-methyl sulfone, D-00021

$C_6H_{15}O_6P$

▷Tetrakis(hydroxymethyl)phosphonium(1+); Acetate, *in* T-00181

$C_6H_{15}P$

Diisopropylphosphine, D-00640
Dipropylphosphine, D-01204
Hexylphosphine, H-00096
▷Triethylphosphine, T-00441

$C_6H_{15}PS$

2-(Diethylphosphino)ethanethiol, D-00326
Diisopropylphosphine; Sulfide, *in* D-00640
Diisopropylphosphinothious acid, D-00647
Dipropylphosphine; Sulfide, *in* D-01204
Ethyl diethylphosphinothioite, *in* D-00332
Triethylphosphine sulfide, T-00444

$C_6H_{15}PSSe$

S-Ethyl diethylphosphinoselenothioate, *in* D-00329
Se-Ethyl diethylphosphinoselenothioate, *in* D-00329

$C_6H_{15}PS_2$

Butyl dimethylphosphinodithioate, *in* D-00834
Diethyl ethylphosphonodithioite, *in* E-00141
Diisopropylphosphinodithioic acid, D-00643
Dipropylphosphinodithioic acid, D-01207
Ethyl diethylphosphinodithioate, *in* D-00325

$C_6H_{15}PS_3$

Diethyl ethylphosphonotrithioate, *in* E-00152
▷Triethyl phosphorotrithioite, T-00461

$C_6H_{15}PS_4$

Triethyl phosphorotetrathionate, T-00456

$C_6H_{15}PSe$

Triethylphosphine; Selenide, *in* T-00441

$C_6H_{15}PSe_2$

Ethyl diethylphosphinodiselenoate, *in* D-00324

$C_6H_{15}PTe$

Diisopropylphosphinotellurous acid, D-00645
Triethylphosphine; Telluride, *in* T-00441

$C_6H_{16}ClN_2OP$

P-(2-Chloroethyl)-*N,N,N',N'*-tetramethyl-phosphonic diamide, *in* C-00075
N,N'-Diisopropylphosphorodiamidic acid; Chloride, *in* D-00661
N,N'-Dipropylphosphorodiamidic acid; Chloride, *in* D-01223

$C_6H_{16}FN_2OP$

N,N'-Dipropylphosphorodiamidic acid; Fluoride, *in* D-01223
▷Mipafox, *in* D-00661

$C_6H_{16}NOP$

N,N-Diethyl-*P,P*-dimethylphosphinic amide, *in* D-00830
P,P-Diisopropylphosphinic amide, D-00642
P,P-Dipropylphosphinic amide, D-01206
N-*tert*-Butyl-*P,P*-dimethylphosphinic amide, *in* D-00830
N,N,P-Triethylphosphinic amide, *in* E-00124
N,P,P-Triethylphosphinic amide, *in* D-00323

$C_6H_{16}NOPS$

O-Ethyl *N*-isopropyl-*P*-methyl-phosphonamidothioate, *in* I-00037
O-Isopropyl trimethylphosphonamidothioate, *in* T-00540
Triethylphosphonamidothioic acid, T-00448

$C_6H_{16}NOPS_2$

S,S-Dipropyl phosphoramidodithioate, *in* D-01227
O-Ethyl *S*-propyl methyl-phosphoramidodithioate, *in* M-00330

$C_6H_{16}NO_2P$

Diisopropylphosphoramidous acid, D-00656
Dipropylphosphoramidous acid, D-01215
Ethyl *N,N*-diethylphosphonamidate, *in* D-00337
Ethyl *N,P*-diethylphosphonamidate, *in* D-00336
Ethyl *N*-isopropyl-*P*-methylphosphonamidate, *in* I-00036
Isopropyl *N*-ethyl-*P*-methylphosphonamidate, *in* E-00088
Isopropyl trimethylphosphonamidate, *in* T-00537
Methyl *N*-*tert*-butyl-*P*-methyl-phosphonamidate, *in* B-00572
Methyl *N,N*-diethyl-*P*-methylphosphonamidate, *in* D-00296
Triethylphosphonamidic acid, T-00446

$C_6H_{16}NO_2PS$

O,O-Diethyl dimethylphosphoramidothioate, *in* D-00871
O,O-Diethyl ethylphosphoramidothioate, *in* E-00159

O,O-Dimethyl diethylphosphoramidothioate, *in* D-00356

$C_6H_{16}NO_2PSe$

O,O-Diethyl dimethylphosphoramidoselenoate, *in* D-00869
Diisopropylphosphoramidoselenoic acid, D-00655
O,O-Dimethyl diethylphosphoramidoselenoate, *in* D-00353

$C_6H_{16}NO_3P$

2-Aminoethylphosphonic acid; Di-Et ester, *in* A-00079
Diethyl (1-aminoethyl)phosphonate, *in* A-00078
Diethyl dimethylphosphoramidate, D-00286
Diethyl ethylphosphoramidate, D-00293
N,N-Diethyl-*P*-(methoxymethyl)-phosphonamidic acid, D-00295
Diisopropyl phosphoramidate, D-00653
Dimethyl butylphosphoramidate, *in* B-00621
Dimethyl *tert*-butylphosphoramidate, *in* B-00622
Dimethyl diethylphosphoramidate, *in* D-00343
Dipropyl phosphoramidate, D-01213
Dipropylphosphoramidic acid, D-01214
Ethyl hydrogen (1-amino-1-methylpropyl)-phosphonate, *in* A-00093
Hexylphosphoramidic acid, H-00102

$C_6H_{16}NO_4P$

Diethyl (2-hydroxyethyl)phosphoramidate, *in* H-00148
2-[(Hydroxymethoxyphosphinyl)oxy]-*N,N,N*-trimethylethanaminium hydroxide, inner salt, *in* C-00198

$C_6H_{16}NP$

P,P-Diethyl-*N,N*-dimethylphosphinous amide, *in* D-00334
P,P-Diisopropylphosphinous amide, D-00649
P,P,P-Triethylphosphine imide, T-00442

$C_6H_{16}NPS$

P,P-Diethyl-*N,N*-dimethylphosphinothioic amide, *in* D-00331

$C_6H_{16}NPS_2$

Triethylphosphonamidodithioic acid, T-00447

$C_6H_{16}N_3OP$

▷*P*-1-Aziridinyl-*N,N,N',N'*-tetramethyl-phosphonic diamide, *in* A-00153
1,3-Dimethyl-2-dimethylamino-1,3,2-diazaphospholidine; 2-Oxide, *in* D-00724

$C_6H_{16}N_3P$

P-1-Aziridinyl-*N,N,N',N'*-tetramethyl-phosphonous diamide, *in* A-00155
1,3-Dimethyl-2-dimethylamino-1,3,2-diazaphospholidine, D-00724

$C_6H_{16}O_2P_2$

1,2-Bis(dimethylphosphinyl)ethane, *in* B-00200

$C_6H_{16}O_3P_2$

Ethylmethylphosphinic acid; Anhydride, *in* E-00085

$C_6H_{16}O_3P_2S_2$

O-Ethyl methylphosphonothioic monoanhydride, *in* D-00925
S-Ethyl methylphosphonothioic monoanhydride, *in* D-00925

$C_6H_{16}O_4P_2$

1,6-Hexanediphosphonous acid, H-00086

$C_6H_{16}O_5P_2$

Diethyl dimethyldiphosphonate, *in* D-00751
Dimethyl diethyldiphosphonate, *in* D-00290
Dipropyldiphosphonic acid, D-01202

$C_6H_{16}O_6P_2$

1,6-Hexanediphosphonic acid, H-00085
Tetramethyl 1,2-ethanediylbisphosphonate, *in* E-00006

$C_6H_{16}O_7P_2$

1-Hydroxy-1,1-hexanediphosphonic acid, H-00153
Tetramethyl (1-hydroxyethylidene)-bisphosphonate, *in* H-00144

$C_6H_{16}P_2$

1,2-Bis(dimethylphosphino)ethane, B-00200
1,6-Diphosphinohexane, D-01177

$C_6H_{17}N_2OP$

N,N'-Dimethyl-*P*-(1,1-dimethylethyl)-phosphonic diamide, *in* B-00597
P-Ethyl-*N,N,N',N'*-tetramethylphosphonic diamide, *in* E-00130
Ethyl tetramethylphosphorodiamidite, *in* T-00227

$C_6H_{17}N_2O_2P$

N,N'-Diisopropylphosphorodiamidic acid, D-00661
N,N'-Dipropylphosphorodiamidic acid, D-01223
Ethyl *N,N'*-diethylphosphorodiamidate, *in* D-00382
Ethyl tetramethylphosphorodiamidate, *in* T-00218

$C_6H_{17}N_2O_3P$

Diethyl (2-aminoethyl)phosphoramidate, *in* A-00080

$C_6H_{17}N_2P$

N,N'-Dimethyl-*P*-(1,1-dimethylethyl)-phosphonous diamide, *in* B-00617

$C_6H_{17}N_2PS$

P-Ethyl-*N,N,N',N'*-tetramethylphosphonothioic diamide, *in* E-00149

$C_6H_{17}N_2PS_2$

Ethyl tetramethylphosphorodiamidodithioate, *in* T-00222

$C_6H_{17}N_2PSe$

Ethylphosphonoselenoic bis(dimethylamide), *in* E-00146

$C_6H_{17}O_3PS$

Diethyl [(methylthio)methyl]phosphonate, *in* M-00415

$C_6H_{17}O_3PSi_2$

4-Hydroxy-2,2,6,6-tetramethyl-1-oxa-4-phospha-2,6-disilacyclohexane 4-oxide, H-00198

$C_6H_{18}ClN_4P$

N,N,N',N',N'',N''-Hexamethylphosphorimidic triamide; *N'''*-Chloro, *in* H-00077

$C_6H_{18}Cl_2NPSi$

Bis(trimethylsilyl)phosphoramidous acid; Dichloride, *in* B-00438

$C_6H_{18}Cl_2N_6O_2P_2$

4,6-Dichloro-1,3,5,7,9,10-hexamethyl-1,3,5,7,9,10-hexaaza-4,6-diphospha(4,6-P^V)-dispiro[3.1.3.1]decane-2,8-dione, D-00170

$C_6H_{18}Cl_3N_6P_3$

2,2,4-Trichloro-4,6,6-tris(dimethylamino)-2,2,4,4,6,6-hexahydro-1,3,5-triaza-2,4,6-triphosphorine, T-00421
2,4,6-Trichloro-2,4,6-tris(dimethylamino)-2,2,4,4,6,6-hexahydro-1,3,5-triaza-2,4,6-triphosphorine, T-00422

$C_6H_{18}F_2NPSi$

Bis(trimethylsilyl)phosphoramidous acid; Difluoride, *in* B-00438

$C_6H_{18}F_2N_3P$

Tris(dimethylamino)difluorophosphorane, T-00760

$C_6H_{18}F_6N_6P_2$

Azidotris(dimethylamino)phosphorus(1+); Hexafluorophosphate, *in* A-00149

$C_6H_{18}NO_3PSi$

Trimethyl (trimethylsilyl)phosphorimidate, *in* T-00557

$C_6H_{18}NO_5PS$

Triethyl methylsulfonylphosphorimidate, *in* M-00402

$C_6H_{18}NO_9P_3$

Triethylamine-2,2',2''-triphosphonic acid, T-00438

$C_6H_{18}NPSi$

P,P,P-Trimethyl-*N*-trimethylsilylphosphine imide, *in* T-00531

$C_6H_{18}N_2O_2P_2S_2$

O,O'-Diethyl hydrazobismethyl-phosphinothioate, *in* H-00105

$C_6H_{18}N_2O_3P_2$

▷*P,P'*-Dimethyl-*N,N,N',N'*-tetramethyl-diphosphonic diamide, *in* D-00751

$C_6H_{18}N_3OP$

▷Hexamethylphosphoric triamide, H-00075
N,N',N''-Triethylphosphoric triamide, *in* T-00462

$C_6H_{18}N_3P$

Hexamethylphosphonimidic diamide, *in* P-00023
▷Hexamethylphosphorous triamide, H-00080
N,N',N''-Triethylphosphorous triamide, T-00462

$C_6H_{18}N_3PS$

▷Hexamethylphosphorothioic triamide, H-00079
N,N',N''-Triethylphosphorothioic triamide, *in* T-00462

$C_6H_{18}N_3PSe$

Hexamethylphosphoroselenoic triamide, H-00078

$C_6H_{18}N_3P_3$

2,2,4,4,6,6-Hexahydro-2,2,4,4,6,6-hexamethyl-1,3,5-triaza-2,4,6-triphosphorine, H-00028

$C_6H_{18}N_6OP_4S_3$

2,4,6,8,9,10-Hexamethyl-2,4,6,8,9,10-hexaaza-1,3,5,7-tetraphosphatricyclo[$3.3.1.1^{3,7}$]-decane; 1-Oxide, 3,5,7-Trisulfide, *in* H-00071

$C_6H_{18}N_6O_2P_2$

2,3,5,6,7,8-Hexamethyl-2,3,5,6,7,8-hexaaza-1,4-diphosphabicyclo[2.2.2]octane; 1,4-Dioxide, *in* H-00070

$C_6H_{18}N_6O_2P_4S_2$

2,4,6,8,9,10-Hexamethyl-2,4,6,8,9,10-hexaaza-1,3,5,7-tetraphosphatricyclo[$3.3.1.1^{3,7}$]-decane; 1,3-Dioxide, 5,7-Disulfide, *in* H-00071

$C_6H_{18}N_6O_3PS$

2,4,6,8,9,10-Hexamethyl-2,4,6,8,9,10-hexaaza-1,3,5,7-tetraphosphatricyclo[$3.3.1.1^{3,7}$]-decane; 1,3,5-Trioxide, 7-Sulfide, *in* H-00071

$C_6H_{18}N_6O_4P_4$

2,4,6,8,9,10-Hexamethyl-2,4,6,8,9,10-hexaaza-1,3,5,7-tetraphosphatricyclo[$3.3.1.1^{3,7}$]-decane; 1,3,5,7-Tetroxide, *in* H-00071

$C_6H_{18}N_6P^{\oplus}$

Azidotris(dimethylamino)phosphorus(1+), A-00149

$C_6H_{18}N_6P_2$

2,3,5,6,7,8-Hexamethyl-2,3,5,6,7,8-hexaaza-1,4-diphosphabicyclo[2.2.2]octane, H-00070

$C_6H_{18}N_6P_2S_2$

2,3,5,6,7,8-Hexamethyl-2,3,5,6,7,8-hexaaza-1,4-diphosphabicyclo[2.2.2]octane; 1,4-Disulfide, *in* H-00070

$C_6H_{18}N_6P_2Se_2$

2,3,5,6,7,8-Hexamethyl-2,3,5,6,7,8-hexaaza-1,4-diphosphabicyclo[2.2.2]octane; 1,4-Diselenide, *in* H-00070

$C_6H_{18}N_6P_4$

2,4,6,8,9,10-Hexamethyl-2,4,6,8,9,10-hexaaza-1,3,5,7-tetraphosphatricyclo[$3.3.1.1^{3,7}$]-decane, H-00071

$C_6H_{18}N_6P_4S$

2,4,6,8,9,10-Hexamethyl-2,4,6,8,9,10-hexaaza-1,3,5,7-tetraphosphatricyclo[$3.3.1.1^{3,7}$]-decane; 1-Sulfide, *in* H-00071

$C_6H_{18}N_6P_4S_2$
2,4,6,8,9,10-Hexamethyl-2,4,6,8,9,10-hexaaza-1,3,5,7-tetraphosphatricyclo[3.3.1.1^{3,7}]-decane; 1,3-Disulfide, *in* H-00071

$C_6H_{18}N_6P_4S_3$
2,4,6,8,9,10-Hexamethyl-2,4,6,8,9,10-hexaaza-1,3,5,7-tetraphosphatricyclo[3.3.1.1^{3,7}]-decane; 1,3,5-Trisulfide, *in* H-00071

$C_6H_{18}N_6P_4S_4$
2,4,6,8,9,10-Hexamethyl-2,4,6,8,9,10-hexaaza-1,3,5,7-tetraphosphatricyclo[3.3.1.1^{3,7}]-decane; 1,3,5,7-Tetrasulfide, *in* H-00071

$C_6H_{18}O_2PSe$
O,O-Neopentylene methylphosphonoselenoate, *in* T-00512

$C_6H_{19}N_4P$
N,N,N′,N′,N″,N″-Hexamethylphosphorimidic triamide, H-00077

$C_6H_{19}O_2PSSi_2$
O,O-Bis(trimethylsilyl) phosphonothioate, B-00437

$C_6H_{19}O_2PSe$
Bis(trimethylsilyl) selenophosphonate, B-00440

$C_6H_{19}O_2PSi$
▷Bis(trimethylsilyl) phosphonite, B-00436

$C_6H_{19}O_2PTe$
Bis(trimethylsilyl) tellurophosphonate, B-00441

$C_6H_{19}O_3PSSi$
O,O-Bis(trimethylsilyl) phosphorothioate, B-00439

$C_6H_{19}O_3PSi_2$
Bis(trimethylsilyl) phosphonate, B-00435

$C_6H_{20}BrN_4P$
N,N,N′,N′,N″,N″-Hexamethyl-tetraaminophosphonium, *in* H-00077

$C_6H_{20}ClN_4P$
N,N,N′,N′,N″,N″-Hexamethyl-tetraaminophosphonium chloride, *in* H-00077

$C_6H_{20}NO_2PSi$
Bis(trimethylsilyl)phosphoramidous acid, B-00438

$C_6H_{20}N_2O_{12}P_4$
[1,2-Ethanediylbis[nitrilobis[methylene]]]-tetrakisphosphonic acid, E-00014

$C_6H_{21}N_6OP$
2,2,2′,2′,2″,2″-Hexamethylphosphoric trihydrazide, H-00076

$C_6H_{24}N_9P_3$
2,2,4,4,6,6-Hexahydro-2,2,4,4,6,6-hexakis(methylamino)-1,3,5-triaza-2,4,6-triphosphorine, H-00027

$C_7F_{15}P_3$
▷2,3-Dihydro-1,2,3,4,5-pentakis(trifluoromethyl)-1*H*-1,2,3-triphosphole, D-00536

$C_7H_4ClO_3P$
2-Chloro-4*H*-1,3,2-benzodioxaphosphorinan-4-one, C-00022

$C_7H_4ClO_4P$
2-Chloro-4*H*-1,3,2-benzodioxaphosphorinan-4-one; 2-Oxide, *in* C-00022

$C_7H_4Cl_2F_3OP$
(4-Trifluoromethylphenyl)phosphonic acid; Dichloride, *in* T-00473

$C_7H_4Cl_3O_2P$
(4-Chlorocarbonyl)phenylphosphonic dichloride, *in* P-00380
(3-Chloroformylphenyl)phosphonic dichloride, *in* P-00379

$C_7H_4Cl_3O_3P$
2-(Chlorocarbonyl)phenyl phosphorodichloridate, C-00027
3-(Chlorocarbonyl)phenyl phosphorodichloridate, C-00028
4-(Chlorocarbonyl)phenyl phosphorodichloridate, C-00029

$C_7H_4Cl_5OP$
(4-Trichloromethylphenyl)phosphonic acid; Dichloride, *in* T-00409

$C_7H_4F_5OP$
(4-Trifluoromethylphenyl)phosphonic acid; Difluoride, *in* T-00473

C_7H_4PS
1,4-Dimethylphosphorinane; Sulfide, *in* D-00883

$C_7H_5ClNO_2P$
▷2-Chloro-5-phenyl-1,3,4,2-dioxazaphosphole, C-00126

$C_7H_5ClNO_3P$
2-Chloro-5-phenyl-1,3,4,2-dioxazaphosphole; 2-Oxide, *in* C-00126
Phenyl phosphorochloridisocyanatidate, P-00270

$C_7H_5Cl_2N_2OP$
2,2-Dichloro-2,2-dihydro-3-phenyl-Δ^1-1,3,2-diazaphosphetin-4-one, *in* D-00557

$C_7H_5Cl_2O_3P$
Benzoyl phosphorodichloridate, *in* M-00442

$C_7H_5Cl_3NOP$
Benzoylphosphorimidic trichloride, B-00040

$C_7H_5Cl_3NO_2P$
2,2,2-Trichloro-2,2-dihydro-5-phenyl-1,3,4,2-dioxazaphosphole, *in* C-00126

$C_7H_5Cl_3NO_4P$
(2-Chloromethyl-4-nitrophenyl) phosphorodichloridate, *in* M-00455

$C_7H_5Cl_4OP$
Phenyl(trichloromethyl)phosphinic acid; Chloride, *in* P-00327

$C_7H_5F_2O_3P$
Benzoyl phosphorodifluoridate, *in* M-00442

$C_7H_5F_6O_3PS_2$
4,5-Bis(trifluoromethyl)spiro[1,3,2-dithiaphosphole-2,1′-[2,6,7]trioxa[1]-phosphabicyclo[2.2.1]heptane, B-00423

C_7H_5P
(Phenylmethylidyne)phosphine, P-00170

C_7H_5PS
1,3-Benzothiaphosphole, B-00025

$C_7H_6Cl_2NO_2P$
Benzoylphosphoramidic dichloride, B-00037

$C_7H_6Cl_3N_2OP$
2,2,2,3-Tetrahydro-5-phenyl-1,3,4,2-oxadiazaphosphole; 2,2,2-Trichloro, *in* T-00131

$C_7H_6Cl_3O_2P$
Phenyl(trichloromethyl)phosphinic acid, P-00327

$C_7H_6Cl_3O_3P$
(4-Trichloromethylphenyl)phosphonic acid, T-00409

$C_7H_6Cl_4NOP$
N-Phenyl-*P*-trichloromethylphosphonamidic acid; Chloride, *in* P-00328

$C_7H_6Cl_4NO_3P$
Dimethyl (2,3,5,6-tetrachloro-4-pyridinyl)-phosphonate, *in* T-00035

$C_7H_6F_3N_2OP$
2,2,2,3-Tetrahydro-5-phenyl-1,3,4,2-oxadiazaphosphole; 2,2,2-Trifluoro, *in* T-00131

$C_7H_6F_3O_2P$
Phenyl(trifluoromethyl)phosphinic acid, P-00329

$C_7H_6F_3O_3P$
(4-Trifluoromethylphenyl)phosphonic acid, T-00473

$C_7H_6NOPS_2$
Phenylphosphonisothiocyanatidothioic acid, P-00227

$C_7H_6NO_2P$
(4-Cyanophenyl)phosphinic acid, C-00222

$C_7H_6NO_3P$
(2-Cyanophenyl)phosphonic acid, C-00223
(3-Cyanophenyl)phosphonic acid, C-00224
(4-Cyanophenyl)phosphonic acid, C-00225

$C_7H_6NO_4P$
6-Benzoxazolephosphonic acid, B-00027
Mono(2-cyanophenyl) phosphate, M-00460

C_7H_6NP
1*H*-1,3-Benzazaphosphole, B-00005

$C_7H_6O_3PS_2$
1,3-Benzodithiol-2-ylphosphonic acid, B-00020

$C_7H_7Br_2P$
Benzyldibromophosphine, *in* B-00068
Dibromo-*m*-tolylphosphine, *in* M-00247
Dibromo-*o*-tolylphosphine, *in* M-00246
Dibromo-*p*-tolylphosphine, *in* M-00248

$C_7H_7Br_2PS$
Benzylphosphonothioic acid; Dibromide, *in* B-00067

$C_7H_7Br_3NO_2PS$
P,P,P-Tribromo-*N*-*p*-toluenesulfonylphosphazene, *in* M-00268

C_7H_7ClFOP
(4-Chlorophenyl)methylphosphinic acid; Fluoride, *in* C-00131

$C_7H_7ClNO_3P$
(2-Nitrophenyl)methylphosphinic acid; Chloride, *in* N-00042

$C_7H_7ClNO_6P$
Mono(2-chloromethyl-4-nitrophenyl) phosphate, M-00455

C_7H_7ClNPS
2-Chloro-2,3-dihydro-3-methyl-1,3,2-benzothiazaphosphole, C-00043

$C_7H_7ClNPS_2$
2-Chloro-2,3-dihydro-3-methyl-1,3,2-benzothiazaphosphole; 2-Sulfide, *in* C-00043

$C_7H_7Cl_2NO_5P$
Hydrogen 4-nitrophenyl (dichloromethyl)-phosphonate, *in* D-00175

$C_7H_7Cl_2N_2O_2P$
(Phenylamino)carbonylphosphoramidic acid; Dichloride, *in* P-00081

$C_7H_7Cl_2OP$
Benzylphosphonic dichloride, B-00062
Benzyl phosphorodichloridite, B-00070
(Chloromethyl)phenylphosphinic acid; Chloride, *in* C-00103
(4-Chlorophenyl)methylphosphinic acid; Chloride, *in* C-00131
(2-Methylphenyl)phosphonic acid; Dichloride, *in* M-00235
(3-Methylphenyl)phosphonic acid; Dichloride, *in* M-00236
(4-Methylphenyl)phosphonic dichloride, M-00239
2-Methylphenyl phosphorodichloridite, M-00258
3-Methylphenyl phosphorodichloridite, M-00259
4-Methylphenyl phosphorodichloridite, M-00260

$C_7H_7Cl_2OPS$
O-(2-Methylphenyl) phosphorodichloridothioate, M-00261
O-(4-Methylphenyl) phosphorodichloridothioate, M-00262

$C_7H_7Cl_2O_2P$
(4-Methoxyphenyl)phosphonic dichloride, M-00073
2-Methylphenyl phosphorodichloridate, *in* M-00485
3-Methylphenyl phosphorodichloridate, *in* M-00486
4-Methylphenyl phosphorodichloridate, M-00257

Phenyl (chloromethyl)phosphonochloridate, *in* C-00111

$C_7H_7Cl_2O_2PS$
O-(4-Methoxyphenyl) phosphorodichloridothioate, M-00082

$C_7H_7Cl_2O_3P$
2-Methoxyphenyl phosphorodichloridate, *in* M-00480
3-Methoxyphenyl phosphorodichloridate, M-00080
4-Methoxyphenyl phosphorodichloridate, M-00081

$C_7H_7Cl_2P$
Benzyldichlorophosphine, *in* B-00068
Dichloro-*m*-tolylphosphine, *in* M-00247
Dichloro-*o*-tolylphosphine, *in* M-00246
(4-Methylphenyl)phosphonous dichloride, M-00249

$C_7H_7Cl_2PS$
Benzylphosphonothioic acid; Dichloride, *in* B-00067

$C_7H_7Cl_3NO_2P$
Phenyl *P*-trichloromethylphosphonamidate, *in* T-00410
N-Phenyl-*P*-trichloromethylphosphonamidic acid, P-00328

$C_7H_7Cl_3NO_2PS$
P,P,P-Trichloro-*N*-*p*-toluenesulfonylphosphazene, *in* M-00268

$C_7H_7Cl_3NO_3PS$
▷Chlorpyrifos Methyl, C-00195

$C_7H_7Cl_3NO_4P$
▷Fospirate, F-00063

$C_7H_7Cl_4NO_2P_2$
Phenyl *P*-chloromethyl-*N*-(dichlorophosphinyl)-phosphonochloridimidate, *in* C-00091

$C_7H_7F_2OP$
(3-Methylphenyl)phosphonic acid; Difluoride, *in* M-00236
(4-Methylphenyl)phosphonic acid; Difluoride, *in* M-00237

$C_7H_7F_2O_2P$
2-Methylphenyl phosphorodifluoridate, *in* M-00485
4-Methylphenyl phosphorodifluoridate, *in* M-00487

$C_7H_7F_2O_3P$
4-Methoxyphenyl phosphorodifluoridate, *in* M-00481

$C_7H_7F_2P$
Benzyldifluorophosphine, *in* B-00068
Difluoro-*p*-tolylphosphine, *in* M-00248

$C_7H_7I_2P$
Diiodo-*p*-tolylphosphine, *in* M-00248

$C_7H_7N_2OP$
2,2-Dihydro-3-phenyl-Δ^1-1,3,2-diazaphosphetin-4-one, D-00557

$C_7H_7N_2O_3P$
5-Benzimidazolephosphonic acid, B-00016
(Diazophenylmethyl)phosphonic acid, D-00037

C_7H_7OPS
1,3,2-Benzoxathiaphosphole; 2-Me, *in* B-00026

$C_7H_7O_2P$
2-Methyl-1,3,2-benzodioxaphosphole, M-00090

$C_7H_7O_2PS$
O,O-Diethyl propylphosphonothioate, *in* P-00480
O,O-*o*-Phenylene thiophosphonate, *in* M-00090

$C_7H_7O_2PS_2$
2-Mercapto-1,3,2-benzodioxaphosphole 2-sulfide; 2-Methylthio (Me ester), *in* M-00005
2-Mercapto-4*H*-1,3,2-benzodioxaphosphorin 2-sulfide, M-00006

$C_7H_7O_2PSe$
O,O-*o*-Phenylene selenophosphonate, *in* M-00090

$C_7H_7O_3P$
1,3-Dihydro-1-hydroxy-2,1-benzoxaphosphole 1-oxide, D-00490
2-Hydroxy-2,3-dihydro-1,2-benzoxaphosphole 2-oxide, H-00119
2-Methoxy-1,3,2-benzodioxaphosphole, M-00026
o-Phenylene methylphosphonate, *in* M-00090

$C_7H_7O_3PS$
2-Hydroxy-4*H*-1,3,2-benzodioxaphosphorin 2-sulfide, H-00110

$C_7H_7O_4P$
Benzoylphosphonic acid, B-00034
2-Hydroxy-1,3,2-benzodioxaphosphole 2-oxide; 2-Methoxy (Me ester), *in* H-00108
2-Hydroxy-4*H*-1,3,2-benzodioxaphosphorin 2-oxide, H-00109
Methyl *o*-phenylene phosphate, *in* M-00026

$C_7H_7O_5P$
(1,3-Benzodioxol-5-yl)phosphonic acid, B-00019
Monobenzoyl phosphate, M-00442
2-Phosphonobenzoic acid, P-00378
3-Phosphonobenzoic acid, P-00379
4-Phosphonobenzoic acid, P-00380

$C_7H_7O_6P$
▷2-(Phosphonooxy)benzoic acid, P-00391
4-(Phosphonoxy)benzoic acid, P-00397

$C_7H_8BrO_2P$
Methyl 3-bromophenylphosphinate, *in* B-00488

$C_7H_8BrO_3P$
(Bromophenylmethyl)phosphonic acid, B-00482
[(2-Bromophenyl)methyl]phosphonic acid, B-00483
[(4-Bromophenyl)methyl]phosphonic acid, B-00484

C_7H_8BrP
Bromomethylphenylphosphine, *in* M-00227

C_7H_8BrPS
Methylphenylphosphinothioic acid; Bromide, *in* M-00225

$C_7H_8BrPS_2$
Methyl phenylphosphonobromidodithioate, *in* P-00228

C_7H_8ClOP
Methylphenylphosphinic acid; Chloride, *in* M-00219

C_7H_8ClOPS
Benzylphosphonochloridothioic acid, B-00064
O-Methyl phenylphosphonochloridothioate, *in* P-00232
S-Methyl phenylphosphonochloridothioate, *in* P-00232
O-Phenyl methylphosphonochloridothioate, *in* M-00298
S-Phenyl methylphosphonochloridothioate, *in* M-00298

$C_7H_8ClO_2P$
Benzylphosphonochloridic acid, B-00063
(Chloromethyl)phenylphosphinic acid, C-00103
(4-Chlorophenyl)methylphosphinic acid, C-00131
Methyl (3-chlorophenyl)phosphinate, *in* C-00143
Methyl phenylphosphonochloridate, *in* P-00229
Phenyl methylphosphonochloridate, *in* M-00296

$C_7H_8ClO_3P$
[(2-Chlorophenyl)methyl]phosphonic acid, C-00132
[(3-Chlorophenyl)methyl]phosphonic acid, C-00133
[(4-Chlorophenyl)methyl]phosphonic acid, C-00134
Methyl phenyl phosphorochloridate, *in* P-00269

C_7H_8ClP
Methylphenylphosphinous chloride, M-00228

C_7H_8ClPS
Methylphenylphosphinothioic acid; Chloride, *in* M-00225
Phenyl methylphosphonochloridothioite, *in* M-00299

$C_7H_8ClPS_2$
Phenyl methylphosphonochloridodithioate, *in* M-00297

C_7H_8ClPSe
Methyl phenylphosphonochloridoselenoite, *in* P-00231

$C_7H_8Cl_2NOP$
(2-Methylphenyl)phosphoramidic acid; Dichloride, *in* M-00251
(4-Methylphenyl)phosphoramidic acid; Dichloride, *in* M-00253

$C_7H_8Cl_2NO_2P$
(4-Methoxyphenyl)phosphoramidic acid; Dichloride, *in* M-00076

$C_7H_8Cl_2O_3P$
Monophenyl (dichloromethyl)phosphonate, *in* D-00175

C_7H_8FOP
Methylphenylphosphinic acid; Fluoride, *in* M-00219
Phenyl methylphosphonofluoridite, *in* M-00307

$C_7H_8FO_2P$
Benzylphosphonofluoridic acid, B-00066
(Fluoromethyl)phenylphosphinic acid, F-00022
Methyl phenylphosphonofluoridate, *in* P-00239

$C_7H_8FO_3P$
[(2-Fluorophenyl)methyl]phosphonic acid, F-00029
[(3-Fluorophenyl)methyl]phosphonic acid, F-00030
[(4-Fluorophenyl)methyl]phosphonic acid, F-00031

C_7H_8FPS
Methylphenylphosphinothioic acid; Fluoride, *in* M-00225

$C_7H_8F_2NOP$
(2-Methylphenyl)phosphoramidic acid; Difluoride, *in* M-00251
(3-Methylphenyl)phosphoramidic acid; Difluoride, *in* M-00252
(4-Methylphenyl)phosphoramidic acid; Difluoride, *in* M-00253

$C_7H_8F_2NOPS$
(2-Methoxyphenyl)phosphoramidothioic acid; Difluoride, *in* M-00077

$C_7H_8F_2NPS$
(3-Methylphenyl)phosphoramidothioic acid; Difluoride, *in* M-00255

$C_7H_8F_2O_2P$
(2-Methoxyphenyl)phosphoramidic acid; Difluoride, *in* M-00074

$C_7H_8IO_2P$
(Iodomethyl)phenylphosphinic acid, I-00012

$C_7H_8IO_3P$
1,3-Dihydro-1-hydroxy-3-methyl-1,2,3-benzodoxaphosphole 3-oxide, *in* D-00488
[(2-Iodophenyl)methyl]phosphonic acid, I-00015

$C_7H_8IO_5P$
1-Hydroxy-3-methoxy-1*H*-1,2,4,3-benziodadioxaphosphorine 3-oxide, *in* D-00616

$C_7H_8NO_3PS$
Benzoylphosphoramidothioic acid, B-00038

$C_7H_8NO_4P$
Benzoylphosphoramidic acid, B-00036
Benzoylphosphorimidic acid, B-00039
(2-Nitrophenyl)methylphosphinic acid, N-00042
(4-Nitrophenyl)methylphosphinic acid, N-00043
4-Phosphonobenzamide, *in* P-00380

$C_7H_8NO_5P$
Methyl hydrogen (2-nitrophenyl)phosphonate, *in* N-00055
Methyl hydrogen (4-nitrophenyl)phosphonate, *in* N-00057
[(2-Nitrophenyl)methyl]phosphonic acid, N-00044
[(3-Nitrophenyl)methyl]phosphonic acid, N-00045
[(4-Nitrophenyl)methyl]phosphonic acid, N-00046
4-(Phosphonoamino)benzoic acid, P-00377

C_7H_8NPS
2,3-Dihydro-2-methyl-1,3,2-benzothiazaphosphole, D-00519

C_7H_8NPSSe
2,3-Dihydro-2-methyl-1,3,2-benzothiazaphosphole; 2-Selenide, *in* D-00519

$C_7H_8NPS_2$
2,3-Dihydro-2-methyl-1,3,2-benzothiazaphosphole; 2-Sulfide, *in* D-00519

$C_7H_8N_3OP$
Methylphenylphosphinic acid; Azide, *in* M-00219

$C_7H_8N_3O_2P$
Methyl phenylphosphonazidate, *in* P-00220
Phenyl methylphosphonazidate, *in* M-00286

$C_7H_9ClNOPS$
O-Phenyl methylphosphoramidochloridothioate, *in* M-00329

$C_7H_9ClNO_2P$
Phenyl *P*-chloromethylphosphonamidate, *in* C-00105

$C_7H_9FN_3O_2P$
▷Flurofamide, F-00051

$C_7H_9F_6O_3PS_2$
2,2-Dihydro-2,2,2-trimethoxy-4,5-bis(trifluoromethyl)-1,3,2-dithiaphosphole, D-00592

$C_7H_9N_2OP$
2,2,2,3-Tetrahydro-5-phenyl-1,3,4,2-oxadiazaphosphole, T-00131

$C_7H_9N_2O_2P$
P-Benzoylphosphonic diamide, B-00035

$C_7H_9N_2O_3P$
N-(1,2-Dimethylethenylenedioxyphosphoryl)-imidazole, D-00753

$C_7H_9N_2O_3PS$
[(Phenylamino)carbonyl]phosphoramidothioic acid, P-00082

$C_7H_9N_2O_4P$
(Phenylamino)carbonylphosphoramidic acid, P-00081

$C_7H_9N_4OP$
P-Methyl-*N*-phenylphosphonazidic amide, *in* M-00287

C_7H_9OP
(4-Methoxyphenyl)phosphine, M-00067
Methylphenylphosphine; Oxide, *in* M-00215
Methylphenylphosphinous acid, M-00227
2,2,2,3-Tetrahydro-1,2-benzoxaphosphole, T-00088

C_7H_9OPS
Methylphenylphosphinothioic acid, M-00225

$C_7H_9OPS_2$
Benzylphosphonodithioic acid, B-00065
O-Phenyl methylphosphonodithioate, P-00172

$C_7H_9O_2P$
Benzyl phosphinate, B-00057
Benzylphosphinic acid, B-00059
Benzylphosphonous acid, B-00068
Methyl phenylphosphinate, *in* P-00210
Methylphenylphosphinic acid, M-00219
(2-Methylphenyl)phosphinic acid, M-00220
(3-Methylphenyl)phosphinic acid, M-00221
▷(4-Methylphenyl)phosphinic acid, M-00222
(2-Methylphenyl)phosphonous acid, M-00246
(3-Methylphenyl)phosphonous acid, M-00247
(4-Methylphenyl)phosphonous acid, M-00248
Phenyl methylphosphinate, *in* M-00278

$C_7H_9O_2PS$
Benzylphosphonothioic acid, B-00067
O-Methyl phenylphosphonothioate, M-00243
S-Methyl phenylphosphonothioate, M-00244

$C_7H_9O_2PS_2$
O-Methyl *O*-phenyl phosphorodithioate, M-00263

$C_7H_9O_3P$
Benzylphosphonic acid, B-00060
(Hydroxymethyl)phenylphosphinic acid, H-00160
(2-Methoxyphenyl)phosphinic acid, M-00068
(4-Methoxyphenyl)phosphinic acid, M-00069
Methyl phenyl phosphonate, M-00234
(2-Methylphenyl)phosphonic acid, M-00235
(3-Methylphenyl)phosphonic acid, M-00236
▷(4-Methylphenyl)phosphonic acid, M-00237
Monobenzyl phosphonate, M-00444
Monomethyl phenylphosphonate, M-00488
Phenyl methylphosphonate, P-00171

$C_7H_9O_3PS$
S-Benzyl phosphorothioate, B-00071
(4-Methylthiophenyl)phosphonic acid, M-00418
[(Phenylthio)methyl]phosphonic acid, P-00326

$C_7H_9O_4P$
(Hydroxyphenylmethyl)phosphonic acid, H-00174
[(2-Hydroxyphenyl)methyl]phosphonic acid, H-00175
[(4-Hydroxyphenyl)methyl]phosphonic acid, H-00176
(2-Methoxyphenyl)phosphonic acid, M-00070
(3-Methoxyphenyl)phosphonic acid, M-00071
(4-Methoxyphenyl)phosphonic acid, M-00072
Monobenzyl phosphate, M-00443
Mono(2-methylphenyl) phosphate, M-00485
Mono(3-methylphenyl) phosphate, M-00486
Mono(4-methylphenyl) phosphate, M-00487

$C_7H_9O_4PS$
[(Phenylsulfinyl)methyl]phosphonic acid, P-00310

$C_7H_9O_5P$
Mono(2-methoxyphenyl) phosphate, M-00480
Mono(4-methoxyphenyl) phosphate, M-00481

$C_7H_9O_5PS$
[(Phenylsulfonyl)methyl]phosphonic acid, P-00311

C_7H_9P
Benzylphosphine, B-00058
2,6-Dimethylphosphorin, D-00881
Methylphenylphosphine, M-00215
(2-Methylphenyl)phosphine, M-00216
(3-Methylphenyl)phosphine, M-00217
(4-Methylphenyl)phosphine, M-00218

$C_7H_9PS_2$
Methylphenylphosphinodithioic acid, M-00224

$C_7H_{10}Cl_3O_3P$
Di-2-propenyl (trichloromethyl)phosphonate, *in* T-00411

$C_7H_{10}NOP$
P-Methyl-*P*-phenylphosphinic amide, M-00223

$C_7H_{10}NOPS$
P-Methyl-*N*-phenylphosphonamidothioic acid, M-00233

$C_7H_{10}NO_2P$
(2-Aminophenyl)methylphosphinic acid, A-00103
(4-Aminophenyl)methylphosphinic acid, A-00104
Methyl *P*-phenylphosphonamidate, *in* P-00218
N-Methyl-*P*-phenylphosphonamidic acid, M-00230
P-Methyl-*N*-phenylphosphonamidic acid, M-00231
N-Methyl-*P*-phenylphosphonimidic acid, M-00240
Phenyl *P*-methylphosphonamidate, *in* M-00283

$C_7H_{10}NO_2PS$
(2-Methylphenyl)phosphoramidothioic acid, M-00254
(3-Methylphenyl)phosphoramidothioic acid, M-00255
(4-Methylphenyl)phosphoramidothioic acid, M-00256

$C_7H_{10}NO_3P$
(Aminophenylmethyl)phosphonic acid, A-00105
[(2-Aminophenyl)methyl]phosphonic acid, A-00106
[(4-Aminophenyl)methyl]phosphonic acid, A-00107
Benzylphosphoramidic acid, B-00069
Dimethyl 2-pyridinylphosphonate, *in* P-00503
Dimethyl 4-pyridinylphosphonate, *in* P-00505
Methyl hydrogen (2-aminophenyl)phosphonate, *in* A-00109
Methyl hydrogen (4-aminophenyl)phosphonate, *in* A-00111
Methylphenylphosphoramidic acid, M-00250
(2-Methylphenyl)phosphoramidic acid, M-00251
(3-Methylphenyl)phosphoramidic acid, M-00252
(4-Methylphenyl)phosphoramidic acid, M-00253
[(Phenylamino)methyl]phosphonic acid, P-00084

$C_7H_{10}NO_3PS$
(2-Methoxyphenyl)phosphoramidothioic acid, M-00077
(3-Methoxyphenyl)phosphoramidothioic acid, M-00078
(4-Methoxyphenyl)phosphoramidothioic acid, M-00079

$C_7H_{10}NO_4P$
(2-Methoxyphenyl)phosphoramidic acid, M-00074
(3-Methoxyphenyl)phosphoramidic acid, M-00075
(4-Methoxyphenyl)phosphoramidic acid, M-00076

$C_7H_{10}NO_5PS$
[(4-Methylphenyl)sulfonyl]phosphorimidic acid, M-00268

$C_7H_{10}NPS$
Methylphenylphosphinothioic amide, M-00226

$C_7H_{10}NPS_2$
P-Methyl-*N*-phenylphosphonamidodithioic acid, M-00232

$C_7H_{10}O_7P_2$
(Hydroxyphenylmethyl)bisphosphonic acid, H-00172

$C_7H_{11}NO_6P_2$
(Aminobenzylidene)bisphosphonic acid, A-00053

$C_7H_{11}N_2OP$
P-Benzylphosphonic diamide, B-00061
P-(4-Methylphenyl)phosphonic diamide, M-00238

$C_7H_{11}N_2OPS$
P-Methyl-N^2-phenylphosphonohydrazidothioic acid, M-00242
O-Phenyl *P*-methylphosphonohydrazidothioate, *in* M-00309

$C_7H_{11}N_2O_2P$
Methyl *P*-phenylphosphonohydrazidate, *in* P-00241
P-Methyl-N^2-phenylphosphonohydrazidic acid, M-00241

$C_7H_{11}N_2O_3P$
Dimethyl 2-pyridylphosphoramidate, *in* P-00510
Dimethyl 3-pyridylphosphoramidate, *in* P-00511

$C_7H_{11}N_2PS$
P-Methyl-*N*-phenylphosphonothioic diamide, M-00245

$C_7H_{11}O_2P$
1-Methoxy-3,4-dimethyl-1*H*-phosphole 1-oxide, *in* H-00130

$C_7H_{11}O_3P$
6-Methyl-4-methylene-4*H*-1,3,2-dioxaphosphorin; 2-Ethoxy, *in* M-00167

$C_7H_{11}O_3PS$
Dimethyl (2-thienylmethyl)phosphonate, *in* T-00304
Dimethyl (3-thienylmethyl)phosphonate, *in* T-00305

$C_7H_{11}O_4P$
6-Methyl-4-methylene-4*H*-1,3,2-dioxaphosphorin; 2-Ethoxy, 2-oxide, *in* M-00167

$C_7H_{11}O_5P$
Dimethyl (2-furanylhydroxymethyl)-phosphonate, *in* F-00071

$C_7H_{11}O_9P$
2-Phosphono-1,2,4-butanetricarboxylic acid, P-00382
3-Phosphono-1,2,3-butanetricarboxylic acid, P-00383
4-Phosphono-1,2,3-butanetricarboxylic acid, P-00384

$C_7H_{11}P$
1,3,4-Trimethyl-1*H*-phosphole, T-00536

$C_7H_{11}PS$
1,3,4-Trimethyl-1*H*-phosphole; Sulfide, *in* T-00536

$C_7H_{12}ClO_2P$
2-Chlorohexahydro-4*H*-1,3,2-benzodioxaphosphorin, C-00085

$C_7H_{12}ClO_3P$
2-Chlorohexahydro-4*H*-1,3,2-benzodioxaphosphorin; 2-Oxide, *in* C-00085
Di-2-propenyl chloromethylphosphonate, *in* C-00106

$C_7H_{12}NO_3P$
2-(Diethoxyphosphinyl)propenoic acid; Nitrile, *in* D-00260
1*H*-Pyrrol-2-ylphosphonic acid; *N*-Me, Mono-Et ester, *in* P-00521

$C_7H_{13}Cl_2N_2O_2P$
Iminocyclophosphamide, I-00004

$C_7H_{13}Cl_2N_2O_3P$
▷4-Oxocyclophosphamide, O-00058

$C_7H_{13}IP$
Triethenylmethylphosphonium iodide, *in* T-00877

$C_7H_{13}N_2O_2P$
4,5-Diethoxy-2-methyl-2*H*-1,3,2-diazaphosphole, D-00235

$C_7H_{13}N_2O_2PS$
4,5-Diethoxy-2-methyl-2*H*-1,3,2-diazaphosphole; 2-Sulfide, *in* D-00235

$C_7H_{13}N_2O_3P$
(Diazocyclohexylmethyl)phosphonic acid, D-00028
4,5-Diethoxy-2-methyl-2*H*-1,3,2-diazaphosphole; 2-Oxide, *in* D-00235
Diethyl 1*H*-imidazol-1-ylphosphonate, *in* I-00003

$C_7H_{13}OP$
2,5-Dihydro-1,3,4-trimethyl-1*H*-phosphole; 1-Oxide, *in* D-00600
1-Ethoxy-2,5-dihydro-3-methyl-1*H*-phosphole, *in* D-00530
1-Phosphabicyclo[2.2.2]octane; 1-Oxide, *in* P-00346

$C_7H_{13}O_2P$
2,3-Dihydro-1-ethoxy-4-methyl-1*H*-phosphole 1-oxide, *in* D-00496
2,5-Dihydro-1-methoxy-3,4-dimethyl-1*H*-phosphole 1-oxide, *in* D-00493
2,3-Dihydro-3,3,5-trimethyl-1,2-oxaphosphole 2-oxide; 2-Me, *in* D-00596
1,4-Dioxa-5-phospha(5-P^V)spiro[4.4]non-7-ene; 5-Me, *in* D-00967
1-Ethoxy-2,3-dihydro-5-methyl-1*H*-phosphole 1-oxide, *in* D-00497
1-Ethoxy-2,5-dihydro-2-methyl-1*H*-phosphole 1-oxide, *in* D-00498
1-Ethoxy-2,5-dihydro-3-methyl-1*H*-phosphole 1-oxide, *in* D-00499
Hexahydro-2*H*-[1,2]oxaphosphorino[2,3-*b*][1,2]oxaphosphorin, H-00037
Methyl di-2-propenylphosphinate, *in* D-01193

$C_7H_{13}O_2PS$
O,O-Dimethyl (2-methyl-1,3-butadienyl)-phosphonothioate, *in* M-00097

$C_7H_{13}O_3P$
Diethyl 1,2-propadienylphosphonate, *in* P-00441
Diethyl 1-propynylphosphonate, *in* P-00494
2,5-Dihydro-5,5-dimethyl-1,2-oxaphosphole 2-oxide; 2-Ethoxy, *in* D-00464
2,5-Dihydro-2-hydroxy-1,2-oxaphosphole 2-oxide; 2-Butoxy, *in* D-00500
▷2,3-Dihydro-3,3,5-trimethyl-1,2-oxaphosphole 2-oxide; 2-Methoxy, *in* D-00596
Dimethyl 1-cyclopenten-1-ylphosphonate, *in* C-00277
Dimethyl (2-methyl-1,3-butadienyl)-phosphonate, *in* M-00093
Dimethyl (3-methyl-1,2-butadienyl)-phosphonate, *in* M-00094
Dimethyl (3-methyl-1,3-butadienyl)-phosphonate, *in* M-00095
Dimethyl 1,3-pentadienylphosphonate, *in* P-00009
Di-2-propenyl methylphosphonate, *in* M-00288
Fosmenic acid, F-00061
Hexahydro-2*H*-[1,2]oxaphosphorino[2,3-*b*][1,2]oxaphosphorin; 9-Oxide, *in* H-00037

$C_7H_{13}O_4P$
Cyclohexanecarbonylphosphonic acid, C-00239
Hexahydro-2-hydroxycyclopenta[*d*]-1,3,2-dioxaphosphorin 2-oxide; 2-Methoxy (Me ester), *in* H-00034
2-Hydroxyhexahydro-4*H*-1,3,2-benzodioxaphosphorin 2-oxide, H-00152
9-Methyl-1,4,6-trioxa-5-phospha(5-P^V)-spiro[4.4]non-7-ene; 5-Methoxy, *in* M-00428

$C_7H_{13}O_5P$
2-(Diethoxyphosphinyl)propenoic acid, D-00260
3-(Diethoxyphosphinyl)propenoic acid, D-00261
4,4-Dimethylspiro[1,3,2-dioxaphospholane-2,1′-[2,6,7]trioxa[1]phosphabicyclo[2.2.1]-heptane], D-00919
2,3-Dimethyl-1,4,6,9-tetraoxa-5-phospha(5-P^V)spiro[4.4]non-2-ene; 5-Methoxy, *in* D-00924

$C_7H_{13}O_5PS$
Methyl 3-[(dimethoxyphosphinothioyl)oxy]-2-methyl-2-propenoate, *in* D-00673

$C_7H_{13}O_6P$
2-[(Diethoxyphosphinyl)oxy]-2-propenoic acid, D-00256
Methyl (diethoxyphosphinyl)oxoacetate, *in* D-00252
▷Mevinphos, M-00440

$C_7H_{13}P$
2,5-Dihydro-1,3,4-trimethyl-1*H*-phosphole, D-00600
1-Phosphabicyclo[2.2.2]octane, P-00346

$C_7H_{13}PS$
1-Ethylthio-1,5-dihydro-3-methyl-1*H*-phosphole, *in* D-00530

$C_7H_{14}BrO_5P$
Methyl bromo(diethoxyphosphinyl)acetate, *in* B-00467

$C_7H_{14}ClN_2O_2P$
1-(2-Chloroethyl)tetrahydro-1*H*,5*H*-[1.3.2]-diazaphospholo[2,1-*b*][1.3.2]-oxazaphosphorine 9-oxide, C-00082

$C_7H_{14}ClN_2O_3P$
2-Amino-3-(2-chloroethyl)tetrahydro-4*H*-1,3,2-oxazaphosphorine 2-oxide; *N*-Ac, *in* A-00063

$C_7H_{14}ClO_2P$
5-*tert*-Butyl-2-chloro-1,3,2-dioxaphosphorinane, B-00547
Methyl cyclohexylphosphonochloridate, *in* C-00259

$C_7H_{14}ClPS$
Cyclohexylmethylphosphinothioic acid; Chloride, *in* C-00248

$C_7H_{14}Cl_3O_3P$
Diisopropyl (trichloromethyl)phosphonate, *in* T-00411

$C_7H_{14}Cl_4O_2P_2$
1,7-Heptanediylbis(phosphonic dichloride), *in* H-00002

$C_7H_{14}FOP$
Cyclohexylmethylphosphinic acid; Fluoride, *in* C-00247

$C_7H_{14}FO_4P$
Methyl (diethoxyphosphinyl)fluoroacetate, *in* D-00250

$C_7H_{14}F_3O_2P$
Diisopropyl (trifluoromethyl)phosphonite, *in* T-00482

$C_7H_{14}NOPS$
Diisopropylphosphinic acid; Isothiocyanate, *in* D-00641

$C_7H_{14}NO_2P$
N,N-Diethyl-*P*-propadienylphosphonamidic acid, D-00399
N,N-Diethyl-*P*-1-propynylphosphonamidic acid, D-00403
Diisopropylphosphinic acid; Isocyanate, *in* D-00641

$C_7H_{14}NO_2PS$
5-Methyl-1,3,2-oxathiaphosphole 2-oxide; 2-Diethylamino, *in* M-00184

$C_7H_{14}NO_3P$
3-(Diethoxyphosphinyl)propanenitrile, *in* D-00259
Diethyl (1-cyanoethyl)phosphonate, *in* C-00210
▷Diethyl (2-cyanoethyl)phosphonate, *in* C-00211
Ethyl butylphosphonisocyanatidate, *in* B-00601

$C_7H_{14}NO_3PS_2$
▷Phosfolan, P-00342

$C_7H_{14}NO_4P$
2-(Diethoxyphosphinyl)propenoic acid; Amide, *in* D-00260

$C_7H_{14}NO_5P$
Monocrotophos, M-00459

$C_7H_{14}NP$
Cyanodiisopropylphosphine, *in* D-00648
Cyanodipropylphosphine, *in* D-01209
2,5-Dihydro-1-dimethylamino-3-methyl-1*H*-phosphole, *in* D-00530

$C_7H_{14}N_3O_3P$
▷Uredepa, U-00001

$C_7H_{14}OP$
1,3-Dimethylphosphorinane; Oxide, *in* D-00882

1,4-Dimethylphosphorinane; Oxide, *in* D-00883

$C_7H_{14}O_4P_2S_2$
3,9-Dimercapto-2,4,8,10-tetraoxa-3,9-diphosphaspiro[5.5]undecane 3,9-disulfide; 3,9-Bis(methylthio) (Di-Me ester), *in* D-00670

$C_7H_{14}P$
1,3-Dimethylphosphorinane, D-00882
1,4-Dimethylphosphorinane, D-00883

$C_7H_{14}PS$
1,3-Dimethylphosphorinane; Sulfide, *in* D-00882

$C_7H_{15}ClNOPS$
O-Methyl cyclohexylphosphoramidochloridothioate, *in* C-00270

$C_7H_{15}Cl_2N_2O_2P$
▷Cyclophosphamide, C-00283
▷Ifosfamide, I-00001

$C_7H_{15}Cl_2N_2O_3P$
Aldophosphamide, A-00050
4-Hydroxycyclophosphamide, H-00115

$C_7H_{15}Cl_2N_2O_4P$
4-Hydroperoxycyclophosphamide, H-00106

$C_7H_{15}Cl_2OP$
Heptylphosphonic acid; Dichloride, *in* H-00006
Heptyl phosphorodichloridite, H-00010

$C_7H_{15}Cl_2O_2P$
Diisopropyl (dichloromethyl)phosphonite, *in* D-00181

$C_7H_{15}Cl_2P$
Dichloroheptylphosphine, *in* H-00008

$C_7H_{15}F_2OP$
Heptylphosphonic acid; Difluoride, *in* H-00006

$C_7H_{15}IN_3OP$
1,3,5-Triaza-7-phosphatricyclo[3.3.1.1^{3,7}]-decane; 7-Oxide; B,MeI, *in* T-00327

$C_7H_{15}IN_3P$
1,3,5-Triaza-7-phosphatricyclo[3.3.1.1^{3,7}]-decane; B,MeI, *in* T-00327

$C_7H_{15}IN_3PS$
1,3,5-Triaza-7-phosphatricyclo[3.3.1.1^{3,7}]-decane; 7-Sulfide; B,MeI, *in* T-00327

$C_7H_{15}IP_2$
Tetrahydro-4-methyl-1*H*,5*H*-[1,2]-diphospholo[1,2-*a*][1,2]diphospholium iodide, *in* D-01170

$C_7H_{15}N_2OP$
Isopropyl bis(1-aziridinyl)phosphinite, *in* B-00095
(2-Methyl-1,3-butadienyl)phosphonic bis(methylamide), *in* M-00096
N,N,N′,N′-Tetramethyl-*P*-1,2-propadienylphosphonic diamide, *in* P-00442
N,N,N′,N′-Tetramethyl-*P*-1-propynylphosphonic diamide, *in* P-00494

$C_7H_{15}N_2OPS$
O-Propyl bis(1-aziridinyl)phosphinothioate, *in* B-00093

$C_7H_{15}N_2O_2P$
▷Isopropyl bis(1-aziridinyl)phosphinate, *in* B-00091

$C_7H_{15}N_2O_3P$
Diethyl (1*H*-pyrazol-3-yl)phosphonate, *in* P-00501

$C_7H_{15}N_2P$
N,N′-Dimethyl-*N,N′*-trimethylene-*P*-methylphosphonous diamide, *in* H-00022

$C_7H_{15}N_4OP$
P-Imidazol-1-yl-*N,N,N′,N′*-tetramethylphosphonic diamide, *in* I-00003

$C_7H_{15}OP$
1-Oxa-5-phospha(5-*P*^{V})spiro[4.4]nonane, O-00051

$C_7H_{15}OPS$
Cyclohexylmethylphosphinothioic acid, C-00248

$C_7H_{15}OPS_2$
S,S-Diethyl 2-propenylphosphonodithioate, *in* P-00457

$C_7H_{15}O_2P$
Butyl 2-propenylphosphinate, *in* P-00452
Cyclohexylmethylphosphinic acid, C-00247
Diethyl 2-propenylphosphonite, *in* P-00463
Ethyl (2-butenyl)methylphosphinate, *in* B-00523
Ethyl pentamethylenephosphinate, *in* H-00188
Methyl cyclohexylphosphinate, *in* C-00253
5,5,6-Trimethyl-4,7-dioxa-3-phospha(3-*P*^{V})-spiro[2.4]heptane, T-00510

$C_7H_{15}O_2PS$
O,O-Diethyl 2-propenylphosphonothioate, *in* P-00461
2-Ethylthio-5,5-dimethyl-1,3,2-dioxaphosphorinane, E-00190

$C_7H_{15}O_2PS_2$
S-Ethyl *O,O*-neopentylene phosphorodithioate, *in* E-00190

$C_7H_{15}O_2PSe$
O,O-Diethyl (2-propenyl)phosphonoselenoate, *in* P-00458

$C_7H_{15}O_2PTe$
Diethyl (2-propenyl)phosphonotelluroate, *in* P-00460

$C_7H_{15}O_3P$
Cyclohexyl hydrogen methylphosphonate, *in* M-00288
Diethyl (1-methylethenyl)phosphonate, *in* M-00149
Diethyl 1-propenylphosphonate, *in* P-00453
Diethyl 2-propenylphosphonate, *in* P-00454
Dimethyl cyclopentylphosphonate, *in* C-00276
Dimethyl (3-methyl-2-butenyl)phosphonate, *in* M-00102
2-Ethoxy-5,5-dimethyl-1,3,2-dioxaphosphorinane, E-00032
2-Ethoxy-1,2-oxaphosphepane 2-oxide, *in* H-00165
2-Methoxy-4,4,5,5-tetramethyl-1,3,2-dioxaphospholane, M-00085
2-Methoxy-3,5,5-trimethyl-1,2-oxaphospholan-3-ol 2-oxide, *in* T-00519
Propyl acetylethylphosphinate, *in* A-00017

$C_7H_{15}O_3PS$
O,O-Diethyl (2-oxopropyl)phosphonothioate, *in* O-00079
O,O-Diethyl (1-oxopropyl)phosphonothioate, *in* O-00078
O-Ethyl *O,O*-neopentylene phosphorothioate, *in* E-00032
S-Ethyl *O,O*-neopentylene phosphorothioate, *in* E-00190

$C_7H_{15}O_4P$
Diethyl (3-methyl-2-oxiranyl)phosphonate, *in* M-00185
Diethyl (1-oxopropyl)phosphonate, *in* O-00074
Diethyl (2-oxopropyl)phosphonate, *in* O-00075
Diethyl (3-oxopropyl)phosphonate, *in* O-00076
Ethyl neopentylene phosphate, *in* E-00032
1-Hydroxy-2,4,6-trimethyl-1,3,5-dioxaphosphorinane 5-oxide; 5-Methoxy (Me ester), *in* H-00203
Methyl tetramethylethylene phosphate, *in* M-00085
(2-Oxoheptyl)phosphonic acid, O-00061
2,2,2,3-Tetrahydro-2,2,2-trimethoxy-5-methyl-1,2-oxaphosphole, T-00157
1,4,6,12-Tetraoxa-5-phospha(5-*P*^{V})spiro[4.7]-dodecane, T-00243

$C_7H_{15}O_4PS_2$
Methylacetophos, M-00086

$C_7H_{15}O_4PSi$
Dimethylvinylene trimethylsilyl phosphate, *in* H-00123

$C_7H_{15}O_5P$
2-(Diethoxyphosphinyl)propanoic acid, D-00258
3-(Diethoxyphosphinyl)propanoic acid, D-00259
Diethyl [(acetyloxy)methyl]phosphonate, *in* A-00007
Diethyl (methoxycarbonylmethyl)phosphonate, *in* D-00241
Diethyl propanoyl phosphate, D-00400
2,2-Dihydro-2,2,2-trimethoxy-4,5-dimethyl-1,3,2-dioxaphosphole, D-00593
Dimethyl (3-acetyloxypropyl)phosphonate, *in* A-00010
Ethyl 2-(dimethoxyphosphinyl)propanoate, *in* D-00684
Methyl 2-(dimethoxyphosphinyl)butanoate, *in* D-00679
Triethyl phosphonoformate, *in* D-00236

$C_7H_{15}O_6P$
2-[(Diethoxyphosphinyl)oxy]propanoic acid, D-00255
Ethyl carbonate-diethyl phosphate anhydride, E-00055
Ethyl 2-[(dimethoxyphosphinyl)oxy]-propanoate, *in* D-00682

$C_7H_{16}BrO_3P$
Diethyl (1-bromopropyl)phosphonate, *in* B-00500
Diethyl (3-bromopropyl)phosphonate, *in* B-00501

$C_7H_{16}BrP$
1,1-Dimethylphosphorinanium bromide, *in* M-00339

$C_7H_{16}ClOP$
tert-Butylisopropylphosphinic acid; Chloride, *in* B-00562
Hexylmethylphosphinic acid; Chloride, *in* H-00094

$C_7H_{16}ClOPS$
O-Propyl butylphosphonochloridothioate, *in* B-00604

$C_7H_{16}ClO_3P$
Diethyl (1-chloropropyl)phosphonate, *in* C-00170
Diethyl (2-chloropropyl)phosphonate, *in* C-00171
Diethyl (3-chloropropyl)phosphonate, *in* C-00172
Diisopropyl chloromethylphosphonate, *in* C-00106

$C_7H_{16}Cl_2NOP$
Heptylphosphoramidic acid; Dichloride, *in* H-00009

$C_7H_{16}FO_2P$
▷Soman, S-00005

$C_7H_{16}NOP$
3,3,4,4-Tetramethyl-1,2-azaphospholidine 2-oxide, T-00191
4,4,5,5-Tetramethyl-1,2-azaphospholidine 2-oxide, T-00192
3,3,5,5-Tetramethyl-1,2-oxaphospholidine 2-oxide, T-00208

$C_7H_{16}NOPS_2$
S,S′-1,3-Propanediyl *N,N*-diethylphosphoramidodithioate, *in* D-01248

$C_7H_{16}NO_2P$
2-Dimethylamino-5,5-dimethyl-1,3,2-dioxaphosphorinane, D-00693

$C_7H_{16}NO_2PS$
O,O-Dimethyl 1-piperidinylphosphonothioate, *in* P-00431
Neopentylene dimethylphosphoramidothioate, *in* D-00693

$C_7H_{16}NO_2PSe$
Neopentylene dimethylphosphoramidoselenoate, *in* D-00693

$C_7H_{16}NO_3P$
Diethyl 2-propenylphosphoramidate, *in* P-00464
Dimethyl 1-piperidinylphosphonate, *in* P-00430
Ethyl *N*-(diethylphosphinyl)carbamate, *in* D-00323
Methyl *P*-acetyl-*N*,*N*-diethylphosphonamidate, *in* A-00014
Neopentylene dimethylphosphoramidate, *in* D-00693

$C_7H_{16}NO_4P$
3-(Diethoxyphosphinyl)propanoic acid; Amide, *in* D-00259
Diethyl (acetamidomethyl)phosphonate, *in* A-00012
Diethyl (1-nitroso-1-methylethyl)phosphonate, *in* N-00066
Diisopropyl aminocarbonylphosphonate, *in* C-00003

$C_7H_{16}NO_4PS_2$
▷Amidithion, A-00051

$C_7H_{16}NO_5P$
Ethyl *N*-(diethoxyphosphinyl)carbamate, *in* D-00249

$C_7H_{16}NO_6P$
Dimethyl [3-(acetylhydroxyimino)-2-hydroxypropyl]phosphonate, *in* A-00138

$C_7H_{16}N_3O_5P$
2-[(Dimethoxyphosphinyl)amino]pentanedioic diamide, *in* P-00367

$C_7H_{16}O_3P$
Diisopropyl fluoromethylphosphonate, *in* F-00023

$C_7H_{16}O_7P_2$
Tetramethyl (2-oxo-1,3-propanediyl)-bisphosphonate, *in* O-00072

$C_7H_{17}AsNP$
1-Ethyl-2-methyl-1,2,3-azaphospharsolidine; 3-Et, *in* E-00083

$C_7H_{17}ClNOP$
N,*N*-Diethyl-*P*-propylphosphonamidic acid; Chloride, *in* D-00401
Methyl diisopropylphosphoramidochloridite, *in* D-00654

$C_7H_{17}ClNO_2P$
Isopropyl diethylphosphoramidochloridate, *in* D-00346

$C_7H_{17}ClNPS$
N,*N*-Diisopropyl-*P*-methylphosphonamidothioic acid; Chloride, *in* D-00637

$C_7H_{17}Cl_2N_2O_4P$
Alcophosphamide, *in* B-00126

$C_7H_{17}FNOP$
N,*N*-Diethyl-*P*-propylphosphonamidic acid; Fluoride, *in* D-00401

$C_7H_{17}FNO_2P$
Isopropyl diethylphosphoramidofluoridate, *in* D-00351

$C_7H_{17}N_2OP$
5,5-Dimethyl-2-dimethylaminotetrahydro-2*H*-1,3,2-oxazaphosphorine, D-00726
Ethyl *N*,*N*′-dimethyl-*N*,*N*′-trimethylene phosphorodiamidite, *in* H-00022
Methyl *N*,*N*′-1,3-butylene-*N*,*N*′-dimethyl-phosphorodiamidite, *in* H-00057
N,*N*,*N*′,*N*′-Tetramethyl-*P*-2-propenylphosphonic diamide, *in* P-00455

$C_7H_{17}N_2O_2P$
Cyclohexyl *P*-methylphosphonohydrazidate, *in* M-00308
5,5-Dimethyl-2-dimethylaminotetrahydro-2*H*-1,3,2-oxazaphosphorine; 2-Oxide, *in* D-00726
2-Propenyl tetramethylphosphorodiamidate, *in* T-00218

$C_7H_{17}N_2O_3P$
Ethyl *P*-aminocarbonyl-*N*,*N*-diethylphosphonamidate, *in* A-00059

$C_7H_{17}N_2O_4P$
Diisopropyl aminocarbonylphosphoramidate, *in* A-00062

$C_7H_{17}N_2P$
N,*N*,*N*′,*N*′-Tetramethyl-*P*-2-propenylphosphonous diamide, *in* P-00463

$C_7H_{17}N_2PS$
N,*N*,*N*′,*N*′-Tetramethyl-*P*-2-propenylphosphonothioic diamide, *in* P-00462

$C_7H_{17}N_2PSe$
N,*N*,*N*′,*N*′-Tetramethyl-*P*-(2-propenyl)-phosphonoselenoic diamide, *in* P-00459

$C_7H_{17}OP$
2,2-Dihydro-2,2,2-trimethyl-1,2-oxaphosphorinane, D-00597
▷Isopropyl diethylphosphinite, *in* D-00333

$C_7H_{17}OPS$
O-Isopropyl diethylphosphinothioate, *in* D-00330
S-Isopropyl diethylphosphinothioate, *in* D-00330
O-Methyl diisopropylphosphinothioate, *in* D-00646
S-Methyl diisopropylphosphinothioate, *in* D-00646
O-Methyl dipropylphosphinothioate, *in* D-01208
S-Methyl dipropylphosphinothioate, *in* D-01208
O-Propyl butylphosphinothioate, *in* B-00593

$C_7H_{17}OPS_2$
O-Ethyl hydrogen pentylphosphonodithioate, *in* P-00051
O-Isopropyl butylphosphonodithioate, *in* B-00606
O-Isopropyl hydrogen (1-methylpropyl)-phosphonodithioate, *in* M-00380
O-Methyl hydrogen hexylphosphonodithioate, *in* H-00099

$C_7H_{17}OPSe$
O-Isopropyl diethylphosphinoselenoate, *in* D-00327

$C_7H_{17}O_2P$
Butyl isopropylphosphinate, *in* I-00049
tert-Butylisopropylphosphinic acid, B-00562
Diethyl propylphosphonite, *in* P-00482
Diisopropyl methylphosphonite, *in* M-00319
Heptylphosphonous acid, H-00008
Hexylmethylphosphinic acid, H-00094
Isopropyl butylphosphinate, *in* B-00590
Isopropyl diethylphosphinate, *in* D-00322
Methyl diisopropylphosphinate, *in* D-00641
2-Methylpropyl ethylmethylphosphinate, *in* E-00085

$C_7H_{17}O_2PS$
▷*S*-Butyl *O*-ethyl methylphosphonothioate, *in* E-00093
O,*O*-Diethyl isopropylphosphonothioate, *in* I-00054
O,*O*-Diisopropyl methylphosphonothioate, *in* M-00313
Heptylphosphonothioic acid, H-00007

$C_7H_{17}O_2PS_3$
O,*O*-Diethyl *SS*-propyl phosphoro(dithioperoxo)thioate, D-00402
▷Isothioate, I-00067
▷Phorate, P-00338

$C_7H_{17}O_3P$
Diethyl isopropylphosphonate, *in* I-00050
Diethyl propylphosphonate, *in* P-00473
Diethyl 2-propynylphosphonate, *in* P-00495
▷Diisopropyl methylphosphonate, *in* M-00288
Dimethyl (2,2-dimethylpropyl)phosphonate, *in* D-00912
Dimethyl pentylphosphonate, *in* P-00050
Dipropyl methylphosphonate, *in* M-00288
Heptylphosphonic acid, H-00006
Isopropyl phenyl methylphosphonate, *in* I-00038

$C_7H_{17}O_3PS$
Diethyl [(ethylthio)methyl]phosphonate, *in* E-00192
Diethyl [1-(methylthio)ethyl]phosphonate, *in* M-00412

$C_7H_{17}O_3PSSi$
O,*O*-2,3-Butylene *O*-trimethylsilyl phosphorothioate, *in* D-00930

$C_7H_{17}O_3PS_2$
Diethyl [bis(methylthio)methyl]phosphonate, *in* B-00364
▷*O*,*O*-Diethyl *S*-[(ethylsulfinyl)methyl] phosphorodithioate, *in* P-00338

$C_7H_{17}O_3PSi$
4,5-Dimethyl-2-trimethylsilyloxy-1,3,2-dioxaphospholane, D-00930

$C_7H_{17}O_4P$
OO-*tert*-Butyl *O*-ethyl methyl-phosphonoperoxoate, *in* M-00310
OO-*tert*-Butyl *O*-methyl ethylphosphonoperoxoate, *in* E-00144
Diethyl (1-hydroxy-2-methylethyl)phosphonate, *in* H-00159
Diethyl (1-hydroxypropyl)phosphonate, *in* H-00190
Diethyl (2-hydroxypropyl)phosphonate, *in* H-00191
Diethyl (3-hydroxypropyl)phosphonate, *in* H-00192
Monoheptyl phosphate, M-00474

$C_7H_{17}O_4PS_2$
▷*OO*-Diethyl *S*-[(ethylsulfonyl)methyl] phosphorodithioate, *in* P-00338
▷Oxydeprofos, O-00099

$C_7H_{17}O_5P$
Diethyl (2,3-dihydroxypropyl)phosphonate, *in* D-00624

$C_7H_{17}O_6P$
Glycerol 1-monophosphate; Di-Me ether, di-Me ester, *in* G-00066

$C_7H_{17}P$
Heptylphosphine, H-00005

$C_7H_{17}PS$
Isopropyl diethylphosphinothioite, *in* D-00332

$C_7H_{17}PSSe$
Se-Propyl diethylphosphinoselenothioate, *in* D-00329

$C_7H_{17}PS_2$
Ethyl ethylpropylphosphinodithioate, *in* E-00185
Isopropyl diethylphosphinodithioate, *in* D-00325
Methyl diisopropylphosphinodithioate, *in* D-00643

$C_7H_{17}PSe_2$
Propyl diethylphosphinodiselenoate, *in* D-00324

$C_7H_{18}Cl_2N_3P$
Dichloromethylenetris(dimethylamino)-phosphorane, D-00173

$C_7H_{18}IP$
Triethylmethylphosphonium iodide, *in* T-00441

$C_7H_{18}IPS_3$
Tri(ethylthio)methylphosphonium iodide, *in* T-00461

$C_7H_{18}NOPS$
N,*N*-Diisopropyl-*P*-methylphosphonamidothioic acid, D-00637
O-Methyl triethylphosphonamidothioate, *in* T-00448

$C_7H_{18}NO_2P$
[1-(Butylamino)-1-methylethyl]phosphinic acid, B-00542
N,*N*-Diethyl-*P*-propylphosphonamidic acid, D-00401
Diethyl propylphosphoramidite, *in* P-00486
N,*N*-Diisopropyl-*P*-methylphosphonamidic acid, D-00636

Diisopropyl methylphosphoramidite, *in* M-00335
Ethyl *N*-*tert*-butyl-*P*-methylphosphonamidate, *in* B-00572
Ethyl *N*,*N*-diethyl-*P*-methylphosphonamidate, *in* D-00296
Isopropyl *N*-isopropyl-*P*-methylphosphonamidate, *in* I-00036
Methyl triethylphosphonamidate, *in* T-00446

$C_7H_{18}NO_2PS$
O,*O*-Diethyl isopropylphosphoramidothioate, *in* I-00059
O,*O*-Diethyl propylphosphoramidothioate, *in* P-00485

$C_7H_{18}NO_2PSe$
O,*O*-Diisopropyl methylphosphoramidoselenoate, *in* M-00332

$C_7H_{18}NO_3P$
Diethyl (1-amino-1-methylethyl)phosphonate, *in* A-00087
Diethyl (2-amino-1-methylethyl)phosphonate, *in* A-00088
Diethyl (1-aminopropyl)phosphonate, *in* A-00125
Diethyl (2-aminopropyl)phosphonate, *in* A-00126
Diethyl (3-aminopropyl)phosphonate, *in* A-00127
Diethyl isopropylphosphoramidate, *in* I-00058
Diethyl propylphosphoramidate, *in* P-00484
Dipropyl methylphosphoramidate, *in* M-00326
Heptylphosphoramidic acid, H-00009
Triethyl methylphosphorimidate, *in* M-00336
Trimethyl *tert*-butylphosphorimidate, *in* B-00625

$C_7H_{18}NO_3PS$
O,*O*-Diethyl (2-methoxyethyl)phosphoramidothioate, *in* M-00041

$C_7H_{18}NO_4P$
2-[(Ethoxyhydroxyphosphinyl)oxy]-*N*,*N*,*N*-trimethylethanaminium hydroxide, inner salt, *in* C-00198

$C_7H_{18}NP$
N-*tert*-Butyl-*P*,*P*,*P*-trimethylphosphine imide, *in* T-00531
P,*P*-Diisopropyl-*N*-methylphosphinous amide, *in* D-00649

$C_7H_{18}NPS$
P,*P*-Diisopropyl-*N*-methylphosphinothioic amide, *in* D-00646

$C_7H_{18}N_3OP$
N,*N*,*N'*,*N'*-Tetramethyl-*N''*-(2-propenyl)phosphoric triamide, *in* T-00216
N,*N'*,*N''*,*N''*-Tetramethyl-*N*,*N'*-trimethylenephosphoric triamide, *in* H-00023

$C_7H_{18}O_2P_2$
1,3-Bis(dimethylphosphinyl)propane, *in* B-00206

$C_7H_{18}O_4P_2$
Diethyl methylenebis[methylphosphinate], *in* M-00140
Diisopropyl methylenebisphosphinate, *in* M-00143

$C_7H_{18}O_6P_2$
1,7-Heptanediphosphonic acid, H-00002

$C_7H_{18}P_2$
1,3-Bis(dimethylphosphino)propane, B-00206
Bis(trimethylphosphoranylidene)methane, B-00429

$C_7H_{18}P_2S_2$
1,3-Bis(dimethylphosphinothioyl)propane, *in* B-00206

$C_7H_{19}BrP$
Trimethyl[(trimethylphosphoranylidene)methyl]phosphonium(1+); Bromide, *in* T-00565

$C_7H_{19}ClP$
Trimethyl[(trimethylphosphoranylidene)methyl]phosphonium(1+); Chloride, *in* T-00565

$C_7H_{19}GeP$
Trimethyl(trimethylgermyl)methylenephosphorane, T-00564

$C_7H_{19}IP$
Trimethyl[(trimethylphosphoranylidene)methyl]phosphonium(1+); Iodide, *in* T-00565

$C_7H_{19}NO_5P_2S$
Trimethyl (diethoxyphosphinothioyl)phosphorimidate, *in* D-00238

$C_7H_{19}NO_6P_2$
Tetramethyl [(dimethylamino)methylene]bisphosphonate, *in* D-00701
Trimethyl (diethoxyphosphinyl)phosphorimidate, *in* D-00257

$C_7H_{19}N_2OP$
P-Isopropyl-*N*,*N*,*N'*,*N'*-tetramethylphosphonic diamide, *in* I-00051
Isopropyl tetramethylphosphorodiamidite, *in* T-00227
N,*N*,*N'*,*N'*-Tetramethyl-*P*-propylphosphonic diamide, *in* P-00474

$C_7H_{19}N_2PS$
P-Methyl-*N*,*N'*-diisopropylphosphonothioic diamide, *in* M-00314

$C_7H_{19}OPSSi$
O-Trimethylsilyl diethylphosphinothioate, *in* D-00330

$C_7H_{19}O_2PSeSi$
O-Ethyl *O*-trimethylsilyl ethylphosphonoselenoate, *in* E-00072

$C_7H_{19}O_2PSi$
Trimethylsilyl diethylphosphinate, *in* D-00322

$C_7H_{19}O_2P_3$
Bis[(dimethylphosphinyl)methyl]methylphosphine, *in* B-00204

$C_7H_{19}O_3PSi$
Diethyl trimethylsilyl phosphite, D-00408

$C_7H_{19}O_4PSi$
Diethyl trimethylsilyl phosphate, D-00407
Dimethyl [1-[(trimethylsilyl)oxy]ethyl]phosphonate, *in* T-00555

$C_7H_{19}O_5PSi_2$
Bis(trimethylsilyloxy)phosphinecarboxylic acid oxide, B-00433

$C_7H_{19}PS_2Si$
Trimethylsilyl diethylphosphinodithioate, *in* D-00325

$C_7H_{19}PSi$
Trimethyl(trimethylsilylmethylene)phosphorane, T-00566

$C_7H_{19}P_2^{\oplus}$
Trimethyl[(trimethylphosphoranylidene)methyl]phosphonium(1+), T-00565

$C_7H_{19}P_3$
Bis[(dimethylphosphino)methyl]methylphosphine, B-00204
Tris(dimethylphosphino)methane, T-00771

$C_7H_{19}P_3S_3$
Tris(dimethylphosphinothioyl)methane, *in* T-00771

$C_7H_{20}NPSSi$
P,*P*-Diethyl-*N*-trimethylsilylphosphinothioic amide, *in* D-00331

$C_7H_{20}N_3P$
N,*N*,*N'*,*N'*,*N''*,*N''*-Hexamethyl-1-methylenephosphoranetriamide, H-00072

$C_7H_{21}N_2PSi$
N,*N*,*P*,*P*-Tetramethyl-*N'*-trimethylsilylphosphinimidic amide, *in* D-00833

$C_7H_{21}N_3PS_2Si$
Trimethylsilyl tetramethylphosphorodiamidodithioate, *in* T-00222

$C_7H_{21}N_4P$
N,*N*,*N'*,*N'*,*N''*,*N''*,*N'''*-Heptamethylphosphorimidic triamide, H-00001

$C_7H_{21}O_3PSi_2$
Methylphosphonic acid; Bis(trimethylsilyl) ester, *in* M-00288

$C_7H_{21}PS_3Si_2$
Methylphosphonotrithioic acid; Bis(trimethylsilyl) ester, *in* M-00318

$C_7H_{22}IN_4P$
Tris(dimethylamino)(methylamino)phosphonium iodide, *in* H-00001

$C_7H_{22}N_3OPSi$
N,*N*,*N'*,*N'*-Tetramethylphosphoric triamide; *N''*-Trimethylsilyl, *in* T-00216

$C_8F_{12}P_2$
1,3,4,6-Tetrakis(trifluoromethyl)-2,5-diphosphatricyclo[3.1.0.0^{2,6}]hex-3-ene, T-00186

$C_8HF_{12}PS_4$
2,3,7,8-Tetrakis(trifluoromethyl)-1,4,6,9-tetrathia-5-phospha(5-P^V)spiro[4.4]nona-2,7-diene, T-00189

$C_8H_4FN_2P$
(3-Fluorophenyl)phosphonous acid; Dicyanide, *in* F-00041
(4-Fluorophenyl)phosphonous acid; Dicyanide, *in* F-00042

$C_8H_5Cl_2OP$
Phenylethynylphosphonic acid; Dichloride, *in* P-00166

$C_8H_5Cl_2PS$
(Phenylethynyl)phosphonothioic acid; Dichloride, *in* P-00167

$C_8H_5Cl_7NP$
N-Phenyl-*P*,*P*-bis(trichloromethyl)phosphinimidic chloride, *in* B-00403

$C_8H_5F_6PSe_2$
Bis(trifluoromethyl) phenylphosphonodiselenoite, *in* P-00235

$C_8H_5N_2OP$
Phenylphosphonic acid; Dicyanide, *in* P-00221

$C_8H_5N_2OPS_2$
Phenylphosphonic acid; Diisothiocyanate, *in* P-00221

$C_8H_5N_2O_2P$
Phenylphosphonous acid; Diisocyanate, *in* P-00251

$C_8H_5N_2O_2PS_2$
Phenyl phosphorodi(isothiocyanatate), P-00294

$C_8H_5N_2O_3P$
Phenylphosphonic diisocyanate, P-00226

$C_8H_5N_2O_3PS$
O-Phenyl phosphorodiisocyanatidothioate, P-00293

$C_8H_5N_2O_4P$
Phenyl phosphorodiisocyanatidate, P-00292

$C_8H_5N_2P$
▷Phenylphosphonous dicyanide, P-00255

$C_8H_5N_2PS_2$
Phenylphosphonous acid; Diisothiocyanate, *in* P-00251

$C_8H_5N_2PS_3$
Phenylphosphonothioic acid; Diisothiocyanate, *in* P-00244

$C_8H_5O_2P$
1*H*-Isophosphindole-1,3(2*H*)-dione, I-00026

$C_8H_6Cl_5O_3P$
2,2,2-Trichloroethyl 2-chlorophenyl phosphorochloridate, T-00405

$C_8H_6F_6NP$
N-Phenyl-*P*,*P*-bis(trifluoromethyl)phosphinous amide, *in* B-00419

$C_8H_6NO_4PS$
(1,3-Dihydro-1,3-dioxo-2*H*-isoindol-2-yl)-phosphonothioic acid, D-00475

$C_8H_6O_{12}P_2$
2,7,9,14,15,16,17,18-Octaoxa-1,8-diphospha(1,8-P^V)-pentacyclo[10.2.1.1^{1,4}.1^{5,8}.1^{8,11}]octadecane-3,6,10,13-tetrone, O-00036

$C_8H_7ClNO_2P$
1,2-Dihydro-1-methyl-4*H*-3,1,2-benzoxazaphosphorin-4-one; 1-Chloro, *in* D-00523

$C_8H_7Cl_2OP$
(2-Phenylethenyl)phosphonic dichloride, P-00157

$C_8H_7Cl_2O_2PS$
O-(2-Acetylphenyl) phosphorodichloridothioate, A-00022
O-(3-Acetylphenyl) phosphorodichloridothioate, A-00023
O-(4-Acetylphenyl) phosphorodichloridothioate, A-00024

$C_8H_7Cl_2P$
(2-Phenylethenyl)phosphonous dichloride, P-00162

$C_8H_7Cl_2PS$
(2-Phenylethenyl)phosphonothioic acid; Dichloride, *in* P-00160

$C_8H_7F_2OP$
(2-Phenylethenyl)phosphonic difluoride, P-00158

$C_8H_7O_2PS$
(Phenylethynyl)phosphonothioic acid, P-00167

$C_8H_7O_2PS_2$
Di-2-thienylphosphinic acid, D-01250

$C_8H_7O_3P$
2-Hydroxy-2*H*-1,2-benzoxaphosphorin 2-oxide, H-00112
Phenylethynylphosphonic acid, P-00166

$C_8H_7O_3PS$
Di-2-furanylphosphinothioic acid, D-00422

$C_8H_7O_4P$
Di-2-furanylphosphinic acid, D-00421
Phenylphosphinedicarboxylic acid, P-00209

$C_8H_7O_5P$
(1,3-Dihydro-3-oxo-1-isobenzofuranyl)-phosphonic acid, D-00535

$C_8H_8BrCl_2O_2PS$
▷Bromophos, B-00498

$C_8H_8Br_3ClNO_2P$
2,2,2-Tribromoethyl phenyl-phosphoramidochloridate, *in* P-00259

$C_8H_8ClN_2OP$
2-Chloro-2,3-dihydro-5-methyl-3-phenyl-1,3,4,2-oxadiazaphosphole, C-00044

$C_8H_8ClN_2OPS$
2-Chloro-2,3-dihydro-5-methyl-3-phenyl-1,3,4,2-oxadiazaphosphole; 2-Sulfide, *in* C-00044

$C_8H_8ClN_2O_2P$
2-Chloro-2,3-dihydro-5-methyl-3-phenyl-1,3,4,2-oxadiazaphosphole; 2-Oxide, *in* C-00044

$C_8H_8ClN_2PS_2$
2,3-Dihydro-5-methyl-3-phenyl-1,3,4,2-thiadiazaphosphole 2-sulfide; 2-Chloro, *in* D-00529

C_8H_8ClOP
Phenylvinylphosphinic acid; Chloride, *in* P-00337

$C_8H_8ClO_2P$
3-Chloro-1,5-dihydro-2,4,3-benzodioxaphosphepin, C-00038

$C_8H_8ClO_3P$
3-Chloro-1,5-dihydro-2,4,3-benzodioxaphosphepin; 3-Oxide, *in* C-00038

$C_8H_8Cl_2IO_3PS$
▷Iodofenphos, I-00011

$C_8H_8Cl_3O_2P$
Methyl phenyl(trichloromethyl)phosphinate, *in* P-00327

$C_8H_8Cl_3O_3PS$
▷Fenchlorphos, F-00003

$C_8H_8Cl_4NO_2P$
2,2,2-Trichloroethyl phenyl-phosphoramidochloridate, *in* P-00259

$C_8H_8Cl_4O_2P_2$
[1,4-Phenylenebis(methylene)]bisphosphonic acid; Tetra-chloride, *in* P-00131

$C_8H_8NOPS_2$
O-Methyl phenyl-phosphonoisothiocyanatidothioate, *in* P-00227

$C_8H_8NO_2P$
1,2-Dihydro-1-methyl-4*H*-3,1,2-benzoxazaphosphorin-4-one, D-00523

$C_8H_8NO_2PS$
Phenyl methylphosphonisothiocyanatidate, *in* M-00294

$C_8H_8NO_3P$
(Cyanophenylmethyl)phosphonic acid, C-00219
[(2-Cyanophenyl)methyl]phosphonic acid, C-00220
[(4-Cyanophenyl)methyl]phosphonic acid, C-00221
Phenyl methylphosphonisocyanatidate, *in* M-00292

C_8H_8NP
1*H*-1,3-Benzazaphosphole; 1-Me, *in* B-00005
Methylphenylphosphinous acid; Cyanide, *in* M-00227

$C_8H_8N_3P$
1-Methyl-5-phenyl-1*H*-1,2,4,3-triazaphosphole, M-00269

$C_8H_8O_3P$
Diethyl 3-buten-1-ynylphosphonate, *in* B-00529

$C_8H_8P^{\oplus}$
1,2,4,5-Tetrakis(methylene)-3-phosphoniaspiro[2.2]pentane, T-00182

C_8H_9ClNOP
2-Chloro-3-phenyl-1,3,2-oxazaphospholidine, C-00138

$C_8H_9ClNOPS$
2-Chloro-3-phenyl-1,3,2-oxazaphospholidine; 2-Sulfide, *in* C-00138

$C_8H_9ClNO_2P$
N-Benzoyl-*P*-methylphosphonamidic chloride, *in* B-00031
2-Chloro-3-phenyl-1,3,2-oxazaphospholidine; 2-Oxide, *in* C-00138

$C_8H_9ClNO_5PS$
▷Chlorthion, C-00196

C_8H_9ClOPS
O-Phenyl bis(chloromethyl)phosphinothioate, *in* B-00129

$C_8H_9Cl_2OP$
(2,5-Dimethylphenyl)phosphonic acid; Dichloride, *in* D-00813
[(4-Methylphenyl)methyl]phosphonic acid; Dichloride, *in* M-00208

$C_8H_9Cl_2O_2P$
2,6-Dimethylphenyl phosphorodichloridate, *in* M-00466
3,5-Dimethylphenyl phosphorodichloridate, D-00822
Phenyl (2-chloroethyl)phosphonochloridate, *in* C-00077
2-Phenylethyl phosphorodichloridate, *in* M-00506

$C_8H_9Cl_3NO_2P$
Methyl *N*-phenyl-*P*-trichloromethyl-phosphonamidate, *in* P-00328

$C_8H_9I_2O_2P$
Phenyl bis(iodomethyl)phosphinate, *in* B-00288

$C_8H_9N_2O_7P$
Dimethyl (2,4-dinitrophenyl)phosphonate, *in* D-00946

$C_8H_9N_2PS_2$
2,3-Dihydro-5-methyl-3-phenyl-1,3,4,2-thiadiazaphosphole 2-sulfide, D-00529

$C_8H_9N_3P$
2-Methyl-5-phenyl-2*H*-1,2,4,3-triazaphosphole, M-00270

$C_8H_9N_4O_6P$
Thalassemine, *in* L-00002

C_8H_9OP
2,3-Dihydro-1*H*-phosphindole 1-oxide, D-00570

C_8H_9OPS
O,S-Ethylene phenylphosphonothioite, *in* O-00056

$C_8H_9OPS_2$
S,S-Ethylene phenylphosphonodithioate, *in* P-00121
(2-Phenylethenyl)phosphonodithioic acid, P-00159

$C_8H_9OPS_3$
2-Phenoxy-1,3,2-dithiaphospholane 2-sulfide, *in* D-01245

C_8H_9OPSe
O-Methyl dimethylphosphinoselenoate, *in* D-00835

$C_8H_9O_2P$
2,3-Dihydro-2-hydroxy-1*H*-isophosphindole 2-oxide, D-00494
2,3-Dihydro-2-methyl-1,4,2-benzodioxaphosphorin, D-00517
1-Hydroxy-2,3-dihydro-1*H*-phosphindole 1-oxide, H-00120
2-Phenyl-1,3,2-dioxaphospholane, P-00115
(2-Phenylethenyl)phosphinic acid, P-00153
(2-Phenylethenyl)phosphonous acid, P-00161
Phenylvinylphosphinic acid, P-00337

$C_8H_9O_2PS$
O,O-Ethylene phenylphosphonothioate, *in* P-00115
2-Methylthio-4*H*-1,3,2-benzodioxaphosphorin, M-00411
2-Phenoxy-1,3,2-oxathiaphospholane, P-00072
(2-Phenylethenyl)phosphonothioic acid, P-00160
Phenyl thiodimethylenephosphinate, *in* H-00201

$C_8H_9O_2PS_2$
2-Mercapto-1,3,2-benzodioxaphosphole 2-sulfide; 2-Ethylthio (Et ester), *in* M-00005
2-Methylthio-4*H*-1,3,2-benzodioxaphosphorin; 2-Sulfide, *in* M-00411

$C_8H_9O_3P$
Acetylphenylphosphinic acid, A-00021
2,3-Dihydro-2-methoxy-1,2-benzoxaphosphole 2-oxide, *in* H-00119
2,3-Dihydro-2-methyl-1,4,2-benzodioxaphosphorin; 2-Oxide, *in* D-00517
2-Ethoxy-1,3,2-benzodioxaphosphole, E-00024
Ethylene phenylphosphonate, *in* P-00115
2-Phenoxy-1,3,2-dioxaphospholane, P-00067
(1-Phenylethenyl)phosphonic acid, P-00154
2-Phenylethenylphosphonic acid, P-00155

$C_8H_9O_3PS$
O,O-Ethylene *O*-phenyl phosphorothioate, *in* P-00067
O,S-Ethylene *O*-phenyl phosphorothioate, *in* P-00072
O-Ethyl *O,O*-*o*-phenylene thiophosphate, *in* E-00024
2-Methylthio-4*H*-1,3,2-benzodioxaphosphorin; 2-Oxide, *in* M-00411

Spiro[1,3,2-benzoxathiaphosphole-2,2′-[1,3,2]-dioxaphospholane], S-00011

$C_8H_9O_3SP$

▷Salithion, S-00001

$C_8H_9O_4P$

Ethylene phenyl phosphate, *in* P-00067
Ethyl *o*-phenylene phosphate, *in* E-00024
(2-Oxo-2-phenylethyl)phosphonic acid, O-00066
Salioxon, *in* H-00109

$C_8H_9O_5P$

α-Phosphonophenylacetic acid, P-00393

$C_8H_9O_6P$

4-Phenyl-1,2,3,5,8-pentaoxa-4-phospha(4-P^V)-spiro[3.4]octane, P-00181

C_8H_9P

2,3-Dihydro-1*H*-isophosphindole, D-00509
(2-Phenylethenyl)phosphine, P-00152
1-Phenylphosphirane, P-00214

$C_8H_9PS_2$

2,5-Dimethyl-1,3,2-benzodithiaphosphole, D-00712
2-Phenyl-1,3,2-dithiaphospholane, P-00121

$C_8H_9PS_3$

Ethylene phenylphosphonotrithioate, *in* P-00121

$C_8H_{10}BrO_2P$

Dimethyl (4-bromophenyl)phosphonite, *in* B-00493
Ethyl (4-bromophenyl)phosphinate, *in* B-00489

$C_8H_{10}BrO_3P$

Dimethyl (2-bromophenyl)phosphonate, *in* B-00490
Dimethyl (3-bromophenyl)phosphonate, *in* B-00491
Dimethyl (4-bromophenyl)phosphonate, *in* B-00492

$C_8H_{10}BrP$

Bromoethylphenylphosphine, *in* E-00108

$C_8H_{10}ClN_2P$

2,3-Dihydro-1,3-dimethyl-1*H*-1,3,2-benzodiazaphosphole; 2-Chloro, *in* D-00456

$C_8H_{10}ClOP$

Ethylphenylphosphinic acid; Chloride, *in* E-00101
Ethyl phenylphosphonochloridite, *in* P-00234
(2-Methylphenyl)methylphosphinic acid; Chloride, *in* M-00203
(4-Methylphenyl)methylphosphinic acid; Chloride, *in* M-00204

$C_8H_{10}ClOPS$

▷*O*-Ethyl phenylphosphonochloridothioate, E-00112
S-Ethyl phenylphosphonochloridothioate, *in* P-00232
O-Phenyl ethylphosphonochloridothioate, *in* E-00137

$C_8H_{10}ClO_2P$

Dimethyl (4-chlorophenyl)phosphonite, *in* C-00153
Ethyl (4-chlorophenyl)phosphinate, *in* C-00144
Ethyl phenylphosphonochloridate, *in* P-00229
Methyl (chloromethyl)phenylphosphinate, *in* C-00103
Methyl (4-chlorophenyl)methylphosphinate, *in* C-00131

$C_8H_{10}ClO_2PS$

O-Ethyl *O*-phenyl phosphorochloridothioate, E-00117
O-Ethyl *S*-phenyl phosphorochloridothioate, E-00118
S-Ethyl *O*-phenyl phosphorochloridothioate, E-00119

$C_8H_{10}ClO_3P$

[(2-Chlorophenyl)methyl]phosphonic acid; Mono-Me ester, *in* C-00132
Dimethyl (2-chlorophenyl)phosphonate, *in* C-00145
Dimethyl (3-chlorophenyl)phosphonate, *in* C-00146
Dimethyl (4-chlorophenyl)phosphonate, *in* C-00147
Ethyl phenyl phosphorochloridate, *in* P-00269

$C_8H_{10}ClP$

Chloroethylphenylphosphine, *in* E-00108

$C_8H_{10}ClPS$

Ethylphenylphosphinothioic acid; Chloride, *in* E-00105
Ethyl phenylphosphonochloridothioite, *in* P-00233

$C_8H_{10}ClPS_2$

Ethyl phenylphosphonochloridodithioate, *in* P-00230

$C_8H_{10}ClPSe$

Ethylphenylphosphinoselenoic acid; Chloride, *in* E-00104

$C_8H_{10}Cl_2NOP$

P,*P*-Bis(chloromethyl)-*N*-phenylphosphinic amide, *in* B-00128
(2,4-Dimethylphenyl)phosphoramidic acid; Dichloride, *in* D-00818

$C_8H_{10}Cl_2NPS$

N-Phenyl-*P*,*P*-bis(chloromethyl)-phosphinothioic amide, *in* B-00129

$C_8H_{10}FOP$

Ethylphenylphosphinic acid; Fluoride, *in* E-00101

$C_8H_{10}FO_2P$

Dimethyl (3-fluorophenyl)phosphonite, *in* F-00041
Dimethyl (4-fluorophenyl)phosphonite, *in* F-00042
Ethyl phenylphosphonofluoridate, *in* P-00239

$C_8H_{10}FO_3P$

Dimethyl (3-fluorophenyl)phosphonate, *in* F-00037
Dimethyl (4-fluorophenyl)phosphonate, *in* F-00038

$C_8H_{10}FPS$

Ethylphenylphosphinothioic acid; Fluoride, *in* E-00105

$C_8H_{10}F_5O_3P$

Dimethyl (pentafluorophenyl)phosphonate, *in* P-00015

$C_8H_{10}F_6N_5P$

1,4-Dimethyl-7,9-bis(trifluoromethyl)-1,4,6,8,10-pentaaza-5-phospha(5-P^V)-spiro[4.5]deca-5,7,9-triene, D-00714

$C_8H_{10}NOP$

3,4-Dihydro-2-methyl-2*H*-1,4,2-benzoxazaphosphorine, D-00522
2-Phenyl-1,3,2-oxazaphospholidine, P-00179
3-Phenyl-1,3,2-oxazaphospholidine, P-00180

$C_8H_{10}NOPS$

2,3-Dihydro-2-methyl-1*H*-4,1,2-benzothiazaphosphorine; 2-Oxide, *in* D-00520
3,4-Dihydro-2-methyl-2*H*-1,4,2-benzoxazaphosphorine; 2-Sulfide, *in* D-00522

$C_8H_{10}NOPS_2$

2-(4-Methoxyphenyl)-1,3,2-thiazaphosphetidine 2-sulfide, M-00083
1,3,2-Thiazaphospholidine 2-sulfide; 2-Phenoxy, *in* T-00303

$C_8H_{10}NO_2P$

3,4-Dihydro-2-methyl-2*H*-1,4,2-benzoxazaphosphorine; 2-Oxide, *in* D-00522
2-Dimethylamino-1,3,2-benzodioxaphosphole, D-00689
2-Ethoxy-2,3-dihydro-1,3,2-benzoxazaphosphole, E-00027
1-Phenylamino-1,3,2-dioxaphospholane, P-00083
3-Phenyl-1,3,2-oxazaphospholidine; Oxide, *in* P-00180

$C_8H_{10}NO_2PS$

Ethylene phenylphosphoramidothioate, *in* P-00083

$C_8H_{10}NO_3P$

N-Benzoyl-*P*-methylphosphonamidic acid, B-00031
2,2-Dihydro-1,3,2-benzoxazaphosphole; 2,2-Dimethoxy, *in* D-00444
Ethylene phenylphosphoramidate, *in* P-00083
o-Phenylene dimethylphosphoramidate, *in* D-00689
Spiro[1,3,2-benzodioxaphosphole-2,2′-[1,3,2]-oxazaphospholidine], S-00009
Spiro[1,3,2-benzoxazaphosphole-2(3*H*),2′-[1,3,2]dioxaphospholane], S-00012

$C_8H_{10}NO_4P$

(*N*-Benzoylaminomethyl)phosphonic acid, *in* A-00091

$C_8H_{10}NO_4PS$

▷*O*,*S*-Dimethyl *O*-4-nitrophenyl phosphorothioate, D-00774

$C_8H_{10}NO_4PSSe$

O,*O*-Dimethyl-*S*,*Se*-2-nitrophenyl-phosphoro(selenothioperoxo)thioate, D-00772

$C_8H_{10}NO_5P$

Dimethyl (2-nitrophenyl)phosphonate, *in* N-00055
Dimethyl (3-nitrophenyl)phosphonate, *in* N-00056
Ethyl hydrogen (4-nitrophenyl)phosphonate, *in* N-00057

$C_8H_{10}NO_5PS$

▷*O*,*O*-Dimethyl *S*-(4-nitrophenyl) phosphorothioate, D-00773
▷Parathion methyl, P-00003

$C_8H_{10}NO_5PSe_2$

O,*O*-Dimethyl *Se*,*Se*-(2-nitrophenyl) phosphoro(diselenoperoxoate), D-00771

$C_8H_{10}NO_6P$

▷Dimethyl 2-nitrophenyl phosphate, D-00768
▷Dimethyl 3-nitrophenyl phosphate, D-00769
▷Dimethyl 4-nitrophenyl phosphate, D-00770
▷Pyridoxal phosphate, P-00508

$C_8H_{10}NP$

3-Phenyl-1,3-azaphosphetidine, P-00087

$C_8H_{10}NPS$

2,3-Dihydro-2-methyl-1*H*-4,1,2-benzothiazaphosphorine, D-00520
3,4-Dihydro-2-methyl-2*H*-1,4,2-benzothiazaphosphorine, D-00521

$C_8H_{10}NPSSe$

3,4-Dihydro-2-methyl-2*H*-1,4,2-benzothiazaphosphorine; 2-Selenide, *in* D-00521

$C_8H_{10}NPS_2$

2,3-Dihydro-2-methyl-1*H*-4,1,2-benzothiazaphosphorine; 2-Sulfide, *in* D-00520
3,4-Dihydro-2-methyl-2*H*-1,4,2-benzothiazaphosphorine; 2-Sulfide, *in* D-00521
1,3,2-Thiazaphospholidine 2-sulfide; 2-Ph, *in* T-00303

$C_8H_{10}N_2O_2P$

(1-Diazoethyl)phenylphosphinic acid, D-00032

$C_8H_{10}N_3OP$

Ethylphenylphosphinic acid; Azide, *in* E-00101

$C_8H_{10}N_3PS_2$

2-Amino-2,3-dihydro-5-methyl-3-phenyl-3*H*-1,3,4,2-thiadiazaphosphole 2-sulfide, A-00072

$C_8H_{10}O_4P_2$

1,8-Dihydroxy-3*a*,4,7,7*a*-tetrahydro-4,7-phosphinidene-1*H*-phosphindole 1,8-dioxide, D-00625

$C_8H_{11}ClNOP$
N,N-Dimethyl-*P*-phenylphosphonamidic acid; Chloride, *in* D-00808
Phenyl dimethylphosphoramidochloridite, *in* D-00862

$C_8H_{11}ClNOPS$
O-Phenyl dimethylphosphoramidochloridothioate, *in* D-00861
O-Phenyl ethylphosphoramidochloridothioate, *in* E-00157

$C_8H_{11}ClNO_2P$
Ethyl phenylphosphoramidochloridate, *in* P-00259
Phenyl dimethylphosphoramidochloridate, *in* D-00860

$C_8H_{11}ClNO_2PS$
O,O-Dimethyl (4-chlorophenyl)phosphoramidothioate, *in* C-00160

$C_8H_{11}ClNO_3P$
Dimethyl (4-chlorophenyl)phosphoramidate, *in* C-00157

$C_8H_{11}ClNO_3PS_2$
N-[(4-Chlorophenyl)sulfonyl]-*P*-ethylphosphonimidothioic acid, C-00165

$C_8H_{11}FNOPS$
O-Phenyl dimethylphosphoramidofluoridothioate, *in* D-00868

$C_8H_{11}FNO_3P$
Dimethyl (3-fluorophenyl)phosphoramidate, *in* F-00043
Dimethyl (4-fluorophenyl)phosphoramidate, *in* F-00044

$C_8H_{11}FNPS_2$
Phenyl dimethylphosphoramidofluoridodithioate, *in* D-00867

$C_8H_{11}N_2OP$
2-Ethoxy-2,3-dihydro-1*H*-1,3,2-benzodiazaphosphole, E-00026
P-(2-Phenylethenyl)phosphonic diamide, P-00156

$C_8H_{11}N_2OPS$
O-Ethyl *N,N'*-*o*-phenylenephosphorodiamidothioate, *in* E-00026

$C_8H_{11}N_2O_2P$
1,3-Dihydrospiro[2*H*-1,3,2-benzodiazaphosphole-2,2'-[1,3,2]-dioxaphosholane], D-00575

$C_8H_{11}N_2O_5P$
Dimethyl (3-nitrophenyl)phosphoramidate, *in* N-00059
Dimethyl (4-nitrophenyl)phosphoramidate, *in* N-00060

$C_8H_{11}N_2O_6P$
Mono[2-(dimethylamino)-4-nitrophenyl] phosphate, M-00464
Pyridoxal phosphate; Oxime, *in* P-00508

$C_8H_{11}N_2P$
2,3-Dihydro-1,3-dimethyl-1*H*-1,3,2-benzodiazaphosphole, D-00456

$C_8H_{11}N_4OP$
N,N-Dimethyl-*P*-phenylphosphonamidic azide, D-00809

$C_8H_{11}OP$
Dimethylphenylphosphine; Oxide, *in* D-00803
Ethylphenylphosphinous acid, E-00108
Methyl methylphenylphosphinite, *in* M-00227
4-Methyl-4-phosphatetracyclo[3.3.0.0^{2,8}.0^{3,6}]-octane; 4-Oxide, *in* M-00275
Phenyl dimethylphosphinite, *in* D-00844
2,3,4,7-Tetrahydro-1*H*-isophosphindole 2-oxide, T-00109

$C_8H_{11}OPS$
O-Ethyl phenylphosphinothioate, *in* P-00213
Ethylphenylphosphinothioic acid, E-00105
O-Methyl methylphenylphosphinothioate, *in* M-00225
S-Methyl methylphenylphosphinothioate, *in* M-00225
Methylphenylphosphinothioic acid; *O*-Me ester, *in* M-00225
O-Phenyl dimethylphosphinothioate, *in* D-00839

$C_8H_{11}OPS_2$
O-Ethyl phenylphosphonodithioate, E-00113

$C_8H_{11}OPSe$
Ethylphenylphosphinoselenoic acid, E-00104

$C_8H_{11}O_2P$
Benzylmethylphosphinic acid, B-00053
Bis(hydroxymethyl)phenylphosphine, B-00281
(2,4-Dimethylphenyl)phosphinic acid, D-00804
(2,5-Dimethylphenyl)phosphinic acid, D-00805
(3,4-Dimethylphenyl)phosphinic acid, D-00806
Dimethyl phenylphosphonite, *in* P-00251
▷Ethyl phenylphosphinate, *in* P-00210
Ethylphenylphosphinic acid, E-00101
Methyl methylphenylphosphinate, *in* M-00219
Methyl (4-methylphenyl)phosphinate, *in* M-00222
(2-Methylphenyl)methylphosphinic acid, M-00203
(4-Methylphenyl)methylphosphinic acid, M-00204
Phenyl dimethylphosphinate, *in* D-00829
Phenyl ethylphosphinate, *in* E-00124
2,3,4,7-Tetrahydro-1*H*-isophosphindole 2-oxide; 2-Hydroxy, *in* T-00109

$C_8H_{11}O_2PS$
O,O-Dimethyl phenylphosphonothioate, *in* P-00244
O,S-Dimethyl phenylphosphonothioate, *in* M-00243
O-Ethyl phenylphosphonothioate, E-00114
O-Methyl *S*-phenyl methylphosphonothioate, *in* M-00171

$C_8H_{11}O_2PS_2$
O-Ethyl *O*-phenyl phosphorodithioate, E-00120
O-Ethyl *S*-phenyl phosphorodithioate, E-00121

$C_8H_{11}O_2PSe$
O,O-Dimethyl phenylphosphonoselenoate, *in* P-00243

$C_8H_{11}O_3P$
Dimethyl phenylphosphonate, D-00810
(2,3-Dimethylphenyl)phosphonic acid, D-00811
(2,4-Dimethylphenyl)phosphonic acid, D-00812
(2,5-Dimethylphenyl)phosphonic acid, D-00813
(2,6-Dimethylphenyl)phosphonic acid, D-00814
(3,4-Dimethylphenyl)phosphonic acid, D-00815
(3,5-Dimethylphenyl)phosphonic acid, D-00816
Ethyl phenyl phosphonate, E-00111
(Methoxymethyl)phenylphosphinic acid, M-00052
Methyl hydrogen benzylphosphonate, *in* B-00060
Methyl (hydroxymethyl)phenylphosphinate, *in* H-00160
Methyl phenyl methylphosphonate, M-00205
[(2-Methylphenyl)methyl]phosphonic acid, M-00206
[(3-Methylphenyl)methyl]phosphonic acid, M-00207
[(4-Methylphenyl)methyl]phosphonic acid, M-00208
Monoethyl phenylphosphonate, M-00471
1-Phenylethylphosphonic acid, P-00164
(2-Phenylethyl)phosphonic acid, P-00165
(Phenylphosphinylidene)bismethanol, *in* B-00281

$C_8H_{11}O_3PS$
[(Benzylthio)methyl]phosphonic acid, B-00072

$C_8H_{11}O_4P$
Dimethyl (2-hydroxyphenyl)phosphonate, *in* H-00179
Dimethyl (3-hydroxyphenyl)phosphonate, *in* H-00180
Dimethyl (4-hydroxyphenyl)phosphonate, *in* H-00181
Ethyl hydrogen (4-hydroxyphenyl)phosphonate, *in* H-00181
[(3-Methoxyphenyl)methyl]phosphonic acid, M-00060
[(4-Methoxyphenyl)methyl]phosphonic acid, M-00061
Methyl hydrogen (α-hydroxybenzyl)-phosphonate, *in* H-00174
Mono-2,6-dimethylphenyl phosphate, M-00466
Mono(2-phenylethyl) phosphate, M-00506

$C_8H_{11}P$
9,9-Dihydro-9-phosphapentacyclo[4.3.0.0^{2,5}.0^{3,8}.0^{4,7}]nonane, D-00569
Dimethylphenylphosphine, D-00803
4-Methyl-4-phosphatetracyclo[3.3.0.0^{2,8}.0^{3,6}]-octane, M-00275
(2-Phenylethyl)phosphine, P-00163

$C_8H_{11}PO_4S$
[[(4-Methylphenyl)sulfinyl]methyl]phosphonic acid, M-00266

$C_8H_{11}PS$
Dimethylphenylphosphine; Sulfide, *in* D-00803
Ethylphenylphosphinothious acid, E-00107
4-Methyl-4-phosphatetracyclo[3.3.0.0^{2,8}.0^{3,6}]-octane; 4-Sulfide, *in* M-00275

$C_8H_{11}PSSe$
Ethylphenylphosphinothioselenoic acid, E-00106

$C_8H_{11}PS_2$
Ethylphenylphosphinodithioic acid, E-00103
Phenyl dimethylphosphinodithioate, *in* D-00834

$C_8H_{11}PSe$
Phenyl dimethylphosphinoselenoite, *in* D-00838

$C_8H_{11}PSe_2$
Dimethyl phenylphosphonodiselenoite, *in* P-00235

$C_8H_{12}BrP$
Tetravinylphosphonium(1+); Bromide, *in* T-00296

$C_8H_{12}ClO_5P$
Ethyl chloro(diethoxyphosphinyl)acetate, *in* C-00036

$C_8H_{12}Cl_6NOP$
N-Cyclohexyl-*P,P*-bis(trichloromethyl)-phosphinic amide, *in* B-00402

$C_8H_{12}Cl_7O_3P$
Bis(2,2,2-trichloro-1,1-dimethylethyl) phosphorochloridate, B-00399

$C_8H_{12}F_4O_6P_2$
Tetramethyl (3,3,4,4-tetrafluoro-1-cyclobutene-1,2-diyl)bisphosphonate, *in* T-00078

$C_8H_{12}NOP$
N,N-Dimethyl-*P,P*-dipropynylphosphinic amide, *in* D-01231
P-Ethyl-*P*-phenylphosphinic amide, E-00102

$C_8H_{12}NOPS$
N-Benzoyl-*P,P*-dimethylphosphinothioic amide, *in* D-00840

$C_8H_{12}NOPS_2$
S,S-Dimethyl phenylphosphoramidodithioate, *in* D-00866
O-Methyl-*S*-phenyl methylphosphoramidodithioate, *in* M-00330

$C_8H_{12}NO_2P$
N,N-Dimethyl-*P*-phenylphosphonamidic acid, D-00808
Ethyl *P*-phenylphosphonamidate, *in* P-00218
N-Ethyl-*P*-phenylphosphonamidic acid, E-00109
P-Ethyl-*N*-phenylphosphonamidic acid, E-00110
Methyl *N*-methyl-*P*-phenylphosphonamidate, *in* M-00230
Methyl *P*-methyl-*N*-phenylphosphonamidate, *in* M-00231
Phenyl *N,P*-dimethylphosphonamidate, *in* D-00849

$C_8H_{12}NO_2PS$
O,O-Dimethyl phenylphosphoramidothioate, *in* P-00262

$C_8H_{12}NO_3P$
(1-Amino-1-phenylethyl)phosphonic acid, A-00100
(1-Amino-2-phenylethyl)phosphonic acid, A-00101
Dimethyl (2-aminophenyl)phosphonate, *in* A-00109
Dimethyl (4-aminophenyl)phosphonate, *in* A-00111
(4-Dimethylaminophenyl)phosphonic acid, D-00706
Dimethyl 2-methyl-4-pyridinephosphonate, *in* M-00388
Dimethyl (3-methyl-4-pyridinyl)phosphonate, *in* M-00390
Dimethyl phenylphosphoramidate, D-00817
(2,4-Dimethylphenyl)phosphoramidic acid, D-00818
Ethyl hydrogen (2-aminophenyl)phosphonate, *in* A-00109
Ethyl hydrogen (4-aminophenyl)phosphonate, *in* A-00111
Ethyl phenyl phosphoramidate, E-00115
Ethylphenylphosphoramidic acid, E-00116

$C_8H_{12}NO_4P$
Phenyl hydrogen 2-aminoethyl phosphate, *in* A-00075

$C_8H_{12}NO_5PS$
Dimethyl phenylsulfonylphosphoramidate, *in* P-00312

$C_8H_{12}NO_5PS_2$
▷*O,O*-Dimethyl *O*-4-sulfamoylphenyl phosphorothioate, D-00920

$C_8H_{12}NO_6$
2-Amino-2-deoxyglucitol; *N*-Ac, *in* A-00066

$C_8H_{12}NPS$
P,P-Dimethyl-*N*-phenylphosphinothioic amide, *in* D-00840

$C_8H_{12}NPS_2$
Methyl *P*-methyl-*N*-phenylphosphonamidodithioate, *in* M-00232

$C_8H_{12}NPSe$
P,P-Dimethyl-*N*-phenylphosphinoselenoic amide, *in* D-00836

$C_8H_{12}N_3O_2P$
Tolfamide, T-00323

$C_8H_{12}N_3O_9P$
6-Azauridine 5′-phosphate, A-00145

$C_8H_{12}N_3P$
Hexahydro-3-phenyl-1,2,4,3-triazaphosphorine, H-00051

$C_8H_{12}N_5OP$
▷Phosphemide, P-00353

$C_8H_{12}O_4P_2$
1,4-Phenylenebis[methylphosphinic acid], P-00137

$C_8H_{12}O_5P_2$
2,2′-Oxybis[4,5-dimethyl-1,3,2-dioxaphosphole], O-00088

$C_8H_{12}O_6P_2$
[1,2-Phenylenebis(methylene)]bisphosphonic acid, P-00129
[1,3-Phenylenebis(methylene)]bisphosphonic acid, P-00130
[1,4-Phenylenebis(methylene)]bisphosphonic acid, P-00131

$C_8H_{12}O_7P_2$
Bis(dimethylvinylene) pyrophosphate, *in* O-00088

$C_8H_{12}P^{\oplus}$
Tetravinylphosphonium(1+), T-00296

$C_8H_{12}P_2S_4$
1,4-Phenylenebis[methylphosphinodithioic acid], P-00138

$C_8H_{13}BrPS$
Bis(2-methylpropyl)phosphinothioic acid; Bromide, *in* B-00351

$C_8H_{13}N_2O_2P$
▷Diamidafos, D-00024
Ethyl *N*-phenylphosphorodiamidate, *in* P-00273

$C_8H_{13}N_2O_2PS$
O,O-Dimethyl 2-phenylphosphorohydrazidothioate, *in* P-00296

$C_8H_{13}N_2O_3P$
Dimethyl (4-aminophenyl)phosphoramidate, *in* A-00115
Phenyl hydrogen (2-aminoethyl)-phosphoramidate, *in* A-00080

$C_8H_{13}N_2O_3PS$
▷Thionazin, T-00314

$C_8H_{13}N_2O_5P$
Pyridoxamine 5′-phosphate, P-00509

$C_8H_{13}N_2PS$
N,N′-Dimethyl-*P*-phenylphosphonothioic diamide, *in* P-00246
N,P-Dimethyl-*N′*-phenylphosphonothioic diamide, *in* M-00245

$C_8H_{13}OP$
2,3,4,5,6,7-Hexahydro-1*H*-isophosphindole 2-oxide, H-00035
8-Methyl-8-phosphabicyclo[3.2.1]oct-6-ene 8-oxide, M-00273

$C_8H_{13}OPS$
10-Phenyl-10*H*-phenothiaphosphine; 10-Oxide, *in* P-00182

$C_8H_{13}O_2P$
2,3,4,5,6,7-Hexahydro-1*H*-isophosphindole 2-oxide; 2-Hydroxy, *in* H-00035

$C_8H_{13}O_2PS$
Diethyl 2-thienylphosphonite, *in* T-00308
10-Phenyl-10*H*-phenothiaphosphine; 5,5-Dioxide, *in* P-00182

$C_8H_{13}O_3PS$
Diethyl 2-thienylphosphonate, *in* T-00307

$C_8H_{13}O_4P$
Diethyl 2-furanylphosphonate, *in* F-00073
Diethyl 3-furanylphosphonate, *in* F-00074

$C_8H_{13}O_5P$
Spiro[1,3,2-dioxaphospholane-2,1′-[2,8,9]-trioxa[1]phosphatricyclo[3.3.1.1^{3,7}]decane], S-00024

$C_8H_{13}PS_2$
10-Phenyl-10*H*-phenothiaphosphine; 10-Sulfide, *in* P-00182

$C_8H_{14}Cl_2N_5OP$
N″-2-Pyrimidinylphosphoric triamide; *N,N′*-Bis(2-chloroethyl), *in* P-00517

$C_8H_{14}Cl_3O_5P$
▷Butonate, B-00530
▷Diethyl 1-acetoxy-2,2,2-trichloroethylphosphonate, *in* A-00011

$C_8H_{14}NO_3P$
Diethyl 4-morpholinylphosphonite, *in* M-00529
Diethyl 1*H*-pyrrol-2-ylphosphonate, *in* P-00521

$C_8H_{14}NO_6P$
2-Diethylamino-1,3,2-dioxaphospholan-4,5-dicarboxylic acid, D-00265

$C_8H_{14}O_4P$
Diisopropyl acetylphosphonate, *in* A-00026

$C_8H_{14}O_4P_2$
Octahydro-1,8-dihydroxy-4,7-phosphinidene-1*H*-phosphindole 1,8-dioxide, *in* D-00625

$C_8H_{14}P^{\oplus}$
1,1,3,4-Tetramethyl-1*H*-phospholium iodide, *in* T-00536

$C_8H_{15}ClN_5OP$
N″-2-Pyrimidinylphosphoric triamide; *N′*-Et-*N*-(2-chloroethyl), *in* P-00517

$C_8H_{15}N_4OP$
1,1′-Dimethyl-*P*-phenylphosphonic dihydrazide, *in* P-00225

$C_8H_{15}N_4PS$
N[1],*N*[1′]-Dimethyl-*P*-phosphonothioic dihydrazide, *in* P-00249

$C_8H_{15}OP$
2,3-Dihydro-1-phenyl-1*H*-benzo[*g*]-phosphindole; 1-Oxide, *in* D-00551
1,4-Dimethyl-2-phosphabicyclo[2.2.1]heptane 2-oxide, D-00825
8-Methyl-8-phosphabicyclo[3.2.1]octane; Oxide, *in* M-00272
1-Phosphabicyclo[3.3.1]nonane; 1-Oxide, *in* P-00345

$C_8H_{15}OPS$
S-Butyl divinylphosphinothioate, *in* D-01259

$C_8H_{15}OPSe$
O-Phenyl diethylphosphinoselenoate, *in* D-00327

$C_8H_{15}O_2P$
1-Ethoxy-2,5-dihydro-3,4-dimethyl-1*H*-phosphole 1-oxide, *in* D-00493
Ethyl di-2-propenylphosphinate, *in* D-01193

$C_8H_{15}O_2PS$
O,O-Diethyl (1,3-butadien-1-yl)-phosphonothioate, *in* B-00511

$C_8H_{15}O_3P$
▷4-*tert*-Butyl-2,6,7-trioxa-1-phosphabicyclo[2.2.2]octane, B-00640
Diethyl (1,2-butadienyl)phosphonate, *in* B-00509
Diethyl (1,3-butadienyl)phosphonate, *in* B-00510
Diethyl 1-butynylphosphonate, *in* B-00643
Diethyl 2-butynylphosphonate, *in* B-00644
2,3-Dihydro-3,3,5-trimethyl-1,2-oxaphosphole 2-oxide; 2-Ethoxy, *in* D-00596
Diisopropyl ethynylphosphonate, *in* E-00203
Dimethyl 2-cyclohexen-1-ylphosphonate, *in* C-00242
8,8-Dimethyl-1,6,10-trioxa-5-phospha(5-P^V)-spiro[4.5]dec-2-ene, D-00933

$C_8H_{15}O_3PS$
▷*tert*-Butyl bicyclic thiophosphate, *in* B-00640

$C_8H_{15}O_4P$
▷*tert*-Butyl bicyclic phosphate, *in* B-00640
tert-Butyl dimethylvinylene phosphate, *in* H-00123
9-Methyl-1,4,6-trioxa-5-phospha(5-P^V)-spiro[4.4]non-7-ene; 5-Ethoxy, *in* M-00428

$C_8H_{15}O_5P$
4-(Diethoxyphosphinyl)-2-butenoic acid, D-00248
Diethyl(tetrahydro-2-oxo-2-furanyl)-phosphonate, *in* T-00121
Methyl 2-(diethoxyphosphinyl)propenoate, *in* D-00260
Methyl 3-(diethoxyphosphinyl)propenoate, *in* D-00261

$C_8H_{15}O_6P$
▷2-(Diethoxyphosphinyl)-3-oxobutanoic acid, D-00253
4-(Diethoxyphosphinyl)-3-oxobutanoic acid, D-00254

Ethyl 3-[(dimethoxyphosphinyl)oxy]crotonate, *in* D-00681
Methyl 2-[(diethoxyphosphinyl)oxy]-2-propenoate, *in* D-00256

$C_8H_{15}O_6PS_2$
[(Diethoxyphosphinothioyl)thio]butanedioic acid, D-00240
▷Dimethyl [(dimethoxyphosphinothioyl)thio]-butanedioate, *in* D-00677
1-Ethyl [(dimethoxyphosphinothioyl)thio]-butanedioate, *in* D-00677
2-Ethyl [(dimethoxyphosphinothioyl)thio]-butanedioate, *in* D-00677

$C_8H_{15}O_7P$
(Diethoxyphosphinyl)butanedioic acid, D-00245
Diethyl phosphonobutanedioate, *in* P-00381

$C_8H_{15}O_8P$
Ethyl (diethoxyphosphinyl)oxoacetate, *in* D-00252

$C_8H_{15}P$
8-Methyl-8-phosphabicyclo[3.2.1]octane, M-00272
1-Phosphabicyclo[3.3.1]nonane, P-00345

$C_8H_{15}PS$
1-Phosphabicyclo[3.3.1]nonane; 1-Sulfide, *in* P-00345

$C_8H_{16}BF_4IP$
5-Phosphoniaspiro[4.4]nonane; Tetrafluoroborate, *in* P-00374

$C_8H_{16}BrOP$
1-Hydroxy-2,2,3,4,4-pentamethylphosphetane 1-oxide; Bromide, *in* H-00168

$C_8H_{16}BrO_5P$
Ethyl bromo(diethoxyphosphinyl)acetate, *in* B-00467

$C_8H_{16}BrP$
2,5-Dihydro-1,1,3,4-tetramethyl-1*H*-phospholium bromide, *in* D-00600
5-Phosphoniaspiro[4.4]nonane; Bromide, *in* P-00374

$C_8H_{16}ClN_2OP$
P,P-Di-1-pyrrolidinylphosphinic acid; Chloride, *in* D-01233

$C_8H_{16}ClN_2O_3P$
Di-4-morpholinylphosphinic acid; Chloride, *in* D-00935

$C_8H_{16}ClN_4P$
Octahydro-8*bH*-2*a*,4*a*,6*a*,8*a*-tetraaza-8*b*-phosphapentaleno[1,6-*cd*]pentalene; 8*b*-Chloro, *in* O-00025

$C_8H_{16}ClOP$
1-Chloro-2,2,3,4,4-pentamethylphosphetane 1-oxide, *in* C-00124

$C_8H_{16}ClO_2P$
2-Chloro-4-isopropyl-5,5-dimethyl-1,3,2-dioxaphosphorinane, C-00090
Ethyl cyclohexylphosphonochloridate, *in* C-00259

$C_8H_{16}ClO_2PS$
2-Chloro-4-isopropyl-5,5-dimethyl-1,3,2-dioxaphosphorinane; 2-Sulfide, *in* C-00090

$C_8H_{16}ClO_2PSe$
2-Chloro-4-isopropyl-5,5-dimethyl-1,3,2-dioxaphosphorinane; 2-Selenide, *in* C-00090

$C_8H_{16}ClO_3P$
2-Chloro-4-isopropyl-5,5-dimethyl-1,3,2-dioxaphosphorinane; 2-Oxide, *in* C-00090

$C_8H_{16}ClO_4P$
Diisopropyl(chloroacetyl)phosphonate, *in* C-00020

$C_8H_{16}ClP$
1-Chloro-2,2,3,4,4-pentamethylphosphetane, C-00124
5-Phosphoniaspiro[4.4]nonane; Chloride, *in* P-00374

$C_8H_{16}ClPS$
1-Chloro-2,2,3,4,4-pentamethylphosphetane 1-sulfide, *in* C-00124

$C_8H_{16}Cl_3O_4P$
Diisopropyl (2,2,2-trichloro-1-hydroxyethyl)-phosphonate, *in* T-00408

$C_8H_{16}Cl_4O_2P_2$
1,8-Octanediphosphonic acid; Tetrachloride, *in* O-00035

$C_8H_{16}FNP$
Octahydro-8*bH*-2*a*,4*a*,6*a*,8*a*-tetraaza-8*b*-phosphapentaleno[1,6-*cd*]pentalene; 8*b*-Fluoro, *in* O-00025

$C_8H_{16}FN_2O_3P$
Di-4-morpholinylphosphinic acid; Fluoride, *in* D-00935

$C_8H_{16}FO_4P$
Ethyl (diethoxyphosphinyl)fluoroacetate, *in* D-00250

$C_8H_{16}IP$
5-Phosphoniaspiro[4.4]nonane; Iodide, *in* P-00374

$C_8H_{16}NOP$
P,P-Diethenyl-*N,N*-diethylphosphinic amide, *in* D-01258
N,N-Dimethyl-*P,P*-di-2-propenylphosphinic amide, *in* D-01193

$C_8H_{16}NO_2P$
P-(1,3-Butadienyl)-*N,N*-diethylphosphonamidic acid, B-00508
2,3-Dihydro-5-methyl-1,2-oxaphosphole 2-oxide; 2-Diethylamino, *in* D-00525

$C_8H_{16}NO_3P$
Diethyl (1-cyanopropyl)phosphonate, *in* C-00229
Diethyl (3-cyanopropyl)phosphonate, *in* C-00230

$C_8H_{16}NO_3PS_2$
▷Mephosfolan, M-00004

$C_8H_{16}NO_4P$
[(5,6-Dihydro-4,4,6-trimethyl-4*H*-1,3-oxazin-2-yl)methyl]phosphonic acid, D-00598
Diisopropyl (isocyanatomethyl)phosphonate, *in* I-00022
N,N,2,3-Tetramethyl-1,4,6,9-tetraoxa-5-phospha(5-P^V)spiro[4.4]non-2-en-5-amine, *in* D-00924

$C_8H_{16}NO_5P$
▷Dicrotophos, D-00204

$C_8H_{16}NO_8P$
2-Amino-2-deoxyglucose 6-(dihydrogen phosphate); *N*-Ac, *in* A-00069

$C_8H_{16}NO_9P$
2-Acetamido-2-deoxy-α-D-glucose 1-(dihydrogen phosphate), *in* A-00067
2-Acetamido-2-deoxy-β-D-glucose 1-(dihydrogen phosphate), *in* A-00067
2-Amino-2-deoxygalactose 1-(dihydrogen phosphate); *N*-Ac, *in* A-00065

$C_8H_{16}N_3OP$
1,1,2,3,3-Pentamethyltrimethylenephosphinic azide, *in* H-00168

$C_8H_{16}N_3P$
Hexahydro-3*H*,6*H*-2*a*,5*a*,8*a*-triaza-8*b*-phosphaacenaphthylene, H-00054

$C_8H_{16}N_4O_2P_2$
[1,4-Phenylenebis(methylene)]bisphosphonic acid; Tetra-amide, *in* P-00131

$C_8H_{16}O_5P_2$
2,2′-Oxybis[4-methyl-1,3,2-dioxaphosphorinane], O-00096

$C_8H_{16}O_5P_2S_2$
Bis(4-methyl-2-thioxo-1,3,2-dioxaphosphorinan-2-yl) oxide, B-00365

$C_8H_{16}O_8P_2$
Tetramethyl (1,4-dioxo-1,4-butanediyl)-bisphosphonate, *in* D-00975

$C_8H_{16}P^{\oplus}$
5-Phosphoniaspiro[4.4]nonane, P-00374

$C_8H_{16}P_2$
1,1′-Biphospholane, B-00084
1,2,3,6-Tetrahydro-1,2,4,5-tetramethyl-1,2-diphosphinine, *in* T-00095

$C_8H_{16}P_2S_2$
1,1′-Biphospholane; Disulfide, *in* B-00084
1,2,3,6-Tetrahydro-4,5-dimethyl-1,2-diphosphorine; 1,2-Di-Me, 1,2-disulfide, *in* T-00095

$C_8H_{17}Br_2P$
Dibromooctylphosphine, *in* O-00044

$C_8H_{17}ClNPS$
P-Cyclohexyl-*N,N*-dimethyl-phosphonamidothioic chloride, *in* C-00255

$C_8H_{17}Cl_2N_2O_2P$
3-Butenyl *N,N′*-bis(2-chloroethyl)-phosphorodiamidate, *in* B-00125
3-Butenyl *N,N*-bis(2-chloroethyl)-phosphorodiamidate, *in* B-00126

$C_8H_{17}Cl_2OP$
Octylphosphonic dichloride, O-00043
Octyl phosphorodichloridite, O-00046

$C_8H_{17}Cl_2P$
Dichlorooctylphosphine, *in* O-00044

$C_8H_{17}F_2OP$
Octylphosphonic acid; Difluoride, *in* O-00042

$C_8H_{17}IP$
1,1,3-Trimethylphosphorinanium iodide, *in* D-00882

$C_8H_{17}N_2O_2P$
Butyl bis(1-aziridinyl)phosphinate, *in* B-00091
P,P-Di-1-pyrrolidinylphosphinic acid, D-01233

$C_8H_{17}N_2O_3P$
Di-4-morpholinylphosphinous acid, D-00936

$C_8H_{17}N_2O_4P$
Di-4-morpholinylphosphinic acid, D-00935

$C_8H_{17}N_2PSe$
P-Ethyl-*N,N′*-diphenylphosphonoselenoic diamide, *in* E-00146

$C_8H_{17}N_4OP$
Octahydro-8*bH*-2*a*,4*a*,6*a*,8*a*-tetraaza-8*b*-phosphapentaleno[1,6-*cd*]pentalene; 8*b*-Hydroxy, *in* O-00025

$C_8H_{17}N_4P$
Octahydro-8*bH*-2*a*,4*a*,6*a*,8*a*-tetraaza-8*b*-phosphapentaleno[1,6-*cd*]pentalene, O-00025

$C_8H_{17}OP$
1-Oxa-5-phospha(5-P^V)spiro[4.5]decane, O-00049
6-Oxa-5-phospha(5-P^V)spiro[4.5]decane, O-00050
1-Oxa-5-phospha(5-P^V)spiro[4.4]nonane; 5-Me, *in* O-00051

$C_8H_{17}OPS$
1-Hydroxy-2,2,3,4,4-pentamethylphosphetane 1-sulfide, H-00169

$C_8H_{17}OPS_2$
O-Ethyl hydrogen cyclohexylphosphonodithioate, *in* C-00261

$C_8H_{17}OPS_3$
2-*tert*-Butyl-1,3,6,2-trithiaphosphocane; 2-Oxide, *in* B-00642

$C_8H_{17}O_2P$
5-*tert*-Butyl-2-methyl-1,3,2-dioxaphosphorinane, B-00564
Dimethyl cyclohexylphosphonite, *in* C-00267
Ethyl cyclohexylphosphinate, *in* C-00253
Ethyl 3-(ethoxymethylphosphinyl)propanoate, *in* H-00161
1-Hydroxy-2,2,3,4,4-pentamethylphosphetane 1-oxide, H-00168
Isopropyl (2-butenyl)methylphosphinate, *in* B-00523
4-Isopropyl-5,5-dimethyl-1,3,2-dioxaphosphorinane, I-00028

Methyl cyclohexylmethylphosphinate, *in* C-00247
Tetramethylene *tert*-butylphosphonite, *in* D-00968

$C_8H_{17}O_2PS$

O,O-Diethyl 2-butenylphosphonothioate, *in* B-00527
O,O-Diethyl (2-methyl-1-propenyl)-phosphonothioate, *in* M-00366
O-Ethyl hydrogen cyclohexylphosphonothioate, *in* C-00264
S-Ethyl *O,O*-tetramethylethylene phosphorodithioate, *in* E-00193
2-Ethylthio-4,4,5,5-tetramethyl-1,3,2-dioxaphospholane, E-00193

$C_8H_{17}O_3P$

2-*tert*-Butoxy-4,5-dimethyl-1,3,2-dioxaphospholane, B-00533
2-*tert*-Butoxy-4-methyl-1,3,2-dioxaphosphorinane, B-00537
5-*tert*-Butyl-2-methoxy-1,3,2-dioxaphosphorinane, B-00563
5-*tert*-Butyl-2-methyl-1,3,2-dioxaphosphorinane; 2-Oxide, *in* B-00564
Diethyl 1-butenylphosphonate, *in* B-00524
Diisopropyl vinylphosphonate, *in* V-00003
Dimethyl cyclohexylphosphonate, *in* C-00256
(1,1-Dimethyl-3-oxobutyl)ethylphosphinic acid, D-00782
2-Ethoxy-4,4,5,5-tetramethyl-1,3,2-dioxaphospholane, E-00052
2-Ethoxy-3,5,5-trimethyl-1,2-oxaphospholan-3-ol 2-oxide, *in* T-00519
4-Isopropyl-5,5-dimethyl-1,3,2-dioxaphosphorinane; 2-Oxide, *in* I-00028
Monoethyl cyclohexylphosphonate, *in* C-00256

$C_8H_{17}O_3PS$

O-Ethyl *O,O*-tetramethylethylene phosphorothioate, *in* E-00052

$C_8H_{17}O_3PS_2$

Diethyl 1,3-dithian-2-ylphosphonate, *in* D-01240

$C_8H_{17}O_4P$

tert-Butyl 1,3-butylene phosphate, *in* B-00537
5-*tert*-Butyl-2-methoxy-1,3,2-dioxaphosphorinane; 2-Oxide, *in* B-00563
(3-Methyl-2-oxoheptyl)phosphonic acid, M-00187
(2-Oxooctyl)phosphonic acid, O-00064
2,2,3,3-Tetramethyl-1,4,6,9-tetraoxa-5-phospha(5-P^V)spiro[4.4]nonane, T-00235

$C_8H_{17}O_4PS$

Ethyl (diethoxyphosphinothioyl)acetate, *in* D-00237

$C_8H_{17}O_4PS_2$

▷Acetophos, A-00004
(Diisopropoxyphosphinothioyl)thioacetic acid, D-00632

$C_8H_{17}O_5P$

2-(Diethoxyphosphinyl)butanoic acid, D-00246
4-(Diethoxyphosphinyl)butanoic acid, D-00247
Diethyl (2-acetyloxyethyl)phosphonate, *in* A-00006
▷Diethyl (ethoxycarbonylmethyl)phosphonate, *in* D-00241
(Diisopropoxyphosphinyl)acetic acid, D-00633
(Dipropoxyphosphinyl)acetic acid, D-01199
Methyl 2-(diethoxyphosphinyl)propanoate, *in* D-00258
Methyl 3-(diethoxyphosphinyl)propanoate, *in* D-00259

$C_8H_{17}O_6P$

Methyl 2-[(diethoxyphosphinyl)oxy]-propanoate, *in* D-00255

$C_8H_{17}P$

2,2,3,4,4-Pentamethylphosphetane, P-00022
5-Phospha(5-P^V)spiro[4.4]nonane, P-00350

$C_8H_{17}P^{\oplus}$

1,1,4-Trimethylphosphoranium iodide, *in* D-00883

$C_8H_{17}PSSi$

5-Methyl-1-phospha-5-silabicyclo[3.3.1]-nonane; 1-Sulfide, *in* M-00274

$C_8H_{17}PS_3$

2-*tert*-Butyl-1,3,6,2-trithiaphosphocane, B-00642

$C_8H_{17}PS_4$

2-*tert*-Butyl-1,3,6,2-trithiaphosphocane; 2-Sulfide, *in* B-00642

$C_8H_{17}PSi$

5-Methyl-1-phospha-5-silabicyclo[3.3.1]-nonane, M-00274

$C_8H_{18}BrOP$

Di-*tert*-butylphosphinic acid; Bromide, *in* D-00110

$C_8H_{18}BrO_2P$

Dibutyl phosphorobromidite, D-00139

$C_8H_{18}BrO_3P$

Di-*tert*-butyl phosphorobromidate, D-00138
Diethyl (4-bromobutyl)phosphonate, *in* B-00466

$C_8H_{18}BrP$

Bromodibutylphosphine, *in* D-00121
Bromodi-*tert*-butylphosphine, *in* D-00122

$C_8H_{18}BrPS$

Dibutylphosphinothioic acid; Bromide, *in* D-00117
Di-*tert*-butylphosphinothioic acid; Bromide, *in* D-00118

$C_8H_{18}Br_2NP$

Di-*tert*-butylphosphoramidous acid; Dibromide, *in* D-00137

$C_8H_{18}ClN_2OPS_2$

2,4-Di-*tert*-butyl-3-chloro-1,2,4,3-thiadiazaphosphetidine; 1-Oxide, 3-Sulfide, *in* D-00079

$C_8H_{18}ClN_2O_2P$

6,9-Dimethyl-1,3-dioxa-6,9-diaza-2-phosphacycloundecane; 2-Chloro, *in* D-00728

$C_8H_{18}ClN_2O_2PS$

2,4-Di-*tert*-butyl-3-chloro-1,2,4,3-thiadiazaphosphetidine; 1,1-Dioxide, *in* D-00079

$C_8H_{18}ClN_2O_5PS$

▷Sufosfamide, S-00026

$C_8H_{18}ClN_2PS$

2,4-Di-*tert*-butyl-3-chloro-1,2,4,3-thiadiazaphosphetidine, D-00079

$C_8H_{18}ClOP$

Bis(1-methylpropyl)phosphinic acid; Chloride, *in* B-00345
Bis(2-methylpropyl)phosphinic acid; Chloride, *in* B-00346
Butyl *tert*-butylphosphonochloridite, *in* D-00130
Butyl butylphosphonochoridite, *in* B-00605
Dibutylphosphinic acid; Chloride, *in* D-00109
Di-*tert*-butylphosphinic acid; Chloride, *in* D-00110

$C_8H_{18}ClOPS_2$

S,S-Dibutyl phosphorochloridodithioate, D-00144

$C_8H_{18}ClO_2P$

Dibutyl phosphorochloridite, D-00142
Di-*tert*-butyl phosphorochloridite, D-00143

$C_8H_{18}ClO_2PS$

O,O-Bis(2-methylpropyl) phosphorochloridothioate, B-00357
O,O-Dibutyl phosphorochloridothioate, D-00145

$C_8H_{18}ClO_2PSe$

O,O-Bis(2-methylpropyl) phosphorochloridoselenoate, B-00356
O,O-Dibutyl phosphorochloridoselenoate, *in* D-00154

$C_8H_{18}ClO_3P$

Bis(1-methylpropyl) phosphorochloridate, *in* B-00341
Bis(2-methylpropyl) phosphorochloridate, *in* B-00342
Dibutyl phosphorochloridate, D-00140
Di-*tert*-butyl phosphorochloridate, D-00141
Diethyl (4-chlorobutyl)phosphonate, *in* C-00025

$C_8H_{18}ClO_3PS$

▷Dibutyl chlorothiophosphonate, D-00080

$C_8H_{18}ClO_4PS$

O,O-Bis(2-ethoxyethyl) phosphorochloridothioate, B-00271

$C_8H_{18}ClP$

Di-*tert*-butylphosphinous chloride, D-00125

$C_8H_{18}ClPS$

Bis(2-methylpropyl)phosphinothioic acid; Chloride, *in* B-00351
Dibutylphosphinothioic acid; Chloride, *in* D-00117
Di-*tert*-butylphosphinothioic acid; Chloride, *in* D-00118

$C_8H_{18}ClPSe$

Dibutylphosphinoselenoic acid; Chloride, *in* D-00114

$C_8H_{18}Cl_2NOP$

Dibutylphosphoramidic acid; Dichloride, *in* D-00134
Octylphosphoramidic acid; Dichloride, *in* O-00045

$C_8H_{18}Cl_2NO_3P$

tert-Butyl hydrogen bis(2-chloroethyl)-phosphoramidate, *in* B-00121

$C_8H_{18}Cl_2NP$

Di-*tert*-butylphosphoramidous acid; Dichloride, *in* D-00137

$C_8H_{18}Cl_2N_2OP_2$

2,4-Dichloro-1,3-di-*tert*-butyl-1,3,2,4-diazadiphosphetidine; 2-Oxide, *in* D-00163

$C_8H_{18}Cl_2N_2OP_2S$

2,4-Dichloro-1,3-di-*tert*-butyl-1,3,2,4-diazadiphosphetidine; 2-Oxide, 4-sulfide, *in* D-00163

$C_8H_{18}Cl_2N_2O_2P_2$

2,4-Dichloro-1,3-di-*tert*-butyl-1,3,2,4-diazadiphosphetidine; 2,4-Dioxide, *in* D-00163

$C_8H_{18}Cl_2N_2P_2$

2,4-Dichloro-1,3-di-*tert*-butyl-1,3,2,4-diazadiphosphetidine, D-00163

$C_8H_{18}Cl_2N_2P_2S$

2,4-Dichloro-1,3-di-*tert*-butyl-1,3,2,4-diazadiphosphetidine; 2-Sulfide, *in* D-00163

$C_8H_{18}FOP$

Bis(2-methylpropyl)phosphinic acid; Fluoride, *in* B-00346
Dibutylphosphinic acid; Fluoride, *in* D-00109
Di-*tert*-butylphosphinic acid; Fluoride, *in* D-00110

$C_8H_{18}FO_3P$

Bis(1-methylpropyl) phosphorofluoridate, *in* B-00341
Bis(2-methylpropyl) phosphorofluoridate, B-00362
Dibutyl phosphorofluoridate, D-00153

$C_8H_{18}FP$

▷Di-*tert*-butylfluorophosphine, *in* D-00122

$C_8H_{18}FPS$

Dibutylphosphinothioic acid; Fluoride, *in* D-00117
Di-*tert*-butylphosphinothioic acid; Fluoride, *in* D-00118

$C_8H_{18}F_2NP$
Di-*tert*-butylphosphoramidous acid; Difluoride, *in* D-00137

$C_8H_{18}IOP$
Di-*tert*-butylphosphinic acid; Iodide, *in* D-00110

$C_8H_{18}IP$
Dibutyliodophosphine, *in* D-00121
Di-*tert*-butyliodophosphine, *in* D-00122

$C_8H_{18}IPS$
Dibutylphosphinothioic acid; Iodide, *in* D-00117

$C_8H_{18}I_2P_2$
1,4-Dimethyl-1,4-diphosphoniabicyclo[2.2.2]-octane diiodide, *in* D-01169

$C_8H_{18}NOP$
2,2,3,4,4-Pentamethylphosphetanamine 1-oxide, P-00021

$C_8H_{18}NO_2P$
2-*tert*-Butylamino-4-methyl-1,3,2-dioxaphosphorinane, B-00541
N-Cyclohexyl-*P*-ethylphosphonamidic acid, C-00246
Diethyl 1-pyrrolidinylphosphonite, *in* P-00520
Diisopropyl 1-aziridinylphosphonite, *in* A-00154
2-Dimethylamino-4,4,5,5-tetramethyl-1,3,2-dioxaphospholane, D-00710
2-Methoxy-3,3,4,4-tetramethyl-1,2-azaphospholidine 2-oxide, *in* T-00191
2-Methoxy-3,3,5,5-tetramethyl-1,2-azaphospholidine 2-oxide, *in* T-00208
2-Methoxy-4,4,5,5-tetramethyl-1,2-azaphospholidine 2-oxide, *in* T-00192

$C_8H_{18}NO_2PS$
O,O-Dimethyl cyclohexylphosphoramidothioate, *in* C-00271
4,4,5,5-Tetramethylethylene dimethylphosphoramidothioate, *in* D-00710

$C_8H_{18}NO_2PSe$
O,O-1,3-Butylene *tert*-butylphosphoramidoselenoate, *in* B-00541

$C_8H_{18}NO_3P$
1,3-Butylene *tert*-butylphosphoramidate, *in* B-00541
Dimethyl cyclohexylphosphoramidate, *in* C-00269
Ethyl *P*-acetyl-*N,N*-diethylphosphonamidate, *in* A-00014
Isopropyl *P*-methyl-4-morpholinylphosphonamidate, *in* M-00175
4,4,5,5-Tetramethylethylene dimethylphosphoramidate, *in* D-00710
2,2,3,3-Tetramethyl-1,4,6-trioxa-9-aza-5-phospha(5-P^V)spiro[4.4]nonane, T-00238
3,5,5-Trimethyl-2-dimethylamino-1,2-oxaphospholan-3-ol 2-oxide, *in* T-00519

$C_8H_{18}NO_4P$
[Bis(1-methylethoxy)phosphinyl]acetamide, *in* D-00633
Bis(1-methylethyl) (2-amino-2-oxoethyl)-phosphonate, *in* A-00060
Diethyl (2-dimethylamino-2-oxoethyl)-phosphonate, *in* D-00703
Diethyl 4-morpholinylphosphonate, *in* M-00523
(Dipropoxyphosphinyl)acetamide, *in* D-01199

$C_8H_{18}NO_4PS_2$
▷Vamidothion, V-00001

$C_8H_{18}NO_5P$
Triethyl phosphonoglycinate, *in* P-00368

$C_8H_{18}NP$
Di-*tert*-butylphosphazene, D-00106

$C_8H_{18}N_2O_6$
Di-4-morpholinyldiphosphonic acid, *in* M-00523

$C_8H_{18}N_3OP$
P,P-Bis(1-aziridinyl)-*N,N*-diethylphosphinic amide, *in* B-00092
Di-*tert*-butylphosphinic acid; Azide, *in* D-00110
2,4,6,7-Tetramethyl-2,6,7-triaza-1-phosphabicyclo[2.2.2]octane; 1-Oxide, *in* T-00237

$C_8H_{18}N_3P$
P,P-Bis(1-aziridinyl)-*N,N*-diethylphosphinous amide, *in* B-00096
Octahydro-1,9-dimethyl-[1,3,2]-diazaphosphorino[1,2-*a*][1,3,2]-diazaphosphorine, O-00005
2,4,6,7-Tetramethyl-2,6,7-triaza-1-phosphabicyclo[2.2.2]octane, T-00237

$C_8H_{18}N_3PS$
2,4,6,7-Tetramethyl-2,6,7-triaza-1-phosphabicyclo[2.2.2]octane; 1-Sulfide, *in* T-00237

$C_8H_{18}N_3PSe$
2,4,6,7-Tetramethyl-2,6,7-triaza-1-phosphabicyclo[2.2.2]octane; 1-Selenide, *in* T-00237

$C_8H_{18}O_4PSi_2$
Bis(trimethylsilyl) acetylphosphonate, *in* A-00026

$C_8H_{18}O_4P_2$
1,3-Bis(1-methylethoxy)-1,3-diphosphetane 1,3-dioxide, *in* D-00620

$C_8H_{18}O_6P_2$
Tetramethyl (2-butene-1,4-diyl)bisphosphonate, *in* B-00519

$C_8H_{18}O_8P_2$
Methyl 3,3-bis(dimethoxyphosphinyl)-propanoate, *in* D-01187
Tetramethyl [1-(acetoxy)ethylidene]-bisphosphonate, *in* A-00005

$C_8H_{18}P_2$
1,6-Diphosphecane, D-01171

$C_8H_{18}P_2S_4$
2,4-Bis(1-methylpropyl)-1,3,2,4-dithiodiphosphetane 2,4-disulfide, *in* M-00382
2,4-Dibutyl-1,3,2,4-dithiaphosphetane 2,4-disulfide, *in* B-00614

$C_8H_{19}ClNOP$
Dibutylphosphoramidochloridous acid, D-00135

$C_8H_{19}N_2O_2P$
6,9-Dimethyl-1,3-dioxa-6,9-diaza-2-phosphacycloundecane, D-00728
2,2,7,7-Tetramethyl-1,6-dioxa-4,9-diaza-5-phospha(5-P^V)spiro[4.4]nonane, T-00195
3,3,8,8-Tetramethyl-1,6-dioxa-4,9-diaza-5-phospha(P^V)spiro[4.4]nonane, T-00196

$C_8H_{19}N_2O_3P$
Ethyl [bis(dimethylamino)phosphinyl]acetate, *in* B-00193

$C_8H_{19}OP$
Bis(2-methylpropyl)phosphine; Oxide, *in* B-00344
Dibutylphosphine; Oxide, *in* D-00107
Di-*tert*-butylphosphine; Oxide, *in* D-00108
Dibutylphosphinous acid, D-00121
Di-*tert*-butylphosphinous acid, D-00122
Ethyl diisopropylphosphinite, *in* D-00648
Octylphosphine; Oxide, *in* O-00041

$C_8H_{19}OPS$
Bis(1-methylpropyl)phosphinothioic acid, B-00350
Bis(2-methylpropyl)phosphinothioic acid, B-00351
S-Butyl diethylphosphinothioate, *in* D-00330
Dibutylphosphinothioic acid, D-00117
Di-*tert*-butylphosphinothioic acid, D-00118

$C_8H_{19}OPSSe_2$
Se,Se-Dibutyl phosphorodiselenothioate, D-00149

$C_8H_{19}OPS_2$
O-Ethyl hydrogen hexylphosphonodithioate, *in* H-00099
O-Isopropyl hydrogen pentylphosphonodithioate, *in* P-00051

$C_8H_{19}OPSe$
Bis(1-methylpropyl)phosphinoselenoic acid, B-00349
Se-Butyl diethylphosphinoselenoate, *in* D-00327
Dibutylphosphinoselenoic acid, D-00114

$C_8H_{19}O_2P$
Bis(1-methylpropyl)phosphinic acid, B-00345
Bis(2-methylpropyl)phosphinic acid, B-00346
Butyl-*tert*-butylphosphinic acid, B-00546
▷Dibutylphosphinic acid, D-00109
Di-*tert*-butylphosphinic acid, D-00110
Dibutyl phosphonite, D-00129
Diethyl butylphosphonite, *in* B-00615
Diethyl *tert*-butylphosphonite, *in* B-00616
Diisopropyl ethylphosphonite, *in* E-00153
Ethyl diisopropylphosphinate, *in* D-00641
Ethyl dipropylphosphinate, *in* D-01205
2-Methylpropyl methylpropylphosphinate, *in* M-00375
Octylphosphonous acid, O-00044

$C_8H_{19}O_2PS$
O,O-Dibutyl phosphonothioate, D-00132
O,O-Diethyl butylphosphonodithioate, *in* B-00611
O,S-Diethyl butylphosphonothioate, *in* B-00611
O,O-Diisopropyl ethylphosphonothioate, *in* E-00148

$C_8H_{19}O_2PS_2$
O,O-Bis(1-methylpropyl) phosphorodithioate, B-00359
O,O-Bis(2-methylpropyl) phosphorodithioate, B-00360
S,S-Bis(2-methylpropyl) phosphorodithioate, B-00361
O,O-Dibutyl phosphorodithioate, D-00150
O,O-Di-*tert*-butyl phosphorodithioate, D-00151
S,S-Dibutyl phosphorodithioate, D-00152
▷Ethoprophos, E-00023

$C_8H_{19}O_2PS_3$
▷Disulfoton, D-01236

$C_8H_{19}O_2PSe$
O,O-Dibutyl phosphonoselenoate, D-00131
O,O-Diethyl butylphosphonoselenoate, *in* B-00609
O,O-Dipropyl phosphonoselenoate, *in* E-00145

$C_8H_{19}O_2PSe_2$
O,O-Dibutyl phosphorodiselenoate, D-00148

$C_8H_{19}O_3P$
Bis(1-methylpropyl) phosphonate, B-00353
Bis(2-methylpropyl) phosphonate, B-00354
▷Dibutyl phosphonate, D-00127
Di-*tert*-butyl phosphonate, D-00128
Diethyl 3-buten-1-ylphosphonate, *in* B-00526
Diethyl butylphosphonate, *in* B-00594
Diethyl (1,1-dimethylethyl)phosphonate, *in* B-00595
Diethyl (1-methylpropyl)phosphonate, *in* M-00377
Diethyl (2-methylpropyl)phosphonate, *in* M-00378
Dimethyl hexylphosphonate, *in* H-00097
Octylphosphonic acid, O-00042

$C_8H_{19}O_3PS$
O,O-Bis(2-methylpropyl) phosphorothioate, B-00363
O,O-Dibutyl phosphorothioate, D-00155
O,O-Di-*tert*-butyl phosphorothioate, D-00156

$C_8H_{19}O_3PS_2$
▷Demeton-*O*, D-00019
O,O-Diethyl *S*-[2-(ethylthio)ethyl] phosphorothioate, D-00294

$C_8H_{19}O_3PSe$

O,O-Dibutyl phosphoroselenoate, D-00154

$C_8H_{19}O_4P$

Bis(1-methylpropyl) phosphate, B-00341
Bis(2-methylpropyl) phosphate, B-00342
OO-tert-Butyl *O*-ethyl ethylphosphonoperoxoate, *in* E-00144
OO-tert-Butyl *O*-isopropyl methylphosphonoperoxoate, *in* M-00310
▷Dibutyl phosphate, D-00104
Di-*tert*-butyl phosphate, D-00105
Diethyl (2-ethoxyethyl)phosphonate, *in* H-00146
Ethyl bis(ethoxymethyl)phosphinate, *in* B-00272
Mono(2-ethylhexyl) phosphate, M-00470
Monooctyl phosphate, M-00501

$C_8H_{19}O_5P$

OO-tert-Butyl *O,O*-diethyl phosphoroperoxoate, B-00551
Diethyl[(2-methoxyethoxy)methyl]phosphonate, *in* M-00040
2,2,2-Triethoxy-2,2-dihydro-1,3,2-dioxaphospholane, T-00436

$C_8H_{19}P$

Bis(1-methylpropyl)phosphine, B-00343
Bis(2-methylpropyl)phosphine, B-00344
Dibutylphosphine, D-00107
Di-*tert*-butylphosphine, D-00108
Octylphosphine, O-00041

$C_8H_{19}PS$

Bis(2-methylpropyl)phosphine; Sulfide, *in* B-00344
Bis(2-methylpropyl)phosphinothious acid, B-00352
Dibutylphosphine; Sulfide, *in* D-00107
Di-*tert*-butylphosphinothious acid, D-00120

$C_8H_{19}PS_2$

Bis(1-methylpropyl)phosphinodithioic acid, B-00347
Bis(2-methylpropyl)phosphinodithioic acid, B-00348
Dibutylphosphinodithioic acid, D-00112
Di-*tert*-butylphosphinodithioic acid, D-00113
Diethyl butylphosphonodithioite, *in* B-00607
Ethyl diisopropylphosphinodithioate, *in* D-00643

$C_8H_{19}PSe$

Bis(2-methylpropyl)phosphine; Selenide, *in* B-00344

$C_8H_{19}PSi$

1,1,3,3-Tetramethyl-1-phospha-3-silacyclohex-1-ene, T-00212

$C_8H_{19}PTe$

Di-*tert*-butylphosphinotellurous acid, D-00116

$C_8H_{20}BrN_2P$

Tetraethylphosphorodiamidous acid; Bromide, *in* T-00066

$C_8H_{20}BrP$

Tetraethylphosphonium(1+); Bromide, *in* T-00056

$C_8H_{20}ClN_2OP$

Tetraethylphosphorodiamidic chloride, T-00060

$C_8H_{20}ClN_2P$

▷Tetraethylphosphorodiamidous chloride, T-00067

$C_8H_{20}ClN_2PS$

Tetraethylphosphorodiamidothioic chloride, T-00065

$C_8H_{20}ClN_2PSe$

Tetraethylphosphorodiamidoselenoic acid; Chloride, *in* T-00061

$C_8H_{20}ClP$

Tetraethylphosphonium(1+); Chloride, *in* T-00056

$C_8H_{20}FN_2OP$

▷Butafox, *in* D-00146
▷Tetraethylphosphorodiamidic acid; Fluoride, *in* T-00059

$C_8H_{20}IN_2P$

Tetraethylphosphorodiamidous acid; Iodide, *in* T-00066

$C_8H_{20}IP$

▷Tetraethylphosphonium(1+); Iodide, *in* T-00056

$C_8H_{20}IPS_3$

Ethyltri(ethylthio)phosphonium iodide, *in* T-00461

$C_8H_{20}NOP$

P,P-Bis(2-methylpropyl)phosphinic amide, *in* B-00346
P,P-Dibutylphosphinic amide, D-00111
P,P-Diisopropyl-*N,N*-dimethylphosphinic amide, *in* D-00642
Tetraethylphosphinic amide, *in* D-00323

$C_8H_{20}NOPS$

O-Ethyl triethylphosphonamidothioate, *in* T-00448
S-Ethyl triethylphosphonamidothioate, *in* T-00448

$C_8H_{20}NOPS_2$

S,S-Dibutyl phosphoramidodithioate, *in* D-00152

$C_8H_{20}NO_2P$

Butyl *N,N*-diethylphosphonamidate, *in* D-00337
N,N-Dibutylphosphonamidic acid, D-00126
Dibutylphosphoramidous acid, D-00136
Di-*tert*-butylphosphoramidous acid, D-00137
Diethyl *tert*-butylphosphoramidite, *in* B-00624
Diethyl diethylphosphoramidite, *in* D-00360
Ethyl triethylphosphonamidate, *in* T-00446
Isopropyl *N,N*-diethyl-*P*-methylphosphonamidate, *in* D-00296
2-Methylpropyl *N,P*-diethylphosphonamidate, *in* D-00336

$C_8H_{20}NO_2PS$

O,O-Diethyl *tert*-butylphosphoramidothioate, *in* B-00623
O,O-Diethyl diethylphosphoramidothioate, *in* D-00356
O,O-Diisopropyl dimethylphosphoramidothioate, *in* D-00871
O,O-Diisopropyl ethylphosphoramidothioate, *in* E-00159

$C_8H_{20}NO_2PSe$

O,O-Diethyl diethylphosphoramidoselenoate, *in* D-00353

$C_8H_{20}NO_3P$

Bis(2-methylpropyl)phosphoramidic acid, B-00355
▷Dibutyl phosphoramidate, D-00133
Di-*tert*-butyl phosphoramidate, *in* D-00105
Dibutylphosphoramidic acid, D-00134
Diethyl (1-aminobutyl)phosphonate, *in* A-00054
Diethyl (2-aminobutyl)phosphonate, *in* A-00055
Diethyl (3-aminobutyl)phosphonate, *in* A-00056
Diethyl (4-aminobutyl)phosphonate, *in* A-00057
Diethyl (1-amino-1-methylpropyl)phosphonate, *in* A-00093
Diethyl (1-amino-2-methylpropyl)phosphonate, *in* A-00094
Diethyl butylphosphoramidate, *in* B-00621
Diethyl *tert*-butylphosphoramidate, *in* B-00622
Diethyl diethylphosphoramidate, D-00281
Diethyl (2-methylpropyl)phosphoramidate, *in* M-00384
Diisopropyl dimethylphosphoramidate, *in* D-00855
Dimethyl dipropylphosphoramidate, *in* D-01214
Dimethyl hexylphosphoramidate, *in* H-00102
Dipropyl ethylphosphoramidate, *in* E-00156
Ethyl *N,N*-diethyl-*P*-(methoxymethyl)phosphonamidate, *in* D-00295
Octylphosphoramidic acid, O-00045
Triethyl ethylphosphorimidate, *in* E-00161

$C_8H_{20}NO_3PS$

O,O-Diethyl *S*-(2-dimethylaminoethyl) phosphorothioate, D-00284

$C_8H_{20}NO_3PSi_2$

Bis(trimethylsilyl) (cyanomethyl)phosphonate, *in* C-00215

$C_8H_{20}NO_4P$

Diethyl 2-dimethylaminoethyl phosphate, D-00283
Diisopropyl 2-aminoethyl phosphate, *in* A-00075
Diisopropyl (2-hydroxyethyl)phosphoramidate, *in* H-00148
2-[(Hydroxyisopropoxyphosphinyl)oxy]-*N,N,N*-trimethylethanaminium hydroxide, inner salt, *in* C-00198

$C_8H_{20}NP$

N-tert-Butyl-*P,P*-diethylphosphinous amide, *in* D-00334
P,P-Dibutylphosphinous amide, D-00123
P,P-Di-*tert*-butylphosphinous amide, D-00124
P,P-Diisopropyl-*N,N*-dimethylphosphinous amide, *in* D-00649
Tetraethylphosphinous amide, *in* D-00334

$C_8H_{20}NPS$

Di-*tert*-butylphosphinothioic amide, D-00119
Tetraethylphosphinothioic amide, *in* D-00331

$C_8H_{20}NPS_2$

Ethyl triethylphosphonamidodithioate, *in* T-00447

$C_8H_{20}NPSe$

P,P-Di-*tert*-butylphosphinoselenoic amide, D-00115
Tetraethylphosphinoselenoic amide, *in* D-00328

$C_8H_{20}N_3OP$

N,N,N',N'-Tetramethyl-*P*-4-morpholinylphosphonous diamide, *in* M-00529

$C_8H_{20}N_3OPS$

N,N,N',N'-Tetramethyl-*P*-4-morpholinylphosphonothioic diamide, *in* M-00528

$C_8H_{20}N_3PS$

3-Amino-2,3-di-*tert*-butyl-1,2,4,3-thiadiazaphosphetidine, A-00071

$C_8H_{20}N_5OP$

Tetraethylphosphorodiamidic acid; Azide, *in* T-00059

$C_8H_{20}N_5PS$

Tetraethylphosphorodiamidothioic azide, T-00064

$C_8H_{20}OP_2S_2$

Diethylphosphinothioic acid; Anhydride, *in* D-00330

$C_8H_{20}O_2P_2$

1,4-Bis(dimethylphosphinyl)butane, *in* B-00199

$C_8H_{20}O_2P_2S$

Diethylphosphinic diethylphosphinothioic anhydride, *in* D-00322

$C_8H_{20}O_2P_2S_3$

O,O'-Diethyl diethylthiodiphosphonate, *in* D-00405

$C_8H_{20}O_3P_2$

1,1-Diethoxy-2,2-diethyldiphosphine 1-oxide, D-00229
▷Diethylphosphinic acid; Anhydride, *in* D-00322

$C_8H_{20}O_4P_2$
Tetraethyl hypodiphosphite, T-00049

$C_8H_{20}O_4P_2S_2$
Tetraethyl thiohypophosphate, T-00072

$C_8H_{20}O_4P_2S_3$
▷*O,O,O′,O′*-Tetraethyl trithiopyrophosphate, T-00076

$C_8H_{20}O_4P_2S_4$
Bis(diethoxyphosphinothioyl) disulfide, B-00170

$C_8H_{20}O_4P_2S_5$
Bis(diethoxyphosphinothioyl) trisulfide, B-00171

$C_8H_{20}O_5P_2$
Dibutyldiphosphonic acid, D-00092
Diethyl diethyldiphosphonate, *in* D-00290
Tetraethyl pyrophosphite, T-00070

$C_8H_{20}O_5P_2S$
O,O,O′,O′-Tetraethyl thiohypophosphate, T-00073

$C_8H_{20}O_5P_2S_2$
▷*O,O,O,O*-Tetraethyl dithiopyrophosphate, T-00045
unsym-O,O,O,O-Tetraethyl dithiopyrophosphate, T-00046

$C_8H_{20}O_6P_2$
1,8-Octanediphosphonic acid, O-00035
Tetraethyl hypophosphate, T-00050

$C_8H_{20}O_6P_2S$
sym-Tetraethyl monothiopyrophosphate, T-00052
▷*unsym*-Tetraethyl monothiopyrophosphate, T-00053

$C_8H_{20}O_6P_2S_2$
▷Bis(diethoxyphosphinyl) disulfide, B-00172

$C_8H_{20}O_7P_3$
▷Tetraethyl pyrophosphate, T-00069

$C_8H_{20}P^{\oplus}$
Tetraethylphosphonium(1+), T-00056

$C_8H_{20}P_2$
1,4-Bis(dimethylphosphino)butane, B-00199
Tetraethyldiphosphine, T-00043

$C_8H_{20}P_2S$
Tetraethyldiphosphine; Monosulfide, *in* T-00043

$C_8H_{20}P_2S_2$
1,4-Bis(dimethylphosphinothioyl)butane, *in* B-00199
Tetraethyldiphosphine disulfide, T-00044

$C_8H_{20}P_2S_3$
Diethylphosphinodithioic acid; Anhydrosulfide, *in* D-00325

$C_8H_{20}P_2Se_3$
Diethylphosphinodiselenoic acid; Anhydroselenide, *in* D-00324

$C_8H_{20}P_4$
Tetraethyltetraphosphetane, T-00071

$C_8H_{20}P_6$
Octahydro-1,2,3,4,5,6-hexamethyl-[1,2,3]-triphospholo[4,5-*d*]-1,2,3-triphosphole, O-00017

$C_8H_{21}NO_5PS$
O,O,O′,O′-Tetraethyl thioimidodiphosphate, T-00074

$C_8H_{21}NO_5P_2S$
Triethyl (dimethoxyphosphinothioyl)-phosphorimidate, *in* D-00674

$C_8H_{21}NO_6P_2$
O,O,O′,O′-Tetraethyl imidodiphosphate, T-00051
Triethyl (dimethoxyphosphinyl)-phosphorimidate, *in* D-00683

$C_8H_{21}NP_2S_2$
N-(Diethylphosphinothioyl)-*P,P*-diethylphosphinothioic amide, *in* D-00331

$C_8H_{21}N_2OP$
P-Butyl-*N,N,N′,N′*-tetramethylphosphonic diamide, *in* B-00596
tert-Butyl tetramethylphosphorodiamidite, *in* T-00227
Tetraethylphosphorodiamidous acid, T-00066

$C_8H_{21}N_2OPS$
Tetraethylphosphorodiamidothioic acid, T-00063

$C_8H_{21}N_2OPSe$
Tetraethylphosphorodiamidoselenoic acid, T-00061

$C_8H_{21}N_2OPTe$
Tetraethylphosphorodiamidotelluroic acid, T-00062

$C_8H_{21}N_2O_2P$
Butyl tetramethylphosphorodiamidate, *in* T-00218
N,N′-Dibutylphosphorodiamidic acid, D-00146
Ethyl *N,N′*-dipropylphosphorodiamidate, *in* D-01223
Tetraethylphosphorodiamidic acid, T-00059

$C_8H_{21}N_2PS$
N,N,N′,N′-Tetraethylphosphonothioic diamide, T-00057

$C_8H_{21}N_2PS_2$
N,N′-Dibutylphosphorodiamidodithioic acid, D-00147

$C_8H_{21}N_2PSe_2$
N,N′-Bis(2-methylpropyl)-phosphorodiamidodiselenoic acid, B-00358

$C_8H_{21}O_2PSSi$
O-Methyl *O*-trimethylsilyl *tert*-butylphosphonothioate, *in* M-00107

$C_8H_{21}O_4PSi$
Diethyl[(trimethylsilyloxy)methyl]phosphonate, *in* T-00556

$C_8H_{21}O_5PSi_2$
[Bis(trimethylsilyloxy)phosphinyl]acetic acid, B-00434
Methyl bis(trimethylsilyloxy)-phosphinecarboxylate oxide, *in* B-00433

$C_8H_{21}PSi_2$
1,1-Dihydro-1,1,3,3,4,4-hexamethyl-1-phospha-3,4-disilacyclopent-1-ene, D-00486

$C_8H_{21}P_3$
[Bis(dimethylphosphino)methylene]trimethyl-phosphorane, B-00203

$C_8H_{22}ClO_3PSi_2$
Bis(trimethylsilyl) (2-chloroethyl)phosphonate, *in* C-00074

$C_8H_{22}N_2O_3P_2$
▷*P,P′*-Diethyl-*N,N,N′,N′*-tetramethyl-diphosphonic diamide, *in* D-00291

$C_8H_{22}N_2O_4P_2S_2$
O,O,O′,O′-Tetraethyl hydrazodiphosphonothioate, T-00048

$C_8H_{22}N_3OP$
N,N,N′,N′-Tetraethylphosphoric triamide, T-00058

$C_8H_{22}O_4P_2$
Tetramethyl 1,4-butanediylbisphosphonite, *in* B-00515

$C_8H_{23}N_3O_4P_2$
[Bis(diethylamino)phosphinyl]phosphorimidic acid, B-00176

$C_8H_{23}O_3PSi_2$
Bis(trimethylsilyl) ethylphosphonate, *in* E-00129

$C_8H_{23}O_4PSi_2$
Bis(trimethylsilyl) (2-hydroxyethyl)-phosphonate, *in* H-00146

$C_8H_{23}PS_2$
Dibutyl ethylphosphonodithioite, *in* E-00141

$C_8H_{24}Cl_2N_7P_3$
2,4-Dichloro-2,4,6,6-tetrakis(dimethylamino)-2,2,4,4,6,6-hexahydro-1,3,5-triaza-2,4,6-triphosphorine, D-00194

$C_8H_{24}IN_4P$
Tetrakis(dimethylamino)iodophosphorane, T-00178

$C_8H_{24}NPSSi_2$
P,P-Dimethyl-*N,N*-bis(trimethylsilyl)-phosphinothioic amide, *in* D-00840

$C_8H_{24}NPSi_2$
P,P-Dimethyl-*N,N*-bis(trimethylsilyl)-phosphinous amide, *in* D-00845

$C_8H_{24}N_4OP_2S_2$
Octamethyl thiodiphosphoramide, O-00034

$C_8H_{24}N_4O_2P_2S$
sym-Octamethylmonothiopyrophosphoramide, O-00030
unsym-Octamethyl-monothiopyrophosphoramide, O-00031

$C_8H_{24}N_4O_3P_2$
▷Octamethyldiphosphoramide, O-00029

$C_8H_{24}N_4P_2$
Tetrakis(dimethylamino)diphosphine, T-00177

$C_8H_{35}O_3P$
Bis(1-methylheptyl) phosphonate, B-00307

$C_8H_{51}O_3P$
▷Tris(2-ethylhexyl) phosphite, T-00780

$C_9Cl_2F_{12}P_2$
7,7-Dichloro-2,3,5,6-tetrakis(trifluoromethyl)-1,4-diphosphabicyclo[2.2.1]hepta-2,5-diene, D-00195

$C_9H_4NO_3P$
[(1-Benzylamino)ethyl]phosphonic acid, *in* A-00078

$C_9H_6N_3PS_3$
Tri(2-thiazolyl)phosphine, T-00873

$C_9H_7F_{12}O_4P$
2,2,3,3-Tetrakis(trifluoromethyl)-1,4,6,10-tetraoxa-5-phospha(5-P^V)spiro[4.5]decane, T-00187

$C_9H_7N_2O_3P$
Benzylphosphonic acid; Diisocyanate, *in* B-00060

C_9H_7P
Isophosphinoline, I-00027

$C_9H_8Cl_2NO_3P$
4-Chlorophenyl 2-cyanoethyl phosphorochloridate, C-00125

$C_9H_8F_{12}NO_3P$
2,2,3,3-Tetrakis(trifluoromethyl)-1,4,6-trioxa-10-aza-5-phospha(5-P^V)spiro[4.5]decane, T-00190

$C_9H_8NO_3P$
2-Quinolinephosphonic acid, Q-00002
4-Quinolinephosphonic acid, Q-00003

$C_9H_8NO_5P$
N-Phthalimidomethylphosphonic acid, P-00426

C_9H_8NP
2-Phenyl-1*H*-1,3-azaphosphole, P-00088

$C_9H_9Cl_2O_2P$
3-(Chlorophenylphosphinyl)propanoyl chloride, *in* H-00178

$C_9H_9F_6O_3PS_2$
4′-Methyl-4,5-bis(trifluoromethyl)spiro[1,3,2-dithiaphosphole-2,1′-[2,6,7]trioxa[1]-phosphabicyclo[2.2.2]octane], M-00092

$C_9H_9F_{12}N_2O_3P$
2,2-Dihydro-3,3-bis(trifluoromethyl)-1-[[2,2,2-trifluoro-1-(trifluoromethyl)ethylidene]-amino]azaphosphiridine; 2,2,2-Trimethoxy, *in* D-00446

$C_9H_9F_{12}O_2P$
2,2-Dihydro-2,2,2-trimethyl-4,4,5,5-tetrakis(trifluoromethyl)-1,3,2-dioxaphospholane, D-00601

$C_9H_9F_{12}O_5P$
2,2-Dihydro-2,2,2-trimethoxy-4,4,5,5-tetrakis(trifluoromethyl)-1,3,2-dioxaphospholane, D-00595

$C_9H_9N_2P$
5-Methyl-2-phenyl-2*H*-1,2,3-diazaphosphole, M-00197

$C_9H_9N_6OP$
1,1′,1″-Phosphinylidynetris(1*H*-imidazole), *in* T-00495
1,1′,1″-Phosphinylidynetris(1*H*-pyrazole), *in* T-00722

$C_9H_9N_6P$
Tri-1-imidazolylphosphine, T-00495
Tri(2-imidazolyl)phosphine, T-00496
Tri(1*H*-pyrazol-1-yl)phosphine, T-00722

$C_9H_9N_6PS$
1,1′,1″-Phosphinothioyldiynetris(1*H*-pyrazole), *in* T-00722

C_9H_9OP
Tri-1-propynylphosphine; Oxide, *in* T-00721

$C_9H_9O_2P$
1,2-Dihydro-2-hydroxyisophosphinoline 2-oxide, D-00495
1,2-Dihydro-1-hydroxyphosphinoline 1-oxide, D-00505
1*H*-Indene-1-phosphonous acid, I-00006

$C_9H_9O_3P$
Diethyl(3-methyl-1-butenyl)phosphonate, *in* M-00101
1*H*-Inden-2-ylphosphonic acid, I-00007
2-Phenyl-1,2-oxaphospholan-2-one 2-oxide, *in* O-00053
(1-Phenyl-1,2-propadienyl)phosphonic acid, P-00298

C_9H_9P
1,2-Dihydroisophosphinoline, D-00510
Tri-1-propynylphosphine, T-00721

C_9H_9PS
Tri-1-propynylphosphine; Sulfide, *in* T-00721

$C_9H_{10}ClN_2O_5PS$
Azemethephos, A-00146

$C_9H_{10}ClP$
1-Chloro-1,2,3,4-tetrahydrophosphinoline, *in* T-00135

$C_9H_{10}Cl_2NO_3P$
(4-Carbethoxyphenyl)phosphoramidic acid; Dichloride, *in* C-00006

$C_9H_{10}Cl_3O_2P$
Ethyl phenyl(trichloromethyl)phosphinate, *in* P-00327

$C_9H_{10}Cl_4NO_3P$
Diethyl (2,3,5,6-tetrachloro-4-pyridinyl)-phosphonate, *in* T-00035

$C_9H_{10}NOP$
1-Phenyl-1,2-azaphospholidin-5-one, P-00092

$C_9H_{10}NOPS_2$
O-Ethyl phenyl-phosphonisothiocyanatidothioate, *in* P-00227

$C_9H_{10}NO_3P$
1,2-Dihydro-1-methyl-4*H*-3,1,2-benzoxazaphosphorin-4-one; 1-Methoxy, *in* D-00523
Dimethyl (4-cyanophenyl)phosphonate, *in* C-00225
Spiro[1,3,2-benzoxazaphosphole-2(3*H*),2′-[1,2]oxaphospholan]-5′-one, S-00013

$C_9H_{10}NO_3PS$
▷Cyanophos, C-00226

$C_9H_{10}NP$
Cyanoethylphenylphosphine, *in* E-00108
4,5-Dihydro-3-phenyl-3*H*-1,3-azaphosphole, D-00549

$C_9H_{10}NPS$
3-Phenyl-1,3-azaphospholidin-2-thione, P-00093

$C_9H_{10}NPS_2$
3-Phenyl-1,3-azaphospholidin-2-thione; 3-Sulfide, *in* P-00093

$C_9H_{11}Br_2O_3P$
(1,2-Dibromo-1-phenylpropyl)phosphonic acid, D-00068

$C_9H_{11}Cl_2OP$
(2,4,5-Trimethylphenyl)phosphonic acid; Dichloride, *in* T-00526

$C_9H_{11}Cl_2O_3PS$
Toclofos-Methyl, T-00321

$C_9H_{11}Cl_3NO_3PS$
▷Chlorpyrifos, C-00194

$C_9H_{11}F_2N_2OP$
2,2-Difluoro-2,2-dihydro-1,3-dimethyl-2-phenyl-1,3,2-diazaphosphetidin-4-one, *in* D-00459

$C_9H_{11}N_2OP$
2,3-Dihydro-1,3-dimethyl-1,3,2-benzodiazaphosphorin-4(1*H*)-one, D-00457

$C_9H_{11}N_2O_2P$
2,3-Dihydro-1,3-dimethyl-1,3,2-benzodiazaphosphorin-4(1*H*)-one; 2-Oxide, *in* D-00457
2,3-Dihydro-2-methoxy-5-methyl-3-phenyl-1,3,4,2-oxadiazaphosphole, D-00513

$C_9H_{11}N_2O_2PS$
2,3-Dihydro-2-methoxy-5-methyl-3-phenyl-1,3,4,2-oxadiazaphosphole; 2-Sulfide, *in* D-00513

$C_9H_{11}N_2O_3P$
Dimethyl (diazophenylmethyl)phosphonate, *in* D-00037

$C_9H_{11}OP$
2-Phenyl-1,2-oxaphospholane, P-00174

$C_9H_{11}OPS$
2-Phenyl-1,2-oxaphospholane; 2-Sulfide, *in* P-00174

$C_9H_{11}O_2P$
2,3-Dihydro-2-methoxy-1*H*-isophosphindole 2-oxide, *in* D-00494
2,3-Dihydro-1-methoxy-1*H*-phosphindole 1-oxide, *in* H-00120
1,5-Dihydro-3-methyl-2,4,3-benzodioxaphosphepin, D-00516
1-Hydroxy-1,2,3,4-tetrahydrophosphinoline 1-oxide, H-00195
Methyl ethenylphenylphosphinate, *in* P-00337
4-Methyl-2-phenyl-1,3,2-dioxaphospholane, M-00198
Methyl (2-phenylethenyl)phosphinate, *in* P-00153
2-Phenyl-1,3,2-dioxaphosphorinane, P-00116
2-Phenyl-1,2-oxaphospholane; 2-Oxide, *in* P-00174
3-Phenyl-1,3-oxaphosphorinan-2-one, P-00177
Phenyl(1-propenyl)phosphinic acid, P-00299
Phenyl(2-propenyl)phosphinic acid, P-00300

$C_9H_{11}O_2PS$
1,5-Dihydro-3-methyl-2,4,3-benzodioxaphosphepin; 4-Sulfide, *in* D-00516
1,3,2-Oxathiaphosphorinane 2-oxide; 2-Ph, *in* O-00057
3-Phenyl-1,3-oxaphosphorinan-2-one; 3-Sulfide, *in* P-00177
O,O-Trimethylene phenylphosphonothioate, *in* P-00116

$C_9H_{11}O_2PS_2$
2-Mercapto-4*H*-1,3,2-benzodioxaphosphorin 2-sulfide; 2-Ethylthio (2-Et ester), *in* M-00006

$C_9H_{11}O_3P$
1,3-Dihydro-1-hydroxy-2,1-benzoxaphosphole 1-oxide; 1-Ethoxy (Et ester), *in* D-00490
2-Ethoxy-2,3-dihydro-1,2-benzoxaphosphole 2-oxide, *in* H-00119
Methyl acetylphenylphosphinate, *in* A-00021
Methyl hydrogen (2-phenylethenyl)-phosphonate, *in* P-00155
4-Methyl-2-phenoxy-1,3,2-dioxaphospholane, M-00190
4-Methyl-2-phenyl-1,3,2-dioxaphospholane; 2-Oxide, *in* M-00198
2-Phenoxy-1,3,2-dioxaphosphorinane, P-00068
(1-Phenyl-1-propenyl)phosphonic acid, P-00301
Trimethylene phenylphosphonate, *in* P-00116

$C_9H_{11}O_3PS$
O-Ethyl saligenin cyclic thiophosphate, *in* H-00110
S-Ethyl saligenin cyclic thiophosphate, *in* H-00110
1,3,2-Oxathiaphosphorinane 2-oxide; 2-Phenoxy, *in* O-00057

$C_9H_{11}O_3PSe$
2-Phenoxy-1,3,2-oxaselenaphosphorinane 2-oxide, P-00071

$C_9H_{11}O_4P$
(3,4-Dihydro-1*H*-2-benzopyran-1-yl)-phosphonic acid, D-00443
Dimethyl benzoylphosphonate, *in* B-00034
2-Hydroxy-4*H*-1,3,2-benzodioxaphosphorin 2-oxide; 2-Ethoxy (Et ester), *in* H-00109
2-Hydroxy-4-phenyl-1,3,2-dioxaphosphorinane 2-oxide, H-00171
3-(Hydroxyphenylphosphinyl)propanoic acid, H-00178
(2-Oxo-3-phenylpropyl)phosphonic acid, O-00068
Phenyl propylene phosphate, *in* M-00190
Trimethylene phenyl phosphate, *in* P-00068

$C_9H_{11}O_5P$
[(Acetyloxy)phenylmethyl]phosphonic acid, A-00020
Ethyl 4-phosphonobenzoate, *in* P-00380
Phosphonoacetic acid; Benzyl ester, *in* P-00376

$C_9H_{11}O_6P$
Ethyl 4-(phosphonoxy)benzoate, *in* P-00397

$C_9H_{11}P$
1,2,3,4-Tetrahydroisophosphinoline, T-00110
1,2,3,4-Tetrahydrophosphinoline, T-00135

$C_9H_{11}PS$
2-Phenyl-1,2-thiaphospholane, P-00324

$C_9H_{11}PS_2$
2,3-Dihydro-2,7-dimethyl-1,4,2-benzodithiaphosphorin, D-00458
4-Methyl-2-phenyl-1,3,2-dithiaphospholane, *in* M-00136
2-Phenyl-1,3,2-dithiaphosphorinane, *in* D-01247
2-Phenyl-1,2-thiaphospholane; 2-Sulfide, *in* P-00324
1,2-Thiaphospholane 2-sulfide; 2-Ph, *in* T-00298

$C_9H_{11}PS_2Se$
2,3-Dihydro-2,7-dimethyl-1,4,2-benzodithiaphosphorin; 2-Selenide, *in* D-00458

$C_9H_{12}BrPS$
Benzyl ethylphosphonobromidothioite, *in* E-00134

$C_9H_{12}ClOP$
Isopropylphenylphosphinic acid; Chloride, *in* I-00043

$C_9H_{12}ClOPS$
O-Isopropyl phenylphosphonochloridothioate, *in* P-00232
S-Isopropyl phenylphosphonochloridothionate, *in* P-00232
S-Propyl phenylphosphonochloridothioate, *in* P-00232

$C_9H_{12}ClO_2P$
Benzyl ethylphosphonochloridate, *in* E-00135

Ethyl benzylphosphonochloridate, *in* B-00063
Ethyl (chloromethyl)phenylphosphinate, *in* C-00103

$C_9H_{12}ClO_2PS_3$

▷Methyl trithion, M-00437

$C_9H_{12}ClO_3P$

Dimethyl [(2-chlorophenyl)methyl]-phosphonate, *in* C-00132
Isopropyl phenyl phosphorochloridate, *in* P-00269
Phenyl propyl phosphorochloridate, *in* P-00269
Tetramethylethylene chlorophosphate, *in* C-00185

$C_9H_{12}ClO_4P$

▷Heptenophos, H-00004

$C_9H_{12}ClPS$

Isopropylphenylphosphinothioic acid; Chloride, *in* I-00046
Isopropyl phenylphosphonochloridothioite, *in* P-00233

$C_9H_{12}Cl_2NOP$

N-Benzyl-*P*,*P*-bis(chloromethyl)phosphinic amide, *in* B-00128

$C_9H_{12}F_6O_6P_2$

Tetramethyl (3,3,4,4,5,5-hexafluoro-1-cyclopentene-1,2-diyl)bisphosphonate, *in* H-00018

$C_9H_{12}IO_2P$

Ethyl (iodomethyl)phenylphosphinate, *in* I-00012

$C_9H_{12}NOP$

2-Methyl-3-phenyl-1,3,2-oxazaphospholidine, M-00212
3-Methyl-2-phenyl-1,3,2-oxazaphospholidine, M-00213
Tetrahydro-2-phenyl-2*H*-1,3,2-oxazaphosphorine, T-00132

$C_9H_{12}NOPS$

2-Methyl-3-phenyl-1,3,2-oxazaphospholidine; 2-Sulfide, *in* M-00212
3-Methyl-2-phenyl-1,3,2-oxazaphospholidine; 2-Sulfide, *in* M-00213
Tetrahydro-2-phenyl-2*H*-1,3,2-oxazaphosphorine; 2-Sulfide, *in* T-00132
Tetrahydro-2*H*-1,3,2-thiazaphosphorine 2-oxide; 2-Ph, *in* T-00154

$C_9H_{12}NOPS_2$

2-(4-Methoxyphenyl)-1,3,2-thiazaphosphetidine 2-sulfide; N-Me, *in* M-00083

$C_9H_{12}NO_2P$

3-Methyl-2-phenoxy-1,3,2-oxazaphospholidine, M-00192
Methyl phenyl 1-aziridinylphosphonite, *in* A-00154
2-Methyl-3-phenyl-1,3,2-oxazaphospholidine; 2-Oxide, *in* M-00212
Tetrahydro-2-phenoxy-2*H*-1,3,2-oxazaphosphorine, T-00126

$C_9H_{12}NO_2PS$

Tetrahydro-2-phenoxy-2*H*-1,3,2-oxazaphosphorine; 2-Sulfide, *in* T-00126
Tetrahydro-2*H*-1,3,2-thiazaphosphorine 2-oxide; 2-Phenoxy, *in* T-00154

$C_9H_{12}NO_3P$

1-(Benzylideneamino)ethylphosphonic acid, B-00049
1,4-Dihydrospiro[2*H*-3,1,2-benzoxazaphosphorine-2,2′-[1,3,2]-dioxaphospholane], D-00576
Methyl *N*-benzoyl-*P*-methylphosphonamidate, *in* B-00031
3-Methyl-2-phenoxy-1,3,2-oxazaphospholidine; 2-Oxide, *in* M-00192
Tetrahydro-2-phenoxy-2*H*-1,3,2-oxazaphosphorine; 2-Oxide, *in* T-00126

$C_9H_{12}NO_4P$

[1-(Benzoylamino)ethyl]phosphonic acid, *in* A-00078
2-(Benzoylamino)ethylphosphonic acid, *in* A-00079
Dimethyl benzoylphosphoramidate, *in* B-00036
Ethyl (2-nitrophenyl)methylphosphinate, *in* N-00042
Phenyl [(methoxycarbonyl)methyl]-phosphonamidate, *in* M-00028

$C_9H_{12}NO_5P$

(2-Carbethoxyphenyl)phosphoramidic acid, C-00004
(3-Carbethoxyphenyl)phosphoramidic acid, C-00005
(4-Carbethoxyphenyl)phosphoramidic acid, C-00006
Dimethyl [(2-nitrophenyl)methyl]phosphonate, *in* N-00044
[(4-Nitrophenyl)methyl]phosphonic acid; Di-Me ester, *in* N-00046
N-Phosphophenylalanine, P-00400

$C_9H_{12}NO_5PS$

▷Fenitrothion, F-00005

$C_9H_{12}NO_6P$

2-Amino-3-[(hydroxyphenoxyphosphinyl)oxy]-propanoic acid, *in* P-00417
▷Fenitrooxon, F-00004
N-Phosphotyrosine, P-00422
O-Phosphotyrosine, P-00423

$C_9H_{12}NP$

2-Phenyl-1,2-azaphospholidine, P-00089
2-Phenyl-1,3-azaphospholidine, P-00090
3-Phenyl-1,3-azaphospholidine, P-00091

$C_9H_{12}N_2O_2P$

Methyl (1-diazoethyl)phenylphosphinate, *in* D-00032

$C_9H_{12}N_3OP$

Isopropylphenylphosphinic acid; Azide, *in* I-00043
3,3′,3″-Phosphinylidynetris[propanenitrile], *in* T-00752

$C_9H_{12}N_3O_7P$

Cytidine 2′,3′-phosphate, C-00294

$C_9H_{12}N_3P$

Tris(2-cyanoethyl)phosphine, T-00752

$C_9H_{12}N_3PS$

▷3,3′,3″-Phosphinothioylidynetris[propanenitrile], *in* T-00752

$C_9H_{12}N_3PSe$

3,3′,3″-Phosphinoselenoylidynetris[propanenitrile], *in* T-00752

$C_9H_{12}OP$

6-Oxa-5-phospha(5-P^V)spiro[4.5]decane; 5-Me, *in* O-00050

$C_9H_{12}O_3P$

3-Phenyl-2-propenylphosphonic acid, P-00302

$C_9H_{12}P_2$

2,2,4,4-Tetrahydro-1*H*-2,4-benzodiphosphepin, T-00087

$C_9H_{13}NO_9P_2$

N,*O*-Diphosphotyrosine, D-01188

$C_9H_{13}N_2OP$

Hexahydro-2-phenoxy-1,3,2-diazaphosphorine, H-00039
N,*N*′-Trimethylene-*P*-phenylphosphonic diamide, *in* H-00020

$C_9H_{13}N_2OPS$

O-Phenyl *N*,*N*′-trimethylenephosphorodiamidothioate, *in* H-00039

$C_9H_{13}N_2O_2P$

N-(Dimethylphosphinyl)-*N*′-phenylurea, *in* D-00830
Phenyl *N*,*N*′-trimethylenephosphorodiamidate, *in* H-00039

$C_9H_{13}N_2O_3P$

Dimethyl phosphorohydrazidate; N^2-Benzylidene, *in* D-00904
Tetrahydro-4-methyl-2*H*-1,3,4,2-oxadiazaphosphorine 2-oxide; 2-Phenoxy, *in* T-00116

$C_9H_{13}N_2O_3PS$

O,*S*-Dimethyl [(phenylamino)carbonyl]-phosphoramidothioate, *in* P-00082

$C_9H_{13}N_2O_4P$

Dimethyl (phenylamino)-carbonylphosphoramidate, *in* P-00081

$C_9H_{13}N_2O_5P$

Trimethyl (4-nitrophenyl)phosphorimidate, *in* N-00063

$C_9H_{13}OP$

Benzyldiphenylphosphine; Oxide, *in* B-00046
Dimethyl(2-methylphenyl)phosphine; Oxide, *in* D-00759
Dimethyl(3-methylphenyl)phosphine; Oxide, *in* D-00760
Dimethyl(4-methylphenyl)phosphine; Oxide, *in* D-00761
Ethylmethylphenylphosphine; Oxide, *in* E-00084
Methyl ethylphenylphosphinite, *in* E-00108
2,3,4,7-Tetrahydro-1*H*-isophosphindole 2-oxide; 2-Me, *in* T-00109

$C_9H_{13}OPS$

O-Ethyl methylphenylphosphinothioate, *in* M-00225
S-Ethyl methylphenylphosphinothioate, *in* M-00225
Isopropylphenylphosphinothioic acid, I-00046
O-Methyl ethylphenylphosphinothioate, *in* E-00105
S-Methyl ethylphenylphosphinothioate, *in* E-00105
O-Propyl phenylphosphinothioate, *in* P-00213

$C_9H_{13}OPS_2$

O-Methyl *S*-phenyl ethylphosphonodithioate, *in* E-00140

$C_9H_{13}OPSe$

O-Methyl ethylphenylphosphinoselenoate, *in* E-00104

$C_9H_{13}O_2P$

Dimethyl (4-methylphenyl)phosphonite, *in* M-00248
Ethyl methylphenylphosphinate, *in* M-00219
Isopropyl phenylphosphinate, *in* P-00210
Isopropylphenylphosphinic acid, I-00043
Methyl benzylmethylphosphinate, *in* B-00053
Methyl (2,5-dimethylphenyl)phosphinate, *in* D-00805
Methyl (3,4-dimethylphenyl)phosphinate, *in* D-00806
Methyl ethylphenylphosphinate, *in* E-00101
Phenylpropylphosphinic acid, P-00303
(2,4,5-Trimethylphenyl)phosphinic acid, T-00524
(2,4,6-Trimethylphenyl)phosphinic acid, T-00525

$C_9H_{13}O_2PS$

O-Ethyl hydrogen benzylphosphonothioate, *in* B-00067
O-Ethyl *O*-methyl phenylphosphonothioate, *in* M-00243
▷*O*-Ethyl *S*-methyl phenylphosphonothioate, *in* M-00244
S-Ethyl *O*-methyl phenylphosphonothioate, *in* P-00244

$C_9H_{13}O_3P$

Benzyl methyl methylphosphonate, *in* M-00168
Diethyl 1,3-pentadiyn-1-ylphosphonate, *in* P-00010
Dimethyl benzylphosphonate, *in* B-00060
Dimethyl (2,6-dimethylphenyl)phosphonate, *in* D-00814
Dimethyl (2-methylphenyl)phosphate, *in* M-00235
Dimethyl (3-methylphenyl)phosphonate, *in* M-00236
Dimethyl (4-methylphenyl)phosphonate, *in* M-00237

Ethyl hydrogen benzylphosphonate, *in* B-00060
Ethyl (4-methoxyphenyl)phosphinate, *in* M-00069
Ethyl phenyl methylphosphonate, E-00100
Isopropyl phenyl phosphonate, I-00047
Monopropyl phenylphosphonate, M-00511
Phenyl propyl phosphonate, P-00304
(3-Phenylpropyl)phosphonic acid, P-00305
3,8-Phosphonanedione, P-00373
(2,4,5-Trimethylphenyl)phosphonic acid, T-00526
(2,4,6-Trimethylphenyl)phosphonic acid, T-00527

$C_9H_{13}O_3PS$

Dimethyl [(phenylthio)methyl]phosphonate, *in* P-00326

$C_9H_{13}O_4P$

Dimethyl (hydroxyphenylmethyl)phosphonate, *in* H-00174
Dimethyl [(2-hydroxyphenyl)methyl]-phosphonate, *in* H-00175
Dimethyl [(4-hydroxyphenyl)methyl]-phosphonate, *in* H-00176
Dimethyl (2-methoxyphenyl)phosphonate, *in* M-00070
Dimethyl (3-methoxyphenyl)phosphonate, *in* M-00071
Dimethyl (4-methoxyphenyl)phosphonate, *in* M-00072
1-Hydroxy-3,8-phosphonanedione 1-oxide, *in* P-00373

$C_9H_{13}O_4PS$

▷Dimethyl 4-(methylthio)phenyl phosphate, D-00767
Dimethyl [(phenylsulfinyl)methyl]phosphonate, *in* P-00310
Methyl hydrogen [[(4-methylphenyl)sulfinyl]-methyl]phosphonate, *in* M-00266

$C_9H_{13}O_5P$

2,2-Dihydro-1,3,2-benzodioxaphosphole; 2,2,2-Trimethoxy, *in* D-00439

$C_9H_{13}P$

Dimethyl(2-methylphenyl)phosphine, D-00759
Dimethyl(3-methylphenyl)phosphine, D-00760
Dimethyl(4-methylphenyl)phosphine, D-00761
Ethylmethylphenylphosphine, E-00084

$C_9H_{13}PS$

Benzyldiphenylphosphine; Sulfide, *in* B-00046
Dimethyl(4-methylphenyl)phosphine; Sulfide, *in* D-00761
Ethylmethylphenylphosphine; Sulfide, *in* E-00084

$C_9H_{13}PS_2$

Isopropylphenylphosphinodithioic acid, I-00045

$C_9H_{14}BrP$

Trimethylphenylphosphonium bromide, *in* D-00803

$C_9H_{14}NOP$

P-Isopropyl-*P*-phenylphosphinic amide, I-00044

$C_9H_{14}NOPS$

S-Ethyl *P*-methyl-*N*-phenyl-phosphonamidothioate, *in* M-00233

$C_9H_{14}NOPS_2$

O-Ethyl-*S*-phenyl methyl-phosphoramidodithioate, *in* M-00330

$C_9H_{14}NO_2P$

(4-Dimethylaminophenyl)methylphosphinic acid, D-00705
Dimethyl *N*-methyl-*P*-phenylphosphonimidate, *in* M-00240
Ethyl *P*-methyl-*N*-phenylphosphonamidate, *in* M-00231
Methyl *N*,*N*-dimethyl-*P*-phenyl-phosphonamidate, *in* D-00808
Methyl *N*-ethyl-*P*-phenylphosphonamidate, *in* E-00109
Methyl *P*-ethyl-*N*-phenylphosphonamidate, *in* E-00110
Phenyl *N*-ethyl-*P*-methylphosphonamidate, *in* E-00088
Toldimfos, T-00322

$C_9H_{14}NO_3P$

Diethyl 2-pyridinylphosphonate, *in* P-00503
Diethyl 4-pyridinylphosphonate, *in* P-00505
Dimethyl [(2-aminophenyl)methyl]-phosphonate, *in* A-00106
Dimethyl [(4-aminophenyl)methyl]-phosphonate, *in* A-00107
Dimethyl benzylphosphoramidate, *in* B-00069
Dimethyl methylphenylphosphoramidate, *in* M-00250
Dimethyl (2-methylphenyl)phosphoramidate, *in* M-00251
Dimethyl (3-methylphenyl)phosphoramidate, *in* M-00252
Dimethyl (4-methylphenyl)phosphoramidate, *in* M-00253
Ethyl hydrogen (α-aminobenzyl)phosphonate, *in* A-00105
[3-(Phenylamino)propyl]phosphonic acid, P-00085
3-Pyridinephosphonic acid; Di-Et ester, *in* P-00504
Trimethyl phenylphosphorimidate, *in* P-00264

$C_9H_{14}NO_3PS$

O-(4-Amino-3-methylphenyl) *O*,*O*-dimethyl phosphorothioate, A-00090

$C_9H_{14}NO_4P$

Dimethyl (2-methoxyphenyl)phosphoramidate, *in* M-00074
Dimethyl (3-methoxyphenyl)phosphoramidate, *in* M-00075
Dimethyl (4-methoxyphenyl)phosphoramidate, *in* M-00076

$C_9H_{14}NO_5PS$

Trimethyl phenysulfonylphosphorimidate, *in* P-00315

$C_9H_{14}NP$

P,*P*,*P*-Trimethyl-*N*-phenylphosphine imide, *in* T-00531

$C_9H_{14}NPS$

N-Benzyl-*P*,*P*-dimethylphosphinothioic amide, *in* D-00840
N,*N*,*P*-Trimethyl-*P*-phenylphosphinothioic amide, *in* M-00226

$C_9H_{14}NPS_2$

▷Phenyl trimethylphosphonoamidodithioate, *in* T-00539

$C_9H_{14}N_3O_8P$

Cytidine 2′-(dihydrogen phosphate), C-00288
Cytidine 3′-(dihydrogen phosphate), C-00289
▷Cytidine 5′-(dihydrogen phosphate), C-00290

$C_9H_{14}N_4P_2S_2$

P,*P*′-Methylenebis[(*N*,*N*,*N*′,*N*′-tetramethyl)-phosphonodiamidothioate], *in* M-00023

$C_9H_{14}N_5O_3P$

Benfosformin, B-00001

$C_9H_{15}N_2O_3P$

Diethyl 2-pyridylphosphoramidate, *in* P-00510
Diethyl 3-pyridylphosphoramidate, *in* P-00511

$C_9H_{15}N_4OP$

Hexahydro-2,4-dimethyl-3-phenoxy-1,2,4,5,3-tetrazaphosphorine, H-00024

$C_9H_{15}N_4OPS$

Hexahydro-2,4-dimethyl-3-phenoxy-1,2,4,5,3-tetrazaphosphorine; 3-Sulfide, *in* H-00024

$C_9H_{15}N_4O_2P$

Hexahydro-2,4-dimethyl-3-phenoxy-1,2,4,5,3-tetrazaphosphorine; 3-Oxide, *in* H-00024

$C_9H_{15}N_4O_8P$

5-Amino-1-ribofuranosylimidazole-4-carboxamide 5′-(dihydrogen phosphate), A-00130

$C_9H_{15}OP$

2,3,4,5,6,7-Hexahydro-1*H*-isophosphindole 2-oxide; 2-Me, *in* H-00035
Octahydro-1*H*-cyclopropa[*c*]phosphindole 3-oxide, O-00004
Tricyclopropylphosphine; Oxide, *in* T-00432
Tri-2-propenylphosphine; Oxide, *in* T-00707

$C_9H_{15}O_2P$

2-Propenyl di-2-propenylphosphinate, *in* D-01193

$C_9H_{15}O_2PSi$

Trimethylsilyl phenylphosphinate, *in* P-00210

$C_9H_{15}O_3P$

Diethyl (3-penten-1-ynyl)phosphonate, *in* P-00045
Diethyl (4-penten-2-yn-1-yl)phosphonate, *in* P-00046
▷Tri-2-propenyl phosphite, T-00708

$C_9H_{15}O_3PS$

Diethyl (2-thienylmethyl)phosphonate, *in* T-00304
Diethyl (3-thienylmethyl)phosphonate, *in* T-00305

$C_9H_{15}O_4P$

Diethyl 2-furanylmethylphosphonate, *in* F-00072
Di-2-propenyl (3-methyl-2-oxiranyl)-phosphonate, *in* M-00185
▷Tri-2-propenyl phosphate, T-00705

$C_9H_{15}O_5P$

Diethyl (2-furanylhydroxymethyl)phosphonate, *in* F-00071

$C_9H_{15}O_6P$

3,3′,3″-Phosphinidynetrispropanoic acid, P-00357
Triethoxycarbonylphosphine, *in* P-00355
Tris(acetoxymethyl)phosphine, *in* T-00784

$C_9H_{15}O_7P$

Tris(hydroxymethyl)phosphine; Oxide, tri-Ac, *in* T-00784

$C_9H_{15}O_8P$

▷Bomyl, B-00461

$C_9H_{15}P$

Tricyclopropylphosphine, T-00432
Tri-1-propenylphosphine, T-00706
Tri-2-propenylphosphine, T-00707

$C_9H_{15}PS$

Tricyclopropylphosphine; Sulfide, *in* T-00432

$C_9H_{15}PSe$

Tricyclopropylphosphine; Selenide, *in* T-00432

$C_9H_{16}ClN_2OP$

P-(3-Chloropropyl)-*N*,*N*′-di-2-propenylphosphonic diamide, *in* C-00173

$C_9H_{16}ClO_3P$

Di-2-propenyl (3-chloropropyl)phosphonate, *in* C-00172

$C_9H_{16}ClO_5PS_2$

Methyl 2,3-di-*O*-methyl-4-thioglucopyranoside 4,6-cyclic phosphonothioate; *P*-Chloro, *in* M-00123

$C_9H_{16}ClO_6PS$

Methyl 2,3-di-*O*-methylglucopyranoside 4,6-cyclic phosphonothioate; *P*-Chloro, *in* M-00120
Methyl 2,3-di-*O*-methyl-4-thioglucopyranoside 4,6-cyclic phosphonate; *P*-Chloro, *in* M-00121

Methyl 2,3-di-*O*-methyl-6-thioglucopyranoside 4,6-cyclic phosphonate; *P*-Chloro, *in* M-00122

$C_9H_{16}ClO_7P$

Methyl 2,3-di-*O*-methylglucopyranoside 4,6-cyclic phosphonate; *P*-Chloro, *in* M-00119

$C_9H_{16}FO_7P$

Methyl 2,3-di-*O*-methylglucopyranoside 4,6-cyclic phosphonate; *P*-Fluoro, *in* M-00119

$C_9H_{16}IO_2PS$

10-Methyl-10-phenyl-10*H*-phenothiaphosphinium 5,5-dioxide iodide, *in* P-00182

$C_9H_{16}NO_2P$

6-Methyl-4-methylene-4*H*-1,3,2-dioxaphosphorin; 2-Diethylamino, *in* M-00167

$C_9H_{16}NO_3P$

Diethyl 1-methyl-1*H*-pyrrol-2-ylphosphonate, *in* P-00521

$C_9H_{16}NO_5P$

Ethyl cyano(diethoxyphosphinyl)acetate, E-00057

$C_9H_{16}P_2$

2,3-Diethyl-2,3-diphosphabicyclo[2.2.1]hept-5-ene, D-00287

$C_9H_{16}P_2S_2$

2,3-Diethyl-2,3-diphosphabicyclo[2.2.1]hept-5-ene; Disulfide, *in* D-00287

$C_9H_{17}ClN_3O_3PS$

▷Isazofos, I-00021

$C_9H_{17}N_2O_3P$

Diisopropyl 1*H*-imidazol-1-ylphosphonate, *in* I-00003
Dimethyl (diazocyclohexylmethyl)phosphonate, *in* D-00028

$C_9H_{17}OP$

1,2,4-Trimethyl-2-phosphabicyclo[2.2.1]-heptane 2-oxide, *in* D-00825

$C_9H_{17}O_2PS$

(2-Methyl-1,3-butadienyl)phosphonothioic acid; *O*,*O*-Di-Et ester, *in* M-00097

$C_9H_{17}O_3P$

Bis(1-methylethyl) 1,2-propadienylphosphonate, *in* P-00441
Diethyl 1-cyclopenten-1-ylphosphonate, *in* C-00277
Diethyl 2-cyclopenten-1-ylphosphonate, *in* C-00278
Diethyl (2-methyl-1,3-butadienyl)phosphonate, *in* M-00093
Diethyl (3-methyl-1,2-butadienyl)phosphonate, *in* M-00094
Diethyl (3-methyl-1,3-butadienyl)phosphonate, *in* M-00095
Diethyl 1,3-pentadienylphosphonate, *in* P-00009
Diisopropyl 1-propynylphosphonate, *in* P-00494

$C_9H_{17}O_4P$

Dimethyl (cyclohexylcarbonyl)phosphonate, *in* C-00239
8,8-Dimethyl-1,6,10-trioxa-5-phospha(5-P^V)-spiro[4.5]dec-2-ene; 5-Methoxy, *in* D-00933

$C_9H_{17}O_5P$

4-(Diethoxyphosphinyl)-3-methyl-2-butenoic acid, D-00251
Ethyl 2-(diethoxyphosphinyl)propenoate, *in* D-00260
Ethyl 3-(diethoxyphosphinyl)propenoate, *in* D-00261
Methyl 4-(diethoxyphosphinyl)-2-butenoate, *in* D-00248

$C_9H_{17}O_5PS_2$

Methyl 2,3-di-*O*-methyl-4-thioglucopyranoside 4,6-cyclic phosphonothioate, M-00123

$C_9H_{17}O_6P$

Ethyl 2-[(diethoxyphosphinyl)oxy]-2-propenoate, *in* D-00256

Methyl 2,3-di-*O*-methylglucopyranoside 4,6-cyclic phosphonothioate, M-00120

$C_9H_{17}O_6PS$

Methyl 2,3-di-*O*-methyl-4-thioglucopyranoside 4,6-cyclic phosphonate, M-00121
Methyl 2,3-di-*O*-methyl-6-thioglucopyranoside 4,6-cyclic phosphonate, M-00122

$C_9H_{17}O_7P$

Methyl 2,3-di-*O*-methylglucopyranoside 4,6-cyclic phosphonate, M-00119

$C_9H_{18}ClN_2OP$

2-Cyanoethyl diisopropylphosphoramidochloridite, *in* D-00654

$C_9H_{18}ClN_4P$

Octahydro-7*H*,9*bH*-2*a*,4*a*,6*a*,9*a*-tetraaza-9*b*-phosphapentaleno[1,6-*cd*]indene; 9*b*-Chloro, *in* O-00023

$C_9H_{18}ClO_2P$

Cyclohexyl propylphosphonochloridate, *in* P-00476

$C_9H_{18}Cl_3N_2O_2P$

▷Trifosfamide, T-00485

$C_9H_{18}Cl_4O_2P_2$

1,9-Nonanediphosphonic acid; Tetrachloride, *in* N-00069

$C_9H_{18}FN_4P$

Octahydro-7*H*,9*bH*-2*a*,4*a*,6*a*,9*a*-tetraaza-9*b*-phosphapentaleno[1,6-*cd*]indene; 9*b*-Fluoro, *in* O-00023

$C_9H_{18}FO_2P$

Cyclohexyl propylphosphonofluoridate, *in* P-00479

$C_9H_{18}IP$

1-Methyl-1-phosphoniabicyclo[3.3.1]nonane iodide, *in* P-00345

$C_9H_{18}NOP$

Dibutylphosphinous acid; Isocyanate, *in* D-00121
1-Diethylamino-2,5-dihydro-2-methyl-1*H*-phosphole 1-oxide, *in* D-00498

$C_9H_{18}NO_2P$

Dibutylphosphinic acid; Isocyanate, *in* D-00109
2,5-Dihydro-5,5-dimethyl-1,2-oxaphosphole 2-oxide; 2-Diethylamino, *in* D-00464
Ethyl *N*,*N*-diethyl-*P*-propadienylphosphonamidate, *in* D-00399
Ethyl *N*,*N*-diethyl-*P*-1-propynylphosphonamidate, *in* D-00403

$C_9H_{18}NO_3P$

Diethyl [(ethylimino)methylethenyl]-phosphonate, *in* E-00081
Diisopropyl (2-cyanoethyl)phosphonate, *in* C-00211
4-Methyl-2,6-dioxa-7-aza-1-phosphabicyclo[2.2.2]octane 1-oxide; 7-*tert*-Butyl, *in* M-00124

$C_9H_{18}NP$

N,*P*-Di-*tert*-butylcarboimidophosphene, D-00077
Dibutylcyanophosphine, *in* D-00121

$C_9H_{18}N_2O_2P$

Cyclohexyl dimethylphosphoramidocyanidate, *in* D-00863

$C_9H_{18}N_3OP$

1,1′,1″-Phosphinylidynetrisazetidine, *in* T-00328
2,8,9-Trimethyl-2,8,9-triaza-1-phosphatricyclo[3.3.1.1[3,7]]decane; Oxide, *in* T-00563
N,*N*′,*N*″-Tri-2-propenylphosphorous triamide; Oxide, *in* T-00709

$C_9H_{18}N_3O_3P$

Tris(2-carbamoylethyl)phosphine, T-00738

$C_9H_{18}N_3O_4P$

3,3′,3″-Phosphinylidynetris[propanamide], *in* T-00738

$C_9H_{18}N_3P$

Tri(1-azetidinyl)phosphine, T-00328
2,8,9-Trimethyl-2,8,9-triaza-1-phosphatricyclo[3.3.1.1[3,7]]decane, T-00563
N,*N*′,*N*″-Tri-2-propenylphosphorous triamide, T-00709

$C_9H_{18}N_3PS$

2,8,9-Trimethyl-2,8,9-triaza-1-phosphatricyclo[3.3.1.1[3,7]]decane; Sulfide, *in* T-00563
N,*N*′,*N*″-Tri-2-propenylphosphorous triamide; Sulfide, *in* T-00709

$C_9H_{18}O_4P_2$

Decahydro-1,5-dihydroxy-1*H*-1,5-benzodiphosphepin 1,5-dioxide, D-00002

$C_9H_{18}O_4P_2S_2$

3,9-Dimercapto-2,4,8,10-tetraoxa-3,9-diphosphaspiro[5.5]undecane 3,9-disulfide; 3,9-Bis(ethylthio) (Di-Et ester), *in* D-00670

$C_9H_{19}Cl_2OP$

Nonylphosphonic acid; Dichloride, *in* N-00076

$C_9H_{19}F_2OP$

Nonylphosphonic acid; Difluoride, *in* N-00076

$C_9H_{19}N_2OP$

(2-Methyl-1,3-butadienyl)phosphonic bis(dimethylamide), *in* M-00096

$C_9H_{19}N_2O_3P$

Methyl di-4-morpholinylphosphinite, *in* D-00936

$C_9H_{19}N_2PS$

N,*N*,*N*′,*N*′-Tetramethyl-*P*-(2-methyl-1,3-butadienyl)phosphonothioic diamide, *in* M-00097

$C_9H_{19}N_4OP$

Octahydro-8*bH*-2*a*,4*a*,6*a*,8*a*-tetraaza-8*b*-phosphapentaleno[1,6-*cd*]pentalene; 8*b*-Methoxy, *in* O-00025

$C_9H_{19}N_4P$

Octahydro-7*H*,9*bH*-2*a*,4*a*,6*a*,9*a*-tetraaza-9*b*-phosphapentaleno[1,6-*cd*]indene, O-00023
Octahydro-8*bH*-2*a*,4*a*,6*a*,8*a*-tetraaza-8*b*-phosphapentaleno[1,6-*cd*]pentalene; 8*b*-Me, *in* O-00025

$C_9H_{19}OP$

1,2,2,3,4,4-Hexamethylphosphetane; Oxide, *in* H-00074
5-Methoxy-5-phospha(5-P^V)spiro[4.4]nonane, *in* P-00350
1-Methylphosphonane; 1-Oxide, *in* M-00285
1-Oxa-5-phospha(5-P^V)spiro[4.5]decane; 5-Me, *in* O-00049
1-Oxa-6-phospha(6-P^V)spiro[5.5]undecane, O-00052

$C_9H_{19}OPS$

1-Methylthio-2,2,3,4,4-pentamethyl-phosphetane 1-oxide, *in* H-00169

$C_9H_{19}OPS_2$

S,*S*-Diisopropyl 2-propenylphosphonodithioate, *in* P-00457

$C_9H_{19}O_2P$

Diisopropyl allylphosphonite, *in* P-00463
Isopropyl cyclohexylphosphinate, *in* C-00253
Methyl 1,1,2,3,3-pentamethyltrimethyl-enephosphinate, *in* H-00168

$C_9H_{19}O_2PS$

O,*O*-Bis(1-methylethyl) 2-propenylphosphonothioate, *in* P-00461

$C_9H_{19}O_2PSe$

Bis(1-methylethyl) (2-propenyl)-phosphonoselenoate, *in* P-00458

$C_9H_{19}O_3P$

2-*tert*-Butoxy-5,5-dimethyl-1,3,2-dioxaphosphorinane, B-00534
Di-*tert*-butyl 1-propenylphosphonate, *in* P-00453

Diethyl (1-methyl-3-butenyl)phosphonate, *in* M-00099
Diethyl (3-methyl-2-butenyl)phosphonate, *in* M-00102
Diethyl (3-methyl-3-butenyl)phosphonate, *in* M-00103
Diisopropyl 1-propenylphosphonate, *in* P-00453
4-Isopropyl-2-methoxy-5,5-dimethyl-1,3,2-dioxaphosphorinane, I-00033
Methyl (1,1-dimethyl-3-oxobutyl)-ethylphosphinate, *in* D-00782
Monoisopropyl cyclohexylphosphonate, *in* C-00256

$C_9H_{19}O_3PSe$

4-Isopropyl-2-methoxy-5,5-dimethyl-1,3,2-dioxaphosphorinane; 2-Selenide, *in* I-00033

$C_9H_{19}O_4P$

Bis(1-methylethyl) (2-oxopropyl)phosphonate, *in* O-00075
tert-Butyl neopentylene phosphate, *in* B-00534
Dimethyl (2-oxoheptyl)phosphonate, *in* O-00061
(3,3-Dimethyl-2-oxoheptyl)phosphonic acid, D-00785
Dipropyl (3-methyl-2-oxiranyl)phosphonate, *in* M-00185
2-Oxononylphosphonic acid, O-00063
2,2,2,3-Tetrahydro-1,2-oxaphosphole; 2,2,2-Triethoxy, *in* T-00119

$C_9H_{19}O_5P$

Diethyl [1-(acetyloxy)propyl]phosphonate, *in* A-00008
Diethyl [2-(acetyloxy)propyl]phosphonate, *in* A-00009
Diethyl (3-acetyloxypropyl)phosphonate, *in* A-00010
Ethyl 2-(diethoxyphosphinyl)propanoate, E-00059
Ethyl 3-(diethoxyphosphinyl)propanoate, *in* D-00259
Methyl 4-(diethoxyphosphinyl)butanoate, *in* D-00247

$C_9H_{19}O_6P$

tert-Butoxycarbonyl diethyl phosphate, B-00531
Ethyl 2-[(diethoxyphosphinyl)oxy]propanoate, *in* D-00255

$C_9H_{19}O_7P$

Glycerol 2-monophosphate; 1-Hexanoyl, *in* G-00067

$C_9H_{19}P$

1,2,2,3,4,4-Hexamethylphosphetane, H-00074
5-Methyl-5-phospha(5-P^V)spiro[4.4]nonane, *in* P-00350
1-Methylphosphonane, M-00285

$C_9H_{20}ClO_2P$

Dibutyl (chloromethyl)phosphonite, *in* C-00115

$C_9H_{20}Cl_2NOP$

Nonylphosphoramidic acid; Dichloride, *in* N-00077

$C_9H_{20}Cl_2O_6P_2$

Tetraethyl dichloromethylenebisphosphonate, *in* D-00172

$C_9H_{20}Cl_3N_2OP$

P-(Trichloromethyl)-*N,N,N',N'*-tetraethylphosphonic diamide, *in* T-00412

$C_9H_{20}Cl_3N_2O_3P$

▷Defosfamide, D-00015

$C_9H_{20}Cl_4N_3P$

N''-Chloro-*N,N,N',N'*-tetraethyl-*P*-(trichloromethyl)phosphonimidic diamide, *in* T-00075

$C_9H_{20}IPSi$

1,5-Dimethyl-1-phosphonia-5-silabicyclo[3.3.1]nonane iodide, *in* M-00274

$C_9H_{20}NOP$

N,2,2,3,4,4-Hexamethylphosphetanamine 1-oxide, *in* P-00021

$C_9H_{20}NO_2P$

2-*tert*-Butylamino-5,5-dimethyl-1,3,2-dioxaphosphorinane, B-00538
5-*tert*-Butyl-2-dimethylamino-1,3,2-dioxaphosphorinane, B-00555
Diethyl 1-piperidinylphosphonite, *in* P-00432
Isopropyl *P*-methyl-1-piperidinylphosphonamidate, *in* M-00361

$C_9H_{20}NO_2PS$

Neopentylene *tert*-butylphosphoramidothioate, *in* B-00538

$C_9H_{20}NO_3P$

2-*tert*-Butyltrimethylene dimethyl-phosphoramidate, *in* B-00555
Diethyl 1-piperidinylphosphonate, *in* P-00430
Neopentylene *tert*-butylphosphoramidate, *in* B-00538

$C_9H_{20}NO_3PS_2$

▷Prothoate, P-00499

$C_9H_{20}NO_4P$

(3-Aminopropyl)phosphonic acid; *N*-Ac, Di-Et ester, *in* A-00127
Diethyl (4-morpholinylmethyl)phosphonate, *in* M-00521
Diisopropyl (acetamidomethyl)phosphonate, *in* A-00012
Diisopropyl (1-nitroso-1-methylethyl)-phosphonate, *in* N-00066
Tetraethyl carbamoylphosphonate, *in* D-00264

$C_9H_{20}N_3OPS$

Tetraethyl phosphor(isothiocyanatidic) diamide, *in* T-00059

$C_9H_{20}N_3O_2P$

Tetrahydro-4-methyl-2*H*-1,3,4,2-oxadiazaphosphorine 2-oxide; 2-Cyclohexylamino, *in* T-00116

$C_9H_{20}N_3O_2PS$

Tetraethyl phosphor(isocyanatidic) diamide, *in* T-00059

$C_9H_{20}N_3P$

4-Ethyl-2,6,7-trimethyl-2,6,7-triaza-1-phosphabicyclo[2.2.2]octane, E-00195

$C_9H_{20}N_3PS$

▷4-Ethyl-2,6,7-trimethyl-2,6,7-triaza-1-phosphabicyclo[2.2.2]octane; 1-Sulfide, *in* E-00195

$C_9H_{20}O_4P_2$

1,3-Diisopropoxy-1,3-diphospholane 1,3-dioxide, *in* D-00621

$C_9H_{20}P_2$

1,2-Diphospholane; 1,2-Diisopropyl, *in* D-01181

$C_9H_{21}Cl_2N_2OP$

P-(Dichloromethyl)phosphonic bis(diethylamide), *in* D-00176

$C_9H_{21}Cl_3N_3P$

N,N,N',N'-Tetraethyl-*P*-(trichloromethyl)-phosphonimidic diamide, T-00075

$C_9H_{21}NO_5P$

2,5-Dihydroxy-3-(1-methylpropyl)-1,3,2,5-oxazadiphospholidine 2,5-dioxide; 2,5-Diethoxy (Di-Et ester), *in* D-00623

$C_9H_{21}N_2OP$

tert-Butyl *N,N'*-dimethyl-*N,N'*-trimethylene phosphorodiamidite, *in* H-00022
5-*tert*-Butyltetrahydro-2-dimethylamino-2*H*-1,3,2-oxazaphosphorine, B-00635

$C_9H_{21}N_2OPS$

5-*tert*-Butyltetrahydro-2-dimethylamino-2*H*-1,3,2-oxazaphosphorine; 2-Sulfide, *in* B-00635

$C_9H_{21}N_2O_2P$

5-*tert*-Butyltetrahydro-2-dimethylamino-2*H*-1,3,2-oxazaphosphorine; 2-Oxide, *in* B-00635
3,3,8,8-Tetramethyl-1,6-dioxa-4,9-diaza-5-phospha(P^V)spiro[4.4]nonane; 5-Me, *in* T-00196

$C_9H_{21}N_2O_3P$

6,9-Dimethyl-1,3-dioxa-6,9-diaza-2-phosphacycloundecane; 2-Methoxy, *in* D-00728

$C_9H_{21}N_2P$

2-Amino-1,3-di-*tert*-butylazaphosphiridine, A-00070

$C_9H_{21}N_6O_9P_3$

Hexamethyl 1,3,5-triazine-2,4,6-triyltriphosphoramidate, *in* T-00330

$C_9H_{21}OP$

Dibutylmethylphosphine; Oxide, *in* D-00098
Di-*tert*-butylmethylphosphine; Oxide, *in* D-00099
2,2-Dihydro-1,2-oxaphospholane; 2,2,2-Tri-Et, *in* D-00534
Methyl di-*tert*-butylphosphinite, *in* D-00122
Triisopropylphosphine; Oxide, *in* T-00500
Tripropylphosphine; Oxide, *in* T-00711

$C_9H_{21}OPS$

O-Methyl di-*tert*-butylphosphinothioate, *in* D-00118
S-Methyl dibutylphosphinothioate, *in* D-00117
S-Methyl di-*tert*-butylphosphinothioate, *in* D-00118
O-Propyl dipropylphosphinothioate, *in* D-01208
O-*tert*-Butyl *tert*-butylmethylphosphinothioate, *in* B-00570

$C_9H_{21}OPS_2$

S,S-Dipropyl propylphosphonodithioate, *in* P-00478

$C_9H_{21}OPS_3$

O,S,S-Triisopropyl phosphorotrithioate, T-00507
S,S,S-Triisopropyl phosphorotrithioate, T-00508
S,S,S-Tripropyl phosphorotrithioate, T-00717

$C_9H_{21}O_2P$

Butyl butylmethylphosphinate, *in* B-00566
Diisopropyl isopropylphosphonite, *in* I-00056
Diisopropyl propylphosphonite, *in* P-00482
Isobutyl butylmethylphosphinate, *in* B-00566
Isopropyl diisopropylphosphinate, *in* D-00641
Methyl dibutylphosphinate, *in* D-00109
Methyl di-*tert*-butylphosphinate, *in* D-00110

$C_9H_{21}O_2PS$

▷*O,S*-Dibutyl methylphosphonothioate, *in* M-00313

$C_9H_{21}O_2PS_2$

O,O,S-Triisopropyl phosphorodithioate, T-00503

$C_9H_{21}O_2PS_3$

▷Terbufos, T-00004

$C_9H_{21}O_2PSe$

O,O-Dibutyl methylphosphonoselenoate, *in* M-00311

$C_9H_{21}O_2PSi$

Trimethylsilyl cyclohexylphosphinate, *in* C-00253

$C_9H_{21}O_3P$

Diethyl (1-methylbutyl)phosphonate, *in* M-00105
Diethyl (3-methylbutyl)phosphonate, *in* M-00106
Diethyl pentylphosphonate, *in* P-00050
Dimethyl heptylphosphonate, *in* H-00006

Dipropyl propylphosphonate, *in* P-00473
Nonylphosphonic acid, N-00076
▷Triisopropyl phosphite, T-00502
▷Tripropyl phosphite, T-00712

$C_9H_{21}O_3PS$
O,O,O-Tripropyl phosphorothioate, T-00715
O,O,S-Tripropyl phosphorothioate, T-00716

$C_9H_{21}O_3PS_3$
▷Aphidan, A-00142
S-[(*tert*-Butylsulfinyl)methyl] *O,O*-diethyl phosphorodithioate, *in* T-00004

$C_9H_{21}O_3PSe$
O,O,O-Triisopropyl phosphoroselenoate, T-00504
O,O,O-Tripropyl phosphoroselenoate, T-00713

$C_9H_{21}O_3P_3$
2,4,6-Tripropyl-1,3,5,2,4,6-trioxatriphosphorinane, T-00720

$C_9H_{21}O_3P_3S_3$
Propylphosphonothioic acid trimer trianhydride, *in* T-00720

$C_9H_{21}O_4P$
O-Butyl *OO-tert*-butyl methylphosphonoperoxoate, *in* M-00310
OO-tert-Butyl *O*-isopropyl ethylphosphonoperoxoate, *in* E-00144
Diisopropyl (2-hydroxypropyl)phosphonate, *in* H-00191
Triisopropyl phosphate, T-00499
Tripropyl phosphate, T-00710

$C_9H_{21}O_4PS_3$
S-[(*tert*-Butylsulfonyl)methyl] *O,O*-diethyl phosphorodithioate, *in* T-00004

$C_9H_{21}O_4P_3$
1,3-Diisopropoxy-2-isopropyltriphosphetane 1,3-dioxide, *in* D-00627

$C_9H_{21}O_5P$
2,2,2-Triethoxy-2,2-dihydro-1,3,2-dioxaphosphorinane, T-00437

$C_9H_{21}O_5PSi$
Diethyl [(trimethylsilyloxy)carbonyl]-phosphonate, *in* D-00241

$C_9H_{21}O_6PS_2$
Methyl 2,3-di-*O*-methyl-4-thioglucopyranoside 4,6-cyclic phosphonothioate; *P*-Ethoxy, *in* M-00123

$C_9H_{21}O_6P_3$
Propylphosphonic acid trimer trianhydride, *in* T-00720

$C_9H_{21}P$
Dibutylmethylphosphine, D-00098
Di-*tert*-butylmethylphosphine, D-00099
Nonylphosphine, N-00075
Triisopropylphosphine, T-00500
▷Tripropylphosphine, T-00711

$C_9H_{21}PS$
Dibutylmethylphosphine; Sulfide, *in* D-00098
Methyl di-*tert*-butylphosphinothioite, *in* D-00120
Triisopropylphosphine; Sulfide, *in* T-00500
Tripropylphosphine; Sulfide, *in* T-00711

$C_9H_{21}PS_2$
Methyl dibutylphosphinodithioate, *in* D-00112

$C_9H_{21}PS_3$
Tripropyl phosphorotrithioite, T-00718

$C_9H_{21}PS_4$
Triisopropyl phosphorotetrathioate, T-00505
Tripropyl phosphorotetrathioate, T-00714

$C_9H_{21}PSe$
Tripropylphosphine; Selenide, *in* T-00711

$C_9H_{21}PTe$
Tripropylphosphine; Telluride, *in* T-00711

$C_9H_{22}ClN_2OP$
P-Chloromethyl-*N,N,N',N'*-tetraethylphosphonic diamide, *in* C-00107

$C_9H_{22}Cl_4N_3P$
Aminobis(diethylamino)(trichloromethyl)-phosphonium chloride, *in* T-00075

$C_9H_{22}NOP$
P-tert-Butyl-*N,N*-diethyl-*P*-methylphosphinic amide, *in* B-00567

$C_9H_{22}NOPS$
O-Isopropyl triethylphosphonamidothioate, *in* T-00448

$C_9H_{22}NO_2P$
Ethyl *N,N*-diisopropyl-*P*-methylphosphonamidate, *in* D-00636
▷Isopropyl triethylphosphonamidate, *in* T-00446

$C_9H_{22}NO_3P$
P-(Butoxymethyl)-*N,N*-diethylphosphonamidic acid, B-00536
Dibutyl methylphosphoramidate, *in* M-00326
Diethyl (1-aminopentyl)phosphonate, *in* A-00096
Diethyl (5-aminopentyl)phosphonate, *in* A-00097
Diethyl pentylphosphoramidate, *in* P-00053
Diisopropyl (3-aminopropyl)phosphonate, *in* A-00127
Diisopropyl isopropylphosphoramidate, *in* I-00058
Dipropyl propylphosphoramidate, *in* P-00484
Nonylphosphoramidic acid, N-00077
Propyl *N,N*-diethyl-*P*-(methoxymethyl)-phosphonamidate, *in* D-00295

$C_9H_{22}NO_5P$
Dibutyl (nitromethyl)phosphonate, *in* N-00037

$C_9H_{22}NP$
P,P-Di-*tert*-butyl-*N*-methylphosphinous amide, *in* D-00124
P,P,P-Triisopropylphosphine imide, T-00501

$C_9H_{22}NPS$
P,P-Di-*tert*-butyl-*N*-methylphosphinothioic amide, *in* D-00119

$C_9H_{22}NPS_2$
Isopropyl triethylphosphonamidodithioate, *in* T-00447

$C_9H_{22}N_2P_2$
1-(Di-*tert*-butyldiphosphinylidene)-methanediamine, D-00090

$C_9H_{22}N_3O_2P$
P-Aminocarbonyl-*N,N,N',N'*-tetraethylphosphonic diamide, *in* C-00003

$C_9H_{22}N_3P$
N,N,N',N'-Tetramethyl-*P*-1-piperidinylphosphonous diamide, *in* P-00432

$C_9H_{22}O_2P_2$
1,5-Bis(dimethylphosphinyl)pentane, *in* B-00205

$C_9H_{22}O_3P$
Dibutyl methylphosphonate, *in* M-00288
Di-*tert*-butyl methylphosphonate, *in* M-00288

$C_9H_{22}O_4P_2$
Diisopropyl methylenebis[methylphosphinate], *in* M-00140
Tetraethyl methylenebisphosphonite, *in* M-00024

$C_9H_{22}O_4P_2S_2$
O,O,O,O-Tetraethyl methylenebisphosphonothioate, *in* M-00023

$C_9H_{22}O_4P_2S_4$
▷Ethion, E-00022

$C_9H_{22}O_6P_2$
1,9-Nonanediphosphonic acid, N-00069
▷Tetraethyl methylenebisphosphonate, *in* M-00022

$C_9H_{22}P_2$
1,5-Bis(dimethylphosphino)pentane, B-00205

$C_9H_{23}ClPSi_2$
1,1,3,3,5,5-Hexamethyl-1-phosphonia-3,5-disilacyclohexane chloride, *in* D-00485

$C_9H_{23}IP_4$
Tetraethylmethyltetraphosphetanium iodide, *in* T-00071

$C_9H_{23}NO_3PS^{\oplus}$
Ecothiopate, E-00001

$C_9H_{23}NO_4PS_2$
O,O,O,O-Tetraethyl methylthioimidodiphosphate, *in* M-00161

$C_9H_{23}NO_4P_2$
Tetraethyl methylimidodiphosphite, *in* M-00157

$C_9H_{23}NO_5P_2S$
O,O-Diethyl (diethoxyphosphinyl)-methylphosphoramidothioate, *in* M-00413

$C_9H_{23}NO_6P_2$
Tetraethyl methylimidodiphosphate, *in* M-00156

$C_9H_{23}N_2OP$
N,N'-Di-*tert*-butyl-*P*-methylphosphonic diamide, *in* M-00289
N,N,N',N'-Tetraethyl-*P*-methylphosphonic diamide, *in* M-00289

$C_9H_{23}N_2OPSe$
O-Methyl tetraethylphosphorodiamidoselenoate, *in* T-00061

$C_9H_{23}N_2O_2P$
Methyl tetraethylphosphorodiamidate, *in* T-00059

$C_9H_{23}N_2P$
N,N'-Di-*tert*-butyl-*P*-methylphosphonous diamide, *in* M-00320
N,N,N',N'-Tetraethyl-*P*-methylphosphonous diamide, *in* M-00320

$C_9H_{23}N_2PS$
Methylphosphonothioic bis(diethylamide), *in* M-00314

$C_9H_{23}N_3O_2P_2$
1-*tert*-Butyl-1,2,4-azadiphosphetidine; 2,4-Bis(dimethylamino), 2,4-dioxide, *in* B-00543

$C_9H_{23}OPSi$
Trimethylsilyl diisopropylphosphinite, *in* D-00648

$C_9H_{23}O_2PSi$
Trimethylsilyl dipropylphosphinate, *in* D-01205

$C_9H_{23}O_4PSi$
Diethyl [1-[(trimethylsilyl)oxy]ethyl]-phosphonate, *in* T-00555

$C_9H_{23}O_4PSi_2$
Bis(trimethylsilyl) (1-oxopropyl)phosphonate, *in* O-00074
Bis(trimethylsilyl) (2-oxopropyl)phosphonate, *in* O-00075

$C_9H_{23}O_5PSi_2$
Ethyl bis(trimethylsilyloxy)-phosphinecarboxylate oxide, *in* B-00433
Methyl [bis(trimethylsilyloxy)phosphinyl]-acetate, *in* B-00434

$C_9H_{23}PSi_2$
1,1-Dihydro-1,1,3,3,5,5-hexamethyl-1-phospha-3,5-disilacyclohex-1-ene, D-00485

$C_9H_{24}NO_3PSi$
Triethyl (trimethylsilyl)phosphorimidate, *in* T-00557

$C_9H_{24}NO_4PSi_2$
Bis(trimethylsilyl) (acetamidomethyl)-phosphonate, *in* A-00012
Bis(trimethylsilyl) (dimethylaminocarbonyl)-phosphonate, *in* D-00691

$C_9H_{24}NPSi$
P,P,P-Triethyl-*N*-trimethylsilylphosphine imide, *in* T-00442

$C_9H_{24}N_3OP$
N,N′,N″-Triisopropylphosphoric triamide, *in* T-00509
N,N′,N″-Tripropylphosphoric triamide, *in* T-00719

$C_9H_{24}N_3O_9P_3$
Hexamethyl 1,3,5-triazine-2,4,6-triyltriphosphonate, *in* T-00329
N,N′,N″-Tris(dihydroxyphosphinylmethyl)-1,4,7-triazacyclononane, T-00757

$C_9H_{24}N_3P$
N,N′,N″-Triisopropylphosphorous triamide, T-00509
N,N′,N″-Tripropylphosphorous triamide, T-00719

$C_9H_{24}N_3PS$
N,N′,N″-Triisopropylphosphorothioic triamide, *in* T-00509
N,N′,N″-Tripropylphosphorothioic triamide, *in* T-00719

$C_9H_{24}O_{10}P_4$
Hexamethyl [phosphinylidynetris(methylene)]-trisphosphonate, *in* P-00360

$C_9H_{26}N_4O_2P_2$
P,P′-Methylenebis(phosphonic diamide); Octa-Me, *in* M-00144

$C_9H_{27}Br_2N_2PSi_3$
P,P-Dibromo-*N*-trimethylsilyl-*P*-[bis(trimethylsilyl)amino]phosphazene, *in* T-00868

$C_9H_{27}Cl_2N_2PSi_3$
P,P-Dichloro-*N*-trimethylsilyl-*P*-[bis(trimethylsilyl)amino]phosphazene, *in* T-00868

$C_9H_{27}F_2N_2PSi_3$
P,P-Difluoro-*N*-trimethylsilyl-*P*-[bis(trimethylsilyl)amino]phosphazene, *in* T-00868

$C_9H_{27}Ge_3P$
Tris(trimethylgermyl)phosphine, T-00860

$C_9H_{27}Ge_3PO_4$
Tris(trimethylgermyl)phosphate, T-00859

$C_9H_{27}I_2N_2PSi_3$
P,P-Diiodo-*N*-trimethylsilyl-*P*-[bis(trimethylsilyl)amino]phosphazene, *in* T-00868

$C_9H_{27}N_4PSi$
N,N,N′,N′,N″,N″-Hexamethyl-*N‴*-trimethylsilylphosphorimidic triamide, H-00082

$C_9H_{27}N_5OP_2S$
N″-Bis(dimethylamino)phosphinyl-*N,N,N′,N′,N″*-pentamethylphosphorotriamidothioate, N-00032

$C_9H_{27}N_5O_2P_2$
Nonamethyl imidodiphosphoramide, N-00068

$C_9H_{27}N_5P_2$
Nonamethylimidodiphosphorous tetraamide, *in* M-00157

$C_9H_{27}N_5P_2S_2$
Bis(dimethylaminophosphinothioyl)-pentamethylthiophosphoric triamide, B-00191

$C_9H_{27}O_3PS$
O,O,O-Triisopropyl phosphorothioate, T-00506

$C_9H_{27}O_3PSi_3$
Tris(trimethylsilyl) phosphite, T-00866

$C_9H_{27}O_4PSi$
▷Tris(trimethylsilyl) phosphate, T-00864

$C_9H_{27}PPb_3$
Tris(trimethylplumbyl)phosphine, T-00863

$C_9H_{27}PS_4Si_3$
Tris(trimethylsilyl) phosphorotetrathioate, T-00869

$C_9H_{27}PSi_3$
▷Tris(trimethylsilyl)phosphine, T-00865

$C_9H_{29}N_2OPSi_3$
NNN′-Tris(trimethylsilyl)phosphonamidimidic acid, T-00867

$C_9H_{29}N_2O_2PSi_3$
N,N,N′-Tris(trimethylsilyl)phosphoramidimidic acid, T-00868

$C_{10}H_4F_7O_4P$
4′,5′-Bis(trifluoromethyl)spiro[1,3,2-benzodioxaphosphole-2,2′-[1,3,2]-dioxaphosphole]; 2-Fluoro, *in* B-00421

$C_{10}H_5F_6O_2PS_2$
4′,5′-Bis(trifluoromethyl)spiro[1,3,2-benzodioxaphosphole-2,2′-[1,3,2]-dithiaphosphole], B-00422

$C_{10}H_5F_6O_4P$
4′,5′-Bis(trifluoromethyl)spiro[1,3,2-benzodioxaphosphole-2,2′-[1,3,2]-dioxaphosphole], B-00421

$C_{10}H_6ClO_2P$
2-Chloronaphtho[1,2-*d*]-1,3,2-dioxaphosphole, C-00118
2-Chloronaphtho[2,3-*d*]-1,3,2-dioxaphosphole, C-00119
2-Chloronaphtho[1,8-*de*]-1,3,2-dioxaphosphorin, C-00120

$C_{10}H_6ClO_2PS$
2,3-Naphthylene phosphorochloridothioate, *in* C-00119

$C_{10}H_6ClO_3P$
2,3-Naphthylene phosphorochloridate, *in* C-00119

$C_{10}H_7Br_2P$
Dibromo-1-naphthylphosphine, *in* N-00025

$C_{10}H_7Cl_2OP$
1-Naphthylphosphonic acid; Dichloride, *in* N-00021
2-Naphthylphosphonic acid; Dichloride, *in* N-00022
1-Naphthyl phosphorodichloridite, N-00030
2-Naphthyl phosphorodichloridite, N-00031

$C_{10}H_7Cl_2O_2P$
1-Naphthyl phosphorodichloridate, *in* M-00494
2-Naphthyl phosphorodichloridate, *in* M-00495

$C_{10}H_7Cl_2P$
Dichloro-1-naphthylphosphine, *in* N-00025
Dichloro-2-naphthylphosphine, *in* N-00026

$C_{10}H_7Cl_2PS$
1-Naphthylphosphonothioic acid; Dichloride, *in* N-00023

$C_{10}H_7F_2O_2P$
1-Naphthyl phosphorodifluoridate, *in* M-00494

$C_{10}H_7O_2P$
Naphtho[1,2-*d*]-1,3,2-dioxaphosphole, N-00009
Naphtho[1,8-*de*]-1,3,2-dioxaphosphorin, N-00010

$C_{10}H_7O_3P$
1,2-Naphthylene phosphonate, *in* N-00009
1,8-Naphthylene phosphonate, *in* N-00010

$C_{10}H_7O_4P$
2-Hydroxynaphtho[2,3-*d*]-1,3,2-dioxaphosphole 2-oxide, H-00164

$C_{10}H_8Cl_2NOP$
1-Naphthylphosphoramidic acid; Dichloride, *in* N-00028
2-Naphthylphosphoramidic acid; Dichloride, *in* N-00029

$C_{10}H_8Cl_2NO_4P$
2-Phthalimidoethyl phosphorodichloridate, P-00425

$C_{10}H_8Cl_2PS$
2-Naphthylphosphonothioic acid; Dichloride, *in* N-00024

$C_{10}H_8F_6NPS$
2,2,2,3-Tetrahydro-5-phenyl-3,3-bis(trifluoromethyl)-1,4,2-thiazaphosphole, T-00129

$C_{10}H_9Cl_4O_4P$
▷Tetrachlorvinphos, T-00038

$C_{10}H_9F_6O_3PS_2$
4,5-Bis(trifluoromethyl)spiro[1,3,2-dithiaphosphole-2,1′-[2,8,9]trioxa[1]-phosphatricyclo[3.3.1.1^{5,7}]decane], B-00424

$C_{10}H_9N_2P$
2,3-Dihydro-1*H*-naphtho[1,8-*de*]-1,3,2-diazaphosphorine, D-00532

$C_{10}H_9OP$
4-Phenyl-4*H*-1,4-oxaphosphorin, P-00175

$C_{10}H_9OPSe$
4-Phenyl-4*H*-1,4-selenaphosphorin; 4-Oxide, *in* P-00307

$C_{10}H_9OPTe$
4-Phenyl-4*H*-1,4-telluraphosphorin; 4-Oxide, *in* P-00317

$C_{10}H_9O_2P$
1-Naphthylphosphinic acid, N-00019
2-Naphthylphosphinic acid, N-00020
1-Naphthylphosphonous acid, N-00025
2-Naphthylphosphonous acid, N-00026
4-Phenyl-4*H*-1,4-oxaphosphorin; 4-Oxide, *in* P-00175

$C_{10}H_9O_2PS$
1-Naphthylphosphonothioic acid, N-00023
2-Naphthylphosphonothioic acid, N-00024

$C_{10}H_9O_3P$
Mono-1-naphthyl phosphonate, M-00496
Mono-2-naphthyl phosphonate, M-00497
1-Naphthylphosphonic acid, N-00021
2-Naphthylphosphonic acid, N-00022

$C_{10}H_9O_4P$
Mono-1-naphthyl phosphate, M-00494
Mono-2-naphthyl phosphate, M-00495

$C_{10}H_9P$
1-Methylisophosphinoline, M-00162
3-Methylisophosphinoline, M-00163
1-Phenyl-1*H*-phosphole, P-00217

$C_{10}H_9PS$
4*H*-1,4-Thiaphosphorin; 4-Ph, *in* T-00299

$C_{10}H_9PSe$
4-Phenyl-4*H*-1,4-selenaphosphorin, P-00307

$C_{10}H_9PTe$
4-Phenyl-4*H*-1,4-telluraphosphorin, P-00317

$C_{10}H_{10}BrCl_2O_4P$
▷Bromfenvinfos-methyl, B-00463

$C_{10}H_{10}BrO_2P$
Spiro[1,3,2-benzodioxaphosphole-2,1′-[3]-phospholene]; 2-Bromo, *in* S-00010

$C_{10}H_{10}Cl_3O_4P$
▷2-Chloro-1-(2,4-dichlorophenyl)vinyl dimethyl phosphate, C-00035

$C_{10}H_{10}Cl_6NO_3P$
Bis(2,2,2-trichloroethyl) phenylphosphoramidate, *in* B-00400

$C_{10}H_{10}FO_2P$
Spiro[1,3,2-benzodioxaphosphole-2,1′-[3]-phospholene]; 2-Fluoro, *in* S-00010

$C_{10}H_{10}F_5O_3P$
Diethyl (pentafluorophenyl)phosphonate, *in* P-00015

$C_{10}H_{10}NO_3P$
1-Naphthylphosphoramidic acid, N-00028
2-Naphthylphosphoramidic acid, N-00029
[(2-Quinolinyl)methyl]phosphonic acid, Q-00004
[(8-Quinolinyl)methyl]phosphonic acid, Q-00005

$C_{10}H_{10}NO_4PS$
O,O-Dimethyl (1,3-dihydro-1,3-dioxo-2*H*-isoindol-2-yl)phosphonothioate, *in* D-00475

$C_{10}H_{11}Cl_3NO_3P$
2,4-Dichlorophenyl 4-morpholinylphosphonochloridate, *in* M-00524

$C_{10}H_{11}F_{12}N_2O_2P$
6,9-Dimethyl-2,2,3,3-tetrakis(trifluoromethyl)-1,4-dioxa-6,9-diaza-5-phospha(5-P^V)-spiro[4.4]nonane, D-00922

$C_{10}H_{11}N_2OPS$
O-2-Naphthyl phosphoradiamidothioate, N-00027

$C_{10}H_{11}N_2O_2PS$
Phenyl *P*-methyl-*N*-2-thiazolylphosphonamidate, *in* M-00409

$C_{10}H_{11}OP$
2,3-Dihydro-1-phenyl-1*H*-phosphole; 1-Oxide, *in* D-00566
2,5-Dihydro-1-phenyl-1*H*-phosphole; 1-Oxide, *in* D-00567
Phenyldivinylphosphine; Oxide, *in* P-00122

$C_{10}H_{11}O_2P$
1,2-Dihydro-1-methoxyphosphinoline 1-oxide, *in* D-00505
2,3-Dihydro-1-phenoxy-1*H*-phosphole 1-oxide, *in* D-00506
4,7-Dihydro-2-phenyl-1,3,2-dioxaphosphepin, D-00561
Divinyl phenylphosphonite, *in* P-00251
4-Phenyl-3,5-dioxa-1-phosphabicyclo[2.2.1]-heptane, P-00113
Spiro[1,3,2-benzodioxaphosphole-2,1′-[3]-phospholene], S-00010

$C_{10}H_{11}O_2PS$
4,7-Dihydro-2-phenyl-1,3,2-dioxaphosphepin; 2-Sulfide, *in* D-00561
O,O-Dimethyl (phenylethynyl)-phosphonothioate, *in* P-00167

$C_{10}H_{11}O_2PS_2$
Ethyl bis(2-thienyl)phosphinate, *in* D-01250

$C_{10}H_{11}O_3P$
4,7-Dihydro-1,3,2-dioxaphosphepin; 2-Phenoxy, *in* D-00472
Dimethyl phenylethynylphosphonate, *in* P-00166
2-Hydroxy-2*H*-1,2-benzoxaphosphorin 2-oxide; 2-Ethoxy (Et ester), *in* H-00112
5′-Methylspiro[1,3,2-benzodioxaphosphole-2,2′(3*H*)-[1,2]oxaphosphole], M-00396
4-Phenyl-3,5-dioxa-1-phosphabicyclo[2.2.1]-heptane; 1-Oxide, *in* P-00113

$C_{10}H_{11}O_3PS$
4,7-Dihydro-1,3,2-dioxaphosphepin 2-sulfide; 2-Phenoxy, *in* D-00473
O-Ethyl di-2-furanylphosphinothioate, *in* D-00422

$C_{10}H_{11}O_4P$
Bis(methoxycarbonyl)phenylphosphine, *in* P-00209
Dimethylvinylene phenyl phosphate, *in* H-00123
Ethyl di-2-furanylphosphinate, *in* D-00421

$C_{10}H_{11}O_5P$
Dimethyl (1,3-dihydro-3-oxo-1-isobenzofuranyl)phosphonate, *in* D-00535

$C_{10}H_{11}P$
2,3-Dihydro-1-phenyl-1*H*-phosphole, D-00566
2,5-Dihydro-1-phenyl-1*H*-phosphole, D-00567
Phenyldivinylphosphine, P-00122

$C_{10}H_{11}PS$
2,3-Dihydro-1-phenyl-1*H*-phosphole; 1-Sulfide, *in* D-00566

$C_{10}H_{11}PS_2$
3,6-Dihydro-2*H*-1,2-thiaphosphorin 2-sulfide; 2-Ph, *in* D-00586

$C_{10}H_{12}BrCl_2O_3PS$
▷Bromophos-Et, B-00499

$C_{10}H_{12}ClN_2O_5P$
4-Nitrophenyl 4-morpholinylphosphonochloridate, *in* M-00524

$C_{10}H_{12}Cl_2IO_2PS$
O-(2,5-Dichloro-4-iodophenyl) *O*-ethyl ethylphosphonothioate, D-00171

$C_{10}H_{12}Cl_3O_2PS$
▷Trichloronate, T-00419

$C_{10}H_{12}NOPS_2$
O-Isopropyl phenyl-phosphonisothiocyanatidothioate, *in* P-00227

$C_{10}H_{12}NO_3P$
1,2-Dihydro-1-methyl-4*H*-3,1,2-benzoxazaphosphorin-4-one; 1-Ethoxy, *in* D-00523
Dimethyl [(2-cyanophenyl)methyl]-phosphonate, *in* C-00220

$C_{10}H_{12}N_3O_3PS_2$
▷Azinphos-Methyl, A-00151

$C_{10}H_{12}N_5O_6P$
▷Cyclic AMP, C-00233

$C_{10}H_{12}N_5O_7P$
Cyclic GMP, C-00234

$C_{10}H_{13}Br_2O_3P$
Diethyl (2,5-dibromophenyl)phosphonate, *in* D-00066
Methyl hydrogen (1,2-dibromo-1-phenyl-propyl)phosphonate, *in* D-00068

$C_{10}H_{13}Br_3NO_3P$
Diethyl (2,4,6-tribromophenyl)-phosphoramidate, *in* T-00359

$C_{10}H_{13}ClNOP$
3-Benzyl-2-chlorotetrahydro-2*H*-1,3,2-oxazaphosphorine, B-00041
2-Chloro-3,4-dimethyl-5-phenyl-1,3,2-oxazaphospholidine, C-00061

$C_{10}H_{13}ClNOPS$
3-Benzyl-2-chlorotetrahydro-2*H*-1,3,2-oxazaphosphorine; 2-Sulfide, *in* B-00041
2-Chloro-3,4-dimethyl-5-phenyl-1,3,2-oxazaphospholidine; 2-Sulfide, *in* C-00061

$C_{10}H_{13}ClNO_2P$
2-Chloro-3,4-dimethyl-5-phenyl-1,3,2-oxazaphospholidine; 2-Oxide, *in* C-00061

$C_{10}H_{13}ClNO_2PS$
O-Phenyl 4-morpholinylphosphonochloridothioate, *in* M-00525

$C_{10}H_{13}ClNO_3P$
Phenyl 4-morpholinylphosphonochloridate, *in* M-00524

$C_{10}H_{13}Cl_2O_2P$
2-*tert*-Butylphenyl phosphorodichloridate, *in* M-00449
4-*tert*-Butylphenyl phosphorodichloridate, *in* M-00450

$C_{10}H_{13}Cl_2O_3P$
Diethyl (2,5-dichlorophenyl)phosphonate, *in* D-00185
Diethyl (3,4-dichlorophenyl)phosphonate, *in* D-00186

$C_{10}H_{13}Cl_2O_3PS$
▷Dichlofenthion, D-00161

$C_{10}H_{13}N_2OP$
2,3-Dihydro-1,3-dimethyl-1,3,2-benzodiazaphosphorin-4(1*H*)-one; 2-Me, 2-oxide, *in* D-00457
Phenyl bis(1-aziridinyl)phosphinite, *in* B-00095

$C_{10}H_{13}N_2OPS_2$
2,3-Dihydro-5-methyl-3-phenyl-1,3,4,2-thiadiazaphosphole 2-sulfide; 2-Ethoxy, *in* D-00529
O-Phenyl *N*-(4,5-dihydro-2-thiazolyl)-*P*-methylphosphonamidothioate, *in* D-00588

$C_{10}H_{13}N_2O_2P$
1,3-Dimethyl-2-phenyl-1,3-diazaphospholidin-4-one 2-oxide, *in* D-00721
Phenyl bis(1-aziridinyl)phosphinate, *in* B-00091

$C_{10}H_{13}N_2O_2PS$
Phenyl *N*-(4,5-dihydro-2-thiazolyl)-*P*-methyl-phosphonamidate, *in* D-00587

$C_{10}H_{13}N_2O_3P$
4′,5′-Dihydro-3′,5′-dimethylspiro[1,3,2-benzoxazaphosph(P^V)ole-2,2′-[1,3,2]-oxazaphospholidin]-4′-one, D-00470
1,3-Dimethyl-2-phenoxy-1,3-diazaphospholidin-4-one 2-oxide, *in* D-00721
Phosphotryptophan, P-00421

$C_{10}H_{13}N_2O_7P$
Diethyl (2,4-dinitrophenyl)phosphonate, *in* D-00946

$C_{10}H_{13}N_3OP$
3-Ethoxy-2,3-dihydro-2-methyl-5-phenyl-1*H*-1,2,4,3-triazaphosphole, E-00028

$C_{10}H_{13}N_4O_9P$
3′-Xanthylic acid, X-00002

$C_{10}H_{13}OP$
Methylphenyl-2-propenylphosphine; Oxide, *in* M-00264
1,2-Oxaphosphorinane; 2-Ph, *in* O-00054
4-Phenyl-1,4-oxaphosphorinane, P-00176
1-Phenylphospholane; 1-Oxide, *in* P-00216

$C_{10}H_{13}OPS$
2-Phenyl-1,3,2-oxathiaphosphepane, P-00178

$C_{10}H_{13}OPS_2$
2-Phenyl-1,3,2-oxathiaphosphepane; 2-Sulfide, *in* P-00178

$C_{10}H_{13}O_2P$
4,5-Dimethyl-2-phenyl-1,3,2-dioxaphospholane, D-00794
Dimethyl (2-phenylethenyl)phosphonite, *in* P-00161
Ethyl ethenylphenylphosphinate, *in* P-00337
Ethyl (2-phenylethenyl)phosphinate, *in* P-00153
5-Methyl-1,2-oxaphospholane 2-oxide; 2-Ph, *in* M-00183
4-Methyl-2-phenyl-1,3,2-dioxaphosphorinane, M-00199
Methyl phenyl(1-propenyl)phosphinate, *in* P-00299
Methyl phenyl(2-propenyl)phosphinate, *in* P-00300
1,2-Oxaphosphorinane; 2-Ph, 2-oxide, *in* O-00054
2-Phenyl-1,3,2-dioxaphosphepane, P-00114
4-Phenyl-1,4-oxaphosphorinane; Oxide, *in* P-00176
1,2,3,4-Tetrahydro-1-methoxyphosphinoline 1-oxide, *in* H-00195

$C_{10}H_{13}O_2PS$
4,5-Dimethyl-2-phenylthio-1,3,2-dioxaphospholane, D-00824
2-Phenyl-1,3,6,2-dioxathiaphosphocane, P-00117
O,O-Tetramethylene phenylphosphonothioate, *in* P-00114

$C_{10}H_{13}O_2PS_2$
S-Benzyl *O,O*-propylene phosphorodithioate, *in* M-00016
O,O-2,3-Butylene *S*-phenyl dithiophosphate, *in* D-00824
2-Mercapto-4*H*-1,3,2-benzodioxaphosphorin 2-sulfide; 2-Isopropylthio (2-Isopropyl ester), *in* M-00006
2-Phenyl-1,3,6,2-dioxathiaphosphocane; 2-Sulfide, *in* P-00117

$C_{10}H_{13}O_3P$
1,3-Butylene phenylphosphonate, *in* M-00199
2,3-Butylene phenylphosphonate, *in* D-00794
4,5-Dimethyl-2-phenoxy-1,3,2-dioxaphospholane, D-00789
Dimethyl (1-phenylethenyl)phosphonate, *in* P-00154
4-Methyl-2-phenoxy-1,3,2-dioxaphosphorinane, M-00191
Monoethyl (2-phenylethenyl)phosphonate, *in* P-00155
2-Phenoxy-1,2-oxaphosphorinane 2-oxide, *in* H-00167

(1-Phenyl-1-propenyl)phosphonic acid; Mono-Me ester, *in* P-00301
Tetramethylene phenyl phosphite, *in* D-00968
Tetramethylene phenylphosphonate, *in* P-00114

$C_{10}H_{13}O_3PS$
O,O-1,3-Butylene *O*-phenyl thiophosphate, *in* M-00191
2-Phenyl-1,3,6,2-dioxathiaphosphocane; 2-Oxide, *in* P-00117
O,O-Tetramethylene *O*-phenyl phosphorothioate, *in* D-00969
1,4-Thiaphosphorinane 1,1,4-trioxide; 4-Ph, *in* T-00300

$C_{10}H_{13}O_4P$
1,3-Butylene phenyl phosphate, *in* M-00191
Dimethyl (2-oxo-2-phenylethyl)phosphonate, *in* O-00066
4,5-Dimethyl-2-phenoxy-1,3,2-dioxaphospholane; 2-Oxide, *in* D-00789
(2-Oxo-4-phenylbutyl)phosphonic acid, O-00065
2-Phenoxy-1,3,2-dioxaphosphepane 2-oxide, *in* H-00132
1,4,6,9-Tetraoxa-5-phospha(5-P^V)spiro[4.4]-nonane; 5-Ph, *in* T-00244

$C_{10}H_{13}O_4PS_2$
α-[(Dimethoxyphosphinothioyl)thio]-benzeneacetic acid, D-00676

$C_{10}H_{13}O_5P$
Phenyl (dimethoxyphosphinyl)acetate, *in* D-00678
1,3,6,2-Trioxaphosphocane; 2-Phenoxy, 2-oxide, *in* T-00592

$C_{10}H_{13}O_6P$
Methyl 2-(dimethoxyphosphinyl)oxybenzoate, *in* P-00391

$C_{10}H_{13}P$
Methylphenyl-2-propenylphosphine, M-00264
1-Phenylphospholane, P-00216

$C_{10}H_{13}PS$
Methylphenyl-2-propenylphosphine; Sulfide, *in* M-00264
1-Phenylphospholane; 1-Sulfide, *in* P-00216
2-Phenyl-1,2-thiaphosphorinane, P-00325

$C_{10}H_{13}PS_2$
5-Methyl-1,2-thiaphospholane 2-sulfide; 2-Ph, *in* M-00407

$C_{10}H_{14}AsP$
1,2,3,4-Tetrahydro-1,4-dimethyl-1,4-benzophospharsenine, T-00091

$C_{10}H_{14}BrOPS_2$
S-4-Bromophenyl *O*-ethyl ethylphosphonodithioate, *in* E-00070

$C_{10}H_{14}BrO_2P$
Diethyl (4-bromophenyl)phosphonite, *in* B-00493

$C_{10}H_{14}BrO_3P$
Diethyl (2-bromophenyl)phosphonate, *in* B-00490
Diethyl (3-bromophenyl)phosphonate, *in* B-00491
Diethyl (4-bromophenyl)phosphonate, *in* B-00492

$C_{10}H_{14}Br_2NO_3P$
Diethyl (2,4-dibromophenyl)phosphoramidate, *in* D-00067

$C_{10}H_{14}ClOP$
tert-Butylphenylphosphinic acid; Chloride, *in* B-00579
tert-Butylphenylphosphinic acid; Chloride, *in* B-00579
(1-Methylpropyl)phenylphosphinic acid; Chloride, *in* M-00372

$C_{10}H_{14}ClOPS_2$
▷*O*-Ethyl *S*-4-chlorophenyl ethylphosphonodithioate, *in* E-00070

$C_{10}H_{14}ClO_2P$
Butyl phenylphosphonochloridate, *in* P-00229
Diethyl (3-chlorophenyl)phosphinate, *in* C-00152
Diethyl (4-chlorophenyl)phosphonite, *in* C-00153

$C_{10}H_{14}ClO_2PS$
O,O-Diethyl (4-chlorophenyl)phosphonothioate, *in* C-00150

$C_{10}H_{14}ClO_3P$
Butyl phenyl phosphorochloridate, *in* P-00269
Diethyl (2-chlorophenyl)phosphonate, *in* C-00145
Diethyl (3-chlorophenyl)phosphonate, *in* C-00146
Diethyl (4-chlorophenyl)phosphonate, *in* C-00147

$C_{10}H_{14}ClPS$
tert-Butylphenylphosphinothioic acid; Chloride, *in* B-00583
tert-Butylphenylphosphinothioic acid; Chloride, *in* B-00583

$C_{10}H_{14}ClPSi$
[Phenyl(trimethylsilyl)methylene]phosphinous chloride, P-00330

$C_{10}H_{14}Cl_2NO_2P$
Phenyl hydrogen bis(2-chloroethyl)-phosphoramidate, *in* B-00121

$C_{10}H_{14}Cl_2NO_2PS$
O-2,4-Dichlorophenyl *S*-methyl isopropylphosphoramidothioate, D-00183
▷Zytron, Z-00002

$C_{10}H_{14}Cl_2NO_3P$
Diethyl (2,4-dichlorophenyl)phosphoramidate, *in* D-00188

$C_{10}H_{14}FOP$
tert-Butylphenylphosphinic acid; Fluoride, *in* B-00579

$C_{10}H_{14}FO_3P$
Diethyl (2-fluorophenyl)phosphonate, *in* F-00036
Diethyl (3-fluorophenyl)phosphonate, *in* F-00037
Diethyl (4-fluorophenyl)phosphonate, *in* F-00038

$C_{10}H_{14}IO_2P$
3-Methyl-2-oxo-3-phenyl-1,3-oxaphosphorinanium iodide, *in* P-00177

$C_{10}H_{14}NO_2P$
1,5-Dihydro-3-dimethylamino-2,4,3-benzodioxaphosphepin, D-00453
4,5-Dihydro-2-dimethylamino-1,3,2-benzodioxaphosphepin, D-00454
2-Ethoxy-3-phenyl-1,3,2-oxazaphospholidine, E-00048
Ethyl phenyl 1-aziridinylphosphonite, *in* A-00154
2,3,5,6,8,8-Hexahydro-8-phenyl-[1,3,2]-oxazaphospholo[2,3-*b*][1,3,2]-oxazaphosphole, H-00049
4-Methyl-2-phenylamino-1,3,2-dioxaphosphorinane, M-00193
3-Methyltetrahydro-2-phenoxy-2*H*-1,3,2-oxazaphosphorine, M-00404

$C_{10}H_{14}NO_2PS$
O,O-1,3-Butylene phenylphosphoramidothioate, *in* M-00193
1,5-Dihydro-3-dimethylamino-2,4,3-benzodioxaphosphepin; 3-Sulfide, *in* D-00453
2-Ethoxy-3-phenyl-1,3,2-oxazaphospholidine; 2-Sulfide, *in* E-00048
3-Methyltetrahydro-2-phenoxy-2*H*-1,3,2-oxazaphosphorine; 2-Sulfide, *in* M-00404

$C_{10}H_{14}NO_2PSe$
O,O-1,3-Butylene phenyl-phosphoramidoselenoate, *in* M-00193
1,5-Dihydro-3-dimethylamino-2,4,3-benzodioxaphosphepin; 3-Selenide, *in* D-00453

$C_{10}H_{14}NO_3P$
1,3-Butylene phenylphosphoramidate, *in* M-00193
2,2-Dihydro-1,3,2-benzoxazaphosphole; 2,2-Diethoxy, *in* D-00444
1,5-Dihydro-3-dimethylamino-2,4,3-benzodioxaphosphepin; 3-Oxide, *in* D-00453
4,5-Dihydro-2-dimethylamino-1,3,2-benzodioxaphosphepin; 2-Oxide, *in* D-00454

$C_{10}H_{14}NO_4P$
(1-Benzoylamino-1-methylethyl)phosphonic acid, *in* A-00087
[1-(*N*-Benzoylamino)propyl]phosphonic acid, *in* A-00125
Trimethyl benzoylphosphorimidate, *in* B-00039

$C_{10}H_{14}NO_5P$
Diethyl (2-nitrophenyl)phosphonate, *in* N-00055
Diethyl (4-nitrophenyl)phosphonate, *in* N-00057

$C_{10}H_{14}NO_5PS$
▷Parathion, P-00002

$C_{10}H_{14}NO_6P$
▷Diethyl 2-nitrophenyl phosphate, D-00303
▷Diethyl 3-nitrophenyl phosphate, D-00304
▷Paraoxon, P-00001

$C_{10}H_{14}NP$
Hexahydro-3-phenyl-1,3-azaphosphorine, H-00041
Hexahydro-4-phenyl-1,4-azaphosphorine, H-00042

$C_{10}H_{14}NPS$
Hexahydro-4-phenyl-1,4-azaphosphorine; 4-Sulfide, *in* H-00042

$C_{10}H_{14}N_2O_2P$
Ethyl (1-diazoethyl)phenylphosphinate, *in* D-00032

$C_{10}H_{14}N_3OP$
P,P-Bis(1-aziridinyl)-*N*-phenylphosphinic amide, *in* B-00092
tert-Butylphenylphosphinic acid; Azide, *in* B-00579
2,3-Dihydro-2-dimethylamino-5-methyl-3-phenyl-1,3,4,2-oxadiazaphosphole, D-00455

$C_{10}H_{14}N_3OPS$
2,3-Dihydro-2-dimethylamino-5-methyl-3-phenyl-1,3,4,2-oxadiazaphosphole; 2-Sulfide, *in* D-00455
3-Ethoxy-2,3-dihydro-2-methyl-5-phenyl-1*H*-1,2,4,3-triazaphosphole; 3-Sulfide, *in* E-00028

$C_{10}H_{14}N_3O_2P$
2,3-Dihydro-2-dimethylamino-5-methyl-3-phenyl-1,3,4,2-oxadiazaphosphole; 2-Oxide, *in* D-00455
3-Ethoxy-2,3-dihydro-2-methyl-5-phenyl-1*H*-1,2,4,3-triazaphosphole; 3-Oxide, *in* E-00028

$C_{10}H_{14}N_3PS$
P,P-Bis(1-aziridinyl)-*N*-phenylphosphinothioic amide, *in* B-00094

$C_{10}H_{14}N_3PS_2$
2,3-Dihydro-*N,N*,5-trimethyl-3-phenyl-1,3,4,2-thiadiazaphosphole-2(3*H*)-amine-2-sulfide, *in* A-00072

$C_{10}H_{14}N_5O_7P$
2′-Adenylic acid, A-00045
3′-Adenylic acid, A-00046

$C_{10}H_{14}N_5O_8P$
3′-Guanylic acid, G-00102
▷5′-Guanylic acid, G-00103

$C_{10}H_{14}O_2P_2$
3*a*,4,7,7*a*-Tetrahydro-1,8-dimethyl-4,7-phosphinidene-1*H*-phosphindole; 1,8-Dioxide, *in* T-00099

$C_{10}H_{14}P_2$
3*a*,4,7,7*a*-Tetrahydro-1,8-dimethyl-4,7-phosphinidene-1*H*-phosphindole, T-00099

$C_{10}H_{14}P_2S_2$
3*a*,4,7,7*a*-Tetrahydro-1,8-dimethyl-4,7-phosphinidene-1*H*-phosphindole; 1,8-Disulfide, *in* T-00099

$C_{10}H_{15}BrNO_3P$
Diethyl (4-bromophenyl)phosphoramidate, *in* B-00496

$C_{10}H_{15}ClNOP$
N,N-Diethyl-*P*-phenylphosphonamidic acid; Chloride, *in* D-00307

$C_{10}H_{15}ClNOPS$
O-Phenyl diethylphosphoramidochloridothioate, *in* D-00348

$C_{10}H_{15}ClNOPSe$
O-Phenyl diethylphosphoramidochloridoselenoate, *in* D-00347

$C_{10}H_{15}ClNO_2P$
P-(4-Chlorobutyl)-*N*-phenylphosphonamidic acid, C-00024
Phenyl diethylphosphoramidochloridate, *in* D-00346

$C_{10}H_{15}ClNO_2PS$
O,O-Diethyl (2-chlorophenyl)-phosphoramidothioate, *in* C-00158
O,O-Diethyl (3-chlorophenyl)-phosphoramidothioate, *in* C-00159
O,O-Diethyl (4-chlorophenyl)-phosphoramidothioate, *in* C-00160

$C_{10}H_{15}ClNO_3P$
Diethyl (4-chlorophenyl)phosphoramidate, *in* C-00157

$C_{10}H_{15}ClNPS$
N,N-Diethyl-*P*-phenylphosphonamidothioic acid; Chloride, *in* D-00309

$C_{10}H_{15}Cl_2N_2O_2P$
Phenyl *N,N*-bis(2-chloroethyl)-phosphorodiamidate, *in* B-00126

$C_{10}H_{15}Cl_2N_2P$
P,P-Dichloro-*P*-diethylamino-*N*-phenylphosphine imide, *in* D-00313

$C_{10}H_{15}Cl_2OP$
1-Adamantylphosphonic acid; Dichloride, *in* A-00034
2-Adamantylphosphonic acid; Dichloride, *in* A-00035

$C_{10}H_{15}FNOP$
N,N-Diethyl-*P*-phenylphosphonamidic acid; Fluoride, *in* D-00307

$C_{10}H_{15}FNO_2PS$
O,O-Diethyl (3-fluorophenyl)-phosphoramidothioate, *in* F-00045
O,O-Diethyl (4-fluorophenyl)-phosphoramidothioate, *in* F-00046

$C_{10}H_{15}FNO_3P$
Diethyl (3-fluorophenyl)phosphoramidate, *in* F-00043
Diethyl (4-fluorophenyl)phosphoramidate, *in* F-00044

$C_{10}H_{15}FNPS_2$
Phenyl diethylphosphoramidofluoridodithioate, *in* D-00352

$C_{10}H_{15}IN_3P$
Tris(2-cyanoethyl)methylphosphonium iodide, *in* T-00752

$C_{10}H_{15}NPSe$
P,P-Diethyl-*N*-phenylphosphinoselenoic amide, *in* D-00328

$C_{10}H_{15}N_2OP$
1,3-Dimethyl-2-phenoxy-1,3,2-diazaphospholidine, D-00788
Hexahydro-2-phenoxy-1*H*-1,3,2-diazaphosphepine, H-00038

$C_{10}H_{15}N_2OPS$
Hexahydro-2-phenoxy-1*H*-1,3,2-diazaphosphepine; 2-Sulfide, *in* H-00038

$C_{10}H_{15}N_2O_2P$
1,6-Dioxa-4,9-diaza-5-phospha(5-P^V)spiro[4.4]-nonane; 5-Ph, *in* D-00963

$C_{10}H_{15}N_2O_4PS$
O-(4-Methyl-2-nitrophenyl) isopropylphosphoramidothioate, M-00180

$C_{10}H_{15}N_2O_5P$
Diethyl (3-nitrophenyl)phosphoramidate, *in* N-00059
Diethyl (4-nitrophenyl)phosphoramidate, *in* N-00060

$C_{10}H_{15}N_2P$
1,2-Dimethyl-4-phenyl-1,2,4-diazaphospholidine, *in* D-00718
1,3-Dimethyl-2-phenyl-1,3,2-diazaphospholidine, D-00793
N,N'-1,2-Ethanediyl-*N,N'*-dimethyl-*P*-phenylphosphonic diamide, *in* D-00793

$C_{10}H_{15}N_5O_{10}P_2$
▷Adenosine diphosphate, A-00038
Adenosine 2',5'-diphosphate, A-00039
Adenosine 3',5'-diphosphate, A-00040

$C_{10}H_{15}N_5O_{11}P_2$
Guanosine 5'-diphosphate, G-00099

$C_{10}H_{15}OP$
Diethylphenylphosphine; Oxide, *in* D-00305
Diethylphosphinous acid; Ph ester, *in* D-00333
Ethyl ethylphenylphosphinite, *in* E-00108
Methylphenylpropylphosphine; Oxide, *in* M-00265

$C_{10}H_{15}OPS$
tert-Butylphenylphosphinothioic acid, B-00583
S-Ethyl ethylphenylphosphinothioate, *in* E-00105
S-Isopropyl methylphenylhosphinothioate, *in* M-00225
O-Isopropyl methylphenylphosphinothioate, *in* M-00225
Isopropylphenylphosphinothioic acid; *S*-Me ester, *in* I-00046
O-Phenyl diethylphosphinothioate, *in* D-00330

$C_{10}H_{15}OPS_2$
O-Butyl phenylphosphonodithioate, *in* P-00236
S,S-Diethyl phenylphosphonodithioate, *in* P-00236
▷Fonofos, F-00052

$C_{10}H_{15}OPSe$
tert-Butylphenylphosphinoselenoic acid, B-00582
O-Ethyl ethylphenylphosphinoselenoate, *in* E-00104

$C_{10}H_{15}O_2P$
Butylphenylphosphinic acid, B-00578
tert-Butylphenylphosphinic acid, B-00579
Diethyl phenylphosphonite, D-00311
(3,4-Dimethylphenyl)phosphinic acid; Et ester, *in* D-00806
Ethyl benzylmethylphosphinate, *in* B-00053
Ethyl (2,5-dimethylphenyl)phosphinate, *in* D-00805
Ethyl ethylphenylphosphinate, *in* E-00101
Isopropyl methylphenylphosphinate, *in* M-00219
Isopropyl (4-methylphenyl)phosphinate, *in* M-00222
Methyl isopropylphenylphosphinate, *in* I-00043
(1-Methylpropyl)phenylphosphinic acid, M-00372
(2-Methylpropyl)phenylphosphinic acid, M-00373
Phenyl butylphosphinate, *in* B-00590
Phenyl *tert*-butylphosphinate, *in* B-00591
Phenyl diethylphosphinate, *in* D-00322

$C_{10}H_{15}O_2PS$
O,O-Diethyl phenylphosphonothioate, *in* P-00244
O,S-Diethyl phenylphosphonothioate, *in* P-00244
O-Ethyl *S*-phenyl ethylphosphonothioate, *in* E-00074
O-Isopropyl *S*-phenyl methylphosphonothioate, *in* I-00041

$C_{10}H_{15}O_2PSSi$
2,2-Dimethyl-1,3-dioxa-5-phospha-2-silacyclohexane; 5-Ph, 5-sulfide, *in* D-00731

$C_{10}H_{15}O_2PS_3$
O,O-Diethyl *SS*-phenyl phosphoro(dithioperoxo)thioate, D-00315

$C_{10}H_{15}O_2PSe$
O,O-Diethyl phenylphosphonoselenoate, *in* P-00243
O,O-Diethyl *Se*-phenyl phosphoroselenoite, D-00316

$C_{10}H_{15}O_2PSi$
2,2-Dimethyl-1,3-dioxa-5-phospha-2-silacyclohexane; 5-Ph, *in* D-00731

$C_{10}H_{15}O_3P$
Butyl phenyl phosphonate, B-00585
▷Diethyl phenylphosphonate, D-00310
Dimethyl 1-phenylethylphosphonate, *in* P-00164
Dimethyl 2-phenylethylphosphonate, *in* P-00165
Ethyl (methoxymethyl)phenylphosphinate, *in* M-00052
1-Methyl-3,8-phosphonanedione 1-oxide, *in* P-00373

$C_{10}H_{15}O_3PS$
O,O-Diethyl *S*-phenyl phosphorothioate, D-00317

$C_{10}H_{15}O_3PS_2$
O,O-Diethyl *SS*-phenyl-phosphoro(dithioperoxoate), D-00314
▷Fenthion, F-00007

$C_{10}H_{15}O_4P$
Diethyl (2-hydroxyphenyl)phosphonate, *in* H-00179
Diethyl (4-hydroxyphenyl)phosphonate, *in* H-00181
Dimethyl [(4-methoxyphenyl)methyl]-phosphonate, *in* M-00061
Mono-2-*tert*-butylphenyl phosphate, M-00449
Mono-4-*tert*-butylphenyl phosphate, M-00450

$C_{10}H_{15}O_4PS$
[[(4-Methylphenyl)sulfinyl]methyl]phosphonic acid; Di-Me ester, *in* M-00266

$C_{10}H_{15}O_4PS_2$
O,O-Dimethyl *O*-[(3-methyl-4-methylsulfinyl)-phenyl] phosphorothioate, *in* F-00007

$C_{10}H_{15}O_5PS_2$
O,O-Dimethyl *O*-[(3-methyl-4-methylsulfonyl)-phenyl]phosphorothioate, *in* F-00007

$C_{10}H_{15}P$
Diethylphenylphosphine, D-00305
Methylphenylpropylphosphine, M-00265

$C_{10}H_{15}PS$
Diethylphenylphosphine; Sulfide, *in* D-00305
Ethyl ethylphenylphosphinothioite, *in* E-00107
Methylphenylpropylphosphine; Sulfide, *in* M-00265
Phenyl diethylphosphinothioite, *in* D-00332

$C_{10}H_{15}PSSe$
tert-Butylphenylphosphinothioselenoic acid, B-00584
S-Ethyl ethylphenylphosphinothioselenoate, *in* E-00106

$C_{10}H_{15}PS_2$
tert-Butylphenylphosphinodithioic acid, B-00581
Diethyl phenylphosphonodithioite, *in* P-00238
Ethyl ethylphenylphosphinodithioate, *in* E-00103
Phenyl diethylphosphinodithioate, *in* D-00325

$C_{10}H_{15}PS_3$
Diethyl phenylphosphonotrithioate, *in* P-00250

$C_{10}H_{15}PSe$
Diethylphenylphosphine; Selenide, *in* D-00305
Methylphenylpropylphosphine; Selenide, *in* M-00265

$C_{10}H_{16}AsP$
[2-(Dimethylarsino)phenyl]dimethylphosphine, D-00711

$C_{10}H_{16}ClN_2OP$
P-4-Chlorophenyl-*N*,*N*,*N'*,*N'*-tetramethylphosphonic diamide, *in* C-00147

$C_{10}H_{16}ClN_2P$
P-3-Chlorophenyl-*N*,*N*,*N'*,*N'*-tetramethylphosphonous diamide, *in* C-00152
P-4-Chlorophenyl-*N*,*N*,*N'*,*N'*-tetramethylphosphonous diamide, *in* C-00153

$C_{10}H_{16}ClN_2PS$
P-4-Chlorophenyl-*N*,*N*,*N'*,*N'*-tetramethylphosphonothioic diamide, *in* C-00150

$C_{10}H_{16}Cl_2NOP$
1-Adamantylphosphoramidic acid; Dichloride, *in* A-00036
2-Adamantylphosphoramidic acid; Dichloride, *in* A-00037

$C_{10}H_{16}FN_2P$
P-4-Fluorophenyl-*N*,*N*,*N'*,*N'*-tetraethylphosphonous diamide, *in* F-00042
P-3-Fluorophenyl-*N*,*N*,*N'*,*N'*-tetramethylphosphonous diamide, *in* F-00041

$C_{10}H_{16}IP$
Dimethylethylphenylphosphonium iodide, *in* E-00084

$C_{10}H_{16}NOP$
P-*tert*-Butyl-*P*-phenylphosphinic amide, B-00580
P,*P*-Diethyl-*N*-phenylphosphinimidic acid, D-00306

$C_{10}H_{16}NOPS$
N,*N*-Diethyl-*P*-phenylphosphonamidothioic acid, D-00309
S-Propyl *P*-methyl-*N*-phenylphosphonamidothioate, *in* M-00233

$C_{10}H_{16}NOPS_2$
O-Isopropyl *S*-phenyl methylphosphoramidodithioate, *in* M-00330

$C_{10}H_{16}NO_2P$
N,*N*-Diethyl-*P*-phenylphosphonamidic acid, D-00307
Diethyl phenylphosphoramidite, *in* P-00263
Ethyl *P*-ethyl-*N*-phenylphosphonamidate, *in* E-00110
Isopropyl *P*-methyl-*N*-phenylphosphonamidate, *in* M-00231
Methyl (4-dimethylaminophenyl)methylphosphinate, *in* D-00705
Phenyl *N*-isopropyl-*P*-methylphosphonamidate, *in* I-00036

$C_{10}H_{16}NO_2PS$
O,*O*-Diethyl phenylphosphoramidothioate, *in* P-00262

$C_{10}H_{16}NO_2PSe$
O,*O*-Diethyl phenylphosphoramidoselenoate, *in* P-00261

$C_{10}H_{16}NO_3P$
(1-Benzylamino-1-methylethyl)phosphonic acid, *in* A-00087
[1-(Benzylamino)propyl]phosphonic acid, *in* A-00125
Diethyl (2-aminophenyl)phosphonate, *in* A-00109
Diethyl (4-aminophenyl)phosphonate, *in* A-00111
Diethyl (3-methyl-2-pyridinyl)phosphonate, *in* M-00389
Diethyl (4-methyl-2-pyridinyl)phosphonate, *in* M-00391
Diethyl (6-methyl-2-pyridinyl)phosphonate, *in* M-00392
Diethyl phenylphosphoramidate, D-00312
Diethyl (2-pyridinylmethyl)phosphonate, *in* P-00506
Diethyl (4-pyridinylmethyl)phosphonate, *in* P-00507

$C_{10}H_{16}NO_3PS$
O-(4-Aminophenyl) *O*,*O*-diethyl phosphorothioate, A-00099

$C_{10}H_{16}NO_4P$
4-Aminophenyl diethyl phosphate, A-00098
(1-Benzoylamino-2-methylpropyl)phosphonic acid, *in* A-00094
(4-Pyridinylmethyl)phosphonic acid; Di-Et ester, *N*-oxide, *in* P-00507

$C_{10}H_{16}NO_5PS$
Diethyl phenylsulfonylphosphoramidate, *in* P-00312
Trimethyl [(4-methylphenyl)sulfonyl]phosphorimidate, *in* M-00268

$C_{10}H_{16}NO_5PS_2$
▷Famphur, F-00001

$C_{10}H_{16}NP$
P,*P*-Diethyl-*N*-phenylphosphinous amide, *in* D-00334

$C_{10}H_{16}NPS$
P,*P*-Diethyl-*N*-phenylphosphinothioic amide, *in* D-00331
P-Ethyl-*N*,*N*-dimethyl-*P*-phenylphosphinothioic amide, *in* E-00105

$C_{10}H_{16}NPS_2$
N,*N*-Diethyl-*P*-phenylphosphonamidodithioic acid, D-00308
Phenyl *P*-ethyl-*N*,*N*-dimethylphosphonamidodithioate, *in* E-00063

$C_{10}H_{16}N_2O_{11}P_2$
Thymidine 5′-pyrophosphate, T-00320

$C_{10}H_{16}N_3OP$
2,2-Dihydro-1,3,2-benzoxazaphosphole; 2,2-Bis(dimethylamino), *in* D-00444

$C_{10}H_{16}N_3P$
Hexahydro-3-phenyl-1,2,4,3-triazaphosphorine; 1,4-Di-Me, *in* H-00051

$C_{10}H_{16}N_5O_{13}P_3$
▷Adenosine triphosphate, A-00043

$C_{10}H_{16}N_5O_{14}P_3$
▷Guanosine 5′-triphosphate, G-00101

$C_{10}H_{16}O_2P_2$
1,3-Bis(dimethylphosphinyl)benzene, *in* P-00124
1,4-Bis(dimethylphosphinyl)benzene, *in* P-00125

$C_{10}H_{16}O_4P_2$
Dimethyl 1,4-phenylenebis[methylphosphinate], *in* P-00137
Tetramethyl 1,4-phenylenebisphosphonite, *in* B-00010

$C_{10}H_{16}O_6P_2$
Tetramethyl 1,2-phenylenebisphosphonate, *in* B-00006
Tetramethyl 1,3-phenylenebisphosphonate, *in* B-00007
Tetramethyl 1,4-phenylenebisphosphonate, *in* B-00008

$C_{10}H_{16}P_2$
1,2-Phenylenebis[dimethylphosphine], P-00123
1,3-Phenylenebis[dimethylphosphine], P-00124
1,4-Phenylenebis[dimethylphosphine], P-00125

$C_{10}H_{16}P_2S_2$
1,3-Bis(dimethylphosphinothioyl)benzene, *in* P-00124
1,4-Bis(dimethylphosphinothioyl)benzene, *in* P-00125

$C_{10}H_{16}P_2S_4$
Dimethyl 1,4-phenylenebis[methylphosphinodithioate], *in* P-00138

$C_{10}H_{16}P_2Se_2$
1,4-Bis(dimethylphosphinoselenoyl)benzene, *in* P-00125

$C_{10}H_{17}N_2OP$
N,*N'*-Diethyl-*P*-phenylphosphonic diamide, *in* P-00222
Phenyl tetramethylphosphorodiamidite, *in* T-00227

$C_{10}H_{17}N_2O_2P$
N,*N*-Diethyl-*N'*-phenylphosphoramidoimidic acid, D-00313
Phenyl *N*,*N'*-diethylphosphorodiamidate, *in* D-00382
Phenyl tetramethylphosphorodiamidate, *in* T-00218

$C_{10}H_{17}N_2O_2PS$
O,*O*-Diethyl 2-phenylphosphorohydrazidothioate, *in* P-00296

$C_{10}H_{17}N_2O_3P$
Diisopropyl (2-pyrimidinyl)phosphonate, *in* P-00515
Diisopropyl (4-pyrimidinyl)phosphonate, *in* P-00516

$C_{10}H_{17}N_2O_4PS$
▷Etrimfos, E-00205

$C_{10}H_{17}N_2P$
Bis(2-aminoethyl)phenylphosphine, B-00086
N,*N'*-Diethyl-*P*-phenylphosphonous diamide, *in* P-00252

$C_{10}H_{17}N_2PS$
N,*N'*-Diethyl-*P*-phenylphosphonothioic diamide, *in* P-00246
P-Methyl-*N*-isopropyl-*N'*-phenylphosphonothioic diamide, *in* M-00245

$C_{10}H_{17}N_2PSe$
N,*N*,*N'*,*N'*-Tetramethyl-*P*-phenylphosphonoselenoic diamide, *in* P-00243

$C_{10}H_{17}OP$
Octahydro-3-methyl-1*H*-cyclopropa[*c*]phosphindole 3-oxide, *in* O-00004

$C_{10}H_{17}O_2P$
Octahydro-3-methoxy-1*H*-cyclopropa[*c*]phosphindole 3-oxide, *in* O-00004

$C_{10}H_{17}O_2PS$
Diisopropyl 2-thienylphosphonite, *in* T-00308

$C_{10}H_{17}O_3P$
1-Adamantylphosphonic acid, A-00034
2-Adamantylphosphonic acid, A-00035

$C_{10}H_{17}O_4P$
2,4-Bis(1-methylethylidene)-1,3,6,9-tetraoxa-5-phospha(5-P^V)spiro[4.4]nonane, B-00306

$C_{10}H_{17}O_4PSi$
[Phenyl[(trimethylsilyl)oxy]methyl]phosphonic acid, P-00331

$C_{10}H_{17}O_7P$
2-Ethoxy-1,3,2-dioxaphospholan-4,5-dicarboxylic acid; Di-Et ester, *in* E-00033

$C_{10}H_{17}P$
1-*tert*-Butyl-3,4-dimethyl-1*H*-phosphole, E-00004

$C_{10}H_{17}PS$
1-*tert*-Butyl-3,4-dimethyl-1*H*-phosphole; 1-Sulfide, *in* E-00004

$C_{10}H_{18}Cl_3O_5P$
Diisopropyl 1-acetoxy-2,2,2-trichloroethylphosphonate, *in* A-00011

$C_{10}H_{18}IP$
Methyltri-1-propenylphosphonium iodide, *in* T-00706
Methyltri-2-propenylphosphonium iodide, *in* T-00707
Tricyclopropylmethylphosphonium iodide, *in* T-00432

$C_{10}H_{18}NO_3P$
1-Adamantylphosphoramidic acid, A-00036

2-Adamantylphosphoramidic acid, A-00037

$C_{10}H_{18}NO_6P$

2-Diethylamino-1,3,2-dioxaphospholan-4,5-dicarboxylic acid; Di-Me ester, *in* D-00265

$C_{10}H_{18}N_3O_3P$

Diisopropyl (2-amino-4-pyrimidinyl)-phosphonate, *in* A-00128

$C_{10}H_{19}ClNOP$

3,5-Di-*tert*-butyl-2-chloro-2,3-dihydro-1,3,2-oxazaphosphole, D-00078

$C_{10}H_{19}ClNOPS$

3,5-Di-*tert*-butyl-2-chloro-2,3-dihydro-1,3,2-oxazaphosphole; 2-Sulfide, *in* D-00078

$C_{10}H_{19}ClNO_2P$

3,5-Di-*tert*-butyl-2-chloro-2,3-dihydro-1,3,2-oxazaphosphole; 2-Oxide, *in* D-00078

$C_{10}H_{19}ClNO_5P$

▷Phosphamidon, P-00347

$C_{10}H_{19}ClNO_5PS$

Methyl 6-deoxy-2,3-di-*O*-methyl-6-methyl-aminoglucopyranoside 4,6-cyclic thiophosphonamide; *P*-Chloro, *in* M-00113

$C_{10}H_{19}ClNO_6P$

Methyl 6-deoxy-2,3-di-*O*-methyl-6-methyl-aminoglucopyranoside 4,6-cyclic phosphonamide; *P*-Chloro, *in* M-00112

$C_{10}H_{19}Cl_2P$

(4-*tert*-Butylcyclohexyl)dichlorophosphine, *in* B-00550

$C_{10}H_{19}N_2O_4PS$

▷Cyanthoate, C-00232

$C_{10}H_{19}N_4OP$

N,N,N',N'-Tetramethyl-2-phenyl-phosphorodiamidic hydrazide, *in* P-00274

$C_{10}H_{19}O_2P$

Dibutyl ethynylphosphonite, *in* E-00204

$C_{10}H_{19}O_3P$

Diethyl 1-cyclohexen-1-ylphosphonate, *in* C-00241

Diethyl 2-cyclohexen-1-ylphosphonate, *in* C-00242

Diethyl 3-cyclohexen-1-ylphosphonate, *in* C-00243

$C_{10}H_{19}O_4P$

3,7-Dimethyl-1,6-octadien-3-yl dihydrogen phosphate, D-00775

3,7-Dimethyl-2,6-octadien-1-yl dihydrogen phosphate, D-00776

$C_{10}H_{19}O_5P$

4-(Diethoxyphosphinyl)-3-methyl-2-butenoic acid; Me ester, *in* D-00251

Ethyl 4-(diethoxyphosphinyl)-2-butenoate, *in* D-00248

$C_{10}H_{19}O_6P$

▷Ethyl 2-(diethoxyphosphinyl)-3-oxobutanoate, *in* D-00253

Ethyl 4-(diethoxyphosphinyl)-3-oxobutanoate, *in* D-00254

$C_{10}H_{19}O_6PS$

Methyl 2,3-di-*O*-methylglucopyranoside 4,6-cyclic phosphonothioate; *P*-Me, *in* M-00120

Methyl 2,3-di-*O*-methyl-4-thioglucopyranoside 4,6-cyclic phosphonate; *P*-Me, *in* M-00121

Methyl 2,3-di-*O*-methyl-6-thioglucopyranoside 4,6-cyclic phosphonate; *P*-Me, *in* M-00122

$C_{10}H_{19}O_6PS_2$

Dimethyl [(diethoxyphosphinothioyl)thio]-butanedioate, *in* D-00240

▷Malathion, M-00001

$C_{10}H_{19}O_7P$

Dimethyl (diethoxyphosphinyl)butanedioate, *in* D-00245

Methyl 2,3-di-*O*-methylglucopyranoside 4,6-cyclic phosphonate; *P*-Me, *in* M-00119

$C_{10}H_{19}P$

1,3,4,5,5,6,7,8,9,9*a*-Decahydro-5-methylene-2*H*-phosphinolizine, D-00005

$C_{10}H_{20}ClN_2OP$

Di-1-piperidinylphosphinic acid; Chloride, *in* D-01190

$C_{10}H_{20}ClN_4OP$

2,4,8,10-Tetramethyl-3,9-dioxo-2,4,8,10-tetraaza-6-phosphoniaspiro[5.5]-undecane(1+); Chloride, *in* T-00198

$C_{10}H_{20}ClN_4P$

Octahydro-5*H*,8*H*,10*bH*-2*a*,4*a*,7*a*,10*a*-tetraaza-10*b*-phosphapentaleno[1,6-*de*]naphthalene; 10*b*-Chloro, *in* O-00024

$C_{10}H_{20}ClOPS$

O-(2-Methylpropyl) cyclohexylphosphonochloridothioate, *in* C-00260

$C_{10}H_{20}ClO_2P$

2-Chloro-1,3-dioxa-2-phosphacyclotridecane, *in* D-00966

Cyclohexyl butylphosphonochloridate, *in* B-00603

$C_{10}H_{20}ClP$

6-Phosphoniaspiro[5.5]undecane; Chloride, *in* P-00375

$C_{10}H_{20}FN_2OP$

Di-1-piperidinylphosphinic acid; Fluoride, *in* D-01190

$C_{10}H_{20}FN_4P$

Octahydro-3*H*,8*H*,10*bH*-2*a*,5*a*,7*a*,10*a*-tetraaza-10*b*-phosphaindeno[7,1-*cd*]indene; 10*b*-Fluoro, *in* O-00022

Octahydro-5*H*,8*H*,10*bH*-2*a*,4*a*,7*a*,10*a*-tetraaza-10*b*-phosphapentaleno[1,6-*de*]naphthalene; 10*b*-Fluoro, *in* O-00024

$C_{10}H_{20}FO_2P$

Cyclohexyl butylphosphonofluoridate, *in* B-00608

2-Fluoro-1,3-dioxa-2-phosphacyclotridecane, *in* D-00966

$C_{10}H_{20}NOP$

N,N-Diethyl-*P,P*-di-2-propenylphosphinic amide, *in* D-01193

Hexahydrospiro[1,3,2-benzoxazaphosphole-2(3*H*),1'-phospholane], H-00052

$C_{10}H_{20}NO_2P$

2,3-Dihydro-3,3,5-trimethyl-1,2-oxaphosphole 2-oxide; 2-Diethylamino, *in* D-00596

Ethyl *P*-(1,3-butadienyl)-*N,N*-diethylphosphonamidate, *in* B-00508

$C_{10}H_{20}NO_3P$

Diethyl di-2-propenylphosphoramidate, *in* D-01195

7-Isopropylidene-4,8,8-trimethyl-1,6-dioxa-4-aza-5-phospha(5-P^V)spiro[4.4]nonan-9-one, I-00032

$C_{10}H_{20}NO_4P$

Diethyl *tert*-butoxycyanomethylphosphonate, D-00273

Dimethyl [(5,6-dihydro-4,4,6-trimethyl-4*H*-1,3-oxazin-2-yl]phosphonate, *in* D-00598

$C_{10}H_{20}NO_4PS$

▷Propetamphos, P-00466

$C_{10}H_{20}NO_5PS$

Methyl 6-deoxy-2,3-di-*O*-methyl-6-methyl-aminoglucopyranoside 4,6-cyclic thiophosphonamide, M-00113

$C_{10}H_{20}NO_5PS_2$

▷Mecarbam, M-00002

$C_{10}H_{20}NO_6P$

Methyl 6-deoxy-2,3-di-*O*-methyl-6-methyl-aminoglucopyranoside 4,6-cyclic phosphonamide, M-00112

$C_{10}H_{20}N_4O_2P^{\oplus}$

2,4,8,10-Tetramethyl-3,9-dioxo-2,4,8,10-tetraaza-6-phosphoniaspiro[5.5]-undecane(1+), T-00198

$C_{10}H_{20}O_4P_2$

2,2'-Bis(5,5-dimethyl-1,3,2-dioxaphosphorinane), B-00197

$C_{10}H_{20}O_4P_2S$

2,2'-Thiobis[5,5-dimethyl-1,3,2-dioxaphosphorinane], T-00309

$C_{10}H_{20}O_4P_2S_2$

2,2'-Dithiobis[5,5-dimethyl-1,3,2-dioxaphosphorinane], D-01251

Thiohypophosphoric acid cyclic *P,P,P',P'*-bis(2,2-dimethyltrimethylene) ester, *in* B-00197

$C_{10}H_{20}O_4P_2S_2Se_2$

Bis(5,5-dimethyl-2-selenoxo-1,3,2-dioxaphosphorinan-2-yl) disulfide, *in* D-01251

2,2'-Diselenobis[5,5-dimethyl-2-thioxo-1,3,2-dioxaphosphorinane], *in* D-01234

$C_{10}H_{20}O_4P_2S_3$

2,2'-Thiobis[5,5-dimethyl-1,3,2-dioxaphosphorinane]; 2,2'-Disulfide, *in* T-00309

$C_{10}H_{20}O_4P_2S_4$

Bis(5,5-dimethyl-2-thioxo-1,3,2-dioxaphosphorinan-2-yl) disulfide, *in* D-01251

$C_{10}H_{20}O_4P_2Se_2$

2,2'-Diselenobis[5,5-dimethyl-1,3,2-dioxaphosphorinane], D-01234

$C_{10}H_{20}O_5P_2$

2,2'-Oxybis[5,5-dimethyl-1,3,2-dioxaphosphorinane], O-00089

$C_{10}H_{20}O_5P_2S$

Thiohypophosphoric acid cyclic *P,P,P',P'*-bis(2,2-dimethyltrimethylene) ester, *in* B-00197

$C_{10}H_{20}O_5P_2S_2$

2-[(5,5-Dimethyl-1,3,2-dioxaphosphorinan-2-yl)thio]-5,5-dimethyl-1,3,2-dioxaphosphorinane *P*-oxide 2-sulfide, *in* T-00309

Sandoflam, *in* O-00089

$C_{10}H_{20}O_6P_2$

2,2'-Bis(5,5-dimethyl-1,3,2-dioxaphosphorinane) 2,2'-dioxide, *in* B-00197

2,2'-Oxybis[5,5-dimethyl-1,3,2-dioxaphosphorinane]; 2-Oxide, *in* O-00089

$C_{10}H_{20}O_6P_2S$

2,2'-Oxybis[5,5-dimethyl-1,3,2-dioxaphosphorinane] 2-oxide 2'-sulfide, O-00090

2,2'-Thiobis[5,5-dimethyl-1,3,2-dioxaphosphorinane] 2,2'-dioxide, T-00310

$C_{10}H_{20}O_6P_2S_2$

Bis(5,5-dimethyl-2-oxo-1,3,2-dioxaphosphorinan-2-yl) disulfide, *in* D-01251

$C_{10}H_{20}O_7P_2$

Mono(3,7-dimethyl-2,6-octadienyl) diphosphate, M-00465

2,2'-Oxybis[5,5-dimethyl-1,3,2-dioxaphosphorinane]; 2,2'-Dioxide, *in* O-00089

$C_{10}H_{20}O_{10}P_2$

Hexamethyl 2,3-diphosphonobutanedioate, *in* D-01182

$C_{10}H_{20}P^{\oplus}$

6-Phosphoniaspiro[5.5]undecane, P-00375

$C_{10}H_{20}P_2$

1,1'-Biphosphorinane, B-00085

Bis(diethylphosphino)acetylene, B-00178

2,3-Diethyl-2,3-diphosphabicyclo[2.2.2]octane, D-00288

$C_{10}H_{20}P_2S_2$

1,1'-Biphosphorinane; Disulfide, *in* B-00085

2,3-Diethyl-2,3-diphosphabicyclo[2.2.2]octane; Disulfide, *in* D-00288

$C_{10}H_{21}ClNPS$

P-Cyclohexyl-*N,N*-diethylphosphonamidothioic chloride, *in* C-00255

$C_{10}H_{21}Cl_2OP$

Decylphosphonic acid; Dichloride, *in* D-00012

(1-Methylnonyl)phosphonic acid; Dichloride, *in* M-00181
(1-Propylheptyl)phosphonic acid; Dichloride, *in* P-00467

$C_{10}H_{21}FNO_4P$
N,N-Diethyl(diethoxyphosphinyl)-fluoroacetamide, *in* D-00250

$C_{10}H_{21}F_2OP$
Decylphosphonic acid; Difluoride, *in* D-00012

$C_{10}H_{21}N_2O_2P$
Di-1-piperidinylphosphinic acid, D-01190

$C_{10}H_{21}N_2O_3P$
Ethyl di-4-morpholinylphosphinite, *in* D-00936

$C_{10}H_{21}N_2O_4P$
Ethyl di-4-morpholinylphosphinate, *in* D-00935

$C_{10}H_{21}N_2P$
4-Cyclohexyl-1,3-dimethyl-1,2,4-diazaphospholidine, *in* D-00718

$C_{10}H_{21}N_2PS_2$
Di-1-piperidinylphosphinodithioic acid, D-01191

$C_{10}H_{21}N_3OP$
N,N-Dimethyl-*P,P*-di-1-pyrrolidinylphosphinic amide, *in* D-01233

$C_{10}H_{21}N_4P$
Octahydro-3*H*,8*H*,10b*H*-2*a*,5*a*,7*a*,10*a*-tetraaza-10*b*-phosphaindeno[7,1-*cd*]indene, O-00022
Octahydro-5*H*,8*H*,10b*H*-2*a*,4*a*,7*a*,10*a*-tetraaza-10*b*-phosphapentaleno[1,6-*de*]naphthalene, O-00024

$C_{10}H_{21}OP$
1-Oxa-6-phospha(6-P^V)spiro[5.5]undecane; 6-Me, *in* O-00052

$C_{10}H_{21}OPS$
1-Ethoxy-2,2,3,4,4-pentamethylphosphetane 1-sulfide, *in* H-00169
1-Ethylthio-2,2,3,4,4-pentamethylphosphetane 1-oxide, *in* H-00169

$C_{10}H_{21}OPS_2$
O,S-Diethyl cyclohexylphosphonodithioate, *in* C-00261
S,S-Diethyl cyclohexylphosphonodithioate, *in* C-00261

$C_{10}H_{21}OP_3$
Triphosphetan-4-one; 1,3-Di-*tert*-butyl, 2-Me, *in* T-00703

$C_{10}H_{21}O_2P$
4-*tert*-Butylcyclohexylphosphonous acid, B-00550
Butyl cyclopentylmethylphosphinate, *in* C-00279
Diethyl cyclohexylphosphonite, *in* C-00267
1,3-Dioxa-2-phosphacyclotridecane, D-00966
Ethyl 1,1,2,3,3-pentamethyltrimethyl-enephosphinate, *in* H-00168

$C_{10}H_{21}O_2PS$
O,O-Diethyl cyclohexylphosphonothioate, *in* C-00264

$C_{10}H_{21}O_3P$
Diethyl cyclohexylphosphonate, *in* C-00256
Ethyl (1,1-dimethyl-3-oxobutyl)-ethylphosphinate, *in* D-00782

$C_{10}H_{21}O_3PS$
O,O-Dibutyl acetylphosphonothioate, *in* A-00028

$C_{10}H_{21}O_4P$
Diethyl (3,3-dimethyl-2-oxobutyl)phosphonate, *in* D-00783
Dimethyl (3-methyl-2-oxoheptyl)phosphonate, *in* M-00187
Dimethyl (2-oxooctyl)phosphonate, *in* O-00064
3,3,9,9-Tetramethyl-1,5,7,11-tetraoxa-6-phospha(6-P^V)spiro[5.5]undecane, T-00236

$C_{10}H_{21}O_4PS_2$
Ethyl (diisopropoxyphosphinothioyl)thioacetate, *in* D-00632

$C_{10}H_{21}O_4PSi_2$
Bis(trimethylsilyl) 2-furanylphosphonate, *in* F-00073

$C_{10}H_{21}O_5P$
(Dibutoxyphosphinyl)acetic acid, D-00073
Diethyl[[(tetrahydro-2*H*-pyran-2-yl)oxy]-methyl]phosphonate, *in* T-00136
Ethyl 2-(diethoxyphosphinyl)butanoate, *in* D-00246
Ethyl 4-(diethoxyphosphinyl)butanoate, *in* D-00247
Ethyl (diisopropoxyphosphinyl)acetate, *in* D-00633
Ethyl (dipropoxyphosphinyl)acetate, *in* D-01199

$C_{10}H_{22}BrP$
1,1,2,2,3,4,4-Heptamethylphosphetanium bromide, *in* H-00074

$C_{10}H_{22}ClOP$
Dipentylphosphinic acid; Chloride, *in* D-00981

$C_{10}H_{22}ClO_2P$
Bis(2,2-dimethylpropyl) phosphorochloridite, B-00209
Dipentyl phosphorochloridite, D-00983

$C_{10}H_{22}ClO_2PS$
O,O-Dipentyl phosphorochloridothioate, D-00984

$C_{10}H_{22}ClO_3P$
Bis(3-methylbutyl) phosphorochloridate, *in* B-00302
Dipentyl phosphorochloridate, *in* D-00979

$C_{10}H_{22}Cl_2NOP$
Decylphosphoramidic acid; Dichloride, *in* D-00013

$C_{10}H_{22}Cl_4O_2P_2$
1,10-Decanediphosphonic acid; Bis(dichloride), *in* D-00008

$C_{10}H_{22}FOP$
Dipentylphosphinic acid; Fluoride, *in* D-00981

$C_{10}H_{22}FO_3P$
Bis(3-methylbutyl) phosphorofluoridate, *in* B-00302
▷Dipentyl phosphorofluoridate, *in* D-00979

$C_{10}H_{22}IP$
1,1-Dimethylphosphonanium iodide, *in* M-00285
1,1,2,2,3,4,4-Heptamethylphosphetanium iodide, *in* H-00074

$C_{10}H_{22}NOP$
P-Cyclohexyl-*N,N*-diethylcyclohexylphosphinic amide, *in* C-00253
N,N,2,2,3,4,4-Heptamethylphosphetanamine 1-oxide, *in* P-00021

$C_{10}H_{22}NO_2P$
Diisopropyl 1-pyrrolidinylphosphonite, *in* P-00520
2-Dimethylamino-4-isopropyl-5,5-dimethyl-1,3,2-dioxaphosphorinane, D-00699

$C_{10}H_{22}NO_2PS$
O,O-Diethyl cyclohexylphosphoramidothioate, *in* C-00271
O,S-Diethyl cyclohexylphosphoramidothioate, *in* C-00271

$C_{10}H_{22}NO_3P$
2-[(Butylamino)-1-hexenyl]phosphonic acid, B-00540
2-Diethylamino-3,5,5-trimethyl-1,2-oxaphospholan-3-ol 2-oxide, *in* T-00519
Diethyl cyclohexylphosphoramidate, *in* C-00269
Diethyl (1-piperidinylmethyl)phosphonate, *in* P-00429
2,2,3,3,7,7-Hexamethyl-1,4,6-trioxa-9-aza-5-phospha(5-P^V)spiro[4.4]nonane, H-00083
2,2,3,3,8,8-Hexamethyl-1,4,6-trioxa-9-aza-5-phospha(5P^V)spiro[4.4]nonane, H-00084

$C_{10}H_{22}NO_4P$
(Dibutoxyphosphinyl)acetamide, *in* D-00073
Diethyl (2-diethylamino-2-oxoethyl)-phosphonate, *in* D-00268
N,N,2,2,3,3-Hexamethyl-1,4,6,9-tetraoxa-5-phospha(5-P^V)spiro[4.4]nonan-5-amine, *in* T-00235

$C_{10}H_{22}NO_5P$
Methyl [3-dimethylamino-3-(diethoxyphosphinyl)]propanoate, *in* A-00122

$C_{10}H_{22}NO_5PS$
Methyl 2-[(diisopropoxyphosphinyl)amino]-3-mercaptopropanoate, *in* P-00365

$C_{10}H_{22}NO_6P$
Methyl 3-hydroxy-2-[(diisoproproxyphosphinyl)amino]-propanoate, *in* P-00418

$C_{10}H_{22}N_3O_3PS$
2-(1-Methyl-2-oxopropylidene)-phosphorohydrazidothioate oxime, M-00189

$C_{10}H_{22}O_2P_2$
1,2-Bis(diethylphosphino)ethylene; Dioxide, *in* B-00181

$C_{10}H_{22}O_4P_2$
1,3-Diisopropoxy-1,3-diphosphorinane 1,3-dioxide, *in* D-00622

$C_{10}H_{22}O_6P_2$
2,10-Dimethyl-1,3,6,9,11,14-hexaoxa-2,10-diphosphacyclohexadecane, D-00757
Tetraethyl ethenylidenebisphosphonate, *in* E-00020

$C_{10}H_{22}P_2$
1,2-Bis(diethylphosphino)ethylene, B-00181

$C_{10}H_{23}ClNOP$
Ethyl dibutylphosphoramidochloridite, *in* D-00135

$C_{10}H_{23}N_2OP$
P-Cyclohexyl-*N,N,N',N'*-tetramethyl-phosphonic diamide, *in* C-00257
P-Ethenyl-*N,N,N',N'*-tetraethylphosphonic diamide, *in* V-00004

$C_{10}H_{23}N_2O_2P$
P-Acetyl-*N,N,N',N'*-tetraethylphosphonic diamide, *in* A-00027

$C_{10}H_{23}N_2O_3P$
[Bis(diethylamino)phosphinyl]acetic acid, B-00175
6,9-Dimethyl-1,3-dioxa-6,9-diaza-2-phosphacycloundecane; 2-Ethoxy, *in* D-00728
N,N,2,2,3,3-Hexamethyl-1,4,6-trioxa-9-aza-5-phospha(5-P^V)spiro[4.4]nonan-5-amine, *in* T-00238

$C_{10}H_{23}N_6O_6P$
Fosfazinomycin *B*, F-00059

$C_{10}H_{23}OP$
Butyl diisopropylphosphinite, *in* D-00648
Butyl dipropylphosphinite, *in* D-01209
Dipentylphosphine; Oxide, *in* D-00980
Ethyl dibutylphosphinite, *in* D-00121
Ethyl di-*tert*-butylphosphinite, *in* D-00122

$C_{10}H_{23}OPS$
S-Butyl dipropylphosphinothioate, *in* D-01208
O-Ethyl dibutylphosphinothioate, *in* D-00117

$C_{10}H_{23}OPSSe_2$
Se,Se-Dibutyl *O*-ethyl phosphorodiselenothioate, *in* D-00149

$C_{10}H_{23}OPS_2$
Ethylphosphonodithioic acid; *S,S*-Dibutyl ester, *in* E-00140

$C_{10}H_{23}O_2P$
Diisopropyl butylphosphonite, *in* B-00615
Diisopropyl *tert*-butylphosphonite, *in* B-00616

Dipentylphosphinic acid, D-00981
Ethyl dibutylphosphinate, *in* D-00109
Ethyl di-*tert*-butylphosphinate, *in* D-00110

$C_{10}H_{23}O_2PS_2$
O,O-Bis(3-methylbutyl) phosphorodithioate, B-00304
O,O-Dipentyl phosphorodithioate, D-00986

$C_{10}H_{23}O_2PSe_2$
O,O-Dipentyl phosphorodiselenoate, D-00985

$C_{10}H_{23}O_3P$
Bis(2,2-dimethylpropyl) phosphonate, B-00208
Bis(3-methylbutyl) phosphonate, B-00303
▷Decylphosphonic acid, D-00012
Dibutyl ethylphosphonate, *in* E-00129
▷Diethyl hexylphosphonate, *in* H-00097
Dimethyl octylphosphonate, *in* O-00042
Dipentyl phosphonate, D-00982
(1-Methylnonyl)phosphonic acid, M-00181
(1-Propylheptyl)phosphonic acid, P-00467

$C_{10}H_{23}O_4P$
Bis(3-methylbutyl) phosphate, B-00302
O-Butyl *OO-tert*-butyl ethylphosphonoperoxoate, *in* E-00144
Dipentyl phosphate, D-00979
Monodecyl phosphate, M-00462

$C_{10}H_{23}O_5P$
OO-tert-Butyl *O,O*-diisopropyl phosphoroperoxoate, B-00554
Diethyl (2,2-diethoxyethyl)phosphonate, *in* D-00234

$C_{10}H_{23}O_5PSi$
Trimethylsilyl 2-(diethoxyphosphinyl)-propanoate, *in* D-00258

$C_{10}H_{23}P$
Decylphosphine, D-00011
Dipentylphosphine, D-00980

$C_{10}H_{23}PS_2$
Ethyl dibutylphosphinodithioate, *in* D-00112

$C_{10}H_{23}PS_3$
Dibutyl ethylphosphonotrithioate, *in* E-00152

$C_{10}H_{24}ClN_2OP$
P-(2-Chloroethyl)-*N,N,N′,N′*-tetraethylphosphonic diamide, *in* C-00075

$C_{10}H_{24}ClN_2PSi$
1,3-Di-*tert*-butyl-4,4-dimethyl-1,3,2-diazaphosphasiletidine; 2-Chloro, *in* D-00087

$C_{10}H_{24}IP$
Methyltripropylphosphonium iodide, *in* T-00711
Triisopropylmethylphosphonium iodide, *in* T-00500

$C_{10}H_{24}IPS_3$
Methyltri(propylthio)phosphonium iodide, *in* T-00718

$C_{10}H_{24}NOP$
P,P-Dibutyl-*N,N*-dimethylphosphinic amide, *in* D-00111
P,P-Dibutyl-*N*-ethylphosphinic amide, *in* D-00111
N,N-Diethyl-*P,P*-dipropylphosphinic amide, *in* D-01206
N-tert-Butyl-*P,P*-diisopropylphosphinic amide, *in* D-00642

$C_{10}H_{24}NO_2P$
Di-*tert*-butyl dimethylphosphoramidite, *in* D-00875
Ethyl *N,N*-dibutylphosphonamidate, *in* D-00126

$C_{10}H_{24}NO_2PS$
O,O-Di-*tert*-butyl dimethyl-phosphoramidothioate, *in* D-00871
O,O-Di-*tert*-butyl ethylphosphoramidothioate, *in* E-00159
O,O-Diisopropyl diethylphosphoramidothioate, *in* D-00356

$C_{10}H_{24}NO_2PSe$
O,O-Diethyl diisopropylphosphoramidoselenoate, *in* D-00655
O,O-Diisopropyl diethylphosphoramidoselenoate, *in* D-00353

$C_{10}H_{24}NO_3P$
(10-Aminodecyl)phosphonic acid, A-00064
Bis(3-methylbutyl) phosphoramidate, *in* B-00302
Decylphosphoramidic acid, D-00013
Dibutyl ethylphosphoramidate, *in* E-00156
Diethyl dipropylphosphoramidate, *in* D-01214
Diisopropyl diethylphosphoramidate, *in* D-00343
Dimethyl dibutylphosphoramidate, *in* D-00134
Dimethyl octylphosphoramidate, *in* O-00045

$C_{10}H_{24}NO_3PS$
▷Amiton, A-00133

$C_{10}H_{24}NP$
P,P-Dibutyl-*N,N*-dimethylphosphinous amide, *in* D-00123
P,P-Dibutyl-*N*-ethylphosphinous amide, *in* D-00123
N,N-Diethyl-*P,P*-diisopropylphosphinous amide, *in* D-00649
P,P-Diisopropyl-*N-tert*-butylphosphinous amide, *in* D-00649
P,P,P-Triisopropyl-*N*-methylphosphine imide, *in* T-00501

$C_{10}H_{24}N_2O_2P$
1,3-Di-*tert*-butyl-2,4-dimethyl-1,3,2,4-diazadiphosphetidine; 2,4-Dioxide, *in* D-00086

$C_{10}H_{24}N_2O_2P_2$
1,3-Di-*tert*-butyl-2,4-dimethoxy-1,3,2,4-diazadiphosphetidine, D-00085

$C_{10}H_{24}N_2O_2P_2S$
1,3-Di-*tert*-butyl-2,4-dimethoxy-1,3,2,4-diazadiphosphetidine; 2-Sulfide, *in* D-00085

$C_{10}H_{24}N_2O_2P_2S_2$
1,3-Di-*tert*-butyl-2,4-dimethoxy-1,3,2,4-diazadiphosphetidine; 2,4-Disulfide, *in* D-00085

$C_{10}H_{24}N_2O_2P_2Se$
1,3-Di-*tert*-butyl-2,4-dimethoxy-1,3,2,4-diazadiphosphetidine; 2-Selenide, *in* D-00085

$C_{10}H_{24}N_2O_2P_2Se_2$
1,3-Di-*tert*-butyl-2,4-dimethoxy-1,3,2,4-diazadiphosphetidine; 2,4-Diselenide, *in* D-00085

$C_{10}H_{24}N_2P_2$
1,3-Di-*tert*-butyl-2,4-dimethyl-1,3,2,4-diazadiphosphetidine, D-00086

$C_{10}H_{24}N_2P_2S_2$
1,3-Di-*tert*-butyl-2,4-dimethyl-1,3,2,4-diazadiphosphetidine; 2,4-Disulfide, *in* D-00086

$C_{10}H_{24}N_2P_2Se_2$
1,3-Di-*tert*-butyl-2,4-dimethyl-1,3,2,4-diazadiphosphetidine; 2,4-Diselenide, *in* D-00086

$C_{10}H_{24}N_2P_2Te_2$
1,3-Di-*tert*-butyl-2,4-dimethyl-1,3,2,4-diazadiphosphetidine; 2,4-Ditelluride, *in* D-00086

$C_{10}H_{24}N_3O_2PS$
3-Amino-2,3-di-*tert*-butyl-1,2,4,3-thiadiazaphosphetidine; *N,N*-Di-Me, 1,1-dioxide, *in* A-00071

$C_{10}H_{24}N_3P$
P-1-Aziridinyl-*N,N,N′,N′*-tetraethylphosphonous amide, *in* A-00155
Hexahydro-1,3,4-trimethyl-1,3,2-diazaphosphorine; 2-Diethylamino, *in* H-00057

$C_{10}H_{24}N_4P_2$
P,P′-1,2-Ethynediylbis[*N,N,N′,N′*-tetramethylphosphonous diamide], E-00200
1,1,4,4-Tetrakis(dimethylamino)-2,3-diphosphabutadiene, *in* T-00005

$C_{10}H_{24}OP_2S_2$
tert-Butylmethylphosphinothioic anhydride, B-00571

$C_{10}H_{24}O_2P_2$
1,2-Bis(diethylphosphinyl)ethane, *in* B-00180
1,6-Bis(dimethylphosphinyl)hexane, *in* B-00201

$C_{10}H_{24}O_2P_2S_3$
O,O′-Diisopropyl diethylthiodiphosphonate, *in* D-00405

$C_{10}H_{24}O_4P_2$
1,2-Ethanediylbis[methylphosphinic acid]; Diisopropyl ester, *in* E-00013
Tetramethyl 1,6-hexanediylbisphosphonite, *in* H-00086

$C_{10}H_{24}O_4P_2S_2$
O,O,O′,O′-Tetraethyl *S,S′*-ethanediyl bisphosphorodithioate, T-00047

$C_{10}H_{24}O_5P_2$
Diethyl dipropyldiphosphonate, *in* D-01202

$C_{10}H_{24}O_6P_2$
1,10-Decanediphosphonic acid, D-00008
Tetraethyl 1,2-ethanediylbisphosphonate, *in* E-00006

$C_{10}H_{24}O_6P_2S$
Tetraethyl [thiobis(methylene)]bisphosphonate, *in* T-00311

$C_{10}H_{24}O_7P_2$
Tetraethyl (1-hydroxyethylidene)-bisphosphonate, *in* H-00144
Tetraethyl [oxybis(methylene)]bisphosphonate, *in* O-00095

$C_{10}H_{24}O_7P_2S$
Thiobis[methanephosphonic acid]; Tetra-Et ester, *S*-oxide, *in* T-00311

$C_{10}H_{24}O_8P_2S$
Thiobis[methanephosphonic acid]; Tetra-Et ester, *S,S*-dioxide, *in* T-00311

$C_{10}H_{24}O_{12}P_4$
2,5-Dihydroxy-3,6-dimethyl-3,6-bis(dihydroxyphosphinyl)-1,4,2,5-dioxadiphosphorinane 2,5-dioxide; Hexa-Me ester, *in* D-00617

$C_{10}H_{24}P_2$
1,2-Bis(diethylphosphino)ethane, B-00180
1,6-Bis(dimethylphosphino)hexane, B-00201
1,2-Di-*tert*-butyl-1,2-dimethyldiphosphine, D-00089
1,10-Diphosphinodecane, D-01175

$C_{10}H_{24}P_2S_2$
1,2-Bis(diethylphosphinothioyl)ethane, *in* B-00180
1,2-Di-*tert*-butyl-1,2-dimethyldiphosphine; Disulfide, *in* D-00089

$C_{10}H_{24}P_2S_3$
tert-Butylmethylphosphinodithioic acid anhydrosulfide, B-00569

$C_{10}H_{25}BrN_3P$
P-Bromo-*P,P*-bis(diethylamino)-*N*-ethylphosphine imide, *in* P-00012

$C_{10}H_{25}ClN_3P$
P-Chloro-*P,P*-bis(diethylamino)-*N*-ethylphosphine imide, *in* P-00012

$C_{10}H_{25}FN_3P$
P,P-Bis(diethylamino)-*N*-ethyl-*P*-fluorophosphine imide, *in* P-00012

$C_{10}H_{25}NO_4P$
Tetraethyl ethylimidodiphosphite, *in* E-00079

$C_{10}H_{25}NO_5P_2S$
▷Triethyl (diethoxyphosphinothioyl)-phosphorimidate, *in* D-00238

$C_{10}H_{25}NO_6P_2$
▷Pentaethyl imidodiphosphate, *in* E-00078
Tetraethyl (1-aminoethylidene)bisphosphonate, *in* A-00073
▷Tetraethyl iminobis(methylene)bisphosphonate, *in* D-00688
Triethyl (diethoxyphosphinyl)phosphorimidate, *in* D-00257

$C_{10}H_{25}N_2OP$
Ethyl tetraethylphosphorodiamidite, *in* T-00066
Pentaethylphosphonic diamide, *in* E-00130

$C_{10}H_{25}N_2OPS$
O-Ethyl tetraethylphosphorodiamidothioate, *in* T-00063
S-Ethyl tetraethylphosphorodiamidothioate, *in* T-00063

$C_{10}H_{25}N_2OPSe$
O-Ethyl tetraethylphosphorodiamidoselenoate, *in* T-00061
Se-Ethyl tetraethylphosphorodiamidoselenoate, *in* T-00061

$C_{10}H_{25}N_2OPTe$
O-Ethyl tetraethylphosphorodiamidotelluroate, *in* T-00062

$C_{10}H_{25}N_2O_2P$
Bis(*N-tert*-butylaminomethyl)phosphinic acid, *in* B-00087
Ethyl *N,N'*-dibutylphosphorodiamidate, *in* D-00146
Ethyl tetraethylphosphorodiamidate, *in* T-00059

$C_{10}H_{25}N_2P$
Pentaethylphosphonous diamide, *in* E-00154

$C_{10}H_{25}N_2PS$
Pentaethylphosphonothioic diamide, *in* E-00149

$C_{10}H_{25}N_2PSe$
Ethylphosphonoselenoic bis(diethylamide), *in* E-00146

$C_{10}H_{25}N_2PSi$
1,3-Di-*tert*-butyl-4,4-dimethyl-1,3,2-diazaphosphasiletidine, D-00087

$C_{10}H_{25}O_3PSi_2$
4-Hydroxy-2,2,6,6-tetramethyl-1-oxa-4-phospha-2,6-disilacyclohexane 4-oxide; 4-Butoxy (butyl ester), *in* H-00198

$C_{10}H_{25}O_5P$
Pentaethoxyphosphorane, P-00011

$C_{10}H_{25}O_5PSi_2$
Bis(trimethylsilyl) (2-acetyloxyethyl)-phosphonate, *in* A-00006
Ethyl [bis(trimethylsilyloxy)phosphinyl]acetate, *in* B-00434

$C_{10}H_{25}O_6P$
1,1,3,3-Tetraethoxy-2-ethyltriphosphine 1,3-dioxide, T-00041

$C_{10}H_{26}Br_2N_3P$
Bromobis(diethylamino)(ethylamino)-phosphonium bromide, *in* P-00012

$C_{10}H_{26}Cl_2N_3P$
Chlorobis(diethylamino)(ethylamino)-phosphonium chloride, *in* P-00012

$C_{10}H_{26}N_2O_3P_2$
P,P'-Diisopropyl-*N,N,N',N'*-tetramethyl-diphosphonic diamide, D-00668

$C_{10}H_{26}N_3OP$
Pentaethylphosphoric triamide, *in* T-00058
Pentaethylphosphorodiamidimidic acid, P-00012

$C_{10}H_{26}N_4P_2$
1,1,3,3-Tetrakis(dimethylamino)-1,3-diphosph(P^V)ete, T-00176

$C_{10}H_{27}N_3O_3P_2S$
Triethyl [bis(dimethylamino)phosphinothioyl]-phosphorimidate, *in* B-00192

$C_{10}H_{27}N_3O_4P_2$
Triethyl [bis(dimethylamino)phosphinyl]-phosphorimidate, *in* B-00194

$C_{10}H_{27}PSi_2$
tert-Butylbis(trimethylsilyl)phosphine, B-00545

$C_{10}H_{28}I_3P_3$
Dimethylbis(trimethylphosphonimethyl)-phosphonium triiodide, *in* B-00204

$C_{10}H_{28}NO_2PSi$
Diethyl bis(trimethylsilyl)phosphoramidite, *in* B-00438

$C_{10}H_{28}NPSi_2$
P,P-Diethyl-*N,N*-bis(trimethylsilyl)phosphinous amide, *in* D-00334

$C_{10}H_{29}N_2PSi_2$
P-tert-Butyl-*N,N'*-bis(trimethylsilyl)-phosphonous diamide, *in* B-00617

$C_{10}H_{29}N_5P_2$
N''''-Ethyl-*N,N,N',N',N'',N'',N''',N'''*-octamethylimidodiphosphorous tetraamide, *in* E-00079

$C_{10}H_{30}O_4PSi$
Methoxytris(trimethylsilyloxy)phosphonium iodide, *in* T-00864

$C_{10}H_{31}N_2OPSi_3$
Methyl *N,N,N'*-tris(trimethylsilyl)-phosphonamidimidate, *in* T-00867

$C_{10}H_{41}O_3P$
Diisopropyl 1-butenylphosphonate, *in* B-00524

$C_{11}H_7F_6O_3PS_2$
4',5'-Bis(trifluoromethyl)spiro[1,3,2-benzodioxaphosphole-2,2'-[1,3,2]-dithiaphosphole]; 2-Methoxy, *in* B-00422

$C_{11}H_7F_6O_5P$
4',5'-Bis(trifluoromethyl)spiro[1,3,2-benzodioxaphosphole-2,2'-[1,3,2]-dioxaphosphole]; 2-Methoxy, *in* B-00421

$C_{11}H_9F_{12}O_5P$
4'-Methyl-4,4,5,5-tetrakis(trifluoromethyl)-spiro[1,3,2-dioxaphospholane-2,1'-[2,6,7]-trioxa[1]phosphabicyclo[2.2.2]octane], M-00405

$C_{11}H_9O_4P$
2-Hydroxynaphtho[2,3-*d*]-1,3,2-dioxaphosphole 2-oxide; 2-Methoxy (Me ester), *in* H-00164

$C_{11}H_{11}O_2P$
6-Methyl-4-methylene-4*H*-1,3,2-dioxaphosphorin; 2-Ph, *in* M-00167
Methyl-1-naphthylphosphinic acid, M-00178

$C_{11}H_{11}O_2PS$
O-Methyl naphthylphosphonothioate, *in* N-00023

$C_{11}H_{11}O_3P$
(1-Naphthylmethyl)phosphonic acid, N-00013
(2-Naphthylmethyl)phosphonic acid, N-00014

$C_{11}H_{11}O_3PS$
O-Methyl *O*-1-naphthalenyl phosphorothioate, M-00176

$C_{11}H_{11}P$
1,3-Dimethylisophosphinoline, D-00758

$C_{11}H_{12}NO_3P$
Dimethyl 4-quinolinephosphonate, *in* Q-00003

$C_{11}H_{12}NO_4PS_2$
▷Phosmet, P-00343

$C_{11}H_{12}NO_5P$
Dimethyl [(1,3-dihydro-1,3-dioxo-2*H*-isoindol-2-yl)methyl]phosphonate, *in* P-00426
Ethyl hydrogen [(1,3-dihydro-1,3-diazo-2*H*-isoindol-2-yl)methyl]phosphonate, *in* P-00426

$C_{11}H_{13}F_{12}N_2O_2P$
6,10-Dimethyl-2,2,3,3-tetrakis(trifluoromethyl)-1,4-dioxa-6,10-diaza-5-phospha(5-P^V)spiro[4.5]decane, D-00923

$C_{11}H_{13}OP$
2-*tert*-Butyl-1,3-benzoxaphosphole, B-00544
2,5-Dihydro-3-methyl-1-phenyl-1*H*-phosphole; Oxide, *in* D-00528
1-Phenyl-2-phosphorinanone, P-00267
1-Phenyl-4-phosphorinanone, P-00268

$C_{11}H_{13}OPS$
1-Phenyl-4-phosphorinanone; Sulfide, *in* P-00268

$C_{11}H_{13}O_2P$
2,3-Dihydro-4-methyl-1-phenoxy-1*H*-phosphole 1-oxide, *in* D-00496
2,5-Dihydro-2-methyl-1-phenoxy-1*H*-phosphole 1-oxide, *in* D-00498
2,5-Dihydro-3-methyl-2-phenoxy-1*H*-phosphole 1-oxide, *in* D-00499
1-Phenyl-4-phosphorinanone; Oxide, *in* P-00268

$C_{11}H_{13}O_2PSi$
1*H*-Isophosphindole-1,3(2*H*)-dione; Trimethylsilyl ether, *in* I-00026

$C_{11}H_{13}O_3P$
2,5-Dihydro-5,5-dimethyl-1,2-oxaphosphole 2-oxide; 2-Phenoxy, *in* D-00464
Spiro[1,3,2-benzodioxaphosphole-2,1'-[3]-phospholene]; 2-Methoxy, *in* S-00010

$C_{11}H_{13}O_4P$
5'-Methylspiro[1,3,2-benzodioxaphosphole-2,2'(3*H*)-[1,2]oxaphosphole]; 2-Methoxy, *in* M-00396

$C_{11}H_{13}P$
1-Benzyl-2,5-dihydro-1*H*-phosphole, *in* D-00572
2,5-Dihydro-3-methyl-1-phenyl-1*H*-phosphole, D-00528
1,2,3,6-Tetrahydro-1-phenylphosphorin, T-00134

$C_{11}H_{14}ClO_3P$
5-Chloromethyl-5-methyl-2-phenoxy-1,3,2-dioxaphosphorinane, C-00099
2-Hydroxy-5,5-dimethyl-4-phenyl-1,3,2-dioxaphoshorinane 2-oxide; Chloride, *in* H-00129

$C_{11}H_{14}ClO_4P$
5-Chloromethyl-5-methyl-2-phenoxy-1,3,2-dioxaphosphorinane; 2-Oxide, *in* C-00099

$C_{11}H_{14}IP$
2,3-Dihydro-1-methyl-1-phenyl-1*H*-phospholium iodide, *in* D-00566

$C_{11}H_{14}NOPS$
tert-Butylphenylphosphinic acid; Isothiocyanate, *in* B-00579
tert-Butylphenylphosphinic acid; Thiocyanate, *in* B-00579

$C_{11}H_{14}NO_2P$
tert-Butylphenylphosphinic acid; Isocyanate, *in* B-00579
1-Phenyl-1,2-azaphospholidin-5-one; 2-Et, 2-oxide, *in* P-00092

$C_{11}H_{14}NO_3P$
Diethyl (2-cyanophenyl)phosphonate, *in* C-00223
Diethyl (3-cyanophenyl)phosphonate, *in* C-00224
Diethyl (4-cyanophenyl)phosphonate, *in* C-00225
1,2-Dihydro-1-methyl-4*H*-3,1,2-benzoxazaphosphorin-4-one; 1-Isopropoxy, *in* D-00523
1,4-Dioxa-6-aza-5-phospha(5-P^V)spiro[4.4]-nonan-7-one; 5-Ph, *in* D-00962
4-Methyl-2,6-dioxa-7-aza-1-phosphabicyclo[2.2.2]octane 1-oxide; 7-Ph, *in* M-00124
Spiro[1,3,2-benzoxazaphosphole-2(3*H*),2'-[1,2]oxaphospholan]-5'-one; 2-Et, *in* S-00013

$C_{11}H_{14}NO_6P$
2-Benzoylamino-4-phosphonobutanoic acid, *in* A-00117

$C_{11}H_{14}N_3O_2P$
▷*N*-Benzoyl-*P*,*P*-bis(1-aziridinyl)phosphinic amide, *in* B-00092

$C_{11}H_{15}BrClO_3PS$
▷Profenofos, P-00440

$C_{11}H_{15}BrO_3P$
Diethyl (bromophenylmethyl)phosphonate, *in* B-00482

$C_{11}H_{15}Br_2O_3P$
Dimethyl (1,2-dibromo-1-phenylpropyl)-phosphonate, *in* D-00068

$C_{11}H_{15}ClNO_2P$
2-Chlorotetrahydro-3-(1-phenylethyl)-2*H*-1,3,2-oxazaphosphorine 2-oxide, C-00181

$C_{11}H_{15}Cl_2O_2PS_2$
▷Prothiofos, P-00498

$C_{11}H_{15}Cl_2O_2PS_3$
▷Phenkapton, P-00062

$C_{11}H_{15}Cl_2O_3PS$
▷Ethafos, E-00005

$C_{11}H_{15}Cl_2O_3PS_2$
Chlorthiophos, C-00197

$C_{11}H_{15}F_6PS_2$
1,1,3,3-Tetramethyl-6,7-bis(trifluoromethyl)-5,8-dithia-4-phospha(4-P^V)spiro[3.4]oct-6-ene, T-00193

$C_{11}H_{15}N_2O_2P$
Benzyl bis(1-aziridinyl)phosphinate, *in* B-00091

$C_{11}H_{15}N_2O_3P$
Methyl hydrogen [[1-amino-2-(1*H*-indol-3-yl)]-ethyl]phosphonate, *in* P-00421

$C_{11}H_{15}OP$
1-Ethyl-1,2,3,4-tetrahydrophosphinoline 1-oxide, *in* T-00135
3-Methyl-1-phenylphospholane; 1-Oxide, *in* M-00229
1-Phenylphosphorinane; Oxide, *in* P-00265
1-Phenyl-2-phosphorinanol, P-00266

$C_{11}H_{15}O_2P$
5,5-Dimethyl-2-phenyl-1,3,2-dioxaphosphorinane, D-00795
1-Ethoxy-1,2,3,4-tetrahydrophosphinoline 1-oxide, *in* H-00195
Ethyl phenyl(2-propenyl)phosphinate, *in* P-00300
Isopropyl styrylphosphinate, *in* P-00153

$C_{11}H_{15}O_2PS$
O,*O*-Neopentylene phenylphosphonothioate, *in* D-00795

$C_{11}H_{15}O_3P$
5,5-Dimethyl-2-phenoxy-1,3,2-dioxaphosphorinane, D-00790
Neopentylene phenylphosphonate, *in* D-00795

$C_{11}H_{15}O_3PS$
O,*O*-Neopentylene *O*-phenyl phosphorothioate, *in* D-00790

$C_{11}H_{15}O_3PS_2$
1,3-Benzodithiol-2-ylphosphonic acid; Di-Et ester, *in* B-00020
O,*O*-Diethyl *S*-benzoyl phosphorodithioate, D-00270

$C_{11}H_{15}O_3PSe$
O,*O*-Neopentylene *O*-phenyl phosphoroselenoate, *in* D-00790

$C_{11}H_{15}O_4P$
Benzoyl diethyl phosphite, B-00028
Diethyl benzoylphosphonate, *in* B-00034
Dimethyl(2-oxo-3-phenylpropyl)phosphonate, *in* O-00068
2-Hydroxy-5,5-dimethyl-4-phenyl-1,3,2-dioxaphoshorinane 2-oxide, H-00129
Methyl 3-(methoxyphenylphosphinyl)-propanoate, *in* H-00178
Neopentylene phenyl phosphate, *in* D-00790

$C_{11}H_{15}O_4PS$
O-Benzoyl *O*,*O*-diethylphosphorothioate, B-00029

$C_{11}H_{15}O_5P$
2-(Diethoxyphosphinyl)benzoic acid, D-00243
4-(Diethoxyphosphinyl)benzoic acid, D-00244
Dimethyl [(acetyloxy)phenylmethyl]-phosphonate, *in* A-00020

$C_{11}H_{15}P$
1-Adamantylmethylidenephosphine, A-00033
2-Ethyl-1,2,3,4-tetrahydroisophosphinoline, *in* T-00110
3-Methyl-1-phenylphospholane, M-00229
1-Phenylphosphorinane, P-00265

$C_{11}H_{15}PS$
1-Phenylphosphorinane; Sulfide, *in* P-00265
1,2,3,4-Tetrahydrophosphinoline; 1-Et, 1-sulfide, *in* T-00135

$C_{11}H_{16}AsP$
2,3,4,5-Tetrahydro-1,5-dimethyl-1*H*-1,5-benzophospharsepine, T-00092

$C_{11}H_{16}BrO_3P$
Diethyl [(2-bromophenyl)methyl]phosphonate, *in* B-00483
Diethyl [(4-bromophenyl)methyl]phosphonate, *in* B-00484

$C_{11}H_{16}ClO_2PS_3$
▷Carbophenothion, C-00008

$C_{11}H_{16}ClO_3P$
Diethyl [(2-chlorophenyl)methyl]phosphonate, *in* C-00132
Diethyl [(3-chlorophenyl)methyl]phosphonate, *in* C-00133
Diethyl [(4-chlorophenyl)methyl]phosphonate, *in* C-00134

$C_{11}H_{16}FO_3P$
Diethyl [(3-fluorophenyl)methyl]phosphonate, *in* F-00030
Diethyl [(4-fluorophenyl)methyl]phosphonate, *in* F-00031
Diethyl [(2-fluorophenyl)methyl]phosphonic acid, *in* F-00029

$C_{11}H_{16}IOP$
4-Methyl-4-phenyl-1,4-oxaphosphorinanium iodide, *in* P-00176

$C_{11}H_{16}IO_3P$
Diethyl [(2-iodophenyl)methyl]phosphonate, *in* I-00015

$C_{11}H_{16}IP$
1-Methyl-1-phenylphospholanium iodide, *in* P-00216

$C_{11}H_{16}NOP$
Hexahydro-1-methyl-4-phenyl-1,4-azaphosphorine; 4-Oxide, *in* H-00036
2,3,4-Trimethyl-5-phenyl-1,3,2-oxazaphospholidine, T-00523

$C_{11}H_{16}NOPS$
N-Benzoyl-*P*,*P*-diethylphosphinothioic amide, *in* D-00331
2,3,4-Trimethyl-5-phenyl-1,3,2-oxazaphospholidine; 2-Sulfide, *in* T-00523

$C_{11}H_{16}NOPSe$
2,3,4-Trimethyl-5-phenyl-1,3,2-oxazaphospholidine; 2-Selenide, *in* T-00523

$C_{11}H_{16}NO_2P$
1-Benzylhexahydro-2-hydroxy-1,2-azaphosphorine 2-oxide, B-00048
2-Dimethylamino-4-phenyl-1,3,2-dioxaphosphorinane, D-00704
5,5-Dimethyl-2-phenylamino-1,3,2-dioxaphosphorinane, D-00792
2-Methoxy-3,4-dimethyl-5-phenyl-1,3,2-oxazaphospholidine, M-00035
Tetrahydro-5,5-dimethyl-2-phenoxy-2*H*-1,3,2-oxazaphosphorine, T-00098
2,3,4-Trimethyl-5-phenyl-1,3,2-oxazaphospholidine; 2-Oxide, *in* T-00523

$C_{11}H_{16}NO_2PS$
Neopentylene phenylphosphoramidothioate, *in* D-00792
Tetrahydro-5,5-dimethyl-2-phenoxy-2*H*-1,3,2-oxazaphosphorine; 2-Sulfide, *in* T-00098

$C_{11}H_{16}NO_2PSe$
Neopentylene phenylphosphoramidoselenoate, *in* D-00792

$C_{11}H_{16}NO_3P$
Neopentylene phenylphosphoramidate, *in* D-00792
Tetrahydro-5,5-dimethyl-2-phenoxy-2*H*-1,3,2-oxazaphosphorine; 2-Oxide, *in* T-00098

$C_{11}H_{16}NO_3PS$
O,*O*-Diethyl benzoylphosphoramidothioate, *in* B-00038

$C_{11}H_{16}NO_4P$
(1-Benzoylaminobutyl)phosphonic acid, *in* A-00054
Diethyl benzoylphosphoramidate, *in* B-00036

$C_{11}H_{16}NO_5P$
Diethyl[(2-nitrophenyl)methyl]phosphonate, *in* N-00044
Diethyl [(4-nitrophenyl)methyl]phosphonate, *in* N-00045
[(4-Nitrophenyl)methyl]phosphonic acid; Di-Et ester, *in* N-00046

$C_{11}H_{16}NP$
1-Ethyl-2-phenyl-1,3-azaphospholidine, *in* P-00090
1-Ethyl-3-phenyl-1,3-azaphospholidine, *in* P-00091
Hexahydro-1-methyl-4-phenyl-1,4-azaphosphorine, H-00036
Hexahydro-3-phenyl-1*H*-1,3-azaphosphepine, H-00040

$C_{11}H_{16}NPS$
Hexahydro-1-methyl-4-phenyl-1,4-azaphosphorine; 4-Sulfide, *in* H-00036

$C_{11}H_{16}N_5O_7PS$
2-Methylthioadenosine 5′-(dihydrogen phosphate), M-00410

$C_{11}H_{17}AsNP$
1-Ethyl-2-methyl-1,2,3-azaphospharsolidine; 3-Ph, *in* E-00083

$C_{11}H_{17}ClNO_3PS_2$
O-Ethyl *S*-methyl *N*-[(4-chlorophenyl)-sulfonyl]-*P*-ethylphosphonamidothioate, *in* C-00165

$C_{11}H_{17}N_2OP$
N,*N*′-Dimethyl-*N*,*N*′-trimethylene-*P*-phenyl-phosphonic diamide, *in* H-00023

$C_{11}H_{17}N_2O_2P$
N-(Diethylphosphinyl)-*N*′-phenylurea, *in* D-00323

$C_{11}H_{17}N_2O_3PS$
O,*O*-Diethyl [(phenylamino)carbonyl]-phosphoramidothioate, *in* P-00082
O,*S*-Diethyl [(phenylamino)carbonyl]-phosphoramidothioate, *in* P-00082

$C_{11}H_{17}N_2O_4P$
Diethyl (phenylamino)-carbonylphosphoramidate, *in* P-00081
2-[(Dimethoxyphosphinyl)amino]-3-phenyl-propanamide, *in* P-00400

$C_{11}H_{17}N_2O_4PS$
▷Amiprophos-Methyl, A-00132
S-Methyl *O*-(4-methyl-2-nitrophenyl) isopropylphosphoramidothioate, *in* M-00180

$C_{11}H_{17}N_2P$
N,*N*′-Dimethyl-*N*,*N*′-trimethylene-*P*-phenyl-phosphonous diamide, *in* H-00022

$C_{11}H_{17}OP$
Benzylmethylpropylphosphine; Oxide, *in* B-00054
tert-Butylmethylphenylphosphine; Oxide, *in* B-00565

$C_{11}H_{17}OPS$
O-Methyl *tert*-butylphenylphosphinothioate, *in* B-00583
S-Methyl *tert*-butylphenylphosphinothioate, *in* B-00583

S-Propyl ethylphenylphosphinothioate, *in* E-00105

$C_{11}H_{17}OPS_2$

S,S-Diethyl benzylphosphonodithioate, *in* B-00065

$C_{11}H_{17}OPSe$

Se-Methyl *tert*-butylphenylphosphinoselenoate, *in* B-00582

$C_{11}H_{17}O_2P$

Benzyl diethylphosphinate, *in* D-00322
tert-Butyl methylphenylphosphinate, *in* M-00219
Diethyl benzylphosphonite, *in* B-00068
Diethyl (4-methylphenyl)phosphonite, *in* M-00248
(3,4-Dimethylphenyl)phosphinic acid; Isopropyl ester, *in* D-00806
Isopropyl (2,5-dimethylphenyl)phosphinate, *in* D-00805
Isopropyl ethylphenylphosphinate, *in* E-00101
Methyl *tert*-butylphenylphosphinate, *in* B-00579
Phenyl *tert*-butylmethylphosphinate, *in* B-00567
Propyl (4-methylphenyl)phenylphosphinate, *in* M-00204

$C_{11}H_{17}O_2PS$

O,O-Diethyl benzylphosphonothioate, *in* B-00067
O,S-Diethyl benzylphosphonothioate, *in* B-00067

$C_{11}H_{17}O_2PS_2$

O,O-Diethyl *S*-[(phenylthio)methyl] phosphorodithioate, D-00318

$C_{11}H_{17}O_3P$

▷Diethyl benzylphosphonate, D-00272
Diethyl (2-methylphenyl)phosphonate, *in* M-00235
Diethyl (3-methylphenyl)phosphonate, *in* M-00236
Diethyl (4-methylphenyl)phosphonate, *in* M-00237
Dimethyl (3-phenylpropyl)phosphonate, *in* P-00305

$C_{11}H_{17}O_3PS$

Diethyl (4-methylthiophenyl)phosphonate, *in* M-00418
Diethyl [(phenylthio)methyl]phosphonate, *in* P-00326
▷Kitazin, K-00001

$C_{11}H_{17}O_3PS_2$

▷*O,O*-Diethyl *O*-[4-(methylthio)phenyl] phosphorothioate, D-00302

$C_{11}H_{17}O_4P$

Diethyl (hydroxyphenylmethyl)phosphonate, *in* H-00174
Diethyl [(2-hydroxyphenyl)methyl]-phosphonate, *in* H-00175
Diethyl [(4-hydroxyphenyl)methyl]-phosphonate, *in* H-00176
Diethyl (2-methoxyphenyl)phosphonate, *in* M-00070
Diethyl (3-methoxyphenyl)phosphonate, *in* M-00071
Diethyl (4-methoxyphenyl)phosphonate, *in* M-00072

$C_{11}H_{17}O_4PS$

Diethyl [(phenylsulfinyl)methyl]phosphonate, *in* P-00310

$C_{11}H_{17}O_4PS_2$

▷Fensulfothion, F-00006

$C_{11}H_{17}O_5PS$

Diethyl [(phenylsulfonyl)methyl]phosphonate, *in* P-00311

$C_{11}H_{17}O_5PS_2$

▷*O,O*-Diethyl *O*-[4-(methylsulfonyl)phenyl] phosphorothioate, *in* D-00302

$C_{11}H_{17}P$

Benzylmethylpropylphosphine, B-00054
tert-Butylmethylphenylphosphine, B-00565
Homocubyltrimethylphosphorane, *in* D-00569

$C_{11}H_{17}PS$

tert-Butylmethylphenylphosphine; Sulfide, *in* B-00565
Isopropyl ethylphenylphosphinothioite, *in* E-00107

$C_{11}H_{17}PS_2$

Butyl methylphenylphosphinodithioate, *in* M-00224
Isopropyl ethylmethylphosphinodithioate, *in* E-00103

$C_{11}H_{17}PSi$

2,2-Dimethyl-1-phenyl-1-phospha-2-silacyclopentane, D-00802

$C_{11}H_{18}IP$

Dimethylphenylpropylphosphonium iodide, *in* M-00265

$C_{11}H_{18}NOPS$

O-Methyl *N,N*-diethyl-*P*-phenyl-phosphonamidothioate, *in* D-00309
S-Methyl *N,N*-diethyl-*P*-phenyl-phosphonamidothioate, *in* D-00309
O-Phenyl *N,N*-diethyl-*P*-methyl-phosphonamidothioate, *in* D-00297
S-Phenyl *N,N*-diethyl-*P*-methyl-phosphonamidothioate, *in* D-00297

$C_{11}H_{18}NO_2P$

Methyl *N,N*-diethyl-*P*-phenylphosphonamidate, *in* D-00307
Phenyl *N,N*-diethyl-*P*-methylphosphonamidate, *in* D-00296

$C_{11}H_{18}NO_2PS$

O,O-Diethyl (2-methylphenyl)-phosphoramidothioate, *in* M-00254
O,O-Diethyl (3-methylphenyl)-phosphoramidothioate, *in* M-00255
O,O-Diethyl (4-methylphenyl)-phosphoramidothioate, *in* M-00256

$C_{11}H_{18}NO_3P$

[1-(Benzylamino)butyl]phosphonic acid, *in* A-00054
Diethyl (aminophenylmethyl)phosphonate, *in* A-00105
▷Diethyl [(4-aminophenyl)methyl]phosphonte, *in* A-00107
Diethyl benzylphosphoramidate, *in* B-00069
Diethyl methylphenylphosphoramidate, *in* M-00250
Diethyl (2-methylphenyl)phosphoramidate, *in* M-00251
Diethyl (3-methylphenyl)phosphoramidate, *in* M-00252
Diethyl (4-methylphenyl)phosphoramidate, *in* M-00253
Diethyl [(phenylamino)methyl]phosphonate, *in* P-00084

$C_{11}H_{18}NO_3PS$

O,O-Diethyl (2-methoxyphenyl)-phosphoramidothioate, *in* M-00077
O,O-Diethyl (3-methoxyphenyl)-phosphoramidothioate, *in* M-00078
O,O-Diethyl (4-methoxyphenyl)-phosphoramidothioate, *in* M-00079

$C_{11}H_{18}NO_4P$

Diethyl (4-methoxyphenyl)phosphoramidate, *in* M-00076
2-[(Hydroxyphenoxyphosphinyl)oxy]-*N,N,N*-trimethylethanaminium hydroxide, inner salt, *in* C-00198

$C_{11}H_{18}NO_5P$

Phosphonoformic acid; *C,P*-di-Et ester, anilinium salt, *in* P-00387

$C_{11}H_{18}NO_{13}PS$

Tagetitoxin, T-00002

$C_{11}H_{18}NP$

N,N-Diethyl-*P*-methyl-*P*-phenylphosphinous amide, *in* M-00227

$C_{11}H_{18}NPS$

N-Butyl-*P*-methyl-*P*-phenylphosphinothioic amide, *in* M-00226
P-Isopropyl-*N,N*-dimethyl-*P*-phenyl-phosphinothioic amide, *in* I-00046

$C_{11}H_{18}O_7P_2$

Tetramethyl (hydroxyphenylmethyl)-bisphosphonate, *in* H-00172

$C_{11}H_{19}AsIP$

[2-(Dimethylarsino)phenyl]trimethyl-phosphonium iodide, *in* D-00711

$C_{11}H_{19}INO_2P$

(4-Dimethylaminophenyl)methylphosphinic acid; Me ester; B,MeI, *in* D-00705

$C_{11}H_{19}IP_2$

Trimethyl(2-dimethylphosphinophenyl)-phosphonium iodide, *in* P-00123

$C_{11}H_{19}N_2OP$

Benzyl tetramethylphosphorodiamidite, *in* T-00227
N,N,N′,N-Tetramethyl-*P*-(phenylmethyl)-phosphonic diamide, *in* B-00061

$C_{11}H_{19}N_2O_3P$

Diisopropyl 2-pyridylphosphoramidate, *in* P-00510
Diisopropyl 3-pyridylphosphoramidate, *in* P-00511

$C_{11}H_{19}OP$

5-Isopropyl-2,6-dimethyl-6-phosphabicyclo[3.1.1]hept-2-ene 6-oxide, I-00029
2,3,5-Trimethyl-3-phosphatricyclo[4.2.1.0^{2,5}]-nonane; Oxide, *in* T-00529

$C_{11}H_{19}O_5P$

Diisopropyl α-hydroxy-2-furanylmethyl-phosphonate, *in* F-00071

$C_{11}H_{19}O_9P$

Galactose 3-dihydrogen phosphate; α-1,2-*O*-Isopropylidene, 4,6-*O*-ethylidene, *in* G-00002

$C_{11}H_{19}P$

2,3,5-Trimethyl-3-phosphatricyclo[4.2.1.0^{2,5}]-nonane, T-00529

$C_{11}H_{20}BrO_2P$

3,5-Di-*tert*-butyl-2,5-dihydro-1,2-oxaphosphole 2-oxide; 2-Bromo, *in* D-00082

$C_{11}H_{20}ClO_2P$

3,5-Di-*tert*-butyl-2,5-dihydro-1,2-oxaphosphole 2-oxide; 2-Chloro, *in* D-00082

$C_{11}H_{20}IP$

1-*tert*-Butyl-1,3,4-trimethyl-1*H*-phospholium iodide, *in* E-00004

$C_{11}H_{20}N_3OP$

N″-Benzyl-*N,N,N′,N′*-tetramethylphosphoric triamide, *in* T-00216

$C_{11}H_{20}N_3O_3PS$

▷Pirimiphos-methyl, P-00435

$C_{11}H_{20}N_3P$

N,N,N′,N′,N″-Pentamethyl-*P*-phenyl-phosphonimidic diamide, N-00079

$C_{11}H_{20}N_4O_{11}P_2$

Cytidine diphosphate ethanolamine, C-00292

$C_{11}H_{21}Cl_2O_4P$

2,2-Dichlorovinyl methyl octyl phosphate, D-00202

$C_{11}H_{21}NO_7P$

Methyl 6-deoxy-2,3-di-*O*-methyl-6-methyl-aminoglucopyranoside 4,6-cyclic phosphonamide; *P*-Methoxy, *in* M-00112

$C_{11}H_{21}O_2P$

3,5-Di-*tert*-butyl-2,5-dihydro-1,2-oxaphosphole 2-oxide, D-00082

$C_{11}H_{21}O_3P$

Bis(1,1-dimethylethyl) 1,2-propadienylphosphonate, *in* P-00441
3,5-Di-*tert*-butyl-2,5-dihydro-1,2-oxaphosphole 2-oxide; 2-Hydroxy, *in* D-00082
Dibutyl 2-propynylphosphonate, *in* P-00495
Diisopropyl 1-cyclopenten-1-ylphosphonate, *in* C-00277

Diisopropyl 2-cyclopenten-1-ylphosphonate, *in* C-00278

$C_{11}H_{21}O_4P$
Diethyl (cyclohexylcarbonyl)phosphonate, *in* C-00239
Dimethyl (3,3-dimethyl-2-oxoheptyl)-phosphonate, *in* D-00785

$C_{11}H_{21}O_7PS$
Methyl 2,3-di-*O*-methylglucopyranoside 4,6-cyclic phosphonothioate; *P*-Ethoxy, *in* M-00120
Methyl 2,3-di-*O*-methyl-4-thioglucopyranoside 4,6-cyclic phosphonate; *P*-Ethoxy, *in* M-00121
Methyl 2,3-di-*O*-methyl-6-thioglucopyranoside 4,6-cyclic phosphonate; *P*-Ethoxy, *in* M-00122

$C_{11}H_{21}O_8P$
Methyl 2,3-di-*O*-methylglucopyranoside 4,6-cyclic phosphonate; *P*-Ethoxy, *in* M-00119

$C_{11}H_{22}NO_2P$
Butyl *N*,*N*-diethyl-*P*-propadienylphosphonamidate, *in* D-00399
Butyl *N*,*N*-diethyl-*P*-1-propynylphosphonamidate, *in* D-00403
3,5-Di-*tert*-butyl-2,3-dihydro-1,3,2-oxazaphosphole; 2-Methoxy, *in* D-00083

$C_{11}H_{22}NO_5PS$
Methyl 6-deoxy-2,3-di-*O*-methyl-6-methyl-aminoglucopyranoside 4,6-cyclic thiophosphonamide; *P*-Me, *in* M-00113

$C_{11}H_{22}NO_6P$
Methyl 6-deoxy-2,3-di-*O*-methyl-6-methyl-aminoglucopyranoside 4,6-cyclic phosphonamide; *P*-Me, *in* M-00112

$C_{11}H_{22}NO_6PS$
Methyl 6-deoxy-2,3-di-*O*-methyl-6-methyl-aminoglucopyranoside 4,6-cyclic thiophosphonamide; *P*-Methoxy, *in* M-00113

$C_{11}H_{22}NO_7P$
Methyl 2,3-di-*O*-methylglucopyranoside 4,6-cyclic phosphonate; *P*-Dimethylamino, *in* M-00119

$C_{11}H_{22}N_3O_3P$
Meturedepa, M-00439

$C_{11}H_{22}N_3O_6P$
▷Antibiotic SF 1293, A-00141

$C_{11}H_{23}NO_6P_2$
Tetraethyl (1-cyano-1,2-ethanediyl)-bisphosphonate, *in* B-00173

$C_{11}H_{23}N_2OP$
(2-Methyl-1,3-butadienyl)phosphonic bis(isopropylamide), *in* M-00096

$C_{11}H_{23}N_2O_2P$
Methyl di-1-piperidinylphosphinate, *in* D-01190

$C_{11}H_{23}N_4O_8P_5$
(*N*-Phosphono)methionine-*S*-sulfoximinylalanylalanine, P-00389

$C_{11}H_{23}O_2P$
Isopropyl 1,1,2,3,3-pentamethyltrimethyl-enephosphinate, *in* H-00168

$C_{11}H_{23}O_3P$
Isopropyl (1,1-dimethyl-3-oxobutyl)-ethylphosphinate, *in* D-00782

$C_{11}H_{23}O_4P$
Diethyl (2-oxoheptyl)phosphonate, *in* O-00061
Dimethyl (2-oxononyl)phosphonate, *in* O-00063

$C_{11}H_{23}O_4PSi$
Bis(trimethylsilyl) 2-furanylmethyl-phosphonate, *in* F-00072

$C_{11}H_{23}O_5P$
Isopropyl (diisopropoxyphosphinyl)acetate, *in* D-00633

$C_{11}H_{24}ClO_6P_2$
Tetraisopropyl dichloromethyl-enebisphosphonate, *in* D-00172

$C_{11}H_{24}NO_2P$
Diisopropyl 1-piperidinylphosphonite, *in* P-00432

$C_{11}H_{24}NO_4P$
Diisopropyl (diethylcarbamoyl)phosphonate, *in* D-00264

$C_{11}H_{24}NO_5P$
Tripropyl phosphonoglycinate, *in* P-00368

$C_{11}H_{24}NO_6P$
Methyl 3-hydroxy-2-[(diisopropoxyphosphinyl)-amino]butanoate, *in* P-00420

$C_{11}H_{24}O_4P_2$
1,3-Diisopropoxy-1,3-diphosphepane 1,3-dioxide, *in* D-00619

$C_{11}H_{24}O_7P_2$
Tetraethyl (1-oxo-1,3-propanediyl)-bisphosphonate, *in* O-00071
Tetraethyl (2-oxo-1,3-propanediyl)-bisphosphonate, *in* O-00072

$C_{11}H_{24}O_8P_2$
2,3-Bis(diethoxyphosphinyl)propanoic acid, B-00173

$C_{11}H_{25}As_2P$
2,3-Di-*tert*-butyl-1-isopropylphosphadiarsirane, D-00096

$C_{11}H_{25}NO_2P_2$
1-*tert*-Butyl-1,2,4-azadiphosphetidine; 2,4-Diisopropoxy, *in* B-00543

$C_{11}H_{25}NO_5P$
2,5-Dihydroxy-3-(1-methylpropyl)-1,3,2,5-oxazadiphospholidine 2,5-dioxide; 2,5-Dipropoxy (Dipropyl ester), *in* D-00623

$C_{11}H_{25}N_2O_2P$
N,*N*,*N′*,*N′*-Tetraethyl-*P*-(2-oxopropyl)-phosphonic diamide, *in* O-00077

$C_{11}H_{25}N_2PS$
N,*N*,*N′*,*N′*-Tetraethyl-*P*-2-propenylphosphonothioic diamide, *in* P-00462

$C_{11}H_{25}N_2PSe$
N,*N*,*N′*,*N′*-Tetraethyl-*P*-(2-propenyl)-phosphonoselenoic diamide, *in* P-00459

$C_{11}H_{25}OPS$
O-Isopropyl dibutylphosphinothioate, *in* D-00117

$C_{11}H_{25}OPSSe_2$
Se,*Se*-Dibutyl *O*-propyl phosphorodiselenothioate, *in* D-00149

$C_{11}H_{25}O_2P$
Decylmethylphosphinic acid, D-00009
Dibutyl isopropylphosphonite, *in* I-00056
Diisopropyl pentylphosphonite, *in* P-00052
2-Methylpropyl hexylmethylphosphinate, *in* H-00094

$C_{11}H_{25}O_2PS$
O,*O*-Diethyl heptylphosphonothioate, *in* H-00007

$C_{11}H_{25}O_3P$
Dibutyl propylphosphonate, *in* P-00473
Diethyl heptylphosphonate, *in* H-00006
Dimethyl nonylphosphonate, *in* N-00076

$C_{11}H_{25}O_4P_3$
2-*tert*-Butyl-1,3-diisopropoxytriphosphetane 1,3-dioxide, *in* D-00627

$C_{11}H_{26}ClN_2OP$
P-(3-Chloropropyl)-*N*,*N*,*N′*,*N′*-tetraethylphosphonic diamide, *in* C-00173

$C_{11}H_{26}NO_2PS$
▷*O*-Ethyl *S*-[2-(diisopropylamino)ethyl] methyl-phosphonothioate, E-00062

$C_{11}H_{26}NO_3P$
Ethyl *P*-(butoxymethyl)-*N*,*N*-diethylphosphonamidate, *in* B-00536

$C_{11}H_{26}N_2P$
N,*N*,*N′*,*N′*-Tetraethyl-*P*-(3-methylphenyl)-phosphonous diamide, *in* M-00247
N,*N*,*N′*,*N′*-Tetraethyl-*P*-(2-methylphenyl)-phosphonous diamide, *in* M-00246
N,*N*,*N′*,*N′*-Tetraethyl-*P*-(4-methylphenyl)-phosphonous diamide, *in* M-00248

$C_{11}H_{26}O_4P_2S_4$
O,*O*,*O′*,*O′*-Tetraethyl *S*,*S′*-propanediyl bisphosphorodithioate, T-00068

$C_{11}H_{26}O_6P_2$
Tetraethyl 1,3-propanediylbisphosphonate, *in* P-00444
Tetraethyl 2,2-propanediyldiphosphonate, *in* P-00445

$C_{11}H_{27}NO_6P_2$
Tetraethyl [(dimethylamino)methylene]-bisphosphonate, *in* D-00701

$C_{11}H_{27}N_2OP$
P-Heptyl-*N*,*N*,*N′*,*N′*-tetramethylphosphonic diamide, *in* H-00006
Isopropyl tetraethylphosphorodiamidite, *in* T-00066
N,*N*,*N′*,*N′*-Tetraethyl-*P*-propylphosphonic diamide, *in* P-00474

$C_{11}H_{27}N_2PSi$
1,3-Di-*tert*-butyl-2,4,4-trimethyl-1,3-diaza-2-phospha-4-silacyclobutane, *in* D-00087

$C_{11}H_{27}OPSSi$
O-Trimethylsilyl dibutylphosphinothioate, *in* D-00117

$C_{11}H_{27}OPSi$
Trimethylsilyl dibutylphosphinite, *in* D-00121
Trimethylsilyl di-*tert*-butylphosphinite, *in* D-00122

$C_{11}H_{27}O_2PSi$
Trimethylsilyl di-*tert*-butylphosphinate, *in* D-00110

$C_{11}H_{28}NPSi$
P,*P*-Di-*tert*-butyl-*N*-trimethylsilylphosphinous amide, *in* D-00124

$C_{11}H_{29}NO_6P_2Si$
Tetraethyl trimethylsilylimidodiphosphate, *in* T-00051

$C_{11}H_{29}N_2OPSi$
Trimethylsilyl tetraethylphosphorodiamidite, *in* T-00066

$C_{11}H_{29}N_3O_4P_2$
Trimethyl [bis(diethylamino)phosphinyl]-phosphorimidate, *in* B-00176

$C_{11}H_{30}N_3OPSi$
N,*N*,*N′*,*N′*-Tetraethylphosphoric triamide; *N″*-Trimethylsilyl, *in* T-00058

$C_{11}H_{30}O_4PSi_3$
Trimethylsilyl bis(trimethylsilyloxymethyl)-phosphinate, *in* B-00282

$C_{11}H_{31}N_2O_2PSi_3$
Trimethylsilyl bis[*N*-(trimethylsilyl)-aminomethyl]phosphinate, *in* B-00087

$C_{11}H_{33}N_2O_2PSi_3$
Dimethyl *N*,*N*,*N′*-tris(trimethylsilyl)-phosphoramidimidate, *in* T-00868

$C_{11}H_{33}N_2PSi_3$
P,*P*-Dimethyl-*N*,*N*,*N′*-tris(trimethylsilyl)-phosphinimidic amide, *in* D-00833

$C_{12}BrF_{10}PS$
Bis(pentafluorophenyl)phosphinothioic acid; Bromide, *in* B-00383

$C_{12}ClF_{10}O_3P$
Bis(pentafluorophenyl) phosphorochloridate, *in* B-00381

$C_{12}ClF_{10}PS$
Bis(pentafluorophenyl)phosphinothioic acid; Chloride, *in* B-00383

$C_{12}Cl_{11}O_3P$
Bis(pentachlorophenyl) phosphorochloridate, *in* B-00379

$C_{12}F_{11}OP$
Bis(pentafluorophenyl)phosphinic acid; Fluoride, *in* B-00382

$C_{12}F_{11}PS$
Bis(pentafluorophenyl)phosphinothioic acid; Fluoride, *in* B-00383

$C_{12}F_{18}P_2$
2,3,5,6,7,8-Hexakis(trifluoromethyl)-1,4-diphosphabicyclo[2.2.2]octa-2,5,7-triene, H-00064

$C_{12}HCl_{10}O_4P$
Bis(pentachlorophenyl) phosphate, B-00379

$C_{12}HF_{10}OPS$
Bis(pentafluorophenyl)phosphinothioic acid, B-00383

$C_{12}HF_{10}O_2P$
Bis(pentafluorophenyl)phosphinic acid, B-00382

$C_{12}HF_{10}O_4P$
Bis(pentafluorophenyl) phosphate, B-00381

$C_{12}H_5Cl_6O_2P$
Bis(2,3,6-trichlorophenyl)phosphinic acid, B-00405
Bis(2,4,5-trichlorophenyl)phosphinic acid, B-00406

$C_{12}H_7Br_4O_2P$
Bis(2,5-dibromophenyl)phosphinic acid, B-00156

$C_{12}H_7Cl_4O_2P$
Bis(2,3-dichlorophenyl)phosphinic acid, B-00163
Bis(2,5-dichlorophenyl)phosphinic acid, B-00164
Bis(3,5-dichlorophenyl)phosphinic acid, B-00165

$C_{12}H_7N_4O_{12}P$
Bis(2,4-dinitrophenyl) phosphate, B-00210

$C_{12}H_8Br_2ClOP$
Bis(3-bromophenyl)phosphinic acid; Chloride, *in* B-00113
Bis(4-bromophenyl)phosphinic acid; Chloride, *in* B-00114

$C_{12}H_8ClF_2OP$
Bis(3-fluorophenyl)phosphinic acid; Chloride, *in* B-00279
Bis(4-fluorophenyl)phosphinic acid; Chloride, *in* B-00280

$C_{12}H_8ClN_2O_6PS$
O,O-Bis(4-nitrophenyl) phosphorochloridothioate, B-00377

$C_{12}H_8ClN_2O_7P$
Bis(4-nitrophenyl) phosphorochloridate, B-00376

$C_{12}H_8ClOP$
6-Chloro-6*H*-dibenzo[*c,e*][1,2]oxaphosphorin, C-00033
5-Chloro-5*H*-dibenzophosphole; 5-Oxide, *in* C-00034

$C_{12}H_8ClOPS$
6-Chloro-6*H*-dibenzo[*c,e*][1,2]oxaphosphorin; 6-Sulfide, *in* C-00033

$C_{12}H_8ClO_2P$
6-Chlorodibenzo[*d,f*][1,3,2]dioxaphosphepin, C-00032
6-Chloro-6*H*-dibenzo[*c,e*][1,2]oxaphosphorin; 6-Oxide, *in* C-00033

$C_{12}H_8ClO_2PS$
O,O-(2,2′-Biphenylyl) phosphorochloridothioate, *in* C-00032

$C_{12}H_8ClO_2PSe$
O,O-(2,2′-Biphenylyl) phosphorochloridoselenoate, *in* C-00032

$C_{12}H_8ClO_3P$
O,O-(2,2′-Biphenylyl) phosphorochloridate, *in* C-00032

$C_{12}H_8ClO_4P$
2-Chloro-2,2′-spirobi[1,3,2-benzodioxaphosphole], C-00174

$C_{12}H_8ClP$
5-Chloro-5*H*-dibenzophosphole, C-00034

$C_{12}H_8ClPS$
5-Chloro-5*H*-dibenzophosphole; 5-Sulfide, *in* C-00034

$C_{12}H_8Cl_2N_3OP$
Bis(4-chlorophenyl)phosphinic acid; Azide, *in* B-00139

$C_{12}H_8Cl_2N_3O_2P$
Bis(3-chlorophenyl)phosphinic acid; Azide, *in* B-00138

$C_{12}H_8Cl_2O_2P_2$
5,10-Dichloro-5,10-dihydrophosphanthrene; 5,10-Dioxide, *in* D-00164

$C_{12}H_8Cl_2P_2$
5,10-Dichloro-5,10-dihydrophosphanthrene, D-00164

$C_{12}H_8Cl_3NO_3P$
Bis(3-chlorophenyl) phosphorochloridate, *in* B-00132

$C_{12}H_8Cl_3OP$
Bis(4-chlorophenyl)phosphinic acid; Chloride, *in* B-00139

$C_{12}H_8Cl_3O_2P$
Bis(2-chlorophenyl)phosphinic acid; Chloride, *in* B-00137
Bis(3-chlorophenyl)phosphinic acid; Chloride, *in* B-00138
Bis(2-chlorophenyl) phosphorochloridite, B-00145
Bis(3-chlorophenyl) phosphorochloridite, B-00146
Bis(4-chlorophenyl) phosphorochloridite, B-00147

$C_{12}H_8Cl_3O_2PS$
O,O-Bis(3-chlorophenyl) phosphorochloridothioate, B-00148
O,O-Bis(4-chlorophenyl) phosphorochloridothioate, B-00149

$C_{12}H_8Cl_3O_3P$
Bis(2-chlorophenyl) phosphorochloridate, *in* B-00131
Bis(4-chlorophenyl) phosphorochloridate, *in* B-00133

$C_{12}H_8Cl_3P$
Chlorobis(2-chlorophenyl)phosphine, *in* B-00140
Chlorobis(3-chlorophenyl)phosphine, *in* B-00141

$C_{12}H_8FO_4P$
2-Fluoro-2,2′-spirobi[1,3,2-benzodioxaphosphole], F-00048

$C_{12}H_8F_2O_4P_2$
2,4-Bis(4-fluorophenyl)-1,3,2,4-dioxaphosphetane 2,4-dioxide, B-00275

$C_{12}H_8F_6P_2S_4$
2,2′-Spirobi[1,3,2-benzodithiaphospholium](1+); Hexafluorophosphate, *in* S-00016

$C_{12}H_8O_5P_2$
2,2′-Oxybis[1,3,2-benzodioxaphosphole], O-00087

$C_{12}H_8O_7P_2$
Catechol cyclic phosphoric acid anhydride, *in* O-00087

$C_{12}H_8PS_4^{\oplus}$
2,2′-Spirobi[1,3,2-benzodithiaphospholium](1+), S-00016

$C_{12}H_8P_2S_4$
2,2′-Bi(1,3,2-benzodithiaphosphole), B-00074

$C_{12}H_9Br_2O_2P$
Bis(2-bromophenyl)phosphinic acid, B-00112
Bis(3-bromophenyl)phosphinic acid, B-00113
Bis(4-bromophenyl)phosphinic acid, B-00114

$C_{12}H_9Br_2O_4P$
Bis(4-bromophenyl) phosphate, B-00111

$C_{12}H_9ClNOP$
10-Chloro-5,10-dihydrophenophosphazine; 10-Oxide, *in* C-00049

$C_{12}H_9ClNO_3P$
Spiro[1,3,2-benzodioxaphosphole-2,2′(3′*H*)-[1,3,2]benzoxazaphosphole]; 2-Chloro, *in* S-00008

$C_{12}H_9ClNO_5P$
4-Nitrophenyl phenyl phosphorochloridate, *in* N-00052

$C_{12}H_9ClNP$
10-Chloro-5,10-dihydrophenophosphazine, C-00049

$C_{12}H_9ClNPS$
10-Chloro-5,10-dihydrophenophosphazine; 10-Sulfide, *in* C-00049

$C_{12}H_9Cl_2OP$
Bis(3-chlorophenyl)phosphine; Oxide, *in* B-00135
Bis(4-chlorophenyl)phosphine; Oxide, *in* B-00136
Bis(2-chlorophenyl)phosphinous acid, B-00140
Bis(3-chlorophenyl)phosphinous acid, B-00141
Bis(4-chlorophenyl)phosphinous acid, B-00142

$C_{12}H_9Cl_2OPS$
O-[(1,1′-Biphenyl)-2-yl] phosphorodichloridothioate, B-00083

$C_{12}H_9Cl_2O_2P$
[1,1′-Biphenyl]-2-yl phosphorodichloridate, *in* M-00445
(1,1′-Biphenyl)-3-yl phosphorodichloridate, B-00082
(1,1′-Biphenyl)-4-yl phosphorodichloridate, *in* M-00446
Bis(2-chlorophenyl)phosphinic acid, B-00137
Bis(3-chlorophenyl)phosphinic acid, B-00138
▷Bis(4-chlorophenyl)phosphinic acid, B-00139

$C_{12}H_9Cl_2O_2PS_2$
O,O-Bis(2-chlorophenyl) phosphorodithioate, B-00150
O,O-Bis(4-chlorophenyl) phosphorodithioate, B-00151

$C_{12}H_9Cl_2O_3P$
Bis(2-chlorophenyl) phosphonate, B-00143
Bis(4-chlorophenyl) phosphonate, B-00144
2-Phenoxyphenyl phosphorodichloridate, *in* M-00505

$C_{12}H_9Cl_2O_4P$
Bis(2-chlorophenyl) phosphate, B-00131
Bis(3-chlorophenyl) phosphate, B-00132
Bis(4-chlorophenyl) phosphate, B-00133

$C_{12}H_9Cl_2P$
Bis(2-chlorophenyl)phosphine, B-00134
Bis(3-chlorophenyl)phosphine, B-00135
Bis(4-chlorophenyl)phosphine, B-00136

$C_{12}H_9F_2O_2P$
Bis(2-fluorophenyl)phosphinic acid, B-00278
Bis(3-fluorophenyl)phosphinic acid, B-00279
Bis(4-fluorophenyl)phosphinic acid, B-00280

$C_{12}H_9F_2P$
Bis(3-fluorophenyl)phosphine, B-00276
Bis(4-fluorophenyl)phosphine, B-00277

$C_{12}H_9I_2O_2P$
Bis(2-iodophenyl)phosphinic acid, B-00289
Bis(3-iodophenyl)phosphinic acid, B-00290
Bis(4-iodophenyl)phosphinic acid, B-00291

$C_{12}H_9N_2O_4P$
Bis(4-nitrophenyl)phosphine, B-00372

$C_{12}H_9N_2O_6P$
▷Bis(3-nitrophenyl)phosphinic acid, B-00373
Bis(4-nitrophenyl)phosphinic acid, B-00374

$C_{12}H_9N_2O_7P$
Bis(4-nitrophenyl) phosphonate, B-00375

$C_{12}H_9N_2O_8P$
Bis(2-nitrophenyl) phosphate, B-00369
Bis(3-nitrophenyl) phosphate, B-00370
Bis(4-nitrophenyl) phosphate, B-00371

$C_{12}H_9OP$
6*H*-Dibenzo[*c*,*e*][1,2]oxaphosphorin, D-00039

$C_{12}H_9OPS$
1,3,2-Benzoxathiaphosphole; 2-Ph, *in* B-00026

$C_{12}H_9OPS_3$
Tri-2-thienylphosphine; Oxide, *in* T-00874
Tri-3-thienylphosphine; Oxide, *in* T-00875

$C_{12}H_9O_2P$
Dibenzo[*d*,*f*][1,3,2]dioxaphosphepin, D-00038
6*H*-Dibenzo[*c*,*e*][1,2]oxaphosphorin; 6-Oxide, *in* D-00039
5-Hydroxy-5*H*-dibenzophosphole 5-oxide, H-00118
2-Phenyl-1,3,2-benzodioxaphosphole, P-00094

$C_{12}H_9O_2PS_2$
6-Mercaptodibenzo[*d*,*f*][1,3,2]dioxaphosphepin 6-sulfide, M-00008
2,2′-Spirobi[1,3,2-benzoxathiaphosphole], S-00019

$C_{12}H_9O_3P$
2,2′-Biphenylene phosphonate, *in* D-00038
10-Hydroxy-10*H*-phenoxaphosphine 10-oxide, H-00170
2-Phenoxy-1,3,2-benzodioxaphosphole, P-00063
o-Phenylene phenylphosphonate, *in* P-00094
Tri-2-furanylphosphine, T-00486
Tri-3-furanylphosphine, T-00487

$C_{12}H_9O_3PS$
2-Mercaptodibenzo[*d*,*f*][1,3,2]dioxaphosphepin 2-oxide, M-00007
O-Phenyl *o*-phenylene thiophosphate, *in* P-00063
Spiro[1,3,2-benzodioxaphosphole-2,2′-[1,3,2]-benzoxathiaphosphole], S-00007
Tri-2-furanylphosphine; Sulfide, *in* T-00486
Tri-3-furanylphosphine; Sulfide, *in* T-00487

$C_{12}H_9O_3PSe$
Tri-2-furanylphosphine; Selenide, *in* T-00486
Tri-3-furanylphosphine; Selenide, *in* T-00487

$C_{12}H_9O_4P$
2-Hydroxydibenzo[*d*,*f*][1,3,2]dioxaphosphepin 2-oxide, H-00116
Phenyl *o*-phenylene phosphate, *in* P-00063
2,2′-Spirobi[1,3,2-benzodioxaphosphole], S-00014
Tri-2-furanylphosphine; Oxide, *in* T-00486
Tri-3-furanylphosphine; Oxide, *in* T-00487

$C_{12}H_9P$
5*H*-Dibenzophosphole, D-00040

$C_{12}H_9PS_3$
Tri-2-thienylphosphine, T-00874
Tri-3-thienylphosphine, T-00875

$C_{12}H_9PS_3Se$
Tri-2-thienylphosphine; Selenide, *in* T-00874
Tri-3-thienylphosphine; Selenide, *in* T-00875

$C_{12}H_9PS_4$
2,2′-Spirobi[1,3,2-benzodithiaphosphole], S-00015
Tri-2-thienylphosphine; Sulfide, *in* T-00874
Tri-3-thienylphosphine; Sulfide, *in* T-00875

$C_{12}H_{10}BrOP$
Diphenylphosphinic acid; Bromide, *in* D-01039

$C_{12}H_{10}BrO_2P$
Diphenyl phosphorobromidite, D-01118

$C_{12}H_{10}BrO_2PS$
O,*O*-Diphenyl phosphorobromidothioate, D-01119

$C_{12}H_{10}BrO_2PSe$
O,*O*-Diphenyl phosphorobromidoselenoate, *in* D-01139

$C_{12}H_{10}BrO_3P$
Diphenyl phosphorobromidate, *in* D-01036

$C_{12}H_{10}BrP$
Bromodiphenylphosphine, *in* D-01089

$C_{12}H_{10}BrPS$
Diphenylphosphinothioic acid; Bromide, *in* D-01082

$C_{12}H_{10}Br_2NP$
Diphenylphosphoramidous acid; Dibromide, *in* D-01111

$C_{12}H_{10}Br_2P_2$
Diphenyl hypodiphosphonous acid; Dibromide, *in* D-01020

$C_{12}H_{10}Br_3O_2P$
Tribromodiphenoxyphosphorane, T-00354

$C_{12}H_{10}Br_3P$
Tribromodiphenylphosphorane, T-00355

$C_{12}H_{10}ClN_2O_3PS$
O-(4-Nitrophenyl) phenylphosphoramidochloridothioate, *in* P-00260

$C_{12}H_{10}ClN_2O_4P$
4-Nitrophenyl phenylphosphoramidochloridate, *in* P-00259

$C_{12}H_{10}ClOP$
4-Chloro-1,4-dihydro-2*H*-naphth[2,1-*c*][1,2]-oxaphosphorin, C-00048
Diphenylphosphinic chloride, D-01041

$C_{12}H_{10}ClOPS$
O-Phenyl phenylphosphonochloridothioate, *in* P-00232
S-Phenyl phenylphosphonochloridothioate, *in* P-00232

$C_{12}H_{10}ClO_2P$
Diphenyl phosphorochloridite, D-01121
▷Phenyl phenylphosphonochloridate, *in* P-00229

$C_{12}H_{10}ClO_2PS$
O,*O*-Diphenyl phosphorochloridothioate, D-01123

$C_{12}H_{10}ClO_2PSe$
O,*O*-Diphenyl phosphorochloridoselenoate, D-01122

$C_{12}H_{10}ClO_3P$
Diphenyl phosphorochloridate, D-01120

$C_{12}H_{10}ClP$
▷Diphenylphosphinous chloride, D-01091

$C_{12}H_{10}ClPS$
Diphenylphosphinothioic chloride, D-01084
Phenyl phenylphosphonochloridothioite, *in* P-00233

$C_{12}H_{10}ClPS_2$
Phenyl phenylphosphonochloridodithioate, *in* P-00230

$C_{12}H_{10}ClPSe$
Diphenylphosphinoselenoic acid; Chloride, *in* D-01078

$C_{12}H_{10}Cl_2NOP$
Bis(4-chlorophenyl)phosphinic acid; Amide, *in* B-00139
Diphenylphosphoramidic acid; Dichloride, *in* D-01108

$C_{12}H_{10}Cl_2NO_2P$
2-Chlorophenyl phenylphosphoramidochloridate, *in* P-00259
4-Chlorophenyl phenylphosphoramidochloridate, *in* P-00259

$C_{12}H_{10}Cl_2NO_3P$
Diphenyl dichlorophosphoramidate, D-01002

$C_{12}H_{10}Cl_2NP$
Diphenylphosphoramidous acid; Dichloride, *in* D-01111

$C_{12}H_{10}Cl_3NO_2P_2S$
P,*P*,*P*-Trichloro-*N*-(diphenoxyphosphinylthioyl)phosphine imide, *in* D-00987

$C_{12}H_{10}Cl_3O_2P$
Trichlorodiphenoxyphosphorane, T-00403

$C_{12}H_{10}Cl_3P$
Trichlorodiphenylphosphorane, T-00404

$C_{12}H_{10}Cl_4N_3P_3$
2,2,4,4-Tetrachloro-2,2,4,4,6,6-hexahydro-6,6-diphenyl-1,3,5-triaza-2,4,6-triphosphorine, T-00031

$C_{12}H_{10}FN_2O_2P$
2,2(3*H*,3′*H*)-Spirobi[1,3,2-benzoxazaphosphole]; 2-Fluoro, *in* S-00020

$C_{12}H_{10}FOP$
Diphenylphosphinic fluoride, D-01042

$C_{12}H_{10}FOPS_2$
S,*S*-Diphenyl phosphorofluoridodithioate, *in* D-01132

$C_{12}H_{10}FO_2P$
Phenyl phenylphosphonofluoridate, *in* P-00239

$C_{12}H_{10}FO_2PS$
O,*O*-Diphenyl phosphorofluoridothioate, D-01134

$C_{12}H_{10}FO_3P$
Diphenyl phosphorofluoridate, D-01133

$C_{12}H_{10}FP$
Diphenylphosphinous fluoride, D-01092

$C_{12}H_{10}FPS$
Diphenylphosphinothioic acid; Fluoride, *in* D-01082

$C_{12}H_{10}FPS_3$
Diphenyl phosphorofluoridotrithioate, D-01135

$C_{12}H_{10}F_2NOP$
Diphenylphosphoramidic acid; Difluoride, *in* D-01108

$C_{12}H_{10}F_2NP$
Diphenylphosphoramidous acid; Difluoride, *in* D-01111

$C_{12}H_{10}F_2P_2S_3$
Phenylphosphonodithioic acid anhydrosulfide; Difluoride, *in* P-00237

$C_{12}H_{10}F_3O_2P$
Trifluorodiphenoxyphosphorane, T-00470

$C_{12}H_{10}F_3P$
Trifluorodiphenylphosphorane, T-00471

$C_{12}H_{10}F_4N_3P_3$
2,2,4,6-Tetrafluoro-2,2,4,4,6,6-hexahydro-4,6-diphenyl-1,3,5-triaza-2,4,6-triphosphorine, T-00079

$C_{12}H_{10}IO_2P$
Diphenyl phosphoriodidite, D-01115

$C_{12}H_{10}IO_3P$
1,3-Dihydro-1-hydroxy-3-phenyl-1,2,3-benziodoxaphosphole 3-oxide, *in* D-00488

$C_{12}H_{10}IP$
Iododiphenylphosphine, *in* D-01089

$C_{12}H_{10}IPS$
Diphenylphosphinothioic acid; Iodide, *in* D-01082

$C_{12}H_{10}I_2NP$
Diphenylphosphoramidous acid; Diiodide, *in* D-01111

$C_{12}H_{10}I_2P_2$
Diphenyl hypodiphosphonous acid; Diiodide, *in* D-01020

$C_{12}H_{10}I_3P$
Triiododiphenylphosphorane, T-00497

$C_{12}H_{10}NOP$
5,10-Dihydrophenophosphazine; 10-Oxide, *in* D-00542

$C_{12}H_{10}NOPS$
5,10-Dihydro-10-hydroxyphenophosphazine 10-sulfide, D-00502
2,3-Dihydro-2-phenoxy-1,3,2-benzothiazaphosphole, D-00545

$C_{12}H_{10}NO_2P$
▷Phenazaphosphinic acid, P-00061

$C_{12}H_{10}NO_2PS$
2,3-Dihydro-2-phenoxy-1,3,2-benzothiazaphosphole; 2-Oxide, *in* D-00545

$C_{12}H_{10}NO_3P$
Spiro[1,3,2-benzodioxaphosphole-2,2′(3′*H*)-[1,3,2]benzoxazaphosphole], S-00008

$C_{12}H_{10}NO_5P$
Phenyl hydrogen (4-nitrophenyl)phosphonate, *in* N-00057

$C_{12}H_{10}NO_6P$
4-Nitrophenyl phenyl phosphate, N-00052

$C_{12}H_{10}NP$
5,10-Dihydrophenophosphazine, D-00542

$C_{12}H_{10}N_3OP$
▷Diphenylphosphinic acid; Azide, *in* D-01039

$C_{12}H_{10}N_3O_2PS$
O,O-Diphenyl phosphorazidothioate, D-01113

$C_{12}H_{10}N_3O_3P$
Diphenyl phosphorazidate, D-01112

$C_{12}H_{10}N_3PS$
Diphenylphosphinothioic acid; Azide, *in* D-01082

$C_{12}H_{10}N_3PSe$
Diphenylphosphinoselenoic acid; Azide, *in* D-01078

$C_{12}H_{10}O_4P_2$
5,10-Dihydro-5,10-dihydroxyphosphanthrene 5,10-dioxide, D-00452

$C_{12}H_{10}P_2S_4$
2,4-Diphenyl-1,3,2,4-dithiadiphosphetane 2,4-disulfide, D-01019

$C_{12}H_{11}Br_2N_2OP$
Bis(3-bromophenyl)phosphinic acid; Hydrazide, *in* B-00113
Bis(4-bromophenyl)phosphinic acid; Hydrazide, *in* B-00114

$C_{12}H_{11}ClNOP$
N,P-Diphenylphosphonamidic acid; Chloride, *in* D-01103

$C_{12}H_{11}ClNOPS$
O-Phenyl phenylphosphoramidochloridothioate, *in* P-00260

$C_{12}H_{11}ClNO_2P$
Phenyl phenylphosphoramidic acid; Chloride, *in* P-00189
Phenyl phenylphosphoramidochloridate, P-00190

$C_{12}H_{11}ClNO_3PS$
N-[(4-Chlorophenyl)sulfonyl]-*P*-phenylphosphonimidothioic acid, C-00166

$C_{12}H_{11}ClNP$
4-Chloro-1,2,3,4-tetrahydronaphth[2,1-*c*][1,2]-azaphosphorine, C-00179

$C_{12}H_{11}ClNPS$
4-Chloro-1,2,3,4-tetrahydronaphth[2,1-*c*][1,2]-azaphosphorine; Sulfide, *in* C-00179

$C_{12}H_{11}Cl_2N_2OP$
Bis(4-chlorophenyl)phosphinic acid; Hydrazide, *in* B-00139

$C_{12}H_{11}FNOP$
N,P-Diphenylphosphonamidic acid; Fluoride, *in* D-01103

$C_{12}H_{11}FNP$
P,P-Diphenylphosphinimidic fluoride, D-01044

$C_{12}H_{11}N_2OP$
2,3-Dihydro-2-phenoxy-1*H*-1,3,2-benzodiazaphosphole, D-00543

$C_{12}H_{11}N_2OPS$
O-Phenyl *N,N′-o*-phenylenephosphorodiamidothioate, *in* D-00543

$C_{12}H_{11}N_2O_2P$
1,3-Dihydrospiro[2*H*-1,3,2-benzodiazaphosphole-2,2′-[1,3,2]-benzodioxaphosphole], D-00574
Phenyl *N,N′-o*-phenylenephosphorodiamidate, *in* D-00543
2,2(3*H*,3′*H*)-Spirobi[1,3,2-benzoxazaphosphole], S-00020

$C_{12}H_{11}N_2P$
6,7-Dihydro-5*H*-dibenzo[*d,f*][1,3,2]-diazaphosphepine, D-00450

$C_{12}H_{11}N_4OP$
Phenylphosphonic acid; Diimidazolide, *in* P-00221

$C_{12}H_{11}N_4O_2P$
Phenyl di-1*H*-imidazol-1-ylphosphinate, *in* D-00628

$C_{12}H_{11}OP$
2,3-Dihydro-1*H*-benzo[*e*]phosphindole 3-oxide, D-00442
Diphenylphosphine; Oxide, *in* D-01037
Diphenylphosphine oxide, D-01038
Diphenylphosphinous acid, D-01089
1-Phenyl-1*H*-phosphepin; Oxide, *in* P-00206

$C_{12}H_{11}OPS$
Diphenylphosphinothioic acid, D-01082

$C_{12}H_{11}OPSe$
Diphenylphosphinoselenoic acid, D-01078

$C_{12}H_{11}O_2P$
2,3-Dihydro-1*H*-benzo[*e*]phosphindole 3-oxide; 3-Hydroxy, *in* D-00442
▷Diphenylphosphinic acid, D-01039
Phenyl phenylphosphinate, *in* P-00210

$C_{12}H_{11}O_2PS$
O,O-Diphenyl phosphonothioate, D-01106

$C_{12}H_{11}O_2PSSe$
O,O-Diphenyl phosphoroselenothioate, D-01140

$C_{12}H_{11}O_2PS_2$
O,O-Diphenyl phosphorodithioate, D-01131
S,S-Diphenyl phosphorodithioate, D-01132

$C_{12}H_{11}O_2PSe_2$
O,O-Diphenyl phosphorodiselenoate, D-01130

$C_{12}H_{11}O_3P$
4-Biphenylphosphonic acid, B-00077
Diphenyl phosphonate, D-01105
2-Ethoxynaphtho[1,2-*d*]-1,3,2-dioxaphosphole, E-00044
2-Ethoxynaphtho[1,8-*de*]-1,3,2-dioxaphosphorin, E-00045
2-Ethoxynaphtho[1,8-*de*]-1,3,2-dioxaphosphorin, E-00046
Phenyl hydrogen phenylphosphonate, *in* P-00221

$C_{12}H_{11}O_3PS$
O,O-Diphenyl phosphorothioate, D-01141
O-Ethyl *O,O*-1,2-naphthylene thiophosphate, *in* E-00044
O-Ethyl *O,O*-1,8-naphthylene thiophosphate, *in* E-00045
O-Ethyl *O,O*-1,8-naphthylene thiophosphate, *in* E-00046

$C_{12}H_{11}O_3PSe$
O,O-Diphenyl phosphoroselenoate, D-01139

$C_{12}H_{11}O_4P$
Bis(2-hydroxyphenyl)phosphinic acid, B-00283
Bis(3-hydroxyphenyl)phosphinic acid, B-00284
Bis(4-hydroxyphenyl)phosphinic acid, B-00285
Diphenyl phosphate, D-01036
Ethyl 1,2-naphthylene phosphate, *in* E-00044
Ethyl 1,8-naphthylene phosphate, *in* E-00045
Ethyl 1,8-naphthylene phosphate, *in* E-00046
2-Hydroxynaphtho[2,3-*d*]-1,3,2-dioxaphosphole 2-oxide; 2-Ethoxy (Et ester), *in* H-00164
Mono(2-biphenylyl) phosphate, M-00445
Mono-(1,1′-biphenyl)-4-yl phosphate, M-00446

$C_{12}H_{11}O_5P$
Mono(2-phenoxyphenyl) phosphate, M-00505

$C_{12}H_{11}P$
4,5-Dihydro-3*H*-benzo[*e*]phosphindole, D-00441
▷Diphenylphosphine, D-01037

$C_{12}H_{11}PS$
Diphenylphosphine; Sulfide, *in* D-01037
Diphenylphosphinothious acid, D-01085

$C_{12}H_{11}PSSe$
Diphenylphosphinoselenothioic acid, D-01080

$C_{12}H_{11}PS_2$
Diphenylphosphinodithioic acid, D-01065

$C_{12}H_{11}PSe$
Diphenylphosphine; Selenide, *in* D-01037
Diphenylphosphinoselenous acid, D-01081

$C_{12}H_{11}PSe_2$
Diphenylphosphinodiselenoic acid, D-01064

$C_{12}H_{12}ClN_2OP$
▷*N,N′*-Diphenylphosphorodiamidic chloride, D-01126

$C_{12}H_{12}FN_2OP$
▷*N,N′*-Diphenylphosphorodiamidic fluoride, D-01127

$C_{12}H_{12}NOP$
P,P-Diphenylphosphinic amide, D-01040

$C_{12}H_{12}NOPS$
N,P-Diphenylphosphonamidothioic acid, D-01104

$C_{12}H_{12}NO_2P$
O-(Diphenylphosphinyl)hydroxylamine, D-01097
N,P-Diphenylphosphonamidic acid, D-01103
Diphenyl phosphoramidite, D-01110
Diphenylphosphoramidous acid, D-01111
N-Hydroxy-*P,P*-diphenylphosphinic amide, H-00141
Phenyl *P*-phenylphosphonamidate, *in* P-00218

$C_{12}H_{12}NO_2PSe$
O,O-Diphenyl phosphoramidoselenoate, *in* D-01139

$C_{12}H_{12}NO_3P$
Diphenyl phosphoramidate, D-01107
Diphenylphosphoramidic acid, D-01108
Phenyl hydrogen (4-aminophenyl)phosphonate, *in* A-00111
Phenyl phenylphosphoramidic acid, P-00189

$C_{12}H_{12}NO_4PS$
P-Phenyl-*N*-phenylsulfonylphosphonamidic acid, P-00191
P-Phenyl-*N*-phenylsulfonylphosphonimidic acid, P-00192

$C_{12}H_{12}NP$
P,P-Diphenylphosphinous amide, D-01090

$C_{12}H_{12}NPS$
P,P-Diphenylphosphinothioic amide, D-01083

$C_{12}H_{12}NPSe$
P,P-Diphenylphosphinoselenoic amide, D-01079

$C_{12}H_{12}N_3OP$
1,1′,1″-Phosphinylidynetrispyrrole, *in* T-00725
2,2′,2″-Phosphinylidynetrispyrrole, *in* T-00726

$C_{12}H_{12}N_3P$
Tri-1-pyrrolylphosphine, T-00725
Tri-2-pyrrolylphosphine, T-00726

$C_{12}H_{12}O_2P_2$
Diphenyl hypodiphosphonous acid, D-01020

$C_{12}H_{12}O_2P_2S_3$
Phenylphosphonodithioic acid anhydrosulfide, P-00237

$C_{12}H_{12}O_3P_2$
Phenylphosphinic acid; Anhydride, *in* P-00210

$C_{12}H_{12}O_3P_2S_2$
Phenylphosphonothioic acid monoanhydride, P-00245

$C_{12}H_{12}O_5P_2$
Diphenyldiphosphonic acid, D-01014

$C_{12}H_{12}O_7P_2$
P,P′-Diphenyl diphosphate, D-01011

$C_{12}H_{12}P_2$
1,2-Diphenyldiphosphine, D-01012

$C_{12}H_{12}P_2S_4$
Diphenylthiohypophosphonic acid, D-01153

$C_{12}H_{13}Cl_4O_4P$
4,5,6,7-Tetrachloro-4′,4′,5′,5′-tetramethylspiro[1,3,2-benzodioxaphosphole-2,2′-[1,3,2]-dioxaphospholane], T-00036

$C_{12}H_{13}NO_5P_2S$
(Diphenoxyphosphinothioyl)phosphorimidic acid, D-00987

$C_{12}H_{13}N_2OP$
Bis(2-cyanoethyl)phenylphosphine; Oxide, *in* B-00152
N,P-Diphenylphosphonic diamide, *in* P-00222

$C_{12}H_{13}N_2OPS$
N,N′-Diphenylphosphorodiamidothioic acid, D-01129

$C_{12}H_{13}N_2O_2P$
▷Bis(3-aminophenyl)phosphinic acid, B-00088
Bis(4-aminophenyl)phosphinic acid, B-00089
N,N′-Diphenylphosphorodiamidic acid, D-01125

$C_{12}H_{13}N_2O_2PS$
O,O-Diphenyl phosphorohydrazidothioate, D-01137

$C_{12}H_{13}N_2O_3P$
Diphenyl phosphorohydrazidate, D-01136
Phenyl hydrogen (2-aminophenyl)-phosphoramidate, *in* A-00113

$C_{12}H_{13}N_2P$
Bis(2-cyanoethyl)phenylphosphine, B-00152
P,P-Diphenylphosphinimidic amide, D-01043

$C_{12}H_{13}N_2PS$
Bis(2-cyanoethyl)phenylphosphine; Sulfide, *in* B-00152

$C_{12}H_{13}N_2PS_2$
N,N′-Diphenylphosphorodiamidodithioic acid, D-01128

$C_{12}H_{13}OP$
2,7-Dihydro-1-phenyl-1*H*-phosphepin; Oxide, *in* D-00564
4,5-Dihydro-1-phenyl-1*H*-phosphepin; Oxide, *in* D-00565
3,4-Dimethyl-1-phenyl-1*H*-phosphole; 1-Oxide, *in* D-00807

$C_{12}H_{13}O_2P$
Dimethyl 1-naphthylphosphonite, *in* N-00025
1-Hydroxy-3,4-dimethyl-1*H*-phosphole 1-oxide; 1-Phenoxy (Ph ester), *in* H-00130
Methyl methyl-1-naphthylphosphinate, *in* M-00178

$C_{12}H_{13}O_3P$
Dimethyl 1-naphthylphosphonate, *in* N-00021

$C_{12}H_{13}P$
2,7-Dihydro-1-phenyl-1*H*-phosphepin, D-00564
4,5-Dihydro-1-phenyl-1*H*-phosphepin, D-00565
3,4-Dimethyl-1-phenyl-1*H*-phosphole, D-00807

$C_{12}H_{13}PS$
3,4-Dimethyl-1-phenyl-1*H*-phosphole; 1-Sulfide, *in* D-00807
3-Phenyl-2-thia-3-phosphabicyclo[2.2.2]oct-5-ene, P-00323

$C_{12}H_{13}PS_2$
3-Phenyl-2-thia-3-phosphabicyclo[2.2.2]oct-5-ene; 3-Sulfide, *in* P-00323

$C_{12}H_{13}PSe$
3,4-Dimethyl-1-phenyl-1*H*-phosphole; 1-Selenide, *in* D-00807

$C_{12}H_{14}BrCl_2O_4P$
▷Bromfenvinfos, B-00462

$C_{12}H_{14}Br_3Cl_2O_3P$
(4-*tert*-Butyl-2-chlorophenyl) 2,2,2-tribromoethyl phosphorochloridate, B-00548

$C_{12}H_{14}ClN_2O_3PS$
Chlorphoxim, C-00193

$C_{12}H_{14}Cl_2NO_3PS_2$
Benoxafos, B-00003

$C_{12}H_{14}Cl_3O_4P$
▷Chlorfenvinphos, C-00018

$C_{12}H_{14}NO_3P$
Ethyl hydrogen[(2-quinolinyl)methyl]-phosphonate, *in* Q-00004

$C_{12}H_{14}NO_4PS$
▷Ditalimfos, D-01237

$C_{12}H_{14}N_3OP$
N,N′-Diphenylphosphoric triamide, D-01114

$C_{12}H_{14}N_5O_8P$
Cyclic GMP; 2′-Ac, *in* C-00234

$C_{12}H_{14}O_6P_2$
[1,5-Naphthalenediylbis[methylene]]-bis[phosphonic acid], N-00002
[1,6-Naphthalenediylbis[methylene]]-bis[phosphonic acid], N-00003
[1,8-Naphthalenediylbis[methylene]]-bis[phosphonic acid], N-00004
[2,3-Naphthalenediylbis[methylene]]-bis[phosphonic acid], N-00005
[2,6-Naphthalenediylbis[methylene]]-bis[phosphonic acid], N-00006
[2,7-Naphthalenediylbis[methylene]]-bis[phosphonic acid], N-00007

$C_{12}H_{15}ClNO_3P$
Diphenyl (3-chlorophenyl)phosphoramidate, *in* C-00156

$C_{12}H_{15}ClNO_4PS_2$
▷Phosalone, P-00340

$C_{12}H_{15}F_5NO_3P$
Triethyl pentafluorophenylphosphorimidate, *in* P-00016

$C_{12}H_{15}F_{12}N_2O_3P$
2,2-Dihydro-3,3-bis(trifluoromethyl)-1-[[2,2,2-trifluoro-1-(trifluoromethyl)ethylidene]-amino]azaphosphiridine; 2,2,2-Triethoxy, *in* D-00446

$C_{12}H_{15}N_2O_2P$
4,5-Diethoxy-1,3-dimethyl-2-phenyl-2*H*-1,3,2-diazaphosphole, D-00230
4,5-Diethoxy-1,3-dimethyl-2-phenyl-2*H*-1,3,2-diazaphosphole; 2-Sulfide, *in* D-00230

$C_{12}H_{15}N_2O_3P$
4,5-Diethoxy-2*H*-1,3,2-diazaphosphole; 2-Phenoxy, *in* D-00228
N-Phenylspiro[1,3,2-benzodioxaphosphole-2,2′(3′*H*)-[1,3,2]benzoxazaphosphol]-2-amine, *in* S-00008

$C_{12}H_{15}N_2O_3PS$
▷Phoxim, P-00424
▷Quinalphos, Q-00001

$C_{12}H_{15}N_2O_6P$
4,4-Diethoxy-2,4,4,5-tetrahydro-2-oxo-5-(4-nitrophenyl)-1,3,4-oxazaphosphole, D-00262

$C_{12}H_{15}N_6P$
2,2′,2″-Phosphinidynetris[1-methylimidazole], *in* T-00496

$C_{12}H_{15}OP$
2,5-Dihydro-3,4-dimethyl-1-phenyl-1*H*-phosphole; 1-Oxide, *in* D-00468
Phenyldi-2-propenylphosphine; Oxide, *in* P-00119

$C_{12}H_{15}O_2P$
2,5-Dihydro-3,4-dimethyl-1-phenoxy-1*H*-phosphole 1-oxide, *in* D-00493
2,3-Dihydro-3,3,5-trimethyl-1,2-oxaphosphole 2-oxide; 2-Ph, *in* D-00596
1,4-Dioxa-5-phospha(5-P^V)spiro[4.4]non-7-ene; 5-Ph, *in* D-00967
Di-2-propenyl phenylphosphonite, *in* P-00251
2-Propenyl phenyl(2-propenyl)phosphinate, *in* P-00300

$C_{12}H_{15}O_2PS$
O,O-Diethyl (phenylethynyl)phosphonothioate, *in* P-00167
O,O-Di-2-propenyl phenylphosphonothioate, *in* P-00244

$C_{12}H_{15}O_3P$
Diethyl phenylethynylphosphonate, *in* P-00166
2,3-Dihydro-3,3,5-trimethyl-1,2-oxaphosphole 2-oxide; 2-Phenoxy, *in* D-00596
Di-2-propenyl phenylphosphonate, *in* P-00221
9-Methyl-1,4,6-trioxa-5-phospha(5-P^V)-spiro[4.4]non-7-ene; 5-Ph, *in* M-00428

$C_{12}H_{15}O_4P$
Bis(acetoxymethyl)phenylphosphine, *in* B-00281
Bis(ethoxycarbonyl)phenylphosphine, *in* P-00209
Hexahydro-2-hydroxycyclopenta[*d*]-1,3,2-dioxaphosphorin 2-oxide; 2-Phenoxy (Ph ester), *in* H-00034
5′-Methylspiro[1,3,2-benzodioxaphosphole-2,2′(3*H*)-[1,2]oxaphosphole]; 2-Ethoxy, *in* M-00396
Phenyl di-2-propenyl phosphate, P-00118

$C_{12}H_{15}O_5P$
(Phenylphosphinyldene)bis[methyl acetate], *in* B-00281

$C_{12}H_{15}O_5PS$
Phenyl 3-[(dimethoxyphosphinothioyl)oxy]-2-methyl-2-propenoate, *in* D-00673

$C_{12}H_{15}P$
2,5-Dihydro-3,4-dimethyl-1-phenyl-1*H*-phosphole, D-00468
Phenyldi-2-propenylphosphine, P-00119

$C_{12}H_{15}PS$
2,5-Dihydro-3,4-dimethyl-1-phenyl-1*H*-phosphole; 1-Sulfide, *in* D-00468
Phenyldi-2-propenylphosphine; Sulfide, *in* P-00119

$C_{12}H_{15}PSe$
2,5-Dihydro-3,4-dimethyl-1-phenyl-1*H*-phosphole; 1-Selenide, *in* D-00468

$C_{12}H_{15}PSi$
4,4-Dimethyl-1-phenyl-1-phospha-4-silacyclohexa-2,5-diene, D-00801

$C_{12}H_{16}Br_3ClNO_2P$
2,2,2-Tribromoethyl 4-*tert*-butylphenylphosphoramidic chloride, T-00356

$C_{12}H_{16}ClOP$
Cyclohexylphenylphosphinic acid; Chloride, *in* C-00250

$C_{12}H_{16}ClO_2P$
Cyclohexyl phenylphosphonochloridate, *in* P-00229

$C_{12}H_{16}FO_2P$
Cyclohexyl phenylphosphonofluoridate, *in* P-00239

$C_{12}H_{16}F_{12}N_3O_2P$
6,9,*N*,*N*-Tetramethyl-2,2,3,3-tetrakis(trifluoromethyl)-1,4-dioxa-6,9-diaza-5-phospha(P^V)spiro[4.4]nonan-5-amine, *in* D-00922

$C_{12}H_{16}IOP$
1-Methyl-4-oxo-1-phenyl-4-phosphorinanium iodide, *in* P-00268

$C_{12}H_{16}NOP$
3,4,5,6-Tetrahydro-4-methyl-2*H*-1,5-methano-4,1-benzazaphosphocine; 1-Oxide, *in* T-00115

$C_{12}H_{16}NO_3P$
Diethyl (cyanophenylmethyl)phosphonate, *in* C-00219
Diethyl [(4-cyanophenyl)methyl]phosphonate, *in* C-00221
4-Methyl-2,6-dioxa-7-aza-1-phosphabicyclo[2.2.2]octane 1-oxide; 7-Benzyl, *in* M-00124

$C_{12}H_{16}NP$
7-Phenyl-6-aza-7-phosphabicyclo[3.2.1]octane, P-00086
3,4,5,6-Tetrahydro-4-methyl-2*H*-1,5-methano-4,1-benzazaphosphocine, T-00115

$C_{12}H_{16}NPS$
3,4,5,6-Tetrahydro-4-methyl-2*H*-1,5-methano-4,1-benzazaphosphocine; 1-Sulfide, *in* T-00115

$C_{12}H_{16}N_3O_3P$
Benzodepa, B-00017

$C_{12}H_{16}N_3O_3PS$
▷Triazophos, T-00333

$C_{12}H_{16}N_3O_3PS_2$
▷Azinphos-ethyl, A-00150

$C_{12}H_{16}N_4O_{11}P_2$
Bis(2-oxo-3-oxazolidinyl)phosphinic acid; Anhydride, *in* B-00378

$C_{12}H_{16}N_5OP$
2,2′-Diphenylphosphoramidic dihydrazide, D-01109

$C_{12}H_{17}ClNO_2P$
Phenyl cyclohexylphosphoramidochloridate, P-00104

$C_{12}H_{17}N_2O_3P$
Dimethyl [[1-amino-2-(1*H*-indol-3-yl)]ethyl]-phosphonate, *in* P-00421

$C_{12}H_{17}N_2O_7P$
Diisopropyl (2,4-dinitrophenyl)phosphonate, *in* D-00946

$C_{12}H_{17}N_2P$
9-Phenyl-1,5-diaza-9-phosphabicyclo[3.3.1]-nonane, P-00105

$C_{12}H_{17}N_4O_2P$
1,3,5,7-Tetramethyl-1,3,5,7-tetraaza-4-phospha(4-P^V)spiro[3.3]heptane-2,6-dione; 4-Ph, *in* T-00234

$C_{12}H_{17}OP$
Cyclohexylphenylphosphine; Oxide, *in* C-00249
2,6-Dimethyl-4-phenyl-1,4-oxaphosphorinane, D-00797
1-Phenylphosphepane; 1-Oxide, *in* P-00205
2,2,5-Trimethyl-3-phenyl-1,3-oxaphospholane, T-00522

$C_{12}H_{17}OPS_2$
S,*S*-Diethyl (2-phenylethenyl)-phosphonodithioate, *in* P-00159

$C_{12}H_{17}O_2P$
Cyclohexylphenylphosphinic acid, C-00250
Diethyl (2-phenylethenyl)phosphonite, *in* P-00161
2-Phenyl-4,4,5,5-tetramethyl-1,3,2-dioxaphospholane, P-00322
2,4,6-Trimethyl-5-phenyl-1,3,5-dioxaphosphorinane, T-00521
2,2,5-Trimethyl-3-phenyl-1,3-oxaphospholane; 3-Oxide, *in* T-00522

$C_{12}H_{17}O_2PS$
O,*O*-Diethyl (2-phenylethenyl)-phosphonothioate, *in* P-00160
2,4,6-Trimethyl-5-phenyl-1,3,5-dioxaphosphorinane; 5-Sulfide, *in* T-00521

$C_{12}H_{17}O_2PSe$
2,4,6-Trimethyl-5-phenyl-1,3,5-dioxaphosphorinane; 5-Selenide, *in* T-00521

$C_{12}H_{17}O_3P$
Diethyl (1-phenylethenyl)phosphonate, *in* P-00154
Diethyl (2-phenylethenyl)phosphonate, *in* P-00155
Monophenyl cyclohexylphosphonate, *in* C-00256
2-Phenoxy-4,4,5,5-tetramethyl-1,3,2-dioxaphospholane, P-00075
Tetramethylethylene phenylphosphonate, *in* P-00322
4,4,6-Trimethyl-2-phenoxy-1,3,2-dioxaphosphorinane, T-00520
2,4,6-Trimethyl-5-phenyl-1,3,5-dioxaphosphorinane; 5-Oxide, *in* T-00521
3,5,5-Trimethyl-2-phenyl-1,2-oxaphospholan-3-ol 2-oxide, *in* T-00519
3,5,5-Trimethyl-2-phenyl-1,2-oxaphospholan-3-ol 2-oxide, *in* T-00519

$C_{12}H_{17}O_4P$
Diethyl (2-oxo-2-phenylethyl)phosphonate, *in* O-00066
Dimethyl (2-oxo-4-phenylbutyl)phosphonate, *in* O-00065
(3,3-Dimethyl-2-oxo-4-phenylbutyl)phosphonic acid, D-00786
2-Phenoxy-1,3,2-dioxaphosphonane 2-oxide, *in* H-00137
4,4,5,5-Tetramethyl phenyl phosphate, *in* P-00075
4′,4′,5′,5′-Tetramethylspiro[1,3,2-benzodioxaphosphole-2,2′-[1,3,2]-dioxaphospholane], T-00230
1,4,6,11-Tetraoxa-5-phospha(5-P^V)spiro[4.6]-undecane; 5-Ph, *in* T-00245
3,5,5-Trimethyl-1,2-oxaphospholan-3-ol 2-oxide; 2-Phenoxy, *in* T-00519

$C_{12}H_{17}O_4PS_2$
▷Phenthoate, P-00078

$C_{12}H_{17}O_5P$
α-(Diethoxyphosphinyl)benzeneacetic acid, D-00242
Methyl 2-(diethoxyphosphinyl)benzoate, *in* D-00243
Methyl 4-(diethoxyphosphinyl)benzoate, *in* D-00244

$C_{12}H_{17}P$
Cyclohexylphenylphosphine, C-00249
1-Phenylphosphepane, P-00205

$C_{12}H_{17}PS$
Cyclohexylphenylphosphine; Sulfide, *in* C-00249
1-Phenylphosphepane; Sulfide, *in* P-00205

$C_{12}H_{18}BrP$
1-Benzyl-1-methylphospholanium bromide, *in* M-00280
1,3-Dimethyl-1-phenylphospholanium bromide, *in* M-00229

$C_{12}H_{18}ClO_2P$
Diisopropyl (4-chlorophenyl)phosphonite, *in* C-00153

$C_{12}H_{18}Cl_2NOPS$
▷*P*-Chloromethyl-*O*-(2-chloro-4-methylphenyl)-*N*-(1-methylpropyl) phosphonamidothioate, *in* C-00101

$C_{12}H_{18}Cl_3P$
Chlorobis(4-chlorophenyl)phosphine, *in* B-00142

$C_{12}H_{18}F_{12}N_5P$
2,2-Dihydro-3,3-bis(trifluoromethyl)-1-[[2,2,2-trifluoro-1-(trifluoromethyl)ethylidene]-amino]azaphosphiridine; 2,2,2-Tris(dimethylamino), *in* D-00446

$C_{12}H_{18}IO_2P$
2,5,5-Trimethyl-2-phenyl-1,3,2-dioxaphosphorinanium iodide, *in* D-00795

$C_{12}H_{18}IP$
1-Methyl-1-phenylphosphorinanium iodide, *in* P-00265

$C_{12}H_{18}NOP$
2-Phenyl-1-oxa-4-aza-5-phospha(5-P^V)-spiro[4.4]nonane, P-00173

$C_{12}H_{18}NO_2P$
N-Cyclohexyl-*P*-phenylphosphonamidic acid, C-00251
2-Ethoxy-1-(2-phenylethyl)-1,2-azaphosphetidine 2-oxide, *in* H-00107

$C_{12}H_{18}NO_3P$
Diethyl [(benzylideneamino)methyl]-phosphonate, D-00271
4′,4′,5′,5′-Tetramethylspiro[1,3,2-benzoxazaphosphole-2(3*H*),2′-[1,3,2]-dioxaphospholane], T-00231

$C_{12}H_{18}NO_4P$
(1-Amino-2-methylpropyl)phosphinic acid; *N*-Benzyloxycarbonyl, *in* A-00092
(1-Benzoylaminopentyl)phosphonic acid, *in* A-00096

$C_{12}H_{18}NO_5P$
Diisopropyl (2-nitrophenyl)phosphonate, *in* N-00055
Diisopropyl (3-nitrophenyl)phosphonate, *in* N-00056

$C_{12}H_{18}NP$
2-Ethyl-2,3,4,5-tetrahydro-2-methyl-1*H*-1,3-benzazaphosphepine, E-00188

$C_{12}H_{18}N_3PS_2$
N,*N*-Diethyl-2,3-dihydro-5-methyl-3-phenyl-1,3,4,2-thiadiazaphosphole-2(3*H*)-amine 2-sulfide, *in* A-00072

$C_{12}H_{18}N_8OP$
▷Fopurine, F-00053

$C_{12}H_{18}O_4P_2$
1,8-Diethoxy-3*a*,4,7,7*a*-tetrahydro-4,7-phosphinidene-1*H*-phosphinidole 1,8-dioxide, *in* D-00625

$C_{12}H_{19}ClNO_3P$
▷Crufomate, C-00208

$C_{12}H_{19}ClN_4O_7P_2S$
▷Thiamine diphosphate, T-00297

$C_{12}H_{19}FNO_2PS$
O,*O*-Diisopropyl (4-fluorophenyl)-phosphoroamidothioate, *in* F-00046

$C_{12}H_{19}FNO_3P$
Diisopropyl (3-fluorophenyl)phosphoramidate, *in* F-00043
Diisopropyl (4-fluorophenyl)phosphoramidate, *in* F-00044

$C_{12}H_{19}N_2OP$
2-Diethylamino-3-phenyl-1,3,2-oxazaphospholidine, D-00269
2-Dimethylamino-3,4-dimethyl-5-phenyl-1,3,2-oxazaphospholidine, D-00694
Phenyl *N*,*N*′-1,3-butylene-*N*,*N*′-dimethylphosphorodiamidite, *in* H-00057
(2-Phenylethenyl)phosphonic bis(dimethylamide), *in* P-00156

$C_{12}H_{19}N_2OPS$

2-Diethylamino-3-phenyl-1,3,2-oxazaphospholidine; 2-Sulfide, *in* D-00269

$C_{12}H_{19}N_2O_2P$

2-Diethylamino-3-phenyl-1,3,2-oxazaphospholidine; 2-Oxide, *in* D-00269
N,N,N′,N′-Tetramethyl-*P*-(2-oxo-2-phenylethyl)phosphonic diamide, *in* O-00066

$C_{12}H_{19}N_2O_4PS$

Amiprophos *M*, A-00131
S-Ethyl *O*-(4-methyl-2-nitrophenyl) isopropylphosphoramidothioate, *in* M-00180

$C_{12}H_{19}N_2O_5P$

Triethyl (2-nitrophenyl)phosphorimidate, *in* N-00061
Triethyl (4-nitrophenyl)phosphorimidate, *in* N-00063

$C_{12}H_{19}N_4O_2P$

Phenyl phosphorodihydrazidate; 2,2′-Diisopropylidene, *in* P-00290

$C_{12}H_{19}N_6OP$

▷Triamiphos, T-00326

$C_{12}H_{19}OP$

Diisopropylphenylphosphine; Oxide, *in* D-00638

$C_{12}H_{19}OPS$

O-Phenyl diisopropylphosphinothioate, *in* D-00646

$C_{12}H_{19}O_2P$

Bis(3-hydroxypropyl)phenylphosphine, B-00286
Diisopropyl phenylphosphonite, *in* P-00251
Isopropyl isopropylphenylphosphinate, *in* I-00043
Propyl phenylpropylphosphinate, *in* P-00303

$C_{12}H_{19}O_2PS$

3,3′-(Phenylphosphinothioylidene)bis[1-propanol], *in* B-00286

$C_{12}H_{19}O_2PS_3$

▷Sulprofos, S-00027

$C_{12}H_{19}O_3P$

Diethyl (2,3-dimethylphenyl)phosphonate, *in* D-00811
Diethyl (2,4-dimethylphenyl)phosphonate, *in* D-00812
Diethyl (2,5-dimethylphenyl)phosphonate, *in* D-00813
Diethyl(3,4-dimethylphenyl)phosphonate, *in* D-00815
Diethyl (3,5-dimethylphenyl)phosphonate, *in* D-00816
Diethyl [(3-methylphenyl)methyl]phosphonate, *in* M-00207
Diethyl [(4-methylphenyl)methyl]phosphonate, *in* M-00208
▷Diethyl 1-phenylethylphosphonate, *in* P-00164
Diethyl 2-phenylethylphosphonate, *in* P-00165
Diisopropyl phenylphosphonate, *in* P-00221
Dipropyl phenylphosphonate, *in* P-00221
3,3′-(Phenylphosphinylidene)bis[1-propanol], *in* B-00286

$C_{12}H_{19}O_3PS$

Diethyl [[(phenylmethyl)thio]methyl]-phosphonate, *in* B-00072

$C_{12}H_{19}O_4P$

OO-tert-Butyl *O*-ethyl phenylphosphonoperoxoate, *in* P-00242
Diethyl [(3-methoxyphenyl)methyl]-phosphonate, *in* M-00060
Diethyl [(4-methoxyphenyl)methyl]-phosphonate, *in* M-00061
Diethyl [(2-methylphenyl)methyl]phosphonate, *in* M-00206

$C_{12}H_{19}P$

Diisopropylphenylphosphine, D-00638
Phenyldipropylphosphine, P-00120

$C_{12}H_{19}PS$

Butyl ethylphenylphosphinothioite, *in* E-00107
Diisopropylphenylphosphine; Sulfide, *in* D-00638

$C_{12}H_{19}PS_3$

Dipropyl phenylphosphonotrithioate, *in* P-00250
Ethyl phenyl (1-methylpropyl)-phosphonotrithioate, *in* M-00382

$C_{12}H_{20}ClO_3P$

Dicyclohexyl phosphorochloridate, *in* D-00209

$C_{12}H_{20}FO_3P$

▷Dicyclohexyl phosphorofluoridate, *in* D-00209

$C_{12}H_{20}F_4O_6P_2$

Tetraethyl (3,3,4,4-tetrafluoro-1-cyclobutene-1,2-diyl)bisphosphonate, *in* T-00078

$C_{12}H_{20}IP$

tert-Butyldimethylphenylphosphonium iodide, *in* B-00565

$C_{12}H_{20}I_2P_2$

3*a*,4,7,7*a*-Tetrahydro-1,1,8,8-tetramethyl-4,7-phosphinidene-1*H*-phosphindolium diiodide, *in* T-00099

$C_{12}H_{20}NOP$

Ethyl *P,P*-diethyl-*N*-phenylphosphinimidate, *in* D-00306

$C_{12}H_{20}NOPS$

S-Ethyl *N,N*-diethyl-*P*-phenylphosphonamidothioate, *in* D-00309

$C_{12}H_{20}NOPS_2$

▷Phosbutyl, P-00341

$C_{12}H_{20}NO_2PS$

O,O-Bis(1-methylethyl) phenylphosphoramidothioate, *in* P-00262
O,O-Dipropyl phenylphosphoramidothioate, *in* D-01230

$C_{12}H_{20}NO_3P$

[1-(Benzylamino)pentyl]phosphonic acid, *in* A-00096
Diethyl (1-amino-1-phenylethyl)phosphonate, *in* A-00100
Diethyl [(benzylamino)methyl]phosphonate, *in* A-00091
Diethyl (4-dimethylaminophenyl)phosphonate, *in* D-00706
Diethyl (2,4-dimethylphenyl)phosphoramidate, *in* D-00818
Diethyl ethylphenylphosphoramidate, *in* E-00116
Diisopropyl (3-methyl-4-pyridinyl)phosphonate, *in* M-00390
Diisopropyl phenylphosphoramidate, *in* P-00257
Dipropyl phenylphosphoramidate, *in* P-00257
Triethyl phenylphosphorimidate, *in* P-00264

$C_{12}H_{20}NO_5PS$

Triethyl phenylsulfonylphosphorimidate, *in* P-00315

$C_{12}H_{20}NO_{11}P$

2-Amino-2-deoxyglucose 6-(dihydrogen phosphate); 1,3,4-Tri-Ac, *in* A-00069

$C_{12}H_{20}NP$

(2-Aminoethyl)butylphenylphosphine, A-00074
P,P-Diisopropyl-*N*-phenylphosphinous amide, *in* D-00649

$C_{12}H_{20}NPS_2$

Ethyl *N,N*-diethyl-*P*-phenylphosphonamidodithioate, *in* D-00308

$C_{12}H_{20}N_3O_9P$

▷Plumbemycin *A*, P-00437

$C_{12}H_{20}O_6P_2$

Tetramethyl [1,4-phenylenebis(methylene)]-bisphosphonate, *in* P-00131

$C_{12}H_{21}N_2OP$

Bis(3-aminopropyl)phenylphosphine; *P*-Oxide, *in* B-00090
N,N′-Diisopropyl-*P*-phenylphosphonic diamide, *in* P-00222

$C_{12}H_{21}N_2O_2P$

Phenyl *N,N′*-diisopropylphosphorodiamidate, *in* D-00661
Phenyl *N,N′*-dipropylphosphorodiamidate, *in* D-01223

$C_{12}H_{21}N_2O_2PS$

O,O-Diisopropyl 2-phenylphosphorohydrazidothioate, *in* P-00296

$C_{12}H_{21}N_2O_3PS$

▷Diazinon, D-00027

$C_{12}H_{21}N_2P$

Bis(3-aminopropyl)phenylphosphine, B-00090

$C_{12}H_{21}N_2PS$

N,N′-Diisopropyl-*P*-phenylphosphonothioic diamide, *in* P-00246
3,3′-(Phenylphosphinothioylidene)bis-1-propanamine, *in* B-00090

$C_{12}H_{21}N_3O_{13}P_2$

Cytidine diphosphate glycerol, C-00293

$C_{12}H_{21}N_4O_8P$

▷Plumbemycin B, *in* P-00437

$C_{12}H_{21}O_2P$

5-Hydroxy-5*H*-dibenzophosphole 5-oxide; Dodecahydro, *in* H-00118

$C_{12}H_{21}O_3P$

Dimethyl 1-adamantylphosphonate, *in* A-00034
Dimethyl 2-adamantylphosphonate, *in* A-00035
Tri-2-butenyl phosphite, T-00361

$C_{12}H_{21}O_4P$

Tris(2-methyl-2-propenyl) phosphate, T-00820

$C_{12}H_{21}O_4PSi$

Dimethyl [phenyl[(trimethylsilyl)oxy]methyl]-phosphonate, *in* P-00331

$C_{12}H_{21}O_6P$

3,3′,3″-Phosphinidynetrispropanoic acid; Tri-Me ester, *in* P-00357

$C_{12}H_{21}O_7P$

3,3′,3″-Phosphinylidynetrispropanoic acid trimethyl ester, *in* P-00357

$C_{12}H_{21}O_9P$

1,2:3,4-Di-*O*-isopropylidene-α-D-galactopyranose 6-dihydrogen phosphate, *in* G-00003
Trimethyl (2-dimethoxyphosphinyl)-1,2,4-butanetrioate, *in* P-00382
Trimethyl 3-(dimethoxyphosphinyl)-1,2,3-butanetrioate, *in* P-00383

$C_{12}H_{21}P$

Tri-3-butenylphosphine, T-00360

$C_{12}H_{21}PSi_2$

2,2,5,5-Tetramethyl-1-phenyl-1-phospha-2,5-disilacyclopentane, T-00209

$C_{12}H_{22}BF_4N_6OP$

(1-Hydroxy-1*H*-benzotriazolato-*O*)-tris(dimethylamino)phosphorus(1+); Tetrafluoroborate, *in* H-00111

$C_{12}H_{22}BrO_3PSi_2$

Bis(trimethylsilyl)(4-bromophenyl)-phosphonate, *in* B-00492

$C_{12}H_{22}BrP$

Bromodicyclohexylphosphine, *in* D-00214

$C_{12}H_{22}ClN_6OP$

(1-Hydroxy-1*H*-benzotriazolato-*O*)-tris(dimethylamino)phosphorus(1+); Chloride, *in* H-00111

$C_{12}H_{22}ClOP$

Dicyclohexylphosphinic acid; Chloride, *in* D-00211

$C_{12}H_{22}ClO_3PSi_2$
Bis(trimethylsilyl)(4-chlorophenyl)-phosphonate, *in* C-00147

$C_{12}H_{22}ClP$
Chlorodicyclohexylphosphine, *in* D-00214

$C_{12}H_{22}ClPS$
Dicyclohexylphosphinothioic acid; Chloride, *in* D-00213

$C_{12}H_{22}FOP$
Dicyclohexylphosphinic acid; Fluoride, *in* D-00211

$C_{12}H_{22}FO_3PSi_2$
Bis(trimethylsilyl)(2-fluorophenyl)phosphonate, *in* F-00036
Bis(trimethylsilyl)(4-fluorophenyl)phosphonate, *in* F-00038

$C_{12}H_{22}FPS$
Dicyclohexylphosphinothioic acid; Fluoride, *in* D-00213

$C_{12}H_{22}F_6N_6OP_2$
(1-Hydroxy-1*H*-benzotriazolato-*O*)-tris(dimethylamino)phosphorus(1+); Hexafluorophosphate, *in* H-00111

$C_{12}H_{22}IP$
Dicyclohexyliodophosphine, *in* D-00214

$C_{12}H_{22}NO_3P$
Dimethyl 1-adamantylphosphoramidate, *in* A-00036

$C_{12}H_{22}NO_4P$
N,N-Dimethyl-2,4-bis(1-methylethylidene)-1,3,6,9-tetraoxa-5-phospha(5-P^V)spiro[4.4]-nonan-5-amine, *in* B-00306

$C_{12}H_{22}NO_6P$
2-Diethylamino-1,3,2-dioxaphospholan-4,5-dicarboxylic acid; Di-Et ester, *in* D-00265

$C_{12}H_{22}NO_6PS$
2-Diethylamino-1,3,2-dioxaphospholan-4,5-dicarboxylic acid; Di-Et ester, 2-sulfide, *in* D-00265

$C_{12}H_{22}N_6OP^{\oplus}$
(1-Hydroxy-1*H*-benzotriazolato-*O*)-tris(dimethylamino)phosphorus(1+), H-00111

$C_{12}H_{22}O_4P_2$
1,8-Diethoxyoctahydro-4,7-phosphinidene-1*H*-phosphindole 1,8-dioxide, *in* D-00625

$C_{12}H_{22}P_2S_4$
2,4-Dicyclohexyl-1,3,2,4-dithiadiphosphetane 2,4-disulfide, D-00207

$C_{12}H_{23}NO_7P$
Methyl 6-deoxy-2,3-di-*O*-methyl-6-methyl-aminoglucopyranoside 4,6-cyclic phosphonamide; *P*-Ethoxy, *in* M-00112

$C_{12}H_{23}N_4P$
N,N,N',N',N'',N''-Hexamethyl-*N'''*-phenyl-phosphorimidic triamide, H-00073

$C_{12}H_{23}OP$
Dicyclohexylphosphine; Oxide, *in* D-00210
Dicyclohexylphosphinous acid, D-00214

$C_{12}H_{23}OPS$
Dicyclohexylphosphinothioic acid, D-00213

$C_{12}H_{23}O_2P$
Dicyclohexylphosphinic acid, D-00211

$C_{12}H_{23}O_2PSSi_2$
O,O-Bis(trimethylsilyl) phenyl-phosphonothioate, *in* P-00244

$C_{12}H_{23}O_2PSeSi_2$
O,O-Bis(trimethylsilyl) phenyl-phosphonoselenoate, *in* P-00243

$C_{12}H_{23}O_2PSe_2$
O,O-Dicyclohexyl phosphorodiselenoate, D-00218

$C_{12}H_{23}O_3P$
Dicyclohexyl phosphonate, D-00215
Diisopropyl 2-cyclohexen-1-ylphosphonate, *in* C-00242

$C_{12}H_{23}O_3PSi_2$
Phenylphosphonic acid; Bis(trimethylsilyl) ester, *in* P-00221

$C_{12}H_{23}O_4P$
Dicyclohexyl phosphate, D-00209

$C_{12}H_{23}O_6PS_2$
Bis(1-methylethyl) [(dimethoxyphosphinothioyl)thio]-butanedioate, *in* D-00677
▷Diethyl [(diethoxyphosphinothioyl)thio]-butanedioate, *in* D-00240

$C_{12}H_{23}O_7P$
Tetraethyl phosphonobutanedioate, *in* D-00245

$C_{12}H_{23}O_7PS$
Methyl 2,3-di-*O*-methylglucopyranoside 4,6-cyclic phosphonate; *P*-(Propylthio), *in* M-00119

$C_{12}H_{23}O_{14}P$
Trehalose 6-(dihydrogen phosphate), T-00324

$C_{12}H_{23}P$
Dicyclohexylphosphine, D-00210

$C_{12}H_{23}PS$
Dicyclohexylphosphine; Sulfide, *in* D-00210

$C_{12}H_{23}PS_2$
Dicyclohexylphosphinodithioic acid, D-00212

$C_{12}H_{23}PS_3Si_2$
Bis(trimethylsilyl) phenylphosphonotrithioate, *in* P-00250

$C_{12}H_{24}ClN_2OP$
1,3-Dibutyl-2,3-dihydro-4,5-dimethyl-1*H*-1,3,2-diazaphosphole 2-oxide; 2-Chloro, *in* D-00081

$C_{12}H_{24}ClO_4PS_2$
2,4,8,10-Tetramethyl-3,9-dithia-6-phosphoniaspiro[5.5]undecane; Perchlorate, *in* T-00202

$C_{12}H_{24}FN_2OP$
▷*N,N'*-Dicyclohexylphosphorodiamidic acid; Fluoride, *in* D-00216

$C_{12}H_{24}NO_2P$
3,5-Di-*tert*-butyl-2,3-dihydro-1,3,2-oxazaphosphole; 2-Ethoxy, *in* D-00083

$C_{12}H_{24}NO_3P$
3,5-Di-*tert*-butyl-2,3-dihydro-1,3,2-oxazaphosphole; 2-Ethoxy, 2-oxide, *in* D-00083
Diethyl 2-(cyclohexylamino)vinylphosphonate, D-00278

$C_{12}H_{24}NO_3PSi_2$
Bis(trimethylsilyl)(4-aminophenyl)-phosphonate, *in* A-00111

$C_{12}H_{24}NO_4P$
Diethyl [(5,6-dihydro-4,4,6-trimethyl-4*H*-1,3-oxazin-2-yl]phosphonate, *in* D-00598

$C_{12}H_{24}N_3OP$
1,1',1''-Phosphinylidynetrispyrrolidine, *in* T-00724

$C_{12}H_{24}N_3O_3P$
Trimorpholinophosphine, T-00569

$C_{12}H_{24}N_3O_3PS$
▷4,4',4''-Phosphinothioylidynetrismorpholine, *in* T-00569

$C_{12}H_{24}N_3O_3PSe$
4,4',4''-Phosphinoselenoylidynetrismorpholine, *in* T-00569

$C_{12}H_{24}N_3O_3PTe$
4,4',4''-Phosphinotelluroylidynetrismorpholine, *in* T-00569

$C_{12}H_{24}N_3O_4P$
4,4',4''-Phosphinylidynetrismorpholine, *in* T-00569

$C_{12}H_{24}N_3P$
Tri-1-pyrrolidylphosphine, T-00724

$C_{12}H_{24}N_3PSe$
1,1',1''-Phosphinoselenoylidynetrispyrrolidine, *in* T-00724

$C_{12}H_{24}N_3PTe$
1,1',1''-Phosphinotelluroylidynetrispyrrolidine, *in* T-00724

$C_{12}H_{24}N_6O_2P_2$
▷Dipin, D-01189

$C_{12}H_{24}N_9P_3$
▷2,2,4,4,6,6-Hexakis(1-aziridinyl)-2,2,4,4,6,6-hexahydro-1,3,5-triaza-2,4,6-triphosphorine, H-00060

$C_{12}H_{24}O_5P_2$
Dicyclohexyldiphosphonic acid, D-00206

$C_{12}H_{24}O_6P_2$
Tetraethyl 1,3-butadiene-1,4-diylbisphosphonate, *in* B-00507

$C_{12}H_{24}PS_2^{\oplus}$
2,4,8,10-Tetramethyl-3,9-dithia-6-phosphoniaspiro[5.5]undecane, T-00202

$C_{12}H_{25}Cl_2P$
Dichlorododecylphosphine, *in* D-01265

$C_{12}H_{25}N_2OP$
1,3-Dibutyl-2,3-dihydro-4,5-dimethyl-1*H*-1,3,2-diazaphosphole 2-oxide, D-00081

$C_{12}H_{25}N_2O_2P$
N,N'-Dicyclohexylphosphorodiamidic acid, D-00216
Ethyl di-1-piperidinylphosphinate, *in* D-01190

$C_{12}H_{25}N_2O_3P$
7-Isopropylidene-4,8,8-trimethyl-1,6-dioxa-4-aza-5-phospha(5-P^V)spiro[4.4]nonan-9-one; 5-Dimethylamino, *in* I-00032

$C_{12}H_{25}N_2PS$
P-(1,3-Butadien-1-yl)-*N,N,N',N*-tetraethylphosphonothioic diamide, *in* B-00511

$C_{12}H_{25}N_2PS_2$
N,N'-Dicyclohexylphosphorodiamidodithioic acid, D-00217

$C_{12}H_{25}N_3OP$
N,N-Diethyl-*P,P*-di-1-pyrrolidinylphosphinic amide, *in* D-01233

$C_{12}H_{25}O_2P$
Dimethyl 4-*tert*-butylcyclohexylphosphonite, *in* B-00550

$C_{12}H_{25}O_2PS_2$
2,4,6-Triisopropyl-5-mercapto-1,3,5-dioxaphosphorinane 5-sulfide, T-00498

$C_{12}H_{25}O_3P$
Diisopropyl cyclohexylphosphonate, *in* C-00256

$C_{12}H_{25}O_4P$
Diethyl (2-oxooctyl)phosphonate, *in* O-00064
2,2,3,3,7,7,8,8-Octamethyl-1,4,6,9-tetraoxa-5-phospha(5-P^V)spiro[4.4]nonane, O-00032

$C_{12}H_{25}O_5P$
Ethyl (dibutoxyphosphinyl)acetate, *in* D-00073
Methyl 3-(dibutoxyphosphinyl)propanoate, *in* P-00396
3,3,9,9-Tetramethyl-1,5,7,11-tetraoxa-6-phospha(6-P^V)spiro[5.5]undecane; 6-Ethoxy, *in* T-00236

$C_{12}H_{26}ClOP$
Dihexylphosphinic acid; Chloride, *in* D-00434

$C_{12}H_{26}ClO_2P$
▷Dihexyl phosphorochloridite, D-00436

$C_{12}H_{26}ClO_3P$
Dibutyl (4-chlorobutyl)phosphonate, *in* C-00025
Dihexyl phosphorochloridate, *in* D-00432

$C_{12}H_{26}FOP$
Dihexylphosphinic acid; Fluoride, *in* D-00434

$C_{12}H_{26}NO_3P$
Dibutyl 4-morpholinylphosphonite, *in* M-00529
Diisopropyl cyclohexylphosphoramidate, *in* C-00269

$C_{12}H_{26}NO_4P$
Diisopropyl (diethylcarbamoylmethyl)-phosphonate, *in* D-00268

$C_{12}H_{26}N_4P_2$
15,16-Dimethyl-1,4,8,11-tetraaza-15,16-diphosphatricyclo[9.3.1.1^{4,8}]hexadecane, D-00921

$C_{12}H_{26}O_4P_2$
2,5,5,8,11,11-Hexamethyl-1,3,7,9-tetraoxa-2,8-diphosphacyclododecane, H-00081

$C_{12}H_{26}O_4P_2S_2$
2,5,5,8,11,11-Hexamethyl-1,3,7,9-tetraoxa-2,8-diphosphacyclododecane; 2,8-Disulfide, *in* H-00081

$C_{12}H_{26}O_6P_2$
Tetraethyl (2-butene-1,4-diyl)bisphosphonate, *in* B-00519

$C_{12}H_{26}O_6P_2S_4$
▷Dioxathion, D-00974

$C_{12}H_{26}O_8P_2$
Methyl 2,3-bis(diethoxyphosphinyl)propanoate, *in* B-00173

$C_{12}H_{27}AsP_2$
Tri-*tert*-butyldiphospharsirane, T-00363

$C_{12}H_{27}ClNOP$
Butyl dibutylphosphoramidochloridite, *in* D-00135

$C_{12}H_{27}F_6N_2OPSi_3$
2-[Bis(trimethylsilyl)amino]-2,2-dihydro-3,3-bis(trifluoromethyl)-2-[(trimethylsilyl)-imino]oxaphosphirane, B-00430

$C_{12}H_{27}INO_6P_2$
[(Dimethylamino)methylene]bisphosphonic acid; Tetra-Et ester; B,MeI, *in* D-00701

$C_{12}H_{27}N_2OP$
1,2,3-Tri-*tert*-butyldiazaphosphiridine; 3-Oxide, *in* T-00362

$C_{12}H_{27}N_2O_3P$
Ethyl [bis(diethylamino)phosphinyl]acetate, *in* B-00175

$C_{12}H_{27}N_2P$
1,2,3-Tri-*tert*-butyldiazaphosphiridine, T-00362

$C_{12}H_{27}N_4OPSSi_2$
O-Phenyl phosphorodihydrazidothioate; *N*^{2},*N*$^{2'}$-Bis(trimethylsilyl), *in* P-00291

$C_{12}H_{27}OP$
Butyl dibutylphosphinite, *in* D-00121
Dihexylphosphine; Oxide, *in* D-00433
Tri-*tert*-butylphosphine; Oxide, *in* T-00373
▷Tributylphosphine oxide, T-00376
Tris(2-methylpropyl)phosphine; Oxide, *in* T-00824

$C_{12}H_{27}OPS$
O-Butyl bis(2-methylpropyl)phosphinothioate, *in* B-00351

$C_{12}H_{27}OPSSe_2$
O,Se,Se-Tributyl phosphorodiselenothioate, *in* D-00149

$C_{12}H_{27}OPS_2$
S,S-Dibutyl butylphosphonodithioate, *in* B-00606

$C_{12}H_{27}OPS_3$
O,S,S-Tributyl phosphorotrithioate, T-00390
S,S,S-Tri-*tert*-butyl phosphorotrithioate, T-00391
▷*S,S,S*-Tributyl phosphorotrithioate, T-00392

$C_{12}H_{27}OPSe_2$
O,Se,Se-Tributyl phosphorodiselenoite, T-00383

$C_{12}H_{27}OPSe_3$
O,Se,Se-Tributyl phosphorotriselenoate, T-00389

$C_{12}H_{27}O_2P$
Bis(2-methylpropyl) (2-methylpropyl)-phosphonite, *in* M-00383
Butyl dibutylphosphinate, *in* D-00109
Dibutyl butylphosphonite, *in* B-00615
Diethyl octylphosphonite, *in* O-00044
▷Dihexylphosphinic acid, D-00434
Diisopropyl hexylphosphonite, *in* H-00101
Dodecylphosphonous acid, D-01265
2-Methylpropyl bis(2-methylpropyl)-phosphinate, *in* B-00346

$C_{12}H_{27}O_2PS$
O,O-Bis(2-methylpropyl) (2-methylpropyl)-phosphonothioate, *in* M-00381
O,O-Dibutyl butylphosphonothioate, *in* B-00611

$C_{12}H_{27}O_2PS_2$
O,O-Dihexyl phosphorodithioate, D-00437
O,O,S-Tributyl phosphorodithioate, T-00384

$C_{12}H_{27}O_3H$
Dihexyl phosphonate, D-00435

$C_{12}H_{27}O_3P$
Bis(2-methylpropyl) (2-methylpropyl)-phosphonate, *in* M-00378
▷Dibutyl butylphosphonate, D-00076
Dibutyl (1-methylpropyl)phosphonate, *in* M-00377
Diethyl octylphosphonate, *in* O-00042
Diisopropyl hexylphosphonate, *in* H-00097
Dimethyl decylphosphonate, *in* D-00012
▷Tributyl phosphite, T-00379
Tri-*tert*-butyl phosphite, T-00380
Tris(1-methylpropyl) phosphite, T-00825
Tris(2-methylpropyl) phosphite, T-00826

$C_{12}H_{27}O_3PS$
O,O,O-Tributyl phosphorothioate, T-00387
O,O,S-Tributyl phosphorothioate, T-00388
O,O,S-Tris(2-methylpropyl) phosphorothioate, T-00828

$C_{12}H_{27}O_3PSe$
O,O,O-Tributyl phosphoroselenoate, T-00385
O,O,O-Tris(2-methylpropyl) phosphoroselenoate, T-00827

$C_{12}H_{27}O_3P_3$
2,4,6-Tributyl-1,3,5,2,4,6-trioxatriphosphorinane, T-00396

$C_{12}H_{27}O_3P_3S_3$
Butylphosphonothioic acid trimer trianhydride, *in* T-00396

$C_{12}H_{27}O_4P$
Dihexyl phosphate, D-00432
Monododecyl phosphate, M-00468
▷Tributyl phosphate, T-00370
Tri-*tert*-butyl phosphate, T-00371
Tris(1-methylpropyl) phosphate, T-00821
Tris(2-methylpropyl) phosphate, T-00822

$C_{12}H_{27}O_7P$
Tris(2-ethoxyethyl) phosphate, T-00778

$C_{12}H_{27}P$
Dihexylphosphine, D-00433
▷Tributylphosphine, T-00372
Tri-*tert*-butylphosphine, T-00373
Tris(1-methylpropyl)phosphine, T-00823
Tris(2-methylpropyl)phosphine, T-00824

$C_{12}H_{27}PS$
Butyl bis(2-methylpropyl)phosphinothioite, *in* B-00352
Tri-*tert*-butylphosphine; Sulfide, *in* T-00373
▷Tributylphosphine sulfide, T-00378
Tris(2-methylpropyl)phosphine; Sulfide, *in* T-00824

$C_{12}H_{27}PS_2$
Butyl diisobutylphosphinodithioate, *in* B-00348
Dibutyl butylphosphonodithioite, *in* B-00607

$C_{12}H_{27}PS_3$
Dibutyl butylphosphonotrithioate, *in* B-00614
▷Tributyl phosphorotrithioite, T-00393

$C_{12}H_{27}PS_4$
Tributyl phosphorotetrathioate, T-00386

$C_{12}H_{27}PSe$
Tri-*tert*-butylphosphine; Selenide, *in* T-00373
Tributylphosphine selenide, T-00377

$C_{12}H_{27}PTe$
Tributylphosphine; Telluride, *in* T-00372
Tri-*tert*-butylphosphine; Telluride, *in* T-00373

$C_{12}H_{27}P_3$
Tri-*tert*-butyltriphosphirane, T-00397

$C_{12}H_{28}BrP$
Tetrapropylphosphonium(1+); Bromide, *in* T-00290

$C_{12}H_{28}ClN_2OP$
Tetrapropylphosphorodiamidic acid; Chloride, *in* T-00291

$C_{12}H_{28}ClP$
Tetrapropylphosphonium(1+); Chloride, *in* T-00290

$C_{12}H_{28}IP$
Tetrapropylphosphonium(1+); Iodide, *in* T-00290

$C_{12}H_{28}NOP$
P,P-Dibutyl-*N,N*-diethylphosphinic amide, *in* D-00111

$C_{12}H_{28}NO_2P$
Butyl *N,N*-dibutylphosphonamidate, *in* D-00126
Diethyl dibutylphosphoramidite, *in* D-00136

$C_{12}H_{28}NO_2PS$
O,O-Dibutyl diethylphosphoramidothioate, *in* D-00356

$C_{12}H_{28}NO_3P$
Dibutyl butylphosphoramidate, *in* B-00621
Dibutyl *tert*-butylphosphoramidate, *in* B-00622
Dibutyl diethylphosphoramidate, *in* D-00343
Dibutyl (2-methylpropyl)phosphoramidate, *in* M-00384
▷Diethyl dibutylphosphoramidate, *in* D-00134
Diethyl octylphosphoramidate, *in* O-00045
Dihexyl phosphoramidate, *in* D-00432
Dipropyl dipropylphosphoramidate, *in* D-01214

$C_{12}H_{28}NP$
P,P-Dibutyl-*N,N*-diethylphosphinous amide, *in* D-00123
P,P,P-Tributylphosphine imide, T-00374
P,P,P-Tri-*tert*-butylphosphine imide, T-00375

$C_{12}H_{28}NPSe$
P,P-Dibutyl-*N,N*-diethylphosphinoselenoic amide, *in* D-00114

$C_{12}H_{28}OP$
Tetraisopropyldiphosphine; Monooxide, *in* T-00167

$C_{12}H_{28}OP_2S_2$
Dipropylphosphinothioic acid; Anhydride, *in* D-01208

$C_{12}H_{28}O_2P_2$
Tetraisopropyldiphosphine; Dioxide, *in* T-00167

$C_{12}H_{28}O_3P_2$
Diisopropylphosphinic acid; Anhydride, *in* D-00641

Dipropylphosphinic acid; Anhydride, *in* D-01205

$C_{12}H_{28}O_4P_2$
Tetraethyl 1,4-butanediylbisphosphonite, *in* B-00515
Tetraisopropyl hypodiphosphite, T-00169
Tetrapropyl hypodiphosphite, T-00287

$C_{12}H_{28}O_4P_2S_2$
O,O,O′,O′-Tetraisopropyl thiohypophosphate, T-00173
O,O,O′,O′-Tetrapropyl thiohypophosphate, T-00293

$C_{12}H_{28}O_4P_2S_3$
O,O,O′,O′-Tetraisopropyl trithiopyrophosphate, T-00174

$C_{12}H_{28}O_4P_2S_4$
Bis(diisopropyoxyphosphinothioyl) disulfide, B-00184

$C_{12}H_{28}O_5P_2$
Diethyl dibutyldiphosphonate, *in* D-00092
Tetrapropyl pyrophosphite, T-00292

$C_{12}H_{28}O_5P_2S_2$
▷Aspon, A-00144
▷*sym*-Tetraisopropyl dithiopyrophosphate, T-00168

$C_{12}H_{28}O_6P_2$
Tetraethyl 1,4-butanediylbisphosphonate, *in* B-00514
Tetraisopropyl hypophosphate, T-00170
Tetrapropyl hypophosphate, T-00288

$C_{12}H_{28}O_6P_2S_2$
Bis(diisopropoxyphosphinyl) disulfide, B-00183

$C_{12}H_{28}O_7P_2$
Tetraethyl oxybis(1,2′-ethanediyl)-bisphosphonate, *in* O-00094
▷Tetraisopropyl diphosphate, T-00166
▷Tetrapropyl diphosphate, T-00285

$C_{12}H_{28}P^{\oplus}$
Tetrapropylphosphonium(1+), T-00290

$C_{12}H_{28}P_2$
Tetraisopropyldiphosphine, T-00167
Tetrapropyldiphosphine, T-00286

$C_{12}H_{28}P_2S$
Diisopropylphosphinothious acid; Anhydrosulfide, *in* D-00647
Tetraisopropyldiphosphine; Monosulfide, *in* T-00167

$C_{12}H_{28}P_2S_2$
Tetraisopropyldiphosphine; Disulfide, *in* T-00167
Tetrapropyldiphosphine; Disulfide, *in* T-00286

$C_{12}H_{28}P_2S_3$
Diisopropylphosphinodithioic acid; Anhydrosulfide, *in* D-00643
Dipropylphosphinodithioic acid; Anhydrosulfide, *in* D-01207

$C_{12}H_{29}N_2OP$
P-Butyl-*N,N,N′,N′*-tetraethylphosphonic diamide, *in* B-00596
N,N,N′,N′-Tetraisopropylphosphonic diamide, T-00171
Tetraisopropylphosphorodiamidous acid, T-00172
N,N,N′,N′-Tetrapropylphosphonic diamide, T-00289
N,N′,P-Tris(1,1-dimethylethyl)phosphonic diamide, *in* B-00597

$C_{12}H_{29}N_2OPSe$
O-Butyl tetraethylphosphorodiamidoselenoate, *in* T-00061
Se-Butyl tetraethylphosphorodiamidoselenoate, *in* T-00061

$C_{12}H_{29}N_2O_2P$
Butyl *N,N′*-dibutylphosphorodiamidate, *in* D-00146
Tetrapropylphosphorodiamidic acid, T-00291

$C_{12}H_{29}O_2PSSi_2$
Cyclohexylphosphonothioic acid; *O,O*-Bis(trimethylsilyl) ester, *in* C-00264

$C_{12}H_{29}O_4PSi_2$
Bis(trimethylsilyl) (3,3-dimethyl-2-oxobutyl)-phosphonate, *in* D-00783

$C_{12}H_{29}O_5PSi_2$
tert-Butyl [bis(trimethylsilyloxy)phosphinyl]-acetate, *in* B-00434

$C_{12}H_{29}O_6PSi_3$
Trimethylsilyl 2-[bis(trimethylsilyloxy)-phosphinyl]oxy-2-propenoate, *in* P-00366

$C_{12}H_{29}O_8P_3$
Tetraethyl [(ethoxyphosphinylidene)bis(methylene)]bisphosphonate, *in* B-00182

$C_{12}H_{30}NO_3PSi$
Triisopropyl (trimethylsilyl)phosphorimidate, *in* T-00557

$C_{12}H_{30}NPSi$
P,P,P-Triisopropyl-*N*-trimethylsilylphosphine imide, *in* T-00501

$C_{12}H_{30}N_2O_3P_2$
▷Hexaethyldiphosphonic diamide, *in* D-00291

$C_{12}H_{30}N_3OP$
Hexaethylphosphoric triamide, *in* H-00015
N,N′,N″-Tributylphosphoric triamide, *in* T-00394
N,N′,N″-Tris(1,1-dimethylethyl)phosphoric triamide, *in* T-00395
N,N′,N″-Tris(1-methylpropyl)phosphoric triamide, *in* T-00829
N,N′,N″-Tris(2-methylpropyl)phosphoric triamide, *in* T-00830

$C_{12}H_{30}N_3O_6$
2,4,6-Tris(diethylamino)-1,3,5,2,4,6-trioxatriphosphorinane 2,4,6-trioxide, T-00756

$C_{12}H_{30}N_3P$
Hexaethylphosphorous triamide, H-00015
N,N′,N″-Tributylphosphorous triamide, T-00394
N,N′,N″-Tri-*tert*-butylphosphorous triamide, T-00395
N,N′,N″-Tris(1-methylpropyl)phosphorous triamide, T-00829
N,N′,N″-Tris(2-methylpropyl)phosphorous triamide, T-00830

$C_{12}H_{30}N_3PS$
Hexaethylphosphorothioic triamide, *in* H-00015
N,N′,N″-Tributylphosphorothioic triamide, *in* T-00394
N,N′,N″-Tris(1-methylpropyl)phosphorothioic triamide, *in* T-00829
N,N′,N″-Tris(2-methylpropyl)phosphorothioic triamide, *in* T-00830

$C_{12}H_{30}N_3PSSi$
1,3-Di-*tert*-butyl-4,4-dimethyl-1,3,2-diazaphosphasiletidine; 2-Dimethylamino, 2-sulfide, *in* D-00087

$C_{12}H_{30}N_3PSe$
Hexaethylphosphoroselenoic triamide, *in* H-00015

$C_{12}H_{30}N_3PSi$
N,N,4,4-Tetramethyl-1,3-bis(1,1-dimethylethyl)-1,3-diaza-2-phosphacyclobutane-2-amine, *in* D-00087

$C_{12}H_{30}N_4O_2P_2$
1,3-Di-*tert*-butyl-2,4-bis(dimethylamino)-1,3,2,4-diazadiphosphetidine; 2,4-Dioxide, *in* D-00074

$C_{12}H_{30}N_4P_2$
1,3-Di-*tert*-butyl-2,4-bis(dimethylamino)-1,3,2,4-diazadiphosphetidine, D-00074

$C_{12}H_{30}N_4P_2S$
1,3-Di-*tert*-butyl-2,4-bis(dimethylamino)-1,3,2,4-diazadiphosphetidine; 2-Sulfide, *in* D-00074

$C_{12}H_{30}N_4P_2S_2$
1,3-Di-*tert*-butyl-2,4-bis(dimethylamino)-1,3,2,4-diazadiphosphetidine; 2,4-Disulfide, *in* D-00074

$C_{12}H_{30}N_4P_2Se$
1,3-Di-*tert*-butyl-2,4-bis(dimethylamino)-1,3,2,4-diazadiphosphetidine; 2-Selenide, *in* D-00074

$C_{12}H_{30}N_4P_2Se_2$
1,3-Di-*tert*-butyl-2,4-bis(dimethylamino)-1,3,2,4-diazadiphosphetidine; 2,4-Diselenide, *in* D-00074

$C_{12}H_{30}N_4P_2Te$
1,3-Di-*tert*-butyl-2,4-bis(dimethylamino)-1,3,2,4-diazadiphosphetidine; 2-Telluride, *in* D-00074

$C_{12}H_{30}P_4$
Tris(*tert*-butylphosphino)phosphine, T-00737

$C_{12}H_{31}O_2PSSi_2$
O,O-Bis(trimethylsilyl) hexylphosphonothioate, *in* H-00100

$C_{12}H_{31}O_3PSi_2$
Bis(triethylsilyl) phosphonate, B-00407

$C_{12}H_{32}N_4O_{12}P_4$
N,N′,N″,N‴-Tetrakis(dihydroxyphosphinylmethyl)-1,4,7,10-tetraazacyclododecane, T-00175

$C_{12}H_{33}O_4PSi_3$
(1-Hydroxy-1-methylethyl)phosphonic acid; Bis(trimethylsilyl) ester, *O*-trimethylsilyl ether, *in* H-00159
(1-Hydroxypropyl)phosphonic acid; Bis(trimethylsilyl)ester, *O*-Trimethylsilyl ether, *in* H-00190

$C_{12}H_{34}N_6O_2P_2$
N,N′-Dimethyl-*N,N′*-bis[[bis(dimethylamino)-phosphinyl]amino]ethane, D-00713

$C_{12}H_{35}N_2OPSi_3$
Isopropyl *N,N,N′*-tris(trimethylsilyl)-phosphonamidimidate, *in* T-00867

$C_{12}H_{36}B_2F_8N_6OP_2$
Hexakis(dimethylamino)-*μ*-oxodiphosphorus(2+); Bis(tetrafluoroborate), *in* H-00062

$C_{12}H_{36}Ge_6P_4$
2,2,4,4,6,6,8,8,9,9,10,10-Dodecamethyl-1,3,5,7-tetraphospha-2,4,6,8,9,10-hexagermatricyclo[3.3.1.1^{3,7}]decane, D-01263

$C_{12}H_{36}IO_4PSi_4$
Tetrakis(trimethylsilyloxy)phosphonium iodide, *in* T-00864

$C_{12}H_{36}N_6OP_2^{\oplus\oplus}$
Hexakis(dimethylamino)-*μ*-oxodiphosphorus(2+), H-00062

$C_{12}H_{36}N_9P_3$
▷2,2,4,4,6,6-Hexakis(dimethylamino)-2,2,4,4,6,6-hexahydro-1,3,5-triaza-2,4,6-triphosphorine, H-00061

$C_{12}H_{37}N_2OPSi_4$
Trimethylsilyl *N,N,N′*-tris(trimethylsilyl)-phosphonamidimidate, *in* T-00867

$C_{13}H_3F_{10}OPS$
O-Methyl bis(pentafluorophenyl)-phosphinothioate, *in* B-00383

$C_{13}H_3F_{18}OP_2$
7-Methoxy-2,3,5,6,7,8-hexakis(trifluoromethyl)-1,4-diphosphabicyclo[2.2.2]octa-2,5-diene, M-00042

$C_{13}H_9ClNO_3P$
2-Chloro-2,3-dihydro-3-phenyl-4*H*-1,3,2-benzoxazaphosphorin-4-one; 2-Oxide, *in* C-00050

$C_{13}H_9P$
Acridophosphine, A-00031
Phosphanthridene, P-00348

$C_{13}H_{10}BrCl_2O_2PS$
▷Leptophos, L-00001

$C_{13}H_{10}BrCl_2O_3P$
(4-Bromo-2,5-dichlorophenyl)methyl phenylphosphonate, *in* M-00488

$C_{13}H_{10}ClN_2OP$
2-Chloro-2,3-dihydro-3,5-diphenyl-1,3,4,2-oxadiazaphosphole, C-00042

$C_{13}H_{10}ClN_2OPS$
2-Chloro-2,3-dihydro-3,5-diphenyl-1,3,4,2-oxadiazaphosphole; 2-Sulfide, *in* C-00042

$C_{13}H_{10}ClN_2O_2P$
2-Chloro-2,3-dihydro-3,5-diphenyl-1,3,4,2-oxadiazaphosphole; 2-Oxide, *in* C-00042

$C_{13}H_{10}ClOP$
5-Chloro-5,6-dihydrophosphanthridine 5-oxide, *in* D-00504

$C_{13}H_{10}ClP$
5-Chloro-5,10-dihydroacridophosphine, C-00037

$C_{13}H_{10}Cl_3N_2OP$
2,2,2,3-Tetrahydro-3,5-diphenyl-1,3,4,2-oxadiazaphosphole; 2,2,2-Trichloro, *in* T-00104

$C_{13}H_{10}Cl_3O_3P$
Diphenyl (trichloromethyl)phosphonate, *in* T-00411

$C_{13}H_{10}NOP$
Diphenylphosphinic acid; Cyanide, *in* D-01039

$C_{13}H_{10}NOPS$
1,2-Dihydro-1-hydroxy-1-phenyl-4*H*-3,1,2-benzothiazaphosphorin-4-one 1-sulfide, D-00503
Diphenylphosphinic acid; Isothiocyanate, *in* D-01039
Diphenylphosphinothioic acid; Isocyanate, *in* D-01082

$C_{13}H_{10}NOPS_2$
O-Phenyl phenylphosphonisothiocyanatidothioate, *in* P-00227

$C_{13}H_{10}NO_2P$
5,10-Dihydro-11*H*-dibenz[*b*,*e*][1,4]-azaphosphepin-11-one 5-oxide, D-00449
2,3-Dihydro-3-phenyl-4*H*-1,3,2-benzoxazaphosphorin-4-one, D-00556
Diphenylphosphinic acid; Isocyanate, *in* D-01039

$C_{13}H_{10}NO_3P$
9-Acridinephosphonic acid, A-00030
Diphenyl phosphorocyanidate, D-01124
6-Phenanthridinephosphonic acid, P-00060

$C_{13}H_{10}NO_3PS$
O,*O*-Diphenyl phosphorisocyanatidothioate, D-01117
Diphenyl phosphoroisothiocyanatidate, D-01138

$C_{13}H_{10}NO_4P$
Diphenyl phosphorisocyanatidate, D-01116

$C_{13}H_{10}NP$
Diphenylphosphinous acid; Cyanide, *in* D-01089

$C_{13}H_{10}NPS$
Diphenylphosphinothioic acid; Cyanide, *in* D-01082
Diphenylphosphinous acid; Isothiocyanate, *in* D-01089

$C_{13}H_{10}NPS_2$
1,2-Dihydro-1-phenyl-4*H*-3,1,2-benzothiazaphosphorin-4-thione, D-00552
Diphenylphosphinothioic acid; Isothiocyanate, *in* D-01082

$C_{13}H_{11}Cl_2OP$
[(1,1′-Biphenyl)-2-ylmethyl]phosphonic acid; Dichloride, *in* B-00079
Methyl bis(4-chlorophenyl)phosphinite, *in* B-00142

$C_{13}H_{11}Cl_2O_2P$
Methyl bis(4-chlorophenyl)phosphinate, *in* B-00139

$C_{13}H_{11}Cl_2O_2PS$
▷*O*-2,5-Dichlorophenyl *O*-methyl phenylphosphonothioate, *in* M-00243

$C_{13}H_{11}Cl_2O_3P$
Diphenyl (dichloromethyl)phosphonate, *in* D-00175

$C_{13}H_{11}N_2OP$
(Diazomethyl)diphenylphosphine oxide, D-00034

$C_{13}H_{11}OP$
1,3-Dihydro-1-phenyl-2,1-benzoxaphosphole, D-00553
5-Methyl-5*H*-dibenzophosphole; 5-Oxide, *in* M-00115

$C_{13}H_{11}OPS$
1,3-Dihydro-1-phenyl-2,1-benzoxaphosphole; 1-Sulfide, *in* D-00553
2-Phenyl-4*H*-1,3,2-benzoxathiaphosphorin, P-00097

$C_{13}H_{11}OPS_2$
5-Methyl-2-phenyl-1,3,2-benzodithiaphosphole; 2-Oxide, *in* M-00195

$C_{13}H_{11}O_2P$
5,10-Dihydro-5-hydroxyacridophosphine 5-oxide, D-00487
5,6-Dihydro-5-hydroxyphosphanthridine 5-oxide, D-00504
1,3-Dihydro-1-phenoxy-2,1-benzoxaphosphole, D-00546
1,3-Dihydro-1-phenyl-2,1-benzoxaphosphole; 1-Oxide, *in* D-00553
5-Methoxy-5*H*-dibenzophosphole 5-oxide, *in* H-00118
2-Phenyl-4*H*-1,3,2-benzodioxaphosphorin, P-00095

$C_{13}H_{11}O_2PS$
1,3-Dihydro-1-phenoxy-2,1-benzoxaphosphole; 1-Sulfide, *in* D-00546
2-Phenyl-4*H*-1,3,2-benzodioxaphosphorin; 2-Sulfide, *in* P-00095
2-Phenyl-4*H*-1,3,2-benzoxathiaphosphorin; 2-Oxide, *in* P-00097

$C_{13}H_{11}O_2PS_2$
2-Mercapto-4*H*-1,3,2-benzodioxaphosphorin 2-sulfide; 2-Phenylthio (2-Ph ester), *in* M-00006

$C_{13}H_{11}O_3P$
Benzoylphenylphosphinic acid, B-00032
(9*H*-Fluoren-9-yl)phosphonic acid, F-00011
6-Methoxydibenzo[*d*,*f*][1,3,2]dioxaphosphepin, M-00032
2-Phenoxy-4*H*-1,3,2-benzodioxaphosphorin, P-00064
2-Phenyl-4*H*-1,3,2-benzodioxaphosphorin; 2-Oxide, *in* P-00095

$C_{13}H_{11}O_3PS$
O-Phenyl saligenin cyclic thiophosphate, *in* P-00064
S-Phenyl saligenin cyclic thiophosphate, *in* H-00110

$C_{13}H_{11}O_4P$
2,2′-Biphenyldiyl methyl phosphate, *in* M-00032
2-Methyl-2,2′-spirobi[1,3,2-benzodioxaphosphole], M-00397
▷Phenyl saligenin cyclic phosphate, *in* P-00064
(9*H*-Xanthen-9-yl)phosphonic acid, X-00001

$C_{13}H_{11}P$
5,6-Dihydrophosphanthridene, D-00568
5-Methyl-5*H*-dibenzophosphole, M-00115

$C_{13}H_{11}PS$
5-Methyl-5*H*-dibenzophosphole; 5-Sulfide, *in* M-00115

$C_{13}H_{11}PS_2$
5-Methyl-2-phenyl-1,3,2-benzodithiaphosphole, M-00195

$C_{13}H_{11}PSe$
5-Methyl-5*H*-dibenzophosphole; 5-Selenide, *in* M-00115

$C_{13}H_{12}ClN_2OP$
6-Chloro-5,6,7,12-tetrahydro-2,5,7,10-tetramethyldibenzo[*d*,*g*][1,3,2]-diazaphosphocine; 6-Oxide, *in* C-00182

$C_{13}H_{12}ClN_2P$
6-Chloro-5,6,7,12-tetrahydro-2,5,7,10-tetramethyldibenzo[*d*,*g*][1,3,2]-diazaphosphocine, C-00182

$C_{13}H_{12}ClN_2PS$
6-Chloro-5,6,7,12-tetrahydro-2,5,7,10-tetramethyldibenzo[*d*,*g*][1,3,2]-diazaphosphocine; 6-Sulfide, *in* C-00182

$C_{13}H_{12}ClOP$
Benzylphenylphosphinic acid; Chloride, *in* B-00056

$C_{13}H_{12}ClO_3P$
Diphenyl chloromethylphosphonate, *in* C-00106

$C_{13}H_{12}Cl_2NO_3P$
2-Chlorophenyl (4-methoxyphenyl)-phosphoramidochloridate, C-00130

$C_{13}H_{12}Cl_3N_2OP$
P-(Trichloromethyl)-*N*,*N*′-diphenylphosphonic diamide, *in* T-00412

$C_{13}H_{12}IO_3P$
Diphenyl (iodomethyl)phosphonate, *in* I-00013
Methyltri-2-furanylphosphonium iodide, *in* T-00486
Methyl(tri-3-furanyl)phosphonium iodide, *in* T-00487

$C_{13}H_{12}IPS_3$
Methyltri-2-thienylphosphonium iodide, *in* T-00874
Methyltri-3-thienylphosphonium iodide, *in* T-00875

$C_{13}H_{12}NOP$
10,11-Dihydro-1*H*-dibenz[*b*,*e*][1,4]-azaphosphepine 5-oxide, D-00448

$C_{13}H_{12}NOPS$
5,10-Dihydro-10-methoxyphenophosphazine 10-sulfide, *in* D-00502
5,10-Dihydro-10-methylthiophenophosphazine 10-oxide, *in* D-00502

$C_{13}H_{12}NO_2P$
Methyl phenazaphosphinate, *in* P-00061

$C_{13}H_{12}NO_2PS$
3′-Methylspiro[1,3,2-benzodioxaphosphole-2,2(3′*H*)-[1,3,2]benzothiazaphosphole], M-00394

$C_{13}H_{12}NO_3P$
(9,10-Dihydro-9-acridinyl)phosphonic acid, D-00438
3′-Methylspiro[1,3,2-benzodioxaphosphole-2,2′(3′*H*)-[1,3,2]benzoxazaphosphole], M-00395

$C_{13}H_{12}NO_4P$
Diphenyl aminocarbonylphosphonate, *in* C-00003
4-Nitrophenyl methylphenylphosphinate, *in* M-00219
Spiro[1,3,2-benzodioxaphosphole-2,2′(3′*H*)-[1,3,2]benzoxazaphosphole]; 2-Methoxy, *in* S-00008

$C_{13}H_{12}NO_4PS$
▷EPN-Methyl, *in* M-00243

$C_{13}H_{12}NO_5P$
N-(Diphenoxyphosphinyl)carbamic acid, D-00989

$C_{13}H_{13}Cl_2N_2OP$
(Dichloromethyl)phosphonic dianilide, *in* D-00176

$C_{13}H_{13}FNP$
N-Methyl-*P*,*P*-diphenylphosphinimidic fluoride, *in* D-01044

$C_{13}H_{13}N_2OP$
6,7-Dihydro-5*H*-dibenzo[*d*,*f*][1,3,2]-diazaphosphepine; 6-Me, 6-oxide, *in* D-00450

2,2,2,3-Tetrahydro-3,5-diphenyl-1,3,4,2-oxadiazaphosphole, T-00104

$C_{13}H_{13}N_2O_2P$

2,2(3*H*,3'*H*)-Spirobi[1,3,2-benzoxazaphosphole]; 2-Me, *in* S-00020

$C_{13}H_{13}N_2O_4P$

Diphenyl aminocarbonylphosphoramidate, *in* A-00062

$C_{13}H_{13}OP$

2,3-Dihydro-1-methyl-1*H*-benzo[*g*]-phosphindole; 1-Oxide, *in* D-00518
(Diphenylphosphino)methanol, D-01069
Methyldiphenylphosphine; Oxide, *in* M-00135
Methyl diphenylphosphinite, *in* D-01089

$C_{13}H_{13}OPS$

(Diphenylphosphino)methanol; Sulfide, *in* D-01069
O-Methyl diphenylphosphinothioate, *in* D-01082
S-Methyl diphenylphosphinothioate, *in* D-01082
O-Phenyl methylphenylphosphinothioate, *in* M-00225
S-Phenyl methylphenylphosphinothioate, *in* M-00225

$C_{13}H_{13}OPS_2$

S,S-Diphenyl methylphosphonodithioate, *in* M-00301

$C_{13}H_{13}OPSe$

O-Methyl diphenylphosphinoselenoate, *in* D-01078

$C_{13}H_{13}O_2P$

Benzylphenylphosphinic acid, B-00056
Diphenyl methylphosphonite, *in* M-00319
(Diphenylphosphino)methanol; Oxide, *in* D-01069
Methyl diphenylphosphinate, *in* D-01039
Phenyl methylphenylphosphinate, *in* M-00219

$C_{13}H_{13}O_2PS$

O,O-Diphenyl methylphosphonothioate, *in* M-00313
O,S-Diphenyl methylphosphonothioate, *in* M-00313

$C_{13}H_{13}O_3P$

[(1,1'-Biphenyl)-2-ylmethyl]phosphonic acid, B-00079
[(1,1'-Biphenyl)-4-ylmethyl]phosphonic acid, B-00080
Diphenyl methylphosphonate, D-01024
(Hydroxyphenylmethyl)phenylphosphinic acid, H-00173
Methyl phenyl phenylphosphonate, *in* M-00488

$C_{13}H_{13}O_4P$

[2-(Benzyloxy)phenyl]phosphonic acid, *in* H-00179

$C_{13}H_{13}P$

2,3-Dihydro-1-methyl-1*H*-benzo[*g*]-phosphindole, D-00518
4,5-Dihydro-3-methyl-3*H*-benzo[*e*]-phosphindole, *in* D-00441
2,6-Dimethyl-4-phenylphosphorin, D-00819
Methyldiphenylphosphine, M-00135

$C_{13}H_{13}PS$

Methyldiphenylphosphine; Sulfide, *in* M-00135
Methyl diphenylphosphinothioite, *in* D-01085

$C_{13}H_{13}PS_2$

Diphenyl methylphosphonodithioite, *in* M-00302
Methyl diphenylphosphinodithioate, *in* D-01065

$C_{13}H_{13}PS_3$

Diphenyl methylphosphonotrithioate, *in* M-00318

$C_{13}H_{13}PSe$

Methyldiphenylphosphine; Selenide, *in* M-00135
Methyl diphenylphosphinoselenoite, *in* D-01081

$C_{13}H_{14}ClN_2OP$

Chloromethylphosphonic dianilide, *in* C-00107

$C_{13}H_{14}ClN_2PS$

P-Chloromethyl-*N,N'*-diphenylphosphonothioic diamide, *in* C-00114

$C_{13}H_{14}F_6NO_3PS$

2,2,2,3-Tetrahydro-5-phenyl-3,3-bis(trifluoromethyl)-1,4,2-thiazaphosphole; 2,2,2-Trimethoxy, *in* T-00129

$C_{13}H_{14}F_6NO_4P$

2,2,2,3-Tetrahydro-2,2,2-trimethoxy-5-phenyl-3,3-bis(trifluoromethyl)-1,4,2-oxazaphosphole, T-00158

$C_{13}H_{14}NOP$

(Aminomethyl)diphenylphosphine oxide, A-00085
N-Methyl-*P,P*-diphenylphosphinic amide, *in* D-01040
P-Methyl-*N,P*-diphenylphosphinic amide, *in* M-00223

$C_{13}H_{14}NO_2P$

(Aminophenylmethyl)phenylphosphinic acid, A-00102
Benzyl *P*-phenylphosphonamidate, *in* P-00218
Methyl *N,P*-diphenylphosphonamidate, *in* D-01103
P-(4-Methylphenyl)-*N*-phenylphosphonamidic acid, M-00214
Phenyl *P*-methyl-*N*-phenylphosphonamidate, *in* M-00231

$C_{13}H_{14}NO_3P$

Diphenyl methylphosphoramidate, *in* M-00326

$C_{13}H_{14}NO_4PS$

Methyl *P*-phenyl-*N*-phenylsulfonylphosphonamidate, *in* P-00191

$C_{13}H_{14}NPS$

N-Methyl-*P,P*-diphenylphosphinothioic amide, *in* D-01083

$C_{13}H_{14}N_3O_3P$

Diphenyl (aminoiminomethyl)phosphoramidate, *in* P-00370

$C_{13}H_{14}O_4P_2$

Methylenebis[phenylphosphinic acid], M-00142

$C_{13}H_{15}ClNOP$

(Aminomethyl)diphenylphosphine oxide; B,HCl, *in* A-00085

$C_{13}H_{15}Cl_4O_5P$

4,5,6,7-Tetrachloro-4',4',5',5'-tetramethyl-spiro[1,3,2-benzodioxaphosphole-2,2'-[1,3,2]-dioxaphospholane]; 2-Methoxy, *in* T-00036

$C_{13}H_{15}N_2OPS$

S-Methyl *N,N'*-diphenyl-phosphorodiamidothioate, *in* D-01129
O-Phenyl *P*-methyl-N^2-phenyl-phosphonohydrazidothioate, *in* M-00242

$C_{13}H_{15}N_2O_2P$

Methyl *N,N'*-diphenylphosphorodiamidate, *in* D-01125
Phenyl *P*-methyl-N^2-phenyl-phosphonohydrazidate, *in* M-00241

$C_{13}H_{15}N_2O_5$

P-Methyl-*N,N'*-diphenylphosphonic diamide, *in* M-00289

$C_{13}H_{15}N_2PS$

P-Methyl-*N,N'*-diphenylphosphonothioic diamide, *in* M-00314

$C_{13}H_{15}OP$

4,5-Dimethyl-2-oxa-3-phenyl-1-phosphabicyclo[2.2.1]hept-5-ene, D-00781
8-Phenyl-8-phosphabicyclo[3.2.1]octan-3-one, P-00199
8-Phenyl-8-phosphabicyclo[3.2.1]oct-6-ene; Oxide, *in* P-00200
3-Phenyl-3-phosphabicyclo[3.2.1]oct-6-ene 3-oxide, *in* P-00198
3,3*a*,4,5-Tetrahydro-3-methyl-2*H*-benzo[*e*]-phosphindole 3-oxide, T-00112

$C_{13}H_{15}OPS$

4,5-Dimethyl-2-oxa-3-phenyl-1-phosphabicyclo[2.2.1]hept-5-ene; 1-Sulfide, *in* D-00781

$C_{13}H_{15}O_2P$

8-Phenyl-8-phosphabicyclo[3.2.1]octan-3-one; Oxide, *in* P-00199

$C_{13}H_{15}O_5P$

2,2-Dihydro-2,2,2-trimethoxynaphtho[1,2-*d*]-1,3,2-dioxaphosphole, D-00594

$C_{13}H_{15}P$

8-Phenyl-8-phosphabicyclo[3.2.1]oct-6-ene, P-00200

$C_{13}H_{16}ClO_4P$

4,5-Dihydro-1-methyl-1-phenyl-1*H*-phosphepinium perchlorate, *in* D-00565

$C_{13}H_{16}IP$

1,3,4-Trimethyl-1-phenyl-1*H*-phospholium iodide, *in* D-00807

$C_{13}H_{16}NO_4PS$

▷Isoxathion, I-00068

$C_{13}H_{16}NO_5P$

Diethyl [(1,3-dihydro-1,3-dioxo-2*H*-isoindol-2-yl)methyl]phosphonate, *in* P-00426

$C_{13}H_{17}N_2O_4P$

Octahydro-2-(4-nitrophenoxy)-2*H*-1,3,2-benzoxazaphosphorine, O-00020

$C_{13}H_{17}N_2O_5P$

Octahydro-2-(4-nitrophenoxy)-2*H*-1,3,2-benzoxazaphosphorine; 2-Oxide, *in* O-00020

$C_{13}H_{17}N_2P$

9-Phenyl-1,5-diaza-9-phosphatricyclo[3.3.1.1^{3,7}]decane, P-00106

$C_{13}H_{17}OP$

1,4-Dimethyl-2-phenyl-7-oxa-2-phosphabicyclo[2.2.1]heptane, D-00796
2,5-Dimethyl-1-phenylphosphorinan-4-one, D-00821
3-Phenyl-3-phosphabicyclo[3.2.1]octane; 3-Oxide, *in* P-00198

$C_{13}H_{17}OPS$

2,5-Dimethyl-1-phenylphosphorinan-4-one; 1-Sulfide, *in* D-00821

$C_{13}H_{17}OPSe$

2,5-Dimethyl-1-phenylphosphorinan-4-one; 1-Selenide, *in* D-00821

$C_{13}H_{17}O_2P$

Diethyl 1*H*-1-indenylphosphonite, *in* I-00006
1,4-Dimethyl-2-phenyl-7-oxa-2-phosphabicyclo[2.2.1]heptane; 2-Oxide, *in* D-00796
2,5-Dimethyl-1-phenylphosphorinan-4-one; 1-Oxide, *in* D-00821

$C_{13}H_{17}O_3P$

Diethyl 1*H*-inden-2-ylphosphonate, *in* I-00007
Diethyl (1-phenyl-1,2-propadienyl)phosphonate, *in* P-00298

$C_{13}H_{17}O_4P$

2-Hydroxyhexahydro-4*H*-1,3,2-benzodioxaphosphorin 2-oxide; 2-Phenoxy (2-Ph ester), *in* H-00152

$C_{13}H_{17}P$

3-Phenyl-3-phosphabicyclo[3.2.1]octane, P-00198

$C_{13}H_{18}ClO_2P$

Cyclohexyl benzylphosphonochloridate, *in* B-00063

$C_{13}H_{18}FO_2P$

Cyclohexyl benzylphosphonofluoridate, *in* B-00066

$C_{13}H_{18}IP$

2,5-Dihydro-1,3,4-trimethyl-1-phenyl-1*H*-phospholium bromide, *in* D-00468

$C_{13}H_{18}NO_2P$
1-Benzyl-2-ethoxyhexahydro-1,2-azaphosphorine 2-oxide, *in* B-00048

$C_{13}H_{18}N_2OP$
N-(*tert*-Butylphenylphosphinyl)imidazole, *in* B-00579

$C_{13}H_{18}OP$
1-Phenylphosphocane; Oxide, *in* P-00215
1,3,4,5-Tetrahydro-4,5-dimethyl-2,5-methano-2*H*-2-benzophosphepin; Oxide, *in* T-00096

$C_{13}H_{18}P$
1-Phenylphosphocane, P-00215
1,3,4,5-Tetrahydro-4,5-dimethyl-2,5-methano-2*H*-2-benzophosphepin, T-00096

$C_{13}H_{18}PI$
1-Ethyl-2,5-dihydro-3-methyl-1-phenyl-1*H*-phospholium iodide, *in* D-00528

$C_{13}H_{19}ClNOP$
5-*tert*-Butyl-2-chlorotetrahydro-3-phenyl-2*H*-1,3,2-oxazaphosphorine, B-00549

$C_{13}H_{19}ClNO_3PS_2$
Anilofos, A-00134

$C_{13}H_{19}NO_8P_2$
2,2,7,7-Tetrahydro-3,4,5,6-tetrakis(methoxycarbonyl)-1*H*-1,2,7-azadiphosphepine, T-00138

$C_{13}H_{19}N_2O_3P$
Diisopropyl (diazophenylmethyl)phosphonate, *in* D-00037

$C_{13}H_{19}N_2O_4PS$
O-Methyl *O*-4-nitrophenyl cyclohexylphosphoramidothioate, M-00179

$C_{13}H_{19}N_2O_6P$
Alafosfalin; *N*-Benzyloxycarbonyl, *in* A-00049

$C_{13}H_{19}N_2P$
3-Methyl-2-phenyl-1,3-diaza-2-phosphabicyclo[3.3.1]nonane, M-00196

$C_{13}H_{19}N_4O_{12}P$
N-(5-Amino-1-*β*-D-ribofuranosylimidazole-4-carbonyl)-L-aspartic acid 5′-phosphate, A-00129

$C_{13}H_{19}OP$
2,5-Dimethyl-1-phenylphosphorinan-4-ol, D-00820

$C_{13}H_{19}OPS$
2,5-Dimethyl-1-phenylphosphorinan-4-ol; 1-Sulfide, *in* D-00820

$C_{13}H_{19}O_2P$
5-*tert*-Butyl-2-phenyl-1,3,2-dioxaphosphorinane, B-00577

$C_{13}H_{19}O_3P$
2-*tert*-Butyltrimethylene phenylphosphonate, *in* B-00577
Diethyl 3-phenyl-2-propenylphosphonate, *in* P-00302

$C_{13}H_{19}O_4P$
Diethyl (3,4-dihydro-1*H*-2-benzopyran-1-yl)-phosphonate, *in* D-00443
Diisopropyl benzoylphosphonate, *in* B-00034
Ethyl 3-(ethoxyphenylphosphinyl)propanoate, *in* H-00178
1,4,6,12-Tetraoxa-5-phospha(5-P^V)spiro[4.7]-dodecane; 5-Ph, *in* T-00243
(9*H*-Xanthen-9-yl)phosphonic acid; Octahydro, *in* X-00001

$C_{13}H_{19}O_5P$
Diethyl [(acetyloxy)phenylmethyl]phosphonate, *in* A-00020
Diethyl (benzyloxycarbonylmethyl)-phosphonate, *in* D-00241
Ethyl [2-(diethoxyphosphinyl)]benzoate, *in* P-00378
Ethyl [3-(diethoxyphosphinyl)]benzoate, *in* P-00379
Ethyl [4-(diethoxyphosphinyl)]benzoate, *in* P-00380
Phenyl (2-diethoxyphosphinyl)propanoate, *in* D-00258
4′,4′,5′,5′-Tetramethylspiro[1,3,2-benzodioxaphosphole-2,2′-[1,3,2]-dioxaphospholane]; 2-Methoxy, *in* T-00230

$C_{13}H_{19}O_6P$
Ethyl 2-(diethoxyphosphinyl)oxybenzoate, *in* P-00391

$C_{13}H_{20}BrP$
Methylphenyl(2-propenyl)propylphosphonium bromide, *in* M-00265

$C_{13}H_{20}ClN_2O_2P$
2-[(2-Chloroethyl)amino]tetrahydro-2*H*-1,3,2-oxazaphosphorine 2-oxide; 3-[(*S*)-1-Phenylethyl], *in* C-00072

$C_{13}H_{20}F_6O_6P_2$
Tetraethyl (3,3,4,4,5,5-hexafluoro-1-cyclopentene-1,2-diyl)bisphosphonate, *in* H-00018

$C_{13}H_{20}NO_2P$
Methyl *N*-cyclohexyl-*P*-phenyl-phosphonamidate, *in* C-00251

$C_{13}H_{20}NO_3P$
Diethyl 1-benzylideneaminophosphonate, *in* B-00049

$C_{13}H_{20}NO_3PS$
O,O-Diisopropyl benzoylphosphoramidothioate, *in* B-00038

$C_{13}H_{20}NO_4P$
Diisopropyl benzoylphosphoramidate, *in* B-00036
4′,4′,5′,5′-Tetramethylspiro[1,3,2-benzoxazaphosphole-2(3*H*),2′-[1,3,2]-dioxaphospholane]; 2-Methoxy, *in* T-00231
Triethyl benzoylphosphorimidate, *in* B-00039

$C_{13}H_{20}NO_5P$
Ethyl 2-[(diethoxyphosphinyl)amino]benzoate, *in* C-00004
Triethyl 3-(*N*-phosphono)benzoate, *in* C-00005

$C_{13}H_{20}NPS$
P-Cyclohexyl-*P*-methyl-*N*-phenyl-phosphinothioic amide, *in* C-00248

$C_{13}H_{20}P_2$
2,2,4,4-Tetrahydro-2,2,4,4-tetramethyl-1*H*-2,4-benzodiphosphepin, *in* T-00087

$C_{13}H_{21}INP$
2-Ethyl-2,3,4,5-tetrahydro-2-methyl-1*H*-1,3-benzazaphosphepine; B,MeI, *in* E-00188

$C_{13}H_{21}N_2OP$
Methyl tetraisopropylphosphorodiamidite, *in* T-00172

$C_{13}H_{21}N_2O_4P$
Diisopropyl (phenylamino)-carbonylphosphoramidate, *in* P-00081

$C_{13}H_{21}N_2O_4PS$
▷Butamifos, B-00513

$C_{13}H_{21}OPS_2$
S,S-Dipropyl benzylphosphonodithioate, *in* B-00065

$C_{13}H_{21}O_2P$
Diisopropyl benzylphosphonite, *in* B-00068
Diisopropyl *p*-tolylphosphonite, *in* M-00248

$C_{13}H_{21}O_2PS_2$
▷Conen, C-00204

$C_{13}H_{21}O_3P$
Diethyl (3-phenylpropyl)phosphonate, *in* P-00305
Diethyl(2,4,6-trimethylphenyl)phosphonate, *in* T-00527

$C_{13}H_{21}O_3PS$
O-Benzyl *O,O*-diisopropyl phosphorothioate, B-00044
▷*S*-Benzyl *O,O*-diisopropyl phosphorothioate, B-00045

$C_{13}H_{21}O_4P$
Bis(1-methylethyl) (hydroxyphenylmethyl)-phosphonate, *in* H-00174
2,2,2,3-Tetrahydro-1,2-benzoxaphosphole; 2,2,2-Triethoxy, *in* T-00088

$C_{13}H_{21}O_4PS$
▷Propaphos, P-00451

$C_{13}H_{22}AsBr_2P$
2,3,4,5-Tetrahydro-1,1,5,5-tetramethyl-1*H*-1,5-benzophospharsepinium dibromide, *in* T-00092

$C_{13}H_{22}Br_2P_2$
2,3,4,5-Tetrahydro-2,2,4,4-tetramethyl-1*H*-2,4-benzodiphosphepinium(2+); Dibromide, *in* T-00139

$C_{13}H_{22}IO_2P$
Bis(3-hydroxypropyl)methylphenyl-phosphonium iodide, *in* B-00286

$C_{13}H_{22}IP$
Diisopropylmethylphenylphosphonium iodide, *in* D-00638

$C_{13}H_{22}NOPS$
S-Isopropyl *N,N*-diethyl-*P*-phenyl-phosphonamidothioate, *in* D-00309

$C_{13}H_{22}NO_3P$
Benzyl triethylphosphonamidate, *in* T-00446
Diisopropyl benzylphosphoramidate, *in* B-00069
Diisopropyl [(phenylamino)methyl]-phosphonate, *in* P-00084

$C_{13}H_{22}NO_3PS$
▷Fenamiphos, F-00002

$C_{13}H_{22}NO_5PS$
Triethyl [(4-methylphenyl)sulfonyl]-phosphorimidate, *in* M-00268

$C_{13}H_{22}NPS_2$
Isopropyl *N,N*-diethyl-*P*-phenyl-phosphonamidodithioate, *in* D-00308

$C_{13}H_{22}O_6P_2$
▷Tetraallyl methylenediphosphonate, *in* M-00022

$C_{13}H_{22}P_2^{\oplus\oplus}$
2,3,4,5-Tetrahydro-2,2,4,4-tetramethyl-1*H*-2,4-benzodiphosphepinium(2+), T-00139

$C_{13}H_{23}O_3PSi_2$
Bis(trimethylsilyl) benzoylphosphonate, *in* B-00034

$C_{13}H_{24}Cl_3N_2OP$
P-(Trichloromethyl)-*N,N*′-dicyclohexylphosphonic diamide, *in* T-00412

$C_{13}H_{24}NO_3PSi$
Diethyl (trimethylsilylphenyl)phosphoramidate, *in* D-00312

$C_{13}H_{24}N_3OP$
N,N,N′,*N*′-Tetraethyl-*P*-2-pyridinylphosphonic diamide, *in* P-00503

$C_{13}H_{24}N_3O_3PS$
▷Pirimiphos-ethyl, P-00434

$C_{13}H_{25}O_2P$
Dibutyl 2,4-cyclopentadienylphosphonite, *in* C-00275
Methyl dicyclohexylphosphinate, *in* D-00211

$C_{13}H_{25}O_3PSi_2$
Bis(trimethylsilyl) benzylphosphonate, *in* B-00060
Bis(trimethylsilyl)(3-methylphenyl)-phosphonate, *in* M-00236
Bis(trimethylsilyl)(4-methylphenyl)-phosphonate, *in* M-00237

$C_{13}H_{25}O_4P$
Diisopropyl (cyclohexylcarbonyl)phosphonate, *in* C-00239

$C_{13}H_{25}O_4PSi_2$
Bis(trimethylsilyl)(2-methoxyphenyl)-phosphonate, *in* M-00070

$C_{13}H_{25}PS_2$
Methyl dicyclohexylphosphinodithioate, *in* D-00212

$C_{13}H_{26}NO_3P$
Diethyl 2-(cyclohexylamino)-2-methylethenyl phosphonate, D-00276

$C_{13}H_{26}N_2OP$
(2-Methyl-1,3-butadienyl)phosphonic bis(diethylamide), *in* M-00096

$C_{13}H_{27}IN_3O_3P$
Methyltris(morpholino)phosphonium iodide, *in* T-00569

$C_{13}H_{27}N_2PS$
N,N,N′,N′-Tetraethyl-*P*-(2-methyl-1,3-butadienyl)phosphonothioic diamide, *in* M-00097

$C_{13}H_{27}OP_3$
Triphosphetan-4-one; 1,2,3-Tri-*tert*-butyl, *in* T-00703

$C_{13}H_{27}O_2P$
Dibutyl cyclopentylphosphonite, *in* C-00281

$C_{13}H_{27}O_2PS$
O,O-Dibutyl cyclopentylphosphonothioate, *in* C-00280

$C_{13}H_{27}O_4P$
Diethyl (2-oxononyl)phosphonate, *in* O-00063
2,2,3,3,7,7,8,8-Octamethyl-1,4,6,9-tetraoxa-5-phospha(5-P^V)spiro[4.4]nonane; 5-Me, *in* O-00032

$C_{13}H_{28}O_8P_2$
Ethyl 2,3-bis(diethoxyphosphinyl)propanoate, *in* B-00173
Ethyl 3,3-bis(diethoxyphosphinyl)propanoate, *in* D-01187
Methyl 4,4-bis(diethoxyphosphinyl)butanoate, *in* D-01183

$C_{13}H_{29}NO_7P_2$
Tetraethyl (4-morpholinylmethylene)-bisphosphonate, *in* M-00520

$C_{13}H_{29}O_3P$
Diisopropyl heptylphosphonate, *in* H-00006

$C_{13}H_{29}P$
Tri-*tert*-butylmethylenephosphorane, T-00365

$C_{13}H_{30}BrP$
Tri-*tert*-butylmethylphosphonium bromide, *in* T-00373

$C_{13}H_{30}IP$
Methyltris(1-methylpropyl)phosphonium iodide, *in* T-00823
Methyltris(2-methylpropyl)phosphonium iodide, *in* T-00824
▷Tributylmethylphosphonium iodide, *in* T-00372
Tri-*tert*-butylmethylphosphonium iodide, *in* T-00373

$C_{13}H_{30}NO_3P$
Butyl *P*-(butoxymethyl)-*N,N*-diethylphosphonamidate, *in* B-00536

$C_{13}H_{30}N_2P_2$
1-(Di-*tert*-butyldiphosphinylidene)-methanediamine; *N,N,N′,N′*-Tetra-Me, *in* D-00090

$C_{13}H_{30}O_4P_2$
Tetraethyl 1,5-pentanediylbisphosphonite, *in* P-00033

$C_{13}H_{30}O_6P_2$
Tetraethyl 1,5-pentanediylbisphosphonate, *in* P-00032
Tetraethyl pentylidenebisphosphonate, *in* P-00031
Tetrakis(1-methylethyl) methyl-enebisphosphonate, *in* M-00022

$C_{13}H_{31}NO_6P_2$
Tetraisopropyl methyliminodiphosphate, *in* M-00156
Triisopropyl (diethoxyphosphinyl)-phosphorimidate, *in* D-00257

$C_{13}H_{31}N_2O_2P$
Methyl tetrapropylphosphorodiamidate, *in* T-00291

$C_{13}H_{33}O_2PSSi_2$
Heptylphosphonothioic acid; *O,O*-Bis(trimethyl-silyl) ester, *in* H-00007

$C_{13}H_{33}O_4PSi_2$
Bis(triethylsilyl) (hydroxymethyl)phosphonate, *in* H-00162

$C_{13}H_{37}N_2OPSi_3$
tert-Butyl *N,N,N′*-tris(trimethylsilyl)-phosphonamidimidate, *in* T-00867

$C_{13}H_{39}N_2PSi_4$
P-Methyl-*N,N,N′,N′*-tetrakis(trimethylsilyl)-phosphonous diamide, *in* M-00320

$C_{14}H_5F_{10}OPS$
O-Ethyl bis(pentafluorophenyl)-phosphinothioate, *in* B-00383

$C_{14}H_5F_{12}NP_2$
7-Phenyl-1,3,4,6-tetrakis(trifluoromethyl)-7-aza-2,5-diphosphatetracyclo[4.1.0.0^{2,4}.0^{3,5}]-heptane, P-00321

$C_{14}H_5F_{12}PS_4$
2,3,7,8-Tetrakis(trifluoromethyl)-1,4,6,9-tetrathia-5-phospha(5-P^V)spiro[4.4]nona-2,7-diene; 5-Ph, *in* T-00189

$C_{14}H_8Cl_3O_3P$
Bis(4-chlorocarbonylphenyl)phosphinic chloride, *in* B-00118

$C_{14}H_9O_2P$
1*H*-Isophosphindole-1,3(2*H*)-dione; 2-Ph, *in* I-00026

$C_{14}H_9O_2PS$
1*H*-Isophosphindole-1,3(2*H*)-dione; 2-Ph, 2-sulfide, *in* I-00026

$C_{14}H_9O_4P$
1,1′(3*H*,3′*H*)-Spirobi[2,1-benzoxaphosphole]-3,3′-dione, S-00018

$C_{14}H_9O_5P$
1-Hydroxy-1,1′(3*H*,3′*H*)-spirobi[2,1-benzoxaphosphole]-3,3′-dione, H-00194

$C_{14}H_{10}ClOP$
5-Chloro-5*H*-dibenzo[*b,f*]phosphepin 5-oxide, *in* H-00117

$C_{14}H_{10}ClO_2P$
2-Chloro-4,5-diphenyl-1,3,2-dioxaphosphole, C-00070

$C_{14}H_{10}Cl_2N_2O_8P$
Bis(4-nitrophenyl)(dichloromethyl)-phosphonate, *in* D-00175

$C_{14}H_{10}Cl_6O_3P_2$
Phenyl(trichloromethyl)phosphinic acid; Anhydride, *in* P-00327

$C_{14}H_{10}F_6O_3P_2$
Phenyl(trifluoromethyl)phosphinic acid; Anhydride, *in* P-00329

$C_{14}H_{10}F_{12}P_2$
3*a*,4,7,7*a*-Tetrahydro-5,6-dimethyl-2,3*a*,7*a*,8-tetrakis(trifluoromethyl)-1,2,3-metheno-2*H*-benzodiphosphole, T-00100

$C_{14}H_{11}Cl_4O_4P$
Bis(2,4-dichlorophenyl) ethyl phosphate, B-00162

$C_{14}H_{11}N_2O_3P$
Diazo(diphenylphosphinyl)acetic acid, D-00030

$C_{14}H_{11}N_2P$
3,5-Diphenyl-1*H*-1,2,4-diazaphosphole, D-01000

$C_{14}H_{11}N_4O_2P$
Bis(1*H*-benzimidazol-1-yl)phosphinic acid, B-00097

$C_{14}H_{11}OP$
Diphenylphosphinylacetylene, *in* E-00201
2-Phenyl-2*H*-isophosphindole 2-oxide, P-00169
9-Phenyl-9-phosphatricyclo[4.2.1.0^{2,5}]nona-3,7-diene; *endo-anti*-9-Oxide, *in* P-00203
1-Phenyl-1*H*-phosphindole; 1-Oxide, *in* P-00207

$C_{14}H_{11}O_2P$
2,2-Dihydrophenanthro[9,10-*d*]-1,3,2-dioxaphosphole, D-00541
5-Hydroxy-5*H*-dibenzo[*b,f*]phosphepin 5-oxide, H-00117
Phenyl(phenylethynyl)phosphinic acid, P-00187

$C_{14}H_{11}O_3P$
2-Hydroxy-2*H*-1,2-benzoxaphosphorin 2-oxide; 2-Phenoxy (Ph ester), *in* H-00112

$C_{14}H_{11}O_4P$
2-Hydroxy-4,5-diphenyl-1,3,2-dioxaphosphole 2-oxide, H-00139

$C_{14}H_{11}O_6P$
Bis(2-carboxyphenyl)phosphinic acid, B-00117
Bis(4-carboxyphenyl)phosphinic acid, B-00118
Dibenzoyl phosphate, D-00042

$C_{14}H_{11}P$
Ethynyldiphenylphosphine, E-00201
10-Methylacridophosphine, M-00087
9-Phenyl-9-phosphatricyclo[4.2.1.0^{2,5}]nona-3,7-diene, P-00203
1-Phenyl-1*H*-phosphindole, P-00207

$C_{14}H_{12}BrCl_2O_2PS$
Leptophos-Ethyl, *in* E-00114

$C_{14}H_{12}ClN_2O_7P$
Bis[(2-nitrophenyl)methyl] phosphorochloridate, *in* B-00366

$C_{14}H_{12}ClN_3O_7P$
Bis[(4-nitrophenyl)methyl] phosphorochloridate, *in* B-00367

$C_{14}H_{12}ClOP$
5-Chloro-10,11-dihydro-5*H*-dibenzo[*b,f*]-phosphepin 5-oxide, *in* D-00492

$C_{14}H_{12}ClO_2P$
2-Chloro-4,5-diphenyl-1,3,2-dioxaphospholane, C-00069

$C_{14}H_{12}ClO_3P$
2-Chloro-4,5-diphenyl-1,3,2-dioxaphospholane; 2-Oxide, *in* C-00069

$C_{14}H_{12}ClP$
5-Chloro-5,10-dihydro-10-methyl-acridophosphine, *in* D-00515

$C_{14}H_{12}NOP$
(Diphenylphosphinyl)acetonitrile, *in* D-01046
2,2,2,3-Tetrahydrophenanthro[9,10-*d*]-1,3,2-oxazaphosphole, T-00125

$C_{14}H_{12}NOPS_3$
1,2-Dihydro-1-phenyl-4*H*-3,1,2-benzothiazaphosphorin-4-thione; 2-Methoxy, 2-sulfide, *in* D-00552

$C_{14}H_{12}NO_2P$
5,10-Dihydro-11*H*-dibenz[*b,e*][1,4]-azaphosphepin-11-one 5-oxide; 5-Me, *in* D-00449

$C_{14}H_{12}NO_3P$
2,3-Dihydro-3-phenyl-4*H*-1,3,2-benzoxazaphosphorin-4-one; 2-Methoxy, *in* D-00556

$C_{14}H_{12}NP$
(Diphenylphosphino)acetonitrile, D-01046

$C_{14}H_{12}NPS$
(Diphenylphosphinothioyl)acetonitrile, *in* D-01046

$C_{14}H_{12}O_4P_2$
4,5,9,10-Tetrahydro-4,9-dihydroxyphosphorino[2,3,4,5-*lmn*]-phosphanthridine 4,9-dioxide, T-00090

$C_{14}H_{13}Br_2O_2P$
Ethyl bis(2-bromophenyl)phosphinate, *in* B-00112

$C_{14}H_{13}Br_2O_3P$
Bis[(4-bromophenyl)methyl] phosphonate, B-00110

$C_{14}H_{13}Br_2O_4P$
Bis[(4-bromophenyl)methyl] phosphate, B-00109

$C_{14}H_{13}Cl_2N_2O_2PS$
▷O,O-Bis(4-chlorophenyl) (1-iminoethyl)-phosphoramidothioate, *in* I-00005

$C_{14}H_{13}Cl_2OP$
Ethyl bis(2-chlorophenyl)phosphinate, *in* B-00140
Ethyl bis(3-chlorophenyl)phosphinite, *in* B-00141
Ethyl bis(4-chlorophenyl)phosphinite, *in* B-00142

$C_{14}H_{13}Cl_2O_2P$
Ethyl bis(3-chlorophenyl)phosphinate, *in* B-00138

$C_{14}H_{13}Cl_2O_2PS$
▷S-Seven, S-00004

$C_{14}H_{13}Cl_2O_3P$
Bis[(4-chlorophenyl)methyl] phosphonate, B-00130

$C_{14}H_{13}N_2OP$
Phenylphosphonic acid; Di-1-pyrrolide, *in* P-00221

$C_{14}H_{13}N_2O_7PS$
▷O-Ethyl O,O-bis(4-nitrophenyl) phosphorothioate, E-00054

$C_{14}H_{13}N_2O_8P$
Bis[(2-nitrophenyl)methyl] phosphate, B-00366
Bis[(4-nitrophenyl)methyl] phosphate, B-00367

$C_{14}H_{13}N_2O_{11}P$
Bis[(4-nitrophenyl)methyl]phosphonate, B-00368

$C_{14}H_{13}OP$
Acetyldiphenylphosphine, A-00016
10,11-Dihydro-5*H*-dibenzo[*b,f*]phosphepin; Oxide, *in* D-00451
5,6-Dihydro-5-methylphosphanthridene 5-oxide, *in* D-00568
3,4-Dihydro-3-phenyl-1*H*-2,3-benzoxaphosphorin, D-00554
2,3-Dihydro-2-phenyl-1*H*-isophosphindole; Oxide, *in* D-00562
2,3-Dihydro-1*H*-phosphindole 1-oxide; 1-Ph, *in* D-00570
Diphenylvinylphosphine; Oxide, *in* D-01168
9-Phenyl-9-phosphabicyclo[4.2.1]nona-2,4,7-triene; Oxide, *in* P-00196
9-Phenyl-9-phosphapentacyclo[4.3.0.0^{2,5}.0^{3,8}.0^{4,7}]nonane; 9-Oxide, *in* P-00201
9-Phenyl-9-phosphatricyclo[3.3.1.0^{2,8}]nona-3,6-diene; 9-Oxide, *in* P-00202

$C_{14}H_{13}OPS$
2,2-Dihydro-4,5-diphenyl-1,3,2-oxathiaphosphole, D-00480
3,4-Dihydro-3-phenyl-1*H*-2,3-benzoxaphosphorin; 3-Sulfide, *in* D-00554
4,5-Dihydro-2-phenyl-1,3,2-benzoxathiaphosphepin, D-00555

$C_{14}H_{13}OPS_2$
S-Acetyl diphenylphosphinodithioate, *in* D-01065

$C_{14}H_{13}O_2P$
10,11-Dihydro-5-hydroxy-5*H*-dibenzo[*b,f*]-phosphepin 5-oxide, D-00492
5,10-Dihydro-5-hydroxy-10-methyl-acridophosphine 5-oxide, *in* D-00515
5,10-Dihydro-5-methoxyacridophosphine 5-oxide, *in* D-00487
1,5-Dihydro-3-phenyl-2,4,3-benzodioxaphosphepin, D-00550
3,4-Dihydro-3-phenyl-1*H*-2,3-benzoxaphosphorin; 3-Oxide, *in* D-00554
(Diphenylphosphino)acetic acid, D-01045
1-(Diphenylphosphinyl)ethanone, *in* A-00016
5-Ethoxy-5*H*-dibenzophosphole 5-oxide, *in* H-00118
5-Methoxy-5,6-dihydrophosphanthridene 5-oxide, *in* D-00504
Phenyl ethenylphenylphosphinate, *in* P-00337
Phenyl(2-phenylethenyl)phosphinic acid, P-00186
1,1′(3*H*,3′*H*)-Spirobi[2,1-benzoxaphosphole], S-00017

$C_{14}H_{13}O_2PS$
1,5-Dihydro-2-phenoxy-2,4,3-benzodioxaphosphepin; 3-Sulfide, *in* D-00544
4,5-Dihydro-2-phenyl-1,3,2-benzoxathiaphosphepin; 2-Oxide, *in* D-00555
(Diphenylphosphinothioyl)acetic acid, D-01086
10-Hydroxy-2,8-dimethyl-10*H*-phenothiaphosphine 10-oxide, H-00127

$C_{14}H_{13}O_3P$
Acetic diphenylphosphinic anhydride, *in* D-01039
1,5-Dihydro-2-phenoxy-2,4,3-benzodioxaphosphepin, D-00544
(Diphenylphosphinyl)acetic acid, D-01093
10-Hydroxy-2,8-dimethyl-10*H*-phenoxaphosphine 10-oxide, H-00128
Methyl benzoylphenylphosphinate, *in* B-00032
2-Phenoxy-1,5-dihydro-2,4,3-benzodioxaphosphepin, P-00066

$C_{14}H_{13}O_3PS$
2-Phenoxy-1,5-dihydro-2,4,3-benzodioxaphosphepin; Sulfide, *in* P-00066
Spiro[1,3,2-benzoxathiaphosphole-2,2′-[1,3,2]-dioxaphospholane]; 2-Ph, *in* S-00011

$C_{14}H_{13}O_3PSe$
2-Phenoxy-1,5-dihydro-2,4,3-benzodioxaphosphepin; Selenide, *in* P-00066

$C_{14}H_{13}O_4P$
1,5-Dihydro-2-phenoxy-2,4,3-benzodioxaphosphepin; 3-Oxide, *in* D-00544
Diphenyl acetylphosphonate, *in* A-00026
2-Hydroxy-4*H*-1,3,2-benzodioxaphosphorin 2-oxide; 2-(2-Methylphenoxy) (*o*-Tolyl ester), *in* H-00109
2-Phenoxy-1,5-dihydro-2,4,3-benzodioxaphosphepin; Oxide, *in* P-00066

$C_{14}H_{13}O_5P$
(Diphenoxyphosphinyl)acetic acid, D-00988

$C_{14}H_{13}P$
5,10-Dihydro-10-methylacridophosphine, D-00515
2,3-Dihydro-2-phenyl-1*H*-isophosphindole, D-00562
Diphenylvinylphosphine, D-01168
9-Phenyl-9-phosphabicyclo[4.2.1]nona-2,4,7-triene, P-00196
9-Phenyl-9-phosphabicyclo[6.1.0]none-2,4,6-triene, P-00197
9-Phenyl-9-phosphapentacyclo[4.3.0.0^{2,5}.0^{3,8}.0^{4,7}]nonane, P-00201
9-Phenyl-9-phosphatricyclo[3.3.1.0^{2,8}]nona-3,6-diene, P-00202

$C_{14}H_{13}PS$
Diphenylvinylphosphine; Sulfide, *in* D-01168
9-Phenyl-9-phosphapentacyclo[4.3.0.0^{2,5}.0^{3,8}.0^{4,7}]nonane; 9-Sulfide, *in* P-00201

$C_{14}H_{14}BO_2P$
2,5-Diphenyl-1,3,5,2-dioxaphosphaborinane, D-01003

$C_{14}H_{14}BO_2PS$
2,5-Diphenyl-1,3,5,2-dioxaphosphaborinane; 5-Sulfide, *in* D-01003

$C_{14}H_{14}BrP$
Dibenzylbromophosphine, *in* D-00058

$C_{14}H_{14}Br_2NO_3P$
Bis[(4-bromophenyl)methyl] phosphoramidate, *in* B-00109

$C_{14}H_{14}ClN_2PS_2$
3,3′-Dimethyl-2,2′-(3*H*,3′*H*)-spirobi[1,3,2-benzothiazaphosphole]; 2-Chloro, *in* D-00916

$C_{14}H_{14}ClOP$
Bis(2-methylphenyl)phosphinic acid; Chloride, *in* B-00318
Bis(3-methylphenyl)phosphinic acid; Chloride, *in* B-00319
Bis(4-methylphenyl)phosphinic acid; Chloride, *in* B-00320
Dibenzylphosphinic acid; Chloride, *in* D-00055

$C_{14}H_{14}ClO_2P$
Bis(2-methylphenyl) phosphorochloridite, B-00330
Bis(3-methylphenyl) phosphorochloridite, B-00331
Bis(4-methylphenyl) phosphorochloridite, B-00332

$C_{14}H_{14}ClO_2PS$
O,O-Bis(2-methylphenyl) phosphorochloridothioate, B-00333
O,O-Bis(3-methylphenyl) phosphorochloridothioate, B-00334
O,O-Bis(4-methylphenyl) phosphorochloridothioate, B-00335

$C_{14}H_{14}ClO_3P$
Bis(4-methoxyphenyl)phosphinic acid; Chloride, *in* B-00301
Bis(2-methylphenyl) phosphorochloridate, *in* B-00312
Bis(3-methylphenyl) phosphorochloridate, *in* B-00313
Bis(4-methylphenyl) phosphorochloridate, B-00329
▷Dibenzyl phosphorochloridate, D-00060
Diphenyl (2-chloroethyl)phosphonate, *in* C-00074

$C_{14}H_{14}ClO_5P$
Bis(2-methoxyphenyl) phosphorochloridate, *in* B-00294
Bis(4-methoxyphenyl) phosphorochloridate, *in* B-00295

$C_{14}H_{14}ClP$
Chlorobis(3-methylphenyl)phosphine, *in* B-00324
Chlorobis(4-methylphenyl)phosphine, *in* B-00325
Chlorodibenzylphosphine, *in* D-00058
Chlorodi-*o*-tolylphosphine, *in* B-00323

$C_{14}H_{14}ClPS$
Bis(4-methylphenyl)phosphinothioic acid; Chloride, *in* B-00322
Dibenzylphosphinothioic acid; Chloride, *in* D-00057

$C_{14}H_{14}Cl_2O_2P_2$
1,2-Ethanediylbis[phenylphosphinic acid]; Dichloride, *in* E-00015

$C_{14}H_{14}Cl_2O_3P_2$
(Chloromethyl)phenylphosphinic acid; Anhydride, *in* C-00103

$C_{14}H_{14}FOP$
Dibenzylphosphinic acid; Fluoride, *in* D-00055

$C_{14}H_{14}FPS$
Bis(4-methylphenyl)phosphinothioic acid; Fluoride, *in* B-00322

$C_{14}H_{14}IP$
5,5-Dimethyl-5*H*-dibenzophospholium iodide, *in* M-00115

$C_{14}H_{14}NOP$
Bis(2-methylphenyl)phosphinic acid; Amide, *in* B-00318
10,11-Dihydro-1*H*-dibenz[*b,e*][1,4]-azaphosphepine 5-oxide; 5-Me, *in* D-00448
5,10-Dihydro-5,10-dimethylphenophosphazine; 10-Oxide, *in* D-00466
2,3-Diphenyl-1,3,2-oxazaphospholidine, D-01029
2-(Diphenylphosphino)acetamide, *in* D-01045

$C_{14}H_{14}NOPS$

N-Acetyl-*P*,*P*-diphenylphosphinothioic amide, *in* D-01083
5,10-Dihydro-10-ethoxyphenophosphazine 10-sulfide, *in* D-00502
2,3-Diphenyl-1,3,2-oxazaphospholidine; 2-Sulfide, *in* D-01029
10-Ethylthio-5,10-dihydrophenophosphazine 10-oxide, *in* D-00502

$C_{14}H_{14}NO_2P$

N-Acetyl-*P*,*P*-diphenylphosphinic amide, *in* D-01040
Diphenyl 1-aziridinylphosphonite, *in* A-00154
2-Phenoxy-3-phenyl-1,3,2-oxazaphospholidine, P-00073

$C_{14}H_{14}NO_2PS$

2-Phenylspiro[1,3,2-benzothiazaphosphole-2(3*H*),2′-[1,3,2]dioxaphospholane], P-00308

$C_{14}H_{14}NO_3P$

2,3′-Dimethylspiro[1,3,2-benzodioxaphosphole-2,2′(3′*H*)-[1,3,2]benzoxazaphosphole], *in* M-00395
O-(Diphenylphosphinyl)hydroxylamine; *N*-Ac, *in* D-01097
N-Hydroxy-*P*,*P*-diphenylphosphinic amide; *O*-Ac, *in* H-00141
2-Phenoxy-3-phenyl-1,3,2-oxazaphospholidine; 2-Oxide, *in* P-00073

$C_{14}H_{14}NO_3PS$

N,*N*-Dimethylspiro[1,3,2-benzodioxaphosphole-2,2′-[1,3,2]benzoxathiaphosphol]-2-amine, *in* S-00007

$C_{14}H_{14}NO_4P$

[(Benzoylamino)phenylmethyl]phosphonic acid, *in* A-00105
Spiro[1,3,2-benzodioxaphosphole-2,2′(3′*H*)-[1,3,2]benzoxazaphosphole]; 2-Ethoxy, *in* S-00008

$C_{14}H_{14}NO_4PS$

▷*O*-Ethyl *O*-4-nitrophenyl phenylphosphonothionate, E-00099

$C_{14}H_{14}NO_5P$

Diphenyl (2-nitroethyl)phosphonate, *in* N-00036
▷Ethyl 4-nitrophenyl phenylphosphonate, *in* M-00471
Methyl *N*-(diphenoxyphosphinyl)carbamate, *in* D-00989

$C_{14}H_{14}NP$

5,10-Dihydro-5,10-dimethylphenophosphazine, D-00466
1,3-Diphenyl-1,3-azaphosphetidine, *in* P-00087

$C_{14}H_{14}NPS$

5,10-Dihydro-5,10-dimethylphenophosphazine; 10-Sulfide, *in* D-00466

$C_{14}H_{14}N_3OP$

Bis(3-methylphenyl)phosphinic acid; Azide, *in* B-00319
Bis(4-methylphenyl)phosphinic acid; Azide, *in* B-00320
2,3-Dihydro-5-methyl-3-phenyl-2-phenylamino-1,3,4,2-oxadiazaphosphole, D-00527

$C_{14}H_{14}N_3OPS$

2,3-Dihydro-5-methyl-3-phenyl-2-phenylamino-1,3,4,2-oxadiazaphosphole; 2-Sulfide, *in* D-00527

$C_{14}H_{14}N_3O_2P$

2,3-Dihydro-5-methyl-3-phenyl-2-phenylamino-1,3,4,2-oxadiazaphosphole; 2-Oxide, *in* D-00527

$C_{14}H_{14}N_3O_3P$

Bis(4-methoxyphenyl)phosphinic acid; Azide, *in* B-00301

$C_{14}H_{14}N_3PS_2$

2,3-Dihydro-5-methyl-*N*,3-diphenyl-1,3,4,2-thiadiazaphosphole-2(3*H*)amine 2-sulfide, *in* A-00072

$C_{14}H_{14}O_2P_2S_4$

2,4-Bis(4-methoxyphenyl)-1,3,2,4-dithiadiphosphetane 2,4-disulfide, B-00293

$C_{14}H_{14}O_4P$

1,2-Ethenediylbis(phenylphosphinic acid), E-00018

$C_{14}H_{15}N_2O_2P$

1,3-Dihydro-1,3-dimethylspiro[2*H*-1,3,2-benzodiazaphosphole-2,2′[1,3,2]-benzodioxaphosphole], D-00469
1,3-Dihydrospiro[2*H*-1,3,2-benzodiazaphosphole-2,2′-[1,3,2]-dioxaphosholane]; 2-Ph, *in* D-00575
3,3-Dimethyl-2,2(3*H*,3′*H*)-spirobi[1,3,2-benzoxazaphosphole], D-00917
Diphenylphosphinylacetic acid hydrazide, D-01094
2,2(3*H*,3′*H*)-Spirobi[1,3,2-benzoxazaphosphole]; 2-Et, *in* S-00020

$C_{14}H_{15}N_2P$

2,3-Dihydro-1,3-dimethyl-2-phenyl-1*H*-benzodiazaphosphole, D-00467

$C_{14}H_{15}N_2PS$

2,3-Dihydro-1,3-dimethyl-2-phenyl-1*H*-benzodiazaphosphole; 2-Sulfide, *in* D-00467
P-Ethenyl-*N*,*N*′-diphenylphosphosphonothioic diamide, *in* V-00007

$C_{14}H_{15}N_2PS_2$

3,3′-Dimethyl-2,2′-(3*H*,3′*H*)-spirobi[1,3,2-benzothiazaphosphole], D-00916

$C_{14}H_{15}OP$

Benzylmethylphenylphosphine; Oxide, *in* B-00051
Bis(2-methylphenyl)phosphine; Oxide, *in* B-00315
Bis(4-methylphenyl)phosphine; Oxide, *in* B-00317
Bis(2-methylphenyl)phosphinous acid, B-00323
Bis(3-methylphenyl)phosphinous acid, B-00324
Bis(4-methylphenyl)phosphinous acid, B-00325
Dibenzylphosphine; Oxide, *in* D-00054
Dibenzylphosphinous acid, D-00058
Ethyldiphenylphosphine; Oxide, *in* E-00066
Ethyl diphenylphosphinite, *in* D-01089
Methoxymethyldiphenylphosphine, *in* D-01069
(2-Methoxyphenyl)methylphenylphosphine, M-00059
2,3,4,7-Tetrahydro-1*H*-isophosphindole 2-oxide; 2-Ph, *in* T-00109

$C_{14}H_{15}OPS$

Bis(4-methylphenyl)phosphinothioic acid, B-00322
Dibenzylphosphinothioic acid, D-00057
O-Ethyl diphenylphosphinothioate, *in* D-01082
S-Ethyl diphenylphosphinothioate, *in* D-01082

$C_{14}H_{15}OPSe$

O-Ethyl diphenylphosphinoselenoate, *in* D-01078

$C_{14}H_{15}O_2P$

Bis(2-methoxyphenyl)phosphine, B-00296
Bis(3-methoxyphenyl)phosphine, B-00297
Bis(4-methoxyphenyl)phosphine, B-00298
Bis(2-methoxyphenyl)phosphinic acid, B-00299
Bis(3-methoxyphenyl)phosphinic acid, B-00300
Bis(2-methylphenyl)phosphinic acid, B-00318
Bis(3-methylphenyl)phosphinic acid, B-00319
Bis(4-methylphenyl)phosphinic acid, B-00320
Dibenzylphosphinic acid, D-00055
Diphenyl ethylphosphonite, *in* E-00153
(Diphenylphosphinyl)dimethyl ether, *in* D-01069
Ethyl diphenylphosphinate, *in* D-01039
(2-Methoxyphenyl)methylphenylphosphine; Oxide, *in* M-00059
Methyl benzylphenylphosphinate, *in* B-00056

$C_{14}H_{15}O_2PS$

O,*O*-Diphenyl ethylphosphonothioate, *in* E-00148

$C_{14}H_{15}O_2PS_2$

O,*O*-Bis(2-methylphenyl) phosphorodithioate, B-00337
▷*O*,*O*-Bis(3-methylphenyl) phosphorodithioate, B-00338
O,*O*-Bis(4-methylphenyl) phosphorodithioate, B-00339
O,*O*-Dibenzyl phosphorodithioate, D-00063
▷Edifenphos, E-00002

$C_{14}H_{15}O_2PSe_2$

O,*O*-Bis(4-methylphenyl) phosphorodiselenoate, B-00336

$C_{14}H_{15}O_3P$

Benzyl methyl phenylphosphonate, *in* M-00488
Bis(2-methoxyphenyl)phosphine; Oxide, *in* B-00296
Bis(3-methoxyphenyl)phosphine; Oxide, *in* B-00297
Bis(4-methoxyphenyl)phosphine; Oxide, *in* B-00298
Bis(2-methylphenyl) phosphonate, B-00326
Bis(3-methylphenyl) phosphonate, B-00327
Bis(4-methylphenyl) phosphonate, B-00328
▷Dibenzyl phosphonate, D-00059
Diphenyl ethylphosphonate, *in* E-00129
Ethyl phenyl phenylphosphonate, *in* M-00471

$C_{14}H_{15}O_4P$

Bis(4-methoxyphenyl)phosphinic acid, B-00301
▷Bis(2-methylphenyl) phosphate, B-00312
Bis(3-methylphenyl) phosphate, B-00313
Bis(4-methylphenyl) phosphate, B-00314
Bis(phenoxymethyl)phosphinic acid, B-00384
Dibenzyl phosphate, D-00053
2,2-Dihydro-1,3,2-benzodioxaphosphole; 2,2-Dimethoxy, 2-Ph, *in* D-00439
Ethyl bis(2-hydroxyphenyl)phosphinate, *in* B-00283

$C_{14}H_{15}O_6P$

Bis(2-methoxyphenyl) phosphate, B-00294
Bis(4-methoxyphenyl) phosphate, B-00295

$C_{14}H_{15}P$

Benzylmethylphenylphosphine, B-00051
▷Bis(2-methylphenyl)phosphine, B-00315
▷Bis(3-methylphenyl)phosphine, B-00316
Bis(4-methylphenyl)phosphine, B-00317
Dibenzylphosphine, D-00054
Ethyldiphenylphosphine, E-00066

$C_{14}H_{15}PS$

Benzylmethylphenylphosphine; Sulfide, *in* B-00051
Ethyldiphenylphosphine; Sulfide, *in* E-00066
Ethyl diphenylphosphinothioite, *in* D-01085

$C_{14}H_{15}PSSe$

S-Ethyl diphenylphosphinoselenothioate, *in* D-01080

$C_{14}H_{15}PS_2$

Dibenzylphosphinodithioic acid, D-00056
Diphenyl ethylphosphonodithioite, *in* E-00141
Ethyl diphenylphosphinodithioate, *in* D-01065

$C_{14}H_{15}PS_3$

Diphenyl ethylphosphonotrithioate, *in* E-00152

$C_{14}H_{15}PSe$

Ethyldiphenylphosphine; Selenide, *in* E-00066

$C_{14}H_{16}BrN_2OP$
P-(2-Bromoethyl)-*N,N′*-diphenylphosphonic diamide, *in* B-00472

$C_{14}H_{16}ClF_{12}O_2P$
1,1,2,3,3-Pentamethyl-6,6,8,8-tetrakis(trifluoromethyl)-5,7-dioxa-4-phospha(4-P^V)spiro[3.4]octane; 4-Chloro, *in* P-00030

$C_{14}H_{16}ClO_5PS$
▷Coumaphos, C-00205

$C_{14}H_{16}FN_2OP$
▷*N,N′*-Dibenzylphosphorodiamidic acid; Fluoride, *in* D-00061

$C_{14}H_{16}IP$
Dibenzylphosphonium iodide, *in* D-00054
2,3-Dihydro-1,1-dimethyl-1*H*-benzo[*g*]-phosphindolinium iodide, *in* D-00518
Dimethyldiphenylphosphonium iodide, *in* M-00135

$C_{14}H_{16}NOP$
Bis(4-methylphenyl)phosphinic acid; Amide, *in* B-00320
Dibenzylphosphinic amide, *in* D-00055
N,N-Dimethyl-*P,P*-diphenylphosphinic amide, *in* D-01040
P-Ethyl-*N,P*-diphenylphosphinic amide, *in* E-00102

$C_{14}H_{16}NO_2P$
2-Diethylaminonaphtho[1,2-*d*]-1,3,2-dioxaphosphole, D-00266
2-Diethylaminonaphtho[1,8-*de*]-1,3,2-dioxaphosphorin, D-00267
Diphenyl dimethylphosphoramidite, *in* D-00875
P-Ethyl-*N,N*-diphenylphosphonamidic acid, E-00067
Methyl *P*-(4-methylphenyl)-*N*-phenylphosphonamidate, *in* M-00214

$C_{14}H_{16}NO_2PS$
O,O-Diphenyl ethylphosphoramidothioate, *in* E-00159
O,O-1,2-Naphthylene diethylphosphoramidothioate, *in* D-00266
O,O-1,8-Naphthylene diethylphosphoramidothioate, *in* D-00267

$C_{14}H_{16}NO_3P$
Bis(4-methoxyphenyl)phosphinic acid; Amide, *in* B-00301
Dibenzyl phosphoramidate, *in* D-00053
Diphenyl (1-aminoethyl)phosphonate, *in* A-00078
Diphenyl dimethylphosphoramidate, *in* D-00855
Diphenyl ethylphosphoramidate, *in* E-00156
1,2-Naphthylene diethylphosphoramidate, *in* D-00266
1,8-Naphthylene diethylphosphoramidate, *in* D-00267
[Phenyl[(phenylmethyl)amino]methyl]-phosphonic acid, P-00188

$C_{14}H_{16}NO_4PS$
Dimethyl *P*-phenyl-*N*-phenylsulfonylphosphonimidate, *in* P-00192
Ethyl *P*-phenyl-*N*-phenylsulfonylphosphonamidate, *in* P-00191

$C_{14}H_{16}NO_5P$
Bis(4-methoxyphenyl) phosphoramidate, *in* B-00295

$C_{14}H_{16}NP$
N,N-Dimethyl-*P,P*-diphenylphosphinous amide, D-00747
P,P-Dimethyl-*N,N*-diphenylphosphinous amide, *in* D-00845
2-(Diphenylphosphino)ethylamine, D-01066

$C_{14}H_{16}NPS$
N-Benzyl-*P*-methyl-*P*-phenylphosphinothioic amide, *in* M-00226
N,N-Dimethyl-*P,P*-diphenylphosphinothioic amide, *in* D-01083

$C_{14}H_{16}NPS_3$
Diphenyl dimethylphosphoramidotrithioate, *in* D-00874

$C_{14}H_{16}NPSe$
N,N-Dimethyl-*P,P*-diphenylphosphinoselenoic amide, *in* D-01079

$C_{14}H_{16}N_3OP$
P,P-Bis(1-aziridinyl)-*N*-2-naphthylphosphinic amide, *in* B-00092

$C_{14}H_{16}O_2P_2$
1,2-Bis(phenylphosphino)ethane; Dioxide, *in* B-00390

$C_{14}H_{16}O_3P_2$
Methylphenylphosphinic acid; Anhydride, *in* M-00219

$C_{14}H_{16}O_3P_2S_3$
O-Phenyl ethylphosphonothioic monoanhydride, *in* D-00925

$C_{14}H_{16}O_4P_2$
1,2-Ethanediylbis[phenylphosphinic acid], E-00015

$C_{14}H_{16}O_6P_2$
1,4-Bis(phosphonomethyl)biphenyl, B-00391

$C_{14}H_{16}P_2$
1,2-Bis(phenylphosphino)ethane, B-00390
1,2-Dimethyl-1,2-diphenyldiphosphine, D-00744

$C_{14}H_{16}P_2S$
1,2-Dimethyl-1,2-diphenyldiphosphine; Monosulfide, *in* D-00744

$C_{14}H_{16}P_2S_2$
1,2-Bis(phenylphosphino)ethane; Disulfide, *in* B-00390
1,2-Dimethyl-1,2-diphenyldiphosphine; Disulfide, *in* D-00744
1,2-Ethanediylbis[phenylphosphinodithioic acid], E-00016

$C_{14}H_{17}ClNO_4PS_2$
▷Dialifos, D-00023

$C_{14}H_{17}F_{12}O_2P$
1,1,2,3,3-Pentamethyl-6,6,7,7-tetrakis(trifluoromethyl)-5,8-dioxa-4-phospha(4-P^V)spiro[3.4]octane, P-00029
1,1,2,3,3-Pentamethyl-6,6,8,8-tetrakis(trifluoromethyl)-5,7-dioxa-4-phospha(4-P^V)spiro[3.4]octane, P-00030

$C_{14}H_{17}N_2OP$
Bis(3-methylphenyl)phosphinic acid; Hydrazide, *in* B-00319
Bis(4-methylphenyl)phosphinic acid; Hydrazide, *in* B-00320
P-Ethyl-*N,N′*-diphenylphosphonic diamide, *in* E-00130

$C_{14}H_{17}N_2OPS$
N,N′-Dibenzylphosphorodiamidothioic acid, D-00062

$C_{14}H_{17}N_2O_2P$
N,N′-Dibenzylphosphorodiamidic acid, D-00061
Ethyl *N,N′*-diphenylphosphorodiamidate, *in* D-01125

$C_{14}H_{17}N_2PS$
P-Ethyl-*N,N′*-diphenylphosphonothioic diamide, *in* E-00149

$C_{14}H_{17}OP$
2,3,4,5,6,7-Hexahydro-1*H*-isophosphindole 2-oxide; 2-Ph, *in* H-00035
9-Phenyl-9-phosphabicyclo[3.3.1]nonan-3-one, P-00195
9-Phenyl-9-phosphatricyclo[4.2.1.0^{2,5}]nonane; 9-Oxide, *in* P-00204

$C_{14}H_{17}O_2P$
Diethyl 1-naphthylphosphonite, *in* N-00025
9-Phenyl-9-phosphabicyclo[3.3.1]nonan-3-one; Oxide, *in* P-00195

$C_{14}H_{17}O_3P$
Diethyl 2-naphthylphosphonate, *in* N-00022

$C_{14}H_{17}O_3PS$
O-Butyl *O*-1-naphthalenyl phosphorothioate, B-00575

$C_{14}H_{17}P$
9-Phenyl-9-phosphatricyclo[4.2.1.0^{2,5}]nonane, P-00204

$C_{14}H_{18}AsOP$
1-*tert*-Butyl-1,4-dihydro-4-phenyl-1,4-phospharsenin; 1-Oxide, *in* B-00553

$C_{14}H_{18}AsP$
1-*tert*-Butyl-1,4-dihydro-4-phenyl-1,4-phospharsenin, B-00553

$C_{14}H_{18}Cl_4NO_4P$
4,5,6,7-Tetrachloro-*N,N*-dimethylspiro[1,3,2-benzodioxaphosphole-2,2′-[1,3,2]-dioxaphospholan]-2-amine, *in* T-00036

$C_{14}H_{18}IOP$
8-Methyl-8-phenyl-8-phosphoniabicyclo[3.2.1]-octan-3-one iodide, *in* P-00199

$C_{14}H_{18}NO_3P$
Diethyl 1-naphthylphosphoramidate, *in* N-00028
Diethyl 2-naphthylphosphoramidate, *in* N-00029
Diethyl [(2-quinolinyl)methyl]phosphonate, *in* Q-00004
Diethyl [(8-quinolinyl)methyl]phosphonate, *in* Q-00005

$C_{14}H_{18}NO_4PS$
O,O-Diisopropyl (1,3-dihydro-1,3-dioxo-2*H*-isoindol-2-yl)phosphonothioate, *in* D-00475

$C_{14}H_{18}N_2O_2P_2$
1,2-Ethanediylbis[phenylphosphinic acid]; Diamide, *in* E-00015

$C_{14}H_{18}N_3PS$
N,N-Dimethyl-*N′,N″*-diphenylphosphorothioic triamide, *in* D-00909

$C_{14}H_{18}N_5O_{11}P$
Adenylosuccinic acid, A-00047

$C_{14}H_{19}N_2OPS$
O-2-Naphthyl tetramethylphosphorodiamidothioate, *in* N-00027

$C_{14}H_{19}OP$
1,4-Dimethyl-2-phenyl-2-phosphabicyclo[2.2.1]heptane 2-oxide, *in* D-00825
9-Phenyl-9-phosphabicyclo[3.3.1]nonane; Oxide, *in* P-00193
9-Phenyl-9-phosphabicyclo[4.2.1]nonane; Oxide, *in* P-00194

$C_{14}H_{19}O_5P$
Diisopropyl (1,3-dihydro-3-oxo-1-isobenzofuranyl)phosphonate, *in* D-00535

$C_{14}H_{19}O_6P$
Crotoxyphos, C-00207

$C_{14}H_{19}P$
Di-3-butenylphenylphosphine, D-00072
9-Phenyl-9-phosphabicyclo[3.3.1]nonane, P-00193
9-Phenyl-9-phosphabicyclo[4.2.1]nonane, P-00194

$C_{14}H_{20}AsBr_2P$
1,1,4,4-Tetramethyl-1,2,3,4-tetrahydro-1,4-benzophospharseninium dibromide, *in* T-00091

$C_{14}H_{20}BrP$
1-Benzyl-2,5-dihydro-1,3,4-trimethyl-1*H*-phospholium bromide, *in* D-00600

$C_{14}H_{20}Br_3N_6O_4P$
O,O-Bis(1-benzotriazolyl) 2,2,2-tribromoethyl phosphate, B-00100

$C_{14}H_{20}F_8O_6P_2$
Tetraethyl (3,3,4,4,5,5,6,6,-octafluoro-1-cyclohexene-1,2-diyl)bisphosphonate, *in* O-00003

$C_{14}H_{20}IOP$
1,2,5-Trimethyl-4-oxo-1-phenylphosphorinanium iodide, *in* D-00821

$C_{14}H_{20}NO_3P$
Dipropyl (cyanophenylmethyl)phosphonate, *in* C-00219

$C_{14}H_{20}NO_6$
2-Amino-2-deoxyglucitol; *N*-Ac, 3,4:5,6-di-*O*-isopropylidene, *in* A-00066

$C_{14}H_{20}N_3O_5PS$
▷Pyrazophos, P-00502

$C_{14}H_{20}N_5OP$
N,*N*-Dimethyl-2,2′-diphenylphosphoramidic dihydrazide, *in* D-01109

$C_{14}H_{21}IP$
1,3,4,5-Tetrahydro-2,4,5-trimethyl-2,5-methano-2*H*-2-benzophosphepinium iodide, *in* T-00096

$C_{14}H_{21}N_2O_2P$
Phenyl bis(1-pyrrolidinyl)phosphinate, *in* D-01233

$C_{14}H_{21}N_2O_3P$
Phenyl di-4-morpholinylphosphinite, *in* D-00936

$C_{14}H_{21}N_2O_4P$
▷Phenyl di-4-morpholinylphosphinate, *in* D-00935

$C_{14}H_{21}N_4P$
Octahydro-8*bH*-2*a*,4*a*,6*a*,8*a*-tetraaza-8*b*-phosphapentaleno[1,6-*cd*]pentalene; 8*b*-Ph, *in* O-00025

$C_{14}H_{21}OP$
2,2,3,4,4-Pentamethyl-1-phenylphosphetane; Oxide, *in* P-00020

$C_{14}H_{21}OPS_2$
S-Ethyl *O*-phenyl cyclohexylphosphonodithioate, *in* C-00261

$C_{14}H_{21}O_2P$
Diisopropyl (2-phenylethenyl)phosphonite, *in* P-00161
Ethyl cyclohexylphenylphosphinate, *in* C-00250
2′,2′,3′,4′,4′-Pentamethylspiro[1,3,2-benzodioxaphosphole-2,1′-phosphetane], P-00027
Phenyl 1,1,2,3,3-pentamethyltrimethylenephosphinate, *in* H-00168

$C_{14}H_{21}O_4P$
Diethyl (2-oxo-4-phenylbutyl)phosphonate, *in* O-00065
Dimethyl (3,3-dimethyl-2-oxo-4-phenylbutyl)-phosphonate, *in* D-00786

$C_{14}H_{21}O_5P$
Ethyl α-(diethoxyphosphinyl)benzeneacetate, *in* D-00242
Phenyl 2-(diethoxyphosphinyl)butanoate, *in* D-00246
2,2,3,3-Tetramethyl-1,4,6,9-tetraoxa-5-phospha(5-P^V)spiro[4.4]nonane; 5-Phenoxy, *in* T-00235

$C_{14}H_{21}P$
2,2,3,4,4-Pentamethyl-1-phenylphosphetane, P-00020

$C_{14}H_{21}PS$
2,2,3,4,4-Pentamethyl-1-phenylphosphetane; Sulfide, *in* P-00020

$C_{14}H_{22}NOP$
N-Phenyl-2,2,3,4,4-pentamethylphosphetanamine 1-oxide, *in* P-00021

$C_{14}H_{22}NO_4P$
Diethyl [1-(*N*-benzoylamino)propyl]-phosphonate, *in* A-00125
N,*N*,4′,4′,5′,5′-Hexamethylspiro[1,3,2-benzodioxaphosphole-2,2′-[1,3,2]-dioxaphospholan]-2-amine, *in* T-00230

$C_{14}H_{22}NO_{12}P$
2-Amino-2-deoxyglucose 6-(dihydrogen phosphate); 1,2*N*,3,4-Tetra-Ac, *in* A-00069

$C_{14}H_{22}O_2P_2$
3*a*,4,7,7*a*-Tetrahydro-1,2,3,5,6,8-hexamethyl-4,7-phosphinidene-1*H*-phosphindole; 1,8-Dioxide, *in* T-00106

$C_{14}H_{22}P_2$
3*a*,4,7,7*a*-Tetrahydro-1,2,3,5,6,8-hexamethyl-4,7-phosphinidene-1*H*-phosphindole, T-00106

$C_{14}H_{22}P_2S_2$
3*a*,4,7,7*a*-Tetrahydro-1,2,3,5,6,8-hexamethyl-4,7-phosphinidene-1*H*-phosphindole; 1,8-Disulfide, *in* T-00106

$C_{14}H_{23}N_2O_3P$
4,4,5,5,*N*,*N*-Hexamethylspiro[1,3,2-benzoxazaphosphole-2(3*H*),2′-[1,3,2]-dioxaphospholan]-2-amine, *in* T-00231

$C_{14}H_{23}OP$
Dibutylphenylphosphine; Oxide, *in* D-00102
Phenyl di-*tert*-butylphosphinite, *in* D-00122

$C_{14}H_{23}OPS$
O-Phenyl di-*tert*-butylphosphinothioate, *in* D-00118

$C_{14}H_{23}O_2P$
Dibutyl phenylphosphonite, *in* P-00251
2-Methylpropyl (2-methylpropyl)-phenylphosphinate, *in* M-00373
Octylphenylphosphinic acid, O-00040
Phenyl di-*tert*-butylphosphinate, *in* D-00110

$C_{14}H_{23}O_3P$
Bis(2-methylpropyl) phenylphosphonate, *in* P-00221
Dibutyl phenylphosphonate, *in* P-00221

$C_{14}H_{23}P$
Bis(1-methylpropyl)phenylphosphine, B-00340
▷Dibutylphenylphosphine, D-00102
Di-*tert*-butylphenylphosphine, D-00103

$C_{14}H_{23}PS$
Dibutylphenylphosphine; Sulfide, *in* D-00102
Phenyl bis(2-methylpropyl)phosphinothioite, *in* B-00352

$C_{14}H_{23}PSe$
Dibutylphenylphosphine; Selenide, *in* D-00102

$C_{14}H_{24}ClN_2P$
P-2-Chlorophenyl-*N*,*N*,*N*′,*N*′-tetraethylphosphonous diamide, *in* C-00151
P-3-Chlorophenyl-*N*,*N*,*N*′,*N*′-tetraethylphosphonous diamide, *in* C-00152

$C_{14}H_{24}NO_3P$
Bis(1-methylpropyl) phenylphosphoramidate, *in* P-00257
Bis(2-methylpropyl) phenylphosphoramidate, *in* P-00257
Dibutyl phenylphosphoramidate, *in* P-00257

$C_{14}H_{24}NO_4PS_3$
▷Bensulide, B-00004

$C_{14}H_{24}NP$
P,*P*-Di-*tert*-butyl-*N*-phenylphosphinous amide, *in* D-00124

$C_{14}H_{24}NPS$
P,*P*-Di-*tert*-butyl-*N*-phenylphosphinothioic amide, *in* D-00119
P,*P*-Dibutyl-*N*-phenylphosphinothioic amide, *in* D-00117

$C_{14}H_{24}NPSe$
P,*P*-Di-*tert*-butyl-*N*-phenylphosphinoselenoic amide, *in* D-00115

$C_{14}H_{24}N_3OP$
▷*N*-(1-Adamantyl)-*P*,*P*-bis(aziridinyl)phosphinic amide, A-00032

$C_{14}H_{24}N_3O_2P$
3,3,8,8-Tetramethyl-1,6-dioxa-4,9-diaza-5-phospha(P^V)spiro[4.4]nonane; 5-Anilino, *in* T-00196

$C_{14}H_{24}N_3P$
2,3-Dihydro-1,3-dimethyl-1*H*-1,3,2-benzodiazaphosphole; 2-Diisopropylamino, *in* D-00456
Hexahydro-3-phenyl-1,2,4,3-triazaphosphorine; 1,4-Diisopropyl, *in* H-00051

$C_{14}H_{24}O_4P_2$
Tetraethyl 1,2-phenylenebisphosphonite, *in* B-00009
Tetraethyl 1,4-phenylenebisphosphonite, *in* B-00010

$C_{14}H_{24}O_6P_2$
Tetraethyl 1,3-phenylenebisphosphonate, *in* B-00007
Tetraethyl 1,4-phenylenebisphosphonate, *in* B-00008

$C_{14}H_{24}O_8P_2$
Tetraethyl 1,2-benzenediol bis(phosphate), T-00042

$C_{14}H_{25}ClN_3P$
P-Chloro-*P*,*P*-bis(diethylamino)-*N*-phenylphosphazene, *in* T-00054

$C_{14}H_{25}FN_3P$
P,*P*-Bis(diethylamino)-*P*-fluoro-*N*-phenylphosphazene, *in* T-00054

$C_{14}H_{25}IN_3P$
P,*P*-Bis(diethylamino)-*P*-iodo-*N*-phenylphosphazene, *in* T-00054

$C_{14}H_{25}NO_6P_2$
Tetraethyl phenylimidodiphosphate, *in* P-00168

$C_{14}H_{25}N_2OP$
N,*N*′-Di-*tert*-butyl-*P*-phenylphosphonic diamide, *in* P-00222
Phenyl tetraethylphosphorodiamidite, *in* T-00066

$C_{14}H_{25}N_2OPS$
O-Phenyl tetraethylphosphorodiamidothioate, *in* T-00063
S-Phenyl tetraethylphosphorodiamidothioate, *in* T-00063

$C_{14}H_{25}N_2OPSe$
O-Phenyl tetraethylphosphorodiamidoselenoate, *in* T-00061

$C_{14}H_{25}N_2O_2P$
Diethyl *N*,*N*-diethyl-*N*′-phenylphosphoramidimidate, *in* D-00313
Phenyl *N*,*N*′-dibutylphosphorodiamidate, *in* D-00146
Phenyl tetraethylphosphorodiamidate, P-00318

$C_{14}H_{25}N_2P$
N,*N*,*N*′,*N*′-Tetraethyl-*P*-phenylphosphonous diamide, *in* P-00252

$C_{14}H_{25}N_2PS$
N,*N*′-Di-*tert*-butyl-*P*-phenylphosphonothioic diamide, *in* P-00246

$C_{14}H_{25}O_3P$
Diethyl 1-adamantylphosphonate, *in* A-00034

$C_{14}H_{25}O_4PSi$
Diethyl [phenyl[(trimethylsilyl)oxy]methyl]-phosphonate, *in* P-00331

$C_{14}H_{25}O_5PSi_2$
Phenyl [bis(trimethylsilyloxy)phosphinyl]-acetate, *in* B-00434

$C_{14}H_{26}NO_3P$
Diethyl 1-adamantylphosphoramidate, *in* A-00036

$C_{14}H_{26}NO_6P$
2-Diethylamino-1,3,2-dioxaphospholan-4,5-dicarboxylic acid; Diisopropyl ester, *in* D-00265

$C_{14}H_{26}N_2O_6P_2$
Tetraethyl 1,4-phenylenebisphosphoramidate, *in* B-00012

$C_{14}H_{26}N_3OP$
N,*N*,*N*′,*N*′-Tetraethyl-*N*″-phenylphosphorodiamidimidic acid, T-00054
Tetraethylphosphorodiamidic anilide, *in* T-00058

$C_{14}H_{26}N_4O_{11}P_2$
▷Cytidine diphosphate choline, C-00291

$C_{14}H_{27}N_2O_4P$
Cyclohexyl di-4-morpholinylphosphinate, *in* D-00935

$C_{14}H_{27}N_4OP$
N,N,N′,N′-Tetraethyl-2-phenylphosphorodiamidic hydrazide, *in* P-00274

$C_{14}H_{27}OP$
Ethyl dicyclohexylphosphinite, *in* D-00214

$C_{14}H_{27}OPS_2$
O-Cyclohexyl *S*-ethyl cyclohexylphosphonodithioate, *in* C-00261

$C_{14}H_{27}O_2P$
Ethyl dicyclohexylphosphinate, *in* D-00211

$C_{14}H_{27}O_6PS_2$
Diisopropyl [(diethoxyphosphinothioyl)thio]-butanedioate, *in* D-00240

$C_{14}H_{28}ClOPS$
O-Octyl cyclohexylphosphonochloridothioate, *in* C-00260

$C_{14}H_{28}Cl_4N_4O_6P_2$
Peroxycyclophosphamide, P-00057

$C_{14}H_{28}NO_4PS$
▷Piperophos, P-00433

$C_{14}H_{28}N_3O_6P$
Phosalacine, P-00339

$C_{14}H_{28}N_4P_2$
N,N,N′,N′,N″,N″,N‴,N‴-Octamethyl-*P,P′*-1,2-phenylenebis[phosphonous diamide], *in* B-00009

$C_{14}H_{28}N_9OP_3$
▷Fotrin, F-00066

$C_{14}H_{28}O_{10}P_2$
Dimethyl 1,2-bis(diethoxyphosphinyl)succinate, *in* D-01182

$C_{14}H_{29}NO_2P$
Cyclohexyl *N*-cyclohexyl-*P*-ethylphosphonamidate, *in* C-00246

$C_{14}H_{29}N_2OP$
N,N′-Dicyclohexyl-*P*-ethylphosphonic diamide, *in* E-00130

$C_{14}H_{29}N_2O_2P$
1,3-Dibutyl-2,3-dihydro-4,5-dimethyl-1*H*-1,3,2-diazaphosphole 2-oxide; 2-Ethoxy, *in* D-00081
Ethyl *N,N′*-dicyclohexylphosphorodiamidate, *in* D-00216

$C_{14}H_{29}O_2PS$
O,O-Dibutyl cyclohexylphosphonothioate, *in* C-00264

$C_{14}H_{29}O_5P$
Dibutyl[[(tetrahydro-2*H*-pyran-2-yl)oxy]-methyl]phosphonate, *in* T-00136
Tributyl phosphonoacetate, *in* D-00073

$C_{14}H_{30}ClO_3P$
Diheptyl phosphorochloridate, *in* D-00424

$C_{14}H_{30}NO_3P$
Di-*tert*-butyl cyclohexylphosphoramidate, *in* C-00269
Diethyl [2-(butylamino)-1-hexenyl]-phosphonate, *in* B-00540

$C_{14}H_{30}NO_4P$
Dibutyl (2-diethylamino-2-oxoethyl)-phosphonate, *in* D-00268

$C_{14}H_{30}O_6P_2$
Tetraethyl 1,2-cyclohexanediylbisphosphonate, *in* C-00240
Tetraisopropyl ethenylidenebisphosphonate, *in* E-00020

$C_{14}H_{30}P_2S_4$
2,8-Di-*tert*-butyl-1,3,7,9-tetrathia-2,8-diphosphacyclododecane, D-00159

$C_{14}H_{30}P_2S_6$
2,8-Di-*tert*-butyl-1,3,7,9-tetrathia-2,8-diphosphacyclododecane; 2,8-Disulfide, *in* D-00159

$C_{14}H_{31}N_2PS$
P-Cyclohexyl-*N,N,N′,N′*-tetraethylphosphonothioic diamide, *in* C-00265

$C_{14}H_{31}OP$
2,2-Dihydro-1,2-oxaphosphetane; 2,2,2-Tributyl, *in* D-00533
Dodecyldimethylphosphine; Oxide, *in* D-01264

$C_{14}H_{31}O_2P$
Diheptylphosphinic acid, D-00425
Diisopropyl octylphosphonite, *in* O-00044
Ethyl dihexylphosphinate, *in* D-00434

$C_{14}H_{31}O_2PS_2$
O,O-Diheptyl phosphorodithioate, D-00427

$C_{14}H_{31}O_3P$
Diethyl decylphosphonate, *in* D-00012
Diheptyl phosphonate, D-00426

$C_{14}H_{31}O_4P$
Diheptyl phosphate, D-00424
Monotetradecyl phosphate, M-00514

$C_{14}H_{31}P$
Dodecyldimethylphosphine, D-01264

$C_{14}H_{32}BrP$
Tri-*tert*-butylethylphosphonium bromide, *in* T-00373

$C_{14}H_{32}NO_3P$
Diisopropyl dibutylphosphoramidate, *in* D-00134
Diisopropyl diisobutylphosphoramidate, *in* B-00355

$C_{14}H_{32}N_3O_2P$
[(*N,N*-Diethylcarbamoyl)methyl]phosphonic bis(diethylamide), *in* B-00175

$C_{14}H_{32}N_3P$
(Di-*tert*-butylimino)(diisopropylamino)-phosphorane, D-00095

$C_{14}H_{32}O_3P_2$
Hexylmethylphosphinic acid; Anhydride, *in* H-00094

$C_{14}H_{32}O_4P_2$
Tetraethyl 1,6-hexanediylbisphosphonite, *in* H-00086

$C_{14}H_{32}O_6P_2$
Tetraethyl 1,6-hexanediylbisphosphonate, *in* H-00085

$C_{14}H_{32}O_6P_2S$
Tetraisopropyl [thiobis(methylene)]-bisphosphonate, *in* T-00311

$C_{14}H_{33}NO_5P_2S$
Tributyl (dimethoxyphosphinothioyl)-phosphorimidate, *in* D-00674

$C_{14}H_{33}NO_6P_2$
Tetraisopropyl ethyliminodiphosphate, *in* E-00078
Tris(2-methylpropyl) (dimethoxyphosphinyl)-phosphorimidate, *in* D-00683

$C_{14}H_{33}N_2O_2P$
Ethyl tetrapropylphosphorodiamidate, *in* T-00291

$C_{14}H_{34}N_2O_4P_2S_2$
O,O,O,O-Tetrapropyl 1,2-ethanediylbisphosphoramidothioate, *in* E-00010

$C_{14}H_{34}N_2O_6P_2$
Tetraisopropyl 1,2-ethanediylbisphosphoramidate, *in* E-00009
Tetrapropyl 1,2-ethanediylbisphosphoramidate, *in* E-00009

$C_{14}H_{35}N_3O_4P_2$
Triethyl [bis(diethylamino)phosphinyl]-phosphorimidate, *in* B-00176

$C_{14}H_{36}ClPSi_4$
[Bis[bis(trimethylsilyl)methylene]]-chlorophosphorane, B-00105

$C_{14}H_{36}F_6N_6O_7P_2S_2$
Hexakis(dimethylamino)-*μ*-oxodiphosphorus(2+); Bis(trifluoromethanesulfonate), *in* H-00062

$C_{14}H_{36}N_4O_2P_2SSi_2$
N,N,N′,N′,6,7-Hexamethyl-1,3-bis(trimethylsilyl)-5,8-dioxa-1,3-diaza-2,4-diphospha(4-P^V)spiro[3.4]oct-6-ene-2,4-diamine; 2-Sulfide, *in* H-00068

$C_{14}H_{36}N_4O_2P_2Si_2$
N,N,N′,N′,6,7-Hexamethyl-1,3-bis(trimethylsilyl)-5,8-dioxa-1,3-diaza-2,4-diphospha(4-P^V)spiro[3.4]oct-6-ene-2,4-diamine, H-00068

$C_{14}H_{41}N_2PSi_4$
P-Ethyl-*N,N,N′,N′*-tetrakis(trimethylsilyl)-phosphonous diamide, *in* E-00154

$C_{14}N_{12}NO_2PS_2$
1,2-Dihydro-1-hydroxy-1-phenyl-4*H*-3,1,2-benzothiazaphosphorin-4-one 1-sulfide; 2-Methoxy (*O*-Me ester), *in* D-00503

$C_{15}H_9Cl_3NO_3P$
4-Chlorophenyl 5-chloro-8-quinolinyl phosphorochloridate, *in* M-00456

$C_{15}H_{10}Cl_2NO_4P$
Mono(4-chlorophenyl) mono(5-chloro-8-quinolinyl) phosphate, M-00456

$C_{15}H_{11}BrF_{12}NO_3P$
2,2,3,3-Tetrakis(trifluoromethyl)-1,4,6-trioxa-10-aza-5-phospha(5-P^V)spiro[4.5]decane; 5-(4-Bromophenoxy), *in* T-00190

$C_{15}H_{11}F_{12}O_5P$
2,2,3,3-Tetrakis(trifluoromethyl)-1,4,6,10-tetraoxa-5-phospha(5-P^V)spiro[4.5]decane; 5-Phenoxy, *in* T-00187

$C_{15}H_{11}NO_3P$
Phenyl 8-quinolyl phosphate, P-00306

$C_{15}H_{11}O_2P$
4,5-Dihydro-4-hydroxybenzo[*lmn*]-phosphanthridine 4-oxide, D-00489

$C_{15}H_{11}O_4P$
1,1′(3*H*,3′*H*)-Spirobi[2,1-benzoxaphosphole]-3,3′-dione; 1-Me, *in* S-00018

$C_{15}H_{11}O_5P$
1-Hydroxy-1,1′(3*H*,3′*H*)-spirobi[2,1-benzoxaphosphole]-3,3′-dione; 1-Methoxy, *in* H-00194

$C_{15}H_{11}P$
2-Phenylphosphinoline, P-00212

$C_{15}H_{12}NO_3P$
(Diphenylphosphinyl)isocyanoacetic acid, D-01098

$C_{15}H_{12}NO_4P$
2,3-Dihydro-3-phenyl-4*H*-1,3,2-benzoxazaphosphorin-4-one; 2-Acetoxy, *in* D-00556

$C_{15}H_{12}N_3OP$
Tri-2-pyridylphosphine; *P*-Oxide, *in* T-00723

$C_{15}H_{12}N_3P$
Tri-2-pyridylphosphine, T-00723

$C_{15}H_{12}N_3PS$
Tri-2-pyridylphosphine; Sulfide, *in* T-00723

$C_{15}H_{13}N_2O_2P$
1-Diazo-1-(diphenylphosphinyl)-2-propanone, D-00031

$C_{15}H_{13}N_2O_3P$
Diphenyl 1*H*-imidazol-1-ylphosphonate, *in* I-00003

$C_{15}H_{13}N_2P$
3,5-Diphenyl-1*H*-1,2,4-diazaphosphole; 1-Me, *in* D-01000

$C_{15}H_{13}OP$
Diphenylphosphinylpropadiene, *in* D-01142
1-(Diphenylphosphinyl)propyne, *in* D-01146
1,2,3,4-Tetrahydro-4-oxo-1-phenylphosphinoline, T-00122

$C_{15}H_{13}O_2P$
5-Methoxy-5*H*-dibenzo[*b,f*]phosphepin 5-oxide, *in* H-00117
1,2,3,4-Tetrahydro-4-oxo-1-phenylphosphinoline; 1-Oxide, *in* T-00122

$C_{15}H_{13}O_3P$
Diphenyl 1-propynylphosphonate, *in* P-00494

$C_{15}H_{13}O_4P$
2-Hydroxy-4,5-diphenyl-1,3,2-dioxaphosphole 2-oxide; 2-Methoxy (Me ester), *in* H-00139

$C_{15}H_{13}P$
Diphenyl-1,2-propadienylphosphine, D-01142
Diphenyl(1-propynyl)phosphine, D-01146

$C_{15}H_{13}PS$
1-(Diphenylphosphinothioyl)propyne, *in* D-01146

$C_{15}H_{14}NOP$
3-(Diphenylphosphinyl)propanenitrile, *in* D-01073

$C_{15}H_{14}NOPS_3$
1,2-Dihydro-1-phenyl-4*H*-3,1,2-benzothiazaphosphorin-4-thione; 2-Ethoxy, 2-sulfide, *in* D-00552

$C_{15}H_{14}NO_2P$
1-Phenyl-1,2-azaphospholidin-5-one; 2-Ph, 2-oxide, *in* P-00092

$C_{15}H_{14}NO_2PS$
▷Cyanofenphos, C-00213

$C_{15}H_{14}NO_2PS_2$
1,2-Dihydro-1-hydroxy-1-phenyl-4*H*-3,1,2-benzothiazaphosphorin-4-one 1-sulfide; 2-Ethoxy (*O*-Et ester), *in* D-00503

$C_{15}H_{14}NO_3P$
4-Cyanophenyl ethyl phenylphosphonate, *in* M-00471
Diethyl methylphosphoramidate, D-00299
2,3-Dihydro-3-phenyl-4*H*-1,3,2-benzoxazaphosphorin-4-one; 2-Ethoxy, *in* D-00556
Diphenyl (1-cyanoethyl)phosphonate, *in* C-00210

$C_{15}H_{14}NP$
3-(Diphenylphosphino)propanenitrile, D-01073

$C_{15}H_{14}NPS$
3-(Diphenylphosphinothioyl)propanenitrile, *in* D-01073

$C_{15}H_{15}NO_3P$
1,4-Dihydrospiro[2*H*-3,1,2-benzoxazaphosphorine-2,2′-[1,3,2]-dioxaphospholane]; 2-Ph, *in* D-00576

$C_{15}H_{15}OP$
10,11-Dihydro-5-methyl-5*H*-dibenzo[*b*,*f*]-phosphepin; Oxide, *in* D-00524
2,3-Dihydro-1*H*-phosphindole 1-oxide; 1-Benzyl, *in* D-00570
2,3-Diphenyl-1,3-oxaphospholane, D-01027
1-(Diphenylphosphino)-2-propanone, D-01075
Diphenyl-1-propenylphosphine; Oxide, *in* D-01143
Diphenyl-1-propenylphosphine; Oxide, *in* D-01143
1-Phenyl-1,2,3,4-tetrahydrophosphinoline; 1-Oxide, *in* P-00320
2-Propenyl diphenylphosphinite, *in* D-01089

$C_{15}H_{15}OPS$
2,3-Diphenyl-1,3-oxaphospholane; 3-Sulfide, *in* D-01027
2,3-Diphenyl-1,3-thiaphospholane; 3-Oxide, *in* D-01151

$C_{15}H_{15}O_2P$
10,11-Dihydro-5-methoxy-5*H*-dibenzo[*b*,*f*]-phosphepin 5-oxide, *in* D-00492
2,3-Diphenyl-1,3-oxaphospholane; 3-Oxide, *in* D-01027
1-(Diphenylphosphinyl)-2-propanone, *in* D-01075
Methyl phenyl(2-phenylethenyl)phosphinate, *in* P-00186

$C_{15}H_{15}O_2PS$
10-Hydroxy-2,8-dimethyl-10*H*-phenothiaphosphine 10-oxide; 10-Methoxy (Me ester), *in* H-00127

$C_{15}H_{15}O_3P$
5,6-Dimethylspiro[1,3,2-benzodioxaphosphole-2,1′(3′*H*)-[2,1]benzoxaphosphole, D-00915
Diphenyl (1-methylethenyl)phosphonate, *in* M-00149
Ethyl benzoylphenylphosphinate, *in* B-00032
2-Methoxy-4,5-diphenyl-1,3,2-dioxaphospholane, M-00039
Methyl (diphenylphosphinyl)acetate, *in* D-01093
Mono(phenylmethyl) (2-phenylethenyl)-phosphonate, *in* P-00155

$C_{15}H_{15}O_3PS$
2-Methoxy-4,5-diphenyl-1,3,2-dioxaphospholane; 2-Sulfide, *in* M-00039

$C_{15}H_{15}O_4P$
Dimethyl (9*H*-xanthen-9-yl)phosphonate, *in* X-00001
Diphenyl (3-methyl-2-oxiranyl)phosphonate, *in* M-00185
Diphenyl (2-oxopropyl)phosphonate, *in* O-00075
2-Methoxy-4,5-diphenyl-1,3,2-dioxaphospholane; 2-Oxide, *in* M-00039

$C_{15}H_{15}P$
10,11-Dihydro-5-methyl-5*H*-dibenzo[*b*,*f*]-phosphepin, D-00524
Diphenyl-1-propenylphosphine, D-01143
Diphenyl-2-propenylphosphine, D-01144
1-Phenyl-1,2,3,4-tetrahydrophosphinoline, P-00320
Phosphanthridene; 5,5-Dihydro, 5,5-di-Me, *in* P-00348
1,2,3,4-Tetrahydro-2-phenylisophosphinoline, *in* T-00110

$C_{15}H_{15}PO$
Diphenyl-2-propenylphosphine; Oxide, *in* D-01144

$C_{15}H_{15}PS$
Diphenyl-1-propenylphosphine; Sulfide, *in* D-01143
Diphenyl-2-propenylphosphine; Sulfide, *in* D-01144
2,3-Diphenyl-1,3-thiaphospholane, D-01151

$C_{15}H_{15}PS_2$
2-Propenyl diphenylphosphinodithioate, *in* D-01065

$C_{15}H_{15}PSe$
Diphenyl-2-propenylphosphine; Selenide, *in* D-01144

$C_{15}H_{16}ClN_2OP$
2-Chloro-1,2,3,4-tetrahydro-6-methyl-3-(4-methylphenyl)-1,3,2-benzodiazaphosphorine 2-oxide, C-00177

$C_{15}H_{16}ClO_3P$
Diphenyl (3-chloropropyl)phosphonate, *in* C-00172

$C_{15}H_{16}ClP$
(3-Chloropropyl)diphenylphosphine, C-00169
5,6-Dihydro-5,5-dimethylphosphanthridinium chloride, *in* D-00568

$C_{15}H_{16}IP$
2,3-Dihydro-2-methyl-2-phenyl-1*H*-isophosphindolium iodide, *in* D-00562
Ethenylmethyldiphenylphosphonium iodide, *in* D-01168
9-Methyl-9-phenyl-9-phosphoniabicyclo[4.2.1]-nona-2,4,7-triene iodide, *in* P-00196
9-Methyl-9-phenyl-9-phosphoniahomocubane iodide, *in* P-00201

$C_{15}H_{16}NOPS$
5,10-Dihydro-10-isopropoxyphenophosphazine 10-sulfide, *in* D-00502

$C_{15}H_{16}NO_2P$
2-[(Diphenylphosphino)oxyimino]propane, *in* D-01097
Tetrahydro-2-phenoxy-3-phenyl-2*H*-1,3,2-oxazaphosphorine, T-00127

$C_{15}H_{16}NO_3P$
Dimethyl (9,10-dihydro-9-acridinyl)-phosphonate, *in* D-00438
Diphenyl 2-propenylphosphoramidate, *in* P-00464
Tetrahydro-2-phenoxy-3-phenyl-2*H*-1,3,2-oxazaphosphorine; 2-Oxide, *in* T-00127

$C_{15}H_{16}NO_5P$
Ethyl *N*-(diphenoxyphosphinyl)carbamate, *in* D-00989

$C_{15}H_{16}NO_6P$
2-Amino-3-[(diphenoxyphosphinyl)oxy]-propanoic acid, *in* P-00417

$C_{15}H_{16}NP$
1-Benzyl-3-phenyl-1,3-azaphosphetidine, *in* P-00087
1,2-Diphenyl-1,2-azaphospholidine, *in* P-00089
2,3-Diphenyl-1,3-azaphospholidine, D-00995

$C_{15}H_{16}NPS$
2,3-Diphenyl-1,3-azaphospholidine; 3-Sulfide, *in* D-00995

$C_{15}H_{16}N_3O_5P$
N-[Imino(diphenoxyphosphinylamino)methyl]-glycine, *in* P-00369

$C_{15}H_{16}P_2$
1,3-Diphenyl-1,3-diphospholane, D-01013
1,2-Diphospholane; 1,2-Di-Ph, *in* D-01181
6,7,8,9-Tetrahydro-5*H*-dibenzo[*f*,*h*][1,5]-diphosphonin, T-00089

$C_{15}H_{16}P_2S_2$
1,2-Diphenyl-1,2-diphospholane 1,2-disulfide, *in* D-01181

$C_{15}H_{17}ClNO_3PS_2$
O-Ethyl *S*-methyl *N*-[(4-chlorophenyl)-sulfonyl]-*P*-phenylphosphonimidothioate, *in* C-00166

$C_{15}H_{17}NP_2$
1-Methyl-2,5-diphenyl-1,2,5-azadiphospholidine, *in* M-00089

$C_{15}H_{17}N_2OP$
N,*N*′-Diphenyl-*P*-2-propenylphosphonic diamide, *in* P-00455
Tetrahydro-3-phenyl-2-(phenylamino)-2*H*-1,3,2-oxazaphosphorine, T-00133

$C_{15}H_{17}N_2O_2P$
1,3-Dihydro-1,2,3-trimethylspiro[2*H*-1,3,2-benzodiazaphosphole-2,2′-[1,3,2]-benzodioxaphosphole], *in* D-00469
Tetrahydro-3-phenyl-2-(phenylamino)-2*H*-1,3,2-oxazaphosphorine; 2-Oxide, *in* T-00133
2,3,3′-Trimethyl-2,2′(3*H*,3′*H*)-spirobi[1,3,2-benzoxazaphosphole], *in* D-00917

$C_{15}H_{17}N_2O_2PS$
3′,*N*,*N*-Trimethylspiro[1,3,2-benzodioxaphosphole-2,2′(3′*H*)-[1,3,2]-benzothiazaphosphol]-2-amine, *in* M-00394

$C_{15}H_{17}N_2O_3P$
Diphenyl phosphorohydrazidate; 2-Isopropylidene deriv., *in* D-01136

$C_{15}H_{17}N_2PS$
N,*N*′-Diphenyl-*P*-2-propenylphosphonothioic diamide, *in* P-00462

$C_{15}H_{17}N_4PS$
N^2,$N^{2'}$-Dibenzylidene-*P*-phenylphosphonothioic dihydrazide, *in* P-00249

$C_{15}H_{17}OP$
Benzylethylphenylphosphine; Oxide, *in* B-00047
3-(Diphenylphosphino)-1-propanol, D-01074
Diphenylpropylphosphine; Oxide, *in* D-01145
Isopropyldiphenylphosphine; Oxide, *in* I-00031
Methyl bis(4-methylphenyl)phosphinite, *in* B-00325

$C_{15}H_{17}OPS_2$
S,*S*-Diphenyl propylphosphonodithioate, *in* P-00478

$C_{15}H_{17}O_2P$
Diphenyl isopropylphosphonite, *in* I-00056
3-(Diphenylphosphino)-1-propanol; Oxide, *in* D-01074
Ethyl benzylphenylphosphinate, *in* B-00056
Isopropyl diphenylphosphinate, *in* D-01039
Methyl bis(3-methoxyphenyl)phosphinate, *in* B-00300
Methyl bis(2-methylphenyl)phosphinate, *in* B-00318
Methyl bis(4-methylphenyl)phosphinate, *in* B-00320
Methyl dibenzylphosphinate, *in* D-00055

$C_{15}H_{17}O_2PS$
Inezin, I-00009

$C_{15}H_{17}O_3P$
Benzyl phenyl phenylphosphonate, *in* M-00471
Diphenyl isopropylphosphonate, *in* I-00050
Ethyl (α-hydroxybenzyl)phenylphosphinate, *in* H-00173
1-Phenyl-3,8-phosphonanedione 1-oxide, *in* P-00373
Phenyl propyl phenylphosphonate, *in* M-00511

$C_{15}H_{17}O_4P$
Diphenyl (1-hydroxy-1-methylethyl)-phosphonate, *in* H-00159
Methyl bis(4-methoxyphenyl)phosphinate, *in* B-00301

$C_{15}H_{17}O_4PSi$
2,2′-Spirobi[1,3,2-benzodioxaphosphole]; 2-Trimethylsilyl, *in* S-00014

$C_{15}H_{17}O_6P$
Glycerol 2-monophosphate; Di-Ph ether, *in* G-00067

$C_{15}H_{17}P$
Benzylethylphenylphosphine, B-00047
Diphenylpropylphosphine, D-01145
Isopropyldiphenylphosphine, I-00031

$C_{15}H_{17}PS$
Benzylethylphenylphosphine; Sulfide, *in* B-00047
Diphenylpropylphosphine; Sulfide, *in* D-01145
Isopropyldiphenylphosphine; Sulfide, *in* I-00031

$C_{15}H_{17}PS_2$
Dibenzyl methylphosphonodithioite, *in* M-00302
Isopropyl diphenylphosphinodithioate, *in* D-01065

$C_{15}H_{18}ClN_2OP$
P-(3-Chloropropyl)-*N*,*N*′-diphenylphosphonic diamide, *in* C-00173

$C_{15}H_{18}IP$
Benzyldimethylphenylphosphonium iodide, *in* B-00051
Ethylmethyldiphenylphosphonium iodide, *in* E-00066

$C_{15}H_{18}NOP$
P,*P*-Diphenyl-*N*-propylphosphinic amide, *in* D-01040
N-Isopropyl-*P*,*P*-diphenylphosphinic amide, *in* D-01040

$C_{15}H_{18}NO_2P$
[1-(Diphenylmethylamino)ethyl]phosphinic acid, *in* A-00077
Ethyl (aminophenylmethyl)phenylphosphinate, *in* A-00102
Ethyl *P*-(4-methylphenyl)-*N*-phenylphosphonamidate, *in* M-00214

$C_{15}H_{18}NO_2PS$
O,*O*-Diphenyl isopropylphosphoramidothioate, *in* I-00059

$C_{15}H_{18}NO_3P$
Dimethyl (9,10-dihydro-5-methyl-9-acridinyl)-phosphonate, *in* D-00438
Diphenyl isopropylphosphoramidate, *in* I-00058
Diphenyl propylphosphoramidate, *in* P-00484

$C_{15}H_{18}NP$
(Dimethylaminomethyl)diphenylphosphine, D-00700

$C_{15}H_{18}N_3OP$
2,2′,2″-Phosphinylidynetris(1-methylpyrrole), *in* T-00726

$C_{15}H_{18}N_3O_3P$
Dibenzyl (aminoiminomethyl)phosphoramidate, *in* P-00370

$C_{15}H_{18}O_4P_2$
1,3-Propanediylbis[phenylphosphinic acid], P-00447

$C_{15}H_{19}FNPSi$
P,*P*-Diphenyl-*N*-trimethylsilylphosphinimidic fluoride, *in* D-01157

$C_{15}H_{19}NO_5P_2S$
Trimethyl (diphenoxyphosphinothioyl)-phosphorimidate, *in* D-00987

$C_{15}H_{19}N_2OP$
N,*N*′-Diphenyl-*P*-propylphosphonic diamide, *in* I-00051

$C_{15}H_{19}N_2O_2P$
Isopropyl *N*,*N*′-diphenylphosphorodiamidate, *in* D-01125

$C_{15}H_{19}N_2PS$
N,*N*′-Diphenyl-*P*-propylphosphonothioic diamide, *in* P-00480
N,*N*′-Diphenyl-*P*-propylphosphonothioic diamide, *in* P-00481

$C_{15}H_{19}OP$
6-Benzyl-2-oxa-6-phosphatricyclo[3.3.1.1^{3,7}]-decane, B-00055
Octahydro-3-phenyl-1*H*-cyclopropa[*c*]-phosphindole 3-oxide, *in* O-00004

$C_{15}H_{19}OPSSi$
O-Trimethylsilyl diphenylphosphinothioate, *in* D-01082

$C_{15}H_{19}O_2P$
6-Benzyl-2-oxa-6-phosphatricyclo[3.3.1.1^{3,7}]-decane; 6-Oxide, *in* B-00055

$C_{15}H_{19}O_3P$
▷Diethyl 1-naphthylmethylphosphonate, *in* N-00013
Diethyl 2-naphthylmethylphosphonate, *in* N-00014

$C_{15}H_{19}PS_2Si$
Trimethylsilyl diphenylphosphinodithioate, *in* D-01065

$C_{15}H_{19}PSi$
Diphenyl(trimethylsilyl)phosphine, D-01156

$C_{15}H_{20}NOPSi$
P,*P*-Diphenyl-*N*-trimethylsilylphosphinimidic acid, D-01157

$C_{15}H_{20}NO_{10}P$
Methyl 2,3-di-*O*-methylglucopyranoside 4,6-cyclic phosphonate; *P*-(4-Nitrophenoxy), *in* M-00119

$C_{15}H_{20}NPSSi$
P,*P*-Diphenyl-*N*-trimethylsilylphosphinothioic amide, *in* D-01083

$C_{15}H_{21}OP$
2,2,6,6-Tetramethyl-1-phenyl-4-phosphorinanone, T-00210

$C_{15}H_{21}OPS$
2,2,6,6-Tetramethyl-1-phenyl-4-phosphorinanone; Sulfide, *in* T-00210

$C_{15}H_{21}O_2P$
Benzyl 1,1,2,3,3-pentamethyltrimethyl-enephosphinate, *in* H-00168
2,2,6,6-Tetramethyl-1-phenyl-4-phosphorinanone; Oxide, *in* T-00210

$C_{15}H_{21}O_2PS$
2,5-Dimethyl-1-phenylphosphorinan-4-ol; 1-Sulfide, 4-*O*-Ac, *in* D-00820

$C_{15}H_{21}O_6PS$
Methyl 2,3-di-*O*-methylglucopyranoside 4,6-cyclic phosphonothioate; *P*-Ph, *in* M-00120

$C_{15}H_{21}O_7P$
Methyl 2,3-di-*O*-methylglucopyranoside 4,6-cyclic phosphonate; *P*-Ph, *in* M-00119

$C_{15}H_{22}IP$
9-Methyl-9-phenyl-9-phosphoniabicyclo[4.2.1]-nonane iodide, *in* P-00194

$C_{15}H_{23}N_5O_{14}P_2$
Adenosine diphosphate ribose, A-00041

$C_{15}H_{23}OP$
4-*tert*-Butyl-1-phenylphosphorinane; Oxide, *in* B-00586

$C_{15}H_{23}OP_3$
Triphosphetan-4-one; 1,3-Di-*tert*-butyl, 2-Ph, *in* T-00703

$C_{15}H_{23}O_5P$
Benzyl 4-(diethoxyphosphinyl)butanoate, *in* D-00247

$C_{15}H_{23}P$
4-*tert*-Butyl-1-phenylphosphorinane, B-00586

$C_{15}H_{24}BrP$
1,2,2,3,4,4-Hexamethyl-1-phenyl-phosphetanium bromide, *in* P-00020

$C_{15}H_{24}NOP$
N-Benzyl-*P*,*P*-(1,1,2,3,3-pentamethyltrimethyl-ene)phosphinic amide, *in* P-00021

$C_{15}H_{24}NO_2P$
Benzyl *N*-cyclohexyl-*P*-ethylphosphonamidate, *in* C-00246

$C_{15}H_{24}NO_4P$
Diethyl (1-*N*-benzoylamino-2-methylpropyl)-phosphonate, *in* A-00094

$C_{15}H_{24}NO_4PS$
▷Isofenphos, I-00025

$C_{15}H_{24}NO_5P$
Dibutyl [(2-nitrophenyl)methyl]phosphonate, *in* N-00044

$C_{15}H_{24}N_2O_{17}P_2$
Uridine diphosphate glucose, U-00002

$C_{15}H_{24}N_5O_6P$
Phosphophenylalanylarginine, P-00401

$C_{15}H_{25}N_2OP$
5-*tert*-Butyltetrahydro-2-dimethylamino-3-phenyl-2*H*-1,3,2-oxazaphosphorine, B-00636

$C_{15}H_{25}N_2OPS$
5-*tert*-Butyltetrahydro-2-dimethylamino-3-phenyl-2*H*-1,3,2-oxazaphosphorine; 2-Sulfide, *in* B-00636

$C_{15}H_{25}N_2O_2P$
P-Benzoyl-*N*,*N*,*N*′,*N*′-tetraethylphosphonic diamide, *in* B-00035
5-*tert*-Butyltetrahydro-2-dimethylamino-3-phenyl-2*H*-1,3,2-oxazaphosphorine; 2-Oxide, *in* B-00636

$C_{15}H_{25}OPS$
O-Benzyl di-*tert*-butylphosphinothioate, *in* D-00118

$C_{15}H_{25}O_2P$
Nonylphenylphosphinic acid, N-00074
Octyl (4-methylphenyl)phosphinate, *in* M-00222
Pentyl (1-methylpropyl)phenylphosphinate, *in* M-00372

$C_{15}H_{25}O_3P$
▷Dibutyl benzylphosphonate, *in* B-00060

$C_{15}H_{25}O_4P$
(3,5-Di-*tert*-butyl-4-hydroxybenzyl)phosphonic acid, D-00094

$C_{15}H_{26}IP$
Dibutylmethylphenylphosphonium iodide, *in* D-00102
Di-*tert*-butylmethylphosphonium iodide, *in* D-00103

$C_{15}H_{26}NO_3P$
Dibutyl [(2-aminophenyl)methyl]phosphonate, *in* A-00106
Di-*tert*-butyl benzylphosphoramidate, *in* B-00069
Triisopropyl phenylphosphorimidate, *in* P-00264

$C_{15}H_{26}NO_6PS$
2,2-Dihydro-2,2,2-triethoxy-4-[(4-methylphenyl)sulfonyl]-1,4,2-oxazaphospholidine, D-00590

$C_{15}H_{27}NO_6P_2$
Tetraethyl (aminophenylmethyl)-bisphosphonate, *in* A-00053

$C_{15}H_{27}N_2OP$
N,N,N',N'-Tetraethyl-*P*-(phenylmethyl)-phosphonic diamide, *in* B-00061

$C_{15}H_{27}O_4P$
3,7,11-Trimethyl-1,6,10-dodecatrien-3-yl dihydrogen phosphate, T-00514
3,7,11-Trimethyl-2,6,10-dodecatrien-1-yl dihydrogen phosphate, T-00515

$C_{15}H_{27}O_5PSi_2$
Benzyl [bis(trimethylsilyloxy)phosphinyl]-acetate, *in* B-00434

$C_{15}H_{27}O_6P$
3,3',3''-Phosphinidynetrispropanoic acid; Tri-Et ester, *in* P-00357
Tributoxycarbonylphosphine, *in* P-00355

$C_{15}H_{27}O_7P$
3,3',3''-Phosphinyldynetrispropanoic acid triethyl ester, *in* P-00357

$C_{15}H_{27}P$
Tricyclopentylphosphine, T-00431

$C_{15}H_{28}O_7P_2$
Mono(3,7,11-trimethyl-2,6,10-dodecatrienyl) diphosphate, M-00517

$C_{15}H_{30}N_3OP$
▷1,1',1''-Phosphinylidynetrispiperidine, *in* T-00704

$C_{15}H_{30}N_3O_9P_3$
Hexaethyl 1,3,5-triazine-2,4,6-triyltriphosphonate, *in* T-00329

$C_{15}H_{30}N_3P$
▷Tri-1-piperidinophosphine, T-00704

$C_{15}H_{30}N_3PS$
1,1',1''-Phosphinothioylidynetrispiperidine, *in* T-00704

$C_{15}H_{30}N_3PSe$
▷1,1',1''-Phosphinoselenoylidynetrispiperidine, *in* T-00704

$C_{15}H_{30}N_3PTe$
1,1',1''-Phosphinotelluroylidynetrispiperidine, *in* T-00704

$C_{15}H_{31}OPSi$
Trimethylsilyl dicyclohexylphosphinite, *in* D-00214

$C_{15}H_{31}O_7P$
Glycerol 2-monophosphate; 1-Dodecanoyl, *in* G-00067

$C_{15}H_{32}N_3OP$
Tetraisopropylphosphorodiamidous acid; 2-Cyanoethyl ester, *in* T-00172

$C_{15}H_{32}N_7O_7P$
Fosfazinomycin *A*, F-00058

$C_{15}H_{32}O_5P_2$
5-*tert*-Butyl-2,2,4,5-tetrahydro-2,2,4-triisopropoxy-1,2,4-oxadiphosphole, B-00637

$C_{15}H_{33}N_2OPSi_3$
Phenyl *N,N,N'*-tris(trimethylsilyl)-phosphonamidimidate, *in* T-00867

$C_{15}H_{33}N_6O_9P_3$
Hexaethyl 1,3,5-triazine-2,4,6-triyltriphosphoramidate, *in* T-00330

$C_{15}H_{33}OP$
▷Tripentylphosphine; Oxide, *in* T-00594

$C_{15}H_{33}O_2P$
Dipentyl pentylphosphonite, *in* P-00052
2-Methylpropyl decylmethylphosphinate, *in* D-00009
Pentyl dipentylphosphinate, *in* D-00981

$C_{15}H_{33}O_2PS$
O,O-Dibutyl heptylphosphonothioate, *in* H-00007

$C_{15}H_{33}O_3P$
Diisopropyl nonylphosphonate, *in* N-00076
Dipentyl pentylphosphonate, *in* P-00050
Tripentyl phosphite, T-00595
Tris(2,2-dimethylpropyl) phosphite, T-00772

$C_{15}H_{33}O_4P$
Tripentyl phosphate, T-00593
Tris(3-methylbutyl) phosphate, T-00796

$C_{15}H_{33}P$
Tripentylphosphine, T-00594

$C_{15}H_{33}PS$
Tripentylphosphine; Sulfide, *in* T-00594

$C_{15}H_{33}PSe$
Tripentylphosphine; Selenide, *in* T-00594

$C_{15}H_{33}PTe$
Tripentylphosphine; Telluride, *in* T-00594

$C_{15}H_{34}BrP$
Tri-*tert*-butylisopropylphosphonium bromide, *in* T-00373

$C_{15}H_{34}IP$
Dodecyltrimethylphosphonium iodide, *in* D-01264

$C_{15}H_{34}N_9O_8PS$
▷Phaseolotoxin, P-00058

$C_{15}H_{34}O_6P_2$
Tetraethyl 1,7-heptanediylbisphosphonate, *in* H-00002
Tetraisopropyl 2,2-propanediphosphonate, *in* P-00445

$C_{15}H_{35}BrN_3P$
P-Bromo-*P,P*-bis(dipropylamino)-*N*-propylphosphazene, *in* P-00043

$C_{15}H_{35}ClN_3P$
P-Chloro-*P,P*-bis(dipropylamino)-*N*-propylphosphazene, *in* P-00043

$C_{15}H_{35}FN_3P$
P,P-Bis(dipropylamino)-*P*-fluoro-*N*-propylphosphazene, *in* P-00043

$C_{15}H_{35}N_2OP$
Pentapropylphosphonic diamide, *in* P-00474

$C_{15}H_{35}N_2O_2P$
Propyl tetrapropylphosphorodiamidate, *in* T-00291

$C_{15}H_{35}N_2P$
Pentapropylphosphonous diamide, *in* P-00482

$C_{15}H_{35}N_4P_3$
4-*tert*-Butyl-1,1,3,3-tetrakis(dimethylamino)-1,3,5-triphosphorin, B-00638

$C_{15}H_{35}O_5PSi_2$
Bis(triethylsilyl) [(acetyloxy)methyl]-phosphonate, *in* A-00007

$C_{15}H_{35}O_9P_3$
Hexaethyl 1,1,3-propanetriyltrisphosphonate, *in* P-00449
Hexaethyl 1,2,3-propanetriyltrisphosphonate, *in* P-00450

$C_{15}H_{36}Br_2N_3P$
Bromobis(dipropylamino)(propylamino)-phosphonium bromide, *in* P-00043

$C_{15}H_{36}NO_9P_3$
Hexaethyl [nitrilotris(methylene)]-trisphosphonate, *in* N-00033

$C_{15}H_{36}NPSi$
P,P,P-Tributyl-*N*-trimethylsilylphosphine imide, *in* T-00374
P,P,P-Tri-*tert*-butyl-*N*-trimethylsilylphosphine imide, *in* T-00375

$C_{15}H_{36}N_2O_2P$
Heptyl tetraethylphosphorodiamidate, *in* T-00059

$C_{15}H_{36}N_3OP$
Pentapropylphosphorodiamidimidic acid, P-00043

$C_{15}H_{36}O_{10}P_4$
Hexaethyl [phosphinylidynetris(methylene)]-trisphosphonate, *in* P-00360

$C_{15}H_{36}P_2Si_2$
2-[Bis(trimethylsilyl)methylene]-1,1-di-*tert*-butyldiphosphine, B-00431

$C_{16}H_4F_{18}P_2$
2,5,7,10,11,12-Hexakis(trifluoromethyl)-1,6-diphosphahexacyclo[4.4.2.$0^{2,5}$.$0^{3,9}$.$0^{4,8}$.$0^{7,10}$]-dodec-11-ene, H-00065
4,5,9,10,11,12-Hexakis(trifluoromethyl)-1,8-diphosphatetracyclo[6.2.2.$0^{2,7}$.$0^{3,6}$]dodecane, H-00066
2,7,9,10,11,12-Hexakis(trifluoromethyl)-1,8-diphosphatetracyclo[6.2.2.$0^{2,7}$.$0^{3,6}$]dodeca-4,9,11-triene, H-00067

$C_{16}H_8NO_5P$
(1-Amino-2-hydroxyethyl)phosphonic acid; *O*-Ac, Di-Ph ester, *in* A-00083

$C_{16}H_9F_6O_5P$
4',5'-Bis(trifluoromethyl)spiro[1,3,2-benzodioxaphosphole-2,2'-[1,3,2]-dioxaphosphole]; 2-Phenoxy, *in* B-00421

$C_{16}H_{11}ClOP$
2-Naphthylphenylphosphinic acid; Chloride, *in* N-00018

$C_{16}H_{11}O_2P$
Bis(phenylethynyl)phosphinic acid, B-00388
Phenyl(phenylbutadiynyl)phosphinic acid, P-00185

$C_{16}H_{11}O_3P$
2-Phenoxynaphtho[1,8-*de*]-1,3,2-dioxaphosphorine, P-00070

$C_{16}H_{11}O_4P$
1,8-Naphthylene phenyl phosphate, *in* P-00070

$C_{16}H_{11}PS$
7*H*-Benzo[*c*]phenothiaphosphine, B-00022
12*H*-Benzo[*c*]phenothiaphosphine, B-00023

$C_{16}H_{12}ClOPS$
4-Chloro-2,6-diphenyl-4*H*-1,4-thiaphosphorin 4-oxide, *in* H-00143

$C_{16}H_{12}ClO_2P$
4-Chloro-2,6-diphenyl-4*H*-1,4-oxaphosphorin 4-oxide, *in* H-00140

$C_{16}H_{12}NOP$
7,12-Dihydrobenzo[*e*]phenophosphazine; 7-Oxide, *in* D-00440

$C_{16}H_{12}NO_2P$
Benzo[*e*]phenazaphosphinic acid, *in* D-00440

$C_{16}H_{12}NP$
7,12-Dihydrobenzo[*e*]phenophosphazine, D-00440
2,6-Diphenyl-1,4-azaphosphorine, D-00996

$C_{16}H_{12}N_3O_4P$
2,3-Dihydro-1*H*-naphtho[1,8-*de*]-1,3,2-diazaphosphorine; 2-(4-Nitrophenoxy), 2-oxide, *in* D-00532

$C_{16}H_{12}O_2P$
2-Naphthylphenylphosphinic acid, N-00018

$C_{16}H_{13}Cl_2O_2PS$
O-(2-Butoxyethyl) phosphorodichloridothioate, *in* H-00149

$C_{16}H_{13}FeP$
1,1'-(Phenylphosphinidene)ferrocene, P-00211

$C_{16}H_{13}FePS$
1,1'-(Phenylphosphinidene)ferrocene; Sulfide, *in* P-00211

$C_{16}H_{13}OP$
3-Phenyl-3*H*-benzophosphepin; 3-Oxide, *in* P-00096

$C_{16}H_{13}O_2P$
1-Hydroxy-3,4-diphenyl-1*H*-phosphole 1-oxide, H-00142
1-Naphthylphenylphosphinic acid, N-00017

$C_{16}H_{13}O_2PS$
Diphenyl 2-thienylphosphonite, *in* T-00308
4-Hydroxy-2,6-diphenyl-4*H*-1,4-thiaphosphorine 2-oxide, H-00143

$C_{16}H_{13}O_3P$
4-Hydroxy-2,6-diphenyl-4*H*-1,4-oxaphosphorin 4-oxide, H-00140

$C_{16}H_{13}O_3PS$
Diphenyl 2-thienylphosphonate, *in* T-00307

$C_{16}H_{13}O_4P$
Diphenyl 2-furanylphosphonate, *in* F-00073

$C_{16}H_{13}P$
3-Phenyl-3*H*-benzophosphepin, P-00096

$C_{16}H_{14}ClOP$
Bis(2-phenylethenyl)phosphinic acid; Chloride, *in* B-00386
Phenyl(4-phenyl-1,3-butadienyl)phosphinic acid; Chloride, *in* P-00184

$C_{16}H_{14}NO_2P$
2,6-Diphenyl-4*H*-1,4-oxaphosphorin-4-amine 4-oxide, *in* H-00140

$C_{16}H_{14}NO_6P$
Diphenyl succinimido phosphate, D-01148

$C_{16}H_{15}F_{12}N_2O_2PS$
6,9-Dimethyl-5-phenylthio-2,2,3,3-tetrakis(trifluoromethyl)-1,4-dioxa-6,9-diaza-5-phospha(5-P^V)spiro[4.4]nonane, *in* D-00922

$C_{16}H_{15}F_{12}N_2O_3P$
6,9-Dimethyl-5-phenoxy-2,2,3,3-tetrakis(trifluoromethyl)-1,4-dioxa-6,9-diaza-5-phospha(5-P^V)spiro[4.4]nonane, *in* D-00922

$C_{16}H_{15}N_2O_3P$
Ethyl diazo(diphenylphosphinyl)acetate, *in* D-00030

$C_{16}H_{15}O_2P$
Bis(2-phenylethenyl)phosphinic acid, B-00386
Phenyl(4-phenyl-1,3-butadienyl)phosphinic acid, P-00184
Spiro[1,3,2-benzodioxaphosphole-2,1′-[3]-phospholene]; 2-Ph, *in* S-00010

$C_{16}H_{15}O_3P$
Diphenyl (1,3-butadienyl)phosphonate, *in* B-00510
5′-Methylspiro[1,3,2-benzodioxaphosphole-2,2′(3*H*)-[1,2]oxaphosphole]; 2-Ph, *in* M-00396

$C_{16}H_{15}O_4P$
4-(Diphenylphosphinyl)-3-oxobutanoic acid, D-01102
2,3-Diphenyl-1,4,6,9-tetraoxa-5-phospha(5-P^V)-spiro[4.4]non-2-ene, D-01150
2-Hydroxy-4,5-diphenyl-1,3,2-dioxaphosphole 2-oxide; 2-Ethoxy (Et ester), *in* H-00139

$C_{16}H_{15}O_4PS_2$
Bis(benzoylthiomethyl)phosphinic acid, *in* B-00292

$C_{16}H_{15}O_6P$
Bis(4-methoxycarbonylphenyl)phosphinic acid, *in* B-00118

$C_{16}H_{16}BrP$
1,1′,3,3′-Tetrahydro-2,2′-spirobi[2*H*-isophosphindolium](1+); Bromide, *in* T-00137

$C_{16}H_{16}ClN_4O_2P$
1,3-Dimethyl-5,7-diphenyl-1,3,5,7-tetraaza-4-phospha(4-P^V)spiro[3.3]heptane-2,6-dione; 4-Chloro, *in* D-00748

$C_{16}H_{16}IP$
Methyldiphenyl-(1-propynyl)phosphonium iodide, *in* D-01146

$C_{16}H_{16}NO_3P$
2,3-Diphenyl-1,4,6-trioxa-9-aza-5-phospha(5-P^V)spiro[4.4]non-2-ene, D-01159

$C_{16}H_{16}NO_6P$
▷Naphthalophos, N-00008

$C_{16}H_{16}N_7O_4P$
O,O-Bis(1-benzotriazolyl) morpholinyl phosphate, B-00099

$C_{16}H_{16}O_4P_2$
4,5,9,10-Tetrahydro-4,9-dihydroxyphosphorino[2,3,4,5-*lmn*]-phosphanthridine 4,9-dioxide; 4,9-Dimethoxy (di-Me ester), *in* T-00090

$C_{16}H_{16}P^{\oplus}$
1,1′,3,3′-Tetrahydro-2,2′-spirobi[2*H*-isophosphindolium](1+), T-00137

$C_{16}H_{16}P_2$
1,2,3,6-Tetrahydro-1,2-diphenyl-1,2-diphosphorin, T-00103

$C_{16}H_{16}P_2S_2$
1,2,3,6-Tetrahydro-1,2-diphenyl-1,2-diphosphorin; 1,2-Disulfide, *in* T-00103

$C_{16}H_{17}Cl_2OP$
Butyl bis(4-chlorophenyl)phosphinite, *in* B-00142

$C_{16}H_{17}Cl_2O_2P$
Butyl bis(4-chlorophenyl)phosphinate, *in* B-00139

$C_{16}H_{17}F_6N_2O_3P$
4′-Ethyl-1,3-dihydro-5-phenyl-3,3-bis(trifluoromethyl)spiro[2*H*-1,4,2-diazaphosphole-2,1′-[2,6,7]trioxa[1]-phosphabicyclo[2.2.2]octane], E-00061

$C_{16}H_{17}N_4O_2P$
1,3-Dimethyl-5,7-diphenyl-1,3,5,7-tetraaza-4-phospha(4-P^V)spiro[3.3]heptane-2,6-dione, D-00748

$C_{16}H_{17}N_4P$
4-Dimethylamino-2,2-dihydro-2,2-diphenyl-1,3,5,2-triazaphosphorine, D-00692

$C_{16}H_{17}OP$
3-Butenyldiphenylphosphine; Oxide, *in* B-00521
(2-Buten-1-yl)diphenylphosphine; Oxide, *in* B-00520
2,3-Diphenyl-1,3-oxaphosphorinane, D-01028
1-Diphenylphosphinyl-2-methylpropene, *in* M-00362
5-Methyl-2,3-diphenyl-1,3-oxaphospholane, M-00133
7-Phenyl-7*H*-dibenzo[*d,f*]phosphonin; 7-Oxide, *in* P-00111

$C_{16}H_{17}OPS$
2,3-Diphenyl-1,3-oxaphosphorinane; 3-Sulfide, *in* D-01028

$C_{16}H_{17}O_2P$
Ethyl (diphenylphosphino)acetate, *in* D-01045
Ethyl phenyl(2-phenylethenyl)phosphinate, *in* P-00186

$C_{16}H_{17}O_2PS$
Ethyl (diphenylphosphinothioyl)acetate, *in* D-01086

$C_{16}H_{17}O_5P$
Acetic acid anhydride with dibenzyl hydrogen phosphate, *in* A-00025
[Bis(phenylmethoxy)phosphinyl]acetic acid, B-00389

$C_{16}H_{17}P$
3-Butenyldiphenylphosphine, B-00521
(2-Buten-1-yl)diphenylphosphine, B-00520
(2-Methyl-1-propenyl)diphenylphosphine, M-00362
7-Phenyl-7*H*-dibenzo[*d,f*]phosphonin, P-00111

$C_{16}H_{17}PS$
2,3-Diphenyl-1,3-thiaphosphorinane, D-01152

$C_{16}H_{18}BrP$
1,1-Diphenylphospholanium bromide, *in* P-00216

$C_{16}H_{18}ClNO_6P$
4-Morpholinylphosphonochloridic acid; Anhydride with diphenyl phosphate, *in* M-00524

$C_{16}H_{18}ClN_2OP$
1,4-Dimethyl-3,5-diphenyl-1,4,2-diazaphospholidine 2-oxide; 2-Chloro, *in* D-00739

$C_{16}H_{18}Cl_2NO_2P$
Diphenyl bis(2-chloroethyl)phosphoramidate, *in* B-00121

$C_{16}H_{18}Cl_2N_2P$
P,P-Bis(3-chlorophenyl)-*N,N*-diethylphosphinous amide, *in* B-00141
P,P-Bis(4-chlorophenyl)-*N,N*-diethylphosphinous amide, *in* B-00142

$C_{16}H_{18}F_2N_2O_4P_2$
6,12-Diphenyl-1,4,8,11-tetraoxa-6,12-diaza-5,7-diphospha(5,7-P^V)dispiro[4.1.4.1]dodecane; 5,7-Difluoro, *in* D-01149

$C_{16}H_{18}IOP$
3-Methyl-2,3-diphenyl-1,3-oxaphospholanium iodide, *in* D-01027

$C_{16}H_{18}IP$
10,11-Dihydro-5,5-dimethyl-5*H*-dibenzo[*b,f*]-phosphepinium iodide, *in* D-00524
Methyldiphenyl-2-propenylphosphonium iodide, *in* D-01144

$C_{16}H_{18}NOP$
Hexahydro-4-phenyl-1,4-azaphosphorine; 1-Ph, 4-oxide, *in* H-00042

$C_{16}H_{18}NO_2P$
3,4-Dimethyl-2-phenoxy-5-phenyl-1,3,2-oxazaphospholidine, D-00791

$C_{16}H_{18}NO_3P$
3,4-Dimethyl-2-phenoxy-5-phenyl-1,3,2-oxazaphospholidine; 2-Oxide, *in* D-00791
Diphenyl 4-morpholinylphosphonite, *in* M-00529

$C_{16}H_{18}NO_3PS$
O,O-Diphenyl 4-morpholinylphosphonothioate, *in* M-00527

$C_{16}H_{18}NO_4P$
▷Diphenyl 4-morpholinylphosphonate, *in* M-00523

$C_{16}H_{18}NO_5P$
Isopropyl *N*-(diphenoxyphosphinyl)carbamate, *in* D-00989

$C_{16}H_{18}NP$
Hexahydro-1,3-diphenyl-1,3-azaphosphorine, *in* H-00041
Hexahydro-1,4-diphenyl-1,4-azaphosphorine, *in* H-00042

$C_{16}H_{18}NPS$
Hexahydro-3-phenyl-1,3-azaphosphorine; 1-Ph, 3-sulfide, *in* H-00041

$C_{16}H_{18}N_3O_5P$
N-[Imino(diphenoxyphosphinylamino)methyl]-*N*-methylglycine, *in* P-00364
Methyl *N*-[imino(diphenoxyphosphinylamino)-methyl]glycinate, *in* P-00369

$C_{16}H_{18}O_2P_2$
1,4-Diphenyl-1,4-diphosphorinane; 1,4-Dioxide, *in* D-01017

$C_{16}H_{18}P_2$
1,2-Diphenyl-1,2-diphosphorinane, D-01015
1,3-Diphenyl-1,3-diphosphorinane, D-01016
1,4-Diphenyl-1,4-diphosphorinane, D-01017
5,6,7,8,9,10-Hexahydrodibenzo[*b,d*][1,6]-diphosphecin, H-00021

$C_{16}H_{18}P_2S_2$
1,2-Diphenyl-1,2-diphosphorinane; 1,2-Disulfide, *in* D-01015

$C_{16}H_{19}Cl_2N_2O_2P$
Phenyl *N,N*-bis(2-chloroethyl)-*N′*-phenyl-phosphorodiamidate, *in* B-00121

$C_{16}H_{19}N_2OP$

1,4-Dimethyl-3,5-diphenyl-1,4,2-diazaphospholidine 2-oxide, D-00739

$C_{16}H_{19}N_2O_2P$

3′,4′-Dimethyl-5′-phenylspiro[1,3,2-benzoxazaphosphole-2(3*H*),2′-[1,3,2]-oxazaphospholidine], D-00823

$C_{16}H_{19}N_2O_4P$

2,2,2,3-Tetrahydro-3,5-diphenyl-1,3,4,2-oxadiazaphosphole; 2,2,2-Trimethoxy, *in* T-00104

$C_{16}H_{19}OP$

Butyl diphenylphosphinite, *in* D-01089
Ethyl bis(2-methylphenyl)phosphinite, *in* B-00323
Ethyl bis(4-methylphenyl)phosphinite, *in* B-00325
Ethyl dibenzylphosphinite, *in* D-00058
(1-Methylpropyl)diphenylphosphine; Oxide, *in* M-00369
(2-Methylpropyl)diphenylphosphine; Oxide, *in* M-00370

$C_{16}H_{19}OPS_2$

S,S-Diphenyl butylphosphonodithioate, *in* B-00606

$C_{16}H_{19}O_2P$

Bis(2,5-dimethylphenyl)phosphinic acid, B-00198
tert-Butyl diphenylphosphinate, *in* D-01039
Dibenzyl ethylphosphonite, *in* E-00153
Diphenyl butylphosphonite, *in* B-00615
Diphenyl *tert*-butylphosphonite, *in* B-00616
Ethyl dibenzylphosphinate, *in* D-00055
Isopropyl benzylphenylphosphinate, *in* B-00056
Phenyl *tert*-butylphenylphosphinate, *in* B-00579

$C_{16}H_{19}O_2PS_2$

Bis(benzylthiomethyl)phosphinic acid, B-00101

$C_{16}H_{19}O_3P$

Benzyl propyl phenylphosphonate, *in* M-00511
Dibenzyl ethylphosphonate, *in* E-00129
Diethyl 4-biphenylylphosphonate, *in* B-00077
Diphenyl butylphosphonate, *in* B-00594
Diphenyl (1,1-dimethylethyl)phosphonate, *in* B-00595
Diphenyl (2-methylpropyl)phosphonate, *in* M-00378

$C_{16}H_{19}O_4P$

Ethyl bis(phenoxymethyl)phosphinate, *in* B-00384

$C_{16}H_{19}O_6PS_2$

Bis(phenylmethylsulfonylmethyl)phosphinic acid, *in* B-00101

$C_{16}H_{19}P$

Bis(2-phenylethyl)phosphine, B-00387
Butyldiphenylphosphine, B-00557
tert-Butyldiphenylphosphine, B-00558
(1-Methylpropyl)diphenylphosphine, M-00369
(2-Methylpropyl)diphenylphosphine, M-00370

$C_{16}H_{19}PO$

Butyldiphenylphosphine; Oxide, *in* B-00557

$C_{16}H_{19}PS$

Butyldiphenylphosphine; Sulfide, *in* B-00557
tert-Butyldiphenylphosphine; Sulfide, *in* B-00558
Butyl diphenylphosphinothioite, *in* D-01085
(2-Methylpropyl)diphenylphosphine; Sulfide, *in* M-00370

$C_{16}H_{19}PS_2$

tert-Butyl diphenylphosphinodithioate, *in* D-01065
Dibenzyl ethylphosphonodithioite, *in* E-00141
Diphenyl butylphosphonodithioite, *in* B-00607

$C_{16}H_{19}PSe$

Butyldiphenylphosphine; Selenide, *in* B-00557

$C_{16}H_{20}BrN_2O_2PS$

O,O-Bis(4-dimethylaminophenyl) phosphorobromidothioate, *in* B-00190

$C_{16}H_{20}ClN_2OP$

(4-Chlorobutyl)phosphonic dianilide, *in* C-00026

$C_{16}H_{20}ClN_2O_2PS$

O,O-Bis(4-dimethylaminophenyl) phosphorochloridothioate, *in* B-00190
O,O-Bis(3-dimethylaminophenyl) phosphorothioate; Chloride, *in* B-00189

$C_{16}H_{20}ClN_2P$

P-tert-Butyl-*N,N′*-diphenylphosphonamidimidic chloride, B-00559

$C_{16}H_{20}Cl_2N_3OP$

N,N-Bis(2-chloroethyl)-*N′,N″*-diphenylphosphoric triamide, *in* B-00121

$C_{16}H_{20}FN_2O_2PS$

O,O-Bis(4-dimethylaminophenyl) phosphorofluoridothioate, *in* B-00190

$C_{16}H_{20}IP$

Benzylethylmethylphenylphosphonium iodide, *in* B-00047
Benzylethylmethylphenylphosphonium iodide, *in* E-00084
Isopropylmethyldiphenylphosphonium iodide, *in* I-00031

$C_{16}H_{20}NOP$

P-tert-Butyl-*N,P*-diphenylphosphinic amide, *in* B-00580
N,N-Diethyl-*P,P*-diphenylphosphinic amide, *in* D-01040
N-tert-Butyl-*P,P*-diphenylphosphinic amide, *in* D-01040

$C_{16}H_{20}NOPS$

S-Phenyl *N,N*-diethyl-*P*-phenylphosphonamidothioate, *in* D-00309

$C_{16}H_{20}NO_2P$

Diphenyl diethylphosphoramidite, *in* D-00360
Ethyl *P*-ethyl-*N,N*-diphenylphosphonamidate, *in* E-00067
Phenyl *N,N*-diethyl-*P*-phenylphosphonamidate, *in* D-00307

$C_{16}H_{20}NO_2PS$

O,O-Diphenyl diethylphosphoramidothioate, *in* D-00356

$C_{16}H_{20}NO_3P$

Diethyl diphenylphosphoramidate, *in* D-01108
Diphenyl (1-amino-2-methylpropyl)-phosphonate, *in* A-00094
Diphenyl butylphosphoramidate, *in* B-00621
Diphenyl *tert*-butylphosphoramidate, *in* B-00622
Diphenyl diethylphosphoramidate, *in* D-00343
Diphenyl (2-methylpropyl)phosphoramidate, *in* M-00384

$C_{16}H_{20}NP$

N,N-Diethyl-*P,P*-diphenylphosphinous amide, *in* D-01090

$C_{16}H_{20}NPS$

N-tert-Butyl-*P,P*-diphenylphosphinothioic amide, *in* D-01083

$C_{16}H_{20}NPS_3$

Diphenyl diethylphosphoramidotrithioate, *in* D-00359

$C_{16}H_{20}N_2O_4P_2$

6,12-Diphenyl-1,4,8,11-tetraoxa-6,12-diaza-5,7-diphospha(5,7-P^V)dispiro[4.1.4.1]dodecane, D-01149

$C_{16}H_{20}N_3OPS$

P-Morpholinyl-*N,N′*-diphenylphosphonothioic diamide, *in* M-00528

$C_{16}H_{20}O_2P_2$

1,2-Ethanediylbis[methylphenylphosphine]; Dioxide, *in* E-00012

$C_{16}H_{20}O_2P_2S_3$

O-Ethyl *S*-(ethoxyphenylphosphinothioyl) phenylphosphonodithioate, *in* P-00237

$C_{16}H_{20}O_4P$

Dimethyl 1,2-ethenediylbis(phenylphosphinate), *in* E-00018

$C_{16}H_{20}O_6P_2S_3$

▷Temephos, T-00003

$C_{16}H_{20}P_2$

1,2-Dibenzyl-1,2-dimethyldiphosphine, D-00043
1,2-Ethanediylbis[methylphenylphosphine], E-00012

$C_{16}H_{20}P_2S_2$

1,4-Butanediylbis[phenylphosphinodithioic acid], B-00516
1,2-Dibenzyl-1,2-dimethyldiphosphine; Disulfide, *in* D-00043

$C_{16}H_{21}INP$

(Dimethylaminomethyl)diphenylphosphine; B,MeI, *in* D-00700

$C_{16}H_{21}N_2OP$

P-(1,1-Dimethylethyl)-*N,N′*-diphenylphosphonic diamide, *in* B-00597
P-(1-Methylpropyl)-*N,N′*-diphenylphosphonic diamide, *in* M-00377

$C_{16}H_{21}N_2O_2P$

Bis(*N*-benzylaminomethyl)phosphinic acid, *in* B-00087
Bis(4-dimethylaminophenyl)phosphinic acid, B-00188

$C_{16}H_{21}N_2O_3PS$

O,O-Bis(3-dimethylaminophenyl) phosphorothioate, B-00189
O,O-Bis(4-dimethylaminophenyl) phosphorothioate, B-00190

$C_{16}H_{21}N_2P$

P,P-Diethyl-*N,N′*-diphenylphosphinimidic amide, *in* D-00306

$C_{16}H_{21}N_2PS$

N,N′-Dibenzyl-*P*-ethylphosphonothioic diamide, *in* E-00149

$C_{16}H_{21}N_4P$

1,1′,3,3′-Tetrahydro-1,1′,3,3′-tetramethyl-2,2′-spirobi[2*H*-1,3,2-benzodiazaphosphole], T-00142

$C_{16}H_{21}O_2P$

Diisopropyl 1-naphthylphosphonite, *in* N-00025

$C_{16}H_{21}O_4P$

2,4-Bis(1-methylethylidene)-1,3,6,9-tetraoxa-5-phospha(5-P^V)spiro[4.4]nonane; 5-Ph, *in* B-00306

$C_{16}H_{22}Cl_2N_5P_3$

2,4-Dichloro-2,4-bis(dimethylamino)-2,2,4,4,6,6-hexahydro-6,6-diphenyl-1,3,5-triaza-2,4,6-triphosphorine, D-00162

$C_{16}H_{22}F_{12}NO_2P$

1,1,2,3,3-Pentamethyl-6,6,7,7-tetrakis(trifluoromethyl)-5,8-dioxa-4-phospha(4-P^V)spiro[3.4]octane; 4-Dimethylamino, *in* P-00029

$C_{16}H_{22}N_3OP$

N,N-Diethyl-*N′,N″*-diphenylphosphoric triamide, *in* D-00364
N-2-Methylpropyl-*N′,N″*-diphenylphosphoric triamide, *in* M-00384

$C_{16}H_{22}N_3PS$

N,N-Diethyl-*N′,N″*-diphenylphosphorothioic triamide, *in* D-00398

$C_{16}H_{22}OP$

1,2,5,6-Tetrahydro-1,1,6,6,9-pentamethyl-4*H*-phospholo[3,2,1-*ij*]phosphinoline; 3-Oxide, *in* T-00123

$C_{16}H_{22}O_6P_2$
Dimethyl[1,8-naphthalenediylbis[methylene]]-bisphosphonate, *in* N-00004

$C_{16}H_{22}P$
1,2,5,6-Tetrahydro-1,1,6,6,9-pentamethyl-4*H*-phospholo[3,2,1-*ij*]phosphinoline, T-00123

$C_{16}H_{23}BOP$
N,N-Diethyl-2,6-dimethyl-1-phenyl-1,4-phosphaborin-4(1*H*)-amine; 1-Oxide, *in* D-00285

$C_{16}H_{23}BP$
N,N-Diethyl-2,6-dimethyl-1-phenyl-1,4-phosphaborin-4(1*H*)-amine, D-00285

$C_{16}H_{23}BPS$
N,N-Diethyl-2,6-dimethyl-1-phenyl-1,4-phosphaborin-4(1*H*)-amine; 1-Sulfide, *in* D-00285

$C_{16}H_{23}BPSe$
N,N-Diethyl-2,6-dimethyl-1-phenyl-1,4-phosphaborin-4(1*H*)-amine; 1-Selenide, *in* D-00285

$C_{16}H_{23}N_2O_9P$
Methyl 6-deoxy-2,3-di-*O*-methyl-6-methyl-aminoglucopyranoside 4,6-cyclic phosphonamide; *P*-(4-Nitrophenoxy), *in* M-00112

$C_{16}H_{23}O_4P$
Bis(3-hydroxypropyl)phenylphosphine; Di-Ac, *in* B-00286

$C_{16}H_{23}O_4PS$
3,3′-(Phenylphosphinothioylidene)bis[1-propyl acetate], *in* B-00286

$C_{16}H_{23}O_5P$
3,3′-(Phenylphosphinylidene)bis[1-propyl acetate], *in* B-00286

$C_{16}H_{24}IOP$
4-Oxo-1,2,2,6,6-pentamethyl-1-phenyl-phosphorinanium iodide, *in* T-00210

$C_{16}H_{24}NOP$
Hexahydrospiro[1,3,2-benzoxazaphosphole-2(3*H*),1′-phospholane]; 2-Ph, *in* H-00052

$C_{16}H_{24}NO_3P$
Dibutyl [(4-cyanophenyl)methyl]phosphonate, *in* C-00221

$C_{16}H_{24}NO_6P$
Methyl 6-deoxy-2,3-di-*O*-methyl-6-methyl-aminoglucopyranoside 4,6-cyclic phosphonamide; *P*-Ph, *in* M-00112

$C_{16}H_{24}N_5OP$
N,N-Diethyl-2,2′-diphenylphosphoramidic dihydrazide, *in* D-01109

$C_{16}H_{25}N_2O_2P$
Phenyl di-1-piperidinylphosphinate, *in* D-01190

$C_{16}H_{25}N_2O_2PS_2$
3,4-Di-*tert*-butyl-2-(4-methylphenyl)-1,3,4,2-thiadiazaphospholidin-5-one 2-sulfide, D-00097

$C_{16}H_{25}N_3OP$
N-Phenyl-*P,P*-di-1-piperidinylphosphinic amide, *in* D-01190

$C_{16}H_{25}N_5O_{15}P_2$
Adenosine 5′-diphosphoglucose, A-00042

$C_{16}H_{25}N_5O_{16}P_2$
Guanosine diphosphate mannose, G-00100

$C_{16}H_{25}O_2P$
Menthyl phenylphosphinate, *in* P-00210

$C_{16}H_{25}O_4P$
Dibutyl (2-oxo-2-phenylethyl)phosphonate, *in* O-00066
3,3,9,9-Tetramethyl-1,5,7,11-tetraoxa-6-phospha(6-P^V)spiro[5.5]undecane; 6-Ph, *in* T-00236

$C_{16}H_{26}BrP$
1-Benzyl-1,2,2,3,4,4-hexamethylphosphetanium bromide, *in* H-00074
4-*tert*-Butyl-1-methyl-1-phenyl-phosphorinanium bromide, *in* B-00586

$C_{16}H_{26}NO_4P$
Triisopropyl benzoylphosphorimidate, *in* B-00039

$C_{16}H_{27}N_2OP$
(2-Phenylethenyl)phosphonic bis(diethylamide), *in* P-00156

$C_{16}H_{27}N_2O_3P$
[Bis(diethylamino)phosphinyl]acetic acid; Ph ester, *in* B-00175

$C_{16}H_{27}N_2P$
N,N,N′,N′-Tetraethyl-*P*-(2-phenylethenyl)-phosphonous diamide, *in* P-00161

$C_{16}H_{27}O_2P$
Decylphenylphosphinic acid, D-00010
Ethyl octylphenylphosphinate, *in* O-00040
Methyl nonylphenylphosphinate, *in* N-00074

$C_{16}H_{27}O_4PSi$
Diethyl[3-phenyl-1-[(trimethylsilyl)oxy]-2-propenyl]phosphonate, D-00319

$C_{16}H_{27}P$
Dipentylphenylphosphine, D-00978

$C_{16}H_{28}NO_3P$
Bis(3-methylbutyl) phenylphosphoramidate, *in* B-00302
Dipentyl phenylphosphoramidate, *in* D-00979

$C_{16}H_{28}N_3P$
Hexahydro-3-phenyl-1,2,4,3-triazaphosphorine; 1,4-Dibutyl, *in* H-00051

$C_{16}H_{28}O_6P_2$
Tetraethyl [1,2-phenylenebis(methylene)]-bisphosphonate, *in* P-00129
Tetraethyl [1,2-phenylenebis(methylene)]-bisphosphonate, *in* P-00130
Tetraethyl [1,4-phenylenebis(methylene)]-bisphosphonate, *in* P-00131

$C_{16}H_{29}N_2P$
Bis[3-(dimethylamino)propyl]phenylphosphine, B-00195

$C_{16}H_{29}O_3P$
Diisopropyl 1-adamantylphosphonate, *in* A-00034

$C_{16}H_{29}O_9P$
O^1,O^2-Diethyl O^3-methyl 3-(diethoxyphosphinyl)-1,2,3-butanetrioate, *in* P-00383

$C_{16}H_{30}IP$
Methyltricyclopentylphosphonium iodide, *in* T-00431

$C_{16}H_{31}OPSn$
4,4-Dibutyl-1-*tert*-butyl-1,4-dihydro-1,4-phosphastannin; 1-Oxide, *in* D-00075

$C_{16}H_{31}PS_2$
Butyl dicyclohexylphosphinodithioate, *in* D-00212

$C_{16}H_{31}PSn$
4,4-Dibutyl-1-*tert*-butyl-1,4-dihydro-1,4-phosphastannin, D-00075

$C_{16}H_{32}N_8P_2$
Bis(cyclenphosphorane), B-00154

$C_{16}H_{32}O_3P_2$
1-Hydroxy-2,2,3,4,4-pentamethylphosphetane 1-oxide; Anhydride, *in* H-00168

$C_{16}H_{33}IN_3P$
Methyltris(piperidino)phosphonium iodide, *in* T-00704

$C_{16}H_{34}ClOP$
Dioctylphosphinic acid; Chloride, *in* D-00958

$C_{16}H_{34}ClO_2P$
Bis(1-methylheptyl) phosphorochloridite, B-00308

$C_{16}H_{34}ClO_3P$
Dioctyl phosphorochloridate, *in* D-00956

$C_{16}H_{34}NO_3P$
Diisopropyl [2-(butylamino)-1-hexenyl)]-phosphonate, *in* B-00540

$C_{16}H_{34}O_2P_2S_6$
2,10-Di-*tert*-butyl-1,3,6,9,11,14-hexathia-2,10-diphosphacyclohexadecane; 2,10-Dioxide, *in* D-00093

$C_{16}H_{34}O_4P_2$
2,9-Di-*tert*-butyl-1,3,8,10-tetraoxa-2,9-diphosphacyclotetradecane, D-00158

$C_{16}H_{34}O_4P_2S_2$
2,9-Di-*tert*-butyl-1,3,8,10-tetraoxa-2,9-diphosphacyclotetradecane; 2,9-Disulfide, *in* D-00158

$C_{16}H_{34}P_2S_6$
2,10-Di-*tert*-butyl-1,3,6,9,11,14-hexathia-2,10-diphosphacyclohexadecane, D-00093

$C_{16}H_{34}P_2S_8$
2,10-Di-*tert*-butyl-1,3,6,9,11,14-hexathia-2,10-diphosphacyclohexadecane; 2,10-Disulfide, *in* D-00093

$C_{16}H_{35}OP$
▷Dioctylphosphine; Oxide, *in* D-00957

$C_{16}H_{35}O_2P$
Dioctylphosphinic acid, D-00958

$C_{16}H_{35}O_2PS_2$
▷*O,O*-Bis(2-ethylhexyl) phosphorodithioate, B-00274
O,O-Dioctyl phosphorodithioate, D-00960

$C_{16}H_{35}O_3P$
▷Bis(2-ethylhexyl) phosphonate, B-00273
Dibutyl octylphosphonate, *in* O-00042
Dioctyl phosphonate, D-00959

$C_{16}H_{35}O_4P$
Dioctyl phosphate, D-00956
Monohexadecyl phosphate, M-00475

$C_{16}H_{35}P$
Dioctylphosphine, D-00957

$C_{16}H_{36}As_2P_2$
Tetra-*tert*-butyl-1,2,3,4-diphosphadiarsetane, T-00013

$C_{16}H_{36}BF_4P$
Tetra-*tert*-butylphosphonium(1+); Tetrafluoroborate, *in* T-00020

$C_{16}H_{36}BrP$
▷Tetrabutylphosphonium(1+); Bromide, *in* T-00019

$C_{16}H_{36}ClN_2OP$
Tetrabutylphosphorodiamidic chloride, T-00022

$C_{16}H_{36}ClP$
▷Tetrabutylphosphonium(1+); Chloride, *in* T-00019

$C_{16}H_{36}IP$
▷Tetrabutylphosphonium(1+); Iodide, *in* T-00019
Tetra-*tert*-butylphosphonium(1+); Iodide, *in* T-00020

$C_{16}H_{36}NOP$
N-Butyl-*P,P*-dihexylphosphinic amide, *in* D-00434

$C_{16}H_{36}NO_2P$
Dibutyl dibutylphosphoramidite, *in* D-00136

$C_{16}H_{36}NO_3P$
Bis(2-methylpropyl) bis(2-methylpropyl)-phosphoramidate, *in* B-00355
Dibutyl dibutylphosphoramidate, *in* D-00134

$C_{16}H_{36}OP_2$
Di-*tert*-butylphosphinous acid; Anhydride, *in* D-00122
Tetrabutyldiphosphine; Monooxide, *in* T-00014
Tetra-*tert*-butyldiphosphine; Monooxide, *in* T-00015

$C_{16}H_{36}OP_2S_2$
Dibutylphosphinothioic acid; Anhydride, *in* D-00117
Di-*tert*-butylphosphinothioic acid; Anhydride, *in* D-00118

$C_{16}H_{36}O_2P_2$
Tetrabutyldiphosphine; Dioxide, *in* T-00014

$C_{16}H_{36}O_3P_2$
Dibutylphosphinic acid; Anhydride, *in* D-00109

$C_{16}H_{36}O_4P_2$
Tetrabutyl hypodiphosphite, T-00016
Tetraisopropyl 1,4-butanediylbisphosphonite, *in* B-00515

$C_{16}H_{36}O_5P_2$
Tetrabutyl pyrophosphite, T-00025

$C_{16}H_{36}O_6P_2$
Tetrabutyl hypophosphate, T-00017
Tetraethyl 1,8-octanediylbisphosphonate, *in* O-00035

$C_{16}H_{36}O_7P_2$
▷Tetrabutyl pyrophosphate, T-00024
Tetraisopropyl oxybis(1,2′-ethanediyl)-bisphosphonate, *in* O-00094

$C_{16}H_{36}P^{\oplus}$
Tetrabutylphosphonium(1+), T-00019
Tetra-*tert*-butylphosphonium(1+), T-00020

$C_{16}H_{36}P_2$
Tetrabutyldiphosphine, T-00014
Tetra-*tert*-butyldiphosphine, T-00015

$C_{16}H_{36}P_2S$
Di-*tert*-butylphosphinothious acid; Anhydrosulfide, *in* D-00120
Tetrabutyldiphosphine; Monosulfide, *in* T-00014

$C_{16}H_{36}P_2S_2$
Tetrabutyldiphosphine; Disulfide, *in* T-00014

$C_{16}H_{36}P_2S_3$
Dibutylphosphinodithioic acid; Anhydrosulfide, *in* D-00112
Di-*tert*-butylphosphinodithioic anhydrosulfide, *in* D-00113

$C_{16}H_{36}P_4$
Tetra-*tert*-butylcyclotetraphosphine, T-00012

$C_{16}H_{37}N_2OP$
N,N,N′,N′-Tetrabutylphosphonic diamide, T-00018

$C_{16}H_{37}N_2O_2P$
Tetrabutylphosphorodiamidic acid, T-00021

$C_{16}H_{37}N_2PS_2$
Tetrabutylphosphorodiamidodithioic acid, T-00023

$C_{16}H_{39}N_3O_3P_2S$
Tributyl [bis(dimethylamino)phosphinothioyl]-phosphorimidate, *in* B-00192

$C_{16}H_{40}N_4O_3P_2$
Octaethyldiphosphoramide, O-00001

$C_{16}H_{41}N_5OP_2S$
N″-[Bis(diethylamino)phosphinothioyl]-*N,N,N′,N′*-tetraethylphosphorodiamidoimidic acid, B-00174

$C_{16}H_{41}N_5O_2P_2$
N″-[Bis(diethylamino)phosphinyl]-*N,N,N′,N′*-tetraethylphosphorodiamidoimidic acid, B-00177

$C_{17}H_{10}Cl_4NO_3P$
Diphenyl (2,3,5,6-tetrachloro-4-pyridinyl)-phosphonate, *in* T-00035

$C_{17}H_{11}F_{10}O_5P$
2,2-Dihydro-2,2,2-trimethoxy-3,5-bis(pentafluorophenyl)-1,4,2-dioxaphospholane, D-00591

$C_{17}H_{12}Cl_4N_3OP$
N,N′-Diphenyl *P*-(2,3,5,6-tetrachloro-4-pyridinyl)phosphonic diamide, *in* T-00035

$C_{17}H_{14}ClN_2O_2P$
2,3-Dihydro-1*H*-naphtho[1,8-*de*]-1,3,2-diazaphosphorine; 2-(4-Chloro-3-methylphenoxy), 2-oxide, *in* D-00532

$C_{17}H_{14}ClOPS$
1-Naphthyl benzylphosphonochloridothioate, *in* B-00064
2-Naphthyl benzylphosphonochloridothioate, *in* B-00064

$C_{17}H_{14}NOP$
2-(Diphenylphosphinyl)pyridine, *in* D-01076
4-(Diphenylphosphinyl)pyridine, *in* D-01077

$C_{17}H_{14}NO_2P$
4-(Diphenylphosphino)pyridine; *N,P*-Dioxide, *in* D-01077

$C_{17}H_{14}NP$
2-(Diphenylphosphino)pyridine, D-01076
4-(Diphenylphosphino)pyridine, D-01077

$C_{17}H_{14}NPS$
2-(Diphenylphosphinothioyl)pyridine, *in* D-01076

$C_{17}H_{14}O_2P$
Methyl 2-naphthylphenylphosphinate, *in* N-00018

$C_{17}H_{14}O_2PS$
4-Methoxy-2,6-diphenyl-4*H*-1,4-thiaphosphorin 4-oxide, *in* H-00143

$C_{17}H_{15}N_2O_2PS$
OO-Diphenyl 2-pyridinylphosphoramidothioate, *in* P-00513
O,O-Diphenyl 4-pyridinylphosphoramidothioate, *in* P-00514

$C_{17}H_{15}N_2O_3P$
Diphenyl 2-pyridylphosphoramidate, *in* P-00510
Diphenyl 3-pyridylphosphoramidate, *in* P-00511
Diphenyl 4-pyridylphosphoramidate, *in* P-00512

$C_{17}H_{15}OP$
Methyl-2-naphthylphenylphosphine; Oxide, *in* M-00177

$C_{17}H_{15}O_2P$
Methyl 1-naphthylphenylphosphinate, *in* N-00017

$C_{17}H_{15}O_3PS$
5,6-Dimethyl-7-phenyl-7-phosphabicyclo[2.2.1]hept-5-ene-2,3-dicarboxylic acid; 7-Sulfide, anhydride, *in* D-00800

$C_{17}H_{15}O_4P$
Spiro[1,3,2-dioxaphosphorinane-2,2′-phenanthro[9,10-*d*][1,3,2]dioxaphosphole], S-00025

$C_{17}H_{15}P$
1,1-Dihydro-1,1-diphenylphosphorin, D-00484
Methyl-2-naphthylphenylphosphine, M-00177

$C_{17}H_{16}NO_2PS$
▷Quintiofos, Q-00006

$C_{17}H_{16}NO_3P$
Diphenyl 1-methyl-1*H*-pyrrol-2-ylphosphonate, *in* P-00521

$C_{17}H_{16}NP$
1,2-Dimethyl-4,5-diphenyl-1*H*-1,3-azaphosphole, D-00736

$C_{17}H_{17}F_{12}N_2O_3P$
6,10-Dimethyl-2,2,3,3-tetrakis(trifluoromethyl)-1,4-dioxa-6,10-diaza-5-phospha(5-P^V)spiro[4.5]decane; 5-Phenoxy, *in* D-00923

$C_{17}H_{17}N_2O_3P$
1,3-Dimethyl-6,7-diphenyl-5,8-dioxa-1,3-diaza-4-phospha(4-P^V)spiro[3.4]oct-6-en-2-one, D-00742

$C_{17}H_{17}N_4OP$
N,N′-Diphenyl-*N″*-2-pyridylphosphoric triamide, *in* P-00510
N,N′-Diphenyl-*N″*-3-pyridylphosphoric triamide, *in* P-00511

$C_{17}H_{17}O_3P$
Diphenyl (2-methyl-1,3-butadienyl)-phosphonate, *in* M-00093

$C_{17}H_{17}O_4P$
5,6-Dimethyl-7-phenyl-7-phosphabicyclo[2.2.1]hept-5-ene-2,3-dicarboxylic acid, D-00800

$C_{17}H_{17}O_5P$
2,2-Dihydrophenanthro[9,10-*d*]-1,3,2-dioxaphosphole; 2,2,2-Trimethoxy, *in* D-00541
2,3-Diphenyl-1,4,6,9-tetraoxa-5-phospha(5-P^V)-spiro[4.4]non-2-ene; 5-Methoxy, *in* D-01150

$C_{17}H_{18}NO_3P$
Diethyl 9-acridinylphosphonate, *in* A-00030
Diethyl 6-phenanthradinylphosphonate, *in* P-00060
2,3-Diphenyl-1,6,10-trioxa-4-aza-5-phospha(5-P^V)spiro[4.5]dec-2-ene, D-01158

$C_{17}H_{18}NO_4P$
2,3-Diphenyl-1,4,6-trioxa-9-aza-5-phospha(5-P^V)spiro[4.4]non-2-ene; 5-Methoxy, *in* D-01159

$C_{17}H_{18}NO_5P$
2-[(Dibenzyloxyphosphinyl)oxy]-2-propenoamide, *in* P-00366

$C_{17}H_{18}O_6P_2$
3,9-Diphenoxy-2,4,8,10-tetraoxa-3,9-diphosphaspiro[5.5]undecane, D-00991

$C_{17}H_{18}O_7P_2$
3,9-Diphenoxy-2,4,8,10-tetraoxa-3,9-diphosphaspiro[5.5]undecane; Monooxide, *in* D-00991

$C_{17}H_{18}O_8P_2$
3,9-Diphenoxy-2,4,8,10-tetraoxa-3,9-diphosphaspiro[5.5]undecane; Dioxide, *in* D-00991

$C_{17}H_{19}F_6PS_2$
1,1,3,3-Tetramethyl-6,7-bis(trifluoromethyl)-5,8-dithia-4-phospha(4-P^V)spiro[3.4]oct-6-ene; 4-Ph, *in* T-00193

$C_{17}H_{19}OP$
2,3-Diphenyl-1,3-oxaphosphepane, D-01026

$C_{17}H_{19}O_2P$
3-Methoxy-18-nor-17-phosphaestra-1,3,5(10),9(11)-tetraen-15-one, M-00054

$C_{17}H_{19}O_3P$
Diethyl (9*H*-fluoren-9-yl)phosphonate, *in* F-00011

$C_{17}H_{19}O_4PS$
2,2-Dihydro-4,5-diphenyl-1,3,2-oxathiaphosphole; 2,2,2-Trimethoxy, *in* D-00480

$C_{17}H_{19}O_5P$
17-Methyl-6,7,9,10-tetrahydrodibenzo[*d,m*][1,3,6,9,12,2]-pentaoxaphosphacyclotetradecin, M-00403

$C_{17}H_{19}O_5PS$
17-Methyl-6,7,9,10-tetrahydrodibenzo[*d,m*][1,3,6,9,12,2]-pentaoxaphosphacyclotetradecin; 17-Sulfide, *in* M-00403

$C_{17}H_{19}O_6P$
17-Methyl-6,7,9,10-tetrahydrodibenzo[*d,m*][1,3,6,9,12,2]-pentaoxaphosphacyclotetradecin; 17-Oxide, *in* M-00403

$C_{17}H_{20}BrP$
1-Benzyl-1-phenylphospholanium bromide, *in* P-00216
1,1-Diphenylphosphorinanium bromide, *in* P-00265
Methylphenyl(phenylmethyl)-2-propenylphosphonium bromide, *in* M-00264

$C_{17}H_{20}IOP$
3-Methyl-2,3-diphenyl-1,3-oxaphosphorinanium iodide, *in* D-01028

$C_{17}H_{20}IP$
7-Methyl-7-phenyl-7*H*-dibenzo[*d,f*]-phosphoninium iodide, *in* P-00111

$C_{17}H_{20}NO_2P$
(Morpholinomethyl)diphenylphosphine oxide, M-00519
N-[Phenyl(phenylmethyl)phosphinyl]-morpholine, *in* B-00056

$C_{17}H_{20}NO_3P$
Diethyl (9,10-dihydro-9-acridinyl)phosphonate, *in* D-00438
Diphenyl 1-piperidinylphosphonate, *in* P-00430
N,*N*,5,6-Tetramethylspiro[1,3,2-benzodioxaphosphole-2,1′(3′*H*)-[2,1]-benzoxaphosphol]-2-amine, *in* D-00915

$C_{17}H_{20}NO_6P$
Ethyl 2-amino-3-[(diphenoxyphosphinyl)oxy]-propanoate, *in* P-00417

$C_{17}H_{20}NP$
1-Ethyl-2,3-diphenyl-1,3-azaphosphacyclopentane, *in* D-00995

$C_{17}H_{20}NPS$
1-(Diphenylphosphinothioyl)piperidine, *in* D-01082

$C_{17}H_{20}N_3O_5P$
Ethyl *N*-[imino(diphenoxyphosphinylamino)-methyl]glycinate, *in* P-00369

$C_{17}H_{20}P_2$
5,7-Diethyl-6,7-dihydro-5*H*-dibenzo[*d,f*][1,3]-diphosphepin, D-00282

$C_{17}H_{21}BrP_2$
5,5,7,7-Tetramethyl[*d,f*][1,λ⁵3]-diphosphepinium bromide, *in* D-00577

$C_{17}H_{21}N_2O_4P$
Dibenzyl [1-(aminocarbonyl)ethyl]-phosphoramidate, *in* P-00362

$C_{17}H_{21}OPSi$
2,2-Dimethyl-5,6-diphenyl-1-oxa-5-phospha-2-silacyclohexane, D-00746

$C_{17}H_{21}O_3P$
Diethyl [(1,1′-biphenyl)-2-ylmethyl]-phosphonate, *in* B-00079
Diethyl[(1,1′-biphenyl)-4-ylmethyl]-phosphonate, *in* B-00080
Diphenyl (2,2-dimethylpropyl)phosphonate, *in* D-00912

$C_{17}H_{21}P$
Methylphenyl(phenylmethylene)-propylphosphorane, *in* B-00052

$C_{17}H_{22}BrP$
Benzylmethylphenylpropylphosphonium(1+); Bromide, *in* B-00052

$C_{17}H_{22}Br_2P_2$
6,7-Dihydro-5,5,7,7-tetramethyl-5*H*-dibenzo[*d,f*][1,3]diphosphepinium(2+); Dibromide, *in* D-00577
1,3-Dimethyl-1,3-diphenyl-1,3-diphospholanium dibromide, *in* D-01013

$C_{17}H_{22}IP$
tert-Butylmethyldiphenylphosphonium iodide, *in* B-00558
Methyl(1-methylpropyl)diphenylphosphonium iodide, *in* M-00369

$C_{17}H_{22}NO_2P$
(1-Amino-2-methylpropyl)phosphinic acid; *N*-(Diphenylmethyl), *in* A-00092

$C_{17}H_{22}NO_3P$
Diethyl (9,10-dihydro-5-methyl-9-acridinyl)-phosphonate, *in* D-00438

$C_{17}H_{22}P^{\oplus}$
Benzylmethylphenylpropylphosphonium(1+), B-00052

$C_{17}H_{22}P_2$
1,3-Bis(methylphenylphosphino)propane, B-00321

$C_{17}H_{22}P_2^{\oplus\oplus}$
6,7-Dihydro-5,5,7,7-tetramethyl-5*H*-dibenzo[*d,f*][1,3]diphosphepinium(2+), D-00577

$C_{17}H_{22}P_2S_2$
1,3-Bis(methylphenylphosphinothioyl)propane, *in* B-00321

$C_{17}H_{23}N_2O_2P$
Methyl bis(4-dimethylaminophenyl)-phosphinate, *in* B-00188

$C_{17}H_{23}N_4P$
1,1′,3,3′-Tetrahydro-1,1′,2,3,3′-pentamethyl-2,2′-spirobi[2*H*-1,3,2-benzodiazaphosphole], *in* T-00142

$C_{17}H_{24}N_3OP$
N-Pentyl-*N*′,*N*″-diphenylphosphoric triamide, *in* P-00053

$C_{17}H_{25}IP$
1,2,5,6-Tetrahydro-1,1,3,6,6,9-hexamethyl-4*H*-phospholo[3,2,1-*i,j*]phosphinolinium iodide, *in* T-00123

$C_{17}H_{25}OP$
2,3,6,7-Tetrahydro-1,1,7,7,9-pentamethyl-1*H*,5*H*-phosphorino[3,2,1-*ij*]phosphinoline; 4-Oxide, *in* T-00124

$C_{17}H_{25}P$
2,3,6,7-Tetrahydro-1,1,7,7,9-pentamethyl-1*H*,5*H*-phosphorino[3,2,1-*ij*]phosphinoline, T-00124

$C_{17}H_{27}NO_8P_2$
2,2,7,7-Tetrahydro-3,4,5,6-tetrakis(methoxycarbonyl)-1,2,2,5,5-pentamethyl-1*H*-1,2,7-azadiphosphepine, *in* T-00138

$C_{17}H_{27}O_2P$
Menthyl methylphenylphosphinate, *in* M-00219

$C_{17}H_{27}O_4P$
3,3,9,9-Tetramethyl-1,5,7,11-tetraoxa-6-phospha(6-P^V)spiro[5.5]undecane; 6-Benzyl, *in* T-00236

$C_{17}H_{27}O_5P$
Benzyl (dibutoxyphosphinyl)acetate, *in* D-00073

$C_{17}H_{29}O_2P$
Ethyl nonylphenylphosphinate, *in* N-00074

$C_{17}H_{29}P$
2,4,6-Tri-*tert*-butylphosphorin, T-00382

$C_{17}H_{30}IP$
Methyldipentylphenylphosphonium iodide, *in* D-00978

$C_{17}H_{30}N_4O_7P_2S_2$
Imcarbofos, I-00002

$C_{17}H_{33}O_2P$
Menthyl cyclohexylmethylphosphinate, *in* C-00247

$C_{17}H_{34}O_{10}P_2$
Diethyl [bis(diethoxyphosphinyl)methyl]-butanedioate, *in* D-01185
Diethyl 2,4-bis(diethoxyphosphinyl)-pentanedioate, *in* D-01186

$C_{17}H_{36}NO_4P$
Dihexyl (diethylcarbamoyl)phosphonate, *in* D-00264

$C_{17}H_{36}P_4$
1,2,4,5-Tetra-*tert*-butyl-1,2,4,5-tetraphosphaspiro[2.2]pentane, T-00026

$C_{17}H_{38}N_2P_2$
1-(Di-*tert*-butyldiphosphinylidene)-methanediamine; *N*,*N*,*N*′,*N*′-Tetra-Et, *in* D-00090

$C_{17}H_{38}O_2P_2$
Bis(dibutylphosphinyl)methane, *in* B-00160

$C_{17}H_{38}O_6P_2$
Tetraethyl 1,9-nonanediylbisphosphonate, *in* N-00069
Tetraisopropyl pentylidenebisphosphonate, *in* P-00031

$C_{17}H_{38}P_2$
Bis(dibutylphosphino)methane, B-00160

$C_{17}H_{39}IP_4$
Tetra-*tert*-butylmethyltetraphosphetanium iodide, *in* T-00012

$C_{17}H_{42}N_4P_2$
Methylenebis[*N*,*N*,*N*′,*N*′-tetraethylphosphonous diamide], *in* M-00024

$C_{18}Cl_{15}O_4P$
Tris(pentachlorophenyl) phosphate, T-00841

$C_{18}F_{15}O_4P$
Tris(pentafluorophenyl) phosphate, T-00842

$C_{18}H_5F_{10}OP$
Bis(pentafluorophenyl)phenylphosphine; Oxide, *in* B-00380

$C_{18}H_5F_{10}OPS$
O-Phenyl bis(pentafluorophenyl)-phosphinothioate, *in* B-00383

$C_{18}H_5F_{10}P$
Bis(pentafluorophenyl)phenylphosphine, B-00380

$C_{18}H_5F_{10}PS$
Bis(pentafluorophenyl)phenylphosphine; Sulfide, *in* B-00380

$C_{18}H_9ClNO_2P$
2-Chloro-2,3-dihydro-3-phenyl-4*H*-1,3,2-benzoxazaphosphorin-4-one, C-00050

$C_{18}H_{10}F_5OP$
(Pentafluorophenyl)diphenylphosphine; Oxide, *in* P-00014

$C_{18}H_{10}F_5P$
(Pentafluorophenyl)diphenylphosphine, P-00014

$C_{18}H_{10}F_5PS$
(Pentafluorophenyl)diphenylphosphine; Sulfide, *in* P-00014

$C_{18}H_{11}O_2PS$
(Phenylphosphinothioylidene)bismethanol, *in* B-00281

$C_{18}H_{12}Br_3OP$
Tris(3-bromophenyl)phosphine; Oxide, *in* T-00734
Tris(4-bromophenyl)phosphine; Oxide, *in* T-00735

$C_{18}H_{12}Br_3P$
Tris(3-bromophenyl)phosphine, T-00734
Tris(4-bromophenyl)phosphine, T-00735

$C_{18}H_{12}Br_3PS$
Tris(4-bromophenyl)phosphine; Sulfide, *in* T-00735

$C_{18}H_{12}ClN_6O_4P$
O,*O*-Bis(1-benzotriazolyl) 2-chlorophenyl phosphate, B-00098

$C_{18}H_{12}ClPS$
Tris(3-chlorophenyl)phosphine; Sulfide, *in* T-00748

$C_{18}H_{12}Cl_3OP$
Tris(2-chlorophenyl)phosphine; Oxide, *in* T-00747
Tris(3-chlorophenyl)phosphine; Oxide, *in* T-00748
▷Tris(4-chlorophenyl)phosphine; Oxide, *in* T-00749

$C_{18}H_{12}Cl_3O_3P$
Tris(2-chlorophenyl) phosphite, T-00750
Tris(4-chlorophenyl) phosphite, T-00751

$C_{18}H_{12}Cl_3O_4P$
Tris(2-chlorophenyl) phosphate, T-00745
Tris(4-chlorophenyl) phosphate, T-00746

$C_{18}H_{12}Cl_3P$
Tris(2-chlorophenyl)phosphine, T-00747
Tris(3-chlorophenyl)phosphine, T-00748
Tris(4-chlorophenyl)phosphine, T-00749

$C_{18}H_{12}Cl_3PS$
Tris(4-chlorophenyl)phosphine; Sulfide, *in* T-00749

$C_{18}H_{12}Cl_3PSe$
Tris(3-chlorophenyl)phosphine; Selenide, *in* T-00748
Tris(4-chlorophenyl)phosphine; Selenide, *in* T-00749

$C_{18}H_{12}F_3OP$
Tris(2-fluorophenyl)phosphine; Oxide, *in* T-00781
Tris(3-fluorophenyl)phosphine; Oxide, *in* T-00782
Tris(4-fluorophenyl)phosphine; Oxide, *in* T-00783

$C_{18}H_{12}F_3P$
Tris(2-fluorophenyl)phosphine, T-00781
Tris(3-fluorophenyl)phosphine, T-00782
Tris(4-fluorophenyl)phosphine, T-00783

$C_{18}H_{12}F_3PS$
Tris(3-fluorophenyl)phosphine; Sulfide, *in* T-00782
Tris(4-fluorophenyl)phosphine; Sulfide, *in* T-00783

$C_{18}H_{12}F_3PSe$
Tris(3-fluorophenyl)phosphine; Selenide, *in* T-00782
Tris(4-fluorophenyl)phosphine; Selenide, *in* T-00783

$C_{18}H_{12}NOP$
5,10[1′,2′]-Benzenophenophosphazine; Oxide, *in* B-00015

$C_{18}H_{12}NP$
5,10[1′,2′]-Benzenophenophosphazine, B-00015

$C_{18}H_{12}N_3O_6P$
Tris(3-nitrophenyl)phosphine, T-00838

$C_{18}H_{12}N_3O_6PS$
Tris(3-nitrophenyl)phosphine; Sulfide, *in* T-00838

$C_{18}H_{12}N_3O_7P$
Tris(3-nitrophenyl)phosphine; Oxide, *in* T-00838

$C_{18}H_{12}N_3O_9P$
Tris(2-nitrophenyl) phosphite, T-00839
Tris(4-nitrophenyl) phosphite, T-00840

$C_{18}H_{12}N_3O_{10}P$
Tris(2-nitrophenyl) phosphate, T-00836
Tris(4-nitrophenyl) phosphate, T-00837

$C_{18}H_{12}O_2P_2$
5,10-[1′,2′]Benzophosphenanthrene; 5,10-Dioxide, *in* B-00024

$C_{18}H_{12}O_6P^{\ominus}$
Tris[1,2-benzenediolato(2−)-*O,O′*] phosphate(1−), T-00730

$C_{18}H_{12}P_2$
5,10-[1′,2′]Benzophosphenanthrene, B-00024

$C_{18}H_{12}P_2S_2$
5,10-[1′,2′]Benzophosphenanthrene; 5,10-Disulfide, *in* B-00024

$C_{18}H_{13}Br_2O_2P$
Phenyl bis(4-bromophenyl)phosphinate, *in* B-00114

$C_{18}H_{13}Cl_3NO_3P$
Diphenyl (2,4,6-tribromophenyl)-phosphoramidate, *in* T-00359
Diphenyl (2,4,6-trichlorophenyl)-phosphoramidate, *in* T-00420

$C_{18}H_{13}N_2O_7P$
Phenylphosphonic acid; Bis(4-nitrophenyl) ester, *in* P-00221

$C_{18}H_{13}OP$
5-Phenyl-5*H*-dibenzophosphole; 5-Oxide, *in* P-00110
10-Phenyl-10*H*-phenoxaphosphine, P-00183

$C_{18}H_{13}OPS$
10-Phenyl-10*H*-phenoxaphosphine; 10-Sulfide, *in* P-00183

$C_{18}H_{13}O_2P$
6-Phenoxy-6*H*-dibenzo[*c,e*][1,2]oxaphosphorin, P-00065
6-Phenyldibenzo[*d,f*][1,3,2]dioxaphosphepin, P-00107
10-Phenyl-10*H*-phenoxaphosphine; 10-Oxide, *in* P-00183

$C_{18}H_{13}O_2PS_2$
10-Phenyl-10*H*-phenothiaphosphine; 5,5-Dioxide, 10-sulfide, *in* P-00182
2,2′-Spirobi[1,3,2-benzoxathiaphosphole]; 2-Ph, *in* S-00019

$C_{18}H_{13}O_3P$
6-Phenyldibenzo[*d,f*][1,3,2]dioxaphosphepin; Oxide, *in* P-00107

$C_{18}H_{13}O_3PS$
10-Phenyl-10*H*-phenothiaphosphine; 5,5,10-Trioxide, *in* P-00182

$C_{18}H_{13}O_4P$
2-Phenyl-2,2′-spirobi[1,3,2-benzodioxaphosphole], P-00309

$C_{18}H_{13}O_4PS$
Spiro[1,3,2-benzodioxaphosphole-2,2′-[1,3,2]-benzoxathiaphosphole]; 2-Phenoxy, *in* S-00007

$C_{18}H_{13}O_5P$
2-Phenoxy-2,2′-spirobi[1,3,2-benzodioxaphosphole], P-00074

$C_{18}H_{13}P$
5-Phenyl-5*H*-dibenzophosphole, P-00110

$C_{18}H_{13}PS$
5-Phenyl-5*H*-dibenzophosphole; 5-Sulfide, *in* P-00110
10-Phenyl-10*H*-phenothiaphosphine, P-00182

$C_{18}H_{13}PS_4$
2,2′-Spirobi[1,3,2-benzodithiaphosphole]; 2-Ph, *in* S-00015

$C_{18}H_{13}PSe$
5-Phenyl-5*H*-dibenzophosphole; 5-Selenide, *in* P-00110

$C_{18}H_{14}BrOP$
(4-Bromophenyl)diphenylphosphine; Oxide, *in* B-00481

$C_{18}H_{14}BrO_2P$
Diphenyl (4-bromophenyl)phosphonite, *in* B-00493

$C_{18}H_{14}BrP$
(2-Bromophenyl)diphenylphosphine, B-00480
(4-Bromophenyl)diphenylphosphine, B-00481

$C_{18}H_{14}BrPS$
(4-Bromophenyl)diphenylphosphine; Sulfide, *in* B-00481

$C_{18}H_{14}Br_2NO_3P$
Diphenyl (2,4-dibromophenyl)-phosphoramidate, *in* D-00067

$C_{18}H_{14}ClOP$
(2-Chlorophenyl)diphenylphosphine; Oxide, *in* C-00127
(3-Chlorophenyl)diphenylphosphine; Oxide, *in* C-00128
(4-Chlorophenyl)diphenylphosphine; Oxide, *in* C-00129

$C_{18}H_{14}ClP$
(2-Chlorophenyl)diphenylphosphine, C-00127
(3-Chlorophenyl)diphenylphosphine, C-00128
(4-Chlorophenyl)diphenylphosphine, C-00129

$C_{18}H_{14}ClPS$
(3-Chlorophenyl)diphenylphosphine; Sulfide, *in* C-00128
(4-Chlorophenyl)diphenylphosphine; Sulfide, *in* C-00129

$C_{18}H_{14}ClPSe$
(3-Chlorophenyl)diphenylphosphine; Selenide, *in* C-00128
(4-Chlorophenyl)diphenylphosphine; Selenide, *in* C-00129

$C_{18}H_{14}Cl_2NO_3P$
Diphenyl (2,4-dichlorophenyl)phosphoramidate, *in* D-00188

$C_{18}H_{14}FOP$
(2-Fluorophenyl)diphenylphosphine; Oxide, *in* F-00026
(3-Fluorophenyl)diphenylphosphine; Oxide, *in* F-00027
(4-Fluorophenyl)diphenylphosphine; Oxide, *in* F-00028

$C_{18}H_{14}FP$
(2-Fluorophenyl)diphenylphosphine, F-00026
(3-Fluorophenyl)diphenylphosphine, F-00027
(4-Fluorophenyl)diphenylphosphine, F-00028

$C_{18}H_{14}FPS$
(3-Fluorophenyl)diphenylphosphine; Sulfide, *in* F-00027
(4-Fluorophenyl)diphenylphosphine; Sulfide, *in* F-00028
(4-Fluorophenyl)diphenylphosphine; Selenide, *in* F-00028

$C_{18}H_{14}FPSe$
(3-Fluorophenyl)diphenylphosphine; Selenide, *in* F-00027

$C_{18}H_{14}NOP$
2,2-Dihydro-1,3,2-benzoxazaphosphole; 2,2-Di-Ph, *in* D-00444
2,3-Dihydro-2,3-diphenyl-1,3,2-benzoxazaphosphole, D-00478
5,10-Dihydro-10-phenylphenophosphazine; 10-Oxide, *in* D-00563

$C_{18}H_{14}NO_2P$
2,3-Dihydro-2,3-diphenyl-1,3,2-benzoxazaphosphole; 2-Oxide, *in* D-00478
(3-Nitrophenyl)diphenylphosphine, N-00039
(4-Nitrophenyl)diphenylphosphine, N-00040

$C_{18}H_{14}NO_2PS$
(3-Nitrophenyl)diphenylphosphine; Sulfide, *in* N-00039
(4-Nitrophenyl)diphenylphosphine; Sulfide, *in* N-00040

$C_{18}H_{14}NO_3P$
2,2-Dihydro-1,3,2-benzoxazaphosphole; 2,2-Diphenoxy, *in* D-00444
(3-Nitrophenyl)diphenylphosphine; Oxide, *in* N-00039
(4-Nitrophenyl)diphenylphosphine; Oxide, *in* N-00040

$C_{18}H_{14}NO_4P$
Spiro[1,3,2-benzodioxaphosphole-2,2′(3′*H*)-[1,3,2]benzoxazaphosphole]; 2-Phenoxy, *in* S-00008

$C_{18}H_{14}NP$
5,10-Dihydro-10-phenylphenophosphazine, D-00563

$C_{18}H_{14}NPS$
5,10-Dihydro-10-phenylphenophosphazine; 10-Sulfide, *in* D-00563

$C_{18}H_{14}N_3O_7P$
Diphenyl (2,4-dinitrophenyl)phosphoramidate, *in* D-00947

$C_{18}H_{15}BrNO_3P$
Diphenyl (2-bromophenyl)phosphoramidate, *in* B-00494
Diphenyl (3-bromophenyl)phosphoramidate, *in* B-00495
Diphenyl (4-bromophenyl)phosphoramidate, *in* B-00496

$C_{18}H_{15}BrNP$
N-Bromo-*P,P,P*-triphenylphosphine imide, *in* T-00626

$C_{18}H_{15}Br_2O_3P$
Dibromotriphenoxyphosphorane, D-00070

$C_{18}H_{15}Br_2P$
Dibromotriphenylphosphorane, D-00071

$C_{18}H_{15}Br_3N_3P_3$
2,4,6-Tribromo-2,2,4,4,6,6-hexahydro-2,4,6-triphenyl-1,3,5-triaza-2,4,6-triphosphorine, T-00358

$C_{18}H_{15}ClNO_2PS$
N-Benzenesulfonyl-*P*,*P*-diphenylphosphinimidic acid; Chloride, *in* B-00013

$C_{18}H_{15}ClNO_3P$
Diphenyl (2-chlorophenyl)phosphoramidate, *in* C-00155
Diphenyl (4-chlorophenyl)phosphoramidate, *in* C-00157

$C_{18}H_{15}ClNP$
P-Chloro-*N*,*P*,*P*-triphenylphosphazene, *in* T-00631
N-Chloro-*P*,*P*,*P*-triphenylphosphine imide, *in* T-00626

$C_{18}H_{15}Cl_2NOP_2$
(Triphenylphosphoranylidene)phosphoramidic dichloride, T-00656

$C_{18}H_{15}Cl_2N_2O_2P$
2,4-Dichlorophenyl *N*,*N'*-diphenylphosphorodiamidate, *in* M-00463

$C_{18}H_{15}Cl_2O_3P$
Dichlorotriphenoxyphosphorane, D-00199

$C_{18}H_{15}Cl_2P$
Dichlorotriphenylphosphorane, D-00200

$C_{18}H_{15}Cl_3N_3P_3$
2,4,6-Trichloro-2,2,4,4,6,6-hexahydro-2,4,6-triphenyl-1,3,5-triaza-2,4,6-triphosphorine, T-00407

$C_{18}H_{15}F_2O_3P$
Difluorotriphenoxyphosphorane, D-00418

$C_{18}H_{15}F_2P$
Difluorotriphenylphosphorane, D-00419

$C_{18}H_{15}I_2O_3P$
Diiodotriphenoxyphosphorane, D-00630

$C_{18}H_{15}I_2P$
▷Diiodotriphenylphosphorane, D-00631

$C_{18}H_{15}I_4P$
Iodotriphenylphosphonium triiodide, *in* D-00631

$C_{18}H_{15}NP_2S$
2,3,4-Triphenyl-1,3,2,4-thiazadiphosphetidine, T-00687

$C_{18}H_{15}NP_2S_3$
2,3,4-Triphenyl-1,3,2,4-thiazadiphosphetidine; 2,4-Disulfide, *in* T-00687

$C_{18}H_{15}N_2OP$
6,7-Dihydro-5*H*-dibenzo[*d*,*f*][1,3,2]-diazaphosphepine; 6-Ph, 6-oxide, *in* D-00450

$C_{18}H_{15}N_2O_2P$
2,2(3*H*,3'*H*)-Spirobi[1,3,2-benzoxazaphosphole]; 2-Ph, *in* S-00020

$C_{18}H_{15}N_2O_3P$
2,2(3*H*,3'*H*)-Spirobi[1,3,2-benzoxazaphosphole]; 2-Phenoxy, *in* S-00020

$C_{18}H_{15}N_2O_5P$
Diphenyl (2-nitrophenyl)phosphoramidate, *in* N-00058
Diphenyl (3-nitrophenyl)phosphoramidate, *in* N-00059
Diphenyl (4-nitrophenyl)phosphoramidate, *in* N-00060

$C_{18}H_{15}OP$
2,3-Dihydro-1*H*-benzo[*e*]phosphindole 3-oxide; 3-Ph, *in* D-00442
2-(Diphenylphosphino)phenol, D-01070
3-(Diphenylphosphino)phenol, D-01071
4-(Diphenylphosphino)phenol, D-01072
Phenyl diphenylphosphinite, *in* D-01089
Triphenylphosphine oxide, T-00627

$C_{18}H_{15}OPS$
2-(Diphenylphosphinothioyl)phenol, *in* D-01070
4-(Diphenylphosphinothioyl)phenol, *in* D-01072
O-Phenyl diphenylphosphinothioate, *in* D-01082
S-Phenyl diphenylphosphinothioate, *in* D-01082

$C_{18}H_{15}OPS_3$
S,*S*,*S*-Triphenyl phosphorotrithioate, T-00675

$C_{18}H_{15}OPSe$
O-Phenyl diphenylphosphinoselenoate, *in* D-01078
Se-Phenyl diphenylphosphinoselenoate, *in* D-01078

$C_{18}H_{15}OPSe_3$
Se,*Se*,*Se*-Triphenyl phosphorotriselenoate, T-00672

$C_{18}H_{15}O_2P$
Diphenyl phenylphosphonite, *in* P-00251
3-(Diphenylphosphino)phenol; Oxide, *in* D-01071
2-(Diphenylphosphinyl)phenol, *in* D-01070
4-(Diphenylphosphinyl)phenol, *in* D-01072
Ethyl bis(phenylethynyl)phosphinate, *in* B-00388
▷Phenyl diphenylphosphinate, *in* D-01039

$C_{18}H_{15}O_2PS$
O,*O*-Diphenyl phenylphosphonothioate, *in* P-00244

$C_{18}H_{15}O_3P$
Diphenyl phenylphosphonate, *in* P-00221
▷Triphenyl phosphite, T-00633
Tris(2-hydroxyphenyl)phosphine, T-00788
Tris(3-hydroxyphenyl)phosphine, T-00789
Tris(4-hydroxyphenyl)phosphine, T-00790

$C_{18}H_{15}O_3PS$
O,*O*,*O*-Triphenyl phosphorothioate, T-00670
O,*O*,*S*-Triphenyl phosphorothioate, T-00671

$C_{18}H_{15}O_3PSe$
O,*O*,*O*-Triphenyl phosphoroselenoate, T-00667

$C_{18}H_{15}O_3P_3$
2,4,6-Triphenyl-1,3,5,2,4,6-trioxatriphosphorinane, T-00694

$C_{18}H_{15}O_3P_3S_3$
Phenylphosphonothioic acid trimer trianhydride, *in* T-00694

$C_{18}H_{15}O_4P$
3,3',3''-Phosphinylidynetrisphenol, *in* T-00788
3,3',3''-Phosphinylidynetrisphenol, *in* T-00789
4,4',4''-Phosphinylidynetrisphenol, *in* T-00790
▷Triphenyl phosphate, T-00623

$C_{18}H_{15}O_4PS$
12*H*-Benzo[*c*]phenothiaphosphine; 12-Ethoxy, 7,7,12-trioxide, *in* B-00023

$C_{18}H_{15}O_6P$
4,4-Dihydro-4,4,4-triphenoxytrioxaphosphetane, D-00603

$C_{18}H_{15}O_6P_3$
Phenylphosphonic acid trimer trianhydride, *in* T-00694

$C_{18}H_{15}O_7P$
Tris(2-hydroxyphenyl) phosphate, T-00785
Tris(3-hydroxyphenyl) phosphate, T-00786
Tris(4-hydroxyphenyl) phosphate, T-00787

$C_{18}H_{15}P$
2,3-Dihydro-1-phenyl-1*H*-benzo[*g*]-phosphindole, D-00551
4,5-Dihydro-3-phenyl-3*H*-benzo[*e*]-phosphindole, *in* D-00441
▷Triphenylphosphine, T-00624

$C_{18}H_{15}PS$
Phenyl diphenylphosphinothioite, *in* D-01085
▷Triphenylphosphine sulfide, T-00629

$C_{18}H_{15}PSSe_3$
Se,*Se*,*Se*-Triphenyl phosphorotriselenothioate, T-00674

$C_{18}H_{15}PS_2$
Diphenyl phenylphosphonodithioite, *in* P-00238
Phenyl diphenylphosphinodithioate, *in* D-01065

$C_{18}H_{15}PS_3$
Triphenyl phosphorotrithioite, T-00676

$C_{18}H_{15}PS_4$
Triphenyl phosphorotetrathioate, T-00669

$C_{18}H_{15}PSe$
Phenyl diphenylphosphinoselenoite, *in* D-01081
Triphenylphosphine selenide, T-00628

$C_{18}H_{15}PSe_3$
Triphenyl phosphorotriselenoite, T-00673

$C_{18}H_{15}PSe_4$
Triphenyl phosphorotetraselenoate, T-00668

$C_{18}H_{15}PTe$
Triphenylphosphine; Telluride, *in* T-00624

$C_{18}H_{15}P_3$
Triphenyltriphosphirane, T-00698

$C_{18}H_{15}P_3S_2$
2,4,5-Triphenyl-1,3,2,4,5-dithiatriphospholane, T-00606

$C_{18}H_{15}P_3S_3$
2,4,5-Triphenyl-1,3,2,4,5-dithiatriphospholane; 4-Sulfide, *in* T-00606

$C_{18}H_{16}Br_2N_3OP$
N-(2,4-Dibromophenyl)-*N'*,*N''*-diphenylphosphoric triamide, *in* D-00067

$C_{18}H_{16}ClN_2O_2P$
4-Chlorophenyl *N*,*N'*-diphenylphosphorodiamidate, *in* M-00458

$C_{18}H_{16}NOP$
2-(Diphenylphosphinyl)aniline, *in* D-01047
3-(Diphenylphosphinyl)aniline, *in* D-01048
4-(Diphenylphosphinyl)aniline, *in* D-01049
N,*P*,*P*-Triphenylphosphinic amide, T-00630
N,*P*,*P*-Triphenylphosphinimidic acid, T-00631

$C_{18}H_{16}NOPS$
O-Phenyl *N*,*P*-diphenylphosphonamidothioate, *in* D-01104

$C_{18}H_{16}NO_2P$
▷Phenyl *N*,*P*-diphenylphosphonamidate, *in* D-01103

$C_{18}H_{16}NO_2PS$
O,*O*-Diphenyl phenylphosphoramidothioate, *in* P-00262

$C_{18}H_{16}NO_3P$
Diphenyl (4-aminophenyl)phosphonate, *in* A-00111
Diphenyl phenylphosphoramidate, D-01035

$C_{18}H_{16}NO_3PS$
N-Benzenesulfonyl-*P*,*P*-diphenylphosphinimidic acid, B-00013
P,*P*-Diphenyl-*N*-phenylsulfonylphosphinic amide, *in* D-01040
Diphenyl phenylsulfonylphosphoramidate, *in* P-00312

$C_{18}H_{16}NO_4PS$
O,*O*-Diphenyl phenylsulfonylphosphoramidothioate, *in* P-00314
Phenyl *P*-phenyl-*N*-phenylsulfonylphosphonamidate, *in* P-00191

$C_{18}H_{16}NP$
2-(Diphenylphosphino)aniline, D-01047
3-(Diphenylphosphino)aniline, D-01048
4-(Diphenylphosphino)aniline, D-01049
P,*P*,*P*-Triphenylphosphine imide, T-00626

$C_{18}H_{16}NPS$
3-(Diphenylphosphinothioyl)aniline, *in* D-01048
4-(Diphenylphosphinothioyl)aniline, *in* D-01049
N,*P*,*P*-Triphenylphosphinothioic amide, *in* D-01083

$C_{18}H_{16}NPSe$
N,P,P-Triphenylphosphinoselenoic amide, *in* D-01079

$C_{18}H_{16}N_3O_3P$
P-(3-Nitrophenyl)-*N,N'*-diphenylphosphonic diamide, *in* N-00056
P-(4-Nitrophenyl)-*N,N'*-diphenylphosphonic diamide, *in* N-00057

$C_{18}H_{16}N_3O_4P$
3-Nitrophenyl *N,N'*-diphenylphosphorodiamidate, *in* M-00499
4-Nitrophenyl *N,N'*-diphenylphosphorodiamidate, *in* M-00500

$C_{18}H_{16}O_2P$
Ethyl 2-naphthylphenylphosphinate, *in* N-00018

$C_{18}H_{16}O_2PS$
4-Ethoxy-2,6-diphenyl-4*H*-1,4-thiaphosphorin 4-oxide, *in* H-00143

$C_{18}H_{16}O_4P_2$
1,2-Phenylenebis[phenylphosphinic acid], P-00139
1,4-Phenylenebis[phenylphosphinic acid], P-00140

$C_{18}H_{17}BrNP$
Methyldiphenyl-4-pyridinylphosphonium bromide, *in* D-01077

$C_{18}H_{17}Cl_4O_4PS$
4,5,6,7-Tetrachloro-4',4',5',5'-tetramethylspiro[1,3,2-benzodioxaphosphole-2,2'-[1,3,2]-dioxaphospholane]; 2-Phenylthio, *in* T-00036

$C_{18}H_{17}Cl_4O_5P$
4,5,6,7-Tetrachloro-4',4',5',5'-tetramethylspiro[1,3,2-benzodioxaphosphole-2,2'-[1,3,2]-dioxaphospholane]; 2-Phenoxy, *in* T-00036

$C_{18}H_{17}INP$
2-(Diphenylphosphino)pyridine; B,MeI, *in* D-01076

$C_{18}H_{17}N_2OP$
4-Cyano-2,2-dihydro-3,5-diphenyl-3*H*-1,2-azaphosphole, *in* D-00461
▷*N,N',P*-Triphenylphosphonic diamide, *in* P-00222

$C_{18}H_{17}N_2OPS$
O-Phenyl *N,N'*-diphenylphosphorodiamidothioate, *in* D-01129

$C_{18}H_{17}N_2O_2P$
Phenyl *N,N'*-diphenylphosphorodiamidate, *in* D-01125

$C_{18}H_{17}N_2O_3P$
Diphenyl (3-aminophenyl)phosphoramidate, *in* A-00114

$C_{18}H_{17}N_2PS$
P-Phenylphosphonothioic dianilide, *in* P-00246

$C_{18}H_{17}N_4O_3P$
N-(3-Nitrophenyl)-*N',N''*-diphenylphosphoric triamide, *in* N-00059
N-(4-Nitrophenyl)-*N',N''*-diphenylphosphoric triamide, *in* N-00060

$C_{18}H_{17}OP$
2,3-Dihydro-1-phenyl-1*H*-benzo[*g*]-phosphindole; 4,5-Dihydro, 1-Oxide, *in* D-00551

$C_{18}H_{17}O_2P$
1-Ethoxy-3,4-diphenyl-1*H*-phosphole 1-oxide, *in* H-00142

$C_{18}H_{17}O_3P$
4-Ethoxy-2,6-diphenyl-4*H*-1,4-oxaphosphorin 4-oxide, *in* H-00140

$C_{18}H_{17}O_8P_2$
Picofosforic acid, P-00428

$C_{18}H_{18}ClN_2O_2P$
Fosazepam, F-00057

$C_{18}H_{18}NO_2P$
2,2-Dihydro-2,2-dimethyl-3,5-diphenyl-3*H*-1,2-azaphosphole-4-carboxylic acid, D-00461

$C_{18}H_{18}N_2O_5P_2S$
[[(Triphenylphosphoranylidene)amino]-sulfonyl]phosphorimidic acid, T-00642

$C_{18}H_{18}N_2O_9P$
Bis(2-oxo-3-oxazolidinyl) diphenyl phosphoric anhydride, *in* B-00378

$C_{18}H_{18}N_3OP$
P-(4-Aminophenyl)-*N,N'*-diphenylphosphonic diamide, *in* A-00112
N,N',N''-Triphenylphosphoric triamide, *in* T-00677

$C_{18}H_{18}N_3O_3P$
Tris(2-aminophenyl) phosphite, T-00728
Tris(3-aminophenyl) phosphite, T-00729

$C_{18}H_{18}N_3O_4P$
Tris(4-aminophenyl) phosphate, T-00727

$C_{18}H_{18}N_3P$
N,N',N''-Triphenylphosphorous triamide, T-00677

$C_{18}H_{19}N_2O_2P$
1,3-Dimethylspiro[1,3,2-diazaphospholidine-2,2'-phenanthro[9,10-*d*][1,3,2]-dioxaphosphole], D-00918

$C_{18}H_{19}N_2O_4P$
1,3-Dimethyl-6,7-diphenyl-5,8-dioxa-1,3-diaza-4-phospha(4-P^V)spiro[3.4]oct-6-en-2-one; 4-Methoxy, *in* D-00742

$C_{18}H_{19}N_2O_5P$
Phosphotryptophan; Di-Me ester, *N*-benzyloxycarbonyl, *in* P-00421

$C_{18}H_{19}N_3PS$
N,N',N''-Triphenylphosphorothioic triamide, *in* T-00677

$C_{18}H_{19}N_4PS$
N^2,$N^{2'}$,*P*-Triphenylphosphonothioic dihydrazide, *in* P-00249

$C_{18}H_{19}OP$
3-Methoxy-15-methyl-18-nor-15-phosphaestra-1,3,5(10),6,8,13-hexaene, M-00048

$C_{18}H_{19}O_2P$
3-Methoxy-15-methyl-18-nor-15-phosphaestra-1,3,5(10),6,8,13-hexaene; 15-Oxide, *in* M-00048

$C_{18}H_{19}O_4P$
Ethyl 4-(diphenylphosphinyl)-3-oxobutanoate, *in* D-01102

$C_{18}H_{19}P$
2,3,3*a*,4,5,9*b*-Hexahydro-4-phenyl-1*H*-cyclopent[*c*]isophosphinoline, H-00048

$C_{18}H_{20}ClO_2P$
3,3,3',3'-Tetramethyl-1,1'(3*H*,3'*H*)-spirobi[2,1-benzoxaphosphole]; 1-Chloro, *in* T-00232
3,3,3',3'-Tetramethyl-1,1'(3*H*,3'*H*)-spirobi[2,1-benzoxaphospholium](1+); Chloride, *in* T-00233

$C_{18}H_{20}NO_3P$
Diphenyl di-2-propenylphosphoramidate, *in* D-01195

$C_{18}H_{20}NO_3PS$
Fostedil, F-00064

$C_{18}H_{20}NO_4P$
N,N-Dimethyl-2,3-diphenyl-1,4,6,9-tetraoxa-5-phospha(5-P^V)spiro[4.4]non-2-en-5-amine, *in* D-01150

$C_{18}H_{20}O_2P^{\oplus}$
3,3,3',3'-Tetramethyl-1,1'(3*H*,3'*H*)-spirobi[2,1-benzoxaphospholium](1+), T-00233

$C_{18}H_{20}PBr$
1-Benzyl-2,5-dihydro-3-methyl-1-phenyl-1*H*-phospholium bromide, *in* D-00528
1,2,3,6-Tetrahydro-1-phenyl-1-phenylmethylphosphininium bromide, *in* T-00134

$C_{18}H_{20}P_2$
1,2,3,6-Tetrahydro-4,5-dimethyl-1,2-diphenyl-1,2-diphosphinine, *in* T-00095

$C_{18}H_{20}P_2S_2$
1,2,3,6-Tetrahydro-4,5-dimethyl-1,2-diphosphorine; 1,2-Di-Ph, 1,2-disulfide, *in* T-00095

$C_{18}H_{21}N_2O_2P$
6,9-Dimethyl-2,3-diphenyl-1,4-dioxa-6,9-diaza-5-phospha(5-P^V)spiro[4.4]non-2-ene, D-00741

$C_{18}H_{21}N_6OP$
2,2',2''-Triphenylphosphoric trihydrazide, T-00664

$C_{18}H_{21}OP$
Diphenylphosphinylcyclohexane, *in* C-00245

$C_{18}H_{21}O_2P$
Diphenyl cyclohexylphosphonite, *in* C-00267
3,3,3',3'-Tetramethyl-1,1'(3*H*,3'*H*)-spirobi[2,1-benzoxaphosphole], T-00232

$C_{18}H_{21}O_3P$
Diphenyl cyclohexylphosphonate, *in* C-00256
1-Hydroxy-3,3,3',3'-tetramethyl-1,1'(3*H*,3'*H*)-spirobi[2,1-benzoxaphosphole], H-00199
3-Methoxy-18-nor-17-phosphaestra-1,3,5(10),9(11)-tetraen-15-one; 17-Me, 17-oxide, *in* M-00054

$C_{18}H_{21}O_4PS$
4',4',5',5'-Tetramethylspiro[1,3,2-benzodioxaphosphole-2,2'-[1,3,2]-dioxaphospholane]; 2-Phenylthio, *in* T-00230

$C_{18}H_{21}O_4PSe$
4',4',5',5'-Tetramethylspiro[1,3,2-benzodioxaphosphole-2,2'-[1,3,2]-dioxaphospholane]; 2-Phenylseleno, *in* T-00230

$C_{18}H_{21}O_5P$
4',4',5',5'-Tetramethylspiro[1,3,2-benzodioxaphosphole-2,2'-[1,3,2]-dioxaphospholane]; 2-Phenoxy, *in* T-00230

$C_{18}H_{21}P$
Cyclohexyldiphenylphosphine, C-00245

$C_{18}H_{21}PS$
Diphenylphosphinothioylcyclohexane, *in* C-00245

$C_{18}H_{21}PS_2$
Diphenyl cyclohexylphosphonodithioite, *in* C-00262
3,4,5,6,7,8-Hexahydro-5-phenyl-2*H*-1,9,5-benzodithiaphosphacycloundecin, H-00044

$C_{18}H_{22}BrP$
1-Benzyl-3-methyl-1-phenylphospholanium bromide, *in* M-00229
1,1-Diphenylphosphepanium bromide, *in* P-00205
1-Phenyl-1-phenylmethylphosphorinanium bromide, *in* P-00265

$C_{18}H_{22}ClOP$
Bis(2,4,6-trimethylphenyl)phosphinic acid; Chloride, *in* B-00428

$C_{18}H_{22}IP$
1-Benzyl-3-methyl-1-phenylphospholanium iodide, *in* M-00229

$C_{18}H_{22}NOP$
N-Cyclohexyl-*P,P*-diphenylphosphinic amide, *in* D-01040
2-Phenyl-1-oxa-4-aza-5-phospha(5-P^V)-spiro[4.4]nonane; 5-Ph, *in* P-00173

$C_{18}H_{22}NO_3P$
Diphenyl cyclohexylphosphoramidate, *in* C-00269

$C_{18}H_{22}NO_4P$
2,2,2,3-Tetrahydro-2,2,2-trimethoxy-3-methyl-4,5-diphenyl-1,3,2-oxazaphosphole, T-00156
4',4',5',5'-Tetramethylspiro[1,3,2-benzoxazaphosphole-2(3*H*),2'-[1,3,2]-dioxaphospholane]; 2-Phenoxy, *in* T-00231

$C_{18}H_{22}NO_5P$
Dibenzyl [1-(methoxycarbonyl)ethyl]-phosphoramidate, *in* P-00362

$C_{18}H_{22}NO_5PS$
Methyl 2-[(dibenzyloxyphosphinyl)amino]-3-mercaptopropanoate, *in* P-00365

$C_{18}H_{22}NO_6P$
Ethyl 2-amino-3-[(diphenoxyphosphinyl)oxy]-butanoate, *in* A-00120
Methyl 3-hydroxy-2-[(dibenzyloxyphosphinyl)-amino]propanoate, *in* P-00418

$C_{18}H_{22}O_4P_2$
Diethyl 1,2-ethenediylbis(phenylphosphinate), *in* E-00018

$C_{18}H_{22}P_2$
5,8-Diethyl-5,6,7,8-tetrahydrobibenzo[*e*,*g*][1,4]diphosphocin, D-00404
2,2-Dimethyl-1,3-diphenyl-1,3-diphosphorinane, D-00745

$C_{18}H_{23}N_2OP$
P-Cyclohexyl-*N*,*N'*-diphenylphosphonic diamide, *in* C-00257

$C_{18}H_{23}N_2O_2P$
3,8-Dimethyl-2,7-diphenyl-1,6-dioxa-4,9-diaza-5-phospha(5-P^V)spiro[4.4]nonane, D-00740

$C_{18}H_{23}N_2O_3P$
1,4-Dimethyl-3,5-diphenyl-1,4,2-diazaphospholidine 2-oxide; 2-Ethoxy, *in* D-00739

$C_{18}H_{23}N_2PS$
P-Cyclohexyl-*N*,*N'*-diphenylphosphonothioic diamide, *in* C-00265

$C_{18}H_{23}OP$
Bis(2,4,6-trimethylphenyl)phosphine; Oxide, *in* B-00426
Butyl bis(2-methylphenyl)phosphinite, *in* B-00323
Butyl dibenzylphosphinite, *in* D-00058
1-(Diphenylphosphinyl)hexane, *in* H-00093

$C_{18}H_{23}O_2P$
Bis(2,4,5-trimethylphenyl)phosphinic acid, B-00427
Bis(2,4,6-trimethylphenyl)phosphinic acid, B-00428

$C_{18}H_{23}O_5P$
Diphenyl (2,2-diethoxyethyl)phosphonate, *in* D-00234

$C_{18}H_{23}P$
Bis(2,4,6-trimethylphenyl)phosphine, B-00426
Hexyldiphenylphosphine, H-00093

$C_{18}H_{23}PS$
1-(Diphenylphosphinothioyl)hexane, *in* H-00093

$C_{18}H_{24}BrP$
Benzyl-*tert*-butylmethylphenylphosphonium bromide, *in* B-00565

$C_{18}H_{24}Br_2P_2$
1,4-Dimethyl-1,4-diphenyl-1,4-diphosphorinanium dibromide, *in* D-01017

$C_{18}H_{24}ClN_2OP$
(4-Chlorobutyl)phosphonic *N*,*N'*-dibenzylamide, *in* C-00026

$C_{18}H_{24}IP$
Butylethyldiphenylphosphonium iodide, *in* B-00557

$C_{18}H_{24}NOP$
P,*P*-Bis(2,4,6-trimethylphenyl)phosphinic amide, *in* B-00428

$C_{18}H_{24}NO_2P$
Diphenyl diisopropylphosphoramidite, *in* D-00656
Diphenyl dipropylphosphoramidite, *in* D-01215

$C_{18}H_{24}NO_3P$
Dibenzyl diethylphosphoramidate, *in* D-00343
Diphenyl dipropylphosphoramidate, *in* D-01214
Diphenyl hexylphosphoramidate, *in* H-00102
[Phenyl[(phenylmethyl)amino]methyl]-phosphonic acid; Di-Et ester, *in* P-00188

$C_{18}H_{24}NP$
P,*P*-Dibenzyl-*N*,*N*-diethylphosphinous amide, *in* D-00058
N,*N*-Diisopropyl-*P*,*P*-diphenylphosphinous amide, *in* D-01090

$C_{18}H_{24}N_2O_2P_2$
Octahydro-3,8-dimethyl-2,7-diphenyl-1,6,3,8,2,7-dioxadiazadiphosphecine, O-00006

$C_{18}H_{24}N_2O_2P_2S_2$
Octahydro-3,8-dimethyl-2,7-diphenyl-1,6,3,8,2,7-dioxadiazadiphosphecine; 2,7-Disulfide (isomer 1), *in* O-00006
Octahydro-3,8-dimethyl-2,7-diphenyl-1,6,3,8,2,7-dioxadiazadiphosphecine; 2,7-Disulfide (isomer 2), *in* O-00006

$C_{18}H_{24}N_2O_6P_2$
6,12-Diphenyl-1,4,8,11-tetraoxa-6,12-diaza-5,7-diphospha(5,7-P^V)dispiro[4.1.4.1]dodecane; 5,7-Dimethoxy, *in* D-01149

$C_{18}H_{24}N_5O_8P$
▷Bucladesine, *in* C-00233

$C_{18}H_{24}O_2P_2S_3$
O-Isopropyl *S*-(isopropoxyphenyl-phosphinothioyl) phenylphosphonodithioate, *in* P-00237

$C_{18}H_{24}O_4P_2$
Diethyl 1,2-ethanediylbis(phenylphosphinate), *in* E-00015

$C_{18}H_{25}F_{12}O_2P$
1,1,2,3,3-Pentamethyl-6,6,7,7-tetrakis(trifluoromethyl)-5,8-dioxa-4-phospha(4-P^V)spiro[3.4]octane; 4-*tert*-Butyl, *in* P-00029

$C_{18}H_{25}NO_5P_2S$
Triethyl (diphenoxyphosphinothioyl)-phosphorimidate, *in* D-00987

$C_{18}H_{25}NP_2$
P,*P*,*P*-Triethyl-*N*-(diphenylphosphino)-phosphine imide, *in* T-00442

$C_{18}H_{25}N_2OP$
Butylphosphonic *N*,*N'*-dibenzyldiamide, *in* B-00596

$C_{18}H_{25}N_2O_2P$
Ethyl bis(4-dimethylaminophenyl)phosphinate, *in* B-00188

$C_{18}H_{26}Br_2P_2$
1,2-Ethanediylbis[dimethyl-phenylphosphonium] dibromide, *in* E-00012

$C_{18}H_{26}NOP$
2,6-Di-*tert*-butyl-1,4-dihydro-4-phenyl-1,4-azaphosphorine; 4-Oxide, *in* D-00084

$C_{18}H_{26}NP$
2,6-Di-*tert*-butyl-1,4-dihydro-4-phenyl-1,4-azaphosphorine, D-00084

$C_{18}H_{26}N_9O_9P$
3,3,3-Tris-(4-morpholinophosphoranylidene)-1-(2,4,6-trinitrophenyl)triazene, T-00833

$C_{18}H_{27}NPS$
7-Phenyl-6-aza-7-phosphabicyclo[3.2.1]octane; 6-Cyclohexyl, 7-sulfide, *in* P-00086

$C_{18}H_{27}OP$
Dicyclohexylphenylphosphine; Oxide, *in* D-00208

$C_{18}H_{27}OPS$
S-Phenyl dicyclohexylphosphinothioate, *in* D-00213

$C_{18}H_{27}O_2P$
Phenyl dicyclohexylphosphinate, *in* D-00211

$C_{18}H_{27}P$
Dicyclohexylphenylphosphine, D-00208

$C_{18}H_{27}PS$
Dicyclohexylphenylphosphine; Sulfide, *in* D-00208

$C_{18}H_{28}NOPSi_2$
Trimethylsilyl *P*,*P*-diphenyl-*N*-trimethyl-silylphosphinimidate, *in* D-01157

$C_{18}H_{28}NO_3P$
Diethyl [2-(cyclohexylamino)-2-phenylethenyl]-phosphonate, D-00277

$C_{18}H_{28}N_4O_2P_2$
N,*N*,*N'*,*N'*,*N''*,*N''*-Hexamethyl-*N'''*-diphenyl-phosphinophosphorimidic triamide; Oxide, *in* H-00069

$C_{18}H_{28}N_4P_2$
N,*N*,*N'*,*N'*,*N''*,*N''*-Hexamethyl-*N'''*-diphenyl-phosphinophosphorimidic triamide, H-00069

$C_{18}H_{29}N_2OP$
N,*N'*-Dicyclohexyl-*P*-phenylphosphonic diamide, *in* P-00222

$C_{18}H_{29}N_2O_2P$
Phenyl *N*,*N'*-dicyclohexylphosphorodiamidate, *in* D-00216

$C_{18}H_{29}N_2PSi_2$
P,*P*-Diphenyl-*N*,*N'*-bis(trimethylsilyl)-phosphinimidic amide, *in* D-01043

$C_{18}H_{29}OPS$
$(S)_P$-*O*-Menthyl ethylphenylphosphinothioate, *in* E-00105

$C_{18}H_{29}O_4P$
2,2,3,3,7,7,8,8-Octamethyl-1,4,6,9-tetraoxa-5-phospha(5-P^V)spiro[4.4]nonane; 5-Ph, *in* O-00032

$C_{18}H_{29}O_4PSe$
2,2,3,3,7,7,8,8-Octamethyl-1,4,6,9-tetraoxa-5-phospha(5-P^V)spiro[4.4]nonane; 5-Phenylselenoxy, *in* O-00032

$C_{18}H_{29}O_5P$
2,2,3,3,7,7,8,8-Octamethyl-1,4,6,9-tetraoxa-5-phospha(5-P^V)spiro[4.4]nonane; 5-Phenoxy, *in* O-00032

$C_{18}H_{29}PS_2$
Dithioxo(2,4,6-tri-*tert*-butylphenyl)-phosphorane, D-01254

$C_{18}H_{29}PSe_2$
Diselenoxo(2,4,6-tri-*tert*-butylphenyl)-phosphorane, D-01235

$C_{18}H_{30}N_4P_2$
1,1,4,4-Tetrakis(diethylamino)-2,3-diphosphabutadiene, *in* T-00005

$C_{18}H_{30}O_2P$
(4-Methylphenyl)methylphosphinic acid; (−)-Menthyl ester, *in* M-00204

$C_{18}H_{31}OP$
Dihexylphenylphosphine; Oxide, *in* D-00431
(2,4,6-Tri-*tert*-butylphenyl)phosphine; Oxide, *in* T-00369

$C_{18}H_{31}O_2P$
Ethyl decylphenylphosphinate, *in* D-00010

$C_{18}H_{31}P$
Dihexylphenylphosphine, D-00431
(2,4,6-Tri-*tert*-butylphenyl)phosphine, T-00369

$C_{18}H_{31}PS$
(2,4,6-Tri-*tert*-butylphenyl)phosphine; Sulfide, *in* T-00369

$C_{18}H_{32}NO_9P_3$
Hexaethyl (nitrilotri-2,1-ethanediyl)-trisphosphonate, *in* T-00438

$C_{18}H_{33}OP$
▷Tricyclohexylphosphine; Oxide, *in* T-00426

$C_{18}H_{33}O_2P$
Cyclohexyl dicyclohexylphosphinate, *in* D-00211

$C_{18}H_{33}O_3P$
Tricyclohexyl phosphite, T-00428

$C_{18}H_{33}O_3P_3$
2,4,6-Tricyclohexyl-1,3,5,2,4,6-trioxatriphosphorinane, T-00430

$C_{18}H_{33}O_3P_3S_3$
Cyclohexylphosphonothioic acid trimer trianhydride, *in* T-00430

$C_{18}H_{33}O_4P$
Tricyclohexyl phosphate, T-00425

$C_{18}H_{33}P$
Tricyclohexylphosphine, T-00426

$C_{18}H_{33}PS$
Tricyclohexylphosphine; Sulfide, *in* T-00426

$C_{18}H_{33}PSe$
Tricyclohexylphosphine; Selenide, *in* T-00426

$C_{18}H_{33}PTe$
Tricyclohexylphosphine; Telluride, *in* T-00426

$C_{18}H_{34}NO_4P$
Dicyclohexyl (2-diethylamino-2-oxoethyl)-phosphonate, *in* D-00268

$C_{18}H_{34}NP$
P,P,P-Tricyclohexylphosphine imide, T-00427

$C_{18}H_{34}PS_2$
Cyclohexyl dicyclohexylphosphinodithioate, *in* D-00212

$C_{18}H_{36}N_2P_2$
1,2-Bis[1-(2,2,6,6-tetramethylpiperidino)]-diphosphene, B-00392

$C_{18}H_{36}N_3OP$
N,N′,N″-Tricyclohexylphosphoric triamide, *in* T-00429

$C_{18}H_{36}N_3P$
N,N′,N″-Tricyclohexylphosphorous triamide, T-00429

$C_{18}H_{36}N_3PS$
▷*N,N′,N″*-Tricyclohexylphosphorothioic triamide, *in* T-00429

$C_{18}H_{36}O_{10}P_2$
Diethyl 2,5-bis(diethoxyphosphinyl)-hexanedioate, *in* D-01184
Diethyl 2-diethoxyphosphinyl-2-[2-(diethoxyphosphinyl)ethyl]butanedioate, *in* P-00394

$C_{18}H_{38}ClO_3P$
Dinonyl phosphorochloridate, *in* D-00948

$C_{18}H_{38}OP_2$
1,2-Bis(dibutylphosphino)ethylene; Monoxide, *in* B-00159

$C_{18}H_{38}O_2P_2$
1,2-Bis(dibutylphosphinyl)ethylene, *in* B-00159

$C_{18}H_{38}P_2$
1,2-Bis(dibutylphosphino)ethylene, B-00159

$C_{18}H_{39}OP$
▷Trihexylphosphine; Oxide, *in* T-00492

$C_{18}H_{39}O_2P$
Diisopropyl dodecylphosphonite, *in* D-01265
Dinonylphosphinic acid, D-00949
Hexyl dihexylphosphinate, *in* D-00434
Tetrabutylphosphonium(1+); Acetate, *in* T-00019

$C_{18}H_{39}O_2PS_2$
O,O-Dinonyl phosphorodithioate, D-00951

$C_{18}H_{39}O_3P$
Dibutyl decylphosphonate, *in* D-00012
Dinonyl phosphonate, D-00950
Trihexyl phosphite, T-00493

$C_{18}H_{39}O_4P$
Dinonyl phosphate, D-00948
Trihexyl phosphate, T-00491

$C_{18}H_{39}O_6P_3$
2,5,5,8,11,11,14,17,17-Nonamethyl-1,3,7,9,13,15-hexaoxa-2,8,14-triphosphacyclooctadecane, N-00067

$C_{18}H_{39}O_6P_3S_3$
2,5,5,8,11,11,14,17,17-Nonamethyl-1,3,7,9,13,15-hexaoxa-2,8,14-triphosphacyclooctadecane; 2,8,14-Trisulfide, *in* N-00067

$C_{18}H_{39}P$
Trihexylphosphine, T-00492

$C_{18}H_{39}PS$
Trihexylphosphine; Sulfide, *in* T-00492

$C_{18}H_{39}PSe$
Trihexylphosphine; Selenide, *in* T-00492

$C_{18}H_{40}NO_3P$
Dibutyl decylphosphoramidate, *in* D-00013

$C_{18}H_{40}O_2P_2$
1,2-Bis(dibutylphosphinyl)ethane, *in* B-00158

$C_{18}H_{40}O_4P_2$
Tetraisopropyl 1,6-hexanediylbisphosphonite, *in* H-00086

$C_{18}H_{40}O_6P_2$
Tetrabutyl 1,2-ethanediylbisphosphonate, *in* E-00006
Tetraethyl 1,10-decanediylbisphosphonate, *in* D-00008

$C_{18}H_{40}O_7P_2$
Tetrabutyl [oxybis(methylene)]bisphosphonate, *in* O-00095

$C_{18}H_{40}P_2$
1,2-Bis(dibutylphosphino)ethane, B-00158

$C_{18}H_{40}P_4$
1,2,4,5-Tetra-*tert*-butyl-1,2,4,5-tetraphosphorinane, T-00027

$C_{18}H_{41}N_2O_2P$
Ethyl tetrabutylphosphorodiamidate, *in* T-00021

$C_{18}H_{42}N_3OP$
Hexapropylphosphoric triamide, *in* H-00092

$C_{18}H_{42}N_3O_6P_3$
2,2,4,4,6,6-Hexahydro-2,2,4,4,6,6-hexapropoxy-1,3,5-triaza-2,4,6-triphosphorine, H-00033

$C_{18}H_{42}N_3P$
Hexapropylphosphorous triamide, H-00092

$C_{18}H_{42}N_3PS$
Hexapropylphosphorothioic triamide, *in* H-00092

$C_{18}H_{42}N_3PSSi$
1,3-Di-*tert*-butyl-4,4-dimethyl-1,3,2-diazaphosphasiletidine; 2-(Di-*tert*-butyl)-amino, 2-sulfide, *in* D-00087

$C_{18}H_{42}N_3PSi$
N,N,1,3-Tetrakis(1,1-dimethylethyl)-4,4-dimethyl-1,3-diaza-2-phosphacyclobutane-2-amine, *in* D-00087

$C_{18}H_{44}N_4P_2$
P,P′-1,2-Ethanediylbis[*N,N,N′,N′*-tetraethylphosphonous diamide], *in* E-00008

$C_{18}H_{45}N_5OP_2S$
Ethyl *N″*-[bis(diethylamino)phosphinothioyl]-*N,N,N′,N′*-tetraethylphosphorodiamidoimidate, *in* B-00174

$C_{18}H_{45}N_5O_2P_2$
Ethyl *N″*-[bis(diethylamino)phosphinyl]-*N,N,N′,N′*-tetraethylphosphorodiamidoimidate, *in* B-00177

$C_{18}H_{45}O_3PSi_3$
Tris(triethylsilyl) phosphite, T-00850

$C_{18}H_{45}O_4PSi$
Tris(triethylsilyl) phosphate, T-00849

$C_{19}H_{13}NP$
P,P,P-Tricyclohexyl-*N*-methylphosphine imide, *in* T-00427

$C_{19}H_{13}OP$
10*H*-5,10[1′,2′]Benzenoacridophosphine; 5-Oxide, *in* B-00014
5-Phenyl-10(5*H*)-acridophosphinone, P-00080

$C_{19}H_{13}O_2P$
5-Phenyl-10(5*H*)-acridophosphinone; 5-Oxide, *in* P-00080

$C_{19}H_{13}P$
10*H*-5,10[1′,2′]Benzenoacridophosphine, B-00014
10-Phenylacridophosphine, P-00079

$C_{19}H_{14}ClP$
5-Chloro-5,10-dihydro-10-phenyl-acridophosphine, *in* D-00548

$C_{19}H_{14}F_3OP$
Diphenyl(3-trifluoromethylphenyl)phosphine; Oxide, *in* D-01155

$C_{19}H_{14}F_3P$
Diphenyl(2-trifluoromethylphenyl)phosphine, D-01154
Diphenyl(3-trifluoromethylphenyl)phosphine, D-01155

$C_{19}H_{14}NOP$
3-(Diphenylphosphinyl)benzonitrile, *in* D-01057
4-(Diphenylphosphinyl)benzonitrile, *in* D-01058

$C_{19}H_{14}NO_2P$
5,10-Dihydro-11*H*-dibenz[*b,e*][1,4]-azaphosphepin-11-one 5-oxide; 5-Ph, *in* D-00449

$C_{19}H_{14}NO_3P$
2,3-Dihydro-3-phenyl-4*H*-1,3,2-benzoxazaphosphorin-4-one; 2-Phenoxy, *in* D-00556

$C_{19}H_{14}NP$
2-(Diphenylphosphino)benzonitrile, D-01056
3-(Diphenylphosphino)benzonitrile, D-01057
4-(Diphenylphosphino)benzonitrile, D-01058

$C_{19}H_{14}NPS$
3-(Diphenylphosphinothioyl)benzonitrile, *in* D-01057
4-(Diphenylphosphinothioyl)benzonitrile, *in* D-01058

$C_{19}H_{15}BrCl_3P$
Tris(4-chlorophenyl)methylphosphonium bromide, *in* T-00749

$C_{19}H_{15}BrF_2P^{\oplus}$
(Bromodifluoromethyl)triphenyl-phosphonium(1+), B-00468

$C_{19}H_{15}Br_2F_2P$
(Bromodifluoromethyl)triphenyl-phosphonium(1+); Bromide, *in* B-00468

$C_{19}H_{15}Br_2P$
(Dibromomethylene)triphenylphosphorane, *in* D-00065

$C_{19}H_{15}Cl_2OP$
Boyd's chloride, *in* T-00608

$C_{19}H_{15}Cl_2P$
(Dichloromethylene)triphenylphosphorane, *in* D-00182

$C_{19}H_{15}Cl_3IP$
Tris(4-chlorophenyl)methylphosphonium iodide, *in* T-00749

$C_{19}H_{15}F_2P$
(Difluoromethylene)triphenylphosphorane, D-00414

$C_{19}H_{15}F_3IP$
Methyltris(3-fluorophenyl)phosphonium iodide, *in* T-00782
Methyltris(4-fluorophenyl)phosphonium iodide, *in* T-00783

$C_{19}H_{15}INP$
10-Methyl-5,10[1′,2′]-benzophenophosphazinium iodide, *in* B-00015

$C_{19}H_{15}N_2OP$

(Diazophenylmethyl)diphenylphosphine oxide, D-00036
2,2-Dihydro-2,2,3-triphenyl-Δ^1-1,3,2-diazaphosphetin-4-one, *in* D-00557

$C_{19}H_{15}N_2O_3P$

2,2-Dihydro-2,2-diphenoxy-3-phenyl-Δ^1-1,3,2-diazaphosphetin-4-one, *in* D-00557
Triphenyl cyanophosphorimidate, *in* C-00228

$C_{19}H_{15}N_2P$

N-Cyano-*P*,*P*,*P*-triphenylphosphine imide, *in* T-00626

$C_{19}H_{15}OP$

Benzoyldiphenylphosphine, B-00030
5,6-Dihydro-5-phenylphosphanthridene 5-oxide, *in* D-00568
2-(Diphenylphosphino)benzaldehyde, D-01050
3-(Diphenylphosphino)benzaldehyde, D-01051
4-(Diphenylphosphino)benzaldehyde, D-01052

$C_{19}H_{15}OPS$

2-(Diphenylphosphinothioyl)benzaldehyde, *in* D-01050
3-(Diphenylphosphinothioyl)benzaldehyde, *in* D-01051

$C_{19}H_{15}OPS_2$

S-Benzoyl diphenylphosphinodithioate, *in* D-01065

$C_{19}H_{15}O_2P$

Benzoyldiphenylphosphine; Oxide, *in* B-00030
5,10-Dihydro-5-hydroxy-10-phenylacridophosphine 5-oxide, *in* D-00548
2-(Diphenylphosphino)benzoic acid, D-01053
3-(Diphenylphosphino)benzoic acid, D-01054
4-(Diphenylphosphino)benzoic acid, D-01055
2-(Diphenylphosphinyl)benzaldehyde, *in* D-01050
3-(Diphenylphosphinyl)benzaldehyde, *in* D-01051
4-(Diphenylphosphinyl)benzaldehyde, *in* D-01052
5-Phenoxy-5,6-dihydrophosphanthridene 5-oxide, *in* D-00504

$C_{19}H_{15}O_2PS$

3-(Diphenylphosphinothioyl)benzoic acid, *in* D-01054
p-(Diphenylphosphinothioyl)benzoic acid, *in* D-01055

$C_{19}H_{15}O_3P$

Benzoic diphenylphosphinic anhydride, *in* D-01039
2-(Diphenylphosphino)benzoic acid; Oxide, *in* D-01053
3-(Diphenylphosphinyl)benzoic acid, *in* D-01054
4-(Diphenylphosphinyl)benzoic acid, *in* D-01055

$C_{19}H_{15}P$

5,10-Dihydro-5-phenylacridophosphine, D-00547
5,10-Dihydro-10-phenylacridophosphine, D-00548

$C_{19}H_{16}BrP$

(Bromomethylene)triphenylphosphorane, *in* B-00477

$C_{19}H_{16}Br_2P^{\oplus}$

(Dibromomethyl)triphenylphosphonium(1+), D-00065

$C_{19}H_{16}Br_3P$

(Dibromomethyl)triphenylphosphonium(1+); Bromide, *in* D-00065

$C_{19}H_{16}ClP$

(Chloromethyl)triphenylphosphonium(1+); Ylide, *in* C-00117

$C_{19}H_{16}Cl_2P^{\oplus}$

(Dichloromethyl)triphenylphosphonium(1+), D-00182

$C_{19}H_{16}Cl_3N_2OP$

N,*N'*-Diphenyl-*P*-[(4-trichloromethyl)phenyl]-phosphonic diamide, *in* T-00409

$C_{19}H_{16}Cl_3P$

(Dichloromethyl)triphenylphosphonium(1+); Chloride, *in* D-00182

$C_{19}H_{16}FP$

(Fluoromethylene)triphenylphosphorane, *in* F-00025

$C_{19}H_{16}IOP$

10-Methyl-10-phenyl-10*H*-phenoxaphosphinium iodide, *in* P-00183

$C_{19}H_{16}IP$

5-Methyl-5-phenyl-5*H*-dibenzophospholium iodide, *in* P-00110

$C_{19}H_{16}IPS$

10-Methyl-10-phenyl-10*H*-phenothiaphosphinium iodide, *in* P-00182

$C_{19}H_{16}NOP$

5-Anilino-5,6-dihydrophosphanthridene 5-oxide, *in* D-00504
10,11-Dihydro-1*H*-dibenz[*b*,*e*][1,4]-azaphosphepine 5-oxide; 5-Ph, *in* D-00448
5,10-Dihydro-5-methyl-10-phenyl-phenophosphazine; 10-Oxide, *in* D-00526
N,1,1-Triphenylphosphinecarbonamide, T-00625

$C_{19}H_{16}NOPS$

N-Benzoyl-*P*,*P*-diphenylphosphinothioic amide, *in* D-01083
N,1,1-Triphenylphosphinecarbonamide; Sulfide, *in* T-00625

$C_{19}H_{16}NO_2P$

N-Benzoyl-*P*,*P*-diphenylphosphinic amide, *in* D-01040
N,1,1-Triphenylphosphinecarbonamide; Oxide, *in* T-00625
(Triphenylphosphoranylidene)carbamic acid, T-00648

$C_{19}H_{16}NO_3PS$

O,*O*-Diphenyl benzoylphosphoramidothioate, *in* B-00038
3'-Methylspiro[1,3,2-benzodioxaphosphole-2,2(3'*H*)-[1,3,2]benzothiazaphosphole]; 2-Phenoxy, *in* M-00394

$C_{19}H_{16}NO_4P$

Diphenyl benzoylphosphoramidate, *in* B-00036

$C_{19}H_{16}NO_5P$

N-(Diphenoxyphosphinyl)-4-aminobenzoic acid, *in* P-00377
Phenyl *N*-(diphenoxyphosphinyl)carbamate, *in* D-00989

$C_{19}H_{16}NP$

5,10-Dihydro-5-methyl-10-phenyl-phenophosphazine, D-00526

$C_{19}H_{16}P_2$

2,3-Dihydro-1,3-diphenyl-1*H*-1,3-benzodiphosphole, D-00476

$C_{19}H_{17}BBrF_4P$

(Bromomethyl)triphenylphosphonium(1+); Tetrafluoroborate, *in* B-00477

$C_{19}H_{17}BClF_4P$

(Chloromethyl)triphenylphosphonium(1+); Tetrafluoroborate, *in* C-00117

$C_{19}H_{17}BrNO_2P$

(Nitromethyl)triphenylphosphonium(1+); Bromide, *in* N-00038

$C_{19}H_{17}BrP^{\oplus}$

(Bromomethyl)triphenylphosphonium(1+), B-00477

$C_{19}H_{17}Br_2P$

(Bromomethyl)triphenylphosphonium(1+); Bromide, *in* B-00477

$C_{19}H_{17}ClIP$

(Chloromethyl)triphenylphosphonium(1+); Iodide, *in* C-00117

$C_{19}H_{17}ClP^{\oplus}$

(Chloromethyl)triphenylphosphonium(1+), C-00117

$C_{19}H_{17}Cl_2P$

▷(Chloromethyl)triphenylphosphonium(1+); Chloride, *in* C-00117

$C_{19}H_{17}FIP$

(Fluoromethyl)triphenylphosponium(1+); Iodide, *in* F-00025

$C_{19}H_{17}FP^{\oplus}$

(Fluoromethyl)triphenylphosponium(1+), F-00025

$C_{19}H_{17}INO_2P$

Methyl(3-nitrophenyl)diphenylphosphonium iodide, *in* N-00039
Methyl(4-nitrophenyl)diphenylphosphonium iodide, *in* N-00040

$C_{19}H_{17}NO_2P^{\oplus}$

(Nitromethyl)triphenylphosphonium(1+), N-00038

$C_{19}H_{17}N_2O_2P$

1,3-Dihydrospiro[2*H*-1,3,2-benzodiazaphosphole-2,2'-[1,3,2]-benzodioxaphosphole]; 1-Benzyl, *in* D-00574

$C_{19}H_{17}N_2O_3P$

Diphenyl phosphorohydrazidate; 2-Benzylidene deriv., *in* D-01136

$C_{19}H_{17}N_2O_3PS$

O,*O*-Diphenyl [(phenylamino)carbonyl]-phosphoramidothioate, *in* P-00082

$C_{19}H_{17}N_2O_4P$

Diphenyl (phenylamino)-carbonylphosphoramidate, *in* P-00081

$C_{19}H_{17}N_2P$

Formaldehyde (triphenylphosphoranylidene)-hydrazone, F-00054

$C_{19}H_{17}OP$

(2-Methoxyphenyl)diphenylphosphine, M-00056
(3-Methoxyphenyl)diphenylphosphine, M-00057
(4-Methoxyphenyl)diphenylphosphine, M-00058
(2-Methylphenyl)diphenylphosphine; Oxide, *in* M-00200
(3-Methylphenyl)diphenylphosphine; Oxide, *in* M-00201
(4-Methylphenyl)diphenylphosphine; Oxide, *in* M-00202
1,2,3,4-Tetrahydro-1-phenylbenzo[*h*]-phosphinoline; 1-Oxide, *in* T-00128

$C_{19}H_{17}OPS$

O-Benzyl diphenylphosphinothioate, *in* D-01082
S-Benzyl diphenylphosphinothioate, *in* D-01082
p-(Diphenylphosphinothioyl)anisole, *in* M-00058
(2-Methoxyphenyl)diphenylphosphine; Sulfide, *in* M-00056

$C_{19}H_{17}O_2P$

Benzyl diphenylphosphinate, *in* D-01039
(Diphenylmethyl)phenylphosphinic acid, D-01023
Diphenyl (4-methylphenyl)phosphonite, *in* M-00248
m-(Diphenylphosphinyl)anisole, *in* M-00057
p-(Diphenylphosphinyl)anisole, *in* M-00058
(2-Methoxyphenyl)diphenylphosphine; Oxide, *in* M-00056

$C_{19}H_{17}O_3P$

2,2-Dihydro-1,3,2-benzodioxaphosphole; 2-Methoxy, 2,2-di-Ph, *in* D-00439
Diphenyl benzylphosphonate, *in* B-00060
Triphenylmethylphosphonic acid, T-00608

$C_{19}H_{17}O_4P$
Diphenyl (4-methoxyphenyl)phosphonate, *in* M-00072
Diphenyl (4-methylphenyl)phosphonate, *in* M-00237

$C_{19}H_{17}P$
Benzyldiphenylphosphine, B-00046
Methylenetriphenylphosphorane, M-00148
(2-Methylphenyl)diphenylphosphine, M-00200
(3-Methylphenyl)diphenylphosphine, M-00201
(4-Methylphenyl)diphenylphosphine, M-00202
1-Phenyl-1,2,3,4-tetrahydrobenzo[*h*]-phosphinoline, P-00319
1,2,3,4-Tetrahydro-1-phenylbenzo[*h*]-phosphinoline, T-00128

$C_{19}H_{17}PS$
Benzyl diphenylphosphinothioite, *in* D-01085
(2-Methylphenyl)diphenylphosphine; Sulfide, *in* M-00200
(4-Methylphenyl)diphenylphosphine; Sulfide, *in* M-00202
[2-(Methylthio)phenyl]diphenylphosphine, M-00417

$C_{19}H_{17}PS_2$
Benzyl diphenylphosphinodithioate, *in* D-01065

$C_{19}H_{17}PSe$
(2-Methylphenyl)diphenylphosphine; Selenide, *in* M-00200
[2-(Methylseleno)phenyl]diphenylphosphine, M-00393

$C_{19}H_{18}BF_4P$
Methyltriphenylphosphonium(1+); Tetrafluoroborate, *in* M-00432

$C_{19}H_{18}BrP$
Methyltriphenylphosphonium(1+); Bromide, *in* M-00432

$C_{19}H_{18}ClOP$
Hydroxymethyltriphenylphosphonium(1+); Chloride, *in* H-00163

$C_{19}H_{18}ClO_4P$
Methyltriphenylphosphonium(1+); Perchlorate, *in* M-00432

$C_{19}H_{18}ClP$
Methyltriphenylphosphonium(1+); Chloride, *in* M-00432

$C_{19}H_{18}IOP$
Hydroxymethyltriphenylphosphonium(1+); Iodide, *in* H-00163
(2-Hydroxyphenyl)methyldiphenyl-phosphonium iodide, *in* D-01070

$C_{19}H_{18}IO_3P$
Triphenylphosphite methiodide, *in* M-00429
Tris(2-hydroxyphenyl)methylphosphonium iodide, *in* T-00788

$C_{19}H_{18}IP$
Methyltriphenylphosphonium(1+); Iodide, *in* M-00432

$C_{19}H_{18}MnO_4P$
▷Methyltriphenylphosphonium(1+); Permanganate, *in* M-00432

$C_{19}H_{18}NOP$
N-Benzyl-*P*,*P*-diphenylphosphinic amide, *in* D-01040

$C_{19}H_{18}NO_2P$
Phenyl *P*-(4-methylphenyl)-*N*-phenyl-phosphonamidate, *in* M-00214

$C_{19}H_{18}NO_2PS$
O,*O*-Diphenyl (2-methylphenyl)-phosphoramidothioate, *in* M-00254
O,*O*-Diphenyl (3-methylphenyl)-phosphoramidothioate, *in* M-00255
O,*O*-Diphenyl (4-methylphenyl)-phosphoramidothioate, *in* M-00256

$C_{19}H_{18}NO_3P$
Diphenyl (aminophenylmethyl)phosphonate, *in* A-00105
Diphenyl benzylphosphoramidate, *in* B-00069
Diphenyl methylphenylphosphoramidate, *in* M-00250
Diphenyl (2-methylphenyl)phosphoramidate, *in* M-00251
Diphenyl (3-methylphenyl)phosphoramidate, *in* M-00252
Diphenyl (4-methylphenyl)phosphoramidate, *in* M-00253

$C_{19}H_{18}NO_3PS$
O,*O*-Diphenyl (4-methoxyphenyl)-phosphoramidothioate, *in* M-00079
Methyl *P*,*P*-diphenyl-*N*-phenyl-sulfonylphosphinimidate, *in* B-00013

$C_{19}H_{18}NO_4P$
Diphenyl (4-methoxyphenyl)phosphoramidate, *in* M-00076

$C_{19}H_{18}NO_5PS$
Triphenyl methylsulfonylphosphorimidate, *in* M-00402

$C_{19}H_{18}NP$
N-Methyl-*P*,*P*,*P*-triphenylphosphine imide, *in* T-00626

$C_{19}H_{18}N_3O_2P$
Benzoylphosphoramidic dianilide, *in* D-01114

$C_{19}H_{18}OP^{\oplus}$
Hydroxymethyltriphenylphosphonium(1+), H-00163

$C_{19}H_{18}O_3P^{\oplus}$
Methyltriphenoxyphosphonium(1+), M-00429

$C_{19}H_{18}P^{\oplus}$
Methyltriphenylphosphonium(1+), M-00432

$C_{19}H_{19}INP$
(3-Aminophenyl)methyldiphenylphosphonium iodide, *in* D-01048
(4-Aminophenyl)methyldiphenylphosphonium iodide, *in* D-01049

$C_{19}H_{19}N_2OP$
Benzylphosphonic dianilide, *in* B-00061
P-(2-Methylphenyl)-*N*,*N'*-diphenylphosphonic diamide, *in* M-00235
P-(4-Methylphenyl)-*N*,*N'*-diphenylphosphonic diamide, *in* M-00238

$C_{19}H_{19}N_2O_2P$
P-(4-Methoxyphenyl)-*N*,*N'*-diphenylphosphonic diamide, *in* M-00072

$C_{19}H_{19}N_2O_3P$
4-Methoxyphenyl *N*,*N'*-diphenyl-phosphorodiamidate, *in* M-00481

$C_{19}H_{19}O_2P$
2,3-Dihydro-4,5-dimethyl-2,7-diphenyl-1,2-oxaphosphepin 2-oxide, D-00462

$C_{19}H_{19}PS$
2,3-Dihydro-4,5-dimethyl-2,2-diphenyl-1,2-thiaphosphepin, D-00463

$C_{19}H_{19}PS_2$
2,3-Dihydro-4,5-dimethyl-2,2-diphenyl-1,2-thiaphosphepin; 2-Sulfide, *in* D-00463

$C_{19}H_{20}BrP$
1-Benzyl-4,5-dichloro-1-phenyl-1*H*-phosphepinium bromide, *in* D-00565
1-Benzyl-3,4-dimethyl-1-phenyl-1*H*-phospholium bromide, *in* D-00807

$C_{19}H_{20}F_3O_5PS$
3,3,3',3'-Tetramethyl-1,1'(3*H*,3'*H*)-spirobi[2,1-benzoxaphospholium](1+); Trifluoromethanesulfonate, *in* T-00233

$C_{19}H_{20}NOP$
2,3-Dihydro-4,5-dimethyl-2,7-diphenyl-1*H*-1,2-azaphosphepine; 2-Oxide, *in* D-00460

$C_{19}H_{20}NO_3P$
(Diphenylphosphinyl)isocyanoacetic acid; *tert*-Butyl ester, *in* D-01098

$C_{19}H_{20}NO_4P$
N,*N*-Dimethylspiro[1,3,2-dioxaphosphorinane-2,2'-phenanthro[9,10-*d*][1,3,2]-dioxaphosphol]-2-amine, *in* S-00025

$C_{19}H_{20}NP$
2,3-Dihydro-4,5-dimethyl-2,7-diphenyl-1*H*-1,2-azaphosphepine, D-00460

$C_{19}H_{20}NPS$
2,3-Dihydro-4,5-dimethyl-2,7-diphenyl-1*H*-1,2-azaphosphepine; 2-Sulfide, *in* D-00460

$C_{19}H_{20}N_3OP$
N-(4-Methylphenyl)-*N'*,*N''*-diphenylphosphoric triamide, *in* M-00253

$C_{19}H_{21}N_2O_3P$
1,3-Dimethylspiro[1,3,2-diazaphospholidine-2,2'-phenanthro[9,10-*d*][1,3,2]-dioxaphosphole]; 2-Methoxy, *in* D-00918

$C_{19}H_{21}N_3O_3P$
1,3-Dimethyl-6,7-diphenyl-5,8-dioxa-1,3-diaza-4-phospha(4-P^V)spiro[3.4]oct-6-en-2-one; 4-Dimethylamino, *in* D-00742

$C_{19}H_{21}O_4P$
8,8-Dimethyl-2,3-diphenyl-1,4,6,10-tetraoxa-5-phospha(5-P^V)spiro[4.5]dec-2-ene, D-00749
5,6-Dimethyl-7-phenyl-7-phosphabicyclo[2.2.1]hept-5-ene-2,3-dicarboxylic acid; Di-Me ester, *in* D-00800

$C_{19}H_{21}O_4PS$
5,6-Dimethyl-7-phenyl-7-phosphabicyclo[2.2.1]hept-5-ene-2,3-dicarboxylic acid; 7-Sulfide, Di-Me ester, *in* D-00800

$C_{19}H_{22}BrP$
1-Benzyl-2,5-dihydro-3,4-dimethyl-1-phenyl-1*H*-phospholium bromide, *in* D-00468
3,3-Diphenyl-3-phosphoniabicyclo[3.2.1]octane bromide, *in* P-00198
1-Ethyl-1,2,3,4-tetrahydro-1-phenylbenzo[*h*]-phosphinolinium bromide, *in* T-00128

$C_{19}H_{22}NOP$
P-Benzyl-*N*,*P*-diphenylphosphinic amide, *in* B-00056

$C_{19}H_{23}BrP$
1,1-Diphenylphosphocanium bromide, *in* P-00215

$C_{19}H_{23}N_2O_3P$
6,9-Dimethyl-2,3-diphenyl-1,4-dioxa-6,9-diaza-5-phospha(5-P^V)spiro[4.4]non-2-ene; 5-Methoxy, *in* D-00741

$C_{19}H_{23}N_4O_6PS$
▷Benfotiamine, B-00002

$C_{19}H_{23}OP$
6-Methyl-1,5-diphenyl-6-phosphabicyclo[3.2.1]octane; Oxide, *in* M-00134

$C_{19}H_{23}O_2P$
1-Hydroxy-3,3,3',3'-tetramethyl-1,1'(3*H*,3'*H*)-spirobi[2,1-benzoxaphosphole]; 1-Methoxy, *in* H-00199

$C_{19}H_{23}P$
6-Methyl-1,5-diphenyl-6-phosphabicyclo[3.2.1]octane, M-00134

$C_{19}H_{24}BrOP$
3-Benzyl-2,2,5-trimethyl-3-phenyl-1,3-oxaphospholanium bromide, *in* T-00522

$C_{19}H_{24}BrP$
1-Benzyl-1-phenylphosphepanium bromide, *in* P-00205

$C_{19}H_{24}NO_2P$
Benzyl *N*-cyclohexyl-*P*-phenyl-phosphonamidate, *in* C-00251

$C_{19}H_{24}NO_4P$
Diethyl (1-benzoylamino-2-phenylethyl)-phosphonate, *in* A-00101

$C_{19}H_{24}NO_6P$
Methyl 3-hydroxy-2-[(dibenzyloxyphosphinyl)-amino]butanoate, *in* P-00420

$C_{19}H_{24}N_7O_{12}P$
Adenosine 5′-uridine 5′-phosphate, A-00044

$C_{19}H_{24}P_2$
1,4-Dibenzyl-1,4-diphosphepane, D-00047

$C_{19}H_{26}NO_3P$
Diisopropyl (9,10-dihydro-5-methyl-9-acridinyl)phosphonate, *in* D-00438

$C_{19}H_{27}N_2O_3PS$
O,O-Bis(4-dimethylaminophenyl) *O*-isopropyl phosphorothioate, *in* B-00190

$C_{19}H_{27}O_8P$
Antibiotic PD 113270, *in* A-00136

$C_{19}H_{27}O_9P$
Antibiotic CI 920, A-00136

$C_{19}H_{27}O_{10}P$
Antibiotic PD 113271, *in* A-00136

$C_{19}H_{28}NOP$
2,6-Di-*tert*-butyl-1,4-dihydro-4-phenyl-1,4-azaphosphorine; 1-Me, 4-oxide, *in* D-00084

$C_{19}H_{29}O_5P$
2,2,3,3,7,7,8,8-Octamethyl-1,4,6,9-tetraoxa-5-phospha(5-P^V)spiro[4.4]nonane; 5-Benzoyl, *in* O-00032

$C_{19}H_{29}P$
(2,4,6-Tri-*tert*-butylphenyl)-methylidynephosphine, T-00368

$C_{19}H_{29}PS_2$
Benzyl dicyclohexylphosphinodithioate, *in* D-00212

$C_{19}H_{31}O_4P$
2,2,3,3,7,7,8,8-Octamethyl-1,4,6,9-tetraoxa-5-phospha(5-P^V)spiro[4.4]nonane; 5-Benzyl, *in* O-00032

$C_{19}H_{32}Cl_2P^{\oplus}$
Chlorphonium, C-00192

$C_{19}H_{32}Cl_3P$
▷Chlorphonium; Chloride, *in* C-00192

$C_{19}H_{32}O_3P$
Phenylpropylphosphinic acid; (−)-Menthyl ester, *in* P-00303

$C_{19}H_{33}O_2PSi_2$
1,2,3,4,7,8-Hexahydro-1-phenyl-5,6-bis[(trimethylsilyl)oxy]phosphocin, H-00046

$C_{19}H_{33}O_4PS_2$
Dimethyl 2-(tributylphosphoranylidene)-1,3-dithiol-4,5-dicarboxylate, D-00927

$C_{19}H_{34}BF_4O_4PS_2$
Dimethyl 2-(tributylphosphoranylidene)-1,3-dithiol-4,5-dicarboxylate; Conj. tetrafluoroborate, *in* D-00927

$C_{19}H_{34}IP$
Dihexylmethylphenylphosphonium iodide, *in* D-00431

$C_{19}H_{35}IP$
1,1,1-Tributyl-1-iodo-*N*-methyl-*N*-phenyl-phosphoranamine, *in* T-00366

$C_{19}H_{35}NP^{\oplus}$
Tributyl(methylphenylamino)-phosphonium(1+), T-00366

$C_{19}H_{36}IP$
Tricyclohexylmethylphosphonium iodide, *in* T-00426

$C_{19}H_{39}OP$
1-(Tributylphosphoranylidene)-2-heptanone, T-00381

$C_{19}H_{41}IOP_2$
Dibutylmethyl[(dibutylphosphinyl)ethenyl]-phosphonium iodide, *in* B-00159

$C_{20}H_{13}O_2P$
6-Phenyl-5*H*-dibenzo[*c,e*]phosphepin-5,7(6*H*)-dione, P-00109

$C_{20}H_{13}O_2PS_2$
4-Mercaptodinaphtho[2,1-*d*:1′,2′-*f*][1,3,2]-dioxaphosphepin 4-sulfide, M-00012

$C_{20}H_{13}O_4P$
4-Hydroxydinaphtho[2,1-*d*:1′,2′-*f*][1,3,2]-dioxaphosphepin 4-oxide, H-00131
1,1′(3*H*,3′*H*)-Spirobi[2,1-benzoxaphosphole]-3,3′-dione; 1-Ph, *in* S-00018

$C_{20}H_{14}ClOP$
Di-1-naphthylphosphinic acid; Chloride, *in* D-00941

$C_{20}H_{14}ClO_2P$
Di-1-naphthyl phosphorochloridite, D-00942
Di-2-naphthyl phosphorochloridite, D-00943

$C_{20}H_{14}ClO_3P$
Di-1-naphthyl phosphorochloridate, *in* D-00937

$C_{20}H_{15}N_2OP$
2-Diazo-2-(diphenylphosphino)-1-phenyl-ethanone, D-00029
2,4,5-Triphenyl-2*H*-1,3,2-diazaphosphole; 2-Oxide, *in* T-00600

$C_{20}H_{15}N_2OPS$
2-Diazo-2-(diphenylphosphinothioyl)-1-phenyl-ethanone, *in* D-00029

$C_{20}H_{15}N_2O_2P$
2-Diazo-2-(diphenylphosphinyl)-1-phenyl-ethanone, *in* D-00029

$C_{20}H_{15}N_2P$
3,5-Diphenyl-1*H*-1,2,4-diazaphosphole; 1-Ph, *in* D-01000
2,4,5-Triphenyl-2*H*-1,3,2-diazaphosphole, T-00600
3,4,5-Triphenyl-4*H*-1,2,4-diazaphosphole, T-00601

$C_{20}H_{15}N_4O_2P$
Phenyl bis(1*H*-benzimidazol-1-yl)phosphinate, *in* B-00097

$C_{20}H_{15}OP$
Di-1-naphthylphosphine; Oxide, *in* D-00939
Di-2-naphthylphosphine; Oxide, *in* D-00940
(Triphenylphosphoranylidene)ethenone, T-00651

$C_{20}H_{15}O_2P$
Di-1-naphthylphosphinic acid, D-00941
2,4,5-Triphenyl-1,3,2-dioxaphosphole, T-00602

$C_{20}H_{15}O_2PS$
2,4,5-Triphenyl-1,3,2-dioxaphosphole; 2-Sulfide, *in* T-00602

$C_{20}H_{15}O_2PS_2$
O,O-Di-1-naphthyl phosphorodithioate, D-00944
O,O-Di-2-naphthyl phosphorodithioate, D-00945

$C_{20}H_{15}O_4P$
Bis(phenoxycarbonyl)phenylphosphine, *in* P-00209
Di-1-naphthyl phosphate, D-00937
Di-2-naphthyl phosphate, D-00938

$C_{20}H_{15}P$
Di-1-naphthylphosphine, D-00939
Di-2-naphthylphosphine, D-00940
5-Phenyl-5*H*-dibenzo[*b,f*]phosphepin, P-00108
1,2,3-Triphenyl-1*H*-phosphirene, T-00632

$C_{20}H_{15}PS$
5-Phenyl-5*H*-dibenzo[*b,f*]phosphepin; Sulfide, *in* P-00108
(Triphenylphosphoranylidene)ethenethione, T-00650

$C_{20}H_{16}BrO_2P$
Bromo(triphenylphosphoranylidene)acetic acid, B-00504

$C_{20}H_{16}ClO_2P$
Chloro(triphenylphosphoranylidene)acetic acid, C-00190

$C_{20}H_{16}IP$
5-Methyl-10*H*-5,10[1′,2′]-benzenoacridophosphonium iodide, *in* B-00014

$C_{20}H_{16}NOP$
α-(Diphenylphosphinyl)benzeneacetonitrile, D-01095

$C_{20}H_{16}NO_2P$
5,10-Dihydro-11*H*-dibenz[*b,e*][1,4]-azaphosphepin-11-one 5-oxide; *N*-Me, 5-Ph, *in* D-00449
10,10-Dihydro-2,3-diphenyl-[1,3,2]-oxazaphospholo[2,3-*b*][1,3,2]-benzoxazaphosphole, D-00481

$C_{20}H_{16}NO_3P$
Di-2-naphthyl phosphoramidate, *in* D-00938

$C_{20}H_{16}NP$
Triphenylphosphoranylideneacetonitrile, T-00641

$C_{20}H_{16}N_5OP$
P,P-Bis(1*H*-benzimidazol-1-yl)-*N*-phenyl-phosphinic amide, *in* B-00097

$C_{20}H_{16}P_2$
1,4-Dihydro-1,4-diphenyl-1,4-benzodiphosphorin, D-00477

$C_{20}H_{17}ClNP$
▷(Cyanomethyl)triphenylphosphonium(1+); Chloride, *in* C-00217

$C_{20}H_{17}F_3IP$
Methyldiphenyl(3-trifluoromethylphenyl)-phosphonium iodide, *in* D-01155

$C_{20}H_{17}INP$
(Cyanomethyl)triphenylphosphonium(1+); Iodide, *in* C-00217

$C_{20}H_{17}NP^{\oplus}$
(Cyanomethyl)triphenylphosphonium(1+), C-00217

$C_{20}H_{17}OP$
10,11-Dihydro-5-phenyl-5*H*-dibenzo[*b,f*]-phosphepin; Oxide, *in* D-00559
7,12-Dihydro-12-phenyl-5*H*-dibenz[*c,f*][1,5]-oxaphosphocin, D-00560
2,8-Dimethyl-10-phenyl-10*H*-phenoxaphosphine, D-00799
Diphenyl(2-phenylethenyl)phosphine; Oxide, *in* D-01033
Triphenylphosphoranylideneacetaldehyde, T-00639

$C_{20}H_{17}OPS$
2,4,5-Triphenyl-1,3,4-oxathiaphospholane, T-00611

$C_{20}H_{17}OPS_2$
S,S-Diphenyl (2-phenylethenyl)-phosphonodithioate, *in* P-00159
2,4,5-Triphenyl-1,3,4-oxathiaphospholane; 4-Sulfide, *in* T-00611

$C_{20}H_{17}O_2P$
8,8-Dihydro-8,8-diphenyl-2*H*,6*H*-[1,2]-oxaphospholo[4,3,2-*hi*][2,1]-benzoxaphosphole, D-00479
7,12-Dihydro-12-phenyl-5*H*-dibenz[*c,f*][1,5]-oxaphosphocin; 12-Oxide, *in* D-00560
2,8-Dimethyl-10-phenyl-10*H*-phenoxaphosphine; 10-Oxide, *in* D-00799
2-(Diphenylphosphino)benzoic acid; Me ester, *in* D-01053
3-(Diphenylphosphino)phenol; Ac, *in* D-01071
Methyl 3-(diphenylphosphino)benzoate, *in* D-01054
Methyl 4-(diphenylphosphino)benzoate, *in* D-01055
1,1′(3*H*,3′*H*)-Spirobi[2,1-benzoxaphosphole]; 1-Ph, *in* S-00017
(Triphenylphosphoranylidene)acetic acid, T-00640

$C_{20}H_{17}O_2PS$
Methyl (4-diphenylphosphinothioyl)benzoate, *in* D-01055

$C_{20}H_{17}O_3P$
Diphenyl (1-phenylethenyl)phosphonate, *in* P-00154
2-(Diphenylphosphino)benzoic acid; Oxide, Me ester, *in* D-01053

3-(Diphenylphosphino)phenol; Oxide, Ac, *in* D-01071
Methyl 4-(diphenylphosphinyl)benzoate, *in* D-01055
2-Phenoxy-4,5-diphenyl-1,3,2-dioxaphospholane, P-00069

$C_{20}H_{17}O_3PS$
2,8-Dimethyl-10-phenyl-10*H*-phenothiaphosphine; 5,5,10-Trioxide, *in* D-00798

$C_{20}H_{17}O_4P$
2-Phenoxy-4,5-diphenyl-1,3,2-dioxaphospholane; 2-Oxide, *in* P-00069

$C_{20}H_{17}O_5P$
Phenyl(diphenoxyphosphinyl)acetate, *in* D-00988

$C_{20}H_{17}O_6P$
Ethyl bis(2-ethoxycarbonylphenyl)phosphinate, *in* B-00117

$C_{20}H_{17}P$
1,1-Dihydro-1,1-diphenyl-3*H*-phosphindole, D-00483
10,11-Dihydro-5-phenyl-5*H*-dibenzo[*b,f*]-phosphepin, D-00559
Diphenyl(2-phenylethenyl)phosphine, D-01033

$C_{20}H_{17}PS$
10,11-Dihydro-5-phenyl-5*H*-dibenzo[*b,f*]-phosphepin; Sulfide, *in* D-00559
2,8-Dimethyl-10-phenyl-10*H*-phenothiaphosphine, D-00798
Diphenyl(2-phenylethenyl)phosphine; Sulfide, *in* D-01033

$C_{20}H_{17}P_3$
2,3-Dihydro-1,2,3-triphenyl-1*H*-1,2,3-triphosphole, *in* D-00615

$C_{20}H_{18}BF_4OP$
(2-Oxoethyl)triphenylphosphonium(1+); Tetrafluoroborate, *in* O-00060

$C_{20}H_{18}BF_4O_2P$
(Carboxymethyl)triphenylphosphonium(1+); Tetrafluoroborate, *in* C-00014

$C_{20}H_{18}BrOP$
(2-Oxoethyl)triphenylphosphonium(1+); Bromide, *in* O-00060

$C_{20}H_{18}BrO_2P$
(Carboxymethyl)triphenylphosphonium(1+); Bromide, *in* C-00014

$C_{20}H_{18}BrP$
2,3-Dihydro-2,2-diphenyl-1*H*-isophosphindolium bromide, *in* D-00562
9,9-Diphenyl-9-phosphoniahomocubane bromide, *in* P-00201
▷Triphenylvinylphosphonium(1+); Bromide, *in* T-00702

$C_{20}H_{18}Br_2NO_3P$
Bis[(4-bromophenyl)methyl] phenylphosphoramidate, *in* B-00109

$C_{20}H_{18}ClOP$
Bis(4-phenyl-1,3-butadienyl)phosphinic acid; Chloride, *in* B-00385
(2-Oxoethyl)triphenylphosphonium(1+); Chloride, *in* O-00060

$C_{20}H_{18}ClO_2P$
▷(Carboxymethyl)triphenylphosphonium(1+); Chloride, *in* C-00014

$C_{20}H_{18}ClO_4P$
2,3-Dihydro-2,2-diphenyl-1*H*-isophosphindolium perchlorate, *in* D-00562

$C_{20}H_{18}IO_2P$
(2-Carboxyphenyl)methyldiphenylphosphonium iodide, *in* D-01053

$C_{20}H_{18}NOP$
N-Acetyl-*P,P,P*-triphenylphosphine imide, *in* T-00626
[(Aminocarbonyl)methylene]triphenylphosphorane, *in* A-00061
3-(Diphenylphosphino)acetanilide, *in* D-01048
5,6,7,12-Tetrahydro-12-phenyldibenz[*c,f*][1,5]-azaphosphocine; 12-Oxide, *in* T-00130

$C_{20}H_{18}NOPS$
3-(Diphenylphosphinothioyl)acetanilide, *in* D-01048

$C_{20}H_{18}NO_2P$
Methyl (triphenylphosphoranylidene)-carbamate, *in* T-00648

$C_{20}H_{18}NO_5P$
Benzyl *N*-(diphenoxyphosphinyl)carbamate, *in* D-00989

$C_{20}H_{18}NP$
5,6,7,12-Tetrahydro-12-phenyldibenz[*c,f*][1,5]-azaphosphocine, T-00130

$C_{20}H_{18}OP^{\oplus}$
(2-Oxoethyl)triphenylphosphonium(1+), O-00060

$C_{20}H_{18}O_2P^{\oplus}$
(Carboxymethyl)triphenylphosphonium(1+), C-00014

$C_{20}H_{18}O_2P_2$
Octahydro-2,4-diphenyl-1,3,5-metheno-1*H*-2,4-diphosphacyclobuta[*c,d*]pentalene; 2,4-Dioxide, *in* O-00016
3*a*,4,7,7*a*-Tetrahydro-1,8-diphenyl-4,7-phosphinidene-1*H*-phosphindole; 1,8-Dioxide, *in* T-00105

$C_{20}H_{18}P^{\oplus}$
Triphenylvinylphosphonium(1+), T-00702

$C_{20}H_{18}P_2$
Octahydro-2,4-diphenyl-1,3,5-metheno-1*H*-2,4-diphosphacyclobuta[*c,d*]pentalene, O-00016
1,2,3,4-Tetrahydro-1,4-diphenyl-1,4-benzodiphosphorin, T-00102
3*a*,4,7,7*a*-Tetrahydro-1,8-diphenyl-4,7-phosphinidene-1*H*-phosphindole, T-00105
9,10,15,16-Tetrahydrotribenzo[*b,d,h*][1,6]-diphosphecin, T-00155

$C_{20}H_{18}P_2S_2$
1,2,3,4-Tetrahydro-1,4-diphenyl-1,4-benzodiphosphorin; 1,4-Disulfide, *in* T-00102
3*a*,4,7,7*a*-Tetrahydro-1,8-diphenyl-4,7-phosphinidene-1*H*-phosphindole; 1,8-Disulfide, *in* T-00105

$C_{20}H_{19}AsP_2$
1,2,3-Triphenyl-1,3,2-diphospharsolane, T-00603

$C_{20}H_{19}BrClP$
(2-Chloroethyl)triphenylphosphonium(1+); Bromide, *in* C-00083

$C_{20}H_{19}ClNOP$
▷[(Aminocarbonyl)methyl]triphenylphosphonium(1+); Chloride, *in* A-00061

$C_{20}H_{19}ClP^{\oplus}$
(2-Chloroethyl)triphenylphosphonium(1+), C-00083

$C_{20}H_{19}NOP^{\oplus}$
[(Aminocarbonyl)methyl]triphenylphosphonium(1+), A-00061

$C_{20}H_{19}N_2O_2P$
[(Triphenylphosphoranylidene)hydrazono]-acetic acid, T-00653

$C_{20}H_{19}N_4O_2P$
Phenyl phosphorodihydrazidate; 2,2′-Dibenzylidene, *in* P-00290

$C_{20}H_{19}OP$
Bis(2-methylphenyl)phenylphosphine; Oxide, *in* B-00309
Bis(3-methylphenyl)phenylphosphine; Oxide, *in* B-00310
Bis(4-methylphenyl)phenylphosphine; Oxide, *in* B-00311
Dibenzylphenylphosphine; Oxide, *in* D-00052
Diphenyl(2-phenylethyl)phosphine; Oxide, *in* D-01034
1,4,4*a*,9,9*a*,10-Hexahydro-11-phenyl-9,10-phosphinideneanthracene; 11-Oxide, *in* H-00050
Methoxymethylenetriphenylphosphorane, *in* M-00053
Phenyl dibenzylphosphinite, *in* D-00058

$C_{20}H_{19}OPS$
7*H*-Benzo[*c*]phenothiaphosphine; 7-Butyl, 7-oxide, *in* B-00022

$C_{20}H_{19}O_2P$
Bis(4-phenyl-1,3-butadienyl)phosphinic acid, B-00385
Dibenzyl phenylphosphonite, *in* P-00251
2,2-Dihydro-2,2,2-triphenyl-1,3,2-dioxaphospholane, D-00611

$C_{20}H_{19}O_2PS$
[(Methylsulfonyl)methylene]triphenylphosphorane, M-00399

$C_{20}H_{19}P$
Bis(2-methylphenyl)phenylphosphine, B-00309
Bis(3-methylphenyl)phenylphosphine, B-00310
Bis(4-methylphenyl)phenylphosphine, B-00311
▷Dibenzylphenylphosphine, D-00052
Diphenyl(2-phenylethyl)phosphine, D-01034
Ethylidenetriphenylphosphorane, E-00076
1,4,4*a*,9,9*a*,10-Hexahydro-11-phenyl-9,10-phosphinideneanthracene, H-00050

$C_{20}H_{19}PS$
Bis(3-methylphenyl)phenylphosphine; Sulfide, *in* B-00310
Bis(4-methylphenyl)phenylphosphine; Sulfide, *in* B-00311
Dibenzylphenylphosphine; Sulfide, *in* D-00052
[(Methylthio)methylene]triphenylphosphorane, *in* M-00416

$C_{20}H_{20}BF_4P$
Ethyltriphenylphosphonium(1+); Tetrafluoroborate, *in* E-00197

$C_{20}H_{20}BrF_{12}O_2P$
1,1,2,3,3-Pentamethyl-6,6,7,7-tetrakis(trifluoromethyl)-5,8-dioxa-4-phospha(4-P^V)spiro[3.4]octane; 4-(4-Bromophenyl), *in* P-00029

$C_{20}H_{20}BrOP$
(2-Hydroxyethyl)triphenylphosphonium(1+); Bromide, *in* H-00151
(Methoxymethyl)triphenylphosphonium(1+); Bromide, *in* M-00053

$C_{20}H_{20}BrP$
Ethyltriphenylphosphonium(1+); Bromide, *in* E-00197

$C_{20}H_{20}ClOP$
(2-Hydroxyethyl)triphenylphosphonium(1+); Chloride, *in* H-00151
(Methoxymethyl)triphenylphosphonium(1+); Chloride, *in* M-00053

$C_{20}H_{20}ClP$
Ethyltriphenylphosphonium(1+); Chloride, *in* E-00197

$C_{20}H_{20}ClPS$
[(Methylthio)methyl]triphenylphosphonium(1+); Chloride, *in* M-00416

$C_{20}H_{20}IOP$
(2-Hydroxyethyl)triphenylphosphonium(1+); Iodide, *in* H-00151
(Methoxymethyl)triphenylphosphonium(1+); Iodide, *in* M-00053
(2-Methoxyphenyl)methyldiphenylphosphonium iodide, *in* M-00056
(4-Methoxyphenyl)methyldiphenylphosphonium iodide, *in* M-00058

$C_{20}H_{20}IP$
Benzylmethyldiphenylphosphonium iodide, *in* B-00046
▷Ethyltriphenylphosphonium(1+); Iodide, *in* E-00197

$C_{20}H_{20}NOP$
P,P-Bis(2-methylphenyl)-*N*-phenylphosphinic amide, *in* B-00318
P,P-Bis(4-methylphenyl)-*N*-phenylphosphinic amide, *in* B-00320

P,P-Dibenzyl-*N*-phenylphosphinic amide, *in* D-00055
2-Diphenylphosphinyl-*N,N*-dimethylaniline, *in* D-01059
3-Diphenylphosphinyl-*N,N*-dimethylaniline, *in* D-01060
4-Diphenylphosphinyl-*N,N*-dimethylaniline, *in* D-01061
N-[(Diphenylphosphinyl)methyl]-*N*-methylaniline, D-01100
Ethyl *N,P,P*-triphenylphosphinimidate, *in* T-00631

$C_{20}H_{20}NO_2P$
[(Phenylmethylamino)phenylmethyl]-phenylphosphinic acid, *in* A-00102

$C_{20}H_{20}NO_3P$
Dibenzyl phenylphosphoramidate, *in* D-00053
Diphenyl (2,4-dimethylphenyl)-phosphoramidate, *in* D-00818
Triphenyl ethylphosphorimidate, *in* E-00161

$C_{20}H_{20}NO_3PS$
Ethyl *P,P*-diphenyl-*N*-phenyl-sulfonylphosphinimidate, *in* B-00013

$C_{20}H_{20}NP$
2-(Diphenylphosphino)-*N,N*-dimethylaniline, D-01059
3-(Diphenylphosphino)-*N,N*-dimethylaniline, D-01060
4-Diphenylphosphino-*N,N*-dimethylaniline, D-01061
N-Ethyl-*P,P,P*-triphenylphosphine imide, *in* T-00626

$C_{20}H_{20}NPS$
3-Diphenylphosphinothioyl-*N,N*-dimethyl-aniline, *in* D-01060
4-Diphenylphosphinothioyl-*N,N*-dimethyl-aniline, *in* D-01061

$C_{20}H_{20}OP$
2,2-Dihydro-1,2-oxaphosphetane; 2,2,2-Tri-Ph, *in* D-00533
(Methoxymethyl)triphenylphosphonium(1+), M-00053

$C_{20}H_{20}OP^{\oplus}$
(2-Hydroxyethyl)triphenylphosphonium(1+), H-00151

$C_{20}H_{20}O_4P_2$
Dimethyl 1,2-phenylenebis(phenylphosphinate), *in* P-00139

$C_{20}H_{20}P^{\oplus}$
▷Ethyltriphenylphosphonium(1+), E-00197

$C_{20}H_{20}PS^{\oplus}$
[(Methylthio)methyl]triphenyl-phosphonium(1+), M-00416

$C_{20}H_{21}F_{12}O_2P$
1,1,2,3,3-Pentamethyl-6,6,7,7-tetrakis(trifluoromethyl)-5,8-dioxa-4-phospha(4-P^V)spiro[3.4]octane; 4-Ph (*cis*-), *in* P-00029
1,1,2,3,3-Pentamethyl-6,6,7,7-tetrakis(trifluoromethyl)-5,8-dioxa-4-phospha(4-P^V)spiro[3.4]octane; 4-Ph (*trans*-), *in* P-00029

$C_{20}H_{21}F_{12}O_3P$
1,1,2,3,3-Pentamethyl-6,6,7,7-tetrakis(trifluoromethyl)-5,8-dioxa-4-phospha(4-P^V)spiro[3.4]octane; 4-Phenoxy, *in* P-00029

$C_{20}H_{21}NO_5P_2S$
Triphenyl (dimethoxyphosphinothioyl)-phosphorimidate, *in* D-00674

$C_{20}H_{21}N_2OPS$
O-Phenyl-*N,N'*-dibenzylphosphorodiamidothioate, *in* D-00062

$C_{20}H_{21}N_2O_2P$
Phenyl *N,N'*-dibenzylphosphorodiamidate, *in* D-00061

$C_{20}H_{21}N_2PS$
N,N'-Dibenzyl-*P*-phenylphosphonothioic diamide, *in* P-00246

$C_{20}H_{21}O_4P$
2,2,2,3,4,5-Hexahydro-2,2,2-trimethoxy-3-phenylnaphth[2,1-*d*]-1,2-oxaphosphole, H-00056

$C_{20}H_{22}BrP$
9,9-Diphenyl-9-phosphoniatricyclo[4.2.1.0^{2,5}]-nonane bromide, *in* P-00204

$C_{20}H_{22}ClOP$
8-Benzyl-8-phenyl-8-phosphoniabicyclo[3.2.1]-octan-3-one chloride, *in* P-00199

$C_{20}H_{22}NO_2P$
2,2-Dihydro-4-ethoxycarbonyl-2,2-dimethyl-3,5-diphenyl-3*H*-1,2-azaphosphole, *in* D-00461

$C_{20}H_{23}O_2P$
9,9-Dimethyl-2,8-diphenyl-3,7-dioxa-1-phosphabicyclo[3.3.1]nonane, D-00743

$C_{20}H_{23}O_2PS$
9,9-Dimethyl-2,8-diphenyl-3,7-dioxa-1-phosphabicyclo[3.3.1]nonane; 1-Sulfide, *in* D-00743

$C_{20}H_{23}O_6P$
Ethyl bis(4-ethoxycarbonylphenyl)phosphinate, *in* B-00118

$C_{20}H_{24}BrP$
9,9-Diphenyl-9-phosphoniabicyclo[4.2.1]nonane bromide, *in* P-00194

$C_{20}H_{24}NO_8P$
Benzyl 2-amino-4,6-*O*-benzylidene-2-deoxy-α-D-glucopyranoside 3-(dihydrogen phosphate), *in* A-00068

$C_{20}H_{24}N_3O_2P$
N,N,1,3-Tetramethylspiro[1,3,2-diazaphospholidine-2,2'-phenanthro[9,10-*d*][1,3,2]dioxaphosphol]-2-amine, *in* D-00918

$C_{20}H_{25}BrP$
1-Benzyl-1-phenylphosphocanium bromide, *in* P-00215

$C_{20}H_{25}O_2P$
2',2',3',4',4'-Pentamethylspiro[1,3,2-benzodioxaphosphole-2,1'-phosphetane]; 2-Ph, *in* P-00027

$C_{20}H_{26}ClO_3P$
Bis(2-*tert*-butylphenyl) phosphorochloridate, *in* B-00115

$C_{20}H_{26}NO_3P$
Dibenzyl cyclohexylphosphoramidate, *in* C-00269
Diphenyl cyclooctylphosphoramidate, D-00998

$C_{20}H_{26}N_3O_2P$
N,N,6,9-Tetramethyl-2,3-diphenyl-1,4-dioxa-6,9-diaza-5-phospha(5-P^V)spiro[4.4]non-2-en-5-amine, *in* D-00741

$C_{20}H_{26}N_4O_5P_2$
1,3-Dimethyl-5,7-diphenyl-1,3,5,7-tetraaza-4-phospha(4-P^V)spiro[3.3]heptane-2,6-dione; 4-(Diethoxyphosphinyl), *in* D-00748

$C_{20}H_{26}O_2P_2$
1,6-Diphosphecane; 1,6-Di-Ph, 1,6-dioxide, *in* D-01171

$C_{20}H_{26}P_2$
1,4-Dibenzyl-1,4-diphosphocane, D-00048
1,5-Dibenzyl-1,5-diphosphocane, D-00049

$C_{20}H_{27}As_2P$
2,3,4,5,6,7,8,9-Octahydro-1,9-dimethyl-5-phenyl-1*H*-5,1,9-benzophosphadiarsacycloundecin, O-00007

$C_{20}H_{27}N_2P$
Hexahydro-1,5-bis(1-phenylethyl)-1*H*-1,5,3-diazaphosphepine, H-00019

$C_{20}H_{27}OP$
1-(Diphenylphosphinyl)octane, *in* O-00039

$C_{20}H_{27}O_3P$
Bis(4-*tert*-butylphenyl) phosphonate, B-00116
Diphenyl octylphosphonate, *in* O-00042

$C_{20}H_{27}O_4P$
Bis(2-*tert*-butylphenyl) phosphate, B-00115
▷2-Ethylhexyl diphenyl phosphate, E-00075

$C_{20}H_{27}P$
2,7-Di-*tert*-butyl-1-phenylphosphepin, D-00101
Octyldiphenylphosphine, O-00039

$C_{20}H_{27}PS$
1-(Diphenylphosphinothioyl)octane, *in* O-00039

$C_{20}H_{28}Cl_2N_2O_3P_2$
P-(4-Chlorobutyl)-*N*-phenylphosphonamidic acid; Anhydride, *in* C-00024

$C_{20}H_{28}NO_3P$
Diphenyl bis(2-methylpropyl)phosphoramidate, *in* B-00355

$C_{20}H_{28}N_2O_4P_2S_2$
6,12-Diphenyl-1,4,8,11-tetraoxa-6,12-diaza-5,7-diphospha(5,7-P^V)dispiro[4.1.4.1]dodecane; 5,7-Bis(ethylthio), *in* D-01149

$C_{20}H_{28}N_2O_6P_2$
6,12-Diphenyl-1,4,8,11-tetraoxa-6,12-diaza-5,7-diphospha(5,7-P^V)dispiro[4.1.4.1]dodecane; 5,7-Diethoxy, *in* D-01149

$C_{20}H_{28}N_{10}O_{21}P_4$
P^1,P^4-Diguanosine 5'-tetraphosphate, D-00423

$C_{20}H_{28}O_2P_2$
2,2-Bis(diethylphosphinyl)biphenyl, *in* B-00179

$C_{20}H_{28}O_4P_2$
Diisopropyl 1,2-ethanediylbis(phenyl-phosphinate), *in* E-00015

$C_{20}H_{28}P_2$
2,2'-Bis(diethylphosphino)biphenyl, B-00179

$C_{20}H_{29}O_5P$
4,6-Di-*tert*-butylspiro[1,3,2-benzodioxaphosphole-2,1'-[2,8,9]-trioxa[1]-phosphaadamantane], D-00157

$C_{20}H_{30}N_3OP$
Phenyl *N,N,N',N'*-tetraethyl-*N''*-phenyl-phosphorodiamidimidate, *in* T-00054

$C_{20}H_{30}O_6P_2$
Diethyl [1,5-naphthalenediylbis[methylene]]-bisphosphonate, *in* N-00002
Diethyl [1,6-naphthalenediylbis[methylene]]-bisphosphonate, *in* N-00003
Tetraethyl [2,3-naphthalenediylbis[methyl-ene]]bisphosphonate, *in* N-00005
Tetraethyl [2,6-naphthalenediylbis[methyl-ene]]bisphosphonate, *in* N-00006
Tetraethyl [2,7-naphthalenediylbis[methyl-ene]]bisphosphonate, *in* N-00007

$C_{20}H_{33}O_2P$
Menthyl *tert*-butylphenylphosphinate, *in* B-00579

$C_{20}H_{33}O_4P$
8,9,10,11,12,13,14,15-Octahydro-4-hydroxydinaphtho[2,1-*d*:1',2'-*f*][1,3,2]-dioxaphosphepin 4-oxide, O-00018

$C_{20}H_{35}N_2OP$
(2-Phenylethenyl)phosphonic bis(dipropylamide), *in* P-00156

$C_{20}H_{35}O_2PSSi_2$
2,3,4,7,8,9-Hexahydro-1-phenyl-5,6-bis[(trimethylsilyl)oxy]-1*H*-phosphonine; Sulfide, *in* H-00047

$C_{20}H_{35}O_2PSi_2$
2,3,4,7,8,9-Hexahydro-1-phenyl-5,6-bis[(trimethylsilyl)oxy]-1*H*-phosphonine, H-00047

$C_{20}H_{36}O_6P_2$
Tetrapropyl [1,4-phenylenebis(methylene)]-bisphosphonate, *in* P-00131

$C_{20}H_{36}O_7P_2$
Mono(3,7,11,15-tetramethyl-2,6,10,14-hexadecatetraenyl) diphosphate, M-00515

$C_{20}H_{41}PSi_2$
Bis[bis(trimethylsilyl)methylene]-phenylphosphorane, B-00107

$C_{20}H_{42}ClOP$
Didecylphosphinic acid; Chloride, *in* D-00221

$C_{20}H_{42}ClO_2PS$
O,O-Didecyl phosphorochloridothioate, D-00223

$C_{20}H_{42}ClO_3P$
Didecyl phosphorochloridate, *in* D-00220

$C_{20}H_{42}NO_4P$
Dioctyl (dimethylcarbamoylmethyl)-phosphonate, *in* D-00703

$C_{20}H_{42}N_8O_8P_2S$
2,4,8,10-Tetramethyl-3,9-dioxo-2,4,8,10-tetraaza-6-phosphoniaspiro[5.5]-undecane(1+); Sulfate, *in* T-00198

$C_{20}H_{42}O_3P$
Didecyl phosphonate, D-00222

$C_{20}H_{43}O_2P$
Butyl dioctylphosphinate, *in* D-00958
Didecylphosphinic acid, D-00221

$C_{20}H_{43}O_2PS_2$
O,O-Didecyl phosphorodithioate, D-00224

$C_{20}H_{43}O_4P$
Didecyl phosphate, D-00220

$C_{20}H_{44}I_2P_2$
1,2-Ethenediylbis[(dibutylmethyl)-phosphonium] diiodide, *in* B-00159

$C_{20}H_{44}O_2P_2$
1,4-Bis(dibutylphosphinyl)butane, *in* B-00157

$C_{20}H_{44}P_2$
1,4-Bis(dibutylphosphino)butane, B-00157

$C_{20}H_{45}BrN_3P$
P-Bromo-*N*-butyl-*P,P*-bis(dibutylamino)-phosphazene, *in* P-00004

$C_{20}H_{45}ClN_3P$
N-Butyl-*P*-chloro-*P,P*-bis(dibutylamino)-phosphazene, *in* P-00004

$C_{20}H_{45}FN_3P$
N-Butyl-*P,P*-bis(dibutylamino)-*P*-fluorophosphazene, *in* P-00004

$C_{20}H_{45}N_2O_2P$
Butyl tetrabutylphosphorodiamidate, *in* T-00021

$C_{20}H_{45}N_2P$
Pentabutylphosphonous diamide, *in* B-00615

$C_{20}H_{46}Br_2N_3P$
Bromo(butylamino)bis(dibutylamino)-phosphonium bromide, *in* P-00004

$C_{20}H_{46}Cl_2N_3P$
(Butylamino)chlorobis(dibutylamino)-phosphonium chloride, *in* P-00004

$C_{20}H_{46}N_3OP$
Pentabutylphosphorodiamidimidic acid, P-00004

$C_{20}H_{46}O_{12}P_4$
Octaethyl 1,2,3,4-butanetetrayltetraphosphonate, *in* B-00518

$C_{20}H_{48}N_4P_2$
P,P'-1,4-Butanediylbis[bis *N,N*-diethylphosphonous diamide], *in* B-00515

$C_{20}H_{49}N_5OP_2S$
Butyl *N''*-[bis(diethylamino)phosphinothioyl]-*N,N,N',N'*-tetraethylphosphorodiamidoimidate, *in* B-00174

$C_{20}H_{49}N_5O_2P_2$
Butyl *N''*-[bis(diethylamino)phosphinyl]-*N,N,N',N'*-tetraethylphosphorodiamidoimidate, *in* B-00177

$C_{21}H_{12}F_9OP$
Tris[2-(trifluoromethyl)phenyl]phosphine; Oxide, *in* T-00852
▷Tris[3-(trifluoromethyl)phenyl]phosphine; Oxide, *in* T-00853
Tris[4-(trifluoromethyl)phenyl]phosphine; Oxide, *in* T-00854

$C_{21}H_{12}F_9P$
Tris[2-(trifluoromethyl)phenyl]phosphine, T-00852
Tris[3-(trifluoromethyl)phenyl]phosphine, T-00853
Tris[4-(trifluoromethyl)phenyl]phosphine, T-00854

$C_{21}H_{12}F_9PSe$
Tris[3-(trifluoromethyl)phenyl]phosphine; Selenide, *in* T-00853

$C_{21}H_{12}N_3OP$
3,3',3''-Phosphinylidynetris[benzonitrile], *in* T-00753
4,4',4''-Phosphinylidynetris[benzonitrile], *in* T-00754

$C_{21}H_{12}N_3O_3P$
Tris(4-cyanophenyl) phosphite, T-00755

$C_{21}H_{12}N_3P$
Tris(3-cyanophenyl)phosphine, T-00753
Tris(4-cyanophenyl)phosphine, T-00754

$C_{21}H_{12}N_3PS$
3,3',3''-Phosphinothioylidynetris[benzonitrile], *in* T-00753
4,4',4''-Phosphinothioylidynetris[benzonitrile], *in* T-00754

$C_{21}H_{12}N_3PS_3$
Tri(2-benzothiazolyl)phosphine, T-00334

$C_{21}H_{12}PS$
Tris(3-methylphenyl)phosphine; Sulfide, *in* T-00807

$C_{21}H_{15}F_6P$
Triphenyl[bis(trifluoromethyl)methylene]-phosphorane, T-00597

$C_{21}H_{15}N_2P$
(Dicyanomethylene)triphenylphosphorane, D-00205

$C_{21}H_{15}O_3P$
Tribenzoylphosphine, T-00335

$C_{21}H_{15}O_4P$
4-Methoxydinaphtho[2,1-*d*,1',2'-*f*][1,3,2]-dioxaphosphepin 4-oxide, *in* H-00131

$C_{21}H_{15}O_6P$
2,5'-Diphenylspiro[4*H*-1,3,2-benzodioxaphosphorin-2,2'-[1,3,2]-dioxaphospholane]-4,4'-dione, D-01147
Tribenzoyl phosphite, T-00336
Triphenoxycarbonylphosphine, *in* P-00355
Tris(3-carboxyphenyl)phosphine, T-00739
Tris(4-carboxyphenyl)phosphine, T-00740

$C_{21}H_{15}O_6PS$
4,4',4''-Phosphinothioylidynetris[benzoic acid], *in* T-00740

$C_{21}H_{15}O_7P$
3,3',3''-Phosphinylidynetris[benzoic acid], *in* T-00739
4,4',4''-Phosphinylidynetris[benzoic acid], *in* T-00740

$C_{21}H_{15}P$
12-Phenyl-10*H*-5,10-ethenoacridophosphine, P-00151

$C_{21}H_{16}NO_2P$
Cyano(triphenylphosphoranylidene)acetic acid, C-00231

$C_{21}H_{17}NO_2P$
2,3-Dihydro-2-hydroxy-1,3,5-triphenyl-1*H*-1,2-azaphosphole 2-oxide, D-00508

$C_{21}H_{17}O_2PS_2$
3-(Ethylthio)-3-thioxo-2-(triphenyl-phosphoranylidene)propanoic acid, E-00194

$C_{21}H_{17}O_4P$
(Triphenylphosphoranylidene)propanedioic acid, T-00658

$C_{21}H_{17}P$
Triphenylpropadienylidenephosphorane, T-00678
Triphenyl(propynylidene)phosphorane, *in* T-00681

$C_{21}H_{18}BrOP$
1-Bromo-3-(triphenylphosphoranylidene)-2-propanone, B-00506

$C_{21}H_{18}BrO_2P$
Methyl bromo(triphenylphosphoranylidene)-acetate, *in* B-00504

$C_{21}H_{18}BrP$
Triphenyl-2-propynylphosphonium(1+); Bromide, *in* T-00681

$C_{21}H_{18}Br_3OP$
Tris[(4-bromophenyl)methyl]phosphine; Oxide, *in* T-00733

$C_{21}H_{18}Br_3O_4P$
Tris[(4-bromophenyl)methyl] phosphate, T-00732

$C_{21}H_{18}Br_3P$
Tris[(4-bromophenyl)methyl]phosphine, T-00733

$C_{21}H_{18}ClOP$
1-Chloro-3-(triphenylphosphoranylidene)-2-propanone, C-00191

$C_{21}H_{18}ClO_2P$
Methyl chloro(triphenylphosphoranylidene)-acetate, *in* C-00190

$C_{21}H_{18}Cl_3OP$
Tris[(4-chlorophenyl)methyl]phosphine; Oxide, *in* T-00744

$C_{21}H_{18}Cl_3P$
Tris[(4-chlorophenyl)methyl]phosphine, T-00744

$C_{21}H_{18}IP$
Methyldi-1-naphthylphosphonium iodide, *in* D-00939

$C_{21}H_{18}NOP$
1,2,3-Triphenyl-1,3-azaphospholidin-5-one, T-00596

$C_{21}H_{18}NO_2P$
1,2,3-Triphenyl-1,3-azaphospholidin-5-one; 3-Oxide, *in* T-00596

$C_{21}H_{18}NO_2PS$
1,2,3-Triphenyl-1,3-azaphospholidin-5-one; 3-Sulfide, *in* T-00596

$C_{21}H_{18}NP$
(2-Cyanoethylidene)triphenylphosphorane, *in* C-00212
2-(Triphenylphosphoranylidene)propanenitrile, T-00659

$C_{21}H_{18}N_3O_6P$
Tris[(4-nitrophenyl)methyl]phosphine, T-00835

$C_{21}H_{18}N_3O_7P$
Tris[(4-nitrophenyl)methyl]phosphine; Oxide, *in* T-00835

$C_{21}H_{18}N_3O_{10}P$
Tris[(4-nitrophenyl)methyl] phosphate, T-00834

$C_{21}H_{18}P^{\oplus}$
Triphenyl-2-propynylphosphonium(1+), T-00681

$C_{21}H_{19}BF_4NP$
(2-Cyanoethyl)triphenylphosphonium(1+); Tetrafluoroborate, *in* C-00212

$C_{21}H_{19}BrNP$
(2-Cyanoethyl)triphenylphosphonium(1+); Bromide, *in* C-00212

$C_{21}H_{19}BrOP^{\oplus}$
(3-Bromo-2-oxopropyl)triphenyl-phosphonium(1+), B-00478

$C_{21}H_{19}Br_2OP$
(3-Bromo-2-oxopropyl)triphenyl-phosphonium(1+); Bromide, *in* B-00478

$C_{21}H_{19}ClOP^{\oplus}$
(3-Chloro-2-oxopropyl)triphenylphosphonium(1+), C-00123

$C_{21}H_{19}ClO_2P^{\oplus}$
2-[(Chloroformyl)oxy]ethyltriphenylphosphonium(1+), C-00084

$C_{21}H_{19}Cl_2OP$
(3-Chloro-2-oxopropyl)triphenylphosphonium(1+); Chloride, *in* C-00123

$C_{21}H_{19}Cl_2O_2P$
2-[(Chloroformyl)oxy]ethyltriphenylphosphonium(1+); Chloride, *in* C-00084

$C_{21}H_{19}NP^{\oplus}$
(2-Cyanoethyl)triphenylphosphonium(1+), C-00212

$C_{21}H_{19}OP$
2-(Triphenylphosphoranylidene)propanal, T-00657
▷1-Triphenylphosphoranylidene-2-propanone, T-00662

$C_{21}H_{19}O_2P$
(2-Carboxyethyl)triphenylphosphonium(1+); Betaine, *in* C-00011
2,2-Dihydro-1,3,2-benzodioxaphosphole; 2,2,2-Tri-Ph, *in* D-00439
Methyl triphenylphosphoranylideneacetate, M-00433
2-(Triphenylphosphoranylidene)propanoic acid, T-00660
3-(Triphenylphosphoranylidene)propanoic acid, T-00661

$C_{21}H_{19}O_3P$
5,6-Dimethylspiro[1,3,2-benzodioxaphosphole-2,1′(3′*H*)-[2,1]benzoxaphosphole; 2-Ph, *in* D-00915
(1,3-Diphenyl-3-oxopropyl)phenylphosphinic acid, D-01030

$C_{21}H_{19}O_4P$
5,6-Dimethylspiro[1,3,2-benzodioxaphosphole-2,1′(3′*H*)-[2,1]benzoxaphosphole; 2-Phenoxy, *in* D-00915
5-Hydroxy-2,4,6-triphenyl-1,3,5-dioxaphosphorinane 5-oxide, H-00204

$C_{21}H_{19}P$
Cyclopropylidenetriphenylphosphorane, C-00284
Triphenyl-2-propenylidenephosphorane, *in* T-00679

$C_{21}H_{19}PS_2$
1,3-Dithiol-2-ylidenetriphenylphosphorane, *in* D-01253

$C_{21}H_{20}BF_4OP$
(2-Oxopropyl)triphenylphosphonium(1+); Tetrafluoroborate, *in* O-00080

$C_{21}H_{20}BF_4O_2P$
(2-Methoxy-2-oxoethyl)triphenylphosphonium(1+); Tetrafluoroborate, *in* M-00055

$C_{21}H_{20}BF_4PS_2$
1,3-Dithiolan-2-yltriphenylphosphonium(1+); Tetrafluoroborate, *in* D-01253

$C_{21}H_{20}BrOP$
(2-Oxopropyl)triphenylphosphonium(1+); Bromide, *in* O-00080

$C_{21}H_{20}BrO_2P$
(2-Carboxyethyl)triphenylphosphonium(1+); Bromide, *in* C-00011
(2-Methoxy-2-oxoethyl)triphenylphosphonium(1+); Bromide, *in* M-00055

$C_{21}H_{20}BrP$
Cyclopropyltriphenylphosphonium(1+); Bromide, *in* C-00287
▷Triphenyl-2-propenylphosphonium(1+); Bromide, *in* T-00679

$C_{21}H_{20}Br_3O_2P$
(2-Carboxyethyl)triphenylphosphonium(1+); Tribromide, *in* C-00011

$C_{21}H_{20}ClOP$
▷(2-Oxopropyl)triphenylphosphonium(1+); Chloride, *in* O-00080

$C_{21}H_{20}ClO_2P$
(2-Carboxyethyl)triphenylphosphonium(1+); Chloride, *in* C-00011
(2-Methoxy-2-oxoethyl)triphenylphosphonium(1+); Chloride, *in* M-00055

$C_{21}H_{20}ClP$
Triphenyl-2-propenylphosphonium(1+); Chloride, *in* T-00679

$C_{21}H_{20}IOP$
7,12-Dihydro-12-methyl-12-phenyl-5*H*-dibenz[*c,f*][1,5]oxaphosphocinium iodide, *in* D-00560
(2-Oxopropyl)triphenylphosphonium(1+); Iodide, *in* O-00080
2,8,10-Trimethyl-10-phenyl-10*H*-phenoxaphosphinium iodide, *in* D-00799

$C_{21}H_{20}IO_2P$
(2-Methoxy-2-oxoethyl)triphenylphosphonium(1+); Iodide, *in* M-00055

$C_{21}H_{20}IP$
10,11-Dihydro-5-methyl-5-phenyl-5*H*-dibenzo[*b,f*]phosphepinium iodide, *in* D-00559
Methyl diphenyl(phenylethenyl)phosphonium iodide, *in* D-01033

$C_{21}H_{20}NO_2P$
Ethyl (triphenylphosphoranylidene)carbamate, *in* T-00648

$C_{21}H_{20}NO_3P$
5,6-Dimethyl-*N*-phenylspiro[1,3,2-benzodioxaphosphole-2,1′(3′*H*)-[2,1]-benzoxaphosphol]-2-amine, *in* D-00915

$C_{21}H_{20}NO_4P$
Dibenzyl benzoylphosphoramidate, *in* B-00036

$C_{21}H_{20}NO_5P$
Diphenyl [4-(ethoxycarbonyl)phenyl]-phosphoramidate, *in* C-00006
Ethyl 2-[(diphenoxyphosphinyl)amino]-benzoate, *in* C-00004

$C_{21}H_{20}NP$
1,2,3-Triphenyl-1,3-azaphosphacyclopentane, *in* D-00995

$C_{21}H_{20}OP^{\oplus}$
(2-Oxopropyl)triphenylphosphonium(1+), O-00080

$C_{21}H_{20}O_2P^{\oplus}$
(1-Carboxyethyl)triphenylphosphonium(1+), C-00010
(2-Carboxyethyl)triphenylphosphonium(1+), C-00011
(2-Methoxy-2-oxoethyl)triphenylphosphonium(1+), M-00055

$C_{21}H_{20}O_4P_2$
[2-Oxo-2-(triphenylphosphoranylidene)propyl]-phosphonic acid, O-00086

$C_{21}H_{20}P^{\oplus}$
Cyclopropyltriphenylphosphonium(1+), C-00287
Triphenyl-2-propenylphosphonium(1+), T-00679

$C_{21}H_{20}PS_2^{\oplus}$
1,3-Dithiolan-2-yltriphenylphosphonium(1+), D-01253

$C_{21}H_{20}P_2$
2,3,4,5-Tetrahydro-1,5-diphenyl-1*H*-1,5-benzodiphosphepin, T-00101

$C_{21}H_{21}BrP^{\oplus}$
▷(3-Bromopropyl)triphenylphosphonium(1+), B-00502

$C_{21}H_{21}Br_2P$
(3-Bromopropyl)triphenylphosphonium(1+); Bromide, *in* B-00502

$C_{21}H_{21}INOP$
(3-Acetylaminophenyl)methyldiphenylphosphonium iodide, *in* D-01048

$C_{21}H_{21}N_2OP$
Hexahydro-1,3,5-triphenyl-1,3,5-diazaphosphorine; 5-Oxide, *in* H-00058

$C_{21}H_{21}N_2P$
Hexahydro-1,3,5-triphenyl-1,3,5-diazaphosphorine, H-00058

$C_{21}H_{21}N_2PS$
Hexahydro-1,3,5-triphenyl-1,3,5-diazaphosphorine; 5-Sulfide, *in* H-00058

$C_{21}H_{21}OP$
2,2-Dihydro-1,2-oxaphospholane; 2,2,2-Tri-Ph, *in* D-00534
▷Tribenzylphosphine; Oxide, *in* T-00338
Tris(2-methylphenyl)phosphine; Oxide, *in* T-00806
Tris(3-methylphenyl)phosphine; Oxide, *in* T-00807
▷Tris(4-methylphenyl)phosphine; Oxide, *in* T-00808

$C_{21}H_{21}OPS_3$
S,S,S-Tribenzyl phosphorotrithioate, T-00342
S,S,S-Tris(2-methylphenyl) phosphorotrithioate, T-00815
S,S,S-Tris(4-methylphenyl) phosphorotrithioate, T-00816

$C_{21}H_{21}O_2P$
Ethyl(diphenylmethyl)phenylphosphinate, *in* D-01023

$C_{21}H_{21}O_3P$
Dimethyl triphenylmethylphosphonate, *in* T-00608
Diphenyl (2,4,5-trimethylphenyl)phosphonate, *in* T-00526
Tribenzyl phosphite, T-00340
Tris(2-methoxyphenyl)phosphine, T-00791
Tris(3-methoxyphenyl)phosphine, T-00792
Tris(4-methoxyphenyl)phosphine, T-00793
▷Tris(2-methylphenyl) phosphite, T-00809
▷Tris(3-methylphenyl) phosphite, T-00810
▷Tris(4-methylphenyl) phosphite, T-00811

$C_{21}H_{21}O_3PS$
Tris(4-methoxyphenyl)phosphine; Sulfide, *in* T-00793
O,O,O-Tris(2-methylphenyl) phosphorothioate, T-00812
O,O,O-Tris(3-methylphenyl) phosphorothioate, T-00813
O,O,O-Tris(4-methylphenyl) phosphorothioate, T-00814

$C_{21}H_{21}O_3PTe_3$
Tris(4-methoxyphenyl) phosphorotritelluroite, T-00795

$C_{21}H_{21}O_4P$
Tribenzyl phosphate, T-00337
Tris(2-methoxyphenyl)phosphine; Oxide, *in* T-00791
Tris(3-methoxyphenyl)phosphine; Oxide, *in* T-00792
Tris(4-methoxyphenyl)phosphine; Oxide, *in* T-00793
▷Tris(2-methylphenyl) phosphate, T-00803
Tris(3-methylphenyl) phosphate, T-00804
Tris(4-methylphenyl) phosphate, T-00805

$C_{21}H_{21}O_6P$
Tris(2-methoxyphenyl) phosphite, T-00794

$C_{21}H_{21}O_7P$
Tris(4-methoxyphenyl) phosphate, *in* T-00787

$C_{21}H_{21}P$
(1-Methylethylidene)triphenylphosphorane, M-00153
Tribenzylphosphine, T-00338
Triphenylpropylphosphonium(1+); Ylide, *in* T-00680
Tris(2-methylphenyl)phosphine, T-00806
Tris(3-methylphenyl)phosphine, T-00807
▷Tris(4-methylphenyl)phosphine, T-00808

$C_{21}H_{21}PS$
Tribenzylphosphine; Sulfide, *in* T-00338
Tris(2-methylphenyl)phosphine; Sulfide, *in* T-00806
Tris(4-methylphenyl)phosphine; Sulfide, *in* T-00808

$C_{21}H_{21}PS_2$
Benzyl dibenzylphosphinodithioate, *in* D-00056

$C_{21}H_{21}PS_3$
Tribenzyl phosphorotrithioite, T-00343

$C_{21}H_{21}PS_4$
Tribenzyl phosphorotetrathioate, T-00341

$C_{21}H_{21}PSe$
Tris(4-methylphenyl)phosphine; Selenide, *in* T-00808

$C_{21}H_{21}S_3P$
Tris[2-(methylthio)phenyl]phosphine, T-00831
▷Tris[4-(methylthio)phenyl]phosphine, T-00832

$C_{21}H_{22}BrOP$
(3-Hydroxypropyl)triphenylphosphonium(1+); Bromide, *in* H-00193

$C_{21}H_{22}BrP$
1-Ethyl-1,2,3,4-tetrahydro-1-phenylbenzo[*h*]-phosphinolinium bromide, *in* P-00319
Isopropyltriphenylphosphonium(1+); Bromide, *in* I-00065
Methylbis(2-methylphenyl)phenylphosphonium bromide, *in* B-00309
Triphenylpropylphosphonium(1+); Bromide, *in* T-00680

$C_{21}H_{22}Br_2P_2$
2,3-Dihydro-1,3-dimethyl-1,3-diphenyl-1*H*-1,3-benzodiphospholium dibromide, *in* D-00476

$C_{21}H_{22}ClOP$
(3-Hydroxypropyl)triphenylphosphonium(1+); Chloride, *in* H-00193

$C_{21}H_{22}ClO_4P$
Isopropyltriphenylphosphonium(1+); Perchlorate, *in* I-00065

$C_{21}H_{22}ClP$
Triphenylpropylphosphonium(1+); Chloride, *in* T-00680

$C_{21}H_{22}IOP$
(3-Hydroxypropyl)triphenylphosphonium(1+); Iodide, *in* H-00193

$C_{21}H_{22}IP$
Isopropyltriphenylphosphonium(1+); Iodide, *in* I-00065
Triphenylpropylphosphonium(1+); Iodide, *in* T-00680

$C_{21}H_{22}NO_3P$
Dibenzyl benzylphosphoramidate, *in* B-00069
Dibenzyl methylphenylphosphoramidate, *in* M-00250
Dibenzyl (4-methylphenyl)phosphoramidate, *in* M-00253

$C_{21}H_{22}NO_4P$
2,2-Dihydro-3,4-bis(methoxycarbonyl)-2,2-dimethyl-5,5-diphenyl-5*H*-1,2-azaphosphole, D-00445

$C_{21}H_{22}NP$
N-Isopropyl-*P,P,P*-triphenylphosphine imide, *in* T-00626
P,P,P-Tribenzylphosphine imide, T-00339
P,P,P-Triphenyl-*N*-propylphosphine imide, *in* T-00626

$C_{21}H_{22}OP^{\oplus}$
(3-Hydroxypropyl)triphenylphosphonium(1+), H-00193

$C_{21}H_{22}P^{\oplus}$
Isopropyltriphenylphosphonium(1+), I-00065
Triphenylpropylphosphonium(1+), T-00680

$C_{21}H_{23}N_2OP$
P-(2,4,5-Trimethylphenyl)-*N,N'*-diphenylphosphonic diamide, *in* T-00526

$C_{21}H_{24}GeNP$
N-Trimethylgermyl-*P,P,P*-triphenylphosphine imide, *in* T-00626

$C_{21}H_{24}NPSi$
N-Trimethylsilyl-*P,P,P*-triphenylphosphine imide, *in* T-00626

$C_{21}H_{24}NPSn$
N-Trimethylstannyl-*P,P,P*-triphenylphosphine imide, *in* T-00626

$C_{21}H_{24}N_2O_5P_2S$
Trimethyl [[(triphenylphosphoranylidene)-amino]sulfonyl]phosphorimidate, *in* T-00642

$C_{21}H_{24}N_3OP$
N,N',N''-Tribenzylphosphoric triamide, *in* T-00344
N,N',N''-Tris(2-methylphenyl)phosphoric triamide, *in* T-00817
N,N',N''-Tris(3-methylphenyl)phosphoric triamide, *in* T-00818
N,N',N''-Tris(4-methylphenyl)phosphoric triamide, *in* T-00819

$C_{21}H_{24}N_3P$
N,N',N''-Tribenzylphosphorous triamide, T-00344
N,N',N''-Tris(2-methylphenyl)phosphorous triamide, T-00817
N,N',N''-Tris(3-methylphenyl)phosphorous triamide, T-00818
N,N',N''-Tris(4-methylphenyl)phosphorous triamide, T-00819

$C_{21}H_{24}N_3PS$
N,N',N''-Tribenzylphosphorothioic triamide, *in* T-00344
N,N',N''-Tris(2-methylphenyl)phosphorothioic triamide, *in* T-00817
N,N',N''-Tris(4-methylphenyl)phosphorothioic triamide, *in* T-00819

$C_{21}H_{24}N_4P_2$
1,3*a*,3*b*,6-Tetrahydro-3,4,8,8-tetramethyl-1,6-diphenyl-8*H*-[1,3]diphospholo[1,5-*e*;3,4-*c'*]-bis[1,2,3]diazaphosphole, T-00140

$C_{21}H_{24}N_4P_2S_2$
1,3*a*,3*b*,6-Tetrahydro-3,4,8,8-tetramethyl-1,6-diphenyl-8*H*-[1,3]diphospholo[1,5-*e*;3,4-*c'*]-bis[1,2,3]diazaphosphole; Disulfide, *in* T-00140

$C_{21}H_{26}IOP$
1,9,9-Trimethyl-2,8-diphenyl-3,7-dioxa-1-phosphoniabicyclo[3.3.1]nonane, *in* D-00743

$C_{21}H_{26}NO_7P$
Diethyl 2-[(diphenoxyphosphinyl)amino]-pentanedioate, *in* P-00367

$C_{21}H_{27}NO_7P$
Dimethyl 2-[(dibenzyloxyphosphinyl)amino]-pentanedioate, *in* P-00367

$C_{21}H_{27}N_7O_{14}P_2$
▷Coenzyme I, C-00202

$C_{21}H_{27}O_2P$
2',2',3',4',4'-Pentamethylspiro[1,3,2-benzodioxaphosphole-2,1'-phosphetane]; 2-Benzyl, *in* P-00027

$C_{21}H_{27}O_7P$
6,7,11,12,19,20,22,23-Octahydro-9-methyl-dibenzo[*g,p*][1,3,6,9,12,15,18,21]-heptaoxaphosphacycloeicosin, O-00019
Tris(2-methoxyphenyl)phosphate, *in* T-00785

$C_{21}H_{27}O_8P$
6,7,11,12,19,20,22,23-Octahydro-9-methyl-dibenzo[*g,p*][1,3,6,9,12,15,18,21]-heptaoxaphosphacycloeicosin; 9-Oxide, *in* O-00019

$C_{21}H_{28}BrP$
1-Benzyl-2,2,3,4,4-pentamethyl-1-phenyl-phosphetanium bromide, *in* P-00020

$C_{21}H_{28}N_7O_{17}P_3$
Coenzyme II, C-00203

$C_{21}H_{29}N_2P$
Hexahydro-1,5-bis(1-phenylethyl)-1*H*-1,5,3-diazaphosphepine; 3-Me, *in* H-00019

$C_{21}H_{30}Br_2P_2$
5,5,7,7-Tetraethyl-6,7-dihydro-5*H*-dibenzo[*d,f*][1,3]diphosphepinium dibromide, *in* D-00282

$C_{21}H_{30}O_4P_2$
1,3-Propanediylbis[phenylphosphinic acid]; Diisopropyl ester, *in* P-00447

$C_{21}H_{35}O_4P$
8,9,10,11,12,13,14,15-Octahydro-4-hydroxydinaphtho[2,1-*d*:1',2'-*f*][1,3,2]-dioxaphosphepin 4-oxide; 4-Methoxy (Me ester), *in* O-00018

$C_{21}H_{36}N_7O_{16}P_3S$
Coenzyme *A*, C-00200

$C_{21}H_{36}N_7O_{16}P_3Se$
Selenocoenzyme *A*, S-00003

$C_{21}H_{37}O_2PSi_2$
1,2,3,4,5,8,9,10-Octahydro-1-phenyl-6,7-bis[(trimethylsilyl)oxy]phosphecine, O-00021

$C_{21}H_{41}O_4PSi_2$
Bis(trimethylsilyl)[3,5-di-*tert*-butyl-4-hydroxyphenyl)methyl]phosphonate, *in* D-00094

$C_{21}H_{42}N_3O_9P_3$
Hexaisopropyl 1,3,5-triazine-2,4,6-triylphosphonate, *in* T-00329

$C_{21}H_{42}P_2$
1,4-Dibenzyl-1,4-diphosphonane, D-00050
1,5-Dibenzyl-1,5-diphosphonane, D-00051

$C_{21}H_{45}OP$
▷Triheptylphosphine; Oxide, *in* T-00489

$C_{21}H_{45}O_2P$
Diheptyl heptylphosphonite, *in* H-00008
Heptyl diheptylphosphinate, *in* D-00425

$C_{21}H_{45}O_3P$
Diheptyl heptylphosphonate, *in* H-00006
Triheptyl phosphite, T-00490

$C_{21}H_{45}O_4P$
Triheptyl phosphate, T-00488

$C_{21}H_{45}P$
Triheptylphosphine, T-00489

$C_{21}H_{45}PS$
Triheptylphosphine; Sulfide, *in* T-00489

$C_{21}H_{45}PSe$
Triheptylphosphine; Selenide, *in* T-00489

$C_{21}H_{46}O_2P_2$
1,5-Bis(dibutylphosphinyl)pentane, *in* B-00161

$C_{21}H_{46}P_2$
1,5-Bis(dibutylphosphino)pentane, B-00161

$C_{21}H_{47}O_9P_3$
Hexaisopropyl 1,2,3-propanetriyltrisphosphonate, *in* P-00450

$C_{21}H_{48}O_{10}P_4$
Hexaisopropyl [phosphinylidynetris(methyl-ene)]trisphosphonate, *in* P-00360

$C_{21}H_{50}N_4P_2$
P,P'-1,5-Pentanediylbis[bis-*N,N*-diethylphosphonous diamide], *in* P-00033

$C_{22}H_{15}F_6P$
(Hexafluorocyclobutylidene)triphenyl-phosphorane, H-00017

$C_{22}H_{15}OP$
Diphenyl(4-phenyl-1,3-butadiynyl)phosphine; Oxide, *in* D-01032
Phenylbis(phenylethynyl)phosphine; Oxide, *in* P-00099

$C_{22}H_{15}O_2P$
Phenyl bis(phenylethynyl)phosphinate, *in* B-00388

$C_{22}H_{15}P$
Diphenyl(4-phenyl-1,3-butadiynyl)phosphine, D-01032
Phenylbis(phenylethynyl)phosphine, P-00099

$C_{22}H_{15}PS$
Diphenyl(4-phenyl-1,3-butadiynyl)phosphine; Sulfide, *in* D-01032
Phenylbis(phenylethynyl)phosphine; Sulfide, *in* P-00099

$C_{22}H_{16}NOP$
N-Phenyl-*P*,*P*-bis(phenylethynyl)phosphinic amide, *in* B-00388

$C_{22}H_{16}NO_3P$
3-Cyano-2-oxo-3-(triphenylphosphoranylidene)-propanoic acid, C-00218

$C_{22}H_{16}N_3O_2PS$
O,*O*-Bis(3-dimethylaminophenyl) phenyl-phosphoramidothioate, *in* B-00189

$C_{22}H_{17}OP$
2,3-Dihydro-2-phenyl-1*H*-dibenz[*e*,*g*]-isophosphindole; 2-Oxide, *in* D-00558
1-(Diphenylphosphinyl)naphthalene, *in* N-00011
2,4,6-Triphenyl-4*H*-1,4-oxaphosphorin, T-00610
1,2,5-Triphenyl-1*H*-phosphole; Oxide, *in* T-00634

$C_{22}H_{17}OPS$
2,4,6-Triphenyl-4*H*-1,4-thiaphosphorin; 4-Oxide, *in* T-00686

$C_{22}H_{17}O_2P$
Diphenyl 1-naphthylphosphonite, *in* N-00025
1-Naphthyl diphenylphosphinate, *in* D-01039
1-Phenoxy-3,4-diphenyl-1*H*-phosphole 1-oxide, *in* H-00142
2,4,6-Triphenyl-4*H*-1,4-oxaphosphorin; 4-Oxide, *in* T-00610

$C_{22}H_{17}O_3P$
Dihydro(triphenylphosphoranylidene)-2,5-furandione, D-00612

$C_{22}H_{17}O_3PS$
2,4,6-Triphenyl-4*H*-1,4-thiaphosphorin; 1,1,4-Trioxide, *in* T-00686

$C_{22}H_{17}P$
2,3-Dihydro-2-phenyl-1*H*-dibenz[*e*,*g*]-isophosphindole, D-00558
1-Naphthyldiphenylphosphine, N-00011
2-Naphthyldiphenylphosphine, N-00012
1,2,5-Triphenyl-1*H*-phosphole, T-00634

$C_{22}H_{17}PS$
1,2,5-Triphenyl-1*H*-phosphole; Sulfide, *in* T-00634
2,4,6-Triphenyl-4*H*-1,4-thiaphosphorin, T-00686

$C_{22}H_{17}PSe$
1,2,5-Triphenyl-1*H*-phosphole; Selenide, *in* T-00634

$C_{22}H_{18}NOP$
1,4-Dihydro-2,4,6-triphenyl-1,4-azaphosphorine; 4-Oxide, *in* D-00604
N-1-Naphthyl-*P*,*P*-diphenylphosphinic amide, *in* D-01040
N-2-Naphthyl-*P*,*P*-diphenylphosphinic amide, *in* D-01040

$C_{22}H_{18}NO_2P$
Methyl cyano(triphenylphosphoranylidene)-acetate, *in* C-00231
3-(Triphenylphosphoranylidene)-2,5-pyrrolidenedione, T-00663

$C_{22}H_{18}NP$
1,4-Dihydro-2,4,6-triphenyl-1,4-azaphosphorine, D-00604

$C_{22}H_{19}F_6O_2P$
Bis(2,2,2-trifluoroethoxy)triphenylphosphorane, B-00408

$C_{22}H_{19}FeOP$
Diphenylphosphinylferrocene, *in* D-01068

$C_{22}H_{19}FeP$
(Diphenylphosphino)ferrocene, D-01068

$C_{22}H_{19}OP$
Diphenyl(4-phenyl-1,3-butadienyl)phosphine; Oxide, *in* D-01031

$C_{22}H_{19}O_2P$
(2-Carboxy-2-propenylidene)triphenyl-phosphorane, *in* C-00016
Ethyl di-1-naphthylphosphinate, *in* D-00941
Phenyl phenyl(4-phenyl-1,3-butadienyl)-phosphinate, *in* P-00184
4-(Triphenylphosphoranylidene)-2-butenoic acid, T-00647

$C_{22}H_{19}O_3P$
6-Methyl-2,3,4-triphenyl-2*H*-1,5,2-dioxaphosphorin 2-oxide, M-00430
3-Oxo-2-(triphenylphosphoranylidene)butanoic acid, O-00084

$C_{22}H_{19}O_4P$
(Triphenylphosphoranylidene)butanedioic acid, T-00646

$C_{22}H_{19}O_5P$
2,3-Diphenyl-1,4,6,9-tetraoxa-5-phospha(5-P^V)-spiro[4.4]non-2-ene; 5-Phenoxy, *in* D-01150

$C_{22}H_{19}P$
Diphenyl(4-phenyl-1,3-butadienyl)phosphine, D-01031

$C_{22}H_{19}PS$
Diphenyl(4-phenyl-1,3-butadienyl)phosphine; Sulfide, *in* D-01031

$C_{22}H_{20}BrF_{12}O_2P$
4-(4-Bromophenyl)-2′,2′,3′,4′,4′-pentamethyl-3,3,6,6-tetrakis(trifluoromethyl)spiro[2,7-dioxa-1-phosphabicyclo[3.2.0]hept-3-ene-1,1′-phosphetane], B-00485

$C_{22}H_{20}BrOP$
1-Bromo-1-(triphenylphosphoranylidene)-2-butanone, B-00505
(3-Methoxy-2-propenyl)triphenyl-phosphonium(1+); Bromide, *in* M-00084

$C_{22}H_{20}BrO_2P$
(2-Carboxy-2-propenyl)triphenyl-phosphonium(1+); Bromide, *in* C-00016
Ethyl bromo(triphenylphosphoranylidene)-acetate, *in* B-00504

$C_{22}H_{20}BrO_4P$
(1,2-Dicarboxyethyl)triphenyl-phosphonium(1+); Bromide, *in* D-00160

$C_{22}H_{20}BrP$
(1,3-Butadienyl)triphenylphosphonium(1+); Bromide, *in* B-00512

$C_{22}H_{20}ClN_2P$
2,3,4,5-Tetrahydro-6-methyl-2,4,4-triphenyl-1,2,3-diazaphosphorine; 3-Chloro, *in* T-00118

$C_{22}H_{20}ClO_2P$
Ethyl chloro(triphenylphosphoranylidene)-acetate, *in* C-00190

$C_{22}H_{20}ClO_4P$
(1,2-Dicarboxyethyl)triphenyl-phosphonium(1+); Chloride, *in* D-00160

$C_{22}H_{20}ClP$
(1,3-Butadienyl)triphenylphosphonium(1+); Chloride, *in* B-00512

$C_{22}H_{20}NOP$
N,*P*-Diphenyl-*P*-(4-phenyl-1,3-butadienyl)-phosphinic amide, *in* P-00184

$C_{22}H_{20}NO_2P$
10,10-Dihydro-2,3-diphenyl-[1,3,2]-oxazaphospholo[2,3-*b*][1,3,2]-benzoxazaphosphole; 10,10-Di-Me, *in* D-00481

$C_{22}H_{20}NO_4P$
10,10-Dihydro-2,3-diphenyl-[1,3,2]-oxazaphospholo[2,3-*b*][1,3,2]-benzoxazaphosphole; 10,10-Dimethoxy, *in* D-00481

$C_{22}H_{20}OP^{\oplus}$
(3-Methoxy-2-propenyl)triphenyl-phosphonium(1+), M-00084

$C_{22}H_{20}O_2P^{\oplus}$
(2-Carboxy-2-propenyl)triphenyl-phosphonium(1+), C-00016

$C_{22}H_{20}O_4P^{\oplus}$
(1,2-Dicarboxyethyl)triphenyl-phosphonium(1+), D-00160

$C_{22}H_{20}P^{\oplus}$
(1,3-Butadienyl)triphenylphosphonium(1+), B-00512

$C_{22}H_{21}INOP$
3-Methyl-1,2,3-triphenyl-5-oxo-1,3-azaphospholidinium iodide, *in* T-00596

$C_{22}H_{21}N_2O_2P$
Ethyl [(triphenylphosphoranylidene)-hydrazono]acetate, *in* T-00653

$C_{22}H_{21}N_2P$
2,3,4,5-Tetrahydro-6-methyl-2,4,4-triphenyl-1,2,3-diazaphosphorine, T-00118

$C_{22}H_{21}OP$
(3-Methoxy-2-propenylidene)triphenyl-phosphorane, *in* M-00084

$C_{22}H_{21}O_2P$
[(1,3-Dioxolan-2-yl)methylene]triphenyl-phosphorane, *in* D-00976
Ethyl triphenylphosphoranylideneacetate, E-00198
Methyl (2-triphenylphosphoranylidene)-propanoate, M-00434
4-(Triphenylphosphoranylidene)butanoic acid, *in* C-00017
3-(Triphenylphosphoranylidene)propanoic acid; Me ester, *in* T-00661

$C_{22}H_{21}O_4P$
5-Hydroxy-2,4,6-triphenyl-1,3,5-dioxaphosphorinane 5-oxide; 5-Methoxy (Me ester), *in* H-00204

$C_{22}H_{21}O_6P$
Benzyl 3-[(diphenoxyphosphinyl)oxy]-propanoate, *in* P-00399

$C_{22}H_{21}P$
2-Butenylidenetriphenylphosphorane, B-00522
Cyclobutylidenetriphenylphosphorane, *in* C-00235
(Cyclopropylmethylene)triphenylphosphorane, *in* C-00286
(Diphenylmethylene)(2,4,6-trimethylphenyl)-phosphine, D-01021
6,6*a*,7,8,9,9*a*-Hexahydro-5-phenyl-1*H*-benzo[*h*]cyclopenta[*c*]phosphinoline, H-00043
(2-Methyl-2-propenylidene)triphenyl-phosphorane, M-00363

$C_{22}H_{21}PS_2$
1,3-Dithian-2-ylidenetriphenylphosphorane, *in* D-01242
(1,3-Dithiolan-2-ylmethylene)triphenyl-phosphorane, *in* D-01252

$C_{22}H_{22}BF_4OP$
(2-Ethoxyethenyl)triphenylphosphonium(1+); Tetrafluoroborate, *in* E-00039

$C_{22}H_{22}BrOP$
(2-Ethoxyethenyl)triphenylphosphonium(1+); Bromide, *in* E-00039

$C_{22}H_{22}BrO_2P$
(3-Carboxypropyl)triphenylphosphonium(1+); Bromide, *in* C-00017
[(1,3-Dioxolan-2-yl)methyl]triphenyl-phosphonium(1+); Bromide, *in* D-00976
(2-Ethoxy-2-oxoethyl)triphenylphosphonium; Bromide, *in* E-00047

$C_{22}H_{22}BrP$
Benzyldiphenyl-2-propenylphosphonium bromide, *in* D-01144
(2-Butenyl)triphenylphosphonium(1+); Bromide, *in* B-00528
Cyclobutyltriphenylphosphonium(1+); Bromide, *in* C-00235
(Cyclopropylmethyl)triphenyl-phosphonium(1+); Bromide, *in* C-00286

$C_{22}H_{22}BrPS$
(1,3-Dithiolan-2-ylmethyl)triphenylphosphonium(1+); Bromide, *in* D-01252

$C_{22}H_{22}ClO_2P$
(3-Carboxypropyl)triphenylphosphonium(1+); Chloride, *in* C-00017
(2-Ethoxy-2-oxoethyl)triphenylphosphonium; Chloride, *in* E-00047

$C_{22}H_{22}ClP$
(2-Methyl-2-propenyl)triphenylphosphonium(1+); Chloride, *in* M-00367

$C_{22}H_{22}ClPS_2$
1,3-Dithian-2-yltriphenylphosphonium(1+); Chloride, *in* D-01242

$C_{22}H_{22}IOP$
(2-Ethoxyethenyl)triphenylphosphonium(1+); Iodide, *in* E-00039

$C_{22}H_{22}IO_2P$
[(1,3-Dioxolan-2-yl)methyl]triphenylphosphonium(1+); Iodide, *in* D-00976
(1-Methoxycarbonylethyl)triphenylphosphonium(1+); Iodide, *in* M-00027

$C_{22}H_{22}NO_2P$
Isopropyl (triphenylphosphoranylidene)carbamate, *in* T-00648
Propyl (triphenylphosphoranylidene)carbamate, *in* T-00648

$C_{22}H_{22}NO_2PS$
2-Phenylspiro[1,3,2-benzothiazaphosphole-2(3*H*),2′-[1,3,2]dioxaphospholane]; 3-(2,6-Dimethylphenyl), *in* P-00308

$C_{22}H_{22}N_3O_5P$
Benzyl *N*-[imino(diphenoxyphosphinylamino)methyl]glycinate, *in* P-00369

$C_{22}H_{22}OP^{\oplus}$
(2-Ethoxyethenyl)triphenylphosphonium(1+), E-00039

$C_{22}H_{22}O_2P^{\oplus}$
(3-Carboxypropyl)triphenylphosphonium(1+), C-00017
[(1,3-Dioxolan-2-yl)methyl]triphenylphosphonium(1+), D-00976
(2-Ethoxy-2-oxoethyl)triphenylphosphonium, E-00047
(1-Methoxycarbonylethyl)triphenylphosphonium(1+), M-00027

$C_{22}H_{22}P^{\oplus}$
(2-Butenyl)triphenylphosphonium(1+), B-00528
Cyclobutyltriphenylphosphonium(1+), C-00235
(Cyclopropylmethyl)triphenylphosphonium(1+), C-00286
(2-Methyl-2-propenyl)triphenylphosphonium(1+), M-00367

$C_{22}H_{22}PS_2^{\oplus}$
1,3-Dithian-2-yltriphenylphosphonium(1+), D-01242
(1,3-Dithiolan-2-ylmethyl)triphenylphosphonium(1+), D-01252

$C_{22}H_{23}P$
Butylidenetriphenylphosphorane, *in* B-00641
(1-Methylpropylidene)triphenylphosphorane, *in* M-00386
(2-Methylpropylidene)triphenylphosphorane, M-00371

$C_{22}H_{24}BrO_3P$
Methyltris(4-methoxyphenyl)phosphonium bromide, *in* T-00793

$C_{22}H_{24}BrP$
▷Butyltriphenylphosphonium(1+); Bromide, *in* B-00641
(1-Methylpropyl)triphenylphosphonium(1+); Bromide, *in* M-00386
(2-Methylpropyl)triphenylphosphonium(1+); Bromide, *in* M-00387
Methyltris(4-methylphenyl)phosphonium bromide, *in* T-00808

$C_{22}H_{24}ClP$
Butyltriphenylphosphonium(1+); Chloride, *in* B-00641

$C_{22}H_{24}IO_3P$
Methyltris(4-methoxyphenyl)phosphonium iodide, *in* T-00793
Tris(2-methoxyphenyl)methylphosphonium iodide, *in* T-00791
Tris(3-methoxyphenyl)methylphosphonium iodide, *in* T-00792

$C_{22}H_{24}IP$
▷Butyltriphenylphosphonium(1+); Iodide, *in* B-00641
(1-Methylpropyl)triphenylphosphonium(1+); Iodide, *in* M-00386
(2-Methylpropyl)triphenylphosphonium(1+); Iodide, *in* M-00387
Methyltris(4-methylphenyl)phosphonium iodide, *in* T-00808
Tribenzylmethylphosphonium iodide, *in* T-00338

$C_{22}H_{24}I_2P_2$
1,2,3,4-Tetrahydro-1,4-dimethyl-1,4-diphenyl-1,4-benzodiphosphorinium diiodide, *in* T-00102

$C_{22}H_{24}I_3P$
Butyltriphenylphosphonium(1+); Triiodide, *in* B-00641

$C_{22}H_{24}I_5P$
Butyltriphenylphosphonium(1+); Pentaiodide, *in* B-00641

$C_{22}H_{24}I_7P$
Butyltriphenylphosphonium(1+); Heptaiodide, *in* B-00641

$C_{22}H_{24}NP$
N-tert-Butyl-*P,P,P*-triphenylphosphine imide, *in* T-00626
(2-Dimethylaminoethylidene)triphenylphosphorane, *in* D-00698
N-Me, *in* T-00339

$C_{22}H_{24}O_4P_2$
Diethyl 1,2-phenylenebis(phenylphosphinate), *in* P-00139

$C_{22}H_{24}P^{\oplus}$
Butyltriphenylphosphonium(1+), B-00641
(1-Methylpropyl)triphenylphosphonium(1+), M-00386
(2-Methylpropyl)triphenylphosphonium(1+), M-00387

$C_{22}H_{25}BrNP$
(2-Dimethylaminoethyl)triphenylphosphonium(1+); Bromide, *in* D-00698

$C_{22}H_{25}F_6N_2O_3P$
2,2,2,3-Tetrahydro-2,2,2-trimethoxy-5-phenyl-3,3-bis(trifluoromethyl)-1-(2,4,6-trimethylphenyl)-1*H*-1,4,2-diazaphosphole, T-00159

$C_{22}H_{25}NO_5P_2S$
Triphenyl (diethoxyphosphinothioyl)phosphorimidate, *in* D-00238

$C_{22}H_{25}NP^{\oplus}$
(2-Dimethylaminoethyl)triphenylphosphonium(1+), D-00698

$C_{22}H_{25}PSi$
Triphenyl[(trimethylsilyl)methylene]phosphorane, T-00692

$C_{22}H_{26}BF_4PSi$
Triphenyl[(trimethylsilyl)methyl]phosphonium(1+); Tetrafluoroborate, *in* T-00693

$C_{22}H_{26}ClOP$
6,6-Dibenzyl-6-phosphonia-2-oxaadamantane chloride, *in* B-00055

$C_{22}H_{26}ClPSi$
Triphenyl[(trimethylsilyl)methyl]phosphonium(1+); Chloride, *in* T-00693

$C_{22}H_{26}IPSi$
Triphenyl[(trimethylsilyl)methyl]phosphonium(1+); Iodide, *in* T-00693

$C_{22}H_{26}NO_3P$
Diphenyl 1-adamantylphosphoramidate, *in* A-00036

$C_{22}H_{26}O_2P_2$
1,6-Diphosphecane; 1,6-Dibenzyl, 1,6-dioxide, *in* D-01171

$C_{22}H_{26}PSi^{\oplus}$
Triphenyl[(trimethylsilyl)methyl]phosphonium(1+), T-00693

$C_{22}H_{26}P_2$
1,6-Dibenzyl-1,6-diphosphecane, *in* D-01171

$C_{22}H_{27}N_3O_4P_2$
Triphenyl [bis(dimethylamino)phosphinyl]phosphorimidate, *in* B-00194

$C_{22}H_{28}BrOP$
1-Benzyl-4-oxo-2,2,6,6-tetramethyl-1-phenylphosphorinanium bromide, *in* T-00210

$C_{22}H_{29}OP$
(5-Methyl-2-isopropylcyclohexyl)diphenylphosphine; Oxide, *in* M-00164

$C_{22}H_{29}O_2P$
4,8-Di-*tert*-butyl-2,10-dimethyl-12*H*-dibenzo[*d,g*][1,3,2]dioxaphosphocin, D-00088

$C_{22}H_{29}O_3P$
4,8-Di-*tert*-butyl-2,10-dimethyl-12*H*-dibenzo[*d,g*][1,3,2]dioxaphosphocin; 6-Oxide, *in* D-00088

$C_{22}H_{29}O_5P$
7,8,10,11,13,14,16,17,19,24-Decahydro-5*H*-dibenzo[*o,r*][1,4,7,10,13,17]-pentaoxaphosphacycloeicosin, D-00001

$C_{22}H_{29}P$
(5-Methyl-2-isopropylcyclohexyl)diphenylphosphine, M-00164

$C_{22}H_{30}BrP$
1-Benzyl-4-*tert*-butyl-1-phenylphosphorinanium bromide, *in* B-00586

$C_{22}H_{31}N_2P$
Hexahydro-1,5-bis(1-phenylethyl)-1*H*-1,5,3-diazaphosphepine; 3-Et, *in* H-00019

$C_{22}H_{31}O_3P$
Diphenyl decylphosphonate, *in* D-00012

$C_{22}H_{32}Br_2P_2$
5,5,8,8-Tetraethyl-5,6,7,8-tetrahydrodibenzo[*e,g*][1,4]diphosphocinium dibromide, *in* D-00404

$C_{22}H_{32}N_4O_5P$
Dibenzyl [[[(4-amino-4-methoxycarbonyl)butyl]amino](imino)methyl]phosphoramidate, *in* P-00363

$C_{22}H_{32}N_5O_{19}P$
Thuringiensin, T-00319

$C_{22}H_{32}O_6P_2$
Tetraethyl [[1,1′-biphenyl]-4,4′-diylbis(methylene)]bisphosphonate, *in* B-00391

$C_{22}H_{32}P_2S_2$
Dimethyl 1,4-butanediylbis[phenylphosphinodithioate], *in* B-00516

$C_{22}H_{34}I_2P_2$
[1,1′-Biphenyl]-2,2′-diylbis(diethylmethylphosphonium)diiodide, *in* B-00179

$C_{22}H_{34}O_2PSSi_2$
1-Phenyl-6,7-bis[(trimethylsilyl)oxy]-phosphacycloundec-6-ene; Sulfide, *in* P-00100

$C_{22}H_{34}O_2PSi_2$
1-Phenyl-6,7-bis[(trimethylsilyl)oxy]-phosphacycloundec-6-ene, P-00100

$C_{22}H_{36}N_6O_{16}P_2$
Agrocin 84, A-00048

$C_{22}H_{38}NOP$
N-tert-Butyl-*P*-(2,4,6-tri-*tert*-butylphenyl)-metaphosphonamidate, *in* B-00639

$C_{22}H_{38}NP$
N-tert-Butyl-*P*-(2,4,6-tri-*tert*-butylphenyl)-metaphosphonimidate, B-00639

$C_{22}H_{38}NPS$
N-tert-Butyl-*P*-(2,4,6-tri-*tert*-butylphenyl)-metaphosphonamidothioate, *in* B-00639

$C_{22}H_{38}NPSe$
N-tert-Butyl-*P*-(2,4,6-tri-*tert*-butylphenyl)-metaphosphonamidoselenoate, *in* B-00639

$C_{22}H_{39}OP$
Dioctylphenylphosphine; Oxide, *in* D-00955

$C_{22}H_{39}O_3P$
▷Dioctyl phenylphosphonate, *in* P-00221

$C_{22}H_{39}P$
Dioctylphenylphosphine, D-00955

$C_{22}H_{44}NOP$
N-Butyl-*P*,*P*-dioctylphosphinic amide, *in* D-00958

$C_{22}H_{47}O_2P$
Diundecylphosphinic acid, D-01256

$C_{22}H_{47}O_4P$
Diundecyl phosphate, D-01255

$C_{22}H_{52}N_4P_2$
P,*P*′-1,6-Hexanediylbis[bis *N*,*N*-diethylphosphonous diamide], *in* H-00086

$C_{23}H_{17}Br_2P_2$
1,1-Dibromo-1,1-dihydro-2,4,6-triphenyl-phosphorin, *in* D-00613

$C_{23}H_{17}Cl_2P_2$
1,1-Dichloro-1,1-dihydro-2,4,6-triphenyl-phosphorin, *in* D-00613

$C_{23}H_{17}F_2P_2$
1,1-Difluoro-1,1-dihydro-2,4,6-triphenyl-phosphorin, *in* D-00613

$C_{23}H_{17}P$
2,4,6-Triphenylphosphorin, T-00665

$C_{23}H_{18}NO_2P$
(2-Nitro-2,4-cyclopentadienylidene)triphenyl-phosphorane, N-00034

$C_{23}H_{18}NO_3P$
Methyl cyano(triphenylphosphoranylidene)-pyruvate, *in* C-00218

$C_{23}H_{19}OP$
(2-Methoxyphenyl)-2-naphthalenylphenyl-phosphine, M-00065
(4-Methoxyphenyl)-1-naphthalenylphenyl-phosphine, M-00066
1,2,6-Triphenyl-4-phosphorinanone, T-00666

$C_{23}H_{19}OPS$
1,2,6-Triphenyl-4-phosphorinanone; Sulfide, *in* T-00666

$C_{23}H_{19}O_2P$
(2-Methoxyphenyl)-2-naphthalenylphenyl-phosphine; Oxide, *in* M-00065
(4-Methoxyphenyl)-1-naphthalenylphenyl-phosphine; Oxide, *in* M-00066
3-Methyl-5-(triphenylphosphoranylidene)-2(5*H*)-furanone, *in* M-00186
1,2,6-Triphenyl-4-phosphorinanone; Oxide, *in* T-00666

$C_{23}H_{19}P$
2,4-Cyclopentadien-1-ylidenetriphenyl-phosphorane, C-00274
2-Penten-4-ynylidenetriphenylphosphorane, *in* P-00047

$C_{23}H_{19}P_2$
1,1-Dihydro-2,4,6-triphenylphosphorin, D-00613

$C_{23}H_{20}BrO_2P$
(3-Methyl-2-oxo-2(5*H*)-furanyl)triphenyl-phosphonium(1+); Bromide, *in* M-00186

$C_{23}H_{20}BrP$
2-Penten-4-ynyltriphenylphosphonium(1+); Bromide, *in* P-00047

$C_{23}H_{20}IP$
1-Methyl-1,2,5-triphenyl-1*H*-phospholium iodide, *in* T-00634

$C_{23}H_{20}O_2P^{\oplus}$
(3-Methyl-2-oxo-2(5*H*)-furanyl)triphenyl-phosphonium(1+), M-00186

$C_{23}H_{20}P^{\oplus}$
2-Penten-4-ynyltriphenylphosphonium(1+), P-00047

$C_{23}H_{21}NO_2P$
2,3-Dihydro-2-ethoxy-1,3,5-triphenyl-1*H*-1,2-azaphosphole 2-oxide, *in* D-00508

$C_{23}H_{21}N_2O_2P$
4-Ethoxy-1,4-dihydro-1,3,5-triphenyl-1,2,4-diazaphosphorine 4-oxide, E-00029

$C_{23}H_{21}O_2P$
Methyl 4-(triphenylphosphoranylidene)-2-butanoate, *in* T-00647
3-(Triphenylphosphoranylidene)-2,4-pentanedione, T-00654

$C_{23}H_{21}O_3P$
Methyl 3-oxo-2-(triphenylphosphoranylidene)-butanoate, *in* O-00084
4-Oxo-5-(triphenylphosphoranylidene)-pentanoic acid, O-00085

$C_{23}H_{21}O_4P$
Dimethyl (triphenylphosphoranylidene)-propanedioate, *in* T-00658
α-Methoxycarbonyltriphenyl-phosphoranylidenepropanoic acid, *in* T-00646

$C_{23}H_{21}P$
4*b*,5,6,10*b*,11,12-Hexahydro-5-phenylbenzo[*c*]-phosphanthridine, H-00045

$C_{23}H_{22}BrO_2P$
(2-Carboxy-2-propenyl)triphenyl-phosphonium(1+); Me ester, bromide, *in* C-00016

$C_{23}H_{22}ClO_2P$
(Diacetylmethyl)triphenylphosphonium(1+); Chloride, *in* D-00022

$C_{23}H_{22}FeIP$
Ferrocenylmethyldiphenylphosphonium iodide, *in* D-01068

$C_{23}H_{22}NO_3P$
2,3-Diphenyl-1,6,10-trioxa-4-aza-5-phospha(5-P^V)spiro[4.5]dec-2-ene; 5-Ph, *in* D-01158

$C_{23}H_{22}O_2P^{\oplus}$
(Diacetylmethyl)triphenylphosphonium(1+), D-00022

$C_{23}H_{23}N_2OP$
2,3,4,5-Tetrahydro-6-methyl-2,4,4-triphenyl-1,2,3-diazaphosphorine; 3-Methoxy, *in* T-00118
2,3,4,5-Tetrahydro-6-methyl-2,4,4-triphenyl-1,2,3-diazaphosphorine; 3-Me, 3-oxide, *in* T-00118

$C_{23}H_{23}N_2OPS$
2,3,4,5-Tetrahydro-6-methyl-2,4,4-triphenyl-1,2,3-diazaphosphorine; 3-Methoxy, 3-sulfide, *in* T-00118

$C_{23}H_{23}N_2O_2P$
4-Ethoxy-1,4,5,6-tetrahydro-1,3,5-triphenyl-1,2,4-diazaphosphorine 4-oxide, *in* E-00029

$C_{23}H_{23}OP$
(2-Ethoxy-2-propenylidene)triphenyl-phosphorane, E-00049
4-Triphenylphosphoranylidene(tetrahydro-2*H*-pyran), *in* T-00684

$C_{23}H_{23}O_2P$
Ethyl 2-triphenylphosphoranylidenepropanoate, E-00199
5-(Triphenylphosphoranylidene)pentanoic acid, T-00655
3-(Triphenylphosphoranylidene)propanoic acid; Et ester, *in* T-00661

$C_{23}H_{23}O_3P$
3-Methoxy-18-nor-17-phosphaestra-1,3,5(10),9(11)-tetraen-15-one; 17-Ph, 17-oxide, *in* M-00054

$C_{23}H_{23}P$
Cyclopentylidenetriphenylphosphorane, *in* C-00282
3-Methyl-2-butenylidenetriphenylphosphorane, M-00098
4-Pentenylidenetriphenylphosphorane, *in* P-00044

$C_{23}H_{23}PS$
4-Triphenylphosphoranylidene(tetrahydro-2*H*-thiopyran), *in* T-00685

$C_{23}H_{23}PS_2$
(1,3-Dithian-2-ylmethylene)triphenyl-phosphorane, *in* D-01241

$C_{23}H_{24}BrOP$
Triphenyl(tetrahydro-2*H*-pyran-4-yl)-phosphonium(1+); Bromide, *in* T-00684

$C_{23}H_{24}BrO_2P$
(4-Carboxybutyl)triphenylphosphonium(1+); Bromide, *in* C-00009
(1-Ethoxycarbonyl)triphenylphosphonium bromide, *in* C-00010

$C_{23}H_{24}BrP$
Cyclopentyltriphenylphosphonium(1+); Bromide, *in* C-00282
(3-Methyl-2-butenyl)triphenyl-phosphonium(1+); Bromide, *in* M-00104
4-Pentenyltriphenylphosphonium(1+); Bromide, *in* P-00044

$C_{23}H_{24}BrPS$
(1,3-Dithian-2-ylmethyl)triphenyl-phosphonium(1+); Bromide, *in* D-01241
Triphenyl(tetrahydro-2*H*-thiopyran-4-yl)-phosphonium(1+); Bromide, *in* T-00685

$C_{23}H_{24}ClO_2P$
(4-Carboxybutyl)triphenylphosphonium(1+); Chloride, *in* C-00009
(1-Carboxyethyl)triphenylphosphonium(1+); Et ester, chloride, *in* C-00010

$C_{23}H_{24}ClP$
Cyclopentyltriphenylphosphonium(1+); Chloride, *in* C-00282
(3-Methyl-2-butenyl)triphenyl-phosphonium(1+); Chloride, *in* M-00104

$C_{23}H_{24}F_6P_2$
2,3,4,5-Tetrahydro-5-methyl-2,2-diphenyl-1*H*-2-benzophosphepinium(1+); Hexafluorophosphate, *in* T-00114

$C_{23}H_{24}IOP$
(2-Ethoxypropenyl)triphenylphosphonium(1+); Iodide, *in* E-00050

$C_{23}H_{24}IO_2P$
(4-Carboxybutyl)triphenylphosphonium(1+); Iodide, *in* C-00009

$C_{23}H_{24}IPS$
(1,3-Dithian-2-ylmethyl)triphenyl-phosphonium(1+); Iodide, *in* D-01241

$C_{23}H_{24}NO_2PS$
2-Phenylspiro[1,3,2-benzothiazaphosphole-2(3*H*),2′-[1,3,2]dioxaphospholane]; 3-(2,4,6-Trimethylphenyl), *in* P-00308

$C_{23}H_{24}NO_5P$
Ethyl 2-[(diphenoxyphosphinyl)amino]-3-phenylpropanoate, *in* P-00400
Tribenzyl phosphonoglycinate, *in* P-00368

$C_{23}H_{24}NO_6P$
Diphenyl [1-ethoxycarbonyl-2-(4-hydroxybenzyl)ethyl]phosphoramidate, *in* P-00422

$C_{23}H_{24}N_3O_5P$
Benzyl *N*-[Imino(diphenoxyphosphinylamino)-methyl]-*N*-methylglycinate, *in* P-00364

$C_{23}H_{24}OP^{\oplus}$
(2-Ethoxypropenyl)triphenylphosphonium(1+), E-00050
Triphenyl(tetrahydro-2*H*-pyran-4-yl)-phosphonium(1+), T-00684

$C_{23}H_{24}O_2P^{\oplus}$
(4-Carboxybutyl)triphenylphosphonium(1+), C-00009

$C_{23}H_{24}P^{\oplus}$
Cyclopentyltriphenylphosphonium(1+), C-00282
(3-Methyl-2-butenyl)triphenylphosphonium(1+), M-00104
4-Pentenyltriphenylphosphonium(1+), P-00044
2,3,4,5-Tetrahydro-5-methyl-2,2-diphenyl-1*H*-2-benzophosphepinium(1+), T-00114

$C_{23}H_{24}PS^{\oplus}$
Triphenyl(tetrahydro-2*H*-thiopyran-4-yl)phosphonium(1+), T-00685

$C_{23}H_{24}PS_2^{\oplus}$
(1,3-Dithian-2-ylmethyl)triphenylphosphonium(1+), D-01241

$C_{23}H_{25}O_3P$
Diethyl triphenylmethylphosphonate, *in* T-00608

$C_{23}H_{25}P$
(2,2-Dimethylpropylidene)triphenylphosphorane, *in* D-00914
3-Methylbutylidenetriphenylphosphorane, *in* M-00111
Pentylidenetriphenylphosphorane, *in* P-00056

$C_{23}H_{26}BrP$
(3-Methylbutyl)triphenylphosphonium(1+); Bromide, *in* M-00111
Pentyltriphenylphosphonium(1+); Bromide, *in* P-00056

$C_{23}H_{26}IP$
(2,2-Dimethylpropyl)triphenylphosphonium(1+); Iodide, *in* D-00914
(3-Methylbutyl)triphenylphosphonium(1+); Iodide, *in* M-00111
Pentyltriphenylphosphonium(1+); Iodide, *in* P-00056

$C_{23}H_{26}NP$
[3-(Dimethylamino)propylidene]triphenylphosphorane, *in* D-00707

$C_{23}H_{26}P^{\oplus}$
(2,2-Dimethylpropyl)triphenylphosphonium(1+), D-00914
(3-Methylbutyl)triphenylphosphonium(1+), M-00111
Pentyltriphenylphosphonium(1+), P-00056

$C_{23}H_{27}BrNP$
[3-(Dimethylamino)propyl]triphenylphosphonium(1+); Bromide, *in* D-00707

$C_{23}H_{27}ClNP$
[3-(Dimethylamino)propyl]triphenylphosphonium(1+); Chloride, *in* D-00707

$C_{23}H_{27}NP^{\oplus}$
[3-(Dimethylamino)propyl]triphenylphosphonium(1+), D-00707

$C_{23}H_{27}PSi$
[2-(Trimethylsilyl)ethylidene]triphenylphosphorane, *in* T-00553

$C_{23}H_{27}PSn$
[2-(Trimethylstannyl)ethylidene]triphenylphosphorane, *in* T-00561

$C_{23}H_{28}IPSi$
[2-(Trimethylsilyl)ethyl]triphenylphosphonium(1+); Iodide, *in* T-00553

$C_{23}H_{28}IPSn$
[2-(Trimethylstannyl)ethyl]triphenylphosphonium; Iodide, *in* T-00561

$C_{23}H_{28}PSi^{\oplus}$
[2-(Trimethylsilyl)ethyl]triphenylphosphonium(1+), T-00553

$C_{23}H_{28}PSn^{\oplus}$
[2-(Trimethylstannyl)ethyl]triphenylphosphonium, T-00561

$C_{23}H_{29}O_5P$
2,2-Dihydrophenanthro[9,10-*d*]-1,3,2-dioxaphosphole; 2,2,2-Triisopropoxy, *in* D-00541

$C_{23}H_{31}O_7P$
7,8,10,11,13,14,16,17,19,24-Decahydro-5*H*-dibenzo[*o,r*][1,4,7,10,13,17]-pentaoxaphosphacycloeicosin; 24-Methoxy, 24-oxide, *in* D-00001

$C_{23}H_{32}N_3O_6P$
Antibiotic I5*B*1, *in* A-00139

$C_{23}H_{32}N_3O_7P$
Antibiotic I5*B*2, *in* A-00139

$C_{23}H_{32}P_2$
1,6-Dibenzyl-1,6-diphosphacycloundecane, D-00046

$C_{23}H_{34}Br_2P_2$
5,5,9,9-Tetraethyl-6,7,8,9-tetrahydro-5*H*-dibenzo[*f,h*][1,5]diphosphoninium dibromide, *in* T-00089

$C_{23}H_{34}N_3O_{10}P$
Phosphoramidon, P-00402

$C_{23}H_{42}IP$
Methyldioctylphenylphosphonium iodide, *in* D-00955

$C_{23}H_{43}N_2PS_2$
Benzyl tetrabutylphosphorodiamidodithioate, *in* T-00023

$C_{24}F_{20}P_2$
Tetrakis(pentafluorophenyl)diphosphine, T-00184

$C_{24}F_{20}P_4$
Tetrakis(pentafluorophenyl)tetraphosphetane, T-00185

$C_{24}H_{15}F_{12}O_2P$
2,2-Dihydro-2,2,2-triphenyl-4,4,5,5-tetrakis(trifluoromethyl)-1,3,2-dioxaphospholane, D-00614

$C_{24}H_{15}OP$
Tris(phenylethynyl)phosphine; Oxide, *in* T-00846

$C_{24}H_{15}P$
Tris(phenylethynyl)phosphine, T-00846

$C_{24}H_{15}PS$
Tris(phenylethynyl)phosphine; Sulfide, *in* T-00846

$C_{24}H_{15}PSe$
Tris(phenylethynyl)phosphine; Selenide, *in* T-00846

$C_{24}H_{16}BrP$
5,5′-Spirobi[5*H*-dibenzophospholium](1+); Bromide, *in* S-00022

$C_{24}H_{16}IP$
5,5′-Spirobi[5*H*-dibenzophospholium](1+); Iodide, *in* S-00022

$C_{24}H_{16}N_4O_{15}P_2$
Tetrakis(4-nitrophenyl) diphosphate, T-00183

$C_{24}H_{16}O_3P_2$
5,5′-Oxybis[5*H*-benzophosphindole] 5,5′-dioxide, *in* H-00118

$C_{24}H_{16}P^{\oplus}$
5,5′-Spirobi[5*H*-dibenzophospholium](1+), S-00022

$C_{24}H_{17}OP$
9*H*-Dibenzo[2,3:4,5]phospholo[1,2-*f*]-phosphanthridene; 17-Oxide, *in* D-00041

$C_{24}H_{17}P$
9*H*-Dibenzo[2,3:4,5]phospholo[1,2-*f*]-phosphanthridene, D-00041
5,5′-Spirobi[5*H*-dibenzophosphole], S-00021

$C_{24}H_{18}AsOP$
5,10-Diphenyldibenzo[1,4]phospharsenin; *P*-Oxide, *in* D-01001

$C_{24}H_{18}AsO_2P$
5,10-Diphenyldibenzo[1,4]phospharsenin; *As,P*-Dioxide, *in* D-01001

$C_{24}H_{18}AsP$
5,10-Diphenyldibenzo[1,4]phospharsenin, D-01001

$C_{24}H_{18}Br_2N_2O_4P_2$
1′,3′-Diphenyldispiro[1,3,2-benzodioxaphosphole-2,2′-[1,3,2,4]-diazaphosphetidine-4′,2″-[1,3,2]-benzodioxaphosphole]; 2,4′-Dibromo, *in* D-01018

$C_{24}H_{18}ClN_2P$
10,10′(5*H*,5′*H*)-Spirobiphenophosphazinium(1+); Chloride, *in* S-00023

$C_{24}H_{18}ClO_3P$
Bis[(1,1′-biphenyl)-2-yl] phosphorochloridate, B-00102

$C_{24}H_{18}Cl_2N_2O_4P_2$
1′,3′-Diphenyldispiro[1,3,2-benzodioxaphosphole-2,2′-[1,3,2,4]-diazaphosphetidine-4′,2″-[1,3,2]-benzodioxaphosphole]; 2,4′-Dichloro, *in* D-01018

$C_{24}H_{18}F_2N_2O_4P_2$
1′,3′-Diphenyldispiro[1,3,2-benzodioxaphosphole-2,2′-[1,3,2,4]-diazaphosphetidine-4′,2″-[1,3,2]-benzodioxaphosphole]; 2,4′-Difluoro, *in* D-01018

$C_{24}H_{18}IOP$
10,10-Diphenyl-10*H*-phenoxaphosphonium iodide, *in* P-00183

$C_{24}H_{18}IP$
5,5-Diphenyl-5*H*-dibenzophospholium iodide, *in* P-00110

$C_{24}H_{18}NP$
Aminobisbiphenylenephosphorane, *in* S-00021

$C_{24}H_{18}N_2P^{\oplus}$
10,10′(5*H*,5′*H*)-Spirobiphenophosphazinium(1+), S-00023

$C_{24}H_{18}OP_2$
5,10-Dihydro-5,10-diphenylphosphanthrene; Monooxide, *in* D-00482

$C_{24}H_{18}O_2P_2$
5,10-Dihydro-5,10-diphenylphosphanthrene; 5,10-Dioxide, *in* D-00482

$C_{24}H_{18}P_2$
5,10-Dihydro-5,10-diphenylphosphanthrene, D-00482

$C_{24}H_{19}AsP_2$
2,3-Dihydro-1,2,3-triphenyl-1*H*-1,3,2-benzodiphospharsole, D-00605

$C_{24}H_{19}AsP_2S_2$
2,3-Dihydro-1,2,3-triphenyl-1*H*-1,3,2-benzodiphospharsole; 1,3-Disulfide, *in* D-00605

$C_{24}H_{19}N_2O_5P$
Triphenyl (2-nitrophenyl)phosphorimidate, *in* N-00061
Triphenyl (3-nitrophenyl)phosphorimidate, *in* N-00062
Triphenyl (4-nitrophenyl)phosphorimidate, *in* N-00063

$C_{24}H_{19}OP$
4-(Diphenylphosphinyl)biphenyl, *in* B-00078
4-Triphenylphosphoranylidene-2,5-cyclohexadien-1-one, T-00649

$C_{24}H_{19}O_4P$
Bis(4-phenoxyphenyl)phosphinic acid, *in* B-00285

$C_{24}H_{19}O_5P$
2,2-Dihydro-2,2,2-triphenoxy-1,3,2-benzodioxaphosphole, D-00602

$C_{24}H_{19}P$
[1,1′-Biphenyl]-4-yldiphenylphosphine, B-00078

$C_{24}H_{19}P_3$
2,3-Dihydro-1,2,3-triphenyl-1*H*-benzotriphosphole, D-00606

$C_{24}H_{19}P_3S$
2,3-Dihydro-1,2,3-triphenyl-1*H*-benzotriphosphole; 1-Sulfide, *in* D-00606

$C_{24}H_{19}P_3S_2$
2,3-Dihydro-1,2,3-triphenyl-1*H*-benzotriphosphole; 1,3-Disulfide, *in* D-00606

$C_{24}H_{20}BrOP$
(2-Hydroxyphenyl)triphenylphosphonium(1+); Bromide, *in* H-00182
(4-Hydroxyphenyl)triphenylphosphonium(1+); Bromide, *in* H-00183

$C_{24}H_{20}BrO_4P$
Bromotetraphenoxyphosphorane, B-00503

$C_{24}H_{20}BrP$
Tetraphenylphosphonium(1+); Bromide, *in* T-00272

$C_{24}H_{20}Br_3P$
Tetraphenylphosphonium(1+); Tribromide, *in* T-00272

$C_{24}H_{20}ClNOP$
P,P-Diphenyl-*N*-(diphenylphosphinyl)-phosphinimidic acid; Chloride, *in* D-01007

$C_{24}H_{20}ClNP_2S$
P,P-Diphenyl-*N*-(diphenylphosphinothioyl)-phosphinimidic acid; Chloride, *in* D-01006

$C_{24}H_{20}ClN_2P$
Tetraphenylphosphorodiamidous acid; Chloride, *in* T-00274

$C_{24}H_{20}ClO_4P$
Chlorotetraphenoxyphosphorane, C-00186

$C_{24}H_{20}ClO_5P$
Phenoxytriphenylphosphorus(1+); Perchlorate, *in* P-00077

$C_{24}H_{20}ClP$
Tetraphenylphosphonium(1+); Chloride, *in* T-00272

$C_{24}H_{20}Cl_2N_3P_3$
2,2-Dichloro-2,2,4,4,6,6-hexahydro-4,4,6,6-tetraphenyl-1,3,5-triaza-2,4,6-triphosphorine, D-00169

$C_{24}H_{20}Cl_6OPSb$
Phenoxytriphenylphosphorus(1+); Hexachloroantimonate, *in* P-00077

$C_{24}H_{20}F_2N_3P_3$
2,2-Difluoro-2,2,4,4,6,6-hexahydro-4,4,6,6-tetraphenyl-1,3,5-triaza-2,4,6-triphosphorine, D-00412

$C_{24}H_{20}IOP$
(2-Hydroxyphenyl)triphenylphosphonium(1+); Iodide, *in* H-00182

$C_{24}H_{20}IO_4P$
Iodotetraphenoxyphosphorane, I-00020

$C_{24}H_{20}IP$
Tetraphenylphosphonium(1+); Iodide, *in* T-00272

$C_{24}H_{20}NOP$
Tetraphenylphosphinic amide, *in* D-01040

$C_{24}H_{20}NO_2P$
Diphenyl diphenylphosphoramidite, *in* D-01111

$C_{24}H_{20}NO_2PS$
P,P,P-Triphenyl-*N*-phenylsulfonylphosphine imide, *in* T-00626

$C_{24}H_{20}NO_3P$
Diphenyl diphenylphosphoramidate, *in* D-01108
Ethyl cyano(triphenylphosphoranylidene)-pyruvate, *in* C-00218
Triphenyl phenylphosphorimidate, *in* P-00264

$C_{24}H_{20}NO_3PS$
Phenyl *P,P*-diphenyl-*N*-phenyl-sulfonylphosphinimidate, *in* B-00013

$C_{24}H_{20}NO_4PS$
Diphenyl *P*-phenyl-*N*-phenyl-sulfonylphosphonimidate, *in* P-00192

$C_{24}H_{20}NO_5PS$
Triphenyl phenylsulfonylphosphorimidate, *in* P-00315

$C_{24}H_{20}NP$
Tetraphenylphosphine imide, T-00269

$C_{24}H_{20}N_2O_2P_2$
1,2,3,4-Tetraphenyl-1,3,2,4-diazadiphosphetidine; 2,4-Dioxide, *in* T-00253

$C_{24}H_{20}N_2O_4P_2$
1′,3′-Diphenyldispiro[1,3,2-benzodioxaphosphole-2,2′-[1,3,2,4]-diazaphosphetidine-4′,2″-[1,3,2]-benzodioxaphosphole], D-01018

$C_{24}H_{20}N_2P_2$
1,2,3,4-Tetraphenyl-1,3,2,4-diazadiphosphetidine, T-00253

$C_{24}H_{20}N_2P_2S_2$
1,2,3,4-Tetraphenyl-1,3,2,4-diazadiphosphetidine; 2,4-Disulfide, *in* T-00253

$C_{24}H_{20}OP^{\oplus}$
(2-Hydroxyphenyl)triphenylphosphonium(1+), H-00182
(4-Hydroxyphenyl)triphenylphosphonium(1+), H-00183
Phenoxytriphenylphosphorus(1+), P-00077

$C_{24}H_{20}OP_2$
Tetraphenyldiphosphine; Mono-oxide, *in* T-00258

$C_{24}H_{20}OP_2S$
Tetraphenyldiphosphine; Mono-oxide, monosulfide, *in* T-00258

$C_{24}H_{20}OP_2S_2$
Diphenylphosphinothioic anhydride, *in* D-01082

$C_{24}H_{20}O_2P_2$
Tetraphenyldiphosphine dioxide, T-00259

$C_{24}H_{20}O_3P_2$
Diphenylphosphinic acid; Anhydride, *in* D-01039

$C_{24}H_{20}O_4P_2$
Dioxybis[diphenylphosphine oxide], D-00977

$C_{24}H_{20}O_4P_2S_2$
O,O,O′,O′-Tetraphenyl thiohypophosphate, T-00283

$C_{24}H_{20}O_4P_2S_3$
O,O,O′,O′-Tetraphenyl trithiopyrophosphate, T-00284

$C_{24}H_{20}O_4P_2S_4$
Bis[(diphenoxyphosphino)thioyl] disulfide, B-00211

$C_{24}H_{20}O_4P_2Se_4$
Bis(*O,O*-diphenyl phosphoroselenoyl) diselenide, B-00269

$C_{24}H_{20}O_5P_2$
Diphenyl diphenyldiphosphonate, *in* D-01014

$C_{24}H_{20}O_7P_2$
Tetraphenyl diphosphate, T-00256

$C_{24}H_{20}P^{\oplus}$
Tetraphenylphosphonium(1+), T-00272

$C_{24}H_{20}P_2$
Tetraphenyldiphosphine, T-00258

$C_{24}H_{20}P_2S$
Tetraphenyldiphosphine; Monosulfide, *in* T-00258

$C_{24}H_{20}P_2S_2$
Tetraphenyldiphosphine disulfide, T-00260

$C_{24}H_{20}P_2S_3$
Diphenylphosphinodithioic acid; Anhydrosulfide, *in* D-01065

$C_{24}H_{20}P_2Se_3$
Diphenylphosphinodiselenoic acid; Anhydroselenide, *in* D-01064

$C_{24}H_{20}P_4$
Tetraphenyltetraphosphetane, T-00278

$C_{24}H_{20}P_4S$
Tetraphenylthiatetraphospholane, T-00282

$C_{24}H_{20}P_4S_4$
Tetraphenyltetraphosphetane; Tetrasulfide, *in* T-00278

$C_{24}H_{21}NOP_2$
P,P-Diphenylphosphinic amide; *N*-Diphenylphosphino, *in* D-01040

$C_{24}H_{21}NOP_2S$
P,P-Diphenyl-*N*-(diphenylphosphinothioyl)-phosphinimidic acid, D-01006
P,P-Diphenylphosphinic amide; *N*-Diphenylphosphinothioyl, *in* D-01040

$C_{24}H_{21}NO_2P_2$
P,P-Diphenyl-*N*-(diphenylphosphinyl)-phosphinimidic acid, D-01007
N-(Diphenylphosphinyl)-*P,P*-diphenylphosphinic amide, D-01096

$C_{24}H_{21}NO_4P_2S_2$
O,O-Diphenyl (diphenoxyphosphinothioyl)-phosphoramidothioate, D-01004

$C_{24}H_{21}NO_6P$
Methyl 2-[(dibenzyloxyphosphinyl)amino]-3-phenylpropanoate, *in* P-00400

$C_{24}H_{21}NO_6P_2$
O,O,O′,O′-Tetraphenyl imidodiphosphate, T-00265

$C_{24}H_{21}NP_2$
N-Diphenylphosphino-*P,P*-diphenylphosphinous amide, D-01063

$C_{24}H_{21}NP_2S_2$
N-Diphenylphosphinothioyl-*P,P*-diphenylphosphinothioic amide, D-01087

$C_{24}H_{21}N_2OP$
2,3-Dihydro-2,4,9-triphenyl-1*H*-1,3,2-diazaphosphonine 2-oxide, D-00609
N,N,N′,N′-Tetraphenylphosphonic diamide, T-00271
Tetraphenylphosphorodiamidous acid, T-00274

$C_{24}H_{21}N_2O_2P$
[1,1′-Biphenyl]-2-yl *N,N′*-diphenylphosphorodiamidate, *in* M-00445

$C_{24}H_{21}N_2P$
▷*N,N′,P,P*-Tetraphenylphosphinimidic amide, *in* T-00631

$C_{24}H_{21}N_2PS$
N,N,N′,N′-Tetraphenylphosphonothioic diamide, T-00273

$C_{24}H_{21}OP$
Tris(2-phenylethenyl)phosphine; Oxide, *in* T-00843

$C_{24}H_{21}O_5PS$
O,O-Diphenyl (diphenoxyphosphinyl)-phosphoramidothioate, D-01005

$C_{24}H_{21}O_6P$
Tris(benzoyloxymethyl)phosphine oxide, *in* T-00784
Tris(3-carboxyphenyl)phosphine; Tri-Me ester, *in* T-00739
Tris(4-carboxyphenyl)phosphine; Tri-Me ester, *in* T-00740
Tris(4-hydroxyphenyl)phosphine; Tri-Ac, *in* T-00790

$C_{24}H_{21}O_6PS$
Tris(4-carboxyphenyl)phosphine; Sulfide, tri-Me ester, *in* T-00740

$C_{24}H_{21}O_7P$
Tris(2-acetyloxyphenyl)phosphine oxide, *in* T-00788
Tris(3-carboxyphenyl)phosphine; Oxide, tri-Me ester, *in* T-00739
Tris(4-carboxyphenyl)phosphine; Oxide, tri-Me ester, *in* T-00740

$C_{24}H_{21}P$
Tris(2-phenylethenyl)phosphine, T-00843

$C_{24}H_{21}PS$
Tris(2-phenylethenyl)phosphine; Sulfide, *in* T-00843

$C_{24}H_{22}BrOP$
(4-Methoxyphenyl)-1-naphthalenylphenylphosphine; B,PhCH$_2$Br, *in* M-00066

$C_{24}H_{22}BrP$
Methyl-2-naphthalenylphenyl(phenylmethyl)phosphonium bromide, *in* M-00177

$C_{24}H_{22}ClOP$
(4-Methoxyphenyl)-1-naphthalenylphenylphosphine; B,PhCH$_2$Cl, *in* M-00066

$C_{24}H_{22}N_2P_2S$
P,P-Diphenyl-*N*-(diphenylphosphinothioyl)phosphinimidic acid; Amide, *in* D-01006

$C_{24}H_{22}N_3OP$
N,N′,N″,N‴-Tetraphenylphosphoric triamide, *in* D-01108

$C_{24}H_{23}O_2P$
Methyl α-(triphenylphosphoranylidene)cyclopropaneacetate, *in* C-00285
4-(Triphenylphosphoranylidene)-2-butenoic acid; Et ester, *in* T-00647

$C_{24}H_{23}O_3P$
Methyl 4-oxo-5-(triphenylphosphoranylidene)pentanoate, *in* O-00085

$C_{24}H_{23}O_4P$
Dimethyl (triphenylphosphoranylidene)butanedioate, *in* T-00646
α-Ethoxycarbonyltriphenylphosphoranylidenepropanoic acid, *in* T-00646

$C_{24}H_{24}BF_4O_3P$
4-Methyl-4-triphenylmethyl-2,6,7-trioxa-1-phosphoniabicyclo[2.2.2]octane tetrafluoroborate, *in* M-00424

$C_{24}H_{24}BrO_2P$
(2-Carboxy-2-propenyl)triphenylphosphonium(1+); Et ester, bromide, *in* C-00016
[α-Cyclopropyl(methoxycarbonyl)methyl]triphenylphosphonium(1+); Bromide, *in* C-00285

$C_{24}H_{24}BrP$
2,3,3*a*,4,5,9*b*-Hexahydro-4,4-diphenyl-1*H*-cyclopent[*c*]isophosphinolinium bromide, *in* H-00048

$C_{24}H_{24}ClO_7P$
4-Methyl-4-triphenylmethyl-2,6,7-trioxa-1-phosphoniabicyclo[2.2.2]octane perchlorate, *in* M-00424

$C_{24}H_{24}F_6P_2$
2,3,3*a*,4,5,9*b*-Hexahydro-4,4-diphenyl-1*H*-cyclopent[*c*]isophosphinolinium hexafluorophosphate, *in* H-00048

$C_{24}H_{24}IO_2P$
[α-Cyclopropyl(methoxycarbonyl)methyl]triphenylphosphonium(1+); Iodide, *in* C-00285

$C_{24}H_{24}N_4O_3P_2$
N,N′-Diphenylphosphorodiamidic acid; Anhydride, *in* D-01125

$C_{24}H_{24}N_5P_3$
2,2-Diamino-2,2,4,4,6,6-hexahydro-4,4,6,6-tetraphenyl-1,3,5-triaza-2,4,6-triphosphorine, D-00025

$C_{24}H_{24}O_2P^{\oplus}$
[α-Cyclopropyl(methoxycarbonyl)methyl]triphenylphosphonium(1+), C-00285
1-(Ethoxycarbonyl)cyclopropyltriphenylphosphonium(1+), E-00025

$C_{24}H_{24}P_2$
3,3′,4,4′-Tetramethyl-1,1′-diphenyl-2,2′-bi-1*H*-phosphole, T-00199

$C_{24}H_{24}P_2S_2$
3,3′,4,4′-Tetramethyl-1,1′-diphenyl-2,2′-bi-1*H*-phosphole; 1,1′-Disulfide, *in* T-00199

$C_{24}H_{25}F_6N_2O_3P$
4′-Ethyl-1,3-dihydro-5-phenyl-3,3-bis(trifluoromethyl)spiro[2*H*-1,4,2-diazaphosphole-2,1′-[2,6,7]trioxa[1]phosphabicyclo[2.2.2]octane]; 1-(2,6-Dimethylphenyl), *in* E-00061

$C_{24}H_{25}OP$
3,3-Dimethyl-1-triphenylphosphoranylidene-2-butanone, D-00934

$C_{24}H_{25}O_2P$
(Diethoxyethenylidene)triphenylphosphorane, D-00232
[2-(1,3-Dioxan-2-yl)ethylidene]triphenylphosphorane, *in* D-00965
3,3,3′,3′-Tetramethyl-1,1′(3*H*,3′*H*)-spirobi[2,1-benzoxaphosphole]; 1-Ph, *in* T-00232
(Triphenylphosphoranylidene)acetic acid; *tert*-Butyl ester, *in* T-00640
6-(Triphenylphosphoranylidene)hexanoic acid, *in* C-00015

$C_{24}H_{25}P$
Cyclohexylidenetriphenylphosphorane, *in* C-00273

$C_{24}H_{26}BrOP$
(3,3-Dimethyl-2-oxobutyl)triphenylphosphonium(1+); Bromide, *in* D-00784
(3,3-Dimethyl-2-oxopropyl)triphenylphosphonium(1+); Bromide, *in* D-00787

$C_{24}H_{26}BrO_2P$
(5-Carboxypentyl)triphenylphosphonium(1+); Bromide, *in* C-00015
[2-(1,3-Dioxan-2-yl)ethyl]triphenylphosphonium(1+); Bromide, *in* D-00965

$C_{24}H_{26}BrP$
Cyclohexyltriphenylphosphonium(1+); Bromide, *in* C-00273

$C_{24}H_{26}ClO_4P$
5-Ethyl-6,6*a*,7,8,9,9*a*-hexahydro-5-phenyl-1*H*-benzo[*h*]cyclopenta[*c*]phosphinolinium perchlorate, *in* H-00043

$C_{24}H_{26}NO_5P$
Dibenzyl [1-(benzyloxycarbonyl)ethyl]phosphoramidate, *in* P-00362

$C_{24}H_{26}N_3O_2P$
10,10-Dihydro-2,3-diphenyl-[1,3,2]oxazaphospholo[2,3-*b*][1,3,2]benzoxazaphosphole; 10,10-Bis(dimethylamino), *in* D-00481

$C_{24}H_{26}OP^{\oplus}$
(3,3-Dimethyl-2-oxobutyl)triphenylphosphonium(1+), D-00784
(3,3-Dimethyl-2-oxopropyl)triphenylphosphonium(1+), D-00787

$C_{24}H_{26}OP_2$
1,2,3,4,6,7,8,9-Octahydro-1,9-diphenyl-5,1,9-benzoxadiphosphacycloundecin, O-00012

$C_{24}H_{26}O_2P^{\oplus}$
(5-Carboxypentyl)triphenylphosphonium(1+), C-00015
[2-(1,3-Dioxan-2-yl)ethyl]triphenylphosphonium(1+), D-00965

$C_{24}H_{26}O_2P_2$
3*a*,4,7,7*a*-Tetrahydro-2,3,5,6-tetramethyl-1,8-diphenyl-4,7-phosphinidene-1*H*-phosphindole; 1,8-Dioxide, *in* T-00141

$C_{24}H_{26}P^{\oplus}$
Cyclohexyltriphenylphosphonium(1+), C-00273

$C_{24}H_{26}P_2$
3*a*,4,7,7*a*-Tetrahydro-2,3,5,6-tetramethyl-1,8-diphenyl-4,7-phosphinidene-1*H*-phosphindole, T-00141

$C_{24}H_{26}P_2S$
1,2,3,4,6,7,8,9-Octahydro-1,9-diphenyl-5,9-benzothiadiphosphacycloundecin, O-00011

$C_{24}H_{27}NP_2$
2,3,4,5,6,7,8,9-Octahydro-1,9-diphenyl-1*H*-5,1,9-benzazadiphosphacycloundecine, O-00009

$C_{24}H_{27}OP$
Tris(2,4-dimethylphenyl)phosphine; Oxide, *in* T-00768
Tris(2,5-dimethylphenyl)phosphine; Oxide, *in* T-00769
Tris(2,6-dimethylphenyl)phosphine; Oxide, *in* T-00770
Tris[(2-methylphenyl)methyl]phosphine; Oxide, *in* T-00800
Tris[(3-methylphenyl)methyl]phosphine; Oxide, *in* T-00801
Tris[(4-methylphenyl)methyl]phosphine; Oxide, *in* T-00802
Tris(2-phenylethyl)phosphine; Oxide, *in* T-00844

$C_{24}H_{27}O_2P$
(2,2-Diethoxyethylidene)triphenylphosphorane, D-00233
2,2-Dihydro-4,4,5,5-tetramethyl-2,2,2-triphenyl-1,3,2-dioxaphospholane, D-00578

$C_{24}H_{27}O_3P$
Bis(2-phenylethyl)(2-phenylethyl)phosphonate, *in* P-00165
Tris(2-phenylethyl) phosphite, T-00845

$C_{24}H_{27}O_4P$
Tris(2,5-dimethylphenyl) phosphate, T-00767

$C_{24}H_{27}O_6P$
Tris(2,4-dimethoxyphenyl)phosphine, T-00758
Tris(3,5-dimethoxyphenyl)phosphine, T-00759

$C_{24}H_{27}O_7P$
Tris(3,5-dimethoxyphenyl)phosphine; Oxide, *in* T-00759

$C_{24}H_{27}P$
Hexylidenetriphenylphosphorane, *in* H-00104
Tris(2,4-dimethylphenyl)phosphine, T-00768
Tris(2,5-dimethylphenyl)phosphine, T-00769
Tris(2,6-dimethylphenyl)phosphine, T-00770
Tris[(2-methylphenyl)methyl]phosphine, T-00800
Tris[(3-methylphenyl)methyl]phosphine, T-00801
Tris[(4-methylphenyl)methyl]phosphine, T-00802
Tris(2-phenylethyl)phosphine, T-00844

$C_{24}H_{27}PS$
Tris(2,4-dimethylphenyl)phosphine; Sulfide, *in* T-00768
Tris(2,5-dimethylphenyl)phosphine; Sulfide, *in* T-00769
Tris[(2-methylphenyl)methyl]phosphine; Sulfide, *in* T-00800
Tris[(3-methylphenyl)methyl]phosphine; Sulfide, *in* T-00801
Tris[(4-methylphenyl)methyl]phosphine; Sulfide, *in* T-00802

$C_{24}H_{27}P_3$
3,3,5,5-Tetramethyl-1,2,4-triphenyl-1,2,4-triphospholane, T-00239

$C_{24}H_{28}BrP$
Hexyltriphenylphosphonium(1+); Bromide, *in* H-00104

$C_{24}H_{28}IP$
Hexyltriphenylphosphonium(1+); Iodide, *in* H-00104

$C_{24}H_{28}NOP$
N-Phenyl-*P,P*-bis(2,4,6-trimethylphenyl)phosphinic amide, *in* B-00428

$C_{24}H_{28}NO_{11}P$
2-Amino-2-deoxyglucose 6-(dihydrogen phosphate); 1,3,4-Tri-Ac, di-Ph ester, *in* A-00069

$C_{24}H_{28}NP$
(4-Dimethylaminobutylidene)triphenylphosphorane, *in* D-00690

$C_{24}H_{28}P^{\oplus}$
▷Hexyltriphenylphosphonium(1+), H-00104

$C_{24}H_{29}BrNP$
[4-(Dimethylamino)butyl]triphenylphosphonium(1+); Bromide, *in* D-00690

$C_{24}H_{29}NP^{\oplus}$
[4-(Dimethylamino)butyl]triphenylphosphonium(1+), D-00690

$C_{24}H_{29}OPSi$
[[2-(Trimethylsilyl)ethoxy]methylene]triphenylphosphorane, *in* T-00552

$C_{24}H_{30}ClOPSi$
[[2-(Trimethylsilyl)ethoxy]methyl]triphenylphosphonium(1+); Chloride, *in* T-00552

$C_{24}H_{30}NO_3P$
Dibenzyl 1-adamantylphosphoramidate, *in* A-00036

$C_{24}H_{30}N_2O_5P_2S$
Triethyl [[(triphenylphosphoranylidene)amino]sulfonyl]phosphorimidate, *in* T-00642

$C_{24}H_{30}N_3OP$
Tris(4-dimethylaminophenyl)phosphine; Oxide, *in* T-00765

$C_{24}H_{30}N_3O_3P$
Tris(4-dimethylaminophenyl) phosphite, T-00766

$C_{24}H_{30}N_3O_4P$
Tris[2-(dimethylamino)phenyl] phosphate, T-00761
Tris[3-(dimethylamino)phenyl] phosphate, T-00762
Tris[4-(dimethylamino)phenyl] phosphate, T-00763

$C_{24}H_{30}N_3P$
Tris(2-dimethylaminophenyl)phosphine, T-00764
Tris(4-dimethylaminophenyl)phosphine, T-00765

$C_{24}H_{30}N_3PS$
Tris(4-dimethylaminophenyl)phosphine; Sulfide, *in* T-00765

$C_{24}H_{30}OPSi^{\oplus}$
[[2-(Trimethylsilyl)ethoxy]methyl]triphenylphosphonium(1+), T-00552

$C_{24}H_{31}BrNP$
2,6-Di-*tert*-butyl-1,4-dihydro-4,4-diphenyl-1,4-azaphosphorinium bromide, *in* D-00084

$C_{24}H_{31}O_3P$
4,8-Di-*tert*-butyl-2,10-dimethyl-12*H*-dibenzo[*d,g*][1,3,2]dioxaphosphocin; 6-Ethoxy, *in* D-00088

$C_{24}H_{32}NO_3P$
Dibenzyl 2-naphthylphosphoramidate, *in* N-00029

$C_{24}H_{32}N_2P_2$
12,14-Dimethyl-4,8-diphenyl-1,11-diaza-4,8-diphosphacyclotetradeca-11,14-diene, D-00738

$C_{24}H_{33}Ge_3P_3$
Hexahydro-2,2,4,4,6,6-hexamethyl-1,3,5-triphenyl-1,3,5,2,4,6-triphosphatrigermanin, H-00029

$C_{24}H_{33}O_6P$
7,8,10,11,13,14,16,17,19,20,22,27-Dodecahydro-5*H*-dibenzo[*r,u*][1,4,7,10,13,16,20]-hexaoxaphosphacyclotricosin, D-01261

$C_{24}H_{36}Br_2P_2$
5,5,10,10-Tetraethyl-5,6,7,8,9,10-hexahydrodibenzo[*b,d*][1,6]diphosphecinium dibromide, *in* H-00021

$C_{24}H_{36}P_2S_2$
Diethyl 1,4-butanediylbis[phenylphosphinodithioate], *in* B-00516

$C_{24}H_{43}N_2OP$
(2-Phenylethenyl)phosphonic bis(dibutylamide), *in* P-00156

$C_{24}H_{44}OP$
Tetracyclohexyldiphosphine; Monoxide, *in* T-00039

$C_{24}H_{44}O_2P_2$
Tetracyclohexyldiphosphine; Dioxide, *in* T-00039

$C_{24}H_{44}P_2$
Tetracyclohexyldiphosphine, T-00039

$C_{24}H_{44}P_2S_2$
Tetracyclohexyldiphosphine; Disulfide, *in* T-00039

$C_{24}H_{50}ClOP$
Didodecylphosphinic acid; Chloride, *in* D-00226

$C_{24}H_{50}O_3P$
Didodecyl phosphonate, D-00227

$C_{24}H_{51}OP$
Trioctylphosphine; Oxide, *in* T-00582

$C_{24}H_{51}O_2P$
Didodecylphosphinic acid, D-00226

$C_{24}H_{51}O_3P$
▷Trioctyl phosphite, T-00583
Tris(1-methylheptyl) phosphite, T-00797
▷Tris(6-methylheptyl) phosphite, T-00798

$C_{24}H_{51}O_4P$
Didodecyl phosphate, D-00225
Trioctyl phosphate, T-00581
▷Tris(2-ethylhexyl) phosphate, T-00779

$C_{24}H_{51}P$
Trioctylphosphine, T-00582

$C_{24}H_{51}PS$
Trioctylphosphine; Sulfide, *in* T-00582

$C_{24}H_{54}N_3OP$
Hexakis(2-methylpropyl)phosphoric triamide, *in* H-00063

$C_{24}H_{54}N_3P$
Hexabutylphosphorous triamide, H-00013
Hexakis(2-methylpropyl)phosphorous triamide, H-00063

$C_{24}H_{54}N_3PO$
Hexabutylphosphoric triamide, *in* H-00013

$C_{24}H_{54}N_3PS$
Hexabutylphosphorothioic triamide, *in* H-00013

$C_{24}H_{60}Ge_6P_4$
2,2,4,4,6,6,8,8,9,9,10,10-Dodecaethyl-1,3,5,7-tetraphospha-2,4,6,8,9,10-hexagermatricyclo[3.3.1.1^{3,7}]decane, D-01260

$C_{25}H_{19}OP$
5,10-Dihydro-5,10-diphenylacridophosphine 5-oxide, *in* D-00548

$C_{25}H_{19}P$
Methylbisbiphenylenephosphorane, *in* S-00021

$C_{25}H_{19}PS_2$
(1,3-Benzodithiol-2-ylidene)triphenylphosphorane, *in* B-00021

$C_{25}H_{20}BF_4PS_2$
1,3-Benzodithiol-2-yltriphenylphosphonium(1+); Tetrafluoroborate, *in* B-00021

$C_{25}H_{20}BrP$
5,6-Dihydro-5,5-diphenylphosphanthridinium bromide, *in* D-00568

$C_{25}H_{20}ClOP$
Phenyl(triphenylmethyl)phosphinic acid; Chloride, *in* P-00332

$C_{25}H_{20}ClP$
5-Benzyl-5-phenyl-5*H*-dibenzophospholium chloride, *in* P-00110
(2-Chlorophenylmethylene)triphenylphosphorane, *in* C-00135
(3-Chlorophenylmethylene)triphenylphosphorane, *in* C-00136
(4-Chlorophenylmethylene)triphenylphosphorane, *in* C-00137

$C_{25}H_{20}ClPS_2$
1,3-Benzodithiol-2-yltriphenylphosphonium(1+); Chloride, *in* B-00021

$C_{25}H_{20}FP$
[(4-Fluorophenyl)methylene]triphenylphosphorane, *in* F-00032

$C_{25}H_{20}NOP$
N-Benzoyl-*P,P,P*-triphenylphosphine imide, *in* T-00626

$C_{25}H_{20}NO_2P$
[(2-Nitrophenyl)methylene]triphenylphosphorane, *in* N-00047
[(3-Nitrophenyl)methylene]triphenylphosphorane, *in* N-00048
[(4-Nitrophenyl)methylene]triphenylphosphorane, N-00041

$C_{25}H_{20}NO_4P$
Triphenyl benzoylphosphorimidate, *in* B-00039

$C_{25}H_{20}OP_2$
Bis(diphenylphosphino)methanone, B-00245

$C_{25}H_{20}PS_2^{\oplus}$
1,3-Benzodithiol-2-yltriphenylphosphonium(1+), B-00021

$C_{25}H_{21}AsIP$
5-Methyl-5,10-diphenyldibenzo[1,4]-phosphoniaarsenin, *in* D-01001

$C_{25}H_{21}BrClP$
Benzyl-4-bromophenyldiphenylphosphonium bromide, *in* B-00481
Benzyl-2-bromophenyldiphenylphosphonium chloride, *in* B-00480
[(2-Chlorophenyl)methyl]triphenylphosphonium(1+); Bromide, *in* C-00135
[(3-Chlorophenyl)methyl]triphenylphosphonium(1+); Bromide, *in* C-00136
[(4-Chlorophenyl)methyl]triphenylphosphonium(1+); Bromide, *in* C-00137

$C_{25}H_{21}BrFP$
[(4-Fluorophenyl)methyl]triphenylphosphonium(1+); Bromide, *in* F-00032

$C_{25}H_{21}BrNO_2P$
[(2-Nitrophenyl)methyl]triphenylphosphonium(1+); Bromide, *in* N-00047
[(3-Nitrophenyl)methyl]triphenylphosphonium(1+); Bromide, *in* N-00048
▷[(4-Nitrophenyl)methyl]triphenylphosphonium(1+); Bromide, *in* N-00049

$C_{25}H_{21}ClFP$
[(4-Fluorophenyl)methyl]triphenylphosphonium(1+); Chloride, *in* F-00032

$C_{25}H_{21}ClNO_2P$
[(2-Nitrophenyl)methyl]triphenylphosphonium(1+); Chloride, *in* N-00047
[(3-Nitrophenyl)methyl]triphenylphosphonium(1+); Chloride, *in* N-00048
[(4-Nitrophenyl)methyl]triphenylphosphonium(1+); Chloride, *in* N-00049

$C_{25}H_{21}ClP^{\oplus}$
[(2-Chlorophenyl)methyl]triphenylphosphonium(1+), C-00135
[(3-Chlorophenyl)methyl]triphenylphosphonium(1+), C-00136
[(4-Chlorophenyl)methyl]triphenylphosphonium(1+), C-00137

$C_{25}H_{21}Cl_2P$
Benzyl(2-chlorophenyl)diphenylphosphonium chloride, *in* C-00127
[(2-Chlorophenyl)methyl]triphenylphosphonium(1+); Chloride, *in* C-00135
[(3-Chlorophenyl)methyl]triphenylphosphonium(1+); Chloride, *in* C-00136
[(4-Chlorophenyl)methyl]triphenylphosphonium(1+); Chloride, *in* C-00137

$C_{25}H_{21}FP^{\oplus}$
[(4-Fluorophenyl)methyl]triphenylphosphonium(1+), F-00032

$C_{25}H_{21}NO_2P^{\oplus}$
[(2-Nitrophenyl)methyl]triphenylphosphonium(1+), N-00047
[(3-Nitrophenyl)methyl]triphenylphosphonium(1+), N-00048
[(4-Nitrophenyl)methyl]triphenylphosphonium(1+), N-00049

$C_{25}H_{21}OP$
[(2-Hydroxyphenyl)methyl]triphenylphosphorane, *in* H-00177

$C_{25}H_{21}O_2P$
Phenyl(triphenylmethyl)phosphinic acid, P-00332

$C_{25}H_{21}P$
Benzylidenetriphenylphosphorane, B-00050

$C_{25}H_{21}PS$
Triphenyl[(phenylthio)methylene]phosphorane, T-00620

$C_{25}H_{21}PS_2$
6,8,10-Triphenyl-1,4-dithia-5-phospha(5-P^V)-spiro[4.5]deca-5,7,9-triene, T-00604

$C_{25}H_{22}BrOP$
[(2-Hydroxyphenyl)methyl]triphenylphosphonium(1+); Bromide, *in* H-00177

$C_{25}H_{22}BrP$
Benzyltriphenylphosphonium(1+); Bromide, *in* B-00073

$C_{25}H_{22}ClP$
Benzyltriphenylphosphonium(1+); Chloride, *in* B-00073

$C_{25}H_{22}ClPS$
Triphenyl(phenylthiomethyl)phosphonium(1+); Chloride, *in* T-00621

$C_{25}H_{22}IP$
▷Benzyltriphenylphosphonium(1+); Iodide, *in* B-00073

$C_{25}H_{22}I_3P$
Benzyltriphenylphosphonium(1+); Triodide, *in* B-00073

$C_{25}H_{22}I_5P$
Benzyltriphenylphosphonium(1+); Pentaiodide, *in* B-00073

$C_{25}H_{22}I_7P$
Benzyltriphenylphosphonium(1+); Heptaiodide, *in* B-00073

$C_{25}H_{22}NOP$
6,8,10-Triphenyl-1-oxa-4-aza-5-phosphaspiro(5-P^V)spiro[4.5]deca-5,7,9-triene, T-00609

$C_{25}H_{22}NO_5PS$
Triphenyl [(4-methylphenyl)sulfonyl]-phosphorimidate, *in* M-00268

$C_{25}H_{22}N_3O_2P$
P-(4-*N*-phenylcarbamoylphenyl)-*N*,*N*′-diphenylphosphonic diamide, *in* P-00380

$C_{25}H_{22}OP^{\oplus}$
[(2-Hydroxyphenyl)methyl]triphenylphosphonium(1+), H-00177

$C_{25}H_{22}O_2P_2$
Bis(diphenylphosphinyl)methane, *in* B-00244

$C_{25}H_{22}O_4P_2$
Tetraphenyl methylenebisphosphonite, *in* M-00024

$C_{25}H_{22}O_6P_2$
Tetraphenyl methylenebisphosphonate, *in* M-00022

$C_{25}H_{22}P^{\oplus}$
Benzyltriphenylphosphonium(1+), B-00073

$C_{25}H_{22}PS^{\oplus}$
Triphenyl(phenylthiomethyl)phosphonium(1+), T-00621

$C_{25}H_{22}P_2$
Bis(diphenylphosphino)methane, B-00244

$C_{25}H_{22}P_2S_2$
Bis(diphenylphosphinothioyl)methane, *in* B-00244

$C_{25}H_{22}P_2Se$
Diphenyl[(diphenylphosphinoselenoyl)methyl]-phosphine, *in* B-00244

$C_{25}H_{22}P_2Se_2$
Bis(diphenylphosphinoselenoyl)methane, *in* B-00244

$C_{25}H_{22}P_4$
1,2,3,4-Tetraphenyltetraphospholane, T-00280

$C_{25}H_{22}P_4S$
1,2,3,4-Tetraphenyltetraphospholane; 1-Sulfide, *in* T-00280

$C_{25}H_{22}P_4S_2$
1,2,3,4-Tetraphenyltetraphospholane; 1,4-Disulfide, *in* T-00280

$C_{25}H_{23}BrNP$
1-Bromo-*N*-methyl-*N*,1,1-tetraphenyl-phosphoranamine, *in* M-00194

$C_{25}H_{23}ClN_3P_3$
2-Chloro-2,2,4,4,6,6-hexahydro-2-methyl-4,4,6,6-tetraphenyl-1,3,5-triaza-2,4,6-triphosphorine, C-00086

$C_{25}H_{23}FO_2P$
1-Fluoro-2,2,3,4,4-pentamethylphosphetane 1-oxide, *in* H-00168

$C_{25}H_{23}INP$
1-Iodo-*N*-methyl-*N*,1,1,1-tetraphenyl-phosphoranamine, *in* M-00194

$C_{25}H_{23}NOP_2S$
O-Methyl *P*,*P*-diphenyl-*N*-(diphenyl-phosphinothioyl)phosphinimidate, *in* D-01006

$C_{25}H_{23}NO_4P_2$
Tetraphenyl methylimidodiphosphite, *in* M-00157

$C_{25}H_{23}NO_6P_2$
Tetraphenyl methylimidodiphosphate, *in* M-00156

$C_{25}H_{23}NP^{\oplus}$
(Methylphenylamino)triphenyl-phosphonium(1+), M-00194

$C_{25}H_{23}N_2OP$
P-[(1,1′-Biphenyl)-2-ylmethyl]-*N*,*N*′-diphenyl-phosphonic diamide, *in* B-00079
Methyl tetraphenylphosphorodiamidite, *in* T-00274

$C_{25}H_{23}N_2P$
6,8,10-Triphenyl-1,4-diaza-5-phospha(5-P^V)-spiro[4.5]deca-5,7,9-triene, T-00599

$C_{25}H_{23}OP$
Tetraphenylphosphonium(1+); Methoxide, *in* T-00272

$C_{25}H_{23}O_2P$
1,1-Dihydro-1,1-dimethoxy-2,4,6-triphenyl-phosphorin, *in* D-00613

$C_{25}H_{23}PS_2$
1,1-Dihydro-1,1-bis(methylthio)-2,4,6-triphenylphosphorin, *in* D-00613

$C_{25}H_{24}IO_6P$
Tris(4-hydroxyphenyl)phosphine; Tri-Ac; B,MeI, *in* T-00790

$C_{25}H_{24}NO_2P$
tert-Butyl cyano(triphenylphosphoranylidene)-acetate, *in* C-00231

$C_{25}H_{25}N_4P_3$
2-Amino-2,2,4,4,6,6-hexahydro-2-methyl-4,4,6,6-tetraphenyl-1,3,5-triaza-2,4,6-triphosphorine, A-00081

$C_{25}H_{25}O_2P$
Diethoxytriphenylphosphorane, D-00263

$C_{25}H_{25}O_2PS_2$
Ethyl 3-(ethylthio)-3-thioxo-2-(triphenyl-phosphoranylidene)propanoate, *in* E-00194

$C_{25}H_{25}O_4P$
Diethyl (triphenylphosphoranylidene)-propanedioate, *in* T-00658

$C_{25}H_{25}O_5P$
8,8-Dimethyl-2,3-diphenyl-1,4,6,10-tetraoxa-5-phospha(5-P^V)spiro[4.5]dec-2-ene; 5-Phenoxy, *in* D-00749

$C_{25}H_{26}I_2NO_6P$
Bis-4-iodobenzyl [1-ethoxycarbonyl-2-(4-hydroxybenzyl)ethyl]phosphoramidate, *in* P-00422

$C_{25}H_{27}ClO_2P$
(6-Carboxyhexyl)triphenylphosphonium(1+); Chloride, *in* C-00013

$C_{25}H_{27}F_6N_2O_3P$
4′-Ethyl-1,3-dihydro-5-phenyl-3,3-bis(trifluoromethyl)spiro[2*H*-1,4,2-diazaphosphole-2,1′-[2,6,7]trioxa[1]-phosphabicyclo[2.2.2]octane]; 1-(2,4,6-Trimethylphenyl), *in* E-00061

$C_{25}H_{27}OP$
3-Methoxy-17-methyl-15-phenyl-15-phosphaestra-1,3,5(10),8,16-pentaene, M-00051
1-(Triphenylphosphoranylidene)-2-heptanone, *in* O-00062

$C_{25}H_{27}OPS$
3-Methoxy-17-methyl-15-phenyl-15-phosphaestra-1,3,5(10),8,16-pentaene; 15-Sulfide, *in* M-00051

$C_{25}H_{27}O_2P$
3-Methoxy-17-methyl-15-phenyl-15-phosphaestra-1,3,5(10),8,16-pentaene; 15-Oxide, *in* M-00051
[3-(2-Methyl-1,3-dioxolan-2-yl)propylidene]-triphenylphosphorane, *in* M-00132
3,3,3′,3′-Tetramethyl-1,1′(3*H*,3′*H*)-spirobi[2,1-benzoxaphosphole]; 1-Benzyl, *in* T-00232
5-(Triphenylphosphoranylidene)pentanoic acid; Et ester, *in* T-00655

$C_{25}H_{27}O_2P^{\oplus}$
(6-Carboxyhexyl)triphenylphosphonium(1+), C-00013

$C_{25}H_{27}P$
Cycloheptylidenetriphenylphosphorane, *in* C-00238

$C_{25}H_{27}PSi$
[4-(Trimethylsilyl)-3-butyn-1-ylidene]-triphenylphosphorane, *in* T-00691

$C_{25}H_{28}BrOP$
(2-Oxoheptyl)triphenylphosphonium(1+); Bromide, *in* O-00062

$C_{25}H_{28}BrO_2P$
[3-(2-Methyl-1,3-dioxolan-2-yl)propyl]-triphenylphosphonium; Bromide, *in* M-00132

$C_{25}H_{28}BrP$
Cycloheptyltriphenylphosphonium(1+); Bromide, *in* C-00238

$C_{25}H_{28}ClOP$
(2-Oxoheptyl)triphenylphosphonium(1+); Chloride, *in* O-00062

$C_{25}H_{28}IO_2P$
(4-Carboxybutyl)triphenylphosphonium(1+); Et ester, iodide, *in* C-00009
[3-(2-Methyl-1,3-dioxolan-2-yl)propyl]-triphenylphosphonium; Iodide, *in* M-00132

$C_{25}H_{28}IPSi$
Triphenyl[4-(trimethylsilyl)-3-butynyl]-phosphonium(1+); Iodide, *in* T-00691

$C_{25}H_{28}NO_6P$
Dibenzyl [1-ethoxycarbonyl-2-(4-hydroxybenzyl)ethyl]phosphoramidate, *in* P-00422

$C_{25}H_{28}OP^{\oplus}$
(2-Oxoheptyl)triphenylphosphonium(1+), O-00062

$C_{25}H_{28}O_2P$
tert-Butyl 2-(triphenylphosphoranylidene)-propanoate, *in* T-00660

$C_{25}H_{28}O_2P^{\oplus}$
[3-(2-Methyl-1,3-dioxolan-2-yl)propyl]-triphenylphosphonium, M-00132

$C_{25}H_{28}O_4P_2$
Diethyl [2-oxo-2-(triphenylphosphoranylidene)-propyl]phosphonate, *in* O-00086

$C_{25}H_{28}P^{\oplus}$
Cycloheptyltriphenylphosphonium(1+), C-00238

$C_{25}H_{28}PSi^{\oplus}$
Triphenyl[4-(trimethylsilyl)-3-butynyl]-phosphonium(1+), T-00691

$C_{25}H_{28}P_2$
2,3,4,5,6,7,8,9-Octahydro-1,9-diphenyl-1*H*-1,9-benzodiphosphacycloundecin, O-00010

$C_{25}H_{29}NP_2$
2,3,4,5,6,7,8,9-Octahydro-1,9-diphenyl-1*H*-5,1,9-benzazadiphosphacycloundecine; 5-Me, *in* O-00009

$C_{25}H_{29}P$
Heptylidenetriphenylphosphorane, *in* H-00011

$C_{25}H_{30}BrP$
Heptyltriphenylphosphonium(1+); Bromide, *in* H-00011

$C_{25}H_{30}IO_6P$
Tris(2,4-dimethoxyphenyl)methylphosphonium iodide, *in* T-00758

$C_{25}H_{30}IP$
Heptyltriphenylphosphonium(1+); Iodide, *in* H-00011
Methyltris(2,4-dimethylphenyl)phosphonium iodide, *in* T-00768
Methyltris(2,5-dimethylphenyl)phosphonium iodide, *in* T-00769

$C_{25}H_{30}P^{\oplus}$
Heptyltriphenylphosphonium(1+), H-00011

$C_{25}H_{33}PSi_2$
[Bis(trimethylsilyl)methylene]triphenylphosphorane, *in* B-00432

$C_{25}H_{34}BrPSi_2$
[Bis(trimethylsilyl)methyl]triphenylphosphonium(1+); Bromide, *in* B-00432

$C_{25}H_{34}N_3O_8P$
Antibiotic K 26, A-00140

$C_{25}H_{34}PSi_2^{\oplus}$
[Bis(trimethylsilyl)methyl]triphenylphosphonium(1+), B-00432

$C_{25}H_{43}NO_6P_2$
Tetrabutyl (aminophenylmethyl)bisphosphonate, *in* A-00053

$C_{25}H_{45}N_6O_{13}P$
Antibiotic BMG 59-R2, A-00135

$C_{25}H_{47}OPSi_2$
1-Trimethylsilyl-2-trimethylsilyloxy-2-(2,4,6-tri-*tert*-butylphenyl)methylenephosphine, T-00560

$C_{25}H_{54}IP$
Methyltrioctylphosphonium iodide, *in* T-00582

$C_{26}H_{19}NOP$
P,P-Di-1-naphthyl-*N*-phenylphosphinic amide, *in* D-00941

$C_{26}H_{20}BrP$
5-Benzyl-10*H*-5,10[1′,2′]-benzenoacridophosphonium bromide, *in* B-00014
Triphenyl(2-phenylethynyl)phosphonium(1+); Bromide, *in* T-00614

$C_{26}H_{20}ClOP$
[(1-Chloro-2-oxo-2-phenyl)ethylidene]-triphenylphosphorane, C-00122
9,10-Dihydro-10-oxo-9-phenyl-9-phosphoniaanthracene chloride, *in* P-00080

$C_{26}H_{20}ClP$
Triphenyl(2-phenylethynyl)phosphonium(1+); Chloride, *in* T-00614

$C_{26}H_{20}IOP$
[(1-Iodo-2-oxo-2-phenyl)ethylidene]triphenylphosphorane, I-00014

$C_{26}H_{20}NO_3P$
(4-Nitrophenyl-2-oxoethylidene)triphenylphosphorane, N-00050

$C_{26}H_{20}NP$
N-[(Triphenylphosphoranylidene)ethylidene]-aniline, T-00652

$C_{26}H_{20}O_2P_2$
1,2-Bis(diphenylphosphino)-1,2-ethanedione, B-00235
1,2-Bis(diphenylphosphinyl)ethyne, *in* B-00216

$C_{26}H_{20}O_3P_2$
5,6-Dihydro-5-hydroxyphosphanthridine 5-oxide; Anhydride, *in* D-00504
2,5-Dihydro-2,3,4,5-tetraphenyl-1,2,5-oxadiphosphole 2,5-dioxide, D-00583

$C_{26}H_{20}P^{\oplus}$
Triphenyl(2-phenylethynyl)phosphonium(1+), T-00614

$C_{26}H_{20}P_2$
Bis(diphenylphosphino)acetylene, B-00216
1,2-Dihydro-1,2,3,4-tetraphenyl-1,2-diphosphete, D-00581

$C_{26}H_{20}P_2S$
2,5-Dihydro-2,3,4,5-tetraphenyl-1,2,5-thiadiphosphole, D-00585

$C_{26}H_{20}P_2S_2$
1,2-Bis(diphenylphosphinothioyl)ethyne, *in* B-00216

$C_{26}H_{20}P_2S_3$
2,5-Dihydro-2,3,4,5-tetraphenyl-1,2,5-thiadiphosphole; 2,5-Disulfide, *in* D-00585

$C_{26}H_{21}BrNO_3P$
[2-(4-Nitrophenyl)-2-oxoethyl]triphenylphosphonium(1+); Bromide, *in* N-00051

$C_{26}H_{21}INO_3P$
[2-(4-Nitrophenyl)-2-oxoethyl]triphenylphosphonium(1+); Iodide, *in* N-00051

$C_{26}H_{21}NO_3P^{\oplus}$
[2-(4-Nitrophenyl)-2-oxoethyl]triphenylphosphonium(1+), N-00051

$C_{26}H_{21}N_2OP$
3,4-Dihydro-2,3,4,5-tetraphenyl-2*H*-1,3,2-diazaphosphole; 2-Oxide, *in* D-00580

$C_{26}H_{21}N_2P$
3,4-Dihydro-2,3,4,5-tetraphenyl-2*H*-1,2,3-diazaphosphole, D-00579
3,4-Dihydro-2,3,4,5-tetraphenyl-2*H*-1,3,2-diazaphosphole, D-00580

$C_{26}H_{21}OP$
2,2-Dihydro-2,2,2-triphenyl-2*H*-1,2-benzoxaphosphorin, D-00607
▷1-Phenyl-2-(triphenylphosphoranylidene)-ethanone, P-00333
α-(Triphenylphosphoranylidene)-benzeneacetaldehyde, T-00643

$C_{26}H_{21}O_2P$
2,3,3,4-Tetraphenyl-1,2-oxaphosphetane 2-oxide, T-00268
α-(Triphenylphosphoranylidene)benzeneacetic acid, T-00644

$C_{26}H_{21}P$
5-Benzyl-5,5-dihydro-5-phenylacridophosphine, B-00042
5-Benzyl-5,6-dihydro-6-phenylphosphanthridine, B-00043

$C_{26}H_{21}PS$
5-Benzyl-5,6-dihydro-6-phenylphosphanthridine; 5-Sulfide, *in* B-00043

$C_{26}H_{22}BrOP$
(α-Formylbenzyl)triphenylphosphonium(1+); Bromide, *in* F-00056
▷(2-Oxo-2-phenylethyl)triphenylphosphonium(1+); Bromide, *in* O-00067

$C_{26}H_{22}BrP$
5-Benzyl-5,10-dihydro-5-phenylacridophosphonium bromide, *in* D-00547

$C_{26}H_{22}ClOP$
(α-Formylbenzyl)triphenylphosphonium(1+); Chloride, *in* F-00056
(2-Oxo-2-phenylethyl)triphenylphosphonium(1+); Chloride, *in* O-00067

$C_{26}H_{22}ClPS$
Triphenyl[1-(phenylthio)ethenyl]phosphonium(1+); Chloride, *in* T-00619

$C_{26}H_{22}IOP$
▷(2-Oxo-2-phenylethyl)triphenylphosphonium(1+); Iodide, *in* O-00067

$C_{26}H_{22}IP$
5,10-Dihydro-5-methyl-5,10-diphenylacridophosphonium iodide, *in* D-00548

$C_{26}H_{22}IPS$
Triphenyl[1-(phenylthio)ethenyl]phosphonium(1+); Iodide, *in* T-00619

$C_{26}H_{22}NOP$
4,5,5,5-Tetrahydro-3,5,5,5-tetraphenyl-1,2,5-oxazaphosphole, T-00153

$C_{26}H_{22}N_2P_2$
2,2,4,4-Tetrahydro-2,2,4,4-tetraphenyl-1,5,2,4-diazadiphosphorine, T-00146

$C_{26}H_{22}OP^{\oplus}$
(α-Formylbenzyl)triphenylphosphonium(1+), F-00056
(2-Oxo-2-phenylethyl)triphenylphosphonium(1+), O-00067

$C_{26}H_{22}O_2P_2$
1,2-Bis(diphenylphosphino)ethylene; *P,P′*-Dioxide, *in* B-00237
1,1-Bis(diphenylphosphinyl)ethylene, *in* B-00236

$C_{26}H_{22}PS^{\oplus}$
Triphenyl[1-(phenylthio)ethenyl]phosphonium(1+), T-00619

$C_{26}H_{22}P_2$
1,1-Bis(diphenylphosphino)ethylene, B-00236
1,2-Bis(diphenylphosphino)ethylene, B-00237

$C_{26}H_{22}P_2S_2$
1,2-Bis(diphenylphosphino)ethylene; *P,P′*-Disulfide, *in* B-00237
1,1-Bis(diphenylphosphinothioyl)ethylene, *in* B-00236

$C_{26}H_{22}P_2Se_2$
1,1-Bis(diphenylphosphinoselenoyl)ethylene, *in* B-00236

$C_{26}H_{23}Fe_2OP$
1,1″-(Phenylphosphinylidene)bisferrocene, *in* D-00410

$C_{26}H_{23}Fe_2P$
Diferrocenylphenylphosphine, D-00410

$C_{26}H_{23}N_2O_2P$
Phenyl [(triphenylphosphoranylidene)-hydrazono]acetate, *in* T-00653

$C_{26}H_{23}OP$
[(2-Methoxyphenyl)methylene]triphenylphosphorane, *in* M-00062
[(3-Methoxyphenyl)methylene]triphenylphosphorane, *in* M-00063
[(4-Methoxyphenyl)methylene]triphenylphosphorane, *in* M-00064

$C_{26}H_{23}O_2P$
Phenyl bis(4-phenyl-1,3-butadienyl)-phosphinate, *in* B-00385

$C_{26}H_{23}O_2PS$
4-Methylphenyl triphenylphosphoranylidenemethyl sulfone, M-00271

$C_{26}H_{23}P$
(1,3,6-Cycloheptatrien-1-ylmethylene)-triphenylphosphorane, C-00236
Homocubyltriphenylphosphorane, *in* D-00569
[(2-Methylphenyl)methylene]triphenylphosphorane, *in* M-00209
[(3-Methylphenyl)methylene]triphenylphosphorane, *in* M-00210
[(4-Methylphenyl)methylene]triphenylphosphorane, *in* M-00211
Triphenyl(1-phenylethylidene)phosphorane, *in* T-00612
Triphenyl(2-phenylethylidene)phosphorane, *in* T-00613

$C_{26}H_{23}PS_2$
7,9,11-Triphenyl-1,5-dithia-6-phospha(6-P^V)-spiro[5.5]undeca-6,8,10-triene, T-00605

$C_{26}H_{24}AsP$
▷[2-(Diphenylarsino)ethyl]diphenylphosphine, D-00993

$C_{26}H_{24}BF_4P$
(2,4,6-Cycloheptatrien-1-ylmethyl)triphenylphosphonium(1+); Tetrafluoroborate, *in* C-00237

$C_{26}H_{24}BrOP$
Benzyl(2-methoxyphenyl)diphenylphosphonium bromide, *in* M-00056
[(2-Methoxyphenyl)methyl]triphenylphosphonium(1+); Bromide, *in* M-00062
[(3-Methoxyphenyl)methyl]triphenylphosphonium(1+); Bromide, *in* M-00063
[(4-Methoxyphenyl)methyl]triphenylphosphonium(1+); Bromide, *in* M-00064

$C_{26}H_{24}BrO_2PS$
[[(4-Methylphenyl)sulfonyl]methyl]triphenylphosphonium(1+); Bromide, *in* M-00267

$C_{26}H_{24}BrP$
▷[(2-Methylphenyl)methyl]triphenylphosphonium(1+); Bromide, *in* M-00209
[(3-Methylphenyl)methyl]triphenylphosphonium(1+); Bromide, *in* M-00210
[(4-Methylphenyl)methyl]triphenylphosphonium(1+); Bromide, *in* M-00211
Triphenyl(1-phenylethyl)phosphonium(1+); Bromide, *in* T-00612
Triphenyl(2-phenylethyl)phosphonium(1+); Bromide, *in* T-00613

$C_{26}H_{24}ClNO_4P_2$
3,4,5,5-Tetrahydro-2,2,5,5-tetraphenyl-1*H*-1,2,5-azadiphospholium(1+); Perchlorate, *in* T-00143

$C_{26}H_{24}ClOP$
Benzyl(3-methoxyphenyl)diphenylphosphonium chloride, *in* M-00057
Benzyl(4-methoxyphenyl)diphenylphosphonium chloride, *in* M-00058
[(2-Methoxyphenyl)methyl]triphenylphosphonium(1+); Chloride, *in* M-00062
[(3-Methoxyphenyl)methyl]triphenylphosphonium(1+); Chloride, *in* M-00063
[(4-Methoxyphenyl)methyl]triphenylphosphonium(1+); Chloride, *in* M-00064

$C_{26}H_{24}ClP$
[(2-Methylphenyl)methyl]triphenylphosphonium(1+); Chloride, *in* M-00209
[(3-Methylphenyl)methyl]triphenylphosphonium(1+); Chloride, *in* M-00210
[(4-Methylphenyl)methyl]triphenylphosphonium(1+); Chloride, *in* M-00211

$C_{26}H_{24}ClPS$
Benzyl(2-methylthiophenyl)diphenylphosphonium chloride, *in* M-00417

$C_{26}H_{24}F_6P_2$
1,2,3,4-Tetrahydro-4-methyl-2,2-diphenylbenz[*h*]isophosphinolinium(1+); Hexafluorophosphate, *in* T-00113

$C_{26}H_{24}IP$
Triphenyl(1-phenylethyl)phosphonium(1+); Iodide, *in* T-00612

$C_{26}H_{24}I_2P_2$
9,10-Dihydro-9,10-dimethyl-9,10-diphenylphosphaanthranium diiodide, *in* D-00482

$C_{26}H_{24}NOP$
N,*N*-Dibenzyl-*P*,*P*-diphenylphosphinic amide, *in* D-01040
N-Phenyl-*P*,*P*-bis(4-phenyl-1,3-butadienyl)phosphinic amide, *in* B-00385
6,8,10-Triphenyl-1-oxa-4-aza-5-phosphaspiro(5-P^V)spiro[4.5]deca-5,7,9-triene; *N*-Me, *in* T-00609

$C_{26}H_{24}NP_2^{\oplus}$
3,4,5,5-Tetrahydro-2,2,5,5-tetraphenyl-1*H*-1,2,5-azadiphospholium(1+), T-00143

$C_{26}H_{24}N_2O_6P_2$
1′,3′-Diphenyldispiro[1,3,2-benzodioxaphosphole-2,2′-[1,3,2,4]-diazaphosphetidine-4′,2″-[1,3,2]-benzodioxaphosphole]; 2,4′-Dimethoxy, *in* D-01018

$C_{26}H_{24}OP^{\oplus}$
[(2-Methoxyphenyl)methyl]triphenylphosphonium(1+), M-00062
[(3-Methoxyphenyl)methyl]triphenylphosphonium(1+), M-00063
[(4-Methoxyphenyl)methyl]triphenylphosphonium(1+), M-00064

$C_{26}H_{24}OP_2$
Diphenyl[(2-diphenylphosphinyl)ethyl]phosphine, *in* B-00234

$C_{26}H_{24}O_2PS^{\oplus}$
[[(4-Methylphenyl)sulfonyl]methyl]triphenylphosphonium(1+), M-00267

$C_{26}H_{24}O_2P_2$
1,2-Bis(diphenylphosphinyl)ethane, *in* B-00234

$C_{26}H_{24}O_4P_2$
Tetraphenyl 1,2-ethanediylbisphosphonite, *in* E-00008

$C_{26}H_{24}O_6P_2$
Tetraphenyl 1,2-ethanediylbisphosphonate, *in* E-00006

$C_{26}H_{24}O_7P_2$
P,*P*-Dibenzyl *P*′,*P*′-diphenyl diphosphate, D-00044

$C_{26}H_{24}P^{\oplus}$
(2,4,6-Cycloheptatrien-1-ylmethyl)triphenylphosphonium(1+), C-00237
[(2-Methylphenyl)methyl]triphenylphosphonium(1+), M-00209
[(3-Methylphenyl)methyl]triphenylphosphonium(1+), M-00210
[(4-Methylphenyl)methyl]triphenylphosphonium(1+), M-00211
1,2,3,4-Tetrahydro-4-methyl-2,2-diphenylbenz[*h*]isophosphinolinium(1+), T-00113
Triphenyl(1-phenylethyl)phosphonium(1+), T-00612
Triphenyl(2-phenylethyl)phosphonium(1+), T-00613

$C_{26}H_{24}P_2$
▷1,2-Bis(diphenylphosphino)ethane, B-00234
1,2-Dibenzyl-1,2-diphenyldiphosphine, D-00045

$C_{26}H_{24}P_2S_2$
1,2-Bis(diphenylphosphinothioyl)ethane, *in* B-00234
1,2-Dibenzyl-1,2-diphenyldiphosphine; Disulfide, *in* D-00045

$C_{26}H_{24}P_2Se_2$
1,2-Bis(diphenylphosphinoselenoyl)ethane, *in* B-00234

$C_{26}H_{24}P_4$
1,2,4,5-Tetraphenyl-1,2,4,5-tetraphosphorinane, T-00281

$C_{26}H_{25}NO_2P$
Ethyl *P*,*P*-diphenyl-*N*-(diphenylphosphinyl)phosphinimidate, *in* D-01007

$C_{26}H_{25}NO_4P_2$
Tetraphenyl ethylimidodiphosphite, *in* E-00079

$C_{26}H_{25}NP_2$
N-(Diphenylphosphino)-*N*-ethyl-*P*,*P*-diphenylphinous amide, D-01067

$C_{26}H_{25}N_2OP$
2,3-Dihydro-2,4,9-triphenyl-1*H*-1,3,2-diazaphosphonine 2-oxide; 1,3-Di-Me, *in* D-00609

$C_{26}H_{26}N_2O_2P_2$
N,*N*-Diphenyl-*P*,*P*-1,2-ethanediylbis[phenylphosphinic amide], *in* E-00015

$C_{26}H_{26}N_2O_4P_2S_2$
O,*O*,*O*,*O*-Tetraphenyl 1,2-ethanediylbisphosphoramidothioate, *in* E-00010

$C_{26}H_{26}N_2O_6P_2$
Tetraphenyl 1,2-ethanediylbisphosphoramidate, *in* E-00009

$C_{26}H_{26}N_3P_3$
2,2,4,4,6,6-Hexahydro-2,2-dimethyl-4,4,6,6-tetraphenyl-1,3,5-triaza-2,4,6-triphosphorine, H-00025

$C_{26}H_{26}O_2P^{\oplus}$
(6-Methoxy-3-methyl-6-oxo-2,4-hexadienyl)triphenylphosphonium(1+), M-00050

$C_{26}H_{27}O_4P$
Diethyl (triphenylphosphoranylidene)butanedioate, *in* T-00646

$C_{26}H_{28}Cl_2FeNPPd$
1-[1-(Dimethylamino)ethyl]-2-(diphenylphosphino)ferrocene; $PdCl_2$ complex, *in* D-00697

$C_{26}H_{28}FeNP$
1-[1-(Dimethylamino)ethyl]-2-(diphenylphosphino)ferrocene, D-00697

$C_{26}H_{28}NOP$
[(5,6-Dihydro-4,4,6-trimethyl-4*H*-1,3-oxazin-2-yl)methylene]triphenylphosphorane, *in* D-00599

$C_{26}H_{28}NO_8P$
Ethyl 2-(benzyloxycarbonyl)amino-3-[(diphenoxyphosphinoyl)oxy]butanoate, *in* A-00120

$C_{26}H_{28}O_4IP$
(1,2-Dicarboxyethyl)triphenylphosphonium(1+); Di-Et ester, iodide, *in* D-00160

$C_{26}H_{29}ClNOP$
[(5,6-Dihydro-4,4,6-trimethyl-4*H*-1,3-oxazin-2-yl)methyl]triphenylphosphonium(1+); Chloride, *in* D-00599

$C_{26}H_{29}NOP^{\oplus}$
[(5,6-Dihydro-4,4,6-trimethyl-4*H*-1,3-oxazin-2-yl)methyl]triphenylphosphonium(1+), D-00599

$C_{26}H_{29}O_2P$
(7-Carboxyheptylidene)triphenylphosphorane, *in* C-00012
Triphenyl[3-[(tetrahydro-2*H*-pyran-2-yl)oxy]propylidene]phosphorane, *in* T-00683

$C_{26}H_{30}BrO_2P$
(7-Carboxyheptyl)triphenylphosphonium(1+); Bromide, *in* C-00012
Triphenyl[[3-(tetrahydro-2*H*-pyran-2-yl)oxy]propyl]phosphonium(1+); Bromide, *in* T-00683

$C_{26}H_{30}IO_2P$
Triphenyl[[3-(tetrahydro-2*H*-pyran-2-yl)oxy]propyl]phosphonium(1+); Iodide, *in* T-00683

$C_{26}H_{30}NO_{12}P$
2-Amino-2-deoxyglucose 6-(dihydrogen phosphate); 1,2*N*,3,4-Tetra-Ac, di-Ph ester, *in* A-00069

$C_{26}H_{30}O_2P$
2-Naphthylphenylphosphinic acid; (−)-Menthyl ester, *in* N-00018

$C_{26}H_{30}O_2P^{\oplus}$
(7-Carboxyheptyl)triphenylphosphonium(1+), C-00012
Triphenyl[[3-(tetrahydro-2*H*-pyran-2-yl)oxy]propyl]phosphonium(1+), T-00683

$C_{26}H_{30}P$
Octylidenetriphenylphosphorane, *in* O-00047

$C_{26}H_{31}N_2P$
Hexahydro-1,5-bis(1-phenylethyl)-1*H*-1,5,3-diazaphosphepine; 3-Ph, *in* H-00019

$C_{26}H_{32}BrP$
Octyltriphenylphosphonium(1+); Bromide, *in* O-00047

$C_{26}H_{32}ClP$
Octyltriphenylphosphonium(1+); Chloride, *in* O-00047

$C_{26}H_{32}P^{\oplus}$
Octyltriphenylphosphonium(1+), O-00047

$C_{26}H_{35}P$
Decylidenetriphenylphosphorane, *in* D-00014

$C_{26}H_{36}BrP$
Decyltriphenylphosphonium(1+); Bromide, *in* D-00014

$C_{26}H_{36}N_2P_2$
2,4-Dimethyl-9,13-diphenyl-1,5-diaza-9,13-diphosphacyclohexadeca-1,4-diene, D-00737

$C_{26}H_{36}P^{\oplus}$
Decyltriphenylphosphonium(1+), D-00014

$C_{26}H_{37}N_2P$
Hexahydro-1,5-bis(1-phenylethyl)-1*H*-1,5,3-diazaphosphepine; 3-Cyclohexyl, *in* H-00019

$C_{26}H_{38}N_2O_6P_2S_4$
Zilantel, Z-00001

$C_{26}H_{38}N_3O_{11}P$
Phosphoramidon; *N*-Ac, Me ester, *in* P-00402

$C_{26}H_{44}O_2P_4$
4,7,13,16-Tetraphenyl-1,10-dioxa-4,7,13,16-tetraphosphacyclooctadecane, T-00255

$C_{26}H_{47}OP$
Didecylphenylphosphine; Oxide, *in* D-00219

$C_{26}H_{47}P$
Didecylphenylphosphine, D-00219

$C_{26}H_{48}O_2P_2$
1,2-Bis(dicyclophosphinyl)ethane, *in* B-00167

$C_{26}H_{48}P_2$
1,2-Bis(dicyclohexylphosphino)ethane, B-00167

$C_{26}H_{48}P_2S_2$
1,2-Bis(dicyclophosphinothioyl)ethane, *in* B-00167

$C_{26}H_{48}P_4$
1,2,3,4-Tetracyclohexyl-1,2,3,4-tetraphosphorinane, T-00040

$C_{26}H_{50}N_6O_{13}P_2S_2$
Hexakis(dimethylamino)-μ-oxodiphosphorus(2+); Bis-(4-methylbenzenesulfonate), *in* H-00062

$C_{27}H_{17}F_6P$
4,6,7-Triphenyl-2,3-bis(trifluoromethyl)-1-phosphabicyclo[2.2.2]octa-2,5,7-triene, T-00598

$C_{27}H_{20}NOP$
Benzoyl(triphenylphosphoranylidene)-acetonitrile, *in* O-00083

$C_{27}H_{21}O_3P$
β-Oxo-α-(triphenylphosphoranylidene)-benzenepropanoic acid, O-00083

$C_{27}H_{21}P$
1-Indenylidenetriphenylphosphorane, *in* I-00008

$C_{27}H_{22}BrP$
1-Indenyltriphenylphosphonium(1+); Bromide, *in* I-00008

$C_{27}H_{22}P^{\oplus}$
1-Indenyltriphenylphosphonium(1+), I-00008

$C_{27}H_{23}OP$
1-Phenyl-1-(triphenylphosphoranylidene)-2-propanone, P-00334
1-Phenyl-2-(triphenylphosphoranylidene)-1-propanone, P-00335
1-Phenyl-3-(triphenylphosphoranylidene)-2-propanone, P-00336

$C_{27}H_{23}O_2P$
[[2-(Methoxycarbonyl)phenyl]methylene]-triphenylphosphorane, *in* M-00029
[[4-(Methoxycarbonyl)phenyl]methylene]-triphenylphosphorane, *in* M-00031
[[3-(3-Methoxycarbonyl)phenyl]methylene]-triphenylphosphorane, *in* M-00030
Methyl α-(triphenylphosphoranylidene)-benzeneacetate, *in* T-00644
Phenyl 2-(triphenylphosphoranylidene)-propanote, *in* T-00660
2,4,5,6-Tetraphenyl-1,3,5-dioxaphosphorinane, T-00254
(Triphenylphosphoranylidene)acetic acid; Benzyl ester, *in* T-00640
α-(Triphenylphosphoranylidene)-benzenepropanoic acid, T-00645

$C_{27}H_{23}O_2PS$
2,4,5,6-Tetraphenyl-1,3,5-dioxaphosphorinane; 5-Sulfide, *in* T-00254

$C_{27}H_{23}O_3P$
2,4,5,6-Tetraphenyl-1,3,5-dioxaphosphorinane; 5-Oxide, *in* T-00254

$C_{27}H_{23}P$
Triphenyl(3-phenyl-2-propenylidene)-phosphorane, T-00615

$C_{27}H_{24}BF_4PS$
Triphenyl[1-(phenylthio)cyclopropyl]-phosphonium(1+); Tetrafluoroborate, *in* T-00618

$C_{27}H_{24}BrOP$
(1-Methyl-2-oxo-2-phenylethyl)triphenyl-phosphonium(1+); Bromide, *in* M-00188
(2-Oxo-1-phenylpropyl)triphenyl-phosphonium(1+); Bromide, *in* O-00069
(3-Oxo-3-phenylpropyl)triphenyl-phosphonium(1+); Bromide, *in* O-00070

$C_{27}H_{24}BrO_2P$
[[2-(Methoxycarbonyl)phenyl]methyl]-triphenylphosphonium(1+); Bromide, *in* M-00029
[[3-(Methoxycarbonyl)phenyl]methyl]-triphenylphosphonium(1+); Bromide, *in* M-00030
[[4-(Methoxycarbonyl)phenyl]methyl]-triphenylphosphonium(1+); Bromide, *in* M-00031

$C_{27}H_{24}BrP$
Triphenyl(3-phenyl-2-propenyl)-phosphonium(1+); Bromide, *in* T-00616

$C_{27}H_{24}BrPS$
Triphenyl[1-(phenylthio)cyclopropyl]-phosphonium(1+); Bromide, *in* T-00618

$C_{27}H_{24}ClP$
Triphenyl(3-phenyl-2-propenyl)-phosphonium(1+); Chloride, *in* T-00616

$C_{27}H_{24}IOP$
(1-Methyl-2-oxo-2-phenylethyl)triphenyl-phosphonium(1+); Iodide, *in* M-00188
(3-Oxo-3-phenylpropyl)triphenyl-phosphonium(1+); Iodide, *in* O-00070

$C_{27}H_{24}NOP$
5,6,7,12-Tetrahydro-12-phenyldibenz[*c,f*][1,5]-azaphosphocine; 6-Benzyl, 12-oxide, *in* T-00130

$C_{27}H_{24}OP^{\oplus}$
(1-Methyl-2-oxo-2-phenylethyl)triphenyl-phosphonium(1+), M-00188
(2-Oxo-1-phenylpropyl)triphenyl-phosphonium(1+), O-00069
(3-Oxo-3-phenylpropyl)triphenyl-phosphonium(1+), O-00070

$C_{27}H_{24}O_2P^{\oplus}$
[[2-(Methoxycarbonyl)phenyl]methyl]-triphenylphosphonium(1+), M-00029
[[3-(Methoxycarbonyl)phenyl]methyl]-triphenylphosphonium(1+), M-00030
[[4-(Methoxycarbonyl)phenyl]methyl]-triphenylphosphonium(1+), M-00031

$C_{27}H_{24}O_2P_2$
1,1-Bis(diphenylphosphinyl)cyclopropane, *in* B-00229
1,2-Bis(diphenylphosphinyl)cyclopropane, *in* B-00230

$C_{27}H_{24}P^{\oplus}$
Triphenyl(3-phenyl-2-propenyl)-phosphonium(1+), T-00616

$C_{27}H_{24}PS^{\oplus}$
Triphenyl[1-(phenylthio)cyclopropyl]-phosphonium(1+), T-00618

$C_{27}H_{24}P_2$
1,1-Bis(diphenylphosphino)cyclopropane, B-00229
1,2-Bis(diphenylphosphino)cyclopropane, B-00230
1,1,3,3-Tetrahydro-1,1,3,3-tetraphenyl-1,3-diphosphacyclopenta-1,2-diene, T-00150

$C_{27}H_{24}P_2S_2$
1,1-Bis(diphenylphosphino)cyclopropane; Disulfide, *in* B-00229

$C_{27}H_{25}OP$
2,2-Dihydro-4-methyl-2,2,2,3-tetraphenyl-1,2-oxaphosphetane, D-00531
6-Methyl-3,5,7-triphenyl-6-phosphatricyclo[3.3.1.0^{2,7}]non-3-ene; 6-Oxide, *in* M-00431

$C_{27}H_{25}P$
6-Methyl-3,5,7-triphenyl-6-phosphatricyclo[3.3.1.0^{2,7}]non-3-ene, M-00431
Triphenyl(3-phenylpropylidene)phosphorane, *in* T-00617

$C_{27}H_{26}BrP$
Triphenyl(3-phenylpropyl)phosphonium(1+); Bromide, *in* T-00617

$C_{27}H_{26}Br_2OP_2$
3,3,5,5-Tetraphenyl-1,3,5-oxadiphosphorinanium(2+); Dibromide, *in* T-00267

$C_{27}H_{26}Br_2P_2$
1,1,3,3-Tetraphenyl-1,3-diphospholanium dibromide, *in* D-01013

$C_{27}H_{26}Cl_2OP_2$
3,3,5,5-Tetraphenyl-1,3,5-oxadiphosphorinanium(2+); Dichloride, *in* T-00267

$C_{27}H_{26}Fe_2IP$
Diferrocenylmethylphenylphosphonium iodide, *in* D-00410

$C_{27}H_{26}OP_2^{\oplus\oplus}$
3,3,5,5-Tetraphenyl-1,3,5-oxadiphosphorinanium(2+), T-00267

$C_{27}H_{26}O_2P_2$
1,3-Bis(diphenylphosphinyl)propane, *in* B-00265

$C_{27}H_{26}P^{\oplus}$
Triphenyl(3-phenylpropyl)phosphonium(1+), T-00617

$C_{27}H_{26}P_2$
1,2-Bis(diphenylphosphino)propane, B-00264
1,3-Bis(diphenylphosphino)propane, B-00265

$C_{27}H_{26}P_2S_2$
1,3-Bis(diphenylphosphinothioyl)propane, *in* B-00265

$C_{27}H_{26}P_2Se_2$
1,3-Bis(diphenylphosphino)propane; Diselenide, *in* B-00265

$C_{27}H_{27}N_2P$
6,8,10-Triphenyl-1,4-diaza-5-phospha(5-P^V)-spiro[4.5]deca-5,7,9-triene; 1,4-Di-Me, *in* T-00599

$C_{27}H_{28}N_2O_4P_2S_2$
O,O,O′,O′-Tetraphenyl 1,3-propanediphosphoramidothioate, *in* P-00446

$C_{27}H_{29}O_4P$
tert-Butyl α-methoxycarbonyltriphenyl-phosphoranylidenepropanoate, *in* T-00646

$C_{27}H_{31}NO_8P_2$
2,2,7,7-Tetrahydro-3,4,5,6-tetrakis(methoxycarbonyl)-1*H*-1,2,7-azadiphosphepine; 2,2-Di-Me, 7,7-di-Ph, *in* T-00138

$C_{27}H_{31}P$
2-Nonenylidenetriphenylphosphorane, N-00071
3-Nonenylidenetriphenylphosphorane, *in* N-00072
4-Nonenylidenetriphenylphosphorane, *in* N-00073

$C_{27}H_{32}BrP$
4-Nonenyltriphenylphosphonium(1+); Bromide, *in* N-00073

$C_{27}H_{32}ClP$
3-Nonenyltriphenylphosphonium(1+); Chloride, *in* N-00072

$C_{27}H_{32}IP$
3-Nonenyltriphenylphosphonium(1+); Iodide, *in* N-00072

$C_{27}H_{32}P^{\oplus}$
3-Nonenyltriphenylphosphonium(1+), N-00072
4-Nonenyltriphenylphosphonium(1+), N-00073

$C_{27}H_{33}OP$
Tris(2,4,6-trimethylphenyl)phosphine; Oxide, *in* T-00861

$C_{27}H_{33}O_9P$
Tris(2,4-6-trimethoxyphenyl)phosphine, T-00858

$C_{27}H_{33}O_9PSe$
Tris(2,4-6-trimethoxyphenyl)phosphine; Selenide, *in* T-00858

$C_{27}H_{33}P$
Nonylidenetriphenylphosphorane, *in* N-00078
Tris(2,4,6-trimethylphenyl)phosphine, T-00861

$C_{27}H_{33}PS$
Tris(2,4,6-trimethylphenyl)phosphine; Sulfide, *in* T-00861

$C_{27}H_{34}BrP$
Nonyltriphenylphosphonium(1+); Bromide, *in* N-00078

$C_{27}H_{34}IP$
Nonyltriphenylphosphonium(1+); Iodide, *in* N-00078

$C_{27}H_{34}P^{\oplus}$
Nonyltriphenylphosphonium(1+), N-00078

$C_{27}H_{44}LiPSi_4$
Bis[bis(trimethylsilyl)methylene]-9*H*-fluoren-9-ylphosphorane; [Li(THF)$_4$]$^+$ complex, *in* B-00106

$C_{27}H_{45}PSi_4$
Bis[bis(trimethylsilyl)methylene]-9*H*-fluoren-9-ylphosphorane, B-00106

$C_{27}H_{50}P_2$
1,3-Bis(dicyclohexylphosphino)propane, B-00169

$C_{27}H_{50}P_2S_2$
1,3-Bis(dicyclohexylphosphinothioyl)propane, *in* B-00169

$C_{27}H_{57}OP$
Trinonylphosphine; Oxide, *in* T-00578

$C_{27}H_{57}O_3P$
Trinonyl phosphite, T-00579

$C_{27}H_{57}P$
Trinonylphosphine, T-00578

$C_{27}H_{57}PS$
Trinonylphosphine; Sulfide, *in* T-00578

$C_{27}H_{57}PSe$
Trinonylphosphine; Selenide, *in* T-00578

$C_{28}H_{20}F_4P_2$
3,3,4,4-Tetrafluoro-1,2-bis(diphenylphosphino)-cyclobutene, T-00077

$C_{28}H_{20}F_6NO_3PS$
2,2,2,3-Tetrahydro-5-phenyl-3,3-bis(trifluoromethyl)-1,4,2-thiazaphosphole; 2,2,2-Triphenoxy, *in* T-00129

$C_{28}H_{20}O_2P_2$
3,4-Bis(diphenylphosphino)-3-cyclobutene-1,2-dione, B-00228

$C_{28}H_{20}P_2$
1,4-Bis(diphenylphosphino)-1,3-butadiyne, B-00223

$C_{28}H_{21}OP$
[1,1'-Biphenyl]-4-yl-1-naphthalenylphenylphosphine; Oxide, *in* B-00081

$C_{28}H_{21}O_2P$
1-Hydroxy-2,3,4,5-tetraphenyl-1*H*-phosphole 1-oxide, H-00200

$C_{28}H_{21}O_4P$
2,3,7,8-Tetraphenyl-1,4,6,9-tetraoxa-5-phospha(5-P^V)spiro[4.4]nona-2,7-diene, T-00275

$C_{28}H_{21}P$
[1,1'-Biphenyl]-4-yl-1-naphthalenylphenylphosphine, B-00081

$C_{28}H_{22}BrOP$
2,4,4,6-Tetraphenyl-4*H*-1,4-oxaphosphorinium bromide, *in* T-00610

$C_{28}H_{22}ClO_5P$
2,4,6-Triphenyl-4*H*-1,4-oxaphosphorin; B,PhClO$_4$, *in* T-00610

$C_{28}H_{22}NP$
N-1-Naphthyl-*P,P,P*-triphenylphosphine imide, *in* T-00626
N-2-Naphthyl-*P,P,P*-triphenylphosphine imide, *in* T-00626

$C_{28}H_{23}BrNO_2P$
[2-(1,3-Dihydro-1,3-dioxo-2*H*-isoindol-2-yl)-ethyl]triphenylphosphonium(1+); Bromide, *in* D-00474

$C_{28}H_{23}ClNO_2P$
[2-(1,3-Dihydro-1,3-dioxo-2*H*-isoindol-2-yl)-ethyl]triphenylphosphonium(1+); Chloride, *in* D-00474

$C_{28}H_{23}NO_2P^{\oplus}$
[2-(1,3-Dihydro-1,3-dioxo-2*H*-isoindol-2-yl)-ethyl]triphenylphosphonium(1+), D-00474

$C_{28}H_{23}O_3P$
γ-Oxo-β-(triphenylphosphoranylidene)-benzenebutanoic acid, O-00082

$C_{28}H_{24}Br_2P_2$
1,4-Dihydro-1,1,4,4-tetraphenyl-1,4-diphosphorinium(2+); Dibromide, *in* D-00582

$C_{28}H_{24}Br_4O_7P_2$
Bis[(4-bromophenyl)methyl] phosphate; Anhydride, *in* B-00109

$C_{28}H_{24}Cl_2P_2$
1,4-Dihydro-1,1,4,4-tetraphenyl-1,4-diphosphorinium(2+); Dichloride, *in* D-00582

$C_{28}H_{24}NO_2P$
3-(Triphenylphosphoranylidene)-2,5-pyrrolidenedione; *N*-Ph, *in* T-00663

$C_{28}H_{24}N_4O_{15}P_2$
Tetrakis[(4-nitrophenyl)methyl] diphosphate, *in* B-00367

$C_{28}H_{24}O_2P_2$
1,4-Bis(diphenylphosphinyl)-2-butyne, *in* B-00227

$C_{28}H_{24}P_2$
1,4-Bis(diphenylphosphino)-2-butyne, B-00227
1,2,3,6-Tetrahydro-1,2,4,5-tetraphenyl-1,2-diphosphorine, T-00151

$C_{28}H_{24}P_2^{\oplus\oplus}$
1,4-Dihydro-1,1,4,4-tetraphenyl-1,4-diphosphorinium(2+), D-00582

$C_{28}H_{24}P_2S_2$
1,4-Bis(diphenylphosphinothioyl)-2-butyne, *in* B-00227
1,2,3,6-Tetrahydro-1,2,4,5-tetraphenyl-1,2-diphosphorine; 1,2-Disulfide, *in* T-00151

$C_{28}H_{25}O_2P$
Benzyl 2-(triphenylphosphoranylidene)-propanoate, *in* T-00660
Methyl α-(triphenylphosphoranylidene)-hydrocinnamate, *in* T-00645

$C_{28}H_{26}BrO_2P$
[[2-[(Acetyloxy)methyl]phenyl]methyl]-triphenylphosphonium(1+); Bromide, *in* A-00019

$C_{28}H_{26}Br_2OP_2$
5-Oxo-1,1,3,3-tetraphenyl-1,3-diphosphorinanium(2+); Dibromide, *in* O-00081

$C_{28}H_{26}Br_2P_2$
1,2,3,4-Tetrahydro-1,1,4,4-tetraphenyl-1,4-diphosphorinium(2+); Dibromide, *in* T-00152

$C_{28}H_{26}Cl_2OP_2$
5-Oxo-1,1,3,3-tetraphenyl-1,3-diphosphorinanium(2+); Dichloride, *in* O-00081

$C_{28}H_{26}NOP$
Tetrahydro-3-methyl-2,4,5,6-tetraphenyl-2*H*-1,3,5-oxazaphosphorine, T-00117

$C_{28}H_{26}NO_5P$
Benzyl 2-[(diphenoxyphosphinyl)amino]-3-phenylpropanoate, *in* P-00400

$C_{28}H_{26}OP_2^{\oplus\oplus}$
5-Oxo-1,1,3,3-tetraphenyl-1,3-diphosphorinanium(2+), O-00081

$C_{28}H_{26}O_2P^{\oplus}$
[[2-[(Acetyloxy)methyl]phenyl]methyl]-triphenylphosphonium(1+), A-00019

$C_{28}H_{26}P_2$
1,1,3,3-Tetrahydro-1,1,3,3-tetraphenyl-1,3-diphosphacyclohexa-1,2-diene, T-00149

$C_{28}H_{26}P_2^{\oplus\oplus}$
1,2,3,4-Tetrahydro-1,1,4,4-tetraphenyl-1,4-diphosphorinium(2+), T-00152

$C_{28}H_{27}NP_2$
3,4-Bis(diphenylphosphino)pyrrolidine, B-00267

$C_{28}H_{28}Br_2OP_2$
3,3,6,6-Tetraphenyl-1,3,6-oxadiphosphepanium(2+); Dibromide, *in* T-00266

$C_{28}H_{28}Br_2P_2$
1,1,3,3-Tetraphenyl-1,3-diphosphorinanium(2+); Dibromide, *in* T-00262
1,1,4,4-Tetraphenyl-1,4-diphosphorinanium(2+); Dibromide, *in* T-00263
1,1,4,4-Tetraphenyl-1,4-diphosphorinanium dibromide, *in* D-01017

$C_{28}H_{28}ClP$
Tetrabenzylphosphonium chloride, *in* T-00338

$C_{28}H_{28}Cl_2P_2$
1,1,3,3-Tetraphenyl-1,3-diphosphorinanium(2+); Dichloride, *in* T-00262

$C_{28}H_{28}I_2P_2$
1,1-Bis(methyldiphenylphosphonio)ethene diiodide, *in* B-00236

$C_{28}H_{28}N_2O_6P_2$
1',3'-Diphenyldispiro[1,3,2-benzodioxaphosphole-2,2'-[1,3,2,4]-diazaphosphetidine-4',2''-[1,3,2]-benzodioxaphosphole]; 2,4'-Diethoxy, *in* D-01018
6,12-Diphenyl-1,4,8,11-tetraoxa-6,12-diaza-5,7-diphospha(5,7-P^V)dispiro[4.1.4.1]dodecane; 5,7-Diphenoxy, *in* D-01149

$C_{28}H_{28}N_2P_2$
Octahydro-1,3,5,7-tetraphenyl-1,5,3,7-diazadiphosphocine, O-00026

$C_{28}H_{28}N_2P_2S_2$
Octahydro-1,3,5,7-tetraphenyl-1,5,3,7-diazadiphosphocine; 3,7-Disulfide, *in* O-00026

$C_{28}H_{28}OP_2$
Oxybis[2-(diphenylphosphino)ethane], O-00093

$C_{28}H_{28}OP_2^{\oplus\oplus}$
3,3,6,6-Tetraphenyl-1,3,6-oxadiphosphepanium(2+), T-00266

$C_{28}H_{28}O_2P_2$
1,4-Bis(diphenylphosphinyl)butane, *in* B-00225
1,2-Ethanediylbis[(2-methoxyphenyl)-phenylphosphine], E-00011
Tetrabenzyldiphosphine; Dioxide, *in* T-00007

$C_{28}H_{28}O_3P$
Bis(2-methylphenyl)phosphinic acid; Anhydride, *in* B-00318

$C_{28}H_{28}O_3P_2$
Bis[(2-diphenylphosphinyl)ethyl]ether, *in* O-00093

$C_{28}H_{28}O_4P_2$
1,2-Ethanediylbis[(2-methoxyphenyl)-phenylphosphine]; Dioxide, *in* E-00011

$C_{28}H_{28}O_4P_2S_4$
Bis(dibenzyloxyphosphinothioyl) disulfide, B-00155

$C_{28}H_{28}O_7P_2$
Bis(3-methylphenyl) phosphate; Anhydride, *in* B-00313
Tetrabenzyl diphosphate, T-00006

$C_{28}H_{28}P_2$
1,3-Bis(diphenylphosphino)butane, B-00224
1,4-Bis(diphenylphosphino)butane, B-00225
2,3-Bis(diphenylphosphino)butane, B-00226
Tetrabenzyldiphosphine, T-00007

$C_{28}H_{28}P_2^{\oplus\oplus}$
1,1,3,3-Tetraphenyl-1,3-diphosphorinanium(2+), T-00262
1,1,4,4-Tetraphenyl-1,4-diphosphorinanium(2+), T-00263

$C_{28}H_{28}P_2S_2$
1,4-Bis(diphenylphosphinothioyl)butane, *in* B-00225
Tetrabenzyldiphosphine; Disulfide, *in* T-00007

$C_{28}H_{28}P_2S_3$
Dibenzylphosphinodithioic acid; Anhydrosulfide, *in* D-00056

$C_{28}H_{28}P_2Se_2$
1,4-Bis(diphenylphosphinoselenoyl)butane, *in* B-00225

$C_{28}H_{28}P_4$
1,3,5,7-Tetraphenyl-1,3,5,7-tetraphosphocane, T-00279

$C_{28}H_{28}P_4Si$
1,4,6,9-Tetraphenyl-1,4,6,9-tetraphospha-5-silaspiro[4.4]nonane, T-00277

$C_{28}H_{29}NO_2P_2$
2,2′-Bis(diphenylphosphino)diethylamine; Dioxide, *in* B-00231

$C_{28}H_{29}NP_2$
2,2′-Bis(diphenylphosphino)diethylamine, B-00231

$C_{28}H_{29}P$
Triphenyl[(2,6,6-trimethyl-1,3-cyclohexadien-1-yl)methylene]phosphorane, *in* T-00688

$C_{28}H_{30}BrP$
Triphenyl[(2,6,6-trimethyl-1,3-cyclohexadien-1-yl)methyl]phosphonium(1+); Bromide, *in* T-00688

$C_{28}H_{30}I_2P_2$
1,2-Ethanediylbis[methyldiphenyl-phosphonium] diiodide, *in* B-00234

$C_{28}H_{30}N_2P_2$
2,3-Bis[(diphenylphosphino)amino]butane, B-00218

$C_{28}H_{30}N_4O_4P_2$
N,N,N′,N′-Tetramethyl-1′,3′-diphenyl-dispiro[1,3,2-benzodioxophosphole-2,2′-[1,3,2,4]diazaphosphetidine-4′,2″-[1,3,2]-benzodioxaphosphol]-2,2′-diamine, *in* D-01018

$C_{28}H_{30}P^{\oplus}$
Triphenyl[(2,6,6-trimethyl-1,3-cyclohexadien-1-yl)methyl]phosphonium(1+), T-00688

$C_{28}H_{31}P$
(3,7-Dimethyl-2,6-octadienylidene)triphenyl-phosphorane, *in* D-00777

$C_{28}H_{32}BrP$
(3,7-Dimethyl-2,6-octadienyl)triphenyl-phosphonium(1+); Bromide, *in* D-00777
Triphenyl[(2,6,6-trimethyl-1-cyclohexen-1-yl)-methyl]phosphonium(1+); Bromide, *in* T-00689

$C_{28}H_{32}ClP$
Triphenyl[(2,6,6-trimethyl-1-cyclohexen-1-yl)-methyl]phosphonium(1+); Chloride, *in* T-00689

$C_{28}H_{32}N_5P_3$
2,2-Diamino-2,2,4,4,6,6-hexahydro-4,4,6,6-tetraphenyl-1,3,5-triaza-2,4,6-triphosphorine; *N,N,N′,N′*-Tetra-Me, *in* D-00025

$C_{28}H_{32}P^{\oplus}$
(3,7-Dimethyl-2,6-octadienyl)triphenyl-phosphonium(1+), D-00777
Triphenyl[(2,6,6-trimethyl-1-cyclohexen-1-yl)-methyl]phosphonium(1+), T-00689

$C_{28}H_{33}O_3P$
4,8-Di-*tert*-butyl-2,10-dimethyl-12*H*-dibenzo[*d,g*][1,3,2]dioxaphosphocin; 6-Phenoxy, *in* D-00088

$C_{28}H_{33}O_6P$
7,8,10,11,13,14,16,17,19,24-Decahydro-5*H*-dibenzo[*o,r*][1,4,7,10,13,17]-pentaoxaphosphacycloeicosin; 24-Ph, 24-oxide, *in* D-00001

$C_{28}H_{36}Br_2P_2$
9,9,16,16-Tetraethyl-9,10,15,16-tetrahydrotribenzo[*b,d,h*][1,6]-diphosphecinium dibromide, *in* T-00155

$C_{28}H_{36}IO_9P$
Tris(2,4,6-trimethoxyphenyl)-methylphosphonium iodide, *in* T-00858

$C_{28}H_{36}IP$
Methyltris(2,4,6-trimethylphenyl)phosphonium iodide, *in* T-00861

$C_{28}H_{37}O_2P$
Bis(2,2-dimethylpropoxy)triphenylphosphorane, B-00207

$C_{28}H_{42}O_2P_2$
1,10-Diphenyl-1,10-diphosphacyclooctadecane; 1,10-Dioxide, *in* D-01009

$C_{28}H_{42}P_2$
1,10-Diphenyl-1,10-diphosphacyclooctadecane, D-01009

$C_{28}H_{52}P_2$
1,4-Bis(dicyclohexylphosphino)butane, B-00166

$C_{28}H_{52}P_2S_2$
1,4-Bis(dicyclohexylphosphinothioyl)butane, *in* B-00166

$C_{28}H_{59}O_2P$
Ditetradecylphosphinic acid, D-01238

$C_{28}H_{59}O_3P$
Ditetradecyl phosphonate, D-01239

$C_{28}H_{60}BrP$
Tributyl(hexadecyl)phosphonium(1+); Bromide, *in* T-00364

$C_{28}H_{60}ClP$
Tributyl(hexadecyl)phosphonium(1+); Chloride, *in* T-00364

$C_{28}H_{60}IP$
Tributyl(hexadecyl)phosphonium(1+); Iodide, *in* T-00364

$C_{28}H_{60}P^{\oplus}$
Tributyl(hexadecyl)phosphonium(1+), T-00364

$C_{29}H_{20}F_6O_2P_2$
1,2-Bis(diphenylphosphinyl)-3,3,4,4,5,5-hexafluorocyclopentene, *in* H-00016

$C_{29}H_{20}F_6P_2$
3,3,4,4,5,5-Hexafluoro-1,2-bis(diphenyl-phosphino)cyclopentene, H-00016

$C_{29}H_{21}P$
2,3,10-Triphenyl-4*H*-1,4-ethenophosphinoline, T-00607

$C_{29}H_{23}OP$
1,2-Dihydro-1,2,4,6-tetraphenylphosphorin; Oxide, *in* D-00584

$C_{29}H_{23}O_2P$
1-Methoxy-2,3,4,5-tetraphenyl-1*H*-phosphole 1-oxide, *in* H-00200

$C_{29}H_{23}P$
(1-Azulenylmethylene)triphenylphosphorane, *in* A-00156
1,2-Dihydro-1,2,4,6-tetraphenylphosphorin, D-00584
1-Naphthalenylmethylenetriphenyl-phosphorane, *in* N-00015
2-Naphthalenylmethylenetriphenyl-phosphorane, *in* N-00016

$C_{29}H_{24}BrP$
[1,1′-Biphenyl]-4-ylmethyl-1-naphthalenylphenylphosphonium bromide, *in* B-00081
(1-Naphthylmethyl)triphenyl-phosphonium(1+); Bromide, *in* N-00015
(2-Naphthylmethyl)triphenyl-phosphonium(1+); Bromide, *in* N-00016

$C_{29}H_{24}ClP$
(1-Naphthylmethyl)triphenyl-phosphonium(1+); Chloride, *in* N-00015

$C_{29}H_{24}IP$
(1-Azulenylmethyl)triphenylphosphonium(1+); Iodide, *in* A-00156
[1,1′-Biphenyl]-4-yl-1-naphthalenylphenyl-phosphine; B,MeI, *in* B-00081

$C_{29}H_{24}NPSSi$
P,P-Dipropyl-*N*-trimethylsilylphosphinothioic amide, *in* D-01208

$C_{29}H_{24}P^{\oplus}$
(1-Azulenylmethyl)triphenylphosphonium(1+), A-00156
(1-Naphthylmethyl)triphenyl-phosphonium(1+), N-00015
(2-Naphthylmethyl)triphenyl-phosphonium(1+), N-00016

$C_{29}H_{25}FeP$
(Ferrocenylmethylene)triphenylphosphorane, *in* F-00008

$C_{29}H_{25}O_3P$
Ethyl β-oxo-α-(triphenylphosphoranylidene)-benzenepropanoate, *in* O-00083
Methyl γ-oxo-β-(triphenylphosphoranylidene)-benzenebutanoate, *in* O-00082

$C_{29}H_{26}BrP$
Benzyldiphenyl(4-phenyl-1,3-butadienyl)-phosphonium bromide, *in* D-01031

$C_{29}H_{26}ClFeP$
Ferrocenyldiphenyl(phenylmethyl)phosphonium chloride, *in* D-01068

$C_{29}H_{26}F_6P_2$
4*b*,5,6,10*b*,11,12-Hexahydro-5,5-diphenyl-benzo[*c*]phosphanthridinium hexafluorophosphate, *in* H-00045

$C_{29}H_{26}FeIP$
(Ferrocenylmethyl)triphenylphosphonium(1+); Iodide, *in* F-00008

$C_{29}H_{26}FeP^{\oplus}$
(Ferrocenylmethyl)triphenylphosphonium(1+), F-00008

$C_{29}H_{27}O_2P$
Ethyl α-(triphenylphosphoranylidene)-hydrocinnamate, *in* T-00645

$C_{29}H_{27}P$
2,7,10,15,17-Pentamethyl-17*H*-tetrabenzo[*b,d,f,h*]phosphonin, P-00028

$C_{29}H_{28}P_2$
1,1,3,3-Tetrahydro-1,1,3,3-tetraphenyl-1,3-diphosphacyclohepta-1,2-diene, T-00148

$C_{29}H_{29}NP_2$
4-(Diphenylphosphino)-2-(diphenyl-phosphinomethyl)pyrrolidine, D-01062

$C_{29}H_{30}Br_2P_2$
1,1,3,3-Tetraphenyl-1,3-diphosphepanium; Dibromide, *in* T-00257

$C_{29}H_{30}O_2P_2$
1,5-Bis(diphenylphosphinyl)pentane, *in* B-00259

$C_{29}H_{30}O_4P_2$
Tetraphenyl 1,5-pentanediylbisphosphonite, *in* P-00033

$C_{29}H_{30}P_2$
1,5-Bis(diphenylphosphino)pentane, B-00259
2,4-Bis(diphenylphosphino)pentane, B-00260

$C_{29}H_{30}P_2^{\oplus\oplus}$
1,1,3,3-Tetraphenyl-1,3-diphosphepanium, T-00257

$C_{29}H_{30}P_2Se_2$
1,5-Bis(diphenylphosphinoselenoyl)pentane, *in* B-00259

$C_{29}H_{33}P$
Triphenyl(2,5-undecadienylidene)phosphorane, T-00700

$C_{29}H_{36}BrP$
Benzylneomenthyldiphenylphosphonium bromide, *in* M-00164

$C_{29}H_{36}N_5O_{18}P$
Coenzyme F_{420}, C-00201

$C_{29}H_{38}BrP$
Triphenylundecylphosphonium(1+); Bromide, *in* T-00701

$C_{29}H_{38}P$
Triphenylundecylidenephosphorane, *in* T-00701

$C_{29}H_{38}P^{\oplus}$
Triphenylundecylphosphonium(1+), T-00701

$C_{29}H_{58}NO_8P$
Glycerol 1,2-didodecanoate 3-phosphoethanolamine, G-00009

$C_{30}H_{20}F_8P_2$
3,3,4,4,5,5,6,6-Octafluoro-1,2-bis(diphenylphosphino)cyclohexene, O-00002

$C_{30}H_{21}OP$
Tri-1-naphthylphosphine; Oxide, *in* T-00572
Tri-2-naphthylphosphine; Oxide, *in* T-00573

$C_{30}H_{21}O_3P$
Tri-1-naphthyl phosphite, T-00574
Tri-2-naphthyl phosphite, T-00575

$C_{30}H_{21}O_4P$
Tri-1-naphthyl phosphate, T-00570
Tri-2-naphthyl phosphate, T-00571

$C_{30}H_{21}O_8P$
5,6,9-Triphenyl-5-phospha(5-P^V)-spiro[4.4]-nona-1,3,6,8-tetraene-1,2,3,4-tetracarboxylic acid, T-00622

$C_{30}H_{21}P$
Phenylbisbiphenylenephosphorane, *in* S-00021
Tri-1-naphthylphosphine, T-00572
Tri-2-naphthylphosphine, T-00573

$C_{30}H_{21}PS$
Tri-1-naphthylphosphine; Sulfide, *in* T-00572

$C_{30}H_{23}P$
1-Acenaphthenylidenetriphenylphosphorane, A-00002
5,5-Dihydro-5,5,5-triphenyl-5*H*-dibenzophosphole, D-00610

$C_{30}H_{24}AsP$
[2-(Diphenylarsino)phenyl]diphenylphosphine, D-00994

$C_{30}H_{24}AsPS$
[2-(Diphenylarsino)phenyl]diphenylphosphine; *P*-Sulfide, *in* D-00994

$C_{30}H_{24}N_3OP$
N,N′,N″-Tri-1-naphthalenylphosphoric triamide, *in* T-00576
N,N′,N″-Tri-2-naphthylphosphoric triamide, *in* T-00577

$C_{30}H_{24}N_3P$
N,N′,N″-Tri-1-naphthylphosphorous triamide, T-00576
N,N′,N″-Tri-2-naphthylphosphorous triamide, T-00577

$C_{30}H_{24}O_2P_2$
1,4-Bis(diphenylphosphinyl)benzene, *in* P-00128

$C_{30}H_{24}P_2$
1,2-Phenylenebis[diphenylphosphine], P-00126
1,3-Phenylenebis[diphenylphosphine], P-00127
1,4-Phenylenebis[diphenylphosphine], P-00128

$C_{30}H_{25}AsNOP$
P,P-Diphenyl-*N*-(triphenylarsoranylidene)-phosphinic amide, *in* D-01160

$C_{30}H_{25}AsNP$
P,P-Diphenyl-*N*-(triphenylarsoranylidene)-phosphinous amide, D-01160

$C_{30}H_{25}ClN_3O_5P$
2-Chloro-2,2,4,4,6,6-hexahydro-2,4,4,6,6-pentaphenoxy-1,3,5-triaza-2,4,6-triphosphorine, C-00087

$C_{30}H_{25}ClN_3P_3$
2-Chloro-2,2,4,4,6,6-hexahydro-2,4,4,6,6-pentaphenyl-1,3,5-triaza-2,4,6-triphosphorine, C-00088

$C_{30}H_{25}GeP$
Diphenyl(triphenylgermyl)phosphine, D-01161

$C_{30}H_{25}NOP_2$
N-Diphenylphosphinyl-*P,P,P*-triphenylphosphine imide, *in* D-01088
P,P-Diphenyl-*N*-(triphenylphosphoranylidene)-phosphinous amide, D-01165

$C_{30}H_{25}NOP_2S$
P,P-Diphenyl-*N*-(triphenylphosphoranylidene)-phosphinothioic amide, *in* D-01165

$C_{30}H_{25}NOP_2Se$
P,P-Diphenyl-*N*-(triphenylphosphoranylidene)-phosphinoselenoic amide, *in* D-01165

$C_{30}H_{25}NOP_2Te$
P,P-Diphenyl-*N*-(triphenylphosphoranylidene)-phosphinotelluroic amide, *in* D-01165

$C_{30}H_{25}NO_2P_2$
P,P-Diphenyl-*N*-(triphenylphosphoranylidene)-phosphinic amide, *in* D-01165

$C_{30}H_{25}NO_5P_2S$
Triphenyl (diphenoxyphosphinothioyl)-phosphorimidate, *in* D-00987

$C_{30}H_{25}NP_2$
N-Diphenylphosphino-*P,P,P*-triphenylphosphine imide, D-01088

$C_{30}H_{25}NP_2S$
N-Diphenylphosphinothioyl-*P,P,P*-triphenylphosphine imide, *in* D-01088

$C_{30}H_{25}N_2OP$
Phenyl tetraphenylphosphorodiamidite, *in* T-00274

$C_{30}H_{25}N_2P$
Pentaphenylphosphonous diamide, *in* P-00252

$C_{30}H_{25}OP$
Phenoxytetraphenylphosphorane, P-00076

$C_{30}H_{25}O_2P$
Diphenoxytriphenylphosphorane, D-00992
1-Ethoxy-2,3,4,5-tetraphenyl-1*H*-phosphole 1-oxide, *in* H-00200

$C_{30}H_{25}O_5P$
Pentaphenoxyphosphorane, P-00035

$C_{30}H_{25}P$
Pentaphenylphosphorane, P-00039

$C_{30}H_{25}PSn$
Diphenyl(triphenylstannyl)phosphine, D-01167

$C_{30}H_{25}P_5$
Pentaphenylpentaphospholane, P-00037

$C_{30}H_{26}IP$
1,2-Dihydro-1-methyl-1,2,4,6-tetraphenylphosphorinium iodide, *in* D-00584

$C_{30}H_{26}N_2O_6P_2$
Tetraphenyl 1,3-phenylenebisphosphoramidate, *in* B-00011
Tetraphenyl 1,4-phenylenebisphosphoramidate, *in* B-00012

$C_{30}H_{26}N_2P_2$
P,P-Diphenyl-*N*-(triphenylphosphoranylidene)-phosphinimidic amide, D-01164

$C_{30}H_{27}Fe_3OP$
1′,1″,1‴-Phosphinylidynetrisferrocene, *in* T-00464

$C_{30}H_{27}Fe_3P$
Triferrocenylphosphine, T-00464

$C_{30}H_{27}Fe_3PS$
1′,1″,1‴-Phosphinothioylidynetrisferrocene, *in* T-00464

$C_{30}H_{27}Fe_3PSe$
1′,1″,1‴-Phosphinoselenoylidynetrisferrocene, *in* T-00464

$C_{30}H_{27}N_4O_5P$
2-Amino-2,2,4,4,6,6-hexahydro-2,4,4,6,6-pentaphenoxy-1,3,5-triaza-2,4,6-triphosphorine, A-00082

$C_{30}H_{28}N_4O_3P$
N,N′,N″-Triphenyl (diphenoxyphosphinothioyl)-phosphorotriamidoimidate, *in* D-00987

$C_{30}H_{28}O_2P_2$
1,4:3,6-Dianhydro-2,5-deoxy-2,5-bis(diphenylphosphino)iditol, D-00026

$C_{30}H_{30}IP$
2,7,10,15,17,17-Hexamethyl-17*H*-tetrabenzo[*b,d,f,h*]phosphoninium iodide, *in* P-00028

$C_{30}H_{30}NO_5P$
Benzyl 2-[(dibenzyloxyphosphinyl)amino]-3-phenylpropanoate, *in* P-00400

$C_{30}H_{30}NO_6P$
Dibenzyl [1-benzyloxycarbonyl-2-(4-hydroxyphenyl)ethyl]phosphoramidate, *in* P-00422

$C_{30}H_{30}O_3P_2$
6,7,9,10,16,17,18,19-Octahydro-16,19-diphenyldibenzo[*h,n*][1,4,7,10,13]-trioxadiphosphacyclopentadecin, O-00015

$C_{30}H_{30}O_5P_2$
6,7,9,10,16,17,18,19-Octahydro-16,19-diphenyldibenzo[*h,n*][1,4,7,10,13]-trioxadiphosphacyclopentadecin; 16,19-Dioxide, *in* O-00015

$C_{30}H_{30}P_2$
1,2-Bis(diphenylphosphinomethyl)cyclobutane, B-00249

$C_{30}H_{30}P_2S_2$
7,8,15,16,17,18-Hexahydro-14,18-diphenyl-6*H*,18*H*-1,6*H*-dibenzo[*b,i*][1,11,4,8]-dithiadiphosphacyclotetradecin, H-00026
6,7,8,9,14,15,16,17-Octahydro-9,14-diphenyl-dibenzo[*b,i*][1,4,8,11]-dithiadiphosphacyclotetradecin, O-00013
6,7,8,9,15,16,17,18-Octahydro-9,18-diphenyl-dibenzo[*b,i*][1,8,4,11]-dithiadiphosphacyclotetradecin, O-00014

$C_{30}H_{31}AsP_3$
2,3,4,5,6,7,8,9-Octahydro-1,5,9-triphenyl-1*H*-1,9,5-benzodiphosphaarsacycloundecin, O-00027

$C_{30}H_{31}NOP_2$
2,2′-Bis(diphenylphosphino)diethylamine; *N*-Ac, *in* B-00231

$C_{30}H_{31}NP_2$
2,3,4,5,6,7,8,9-Octahydro-1,9-diphenyl-1*H*-5,1,9-benzazadiphosphacycloundecine; 5-Ph, *in* O-00009

$C_{30}H_{31}P_3$
2,3,4,5,6,7,8,9-Octahydro-1,5,9-triphenyl-1*H*-1,5,9-benzotriphosphacycloundecin, O-00028

$C_{30}H_{32}Br_2P_2$
1,1,5,5-Tetraphenyl-1,5-diphosphocanium(2+); Dibromide, *in* T-00261

$C_{30}H_{32}I_2P_2$
1,1,5,5-Tetraphenyl-1,5-diphosphocanium(2+); Diiodide, *in* T-00261

$C_{30}H_{32}N_2O_6P_2$
1′,3′-Diphenyldispiro[1,3,2-benzodioxaphosphole-2,2′-[1,3,2,4]-diazaphosphetidine-4′,2″-[1,3,2]-benzodioxaphosphole]; 2,4′-Diisopropoxy, *in* D-01018

$C_{30}H_{32}N_2P_2$
1,2-Bis[(*N*-diphenylphosphino)amino]-cyclohexane, B-00219

$C_{30}H_{32}O_2P_2$
1,6-Bis(diphenylphosphinyl)hexane, *in* B-00243

$C_{30}H_{32}P_2$
1,2-Bis[(bis-2-methylphenyl)phosphino]ethane, B-00103
1,2-Bis[(bis-4-methylphenyl)phosphino]ethane, B-00104
1,6-Bis(diphenylphosphino)hexane, B-00243

$C_{30}H_{32}P_2^{\oplus\oplus}$
1,1,5,5-Tetraphenyl-1,5-diphosphocanium(2+), T-00261

$C_{30}H_{32}P_2S_2$
1,6-Bis(diphenylphosphinothioyl)hexane, *in* B-00243

$C_{30}H_{32}P_2Se_2$
1,6-Bis(diphenylphosphinoselenoyl)hexane, *in* B-00243

$C_{30}H_{33}NO_2P_2$
N,*N*-Bis[(2-diphenylphosphino)ethyl]-ethylamine; Dioxide, *in* B-00238

$C_{30}H_{33}NP_2$
N,*N*-Bis[(2-diphenylphosphino)ethyl]-ethylamine, B-00238

$C_{30}H_{34}Br_2P_2$
1,2-Ethanediylbis[methylphenylphosphine]; B,2PhCH$_2$Br, *in* E-00012

$C_{30}H_{34}I_2N_2P_2$
Octahydro-3,7-dimethyl-1,3,5,7-tetraphenyl-1,5,3,7-diazadiphosphocinium iodide, *in* O-00026

$C_{30}H_{35}O_4P$
Di-*tert*-butyl (triphenylphosphoranylidene)-butanedioate, *in* T-00646

$C_{30}H_{37}O_7P$
7,8,10,11,13,14,16,17,19,20,22,27-Dodecahydro-5*H*-dibenzo[*r*,*u*][1,4,7,10,13,16,20]-hexaoxaphosphacyclotricosin; 27-Ph, 27-oxide, *in* D-01261

$C_{30}H_{39}O_3P$
Tris(4-*tert*-butylphenyl) phosphite, T-00736

$C_{30}H_{39}P$
Dodecylidenetriphenylphosphorane, *in* D-01266

$C_{30}H_{40}BrP$
Dodecyltriphenylphosphonium(1+); Bromide, *in* D-01266

$C_{30}H_{40}P^{\oplus}$
Dodecyltriphenylphosphonium(1+), D-01266

$C_{30}H_{42}P_2$
4,4′-Di-*tert*-butyl-1,1′,2,2′,5,5′,6,6′-octahydro-1,1′-diphenyl-2,2′-biphosphorin, D-00100

$C_{30}H_{42}P_2S_2$
4,4′-Di-*tert*-butyl-1,1′,2,2′,5,5′,6,6′-octahydro-1,1′-diphenyl-2,2′-biphosphorin; *P*,*P*′-Disulfide, *in* D-00100

$C_{30}H_{63}OP$
Tri(decyl)phosphine; Oxide, *in* T-00433

$C_{30}H_{63}O_3P$
Tridecyl phosphite, T-00434
Tris(8-methylnonyl) phosphite, T-00799

$C_{30}H_{63}P$
Tri(decyl)phosphine, T-00433

$C_{30}H_{63}PS$
Tri(decyl)phosphine; Selenide, *in* T-00433

$C_{31}H_{23}P$
9*H*-Fluoren-9-ylidenetriphenylphosphorane, F-00010

$C_{31}H_{24}BF_4P$
9*H*-Fluoren-9-yltriphenylphosphonium(1+); Tetrafluoroborate, *in* F-00012

$C_{31}H_{24}BrP$
9*H*-Fluoren-9-yltriphenylphosphonium(1+); Bromide, *in* F-00012

$C_{31}H_{24}P^{\oplus}$
9*H*-Fluoren-9-yltriphenylphosphonium(1+), F-00012

$C_{31}H_{25}AsBrP$
5-Phenyl-5,10-diphenyldibenzo[1,4]-phosphoniaarsenin, *in* D-01001

$C_{31}H_{25}N_2O_4P$
2,2,2,3-Tetrahydro-3,5-diphenyl-1,3,4,2-oxadiazaphosphole; 2,2,2-Triphenoxy, *in* T-00104

$C_{31}H_{25}P$
(Diphenylmethylene)triphenylphosphorane, D-01022

$C_{31}H_{26}BrP$
(Diphenylmethyl)triphenylphosphonium(1+); Bromide, *in* D-01025

$C_{31}H_{26}OP_2$
(Diphenylphosphinylmethylene)triphenyl-phosphorane, D-01099

$C_{31}H_{26}O_3P_2$
Diphenyl [(triphenylphosphoranylidene)-methyl]phosphonate, D-01163

$C_{31}H_{26}P^{\oplus}$
(Diphenylmethyl)triphenylphosphonium(1+), D-01025

$C_{31}H_{27}BrOP_2$
[(Diphenylphosphinyl)methyl]triphenyl-phosphonium(1+); Bromide, *in* D-01101

$C_{31}H_{27}ClOP_2$
[(Diphenoxyphosphinyl)methyl]triphenyl-phosphonium(1+); Chloride, *in* D-00990
[(Diphenylphosphinyl)methyl]triphenyl-phosphonium(1+); Chloride, *in* D-01101

$C_{31}H_{27}IOP_2$
[(Diphenoxyphosphinyl)methyl]triphenyl-phosphonium(1+); Iodide, *in* D-00990

$C_{31}H_{27}OP_2^{\oplus}$
[(Diphenoxyphosphinyl)methyl]triphenyl-phosphonium(1+), D-00990
[(Diphenylphosphinyl)methyl]triphenyl-phosphonium(1+), D-01101

$C_{31}H_{28}O_2P_2$
5,6-Bis(diphenylphosphinyl)bicyclo[2.2.1]hept-2-ene, *in* B-00220

$C_{31}H_{28}P_2$
5,6-Bis(diphenylphosphino)bicyclo[2.2.1]hept-2-ene, B-00220

$C_{31}H_{30}P_2$
2,3-Bis(diphenylphosphino)bicyclo[2.2.1]-heptane, *in* B-00220

$C_{31}H_{31}NOP_2$
4-(Diphenylphosphino)-2-(diphenyl-phosphinomethyl)pyrrolidine; *N*-Ac, *in* D-01062

$C_{31}H_{32}O_2P_2$
4,5-Bis[(diphenylphosphino)methyl]-2,2-dimethyl-1,3-dioxolane, B-00251

$C_{31}H_{32}O_3PS$
Cyclohexyltriphenylphosphonium(1+); 4-Methylbenzenesulfonate, *in* C-00273

$C_{31}H_{34}O_2P_2$
1,7-Bis(diphenylphosphinyl)heptane, *in* B-00242

$C_{31}H_{34}P_2$
1,7-Bis(diphenylphosphino)heptane, B-00242

$C_{31}H_{35}P$
β-Ionylidenetriphenylphosphorane, *in* M-00422

$C_{31}H_{36}BrP$
[1-Methyl-3-(2,6,6-trimethyl-1-cyclohexen-1-yl)-2-propenyl]triphenylphosphonium(1+); Bromide, *in* M-00422

$C_{31}H_{36}ClP$
[1-Methyl-3-(2,6,6-trimethyl-1-cyclohexen-1-yl)-2-propenyl]triphenylphosphonium(1+); Chloride, *in* M-00422

$C_{31}H_{36}P^{\oplus}$
[1-Methyl-3-(2,6,6-trimethyl-1-cyclohexen-1-yl)-2-propenyl]triphenylphosphonium(1+), M-00422

$C_{31}H_{44}NaO_7P$
▷Proticin, P-00500

$C_{31}H_{45}O_6P$
Difficidin, D-00411

$C_{31}H_{45}O_7P$
Oxydifficidin, *in* D-00411

$C_{31}H_{46}O_3$
Difficol, *in* D-00411

$C_{31}H_{46}O_4$
Oxydifficol, *in* D-00411

$C_{31}H_{50}O_8P_2$
Phytonadiol diphosphate, P-00427

$C_{31}H_{56}O_2P_2$
4,5-Bis[(dicyclohexylphosphino)methyl]-2,2-dimethyl-1,3-dioxolane, B-00168

$C_{31}H_{61}O_8P$
Glycerol 1,2-ditetradecanoate 3-phosphate, G-00047
Glycerol 2-monophosphate; 1,3-Ditetradecanoyl, *in* G-00067

$C_{32}H_{23}O_2P$
2,2-Dihydrophenanthro[9,10-*d*]-1,3,2-dioxaphosphole; 2,2,2-Tri-Ph, *in* D-00541

$C_{32}H_{23}O_5P$
2,2-Dihydrophenanthro[9,10-*d*]-1,3,2-dioxaphosphole; 2,2,2-Triphenoxy, *in* D-00541

$C_{32}H_{24}NO_2P$
10,10-Dihydro-2,3-diphenyl-[1,3,2]-oxazaphospholo[2,3-*b*][1,3,2]-benzoxazaphosphole; 10,10-Di-Ph, *in* D-00481

$C_{32}H_{24}NO_4P$
10,10-Dihydro-2,3-diphenyl-[1,3,2]-oxazaphospholo[2,3-*b*][1,3,2]-benzoxazaphosphole; 10,10-Diphenoxy, *in* D-00481

$C_{32}H_{24}O_5P_2$
Oxybis(2,6-diphenyl-4*H*-1,4-oxaphosphorin)-4,4′-dioxide, *in* H-00140

$C_{32}H_{24}P_2$
2,2′,5,5′-Tetraphenyl-1*H*,1′*H*-biphosphole, T-00248

$C_{32}H_{25}P$
[(9,10-Dihydro-9-phenanthrenenyl)methylene]-triphenylphosphorane, *in* D-00540

$C_{32}H_{25}P_3$
2,3-Dihydro-1,2,3,4,5-pentaphenyl-1*H*-1,2,3-triphosphole, D-00539

$C_{32}H_{25}P_3S_2$
2,3-Dihydro-1,2,3,4,5-pentaphenyl-1*H*-1,2,3-triphosphole; 1,3-Disulfide, *in* D-00539

$C_{32}H_{26}BrP$
(9,10-Dihydro-9-phenanthreneyl)triphenyl-phosphonium(1+); Bromide, *in* D-00540

$C_{32}H_{26}NOP$
2,2,2,3,3*a*,11*b*-Hexahydro-2,2,2-triphenyl-phenanthro[9,10-*d*]-1,3,2-oxazaphosphole, H-00059

$C_{32}H_{26}P^{\oplus}$
(9,10-Dihydro-9-phenanthreneyl)triphenyl-phosphonium(1+), D-00540

$C_{32}H_{27}O_2P$
2,2-Dihydro-2,2,2,4,5-pentaphenyl-1,3,2-dioxaphospholane, D-00537

$C_{32}H_{28}O_2P_2$
(1-Phenyl-1,2-ethanediyl)bis[diphenylphosphine]; Dioxide, *in* P-00150

$C_{32}H_{28}O_6P_2$
Tetraphenyl [1,4-phenylenebis(methylene)]bisphosphonate, *in* P-00131

$C_{32}H_{28}P_2$
(1-Phenyl-1,2-ethanediyl)bis[diphenylphosphine], P-00150

$C_{32}H_{29}O_3P_3$
Bis[(diphenylphosphinyl)methyl]phenylphosphine oxide, *in* B-00253

$C_{32}H_{29}O_8P$
7-Butyl-2,3-dihydro-1,6-diphenyl-1*H*-1-benzophosphonin-2,3,4,5-tetracarboxylic acid, B-00552

$C_{32}H_{29}P_3$
Bis[(diphenylphosphino)methyl]phenylphosphine, B-00253

$C_{32}H_{29}P_3S_3$
Bis[(diphenylphosphinothioyl)methyl]phenylphosphine sulfide, *in* B-00253

$C_{32}H_{30}P_2$
5,6-Bis(diphenylphosphino)-2-methylbicyclo[2.2.1]hept-2-ene, B-00247

$C_{32}H_{34}P_2$
1,2-Bis(diphenylphosphinomethyl)cyclohexane, B-00250
1-Cyclohexyl-1,2-bis(diphenylphosphino)ethane, C-00244

$C_{32}H_{35}O_3P$
4,8-Di-*tert*-butyl-2,10-dimethyl-12*H*-dibenzo[*d,g*][1,3,2]dioxaphosphocin; 6-(1-Naphthoxy), *in* D-00088

$C_{32}H_{36}As_2P_2$
5,6,7,8,9,14,15,16,17,18-Decahydro-9,14-dimethyl-5,18-diphenyl-dibenzo[*b,i*][1,4,8,11]-diphosphadiarsacyclotetradecin, D-00004

$C_{32}H_{36}N_2P_2$
1,2-Bis[(*N*-diphenylphosphino)amino]cyclohexane; *N,N'*-Di-Me, *in* B-00219

$C_{32}H_{36}O_2P_2$
1,8-Bis(diphenylphosphinyl)octane, *in* B-00258

$C_{32}H_{36}P_2$
1,8-Bis(diphenylphosphino)octane, B-00258

$C_{32}H_{38}O_2P_4Si_2$
2,3,5,6-Tetraphenyl-1,4-bis[(trimethylsilyl)oxy]-2,3,5,6-tetraphosphabicyclo[2.2.0]hexane, T-00251

$C_{32}H_{43}P$
Triphenyltetradecylidenephosphorane, *in* T-00682

$C_{32}H_{44}BrP$
Triphenyl(tetradecyl)phosphonium(1+); Bromide, *in* T-00682

$C_{32}H_{44}P^{\oplus}$
Triphenyl(tetradecyl)phosphonium(1+), T-00682

$C_{32}H_{50}O_2P_2$
1,12-Diphenyl-1,12-diphosphacyclodocosane; 1,12-Dioxide, *in* D-01008

$C_{32}H_{50}P_2$
1,12-Diphenyl-1,12-diphosphacyclodocosane, D-01008

$C_{32}H_{64}NO_8P$
Glycerol 1,2-didodecanoate 3-phosphocholine, G-00008
Glycerol 1,2-di-10-methylundecanoate 3-phosphocholine, G-00025

$C_{32}H_{67}O_2P$
Dihexadecylphosphinic acid, D-00429

$C_{32}H_{67}O_3P$
▷Dihexadecyl phosphonate, D-00430

$C_{32}H_{67}O_4P$
Dihexadecyl phosphate, D-00428

$C_{33}H_{23}P$
(9*H*-Fluoren-9-ylideneethenylidene)triphenylphosphorane, F-00009

$C_{33}H_{25}O_2P$
1,3-Diphenyl-2-(triphenylphosphoranylidene)-1,3-propanedione, D-01166

$C_{33}H_{25}O_3P$
2,2,3,4,6-Pentaphenyl-2*H*-1,5,2-dioxaphosphorin 2-oxide, P-00036

$C_{33}H_{25}P$
(3-Phenanthrenylmethylene)triphenylphosphorane, *in* T-00637
(9-Phenanthrenylmethylene)triphenylphosphorane, *in* T-00638

$C_{33}H_{26}BrP$
3-(Triphenylphosphoniomethyl)phenanthrene(1+); Bromide, *in* T-00637
9-(Triphenylphosphoniomethyl)phenanthrene(1+); Bromide, *in* T-00638

$C_{33}H_{26}ClP$
9-(Triphenylphosphoniomethyl)phenanthrene(1+); Chloride, *in* T-00638

$C_{33}H_{26}P^{\oplus}$
3-(Triphenylphosphoniomethyl)phenanthrene(1+), T-00637
9-(Triphenylphosphoniomethyl)phenanthrene(1+), T-00638

$C_{33}H_{28}P_2$
2,2,4,4-Tetrahydro-2,2,4,4-tetraphenyl-1*H*-2,4-benzodiphosphepin, *in* T-00087

$C_{33}H_{29}ClP_2$
[1-(Diphenylphosphino)cyclopropyl]triphenylphosphonium chloride, *in* B-00229

$C_{33}H_{30}Br_2P_2$
2,3,4,5-Tetrahydro-2,2,4,4-tetraphenyl-1*H*-2,4-benzodiphosphonium(2+); Dibromide, *in* T-00144

$C_{33}H_{30}ClFe_2P$
Benzyldiferrocenylphenylphosphonium chloride, *in* D-00410

$C_{33}H_{30}P_2^{\oplus\oplus}$
2,3,4,5-Tetrahydro-2,2,4,4-tetraphenyl-1*H*-2,4-benzodiphosphonium(2+), T-00144

$C_{33}H_{32}P_2$
Bicyclo[2.2.1]hept-5-ene-2,3-diylbis(methylene)bis[diphenylphosphine], B-00075

$C_{33}H_{34}NO_7P$
Dibenzyl 2-[(dibenzyloxyphosphinyl)amino]pentanedioate, *in* P-00367

$C_{33}H_{37}P$
[3-Methyl-5-(2,6,6-trimethyl-1-cyclohexen-1-yl)-2,4-pentadienylidene]triphenylphosphorane, *in* M-00421

$C_{33}H_{38}BrOP$
[5-(4-Hydroxy-2,6,6-trimethyl-1-cyclohexen-1-yl)-3-methyl-2,4-pentadienyl]triphenylphosphonium(1+); Bromide, *in* H-00202

$C_{33}H_{38}BrP$
[3-Methyl-5-(2,6,6-trimethyl-1-cyclohexen-1-yl)-2,4-pentadienyl]triphenylphosphonium(1+); Bromide, *in* M-00421

$C_{33}H_{38}Br_2P_2$
1,1,4,4-Tetrabenzyl-1,4-diphosphepanium dibromide, *in* D-00047

$C_{33}H_{38}ClOP$
[5-(4-Hydroxy-2,6,6-trimethyl-1-cyclohexen-1-yl)-3-methyl-2,4-pentadienyl]triphenylphosphonium(1+); Chloride, *in* H-00202

$C_{33}H_{38}ClP$
[3-Methyl-5-(2,6,6-trimethyl-1-cyclohexen-1-yl)-2,4-pentadienyl]triphenylphosphonium(1+); Chloride, *in* M-00421

$C_{33}H_{38}OP^{\oplus}$
[5-(4-Hydroxy-2,6,6-trimethyl-1-cyclohexen-1-yl)-3-methyl-2,4-pentadienyl]triphenylphosphonium(1+), H-00202

$C_{33}H_{38}P^{\oplus}$
[3-Methyl-5-(2,6,6-trimethyl-1-cyclohexen-1-yl)-2,4-pentadienyl]triphenylphosphonium(1+), M-00421

$C_{33}H_{39}P$
1-(2,4,6-Tri-*tert*-butylphenyl)-4,4-diphenyl-1-phospha-1,2,3-butatriene, T-00367

$C_{33}H_{45}P$
Triphenyl(3,7,11-trimethyldodecylidene)phosphorane, *in* T-00690

$C_{33}H_{46}BrP$
Triphenyl(3,7,11-trimethyldodecyl)phosphonium(1+); Bromide, *in* T-00690

$C_{33}H_{46}P^{\oplus}$
Triphenyl(3,7,11-trimethyldodecyl)phosphonium(1+), T-00690

$C_{33}H_{66}NO_8P$
Glycerol 1,2-ditetradecanoate 3-phosphoethanolamine, G-00049

$C_{34}H_{23}OP$
Phenylbis[2-(phenylethynyl)phenyl]phosphine; Oxide, *in* P-00098

$C_{34}H_{23}P$
Phenylbis[2-(phenylethynyl)phenyl]phosphine, P-00098

$C_{34}H_{24}P_2$
2,2',6,6'-Tetraphenyl-4,4'-biphosphorine, T-00249

$C_{34}H_{25}OP$
Pentaphenyl-1*H*-phosphole; Oxide, *in* P-00038

$C_{34}H_{25}O_2P$
1-Phenoxy-2,3,4,5-tetraphenyl-1*H*-phosphole 1-oxide, *in* H-00200

$C_{34}H_{25}O_4P$
2,3,7,8-Tetraphenyl-1,4,6,9-tetraoxa-5-phospha(5-P^V)spiro[4.4]nona-2,7-diene; 5-Ph, *in* T-00275

$C_{34}H_{25}P$
Pentaphenyl-1*H*-phosphole, P-00038

$C_{34}H_{25}PS$
Pentaphenyl-1*H*-phosphole; Sulfide, *in* P-00038

$C_{34}H_{26}O_2P_2$
1,4-Bis(diphenylphosphinyl)naphthalene, *in* B-00254
1,5-Bis(diphenylphosphinyl)naphthalene, *in* B-00255
2,6-Bis(diphenylphosphinyl)naphthalene, *in* B-00256
2,7-Bis(diphenylphosphinyl)naphthalene, *in* B-00257

$C_{34}H_{26}P_2$
1,4-Bis(diphenylphosphino)naphthalene, B-00254
1,5-Bis(diphenylphosphino)naphthalene, B-00255
2,6-Bis(diphenylphosphino)naphthalene, B-00256
2,7-Bis(diphenylphosphino)naphthalene, B-00257
3,5,8,10-Tetraphenyl-1*H*,5*H*-bisphospholo[1,2-*a*:1',2'-*d*][1,4]diphosphorin, T-00250

$C_{34}H_{27}O_2P$
1,4-Diphenyl-2-(triphenylphosphoranylidene)-1,4-butanedione, D-01162

$C_{34}H_{28}FeO_2P_2$
1,1'-Bis(diphenylphosphinyl)ferrocene, *in* B-00241

$C_{34}H_{28}FeP_2$
1,1'-Bis(diphenylphosphino)ferrocene, B-00241

$C_{34}H_{28}FeP_2S_2$
1,1'-Bis(diphenylphosphinothioyl)ferrocene, *in* B-00241

$C_{34}H_{29}O_8P$
Tetramethyl 1,2,5-triphenyl-5-phospha(5-P^V)-spiro[4.4]nona-1,3,6,8-tetraene-1,2,3,4-tetracarboxylate, *in* T-00622

$C_{34}H_{29}P$
2,7,10,15-Tetramethyl-17-phenyl-17*H*-tetrabenzo[*b,d,f,h*]phosphonin, T-00211

$C_{34}H_{30}Br_2P_2$
1,2,5,6-Tetrahydro-2,2,5,5-tetraphenyl-2,5-benzodiphocinium dibromide, *in* H-00053

$C_{34}H_{32}Br_2P_2$
Tetraphenyl-1,4-diphosphonia-6,7-benzocyclooct-6-ene, *in* H-00053

$C_{34}H_{32}NOP$
5,5-Dihydro-4,4-dimethyl-2,3,5,5,5-pentaphenyl-1,2,5-oxazaphospholidine, D-00465

$C_{34}H_{32}P_2^{\oplus\oplus}$
1,2,3,4,5,6-Hexahydro-2,2,5,5-tetraphenyl-2,5-benzodiphosphocinium(2+), H-00053

$C_{34}H_{33}As_2O_3P$
Bis[2-(diphenylarsino)ethyl]phenylphosphine; *As,As,P*-Trioxide, *in* B-00212

$C_{34}H_{33}As_2P$
▷Bis[2-(diphenylarsino)ethyl]phenylphosphine, B-00212

$C_{34}H_{33}As_2PS$
Bis[2-(diphenylarsino)ethyl]phenylphosphine; *P*-Sulfide, *in* B-00212

$C_{34}H_{33}N_2P$
N,N,N′,N′-Tetrabenzyl-*P*-phenylphosphonous diamide, *in* P-00252

$C_{34}H_{33}O_3P_3$
Bis[2-(diphenylphosphino)ethyl]-phenylphosphine; Trioxide, *in* B-00239

$C_{34}H_{33}P_3$
Bis[2-(diphenylphosphino)ethyl]-phenylphosphine, B-00239

$C_{34}H_{33}P_3S_3$
Bis[2-(diphenylphosphino)ethyl]-phenylphosphine; Trisulfide, *in* B-00239

$C_{34}H_{33}P_3Se_3$
Bis[2-(diphenylphosphino)ethyl]-phenylphosphine; Triselenide, *in* B-00239

$C_{34}H_{37}NO_2P_2$
N-tert-Butoxycarbonyl-4-diphenylphosphino-2-diphenylphosphinomethylpyrrolidine, B-00532

$C_{34}H_{38}O_{10}P_2$
13,14,16,17,31,32,34,35-Octahydro-6,24-dimethyltetrabenzo[*d,m,r,a′*]-[1,3,6,9,12,15,20,23,26,2,16]decaoxadiphosphacyclooctacosin, O-00008

$C_{34}H_{38}O_{10}P_2S_2$
13,14,16,17,31,32,34,35-Octahydro-6,24-dimethyltetrabenzo[*d,m,r,a′*]-[1,3,6,9,12,15,20,23,26,2,16]decaoxadiphosphacyclooctacosin; 6,24-Disulfide, *in* O-00008

$C_{34}H_{38}O_{12}P_2$
13,14,16,17,31,32,34,35-Octahydro-6,24-dimethyltetrabenzo[*d,m,r,a′*]-[1,3,6,9,12,15,20,23,26,2,16]decaoxadiphosphacyclooctacosin; 6,24-Dioxide, *in* O-00008

$C_{34}H_{40}Br_2P_2$
1,6-Dibenzyl-1,6-diphenyl-1,6-diphosphecanium dibromide, *in* D-01171
1,1,4,4-Tetrabenzyl-1,4-diphosphocanium dibromide, *in* D-00048
1,1,5,5-Tetrabenzyl-1,5-diphosphocanium dibromide, *in* D-00049

$C_{34}H_{52}BrP$
Hexadecyltriphenylphosphonium(1+); Bromide, *in* H-00014

$C_{34}H_{52}P^{\oplus}$
Hexadecyltriphenylphosphonium(1+), H-00014

$C_{34}H_{66}NO_{10}P$
Glycerol 1,2-ditetradecanoate 3-phosphoserine, G-00051

$C_{34}H_{68}NO_8P$
Glycerol 1,2-di-11-methyldodecanoate 3-phosphocholine, G-00019
Glycerol 1,2-ditetradecanoate 3-phosphono-2-amino-1-propanol, G-00050
Glycerol 1,2-ditridecanoate 3-phosphocholine, G-00052

$C_{35}H_{25}P$
2,3,4,5,6-Pentaphenylphosphorin, P-00040

$C_{35}H_{26}BrO_3P$
2-(4-Bromophenyl)-8,8-dihydro-4,6,8,8-tetraphenyl[1,2]oxaphospholo[2,3-*b*][1,2,5]-dioxaphosphorin, B-00479

$C_{35}H_{27}O_2P$
1,1-Dihydro-1,1-diphenoxy-2,4,6-triphenylphosphorin, *in* D-00613

$C_{35}H_{27}P$
(2,4-Diphenyl-2,4-cyclopentadien-1-ylidene)-triphenylphosphorane, D-00999

$C_{35}H_{27}PS_2$
1,1-Dihydro-1,1-bis(phenylthio)-2,4,6-triphenylphosphorin, *in* D-00613

$C_{35}H_{31}NOP_2$
3,4-Bis(diphenylphosphino)pyrrolidine; *N*-Benzoyl, *in* B-00267

$C_{35}H_{33}NP_2$
3,4-Bis(diphenylphosphino)pyrrolidine; *N*-Benzyl, *in* B-00267

$C_{35}H_{56}Br_2P_2$
1,1,4,4-Tetrabenzyl-1,4-diphosphonanium dibromide, *in* D-00050
1,1,5,5-Tetrabenzyl-1,5-diphosphonanium dibromide, *in* D-00051

$C_{35}H_{67}O_8P$
Glycerol 1-9-octadecenoate 2-tetradecanoate 3-phosphate, G-00094

$C_{35}H_{69}O_8P$
Glycerol 1,2-dihexadecanoate 3-phosphate, G-00011

$C_{36}H_{24}KP$
Tris[[1,1′-biphenyl]-2,2′-diyl]phosphate(1−); K salt, *in* T-00731

$C_{36}H_{24}LiP$
Tris[[1,1′-biphenyl]-2,2′-diyl]phosphate(1−); Li salt, *in* T-00731

$C_{36}H_{24}NaP$
Tris[[1,1′-biphenyl]-2,2′-diyl]phosphate(1−); Na salt, *in* T-00731

$C_{36}H_{24}P^{\ominus}$
Tris[[1,1′-biphenyl]-2,2′-diyl]phosphate(1−), T-00731

$C_{36}H_{27}OP$
Tri-2-biphenylylphosphine; Oxide, *in* T-00348
Tri-4-biphenylylphosphine; Oxide, *in* T-00350

$C_{36}H_{27}O_2P$
Pentaphenyl-1*H*-phosphole; Maleic anhydride adduct, *in* P-00038

$C_{36}H_{27}O_3P$
Tri([1,1-biphenyl]-2-yl) phosphite, T-00351

$C_{36}H_{27}O_4P$
Tri-2-biphenylyl phosphate, T-00345
Tri-3-biphenylyl phosphate, T-00346
Tri-4-biphenylyl phosphate, T-00347

$C_{36}H_{27}P$
Tri-2-biphenylylphosphine, T-00348
Tri-3-biphenylylphosphine, T-00349
Tri-4-biphenylylphosphine, T-00350

$C_{36}H_{27}PS$
Tri-4-biphenylylphosphine; Sulfide, *in* T-00350

$C_{36}H_{28}N_2O_4P_2$
1′,3′-Diphenyldispiro[1,3,2-benzodioxaphosphole-2,2′-[1,3,2,4]-diazaphosphetidine-4′,2″-[1,3,2]-benzodioxaphosphole]; 2,4′-Di-Ph, *in* D-01018

$C_{36}H_{28}N_2O_6P_2$
1′,3′-Diphenyldispiro[1,3,2-benzodioxaphosphole-2,2′-[1,3,2,4]-diazaphosphetidine-4′,2″-[1,3,2]-benzodioxaphosphole]; 2,4′-Diphenoxy, *in* D-01018

$C_{36}H_{28}P_2$
3,3′-Bis(diphenylphosphino)biphenyl, B-00222

$C_{36}H_{28}P_4$
5,6,11,12-Tetrahydro-5,6,11,12-tetraphenyl-dibenzo[*c,g*][1,2,5,6]tetraphosphocin, T-00147

$C_{36}H_{28}P_4S_2$
5,6,11,12-Tetrahydro-5,6,11,12-tetraphenyl-dibenzo[*c,g*][1,2,5,6]tetraphosphocin; 5,11-(or 5,6)-Disulfide, *in* T-00147

$C_{36}H_{29}FN_2P_2$
N′-(4-Fluorophenyl)-*P,P*-diphenyl-*N*-(triphenylphosphoranylidene)phosphinimidic amide, *in* D-01164

$C_{36}H_{29}N_3O_2P_2$
N′-(4-Nitrophenyl)-*P,P*-diphenyl-*N*-(triphenylphosphoranylidene)phosphinimidic amide, *in* D-01164

$C_{36}H_{30}BrNP_2$
Bis(triphenylphosphine)iminium(1+); Bromide, *in* B-00442

$C_{36}H_{30}ClNP_2$
N-Chlorotriphenylphosphoranyl-*P,P,P*-triphenylphosphine imide, *in* B-00442

$C_{36}H_{30}FNP_2$
Bis(triphenylphosphine)iminium(1+); Fluoride, *in* B-00442

$C_{36}H_{30}GeNP$
Triphenyl(triphenylgermyl)phosphinimine, T-00695

$C_{36}H_{30}GeP_2$
Bis(diphenylphosphino)diphenylgermane, B-00232

$C_{36}H_{30}Ge_2P_2$
Hexaphenyl-1,3-diphospha-2,4-digermacyclobutane, H-00087

$C_{36}H_{30}INP_2$
Bis(triphenylphosphine)iminium(1+); Iodide, *in* B-00442

$C_{36}H_{30}NP_2^{\oplus}$
Bis(triphenylphosphine)iminium(1+), B-00442

$C_{36}H_{30}NP_3$
N,N-Bis(diphenylphosphino)-*P,P*-diphenylphosphinous amide, B-00233

$C_{36}H_{30}N_2O_5P_2S$
Triphenyl [[(triphenylphosphoranylidene)-amino]sulfonyl]phosphorimidate, *in* T-00642

$C_{36}H_{30}N_2P_2$
Bis(triphenylphosphoranylidene)hydrazine, B-00458
N′,P,P-Triphenyl-*N*-(triphenylphosphoranylidene)phosphinimidic amide, *in* D-01164

$C_{36}H_{30}N_3OP$
Hexaphenylphosphoric triamide, *in* H-00090

$C_{36}H_{30}N_3O_6P_3$
2,2,4,4,6,6-Hexahydro-2,2,4,4,6,6-hexaphenoxy-1,3,5-triaza-2,4,6-triphosphorine, H-00030

$C_{36}H_{30}N_3P$
Hexaphenylphosphorous triamide, H-00090

$C_{36}H_{30}N_3P_3$
2,2,4,4,6,6-Hexahydro-2,2,4,4,6,6-hexaphenyl-1,3,5-triaza-2,4,6-triphosphorine, H-00032

$C_{36}H_{30}N_4O_4P_2$
1′,3′-Diphenyldispiro[1,3,2-benzodioxaphosphole-2,2′-[1,3,2,4]-diazaphosphetidine-4′,2″-[1,3,2]-benzodioxaphosphole]; 2,4′-Bis(phenylamino), *in* D-01018

$C_{36}H_{30}P_6$
Hexaphenylhexaphosphorinane, H-00089

$C_{36}H_{32}Cl_2N_2P_2$
N,N′-Bis(triphenylphosphonio)hydrazine dichloride, *in* B-00458

$C_{36}H_{32}FeI_2P_2$
1,1′-Bis(methyldiphenylphosphonio)ferrocene diiodide, *in* B-00241

$C_{36}H_{32}N_5P_3$
2,2-Diamino-2,2,4,4,6,6-hexahydro-4,4,6,6-tetraphenyl-1,3,5-triaza-2,4,6-triphosphorine; *N,N′*-Di-Ph, *in* D-00025

$C_{36}H_{34}IP$
17-Benzyl-2,7,10,15,17-pentamethyl-17*H*-tetrabenzo[*b,d,f,h*]phosphoninium iodide, *in* P-00028

$C_{36}H_{37}OP_3$
Bis[3-(diphenylphosphino)propyl]-phenylphosphine; 5-Oxide, *in* B-00266

$C_{36}H_{37}O_9P$
7-Butyl-2,3-dihydro-1,6-diphenyl-1*H*-1-benzophosphonin-2,3,4,5-tetracarboxylic acid; Tetra-Me ester, oxide, *in* B-00552

$C_{36}H_{37}P_3$
Bis[3-(diphenylphosphino)propyl]-phenylphosphine, B-00266

$C_{36}H_{40}Br_2P_2$
1,1,6,6-Tetrabenzyl-1,6-diphosphecanium dibromide, *in* D-01171

$C_{36}H_{44}O_3P_2$
Bis(2,4,6-trimethylphenyl)phosphinic acid; Anhydride, *in* B-00428

$C_{36}H_{44}O_4P_4$
1,5,9,13-Tetraphenyl-1,5,9,13-tetraphosphacyclohexadecene; Tetraoxide, *in* T-00276

$C_{36}H_{44}P_4$
1,5,9,13-Tetraphenyl-1,5,9,13-tetraphosphacyclohexadecene, T-00276

$C_{36}H_{48}P_2$
Phellanphos, P-00059

$C_{36}H_{48}P_2S_2$
Phellanphos; Disulfide, *in* P-00059

$C_{36}H_{58}OP_2$
Bis(2,4,6-tri-*tert*-butylphenyl)diphosphene; Oxide, *in* B-00395

$C_{36}H_{58}O_2P_2$
1,12-Diphenyl-1,12-diphosphacyclotetracosane; 1,12-Dioxide, *in* D-01010

$C_{36}H_{58}P_2$
Bis(2,4,6-tri-*tert*-butylphenyl)diphosphene, B-00395
1,12-Diphenyl-1,12-diphosphacyclotetracosane, D-01010

$C_{36}H_{58}P_2S$
Bis(2,4,6-tri-*tert*-butylphenyl)diphosphene; Sulfide, *in* B-00395
2,3-Bis(2,4,6-tri-*tert*-butylphenyl)-thiadiphosphirane, B-00397

$C_{36}H_{58}P_2Se$
2,3-Bis(2,4,6-tri-*tert*-butylphenyl)-1,2,3-selenadiphosphirane, B-00396

$C_{36}H_{72}NO_8P$
Glycerol 1,2-di-12-methyltridecanoate 3-phosphocholine, G-00024
Glycerol 1,2-ditetradecanoate 3-phosphocholine, G-00048

$C_{36}H_{75}OP$
Tridodecylphosphine; Oxide, *in* T-00435

$C_{36}H_{75}O_2P$
Dioctadecylphosphinic acid, D-00953

$C_{36}H_{75}O_3P$
Dioctadecyl phosphonate, D-00954

$C_{36}H_{75}O_4P$
Dioctadecyl phosphate, D-00952

$C_{36}H_{75}P$
Tridodecylphosphine, T-00435

$C_{37}H_{27}P$
(Benzo[*e*]phenanthren-2-ylmethylene)-triphenylphosphorane, *in* T-00636

$C_{37}H_{28}BrP$
2-(Triphenylphosphoniomethyl)benzo[*c*]-phenanthrene(1+); Bromide, *in* T-00636

$C_{37}H_{28}P^{\oplus}$
2-(Triphenylphosphoniomethyl)benzo[*c*]-phenanthrene(1+), T-00636

$C_{37}H_{30}BrP$
1,2,3,4-Tetraphenyl-5-phosphoniaspiro[4.5]-deca-1,3-diene; Bromide, *in* T-00270

$C_{37}H_{30}P^{\oplus}$
1,2,3,4-Tetraphenyl-5-phosphoniaspiro[4.5]-deca-1,3-diene, T-00270

$C_{37}H_{30}P_2$
Bis(triphenylphosphoranylidene)methane, B-00459

$C_{37}H_{31}BrP_2$
Triphenyl[(triphenylphosphoranylidene)-methyl]phosphonium(1+); Bromide, *in* T-00697

$C_{37}H_{31}ClP_2$
Triphenyl[(triphenylphosphoranylidene)-methyl]phosphonium(1+); Chloride, *in* T-00697

$C_{37}H_{31}IP_2$
Triphenyl[(triphenylphosphoranylidene)-methyl]phosphonium(1+); Iodide, *in* T-00697

$C_{37}H_{31}N_2O_4P$
3,3,3,6,6-Pentaphenyl-3,3,3,6-tetrahydro-4,5-bis(methoxycarbonyl)-1,2,3-diazaphosphorine, P-00041

$C_{37}H_{31}OP_3S_2$
Tris(diphenylphosphino)methane; Oxide, disulfide, *in* T-00775

$C_{37}H_{31}O_2P_3$
Tris(diphenylphosphino)methane; Dioxide, *in* T-00775

$C_{37}H_{31}O_2P_3S$
Tris(diphenylphosphino)methane; Dioxide, monosulfide, *in* T-00775

$C_{37}H_{31}O_3P_3$
Tris(diphenylphosphino)methane; Trioxide, *in* T-00775

$C_{37}H_{31}P_2^{\oplus}$
Triphenyl[(triphenylphosphoranylidene)-methyl]phosphonium(1+), T-00697

$C_{37}H_{31}P_3$
Tris(diphenylphosphino)methane, T-00775

$C_{37}H_{31}P_3S$
Tris(diphenylphosphino)methane; Monosulfide, *in* T-00775

$C_{37}H_{31}P_3S_2$
Tris(diphenylphosphino)methane; Disulfide, *in* T-00775

$C_{37}H_{31}P_3S_3$
Tris(diphenylphosphino)methane; Trisulfide, *in* T-00775

$C_{37}H_{32}Br_2P_2$
Methylenebis[triphenylphosphonium](2+); Dibromide, *in* M-00147

$C_{37}H_{32}Cl_2P_2$
Methylenebis[triphenylphosphonium](2+); Dichloride, *in* M-00147

$C_{37}H_{32}I_2P_2$
Methylenebis[triphenylphosphonium](2+); Diiodide, *in* M-00147

$C_{37}H_{32}P_2^{\oplus\oplus}$
Methylenebis[triphenylphosphonium](2+), M-00147

$C_{37}H_{35}NO_8P_2$
2,2,7,7-Tetrahydro-3,4,5,6-tetrakis(methoxycarbonyl)-1*H*-1,2,7-azadiphosphepine; 2,2,7,7-Tetra-Ph, *in* T-00138

$C_{37}H_{36}F_6FeNP_2Rh$
1-[1-(Dimethylamino)ethyl]-2-(diphenyl-phosphino)ferrocene; (Norbornadiene)-rhodiumhexafluorophosphate complex, *in* D-00697

$C_{37}H_{37}NO_9P_2$
Dibenzyl [1-carboxy-2-[4-(dibenzyloxyphosphinyloxy)phenyl]ethyl]-phosphoramidate, *in* D-01188

$C_{37}H_{42}I_3P_3$
Bis[2-(methyldiphenylphosphonio)ethyl]-methylphenylphosphonium triiodide, *in* B-00239

$C_{37}H_{46}Br_2P_2$
1,1,6,6-Tetrabenzyl-1,6-diphosphoniacycloundecane dibromide, *in* D-00046

$C_{37}H_{58}P_2$
P,P′-Bis(2,4,6-tri-*tert*-butylphenyl)-1,3-diphosphaallene, B-00394

$C_{37}H_{71}O_8P$
Glycerol 1-hexadecanoate 2-9-octadecenoate 3-phosphate, G-00060
Glycerol 1-9-octadecenoate 2-hexadecanoate 3-phosphate, G-00086

$C_{37}H_{74}NO_8P$
Glycerol 1,2-dihexadecanoate 3-phosphoethanolamine, G-00013

$C_{38}H_{28}P_2$
9,10-Bis(diphenylphosphino)phenanthrene, B-00261

$C_{38}H_{30}NP^{\oplus}$
[(Cyanomethylene)triphenylphosphoranyl]-triphenylphosphonium, *in* C-00214

$C_{38}H_{30}P_2$
1,2,2,3,4,4-Hexaphenyl-1,3-diphosphetane, H-00088

$C_{38}H_{31}Br_2NP_2$
Cyanomethylenebis(triphenyl-phosphonium)(2+); Dibromide, *in* C-00214

$C_{38}H_{31}NP_2^{\oplus\oplus}$
Cyanomethylenebis(triphenyl-phosphonium)(2+), C-00214

$C_{38}H_{32}Br_2P_2$
1,2-Ethenediylbis[triphenylphosphonium](2+); Dibromide, *in* E-00019

$C_{38}H_{32}I_2P_2$
1,2-Ethenediylbis[triphenylphosphonium](2+); Diiodide, *in* E-00019

$C_{38}H_{32}OP_2$
[Oxybis[methylene]]bis[triphenyl-phosphonium](2+); Bis-ylide, *in* O-00097

$C_{38}H_{32}PS$
[Thiobis[methylene]]bis[triphenyl-phosphorane], *in* T-00312

$C_{38}H_{32}P_2^{\oplus\oplus}$
1,2-Ethenediylbis[triphenylphosphonium](2+), E-00019

$C_{38}H_{34}Br_2OP_2$
[Oxybis[methylene]]bis[triphenyl-phosphonium](2+); Dibromide, *in* O-00097

$C_{38}H_{34}Br_2P_2$
▷1,2-Ethanediylbis[triphenylphosphonium](2+); Dibromide, *in* E-00017

$C_{38}H_{34}Br_2P_2S$
[Thiobis[methylene]]bis[triphenyl-phosphonium](2+); Dibromide, *in* T-00312

$C_{38}H_{34}Cl_2OP_2$
[Oxybis[methylene]]bis[triphenyl-phosphonium](2+); Dichloride, *in* O-00097

$C_{38}H_{34}Cl_2P_2S$
[Thiobis[methylene]]bis[triphenyl-phosphonium](2+); Dichloride, *in* T-00312

$C_{38}H_{34}IO_3P_2$
Tris(diphenylphosphino)methane; Dioxide, monomethiodide, *in* T-00775

$C_{38}H_{34}N_2P_2$
2,2′-Bis(diphenylphosphinamido)-6,6′-dimethyl-biphenyl, B-00215

$C_{38}H_{34}OP_2^{\oplus\oplus}$
[Oxybis[methylene]]bis[triphenyl-phosphonium](2+), O-00097

$C_{38}H_{34}O_3P_2S$
[(Diphenylphosphinyl)methyl]triphenyl-phosphonium(1+); 4-Methylbenzenesulfonate, *in* D-01101

$C_{38}H_{34}P_2^{\oplus\oplus}$
1,2-Ethanediylbis[triphenylphosphonium](2+), E-00017

$C_{38}H_{34}P_2S^{\oplus\oplus}$
[Thiobis[methylene]]bis[triphenylphosphonium](2+), T-00312

$C_{38}H_{36}O_6P_2$
Methyl 2,3-bis-*O*-diphenylphosphino-4,6-*O*-benzylideneglucopyranoside, M-00091

$C_{38}H_{43}P$
[3,7-Dimethyl-9-(2,6,6-trimethyl-1-cyclohexen-1-yl)-2,4,6,8-nonatetraenyl]triphenylphosphonium(1+); Ylide, *in* D-00929

$C_{38}H_{44}IO_4P$
[3,7-Dimethyl-9-(2,6,6-trimethyl-1-cyclohexen-1-yl)-2,4,6,8-nonatetraenyl]triphenylphosphonium(1+); Periodate, *in* D-00929

$C_{38}H_{44}IP$
[3,7-Dimethyl-9-(2,6,6-trimethyl-1-cyclohexen-1-yl)-2,4,6,8-nonatetraenyl]triphenylphosphonium(1+); Iodide, *in* D-00929

$C_{38}H_{44}P^{\oplus}$
[3,7-Dimethyl-9-(2,6,6-trimethyl-1-cyclohexen-1-yl)-2,4,6,8-nonatetraenyl]triphenylphosphonium(1+), D-00929

$C_{38}H_{74}NO_{10}P$
Glycerol 1,2-dihexadecanoate 3-phosphoserine, G-00016

$C_{38}H_{74}O_{13}P_2$
Glycerol 1-9-octadecenoate 2-tetradecanoate 3-phosphoglycerophosphate, G-00095

$C_{38}H_{75}O_{10}P$
Glycerol 1,2-dihexadecanoate 3-phospho-1′-glycerol, G-00014

$C_{38}H_{76}NO_8P$
Glycerol 1,2-di-13-methyltetradecanoate 3-phosphocholine, G-00023
Glycerol 1-dodecanoate 2-octadecanoate 3-phosphocholine, G-00053
Glycerol 1-hexadecanoate 2-tetradecanoate 3-phosphocholine, G-00064
Glycerol 1-octadecanoate 2-dodecanoate 3-phosphocholine, G-00074
Glycerol 1-tetradecanoate 2-hexadecanoate 3-phosphocholine, G-00096

$C_{38}H_{76}O_8P$
Glycerol 1,2-dipentadecanoate 3-phosphocholine, G-00046

$C_{39}H_{30}N_3O_9P_3$
Hexaphenyl 1,3,5-triazine-2,4,6-triylphosphonate, *in* T-00329

$C_{39}H_{32}OP_2$
1,3-Bis(triphenylphosphoranylidene)-2-propanone, *in* O-00073

$C_{39}H_{34}Br_2OP_2$
(2-Oxo-1,3-propanediyl)bis(triphenylphosphonium)(2+); Dibromide, *in* O-00073

$C_{39}H_{34}Cl_2OP_2$
(2-Oxo-1,3-propanediyl)bis(triphenylphosphonium)(2+); Dichloride, *in* O-00073

$C_{39}H_{34}OP_2^{\oplus\oplus}$
(2-Oxo-1,3-propanediyl)bis(triphenylphosphonium)(2+), O-00073

$C_{39}H_{34}P_2$
1,3-Propanediylidenebis[triphenylphosphorane], *in* P-00448

$C_{39}H_{36}Br_2P_2$
▷1,3-Propanediylbis[triphenylphosphonium](2+); Dibromide, *in* P-00448

$C_{39}H_{36}P_2^{\oplus\oplus}$
1,3-Propanediylbis[triphenylphosphonium](2+), P-00448

$C_{39}H_{42}P_4$
2,3,4,5,6,7,8,9,10,11,12,13-Dodecahydro-2,5,9,12-tetraphenyl-1*H*-2,5,9,12-benzotetraphosphacyclopentadecin, D-01262

$C_{39}H_{65}O_8P$
Glycerol 1,2-di-9,12,15-octadecatrienoate 3-phosphate, G-00039

$C_{39}H_{69}O_8P$
Glycerol 1,2-di-9,12-octadecadienoate 3-phosphate, G-00027

$C_{39}H_{72}NO_8P$
Glycerol 1-hexadecanoate 2-9,12,15-octadecatrienoate 3-phosphoethanolamine, G-00059
Glycerol 1-9,12,15-octadecatrienoate 2-hexadecanoate 3-phosphoethanolamine, G-00085

$C_{39}H_{73}O_{13}P$
Glycerol 1-dodecanoate 2-9-octadecenoate 3-phosphoinositol, G-00054

$C_{39}H_{74}NO_8P$
Glycerol 1-9,12-octadecadienoate 2-hexadecanoate 3-phosphoethanolamine, G-00069

$C_{39}H_{76}NO_8P$
Glycerol 1-hexadecanoate 2-9-octadecenoate 3-phosphoethanolamine, G-00062

$C_{39}H_{77}O_8P$
Glycerol 1,2-dioctadecanoate 3-phosphate, G-00032
Glycerol 2-monophosphate; 1,3-Dioctadecanoyl, *in* G-00067

$C_{40}H_{28}O_3P$
Di-1-naphthylphosphinic acid; Anhydride, *in* D-00941

$C_{40}H_{29}OP$
2,3-Benzo-1,4,5,6,7-pentaphenyl-7-phosphabicyclo[2.2.1]hept-5-ene 7-oxide, *in* D-00538

$C_{40}H_{29}P$
1,4-Dihydro-1,2,3,4,9-pentaphenyl-1,4-phosphinidenenaphthalene, D-00538

$C_{40}H_{30}F_6OP_2$
2,2-Dihydro-2,2,2-triphenyl-4,4-bis(trifluoromethyl)-3-(triphenylphosphoranylidene)-1,2-oxaphosphetane, D-00608

$C_{40}H_{32}N_6O_{15}P_2$
3,3,6,6-Tetraphenyl-1,3,6-oxadiphosphepanium(2+); Dipicrate, *in* T-00266

$C_{40}H_{36}P_2$
1,4-Butanediylidenebis[triphenylphosphorane], *in* B-00517

$C_{40}H_{38}Br_2P_2$
1,4-Butanediylbis[triphenylphosphonium](2+); Dibromide, *in* B-00517

$C_{40}H_{38}Br_6P_2$
1,4-Butanediylbis[triphenylphosphonium](2+); Perbromide, *in* B-00517

$C_{40}H_{38}P_2^{\oplus\oplus}$
1,4-Butanediylbis[triphenylphosphonium](2+), B-00517

$C_{40}H_{39}BClP$
(3-Chloropropyl)methyldiphenylphosphonium tetraphenylborate, *in* C-00169

$C_{40}H_{67}OP_3Si$
1,4-Bis(2,4,6-tri-*tert*-butylphenyl)-3-trimethylsilyloxy-1,2,4-triphospha-1,3-butadiene, B-00398

$C_{40}H_{77}O_{10}P$
Glycerol 1-9-octadecenoate 2-hexadecanoate 3-phospho-1′-glycerol, G-00088

$C_{40}H_{80}NO_8P$
Glycerol 1,2-dihexadecanoate 3-phosphocholine, G-00012
Glycerol 1,2-di-14-methylpentadecanoate 3-phosphocholine, G-00022

$C_{41}H_{29}P$
(Benzo[*g*]chrysen-10-ylmethylene)triphenylphosphorane, *in* T-00635

$C_{41}H_{30}BrP$
10-(Triphenylphosphoniomethyl)benzo[*g*]chrysene(1+); Bromide, *in* T-00635

$C_{41}H_{30}P^{\oplus}$
10-(Triphenylphosphoniomethyl)benzo[*g*]chrysene(1+), T-00635

$C_{41}H_{36}IP$
17-Benzyl-2,7,10,15-tetramethyl-17-phenyl-17*H*-tetrabenzo[*b,d,f,h*]phosphoninium iodide, *in* T-00211

$C_{41}H_{38}P_2$
1,5-Pentanediylidenebis[triphenylphosphorane], *in* P-00034

$C_{41}H_{39}P_3$
1,1,1-Tris(diphenylphosphinomethyl)ethane, T-00776

$C_{41}H_{40}Br_2P_2$
1,5-Pentanediylbis[triphenylphosphonium](2+); Dibromide, *in* P-00034

$C_{41}H_{40}P_2^{\oplus\oplus}$
▷1,5-Pentanediylbis[triphenylphosphonium](2+), P-00034

$C_{41}H_{70}NO_8P$
Glycerol 1,2-di-9,12,15-octadecatrienoate 3-phosphoethanolamine, G-00041

$C_{41}H_{74}NO_8P$
Glycerol 1,2-di-9,12-octadecadienoate 3-phosphoethanolamine, G-00029

$C_{41}H_{76}NO_8P$
Glycerol 1-octadecanoate 2-9,12,15-octadecatrienoate 3-phosphoethanolamine, G-00081
Glycerol 1-9-octadecenoate 2-9,12-octadecadienoate 3-phosphoethanolamine, G-00089

$C_{41}H_{78}NO_8P$
Glycerol 1,2-di-9-octadecenoate 3-phosphoethanolamine, G-00043
Glycerol 1-9,12-octadecadienoate 2-octadecanoate 3-phosphoethanolamine, G-00071
Glycerol 1-octadecanoate 2-9,12-octadecadienoate 3-phosphoethanolamine, G-00079

$C_{41}H_{79}O_{13}P$
Glycerol 1,2-dihexadecanoate 3-phosphoinositol, G-00015

$C_{41}H_{80}NO_8P$
Glycerol 1-octadecanoate 2-9-octadecenoate 3-phosphoethanolamine, G-00083
Glycerol 1-9-octadecenoate 2-octadecanoate 3-phosphoethanolamine, G-00092

$C_{41}H_{82}NO_8P$
Glycerol 1,2-dioctadecanoate 3-phosphonoethanolamine, G-00037

$C_{42}H_{25}F_8P_3$
Bis[2-(diphenylphosphino)-3,4,5,6-tetrafluorophenyl]phenylphosphine, B-00268

$C_{42}H_{33}As_2P$
Bis[2-(diphenylarsino)phenyl]phenylphosphine, B-00214

$C_{42}H_{33}O_3P_3$
Bis[4-(diphenylphosphino)phenyl]-phenylphosphine; Trioxide, *in* B-00263

$C_{42}H_{33}P_3$
Bis[2-(diphenylphosphino)phenyl]-phenylphosphine, B-00262
Bis[4-(diphenylphosphino)phenyl]-phenylphosphine, B-00263

$C_{42}H_{33}P_3S_3$
Bis[4-(diphenylphosphino)phenyl]-phenylphosphine; Trisulfide, *in* B-00263

$C_{42}H_{34}N_2P_2$
N,N′-1,2-Phenylenebis[*P,P,P*-triphenylphosphine imide], P-00143
N,N′-1,3-Phenylenebis[*P,P,P*-triphenylphosphine imide], P-00144
N,N′-1,4-Phenylenebis[*P,P,P*-triphenylphosphine imide], P-00145
P,P′-1,4-Phenylenebis[*N,P,P*-triphenylphosphine imide], P-00146

$C_{42}H_{34}OP_2$
[2,5-Furandiylbis[methylene]]bis[triphenylphosphorane], *in* B-00448

[3,4-Furandiylbis[methylene]]bis[triphenylphosphorane], *in* B-00449

$C_{42}H_{34}P_2S$
[2,5-Thiophenediylbis[methylene]]-bis[triphenylphosphorane], *in* B-00456
[3,4-Thiophenediylbis[methylene]]-bis[triphenylphosphorane], *in* B-00457

$C_{42}H_{35}Ge_2P$
Phenylbis(triphenylgermyl)phosphine, P-00101

$C_{42}H_{35}N_2OP_3$
P-Phenyl-*N*,*N'*-bis(triphenylphosphoranylidene)phosphonic diamide, *in* P-00102

$C_{42}H_{35}N_2P_3$
P-Phenyl-*N*,*N'*-bis(triphenylphosphoranylidene)phosphonous diamide, P-00102

$C_{42}H_{35}N_2P_3S$
P-Phenyl-*N*,*N'*-bis(triphenylphosphoranylidene)phosphonothioic diamide, *in* P-00102

$C_{42}H_{36}Br_2OP_2$
3,4-Bis[(triphenylphosphonio)methyl]-furan(2+); Dibromide, *in* B-00449

$C_{42}H_{36}Br_2P_2S$
3,4-Bis[(triphenylphosphonio)methyl]-thiophene(1+); Dibromide, *in* B-00457

$C_{42}H_{36}Cl_2OP_2$
2,5-Bis[(triphenylphosphonio)methyl]-furan(2+); Dichloride, *in* B-00448
3,4-Bis[(triphenylphosphonio)methyl]-furan(2+); Dichloride, *in* B-00449

$C_{42}H_{36}Cl_2P_2S$
2,5-Bis[(triphenylphosphonio)methyl]-thiophene(2+); Dichloride, *in* B-00456

$C_{42}H_{36}OP_2^{\oplus\oplus}$
2,5-Bis[(triphenylphosphonio)methyl]-furan(2+), B-00448
3,4-Bis[(triphenylphosphonio)methyl]-furan(2+), B-00449

$C_{42}H_{36}P_2S^{\oplus\oplus}$
2,5-Bis[(triphenylphosphonio)methyl]-thiophene(2+), B-00456
3,4-Bis[(triphenylphosphonio)methyl]-thiophene(1+), B-00457

$C_{42}H_{40}P_4$
5,6,7,8,9,14,15,16,17,18-Decahydro-5,9,14,18-tetraphenyldibenzo[*b*,*i*][1,4,8,11]-tetraphosphacyclotetradecin, D-00006

$C_{42}H_{42}NP_3$
Tris[(2-diphenylphosphino)ethyl]amine, T-00773

$C_{42}H_{42}N_3P$
Hexabenzylphosphorous triamide, H-00012

$C_{42}H_{42}O_4P_4$
1,1,4,7,10,10-Hexaphenyl-1,4,7,10-tetraphosphadecane; Tetraoxide, *in* H-00091
Tris[2-(diphenylphosphino)ethyl]phosphine; Tetraoxide, *in* T-00774

$C_{42}H_{42}P_4$
1,1,4,7,10,10-Hexaphenyl-1,4,7,10-tetraphosphadecane, H-00091
Tris[2-(diphenylphosphino)ethyl]phosphine, T-00774

$C_{42}H_{42}P_4S_4$
Tris[2-(diphenylphosphino)ethyl]phosphine; Tetrasulfide, *in* T-00774

$C_{42}H_{56}O_4P_4$
1,5,10,14-Tetrabenzyl-1,5,10,14-tetraphosphacyclooctadecane; 1,5,10,14-Tetraoxide, *in* T-00009

$C_{42}H_{56}P_4$
1,5,10,14-Tetrabenzyl-1,5,10,14-tetraphosphacyclooctadecane, T-00009

$C_{42}H_{58}N_2P_4$
4,7,13,16-Tetraphenyl-1,10-dipropyl-1,10-diaza-4,7,13,16-tetraphosphacyclooctadecane, T-00264

$C_{42}H_{74}NO_{10}P$
Glycerol 1,2-di-9,12-octadecadienoate 3-phosphoserine, G-00031

$C_{42}H_{76}NO_{10}P$
Glycerol 1-9-octadecenoate 2-9,12-octadecadienoate 3-phosphoserine, G-00090

$C_{42}H_{78}NO_8P$
Glycerol 1-hexadecanoate 2-9,12,15-octadecatrienoate 3-phosphocholine, G-00058

$C_{42}H_{78}NO_{10}P$
Glycerol 1,2-di-9-octadecenoate 3-phosphoserine, G-00045

$C_{42}H_{78}O_{10}P$
Glycerol 1,2-di-9-octadecenoate 3-phospho-1'-glycerol, G-00044

$C_{42}H_{80}NO_8P$
Glycerol 1-hexadecanoate 2-9,12-octadecadienoate 3-phosphocholine, G-00056
Glycerol 1-9,12-octadecadienoate 2-hexadecanoate 3-phosphocholine, G-00068

$C_{42}H_{80}NO_{10}P$
Glycerol 1-9-octadecenoate 2-octadecanoate 3-phosphoserine, G-00093

$C_{42}H_{82}NO_8P$
Glycerol 1-hexadecanoate 2-9-octadecenoate 3-phosphocholine, G-00061
Glycerol 1-9-octadecenoate 2-hexadecanoate 3-phosphocholine, G-00087

$C_{42}H_{82}NO_{10}P$
Glycerol 1,2-dioctadecanoate 3-phosphoserine, G-00038

$C_{42}H_{83}O_{10}P$
Glycerol 1,2-dioctadecanoate 3-phospho-1'-glycerol, G-00035

$C_{42}H_{84}NO_8P$
Glycerol 1,2-diheptadecanoate 3-phosphocholine, G-00010
Glycerol 1,2-di-15-methylhexadecanoate 3-phosphocholine, G-00021
Glycerol 1,2-dioctadecanoate 3-phospho-2-amino-1-propanol, G-00033
Glycerol 1-hexadecanoate 2-octadecanoate 3-phosphocholine, G-00057
Glycerol 1-octadecanoate 2-hexadecanoate 3-phosphocholine, G-00075

$C_{43}H_{35}As_2P$
[Bis(diphenylarsino)methylene]triphenylphosphorane, B-00213

$C_{43}H_{35}PSb_2$
[Bis(diphenylstibino)methylene]triphenylphosphorane, B-00270

$C_{43}H_{35}P_3$
[Bis(diphenylphosphino)methylene]triphenylphosphorane, B-00252

$C_{43}H_{37}BBrP$
(Bromomethyl)triphenylphosphonium(1+); Tetraphenylborate, *in* B-00477

$C_{43}H_{37}BClP$
(Chloromethyl)triphenylphosphonium(1+); Tetraphenylborate, *in* C-00117

$C_{43}H_{38}BP$
Methyltriphenylphosphonium(1+); Tetraphenylborate, *in* M-00432

$C_{43}H_{78}NO_8P$
Glycerol 1-5,8,11,14-icosatetraenoate 2-octadecanoate 3-phosphoethanolamine, G-00065
Glycerol 1-octadecanoate 2-5,8,11,14-icosatetraenoate 3-phosphoethanolamine, G-00077

$C_{43}H_{81}O_{13}P$
Glycerol-1-hexadecanoate 2-9-octadecenoate 3-phosphoinositol, G-00063

$C_{44}H_{32}O_2P_2$
2,2'-Bis(diphenylphosphino)-1,1'-binaphthyl; Dioxide, *in* B-00221

$C_{44}H_{32}P_2$
2,2'-Bis(diphenylphosphino)-1,1'-binaphthyl, B-00221

$C_{44}H_{34}N_2P_2$
2,2'-Bis[(diphenylphosphino)amino]-1,1'-binaphthyl, B-00217

$C_{44}H_{34}O_2P_2$
2,11-Bis[(diphenylphosphino)methyl]benzo[*c*]-phenanthrene; Dioxide, *in* B-00246

$C_{44}H_{34}P_2$
2,11-Bis[(diphenylphosphino)methyl]benzo[*c*]-phenanthrene, B-00246
1,2-Bis(triphosphoranylidene)-benzocyclobutene, B-00460
1,3*a*,3*b*,4,6*a*,6*b*-Hexahydro-1,2,3*b*,4,5,6*b*-hexaphenylcyclobuta[1,2-*b*:3,4-*b'*]-diphosphole, H-00031

$C_{44}H_{36}Br_2P_2$
1,2-Dihydro-1,2-bis(triphenylphosphonio)-benzocyclobutene(2+); Dibromide, *in* D-00447

$C_{44}H_{36}Cl_2O_8P_2$
1,2-Dihydro-1,2-bis(triphenylphosphonio)-benzocyclobutene(2+); Diperchlorate, *in* D-00447

$C_{44}H_{36}P_2$
[1,2-Phenylenebis[methylene]]bis[triphenylphosphorane], P-00135
[1,3-Phenylenebis[methylene]]bis[triphenylphosphorane], *in* P-00133
[1,4-Phenylenebis(methylene)]bis[triphenylphosphorane], P-00136

$C_{44}H_{36}P_2^{\oplus\oplus}$
1,2-Dihydro-1,2-bis(triphenylphosphonio)-benzocyclobutene(2+), D-00447

$C_{44}H_{38}Br_2P_2$
1,2-[Phenylenebis[methylene]]bis[triphenylphosphonium](2+); Dibromide, *in* P-00132
[1,3-Phenylenebis[methylene]]bis[triphenylphosphonium](2+); Dibromide, *in* P-00133
1,4-[Phenylenebis[methylene]]bis[triphenylphosphonium](2+); Dibromide, *in* P-00134

$C_{44}H_{38}Cl_2P_2$
[1,3-Phenylenebis[methylene]]bis[triphenylphosphonium](2+); Dichloride, *in* P-00133
▷1,4-[Phenylenebis[methylene]]bis[triphenylphosphonium](2+); Dichloride, *in* P-00134

$C_{44}H_{38}P_2^{\oplus\oplus}$
1,2-[Phenylenebis[methylene]]bis[triphenylphosphonium](2+), P-00132
[1,3-Phenylenebis[methylene]]bis[triphenylphosphonium](2+), P-00133
1,4-[Phenylenebis[methylene]]bis[triphenylphosphonium](2+), P-00134

$C_{44}H_{40}BOP$
(Methoxymethyl)triphenylphosphonium(1+); Tetraphenylborate, *in* M-00053

$C_{44}H_{40}BP$
Ethyltriphenylphosphonium(1+); Tetraphenylborate, *in* E-00197

$C_{44}H_{40}BPS$
[(Methylthio)methyl]triphenylphosphonium(1+); Tetraphenylborate, *in* M-00416

$C_{44}H_{48}I_3P_3$
1,1,1-Tris(methyldiphenylphosphoniomethyl)-ethane, *in* T-00776

$C_{44}H_{60}P_4$
1,6,11,16-Tetrabenzyl-1,6,11,16-tetraphosphacycloeicosane, T-00008

$C_{44}H_{76}NO_8P$
Glycerol 1,2-di-9,12,15-octadecatrienoate 3-phosphocholine, G-00040

$C_{44}H_{76}O_2P_2Si_2$
1,4-Bis(2,4,6-tri-*tert*-butylphenyl)-1,4-bis(trimethylsilyloxy)-2,3-diphosphabutadiene, B-00393

$C_{44}H_{80}NO_8P$
Glycerol 1,2-di-9,12-octadecadienoate 3-phosphocholine, G-00028
Glycerol 1-hexadecanoate 2-5,8,11,14-icosatetraenoate 3-phosphocholine, G-00055

$C_{44}H_{82}NO_8P$
Glycerol 1-9,12-octadecadienoate 2-9-octadecenoate 3-phosphocholine, G-00072
Glycerol 1-octadecanoate 2-9,12,15-octadecatrienoate 3-phosphocholine, G-00080

$C_{44}H_{84}NO_8P$
Glycerol 1,2-di-9-octadecenoate 3-phosphocholine, G-00042
Glycerol 1-9,12-octadecadienoate 2-octadecanoate 3-phosphocholine, G-00070
Glycerol 1-octadecanoate 2-9,12-octadecadienoate 3-phosphocholine, G-00078

$C_{44}H_{86}NO_8P$
Glycerol 1-octadecanoate 2-9-octadecenoate 3-phosphocholine, G-00082
Glycerol 1-9-octadecenoate 2-octadecanoate 3-phosphocholine, G-00091

$C_{44}H_{88}NO_8P$
Glycerol 1,2-di-16-methylheptadecanoate 3-phosphocholine, G-00020
Glycerol 1,2-dioctadecanoate 3-phosphocholine, G-00034

$C_{45}H_{38}P_2$
[2-[2-(Triphenylphosphoranylidenemethyl)-phenyl]ethylidenetriphenylphosphorane, *in* T-00696

$C_{45}H_{40}Br_2P_2$
Triphenyl[[2-[2-(triphenylphosphonio)ethyl]-phenyl]methyl]phosphonium(2+); Dibromide, *in* T-00696

$C_{45}H_{40}I_2O_8P_2$
Triphenyl[[2-[2-(triphenylphosphonio)ethyl]-phenyl]methyl]phosphonium(2+); Diperiodate, *in* T-00696

$C_{45}H_{40}P_2^{\oplus\oplus}$
Triphenyl[[2-[2-(triphenylphosphonio)ethyl]-phenyl]methyl]phosphonium(2+), T-00696

$C_{45}H_{74}NO_8P$
Glycerol 1,2-di-5,8,11,14-icosatetraenoate 3-phosphoethanolamine, G-00018

$C_{45}H_{79}O_{13}P$
Glycerol 1,2-di-9,12-octadecadienoate 3-phosphoinositol, G-00030

$C_{45}H_{81}O_{13}P$
Glycerol 1-9,12-octadecadienoate 2,9-octadecenoate 3-phosphoinositol, G-00073

$C_{45}H_{85}O_{13}P$
Glycerol 1-octadecanoate 2-9-octadecenoate 3-phosphoinositol, G-00084

$C_{45}H_{87}O_{13}P$
Glycerol 1,2-dioctadecanoate 3-phosphoinositol, G-00036

$C_{46}H_{36}O_2P_2$
2,2′-Bis[(diphenylphosphino)methyl]-1,1′-binaphthyl; Dioxide, *in* B-00248

$C_{46}H_{36}P_2$
2,2′-Bis[(diphenylphosphino)methyl]-1,1′-binaphthyl, B-00248

$C_{46}H_{42}BOP$
(2-Ethoxyethenyl)triphenylphosphonium(1+); Tetraphenylborate, *in* E-00039

$C_{46}H_{42}P_2$
(2,7-Dimethyl-2,4,6-octatrien-1,8-diylidene)-bis[triphenylphosphorane], *in* D-00778

$C_{46}H_{44}Br_2P_2$
(2,7-Dimethyl-2,4,6-octatrien-1,8-diyl)-bis[triphenylphosphonium](2+); Dibromide, *in* D-00778

$C_{46}H_{44}P_2^{\oplus\oplus}$
(2,7-Dimethyl-2,4,6-octatrien-1,8-diyl)-bis[triphenylphosphonium](2+), D-00778

$C_{46}H_{84}NO_8P$
Glycerol 1-octadecanoate 2-5,8,11,14-icosatetraenoate 3-phosphocholine, G-00076

$C_{46}H_{92}NO_8P$
Glycerol 1,2-dinonadecanoate 3-phosphocholine, G-00026

$C_{47}H_{35}P$
(Tetraphenyl-2,4-cyclopentadien-1-ylidene)-triphenylphosphorane, T-00252

$C_{48}H_{33}OP$
1,2,3,4,9-Pentaphenyl-1*H*-tribenzo[*b*,*d*,*f*]-phosphepin; 9-Oxide, *in* P-00042

$C_{48}H_{33}P$
1,2,3,4,9-Pentaphenyl-1*H*-tribenzo[*b*,*d*,*f*]-phosphepin, P-00042

$C_{48}H_{38}P_2$
[1,5-Naphthalenediylbis[methylene]]-bis[triphenylphosphorane], *in* B-00451
[1,8-Naphthalenediylbis[methylene]]-bis[triphenylphosphorane], *in* B-00452
[(2,3-Naphthalenediylbis[methylene]]-bis[triphenylphosphorane], *in* B-00453
[2,7-Naphthalenediylbis[methylene]]-bis[triphenylphosphorane], *in* B-00455

$C_{48}H_{40}Br_2P_2$
1,4-Bis(triphenylphosphoniomethyl)-naphthalene(2+); Dibromide, *in* B-00450
1,5-Bis(triphenylphosphoniomethyl)-naphthalene(2+); Dibromide, *in* B-00451
1,8-Bis(triphenylphosphoniomethyl)-naphthalene(2+); Dibromide, *in* B-00452
2,3-Bis(triphenylphosphoniomethyl)-naphthalene(2+); Dibromide, *in* B-00453
2,6-Bis(triphenylphosphoniomethyl)-naphthalene(2+); Dibromide, *in* B-00454
2,7-Bis(triphenylphosphoniomethyl)-naphthalene(2+); Dibromide, *in* B-00455

$C_{48}H_{40}P_2$
(1,2-Phenylenedi-2-propene-3,1-diylidene)-bis[triphenylphosphorane], *in* P-00149

$C_{48}H_{40}P_2^{\oplus\oplus}$
1,4-Bis(triphenylphosphoniomethyl)-naphthalene(2+), B-00450
1,5-Bis(triphenylphosphoniomethyl)-naphthalene(2+), B-00451
1,8-Bis(triphenylphosphoniomethyl)-naphthalene(2+), B-00452
2,3-Bis(triphenylphosphoniomethyl)-naphthalene(2+), B-00453
2,6-Bis(triphenylphosphoniomethyl)-naphthalene(2+), B-00454
2,7-Bis(triphenylphosphoniomethyl)-naphthalene(2+), B-00455

$C_{48}H_{41}N_2PSi_2$
P,*P*-Diphenyl-*N*,*N*′-bis(triphenylsilyl)-phosphinimidic amide, *in* D-01043

$C_{48}H_{41}P_3$
Phenylbis[3-(triphenylphosphoranylidene)-propenyl]phosphine, P-00103

$C_{48}H_{42}B_2F_8P_2$
(1,2-Phenylenedi-2-propene-3,1-diyl)-bis[triphenylphosphonium](2+); Bis(tetrafluoroborate), *in* P-00149

$C_{48}H_{42}Br_2P_2$
(1,2-Phenylenedi-2-propene-3,1-diyl)-bis[triphenylphosphonium](2+); Dibromide, *in* P-00149

$C_{48}H_{42}P_2^{\oplus\oplus}$
(1,2-Phenylenedi-2-propene-3,1-diyl)-bis[triphenylphosphonium](2+), P-00149

$C_{48}H_{44}P_4$
1,2,7,8,9,10,15,16-Octamethyl-3,6,11,14-tetraphenyltetrakisphospholo[1,2-*b*:2′,1′-*d*:1″,2″-*f*:2‴,1‴-*h*][1,2,5,6]tetraphosphocin, O-00033

$C_{48}H_{80}NO_8P$
Glycerol 1,2-di-5,8,11,14-icosatetraenoate 3-phosphocholine, G-00017

$C_{49}H_{36}IP$
9-Methyl-1,2,3,4,9-pentaphenyl-1*H*-tribenzo[*b*,*d*,*f*]phosphepinium iodide, *in* P-00042

$C_{49}H_{37}P$
2,4,6,7,7,8,8-Heptaphenyl-1-phosphabicyclo[4.2.0]octa-2,4-diene, H-00003

$C_{50}H_{38}P_2$
(1,8-Biphenylenediyldimethylidyne)-bis[triphenylphosphorane], B-00076

$C_{50}H_{40}P_2$
[[1,1′-Biphenyl]-2,2′-diylbis[methylene]]-bis[triphenylphosphorane], *in* B-00445
[[1,1′-Biphenyl]-3,3′-diylbis[methylene]]-bis[triphenylphosphorane], *in* B-00446
[[1,1′-Biphenyl]-4,4′-diylbis[methylene]]-bis[triphenylphosphorane], *in* B-00447

$C_{50}H_{42}Br_2P_2$
(5,6-Acenaphthenylenedimethylene)-bis[triphenylphosphonium](2+); Dibromide, *in* A-00001
2,2′-Bis[(triphenylphosphonio)methyl]-biphenyl(2+); Dibromide, *in* B-00445
3,3′-Bis[(triphenylphosphonio)methyl]-biphenyl(2+); Dibromide, *in* B-00446

$C_{50}H_{42}Cl_2P_2$
4,4′-Bis[(triphenylphosphonio)methyl]-biphenyl(2+); Dichloride, *in* B-00447

$C_{50}H_{42}I_2O_8P_2$
(5,6-Acenaphthenylenedimethylene)-bis[triphenylphosphonium](2+); Diperiodate, *in* A-00001
2,2′-Bis[(triphenylphosphonio)methyl]-biphenyl(2+); Diperiodate, *in* B-00445

$C_{50}H_{42}P_2^{\oplus\oplus}$
(5,6-Acenaphthenylenedimethylene)-bis[triphenylphosphonium](2+), A-00001
2,2′-Bis[(triphenylphosphonio)methyl]-biphenyl(2+), B-00445
3,3′-Bis[(triphenylphosphonio)methyl]-biphenyl(2+), B-00446
4,4′-Bis[(triphenylphosphonio)methyl]-biphenyl(2+), B-00447

$C_{50}H_{81}N_4O_{15}P$
Calyculin *A*, C-00001

$C_{51}H_{10}O_4P_4$
1,1,3,3-Tetrakis(diphenylphosphino)-1,2-propadiene; Tetraoxide, *in* T-00180

$C_{51}H_{10}P_4$
1,1,3,3-Tetrakis(diphenylphosphino)-1,2-propadiene, T-00180

$C_{51}H_{42}P_2$
[Methylenebis(2,1-phenylenemethylene)]-bis[triphenylphosphorane], *in* M-00141

$C_{51}H_{44}Br_2P_2$
[Methylenebis(2,1-phenylenemethylene)]-bis[triphenylphosphonium](2+); Dibromide, *in* M-00141

$C_{51}H_{44}I_2O_8P_2$
[Methylenebis(2,1-phenylenemethylene)]-bis[triphenylphosphonium](2+); Diperiodate, *in* M-00141

$C_{51}H_{44}P_2^{\oplus\oplus}$
[Methylenebis(2,1-phenylenemethylene)]-bis[triphenylphosphonium](2+), M-00141

$C_{52}H_{13}IP_4$
1,1,3,3-Tetrakis(diphenylphosphino)-1,2-propadiene; B,MeI, *in* T-00180

$C_{52}H_{40}O_2P_2$
1,4-Diphenyl-2,3-bis(triphenyl-phosphoranylidene)-1,4-butanedione, D-00997
5*a*,6,11,11*a*-Tetrahydro-1,2,3,6,11,12-hexaphenyl-6,11-phosphinidene-1*H*-naphth[2,3-*e*]isophosphindole; 2,12-Dioxide, *in* T-00107

$C_{52}H_{40}P_2$
[1,8-Anthracenediylbis[methylene]]-bis[triphenylphosphorane], *in* B-00443
5*a*,6,11,11*a*-Tetrahydro-1,2,3,6,11,12-hexaphenyl-6,11-phosphinidene-1*H*-naphth[2,3-*e*]isophosphindole, T-00107

$C_{52}H_{42}Br_2P_2$
1,8-Bis(triphenylphosphoniomethyl)-anthracene(2+); Dibromide, *in* B-00443

$C_{52}H_{42}P_2^{\oplus\oplus}$
1,8-Bis(triphenylphosphoniomethyl)-anthracene(2+), B-00443

$C_{53}H_{48}O_4P_4$
Tetrakis[(diphenylphosphinyl)methyl]methane, *in* T-00179

$C_{53}H_{48}P_4$
Tetrakis[(diphenylphosphino)methyl]methane, T-00179

$C_{53}H_{48}P_4S_4$
Tetrakis[(diphenylphosphinothioyl)methyl]-methane, *in* T-00179

$C_{53}H_{48}P_4Se_4$
Tetrakis[(diphenylphosphinoselenoyl)methyl]-methane, *in* T-00179

$C_{54}H_{42}P_4$
Tris(2-diphenylphosphinophenyl)phosphine, T-00777

$C_{54}H_{45}Ge_3P$
Tris(triphenylgermyl)phosphine, T-00870

$C_{54}H_{45}Ge_3P_3$
Nonaphenyl-1,3,5-triphospha-2,4,6-trigermacyclohexane, N-00070

$C_{54}H_{45}N_3OP_4$
Tris(triphenylphosphoranylidene)phosphoric triamide, *in* T-00872

$C_{54}H_{45}N_3P_4$
Tris(triphenylphosphoranylidene)phosphorous triamide, T-00872

$C_{54}H_{87}O_6P_3$
2,4,6-Tris(2,4,6-tri-*tert*-butylphenoxy)-1,3,5,2,4,6-trioxatriphosphorine, T-00847

$C_{54}H_{87}P_3S_3$
2,4,6-Tris(2,4,6-trimethylphenyl)-1,3,5,2,4,6-trithiatriphosphorinane, T-00862

$C_{54}H_{101}O_3P$
Trioctadecyl phosphite, T-00580

$C_{55}H_{47}BOP_2$
[(Diphenylphosphinyl)methyl]triphenyl-phosphonium(1+); Tetraphenylborate, *in* D-01101

$C_{55}H_{48}CuP_3$
Methyltris(triphenylphosphine)copper, M-00435

$C_{57}H_{48}OP_4$
Tris[(triphenylphosphoranylidene)methyl]-phosphine; Oxide, *in* T-00871

$C_{57}H_{48}P_4$
Tris[(triphenylphosphoranylidene)methyl]-phosphine, T-00871

$C_{57}H_{48}P_4S$
Tris[(triphenylphosphoranylidene)methyl]-phosphine; Sulfide, *in* T-00871

$C_{58}H_{46}Br_2P_2$
2,2′-Bis[(triphenylphosphonio)methyl]-1,1-binaphthyl(2+); Dibromide, *in* B-00444

$C_{58}H_{46}I_2O_8P_2$
2,2′-Bis[(triphenylphosphonio)methyl]-1,1-binaphthyl(2+); Diperiodate, *in* B-00444

$C_{58}H_{46}P_2^{\oplus\oplus}$
2,2′-Bis[(triphenylphosphonio)methyl]-1,1-binaphthyl(2+), B-00444

$C_{58}H_{60}P_6$
4,7-Bis[2-(diphenylphosphino)ethyl]-1,1,10,10-tetraphenyl-1,4,7,10-tetraphosphadecane, B-00240

$C_{60}H_{40}P_2$
Tris[[1,1′-biphenyl]-2,2′-diyl]phosphate(1−); 5,5′-Spirobi[5*H*-dibenzophospholium] salt, *in* T-00731

$C_{60}H_{48}Br_2P_2$
1,2,3,4,9,10,11,12-Octaphenyl-5,8-diphosphoniadispiro[4.2.4.2]deca-1,3,9,11-tetraene(2+); Dibromide, *in* O-00037

$C_{60}H_{48}P_2^{\oplus\oplus}$
1,2,3,4,9,10,11,12-Octaphenyl-5,8-diphosphoniadispiro[4.2.4.2]deca-1,3,9,11-tetraene(2+), O-00037

$C_{60}H_{53}N_2O_4P$
Tris[[1,1′-biphenyl]-2,2′-diyl]phosphate(1−); 19-Methylbrucinium salt, *in* T-00731

$C_{61}H_{51}BP_2$
Triphenyl[(triphenylphosphoranylidene)-methyl]phosphonium(1+); Tetraphenyl-borate, *in* T-00697

$C_{62}H_{50}Cl_2P_2$
1,1,3,3-Tetrahydro-1,1,3,3-tetraphenyl-2,4-bis(triphenylphosphonio)-1,3-diphosphete; Dichloride, *in* T-00145

$C_{62}H_{50}P_2^{\oplus\oplus}$
1,1,3,3-Tetrahydro-1,1,3,3-tetraphenyl-2,4-bis(triphenylphosphonio)-1,3-diphosphete, T-00145

$C_{62}H_{92}CoN_{13}O_{15}P$
Vitamin B_{12a}, V-00010

$C_{62\text{-}74}H_{96\text{-}103}N_{5\text{-}6}O_{32\text{-}40}P$
Prasinomycin, P-00438

$C_{63}H_{88}CoN_{14}O_{14}P$
▷Vitamin B_{12}, V-00009

$C_{63}H_{110}N_9O_{36}P$
Ensanchomycin, E-00003

$C_{64}H_{72}Br_4P_4$
1,5,9,13-Tetrabenzyl-1,5,9,13-tetraphenyl-1,5,9,13-tetraphosphoniacyclohexadecane tetrabromide, *in* T-00276

$C_{65}H_{126}O_{17}P_2$
Tetramyristoyldiphosphatidylglycerol, T-00241

$C_{69}H_{107}N_4O_{35}P$
Moenomycin *A*, *in* M-00441

$C_{70}H_{84}Br_4P_4$
1,1,5,5,10,10,14,14-Octabenzyl-1,5,10,14-tetraphosphoniacyclooctadecane tetrabromide, *in* T-00009

$C_{70}H_{135}O_{12}P$
Tetrapalmitoylphosphatidylglycerol, T-00247

$C_{72}H_{90}Br_4P_4$
1,1,6,6,11,11,16,16-Octabenzyl-1,6,11,16-tetraphosphoniacycloeicosane tetrabromide, *in* T-00008

$C_{72}H_{125}N_6O_{42}P$
Prenomycin, P-00439

$C_{73}H_{142}O_{17}P_2$
Tetrapalmitoyldiphosphatidylglycerol, T-00246

$C_{75}H_{135}N_7O_{42}P$
Moenomycin *C*, *in* M-00441

$C_{78}H_{143}O_{12}P$
Tetra-9-octadecenoylphosphatidylglycerol, T-00242

$C_{78}H_{151}O_{12}P$
Tetrastearoylphosphatidylglycerol, T-00295

$C_{81}H_{158}O_{17}P_2$
Tetrastearoyldiphosphatidylglycerol, T-00294

CAS Registry Number Index

CAS Registry Number Index

50-18-0 ▷Cyclophosphamide, C-00283
52-24-4 ▷Thiotepa, T-00317
52-46-0 ▷2,2,4,4,6,6-Hexakis(1-aziridinyl)-2,2,4,4,6,6-hexahydro-1,3,5-triaza-2,4,6-triphosphorine, H-00060
52-68-6 ▷Trichlorphon, T-00423
52-85-7 ▷Famphur, F-00001
53-59-8 Coenzyme II, C-00203
53-84-9 ▷Coenzyme I, C-00202
54-47-7 ▷Pyridoxal phosphate, P-00508
55-38-9 ▷Fenthion, F-00007
55-91-4 ▷Diisopropyl phosphorofluoridate, D-00665
56-38-2 ▷Parathion, P-00002
56-65-5 ▷Adenosine triphosphate, A-00043
56-72-4 ▷Coumaphos, C-00205
56-73-5 Glucose 6-dihydrogen phosphate; D-*form*, *in* G-00007
58-64-0 ▷Adenosine diphosphate, A-00038
60-51-5 ▷Dimethoate, D-00671
60-92-4 ▷Cyclic AMP, C-00233
62-73-7 ▷Dichlorvos, D-00203
63-37-6 ▷Cytidine 5′-(dihydrogen phosphate), C-00290
63-89-8 Glycerol 1,2-dihexadecanoate 3-phosphocholine; (*R*)-L-*form*, *in* G-00012
67-07-2 *N*-Phosphocreatine, P-00364
68-19-9 ▷Vitamin B_{12}, V-00009
77-81-6 ▷Tabun, T-00001
78-32-0 Tris(4-methylphenyl) phosphate, T-00805
78-34-2 ▷Dioxathion, D-00974
78-38-6 Diethyl ethylphosphonate, D-00292
78-40-0 ▷Triethyl phosphate, T-00440
78-42-2 ▷Tris(2-ethylhexyl) phosphate, T-00779
78-46-6 ▷Dibutyl butylphosphonate, D-00076
78-47-7 *O,O,O*-Tributyl phosphorothioate, T-00387
78-48-8 ▷*S,S,S*-Tributyl phosphorotrithioate, T-00392
78-53-5 ▷Amiton, A-00133
78-57-9 ▷Menazon, M-00003
78-64-8 *O,O*-Dihexyl phosphorodithioate, D-00437
84-52-6 Cytidine 3′-(dihydrogen phosphate), C-00289
85-32-5 ▷5′-Guanylic acid, G-00103
85-61-0 Coenzyme *A*, C-00200
85-94-9 Cytidine 2′-(dihydrogen phosphate), C-00288
86-01-1 ▷Guanosine 5′-triphosphate, G-00101
86-50-0 ▷Azinphos-Methyl, A-00151
96-64-0 ▷Soman, S-00005
97-17-6 ▷Dichlofenthion, D-00161
101-02-0 ▷Triphenyl phosphite, T-00633
102-84-1 ▷Tri-2-propenyl phosphite, T-00708
102-85-2 ▷Tributyl phosphite, T-00379
107-44-8 ▷Sarin, S-00002
107-49-3 ▷Tetraethyl pyrophosphate, T-00069
107-55-1 *O,O*-Bis(1-methylpropyl) phosphorodithioate, B-00359
107-56-2 *O,O*-Diisopropyl phosphorodithioate, D-00663
107-66-4 ▷Dibutyl phosphate, D-00104
115-26-4 ▷Dimefox, D-00669
115-78-6 ▷Chlorphonium; Chloride, *in* C-00192
115-86-6 ▷Triphenyl phosphate, T-00623
115-90-2 ▷Fensulfothion, F-00006
115-93-5 ▷*O,O*-Dimethyl *O*-4-sulfamoylphenyl phosphorothioate, D-00920
116-17-6 ▷Triisopropyl phosphite, T-00502
117-68-0 3′-Guanylic acid, G-00102
121-45-9 ▷Trimethyl phosphite, T-00535
121-70-0 Phenylphosphonous acid, P-00251
121-75-5 ▷Malathion, M-00001
122-10-1 ▷Bomyl, B-00461
122-14-5 ▷Fenitrothion, F-00005
122-52-1 ▷Triethyl phosphite, T-00445
124-64-1 ▷Tetrakis(hydroxymethyl)-phosphonium(1+); Chloride, *in* T-00181
126-22-7 ▷Butonate, B-00530
126-68-1 ▷*O,O,O*-Triethyl phosphorothioate, T-00458
126-71-6 Tris(2-methylpropyl) phosphate, T-00822
126-73-8 ▷Tributyl phosphate, T-00370
127-88-8 ▷Bis(2-chloroethyl)phosphoramidic dichloride, B-00122
130-49-4 2′-Adenylic acid, A-00045
133-89-1 Uridine diphosphate glucose, U-00002
138-08-9 Phosphoenolpyruvic acid, P-00366
141-66-2 ▷Dicrotophos; (*E*)-*form*, *in* D-00204
142-10-9 2-Hydroxy-3-(phosphonoxy)propanal, H-00187
144-35-4 3,9-Diphenoxy-2,4,8,10-tetraoxa-3,9-diphosphaspiro[5.5]-undecane, D-00991
146-91-8 Guanosine 5′-diphosphate, G-00099
150-50-5 ▷Tributyl phosphorotrithioite, T-00393
152-16-9 ▷Octamethyldiphosphoramide, O-00029
152-18-1 ▷*O,O,O*-Trimethyl phosphorothioate, T-00545
152-19-2 *O,O,O*-Trimethyl phosphoroselenoate, T-00543
152-20-5 ▷*O,O,S*-Trimethyl phosphorothioate, T-00546
154-87-0 ▷Thiamine diphosphate, T-00297
161-95-5 Phosphanthridene, P-00348
176-50-1 5-Phosphoniaspiro[4.4]nonane, P-00374
181-07-7 6-Phosphoniaspiro[5.5]undecane, P-00375
181-85-1 2,2′-Spirobi[1,3,2-benzodioxaphosphole], S-00014
188-36-3 4-Mercaptodinaphtho[2,1-*d*:1′,2′-*f*][1,3,2]dioxaphosphepin 4-sulfide, M-00012
234-23-1 Naphtho[1,2-*d*]-1,3,2-dioxaphosphole, N-00009
244-87-1 5*H*-Dibenzophosphole, D-00040
253-37-2 Isophosphinoline, I-00027
279-53-8 2,6,7-Trioxa-1-phosphabicyclo[2.2.1]-heptane, T-00587
280-35-3 1-Phosphabicyclo[2.2.2]octane, P-00346
280-45-5 2,6,7-Trioxa-1-phosphabicyclo[2.2.2]-octane, T-00588
280-61-5 1,4-Diphosphabicyclo[2.2.2]octane, D-01169
281-33-4 2,8,9-Trioxa-1-phosphatricyclo[3.3.1.1^{3,7}]decane, T-00591
288-01-7 1*H*-Phosphole, P-00372
288-78-8 1,3,2-Dithiaphosphole, D-01246
289-68-9 Phosphorin, P-00403
297-97-2 ▷Thionazin, T-00314
297-99-4 ▷Phosphamidon; (*E*)-*form*, *in* P-00347
298-00-0 ▷Parathion methyl, P-00003
298-01-1 Mevinphos; (*E*)-*form*, *in* M-00440
298-02-2 ▷Phorate, P-00338
298-03-3 ▷Demeton-*O*, D-00019
298-04-4 ▷Disulfoton, D-01236
298-06-6 ▷*O,O*-Diethyl phosphorodithioate, D-00384
299-84-3 ▷Fenchlorphos, F-00003
299-85-4 ▷Zytron, Z-00002
299-86-5 ▷Crufomate, C-00208
300-76-5 ▷Naled, N-00001
301-12-2 ▷Oxydemeton-methyl, O-00098
301-13-3 ▷Tris(2-ethylhexyl) phosphite, T-00780
302-48-7 ▷*P,P*-Bis(1-aziridinyl)-*N*-ethylphosphinic amide, *in* B-00092
302-49-8 ▷Uredepa, U-00001
310-40-7 Diethyl (4-fluorophenyl)-phosphonate, *in* F-00038
311-33-1 Spiro[1,3,2-benzodioxaphosphole-2,1′-[3]phospholene], S-00010
311-45-5 ▷Paraoxon, P-00001
311-46-6 Ethyl (dimethoxyphosphinyl)acetate, *in* D-00678
312-05-0 Diethyl (2-fluorophenyl)-phosphonate, *in* F-00036
326-12-5 Crotoxyphos, C-00207
327-98-0 ▷Trichloronate, T-00419
330-08-5 ▷*N,N′*-Diphenylphosphorodiamidic fluoride, D-01127
330-13-2 Mono(4-nitrophenyl) phosphate, M-00500
333-41-5 ▷Diazinon, D-00027
338-45-4 Mevinphos; (*Z*)-*form*, *in* M-00440
348-71-0 *N*-Phenyl-*P,P*-bis(trifluoromethyl)-phosphinous amide, *in* B-00419
349-87-1 (4-Fluorophenyl)phosphonic acid, F-00038
350-65-2 ▷*N,N′*-Dicyclohexylphosphorodiamidic acid; Fluoride, *in* D-00216
353-88-8 Methyl methylphosphonofluoridate, *in* M-00303
354-43-8 Dimethylphosphoramidic difluoride, D-00858
358-11-7 *O,O*-Diethyl phosphorofluoridothioate, D-00389
358-29-2 ▷Ethyl dimethyl-phosphoramidofluoridate, E-00064
358-71-4 Mono(3-methyl-3-butenyl) diphosphate, M-00483
358-72-5 Mono(3-methyl-2-butenyl) diphosphate, M-00482
358-74-7 ▷Diethyl phosphorofluoridate, D-00387
359-64-8 Iodobis(trifluoromethyl)phosphine, *in* B-00418
359-65-9 Bis(trifluoromethyl)phosphinous acid, B-00418
359-94-4 Diethylphosphoramidic difluoride, D-00345
359-95-5 Diethylphosphoramidothioic difluoride, D-00358
362-74-3 ▷Bucladesine, *in* C-00233
363-84-8 Diethylphosphoramidous acid; Difluoride, *in* D-00360
371-68-6 Ethyl phosphorofluoridate, E-00179
371-86-8 ▷Mipafox, *in* D-00661
372-97-4 Mono(3,7,11-trimethyl-2,6,10-dodecatrienyl) diphosphate; (*E,E*)-*form*, *in* M-00517
374-09-4 (Trifluoromethyl)phosphonic acid, T-00478
381-45-3 ▷Dipropyl phosphorofluoridate, D-01228
381-65-7 Methyl phosphorodifluoridite, M-00353
393-02-2 ▷Tetrakis(trifluoromethyl)-tetraphosphetane, T-00188
398-14-1 Acridophosphine, A-00031
403-65-6 Diphenyl phosphorofluoridate, D-01133
407-41-0 Phosphoserine; (*S*)-*form*, *in* P-00417
420-52-0 (Trifluoromethyl)phosphine, T-00474
420-64-4 Tetrafluoromethylphosphorane, T-00081
421-58-9 (Trifluoromethyl)phosphonous dichloride, T-00483
421-59-0 (Trifluoromethyl)diiodophosphine, *in* T-00482

422-94-6 Bis(trifluoromethyl)phosphinic acid, B-00411
423-01-8 Tris(trifluoromethyl)phosphine oxide, T-00856
428-67-1 *O,O*-Dimethyl phosphorofluoridothioate, D-00902
430-78-4 Ethyldifluorophosphine, *in* E-00153
431-95-8 ▷*P,P*-Bis(trifluoromethyl)-phosphinous amide, B-00419
431-97-0 ▷Cyanobis(trifluoromethyl)-phosphine, *in* B-00418
431-98-1 *N*-Methyl-*P,P*-bis(trifluoromethyl)-phosphinous amide, *in* B-00419
432-01-9 *N,N*-Dimethyl-*P,P*-bis(trifluoromethyl)-phosphinous amide, *in* B-00419
432-04-2 Tris(trifluoromethyl)phosphine, T-00855
460-52-6 ▷Ethyl phosphorodifluoridate, E-00173
460-53-7 *O*-Ethyl phosphorodifluoridothioate, E-00175
462-57-7 1,2-Bis(trifluoromethyl)-diphosphine, B-00410
466-15-9 ▷Methyl bis(1-aziridinyl)-phosphinate, *in* B-00091
469-35-2 ▷Ethyl bis(1-aziridinyl)-phosphinate, *in* B-00091
469-47-6 Dimethyl 1-aziridinylphosphonate, *in* A-00153
470-27-9 Diethyl 1-aziridinylphosphonate, *in* A-00153
470-90-6 ▷Chlorfenvinphos, C-00018
470-94-0 ▷*O,O*-Diethyl *S*-(2-dimethylaminoethyl) phosphorothioate; Hydrogen oxalate, *in* D-00284
472-43-5 ▷Phenazaphosphinic acid, P-00061
488-69-7 Fructose 1,6-bis(dihydrogen phosphate); D-*form*, *in* F-00067
491-97-4 Thymidine 5′-pyrophosphate, T-00320
500-28-7 ▷Chlorthion, C-00196
507-15-3 ▷Dimethylphosphinous fluoride, D-00848
512-56-1 ▷Trimethyl phosphate, T-00528
513-02-0 Triisopropyl phosphate, T-00499
513-08-6 Tripropyl phosphate, T-00710
513-10-0 ▷Ecothiopate; Iodide, *in* E-00001
524-49-2 5-Hydroxy-5*H*-dibenzophosphole 5-oxide, H-00118
529-96-4 Pyridoxamine 5′-phosphate, P-00509
538-37-4 ▷Dibenzyl phosphorochloridate, D-00060
545-55-1 ▷1,1′,1″-Phosphinylidynetrisaziridine, *in* T-00331
554-70-1 ▷Triethylphosphine, T-00441
562-17-4 ▷Tetraethylphosphorodiamidic acid; Fluoride, *in* T-00059
563-12-2 ▷Ethion, E-00022
563-22-4 Bis(2-methylpropyl) phosphorofluoridate, B-00362
575-75-7 Tonophosphan, *in* T-00322
589-57-1 ▷Diethyl phosphorochloridite, D-00374
590-54-5 Acetyl phosphate, A-00025
590-55-6 ▷Carbamoyl dihydrogen phosphate, C-00002
590-69-2 ▷Butafox, *in* D-00146
593-54-4 ▷Methylphosphine, M-00277
593-68-0 ▷Ethylphosphine, E-00123
594-09-2 ▷Trimethylphosphine, T-00530
597-07-9 Dimethyl dimethylphosphoramidate, D-00727
597-24-0 Dimethyl 1-piperidinylphosphonate, *in* P-00430
597-25-1 ▷Dimethyl 4-morpholinylphosphonate, *in* M-00523
597-50-2 Triethylphosphine oxide, T-00443
597-51-3 Triethylphosphine sulfide, T-00444
597-82-0 *O,O,O*-Triphenyl phosphorothioate, T-00670
597-83-1 *S,S,S*-Triphenyl phosphorotrithioate, T-00675
597-84-2 *O,O,O*-Tris(4-methylphenyl) phosphorothioate, T-00814
597-89-7 ▷*O,S*-Dimethyl *O*-4-nitrophenyl phosphorothioate, D-00774
598-02-7 Diethyl phosphate, D-00320
603-35-0 ▷Triphenylphosphine, T-00624
607-01-2 Ethyldiphenylphosphine, E-00066
618-21-3 4-Phosphonobenzoic acid, P-00380
620-38-2 ▷Tris(3-methylphenyl) phosphite, T-00810
620-42-8 ▷Tris(4-methylphenyl) phosphite, T-00811
627-49-6 ▷Diethylphosphine, D-00321
631-44-7 Tris(2-chlorophenyl) phosphate, T-00745
631-45-8 *O,O,O*-Tris(2-methylphenyl) phosphorothioate, T-00812
633-90-9 Cytidine 2′,3′-phosphate, C-00294
638-21-1 ▷Phenylphosphine, P-00208
640-15-3 ▷Thiometon, T-00313
643-13-0 Fructose 6-dihydrogen phosphate; D-*form*, *in* F-00070
644-97-3 ▷Phenylphosphonous dichloride, P-00254
645-15-8 Bis(4-nitrophenyl) phosphate, B-00371
645-78-3 ▷*unsym*-Tetraethyl monothiopyrophosphate, T-00053
646-44-6 1,1,1-Trifluoro-1,1-dihydrophospholane, T-00466
646-71-9 Bis(trifluoromethyl)phosphinic acid; Chloride, *in* B-00411
650-09-9 Diethyl trifluoroacetyl phosphate, D-00406
650-20-4 Ethyl ethylphosphonofluoridate, *in* E-00142
650-52-2 ▷Bis(trifluoromethyl)phosphinous chloride, B-00420
650-99-7 Methyl phenylphosphonofluoridate, *in* P-00239
657-38-5 Methylphenylphosphinothioic acid; (±)-*form*, Fluoride, *in* M-00225
657-39-6 Phenylphosphonic difluoride, P-00224
657-40-9 Phenylphosphonothioic difluoride, P-00248
657-81-8 (4-Fluorophenyl)phosphonic dichloride, F-00039
657-97-6 Phenylphosphonous difluoride, P-00256
658-35-5 *O*-Phenyl phosphorodifluoridothioate, P-00287
661-42-7 Difluorotrimethylphosphorane, D-00417
661-45-0 Difluorotris(trifluoromethyl)-phosphorane, D-00420
661-60-9 ▷Trimethylphosphonamidic fluoride, T-00538
661-61-0 ▷Trimethylphosphonamidothioic acid; Fluoride, *in* T-00540
665-03-2 Methyl ethylphosphonofluoridate, *in* E-00142
665-92-9 Dipropylphosphinic acid; Fluoride, *in* D-01205
666-00-2 Dibutylphosphinic acid; Fluoride, *in* D-00109
666-23-9 Tetrafluorophenylphosphorane, T-00084
667-21-0 Ethyl diethylphosphoramidofluoridate, *in* D-00351
671-64-7 Trifluoromethyl bis(trifluoromethyl)-phosphinothioite, *in* B-00417
671-65-8 Trifluoromethyl bis(trifluoromethyl)-phosphinoselenoite, *in* B-00413
672-66-2 Dimethylphenylphosphine, D-00803
673-97-2 Ethyl methylphosphonofluoridate, *in* M-00303
673-98-3 *S*-Ethyl methylphosphonofluoridothioate, *in* M-00305
674-48-6 Dibutyl phosphorofluoridate, D-00153
676-59-5 ▷Dimethylphosphine, D-00827
676-83-5 ▷Methylphosphonous dichloride, M-00322
676-96-0 Trimethylphosphine oxide, T-00532
676-97-1 ▷Methylphosphonic dichloride, M-00290
676-98-2 ▷Methylphosphonothioic dichloride, M-00316
676-99-3 ▷Methylphosphonic difluoride, M-00291
677-24-7 Methyl phosphorodichloridate, M-00345
677-42-9 Isopropylphosphonic acid; Difluoride, *in* I-00050
677-43-0 ▷Dimethylphosphoramidic dichloride, D-00856
677-72-5 Di-*tert*-butylphosphinic acid; Bromide, *in* D-00110
677-74-7 Di-*tert*-butylphosphinic acid; Chloride, *in* D-00110
677-76-9 Di-*tert*-butylphosphinic acid, D-00110
679-37-8 Tetraethyl hypophosphate, T-00050
679-38-9 Tetrapropyl hypophosphate, T-00288
679-39-0 Tetrabutyl hypophosphate, T-00017
680-03-5 Diisopropylphosphinic acid, D-00641
680-31-9 ▷Hexamethylphosphoric triamide, H-00075
681-02-7 Diethylphosphinothioic acid; Fluoride, *in* D-00330
681-06-1 *O,O*-Dimethyl methylphosphonothioate, *in* M-00313
681-71-0 ▷*S,S,S*-Trimethyl phosphorotrithioate, T-00549
682-30-4 Diethyl ethenylphosphonate, *in* V-00003
682-34-8 2-Butenylphosphonic acid, B-00525
682-80-4 Demephion-*O*, D-00017
683-08-9 Diethyl methylphosphonate, D-00298
683-16-9 Dibutylphosphinic acid; Chloride, *in* D-00109
683-19-2 Dioctylphosphinic acid, D-00958
683-84-1 Tetramethylphosphinous amide, T-00213
683-85-2 ▷Dimethylphosphoramidous dichloride, D-00876
684-17-3 (2,2,2-Trichloro-1-hydroxyethyl)-phosphonic acid, T-00408
684-19-5 Di-*tert*-butylphosphine; Oxide, *in* D-00108
684-25-3 Methyl bis(trifluoromethyl)-phosphinite, *in* B-00418
684-56-0 Dimethyl (trifluoromethyl)-phosphonite, *in* T-00482
685-83-6 ▷Tetraethylphosphorodiamidous chloride, T-00067
686-20-4 Tetraethylphosphinous amide, *in* D-00334
686-69-1 Diethylphosphinous chloride, D-00335
688-62-0 ▷Triethyl phosphorotrithioite, T-00461
689-62-3 Dibutylphosphine; Sulfide, *in* D-00107
690-12-0 (2-Chloroethyl)phosphonic dichloride, C-00076
690-49-3 ▷Tetramethyl pyrophosphate, T-00229
690-64-2 Isopropyl methyl methylphosphonate, *in* M-00168
690-97-1 Butylphosphonic acid; Difluoride, *in* B-00594
693-00-5 Butyl phosphorodifluoridite, B-00633
694-24-6 2,3-Dihydro-1-hydroxy-1*H*-phosphole 1-oxide, D-00506
694-65-5 ▷2,5-Dihydro-1-methoxy-1*H*-phosphole 1-oxide, *in* D-00507
694-66-6 ▷2,3-Dihydro-1-methoxy-1*H*-phosphole 1-oxide, *in* D-00506
695-11-4 ▷2-Ethoxy-1,3,2-dioxaphospholane, E-00034
695-31-8 ▷Trimethylene fluorophosphate, *in* F-00018
695-59-0 ▷2,5-Dihydro-1-methoxy-3-methyl-1*H*-phosphole 1-oxide, *in* D-00499

695-62-5 ▷1-Ethoxy-2,5-dihydro-1*H*-phosphole 1-oxide, *in* D-00507
695-63-6 1-Ethoxy-2,3-dihydro-1*H*-phosphole 1-oxide, *in* D-00506
695-67-0 ▷2,5-Dihydro-1-hydroxy-3,4-dimethyl-1*H*-phosphole 1-oxide, D-00493
695-68-1 2-Mercapto-4,5-dimethyl-1,3,2-dioxaphospholane 2-sulfide, M-00009
696-50-4 2-Ethoxy-4-methyl-1,3,2-dioxaphospholane, E-00040
696-58-2 2-Ethoxy-1,3,2-dioxaphosphorinane, E-00037
697-29-0 2,5-Dihydro-1-methoxy-3,4-dimethyl-1*H*-phosphole 1-oxide, *in* D-00493
697-31-4 1-Ethoxy-2,5-dihydro-3-methyl-1*H*-phosphole 1-oxide, *in* D-00499
697-32-5 2,3-Dihydro-1-ethoxy-4-methyl-1*H*-phosphole 1-oxide, *in* D-00496
697-39-2 Ethyl trimethylene phosphate, *in* E-00037
697-45-0 2-Mercapto-5,5-dimethyl-1,3,2-dioxaphosphorinane 2-sulfide, M-00012
698-89-5 Benzyldibromophosphine, *in* B-00068
699-36-5 2-Mercapto-4,4,5,5-tetramethyl-1,3,2-dioxaphospholane 2-sulfide, M-00019
700-24-3 (2-Fluorophenyl)phosphonic acid, F-00036
701-03-1 Methylphenylpropylphosphine; (*S*)-*form*, *in* M-00265
701-32-6 (4-Methylphenyl)phosphonic acid; Difluoride, *in* M-00237
701-61-1 Phenylphosphoramidic acid; Difluoride, *in* P-00257
701-64-4 Monophenyl phosphate, M-00507
702-86-3 Methyl *o*-phenylene phosphate, *in* M-00026
703-03-7 2,3-Dihydro-1-phenyl-1*H*-phosphole; 1-Oxide, *in* D-00566
703-06-0 Ethyl phenylphosphonofluoridate, *in* P-00239
705-90-8 (2-Phenylethenyl)phosphonic difluoride, P-00158
710-09-8 Benzyl methyl methylphosphonate, *in* M-00168
710-89-4 2,5-Dihydro-3,4-dimethyl-1-phenyl-1*H*-phosphole; 1-Oxide, *in* D-00468
711-07-9 Trimethylene phenyl phosphate, *in* P-00068
714-87-4 3,9-Dichloro-2,4,8,10-tetraoxa-3,9-diphosphaspiro[5.5]-undecane; 3,9-Dioxide, *in* D-00196
716-85-8 *N,N′*-Diisopropyl-*P*-phenylphosphonous diamide, *in* P-00252
719-80-2 Ethyl diphenylphosphinite, *in* D-01089
721-74-4 Benzylmethylphenylphosphine; (*S*)-*form*, Oxide, *in* B-00051
732-11-6 ▷Phosmet, P-00343
734-59-8 (4-Bromophenyl)diphenylphosphine, B-00481
734-60-1 (4-Chlorophenyl)diphenylphosphine, C-00129
738-99-8 ▷Dipin, D-01189
739-58-2 4-Diphenylphosphino-*N,N*-dimethylaniline, D-01061
741-58-2 ▷Bensulide, B-00004
752-23-8 Bis(triphenylphosphoranylidene)-hydrazine, B-00458
753-59-3 Methylphosphonous difluoride, M-00324
753-70-8 Dimethylphosphinic fluoride, D-00832
753-72-0 Methylphosphonothioic difluoride, M-00317
753-98-0 Ethylphosphonic difluoride, E-00132
753-99-1 Ethylphosphonothioic difluoride, E-00151
754-01-8 Methyl phosphorochloridofluoridate, M-00341
754-24-5 *tert*-Butylphosphonic difluoride, B-00600
754-37-0 Bis(trifluoromethyl)phosphinic acid; Bromide, *in* B-00411
756-78-5 Diethylphosphinic acid; Fluoride, *in* D-00322
756-79-6 ▷Dimethyl methylphosphonate, D-00762
756-80-9 ▷*O,O*-Dimethyl phosphorodithioate, D-00898
757-54-0 Dimethyl 2-propenylphosphonate, *in* P-00454
757-71-1 Ethyl di-2-propenylphosphinate, *in* D-01193
757-86-8 Methyl [(dimethoxyphosphinothioyl)-thio]acetate, *in* D-00675
758-45-2 Bromobis(trifluoromethyl)phosphine, *in* B-00418
762-04-9 ▷Diethyl phosphonate, D-00338
762-77-6 Ethyl phosphorochloridofluoridate, E-00164
763-10-0 Mono(3,7-dimethyl-2,6-octadienyl) diphosphate; (*E*)-*form*, *in* M-00465
764-11-4 Heptylphosphonic acid; Dichloride, *in* H-00006
765-40-2 2-Fluoro-1,3,2-dioxaphospholane, F-00017
768-87-6 2,6,7-Trioxa-1-phosphabicyclo[2.2.2]-octane-4-methanol; Sulfide, *in* T-00590
769-91-5 (3-Methylphenyl)phosphonic acid; Difluoride, *in* M-00236
770-12-7 Phenyl phosphorodichloridate, P-00279
772-79-2 4-Chlorophenyl phosphorodichloridate, C-00162
773-47-7 Dimethyl benzylphosphonate, *in* B-00060
774-49-2 *N,N′*-Diethyl-*P*-phenylphosphonous diamide, *in* P-00252
775-08-6 3,5-Dimethylphenyl phosphorodichloridate, D-00822
777-52-6 4-Nitrophenyl phosphorodichloridate, N-00064
786-19-6 ▷Carbophenothion, C-00008
791-28-6 Triphenylphosphine oxide, T-00627
791-36-6 *S*-Phenyl *N,N*-diethyl-*P*-phenylphosphonamidothioate, *in* D-00309
791-43-5 2,4-Bis(4-fluorophenyl)-1,3,2,4-dioxaphosphetane 2,4-dioxide, B-00275
793-10-2 4-Nitrophenyl phenyl phosphorochloridate, *in* N-00052
793-43-1 4-(Diphenylphosphinyl)phenol, *in* D-01072
794-39-8 Diphenyl(2-phenylethenyl)phosphine, D-01033
795-44-8 *p*-(Diphenylphosphinyl)anisole, *in* M-00058
797-70-6 ▷Tris(4-methylphenyl)phosphine; Oxide, *in* T-00808
797-71-7 4,4′,4″-Phosphinylidynetrisphenol, *in* T-00790
797-72-8 4-Diphenylphosphinyl-*N,N*-dimethylaniline, *in* D-01061
803-14-5 Ethyl bromo(triphenyl-phosphoranylidene)acetate, *in* B-00504
803-17-8 Tris(4-methoxyphenyl)phosphine; Oxide, *in* T-00793
807-19-2 4,4′,4″-Phosphinylidynetris[benzoic acid], *in* T-00740
809-44-9 Tris(4-carboxyphenyl)phosphine; Oxide, tri-Me ester, *in* T-00740
811-62-1 ▷Dimethylphosphinous chloride, D-00846
811-71-2 Dimethylphosphinothioic acid; Fluoride, *in* D-00839
811-79-0 Trifluorodimethylphosphorane, T-00469
811-99-4 ▷*N,P*-Dimethylphosphonamidothioic fluoride, *in* D-00850
812-00-0 ▷Monomethyl phosphate, M-00489
812-13-5 Dimethylphosphoramidothioic difluoride, D-00873
813-76-3 Diethylphosphinic acid, D-00322
813-77-4 Dimethyl phosphorochloridate, D-00887
813-78-5 Dimethyl phosphate, D-00826
814-16-4 Diethyl (3,3-dimethyl-2-oxobutyl)phosphonate, *in* D-00783
814-29-9 ▷Tributylphosphine oxide, T-00376
814-49-3 ▷Diethyl phosphorochloridate, D-00373
814-97-1 Dimethylphosphoramidous difluoride, D-00877
816-80-8 ▷Trimethyl phosphorotrithioite, T-00550
816-94-4 Glycerol 1,2-dioctadecanoate 3-phosphocholine; (*R*)-*form*, *in* G-00034
819-04-5 Ethylphosphoramidic acid; Difluoride, *in* E-00156
819-19-2 Di-*tert*-butylphosphine, D-00108
819-43-2 Dibutyl phosphorochloridate, D-00140
822-39-9 2-Chloro-1,3,2-dioxaphospholane, C-00063
822-47-9 1-Chloro-2,5-dihydro-1*H*-phosphole; 1-Oxide, *in* C-00052
822-68-4 ▷Cyclohexylphosphine, C-00252
823-14-3 1-Chloro-2,3-dihydro-4-methyl-1*H*-phosphole; 1-Oxide, *in* C-00045
823-31-4 2-Ethoxy-1,3,2-dioxaphospholane 2-oxide, E-00035
824-11-3 ▷4-Ethyl-2,6,7-trioxa-1-phosphabicyclo[2.2.2]-octane, E-00196
824-71-5 Cyclohexylmethylphosphinic acid; Fluoride, *in* C-00247
824-72-6 Phenylphosphonic dichloride, P-00223
829-84-5 Dicyclohexylphosphine, D-00210
829-85-6 ▷Diphenylphosphine, D-01037
832-39-3 Dicyclohexylphosphinic acid, D-00211
833-57-8 Di-2-propenyl phenylphosphonite, *in* P-00251
838-85-7 Diphenyl phosphate, D-01036
839-28-1 Benzylmethyl-phenylpropylphosphonium(1+); (*R*)-*form*, Bromide, *in* B-00052
843-24-3 Bis(4-methylphenyl) phosphate, B-00314
845-64-7 Difluorotriphenylphosphorane, D-00419
855-38-9 Tris(4-methoxyphenyl)phosphine, T-00793
859-65-4 ▷1-Phenyl-2-(triphenylphosphoranylidene)-ethanone, P-00333
866-23-9 ▷Diethyl (trichloromethyl)-phosphonate, *in* T-00411
866-32-0 ▷Dibutylphosphinic acid, D-00109
866-53-5 Diethylphosphinothioic acid, D-00330
866-54-6 Diethylphosphinodithioic acid, D-00325
866-68-2 Diisopropylphosphinic acid; Isocyanate, *in* D-00641
867-13-0 ▷Diethyl (ethoxycarbonylmethyl)-phosphonate, *in* D-00241
867-33-4 Dipropylphosphinic acid, D-01205
867-34-5 Dipropylphosphinothioic acid, D-01208
868-85-9 ▷Dimethyl phosphonate, D-00851
869-56-7 Tripropyl phosphorotrithioite, T-00718
870-30-4 Phosphorisocyanatidic dichloride, P-00405
870-52-0 ▷Dibutyl phosphoramidate, D-00133
872-99-1 2-Chloro-1,3,2-dioxaphosphorinane 2-oxide, C-00067
873-13-2 Bicyclic phosphate, *in* T-00588
873-16-5 1-Chloro-2,5-dihydro-3,4-dimethyl-1*H*-phosphole; 1-Oxide, *in* C-00041
873-93-8 2,6,7-Trioxa-1-phosphabicyclo[2.2.2]-octane-4-methanol, T-00590

873-97-2 Neopentylene methylphosphonate, *in* T-00512
873-98-3 2-Chloro-5,5-dimethyl-1,3,2-dioxaphosphorinane 2-sulfide, C-00060
874-65-7 2-Ethoxy-4-methyl-1,3,2-dioxaphosphorinane, E-00041
875-12-7 ▷2,8,9-Trioxa-1-phosphatricyclo[3.3.1.1^{3,7}]decane; 1-Oxide, *in* T-00591
876-48-2 Neopentylene dimethylphosphoramidate, *in* D-00693
876-95-4 Dodecyldimethylphosphine; Oxide, *in* D-01264
878-17-1 4-Methylphenyl phosphorodichloridate, M-00257
880-68-2 1,4-Benzenediphosphonic acid, B-00008
882-58-6 ▷Phosphemide, P-00353
882-69-9 Neopentylene phenylphosphonate, *in* D-00795
884-74-2 (Diphenylphosphino)methanol; Oxide, *in* D-01069
884-89-9 Neopentylene phenyl phosphate, *in* D-00790
889-57-6 3-(Diphenylphosphino)-1-propanol; Oxide, *in* D-01074
896-33-3 Ethyltriphenylphosphonium(1+); Chloride, *in* E-00197
896-84-4 5,10-Dihydro-5-methyl-10-phenylphenophosphazine, D-00526
896-89-9 (4-Methoxyphenyl)diphenyl-phosphine, M-00058
900-99-2 Tris(4-bromophenyl)phosphine; Oxide, *in* T-00735
903-22-0 Hexamethyl 1,3,5-triazine-2,4,6-triyltriphosphonate, *in* T-00329
906-65-0 Dihydro(triphenyl-phosphoranylidene)-2,5-furandione, D-00612
912-40-3 Hexaethyl 1,3,5-triazine-2,4,6-triyltriphosphonate, *in* T-00329
918-47-8 ▷Tetramethylphosphorodiamidothioic fluoride, T-00226
919-19-7 Diethyl acetylphosphonate, *in* A-00026
919-44-8 Monocrotophos; (*Z*)-*form*, *in* M-00459
919-48-2 Tetrabutyl 1,2-ethanediylbisphosphonate, *in* E-00006
919-54-0 ▷Acetophos, A-00004
919-62-0 Tris(3-methylbutyl) phosphate, T-00796
919-76-6 ▷Amidithion, A-00051
919-86-8 ▷Demeton-*S*-methyl, D-00020
921-26-6 Diisopropylphosphoramidous acid; Dichloride, *in* D-00656
921-27-7 Diisopropylphosphoramidous acid; Difluoride, *in* D-00656
923-61-5 Glycerol 1,2-dihexadecanoate 3-phosphoethanolamine; (*R*)-*form*, *in* G-00013
923-99-9 ▷Tripropyl phosphite, T-00712
926-46-5 Pentylphosphonic acid; Dichloride, *in* P-00050
928-64-3 Hexylphosphonic dichloride, H-00098
932-74-1 4-Morpholinylphosphonous acid; Dichloride, *in* M-00529
933-43-7 Methyl dimethylvinylene phosphate, *in* H-00123
934-35-0 *o*-Phenylene phosphite, *in* B-00018
935-52-4 ▷Ethyl bicyclic thiophosphate, *in* E-00196
935-59-1 *S*-Methyl *O,O*-neopentylene phosphorodithioate, *in* D-00766
940-18-1 3-Methylphenyl phosphorodichloridate, *in* M-00486
941-39-9 *O,O*-Dimethyl cyclohexylphosphoramidothioate, *in* C-00271
944-21-8 ▷Fonofos oxon, *in* E-00074
944-22-9 ▷Fonofos, F-00052
944-23-0 Neopentylene *tert*-butylphosphoramidate, *in* B-00538
947-02-4 ▷Phosfolan, P-00342
947-28-4 3,9-Dihydroxy-2,4,8,10-Tetraoxa-3,9-diphosphaspiro[5.5]-undecane 3,9-dioxide, D-00626
950-10-7 ▷Mephosfolan, M-00004
950-35-6 ▷Dimethyl 4-nitrophenyl phosphate, D-00770
950-37-8 ▷Methidathion, M-00025
953-17-3 ▷Methyl trithion, M-00437
953-60-6 2,2′,2″-Phosphinylidynetrispyrrole, *in* T-00726
957-07-3 2,2,4-Trichloro-4,6,6-tris(dimethylamino)-2,2,4,4,6,6-hexahydro-1,3,5-triaza-2,4,6-triphosphorine, T-00421
957-10-8 2,4,6-Trichloro-2,4,6-tris(dimethylamino)-2,2,4,4,6,6-hexahydro-1,3,5-triaza-2,4,6-triphosphorine; (2α,4α,6β)-*form*, *in* T-00422
961-11-5 ▷Tetrachlorvinphos, T-00038
963-05-3 2,4-Dichloro-2,4,6,6-tetrakis(dimethylamino)-2,2,4,4,6,6-hexahydro-1,3,5-triaza-2,4,6-triphosphorine; (4*RS*,6*SR*)-*form*, *in* D-00194
974-68-5 ▷2,2,4,4,6,6-Hexakis(dimethylamino)-2,2,4,4,6,6-hexahydro-1,3,5-triaza-2,4,6-triphosphorine, H-00061
976-56-7 Diethyl [(3,5-di-*tert*-butyl-4-hydroxyphenyl)methyl]-phosphonate, *in* D-00094
983-81-3 1,2-Bis(diphenylphosphino)ethylene; (*E*)-*form*, *in* B-00237
983-90-2 1,2-Bis(diphenylphosphino)ethylene; (*Z*)-*form*, *in* B-00237
986-06-1 1,2-Bis(diphenylphosphino)ethylene; (*Z*)-*form*, *P,P′*-Disulfide, *in* B-00237
986-07-2 1,2-Bis(diphenylphosphino)ethylene; (*E*)-*form*, *P,P′*-Disulfide, *in* B-00237
987-78-0 ▷Cytidine diphosphate choline, C-00291
990-91-0 Tetrabenzyl diphosphate, T-00006
993-11-3 ▷Tetramethylphosphonium(1+); Iodide, *in* T-00215
993-12-4 Dimethylphosphinothioic chloride, D-00842
993-13-5 ▷Methylphosphonic acid, M-00288
993-43-1 Ethylphosphonothioic dichloride, E-00150
993-44-2 *O*-Methyl methylphosphonodithioate, M-00170
993-86-2 ▷Diethyl (2,2,2-trichloro-1-hydroxyethyl)phosphonate, *in* T-00408
993-88-4 Methyl 2-(diethoxyphosphinyl)-propenoate, *in* D-00260
995-17-5 Methyl *N*-(dimethoxyphosphinyl)-carbamate, *in* D-00680
995-32-4 Tetraethyl 1,2-ethanediylbisphosphonate, *in* E-00006
995-37-9 Ethyl 3-(diethoxyphosphinyl)-propenoate, *in* D-00261
995-79-9 *O*-Ethyl ethylphosphonodithioate, E-00070
995-87-9 Ethyl phosphorochloridisocyanatidate, E-00163
995-88-0 *S,S*-Diethyl methylphosphonodithioate, *in* M-00301
996-42-9 Diisopropyl (2,2,2-trichloro-1-hydroxyethyl)phosphonate, *in* T-00408
998-06-1 Glycerol 1,2-di-9,12-octadecadienoate 3-phosphocholine; (*R*)-L-(*all*-*Z*)-*form*, *in* G-00028
998-07-2 Glycerol 1,2-ditetradecanoate 3-phosphoethanolamine; (*R*)-*form*, *in* G-00049
998-40-3 ▷Tributylphosphine, T-00372
998-80-1 Ethyl ethylphosphinate, E-00068
999-01-9 *O,O*-Diethyl phosphonothioate, D-00341
999-02-0 Diethyl methylphosphonodithioite, *in* M-00302
999-82-6 (2-Bromoethyl)phosphonic acid, B-00472
999-83-7 *O*-Ethyl methylphosphonodithioate, E-00091
999-87-1 *O*-Isopropyl methylphosphonodithioate; (±)-*form*, *in* I-00040
1000-53-9 *O*-Propyl methylphosphonodithioate, P-00469
1003-11-8 ▷2-Hydroxy-1,3,2-dioxaphospholane, *in* D-00970
1003-11-8 ▷2-Hydroxy-1,3,2-dioxaphospholane, H-00133
1003-18-5 1-Chloro-2,3-dihydro-1*H*-phosphole; 1-Oxide, *in* C-00051
1004-78-0 Dichloro(2-chlorophenyl)phosphine, *in* C-00151
1005-22-7 Cyclohexylphosphonic dichloride, C-00258
1005-23-8 Cyclohexylphosphonic acid, C-00256
1005-32-9 (4-Methylphenyl)phosphonous dichloride, M-00249
1005-33-0 (4-Chlorophenyl)phosphonous dichloride, C-00154
1005-69-2 2-Methoxy-5,5-dimethyl-1,3,2-dioxaphosphorinane, M-00034
1005-93-2 ▷Ethyl bicyclic phosphate, *in* E-00196
1005-95-4 1-Ethoxy-2,5-dihydro-3,4-dimethyl-1*H*-phosphole 1-oxide, *in* D-00493
1005-96-5 Methyl neopentylene phosphate, *in* M-00034
1005-97-6 *O*-Methyl *O,O*-neopentylene phosphorothioate, *in* M-00034
1005-98-7 *S*-Methyl *O,O*-neopentylene phosphorothioate, *in* D-00766
1006-83-3 2-Phenyl-1,3,2-dioxaphospholane, P-00115
1007-57-4 2-Ethoxy-5,5-dimethyl-1,3,2-dioxaphosphorinane, E-00032
1007-80-3 Ethyl neopentylene phosphate, *in* E-00032
1007-92-7 *O*-Ethyl hydrogen cyclohexylphosphonodithioate, *in* C-00261
1007-94-9 *O*-Ethyl phenylphosphonodithioate, E-00113
1012-43-7 Diethyl 1-methyl-1*H*-pyrrol-2-ylphosphonate, *in* P-00521
1015-28-7 Dimethyl (2-oxo-2-phenylethyl)-phosphonate, *in* O-00066
1015-37-8 Diphenylphosphinothioic chloride, D-01084
1015-38-9 Diphenylphosphinodithioic acid, D-01065
1017-60-3 Bis(4-methylphenyl)phosphine, B-00317
1017-88-5 Dimethyldiphenylphosphonium iodide, *in* M-00135
1017-89-6 Trichlorodiphenylphosphorane, T-00404
1017-98-7 Ethyldiphenylphosphine; Sulfide, *in* E-00066
1018-07-1 2,2,4,6-Tetrachloro-4,6-bis(dimethylamino)-2,2,4,4,6,6-hexahydro-1,3,5-triaza-2,4,6-triphosphorine; (4*RS*, 6*SR*)-*form*, *in* T-00028
1018-24-2 Diethyl (2-phenylethenyl)-phosphonate, *in* P-00155
1019-71-2 Chlorobis(4-methylphenyl)-phosphine, *in* B-00325

1021-20-1 Tri-2-furanylphosphine; Oxide, *in* T-00486
1021-21-2 Tri-2-thienylphosphine; Oxide, *in* T-00874
1024-06-2 2,4-Dicyclohexyl-1,3,2,4-dithiadiphosphetane 2,4-disulfide, D-00207
1024-07-3 2,4-Diphenyl-1,3,2,4-dithiadiphosphetane 2,4-disulfide, D-01019
1031-13-6 5-Phenyl-5*H*-dibenzophosphole; 5-Oxide, *in* P-00110
1031-15-8 Methyltriphenylphosphonium(1+); Chloride, *in* M-00432
1031-47-6 ▷Triamiphos, T-00326
1031-93-2 (4-Methylphenyl)diphenylphosphine, M-00202
1034-39-5 Dibromotriphenylphosphorane, D-00071
1034-49-7 (Bromomethyl)-triphenylphosphonium(1+); Bromide, *in* B-00477
1034-94-2 Dibutyl (2-oxo-2-phenylethyl)-phosphonate, *in* O-00066
1038-95-5 ▷Tris(4-methylphenyl)phosphine, T-00808
1041-99-2 4-Diphenylphosphinothioyl-*N*,*N*-dimethylaniline, *in* D-01061
1053-73-2 Adenosine 3′,5′-diphosphate, A-00040
1054-59-7 Tetraphenyldiphosphine dioxide, T-00259
1054-60-0 Tetraphenyldiphosphine disulfide, T-00260
1066-34-8 Methylphosphonous dibromide, M-00321
1066-50-8 ▷Ethylphosphonic dichloride, E-00131
1066-51-9 (Aminomethyl)phosphonic acid, A-00091
1066-52-0 Methyl methylphosphonochloridate, *in* M-00296
1066-53-1 Methyl methylphosphonate, M-00168
1066-97-3 *O*,*O*-Dimethyl phosphorodithioate; NH_4 salt, *in* D-00898
1067-01-2 (Pentachloroethyl)phosphorimidic trichloride, P-00005
1067-12-5 Phosphinylidynetrismethanol, *in* T-00784
1067-70-5 Diisopropyl allylphosphonate, *in* P-00454
1067-71-6 Diethyl (2-oxopropyl)phosphonate, *in* O-00075
1067-72-7 *O*,*O*-Diethyl (2-oxopropyl)-phosphonothioate, *in* O-00079
1067-74-9 Diethyl (methoxycarbonylmethyl)-phosphonate, *in* D-00241
1067-87-4 Diethyl 2-propenylphosphonate, *in* P-00454
1067-88-5 Diethyl 1-propynylphosphonate, *in* P-00494
1068-07-1 Diethyl octylphosphonate, *in* O-00042
1068-13-9 Ethyl [(dimethoxyphosphinothioyl)-thio]acetate, *in* D-00675
1068-20-8 Methylphosphonic acid; Diisocyanate, *in* M-00288
1068-21-9 Diethyl phosphoramidate, D-00342
1068-22-0 ▷*O*,*O*-Diethyl phosphorodithioate; NH_4 salt, *in* D-00384
1068-59-3 Dibromoethylphosphine, *in* E-00153
1069-08-5 Diethylphosphoramidous dichloride, D-00361
1070-03-7 Mono(2-ethylhexyl) phosphate, M-00470
1070-42-4 ▷Bis(2-chloroethyl) phosphonate, B-00120
1070-90-2 Dimethyldiphosphonic acid, D-00751
1071-83-6 ▷Glyphosate, G-00097
1073-47-8 Phenylphosphonous dibromide, P-00253
1073-52-5 1-Piperidinylphosphonous acid; Difluoride, *in* P-00432
1073-98-9 Pentamethylpentaphospholane, P-00019
1076-00-2 *O*-Ethyl *O*,*O*-neopentylene phosphorothioate, *in* E-00032
1077-05-0 2-Phenoxy-1,3,2-dioxaphospholane, P-00067
1078-57-5 2-Phenoxy-1,3,2-dioxaphosphorinane, P-00068
1078-84-8 Phenylphosphonic diisocyanate, P-00226
1079-65-8 Bromodiphenylphosphine, *in* D-01089
1079-66-9 ▷Diphenylphosphinous chloride, D-01091
1080-32-6 ▷Diethyl benzylphosphonate, D-00272
1081-87-4 2,2,4,6-Tetrachloro-4,6-bis(dimethylamino)-2,2,4,4,6,6-hexahydro-1,3,5-triaza-2,4,6-triphosphorine; (4*RS*, 6*RS*)-*form*, *in* T-00028
1083-44-9 4,4,6-Trimethyl-2-phenoxy-1,3,2-dioxaphosphorinane, T-00520
1084-11-3 Bis(4-methylphenyl)phosphinic acid, B-00320
1088-00-2 5-Phenyl-5*H*-dibenzophosphole, P-00110
1089-77-6 1,2-Ethanediylbis[phenylphosphinic acid], E-00015
1091-27-6 10-Phenyl-10*H*-phenoxaphosphine; 10-Oxide, *in* P-00183
1095-04-1 Triphenyl phosphorotrithioite, T-00676
1097-84-3 2,2,4,4,6,6-Hexakis(dimethylamino)-2,2,4,4,6,6-hexahydro-1,3,5-triaza-2,4,6-triphosphorine; B, HCl, *in* H-00061
1099-45-2 Ethyl triphenyl-phosphoranylideneacetate, E-00198
1100-88-5 Benzyltriphenylphosphonium(1+); Chloride, *in* B-00073
1101-41-3 Tetraphenyldiphosphine, T-00258
1104-52-5 Tetraphenyltetraphosphetane, T-00278
1104-78-5 Dimethyl (triphenylphosphoranylidene)-butanedioate, *in* T-00646
1106-06-5 Methyl α-(triphenylphosphoranylidene)-benzeneacetate, *in* T-00644
1110-78-7 2,2,4,4,6,6-Hexahydro-2,2,4,4,6,6-hexaphenyl-1,3,5-triaza-2,4,6-triphosphorine, H-00032
1111-92-8 Dimethylphosphinic chloride, D-00831
1111-98-4 (Chloromethyl)phosphonic difluoride, C-00109
1112-03-4 (Trichloromethyl)phosphonous acid; Difluoride, *in* T-00416
1112-04-5 Trifluoromethylphosphonous difluoride, T-00484
1112-15-8 Diisopropylphosphinic acid; Chloride, *in* D-00641
1112-37-4 Diethylphosphinic acid; Chloride, *in* D-00322
1112-38-5 *O*,*O*-Dimethyl phosphorothioate, D-00907
1112-94-3 Methyl 3-(diethoxyphosphinyl)-propanoate, *in* D-00259
1112-96-5 Tetraethyl (2-butene-1,4-diyl)-bisphosphonate, *in* B-00519
1113-01-5 ▷[(Dimethoxyphosphinothioyl)-thio]acetic acid, D-00675
1113-02-6 ▷Omethoate, O-00048
1113-11-7 Dipropylphosphinic acid; Chloride, *in* D-01205
1114-81-4 2-Amino-3-phosphonooxybutanoic acid; (2*S*,3*R*)-*form*, *in* A-00120
1118-29-2 Bis(2-methylpropyl)phosphinic acid; Chloride, *in* B-00346
1121-81-9 2,4-Dimethyl-1,3,2,4-dithiadiphosphetane 2,4-disulfide, D-00752
1124-70-5 2,4-Diethyl-1,3,2,4-dithiadiphosphetane 2,4-disulfide, *in* E-00152
1126-52-9 Phenyl phosphorodifluoridate, P-00285
1127-41-9 (Hydroxyphenylmethyl)phosphonic acid, H-00174
1135-57-5 2,4-Dibutyl-1,3,2,4-dithiaphosphetane 2,4-disulfide, *in* B-00614
1135-98-4 Diphenylphosphinic fluoride, D-01042
1138-99-4 Trifluorodiphenylphosphorane, T-00471
1145-93-3 Diethyl [(4-methoxyphenyl)-methyl]phosphonate, *in* M-00061
1159-54-2 Tris(4-chlorophenyl)phosphine, T-00749
1162-70-5 1,2,5-Triphenyl-1*H*-phosphole, T-00634
1162-90-9 1-Naphthyldiphenylphosphine, N-00011
1174-61-4 1-Phenyl-3-(triphenylphosphoranylidene)-2-propanone, P-00336
1179-05-1 1,3-Phenylenebis[diphenyl-phosphine], P-00127
1179-06-2 1,4-Phenylenebis[diphenyl-phosphine], P-00128
1181-62-0 Pentaphenyl-1*H*-phosphole, P-00038
1183-90-0 1,2-Dihydro-1,2-bis(triphenylphosphonio)-benzocyclobutene(2+); (1*RS*,2*RS*)-*form*, Dibromide, *in* D-00447
1184-10-7 2,2,4,4,6,6-Hexahydro-2,2,4,4,6,6-hexaphenoxy-1,3,5-triaza-2,4,6-triphosphorine, H-00030
1184-81-2 Tetrafluoro(trifluoromethyl)-phosphorane, T-00086
1184-82-3 Trifluorobis(trifluoromethyl)-phosphorane, T-00465
1186-08-9 ▷Diethyl chlorothiophosphonate, D-00274
1186-09-0 ▷*O*,*O*,*S*-Triethyl phosphorothioate, T-00457
1186-10-3 Diethyl (3-bromopropyl)-phosphonate, *in* B-00501
1186-17-0 Diethyl pentylphosphonate, *in* P-00050
1186-23-8 Diethyl (3-acetyloxypropyl)-phosphonate, *in* A-00010
1189-11-3 Phosphoarginine; (*S*)-*form*, *in* P-00363
1189-24-8 Bis(2-methylpropyl) phosphonate, B-00354
1189-87-3 ▷Isopropyl ethylphosphonofluoridate, *in* E-00142
1190-09-6 (3-Bromopropyl)phosphonic acid, B-00501
1190-28-9 [(Dimethoxyphosphinothioyl)-thio]butanedioic acid, D-00677
1193-39-1 2-Ethoxy-1,2-oxaphospholane 2-oxide, *in* H-00166
1194-41-8 2-Ethoxy-1,2-oxaphosphorinane 2-oxide, *in* H-00167
1194-52-1 *P*,*P*-Bis(1-aziridinyl)-*N*,*N*-dimethylphosphinous amide, *in* B-00096
1194-53-2 ▷Tri-(1-aziridinyl)phosphine, T-00331
1195-18-2 *P*-1-Aziridinyl-*N*,*N*,*N*′,*N*′-tetramethylphosphonous diamide, *in* A-00155
1195-67-1 ▷*P*-1-Aziridinyl-*N*,*N*,*N*′,*N*′-tetramethylphosphonic diamide, *in* A-00153
1195-69-3 ▷*P*,*P*-Bis(1-aziridinyl)-*N*,*N*-dimethylphosphinic amide, *in* B-00092
1205-62-5 [(4-Nitrophenyl)methyl]phosphonic acid, N-00046
1212-80-2 (Hydroxyphenylmethyl)phosphonic acid; (±)-*form*, Monoanilinium salt, *in* H-00174
1216-38-2 2,2,6,6-Tetramethyl-1-phenyl-4-phosphorinanone; Sulfide, *in* T-00210
1216-69-9 2,4,6-Trichloro-2,4,6-tris(dimethylamino)-2,2,4,4,6,6-hexahydro-1,3,5-triaza-2,4,6-triphosphorine; (2α,4α,6α)-*form*, *in* T-00422

1225-16-7 10-Phenyl-10*H*-phenoxaphosphine, P-00183
1231-03-4 *P,P'*-Diphenyl diphosphate, D-01011
1235-21-8 ▷(2-Oxopropyl)-triphenylphosphonium(1+); Chloride, *in* O-00080
1241-94-7 ▷2-Ethylhexyl diphenyl phosphate, E-00075
1243-97-6 ▷Benzyltriphenylphosphonium(1+); Iodide, *in* B-00073
1253-46-9 [[4-(Methoxycarbonyl)phenyl]-methyl]triphenyl-phosphonium(1+); Bromide, *in* M-00031
1277-92-5 1',1'',1'''-Phosphinylidynetrisferrocene, *in* T-00464
1292-82-6 Triferrocenylphosphine, T-00464
1426-00-2 (Chloromethyl)phosphonothioic acid; Difluoride, *in* C-00113
1426-40-0 Fluorobis(trifluoromethyl)-phosphine, *in* B-00418
1426-81-9 Cyclohexyl phenylphosphonofluoridate, *in* P-00239
1429-50-1 [1,2-Ethanediylbis[nitrilobis[methyl-ene]]]tetrakisphosphonic acid, E-00014
1438-74-0 Vinylphosphonic dichloride, V-00005
1439-36-7 ▷1-Triphenylphosphoranylidene-2-propanone, T-00662
1439-41-4 3,3',3''-Phosphinylidynetris[propaneni-trile], *in* T-00752
1439-43-6 (4-Nitrophenyl-2-oxoethylidene)-triphenylphosphorane, N-00050
1443-62-5 5,5'-Spirobi[5*H*-dibenzophosphole], S-00021
1445-36-9 Phenylphosphoramidic acid, P-00257
1445-38-1 Diethyl phenylphosphoramidate, D-00312
1445-75-6 ▷Diisopropyl methylphosphonate, *in* M-00288
1445-76-7 Methylphosphonochloridic acid; (±)-*form*, Isopropyl ester, *in* M-00296
1447-25-2 Tetraphenyltetraphosphetane; Tetrasulfide, *in* T-00278
1449-46-3 Benzyltriphenylphosphonium(1+); Bromide, *in* B-00073
1449-89-4 ▷Trimethylolethane phosphate, *in* M-00424
1449-91-8 ▷4-Methyl-2,6,7-trioxa-1-phosphabicyclo[2.2.2]-octane, M-00424
1450-07-3 1-Phenyl-2-(triphenylphosphoranylidene)-1-propanone, P-00335
1455-05-5 2-Chloroethyl phosphorodichloridate, *in* M-00454
1455-07-8 Pentamethoxyphosphorane, P-00018
1467-28-3 Diethyl phenylsulfonylphosphoramidate, *in* P-00312
1474-08-4 (4-Chlorophenyl)diphenylphosphine; Sulfide, *in* C-00129
1474-09-5 (4-Bromophenyl)diphenylphosphine; Sulfide, *in* B-00481
1474-24-4 1,3-Diphenyl-2-(triphenylphosphoranylidene)-1,3-propanedione, D-01166
1474-31-3 Ethyl *β*-oxo-*α*-(triphenylphosphoranylidene)-benzenepropanoate, *in* O-00083
1474-32-4 3-(Triphenylphosphoranylidene)-2,4-pentanedione, T-00654
1474-75-5 ▷Tetrabutyl pyrophosphate, T-00024
1474-78-8 Triethyl phosphonoformate, *in* D-00236
1478-39-3 1,1,1-Trifluoro-1,1-dihydrophosphorinane, *in* D-00573
1485-88-7 (2-Methoxyphenyl)-methylphenylphosphine, M-00059
1486-18-6 Methyl bis(trifluoromethyl)-phosphinothioite, *in* B-00417
1486-19-7 Bis(trifluoromethyl)-phosphinothious acid, B-00417
1486-20-0 Bis(trifluoromethyl)-phosphinothious acid; Anhydrosulfide, *in* B-00417
1486-28-8 Methyldiphenylphosphine, M-00135
1486-37-9 Diethyl phenylphosphonodithioite, *in* P-00238
1486-38-0 Butyl diphenylphosphinothioite, *in* D-01085
1486-39-1 ▷*S,S,S*-Triethyl phosphorotrithioate, T-00460
1486-40-4 *S*-Ethyl phosphorodichloridothioate, E-00171
1486-42-6 Diethyl phosphorochloridodithioite, D-00376
1486-43-7 Ethyl phosphorodichloridothioite, E-00172
1490-12-6 Dimethyl (cyclohexylcarbonyl)-phosphonate, *in* C-00239
1490-14-8 Diethyl (cyclohexylcarbonyl)-phosphonate, *in* C-00239
1491-22-1 1,6-Dioxa-4,9-diaza-5-phospha(5-P^V)spiro[4.4]-nonane, D-00963
1491-41-4 ▷Naphthalophos, N-00008
1492-08-6 1,1,3,7-Tetrahydroxy-7-methyl-2,4-dioxa-1,3-diphosphanonan-9-oic acid 1,3-dioxide, T-00162
1495-32-5 Trifluoromethanethiol phosphorodichloridite, T-00472
1495-54-1 Phosphorisocyanatidic difluoride, P-00406
1495-59-6 Bis(3-fluorophenyl)phosphinic acid, B-00279
1496-94-2 Tripropylphosphine; Oxide, *in* T-00711
1497-32-1 *O,O*-Diethyl *S*-benzoyl phosphorodithioate, D-00270
1497-67-2 *O*-Propyl ethylphosphonochloridothioate, *in* E-00137
1497-68-3 *O*-Ethyl ethylphosphonochloridothioate, E-00069
1497-69-4 *O*-Methyl ethylphosphonochloridothioate, M-00154
1498-40-4 ▷Ethylphosphonous dichloride, E-00155
1498-42-6 ▷Ethyl phosphorodichloridite, E-00168
1498-46-0 Isopropylphosphonic dichloride, I-00052
1498-47-1 2-Propenylphosphonic dichloride, P-00456
1498-51-7 ▷Ethyl phosphorodichloridate, E-00167
1498-52-8 ▷Butyl phosphorodichloridate, B-00628
1498-54-0 Diethylphosphoramidic dichloride, D-00344
1498-56-2 ▷1-Piperidinylphosphonic acid; Dichloride, *in* P-00430
1498-57-3 4-Morpholinylphosphonic acid; Dichloride, *in* M-00523
1498-60-8 Isopropylphosphonothioic dichloride, I-00055
1498-61-9 2-Propenylphosphonothioic acid; Dichloride, *in* P-00461
1498-62-0 Cyclopentylphosphonothioic acid; Dichloride, *in* C-00280
1498-63-1 Cyclohexylphosphonothioic dichloride, C-00266
1498-64-2 *O*-Ethyl phosphorodichloridothioate, E-00170
1498-65-3 Dimethylphosphoramidothioic dichloride, D-00872
1499-17-8 2-Chloro-1,3,2-benzodioxaphosphole; Oxide, *in* C-00021
1499-19-0 Benzylphosphonic dichloride, B-00062
1499-21-4 Diphenylphosphinic chloride, D-01041
1499-29-2 Methylenebis(phosphonic dichloride), M-00145
1499-30-5 1,2-Ethanediylbis[phosphonic dichloride], *in* E-00006
1499-31-6 1,3-Propanediphosphonic acid; Bis-dichloride, *in* P-00444
1499-32-7 Methanediphosphonothioic acid; Tetrachloride, *in* M-00023
1509-81-5 Glycerol 1-monophosphate; (±)-*form*, *in* G-00066
1511-67-7 ▷Methylphosphonofluoridic acid, M-00303
1513-46-8 Phenyl(trifluoromethyl)phosphinic acid, P-00329
1515-99-7 Methylphenylpropylphosphine; (*R*)-*form*, Oxide, *in* M-00265
1516-67-2 ▷*O,O*-Diethyl phosphorazidothioate, D-00363
1516-68-3 ▷Diethyl phosphorazidate, D-00362
1516-96-8 *N,N'*-Di-*tert*-butyl-*P*-phenylphosphonous diamide, *in* P-00252
1519-45-5 ▷1,2-Ethanediylbis[triphenyl-phosphonium](2+); Dibromide, *in* E-00017
1519-46-6 1,2-[Phenylenebis[methylene]]-bis[triphenylphosphonium](2+); Dibromide, *in* P-00132
1519-47-7 ▷1,4-[Phenylenebis[methylene]]-bis[triphenylphosphonium](2+); Dichloride, *in* P-00134
1523-68-8 Diethyl (1-oxopropyl)phosphonate, *in* O-00074
1526-24-5 2-Fluoro-1,3,2-benzodioxaphosphole, F-00013
1527-34-0 4-(Diethoxyphosphinyl)benzoic acid, D-00244
1530-32-1 Ethyltriphenylphosphonium(1+); Bromide, *in* E-00197
1530-33-2 Isopropyltriphenylphosphonium(1+); Bromide, *in* I-00065
1530-34-3 (3-Methyl-2-butenyl)-triphenylphosphonium(1+); Bromide, *in* M-00104
1530-36-5 ▷[(2-Methylphenyl)methyl]-triphenylphosphonium(1+); Bromide, *in* M-00209
1530-37-6 [(4-Methylphenyl)methyl]-triphenylphosphonium(1+); Chloride, *in* M-00211
1530-38-7 [(4-Methoxyphenyl)methyl]-triphenylphosphonium(1+); Bromide, *in* M-00064
1530-39-8 [(4-Chlorophenyl)methyl]-triphenylphosphonium(1+); Chloride, *in* C-00137
1530-41-2 [(3-Nitrophenyl)methyl]-triphenylphosphonium(1+); Bromide, *in* N-00048
1530-42-3 [(4-Nitrophenyl)methyl]-triphenylphosphonium(1+); Chloride, *in* N-00049
1530-44-5 (Carboxymethyl)-triphenylphosphonium(1+); Bromide, *in* C-00014
1530-45-6 (2-Ethoxy-2-oxoethyl)-triphenylphosphonium; Bromide, *in* E-00047
1535-37-1 (4-Chlorophenyl)phosphonic difluoride, C-00149
1536-92-1 Diisopropylphosphinic acid; Fluoride, *in* D-00641
1538-69-8 Diethyl isopropylphosphonate, *in* I-00050
1538-72-3 Diphenyl isopropylphosphonate, *in* I-00050
1552-41-6 Diethyl [(4-cyanophenyl)methyl]-phosphonate, *in* C-00221
1559-55-3 Diethyl acetylphosphoramidate, *in* A-00029
1560-54-9 ▷Triphenyl-2-propenylphosphonium(1+); Bromide, *in* T-00679
1566-15-0 ▷*N,N*-Bis(2-chloroethyl)-phosphorodiamidic acid; Mono(cyclohexylammonium) salt, *in* B-00126

1571-33-1 ▷Phenylphosphonic acid, P-00221
1604-65-5 Dipentylphosphinic acid; Chloride, *in* D-00981
1605-53-4 Diethylphenylphosphine, D-00305
1605-65-8 Tetramethylphosphorodiamidic chloride, T-00220
1606-75-3 Diethyl (2-oxoethyl)phosphonate, *in* O-00059
1606-96-8 Tris(2-methylpropyl) phosphite, T-00826
1608-26-0 ▷Hexamethylphosphorous triamide, H-00080
1609-72-9 Diethyl 1,2-propadienylphosphonate, *in* P-00441
1623-06-9 Monopropyl phosphate, M-00512
1623-07-0 Monobenzyl phosphate, M-00443
1623-08-1 Dibenzyl phosphate, D-00053
1623-12-7 Phenyl di-2-propenyl phosphate, P-00118
1623-14-9 Monoethyl phosphate, M-00472
1623-15-0 Monobutyl phosphate, M-00451
1623-19-4 ▷Tri-2-propenyl phosphate, T-00705
1623-24-1 ▷Monoisopropyl phosphate, M-00477
1635-63-8 2,2,4,4,6,6-Hexahydro-2,2,4,4,6,6-hexakis(methylamino)-1,3,5-triaza-2,4,6-triphosphorine, H-00027
1636-14-2 *N,N,N′,N′*-Tetraethyl-*P*-phenylphosphonous diamide, *in* P-00252
1636-15-3 *N,N*-Diethyl-*P,P*-diphenylphosphinous amide, *in* D-01090
1638-86-4 Diethyl phenylphosphonite, D-00311
1639-18-5 ▷*O,O*-Diethyl chlorothiophosphonothioate, D-00275
1641-40-3 2-Chloro-1,3,2-benzodioxaphosphole, C-00021
1641-57-2 Bis(trimethylsilyl) ethylphosphonate, *in* E-00129
1641-61-8 Dimethyl cyclohexylphosphonate, *in* C-00256
1641-63-0 Pentaphenyl-1*H*-phosphole; Oxide, *in* P-00038
1642-43-9 Triethyl phosphorotetrathionate, T-00456
1642-44-0 *S,S,S*-Tripropyl phosphorotrithioate, T-00717
1642-47-3 Tributyl phosphorotetrathioate, T-00386
1642-48-4 Tripropyl phosphorotetrathioate, T-00714
1648-37-9 *O,O*-Diphenyl phosphorofluoridothioate, D-01134
1648-39-1 Diphenylphosphinothioic acid; Fluoride, *in* D-01082
1650-91-1 Benzyldiphenylphosphine, B-00046
1656-65-1 Tribenzyl phosphorotrithioite, T-00343
1660-94-2 ▷Tetraethyl methylenebisphosphonate, *in* M-00022
1660-95-3 Tetrakis(1-methylethyl) methylenebisphosphonate, *in* M-00022
1660-97-5 1,4-Butanediphosphonic acid; Bisdichloride, *in* B-00514
1661-12-7 1,2-Ethanediphosphonothioic acid; Tetrachloride, *in* E-00007
1661-16-1 1-Methylphosphorinane; Sulfide, *in* M-00339
1661-17-2 1-Methylphospholane; Sulfide, *in* M-00280
1661-29-6 Meturedepa, M-00439
1663-45-2 ▷1,2-Bis(diphenylphosphino)ethane, B-00234
1665-79-8 2,2-Dihydro-2,2,2-trimethoxy-4,5-dimethyl-1,3,2-dioxaphosphole, D-00593
1666-10-0 Diisopropyl phenylphosphoramidate, *in* P-00257
1666-79-8 Hexylidenetriphenylphosphorane, *in* H-00104
1678-18-8 (2-Oxo-2-phenylethyl)-triphenylphosphonium(1+); Chloride, *in* O-00067
1679-91-0 2,4-Dichloro-1,3-dimethyl-1,3,2,4-diazadiphosphetidine, D-00166
1698-62-0 Dimethyl (trifluoromethyl)-phosphonodithioite, *in* T-00480
1702-41-6 [(3-Methylphenyl)methyl]-triphenylphosphonium(1+); Bromide, *in* M-00210
1706-90-7 Methyl diphenylphosphinate, *in* D-01039
1706-91-8 Isopropyl diphenylphosphinate, *in* D-01039
1706-92-9 *tert*-Butyl diphenylphosphinate, *in* D-01039
1706-96-3 ▷Phenyl diphenylphosphinate, *in* D-01039
1706-99-6 (Diphenylphosphinothioyl)acetic acid, D-01086
1707-00-2 Dimethylphenylphosphine; Sulfide, *in* D-00803
1707-03-5 ▷Diphenylphosphinic acid, D-01039
1707-08-0 2-Phenylethenylphosphonic acid, P-00155
1707-71-7 *P,P′*-Diethyl diphosphate, D-00289
1707-92-2 Tribenzyl phosphate, T-00337
1718-21-4 Dichloro(3-chlorophenyl)phosphine, *in* C-00152
1725-19-5 Diethyl 2-thienylphosphonite, *in* T-00308
1726-72-3 Dichloro-2-thienylphosphine, *in* T-00308
1732-66-7 Bis(2,4,6-trimethylphenyl)-phosphine, B-00426
1732-72-5 Dibutylphosphine, D-00107
1732-74-7 Butylphosphine, B-00588
1733-46-6 *tert*-Butyl bis(trifluoromethyl)-phosphinothioite, *in* B-00417
1733-52-4 1-(Diphenylphosphinyl)-2-propanone, *in* D-01075
1733-55-7 Ethyl diphenylphosphinate, *in* D-01039
1733-57-9 Ethyldiphenylphosphine; Oxide, *in* E-00066
1735-82-6 Tetramethylphosphorodiamidous acid; Fluoride, *in* T-00227
1735-83-7 Bis(dimethylamino)-trifluorophosphorane, B-00196
1743-62-0 Methyl 3-oxo-2-(triphenylphosphoranylidene)-butanoate, *in* O-00084
1746-03-8 Vinylphosphonic acid, V-00003
1754-42-3 Diethyl (4-nitrophenyl)-phosphonate, *in* N-00057
1754-43-4 Diethyl (4-dimethylaminophenyl)-phosphonate, *in* D-00706
1754-46-7 Diethyl (4-methylphenyl)-phosphonate, *in* M-00237
1754-47-8 ▷Dioctyl phenylphosphonate, *in* P-00221
1754-49-0 ▷Diethyl phenylphosphonate, D-00310
1754-58-1 ▷Diamidafos, D-00024
1754-60-5 *N,N′*-Dipropylphosphorodiamidic acid; Chloride, *in* D-01223
1754-88-7 Ethylidenetriphenylphosphorane, E-00076
1769-52-4 2,3-Dihydro-1*H*-phosphole, D-00571
1776-83-6 ▷Quintiofos, Q-00006
1779-48-2 Phenylphosphinic acid, P-00210
1779-49-3 Methyltriphenylphosphonium(1+); Bromide, *in* M-00432
1779-51-7 ▷Butyltriphenylphosphonium(1+); Bromide, *in* B-00641
1779-54-0 [(Methylthio)methyl]-triphenylphosphonium(1+); Chloride, *in* M-00416
1779-58-4 (2-Methoxy-2-oxoethyl)-triphenylphosphonium(1+); Bromide, *in* M-00055
1789-95-3 Dipropyl propylphosphonate, *in* P-00473
1794-24-7 Tetraethylphosphorodiamidic chloride, T-00060
1804-93-9 Diisopropyl phosphate, D-00639
1804-93-9 Dipropyl phosphate, D-01203
1806-54-8 Trioctyl phosphate, T-00581
1808-64-6 *N,N′,N″*-Tris(1,1-dimethylethyl)-phosphoric triamide, *in* T-00395
1809-14-9 Dioctyl phosphonate, D-00959
1809-17-2 Dipentyl phosphonate, D-00982
1809-19-4 ▷Dibutyl phosphonate, D-00127
1809-20-7 ▷Diisopropyl phosphonate, D-00651
1809-21-8 Dipropyl phosphonate, D-01210
1817-85-2 Diphenyl aminocarbonylphosphoramidate, *in* A-00062
1817-86-3 Diphenyl (phenylamino)-carbonylphosphoramidate, *in* P-00081
1820-39-9 1,2-Bis(triphosphoranylidene)-benzocyclobutene, B-00460
1831-25-0 Ethylene methylphosphonate, *in* M-00126
1831-31-8 4-Methyl-2-phenyl-1,3,2-dioxaphospholane; 2-Oxide, *in* M-00198
1831-37-4 (3-Methyl-1,2-butadienyl)-phosphonic acid, M-00094
1831-63-6 (Diphenylphosphinyl)acetic acid, D-01093
1832-53-7 Ethyl methylphosphonate, E-00089
1832-54-8 Isopropyl methylphosphonate, I-00038
1844-12-8 Phenyl phosphorodiisocyanatidate, P-00292
1858-24-8 Phosphoryl isocyanate, P-00415
1858-25-9 Phosphoryl isothiocyanate, P-00416
1858-26-0 Thiophosphoryl isothiocyanate, T-00316
1858-37-3 Phosphorodiisocyanatidic chloride, P-00412
1858-38-4 Phosphor(isothiocyanatidic) dichloride, P-00407
1866-68-8 Carbamoyl dihydrogen phosphate; Di-Li salt, *in* C-00002
1883-26-7 Phenylbis(phenylethynyl)phosphine; Sulfide, *in* P-00099
1885-79-6 3-Methyl-2-phenyl-1,3,2-oxazaphospholidine, M-00213
1885-80-9 3-Methyl-2-phenyl-1,3,2-oxazaphospholidine; 2-Sulfide, *in* M-00213
1892-18-8 (4-Chlorophenyl)phosphoramidic acid, C-00157
1907-75-1 *P,P*-Bis(1-aziridinyl)-*N,N*-diethylphosphinic amide, *in* B-00092
1929-41-5 Methyl hydrogen chloromethylphosphonate, *in* C-00106
1932-60-1 Cyclohexyl hydrogen methylphosphonate, *in* M-00288
1935-19-9 Threonine ethanolamine phosphate; *(S)-form*, *in* T-00318
1941-19-1 Tetramethylphosphonium(1+); Chloride, *in* T-00215
1946-09-4 Diethyl ethylphosphoramidate, D-00293
1950-04-5 Diphenyl bis(2-chloroethyl)-phosphoramidate, *in* B-00121
1980-45-6 Benzodepa, B-00017
1982-13-4 Ethyl dimethoxyphosphinylformate, *in* D-00672
1983-26-2 (Chloromethyl)phosphonic dichloride, C-00108
1983-27-3 (Chloromethyl)phosphonothioic acid; Dichloride, *in* C-00113
1984-15-2 Methanediphosphonic acid, M-00022
1990-22-3 Tripentyl phosphite, T-00595
1991-30-6 Ethyl 2-[(diethoxyphosphinyl)-oxy]-2-propenoate, *in* D-00256
1992-39-8 Monophenyl phosphate; Monoanilinium salt, *in* M-00507
1992-41-2 Monoisopropyl phosphate; Monoanilinium salt, *in* M-00477
2001-45-8 Tetraphenylphosphonium(1+); Chloride, *in* T-00272
2007-97-8 2,2,2-Trichloro-2,2-dihydro-1,3,2-benzodioxaphosphole, T-00400
2012-00-2 ▷Ethyl 4-nitrophenyl phenylphosphonate, *in* M-00471
2014-79-1 Diphenyl methylphosphoramidate, *in* M-00326

2014-81-5 Dibutyl methylphosphoramidate, *in* M-00326
2015-56-7 Diphenyl phosphoramidate, D-01107
2018-19-1 6-Azauridine 5′-phosphate, A-00145
2025-08-3 Tris(trifluoromethyl)phosphine; Sulfide, *in* T-00855
2025-12-9 Tris(phenylethynyl)phosphine; Sulfide, *in* T-00846
2026-42-8 Diethyl (2-thienylmethyl)-phosphonate, *in* T-00304
2035-84-9 Ethenyl phosphorodichloridate, *in* M-00518
2041-14-7 2-Aminoethylphosphonic acid, A-00079
2047-14-5 ▷*O,O*-Diethyl *S*-(2-amino-2-oxoethyl) phosphorodithioate, *in* D-00239
2049-62-9 *Se*-Phenyl diphenylphosphinoselenoate, *in* D-01078
2065-66-9 Methyltriphenylphosphonium(1+); Iodide, *in* M-00432
2065-67-0 Tetraphenylphosphonium(1+); Iodide, *in* T-00272
2071-20-7 Bis(diphenylphosphino)methane, B-00244
2071-21-8 Bis(diphenylphosphinyl)methane, *in* B-00244
2071-24-1 Bis[(diphenylphosphino)methyl]-phenylphosphine, B-00253
2074-67-1 Bis(hydroxymethyl)phosphinic acid, B-00282
2082-80-6 Trioctadecyl phosphite, T-00580
2088-72-4 ▷Ethyl (dimethoxyphosphinyl)-thioacetate, *in* D-00687
2091-46-5 Triphenyl-2-propynylphosphonium(1+); Bromide, *in* T-00681
2096-78-8 Diphenylvinylphosphine; Oxide, *in* D-01168
2096-83-5 Tetraphenyldiphosphine; Mono-oxide, *in* T-00258
2104-64-5 ▷*O*-Ethyl *O*-4-nitrophenyl phenylphosphonothionate, E-00099
2104-96-3 ▷Bromophos, B-00498
2129-29-5 3-(Diphenylphosphinyl)benzoic acid, *in* D-01054
2129-30-8 3-(Diphenylphosphino)benzoic acid, D-01054
2129-31-9 4-(Diphenylphosphino)benzoic acid, D-01055
2129-79-5 Methyl benzylphenylphosphinate, *in* B-00056
2129-89-7 Methyldiphenylphosphine; Oxide, *in* M-00135
2136-75-6 Triphenyl-phosphoranylideneacetaldehyde, T-00639
2140-58-1 Adenosine 5′-diphosphoglucose, A-00042
2152-00-3 Phenyl phenylphosphoramidic acid; Cyclohexylammonium salt, *in* P-00189
2155-78-4 (Chloromethyl)phosphonous dichloride, C-00116
2155-96-6 Diphenylvinylphosphine, D-01168
2156-68-5 *N*-Chlorotriphenylphosphoranyl-*P,P,P*-triphenylphosphine imide, *in* B-00442
2156-69-6 *P,P*-Diphenyl-*N*-(triphenylphosphoranylidene)-phosphinic amide, *in* D-01165
2157-98-4 Monocrotophos, M-00459
2171-87-1 Ethylphosphonofluoridic acid, E-00142
2175-86-2 ▷(4-Nitrophenyl)phosphonic acid, N-00057
2180-41-8 Monopropyl phosphate; Monoanilinium salt, *in* M-00512
2180-42-9 Monoethyl phosphate; Monoanilinium salt, *in* M-00472
2181-97-7 (2-Methoxy-2-oxoethyl)-triphenylphosphonium(1+); Chloride, *in* M-00055
2196-04-5 2-Methoxy-1,3,2-dioxaphospholane 2-oxide, M-00037
2203-74-9 2,2,4,6-Tetrachloro-4,6-bis(dimethylamino)-2,2,4,4,6,6-hexahydro-1,3,5-triaza-2,4,6-triphosphorine, T-00028
2212-88-6 3-Phosphonoxypropanenitrile, P-00398
2234-97-1 ▷Tripropylphosphine, T-00711
2236-01-3 (2-Oxopropyl)-triphenylphosphonium(1+); Bromide, *in* O-00080
2237-89-0 *P,P′*-Diphenyl diphosphate; Biscyclohexylammonium salt, *in* D-01011
2240-31-5 Bromodimethylphosphine, *in* D-00844
2240-32-6 Chloroethylmethylphosphine, *in* E-00087
2240-33-7 Bromoethylmethylphosphine, *in* E-00087
2240-41-7 Dimethyl phenylphosphonate, D-00810
2240-45-1 Bis(1-methylpropyl)phosphine, B-00343
2240-47-3 *P,P,P*-Triphenylphosphine imide, T-00626
2253-43-2 *O,O*-Dipropyl phosphorodithioate, D-01225
2253-44-3 *O,O*-Dibutyl phosphorodithioate, D-00150
2253-52-3 *O,O*-Bis(2-methylpropyl) phosphorodithioate, B-00360
2253-54-5 *O,O*-Dipentyl phosphorodithioate, D-00986
2253-57-8 *O,O*-Dioctyl phosphorodithioate, D-00960
2253-58-9 *O,O*-Dinonyl phosphorodithioate, D-00951
2253-59-0 *O,O*-Didecyl phosphorodithioate, D-00224
2253-60-3 *O,O*-Diphenyl phosphorodithioate, D-01131
2253-62-5 *O,O*-Dibenzyl phosphorodithioate, D-00063
2255-14-3 Galactose 1-dihydrogen phosphate; α-D-*Pyranose-form*, *in* G-00001
2255-17-6 ▷Fenitrooxon, F-00004
2259-81-6 1,2-Ethanediylbis[phenylphosphinic acid]; Dichloride, *in* E-00015
2272-04-0 4-(Diphenylphosphinyl)benzoic acid, *in* D-01055
2272-08-4 *O,O,O*-Tripropyl phosphorothioate, T-00715
2274-67-1 ▷2-Chloro-1-(2,4-dichlorophenyl)vinyl dimethyl phosphate, C-00035
2275-14-1 ▷Phenkapton, P-00062
2275-18-5 ▷Prothoate, P-00499
2275-23-2 ▷Vamidothion, V-00001
2275-61-8 ▷*P,P*-Bis(1-aziridinyl)-*N*-methylphosphinic amide, *in* B-00092
2283-11-6 Hexaethylphosphorous triamide, H-00015
2283-25-2 Bis(1-methylpropyl) phosphonate, B-00353
2302-70-7 1,2,5-Triphenyl-1*H*-phosphole; Selenide, *in* T-00634
2302-80-9 Butylphosphonic dichloride, B-00598
2303-77-7 ▷Ethyl diethylphosphinite, *in* D-00333
2304-30-5 ▷Tetrabutylphosphonium(1+); Chloride, *in* T-00019
2310-17-0 ▷Phosalone, P-00340
2310-66-9 Cyclohexylphenylphosphinic acid, C-00250
2310-71-6 Cyclohexylphosphinic acid, C-00253
2310-87-4 Phenyl hydrogen phenylphosphonate, *in* P-00221
2310-89-6 Monophenyl phosphonate, M-00508
2316-28-1 1,1,4,4-Tetraphenyl-1,4-diphosphorinanium dibromide, *in* D-01017
2316-28-1 1,1,4,4-Tetraphenyl-1,4-diphosphorinanium(2+); Dibromide, *in* T-00263
2318-47-0 3,3,8,8-Tetramethyl-1,6-dioxa-4,9-diaza-5-phospha(P^{V})-spiro[4.4]nonane, T-00196
2325-27-1 Tetraphenylphosphine imide, T-00269
2327-69-7 Ethyl 4-(diethoxyphosphinyl)-butanoate, *in* D-00247
2328-23-6 Methylphenylpropylphosphine; (±)-*form*, Oxide, *in* M-00265
2328-95-2 Dibenzylphosphinous acid, D-00058
2353-98-2 (Dimethylamino)-tetrafluorophosphorane, D-00708
2359-99-1 Tetracyclohexyldiphosphine, T-00039
2360-09-0 3-(Diphenylphosphino)-1-propanol, D-01074
2369-26-8 (3-Fluorophenyl)phosphonic acid, F-00037
2373-43-5 Diethyl (4-chlorophenyl)-phosphonate, *in* C-00147
2375-06-6 2-[(Hydroxymethoxyphosphinyl)-oxy]-*N,N,N*-trimethyl-ethanaminium hydroxide, inner salt, *in* C-00198
2378-86-1 [(4-Methylphenyl)methyl]-triphenylphosphonium(1+); Bromide, *in* M-00211
2381-38-1 Trimethyl phosphorotetrathioate, T-00544
2382-75-4 Mono-*tert*-butyl phosphate, M-00452
2382-76-5 Monopentyl phosphate, M-00504
2382-77-6 Mono(1-methylpropyl) phosphate, M-00492
2386-41-6 Triisopropyl phosphorotetrathioate, T-00505
2397-47-9 Triethyl phenylphosphorimidate, *in* P-00264
2397-48-0 Triethyl (diethoxyphosphinyl)-phosphorimidate, *in* D-00257
2404-03-7 Diethyl dimethylphosphoramidate, D-00286
2404-04-8 Diisopropyl dimethylphosphoramidate, *in* D-00855
2404-55-9 Trimethylphosphine sulfide, T-00534
2404-58-2 Dibutyl ethylphosphonate, *in* E-00129
2404-65-1 Ethyl tetramethyl-phosphorodiamidate, *in* T-00218
2404-73-1 Dibutyl methylphosphonate, *in* M-00288
2404-75-3 Diethyl butylphosphonate, *in* B-00594
2404-78-6 ▷*O,S,S*-Triethyl phosphorodithioate, T-00450
2404-81-1 Ethyl *N,N*-diethyl-*P*-methylphosphonamidate, *in* D-00296
2408-96-0 Diethylphosphinodiselenoic acid; Anhydroselenide, *in* D-00324
2409-61-2 Bis(4-methylphenyl)phosphine; Oxide, *in* B-00317
2409-69-8 (Hydroxyphenylmethyl)-phenylphosphinic acid, H-00173
2423-99-6 *O,O,O′,O′*-Tetraethyl imidodiphosphate, T-00051
2428-06-0 2-Chloro-5,5-dimethyl-1,3,2-dioxaphosphorinane, C-00058
2428-07-1 *O,O*-Neopentylene phosphonothioate, *in* D-00735
2439-99-8 ▷Glyphosine, G-00098
2455-45-0 *O,O*-Diethyl ethylphosphonothioate, *in* E-00148
2462-63-7 Glycerol 1,2-di-9-octadecenoate 3-phosphoethanolamine; (±)-(*all-Z*)-*form*, *in* G-00043
2464-03-1 *O,O,O*-Triisopropyl phosphorothioate, T-00506
2465-65-8 *O,O*-Diethyl phosphorothioate, D-00396
2466-73-1 Mono(2-methylpropyl) phosphate, M-00493
2468-05-5 Bis(3-hydroxypropyl)-phenylphosphine, B-00286
2487-04-9 Diphenyl phosphorisocyanatidate, D-01116
2501-94-2 *tert*-Butylphosphine, B-00589
2502-20-7 Hexylphosphine, H-00096
2510-89-6 Dipropyl phosphorochloridate, D-01217

2511-09-3 ▷Ethyl phenylphosphinate, *in* P-00210
2511-13-9 *O,S*-Diethyl propylphosphonothioate, *in* P-00480
2511-17-3 Pentamethylphosphonic diamide, *in* M-00289
2511-19-5 Ethyl dimethylphosphinate, *in* D-00829
2512-58-5 Dicyclohexylphosphinodithioic acid, D-00212
2520-52-7 Galactose 1-dihydrogen phosphate; *β-D-Pyranose-form*, *in* G-00001
2523-94-6 *O*-Methyl phosphorodichloridothioate, M-00349
2523-96-8 *O*-Propyl phosphorodichloridothioate, P-00489
2523-98-0 Diethylphosphoramidothioic dichloride, D-00357
2524-01-8 Propylphosphonothioic acid; Dichloride, *in* P-00480
2524-03-0 ▷*O,O*-Dimethyl phosphorochloridothioate, D-00891
2524-04-1 ▷*O,O*-Diethyl phosphorochloridothioate, D-00378
2524-05-2 *O,O*-Dipropyl phosphorochloridothioate, D-01221
2524-06-3 ▷*O,O*-Diisopropyl phosphorochloridothioate, D-00659
2524-07-4 *O,O*-Dibutyl phosphorochloridothioate, D-00145
2524-09-6 *O,O,S*-Triethyl phosphorodithioate, T-00451
2524-15-4 *O*-Methyl methylphosphonochloridothioate, M-00169
2524-16-5 *O*-Ethyl methylphosphonochloridothioate, E-00090
2524-17-6 *O*-Isopropyl methylphosphonochloridothioate, I-00039
2524-18-7 Dipropylphosphinothioic acid; Chloride, *in* D-01208
2524-64-3 Diphenyl phosphorochloridate, D-01120
2526-64-9 Dichlorotriphenylphosphorane, D-00200
2528-38-3 Tripentyl phosphate, T-00593
2528-39-4 Trihexyl phosphate, T-00491
2528-40-7 Tricyclohexyl phosphate, T-00425
2528-45-2 Tris(1-methylpropyl) phosphate, T-00821
2536-16-3 Ethyl (diethoxyphosphinyl)-fluoroacetate, *in* D-00250
2537-48-6 Diethyl (cyanomethyl)phosphonate, *in* C-00215
2540-82-1 ▷Formothion, F-00055
2549-63-5 Trinonyl phosphite, T-00579
2551-53-3 Trioctylphosphine; Sulfide, *in* T-00582
2565-58-4 (Chloromethyl)phosphonic acid, C-00106
2572-40-9 2-(Diphenylphosphino)benzoic acid; Oxide, *in* D-01053
2573-70-8 Dibutylphosphinodithioic acid; NH_4 salt, *in* D-00112
2574-25-6 Diisopropyl phosphorochloridate, D-00657
2587-90-8 ▷Demephion-*S*, D-00018
2588-03-6 ▷*O,O*-Diethyl *S*-[(ethylsulfinyl)-methyl] phosphorodithioate, *in* P-00338
2588-04-7 ▷*OO*-Diethyl *S*-[(ethylsulfonyl)-methyl] phosphorodithioate, *in* P-00338
2588-88-7 Pentaphenylphosphorane, P-00039
2590-44-5 *O*-Propyl bis(1-aziridinyl)-phosphinothioate, *in* B-00093
2595-54-2 ▷Mecarbam, M-00002
2596-41-0 Monophenyl cyclohexylphosphonate, *in* C-00256
2597-03-7 ▷Phenthoate, P-00078
2605-67-6 Methyl triphenyl-phosphoranylideneacetate, M-00433
2605-68-7 Methyl (2-triphenylphosphoranylidene)-propanoate, M-00434
2609-49-6 [(4-Nitrophenyl)methyl]phosphonic acid; Di-Et ester, *in* N-00046
2617-47-2 (Hydroxymethyl)phosphonic acid, H-00162
2622-07-3 Hexaethylphosphoric triamide, *in* H-00015
2622-08-4 ▷Tris(2-methylphenyl) phosphite, T-00809
2622-14-2 Tricyclohexylphosphine, T-00426
2627-35-2 Monododecyl phosphate, M-00468
2629-35-8 Dichlorododecylphosphine, *in* D-01265
2632-88-4 Ethyl tetraethylphosphorodiamidite, *in* T-00066
2636-26-2 ▷Cyanophos, C-00226
2642-71-9 ▷Azinphos-ethyl, A-00150
2651-85-6 Diethyl ethylphosphonite, *in* E-00153
2651-89-0 *O,O,O*-Triethyl phosphoroselenoate, T-00454
2666-14-0 ▷(1-Hydroxyethylidene)-bisphosphonic acid; Tri-Na salt, *in* H-00144
2672-32-4 Diethyl propylphosphoramidate, *in* P-00484
2674-91-1 ▷Oxydeprofos, O-00099
2697-42-9 *O,O*-Dimethyl phosphoramidate, D-00854
2708-87-4 Diethyl (trifluoromethyl)-phosphonate, *in* T-00478
2708-92-1 Methyl(2-methylpropyl)phosphinic acid; Fluoride, *in* M-00172
2712-92-7 Dibromo(trifluoromethyl)phosphine, *in* T-00482
2725-66-8 Dichlorotrimethylphosphorane, D-00198
2725-67-9 Trichlorodimethylphosphorane, T-00402
2725-68-0 Tetrachloromethylphosphorane, T-00032
2728-67-8 Bis(trifluoromethyl)phosphinous acid; Anhydride, *in* B-00418
2736-46-1 Ethylphosphonic acid; Diisocyanate, *in* E-00129
2736-47-2 Isopropylphosphonic acid; Diisocyanate, *in* I-00050
2736-49-4 Phenylphosphonous acid; Diisocyanate, *in* P-00251
2741-38-0 Diphenyl-2-propenylphosphine, D-01144
2751-90-8 Tetraphenylphosphonium(1+); Bromide, *in* T-00272
2767-70-6 ▷[(4-Nitrophenyl)methyl]-triphenylphosphonium(1+); Bromide, *in* N-00049
2767-80-8 Tris(hydroxymethyl)phosphine, T-00784
2783-48-4 Bis(2-methylpropyl) phenylphosphonate, *in* P-00221
2797-54-8 *S,S*-Dibutyl butylphosphonodithioate, *in* B-00606
2797-55-9 *S,S*-Dibutyl phosphorochloridodithioate, D-00144
2809-21-4 (1-Hydroxyethylidene)-bisphosphonic acid, H-00144
2809-26-9 (Hydroxyphenylmethyl)-bisphosphonic acid, H-00172
2816-34-4 Diphenyl(triphenylgermyl)-phosphine, D-01161
2844-89-5 Cyclohexylphosphonous dichloride, C-00268
2845-09-2 Dioctylphenylphosphine; Oxide, *in* D-00955
2846-32-4 2,2,4,4-Tetrachloro-2,2,4,4,6,6-hexahydro-6,6-diphenyl-1,3,5-triaza-2,4,6-triphosphorine, T-00031
2851-75-4 Methylphosphoramidous acid; Difluoride, *in* M-00335
2901-90-8 Bis(diethoxyphosphinothioyl) disulfide, B-00170
2921-71-3 2,3-Dihydro-4-methyl-1-phenoxy-1*H*-phosphole 1-oxide, *in* D-00496
2921-71-3 2,5-Dihydro-3-methyl-2-phenoxy-1*H*-phosphole 1-oxide, *in* D-00499
2921-88-2 ▷Chlorpyrifos, C-00194
2925-91-9 2,3,5,6,7,8-Hexakis(trifluoromethyl)-1,4-diphosphabicyclo[2.2.2]octa-2,5,7-triene, H-00064
2935-46-8 Tributylphosphine; Telluride, *in* T-00372
2946-59-0 ▷Phenylphosphonous dicyanide, P-00255
2946-60-3 Tri-2-propenylphosphine; Oxide, *in* T-00707
2946-61-4 Dimethyl phenylphosphonite, *in* P-00251
2948-89-2 Di-2-propenyl phenylphosphonate, *in* P-00221
2950-47-2 Butyl dibutylphosphinate, *in* D-00109
2953-29-9 ▷*O,O,S*-Trimethyl phosphorodithioate, T-00542
2959-65-1 Tris(1-methylpropyl)phosphine; Oxide, *in* T-00823
2959-74-2 Benzyldiphenylphosphine; Oxide, *in* B-00046
2959-75-3 Isopropyldiphenylphosphine; Oxide, *in* I-00031
2960-37-4 *N*-Diphenylphosphino-*P,P*-diphenylphosphinous amide, D-01063
2960-41-0 *N*-(Diphenylphosphino)-*N*-ethyl-*P,P*-diphenylphinous amide, D-01067
2984-64-7 ▷*O*-Ethyl *S*-4-chlorophenyl ethylphosphonodithioate, *in* E-00070
2994-53-8 Ethyltriphenylphosphonium(1+); Tetrafluoroborate, *in* E-00197
3004-47-5 Ethyl bis(1-aziridinyl)-phosphinite, *in* B-00095
3006-66-4 2,4,6-Trichloro-2,2,4,4,6,6-hexahydro-2,4,6-triphenyl-1,3,5-triaza-2,4,6-triphosphorine, T-00407
3009-75-4 2,2-Dihydrophenanthro[9,10-*d*]-1,3,2-dioxaphosphole; 2,2,2-Triisopropoxy, *in* D-00541
3011-76-5 Diheptylphosphinic acid, D-00425
3011-82-3 ▷Dioctylphosphine; Oxide, *in* D-00957
3019-36-1 Phosphoguanidine, P-00370
3026-45-7 Glycerol 1,2-dihexadecanoate 3-phosphoethanolamine, G-00013
3028-88-4 ▷Trioctyl phosphite, T-00583
3030-90-8 Dibutyl phenylphosphonite, *in* P-00251
3031-21-8 Bis(diisopropyoxyphosphinothioyl) disulfide, B-00184
3031-94-5 5-Amino-1-ribofuranosylimidazole-4-carboxamide 5′-(dihydrogen phosphate); *β-D-form*, *in* A-00130
3031-95-6 *N*-(5-Amino-1-*β*-D-ribofuranosylimidazole-4-carbonyl)-L-aspartic acid 5′-phosphate, A-00129
3036-18-8 Cytidine diphosphate ethanolamine, C-00292
3040-63-9 Tetraethyldiphosphine, T-00043
3040-70-8 Tetraethyltetraphosphetane, T-00071
3049-19-2 Diphenyl butylphosphonate, *in* B-00594
3049-24-9 Diphenyl phenylphosphonate, *in* P-00221
3054-93-1 [3-(2-Methyl-1,3-dioxolan-2-yl)-propylidene]triphenylphosphorane, *in* M-00132

3057-08-7 5,5-Dimethyl-2-phenoxy-1,3,2-dioxaphosphorinane, D-00790
3058-18-2 Pentyl dipentylphosphinate, *in* D-00981
3058-19-3 Heptyl diheptylphosphinate, *in* D-00425
3058-20-6 Dibutyl butylphosphonite, *in* B-00615
3058-21-7 Diheptyl heptylphosphonite, *in* H-00008
3064-55-9 (4-Chlorophenyl)phosphonothioic acid; Dichloride, *in* C-00150
3064-56-0 (Diphenylphosphino)acetic acid, D-01045
3070-15-3 ▷*O,O*-Diethyl *O*-[4-(methylthio)-phenyl] phosphorothioate, D-00302
3084-40-0 Diethyl (hydroxymethyl)-phosphonate, *in* H-00162
3084-47-7 ▷Tripentylphosphine; Oxide, *in* T-00594
3084-48-8 ▷Trihexylphosphine; Oxide, *in* T-00492
3084-50-2 ▷Tributylphosphine sulfide, T-00378
3095-05-4 (1,2-Butadienyl)phosphonic acid, B-00509
3095-05-4 Diethyl (1,2-butadienyl)-phosphonate, *in* B-00509
3095-06-5 2-Butynylphosphonic acid, B-00644
3095-06-5 Diethyl 2-butynylphosphonate, *in* B-00644
3095-09-8 Diphenyl 1-propynylphosphonate, *in* P-00494
3095-10-1 Diethyl (1-phenyl-1,2-propadienyl)phosphonate, *in* P-00298
3095-33-8 1-(Diphenylphosphinyl)naphthalene, *in* N-00011
3095-90-7 Octylphosphine, O-00041
3095-94-1 Octylphosphonic dichloride, O-00043
3095-95-2 (Diethoxyphosphinyl)acetic acid, D-00241
3095-96-3 3-(Diethoxyphosphinyl)propanoic acid, D-00259
3096-04-6 *S*-Ethyl diphenylphosphinothioate, *in* D-01082
3096-05-7 *S*-Benzyl diphenylphosphinothioate, *in* D-01082
3096-08-0 *O*-Methyl diphenylphosphinothioate, *in* D-01082
3096-09-1 Diphenylphosphinothioic anhydride, *in* D-01082
3096-30-5 *S*-Methyl diphenylphosphinothioate, *in* D-01082
3101-74-4 Tris(triphenylphosphoranylidene)-phosphoric triamide, *in* T-00872
3115-39-7 Dioctyl phosphate, D-00956
3115-66-0 ▷Tetrabutylphosphonium(1+); Iodide, *in* T-00019
3115-68-2 ▷Tetrabutylphosphonium(1+); Bromide, *in* T-00019
3123-67-9 Guanosine diphosphate mannose, G-00100
3127-08-0 Bis(hydroxymethyl)-phenylphosphine, B-00281
3132-52-3 1,1′,1″-Phosphinylidynetrisazetidine, *in* T-00328
3133-27-5 *O*-Ethyl diphenylphosphinothioate, *in* D-01082
3135-77-1 Dimethyl aminocarbonylphosphoramidate, *in* A-00062
3135-78-2 Diethyl aminocarbonylphosphoramidate, *in* A-00062
3138-42-9 Dipentyl phosphate, D-00979
3138-48-0 Dinonyl phosphate, D-00948
3138-57-6 Tetraphenylphosphonium(1+); Tribromide, *in* T-00272
3151-19-7 5,5′-Spirobi[5*H*-dibenzophospholium](1+); Iodide, *in* S-00022
3151-21-1 Methylbisbiphenylenephosphorane, *in* S-00021
3167-62-2 Diethyl (dichloromethyl)-phosphonate, *in* D-00175
3167-63-3 Diethyl chloromethylphosphonate, *in* C-00106
3167-69-9 Diethyl diethylphosphoramidate, D-00281
3173-38-4 Diethyl [(4-hydroxyphenyl)-methyl]phosphonate, *in* H-00176
3182-66-9 2,3,5,6,7,8-Hexamethyl-2,3,5,6,7,8-hexaaza-1,4-diphosphabicyclo[2.2.2]octane; 1,4-Dioxide, *in* H-00070
3185-65-7 *P,P*-Dichloro-*P*-diethylamino-*N*-phenylphosphine imide, *in* D-00313
3185-66-8 Tetraethylphosphorodiamidic anilide, *in* T-00058
3185-83-9 *P*-Chloro-*P,P*-bis(diethylamino)-*N*-phenylphosphazene, *in* T-00054
3190-77-0 2-(Diethylphosphino)ethanethiol, D-00326
3196-56-3 ▷Trimethylolethane thiophosphate, *in* M-00424
3201-73-8 Diethyl 2-propynylphosphonate, *in* P-00495
3201-76-1 Bis(1-methylethyl) 1,2-propadienylphosphonate, *in* P-00441
3201-79-4 Diisopropyl 1-propynylphosphonate, *in* P-00494
3201-84-1 Diethyl (3-methyl-1,2-butadienyl)phosphonate, *in* M-00094
3220-50-6 (1-Phenylethenyl)phosphonic acid, P-00154
3244-90-4 ▷Aspon, A-00144
3245-78-1 Dipentyl pentylphosphonite, *in* P-00052
3253-29-0 *O,O,O′,O′*-Tetraisopropyl trithiopyrophosphate, T-00174
3254-63-5 ▷Dimethyl 4-(methylthio)phenyl phosphate, D-00767
3266-66-8 Acetyl diethyl phosphite, A-00013
3272-86-4 Trimethyl(trimethylsilylmethylene)-phosphorane, T-00566
3277-27-8 Diethyl benzoylphosphonate, *in* B-00034
3279-26-3 ▷Methyl phosphorodichloridite, M-00346
3283-12-3 Dimethylphosphinic acid, D-00829
3287-84-1 *O,O*-Dipropyl phosphorodithioate; K salt, *in* D-01225
3287-85-2 *O,O*-Bis(1-methylpropyl) phosphorodithioate; K salt, *in* B-00359
3287-86-3 *O,O*-Dipentyl phosphorodithioate; K salt, *in* D-00986
3287-87-4 *O,O*-Dihexyl phosphorodithioate; K salt, *in* D-00437
3287-88-5 *O,O*-Diheptyl phosphorodithioate; K salt, *in* D-00427
3287-89-6 *O,O*-Dioctyl phosphorodithioate; K salt, *in* D-00960
3287-90-9 *O,O*-Dinonyl phosphorodithioate; K salt, *in* D-00951
3296-60-4 (Phenylphosphinothioylidene)-bismethanol, *in* B-00281
3302-83-8 1-Phenylphosphorinane, P-00265
3302-87-2 1-Phenylphospholane, P-00216
3302-93-0 1-Methyl-1-phenylphospholanium iodide, *in* P-00216
3302-94-1 1-Methyl-1-phenylphosphorinanium iodide, *in* P-00265
3321-64-0 Butylphosphonic acid, B-00594
3335-12-4 Diphenyl *tert*-butylphosphoramidate, *in* B-00622
3335-14-6 2,3,5,6,7,8-Hexamethyl-2,3,5,6,7,8-hexaaza-1,4-diphosphabicyclo[2.2.2]octane; 1,4-Disulfide, *in* H-00070
3338-24-7 ▷Hostaflot, *in* D-00384
3347-28-2 *O,S,S*-Trimethyl phosphorotrithioate, T-00548
3347-30-6 ▷*O,S,S*-Triethyl phosphorotrithioate, T-00459
3347-31-7 *O,S*-Diethyl methylphosphonodithioate, *in* M-00301
3347-32-8 *O,S*-Diethyl ethylphosphonodithioate, *in* E-00070
3348-40-1 *P*-Cyclohexyl-*N,N,N′,N′*-tetraethylphosphonous diamide, *in* C-00267
3348-43-4 ▷2,2′-Oxybis[1,3,2-dioxaphospholane], O-00091
3348-44-5 ▷Tetramethylphosphorodiamidous chloride, T-00228
3355-27-9 Glycerol 1,2-dipentadecanoate 3-phosphocholine; (*R*)-L-*form*, *in* G-00046
3356-37-4 *N,N,N′,N′*-Tetramethyl-*P*-1,2-propadienylphosphonic diamide, *in* P-00442
3356-46-5 *N,N′*-1,4-Phenylenebis[*P,P,P*-triphenylphosphine imide], P-00145
3366-72-1 ▷(4-Methylphenyl)phosphonic acid, M-00237
3376-52-1 Pentaphenylpentaphospholane, P-00037
3383-96-8 ▷Temephos, T-00003
3393-57-5 Isopropylmethylphosphinic acid; Chloride, *in* I-00035
3402-24-2 Ethyl tetramethyl-phosphorodiamidite, *in* T-00227
3402-28-6 Isopropyl tetraethylphosphorodiamidite, *in* T-00066
3411-48-1 Tri-1-naphthylphosphine, T-00572
3419-34-9 *O,O*-Diisopropyl phosphorodithioate; K salt, *in* D-00663
3419-35-0 *O,O*-Didecyl phosphorodithioate; K salt, *in* D-00224
3426-89-9 ▷Phenyl phosphorodichloridite, P-00280
3450-34-8 Trimethylphosphonamidothioic acid; Chloride, *in* T-00540
3450-35-9 *N,N*-Diethyl-*P*-methylphosphonamidothioic acid; Chloride, *in* D-00297
3450-65-5 Diethyl (2-oxoheptyl)phosphonate, *in* O-00061
3450-67-7 Diethyl phenylethynylphosphonate, *in* P-00166
3452-99-1 Diethyl (2-oxooctyl)phosphonate, *in* O-00064
3453-00-7 Diethyl (2-oxo-2-phenylethyl)-phosphonate, *in* O-00066
3454-66-8 *O,O*-Diethyl phosphorodithioate; K salt, *in* D-00384
3456-71-1 *tert*-Butylphosphoramidic acid; Dichloride, *in* B-00622
3462-95-1 [(4-Fluorophenyl)methyl]-triphenylphosphonium(1+); Chloride, *in* F-00032
3462-97-3 [(4-Methoxyphenyl)methyl]-triphenylphosphonium(1+); Chloride, *in* M-00064
3466-00-0 Phospholane, P-00371
3478-74-8 2,3,5,6,7,8-Hexamethyl-2,3,5,6,7,8-hexaaza-1,4-diphosphabicyclo[2.2.2]octane, H-00070
3479-30-9 Di-2-propenyl 2-propenylphosphonate, *in* P-00454
3487-44-3 Methylenetriphenylphosphorane, M-00148
3492-46-4 *o*-Phenylene phosphorobromidate, *in* B-00464
3497-00-5 ▷Phenylphosphonothioic dichloride, P-00247
3514-82-7 *O,O*-Diphenyl phosphorodithioate; K salt, *in* D-01131
3539-43-3 Monohexadecyl phosphate, M-00475
3541-77-3 Dioctylphosphine, D-00957
3549-51-7 *O,O*-Dibutyl phosphorodithioate; K salt, *in* D-00150
3549-52-8 *O,O*-Bis(2-methylpropyl) phosphorodithioate; K salt, *in* B-00360
3560-15-4 *N,N,N′,N′*-Tetraethylphosphonic diamide, T-00055

3572-91-6 Phenylbisbiphenylenephosphorane, *in* S-00021
3576-20-3 2,2-Dihydro-1,3-dimethyl-1,3,2-diazaphosphetidin-4-one; 2,2,2-Trichloro, *in* D-00459
3576-26-9 *P,P,P*-Trichloro-*N-p*-toluenesulfonylphosphazene, *in* M-00268
3582-10-3 Tetrachloro(trichloromethyl)-phosphorane, T-00037
3582-11-4 Trichloromethylphosphonous dichloride, T-00417
3582-82-9 Diphenyl(2-phenylethenyl)-phosphine; (*E*)-*form*, Oxide, *in* D-01033
3582-83-0 Diphenyl(2-phenylethenyl)-phosphine; (*E*)-*form*, Sulfide, *in* D-01033
3582-84-1 Diphenyl(2-phenylethyl)phosphine; Oxide, *in* D-01034
3583-02-6 2-Bromo-1,3,2-benzodioxaphosphole, B-00464
3583-94-6 ▷Tetrapropyl diphosphate, T-00285
3587-00-6 2,4,6-Trichloro-2,2,4,4,6,6-hexahydro-2,4,6-triphenyl-1,3,5-triaza-2,4,6-triphosphorine; (2α,4α,6β)-*form*, *in* T-00407
3606-84-6 2-Chloro-2,2,4,4,6,6-hexahydro-2,4,4,6,6-pentaphenyl-1,3,5-triaza-2,4,6-triphosphorine, C-00088
3606-85-7 2,4,6-Trichloro-2,2,4,4,6,6-hexahydro-2,4,6-triphenyl-1,3,5-triaza-2,4,6-triphosphorine; (2α,4α,6α)-*form*, *in* T-00407
3606-94-8 2,2-Dichloro-2,2,4,4,6,6-hexahydro-4,4,6,6-tetraphenyl-1,3,5-triaza-2,4,6-triphosphorine, D-00169
3607-17-8 (3-Bromopropyl)-triphenylphosphonium(1+); Bromide, *in* B-00502
3616-42-0 2-Amino-2-deoxyglucose 6-(dihydrogen phosphate); D-*form*, *in* A-00069
3619-91-8 1,3-Diphosphinopropane, D-01180
3643-70-7 3,9-Dichloro-2,4,8,10-tetraoxa-3,9-diphosphaspiro[5.5]-undecane, D-00196
3644-89-1 Ethyl tetraethylphosphorodiamidate, *in* T-00059
3646-10-4 1,4,6,9-Tetraoxa-5-phospha(5-P^{V})spiro[4.4]nonane, T-00244
3654-42-0 ▷Tetraethyl iminobis(methylene)-bisphosphonate, *in* D-00688
3654-42-0 ▷Pentaethyl imidodiphosphate, *in* E-00078
3658-48-8 ▷Bis(2-ethylhexyl) phosphonate, B-00273
3663-51-2 Bis(triethylsilyl) phosphonate, B-00407
3663-52-3 Bis(trimethylsilyl) phosphonate, B-00435
3666-89-5 Cyclobutyltriphenyl-phosphonium(1+); Bromide, *in* C-00235
3676-91-3 ▷Tetramethyldiphosphine, T-00200
3676-96-8 1,2-Dimethyl-1,2-diphenyldiphosphine, D-00744
3676-97-9 Tetramethyldiphosphine disulfide, T-00201
3676-98-0 Tetracyclohexyldiphosphine; Disulfide, *in* T-00039
3677-97-2 *O*-Ethyl bis(1-aziridinyl)-phosphinothioate, *in* B-00093
3689-24-5 ▷*O,O,O,O*-Tetraethyl dithiopyrophosphate, T-00045
3692-87-3 *O,O*-Dimethyl *S*-(2-amino-2-oxoethyl) phosphorodithioate, *in* D-00675
3694-54-0 *O,O*-Diethyl phenylphosphoramidothioate, *in* P-00262
3699-66-9 Ethyl 2-(diethoxyphosphinyl)-propanoate, E-00059
3699-67-0 Ethyl 3-(diethoxyphosphinyl)-propanoate, *in* D-00259
3699-76-1 Diethyl (2-diethylamino-2-oxoethyl)phosphonate, *in* D-00268
3700-86-5 ▷Diethyl [(diethoxyphosphinothioyl)thio]-butanedioate, *in* D-00240
3700-89-8 ▷Dimethyl [(dimethoxyphosphinothioyl)thio]-butanedioate, *in* D-00677
3711-50-0 *O,S*-Dimethyl phosphorochloridothioate, D-00892
3711-51-1 *O,S*-Diethyl phosphorochloridothioate, D-00379
3712-44-5 2,2,2-Tribromo-2,2-dihydro-1,3,2-benzodioxaphosphole, T-00352
3721-13-9 2,4,6-Trichloro-2,4,6-tris(dimethylamino)-2,2,4,4,6,6-hexahydro-1,3,5-triaza-2,4,6-triphosphorine, T-00422
3728-50-5 Butylidenetriphenylphosphorane, *in* B-00641
3730-54-9 ▷2-(Diethoxyphosphinyl)-3-oxobutanoic acid, D-00253
3730-54-9 ▷Ethyl 2-(diethoxyphosphinyl)-3-oxobutanoate, *in* D-00253
3732-81-8 Tetramethylphosphorodiamidothioic chloride, T-00225
3732-82-9 ▷Hexamethylphosphorothioic triamide, H-00079
3732-84-1 *N,N*-Dimethyl-*P,P*-diphenylphosphinic amide, *in* D-01040
3732-86-3 *N,N,N′,N′*-Tetramethylphosphoric triamide, T-00216
3733-81-1 ▷Defosfamide, D-00015
3734-95-0 ▷Cyanthoate, C-00232
3734-97-2 ▷Amiton; Hydrogen oxalate, *in* A-00133
3735-01-1 *O*-(4-Aminophenyl) *O,O*-diethyl phosphorothioate, A-00099
3735-78-2 ▷Dicrotophos, D-00204
3735-80-6 Salioxon, *in* H-00109
3736-70-7 Boyd's chloride, *in* T-00608
3739-96-6 (2,2-Dimethylpropylidene)-triphenylphosphorane, *in* D-00914
3739-97-7 Triphenyl[(trimethylsilyl)-methylene]phosphorane, T-00692
3739-98-8 Triphenyl[(trimethylsilyl)-methyl]phosphonium(1+); Iodide, *in* T-00693
3741-36-4 ▷2-Methoxy-1,3,2-dioxaphospholane, M-00036
3743-07-5 ▷Dimethyl chlorophosphite, D-00715
3746-01-8 Trivinylphosphine, T-00877
3762-25-2 Diethyl [(4-methylphenyl)-methyl]phosphonate, *in* M-00208
3762-27-4 ▷Dibutyl benzylphosphonate, *in* B-00060
3762-33-2 Diethyl (4-methoxyphenyl)-phosphonate, *in* M-00072
3778-73-2 ▷Ifosfamide, I-00001
3790-23-6 Tetraethyldiphosphine disulfide, T-00044
3792-59-4 ▷*S*-Seven, S-00004
3805-37-6 Adenosine 2′,5′-diphosphate, A-00039
3808-08-0 Diisopropyl benzoylphosphoramidate, *in* B-00036
3808-22-8 Dicyclohexyl phosphonate, D-00215
3811-49-2 ▷Salithion, S-00001
3820-53-9 ▷*O,O*-Dimethyl *S*-(4-nitrophenyl) phosphorothioate, D-00773
3820-71-1 Triphenyl phosphorotetrathioate, T-00669
3824-48-4 Methylphosphoramidic difluoride, M-00328
3842-86-2 Diethyl (2-dimethylamino-2-oxoethyl)phosphonate, *in* D-00703
3848-51-9 Diphenyl phenylphosphoramidate, D-01035
3848-53-1 *O,O,O′,O′*-Tetraphenyl imidodiphosphate, T-00265
3853-90-2 Dimethyl phosphorocyanidate, D-00894
3858-16-0 Tetramethyl (2-butene-1,4-diyl)-bisphosphonate, *in* B-00519
3858-24-0 ▷2,3-Dihydro-1-hydroxy-4-methyl-1*H*-phosphole 1-oxide, D-00496
3858-24-0 ▷2,3-Dihydro-1-hydroxy-5-methyl-1*H*-phosphole 1-oxide, D-00497
3871-20-3 Tris(4-nitrophenyl) phosphate, T-00837
3871-31-6 Tris(4-chlorophenyl) phosphate, T-00746
3878-44-2 Triphenylphosphine selenide, T-00628
3878-45-3 ▷Triphenylphosphine sulfide, T-00629
3881-20-7 Dibenzyl diethylphosphoramidate, *in* D-00343
3883-97-4 Ethyl bicyclic selenophosphate, *in* E-00196
3900-03-6 Monoheptyl phosphate, M-00474
3900-04-7 Monohexyl phosphate, M-00476
3900-12-7 Diheptyl phosphate, D-00424
3900-13-8 Dihexyl phosphate, D-00432
3905-76-8 Dibenzyl phosphoramidate, *in* D-00053
3921-30-0 Monodecyl phosphate, M-00462
3944-25-0 Methyl 3-(diethoxyphosphinyl)-propenoate, *in* D-00261
3944-27-2 (1-Methylethenyl)phosphonic dichloride, M-00150
3956-95-6 Diethyl (3-oxopropyl)phosphonate, *in* O-00076
3957-64-0 Tris(4-hydroxyphenyl) phosphate, T-00787
3958-00-7 *O*-Ethyl ethylphosphonoselenoate, E-00072
3958-21-2 Diethyl 2-dimethylaminoethyl phosphate, D-00283
3964-95-2 Propyl phosphorodifluoridite, P-00492
3965-01-3 Phenyl phosphorodifluoridite, P-00286
3968-92-1 (1-Methylpropyl)-triphenylphosphonium(1+); Bromide, *in* M-00386
3969-46-8 *P*-Phenylphosphonothioic diamide, P-00246
3969-50-4 *O*-Phenyl phosphorodiamidothioate, P-00276
3981-45-1 *P*-Ethyl-*N,N,N′,N′*-tetramethylphosphonothioic diamide, *in* E-00149
3981-46-2 Diethylphosphinothioic acid; Bromide, *in* D-00330
3982-87-4 Tris(2-methylpropyl)phosphine; Sulfide, *in* T-00824
3982-89-6 Diethylphosphinothioic acid; Chloride, *in* D-00330
3991-73-9 Monooctyl phosphate, M-00501
4002-41-9 Diethylphosphinoselenoic acid, D-00327
4004-05-1 Glycerol 1,2-di-9-octadecenoate 3-phosphoethanolamine; (*R*)-(*all-Z*)-*form*, *in* G-00043
4006-38-6 Bis(2-methylpropyl)phosphine, B-00344
4009-98-7 (Methoxymethyl)-triphenylphosphonium(1+); Chloride, *in* M-00053
4020-99-9 Methyl diphenylphosphinite, *in* D-01089
4023-49-8 Bis(2-cyanoethyl)phosphine, B-00153
4023-52-3 (2-Methylpropyl)phosphine, M-00374
4023-53-4 Tris(2-cyanoethyl)phosphine, T-00752
4037-11-0 Dimethyl (1,3-butadienyl)-phosphonate, *in* B-00510
4037-12-1 Dimethyl (2-methyl-1,3-butadienyl)phosphonate, *in* M-00093

4042-59-5 (4-Iodophenyl)phosphonic acid, I-00018
4042-71-1 *P*-Phenyl-*N,N'*-bis(triphenylphosphoranylidene)-phosphonothioic diamide, *in* P-00102
4073-55-6 2-Mercapto-5,5-dimethyl-1,3,2-dioxaphosphorinane 2-oxide; Cyclohexylammonium salt, *in* M-00010
4073-59-0 Bis(5,5-dimethyl-2-thioxo-1,3,2-dioxaphosphorinan-2-yl) disulfide, *in* D-01251
4081-23-6 ▷Phenyl saligenin cyclic phosphate, *in* P-00064
4090-51-1 Sandoflam, *in* O-00089
4090-52-2 2,2'-Oxybis[5,5-dimethyl-1,3,2-dioxaphosphorinane]; 2,2'-Dioxide, *in* O-00089
4090-53-3 2-[(5,5-Dimethyl-1,3,2-dioxaphosphorinan-2-yl)-thio]-5,5-dimethyl-1,3,2-dioxaphosphorinane *P*-oxide 2-sulfide, *in* T-00309
4090-54-4 2,2'-Thiobis[5,5-dimethyl-1,3,2-dioxaphosphorinane]; 2,2'-Disulfide, *in* T-00309
4090-55-5 2-Chloro-5,5-dimethyl-1,3,2-dioxaphosphorinane 2-oxide, C-00059
4090-60-2 Neopentylene phosphonate, *in* D-00735
4090-61-3 2-Mercapto-5,5-dimethyl-1,3,2-dioxaphosphorinane 2-oxide; Triethylammonium salt, *in* M-00010
4104-04-5 Dipropyl phosphorochloridodithioite, D-01220
4104-14-7 ▷*O,O*-Bis(4-chlorophenyl) (1-iminoethyl)-phosphoramidothioate, *in* I-00005
4104-65-8 Bis(2-methylpropyl)-phosphinothioic acid; Chloride, *in* B-00351
4110-66-1 ▷*N*-Benzoyl-*P,P*-bis(1-aziridinyl)phosphinic amide, *in* B-00092
4115-99-5 3,3',3''-Phosphinyldynetrispropanoic acid triethyl ester, *in* P-00357
4116-00-1 3,3',3''-Phosphinylidynetris[propanamide], *in* T-00738
4124-07-6 3-Butenyldiphenylphosphine, B-00521
4124-92-9 Dibutyl phosphorochloridite, D-00142
4125-25-1 Tris(2-methylpropyl)phosphine, T-00824
4129-17-3 ▷Diphenylphosphinic acid; Azide, *in* D-01039
4129-18-4 Bis(4-methylphenyl)phosphinic acid; Azide, *in* B-00320
4129-19-5 Bis(4-chlorophenyl)phosphinic acid; Azide, *in* B-00139
4129-20-8 Diphenylphosphinothioic acid; Azide, *in* D-01082
4129-24-2 *P,P*-Diphenyl-*N*-(triphenylphosphoranylidene)-phosphinothioic amide, *in* D-01165
4129-39-9 Bis(4-chlorophenyl)phosphinic acid; Chloride, *in* B-00139
4129-40-2 Bis(4-methylphenyl)phosphinic acid; Chloride, *in* B-00320
4129-41-3 *N,P,P*-Triphenylphosphinothioic amide, *in* D-01083
4129-47-9 *P*-Phenyl-*N,N'*-bis(triphenylphosphoranylidene)-phosphonic diamide, *in* P-00102
4130-19-2 *P*[1],*P*[4]-Diguanosine 5'-tetraphosphate, D-00423
4140-56-0 Dimethyl phenylsulfonylphosphoramidate, *in* P-00312
4141-48-4 Diphenyl-2-propenylphosphine; Oxide, *in* D-01144
4141-50-8 1,2-Bis(diphenylphosphinyl)ethane, *in* B-00234
4141-59-7 1,2-Bis(dibutylphosphino)ethane, B-00158
4141-63-3 1,2-Bis(dibutylphosphinyl)ethane, *in* B-00158
4145-68-0 Tris[(4-chlorophenyl)methyl]-phosphine; Oxide, *in* T-00744
4151-26-2 1,4-Bis(dibutylphosphinyl)butane, *in* B-00157
4151-27-3 1,4-Bis(diphenylphosphinyl)butane, *in* B-00225
4154-77-2 Hexaethylphosphorothioic triamide, *in* H-00015
4154-80-7 Diethylphosphoramidous acid; Dibromide, *in* D-00360
4167-02-6 2-Bromoethyl phosphorodichloridate, *in* M-00447
4167-52-6 *O,O*-Diethyl dimethylphosphoramidothioate, *in* D-00871
4168-73-4 Trihexylphosphine, T-00492
4188-33-4 1-Methylethenyl phosphorodichloridate, M-00151
4202-14-6 Dimethyl (2-oxopropyl)phosphonate, *in* O-00075
4205-52-1 ▷Phosbutyl, P-00341
4206-94-4 Methylphosphinic acid, M-00278
4214-38-4 (Diphenylmethylene)-triphenylphosphorane, D-01022
4225-51-8 *O*-(4-Nitrophenyl) phosphorodichloridothioate, N-00065
4233-13-0 Butyldiphenylphosphine; Oxide, *in* B-00557
4234-61-1 Tetraethylphosphorodiamidothioic chloride, T-00065
4235-95-4 Glycerol 1,2-di-9-octadecenoate 3-phosphocholine; (*R*)-(*Z,Z*)-*form*, *in* G-00042
4241-37-6 *O*-Isopropyl methylphosphonofluoridothioate, *in* M-00305
4242-21-1 2-Phenyl-4*H*-1,3,2-benzodioxaphosphorin; 2-Oxide, *in* P-00095
4242-25-5 2-Phenyl-4*H*-1,3,2-benzodioxaphosphorin; 2-Sulfide, *in* P-00095
4250-08-2 Dioxybis[diphenylphosphine oxide], D-00977
4252-61-3 (1-Methylpropyl)diphenylphosphine; (±)-*form*, Oxide, *in* M-00369
4252-88-4 Diphenylpropylphosphine; Oxide, *in* D-01145
4271-13-0 Methylphenylphosphinic acid, M-00219
4283-98-1 2,2'-Bis[(triphenylphosphonio)-methyl]biphenyl(2+); Dibromide, *in* B-00445
4297-95-4 Sodium phenylphosphinite, *in* P-00251
4303-59-7 (2-Methyl-2-propenyl)-triphenylphosphonium(1+); Chloride, *in* M-00367
4317-06-0 ▷Tetraethylphosphonium(1+); Iodide, *in* T-00056
4317-07-1 Tetraethylphosphonium(1+); Bromide, *in* T-00056
4319-13-5 Tris(2-phenylethenyl)phosphine; Sulfide, *in* T-00843
4323-64-2 Dibutylchlorophosphine, *in* D-00121
4328-22-7 ▷*O,O,O',O'*-Tetraethyl trithiopyrophosphate, T-00076
4335-89-1 ▷2,3-Dihydro-3,3,5-trimethyl-1,2-oxaphosphole 2-oxide; 2-Methoxy, *in* D-00596
4336-70-3 ▷(Cyanomethyl)-triphenylphosphonium(1+); Chloride, *in* C-00217
4343-60-6 Tris(2-carbamoylethyl)phosphine, T-00738
4358-50-3 2-(Diphenylphosphino)-*N,N*-dimethylaniline, D-01059
4363-06-8 Ethylphosphinic acid, E-00124
4378-43-2 4-Phosphonobutanoic acid, P-00386
4402-24-8 Diethyl (3-aminopropyl)-phosphonate, *in* A-00127
4403-51-4 ▷Bis(diethoxyphosphinyl) disulfide, B-00172
4403-52-5 Triethyl phosphoro(dithioperoxo)thioate, T-00453
4408-78-0 Phosphonoacetic acid, P-00376
4414-16-8 1-Chloro-2,5-dihydro-3-methyl-1*H*-phosphole; 1-Sulfide, *in* C-00047
4414-25-9 Trimethylsilyl phosphorodifluoridate, T-00559
4422-66-6 3-Phosphonobutanoic acid, P-00385
4428-95-9 Phosphonoformic acid, P-00387
4441-12-7 4,4',4''-Phosphinylidynetrismorpholine, *in* T-00569
4441-17-2 ▷1,1',1''-Phosphinylidynetrispiperidine, *in* T-00704
4456-11-5 Monomethoxycarbonyl phosphate, M-00479
4456-15-9 Ethyl carbonate-diethyl phosphate anhydride, E-00055
4478-20-0 2,3-Dihydro-1*H*-isophosphindole, D-00509
4484-88-2 Trehalose 6-(dihydrogen phosphate), T-00324
4486-43-5 *O,O*-Dipropyl phosphorothioate, D-01230
4486-44-6 *O,O*-Diisopropyl phosphorothioate, D-00667
4519-28-2 Tetramethylphosphonium(1+); Bromide, *in* T-00215
4519-33-9 Phenyl tetraethylphosphorodiamidate, P-00318
4519-65-7 Methylphosphonic acid; Diisothiocyanate, *in* M-00288
4526-20-9 Acetic acid anhydride with diethyl hydrogen phosphate, *in* A-00025
4529-76-4 2,3-Dihydro-3,3,5-trimethyl-1,2-oxaphosphole 2-oxide; 2-Ph, *in* D-00596
4532-02-9 ▷Diethyl 2-nitrophenyl phosphate, D-00303
4532-05-2 ▷Dimethyl 3-nitrophenyl phosphate, D-00769
4532-06-3 ▷Diethyl 3-nitrophenyl phosphate, D-00304
4537-76-2 Glycerol 1,2-dioctadecanoate 3-phosphonoethanolamine, G-00037
4537-77-3 Glycerol 1,2-dihexadecanoate 3-phospho-1'-glycerol, G-00014
4537-78-4 Glycerol 1,2-dioctadecanoate 3-phospho-1'-glycerol, G-00035
4538-29-8 Isopropylphosphine, I-00048
4538-32-3 (4-Chlorophenyl)phosphine, C-00141
4538-55-0 ▷Tribenzylphosphine; Oxide, *in* T-00338
4539-70-2 Glycerol 1,2-dioctadecanoate 3-phosphocholine, G-00034
4544-86-9 2,5-Dihydro-1*H*-phosphole, D-00572
4545-85-1 Benzyldichlorophosphine, *in* B-00068
4545-88-4 Tetramethyl (3,3,4,4,5,5-hexafluoro-1-cyclopentene-1,2-diyl)bisphosphonate, *in* H-00018
4545-92-0 Tetraethyl (3,3,4,4-tetrafluoro-1-cyclobutene-1,2-diyl)bisphosphonate, *in* T-00078
4545-93-1 Tetraethyl (3,3,4,4,5,5,6,6,-octafluoro-1-cyclohexene-1,2-diyl)-bisphosphonate, *in* O-00003
4545-95-3 1,2-Bis(diphenylphosphinyl)-3,3,4,4,5,5-hexafluorocyclopentene, *in* H-00016

4546-04-7 Tetraethyl [1,4-phenylenebis(methylene)]-bisphosphonate, *in* P-00131
4546-06-9 [1,4-Phenylenebis(methylene)]-bisphosphonic acid, P-00131
4546-11-6 Hydrogen propyl methyl-phosphonate, *in* M-00288
4546-15-0 Monoethyl cyclohexylphosphonate, *in* C-00256
4546-19-4 Monoethyl phenylphosphonate, M-00471
4547-77-7 Tris(phenylethynyl)phosphine, T-00846
4552-71-0 Hexaphenylhexaphosphorinane, H-00089
4552-75-4 1-Chloro-2,5-dihydro-2-methyl-1*H*-phosphole; 1-Sulfide, *in* C-00046
4559-70-0 Diphenylphosphine; Oxide, *in* D-01037
4559-70-0 Diphenylphosphine oxide, D-01038
4569-35-1 Diphenyl-1-propenylphosphine; (*Z*)-*form*, Oxide, *in* D-01143
4569-37-3 Diphenylphosphinothioic acid; Cyanide, *in* D-01082
4575-86-4 *O*-Phenyl saligenin cyclic thiophosphate, *in* P-00064
4576-56-1 ▷Tris(4-chlorophenyl)phosphine; Oxide, *in* T-00749
4579-03-7 2,6,7-Trioxa-1,4-diphosphabicyclo[2.2.2]-octane, T-00585
4591-40-6 2-Phenoxy-1,3,2-benzodioxaphosphole, P-00063
4600-12-8 2,3-Dihydro-1,3-dimethyl-2-phenyl-1*H*-benzodiazaphosphole; 2-Sulfide, *in* D-00467
4608-06-4 Diphenyl-1-propenylphosphine; (*E*)-*form*, Oxide, *in* D-01143
4614-90-8 Diphenylphosphoramidous acid; Dichloride, *in* D-01111
4621-50-5 Triheptyl phosphate, T-00488
4631-50-9 Diethyl[(trimethylsilyloxy)-methyl]phosphonate, *in* T-00556
4632-37-5 Diphenyl(triphenylstannyl)-phosphine, D-01167
4645-32-3 Dimethyl ethenylphosphonate, *in* V-00003
4645-89-0 Phenyl *N,P*-dimethylphosphonamidate, *in* D-00849
4645-91-4 Phenyl *N*-isopropyl-*P*-methylphosphonamidate, *in* I-00036
4645-95-8 Phenyl *N,N*-diethyl-*P*-methylphosphonamidate, *in* D-00296
4646-06-4 2,6,7-Trioxa-1,4-diphosphabicyclo[2.2.2]-octane; 1-Sulfide, *in* T-00585
4648-58-2 (4-Bromophenyl)phosphonic acid; Dichloride, *in* B-00492
4652-20-4 Bis(2-methylpropyl)-phosphinothioic acid; Bromide, *in* B-00351
4653-50-3 *P*-Methyl-*N,N'*-diphenylphosphonic diamide, *in* M-00289
4653-62-7 Methylphenylpropylphosphine, M-00265
4655-86-1 Tetraethyl (3,3,4,4,5,5-hexafluoro-1-cyclopentene-1,2-diyl)bisphosphonate, *in* H-00018
4669-51-6 2-Chloro-1,3,2-dithiaphospholane, C-00071
4669-52-7 *S,S*-Ethylene phenylphosphonodithioate, *in* P-00121
4669-54-9 2-Phenyl-1,3,2-dithiaphospholane, P-00121
4669-69-6 Diphenylphosphinic acid; Cyanide, *in* D-01039
4671-77-6 1,4-Butanediphosphonic acid, B-00514
4671-82-3 1,3-Propanediphosphonic acid, P-00444
4672-25-7 1,5-Pentanediphosphonic acid, P-00032
4672-26-8 Pentylphosphonic acid, P-00050
4672-27-9 (1,1-Dimethylpropyl)phosphonic acid, D-00911
4672-28-0 1,1-Pentanediphosphonic acid, P-00031
4672-30-4 (2-Phenylethyl)phosphonic acid, P-00165
4672-35-9 (1-Piperidinylmethyl)phosphonic acid, P-00429
4672-37-1 1-Propenylphosphonic acid, P-00453
4672-38-2 ▷Propylphosphonic acid, P-00473
4672-45-1 (2-Chlorophenyl)phosphonic acid; Dichloride, *in* C-00145
4693-63-4 Benzoic diphenylphosphinic anhydride, *in* D-01039
4707-88-4 ▷*P*-Phenylphosphonic diamide, P-00222
4707-91-9 ▷*N,N',P*-Triphenylphosphonic diamide, *in* P-00222
4707-93-1 (1,3-Butadienyl)phosphonic acid; Dichloride, *in* B-00510
4707-94-2 (1-Methylpropyl)phosphonic dichloride, M-00379
4707-95-3 *tert*-Butylphosphonic dichloride, B-00599
4708-01-4 (2-Methyl-1-propenyl)phosphonic dichloride, M-00365
4708-04-7 Propylphosphonic dichloride, P-00475
4708-07-0 (2-Phenylethenyl)phosphonic dichloride, P-00157
4712-06-5 (Aminobenzylidene)bisphosphonic acid, A-00053
4712-55-4 Diphenyl phosphonate, D-01105
4721-16-8 Heptylphosphonic acid, H-00006
4721-22-6 1,6-Hexanediphosphonic acid, H-00085
4721-24-8 Hexylphosphonic acid, H-00097
4721-34-0 (2-Methylpropyl)phosphonic acid, M-00378
4721-37-3 Isopropylphosphonic acid, I-00050
4722-60-5 2,4,5,6-Tetraphenyl-1,3,5-dioxaphosphorinane, T-00254
4724-48-5 Octylphosphonic acid, O-00042
4724-50-9 (2-Oxo-2-phenylethyl)phosphonic acid, O-00066
4726-80-1 ▷2,6,7-Trioxa-1,4-diphosphabicyclo[2.2.2]-octane; 1,4-Dioxide, *in* T-00585
4730-74-9 4-Morpholinylphosphonic acid, M-00523
4730-75-0 (4-Morpholinylmethyl)phosphonic acid, M-00521
4730-79-4 Nonylphosphonic acid, N-00076
4731-53-7 Trioctylphosphine, T-00582
4731-62-8 Iododimethylphosphine, *in* D-00844
4731-65-1 Tris(2-methoxyphenyl)phosphine, T-00791
4736-60-1 ▷Ethyltriphenylphosphonium(1+); Iodide, *in* E-00197
4736-97-4 Ethylphosphonous acid, E-00153
4737-12-6 Phenylsulfonylphosphoramidic acid, P-00312
4737-14-8 Benzoylphosphoramidic dichloride, B-00037
4743-38-8 *N,N',N''*-Triphenylphosphorothioic triamide, *in* T-00677
4743-40-2 Phosphorinane, P-00404
4743-42-4 *N,N'*-Diphenylphosphorodiamidic acid, D-01125
4756-25-6 9*H*-Fluoren-9-ylidenetriphenylphosphorane, F-00010
4757-24-8 1,3,2-Dioxaphosphepane, D-00968
4757-26-0 1,3,2-Dioxaphospholane, D-00970
4757-27-1 1,3,2-Dioxaphosphorinane, D-00972
4759-28-8 Triphenylmethylphosphonic acid, T-00608
4759-30-2 *P*-Methylphosphonic diamide, M-00289
4759-37-9 2-Phenyl-1,3,2-benzodioxaphosphole, P-00094
4762-26-9 Hexyltriphenylphosphonium(1+); Bromide, *in* H-00104
4762-30-5 (1-Methylpropyl)-triphenylphosphonium(1+); Iodide, *in* M-00386
4762-31-6 4-Bromophenyldichlorophosphine, *in* B-00493
4762-37-2 Diphenyl (4-bromophenyl)-phosphonite, *in* B-00493
4762-64-5 Dibutyl 2-propenylphosphonate, *in* P-00454
4764-18-5 1-Piperidinylphosphonic acid, P-00430
4775-09-1 Ethyl diethylphosphinate, *in* D-00322
4775-92-2 3,3-Diphosphonopropanoic acid, D-01187
4779-09-3 Triethyl [(4-methylphenyl)-sulfonyl]phosphorimidate, *in* M-00268
4789-36-0 *O*-Ethyl ethylphosphonothioate; (*R*)-*form*, *in* E-00074
4789-37-1 *O*-Ethyl ethylphosphonochloridothioate; (*R*)-*form*, *in* E-00069
4791-48-4 Diphenylphosphinous acid; Cyanide, *in* D-01089
4795-31-7 Bis(4-chlorophenyl) phosphate, B-00133
4824-78-6 ▷Bromophos-Et, B-00499
4836-72-0 *O,O*-Diphenyl phosphorothioate; K salt, *in* D-01141
4844-39-7 Dimethyl (1-phenylethenyl)-phosphonate, *in* P-00154
4846-23-5 2-Hydroxy-1,3,2-benzodioxaphosphole 2-oxide, H-00108
4848-43-5 2-(Diphenylphosphino)ethylamine, D-01066
4850-60-6 2,2-Dihydrophenanthro[9,10-*d*]-1,3,2-dioxaphosphole; 2,2,2-Triphenoxy, *in* D-00541
4851-51-8 Diethyl ethynylphosphonate, *in* E-00203
4851-54-1 1-Butynylphosphonic acid, B-00643
4851-54-1 Diethyl 1-butynylphosphonate, *in* B-00643
4851-69-8 Methylenebis[phenylphosphinic acid], M-00142
4851-72-3 1,3-Propanediylbis[phenylphosphinic acid], P-00447
4851-97-2 Tris(hydroxymethyl)phosphine; Oxide, tri-Ac, *in* T-00784
4881-11-2 Ethyl di-4-morpholinylphosphinite, *in* D-00936
4881-12-3 Ethyl di-4-morpholinylphosphinate, *in* D-00935
4881-17-8 ▷Phenyl di-4-morpholinylphosphinate, *in* D-00935
4885-47-6 *S*-Ethyl diethylphosphinothioate, *in* D-00330
4885-49-8 *O*-Ethyl diethylphosphinothioate, *in* D-00330
4885-52-3 *O*-Acetyl diethylphosphinothioate, *in* D-00330
4885-53-4 Diethylphosphinic diethylphosphinothioic anhydride, *in* D-00322
4885-54-5 *O*-Trimethylsilyl diethylphosphinothioate, *in* D-00330
4885-57-8 (2-Oxo-1,3-propanediyl)-bis(triphenylphosphonium)(2+); Dichloride, *in* O-00073
4895-48-1 Diethylphosphinothioic acid; Anhydride, *in* D-00330
4895-54-9 Phenyl(phenylethynyl)phosphinic acid, P-00187
4895-65-2 Tetrachlorophenylphosphorane, T-00034
4895-67-4 Phenyl(2-phenylethenyl)phosphinic acid, P-00186
4895-75-4 Bis(2-phenylethenyl)phosphinic acid, B-00386
4895-86-7 Bis(phenylethynyl)phosphinic acid, B-00388
4902-59-4 Monoethyl phosphate; Bis(cyclohexylammonium) salt, *in* M-00472
4903-06-4 2,2-Dihydrophenanthro[9,10-*d*]-1,3,2-dioxaphosphole; 2,2,2-Trimethoxy, *in* D-00541

4923-84-6 *tert*-Butylphosphonic acid, B-00595
4923-86-8 *tert*-Butylphenylphosphinic acid, B-00579
4938-89-0 ▷Dimethyl 2-nitrophenyl phosphate, D-00768
4963-91-1 1-Phenylphospholane; 1-Oxide, *in* P-00216
4963-94-4 1-Phenylphosphorinane; Sulfide, *in* P-00265
4963-95-5 1-Phenylphosphorinane; Oxide, *in* P-00265
4963-96-6 1-Phenylphosphepane, P-00205
4972-36-5 Diethyl 1-piperidinylphosphonate, *in* P-00430
4972-39-8 Diethyl 1-piperidinylphosphonite, *in* P-00432
4972-40-1 Diethyl (1-piperidinylmethyl)-phosphonate, *in* P-00429
4973-47-1 [Thiobis[methylene]]-bis[triphenylphosphonium](2+); Dibromide, *in* T-00312
4981-25-3 1,3-Pentadienylphosphonic acid; Dichloride, *in* P-00009
4981-27-5 (2-Methyl-1,3-butadienyl)-phosphonic acid; Dichloride, *in* M-00093
4981-29-7 (2-Methyl-1,3-butadienyl)-phosphonothioic acid; Dichloride, *in* M-00097
4981-30-0 1-Propynylphosphonic dichloride, P-00496
4981-31-1 3-Buten-1-ynylphosphonic acid; Dichloride, *in* B-00529
4981-32-2 Phenylethynylphosphonic acid; Dichloride, *in* P-00166
5003-99-6 Bis(2-phenylethenyl)phosphinic acid; Chloride, *in* B-00386
5021-92-1 2,5-Diethoxy-1,4,2,5-dioxaphosphorinane 2,5-dioxide, *in* D-00618
5021-98-7 (2-Methylpropyl)phosphonic acid; Dichloride, *in* M-00378
5022-30-0 *O,O*-Dimethyl (1,3-dihydro-1,3-dioxo-2*H*-isoindol-2-yl)-phosphonothioate, *in* D-00475
5022-56-0 *P,P*-Diethylphosphinothioic amide, D-00331
5032-39-3 2-Chloro-2,2,4,4,6,6-hexahydro-2,4,4,6,6-pentaphenoxy-1,3,5-triaza-2,4,6-triphosphorine, C-00087
5032-51-9 Methyl 4-(diphenylphosphino)-benzoate, *in* D-01055
5032-54-2 4-(Diphenylphosphinyl)-benzonitrile, *in* D-01058
5032-60-0 4-(Diphenylphosphinothioyl)-benzonitrile, *in* D-01058
5032-62-2 Tris(4-chlorophenyl)phosphine; Sulfide, *in* T-00749
5032-63-3 (4-Nitrophenyl)diphenylphosphine, N-00040
5032-67-7 3-(Diphenylphosphinyl)-propanenitrile, *in* D-01073
5032-78-0 Tetrakis[(diphenylphosphino)-methyl]methane, T-00179
5032-98-4 Benzoyl(triphenyl-phosphoranylidene)acetonitrile, *in* O-00083
5035-79-0 (1-Amino-1-methylethyl)phosphonic acid, A-00087
5044-52-0 ▷Triphenylvinylphosphonium(1+); Bromide, *in* T-00702
5054-42-2 Phenylphosphinic acid; Anhydride, *in* P-00210
5068-16-6 4-(Diphenylphosphino)benzonitrile, D-01058
5068-18-8 4-(Diphenylphosphino)benzaldehyde, D-01052
5068-21-3 4-(Diphenylphosphino)phenol, D-01072
5068-23-5 4-(Diphenylphosphinyl)-benzaldehyde, *in* D-01052
5068-24-6 *p*-(Diphenylphosphinothioyl)-benzoic acid, *in* D-01055
5074-71-5 Bis(pentafluorophenyl)-phenylphosphine, B-00380
5074-74-8 (Pentafluorophenyl)-diphenylphosphine; Oxide, *in* P-00014
5074-77-1 *O*-Ethyl butylphosphonodithioate, *in* B-00606
5075-13-8 ▷*O*-Ethyl phenylphosphonochloridothioate, E-00112
5075-15-0 Ethylphenylphosphinothioic acid; (±)-*form*, Chloride, *in* E-00105
5101-78-1 *S*-Phenyl diphenylphosphinothioate, *in* D-01082
5110-51-0 Diethyl 1-aziridinylphosphonite, *in* A-00154
5112-95-8 Bis(diphenylphosphino)acetylene, B-00216
5115-19-5 Phosphoglycocyamine, P-00369
5116-77-8 2,2,4,4,6,6-Hexahydro-2,2,4,4,6,6-hexapropoxy-1,3,5-triaza-2,4,6-triphosphorine, H-00033
5120-49-0 Ethyl phenyl-phosphonochloridodithioate, *in* P-00230
5120-50-3 *N,N*-Diethyl-*P*-phenylphosphonamidothioic acid; Chloride, *in* D-00309
5129-68-0 Glycerol 1,2-dihexadecanoate 3-phosphate; (±)-*form*, *in* G-00011
5131-05-5 Dimethyl 1-propynylphosphonate, *in* P-00494
5131-24-8 ▷Ditalimfos, D-01237
5137-89-3 Benzylmethyl-phenylpropylphosphonium(1+); (*S*)-*form*, Bromide, *in* B-00052
5152-74-9 *O*-Ethyl ethylphosphonothioate; (*S*)-*form*, *in* E-00074
5156-98-9 Diethyl phenylphosphoramidite, *in* P-00263
5186-73-2 2,5-Dihydro-1-phenyl-1*H*-phosphole; 1-Oxide, *in* D-00567
5187-31-5 ▷*O,O*-Dipropyl chlorothiophosphonothioate, D-01201
5191-39-5 Diethyl (2-ethoxyethyl)-phosphonate, *in* H-00146
5234-91-3 2,5-Dihydro-1-phenoxy-1*H*-phosphole 1-oxide, *in* D-00507
5259-81-4 *N*-Phosphoglycine, P-00368
5274-20-4 5,5,8,8-Tetraethyl-5,6,7,8-tetrahydrodibenzo[*e,g*]-[1,4]diphosphocinium dibromide, *in* D-00404
5274-21-5 5,5,9,9-Tetraethyl-6,7,8,9-tetrahydro-5*H*-dibenzo[*f,h*][1,5]-diphosphoninium dibromide, *in* T-00089
5274-22-6 5,5-Dimethyl-5*H*-dibenzophospholium iodide, *in* M-00115
5276-94-8 (Aminomethyl)diphenylphosphine oxide, A-00085
5277-10-1 *P*-Butyl-*N,N,N',N'*-tetramethylphosphonic diamide, *in* B-00596
5277-12-3 *P*-Heptyl-*N,N,N',N'*-tetramethylphosphonic diamide, *in* H-00006
5284-09-3 Ethyl methylphosphonochloridate, *in* M-00296
5284-11-7 Monoethyl butylphosphonate, *in* B-00594
5284-12-8 Ethyl phenylphosphonochloridate, *in* P-00229
5290-43-7 *P,P,P*-Trichloro-*N*-phenylphosphine imide, *in* P-00264
5290-45-9 *P*-Chloro-*N,P,P*-triphenylphosphazene, *in* T-00631
5290-57-3 Phenylvinylphosphinic acid; Chloride, *in* P-00337
5293-84-4 Hydroxymethyltriphenyl-phosphonium(1+); Chloride, *in* H-00163
5293-84-5 ▷(Chloromethyl)-triphenylphosphonium(1+); Chloride, *in* C-00117
5301-78-0 2,6,7-Trioxa-1-phosphabicyclo[2.2.2]-octane-4-methanol; Oxide, *in* T-00590
5305-81-7 3,9-Dichloro-2,4,8,10-tetraoxa-3,9-diphosphaspiro[5.5]-undecane; 3,9-Disulfide, *in* D-00196
5310-87-2 Dichloro-*o*-tolylphosphine, *in* M-00246
5324-30-1 Diethyl (2-bromoethyl)phosphonate, *in* B-00472
5326-06-7 Ethyl *P*-phenylphosphonamidate, *in* P-00218
5326-10-3 *N,N',N''*-Triphenylphosphoric triamide, *in* T-00677
5332-68-3 ▷*N,N',N''*-Tricyclohexylphosphorothioic triamide, *in* T-00429
5337-17-7 ▷(4-Aminophenyl)phosphonic acid, A-00111
5337-19-3 (3-Nitrophenyl)phosphonic acid, N-00056
5353-22-0 5,5,7,7-Tetraethyl-6,7-dihydro-5*H*-dibenzo[*d,f*][1,3]-diphosphepinium dibromide, *in* D-00282
5353-66-2 (Diazomethyl)diphenylphosphine oxide, D-00034
5368-60-5 [Oxybis[methylene]]-bis[triphenylphosphonium](2+); Dibromide, *in* O-00097
5381-98-6 2-Chloro-4*H*-1,3,2-benzodioxaphosphorinan-4-one; 2-Oxide, *in* C-00022
5382-00-3 Diphenyl phosphorochloridite, D-01121
5390-60-3 Methyl phosphorodichloridodithioate, M-00347
5391-92-4 Tetraethyl 1,6-hexanediylbisphosphonate, *in* H-00085
5395-21-1 *P*-Phenylphosphonothioic dihydrazide, P-00249
5395-86-8 *N,N',N''*-Tripropylphosphorothioic triamide, *in* T-00719
5427-30-5 ▷(3-Aminophenyl)phosphonic acid, A-00110
5431-19-6 ▷(2-Chlorophenyl)phosphonic acid, C-00145
5431-34-5 ▷(3-Chlorophenyl)phosphonic acid, C-00146
5431-35-6 ▷(4-Chlorophenyl)phosphonic acid, C-00147
5464-68-6 Diethyl (2-amino-2-oxoethyl)-phosphonate, *in* A-00060
5467-82-3 Phenyl *P*-phenylphosphonamidate, *in* P-00218
5510-88-3 Dichloro-*m*-tolylphosphine, *in* M-00247
5510-93-0 Dichloro(4-fluorophenyl)phosphine, *in* F-00042
5510-94-1 Dichloro(3-fluorophenyl)phosphine, *in* F-00041
5518-52-5 Tri-2-furanylphosphine, T-00486
5518-61-6 Diphosphinomethane, D-01178
5518-62-7 ▷1,2-Diphosphinoethane, D-01176
5518-64-9 1,4-Diphosphinobutane, D-01174
5518-69-4 1-Propynylphosphonic acid, P-00494
5518-70-7 Tri-2-furanylphosphine; Sulfide, *in* T-00486
5525-95-1 (Pentafluorophenyl)-diphenylphosphine, P-00014
5553-01-5 *O*-Phenyl dimethylphosphinothioate, *in* D-00839
5554-81-4 4-Methylphenyl triphenylphosphoranylidenemethyl sulfone, M-00271
5554-83-6 [(Methylsulfonyl)methylene]-triphenylphosphorane, M-00399
5559-92-2 2-Propenyl di-2-propenylphosphinate, *in* D-01193
5572-32-7 ▷*O,O*-Diisopropyl chlorothiophosphonothioate, D-00635
5573-42-2 Benzyl diphenylphosphinate, *in* D-01039
5586-04-9 *P,P*-Diethyl-*N*-phenylphosphinous amide, *in* D-00334
5586-05-0 *P,P*-Diethyl-*N,N'*-diphenylphosphinimidic amide, *in* D-00306
5586-09-4 Ethyl *N,N'*-diphenylphosphorodiamidate, *in* D-01125

5587-39-3 (4-Methylphenyl)diphenylphosphine; Sulfide, *in* M-00202
5591-07-1 2,4-Dichloro-2,4,6,6-tetrakis(dimethylamino)-2,2,4,4,6,6-hexahydro-1,3,5-triaza-2,4,6-triphosphorine; (4*RS*,6*RS*)-*form*, *in* D-00194
5592-65-4 Dichloro(2-fluorophenyl)phosphine, *in* F-00040
5593-72-6 Tri-2-furanylphosphine; Selenide, *in* T-00486
5594-90-1 Bis(pentafluorophenyl)-phenylphosphine; Oxide, *in* B-00380
5598-13-0 ▷Chlorpyrifos Methyl, C-00195
5598-52-7 ▷Fospirate, F-00063
5625-99-0 ▷*N,N*′-Diphenylphosphorodiamidic chloride, D-01126
5638-58-4 2-Chlorotetrahydro-2*H*-1,3,2-oxazaphosphorine; 2-Oxide, *in* C-00180
5652-28-8 2-Amino-3-phosphonopropanoic acid, A-00121
5652-40-4 3-Amino-3-phosphonopropanoic acid, A-00122
5666-12-6 Tri-1-pyrrolidylphosphine, T-00724
5666-55-7 Phenylsulfonylphosphorimidic trichloride, P-00316
5675-55-8 Ethyl 3-[(dimethoxyphosphinyl)-oxy]crotonate, *in* D-00681
5679-61-8 Tris(4-chlorophenyl) phosphite, T-00751
5679-72-1 2-Ethoxy-3-phenyl-1,3,2-oxazaphospholidine, E-00048
5679-73-2 2,3-Diphenyl-1,3,2-oxazaphospholidine, D-01029
5681-36-7 Glycerol 1,2-dihexadecanoate 3-phosphoethanolamine; (±)-*form*, *in* G-00013
5681-42-5 [[(4-Methylphenyl)sulfonyl]-methyl]triphenyl-phosphonium(1+); Bromide, *in* M-00267
5683-46-5 Glycerol 1,2-ditetradecanoate 3-phosphoethanolamine; (±)-*form*, *in* G-00049
5689-41-8 Methyl diethylphosphinate, *in* D-00322
5689-46-3 1,4-Butanediylbis[phenyl-phosphinodithioic acid], B-00516
5693-11-8 2,3-Diphosphonobutanedioic acid, D-01182
5717-37-3 Ethyl 2-triphenyl-phosphoranylidenepropanoate, E-00199
5745-32-4 Ethyl diethylphosphinodithioate, *in* D-00325
5745-37-9 Phenyl dimethylphosphinodithioate, *in* D-00834
5746-57-6 Glycerol 1-monophosphate; (*R*)-*form*, *in* G-00066
5756-02-5 Diphenyl ethylphosphoramidate, *in* E-00156
5756-03-6 Diphenyl propylphosphoramidate, *in* P-00484
5756-04-7 Diphenyl isopropylphosphoramidate, *in* I-00058
5756-05-8 Diphenyl butylphosphoramidate, *in* B-00621
5756-07-0 Dibutyl butylphosphoramidate, *in* B-00621
5756-10-5 Dibutyl *tert*-butylphosphoramidate, *in* B-00622
5761-95-5 Dimethylphosphinothioic acid, D-00839
5768-48-9 Phosphonobutanedioic acid, P-00381
5786-71-0 Fosfocreatinine, F-00060
5794-87-6 1-Methylphospholane; Oxide, *in* M-00280
5810-88-8 ▷*O,O*-Bis(2-ethylhexyl) phosphorodithioate, B-00274
5815-61-2 Trimorpholinophosphine, T-00569
5827-05-4 ▷Aphidan, A-00142
5827-17-8 Diphenylphosphinodithioic acid; Na salt, *in* D-01065
5827-24-7 Methylphosphonothioic dibromide, M-00315
5836-28-2 ▷Tetraisopropyl diphosphate, T-00166
5843-26-5 *N,N,N*′,*N*′-Tetramethylphosphonic diamide, T-00214
5843-28-7 *N,N,N*′,*N*′-Tetrapropylphosphonic diamide, T-00289
5843-29-8 *N,N,N*′,*N*′-Tetrabutylphosphonic diamide, T-00018
5848-64-6 Hexapropylphosphorous triamide, H-00092
5848-65-7 Hexabutylphosphorous triamide, H-00013
5849-35-4 Dioctylphosphinic acid; Chloride, *in* D-00958
5849-36-5 Diphenylphosphinic acid; Anhydride, *in* D-01039
5849-47-8 Di-2-thienylphosphinic acid, D-01250
5851-14-9 *O,O*-Di-2-propenyl phosphorodithioate, D-01197
5851-18-3 *O,O*-Diethyl diethylphosphoramidothioate, *in* D-00356
5852-63-1 *O,O*-Diethyl phosphorothioate; Na salt, *in* D-00396
5853-64-5 Diphenylphosphine; Selenide, *in* D-01037
5853-65-6 Bis(2-methylpropyl)phosphine; Selenide, *in* B-00344
5871-16-9 *O,O*-Diethyl phosphorothioate; NH_4 salt, *in* D-00396
5871-17-0 *O,O*-Diethyl phosphorothioate; K salt, *in* D-00396
5927-18-4 Methyl (dimethoxyphosphinyl)-acetate, *in* D-00678
5927-50-4 Tetramethyl 1,2-ethanediylbisphosphonate, *in* E-00006
5930-71-2 ▷Bis(dimethoxyphosphinothioyl) disulfide, B-00185
5930-72-3 *O,O*-Dimethyl phosphonothioate, D-00853
5930-73-4 *O,O,O*′,*O*′-Tetramethyl trithiopyrophosphate, T-00240
5931-53-3 (2-Methylphenyl)diphenylphosphine, M-00200
5931-54-4 3-(Diphenylphosphino)-*N,N*-dimethylaniline, D-01060
5935-34-2 *O,O,O*′,*O*′-Tetraethyl thiohypophosphate, T-00073
5935-39-7 Tetraethyl thiohypophosphate, T-00072
5942-80-3 3,3,4,4,5,5-Hexafluoro-1,2-cyclopentenediphosphonic acid, H-00018
5943-21-5 1,10-Decanediphosphonic acid, D-00008
5943-62-4 Tetraethyl 1,10-decanediylbisphosphonate, *in* D-00008
5944-33-2 [3-(2-Methyl-1,3-dioxolan-2-yl)-propyl]triphenylphosphonium; Bromide, *in* M-00132
5944-50-3 2,4-Dichloro-1,3-dimethyl-1,3,2,4-diazadiphosphetidine; 2,4-Disulfide, *in* D-00166
5944-59-2 2,4-Dichloro-1,3-dimethyl-1,3,2,4-diazadiphosphetidine; 2,4-Dioxide, *in* D-00166
5952-41-0 ▷Dimethyl 1-acetoxy-2,2,2-trichloroethylphosphonate, *in* A-00011
5952-42-1 ▷Diethyl 1-acetoxy-2,2,2-trichloroethylphosphonate, *in* A-00011
5952-47-6 (2-Methylpropyl)diphenylphosphine, M-00370
5952-49-8 Diphenyl(2-phenylethyl)phosphine, D-01034
5954-50-7 ▷Dimethyl phosphorofluoridate, D-00901
5958-44-1 (Diphenylphosphino)methanol, D-01069
5958-53-2 Tetrabutyldiphosphine; Disulfide, *in* T-00014
5958-55-4 1,1′-Biphospholane; Disulfide, *in* B-00084
5958-56-5 1,1′-Biphosphorinane; Disulfide, *in* B-00085
5961-85-3 3,3′,3″-Phosphinidynetrispropanoic acid, P-00357
5962-41-4 2-Phosphonopropanoic acid, P-00395
5962-42-5 ▷3-Phosphonopropanoic acid, P-00396
5988-22-7 Phytonadiol sodium diphosphate, *in* P-00427
5994-41-2 (Trichloromethyl)phosphonic acid, T-00411
5994-56-9 1,2-Phenylenebis[phenylphosphinic acid], P-00139
5994-74-1 *P*-Methyl-*N*-isopropyl-*N*′-phenylphosphonothioic diamide, *in* M-00245
5994-75-2 *N,P*-Dimethyl-*N*′-phenylphosphonothioic diamide, *in* M-00245
5994-87-6 *P,P*-Diphenylphosphinic amide, D-01040
5995-06-2 Tetra-*tert*-butyldiphosphine, T-00015
5995-07-3 Tetra-*tert*-butylcyclotetraphosphine, T-00012
5995-10-8 Di-*tert*-butylphosphinodithioic anhydrosulfide, *in* D-00113
5996-17-8 Glucose 6-dihydrogen phosphate; D-*form*, Di-K salt, *in* G-00007
6002-26-2 *N*′,*P,P*-Triphenyl-*N*-(triphenylphosphoranylidene)-phosphinimidic amide, *in* D-01164
6002-34-2 *tert*-Butyldiphenylphosphine, B-00558
6002-40-0 Di-*tert*-butylmethylphosphine, D-00099
6002-45-5 *tert*-Butylphenylphosphinothioic acid, B-00583
6046-15-7 Dimethylphosphinothioic acid; NH_4 salt, *in* D-00839
6048-29-9 ▷(2-Oxo-2-phenylethyl)-triphenylphosphonium(1+); Bromide, *in* O-00067
6064-83-1 ▷2-(Phosphonooxy)benzoic acid, P-00391
6064-84-2 Mono(2-nitrophenyl) phosphate, M-00498
6068-99-1 Pentyl phosphorodichloridite, P-00054
6069-36-9 2-Chloro-1,3-dimethyl-1,3,2-diazaphospholidine, C-00055
6069-37-0 2-Chloro-1,3-dimethyl-1,3,2-diazaphospholidine; 2-Oxide, *in* C-00055
6069-38-1 1,3-Dimethyl-2-dimethylamino-1,3,2-diazaphospholidine, D-00724
6079-78-3 Diphenylphosphinodithioic acid; Anhydrosulfide, *in* D-01065
6087-94-1 ▷Bis(2-chloroethyl) phosphorochloridate, *in* B-00119
6095-42-7 Trihexyl phosphite, T-00493
6099-41-8 2-(Chlorocarbonyl)phenyl phosphorodichloridate, C-00027
6104-47-8 Ethynyldiphenylphosphine, E-00201
6104-48-9 Diphenylphosphinylacetylene, *in* E-00201
6131-74-4 Dimethyl propylphosphonite, *in* P-00482
6141-77-1 *N,N*′,*N*″-Tris(2-methylpropyl)-phosphoric triamide, *in* T-00830
6141-78-2 *N,N*′,*N*″-Trimethylphosphorothioic triamide, *in* T-00551
6141-79-3 *N,N*′,*N*″-Tris(2-methylpropyl)-phosphorothioic triamide, *in* T-00830
6141-80-6 Propylphosphoramidothioic acid; Dichloride, *in* P-00485
6145-31-9 1,2-Ethanediphosphonic acid, E-00006
6145-32-0 2,2-Propanediphosphonic acid, P-00445
6151-88-8 Tris(3-methylphenyl)phosphine; Oxide, *in* T-00807
6151-90-2 Dihexyl phosphonate, D-00435
6156-15-6 2-Methoxy-4-methyl-1,3,2-dioxaphospholane, M-00043

6163-58-2 Tris(2-methylphenyl)phosphine, T-00806
6163-61-7 Tris(2-methylphenyl)phosphine; Sulfide, *in* T-00806
6163-62-8 Tris(3-methylphenyl)phosphine; Sulfide, *in* T-00807
6163-63-9 Tris(2-methylphenyl)phosphine; Oxide, *in* T-00806
6163-75-3 Dimethyl ethylphosphonate, D-00754
6163-90-2 Diheptyl phosphonate, D-00426
6168-27-0 1,2-Diphenyl-1,2-diphosphorinane; 1,2-Disulfide, *in* D-01015
6171-94-4 Sarin; (+)-*form*, *in* S-00002
6172-92-5 Dimethyl hexylphosphonate, *in* H-00097
6172-97-0 Dimethyl octylphosphonate, *in* O-00042
6190-11-0 1,2-Diphenyl-1,2-diphosphorinane, D-01015
6190-28-9 *N,P,P*-Triphenylphosphinic amide, T-00630
6196-71-0 Didodecylphosphinic acid, D-00226
6200-26-6 2,4,6-Triisopropyl-5-mercapto-1,3,5-dioxaphosphorinane 5-sulfide; NH_4 salt, *in* T-00498
6214-04-6 Diphenyl diethylphosphoramidate, *in* D-00343
6214-09-1 Diphenyl 1-piperidinylphosphonate, *in* P-00430
6224-63-1 Tris(3-methylphenyl)phosphine, T-00807
6224-65-3 Tris(4-methylphenyl)phosphine; Sulfide, *in* T-00808
6224-94-8 Diphenyl(1-propynyl)phosphine, D-01146
6224-95-9 1-(Diphenylphosphinyl)propyne, *in* D-01146
6225-99-6 Ethyl diisopropylphosphinite, *in* D-00648
6226-05-7 *N,N′*-1,2-Ethanediyl-*N,N′*-dimethyl-*P*-phenylphosphonic diamide, *in* D-00793
6228-47-3 Triphenylpropylphosphonium(1+); Bromide, *in* T-00680
6230-82-6 ▷(2-Oxo-2-phenylethyl)-triphenylphosphonium(1+); Iodide, *in* O-00067
6230-93-9 *O*-Ethyl phenylphosphonothioate, E-00114
6230-97-3 *O*-Butyl phenylphosphonodithioate, *in* P-00236
6231-02-3 Phenylphosphonothioic acid; Dibromide, *in* P-00244
6231-03-4 *O,O*-Diethyl phenylphosphonothioate, *in* P-00244
6249-37-2 *P,P*-Diphenyl-*N*-(triphenylarsoranylidene)-phosphinic amide, *in* D-01160
6254-02-0 Phenyl phenylphosphoramidic acid, P-00189
6259-97-8 (Bromomethyl)phosphonic acid; Dibromide, *in* B-00475
6278-47-3 *N,N′*-Diethyl-*P*-phenylphosphonothioic diamide, *in* P-00246
6294-34-4 Bis(2-chloroethyl) (2-chloroethyl)phosphonate, *in* C-00074
6303-30-6 Bis(2-methylpropyl) phosphate, B-00342
6323-97-3 (1-Aminoethyl)phosphonic acid, A-00078
6323-99-5 2-Amino-4-phosphonobutanoic acid, A-00117
6324-00-1 (1-Amino-2-phenylethyl)phosphonic acid, A-00101
6326-70-1 *P*-Chloromethylphosphonic diamide, C-00107
6326-72-3 *N,N′,N″*-Trimethylphosphoric triamide, *in* T-00551
6326-73-4 Diethyl methylphosphoramidate, D-00299
6329-46-0 Dimethyl (hydroxyphenylmethyl)-phosphonate, *in* H-00174
6346-15-2 Dimethyl chloromethylphosphonate, *in* C-00106
6362-86-3 2-Chloro-4-methyl-1,3,2-dioxaphospholane, C-00092
6362-87-4 2-Chloro-4-methyl-1,3,2-dioxaphosphorinane, C-00093
6362-89-6 ▷2-Chloro-1,3,2-dioxaphosphorinane, C-00066
6372-21-0 Diphenyl cyclohexylphosphoramidate, *in* C-00269
6372-40-3 Isopropyldiphenylphosphine, I-00031
6372-41-4 Butyldiphenylphosphine, B-00557
6372-42-5 Cyclohexyldiphenylphosphine, C-00245
6372-43-6 Diisopropylphenylphosphine, D-00638
6372-44-7 ▷Dibutylphenylphosphine, D-00102
6372-48-1 Methylphenylphosphine, M-00215
6374-26-1 Diethyl phosphor(isothiocyanatidate), D-00369
6395-52-4 Diethylphosphinoselenoic acid; Chloride, *in* D-00327
6396-07-2 ▷Diiodotriphenylphosphorane, D-00631
6396-08-3 Triphenylphosphine; B,HI, *in* T-00624
6399-81-1 Triphenylphosphine; B,HBr, *in* T-00624
6410-56-6 Dipropyl methylphosphonate, *in* M-00288
6411-21-8 1,2-Bis(diethylphosphino)ethane, B-00180
6411-24-1 Tridodecylphosphine, T-00435
6415-07-2 1,1′,1″-Phosphinylidynetrispyrrolidine, *in* T-00724
6415-20-9 Diisopropyl phosphoramidate, D-00653
6415-21-0 Diphenyl dimethylphosphoramidate, *in* D-00855
6418-33-7 Butyl dibutylphosphinite, *in* D-00121
6418-56-0 Dipentyl pentylphosphonate, *in* P-00050
6418-95-7 Glycerol 1-octadecanoate 2-9-octadecenoate 3-phosphoethanolamine; (*R*)-(*Z*)-*form*, *in* G-00083
6419-19-8 ▷[Nitrilotris(methylene)]-trisphosphonic acid, N-00033
6460-26-0 Dibromo-*o*-tolylphosphine, *in* M-00246
6460-27-1 Butylphosphonous dichloride, B-00618
6460-28-2 Dichlorohexylphosphine, *in* H-00101
6472-98-6 1,3-Bis(triphenyl-phosphoranylidene)-2-propanone, *in* O-00073
6476-36-4 Triisopropylphosphine, T-00500
6476-37-5 Dicyclohexylphenylphosphine, D-00208
6509-85-9 2,2-Dihydro-2,2,2-triphenyl-4,4,5,5-tetrakis(trifluoromethyl)-1,3,2-dioxaphospholane, D-00614
6509-88-2 2,2-Dihydro-2,2,2-trimethoxy-4,4,5,5-tetrakis(trifluoromethyl)-1,3,2-dioxaphospholane, D-00595
6546-78-7 2,2-Dihydrophenanthro[9,10-*d*]-1,3,2-dioxaphosphole; 2,2,2-Tri-Ph, *in* D-00541
6546-97-0 Bis(4-nitrophenyl) phosphorochloridate, B-00376
6566-81-0 1,3,2-Dioxaphosphonane, D-00971
6569-77-3 1,4-Oxaphosphorinane, O-00055
6569-82-0 Phosphirane, P-00361
6573-35-9 1,3,6,2-Trioxaphosphocane, T-00592
6587-25-3 Ethylene phenylphosphoramidate, *in* P-00083
6587-98-0 Ethylphenylphosphinothious acid; (±)-*form*, Et ester, *in* E-00107
6588-00-7 Ethyl diethylphosphinothioite, *in* D-00332
6588-02-9 Dibutyl ethylphosphonodithioite, *in* E-00141
6588-07-4 *N*-Diphenylphosphinothioyl-*P,P*-diphenylphosphinothioic amide, D-01087
6588-19-8 Benzylphosphonothioic acid; Dichloride, *in* B-00067
6588-21-2 (2-Methylpropyl)phosphonothioic acid; Dichloride, *in* M-00381
6588-22-3 Butylphosphonothioic acid; Dichloride, *in* B-00611
6588-28-9 Ethyl ethylphosphonochloridothioite, *in* E-00138
6588-35-8 Ethyl ethylphenylphosphinodithioate, *in* E-00103
6588-46-1 Isopropyl ethylphenylphosphinothioite, *in* E-00107
6591-05-5 Dimethylphosphine; Sulfide, *in* D-00827
6591-06-6 Diethylphosphine; Sulfide, *in* D-00321
6591-07-7 Diphenylphosphine; Sulfide, *in* D-01037
6591-08-8 *O*-Ethyl phenylphosphinothioate, *in* P-00213
6591-16-8 1-(Diphenylphosphinothioyl)hexane, *in* H-00093
6591-42-0 Diphenyl-phosphinothioylcyclohexane, *in* C-00245
6607-30-3 2,2,4,4,6,6-Hexahydro-2,2,4,4,6,6-hexamethyl-1,3,5-triaza-2,4,6-triphosphorine, H-00028
6609-64-9 2-Chloro-1,3,2-dioxaphospholane 2-oxide, C-00064
6619-48-3 Dimethyl pentylphosphonate, *in* P-00050
6626-39-7 Dibutyl diethylphosphoramidate, *in* D-00343
6630-13-3 Bis(2-methylphenyl) phosphorochloridate, *in* B-00312
6630-14-4 Bis(3-methylphenyl) phosphorochloridate, *in* B-00313
6630-15-5 Bis(4-methylphenyl) phosphorochloridate, B-00329
6632-88-8 Diethyl (1-hydroxy-2-methylethyl)phosphonate, *in* H-00159
6659-26-3 Monobenzoyl phosphate, M-00442
6665-99-2 Cytidine diphosphate glycerol, C-00293
6669-37-0 1,3,2-Dithiaphospholane, D-01243
6680-55-3 1,2-Diphospholane, D-01181
6680-67-7 1,3,2-Dithiaphosphorinane, D-01247
6699-20-3 Mono(3,7,11,15-tetramethyl-2,6,10,14-hexadecatetraenyl) diphosphate; (*E,E,E*)-*form*, *in* M-00515
6711-47-3 2-Hydroxy-1,3,2-dioxaphospholane 2-oxide, H-00134
6729-45-9 Mono(4-methylphenyl) phosphate, M-00487
6736-03-4 *O,O*-Diethyl *S*-(2-dimethylaminoethyl) phosphorothioate, D-00284
6736-03-4 Ecothiopate, E-00001
6737-25-3 *O,O*-Diethyl isopropylphosphoramidothioate, *in* I-00059
6737-42-4 1,3-Bis(diphenylphosphino)propane, B-00265
6737-43-5 Octyldiphenylphosphine, O-00039
6753-56-6 Glycerol 1-octadecanoate 2-9-octadecenoate 3-phosphocholine, G-00082
6775-01-5 Bis(3-aminopropyl)phenylphosphine, B-00090
6778-87-6 (1-Methylpropyl)phosphonic acid, M-00377
6779-08-4 (Dichloromethylene)-triphenylphosphorane, *in* D-00182
6779-09-5 Ethylphosphonic acid, E-00129
6779-49-3 3,3′-(Phenylphosphinothioylidene)-bis-1-propanamine, *in* B-00090
6779-62-0 Diphenylphosphinic acid; Isocyanate, *in* D-01039
6783-71-7 3-Phosphinopropanenitrile, P-00358
6783-73-9 ▷3,3′,3″-Phosphinothioylidynetris[propane-nitrile], *in* T-00752

6784-51-6 ▷*P,P*-Bis(1-aziridinyl)phosphinic amide, B-00092
6784-53-8 *P,P*-Bis(1-aziridinyl)-*N*-phenylphosphinic amide, *in* B-00092
6787-46-8 1-Hydroxyphospholane 1-oxide, H-00185
6789-42-4 1,1′,1″-Phosphinothioylidynetrispiperidine, *in* T-00704
6795-01-3 *OO-tert*-Butyl *O*-methyl methylphosphonoperoxoate, *in* M-00310
6795-02-4 *OO-tert*-Butyl *O*-methyl ethylphosphonoperoxoate, *in* E-00144
6830-45-1 Tetrapropyldiphosphine; Disulfide, *in* T-00286
6833-67-6 2-Propenylphosphonic acid, P-00454
6838-97-7 Dinonyl phosphonate, D-00950
6839-93-6 Dimethylphosphinothioic bromide, D-00841
6839-95-8 Tetramethylphosphinothioic amide, *in* D-00840
6840-01-3 *N,N*-Dimethyl-*P,P*-diphenylphosphinous amide, D-00747
6840-02-4 *N,N*-Dimethyl-*P*-phenylphosphonamidic acid; Chloride, *in* D-00808
6840-11-5 *O,O*-Dimethyl phenylphosphonothioate, *in* P-00244
6840-23-9 Dimethyl (2-methylphenyl)-phosphate, *in* M-00235
6840-25-1 Dimethyl (4-methylphenyl)-phosphonate, *in* M-00237
6840-26-2 (2-Methylphenyl)diphenylphosphine; Oxide, *in* M-00200
6840-27-3 (3-Methylphenyl)diphenylphosphine; Oxide, *in* M-00201
6840-28-4 (4-Methylphenyl)diphenylphosphine; Oxide, *in* M-00202
6851-71-4 *P,P*-Dimethylphosphinothioic amide, D-00840
6857-81-4 2-Chloro-2,2′-spirobi[1,3,2-benzodioxaphosphole], C-00174
6866-70-2 Tri-*tert*-butylphosphine; Oxide, *in* T-00373
6869-04-1 Cyclopentanephosphonic acid, C-00276
6874-57-3 Carbamoylphosphonic acid, C-00003
6874-60-8 ▷Decylphosphonic acid, D-00012
6881-54-5 Acetylphosphonic acid, A-00026
6881-57-8 Benzylphosphonic acid, B-00060
6881-61-4 Benzoylphosphonic acid, B-00034
6903-79-3 ▷Creatinolfosfate, C-00206
6909-62-2 Demanyl phosphate, D-00016
6913-02-6 (2-Oxopropyl)phosphonic acid, O-00075
6923-22-4 ▷Monocrotophos; (*E*)-*form*, *in* M-00459
6926-73-4 Bis(diethoxyphosphinothioyl) trisulfide, B-00171
6933-17-1 [(4-Nitrophenyl)methylene]-triphenylphosphorane, N-00041
6941-20-4 *N*-Cyclohexyl-*P,P*-diphenylphosphinic amide, *in* D-01040
6954-83-2 Diisopropyl chloromethylphosphonate, *in* C-00106
6955-57-3 Phenylphosphoramidic dichloride, P-00258
6959-02-0 (3-Bromophenyl)phosphonic acid, B-00491
6962-87-4 Tris(2-chlorophenyl)phosphine, T-00747
6964-36-9 2-Methylphenyl phosphorodichloridate, *in* M-00485
6996-81-2 *O,O*-Diethyl methylphosphonothioate, *in* M-00313
7000-66-0 Didecyl phosphonate, D-00222
7006-95-3 Dimethyl methylphenylphosphoramidate, *in* M-00250
7016-64-0 Ethyl acetylphenylphosphinate, *in* A-00021
7016-78-6 Diethyl [(acetyloxy)methyl]-phosphonate, *in* A-00007
7051-22-1 2-Ethoxy-2,3-dihydro-1,3,2-benzoxazaphosphole, E-00027
7057-92-3 Didodecyl phosphate, D-00225
7067-72-3 *N*,1,1-Triphenyl-phosphinecarbonamide, T-00625
7067-73-4 *N*,1,1-Triphenyl-phosphinecarbonamide; Oxide, *in* T-00625
7071-12-7 Ethyl chloro(diethoxyphosphinyl)-acetate, *in* C-00036
7077-62-5 Trimethyl phenylphosphorimidate, *in* P-00264
7078-91-3 Methyl acetylphenylphosphinate, *in* A-00021
7091-44-3 Glycerol 1,2-dihexadecanoate 3-phosphate; (*R*)-*form*, *in* G-00011
7092-52-6 1,3,2-Oxathiaphospholane, O-00056
7100-92-7 Ethyl dibutylphosphinate, *in* D-00109
7108-85-2 Diethyl phosphate; Anilinum salt, *in* D-00320
7109-06-0 Trimethyl methylsulfonylphosphorimidate, *in* M-00402
7109-08-2 *O,O,O′,O′*-Tetraethyl thioimidodiphosphate, T-00074
7109-15-1 Dimethyl methylsulfonylphosphoramidate, *in* M-00401
7114-39-8 2-Dimethylamino-1,3,2-dioxaphospholane, D-00695
7114-42-3 4-Methyl-2-dimethylamino-1,3,2-dioxaphospholane, M-00116
7114-53-6 Ethylene dimethylphosphoramidothioate, *in* D-00695
7114-66-1 Ethylene dimethylphosphoramidate, *in* D-00695
7137-86-2 Methyl *N,N′*-ethanediyl-*N,N′*-dimethylphosphorodiamidite, *in* D-00719
7137-89-5 *N,N*,6,9-Tetramethyl-2,3-diphenyl-1,4-dioxa-6,9-diaza-5-phospha(5-P^{V})-spiro[4.4]non-2-en-5-amine, *in* D-00741
7142-76-9 Diethyl (9*H*-fluoren-9-yl)-phosphonate, *in* F-00011
7158-34-1 Diethyl (2-methyl-1,3-butadienyl)phosphonate, *in* M-00093
7158-35-2 Diethyl (1,3-butadienyl)-phosphonate, *in* B-00510
7162-15-4 Methyl phenylphosphinate, *in* P-00210
7163-67-9 Methyl dibutylphosphinate, *in* D-00109
7166-10-1 ▷Triethyl (diethoxyphosphinothioyl)-phosphorimidate, *in* D-00238
7191-13-1 Trimethylene phenylphosphonate, *in* P-00116
7191-20-0 Tetramethylene phenylphosphonate, *in* P-00114
7191-26-6 2-Phenoxy-1,3,2-dioxaphosphepane 2-oxide, *in* H-00132
7203-67-0 Tetraethyl 1,4-butanediylbisphosphonate, *in* B-00514
7211-39-4 Dimethylphosphine oxide, D-00828
7215-33-0 Diethylphosphine; Oxide, *in* D-00321
7219-78-5 Tetramethylphosphorodiamidic azide, T-00219
7234-22-2 2-Methylthio-4*H*-1,3,2-benzodioxaphosphorin; 2-Sulfide, *in* M-00411
7237-07-2 1,2-Phenylenebis[dimethyl-phosphine], P-00123
7237-08-3 (Chloromethyl)phosphine, C-00104
7237-16-3 Diisopropyl phenylphosphonate, *in* P-00221
7246-20-0 ▷Triclofos sodium, *in* T-00424
7248-72-8 *O,O,O*-Triphenyl phosphoroselenoate, T-00667
7253-07-8 9*H*-Fluoren-9-yltriphenylphosphonium(1+); Bromide, *in* F-00012
7259-36-1 Tetrapropylphosphonium(1+); Iodide, *in* T-00290
7264-90-6 Di-4-morpholinylphosphinic acid; Chloride, *in* D-00935
7264-92-8 Mono-2-propenyl phosphate; Biscyclohexylammonium salt, *in* M-00509
7266-53-7 Glycerol 1-octadecanoate 2-9,12-octadecadienoate 3-phosphoethanolamine; (*R*)-(*all-Z*)-*form*, *in* G-00079
7282-98-6 Benzylphenylphosphinic acid, B-00056
7292-16-2 ▷Propaphos, P-00451
7305-61-5 Ethyl hydrogen ethylphosphonate, *in* E-00129
7309-49-1 Ethylmethylphenylphosphine; (±)-*form*, Oxide, *in* E-00084
7319-55-3 Glycerol 1-9-octadecenoate 2-octadecanoate 3-phosphocholine; (*R*)-(*Z*)-*form*, *in* G-00091
7322-77-2 *O,O,O*-Tripropyl phosphoroselenoate, T-00713
7331-08-0 Phosphoserine, P-00417
7333-51-9 Cyclohexyltriphenyl-phosphonium(1+); Bromide, *in* C-00273
7333-52-0 Cyclopentyltriphenyl-phosphonium(1+); Bromide, *in* C-00282
7333-65-5 (Diphenylmethyl)-triphenylphosphonium(1+); Bromide, *in* D-01025
7333-67-7 ▷1,3-Propanediylbis[triphenyl-phosphonium](2+); Dibromide, *in* P-00448
7342-94-1 *sym*-Tetraethyl monothiopyrophosphate, T-00052
7343-26-2 ▷(Carboxymethyl)-triphenylphosphonium(1+); Chloride, *in* C-00014
7348-85-8 *O,S*-Diethyl ethylphosphonothioate, *in* E-00074
7351-34-0 Methyl trimethylphosphonamidate, *in* T-00537
7351-35-1 Methyl *N,P*-dimethylphosphonamidate, *in* D-00849
7355-27-3 Diethyl 2-propenylphosphoramidate, *in* P-00464
7368-65-2 Tetraethylphosphonium(1+); Chloride, *in* T-00056
7369-43-9 Diethyl diethyldiphosphonate, *in* D-00290
7369-51-9 Dibenzylphosphinic acid, D-00055
7393-11-5 Methyl tetramethylphosphorodiamidate, M-00406
7393-12-6 *P*-Chloromethyl-*N,N,N′,N′*-tetramethylphosphonic diamide, *in* C-00107
7393-13-7 Phenyl tetramethylphosphorodiamidate, *in* T-00218
7396-44-3 Diethyl (benzyloxycarbonylmethyl)-phosphonate, *in* D-00241
7412-25-1 ▷Diphenyl 4-morpholinylphosphonate, *in* M-00523
7413-09-4 Diethyl cyclohexylphosphonate, *in* C-00256
7414-83-7 ▷HEDSPA, *in* H-00144
7422-64-2 Hexaphenylphosphoric triamide, *in* H-00090
7422-73-3 Hexamethylphosphoroselenoic triamide, H-00078
7439-48-7 *O,O*-Diethyl *SS*-phenylphosphoro(dithioperoxoate), D-00314
7439-49-8 Bis(dimethoxyphosphinyl) disulfide, B-00186
7439-68-1 Dibutyl (2-diethylamino-2-oxoethyl)phosphonate, *in* D-00268

7441-04-5 Tri-*tert*-butylphosphine; Sulfide, *in* T-00373
7441-05-6 *O,O,O*-Tributyl phosphoroselenoate, T-00385
7443-26-7 Methyl tetramethylethylene phosphate, *in* M-00085
7450-69-3 Phenyl phosphorodiamidate, P-00272
7450-70-6 Phenyl *N,N'*-diethylphosphorodiamidate, *in* D-00382
7452-28-0 *O,O*-Diethyl phosphoroselenoate, D-00393
7452-97-2 *O,O*-Diethyl phosphoroselenoate; Na salt, *in* D-00393
7472-16-4 ▷(2-Aminophenyl)phosphonic acid, A-00109
7484-37-9 Triphenyl(3-phenylpropyl)-phosphonium(1+); Bromide, *in* T-00617
7494-75-9 Dibenzyl cyclohexylphosphoramidate, *in* C-00269
7495-97-8 ▷Diethylphosphinic acid; Anhydride, *in* D-00322
7504-61-2 Tris(1-methylpropyl) phosphite, T-00825
7506-85-6 (2-Methoxyphenyl)phosphonic acid, M-00070
7507-08-6 Didecylphosphinic acid, D-00221
7508-73-8 ▷*O*-Ethyl *O,O*-bis(4-nitrophenyl) phosphorothioate, E-00054
7509-48-0 Diethyl (triphenylphosphoranylidene)-propanedioate, *in* T-00658
7526-25-2 Methyl phenyl methylphosphonate, M-00205
7526-26-3 Diphenyl methylphosphonate, D-01024
7526-29-6 Diphenyl ethylphosphonate, *in* E-00129
7526-31-0 5,5-Dimethyl-2-phenyl-1,3,2-dioxaphosphorinane, D-00795
7526-32-1 2-Phenyl-1,3,2-dioxaphosphorinane, P-00116
7526-37-6 2-Phenyl-1,3,2-dioxaphosphepane, P-00114
7533-52-0 Bis(triphenylphosphoranylidene)-methane, B-00459
7549-83-9 Tris(dimethylamino)-difluorophosphorane, T-00760
7563-17-9 2,3-Dihydro-3,3,5-trimethyl-1,2-oxaphosphole 2-oxide; 2-Phenoxy, *in* D-00596
7563-70-4 Tetraethyl (1-cyano-1,2-ethanediyl)bisphosphonate, *in* B-00173
7575-26-0 Bis(dibenzyloxyphosphinothioyl) disulfide, B-00155
7579-70-6 (3-Methylphenyl)diphenylphosphine, M-00201
7580-37-2 ▷Tetrakis(hydroxymethyl)-phosphonium(1+); Acetate, *in* T-00181
7582-40-3 (Bromomethyl)phosphonic acid, B-00475
7594-78-7 Acetic diphenylphosphinic anhydride, *in* D-01039
7595-89-3 ▷*O,O*-Bis(3-methylphenyl) phosphorodithioate, B-00338
7598-61-0 Diethyl (2,2-diethoxyethyl)-phosphonate, *in* D-00234
7608-32-4 Diphenyl(4-phenyl-1,3-butadiynyl)phosphine; Oxide, *in* D-01032
7614-58-6 2,3-Dihydro-3,3,5-trimethyl-1,2-oxaphosphole 2-oxide; 2-Ethoxy, *in* D-00596
7615-15-1 1,2-Bis(diphenylphosphinothioyl)-ethane, *in* B-00234
7621-16-1 *tert*-Butylmethylphenylphosphine, B-00565
7628-15-1 *N,N*-Bis[(2-diphenylphosphino)-ethyl]ethylamine, B-00238
7646-81-3 ▷Dihexylphosphinic acid, D-00434
7648-76-2 *P,P*-Diphenylphosphinic amide; *N*-Diphenylphosphino, *in* D-01040
7650-79-5 (1-Methylpropyl)diphenylphosphine, M-00369
7650-80-8 Bis(1-methylpropyl)-phenylphosphine, B-00340
7650-83-1 Phenyldipropylphosphine, P-00120
7650-84-2 Diphenylpropylphosphine, D-01145
7650-88-6 Tricyclopentylphosphine, T-00431
7650-89-7 Tribenzylphosphine, T-00338
7650-90-0 ▷Dibenzylphenylphosphine, D-00052
7665-99-8 Cyclic GMP, C-00234
7670-93-1 Monopropyl phenylphosphonate, M-00511
7682-87-3 Tris(2-methylpropyl)phosphine; Oxide, *in* T-00824
7688-25-7 1,4-Bis(diphenylphosphino)butane, B-00225
7697-72-5 Diphenyl cyclohexylphosphonodithioite, *in* C-00262
7697-73-6 Diphenyl butylphosphonodithioite, *in* B-00607
7697-74-7 Dibutyl butylphosphonodithioite, *in* B-00607
7697-75-8 Butyl dicyclohexylphosphinodithioate, *in* D-00212
7697-76-9 Butyl diisobutylphosphinodithioate, *in* B-00348
7697-78-1 Dibutyl butylphosphonotrithioate, *in* B-00614
7697-79-2 *S*-Phenyl dicyclohexylphosphinothioate, *in* D-00213
7697-80-5 Butyl bis(2-methylpropyl)-phosphinothioite, *in* B-00352
7697-83-8 Phenyl bis(2-methylpropyl)-phosphinothioite, *in* B-00352
7700-17-6 ▷Crotoxyphos; (±)-(*E*)-*form*, *in* C-00207
7704-44-1 Monomethyl phosphate; Monoanilinium salt, *in* M-00489
7704-50-9 Mono(2-methylpropyl) phosphate; Monoanilinium salt, *in* M-00493
7704-54-3 Monooctyl phosphate; Monoanilinium salt, *in* M-00501
7735-81-1 2-Methoxy-4-methyl-1,3,2-dioxaphosphorinane; (2*RS*,4*RS*)-*form*, *in* M-00044
7735-85-5 2-Methoxy-4-methyl-1,3,2-dioxaphosphorinane; (2*RS*,4*SR*)-*form*, *in* M-00044
7735-87-7 Pentaethoxyphosphorane, P-00011
7748-09-6 Di-2-propenyl phosphate, D-01192
7766-52-1 Ethyl ethenylphenylphosphinate, *in* P-00337
7776-66-1 *O*-Ethyl ethylphosphonothioate, E-00074
7778-06-5 1,3-Dimethyl-2-dimethylamino-1,3,2-diazaphospholidine; 2-Oxide, *in* D-00724
7778-09-8 *N,N',N'',N''*-Tetramethyl-*N,N'*-trimethylenephosphoric triamide, *in* H-00023
7784-90-9 *N,N'*-Dimethyl-*N,N'*-trimethylene-*P*-phenylphosphonic diamide, *in* H-00023
7786-34-7 ▷Mevinphos, M-00440
7795-87-1 Didecyl phosphate, D-00220
10015-87-9 Glycerol 1-9-octadecenoate 2-hexadecanoate 3-phosphate, G-00086
10022-55-6 2,8,9-Trioxa-5-aza-1-phosphabicyclo[3.3.3]-undecane; 1-Oxide, *in* T-00584
10026-27-4 Ethyl *N,N'*-1,2-ethanediyl-*N,N'*-dimethyl-phosphorodiamidate, *in* E-00030
10038-36-5 1,8-Bis(triphenyl-phosphoniomethyl)-naphthalene(2+); Dibromide, *in* B-00452
10038-37-6 Triphenyl[[2-[2-(triphenylphosphonio)-ethyl]phenyl]methyl]-phosphonium(2+); Dibromide, *in* T-00696
10038-58-1 Tetraisopropyl 2,2-propanediphosphonate, *in* P-00445
10038-65-0 Dimethyl (2-oxoethyl)phosphonate, *in* O-00059
10052-96-7 Phenylphosphine; Oxide, *in* P-00208
10061-87-7 Diphenyl-2-propenylphosphine; Sulfide, *in* D-01144
10072-02-3 2-Ethoxy-1,3,2-benzodioxaphosphole, E-00024
10088-45-6 Monomethyl phenylphosphonate, M-00488
10089-43-7 Tris(4-methylphenyl)phosphine; Selenide, *in* T-00808
10123-62-3 ▷Diethyl (2-cyanoethyl)-phosphonate, *in* C-00211
10159-46-3 ▷Pentamethylphosphoric triamide, P-00026
10159-53-2 *N,N*-Bis(2-chloroethyl)-phosphorodiamidic acid, B-00126
10160-45-9 *OO-tert*-Butyl *O,O*-diethyl phosphoroperoxoate, B-00551
10160-46-0 *OO-tert*-Butyl *O,O*-diisopropyl phosphoroperoxoate, B-00554
10163-62-9 *O,O*-Dibutyl phosphorothioate, D-00155
10173-43-0 ▷Propyl phosphorodichloridate, P-00487
10175-90-3 (3,5-Di-*tert*-butyl-4-hydroxybenzyl)phosphonic acid, D-00094
10177-78-3 ▷Bis(3-methylphenyl)phosphine, B-00316
10184-66-4 Dimethyl (1-hydroxyethyl)-phosphonate, *in* H-00145
10212-05-2 1,4-Phenylenebis[phenylphosphinic acid], P-00140
10235-09-3 ▷*O,O,O*-Tris(2-chloroethyl) phosphorothioate, T-00743
10258-61-4 Phenyl phosphorodichloridodithioate, P-00281
10265-92-6 ▷Methamidophos, M-00021
10273-74-2 [1,3-Phenylenebis[methylene]]-bis[triphenylphosphonium](2+); Dibromide, *in* P-00133
10311-08-7 Dimethylphenylphosphine; Oxide, *in* D-00803
10311-84-9 ▷Dialifos, D-00023
10369-17-2 2,4,6,8,9,10-Hexamethyl-2,4,6,8,9,10-hexaaza-1,3,5,7-tetraphosphatricyclo[3.3.1.1^{3,7}]-decane, H-00071
10419-77-9 Diethyl (iodomethyl)phosphonate, *in* I-00013
10419-78-0 Diethyl (1-chloroethyl)-phosphonate, *in* C-00073
10419-79-1 ▷Diethyl (2-chloroethyl)-phosphonate, *in* C-00074
10419-85-9 Diphenyl chloromethylphosphonate, *in* C-00106
10419-87-1 Diphenyl benzylphosphonate, *in* B-00060
10448-49-4 Tetraphenyl diphosphate, T-00256
10494-05-0 *N*,1,1-Triphenylphosphinecarbonamide; Sulfide, *in* T-00625
10496-10-3 Tripentylphosphine, T-00594
10496-13-6 Butyl phosphorodichloridite, B-00629
10496-14-7 Hexyl phosphorodichloridite, H-00103
10497-05-9 ▷Tris(trimethylsilyl) phosphate, T-00864
10498-56-8 1,4-Benzenediphosphonous acid; Bis(dichloride), *in* B-00010
10498-57-4 1,4-Phenylenebis[dimethyl-phosphine], P-00125
10498-59-6 Tetramethyl 1,4-phenylenebisphosphonite, *in* B-00010
10508-76-6 Ethyl *o*-phenylene phosphate, *in* E-00024
10526-80-4 Phosphoenolpyruvic acid; Monocyclohexylammonium salt, *in* P-00366
10537-55-0 Phenyl bis(1-aziridinyl)-phosphinate, *in* B-00091
10542-07-1 Monobenzyl phosphonate, M-00444

10545-19-4 (Trichloromethyl)phosphorimidic trichloride, T-00418
10545-61-6 Ethyl tetramethylenephosphinate, *in* H-00185
10548-10-4 *S*-[(*tert*-Butylsulfinyl)methyl] *O,O*-diethyl phosphorodithioate, *in* T-00004
10557-66-1 Dibutylphenylphosphine; Oxide, *in* D-00102
10562-82-0 (1-Phenyl-1-propenyl)phosphonic acid; (*E*)-*form*, *in* P-00301
10562-87-5 (1-Phenyl-1-propenyl)phosphonic acid; (*Z*)-*form*, *in* P-00301
10562-88-6 1-Cyclohexenephosphonic acid, C-00241
10565-43-2 *O,O*-Diethyl phosphoroselenothioate, D-00394
10568-62-4 Difluoro-*p*-tolylphosphine, *in* M-00248
10574-82-0 *O,S*-Dimethyl phosphorothioate; NH_4 salt, *in* D-00908
10580-44-6 Dimethyl 1,4-phenylenebis[methyl-phosphinate], *in* P-00137
10589-50-1 Glycerol 1-tetradecanoate 2-hexadecanoate 3-phosphocholine, G-00096
10596-23-3 Dichloromethylenebisphosphonic acid, D-00172
11021-54-8 Prasinomycin; Prasinomycin *A*, *in* P-00438
11021-55-9 Prasinomycin; Prasinomycin *B*, *in* P-00438
11021-56-0 Prasinomycin; Prasinomycin *C*, *in* P-00438
12098-17-8 (Diphenylphosphino)ferrocene, D-01068
12150-46-8 1,1′-Bis(diphenylphosphino)-ferrocene, B-00241
12204-30-7 1,1″-(Phenylphosphinylidene)-bisferrocene, *in* D-00410
12278-69-2 Diferrocenylphenylphosphine, D-00410
12687-95-5 Prasinomycin, P-00438
12689-28-0 ▷Proticin, P-00500
13024-84-5 Dibutyl phenylphosphoramidate, *in* P-00257
13030-37-0 *O,O*-Diethyl phosphorochloridoselenoate, D-00377
13040-68-1 Propyl phosphorodichloridite, P-00488
13041-11-7 4-Methyl-2-dimethylamino-1,3,2-dioxaphosphorinane, M-00117
13058-04-3 Mono(3,7,11-trimethyl-2,6,10-dodecatrienyl) diphosphate, M-00517
13067-93-1 ▷Cyanofenphos, C-00213
13071-79-9 ▷Terbufos, T-00004
13081-74-8 [(4-Methylphenyl)methyl]-phosphonic acid, M-00208
13086-84-5 Di-*tert*-butyl phosphonate, D-00128
13089-54-8 Cyclohexylphosphoramidic acid; Dichloride, *in* C-00269
13091-13-9 Phenyl methylphosphonate, P-00171
13104-21-7 ▷Bromfenvinfos-methyl, B-00463
13106-27-9 *O,O*-Diethyl phosphoroselenothioate; Cyclohexylammonium salt, *in* D-00394
13113-88-7 (Dichloromethyl)phosphonic acid, D-00175
13113-89-8 *S*-Methyl methylphosphonochloridothioate, *in* M-00298
13113-90-1 *S*-Ethyl methylphosphonochloridothioate, *in* M-00298
13113-91-2 *S*-Propyl methylphosphonochloridothioate, *in* M-00298
13113-92-3 *S*-Methyl ethylphosphonochloridothioate, *in* E-00137
13113-93-4 *S*-Ethyl ethylphosphonochloridothioate, *in* E-00137
13113-94-5 *S*-Propyl ethylphosphonochloridothioate, *in* E-00137
13113-95-6 *S*-Methyl phenylphosphonochloridothioate, *in* P-00232
13113-96-7 *S*-Ethyl phenylphosphonochloridothioate, *in* P-00232
13113-98-9 *S*-Isopropyl methylphosphonochloridothioate, *in* M-00298
13113-99-0 *S*-Isopropyl ethylphosphonochloridothioate, *in* E-00137
13114-00-6 *S*-Isopropyl phenyl-phosphonochloridothionate, *in* P-00232
13114-07-3 Dimethyl (4-chlorophenyl)-phosphonate, *in* C-00147
13114-08-4 Methyl (4-chlorophenyl)-methylphosphinate, *in* C-00131
13114-09-5 (4-Chlorophenyl)methylphosphinic acid; Chloride, *in* C-00131
13114-10-8 (4-Chlorophenyl)methylphosphinic acid, C-00131
13118-24-6 Tricyclopropylphosphine, T-00432
13119-01-2 ▷Bis(4-chlorophenyl)phosphinic acid, B-00139
13119-11-4 3,3,6,6-Tetraphenyl-1,3,6-oxadiphosphepanium(2+); Dibromide, *in* T-00266
13119-12-5 3,3,5,5-Tetraphenyl-1,3,5-oxadiphosphorinanium(2+); Dichloride, *in* T-00267
13119-13-6 3,3,5,5-Tetraphenyl-1,3,5-oxadiphosphorinanium(2+); Dibromide, *in* T-00267
13119-19-2 (Dimethylaminomethyl)-diphenylphosphine, D-00700
13125-82-1 1-Indenylidenetriphenyl-phosphorane, *in* I-00008
13138-33-5 (3-Aminopropyl)phosphonic acid, A-00127
13138-36-8 (1-Aminobutyl)phosphonic acid, A-00054
13138-37-9 (1-Aminopentyl)phosphonic acid, A-00096
13143-83-4 3,3,6,6-Tetraphenyl-1,3,6-oxadiphosphepanium(2+); Dipicrate, *in* T-00266
13145-84-1 (3-Methoxyphenyl)-diphenylphosphine, M-00057
13147-57-4 Phosphonooxyacetic acid, P-00390
13153-12-3 (4-Methoxyphenyl)phosphine, M-00067
13153-89-4 Methylphenylpropylphosphine; (*R*)-*form*, *in* M-00265
13153-91-8 Methylphenylpropylphosphine; (*R*)-*form*, Sulfide, *in* M-00265
13163-99-0 *O*-Methyl bis(1-aziridinyl)-phosphinothioate, *in* B-00093
13171-21-6 ▷Phosphamidon, P-00347
13172-76-4 Diphenylphosphinylpropadiene, *in* D-01142
13176-23-3 Tridodecylphosphine; Oxide, *in* T-00435
13194-48-4 ▷Ethoprophos, E-00023
13213-38-2 Ethylmethylphosphinic acid; Chloride, *in* E-00085
13213-39-3 Methylpropylphosphinic acid; Chloride, *in* M-00375
13213-41-7 Butylmethylphosphinic acid; Chloride, *in* B-00566
13213-42-8 Methyl(2-methylpropyl)phosphinic acid; Chloride, *in* M-00172
13213-43-9 Isopropylphenylphosphinic acid; (±)-*form*, Chloride, *in* I-00043
13223-91-1 4-Methyl-2-phenyl-1,3,2-dioxaphosphorinane, M-00199
13231-84-0 *O*-Isopropyl phenylphosphonochloridothioate, *in* P-00232
13232-18-3 2,7,8-Trioxa-1-phosphabicyclo[3.2.1]-octane, T-00589
13237-70-2 Fosmenic acid, F-00061
13238-16-9 Dibenzylphosphine; Oxide, *in* D-00054
13249-95-1 [(2-Chlorophenyl)methyl]-phosphonic acid, C-00132
13274-83-4 Bis(chloromethyl)phosphinic acid, B-00127
13274-84-5 Ethyl bis(chloromethyl)-phosphinate, *in* B-00127
13274-91-4 Benzyl diethylphosphinate, *in* D-00322
13274-97-0 1,2,3,4-Tetrahydro-1,1,4,4-tetraphenyl-1,4-diphosphorinium(2+); Dibromide, *in* T-00152
13275-01-9 1,1,4,4-Tetraphenyl-1,4-diphosphorinanium(2+), T-00263
13275-02-0 1,2-Ethanediylbis[triphenyl-phosphonium](2+), E-00017
13286-32-3 ▷Kitazin, K-00001
13294-07-0 Trimethyl (dimethoxyphosphinyl)-phosphorimidate, *in* D-00683
13294-09-2 Tetraethyl methylimidodiphosphate, *in* M-00156
13294-40-1 Diethyl (2-nitrophenyl)-phosphonate, *in* N-00055
13297-93-3 Ethyl ethylpropylphosphinodithioate, *in* E-00185
13297-95-5 Diethyl ethylphosphonotrithioate, *in* E-00152
13298-01-6 Dibenzylphenylphosphine; Sulfide, *in* D-00052
13298-02-7 (Iodomethyl)phosphonic acid, I-00013
13298-14-1 Methyl methyl-phosphonisothiocyanatidate, *in* M-00294
13298-15-2 Ethyl methyl-phosphonisothiocyanatidate, *in* M-00294
13298-19-4 Isopropyl methyl-phosphonisothiocyanatidate, *in* M-00294
13298-37-8 Ethyl ethylmethylphosphinodithioate, *in* E-00086
13306-69-9 *O*-(4-Amino-3-methylphenyl) *O,O*-dimethyl phosphorothioate, A-00090
13317-09-4 (3-Chloropropyl)phosphonic acid, C-00172
13317-44-7 Ethylphenylphosphinic acid, E-00101
13336-50-6 Isopropyl phenylphosphinate, *in* P-00210
13336-55-5 Methyl (4-methylphenyl)-phosphinate, *in* M-00222
13336-57-7 Isopropyl (4-methylphenyl)-phosphinate, *in* M-00222
13337-10-5 1,2-Bis(diethylphosphinyl)ethane, *in* B-00180
13346-70-8 2-Ethoxy-1,3,2-thiazaphospholidine; 2-Sulfide, *in* E-00053
13346-73-1 1,3,2-Thiazaphospholidine 2-sulfide; 2-Ph, *in* T-00303
13346-75-3 1,3,2-Thiazaphospholidine 2-sulfide; 2-Phenoxy, *in* T-00303
13346-77-5 1,3,2-Thiazaphospholidine 2-sulfide; 2-Methoxy, *in* T-00303
13346-80-0 2-Ethoxy-1,3,2-thiazaphospholidine; 2-Oxide, *in* E-00053
13351-39-8 *O*-Methyl *O*-4-nitrophenyl cyclohexylphosphoramidothioate; (±)-*form*, *in* M-00179
13355-82-3 Tris(trimethylplumbyl)phosphine, T-00863

13360-92-4 Phenyl diphenylphosphinite, *in* D-01089
13360-94-6 Butyl diphenylphosphinite, *in* D-01089
13362-44-2 *O,S,S*-Tributyl phosphorotrithioate, T-00390
13371-17-0 Butyltriphenylphosphonium(1+); Chloride, *in* B-00641
13371-29-4 Phenylbis(triphenylgermyl)-phosphine, P-00101
13371-32-9 Tris(triphenylgermyl)phosphine, T-00870
13376-78-8 α-[(Dimethoxyphosphinothioyl)-thio]benzeneacetic acid, D-00676
13388-88-0 Mono(4-chlorophenyl) phosphate, M-00458
13388-91-5 Mono(3-nitrophenyl) phosphate, M-00499
13406-27-4 Tris[4-(trifluoromethyl)-phenyl]phosphine; Oxide, *in* T-00854
13406-29-6 Tris[4-(trifluoromethyl)-phenyl]phosphine, T-00854
13407-03-9 Trimethylene methylphosphonate, *in* M-00128
13410-61-2 Diphenyl phenylphosphonite, *in* P-00251
13411-65-9 (Diphenylphosphinylmethylene)-triphenylphosphorane, D-01099
13422-52-1 Vitamin B_{12a}, V-00010
13423-48-8 Heptyltriphenylphosphonium(1+); Bromide, *in* H-00011
13428-19-8 Mono(2-chlorophenyl) phosphate, M-00457
13435-44-4 Cytidine 5′-(dihydrogen phosphate); Ba salt, *in* C-00290
13440-07-8 Di-1-naphthylphosphine; Oxide, *in* D-00939
13457-18-6 ▷Pyrazophos, P-00502
13468-89-8 Ethylene phenylphosphonate, *in* P-00115
13482-64-9 Bis(chloromethyl)phosphinic acid; Chloride, *in* B-00127
13497-36-4 2,4,6-Triphenylphosphorin, T-00665
13504-71-7 Methyl cyano(triphenyl-phosphoranylidene)acetate, *in* C-00231
13504-77-3 Methyl bromo(triphenyl-phosphoranylidene)acetate, *in* B-00504
13504-83-1 Ethyl cyano(diethoxyphosphinyl)-acetate, E-00057
13507-10-3 2-Hydroxy-1,3,2-dioxaphosphorinane 2-oxide, H-00138
13534-39-9 Ethyl ethylphosphonisocyanatidate, *in* E-00133
13547-40-5 *O*-Ethyl ethylphosphonochloridothioate; (±)-*form*, *in* E-00069
13547-42-7 *O*-Ethyl ethylphosphonochloridothioate; (*S*)-*form*, *in* E-00069
13557-07-8 *O,O*-Diethyl [(phenylamino)-carbonyl]phosphoramidothioate, *in* P-00082
13561-75-6 *O,O*-Diphenyl phosphorisocyanatidothioate, D-01117
13563-34-3 (2,3-Dihydroxypropyl)phosphonic acid, D-00624
13563-93-4 Glycerol 1,2-dioctadecanoate 3-phosphate; (±)-*form*, *in* G-00032
13590-71-1 ▷Monomethyl phosphonate, M-00490
13593-03-8 ▷Quinalphos, Q-00001
13604-50-7 *O,O*-Diethyl benzoylphosphoramidothioate, *in* B-00038
13604-51-8 *O,O*-Diisopropyl benzoylphosphoramidothioate, *in* B-00038
13604-54-1 *O,O*-Diphenyl benzoylphosphoramidothioate, *in* B-00038
13605-65-7 (3-Chloro-2-oxopropyl)-triphenylphosphonium(1+); Chloride, *in* C-00123
13605-66-8 1-Chloro-3-(triphenylphosphoranylidene)-2-propanone, C-00191
13620-62-7 *O,O*-Diethyl phosphorisocyanatidothioate, D-00367
13639-67-8 Methylphenylphosphinothioic acid; (±)-*form*, Chloride, *in* M-00225
13639-72-0 Tripropylphosphine; Sulfide, *in* T-00711
13639-75-3 1,2-Dimethyl-1,2-diphenyldiphosphine; (*RS,RS*)-*form*, Disulfide, *in* D-00744
13640-94-3 Dimethyl 2-thienylphosphonate, *in* T-00307
13640-95-4 Diethyl 2-thienylphosphonate, *in* T-00307
13640-97-6 Dimethyl 2-furanylphosphonate, *in* F-00073
13648-75-4 Ethyl bis(4-methylphenyl)-phosphinite, *in* B-00325
13683-02-8 Trimethylsilyl dibutylphosphinite, *in* D-00121
13685-23-9 Chlorobis(3-methylphenyl)-phosphine, *in* B-00324
13685-26-2 Chlorobis(4-chlorophenyl)-phosphine, *in* B-00142
13685-27-3 Chlorobis(3-chlorophenyl)-phosphine, *in* B-00141
13685-43-3 Ethyl bis(2-methylphenyl)-phosphinite, *in* B-00323
13685-46-6 Ethyl bis(4-chlorophenyl)-phosphinite, *in* B-00142
13685-47-7 Ethyl bis(3-chlorophenyl)-phosphinite, *in* B-00141
13685-74-0 *O*-Methyl butylphosphonodithioate, *in* B-00606
13685-75-1 *O*-Isopropyl butylphosphonodithioate, *in* B-00606
13685-79-5 *O*-Methyl hydrogen pentylphosphonodithioate, *in* P-00051
13685-85-3 *O*-Ethyl hydrogen hexylphosphonodithioate, *in* H-00099
13685-87-5 *O,S*-Diethyl cyclohexylphosphonodithioate, *in* C-00261
13687-09-7 *P,P*-Bis(1-aziridinyl)-*N*-methylphosphinothioic amide, *in* B-00094
13689-16-2 Diphenyl cyclohexylphosphonate, *in* C-00256
13689-17-3 Ethyl cyclohexylphenylphosphinate, *in* C-00250
13689-19-5 ▷Tricyclohexylphosphine; Oxide, *in* T-00426
13689-20-8 Diphenylphosphinylcyclohexane, *in* C-00245
13689-21-9 Phenyl dicyclohexylphosphinate, *in* D-00211
13689-23-1 1,2-Dihydro-1,2,4,6-tetraphenylphosphorin, D-00584
13689-26-4 1,2-Dihydro-1,2,4,6-tetraphenylphosphorin; Oxide, *in* D-00584
13699-67-7 Trivinylphosphine; Oxide, *in* T-00877
13703-06-5 Methyl *N*-ethyl-*P*-methylphosphonamidate, *in* E-00088
13703-07-6 Methyl *N*-isopropyl-*P*-methylphosphonamidate, *in* I-00036
13703-09-8 Ethyl *N*-ethyl-*P*-methylphosphonamidate, *in* E-00088
13703-11-2 Isopropyl *N,P*-dimethylphosphonamidate, *in* D-00849
13703-12-3 Isopropyl *N*-ethyl-*P*-methylphosphonamidate, *in* E-00088
13703-13-4 Isopropyl *N*-isopropyl-*P*-methylphosphonamidate, *in* I-00036
13703-32-7 Ethyl *N,P*-dimethylphosphonamidate, *in* D-00849
13716-10-4 Di-*tert*-butylphosphinous chloride, D-00125
13716-12-6 Tri-*tert*-butylphosphine, T-00373
13716-45-5 Diethyl trimethylsilyl phosphite, D-00408
13788-50-6 Ethyl dicyclohexylphosphinate, *in* D-00211
13789-63-4 *P*-Methyl-*N,N′*-diphenylphosphonothioic diamide, *in* M-00314
13789-68-9 *P*-Methyl-*N,N′*-diisopropylphosphonothioic diamide, *in* M-00314
13789-70-3 *N,N′*-Diisopropyl-*P*-phenylphosphonothioic diamide, *in* P-00246
13789-71-4 1,2,3,4-Tetramethyl-1,3,2,4-diazadiphosphetidine; 2,4-Disulfide, *in* T-00194
13789-75-8 1,2,3,4-Tetraphenyl-1,3,2,4-diazadiphosphetidine; 2,4-Disulfide, *in* T-00253
13792-55-7 Ethyl *N*-isopropyl-*P*-methylphosphonamidate, *in* I-00036
13795-84-1 Phenyl hydrogen (2-aminophenyl)-phosphoramidate, *in* A-00113
13798-39-5 Monophenyl phosphate; Bis(cyclohexylammonium) salt, *in* M-00507
13798-41-9 Diphenyl phosphate; Cyclohexylammonium salt, *in* D-01036
13799-87-6 *S,S,S*-Tris(4-methylphenyl) phosphorotrithioate, T-00816
13815-96-8 Phenyldivinylphosphine; Oxide, *in* P-00122
13824-09-4 3,9-Dimercapto-2,4,8,10-tetraoxa-3,9-diphosphaspiro[5.5]-undecane 3,9-disulfide, D-00670
13849-00-8 2,6,7-Trioxa-1-phosphabicyclo[2.2.1]-heptane; 1-Oxide, *in* T-00587
13868-71-8 Methylphosphonous diiodide, M-00325
13868-72-9 Ethyldiiodophosphine, *in* E-00153
13869-19-7 Bis(4-methylphenyl) phosphonate, B-00328
13879-80-6 Glycerol 1-9-octadecenoate 2-hexadecanoate 3-phospho-1′-glycerol; (*Z*)-*form*, Na salt, *in* G-00088
13884-92-9 Triphenyl(phenylthiomethyl)-phosphonium(1+); Chloride, *in* T-00621
13885-05-7 Tri-4-biphenylylphosphine, T-00350
13885-11-5 Tri-2-biphenylylphosphine, T-00348
13887-05-3 2,2,6,6-Tetramethyl-1-phenyl-4-phosphorinanone, T-00210
13887-07-5 9-Phenyl-9-phosphabicyclo[4.2.1]-nona-2,4,7-triene, P-00196
13891-63-9 *O,O*-Bis(2-ethoxyethyl) phosphorochloridothioate, B-00271
13891-64-0 *O*-(2-Ethoxyethyl) phosphorodichloridothioate, *in* H-00149
13891-84-4 (Phenylethynyl)phosphonothioic acid; Dichloride, *in* P-00167
13894-35-4 ▷Dimethyl chlorothiophosphonate, D-00716

13904-36-4 Tris(trimethylgermyl)phosphine, T-00860
13904-54-6 Tetrabutyldiphosphine, T-00014
13913-34-3 Bis(1-aziridinyl)phosphinic acid, B-00091
13917-52-7 Methyl bis(4-methylphenyl)-phosphinite, *in* B-00325
13941-61-2 *O,O*-Dimethyl phosphorothioate; Dicyclohexylammonium salt, *in* D-00907
13941-62-3 *O,O*-Diethyl phosphorothioate; Dicyclohexylammonium salt, *in* D-00396
13945-86-3 9-Phenyl-9-phosphabicyclo[4.2.1]nonane, P-00194
13953-50-9 *O*-Isopropyl hydrogen (1-methylpropyl)-phosphonodithioate, *in* M-00380
13954-38-6 ▷Tri-1-piperidinophosphine, T-00704
13968-96-2 2,3-Dihydro-2-methyl-1,3,2-benzothiazaphosphole, D-00519
13968-97-3 1,3,2-Benzoxathiaphosphole; 2-Me, *in* B-00026
13968-98-4 2-Methyl-1,3,2-benzodioxaphosphole, M-00090
13969-05-6 *o*-Phenylene methylphosphonate, *in* M-00090
13969-07-8 2,3-Dihydro-2-methyl-1,3,2-benzothiazaphosphole; 2-Sulfide, *in* D-00519
13969-32-9 2,5-Dimethyl-1,3,2-benzodithiaphosphole, D-00712
13983-95-4 Tetraethylphosphonium(1+), T-00056
13987-57-0 Hexapropylphosphoric triamide, *in* H-00092
13989-64-5 *N-tert*-Butyl-*P,P,P*-triphenylphosphine imide, *in* T-00626
13991-08-7 1,2-Phenylenebis[diphenyl-phosphine], P-00126
13991-98-5 Ethylphosphonothioic acid, E-00148
13992-06-8 Ethylphosphonoselenoic acid, E-00145
14047-23-5 (1-Aminopropyl)phosphonic acid, A-00125
14061-49-5 Dimethyl phosphorochloridodithioite, D-00889
14062-37-4 Dibutylmethylphosphine; Oxide, *in* D-00098
14074-89-6 6-Methoxydibenzo[*d,f*][1,3,2]-dioxaphosphepin, M-00032
14076-14-3 Dimethyl 3-pyridylphosphoramidate, *in* P-00511
14076-16-5 Diisopropyl 3-pyridylphosphoramidate, *in* P-00511
14086-46-5 Dioctylphenylphosphine, D-00955
14090-06-3 Diphenyl(2-phenylethenyl)-phosphine; (*E*)-*form*, *in* D-01033
14090-07-4 Diphenyl(2-phenylethenyl)-phosphine; (*Z*)-*form*, *in* D-01033
14109-35-4 9-Phenyl-9-phosphabicyclo[3.3.1]nonane, P-00193
14110-88-4 Dimethyl (9*H*-xanthen-9-yl)-phosphonate, *in* X-00001
14114-05-7 Cyclopropyltriphenyl-phosphonium(1+); Bromide, *in* C-00287
14114-67-1 Methyl *P*-acetyl-*N,N*-diethylphosphonamidate, *in* A-00014
14114-73-9 *P*-Benzoyl-*N,N,N′,N′*-tetraethylphosphonic diamide, *in* B-00035
14114-74-0 *P*-Acetyl-*N,N,N′,N′*-tetraethylphosphonic diamide, *in* A-00027
14114-77-3 Ethyl diethylphosphoramidochloridite, E-00060
14129-98-7 ▷4,4′,4″-Phosphinothioylidynetrismorpholine, *in* T-00569
14156-07-1 *O,O*-Diphenyl phosphorothioate, D-01141
14173-53-6 Diethyl 3-pyridylphosphoramidate, *in* P-00511
14180-49-5 *p*-(Diphenylphosphinothioyl)-anisole, *in* M-00058
14180-55-3 Tris(4-methoxyphenyl)phosphine; Sulfide, *in* T-00793
14181-19-2 2,2-Dihydro-2,2,2-triphenyl-4,4-bis(trifluoromethyl)-3-(triphenylphosphoranylidene)-1,2-oxaphosphetane, D-00608
14212-91-0 Methyl bis(chloromethyl)-phosphinate, *in* B-00127
14212-98-7 Phenyl bis(chloromethyl)-phosphinate, *in* B-00127
14235-74-6 Phenyl methylphosphonochloridate, *in* M-00296
14254-40-1 Dipentyl phosphorochloridate, *in* D-00979
14255-61-9 Hydroxymethanediphosphonic acid; Di-Na salt, *in* H-00154
14255-72-2 ▷*O,O*-Diethyl *O*-[4-(methylsulfonyl)phenyl] phosphorothioate, *in* D-00302
14264-70-1 4,5,5,5-Tetrahydro-3,5,5,5-tetraphenyl-1,2,5-oxazaphosphole, T-00153
14274-25-0 ▷2-Ethylthio-1,3,2-dioxaphospholane, E-00191
14274-39-6 ▷2-Ethylthio-4,5-dimethyl-1,3,2-dioxaphospholane, E-00189
14274-40-9 2-Ethylthio-4,4,5,5-tetramethyl-1,3,2-dioxaphospholane, E-00193
14274-42-1 2-Dimethylamino-4,4,5,5-tetramethyl-1,3,2-dioxaphospholane, D-00710
14274-45-4 *O*-Ethyl *O,O*-tetramethylethylene phosphorothioate, *in* E-00052
14274-46-5 *S*-Ethyl *O,O*-ethylene dithiophosphate, *in* E-00191
14274-48-7 2-Ethylthio-4,5-dimethyl-1,3,2-dioxaphospholane; 2-Sulfide, *in* E-00189
14274-49-8 *S*-Ethyl *O,O*-tetramethylethylene phosphorodithioate, *in* E-00193
14274-50-1 *O,O*-2,3-Butylene dimethylphosphoramidothioate, *in* D-00725
14274-51-2 4,4,5,5-Tetramethylethylene dimethylphosphoramidothioate, *in* D-00710
14278-36-5 2-Phenylphosphorodifluoridothioic hydrazide, P-00289
14278-72-9 Diphenylphosphinothioic acid, D-01082
14295-11-5 2-Ethoxy-1,3,2-dioxaphospholane 2-sulfide, E-00036
14311-22-9 Phenyl diphenylphosphinothioite, *in* D-01085
14335-48-9 (Trichloroacetyl)phosphorimidic trichloride, T-00398
14337-77-0 Methyl dimethylphosphinate, *in* D-00829
14337-82-7 Dimethylphosphinic acid; Anhydride, *in* D-00829
14350-50-6 Triphenylpropylphosphonium(1+); Iodide, *in* T-00680
14360-85-1 *N,N′*-Diethyl-*P*-phenylphosphonic diamide, *in* P-00222
14366-43-9 *O,O*-Bis(4-methylphenyl) phosphorodithioate, B-00339
14366-45-1 *O,O*-Bis(4-chlorophenyl) phosphorodithioate, B-00151
14378-87-1 3-(Diphenylphosphinothioyl)-benzoic acid, *in* D-01054
14378-89-3 3-(Diphenylphosphino)-benzonitrile, D-01057
14409-95-1 2,5-Dihydro-3,4-dimethyl-1-phenyl-1*H*-phosphole, D-00468
14410-00-5 1,2,3,6-Tetrahydro-4,5-dimethyl-1,2-diphosphorine; 1,2-Di-Me, 1,2-disulfide, *in* T-00095
14410-05-0 2,5-Dihydro-1,3,4-trimethyl-1*H*-phosphole, D-00600
14410-06-1 1,2,3,6-Tetrahydro-1,2,4,5-tetramethyl-1,2-diphosphinine, *in* T-00095
14410-07-2 *O*-Phenyl methylphosphonochloridothioate, *in* M-00298
14418-26-9 2,4,6,7-Tetramethyl-2,6,7-triaza-1-phosphabicyclo[2.2.2]-octane, T-00237
14438-23-4 (Triphenylphosphoranylidene)-butanedioic acid, T-00646
14445-60-4 Tetramethylphosphorodiamidic cyanide, T-00221
14463-04-8 1,2,3,6-Tetrahydro-4,5-dimethyl-1,2-diphosphorine; 1,2-Di-Ph, 1,2-disulfide, *in* T-00095
14463-05-9 1,2,3,6-Tetrahydro-4,5-dimethyl-1,2-diphenyl-1,2-diphosphinine, *in* T-00095
14465-96-4 ▷*P,P*-Bis(1-aziridinyl)-phosphinothioic amide, B-00094
14491-87-3 *P*-Ethenyl-*N,N,N′,N′*-tetraethylphosphonic diamide, *in* V-00004
14496-36-7 Phenyl bis(1-aziridinyl)-phosphinite, *in* B-00095
14496-37-8 Diphenyl 1-aziridinylphosphonite, *in* A-00154
14518-01-5 *P*-(2-Chloroethyl)-*N,N,N′,N′*-tetramethylphosphonic diamide, *in* C-00075
14526-13-7 Phosphor(isothiocyanatidic) difluoride, P-00408
14526-14-8 Phosphorodi(isothiocyanatidic) fluoride, P-00414
14529-09-0 Methylenebis[triphenyl-phosphonium](2+); Dibromide, *in* M-00147
14540-45-5 *tert*-Butyl ethyl phosphonate, B-00561
14540-49-9 *O,O,O*-Triisopropyl phosphoroselenoate, T-00504
14557-33-6 Methyl dichlorophosphinylcarbamate, *in* D-00191
14561-18-3 Diisopropyl benzylphosphonite, *in* B-00068
14561-29-6 5,5-Dihydro-4,4-dimethyl-2,3,5,5,5-pentaphenyl-1,2,5-oxazaphospholidine, D-00465
14572-73-7 Benzyl *P*-phenylphosphonamidate, *in* P-00218
14576-59-1 Hexylphosphonic acid; Difluoride, *in* H-00097
14576-61-5 Heptylphosphonic acid; Difluoride, *in* H-00006
14576-63-7 Octylphosphonic acid; Difluoride, *in* O-00042
14576-64-8 Nonylphosphonic acid; Dichloride, *in* N-00076
14576-66-0 Decylphosphonic acid; Dichloride, *in* D-00012
14579-57-8 Tris(triethylsilyl) phosphate, T-00849
14580-91-7 Trimethylmethylenephosphorane, T-00518
14580-93-9 2,2-Dihydro-1,2-oxaphospholane; 2,2,2-Tri-Ph, *in* D-00534
14596-55-5 2-Methylaminoethylphosphonic acid, *in* A-00079
14605-34-6 *P*-(2-Chloroethyl)-*N,N,N′,N′*-tetraethylphosphonic diamide, *in* C-00075
14609-90-6 Isopropyl phenyl phosphonate, I-00047
14609-91-7 Butyl phenyl phosphonate, B-00585
14609-96-2 *O,O*-Dipropyl phosphonothioate, D-01212
14610-01-6 Ethyl cyclohexylphosphinate, *in* C-00253
14610-03-8 Dicyclohexylphosphine; Sulfide, *in* D-00210
14618-58-7 *tert*-Butoxycarbonyl diethyl phosphate, B-00531

14633-92-2 Bis(diphenylphosphinothioyl)methane, *in* B-00244
14633-95-5 Cyclopropylidenetriphenylphosphorane, C-00284
14646-04-9 Diisopropyl 2-aminoethyl phosphate, *in* A-00075
14646-14-1 2-Quinolinephosphonic acid, Q-00002
14651-62-8 2-Phenyl-1,2-oxaphospholan-2-one 2-oxide, *in* O-00053
14655-46-0 2-Propenyl phenyl(2-propenyl)phosphinate, *in* P-00300
14655-66-4 Bis(4-methylphenyl)phosphinous acid, B-00325
14655-69-7 *P*-Ethyl-*N,N,N′,N′*-tetramethylphosphonic diamide, *in* E-00130
14655-70-0 *N,N,N′,N′*-Tetramethyl-*P*-propylphosphonic diamide, *in* P-00474
14655-73-3 *N,N,N′,N*-Tetramethyl-*P*-(phenylmethyl)phosphonic diamide, *in* B-00061
14655-81-3 1,2-Cyclohexanediphosphonic acid, C-00240
14655-88-0 Tetrakis(pentafluorophenyl)tetraphosphetane, T-00185
14657-64-8 3-(Hydroxyphenylphosphinyl)propanoic acid, H-00178
14657-87-5 2,3,4,5,6-Pentaphenylphosphorin, P-00040
14662-78-3 Diethyl (2-hydroxyethyl)phosphoramidate, *in* H-00148
14684-24-3 Methyl phosphorodichloridothioite, M-00350
14684-25-4 ▷Phenyl phosphorodichloridothioite, P-00284
14684-28-7 Phenyl tetraethylphosphorodiamidite, *in* T-00066
14684-31-2 Methyl phosphinate, M-00276
14684-32-3 ▷Ethyl phosphinate, E-00122
14684-35-0 Diethylphenylphosphine; Sulfide, *in* D-00305
14703-66-3 Monobutyl phosphate; Monocyclohexylammonium salt, *in* M-00451
14705-46-5 Ethylphosphonoselenoic dichloride, E-00147
14707-40-5 2,3-Dihydro-2-methoxy-1,2-benzoxaphosphole 2-oxide, *in* H-00119
14707-41-6 2-Ethoxy-2,3-dihydro-1,2-benzoxaphosphole 2-oxide, *in* H-00119
14717-27-2 *O,O*-Diisopropyl phosphonothioate, D-00652
14717-29-4 Dicyclohexylphosphine; Oxide, *in* D-00210
14744-12-8 Diphenyl ethylphosphonite, *in* E-00153
14747-96-7 Diphenyl diethylphosphoramidite, *in* D-00360
14791-94-7 [2-(Methylthio)phenyl]diphenylphosphine, M-00417
14799-82-7 (Cyclopropylmethyl)triphenylphosphonium(1+); Bromide, *in* C-00286
14806-54-3 *O*-Methyl diethylphosphinothioate, *in* D-00330
14806-66-7 Dimethyl methylphosphonotrithioate, *in* M-00318
14812-59-0 2-Chloro-4,4,5,5-tetramethyl-1,3,2-dioxaphospholane, C-00185
14812-60-3 2-Methoxy-4,4,5,5-tetramethyl-1,3,2-dioxaphospholane, M-00085
14812-61-4 2-Phenoxy-4,4,5,5-tetramethyl-1,3,2-dioxaphospholane, P-00075
14812-63-6 2-Phenyl-4,4,5,5-tetramethyl-1,3,2-dioxaphospholane, P-00322
14816-18-3 ▷Phoxim, P-00424
14866-43-4 Hexadecyltriphenylphosphonium(1+); Bromide, *in* H-00014
14899-31-1 3-Phosphonobenzoic acid, P-00379
14902-12-6 (Cyclopropylmethylene)triphenylphosphorane, *in* C-00286
14902-14-8 Methyl α-(triphenylphosphoranylidene)cyclopropaneacetate, *in* C-00285
14923-49-0 2-Ethoxy-1,2-oxaphosphepane 2-oxide, *in* H-00165
14923-50-3 Ethyl pentamethylenephosphinate, *in* H-00188
14937-39-4 Pentamethylphosphonous diamide, *in* M-00320
14937-45-2 Tributyl(hexadecyl)phosphonium(1+); Bromide, *in* T-00364
14966-25-7 2-Naphthyl phosphorodichloridite, N-00031
14967-93-2 2,3-Dihydro-1-phenoxy-1*H*-phosphole 1-oxide, *in* D-00506
14974-76-6 *S,S,S*-Tribenzyl phosphorotrithioate, T-00342
14984-58-8 4-Aminophenyl diethyl phosphate, A-00098
14990-01-3 Benzylphosphine, B-00058
14994-07-1 Glycerol 1,2-di-5,8,11,14-icosatetraenoate 3-phosphoethanolamine, G-00018
15004-44-1 *O,O-o*-Phenylene thiophosphonate, *in* M-00090
15040-65-0 10-Methyl-10-phenyl-10*H*-phenoxaphosphinium iodide, *in* P-00183
15042-79-2 10-Hydroxy-10*H*-phenoxaphosphine 10-oxide, H-00170
15049-10-2 Tris[(4-methylphenyl)methyl]phosphine; Oxide, *in* T-00802
15049-85-1 1-Amino-1,1-ethanediphosphonic acid, A-00073
15049-86-2 (1-Aminopropylidene)bisphosphonic acid, A-00124
15052-46-7 2,2-Dihydro-2,2,2-trimethoxynaphtho[1,2-*d*]-1,3,2-dioxaphosphole, D-00594
15071-88-2 Glycerol 1-octadecanoate 2-dodecanoate 3-phosphocholine, G-00074
15074-53-0 Bis(4-chlorophenyl) phosphorochloridate, *in* B-00133
15074-54-1 2-Chlorophenyl phosphorodichloridate, C-00161
15090-23-0 3-(Hydroxymethylphosphinyl)propanoic acid, H-00161
15096-31-8 2,2,2,3-Tetrahydro-1,2-oxaphosphole; 2,2,2-Trimethoxy, *in* T-00119
15096-33-0 3-Phenyl-1,3-azaphospholidine, P-00091
15096-99-8 (Tetraphenyl-2,4-cyclopentadien-1-ylidene)triphenylphosphorane, T-00252
15103-99-8 Tetramethyl hypophosphate, T-00204
15107-02-5 ▷*P,P,P*-Trimethylphosphine imide, T-00531
15107-69-4 2,3-Diphenyl-1,3-azaphospholidine; 3-Sulfide, *in* D-00995
15108-81-3 *unsym-O,O,O,O*-Tetraethyl dithiopyrophosphate, T-00046
15110-15-3 2,2,2,3-Tetrahydro-1,2-oxaphosphole; 2,2,2-Triethoxy, *in* T-00119
15114-55-3 Tris[(2-diphenylphosphino)ethyl]amine, T-00773
15120-59-9 1,3,2-Benzoxathiaphosphole, B-00026
15159-51-0 *N,N′,N″*-Triphenylphosphorous triamide, T-00677
15171-48-9 2-Methyl-1,2-oxaphospholan-5-one 2-oxide, *in* O-00053
15199-21-0 2,4,6,7-Tetramethyl-2,6,7-triaza-1-phosphabicyclo[2.2.2]octane; 1-Oxide, *in* T-00237
15199-22-1 2,4,6,7-Tetramethyl-2,6,7-triaza-1-phosphabicyclo[2.2.2]octane; 1-Sulfide, *in* T-00237
15205-57-9 Tribenzyl phosphite, T-00340
15205-58-0 Tricyclohexyl phosphite, T-00428
15205-62-6 Tri-*tert*-butyl phosphite, T-00380
15207-49-5 2,3-Diphenyl-1,3-azaphospholidine, D-00995
15207-88-2 Tetramethyl (1-hydroxyethylidene)bisphosphonate, *in* H-00144
15233-97-3 4,8-Di-*tert*-butyl-2,10-dimethyl-12*H*-dibenzo[*d,g*][1,3,2]dioxaphosphocin; 6-Oxide, *in* D-00088
15247-35-5 Trichlorodiphenoxyphosphorane, T-00403
15272-12-5 *P,P*-Diethyl-*N,N*-dimethylphosphinous amide, *in* D-00334
15272-77-2 Bomyl; (*E*)-*form*, *in* B-00461
15272-78-3 Bomyl; (*Z*)-*form*, *in* B-00461
15286-11-0 Diethyl (2-methylphenyl)phosphonate, *in* M-00235
15286-13-2 Diethyl (3-methylphenyl)phosphonate, *in* M-00236
15286-16-5 Dimethyl (2-methoxyphenyl)phosphonate, *in* M-00070
15286-17-6 Diethyl (2-methoxyphenyl)phosphonate, *in* M-00070
15286-19-8 Dimethyl (4-methoxyphenyl)phosphonate, *in* M-00072
15288-70-7 Methyl diphenylphosphinodithioate, *in* D-01065
15300-97-7 Diethyl (2-acetyloxyethyl)phosphonate, *in* A-00006
15303-25-0 1,2,5,6-Tetrahydro-2,2,5,5-tetraphenyl-2,5-benzodiphocinium dibromide, *in* H-00053
15309-56-5 4-Dimethylamino-2,2-dihydro-2,2-diphenyl-1,3,5,2-triazaphosphorine, D-00692
15309-64-5 5-Chloro-5,10-dihydroacridophosphine, C-00037
15323-49-6 *O*-2-Naphthyl phosphoradiamidothioate, N-00027
15336-73-9 Diethyl (1-hydroxyethyl)phosphonate, *in* H-00145
15352-69-9 Tetraphenyl-1,4-diphosphonia-6,7-benzocyclooct-6-ene, *in* H-00053
15367-52-9 Butyldiphenylphosphine; Sulfide, *in* B-00557
15367-75-6 Benzyldiphenylphosphine; Sulfide, *in* B-00046
15374-50-2 *S*-Ethyl *O*-phenyl cyclohexylphosphonodithioate, *in* C-00261
15383-58-1 1,2-Bis(diphenylphosphino)propane, B-00264
15416-86-1 3,7,11-Trimethyl-2,6,10-dodecatrien-1-yl dihydrogen phosphate, T-00515
15435-17-3 2,3,4-Triphenyl-1,3,2,4-thiazadiphosphetidine; 2,4-Disulfide, *in* T-00687
15450-93-8 2,5-Dihydro-3-methyl-1-phenyl-1*H*-phosphole, D-00528
15468-10-7 Hydroxymethanediphosphonic acid, H-00154
15479-16-0 2-Chloro-4,5-dimethyl-1,3,2-dioxaphospholane; (4*RS*,5*RS*)-*form*, *in* C-00056
15493-07-9 Dichlorotriphenoxyphosphorane, D-00199
15494-45-8 2-Chloro-2,3-dihydro-3-phenyl-4*H*-1,3,2-benzoxazaphosphorin-4-one, C-00050

15494-46-9 2,3-Dihydro-3-phenyl-4*H*-1,3,2-benzoxazaphosphorin-4-one; 2-Methoxy, *in* D-00556

15494-47-0 2,3-Dihydro-3-phenyl-4*H*-1,3,2-benzoxazaphosphorin-4-one; 2-Ethoxy, *in* D-00556

15494-49-2 2,3-Dihydro-3-phenyl-4*H*-1,3,2-benzoxazaphosphorin-4-one; 2-Phenoxy, *in* D-00556

15496-31-8 1*H*-Imidazol-1-ylphosphonic acid, I-00003

15510-55-1 Dodecyltriphenylphosphonium(1+); Bromide, *in* D-01266

15516-38-8 Bis(2-methylphenyl) phosphonate, B-00326

15516-41-3 Bis(4-chlorophenyl) phosphonate, B-00144

15520-16-8 Bis(4-chlorophenyl) phosphorochloridite, B-00147

15536-25-1 ▷*O,S*-Dibutyl methylphosphonothioate, *in* M-00313

15546-42-6 1,4-Butanediylbis[triphenyl-phosphonium](2+); Dibromide, *in* B-00517

15573-31-3 Propylphosphonous dichloride, P-00483

15573-32-7 Dichloropentylphosphine, *in* P-00052

15573-33-8 Dichloroheptylphosphine, *in* H-00008

15573-34-9 Dichlorooctylphosphine, *in* O-00044

15573-37-2 Nonylphosphine, N-00075

15573-38-3 ▷Tris(trimethylsilyl)phosphine, T-00865

15596-07-3 (Triphenylphosphoranylidene)-ethenone, T-00651

15607-00-8 2,3-Diphenyl-1,4,6,9-tetraoxa-5-phospha(5-P^V)spiro[4.4]non-2-ene; 5-Methoxy, *in* D-01150

15607-05-3 *N,N*,1,3-Tetramethylspiro[1,3,2-diazaphospholidine-2,2′-phenanthro[9,10-*d*]-[1,3,2]dioxaphosphol]-2-amine, *in* D-00918

15607-06-4 1,3-Dimethylspiro[1,3,2-diazaphospholidine-2,2′-phenanthro[9,10-*d*]-[1,3,2]dioxaphosphole]; 2-Methoxy, *in* D-00918

15661-97-9 2,3-Dimethyl-1,4,6,9-tetraoxa-5-phospha(5-P^V)spiro[4.4]non-2-ene; 5-Methoxy, *in* D-00924

15677-02-8 (Triphenylphosphoranylidene)-acetic acid, T-00640

15704-45-7 Dimethyl (4-fluorophenyl)-phosphonate, *in* F-00038

15706-68-0 Phenyl di-1*H*-imidazol-1-ylphosphinate, *in* D-00628

15714-21-3 Diethyl ethylphosphonodithioite, *in* E-00141

15715-41-0 ▷Diethyl methylphosphonite, *in* M-00319

15715-42-1 ▷Dimethyl ethylphosphonite, *in* E-00153

15754-51-5 Bis(4-methoxyphenyl)phosphine, B-00298

15754-51-5 Bis(4-methoxyphenyl)phosphine; Oxide, *in* B-00298

15754-54-8 Dibutylphosphine; Oxide, *in* D-00107

15762-00-2 *tert*-Butyl neopentylene phosphate, *in* B-00534

15762-04-6 2,2′-Oxybis[5,5-dimethyl-1,3,2-dioxaphosphorinane] 2-oxide 2′-sulfide, O-00090

15790-43-9 Dimethylphosphinic acid; Isocyanate, *in* D-00829

15790-45-1 Dibutylphosphinic acid; Isocyanate, *in* D-00109

15832-91-4 *O,O*-Diethyl (2-methylphenyl)-phosphoramidothioate, *in* M-00254

15832-92-5 *O,O*-Diethyl (3-methylphenyl)-phosphoramidothioate, *in* M-00255

15832-93-6 *O,O*-Diethyl (4-methylphenyl)-phosphoramidothioate, *in* M-00256

15832-94-7 *O,O*-Diethyl (4-methoxyphenyl)-phosphoramidothioate, *in* M-00079

15832-95-8 *O,O*-Diethyl (2-methoxyphenyl)-phosphoramidothioate, *in* M-00077

15833-50-8 *O,O*-Diethyl (3-methoxyphenyl)-phosphoramidothioate, *in* M-00078

15845-66-6 Monoethyl phosphonate, M-00473

15849-83-9 Chloroethylphenylphosphine, *in* E-00108

15849-84-0 Ethylmethylphenylphosphine, E-00084

15849-86-2 Methylphenylphosphinous chloride, M-00228

15849-93-1 Pentamethylphosphonothioic diamide, *in* M-00314

15849-99-7 Vinylphosphonothioic acid; Dichloride, *in* V-00007

15850-00-7 Divinylphosphinothioic acid; Chloride, *in* D-01259

15853-35-7 Benzyltriphenylphosphonium(1+), B-00073

15853-37-9 Tetrabutylphosphonium(1+), T-00019

15873-69-5 Ethyl cyclohexylphosphonochloridate, *in* C-00259

15873-72-0 Dicyclohexylphosphinic acid; Chloride, *in* D-00211

15887-56-6 *O,O*-Diethyl phosphoroselenothioate; K salt, *in* D-00394

15899-82-8 *O,O*-Diphenyl phosphoroselenothioate; K salt, *in* D-01140

15901-11-8 (Aminomethyl)methylphosphinic acid, A-00089

15905-28-9 Neopentylene dimethylphosphoramidothioate, *in* D-00693

15909-92-9 Bis(2-cyanoethyl)phenylphosphine, B-00152

15912-74-0 Methyltriphenylphosphonium(1+), M-00432

15912-75-1 Triphenylpropylphosphonium(1+), T-00680

15912-76-2 Triphenyl-2-propenylphosphonium(1+), T-00679

15916-61-7 Hexahydro-4-phenyl-1,4-azaphosphorine, H-00042

15916-62-8 Hexahydro-4-phenyl-1,4-azaphosphorine; 4-Sulfide, *in* H-00042

15916-99-1 *N,N*′-Di-*tert*-butyl-*P*-phenylphosphonic diamide, *in* P-00222

15917-00-7 *N,N*′-Di-*tert*-butyl-*P*-phenylphosphonothioic diamide, *in* P-00246

15924-57-9 Bis(2-methylpropyl)phosphinic acid, B-00346

15924-61-5 1,4-Dihydro-1,1,4,4-tetraphenyl-1,4-diphosphorinium(2+); Dibromide, *in* D-00582

15935-94-1 Triphenyl-2-propenylidenephosphorane, *in* T-00679

15948-60-4 Bis(4-chlorophenyl)phosphine; Oxide, *in* B-00136

15950-60-4 *O,O*-Diethyl (2-chlorophenyl)-phosphoramidothioate, *in* C-00158

15950-61-5 *O,O*-Diethyl (3-chlorophenyl)-phosphoramidothioate, *in* C-00159

15978-08-2 Fructose 1-dihydrogen phosphate; D-*form*, *in* F-00068

15990-54-2 Formaldehyde (triphenylphosphoranylidene)-hydrazone, F-00054

15995-44-5 Bis(5,5-dimethyl-2-oxo-1,3,2-dioxaphosphorinan-2-yl) disulfide, *in* D-01251

16001-01-7 ▷Diethyl phosphorochloridotrithioate, D-00380

16001-03-9 Dimethyl phosphorochloridotrithioate, D-00893

16001-05-1 Ethyl phosphorodichloridodithioate, E-00169

16001-93-7 Tetramethyl methylenebisphosphonate, *in* M-00022

16051-76-6 (2-Oxoethyl)phosphonic acid, O-00059

16062-77-4 1,1′,1″-Phosphinylidynetris(1*H*-imidazole), *in* T-00495

16079-34-8 Diphenyl phenylsulfonylphosphoramidate, *in* P-00312

16083-91-3 2,2,3,4,4-Pentamethyl-1-phenylphosphetane; *cis-form*, Oxide, *in* P-00020

16083-94-6 1,2,2,3,4,4-Hexamethylphosphetane; Oxide, *in* H-00074

16083-95-7 2,2,3,4,4-Pentamethyl-1-phenylphosphetane; *trans-form*, *in* P-00020

16084-01-8 1,1,2,2,3,4,4-Heptamethylphosphetanium bromide, *in* H-00074

16088-56-5 ▷Dioxathion; *Cis-form*, *in* D-00974

16102-45-7 Diethyl benzoylphosphoramidate, *in* B-00036

16109-84-5 1,2,2,3,4,4-Hexamethylphosphetane, H-00074

16139-82-5 Methyl cyclohexylphosphonochloridate, *in* C-00259

16139-87-0 *S,S*-Diethyl cyclohexylphosphonodithioate, *in* C-00261

16141-66-5 Benzyldifluorophosphine, *in* B-00068

16141-72-3 Ethyl *N,N*-diethylphosphonamidate, *in* D-00337

16153-59-6 Diethyl [(acetyloxy)-phenylmethyl]phosphonate, *in* A-00020

16165-66-5 ▷Diethyl hexylphosphonate, *in* H-00097

16165-68-7 Diethyl decylphosphonate, *in* D-00012

16167-31-0 *O*-(Pentachlorophenyl) phosphorodichloridothioate, P-00007

16182-83-5 Butyl ethylphosphonochloridothioite, *in* E-00138

16182-89-1 1,8-Diethoxy-3*a*,4,7,7*a*-tetrahydro-4,7-phosphinidene-1*H*-phosphinidole 1,8-dioxide, *in* D-00625

16182-90-4 1,8-Diethoxyoctahydro-4,7-phosphinidene-1*H*-phosphindole 1,8-dioxide, *in* D-00625

16195-98-5 Dimethyl cyclohexylphosphonite, *in* C-00267

16195-99-6 Diethyl cyclohexylphosphonite, *in* C-00267

16196-03-5 Methyl cyclohexylphosphinate, *in* C-00253

16243-48-4 *O*-(2-Methylpropyl) cyclohexylphosphonochloridothioate, *in* C-00260

16259-87-2 Diethyl (3,4-dihydro-1*H*-2-benzopyran-1-yl)-phosphonate, *in* D-00443

16270-86-3 ▷Dioxathion; *Trans-form*, *in* D-00974

16271-10-6 *O,S*-Diethyl phosphoramidothioate, D-00355

16276-72-5 Butyl *N,N*-diethylphosphonamidate, *in* D-00337

16276-74-7 Ethyl 4-morpholinylphosphonochloridite, *in* M-00526
16276-75-8 Ethyl 4-morpholinylphosphinate, *in* M-00522
16284-71-2 ▷Methyl methylphosphonochloridodithioate, *in* M-00297
16284-72-3 Methyl *P*-methyl-*N*-phenylphosphonamidodithioate, *in* M-00232
16284-88-1 *O,O*-Dimethyl phosphorothioate; Trimethylammonium salt, *in* D-00907
16324-17-7 2-Phenyl-1,2-oxaphospholane, P-00174
16324-18-8 2-Phenyl-1,2-oxaphospholane; 2-Sulfide, *in* P-00174
16324-19-9 2-Phenyl-1,2-oxaphospholane; 2-Oxide, *in* P-00174
16327-48-3 1-Methyl-4-phosphorinanone, M-00340
16352-17-3 2,3-Butylene phosphonate, *in* D-00733
16352-18-4 Tetramethylethylene phosphonate, *in* T-00197
16352-21-9 Trimethylene phosphonate, *in* D-00972
16352-25-3 Tetramethylene phosphorochloridite, *in* D-00968
16352-28-6 2-Chloro-4,5-dimethyl-1,3,2-dioxaphospholane, C-00056
16365-56-3 *O*-Methyl phosphorofluoridoisothiocyanatidothioate, M-00358
16367-68-3 Dimethylphosphinodithioic acid, D-00834
16368-06-2 2,2′-Bis(5,5-dimethyl-1,3,2-dioxaphosphorinane) 2,2′-dioxide, *in* B-00197
16368-07-3 Thiohypophosphoric acid cyclic *P,P,P′,P′*-bis(2,2-dimethyltrimethylene) ester, *in* B-00197
16368-08-4 Thiohypophosphoric acid cyclic *P,P,P′,P′*-bis(2,2-dimethyltrimethylene) ester, *in* B-00197
16368-09-5 2,2′-Oxybis[5,5-dimethyl-1,3,2-dioxaphosphorinane]; 2-Oxide, *in* O-00089
16368-12-0 Neopentylene phosphorobromidate, *in* B-00471
16368-13-1 Neopentylene phosphorobromidothioate, *in* B-00471
16368-15-3 *O,O*-2,3-Butylene phosphonothioate, *in* D-00733
16368-16-4 *O,O*-Trimethylene phosphonothioate, *in* D-00972
16368-17-5 4-Methyl-1,3,2-dioxaphosphorinane 2-sulfide, M-00131
16368-20-0 4-Methyl-1,3,2-dioxaphosphorinane 2-oxide, M-00130
16383-57-6 Di-2-propenyl phosphorochloridate, *in* D-01192
16389-70-1 [(Diphenylphosphinyl)methyl]-triphenylphosphonium(1+); Tetraphenylborate, *in* D-01101
16390-99-1 Phenyl propyl phosphonate, P-00304
16391-06-3 Methyl methylphosphinate, *in* M-00278
16391-07-4 Ethyl methylphosphinate, *in* M-00278
16391-15-4 Methyl (2,5-dimethylphenyl)-phosphinate, *in* D-00805
16391-16-5 Ethyl (2,5-dimethylphenyl)-phosphinate, *in* D-00805
16391-19-8 Methyl (3,4-dimethylphenyl)-phosphinate, *in* D-00806
16391-20-1 (3,4-Dimethylphenyl)phosphinic acid; Et ester, *in* D-00806
16391-21-2 (3,4-Dimethylphenyl)phosphinic acid; Isopropyl ester, *in* D-00806
16393-52-5 2-Mercapto-1,3,2-dioxaphospholane, *in* D-00970
16400-19-4 *P*-Ethynyl-*N,N,N′,N′*-tetramethylphosphonic diamide, *in* E-00203
16400-23-0 *N,N,N′,N′*-Tetramethyl-*P*-1-propynylphosphonic diamide, *in* P-00494
16400-31-0 2,2,4,6-Tetrachloro-2,2,4,4,6,6-hexahydro-4,6-bis(methylamino)-1,3,5-triaza-2,4,6-triphosphorine; (4*RS*, 6*SR*)-*form*, *in* T-00030
16401-29-9 3,4-Bis[(triphenylphosphonio)-methyl]furan(2+); Dichloride, *in* B-00449
16403-36-4 [(Diphenylphosphinyl)methyl]-triphenylphosphonium(1+); Chloride, *in* D-01101
16415-09-1 2-Fluoro-4-methyl-1,3,2-dioxaphospholane, F-00020
16416-60-7 2,2,4,6-Tetrachloro-2,2,4,4,6,6-hexahydro-4,6-bis(methylamino)-1,3,5-triaza-2,4,6-triphosphorine; (4*RS*, 6*RS*)-*form*, *in* T-00030
16421-86-6 2,2′-Oxybis[1,3,2-benzodioxaphosphole], O-00087
16456-50-1 *N,N′*-Trimethylene-*P*-phenylphosphonic diamide, *in* H-00020
16456-51-2 *P*-Methyl-*N,N′*-trimethylenephosphonic diamide, *in* H-00020
16456-52-3 Phenyl *N,N′*-trimethylenephosphorodiamidate, *in* H-00039
16456-53-4 Methyl *N,N′*-trimethylenephosphorodiamidate, *in* H-00020
16456-56-7 ▷Monobutyl phosphonate, M-00453
16474-65-0 (3-Methylbutyl)phosphonic acid; Dichloride, *in* M-00106
16474-69-4 (1-Methylnonyl)phosphonic acid; (±)-*form*, Dichloride, *in* M-00181
16474-70-7 (1-Propylheptyl)phosphonic acid; (±)-*form*, Dichloride, *in* P-00467
16492-16-3 Ethylene phenyl phosphate, *in* P-00067
16492-17-4 2,3-Butylene ethyl phosphate, *in* E-00031
16492-44-7 4,4,5,5-Tetramethylethylene dimethylphosphoramidate, *in* D-00710
16498-14-9 *N″,N″*-Dimethyl-*N,N′*-trimethylenephosphoric triamide, *in* H-00020
16522-04-6 1,2-Dihydro-1,2-bis(triphenylphosphonio)-benzocyclobutene(2+); (1*RS*,2*RS*)-*form*, Diperchlorate, *in* D-00447
16522-52-4 2-(Diphenylphosphinyl)phenol, *in* D-01070
16522-54-6 2-(Diphenylphosphinothioyl)-phenol, *in* D-01070
16523-54-9 Chlorodicyclohexylphosphine, *in* D-00214
16523-61-8 *N*-Methyl-*P,P*-diphenylphosphinothioic amide, *in* D-01083
16523-75-4 5,5,10,10-Tetraethyl-5,6,7,8,9,10-hexahydrodibenzo[*b,d*][1,6]-diphosphecinium dibromide, *in* H-00021
16523-79-8 9,9,16,16-Tetraethyl-9,10,15,16-tetrahydrotribenzo[*b,d,h*][1,6]-diphosphecinium dibromide, *in* T-00155
16523-89-0 Tri-2-propenylphosphine, T-00707
16524-41-7 1,3-Bis(diphenylphosphinyl)-propane, *in* B-00265
16527-12-1 1,2-Bis(dicyclophosphinyl)ethane, *in* B-00167
16534-61-5 Bis(4-methylphenyl)-phosphinothioic acid; Chloride, *in* B-00322
16534-67-1 Diphenylphosphinothioic acid; Bromide, *in* D-01082
16540-25-3 Ethyl (diethoxyphosphinyl)-oxoacetate, *in* D-00252
16543-10-5 Diphenylphosphinylacetic acid hydrazide, D-01094
16543-38-7 Diisopropylphenylphosphine; Oxide, *in* D-00638
16543-42-3 *tert*-Butylethylphosphinic acid, B-00560
16543-43-4 Isopropylphenylphosphinic acid, I-00043
16543-45-6 Isopropyl ethylphosphonochloridothioite, *in* E-00138
16546-79-5 5-Methyl-5*H*-dibenzophosphole, M-00115
16566-17-9 [2-(Methylseleno)phenyl]-diphenylphosphine, M-00393
16604-64-1 *O*-Phenyl tetraethylphosphorodiamidothioate, *in* T-00063
16604-66-3 *O*-Phenyl tetraethylphosphorodiamidoselenoate, *in* T-00061
16606-64-7 (1-Aminopropyl)phosphonic acid; (±)-*form*, *in* A-00125
16606-65-8 (1-Aminoethyl)phosphonic acid; (±)-*form*, *in* A-00078
16611-68-0 6-Chlorodibenzo[*d,f*][1,3,2]-dioxaphosphepin, C-00032
16640-68-9 Triphenyl-phosphoranylideneacetonitrile, T-00641
16644-61-4 Butyl dioctylphosphinate, *in* D-00958
16650-20-7 Tetraethyl 1,3-butadiene-1,4-diylbisphosphonate, *in* B-00507
16666-78-7 Triphenylpropylphosphonium(1+); Ylide, *in* T-00680
16666-80-1 (1-Methylethylidene)-triphenylphosphorane, M-00153
16666-81-2 Cyclohexylidenetriphenyl-phosphorane, *in* C-00273
16672-87-0 ▷(2-Chloroethyl)phosphonic acid, C-00074
16721-43-0 Triphenylpropylphosphonium(1+); Chloride, *in* T-00680
16721-45-2 Benzylidenetriphenylphosphorane, B-00050
16726-81-1 *O,O*-Diethyl (1,3-butadien-1-yl)phosphonothioate, *in* B-00511
16727-61-0 2-Hydroxy-4-methyl-1,3,2-dioxaphosphorinane 2-oxide, H-00156
16751-02-3 Mono(3,7-dimethyl-2,6-octadienyl) diphosphate, M-00465
16751-02-3 Mono(3,7-dimethyl-2,6-octadienyl) diphosphate; (*Z*)-*form*, *in* M-00465
16759-59-4 Benoxafos, B-00003
16764-06-0 Ethyl dimethylvinylene phosphate, *in* H-00123
16767-55-8 Phenylsulfonylphosphoramidic dichloride, P-00313
16777-05-2 Phosphinecarbonitrile, P-00354
16777-83-6 Glycerol 1,2-di-9-octadecenoate 3-phosphoethanolamine; (±)-(*all-E*)-*form*, *in* G-00043
16779-00-3 Spiro[1,3,2-benzoxazaphosphole-2(3*H*),2′-[1,3,2]-dioxaphospholane], S-00012
16814-09-8 Diethyl (1-amino-1-methylethyl)phosphonate, *in* A-00087
16822-31-4 Bis(4-methylphenyl)-phenylphosphine; Sulfide, *in* B-00311

16828-78-7 [1,4-Phenylenebis(methylene)]-bis[triphenylphosphorane], P-00136
16830-69-6 2-Hydroxy-4,5-diphenyl-1,3,2-dioxaphosphole 2-oxide; 2-Ethoxy (Et ester), *in* H-00139
16839-13-7 ▷(4-Bromophenyl)phosphonic acid, B-00492
16841-91-1 2,2(3*H*,3′*H*)-Spirobi[1,3,2-benzoxazaphosphole], S-00020
16853-36-4 *N*,*N*,*N*′,*N*″-Tetramethylphosphoric triamide, *in* D-00880
16856-53-4 1-Hydroxy-1,1-butanediphosphonic acid, H-00113
16871-93-5 Tris[[1,1′-biphenyl]-2,2′-diyl]phosphate(1−); (+)-*form*, 5,5′-Spirobi[5*H*-dibenzophospholium] salt, *in* T-00731
16913-98-7 Diethyl 1*H*-imidazol-1-ylphosphonate, *in* I-00003
16934-90-0 Phenylpropylphosphinic acid; (*R*)P-*form*, (−)-Menthyl ester, *in* P-00303
16934-91-1 Phenylpropylphosphinic acid; (*S*)P-*form*, (−)-Menthyl ester, *in* P-00303
16934-92-2 Menthyl methylphenylphosphinate, *in* M-00219
16934-93-3 Methylphenylphosphinic acid; (*S*)-*form*, (−)-Menthyl ester, *in* M-00219
16954-60-2 Isopropyl cyclohexylphosphinate, *in* C-00253
16956-55-1 2,2′-Thiobis[5,5-dimethyl-1,3,2-dioxaphosphorinane] 2,2′-dioxide, T-00310
16958-43-3 3-Butenyldiphenylphosphine; Oxide, *in* B-00521
16965-72-3 Dimethyl (diazophenylmethyl)-phosphonate, *in* D-00037
16965-84-7 Dimethyl [(acetyloxy)-phenylmethyl]phosphonate, *in* A-00020
16979-13-8 (2-Methyl-2-bromomethyl)-trimethylene phosphoryl bromide, *in* B-00465
17023-23-3 *O*,*O*-Diethyl (trichloromethyl)-phosphonothioate, *in* T-00415
17040-19-6 ▷Demeton-*S*-methyl sulfone, D-00021
17043-52-6 2-Mercapto-4-methyl-1,3,2-dioxaphospholane 2-sulfide, M-00016
17045-33-9 Dichloroisobutylphosphine, *in* M-00383
17045-47-5 Ethylmethylphenylphosphine; (*S*)-*form*, Oxide, *in* E-00084
17051-68-2 (2-Methyl-1-butenyl)phosphonic acid; Dichloride, *in* M-00100
17051-69-3 (3-Methyl-1-butenyl)phosphonic acid; Dichloride, *in* M-00101
17051-95-5 Diethyl (trichloromethyl)-phosphonite, *in* T-00416
17052-16-3 Bis(iodomethyl)phosphinic acid, B-00288
17052-17-4 Ethyl bis(iodomethyl)phosphinate, *in* B-00288
17052-19-6 Phenyl bis(iodomethyl)-phosphinate, *in* B-00288
17056-34-7 (4-Chlorophenyl)-difluorophosphine, *in* C-00153
17067-92-4 Ethyl [4-(diethoxyphosphinyl)]-benzoate, *in* P-00380
17105-65-6 Dimethyl [(4-methoxyphenyl)-methyl]phosphonate, *in* M-00061
17106-25-1 Methyl bis(4-chlorophenyl)-phosphinite, *in* B-00142
17106-28-4 Butyl bis(4-chlorophenyl)-phosphinite, *in* B-00142
17109-49-8 ▷Edifenphos, E-00002
17118-61-5 Glycerol 1-hexadecanoate 2-9,12,15-octadecatrienoate 3-phosphocholine, G-00058
17123-05-6 Di-*tert*-butyl methylphosphonate, *in* M-00288
17123-09-0 Dipropyl phosphoramidate, D-01213
17123-10-3 Dipropyl dipropylphosphoramidate, *in* D-01214
17145-91-4 Ethyl 2-(diethoxyphosphinyl)-butanoate, *in* D-00246
17151-43-8 Ethyl ethylphosphonobromidothioite, *in* E-00134
17154-34-6 Diphenyl(trimethylsilyl)-phosphine, D-01156
17158-87-1 Bis(2-methylpropyl) phosphorochloridate, *in* B-00342
17162-55-9 Ethyl ethylphosphonochloridodithioate, *in* E-00136
17162-57-1 Isopropyl ethylphosphonochloridodithioate, *in* E-00136
17162-81-1 *P*,*P*-Diphenyl-*N*-(diphenylphosphinothioyl)-phosphinimidic acid; Chloride, *in* D-01006
17162-82-2 *O*-Methyl *P*,*P*-diphenyl-*N*-(diphenylphosphinothioyl)-phosphinimidate, *in* D-01006
17162-83-3 *P*,*P*-Diphenylphosphinic amide; *N*-Diphenylphosphinothioyl, *in* D-01040
17166-16-4 Methyl tetramethylphosphorodiamidite, *in* T-00227
17166-36-8 Propadienylphosphonic dichloride, P-00443
17166-37-9 1,2-Propadienylphosphonic acid; Difluoride, *in* P-00441
17166-43-7 Dimethyl (3-methyl-1,2-butadienyl)phosphonate, *in* M-00094
17166-90-4 *P*,*P*-Bis(chloromethyl)-*N*,*N*-dimethylphosphinic amide, *in* B-00128
17166-94-8 *P*,*P*-Bis(chloromethyl)-*N*-methylphosphinic amide, *in* B-00128
17166-98-2 *P*,*P*-Bis(chloromethyl)-*N*-phenylphosphinic amide, *in* B-00128
17167-23-6 1-Methyl-1*H*-phosphole, M-00282
17167-30-5 Trimethyl cyanophosphorimidate, *in* C-00228
17167-31-6 Difluorotrimethoxyphosphorane, D-00416
17167-32-7 Difluorotriphenoxyphosphorane, D-00418
17170-48-8 Methylphenylpropylphosphine; (*S*)-*form*, Oxide, *in* M-00265
17176-77-1 ▷Dibenzyl phosphonate, D-00059
17181-54-3 Glycerol 2-monophosphate, G-00067
17195-46-9 Diethyl heptylphosphonate, *in* H-00006
17196-88-2 2,2-Dichlorovinyl methyl octyl phosphate, D-00202
17222-03-6 Dipropyl propylphosphoramidate, *in* P-00484
17233-33-9 Triphenyl cyanophosphorimidate, *in* C-00228
17242-13-6 2,4-Dichloro-2,4-bis(dimethylamino)-2,2,4,4,6,6-hexahydro-6,6-diphenyl-1,3,5-triaza-2,4,6-triphosphorine, D-00162
17244-54-1 2,2-Dihydro-2,2,2-trimethyl-4,4,5,5-tetrakis(trifluoromethyl)-1,3,2-dioxaphospholane, D-00601
17244-68-7 6,9,*N*,*N*-Tetramethyl-2,2,3,3-tetrakis(trifluoromethyl)-1,4-dioxa-6,9-diaza-5-phospha(P^V)-spiro[4.4]nonan-5-amine, *in* D-00922
17261-28-8 2-(Diphenylphosphino)benzoic acid, D-01053
17261-34-6 Dimethylamine-1,1′-diphosphonic acid, D-00688
17262-51-0 ▷Triheptylphosphine; Oxide, *in* T-00489
17262-53-2 Trinonylphosphine; Oxide, *in* T-00578
17262-57-6 Didecylphenylphosphine; Oxide, *in* D-00219
17316-54-0 Dimethyl (2-phenylethenyl)-phosphonite, *in* P-00161
17316-55-1 Diethyl (2-phenylethenyl)-phosphonite, *in* P-00161
17316-57-3 Diisopropyl (2-phenylethenyl)-phosphonite, *in* P-00161
17316-67-5 [1-(Butylamino)-1-methylethyl]-phosphinic acid, B-00542
17321-47-0 *O*,*O*-Dimethyl phosphoramidothioate, D-00870
17321-48-1 *O*,*O*-Diethyl phosphoramidothioate, D-00354
17321-49-2 *O*,*O*-Diethyl methylphosphoramidothioate, *in* M-00333
17325-53-0 1,2-Ethanediylbis[methyl-phenylphosphine]; (*RS*,*SR*)-*form*, *in* E-00012
17325-54-1 1,2-Ethanediylbis[methyl-phenylphosphine]; (*RS*,*RS*)-*form*, *in* E-00012
17366-80-2 *P*,*P*-Diphenylphosphinothioic amide, D-01083
17377-20-7 2-Chloro-4-methyl-1,3,2-dioxaphosphorinane 2-sulfide, C-00095
17382-94-4 Thiophosphoryl isocyanate, T-00315
17383-44-7 Dimethyl butylphosphonite, *in* B-00615
17383-46-9 Diphenyl butylphosphonite, *in* B-00615
17390-19-1 *N*,*N*′-Dicyclohexylphosphorodiamidic acid, D-00216
17391-48-9 Methyl 3-dichlorophosphinylpropanoate, *in* D-00192
17391-49-0 Ethyl 3-dichlorophosphinylpropanoate, *in* D-00192
17391-53-6 (2-Phenylethenyl)phosphonous dichloride, P-00162
17405-93-5 1-Hydroxy-2,2,3,4,4-pentamethylphosphetane 1-oxide; *trans-form*, *in* H-00168
17405-94-6 1-Hydroxy-2,2,3,4,4-pentamethylphosphetane 1-oxide; *cis-form*, *in* H-00168
17405-95-7 1-Hydroxy-2,2,3,4,4-pentamethylphosphetane 1-oxide; *trans-form*, Et ester, *in* H-00168
17405-96-8 Ethyl 1,1,2,3,3-pentamethyltrimethyl-enephosphinate, *in* H-00168
17420-29-0 2,4,6-Tri-*tert*-butylphosphorin, T-00382
17427-91-7 1,2-Bis(diphenylphosphino)-ethylene, B-00237
17430-39-6 [1,1′-Biphenyl]-4-yl-1-naphthalenylphenylphosphine; (*R*)-*form*, *in* B-00081
17436-27-0 Triphenyl [(4-methylphenyl)-sulfonyl]phosphorimidate, *in* M-00268
17436-30-5 Methyl *P*,*P*-diphenyl-*N*-phenylsulfonylphosphinimidate, *in* B-00013
17437-51-3 Ethyl (triphenylphosphoranylidene)-carbamate, *in* T-00648
17437-62-6 Benzoylphosphorimidic trichloride, B-00040
17454-08-9 *N*,*N*-Dimethyl-2,3-diphenyl-1,4,6,9-tetraoxa-5-phospha(5-P^V)spiro[4.4]-non-2-en-5-amine, *in* D-01150
17454-09-0 *N*,*N*,2,3-Tetramethyl-1,4,6,9-tetraoxa-5-phospha(5-P^V)spiro[4.4]non-2-en-5-amine, *in* D-00924

17454-11-4 *N,N*-Dimethylspiro[1,3,2-dioxaphosphorinane-2,2′-phenanthro[9,10-*d*]-[1,3,2]dioxaphosphol]-2-amine, *in* S-00025
17454-25-0 2-Dimethylamino-1,3,2-dioxaphosphorinane, D-00696
17506-98-8 Diphenyl diphenylphosphoramidite, *in* D-01111
17507-47-0 (Triphenylphosphoranylidene)-ethenethione, T-00650
17507-54-9 2-Diazo-2-(diphenylphosphinyl)-1-phenylethanone, *in* D-00029
17507-55-0 Ethyl diazo(diphenylphosphinyl)-acetate, *in* D-00030
17507-57-2 (Diazophenylmethyl)-diphenylphosphine oxide, D-00036
17513-58-5 Triisopropylphosphine; Oxide, *in* T-00500
17513-59-6 Triisopropylphosphine; Sulfide, *in* T-00500
17513-66-5 *P,P*-Diethyl-*N,N*-dimethylphosphinothioic amide, *in* D-00331
17513-68-7 *N,N*-Dimethyl-*P,P*-diphenylphosphinothioic amide, *in* D-01083
17529-42-9 Dihexylphosphine; Oxide, *in* D-00433
17529-47-4 *O,O*-Dibutyl phosphonothioate, D-00132
17534-72-4 10-Chloro-5,10-dihydrophenophosphazine; 10-Oxide, *in* C-00049
17534-85-9 *O*-Phenyl diphenylphosphinothioate, *in* D-01082
17534-96-2 Dimethylphosphinic acid; Cyanide, *in* D-00829
17535-01-2 Bis(dimethylphosphinyl)methane, *in* B-00202
17544-45-5 (Trichloromethyl)phosphonothioic acid; Dichloride, *in* T-00415
17566-84-6 (4-Methylphenyl)phosphonic dichloride, M-00239
17577-28-5 (2-Ethoxy-2-oxoethyl)-triphenylphosphonium; Chloride, *in* E-00047
17579-99-6 Triphenylphosphite methiodide, *in* M-00429
17586-49-1 Tris(1-methylpropyl)phosphine, T-00823
17586-54-0 Methyltri-2-propenylphosphonium iodide, *in* T-00707
17604-67-0 Phenyl (dimethoxyphosphinyl)-acetate, *in* D-00678
17610-56-9 *O,O,S*-Tributyl phosphorodithioate, T-00384
17615-05-3 Methyl γ-oxo-β-(triphenylphosphoranylidene)-benzenebutanoate, *in* O-00082
17617-66-2 Tris[2-(methylthio)phenyl]-phosphine, T-00831
17620-21-2 Di-2-propenylphosphinic acid; Chloride, *in* D-01193
17620-69-8 *S*-Phenyl phosphorodifluoridothioate, P-00288
17620-71-2 Diphenyl phosphorofluoridotrithioate, D-01135
17620-74-5 Phenyl dimethyl-phosphoramidofluoridodithioate, *in* D-00867
17620-75-6 Phenyl diethylphosphoramidofluoridodithioate, *in* D-00352
17620-78-9 Ethylphosphonothioic acid trimer trianhydride, *in* T-00463
17621-04-4 Triheptylphosphine, T-00489
17621-06-6 Trinonylphosphine, T-00578
17621-07-7 Tri(decyl)phosphine, T-00433
17622-57-0 [1,1′-Biphenyl]-4-yl-1-naphthalenylphenylphosphine; (*S*)-*form*, *in* B-00081
17638-55-0 (2-Oxo-2-phenylethyl)-triphenylphosphonium(1+), O-00067
17643-83-3 *O*-Butyl phosphorodichloridothioate, B-00630
17643-87-7 *O,O*-Dibutyl butylphosphonothioate, *in* B-00611
17643-91-3 *O*-Propyl dipropylphosphinothioate, *in* D-01208
17643-99-1 Tripentylphosphine; Sulfide, *in* T-00594
17648-16-7 Methylimidodiphosphorous tetrachloride, M-00158
17648-17-8 Ethylimidodiphosphorous tetrachloride, E-00080
17648-18-9 Methylimidodiphosphorous tetrafluoride, M-00159
17663-89-7 Diphenoxytriphenylphosphorane, D-00992
17668-60-9 (2-Buten-1-yl)diphenylphosphine; (*E*)-*form*, Oxide, *in* B-00520
17672-53-6 Bis(2,2,2-trichloroethyl) phosphorochloridate, *in* B-00400
17674-28-1 Dimethyl acetylphosphonate, *in* A-00026
17677-92-8 Bis(2,2,2-trichloro-1,1-dimethylethyl) phosphorochloridate, B-00399
17708-90-6 Glycerol 1-hexadecanoate 2-9,12-octadecadienoate 3-phosphocholine; (*R*)-L-(*Z,Z*)-*form*, *in* G-00056
17708-93-9 Tetra-9-octadecenoylphosphatidylglycerol; (*all-Z*)-*form*, *in* T-00242
17708-99-5 Glycerol 1-9-octadecenoate 2-tetradecanoate 3-phosphoglycerophosphate; (*Z*)-*form*, Ba salt, *in* G-00095
17718-78-4 2,3-Dihydro-2-methyl-1,4,2-benzodioxaphosphorin; 2-Oxide, *in* D-00517
17718-79-5 3,4-Dihydro-2-methyl-2*H*-1,4,2-benzoxazaphosphorine; 2-Oxide, *in* D-00522
17718-80-8 3,4-Dihydro-2-methyl-2*H*-1,4,2-benzoxazaphosphorine; 2-Sulfide, *in* D-00522
17718-81-9 2,3-Dihydro-2-methyl-1*H*-4,1,2-benzothiazaphosphorine; 2-Oxide, *in* D-00520
17718-82-0 2,3-Dihydro-2-methyl-1*H*-4,1,2-benzothiazaphosphorine; 2-Sulfide, *in* D-00520
17725-01-8 Pentachlorophenyl phosphorodichloridate, P-00006
17726-41-9 Methylphosphonothioic acid trimer trianhydride, *in* T-00567
17729-74-7 1-(Diphenylphosphino)-2-propanone, D-01075
17730-93-7 [2-(4-Nitrophenyl)-2-oxoethyl]-triphenylphosphonium(1+); Bromide, *in* N-00051
17776-78-2 Bis(2-chlorophenyl) phosphorochloridate, *in* B-00131
17788-07-7 Pentafluorophenyl phosphorodichloridate, P-00017
17788-08-8 Bis(pentafluorophenyl) phosphorochloridate, *in* B-00381
17802-67-4 *N″*-2-Pyrimidinylphosphoric triamide; *N,N′*-Bis(2-chloroethyl), *in* P-00517
17814-85-6 (4-Carboxybutyl)-triphenylphosphonium(1+); Bromide, *in* C-00009
17833-37-3 *P*-Cyclohexyl-*N,N*-dimethylphosphonamidothioic chloride, *in* C-00255
17833-38-4 *P*-Cyclohexyl-*N,N*-diethylphosphonamidothioic chloride, *in* C-00255
17833-40-8 *N,N*-Diethyl-*P*-phenylphosphonamidic acid; Chloride, *in* D-00307
17847-94-8 5,6-Dihydro-5-phenylphosphanthridene 5-oxide, *in* D-00568
17847-95-9 5,6-Dihydro-5-methylphosphanthridene 5-oxide, *in* D-00568
17848-02-1 Phenyl methylphosphonisocyanatidate, *in* M-00292
17848-04-3 *N,N,P*-Trimethylphosphonisocyanatidic amide, *in* M-00293
17848-05-4 *N,N*-Diethyl-*P*-methylphosphonisocyanatidic amide, *in* M-00293
17848-06-5 Phosphorodiisocyanatidic fluoride, P-00413
17848-08-7 Phosphorodiisocyanatidic bromide, P-00411
17850-02-1 Chlorodibenzylphosphine, *in* D-00058
17850-04-3 Dibenzylbromophosphine, *in* D-00058
17857-14-6 (3-Carboxypropyl)-triphenylphosphonium(1+); Bromide, *in* C-00017
17858-28-5 *P,P*-Di-*tert*-butylphosphinous amide, D-00124
17858-29-6 *P,P*-Di-*tert*-butyl-*N*-trimethylsilylphosphinous amide, *in* D-00124
17885-08-4 Phosphoserine; (±)-*form*, *in* P-00417
17919-24-3 (Trichloromethyl)phosphonic difluoride, T-00414
17966-16-4 Glycerol 1,2-dioctadecanoate 3-phosphate; (*R*)-*form*, *in* G-00032
17973-39-6 *O*-Vinyl phosphorodichloridothioate, V-00008
17985-95-4 Ethyl *N,P,P*-triphenylphosphinimidate, *in* T-00631
17985-98-7 ▷*N,N′,P,P*-Tetraphenylphosphinimidic amide, *in* T-00631
17986-01-5 *N*-Methyl-*P,P,P*-triphenylphosphine imide, *in* T-00626
17986-07-1 Trimethyl [(4-methylphenyl)-sulfonyl]phosphorimidate, *in* M-00268
17989-41-2 Glycerol 1-monophosphate; (*S*)-*form*, *in* G-00066
18005-37-3 *O*-Propyl methylphosphonochloridothioate, P-00468
18005-38-4 ▷*O*-Butyl methylphosphonochloridothioate, B-00573
18005-39-5 *O*-Methyl methylphosphonothioate, M-00171
18005-40-8 *O*-Ethyl methylphosphonothioate, E-00093
18005-43-1 2,3,7,8-Tetraphenyl-1,4,6,9-tetraoxa-5-phospha(5-P^{V})spiro[4.4]nona-2,7-diene; 5-Ph, *in* T-00275
18005-69-1 2,3-Diethyl-2,3-diphosphabicyclo[2.2.1]-hept-5-ene; Disulfide, *in* D-00287
18005-70-4 2,3-Diethyl-2,3-diphosphabicyclo[2.2.2]-octane; Disulfide, *in* D-00288
18025-89-3 Tetraethyl phosphor(isocyanatidic) diamide, *in* T-00059
18026-39-6 (1,3-Butadienyl)phosphonic acid; Difluoride, *in* B-00510
18026-42-1 1-Propynylphosphonic acid; Difluoride, *in* P-00494

18026-83-0 Trimethylsilyl phosphorodichloridate, T-00558
18032-85-4 Ethyl triethylphosphonamidodithioate, *in* T-00447
18032-89-8 Tetraethylphosphinothioic amide, *in* D-00331
18032-95-6 ▷Ethylphosphonodithioic acid; *S,S*-Di-Et ester, *in* E-00140
18032-97-8 *S*-Butyl diethylphosphinothioate, *in* D-00330
18033-91-5 Tetraethyl [(ethoxyphosphinylidene)-bis(methylene)]bisphosphonate, *in* B-00182
18054-52-9 2,4,5-Triphenyl-2*H*-1,3,2-diazaphosphole; 2-Oxide, *in* T-00600
18106-71-3 Dimethyl benzoylphosphonate, *in* B-00034
18106-76-8 Methyl benzoylphenylphosphinate, *in* B-00032
18108-08-2 Propyl phosphinate, P-00470
18108-09-3 Butyl phosphinate, B-00587
18108-11-7 Pentyl phosphinate, P-00048
18108-16-2 Benzyl phosphinate, B-00057
18108-22-0 (Aminophenylmethyl)phosphonic acid, A-00105
18108-24-2 (1-Amino-2-methylpropyl)-phosphonic acid, A-00094
18108-37-7 5-Methyl-2-phenyl-2*H*-1,2,3-diazaphosphole, M-00197
18110-43-5 Mono(2-phenylethyl) phosphate, M-00506
18117-85-6 Methylphenylphosphinothioic acid; (±)-*form*, Dicyclohexylammonium salt, *in* M-00225
18117-86-7 Methylphenylphosphinothioic acid; (*S*)-*form*, Dicyclohexylammonium salt, *in* M-00225
18117-87-8 Methylphenylphosphinothioic acid; (*R*)-*form*, Dicyclohexylammonium salt, *in* M-00225
18135-13-2 Dimethyl trimethylsilyl phosphate, D-00931
18181-70-9 ▷Iodofenphos, I-00011
18194-24-6 Glycerol 1,2-ditetradecanoate 3-phosphocholine; (*R*)-*form*, *in* G-00048
18194-25-7 Glycerol 1,2-didodecanoate 3-phosphocholine; (*R*)-L-*form*, *in* G-00008
18198-39-5 Tetraphenylphosphonium(1+), T-00272
18226-72-7 *O,S*-Dimethyl phosphorodithioate; (±)-*form*, K salt, *in* D-00899
18250-63-0 Dicrotophos; (*Z*)-*form*, *in* D-00204
18255-62-4 Dibenzylphosphinic acid; Chloride, *in* D-00055
18279-83-9 Methylphosphonic acid; Bis(trimethylsilyl) ester, *in* M-00288
18281-76-0 *S*-Methyl phosphorodichlorodothioate, M-00351
18281-78-2 *S*-Propyl phosphorodichloridothioate, P-00490
18297-98-8 Dihexylphenylphosphine, D-00431
18298-00-5 Hexyldiphenylphosphine, H-00093
18306-68-8 Diethyl trimethylsilyl phosphate, D-00407
18313-77-4 Triethyl (dimethoxyphosphinyl)-phosphorimidate, *in* D-00683
18313-80-9 Triisopropyl (diethoxyphosphinyl)-phosphorimidate, *in* D-00257
18313-83-2 Tetramethyl methylimidodiphosphate, *in* M-00156
18334-55-9 2,2,2,3-Tetrahydro-3,5-diphenyl-1,3,4,2-oxadiazaphosphole; 2,2,2-Trimethoxy, *in* T-00104
18350-98-6 2,6-Dimethylphenyl phosphorodichloridate, *in* M-00466
18351-08-1 *O*-(2-Butoxyethyl) phosphorodichloridothioate, *in* H-00149
18351-11-6 Hexylphosphonothioic acid; Dichloride, *in* H-00100
18351-13-8 *O*-(3-Methylbutyl) phosphorodichloridothioate, M-00109
18351-14-9 *O*-Pentyl phosphorodichloridothioate, P-00055
18351-20-7 *O*-Ethyl *S*-phenyl phosphorochloridothioate, E-00118
18351-25-2 Butyl phenylphosphonochloridate, *in* P-00229
18351-33-2 2-*tert*-Butylphenyl phosphorodichloridate, *in* M-00449
18351-36-5 4-*tert*-Butylphenyl phosphorodichloridate, *in* M-00450
18351-80-9 *tert*-Butylmethylphosphinic acid, B-00567
18351-81-0 Di-*tert*-butylmethylphosphine; Oxide, *in* D-00099
18356-17-7 Dimethyl 1,2-propadienylphosphonate, *in* P-00441
18358-21-9 Butylmethylphosphinic acid; Fluoride, *in* B-00566
18358-25-3 Bis(2-methylpropyl)phosphinic acid; Fluoride, *in* B-00346
18359-32-5 1-Chloro-2,5-dihydro-1*H*-phosphole; 1-Sulfide, *in* C-00052
18365-82-7 2-Phosphonooxypropanoic acid, P-00392
18368-11-1 ▷Diethyl dichlorophosphoramidate, D-00280
18370-12-2 Monomethyl 4-morpholinylphosphonothioate, *in* M-00527
18374-36-2 Phenyl *N,N*-bis(2-chloroethyl)-phosphorodiamidate, *in* B-00126
18389-60-1 2-Dimethylamino-1,3,2-benzodioxaphosphole, D-00689
18389-61-2 2,2,3,3,7,7,8,8-Octamethyl-1,4,6,9-tetraoxa-5-phospha(5-P^V)spiro[4.4]-nonane, O-00032
18389-63-4 4,9-Dimethyl-1,6-dioxa-4,9-diaza-5-phospha(5-P^v)-spiro[4.4]nonane, D-00729
18389-64-5 2,2,3,3-Tetramethyl-1,4,6,9-tetraoxa-5-phospha(5-P^V)spiro[4.4]nonane, T-00235
18389-65-6 2,2,3,3-Tetramethyl-1,4,6-trioxa-9-aza-5-phospha(5-P^V)-spiro[4.4]nonane, T-00238
18406-56-9 Bis(trimethylsilyl) benzylphosphonate, *in* B-00060
18416-85-8 Lombricine; (*S*)-*form*, *in* L-00002
18437-71-3 (4-Fluorophenyl)phosphine, F-00034
18437-72-4 (4-Fluorophenyl)-diphenylphosphine, F-00028
18437-73-5 (4-Fluorophenyl)-diphenylphosphine; Oxide, *in* F-00028
18437-76-8 (4-Fluorophenyl)-diphenylphosphine; Sulfide, *in* F-00028
18437-78-0 Tris(4-fluorophenyl)phosphine, T-00783
18437-80-4 Tris(4-fluorophenyl)phosphine; Sulfide, *in* T-00783
18457-65-3 *O*-Methyl *O*-4-nitrophenyl cyclohexylphosphoramidothioate; (+)-*form*, *in* M-00179
18468-68-3 Glycerol 1-9,12-octadecadienoate 2-octadecanoate 3-phosphoethanolamine; (±)-(*all-Z*)-*form*, *in* G-00071
18480-23-4 Triphenyl-2-propenylphosphonium(1+); Chloride, *in* T-00679
18480-27-8 (1-Carboxyethyl)-triphenylphosphonium(1+); Et ester, chloride, *in* C-00010
18509-25-6 Diethoxytriphenylphosphorane, D-00263
18509-37-0 *P,P*-Dimethyl-*N*-(dimethylphosphinothioyl)-phosphinothioic amide, *in* D-00840
18551-98-9 *S*-Phenyl tetraethylphosphorodiamidothioate, *in* T-00063
18582-73-5 1,4-Dimethyl-1,4-diphenyl-1,4-diphosphorinanium dibromide, *in* D-01017
18582-74-6 1,4-Diphenyl-1,4-diphosphorinane; *trans-form*, B,2MeBr, *in* D-01017
18582-75-7 1,3-Dimethyl-1,3-diphenyl-1,3-diphospholanium dibromide, *in* D-01013
18583-38-5 [(4-Methylphenyl)methyl]-triphenylphosphonium(1+), M-00211
18583-53-6 [(2-Chlorophenyl)methyl]-triphenylphosphonium(1+); Chloride, *in* C-00135
18593-19-6 Bis(2-methylphenyl)phosphinic acid, B-00318
18593-20-9 ▷Bis(3-nitrophenyl)phosphinic acid, B-00373
18593-23-2 Methyl bis(4-dimethylaminophenyl)-phosphinate, *in* B-00188
18593-24-3 5,10-Dihydro-5-hydroxyacridophosphine 5-oxide, D-00487
18593-25-4 5,10-Dihydro-5-methoxyacridophosphine 5-oxide, *in* D-00487
18620-05-8 4-Methyl-3,5,8-trioxa-1-phosphabicyclo[2.2.2]-octane, M-00425
18628-83-6 *N,N*-Diethyl-*P*-phenylphosphonamidothioic acid, D-00309
18628-85-8 *O,O*-Dibutyl diethylphosphoramidothioate, *in* D-00356
18629-24-8 Butylphenylphosphinic acid, B-00578
18629-29-3 Butylphenylphosphinic acid; Dicyclohexylammonium salt, *in* B-00578
18629-48-6 Ethyl *N,N*-diethyl-*P*-propadienylphosphonamidate, *in* D-00399
18629-49-7 Butyl *N,N*-diethyl-*P*-propadienylphosphonamidate, *in* D-00399
18629-57-7 Acetyldiphenylphosphine, A-00016
18629-73-7 Butyl *N,N*-diethyl-*P*-1-propynylphosphonamidate, *in* D-00403
18632-46-7 Methyl diisopropylphosphinate, *in* D-00641
18632-47-8 Isopropyl diethylphosphinate, *in* D-00322
18639-74-2 Methyldiphenylphosphine; Sulfide, *in* M-00135
18655-72-6 2-Chloro-2,3-dihydro-5-methyl-3-phenyl-1,3,4,2-oxadiazaphosphole; 2-Sulfide, *in* C-00044
18689-31-1 1-Butenylphosphonic acid; (*E*)-*form*, Di-Et ester, *in* B-00524
18689-32-2 Diethyl 1-propenylphosphonate, *in* P-00453
18689-34-4 (3-Methyl-1-butenyl)phosphonic acid; (*Z*)-*form*, Di-Et ester, *in* M-00101
18689-36-6 1-Propenylphosphonic acid; (*Z*)-*form*, Di-Et ester, *in* P-00453
18720-17-7 2-Chloro-2,3-dihydro-3,5-diphenyl-1,3,4,2-oxadiazaphosphole; 2-Sulfide, *in* C-00042

18724-97-5 1,2-Bis(dimethylphosphinyl)-ethane, *in* B-00200
18733-15-8 Methyl 3-(dimethoxyphosphinyl)-propanoate, *in* D-00685
18748-21-5 2-Ethoxy-1,3,2-dioxaphosphepane 2-oxide, *in* H-00132
18755-43-6 Dimethyl propylphosphonate, *in* P-00473
18764-12-0 ▷*unsym*-Tetramethyl monothiopyrophosphate, T-00207
18788-02-8 Tris(acetoxymethyl)phosphine, *in* T-00784
18788-05-1 Phenyl diethylphosphinothioite, *in* D-00332
18788-27-7 Ethyl benzoylphenylphosphinate, *in* B-00032
18788-35-7 Hexaethyl [phosphinylidynetris(methylene)]trisphosphonate, *in* P-00360
18788-36-8 Hexamethyl [phosphinylidynetris(methylene)]trisphosphonate, *in* P-00360
18788-65-3 Ethyl acetylethylphosphinate, *in* A-00017
18788-83-5 Isopropyl styrylphosphinate, *in* P-00153
18788-85-7 Ethyl (2-phenylethenyl)-phosphinate, *in* P-00153
18788-86-8 Methyl (2-phenylethenyl)-phosphinate, *in* P-00153
18788-99-3 *O*-Methyl hydrogen hexylphosphonodithioate, *in* H-00099
18789-04-3 *O*-Isopropyl hydrogen pentylphosphonodithioate, *in* P-00051
18789-05-4 *O*-Ethyl hydrogen pentylphosphonodithioate, *in* P-00051
18789-43-0 Ethylene methylphosphonotrithioate, *in* M-00139
18792-78-4 2-Naphthalenylmethylenetriphenylphosphorane, *in* N-00016
18796-57-1 *P,P*-Diphenyl-*N*-trimethylsilylphosphinothioic amide, *in* D-01083
18799-59-2 *P*-(4-Methylphenyl)phosphonic diamide, M-00238
18799-74-1 *N,N*-Dimethyl-*P*-trifluoromethyl-phosphonamidothioic fluoride, D-00928
18799-75-2 Bis(trifluoromethyl)-phosphinodithioic acid, B-00412
18799-76-3 Bis(trifluoromethyl)-phosphinothioic acid; Iodide, *in* B-00415
18799-77-4 Bis(trifluoromethyl)-phosphinothioic acid; Bromide, *in* B-00415
18799-78-5 (Trifluoromethyl)phosphonothioic acid; Dichloride, *in* T-00481
18799-79-6 Methyl bis(trifluoromethyl)-phosphinodithioate, *in* B-00412
18799-82-1 Bis(trifluoromethyl)-phosphinothioic acid; Chloride, *in* B-00415
18803-08-2 Bis(3-methylphenyl)-phenylphosphine, B-00310
18803-09-3 Bis(2-methylphenyl)-phenylphosphine, B-00309
18803-10-6 Bis(3-methylphenyl)-phenylphosphine; Oxide, *in* B-00310
18803-11-7 Bis(2-methylphenyl)-phenylphosphine; Oxide, *in* B-00309
18803-12-8 Trihexylphosphine; Sulfide, *in* T-00492
18812-51-6 Diethyl propylphosphonate, *in* P-00473
18818-32-1 4-Methyl-2,6,7-trithia-1-phosphabicyclo[2.2.2]-octane, M-00436
18818-33-2 ▷4-Methyl-2,6,7-trithia-1-phosphabicyclo[2.2.2]-octane; 1-Sulfide, *in* M-00436
18818-34-3 4-Methyl-2,6,7-trithia-1-phosphabicyclo[2.2.2]-octane; 1-Oxide, *in* M-00436
18852-83-0 *O,O*-Diethyl *S*-phenyl phosphorothioate, D-00317
18852-83-0 *S*-Phenyl phosphorothioate, P-00297
18854-01-8 ▷Isoxathion, I-00068
18854-80-3 *tert*-Butylphosphorimidic trichloride, B-00626
18855-52-2 Tetraethyl [(dimethylamino)-methylene]bisphosphonate, *in* D-00701
18865-25-3 2-Methylthio-4*H*-1,3,2-benzodioxaphosphorin; 2-Oxide, *in* M-00411
18865-31-1 4-Amino-4-phosphonobutanoic acid, A-00119
18868-46-7 2,2,2-Trichloroethyl phosphorodichloridate, *in* T-00424
18872-30-5 Methylphenylphosphinothioic acid; (*S*)-*form*, *S*-Me ester, *in* M-00225
18874-05-0 1-Diethylamino-2,5-dihydro-2-methyl-1*H*-phosphole 1-oxide, *in* D-00498
18874-12-4 1-Ethoxy-2,5-dihydro-2-methyl-1*H*-phosphole 1-oxide, *in* D-00498
18874-12-9 2,5-Dihydro-2-methyl-1-phenoxy-1*H*-phosphole 1-oxide, *in* D-00498
18874-22-1 1-Chloro-2,5-dihydro-3-methyl-1*H*-phosphole; 1-Oxide, *in* C-00047
18880-05-2 [(3-Methoxyphenyl)methyl]-triphenylphosphonium(1+); Chloride, *in* M-00063
18882-24-1 Trimethylene methylphosphonotrithioate, *in* M-00137
18882-25-2 *S,S'*-Propanediyl methylphosphonodithioate, *in* M-00137
18882-26-3 *O,O*-Trimethylene methylphosphonoselenoate, *in* M-00128
18882-27-4 *O,O*-Trimethylene methylphosphonothioate, *in* M-00128
18894-82-1 *P*-Trifluoromethyl-*N,N,N',N'*-tetramethylphosphonothioic diamide, *in* T-00481
18895-84-6 Hexamethyl 1,3,5-triazine-2,4,6-triyltriphosphoramidate, *in* T-00330
18895-85-7 Hexaethyl 1,3,5-triazine-2,4,6-triyltriphosphoramidate, *in* T-00330
18896-56-5 [(2-Methylphenyl)methyl]-phosphonic acid, M-00206
18899-64-4 1,2-Bis(phenylphosphino)ethane, B-00390
18899-86-4 1,4-Diphenyl-1,4-diphosphorinane, D-01017
18904-52-4 *P,P*-Bis(trifluoromethyl)-phosphinothioic amide, B-00416
18906-31-5 *S,S*-Ethylene methylphosphonodithioate, *in* M-00139
18906-32-6 Ethylene methylphosphonothioate, *in* M-00126
18933-98-7 Ethyl hydrogen benzylphosphonate, *in* B-00060
18938-26-6 1-(Dimethylphosphinyl)ethanone, *in* A-00015
18943-90-3 1-Chloro-2,5-dihydro-2-methyl-1*H*-phosphole; 1-Oxide, *in* C-00046
18945-65-8 [(3-Methylphenyl)methyl]-phosphonic acid, M-00207
18945-68-1 Tris[(3-methylphenyl)methyl]-phosphine; Oxide, *in* T-00801
18945-69-2 Tris[(2-methylphenyl)methyl]-phosphine; Oxide, *in* T-00800
18957-70-5 Bis(4-methylphenyl)-phenylphosphine; Oxide, *in* B-00311
18961-94-9 *O*-(4-Methoxyphenyl) phosphorodichloridothioate, M-00082
18961-95-0 *O*-(4-Methylphenyl) phosphorodichloridothioate, M-00262
18961-96-1 *O*-Phenyl phosphorodichloridothioate, P-00282
18962-89-5 Methylphenylphosphinothioic acid; (±)-*form*, *in* M-00225
18962-97-5 Bis(2,4-dinitrophenyl) phosphate, B-00210
18975-70-7 Bis(4-methylphenyl)-phosphinothioic acid; Fluoride, *in* B-00322
18983-86-3 Acetyldimethylphosphine, A-00015
18994-99-5 *N,N'*-Dimethyl-*P*-phenylphosphonothioic diamide, *in* P-00246
18995-01-2 *P*-Phenylphosphonothioic dianilide, *in* P-00246
18995-02-3 Phenyl *N,N'*-diphenylphosphorodiamidate, *in* D-01125
18995-03-4 Phenyl *N,N'*-dibenzylphosphorodiamidate, *in* D-00061
18995-05-6 Phenyl *N,N'*-diisopropylphosphorodiamidate, *in* D-00661
18995-06-7 Phenyl *N,N'*-dibutylphosphorodiamidate, *in* D-00146
19014-66-5 Methylenebis[methylphosphinic acid], M-00140
19016-52-5 2,3-Dihydro-5-methyl-3-phenyl-2-phenylamino-1,3,4,2-oxadiazaphosphole, D-00527
19021-61-5 *O*-Isopropyl phosphorodichloridothioate, I-00062
19046-60-7 Galactose 1-dihydrogen phosphate; α-D-*Pyranose-form*, Di-K salt, *in* G-00001
19046-78-7 Adenylosuccinic acid, A-00047
19047-85-9 Dioctadecyl phosphonate, D-00954
19049-13-9 Tris[[1,1'-biphenyl]-2,2'-diyl]phosphate(1−); (±)-*form*, 5,5'-Spirobi[5*H*-dibenzophospholium] salt, *in* T-00731
19057-03-5 *S*-Methyl propylphosphonochloridothioate, *in* P-00477
19057-04-6 *S*-Propyl propylphosphonochloridothioate, *in* P-00477
19057-06-8 *S*-Ethyl isopropylphosphonochloridothioate, *in* I-00053
19066-48-9 Tris[[1,1'-biphenyl]-2,2'-diyl]phosphate(1−); (+)-*form*, 19-Methylbrucinium salt, *in* T-00731
19073-47-3 Diethylphosphinodithioic acid; NH_4 salt, *in* D-00325
19081-02-8 Ethyl methylphosphonisocyanatidate, *in* M-00292
19081-37-9 *O*-(4-Chlorophenyl) phosphorodichloridothioate, C-00163
19085-97-3 (Triphenylphosphoranylidene)-phosphoramidic dichloride, T-00656
19093-80-2 Diethylphosphine; Li deriv., *in* D-00321
19100-59-5 Bis(pentafluorophenyl)-phenylphosphine; Sulfide, *in* B-00380

19100-62-0 *O*-Methyl bis(pentafluorophenyl)-phosphinothioate, *in* B-00383
19100-63-1 *O*-Ethyl bis(pentafluorophenyl)-phosphinothioate, *in* B-00383
19100-64-2 *O*-Phenyl bis(pentafluorophenyl)-phosphinothioate, *in* B-00383
19107-17-6 Tris[[1,1′-biphenyl]-2,2′-diyl]phosphate(1−); (−)-*form*, K salt, *in* T-00731
19108-57-7 1,1-Dihydro-1,1-diphenylphosphorin, D-00484
19162-94-8 (Trifluoromethyl)phosphonic acid; Difluoride, *in* T-00478
19172-47-5 2,4-Bis(4-methoxyphenyl)-1,3,2,4-dithiadiphosphetane 2,4-disulfide, B-00293
19190-40-0 5-Methyl-5*H*-dibenzophosphole; 5-Oxide, *in* M-00115
19201-23-1 Bis(3-methylbutyl) phosphonate, B-00303
19219-95-5 1,3-Butylene phenyl phosphate, *in* M-00191
19229-33-0 Ethyl ethylmethylphosphinate, *in* E-00085
19233-61-5 *O*-Ethyl *P*-methyl-phosphonohydrazidothioate, *in* M-00309
19233-65-9 Cyclohexyl *P*-methylphosphonohydrazidate, *in* M-00308
19233-68-2 2-Methylpropyl *P*-methylphosphonohydrazidate, *in* M-00308
19233-70-6 Isopropyl *P*-methylphosphonohydrazidate, *in* M-00308
19233-71-7 Ethyl *P*-methylphosphonohydrazidate, *in* M-00308
19259-62-2 Dihexylphenylphosphine; Oxide, *in* D-00431
19259-63-3 Hexyl dihexylphosphinate, *in* D-00434
19259-70-2 1-(Diphenylphosphinyl)hexane, *in* H-00093
19270-90-7 (1,1-Dimethyl-3-oxobutyl)-ethylphosphinic acid, D-00782
19270-91-8 Methyl (1,1-dimethyl-3-oxobutyl)ethylphosphinate, *in* D-00782
19280-60-5 Phenyl *P*-chloromethylphosphonamidate, *in* C-00105
19280-61-6 Ethyl *P*-chloromethylphosphonamidate, *in* C-00105
19280-64-9 Phenyl *P*-methylphosphonamidate, *in* M-00283
19280-65-0 Ethyl *P*-methylphosphonamidate, *in* M-00283
19315-13-0 Methylphenylphosphine; Oxide, *in* M-00215
19316-37-1 *N*,*N*-Dimethylphosphoric triamide, D-00880
19322-22-6 ▷2,4,6-Tribromo-2,2,4,4,6,6-hexahydro-2,4,6-triphenyl-1,3,5-triaza-2,4,6-triphosphorine; (2α,4α,6α)-*form*, *in* T-00358
19329-39-6 2,4,6-Tribromo-2,2,4,4,6,6-hexahydro-2,4,6-triphenyl-1,3,5-triaza-2,4,6-triphosphorine; (2α,4α,6β)-*form*, *in* T-00358
19357-87-0 Dipropylphosphine, D-01204
19430-64-9 Methylphosphonic acid; Dibromide, *in* M-00288
19430-75-2 2,4-Dichlorophenyl phosphorodichloridate, D-00189
19483-45-5 *O*,*O*-Diethyl phosphorodiselenoate; K salt, *in* D-00383
19483-46-6 *O*,*O*-Dipropyl phosphorodiselenoate; K salt, *in* D-01224
19483-48-8 *O*,*O*-Dicyclohexyl phosphorodiselenoate; K salt, *in* D-00218
19483-49-9 *O*,*O*-Dipentyl phosphorodiselenoate; K salt, *in* D-00985
19483-58-0 *N*,*N*′-Bis(2-methylpropyl)-phosphorodiamidodiselenoic acid; Isobutylammonium salt, *in* B-00358
19488-90-5 Benzylmethylphenylphosphine; (*R*)-*form*, Sulfide, *in* B-00051
19491-23-7 Dimethyl (triphenylphosphoranylidene)-propanedioate, *in* T-00658
19503-07-2 2-Chloro-2,3-dihydro-3,5-diphenyl-1,3,4,2-oxadiazaphosphole, C-00042
19525-44-1 2-Chloro-2,3-dihydro-5-methyl-3-phenyl-1,3,4,2-oxadiazaphosphole, C-00044
19535-24-1 Spiro[1,3,2-dioxaphosphorinane-2,2′-phenanthro[9,10-*d*]-[1,3,2]dioxaphosphole], S-00025
19549-57-6 5-Phospha(5-P^V)spiro[4.4]nonane, P-00350
19579-02-3 2-Phenoxy-2,2′-spirobi[1,3,2-benzodioxaphosphole], P-00074
19579-04-5 Tetrachlorophenoxyphosphorane, T-00033
19604-05-8 Fosfocreatinine; Di-Na salt, *in* F-00060
19613-06-0 Pentaphenoxyphosphorane, P-00035
19620-78-1 Diphenyl dimethylphosphoramidite, *in* D-00875
19620-83-8 Diphenyl 4-morpholinylphosphonite, *in* M-00529
19622-51-6 *N*,*N*′,*N*″-Triethylphosphoric triamide, *in* T-00462
19622-52-7 *N*,*N*′,*N*″-Tripropylphosphoric triamide, *in* T-00719
19656-40-7 Diethyl 9-acridinylphosphonate, *in* A-00030
19656-41-8 9-Acridinephosphonic acid, A-00030
19667-83-5 2,2-Dihydro-2,2,2-triphenoxy-1,3,2-benzodioxaphosphole, D-00602
19681-77-7 Tris[[1,1′-biphenyl]-2,2′-diyl]phosphate(1−); (−)-*form*, 19-Methylbrucinium salt, *in* T-00731
19698-29-4 Glycerol 1,2-dihexadecanoate 3-phosphate, G-00011
19714-34-2 Cyclohexylphosphoramidochlorido-thioic acid, C-00270
19732-22-0 6-Chloro-5,6,7,12-tetrahydro-2,5,7,10-tetramethyl-dibenzo[*d*,*g*][1,3,2]-diazaphosphocine; 6-Oxide, *in* C-00182
19733-70-1 2-Chlorotetrahydro-2*H*-1,3,2-oxazaphosphorine; 2-Sulfide, *in* C-00180
19733-71-2 2-Ethoxytetrahydro-2*H*-1,3,2-oxazaphosphorine; 2-Sulfide, *in* E-00051
19753-63-0 (2-Oxoethyl)-triphenylphosphonium(1+); Bromide, *in* O-00060
19753-64-1 (3-Bromo-2-oxopropyl)-triphenylphosphonium(1+); Bromide, *in* B-00478
19753-68-5 1-Bromo-3-(triphenylphosphoranylidene)-2-propanone, B-00506
19843-47-7 *O*,*O*-Dibutyl phosphorodiselenoate; K salt, *in* D-00148
19845-69-3 1,6-Bis(diphenylphosphino)hexane, B-00243
19847-57-5 1-Diazo-1-(diphenylphosphinyl)-2-propanone, D-00031
19858-95-8 2-Phenyl-1,3,2-oxazaphospholidine, P-00179
19874-73-8 (3-Oxo-3-phenylpropyl)-triphenylphosphonium(1+); Bromide, *in* O-00070
19909-82-1 Dimethyl (4-chlorophenyl)-phosphonite, *in* C-00153
19911-11-6 *tert*-Butylphosphonous diamide, B-00617
19911-12-7 *P*-*tert*-Butyl-*N*,*N*′-bis(trimethylsilyl)-phosphonous diamide, *in* B-00617
19928-97-3 Tetraethyl dichloromethyl-enebisphosphonate, *in* D-00172
19930-80-4 *N*,*N*,*N*′,*N*′-Tetraethyl-*P*-(2-oxopropyl)phosphonic diamide, *in* O-00077
19934-95-3 Bis(4-methylphenyl)-phenylphosphine, B-00311
19935-93-4 Diethyl (1,1-dimethylethyl)-phosphonate, *in* B-00595
19935-94-5 Ethyl diisopropylphosphinate, *in* D-00641
19935-95-6 Ethyl di-*tert*-butylphosphinate, *in* D-00110
19945-48-3 4-(Diethoxyphosphinyl)-3-methyl-2-butenoic acid; (*Z*)-*form*, Me ester, *in* D-00251
19945-56-3 4-(Diethoxyphosphinyl)-3-methyl-2-butenoic acid; (*E*)-*form*, Me ester, *in* D-00251
19946-96-4 *O*-Ethyl tetraethylphosphorodiamidothio-ate, *in* T-00063
19948-00-6 2-Fluoro-1,3-dioxa-2-phosphacyclotridecane, *in* D-00966
19948-32-4 Tetramethylene phosphorofluoridite, *in* D-00968
19948-33-5 2-Chloro-1,3,2-dioxaphosphonane, *in* D-00971
19948-34-6 2-Fluoro-1,3,2-dioxaphosphonane, *in* D-00971
19948-35-7 2-Chloro-1,3-dioxa-2-phosphacyclotridecane, *in* D-00966
19952-56-8 2-Fluoro-4-methyl-1,3,2-dioxaphosphorinane, F-00021
19952-57-9 2-Fluoro-4,5-dimethyl-1,3,2-dioxaphospholane, F-00015
19966-81-5 Dicyclohexylphosphine; Li salt, *in* D-00210
20001-73-4 Ethyl phosphorofluoridisocyanatidate, E-00180
20003-39-8 Dipropyl phosphorochloridite, D-01218
20003-41-2 Bis(4-methylphenyl) phosphorochloridite, B-00332
20039-32-1 *O*,*O*-Dimethyl phosphorisocyanatidothioate, D-00884
20039-33-2 Diethyl phosphorisocyanatidate, D-00366
20039-34-3 *O*-Phenyl phosphorodiisocyanatidothioate, P-00293
20043-24-7 Tetrahydro-2*H*-1,3,2-thiazaphosphorine 2-oxide; 2-Phenoxy, *in* T-00154
20043-26-9 Tetrahydro-2*H*-1,3,2-thiazaphosphorine 2-oxide; 2-Ethoxy, *in* T-00154
20043-27-0 1,3,2-Thiazaphospholidine 2-oxide; 2-Butoxy, *in* T-00302
20047-46-5 2,2,3,4,4-Pentamethyl-1-phenylphosphetane; *trans*-*form*, Oxide, *in* P-00020
20056-38-6 2-Nitrophenyl phosphorodichloridate, *in* M-00498
20056-44-4 2,4-Dinitrophenyl phosphorodichloridate, *in* M-00467
20061-84-1 Diethyl 2-(cyclohexylamino)-vinylphosphonate, D-00278
20074-79-7 ▷Diethyl [(4-aminophenyl)-methyl]phosphonte, *in* A-00107

20076-68-0 2,3-Dihydro-2-methoxy-5-methyl-3-phenyl-1,3,4,2-oxadiazaphosphole; 2-Sulfide, *in* D-00513
20095-28-7 1-Cyclohexenephosphonic acid; Dichloride, *in* C-00241
20107-85-1 Tetrafluoromethoxyphosphorane, T-00080
20108-75-2 Methylphenylpropylphosphine; (±)-*form*, *in* M-00265
20115-21-3 *O*-Ethyl *O*-methyl phosphorodithioate, E-00095
20145-77-1 *O*-Methyl phenylphosphonoisothiocyanatidothioate, *in* P-00227
20145-78-2 *O*-Ethyl phenyl-phosphonisothiocyanatidothioate, *in* P-00227
20145-82-8 *O*-Phenyl phenyl-phosphonisothiocyanatidothioate, *in* P-00227
20147-96-0 *O*-Methyl phenylphosphonochloridothioate, *in* P-00232
20148-06-5 *O*-Phenyl phenylphosphonochloridothioate, *in* P-00232
20148-17-8 2,3-Dihydro-2-hydroxy-1*H*-isophosphindole 2-oxide, D-00494
20157-29-3 (Dichloromethyl)phosphonothioic acid; Difluoride, *in* D-00180
20170-34-7 Diethyl (1-methylethenyl)-phosphonate, *in* M-00149
20180-11-4 *O*-Methyl diphenylphosphinoselenoate, *in* D-01078
20180-12-5 *O,O*-Dimethyl phenylphosphonoselenoate, *in* P-00243
20188-02-7 (1-Hydroxyethyl)phosphonic acid, H-00145
20202-72-6 2,2-Dichloroethenyl phosphorodichloridate, D-00168
20217-54-3 Dimethyl dimethylphosphoramidite, *in* D-00875
20217-86-1 *S*-Methyl phosphorodiamidothioate, M-00343
20224-50-4 Tri-*tert*-butyl phosphate, T-00371
20255-95-2 Glycerol 1,2-ditetradecanoate 3-phosphoethanolamine, G-00049
20262-87-7 Diethyl diethylphosphoramidite, *in* D-00360
20263-06-3 2-Amino-3-phosphonopropanoic acid; (±)-*form*, *in* A-00121
20263-07-4 2-Amino-4-phosphonobutanoic acid; (±)-*form*, *in* A-00117
20263-10-9 (1-Aminobutyl)phosphonic acid; (±)-*form*, *in* A-00054
20278-51-7 Dimethyl methylphosphonite, *in* M-00319
20335-88-0 2,3-Dihydro-3,3,5-trimethyl-1,2-oxaphosphole 2-oxide; 2-Diethylamino, *in* D-00596
20335-94-8 5,6-Dihydrophosphanthridene, D-00568
20337-67-1 Ethyl methylphosphonofluoridite, *in* M-00307
20337-70-6 Ethylphenylphosphinous acid; (±)-*form*, Me ester, *in* E-00108
20337-75-1 Phenyl ethylphenylphosphinite, *in* E-00108
20339-23-5 *N,N'*-1,3-Phenylenebis[*P,P,P*-triphenylphosphine imide], P-00144
20341-75-7 Ethyl *P*-methyl-*N*-phenylphosphonamidate, *in* M-00231
20341-81-5 Phenyl *P*-methyl-*N*-phenylphosphonamidate, *in* M-00231
20342-00-1 1-Phenyl-1*H*-phosphole, P-00217
20342-03-4 2-Chloro-2,3-dihydro-4,5-dimethyl-1,2-oxaphosphole; 2-Oxide, *in* C-00040
20342-05-6 3*a*,4,7,7*a*-Tetrahydro-1,8-diphenyl-4,7-phosphinidene-1*H*-phosphindole; 1,8-Dioxide, *in* T-00105
20345-61-3 Ethyl 2-(diethoxyphosphinyl)-propenoate, *in* D-00260
20354-32-9 2-Chloro-1,3,2-oxathiaphospholane, *in* O-00056
20354-83-0 Dibenzyl ethylphosphonodithioite, *in* E-00141
20355-97-9 Diethyl (4-chlorophenyl)-phosphonite, *in* C-00153
20384-84-3 Methylphenylphosphinodithioic acid, M-00224
20384-85-4 Ethylphenylphosphinodithioic acid, E-00103
20384-86-5 Isopropylphenylphosphinodithioic acid, I-00045
20394-86-9 Benzylphosphinic acid, B-00059
20395-32-8 Ethyl *N,P*-diethylphosphonamidate, *in* D-00336
20408-23-5 Diethyl 1,2-ethenediylbis(phenyl-phosphinate), *in* E-00018
20408-30-4 1-Propenylphosphonic acid; (*E*)-*form*, Dichloride, *in* P-00453
20408-31-5 (2-Phenylethenyl)phosphonic dichloride; (*E*)-*form*, *in* P-00157
20408-33-7 2-Phenylethenylphosphonic acid; (*E*)-*form*, Di-Et ester, *in* P-00155
20416-73-3 (Trichloromethyl)phosphonic acid; Diisocyanate, *in* T-00411
20417-42-9 *N,N,N',N'*-Tetraethyl-*P*-heptylphosphonous diamide, *in* H-00008
20420-12-6 *O,O*-Diethyl 2-butenylphosphonothioate, *in* B-00527
20420-16-0 Ethyl (2-butenyl)-methylphosphinate, *in* B-00523
20427-93-4 Tetraethyl (1-hydroxyethylidene)-bisphosphonate, *in* H-00144
20428-20-0 (2-Chloroethyl)phosphonothioic acid; Dichloride, *in* C-00078
20428-23-3 Bis(mercaptomethyl)phosphinic acid, B-00292
20433-63-0 *S*-Phenyl methylphosphonochloridothioate, *in* M-00298
20433-65-2 *S*-Phenyl phenylphosphonochloridothioate, *in* P-00232
20434-05-3 Bis(4-methoxyphenyl)phosphinic acid, B-00301
20434-06-4 Bis(4-methoxyphenyl)phosphinic acid; Chloride, *in* B-00301
20434-08-6 Diethylphosphinic acid; Isocyanate, *in* D-00322
20434-13-4 Diethylphosphinic acid; Isothiocyanate, *in* D-00322
20435-15-8 Tris(2-phenylethenyl)phosphine; Oxide, *in* T-00843
20442-31-3 *O,O,S*-Triisopropyl phosphorodithioate, T-00503
20442-45-9 *O,O*-Di-2-propenyl phosphorodithioate; K salt, *in* D-01197
20442-56-2 Ethyl phenyl phosphonate, E-00111
20443-12-3 Dimethylphosphinic acid; Isothiocyanate, *in* D-00829
20443-14-5 Diphenylphosphinic acid; Isothiocyanate, *in* D-01039
20443-19-0 Phenylphosphonic acid; Diisothiocyanate, *in* P-00221
20459-30-7 *Se,Se,Se*-Triphenyl phosphorotriselenoate, T-00672
20459-69-2 *N*-Phenyl-*P,P*-bis(chloromethyl)-phosphinothioic amide, *in* B-00129
20464-67-9 3-Methoxyphenyl phosphorodichloridate, M-00080
20464-68-0 4-Methoxyphenyl phosphorodichloridate, M-00081
20464-81-7 Bis(2-methoxyphenyl) phosphorochloridate, *in* B-00294
20464-82-8 Bis(4-methoxyphenyl) phosphorochloridate, *in* B-00295
20464-99-7 Dimethyl ethylphosphoramidate, *in* E-00156
20465-00-3 Dimethyl propylphosphoramidate, *in* P-00484
20465-01-4 Dimethyl butylphosphoramidate, *in* B-00621
20465-03-6 Diethyl butylphosphoramidate, *in* B-00621
20465-04-7 Dipropyl ethylphosphoramidate, *in* E-00156
20465-06-9 Dibutyl ethylphosphoramidate, *in* E-00156
20472-19-9 Diiodophenylphosphine, *in* P-00251
20472-46-2 Bromodiethylphosphine, *in* D-00333
20472-47-3 Diethyliodophosphine, *in* D-00333
20472-49-5 Ethyl diphenylphosphinothioite, *in* D-01085
20472-52-0 Iododiphenylphosphine, *in* D-01089
20472-53-1 Diphenylphosphinous fluoride, D-01092
20477-07-0 Trimethyl[(trimethyl-phosphoranylidene)methyl]-phosphonium(1+); Iodide, *in* T-00565
20477-35-4 4,6,7-Triphenyl-2,3-bis(trifluoromethyl)-1-phosphabicyclo[2.2.2]-octa-2,5,7-triene, T-00598
20477-51-4 2-Phenyl-1,3-azaphospholidine, P-00090
20486-23-1 Cytidine 2',3'-phosphate; K salt, *in* C-00294
20490-50-0 1-Ethyl-3-phenyl-1,3-azaphospholidine, *in* P-00091
20491-26-3 $N^2,N^{2'}$,*P*-Triphenylphosphonothioic dihydrazide, *in* P-00249
20491-53-6 Diisopropylphosphine, D-00640
20491-55-8 Tetraisopropyldiphosphine, T-00167
20491-56-9 Tetraisopropyldiphosphine; Disulfide, *in* T-00167
20491-60-5 Tetrapropyldiphosphine, T-00286
20494-66-0 *O,S*-Diethyl phosphorofluoridodithioate, D-00388
20494-69-3 Methyl diethylphosphoramidofluoridodithioate, *in* D-00352
20494-70-6 Ethyl diethylphosphoramidofluoridodithioate, *in* D-00352
20494-70-6 Ethyl dimethyl-phosphoramidofluoridodithioate, *in* D-00867
20494-71-7 Dimethyl phosphorofluoridotrithioate, D-00903
20494-72-8 Diethyl phosphorofluoridotrithioate, D-00390
20495-30-1 Dimethyl (acetamidomethyl)-phosphonate, *in* A-00012
20495-31-2 Diethyl (acetamidomethyl)-phosphonate, *in* A-00012
20495-46-9 Diethylphosphinic acid; Azide, *in* D-00322
20502-36-7 Tetramethylphosphorodiamidous acid; Bromide, *in* T-00227
20502-37-8 Dimethylphosphoramidous acid; Dibromide, *in* D-00875
20502-39-0 Dimethylphosphoramidous acid; Diiodide, *in* D-00875
20502-44-7 Methyl phosphorodibromidite, M-00344
20502-48-1 Diethyl phosphorobromidite, D-00372
20502-49-2 Ethyl phosphorodibromidite, E-00166

20502-50-5 Diethyl phosphoriodidite, D-00365
20502-63-0 Dimethyl phosphonite, D-00852
20502-85-6 Diethyl phosphonite, D-00339
20502-88-9 Methyl dimethylphosphinite, *in* D-00844
20505-16-2 Diethyl ethynylphosphonite, *in* E-00204
20518-03-0 *O*-Methyl methylphosphonofluoridothioate, *in* M-00305
20519-92-0 Trimethylgermylphosphine, T-00516
20519-93-1 Dimethyldiphosphinogermane, D-00750
20536-00-9 *N,N′*-Dimethyl-phosphorodiamidodithioic acid; Methylammonium salt, *in* D-00896
20537-00-2 Diethyl (3-penten-1-ynyl)-phosphonate, *in* P-00045
20537-88-6 ▷Amifostine, A-00052
20537-88-6 ▷Ethiofos, E-00021
20540-86-7 1-Hydroxy-2,4,6-trimethyl-1,3,5-dioxaphosphorinane 5-oxide, H-00203
20540-88-9 1-Hydroxy-2,4,6-trimethyl-1,3,5-dioxaphosphorinane 5-oxide; 5-Methoxy (Me ester), *in* H-00203
20543-86-6 Methyl(trichloromethyl)-phosphinic acid; Chloride, *in* M-00419
20543-88-8 Ethyl methyl(trichloromethyl)-phosphinate, *in* M-00419
20544-49-4 7-Methyl-1,4,6-trioxa-5-phospha(5-P^{V})spiro[4.4]-nonan-8-one; 5-Chloro, *in* M-00427
20553-55-3 7-Methyl-1,4,6-trioxa-5-phospha(5-P^{V})spiro[4.4]-nonan-8-one; 5-Hydroxy, *in* M-00427
20553-92-8 5-Phosphoniaspiro[4.4]nonane; Bromide, *in* P-00374
20554-01-2 5-Phosphoniaspiro[4.4]nonane; Iodide, *in* P-00374
20570-25-6 2-Methoxy-1,3,2-benzodioxaphosphole, M-00026
20580-36-3 Dimethyl (2-cyanoethyl)-phosphonate, *in* C-00211
20612-71-9 Trihexylphosphine; Selenide, *in* T-00492
20612-72-0 Triheptylphosphine; Selenide, *in* T-00489
20612-74-2 Trinonylphosphine; Selenide, *in* T-00578
20612-75-3 Tri(decyl)phosphine; Selenide, *in* T-00433
20626-67-9 Butyl ethylphosphonobromidothioite, *in* E-00134
20626-85-1 *P,P*-Dimethyl-*N,N*-diphenylphosphinous amide, *in* D-00845
20626-86-2 Phenyl dimethylphosphinoselenoite, *in* D-00838
20627-00-3 *O*-Isopropyl methylphosphonothioate, I-00041
20627-09-2 Diethyl (2-furanylhydroxymethyl)-phosphonate, *in* F-00071
20636-79-7 2-Hydroxy-4-methyl-1,3,2-dioxaphospholane 2-oxide, H-00155
20639-78-5 2-Mercapto-4-methyl-1,3,2-dioxaphosphorinane 2-sulfide, M-00017
20650-56-0 4-(Diphenylphosphinothioyl)-phenol, *in* D-01072
20663-39-2 Methyl-2-naphthylphenyl-phosphine; (*S*)-*form*, Oxide, *in* M-00177
20665-23-0 Methylphosphoramidothioic dichloride, M-00334
20676-63-5 (4-Methylphenyl)phosphonous acid, M-00248
20676-64-6 Dimethyl(4-methylphenyl)-phosphine, D-00761
20676-76-0 Bis(3-methylphenyl)-phenylphosphine; Sulfide, *in* B-00310
20677-03-6 Dipropyl phenylphosphonate, *in* P-00221
20677-12-7 Diethyl (4-bromophenyl)-phosphonate, *in* B-00492
20682-72-8 2-Hydroxy-4,5-dimethyl-1,3,2-dioxaphosphole 2-oxide, H-00123
20684-76-8 (Morpholinomethyl)-diphenylphosphine oxide, M-00519
20698-87-7 *O*-Ethyl triethylphosphonamidothioate, *in* T-00448
20701-41-1 Glucose 3-dihydrogen phosphate; D-*form*, *in* G-00005
20701-99-9 3*a*,4,7,7*a*-Tetrahydro-1,8-diphenyl-4,7-phosphinidene-1*H*-phosphindole; 1,8-Disulfide, *in* T-00105
20702-05-0 5,6-Dihydro-5-hydroxyphosphanthridine 5-oxide, D-00504
20707-71-5 Glycerol 1,2-di-9,12-octadecadienoate 3-phosphoethanolamine; (*R*)-(*all-Z*)-*form*, *in* G-00029
20758-93-4 Trifluorodiphenoxyphosphorane, T-00470
20762-30-5 Adenosine diphosphate ribose, A-00041
20763-19-3 Methoxymethylenetriphenyl-phosphorane, *in* M-00053
20783-50-0 ▷(4-Methylphenyl)phosphinic acid, M-00222
20809-97-6 Diethyl (4-methylphenyl)-phosphoramidate, *in* M-00253
20819-54-9 ▷Trimethylphosphine selenide, T-00533
20846-04-2 *P,P*-Bis(4-methylphenyl)-*N*-phenylphosphinic amide, *in* B-00320
20868-28-4 *O*-Ethyl *O*-phenyl phosphorodithioate; K salt, *in* E-00120
20868-30-8 *O*-Methyl *O*-phenyl phosphorodithioate; K salt, *in* M-00263
20920-23-4 Methyltriphenylphosphonium(1+); Perchlorate, *in* M-00432
20995-73-7 1,1-Dihydro-1,1-diphenoxy-2,4,6-triphenylphosphorin, *in* D-00613
20995-81-7 10-Phenylacridophosphine, P-00079
20995-83-9 5-Chloro-5,10-dihydro-10-phenylacridophosphine, *in* D-00548
21047-79-0 3-Methyl-2-phenoxy-1,3,2-oxazaphospholidine; 2-Oxide, *in* M-00192
21049-95-6 Phenyl phosphorodi(isothiocyanatate), P-00294
21050-06-6 Phenyl phosphorochloridisocyanatidate, P-00270
21050-24-8 2,2-Difluoro-2,2,4,4,6,6-hexahydro-4,4,6,6-tetraphenyl-1,3,5-triaza-2,4,6-triphosphorine, D-00412
21063-51-4 Galactose 3-dihydrogen phosphate; D-*form*, Ba salt, *in* G-00002
21079-47-0 2,2,4,6-Tetrafluoro-2,2,4,4,6,6-hexahydro-4,6-diphenyl-1,3,5-triaza-2,4,6-triphosphorine; (4*RS*,6*SR*)-*form*, *in* T-00079
21088-66-4 4-Phenyl-3,5-dioxa-1-phosphabicyclo[2.2.1]-heptane, P-00113
21089-13-4 1-Hydroxy-1,1-propanediphosphonic acid, H-00189
21089-18-9 Trichlorobis(trichloromethyl)-phosphorane, T-00399
21089-19-0 Bis(trichloromethyl)phosphinic acid, B-00402
21089-32-7 3′-Xanthylic acid, X-00002
21100-88-9 Phenyl methyl-phosphonisothiocyanatidate, *in* M-00294
21124-98-1 3-Methyl-1,3,2-thiazaphospholidine; 2-Methoxy, 2-oxide, *in* M-00408
21124-99-2 2-Ethoxy-3-methyl-1,3,2-thiazaphospholidine; 2-Oxide, *in* E-00043
21124-99-2 3-Methyl-1,3,2-thiazaphospholidine; 2-Ethoxy, 2-oxide, *in* M-00408
21135-08-0 5-*tert*-Butyl-2-chloro-1,3,2-dioxaphosphorinane, B-00547
21162-06-1 Phenyl ethylphosphinate, *in* E-00124
21162-48-1 5-Phosphoniaspiro[4.4]nonane; Tetrafluoroborate, *in* P-00374
21163-77-9 Tetrakis(dimethylamino)-iodophosphorane, T-00178
21183-89-1 *P,P*-Di-*tert*-butyl-*N*-methylphosphinous amide, *in* D-00124
21186-90-3 *S*-Phenyl phosphorodichloridothioate, P-00283
21186-91-4 *S*-(4-Chlorophenyl) phosphorodichloridothioate, C-00164
21187-15-5 Tribenzylphosphine; Sulfide, *in* T-00338
21187-16-6 Dibenzylphosphinothioic acid, D-00057
21187-18-8 *tert*-Butylphosphonothioic dichloride, B-00613
21199-96-2 Bis(benzylthiomethyl)phosphinic acid, B-00101
21199-97-3 Bis(benzoylthiomethyl)phosphinic acid, *in* B-00292
21200-06-6 Hexahydro-3-phenyl-1,3-azaphosphorine, H-00041
21204-48-8 Isopropyl methylphosphinate, I-00034
21204-52-4 Ethyl *N,N*-diethyl-*P*-1-propynylphosphonamidate, *in* D-00403
21207-77-2 Methylphosphonofluoridodithioic acid, M-00304
21207-78-3 Ethylphosphonofluoridodithioic acid, E-00143
21207-81-8 Dimethylthiodiphosphonic acid difluoride, *in* M-00304
21228-96-6 Methyl *N*-benzoyl-*P*-methylphosphonamidate, *in* B-00031
21229-04-9 2-Methyl-2,2′-spirobi[1,3,2-benzodioxaphosphole], M-00397
21229-05-0 2-Phenyl-2,2′-spirobi[1,3,2-benzodioxaphosphole], P-00309
21229-79-8 2-Fluoro-2,2′-spirobi[1,3,2-benzodioxaphosphole], F-00048
21229-87-8 *N*-Benzoyl-*P*-methyl-phosphonamidic chloride, *in* B-00031
21230-89-7 2,2,6,6-Tetramethyl-1-phenyl-4-phosphorinanone; Oxide, *in* T-00210
21230-91-1 (2-Hydroxyphenyl)-triphenylphosphonium(1+); Iodide, *in* H-00182
21232-91-7 2-Naphthylphenylphosphinic acid; (*R*)P-*form*, (−)-Menthyl ester, *in* N-00018
21232-92-8 2-Naphthylphenylphosphinic acid; (*S*)P-*form*, (−)-Menthyl ester, *in* N-00018
21245-07-8 Bis(dibutylphosphinyl)methane, *in* B-00160
21248-23-7 Zytron; (+)-*form*, *in* Z-00002
21248-24-8 Zytron; (−)-*form*, *in* Z-00002
21267-11-8 Tetraethyl 1,4-phenylenebisphosphonite, *in* B-00010

21267-14-1 Tetraethyl 1,4-phenylenebisphosphonate, *in* B-00008
21298-94-2 Diphenyl-1-propenylphosphine; (*E*)-*form*, Sulfide, *in* D-01143
21300-86-7 3-[(Dimethoxyphosphinyl)oxy]-2-butenoic acid; (*E*)-*form*, *in* D-00681
21302-09-0 Didodecyl phosphonate, D-00227
21302-85-2 *S*-Methyl phosphorothioate, M-00360
21309-23-9 (1-Aminoethyl)phosphinic acid; (±)-*form*, *in* A-00077
21310-06-5 Phenyl(triphenylmethyl)-phosphinic acid; Chloride, *in* P-00332
21310-07-6 (1,3-Butadienyl)-triphenylphosphonium(1+); Bromide, *in* B-00512
21331-80-6 (2-Dimethylaminoethyl)-triphenylphosphonium(1+); Bromide, *in* D-00698
21341-80-0 *N*-Acetyl-*P*,*P*-diphenylphosphinic amide, *in* D-01040
21348-12-9 *O*-Methyl phosphorodifluoridothioate, M-00354
21382-79-6 Diethyl (3-thienylmethyl)-phosphonate, *in* T-00305
21385-80-8 *N*-[(Triphenylphosphoranylidene)-ethylidene]aniline, T-00652
21385-93-3 *P*,*P*-Dimethyl-*N*,*N*,*N'*-tris(trimethylsilyl)-phosphinimidic amide, *in* D-00833
21388-27-2 *P*-3-Fluorophenyl-*N*,*N*,*N'*,*N'*-tetramethylphosphonous diamide, *in* F-00041
21388-28-3 Dimethyl (3-fluorophenyl)-phosphonite, *in* F-00041
21388-29-4 (3-Fluorophenyl)-diphenylphosphine, F-00027
21388-30-7 Difluoro(3-fluorophenyl)-phosphine, *in* F-00041
21388-33-0 (3-Fluorophenyl)phosphonous acid; Dicyanide, *in* F-00041
21388-35-2 *P*-4-Fluorophenyl-*N*,*N*,*N'*,*N'*-tetraethylphosphonous diamide, *in* F-00042
21388-36-3 Dimethyl (4-fluorophenyl)-phosphonite, *in* F-00042
21388-38-5 Difluoro(4-fluorophenyl)-phosphine, *in* F-00042
21388-41-0 (4-Fluorophenyl)phosphonous acid; Dicyanide, *in* F-00042
21398-17-4 2,3-Dihydro-5-methyl-3-phenyl-2-phenylamino-1,3,4,2-oxadiazaphosphole; 2-Sulfide, *in* D-00527
21398-21-0 2,3-Dihydro-2-dimethylamino-5-methyl-3-phenyl-1,3,4,2-oxadiazaphosphole; 2-Sulfide, *in* D-00455
21406-61-1 Pentyltriphenylphosphonium(1+); Bromide, *in* P-00056
21448-79-3 *tert*-Butylmethylphenylphosphine; (*S*)-*form*, Oxide, *in* B-00565
21458-56-0 Trimethyl(trimethylgermyl)-methylenephosphorane, T-00564
21458-74-2 2-Fluoro-5,5-dimethyl-1,3,2-dioxaphosphorinane, F-00016
21481-98-1 (1-Methylpropylidene)-triphenylphosphorane, *in* M-00386
21482-00-8 Cyclopentylidenetriphenyl-phosphorane, *in* C-00282
21482-16-6 Ditetradecyl phosphonate, D-01239
21510-59-8 (Trichloromethyl)phosphonic dichloride, T-00413
21510-86-1 (3-Chloropropyl)phosphonic acid; Dichloride, *in* C-00172
21522-01-0 Triethylphosphine; Selenide, *in* T-00441
21548-32-3 ▷Fosthietan, F-00065
21609-90-5 ▷Leptophos, L-00001
21610-28-6 2,2,2,3-Tetrahydro-3,5-diphenyl-1,3,4,2-oxadiazaphosphole; 2,2,2-Triphenoxy, *in* T-00104
21612-70-4 Diphenyl octylphosphonate, *in* O-00042
21646-99-1 Tetraethyl pyrophosphite, T-00070
21655-99-2 Phenyl butylphosphinate, *in* B-00590
21658-91-3 1-Methylphosphirane, *in* P-00361
21661-51-8 Methyl butylphosphinate, *in* B-00590
21661-52-9 Ethyl butylphosphinate, *in* B-00590
21661-54-1 Isopropyl butylphosphinate, *in* B-00590
21669-30-7 Tetramethyl (3,3,4,4-tetrafluoro-1-cyclobutene-1,2-diyl)bisphosphonate, *in* T-00078
21680-84-2 2,4,4,6-Tetraphenyl-4*H*-1,4-oxaphosphorinium bromide, *in* T-00610
21680-93-3 2,4,6-Triphenyl-4*H*-1,4-oxaphosphorin; B,$PhClO_4$, *in* T-00610
21693-81-2 *P*-Methylphosphonamidothioic acid; Fluoride, *in* M-00284
21693-82-3 *P*-Ethylphosphonamidothioic acid; Fluoride, *in* E-00128
21713-62-2 Methyl bis(4-chlorophenyl)-phosphinate, *in* B-00139
21713-63-3 Methyl dibenzylphosphinate, *in* D-00055
21719-85-7 4-Methylphenyl phosphorodichloridite, M-00260
21720-16-1 3-Methylphenyl phosphorodichloridite, M-00259
21722-49-6 Triheptylphosphine; Sulfide, *in* T-00489
21722-52-1 Trinonylphosphine; Sulfide, *in* T-00578
21722-85-0 Inezin, I-00009
21758-82-7 2,2'-Oxybis[1,3,2-dioxaphosphorinane]; 2,2'-Dioxide, *in* O-00092
21758-85-0 Phenyl phosphorofluoridic acid; NH_4 salt, *in* P-00295
21776-15-8 Diphenylvinylphosphine; Sulfide, *in* D-01168
21776-17-0 *P*-Ethenyl-*N*,*N*,*N'*,*N'*-tetramethylphosphonic diamide, *in* V-00004
21778-19-8 (4-Methoxyphenyl)phosphonic acid, M-00072
21783-55-1 (Chloromethyl)phosphonic diisocyanate, C-00110
21783-56-2 (Dichloromethyl)phosphonic diisocyanate, D-00179
21820-51-9 *O*-Phosphotyrosine; (*S*)-*form*, *in* P-00423
21843-98-1 *O*-Ethyl *N*-isopropyl-*P*-methylphosphonamidothioate, *in* I-00037
21844-11-1 *O*-Ethyl *N*,*P*-dimethylphosphonamidothioate, *in* D-00850
21851-89-8 Bis[(diphenylphosphinyl)-methyl]phenylphosphine oxide, *in* B-00253
21852-02-8 [Phosphinylidynetris(methylene)]-trisphosphonic acid, P-00360
21852-18-6 Tetrahydro-2*H*-1,3,2-thiazaphosphorine 2-oxide; 2-Me, *in* T-00154
21852-19-7 Tetrahydro-2*H*-1,3,2-thiazaphosphorine 2-oxide; 2-Ph, *in* T-00154
21857-33-0 4-Methyl-2-dimethylamino-1,3,2-dioxaphosphorinane 2-oxide, M-00118
21869-93-2 N^5-Phosphonomethionine sulfoximine; (*S*,*S*)-*form*, *in* P-00388
21882-77-9 (Diethoxyethenylidene)-triphenylphosphorane, D-00232
21890-13-1 Phenyl phenyl-phosphonochloridodithioate, *in* P-00230
21915-78-6 Diphenyl 2-pyridylphosphoramidate, *in* P-00510
21915-79-7 *OO*-Diphenyl 2-pyridinylphosphoramidothioate, *in* P-00513
21923-29-5 *O*,*O*-Diphenyl 4-pyridinylphosphoramidothioate, *in* P-00514
21923-73-9 2,4,6,8-Tetrahydroxy-1,5-dimethyl-3,7,9-trioxa-2,4,6,8-tetraphosphabicyclo[3.3.1]-nonane; Tetra-Na salt, *in* T-00161
21940-87-4 Spiro[1,3,2-benzodioxaphosphole-2,1'-[3]phospholene]; 2-Bromo, *in* S-00010
21940-92-1 Spiro[1,3,2-benzodioxaphosphole-2,1'-[3]phospholene]; 2-Fluoro, *in* S-00010
21940-99-8 Spiro[1,3,2-benzodioxaphosphole-2,1'-[3]phospholene]; 2-Methoxy, *in* S-00010
21949-38-2 Dimethylvinylene chlorophosphate, *in* C-00057
21955-18-0 Diethyl (isocyanatomethyl)-phosphonate, *in* I-00022
21955-58-8 [3-(2-Methyl-1,3-dioxolan-2-yl)propyl]triphenylphosphonium; Iodide, *in* M-00132
21955-74-8 *P*,*P*-Diphenyl-*N*,*N'*-bis(trimethylsilyl)-phosphinimidic amide, *in* D-01043
21955-86-2 5,5'-Spirobi[5*H*-dibenzophospholium](1+); Bromide, *in* S-00022
21959-85-3 Ethyl butylphosphonisocyanatidate, *in* B-00601
21960-26-9 [(4-Methoxyphenyl)methylene]-triphenylphosphorane, *in* M-00064
21960-27-0 (2-Methylpropylidene)-triphenylphosphorane, M-00371
21981-56-6 2,8-Dimethyl-10-phenyl-10*H*-phenoxaphosphine, D-00799
21989-72-0 Benzylethylphenylphosphine; (±)-*form*, Sulfide, *in* B-00047
21990-63-6 2,8-Dimethyl-10-phenyl-10*H*-phenoxaphosphine; 10-Oxide, *in* D-00799
21991-66-2 2-[(Hydroxyisopropoxyphosphinyl)-oxy]-*N*,*N*,*N*-trimethyl-ethanaminium hydroxide, inner salt, *in* C-00198
21992-79-0 2-Bromo-4-methyl-1,3,2-dioxaphospholane, B-00473
21992-80-3 2-Bromo-4,5-dimethyl-1,3,2-dioxaphospholane, B-00470
21993-10-2 Bis(phenoxymethyl)phosphinic acid, B-00384
21993-11-3 Ethyl bis(phenoxymethyl)-phosphinate, *in* B-00384
21993-16-8 Methyl (diphenylphosphinyl)-acetate, *in* D-01093
21996-99-6 Pentaphenyl-1*H*-phosphole; Sulfide, *in* P-00038
22012-72-2 Zilantel, Z-00001
22031-12-5 1,1,1-Tris(diphenylphosphinomethyl)-ethane, T-00776
22063-07-6 Ethylene glycol cyclic *P*,*P*:*P'*,*P'*-pyrophosphate, *in* O-00091
22065-56-1 Ethyl 2-(phosphonoxy)-2-propenoate, *in* P-00366
22069-91-6 *tert*-Butylmethylphosphinothioic acid; (±)-*form*, Bromide, *in* B-00570
22069-92-7 *tert*-Butylmethylphosphinodithioic acid, B-00568

22069-94-9 *tert*-Butylmethylphosphinothioic acid, B-00570
22077-39-0 *O*-Methyl *O*-4-nitrophenyl cyclohexylphosphoramidothioate, M-00179
22077-43-6 *O,O*-Diphenyl 4-morpholinylphosphonothioate, *in* M-00527
22077-44-7 *O,O*-Diphenyl phosphorochloridothioate, D-01123
22077-83-4 *S*-(2-Chloroethyl) phosphorodichloridothioate, C-00081
22080-28-0 2-[(Ethoxyhydroxyphosphinyl)-oxy]-*N,N,N*-trimethyl-ethanaminium hydroxide, inner salt, *in* C-00198
22082-39-9 *S,S*-Diethyl benzylphosphonodithioate, *in* B-00065
22082-71-9 2-Chloro-3-methyl-1,3,2-oxazaphospholidine, C-00102
22089-22-1 ▷Trifosfamide, T-00485
22122-79-8 Phenylphosphonic acid; Dicyanide, *in* P-00221
22138-53-0 *β*-Aspartyl phosphate, A-00143
22140-20-1 2-Methylthioadenosine 5′-(dihydrogen phosphate), M-00410
22151-84-4 *O*-2,4-Dichlorophenyl *S*-methyl isopropylphosphoramidothioate; (+)-*form, in* D-00183
22151-85-5 *O*-2,4-Dichlorophenyl *S*-methyl isopropylphosphoramidothioate; (±)-*form, in* D-00183
22208-82-8 2,6-Dimethyl-4-phenylphosphorin, D-00819
22224-92-6 ▷Fenamiphos, F-00002
22227-09-4 Phenyl propylene phosphate, *in* M-00190
22230-88-2 *O,O,O*-Tris(2-methylpropyl) phosphoroselenoate, T-00827
22248-79-9 ▷Tetrachlorvinphos; (*Z*)-*form, in* T-00038
22259-33-2 Glycerol 1-5,8,11,14-icosatetraenoate 2-octadecanoate 3-phosphoethanolamine, G-00065
22260-28-2 *O*-2,4-Dichlorophenyl *S*-methyl isopropylphosphoramidothioate; (−)-*form, in* D-00183
22289-00-5 Bis(2,2-dimethylpropyl) phosphonate, B-00208
22298-05-1 1,3-Dimethyl-1,3,2-diazaphosphetidin-4-one; 2-Dimethylamino, *in* D-00717
22307-83-1 Diisopropylphosphinothioic acid, D-00646
22336-21-6 (4-Chlorophenyl)phosphinic acid, C-00144
22341-00-0 2,2,4,6-Tetrafluoro-2,2,4,4,6,6-hexahydro-4,6-diphenyl-1,3,5-triaza-2,4,6-triphosphorine; (4*RS*,6*RS*)-*form, in* T-00079
22345-03-5 4,4,5,5-Tetramethyl phenyl phosphate, *in* P-00075
22350-76-1 Tetrachlorvinphos; (*E*)-*form, in* T-00038
22382-13-4 ▷Methyl phosphorodifluoridate, M-00352
22400-41-5 Diphenyl [(triphenylphosphoranylidene)-methyl]phosphonate, D-01163
22401-25-8 Tetraethyl 1,3-propanediylbisphosphonate, *in* P-00444
22403-60-7 2,2-Dichloro-2,2-dihydro-3-phenyl-Δ^1-1,3,2-diazaphosphetin-4-one, *in* D-00557
22403-61-8 2,2-Dihydro-2,2,3-triphenyl-Δ^1-1,3,2-diazaphosphetin-4-one, *in* D-00557
22403-62-9 2,2-Dihydro-2,2-diphenoxy-3-phenyl-Δ^1-1,3,2-diazaphosphetin-4-one, *in* D-00557
22421-85-8 Hexabutylphosphoric triamide, *in* H-00013
22428-64-4 1,2-Bis(diphenylphosphinyl)-ethyne, *in* B-00216
22428-65-5 1,2-Bis(diphenylphosphinothioyl)-ethyne, *in* B-00216
22429-12-5 1,3-Dimethyl-2-phenyl-1,3,2-diazaphospholidine, D-00793
22434-51-1 2,2,3,4,4-Pentamethyl-1-phenylphosphetane; *cis-form, in* P-00020
22457-89-2 ▷Benfotiamine, B-00002
22474-68-6 Pentafluorophenylphosphonic acid; Difluoride, *in* P-00015
22474-69-7 Bis(pentafluorophenyl)phosphinic acid; Fluoride, *in* B-00382
22491-81-2 Butyl methylphosphonofluoridothioite, *in* M-00306
22514-10-9 2,4-Dichloro-2,4,6,6-tetrakis(dimethylamino)-2,2,4,4,6,6-hexahydro-1,3,5-triaza-2,4,6-triphosphorine, D-00194
22537-93-5 2-(Diethoxyphosphinyl)benzoic acid, D-00243
22585-74-6 3-Hydroxy-1,3-thiaphosphetane 3-oxide, H-00201
22585-75-7 Ethyl thiodimethylenephosphinate, *in* H-00201
22585-76-8 3-Chloro-1,3-thiaphosphetane; 3-Oxide, *in* C-00188
22585-79-1 *O,O*-Diethyl (4-chlorophenyl)-phosphonothioate, *in* C-00150
22585-81-5 (4-Chlorophenyl)phosphonic dichloride, C-00148
22585-87-1 *P*-4-Chlorophenyl-*N,N,N′,N′*-tetramethylphosphonous diamide, *in* C-00153
22601-77-0 *O*-Methyl *O*-1-naphthalenyl phosphorothioate; (±)-*form, in* M-00176
22606-68-4 Tetrafluoromethylthiophosphorane, T-00082
22608-53-3 ▷*O,S,S*-Trimethyl phosphorodithioate, T-00541
22608-55-5 *S,S*-Dimethyl phosphorodithioate; K salt, *in* D-00900
22610-15-9 Ethyl [(triphenylphosphoranylidene)-hydrazono]acetate, *in* T-00653
22685-19-4 Diethyl isopropylphosphoramidate, *in* I-00058
22685-20-7 Diethyl *tert*-butylphosphoramidate, *in* B-00622
22689-71-0 Dipropylphosphinodithioic acid, D-01207
22716-98-9 3-[(Dimethoxyphosphinyl)oxy]-2-butenoic acid, D-00681
22737-09-3 Dibutylphosphinodithioic acid; Anhydrosulfide, *in* D-00112
22737-43-5 Diethylphosphinodithioic acid; Anhydrosulfide, *in* D-00325
22737-44-6 Dipropylphosphinodithioic acid; Anhydrosulfide, *in* D-01207
22737-45-7 Dibenzylphosphinodithioic acid; Anhydrosulfide, *in* D-00056
22740-38-1 Phenyl methyl-phosphonochloridodithioate, *in* M-00297
22749-43-5 6-Chloro-6*H*-dibenzo[*c,e*][1,2]-oxaphosphorin, C-00033
22788-91-6 Phenyl methylphosphonofluoridite, *in* M-00307
22794-04-3 Methyl methylphosphonofluoridite, *in* M-00307
22794-06-5 Isopropyl methylphosphonofluoridite, *in* M-00307
22836-03-9 Triphenyl(tetrahydro-2*H*-pyran-4-yl)phosphonium(1+); Bromide, *in* T-00684
22836-04-0 Triphenyl(tetrahydro-2*H*-thiopyran-4-yl)-phosphonium(1+); Bromide, *in* T-00685
22836-06-2 Cycloheptyltriphenyl-phosphonium(1+); Bromide, *in* C-00238
22836-20-0 3-(Diphenylphosphinyl)-benzonitrile, *in* D-01057
22836-23-3 4,4′,4″-Phosphinylidynetris[benzonitrile], *in* T-00754
22836-27-7 Tris(4-carboxyphenyl)phosphine, T-00740
22836-30-2 Tris(4-cyanophenyl)phosphine, T-00754
22836-36-8 Tris(4-carboxyphenyl)phosphine; Sulfide, tri-Me ester, *in* T-00740
22846-16-8 1-Phenylphosphirane, P-00214
22864-69-3 Trimethyl (4-nitrophenyl)-phosphorimidate, *in* N-00063
22880-57-5 2,2,4,4,6,6-Hexahydro-2,2-dimethyl-4,4,6,6-tetraphenyl-1,3,5-triaza-2,4,6-triphosphorine; B, HCl, *in* H-00025
22880-68-8 Bromoethylphenylphosphine, *in* E-00108
22883-70-1 (4-Hydroxyphenyl)-triphenylphosphonium(1+); Bromide, *in* H-00183
22884-29-3 (2-Methylpropyl)-triphenylphosphonium(1+); Bromide, *in* M-00387
22884-31-7 1,5-Pentanediylbis[triphenyl-phosphonium](2+); Dibromide, *in* P-00034
22922-25-4 2,3-Naphthylene phosphorochloridothioate, *in* C-00119
22949-84-4 ▷Butyltriphenylphosphonium(1+); Iodide, *in* B-00641
22950-57-8 Diisopropyl benzoylphosphonate, *in* B-00034
22950-58-9 Diphenyl acetylphosphonate, *in* A-00026
22965-10-2 *N,N′,N″*-Triethylphosphorothioic triamide, *in* T-00462
22987-21-9 (2-Hydroxyethyl)phosphonic acid, H-00146
22987-28-6 Mono(3-methylphenyl) phosphate, M-00486
23039-94-3 Tris(3-fluorophenyl)phosphine, T-00782
23040-22-4 (Diphenylphosphinyl)acetonitrile, *in* D-01046
23040-94-0 *P*-(3-Chloropropyl)-*N,N,N′,N′*-tetraethylphosphonic diamide, *in* C-00173
23040-98-4 *P*-(3-Chloropropyl)-*N,N′*-diphenylphosphonic diamide, *in* C-00173
23041-18-1 (Isocyanatomethyl)phosphonic acid; Dichloride, *in* I-00022
23041-25-0 Bis(trichloromethyl)phosphinic acid; Chloride, *in* B-00402
23041-38-5 2,2,3,4,4-Pentamethyl-1-phenylphosphetane, P-00020
23052-81-5 2-Amino-4-phosphonobutanoic acid; (*S*)-*form, in* A-00117
23055-15-4 (5,6-Acenaphthenylenedimethyl-ene)bis[triphenyl-phosphonium](2+); Dibromide, *in* A-00001
23069-03-6 Triphenyl[(2,6,6-trimethyl-1,3-cyclohexadien-1-yl)-methyl]phosphonium(1+); Bromide, *in* T-00688
23081-47-2 Dimethyl (2-nitrophenyl)-phosphonate, *in* N-00055
23081-49-4 Ethyl (2-nitrophenyl)-methylphosphinate, *in* N-00042
23081-50-7 (2-Nitrophenyl)methylphosphinic acid, N-00042
23081-65-4 Ethyl bis(ethoxymethyl)-phosphinate, *in* B-00272
23081-69-8 *O,S*-Ethylene *O*-methyl phosphorothioite, *in* O-00056

23081-71-2 2,3-Dihydro-2-methoxy-5-methyl-1,2-thiaphosphole; 2-Oxide, *in* D-00514

23081-74-5 2-(Diphenylphosphinyl)aniline, *in* D-01047

23081-75-6 Diethyl (2,4-dinitrophenyl)-phosphonate, *in* D-00946

23081-76-7 Dimethyl (2,4-dinitrophenyl)-phosphonate, *in* D-00946

23081-77-8 Diisopropyl (2,4-dinitrophenyl)phosphonate, *in* D-00946

23081-78-9 Diethyl 2-pyridinylphosphonate, *in* P-00503

23092-11-7 Dimethylphosphinodithioic anhydrosulfide, *in* D-00834

23112-90-5 (3-Methyloxiranyl)phosphonic acid, M-00185

23124-79-0 *O*-Ethyl ethylphosphonodithioate; (*S*)-*form*, Et ester, *in* E-00070

23155-02-4 ▷(3-Methyloxiranyl)phosphonic acid; (2*R*,3*S*)-*form*, *in* M-00185

23168-88-9 2-Methoxy-4-methyl-1,3,2-dioxaphosphorinane 2-sulfide; (2*RS*,4*RS*)-*form*, *in* M-00047

23168-89-0 2-Methoxy-4-methyl-1,3,2-dioxaphosphorinane 2-sulfide; (2*RS*,4*SR*)-*form*, *in* M-00047

23201-70-9 5-*tert*-Butyl-2-methoxy-1,3,2-dioxaphosphorinane; *cis-form*, *in* B-00563

23201-71-0 5-*tert*-Butyl-2-methoxy-1,3,2-dioxaphosphorinane; *trans-form*, *in* B-00563

23206-00-0 Ethyl phenyl 1-aziridinylphosphonite, *in* A-00154

23233-31-0 2-Hydroxy-1,3,2-dioxaphospholane 2-sulfide; 1*H*-Imidazolium salt, *in* H-00135

23233-32-1 2-Hydroxy-4,4,5,5-tetramethyl-1,3,2-dioxaphospholane 2-sulfide; Imidazolium salt, *in* H-00197

23269-98-9 Diethyl (3-chloropropyl)-phosphonate, *in* C-00172

23270-01-1 Diphenyl (3-chloropropyl)-phosphonate, *in* C-00172

23270-16-8 Tricyclopropylphosphine; Sulfide, *in* T-00432

23270-18-0 Tricyclopropylphosphine; Selenide, *in* T-00432

23275-37-8 Benzylmethylphenylphosphine, B-00051

23277-00-1 (1-Naphthylmethyl)-triphenylphosphonium(1+); Chloride, *in* N-00015

23282-01-1 *S*,*S*-Diphenyl phosphorodithioate, D-01132

23308-83-0 [(2-Nitrophenyl)methyl]-triphenylphosphonium(1+); Bromide, *in* N-00047

23344-69-6 *N*,*N'*,*N''*-Tributylphosphoric triamide, *in* T-00394

23358-83-0 1,3,2-Benzoxathiaphosphole; 2-Chloro, *in* B-00026

23360-63-6 Diethyl (triphenylphosphoranylidene)-butanedioate, *in* T-00646

23389-78-8 *N*,*N*,*N'*,*N'*-Tetramethyl-*P*-phenylphosphonoselenoic diamide, *in* P-00243

23415-53-4 Tri-3-thienylphosphine, T-00875

23415-66-9 (3-Chlorophenyl)-diphenylphosphine, C-00128

23415-68-1 (3-Chlorophenyl)-diphenylphosphine; Oxide, *in* C-00128

23415-69-2 (3-Chlorophenyl)-diphenylphosphine; Sulfide, *in* C-00128

23415-70-5 (3-Chlorophenyl)phosphonic acid; Dichloride, *in* C-00146

23415-71-6 Diethyl (3-chlorophenyl)-phosphonate, *in* C-00146

23415-73-8 (3-Chlorophenyl)phosphine, C-00140

23415-85-2 Dichloro(dichloromethyl)-phosphine, *in* D-00181

23416-28-6 *O*,*O*-Diethyl phosphonoselenoate, D-00340

23453-30-7 *P*,*P*,*P*-Trichloro-*N*-methylphosphazene, *in* M-00336

23462-73-9 [(Methylthio)methylene]-triphenylphosphorane, *in* M-00416

23472-22-2 Vinylphosphonic acid; Diisocyanate, *in* V-00003

23485-35-0 Tris(4-nitrophenyl) phosphite, T-00840

23486-86-4 *N*,*N*-Dimethyl-*P*,*P*-diphenylphosphinoselenoic amide, *in* D-01079

23505-41-1 ▷Pirimiphos-ethyl, P-00434

23526-02-5 Thuringiensin, T-00319

23545-86-0 (2-Hydroxyethyl)phosphoramidic acid, H-00148

23560-59-0 ▷Heptenophos, H-00004

23561-01-5 Decylphosphine, D-00011

23561-36-6 (2-Chloromethyl-4-nitrophenyl) phosphorodichloridate, *in* M-00455

23562-59-6 Monobenzyl phosphate; Monoanilinium salt, *in* M-00443

23575-76-0 *PP*-Bis(trichloromethyl)-phosphinimidic chloride, B-00403

23582-00-5 4,7-Bis[2-(diphenylphosphino)-ethyl]-1,1,10,10-tetraphenyl-1,4,7,10-tetraphosphadecane, B-00240

23582-02-7 Bis[2-(diphenylphosphino)-ethyl]phenylphosphine, B-00239

23582-03-8 Tris[2-(diphenylphosphino)-ethyl]phosphine, T-00774

23582-04-9 1,1,4,7,10,10-Hexaphenyl-1,4,7,10-tetraphosphadecane, H-00091

23582-05-0 ▷Bis[2-(diphenylarsino)ethyl]-phenylphosphine, B-00212

23582-06-1 ▷[2-(Diphenylarsino)ethyl]-diphenylphosphine, D-00993

23587-34-0 Bis(3-fluorophenyl)phosphine, B-00276

23588-02-5 Ethyl phenylphosphonochloridothioite, *in* P-00233

23588-03-6 Phenyl methylphosphonochloridothioite, *in* M-00299

23588-06-9 (3-Fluorophenyl)-diphenylphosphine; Sulfide, *in* F-00027

23588-07-0 Diethyl (3-fluorophenyl)-phosphonate, *in* F-00037

23588-09-2 (3-Fluorophenyl)phosphine, F-00033

23588-13-8 (3-Fluorophenyl)phosphonic acid; Dichloride, *in* F-00037

23588-23-0 Bis(3-fluorophenyl)phosphinic acid; Chloride, *in* B-00279

23603-71-6 Diethyl (3-chlorophenyl)-phosphinate, *in* C-00152

23614-63-3 Diphenyl (trichloromethyl)-phosphonate, *in* T-00411

23654-71-9 Ethyl butylphosphonochloridite, *in* B-00605

23669-19-4 2,2-Diamino-2,2,4,4,6,6-hexahydro-4,4,6,6-tetraphenyl-1,3,5-triaza-2,4,6-triphosphorine; *N*,*N'*-Di-Ph, *in* D-00025

23679-20-1 ▷Di-2-propenyl phosphonate, D-01194

23685-56-5 *N*,*N*-Dimethyl-*P*-(1-methylpropyl)phosphonamidic chloride, D-00763

23691-99-8 *P*,*P*-Bis(trichloromethyl)-phosphinic amide, *in* B-00402

23725-53-3 Tris[2-(diphenylphosphino)-ethyl]phosphine; Tetraoxide, *in* T-00774

23728-60-1 2,4-Dichloro-2,4-bis(dimethylamino)-2,2,4,4,6,6-hexahydro-6,6-diphenyl-1,3,5-triaza-2,4,6-triphosphorine; (2*RS*,4*RS*)-*form*, *in* D-00162

23743-26-2 1,2-Bis(dicyclohexylphosphino)-ethane, B-00167

23754-87-2 *O*,*O*-Dimethyl phosphorothioate; Na salt, *in* D-00907

23755-72-8 Methyl bromo(diethoxyphosphinyl)-acetate, *in* B-00467

23755-73-9 Ethyl bromo(diethoxyphosphinyl)-acetate, *in* B-00467

23781-24-0 10-Methyl-10-phenyl-10*H*-phenothiaphosphinium iodide, *in* P-00182

23781-25-1 10-Methyl-10-phenyl-10*H*-phenothiaphosphinium 5,5-dioxide iodide, *in* P-00182

23781-30-8 10-Phenyl-10*H*-phenothiaphosphine; 5,5,10-Trioxide, *in* P-00182

23781-31-9 10-Phenyl-10*H*-phenothiaphosphine; 5,5-Dioxide, 10-sulfide, *in* P-00182

23783-98-4 Phosphamidon; (*Z*)-*form*, *in* P-00347

23808-01-7 1,2-Ethanediylbis[methyl-phenylphosphine], E-00012

23834-60-8 Dibutylphosphinothioic acid; Chloride, *in* D-00117

23834-61-9 Diisopropylphosphinothioic acid; Chloride, *in* D-00646

23844-89-5 10-Phenyl-10*H*-phenothiaphosphine; 5,5-Dioxide, *in* P-00182

23844-90-8 10-Phenyl-10*H*-phenothiaphosphine, P-00182

23853-25-0 Ethyl cyano(triphenyl-phosphoranylidene)pyruvate, *in* C-00218

23855-75-6 10-Hydroxy-2,8-dimethyl-10*H*-phenothiaphosphine 10-oxide; 10-Methoxy (Me ester), *in* H-00127

23855-87-0 1-Phenyl-4-phosphorinanone, P-00268

23855-91-6 1,2,3,6-Tetrahydro-1-phenylphosphorin, T-00134

23855-92-7 1,2,3,6-Tetrahydro-1-phenyl-1-phenylmethylphosphininium bromide, *in* T-00134

23856-98-6 1,2-Dihydro-1-hydroxy-1-phenyl-4*H*-3,1,2-benzothiazaphosphorin-4-one 1-sulfide; 2-Methoxy (*O*-Me ester), *in* D-00503

23857-21-8 1,2-Dihydro-1-phenyl-4*H*-3,1,2-benzothiazaphosphorin-4-thione; 2-Methoxy, 2-sulfide, *in* D-00552

23857-22-9 1,2-Dihydro-1-phenyl-4*H*-3,1,2-benzothiazaphosphorin-4-thione; 2-Ethoxy, 2-sulfide, *in* D-00552

23897-15-6 Tris(2,4,6-trimethylphenyl)-phosphine, T-00861

23897-16-7 Bis(2,4,6-trimethylphenyl)-phosphine; Oxide, *in* B-00426

23897-17-8 Tris(2,4,6-trimethylphenyl)-phosphine; Oxide, *in* T-00861

23900-29-0 1,2-Dihydro-1-hydroxy-1-phenyl-4*H*-3,1,2-benzothiazaphosphorin-4-one 1-sulfide; 2-Ethoxy (*O*-Et ester), *in* D-00503

23906-87-8 1-Adamantylphosphonic acid; Dichloride, *in* A-00034

23906-88-9 1-Adamantylphosphonic acid, A-00034

23936-60-9 1,2-Bis(dimethylphosphino)ethane, B-00200

23973-60-6 Tetraethyl [2,6-naphthalenediylbis[methylene]]-bisphosphonate, *in* N-00006

23988-36-5 2,2,2-Triethoxy-2,2-dihydro-1,3,2-dioxaphosphorinane, T-00437
23988-42-3 2-Ethoxynaphtho[1,2-*d*]-1,3,2-dioxaphosphole, E-00044
23995-96-2 (2,3,5,6-Tetrachloro-4-pyridinyl)phosphonic acid, T-00035
24017-47-8 ▷Triazophos, T-00333
24033-37-2 *P,P*-Diphenyl-*N*-(triphenylphosphoranylidene)-phosphinimidic amide; B, HCl, *in* D-01164
24037-30-7 Diethyl 1-pyrrolidinylphosphonite, *in* P-00520
24058-95-5 Dimethyl 1-pyrrolidinylphosphonate, *in* P-00518
24082-55-1 Tris(2-phenylethenyl)phosphine, T-00843
24090-56-0 *N,N'*-Dimethyl-*P*-(1,1-dimethylethyl)phosphonous diamide, *in* B-00617
24091-50-7 Phenyl diphenylphosphinoselenoite, *in* D-01081
24102-75-8 Phenyl *N,N*-diethyl-*P*-phenylphosphonamidate, *in* D-00307
24102-76-9 Ethyl *P*-ethyl-*N,N*-diphenylphosphonamidate, *in* E-00067
24108-60-9 *O*-(2-Methoxyethyl) phosphorodichloridothioate, *in* H-00149
24110-33-6 1,6-Hexanediphosphonous acid; Bis(dichloride), *in* H-00086
24138-46-3 Diethyl (2,3,5,6-tetrachloro-4-pyridinyl)phosphonate, *in* T-00035
24151-93-7 ▷Piperophos, P-00433
24167-74-6 Dimethyl phosphorobromidate, D-00886
24171-89-9 Tri-2-thienylphosphine, T-00874
24304-64-1 Tetraethylphosphinic amide, *in* D-00323
24310-72-3 Diethyl propylphosphoramidite, *in* P-00486
24323-92-0 Diethylphenylphosphine; Oxide, *in* D-00305
24330-37-8 *O,O*-Dimethyl diethylphosphoramidoselenoate, *in* D-00353
24330-38-9 *O,O*-Diethyl diethylphosphoramidoselenoate, *in* D-00353
24330-40-3 *O,O*-Diisopropyl diethylphosphoramidoselenoate, *in* D-00353
24330-42-5 *O,O*-Diethyl dimethylphosphoramidoselenoate, *in* D-00869
24335-35-1 2,4-Dichloro-1,3-di-*tert*-butyl-1,3,2,4-diazadiphosphetidine, D-00163
24335-36-2 *tert*-Butylphosphoramidous acid; Dichloride, *in* B-00624
24349-90-4 (3-Methyloxiranyl)phosphonic acid; (2*R*,3*S*)-*form*, Monobenzylammonium salt, *in* M-00185
24442-23-7 Phenylbis[3-(triphenylphosphoranylidene)-propenyl]phosphine, P-00103
24442-28-2 Tris[(triphenylphosphoranylidene)-methyl]phosphine; Oxide, *in* T-00871
24442-30-6 Tris[(triphenylphosphoranylidene)-methyl]phosphine; Sulfide, *in* T-00871
24442-45-3 Dibenzylphenylphosphine; Oxide, *in* D-00052
24453-83-6 *O,O*-Ethylene *O*-phenyl phosphorothioate, *in* P-00067
24460-31-9 Tris(2-chlorophenyl) phosphite, T-00750
24470-78-8 Isopropyltriphenyl-phosphonium(1+); Iodide, *in* I-00065
24475-23-8 Dimethyl butylphosphonate, *in* B-00594
24490-39-9 Methyl dimethylphosphinoselenoite, *in* D-00838
24490-40-2 *Se*-Methyl dimethylphosphinoselenothioate, *in* D-00837
24505-49-5 Tris[(triphenylphosphoranylidene)-methyl]phosphine, T-00871
24547-24-8 Cyclohexyldiiodophosphine, *in* C-00267
24576-92-9 Octahydro-1,8-dihydroxy-4,7-phosphinidene-1*H*-phosphindole 1,8-dioxide, *in* D-00625
24590-63-4 (2-Methyl-1,3-butadienyl)-phosphonic acid, M-00093
24610-81-9 *N,N*-Dimethyl-*P,P*-di-2-propenylphosphinic amide, *in* D-01193
24610-82-0 *N,N*-Diethyl-*P,P*-di-2-propenylphosphinic amide, *in* D-01193
24610-83-1 Methyl di-2-propenylphosphinate, *in* D-01193
24611-05-0 *O*-Ethyl dibutylphosphinothioate, *in* D-00117
24623-72-1 *P*-Phenylphosphonamidothioic acid; Fluoride, *in* P-00219
24630-57-7 Dimethyl [(acetyloxy)methyl]-phosphonate, *in* A-00007
24630-67-9 Dimethyl (hydroxymethyl)-phosphonate, *in* H-00162
24630-80-6 Diphenylphosphinous acid, D-01089
24635-39-0 *S*-Ethyl tetraethylphosphorodiamidothioate, *in* T-00063
24647-30-1 2,6,7-Trioxa-4-phospha-1-arsabicyclo[2.2.2]octane, T-00586
24647-32-3 1-Methyl-2,6,7-trioxa-4-phospha-1-silabicyclo[2.2.2]-octane, M-00426
24674-42-8 Dipentylphosphine, D-00980
24674-43-9 Dihexylphosphine, D-00433
24676-98-0 4-(Diethoxyphosphinyl)-2-butenoic acid, D-00248
24676-98-0 Methyl 4-(diethoxyphosphinyl)-2-butenoate, *in* D-00248
24680-53-3 *O*-Ethyl ethylphosphonofluoridothioate; (+)-*form*, *in* E-00071
24680-56-6 Triethylphosphonamidothioic acid; (±)-*form*, *O*-Et ester, *in* T-00448
24680-57-7 *O*-Ethyl ethylphosphonofluoridothioate; (−)-*form*, *in* E-00071
24720-64-7 2-(Triphenylphosphoranylidene)-propanal, T-00657
24838-92-4 *S,S*-Dimethyl 2-propenylphosphonodithioate, *in* P-00457
24838-93-5 *S,S*-Diethyl 2-propenylphosphonodithioate, *in* P-00457
24856-23-3 Dimethyl benzoylphosphoramidate, *in* B-00036
24901-16-4 1,4-Dioxa-5-phospha(5-P^V)-spiro[4.4]non-7-ene; 5-Ph, *in* D-00967
24901-17-5 Spiro[1,3,2-benzodioxaphosphole-2,1'-[3]phospholene]; 2-Ph, *in* S-00010
24901-23-3 Diiodo-*p*-tolylphosphine, *in* M-00248
24901-24-4 (4-Fluorophenyl)diiodophosphine, *in* F-00042
24901-25-5 (4-Chlorophenyl)diiodophosphine, *in* C-00153
24901-29-9 3-Methyl-1-phenylphospholane, M-00229
24933-27-5 Ethyl phosphorodifluoridite, E-00174
24934-91-6 ▷Chlormephos, C-00019
24935-94-2 Dipentylphosphinic acid, D-00981
24942-19-6 5,10[1',2']-Benzenophenophosphazine, B-00015
24942-20-9 5,10[1',2']-Benzenophenophosphazine; Oxide, *in* B-00015
24949-50-6 Tetraethyl phenylimidodiphosphate, *in* P-00168
24958-79-0 Methyl phosphorodiisocyanatidate, M-00356
24958-80-3 Ethyl phosphorodiisocyanatidate, E-00177
25003-28-5 5,5'-Spirobi[5*H*-dibenzophospholium](1+), S-00022
25005-76-9 *O,O*-Diethyl phosphorohydrazidothioate, D-00392
25006-32-0 (4-Bromo-2,5-dichlorophenyl)-methyl phenylphosphonate, *in* M-00488
25022-72-4 ▷Mono-2-propenyl phosphate, M-00509
25030-57-3 Diethyl (3-methyl-2-oxiranyl)-phosphonate, *in* M-00185
25030-78-8 (3-Methyloxiranyl)phosphonic acid; (2*RS*,3*SR*)-*form*, Di-Me ester, *in* M-00185
25033-34-5 *P,P*-Bis(1-aziridinyl)-*N*-phenylphosphinothioic amide, *in* B-00094
25075-80-3 Diethyl [1,5-naphthalenediylbis[methylene]]-bisphosphonate, *in* N-00002
25075-81-4 2,6-Bis(triphenyl-phosphoniomethyl)-naphthalene(2+); Dibromide, *in* B-00454
25076-73-7 [1,1'-Biphenyl]-4-yl-1-naphthalenylphenylphosphine, B-00081
25091-07-0 Hexaethyl 1,2,3-propanetriyltrisphosphonate, *in* P-00450
25091-08-1 Octaethyl 1,2,3,4-butanetetrayltetraphosphonate, *in* B-00518
25103-12-2 ▷Tris(6-methylheptyl) phosphite, T-00798
25130-52-3 2-Chloro-2,3-dihydro-3,5-dimethyl-1,3,4,2-oxadiazaphosphole, C-00039
25132-69-8 1,4-Diphenyl-2,3-bis(triphenylphosphoranylidene)-1,4-butanedione, D-00997
25137-74-0 Zytron; (±)-*form*, *in* Z-00002
25140-24-3 *tert*-Butylmethylphenylphosphine; (*R*)-*form*, *in* B-00565
25140-53-8 Benzylmethylphenylphosphine; (*R*)-*form*, *in* B-00051
25145-23-7 1-Chloro-2,2,3,4,4-pentamethylphosphetane; *cis-form*, *in* C-00124
25145-24-8 1-Chloro-2,2,3,4,4-pentamethylphosphetane; *trans-form*, *in* C-00124
25145-33-9 1-Chloro-2,2,3,4,4-pentamethylphosphetane 1-oxide, *in* C-00124
25145-58-8 (1,3-Butadienyl)phosphonic acid; (*E*)-*form*, Di-Et ester, *in* B-00510
25154-95-4 5-Hydroxy-2,4,6-triphenyl-1,3,5-dioxaphosphorinane 5-oxide, H-00204
25177-27-9 *O*-(2,5-Dichloro-4-iodophenyl) *O*-ethyl ethylphosphonothioate, D-00171
25186-17-8 Bis(4-fluorophenyl)phosphine, B-00277
25196-01-4 (2-Bromoethyl)phosphonic acid; Dibromide, *in* B-00472
25196-02-5 Diethyl(chloroacetyl)phosphonate, *in* C-00020

25196-14-9 1,2-Di-*tert*-butyl-1,2-dimethyldiphosphine, D-00089
25235-15-8 Isopropylphosphonous dichloride, I-00057
25236-29-7 5,5-Dimethyl-1,3,2-dioxaphosphorinane, D-00735
25237-22-3 *S*,*S*-Dimethyl phosphorofluoridodithioate, *in* D-00900
25237-23-4 *S*,*S*-Diphenyl phosphorofluoridodithioate, *in* D-01132
25237-37-0 *S*-Methyl phosphorodifluoridothioate, M-00355
25237-38-1 *S*-Ethyl phosphorodifluoridothioate, E-00176
25239-79-6 *N*-Benzoyl-*P*-phenylphosphonimidic dichloride, B-00033
25246-48-4 *O*-Methyl phosphoramidofluoridothioate, M-00331
25246-49-5 *O*-Ethyl phosphoramidofluoridothioate, E-00158
25264-54-4 2,5-Bis[(triphenylphosphonio)-methyl]thiophene(2+); Dichloride, *in* B-00456
25311-71-1 ▷Isofenphos, I-00025
25316-38-5 *N*,*N*-Diethylphosphoric triamide, D-00364
25317-88-8 10-Phenyl-10*H*-phenothiaphosphine; 10-Sulfide, *in* P-00182
25324-36-1 Bis(4-fluorophenyl)phosphinic acid; Chloride, *in* B-00280
25331-57-1 Phenylphosphonothioic acid, P-00244
25336-93-0 Spiro[1,3,2-benzodioxaphosphole-2,2′-[1,3,2]oxazaphospholidine], S-00009
25359-60-8 *O*,*S*-Dimethyl phosphorisocyanatidothioate, D-00885
25359-61-9 *O*,*S*-Diethyl phosphorisocyanatidothioate, D-00368
25359-63-1 *O*,*S*-Diethyl [(phenylamino)-carbonyl]phosphoramidothioate, *in* P-00082
25361-54-0 ▷[(Aminocarbonyl)methyl]-triphenylphosphonium(1+); Chloride, *in* A-00061
25362-02-1 (2-Phenylethenyl)phosphonic dichloride; (*Z*)-*form*, *in* P-00157
25362-06-5 Dimethyl 1-propenylphosphonate, *in* P-00453
25383-05-5 Di-*tert*-butyl 1-propenylphosphonate, *in* P-00453
25383-07-7 (3-Methyloxiranyl)phosphonic acid; (2*R*,3*S*)-*form*, Mono-(+)-1-phenylethylammonium salt, *in* M-00185
25383-48-6 Bis(1,1-dimethylethyl) 1,2-propadienylphosphonate, *in* P-00441
25404-72-2 1,2,3-Propanetriphosphonic acid, P-00450
25404-73-3 1,2,3,4-Butanetetraphosphonic acid, B-00518
25408-76-8 Tetramethyl-phosphorodiamidoselenoic acid; Chloride, *in* T-00223
25408-77-9 Tetraethylphosphorodiamidoselenoic acid; Chloride, *in* T-00061
25411-71-6 Ethyl bis(phenylethynyl)-phosphinate, *in* B-00388
25411-73-8 Diethyl (diazomethyl)phosphonate, *in* D-00035
25439-11-6 Methyl bis(trifluoromethyl)-phosphinate, *in* B-00411
25448-25-3 Tris(8-methylnonyl) phosphite, T-00799
25460-63-3 (3-Methyloxiranyl)phosphonic acid; (2*S*,3*R*)-*form*, Di-Me ester, *in* M-00185
25484-41-7 Dimethyl (3-methyloxiranyl)-phosphonate, *in* M-00185
25489-13-8 (3-Methyloxiranyl)phosphonic acid; (2*RS*,3*SR*)-*form*, Mono(benzylammonium) salt, *in* M-00185
25508-32-1 Dimethyl [(methylthio)methyl]-phosphonate, *in* M-00415
25508-33-2 Dimethyl [(methylsulfonyl)-methyl]phosphonate, *in* M-00400
25522-46-7 1-Propenylphosphonic acid; (*Z*)-*form*, Dichloride, *in* P-00453
25522-75-2 *N*,*N*′-Dicyclohexylphosphorodiamidodithioic acid; Cyclohexylammonium salt, *in* D-00217
25522-77-4 *N*,*N*′-Dibutylphosphorodiamidodithioic acid; Butylammonium salt, *in* D-00147
25522-78-5 Di-1-piperidinylphosphinodithioic acid; Piperidinium salt, *in* D-01191
25598-80-5 *P*,*P*′-Methylenebis(phosphonic diamide), M-00144
25598-80-5 *P*,*P*′-Methylenebis(phosphonic diamide); Octa-Me, *in* M-00144
25626-00-3 Diethyl (3-methylphenyl)-phosphoramidate, *in* M-00252
25626-89-5 Diethyl ethylphenylphosphoramidate, *in* E-00116
25626-98-6 Dimethyl (2-methylphenyl)-phosphoramidate, *in* M-00251
25626-99-7 Dimethyl (3-methylphenyl)-phosphoramidate, *in* M-00252
25627-01-4 Dimethyl (4-methylphenyl)-phosphoramidate, *in* M-00253
25627-05-8 Dimethyl (4-methoxyphenyl)-phosphoramidate, *in* M-00076
25674-41-3 4,6-Dichloro-1,3,5,7,9,10-hexamethyl-1,3,5,7,9,10-hexaaza-4,6-diphospha(4,6-P^V)-dispiro[3.1.3.1]decane-2,8-dione, D-00170
25688-42-0 Tris[2-(trifluoromethyl)-phenyl]phosphine, T-00852
25688-44-2 Diphenyl(2-trifluoromethylphenyl)phosphine, D-01154
25688-46-4 Tris[3-(trifluoromethyl)-phenyl]phosphine, T-00853
25743-68-4 Triethenylmethylphosphonium iodide, *in* T-00877
25756-78-9 Diphenylphosphinodiselenoic acid, D-01064
25781-00-4 Diphenyl dipropylphosphoramidite, *in* D-01215
25781-02-6 Phenyl dimethylphosphinite, *in* D-00844
25781-03-7 Diethylphosphinous acid; Ph ester, *in* D-00333
25781-13-9 Diethyl *tert*-butylphosphonite, *in* B-00616
25781-14-0 *tert*-Butylethylphosphinic acid; Chloride, *in* B-00560
25781-15-1 *tert*-Butylisopropylphosphinic acid; Chloride, *in* B-00562
25788-98-1 *tert*-Butylmethylphosphinic acid; Chloride, *in* B-00567
25788-99-2 *tert*-Butylisopropylphosphinic acid, B-00562
25791-20-2 Triphenyl(tetradecyl)-phosphonium(1+); Bromide, *in* T-00682
25795-00-0 *O*,*O*-Diethyl *S*-[(phenylthio)-methyl] phosphorodithioate, D-00318
25806-71-7 ▷Bis(3-aminophenyl)phosphinic acid, B-00088
25836-59-3 Phenylphosphonoperoxoic acid; Na salt, *in* P-00242
25836-60-6 Phenylphosphonoperoxoic acid, P-00242
25881-29-2 9-Phenyl-9-phosphatricyclo[3.3.1.$0^{2,8}$]nona-3,6-diene, P-00202
25881-30-5 9-Phenyl-9-phosphatricyclo[3.3.1.$0^{2,8}$]nona-3,6-diene; 9-Oxide, *in* P-00202
25881-31-6 9-Phenyl-9-phosphapentacyclo-[4.3.0.$0^{2,5}$.$0^{3,8}$.$0^{4,7}$]nonane, P-00201
25881-35-0 [Phenyl[(phenylmethyl)amino]-methyl]phosphonic acid, P-00188
25891-89-8 (Aminophenylmethyl)-phenylphosphinic acid, A-00102
25921-17-9 1-Propenylphosphonic acid; (*Z*)-*form*, Difluoride, *in* P-00453
25921-18-0 1-Propenylphosphonic acid; (*Z*)-*form*, Di-Me ester, *in* P-00453
25944-64-3 Diethyl (1-phenylethenyl)-phosphonate, *in* P-00154
25944-79-0 Tetraethyl 1,3-phenylenebisphosphonate, *in* B-00007
25944-84-7 1-Naphthylphenylphosphinic acid, N-00017
25945-92-0 Benzylethylphenylphosphine, B-00047
25954-13-6 ▷Fosamine ammonium, *in* M-00469
25959-38-0 *N*-Methyl-*P*,*P*-diphenylphosphinimidic fluoride, *in* D-01044
25979-07-1 *tert*-Butylphosphonous dichloride, B-00619
25992-25-0 Diphenyl (3-methyl-2-oxiranyl)-phosphonate, *in* M-00185
26003-21-9 9-Phenyl-9-phosphatricyclo[4.2.1.$0^{2,5}$]nona-3,7-diene; *endo-syn*-9-Oxide, *in* P-00203
26003-22-5 9-Phenyl-9-phosphatricyclo[4.2.1.$0^{2,5}$]nona-3,7-diene; *endo-anti*-9-Oxide, *in* P-00203
26016-89-7 (3-Methyloxiranyl)phosphonic acid; (2*S*,3*R*)-*form*, Mono-(−)-α-phenylethylamine salt, *in* M-00185
26016-99-9 ▷(3-Methyloxiranyl)phosphonic acid; (2*R*,3*S*)-*form*, Di-Na salt, *in* M-00185
26017-03-8 (3-Methyloxiranyl)phosphonic acid; (2*S*,3*R*)-*form*, *in* M-00185
26074-17-9 *O*-Ethyl phosphorodiisocyanatidothioate, E-00178
26076-14-2 1-Propenylphosphonic acid; (*E*)-*form*, Difluoride, *in* P-00453
26087-47-8 ▷*S*-Benzyl *O*,*O*-diisopropyl phosphorothioate, B-00045
26092-87-5 2,2′,6,6′-Tetraphenyl-4,4′-biphosphorine, T-00249
26119-41-5 Dimethyl (2-chloroethyl)-phosphonate, *in* C-00074
26120-40-0 Methyl bis(1-aziridinyl)-phosphinite, *in* B-00095
26121-97-1 *S*-Isopropyl phosphorodichloridothioate, *in* I-00064
26155-88-4 *S*-Butyl phosphorodichloridothioate, B-00631
26177-85-5 Fructose 1,6-bis(dihydrogen phosphate); D-*form*, Di-Na salt, *in* F-00067
26190-34-1 ▷*O*,*O*-Diethyl phosphor(isothiocyanatido)-thioate, D-00370
26192-22-3 2,2,2,3-Tetrahydro-2,2,2-trimethoxy-5-methyl-1,2-oxaphosphole, T-00157
26227-69-0 2,4,6-Tris(diethylamino)-1,3,5,2,4,6-trioxatriphosphorinane 2,4,6-trioxide, T-00756
26236-17-9 2,2,4,6-Tetrachloro-2,2-dihydro-1,3,5,2-triazaphosphorine, T-00029
26245-77-2 Dipropyl phenylphosphoramidate, *in* P-00257

26245-90-9 (1-Hydroxybutyl)phosphonic acid, H-00114
26255-46-9 Methylphenyl(phenylmethylene)-propylphosphorane, *in* B-00052
26306-14-9 Cyanodiethylphosphine, *in* D-00333
26327-86-6 Diethyl 2-butenylphosphonate, *in* B-00525
26327-87-7 2-Butenylphosphonic acid; (*Z*)-*form*, Di-Et ester, *in* B-00525
26339-67-3 4-Methyl-1,3,2-dioxaphosphorinane 2-oxide; (2*RS*,4*RS*)-*form*, *in* M-00130
26339-68-4 4-Methyl-1,3,2-dioxaphosphorinane 2-oxide; (2*RS*,4*SR*)-*form*, *in* M-00130
26342-16-5 Ethyl [3-(diethoxyphosphinyl)]-benzoate, *in* P-00379
26343-70-4 Methylphenyl-2-propenylphosphine; (*R*)-*form*, Sulfide, *in* M-00264
26344-06-9 5-*tert*-Butyl-2-methoxy-1,3,2-dioxaphosphorinane; *trans-form*, 2-Oxide, *in* B-00563
26344-07-0 5-*tert*-Butyl-2-methoxy-1,3,2-dioxaphosphorinane; *cis-form*, 2-Oxide, *in* B-00563
26344-10-5 5-*tert*-Butyl-2-methyl-1,3,2-dioxaphosphorinane; *cis-form*, 2-Oxide, *in* B-00564
26344-11-6 5-*tert*-Butyl-2-methyl-1,3,2-dioxaphosphorinane; *trans-form*, 2-Oxide, *in* B-00564
26348-78-7 *P*-Butyl-*N*,*N*,*N'*,*N'*-tetraethylphosphonic diamide, *in* B-00596
26349-88-2 2-Hydroxy-4-methyl-1,3,2-dioxaphosphorinane 2-selenide; (2*RS*,4*RS*)-*form*, Dicyclohexylammonium salt, *in* H-00157
26350-26-5 (Chloromethyl)methylphosphinic acid; Chloride, *in* C-00100
26350-28-7 *S*-Ethyl *P*-ethylphosphonamidothioate, *in* E-00128
26350-29-8 *S*-Methyl *P*-ethylphosphonamidothioate, *in* E-00128
26350-31-2 Methyl *P*-ethylphosphonamidodithioate, *in* E-00127
26384-81-6 Diethyl (3-methyl-2-pyridinyl)-phosphonate, *in* M-00389
26384-86-1 2-Pyridinephosphonic acid, P-00503
26384-87-2 Diethyl (6-methyl-2-pyridinyl)-phosphonate, *in* M-00392
26384-88-3 (6-Methyl-2-pyridinyl)phosphonic acid, M-00392
26384-89-4 Diethyl (4-methyl-2-pyridinyl)-phosphonate, *in* M-00391
26384-90-7 (4-Methyl-2-pyridinyl)phosphonic acid, M-00391
26384-91-8 (3-Methyl-2-pyridinyl)phosphonic acid, M-00389
26386-88-9 Diphenyl phosphorazidate, D-01112
26387-47-3 *O*-Phenyl *N*,*N'*-trimethyl-enephosphorodiamidothioate, *in* H-00039
26387-48-4 Hexahydro-2-phenoxy-1*H*-1,3,2-diazaphosphepine; 2-Sulfide, *in* H-00038
26404-96-6 Phenyl hydrogen (2-aminoethyl)phosphoramidate, *in* A-00080
26437-48-9 Tri-2-pyridylphosphine, T-00723
26447-90-5 Pentaphenylphosphonous diamide, *in* P-00252
26452-49-3 *O*-Methyl tetraethylphosphorodiamidose-lenoate, *in* T-00061
26452-50-6 *O*-Ethyl tetraethylphosphorodiamidose-lenoate, *in* T-00061
26452-52-8 *O*-Butyl tetraethylphosphorodiamidose-lenoate, *in* T-00061
26464-99-3 ▷Dimethyl(trimethylsilyl)-phosphine, D-00932
26465-28-1 Dimethylgermylphosphine, D-00756
26472-47-9 (3-Methyloxiranyl)phosphonic acid; (2*R*,3*S*)-*form*, Ca salt, *in* M-00185
26480-92-2 Methyl α-(triphenylphosphoranylidene)-hydrocinnamate, *in* T-00645
26487-84-3 (3,3-Dimethyl-2-oxobutyl)-triphenylphosphonium(1+); Bromide, *in* D-00784
26487-93-4 3,3-Dimethyl-1-triphenylphosphoranylidene-2-butanone, D-00934
26490-21-1 1-Hydroxy-2,2,3,4,4-pentamethylphosphetane 1-oxide; *trans-form*, Me ester, *in* H-00168
26515-05-9 Ethylmethylphenylphosphine; (*R*)-*form*, Oxide, *in* E-00084
26546-68-9 Pentaethylphosphonous diamide, *in* E-00154
26546-75-8 Phenyl tetramethylphosphorodiamidite, *in* T-00227
26547-89-7 *O*-Isopropyl methylphosphonothioate; (*R*)-*form*, *in* I-00041
26580-25-6 Dimethyl cyclopentylphosphonate, *in* C-00276
26584-15-6 Dimethyl (1-diazoethyl)-phosphonate, *in* D-00033
26662-95-3 Glycerol 1-9,12-octadecadienoate 2-hexadecanoate 3-phosphoethanolamine; (*R*)-(*all*-*Z*)-*form*, *in* G-00069
26674-18-0 1-Chloro-2,2,3,4,4-pentamethylphosphetane; *trans-form*, Oxide, *in* C-00124
26681-85-6 Di-3-butenylphenylphosphine, D-00072
26681-88-9 Phenyldivinylphosphine, P-00122
26707-09-5 Tris(4-hydroxyphenyl)phosphine, T-00790
26729-68-0 3-[(Dimethoxyphosphinyl)oxy]-2-butenoic acid; (*E*)-*form*, Et ester, *in* D-00681
26756-19-4 (2-Phenylethenyl)phosphonothioic acid; Dichloride, *in* P-00160
26818-24-6 *O*,*O*,*S*-Tributyl phosphorothioate, T-00388
26819-90-9 ▷*O*,*O*-Di-2-propynyl phosphorodithioate, D-01232
26852-27-7 Tetraiodophenylphosphorane, T-00165
26853-31-6 Glycerol 1-hexadecanoate 2-9-octadecenoate 3-phosphocholine; (*R*)-L-(*Z*)-*form*, *in* G-00061
26904-52-9 Bis(bromomethyl)phosphinic acid, B-00108
26905-08-8 2,2'-Diselenobis[5,5-dimethyl-2-thioxo-1,3,2-dioxaphosphorinane], *in* D-01234
26905-09-9 Bis(5,5-dimethyl-2-selenoxo-1,3,2-dioxaphosphorinan-2-yl) disulfide, *in* D-01251
26910-74-7 Tri-2-thienylphosphine; Selenide, *in* T-00874
26920-54-7 *tert*-Butylphosphinic acid, B-00591
26943-92-0 4-Bromophenyldiiodophosphine, *in* B-00493
26944-07-0 Tri-3-thienylphosphine; Selenide, *in* T-00875
26978-37-0 Tetraphenyldiphosphine; Monosulfide, *in* T-00258
26978-38-1 Tetramethyldiphosphine; Monosulfide, *in* T-00200
26978-43-8 Tetrabenzyldiphosphine; Disulfide, *in* T-00007
26990-23-8 Isopropyl phenylphosphonochloridothioite, *in* P-00233
26990-26-1 Phenyl phenylphosphonochloridothioite, *in* P-00233
26990-33-0 *S*-Ethyl *N*,*N*-diethyl-*P*-phenylphosphonamidothioate, *in* D-00309
26990-39-6 Ethyl *N*,*N*-diethyl-*P*-phenylphosphonamidodithioate, *in* D-00308
27003-03-8 (3-Fluorophenyl)phosphinic acid, F-00035
27046-19-1 ▷4-Oxocyclophosphamide, O-00058
27098-24-4 Glycerol 1-octadecanoate 2-9,12-octadecadienoate 3-phosphocholine; (*R*)-L-(*all*-*Z*)-*form*, *in* G-00078
27125-51-5 Diphenyl dichlorophosphoramidate, D-01002
27127-08-8 *N*-Benzyl-*P*,*P*-diphenylphosphinic amide, *in* D-01040
27127-09-9 *N*,*N*-Dibenzyl-*P*,*P*-diphenylphosphinic amide, *in* D-01040
27127-15-7 Benzyl ethylphosphonobromidothioite, *in* E-00134
27230-13-3 5-Methyl-1,2-oxaphospholane 2-oxide; 2-Methoxy, *in* M-00183
27255-30-7 Catechol cyclic phosphoric acid anhydride, *in* O-00087
27258-71-5 Tri-1-propynylphosphine; Oxide, *in* T-00721
27258-72-6 Tris(phenylethynyl)phosphine; Selenide, *in* T-00846
27258-73-7 Phenylbis(phenylethynyl)-phosphine, P-00099
27260-90-8 Dimethylphosphinothioic acid; Azide, *in* D-00839
27262-80-2 Trimethylsilyl phenylphosphinate, *in* P-00210
27275-47-4 2-*tert*-Butoxy-5,5-dimethyl-1,3,2-dioxaphosphorinane, B-00534
27286-05-1 Phenyl di-*tert*-butylphosphinate, *in* D-00110
27286-18-6 Phenyl di-*tert*-butylphosphinite, *in* D-00122
27303-84-0 2-Methoxy-4,5-dimethyl-1,3,2-dioxaphospholane; (4*RS*,5*RS*)-*form*, *in* M-00033
27303-86-2 2,3-Butylene methyl phosphate, *in* M-00033
27318-83-8 1,4-Bis(diphenylphosphino)-2-butyne, B-00227
27325-49-1 1-Piperidinylphosphonous acid; Dichloride, *in* P-00432
27353-29-3 Diethyl (4-morpholinylmethyl)-phosphonate, *in* M-00521
27357-50-2 (3-Methyloxiranyl)phosphonic acid; (2*RS*,3*RS*)-*form*, *in* M-00185
27384-09-4 1-(Diphenylphosphinyl)ethanone, *in* A-00016
27384-35-6 Bis[2-(diphenylphosphino)-3,4,5,6-tetrafluorophenyl]-phenylphosphine, B-00268
27387-65-1 *O*-Trimethylsilyl diphenylphosphinothioate, *in* D-01082
27388-03-0 Methylphosphonous dicyanide, M-00323
27443-18-1 Dipropylphosphine; Oxide, *in* D-01204
27445-88-1 Bis(bromomethyl)phosphinic acid; Anhydride, *in* B-00108
27445-88-1 Bis(chloromethyl)phosphinic acid; Anhydride, *in* B-00127
27453-00-5 *P*-(4-Aminophenyl)-*N*,*N'*-dimethylphosphonic diamide, *in* A-00112

27490-04-6 *tert*-Butylmethylphosphinothioic acid; (±)-*form*, Quinine salt, *in* B-00570
27491-70-9 Dimethyl (diazomethyl)-phosphonate, *in* D-00035
27502-54-1 *O*-Trimethylsilyl dibutylphosphinothioate, *in* D-00117
27502-55-2 Tetrabutyldiphosphine; Monooxide, *in* T-00014
27503-04-4 2-Methoxy-4-methyl-1,2-oxaphospholane; 2-Oxide, *in* M-00049
27509-01-9 *tert*-Butylmethylphosphinodithioic acid; Na salt, *in* B-00568
27509-07-5 Di-*tert*-butylphosphinothioic acid; Chloride, *in* D-00118
27509-08-6 *tert*-Butylphosphonothioic acid; Dibromide, *in* B-00612
27509-09-7 Di-*tert*-butylphosphinothioic acid; Bromide, *in* D-00118
27509-18-8 *tert*-Butylmethylphosphinothioic acid; (±)-*form*, Chloride, *in* B-00570
27518-86-1 Uridine diphosphate glucose; Di-Li salt, *in* U-00002
27530-80-9 2-Amino-3-phosphonooxybutanoic acid; (2*RS*,3*SR*)-*form*, *in* A-00120
27721-02-4 1,5-Bis(diphenylphosphino)-pentane, B-00259
27744-98-5 Acetic acid anhydride with dimethyl hydrogen phosphate, *in* A-00025
27771-03-5 1,4-Diphenyl-1,4-diphosphorinane; *cis-form*, 1,4-Dioxide, *in* D-01017
27771-04-6 1,4-Diphenyl-1,4-diphosphorinane; *trans-form*, 1,4-Dioxide, *in* D-01017
27852-48-8 Methyl ethylphosphinate, *in* E-00124
27852-49-9 Isopropyl ethylphosphinate, *in* E-00124
27852-53-5 *tert*-Butyl ethylphosphinate, *in* E-00124
27930-69-4 *N*,*N*-Diethyl-*P*-methylphosphonamidic acid; Chloride, *in* D-00296
27933-10-4 Ethyl *N*,*N*'-dibutylphosphorodiamidate, *in* D-00146
27933-13-7 Ethyl *N*,*N*'-diethylphosphorodiamidate, *in* D-00382
27933-15-9 Butyl *N*,*N*'-dibutylphosphorodiamidate, *in* D-00146
27949-52-6 ▷Conen, C-00204
27971-26-2 (Dimethylamino)-tetraiodophosphorane, D-00709
27972-73-2 Tetramethylphosphorodiamidic acid, T-00218
27976-73-4 *O*,*O*-Diphenyl methylphosphonothioate, *in* M-00313
28003-13-6 *P*-Imidazol-1-yl-*N*,*N*,*N*',*N*'-tetramethylphosphonic diamide, *in* I-00003
28004-49-1 3-Methyl-1,3,2-thiazaphospholidine; 2-Chloro, 2-oxide, *in* M-00408
28004-51-5 3-Methyl-1,3,2-thiazaphospholidine; 2-Chloro, 2-sulfide, *in* M-00408
28022-29-9 Bis(3-nitrophenyl) phosphate, B-00370
28036-01-3 Diethyl(2,4,6-trimethylphenyl)-phosphonate, *in* T-00527
28036-11-5 Ethyl [2-(diethoxyphosphinyl)]-benzoate, *in* P-00378
28036-18-2 Diethyl (2-chlorophenyl)-phosphonate, *in* C-00145
28051-32-3 9-Phenyl-9-phosphapentacyclo-[$4.3.0.0^{2,5}.0^{3,8}.0^{4,7}$]nonane; 9-Oxide, *in* P-00201
28053-08-9 Uridine diphosphate glucose; Di-Na salt, *in* U-00002
28084-39-4 Heptyl tetraethylphosphorodiamidate, *in* T-00059
28096-32-4 (Fluoromethyl)-triphenylphosponium(1+); Iodide, *in* F-00025
28096-33-5 (Fluoromethylene)-triphenylphosphorane, *in* F-00025
28118-79-8 3-(Triphenylphosphoranylidene)-2,5-pyrrolidenedione, T-00663
28124-22-3 1,3*a*,3*b*,4,6*a*,6*b*-Hexahydro-1,2,3*b*,4,5,6*b*-hexaphenyl-cyclobuta[1,2-*b*:3,4-*b*']-diphosphole, H-00031
28133-43-9 2,2,4,4,6,6,8,8,9,9,10,10-Dodecamethyl-1,3,5,7-tetraphospha-2,4,6,8,9,10-hexagermatricyclo[$3.3.1.1^{3,7}$]-decane, D-01263
28167-51-3 *O*,*O*-Dimethyl dimethylphosphoramidothioate, *in* D-00871
28225-41-4 *N*-Methyl-*P*,*P*-bis(trichloromethyl)-phosphinimidic chloride, *in* B-00403
28240-68-8 Methanediphosphonous acid; Bis(dichloride), *in* M-00024
28240-69-9 1,2-Ethanediphosphonous acid; Bis(dichloride), *in* E-00008
28240-71-3 1,4-Butanediphosphonous acid; Bis(dichloride), *in* B-00515
28255-39-2 Diethyl (4-hydroxyphenyl)-phosphonate, *in* H-00181
28255-72-3 Diethyl (4-cyanophenyl)-phosphonate, *in* C-00225
28273-34-9 1-Bromo-2,5-dihydro-3,4-dimethyl-1*H*-phosphole, B-00469
28273-35-0 1-Chloro-2,5-dihydro-3-methyl-1*H*-phosphole, C-00047
28273-36-1 1-Chloro-2,3-dihydro-1*H*-phosphole, C-00051
28273-37-2 1-Chloro-2,3-dihydro-4-methyl-1*H*-phosphole, C-00045
28278-54-8 2,5-Dihydro-1-phenyl-1*H*-phosphole, D-00567
28278-55-9 2,3-Dihydro-1-phenyl-1*H*-phosphole, D-00566
28278-56-0 1-Chloro-2,5-dihydro-1*H*-phosphole, C-00052
28304-87-2 Tris[1,2-benzenediolato(2−)-*O*,*O*'] phosphate(1−); Tetramethylammonium salt, *in* T-00730
28322-40-9 (3-Methylbutyl)-triphenylphosphonium(1+); Bromide, *in* M-00111
28343-42-2 Bis(4-chlorophenyl)phosphinous acid, B-00142
28364-21-8 Cyclohexyl propylphosphonofluoridate, *in* P-00479
28364-22-9 Cyclohexyl butylphosphonofluoridate, *in* B-00608
28364-26-3 Cyclohexyl benzylphosphonofluoridate, *in* B-00066
28387-30-6 *N*-Methyl-*P*,*P*-bis(trichloromethyl)-phosphinic amide, *in* B-00402
28397-15-1 Diethyldiphosphonic acid, D-00290
28447-26-9 Dimethyl [(1,3-dihydro-1,3-dioxo-2*H*-isoindol-2-yl)-methyl]phosphonate, *in* P-00426
28459-97-4 Methyl thiodimethylenephosphinate, *in* H-00201
28460-01-7 Diethyl [(methylthio)methyl]-phosphonate, *in* M-00415
28482-01-1 (2-Bromoethyl)phosphonic acid; Dichloride, *in* B-00472
28486-27-8 (Carboxymethyl)-triphenylphosphonium(1+), C-00014
28522-96-5 *S*,*S*-Dimethyl phosphorochloridodithioate, D-00888
28522-97-6 ▷*S*,*S*-Diethyl phosphorochloridodithioate, D-00375
28522-98-7 *S*,*S*-Dipropyl phosphorochloridodithioate, D-01219
28540-72-9 [(3-Chlorophenyl)methyl]-triphenylphosphonium(1+); Bromide, *in* C-00136
28660-33-5 (2-Aminopropyl)phosphonic acid, A-00126
28660-34-6 (2-Amino-1-methylethyl)-phosphonic acid, A-00088
28691-78-3 Diphenyl-1-propenylphosphine; (*Z*)-*form*, *in* D-01143
28706-85-6 *N*,*N*,*N*',*N*',*N*'',*N*''-Hexamethyl-1-methylenephosphoranetriamide, H-00072
28772-00-1 1,1,2,2,3,4,4-Heptamethylphosphetanium iodide, *in* H-00074
28874-52-4 Glycerol 1,2-ditetradecanoate 3-phosphate; (*R*)-*form*, *in* G-00047
28896-84-6 2-Choro-1,3,2-dithiaphosphorinane, C-00199
28926-65-0 Tris(diphenylphosphino)methane, T-00775
28926-66-1 Tris(diphenylphosphino)methane; Trisulfide, *in* T-00775
28950-17-6 2-*tert*-Butoxy-1,3,2-dioxaphospholane, B-00535
29021-62-3 Ethylene phenylphosphonotrithioate, *in* P-00121
29049-29-4 *tert*-Butylmethylphosphinodithioic acid; Dimethylammonium salt, *in* B-00568
29070-40-4 *P*-Phenylphosphonamidic acid; Fluoride, *in* P-00218
29070-41-5 *P*-Cyclohexylphosphonamidic fluoride, C-00254
29070-42-6 *P*-Ethylphosphonamidic fluoride, *in* E-00126
29070-43-7 *P*-Methylphosphonamidic fluoride, *in* M-00283
29074-98-4 Diethyl [(2-chlorophenyl)-methyl]phosphonate, *in* C-00132
29146-24-5 ▷Di-*tert*-butylfluorophosphine, *in* D-00122
29149-32-4 ▷*tert*-Butylphosphonous difluoride, B-00620
29149-37-9 Di-*tert*-butylphosphinic acid; Fluoride, *in* D-00110
29149-40-4 Di-*tert*-butylphosphinothioic acid; Fluoride, *in* D-00118
29207-36-1 1,1-Bis(diphenylphosphinyl)-ethylene, *in* B-00236
29219-35-0 (2-Methyl-2-propenylidene)-triphenylphosphorane, M-00363
29232-93-7 ▷Pirimiphos-methyl, P-00435
29238-81-1 Dimethyl (trichloromethyl)-phosphonate, *in* T-00411
29259-72-1 8-Phenyl-8-phosphabicyclo[3.2.1]-octan-3-one, P-00199
29259-72-1 8-Phenyl-8-phosphabicyclo[3.2.1]-octan-3-one; *anti-form*, *in* P-00199
29259-76-5 8-Phenyl-8-phosphabicyclo[3.2.1]-octan-3-one; *syn-form*, Oxide, *in* P-00199
29259-77-6 8-Phenyl-8-phosphabicyclo[3.2.1]-octan-3-one; *anti-form*, Oxide, *in* P-00199
29269-17-8 3,3',3''-Phosphinidynetrispropanoic acid; Tri-Me ester, *in* P-00357
29276-10-6 *P*,*P*-Diisopropyl-*N*,*N*-dimethylphosphinic amide, *in* D-00642

29310-25-6 1-Phosphabicyclo[2.2.1]heptane, P-00344
29373-96-4 *O*-2,4-Dichlorophenyl *S*-methyl isopropylphosphoramidothioate, D-00183
29415-54-1 Methyl-2-naphthylphenylphosphine; (*S*)-*form*, *in* M-00177
29473-30-1 2,4-Cyclopentadien-1-ylidenetriphenylphosphorane, C-00274
29476-24-2 1,3-Dimethyl-1,3,2-diazaphosphetidin-4-one; 2-Methoxy, *in* D-00717
29476-26-4 1,3-Dimethyl-6,7-diphenyl-5,8-dioxa-1,3-diaza-4-phospha(4-P^V)spiro[3.4]-oct-6-en-2-one; 4-Dimethylamino, *in* D-00742
29476-27-5 1,3-Dimethyl-6,7-diphenyl-5,8-dioxa-1,3-diaza-4-phospha(4-P^V)spiro[3.4]-oct-6-en-2-one; 4-Methoxy, *in* D-00742
29516-95-8 *O,O,O′,O′*-Tetraphenyl trithiopyrophosphate, T-00284
29541-98-8 Pentylidenetriphenylphosphorane, *in* P-00056
29544-53-4 Diphenylphosphinic acid; Dibenzylammonium salt, *in* D-01039
29634-03-5 1-Phenyl-1*H*-phosphepin; Oxide, *in* P-00206
29634-20-6 2,7-Dihydro-1-phenyl-1*H*-phosphepin; Oxide, *in* D-00564
29651-24-9 *P,P,P*-Trichloro-*N*-methylsulfonylphosphazene, *in* M-00402
29651-27-2 Methylsulfonylphosphoramidic acid; Dichloride, *in* M-00401
29701-83-5 Tris(2-phenylethyl)phosphine; Oxide, *in* T-00844
29701-85-7 1-(Diphenylphosphinyl)octane, *in* O-00039
29703-21-7 *S,S*-Diphenyl propylphosphonodithioate, *in* P-00478
29703-23-9 *S,S*-Diphenyl butylphosphonodithioate, *in* B-00606
29712-28-5 (Aminomethylene)bisphosphonic acid, A-00086
29712-30-9 [(Dimethylamino)methylene]-bisphosphonic acid, D-00701
29727-86-4 Dimethyl dichlorophosphoramidate, D-00723
29749-84-6 (2-Ethoxyethenyl)-triphenylphosphonium(1+); (*Z*)-*form*, Iodide, *in* E-00039
29777-24-0 Methylimidodiphosphoryl chloride, M-00160
29809-48-1 *O,S*-Dimethyl phosphoramidodithioate, D-00865
29809-53-8 *O,S*-Diethyl phosphoramidodithioate, D-00349
29833-83-8 4,4-Dihydro-4,4,4-triphenoxytrioxaphosphetane, D-00603
29918-57-8 *O,O*-Diisopropyl phosphorothioate; NH_4 salt, *in* D-00667
29936-20-7 *P,P′*-1,2-Ethynediylbis[*N,N,N′,N′*-tetramethylphosphonous diamide], E-00200
29936-33-2 Tetraethyl 1,5-pentanediylbisphosphonite, *in* P-00033
29936-34-3 *P,P′*-1,5-Pentanediylbis[bis-*N,N*-diethylphosphonous diamide], *in* P-00033
29936-35-4 1,5-Pentanediphosphonous acid; Bis(dichloride), *in* P-00033
29936-36-5 1,5-Diphosphinopentane, D-01179
29941-08-0 (Dichloromethyl)phosphonic dichloride, D-00177
29949-64-2 ▷Bis(2-methylphenyl)phosphine, B-00315
29949-75-5 Phenyldi-2-propenylphosphine, P-00119
29949-80-2 ▷Tris[4-(methylthio)phenyl]-phosphine, T-00832
29949-81-3 Tris(4-bromophenyl)phosphine, T-00735
29949-84-6 Tris(3-methoxyphenyl)phosphine, T-00792
29949-85-7 Tris(3-chlorophenyl)phosphine, T-00748
29949-92-6 (Chloromethyl)-triphenylphosphonium(1+); Ylide, *in* C-00117
29952-66-7 *O*-Methyl dimethylphosphinothioate, *in* D-00839
29952-67-8 *S*-Methyl dimethylphosphinothioate, *in* D-00839
29952-68-9 Methyl dimethylphosphinodithioate, *in* D-00834
29954-85-6 1,2,3-Triphenyl-1*H*-phosphirene; Oxide, *in* T-00632
30012-71-6 [[1,1′-Biphenyl]-2,2′-diylbis[methylene]]-bis[triphenylphosphorane], *in* B-00445
30018-16-7 (1-Ethoxycarbonyl)-triphenylphosphonium bromide, *in* C-00010
30043-15-3 *P*-Ethylphosphonothioic diamide, E-00149
30092-40-1 Homocubyltriphenylphosphorane, *in* D-00569
30094-28-1 2-(Diethoxyphosphinyl)propanoic acid, D-00258
30094-33-8 Chloro(diethoxyphosphinyl)acetic acid, C-00036
30148-50-6 ▷Bis(trimethylsilyl) phosphonite, B-00436
30170-00-4 Glycerol 1,2-ditetradecanoate 3-phosphate, G-00047
30238-12-1 *O,O*-Diethyl (phenylethynyl)-phosphonothioate, *in* P-00167
30238-13-2 *O,O*-Dimethyl (phenylethynyl)-phosphonothioate, *in* P-00167
30292-78-5 1-Chlorophosphorinane, C-00168
30309-73-0 10,11-Dihydro-5-hydroxy-5*H*-dibenzo[*b,f*]phosphepin 5-oxide, D-00492
30309-74-1 10,11-Dihydro-5-phenyl-5*H*-dibenzo[*b,f*]phosphepin, D-00559
30309-75-2 10,11-Dihydro-5-phenyl-5*H*-dibenzo[*b,f*]phosphepin; Oxide, *in* D-00559
30309-80-9 Bis(2-methylphenyl)phosphine; Oxide, *in* B-00315
30318-19-5 3′,4′-Dimethyl-5′-phenylspiro[1,3,2-benzoxazaphosphole-2(3*H*),2′-[1,3,2]-oxazaphospholidine], D-00823
30337-25-8 2-Phenyl-4*H*-1,3,2-benzoxathiaphosphorin; 2-Oxide, *in* P-00097
30338-49-9 3,5-Di-*tert*-butyl-2,5-dihydro-1,2-oxaphosphole 2-oxide; 2-Bromo, *in* D-00082
30342-92-8 1,3,2-Oxathiaphosphorinane 2-oxide; 2-Phenoxy, *in* O-00057
30342-93-9 1,3,2-Oxathiaphosphorinane 2-oxide; 2-Isopropoxy, *in* O-00057
30342-94-0 *O,O*-Trimethylene phenylphosphonothioate, *in* P-00116
30342-95-1 1,3,2-Oxathiaphosphorinane 2-oxide; 2-Ph, *in* O-00057
30343-12-5 6-Chloro-6*H*-dibenzo[*c,e*][1,2]-oxaphosphorin; 6-Sulfide, *in* C-00033
30345-49-4 Tetrabutylphosphonium(1+); Acetate, *in* T-00019
30354-22-4 Aminobisbiphenylenephosphorane, *in* S-00021
30382-05-9 *S,S′*-1,3-Butylene phosphorochloridodithioite, *in* M-00138
30404-86-5 Hexahydro-2,2,4,4,6,6-hexamethyl-1,3,5-triphenyl-1,3,5,2,4,6-triphosphatrigermanin, H-00029
30404-88-7 Hexaphenyl-1,3-diphospha-2,4-digermacyclobutane, H-00087
30404-89-8 Nonaphenyl-1,3,5-triphospha-2,4,6-trigermacyclohexane, N-00070
30537-09-8 Triphenyl(1-phenylethyl)-phosphonium(1+); Bromide, *in* T-00612
30540-36-4 3,4-Dimethyl-1-phenyl-1*H*-phosphole, D-00807
30541-74-3 1-Ethyl-1,2,3,4-tetrahydro-1-phenylbenzo[*h*]phosphinolinium bromide, *in* T-00128
30541-81-2 1,2,3,4-Tetrahydro-1-phenylbenzo[*h*]phosphinoline; 1-Oxide, *in* T-00128
30546-51-1 *N,N′*-Dibenzylphosphorodiamidic acid, D-00061
30560-19-1 ▷Acephate, A-00003
30594-15-1 Ethyl diethenylphosphinate, *in* D-01258
30594-16-2 Phenyl diethenylphosphinate, *in* D-01258
30594-17-3 *O*-Ethyl divinylphosphinothioate, *in* D-01259
30596-36-2 4,5-Dihydro-3-phenyl-3*H*-1,3-azaphosphole, D-00549
30597-71-8 1,2,3,4-Tetrahydro-1-phenylbenzo[*h*]phosphinoline, T-00128
30610-36-7 Pentaethylphosphonothioic diamide, *in* E-00149
30653-71-5 Dibutyl phosphonite, D-00129
30664-59-6 2,2,3,4,4-Pentamethyl-1-phenylphosphetane; *cis-form*, Sulfide, *in* P-00020
30664-60-9 2,2,3,4,4-Pentamethyl-1-phenylphosphetane; *trans-form*, Sulfide, *in* P-00020
30779-27-2 1-Fluorophosphorinane; Sulfide, *in* F-00047
30779-42-1 Dibutylphosphinothioic acid; Fluoride, *in* D-00117
30798-15-8 2,4,6-Trihydroxy-1,2,4,6-oxatriphosphorinane 2,4,6-trioxide; Tricyclohexylammonium salt, *in* T-00494
30818-70-3 Diethyl[(1,1′-biphenyl)-4-ylmethyl]phosphonate, *in* B-00080
30884-93-6 Tetrabutyl hypodiphosphite, T-00016
30903-12-9 1-Fluorophosphorinane; Oxide, *in* F-00047
30983-71-2 *O*-Isopropyl methylphosphinothioate, *in* M-00279
30993-94-3 *S,S*-Diethyl phosphoramidodithioate, D-00350
31053-08-4 2,2-Difluoro-2,2-dihydro-1,2,3-trimethyl-1,3,2-diazaphosphetidin-4-one, *in* D-00459
31066-81-6 Octylphenylphosphinic acid, O-00040
31078-90-7 Diisopropyl propylphosphonite, *in* P-00482
31078-91-8 Allyldichlorophosphine, *in* P-00463
31080-60-1 *N,N,N′,N′*-Tetramethyl-*P*-2-propenylphosphonous diamide, *in* P-00463
31080-62-3 Diisopropyl allylphosphonite, *in* P-00463

31081-90-0 Triethoxycarbonylphosphine, *in* P-00355
31082-04-9 1-Benzyl-1-phenylphospholanium bromide, *in* P-00216
31082-05-0 1-Phenyl-1-phenylmethylphosphorinanium bromide, *in* P-00265
31120-32-8 Ethyl 2-[(diethoxyphosphinyl)-amino]benzoate, *in* C-00004
31121-06-9 ▷2-Methoxy-1,3,2-dioxaphosphorinane, M-00038
31128-88-8 Tributoxycarbonylphosphine, *in* P-00355
31128-89-9 Triphenoxycarbonylphosphine, *in* P-00355
31139-62-5 *O*-[(1,1′-Biphenyl)-2-yl] phosphorodichloridothioate, B-00083
31142-23-1 Trimethyl phosphonoformate, *in* D-00672
31142-29-7 Diethyl aminocarbonylphosphonate, *in* C-00003
31149-52-7 4-Phenyl-3,5-dioxa-1-phosphabicyclo[2.2.1]-heptane; 1-Oxide, *in* P-00113
31160-09-5 *N,N′,N″*-Tricyclohexylphosphoric triamide, *in* T-00429
31160-10-8 *N,N′,N″*-Tris(4-methylphenyl)-phosphoric triamide, *in* T-00819
31160-11-9 *N,N′,N″*-Tris(2-methylphenyl)-phosphoric triamide, *in* T-00817
31160-12-0 *N,N′,N″*-Tris(3-methylphenyl)-phosphoric triamide, *in* T-00818
31188-53-1 3-Methyl-2-butenylidenetriphenyl-phosphorane, M-00098
31201-88-4 (Diphenylphosphino)acetonitrile, D-01046
31202-14-9 2-Mercapto-1,3,2-benzodioxaphosphole 2-sulfide; Triethylammonium salt, *in* M-00005
31218-83-4 Propetamphos; (*E*)-*form*, *in* P-00466
31236-96-1 2,3-Dihydro-1*H*-phosphindole 1-oxide; 1-Ph, *in* D-00570
31236-97-2 1-Phenyl-1*H*-phosphindole; 1-Oxide, *in* P-00207
31236-98-3 1-Phenyl-1*H*-phosphindole, P-00207
31238-20-7 (2-Chloroethyl)-triphenylphosphonium(1+); Bromide, *in* C-00083
31238-29-6 *OO-tert*-Butyl *O*-ethyl methylphosphonoperoxoate, *in* M-00310
31238-30-9 *OO-tert*-Butyl *O*-isopropyl methylphosphonoperoxoate, *in* M-00310
31238-31-0 *OO-tert*-Butyl *O*-isopropyl ethylphosphonoperoxoate, *in* E-00144
31238-32-1 *O*-Butyl *OO-tert*-butyl methylphosphonoperoxoate, *in* M-00310
31238-50-3 Diethyl (2-aminophenyl)-phosphonate, *in* A-00109
31239-06-2 *P,P*-Diphenyl-*N*-(diphenylphosphinyl)-phosphinimidic acid, D-01007
31246-33-0 *O*-Butyl *OO-tert*-butyl ethylphosphonoperoxoate, *in* E-00144
31335-59-8 Tetraethyl hypodiphosphite, T-00049
31355-98-3 Ethyl ethylphosphinate; (+)-*form*, *in* E-00068
31398-95-5 Phenylbis(phenylethynyl)-phosphine; Oxide, *in* P-00099
31398-96-6 Tris(phenylethynyl)phosphine; Oxide, *in* T-00846
31398-97-7 2,4,6-Triphenyl-4*H*-1,4-oxaphosphorin; 4-Oxide, *in* T-00610
31430-66-7 Isopropyl phosphorodichloridite, I-00061
31459-97-9 *OO-tert*-Butyl *O*-ethyl phenylphosphonoperoxoate, *in* P-00242
31459-98-0 Methyl chloro(triphenyl-phosphoranylidene)acetate, *in* C-00190
31460-11-4 Methyl dichlorophosphinylacetate, *in* D-00190
31460-21-6 Ethyl [(methoxycarbonyl)-methyl]phosphonamidate, *in* M-00028
31460-22-7 Phenyl [(methoxycarbonyl)-methyl]phosphonamidate, *in* M-00028
31464-99-0 *O,O*-Dimethyl methylphosphoramidothioate, *in* M-00333
31480-11-2 *N,N,N′,N′*-Tetramethyl-*P*-2-propenylphosphonothioic diamide, *in* P-00462
31481-91-1 2-Hydroxy-4,5-dimethyl-1,3,2-dioxaphospholane 2-oxide; Anilinium salt, *in* H-00121
31574-96-6 [1,2-Phenylenebis[methylene]]-bis[triphenylphosphorane], P-00135
31581-34-7 Diphenylpropylphosphine; Sulfide, *in* D-01145
31596-41-5 2,3-Dihydro-2-dimethylamino-5-methyl-3-phenyl-1,3,4,2-oxadiazaphosphole; 2-Oxide, *in* D-00455
31599-85-6 *O,O*-Dimethyl diethylphosphoramidothioate, *in* D-00356
31614-36-5 4-Phenyl-1,4-oxaphosphorinane, P-00176
31634-70-5 5,10-[1′,2′]-Benzophosphenanthrene, B-00024
31634-72-7 5,10-[1′,2′]-Benzophosphenanthrene; 5,10-Dioxide, *in* B-00024
31634-79-4 (3-Nitrophenyl)diphenylphosphine, N-00039
31638-73-0 Tris(3-nitrophenyl)phosphine, T-00838
31638-89-8 Tris(3-nitrophenyl)phosphine; Oxide, *in* T-00838
31639-12-0 Bis(2,4,6-trimethylphenyl)-phosphinic acid; Chloride, *in* B-00428
31641-57-3 ▷Dimethylphosphinous cyanide, D-00847
31641-59-5 Methylphosphonic acid; Dicyanide, *in* M-00288
31645-39-3 *N,N′*-Bis(2-chloroethyl)-phosphorodiamidic acid, B-00125
31650-50-7 *S,S*-Diphenyl methylphosphonodithioate, *in* M-00301
31650-51-8 *O,S*-Diphenyl methylphosphonothioate, *in* M-00313
31650-52-9 *S*-Phenyl *N,N*-diethyl-*P*-methylphosphonamidothioate, *in* D-00297
31650-57-4 Diethyl methylphosphonotrithioate, *in* M-00318
31650-59-6 Diphenyl methylphosphonotrithioate, *in* M-00318
31650-60-9 *O*-Phenyl *N,N*-diethyl-*P*-methylphosphonamidothioate, *in* D-00297
31650-61-0 Methylphosphonothioic bis(diethylamide), *in* M-00314
31650-79-0 Ethyl methylphosphonazidate, *in* M-00286
31651-16-8 Diethyl 1-cyclohexen-1-ylphosphonate, *in* C-00241
31651-74-8 2-Naphthyl phosphorodichloridate, *in* M-00495
31651-76-0 1-Naphthyl phosphorodichloridate, *in* M-00494
31662-55-2 (2-Cyanoethyl)-triphenylphosphonium(1+); Tetrafluoroborate, *in* C-00212
31675-43-1 [Phenyl[(trimethylsilyl)oxy]-methyl]phosphonic acid, P-00331
31675-49-7 Tris[(2-methylphenyl)methyl]-phosphine; Sulfide, *in* T-00800
31675-50-0 Tris[(3-methylphenyl)methyl]-phosphine; Sulfide, *in* T-00801
31675-51-1 Tris[(4-methylphenyl)methyl]-phosphine; Sulfide, *in* T-00802
31675-59-9 Trimethylsilyl bis(chloromethyl)phosphinate, *in* B-00127
31683-58-6 Phenyl methylphosphonazidate, *in* M-00286
31700-79-5 2-Ethoxy-3-phenyl-1,3,2-oxazaphospholidine; 2-Sulfide, *in* E-00048
31707-09-2 2-Diethylamino-3-phenyl-1,3,2-oxazaphospholidine, D-00269
31755-40-5 2-Chloro-1,3-dimethyl-1,3,2-diazaphospholidine; 2-Sulfide, *in* C-00055
31860-10-3 2-Methylphenyl phosphorodichloridite, M-00258
31860-18-0 *O,O*-Diphenyl diethylphosphoramidothioate, *in* D-00356
31860-23-8 *O,O*-1,3-Butylene methylphosphonothioate, *in* D-00734
31951-84-5 *O,O*-Neopentylene *O*-phenyl phosphorothioate, *in* D-00790
31951-89-0 *O,O*-1,3-Butylene *O*-phenyl thiophosphate, *in* M-00191
31951-90-3 2-Chloro-4-methyl-1,3,2-dioxaphosphorinane 2-oxide, C-00094
32018-63-6 2-Dimethylamino-3,4-dimethyl-5-phenyl-1,3,2-oxazaphospholidine, D-00694
32045-17-3 Dimethyl *tert*-butylphosphonite, *in* B-00616
32084-79-0 Tetramethylethylene phenylphosphonate, *in* P-00322
32101-77-2 4,5-Dihydro-2-phenyl-1,3,2-benzoxathiaphosphepin; 2-Oxide, *in* D-00555
32102-01-5 2-Phenoxy-1,3,2-oxaselenaphosphorinane 2-oxide, P-00071
32119-41-8 *O*-Benzoyl *O,O*-diethylphosphorothioate, B-00029
32186-89-3 Chlorobis(2-chlorophenyl)-phosphine, *in* B-00140
32186-90-6 Dibromo-1-naphthylphosphine, *in* N-00025
32186-92-8 6-Chloro-6*H*-dibenzo[*c,e*][1,2]-oxaphosphorin; 6-Oxide, *in* C-00033
32187-40-9 Diphenyl (1-phenylethenyl)-phosphonate, *in* P-00154
32249-59-5 Tetramethyl (hydroxyphenylmethyl)-bisphosphonate, *in* H-00172
32271-99-1 Ethyl *P*-(1,3-butadienyl)-*N,N*-diethylphosphonamidate, *in* B-00508
32287-51-7 2-Methyl-3-phenyl-1,3,2-oxazaphospholidine, M-00212
32287-54-0 2-Methyl-3-phenyl-1,3,2-oxazaphospholidine; 2-Oxide, *in* M-00212
32288-17-8 Diethyl dimethyldiphosphonate, *in* D-00751
32294-62-5 1,3-Dimethyl-1,3,2-diazaphospholidine; 2-Me, *in* D-00719
32305-98-9 4,5-Bis[(diphenylphosphino)-methyl]-2,2-dimethyl-1,3-dioxolane; (4*R*,5*R*)-*form*, *in* B-00251
32306-56-2 *P*-(1,1-Dimethylethyl)-*P*-methylphosphinic amide, *in* B-00567

32309-30-1 3,6-Dihydro-2*H*-1,2-thiaphosphorin 2-sulfide; 2-Ph, *in* D-00586
32338-34-4 Diisopropylphosphinodithioic acid, D-00643
32339-43-8 Decyltriphenylphosphonium(1+); Bromide, *in* D-00014
32341-34-7 3-(Diphenylphosphino)phenol, D-01071
32357-32-7 Bis[3-(dimethylamino)propyl]-phenylphosphine, B-00195
32364-85-5 *P,P*-Bis(1-aziridinyl)-*N*-ethylphosphinothioic amide, *in* B-00094
32394-39-1 Chlorotetraphenoxyphosphorane, C-00186
32395-29-2 Benzylphenylphosphinic acid; Chloride, *in* B-00056
32400-38-7 *P*-Cyclohexyl-*N,N,N′,N′*-tetramethylphosphonic diamide, *in* C-00257
32405-87-1 Dimethyl cyclohexylphosphoramidate, *in* C-00269
32405-88-2 Diethyl cyclohexylphosphoramidate, *in* C-00269
32429-65-5 *S*-Ethyl bis(1-aziridinyl)-phosphinothioate, *in* B-00093
32435-35-1 Dibutylphosphinodithioic acid, D-00112
32450-46-7 2-Phosphonooxypropanoic acid; (±)-*form*, *in* P-00392
32488-63-4 Tri-3-furanylphosphine; Oxide, *in* T-00487
32488-65-6 Tri-3-furanylphosphine; Selenide, *in* T-00487
32488-66-7 Methyl(tri-3-furanyl)phosphonium iodide, *in* T-00487
32503-56-3 2,3-Dihydro-3,3,5-trimethyl-1,2-oxaphosphole 2-oxide; 2-Me, *in* D-00596
32503-86-9 2-Methoxy-4,4,5,5-tetramethyl-1,2-azaphospholidine 2-oxide, *in* T-00192
32503-87-0 2-Methoxy-3,3,4,4-tetramethyl-1,2-azaphospholidine 2-oxide, *in* T-00191
32511-22-1 (5-Methyl-2-isopropylcyclohexyl)diphenyl-phosphine, M-00164
32511-23-2 5-*tert*-Butyl-2-methyl-1,3,2-dioxaphosphorinane; *cis-form*, *in* B-00564
32511-62-9 5-*tert*-Butyl-2-methyl-1,3,2-dioxaphosphorinane; *trans-form*, *in* B-00564
32541-80-3 (2-Acetoxyethyl)phosphonic acid, A-00006
32545-75-8 (4-Morpholinylmethylene)-bisphosphonic acid, M-00520
32564-13-9 2-Methyl-1,3,2-dioxaphospholane, M-00126
32565-98-3 Ethyl 2-(diethoxyphosphinyl)-oxybenzoate, *in* P-00391
32579-62-7 2-*tert*-Butoxy-4-methyl-1,3,2-dioxaphosphorinane, B-00537
32589-80-3 Tetramethylphosphonium(1+), T-00215
32596-67-1 Dibutyl dibutylphosphoramidite, *in* D-00136
32596-69-3 Pentapropylphosphonous diamide, *in* P-00482
32596-70-6 Pentabutylphosphonous diamide, *in* B-00615
32597-23-2 Dipropylphosphoramidous acid; Dichloride, *in* D-01215
32597-92-5 [(3-Chlorophenyl)methyl]-triphenylphosphonium(1+); Chloride, *in* C-00136
32627-64-8 Cyanoneocobalamin, *in* V-00009
32673-25-9 Di-*tert*-butylphenylphosphine, D-00103
32707-12-3 2,2-Dihydro-1,3-dimethyl-1,3,2-diazaphosphetidin-4-one; 2,2,2-Trifluoro, *in* D-00459
32707-15-6 2,2-Difluoro-2,2-dihydro-1,3-dimethyl-2-phenyl-1,3,2-diazaphosphetidin-4-one, *in* D-00459
32743-80-9 *O,O*-Dimethyl 4-morpholinylphosphonothioate, *in* M-00527
32755-13-8 Diethyl dibromophosphoramidate, D-00279
32755-55-8 Ethyl (chlorosulfonyl)-phosphorodichlorimidate, E-00056
32847-69-1 2-Chloro-1,3,2-dioxaphospholane 2-sulfide, C-00065
32847-70-4 2-Chloro-3-phenyl-1,3,2-oxazaphospholidine; 2-Sulfide, *in* C-00138
32881-50-8 1*H*-1,3-Benzazaphosphole, B-00005
32914-67-3 (Ferrocenylmethyl)-triphenylphosphonium(1+); Iodide, *in* F-00008
32966-78-2 1,4-Dioxa-5-phospha(5-P^{V})-spiro[4.4]non-7-ene; 5-Me, *in* D-00967
32979-55-8 *S,S*-Dimethyl phosphoramidodithioate, D-00866
32997-75-4 *O,O*-Diisopropyl phosphorodithioate; Anilinium salt, *in* D-00663
33063-22-8 Tetrahydro-4-methyl-2*H*-1,3,4,2-oxadiazaphosphorine 2-oxide; 2-Cyclohexylamino, *in* T-00116
33078-07-8 α-(Triphenylphosphoranylidene)-benzeneacetaldehyde, T-00643
33078-24-9 Methylphosphonic acid; Diazide, *in* M-00288
33078-25-0 Methylphosphonisocyanatidic acid; Azide, *in* M-00292
33078-26-1 *N,N*-Diethyl-*P*-methylphosphonazidic amide, *in* M-00287
33148-57-1 *O*-Methyl *O,O*-trimethylene phosphorothioate, *in* M-00038
33148-58-2 *O*-Ethyl *O,O*-trimethylene thiophosphate, *in* E-00037
33185-66-9 Methylphosphonazidic acid; Chloride, *in* M-00286
33202-91-4 Diethyl (phenylamino)-carbonylphosphoramidate, *in* P-00081
33300-85-5 5-Chloro-5*H*-dibenzophosphole, C-00034
33312-92-4 2,2,3,3,8,8-Hexamethyl-1,4,6-trioxa-9-aza-5-phospha(5P^{V})-spiro[4.4]nonane, H-00084
33313-49-4 2-[(Dimethoxyphosphinyl)oxy]-propanoic acid, D-00682
33327-82-1 4-Methyl-2-phenyl-1,3,2-dioxaphospholane, M-00198
33329-02-1 Ethyl diphenylphosphinodithioate, *in* D-01065
33345-04-9 Ethyl *P,P*-diethyl-*N*-phenylphosphinimidate, *in* D-00306
33374-48-0 Dibutylmethylphosphine, D-00098
33399-00-7 ▷Bromfenvinfos, B-00462
33439-83-7 (1-Aminopentyl)phosphonic acid; (±)-*form*, *in* A-00096
33446-89-8 1-Naphthylphosphonothioic acid; Dichloride, *in* N-00023
33452-31-2 2,2,2,3-Tetrahydro-2,2,2-trimethoxy-3-methyl-4,5-diphenyl-1,3,2-oxazaphosphole, T-00156
33493-42-4 Dihexylphosphinic acid; Chloride, *in* D-00434
33493-44-6 Didecylphosphinic acid; Chloride, *in* D-00221
33494-77-8 Mono(1-methylpropyl) phosphate; (±)-*form*, Monoanilinium salt, *in* M-00492
33494-81-4 Di-*tert*-butyl phosphate, D-00105
33512-26-4 Diethyl [(1,3-dihydro-1,3-dioxo-2*H*-isoindol-2-yl)-methyl]phosphonate, *in* P-00426
33530-51-7 1,2,2,3,4,4-Hexamethylphosphetane; *cis-form*, Oxide, *in* H-00074
33530-51-7 1,2,2,3,4,4-Hexamethylphosphetane; *trans-form*, Oxide, *in* H-00074
33534-83-7 Dimethyl aminocarbonylphosphonate, *in* C-00003
33536-07-1 2,2-Dihydro-3,4-bis(methoxycarbonyl)-2,2-dimethyl-5,5-diphenyl-5*H*-1,2-azaphosphole, D-00445
33536-50-4 Diethyl(3-methyl-1-butenyl)-phosphonate, *in* M-00101
33550-42-4 2,2,2,3-Tetrahydro-2,2,2-trimethoxy-5-phenyl-3,3-bis(trifluoromethyl)-1,4,2-oxazaphosphole, T-00158
33554-05-1 Methyl trimethylene phosphate, *in* M-00038
33558-14-4 (Difluoromethylene)-triphenylphosphorane, D-00414
33625-82-0 *S*-Methyl phenylphosphonothioate; (*S*)-*form*, *O*-Menthyl ester, *in* M-00244
33626-88-9 *S*-Ethyl ethylphenylphosphinothioate, *in* E-00105
33642-98-7 Methylphenylphosphinic acid; (*S*)-*form*, Et ester, *in* M-00219
33670-32-5 (Methoxymethyl)-triphenylphosphonium(1+); Bromide, *in* M-00053
33672-91-2 2-Fluoro-1,3-dimethyl-1,3,2-diazaphospholidine, F-00014
33672-92-3 2-Fluoro-1,3,2-dithiaphospholane, *in* D-01243
33706-30-8 2,3-Dihydro-5-methyl-3-phenyl-2-phenylamino-1,3,4,2-oxadiazaphosphole; 2-Oxide, *in* D-00527
33706-78-4 4′,4′,5′,5′-Tetramethylspiro[1,3,2-benzodioxaphosphole-2,2′-[1,3,2]dioxaphospholane], T-00230
33711-90-9 2-Chloro-2,3-dihydro-5-methyl-3-phenyl-1,3,4,2-oxadiazaphosphole; 2-Oxide, *in* C-00044
33711-91-0 2-Chloro-2,3-dihydro-3,5-diphenyl-1,3,4,2-oxadiazaphosphole; 2-Oxide, *in* C-00042
33712-82-2 Hexapropylphosphorothioic triamide, *in* H-00092
33730-82-4 2-Phenyl-1,2-azaphospholidine, P-00089
33733-31-2 ▷(3-Hydroxyphenyl)phosphonic acid, H-00180
33733-60-7 Dimethyl(3-methylphenyl)-phosphine, D-00760
33771-51-6 5-Chloro-5*H*-dibenzophosphole; 5-Oxide, *in* C-00034
33771-53-8 5-Methyl-5*H*-dibenzophosphole; 5-Sulfide, *in* M-00115
33771-54-9 5-Phenyl-5*H*-dibenzophosphole; 5-Sulfide, *in* P-00110
33771-55-0 5-Chloro-5*H*-dibenzophosphole; 5-Sulfide, *in* C-00034
33795-18-5 ▷(4-Hydroxyphenyl)phosphonic acid, H-00181
33795-32-3 1,2-Diphenyl-1,2-azaphospholidine, *in* P-00089
33801-17-1 2-Naphthyl benzylphosphonochloridothioate, *in* B-00064
33801-18-2 1-Naphthyl benzylphosphonochloridothioate, *in* B-00064
33802-53-8 Dimethyl phenylethynylphosphonate, *in* P-00166
33803-58-6 1-(Triphenylphosphoranylidene)-2-heptanone, *in* O-00062
33829-64-0 Tri-3-furanylphosphine, T-00487
33834-35-4 4-(Diphenylphosphino)aniline, D-01049
33834-36-5 4-(Diphenylphosphinyl)aniline, *in* D-01049
33834-37-6 4-(Diphenylphosphinothioyl)-aniline, *in* D-01049

33835-25-5 2-*tert*-Butoxy-4,5-dimethyl-1,3,2-dioxaphospholane, B-00533
33845-96-4 Homocubyltrimethylphosphorane, *in* D-00569
33857-23-7 Amiprophos *M*, A-00131
33862-44-1 Diphenyl phosphorohydrazidate, D-01136
33876-53-8 Isopropylphosphoramidic acid, I-00058
33876-55-0 1-Pyrrolidinylphosphonic acid, P-00518
33876-58-3 Isopropylphosphoramidic acid; Dichloride, *in* I-00058
33876-59-4 Dibutylphosphoramidic acid; Dichloride, *in* D-00134
33876-85-6 Bis(trimethylsilyl) benzoylphosphonate, *in* B-00034
33892-95-4 2-Methoxy-4-methyl-1,3,2-dioxaphosphorinane, M-00044
33895-27-1 3-(Triphenylphosphoniomethyl)-phenanthrene(1+); Bromide, *in* T-00637
33895-28-2 9-(Triphenylphosphoniomethyl)-phenanthrene(1+); Bromide, *in* T-00638
33909-73-8 Tri-1-propynylphosphine, T-00721
33921-08-3 Tetraethyl [oxybis(methylene)]-bisphosphonate, *in* O-00095
33945-42-5 Methyl acetylmethylphosphinate, *in* A-00018
33965-78-5 2-Methoxyphenyl phosphorodichloridate, *in* M-00480
33985-75-0 Diphenyl benzylphosphoramidate, *in* B-00069
33992-85-7 Diethyl *N*,*N*-diethyl-*N'*-phenylphosphoramidimidate, *in* D-00313
33996-01-9 2-Methoxy-4-methyl-1,3,2-dioxaphosphorinane 2-selenide; (2*RS*,4*RS*)-*form*, *in* M-00046
33996-02-0 2-Methoxy-4-methyl-1,3,2-dioxaphosphorinane 2-selenide; (2*RS*,4*SR*)-*form*, *in* M-00046
33996-04-2 2-Methoxy-4-methyl-1,3,2-dioxaphosphorinane 2-oxide; (2*RS*,4*SR*)-*form*, *in* M-00045
34005-83-9 Bis(trifluoromethyl)phosphinic acid; Fluoride, *in* B-00411
34043-22-6 Bis(trifluoromethyl)phosphinic acid; Anhydride, *in* B-00411
34044-22-9 Diethyl diphenylphosphoramidate, *in* D-01108
34109-34-7 *OO*-*tert*-Butyl *O*-ethyl ethylphosphonoperoxoate, *in* E-00144
34159-46-1 (Dimethoxyphosphinyl)acetic acid, D-00678
34159-52-9 (Diphenoxyphosphinyl)acetic acid, D-00988
34163-96-7 1,2-Propadienylphosphonic acid, P-00441
34170-54-2 Phenyl *P*-methyl-N^2-phenylphosphonohydrazidate, *in* M-00241
34170-57-5 *O*-Phenyl *P*-methyl-N^2-phenyl-phosphonohydrazidothioate, *in* M-00242
34193-25-4 (2-Methyl-1-propenyl)-diphenylphosphine, M-00362
34208-79-2 *O*-Methyl *tert*-butylphosphonothioate; (*S*)-*form*, *in* M-00107
34239-27-5 *O*-Methyl *tert*-butylphosphonothioate; (*R*)-*form*, *in* M-00107
34244-04-7 Nonamethylimidodiphosphorous tetraamide, *in* M-00157
34244-05-8 Tetramethyl methylimidodiphosphite, *in* M-00157
34244-07-0 Tetraethyl methylimidodiphosphite, *in* M-00157
34244-09-2 Bis(dimethyl-aminophosphinothioyl)-pentamethylthiophosphoric triamide, B-00191
34244-10-5 *O*,*O*,*O*,*O*-Tetramethyl methylthioimidodiphosphate, *in* M-00161
34250-29-8 2,4,5-Triphenyl-1,3,2-dioxaphosphole; 2-Sulfide, *in* T-00602
34255-82-8 *O*-Ethyl methylphosphonothioic monoanhydride, *in* D-00925
34255-83-9 *O*-Phenyl ethylphosphonothioic monoanhydride, *in* D-00925
34255-89-5 *O*-Ethyl *P*-methylphosphonamidothioate, *in* M-00284
34257-63-1 1-Iodo-*N*-methyl-*N*,1,1,1-tetraphenylphosphoranamine, *in* M-00194
34260-92-9 *O*-Methyl *tert*-butylphosphonothioate; (±)-*form*, *in* M-00107
34283-79-9 10,10'(5*H*,5'*H*)-Spirobiphenophosphazinium(1+); Chloride, *in* S-00023
34284-47-4 *O*,*O*-Neopentylene phenylphosphonothioate, *in* D-00795
34295-11-9 1-Diphenylphosphinyl-2-methylpropene, *in* M-00362
34303-18-9 (4-Chlorophenyl)-diphenylphosphine; Oxide, *in* C-00129
34303-19-0 Ethylene phosphorochloridotrithioate, *in* C-00071
34306-15-5 2-(4-Bromophenyl)-8,8-dihydro-4,6,8,8-tetraphenyl[1,2]-oxaphospholo[2,3-*b*][1,2,5]-dioxaphosphorin, B-00479
34309-87-0 Diethyl phenylphosphonotrithioate, *in* P-00250
34378-77-3 Candiolin, *in* F-00067
34381-74-3 *O*,*O*-Bis(2-methylphenyl) phosphorodithioate, B-00337
34384-99-1 Hexahydro-2-hydroxycyclopenta[*d*]-1,3,2-dioxaphosphorin 2-oxide; (4*aRS*,7*aSR*)-*form*, Chloride, *in* H-00034
34387-64-9 Triphenyl(2-phenylethynyl)-phosphonium(1+); Bromide, *in* T-00614
34392-13-7 10,10'(5*H*,5'*H*)-Spirobiphenophosphazinium(1+), S-00023
34421-23-3 *O*-Phenyl *P*-methyl-phosphonohydrazidothioate, *in* M-00309
34422-49-6 1-Methoxy-3,4-dimethyl-1*H*-phosphole 1-oxide, *in* H-00130
34442-16-5 Spiro[1,3,2-dioxaphospholane-2,1'-[2,8,9]trioxa[1]-phosphatricyclo[3.3.1.$1^{3,7}$]-decane], S-00024
34478-62-1 1,2-Diphenyldiphosphine, D-01012
34491-76-4 Diethyl (1-cyanopropyl)-phosphonate, *in* C-00229
34492-20-1 Heptylphosphoramidic acid; Dichloride, *in* H-00009
34492-32-5 ▷Bis(2-chloroethyl)-phosphoramidothioic dichloride, B-00123
34549-99-0 (2-Cyanoethyl)phosphonic acid, C-00211
34573-97-2 4,8-Di-*tert*-butyl-2,10-dimethyl-12*H*-dibenzo[*d*,*g*][1,3,2]-dioxaphosphocin; 6-Ethoxy, *in* D-00088
34585-19-8 Diphenyl 2-thienylphosphonite, *in* T-00308
34585-91-6 *O*,*O*-Diphenyl phosphorodiselenoate, D-01130
34585-92-7 *O*,*O*-Dimethyl phosphorodiselenoate; K salt, *in* D-00897
34585-95-0 *O*,*O*-Diisopropyl phosphorodiselenoate; K salt, *in* D-00662
34585-97-2 *O*,*O*-Diphenyl phosphorodiselenoate; K salt, *in* D-01130
34595-07-8 Diethyl (2-cyanophenyl)-phosphonate, *in* C-00223
34595-52-3 Triethylamine-2,2',2''-triphosphonic acid, T-00438
34605-98-6 Methyl *N*,*N*-diethyl-*P*-methylphosphonamidate, *in* D-00296
34608-90-7 Diethyl phosphate; Dicyclohexylammonium salt, *in* D-00320
34637-96-2 Dimethyl diethyldiphosphonate, *in* D-00290
34638-79-4 Ethyl methylphenylphosphinate, *in* M-00219
34643-46-4 ▷Prothiofos, P-00498
34647-06-8 Methylphenylphosphinic acid; (*S*)-*form*, Me ester, *in* M-00219
34647-07-9 Methyl methylphenylphosphinate, *in* M-00219
34664-65-8 [2-(Dimethylarsino)phenyl]-dimethylphosphine, D-00711
34666-29-0 Tetraethyl (1-oxo-1,3-propanediyl)bisphosphonate, *in* O-00071
34666-63-2 *S*-Ethyl *O*-isopropyl methylphosphonothioate, *in* I-00040
34666-66-5 Methylphenylpropylphosphine; (±)-*form*, Sulfide, *in* M-00265
34667-31-7 1,2,3,4-Tetraphenyl-5-phosphoniaspiro[4.5]-deca-1,3-diene; Bromide, *in* T-00270
34667-32-8 1,2,3,4,9,10,11,12-Octaphenyl-5,8-diphosphoniadispiro[4.2.4.2]-deca-1,3,9,11-tetraene(2+); Dibromide, *in* O-00037
34669-04-0 Dimethylphosphinodithioic acid; Na salt, *in* D-00834
34670-52-5 Tetraphenyl 1,2-ethanediylbisphosphoramidate, *in* E-00009
34670-59-2 *O*,*O*,*O*,*O*-Tetraphenyl 1,2-ethanediylbisphosphoramidothioate, *in* E-00010
34670-60-5 1,3-Propanediphosphoramidothioic acid, P-00446
34670-60-5 *O*,*O*,*O'*,*O'*-Tetraphenyl 1,3-propanediphosphoramidothioate, *in* P-00446
34670-66-1 Tetraphenyl 1,4-phenylenebisphosphoramidate, *in* B-00012
34674-30-1 1-Hydroxy-2,3,4,5-tetraphenyl-1*H*-phosphole 1-oxide, H-00200
34679-31-7 Dibenzylphosphinic amide, *in* D-00055
34679-43-1 Ethyl dibenzylphosphinate, *in* D-00055
34714-88-0 Tetraphenyl 1,3-phenylenebisphosphoramidate, *in* B-00011
34734-82-2 *O*,*O*-Bis(trimethylsilyl) phosphonothioate, B-00437
34735-02-9 1,4-Dihydro-2,4,6-triphenyl-1,4-azaphosphorine; 4-Oxide, *in* D-00604
34735-03-0 2,6-Di-*tert*-butyl-1,4-dihydro-4-phenyl-1,4-azaphosphorine; 4-Oxide, *in* D-00084
34736-69-1 2,2-Dihydro-2,2,2-triphenyl-1,3,2-dioxaphospholane, D-00611
34736-77-1 3,3,9,9-Tetramethyl-1,5,7,11-tetraoxa-6-phospha(6-P^V)spiro[5.5]undecane; 6-Ethoxy, *in* T-00236
34736-78-2 3,3,9,9-Tetramethyl-1,5,7,11-tetraoxa-6-phospha(6-P^V)spiro[5.5]undecane; 6-Ph, *in* T-00236

34764-11-9 Bis(*O,O*-diphenyl phosphoroselenoyl) diselenide, B-00269

34813-40-6 Glycerol 1,2-di-9,12,15-octadecatrienoate 3-phosphoethanolamine; (*R*)-(*all-Z*)-*form*, *in* G-00041

34819-50-6 *O,O*-Di-2-propenyl phosphorochloridothioate, D-01196

34819-51-7 *O,O*-Bis(2-chloroethyl) phosphorochloridothioate, B-00124

34824-23-2 1,8-Bis(triphenyl-phosphoniomethyl)-anthracene(2+); Dibromide, *in* B-00443

34825-99-5 2-(Diphenylphosphino)-benzonitrile, D-01056

34833-61-9 Divinylphosphinic acid; Chloride, *in* D-01258

34833-62-0 Diethynylphosphinic acid, D-00409

34834-03-2 Nonamethyl imidodiphosphoramide, N-00068

34868-25-2 Benzylmethylphenylphosphine; (*S*)-*form*, *in* B-00051

34875-40-6 4-Methyl-2-phenylamino-1,3,2-dioxaphosphorinane, M-00193

34875-42-8 1-Phenylamino-1,3,2-dioxaphospholane, P-00083

34882-98-9 4,4,6-Trimethyl-1,3,2-dioxaphosphorinane; (2*RS*,4*SR*)-*form*, 2-Oxide, *in* T-00513

34882-99-0 4,4,6-Trimethyl-1,3,2-dioxaphosphorinane; (2*RS*,4*RS*)-*form*, 2-Oxide, *in* T-00513

34884-33-8 Triphenyl[(phenylthio)-methylene]phosphorane, T-00620

34906-44-0 [(3-Nitrophenyl)methyl]-triphenylphosphonium(1+); Chloride, *in* N-00048

34909-17-6 (3-Nitrophenyl)phosphonic acid; Dichloride, *in* N-00056

34909-18-7 (4-Nitrophenyl)phosphonic acid; Dichloride, *in* N-00057

34979-30-1 Dipropylphosphinic acid; Anhydride, *in* D-01205

34979-31-2 Dibutylphosphinic acid; Anhydride, *in* D-00109

34982-17-7 2,3-Dihydro-2,2-diphenyl-1*H*-isophosphindolium bromide, *in* D-00562

34989-06-5 Bis[3-(diphenylphosphino)-propyl]phenylphosphine, B-00266

35029-38-0 *S,S,S*-Tris(2-methylphenyl) phosphorotrithioate, T-00815

35035-62-2 (2-Chlorophenyl)-diphenylphosphine, C-00127

35060-65-2 *O,O*-Diethyl diisopropylphosphoramidoseleno-ate, *in* D-00655

35099-14-0 Diphenyl(3-trifluoromethylphenyl)phosphine, D-01155

35107-68-7 2,4-Dichloro-1,3-di-*tert*-butyl-1,3,2,4-diazadiphosphetidine; *cis-form*, *in* D-00163

35125-65-6 Diethyl (3-bromophenyl)-phosphonate, *in* B-00491

35144-64-0 Aldophosphamide, A-00050

35160-95-3 (2-Naphthylmethyl)-triphenylphosphonium(1+); Bromide, *in* N-00016

35160-98-6 2-(Triphenylphosphoniomethyl)-benzo[*c*]phenanthrene(1+); Bromide, *in* T-00636

35169-09-6 2,3-Dihydro-5-methyl-3-phenyl-1,3,4,2-thiadiazaphosphole 2-sulfide; 2-Chloro, *in* D-00529

35169-11-0 2,3-Dihydro-3,5-dimethyl-1,3,4,2-thiadiazaphosphole 2-sulfide; 2-Chloro, *in* D-00471

35169-12-1 2,3-Dihydro-*N,N*,5-trimethyl-3-phenyl-1,3,4,2-thiadiazaphosphole-2(3*H*)-amine-2-sulfide, *in* A-00072

35169-13-2 *N,N*-Diethyl-2,3-dihydro-5-methyl-3-phenyl-1,3,4,2-thiadiazaphosphole-2(3*H*)-amine 2-sulfide, *in* A-00072

35169-16-5 2,3-Dihydro-5-methyl-*N*,3-diphenyl-1,3,4,2-thiadiazaphosphole-2(3*H*)amine 2-sulfide, *in* A-00072

35169-18-5 2,3-Dihydro-3,5-dimethyl-1,3,4,2-thiadiazaphosphole 2-sulfide; 2-Ethoxy, *in* D-00471

35169-20-7 2,3-Dihydro-5-methyl-3-phenyl-1,3,4,2-thiadiazaphosphole 2-sulfide; 2-Ethoxy, *in* D-00529

35171-55-2 Pentyltriphenylphosphonium(1+); Iodide, *in* P-00056

35172-21-5 3-Chloro-1,3-thiaphosphetane; 3-Sulfide, *in* C-00188

35193-63-6 4-Hydroxydinaphtho[2,1-*d*:1′,2′-*f*][1,3,2]-dioxaphosphepin 4-oxide, H-00131

35193-64-7 4-Hydroxydinaphtho[2,1-*d*:1′,2′-*f*][1,3,2]-dioxaphosphepin 4-oxide; (*S*)-*form*, *in* H-00131

35193-65-8 4-Hydroxydinaphtho[2,1-*d*:1′,2′-*f*][1,3,2]-dioxaphosphepin 4-oxide; (*S*)-*form*, 2-Methoxy (Me ester), *in* H-00131

35193-66-9 4-Hydroxydinaphtho[2,1-*d*:1′,2′-*f*][1,3,2]-dioxaphosphepin 4-oxide; (±)-*form*, 2-Methoxy (Me ester), *in* H-00131

35196-92-0 1,2-Dimethyl-1,2-diphenyldiphosphine; (*RS,SR*)-*form*, *in* D-00744

35196-93-1 1,2-Dimethyl-1,2-diphenyldiphosphine; (*RS,RS*)-*form*, *in* D-00744

35210-25-4 1-Hydroxy-2,2,3,4,4-pentamethylphosphetane 1-oxide, H-00168

35210-27-6 Bis(1-methylpropyl)phosphinic acid, B-00345

35227-84-0 2-Hydroxydibenzo[*d,f*][1,3,2]-dioxaphosphepin 2-oxide, H-00116

35236-60-3 (Cyanomethyl)phosphonic acid, C-00215

35282-33-8 JAV 852, *in* B-00001

35298-44-3 3-Methyl-5-(triphenylphosphoranylidene)-2(5*H*)-furanone, *in* M-00186

35314-67-1 (3-Chlorophenyl)phosphoramidic acid; Bis(cyclohexylammonium) salt, *in* C-00156

35314-68-2 (4-Bromophenyl)phosphoramidic acid; Bis(cyclohexylammonium) salt, *in* B-00496

35322-07-7 Fosazepam, F-00057

35329-22-7 *O,O*-Dibutyl phosphorothioate; NH_4 salt, *in* D-00155

35329-37-4 *O,S*-Dipropyl phosphorodithioate; NH_4 salt, *in* D-01226

35329-38-5 *S,S*-Dipropyl phosphorodithioate; NH_4 salt, *in* D-01227

35376-07-9 *O,S*-Diethyl phosphorodithioate; Na salt, *in* D-00385

35387-44-1 1,6-Bis(diphenylphosphinyl)-hexane, *in* B-00243

35400-43-2 ▷Sulprofos, S-00027

35412-81-1 ▷*P,P′*-Diethyl-*N,N,N′,N′*-tetramethyldiphosphonic diamide, *in* D-00291

35418-57-6 Glycerol 1-octadecanoate 2-9,12,15-octadecatrienoate 3-phosphocholine; (*R*)-L-(*all-Z*)-*form*, *in* G-00080

35418-58-7 Glycerol 1-hexadecanoate 2-5,8,11,14-icosatetraenoate 3-phosphocholine; (*R*)-L-(*all-Z*)-*form*, *in* G-00055

35418-59-8 Glycerol 1-octadecanoate 2-5,8,11,14-icosatetraenoate 3-phosphocholine; (*R*)-(*all-Z*)-*form*, *in* G-00076

35437-28-6 2-Methyl-1,3,2-dithiophospholane, M-00139

35437-30-0 2-Methoxy-1,3,2-dithiaphospholane, *in* D-01243

35437-31-1 2-(Methylthio)-1,3,2-dithiaphospholane, *in* D-01243

35437-33-3 *S,S*-Ethylene phosphorochloridodithioate, *in* C-00071

35437-36-6 2-Methoxy-1,3,2-dithiaphospholane 2-oxide, *in* D-01244

35456-77-0 5-Ethyl-6,6*a*,7,8,9,9*a*-hexahydro-5-phenyl-1*H*-benzo[*h*]cyclopenta[*c*]-phosphinolinium perchlorate, *in* H-00043

35469-52-4 (4-Pyridinylmethyl)phosphonic acid; Di-Et ester, *N*-oxide, *in* P-00507

35469-53-5 (2-Pyridinylmethyl)phosphonic acid; *N*-Oxide, *in* P-00506

35497-01-9 [2-(4-Nitrophenyl)-2-oxoethyl]-triphenylphosphonium(1+), N-00051

35497-19-9 5-Oxo-1,1,3,3-tetraphenyl-1,3-diphosphorinanium(2+); Dibromide, *in* O-00081

35497-20-2 5-Oxo-1,1,3,3-tetraphenyl-1,3-diphosphorinanium(2+); Dichloride, *in* O-00081

35525-41-8 Pentamethylphosphonoselenoic diamide, P-00024

35525-42-9 *P,P*-Diethyl-*N,N*-dimethylphosphinoselenoic amide, *in* D-00328

35529-13-4 *O,O,O′,O′*-Tetraethyl hydrazodiphosphonothioate, T-00048

35542-86-0 *N,N′,N″*-Tri-2-propenylphosphorous triamide; Oxide, *in* T-00709

35556-64-0 Methylphenylphosphinothioic acid; (*S*)-*form*, *in* M-00225

35556-70-8 Phosphoenolpyruvic acid; Tris(cyclohexylammonium) salt, *in* P-00366

35563-52-1 1-(Tributylphosphoranylidene)-2-heptanone, T-00381

35575-96-3 Azemethephos, A-00146

35597-43-4 ▷Antibiotic SF 1293, A-00141

35597-44-5 Phosphinothricin; (*S*)-*form*, *in* P-00359

35621-97-7 1,2,2,3,4,4-Hexamethylphosphetane; *trans-form*, *in* H-00074

35622-00-5 1,2,2,3,4,4-Hexamethylphosphetane; *cis-form*, *in* H-00074

35622-27-6 (4-Aminobutyl)phosphonic acid, A-00057

35622-32-3 Di-2-propenylphosphinic acid, D-01193

35623-66-6 1-Chloro-2,2,3,4,4-pentamethylphosphetane; *trans-form*, Sulfide, *in* C-00124

35638-72-3 7-Phenyl-6-aza-7-phosphabicyclo[3.2.1]-octane; 6-Cyclohexyl, 7-sulfide, *in* P-00086

35661-65-5 3-Methyl-2-phenyl-1,3-diaza-2-phosphabicyclo[3.3.1]-nonane, M-00196

35661-66-6 9-Phenyl-1,5-diaza-9-phosphabicyclo[3.3.1]-nonane, P-00105
35696-22-1 Tribenzoylphosphine, T-00335
35722-10-2 Spiro[1,3,2-benzodioxaphosphole-2,2′(3′H)-[1,3,2]-benzoxazaphosphole], S-00008
35726-77-3 2-Ethoxytetrahydro-2*H*-1,3,2-oxazaphosphorine, E-00051
35752-46-6 Ethyl dihydroxyphosphinylacetate, *in* P-00376
35776-11-5 2-Diethylamino-3-phenyl-1,3,2-oxazaphospholidine; 2-Sulfide, *in* D-00269
35787-74-7 ▷Bis(2-methylphenyl) phosphate, B-00312
35812-40-9 *O,O*-Dimethyl (2-methoxyethyl)-phosphoramidothioate, *in* M-00041
35812-41-0 *O,O*-Diethyl (2-methoxyethyl)-phosphoramidothioate, *in* M-00041
35814-49-4 Bis(trifluoromethyl)-phosphinothioic acid, B-00415
35814-50-7 Bis(trifluoromethyl)-phosphinothioic acid; Anhydride, *in* B-00415
35820-68-9 9-Phenyl-1,5-diaza-9-phosphatricyclo[3.3.1.1^{3,7}]decane, P-00106
35823-41-7 Methylenebis[triphenyl-phosphonium](2+), M-00147
35823-43-9 1,3-Propanediylbis[triphenyl-phosphonium](2+), P-00448
35852-07-4 Ethyl *N*-(diethoxyphosphinyl)-carbamate, *in* D-00249
35854-55-8 7-Isopropylidene-4,8,8-trimethyl-1,6-dioxa-4-aza-5-phospha(5-P^V)-spiro[4.4]nonan-9-one; 5-Dimethylamino, *in* I-00032
35946-79-3 2-Amino-2-deoxygalactose 1-(dihydrogen phosphate); α-D-*Pyranose-form*, *in* A-00065
35948-25-5 6*H*-Dibenzo[*c,e*][1,2]-oxaphosphorin; 6-Oxide, *in* D-00039
35948-27-7 6-Phenoxy-6*H*-dibenzo[*c,e*][1,2]-oxaphosphorin, P-00065
35989-04-9 *N,N,N′,N′,N″,N″*-Hexamethyl-*N‴*-phenylphosphorimidic triamide, H-00073
35996-72-6 ▷Isopropyl bis(1-aziridinyl)-phosphinate, *in* B-00091
36001-88-4 ▷Amiprophos-Methyl, A-00132
36011-88-8 1,3-Phenylene diphosphoric acid, P-00147
36042-94-1 Chlorodi-*o*-tolylphosphine, *in* B-00323
36043-00-2 Dichloro-1-naphthylphosphine, *in* N-00025
36044-89-0 2,2,3,4,4-Pentamethylphosphetane, P-00022
36050-78-9 [Bis(trimethylsilyl)methylene]-triphenylphosphorane, *in* B-00432
36055-83-1 Isopropyl tetramethylphosphorodiamidite, *in* T-00227
36097-59-3 Phenylphosphoramidic acid; Monoanilinium salt, *in* P-00257
36097-61-7 1-Naphthylphosphoramidic acid, N-00028
36097-62-8 2-Naphthylphosphoramidic acid, N-00029
36097-63-9 Benzoylphosphoramidic acid, B-00036
36103-10-3 *O,O*-Ethylene phenylphosphonothioate, *in* P-00115
36103-43-2 Diethyl benzylphosphonite, *in* B-00068
36120-75-9 Methylvinylphosphinic acid; Chloride, *in* M-00438
36120-86-2 (Mercaptomethyl)phosphonic acid, M-00018
36121-18-3 Fluorotetramethylphosphorane, F-00049
36126-86-0 1-Phenylphosphepane; 1-Oxide, *in* P-00205
36156-14-6 1,1,4,7,10,10-Hexaphenyl-1,4,7,10-tetraphosphadecane; High-melting-*form*, Tetraoxide, *in* H-00091
36175-06-1 Picofosforic acid, P-00428
36190-30-4 *N*-Benzyl-*P,P*-dimethylphosphinothioic amide, *in* D-00840
36206-23-2 3,3,4,4,5,5,6,6-Octafluoro-1,2-cyclohexenediphosphonic acid, O-00003
36219-46-2 Diethenyl phosphorobromidate, *in* D-01257
36230-19-0 (1,8-Biphenylenediyldimethyl-idyne)bis[triphenylphosphorane], B-00076
36231-69-3 6,8,10-Triphenyl-1,4-diaza-5-phospha(5-P^V)spiro[4.5]-deca-5,7,9-triene; 1,4-Di-Me, *in* T-00599
36231-70-6 6,8,10-Triphenyl-1,4-diaza-5-phospha(5-P^V)spiro[4.5]-deca-5,7,9-triene, T-00599
36231-71-7 6,8,10-Triphenyl-1-oxa-4-aza-5-phosphaspiro(5-P^V)-spiro[4.5]deca-5,7,9-triene; *N*-Me, *in* T-00609
36238-99-0 Diisopropyl phenylphosphonite, *in* P-00251
36240-22-9 1,1-Dihydro-1,1-bis(methylthio)-2,4,6-triphenylphosphorin, *in* D-00613
36240-28-5 [1,1′-Biphenyl]-2-yl phosphorodichloridate, *in* M-00445
36267-33-1 3-(Diphenylphosphino)aniline, D-01048
36267-37-5 3-(Diphenylphosphinothioyl)-aniline, *in* D-01048
36296-79-4 4′-Methyl-4,4,5,5-tetrakis(trifluoromethyl)-spiro[1,3,2-dioxaphospholane-2,1′-[2,6,7]trioxa[1]-phosphabicyclo[2.2.2]octane], M-00405
36305-33-6 1-Chloro-2,3-dihydro-1*H*-phosphole; 1-Sulfide, *in* C-00051
36315-13-6 2,5,5-Trimethyl-1,3,2-dioxaphosphorinane, T-00512
36335-67-8 ▷Butamifos, B-00513
36357-23-0 Bis[(1,1′-biphenyl)-2-yl] phosphorochloridate, B-00102
36357-48-9 3-(Diphenylphosphinyl)aniline, *in* D-01048
36357-77-4 Phosphoramidon, P-00402
36366-55-9 Diethyl 2-furanylphosphonate, *in* F-00073
36383-22-9 Bis[(diphenoxyphosphino)thioyl] disulfide, B-00211
36384-93-7 *O*-Methyl phosphorodiisocyanatidothioate, M-00357
36400-46-1 Bis(3-methylphenyl) phosphate, B-00313
36400-49-4 Bis(2-chlorophenyl) phosphate, B-00131
36519-00-3 Bis(2,4-dichlorophenyl) ethyl phosphate, B-00162
36585-29-2 *O*-Ethyl ethylphosphonothioate; (±)-*form*, *in* E-00074
36585-69-0 *O*-Methyl methylphosphonothioate; (±)-*form*, *in* M-00171
36585-70-3 *O*-Ethyl methylphosphonothioate; (±)-*form*, *in* E-00093
36585-72-5 *O*-Isopropyl methylphosphonothioate; (±)-*form*, *in* I-00041
36585-74-7 *O*-Methyl ethylphosphonothioate; (±)-*form*, *in* M-00155
36585-75-8 *O*-Ethyl isopropylphosphonothioate; (±)-*form*, *in* E-00082
36592-32-2 *O,O*-Diethyl ethylphosphoramidothioate, *in* E-00159
36592-34-4 *O,O*-Diethyl propylphosphoramidothioate, *in* P-00485
36598-86-4 Methylphosphoramidic dichloride, M-00327
36614-38-7 ▷Isothioate, I-00067
36626-29-6 (2-Carboxyethyl)-triphenylphosphonium(1+); Chloride, *in* C-00011
36696-09-0 Diisopropyl phosphorochloridothioite, D-00660
36696-24-9 Propyl phosphorodichloridothioite, P-00491
36696-25-0 Isopropyl phosphorodichloridothioite, I-00063
36696-26-1 Butyl phosphorodichloridothioite, B-00632
36712-24-0 10-Phenyl-10*H*-phenoxaphosphine; 10-Sulfide, *in* P-00183
36712-66-0 2,8,9-Trioxa-5-aza-1-phosphabicyclo[3.3.3]-undecane, T-00584
36761-83-8 2-[(2-Chloroethyl)amino]-tetrahydro-2*H*-1,3,2-oxazaphosphorine 2-oxide, C-00072
36792-53-7 2,4,6-Trihydroxy-1,2,4,6-oxatriphosphorinane 2,4,6-trioxide, T-00494
36838-04-7 Benzoyldiphenylphosphine, B-00030
36838-17-2 Diphenyl (4-methylphenyl)-phosphonite, *in* M-00248
36885-49-1 ▷Monovinyl phosphate, M-00518
36951-54-7 Ethyl dichlorophosphinylacetate, *in* D-00190
36952-24-6 (9*H*-Xanthen-9-yl)phosphonic acid, X-00001
36952-30-4 2,2-Dihydro-1,3,2-benzoxazaphosphole; 2,2-Dichloro, *in* D-00444
36952-32-6 2,2-Dihydro-1,3,2-benzoxazaphosphole; 2,2-Diethoxy, *in* D-00444
36952-33-7 2,2-Dihydro-1,3,2-benzoxazaphosphole; 2,2-Dimethoxy, *in* D-00444
36952-34-8 2,2-Dihydro-1,3,2-benzoxazaphosphole; 2,2-Diphenoxy, *in* D-00444
36969-89-8 Dimethyl (2-oxoheptyl)-phosphonate, *in* O-00061
36989-61-4 Monomethyl phosphate; Di-NH_4 salt, *in* M-00489
37032-33-0 ▷Dihexadecyl phosphonate, D-00430
37037-60-8 Diphenyl (1-methylethenyl)-phosphonate, *in* M-00149
37084-64-3 Phenoxytetraphenylphosphorane, P-00076
37097-43-1 Diethyl 4-morpholinylphosphonate, *in* M-00523
37100-67-7 (1-Amino-2-methylpropyl)-phosphonic acid; (±)-*form*, *in* A-00094
37132-72-2 ▷Fotrin, F-00066
37173-14-1 *O,O*-Di-*tert*-butyl phosphorothioate; Na salt, *in* D-00156
37175-34-1 Diethyl 4-pyridinylphosphonate, *in* P-00505
37181-95-6 2,3-Diphenyl-1,3-oxaphospholane, D-01027
37181-98-9 5-Methyl-2,3-diphenyl-1,3-oxaphospholane, M-00133
37182-02-8 2,3-Diphenyl-1,3-oxaphospholane; 3-Oxide, *in* D-01027
37182-04-0 2,3-Diphenyl-1,3-oxaphospholane; 3-Sulfide, *in* D-01027
37385-42-5 4-Cyano-2,2-dihydro-3,5-diphenyl-3*H*-1,2-azaphosphole, *in* D-00461
37385-44-7 2,2-Dihydro-4-ethoxycarbonyl-2,2-dimethyl-3,5-diphenyl-3*H*-1,2-azaphosphole, *in* D-00461

37442-78-7 2-Choro-1,3,2-dithiaphosphorinane; 2-Oxide, *in* C-00199
37443-15-5 *O,O*-Bis(4-methylphenyl) phosphorodiselenoate; K salt, *in* B-00336
37443-16-6 *O*-Ethyl phosphorotriselenoate; Di-K salt, *in* E-00184
37443-71-3 1,3,2-Dithiaphosphorinane 2-oxide, D-01248
37497-25-9 Dimethyl (2-oxononyl)phosphonate, *in* O-00063
37505-84-3 2-Mercapto-1,3,2-dithiaphosphorinane 2-sulfide; Triethylammonium salt, *in* M-00014
37505-86-5 Methyl trimethylene phosphorotetrathioate, *in* M-00014
37515-81-4 2-Adamantylphosphonic acid; Monoanilinum salt, *in* A-00035
37516-01-1 2-Adamantylphosphonic acid; Dichloride, *in* A-00035
37516-02-2 2-Adamantylphosphonic acid, A-00035
37516-03-3 Dimethyl 2-adamantylphosphonate, *in* A-00035
37516-07-7 Dimethyl 1-adamantylphosphonate, *in* A-00034
37516-08-8 Diethyl 1-adamantylphosphonate, *in* A-00034
37516-09-9 Diisopropyl 1-adamantylphosphonate, *in* A-00034
37516-10-2 1-Adamantylphosphonic acid; Monoanilinium salt, *in* A-00034
37521-13-4 *N,N*-Dimethyl-2,4-bis(1-methylethylidene)-1,3,6,9-tetraoxa-5-phospha(5-P^V)spiro[4.4]-nonan-5-amine, *in* B-00306
37521-15-6 2,4-Bis(1-methylethylidene)-1,3,6,9-tetraoxa-5-phospha(5-P^V)spiro[4.4]-nonane; 5-Ph, *in* B-00306
37521-98-5 Tetraethyl 1,2-benzenediol bis(phosphate), T-00042
37555-31-0 5-*tert*-Butyl-2-phenyl-1,3,2-dioxaphosphorinane; *trans-form*, 2-Oxide, *in* B-00577
37605-43-9 Methyl 2,3-bis-*O*-diphenylphosphino-4,6-*O*-benzylideneglucopyranoside; α-D-*form*, *in* M-00091
37624-66-1 *N,N′,N″*-Tribenzylphosphoric triamide, *in* T-00344
37624-73-0 *P*-Cyclohexyl-*N,N′*-diphenylphosphonic diamide, *in* C-00257
37627-97-7 [(Diphenylphosphinyl)methyl]-triphenylphosphonium(1+); 4-Methylbenzenesulfonate, *in* D-01101
37628-15-2 4-Ethoxy-1,4,5,6-tetrahydro-1,3,5-triphenyl-1,2,4-diazaphosphorine 4-oxide, *in* E-00029
37632-32-9 Dimethyl 3-furanylphosphonate, *in* F-00074
37632-33-0 Diethyl 3-furanylphosphonate, *in* F-00074
37714-05-9 (Aminophenylmethyl)phosphonic acid; (*R*)-*form*, *in* A-00105
37714-06-0 (Aminophenylmethyl)phosphonic acid; (*S*)-*form*, *in* A-00105
37736-69-9 Methyl (diethoxyphosphinyl)-oxoacetate, *in* D-00252
37737-13-6 1,1,3,4-Tetramethyl-1*H*-phospholium iodide, *in* T-00536
37739-99-4 1,3,4-Trimethyl-1*H*-phosphole, T-00536
37740-00-4 3*a*,4,7,7*a*-Tetrahydro-2,3,5,6-tetramethyl-1,8-diphenyl-4,7-phosphinidene-1*H*-phosphindole; 1,8-Dioxide, *in* T-00141
37747-07-2 2,4,6,8,9,10-Hexamethyl-2,4,6,8,9,10-hexaaza-1,3,5,7-tetraphosphatricyclo[3.3.1.1^{3,7}]-decane; 1,3,5,7-Tetrasulfide, *in* H-00071
37753-10-9 ▷Sufosfamide, S-00026
37755-70-7 4-Phenyl-4*H*-1,4-oxaphosphorin; 4-Oxide, *in* P-00175
37755-73-0 2,3-Benzo-1,4,5,6,7-pentaphenyl-7-phosphabicyclo[2.2.1]hept-5-ene 7-oxide, *in* D-00538
37759-12-9 6-Benzyl-2-oxa-6-phosphatricyclo[3.3.1.1^{3,7}]decane, B-00055
37803-03-5 2,2′-Bis[(triphenylphosphonio)-methyl]-1,1-binaphthyl(2+); (*S*)-*form*, Dibromide, *in* B-00444
37812-59-2 2,2′-Bis[(triphenylphosphonio)-methyl]-1,1-binaphthyl(2+); (*S*)-*form*, Diperiodate, *in* B-00444
37835-11-3 6-Benzyl-2-oxa-6-phosphatricyclo[3.3.1.1^{3,7}]decane; 6-Oxide, *in* B-00055
37839-81-9 Cyclic AMP; Na salt, *in* C-00233
37862-86-5 Cyclohexylphosphonothioic acid trimer trianhydride, *in* T-00430
37893-34-8 4′,5′-Bis(trifluoromethyl)-spiro[1,3,2-benzodioxaphosphole-2,2′-[1,3,2]dithiaphosphole]; 2-Methoxy, *in* B-00422
37895-63-9 2,2-Dihydro-2,2,2-trimethoxy-4,5-bis(trifluoromethyl)-1,3,2-dithiaphosphole, D-00592
37909-36-7 Dimethyl 1,2-phenylenebis(phenylphosphinate), *in* P-00139
37912-98-4 *O*-Ethyl isopropylphosphonothioate, E-00082
37913-15-8 2,3-Dihydro-1,2,3-triphenyl-1*H*-benzotriphosphole; (1α,2β,3α)-*form*, 1,3-Disulfide, *in* D-00606
37943-90-1 2-(Diphenylphosphino)pyridine, D-01076
37971-96-1 2-Phosphono-1,2,4-butanetricarboxylic acid, P-00382
37983-59-8 Triethyl ethylphosphorimidate, *in* E-00161
38019-09-9 Tris(3-bromophenyl)phosphine; Oxide, *in* T-00734
38021-01-1 Dinonylphosphinic acid, D-00949
38021-51-1 4,5-Dihydro-4-hydroxybenzo[*lmn*]-phosphanthridine 4-oxide, D-00489
38033-18-0 5-Methoxy-5,6-dihydrophosphanthridene 5-oxide, *in* D-00504
38046-64-9 1,1,3,3-Tetramethyl-6,7-bis(trifluoromethyl)-5,8-dithia-4-phospha(4-P^V)spiro[3.4]oct-6-ene; 4-Ph, *in* T-00193
38047-30-2 *N*-Chloro-*P,P*-bis(trichloromethyl)-phosphinimidic chloride, *in* B-00403
38047-33-5 [Bis(trichloromethyl)-phosphinyl]phosphorimidic trichloride, B-00404
38052-05-0 *O*-Ethyl *O*-phenyl phosphorochloridothioate, E-00117
38055-44-6 Dimethylphosphinothioic acid; Anhydride, *in* D-00839
38057-89-5 3,3-Dimethyl-2,2(3*H*,3′*H*)-spirobi[1,3,2-benzoxazaphosphole], D-00917
38066-16-9 Diethyl [(phenylthio)methyl]-phosphonate, *in* P-00326
38066-23-8 3,4-Dimethyl-1-phenyl-1*H*-phosphole; 1-Selenide, *in* D-00807
38066-25-0 1-*tert*-Butyl-3,4-dimethyl-1*H*-phosphole, E-00004
38066-26-1 1-*tert*-Butyl-3,4-dimethyl-1*H*-phosphole; 1-Sulfide, *in* E-00004
38066-33-0 Triphenylvinylphosphonium(1+), T-00702
38074-88-3 Ethyl phenyl methylphosphonate, E-00100
38104-00-6 (2-Carboxy-2-propenyl)-triphenylphosphonium(1+); Et ester, bromide, *in* C-00016
38135-34-1 1,3-Phenylene bisphosphorodichloridate, *in* P-00147
38139-02-5 Ethyl (2-chloroethyl)-phosphonochloridate, *in* C-00077
38151-54-1 6-Chloro-5,6,7,12-tetrahydro-2,5,7,10-tetramethyl-dibenzo[*d,g*][1,3,2]-diazaphosphocine; 6-Sulfide, *in* C-00182
38167-29-2 2,3,5,6,7,8-Hexamethyl-2,3,5,6,7,8-hexaaza-1,4-diphosphabicyclo[2.2.2]-octane; 1,4-Diselenide, *in* H-00070
38169-15-2 (4-Chlorophenyl)methylphosphinic acid; Fluoride, *in* C-00131
38169-17-4 Dicyclohexylphosphinic acid; Fluoride, *in* D-00211
38169-27-6 Ethylphenylphosphinothioic acid; (±)-*form*, Fluoride, *in* E-00105
38169-28-7 Dipropylphosphinothioic acid; Fluoride, *in* D-01208
38169-29-8 Dicyclohexylphosphinothioic acid; Fluoride, *in* D-00213
38169-30-1 Dicyclohexylphosphinothioic acid; Chloride, *in* D-00213
38186-51-5 Diethyl [(4-bromophenyl)-methyl]phosphonate, *in* B-00484
38206-24-5 2-Ethoxy-4,4,5,5-tetramethyl-1,3,2-dioxaphospholane, E-00052
38234-83-2 2,3-Dihydro-1,2,3-triphenyl-1*H*-1,3,2-benzodiphospharsole; (1α,2β,3α)-*form*, *in* D-00605
38234-85-4 2,3-Dihydro-1,3-diphenyl-1*H*-1,3-benzodiphosphole, D-00476
38234-86-5 1,2,3,4-Tetrahydro-1,4-diphenyl-1,4-benzodiphosphorin; (1*RS*,4*SR*)-*form*, *in* T-00102
38234-87-6 1,2,3,4-Tetrahydro-1,4-diphenyl-1,4-benzodiphosphorin; (1*RS*,4*SR*)-*form*, 1,4-Disulfide, *in* T-00102
38234-89-8 1,2-Ethanediylbis[methyl-phenylphosphine]; (*RS,SR*)-*form*, Dioxide, *in* E-00012
38234-90-1 1,2-Ethanediylbis[methyl-phenylphosphine]; (*RS,RS*)-*form*, Dioxide, *in* E-00012
38234-91-2 5,6,11,12-Tetrahydro-5,6,11,12-tetraphenyl-dibenzo[*c,g*][1,2,5,6]-tetraphosphocin, T-00147
38240-96-9 5,6,11,12-Tetrahydro-5,6,11,12-tetraphenyl-dibenzo[*c,g*][1,2,5,6]-tetraphosphocin; 5,11-(or 5,6)-Disulfide, *in* T-00147
38240-97-0 1,4-Dihydro-1,4-diphenyl-1,4-benzodiphosphorin, D-00477
38260-54-7 ▷Etrimfos, E-00205
38281-98-0 Dimethyl (1-methylethenyl)-phosphonate, *in* M-00149
38289-60-0 Phenyl 3-[(dimethoxyphosphinothioyl)-oxy]-2-methyl-2-propenoate, *in* D-00673
38305-79-2 2,3-Dihydro-1,2,3-triphenyl-1*H*-1,3,2-benzodiphospharsole; (1α,2β,3α)-*form*, 1,3-Disulfide, *in* D-00605

38315-71-8 *O*-Methyl methylphosphonothioate; (*R*)-*form*, *in* M-00171
38315-72-9 *O*-Ethyl methylphosphonothioate; (*R*)-*form*, *in* E-00093
38315-75-2 *O*-Ethyl isopropylphosphonothioate; (*R*)-*form*, *in* E-00082
38315-77-4 *O*-Ethyl methylphosphonothioate; (*S*)-*form*, *in* E-00093
38315-80-9 *O*-Methyl methylphosphonochloridothioate; (*S*)-*form*, *in* M-00169
38315-81-0 *O*-Ethyl methylphosphonochloridothioate; (*R*)-*form*, *in* E-00090
38315-82-1 *O*-Propyl methylphosphonochloridothioate; (*S*)-*form*, *in* P-00468
38315-83-2 *O*-Butyl methylphosphonochloridothioate; (*S*)-*form*, *in* B-00573
38315-84-3 *O*-Methyl ethylphosphonochloridothioate; (*R*)-*form*, *in* M-00154
38315-86-5 *O*-Methyl isopropylphosphonochloridothioate; (*S*)-*form*, *in* M-00165
38315-90-1 *O*-Methyl isopropylphosphonochloridothioate; (*R*)-*form*, *in* M-00165
38315-91-2 *O*-Methyl methylphosphonothioate; (*S*)-*form*, *O*-Isopropyl ester, *in* M-00171
38315-92-3 *O*-Isopropyl *O*-methyl methylphosphonothioate, *in* M-00171
38316-42-6 Diphenyl methylphosphonite, *in* M-00319
38319-29-8 3-Nitrophenyl phosphorodichloridate, *in* M-00499
38344-09-1 *O*-Methyl ethylphosphonothioate; (*S*)-*form*, *in* M-00155
38344-09-1 *O*-Ethyl *O*-methyl ethylphosphonothioate, *in* M-00155
38416-66-9 Ethyl hydrogen [(1,3-dihydro-1,3-diazo-2*H*-isoindol-2-yl)methyl]phosphonate, *in* P-00426
38421-29-3 2,3-Dihydro-4,5-dimethyl-2,7-diphenyl-1*H*-1,2-azaphosphepine; 2-Oxide, *in* D-00460
38443-57-1 *O,S*-Ethylene phenylphosphonothioite, *in* O-00056
38448-55-4 2,4,6,8,9,10-Hexamethyl-2,4,6,8,9,10-hexaaza-1,3,5,7-tetraphosphatricyclo[3.3.1.1^{3,7}]-decane; 1,3,5-Trisulfide, *in* H-00071
38448-56-5 2,4,6,8,9,10-Hexamethyl-2,4,6,8,9,10-hexaaza-1,3,5,7-tetraphosphatricyclo[3.3.1.1^{3,7}]-decane; 1,3-Disulfide, *in* H-00071
38448-57-6 2,4,6,8,9,10-Hexamethyl-2,4,6,8,9,10-hexaaza-1,3,5,7-tetraphosphatricyclo[3.3.1.1^{3,7}]-decane; 1-Sulfide, *in* H-00071
38450-57-6 Octaethyldiphosphoramide, O-00001
38451-19-3 1,5-Pentanediylidenebis[triphenylphosphorane], *in* P-00034
38452-89-0 3,4,5,5-Tetrahydro-2,2,5,5-tetraphenyl-1*H*-1,2,5-azadiphospholium(1+); Perchlorate, *in* T-00143
38453-00-8 2,2,4,4-Tetrahydro-2,2,4,4-tetraphenyl-1,5,2,4-diazadiphosphorine, T-00146
38476-60-7 Diphenyl phenylphosphonodithioite, *in* P-00238
38476-63-0 Diethyl butylphosphonodithioite, *in* B-00607
38476-65-2 Diphenyl methylphosphonodithioite, *in* M-00302
38500-29-7 Tetramethyl phosphor(isothiocyanatidic) diamide, *in* T-00218
38527-91-2 ▷Ethafos, E-00005
38555-73-6 Monomethyl phenylphosphonate; (±)-*form*, Mono-cyclohexylammonium salt, *in* M-00488
38562-84-4 *O*-Trimethylsilyl bis(trifluoromethyl)-phosphinothioate, *in* B-00415
38568-67-1 (Methylthio)imidodiphosphoryl chloride, M-00414
38568-76-2 Phenylphosphoramidothioic acid; Dichloride, *in* P-00262
38580-66-4 2-Phenoxy-1,3,2-oxathiaphospholane, P-00072
38590-11-3 *N,N,N′,N′*-Tetraethylphosphoric triamide, T-00058
38605-11-7 *S*-Methyl methylphenylphosphinothioate, *in* M-00225
38605-13-9 Methylphenylphosphinothioic acid; (*S*)-*form*, *O*-Me ester, *in* M-00225
38605-14-0 *O*-Methyl ethylphenylphosphinothioate, *in* E-00105
38607-71-5 *O*-Ethyl isopropylphosphonothioate; (*S*)-*form*, *in* E-00082
38607-73-7 Methylphenylphosphinothioic acid; (*R*)-*form*, *in* M-00225
38607-74-8 Ethylphenylphosphinothioic acid; (*S*)-*form*, *in* E-00105
38633-40-8 Triphenyl(3-phenyl-2-propenyl)-phosphonium(1+); (*E*)-*form*, Bromide, *in* T-00616
38641-40-6 Tris[1,2-benzenediolato(2−)-*O,O′*] phosphate(1−); Dimethyl-ammonium salt, *in* T-00730
38641-94-0 ▷Roundup, *in* G-00097
38654-91-0 α-(Diethoxyphosphinyl)-benzeneacetic acid, D-00242
38654-93-2 α-Phosphonophenylacetic acid, P-00393
38694-48-3 4-(Diethoxyphosphinyl)butanoic acid, D-00247
38696-09-2 (2-Nitrophenyl)phosphonic acid, N-00055
38704-10-8 Methyltris(triphenylphosphine)-copper, M-00435
38706-86-4 2,3-Diphenyl-1,3-oxaphosphorinane, D-01028
38706-92-2 3-Methyl-2,3-diphenyl-1,3-oxaphosphorinanium iodide, *in* D-01028
38707-15-2 1-Phenyl-4-phosphorinanone; Oxide, *in* P-00268
38736-54-8 *tert*-Butylmethylphenylphosphine; (*S*)-*form*, *in* B-00565
38749-18-7 Diisopropyl (diazophenylmethyl)phosphonate, *in* D-00037
38751-14-3 *S*-Butyl methylphosphonochloridothioate, *in* M-00298
38766-19-7 Ethyl (4-bromophenyl)phosphinate, *in* B-00489
38766-20-8 Ethyl (4-chlorophenyl)-phosphinate, *in* C-00144
38766-21-1 Ethyl (4-methoxyphenyl)-phosphinate, *in* M-00069
38802-08-3 *tert*-Butylmethylphenylphosphine; (*R*)-*form*, Oxide, *in* B-00565
38868-17-6 Diisopropyl 1-cyclopenten-1-ylphosphonate, *in* C-00277
38873-91-5 Phenylphosphonic acid; Bis(4-nitrophenyl) ester, *in* P-00221
38874-30-5 (4-Bromophenyl)phosphoramidic acid, B-00496
38874-31-6 (3-Chlorophenyl)phosphoramidic acid, C-00156
38886-39-4 *O,O*-Dibenzyl phosphorodithioate; K salt, *in* D-00063
38897-99-3 (4-Chlorophenylmethylene)-triphenylphosphorane, *in* C-00137
38938-31-7 Methyl *N,P*-diphenylphosphonamidate, *in* D-01103
38945-95-6 *O,O*-Diphenyl phenylphosphoramidothioate, *in* P-00262
38964-67-9 1,4-Dioxa-6,9-dithia-5-phospha(5-P^V)spiro[4.4]-nonane, D-00964
39013-59-7 Methyl (4-methylphenyl)-methylphosphinate, *in* M-00204
39013-98-4 2,3-Bis(triphenyl-phosphoniomethyl)-naphthalene(2+); Dibromide, *in* B-00453
39014-78-4 *O,O*-Diphenyl phosphorochloridoselenoate, D-01122
39030-35-8 Methyl *N,N*-diethyl-*P*-phenylphosphonamidate, *in* D-00307
39043-29-3 Tris[1,2-benzenediolato(2−)-*O,O′*] phosphate(1−); Triethylammonium salt, *in* T-00730
39055-19-1 5′-Methylspiro[1,3,2-benzodioxaphosphole-2,2′(3*H*)-[1,2]oxaphosphole]; 2-Ph, *in* M-00396
39063-55-3 2,2(3*H*,3′*H*)-Spirobi[1,3,2-benzoxazaphosphole]; 2-Et, *in* S-00020
39063-58-6 1,2-Ethanediphosphonous acid; Bis(dibromide), *in* E-00008
39063-70-2 2,5-Dihydro-1-hydroxy-1*H*-phosphole 1-oxide, D-00507
39071-56-2 Triethyl (trimethylsilyl)-phosphorimidate, *in* T-00557
39078-28-9 *O*-Ethyl diethylphosphinoselenoate, *in* D-00327
39078-29-0 *S*-Ethyl diethylphosphinoselenothioate, *in* D-00329
39078-30-7 Phenylphosphonoselenoic acid; Dichloride, *in* P-00243
39096-95-2 *N,N*-Diethylphosphorothioic triamide, D-00398
39106-95-1 Bromodi-*tert*-butylphosphine, *in* D-00122
39110-21-9 [(4-Methylphenyl)methylene]-triphenylphosphorane, *in* M-00211
39110-24-2 3-Methylbutylidenetriphenyl-phosphorane, *in* M-00111
39110-26-4 Triphenyl(3-phenylpropylidene)-phosphorane, *in* T-00617
39118-50-8 Dimethyl (2-acetyloxyethyl)-phosphonate, *in* A-00006
39118-51-9 Dimethyl (3-acetyloxypropyl)-phosphonate, *in* A-00010
39118-56-4 (Phenylphosphinylidene)-bismethanol, *in* B-00281
39124-58-8 2,2-Dichloro-1,1,1-triethyldiphosphinium 2-oxide(1+); Chloride, *in* D-00197
39148-16-8 Monoethyl phosphonate; Na salt, *in* M-00473
39148-24-8 Fosetyl, *in* M-00473
39171-65-8 (1-Naphthylmethyl)-triphenylphosphonium(1+); Bromide, *in* N-00015
39177-91-8 *O*-Ethyl *S*-(ethoxyphenylphosphinothioyl) phenylphosphonodithioate, *in* P-00237
39177-93-0 *O*-Isopropyl *S*-(isopropoxyphenyl-phosphinothioyl) phenyl-phosphonodithioate, *in* P-00237

39181-15-2 Ethylphosphonoselenoic acid; *O,O*-Di-Et ester, *in* E-00145

39181-18-5 *O,O*-Diethyl phenylphosphonoselenoate, *in* P-00243

39181-19-6 *O*-Ethyl diphenylphosphinoselenoate, *in* D-01078

39181-26-5 Tributylphosphine selenide, T-00377

39181-29-8 Hexaethylphosphoroselenoic triamide, *in* H-00015

39181-30-1 Ethylphosphonoselenoic bis(diethylamide), *in* E-00146

39181-31-2 Tetraethylphosphinoselenoic amide, *in* D-00328

39181-35-6 *O,O,Se*-Triethyl phosphoroselenoate, T-00455

39200-99-2 Phenyl (chloromethyl)-phosphonochloridate, *in* C-00111

39222-25-8 Dibutyl ethynylphosphonite, *in* E-00204

39225-05-3 [(4-Chlorophenyl)methyl]-phosphonic acid, C-00134

39225-17-7 Diethyl [(4-chlorophenyl)-methyl]phosphonate, *in* C-00134

39230-41-6 Dimethyl phosphoramidite, D-00859

39230-44-9 Dimethyl *N,N,N'*-tris(trimethylsilyl)-phosphoramidimidate, *in* T-00868

39238-95-4 (4-Bromophenyl)phosphinic acid, B-00489

39480-85-8 Dimethyl methylphosphoramidite, *in* M-00335

39500-93-1 *P,P,P*-Tribromo-*N-tert*-butylphosphazene, *in* B-00625

39528-26-2 Tris(2-methylpropyl) (dimethoxyphosphinyl)-phosphorimidate, *in* D-00683

39536-49-7 5-*tert*-Butyl-2-dimethylamino-1,3,2-dioxaphosphorinane; *trans-form*, 2-Oxide, *in* B-00555

39536-50-0 2-*tert*-Butyltrimethylene dimethylphosphoramidate, *in* B-00555

39536-56-6 5-*tert*-Butyl-2-dimethylamino-1,3,2-dioxaphosphorinane; *trans-form*, *in* B-00555

39536-73-7 5-*tert*-Butyl-2-dimethylamino-1,3,2-dioxaphosphorinane; *cis-form*, *in* B-00555

39564-18-6 1,3-Bis(dimethylphosphino)-propane, B-00206

39589-66-7 Nonyltriphenylphosphonium(1+); Iodide, *in* N-00078

39595-20-5 *O,O*-Diethyl phosphorodiselenoate, D-00383

39598-55-5 (Bromomethylene)-triphenylphosphorane, *in* B-00477

39616-20-1 (2-Butenyl)-triphenylphosphonium(1+); (*Z*)-*form*, Bromide, *in* B-00528

39648-63-0 8,9,10,11,12,13,14,15-Octahydro-4-hydroxydinaphtho[2,1-*d*:1',2'-*f*][1,3,2]dioxaphosphepin 4-oxide, O-00018

39648-64-1 8,9,10,11,12,13,14,15-Octahydro-4-hydroxydinaphtho[2,1-*d*:1',2'-*f*][1,3,2]dioxaphosphepin 4-oxide; 4-Methoxy (Me ester), *in* O-00018

39648-67-4 4-Hydroxydinaphtho[2,1-*d*:1',2'-*f*][1,3,2]-dioxaphosphepin 4-oxide; (*R*)-*form*, *in* H-00131

39695-34-6 *O*-Phenyl *N,N'*-*o*-phenyl-enephosphorodiamidothioate, *in* D-00543

39705-76-5 9,10-Bis(diphenylphosphino)-phenanthrene, B-00261

39720-65-5 (2-Methoxy-2-oxoethyl)-triphenylphosphonium(1+); Iodide, *in* M-00055

39741-81-6 (2-Butenyl)-triphenylphosphonium(1+); (*E*)-*form*, Bromide, *in* B-00528

39746-02-6 (3-Methyl-2-oxoheptyl)phosphonic acid; (±)-*form*, Me ester, *in* M-00187

39746-15-1 Dimethyl (3,3-dimethyl-2-oxoheptyl)phosphonate, *in* D-00785

39749-44-5 Glycerol 1-hexadecanoate 2-9-octadecenoate 3-phosphoethanolamine; (*R*)-(*Z*)-*form*, *in* G-00062

39763-49-0 1-Methylphosphorinane; Oxide, *in* M-00339

39763-50-3 1-Methylphosphorinane, M-00339

39768-04-2 2-Phenylphosphinoline, P-00212

39768-11-1 2,6-Diphenyl-1,4-azaphosphorine, D-00996

39768-15-5 1,4-Dihydro-2,4,6-triphenyl-1,4-azaphosphorine, D-00604

39768-52-0 1,2-Diphenyldiphosphine; (*RS,SR*)-*form*, *in* D-01012

39768-63-3 1,2-Diphenyldiphosphine; (*RS,RS*)-*form*, *in* D-01012

39770-51-9 3,5,5-Trimethyl-1,2-oxaphospholan-3-ol 2-oxide; (*2RS,3SR*)-*form*, 2-Diethylamino, *in* T-00519

39770-55-3 2-Diethylamino-3,5,5-trimethyl-1,2-oxaphospholan-3-ol 2-oxide, *in* T-00519

39770-56-4 3,5,5-Trimethyl-2-phenyl-1,2-oxaphospholan-3-ol 2-oxide, *in* T-00519

39770-56-4 3,5,5-Trimethyl-2-phenyl-1,2-oxaphospholan-3-ol 2-oxide, *in* T-00519

39776-40-4 (2,7-Dimethyl-2,4,6-octatrien-1,8-diylidene)bis[triphenyl-phosphorane], *in* D-00778

39826-54-5 2-Hydroxy-4-methyl-1,3,2-dioxaphosphorinane 2-sulfide, H-00158

39826-69-2 2-Hydroxy-4-methyl-1,3,2-dioxaphosphorinane 2-selenide; (*2RS,4SR*)-*form*, Dicyclohexylammonium salt, *in* H-00157

39826-70-5 2-Hydroxy-4-methyl-1,3,2-dioxaphosphorinane 2-selenide; (*2RS,4SR*)-*form*, *in* H-00157

39826-73-8 2-Hydroxy-4-methyl-1,3,2-dioxaphosphorinane 2-selenide; (*2RS,4RS*)-*form*, *in* H-00157

39834-55-4 1-Methylphospholane, M-00280

39846-24-7 5,5-Dimethyl-2-methylthio-1,3,2-dioxaphosphorinane, D-00766

39846-28-1 Neopentylene phosphorofluoridate, *in* F-00016

39850-00-5 Dimethyl (3-methyl-2-oxoheptyl)phosphonate, *in* M-00187

39850-01-6 (3-Methyl-2-oxoheptyl)phosphonic acid; (*S*)-*form*, Me ester, *in* M-00187

39857-88-0 *O,O*-Diethyl phosphorodithioate; Diethylammonium salt, *in* D-00384

39864-75-0 Di-1-naphthylphosphine, D-00939

39895-79-9 ▷Ethyltriphenylphosphonium(1+), E-00197

39902-05-1 Methyl vinylene phosphate, *in* H-00136

39943-76-5 Dibromotriphenoxyphosphorane, D-00070

39968-54-2 Dibutyl (4-chlorobutyl)-phosphonate, *in* C-00025

39968-97-3 5-(Triphenylphosphoranylidene)-pentanoic acid, T-00655

39980-20-6 [(4-Nitrophenyl)methyl]-phosphonic acid; Di-Me ester, *in* N-00046

39980-64-8 1,1-Dihydro-1,1,3,3,4,4-hexamethyl-1-phospha-3,4-disilacyclopent-1-ene, D-00486

39996-87-7 Diethyl (2-pyridinylmethyl)-phosphonate, *in* P-00506

39997-40-5 Diethyl (2-hydroxyethyl)-phosphonate, *in* H-00146

39997-45-0 2,5-Dihydro-3,4-dimethyl-1-phenyl-1*H*-phosphole; 1-Sulfide, *in* D-00468

40012-64-4 1,2,3-Triphenyl-1,2,3-triphospholane, T-00699

40074-54-2 10-Chloro-5,10-dihydrophenophosphazine; 10-Sulfide, *in* C-00049

40103-72-8 Phenyl(trichloromethyl)-phosphinic acid, P-00327

40110-50-7 2,2-Dihydro-1,2-oxaphosphetane, D-00533

40137-11-9 Diethyl [(methylsulfonyl)-methyl]phosphonate, *in* M-00400

40145-83-3 *S,S*-Dimethyl methylphosphonodithioate, *in* M-00301

40156-84-1 2-Hydroxy-4*H*-1,3,2-benzodioxaphosphorin 2-oxide, H-00109

40156-85-2 *o*-Phenylene phenylphosphonate, *in* P-00094

40168-00-1 2,3-Dihydro-4,5-dimethyl-2,7-diphenyl-1*H*-1,2-azaphosphepine; 2-Sulfide, *in* D-00460

40195-10-6 5,10-Dihydro-10-methoxyphenophosphazine 10-sulfide, *in* D-00502

40200-59-7 Propylphosphine, P-00471

40201-80-7 *N,N'*-Dimethyl-*N,N'*-trimethylene-*P*-phenyl-phosphonous diamide, *in* H-00022

40201-85-2 *N,N'*-Dimethyl-*N,N'*-trimethylene phosphorodiamidous chloride, *in* H-00022

40201-86-3 Methyl *N,N'*-dimethyl-*N,N'*-trimethylene phosphorodiamidite, *in* H-00022

40203-43-8 Diethyldiphosphonic acid; Bis(cyclohexylamine) salt, *in* D-00290

40207-47-4 (Chloromethyl)methylphosphinic acid, C-00100

40215-39-2 5,10-Dihydro-10-hydroxyphenophosphazine 10-sulfide, D-00502

40244-90-4 Diisopropylphosphinous chloride, D-00650

40277-05-2 4-Hydroxycyclophosphamide, H-00115

40290-42-4 Glycerol 1,2-dihexadecanoate 3-phosphoserine; (*R*)-L-*form*, *in* G-00016

40299-58-9 [(3-Nitrophenyl)methyl]-phosphonic acid, N-00045

40299-61-4 [(4-Methoxyphenyl)methyl]-phosphonic acid, M-00061

40299-92-1 2-Naphthylphosphonic acid; Dichloride, *in* N-00022

40329-00-8 1,3-Dihydroxy-1*H*-1,2,4,3-benziodadioxaphosphorin 3-oxide, D-00616

40329-01-9 1-Hydroxy-3-methoxy-1*H*-1,2,4,3-benziodadioxaphosphorine 3-oxide, *in* D-00616

40331-46-2 Tris(3-methoxyphenyl)phosphine; Oxide, *in* T-00792

40334-15-4 2,3,4,7,8,9-Hexahydro-1-phenyl-5,6-bis[(trimethylsilyl)-oxy]-1*H*-phosphonine, H-00047

40334-16-5 1-Phenyl-6,7-bis[(trimethylsilyl)-oxy]phosphacycloundec-6-ene, P-00100

40334-32-5 Methyl *P*-phenylphosphonamidate, *in* P-00218

40334-37-0 *O*-Methyl *P*-methylphosphonamidothioate, *in* M-00284

40334-49-4 *O*-Ethyl phosphorodiamidothioate, *in* E-00182

40334-50-7 *O*-Methyl phosphorodiamidothioate, *in* M-00359

40425-71-6 1,1-Dichloro-1,1-dihydro-2,4,6-triphenylphosphorin, *in* D-00613

40425-79-4 1,1-Difluoro-1,1-dihydro-2,4,6-triphenylphosphorin, *in* D-00613

40425-94-3 1,2,3,4-Tetraphenyltetraphospholane, T-00280

40438-07-1 1,2,3,4,7,8-Hexahydro-1-phenyl-5,6-bis[(trimethylsilyl)-oxy]phosphocin, H-00046

40438-08-2 1,2,3,4,5,8,9,10-Octahydro-1-phenyl-6,7-bis[(trimethylsilyl)-oxy]phosphecine, O-00021

40438-23-1 Methyl (triphenylphosphoranylidene)-carbamate, *in* T-00648

40450-19-9 (1-Azulenylmethyl)-triphenylphosphonium(1+); Iodide, *in* A-00156

40450-24-6 (1-Azulenylmethylene)-triphenylphosphorane, *in* A-00156

40468-55-1 1,2-Bis(diphenylphosphino)-ethylene; (*Z*)-*form*, *P*,*P'*-Dioxide, *in* B-00237

40468-93-7 9-Phenyl-9-phosphabicyclo[4.2.1]nonane; Oxide, *in* P-00194

40524-74-1 Thalassemine, *in* L-00002

40535-35-1 Tetrahydro-4-methyl-2*H*-1,3,4,2-oxadiazaphosphorine 2-oxide; 2-Phenoxy, *in* T-00116

40561-09-9 (Chloromethyl)phenylphosphinic acid; Chloride, *in* C-00103

40585-73-7 2,2'-Spirobi[1,3,2-benzodithiaphosphole]; 2-Ph, *in* S-00015

40601-40-9 Diethyl (2-oxononyl)phosphonate, *in* O-00063

40601-45-4 Diethyl (2-oxo-4-phenylbutyl)-phosphonate, *in* O-00065

40612-16-6 Dimethyl 1,2-ethenediylbis(phenyl-phosphinate), *in* E-00018

40612-18-8 1,2-Bis(diphenylphosphino)-ethylene; (*E*)-*form*, *P*,*P'*-Dioxide, *in* B-00237

40612-55-3 Diisopropyl(chloroacetyl)-phosphonate, *in* C-00020

40614-39-9 1-Phosphabicyclo[2.2.1]heptane; *p*-Oxide, *in* P-00344

40618-53-9 *S*-Ethyl *O*-methyl phenylphosphonothioate, *in* P-00244

40618-55-1 Ethyl phenylphosphonochloridite, *in* P-00234

40633-14-5 *O*,*O*-Dimethyl phosphorothioate; NH_4 salt, *in* D-00907

40633-23-6 Triphenyltriphosphirane, T-00698

40648-79-1 Benzoyl diethyl phosphite, B-00028

40651-48-7 4,4,6-Trimethyl-1,3,2-dioxaphosphorinane; 2-Sulfide, *in* T-00513

40651-74-9 *O*,*O*,*O'*,*O'*-Tetraisopropyl thiohypophosphate, T-00173

40651-78-3 Tetrapropyl hypodiphosphite, T-00287

40651-79-4 Tetraisopropyl hypodiphosphite, T-00169

40657-94-1 (4-Chlorobutyl)phosphonic acid; Dichloride, *in* C-00025

40678-60-2 *P*,*P*,*P*-Trichloro-*N*-trimethylsilylphosphazene, *in* T-00557

40760-48-3 1,2,3,4-Tetraphenyl-1,3,2,4-diazadiphosphetidine; *trans-form*, 2,4-Disulfide, *in* T-00253

40817-03-6 1,4-[Phenylenebis[methylene]]-bis[triphenylphosphonium](2+); Dibromide, *in* P-00134

40881-95-6 Tetrapropylphosphorodiamidic acid; Chloride, *in* T-00291

40881-96-7 Tetrabutylphosphorodiamidic chloride, T-00022

40881-98-9 Dipropylphosphoramidic acid; Dichloride, *in* D-01214

40882-05-1 Propyl tetrapropylphosphorodiamidate, *in* T-00291

40882-06-2 Butyl tetrabutylphosphorodiamidate, *in* T-00021

40894-65-3 9*H*-Dibenzo[2,3:4,5]-phospholo[1,2-*f*]-phosphanthridene, D-00041

40894-65-3 9*H*-Dibenzo[2,3:4,5]-phospholo[1,2-*f*]-phosphanthridene; 17-Oxide, *in* D-00041

40911-36-2 Diphenyl (dichloromethyl)-phosphonate, *in* D-00175

40925-11-9 2-(2-Methylpropoxy)-1,3,2-dioxaphospholane, M-00368

40928-00-5 2-(1-Methylethoxy)-1,3,2-dioxaphospholane, M-00152

40932-00-1 5-Methoxy-5*H*-dibenzophosphole 5-oxide, *in* H-00118

40955-14-4 3-(Triphenylphosphoranylidene)-propanoic acid; Me ester, *in* T-00661

40962-34-3 [(4-Bromophenyl)methyl]-phosphonic acid, B-00484

40962-35-4 (Bromophenylmethyl)phosphonic acid, B-00482

40962-37-6 Methyl dihydroxyphosphinylacetate, *in* P-00376

40964-81-6 4,5,9,10-Tetrahydro-4,9-dihydroxyphosphorino[2,3,4,5-*lmn*]phosphanthridine 4,9-dioxide, T-00090

40964-82-7 6,7-Dihydro-6-hydroxy-5*H*-dibenzo[*c*,*e*]phosphepin 6-oxide, D-00491

40965-68-2 1-Chloro-2,5-dihydro-3,4-dimethyl-1*H*-phosphole, C-00041

40985-25-9 Tribenzylmethylphosphonium iodide, *in* T-00338

40986-03-6 4-Methyl-1,3,2-dioxaphosphorinane; (2*RS*,4*RS*)-*form*, *in* M-00129

41003-33-2 2-Propenyl phosphorodichloridite, P-00465

41003-94-5 Diethyl (isocyanomethyl)-phosphonate, *in* I-00024

41006-23-9 1,2-Bis(diphenylphosphino)-1,2-ethanedione, B-00235

41006-27-3 3,4-Bis(diphenylphosphino)-3-cyclobutene-1,2-dione, B-00228

41012-80-0 *O*,*S*-Ethylene *S*-methyl phosphorodithioite, *in* O-00056

41029-31-6 2,2'-Spirobi[1,3,2-benzodithiaphosphole], S-00015

41044-91-1 Butyl bis(4-chlorophenyl)-phosphinate, *in* B-00139

41064-81-7 Diethyl dibutylphosphoramidite, *in* D-00136

41097-27-2 ▷2,6,7-Trioxa-1,4-diphosphabicyclo[2.2.2]-octane; 4-Oxide, *in* T-00585

41097-29-4 2,6,7-Trioxa-1,4-diphosphabicyclo[2.2.2]-octane; 1,4-Disulfide, *in* T-00585

41105-12-8 1,4-Phenylenebis[phosphorodi-chloridite], P-00142

41116-69-2 *S*-Methyl diethylphosphinothioate, *in* D-00330

41118-97-2 *O*,*O*-Diethyl phosphoroselenoate; K salt, *in* D-00393

41136-82-7 2,2'-Spirobi[1,3,2-benzodithiaphospholium]-(1+); Hexafluorophosphate, *in* S-00016

41157-34-0 Chlorodipropylphosphine, *in* D-01209

41157-86-2 3,5,5-Trimethyl-1,2-oxaphospholan-3-ol 2-oxide; (2*RS*,3*SR*)-*form*, 2-Dimethylamino, *in* T-00519

41162-19-0 Dimethyl (2-oxo-4-phenylbutyl)-phosphonate, *in* O-00065

41172-19-4 (2-Nitro-2,4-cyclopentadienylidene)-triphenylphosphorane, N-00034

41198-08-7 ▷Profenofos, P-00440

41222-85-9 Di-*tert*-butyl phosphoramidate, *in* D-00105

41227-61-6 Phenyl *N*,*N'*-*o*-phenylenephosphorodiamidate, *in* D-00543

41240-73-7 1,4-Phenylenebis(phosphorodi-chloridate), *in* P-00148

41255-51-0 Phenyl 8-quinolyl phosphate, P-00306

41264-12-4 (4-Carboxybutyl)-triphenylphosphonium(1+), C-00009

41265-74-1 Ethyl *N*,*N'*-dimethylphosphorodiamidate, *in* D-00895

41272-12-2 Tributyl(hexadecyl)-phosphonium(1+); Chloride, *in* T-00364

41273-34-1 (3,7-Dimethyl-2,6-octadienyl)-triphenylphosphonium(1+); (*E*)-*form*, Bromide, *in* D-00777

41273-83-0 *N*,*N'*-Diisopropylphosphorodiamidic acid; Chloride, *in* D-00661

41274-67-9 3,6-Dihydro-2*H*-1,2-thiaphosphorin 2-sulfide; 2-Et, *in* D-00586

41304-05-6 (Dimethoxyphosphinyl)thioacetic acid, D-00687

41309-18-6 Triphenyl [[(triphenylphosphoranylidene)-amino]sulfonyl]phosphorimidate, *in* T-00642

41343-62-8 Phenyl 2-(triphenylphosphoranylidene)-propanote, *in* T-00660

41345-58-8 Triphenyl[4-(trimethylsilyl)-3-butynyl]phosphonium(1+); Iodide, *in* T-00691

41392-00-1 Ethyl *P*-acetyl-*N*,*N*-diethylphosphonamidate, *in* A-00014

41392-17-0 5-Methyl-1,2-oxaphospholane 2-oxide; 2-Ph, *in* M-00183

41392-21-6 5-Methyl-1,2-thiaphospholane 2-sulfide; 2-Ph, *in* M-00407

41392-23-8 5-Methyl-1,2-thiaphospholane 2-sulfide; 2-Et, *in* M-00407

41398-45-2 Tris(4-chlorophenyl)phosphine; Selenide, *in* T-00749

41429-69-0 Triphenyl(3-phenyl-2-propenylidene)phosphorane, T-00615

41458-68-8 2,2-Dihydro-1,3,2-benzoxazaphosphole; 2,2-Di-Ph, *in* D-00444

41468-36-4 2-Aminoethylphosphonic acid; Di-Et ester, *in* A-00079

41625-30-3 1,8-Bis(diphenylphosphino)octane, B-00258

41640-08-8 Dimethyl (3,3-dimethyl-2-oxo-4-phenylbutyl)phosphonate, *in* D-00786

41662-41-3 Bis(2,2,2-trichloroethyl) phosphorochloridite, B-00401

41662-51-5 Diisopropyl phosphorochloridite, D-00658
41760-64-9 Diethyl [[(phenylmethyl)thio]-methyl]phosphonate, *in* B-00072
41760-80-9 Diethyl 1-butenylphosphonate, *in* B-00524
41760-84-3 Diethyl methyl phosphonoformate, *in* D-00236
41760-95-6 Ethyl benzylphosphonochloridate, *in* B-00063
41761-00-6 Methyl phenylphosphonochloridate, *in* P-00229
41773-68-6 2-Ethoxy-1,3,2-dioxaphospholan-4,5-dicarboxylic acid; (4*R*,5*R*)-*form*, Di-Et ester, *in* E-00033
41773-69-7 2-Ethoxy-1,3,2-dioxaphospholan-4,5-dicarboxylic acid; (4*RS*,5*RS*)-*form*, Di-Et ester, *in* E-00033
41773-70-0 2-Ethoxy-1,3,2-dioxaphospholan-4,5-dicarboxylic acid; (4*RS*,5*SR*)-*form*, Di-Et ester, *in* E-00033
41784-92-3 2,7-Bis(triphenyl-phosphoniomethyl)-naphthalene(2+); Dibromide, *in* B-00455
41809-52-3 1-Phosphabicyclo[2.2.2]octane; 1-Oxide, *in* P-00346
41821-75-4 2-Hydroxy-4,5-dimethyl-1,3,2-dioxaphospholane 2-oxide; Trimethylammonium salt, *in* H-00121
41821-76-5 2-Hydroxy-4,4,5,5-tetramethyl-1,3,2-dioxaphospholane 2-oxide, H-00196
41821-77-6 2-Hydroxy-1,3,2-dioxaphosphorinane 2-oxide; Tetramethylammonium salt, *in* H-00138
41821-80-1 2-Hydroxy-5,5-dimethyl-1,3,2-dioxaphosphorinane 2-oxide; Tetramethylammonium salt, *in* H-00124
41821-87-8 2-Methoxy-4,5-dimethyl-1,3,2-dioxaphospholane, M-00033
41839-48-9 Butyl butylphosphonochoridite, *in* B-00605
41839-63-8 Diphenyl aminocarbonylphosphonate, *in* C-00003
41845-30-1 Dimethyl(2-methylphenyl)-phosphine, D-00759
41855-87-2 *S*,*S*-Ethylene diethylphosphoramidodithioite, *in* D-01243
41892-64-2 2-Butenylidenetriphenyl-phosphorane, B-00522
41899-40-5 Ethylmethylphenylphosphine; (*S*)-*form*, Sulfide, *in* E-00084
41902-21-0 2,3-Dihydro-4,5-dimethyl-2,7-diphenyl-1,2-oxaphosphepin 2-oxide, D-00462
41924-72-5 Methyl diethenylphosphinate, *in* D-01258
41924-81-6 *S*-Butyl divinylphosphinothioate, *in* D-01259
41928-08-9 (*N*-Phosphono)methionine-*S*-sulfoximinylalanylalanine; all-L-*form*, *in* P-00389
41998-90-7 1-Chloroethyl phosphorodichloridate, C-00079
41999-14-8 *N*,*N*′,*N*″-Tri-2-propenylphosphorous triamide, T-00709
42003-30-5 Ethyl *P*-dichloromethylphosphonamidate, *in* D-00174
42003-31-6 Ethyl *P*-trichloromethylphosphonamidate, *in* T-00410
42003-32-7 Phenyl *P*-trichloromethylphosphonamidate, *in* T-00410
42003-54-3 5*a*,6,11,11*a*-Tetrahydro-1,2,3,6,11,12-hexaphenyl-6,11-phosphinidene-1*H*-naphth[2,3-*e*]-isophosphindole; 2,12-Dioxide, *in* T-00107
42003-78-1 *P*,*P*′-1,4-Phenylenebis[*N*,*P*,*P*-triphenylphosphine imide], P-00146
42023-31-4 Monopropyl phosphonate, M-00513
42023-33-6 Mono-2-propenyl phosphonate, M-00510
42036-78-2 Octyltriphenylphosphonium(1+); Bromide, *in* O-00047
42061-52-9 ▷Fopurine, F-00053
42070-34-8 *O*,*O*-Didecyl phosphorochloridothioate, D-00223
42077-94-1 (Aminophenylmethyl)phosphonic acid; (±)-*form*, Di-Et ester; B,HCl, *in* A-00105
42077-97-4 Diethyl (aminophenylmethyl)-phosphonate, *in* A-00105
42087-74-1 3,5-Di-*tert*-butyl-2,5-dihydro-1,2-oxaphosphole 2-oxide; 2-Chloro, *in* D-00082
42087-75-2 3,5-Di-*tert*-butyl-2,5-dihydro-1,2-oxaphosphole 2-oxide; 2-Hydroxy, *in* D-00082
42092-05-7 Tetraethyl [1,2-phenylenebis(methylene)]-bisphosphonate, *in* P-00129
42092-11-5 1,1′,3,3′-Tetrahydro-2,2′-spirobi[2*H*-isophosphindolium](1+); Bromide, *in* T-00137
42104-58-5 [1,2-Phenylenebis(methylene)]-bisphosphonic acid, P-00129
42104-62-1 2,3-Dihydro-2-methoxy-1*H*-isophosphindole 2-oxide, *in* D-00494
42148-25-4 Methylenebisphosphonic difluoride, M-00146
42201-98-9 Tricyclohexylphosphine; Sulfide, *in* T-00426
42202-12-0 1-Benzyl-4,5-dichloro-1-phenyl-1*H*-phosphepinium bromide, *in* D-00565
42202-33-5 1-Bromo-2,5-dihydro-3,4-dimethyl-1*H*-phosphole; 1-Sulfide, *in* B-00469
42202-49-3 4,5-Dihydro-1-phenyl-1*H*-phosphepin, D-00565
42202-53-9 4,5-Dihydro-1-methyl-1-phenyl-1*H*-phosphepinium perchlorate, *in* D-00565
42202-57-3 4,5-Dihydro-1-phenyl-1*H*-phosphepin; Oxide, *in* D-00565
42255-08-3 *O*-Benzyl *O*,*O*-diisopropyl phosphorothioate, B-00044
42282-06-4 1,1,2,3,3-Pentamethyl-6,6,8,8-tetrakis(trifluoromethyl)-5,7-dioxa-4-phospha(4-P^V)spiro[3.4]-octane; 4-Chloro, *in* P-00030
42295-63-6 2-*tert*-Butyltrimethylene phenylphosphonate, *in* B-00577
42295-71-6 *O*-Ethyl methylphenylphosphinothioate, *in* M-00225
42295-72-7 *O*-Isopropyl methylphenylphosphinothioate, *in* M-00225
42295-74-9 *S*-Ethyl methylphenylphosphinothioate, *in* M-00225
42295-75-0 *S*-Isopropyl methylphenylhosphinothioate, *in* M-00225
42333-51-7 Diethyl [(2-quinolinyl)methyl]-phosphonate, *in* Q-00004
42336-88-9 1-Chloro-2,2,3,4,4-pentamethylphosphetane, C-00124
42346-36-1 Trimethylsilyl dimethylphosphinate, *in* D-00829
42346-37-2 Dimethylphosphinothioic acid; *O*-Trimethylsilyl ester, *in* D-00839
42346-39-4 Trimethylsilyl diethylphosphinate, *in* D-00322
42346-42-9 Trimethylsilyl di-*tert*-butylphosphinate, *in* D-00110
42346-43-0 *O*-Trimethylsilyl di-*tert*-butylphosphinothioate, *in* D-00118
42346-45-2 *P*,*P*-Dimethyl-*N*-trimethylsilylphosphinothioic amide, *in* D-00840
42366-28-9 2,4-Dichloro-1,3-di-*tert*-butyl-1,3,2,4-diazadiphosphetidine; *trans-form*, 2,4-Dioxide, *in* D-00163
42405-96-9 3,3′,3″-Phosphinylidynetrisphenol, *in* T-00789
42408-72-0 Methylphosphonochloridic acid, M-00296
42409-57-4 *S*,*S*-Diisopropyl phosphorodithioate, D-00664
42436-56-6 Glycerol 1,2-didodecanoate 3-phosphoethanolamine; (±)-*form*, *in* G-00009
42437-75-2 Tetramethylphosphine imide, *in* T-00531
42449-24-1 Phenylphosphonic acid; Bis(trimethylsilyl) ester, *in* P-00221
42451-26-3 Tetraphenyl 1,2-ethanediylbisphosphonate, *in* E-00006
42451-87-6 Hexahydro-3-phenyl-1*H*-1,3-azaphosphepine, H-00040
42451-95-6 1,2-Dihydro-1,2,3,4-tetraphenyl-1,2-diphosphete, D-00581
42451-96-7 2,5-Dihydro-2,3,4,5-tetraphenyl-1,2,5-oxadiphosphole 2,5-dioxide, D-00583
42451-97-8 2,3-Dihydro-1,2,3,4,5-pentaphenyl-1*H*-1,2,3-triphosphole; 1,3-Disulfide, *in* D-00539
42451-98-9 2,5-Dihydro-2,3,4,5-tetraphenyl-1,2,5-thiadiphosphole; 2,5-Disulfide, *in* D-00585
42475-16-1 2-Phenoxy-1,3,2-dithiaphospholane 2-sulfide, *in* D-01245
42491-33-8 *tert*-Butylbis(trimethylsilyl)-phosphine, B-00545
42495-86-3 Trivinylphosphine; Sulfide, *in* T-00877
42509-80-8 ▷Isazofos, I-00021
42516-28-9 Ethyl 4-(diethoxyphosphinyl)-2-butenoate, *in* D-00248
42534-57-6 1-Chloro-2,5-dihydro-3,4-dimethyl-1*H*-phosphole; 1-Sulfide, *in* C-00041
42546-50-9 [(2-Nitrophenyl)methylene]-triphenylphosphorane, *in* N-00047
42563-95-1 2,2-Dichloro-2,2-dihydro-4-trichloromethyl-1,3,2-diazaphosphete, D-00165
42576-53-4 *O*,*S*-Dimethyl phosphorothioate, D-00908
42581-74-8 *N*′-(4-Nitrophenyl)-*P*,*P*-diphenyl-*N*-(triphenyl-phosphoranylidene)-phosphinimidic amide, *in* D-01164
42585-08-0 ▷*P*-Chloromethyl-*O*-(2-chloro-4-methylphenyl)-*N*-(1-methylpropyl) phosphonamidothioate, *in* C-00101
42585-47-7 Tri-3-butenylphosphine, T-00360
42591-68-4 Diethyl (2-amino-1-methylethyl)phosphonate, *in* A-00088
42800-31-7 Monoisopropyl phosphonate, M-00478
42809-80-3 (2-Ethoxy-2-oxoethyl)-triphenylphosphonium, E-00047
42822-57-1 Diethyl (4-aminophenyl)-phosphonate, *in* A-00111

42844-49-5 Tris[(4-bromophenyl)methyl]-phosphine, T-00733
42847-72-3 *P,P*-Dimethyl-*N*-phenylphosphinothioic amide, *in* D-00840
42847-83-6 *P,P*-Diethyl-*N*-phenylphosphinothioic amide, *in* D-00331
42847-86-7 *N*-(Diethylphosphinothioyl)-*P,P*-diethylphosphinothioic amide, *in* D-00331
42867-45-8 (Dibromomethylene)-triphenylphosphorane, *in* D-00065
42976-67-0 *O*-Methyl phenylphosphonothioate, M-00243
43017-14-7 9-Phenyl-9-phosphabicyclo[6.1.0]-none-2,4,6-triene, P-00197
43017-17-0 9-Phenyl-9-phosphatricyclo[4.2.1.$0^{2,5}$]-nonane; $(1\alpha,2\alpha,5\alpha,6\alpha)$-*form*, 9-Oxide, *in* P-00204
43017-36-3 1,1-Diphenylphospholanium bromide, *in* P-00216
43055-47-6 2-(Triphenylphosphoranylidene)-propanenitrile, T-00659
43055-48-7 Diethyl (cyanophenylmethyl)-phosphonate, *in* C-00219
43077-29-8 (5-Methyl-2-isopropylcyclohexyl)diphenyl-phosphine; (1*S*,2*S*,5*R*)-*form*, *in* M-00164
43077-31-2 (5-Methyl-2-isopropylcyclohexyl)diphenyl-phosphine; (1*R*,2*S*,5*R*)-*form*, *in* M-00164
43081-11-4 Didecylphenylphosphine, D-00219
43207-24-5 [3,7-Dimethyl-9-(2,6,6-trimethyl-1-cyclohexen-1-yl)-2,4,6,8-nonatetraenyl]-triphenylphosphonium(1+); (*E,E,E,E*)-*form*, Periodate, *in* D-00929
44252-11-1 *O*-Methyl methylphosphonothioate; (*S*)-*form*, *in* M-00171
44657-29-6 *O*-Isopropyl methylphosphonothioate; (*S*)-*form*, *in* I-00041
44657-29-6 *O*-Isopropyl methylphosphonothioate; (*S*)-*form*, *S*-Me ester, *in* I-00041
44991-95-9 Oxybis[methanephosphonic acid], O-00095
45098-72-4 *O,O*-Di-*tert*-butyl phosphorothioate, D-00156
45734-11-0 2-Mercapto-5,5-dimethyl-1,3,2-dioxaphosphorinane 2-oxide, M-00010
45778-98-1 Phenylphosphonofluoridic acid, P-00239
45840-49-1 2-Hydroxy-4,4,5,5-tetramethyl-1,3,2-dioxaphospholane 2-sulfide, H-00197
45893-59-2 2,3-Dihydro-2-mercapto-1,3,2-benzoxazaphosphole 2-sulfide, D-00511
46061-42-1 (4-Methylthiophenyl)phosphonic acid, M-00418
46141-44-0 *tert*-Butylphenylphosphinoselenoic acid, B-00582
46338-54-9 *P,P,P*-Trichloro-*N*-(4-nitrophenyl)phosphine imide, *in* N-00063
46501-98-8 *O*-Methyl *O*-1-naphthalenyl phosphorothioate; (−)-*form*, *in* M-00176
47252-15-3 (2-Methyl-2-propenyl)-triphenylphosphonium(1+), M-00367
47252-66-4 (2-Cyanoethyl)-triphenylphosphonium(1+), C-00212
47383-70-0 Tris[1,2-benzenediolato(2−)-*O,O′*] phosphate(1−), T-00730
47467-89-0 Tris(2-methoxyphenyl)phosphine; Oxide, *in* T-00791
47522-13-4 [(4-Chlorophenyl)methyl]-triphenylphosphonium(1+), C-00137
47562-49-2 Triphenyl(3-phenylpropyl)-phosphonium(1+), T-00617
48154-37-6 *O*-Methyl *O*-1-naphthalenyl phosphorothioate; (+)-*form*, *in* M-00176
48195-46-6 (2-Butenyl)-triphenylphosphonium(1+), B-00528
49594-18-5 *N*-Phthalimidomethylphosphonic acid, P-00426
49595-63-3 2,2-Dihydro-4,4,5,5-tetramethyl-2,2,2-triphenyl-1,3,2-dioxaphospholane, D-00578
49622-63-1 3-Methylisophosphinoline, M-00163
49629-23-4 2,2-Dihydro-3,3-bis(trifluoromethyl)-1-[[2,2,2-trifluoro-1-(trifluoromethyl)-ethylidene]amino]-azaphosphiridine; 2,2,2-Trimethoxy, *in* D-00446
49629-24-5 2,2-Dihydro-3,3-bis(trifluoromethyl)-1-[[2,2,2-trifluoro-1-(trifluoromethyl)-ethylidene]amino]-azaphosphiridine; 2,2,2-Triethoxy, *in* D-00446
49661-82-7 2,2-Dihydro-3,3-bis(trifluoromethyl)-1-[[2,2,2-trifluoro-1-(trifluoromethyl)-ethylidene]amino]-azaphosphiridine; 2,2,2-Tris(dimethylamino), *in* D-00446
49676-42-8 Tris(2,4-dimethylphenyl)-phosphine, T-00768
49749-59-9 Cyclohexylphenylphosphine; Sulfide, *in* C-00249
49774-26-7 2,4-Dichloro-1,3-di-*tert*-butyl-1,3,2,4-diazadiphosphetidine; *cis*-*form*, 2-Oxide, *in* D-00163
49774-27-8 2,4-Dichloro-1,3-di-*tert*-butyl-1,3,2,4-diazadiphosphetidine; *cis*-*form*, 2-Sulfide, *in* D-00163
49774-28-9 2,4-Dichloro-1,3-di-*tert*-butyl-1,3,2,4-diazadiphosphetidine; *cis*-*form*, 2-Oxide, 4-sulfide, *in* D-00163
49778-01-0 *N,N,N′,N′,N″,N″*-Hexamethylphosphorimidic triamide, H-00077
49778-02-1 *N,N,N′,N′,P*-Pentamethylphosphonimidic diamide, P-00023
49778-03-2 *N,N,P,P*-Tetramethylphosphinimidic amide, *in* D-00833
49778-04-3 *N,N,N′,N′,N″,N″,N‴*-Heptamethylphosphorimidic triamide, H-00001
49778-05-4 Hexamethylphosphonimidic diamide, *in* P-00023
49778-06-5 Pentamethylphosphinimidic amide, *in* D-00833
49785-01-5 3-Chloro-1,5-dihydro-2,4,3-benzodioxaphosphepin; 3-Oxide, *in* C-00038
49785-03-7 1,5-Dihydro-2-phenoxy-2,4,3-benzodioxaphosphepin; 3-Oxide, *in* D-00544
49789-34-6 Hexahydro-1,3-diphenyl-1,3-azaphosphorine, *in* H-00041
49789-43-7 Hexahydro-3-phenyl-1,3-azaphosphorine; 1-Ph, 3-sulfide, *in* H-00041
49802-15-5 Diethyl (4-chlorophenyl)-phosphoramidate, *in* C-00157
49802-17-7 Diethyl 1-naphthylphosphoramidate, *in* N-00028
49802-18-8 Diethyl 1-adamantylphosphoramidate, *in* A-00036
49802-22-4 Diphenyl (4-chlorophenyl)-phosphoramidate, *in* C-00157
49802-24-6 Diphenyl 1-adamantylphosphoramidate, *in* A-00036
49868-41-9 (3-Methyl-2-butenyl)-triphenylphosphonium(1+), M-00104
49873-27-0 Methyl diethylphosphinodithioate, *in* D-00325
49873-30-5 *S*-Methyl di-*tert*-butylphosphinothioate, *in* D-00118
49873-30-5 *S*-Methyl diisopropylphosphinothioate, *in* D-00646
49873-31-6 Methyl diisopropylphosphinodithioate, *in* D-00643
49873-32-7 *S*-Methyl dipropylphosphinothioate, *in* D-01208
49873-34-9 Methyl dibutylphosphinodithioate, *in* D-00112
50259-42-2 [(5,6-Dihydro-4,4,6-trimethyl-4*H*-1,3-oxazin-2-yl)-methyl]triphenyl-phosphonium(1+); Chloride, *in* D-00599
50281-51-1 Azidotris(dimethylamino)-phosphorus(1+); Hexafluorophosphate, *in* A-00149
50329-34-5 *O,O*-Diisopropyl phosphorodithioate; Cyclohexylammonium salt, *in* D-00663
50341-15-6 Tris(2,6-dimethylphenyl)-phosphine, T-00770
50351-54-7 *O*-Phenyl diethylphosphinoselenoate, *in* D-00327
50351-61-1 Hexaethylphosphorous triamide; Telluride, *in* H-00015
50375-72-9 Diethyl [(2-hydroxyphenyl)-methyl]phosphonate, *in* H-00175
50375-97-8 2-Butenylidenetriphenyl-phosphorane; (*Z*)-*form*, *in* B-00522
50376-02-8 2-Nonenylidenetriphenyl-phosphorane, N-00071
50387-23-0 *P,P*-Dibutyl-*N,N*-dimethylphosphinic amide, *in* D-00111
50396-26-4 1,2-Bis[(bis-2-methylphenyl)-phosphino]ethane, B-00103
50431-01-1 Diethyl [(5,6-dihydro-4,4,6-trimethyl-4*H*-1,3-oxazin-2-yl]phosphonate, *in* D-00598
50457-19-7 2,2-Diamino-2,2,4,4,6,6-hexahydro-4,4,6,6-tetraphenyl-1,3,5-triaza-2,4,6-triphosphorine, D-00025
50457-20-0 2-Chloro-2,2,4,4,6,6-hexahydro-2-methyl-4,4,6,6-tetraphenyl-1,3,5-triaza-2,4,6-triphosphorine, C-00086
50457-21-1 2-Amino-2,2,4,4,6,6-hexahydro-2-methyl-4,4,6,6-tetraphenyl-1,3,5-triaza-2,4,6-triphosphorine, A-00081
50518-33-7 1,3-Bis(dimethylphosphinothioyl)-propane, *in* B-00206
50538-13-1 Diisopropylphenylphosphine; Sulfide, *in* D-00638
50566-50-2 *O,O*-Di-1-naphthyl phosphorodithioate, D-00944
50577-84-9 (3-Phenylpropyl)phosphonic acid, P-00305
50577-97-4 2-Mercapto-1,3,2-benzodioxaphosphole 2-sulfide, M-00005
50597-14-3 1,3-Dihydrospiro[2*H*-1,3,2-benzodiazaphosphole-2,2′-[1,3,2]benzodioxaphosphole], D-00574
50597-18-7 2-Chloro-4,5-diphenyl-1,3,2-dioxaphospholane, C-00069

50602-70-5 Dibutylphosphinous acid, D-00121
50614-16-8 *O,O*-Bis(2-methylpropyl) phosphorothioate, B-00363
50615-91-3 (Hexafluorocyclobutylidene)-triphenylphosphorane, H-00017
50636-18-5 2,8,9-Trimethyl-2,8,9-triaza-1-phosphatricyclo[3.3.1.1^{3,7}]-decane; Oxide, *in* T-00563
50636-19-6 2,8,9-Trimethyl-2,8,9-triaza-1-phosphatricyclo[3.3.1.1^{3,7}]-decane, T-00563
50636-20-9 2,8,9-Trimethyl-2,8,9-triaza-1-phosphatricyclo[3.3.1.1^{3,7}]-decane; Sulfide, *in* T-00563
50636-30-1 2,3-Dihydro-2,7-dimethyl-1,4,2-benzodithiaphosphorin, D-00458
50636-31-2 3,4-Dihydro-2-methyl-2*H*-1,4,2-benzothiazaphosphorine, D-00521
50636-32-3 2,3-Dihydro-2,7-dimethyl-1,4,2-benzodithiaphosphorin; 2-Selenide, *in* D-00458
50636-33-4 3,4-Dihydro-2-methyl-2*H*-1,4,2-benzothiazaphosphorine; 2-Sulfide, *in* D-00521
50636-34-5 3,4-Dihydro-2-methyl-2*H*-1,4,2-benzothiazaphosphorine; 2-Selenide, *in* D-00521
50636-74-3 *O*-Methyl isopropylphosphonochloridothioate, M-00165
50636-75-4 *O*-Methyl isopropylphosphonothioate, M-00166
50651-75-7 Dibenzyl phosphate; Ag salt, *in* D-00053
50655-63-5 Diethyl (2-methylpropyl)-phosphonate, *in* M-00378
50663-05-3 Tetramethylphosphinic amide, *in* D-00830
50669-78-8 *O*-Ethyl *S*-propyl phosphorodithioate, E-00186
50672-16-7 Diethyl (3-fluorophenyl)-phosphoramidate, *in* F-00043
50672-17-8 *O,O*-Diethyl (3-fluorophenyl)-phosphoramidothioate, *in* F-00045
50672-18-9 Diethyl (4-fluorophenyl)-phosphoramidate, *in* F-00044
50672-19-0 *O,O*-Diethyl (4-fluorophenyl)-phosphoramidothioate, *in* F-00046
50672-29-2 Diisopropyl (4-fluorophenyl)-phosphoramidate, *in* F-00044
50672-30-5 *O,O*-Diisopropyl (4-fluorophenyl)-phosphoroamidothioate, *in* F-00046
50682-94-5 *P*-Methyl-*P*-phenylphosphinic amide; (*S*)-*form*, *N*-Ph, *in* M-00223
50687-78-0 *O,O*-Dimethyl vinylphosphonothioate, *in* V-00007
50694-63-8 2,4,6-Triphenyl-4*H*-1,4-thiaphosphorin; 4-Oxide, *in* T-00686
50694-65-0 2,4,6-Triphenyl-4*H*-1,4-thiaphosphorin; 1,1,4-Trioxide, *in* T-00686
50696-02-1 Tetraphenylthiatetraphospholane; (2α,3β,4α,5β)-*form*, *in* T-00282
50704-15-9 Triethyl phosphoro(dithioperoxoate), T-00452
50704-30-8 *O,O*-Diethyl *SS*-methyl phosphoro(dithioperoxo)-thioate, D-00300
50731-87-8 3-Methoxy-18-nor-17-phosphaestra-1,3,5(10),-9(11)-tetraen-15-one; 13α-*form*, 17-Ph, 17-oxide, *in* M-00054
50731-88-9 3-Methoxy-18-nor-17-phosphaestra-1,3,5(10),-9(11)-tetraen-15-one; 13α-*form*, 17-Me, 17-oxide, *in* M-00054
50732-22-4 Bis(trimethylsilyl)-phosphoramidous acid; Difluoride, *in* B-00438
50732-24-6 Methyl *N,N,N'*-tris(trimethylsilyl)-phosphonamidimidate, *in* T-00867
50738-79-9 2,5-Bis[(triphenylphosphonio)-methyl]furan(2+); Dichloride, *in* B-00448
50742-51-3 3-Benzyl-2-chlorotetrahydro-2*H*-1,3,2-oxazaphosphorine; 2-Sulfide, *in* B-00041
50742-52-4 Tetrahydro-2-phenoxy-2*H*-1,3,2-oxazaphosphorine; 2-Oxide, *in* T-00126
50742-53-5 Tetrahydro-2-phenyl-2*H*-1,3,2-oxazaphosphorine; 2-Sulfide, *in* T-00132
50775-60-5 2-Propenyl tetramethylphosphorodiamidate, *in* T-00218
50777-69-0 3-(Diphenylphosphino)-benzaldehyde, D-01051
50777-70-3 3-(Diphenylphosphinothioyl)-benzaldehyde, *in* D-01051
50777-71-4 3-(Diphenylphosphinyl)-benzaldehyde, *in* D-01051
50777-76-9 2-(Diphenylphosphino)-benzaldehyde, D-01050
50777-77-0 2-(Diphenylphosphinyl)-benzaldehyde, *in* D-01050
50777-78-1 2-(Diphenylphosphinothioyl)-benzaldehyde, *in* D-01050
50782-69-9 ▷*O*-Ethyl *S*-[2-(diisopropylamino)ethyl] methylphosphonothioate, E-00062
50869-62-0 2,3-Dihydro-2-phenyl-1*H*-isophosphindole; Oxide, *in* D-00562
50889-29-7 (5-Carboxypentyl)-triphenylphosphonium(1+); Bromide, *in* C-00015
50897-31-9 3-Methoxy-18-nor-17-phosphaestra-1,3,5(10),-9(11)-tetraen-15-one; 13β-*form*, 17-Ph, 17-oxide, *in* M-00054
50902-83-5 4-Methyl-2-methylthio-1,3,2-dioxaphosphorinane; (2*RS*,4*SR*)-*form*, 2-Oxide, *in* M-00174
50902-84-6 *O,O*-1,3-Butylene *S*-methyl thiophosphate, *in* M-00174
50915-72-1 Diethyl (aminomethyl)-phosphonate, *in* A-00091
50917-73-2 Diethyl [(benzylideneamino)-methyl]phosphonate, D-00271
50964-72-2 2,2,4,4,6,6-Hexahydro-2,2-dimethyl-4,4,6,6-tetraphenyl-1,3,5-triaza-2,4,6-triphosphorine, H-00025
50964-97-1 1,2,4,5-Tetraphenyl-1,2,4,5-tetraphosphorinane, T-00281
51008-95-8 (2-Oxo-4-phenylbutyl)phosphonic acid, O-00065
51010-76-5 Dibenzylphosphinic acid; Fluoride, *in* D-00055
51021-79-5 1,12-Diphenyl-1,12-diphosphacyclotetracosane; *cis-form*, 1,12-Dioxide, *in* D-01010
51021-80-8 1,10-Diphenyl-1,10-diphosphacyclooctadecane; *trans-form*, *in* D-01009
51021-81-9 1,12-Diphenyl-1,12-diphosphacyclodocosane; *trans-form*, *in* D-01008
51021-82-0 1,12-Diphenyl-1,12-diphosphacyclotetracosane; *trans-form*, *in* D-01010
51021-83-1 1,10-Diphenyl-1,10-diphosphacyclooctadecane; *trans-form*, 1,10-Dioxide, *in* D-01009
51028-07-0 *P*-Ethyl-*N,P*-diphenylphosphinic amide, *in* E-00102
51028-11-6 *P-tert*-Butyl-*N,P*-diphenylphosphinic amide, *in* B-00580
51028-14-9 *P*-Ethyl-*P*-phenylphosphinic amide, E-00102
51028-16-1 *P*-Isopropyl-*P*-phenylphosphinic amide, I-00044
51028-18-3 *P-tert*-Butyl-*P*-phenylphosphinic amide, B-00580
51044-11-2 [(4-Fluorophenyl)methyl]-triphenylphosphonium(1+); Bromide, *in* F-00032
51044-12-3 [(4-Chlorophenyl)methyl]-triphenylphosphonium(1+); Bromide, *in* C-00137
51049-70-8 *S*-Methyl dibutylphosphinothioate, *in* D-00117
51064-83-6 1,12-Diphenyl-1,12-diphosphacyclodocosane; *cis-form*, *in* D-01008
51064-84-7 1,12-Diphenyl-1,12-diphosphacyclotetracosane; *cis-form*, *in* D-01010
51066-52-5 Dimethyl (2,3,5,6-tetrachloro-4-pyridinyl)phosphonate, *in* T-00035
51072-22-1 *O*-Methyl dimethylphosphinoselenoate, *in* D-00835
51072-23-2 *O,O*-Dimethyl methylphosphonoselenoate, *in* M-00311
51090-79-0 *O,O*-Bis(2-methylpropyl) phosphorochloridothioate, B-00357
51090-80-3 1-Chloro-2,5-dihydro-2-methyl-1*H*-phosphole, C-00046
51096-67-4 1,12-Diphenyl-1,12-diphosphacyclodocosane; *trans-form*, 1,12-Dioxide, *in* D-01008
51096-67-4 1,10-Diphenyl-1,10-diphosphacyclooctadecane; *cis-form*, 1,10-Dioxide, *in* D-01009
51096-68-5 1,12-Diphenyl-1,12-diphosphacyclodocosane; *cis-form*, 1,12-Dioxide, *in* D-01008
51096-69-6 1,12-Diphenyl-1,12-diphosphacyclotetracosane; *trans-form*, 1,12-Dioxide, *in* D-01010
51101-58-7 Bis(trifluoromethyl)-phosphinoselenous acid; Anhydroselenide, *in* B-00413
51103-89-0 Bis(3-methylphenyl)phosphinic acid; Chloride, *in* B-00319
51103-90-3 Bis(4-bromophenyl)phosphinic acid; Chloride, *in* B-00114
51103-92-5 Bis(3-chlorophenyl) phosphorochloridate, *in* B-00132
51103-93-6 Bis(3-chlorophenyl) phosphorochloridite, B-00146
51103-94-7 *O,O*-Bis(3-chlorophenyl) phosphorochloridothioate, B-00148
51104-85-9 3-Phenyl-1,3,2-oxazaphospholidine; Oxide, *in* P-00180
51114-94-4 (2-Carboxyethyl)-triphenylphosphonium(1+); Bromide, *in* C-00011
51120-35-5 ▷*sym-O,O,O,O*-Tetramethyl dithiopyrophosphate, T-00203
51145-33-6 1,5-Bis(diphenylphosphinyl)-pentane, *in* B-00259
51153-50-5 Benzylmethylphenylphosphine; (±)-*form*, Oxide, *in* B-00051
51154-91-7 1,10-Diphenyl-1,10-diphosphacyclooctadecane; *cis-form*, *in* D-01009
51168-02-6 Tetramethylphosphinoselenoic amide, *in* D-00836

51274-71-6 Peroxycyclophosphamide, P-00057

51276-47-2 Phosphinothricin, P-00359

51303-82-3 Diisopropyl *p*-tolylphosphonite, *in* M-00248

51321-79-0 ▷Sparfosic acid; (*S*)-*form*, *in* S-00006

51329-19-2 2,3-Diphenyl-1,3-thiaphosphorinane, D-01152

51342-28-0 Bis(diphenylphosphino)methanone, B-00245

51346-85-1 Tetraethyl 2,2-propanediyldiphosphonate, *in* P-00445

51354-18-8 Diphenyl phosphorocyanidate, D-01124

51371-50-7 Phenylphosphonothioic acid trimer trianhydride, *in* T-00694

51374-71-1 2-Hydroxy-1,3,2-dioxaphosphepane 2-oxide, H-00132

51395-42-7 (Diphosphonomethyl)butanedioic acid, D-01185

51395-43-8 4-Phosphono-1,2,3-butanetricarboxylic acid, P-00384

51439-10-2 2-(*tert*-Butylamino)-1,3,2-dioxaphospholane, B-00539

51446-62-9 Glycerol 1,2-dioctadecanoate 3-phosphoserine; (*R*)-L-*form*, *in* G-00038

51452-84-7 1,3-Dimethyl-1,3,2-diazaphospholidine 2-oxide, D-00720

51461-98-4 Pentamethylphosphonotelluroic diamide, P-00025

51462-00-1 *O*-Ethyl tetraethylphosphorodiamidotelluroate, *in* T-00062

51463-65-1 Dimethyl (1-oxopropyl)-phosphonate, *in* O-00074

51500-74-4 3-Fluoro-1,3-thiaphosphetane; 3-Sulfide, *in* F-00050

51501-23-6 *P,P,P*-Tribenzylphosphine imide, T-00339

51503-25-4 Diethyl butylphosphonite, *in* B-00615

51528-28-0 Hexylmethylphosphinic acid; Chloride, *in* H-00094

51528-32-6 Ethylmethylphosphinic acid, E-00085

51528-33-7 Dimethylphosphinic acid; NH_4 salt, *in* D-00829

51528-37-1 Ethylmethylphosphinic acid; Anhydride, *in* E-00085

51528-98-4 Hexaphenylphosphorous triamide, H-00090

51531-26-1 1,2,3-Triphenyl-1,3-azaphospholidin-5-one, T-00596

51572-21-5 3,3,8,8-Tetramethyl-1,6-dioxa-4,9-diaza-5-phospha(P^V)-spiro[4.4]nonane; (±)-*form*, *in* T-00196

51576-47-7 Triphenyl (diethoxyphosphinothioyl)-phosphorimidate, *in* D-00238

51576-53-5 Tributyl (dimethoxyphosphinothioyl)-phosphorimidate, *in* D-00674

51576-54-6 Triphenyl (dimethoxyphosphinothioyl)-phosphorimidate, *in* D-00674

51576-55-7 Triethyl (diphenoxyphosphinothioyl)-phosphorimidate, *in* D-00987

51576-56-8 *O,O*-Dipropyl phosphorazidothioate, D-01216

51576-57-9 *O,O*-Dimethyl phosphorazidothioate, D-00879

51576-61-5 *O,O*-Diphenyl phosphorazidothioate, D-01113

51576-73-9 *O,O*-Diphenyl phosphorodithioate; Diethylammonium salt, *in* D-01131

51584-27-1 *tert*-Butylphenylphosphinoselenoic acid; (*R*)-*form*, *in* B-00582

51584-28-2 *tert*-Butylphenylphosphinoselenoic acid; (*S*)-*form*, *in* B-00582

51584-29-3 *S*-Methyl *tert*-butylphenylphosphinothioate, *in* B-00583

51584-30-6 *tert*-Butylphenylphosphinothioic acid; (*S*)-*form*, *S*-Me ester, *in* B-00583

51593-74-9 [Thiobis[methylene]]-bis[triphenylphosphorane], *in* T-00312

51639-96-4 4,4′,4″-Phosphinoselenoylidynetrismorpholine, *in* T-00569

51656-53-2 1,3,2-Benzoxathiaphosphole; 2-Ph, *in* B-00026

51672-37-8 Dicyclohexyl phosphorochloridate, *in* D-00209

51675-70-8 4′,4′,5′,5′-Tetramethylspiro[1,3,2-benzoxazaphosphole-2(3*H*),2′-[1,3,2]-dioxaphospholane], T-00231

51675-92-4 3′-Methylspiro[1,3,2-benzodioxaphosphole-2,2′(3′*H*)-[1,3,2]-benzoxazaphosphole], M-00395

51675-96-8 2,2,7,7-Tetramethyl-1,6-dioxa-4,9-diaza-5-phospha(5-P^V)spiro[4.4]nonane, T-00195

51676-08-5 4′,5′-Dihydro-3′,5′-dimethylspiro[1,3,2-benzoxazaphosph(P^V)ole-2,2′-[1,3,2]oxazaphospholidin]-4′-one, D-00470

51676-09-6 4′,5′-Dihydro-3′,5′-dimethylspiro[1,3,2-benzoxazaphosph(P^V)ole-2,2′-[1,3,2]oxazaphospholidin]-4′-one; (5′*S*)-*form*, 3-Me, *in* D-00470

51703-89-0 *P*-Methyl-*P*-phenylphosphinic amide; (±)-*form*, *N*-Ph, *in* M-00223

51747-25-2 *N,N*-Diisopropyl-*P*-methylphosphonamidothioic acid; Chloride, *in* D-00637

51756-01-5 ▷Diisopropyl chlorothiophosphonate, D-00634

51757-41-6 3-Phosphono-1,2,3-butanetricarboxylic acid, P-00383

51761-27-4 Diethyl phosphorobromidate, D-00371

51761-80-9 *O,O,O′,O′*-Tetrapropyl thiohypophosphate, T-00293

51766-21-3 Phenyl phenylphosphoramidochloridate, P-00190

51771-83-6 *P*-*tert*-Butyl-*N,N′*-diphenylphosphonamidimidic chloride, B-00559

51777-80-1 2,2,3,3,7,7-Hexamethyl-1,4,6-trioxa-9-aza-5-phospha(5-P^V)-spiro[4.4]nonane, H-00083

51795-72-3 Diethyl (3-methyl-2-butenyl)-phosphonate, *in* M-00102

51805-04-0 *S,S*-Diethyl phenylphosphonodithioate, *in* P-00236

51807-11-5 1,2-Ethanediylbis[methylphosphinic acid], E-00013

51825-26-4 Diethyl propylphosphonite, *in* P-00482

51825-87-7 *O,O*-Dibutyl phosphorothioate; K salt, *in* D-00155

51833-57-9 *N,N′*-Dimethyl-*N,N′*-bis[[bis(dimethylamino)-phosphinyl]amino]ethane, D-00713

51833-60-4 *unsym*-Octamethyl-monothiopyrophosphoramide, O-00031

51835-37-1 4-Quinolinephosphonic acid, Q-00003

51860-45-8 (3-Hydroxypropyl)-triphenylphosphonium(1+); Bromide, *in* H-00193

51863-54-8 Dipropyl (cyanophenylmethyl)-phosphonate, *in* C-00219

51870-63-4 *N′*-(4-Fluorophenyl)-*P,P*-diphenyl-*N*-(triphenyl-phosphoranylidene)-phosphinimidic amide, *in* D-01164

51922-82-8 8,8-Dihydro-8,8-diphenyl-2*H*,6*H*-[1,2]oxaphospholo[4,3,2-*hi*][2,1]benzoxaphosphole, D-00479

51926-09-1 *O*-Ethyl hydrogen propylphosphonodithioate, *in* P-00478

51934-48-6 Methyl methylphosphonochloridite, *in* M-00300

51945-70-1 Trimethyl (diethoxyphosphinothioyl)-phosphorimidate, *in* D-00238

51965-64-1 (Trifluoromethyl)phosphonic dichloride, T-00479

51986-54-0 (2-Methoxyphenyl)-diphenylphosphine; Oxide, *in* M-00056

51992-59-7 Tetrahydro-2-phenyl-2*H*-1,3,2-oxazaphosphorine, T-00132

51992-60-0 *N,N,N′*-Triethylphosphoramidimidic dichloride, T-00449

51993-67-0 *O*-Isopropyl diethylphosphinothioate, *in* D-00330

51998-39-1 2-Phenyl-2*H*-isophosphindole 2-oxide, P-00169

52038-87-6 *O,O*-Diethyl isopropylphosphonothioate, *in* I-00054

52041-10-8 *O*-(2-Chloroethyl) phosphorodichloridothioate, C-00080

52045-12-2 1,5-Bis(triphenyl-phosphoniomethyl)-naphthalene(2+); Dibromide, *in* B-00451

52045-25-7 [(2-Methoxyphenyl)methyl]-triphenylphosphonium(1+); Chloride, *in* M-00062

52057-41-7 2-Methylpropyl phosphorodichloridite, M-00385

52103-13-6 Triphenyl[[3-(tetrahydro-2*H*-pyran-2-yl)oxy]propyl]-phosphonium(1+); Iodide, *in* T-00683

52111-99-6 1,2,3-Triphenyl-1,3-azaphospholidin-5-one; 3-Oxide, *in* T-00596

52112-01-3 1,2,3-Triphenyl-1,3-azaphospholidin-5-one; 3-Sulfide, *in* T-00596

52112-04-6 1,2,3-Triphenyl-1,3-azaphosphacyclopentane, *in* D-00995

52119-19-4 Ethylmethylphenylphosphine; (*R*)-*form*, *in* E-00084

52125-84-5 2,3-Diphenyl-1,3-thiaphospholane, D-01151

52134-55-1 4-(Triphenylphosphoranylidene)-butanoic acid, *in* C-00017

52175-06-1 *N,N,N′,N′*-Tetraethyl-*P*-(phenylmethyl)phosphonic diamide, *in* B-00061

52186-87-7 1-(Ethoxycarbonyl)-cyclopropyltriphenyl-phosphonium(1+), E-00025

52186-89-7 1-(Ethoxycarbonyl)-cyclopropyltriphenyl-phosphonium(1+); Tetrafluoroborate, *in* E-00025

52198-45-5 2-Phthalimidoethyl phosphorodichloridate, P-00425

52199-87-8 2-Chloro-1,3,2-benzodithiaphosphole, C-00023

52202-92-3 Tetrahydro-2,6-dimethyl-4*H*-1,3,6,2-dioxazaphosphocine; 2-Oxide, *in* T-00094

52221-68-8 1-Phenyl-1,2,3,4-tetrahydrophosphinoline, P-00320

52221-86-0 1,2,3,4-Tetrahydro-1-methoxyphosphinoline 1-oxide, *in* H-00195
52221-87-1 1-Ethoxy-1,2,3,4-tetrahydrophosphinoline 1-oxide, *in* H-00195
52221-88-2 1,2,3,4-Tetrahydrophosphinoline; 1-Et, 1-sulfide, *in* T-00135
52221-89-3 1-Phenyl-1,2,3,4-tetrahydrophosphinoline; 1-Oxide, *in* P-00320
52221-91-7 Diethyl (3-phenylpropyl)-phosphonate, *in* P-00305
52222-65-8 *P*-(4-Methylphenyl)-*N,N′*-diphenylphosphonic diamide, *in* M-00238
52258-06-7 *O,O*-(2,2′-Biphenylyl) phosphorochloridate, *in* C-00032
52293-07-9 1-Hydroxy-1,2,3,4-tetrahydrophosphinoline 1-oxide, H-00195
52293-08-0 1-Ethyl-1,2,3,4-tetrahydrophosphinoline 1-oxide, *in* T-00135
52330-21-9 Diethyl 2-(cyclohexylamino)-2-methylethenyl phosphonate, D-00276
52336-54-6 Alcophosphamide, *in* B-00126
52343-38-1 Dimethyl(2-oxo-3-phenylpropyl)-phosphonate, *in* O-00068
52364-38-2 2,3,3,4-Tetraphenyl-1,2-oxaphosphetane 2-oxide, T-00268
52364-40-6 2,2,3,4,6-Pentaphenyl-2*H*-1,5,2-dioxaphosphorin 2-oxide, P-00036
52420-88-9 Dimethyl methylphosphoramidate, *in* M-00326
52427-40-4 Hexahydro-1-methyl-4-phenyl-1,4-azaphosphorine; 4-Oxide, *in* H-00036
52427-43-7 4-Phenyl-1,4-oxaphosphorinane; Oxide, *in* P-00176
52427-44-8 Hexahydro-1-methyl-4-phenyl-1,4-azaphosphorine, H-00036
52427-45-9 Hexahydro-1-methyl-4-phenyl-1,4-azaphosphorine; 4-Sulfide, *in* H-00036
52427-49-3 1-Hydroxy-2,3-dihydro-1*H*-phosphindole 1-oxide, H-00120
52427-50-7 2,3-Dihydro-1-methoxy-1*H*-phosphindole 1-oxide, *in* H-00120
52427-76-6 1,2-Dihydro-1-hydroxyphosphinoline 1-oxide, D-00505
52427-77-7 1,2-Dihydro-1-methoxyphosphinoline 1-oxide, *in* D-00505
52427-85-7 3,4,5,6-Tetrahydro-4-methyl-2*H*-1,5-methano-4,1-benzazaphosphocine; 1-Oxide, *in* T-00115
52427-86-8 3,4,5,6-Tetrahydro-4-methyl-2*H*-1,5-methano-4,1-benzazaphosphocine; 1-Oxide; B,HCl, *in* T-00115
52427-87-9 3,4,5,6-Tetrahydro-4-methyl-2*H*-1,5-methano-4,1-benzazaphosphocine, T-00115
52427-88-0 3,4,5,6-Tetrahydro-4-methyl-2*H*-1,5-methano-4,1-benzazaphosphocine; 1-Sulfide, *in* T-00115
52427-89-1 3,4,5,6-Tetrahydro-4-methyl-2*H*-1,5-methano-4,1-benzazaphosphocine; 1-Sulfide; B,HCl, *in* T-00115
52463-56-6 Tetrahydro-3-phenyl-2-(phenylamino)-2*H*-1,3,2-oxazaphosphorine; 2-Oxide, *in* T-00133
52483-84-8 Mono(4-nitrophenyl) phosphate; Bis(cyclohexylammonium) salt, *in* M-00500
52509-14-5 [(1,3-Dioxolan-2-yl)methyl]-triphenylphosphonium(1+); Bromide, *in* D-00976
52513-28-7 [(2-Nitrophenyl)methyl]-triphenylphosphonium(1+); Chloride, *in* N-00047
52551-78-7 Diethyl (2-aminoethyl)-phosphoramidate, *in* A-00080
52553-42-7 Bis(2-*tert*-butylphenyl) phosphorochloridate, *in* B-00115
52577-04-5 Tetramethyl [1,4-phenylenebis(methylene)]-bisphosphonate, *in* P-00131
52604-86-1 Butyl tetramethylphosphorodiamidate, *in* T-00218
52625-61-3 Tetrakis(4-nitrophenyl) diphosphate, T-00183
52670-78-7 Diethyl methylphenylphosphoramidate, *in* M-00250
52670-92-5 Diphenyl methylphenylphosphoramidate, *in* M-00250
52678-80-5 *O,O*-Bis(4-nitrophenyl) phosphorochloridothioate, B-00377
52692-02-1 Benzyl phosphorodichloridate, *in* M-00443
52710-37-9 (3-Methylbutyl)-triphenylphosphonium(1+); Iodide, *in* M-00111
52713-97-0 2-Methyl-5-phenyl-2*H*-1,2,4,3-triazaphosphole, M-00270
52718-98-1 1-Methyl-5-phenyl-1*H*-1,2,4,3-triazaphosphole, M-00269
52726-83-7 Diisopropyl (2-cyanoethyl)-phosphonate, *in* C-00211
52744-21-5 Phenyl phenylphosphinate, *in* P-00210
52750-95-5 (3-Methyl-2-butenyl)-triphenylphosphonium(1+); Chloride, *in* M-00104
52784-98-2 Tricyclohexylphosphine; Selenide, *in* T-00426
52809-04-8 Di-*tert*-butylphosphinous acid, D-00122
52810-61-4 2,5-Difluoro-1-methyl-1,2,5-azadiphospholidine, *in* M-00089
52912-57-9 Diethyl (4-bromophenyl)-phosphoramidate, *in* B-00496
52912-90-0 *O*-Methyl *O,O*-trimethylene phosphoroselenoate, *in* M-00038
52928-43-5 Bis(2-methylpropyl) (2-methylpropyl)-phosphonate, *in* M-00378
52940-07-5 Phenyl(trichloromethyl)-phosphinic acid; Chloride, *in* P-00327
52944-44-0 Tris(2,4-dimethylphenyl)-phosphine; Oxide, *in* T-00768
52944-83-9 Tris(2,6-dimethylphenyl)-phosphine; Oxide, *in* T-00770
52956-93-1 (7-Carboxyheptyl)-triphenylphosphonium(1+); Bromide, *in* C-00012
52961-94-1 2,2(3*H*,3′*H*)-Spirobi[1,3,2-benzoxazaphosphole]; 2-Phenoxy, *in* S-00020
52961-95-2 Phenyl *o*-phenylene phosphate, *in* P-00063
52988-62-2 2,3-Dihydro-1-phenyl-1*H*-phosphole; 1-Sulfide, *in* D-00566
53036-80-9 6-(Triphenylphosphoranylidene)-hexanoic acid, *in* C-00015
53054-21-0 (3-Hydroxypropyl)phosphonic acid, H-00192
53056-49-8 *O*-Phenyl ethylphosphonochloridothioate, *in* E-00137
53088-59-8 1,3-Diphenyl-1,3-diphospholane, D-01013
53097-98-6 Octahydro-3-phenyl-1*H*-cyclopropa[*c*]phosphindole 3-oxide, *in* O-00004
53098-45-6 *P,P,P*-Triethylphosphine imide, T-00442
53103-03-0 Bis[2-(diphenylphosphino)-phenyl]phenylphosphine, B-00262
53104-46-4 (2-Hydroxyphenyl)phosphonic acid, H-00179
53111-20-9 (2-Methoxyphenyl)-diphenylphosphine, M-00056
53121-78-1 *O,O*-Neopentylene methylphosphonothioate, *in* T-00512
53144-23-3 *O,O*-Diphenyl phosphorohydrazidothioate, D-01137
53156-59-5 10-(Triphenylphosphoniomethyl)-benzo[*g*]chrysene(1+); Bromide, *in* T-00635
53159-02-7 Isopropylphenylphosphinothioic acid, I-00046
53159-03-8 Di-*tert*-butylphosphinothioic acid, D-00118
53167-50-3 *N,N,N′,N′,N″,N″*-Hexamethyl-*N‴*-trimethylsilylphosphorimidic triamide, H-00082
53173-36-7 Octahydro-3-methoxy-1*H*-cyclopropa[*c*]phosphindole 3-oxide, *in* O-00004
53213-05-1 Cyclohexylphosphonoselenoic acid; Dichloride, *in* C-00263
53213-06-2 Cyclobutylidenetriphenyl-phosphorane, *in* C-00235
53213-07-3 Cycloheptylidenetriphenyl-phosphorane, *in* C-00238
53213-08-4 Triphenyl(2-phenylethylidene)-phosphorane, *in* T-00613
53213-26-6 Triphenyl(2-phenylethyl)-phosphonium(1+); Bromide, *in* T-00613
53213-42-6 Dimethylphosphoramidoselenoic acid; Difluoride, *in* D-00869
53213-43-7 Diethylphosphoramidoselenoic acid; Difluoride, *in* D-00353
53218-00-1 Tris(2-diphenylphosphinophenyl)-phosphine, T-00777
53227-37-5 *O,O*-Diethyl (diethoxyphosphinyl)-methylphosphoramidothioate, *in* M-00413
53228-52-7 *O*-Ethyl ethylphosphonoselenoate; (*S*)-*form*, *in* E-00072
53228-53-8 *O*-Ethyl ethylphosphonoselenoate; (*R*)-*form*, *in* E-00072
53228-54-9 *O*-Ethyl ethylphosphonoselenoate; (±)-*form*, *in* E-00072
53235-71-5 Diphenyl (2-methylpropyl)-phosphonate, *in* M-00378
53236-29-6 Bis(2,2-dimethylpropyl) phosphorochloridite, B-00209
53243-58-6 [Bis(phenylmethoxy)phosphinyl]-acetic acid, B-00389
53246-95-0 Diethyl octylphosphoramidate, *in* O-00045
53253-54-6 Diethyl (4-aminobutyl)-phosphonate, *in* A-00057
53253-55-7 Diethyl (5-aminopentyl)-phosphonate, *in* A-00097
53253-67-1 Diethyl (3-cyanopropyl)-phosphonate, *in* C-00230
53255-91-7 2-Ethoxy-4,5-dimethyl-1,3,2-dioxaphospholane, E-00031
53282-27-2 [2-(4-Nitrophenyl)-2-oxoethyl]-triphenylphosphonium(1+); Iodide, *in* N-00051
53282-28-3 [3-Methyl-5-(2,6,6-trimethyl-1-cyclohexen-1-yl)-2,4-pentadienyl]triphenyl-phosphonium(1+); (*E,E*)-*form*, Chloride, *in* M-00421
53314-51-5 2-Methylpropyl methyl(2-methylpropyl)phosphinate, *in* M-00172
53314-64-0 Methylvinylphosphinic acid, M-00438
53314-66-2 2-Methyl-1,2-oxaphospholane; 2-Oxide, *in* M-00182
53327-22-3 Bis(pentafluorophenyl)-phosphinothioic acid; Fluoride, *in* B-00383

53327-28-9 Bis(pentafluorophenyl)phosphinothioic acid; Chloride, *in* B-00383

53327-49-4 Phenyl 2-(diethoxyphosphinyl)butanoate, *in* D-00246

53332-51-7 1,2-Ethenediylbis[triphenylphosphonium](2+); (*E*)-*form*, Dibromide, *in* E-00019

53332-52-8 1,2-Ethenediylbis[triphenylphosphonium](2+); (*E*)-*form*, Diiodide, *in* E-00019

53340-10-6 3-Pyridinephosphonic acid; Di-Et ester, *in* P-00504

53340-11-7 3-Pyridinephosphonic acid, P-00504

53364-04-8 *N,N′,N″*-Tributylphosphorothioic triamide, *in* T-00394

53369-07-6 Phosphinothricin; (±)-*form*, *in* P-00359

53378-32-8 Bis(3-methylbutyl) phosphorochloridate, *in* B-00302

53379-57-0 *P,P,P*-Trichloro-*N*-(3-nitrophenyl)phosphazene, *in* N-00062

53426-77-0 Phenyl phosphorodihydrazidate, P-00290

53439-65-9 *N,N*-Dimethyl-*P,P*-di-1-pyrrolidinylphosphinic amide, *in* D-01233

53459-47-5 Tetraethyl pentylidenebisphosphonate, *in* P-00031

53459-55-5 2-Amino-3-(2-chloroethyl)-tetrahydro-4*H*-1,3,2-oxazaphosphorine 2-oxide, A-00063

53482-97-6 Dipropylphosphinodithioic acid; Na salt, *in* D-01207

53483-28-6 Trimethylsilyl di-*tert*-butylphosphinite, *in* D-00122

53483-29-7 Trimethylsilyl dipropylphosphinate, *in* D-01205

53497-47-5 4-(Phosphonoxy)benzoic acid, P-00397

53518-61-9 *O*-Ethyl methylphosphonothioate; (*R*)-*form*, (*R*)-1-Phenylethylammonium salt, *in* E-00093

53518-62-0 *O*-Ethyl methylphosphonothioate; (*S*)-*form*, (*S*)-1-Phenylethylammonium salt, *in* E-00093

53530-44-2 5,5-Diphenyl-5*H*-dibenzophospholium iodide, *in* P-00110

53531-20-7 4,5-Bis[(diphenylphosphino)-methyl]-2,2-dimethyl-1,3-dioxolane, B-00251

53534-28-4 *P,P*-Dibutylphosphinic amide, D-00111

53534-65-9 (4-Methoxyphenyl)phosphinic acid, M-00069

53554-11-3 2,2′-Biphenylene phosphonate, *in* D-00038

53561-53-8 *P,P,P*-Tri-*tert*-butyl-*N*-trimethylsilylphosphine imide, *in* T-00375

53575-08-9 ▷Diethyl 1-naphthylmethylphosphonate, *in* N-00013

53589-30-3 (2-Chloropropyl)phosphonic acid, C-00171

53596-32-0 2,3-Dihydro-1*H*-naphtho[1,8-*de*]-1,3,2-diazaphosphorine; 2-(4-Chloro-3-methylphenoxy), 2-oxide, *in* D-00532

53597-69-6 1,3,5-Triaza-7-phosphatricyclo[3.3.1.1^{3,7}]decane, T-00327

53597-70-9 1,3,5-Triaza-7-phosphatricyclo[3.3.1.1^{3,7}]decane; 7-Oxide, *in* T-00327

53621-85-5 (1-Hydroxypropyl)phosphonic acid, H-00190

53621-86-6 (1-Acetoxypropyl)phosphonic acid, A-00008

53640-96-3 Diethyl benzylphosphoramidate, *in* B-00069

53648-94-5 2-Methoxy-3,4-dimethyl-5-phenyl-1,3,2-oxazaphospholidine; (2*S*,4*S*,5*R*)-*form*, *in* M-00035

53648-95-6 2-Methoxy-3,4-dimethyl-5-phenyl-1,3,2-oxazaphospholidine; (2*R*,4*S*,5*R*)-*form*, *in* M-00035

53648-96-7 2-Methoxy-3,4-dimethyl-5-phenyl-1,3,2-oxazaphospholidine; (2*S*,4*R*,5*S*)-*form*, *in* M-00035

53648-97-8 2-Methoxy-3,4-dimethyl-5-phenyl-1,3,2-oxazaphospholidine; (2*R*,4*R*,5*S*)-*form*, *in* M-00035

53666-65-2 6,9-Dimethyl-5-phenoxy-2,2,3,3-tetrakis(trifluoromethyl)-1,4-dioxa-6,9-diaza-5-phospha(5-P^{V})spiro[4.4]nonane, *in* D-00922

53676-22-5 2,2,2-Tribromoethyl phosphorodichloridate, *in* M-00516

53710-27-3 (2-Hydroxyethyl)-triphenylphosphonium(1+); Bromide, *in* H-00151

53711-17-4 (2-Chloroethyl)-phosphonochloridic acid, C-00077

53718-38-0 *O,O,O′,O′*-Tetramethyl *S,S′*-methylene bisphosphorodithioate, T-00205

53742-11-3 Benzylethylphenylphosphine; (±)-*form*, Oxide, *in* B-00047

53743-43-4 ▷*N*-(1-Adamantyl)-*P,P*-bis(aziridinyl)phosphinic amide, A-00032

53753-37-0 Diethyl (nitromethyl)phosphonate, *in* N-00037

53753-40-5 Dimethyl (nitromethyl)-phosphonate, *in* N-00037

53753-41-6 Dimethyl (1-amino-1-methylethyl)phosphonate, *in* A-00087

53753-58-5 Diisopropylphosphine; Oxide, *in* D-00640

53764-94-6 Dibutyl phosphorobromidite, D-00139

53764-96-8 Butyl phosphorodibromidite, B-00627

53764-97-9 Tetraethylphosphorodiamidous acid; Bromide, *in* T-00066

53772-54-6 (4-Methylphenyl)phosphine, M-00218

53772-55-7 (4-Bromophenyl)phosphine, B-00487

53772-57-9 (2-Chlorophenyl)phosphine, C-00139

53772-58-0 (2-Bromophenyl)phosphine, B-00486

53772-59-1 (2-Methylphenyl)phosphine, M-00216

53778-28-2 5,10-Dihydrophenophosphazine; 10-Oxide, *in* D-00542

53782-53-9 *P*-Cyclohexylphosphonic diamide, C-00257

53796-00-2 Dibutyl dibutylphosphoramidate, *in* D-00134

53798-59-7 7,12-Dihydrobenzo[*e*]-phenophosphazine; 7-Oxide, *in* D-00440

53798-62-2 Benzo[*e*]phenazaphosphinic acid, *in* D-00440

53799-42-1 4′,5′-Bis(trifluoromethyl)-spiro[1,3,2-benzodioxaphosphole-2,2′-[1,3,2]dioxaphosphole]; 2-Phenoxy, *in* B-00421

53799-43-2 4′,5′-Bis(trifluoromethyl)-spiro[1,3,2-benzodioxaphosphole-2,2′-[1,3,2]dioxaphosphole]; 2-Methoxy, *in* B-00421

53799-86-3 2-Propynyl phosphorodichloridate, P-00497

53813-58-4 3,3,3,6,6-Pentaphenyl-3,3,3,6-tetrahydro-4,5-bis(methoxycarbonyl)-1,2,3-diazaphosphorine, P-00041

53820-05-6 *N,N′,N″*-Tribenzylphosphorothioic triamide, *in* T-00344

53820-06-7 *N,N′,N″*-Tri-2-propenylphosphorous triamide; Sulfide, *in* T-00709

53827-16-0 *S,S*-1,2-Propylene *N,N*-dimethylphosphoramidodithioite, *in* M-00136

53882-44-5 *O,S*-Diethyl phosphorothioate, D-00397

53888-89-4 Dimethyl(4-methylphenyl)-phosphine; Oxide, *in* D-00761

53888-90-7 Dimethyl(3-methylphenyl)-phosphine; Oxide, *in* D-00760

53889-34-2 Bis[2-(diphenylphosphino)-ethyl]phenylphosphine; Trioxide, *in* B-00239

53910-02-4 3,6-Dihydro-2-hydroxy-2*H*-1,2-oxaphosphorin 2-oxide; 2-Methoxy (Me ester), *in* D-00501

53931-67-2 Propylphosphoramidic acid; Dichloride, *in* P-00484

53986-90-6 Diphenyl (2-chloroethyl)-phosphonate, *in* C-00074

54000-84-9 Trimethyl methylphosphorimidate, *in* M-00336

54006-08-5 Dimethyl [(2-nitrophenyl)-methyl]phosphonate, *in* N-00044

54006-10-9 Dimethyl [(2-aminophenyl)-methyl]phosphonate, *in* A-00106

54017-36-6 2-Methoxy-4,5-diphenyl-1,3,2-dioxaphospholane; (2α,4β,5β)-*form*, 2-Oxide, *in* M-00039

54036-90-7 Bis(trimethylsilyl)-phosphoramidous acid; Dichloride, *in* B-00438

54044-00-7 Dipropylphosphine; Sulfide, *in* D-01204

54050-02-1 4,4′-Bis[(triphenylphosphonio)-methyl]biphenyl(2+); Dichloride, *in* B-00447

54054-03-4 2-Methoxy-4,5-diphenyl-1,3,2-dioxaphospholane; (2α,4β,5β)-*form*, *in* M-00039

54060-24-1 Diphenylphosphinylferrocene, *in* D-01068

54061-79-9 2-Dimethylamino-3,4-dimethyl-5-phenyl-1,3,2-oxazaphospholidine; (2*R*,4*S*,5*R*)-*form*, *in* D-00694

54061-80-2 2-Dimethylamino-3,4-dimethyl-5-phenyl-1,3,2-oxazaphospholidine; (2*S*,4*R*,5*S*)-*form*, *in* D-00694

54086-38-3 5-Phenyl-10(5*H*)-acridophosphinone; 5-Oxide, *in* P-00080

54086-39-4 5-Phenyl-10(5*H*)-acridophosphinone, P-00080

54091-78-0 Diethyl [(ethylthio)methyl]-phosphonate, *in* E-00192

54098-11-2 *P,P,P*-Tricyclohexylphosphine imide, T-00427

54100-43-5 Diisopropylphosphinic acid; Isothiocyanate, *in* D-00641

54100-47-9 *tert*-Butylphenylphosphinothioic acid; (*R*)-*form*, *in* B-00583

54100-67-3 Phenylbis[2-(phenylethynyl)-phenyl]phosphine, P-00098

54185-79-4 1,3-Dihydro-1-hydroxy-3-phenyl-1,2,3-benziodoxaphosphole 3-oxide, *in* D-00488

54185-80-7 1,3-Dihydro-1,3-dihydroxy-1,2,3-benziodoxaphosphole 3-oxide, *in* D-00488

54185-82-9 (2-Iodophenyl)phosphonic acid, I-00016

54208-04-7 Dodecylidenetriphenylphosphorane, *in* D-01266

54208-05-8 Nonylidenetriphenylphosphorane, *in* N-00078

54229-99-1 2,3,4,5-Tetrahydro-5-methyl-2,2-diphenyl-1*H*-2-benzophosphepinium(1+), T-00114

54230-00-1 2,3,4,5-Tetrahydro-5-methyl-2,2-diphenyl-1*H*-2-benzophosphepinium(1+); Hexafluorophosphate, *in* T-00114
54230-05-6 2,3,3*a*,4,5,9*b*-Hexahydro-4,4-diphenyl-1*H*-cyclopent[*c*]-isophosphinolinium bromide, *in* H-00048
54293-26-4 1,2,3,4-Tetrahydro-4-methyl-2,2-diphenylbenz[*h*]-isophosphinolinium(1+), T-00113
54293-27-5 1,2,3,4-Tetrahydro-4-methyl-2,2-diphenylbenz[*h*]-isophosphinolinium(1+); Hexafluorophosphate, *in* T-00113
54300-34-4 Tris(3-fluorophenyl)phosphine; Oxide, *in* T-00782
54300-35-5 (3-Fluorophenyl)-diphenylphosphine; Oxide, *in* F-00027
54300-36-6 Tris(3-chlorophenyl)phosphine; Oxide, *in* T-00748
54300-42-4 Tris(3-fluorophenyl)phosphine; Sulfide, *in* T-00782
54300-45-7 Tris(3-chlorophenyl)phosphine; Sulfide, *in* T-00748
54300-47-9 Tris(4-fluorophenyl)phosphine; Selenide, *in* T-00783
54300-50-4 Tris(3-fluorophenyl)phosphine; Selenide, *in* T-00782
54300-56-0 Tris(3-chlorophenyl)phosphine; Selenide, *in* T-00748
54305-75-8 Diethylphosphoramidous acid; Diiodide, *in* D-00360
54305-77-0 Diphenylphosphoramidous acid; Diiodide, *in* D-01111
54305-86-1 Tetrabutyl pyrophosphite, T-00025
54356-04-6 3-(Triphenylphosphoranylidene)-propanoic acid; Et ester, *in* T-00661
54377-81-0 *O*-Ethyl phosphorofluoridoisocyanatido-thioate, E-00181
54434-92-3 Tetramethyl phosphor(isothiocyanatido)thioic diamide, T-00217
54458-92-3 [1,1′-Biphenyl]-4-yl-1-naphthalenylphenylphosphine; (±)-*form*, *in* B-00081
54480-52-3 Diisopropyl cyclohexylphosphoramidate, *in* C-00269
54529-67-8 1,1′-Dimethyl-*P*-phenylphosphonic dihydrazide, *in* P-00225
54529-68-9 N^1,$N^{1'}$-Dimethyl-*P*-phosphonothioic dihydrazide, *in* P-00249
54529-69-0 *P*,2,2′-Triphenylphosphonic dihydrazide, *in* P-00225
54552-71-5 (2,2-Dimethylpropyl)phosphonic acid; Dichloride, *in* D-00912
54552-75-9 Diphenyl (2,2-dimethylpropyl)-phosphonate, *in* D-00912
54552-76-0 Dimethyl (2,2-dimethylpropyl)-phosphonate, *in* D-00912
54552-77-1 Dimethyl isopropylphosphonate, *in* I-00050
54552-79-3 1*H*-Isophosphindole-1,3(2*H*)-dione; 2-Ph, 2-sulfide, *in* I-00026
54552-87-3 1-Phenyl-2-phosphorinanol, P-00266
54552-95-3 1-Phenyl-2-phosphorinanone, P-00267
54553-21-8 Diethyl 2-phenylethylphosphonate, *in* P-00165
54560-62-2 Tris(4-bromophenyl)phosphine; Sulfide, *in* T-00735
54565-44-5 *O*,*S*-Dimethyl methylphosphonodithioate, *in* M-00301
54565-51-4 Ethyl diisopropylphosphinodithioate, *in* D-00643
54622-61-6 1,3-Dimethyl-2-phenoxy-1,3,2-diazaphospholidine, D-00788
54622-67-2 4,5,6,7-Tetrachloro-4′,4′,5′,5′-tetramethylspiro[1,3,2-benzodioxaphosphole-2,2′-[1,3,2]-dioxaphospholane]; 2-Phenoxy, *in* T-00036
54622-68-3 4,5,6,7-Tetrachloro-4′,4′,5′,5′-tetramethylspiro[1,3,2-benzodioxaphosphole-2,2′-[1,3,2]-dioxaphospholane]; 2-Phenylthio, *in* T-00036
54622-70-7 4,5,6,7-Tetrachloro-4′,4′,5′,5′-tetramethylspiro[1,3,2-benzodioxaphosphole-2,2′-[1,3,2]-dioxaphospholane]; 2-Methoxy, *in* T-00036
54622-74-1 4′,4′,5′,5′-Tetramethylspiro[1,3,2-benzodioxaphosphole-2,2′-[1,3,2]dioxaphospholane]; 2-Phenoxy, *in* T-00230
54622-76-3 4′,4′,5′,5′-Tetramethylspiro[1,3,2-benzodioxaphosphole-2,2′-[1,3,2]dioxaphospholane]; 2-Phenylthio, *in* T-00230
54632-62-1 Methyl 2-naphthylphenylphosphinate, *in* N-00018
54632-63-2 Ethyl 2-naphthylphenylphosphinate, *in* N-00018
54638-48-1 *P*,*P*-Dibutyl-*N*,*N*-diethylphosphinoselenoic amide, *in* D-00114
54655-13-9 2-Methoxy-3,4-dimethyl-5-phenyl-1,3,2-oxazaphospholidine; (2*R*,4*S*,5*R*)-*form*, 2-Oxide, *in* M-00035
54655-28-6 5-*tert*-Butyl-2-phenyl-1,3,2-dioxaphosphorinane; *cis-form*, *in* B-00577
54655-29-7 5-*tert*-Butyl-2-phenyl-1,3,2-dioxaphosphorinane; *trans-form*, *in* B-00577
54662-09-8 1-Methyl-4-phosphorinanone; Oxide, *in* M-00340
54674-84-9 (3-Hydroxypropyl)-triphenylphosphonium(1+); Chloride, *in* H-00193
54710-83-7 (2,4-Diphenyl-2,4-cyclopentadien-1-ylidene)triphenylphosphorane, D-00999
54731-72-5 Dimethyl (2-hydroxyethyl)-phosphonate, *in* H-00146
54731-74-7 Dimethyl (3-hydroxypropyl)-phosphonate, *in* H-00192
54750-12-8 2-Chloro-3,4-dimethyl-5-phenyl-1,3,2-oxazaphospholidine; (2*S*,4*S*,5*R*)-*form*, 2-Oxide, *in* C-00061
54750-13-9 2-Chloro-3,4-dimethyl-5-phenyl-1,3,2-oxazaphospholidine; (2*R*,4*S*,5*R*)-*form*, 2-Oxide, *in* C-00061
54750-98-0 4-(Diphenylphosphino)pyridine, D-01077
54750-99-1 4-(Diphenylphosphinyl)pyridine, *in* D-01077
54757-38-9 Bis(1-methylpropyl) phosphorochloridate, *in* B-00341
54770-04-6 2,2-Dimethyl-1-phenyl-1-phospha-2-silacyclopentane, D-00802
54772-83-7 Alafosfalin; $(R)_C(S)_P$-*form*, *in* A-00049
54788-43-1 Alafosfalin, A-00049
54799-74-5 2,2,2-Trifluoro-2,2-dihydro-4,4,5,5-tetrakis(trifluoromethyl)-1,3,2-dioxaphospholane, T-00467
54807-86-2 Methylphenyl-2-propenylphosphine, M-00264
54835-97-1 *P*-Methyl-*P*-phenylphosphinic amide; (*S*)-*form*, *in* M-00223
54844-87-0 Dimethyl(4-methylphenyl)-phosphine; Sulfide, *in* D-00761
54854-20-5 1,2-Ethenediylbis[triphenyl-phosphonium](2+), E-00019
54857-00-0 2-Phosphonooxypropanoic acid; (±)-*form*, Me ester, *in* P-00392
54862-45-2 Tribenzoyl phosphite, T-00336
54877-14-4 2,5-Dimethyl-1-phenylphosphorinan-4-one, D-00821
54877-55-3 Tri(1*H*-pyrazol-1-yl)phosphine, T-00722
54882-43-8 2-Methoxy-3,3,5,5-tetramethyl-1,2-azaphospholidine 2-oxide, *in* T-00208
54921-67-4 Cyclohexyl phosphorodichloridite, C-00272
54921-72-1 Monophenyl phosphonate; NH_4 salt, *in* M-00508
54932-28-4 3-Methyl-1-phenylphospholane; (1*S*,3*R*)-*form*, *in* M-00229
54944-19-3 Ethyl phenyl(trichloromethyl)-phosphinate, *in* P-00327
54948-89-9 Triethyl (2-nitrophenyl)-phosphorimidate, *in* N-00061
54970-66-0 2,4,8,10-Tetramethyl-3,9-dithia-6-phosphoniaspiro[5.5]-undecane; Perchlorate, *in* T-00202
54970-71-7 Phenyldi-2-propenylphosphine; Sulfide, *in* P-00119
54975-79-0 Triethyl (dimethoxyphosphinothioyl)-phosphorimidate, *in* D-00674
54975-91-6 Methyl tetramethyl-phosphorodiamidodithioate, *in* T-00222
54975-93-8 Ethyl tetramethyl-phosphorodiamidodithioate, *in* T-00222
54975-97-2 2,2′-Bis(5,5-dimethyl-1,3,2-dioxaphosphorinane), B-00197
54981-26-9 Diphenylphosphoramidic acid; Difluoride, *in* D-01108
54998-94-6 *P*,*P*,*P*-Trichloro-*N*-(2-nitrophenyl)phosphazene, *in* N-00061
55003-02-6 *O*-Phenyl phosphorodihydrazidothioate, P-00291
55042-00-7 2,4-Bis(dimethylamino)-3-methyl-1,3,2,4-thiazadiphosphetidine; 2,4-Disulfide, *in* B-00187
55042-06-3 N″-Bis(dimethylamino)-phosphinyl-*N*,*N*,*N*′,*N*′,*N*″-pentamethyl-phosphorotriamidothioate, N-00032
55043-98-6 3-Methyl-1-phenylphospholane; (1*RS*,3*RS*)-*form*, 1-Oxide, *in* M-00229
55043-99-7 3-Methyl-1-phenylphospholane; (1*RS*,3*SR*)-*form*, 1-Oxide, *in* M-00229
55054-73-4 Hexahydro-3-phenyl-1,2,4,3-triazaphosphorine; 1,4-Di-Me, *in* H-00051
55055-14-6 2-Mercapto-1,3,2-dioxaphosphorinane 2-sulfide, M-00013
55055-15-7 2-Mercapto-1,3,2-dioxaphosphorinane 2-sulfide; Triethylammonium salt, *in* M-00013
55055-16-8 2-Mercapto-4-methyl-1,3,2-dioxaphosphorinane 2-sulfide; Triethylammonium salt, *in* M-00017
55055-20-4 5′-Methylspiro[1,3,2-benzodioxaphosphole-2,2′(3*H*)-[1,2]oxaphosphole]; 2-Methoxy, *in* M-00396
55055-21-5 5′-Methylspiro[1,3,2-benzodioxaphosphole-2,2′(3*H*)-[1,2]oxaphosphole]; 2-Ethoxy, *in* M-00396
55102-61-9 4-Methyl-2-methylthio-1,3,2-dioxaphosphorinane; (2*RS*,4*SR*)-*form*, 2-Sulfide, *in* M-00174
55102-62-0 *O*,*O*-1,3-Butylene *S*-methyl dithiophosphate, *in* M-00174
55108-80-0 Bis(trimethylsilyl) (acetamidomethyl)-phosphonate, *in* A-00012

55120-33-7 3,9-Diphenoxy-2,4,8,10-tetraoxa-3,9-diphosphaspiro[5.5]undecane; Dioxide, *in* D-00991
55134-66-2 2-Methoxy-1,3,2-dithiaphosphorinane, *in* D-01247
55136-69-1 Diiodotriphenoxyphosphorane, D-00630
55137-99-0 Benzylphosphonofluoridic acid, B-00066
55153-57-6 3,5,8,10-Tetraphenyl-1*H*,5*H*-bisphospholo[1,2-*a*:1′,2′-*d*][1,4]diphosphorin, T-00250
55157-75-0 2-Ethoxy-1,3,2-dithiaphosphorinane, *in* D-01247
55176-55-1 Glycerol 1-9-octadecenoate 2-octadecanoate 3-phosphoethanolamine; (±)-(*Z*)-*form*, *in* G-00092
55205-95-3 *O*,*O*-Bis(2-methylpropyl) phosphorochloridoselenoate, B-00356
55205-96-4 Ethylphosphonoselenoic bis(dimethylamide), *in* E-00146
55215-24-2 Trimethyl ethylphosphorimidate, *in* E-00161
55215-26-4 Dimethyl *N*-methyl-*P*-phenylphosphonimidate, *in* M-00240
55215-27-5 *N*,*N*,*N*′,*N*′,N″-Pentamethyl-*P*-phenylphosphonimidic diamide, N-00079
55215-28-6 Methyl *N*,*N*-dimethyl-*P*-phenylphosphonamidate, *in* D-00808
55231-79-3 Mono-(1,1′-biphenyl)-4-yl phosphate, M-00446
55231-79-3 (1,1′-Biphenyl)-4-yl phosphorodichloridate, *in* M-00446
55249-21-3 Butylphosphonoselenoic acid; Dichloride, *in* B-00609
55249-22-4 Dibutylphosphinoselenoic acid; Chloride, *in* D-00114
55249-23-5 Diphenylphosphinoselenoic acid; Chloride, *in* D-01078
55274-09-4 Trimethylsilyl tetraethylphosphorodiamidite, *in* T-00066
55277-64-0 5-Ethoxy-5*H*-dibenzophosphole 5-oxide, *in* H-00118
55287-82-6 *P*,*P*-Dipropyl-*N*-trimethylsilylphosphinothioic amide, *in* D-01208
55287-83-7 Dipropylphosphinothioic acid; Bromide, *in* D-01208
55329-13-0 12*H*-Benzo[*c*]phenothiaphosphine; 12-Ethoxy, 7,7,12-trioxide, *in* B-00023
55337-91-2 [Thiobis[methylene]]-bis[triphenylphosphonium](2+); Dichloride, *in* T-00312
55338-02-8 [Oxybis[methylene]]-bis[triphenylphosphonium](2+); Dichloride, *in* O-00097
55339-99-6 Dimethyl nonylphosphonate, *in* N-00076
55340-00-6 Dimethyl (3-phenylpropyl)-phosphonate, *in* P-00305
55343-32-3 (Dichloromethyl)-difluorophosphine, *in* D-00181
55364-12-0 5,10-Dihydro-5-hydroxy-10-phenylacridophosphine 5-oxide, *in* D-00548
55364-14-2 10*H*-5,10[1′,2′]-Benzenoacridophosphine, B-00014
55364-15-3 5-Methyl-10*H*-5,10[1′,2′]-benzenoacridophosphonium iodide, *in* B-00014
55364-18-6 10*H*-5,10[1′,2′]-Benzenoacridophosphine; 5-Oxide, *in* B-00014
55365-20-3 Diethyl [2-(cyclohexylamino)-2-phenylethenyl]-phosphonate, D-00277
55367-56-1 Heptylidenetriphenylphosphorane, *in* H-00011
55370-43-9 *N*,*N*′-Bis(triphenylphosphonio)-hydrazine dichloride, *in* B-00458
55379-57-2 2-Isoselenocyanato-5,5-dimethyl-1,3,2-dioxaphosphorinane; 2-Oxide, *in* I-00066
55379-58-3 2-Isocyano-5,5-dimethyl-1,3,2-dioxaphosphorinane; 2-Oxide, *in* I-00023
55379-59-4 2-Cyano-5,5-dimethyl-1,3,2-dioxaphosphorinane; 2-Oxide, *in* C-00209
55382-99-5 *P*,*P*-Di-*tert*-butyl-*N*-methylphosphinothioic amide, *in* D-00119
55383-01-2 Bromodiisopropylphosphine, *in* D-00648
55402-79-4 2,4,8,10-Tetramethyl-3,9-dithia-6-phosphoniaspiro[5.5]undecane, T-00202
55422-17-8 (Aminomethyl)diphenylphosphine oxide; B,HCl, *in* A-00085
55453-25-3 Methylphosphonotrithioic acid, M-00318
55482-36-5 Dibutylmethylphosphine; Sulfide, *in* D-00098
55498-89-0 2,2,2,3-Tetrahydro-5-phenyl-1,3,4,2-oxadiazaphosphole; 2,2,2-Trichloro, *in* T-00131
55499-33-7 2,4,5-Triphenyl-1,3,2,4,5-dithiatriphospholane; 4-Sulfide, *in* T-00606
55500-03-3 4*H*-1,4-Thiaphosphorin; 4-Ph, *in* T-00299
55520-53-1 3,4-Dihydro-2,3,4,5-tetraphenyl-2*H*-1,2,3-diazaphosphole; (3*RS*,4*RS*)-*form*, *in* D-00579
55520-54-2 3,4-Dihydro-2,3,4,5-tetraphenyl-2*H*-1,2,3-diazaphosphole; (3*RS*,4*SR*)-*form*, *in* D-00579
55523-19-8 12-Phenyl-10*H*-5,10-ethenoacridophosphine, P-00151
55523-21-2 2,3,10-Triphenyl-4*H*-1,4-ethenophosphinoline, T-00607
55523-58-5 Octahydro-2,4-diphenyl-1,3,5-metheno-1*H*-2,4-diphosphacyclobuta[*c*,*d*]-pentalene; 2,4-Dioxide, *in* O-00016
55526-69-7 *O*,*O*-Bis(4-methylphenyl) phosphorochloridothioate, B-00335
55526-70-0 *O*,*O*-Bis(4-chlorophenyl) phosphorochloridothioate, B-00149
55549-39-8 1,2-Oxaphosphorinane; 2-Ph, 2-oxide, *in* O-00054
55549-40-1 *O*-Ethyl hydrogen (1-methylpropyl)-phosphonodithioate, *in* M-00380
55549-60-5 2-Phenoxy-1,2-oxaphosphorinane 2-oxide, *in* H-00167
55549-63-8 2-Chloro-1,2-oxaphosphorinane 2-oxide, *in* H-00167
55578-32-0 *O*,*O*-Dipropyl phosphorochloridoselenoate, *in* D-01229
55578-33-1 *O*,*O*-Dibutyl phosphorochloridoselenoate, *in* D-00154
55590-36-8 10,10-Dihydro-2,3-diphenyl-[1,3,2]oxazaphospholo[2,3-*b*][1,3,2]benzoxazaphosphole; 10,10-Di-Me, *in* D-00481
55590-37-9 10,10-Dihydro-2,3-diphenyl-[1,3,2]oxazaphospholo[2,3-*b*][1,3,2]benzoxazaphosphole; 10,10-Di-Ph, *in* D-00481
55590-38-0 10,10-Dihydro-2,3-diphenyl-[1,3,2]oxazaphospholo[2,3-*b*][1,3,2]benzoxazaphosphole; 10,10-Bis(dimethylamino), *in* D-00481
55590-39-1 10,10-Dihydro-2,3-diphenyl-[1,3,2]oxazaphospholo[2,3-*b*][1,3,2]benzoxazaphosphole; 10,10-Dimethoxy, *in* D-00481
55602-33-6 (1-Hydroxy-1*H*-benzotriazolato-*O*)tris(dimethylamino)-phosphorus(1+); Hexafluorophosphate, *in* H-00111
55629-68-0 1,2-Bis(diphenylphosphinomethyl)-cyclobutane; (1*R*,2*R*)-*form*, *in* B-00249
55630-58-3 1-[1-(Dimethylamino)ethyl]-2-(diphenylphosphino)-ferrocene; $(S)_C(R)_{planar}$-*form*, *in* D-00697
55655-35-1 *O*,*O*-Diisopropyl diethylphosphoramidothioate, *in* D-00356
55655-36-2 ▷Dipropyl chlorothiophosphonate, D-01200
55656-88-7 Dibutylphosphinothioic acid; Bromide, *in* D-00117
55658-70-3 Tetraphenylthiatetraphospholane, T-00282
55660-58-7 Acetyl phosphate; Di-NH_4 salt, *in* A-00025
55660-59-8 Acetyl phosphate; Dianilinium salt, *in* A-00025
55666-79-0 4,5-Dimethyl-2-dimethylamino-1,3,2-dioxaphospholane, D-00725
55666-82-5 *tert*-Butyl tetramethylphosphorodiamidite, *in* T-00227
55666-86-9 Ethyl *N*,*N*′-dimethyl-*N*,*N*′-trimethylene phosphorodiamidite, *in* H-00022
55677-75-3 [2-(Diphenylarsino)phenyl]-diphenylphosphine, D-00994
55700-44-2 1-[1-(Dimethylamino)ethyl]-2-(diphenylphosphino)-ferrocene; $(R)_C(S)_{planar}$-*form*, *in* D-00697
55701-23-0 Glycerol 2-monophosphate; Ca salt, *in* G-00067
55702-27-7 *N*,*N*′,*P*-Tris(1,1-dimethylethyl)-phosphonic diamide, *in* B-00597
55705-77-6 *tert*-Butylphenylphosphinothioic acid; (*S*)-*form*, *in* B-00583
55705-78-7 *O*-Methyl *tert*-butylphenylphosphinothioate, *in* B-00583
55739-97-4 *O*,*O*-Diethyl cyclohexylphosphoramidothioate, *in* C-00271
55743-26-5 Phenylvinylphosphinic acid, P-00337
55776-57-3 Methyl diphenylphosphinoselenoite, *in* D-01081
55776-58-4 Dimethyl phenylphosphonodiselenoite, *in* P-00235
55776-59-5 Trimethyl phosphorotriselenoite, T-00547
55776-63-1 Methyl phosphorodichloridoselenoite, M-00348
55776-64-2 *O*-Ethyl *Se*-methyl phosphorochloridoselenoite, E-00094
55776-66-4 Dimethyl phosphorochloridoselenoite, D-00890
55776-68-6 Methyl phenyl-phosphonochloridoselenoite, *in* P-00231
55776-69-7 2-Thia-1,3,5-triaza-7-phosphatricyclo[3.3.1.$1^{3,7}$]decane; 2,2-Dioxide, *in* T-00301
55776-70-0 2-Thia-1,3,5-triaza-7-phosphatricyclo[3.3.1.$1^{3,7}$]decane; 2,2,7-Trioxide, *in* T-00301

55776-71-1 7-Methyl-2-thia-1,3,5-triaza-7-phosphoniaadamantane iodide, *in* T-00301
55781-95-8 1,2,3,6-Tetrahydro-1,2,4,5-tetraphenyl-1,2-diphosphorine; 1,2-Disulfide, *in* T-00151
55781-98-1 2-Phenyl-1,3,2-oxathiaphosphepane; 2-Sulfide, *in* P-00178
55816-32-5 8-Methyl-8-phosphabicyclo[3.2.1]-oct-6-ene 8-oxide; *syn-form, in* M-00273
55816-33-6 8-Methyl-8-phosphabicyclo[3.2.1]-oct-6-ene 8-oxide; *anti-form, in* M-00273
55816-41-6 8-Methyl-8-phosphabicyclo[3.2.1]octane; *syn-form*, Oxide, *in* M-00272
55816-42-7 8-Methyl-8-phosphabicyclo[3.2.1]octane; *anti-form*, Oxide, *in* M-00272
55822-08-7 Glycerol 1-octadecanoate 2-9,12,15-octadecatrienoate 3-phosphoethanolamine; (±)-(*all-Z*)-*form, in* G-00081
55849-69-9 Diethyl (3-hydroxypropyl)-phosphonate, *in* H-00192
55874-27-2 Cyclohexyltriphenyl-phosphonium(1+); 4-Methylbenzenesulfonate, *in* C-00273
55878-19-8 Hexakis(dimethylamino)-μ-oxodiphosphorus(2+); Bis-(4-methylbenzenesulfonate), *in* H-00062
55881-03-3 Hexakis(dimethylamino)-μ-oxodiphosphorus(2+); Bis(tetrafluoroborate), *in* H-00062
55894-94-5 Bis(dimethylvinylene) pyrophosphate, *in* O-00088
55895-03-9 Dimethylvinylene phenyl phosphate, *in* H-00123
55905-05-0 Diphenylphosphinothious acid, D-01085
55916-42-2 Dipropyl phosphorochloridotrithioate, D-01222
55920-68-8 Methyl (dihydroxyphosphinyl)-formate, *in* P-00387
55920-71-3 Ethyl (dihydorxyphosphinyl)-formate, *in* P-00387
55927-32-7 *O*-Propyl hydrogen phosphonoselenoate, *in* E-00145
55983-40-9 2,2,3,3,7,7,8,8-Octamethyl-1,4,6,9-tetraoxa-5-phospha(5-P^V)spiro[4.4]-nonane; 5-Ph, *in* O-00032
55983-41-0 1,6-Dioxa-4,9-diaza-5-phospha(5-P^V)spiro[4.4]-nonane; 5-Ph, *in* D-00963
55983-44-3 2,2(3*H*,3'*H*)-Spirobi[1,3,2-benzoxazaphosphole]; 2-Ph, *in* S-00020
55985-85-8 (2-Carboxyethyl)-triphenylphosphonium(1+); Tribromide, *in* C-00011
55987-89-8 *O,O*-Diethyl *Se*-methyl phosphoroselenoite, D-00301
56013-01-5 Triphenyl[(2,6,6-trimethyl-1-cyclohexen-1-yl)methyl]-phosphonium(1+); Bromide, *in* T-00689
56031-25-5 Triethyl (4-nitrophenyl)-phosphorimidate, *in* N-00063
56042-55-8 Methyl (1-diazoethyl)-phenylphosphinate, *in* D-00032
56045-72-8 Tetraisopropyl 1,2-ethanediylbisphosphoramidate, *in* E-00009
56069-39-7 Diethyl [(phenylsulfonyl)-methyl]phosphonate, *in* P-00311
56070-16-7 *S*-[(*tert*-Butylsulfonyl)methyl] *O,O*-diethyl phosphorodithioate, *in* T-00004
56080-45-6 *N,N*-Diethyl-*P,P*-dimethylphosphinic amide, *in* D-00830
56087-62-8 12*H*-Benzo[*c*]phenothiaphosphine, B-00023
56119-60-9 Di-*tert*-butyl phosphorochloridate, D-00141
56148-84-6 Octyl phosphorodichloridite, O-00046
56152-43-3 Tetrahydro-2,6-dimethyl-4*H*-1,3,6,2-dioxazaphosphocine; 2-Sulfide, *in* T-00094
56153-45-8 4-Hydroxy-2,6-diphenyl-4*H*-1,4-oxaphosphorin 4-oxide, H-00140
56168-00-8 Diphenyl (2-methylphenyl)-phosphoramidate, *in* M-00251
56183-69-8 Diethyl phosphorohydrazidate, D-00391
56185-09-2 *o*-Phenylene dimethylphosphoramidate, *in* D-00689
56299-23-1 2-Thia-1,3,5-triaza-7-phosphatricyclo[3.3.1.1^{3,7}]decane, T-00301
56317-58-9 Monobutyl phosphonate; Na salt, *in* M-00453
56317-74-9 Monobenzyl phosphonate; NH_4 salt, *in* M-00444
56318-16-2 Monoisopropyl phosphonate; Al salt, *in* M-00478
56341-78-7 Bis(diisopropoxyphosphinyl) disulfide, B-00183
56341-79-8 *O,O*-Diethyl methylphosphoramidoselenoate, *in* M-00332
56341-80-1 *O,O*-Diisopropyl methylphosphoramidoselenoate, *in* M-00332
56374-57-3 2-Butenylidenetriphenyl-phosphorane; (*E*)-*form, in* B-00522
56374-63-1 2-Nonenylidenetriphenyl-phosphorane; (*Z*)-*form, in* N-00071
56374-64-2 2-Nonenylidenetriphenyl-phosphorane; (*E*)-*form, in* N-00071
56374-74-4 Triphenyl(3-phenyl-2-propenylidene)phosphorane; (*E*)-*form, in* T-00615
56376-11-5 Isopropyl phosphorodichloridate, I-00060
56421-10-4 Glycerol 1-octadecanoate 2-9-octadecenoate 3-phosphocholine; (*R*)-*form, in* G-00082
56436-86-3 Diethyl (1-chloropropyl)-phosphonate, *in* C-00170
56465-64-6 2-Dimethylamino-5,5-dimethyl-1,3,2-dioxaphosphorinane, D-00693
56506-90-2 (Dibromomethyl)-triphenylphosphonium(1+), D-00065
56522-04-4 Dibenzylphosphine, D-00054
56530-33-7 *P*-Ethyl-*N,N'*-diphenylphosphonic diamide, *in* E-00130
56562-96-0 2,3-Dihydro-2,4,9-triphenyl-1*H*-1,3,2-diazaphosphonine 2-oxide, D-00609
56562-99-3 2,3-Dihydro-2,4,9-triphenyl-1*H*-1,3,2-diazaphosphonine 2-oxide; 1,3-Di-Me, *in* D-00609
56577-92-5 2-Methylphosphorin, M-00337
56577-95-8 2,6-Dimethylphosphorin, D-00881
56583-26-7 Methyl(2-propenyl)phosphinic acid; Chloride, *in* M-00364
56583-33-6 Diethyl 6-phenanthradinylphosphonate, *in* P-00060
56583-34-7 6-Phenanthridinephosphonic acid, P-00060
56595-15-4 *O,O*-Dimethyl dimethylphosphoramidoselenoate, *in* D-00869
56595-16-5 *O*-Methyl tetramethyl-phosphorodiamidoselenoate, *in* T-00223
56597-20-7 2-*tert*-Butylamino-5,5-dimethyl-1,3,2-dioxaphosphorinane, B-00538
56598-35-7 *tert*-Butyldiphenylphosphine; Oxide, *in* B-00558
56598-40-4 Ethenylmethyldiphenyl-phosphonium iodide, *in* D-01168
56607-96-6 Di-*tert*-butylphosphoramidous acid; Dichloride, *in* D-00137
56634-29-8 Hexahydro-2,4-dimethyl-3-phenoxy-1,2,4,5,3-tetrazaphosphorine; 3-Sulfide, *in* H-00024
56634-34-5 2,4,10-Trimethyl-1,3,4,5,7,10-hexaaza-3-phosphatricyclo[3.3.1.1^{3,7}]decane; 3-Oxide, *in* T-00517
56634-35-6 2,4,10-Trimethyl-1,3,4,5,7,10-hexaaza-3-phosphatricyclo[3.3.1.1^{3,7}]decane; 3-Sulfide, *in* T-00517
56660-55-0 Ethyl dibutylphosphinite, *in* D-00121
56705-27-2 2,2,3,3,7,7,8,8-Octamethyl-1,4,6,9-tetraoxa-5-phospha(5-P^V)spiro[4.4]-nonane; 5-Me, *in* O-00032
56726-81-9 Tris(triphenylphosphoranylidene)-phosphorous triamide, T-00872
56727-72-1 *N,N,N',N',N'',N''*-Hexamethyl-*N'''*-diphenylphosphinophosphorimidic triamide; Oxide, *in* H-00069
56748-23-3 Ensanchomycin, E-00003
56748-95-9 Prenomycin, P-00439
56764-36-4 Phosphinidynemethylamine, P-00356
56771-29-0 4-Pentenyltriphenyl-phosphonium(1+); Bromide, *in* P-00044
56772-64-6 Tributyl(hexadecyl)-phosphonium(1+); Iodide, *in* T-00364
56782-48-0 Glycerol 1,2-di-5,8,11,14-icosatetraenoate 3-phosphocholine; (*R*)-(*all-Z*)-*form, in* G-00017
56783-21-2 5,10-[1',2']-Benzophosphenanthrene; 5,10-Disulfide, *in* B-00024
56796-56-6 1,3,5-Triaza-7-phosphatricyclo[3.3.1.1^{3,7}]decane; 7-Sulfide, *in* T-00327
56843-02-8 Trimethyl [[(triphenylphosphoranylidene)-amino]sulfonyl]phosphorimidate, *in* T-00642
56843-03-9 Triethyl [[(triphenylphosphoranylidene)-amino]sulfonyl]phosphorimidate, *in* T-00642
56858-93-6 Triacetyl phosphate, T-00325
56875-30-0 Diethyl [(phenylamino)methyl]-phosphonate, *in* P-00084
56875-38-5 Tetraethyl [1,2-phenylenebis(methylene)]-bisphosphonate, *in* P-00130
56879-79-9 Cyclic GMP; 2'-Ac, *in* C-00234
56883-17-1 Bis[(2-nitrophenyl)methyl] phosphorochloridate, *in* B-00366
56883-96-6 Dibenzyl phenylphosphoramidate, *in* D-00053
56883-97-7 Dibenzyl benzylphosphoramidate, *in* B-00069
56884-00-5 Di-*tert*-butyl benzylphosphoramidate, *in* B-00069
56888-24-5 2-Fluoro-4-methyl-1,3,2-dioxaphosphorinane; (2*RS*,4*RS*)-*form, in* F-00021
56888-26-7 *O,O*-1,3-Butylene fluorothiophosphate, *in* F-00021

56898-58-9 *P*,*P*-Dimethyl-*N*-phenylphosphinoselenoic amide, *in* D-00836

56898-59-0 *P*,*P*-Di-*tert*-butyl-*N*-phenylphosphinous amide, *in* D-00124

56898-61-4 *P*,*P*-Di-*tert*-butyl-*N*-phenylphosphinothioic amide, *in* D-00119

56898-62-5 *P*,*P*-Di-*tert*-butyl-*N*-phenylphosphinoselenoic amide, *in* D-00115

56904-86-0 *tert*-Butyl 2-(triphenylphosphoranylidene)-propanoate, *in* T-00660

56908-15-7 1,2,3-Tri-*tert*-butyldiazaphosphiridine; (1α,2β,3α)-*form*, 3-Oxide, *in* T-00362

56981-97-6 [[3-(Methoxycarbonyl)phenyl]-methyl]triphenyl-phosphonium(1+); Bromide, *in* M-00030

57018-04-9 Toclofos-Methyl, T-00321

57027-30-2 1,1,3,3-Tetramethyl-1-phospha-3-silacyclohex-1-ene, T-00212

57028-20-3 1,2-Bis(trifluoromethyl)-diphosphine; (1*RS*,2*SR*)-*form*, *in* B-00410

57028-24-7 1,2-Bis(trifluoromethyl)-diphosphine; (1*RS*,2*RS*)-*form*, *in* B-00410

57039-78-8 *N*,*N*,*N*′,*N*′-Tetraethylphosphonothioic diamide, T-00057

57043-19-3 2,2-Dihydro-1,2-oxaphosphetane; 2,2,2-Tri-Ph, *in* D-00533

57044-92-5 4-Phenyl-4*H*-1,4-selenaphosphorin; 4-Oxide, *in* P-00307

57044-96-9 4-Phenyl-4*H*-1,4-telluraphosphorin; 4-Oxide, *in* P-00317

57055-05-7 Bis(4-methylphenyl)phosphinic acid; Amide, *in* B-00320

57065-70-0 3,3*a*,4,5-Tetrahydro-3-methyl-2*H*-benzo[*e*]phosphindole 3-oxide; (3*RS*,3*aRS*)-*form*, *in* T-00112

57084-22-7 2,2-Dihydro-4,5-diphenyl-1,3,2-oxathiaphosphole; 2,2,2-Trimethoxy, *in* D-00480

57090-22-9 2-Diethylamino-1,3,2-dioxaphospholan-4,5-dicarboxylic acid; (4*R*,5*R*)-*form*, Di-Me ester, *in* D-00265

57093-51-3 2-Methoxy-4-methyl-1,3,2-dithiaphospholane, *in* M-00136

57115-47-6 *P*,*P*-Diisopropylphosphinic amide, D-00642

57115-66-9 2-Phenyl-1,3,2-dithiaphosphorinane, *in* D-01247

57115-76-1 2-Choro-1,3,2-dithiaphosphorinane; 2-Sulfide, *in* C-00199

57115-77-2 Trimethylene phenylphosphonotrithioate, *in* D-01249

57137-55-0 (3-Chloropropyl)-diphenylphosphine, C-00169

57150-67-1 Dichloro-2-naphthylphosphine, *in* N-00026

57154-91-3 Methyl *N*,*N*-bis(2-chloroethyl)-phosphorodiamidate, *in* B-00126

57156-84-0 Phenylphosphonic acid trimer trianhydride, *in* T-00694

57188-46-2 Bis[(4-nitrophenyl)methyl] phosphorochloridate, *in* B-00367

57193-50-7 Phenyl *N*,*N*′-dipropylphosphorodiamidate, *in* D-01223

57194-41-9 2,3-Dihydro-1,2,3-triphenyl-1*H*-benzotriphosphole, D-00606

57212-38-1 (Dichloromethyl)-triphenylphosphonium(1+); Chloride, *in* D-00182

57237-61-3 *P*-Ethyl-*N*-phenylphosphonamidic acid; (*S*)-*form*, Et ester, *in* E-00110

57237-62-4 Ethyl *P*-ethyl-*N*-phenylphosphonamidate, *in* E-00110

57237-67-9 *O*,*O*-Diethyl phenylphosphoramidoselenoate, *in* P-00261

57237-68-0 Neopentylene phenylphosphoramidate, *in* D-00792

57237-69-1 Neopentylene phenylphosphoramidothioate, *in* D-00792

57237-70-4 Neopentylene phenylphosphoramidoselenoate, *in* D-00792

57241-88-0 [2-(Trimethylstannyl)-ethylidene]triphenylphosphorane, *in* T-00561

57242-06-5 4-Methylphosphorin, M-00338

57244-57-2 2,2,3,3,7,7,8,8-Octamethyl-1,4,6,9-tetraoxa-5-phospha(5-P^{V})spiro[4.4]-nonane; 5-Benzoyl, *in* O-00032

57244-61-8 Phenyl dimethylphosphinate, *in* D-00829

57270-64-1 4-Methyl-2-methylthio-1,3,2-dioxaphosphorinane; (2*RS*,4*RS*)-*form*, *in* M-00174

57270-65-2 4-Methyl-2-methylthio-1,3,2-dioxaphosphorinane; (2*RS*,4*SR*)-*form*, 2-Selenide, *in* M-00174

57270-66-3 *O*,*O*-1,3-Butylene *S*-methyl phosphoroselenothioate, *in* M-00174

57277-25-5 Diethyl 2-naphthylmethylphosphonate, *in* N-00014

57301-49-2 3-Methyl-2-phenoxy-1,3,2-oxazaphospholidine, M-00192

57301-51-6 4′,4′,5′,5′-Tetramethylspiro[1,3,2-benzodioxaphosphole-2,2′-[1,3,2]dioxaphospholane]; 2-Methoxy, *in* T-00230

57301-52-7 2,2,3,3,7,7,8,8-Octamethyl-1,4,6,9-tetraoxa-5-phospha(5-P^{V})spiro[4.4]-nonane; 5-Phenoxy, *in* O-00032

57301-53-8 2,2,3,3-Tetramethyl-1,4,6,9-tetraoxa-5-phospha(5-P^{V})spiro[4.4]nonane; 5-Phenoxy, *in* T-00235

57321-50-3 4-Methyl-2-methylthio-1,3,2-dioxaphosphorinane; (2*RS*,4*SR*)-*form*, *in* M-00174

57322-08-4 Methyl ethylphenylphosphinite, *in* E-00108

57322-08-4 Methyl methylphenylphosphinite, *in* M-00227

57322-10-8 *S*-Methyl ethylphenylphosphinothioate, *in* E-00105

57322-18-6 1,1′(3*H*,3′*H*)-Spirobi[2,1-benzoxaphosphole]-3,3′-dione; 1-Ph, *in* S-00018

57322-19-7 1,1′(3*H*,3′*H*)-Spirobi[2,1-benzoxaphosphole]-3,3′-dione; 1-Me, *in* S-00018

57328-54-8 1,2-Dihydro-2-hydroxyisophosphinoline 2-oxide, D-00495

57328-60-6 1-Methylisophosphinoline, M-00162

57328-61-7 1,3-Dimethylisophosphinoline, D-00758

57350-23-9 *O*-Ethyl *S*-phenyl phosphorodithioate, E-00121

57351-96-9 2-Chloro-4-methyl-1,3,2-dithiaphospholane, *in* M-00136

57367-54-1 α-Methoxycarbonyltriphenyl-phosphoranylidenepropanoic acid, *in* T-00646

57367-56-3 *tert*-Butyl α-methoxycarbonyltriphenyl-phosphoranylidenepropanoate, *in* T-00646

57368-24-8 Tris(2,4,6-trimethylphenyl)-phosphine; Sulfide, *in* T-00861

57422-79-4 10-Methylacridophosphine, M-00087

57432-16-3 Trimethyl[(trimethyl-phosphoranylidene)methyl]-phosphonium(1+); Chloride, *in* T-00565

57436-71-2 2-Isoselenocyanato-5,5-dimethyl-1,3,2-dioxaphosphorinane, I-00066

57436-72-3 2-Cyano-5,5-dimethyl-1,3,2-dioxaphosphorinane; 2-Selenide, *in* C-00209

57436-73-4 2-Cyano-5,5-dimethyl-1,3,2-dioxaphosphorinane, C-00209

57437-91-9 Bis(trimethylphosphoranylidene)-methane, B-00429

57454-28-1 1,3-Butylene phenylphosphonate, *in* M-00199

57458-74-9 9-Phenyl-9-phosphabicyclo[3.3.1]nonane; Oxide, *in* P-00193

57468-68-5 ▷Dimethyl phosphorazidate, D-00878

57519-80-9 (2-Cyanoethyl)-triphenylphosphonium(1+); Bromide, *in* C-00212

57545-56-9 (2,7-Dimethyl-2,4,6-octatrien-1,8-diyl)bis[triphenyl-phosphonium](2+); (*E*,*E*,*E*)-*form*, Dibromide, *in* D-00778

57551-08-3 1,1′(3*H*,3′*H*)-Spirobi[2,1-benzoxaphosphole], S-00017

57557-25-2 *O*-Ethyl *O*,*S*-dimethyl phosphorothioate; (*S*)-*form*, *in* E-00065

57557-26-3 *O*-Ethyl *O*,*S*-dimethyl phosphorothioate; (*R*)-*form*, *in* E-00065

57557-27-4 *O*-Isopropyl *O*,*S*-dimethyl phosphorothioate; (*S*)-*form*, *in* I-00030

57557-28-5 *O*-Isopropyl *O*,*S*-dimethyl phosphorothioate; (*R*)-*form*, *in* I-00030

57557-80-9 *O*,*S*-Diethyl phenylphosphonothioate, *in* P-00244

57573-32-7 2-Chloro-3,4-dimethyl-5-phenyl-1,3,2-oxazaphospholidine; (2*R*,4*S*,5*R*)-*form*, 2-Sulfide, *in* C-00061

57575-15-2 2,2,2-Tribromoethyl 4-morpholinylphosphonochloridate, T-00357

57584-33-5 2,2,4,4,6,6,8,8,9,9,10,10-Dodecaethyl-1,3,5,7-tetraphospha-2,4,6,8,9,10-hexagermatricyclo[3.3.1.$1^{3,7}$]-decane, D-01260

57590-54-2 *N*,*N*-Diethyl-2,6-dimethyl-1-phenyl-1,4-phosphaborin-4(1*H*)-amine, D-00285

57602-63-8 2,2(3*H*,3′*H*)-Spirobi[1,3,2-benzoxazaphosphole]; 2-Me, *in* S-00020

57605-13-7 *tert*-Butylphosphonothioic acid, B-00612

57637-97-5 [(Acetylamino)methyl]phosphonic acid, A-00012

57644-86-7 Phenyl diphenylphosphinodithioate, *in* D-01065

57644-87-8 Benzyl diphenylphosphinodithioate, *in* D-01065

57648-42-7 *P*-Cyanomethyl-*N*,*N*,*N*′,*N*′-tetramethylphosphonic diamide, C-00216

57648-76-7 *N*-(1,2-Dimethyl-ethenylenedioxyphosphoryl)-imidazole, D-00753

57651-34-0 2-Chloro-3,4-dimethyl-5-phenyl-1,3,2-oxazaphospholidine; (2*S*,4*S*,5*R*)-*form*, 2-Sulfide, *in* C-00061

57661-74-2 2,3-Diphenyl-1,4,6-trioxa-9-aza-5-phospha(5-P^V)-spiro[4.4]non-2-ene; 5-Methoxy, *in* D-01159

57664-94-5 3-Methyl-1-phenylphospholane; (1*RS*,3*SR*)-*form*, *in* M-00229

57668-23-2 ▷Phenyl *N*,*P*-diphenylphosphonamidate, *in* D-01103

57673-93-5 Ethyl *N*,*N′*-dicyclohexylphosphorodiamidate, *in* D-00216

57673-94-6 *O*,*O*-Diethyl *tert*-butylphosphoramidothioate, *in* B-00623

57674-98-3 (9*H*-Fluoren-9-ylideneethenylidene)-triphenylphosphorane, F-00009

57680-64-5 2,3,5,6,8,8-Hexahydro-8-phenyl-[1,3,2]-oxazaphospholo[2,3-*b*][1,3,2]-oxazaphosphole, H-00049

57733-31-0 2-*tert*-Butylamino-4-methyl-1,3,2-dioxaphosphorinane; (2*RS*,4*SR*)-*form*, *in* B-00541

57733-32-1 2-*tert*-Butylamino-4-methyl-1,3,2-dioxaphosphorinane; (2*RS*,4*RS*)-*form*, *in* B-00541

57733-33-2 *O*,*O*-1,3-Butylene *tert*-butylphosphoramidoselenoate, *in* B-00541

57734-11-9 Diisopropyl phosphate; Cyclohexylammonium salt, *in* D-00639

57745-06-9 2,4-Dimethyl-1,3,2-dioxaphosphorinane; (2*RS*,4*RS*)-*form*, *in* D-00734

57745-06-9 1,3-Butylene methylphosphonate, *in* D-00734

57756-98-6 2-Fluoro-4-methyl-1,3,2-dioxaphosphorinane; (2*RS*,4*RS*)-*form*, 2-Oxide, *in* F-00021

57761-68-9 2,4-Dimethyl-1,3,2-dioxaphosphorinane; (2*RS*,4*SR*)-*form*, *in* D-00734

57761-68-9 2,4-Dimethyl-1,3,2-dioxaphosphorinane; (2*RS*,4*SR*)-*form*, 2-Oxide, *in* D-00734

57767-19-8 1-*tert*-Butyl-1,4-dihydro-4-phenyl-1,4-phospharsenin; *trans*-*form*, *in* B-00553

57767-20-1 1-*tert*-Butyl-1,4-dihydro-4-phenyl-1,4-phospharsenin; *cis*-*form*, *in* B-00553

57767-21-2 1-*tert*-Butyl-1,4-dihydro-4-phenyl-1,4-phospharsenin; *trans*-*form*, 1-Oxide, *in* B-00553

57767-22-3 1-*tert*-Butyl-1,4-dihydro-4-phenyl-1,4-phospharsenin; *cis*-*form*, 1-Oxide, *in* B-00553

57788-18-8 3-Hydroxy-2,2-dimethyl-1,3-oxaphosphetane 3-oxide; Na salt, *in* H-00126

57808-64-7 Toldimfos, T-00322

57831-27-3 *P*,*P*,*P*-Tributylphosphine imide, T-00374

57857-79-1 Tetraphenyl ethylimidodiphosphite, *in* E-00079

57858-63-6 *N*,*N′*,*N″*-Tris(4-methylphenyl)-phosphorothioic triamide, *in* T-00819

57858-64-7 *N*,*N′*,*N″*-Tris(2-methylphenyl)-phosphorothioic triamide, *in* T-00817

57858-66-9 *N*,*N*-Dimethyl-*N′*,*N″*-diphenylphosphorothioic triamide, *in* D-00909

57906-69-1 Bis(3-methylphenyl)phosphinic acid; Azide, *in* B-00319

57906-70-4 Bis(3-chlorophenyl)phosphinic acid; Azide, *in* B-00138

57906-75-9 Bis(3-methylphenyl)phosphinic acid, B-00319

57906-77-1 Bis(3-chlorophenyl)phosphinic acid; Chloride, *in* B-00138

57945-51-4 9*H*-Fluoren-9-yltriphenylphosphonium(1+); Tetrafluoroborate, *in* F-00012

57978-07-1 1,5,9,13-Tetrabenzyl-1,5,9,13-tetraphenyl-1,5,9,13-tetraphosphoniacyclohexadecane tetrabromide, *in* T-00276

57978-08-2 1,5,9,13-Tetraphenyl-1,5,9,13-tetraphosphacyclohexadecene; Tetraoxide, *in* T-00276

57978-09-3 1,5,9,13-Tetraphenyl-1,5,9,13-tetraphosphacyclohexadecene, T-00276

57978-15-1 1,1,6,6,11,11,16,16-Octabenzyl-1,6,11,16-tetraphosphoniacycloeicosane tetrabromide, *in* T-00008

57978-17-3 1,6,11,16-Tetrabenzyl-1,6,11,16-tetraphosphacycloeicosane, T-00008

58045-33-3 *O*,*O*-Diphenyl phosphonothioate, D-01106

58046-12-1 Dimethyl phenylphosphoramidate, D-00817

58068-62-5 2-(Trimethylsilyloxy)-1,3,2-dioxaphospholane, T-00554

58078-24-3 *O*-Ethyl hydrogen cyclohexylphosphonothioate, *in* C-00264

58116-17-9 Glycerol 1-octadecanoate 2-9-octadecenoate 3-phosphoinositol; (*R*)-(*Z*)-*form*, *in* G-00084

58141-57-4 Dibutyl 2,4-cyclopentadienylphosphonite, *in* C-00275

58156-53-9 Methyl di-*tert*-butylphosphinate, *in* D-00110

58159-86-7 2,4,6,7,7,8,8-Heptaphenyl-1-phosphabicyclo[4.2.0]-octa-2,4-diene, H-00003

58159-96-9 *tert*-Butylmethylphenylphosphine; (*R*)-*form*, Sulfide, *in* B-00565

58160-80-8 Butylphosphonothioic acid trimer trianhydride, *in* T-00396

58201-66-4 (Bromodifluoromethyl)-triphenylphosphonium(1+); Bromide, *in* B-00468

58228-19-6 *O*,*O*-Dimethyl-*S*,*Se*-2-nitrophenyl-phosphoro(selenothioperoxo)-thioate, D-00772

58228-22-1 *O*,*O*-Dimethyl *Se*,*Se*-(2-nitrophenyl) phosphoro(diselenoperoxoate), D-00771

58228-28-7 *O*,*O*-Dimethyl phosphoroselenoate; Na salt, *in* D-00906

58228-29-8 *O*,*O*-Diisopropyl phosphoroselenoate; Na salt, *in* D-00666

58245-39-9 Methyl *N*-cyclohexyl-*P*-phenylphosphonamidate, *in* C-00251

58293-33-7 ▷Dibutyl chlorothiophosphonate, D-00080

58293-37-1 *O*-Methyl *O*-1-naphthalenyl phosphorothioate; (+)-*form*, Dicyclohexylammonium salt, *in* M-00176

58293-38-2 *O*-Methyl *O*-1-naphthalenyl phosphorothioate; (−)-*form*, Dicyclohexylammonium salt, *in* M-00176

58299-37-9 2,8,9-Trioxa-1-phosphatricyclo[3.3.1.1^{3,7}]decane; 1-Sulfide, *in* T-00591

58309-95-8 Ethyl di-*tert*-butylphosphinite, *in* D-00122

58310-40-0 (Trifluoromethyl)phosphinous bromide, T-00476

58310-41-1 (Trifluoromethyl)phosphinous chloride, T-00477

58310-46-6 (Trifluoromethyl)phosphonous acid; Dicyanide, *in* T-00482

58322-08-0 (2-Buten-1-yl)diphenylphosphine; (*Z*)-*form*, Oxide, *in* B-00520

58347-13-0 Dimethylphosphinic acid; Azide, *in* D-00829

58347-14-1 Diethylphosphinothioic acid; Azide, *in* D-00330

58359-96-9 Dimethyl 4-*tert*-butylcyclohexylphosphonite, *in* B-00550

58359-97-0 4-*tert*-Butylcyclohexylphosphonous acid; *trans*-*form*, Di-Me ester, *in* B-00550

58498-96-7 2-Phenylspiro[1,3,2-benzothiazaphosphole-2(3*H*),2′-[1,3,2]-dioxaphospholane]; 3-(2,4,6-Trimethylphenyl), *in* P-00308

58498-98-9 2-Phenylspiro[1,3,2-benzothiazaphosphole-2(3*H*),2′-[1,3,2]-dioxaphospholane]; 3-(2,6-Dimethylphenyl), *in* P-00308

58499-02-8 1,4,4*a*,9,9*a*,10-Hexahydro-11-phenyl-9,10-phosphinideneanthracene; 11-Oxide, *in* H-00050

58513-98-7 Triphenyl[(triphenyl-phosphoranylidene)methyl]-phosphonium(1+); Chloride, *in* T-00697

58521-13-4 Tetraphenylphosphorodiamidous acid; Chloride, *in* T-00274

58521-14-5 Diphenylphosphoramidous acid; Difluoride, *in* D-01111

58521-17-8 *N*,*N*,*N′*,*N′*-Tetraphenylphosphonic diamide, T-00271

58521-18-9 Methyl tetraphenylphosphorodiamidite, *in* T-00274

58521-19-0 Phenyl tetraphenylphosphorodiamidite, *in* T-00274

58521-21-4 *N*,*N*,*N′*,*N′*-Tetraphenylphosphonothioic diamide, T-00273

58525-84-1 6-Methyl-3,5,7-triphenyl-6-phosphatricyclo[3.3.1.0^{2,7}]non-3-ene; 6-Oxide, *in* M-00431

58529-35-4 6-Phenyl-5*H*-dibenzo[*c*,*e*]-phosphepin-5,7(6*H*)-dione, P-00109

58558-73-9 Triphenyl phosphorotriselenoite, T-00673

58558-74-0 *Se*,*Se*,*Se*-Triphenyl phosphorotriselenothioate, T-00674

58558-75-1 Tris(4-methoxyphenyl) phosphorotritelluroite, T-00795

58580-13-5 Bromfenvinfos; (*Z*)-*form*, *in* B-00462

58580-14-6 Bromfenvinfos; (*E*)-*form*, *in* B-00462

58594-17-5 4-Ethylspiro[2,8,9-trioxa-1-phosphatricyclo[3.3.1.1^{3,7}]-decane-1,4′-trioxaphosphetane], E-00187

58594-19-7 Triphenyl(1-phenylethylidene)-phosphorane, *in* T-00612

58608-01-8 2,5-Dihydro-3,4-dimethyl-1-phenyl-1*H*-phosphole; 1-Selenide, *in* D-00468

58621-89-9 1,2-Di-*tert*-butyl-1,2-dimethyldiphosphine; (*RS*,*RS*)-*form*, *in* D-00089

58656-41-0 2,3-Dihydro-2,3-diphenyl-1,3,2-benzoxazaphosphole; 2-Oxide, *in* D-00478

58661-92-0 *O*-Ethyl *O*,*O*-*o*-phenylene thiophosphate, *in* E-00024

58673-12-4 Diethyl (1*H*-1,2,4-triazol-1-yl)phosphonite, *in* T-00332

58722-79-5 *O*-Methyl diethylphosphinoselenoate, *in* D-00327

58751-94-3 5-Chloro-5,10-dihydro-10-methylacridophosphine, *in* D-00515

58751-95-4 5,10-Dihydro-5-hydroxy-10-methylacridophosphine 5-oxide, *in* D-00515

58751-96-5 5,10-Dihydro-10-methylacridophosphine, D-00515

58776-19-5 3,4-Bis[(triphenylphosphonio)-methyl]thiophene(1+); Dibromide, *in* B-00457

58776-20-8 3,4-Bis[(triphenylphosphonio)-methyl]furan(2+); Dibromide, *in* B-00449

58802-35-0 1,1-Dihydro-1,1,3,3,5,5-hexamethyl-1-phospha-3,5-disilacyclohex-1-ene, D-00485

58809-19-1 Phenyl cyclohexylphosphoramidochloridate, P-00104

58809-22-6 2,4-Dichlorophenyl 4-morpholinylphosphonochloridate, *in* M-00524

58815-97-7 Diisopropyl (3-methyl-4-pyridinyl)phosphonate, *in* M-00390

58816-01-6 4-Pyridinephosphonic acid, P-00505

58816-02-7 (3-Methyl-4-pyridinyl)phosphonic acid, M-00390

58816-61-8 Dimethyl phosphorohydrazidate, D-00904

58816-62-9 Methyl phenylphosphonazidate, *in* P-00220

58823-95-3 Glucose 6-dihydrogen phosphate; D-*form*, Ba salt, *in* G-00007

58825-45-9 2-Ethoxy-2,3-dihydro-1*H*-1,3,2-benzodiazaphosphole, E-00026

58825-46-0 *O*-Ethyl *N,N'*-*o*-phenylenephosphorodiamidothioate, *in* E-00026

58887-04-0 (2-Hydroxyethyl)-trimethylphosphonium(1+); Chloride, *in* H-00150

58887-04-0 (2-Hydroxyethyl)-triphenylphosphonium(1+); Chloride, *in* H-00151

58898-18-3 Diethyl (dibromomethyl)-phosphonate, *in* D-00064

58910-86-4 Methyl ethylmethylphosphinite, *in* E-00087

58921-82-7 2-Diazo-2-(diphenylphosphinothioyl)-1-phenylethanone, *in* D-00029

58922-31-9 3-Phenyl-2-propenylphosphonic acid, P-00302

58922-31-9 Diethyl 3-phenyl-2-propenylphosphonate, *in* P-00302

58942-11-3 Trimethyl (trimethylsilyl)-phosphorimidate, *in* T-00557

58943-95-6 5,10-Dihydro-5,10-dimethylphenophosphazine, D-00466

58943-96-7 5,10-Dihydro-5-methyl-10-phenylphenophosphazine; 10-Oxide, *in* D-00526

58943-97-8 5,10-Dihydro-5,10-dimethylphenophosphazine; 10-Oxide, *in* D-00466

58972-01-3 *P,P*-Dichloro-*N*-trimethylsilyl-*P*-[bis(trimethylsilyl)-amino]phosphazene, *in* T-00868

58972-02-4 *P,P*-Difluoro-*N*-trimethylsilyl-*P*-[bis(trimethylsilyl)-amino]phosphazene, *in* T-00868

58972-03-5 *P,P*-Dibromo-*N*-trimethylsilyl-*P*-[bis(trimethylsilyl)-amino]phosphazene, *in* T-00868

58972-04-6 *P,P*-Diiodo-*N*-trimethylsilyl-*P*-[bis(trimethylsilyl)-amino]phosphazene, *in* T-00868

58979-11-6 2,4,6,8,9,10-Hexamethyl-2,4,6,8,9,10-hexaaza-1,3,5,7-tetraphosphatricyclo[3.3.1.$1^{3,7}$]-decane; 1,3,5,7-Tetroxide, *in* H-00071

58983-18-9 Diethyl (2,5-dimethylphenyl)-phosphonate, *in* D-00813

58983-19-0 (2,5-Dimethylphenyl)phosphonic acid, D-00813

58983-20-3 Diethyl (2,4-dimethylphenyl)-phosphonate, *in* D-00812

58992-24-8 Triphenyl[1-(phenylthio)-cyclopropyl]phosphonium(1+); Bromide, *in* T-00618

58992-26-0 Triphenyl[1-(phenylthio)-cyclopropyl]phosphonium(1+); Tetrafluoroborate, *in* T-00618

58993-56-9 Dimethyl (dichloromethyl)-phosphonate, *in* D-00175

58993-59-2 *P*-(Dichloromethyl)phosphonic bis(dimethylamide), *in* D-00176

58995-37-2 ▷Propetamphos, P-00466

58995-93-0 *O,O*-1,3-Butylene *O*-ethyl thiophosphate, *in* E-00041

59007-36-2 3-Ethoxy-2,3-dihydro-2-methyl-5-phenyl-1*H*-1,2,4,3-triazaphosphole; 3-Sulfide, *in* E-00028

59022-97-8 *N*-Benzoyl-*P,P*-diphenylphosphinic amide, *in* D-01040

59081-07-1 Tetraethyl (1-aminoethylidene)-bisphosphonate, *in* A-00073

59085-24-4 *O*-Isopropyl diethylphosphinoselenoate, *in* D-00327

59085-25-5 *P,P*-Diethyl-*N*-methylphosphinoselenoic amide, *in* D-00328

59085-26-6 *P,P*-Diethyl-*N*-phenylphosphinoselenoic amide, *in* D-00328

59085-32-4 2,3-Diphenyl-1,3,2-oxazaphospholidine; 2-Sulfide, *in* D-01029

59111-78-3 Agrocin 84, A-00048

59205-19-5 1,3-Benzothiaphosphole, B-00025

59273-35-7 5,10-Dihydro-5-phenylacridophosphine, D-00547

59274-19-0 Diethyl (1-nitroso-1-methylethyl)phosphonate, *in* N-00066

59284-70-7 (Aminophenylmethyl)phosphonic acid; (±)-*form*, *in* A-00105

59313-80-3 2,7,10,15,17-Pentamethyl-17*H*-tetrabenzo[*b,d,f,h*]-phosphonin, P-00028

59319-64-1 *O,O*-1,3-Butylene fluoroselenophosphate, *in* F-00021

59325-05-2 2,7,10,15,17,17-Hexamethyl-17*H*-tetrabenzo[*b,d,f,h*]-phosphoninium iodide, *in* P-00028

59341-95-6 2,7,10,15-Tetramethyl-17-phenyl-17*H*-tetrabenzo[*b,d,f,h*]-phosphonin, T-00211

59344-69-3 Dimethyl (2-nitroethyl)-phosphonate, *in* N-00036

59346-65-5 Di-*tert*-butyl phosphorobromidate, D-00138

59357-76-5 Triiododiphenylphosphorane, T-00497

59360-03-1 *N*-Phosphoglutamic acid; (*S*)-*form*, *in* P-00367

59360-59-7 *N*-Phenyl-*P*-trichloromethylphosphonamidic acid; Chloride, *in* P-00328

59361-27-2 Diphenyl (1,1-dimethylethyl)-phosphonate, *in* B-00595

59375-25-6 Ethyl *N,N*-diethyl-*P*-(methoxymethyl)-phosphonamidate, *in* D-00295

59375-25-7 Propyl *N,N*-diethyl-*P*-(methoxymethyl)-phosphonamidate, *in* D-00295

59375-32-5 Ethyl *P*-(butoxymethyl)-*N,N*-diethylphosphonamidate, *in* B-00536

59378-88-0 Heptyltriphenylphosphonium(1+); Iodide, *in* H-00011

59384-67-7 1,4-Phenylenebis[methylphosphinic acid], P-00137

59386-52-6 1-Phenylphosphocane, P-00215

59386-55-9 1,1-Diphenylphosphepanium bromide, *in* P-00205

59386-61-7 1-Phenylphosphocane; Oxide, *in* P-00215

59413-41-1 Tetramethylene phenyl phosphite, *in* D-00968

59413-42-2 4,7-Dihydro-1,3,2-dioxaphosphepin; 2-Phenoxy, *in* D-00472

59413-43-3 1,5-Dihydro-2-phenoxy-2,4,3-benzodioxaphosphepin, D-00544

59432-47-2 1,1-Diphenylphosphorinanium bromide, *in* P-00265

59432-49-4 1-Benzyl-1-phenylphosphepanium bromide, *in* P-00205

59452-79-8 Bis(2-methylphenyl)phosphinic acid; Amide, *in* B-00318

59452-81-2 *P,P*-Bis(2-methylphenyl)-*N*-phenylphosphinic amide, *in* B-00318

59463-49-9 Diethyl *tert*-butoxycyanomethylphosphonate, D-00273

59466-62-5 1,4-Thiaphosphorinane 1,1,4-trioxide; 4-Ph, *in* T-00300

59466-63-6 1,4-Thiaphosphorinane 1,1,4-trioxide; 4-Me, *in* T-00300

59466-94-3 Diethyl *tert*-butylphosphoramidite, *in* B-00624

59470-47-2 10,11-Dihydro-5-methyl-5*H*-dibenzo[*b,f*]phosphepin; Oxide, *in* D-00524

59472-84-3 Bis(2-methylphenyl)phosphinic acid; Chloride, *in* B-00318

59474-16-7 2,5-Dihydro-5,5-dimethyl-1,2-oxaphosphole 2-oxide, D-00464

59474-16-7 2,5-Dihydro-5,5-dimethyl-1,2-oxaphosphole 2-oxide; 2-Hydroxy, *in* D-00464

59491-62-2 Glycerol 1-9-octadecenoate 2-hexadecanoate 3-phosphocholine; (*R*)-L-(*Z*)-*form*, *in* G-00087

59499-31-9 4-Nonenyltriphenyl-phosphonium(1+); (*Z*)-*form*, Ylide, *in* N-00073

59499-32-0 4-Nonenylidenetriphenyl-phosphorane, *in* N-00073

59502-08-8 7-Butyl-2,3-dihydro-1,6-diphenyl-1*H*-1-benzophosphonin-2,3,4,5-tetracarboxylic acid; Tetra-Me ester, oxide, *in* B-00552

59506-93-3 10,11-Dihydro-5*H*-dibenzo[*b,f*]-phosphepin; Oxide, *in* D-00451

59512-89-9 4-Nonenyltriphenyl-phosphonium(1+); (*Z*)-*form*, Bromide, *in* N-00073

59512-90-2 4-Nonenyltriphenyl-phosphonium(1+); (*E*)-*form*, Bromide, *in* N-00073

59515-30-9 2-Dimethylamino-1,3,2-oxazaphospholidine, D-00702

59523-65-8 Tetramethylethylene chlorothiophosphate, *in* C-00185

59523-66-9 2-Hydroxy-4,4,5,5-tetramethyl-1,3,2-dioxaphospholane 2-sulfide; Dicyclohexylammonium salt, *in* H-00197

59547-94-3 Diphenyl phosphoriodidite, D-01115

59556-98-8 2-Hydroxy-4-phenyl-1,3,2-dioxaphosphorinane 2-oxide, H-00171

59568-74-0 Dipropyl phenylphosphonotrithioate, *in* P-00250

59568-76-2 Isopropyl diphenylphosphinodithioate, *in* D-01065

59578-17-5 Dichloromethylenetris(dimethylamino)phosphorane, D-00173

59590-74-8 5-Benzyl-5,5-dihydro-5-phenylacridophosphine, B-00042

59611-88-0 Ethyl phenyl(2-propenyl)-phosphinate, *in* P-00300

59611-97-1 Diphenyl (2-methyl-1,3-butadienyl)phosphonate, *in* M-00093

59612-01-0 Tetraethylphosphorodiamidous acid; Iodide, *in* T-00066

59612-04-3 Iododiisopropylphosphine, *in* D-00648

59624-92-9 Ethyl ethylphosphinate; (−)-*form*, *in* E-00068

59625-55-7 [(3-Methylphenyl)methylene]-triphenylphosphorane, *in* M-00210

59625-56-8 (2-Chlorophenylmethylene)-triphenylphosphorane, *in* C-00135

59625-60-4 [(4-Fluorophenyl)methylene]-triphenylphosphorane, *in* F-00032

59646-46-7 Tetraethyl (4-morpholinylmethylene)-bisphosphonate, *in* M-00520

59658-74-1 Diisopropyl benzylphosphoramidate, *in* B-00069

59658-82-1 Triethyl benzoylphosphorimidate, *in* B-00039

59658-83-2 Trimethyl benzoylphosphorimidate, *in* B-00039

59658-86-5 Triphenyl benzoylphosphorimidate, *in* B-00039

59659-68-6 [(2-Methoxyphenyl)methylene]-triphenylphosphorane, *in* M-00062

59675-46-6 Carbonylbis[phosphorimidic trichloride], C-00007

59682-40-5 Diethyl (methylaminocarbonyl)-phosphonate, *in* M-00088

59694-43-8 (2-Hydroxyethyl)-trimethylphosphonium(1+); Iodide, *in* H-00150

59694-43-8 (2-Hydroxyethyl)-triphenylphosphonium(1+); Iodide, *in* H-00151

59725-15-4 1,2,4-Trimethyl-1,2,4,3,5-triazadiphospholidine; 3,5-Dichloro, *in* T-00562

59725-16-5 1,2,3,4,5-Pentamethyl-1,2,4,3,5-triazadiphospholidine, *in* T-00562

59725-17-6 1,2,4-Trimethyl-1,2,4,3,5-triazadiphospholidine; 3,5-Dimethoxy, *in* T-00562

59725-18-7 1,2,4-Trimethyl-1,2,4,3,5-triazadiphospholidine; 3,5-Bis(methylthio), *in* T-00562

59725-20-1 ▷2,5-Dichloro-3,4-dimethyl-1,3,4,2,5-thiadiazadiphospholidine, D-00167

59738-68-0 (2-Hydroxyethyl)-trimethylphosphonium(1+), H-00150

59740-56-6 Trimethyl [bis(diethylamino)-phosphinyl]phosphorimidate, *in* B-00176

59740-57-7 Triethyl [bis(diethylamino)-phosphinyl]phosphorimidate, *in* B-00176

59740-66-8 Tetraethylphosphorodiamidic acid; Azide, *in* T-00059

59745-41-4 *O,O*-Dipropyl phosphoroselenoate; K salt, *in* D-01229

59758-17-7 3-Benzyl-2-chlorotetrahydro-2*H*-1,3,2-oxazaphosphorine, B-00041

59758-28-0 Hexabenzylphosphorous triamide, H-00012

59819-52-2 2,2,2-Trichloroethyl 2-chlorophenyl phosphorochloridate, T-00405

59825-34-2 2-Mercapto-4,4,5,5-tetramethyl-1,3,2-dioxaphospholane 2-sulfide; Na salt, *in* M-00019

59833-34-0 Triethyl [bis(dimethylamino)-phosphinyl]phosphorimidate, *in* B-00194

59857-18-0 Cyanophosphoramidic dichloride, C-00227

59892-97-6 2-Hydroxy-1,3,2-dioxaphosphole 2-oxide, H-00136

59895-94-4 *O,O*-Dimethyl phosphorohydrazidothioate, D-00905

59901-84-7 2,3-Dihydro-1,3-dimethyl-2-phenyl-1*H*-benzodiazaphosphole, D-00467

59901-85-8 2,3-Dihydro-1,3-dimethyl-1*H*-1,3,2-benzodiazaphosphole; 2-Chloro, *in* D-00456

59917-79-2 1-Methoxy-1,3-dioxo-2,4-dimethyl-1-phospha-2,4-diazacyclopentane, *in* D-00722

59924-20-8 Tris(2-phenylethyl) phosphite, T-00845

59925-00-7 Diethyl [2-oxo-2-(triphenylphosphoranylidene)-propyl]phosphonate, *in* O-00086

59950-71-9 *N*″-Benzyl-*N,N,N*′,*N*′-tetramethylphosphoric triamide, *in* T-00216

59992-10-8 *N,N,N*′,*N*′,6,7-Hexamethyl-1,3-bis(trimethylsilyl)-5,8-dioxa-1,3-diaza-2,4-diphospha(4-P^{V})-spiro[3.4]oct-6-ene-2,4-diamine, H-00068

59992-11-9 *N,N,N*′,*N*′,6,7-Hexamethyl-1,3-bis(trimethylsilyl)-5,8-dioxa-1,3-diaza-2,4-diphospha(4-P^{V})-spiro[3.4]oct-6-ene-2,4-diamine; 2-Sulfide, *in* H-00068

59998-70-8 Ethyl *N*″-[bis(diethylamino)-phosphinyl]-*N,N,N*′,*N*′-tetraethylphosphorodiamidoimidate, *in* B-00177

59998-72-0 Butyl *N*″-[bis(diethylamino)-phosphinyl]-*N,N,N*′,*N*′-tetraethylphosphorodiamidoimidate, *in* B-00177

59998-75-3 Triethyl [bis(dimethylamino)-phosphinothioyl]-phosphorimidate, *in* B-00192

59998-77-5 Tributyl [bis(dimethylamino)-phosphinothioyl]-phosphorimidate, *in* B-00192

59998-80-0 Ethyl *N*″-[bis(diethylamino)-phosphinothioyl]-*N,N,N*′,*N*′-tetraethylphosphorodiamidoimidate, *in* B-00174

59998-81-1 Butyl *N*″-[bis(diethylamino)-phosphinothioyl]-*N,N,N*′,*N*′-tetraethylphosphorodiamidoimidate, *in* B-00174

59998-82-2 Tetramethylphosphorodiamidothioic azide, T-00224

59998-83-3 Tetraethylphosphorodiamidothioic azide, T-00064

60007-95-6 Cyclophosphamide; (±)-*form*, *in* C-00283

60007-96-7 Cyclophosphamide; (*S*)-*form*, *in* C-00283

60010-51-7 2,2,2-Trichloroethyl phosphorodichloridite, T-00406

60028-25-3 2,8,9-Trioxa-5-aza-1-phosphabicyclo[3.3.3]-undecane; 1-Sulfide, *in* T-00584

60030-06-0 Diethyl methylphosphoramidite, *in* M-00335

60030-72-0 Cyclophosphamide; (*R*)-*form*, *in* C-00283

60049-48-1 4,5-Bis(trifluoromethyl)-spiro[1,3,2-dithiaphosphole-2,1′-[2,8,9]trioxa[1]-phosphatricyclo[3.3.1.$1^{5,7}$]-decane], B-00424

60049-49-2 4′-Methyl-4,5-bis(trifluoromethyl)-spiro[1,3,2-dithiaphosphole-2,1′-[2,6,7]trioxa[1]-phosphabicyclo[2.2.2]octane], M-00092

60049-50-5 4,5-Bis(trifluoromethyl)-spiro[1,3,2-dithiaphosphole-2,1′-[2,6,7]trioxa[1]-phosphabicyclo[2.2.1]heptane, B-00423

60090-34-8 1,1′,1″-Phosphinylidynetrispyrrole, *in* T-00725

60106-53-8 Hexyltriphenylphosphonium(1+); Iodide, *in* H-00104

60146-70-5 Tetramethylethylene chlorophosphate, *in* C-00185

60158-23-8 (3-Oxopropyl)phosphonic acid, O-00076

60190-89-8 2-Phenylethenylphosphonic acid; (*E*)-*form*, Di-Me ester, *in* P-00155

60212-86-4 Phosphorocyanidous dibromide, P-00410

60212-92-2 Phosphorocyanidous bromide iodide, P-00409

60249-40-3 *S*-Ethyl *O*-phenyl phosphorochloridothioate, E-00119

60254-10-6 2-(Diphenylphosphino)phenol, D-01070

60254-11-7 (2-Methoxyphenyl)-diphenylphosphine; Sulfide, *in* M-00056

60259-30-5 Tri-1-pyrrolylphosphine, T-00725

60331-23-9 17-Methyl-6,7,9,10-tetrahydrodibenzo[*d,m*]-[1,3,6,9,12,2]-pentaoxaphosphacyclotetradecin; 17-Oxide, *in* M-00403

60346-00-1 (3,7-Dimethyl-2,6-octadienyl)-triphenylphosphonium(1+); (*Z*)-*form*, Bromide, *in* D-00777

60354-95-2 Bis(pentafluorophenyl)phosphinic acid, B-00382

60414-60-0 *N,N*′,*P,P*-Tetramethylphosphinimidic amide, *in* D-00833

60492-83-3 2,4-Dichloro-2,4-bis(dimethylamino)-2,2,4,4,6,6-hexahydro-6,6-diphenyl-1,3,5-triaza-2,4,6-triphosphorine; (2*RS*,4*SR*)-*form*, *in* D-00162

60494-73-1 [[2-(Methoxycarbonyl)phenyl]-methyl]triphenyl-phosphonium(1+); Bromide, *in* M-00029

60558-59-0 [(Phenylamino)methyl]phosphonic acid, P-00084

60593-25-1 Dimethyl (1-nitroethyl)-phosphonate, *in* N-00035

60593-26-2 Diethyl (1-nitroethyl)-phosphonate, *in* N-00035

60600-13-7 1,2,3-Triphenyl-1,3,2-diphospharsolane, T-00603

60600-14-8 1-Methyl-2,5-diphenyl-1,2,5-azadiphospholidine, *in* M-00089

60600-20-6 2,2-Dimethyl-1,3-diphenyl-1,3-diphosphorinane, D-00745

60609-88-3 1-*tert*-Butyl-1,2,4-azadiphosphetidine; 2,4-Bis(dimethylamino), 2,4-dioxide, *in* B-00543

60609-91-8 1-*tert*-Butyl-1,2,4-azadiphosphetidine; 2,4-Dichloro, 2,4-dioxide, *in* B-00543

60610-05-1 (2-Methylpropyl)-triphenylphosphonium(1+); Iodide, *in* M-00387

60633-18-3 (3-Carboxypropyl)-triphenylphosphonium(1+); Chloride, *in* C-00017

60633-19-4 (4-Carboxybutyl)-triphenylphosphonium(1+); Chloride, *in* C-00009

60668-24-8 Alafosfalin; $(S)_C,(R)_P$-*form*, *in* A-00049

60668-26-0 Alafosfalin; $(S)_C,(R)_P$-*form*, *N*-Benzyloxycarbonyl, *in* A-00049

60668-66-8 Alafosfalin; $(S)_C(S)_P$-*form*, *N*-Benzyloxycarbonyl, *in* A-00049

60669-22-9 Triphenyl-undecylphosphonium(1+); Bromide, *in* T-00701

60678-61-7 2,3-Dihydro-2-ethoxy-1,3,5-triphenyl-1*H*-1,2-azaphosphole 2-oxide, *in* D-00508

60680-68-4 2-Ethyl-2,3,4,5-tetrahydro-2-methyl-1*H*-1,3-benzazaphosphepine, E-00188

60680-72-0 2-Ethyl-2,3,4,5-tetrahydro-2-methyl-1*H*-1,3-benzazaphosphepine; B,HCl, *in* E-00188

60680-84-4 1-Ethyl-2-methyl-1,2,3-azaphospharsolidine; 3-Et, *in* E-00083

60680-85-5 1-Ethyl-2-methyl-1,2,3-azaphospharsolidine; 3-Ph, *in* E-00083

60683-79-6 Glycerol 1,2-di-16-methylheptadecanoate 3-phosphocholine; (*R*)-L-*form*, *in* G-00020

60686-35-3 1,2,3,4,9-Pentaphenyl-1*H*-tribenzo[*b,d,f*]phosphepin, P-00042

60687-36-7 (1-Aminoethyl)phosphonic acid; (*R*)-*form*, *in* A-00078

60711-68-4 Bis(4-methylphenyl)-phosphinothioic acid, B-00322

60714-29-6 Di-*tert*-butylphosphinous acid; Anhydride, *in* D-00122

60714-30-9 Tetra-*tert*-butyldiphosphine; Monooxide, *in* T-00015

60749-96-4 2-Phenoxy-3-phenyl-1,3,2-oxazaphospholidine; 2-Oxide, *in* P-00073

60766-82-7 Diethyl phosphate; Cyclohexylammonium salt, *in* D-00320

60797-48-0 4-(4-Bromophenyl)-2′,2′,3′,4′,4′-pentamethyl-3,3,6,6-tetrakis(trifluoromethyl)-spiro[2,7-dioxa-1-phosphabicyclo[3.2.0]hept-3-ene-1,1′-phosphetane], B-00485

60815-18-2 Diethyl [(3-methoxyphenyl)-methyl]phosphonate, *in* M-00060

60816-98-0 1-[1-(Dimethylamino)ethyl]-2-(diphenylphosphino)-ferrocene, D-00697

60819-73-0 *Se*-Butyl diethylphosphinoselenoate, *in* D-00327

60824-80-8 1-Naphthalenylmethylenetriphenyl-phosphorane, *in* N-00015

60839-30-7 Methanediphosphonous acid; Bis(difluoride), *in* M-00024

60890-66-6 1,4-Dioxa-6-aza-5-phospha(5-P^V)spiro[4.4]nonan-7-one; 5-Ph, *in* D-00962

60902-45-2 Nonyltriphenylphosphonium(1+); Bromide, *in* N-00078

60919-78-0 Triphenyl(3,7,11-trimethyldodecyl)-phosphonium(1+); (3*R*,7*R*)-*form*, Bromide, *in* T-00690

60931-61-5 Tetrapropylphosphonium(1+); Chloride, *in* T-00290

60978-99-6 2-Amino-5-phospho-3-pentenoic acid; (*R*)-*form*, *in* A-00123

60991-59-5 2-Methyl-1,3,2-dithiaphosphorinane, M-00137

61009-58-3 2,2,5-Trimethyl-3-phenyl-1,3-oxaphospholane; (3*RS*,5*SR*)-*form*, *in* T-00522

61009-60-7 2,2,5-Trimethyl-3-phenyl-1,3-oxaphospholane; (3*RS*,5*RS*)-*form*, *in* T-00522

61009-61-8 2,2,5-Trimethyl-3-phenyl-1,3-oxaphospholane; (3*RS*,5*SR*)-*form*, 3-Oxide, *in* T-00522

61009-62-9 2,2,5-Trimethyl-3-phenyl-1,3-oxaphospholane; (3*RS*,5*SR*)-*form*, B,$PhCH_2Br$, *in* T-00522

61009-63-0 3-Benzyl-2,2,5-trimethyl-3-phenyl-1,3-oxaphospholanium bromide, *in* T-00522

61056-26-6 Ethylphosphoramidic acid; Dichloride, *in* E-00156

61083-59-8 2-[(Chloroformyl)oxy]-ethyltriphenylphosphonium(1+); Chloride, *in* C-00084

61102-69-0 2-Diphenylphosphinyl-*N,N*-dimethylaniline, *in* D-01059

61102-87-2 (2-Chlorophenyl)-diphenylphosphine; Oxide, *in* C-00127

61102-88-3 Tris(2-chlorophenyl)phosphine; Oxide, *in* T-00747

61107-65-1 Ethyl hydrogen (2-aminophenyl)-phosphonate, *in* A-00109

61107-67-3 Methyl hydrogen (2-aminophenyl)phosphonate, *in* A-00109

61110-97-2 [(3-Nitrophenyl)methylene]-triphenylphosphorane, *in* N-00048

61142-49-2 1,6-Dibenzyl-1,6-diphosphecane, *in* D-01171

61152-20-3 2,2-Dihydro-1,2-oxaphospholane; 2,2,2-Tri-Me, *in* D-00534

61152-21-4 2,2-Dihydro-1,2-oxaphospholane; 2,2,2-Tri-Et, *in* D-00534

61152-24-7 *N,N′*-Di-*tert*-butyl-*P*-methylphosphonous diamide, *in* M-00320

61152-26-9 1,3-Di-*tert*-butyl-2,4-dimethyl-1,3,2,4-diazadiphosphetidine, D-00086

61152-30-5 1,3-Di-*tert*-butyl-2,4-dimethyl-1,3,2,4-diazadiphosphetidine; 2,4-Dioxide, *in* D-00086

61152-31-6 1,3-Di-*tert*-butyl-2,4-dimethyl-1,3,2,4-diazadiphosphetidine; 2,4-Disulfide, *in* D-00086

61152-33-8 1,3-Di-*tert*-butyl-2,4-dimethyl-1,3,2,4-diazadiphosphetidine; 2,4-Ditelluride, *in* D-00086

61153-56-8 Ethyl bis(4-dimethylaminophenyl)-phosphinate, *in* B-00188

61157-01-5 2,3-Dihydro-4,5-dimethyl-2,2-diphenyl-1,2-thiaphosphepin; 2-Sulfide, *in* D-00463

61157-03-7 2-Phenyl-1,2-thiaphospholane; 2-Sulfide, *in* P-00324

61157-03-7 1,2-Thiaphospholane 2-sulfide; 2-Ph, *in* T-00298

61168-05-6 (4-Carboxybutyl)-triphenylphosphonium(1+); Iodide, *in* C-00009

61173-01-1 *P*-Ethyl-*N,N′*-diphenylphosphonothioic diamide, *in* E-00149

61183-55-9 4-Ethoxy-2,6-diphenyl-4*H*-1,4-oxaphosphorin 4-oxide, *in* H-00140

61185-81-7 *P*-Methyl-*P*-phenylphosphinic amide; (*R*)-*form*, *in* M-00223

61187-71-1 Dimethyl [[(4-methylphenyl)-sulfonyl]methyl]-phosphonate, *in* M-00266

61217-78-5 *P*-Methyl-*N,P*-diphenylphosphinic amide, *in* M-00223

61260-15-9 Dimethyl (1,3-dihydro-3-oxo-1-isobenzofuranyl)-phosphonate, *in* D-00535

61260-18-2 2-Naphthylphosphinic acid, N-00020

61274-57-5 ▷Phenyl phenylphosphonochloridate, *in* P-00229

61324-13-8 1,1′,1″-Phosphinylidynetris(1*H*-pyrazole), *in* T-00722

61324-15-0 1,1′,1″-Phosphinothioyldiynetris(1*H*-pyrazole), *in* T-00722

61332-72-7 4-*tert*-Butyl-1-phenylphosphorinane; *trans-form*, *in* B-00586

61332-73-8 4-*tert*-Butyl-1-phenylphosphorinane; *cis-form*, *in* B-00586

61332-81-8 4-*tert*-Butyl-1-phenylphosphorinane; *trans-form*, Oxide, *in* B-00586

61332-82-9 4-*tert*-Butyl-1-phenylphosphorinane; *cis-form*, Oxide, *in* B-00586

61335-18-0 *O,O*-(2,2′-Biphenylyl) phosphorochloridothioate, *in* C-00032

61335-47-5 4,5-Dimethyl-2-trimethylsilyloxy-1,3,2-dioxaphospholane; $(2\alpha,4\alpha,5\alpha)$-*form*, *in* D-00930

61351-26-6 *O*-Butyl phosphinothioate, B-00592

61358-48-3 3,3′-Bis[(triphenylphosphonio)-methyl]biphenyl(2+); Dibromide, *in* B-00446

61361-04-4 2-Hydroxy-4,5-dimethyl-1,3,2-dioxaphospholane 2-sulfide; $(2\alpha,4\alpha,5\beta)$-*form*, 1*H*-Imidazolium salt, *in* H-00122

61361-99-7 Phenthoate; (±)-*form*, *in* P-00078

61362-00-3 Phenthoate; (−)-*form*, *in* P-00078

61372-42-7 *O*-Ethyl methylphosphonothioate; (±)-*form*, Benzylammonium salt, *in* E-00093

61388-10-1 Ethyl ethylphenylphosphinite, *in* E-00108

61391-87-5 Phenthoate; (+)-*form*, *in* P-00078

61408-88-6 Dimethyl (2-oxooctyl)phosphonate, *in* O-00064

61470-40-4 1-Phenylethylphosphonic acid, P-00164

61478-28-2 *N*-*tert*-Butoxycarbonyl-4-diphenylphosphino-2-diphenylphosphinomethyl-pyrrolidine; (2*S*,4*S*)-*form*, *in* B-00532

61478-29-3 4-(Diphenylphosphino)-2-(diphenylphosphinomethyl)-pyrrolidine; (2*S*,4*S*)-*form*, *in* D-01062

61481-19-4 ▷*tert*-Butyl bicyclic phosphate, *in* B-00640

61500-29-6 *P,P*-Diphenylphosphinimidic amide; *N,N′*-Bis(trimethylsilyl), Na salt, *in* D-01043

61500-34-3 1-Phosphabicyclo[3.3.1]nonane, P-00345

61500-36-5 1-Phosphabicyclo[3.3.1]nonane; 1-Sulfide, *in* P-00345

61500-41-2 1-Phosphabicyclo[3.3.1]nonane; 1-Oxide, *in* P-00345

61509-77-1 Dimethylphosphinoselenoic acid; Chloride, *in* D-00835

61558-36-9 2-Methyl-1,3,2-dioxaphosphorinane, M-00128

61561-85-1 Di-1*H*-imidazol-1-ylphosphinic acid; Chloride, *in* D-00628

61564-31-6 3-Diphenylphosphinyl-*N,N*-dimethylaniline, *in* D-01060

61564-32-7 3-Diphenylphosphinothioyl-*N,N*-dimethylaniline, *in* D-01060

61570-43-2 2,2-Dihydro-2,2,2,4,5-pentaphenyl-1,3,2-dioxaphospholane; (4*RS*,5*RS*)-*form, in* D-00537

61580-09-4 4-Methyl-2,6,7-trioxa-1-phosphabicyclo[2.2.1]-heptane, M-00423

61582-41-0 Diphenylphosphinous acid; Isothiocyanate, *in* D-01089

61614-67-3 Octamethyl thiodiphosphoramide, O-00034

61614-71-9 ▷*sym*-Tetraisopropyl dithiopyrophosphate, T-00168

61617-39-8 2-Hydroxy-4,5-dimethyl-1,3,2-dioxaphospholane 2-sulfide, H-00122

61665-57-4 2-Penten-4-ynyltriphenylphosphonium(1+); Bromide, *in* P-00047

61670-34-6 Ethyl dichlorophosphinylcarbamate, *in* D-00191

61695-12-3 Tri-*tert*-butyltriphosphirane, T-00397

61701-85-7 *P,P*-Diphenyl-*N*-trimethylsilylphosphinimidic fluoride, *in* D-01157

61704-76-5 2,2,2,3,4,5-Hexahydro-2,2,2-trimethoxy-3-phenylnaphth[2,1-*d*]-1,2-oxaphosphole, H-00056

61707-02-6 *O*-Ethyl hydrogen benzylphosphonothioate, *in* B-00067

61739-41-1 2-Chloro-3,4-dimethyl-5-phenyl-1,3,2-oxazaphospholidine; (2*R*,4*S*,5*R*)-*form, in* C-00061

61739-43-3 *O*-Ethyl methylphosphonoselenoate; (*S*)-*form, in* E-00092

61754-87-8 Octahydro-8*bH*-2*a*,4*a*,6*a*,8*a*-tetraaza-8*b*-phosphapentaleno[1,6-*cd*]-pentalene; 8*b*-Fluoro, *in* O-00025

61754-88-9 Octahydro-8*bH*-2*a*,4*a*,6*a*,8*a*-tetraaza-8*b*-phosphapentaleno[1,6-*cd*]-pentalene; 8*b*-Me, *in* O-00025

61754-89-0 Octahydro-8*bH*-2*a*,4*a*,6*a*,8*a*-tetraaza-8*b*-phosphapentaleno[1,6-*cd*]-pentalene; 8*b*-Ph, *in* O-00025

61768-32-9 2-Chlorohexahydropyrano[3,2-*d*]-1,3,2-dioxaphosphorin; (2*RS*,4*aSR*,8*aRS*)-*form*, 2-Sulfide, *in* C-00089

61777-92-2 3-Phosphonoxypropanoic acid, P-00399

61779-87-1 Dimethyl [bis(methylthio)-methyl]phosphonate, *in* B-00364

61820-24-4 (2-Methylphenyl)methylphosphinic acid, M-00203

61826-18-4 2-Chlorohexahydropyrano[3,2-*d*]-1,3,2-dioxaphosphorin; (2*RS*,4*aRS*,8*aSR*)-*form*, 2-Sulfide, *in* C-00089

61864-97-9 Dimethyl 2-pyridinylphosphonate, *in* P-00503

61890-82-2 1,4,6,9-Tetraoxa-5-phospha(5-P^V)spiro[4.4]nonane; 5-Methoxy, *in* T-00244

61892-31-7 2,11-Bis[(diphenylphosphino)-methyl]benzo[*c*]phenanthrene, B-00246

61894-25-5 1,3-Bis(diphenylphosphinothioyl)-propane, *in* B-00265

61894-26-6 1,4-Bis(diphenylphosphinothioyl)-butane, *in* B-00225

61894-27-7 1,6-Bis(diphenylphosphinothioyl)-hexane, *in* B-00243

62013-25-6 2,2,3,3-Tetrakis(trifluoromethyl)-1,4,6,10-tetraoxa-5-phospha(5-P^V)spiro[4.5]decane; 5-Phenoxy, *in* T-00187

62013-28-9 6,10-Dimethyl-2,2,3,3-tetrakis(trifluoromethyl)-1,4-dioxa-6,10-diaza-5-phospha(5-P^V)spiro[4.5]decane; 5-Phenoxy, *in* D-00923

62051-25-6 Hexahydro-3*H*,6*H*-2*a*,5*a*,8*a*-triaza-8*b*-phosphaacenaphthylene, H-00054

62051-26-7 Tetrahydro-1,7-dimethyl-1*H*,5*H*-[1,3,2]diazaphospholo[1,2-*a*][1,3,2]diazaphosphole, T-00093

62128-53-4 2-Chloro-4,5-diphenyl-1,3,2-dioxaphosphole, C-00070

62128-57-8 2-Chloro-2,3-dihydro-3-methyl-1,3,2-benzothiazaphosphole, C-00043

62128-60-3 2-Chloro-2,3-dihydro-3-methyl-1,3,2-benzothiazaphosphole; 2-Sulfide, *in* C-00043

62157-09-9 (1-Hydroxy-1*H*-benzotriazolato-*O*)tris(dimethylamino)-phosphorus(1+); Chloride, *in* H-00111

62217-21-4 1,3-Benzodithiol-2-ylphosphonic acid; Di-Et ester, *in* B-00020

62217-34-9 1,3-Benzodithiol-2-yltriphenylphosphonium(1+); Tetrafluoroborate, *in* B-00021

62217-36-1 (1,3-Benzodithiol-2-ylidene)-triphenylphosphorane, *in* B-00021

62218-18-2 7-Methoxy-2,3,5,6,7,8-hexakis(trifluoromethyl)-1,4-diphosphabicyclo[2.2.2]octa-2,5-diene, M-00042

62241-68-3 3-Methoxy-17-methyl-15-phenyl-15-phosphaestra-1,3,5(10),8,16-pentaene; (±)-*form*, 15-Sulfide, *in* M-00051

62241-69-4 3-Methoxy-17-methyl-15-phenyl-15-phosphaestra-1,3,5(10),8,16-pentaene; (±)-*form*, 15-Oxide, *in* M-00051

62246-59-7 *O*-Methyl di-*tert*-butylphosphinothioate, *in* D-00118

62246-59-7 *O*-Methyl diisopropylphosphinothioate, *in* D-00646

62246-61-1 *S*-Methyl isopropylphenylphosphinothioate, *in* I-00046

62249-77-8 ▷Phaseolotoxin, P-00058

62266-03-9 4-Cyanophenyl ethyl phenylphosphonate, *in* M-00471

62278-16-9 Diethyl [(2-methylphenyl)-methyl]phosphonate, *in* M-00206

62285-48-7 Dimethyl (dimethylcarbamoylmethyl)-phosphonate, *in* D-00703

62285-98-7 [3-Methyl-5-(2,6,6-trimethyl-1-cyclohexen-1-yl)-2,4-pentadienyl]triphenyl-phosphonium(1+); (*E,E*)-*form*, Bromide, *in* M-00421

62290-26-0 2,3-Naphthylene phosphorochloridate, *in* C-00119

62290-27-1 2-Hydroxynaphtho[2,3-*d*]-1,3,2-dioxaphosphole 2-oxide; 2-Ethoxy (Et ester), *in* H-00164

62292-07-3 Dibenzyl phenylphosphonite, *in* P-00251

62336-24-7 (2-Bromophenyl)diphenyl-phosphine, B-00480

62337-53-5 1,1,3,3-Tetrahydro-1,1,3,3-tetraphenyl-2,4-bis(triphenylphosphonio)-1,3-diphosphete; Dichloride, *in* T-00145

62386-52-1 (2-Methylphenyl)phosphonic acid; Dichloride, *in* M-00235

62395-00-9 Bromotetraphenoxyphosphorane, B-00503

62421-62-9 ▷Cyanofenphos; (±)-*form, in* C-00213

62461-25-0 Trimethyl phenysulfonylphosphorimidate, *in* P-00315

62486-05-9 1,4-Butanediylidenebis[triphenyl-phosphorane], *in* B-00517

62496-79-1 6,8,10-Triphenyl-1,4-dithia-5-phospha(5-P^V)spiro[4.5]-deca-5,7,9-triene, T-00604

62496-80-4 7,9,11-Triphenyl-1,5-dithia-6-phospha(6-P^V)spiro[5.5]-undeca-6,8,10-triene, T-00605

62510-54-7 *N″*-2-Pyrimidinylphosphoric triamide; *N′*-Et-*N*-(2-chloroethyl), *in* P-00517

62558-54-7 *O,O*-Dibutyl phosphoroselenoate, D-00154

62569-07-7 Diphenyl (4-methylphenyl)-phosphoramidate, *in* M-00253

62569-08-8 Diphenyl (3-chlorophenyl)-phosphoramidate, *in* C-00156

62569-09-9 Diphenyl (4-bromophenyl)-phosphoramidate, *in* B-00496

62569-10-2 Diphenyl (3-nitrophenyl)-phosphoramidate, *in* N-00059

62591-42-8 3,5-Di-*tert*-butyl-2-chloro-2,3-dihydro-1,3,2-oxazaphosphole; 2-Oxide, *in* D-00078

62608-26-8 2,2,3,3-Tetrakis(trifluoromethyl)-1,4,6-trioxa-10-aza-5-phospha(5-P^V)-spiro[4.5]decane; 5-(4-Bromophenoxy), *in* T-00190

62610-77-9 Methyl 3-[(dimethoxyphosphinothioyl)-oxy]-2-methyl-2-propenoate, *in* D-00673

62621-04-9 Ethyl ethylphenylphosphinothioite, *in* E-00107

62621-05-0 Ethylmethylphenylphosphine; (*R*)-*form*, Sulfide, *in* E-00084

62621-07-2 Ethylphenylphosphinothious acid; (*S*)-*form*, Et ester, *in* E-00107

62639-98-9 (2-Ethoxy-2-propenylidene)-triphenylphosphorane, E-00049

62640-67-9 [(2-Chlorophenyl)methyl]-triphenylphosphonium(1+); Bromide, *in* C-00135

62680-03-9 ▷Fonofos; (*S*)-*form, in* F-00052

62680-68-6 Diethyl [(2-iodophenyl)methyl]-phosphonate, *in* I-00015

62705-71-9 Fonofos; (*R*)-*form, in* F-00052

62729-72-0 *N,N*-Dimethyl-2,2′-diphenylphosphoramidic dihydrazide, *in* D-01109

62753-76-8 Diphenyl (4-methoxyphenyl)-phosphoramidate, *in* M-00076

62757-64-6 1,3-Di-*tert*-butyl-2,4-bis(dimethylamino)-1,3,2,4-diazadiphosphetidine, D-00074

62761-27-7 (1,2-Phenylenedi-2-propene-3,1-diyl)bis[triphenyl-phosphonium](2+); (*E,E*)-*form*, Dibromide, *in* P-00149

62785-51-7 2,2-Dihydro-1,3,2-benzodioxaphosphole; 2,2-Dimethoxy, 2-Ph, *in* D-00439

62785-52-8 2,2-Dihydro-1,3,2-benzodioxaphosphole; 2,2,2-Trimethoxy, *in* D-00439

62808-39-3 Phenylphosphonochloridic acid, P-00229

62839-77-4 2,7,9,10,11,12-Hexakis(trifluoromethyl)-1,8-di-phosphatetracyclo[6.2.2.0^{2,7}.0^{3,6}]-dodeca-4,9,11-triene; (2α,3β,6β,7α)-*form, in* H-00067

62839-78-5 2,5,7,10,11,12-Hexakis(trifluoromethyl)-1,6-di-phosphahexacyclo-[4.4.2.0^{2,5},0^{3,9},0^{4,8},0^{7,10}]dodec-11-ene, H-00065

62839-81-0 2,2-Dihydro-2,2,2-triphenyl-2*H*-1,2-benzoxaphosphorin, D-00607
62839-84-3 *tert*-Butylphenylphosphinothioic acid; (±)-*form*, Chloride, *in* B-00583
62879-51-0 (1-Amino-1-methylpropyl)-phosphonic acid; (+)-*form*, *in* A-00093
62896-17-7 ▷Plumbemycin B, *in* P-00437
62896-18-8 ▷Plumbemycin *A*, P-00437
62920-97-2 *O,O*-Dipropyl phosphorodiselenoate, D-01224
62920-98-3 *O,O*-Diisopropyl phosphorodiselenoate, D-00662
62920-99-4 *O,O*-Dibutyl phosphorodiselenoate, D-00148
62929-10-6 2,7,9,10,11,12-Hexakis(trifluoromethyl)-1,8-di-phosphatetracyclo[6.2.2.$0^{2,7}$.$0^{3,6}$]-dodeca-4,9,11-triene; (2α,3α,6α,7α)-*form*, *in* H-00067
62935-09-5 Dipentyl phosphorochloridite, D-00983
62942-43-2 (2-Oxoethyl)-triphenylphosphonium(1+); Chloride, *in* O-00060
62969-12-4 Tetraisopropyldiphosphine; Monosulfide, *in* T-00167
62969-13-5 Diisopropylphosphinothious acid; Anhydrosulfide, *in* D-00647
62969-14-6 Tetrabutyldiphosphine; Monosulfide, *in* T-00014
62969-15-7 Di-*tert*-butylphosphinothious acid; Anhydrosulfide, *in* D-00120
62974-08-7 Spiro[1,3,2-benzoxathiaphosphole-2,2′-[1,3,2]dioxaphospholane], S-00011
62992-28-3 Ethyl *P*-ethylphosphonamidate, *in* E-00126
62992-30-7 *N,N′*-Dibenzyl-*P*-ethylphosphonothioic diamide, *in* E-00149
62999-70-6 Diethyl [bis(methylthio)-methyl]phosphonate, *in* B-00364
63013-65-0 Trimethyl (diphenoxyphosphinothioyl)-phosphorimidate, *in* D-00987
63023-44-9 3,7,11-Trimethyl-2,6,10-dodecatrien-1-yl dihydrogen phosphate; (*E,E*)-*form*, *in* T-00515
63027-78-1 Phenyl *tert*-butylmethylphosphinate, *in* B-00567
63027-79-2 Phenyl diethylphosphinate, *in* D-00322
63027-81-6 *O*-Phenyl diethylphosphinothioate, *in* D-00330
63027-82-7 *O*-Phenyl di-*tert*-butylphosphinothioate, *in* D-00118
63027-82-7 *O*-Phenyl diisopropylphosphinothioate, *in* D-00646
63067-80-7 *N,N,N′,N′*-Tetrabenzyl-*P*-phenylphosphonous diamide, *in* P-00252
63075-66-1 Diethyl (4-bromobutyl)-phosphonate, *in* B-00466
63075-69-4 1-Benzyl-2-ethoxyhexahydro-1,2-azaphosphorine 2-oxide, *in* B-00048
63075-71-8 1-Benzylhexahydro-2-hydroxy-1,2-azaphosphorine 2-oxide, B-00048
63103-76-4 (2-Methylpropyl)-diphenylphosphine; Oxide, *in* M-00370
63126-94-3 Diethyl (1-amino-1-methylpropyl)phosphonate, *in* A-00093
63129-91-9 3-(Triphenylphosphoranylidene)-propanoic acid, T-00661
63129-92-0 5-(Triphenylphosphoranylidene)-pentanoic acid; Et ester, *in* T-00655
63135-71-7 *P*-Cyclohexyl-*N,N,N′,N′*-tetraethylphosphonothioic diamide, *in* C-00265
63139-09-3 5-Methyl-1*H*-1,2,3-diazaphosphole, M-00114
63147-97-7 (Diazophenylmethyl)phosphonic acid, D-00037
63148-51-6 1,5-Dimethyl-1*H*-1,2,4,3-triazaphosphole, D-00926
63231-19-6 [[(4-Methylphenyl)sulfinyl]-methyl]phosphonic acid; (±)-*form*, Di-Me ester, *in* M-00266
63246-15-1 Menthyl *tert*-butylphenylphosphinate, *in* B-00579
63246-16-2 *tert*-Butylphenylphosphinic acid; (±)-*form*, Chloride, *in* B-00579
63263-75-2 (2-Phenylethenyl)phosphinic acid, P-00153
63268-43-9 [[(4-Methylphenyl)sulfinyl]-methyl]phosphonic acid; (*S*)-*form*, Di-Me ester, *in* M-00266
63283-74-9 *tert*-Butylphenylphosphinic acid; (*S*)-*form*, (−)-Menthyl ester, *in* B-00579
63295-53-4 1,2,3,4-Tetraphenyltetraphospholane; (1α,2β,3α,4β)-*form*, 1,4-Disulfide, *in* T-00280
63299-52-5 Tetramethylene methyl-phosphonate, *in* M-00125
63314-88-5 Methyl methylvinylphosphinate, *in* M-00438
63322-25-8 (2-Buten-1-yl)diphenylphosphine, B-00520
63347-76-2 3-Methoxy-15-methyl-18-nor-15-phosphaestra-1,3,5(10),-6,8,13-hexaene; 15-Oxide, *in* M-00048
63366-48-3 Methanediphosphonous acid; Bis(dibromide), *in* M-00024
63366-59-6 *O,O,O,O*-Tetraethyl methylenebisphosphonothioate, *in* M-00023
63366-61-0 *P,P′*-Methylenebis[(*N,N,N′,N′*-tetramethyl)-phosphonodiamidothioate], *in* M-00023
63368-36-5 [(2-Methylphenyl)methyl]-triphenylphosphonium(1+); Chloride, *in* M-00209
63368-37-6 [(3-Methylphenyl)methyl]-triphenylphosphonium(1+); Chloride, *in* M-00210
63382-79-6 4-Hydroxy-2,2,6,6-tetramethyl-1-oxa-4-phospha-2,6-disilacyclohexane 4-oxide; 4-Butoxy (butyl ester), *in* H-00198
63387-15-5 *P,P,P*-Triisopropylphosphine imide, T-00501
63392-65-4 (3-Methylphenyl)phosphonic acid; Dichloride, *in* M-00236
63462-98-6 Tetrapropylphosphonium(1+); Bromide, *in* T-00290
63469-88-5 Ethyl (1-diazoethyl)-phenylphosphinate, *in* D-00032
63503-62-8 3,3′-Dimethyl-2,2′-(3*H*,3′*H*)-spirobi[1,3,2-benzothiazaphosphole]; 2-Chloro, *in* D-00916
63507-03-9 Dimethyl (4-methylphenyl)-phosphonite, *in* M-00248
63542-12-1 Diethyl (4-nitrophenyl)-phosphoramidate, *in* N-00060
63577-86-6 *O,O*-Diphenyl isopropylphosphoramidothioate, *in* I-00059
63577-87-7 *N-tert*-Butyl-*P,P*-diphenylphosphinothioic amide, *in* D-01083
63581-66-8 Methyl hydrogen benzylphosphonate, *in* B-00060
63585-09-1 ▷Foscarnet sodium, *in* P-00387
63586-83-4 5,10-Dichloro-5,10-dihydrophosphanthrene, D-00164
63586-87-8 5,10-Dichloro-5,10-dihydrophosphanthrene; 5,10-Dioxide, *in* D-00164
63586-90-3 5,10-Dihydro-5,10-dihydroxyphosphanthrene 5,10-dioxide, D-00452
63591-90-2 1,3-Propanediylidenebis[triphenyl-phosphorane], *in* P-00448
63702-96-5 5-Phosphoniaspiro[4.4]nonane; Chloride, *in* P-00374
63702-97-6 5-Methyl-5-phospha(5-P^V)-spiro[4.4]nonane, *in* P-00350
63708-56-5 Decahydro-1,4-dihydroxy-1,4-benzodiphosphorin 1,4-dioxide, D-00003
63708-58-7 Decahydro-1,3,5-trihydroxy-1*H*-1,3,5-benzotriphosphepin 1,3,5-trioxide, D-00007
63708-67-6 Decahydro-1,5-dihydroxy-1*H*-1,5-benzodiphosphepin 1,5-dioxide, D-00002
63744-11-6 *P,P*-Dimethyl-*N,N*-bis(trimethylsilyl)-phosphinous amide, *in* D-00845
63746-57-6 2,2,5,5-Tetramethyl-1-phenyl-1-phospha-2,5-disilacyclopentane, T-00209
63746-61-2 1,4,6,9-Tetraphenyl-1,4,6,9-tetraphospha-5-silaspiro[4.4]-nonane, T-00277
63790-94-3 Tetramethylene *tert*-butylphosphonite, *in* D-00968
63814-77-7 2,4,6-Tris(diethylamino)-1,3,5,2,4,6-trioxatriphosphorinane 2,4,6-trioxide; (2α,4α,6α)-*form*, *in* T-00756
63814-78-8 2,4,6-Tris(diethylamino)-1,3,5,2,4,6-trioxatriphosphorinane 2,4,6-trioxide; (2α,4α,4β)-*form*, *in* T-00756
63818-45-1 Diphenyl 2-furanylphosphonate, *in* F-00073
63818-46-2 Diphenyl 2-thienylphosphonate, *in* T-00307
63822-67-3 1,3-Dithian-2-yltriphenylphosphonium(1+); Chloride, *in* D-01242
63822-68-4 1,3-Dithian-2-ylidenetriphenylphosphorane, *in* D-01242
63832-89-3 Benzyl diphenylphosphinothioite, *in* D-01085
63842-88-6 Triethylphosphonamidic acid; (*S*)-*form*, Et ester, *in* T-00446
63842-89-7 Triethylphosphonamidic acid; (*R*)-*form*, Et ester, *in* T-00446
63842-91-1 Ethyl triethylphosphonamidate, *in* T-00446
63853-21-4 Trimethylsilyl tetramethyl-phosphorodiamidodithioate, *in* T-00222
63853-23-6 Tris(trimethylsilyl) phosphorotetrathioate, T-00869
63853-25-8 Bis(trimethylsilyl) phenylphosphonotrithioate, *in* P-00250
63873-24-5 *S*-Isopropyl diethylphosphinothioate, *in* D-00330
63909-50-2 Diethyl [(3-methylphenyl)-methyl]phosphonate, *in* M-00207
63909-54-6 Diethyl [(2-fluorophenyl)-methyl]phosphonic acid, *in* F-00029
63909-55-7 Diethyl [(2-bromophenyl)-methyl]phosphonate, *in* B-00483
63909-57-9 Diethyl [(3-fluorophenyl)-methyl]phosphonate, *in* F-00030

63909-58-0 Diethyl [(4-fluorophenyl)-methyl]phosphonate, *in* F-00031

63922-69-0 [2-(Trimethylsilyl)ethylidene]-triphenylphosphorane, *in* T-00553

63922-83-8 [2-(Trimethylstannyl)ethyl]-triphenylphosphonium; Iodide, *in* T-00561

63922-84-9 [2-(Trimethylsilyl)ethyl]-triphenylphosphonium(1+); Iodide, *in* T-00553

63923-74-0 1,2,3,4-Tetracyclohexyl-1,2,3,4-tetraphosphorinane, T-00040

64001-72-5 1,2,3,4-Tetraphenyltetraphospholane; $(1\alpha,2\beta,3\alpha,4\beta)$-*form*, 1-Sulfide, *in* T-00280

64008-12-4 *O,O,S*-Tripropyl phosphorothioate, T-00716

64023-57-0 2-Chloro-3,4-dimethyl-5-phenyl-1,3,2-oxazaphospholidine, C-00061

64065-08-3 Bis(dimethylphosphino)methane, B-00202

64075-33-8 2,2′-Bis[(diphenylphosphino)-methyl]-1,1′-binaphthyl; (±)-*form*, *in* B-00248

64091-23-2 2,2′-Bis[(diphenylphosphino)-methyl]-1,1′-binaphthyl; (*S*)-*form*, *in* B-00248

64091-24-3 2,2′-Bis[(diphenylphosphino)-methyl]-1,1′-binaphthyl; (*S*)-*form*, Dioxide, *in* B-00248

64117-64-2 Methyl phenazaphosphinate, *in* P-00061

64128-97-8 Methyl (hydroxymethyl)-phenylphosphinate, *in* H-00160

64149-08-2 Diethyl (trifluoromethyl)-phosphonite, *in* T-00482

64151-43-5 Methylenebis[*N,N,N′,N′*-tetraethylphosphonous diamide], *in* M-00024

64202-92-2 *O,O*-Diethyl *Se*-phenyl phosphoroselenoite, D-00316

64202-98-8 *O,O*-Diisopropyl phosphoroselenoate; K salt, *in* D-00666

64228-49-5 (1-Amino-1-methylpropyl)-phosphonic acid; (±)-*form*, Di-Et ester, *in* A-00093

64249-01-0 Anilofos, A-00134

64278-33-7 7*H*-Benzo[*c*]phenothiaphosphine; 7-Butyl, 7-oxide, *in* B-00022

64286-40-6 Tri-*tert*-butylmethylenephosphorane, T-00365

64317-97-1 Octahydro-8*bH*-2*a*,4*a*,6*a*,8*a*-tetraaza-8*b*-phosphapentaleno[1,6-*cd*]-pentalene, O-00025

64317-98-2 Octahydro-7*H*,9*bH*-2*a*,4*a*,6*a*,9*a*-tetraaza-9*b*-phosphapentaleno[1,6-*cd*]indene, O-00023

64317-99-3 Octahydro-3*H*,8*H*,10*bH*-2*a*,5*a*,7*a*,10*a*-tetraaza-10*b*-phosphaindeno[7,1-*cd*]indene, O-00022

64395-80-8 Benzylmethylphenylphosphine; (±)-*form*, *in* B-00051

64470-11-7 2,2-Dihydro-1,2-oxaphosphetane; 2,2,2-Tributyl, *in* D-00533

64509-14-4 7*H*-Benzo[*c*]phenothiaphosphine, B-00022

64516-41-2 Tetramethyldiphosphine; 1-Oxide 2-sulfide, *in* T-00200

64518-03-2 1,2-Ethenediylbis[triphenyl-phosphonium](2+); (*Z*)-*form*, Dibromide, *in* E-00019

64595-13-7 2,2,2-Trichloro-2,2-dihydro-1,3-dimethyl-6,8-bis(trifluoromethyl)-1,3,5,7,9-pentaaza-2,4-diphospha(4-P^{V})spiro[3.5]nona-4,6,8-triene, T-00401

64595-14-8 1,4-Dimethyl-7,9-bis(trifluoromethyl)-1,4,6,8,10-pentaaza-5-phospha(5-P^{V})spiro[4.5]-deca-5,7,9-triene, D-00714

64598-17-0 (1,2-Dicarboxyethyl)-triphenylphosphonium(1+); Bromide, *in* D-00160

64598-21-6 (1,2-Dicarboxyethyl)-triphenylphosphonium(1+); Chloride, *in* D-00160

64599-80-4 Triphenyltriphosphirane; $(1\alpha,2\alpha,3\beta)$-*form*, *in* T-00698

64642-25-7 1,3-Dimethyl-2-phenyl-1,3-diazaphospholidin-4-one 2-oxide, *in* D-00721

64642-30-4 1,2,3-Trimethyl-1,3-diazaphospholidin-4-one 2-oxide, *in* D-00721

64642-31-5 1,3-Dimethyl-2-phenoxy-1,3-diazaphospholidin-4-one 2-oxide, *in* D-00721

64643-66-9 *O,O*-Diisopropyl phosphoroselenoate, D-00666

64649-14-5 Diethyl [1,6-naphthalenediylbis[methylene]]-bisphosphonate, *in* N-00003

64649-15-6 Tetraethyl [2,7-naphthalenediylbis[methylene]]-bisphosphonate, *in* N-00007

64649-16-7 Tetraethyl [2,3-naphthalenediylbis[methylene]]-bisphosphonate, *in* N-00005

64673-58-1 Dipropylphosphoramidous acid; Diiodide, *in* D-01215

64673-73-0 *Se*-Ethyl tetraethylphosphorodiamidoselenoate, *in* T-00061

64723-48-4 *O,O*-Bis(1-methylethyl) phenylphosphoramidothioate, *in* P-00262

64724-10-3 1-(2-Chloroethyl)tetrahydro-1*H*,5*H*-[1.3.2]-diazaphospholo[2,1-*b*][1.3.2]-oxazaphosphorine 9-oxide, C-00082

64730-63-8 1,7-Bis(diphenylphosphino)-heptane, B-00242

64741-30-6 2-(Diphenylphosphinyl)pyridine, *in* D-01076

64762-31-8 6,9-Dimethyl-1,3-dioxa-6,9-diaza-2-phosphacycloundecane; 2-Methoxy, *in* D-00728

64820-07-1 [(2-Methoxyphenyl)methyl]-triphenylphosphonium(1+); Bromide, *in* M-00062

64822-27-1 (4-Methoxyphenyl)-1-naphthalenylphenylphosphine; (±)-*form*, *in* M-00066

64822-31-7 (4-Methoxyphenyl)-1-naphthalenylphenylphosphine; (+)-*form*, *in* M-00066

64839-73-2 *tert*-Butylmethylphenylphosphine; (±)-*form*, *in* B-00565

64853-94-7 *S*-Methyl *N,N′*-diphenyl-phosphorodiamidothioate, *in* D-01129

64885-97-8 Coenzyme F_{420}, C-00201

64899-38-3 1,3-Di-*tert*-butyl-2,4-dimethoxy-1,3,2,4-diazadiphosphetidine, D-00085

64931-13-1 8,8-Dimethyl-1,6,10-trioxa-5-phospha(5-P^{V})spiro[4.5]-dec-2-ene; 5-Methoxy, *in* D-00933

64970-91-8 [(3-Methoxyphenyl)methylene]-triphenylphosphorane, *in* M-00063

65038-36-0 1,4-Bis(dicyclohexylphosphino)-butane, B-00166

65043-08-5 Ethyl 4-(diethoxyphosphinyl)-3-oxobutanoate, *in* D-00254

65094-30-6 (2-Phenylethenyl)phosphine, P-00152

65111-15-1 Methyl *N*-methyl-*P*-phenylphosphonamidate, *in* M-00230

65111-16-2 Methyl *N*-ethyl-*P*-phenylphosphonamidate, *in* E-00109

65111-20-8 Methyl *P*-phenylphosphonohydrazidate, *in* P-00241

65113-29-3 2,3,4,5,6,7,8,9-Octahydro-1,5,9-triphenyl-1*H*-1,5,9-benzotriphosphacycloundecin, O-00028

65113-30-6 5,6,7,8,9,14,15,16,17,18-Decahydro-5,9,14,18-tetraphenyldibenzo[*b*,*i*]-[1,4,8,11]-tetraphosphacyclotetradecin, D-00006

65114-88-7 1-Methyl-3,8-phosphonanedione 1-oxide, *in* P-00373

65114-89-8 1-Hydroxy-3,8-phosphonanedione 1-oxide, *in* P-00373

65114-90-1 1,3,4,6-Tetrakis(trifluoromethyl)-2,5-diphosphatricyclo[3.1.0.$0^{2,6}$]hex-3-ene, T-00186

65164-90-1 α-(Diphenylphosphinyl)-benzeneacetonitrile, D-01095

65181-55-7 Bis(2-phenylethenyl)phosphinic acid; Butylammonium salt, *in* B-00386

65200-23-9 Tribromodiphenylphosphorane, T-00355

65213-16-3 1,3-Di-*tert*-butyl-2,4-bis(dimethylamino)-1,3,2,4-diazadiphosphetidine; *Trans-form*, *in* D-00074

65213-17-4 1,3-Di-*tert*-butyl-2,4-bis(dimethylamino)-1,3,2,4-diazadiphosphetidine; *Cis-form*, 2-Selenide, *in* D-00074

65213-18-5 1,3-Di-*tert*-butyl-2,4-bis(dimethylamino)-1,3,2,4-diazadiphosphetidine; *Trans-form*, 2-Telluride, *in* D-00074

65253-76-1 (2-Hydroxyethyl)phosphonothioic acid, H-00147

65257-79-6 2,3,4,5,6,7,8,9,10,11,12,13-Dodecahydro-2,5,9,12-tetraphenyl-1*H*-2,5,9,12-benzotetraphosphacyclopentadecin, D-01262

65268-60-2 1,3-Di-*tert*-butyl-2,4-bis(dimethylamino)-1,3,2,4-diazadiphosphetidine; *Cis-form*, *in* D-00074

65268-61-3 1,3-Di-*tert*-butyl-2,4-bis(dimethylamino)-1,3,2,4-diazadiphosphetidine; *Trans-form*, 2-Selenide, *in* D-00074

65273-60-1 [2-(1,3-Dihydro-1,3-dioxo-2*H*-isoindol-2-yl)ethyl]-triphenylphosphonium(1+); Chloride, *in* D-00474

65273-64-5 [2-(1,3-Dihydro-1,3-dioxo-2*H*-isoindol-2-yl)ethyl]-triphenylphosphonium(1+); Bromide, *in* D-00474

65289-19-2 Methyl *N*-(diethoxyphosphinyl)-carbamate, *in* D-00249

65289-20-5 Ethyl *N*-(dimethoxyphosphinyl)-carbamate, *in* D-00680

65331-55-7 *S*-Ethyl *P*-methylphosphonamidothioate, *in* M-00284

65337-14-6 (2-Methoxyphenyl)-methylphenylphosphine; (*S*)-*form*, *in* M-00059

65388-38-7 1,3-Di-*tert*-butyl-2,4-bis(dimethylamino)-1,3,2,4-diazadiphosphetidine; *Cis-form*, 2,4-Disulfide, *in* D-00074

65395-34-8 Tetraethyl ethylimidodiphosphite, *in* E-00079

65402-32-6 2-Phosphono-2-(2-phosphonoethyl)butanedioic acid, P-00394

65423-44-1 2-(Diphenylphosphino)aniline, D-01047
65426-08-6 *O,Se*-Diethyl ethylphosphonoselenoate, *in* E-00145
65426-09-7 *O*-Ethyl *Se*-methyl ethylphosphonoselenoate, *in* E-00072
65438-83-7 Ethylene phenylphosphoramidothioate, *in* P-00083
65438-85-9 *N,P,P*-Triphenylphosphinoselenoic amide, *in* D-01079
65442-15-1 (3-Bromophenyl)phosphonic acid; Dichloride, *in* B-00491
65442-22-0 Diethyl (3-methoxyphenyl)-phosphonate, *in* M-00071
65463-54-9 Bis[(4-nitrophenyl)methyl]-phosphonate, B-00368
65463-55-0 Bis[(4-chlorophenyl)methyl] phosphonate, B-00130
65463-64-1 Bis(4-nitrophenyl) phosphonate, B-00375
65475-24-3 Bis(2-chlorophenyl) phosphonate, B-00143
65482-10-2 2,3,4,5,6,7-Hexahydro-1*H*-isophosphindole 2-oxide; 2-Me, *in* H-00035
65482-11-3 2,3,4,5,6,7-Hexahydro-1*H*-isophosphindole 2-oxide; 2-Hydroxy, *in* H-00035
65534-45-4 7,7-Dichloro-2,3,5,6-tetrakis(trifluoromethyl)-1,4-diphosphabicyclo[2.2.1]hepta-2,5-diene, D-00195
65534-50-1 Bis[(2-diphenylphosphinyl)-ethyl]ether, *in* O-00093
65580-79-2 ▷*O*-Ethyl *O*-4-nitrophenyl phenylphosphonothionate; (+)-*form*, *in* E-00099
65580-80-5 ▷*O*-Ethyl *O*-4-nitrophenyl phenylphosphonothionate; (−)-*form*, *in* E-00099
65617-68-7 1,3-Dihydroxy-1,3-diphosphepane 1,3-dioxide, D-00619
65634-93-7 *P*-Methyl-*P*-phenylphosphinic amide; (±)-*form*, *in* M-00223
65650-33-1 1,3-Dihydroxy-1,3-diphosphorinane 1,3-dioxide, D-00622
65659-19-0 Dimethyl diethylphosphoramidate, *in* D-00343
65665-33-0 Ethylphenylphosphinic acid; (*S*)-*form*, Me ester, *in* E-00101
65669-35-4 1,3,2-Benzoxathiaphosphole; 2-Bromo, *in* B-00026
65670-22-6 Dimethyl (3-methyl-1,3-butadienyl)phosphonate, *in* M-00095
65674-21-7 Dimethyl (9,10-dihydro-9-acridinyl)phosphonate, *in* D-00438
65693-26-7 2,8-Dioxa-5-aza-1-phosphabicyclo[3.3.0]-octane, D-00961
65715-52-8 1,1′(3*H*,3′*H*)-Spirobi[2,1-benzoxaphosphole]; 1-Ph, *in* S-00017
65728-06-5 Methyl dicyclohexylphosphinodithioate, *in* D-00212
65728-16-7 Benzyl dicyclohexylphosphinodithioate, *in* D-00212
65736-00-7 4,4-Dimethyl-1-phenyl-1-phospha-4-silacyclohexa-2,5-diene, D-00801
65747-62-8 Diethyl methylenebis[methyl-phosphinate], *in* M-00140
65747-63-9 Diisopropyl methylenebis[methyl-phosphinate], *in* M-00140
65747-71-9 Diisopropyl methylenebisphosphinate, *in* M-00143
65753-94-8 1,3-Di-*tert*-butyl-2,4-bis(dimethylamino)-1,3,2,4-diazadiphosphetidine; *Trans-form*, 2,4-Dioxide, *in* D-00074
65753-95-9 1,3-Di-*tert*-butyl-2,4-bis(dimethylamino)-1,3,2,4-diazadiphosphetidine; *Trans-form*, 2,4-Diselenide, *in* D-00074
65753-96-0 1,3-Di-*tert*-butyl-2,4-bis(dimethylamino)-1,3,2,4-diazadiphosphetidine; *Cis-form*, 2,4-Diselenide, *in* D-00074
65753-97-1 1,3-Di-*tert*-butyl-2,4-bis(dimethylamino)-1,3,2,4-diazadiphosphetidine; *Cis-form*, 2-Sulfide, *in* D-00074
65753-98-2 1,3-Di-*tert*-butyl-2,4-bis(dimethylamino)-1,3,2,4-diazadiphosphetidine; *Trans-form*, 2-Sulfide, *in* D-00074
65768-04-9 *N,N*-Diethyl-*P,P*-diisopropylphosphinous amide, *in* D-00649
65789-46-0 1,3-Di-*tert*-butyl-2,4-bis(dimethylamino)-1,3,2,4-diazadiphosphetidine; *Cis-form*, 2,4-Dioxide, *in* D-00074
65789-47-1 1,3-Di-*tert*-butyl-2,4-bis(dimethylamino)-1,3,2,4-diazadiphosphetidine; *Trans-form*, 2,4-Disulfide, *in* D-00074
65824-64-8 Tetrabutyl [oxybis(methylene)]-bisphosphonate, *in* O-00095
65831-34-7 2,5-Dimethyl-1-phenylphosphorinan-4-one; (1*SR*,2*RS*,5*RS*)-*form*, 1-Sulfide, *in* D-00821
65831-35-8 2,5-Dimethyl-1-phenylphosphorinan-4-one; (1*SR*,2*RS*,5*RS*)-*form*, 1-Oxide, *in* D-00821
65831-36-9 2,5-Dimethyl-1-phenylphosphorinan-4-one; (1*RS*,2*RS*,5*RS*)-*form*, 1-Sulfide, *in* D-00821
65845-21-8 1,5-Dihydro-3-methyl-2,4,3-benzodioxaphosphepin; 4-Sulfide, *in* D-00516
65857-68-3 *O,O*-Diethyl phosphorotelluroate; Na salt, *in* D-00395
65866-00-4 Methyl 4-(triphenylphosphoranylidene)-2-butanoate, *in* T-00647
65882-82-8 Tetraethyl 1,2-phenylenebisphosphonite, *in* B-00009
65884-57-3 5,10-Dihydro-11*H*-dibenz[*b,e*]-[1,4]azaphosphepin-11-one 5-oxide; 5-Ph, *in* D-00449
65884-58-4 5,10-Dihydro-11*H*-dibenz[*b,e*]-[1,4]azaphosphepin-11-one 5-oxide; 5-Me, *in* D-00449
65884-59-5 10,11-Dihydro-1*H*-dibenz[*b,e*]-[1,4]azaphosphepine 5-oxide; 5-Ph, *in* D-00448
65884-60-8 10,11-Dihydro-1*H*-dibenz[*b,e*]-[1,4]azaphosphepine 5-oxide; 5-Me, *in* D-00448
65887-64-1 Di-2-furanylphosphinic acid, D-00421
65887-65-2 Ethyl bis(2-thienyl)phosphinate, *in* D-01250
65887-66-3 Ethyl di-2-furanylphosphinate, *in* D-00421
65894-65-7 13,14,16,17,31,32,34,35-Octahydro-6,24-dimethyltetra-benzo[*d,m,r,a′*]-[1,3,6,9,12,15,20,23,26,2,16]-decaoxadiphosphacyclooctacosin; 6,24-Dioxide, *in* O-00008
65915-23-3 Dimethyl [(methylsulfinyl)-methyl]phosphonate, *in* M-00398
65915-24-4 Diethyl [(methylsulfinyl)-methyl]phosphonate, *in* M-00398
65915-25-5 Diethyl [(phenylsulfinyl)-methyl]phosphonate, *in* P-00310
65924-29-0 Phenyl bis(4-bromophenyl)-phosphinate, *in* B-00114
65960-95-4 Methamidophos; (*R*)-*form*, *in* M-00021
65972-11-4 *P,P*-Dimethylphosphinic amide, D-00830
66023-94-7 Alafosfalin; $(S)_C(S)_P$-*form*, *in* A-00049
66055-07-0 1,4-Phenylenebis[methyl-phosphinodithioic acid], P-00138
66055-08-1 Dimethyl 1,4-phenylenebis[methyl-phosphinodithioate], *in* P-00138
66076-07-1 Bis(3-methylphenyl) phosphonate, B-00327
66089-08-5 3,3,5,5-Tetramethyl-1,2,4-triphenyl-1,2,4-triphospholane, T-00239
66093-69-4 1,2,4-Trimethyl-1,2,4-triphospholane, T-00568
66130-90-3 Diethyl [(trimethylsilyloxy)-carbonyl]phosphonate, *in* D-00241
66181-02-0 2,2,2-Trichloro-2,2-dihydro-5-phenyl-1,3,4,2-dioxazaphosphole, *in* C-00126
66181-06-4 2-Chloro-5-phenyl-1,3,4,2-dioxazaphosphole; 2-Oxide, *in* C-00126
66181-07-5 ▷2-Chloro-5-phenyl-1,3,4,2-dioxazaphosphole, C-00126
66191-00-2 Ethyl bis(trimethylsilyloxy)-phosphinecarboxylate oxide, *in* B-00433
66193-16-6 1,1-Diethoxy-2,2-diethyldiphosphine 1-oxide, D-00229
66193-23-5 Tetraisopropyldiphosphine; Monooxide, *in* T-00167
66193-25-7 Tetracyclohexyldiphosphine; Monoxide, *in* T-00039
66193-26-8 Dipropylphosphinous acid, D-01209
66197-72-6 Diethyl (bromomethyl)-phosphonate, *in* B-00475
66220-54-0 Ethylmethylphosphinodithioic acid, E-00086
66220-57-3 Methylphenylphosphinodithioic acid; NH_4 salt, *in* M-00224
66221-32-7 1,3-Benzodithiol-2-yltriphenylphosphonium(1+); Chloride, *in* B-00021
66224-11-1 2,6-Dimethyl-4,4-diphenyl-1,4-oxaphosphorinanium bromide, *in* D-00797
66242-28-2 *O,O*-2,3-Butylene *O*-methyl thiophosphate, *in* M-00033
66242-30-6 *O,O*-2,3-Butylene *S*-methyl thiophosphate, *in* D-00765
66242-31-7 *O,O*-2,3-Butylene bromothiophosphate, *in* B-00470
66250-56-4 3,3′,3″-Phosphinylidynetrispropanoic acid trimethyl ester, *in* P-00357
66254-55-5 (1-Amino-2-methylpropyl)-phosphonic acid; (*S*)-*form*, *in* A-00094
66254-56-6 (1-Amino-2-methylpropyl)-phosphonic acid; (*R*)-*form*, *in* A-00094
66288-90-2 4,5-Dimethyl-1,3,2-dioxaphospholane; (4*RS*,5*RS*)-*form*, 2-Sulfide, *in* D-00733
66288-94-6 4,5-Dimethyl-2-trimethylsilyloxy-1,3,2-dioxaphospholane; (2α,4α,5β)-(±)-*form*, *in* D-00930
66288-96-8 4,5-Dimethyl-1,3,2-dioxaphospholane; (4*RS*,5*SR*)-*form*, 2-Oxide, *in* D-00733

66289-05-2 2-Methoxy-4,5-diphenyl-1,3,2-dioxaphospholane; (2α,4β,5β)-*form*, 2-Sulfide, *in* M-00039

66289-11-0 2-Bromo-4,5-dimethyl-1,3,2-dioxaphospholane; (2α,4β,5β)-*form*, 2-Sulfide, *in* B-00470

66289-12-1 ▷2-Bromo-4,5-dimethyl-1,3,2-dioxaphospholane; (2α,4β,5β)-*form*, *in* B-00470

66295-37-2 1,3,6,2-Trioxaphosphocane; 2-Me, *in* T-00592

66295-38-3 1,3,6,2-Trioxaphosphocane; 2-Me, 2-sulfide, *in* T-00592

66295-39-4 2-Methyl-1,3,2-dioxaphosphepane, M-00125

66295-41-8 *O,O*-Tetramethylene methylphosphonothioate, *in* M-00125

66295-44-1 Diisopropyl methylphosphonite, *in* M-00319

66295-77-0 *tert*-Butyldiphenylphosphine; Sulfide, *in* B-00558

66295-79-2 Isopropyldiphenylphosphine; Sulfide, *in* I-00031

66298-61-1 2,2,2,3-Tetrahydro-5-phenyl-3,3-bis(trifluoromethyl)-1,4,2-thiazaphosphole; 2,2,2-Trimethoxy, *in* T-00129

66322-40-5 4,5-Dimethyl-1,3,2-dioxaphospholane; (4*RS*,5*RS*)-*form*, 2-Oxide, *in* D-00733

66322-41-6 4,5-Dimethyl-2-methylthio-1,3,2-dioxaphospholane; (4*RS*,5*SR*)-*form*, 2-Oxide, *in* D-00765

66341-36-4 Hexahydro-2*H*-[1,2]-oxaphosphorino[2,3-*b*]-[1,2]oxaphosphorin; 9-Oxide, *in* H-00037

66376-30-5 *O*-Ethyl methylphosphonoselenoate, E-00092

66379-66-6 ▷Dihexyl phosphorochloridite, D-00436

66379-68-8 *P,P*-Dibutyl-*N,N*-diethylphosphinic amide, *in* D-00111

66392-89-0 Diphenyl (2,4,6-tribromophenyl)-phosphoramidate, *in* T-00359

66407-89-4 6-Methyl-2,3,4-triphenyl-2*H*-1,5,2-dioxaphosphorin 2-oxide; (2*RS*,6*RS*)-*form*, *in* M-00430

66407-90-7 6-Methyl-2,3,4-triphenyl-2*H*-1,5,2-dioxaphosphorin 2-oxide; (2*RS*,6*SR*)-*form*, *in* M-00430

66416-58-8 Trimethylsilyl *P,P*-diphenyl-*N*-trimethylsilylphosphinimidate, *in* D-01157

66417-54-7 Tris(4-carboxyphenyl)phosphine; Tri-Me ester, *in* T-00740

66436-14-4 Tetramethyldiphosphine; Monooxide, *in* T-00200

66441-92-7 Diphenyl *tert*-butylphosphonite, *in* B-00616

66442-24-8 Ethyl dimethyl-phosphoramidochloridite, *in* D-00862

66481-71-8 *O,O*-Bis(trimethylsilyl) phenylphosphonothioate, *in* P-00244

66481-72-9 *O,O*-Bis(trimethylsilyl) phenylphosphonoselenoate, *in* P-00243

66498-95-1 *O,Se,Se*-Tributyl phosphorodiselenoite, T-00383

66498-99-5 *Se,Se*-Dibutyl *O*-ethyl phosphorodiselenothioate, *in* D-00149

66499-00-1 *Se,Se*-Dibutyl *O*-propyl phosphorodiselenothioate, *in* D-00149

66499-01-2 *O,Se,Se*-Tributyl phosphorodiselenothioate, *in* D-00149

66499-12-5 Diphenylphosphinoselenothioic acid, D-01080

66499-13-6 *O*-Ethyl ethylphosphonoselenothioate, E-00073

66499-14-7 *tert*-Butylphenyl-phosphinothioselenoic acid, B-00584

66508-37-0 Fosmidomycin; Mono-Na salt, *in* F-00062

66508-52-9 Antibiotic FR 32863; K salt, *in* A-00137

66508-53-0 Fosmidomycin, F-00062

66508-88-1 Antibiotic FR 32863, A-00137

66508-89-2 Antibiotic FR 33289, A-00138

66517-44-0 *tert*-Butyldiiodophosphine, *in* B-00616

66521-47-9 Diisopropylphosphinodithioic acid; Anhydrosulfide, *in* D-00643

66534-96-1 2,2′-Bis(diphenylphosphino)-diethylamine, B-00231

66534-97-2 2,2′-Bis(diphenylphosphino)-diethylamine; B,HCl, *in* B-00231

66556-69-2 [1-Methyl-3-(2,6,6-trimethyl-1-cyclohexen-1-yl)-2-propenyl]triphenyl-phosphonium(1+); (*E*)-*form*, Bromide, *in* M-00422

66568-25-0 Bis(2-carboxyphenyl)phosphinic acid, B-00117

66608-32-0 Imcarbofos, I-00002

66628-81-7 Diethyl bis(trimethylsilyl)-phosphoramidite, *in* B-00438

66685-93-6 1-Phenyl-4-phosphorinanone; Sulfide, *in* P-00268

66686-22-4 1,3-Dihydroxy-1,3-diphosphetane 1,3-dioxide, D-00620

66726-75-8 [1,3-Phenylenebis[methylene]]-bis[triphenylphosphonium](2+); Dichloride, *in* P-00133

66731-80-4 Diethyl[3-phenyl-1-[(trimethylsilyl)oxy]-2-propenyl]phosphonate, D-00319

66767-39-3 Fonofos; (±)-*form*, *in* F-00052

66777-93-3 Tris[(4-nitrophenyl)methyl] phosphate, T-00834

66778-06-1 Diphenyl 1*H*-imidazol-1-ylphosphonate, *in* I-00003

66783-70-8 Di-*tert*-butyliodophosphine, *in* D-00122

66821-75-8 *O*-Methyl *S,S*-trimethylene phosphonotrithioate, *in* D-01249

66830-60-2 (Dimethylaminomethyl)-diphenylphosphine; B,HCl, *in* D-00700

66830-78-2 3,5,5-Trimethyl-2-dimethylamino-1,2-oxaphospholan-3-ol 2-oxide, *in* T-00519

66849-33-0 Ifosfamide; (*S*)-*form*, *in* I-00001

66849-34-1 Ifosfamide; (*R*)-*form*, *in* I-00001

66850-41-7 2-Diethylamino-3-phenyl-1,3,2-oxazaphospholidine; 2-Oxide, *in* D-00269

66872-86-4 1,5-Diphosphabicyclo[3.3.0]-octane, D-01170

66872-89-7 1,5-Diphosphabicyclo[3.3.0]-octane; Disulfide, *in* D-01170

66872-92-2 Tetrahydro-4-methyl-1*H*,5*H*-[1,2]diphospholo[1,2-*a*]-[1,2]diphospholium iodide, *in* D-01170

66881-05-8 2,2,7,7-Tetrahydro-3,4,5,6-tetrakis(methoxycarbonyl)-1,2,2,5,5-pentamethyl-1*H*-1,2,7-azadiphosphepine, *in* T-00138

66881-06-9 2,2,7,7-Tetrahydro-3,4,5,6-tetrakis(methoxycarbonyl)-1*H*-1,2,7-azadiphosphepine; 2,2,7,7-Tetra-Ph, *in* T-00138

66881-07-0 2,2,7,7-Tetrahydro-3,4,5,6-tetrakis(methoxycarbonyl)-1*H*-1,2,7-azadiphosphepine; 2,2-Di-Me, 7,7-di-Ph, *in* T-00138

66899-05-6 2-Propenylphosphinic acid, P-00452

66911-81-7 *P*-Chloro-*P,P*-bis(diethylamino)-*N*-ethylphosphine imide, *in* P-00012

66911-82-8 *N*-Butyl-*P*-chloro-*P,P*-bis(dibutylamino)-phosphazene, *in* P-00004

66918-48-7 9-Methyl-1,4,6-trioxa-5-phospha(5-P^V)spiro[4.4]-non-7-ene; 5-Ethoxy, *in* M-00428

66959-05-5 (2-Diethylamino-2-oxoethyl)-phosphonic acid, D-00268

66964-03-2 *P*-Chloro-*P,P*-bis(dipropylamino)-*N*-propylphosphazene, *in* P-00043

66997-36-2 Tributyl(hexadecyl)-phosphonium(1+), T-00364

67003-66-1 Dipropylphosphinothioic acid; Anhydride, *in* D-01208

67003-67-2 Dibutylphosphinothioic acid; Anhydride, *in* D-00117

67057-49-2 1,3-Butylene phenylphosphoramidate, *in* M-00193

67057-50-5 4-Methyl-2-phenylamino-1,3,2-dioxaphosphorinane; (2*RS*,4*SR*)-*form*, 2-Oxide, *in* M-00193

67057-51-6 *O,O*-1,3-Butylene phenylphosphoramidothioate, *in* M-00193

67057-52-7 4-Methyl-2-phenylamino-1,3,2-dioxaphosphorinane; (2*RS*,4*SR*)-*form*, 2-Sulfide, *in* M-00193

67057-53-8 *O,O*-1,3-Butylene phenylphosphoramidoselenoate, *in* M-00193

67057-54-9 4-Methyl-2-phenylamino-1,3,2-dioxaphosphorinane; (2*RS*,4*SR*)-*form*, 2-Selenide, *in* M-00193

67057-61-8 4-Methyl-2-dimethylamino-1,3,2-dioxaphosphorinane 2-oxide; (2*RS*,4*SR*)-*form*, *in* M-00118

67057-62-9 4-Methyl-2-dimethylamino-1,3,2-dioxaphosphorinane 2-oxide; (2*RS*,4*RS*)-*form*, *in* M-00118

67057-63-0 4-Methyl-2-dimethylamino-1,3,2-dioxaphosphorinane; (2*RS*,4*RS*)-*form*, 2-Sulfide, *in* M-00117

67057-64-1 *O,O*-1,3-Butylene dimethylphosphoramidothioate, *in* M-00117

67057-65-2 *O,O*-1,3-Butylene dimethylphosphoramidoselenoate, *in* M-00117

67059-56-7 Phenyl bis(1*H*-benzimidazol-1-yl)phosphinate, *in* B-00097

67071-69-6 Tetraphenylphosphinic amide, *in* D-01040

67105-50-4 2-Chlorotetrahydro-3-methyl-2*H*-1,3,2-oxazaphosphorine, C-00178

67106-81-4 (1,2-Dibromo-1-phenylpropyl)-phosphonic acid; (1*RS*,2*RS*)-*form*, *in* D-00068

67106-82-5 Methyl hydrogen (1,2-dibromo-1-phenylpropyl)-phosphonate, *in* D-00068

67106-83-6 (1,2-Dibromo-1-phenylpropyl)-phosphonic acid; (1*RS*,2*SR*)-*form*, Mono-Me ester, *in* D-00068

67106-84-7 (1,2-Dibromo-1-phenylpropyl)-phosphonic acid; (1*RS*,2*SR*)-*form*, *in* D-00068

67106-85-8 Dimethyl (1,2-dibromo-1-phenylpropyl)phosphonate, *in* D-00068

67106-86-9 (1,2-Dibromo-1-phenylpropyl)-phosphonic acid; (1*RS*,2*SR*)-*form*, Di-Me ester, *in* D-00068

67123-93-7 4-Methyl-2-dimethylamino-1,3,2-dioxaphosphorinane; (2*RS*,4*RS*)-*form*, 2-Selenide, *in* M-00117

67173-22-3 5-Isopropyl-2,6-dimethyl-6-phosphabicyclo[3.1.1]-hept-2-ene 6-oxide, I-00029

67201-32-5 2,2-Dihydro-1,3,2-benzoxazaphosphole; 2,2-Bis(dimethylamino), *in* D-00444

67213-42-7 *O*-Methyl phenylphosphonothioate; (±)-*form*, *in* M-00243

67213-43-8 *P*-*tert*-Butyl-*P*-phenylphosphinic amide; (±)-*form*, *in* B-00580

67219-46-9 [[2-[(Acetyloxy)methyl]phenyl]-methyl]triphenyl-phosphonium(1+); Bromide, *in* A-00019

67221-20-9 Diethyl 1,3-pentadienylphosphonate, *in* P-00009

67242-35-7 *S*-Methyl *N*,*P*-dimethylphosphonamidothioate, *in* D-00850

67242-36-8 *S*-Methyl trimethylphosphonamidothioate, *in* T-00540

67242-51-7 Methyl *P*-methyl-*N*,*N*-dimethylphosphonamidodithioate, *in* E-00063

67242-52-8 *S*-Methyl *P*-methylphosphonamidothioate, *in* M-00284

67253-70-7 *O*-Methyl phenylphosphonothioate; (*R*)-*form*, *in* M-00243

67253-71-8 *O*-Methyl phenylphosphonothioate; (*S*)-*form*, *in* M-00243

67257-36-7 Bis(1-methylethyl) (2-oxopropyl)phosphonate, *in* O-00075

67260-61-1 Methylenebis[triphenyl-phosphonium](2+); Diiodide, *in* M-00147

67264-34-0 (2-Aminobutyl)phosphonic acid, A-00055

67265-07-4 Tris[(4-nitrophenyl)methyl]-phosphine; Oxide, *in* T-00835

67285-74-9 1,2-Diphenyl-1,2-diphospholane 1,2-disulfide, *in* D-01181

67309-03-9 Methyl bis(hydroxymethyl)-phosphinate, *in* B-00282

67309-04-0 Ethyl bis(hydroxymethyl)-phosphinate, *in* B-00282

67314-76-5 *tert*-Butylphenylphosphinothioic acid; (±)-*form*, *in* B-00583

67333-92-0 *O*,*O*-Dipropyl phosphorodithioate; Anilinium salt, *in* D-01225

67344-36-9 Bis(trimethylsilyl) (2-chloroethyl)phosphonate, *in* C-00074

67347-97-1 10,10-Dihydro-2,3-diphenyl-[1,3,2]oxazaphospholo[2,3-*b*][1,3,2]benzoxazaphosphole; 10,10-Diphenoxy, *in* D-00481

67348-17-8 4′,5′-Bis(trifluoromethyl)-spiro[1,3,2-benzodioxaphosphole-2,2′-[1,3,2]dioxaphosphole]; 2-Fluoro, *in* B-00421

67353-93-9 2,4-Di-*tert*-butyl-3-chloro-1,2,4,3-thiadiazaphosphetidine; 1-Oxide, 3-Sulfide, *in* D-00079

67353-94-0 Trimethyl *tert*-butylphosphorimidate, *in* B-00625

67364-34-5 Hexahydro-1,3-dimethyl-1,3,2-diazaphosphorine 2-oxide, H-00023

67364-36-7 Hexahydro-1,3,4-trimethyl-1,3,2-diazaphosphorine, H-00057

67364-37-8 *N*,*N*′-1,3-Butylene-*N*,*N*′-dimethylphosphorodiamidous chloride, *in* H-00057

67374-21-4 6,12-Diphenyl-1,4,8,11-tetraoxa-6,12-diaza-5,7-diphospha(5,7-P^V)-dispiro[4.1.4.1]dodecane; 5,7-Dimethoxy, *in* D-01149

67374-25-8 6,12-Diphenyl-1,4,8,11-tetraoxa-6,12-diaza-5,7-diphospha(5,7-P^V)-dispiro[4.1.4.1]dodecane; 5,7-Difluoro, *in* D-01149

67384-54-7 2,5-Dimethyl-1-phenylphosphorinan-4-ol; (1*RS*,2*SR*,4*SR*,5*SR*)-*form*, 1-Sulfide, *in* D-00820

67384-55-8 2,5-Dimethyl-1-phenylphosphorinan-4-ol; (1*RS*,2*SR*,4*SR*,5*SR*)-*form*, 1-Sulfide, 4-*O*-Ac, *in* D-00820

67395-78-2 (2-Bromophenyl)phosphonic acid; Dichloride, *in* B-00490

67397-51-7 *O*,*O*′-Diethyl diethylthiodiphosphonate, *in* D-00405

67398-15-6 (1-Aminoethyl)methylphosphinic acid, A-00076

67424-66-2 2,5-Dimethyl-1-phenylphosphorinan-4-one; (1*SR*,2*RS*,5*RS*)-*form*, 1-Selenide, *in* D-00821

67424-67-3 2,5-Dimethyl-1-phenylphosphorinan-4-one; (1*RS*,2*RS*,5*RS*)-*form*, 1-Selenide, *in* D-00821

67424-69-5 2,5-Dimethyl-1-phenylphosphorinan-4-one; (1*RS*,2*SR*,5*RS*)-*form*, 1-Selenide, *in* D-00821

67424-72-0 2,5-Dimethyl-1-phenylphosphorinan-4-ol; (1*RS*,2*SR*,4*RS*,5*SR*)-*form*, 1-Sulfide, *in* D-00820

67424-73-1 2,5-Dimethyl-1-phenylphosphorinan-4-ol; (1*RS*,2*SR*,4*RS*,5*SR*)-*form*, 1-Sulfide, 4-*O*-Ac, *in* D-00820

67424-74-2 2,5-Dimethyl-1-phenylphosphorinan-4-ol; (1*RS*,2*RS*,4*RS*,5*RS*)-*form*, 1-Sulfide, *in* D-00820

67424-75-3 2,5-Dimethyl-1-phenylphosphorinan-4-ol; (1*RS*,2*RS*,4*RS*,5*RS*)-*form*, 1-Sulfide, 4-*O*-Ac, *in* D-00820

67424-76-4 2,5-Dimethyl-1-phenylphosphorinan-4-ol; (1*RS*,2*RS*,4*SR*,5*RS*)-*form*, 1-Sulfide, *in* D-00820

67424-77-5 2,5-Dimethyl-1-phenylphosphorinan-4-ol; (1*RS*,2*RS*,4*SR*,5*RS*)-*form*, 1-Sulfide, 4-*O*-Ac, *in* D-00820

67424-79-7 2,5-Dimethyl-1-phenylphosphorinan-4-one; (1*RS*,2*RS*,5*RS*)-*form*, 1-Oxide, *in* D-00821

67452-76-0 9-Phenyl-9-phosphapentacyclo-[4.3.0.$0^{2,5}$.$0^{3,8}$.$0^{4,7}$]nonane; 9-Sulfide, *in* P-00201

67471-54-9 Trimethylol selenophosphate, *in* M-00424

67472-25-7 Bis(trimethylsilyl) (1-oxopropyl)phosphonate, *in* O-00074

67472-26-8 Benzoylphosphonic acid; Monoanilinium salt, *in* B-00034

67472-30-4 (1-Oxopropyl)phosphonic acid, O-00074

67504-68-1 9-Methyl-1,4,6-trioxa-5-phospha(5-P^V)spiro[4.4]-non-7-ene; 5-Ph, *in* M-00428

67515-40-6 (1-Diazoethyl)phenylphosphinic acid, D-00032

67517-97-9 Ethylidynephosphine, E-00077

67535-53-9 (3-Chlorophenylmethylene)-triphenylphosphorane, *in* C-00136

67565-40-6 2,2,2,3-Tetrahydro-2,2,2-trimethoxy-5-phenyl-3,3-bis(trifluoromethyl)-1-(2,4,6-trimethylphenyl)-1*H*-1,4,2-diazaphosphole, T-00159

67565-91-7 (Diphenylmethylene)(2,4,6-trimethylphenyl)phosphine, D-01021

67580-24-9 8-Phenyl-8-phosphabicyclo[3.2.1]-octan-3-one; *syn*-*form*, *in* P-00199

67583-31-7 6,12-Diphenyl-1,4,8,11-tetraoxa-6,12-diaza-5,7-diphospha(5,7-P^V)-dispiro[4.1.4.1]dodecane; 5,7-Diphenoxy, *in* D-01149

67590-60-7 4-Ethyl-2,6,7-trimethyl-2,6,7-triaza-1-phosphabicyclo[2.2.2]-octane, E-00195

67590-61-8 ▷4-Ethyl-2,6,7-trimethyl-2,6,7-triaza-1-phosphabicyclo[2.2.2]octane; 1-Sulfide, *in* E-00195

67598-46-3 Benzoyl phosphorodichloridate, *in* M-00442

67598-47-4 Benzoyl phosphorodifluoridate, *in* M-00442

67605-36-1 Methyl bis(trimethylsilyloxy)-phosphinecarboxylate oxide, *in* B-00433

67628-93-7 2-Chloro-1-(2,4-dichlorophenyl)vinyl dimethyl phosphate; (*Z*)-*form*, *in* C-00035

67660-23-5 1,1,1-Tributyl-1-iodo-*N*-methyl-*N*-phenyl-phosphoranamine, *in* T-00366

67673-30-7 1,3-Dihydro-1-hydroxy-3-methyl-1,2,3-benzodoxaphosphole 3-oxide, *in* D-00488

67696-25-7 Bis(2,2,2-trifluoroethoxy)-triphenylphosphorane, B-00408

67704-60-3 *N*,*N*′,*P*-Trimethylphosphonic diamide, *in* M-00289

67711-52-8 Diisopropyl 1*H*-imidazol-1-ylphosphonate, *in* I-00003

67723-07-3 Dimethyl 1*H*-imidazol-1-ylphosphonate, *in* I-00003

67747-67-5 (2-Aminoethyl)-butylphenylphosphine; (±)-*form*, *in* A-00074

67747-68-6 (2-Aminoethyl)-butylphenylphosphine; (+)-*form*, *in* A-00074

67747-69-7 (2-Aminoethyl)-butylphenylphosphine; (−)-*form*, *in* A-00074

67754-89-6 1-Hydroxy-1,1′(3*H*,3′*H*)-spirobi[2,1-benzoxaphosphole]-3,3′-dione, H-00194

67759-42-6 1-Hydroxy-3,3,3′,3′-tetramethyl-1,1′(3*H*,3′*H*)-spirobi[2,1-benzoxaphosphole], H-00199

67759-43-7 1-Hydroxy-3,3,3′,3′-tetramethyl-1,1′(3*H*,3′*H*)-spirobi[2,1-benzoxaphosphole]; Na salt, *in* H-00199

67761-25-5 *O*,*O*-Ethylene *O*-methyl phosphoroselenoate, *in* M-00036

67773-68-6 4-Pentenylidenetriphenyl-phosphorane, *in* P-00044

67777-11-1 *O*,*O*-Diethyl (1-iminoethyl)-phosphoramidothioate, *in* I-00005

67826-78-2 2,2′-Spirobi[1,3,2-benzoxathiaphosphole]; 2-Ph, *in* S-00019

67828-17-5 ▷Diethyl dibutylphosphoramidate, *in* D-00134

67860-71-3 Octahydro-1,2,3,4,5,6-hexamethyl-[1,2,3]-triphospholo[4,5-*d*]-1,2,3-triphosphole; (3*aα*,6*aβ*)-*form*, *in* O-00017

67864-83-9 Acetylphenylphosphinic acid, A-00021

67884-32-6 1,2-Bis(diphenylphosphino)-propane; (*R*)-*form*, *in* B-00264

67884-33-7 1,2-Bis(diphenylphosphino)-propane; (*S*)-*form*, *in* B-00264

67896-52-0 (1-Amino-2-methylpropyl)-phosphinic acid, A-00092

67902-07-2 1,3-Di-*tert*-butyl-2,4-dimethoxy-1,3,2,4-diazadiphosphetidine; *cis-form*, *in* D-00085

67902-08-3 1,3-Di-*tert*-butyl-2,4-dimethoxy-1,3,2,4-diazadiphosphetidine; *trans-form*, *in* D-00085

67902-10-7 1,3-Di-*tert*-butyl-2,4-dimethoxy-1,3,2,4-diazadiphosphetidine; *cis-form*, 2-Sulfide, *in* D-00085

67902-12-9 1,3-Di-*tert*-butyl-2,4-dimethoxy-1,3,2,4-diazadiphosphetidine; *cis-form*, 2-Selenide, *in* D-00085

67941-86-0 *S*,*S*-Diethyl phosphorodithioate, D-00386

67941-87-1 *S*,*S*-Diphenyl phosphorodithioate; Cyclohexylammonium salt, *in* D-01132

67949-88-6 Tetraisopropyldiphosphine; Dioxide, *in* T-00167

67949-89-7 Diisopropylphosphinic acid; Anhydride, *in* D-00641

67964-16-3 *O*-Ethyl phenylphosphonothioate; (±)-*form*, *in* E-00114

67964-17-4 *O*-Ethyl phenylphosphonothioate; (*S*)-*form*, *in* E-00114

67991-12-2 3,3′,3″-Phosphinidynetrispropanoic acid; Tri-Et ester, *in* P-00357

67998-60-1 Diphenylphosphinoselenoic acid, D-01078

67998-65-6 Diphenylphosphinoselenothioic acid; NEt_3 salt, *in* D-01080

68036-30-6 *N*-*tert*-Butyl-*P*,*P*-dimethylphosphinic amide, *in* D-00830

68036-31-7 *N*-*tert*-Butyl-*P*,*P*-diphenylphosphinic amide, *in* D-01040

68036-32-8 Dimethyl *tert*-butylphosphoramidate, *in* B-00622

68064-15-3 1,1′,1″-Phosphinoselenoylidynetrisaziridine, *in* T-00331

68064-25-5 Bis(trimethylsilyl) (2-oxopropyl)phosphonate, *in* O-00075

68064-28-8 Bis(trimethylsilyl) (3,3-dimethyl-2-oxobutyl)phosphonate, *in* D-00783

68064-33-5 (2-Oxopropyl)phosphonic acid; Mono-dicyclohexylammonium salt, *in* O-00075

68064-39-1 (3,3-Dimethyl-2-oxobutyl)-phosphonic acid; Dicyclohexylammonium salt, *in* D-00783

68084-38-0 (3,3-Dimethyl-2-oxobutyl)-phosphonic acid, D-00783

68089-86-1 (Chloromethyl)-triphenylphosphonium(1+); Iodide, *in* C-00117

68090-10-8 1,3,2-Dithiaphosphole; 2-Chloro, *in* D-01246

68090-11-9 1,3,2-Dithiaphosphole; 2-Methoxy, *in* D-01246

68090-74-4 2,3-Dihydro-1,2,3-trimethyl-1*H*-1,2,3-triphosphole, *in* D-00615

68100-16-3 2,4,6-Trimethyl-5-phenyl-1,3,5-dioxaphosphorinane, T-00521

68100-17-4 2,4,6-Trimethyl-5-phenyl-1,3,5-dioxaphosphorinane; (2*α*,4*α*,5*β*,6*α*)-*form*, 5-Sulfide, *in* T-00521

68107-00-6 Bromfenvinfos-methyl; (*E*)-*form*, *in* B-00463

68107-01-7 Bromfenvinfos-methyl; (*Z*)-*form*, *in* B-00463

68108-86-1 *tert*-Butyloxaphosphine, B-00576

68128-37-0 2,4,6-Trimethyl-5-phenyl-1,3,5-dioxaphosphorinane; (2*α*,4*α*,5*α*,6*α*)-*form*, 5-Sulfide, *in* T-00521

68129-82-8 2,2′-Spirobi[1,3,2-benzoxathiaphosphole], S-00019

68144-09-2 *O*-Butyl *O*-1-naphthalenyl phosphorothioate; (+)-*form*, *in* B-00575

68198-84-5 *O*-Butyl *O*-1-naphthalenyl phosphorothioate; (+)-*form*, Tetramethylammonium salt, *in* B-00575

68222-38-8 2-[Bis(trimethylsilyl)amino]-2,2-dihydro-3,3-bis(trifluoromethyl)-2-[(trimethylsilyl)imino]-oxaphosphirane, B-00430

68236-47-5 Phenyl *N*-(4,5-dihydro-2-thiazolyl)-*P*-methyl-phosphonamidate, *in* D-00587

68236-57-7 Phenyl *P*-methyl-*N*-2-thiazolylphosphonamidate, *in* M-00409

68236-58-8 Ethyl *N*-(4,5-dihydro-2-thiazolyl)-*P*-methyl-phosphonamidate, *in* D-00587

68257-69-2 4-Chloro-1,4-dihydro-2*H*-naphth[2,1-*c*][1,2]-oxaphosphorin, C-00048

68280-30-8 *O*-Ethyl phenylphosphonochloridothioate; (*S*)-*form*, *in* E-00112

68286-92-0 3-Phenyl-1,3-azaphospholidin-2-thione, P-00093

68287-03-6 3-Phenyl-1,3-azaphospholidin-2-thione; 3-Sulfide, *in* P-00093

68305-42-0 6-Methyl-1,5-diphenyl-6-phosphabicyclo[3.2.1]-octane; Oxide, *in* M-00134

68305-75-9 (Methoxymethyl)-triphenylphosphonium(1+); Iodide, *in* M-00053

68351-49-5 3,4,5,6,7,8-Hexahydro-5-phenyl-2*H*-1,9,5-benzodithiaphosphacycloundecin, H-00044

68351-50-8 6,7,8,9,14,15,16,17-Octahydro-9,14-diphenyl-dibenzo[*b*,*i*][1,4,8,11]-dithiadiphosphacyclotetradecin, O-00013

68358-05-4 Bis(aminomethyl)phosphinic acid; B,HCl, *in* B-00087

68358-06-5 Bis(aminomethyl)phosphinic acid; B,HBr, *in* B-00087

68358-09-8 Bis(aminomethyl)phosphinic acid, B-00087

68373-17-1 Diethyl (2,5-dichlorophenyl)-phosphonate, *in* D-00185

68374-69-6 [Phenyl[(phenylmethyl)amino]-methyl]phosphonic acid; (±)-*form*, Di-Et ester, *in* P-00188

68378-98-3 Methylglycerol bicyclic selenophosphate, *in* M-00423

68378-99-4 2,4,6,7-Tetramethyl-2,6,7-triaza-1-phosphabicyclo[2.2.2]-octane; 1-Selenide, *in* T-00237

68402-79-9 6,7,9,10,16,17,18,19-Octahydro-16,19-diphenyldibenzo[*h*,*n*]-[1,4,7,10,13]-trioxadiphosphacyclopentadecin; 16,19-Dioxide, *in* O-00015

68434-72-2 *α*-Ethoxycarbonyltriphenyl-phosphoranylidenepropanoic acid, *in* T-00646

68492-54-6 2,5-Dihydro-2-hydroxy-1,2-oxaphosphole 2-oxide; 2-Butoxy, *in* D-00500

68492-55-7 2,5-Dihydro-2-hydroxy-1,2-oxaphosphole 2-oxide, D-00500

68521-29-9 6,7-Dihydro-5*H*-dibenzo[*d*,*f*]-[1,3,2]diazaphosphepine; 6-Me, 6-oxide, *in* D-00450

68521-33-5 6,7-Dihydro-5*H*-dibenzo[*d*,*f*]-[1,3,2]diazaphosphepine; 6-Ph, 6-oxide, *in* D-00450

68539-80-0 Diethyl 2,4-bis(diethoxyphosphinyl)-pentanedioate, *in* D-01186

68541-88-8 ▷1,1′,1″-Phosphinoselenoylidynetrispiperidine, *in* T-00704

68560-85-0 6-Mercaptodibenzo[*d*,*f*][1,3,2]-dioxaphosphepin 6-sulfide; NH_4 salt, *in* M-00008

68560-86-1 6-Mercaptodibenzo[*d*,*f*][1,3,2]-dioxaphosphepin 6-sulfide, M-00008

68593-85-1 Di-1*H*-imidazol-1-ylphosphinic acid, D-00628

68613-50-3 Benzyl 2-(triphenylphosphoranylidene)-propanoate, *in* T-00660

68614-93-7 2,3-Dihydro-1,3-dimethyl-1,3,2-benzodiazaphosphorin-4(1*H*)-one; 2-Oxide, *in* D-00457

68629-93-6 Dimethyl 2-(tributylphosphoranylidene)-1,3-dithiol-4,5-dicarboxylate, D-00927

68641-49-6 Bis(2-oxo-3-oxazolidinyl)-phosphinic acid; Chloride, *in* B-00378

68669-00-1 7,12-Dihydro-12-phenyl-5*H*-dibenz[*c*,*f*][1,5]-oxaphosphocin; 12-Oxide, *in* D-00560

68669-04-5 5,6,7,12-Tetrahydro-12-phenyldibenz[*c*,*f*][1,5]-azaphosphocine; 6-Benzyl, 12-oxide, *in* T-00130

68669-06-7 7,12-Dihydro-12-methyl-12-phenyl-5*H*-dibenz[*c*,*f*]-[1,5]oxaphosphocinium iodide, *in* D-00560

68669-07-8 5,6,7,12-Tetrahydro-12-phenyldibenz[*c*,*f*][1,5]-azaphosphocine; 12-Oxide, *in* T-00130

68669-55-6 (2-Hydroxyethyl)phosphonothioic acid; Monoanilinium salt, *in* H-00147

68669-95-4 Triphenyl(triphenylgermyl)-phosphinimine, T-00695

68683-75-0 1,2-Bis[(*N*-diphenylphosphino)-amino]cyclohexane; (1*R*,2*R*)-*form*, *N*,*N*′-Di-Me, *in* B-00219

68694-30-4 Spiro[1,3,2-benzoxazaphosphole-2(3*H*),2′-[1,2]oxaphospholan]-5′-one; 2-Et, *in* S-00013

68698-15-7 Di-1*H*-imidazol-1-ylphosphinic acid; Tributylammonium salt, *in* D-00628

68726-16-9 2,2-Dimethyl-1,3-dioxa-5-phospha-2-silacyclohexane; 5-Me, *in* D-00731

68755-15-7 2-Phenoxy-1,3,2-dioxaphosphonane 2-oxide, *in* H-00137

68755-16-8 1,3,6,2-Trioxaphosphocane; 2-Phenoxy, 2-oxide, *in* T-00592

68755-19-1 Hexahydro-2-hydroxycyclopenta[*d*]-1,3,2-dioxaphosphorin 2-oxide; (*4aRS,7aSR*)-*form*, 2-Phenoxy (Ph ester), *in* H-00034

68755-20-4 2-Hydroxy-1,3,2-dioxaphosphonane 2-oxide, H-00137

68755-21-5 2-Hydroxy-4-methyl-1,3,2-dioxaphosphorinane 2-oxide; Benzylammonium salt, *in* H-00156

68823-56-3 2,3,7,8-Tetrakis(trifluoromethyl)-1,4,6,9-tetrathia-5-phospha(5-P^{V})-spiro[4.4]nona-2,7-diene; 5-Ph, *in* T-00189

68823-62-1 3,3,3′,3′-Tetramethyl-1,1′(3*H*,3′*H*)-spirobi[2,1-benzoxaphospholium](1+), T-00233

68823-63-2 3,3,3′,3′-Tetramethyl-1,1′(3*H*,3′*H*)-spirobi[2,1-benzoxaphospholium](1+); Trifluoromethanesulfonate, *in* T-00233

68823-65-4 3,3,3′,3′-Tetramethyl-1,1′(3*H*,3′*H*)-spirobi[2,1-benzoxaphosphole], T-00232

68836-92-0 Diisopropylphosphine; Sulfide, *in* D-00640

68896-35-5 Phenyl(2-propenyl)phosphinic acid, P-00300

68897-36-9 *O,S*-Dimethyl phosphorodithioate; (±)-*form*, *in* D-00899

68908-02-1 Ethylphenylphosphinothioic acid; (±)-*form*, *in* E-00105

68957-94-8 Propylphosphonic acid trimer trianhydride, *in* T-00720

68969-73-7 1,2-Di-*tert*-butyldiphosphirane, D-00091

68969-74-4 1,2,4,5-Tetra-*tert*-butyl-1,2,4,5-tetraphosphorinane, T-00027

68997-87-5 Dimethyl [(2-hydroxyphenyl)-methyl]phosphonate, *in* H-00175

68997-88-6 Dimethyl [(4-hydroxyphenyl)-methyl]phosphonate, *in* H-00176

69019-76-7 Tetrahydro-1*H*,5*H*-[1,2,3]-triphospholo[2,1-*a*]-[1,2,3]triphosphole, T-00160

69039-09-4 Ethyl (diphenylphosphinothioyl)acetate, *in* D-01086

69039-10-7 (Diphenylphosphinothioyl)-acetonitrile, *in* D-01046

69052-20-6 Triphenyl(3-phenyl-2-propenyl)-phosphonium(1+); (*E*)-*form*, Chloride, *in* T-00616

69084-18-0 5,5,6-Trimethyl-4,7-dioxa-3-phospha(3-P^{V})spiro[2.4]-heptane, T-00510

69093-80-7 Tripropyl phosphonoglycinate, *in* P-00368

69098-18-6 *P*-(Trichloromethyl)phosphonic diamide, T-00412

69141-62-4 (1,3-Dithiolan-2-ylmethyl)-triphenylphosphonium(1+); Bromide, *in* D-01252

69141-63-5 (1,3-Dithian-2-ylmethyl)-triphenylphosphonium(1+); Bromide, *in* D-01241

69141-64-6 (1,3-Dithian-2-ylmethyl)-triphenylphosphonium(1+); Iodide, *in* D-01241

69162-08-9 *P,P,P*-Trichloro-*N*-(pentafluorophenyl)-phosphazene, *in* P-00016

69174-43-2 5-Methyl-1,3,2-oxathiaphosphole 2-oxide; 2-Ethoxy, *in* M-00184

69209-32-1 Methyl 3-(diphenylphosphino)-benzoate, *in* D-01054

69216-75-7 [(1,3-Dioxolan-2-yl)methyl]-triphenylphosphonium(1+); Iodide, *in* D-00976

69220-11-7 1,4-Bis(dimethylphosphinothioyl)-benzene, *in* P-00125

69220-13-9 4-Isopropyl-5,5-dimethyl-1,3,2-dioxaphosphorinane, I-00028

69220-15-1 4,4,6-Trimethyl-1,3,2-dioxaphosphorinane; (*2RS,4RS*)-*form*, *in* T-00513

69220-16-2 4,4,6-Trimethyl-1,3,2-dioxaphosphorinane; (*2RS,4SR*)-*form*, *in* T-00513

69277-58-3 *P,P,P*-Tri-*tert*-butylphosphine imide, T-00375

69284-88-4 2-Fluoro-1,3-dimethyl-1,3,2-diazaphospholidine; 2-Sulfide, *in* F-00014

69310-55-0 Ethynylphosphonic acid, E-00203

69381-90-4 (1-Phenyl-1,2-ethanediyl)-bis[diphenylphosphine]; (*S*)-*form*, *in* P-00150

69381-91-5 (1-Phenyl-1,2-ethanediyl)-bis[diphenylphosphine]; (*R*)-*form*, *in* P-00150

69387-00-4 Phenylpropylphosphinic acid, P-00303

69387-01-5 Cyclopentylmethylphosphinic acid, C-00279

69404-21-3 Methyl(trichloromethyl)-phosphinic acid, M-00419

69404-44-0 Dimethyl 2-phenylethylphosphonate, *in* P-00165

69404-52-0 Tetraethyl [thiobis(methylene)]-bisphosphonate, *in* T-00311

69404-53-1 Thiobis[methanephosphonic acid], T-00311

69412-49-3 2-Ethoxy-1-(2-phenylethyl)-1,2-azaphosphetidine 2-oxide, *in* H-00107

69423-57-0 Methyl ethylphenylphosphinate, *in* E-00101

69442-51-9 Triphenyl[1-(phenylthio)-ethenyl]phosphonium(1+); Chloride, *in* T-00619

69442-52-0 Triphenyl[1-(phenylthio)-ethenyl]phosphonium(1+); Iodide, *in* T-00619

69460-42-0 Methyl *tert*-butylphenylphosphinate, *in* B-00579

69507-62-6 4-Chlorophenyl 2-cyanoethyl phosphorochloridate, C-00125

69572-88-9 1,1,4,4-Tetrabenzyl-1,4-diphosphocanium dibromide, *in* D-00048

69572-88-9 1,1,5,5-Tetrabenzyl-1,5-diphosphocanium dibromide, *in* D-00049

69572-89-0 1,1,5,5-Tetrabenzyl-1,5-diphosphonanium dibromide, *in* D-00051

69572-90-3 1,1,4,4-Tetrabenzyl-1,4-diphosphonanium dibromide, *in* D-00050

69572-97-0 1,4-Dibenzyl-1,4-diphosphepane, D-00047

69572-97-0 1,1,4,4-Tetrabenzyl-1,4-diphosphepanium dibromide, *in* D-00047

69572-98-1 1,4-Dibenzyl-1,4-diphosphocane, D-00048

69572-99-2 1,5-Dibenzyl-1,5-diphosphocane, D-00049

69573-00-8 1,5-Dibenzyl-1,5-diphosphonane, D-00051

69573-01-9 1,4-Dibenzyl-1,4-diphosphonane, D-00050

69573-04-2 1,6-Dibenzyl-1,6-diphosphacycloundecane, D-00046

69573-06-4 1,5,10,14-Tetrabenzyl-1,5,10,14-tetraphosphacyclooctadecane, T-00009

69576-78-9 Methyl tetramethylene phosphite, *in* D-00968

69639-73-2 (2-Oxoheptyl)phosphonic acid, O-00061

69639-74-3 [(Ethylthio)methyl]phosphonic acid, E-00192

69646-14-6 Diethyl (2-hydroxyphenyl)-phosphonate, *in* H-00179

69646-20-4 Bis(trifluoromethyl) phenylphosphonodiselenoite, *in* P-00235

69695-11-0 1′,3′-Diphenyldispiro[1,3,2-benzodioxaphosphole-2,2′-[1,3,2,4]diazaphosphetidine-4′,2″-[1,3,2]-benzodioxaphosphole]; 2,4′-Dichloro, *in* D-01018

69695-70-1 *O,O*-Diethyl 2-propenylphosphonothioate, *in* P-00461

69695-90-5 *N,N′,N″*-Triisopropylphosphorothioic triamide, *in* T-00509

69695-91-6 *N,N′,N″*-Tris(1-methylpropyl)-phosphorothioic triamide, *in* T-00829

69741-76-0 *S*-Ethyl diphenylphosphinoselenothioate, *in* D-01080

69741-80-6 *S*-Ethyl ethylphenyl-phosphinothioselenoate, *in* E-00106

69750-15-8 2-Hydroxy-2*H*-1,2-benzoxaphosphorin 2-oxide; 2-Phenoxy (Ph ester), *in* H-00112

69750-16-9 2-Hydroxy-2*H*-1,2-benzoxaphosphorin 2-oxide; 2-Ethoxy (Et ester), *in* H-00112

69774-92-1 Spiro[1,3,2-benzodioxaphosphole-2,2′-[1,3,2]-benzoxathiaphosphole]; 2-Phenoxy, *in* S-00007

69774-94-3 3′-Methylspiro[1,3,2-benzodioxaphosphole-2,2(3′*H*)-[1,3,2]-benzothiazaphosphole]; 2-Phenoxy, *in* M-00394

69774-96-5 3′,*N,N*-Trimethylspiro[1,3,2-benzodioxaphosphole-2,2′(3′*H*)-[1,3,2]-benzothiazaphosphol]-2-amine, *in* M-00394

69775-01-5 4,4,5,5,*N,N*-Hexamethylspiro[1,3,2-benzoxazaphosphole-2(3*H*),2′-[1,3,2]-dioxaphospholan]-2-amine, *in* T-00231

69775-02-6 4′,4′,5′,5′-Tetramethylspiro[1,3,2-benzoxazaphosphole-2(3*H*),2′-[1,3,2]-dioxaphospholane]; 2-Methoxy, *in* T-00231

69775-03-7 4′,4′,5′,5′-Tetramethylspiro[1,3,2-benzoxazaphosphole-2(3*H*),2′-[1,3,2]-dioxaphospholane]; 2-Phenoxy, *in* T-00231

69783-46-6 1-Oxa-5-phospha(5-P^{V})-spiro[4.4]nonane; 5-Me, *in* O-00051

69783-47-7 1-Oxa-5-phospha(5-P^{V})-spiro[4.5]decane; 5-Me, *in* O-00049

69783-48-8 2,2-Dihydro-2,2,2-trimethyl-1,2-oxaphosphorinane, D-00597

69783-50-2 6-Oxa-5-phospha(5-P^{V})-spiro[4.5]decane; 5-Me, *in* O-00050

69783-51-3 1-Oxa-6-phospha(6-P^{V})-spiro[5.5]undecane; 6-Me, *in* O-00052

69803-89-4 4-Morpholinylphosphonochloridic acid; Anhydride with diphenyl phosphate, *in* M-00524

69805-25-0 Ethyl ethylphosphonochloridite, *in* E-00139

69813-37-2 4,7-Dihydro-1,3,2-dioxaphosphepin; 2-Dimethyl-amino, *in* D-00472

69813-38-3 4,7-Dihydro-2-phenyl-1,3,2-dioxaphosphepin, D-00561

69813-41-8 1,5-Dihydro-2-phenoxy-2,4,3-benzodioxaphosphepin; 3-Sulfide, *in* D-00544

69813-43-0 4,7-Dihydro-1,3,2-dioxaphosphepin 2-sulfide; 2-Chloro, *in* D-00473

69813-44-1 4,7-Dihydro-1,3,2-dioxaphosphepin 2-sulfide; 2-Phenoxy, *in* D-00473

69813-45-2 4,7-Dihydro-2-phenyl-1,3,2-dioxaphosphepin; 2-Sulfide, *in* D-00561

69813-46-3 4,7-Dihydro-1,3,2-dioxaphosphepin 2-sulfide; 2-Me, *in* D-00473

69813-49-6 *O,O*-Tetramethylene *O*-phenyl phosphorothioate, *in* D-00969

69813-50-9 *O,O*-Tetramethylene phenylphosphonothioate, *in* P-00114

69813-52-1 1,5-Dihydro-3-methyl-2,4,3-benzodioxaphosphepin, D-00516

69813-53-2 1,5-Dihydro-3-dimethylamino-2,4,3-benzodioxaphosphepin, D-00453

69813-54-3 3-Chloro-1,5-dihydro-2,4,3-benzodioxaphosphepin, C-00038

69822-97-5 Diphenyl-1-propenylphosphine, D-01143

69844-21-9 Tetramethyldiphosphine; Dioxide, *in* T-00200

69891-57-2 [2-(1,3-Dioxan-2-yl)-ethylidene]triphenylphosphorane, *in* D-00965

69891-92-5 [2-(1,3-Dioxan-2-yl)ethyl]-triphenylphosphonium(1+); Bromide, *in* D-00965

69914-34-7 Dimethyl 1-phenylethylphosphonate, *in* P-00164

69919-18-2 (4-*tert*-Butyl-2-chlorophenyl) 2,2,2-tribromoethyl phosphorochloridate, B-00548

69923-44-0 (Diisopropoxyphosphinothioyl)-thioacetic acid, D-00632

69928-10-5 Methyl bis(2-methylphenyl)-phosphinate, *in* B-00318

69928-15-0 7,8,10,11,13,14,16,17,19,24-Decahydro-5*H*-dibenzo[*o,r*][1,4,7,10,13,17]-pentaoxaphosphacycloeicosin; 24-Ph, 24-oxide, *in* D-00001

69928-16-1 7,8,10,11,13,14,16,17,19,20,22,27-Dodecahydro-5*H*-dibenzo[*r,u*][1,4,7,10,13,16,20]-hexaoxaphosphacyclotricosin; 27-Ph, 27-oxide, *in* D-01261

69928-19-4 7,8,10,11,13,14,16,17,19,24-Decahydro-5*H*-dibenzo[*o,r*][1,4,7,10,13,17]-pentaoxaphosphacycloeicosin; 24-Methoxy, 24-oxide, *in* D-00001

69981-37-9 *N,N*-Diethyl-*P,P*-di-1-pyrrolidinylphosphinic amide, *in* D-01233

69991-36-2 1,5-Dimethyl-1*H*-1,2,3-diazaphosphole, *in* M-00114

70035-95-9 3,5-Di-*tert*-butyl-2,3-dihydro-1,3,2-oxazaphosphole; 2-Methoxy, *in* D-00083

70058-90-1 Tris(trifluoromethyl) phosphorotriselenoite, T-00857

70058-99-0 *N,P,P*-Trimethylphosphinoselenoic amide, *in* D-00836

70063-12-6 Methyl dimethyl-phosphoramidochloridite, *in* D-00862

70069-20-4 Bis(1-methylpropyl) phosphate, B-00341

70073-11-9 Methyl di-*tert*-butylphosphinite, *in* D-00122

70101-69-8 4-Mercaptodinaphtho[2,1-*d*:1′,2′-*f*][1,3,2]-dioxaphosphepin 4-sulfide; (±)-*form*, *in* M-00012

70111-23-8 1,4-Bis(dimethylphosphinothioyl)-butane, *in* B-00199

70111-25-0 Ethyl dibutylphosphinodithioate, *in* D-00112

70111-27-2 2,3,4,5-Tetrahydro-6-methyl-2,4,4-triphenyl-1,2,3-diazaphosphorine; 3-Methoxy, *in* T-00118

70111-30-7 2,3,4,5-Tetrahydro-6-methyl-2,4,4-triphenyl-1,2,3-diazaphosphorine; 3-Chloro, *in* T-00118

70111-32-9 2,3,4,5-Tetrahydro-6-methyl-2,4,4-triphenyl-1,2,3-diazaphosphorine; 3-Me, 3-oxide, *in* T-00118

70111-33-0 2,3,4,5-Tetrahydro-6-methyl-2,4,4-triphenyl-1,2,3-diazaphosphorine; 3-Methoxy, 3-sulfide, *in* T-00118

70117-03-2 Spiro[1,3,2-benzodioxaphosphole-2,2′-[1,3,2]-benzoxathiaphosphole], S-00007

70144-34-2 4-Mercaptodinaphtho[2,1-*d*:1′,2′-*f*][1,3,2]-dioxaphosphepin 4-sulfide; (*S*)-*form*, *in* M-00012

70144-36-4 4-Mercaptodinaphtho[2,1-*d*:1′,2′-*f*][1,3,2]-dioxaphosphepin 4-sulfide; (*S*)-*form*, (+)-1-Phenylethylammonium salt, *in* M-00012

70179-63-4 2,3,4,7-Tetrahydro-1*H*-isophosphindole 2-oxide; 2-Ph, *in* T-00109

70179-64-5 2,3,4,7-Tetrahydro-1*H*-isophosphindole 2-oxide; 2-Me, *in* T-00109

70179-65-6 2,3,4,7-Tetrahydro-1*H*-isophosphindole 2-oxide; 2-Hydroxy, *in* T-00109

70219-43-1 5-*tert*-Butyltetrahydro-2-dimethylamino-3-phenyl-2*H*-1,3,2-oxazaphosphorine; (2*RS*,5*RS*)-*form*, 2-Oxide, *in* B-00636

70219-44-2 5-*tert*-Butyltetrahydro-2-dimethylamino-3-phenyl-2*H*-1,3,2-oxazaphosphorine; (2*RS*,5*SR*)-*form*, 2-Oxide, *in* B-00636

70223-77-7 1,2-Bis(diphenylphosphinomethyl)-cyclohexane; (1*S*,2*S*)-*form*, *in* B-00250

70225-00-2 6-Oxa-5-phospha(5-P^V)-spiro[4.5]decane, O-00050

70225-02-4 1-Oxa-5-phospha(5-P^V)-spiro[4.5]decane, O-00049

70225-03-5 1-Oxa-5-phospha(5-P^V)-spiro[4.4]nonane, O-00051

70236-64-5 1-Oxa-6-phospha(6-P^V)-spiro[5.5]undecane, O-00052

70307-60-7 (2-Oxo-1,3-propanediyl)-bis(triphenylphosphonium)(2+); Dibromide, *in* O-00073

70311-64-7 2-Chloro-4,4,5,5-tetrakis(trifluoromethyl)-1,3,2-dioxaphospholane, C-00184

70311-67-0 2,2-Dihydro-2,2,2-trichloro-4,4,5,5-tetrakis(trifluoromethyl)-1,3,2-dioxaphospholane, D-00589

70311-69-2 2,2,2-Tribromo-2,2-dihydro-4,4,5,5-tetrakis(trifluoromethyl)-1,3,2-dioxaphospholane, T-00353

70320-30-8 1,2-Bis[(bis-4-methylphenyl)-phosphino]ethane, B-00104

70340-04-4 [(2-Hydroxyphenyl)methyl]-triphenylphosphonium(1+); Bromide, *in* H-00177

70346-98-4 5-*tert*-Butyl-2,2,4,5-tetrahydro-2,2,4-triisopropoxy-1,2,4-oxadiphosphole, B-00637

70355-41-8 Tris(dimethylphosphino)methane, T-00771

70369-42-5 Dimethyl [(phenylthio)methyl]-phosphonate, *in* P-00326

70372-09-7 *O,O*-Tetramethylethylene phosphonothioate, *in* T-00197

70385-78-3 2,2,2,3-Tetrahydro-1,2-benzoxaphosphole; 2,2,2-Triethoxy, *in* T-00088

70403-02-0 *N,N,P*-Triethylphosphinic amide, *in* E-00124

70403-04-2 *N,N*-Diethyl-*P*-phenylphosphinic amide, *in* P-00210

70430-42-1 Dibromo-4-bromophenylphosphine, *in* B-00493

70430-43-2 Dibromo(4-chlorophenyl)-phosphine, *in* C-00153

70430-44-3 Tetrabromophenylphosphorane, T-00011

70430-51-2 2,5′-Diphenylspiro[4*H*-1,3,2-benzodioxaphosphorin-2,2′-[1,3,2]dioxaphospholane]-4,4′-dione, D-01147

70442-86-3 Dimethylphosphinoselenoic acid; Azide, *in* D-00835

70445-76-0 Diphenyl phosphorobromidite, D-01118

70445-77-1 Phenyl phosphorodibromidite, P-00277

70446-63-8 Butyl *tert*-butylphosphonochloridite, *in* D-00130

70446-65-0 Dibutyl isopropylphosphonite, *in* I-00056

70446-71-8 Butyl isopropylphosphinate, *in* I-00049

70448-73-6 Diisopropyl (1-nitroso-1-methylethyl)phosphonate, *in* N-00066

70519-56-1 *O*-Ethyl di-2-furanylphosphinothioate, *in* D-00422

70519-90-3 (3-Aminobutyl)phosphonic acid, A-00056

70525-48-3 Dimethyl (isocyanatomethyl)-phosphonate, *in* I-00022

70532-50-2 Neopentylene phosphorochloridoselenoate, *in* C-00058

70562-38-8 *O,O,S*-Triphenyl phosphorothioate, T-00671

70575-81-4 *O,O*-Diphenyl phenyl-sulfonylphosphoramidothioate, *in* P-00314

70578-12-0 *N,N′*-Diphenyl-phosphorodiamidodithioic acid, D-01128

70589-60-5 *P,P*-Diisopropyl-*N*-phenylphosphinous amide, *in* D-00649

70589-76-3 (3-Methyl-1,3-butadienyl)-phosphonic acid; (*E*)-*form*, Dichloride, *in* M-00095

70602-72-1 Bis(acetoxymethyl)-phenylphosphine, *in* B-00281

70609-27-7 *P,P*-Diisopropylphosphinous amide, D-00649

70610-32-1 2,3-Dihydro-1*H*-benzo[*e*]-phosphindole 3-oxide; 3-Ph, *in* D-00442

70610-33-2 2,3-Dihydro-1*H*-benzo[*e*]-phosphindole 3-oxide; 3-Hydroxy, *in* D-00442

70610-37-6 2,3-Dihydro-1-phenyl-1*H*-benzo[*g*]phosphindole; 4,5-Dihydro, 1-Oxide, *in* D-00551

70610-38-7 2,3-Dihydro-1-phenyl-1*H*-benzo[*g*]phosphindole; 1-Oxide, *in* D-00551

70629-41-3 Diethylphosphinoselenoic acid; Azide, *in* D-00327

70629-46-8 Diphenylphosphinoselenoic acid; Azide, *in* D-01078

70629-47-9 Dimethylphosphinoselenoic acid; Bromide, *in* D-00835

70636-86-1 ▷*tert*-Butyl bicyclic thiophosphate, *in* B-00640

70641-33-7 4,4′-Di-*tert*-butyl-1,1′,2,2′,5,5′,6,6′-octahydro-1,1′-diphenyl-2,2′-biphosphorin; *P,P′*-Disulfide, *in* D-00100

70660-05-8 Diethyl (mercaptomethyl)phosphonate, *in* M-00018
70665-02-0 Triphenyl[[3-(tetrahydro-2*H*-pyran-2-yl)oxy]propyl]phosphonium(1+); Bromide, *in* T-00683
70677-19-9 Diphenyl phosphorobromidate, *in* D-01036
70708-37-1 1,2-Bis[(*N*-diphenylphosphino)amino]cyclohexane; (1*R*,2*R*)-*form*, *in* B-00219
70723-42-1 *O*,*O*-Diisopropyl phosphorodithioate; Dimethylammonium salt, *in* D-00663
70741-64-9 Ethyl phenyl methylphosphonate; (*R*)-*form*, *in* E-00100
70741-65-0 Ethyl phenyl methylphosphonate; (*S*)-*form*, *in* E-00100
70741-68-3 Methyl phenyl methylphosphonate; (*R*)-*form*, *in* M-00205
70741-69-4 Methyl phenyl methylphosphonate; (*S*)-*form*, *in* M-00205
70774-28-6 1,2-Bis(diphenylphosphinomethyl)cyclohexane; (1*R*,2*R*)-*form*, *in* B-00250
70788-28-2 ▷Flurofamide, F-00051
70788-29-3 Tolfamide, T-00323
70869-67-9 2-*tert*-Butylamino-4-methyl-1,3,2-dioxaphosphorinane, B-00541
70870-36-9 4,5-Dimethyl-2-phenoxy-1,3,2-dioxaphospholane, D-00789
70886-00-9 4-Hydroxy-2,6-diphenyl-4*H*-1,4-oxaphosphorin 4-oxide; Triethylammonium salt, *in* H-00140
70886-01-0 4-Chloro-2,6-diphenyl-4*H*-1,4-oxaphosphorin 4-oxide, *in* H-00140
70886-02-1 2,6-Diphenyl-4*H*-1,4-oxaphosphorin-4-amine 4-oxide, *in* H-00140
70886-04-3 Oxybis(2,6-diphenyl-4*H*-1,4-oxaphosphorin)-4,4′-dioxide, *in* H-00140
70897-27-7 Glycerol 1,2-diheptadecanoate 3-phosphocholine; (*R*)-L-*form*, *in* G-00010
71009-87-5 *S*-Methyl bis(trifluoromethyl)phosphinothioate, *in* B-00415
71009-88-6 *S*-Ethyl bis(trifluoromethyl)phosphinothioate, *in* B-00415
71009-89-7 *S*-Isopropyl bis(trifluoromethyl)phosphinothioate, *in* B-00415
71009-90-0 *O*-Ethyl bis(trifluoromethyl)phosphinothioate, *in* B-00415
71009-91-1 *O*-Isopropyl bis(trifluoromethyl)phosphinothioate, *in* B-00415
71040-58-9 *O*-Methyl bis(trifluoromethyl)phosphinothioate, *in* B-00415
71042-54-1 5,6-Bis(diphenylphosphino)bicyclo[2.2.1]hept-2-ene; (1*R*,5*R*,6*R*)-*form*, *in* B-00220
71042-55-2 5,6-Bis(diphenylphosphino)bicyclo[2.2.1]hept-2-ene; (1*S*,5*S*,6*S*)-*form*, *in* B-00220
71071-55-1 Diethyl (3-methyl-3-butenyl)phosphonate, *in* M-00103
71075-22-4 5,6-Bis(diphenylphosphino)bicyclo[2.2.1]hept-2-ene; (1*RS*,5*RS*,6*RS*)-*form*, *P*,*P*′-Dioxide, *in* B-00220
71075-23-5 5,6-Bis(diphenylphosphinyl)bicyclo[2.2.1]hept-2-ene, *in* B-00220
71075-24-6 5,6-Bis(diphenylphosphino)bicyclo[2.2.1]hept-2-ene; (1*S*,5*S*,6*S*)-*form*, *P*,*P*′-Dioxide, *in* B-00220
71089-75-3 Octahydro-5*H*,8*H*,10*bH*-2*a*,4*a*,7*a*,10*a*-tetraaza-10*b*-phosphapentaleno[1,6-*de*]naphthalene, O-00024
71093-77-1 Tetrahydro-2-phenoxy-2*H*-1,3,2-oxazaphosphorine; 2-Sulfide, *in* T-00126
71093-78-2 3-Methyltetrahydro-2-phenoxy-2*H*-1,3,2-oxazaphosphorine; 2-Sulfide, *in* M-00404
71156-57-5 2-*tert*-Butyl-1,3,6,2-trithiaphosphocane; 2-Sulfide, *in* B-00642
71162-24-8 *N*,*N*,*P*-Triethyl-*P*-phenylphosphinous amide, *in* E-00108
71162-25-9 Ethylphenylphosphinous acid; (*S*)-*form*, Diethylamide, *in* E-00108
71182-73-5 1-Hydroxy-1,1′(3*H*,3′*H*)-spirobi[2,1-benzoxaphosphole]-3,3′-dione; 1-Methoxy, *in* H-00194
71242-28-9 Glycerol 1,2-ditridecanoate 3-phosphocholine; (*R*)-L-*form*, *in* G-00052
71256-07-0 1,3,6,2-Trioxaphosphocane; 2-Chloro, *in* T-00592
71256-08-1 1,3,6,2-Trioxaphosphocane; 2-Methoxy, *in* T-00592
71260-66-7 2,2,2-Tribromoethyl 4-*tert*-butylphenylphosphoramidic chloride, T-00356
71276-93-2 (2-Ethoxyethenyl)triphenylphosphonium(1+); Bromide, *in* E-00039
71276-94-3 (2,2-Diethoxyethylidene)triphenylphosphorane, D-00233
71309-04-1 3,3′,3″-Phosphinoselenoylidynetris[propanenitrile], *in* T-00752
71328-66-0 *N*-*tert*-Butyl-*P*,*P*,*P*-trimethylphosphine imide, *in* T-00531
71337-04-7 (3-Methylphenyl)phosphine, M-00217
71348-05-5 *O*-Ethyl *O*-methyl phosphorothioate; (*R*)-*form*, *in* E-00098
71348-18-0 *O*-Ethyl *O*-methyl phosphorothioate; (*S*)-*form*, *in* E-00098
71354-74-0 Phenylphosphonous acid; Diisothiocyanate, *in* P-00251
71360-01-5 Bis(3-methoxyphenyl)phosphine; Oxide, *in* B-00297
71360-03-7 Bis(3-chlorophenyl)phosphine; Oxide, *in* B-00135
71360-04-8 Bis(2-methoxyphenyl)phosphine; Oxide, *in* B-00296
71363-52-5 2-Chloro-1-(2,4-dichlorophenyl)vinyl dimethyl phosphate; (*E*)-*form*, *in* C-00035
71368-21-3 Glycerol 1,2-di-15-methylhexadecanoate 3-phosphocholine; (*R*)-L-*form*, *in* G-00021
71368-22-4 Glycerol 1,2-di-14-methylpentadecanoate 3-phosphocholine; (*R*)-L-*form*, *in* G-00022
71368-23-5 Glycerol 1,2-di-13-methyltetradecanoate 3-phosphocholine; (*R*)-L-*form*, *in* G-00023
71368-24-6 Glycerol 1,2-di-12-methyltridecanoate 3-phosphocholine; (*R*)-L-*form*, *in* G-00024
71368-25-7 Glycerol 1,2-di-11-methyldodecanoate 3-phosphocholine; (*R*)-L-*form*, *in* G-00019
71368-26-8 Glycerol 1,2-di-10-methylundecanoate 3-phosphocholine; (*R*)-L-*form*, *in* G-00025
71401-87-1 3,3,3′,3′-Tetramethyl-1,1′(3*H*,3′*H*)-spirobi[2,1-benzoxaphosphole]; 1-Chloro, *in* T-00232
71402-66-9 Tetraethyl 1,4-phenylenebisphosphoramidate, *in* B-00012
71433-47-1 3,3,3′,3′-Tetramethyl-1,1′(3*H*,3′*H*)-spirobi[2,1-benzoxaphosphole]; 1-Ph, *in* T-00232
71436-82-3 Triphenyl[3-[(tetrahydro-2*H*-pyran-2-yl)oxy]propylidene]phosphorane, *in* T-00683
71444-56-9 2,4,6-Trimethyl-5-phenyl-1,3,5-dioxaphosphorinane; (2α,4α,5β,6α)-*form*, 5-Oxide, *in* T-00521
71456-84-3 3,5-Di-*tert*-butyl-2-chloro-2,3-dihydro-1,3,2-oxazaphosphole; 2-Sulfide, *in* D-00078
71456-85-4 3,5-Di-*tert*-butyl-2-chloro-2,3-dihydro-1,3,2-oxazaphosphole, D-00078
71466-47-2 1,2,4,5-Tetrakis(methylene)-3-phosphoniaspiro[2.2]pentane, T-00182
71466-49-4 5-Phospha(5-P^{V})spiro[4.5]deca-1,3,5,7,9-pentaene, P-00349
71476-14-7 2,3-Dihydro-1,3-dimethyl-1,3,2-benzodiazaphosphorin-4(1*H*)-one; 2-Me, 2-oxide, *in* D-00457
71485-27-3 2,4,6-Trimethyl-5-phenyl-1,3,5-dioxaphosphorinane; (2α,4α,5α,6α)-*form*, *in* T-00521
71485-28-4 2,4,6-Trimethyl-5-phenyl-1,3,5-dioxaphosphorinane; (2α,4α,5β,6α)-*form*, *in* T-00521
71485-34-2 2,4,6-Trimethyl-5-phenyl-1,3,5-dioxaphosphorinane; (2α,4α,5α,6α)-*form*, 5-Oxide, *in* T-00521
71544-98-4 Diisopropyl (trifluoromethyl)phosphonite, *in* T-00482
71545-04-5 Isopropyl (trifluoromethyl)phosphinate, *in* T-00475
71550-48-6 Dibutylphosphinodithioic acid; Na salt, *in* D-00112
71559-18-7 2-Phenyl-1-oxa-4-aza-5-phospha(5-P^{V})spiro[4.4]nonane; 5-Ph, *in* P-00173
71559-20-1 2,3-Diphenyl-1,6,10-trioxa-4-aza-5-phospha(5-P^{V})spiro[4.5]dec-2-ene; 5-Ph, *in* D-01158
71559-28-9 1,4-Dihydrospiro[2*H*-3,1,2-benzoxazaphosphorine-2,2′-[1,3,2]dioxaphospholane]; 2-Ph, *in* D-00576
71559-29-0 1,3-Dihydrospiro[2*H*-1,3,2-benzodiazaphosphole-2,2′-[1,3,2]dioxaphosholane]; 2-Ph, *in* D-00575
71559-33-6 1,4,6,11-Tetraoxa-5-phospha(5-P^{V})spiro[4.6]undecane; 5-Ph, *in* T-00245
71559-34-7 1,4,6,12-Tetraoxa-5-phospha(5-P^{V})spiro[4.7]dodecane; 5-Ph, *in* T-00243
71559-37-0 Spiro[1,3,2-benzoxathiaphosphole-2,2′-[1,3,2]dioxaphospholane]; 2-Ph, *in* S-00011
71640-53-4 Tetraiodomethylphosphorane, T-00163
71761-33-6 2,6-Dimethyl-4-phenyl-1,4-oxaphosphorinane; *cis*-*form*, *in* D-00797
71771-33-0 2-Hydroxyhexahydro-4*H*-1,3,2-benzodioxaphosphorin 2-oxide; (2α,4*a*β,8*a*α)-*form*, 2-(4-Nitrophenoxy) (2-(4-Nitrophenyl)ester), *in* H-00152
71771-35-2 Octahydro-2-(4-nitrophenoxy)-2*H*-1,3,2-benzoxazaphosphorine; (2*RS*,4*aSR*,8*aRS*)-*form*, 2-Oxide, *in* O-00020
71771-36-3 Hexahydro-2*a*,4*a*,6*a*-triaza-6*b*-phosphacyclopenta[*cd*]pentalene, H-00055

71771-37-4 Hexahydro-2*a*,4*a*,6*a*-triaza-6*b*-phosphacyclopenta[*cd*]-pentalene; 6*b*-Oxide, *in* H-00055

71771-38-5 Hexahydro-2*a*,4*a*,6*a*-triaza-6*b*-phosphacyclopenta[*cd*]-pentalene; 6*b*-Sulfide, *in* H-00055

71771-39-6 Hexahydro-2*a*,4*a*,6*a*-triaza-6*b*-phosphacyclopenta[*cd*]-pentalene; 6*b*-Selenide, *in* H-00055

71778-99-9 (Dibromomethyl)phosphonic acid, D-00064

71783-12-5 1,3-Di-*tert*-butyl-2,4-dimethoxy-1,3,2,4-diazadiphosphetidine; *trans-form*, 2,4-Disulfide, *in* D-00085

71783-13-6 1,3-Di-*tert*-butyl-2,4-dimethoxy-1,3,2,4-diazadiphosphetidine; *trans-form*, 2-Sulfide, *in* D-00085

71783-14-7 1,3-Di-*tert*-butyl-2,4-dimethoxy-1,3,2,4-diazadiphosphetidine; *cis-form*, 2,4-Diselenide, *in* D-00085

71783-15-8 1,3-Di-*tert*-butyl-2,4-dimethoxy-1,3,2,4-diazadiphosphetidine; *trans-form*, 2,4-Diselenide, *in* D-00085

71783-16-9 1,3-Di-*tert*-butyl-2,4-dimethoxy-1,3,2,4-diazadiphosphetidine; *trans-form*, 2-Selenide, *in* D-00085

71787-50-3 2,2-Dimethyl-1,3-dioxa-5-phospha-2-silacyclohexane; 5-Ph, *in* D-00731

71787-57-0 17-Methyl-6,7,9,10-tetrahydrodibenzo[*d,m*]-[1,3,6,9,12,2]-pentaoxaphosphacyclotetradecin; 17-Sulfide, *in* M-00403

71802-06-7 1,3-Di-*tert*-butyl-2,4-dimethoxy-1,3,2,4-diazadiphosphetidine; *cis-form*, 2,4-Disulfide, *in* D-00085

71806-76-3 2-Hydroxyhexahydro-4*H*-1,3,2-benzodioxaphosphorin 2-oxide; ($2\alpha,4a\alpha,8a\beta$)-*form*, 2-(4-Nitrophenoxy), *in* H-00152

71806-77-4 Octahydro-2-(4-nitrophenoxy)-2*H*-1,3,2-benzoxazaphosphorine; (2*RS*,4*aRS*,8*aSR*)-*form*, 2-Oxide, *in* O-00020

71817-08-8 6,7,11,12,19,20,22,23-Octahydro-9-methyldibenzo[*g,p*]-[1,3,6,9,12,15,18,21]heptaoxaphosphacycloeicosin; 9-Oxide, *in* O-00019

71838-20-5 2,3,5,6-Tetraphenyl-1,4-bis[(trimethylsilyl)-oxy]-2,3,5,6-tetraphosphabicyclo[2.2.0]-hexane, T-00251

71847-21-7 *N*-Isopropyl-*P,P*-diphenylphosphinic amide, *in* D-01040

71861-22-8 *sym*-Tetramethyl monothiopyrophosphate, T-00206

71867-67-9 Triethyl methylphosphorimidate, *in* M-00336

71877-80-0 *O,S*-Dimethyl phosphorochloridothioate; (*R*)-*form*, *in* D-00892

71877-81-1 *O,S*-Dimethyl phosphorochloridothioate; (*S*)-*form*, *in* D-00892

71877-82-2 Methyl dimethylphosphoramidochloridate, *in* D-00860

71901-72-9 4,5,9,10,11,12-Hexakis(trifluoromethyl)-1,8-diphosphatetracyclo[$6.2.2.0^{2,7}.0^{3,6}$]-dodecane; ($2\alpha,3\beta,6\beta,7\alpha$)-*form*, *in* H-00066

71918-67-7 Dimethyl (dibromomethyl)-phosphonate, *in* D-00064

71934-28-6 Hexahydrospiro[1,3,2-benzoxazaphosphole-2(3*H*),1′-phospholane]; 2-Ph, *in* H-00052

71987-74-1 [3-Methyl-5-(2,6,6-trimethyl-1-cyclohexen-1-yl)-2,4-pentadienylidene]-triphenylphosphorane, *in* M-00421

71990-10-8 1,4,6,11-Tetraoxa-5-phospha(5-P^{V})spiro[4.6]undecane, T-00245

72021-47-7 Phellanphos, P-00059

72054-43-4 4,5-Dihydro-3-phenyl-3*H*-benzo[*e*]phosphindole, *in* D-00441

72090-83-6 1,2-Bis[(*N*-diphenylphosphino)-amino]cyclohexane; (1*S*,2*S*)-*form*, *in* B-00219

72091-04-4 2,3-Dihydro-1,3-diphenyl-1*H*-1,3-benzodiphosphole; (1*RS*,3*SR*)-*form*, B,2MeBr, *in* D-00476

72110-12-4 1,2,3,4,6,7,8,9-Octahydro-1,9-diphenyl-5,9-benzothiadiphosphacycloundecin, O-00011

72110-13-5 1,2,3,4,6,7,8,9-Octahydro-1,9-diphenyl-5,1,9-benzoxadiphosphacycloundecin, O-00012

72110-14-6 2,3,4,5,6,7,8,9-Octahydro-1,9-diphenyl-1*H*-5,1,9-benzazadiphosphacycloundecine; (1*RS*,9*SR*)-*form*, 5-Me, *in* O-00009

72110-15-7 2,3,4,5,6,7,8,9-Octahydro-1,9-diphenyl-1*H*-5,1,9-benzazadiphosphacycloundecine; (1*RS*,9*SR*)-*form*, 5-Ph, *in* O-00009

72110-16-8 2,3,4,5,6,7,8,9-Octahydro-1,9-diphenyl-1*H*-1,9-benzodiphosphacycloundecin, O-00010

72138-53-7 Diphenyl-1-propenylphosphine; (*E*)-*form*, *in* D-01143

72151-42-9 1,2-Bis(diphenylphosphinomethyl)-cyclobutane; (1*RS*,2*RS*)-*form*, *in* B-00249

72151-45-2 1,2-Bis(diphenylphosphinomethyl)-cyclohexane; (1*RS*,2*SR*)-*form*, *in* B-00250

72163-96-3 Ethyl phenyl phosphoramidate; (+)-*form*, *in* E-00115

72163-97-4 Ethyl phenyl phosphoramidate; (−)-*form*, *in* E-00115

72170-80-0 Di-*tert*-butylphosphinothioic acid; Anhydride, *in* D-00118

72184-68-0 Triphenyl(propynylidene)-phosphorane, *in* T-00681

72217-80-2 2-Amino-3-phosphonobutanoic acid, A-00116

72236-74-9 3-Phosphonoxypropanenitrile; Monocyclohexylammonium salt, *in* P-00398

72250-10-3 Triethyl (azidosulfonyl)-phosphorimidate, T-00439

72282-84-9 Trifosfamide; (*R*)-*form*, *in* T-00485

72282-85-0 Trifosfamide; (*S*)-*form*, *in* T-00485

72284-31-2 *S,S*-Diethyl phosphorodithioate; Cyclohexylammonium salt, *in* D-00386

72284-32-3 *S,S*-Diisopropyl phosphorodithioate; Cyclohexylammonium salt, *in* D-00664

72284-34-5 *S,S*-Dibutyl phosphorodithioate; Cyclohexylammonium salt, *in* D-00152

72284-36-7 *S,S*-Bis(2-methylpropyl) phosphorodithioate; Cyclohexylammonium salt, *in* B-00361

72297-07-5 3-Nonenyltriphenyl-phosphonium(1+); (*Z*)-*form*, Iodide, *in* N-00072

72310-28-2 2-Chloronaphtho[1,8-*de*]-1,3,2-dioxaphosphorin, C-00120

72310-30-6 2-Ethoxynaphtho[1,8-*de*]-1,3,2-dioxaphosphorin, E-00045

72310-30-6 2-Ethoxynaphtho[1,8-*de*]-1,3,2-dioxaphosphorin, E-00046

72310-31-7 2-Phenoxynaphtho[1,8-*de*]-1,3,2-dioxaphosphorine, P-00070

72310-32-8 2-Diethylaminonaphtho[1,8-*de*]-1,3,2-dioxaphosphorin, D-00267

72311-12-7 [(3-Methoxyphenyl)methyl]-triphenylphosphonium(1+); Bromide, *in* M-00063

72344-34-4 (Difluoromethylene)-(trifluoromethyl)-phosphine, D-00413

72374-39-1 4,5-Dihydro-2-dimethylamino-1,3,2-benzodioxaphosphepin; 2-Oxide, *in* D-00454

72385-59-0 Cyclopentyltriphenyl-phosphonium(1+); Chloride, *in* C-00282

72450-49-8 Hexakis(dimethylamino)-μ-oxodiphosphorus(2+); Bis(trifluoromethanesulfonate), *in* H-00062

72492-61-6 2,6-Dimethyl-1,2,6-oxadiphosphorinane 2,6-dioxide, D-00780

72492-62-7 2,7-Dimethyl-1,2,7-oxadiphosphepane 2,7-dioxide, D-00779

72542-64-4 Neopentylene *tert*-butylphosphoramidothioate, *in* B-00538

72563-41-8 Bis(trimethylsilyl) (2-acetyloxyethyl)-phosphonate, *in* A-00006

72563-46-3 Ethyl 3-phosphonopropanoate, *in* P-00396

72564-40-0 2,4,5,6-Tetraphenyl-1,3,5-dioxaphosphorinane; 5-Oxide, *in* T-00254

72564-41-1 2,4,5,6-Tetraphenyl-1,3,5-dioxaphosphorinane; 5-Sulfide, *in* T-00254

72564-42-2 2,4,5-Triphenyl-1,3,4-oxathiaphospholane, T-00611

72564-42-2 2,4,5-Triphenyl-1,3,4-oxathiaphospholane; 4-Sulfide, *in* T-00611

72578-62-2 2-Chlorotetrahydro-3-(1-phenylethyl)-2*H*-1,3,2-oxazaphosphorine 2-oxide; (S_C,R_P)-*form*, *in* C-00181

72578-63-3 2-Chlorotetrahydro-3-(1-phenylethyl)-2*H*-1,3,2-oxazaphosphorine 2-oxide; (S_C,S_P)-*form*, *in* C-00181

72578-69-9 2-[(2-Chloroethyl)amino]-tetrahydro-2*H*-1,3,2-oxazaphosphorine 2-oxide; (*R*)-*form*, *in* C-00072

72578-70-2 2-[(2-Chloroethyl)amino]-tetrahydro-2*H*-1,3,2-oxazaphosphorine 2-oxide; (*S*)-*form*, *in* C-00072

72596-30-6 Diethyl(3,4-dimethylphenyl)-phosphonate, *in* D-00815

72596-34-0 Diethyl (2,3-dimethylphenyl)-phosphonate, *in* D-00811

72598-28-8 1-(2-Chloroethyl)tetrahydro-1*H*,5*H*-[1.3.2]-diazaphospholo[2,1-*b*][1.3.2]-oxazaphosphorine 9-oxide; (±)-*form*, *in* C-00082

72616-61-6 Diphenyl (2,2-diethoxyethyl)-phosphonate, *in* D-00234

72617-25-5 2,2-Dimethyl-1,3-dioxa-5-phospha-2-silacyclohexane; 5-Me, 5-sulfide, *in* D-00731

72649-11-7 Di-*tert*-butyl (triphenylphosphoranylidene)-butanedioate, *in* T-00646

72707-67-6 (2-Carboxy-2-propenyl)-triphenylphosphonium(1+); Bromide, *in* C-00016

72707-68-7 (2-Carboxy-2-propenyl)-triphenylphosphonium(1+); Me ester, bromide, *in* C-00016

72729-56-7 2,5-Diphenyl-1,3,5,2-dioxaphosphaborinane, D-01003

72740-04-6 Trimethylsilyl diisopropylphosphinite, *in* D-00648

72740-05-7 *P,P*-Diisopropyl-*N,N*-dimethylphosphinous amide, *in* D-00649

72741-17-4 Tri(1,2,4-triazol-1-yl)phosphine, T-00876

72741-18-5 1,1′,1″-Phosphinylidynetris(1*H*-1,2,4-triazole), *in* T-00876

72804-56-7 *O*-(Diphenylphosphinyl)-hydroxylamine, D-01097

72807-24-0 Triphenyl(1-phenylethyl)-phosphonium(1+); Iodide, *in* T-00612

72862-30-7 1,4-Bis(triphenyl-phosphoniomethyl)-naphthalene(2+); Dibromide, *in* B-00450

72897-05-3 Hexahydro-1,3,5-triphenyl-1,3,5-diazaphosphorine, H-00058

72918-87-7 (Carboxymethyl)-triphenylphosphonium(1+); Tetrafluoroborate, *in* C-00014

72954-06-4 1,1′-(Phenylphosphinidene)-ferrocene, P-00211

72954-09-7 1,1′-(Phenylphosphinidene)-ferrocene; Sulfide, *in* P-00211

72974-35-7 Ethylmethylphenylphosphine; (*S*)-*form*, *in* E-00084

72974-38-0 Ethylphenylphosphinothioic acid; (*S*)-*form*, Dicyclohexylammonium salt, *in* E-00105

72974-42-6 Methyl 2-(diethoxyphosphinyl)-benzoate, *in* D-00243

72974-99-3 Benzyltriphenylphosphonium(1+); Triodide, *in* B-00073

72974-99-3 Benzyltriphenylphosphonium(1+); Heptaiodide, *in* B-00073

72975-00-9 Benzyltriphenylphosphonium(1+); Pentaiodide, *in* B-00073

72975-01-0 Tetraiodophenoxyphosphorane, T-00164

72978-05-3 10,11-Dihydro-5-methyl-5*H*-dibenzo[*b,f*]phosphepin, D-00524

72978-08-6 10,11-Dihydro-5,5-dimethyl-5*H*-dibenzo[*b,f*]phosphepinium iodide, *in* D-00524

73013-46-4 Methyl (chloromethyl)-methylphosphinate, *in* C-00100

73075-73-7 Ethylene trimethylsilyl phosphate, *in* T-00554

73076-90-1 Methyl bis(trifluoromethyl)-phosphinoselenoite, *in* B-00413

73076-91-2 Methyl bis(trifluoromethyl)-phosphinotelluroite, *in* B-00414

73086-51-8 2-Chloro-1,2,3,4-tetrahydro-6-methyl-3-(4-methylphenyl)-1,3,2-benzodiazaphosphorine 2-oxide, C-00177

73098-19-8 *Se*-Methyl *O,O*-neopentylene phosphorodiselenoate, *in* D-00764

73125-47-0 Tetramethyl (2-oxo-1,3-propanediyl)bisphosphonate, *in* O-00072

73144-64-6 Ethyl 3-(ethylthio)-3-thioxo-2-(triphenylphosphoranylidene)-propanoate, *in* E-00194

73176-32-6 Ethylphenylphosphinothioic acid; (*R*)-*form*, *in* E-00105

73178-33-3 *O,O*-Diethyl benzylphosphonothioate, *in* B-00067

73178-34-4 *O,O*-Diethyl cyclohexylphosphonothioate, *in* C-00264

73186-67-1 3-Acetoxypropylphosphonic acid, A-00010

73210-33-0 Diphenyl(4-phenyl-1,3-butadienyl)phosphine; (*E,E*)-*form*, *in* D-01031

73210-72-7 1,2-Bis(phenylphosphino)ethane; (*RS,SR*)-*form*, Dioxide, *in* B-00390

73240-14-9 Antibiotic FR 33289; (*R*)-*form*, Na salt, *in* A-00138

73240-15-0 Antibiotic FR 33289; (*R*)-*form*, *in* A-00138

73240-16-1 Dimethyl [3-(acetylhydroxyimino)-2-hydroxypropyl]phosphonate, *in* A-00138

73270-05-0 *P,P*-Diethyl-*N,N*-bis(trimethylsilyl)-phosphinous amide, *in* D-00334

73270-42-5 Diphenyl (1-amino-2-methylpropyl)phosphonate, *in* A-00094

73270-51-6 *O*-Methyl phenylphosphonothioate; (*S*)-*form*, 1-Phenylethylammonium salt, *in* M-00243

73296-42-1 *N,N,P,P*-Tetramethyl-*N*′-trimethylsilylphosphinimidic amide, *in* D-00833

73326-54-2 (1,2-Phenylenedi-2-propene-3,1-diyl)bis[triphenyl-phosphonium](2+); (*E,E*)-*form*, Bis(tetrafluoroborate), *in* P-00149

73342-45-7 Methylpropylphosphinic acid, M-00375

73373-57-6 Bis(1*H*-benzimidazol-1-yl)-phosphinic acid, B-00097

73410-08-9 Triphenyl[(2,6,6-trimethyl-1-cyclohexen-1-yl)methyl]-phosphonium(1+); Chloride, *in* T-00689

73410-59-0 Hexahydro-2-hydroxycyclopenta[*d*]-1,3,2-dioxaphosphorin 2-oxide; (4*aRS*,7*aSR*)-*form*, *in* H-00034

73410-82-9 5-Methoxy-5-phospha(5-P^V)-spiro[4.4]nonane, *in* P-00350

73424-01-8 Hexahydro-2-hydroxycyclopenta[*d*]-1,3,2-dioxaphosphorin 2-oxide; (4*aRS*,7*aRS*)-*form*, *in* H-00034

73424-19-8 2-*tert*-Butyl-1,3-diisopropoxytriphosphetane 1,3-dioxide, *in* D-00627

73452-52-5 *N*-Hydroxy-*P,P*-diphenylphosphinic amide, H-00141

73469-37-1 1,1,3,3-Tetraethoxy-2-ethyltriphosphine 1,3-dioxide, T-00041

73473-51-5 Diethyl [(ethylimino)-methylethenyl]phosphonate, *in* E-00081

73502-97-3 2,2,4,6-Tetrafluoro-2,2,4,4,6,6-hexahydro-4,6-diphenyl-1,3,5-triaza-2,4,6-triphosphorine, T-00079

73551-19-6 *N*⁗-Ethyl-*N,N,N*′,*N*′,*N*″,*N*″,*N*‴,*N*‴-octamethylimidodiphosphorous tetraamide, *in* E-00079

73577-40-9 *O*-Methyl *N,N*-diethyl-*P*-phenylphosphonamidothioate, *in* D-00309

73577-42-1 *S*-Methyl *N,N*-diethyl-*P*-phenylphosphonamidothioate, *in* D-00309

73607-10-0 Triphenyl[bis(trifluoromethyl)-methylene]phosphorane, T-00597

73608-45-4 *O,O,O*′,*O*′-Tetraphenyl thiohypophosphate, T-00283

73627-86-8 2-Methyl-1,2-oxaphospholane; 2-Sulfide, *in* M-00182

73627-87-9 1,2-Thiaphospholane 2-sulfide; 2-Me, *in* T-00298

73678-99-6 3,4-Di-*tert*-butyl-2-(4-methylphenyl)-1,3,4,2-thiadiazaphospholidin-5-one 2-sulfide; (2α,3β,4α)-*form*, *in* D-00097

73719-56-9 1,2-Dimethyl-1,2-diphenyldiphosphine; (*RS,RS*)-*form*, Monosulfide, *in* D-00744

73719-57-0 1,2-Dimethyl-1,2-diphenyldiphosphine; (*RS,SR*)-*form*, Monosulfide, *in* D-00744

73719-60-5 1,2-Di-*tert*-butyl-1,2-dimethyldiphosphine; (*RS,SR*)-*form*, Disulfide, *in* D-00089

73720-00-0 Bis(2-methylphenyl)phosphinic acid; Anhydride, *in* B-00318

73726-76-8 Bis(4-*tert*-butylphenyl) phosphonate, B-00116

73731-13-2 Tetramethyldiphosphine; 1-Selenide 2-sulfide, *in* T-00200

73731-14-3 Tetramethyldiphosphine; Diselenide, *in* T-00200

73731-22-3 Bis(2,4,6-trimethylphenyl)-phosphinic acid; Anhydride, *in* B-00428

73749-41-4 2-Hydroxynaphtho[2,3-*d*]-1,3,2-dioxaphosphole 2-oxide, H-00164

73785-73-6 5,10-Dihydro-10-phenylphenophosphazine; 10-Oxide, *in* D-00563

73790-51-9 ▷*O*-Ethyl methylphosphonothioate; (±)-*form*, Dicyclohexylammonium salt, *in* E-00093

73834-61-4 2-Chlorotetrahydro-3-(1-phenylethyl)-2*H*-1,3,2-oxazaphosphorine 2-oxide; (R_C,R_P)-*form*, *in* C-00181

73834-62-5 2-Chlorotetrahydro-3-(1-phenylethyl)-2*H*-1,3,2-oxazaphosphorine 2-oxide; (R_C,S_P)-*form*, *in* C-00181

73870-70-9 3-Amino-3-phosphonobutanoic acid, A-00118

73892-37-2 1,2-Bis(diphenylphosphinomethyl)-cyclohexane, B-00250

73913-63-0 Phosphoserine; (*R*)-*form*, *in* P-00417

73946-92-6 Tri-1-imidazolylphosphine, T-00495

73954-60-6 *P*-Bromo-*P,P*-bis(diethylamino)-*N*-ethylphosphine imide, *in* P-00012

73954-61-7 *P*-Bromo-*P,P*-bis(dipropylamino)-*N*-propylphosphazene, *in* P-00043

73958-02-8 Triphenyl(2,5-undecadienylidene)phosphorane; (*Z,Z*)-*form*, *in* T-00700

73958-09-5 3-Nonenylidenetriphenyl-phosphorane, *in* N-00072

73992-81-1 3,5-Di-*tert*-butyl-2,3-dihydro-1,3,2-oxazaphosphole; 2-Ethoxy, *in* D-00083

73992-82-2 3,5-Di-*tert*-butyl-2,3-dihydro-1,3,2-oxazaphosphole; 2-Ethoxy, 2-oxide, *in* D-00083

74038-21-4 ▷Tris[3-(trifluoromethyl)-phenyl]phosphine; Oxide, *in* T-00853

74044-77-2 1,3-Bis(diphenylphosphino)butane; (*R*)-*form*, *in* B-00224

74078-07-2 2,3-Dihydro-2-phenyl-1*H*-dibenz[*e,g*]isophosphindole; 2-Oxide, *in* D-00558

74078-10-7 7-Phenyl-7*H*-dibenzo[*d,f*]-phosphonin; (*E,Z*)-*form*, 7-Oxide, *in* P-00111

74078-11-8 7-Phenyl-7*H*-dibenzo[*d,f*]-phosphonin; (*E,Z*)-*form*, *in* P-00111

74078-12-9 7-Methyl-7-phenyl-7*H*-dibenzo[*d*,*f*]phosphoninium iodide, *in* P-00111

74124-43-9 Dimethyl isopropylphosphoramidate, *in* I-00058

74124-44-0 Dimethyl benzylphosphoramidate, *in* B-00069

74124-46-2 Diisopropyl isopropylphosphoramidate, *in* I-00058

74124-48-4 Diisopropyl diethylphosphoramidate, *in* D-00343

74130-08-8 Dimethyl dibutylphosphoramidate, *in* D-00134

74144-31-3 1,1,3,3-Tetraphenyl-1,3-diphospholanium dibromide, *in* D-01013

74144-32-4 1,1,3,3-Tetraphenyl-1,3-diphosphorinanium(2+); Dibromide, *in* T-00262

74144-33-5 1,1,3,3-Tetraphenyl-1,3-diphosphepanium; Dibromide, *in* T-00257

74144-34-6 1,1,3,3-Tetrahydro-1,1,3,3-tetraphenyl-1,3-diphosphacyclopenta-1,2-diene, T-00150

74144-35-7 1,1,3,3-Tetrahydro-1,1,3,3-tetraphenyl-1,3-diphosphacyclohexa-1,2-diene, T-00149

74144-36-8 1,1,3,3-Tetrahydro-1,1,3,3-tetraphenyl-1,3-diphosphacyclohepta-1,2-diene, T-00148

74206-36-3 *O*,*O*-Diphenyl ethylphosphonothioate, *in* E-00148

74289-41-1 Phenoxytriphenylphosphorus(1+); Perchlorate, *in* P-00077

74292-99-2 Diethyl (1-aminoethyl)-phosphonate, *in* A-00078

74308-16-0 Ethyl (dichlorophosphinyl)-phosphorodichloroimidate, E-00058

74311-54-9 1-[1-(Dimethylamino)ethyl]-2-(diphenylphosphino)-ferrocene; $(R)_C(R)_{planar}$-*form*, *in* D-00697

74333-44-1 (1-Aminoethyl)phosphinic acid, A-00077

74378-76-0 2-Hydroxyhexahydro-4*H*-1,3,2-benzodioxaphosphorin 2-oxide; (2α,4aβ,8aα)-*form*, 2-(4-Methoxyphenoxy) (2-Methoxyphenyl ester), *in* H-00152

74378-77-1 2-Hydroxyhexahydro-4*H*-1,3,2-benzodioxaphosphorin 2-oxide; (2α,4aβ,8aα)-*form*, 2-Phenoxy (2-Ph ester), *in* H-00152

74378-78-2 2-Hydroxyhexahydro-4*H*-1,3,2-benzodioxaphosphorin 2-oxide; (2α,4aα,8aβ)-*form*, 2-(2,4-Dinitrophenoxy), *in* H-00152

74378-80-6 Tetrahydro-2*H*,6*H*-[1.3.2]-oxazaphosphorino[2,3-*b*]-[1.3.2]oxazaphosphorine, T-00120

74410-65-4 2-Hydroxyhexahydro-4*H*-1,3,2-benzodioxaphosphorin 2-oxide; (2α,4aα,8aβ)-*form*, 2-(4-Methoxyphenoxy), *in* H-00152

74410-66-5 2-Hydroxyhexahydro-4*H*-1,3,2-benzodioxaphosphorin 2-oxide; (2α,4aα,8aβ)-*form*, 2-Phenoxy, *in* H-00152

74410-72-3 2-Chlorohexahydro-4*H*-1,3,2-benzodioxaphosphorin; (2*RS*,4a*SR*,8a*SR*)-*form*, 2-Oxide, *in* C-00085

74411-89-5 1,3,5,7-Tetramethyl-1,3,5,7-tetraaza-4-phospha(4-P^V)spiro[3.3]heptane-2,6-dione; 4-Ph, *in* T-00234

74431-09-7 2-Hydroxyhexahydro-4*H*-1,3,2-benzodioxaphosphorin 2-oxide; (2α,4aβ,8aα)-*form*, 2-(2,4-Dinitrophenoxy) (2-(2,4-Dinitrophenyl)ester), *in* H-00152

74483-17-3 [Phenyl(trimethylsilyl)-methylene]phosphinous chloride, P-00330

74514-55-9 4,4-Diphosphonobutanoic acid, D-01183

74536-91-7 *O*-Ethyl *O*,*O*-1,8-naphthylene thiophosphate, *in* E-00045

74536-91-7 *O*-Ethyl *O*,*O*-1,8-naphthylene thiophosphate, *in* E-00046

74560-09-1 2-Hydroxy-4,5-dimethyl-1,3,2-dioxaphospholane 2-sulfide; (2α,4β,5β)-*form*, 1*H*-Imidazolium salt, *in* H-00122

74605-47-3 4,4-Diethoxy-2,4,4,5-tetrahydro-2-oxo-5-(4-nitrophenyl)-1,3,4-oxazaphosphole, D-00262

74607-62-8 Hexahydro-1,3,5-triphenyl-1,3,5-diazaphosphorine; 5-Oxide, *in* H-00058

74607-65-1 Hexahydro-1,3,5-triphenyl-1,3,5-diazaphosphorine; 5-Sulfide, *in* H-00058

74633-13-9 1,3-Diphenyl-1,3-azaphosphetidine, *in* P-00087

74633-16-2 1-Benzyl-3-phenyl-1,3-azaphosphetidine, *in* P-00087

74659-21-5 2,5-Diphenyl-1,3,5,2-dioxaphosphaborinane; 5-Sulfide, *in* D-01003

74676-24-7 1-Ethoxy-2,3,4,5-tetraphenyl-1*H*-phosphole 1-oxide, *in* H-00200

74676-25-8 1-Methoxy-2,3,4,5-tetraphenyl-1*H*-phosphole 1-oxide, *in* H-00200

74676-26-9 1-Phenoxy-2,3,4,5-tetraphenyl-1*H*-phosphole 1-oxide, *in* H-00200

74700-09-7 4-Morpholinylphosphonothioic acid; Dichloride, *in* M-00527

74700-10-0 Phenyl 4-morpholinylphosphonochloridate, *in* M-00524

74700-13-3 *O*-Phenyl 4-morpholinylphosphonochloridothioate, *in* M-00525

74720-40-4 2-Phenyl-1,3,6,2-dioxathiaphosphocane; 2-Oxide, *in* P-00117

74720-44-8 2-Phenyl-1,3,6,2-dioxathiaphosphocane; 2-Sulfide, *in* P-00117

74722-07-9 Di-*tert*-butylphosphinic acid; Azide, *in* D-00110

74722-24-0 3,4-Dihydro-3-phenyl-1*H*-2,3-benzoxaphosphorin; 3-Oxide, *in* D-00554

74722-25-1 3,4-Dihydro-3-phenyl-1*H*-2,3-benzoxaphosphorin; 3-Sulfide, *in* D-00554

74737-35-2 Hexahydro-2,4-dimethyl-3-phenoxy-1,2,4,5,3-tetrazaphosphorine; 3-Oxide, *in* H-00024

74783-07-6 4,7,13,16-Tetraphenyl-1,10-dioxa-4,7,13,16-tetraphosphacyclooctadecane; (4*RS*,7*RS*,13*RS*,16*SR*)-*form*, *in* T-00255

74785-77-6 Acetyl phosphate; Monoanilinium salt, *in* A-00025

74806-16-9 (Aminophenylmethyl)-phenylphosphinic acid; (±)-*form*, *in* A-00102

74807-53-7 4,7,13,16-Tetraphenyl-1,10-dioxa-4,7,13,16-tetraphosphacyclooctadecane; (4*RS*,7*RS*,13*SR*,16*SR*)-*form*, *in* T-00255

74841-97-7 (Aminophenylmethyl)-phenylphosphinic acid; $(R)_C$-*form*, Et ester, *in* A-00102

74841-98-8 (Aminophenylmethyl)-phenylphosphinic acid; $(R)_C$-*form*, *in* A-00102

74860-85-8 Tris[2-(trifluoromethyl)-phenyl]phosphine; Oxide, *in* T-00852

74891-95-5 (1-Aminoethyl)methylphosphinic acid; (±)-*form*, *in* A-00076

74927-63-2 1′,3′-Diphenyldispiro[1,3,2-benzodioxaphosphole-2,2′-[1,3,2,4]diazaphosphetidine-4′,2″-[1,3,2]-benzodioxaphosphole]; 2,4′-Dimethoxy, *in* D-01018

74927-64-3 1′,3′-Diphenyldispiro[1,3,2-benzodioxaphosphole-2,2′-[1,3,2,4]diazaphosphetidine-4′,2″-[1,3,2]-benzodioxaphosphole]; 2,4′-Diethoxy, *in* D-01018

74927-66-5 1′,3′-Diphenyldispiro[1,3,2-benzodioxaphosphole-2,2′-[1,3,2,4]diazaphosphetidine-4′,2″-[1,3,2]-benzodioxaphosphole]; 2,4′-Diphenoxy, *in* D-01018

74930-70-4 3*a*,4,7,7*a*-Tetrahydro-5,6-dimethyl-2,3*a*,7*a*,8-tetrakis(trifluoromethyl)-1,2,3-metheno-2*H*-benzodiphosphole, T-00100

74930-73-7 7-Phenyl-1,3,4,6-tetrakis(trifluoromethyl)-7-aza-2,5-diphosphatetracyclo-[4.1.0.$0^{2,4}$.$0^{3,5}$]heptane, P-00321

74974-14-4 2,2′-Bis[(diphenylphosphino)-amino]-1,1′-binaphthyl; (*R*)-*form*, *in* B-00217

74974-15-5 2,2′-Bis[(diphenylphosphino)-amino]-1,1′-binaphthyl; (*S*)-*form*, *in* B-00217

75045-09-9 Dimethyl 2-chloro-1,3,2-dioxaphospholan-4,5-dicarboxylate, *in* C-00062

75080-72-7 Spiro[1,3,2-benzodioxaphosphole-2,2′(3′*H*)-[1,3,2]-benzoxazaphosphole]; 2-Chloro, *in* S-00008

75080-73-8 *N*-Phenylspiro[1,3,2-benzodioxaphosphole-2,2′(3′*H*)-[1,3,2]-benzoxazaphosphol]-2-amine, *in* S-00008

75080-74-9 Spiro[1,3,2-benzodioxaphosphole-2,2′(3′*H*)-[1,3,2]-benzoxazaphosphole]; 2-Phenoxy, *in* S-00008

75080-75-0 Spiro[1,3,2-benzodioxaphosphole-2,2′(3′*H*)-[1,3,2]-benzoxazaphosphole]; 2-Methoxy, *in* S-00008

75135-39-6 1,3-Dibutyl-2,3-dihydro-4,5-dimethyl-1*H*-1,3,2-diazaphosphole 2-oxide; 2-Ethoxy, *in* D-00081

75135-40-9 1,3-Dibutyl-2,3-dihydro-4,5-dimethyl-1*H*-1,3,2-diazaphosphole 2-oxide; 2-Chloro, *in* D-00081

75213-01-3 *tert*-Butylphenylphosphinic acid; (*S*)-*form*, Chloride, *in* B-00579

75213-02-4 *tert*-Butylphenylphosphinic acid; (*R*)-*form*, Chloride, *in* B-00579

75231-53-7 Methyl diphenylphosphinothioite, *in* D-01085

75231-66-2 5-Chloro-10,11-dihydro-5*H*-dibenzo[*b,f*]phosphepin 5-oxide, *in* D-00492

75231-67-3 10,11-Dihydro-5-methoxy-5*H*-dibenzo[*b,f*]phosphepin 5-oxide, *in* D-00492

75231-73-1 5-Hydroxy-5*H*-dibenzo[*b,f*]-phosphepin 5-oxide, H-00117

75231-74-2 5-Chloro-5*H*-dibenzo[*b,f*]-phosphepin 5-oxide, *in* H-00117

75231-75-3 5-Methoxy-5*H*-dibenzo[*b,f*]-phosphepin 5-oxide, *in* H-00117

75238-07-2 5-Phenyl-5*H*-dibenzo[*b,f*]-phosphepin; Sulfide, *in* P-00108

75263-36-4 2-Thienylphosphinic acid, T-00306

75271-90-8 Dibutyliodophosphine, *in* D-00121

75281-59-3 Ethylphenylphosphinous acid; (±)-*form*, Diethylamide, *in* E-00108

75292-68-1 *N,N,N′,N′*-Tetraisopropylphosphonic diamide, T-00171

75340-94-2 Tetraisopropyl hypophosphate, T-00170

75354-35-7 Cyclohexylphenylphosphine; Oxide, *in* C-00249

75355-36-1 Diethyl [(8-quinolinyl)methyl]-phosphonate, *in* Q-00005

75355-39-4 [(8-Quinolinyl)methyl]phosphonic acid, Q-00005

75373-19-2 1,4-Dioxa-6,9-dithia-5-phospha(5-P^{V})spiro[4.4]-nonane; 5-Methoxy, *in* D-00964

75373-22-7 *S,S*-Ethylene *N,N*-dimethyl phosphoramidodithioate, *in* D-01244

75374-08-2 2,2,2,3-Tetrahydro-5-phenyl-3,3-bis(trifluoromethyl)-1,4,2-thiazaphosphole; 2,2,2-Triphenoxy, *in* T-00129

75378-51-7 Ethyl 4-(phosphonoxy)benzoate, *in* P-00397

75386-25-3 *N,N*,2,2,3,3-Hexamethyl-1,4,6-trioxa-9-aza-5-phospha(5-P^{V})-spiro[4.4]nonan-5-amine, *in* T-00238

75401-31-9 1-Methylphosphonane; 1-Oxide, *in* M-00285

75401-43-3 7-Phenyl-7*H*-dibenzo[*d,f*]-phosphonin, P-00111

75401-44-4 1,1-Dimethylphosphonanium iodide, *in* M-00285

75403-51-9 Octahydro-1,9-dimethyl-[1,3,2]-diazaphosphorino[1,2-*a*]-[1,3,2]diazaphosphorine, O-00005

75404-54-5 4,4′,4″-Phosphinotelluroylidynetrismorpholine, *in* T-00569

75404-55-6 1,1′,1″-Phosphinotelluroylidynetrispiperidine, *in* T-00704

75404-56-7 1,1′,1″-Phosphinotelluroylidynetrispyrrolidine, *in* T-00724

75404-57-8 1,1′,1″-Phosphinoselenoylidynetrispyrrolidine, *in* T-00724

75415-25-7 3-Nonenyltriphenylphosphonium(1+); (*Z*)-*form*, Chloride, *in* N-00072

75421-31-7 1-Cyclohexyl-1,2-bis(diphenylphosphino)-ethane; (*R*)-*form*, *in* C-00244

75482-41-6 4-Ethoxy-1,4-dihydro-1,3,5-triphenyl-1,2,4-diazaphosphorine 4-oxide, E-00029

75492-61-4 (2-Methylphenyl)-diphenylphosphine; Selenide, *in* M-00200

75502-78-2 Bis(trimethylsilyl) (2-hydroxyethyl)-phosphonate, *in* H-00146

75502-79-3 (1-Hydroxyethyl)phosphonic acid; (±)-*form*, Monoanilinium salt, *in* H-00145

75502-79-3 (2-Hydroxyethyl)phosphonic acid; Monoanilinium salt, *in* H-00146

75502-80-6 (2-Hydroxyethyl)phosphonic acid; Mono-dicyclohexylammonium salt, *in* H-00146

75502-81-7 (2-Acetoxyethyl)phosphonic acid; Monoanilinium salt, *in* A-00006

75502-82-8 (2-Acetoxyethyl)phosphonic acid; Mono-dicyclohexylammonium salt, *in* A-00006

75511-33-0 *O*-Isopropyl *O,S*-dimethyl phosphorothioate, I-00030

75513-55-2 Diphenyl succinimido phosphate, D-01148

75543-48-5 5-Benzyl-5,6-dihydro-6-phenylphosphanthridine; (5*RS*,6*SR*)-*form*, 5-Sulfide, *in* B-00043

75543-49-6 5-Benzyl-5,6-dihydro-6-phenylphosphanthridine; (5*RS*,6*RS*)-*form*, 5-Sulfide, *in* B-00043

75543-51-0 5-Benzyl-5,6-dihydro-6-phenylphosphanthridine; (5*RS*,6*RS*)-*form*, *in* B-00043

75543-52-1 5-Benzyl-5,6-dihydro-6-phenylphosphanthridine; (5*RS*,6*SR*)-*form*, *in* B-00043

75543-58-7 2,4,10-Trimethyl-1,3,4,5,7,10-hexaaza-3-phosphatricyclo[3.3.1.$1^{3,7}$]decane, T-00517

75550-55-9 4,6-Di-*tert*-butylspiro[1,3,2-benzodioxaphosphole-2,1′-[2,8,9]-trioxa[1]-phosphaadamantane], D-00157

75558-19-9 1,2,5,6-Tetrahydro-1,1,6,6,9-pentamethyl-4*H*-phospholo[3,2,1-*ij*]phosphinoline, T-00123

75558-20-2 2,3,6,7-Tetrahydro-1,1,7,7,9-pentamethyl-1*H*,5*H*-phosphorino[3,2,1-*ij*]-phosphinoline, T-00124

75558-26-8 1,2,5,6-Tetrahydro-1,1,6,6,9-pentamethyl-4*H*-phospholo[3,2,1-*ij*]phosphinoline; 3-Oxide, *in* T-00123

75558-27-9 1,2,5,6-Tetrahydro-1,1,3,6,6,9-hexamethyl-4*H*-phospholo[3,2,1-*ij*]-phosphinolinium iodide, *in* T-00123

75558-32-6 2,3,6,7-Tetrahydro-1,1,7,7,9-pentamethyl-1*H*,5*H*-phosphorino[3,2,1-*ij*]-phosphinoline; 4-Oxide, *in* T-00124

75593-74-7 Octahydro-1,3,5,7-tetraphenyl-1,5,3,7-diazadiphosphocine, O-00026

75600-55-4 2,3-Dihydro-1,2,3,4,5-pentaphenyl-1*H*-1,2,3-triphosphole, D-00539

75600-57-6 2,3-Dihydro-1,2,3-triphenyl-1*H*-1,2,3-triphosphole, *in* D-00615

75619-23-7 4′-Ethyl-1,3-dihydro-5-phenyl-3,3-bis(trifluoromethyl)spiro[2*H*-1,4,2-diazaphosphole-2,1′-[2,6,7]-trioxa[1]phosphabicyclo[2.2.2]-octane]; 1-(2,4,6-Trimethyl-phenyl), *in* E-00061

75620-16-5 Monomethyl phenylphosphonate; (±)-*form*, Mono-dicyclohexylammonium salt, *in* M-00488

75716-68-6 2-Mercapto-5,5-dimethyl-1,3,2-dioxaphosphorinane 2-oxide; Methyltriethylammonium salt, *in* M-00010

75770-58-0 Diphenylphosphinic acid; Bromide, *in* D-01039

75777-29-6 1,3-Dihydro-1-phenyl-2,1-benzoxaphosphole; 1-Oxide, *in* D-00553

75777-31-0 1,3-Dihydro-1-hydroxy-2,1-benzoxaphosphole 1-oxide; 1-Ethoxy (Et ester), *in* D-00490

75777-32-1 1,3-Dihydro-1-hydroxy-2,1-benzoxaphosphole 1-oxide, D-00490

75779-67-8 2,5-Dihydro-5,5-dimethyl-1,2-oxaphosphole 2-oxide; 2-Chloro, *in* D-00464

75779-68-9 (3-Methyl-1,3-butadienyl)-phosphonic acid; (*Z*)-*form*, Dichloride, *in* M-00095

75779-78-1 Propylphosphinic acid, P-00472

75780-25-5 6-Phospha(6-P^{V})spiro[5.5]-undeca-1(6)-ene, P-00351

75780-26-6 1,3,4,5,5,6,7,8,9,9*a*-Decahydro-5-methylene-2*H*-phosphinolizine, D-00005

75780-28-8 6-Phosphoniaspiro[5.5]undecane; Chloride, *in* P-00375

75794-09-1 *N*-Me, *in* T-00339

75794-10-4 *P,P,P*-Tricyclohexyl-*N*-methylphosphine imide, *in* T-00427

75860-43-4 (2-Cyanoethylidene)-triphenylphosphorane, *in* C-00212

75861-17-5 2,3-Dihydro-1*H*-naphtho[1,8-*de*]-1,3,2-diazaphosphorine; 2-(4-Nitrophenoxy), 2-oxide, *in* D-00532

75889-62-2 Fostedil, F-00064

75905-80-5 Diphenyl [4-(ethoxycarbonyl)-phenyl]phosphoramidate, *in* C-00006

75918-60-4 Octahydro-3,8-dimethyl-2,7-diphenyl-1,6,3,8,2,7-dioxadiazadiphosphecine, O-00006

75944-12-6 Triphenyl [bis(dimethylamino)-phosphinyl]phosphorimidate, *in* B-00194

75956-77-3 Dimethyl methylphosphonodithioite, *in* M-00302

75956-78-4 Methyl di-*tert*-butylphosphinothioite, *in* D-00120

76001-34-8 3,3-Dichloro-1,1,2,2-tetrahydro-1,1,1,2,2,2-hexaphenyldiphosphirane, D-00193

76010-93-0 5,6,7,8,9,14,15,16,17,18-Decahydro-9,14-dimethyl-5,18-diphenyldibenzo[*b,i*][1,4,8,11]-diphosphadiarsacyclotetradecin; (5*RS*,9*RS*,14*RS*,18*SR*)-*form*, *in* D-00004

76037-02-0 2,4-Di-*tert*-butyl-3-chloro-1,2,4,3-thiadiazaphosphetidine; 1,1-Dioxide, *in* D-00079

76037-04-2 3-Amino-2,3-di-*tert*-butyl-1,2,4,3-thiadiazaphosphetidine; *N,N*-Di-Me, 1,1-dioxide, *in* A-00071

76045-08-4 2-Mercaptodibenzo[*d,f*][1,3,2]-dioxaphosphepin 2-oxide; NH_4 salt, *in* M-00007

76045-15-3 2,2′-Biphenyldiyl methyl phosphate, *in* M-00032

76062-43-6 4*b*,5,6,10*b*,11,12-Hexahydro-5,5-diphenylbenzo[*c*]-phosphanthridinium hexafluorophosphate, *in* H-00045

76065-67-3 4-Chloro-1,2,3,4-tetrahydronaphth[2,1-*c*]-[1,2]azaphosphorine; Sulfide, *in* C-00179

76068-72-9 1-Phenyl-1-(triphenylphosphoranylidene)-2-propanone, P-00334

76095-39-1 Moenomycin; Moenomycin *A*, *in* M-00441

76101-29-6 Benzyl phosphorodichloridite, B-00070

76144-87-1 2,2′-Bis(diphenylphosphino)-1,1′-binaphthyl; (±)-*form*, *in* B-00221
76152-72-2 3-Phenyl-3-phosphabicyclo[3.2.1]octane; *endo-form*, 3-Oxide, *in* P-00198
76152-73-0 3-Phenyl-3-phosphabicyclo[3.2.1]octane; *endo-form*, *in* P-00198
76156-74-6 1,2,6-Triphenyl-4-phosphorinanone; (1α,2α,6β)-*form*, Oxide, *in* T-00666
76156-75-7 1,2,6-Triphenyl-4-phosphorinanone; (1α,2α,6β)-*form*, Sulfide, *in* T-00666
76164-17-5 2-Hydroxy-2*H*-1,2-benzoxaphosphorin 2-oxide, H-00112
76168-01-9 Diphenyl (3-methylphenyl)-phosphoramidate, *in* M-00252
76174-11-3 2,2,2,3-Tetrahydro-5-phenyl-1,3,4,2-oxadiazaphosphole; 2,2,2-Trifluoro, *in* T-00131
76179-26-5 Methyl 2-[(diethoxyphosphinyl)-oxy]-2-propenoate, *in* D-00256
76189-55-4 2,2′-Bis(diphenylphosphino)-1,1′-binaphthyl; (*R*)-*form*, *in* B-00221
76189-56-5 2,2′-Bis(diphenylphosphino)-1,1′-binaphthyl; (*S*)-*form*, *in* B-00221
76189-76-9 1,2,6-Triphenyl-4-phosphorinanone; (1α,2α,6β)-*form*, *in* T-00666
76189-77-0 1,2,6-Triphenyl-4-phosphorinanone; (1α,2β,6β)-*form*, *in* T-00666
76195-19-2 1-Bromo-*N*-methyl-*N*,1,1,1-tetraphenylphosphoranamine, *in* M-00194
76232-77-4 1,3,4,5-Tetrahydro-4,5-dimethyl-2,5-methano-2*H*-2-benzophosphepin; Oxide, *in* T-00096
76232-78-5 1,3,4,5-Tetrahydro-4,5-dimethyl-2,5-methano-2*H*-2-benzophosphepin, T-00096
76232-79-6 1,3,4,5-Tetrahydro-2,4,5-trimethyl-2,5-methano-2*H*-2-benzophosphepinium iodide, *in* T-00096
76254-55-2 1-Hydroxy-1,1-hexanediphosphonic acid, H-00153
76257-42-6 Diisopropyl *tert*-butylphosphonite, *in* B-00616
76274-77-6 (2-Hydroxypropyl)phosphonic acid, H-00191
76287-39-3 2,3,5-Trimethyl-3-phosphatricyclo[4.2.1.$0^{2,5}$]-nonane; Oxide, *in* T-00529
76297-11-5 Diisopropyl butylphosphonite, *in* B-00615
76373-02-9 (1,3-Butadienyl)-triphenylphosphonium(1+); Chloride, *in* B-00512
76374-09-9 1-[1-(Dimethylamino)ethyl]-2-(diphenylphosphino)-ferrocene; $(R)_C(S)_{planar}$-*form*, $PdCl_2$ complex, *in* D-00697
76380-84-2 Methyl 1-naphthylphenylphosphinate, *in* N-00017
76380-85-3 Isopropylphenylphosphinothioic acid; (±)-*form*, *S*-Me ester, *in* I-00046
76380-87-5 *O*-Phenyl methylphenylphosphinothioate, *in* M-00225
76380-88-6 Isopropylphenylphosphinothioic acid; (±)-*form*, *in* I-00046
76380-89-7 Isopropylphenylphosphinothioic acid; (±)-*form*, Dicyclohexylammonium salt, *in* I-00046
76380-91-1 Isopropylphenylphosphinothioic acid; (+)-*form*, Dicyclohexylammonium salt, *in* I-00046
76380-92-2 Isopropylphenylphosphinothioic acid; (+)-*form*, *in* I-00046
76420-29-6 Methyl isopropylphenylphosphinate, *in* I-00043
76420-32-1 Phenyl methylphenylphosphinate, *in* M-00219
76420-33-2 Methylphenylphosphinothioic acid; (±)-*form*, *S*-Me ester, *in* M-00225
76420-34-3 Ethylphenylphosphinothioic acid; (±)-*form*, *S*-Me ester, *in* E-00105
76420-35-4 *S*-Phenyl methylphenylphosphinothioate, *in* M-00225
76420-36-5 Methylphenylphosphinothioic acid; (±)-*form*, *O*-Me ester, *in* M-00225
76420-37-6 *tert*-Butylphenylphosphinothioic acid; (±)-*form*, *O*-Me ester, *in* B-00583
76442-57-4 7-Chloro-6-thia-7-phosphabicyclo[3.2.1]-octane; 7-Sulfide, *in* C-00187
76442-61-0 2-Chloro-3,6-dihydro-2*H*-1,2-thiaphosphorin 2-sulfide, C-00053
76442-64-3 5-Methyl-1,2-thiaphospholane 2-sulfide; 2-Chloro, *in* M-00407
76457-37-9 2-Fluoro-1,3-dimethyl-1,3,2-diazaphospholidine; 2-Oxide, *in* F-00014
76509-66-5 Diisopropyl ethylphosphonite, *in* E-00153
76527-75-8 *N*-[(Diphenylphosphinyl)methyl]-*N*-methylaniline, D-01100
76640-63-4 (2,4,6-Cycloheptatrien-1-ylmethyl)triphenyl-phosphonium(1+); Tetrafluoroborate, *in* C-00237
76659-90-0 Dimethyl (4-cyanophenyl)-phosphonate, *in* C-00225
76682-26-3 [5-(4-Hydroxy-2,6,6-trimethyl-1-cyclohexen-1-yl)-3-methyl-2,4-pentadienyl]-triphenylphosphonium(1+); (*R*)-(*E,E*)-*form*, Bromide, *in* H-00202
76684-21-4 (Phenylmethylidyne)phosphine, P-00170
76686-23-2 [5-(4-Hydroxy-2,6,6-trimethyl-1-cyclohexen-1-yl)-3-methyl-2,4-pentadienyl]-triphenylphosphonium(1+); (*S*)-(*E,E*)-*form*, Chloride, *in* H-00202
76686-31-2 [5-(4-Hydroxy-2,6,6-trimethyl-1-cyclohexen-1-yl)-3-methyl-2,4-pentadienyl]-triphenylphosphonium(1+); (*R*)-(*E,E*)-*form*, Chloride, *in* H-00202
76740-45-9 5,6-Bis(diphenylphosphino)-bicyclo[2.2.1]hept-2-ene; (1*RS*,5*RS*,6*RS*)-*form*, *in* B-00220
76773-16-5 Trimethylphosphine; Telluride, *in* T-00530
76795-74-9 6-Methyl-4-methylene-4*H*-1,3,2-dioxaphosphorin; 2-Ph, *in* M-00167
76819-59-5 *tert*-Butyl 1,3-butylene phosphate, *in* B-00537
76819-63-1 *tert*-Butyl ethylene phosphate, *in* B-00535
76835-82-0 Butyltriphenylphosphonium(1+); Triiodide, *in* B-00641
76835-83-1 Butyltriphenylphosphonium(1+); Pentaiodide, *in* B-00641
76835-84-2 Butyltriphenylphosphonium(1+); Heptaiodide, *in* B-00641
76842-91-6 Ethyl 4-(diphenylphosphinyl)-3-oxobutanoate, *in* D-01102
76905-66-3 Methyl dicyclohexylphosphinate, *in* D-00211
77013-89-9 Tris(2-hydroxyphenyl)phosphine, T-00788
77013-92-4 2-*tert*-Butyl-1,3-benzoxaphosphole, B-00544
77047-42-8 Diethyl (4-pyridinylmethyl)-phosphonate, *in* P-00507
77075-77-5 2-*tert*-Butyl-1,3,6,2-trithiaphosphocane, B-00642
77075-78-6 2-*tert*-Butyl-1,3,6,2-trithiaphosphocane; 2-Oxide, *in* B-00642
77075-96-8 2,10-Di-*tert*-butyl-1,3,6,9,11,14-hexathia-2,10-diphosphacyclohexadecane; *cis-form*, *in* D-00093
77075-97-9 2,10-Di-*tert*-butyl-1,3,6,9,11,14-hexathia-2,10-diphosphacyclohexadecane; *trans-form*, 2,10-Dioxide, *in* D-00093
77075-98-0 2,10-Di-*tert*-butyl-1,3,6,9,11,14-hexathia-2,10-diphosphacyclohexadecane; *cis-form*, 2,10-Disulfide, *in* D-00093
77113-17-8 2-Furanylphosphonic acid, F-00073
77113-29-2 Bis(trimethylsilyl) 2-furanylphosphonate, *in* F-00073
77121-94-9 2,10-Di-*tert*-butyl-1,3,6,9,11,14-hexathia-2,10-diphosphacyclohexadecane; *trans-form*, *in* D-00093
77121-95-0 2,10-Di-*tert*-butyl-1,3,6,9,11,14-hexathia-2,10-diphosphacyclohexadecane; *trans-form*, 2,10-Disulfide, *in* D-00093
77123-34-3 2,10-Di-*tert*-butyl-1,3,6,9,11,14-hexathia-2,10-diphosphacyclohexadecane; *cis-form*, 2,10-Dioxide, *in* D-00093
77145-22-5 (4-Methylphenyl)methylphosphinic acid; $(S)_P$-*form*, (−)-Menthyl ester, *in* M-00204
77145-23-4 (4-Methylphenyl)methylphosphinic acid; (±)-*form*, Chloride, *in* M-00204
77173-37-6 Ethyl hydrogen (4-hydroxyphenyl)phosphonate, *in* H-00181
77173-40-1 Monoethyl phenylphosphonate; (±)-*form*, Dicyclohexylammonium salt, *in* M-00471
77173-41-2 *S*-Methyl phenylphosphonothioate, M-00244
77181-80-7 4-Chlorophenyl 5-chloro-8-quinolinyl phosphorochloridate, *in* M-00456
77205-54-0 5,10-Dihydro-5,10-dimethylphenophosphazine; 10-Sulfide, *in* D-00466
77318-59-3 Tetramethyl-phosphorodiamidodithioic acid; Dimethylammonium salt, *in* T-00222
77326-05-7 *O*-Methyl phenylphosphonothioate; (*S*)-*form*, *S*-Me ester, *in* M-00243
77326-10-4 Hexylphosphonothioic acid, H-00100
77339-53-8 *P*-(Trichloromethyl)-*N*,*N*,*N*′,*N*′-tetraethylphosphonic diamide, *in* T-00412
77339-54-9 *N*,*N*,*N*′,*N*′-Tetraethyl-*P*-(trichloromethyl)-phosphonimidic diamide, T-00075
77339-55-0 Aminobis(diethylamino)-(trichloromethyl)-phosphonium chloride, *in* T-00075

77339-59-4 N″-Chloro-N,N,N′,N′-tetraethyl-P-(trichloromethyl)-phosphonimidic diamide, *in* T-00075
77339-64-1 Cyclohexylphosphonothioic acid; O,O-Bis(trimethylsilyl) ester, *in* C-00264
77339-65-2 O,O-Bis(trimethylsilyl) hexylphosphonothioate, *in* H-00100
77339-66-3 Heptylphosphonothioic acid; O,O-Bis(trimethylsilyl) ester, *in* H-00007
77339-68-5 Heptylphosphonothioic acid; Monoanilinium salt, *in* H-00007
77339-71-0 Trimethylsilyl cyclohexylphosphinate, *in* C-00253
77382-31-1 1,3-Di-*tert*-butyl-4,4-dimethyl-1,3,2-diazaphosphasiletidine; 2-Chloro, *in* D-00087
77432-49-6 1,3-Dithiolan-2-yltriphenylphosphonium(1+); Tetrafluoroborate, *in* D-01253
77490-41-6 1,1′,3,3′-Tetrahydro-1,1′,2,3,3′-pentamethyl-2,2′-spirobi[2H-1,3,2-benzodiazaphosphole], *in* T-00142
77507-70-1 1,3,5,7-Tetramethyl-1,3,5,7-tetraaza-4-phospha(4-P^V)spiro[3.3]heptane-2,6-dione; 4-Chloro, *in* T-00234
77526-90-1 Diethyl (2-bromophenyl)-phosphonate, *in* B-00490
77547-92-3 Monoisopropyl phosphonate; Monocyclohexylammonium salt, *in* M-00478
77547-94-5 O,S-Dimethyl phosphorodithioate; (±)-*form*, Cyclohexylammonium salt, *in* D-00899
77585-91-2 O,O-Trimethylene phosphorochloridoselenoate, *in* C-00066
77614-73-4 Tri-*tert*-butyldiphospharsirane, T-00363
77655-19-7 1,3-Dimethyl-5,7-diphenyl-1,3,5,7-tetraaza-4-phospha(4-P^V)spiro[3.3]-heptane-2,6-dione; 4-Chloro, *in* D-00748
77659-30-4 O,O-Diethyl (2-propenyl)-phosphonoselenoate, *in* P-00458
77671-18-2 O,O-(2,2′-Biphenylyl) phosphorochloridoselenoate, *in* C-00032
77706-35-5 (3,7-Dimethyl-2,6-octadienylidene)-triphenylphosphorane, *in* D-00777
77713-27-0 Phenyl N,N,N′,N′-tetraethyl-N″-phenyl-phosphorodiamidimidate, *in* T-00054
77748-20-0 O,O-Diethyl SS-phenyl phosphoro(dithioperoxo)-thioate, D-00315
77761-66-1 Dicyclohexyl (2-diethylamino-2-oxoethyl)phosphonate, *in* D-00268
77785-52-5 (Cyanomethyl)-triphenylphosphonium(1+); Iodide, *in* C-00217
77801-12-8 9,9-Dimethyl-2,8-diphenyl-3,7-dioxa-1-phosphabicyclo[3.3.1]-nonane; 1-Sulfide, *in* D-00743
77815-22-6 5-*tert*-Butyltetrahydro-2-dimethylamino-2H-1,3,2-oxazaphosphorine; (2RS,5SR)-*form*, 2-Oxide, *in* B-00635
77815-23-7 5-*tert*-Butyltetrahydro-2-dimethylamino-2H-1,3,2-oxazaphosphorine; (2RS,5RS)-*form*, 2-Oxide, *in* B-00635
77837-61-7 [1-Methyl-3-(2,6,6-trimethyl-1-cyclohexen-1-yl)-2-propenyl]triphenyl-phosphonium(1+); (E)-*form*, Chloride, *in* M-00422
77876-46-1 1,3-Bis(diphenylphosphino)butane; (S)-*form*, *in* B-00224
77876-82-5 1,4-Bis(dimethylphosphinyl)-benzene, *in* P-00125
77891-20-4 (Diphenylphosphinyl)-isocyanoacetic acid; *tert*-Butyl ester, *in* D-01098
77918-49-1 (3-Methoxyphenyl)phosphonic acid, M-00071
77929-80-7 N,P-Diphenylphosphonamidic acid; Chloride, *in* D-01103
78037-58-8 O-Methyl methylphenylphosphinothioate, *in* M-00225
78045-07-5 Diphenyl-1-propenylphosphine; (Z)-*form*, Sulfide, *in* D-01143
78045-10-0 Diphenyl(2-phenylethenyl)-phosphine; (Z)-*form*, Oxide, *in* D-01033
78050-96-1 N,N,N′,N′,N″,N″-Hexamethylphosphorimidic triamide; N‴-Chloro, *in* H-00077
78055-64-8 Diethyl [(3-chlorophenyl)-methyl]phosphonate, *in* C-00133
78089-64-2 (2-Methylphenyl)methylphosphinic acid; Chloride, *in* M-00203
78129-68-7 (2,2-Dimethylpropylidyne)-phosphine, D-00910
78133-46-7 Dimethyl 4-pyridinylphosphonate, *in* P-00505
78133-47-8 Dimethyl 2-methyl-4-pyridinephosphonate, *in* M-00388
78133-49-0 Dimethyl (3-methyl-4-pyridinyl)phosphonate, *in* M-00390
78133-51-4 Dimethyl 4-quinolinephosphonate, *in* Q-00003
78172-51-7 3,4-Bis(trifluoromethyl)-1,2,5-thiadiphosphole, B-00425
78177-25-0 5,6-Dimethyl-7-phenyl-7-phosphabicyclo[2.2.1]-hept-5-ene-2,3-dicarboxylic acid; *endo,endo,anti-form*, Di-Me ester, *in* D-00800
78177-26-7 5,6-Dimethyl-7-phenyl-7-phosphabicyclo[2.2.1]-hept-5-ene-2,3-dicarboxylic acid; *endo,endo,syn-form*, Di-Me ester, *in* D-00800
78211-40-2 1,2-Dimethyl-4-phenyl-1,2,4-diazaphospholidine, *in* D-00718
78227-61-9 3-Phenyl-1,3-azaphosphetidine, P-00087
78258-13-6 Dimethyl (4-nitrophenyl)-phosphoramidate, *in* N-00060
78259-08-2 2-Methyl-3-phenyl-1,3,2-oxazaphospholidine; 2-Sulfide, *in* M-00212
78269-37-1 2,3,4,5-Tetrahydro-1,5-diphenyl-1H-1,5-benzodiphosphepin; *cis-form*, *in* T-00101
78271-46-2 Tetramethyl 1,3-phenylenebisphosphonate, *in* B-00007
78300-89-7 Diisopropylphosphinic acid; Azide, *in* D-00641
78339-70-5 6,7,8,9,14,15,16,17-Octahydro-9,14-diphenyl-dibenzo[b,i][1,4,8,11]-dithiadiphosphacyclotetradecin; (9RS,14SR)-*form*, *in* O-00013
78342-55-9 2-Amino-1,3-di-*tert*-butylazaphosphiridine; (2RS,3SR)-*form*, N,N-Diisopropyl, *in* A-00070
78463-56-6 Dimethyl methylenebis(methyl-phosphinate), *in* M-00140
78504-35-5 Diethyl (2-methylphenyl)-phosphoramidate, *in* M-00251
78532-89-5 2,3,4,5-Tetrahydro-2,2,4,4-tetramethyl-1H-2,4-benzodiphosphepinium(2+); Dibromide, *in* T-00139
78532-90-8 2,3,4,5-Tetrahydro-2,2,4,4-tetraphenyl-1H-2,4-benzodiphosphonium(2+); Dibromide, *in* T-00144
78532-93-1 2,2,4,4-Tetrahydro-2,2,4,4-tetramethyl-1H-2,4-benzodiphosphepin, *in* T-00087
78532-94-2 2,2,4,4-Tetrahydro-2,2,4,4-tetraphenyl-1H-2,4-benzodiphosphepin, *in* T-00087
78543-77-8 Di-*tert*-butyl phosphorochloridite, D-00143
78550-26-0 3-[(Dimethoxyphosphinyl)oxy]-2-butenoic acid; (Z)-*form*, Et ester, *in* D-00681
78550-67-1 1,4-Diphosphinobenzene, D-01173
78554-61-7 Diisopropyl [2-(butylamino)-1-hexenyl)]phosphonate, *in* B-00540
78570-13-5 *tert*-Butyl ethylphosphonochloridodithioate, *in* E-00136
78605-34-2 Diethyl 1,2-phenylenebis(phenylphosphinate), *in* P-00139
78721-85-4 2,5-Dimethyl-1-phenylphosphorinan-4-ol, D-00820
78725-78-7 Octahydro-7H,9bH-2a,4a,6a,9a-tetraaza-9b-phosphapentaleno[1,6-cd]indene; 9b-Fluoro, *in* O-00023
78725-79-8 Octahydro-5H,8H,10bH-2a,4a,7a,10a-tetraaza-10b-phosphapentaleno[1,6-de]-naphthalene; 10b-Fluoro, *in* O-00024
78725-80-1 Octahydro-3H,8H,10bH-2a,5a,7a,10a-tetraaza-10b-phosphaindeno[7,1-cd]indene; 10b-Fluoro, *in* O-00022
78739-01-2 2-Amino-4-phosphonobutanoic acid; (R)-*form*, *in* A-00117
78788-15-5 S,S,S-Triisopropyl phosphorotrithioate, T-00508
78788-16-6 S,S,S-Tri-*tert*-butyl phosphorotrithioate, T-00391
78819-34-8 Tetraphenyl 1,2-ethanediylbisphosphonite, *in* E-00008
78871-05-3 Di-2-naphthylphosphine; Oxide, *in* D-00940
78871-06-4 Di-2-naphthylphosphine, D-00940
78928-19-5 2-Mercapto-4-methyl-1,3,2-dioxaphospholane 2-oxide, M-00015
78944-60-2 Mono-*tert*-butyl phosphate; Monoanilinium salt, *in* M-00452
78949-15-2 Di-1-piperidinylphosphinic acid, D-01190
78950-65-9 [(1,3-Dioxolan-2-yl)methylene]-triphenylphosphorane, *in* D-00976
78961-27-0 O,S-Ethylene O-phenyl phosphorothioate, *in* P-00072
78961-32-7 O,O-Diphenyl phosphoroselenoate, D-01139
78993-09-6 Bis(2-nitrophenyl) phosphate, B-00369
79014-65-6 (1-Amino-1-methylpropyl)-phosphonic acid, A-00093
79014-69-0 (1-Amino-1-phenylethyl)-phosphonic acid, A-00100
79117-07-0 3-Ethoxy-2,3-dihydro-2-methyl-5-phenyl-1H-1,2,4,3-triazaphosphole; 3-Oxide, *in* E-00028

79129-16-1 2,3′-Dimethylspiro[1,3,2-benzodioxaphosphole-2,2′(3′H)-[1,3,2]-benzoxazaphosphole], *in* M-00395
79129-17-2 1,3-Dihydro-1,2,3-trimethylspiro[2H-1,3,2-benzodiazaphosphole-2,2′-[1,3,2]benzodioxaphosphole], *in* D-00469
79129-18-3 2,3,3′-Trimethyl-2,2′(3H,3′H)-spirobi[1,3,2-benzoxazaphosphole], *in* D-00917
79133-84-9 5-Methyl-5H-dibenzophosphole; 5-Selenide, *in* M-00115
79138-42-4 Monopropyl phosphate; Monocyclohexylammonium salt, *in* M-00512
79138-43-5 Mono(1-methylpropyl) phosphate; (±)-*form*, Monocyclohexylammonium salt, *in* M-00492
79150-96-2 3-Phenyl-3-phosphabicyclo[3.2.1]-oct-6-ene 3-oxide, *in* P-00198
79157-74-7 5,6-Dimethylspiro[1,3,2-benzodioxaphosphole-2,1′(3′H)-[2,1]benzoxaphosphole, D-00915
79157-75-8 5,6-Dimethyl-N-phenylspiro[1,3,2-benzodioxaphosphole-2,1′(3′H)-[2,1]benzoxaphosphol]-2-amine, *in* D-00915
79157-76-9 5,6-Dimethylspiro[1,3,2-benzodioxaphosphole-2,1′(3′H)-[2,1]benzoxaphosphole; 2-Phenoxy, *in* D-00915
79157-77-0 N,N,5,6-Tetramethylspiro[1,3,2-benzodioxaphosphole-2,1′(3′H)-[2,1]benzoxaphosphol]-2-amine, *in* D-00915
79157-78-1 5,6-Dimethylspiro[1,3,2-benzodioxaphosphole-2,1′(3′H)-[2,1]benzoxaphosphole; 2-Ph, *in* D-00915
79157-80-5 1,3-Dihydro-1-phenoxy-2,1-benzoxaphosphole, D-00546
79157-82-7 1,3-Dihydro-1-phenyl-2,1-benzoxaphosphole, D-00553
79157-84-9 1,3-Dihydro-1-phenoxy-2,1-benzoxaphosphole; 1-Sulfide, *in* D-00546
79157-86-1 1,3-Dihydro-1-phenyl-2,1-benzoxaphosphole; 1-Sulfide, *in* D-00553
79157-90-7 Methyl *tert*-butylphosphinate, *in* B-00591
79157-92-9 *tert*-Butylphenylphosphinic acid; (*R*)-*form*, Isocyanate, *in* B-00579
79157-96-3 *tert*-Butylphenylphosphinic acid; (*S*)-*form*, Isocyanate, *in* B-00579
79157-98-5 Methyl *tert*-butylphosphon(isocyanatidate), *in* B-00602
79158-00-2 *tert*-Butylphenylphosphinic acid; (*S*)-*form*, Isothiocyanate, *in* B-00579
79184-79-5 1,4-Dimethyl-2-phenyl-7-oxa-2-phosphabicyclo[2.2.1]-heptane; 2-Oxide, *in* D-00796
79220-12-5 S-*tert*-Butyl methyl-phosphonochloridodithioate, *in* M-00297
79227-86-4 Mono(4-methylphenyl) phosphate; Monocyclohexylammonium salt, *in* M-00487
79227-87-5 Mono(4-chlorophenyl) phosphate; Monocyclohexylammonium salt, *in* M-00458
79238-62-3 Diethyl (3,5-dimethylphenyl)-phosphonate, *in* D-00816
79251-52-8 2,6-Dimethyl-1,3,2,6-dioxadiphosphocane; *cis*-*form*, 2,6-Diselenide, *in* D-00730
79251-53-9 2,6-Dimethyl-1,3,2,6-dioxadiphosphocane; *trans*-*form*, 2,6-Diselenide, *in* D-00730
79251-54-0 2,6-Dimethyl-1,3,2,6-dioxadiphosphocane; *cis*-*form*, 2,6-Disulfide, *in* D-00730
79251-55-1 2,6-Dimethyl-1,3,2,6-dioxadiphosphocane; *trans*-*form*, 2,6-Disulfide, *in* D-00730
79251-56-2 2,6-Dimethyl-1,3,2,6-dioxadiphosphocane; *cis*-*form*, *in* D-00730
79251-57-3 2,6-Dimethyl-1,3,2,6-dioxadiphosphocane; *trans*-*form*, *in* D-00730
79272-33-6 Phenyl (dichlorophosphinothioyl)-phosphorodichlorimidate, P-00112
79272-36-9 Triphenyl (diphenoxyphosphinothioyl)-phosphorimidate, *in* D-00987
79296-50-7 Methyl hydrogen (α-hydroxybenzyl)phosphonate, *in* H-00174
79303-16-5 Tetrafluorophenoxyphosphorane, T-00083
79312-13-3 Monohexyl phosphate; Monoanilinium salt, *in* M-00476
79317-54-7 Dimethyl(2-methylphenyl)-phosphine; Oxide, *in* D-00759
79317-63-8 2-(Diphenylphosphino)benzoic acid; Oxide, Me ester, *in* D-01053
79343-17-2 2-Azido-1,3,2-benzodioxaphosphole, A-00148
79356-79-9 2-Chloro-4,5-diphenyl-1,3,2-dioxaphospholane; (2α,4β,5β)-*form*, 2-Oxide, *in* C-00069
79426-24-7 Bicyclo[2.2.1]hept-5-ene-2,3-diylbis(methylene)-bis[diphenylphosphine]; (2*RS*,3*RS*)-*form*, *in* B-00075
79426-25-8 Bicyclo[2.2.1]hept-5-ene-2,3-diylbis(methylene)-bis[diphenylphosphine]; (2*RS*,3*RS*)-*form*, Dioxide, *in* B-00075
79426-26-9 Bicyclo[2.2.1]hept-5-ene-2,3-diylbis(methylene)-bis[diphenylphosphine]; (2*S**,3*R**)-*form*, Dioxide, *in* B-00075
79426-27-0 Bicyclo[2.2.1]hept-5-ene-2,3-diylbis(methylene)-bis[diphenylphosphine]; (2*R**,3*R**)-*form*, Dioxide, *in* B-00075
79426-28-1 Bicyclo[2.2.1]hept-5-ene-2,3-diylbis(methylene)-bis[diphenylphosphine]; (2*S**,3*R**)-*form*, *in* B-00075
79426-29-2 Bicyclo[2.2.1]hept-5-ene-2,3-diylbis(methylene)-bis[diphenylphosphine]; (2*R**,3*R**)-*form*, *in* B-00075
79521-54-3 Diethyl (3-methyl-1,3-butadienyl)phosphonate, *in* M-00095
79543-93-4 1-(Diphenylphosphino)-2-propanone; Oxide, 2,4-dinitrophenylhydrazone, *in* D-01075
79563-56-7 4-(Diethoxyphosphinyl)-2-butenoic acid; (*E*)-*form*, *in* D-00248
79564-52-6 Bis(methoxycarbonyl)-phenylphosphine, *in* P-00209
79564-54-8 Bis(ethoxycarbonyl)-phenylphosphine, *in* P-00209
79564-58-2 Bis(phenoxycarbonyl)-phenylphosphine, *in* P-00209
79574-60-0 2-Dimethylamino-1,3,2-oxazaphospholidine; 2-Oxide, *in* D-00702
79629-50-8 *tert*-Butyl N,N,N′-tris(trimethylsilyl)-phosphonamidimidate, *in* T-00867
79629-54-2 Phenyl N,N,N′-tris(trimethylsilyl)-phosphonamidimidate, *in* T-00867
79639-92-2 Dimethyl (3-fluorophenyl)-phosphoramidate, *in* F-00043
79639-93-3 Dimethyl (4-fluorophenyl)-phosphoramidate, *in* F-00044
79639-95-5 Dimethyl (3-nitrophenyl)-phosphoramidate, *in* N-00059
79673-93-1 Bis(2-oxo-3-oxazolidinyl)-phosphinic acid, B-00378
79691-65-9 5,10-Dihydro-10-phenylphenophosphazine; 10-Sulfide, *in* D-00563
79691-79-5 Ethyl (diisopropoxyphosphinothioyl)-thioacetate, *in* D-00632
79722-69-3 5,10-Dihydro-10-phenylphenophosphazine, D-00563
79734-41-1 [5-(4-Hydroxy-2,6,6-trimethyl-1-cyclohexen-1-yl)-3-methyl-2,4-pentadienyl]-triphenylphosphonium(1+); (±)-(*E*,*E*)-*form*, Bromide, *in* H-00202
79735-27-6 10-Chloro-5,10-dihydrophenophosphazine, C-00049
79750-11-1 *O*,*O*-1,8-Naphthylene diethylphosphoramidothioate, *in* D-00267
79750-12-2 1,8-Naphthylene diethylphosphoramidate, *in* D-00267
79768-77-7 1,2-Bis(diphenylphosphinomethyl)-cyclobutane; (1*RS*,2*SR*)-*form*, *in* B-00249
79772-19-3 1,8-Naphthylene phosphonate, *in* N-00010
79788-73-1 2-Chloronaphtho[1,2-*d*]-1,3,2-dioxaphosphole, C-00118
79814-92-9 (1-Phenyl-1,2-ethanediyl)-bis[diphenylphosphine]; (±)-*form*, *in* P-00150
79827-34-2 Decylidenetriphenylphosphorane, *in* D-00014
79838-05-4 4-Nitrophenyl 4-morpholinylphosphonochloridate, *in* M-00524
79922-17-1 2-(4-Methoxyphenyl)-1,3,2-thiazaphosphetidine 2-sulfide; N-Me, *in* M-00083
79926-86-6 9-(Triphenylphosphoniomethyl)-phenanthrene(1+); Chloride, *in* T-00638
79932-99-3 2-(Diphenylphosphino)benzoic acid; Me ester, *in* D-01053
80034-71-5 5,10-Dihydro-5,10-diphenylacridophosphine 5-oxide, *in* D-00548
80049-55-4 1,3-Dimethyl-5,7-diphenyl-1,3,5,7-tetraaza-4-phospha(4-P^{V})spiro[3.3]-heptane-2,6-dione; 4-(Diethoxyphosphinyl), *in* D-00748
80093-92-1 2,4,5-Trimethyl-1,3,2-dioxaphospholane; (4*R*,5*R*)-*form*, *in* T-00511
80125-96-8 2,3-Butylene methylphosphonate, *in* T-00511
80138-63-2 *O*,*O*-Diethyl phosphoroselenoate; Dicyclohexylammonium salt, *in* D-00393
80166-20-7 N-Phenyl-P,P-bis(trichloromethyl)-phosphinimidic chloride, *in* B-00403
80173-16-6 (1-Methylpropyl)-diphenylphosphine; (*R*)-*form*, *in* M-00369
80173-17-2 (1-Methylpropyl)-diphenylphosphine; (*R*)-*form*, Oxide, *in* M-00369

80202-58-0 2,2-Dimethyl-1,3-dioxa-5-phospha-2-silacyclohexane; 5-Ph, 5-sulfide, *in* D-00731

80226-61-5 (1-Methylpropyl)-diphenylphosphine; (±)-*form*, *in* M-00369

80241-43-6 (4-Pyridinylmethyl)phosphonic acid, P-00507

80241-45-8 (2-Pyridinylmethyl)phosphonic acid, P-00506

80248-48-2 (Pentafluoroethyl)phosphorimidic trichloride, P-00013

80249-44-1 (Bromomethyl)-triphenylphosphonium(1+); Tetrafluoroborate, *in* B-00477

80325-32-2 (1,3,6-Cycloheptatrien-1-ylmethylene)triphenyl-phosphorane, C-00236

80354-12-7 ▷Diisopropyl phosphorobromidate, *in* D-00639

80358-18-5 1,2-Bis(phenylphosphino)ethane; (*RS,SR*)-*form*, Disulfide, *in* B-00390

80358-19-6 1,2-Bis(phenylphosphino)ethane; (*RS,RS*)-*form*, Disulfide, *in* B-00390

80358-25-4 1,2-Bis(phenylphosphino)ethane; (*RS,RS*)-*form*, Dioxide, *in* B-00390

80388-56-3 (3,7-Dimethyl-2,6-octadienyl)-triphenylphosphonium(1+); (*Z*)-*form*, Ylide, *in* D-00777

80395-11-5 [(3-Chlorophenyl)methyl]-phosphonic acid, C-00133

80395-13-7 [(2-Bromophenyl)methyl]-phosphonic acid, B-00483

80395-14-8 [(4-Fluorophenyl)methyl]-phosphonic acid, F-00031

80395-15-9 [(2-Fluorophenyl)methyl]-phosphonic acid, F-00029

80395-16-0 [(3-Fluorophenyl)methyl]-phosphonic acid, F-00030

80413-46-3 Ethyl dicyclohexylphosphinite, *in* D-00214

80454-59-7 2,7,9,14,15,16,17,18-Octaoxa-1,8-diphospha(1,8-P^V)-pentacyclo[10.2.1.1^{1,4}.1^{5,8}.1^{8,11}]-octadecane-3,6,10,13-tetrone; (*R,R,R,R*)-*form*, Bis(triethylammonium) salt, *in* O-00036

80454-84-8 2,7,9,14,15,16,17,18-Octaoxa-1,8-diphospha(1,8-P^V)-pentacyclo[10.2.1.1^{1,4}.1^{5,8}.1^{8,11}]-octadecane-3,6,10,13-tetrone, O-00036

80461-84-3 1-Methylphosphonane, M-00285

80483-64-3 2,7,9,14,15,16,17,18-Octaoxa-1,8-diphospha(1,8-P^V)-pentacyclo[10.2.1.1^{1,4}.1^{5,8}.1^{8,11}]-octadecane-3,6,10,13-tetrone; (*R,R,R,R*)-*form*, 1,8-Dihydroxy, bistriethylammonium salt, *in* O-00036

80510-04-9 1,2-Diphosphinobenzene, D-01172

80518-21-4 Dibromo-*tert*-butylphosphine, *in* B-00616

80625-78-1 Octylidenetriphenylphosphorane, *in* O-00047

80679-23-8 Tri(2-thiazolyl)phosphine, T-00873

80679-24-9 Tri(2-benzothiazolyl)phosphine, T-00334

80679-27-2 2,2′,2″-Phosphinidynetris[1-methylimidazole], *in* T-00496

80737-80-0 1,2,7,8,9,10,15,16-Octamethyl-3,6,11,14-tetraphenyl-tetrakisphospholo[1,2-*b*:2′,1′-*d*:1″,2″-*f*:2‴,1‴-*h*][1,2,5,6]-tetraphosphocin, O-00033

80754-56-9 2,3,4,5,6,7-Hexahydro-1*H*-isophosphindole 2-oxide; 2-Ph, *in* H-00035

80754-60-5 1-Phenyl-3,8-phosphonanedione 1-oxide, *in* P-00373

80799-69-5 *O*-Isopropyl methylphosphonochloridothioate; (*R*)-*form*, *in* I-00039

80812-96-0 4-Methyl-1,3,2-dioxaphosphorinane; (2*RS*,4*SR*)-*form*, *in* M-00129

80812-97-1 2-Chloro-4,4,6-trimethyl-1,3,2-dioxaphosphorinane, C-00189

80817-46-5 *O,O*-Bis(1-benzotriazolyl) 2-chlorophenyl phosphate, B-00098

80821-21-2 *O,Se,Se*-Tributyl phosphorotriselenoate, T-00389

80920-38-3 Pentaethylphosphoric triamide, *in* T-00058

80920-97-4 Dibromo(dichloromethyl)-phosphine, *in* D-00181

80920-98-5 Diisopropyl (dichloromethyl)-phosphonite, *in* D-00181

80959-91-7 2,3-Diphenyl-1,3-thiaphospholane; 3-Oxide, *in* D-01151

80973-45-1 *O,O*-Bis(1-benzotriazolyl) morpholinyl phosphate, B-00099

80980-28-5 *O,O*-Bis(1-benzotriazolyl) 2,2,2-tribromoethyl phosphate, B-00100

81048-09-1 5-Hydroxy-2,4,6-triphenyl-1,3,5-dioxaphosphorinane 5-oxide; 5-Methoxy (Me ester), *in* H-00204

81050-37-5 Methylenebis[phosphinic acid], M-00143

81123-19-5 Profenofos; (*S*)-*form*, *in* P-00440

81236-66-0 4′,4′,5′,5′-Tetramethylspiro[1,3,2-benzodioxaphosphole-2,2′-[1,3,2]dioxaphospholane]; 2-Phenylseleno, *in* T-00230

81349-02-2 (4-Nitrophenyl)methylphosphinic acid, N-00043

81358-38-5 Tris[3-(trifluoromethyl)-phenyl]phosphine; Selenide, *in* T-00853

81366-72-5 Mono(4-chlorophenyl) mono(5-chloro-8-quinolinyl) phosphate, M-00456

81373-55-9 Diethylphosphinothioic acid; Iodide, *in* D-00330

81373-56-0 Dipropylphosphinothioic acid; Iodide, *in* D-01208

81373-57-1 Dibutylphosphinothioic acid; Iodide, *in* D-00117

81373-58-2 Iododipropylphosphine, *in* D-01209

81431-68-7 Tris(dimethylphosphinothioyl)-methane, *in* T-00771

81439-66-9 2,4,6-Trimethyl-5-phenyl-1,3,5-dioxaphosphorinane; (2α,4α,5α,6α)-*form*, 5-Selenide, *in* T-00521

81485-04-3 2-Chloro-3-(2-chloroethyl)-tetrahydro-2*H*-1,3,2-oxazaphosphorine; (±)-*form*, 2-Oxide, *in* C-00030

81487-72-1 Trimethyl[(trimethyl-phosphoranylidene)methyl]-phosphonium(1+); Bromide, *in* T-00565

81487-79-8 2,3-Dihydro-1-methyl-1*H*-benzo[*g*]phosphindole; 1-Oxide, *in* D-00518

81487-80-1 2,3-Dihydro-1-methyl-1*H*-benzo[*g*]phosphindole, D-00518

81488-03-1 2,3-Dihydro-1,1-dimethyl-1*H*-benzo[*g*]phosphindolinium iodide, *in* D-00518

81492-17-3 Dibromotrimethylphosphorane, D-00069

81495-87-6 2,4,6-Trimethyl-5-phenyl-1,3,5-dioxaphosphorinane; (2α,4α,5β,6α)-*form*, 5-Selenide, *in* T-00521

81500-73-4 1,2,3,6-Tetrahydro-1,2-diphenyl-1,2-diphosphorin; (1*RS*,2*SR*)-*form*, 1,2-Disulfide, *in* T-00103

81500-74-5 1,2,3,6-Tetrahydro-1,2-diphenyl-1,2-diphosphorin; (1*RS*,2*RS*)-*form*, 1,2-Disulfide, *in* T-00103

81532-95-8 *O,O*-Bis(2-methylphenyl) phosphorochloridothioate, B-00333

81532-96-9 *O,O*-Bis(3-methylphenyl) phosphorochloridothioate, B-00334

81592-48-5 (2,4-Dimethylphenyl)phosphinic acid, D-00804

81626-09-7 Bis[(dimethylphosphino)methyl]-methylphosphine, B-00204

81675-82-3 *P*-Bromo-*N*-butyl-*P,P*-bis(dibutylamino)-phosphazene, *in* P-00004

81739-37-9 Tri-*tert*-butyltriphosphirane; (1α,2α,3β)-*form*, *in* T-00397

81928-62-3 Tetraethyl (aminophenylmethyl)-bisphosphonate, *in* A-00053

81928-64-5 Tetrabutyl (aminophenylmethyl)-bisphosphonate, *in* A-00053

82064-34-4 Phosphophenylalanylarginine, P-00401

82134-83-6 *P,P*-Diethyl-*N*-methylphosphinic amide, *in* D-00323

82159-11-3 [Bis(dimethylphosphino)-methylene]trimethylphosphorane, B-00203

82159-17-9 1,5-Pentanediphosphonous acid; Bis(dibromide), *in* P-00033

82159-18-0 1,6-Hexanediphosphonous acid; Bis(dibromide), *in* H-00086

82159-30-6 1,2-Ethanediphosphonothioic acid; Tetrabromide, *in* E-00007

82159-33-9 1,2-Diphospholane; 1,2-Diisopropyl, *in* D-01181

82159-40-8 *P,P′*-1,6-Hexanediylbis[bis *N,N*-diethylphosphonous diamide], *in* H-00086

82224-12-2 (1-Methylpropyl)phenylphosphinic acid, M-00372

82237-09-0 Phenyl dimethyl-phosphoramidochloridite, *in* D-00862

82239-68-7 4,5-Bis[(dicyclohexylphosphino)-methyl]-2,2-dimethyl-1,3-dioxolane; (4*R*,5*R*)-*form*, *in* B-00168

82340-10-1 1,3-Bis(dimethylphosphinothioyl)-benzene, *in* P-00124

82359-29-3 4*H*-1,4-Thiaphosphorin; 4-Chloro, 4-oxide, *in* T-00299

82403-26-7 Methyl methyl(trifluoromethyl)-phosphinate, *in* M-00420

82403-28-9 Butyl methyl(trifluoromethyl)-phosphinate, *in* M-00420

82404-41-9 6,7-Dihydro-5,5,7,7-tetramethyl-5*H*-dibenzo[*d,f*][1,3]-diphosphepinium(2+); Dibromide, *in* D-00577

82404-42-0 5,5,7,7-Tetramethyl[*d,f*]-[1,λ^{5}3]diphosphepinium bromide, *in* D-00577

82404-44-2 Phosphanthridene; 5,5-Dihydro, 5,5-di-Me, *in* P-00348

82404-45-3 5,6-Dihydro-5,5-dimethylphosphanthridinium chloride, *in* D-00568

82495-65-6 *N,N,N′,N′,N″,N″,N‴,N‴*-Octamethyl-*P,P′*-1,2-phenylenebis[phosphonous diamide], *in* B-00009

82495-67-8 1,2-Benzenediphosphonous acid; Bis(dichloride), *in* B-00009

82556-11-4 2-Chlorophenyl (4-methoxyphenyl)-phosphoramidochloridate, C-00130

82564-88-3 2-Phenoxy-3-phenyl-1,3,2-oxazaphospholidine, P-00073

82568-46-5 2-Iodo-5,5-dimethyl-1,3,2-dioxaphosphorinane, I-00010

82581-87-1 *P*-Methyl-*N,N,N′,N′*-tetrakis(trimethylsilyl)-phosphonous diamide, *in* M-00320

82631-91-2 (1-Phenyl-1,2-ethanediyl)bis[diphenylphosphine]; (*S*)-*form*, Dioxide, *in* P-00150
82638-81-1 2-(1-Methyl-2-oxopropylidene)phosphorohydrazidothioate oxime; (*E,E*)-*form*, *in* M-00189
82660-49-9 (1-Phenyl-1,2-ethanediyl)bis[diphenylphosphine]; (±)-*form*, Dioxide, *in* P-00150
82685-97-0 1,3*a*,3*b*,6-Tetrahydro-3,4,8,8-tetramethyl-1,6-diphenyl-8*H*-[1,3]-diphospholo[1,5-*e*;3,4-*c'*]bis[1,2,3]diazaphosphole, T-00140
82756-86-3 Dibutyl 2-propynylphosphonate, *in* P-00495
82757-16-2 5-*tert*-Butyltetrahydro-2-dimethylamino-2*H*-1,3,2-oxazaphosphorine; (2*RS*,5*SR*)-*form*, 2-Sulfide, *in* B-00635
82784-34-7 (3-Methoxy-2-propenylidene)triphenylphosphorane, *in* M-00084
82881-16-1 *N,N*,4,4-Tetramethyl-1,3-bis(1,1-dimethylethyl)-1,3-diaza-2-phosphacyclobutane-2-amine, *in* D-00087
82881-21-8 *N,N*,1,3-Tetrakis(1,1-dimethylethyl)-4,4-dimethyl-1,3-diaza-2-phosphacyclobutane-2-amine, *in* D-00087
82881-24-1 1,3-Di-*tert*-butyl-4,4-dimethyl-1,3,2-diazaphosphasiletidine; 2-Dimethylamino, 2-sulfide, *in* D-00087
82881-29-6 1,3-Di-*tert*-butyl-4,4-dimethyl-1,3,2-diazaphosphasiletidine; 2-(Di-*tert*-butyl)amino, 2-sulfide, *in* D-00087
82943-05-3 Ethyl phenyl(2-phenylethenyl)phosphinate, *in* P-00186
82945-12-8 *tert*-Butylphenylphosphinic acid; (*S*)-*form*, Thiocyanate, *in* B-00579
83037-86-9 Phenyl(diphenoxyphosphinyl)acetate, *in* D-00988
83085-52-3 2,2',5,5'-Tetraphenyl-1*H*,1'*H*-biphosphole, T-00248
83096-33-7 5-*tert*-Butyltetrahydro-2-dimethylamino-2*H*-1,3,2-oxazaphosphorine; (2*RS*,5*RS*)-*form*, 2-Sulfide, *in* B-00635
83096-34-8 5-*tert*-Butyltetrahydro-2-dimethylamino-3-phenyl-2*H*-1,3,2-oxazaphosphorine; (2*RS*,5*RS*)-*form*, 2-Sulfide, *in* B-00636
83096-35-9 5-*tert*-Butyltetrahydro-2-dimethylamino-3-phenyl-2*H*-1,3,2-oxazaphosphorine; (2*RS*,5*SR*)-*form*, 2-Sulfide, *in* B-00636
83096-40-6 5-*tert*-Butyltetrahydro-2-dimethylamino-3-phenyl-2*H*-1,3,2-oxazaphosphorine; (2*RS*,5*SR*)-*form*, *in* B-00636
83096-41-7 5-*tert*-Butyltetrahydro-2-dimethylamino-3-phenyl-2*H*-1,3,2-oxazaphosphorine; (2*RS*,5*RS*)-*form*, *in* B-00636
83096-42-8 5-*tert*-Butyl-2-chlorotetrahydro-3-phenyl-2*H*-1,3,2-oxazaphosphorine, B-00549
83115-12-2 (2,4,6-Tri-*tert*-butylphenyl)phosphine, T-00369
83245-93-6 Diethyl (1-aminobutyl)phosphonate, *in* A-00054
83299-97-2 [4-(Dimethylamino)butyl]triphenylphosphonium(1+); Bromide, *in* D-00690
83318-64-3 *N,N,N',N',N'',N''*-Hexamethyl-*N'''*-diphenylphosphinophosphorimidic triamide, H-00069
83436-57-1 *O,O*-Dimethyl phenylphosphoramidothioate, *in* P-00262
83438-72-6 [Bis[bis(trimethylsilyl)methylene]]chlorophosphorane, B-00105
83462-66-2 5,6-Bis(diphenylphosphino)-2-methylbicyclo[2.2.1]hept-2-ene; (1*S**,5*S**,6*S**)-*form*, *in* B-00247
83466-54-0 Bis(2,4,6-tri-*tert*-butylphenyl)diphosphene; (*E*)-*form*, *in* B-00395
83470-25-1 Bis(4-methoxyphenyl)phosphinic acid; Azide, *in* B-00301
83470-29-5 Dimethyl (4-chlorophenyl)phosphoramidate, *in* C-00157
83470-30-8 Methyl bis(4-methylphenyl)phosphinate, *in* B-00320
83470-31-9 Methyl bis(4-methoxyphenyl)phosphinate, *in* B-00301
83470-32-0 Bis(4-methoxyphenyl)phosphinic acid; Amide, *in* B-00301
83470-33-1 Bis(4-chlorophenyl)phosphinic acid; Amide, *in* B-00139
83582-59-6 ▷Isopropyl diethylphosphinite, *in* D-00333
83587-40-0 *P*-Ethyl-*N,N*-dimethyl-*P*-phenylphosphinothioic amide, *in* E-00105
83587-43-3 *P*-Isopropyl-*N,N*-dimethyl-*P*-phenylphosphinothioic amide, *in* I-00046
83622-76-8 5-Methyl-1-phospha-5-silabicyclo[3.3.1]nonane, M-00274
83622-80-4 5-Methyl-1-phospha-5-silabicyclo[3.3.1]nonane; 1-Sulfide, *in* M-00274
83627-25-2 2-Ethoxy-3,5,5-trimethyl-1,2-oxaphospholan-3-ol 2-oxide, *in* T-00519
83630-54-0 *N,N,N',N'*-Tetramethyl-*P*-(2-methyl-1,3-butadienyl)phosphonothioic diamide, *in* M-00097
83631-23-6 Dioctyl (dimethylcarbamoylmethyl)phosphonate, *in* D-00703
83646-38-2 Diethyl (3-nitrophenyl)phosphoramidate, *in* N-00059
83692-00-6 2,4,6-Tris(2,4,6-trimethylphenyl)-1,3,5,2,4,6-trithiatriphosphorinane; (2*α*,4*α*,6*β*)-*form*, *in* T-00862
83802-21-5 2-Amino-3-(2-chloroethyl)tetrahydro-4*H*-1,3,2-oxazaphosphorine 2-oxide; (*R*)-*form*, *in* A-00063
83802-22-6 2-Amino-3-(2-chloroethyl)tetrahydro-4*H*-1,3,2-oxazaphosphorine 2-oxide; (*S*)-*form*, *in* A-00063
84050-72-6 *P*-Ethyl-*N,N,N',N'*-tetrakis(trimethylsilyl)phosphonous diamide, *in* E-00154
84251-28-5 Bis(3-methoxyphenyl)phosphinic acid, B-00300
84251-29-6 Bis(3-hydroxyphenyl)phosphinic acid, B-00284
84350-73-2 Tris(2-fluorophenyl)phosphine, T-00781
84372-47-4 Diethyl 1-naphthylphosphonite, *in* N-00025
84372-54-3 (2,5-Dimethylphenyl)phosphinic acid, D-00805
84382-98-9 2-Hydroxy-4,5-diphenyl-1,3,2-dioxaphosphole 2-oxide; 2-Methoxy (Me ester), *in* H-00139
84383-01-7 2-Chloro-4,5-dimethyl-1,3,2-dioxaphosphole, C-00057
84441-06-5 3-Phenyl-1,3-oxaphosphorinan-2-one; 3-Sulfide, *in* P-00177
84441-53-2 Tetraphenyl methylenebisphosphonate, *in* M-00022
84441-54-3 Diphenyl (iodomethyl)phosphonate, *in* I-00013
84451-63-8 Tetrahydro-2-methoxy-2*H*-1,3,2-oxazaphosphorine; 2-Sulfide, *in* T-00111
84489-09-8 Iminocyclophosphamide, I-00004
84494-89-3 1,1-Bis(diphenylphosphino)ethylene, B-00236
84494-91-7 1,1-Bis(diphenylphosphinoselenoyl)ethylene, *in* B-00236
84530-50-7 5,5'-Oxybis[5*H*-benzophosphindole] 5,5'-dioxide, *in* H-00118
84534-45-2 Methanediphosphonothioic acid; Tetrafluoride, *in* M-00023
84549-28-0 2-Isopropyl-1,3,2-oxathiaphospholane 2-oxide, *in* I-00042
84549-29-1 2-Isopropyl-1,3,2-oxathiaphospholane 2-sulfide, *in* I-00042
84589-83-3 Diphenylphosphinothioic acid; Iodide, *in* D-01082
84591-34-4 Bromobis(diethylamino)(ethylamino)phosphonium bromide, *in* P-00012
84591-35-5 Bromobis(dipropylamino)(propylamino)phosphonium bromide, *in* P-00043
84591-36-6 Bromo(butylamino)bis(dibutylamino)phosphonium bromide, *in* P-00004
84591-37-7 Chlorobis(diethylamino)(ethylamino)phosphonium chloride, *in* P-00012
84591-39-9 (Butylamino)chlorobis(dibutylamino)phosphonium chloride, *in* P-00004
84635-68-7 1',3'-Diphenyldispiro[1,3,2-benzodioxaphosphole-2,2'-[1,3,2,4]diazaphosphetidine-4',2''-[1,3,2]-benzodioxaphosphole]; 2,4'-Difluoro, *in* D-01018
84635-69-8 1',3'-Diphenyldispiro[1,3,2-benzodioxaphosphole-2,2'-[1,3,2,4]diazaphosphetidine-4',2''-[1,3,2]-benzodioxaphosphole]; 2,4'-Dibromo, *in* D-01018
84635-70-1 1',3'-Diphenyldispiro[1,3,2-benzodioxaphosphole-2,2'-[1,3,2,4]diazaphosphetidine-4',2''-[1,3,2]-benzodioxaphosphole]; 2,4'-Diisopropoxy, *in* D-01018
84681-48-1 Benzyl bis(1-aziridinyl)phosphinate, *in* B-00091
84711-20-6 Ifosfamide; (±)-*form*, *in* I-00001
84759-25-1 1*H*-1,3-Benzazaphosphole; 1-Me, *in* B-00005
84834-74-2 Octyltriphenylphosphonium(1+); Chloride, *in* O-00047
84890-90-4 Antibiotic I5*B*; Antibiotic I5*B*1, *in* A-00139
84890-91-5 Antibiotic K 26, A-00140
84905-98-6 Tetrapalmitoylphosphatidylglycerol, T-00247
84923-95-5 Dipentylphosphinic acid; Fluoride, *in* D-00981
84923-96-6 Dihexylphosphinic acid; Fluoride, *in* D-00434
85000-72-2 2,8,9-Trioxa-1-phosphatricyclo[3.3.1.1^{3,7}]decane; 1-Selenide, *in* T-00591
85046-80-6 Dimethyl acetylphosphoramidate, *in* A-00029
85152-45-0 2,5-Dihydro-5,5-dimethyl-1,2-oxaphosphole 2-oxide; 2-Phenoxy, *in* D-00464
85152-46-1 2,5-Dihydro-5,5-dimethyl-1,2-oxaphosphole 2-oxide; 2-Diethylamino, *in* D-00464

85185-99-5 Methyl benzylmethylphosphinate, *in* B-00053

85207-39-2 Methyl 3-amino-3-phosphonopropanoate, *in* A-00122

85231-81-8 Diethyl pentylphosphoramidate, *in* P-00053

85249-44-1 3-Amino-3-phosphonopropanoic acid; (*R*)*-form*, *in* A-00122

85272-73-7 1,3*a*,3*b*,6-Tetrahydro-3,4,8,8-tetramethyl-1,6-diphenyl-8*H*-[1,3]-diphospholo[1,5-*e*;3,4-*c*′]bis[1,2,3]diazaphosphole; (3*aβ*,3*bα*,7*α*,9*α*)*-form*, *in* T-00140

85289-24-3 Tetrahydro-5,5-dimethyl-2-phenoxy-2*H*-1,3,2-oxazaphosphorine; 2-Oxide, *in* T-00098

85289-25-4 Tetrahydro-5,5-dimethyl-2-phenoxy-2*H*-1,3,2-oxazaphosphorine; 2-Sulfide, *in* T-00098

85289-31-2 2-Chlorotetrahydro-3,5,5-trimethyl-2*H*-1,3,2-oxazaphosphorine; 2-Oxide, *in* C-00183

85289-33-4 2-Chlorotetrahydro-3,5,5-trimethyl-2*H*-1,3,2-oxazaphosphorine; 2-Sulfide, *in* C-00183

85290-04-6 1,4-Dimethyl-3,5-diphenyl-1,4,2-diazaphospholidine 2-oxide; 2-Chloro, *in* D-00739

85290-05-7 1,4-Dimethyl-3,5-diphenyl-1,4,2-diazaphospholidine 2-oxide; 2-Ethoxy, *in* D-00739

85290-14-8 *O*,*O*-Dibutyl phosphoroselenoate; K salt, *in* D-00154

85290-15-9 *O*,*O*-Dimethyl phosphoroselenoate; K salt, *in* D-00906

85357-61-5 Methyl cyclohexylphenylphosphinate, *in* C-00250

85414-46-6 1,3-Di-*tert*-butyl-2,4,4-trimethyl-1,3-diaza-2-phospha-4-silacyclobutane, *in* D-00087

85421-75-6 2,3-Bis(2,4,6-tri-*tert*-butylphenyl)thiadiphosphirane; (2*RS*,3*RS*)*-form*, *in* B-00397

85430-61-1 Bis(2-oxo-3-oxazolidinyl)-phosphinic acid; Anhydride, *in* B-00378

85437-50-9 Methyl *P*-dichloromethylphosphonamidate, *in* D-00174

85501-46-8 Ethyl hydrogen (4-nitrophenyl)-phosphonate, *in* N-00057

85505-00-6 2,4,6-Tris(2,4,6-trimethylphenyl)-1,3,5,2,4,6-trithiatriphosphorinane, T-00862

85556-99-6 2-Mercapto-1,3,2-dioxaphosphorinane 2-sulfide; K salt, *in* M-00013

85599-22-0 Diphenyl (4-methoxyphenyl)-phosphonate, *in* M-00072

85599-24-2 Diphenyl (4-aminophenyl)-phosphonate, *in* A-00111

85653-24-3 (2-Aminopropyl)phosphonic acid; (+)*-form*, *in* A-00126

85656-05-9 Isopropylphenylphosphinic acid; (±)*-form*, Azide, *in* I-00043

85656-06-0 Methyl *P*-methyl-*N*-phenylphosphonamidate, *in* M-00231

85656-09-3 Methyl *N*-*tert*-butyl-*P*-methylphosphonamidate, *in* B-00572

85684-40-8 Octahydro-1,3,5,7-tetraphenyl-1,5,3,7-diazadiphosphocine; 3,7-Disulfide, *in* O-00026

85684-42-0 Octahydro-3,7-dimethyl-1,3,5,7-tetraphenyl-1,5,3,7-diazadiphosphocinium iodide, *in* O-00026

85685-74-1 1-*tert*-Butyl-1,2,4-azadiphosphetidine; 2,4-Diisopropoxy, *in* B-00543

85741-82-8 *o*-Phenylene phosphoryl azide, *in* A-00148

85913-94-6 *O*,*O*-Diphenyl phosphorothioate; Triethylammonium salt, *in* D-01141

85915-09-9 Diethyl (3-cyanophenyl)-phosphonate, *in* C-00224

86012-31-9 Dibromo(4-fluorophenyl)phosphine, *in* F-00042

86012-32-0 Dibromocyclohexylphosphine, *in* C-00267

86012-34-2 Dibromo(chloromethyl)phosphine, *in* C-00115

86030-42-4 Methyl 4-morpholinylphosphonochloridite, *in* M-00526

86030-43-5 Methyl diisopropylphosphoramidochloridite, *in* D-00654

86030-44-6 Methyl 1-pyrrolidinylphosphonochloridite, *in* P-00519

86126-37-6 Diphenyl cyclooctylphosphoramidate, D-00998

86334-02-3 4-Methoxydinaphtho[2,1-*d*,1′,2′-*f*][1,3,2]-dioxaphosphepin 4-oxide, *in* H-00131

86343-95-5 2-Hydroxy-4-methyl-1,3,2-dioxaphosphorinane 2-selenide, H-00157

86428-81-1 2-Mercapto-4,4,5,5-tetramethyl-1,3,2-dioxaphospholane 2-sulfide; NH_4 salt, *in* M-00019

86507-04-2 2-Hydroxynaphtho[2,3-*d*]-1,3,2-dioxaphosphole 2-oxide; 2-Methoxy (Me ester), *in* H-00164

86539-32-4 (2,4,6-Tri-*tert*-butylphenyl)-phosphine; Oxide, *in* T-00369

86569-32-6 2-Chloro-4-methyl-1,3,2-dioxaphosphorinane 2-sulfide; (2*RS*,4*RS*)*-form*, *in* C-00095

86600-99-9 *P*,*P*-Bis(diethylamino)-*N*-ethyl-*P*-fluorophosphine imide, *in* P-00012

86601-00-5 *P*,*P*-Bis(dipropylamino)-*P*-fluoro-*N*-propylphosphazene, *in* P-00043

86601-01-6 *N*-Butyl-*P*,*P*-bis(dibutylamino)-*P*-fluorophosphazene, *in* P-00004

86601-02-7 *P*,*P*-Bis(diethylamino)-*P*-fluoro-*N*-phenylphosphazene, *in* T-00054

86703-68-6 3,5,5-Trimethyl-1,2-oxaphospholan-3-ol 2-oxide; (2*RS*,3*SR*)*-form*, 2-Methoxy, *in* T-00519

86703-69-7 2-Methoxy-3,5,5-trimethyl-1,2-oxaphospholan-3-ol 2-oxide, *in* T-00519

86919-76-8 2-Hydroxy-4-methyl-1,3,2-dioxaphosphorinane 2-sulfide; (2*RS*,4*RS*)*-form*, Dicyclohexylammonium salt, *in* H-00158

86926-28-5 *P*,*P*′-1,2-Ethanediylbis[*N*,*N*,*N*′,*N*′-tetraethylphosphonous diamide], *in* E-00008

86926-30-9 *P*,*P*′-1,4-Butanediylbis[bis *N*,*N*-diethylphosphonous diamide], *in* B-00515

86936-48-3 *N*,*N*,*N*′,*N*′-Tetraethyl-*P*-(2-phenylethenyl)phosphonous diamide, *in* P-00161

86947-46-8 4-Methyl-2,6-dioxa-7-aza-1-phosphabicyclo[2.2.2]-octane 1-oxide; 7-Benzyl, *in* M-00124

86947-47-9 4-Methyl-2,6-dioxa-7-aza-1-phosphabicyclo[2.2.2]-octane 1-oxide; 7-Ph, *in* M-00124

86947-48-0 4-Methyl-2,6-dioxa-7-aza-1-phosphabicyclo[2.2.2]-octane 1-oxide; 7-*tert*-Butyl, *in* M-00124

87079-91-2 1,2-Oxaphosphorinane; 2-Ph, *in* O-00054

87363-91-5 6-Methyl-4-methylene-4*H*-1,3,2-dioxaphosphorin; 2-Ethoxy, *in* M-00167

87363-94-8 6-Methyl-4-methylene-4*H*-1,3,2-dioxaphosphorin; 2-Me, 2-oxide, *in* M-00167

87367-43-9 *N*,*N*-Diethyl-2,6-dimethyl-1-phenyl-1,4-phosphaborin-4(1*H*)-amine; 1-Oxide, *in* D-00285

87367-44-0 *N*,*N*-Diethyl-2,6-dimethyl-1-phenyl-1,4-phosphaborin-4(1*H*)-amine; 1-Sulfide, *in* D-00285

87423-10-7 Fosfazinomycin *A*, F-00058

87423-11-8 Fosfazinomycin *B*, F-00059

87453-82-5 *N*,*N*-Diethyl-2,6-dimethyl-1-pyrrenyl-1,4-phosphaborin-4(1*H*)-amine; 1-Selenide, *in* D-00285

87482-32-4 [Bis(diphenylstibino)-methylene]triphenylphosphorane, B-00270

87648-02-0 Tetraphenyl methylenebisphosphonite, *in* M-00024

87810-56-8 Antibiotic CI 920, A-00136

87860-37-5 Antibiotic PD 113270, *in* A-00136

87860-38-6 Antibiotic PD 113271, *in* A-00136

87863-65-8 *N*,*N*,*N*′,*N*′,*N*″,*N*″-Hexamethyl-tetraaminophosphonium chloride, *in* H-00077

87913-21-1 Tagetitoxin, T-00002

87934-75-6 *N*,*N*,*N*′,*N*′-Tetramethyl-*P*-1-piperidinylphosphonous diamide, *in* P-00432

87992-68-5 Trimethyl (diethoxyphosphinyl)-phosphorimidate, *in* D-00257

88001-79-0 Dithioxo(2,4,6-tri-*tert*-butylphenyl)phosphorane, D-01254

88017-79-2 Diphenyl hypodiphosphonous acid; Dibromide, *in* D-01020

88133-77-1 1,4:3,6-Dianhydro-2,5-deoxy-2,5-bis(diphenylphosphino)iditol; L*-form*, *in* D-00026

88142-41-0 Tetrahydro-2-methoxy-2*H*-1,3,2-oxazaphosphorine, T-00111

88239-51-4 *O*,*O*-Diphenyl phenylphosphonothioate, *in* P-00244

88286-55-9 N^{δ}-Phosphosulfamylornithine, P-00419

88515-16-6 Triethyl pentafluorophenyl-phosphorimidate, *in* P-00016

88576-13-0 4,5-Dimethyl-2-trimethylsilyloxy-1,3,2-dioxaphospholane, D-00930

88652-78-2 *N*-*tert*-Butyl-*P*-methylphosphonamidic acid; Chloride, *in* B-00572

88766-69-2 Mono(4-chlorophenyl) phosphate; Bis(cyclohexylammonium) salt, *in* M-00458

88834-10-0 *P*,*P*-Diethyl-*N*-methylphosphinous amide, *in* D-00334

88834-11-1 *P*,*P*-Diisopropyl-*N*-methylphosphinous amide, *in* D-00649

88834-15-5 *N*-*tert*-Butyl-*P*,*P*-diethylphosphinous amide, *in* D-00334

88946-46-7 5,5-Dimethyl-2-dimethylaminotetrahydro-2*H*-1,3,2-oxazaphosphorine; 2-Oxide, *in* D-00726

89104-48-3 2,3-Butylene chlorophosphate, *in* C-00056
89104-49-4 2-Chloro-4,5-dimethyl-1,3,2-dioxaphospholane; (4*R*,5*R*)-*form*, *in* C-00056
89176-09-0 (2,4,6-Tri-*tert*-butylphenyl)-phosphine; Sulfide, *in* T-00369
89262-61-3 Butyldiiodophosphine, *in* B-00615
89262-65-7 *O,O*-Diethyl (trifluoromethyl)-phosphonothioate, *in* T-00481
89262-66-8 *O,S*-Diethyl (trifluoromethyl)-phosphonothioate, *in* T-00481
89596-83-8 Trimethylsilyl *N,N,N'*-tris(trimethylsilyl)-phosphonamidimidate, *in* T-00867
89699-33-2 (Chloroacetyl)phosphonic acid, C-00020
89767-35-1 1,3-Dihydrospiro[2*H*-1,3,2-benzodiazaphosphole-2,2'-[1,3,2]benzodioxaphosphole]; 1-Benzyl, *in* D-00574
89777-65-7 Cyclopentadienylphosphonous acid; Dichloride, *in* C-00275
89915-88-8 Tris(diphenylphosphino)methane; Dioxide, monosulfide, *in* T-00775
89915-89-9 Tris(diphenylphosphino)methane; Trioxide, *in* T-00775
89982-28-5 1,3,5,7-Tetramethyl-1,3,5,7-tetraaza-4-phospha(4-P^V)spiro[3.3]heptane-2,6-dione; 4-Azido, *in* T-00234
89982-50-3 1,2-Bis[1-(2,2,6,6-tetramethylpiperidino)]-diphosphene; (*E*)-*form*, *in* B-00392
89992-69-8 2-Cyanoethyl dimethyl-phosphoramidochloridite, *in* D-00862
89992-70-1 2-Cyanoethyl diisopropylphosphoramidochloridite, *in* D-00654
90119-88-3 Antibiotic BMG 59-R2, A-00135
90285-16-8 Cyclopentadienylphosphonous acid; Bromide, *in* C-00275
90293-10-0 Salithion; (*S*)-*form*, *in* S-00001
90293-16-6 Salithion; (*R*)-*form*, *in* S-00001
90430-77-6 *O,O*-Diphenyl (diphenoxyphosphinothioyl)-phosphoramidothioate, D-01004
90501-97-6 Tris[1,2-benzenediolato(2−)-*O,O'*] phosphate(1−); 4-Methylmorpholinium salt, *in* T-00730
90599-65-8 2,3-Bis(2,4,6-tri-*tert*-butylphenyl)-1,2,3-selenadiphosphirane, B-00396
90599-66-9 Diselenoxo(2,4,6-tri-*tert*-butylphenyl)phosphorane, D-01235
90696-72-3 15,16-Dimethyl-1,4,8,11-tetraaza-15,16-diphosphatricyclo[9.3.1.$1^{4,8}$]-hexadecane, D-00921
91043-34-4 Diethyl dipropylphosphoramidate, *in* D-01214
91425-19-3 *P,P'*-Bis(2,4,6-tri-*tert*-butylphenyl)-1,3-diphosphaallene, B-00394
91670-70-1 Neopentylene dimethylphosphoramidoselenoate, *in* D-00693
91740-93-1 2-Phenoxy-4,5-diphenyl-1,3,2-dioxaphospholane; (4*RS*,5*RS*)-*form*, *in* P-00069
92136-94-2 2-Hydroxy-1,3,2-dioxaphospholane 2-oxide; Cyclohexylammonium salt, *in* H-00134
92470-71-8 ▷Salithion; (±)-*form*, *in* S-00001
92567-89-0 Phosalacine, P-00339
92611-10-4 Methyl tetraisopropylphosphorodiamidite, *in* T-00172
92760-41-3 Profenofos; (±)-*form*, *in* P-00440
93032-39-4 2-Adamantylphosphoramidic acid; Dichloride, *in* A-00037
93183-35-8 Methyl di-4-morpholinylphosphinite, *in* D-00936
93214-19-8 2,2,2,3,3*a*,11*b*-Hexahydro-2,2,2-triphenylphenanthro[9,10-*d*]-1,3,2-oxazaphosphole; (9*RS*,10*RS*)-*form*, *in* H-00059
93289-64-6 Octicidin, O-00038
93374-14-2 Phenyl di-4-morpholinylphosphinite, *in* D-00936
93381-80-7 2-Chlorohexahydro-4*H*-1,3,2-benzodioxaphosphorin; (2*RS*,4*aSR*,8*aSR*)-*form*, *in* C-00085
93523-22-9 1,2-Dihydro-1-methyl-4*H*-3,1,2-benzoxazaphosphorin-4-one; 1-Methoxy, *in* D-00523
93523-23-0 1,2-Dihydro-1-methyl-4*H*-3,1,2-benzoxazaphosphorin-4-one; 1-Ethoxy, *in* D-00523
93523-24-1 1,2-Dihydro-1-methyl-4*H*-3,1,2-benzoxazaphosphorin-4-one; 1-Isopropoxy, *in* D-00523
93633-03-5 2-*tert*-Butylamino-4-methyl-1,3,2-dioxaphosphorinane; (2*RS*,4*SR*)-*form*, 2-Oxide, *in* B-00541
93633-04-6 1,3-Butylene *tert*-butylphosphoramidate, *in* B-00541
93646-66-3 3,5-Diphenyl-1*H*-1,2,4-diazaphosphole, D-01000
93646-72-1 3,5-Diphenyl-1*H*-1,2,4-diazaphosphole; 1-Me, *in* D-01000
93701-60-1 3,9-Diphenoxy-2,4,8,10-tetraoxa-3,9-diphosphaspiro[5.5]-undecane; Monooxide, *in* D-00991
93714-99-9 3,5-Diphenyl-1*H*-1,2,4-diazaphosphole; 1-Ph, *in* D-01000
93768-49-1 Antibiotic I5*B*; Antibiotic I5*B*2, *in* A-00139
93772-86-2 1,2-Dihydro-1-methyl-4*H*-3,1,2-benzoxazaphosphorin-4-one; 1-Chloro, *in* D-00523
94721-88-7 3,3,3-Tris-(4-morpholinophosphoranylidene)-1-(2,4,6-trinitrophenyl)triazene; (*E*)-*form*, *in* T-00833
94799-21-0 1,3*a*,3*b*,6-Tetrahydro-3,4,8,8-tetramethyl-1,6-diphenyl-8*H*-[1,3]-diphospholo[1,5-*e*;3,4-*c'*]bis[1,2,3]diazaphosphole; (3*aα*,3*bα*,7*α*,9*α*)-*form*, *in* T-00140
94890-89-8 6,7,8,9,15,16,17,18-Octahydro-9,18-diphenyl-dibenzo[*b,i*][1,8,4,11]-dithiadiphosphacyclotetradecin; (9*RS*,18*RS*)-*form*, *in* O-00014
95115-80-3 2-Chloro-4-isopropyl-5,5-dimethyl-1,3,2-dioxaphosphorinane; (2*RS*,4*RS*)-*form*, *in* C-00090
95115-81-4 2-Chloro-4-isopropyl-5,5-dimethyl-1,3,2-dioxaphosphorinane; (2*RS*,4*RS*)-*form*, 2-Oxide, *in* C-00090
95115-82-5 4-Isopropyl-2-methoxy-5,5-dimethyl-1,3,2-dioxaphosphorinane; (2*RS*,4*SR*)-*form*, *in* I-00033
95115-83-6 4-Isopropyl-2-methoxy-5,5-dimethyl-1,3,2-dioxaphosphorinane; (2*RS*,4*RS*)-*form*, *in* I-00033
95115-84-7 4-Isopropyl-2-methoxy-5,5-dimethyl-1,3,2-dioxaphosphorinane; (2*RS*,4*RS*)-*form*, 2-Selenide, *in* I-00033
95115-86-9 2-Chloro-4-isopropyl-5,5-dimethyl-1,3,2-dioxaphosphorinane; (2*RS*,4*SR*)-*form*, 2-Oxide, *in* C-00090
95152-88-8 Difficidin, D-00411
95152-89-9 Oxydifficidin, *in* D-00411
95531-55-8 [Bis(diphenylarsino)methylene]-triphenylphosphorane, B-00213
95539-24-5 *O,O*-Neopentylene *O*-phenyl phosphoroselenoate, *in* D-00790
95552-76-4 Di-*tert*-butylphosphazene, D-00106
95792-75-9 *N*-*tert*-Butyl-*P*-(2,4,6-tri-*tert*-butylphenyl)-metaphosphonimidate; (*E*)-*form*, *in* B-00639
95837-80-2 1,2-Di-*tert*-butyl-1,2-dimethyldiphosphine; (*RS,RS*)-*form*, Disulfide, *in* D-00089
95837-84-6 *tert*-Butylmethylphosphinodithioic acid anhydrosulfide; (*RS,RS*)-*form*, *in* B-00569
95888-14-5 *tert*-Butylmethylphosphinothioic anhydride; (*RS,SR*)-*form*, *in* B-00571
96688-95-8 Bis[bis(trimethylsilyl)-methylene]-9*H*-fluoren-9-ylphosphorane, B-00106
96705-27-0 Bis[bis(trimethylsilyl)-methylene]-9*H*-fluoren-9-ylphosphorane; $[Li(THF)_4]^+$ complex, *in* B-00106
96921-88-9 1,1-Bis(diphenylphosphino)-cyclopropane, B-00229
96921-89-0 1,1-Bis(diphenylphosphinyl)-cyclopropane, *in* B-00229
96921-90-3 1,1-Bis(diphenylphosphino)-cyclopropane; Disulfide, *in* B-00229
97106-56-4 1,1-Bis(diphenylphosphinothioyl)-ethylene, *in* B-00236
97322-65-1 *N,N*-Bis(diphenylphosphino)-*P,P*-diphenylphosphinous amide, B-00233
98060-75-4 2,3-Bis(2,4,6-tri-*tert*-butylphenyl)thiadiphosphirane, B-00397
98188-41-1 Diphenylphosphinodiselenoic acid; Anhydroselenide, *in* D-01064
98820-75-8 Alafosfalin; $(R)_C(R)_P$-*form*, *N*-Benzyloxycarbonyl, *in* A-00049
98820-76-9 Alafosfalin; $(R)_C(S)_P$-*form*, *N*-Benzyloxycarbonyl, *in* A-00049
98857-06-8 Alafosfalin; $(R)_C(R)_P$-*form*, *in* A-00049
99136-00-2 Bis(trimethylsilyl)(2-fluorophenyl)phosphonate, *in* F-00036
99136-02-4 Bis(trimethylsilyl)(2-methoxyphenyl)phosphonate, *in* M-00070
99136-05-7 Bis(trimethylsilyl)(3-methylphenyl)phosphonate, *in* M-00236
99136-07-9 Bis(trimethylsilyl)(4-aminophenyl)phosphonate, *in* A-00111
99136-08-0 Bis(trimethylsilyl)(4-bromophenyl)phosphonate, *in* B-00492
99632-61-8 2-Phenyl-1,2-thiaphospholane, P-00324
99632-63-0 2-Phenyl-1,2-thiaphosphorinane, P-00325
99706-39-5 Ethyl bis(2-hydroxyphenyl)-phosphinate, *in* B-00283
99706-43-1 Bis(2-hydroxyphenyl)phosphinic acid, B-00283
101409-46-5 1-(Di-*tert*-butyldiphosphinylidene)-methanediamine; *N,N,N',N'*-Tetra-Me, *in* D-00090

101409-48-7 1-(Di-*tert*-butyldiphosphinylidene)-methanediamine; *N,N,N′,N′*-Tetra-Et, *in* D-00090

102615-39-4 Tris(diphenylphosphino)methane; Oxide, disulfide, *in* T-00775

102691-36-1 Tetraisopropylphosphorodiamidous acid; 2-Cyanoethyl ester, *in* T-00172

103678-18-8 1,1,3,3-Tetrakis(dimethylamino)-1,3-diphosph(P^V)ete, T-00176

104189-33-5 Bis(trimethylsilyl) selenophosphonate, B-00440

104189-34-6 Bis(trimethylsilyl) tellurophosphonate, B-00441

104412-65-9 Bis(trimethylsilyl)(4-chlorophenyl)phosphonate, *in* C-00147

104412-66-0 Bis(trimethylsilyl)(4-methylphenyl)phosphonate, *in* M-00237

104412-67-1 Bis(trimethylsilyl)(4-fluorophenyl)phosphonate, *in* F-00038

Type of Compound Index

Sections appear in the following order:

Primary phosphines and derivatives
Secondary phosphines and derivatives
Tertiary phosphines and derivatives
Polyphosphines and derivatives
Phosphenes and derivatives
Phosphonium salts
Phosphoranes
Methylenephosphoranes and derivatives
Methylenephosphines
Iminophosphines
Hexacoordinate compounds
Phosphinidynes
Phosphorous acids and derivatives
Thiophosphorous acids and derivatives
Selenophosphorous acids and derivatives
Tellurophosphorous acids and derivatives
Phosphonous acids and derivatives
Thiophosphonous acids and derivatives
Selenophosphonous acids and derivatives
Tellurophosphonous acids and derivatives
Phosphinous acids and derivatives
Thiophosphinous acids and derivatives
Selenophosphinous acids and derivatives
Tellurophosphinous acids and derivatives
Phosphoric acids and derivatives
Thiophosphoric acids and derivatives
Selenophosphoric acids and derivatives
Tellurophosphoric acids and derivatives
Imidophosphoric acids and derivatives
Phosphonic acids and derivatives
Thiophosphonic acids and derivatives
Selenophosphonic acids and derivatives
Tellurophosphonic acids and derivatives
Imidophosphonic acids and derivatives
Phosphinic acids and derivatives
Thiophosphinic acids and derivatives
Selenophosphinic acids and derivatives
Tellurophosphinic acids and derivatives
Imidophosphinic acids and derivatives
Pentaoxyphosphoranes and derivatives
Phosphenous and phosphenic acids and derivatives
Thiophosphenous and thiophosphenic acids and derivatives
Selenophosphenous and selenophosphenic acids and derivatives
Tellurophosphenous and tellurophosphenic acids and derivatives

Followed by a classification of compounds according to the number and size of phosphorus-containing rings present.

Type of Compound Index

Primary phosphines, RPH_2

Benzylphosphine, B-00058
(2-Bromophenyl)phosphine, B-00486
(4-Bromophenyl)phosphine, B-00487
Butylphosphine, B-00588
tert-Butylphosphine, B-00589
(Chloromethyl)phosphine, C-00104
(2-Chlorophenyl)phosphine, C-00139
(3-Chlorophenyl)phosphine, C-00140
(4-Chlorophenyl)phosphine, C-00141
▷Cyclohexylphosphine, C-00252
Decylphosphine, D-00011
1,2-Diphosphinobenzene, D-01172
1,4-Diphosphinobenzene, D-01173
1,4-Diphosphinobutane, D-01174
1,10-Diphosphinodecane, D-01175
▷1,2-Diphosphinoethane, D-01176
1,6-Diphosphinohexane, D-01177
Diphosphinomethane, D-01178
1,5-Diphosphinopentane, D-01179
1,3-Diphosphinopropane, D-01180
▷Ethylphosphine, E-00123
(3-Fluorophenyl)phosphine, F-00033
(4-Fluorophenyl)phosphine, F-00034
Heptylphosphine, H-00005
Hexylphosphine, H-00096
Isopropylphosphine, I-00048
(4-Methoxyphenyl)phosphine, M-00067
(2-Methylphenyl)phosphine, M-00216
(3-Methylphenyl)phosphine, M-00217
(4-Methylphenyl)phosphine, M-00218
▷Methylphosphine, M-00277
(2-Methylpropyl)phosphine, M-00374
Nonylphosphine, N-00075
Octylphosphine, O-00041
Pentylphosphine, P-00049
(2-Phenylethenyl)phosphine, P-00152
(2-Phenylethyl)phosphine, P-00163
▷Phenylphosphine, P-00208
Phosphinecarbonitrile, P-00354
3-Phosphinopropanenitrile, P-00358
Propylphosphine, P-00471
(2,4,6-Tri-*tert*-butylphenyl)phosphine, T-00369
(Trifluoromethyl)phosphine, T-00474

Primary phosphine oxides, RPH_2O

Octylphosphine; Oxide, *in* O-00041
Phenylphosphine; Oxide, *in* P-00208
(2,4,6-Tri-*tert*-butylphenyl)phosphine; Oxide, *in* T-00369

Primary phosphine sulfides, RPH_2S

Phenylphosphine; Sulfide, *in* P-00208
(2,4,6-Tri-*tert*-butylphenyl)phosphine; Sulfide, *in* T-00369

Primary phosphine imides, $RPH_2{=}NR$

N-Acetyl-*P,P,P*-triphenylphosphine imide, *in* T-00626
N-Benzoyl-*P,P,P*-triphenylphosphine imide, *in* T-00626
N-Bromo-*P,P,P*-triphenylphosphine imide, *in* T-00626
N-tert-Butyl-*P,P,P*-triphenylphosphine imide, *in* T-00626
N-Chloro-*P,P,P*-triphenylphosphine imide, *in* T-00626
N-Cyano-*P,P,P*-triphenylphosphine imide, *in* T-00626
N-Ethyl-*P,P,P*-triphenylphosphine imide, *in* T-00626
N-Isopropyl-*P,P,P*-triphenylphosphine imide, *in* T-00626
N-Methyl-*P,P,P*-triphenylphosphine imide, *in* T-00626
N-1-Naphthyl-*P,P,P*-triphenylphosphine imide, *in* T-00626
N-2-Naphthyl-*P,P,P*-triphenylphosphine imide, *in* T-00626
N-Trimethylgermyl-*P,P,P*-triphenylphosphine imide, *in* T-00626
N-Trimethylsilyl-*P,P,P*-triphenylphosphine imide, *in* T-00626
N-Trimethylstannyl-*P,P,P*-triphenylphosphine imide, *in* T-00626
P,P,P-Triphenyl-*N*-phenylsulfonylphosphine imide, *in* T-00626
P,P,P-Triphenylphosphine imide, T-00626
P,P,P-Triphenyl-*N*-propylphosphine imide, *in* T-00626

Secondary phosphines, R_2PH

Bis(2-chlorophenyl)phosphine, B-00134
Bis(3-chlorophenyl)phosphine, B-00135
Bis(4-chlorophenyl)phosphine, B-00136
Bis(2-cyanoethyl)phosphine, B-00153
Bis(3-fluorophenyl)phosphine, B-00276
Bis(4-fluorophenyl)phosphine, B-00277
Bis(2-methoxyphenyl)phosphine, B-00296
Bis(3-methoxyphenyl)phosphine, B-00297
Bis(4-methoxyphenyl)phosphine, B-00298
▷Bis(2-methylphenyl)phosphine, B-00315
▷Bis(3-methylphenyl)phosphine, B-00316
Bis(4-methylphenyl)phosphine, B-00317
Bis(1-methylpropyl)phosphine, B-00343
Bis(2-methylpropyl)phosphine, B-00344
Bis(4-nitrophenyl)phosphine, B-00372
Bis(2-phenylethyl)phosphine, B-00387
1,2-Bis(phenylphosphino)ethane, B-00390
Bis(2,4,6-trimethylphenyl)phosphine, B-00426
Cyclohexylphenylphosphine, C-00249
5*H*-Dibenzophosphole, D-00040
Dibenzylphosphine, D-00054
Dibutylphosphine, D-00107
Di-*tert*-butylphosphine, D-00108
Dicyclohexylphosphine, D-00210
▷Diethylphosphine, D-00321
Dihexylphosphine, D-00433
2,3-Dihydro-1*H*-isophosphindole, D-00509
1,2-Dihydroisophosphinoline, D-00510
5,10-Dihydro-10-methylacridophosphine, D-00515
2,5-Dihydro-3-methyl-1*H*-phosphole, D-00530
5,10-Dihydrophenophosphazine; 10-Oxide, *in* D-00542
5,6-Dihydrophosphanthridene, D-00568
2,3-Dihydro-1*H*-phosphole, D-00571
2,5-Dihydro-1*H*-phosphole, D-00572
Diisopropylphosphine, D-00640
▷Dimethylphosphine, D-00827
Di-1-naphthylphosphine, D-00939
Di-2-naphthylphosphine, D-00940
Dioctylphosphine, D-00957
Dipentylphosphine, D-00980
2,6-Diphenyl-1,4-azaphosphorine, D-00996
▷Diphenylphosphine, D-01037
1,2-Diphospholane, D-01181
Dipropylphosphine, D-01204
1-Ethyl-2-phenyl-1,3-azaphospholidine, *in* P-00090
2-Ethyl-2,3,4,5-tetrahydro-2-methyl-1*H*-1,3-benzazaphosphepine, E-00188
Methylphenylphosphine, M-00215
1,4-Oxaphosphorinane, O-00055
2,2,3,4,4-Pentamethylphosphetane, P-00022
2-Phenyl-1,3-azaphospholidine, P-00090
Phosphirane, P-00361
Phospholane, P-00371
1*H*-Phosphole, P-00372
Phosphorinane, P-00404

Secondary phosphine oxides, R_2PHO

Bis(3-chlorophenyl)phosphine; Oxide, *in* B-00135
Bis(4-chlorophenyl)phosphine; Oxide, *in* B-00136
Bis(2-cyanoethyl)phosphine; Oxide, *in* B-00153
Bis(2-methoxyphenyl)phosphine; Oxide, *in* B-00296

Bis(3-methoxyphenyl)phosphine; Oxide, *in* B-00297
Bis(4-methoxyphenyl)phosphine; Oxide, *in* B-00298
Bis(2-methylphenyl)phosphine; Oxide, *in* B-00315
Bis(4-methylphenyl)phosphine; Oxide, *in* B-00317
Bis(2-methylpropyl)phosphine; Oxide, *in* B-00344
1,2-Bis(phenylphosphino)ethane; (*RS,RS*)-*form*, Dioxide, *in* B-00390
1,2-Bis(phenylphosphino)ethane; (*RS,SR*)-*form*, Dioxide, *in* B-00390
Bis(2,4,6-trimethylphenyl)phosphine; Oxide, *in* B-00426
Cyclohexylphenylphosphine; Oxide, *in* C-00249
Dibenzylphosphine; Oxide, *in* D-00054
Dibutylphosphine; Oxide, *in* D-00107
Di-*tert*-butylphosphine; Oxide, *in* D-00108
Dicyclohexylphosphine; Oxide, *in* D-00210
Diethylphosphine; Oxide, *in* D-00321
Dihexylphosphine; Oxide, *in* D-00433
7,12-Dihydrobenzo[*e*]phenophosphazine; 7-Oxide, *in* D-00440
10,11-Dihydro-5*H*-dibenzo[*b,f*]phosphepin; Oxide, *in* D-00451
5,10-Dihydro-5,10-diphenylacridophosphine 5-oxide, *in* D-00548
Diisopropylphosphine; Oxide, *in* D-00640
Dimethylphosphine oxide, D-00828
Di-1-naphthylphosphine; Oxide, *in* D-00939
Di-2-naphthylphosphine; Oxide, *in* D-00940
▷Dioctylphosphine; Oxide, *in* D-00957
Dipentylphosphine; Oxide, *in* D-00980
Diphenylphosphine; Oxide, *in* D-01037
Diphenylphosphine oxide, D-01038
Diphenylphosphinous acid, D-01089
Dipropylphosphine; Oxide, *in* D-01204
Methylphenylphosphine; Oxide, *in* M-00215

Secondary phosphine sulfides, R_2PHS

Bis(2-methylpropyl)phosphine; Sulfide, *in* B-00344
1,2-Bis(phenylphosphino)ethane; (*RS,RS*)-*form*, Disulfide, *in* B-00390
1,2-Bis(phenylphosphino)ethane; (*RS,SR*)-*form*, Disulfide, *in* B-00390
Cyclohexylphenylphosphine; Sulfide, *in* C-00249
Dibutylphosphine; Sulfide, *in* D-00107
Dicyclohexylphosphine; Sulfide, *in* D-00210
Diethylphosphine; Sulfide, *in* D-00321
Diisopropylphosphine; Sulfide, *in* D-00640
Dimethylphosphine; Sulfide, *in* D-00827
Diphenylphosphine; Sulfide, *in* D-01037
Dipropylphosphine; Sulfide, *in* D-01204
1-Methylphosphonane; 1-Oxide, *in* M-00285

Secondary phosphine selenides, R_2PHSe

Bis(2-methylpropyl)phosphine; Selenide, *in* B-00344
Diphenylphosphine; Selenide, *in* D-01037

Secondary phosphine tellurides, R_2PHTe

Di-*tert*-butylphosphinotellurous acid, D-00116
Diisopropylphosphinotellurous acid, D-00645

Tertiary phosphines, R_3P

Acetyldimethylphosphine, A-00015
Acetyldiphenylphosphine, A-00016
Acridophosphine, A-00031
(2-Aminoethyl)butylphenylphosphine; (+)-*form*, *in* A-00074
(2-Aminoethyl)butylphenylphosphine; (−)-*form*, *in* A-00074
(2-Aminoethyl)butylphenylphosphine; (±)-*form*, *in* A-00074
5,10[1′,2′]-Benzenophenophosphazine, B-00015
5,10-[1′,2′]Benzophosphenanthrene, B-00024
Benzoyldiphenylphosphine, B-00030
5-Benzyl-5,6-dihydro-6-phenylphosphanthridine, B-00043
1-Benzyl-2,5-dihydro-1*H*-phosphole, *in* D-00572
Benzyldiphenylphosphine, B-00046
Benzylethylphenylphosphine; (±)-*form*, *in* B-00047
Benzylmethylphenylphosphine; (*R*)-*form*, *in* B-00051
Benzylmethylphenylphosphine; (*S*)-*form*, *in* B-00051
Benzylmethylphenylphosphine; (±)-*form*, *in* B-00051
Benzylmethylpropylphosphine, B-00054
6-Benzyl-2-oxa-6-phosphatricyclo[3.3.1.1^{3,7}]decane, B-00055
1-Benzyl-3-phenyl-1,3-azaphosphetidine, *in* P-00087
Bicyclo[2.2.1]hept-5-ene-2,3-diylbis(methylene)-bis[diphenylphosphine]; (2*R**,3*R**)-*form*, *in* B-00075
Bicyclo[2.2.1]hept-5-ene-2,3-diylbis(methylene)-bis[diphenylphosphine]; (2*S**,3*R**)-*form*, *in* B-00075
[1,1′-Biphenyl]-4-yldiphenylphosphine, B-00078
[1,1′-Biphenyl]-4-yl-1-naphthalenylphenylphosphine; (*R*)-*form*, *in* B-00081
[1,1′-Biphenyl]-4-yl-1-naphthalenylphenylphosphine; (*S*)-*form*, *in* B-00081
[1,1′-Biphenyl]-4-yl-1-naphthalenylphenylphosphine; (±)-*form*, *in* B-00081
Bis(acetoxymethyl)phenylphosphine, *in* B-00281
Bis(2-aminoethyl)phenylphosphine, B-00086
Bis(3-aminopropyl)phenylphosphine, B-00090
1,2-Bis[(bis-2-methylphenyl)phosphino]ethane, B-00103
1,2-Bis[(bis-4-methylphenyl)phosphino]ethane, B-00104
Bis(2-cyanoethyl)phenylphosphine, B-00152
1,4-Bis(dibutylphosphino)butane, B-00157
1,2-Bis(dibutylphosphino)ethylene; (*E*)-*form*, *in* B-00159
1,2-Bis(dibutylphosphino)ethylene; (*E*)-*form*, Monoxide, *in* B-00159
1,2-Bis(dicyclohexylphosphino)ethane, B-00167
4,5-Bis[(dicyclohexylphosphino)methyl]-2,2-dimethyl-1,3-dioxolane, B-00168
1,3-Bis(dicyclohexylphosphino)propane, B-00169
Bis(diethylphosphino)acetylene, B-00178
2,2′-Bis(diethylphosphino)biphenyl, B-00179
1,2-Bis(diethylphosphino)ethane, B-00180
1,2-Bis(diethylphosphino)ethylene, B-00181
Bis[3-(dimethylamino)propyl]phenylphosphine, B-00195
1,4-Bis(dimethylphosphino)butane, B-00199
1,2-Bis(dimethylphosphino)ethane, B-00200
Bis(dimethylphosphino)methane, B-00202
[Bis(dimethylphosphino)methylene]trimethylphosphorane, B-00203
Bis[(dimethylphosphino)methyl]methylphosphine, B-00204
1,5-Bis(dimethylphosphino)pentane, B-00205
1,3-Bis(dimethylphosphino)propane, B-00206
Bis[(dimethylphosphinyl)methyl]methylphosphine, *in* B-00204
▷Bis[2-(diphenylarsino)ethyl]phenylphosphine, B-00212
Bis[2-(diphenylarsino)phenyl]phenylphosphine, B-00214
Bis(diphenylphosphino)acetylene, B-00216
2,3-Bis(diphenylphosphino)bicyclo[2.2.1]heptane, *in* B-00220
5,6-Bis(diphenylphosphino)bicyclo[2.2.1]hept-2-ene; (1*R*,5*R*,6*R*)-*form*, *in* B-00220
5,6-Bis(diphenylphosphino)bicyclo[2.2.1]hept-2-ene; (1*S*,5*S*,6*S*)-*form*, *in* B-00220
5,6-Bis(diphenylphosphino)bicyclo[2.2.1]hept-2-ene; (1*RS*,5*RS*,6*RS*)-*form*, *in* B-00220
2,2′-Bis(diphenylphosphino)-1,1′-binaphthyl; (*R*)-*form*, *in* B-00221
2,2′-Bis(diphenylphosphino)-1,1′-binaphthyl; (*S*)-*form*, *in* B-00221
2,2′-Bis(diphenylphosphino)-1,1′-binaphthyl; (±)-*form*, *in* B-00221
3,3′-Bis(diphenylphosphino)biphenyl, B-00222
1,4-Bis(diphenylphosphino)-1,3-butadiyne, B-00223
1,3-Bis(diphenylphosphino)butane; (*R*)-*form*, *in* B-00224
1,3-Bis(diphenylphosphino)butane; (*S*)-*form*, *in* B-00224
1,4-Bis(diphenylphosphino)butane, B-00225
2,3-Bis(diphenylphosphino)butane; (2*R*,3*R*)-*form*, *in* B-00226
2,3-Bis(diphenylphosphino)butane; (2*S*,3*S*)-*form*, *in* B-00226
1,4-Bis(diphenylphosphino)-2-butyne, B-00227
3,4-Bis(diphenylphosphino)-3-cyclobutene-1,2-dione, B-00228
1,1-Bis(diphenylphosphino)cyclopropane, B-00229
1,2-Bis(diphenylphosphino)cyclopropane; (1*RS*,2*RS*)-*form*, *in* B-00230
2,2′-Bis(diphenylphosphino)diethylamine, B-00231
▷1,2-Bis(diphenylphosphino)ethane, B-00234
1,2-Bis(diphenylphosphino)-1,2-ethanedione, B-00235
1,1-Bis(diphenylphosphino)ethylene, B-00236
1,2-Bis(diphenylphosphino)ethylene, B-00237
N,N-Bis[(2-diphenylphosphino)ethyl]ethylamine, B-00238
Bis[2-(diphenylphosphino)ethyl]phenylphosphine, B-00239

4,7-Bis[2-(diphenylphosphino)ethyl]-1,1,10,10-tetraphenyl-1,4,7,10-tetraphosphadecane, B-00240
1,1′-Bis(diphenylphosphino)ferrocene, B-00241
1,7-Bis(diphenylphosphino)heptane, B-00242
1,6-Bis(diphenylphosphino)hexane, B-00243
2,11-Bis[(diphenylphosphino)methyl]benzo[*c*]phenanthrene, B-00246
5,6-Bis(diphenylphosphino)-2-methylbicyclo[2.2.1]hept-2-ene; (1*R**,5*R**,6*R**)-*form*, *in* B-00247
5,6-Bis(diphenylphosphino)-2-methylbicyclo[2.2.1]hept-2-ene; (1*S**,5*S**,6*S**)-*form*, *in* B-00247
2,2′-Bis[(diphenylphosphino)methyl]-1,1′-binaphthyl; (*S*)-*form*, *in* B-00248
2,2′-Bis[(diphenylphosphino)methyl]-1,1′-binaphthyl; (±)-*form*, *in* B-00248
1,2-Bis(diphenylphosphinomethyl)cyclobutane; (1*R*,2*R*)-*form*, *in* B-00249
1,2-Bis(diphenylphosphinomethyl)cyclobutane; (1*RS*,2*RS*)-*form*, *in* B-00249
1,2-Bis(diphenylphosphinomethyl)cyclobutane; (1*RS*,2*SR*)-*form*, *in* B-00249
1,2-Bis(diphenylphosphinomethyl)cyclohexane; (1*R*,2*R*)-*form*, *in* B-00250
1,2-Bis(diphenylphosphinomethyl)cyclohexane; (1*S*,2*S*)-*form*, *in* B-00250
1,2-Bis(diphenylphosphinomethyl)cyclohexane; (1*RS*,2*SR*)-*form*, *in* B-00250
[Bis(diphenylphosphino)methylene]triphenylphosphorane, B-00252
Bis[(diphenylphosphino)methyl]phenylphosphine, B-00253
1,4-Bis(diphenylphosphino)naphthalene, B-00254
1,5-Bis(diphenylphosphino)naphthalene, B-00255
2,6-Bis(diphenylphosphino)naphthalene, B-00256
2,7-Bis(diphenylphosphino)naphthalene, B-00257
1,8-Bis(diphenylphosphino)octane, B-00258
2,4-Bis(diphenylphosphino)pentane, B-00260
9,10-Bis(diphenylphosphino)phenanthrene, B-00261
Bis[2-(diphenylphosphino)phenyl]phenylphosphine, B-00262
Bis[4-(diphenylphosphino)phenyl]phenylphosphine, B-00263
1,2-Bis(diphenylphosphino)propane; (*R*)-*form*, *in* B-00264
1,2-Bis(diphenylphosphino)propane; (*S*)-*form*, *in* B-00264
1,3-Bis(diphenylphosphino)propane, B-00265
Bis[3-(diphenylphosphino)propyl]phenylphosphine, B-00266
Bis[3-(diphenylphosphino)propyl]phenylphosphine; 5-Oxide, *in* B-00266
3,4-Bis(diphenylphosphino)pyrrolidine; (3*R*,4*R*)-*form*, *in* B-00267
Bis[2-(diphenylphosphino)-3,4,5,6-tetrafluorophenyl]-phenylphosphine, B-00268
Bis(ethoxycarbonyl)phenylphosphine, *in* P-00209
Bis(hydroxymethyl)phenylphosphine, B-00281
Bis(3-hydroxypropyl)phenylphosphine, B-00286
Bis(3-hydroxypropyl)phenylphosphine; Di-Ac, *in* B-00286
Bis(methoxycarbonyl)phenylphosphine, *in* P-00209
Bis(2-methylphenyl)phenylphosphine, B-00309
Bis(3-methylphenyl)phenylphosphine, B-00310
Bis(4-methylphenyl)phenylphosphine, B-00311
1,3-Bis(methylphenylphosphino)propane, B-00321
Bis(1-methylpropyl)phenylphosphine, B-00340
Bis(pentafluorophenyl)phenylphosphine, B-00380
Bis(phenoxycarbonyl)phenylphosphine, *in* P-00209
(2-Bromophenyl)diphenylphosphine, B-00480
(4-Bromophenyl)diphenylphosphine, B-00481
3-Butenyldiphenylphosphine, B-00521
(2-Buten-1-yl)diphenylphosphine, B-00520
N-*tert*-Butoxycarbonyl-4-diphenylphosphino-2-diphenylphosphinomethylpyrrolidine, B-00532
tert-Butylbis(trimethylsilyl)phosphine, B-00545
1-*tert*-Butyl-1,4-dihydro-4-phenyl-1,4-phospharsenin; *cis*-*form*, *in* B-00553
1-*tert*-Butyl-1,4-dihydro-4-phenyl-1,4-phospharsenin; *trans*-*form*, *in* B-00553
Butyldiphenylphosphine, B-00557
tert-Butyldiphenylphosphine, B-00558
tert-Butylmethylphenylphosphine; (*R*)-*form*, *in* B-00565
tert-Butylmethylphenylphosphine; (*S*)-*form*, *in* B-00565
tert-Butylmethylphenylphosphine; (±)-*form*, *in* B-00565
4-*tert*-Butyl-1-phenylphosphorinane; *cis*-*form*, *in* B-00586
4-*tert*-Butyl-1-phenylphosphorinane; *trans*-*form*, *in* B-00586
(2-Chlorophenyl)diphenylphosphine, C-00127
(3-Chlorophenyl)diphenylphosphine, C-00128
(4-Chlorophenyl)diphenylphosphine, C-00129
(3-Chloropropyl)diphenylphosphine, C-00169
1-Cyclohexyl-1,2-bis(diphenylphosphino)ethane, C-00244
4-Cyclohexyl-1,3-dimethyl-1,2,4-diazaphospholidine, *in* D-00718
Cyclohexyldiphenylphosphine, C-00245
5,6,7,8,9,14,15,16,17,18-Decahydro-9,14-dimethyl-5,18-diphenyldibenzo[*b*,*i*][1,4,8,11]-diphosphadiarsacyclotetradecin, D-00004
5,6,7,8,9,14,15,16,17,18-Decahydro-5,9,14,18-tetraphenyldibenzo[*b*,*i*][1,4,8,11]-tetraphosphacyclotetradecin; (1α,4α,8α,11α)-*form*, *in* D-00006
5,6,7,8,9,14,15,16,17,18-Decahydro-5,9,14,18-tetraphenyldibenzo[*b*,*i*][1,4,8,11]-tetraphosphacyclotetradecin; (1α,4α,8β,11β)-*form*, *in* D-00006
1,4:3,6-Dianhydro-2,5-deoxy-2,5-bis(diphenylphosphino)-iditol, D-00026
1,6-Dibenzyl-1,6-diphosphacycloundecane, D-00046
1,6-Dibenzyl-1,6-diphosphecane, *in* D-01171
1,4-Dibenzyl-1,4-diphosphepane, D-00047
1,4-Dibenzyl-1,4-diphosphocane, D-00048
1,5-Dibenzyl-1,5-diphosphocane, D-00049
1,4-Dibenzyl-1,4-diphosphonane, D-00050
1,5-Dibenzyl-1,5-diphosphonane, D-00051
▷Dibenzylphenylphosphine, D-00052
Di-3-butenylphenylphosphine, D-00072
4,4-Dibutyl-1-*tert*-butyl-1,4-dihydro-1,4-phosphastannin, D-00075
N,*P*-Di-*tert*-butylcarboimidophosphene, D-00077
Dibutylmethylphosphine, D-00098
Di-*tert*-butylmethylphosphine, D-00099
▷Dibutylphenylphosphine, D-00102
Di-*tert*-butylphenylphosphine, D-00103
7,7-Dichloro-2,3,5,6-tetrakis(trifluoromethyl)-1,4-diphosphabicyclo[2.2.1]hepta-2,5-diene, D-00195
Dicyclohexylphenylphosphine, D-00208
Didecylphenylphosphine, D-00219
N,*N*-Diethyl-2,6-dimethyl-1-phenyl-1,4-phosphaborin-4(1*H*)-amine, D-00285
Diethylphenylphosphine, D-00305
2-(Diethylphosphino)ethanethiol, D-00326
Diferrocenylphenylphosphine, D-00410
Dihexylphenylphosphine, D-00431
5,10-Dihydro-5,10-dimethylphenophosphazine, D-00466
2,5-Dihydro-3,4-dimethyl-1-phenyl-1*H*-phosphole, D-00468
2,3-Dihydro-1,3-diphenyl-1*H*-1,3-benzodiphosphole, D-00476
1,4-Dihydro-1,4-diphenyl-1,4-benzodiphosphorin, D-00477
5,10-Dihydro-5,10-diphenylphosphanthrene, D-00482
2,3-Dihydro-1-methyl-1*H*-benzo[*g*]phosphindole, D-00518
4,5-Dihydro-3-methyl-3*H*-benzo[*e*]phosphindole, *in* D-00441
10,11-Dihydro-5-methyl-5*H*-dibenzo[*b*,*f*]phosphepin, D-00524
5,10-Dihydro-5-methyl-10-phenylphenophosphazine, D-00526
2,5-Dihydro-3-methyl-1-phenyl-1*H*-phosphole, D-00528
5,10-Dihydro-5-phenylacridophosphine, D-00547
4,5-Dihydro-3-phenyl-3*H*-1,3-azaphosphole, D-00549
4,5-Dihydro-3-phenyl-3*H*-benzo[*e*]phosphindole, *in* D-00441
10,11-Dihydro-5-phenyl-5*H*-dibenzo[*b*,*f*]phosphepin, D-00559
2,3-Dihydro-2-phenyl-1*H*-isophosphindole, D-00562
5,10-Dihydro-10-phenylphenophosphazine, D-00563
2,7-Dihydro-1-phenyl-1*H*-phosphepin, D-00564
4,5-Dihydro-1-phenyl-1*H*-phosphepin, D-00565
2,3-Dihydro-1-phenyl-1*H*-phosphole, D-00566
1,2-Dihydro-1,2,4,6-tetraphenylphosphorin, D-00584
2,5-Dihydro-1,3,4-trimethyl-1*H*-phosphole, D-00600
1,4-Dihydro-2,4,6-triphenyl-1,4-azaphosphorine, D-00604
Diisopropylphenylphosphine, D-00638
(Dimethylaminomethyl)diphenylphosphine, D-00700
[2-(Dimethylarsino)phenyl]dimethylphosphine, D-00711
2,2-Dimethyl-1,3-dioxa-5-phospha-2-silacyclohexane; 5-Me, *in* D-00731
2,2-Dimethyl-1,3-dioxa-5-phospha-2-silacyclohexane; 5-Ph, *in* D-00731

2,4-Dimethyl-9,13-diphenyl-1,5-diaza-9,13-diphosphacyclohexadeca-1,4-diene, D-00737
12,14-Dimethyl-4,8-diphenyl-1,11-diaza-4,8-diphosphacyclotetradeca-11,14-diene, D-00738
2,2-Dimethyl-1,3-diphenyl-1,3-diphosphorinane, D-00745
2,2-Dimethyl-5,6-diphenyl-1-oxa-5-phospha-2-silacyclohexane, D-00746
1,3-Dimethylisophosphinoline, D-00758
Dimethyl(2-methylphenyl)phosphine, D-00759
Dimethyl(3-methylphenyl)phosphine, D-00760
Dimethyl(4-methylphenyl)phosphine, D-00761
1,2-Dimethyl-4-phenyl-1,2,4-diazaphospholidine, *in* D-00718
2,8-Dimethyl-10-phenyl-10*H*-phenothiaphosphine, D-00798
2,8-Dimethyl-10-phenyl-10*H*-phenoxaphosphine, D-00799
5,6-Dimethyl-7-phenyl-7-phosphabicyclo[2.2.1]hept-5-ene-2,3-dicarboxylic acid; *endo,endo,syn-form*, Di-Me ester, *in* D-00800
5,6-Dimethyl-7-phenyl-7-phosphabicyclo[2.2.1]hept-5-ene-2,3-dicarboxylic acid; *endo,endo,anti-form*, Di-Me ester, *in* D-00800
4,4-Dimethyl-1-phenyl-1-phospha-4-silacyclohexa-2,5-diene, D-00801
2,2-Dimethyl-1-phenyl-1-phospha-2-silacyclopentane, D-00802
Dimethylphenylphosphine, D-00803
3,4-Dimethyl-1-phenyl-1*H*-phosphole, D-00807
2,5-Dimethyl-1-phenylphosphorinan-4-one, D-00821
2,6-Dimethylphosphorin, D-00881
1,3-Dimethylphosphorinane, D-00882
1,4-Dimethylphosphorinane, D-00883
1,4-Dimethylphosphorinane; Sulfide, *in* D-00883
▷Dimethyl(trimethylsilyl)phosphine, D-00932
Dioctylphenylphosphine, D-00955
Dipentylphenylphosphine, D-00978
[(Diphenoxyphosphinyl)methyl]triphenylphosphonium(1+); Chloride, *in* D-00990
▷[2-(Diphenylarsino)ethyl]diphenylphosphine, D-00993
1,3-Diphenyl-1,3-azaphosphetidine, *in* P-00087
2,3-Diphenyl-1,3-azaphospholidine, D-00995
5,10-Diphenyldibenzo[1,4]phospharsenin, D-01001
2,5-Diphenyl-1,3,5,2-dioxaphosphaborinane, D-01003
Diphenyl[(diphenylphosphinoselenoyl)methyl]phosphine, *in* B-00244
Diphenyl[(2-diphenylphosphinyl)ethyl]phosphine, *in* B-00234
1,12-Diphenyl-1,12-diphosphacyclodocosane; *cis-form*, *in* D-01008
1,12-Diphenyl-1,12-diphosphacyclodocosane; *trans-form*, *in* D-01008
1,10-Diphenyl-1,10-diphosphacyclooctadecane; *cis-form*, *in* D-01009
1,10-Diphenyl-1,10-diphosphacyclooctadecane; *trans-form*, *in* D-01009
1,12-Diphenyl-1,12-diphosphacyclotetracosane; *cis-form*, *in* D-01010
1,12-Diphenyl-1,12-diphosphacyclotetracosane; *trans-form*, *in* D-01010
1,3-Diphenyl-1,3-diphospholane, D-01013
1,3-Diphenyl-1,3-diphosphorinane, D-01016
1,4-Diphenyl-1,4-diphosphorinane, D-01017
(Diphenylmethylene)(2,4,6-trimethylphenyl)phosphine, D-01021
2,3-Diphenyl-1,3-oxaphosphepane, D-01026
2,3-Diphenyl-1,3-oxaphospholane, D-01027
2,3-Diphenyl-1,3-oxaphosphorinane, D-01028
Diphenyl(4-phenyl-1,3-butadienyl)phosphine, D-01031
Diphenyl(4-phenyl-1,3-butadiynyl)phosphine, D-01032
Diphenyl(2-phenylethenyl)phosphine; (*E*)*-form*, *in* D-01033
Diphenyl(2-phenylethenyl)phosphine; (*Z*)*-form*, *in* D-01033
Diphenyl(2-phenylethyl)phosphine, D-01034
2-(Diphenylphosphino)acetamide, *in* D-01045
3-(Diphenylphosphino)acetanilide, *in* D-01048
(Diphenylphosphino)acetic acid, D-01045
(Diphenylphosphino)acetonitrile, D-01046
2-(Diphenylphosphino)aniline, D-01047
3-(Diphenylphosphino)aniline, D-01048
4-(Diphenylphosphino)aniline, D-01049
2-(Diphenylphosphino)benzaldehyde, D-01050
3-(Diphenylphosphino)benzaldehyde, D-01051
4-(Diphenylphosphino)benzaldehyde, D-01052
2-(Diphenylphosphino)benzoic acid, D-01053
2-(Diphenylphosphino)benzoic acid; Me ester, *in* D-01053
3-(Diphenylphosphino)benzoic acid, D-01054
4-(Diphenylphosphino)benzoic acid, D-01055
2-(Diphenylphosphino)benzonitrile, D-01056
3-(Diphenylphosphino)benzonitrile, D-01057
4-(Diphenylphosphino)benzonitrile, D-01058
2-(Diphenylphosphino)-*N*,*N*-dimethylaniline, D-01059
3-(Diphenylphosphino)-*N*,*N*-dimethylaniline, D-01060
4-Diphenylphosphino-*N*,*N*-dimethylaniline, D-01061
4-(Diphenylphosphino)-2-(diphenylphosphinomethyl)-pyrrolidine; (2*S*,4*S*)*-form*, *in* D-01062
4-(Diphenylphosphino)-2-(diphenylphosphinomethyl)-pyrrolidine; (2*S*,4*S*)*-form*, *N*-Ac, *in* D-01062
2-(Diphenylphosphino)ethylamine, D-01066
(Diphenylphosphino)methanol, D-01069
2-(Diphenylphosphino)phenol, D-01070
3-(Diphenylphosphino)phenol, D-01071
3-(Diphenylphosphino)phenol; Ac, *in* D-01071
3-(Diphenylphosphino)phenol; Me ether, *in* D-01071
4-(Diphenylphosphino)phenol, D-01072
3-(Diphenylphosphino)propanenitrile, D-01073
3-(Diphenylphosphino)-1-propanol, D-01074
1-(Diphenylphosphino)-2-propanone, D-01075
2-(Diphenylphosphino)pyridine, D-01076
4-(Diphenylphosphino)pyridine, D-01077
Diphenyl-1,2-propadienylphosphine, D-01142
Diphenyl-1-propenylphosphine, D-01143
Diphenyl-2-propenylphosphine, D-01144
Diphenylpropylphosphine, D-01145
Diphenyl(1-propynyl)phosphine, D-01146
2,3-Diphenyl-1,3-thiaphospholane, D-01151
2,3-Diphenyl-1,3-thiaphosphorinane, D-01152
Diphenyl(2-trifluoromethylphenyl)phosphine, D-01154
Diphenyl(3-trifluoromethylphenyl)phosphine, D-01155
Diphenyl(trimethylsilyl)phosphine, D-01156
Diphenylvinylphosphine, D-01168
1,4-Diphosphabicyclo[2.2.2]octane, D-01169
1,5-Diphosphabicyclo[3.3.0]octane, D-01170
1,2-Diphospholane; 1,2-Diisopropyl, *in* D-01181
1,2-Diphospholane; 1,2-Di-Ph, *in* D-01181
2,3,4,5,6,7,8,9,10,11,12,13-Dodecahydro-2,5,9,12-tetraphenyl-1*H*-2,5,9,12-benzotetraphosphacyclopentadecin, D-01262
Dodecyldimethylphosphine, D-01264
1-*tert*-Butyl-3,4-dimethyl-1*H*-phosphole, E-00004
1,2-Ethanediylbis[(2-methoxyphenyl)phenylphosphine]; (*R**,*R**)*-form*, *in* E-00011
1,2-Ethanediylbis[(2-methoxyphenyl)phenylphosphine]; (*RS*,*SR*)*-form*, *in* E-00011
1,2-Ethanediylbis[methylphenylphosphine]; (*RS*,*RS*)*-form*, *in* E-00012
1,2-Ethanediylbis[methylphenylphosphine]; (*RS*,*SR*)*-form*, *in* E-00012
Ethyldiphenylphosphine, E-00066
Ethyl (diphenylphosphino)acetate, *in* D-01045
Ethylmethylphenylphosphine; (*S*)*-form*, *in* E-00084
Ethylmethylphenylphosphine; (±)*-form*, *in* E-00084
1-Ethyl-3-phenyl-1,3-azaphospholidine, *in* P-00091
2-Ethyl-1,2,3,4-tetrahydroisophosphinoline, *in* T-00110
Ethynyldiphenylphosphine, E-00201
(2-Fluorophenyl)diphenylphosphine, F-00026
(3-Fluorophenyl)diphenylphosphine, F-00027
(4-Fluorophenyl)diphenylphosphine, F-00028
2,4,6,7,7,8,8-Heptaphenyl-1-phosphabicyclo[4.2.0]octa-2,4-diene, H-00003
3,3,4,4,5,5-Hexafluoro-1,2-bis(diphenylphosphino)-cyclopentene, H-00016
Hexahydro-1,5-bis(1-phenylethyl)-1*H*-1,5,3-diazaphosphepine; (*R*,*R*)*-form*, 3-Me, *in* H-00019
Hexahydro-1,5-bis(1-phenylethyl)-1*H*-1,5,3-diazaphosphepine; (*R*,*R*)*-form*, 3-Ph, *in* H-00019
Hexahydro-1,5-bis(1-phenylethyl)-1*H*-1,5,3-diazaphosphepine; (*S*,*S*)*-form*, 3-Et, *in* H-00019

Hexahydro-1,5-bis(1-phenylethyl)-1*H*-1,5,3-diazaphosphepine; (*S*,*S*)-*form*, 3-Cyclohexyl, *in* H-00019
Hexahydro-1,3-diphenyl-1,3-azaphosphorine, *in* H-00041
Hexahydro-1,4-diphenyl-1,4-azaphosphorine, *in* H-00042
7,8,15,16,17,18-Hexahydro-14,18-diphenyl-6*H*,18*H*-1,6*H*-dibenzo[*b*,*i*][1,11,4,8]dithiadiphosphacyclotetradecin, H-00026
1,3*a*,3*b*,4,6*a*,6*b*-Hexahydro-1,2,3*b*,4,5,6*b*-hexaphenylcyclobuta[1,2-*b*:3,4-*b*′]diphosphole, H-00031
Hexahydro-1-methyl-4-phenyl-1,4-azaphosphorine, H-00036
Hexahydro-3-phenyl-1*H*-1,3-azaphosphepine, H-00040
Hexahydro-3-phenyl-1,3-azaphosphorine, H-00041
Hexahydro-4-phenyl-1,4-azaphosphorine, H-00042
3,4,5,6,7,8-Hexahydro-5-phenyl-2*H*-1,9,5-benzodithiaphosphacycloundecin, H-00044
1,2,3,4,7,8-Hexahydro-1-phenyl-5,6-bis[(trimethylsilyl)oxy]-phosphocin, H-00046
2,3,4,7,8,9-Hexahydro-1-phenyl-5,6-bis[(trimethylsilyl)oxy]-1*H*-phosphonine, H-00047
Hexahydro-1,3,5-triphenyl-1,3,5-diazaphosphorine, H-00058
2,3,5,6,7,8-Hexakis(trifluoromethyl)-1,4-diphosphabicyclo[2.2.2]octa-2,5,7-triene, H-00064
2,5,7,10,11,12-Hexakis(trifluoromethyl)-1,6-diphosphahexacyclo[$4.4.2.0^{2,5}.0^{3,9}.0^{4,8}.0^{7,10}$]dodec-11-ene, H-00065
4,5,9,10,11,12-Hexakis(trifluoromethyl)-1,8-diphosphatetracyclo[$6.2.2.0^{2,7}.0^{3,6}$]dodecane, H-00066
2,7,9,10,11,12-Hexakis(trifluoromethyl)-1,8-diphosphatetracyclo[$6.2.2.0^{2,7}.0^{3,6}$]dodeca-4,9,11-triene; ($2\alpha,3\alpha,6\alpha,7\alpha$)-*form*, *in* H-00067
2,7,9,10,11,12-Hexakis(trifluoromethyl)-1,8-diphosphatetracyclo[$6.2.2.0^{2,7}.0^{3,6}$]dodeca-4,9,11-triene; ($2\alpha,3\beta,6\beta,7\alpha$)-*form*, *in* H-00067
1,2,2,3,4,4-Hexaphenyl-1,3-diphosphetane, H-00088
1,1,4,7,10,10-Hexaphenyl-1,4,7,10-tetraphosphadecane, H-00091
Hexyldiphenylphosphine, H-00093
1*H*-Isophosphindole-1,3(2*H*)-dione; 2-Ph, *in* I-00026
1*H*-Isophosphindole-1,3(2*H*)-dione; *Enol-form*, Trimethylsilyl ether, *in* I-00026
Isophosphinoline, I-00027
Isopropyldiphenylphosphine, I-00031
Menthyldiphenylphosphine, *in* M-00164
7-Methoxy-2,3,5,6,7,8-hexakis(trifluoromethyl)-1,4-diphosphabicyclo[2.2.2]octa-2,5-diene, M-00042
Methoxymethyldiphenylphosphine, *in* D-01069
(2-Methoxyphenyl)diphenylphosphine, M-00056
(3-Methoxyphenyl)diphenylphosphine, M-00057
(4-Methoxyphenyl)diphenylphosphine, M-00058
(2-Methoxyphenyl)methylphenylphosphine, M-00059
(2-Methoxyphenyl)-2-naphthalenylphenylphosphine; (*R*)-*form*, *in* M-00065
(2-Methoxyphenyl)-2-naphthalenylphenylphosphine; (*S*)-*form*, *in* M-00065
(2-Methoxyphenyl)-2-naphthalenylphenylphosphine; (±)-*form*, *in* M-00065
(4-Methoxyphenyl)-1-naphthalenylphenylphosphine; (+)-*form*, *in* M-00066
(4-Methoxyphenyl)-1-naphthalenylphenylphosphine; (±)-*form*, *in* M-00066
10-Methylacridophosphine, M-00087
Methyl 2,3-bis-*O*-diphenylphosphino-4,6-*O*-benzylideneglucopyranoside; α-D-*form*, *in* M-00091
5-Methyl-5*H*-dibenzophosphole, M-00115
5-Methyl-2,3-diphenyl-1,3-oxaphospholane, M-00133
Methyldiphenylphosphine, M-00135
Methyl 3-(diphenylphosphino)benzoate, *in* D-01054
Methyl 4-(diphenylphosphino)benzoate, *in* D-01055
1-Methylisophosphinoline, M-00162
3-Methylisophosphinoline, M-00163
Methyl-2-naphthylphenylphosphine, M-00177
Methyl-2-naphthylphenylphosphine; (*S*)-*form*, *in* M-00177
3-Methyl-2-phenyl-1,3-diaza-2-phosphabicyclo[3.3.1]nonane, M-00196
(2-Methylphenyl)diphenylphosphine, M-00200
(3-Methylphenyl)diphenylphosphine, M-00201
(4-Methylphenyl)diphenylphosphine, M-00202
3-Methyl-1-phenylphospholane; (1*S*,3*R*)-*form*, *in* M-00229
Methylphenyl-2-propenylphosphine; (*R*)-*form*, *in* M-00264
Methylphenyl-2-propenylphosphine; (*S*)-*form*, *in* M-00264
Methylphenyl-2-propenylphosphine; (±)-*form*, *in* M-00264
Methylphenylpropylphosphine, M-00265
5-Methyl-1-phospha-5-silabicyclo[3.3.1]nonane, M-00274
4-Methyl-4-phosphatetracyclo[$3.3.0.0^{2,8}.0^{3,6}$]octane, M-00275
1-Methylphosphirane, *in* P-00361
1-Methylphospholane, M-00280
3-Methylphospholane, M-00281
1-Methyl-1*H*-phosphole, M-00282
2-Methylphosphorin, M-00337
4-Methylphosphorin, M-00338
1-Methylphosphorinane, M-00339
1-Methyl-4-phosphorinanone, M-00340
(2-Methyl-1-propenyl)diphenylphosphine, M-00362
(1-Methylpropyl)diphenylphosphine; (*R*)-*form*, *in* M-00369
(1-Methylpropyl)diphenylphosphine; (±)-*form*, *in* M-00369
[2-(Methylseleno)phenyl]diphenylphosphine, M-00393
[2-(Methylthio)phenyl]diphenylphosphine, M-00417
4-Methyl-3,5,8-trioxa-1-phosphabicyclo[2.2.2]octane, M-00425
1-Methyl-2,6,7-trioxa-4-phospha-1-silabicyclo[2.2.2]octane, M-00426
1-Naphthyldiphenylphosphine, N-00011
2-Naphthyldiphenylphosphine, N-00012
Neomenthydiphenylphosphine, *in* M-00164
(3-Nitrophenyl)diphenylphosphine, N-00039
(4-Nitrophenyl)diphenylphosphine, N-00040
3,3,4,4,5,5,6,6-Octafluoro-1,2-bis(diphenylphosphino)-cyclohexene, O-00002
2,3,4,5,6,7,8,9-Octahydro-1,9-dimethyl-5-phenyl-1*H*-5,1,9-benzophosphadiarsacycloundecin, O-00007
2,3,4,5,6,7,8,9-Octahydro-1,9-diphenyl-1*H*-5,1,9-benzazadiphosphacycloundecine; (1*RS*,9*SR*)-*form*, *in* O-00009
2,3,4,5,6,7,8,9-Octahydro-1,9-diphenyl-1*H*-5,1,9-benzazadiphosphacycloundecine; (1*RS*,9*SR*)-*form*, 5-Me, *in* O-00009
2,3,4,5,6,7,8,9-Octahydro-1,9-diphenyl-1*H*-5,1,9-benzazadiphosphacycloundecine; (1*RS*,9*SR*)-*form*, 5-Ph, *in* O-00009
2,3,4,5,6,7,8,9-Octahydro-1,9-diphenyl-1*H*-1,9-benzodiphosphacycloundecin, O-00010
1,2,3,4,6,7,8,9-Octahydro-1,9-diphenyl-5,9-benzothiadiphosphacycloundecin, O-00011
1,2,3,4,6,7,8,9-Octahydro-1,9-diphenyl-5,1,9-benzoxadiphosphacycloundecin, O-00012
6,7,8,9,14,15,16,17-Octahydro-9,14-diphenyldibenzo[*b*,*i*]-[1,4,8,11]dithiadiphosphacyclotetradecin, O-00013
6,7,8,9,15,16,17,18-Octahydro-9,18-diphenyldibenzo[*b*,*i*]-[1,8,4,11]dithiadiphosphacyclotetradecin, O-00014
1,2,3,4,5,8,9,10-Octahydro-1-phenyl-6,7-bis[(trimethylsilyl)oxy]phosphecine, O-00021
2,3,4,5,6,7,8,9-Octahydro-1,5,9-triphenyl-1*H*-1,9,5-benzodiphosphaarsacycloundecin, O-00027
2,3,4,5,6,7,8,9-Octahydro-1,5,9-triphenyl-1*H*-1,5,9-benzotriphosphacycloundecin, O-00028
1,2,7,8,9,10,15,16-Octamethyl-3,6,11,14-tetraphenyltetrakisphospholo[1,2-*b*:2′,1′-*d*:1″,2″-*f*:2‴,1‴-*h*][1,2,5,6]tetraphosphocin, O-00033
Octyldiphenylphosphine, O-00039
Oxybis[2-(diphenylphosphino)ethane], O-00093
(Pentafluorophenyl)diphenylphosphine, P-00014
2,2,3,4,4-Pentamethyl-1-phenylphosphetane; *cis*-*form*, *in* P-00020
2,2,3,4,4-Pentamethyl-1-phenylphosphetane; *trans*-*form*, *in* P-00020
Pentaphenyl-1*H*-phosphole, P-00038
Pentaphenyl-1*H*-phosphole; Maleic anhydride adduct, *in* P-00038
2,3,4,5,6-Pentaphenylphosphorin, P-00040
1,2,3,4,9-Pentaphenyl-1*H*-tribenzo[*b*,*d*,*f*]phosphepin, P-00042
Phellanphos, P-00059
10-Phenylacridophosphine, P-00079
5-Phenyl-10(5*H*)-acridophosphinone, P-00080
3-Phenyl-1,3-azaphospholidine, P-00091

3-Phenyl-1,3-azaphospholidin-2-thione, P-00093
3-Phenyl-3*H*-benzophosphepin, P-00096
Phenylbis[2-(phenylethynyl)phenyl]phosphine, P-00098
Phenylbis(phenylethynyl)phosphine, P-00099
1-Phenyl-6,7-bis[(trimethylsilyl)oxy]phosphacycloundec-6-ene, P-00100
Phenylbis[3-(triphenylphosphoranylidene)propenyl]phosphine, P-00103
5-Phenyl-5*H*-dibenzo[*b,f*]phosphepin, P-00108
6-Phenyl-5*H*-dibenzo[*c,e*]phosphepin-5,7(6*H*)-dione, P-00109
5-Phenyl-5*H*-dibenzophosphole, P-00110
7-Phenyl-7*H*-dibenzo[*d,f*]phosphonin, P-00111
4-Phenyl-3,5-dioxa-1-phosphabicyclo[2.2.1]heptane, P-00113
Phenyldi-2-propenylphosphine, P-00119
Phenyldipropylphosphine, P-00120
Phenyldivinylphosphine, P-00122
1,2-Phenylenebis[dimethylphosphine], P-00123
1,3-Phenylenebis[dimethylphosphine], P-00124
1,4-Phenylenebis[dimethylphosphine], P-00125
1,2-Phenylenebis[diphenylphosphine], P-00126
1,3-Phenylenebis[diphenylphosphine], P-00127
(1-Phenyl-1,2-ethanediyl)bis[diphenylphosphine]; (*R*)-*form*, *in* P-00150
(1-Phenyl-1,2-ethanediyl)bis[diphenylphosphine]; (*S*)-*form*, *in* P-00150
12-Phenyl-10*H*-5,10-ethenoacridophosphine, P-00151
4-Phenyl-1,4-oxaphosphorinane, P-00176
3-Phenyl-1,3-oxaphosphorinan-2-one, P-00177
10-Phenyl-10*H*-phenothiaphosphine, P-00182
10-Phenyl-10*H*-phenothiaphosphine; 5,5-Dioxide, *in* P-00182
10-Phenyl-10*H*-phenoxaphosphine, P-00183
9-Phenyl-9-phosphabicyclo[3.3.1]nonane, P-00193
9-Phenyl-9-phosphabicyclo[4.2.1]nonane, P-00194
9-Phenyl-9-phosphabicyclo[4.2.1]nona-2,4,7-triene; *syn-form*, *in* P-00196
9-Phenyl-9-phosphabicyclo[4.2.1]nona-2,4,7-triene; *anti-form*, *in* P-00196
9-Phenyl-9-phosphabicyclo[6.1.0]none-2,4,6-triene, P-00197
8-Phenyl-8-phosphabicyclo[3.2.1]octan-3-one; *anti-form*, *in* P-00199
9-Phenyl-9-phosphapentacyclo[4.3.0.$0^{2,5}$.$0^{3,8}$.$0^{4,7}$]nonane, P-00201
9-Phenyl-9-phosphatricyclo[3.3.1.$0^{2,8}$]nona-3,6-diene, P-00202
9-Phenyl-9-phosphatricyclo[4.2.1.$0^{2,5}$]nona-3,7-diene, P-00203
1-Phenylphosphepane, P-00205
1-Phenyl-1*H*-phosphepin, P-00206
1-Phenyl-1*H*-phosphindole, P-00207
2-Phenylphosphinoline, P-00212
1-Phenylphosphirane, P-00214
1-Phenylphosphocane, P-00215
1-Phenylphospholane, P-00216
1-Phenyl-1*H*-phosphole, P-00217
1-Phenylphosphorinane, P-00265
1-Phenyl-2-phosphorinanol, P-00266
1-Phenyl-2-phosphorinanone, P-00267
1-Phenyl-4-phosphorinanone, P-00268
1-Phenyl-1,2,3,4-tetrahydrophosphinoline, P-00320
7-Phenyl-1,3,4,6-tetrakis(trifluoromethyl)-7-aza-2,5-diphosphatetracyclo[4.1.0.$0^{2,4}$.$0^{3,5}$]heptane, P-00321
1-Phosphabicyclo[2.2.1]heptane, P-00344
1-Phosphabicyclo[3.3.1]nonane, P-00345
Phosphanthridene, P-00348
2,2′,2″-Phosphinidynetris[1-methylimidazole], *in* T-00496
3,3′,3″-Phosphinidynetrispropanoic acid; Tri-Me ester, *in* P-00357
3,3′,3″-Phosphinidynetrispropanoic acid; Tri-Et ester, *in* P-00357
Phosphorin, P-00403
1,6,11,16-Tetrabenzyl-1,6,11,16-tetraphosphacycloeicosane, T-00008
1,5,10,14-Tetrabenzyl-1,5,10,14-tetraphosphacyclooctadecane, T-00009
1,2,4,5-Tetra-*tert*-butyl-1,2,4,5-tetraphosphorinane, T-00027
3,3,4,4-Tetrafluoro-1,2-bis(diphenylphosphino)cyclobutene, T-00077
1,3,4,5-Tetrahydro-4,5-dimethyl-2,5-methano-2*H*-2-benzophosphepin, T-00096
3*a*,4,7,7*a*-Tetrahydro-1,8-dimethyl-4,7-phosphinidene-1*H*-phosphindole, T-00099
3*a*,4,7,7*a*-Tetrahydro-5,6-dimethyl-2,3*a*,7*a*,8-tetrakis(trifluoromethyl)-1,2,3-metheno-2*H*-benzodiphosphole, T-00100
2,3,4,5-Tetrahydro-1,5-diphenyl-1*H*-1,5-benzodiphosphepin, T-00101
1,2,3,4-Tetrahydro-1,4-diphenyl-1,4-benzodiphosphorin, T-00102
3*a*,4,7,7*a*-Tetrahydro-1,8-diphenyl-4,7-phosphinidene-1*H*-phosphindole, T-00105
3*a*,4,7,7*a*-Tetrahydro-1,2,3,5,6,8-hexamethyl-4,7-phosphinidene-1*H*-phosphindole, T-00106
3,4,5,6-Tetrahydro-4-methyl-2*H*-1,5-methano-4,1-benzazaphosphocine, T-00115
Tetrahydro-3-methyl-2,4,5,6-tetraphenyl-2*H*-1,3,5-oxazaphosphorine, T-00117
1,2,3,4-Tetrahydro-4-oxo-1-phenylphosphinoline, T-00122
1,2,5,6-Tetrahydro-1,1,6,6,9-pentamethyl-4*H*-phospholo[3,2,1-*ij*]phosphinoline, T-00123
2,3,6,7-Tetrahydro-1,1,7,7,9-pentamethyl-1*H*,5*H*-phosphorino[3,2,1-*ij*]phosphinoline, T-00124
1,2,3,4-Tetrahydro-1-phenylbenzo[*h*]phosphinoline, T-00128
5,6,7,12-Tetrahydro-12-phenyldibenz[*c,f*][1,5]-azaphosphocine, T-00130
1,2,3,4-Tetrahydro-2-phenylisophosphinoline, *in* T-00110
1,2,3,6-Tetrahydro-1-phenylphosphorin, T-00134
Tetrakis[(diphenylphosphino)methyl]methane, T-00179
1,1,3,3-Tetrakis(diphenylphosphino)-1,2-propadiene, T-00180
1,3,4,6-Tetrakis(trifluoromethyl)-2,5-diphosphatricyclo[3.1.0.$0^{2,6}$]hex-3-ene, T-00186
3,3′,4,4′-Tetramethyl-1,1′-diphenyl-2,2′-bi-1*H*-phosphole, T-00199
2,2,5,5-Tetramethyl-1-phenyl-1-phospha-2,5-disilacyclopentane, T-00209
2,2,6,6-Tetramethyl-1-phenyl-4-phosphorinanone, T-00210
2,7,10,15-Tetramethyl-17-phenyl-17*H*-tetrabenzo[*b,d,f,h*]-phosphonin, T-00211
3,3,5,5-Tetramethyl-1,2,4-triphenyl-1,2,4-triphospholane, T-00239
2,2′,6,6′-Tetraphenyl-4,4′-biphosphorine, T-00249
3,5,8,10-Tetraphenyl-1*H*,5*H*-bisphospholo[1,2-*a*:1′,2′-*d*][1,4]-diphosphorin, T-00250
2,3,5,6-Tetraphenyl-1,4-bis[(trimethylsilyl)oxy]-2,3,5,6-tetraphosphabicyclo[2.2.0]hexane, T-00251
2,4,5,6-Tetraphenyl-1,3,5-dioxaphosphorinane, T-00254
4,7,13,16-Tetraphenyl-1,10-dioxa-4,7,13,16-tetraphosphacyclooctadecane; (4*RS*,7*RS*,13*SR*,16*SR*)-*form*, *in* T-00255
4,7,13,16-Tetraphenyl-1,10-dioxa-4,7,13,16-tetraphosphacyclooctadecane; (4*RS*,7*RS*,13*RS*,16*SR*)-*form*, *in* T-00255
4,7,13,16-Tetraphenyl-1,10-dipropyl-1,10-diaza-4,7,13,16-tetraphosphacyclooctadecane, T-00264
1,5,9,13-Tetraphenyl-1,5,9,13-tetraphosphacyclohexadecene, T-00276
1,4,6,9-Tetraphenyl-1,4,6,9-tetraphospha-5-silaspiro[4.4]-nonane, T-00277
1,3,5,7-Tetraphenyl-1,3,5,7-tetraphosphocane, T-00279
1,2,4,5-Tetraphenyl-1,2,4,5-tetraphosphorinane, T-00281
4*H*-1,4-Thiaphosphorin; 4-Ph, *in* T-00299
1,3,5-Triaza-7-phosphatricyclo[3.3.1.$1^{3,7}$]decane, T-00327
Tribenzoylphosphine, T-00335
Tribenzylphosphine, T-00338
Tri-2-biphenylylphosphine, T-00348
Tri-3-biphenylylphosphine, T-00349
Tri-4-biphenylylphosphine, T-00350
Tri-3-butenylphosphine, T-00360
Tributoxycarbonylphosphine, *in* P-00355
▷Tributylphosphine, T-00372
Tri-*tert*-butylphosphine, T-00373
2,4,6-Tri-*tert*-butylphosphorin, T-00382
Tricyclohexylphosphine, T-00426
Tricyclopentylphosphine, T-00431
Tricyclopropylphosphine, T-00432

Tri(decyl)phosphine, T-00433
Tridodecylphosphine, T-00435
Triethoxycarbonylphosphine, *in* P-00355
▷Triethylphosphine, T-00441
Tri-2-furanylphosphine, T-00486
Tri-3-furanylphosphine, T-00487
Triheptylphosphine, T-00489
Trihexylphosphine, T-00492
Tri(2-imidazolyl)phosphine, T-00496
Triisopropylphosphine, T-00500
2,4,6-Trimethyl-5-phenyl-1,3,5-dioxaphosphorinane, T-00521
2,2,5-Trimethyl-3-phenyl-1,3-oxaphospholane; (3*RS*,5*RS*)-*form*, *in* T-00522
2,2,5-Trimethyl-3-phenyl-1,3-oxaphospholane; (3*RS*,5*SR*)-*form*, *in* T-00522
▷Trimethylphosphine, T-00530
1,3,4-Trimethyl-1*H*-phosphole, T-00536
1,2,4-Trimethyl-1,2,4-triphospholane, T-00568
Tri-1-naphthylphosphine, T-00572
Tri-2-naphthylphosphine, T-00573
Trinonylphosphine, T-00578
Trioctylphosphine, T-00582
2,6,7-Trioxa-1,4-diphosphabicyclo[2.2.2]octane, T-00585
2,6,7-Trioxa-1,4-diphosphabicyclo[2.2.2]octane; 1-Oxide, *in* T-00585
2,6,7-Trioxa-1,4-diphosphabicyclo[2.2.2]octane; 1-Sulfide, *in* T-00585
2,6,7-Trioxa-4-phospha-1-arsabicyclo[2.2.2]octane, T-00586
Tripentylphosphine, T-00594
Triphenoxycarbonylphosphine, *in* P-00355
1,2,3-Triphenyl-1,3-azaphospholidin-5-one, T-00596
4,6,7-Triphenyl-2,3-bis(trifluoromethyl)-1-phosphabicyclo[2.2.2]octa-2,5,7-triene, T-00598
3,4,5-Triphenyl-4*H*-1,2,4-diazaphosphole, T-00601
2,3,10-Triphenyl-4*H*-1,4-ethenophosphinoline, T-00607
▷Triphenylphosphine, T-00624
N,1,1-Triphenylphosphinecarbonamide, T-00625
1,2,3-Triphenyl-1*H*-phosphirene, T-00632
1,2,5-Triphenyl-1*H*-phosphole, T-00634
2,4,6-Triphenylphosphorin, T-00665
1,2,6-Triphenyl-4-phosphorinanone, T-00666
Tri-1-propenylphosphine; (*E*,*E*,*E*)-*form*, *in* T-00706
Tri-1-propenylphosphine; (*Z*,*Z*,*Z*)-*form*, *in* T-00706
Tri-2-propenylphosphine, T-00707
▷Tripropylphosphine, T-00711
Tri-1-propynylphosphine, T-00721
Tri-2-pyridylphosphine, T-00723
Tri-2-pyrrolylphosphine, T-00726
Tris(acetoxymethyl)phosphine, *in* T-00784
Tris[(4-bromophenyl)methyl]phosphine, T-00733
Tris(3-bromophenyl)phosphine, T-00734
Tris(4-bromophenyl)phosphine, T-00735
Tris(2-carbamoylethyl)phosphine, T-00738
Tris(3-carboxyphenyl)phosphine, T-00739
Tris(3-carboxyphenyl)phosphine; Tri-Me ester, *in* T-00739
Tris(4-carboxyphenyl)phosphine, T-00740
Tris(4-carboxyphenyl)phosphine; Tri-Me ester, *in* T-00740
Tris(2-chlorophenyl)phosphine, T-00747
Tris(3-chlorophenyl)phosphine, T-00748
Tris(4-chlorophenyl)phosphine, T-00749
Tris(2-cyanoethyl)phosphine, T-00752
Tris(3-cyanophenyl)phosphine, T-00753
Tris(4-cyanophenyl)phosphine, T-00754
Tris(2,4-dimethoxyphenyl)phosphine, T-00758
Tris(3,5-dimethoxyphenyl)phosphine, T-00759
Tris(2-dimethylaminophenyl)phosphine, T-00764
Tris(4-dimethylaminophenyl)phosphine, T-00765
Tris(2,4-dimethylphenyl)phosphine, T-00768
Tris(2,5-dimethylphenyl)phosphine, T-00769
Tris(2,6-dimethylphenyl)phosphine, T-00770
Tris(dimethylphosphino)methane, T-00771
Tris[(2-diphenylphosphino)ethyl]amine, T-00773
Tris[2-(diphenylphosphino)ethyl]phosphine, T-00774
Tris(diphenylphosphino)methane, T-00775
Tris(diphenylphosphino)methane; Dioxide, *in* T-00775
Tris(diphenylphosphino)methane; Monosulfide, *in* T-00775
Tris(diphenylphosphino)methane; Disulfide, *in* T-00775
1,1,1-Tris(diphenylphosphinomethyl)ethane, T-00776
Tris(2-diphenylphosphinophenyl)phosphine, T-00777
Tris(2-fluorophenyl)phosphine, T-00781
Tris(3-fluorophenyl)phosphine, T-00782
Tris(4-fluorophenyl)phosphine, T-00783
Tris(hydroxymethyl)phosphine, T-00784
Tris(2-hydroxyphenyl)phosphine, T-00788
Tris(4-hydroxyphenyl)phosphine, T-00790
Tris(4-hydroxyphenyl)phosphine; Tri-Ac, *in* T-00790
Tris(2-methoxyphenyl)phosphine, T-00791
Tris(3-methoxyphenyl)phosphine, T-00792
Tris(4-methoxyphenyl)phosphine, T-00793
Tris(2-methylphenyl)phosphine, T-00806
Tris(3-methylphenyl)phosphine, T-00807
▷Tris(4-methylphenyl)phosphine, T-00808
Tris(1-methylpropyl)phosphine, T-00823
Tris(2-methylpropyl)phosphine, T-00824
Tris[2-(methylthio)phenyl]phosphine, T-00831
▷Tris[4-(methylthio)phenyl]phosphine, T-00832
Tris[(4-nitrophenyl)methyl]phosphine, T-00835
Tris(3-nitrophenyl)phosphine, T-00838
Tris(2-phenylethenyl)phosphine, T-00843
Tris(2-phenylethyl)phosphine, T-00844
Tris(phenylethynyl)phosphine, T-00846
Tris[2-(trifluoromethyl)phenyl]phosphine, T-00852
Tris[3-(trifluoromethyl)phenyl]phosphine, T-00853
Tris[4-(trifluoromethyl)phenyl]phosphine, T-00854
Tris(trifluoromethyl)phosphine, T-00855
Tris(2,4-6-trimethoxyphenyl)phosphine, T-00858
Tris(2,4,6-trimethylphenyl)phosphine, T-00861
▷Tris(trimethylsilyl)phosphine, T-00865
Tris[(triphenylphosphoranylidene)methyl]phosphine, T-00871
Tri(2-thiazolyl)phosphine, T-00873
Tri-2-thienylphosphine, T-00874
Tri-3-thienylphosphine, T-00875
Trivinylphosphine, T-00877

Tertiary phosphine oxides, R_3PO

(Aminomethyl)diphenylphosphine oxide, A-00085
10*H*-5,10[1′,2′]Benzenoacridophosphine; 5-Oxide, *in* B-00014
5,10[1′,2′]-Benzenophenophosphazine; Oxide, *in* B-00015
2,3-Benzo-1,4,5,6,7-pentaphenyl-7-phosphabicyclo[2.2.1]-hept-5-ene 7-oxide, *in* D-00538
7*H*-Benzo[*c*]phenothiaphosphine; 7-Butyl, 7-oxide, *in* B-00022
5,10-[1′,2′]Benzophosphenanthrene; 5,10-Dioxide, *in* B-00024
Benzoyldiphenylphosphine; Oxide, *in* B-00030
Benzyldiphenylphosphine; Oxide, *in* B-00046
Benzylethylphenylphosphine; (±)-*form*, Oxide, *in* B-00047
Benzylmethylphenylphosphine; (*S*)-*form*, Oxide, *in* B-00051
Benzylmethylphenylphosphine; (±)-*form*, Oxide, *in* B-00051
Benzylmethylpropylphosphine; Oxide, *in* B-00054
6-Benzyl-2-oxa-6-phosphatricyclo[3.3.1.1^{3,7}]decane; 6-Oxide, *in* B-00055
Bicyclo[2.2.1]hept-5-ene-2,3-diylbis(methylene)-bis[diphenylphosphine]; (2*R**,3*R**)-*form*, Dioxide, *in* B-00075
Bicyclo[2.2.1]hept-5-ene-2,3-diylbis(methylene)-bis[diphenylphosphine]; (2*S**,3*R**)-*form*, Dioxide, *in* B-00075
Bicyclo[2.2.1]hept-5-ene-2,3-diylbis(methylene)-bis[diphenylphosphine]; (2*RS*,3*RS*)-*form*, Dioxide, *in* B-00075
[1,1′-Biphenyl]-4-yl-1-naphthalenylphenylphosphine; (*R*)-*form*, Oxide, *in* B-00081
[1,1′-Biphenyl]-4-yl-1-naphthalenylphenylphosphine; (*S*)-*form*, Oxide, *in* B-00081
[1,1′-Biphenyl]-4-yl-1-naphthalenylphenylphosphine; (±)-*form*, Oxide, *in* B-00081
Bis(3-aminopropyl)phenylphosphine; *P*-Oxide, *in* B-00090
Bis(2-cyanoethyl)phenylphosphine; Oxide, *in* B-00152
1,2-Bis(dibutylphosphino)ethylene; (*E*)-*form*, Monoxide, *in* B-00159
1,4-Bis(dibutylphosphinyl)butane, *in* B-00157
1,2-Bis(dibutylphosphinyl)ethane, *in* B-00158
1,2-Bis(dibutylphosphinyl)ethylene, *in* B-00159
Bis(dibutylphosphinyl)methane, *in* B-00160

1,5-Bis(dibutylphosphinyl)pentane, *in* B-00161
1,2-Bis(dicyclophosphinyl)ethane, *in* B-00167
1,2-Bis(diethylphosphino)ethylene; (*E*)-*form*, Dioxide, *in* B-00181
2,2-Bis(diethylphosphinyl)biphenyl, *in* B-00179
1,2-Bis(diethylphosphinyl)ethane, *in* B-00180
1,3-Bis(dimethylphosphinyl)benzene, *in* P-00124
1,4-Bis(dimethylphosphinyl)benzene, *in* P-00125
1,4-Bis(dimethylphosphinyl)butane, *in* B-00199
1,2-Bis(dimethylphosphinyl)ethane, *in* B-00200
1,6-Bis(dimethylphosphinyl)hexane, *in* B-00201
Bis(dimethylphosphinyl)methane, *in* B-00202
Bis[(dimethylphosphinyl)methyl]methylphosphine, *in* B-00204
1,5-Bis(dimethylphosphinyl)pentane, *in* B-00205
1,3-Bis(dimethylphosphinyl)propane, *in* B-00206
Bis[2-(diphenylarsino)ethyl]phenylphosphine; *As*,*As*,*P*-Trioxide, *in* B-00212
5,6-Bis(diphenylphosphino)bicyclo[2.2.1]hept-2-ene; (1*S*,5*S*,6*S*)-*form*, *P*,*P*′-Dioxide, *in* B-00220
5,6-Bis(diphenylphosphino)bicyclo[2.2.1]hept-2-ene; (1*RS*,5*RS*,6*RS*)-*form*, *P*,*P*′-Dioxide, *in* B-00220
2,2′-Bis(diphenylphosphino)-1,1′-binaphthyl; (*R*)-*form*, Dioxide, *in* B-00221
2,2′-Bis(diphenylphosphino)-1,1′-binaphthyl; (*S*)-*form*, Dioxide, *in* B-00221
2,2′-Bis(diphenylphosphino)-1,1′-binaphthyl; (±)-*form*, Dioxide, *in* B-00221
2,2′-Bis(diphenylphosphino)diethylamine; Dioxide, *in* B-00231
2,2′-Bis(diphenylphosphino)diethylamine; *N*-Ac, *in* B-00231
1,2-Bis(diphenylphosphino)ethylene; (*E*)-*form*, *P*,*P*′-Dioxide, *in* B-00237
1,2-Bis(diphenylphosphino)ethylene; (*Z*)-*form*, *P*,*P*′-Dioxide, *in* B-00237
N,*N*-Bis[(2-diphenylphosphino)ethyl]ethylamine; Dioxide, *in* B-00238
Bis[2-(diphenylphosphino)ethyl]phenylphosphine; Trioxide, *in* B-00239
2,11-Bis[(diphenylphosphino)methyl]benzo[*c*]phenanthrene; Dioxide, *in* B-00246
5,6-Bis(diphenylphosphino)-2-methylbicyclo[2.2.1]hept-2-ene; (1*R**,5*R**,6*R**)-*form*, *P*,*P*′-Dioxide, *in* B-00247
5,6-Bis(diphenylphosphino)-2-methylbicyclo[2.2.1]hept-2-ene; (1*S**,5*S**,6*S**)-*form*, P,P′-Dioxide, *in* B-00247
2,2′-Bis[(diphenylphosphino)methyl]-1,1′-binaphthyl; (*S*)-*form*, Dioxide, *in* B-00248
Bis[4-(diphenylphosphino)phenyl]phenylphosphine; Trioxide, *in* B-00263
Bis[3-(diphenylphosphino)propyl]phenylphosphine; 5-Oxide, *in* B-00266
1,4-Bis(diphenylphosphinyl)benzene, *in* P-00128
5,6-Bis(diphenylphosphinyl)bicyclo[2.2.1]hept-2-ene, *in* B-00220
1,4-Bis(diphenylphosphinyl)butane, *in* B-00225
1,4-Bis(diphenylphosphinyl)-2-butyne, *in* B-00227
1,1-Bis(diphenylphosphinyl)cyclopropane, *in* B-00229
1,2-Bis(diphenylphosphinyl)cyclopropane, *in* B-00230
1,2-Bis(diphenylphosphinyl)ethane, *in* B-00234
1,1-Bis(diphenylphosphinyl)ethylene, *in* B-00236
Bis[(2-diphenylphosphinyl)ethyl]ether, *in* O-00093
1,2-Bis(diphenylphosphinyl)ethyne, *in* B-00216
1,1′-Bis(diphenylphosphinyl)ferrocene, *in* B-00241
1,7-Bis(diphenylphosphinyl)heptane, *in* B-00242
1,2-Bis(diphenylphosphinyl)-3,3,4,4,5,5-hexafluorocyclopentene, *in* H-00016
1,6-Bis(diphenylphosphinyl)hexane, *in* B-00243
Bis(diphenylphosphinyl)methane, *in* B-00244
Bis[(diphenylphosphinyl)methyl]phenylphosphine oxide, *in* B-00253
1,4-Bis(diphenylphosphinyl)naphthalene, *in* B-00254
1,5-Bis(diphenylphosphinyl)naphthalene, *in* B-00255
2,6-Bis(diphenylphosphinyl)naphthalene, *in* B-00256
2,7-Bis(diphenylphosphinyl)naphthalene, *in* B-00257
1,8-Bis(diphenylphosphinyl)octane, *in* B-00258
1,5-Bis(diphenylphosphinyl)pentane, *in* B-00259
1,3-Bis(diphenylphosphinyl)propane, *in* B-00265
Bis(2-methylphenyl)phenylphosphine; Oxide, *in* B-00309
Bis(3-methylphenyl)phenylphosphine; Oxide, *in* B-00310
Bis(4-methylphenyl)phenylphosphine; Oxide, *in* B-00311
Bis(pentafluorophenyl)phenylphosphine; Oxide, *in* B-00380
(4-Bromophenyl)diphenylphosphine; Oxide, *in* B-00481
3-Butenyldiphenylphosphine; Oxide, *in* B-00521
(2-Buten-1-yl)diphenylphosphine; (*E*)-*form*, Oxide, *in* B-00520
(2-Buten-1-yl)diphenylphosphine; (*Z*)-*form*, *in* B-00520
(2-Buten-1-yl)diphenylphosphine; (*Z*)-*form*, Oxide, *in* B-00520
7-Butyl-2,3-dihydro-1,6-diphenyl-1*H*-1-benzophosphonin-2,3,4,5-tetracarboxylic acid; Tetra-Me ester, oxide, *in* B-00552
1-*tert*-Butyl-1,4-dihydro-4-phenyl-1,4-phospharsenin; *cis*-*form*, 1-Oxide, *in* B-00553
1-*tert*-Butyl-1,4-dihydro-4-phenyl-1,4-phospharsenin; *trans*-*form*, 1-Oxide, *in* B-00553
Butyldiphenylphosphine; Oxide, *in* B-00557
tert-Butyldiphenylphosphine; Oxide, *in* B-00558
tert-Butylmethylphenylphosphine; (*R*)-*form*, Oxide, *in* B-00565
tert-Butylmethylphenylphosphine; (*S*)-*form*, Oxide, *in* B-00565
tert-Butylmethylphenylphosphine; (±)-*form*, Oxide, *in* B-00565
4-*tert*-Butyl-1-phenylphosphorinane; *cis*-*form*, Oxide, *in* B-00586
4-*tert*-Butyl-1-phenylphosphorinane; *trans*-*form*, Oxide, *in* B-00586
(2-Carboxyphenyl)methyldiphenylphosphonium iodide, *in* D-01053
(2-Chlorophenyl)diphenylphosphine; Oxide, *in* C-00127
(3-Chlorophenyl)diphenylphosphine; Oxide, *in* C-00128
(4-Chlorophenyl)diphenylphosphine; Oxide, *in* C-00129
7,8,10,11,13,14,16,17,19,24-Decahydro-5*H*-dibenzo[*o*,*r*]-[1,4,7,10,13,17]pentaoxaphosphacycloeicosin; 24-Ph, 24-oxide, *in* D-00001
2-Diazo-2-(diphenylphosphinyl)-1-phenylethanone, *in* D-00029
1-Diazo-1-(diphenylphosphinyl)-2-propanone, D-00031
(Diazomethyl)diphenylphosphine oxide, D-00034
(Diazophenylmethyl)diphenylphosphine oxide, D-00036
9*H*-Dibenzo[2,3:4,5]phospholo[1,2-*f*]phosphanthridene; 17-Oxide, *in* D-00041
Dibenzylphenylphosphine; Oxide, *in* D-00052
4,4-Dibutyl-1-*tert*-butyl-1,4-dihydro-1,4-phosphastannin; 1-Oxide, *in* D-00075
2,6-Di-*tert*-butyl-1,4-dihydro-4-phenyl-1,4-azaphosphorine; 4-Oxide, *in* D-00084
2,6-Di-*tert*-butyl-1,4-dihydro-4-phenyl-1,4-azaphosphorine; 1-Me, 4-oxide, *in* D-00084
Dibutylmethyl[(dibutylphosphinyl)ethenyl]phosphonium iodide, *in* B-00159
Dibutylmethylphosphine; Oxide, *in* D-00098
Di-*tert*-butylmethylphosphine; Oxide, *in* D-00099
Dibutylphenylphosphine; Oxide, *in* D-00102
Dicyclohexylphenylphosphine; Oxide, *in* D-00208
Didecylphenylphosphine; Oxide, *in* D-00219
N,*N*-Diethyl-2,6-dimethyl-1-phenyl-1,4-phosphaborin-4(1*H*)-amine; 1-Oxide, *in* D-00285
Diethylphenylphosphine; Oxide, *in* D-00305
Dihexylphenylphosphine; Oxide, *in* D-00431
2,3-Dihydro-1*H*-benzo[*e*]phosphindole 3-oxide; 3-Ph, *in* D-00442
10,11-Dihydro-1*H*-dibenz[*b*,*e*][1,4]azaphosphepine 5-oxide; 5-Me, *in* D-00448
10,11-Dihydro-1*H*-dibenz[*b*,*e*][1,4]azaphosphepine 5-oxide; 5-Ph, *in* D-00448
5,10-Dihydro-11*H*-dibenz[*b*,*e*][1,4]azaphosphepin-11-one 5-oxide; 5-Me, *in* D-00449
5,10-Dihydro-11*H*-dibenz[*b*,*e*][1,4]azaphosphepin-11-one 5-oxide; 5-Ph, *in* D-00449
5,10-Dihydro-11*H*-dibenz[*b*,*e*][1,4]azaphosphepin-11-one 5-oxide; *N*-Me, 5-Ph, *in* D-00449
5,10-Dihydro-5,10-dimethylphenophosphazine; 10-Oxide, *in* D-00466
2,5-Dihydro-3,4-dimethyl-1-phenyl-1*H*-phosphole; 1-Oxide, *in* D-00468
5,10-Dihydro-5,10-diphenylphosphanthrene; Monooxide, *in* D-00482
5,10-Dihydro-5,10-diphenylphosphanthrene; 5,10-Dioxide, *in* D-00482
2,3-Dihydro-1-methyl-1*H*-benzo[*g*]phosphindole; 1-Oxide, *in* D-00518

10,11-Dihydro-5-methyl-5*H*-dibenzo[*b,f*]phosphepin; Oxide, *in* D-00524
5,10-Dihydro-5-methyl-10-phenylphenophosphazine; 10-Oxide, *in* D-00526
2,5-Dihydro-3-methyl-1-phenyl-1*H*-phosphole; Oxide, *in* D-00528
5,6-Dihydro-5-methylphosphanthridene 5-oxide, *in* D-00568
2,3-Dihydro-1-phenyl-1*H*-benzo[*g*]phosphindole; 1-Oxide, *in* D-00551
2,3-Dihydro-1-phenyl-1*H*-benzo[*g*]phosphindole; 4,5-Dihydro, 1-Oxide, *in* D-00551
2,3-Dihydro-2-phenyl-1*H*-dibenz[*e,g*]isophosphindole; 2-Oxide, *in* D-00558
10,11-Dihydro-5-phenyl-5*H*-dibenzo[*b,f*]phosphepin; Oxide, *in* D-00559
7,12-Dihydro-12-phenyl-5*H*-dibenz[*c,f*][1,5]oxaphosphocin; 12-Oxide, *in* D-00560
2,3-Dihydro-2-phenyl-1*H*-isophosphindole; Oxide, *in* D-00562
5,10-Dihydro-10-phenylphenophosphazine; 10-Oxide, *in* D-00563
5,6-Dihydro-5-phenylphosphanthridene 5-oxide, *in* D-00568
2,7-Dihydro-1-phenyl-1*H*-phosphepin; Oxide, *in* D-00564
4,5-Dihydro-1-phenyl-1*H*-phosphepin; Oxide, *in* D-00565
2,3-Dihydro-1-phenyl-1*H*-phosphole; 1-Oxide, *in* D-00566
2,5-Dihydro-1-phenyl-1*H*-phosphole; 1-Oxide, *in* D-00567
2,3-Dihydro-1*H*-phosphindole 1-oxide; 1-Ph, *in* D-00570
2,3-Dihydro-1*H*-phosphindole 1-oxide; 1-Benzyl, *in* D-00570
1,2-Dihydro-1,2,4,6-tetraphenylphosphorin; Oxide, *in* D-00584
2,5-Dihydro-1,3,4-trimethyl-1*H*-phosphole; 1-Oxide, *in* D-00600
1,4-Dihydro-2,4,6-triphenyl-1,4-azaphosphorine; 4-Oxide, *in* D-00604
Diisopropylphenylphosphine; Oxide, *in* D-00638
Dimethyl(2-methylphenyl)phosphine; Oxide, *in* D-00759
Dimethyl(3-methylphenyl)phosphine; Oxide, *in* D-00760
Dimethyl(4-methylphenyl)phosphine; Oxide, *in* D-00761
1,4-Dimethyl-2-phenyl-7-oxa-2-phosphabicyclo[2.2.1]heptane; 2-Oxide, *in* D-00796
2,8-Dimethyl-10-phenyl-10*H*-phenothiaphosphine; 5,5,10-Trioxide, *in* D-00798
2,8-Dimethyl-10-phenyl-10*H*-phenoxaphosphine; 10-Oxide, *in* D-00799
1,4-Dimethyl-2-phenyl-2-phosphabicyclo[2.2.1]heptane 2-oxide, *in* D-00825
Dimethylphenylphosphine; Oxide, *in* D-00803
3,4-Dimethyl-1-phenyl-1*H*-phosphole; 1-Oxide, *in* D-00807
2,5-Dimethyl-1-phenylphosphorinan-4-one; (1*RS*,2*RS*,5*RS*)-*form*, 1-Oxide, *in* D-00821
2,5-Dimethyl-1-phenylphosphorinan-4-one; (1*SR*,2*RS*,5*RS*)-*form*, 1-Oxide, *in* D-00821
1-(Dimethylphosphinyl)ethanone, *in* A-00015
1,3-Dimethylphosphorinane; Oxide, *in* D-00882
1,4-Dimethylphosphorinane; Oxide, *in* D-00883
Dioctylphenylphosphine; Oxide, *in* D-00955
2,3-Diphenyl-1,3-azaphospholidine; 3-Sulfide, *in* D-00995
5,10-Diphenyldibenzo[1,4]phospharsenin; *P*-Oxide, *in* D-01001
5,10-Diphenyldibenzo[1,4]phospharsenin; *As,P*-Dioxide, *in* D-01001
Diphenyl[(2-diphenylphosphinyl)ethyl]phosphine, *in* B-00234
1,12-Diphenyl-1,12-diphosphacyclodocosane; *cis-form*, 1,12-Dioxide, *in* D-01008
1,12-Diphenyl-1,12-diphosphacyclodocosane; *trans-form*, 1,12-Dioxide, *in* D-01008
1,10-Diphenyl-1,10-diphosphacyclooctadecane; *cis-form*, 1,10-Dioxide, *in* D-01009
1,10-Diphenyl-1,10-diphosphacyclooctadecane; *trans-form*, 1,10-Dioxide, *in* D-01009
1,12-Diphenyl-1,12-diphosphacyclotetracosane; *cis-form*, 1,12-Dioxide, *in* D-01010
1,12-Diphenyl-1,12-diphosphacyclotetracosane; *trans-form*, 1,12-Dioxide, *in* D-01010
1,4-Diphenyl-1,4-diphosphorinane; *cis-form*, 1,4-Dioxide, *in* D-01017
1,4-Diphenyl-1,4-diphosphorinane; *trans-form*, 1,4-Dioxide, *in* D-01017
2,3-Diphenyl-1,3-oxaphospholane; 3-Oxide, *in* D-01027
Diphenyl(4-phenyl-1,3-butadienyl)phosphine; Oxide, *in* D-01031
Diphenyl(4-phenyl-1,3-butadiynyl)phosphine; Oxide, *in* D-01032
Diphenyl(2-phenylethenyl)phosphine; (*E*)-*form*, Oxide, *in* D-01033
Diphenyl(2-phenylethenyl)phosphine; (*Z*)-*form*, Oxide, *in* D-01033
Diphenyl(2-phenylethyl)phosphine; Oxide, *in* D-01034
2-(Diphenylphosphino)benzoic acid; Oxide, *in* D-01053
2-(Diphenylphosphino)benzoic acid; Oxide, Me ester, *in* D-01053
(Diphenylphosphino)methanol; Oxide, *in* D-01069
3-(Diphenylphosphino)phenol; Oxide, *in* D-01071
3-(Diphenylphosphino)phenol; Oxide, Ac, *in* D-01071
3-(Diphenylphosphino)-1-propanol; Oxide, *in* D-01074
4-(Diphenylphosphino)pyridine; *N,P*-Dioxide, *in* D-01077
(Diphenylphosphinyl)acetic acid, D-01093
Diphenylphosphinylacetic acid hydrazide, D-01094
(Diphenylphosphinyl)acetonitrile, *in* D-01046
Diphenylphosphinylacetylene, *in* E-00201
2-(Diphenylphosphinyl)aniline, *in* D-01047
3-(Diphenylphosphinyl)aniline, *in* D-01048
4-(Diphenylphosphinyl)aniline, *in* D-01049
m-(Diphenylphosphinyl)anisole, *in* M-00057
p-(Diphenylphosphinyl)anisole, *in* M-00058
2-(Diphenylphosphinyl)benzaldehyde, *in* D-01050
3-(Diphenylphosphinyl)benzaldehyde, *in* D-01051
4-(Diphenylphosphinyl)benzaldehyde, *in* D-01052
α-(Diphenylphosphinyl)benzeneacetonitrile, D-01095
3-(Diphenylphosphinyl)benzoic acid, *in* D-01054
4-(Diphenylphosphinyl)benzoic acid, *in* D-01055
3-(Diphenylphosphinyl)benzonitrile, *in* D-01057
4-(Diphenylphosphinyl)benzonitrile, *in* D-01058
4-(Diphenylphosphinyl)biphenyl, *in* B-00078
Diphenylphosphinylcyclohexane, *in* C-00245
2-Diphenylphosphinyl-*N,N*-dimethylaniline, *in* D-01059
3-Diphenylphosphinyl-*N,N*-dimethylaniline, *in* D-01060
4-Diphenylphosphinyl-*N,N*-dimethylaniline, *in* D-01061
(Diphenylphosphinyl)dimethyl ether, *in* D-01069
1-(Diphenylphosphinyl)ethanone, *in* A-00016
1-(Diphenylphosphinyl)hexane, *in* H-00093
(Diphenylphosphinyl)isocyanoacetic acid; *tert*-Butyl ester, *in* D-01098
(Diphenylphosphinylmethylene)triphenylphosphorane, D-01099
N-[(Diphenylphosphinyl)methyl]-*N*-methylaniline, D-01100
1-Diphenylphosphinyl-2-methylpropene, *in* M-00362
[(Diphenylphosphinyl)methyl]triphenylphosphonium(1+); Chloride, *in* D-01101
[(Diphenylphosphinyl)methyl]triphenylphosphonium(1+); Bromide, *in* D-01101
[(Diphenylphosphinyl)methyl]triphenylphosphonium(1+); Tetraphenylborate, *in* D-01101
[(Diphenylphosphinyl)methyl]triphenylphosphonium(1+); 4-Methylbenzenesulfonate, *in* D-01101
1-(Diphenylphosphinyl)naphthalene, *in* N-00011
1-(Diphenylphosphinyl)octane, *in* O-00039
2-(Diphenylphosphinyl)phenol, *in* D-01070
4-(Diphenylphosphinyl)phenol, *in* D-01072
Diphenylphosphinylpropadiene, *in* D-01142
3-(Diphenylphosphinyl)propanenitrile, *in* D-01073
1-(Diphenylphosphinyl)-2-propanone, *in* D-01075
1-(Diphenylphosphinyl)propyne, *in* D-01146
2-(Diphenylphosphinyl)pyridine, *in* D-01076
4-(Diphenylphosphinyl)pyridine, *in* D-01077
Diphenyl-1-propenylphosphine; (*E*)-*form*, Oxide, *in* D-01143
Diphenyl-2-propenylphosphine; Oxide, *in* D-01144
Diphenylpropylphosphine; Oxide, *in* D-01145
2,3-Diphenyl-1,3-thiaphospholane; 3-Oxide, *in* D-01151
Diphenyl(3-trifluoromethylphenyl)phosphine; Oxide, *in* D-01155
Diphenylvinylphosphine; Oxide, *in* D-01168
1,4-Diphosphabicyclo[2.2.2]octane; Dioxide, *in* D-01169
1,6-Diphosphecane; 1,6-Di-Ph, 1,6-dioxide, *in* D-01171
1,6-Diphosphecane; 1,6-Dibenzyl, 1,6-dioxide, *in* D-01171
7,8,10,11,13,14,16,17,19,20,22,27-Dodecahydro-5*H*-dibenzo[*r,u*][1,4,7,10,13,16,20]-hexaoxaphosphacyclotricosin; 27-Ph, 27-oxide, *in* D-01261

Dodecyldimethylphosphine; Oxide, *in* D-01264
1,2-Ethanediylbis[(2-methoxyphenyl)phenylphosphine]; (*R**,*R**)-*form*, Dioxide, *in* E-00011
1,2-Ethanediylbis[(2-methoxyphenyl)phenylphosphine]; (*RS*,*SR*)-*form*, Dioxide, *in* E-00011
1,2-Ethanediylbis[methylphenylphosphine]; (*RS*,*RS*)-*form*, Dioxide, *in* E-00012
1,2-Ethanediylbis[methylphenylphosphine]; (*RS*,*SR*)-*form*, Dioxide, *in* E-00012
Ethyl diazo(diphenylphosphinyl)acetate, *in* D-00030
Ethyl (dihydorxyphosphinyl)formate, *in* P-00387
1-Ethyl-2,3-diphenyl-1,3-azaphosphacyclopentane, *in* D-00995
Ethyldiphenylphosphine; Oxide, *in* E-00066
Ethyl 4-(diphenylphosphinyl)-3-oxobutanoate, *in* D-01102
Ethylmethylphenylphosphine; (*R*)-*form*, Oxide, *in* E-00084
Ethylmethylphenylphosphine; (*S*)-*form*, Oxide, *in* E-00084
Ethylmethylphenylphosphine; (±)-*form*, Oxide, *in* E-00084
1-Ethyl-1,2,3,4-tetrahydrophosphinoline 1-oxide, *in* T-00135
(2-Fluorophenyl)diphenylphosphine; Oxide, *in* F-00026
(3-Fluorophenyl)diphenylphosphine; Oxide, *in* F-00027
(4-Fluorophenyl)diphenylphosphine; Oxide, *in* F-00028
Hexaethyl [phosphinylidynetris(methylene)]trisphosphonate, *in* P-00360
2,3,4,5,6,7-Hexahydro-1*H*-isophosphindole 2-oxide; 2-Me, *in* H-00035
2,3,4,5,6,7-Hexahydro-1*H*-isophosphindole 2-oxide; 2-Ph, *in* H-00035
Hexahydro-1-methyl-4-phenyl-1,4-azaphosphorine; 4-Oxide, *in* H-00036
Hexahydro-4-phenyl-1,4-azaphosphorine; 1-Ph, 4-oxide, *in* H-00042
1,4,4*a*,9,9*a*,10-Hexahydro-11-phenyl-9,10-phosphinideneanthracene; 11-Oxide, *in* H-00050
Hexahydro-1,3,5-triphenyl-1,3,5-diazaphosphorine; 5-Oxide, *in* H-00058
Hexaisopropyl [phosphinylidynetris(methylene)]-trisphosphonate, *in* P-00360
1,2,2,3,4,4-Hexamethylphosphetane; Oxide, *in* H-00074
1,2,2,3,4,4-Hexamethylphosphetane; *cis-form*, Oxide, *in* H-00074
1,2,2,3,4,4-Hexamethylphosphetane; *trans-form*, Oxide, *in* H-00074
Hexamethyl [phosphinylidynetris(methylene)]trisphosphonate, *in* P-00360
1,1,4,7,10,10-Hexaphenyl-1,4,7,10-tetraphosphadecane; High-melting-*form*, Tetraoxide, *in* H-00091
5-Isopropyl-2,6-dimethyl-6-phosphabicyclo[3.1.1]hept-2-ene 6-oxide, I-00029
Isopropyldiphenylphosphine; Oxide, *in* I-00031
3-Methoxy-15-methyl-18-nor-15-phosphaestra-1,3,5(10),-6,8,13-hexaene; 15-Oxide, *in* M-00048
3-Methoxy-17-methyl-15-phenyl-15-phosphaestra-1,3,5(10),-8,16-pentaene; (±)-*form*, 15-Oxide, *in* M-00051
3-Methoxy-18-nor-17-phosphaestra-1,3,5(10),9(11)-tetraen-15-one; 13α-*form*, 17-Me, 17-oxide, *in* M-00054
3-Methoxy-18-nor-17-phosphaestra-1,3,5(10),9(11)-tetraen-15-one; 13α-*form*, 17-Ph, 17-oxide, *in* M-00054
3-Methoxy-18-nor-17-phosphaestra-1,3,5(10),9(11)-tetraen-15-one; 13β-*form*, 17-Ph, 17-oxide, *in* M-00054
(2-Methoxyphenyl)diphenylphosphine; Oxide, *in* M-00056
(2-Methoxyphenyl)methylphenylphosphine; (*S*)-*form*, Oxide, *in* M-00059
(2-Methoxyphenyl)-2-naphthalenylphenylphosphine; (*R*)-*form*, Oxide, *in* M-00065
(2-Methoxyphenyl)-2-naphthalenylphenylphosphine; (*S*)-*form*, Oxide, *in* M-00065
(2-Methoxyphenyl)-2-naphthalenylphenylphosphine; (±)-*form*, Oxide, *in* M-00065
(4-Methoxyphenyl)-1-naphthalenylphenylphosphine; (±)-*form*, Oxide, *in* M-00066
5-Methyl-5*H*-dibenzophosphole; 5-Oxide, *in* M-00115
Methyl (dihydroxyphosphinyl)formate, *in* P-00387
6-Methyl-1,5-diphenyl-6-phosphabicyclo[3.2.1]octane; Oxide, *in* M-00134
Methyldiphenylphosphine; Oxide, *in* M-00135
Methyl (diphenylphosphinyl)acetate, *in* D-01093
Methyl 4-(diphenylphosphinyl)benzoate, *in* D-01055
(5-Methyl-2-isopropylcyclohexyl)diphenylphosphine; (1*R*,2*S*,5*R*)-*form*, Oxide, *in* M-00164
(5-Methyl-2-isopropylcyclohexyl)diphenylphosphine; (1*S*,2*S*,5*R*)-*form*, Oxide, *in* M-00164
Methyl-2-naphthylphenylphosphine; (*R*)-*form*, Oxide, *in* M-00177
Methyl-2-naphthylphenylphosphine; (*S*)-*form*, Oxide, *in* M-00177
(2-Methylphenyl)diphenylphosphine; Oxide, *in* M-00200
(3-Methylphenyl)diphenylphosphine; Oxide, *in* M-00201
(4-Methylphenyl)diphenylphosphine; Oxide, *in* M-00202
3-Methyl-1-phenylphospholane; (1*S*,3*R*)-*form*, 1-Oxide, *in* M-00229
3-Methyl-1-phenylphospholane; (1*RS*,3*RS*)-*form*, 1-Oxide, *in* M-00229
3-Methyl-1-phenylphospholane; (1*RS*,3*SR*)-*form*, 1-Oxide, *in* M-00229
Methylphenyl-2-propenylphosphine; (*S*)-*form*, Oxide, *in* M-00264
Methylphenylpropylphosphine; (*R*)-*form*, Oxide, *in* M-00265
Methylphenylpropylphosphine; (*S*)-*form*, Oxide, *in* M-00265
Methylphenylpropylphosphine; (±)-*form*, Oxide, *in* M-00265
8-Methyl-8-phosphabicyclo[3.2.1]octane; *syn-form*, Oxide, *in* M-00272
8-Methyl-8-phosphabicyclo[3.2.1]octane; *anti-form*, Oxide, *in* M-00272
8-Methyl-8-phosphabicyclo[3.2.1]oct-6-ene 8-oxide; *syn-form*, *in* M-00273
8-Methyl-8-phosphabicyclo[3.2.1]oct-6-ene 8-oxide; *anti-form*, *in* M-00273
4-Methyl-4-phosphatetracyclo[3.3.0.0^{2,8}.0^{3,6}]octane; *endo-form*, 4-Oxide, *in* M-00275
4-Methyl-4-phosphatetracyclo[3.3.0.0^{2,8}.0^{3,6}]octane; *exo-form*, 4-Oxide, *in* M-00275
1-Methylphospholane; Oxide, *in* M-00280
1-Methyl-3,8-phosphonanedione 1-oxide, *in* P-00373
1-Methylphosphorinane; Oxide, *in* M-00339
1-Methyl-4-phosphorinanone; Oxide, *in* M-00340
(1-Methylpropyl)diphenylphosphine; (*R*)-*form*, Oxide, *in* M-00369
(1-Methylpropyl)diphenylphosphine; (±)-*form*, Oxide, *in* M-00369
(2-Methylpropyl)diphenylphosphine; Oxide, *in* M-00370
7-Methyl-2-thia-1,3,5-triaza-7-phosphoniaadamantane iodide, *in* T-00301
6-Methyl-3,5,7-triphenyl-6-phosphatricyclo[3.3.1.0^{2,7}]non-3-ene; 6-Oxide, *in* M-00431
(Morpholinomethyl)diphenylphosphine oxide, M-00519
(3-Nitrophenyl)diphenylphosphine; Oxide, *in* N-00039
(4-Nitrophenyl)diphenylphosphine; Oxide, *in* N-00040
6,7,9,10,16,17,18,19-Octahydro-16,19-diphenyldibenzo[*h*,*n*]-[1,4,7,10,13]trioxadiphosphacyclopentadecin; 16,19-Dioxide, *in* O-00015
Octahydro-2,4-diphenyl-1,3,5-metheno-1*H*-2,4-diphosphacyclobuta[*c*,*d*]pentalene; 2,4-Dioxide, *in* O-00016
Octahydro-3-methyl-1*H*-cyclopropa[*c*]phosphindole 3-oxide, *in* O-00004
Octahydro-3-phenyl-1*H*-cyclopropa[*c*]phosphindole 3-oxide, *in* O-00004
(Pentafluorophenyl)diphenylphosphine; Oxide, *in* P-00014
2,2,3,4,4-Pentamethyl-1-phenylphosphetane; *cis-form*, Oxide, *in* P-00020
2,2,3,4,4-Pentamethyl-1-phenylphosphetane; *trans-form*, Oxide, *in* P-00020
2,7,10,15,17-Pentamethyl-17*H*-tetrabenzo[*b*,*d*,*f*,*h*]phosphonin, P-00028
Pentaphenyl-1*H*-phosphole; Oxide, *in* P-00038
1,2,3,4,9-Pentaphenyl-1*H*-tribenzo[*b*,*d*,*f*]phosphepin; 9-Oxide, *in* P-00042
5-Phenyl-10(5*H*)-acridophosphinone; 5-Oxide, *in* P-00080
3-Phenyl-3*H*-benzophosphepin; 3-Oxide, *in* P-00096
Phenylbis[2-(phenylethynyl)phenyl]phosphine; Oxide, *in* P-00098
Phenylbis(phenylethynyl)phosphine; Oxide, *in* P-00099
5-Phenyl-5*H*-dibenzophosphole; 5-Oxide, *in* P-00110
7-Phenyl-7*H*-dibenzo[*d*,*f*]phosphonin; (*E*,*Z*)-*form*, 7-Oxide, *in* P-00111

4-Phenyl-3,5-dioxa-1-phosphabicyclo[2.2.1]heptane; 1-Oxide, *in* P-00113
Phenyldi-2-propenylphosphine; Oxide, *in* P-00119
Phenyldivinylphosphine; Oxide, *in* P-00122
(1-Phenyl-1,2-ethanediyl)bis[diphenylphosphine]; (*S*)-*form*, Dioxide, *in* P-00150
(1-Phenyl-1,2-ethanediyl)bis[diphenylphosphine]; (±)-*form*, Dioxide, *in* P-00150
2-Phenyl-2*H*-isophosphindole 2-oxide, P-00169
4-Phenyl-4*H*-1,4-oxaphosphorin; 4-Oxide, *in* P-00175
4-Phenyl-1,4-oxaphosphorinane; Oxide, *in* P-00176
10-Phenyl-10*H*-phenothiaphosphine; 10-Oxide, *in* P-00182
10-Phenyl-10*H*-phenothiaphosphine; 5,5,10-Trioxide, *in* P-00182
10-Phenyl-10*H*-phenoxaphosphine; 10-Oxide, *in* P-00183
9-Phenyl-9-phosphabicyclo[3.3.1]nonane; Oxide, *in* P-00193
9-Phenyl-9-phosphabicyclo[4.2.1]nonane; Oxide, *in* P-00194
9-Phenyl-9-phosphabicyclo[3.3.1]nonan-3-one; *syn-form*, Oxide, *in* P-00195
9-Phenyl-9-phosphabicyclo[3.3.1]nonan-3-one; *anti-form*, Oxide, *in* P-00195
9-Phenyl-9-phosphabicyclo[4.2.1]nona-2,4,7-triene; *syn-form*, Oxide, *in* P-00196
9-Phenyl-9-phosphabicyclo[4.2.1]nona-2,4,7-triene; *anti-form*, Oxide, *in* P-00196
3-Phenyl-3-phosphabicyclo[3.2.1]octane; *endo-form*, 3-Oxide, *in* P-00198
3-Phenyl-3-phosphabicyclo[3.2.1]octane; *exo-form*, 6,7-Didehydro, 3-oxide, *in* P-00198
8-Phenyl-8-phosphabicyclo[3.2.1]octan-3-one; *syn-form*, Oxide, *in* P-00199
8-Phenyl-8-phosphabicyclo[3.2.1]octan-3-one; *anti-form*, Oxide, *in* P-00199
8-Phenyl-8-phosphabicyclo[3.2.1]oct-6-ene; Oxide, *in* P-00200
3-Phenyl-3-phosphabicyclo[3.2.1]oct-6-ene 3-oxide, *in* P-00198
9-Phenyl-9-phosphapentacyclo[$4.3.0.0^{2,5}.0^{3,8}.0^{4,7}$]nonane; 9-Oxide, *in* P-00201
9-Phenyl-9-phosphatricyclo[$3.3.1.0^{2,8}$]nona-3,6-diene; 9-Oxide, *in* P-00202
9-Phenyl-9-phosphatricyclo[$4.2.1.0^{2,5}$]nona-3,7-diene; *endo-syn*-9-Oxide, *in* P-00203
9-Phenyl-9-phosphatricyclo[$4.2.1.0^{2,5}$]nona-3,7-diene; *endo-anti*-9-Oxide, *in* P-00203
9-Phenyl-9-phosphatricyclo[$4.2.1.0^{2,5}$]nonane; ($1\alpha,2\alpha,5\alpha,6\alpha$)-*form*, 9-Oxide, *in* P-00204
1-Phenylphosphepane; 1-Oxide, *in* P-00205
1-Phenyl-1*H*-phosphepin; Oxide, *in* P-00206
1-Phenyl-1*H*-phosphindole; 1-Oxide, *in* P-00207
(Phenylphosphinyldene)bis[methyl acetate], *in* B-00281
1,1″-(Phenylphosphinylidene)bisferrocene, *in* D-00410
(Phenylphosphinylidene)bismethanol, *in* B-00281
3,3′-(Phenylphosphinylidene)bis[1-propanol], *in* B-00286
3,3′-(Phenylphosphinylidene)bis[1-propyl acetate], *in* B-00286
1-Phenylphosphocane; Oxide, *in* P-00215
1-Phenylphospholane; 1-Oxide, *in* P-00216
1-Phenyl-3,8-phosphonanedione 1-oxide, *in* P-00373
1-Phenylphosphorinane; Oxide, *in* P-00265
1-Phenyl-4-phosphorinanone; Oxide, *in* P-00268
4-Phenyl-4*H*-1,4-selenaphosphorin; 4-Oxide, *in* P-00307
4-Phenyl-4*H*-1,4-telluraphosphorin; 4-Oxide, *in* P-00317
1-Phenyl-1,2,3,4-tetrahydrophosphinoline; 1-Oxide, *in* P-00320
1-Phosphabicyclo[2.2.1]heptane; *p*-Oxide, *in* P-00344
1-Phosphabicyclo[3.3.1]nonane; 1-Oxide, *in* P-00345
1-Phosphabicyclo[2.2.2]octane; 1-Oxide, *in* P-00346
3,3′,3″-Phosphinyldynetrispropanoic acid triethyl ester, *in* P-00357
4,4′,4″-Phosphinylidynetris[benzoic acid], *in* T-00740
3,3′,3″-Phosphinylidynetris[benzonitrile], *in* T-00753
4,4′,4″-Phosphinylidynetris[benzonitrile], *in* T-00754
Phosphinylidynetrismethanol, *in* T-00784
[Phosphinylidynetris(methylene)]trisphosphonic acid, P-00360
2,2′,2″-Phosphinylidynetris(1-methylpyrrole), *in* T-00726
3,3′,3″-Phosphinylidynetrisphenol, *in* T-00788
3,3′,3″-Phosphinylidynetrisphenol, *in* T-00789
4,4′,4″-Phosphinylidynetrisphenol, *in* T-00790
3,3′,3″-Phosphinylidynetris[propanamide], *in* T-00738
3,3′,3″-Phosphinylidynetris[propanenitrile], *in* T-00752
3,3′,3″-Phosphinylidynetrispropanoic acid trimethyl ester, *in* P-00357
2,2′,2″-Phosphinylidynetrispyrrole, *in* T-00726
Phosphonoformic acid, P-00387
Phosphonoformic acid; *C*-Et ester, dianilinium salt, *in* P-00387
Phosphonoformic acid; *C*,*P*-di-Et ester, anilinium salt, *in* P-00387
1,5,10,14-Tetrabenzyl-1,5,10,14-tetraphosphacyclooctadecane; 1,5,10,14-Tetraoxide, *in* T-00009
5,5,8,8-Tetraethyl-5,6,7,8-tetrahydrodibenzo[*e*,*g*][1,4]-diphosphocinium dibromide, *in* D-00404
1,3,4,5-Tetrahydro-4,5-dimethyl-2,5-methano-2*H*-2-benzophosphepin; Oxide, *in* T-00096
3*a*,4,7,7*a*-Tetrahydro-1,8-dimethyl-4,7-phosphinidene-1*H*-phosphindole; 1,8-Dioxide, *in* T-00099
3*a*,4,7,7*a*-Tetrahydro-1,8-diphenyl-4,7-phosphinidene-1*H*-phosphindole; 1,8-Dioxide, *in* T-00105
3*a*,4,7,7*a*-Tetrahydro-1,2,3,5,6,8-hexamethyl-4,7-phosphinidene-1*H*-phosphindole; 1,8-Dioxide, *in* T-00106
5*a*,6,11,11*a*-Tetrahydro-1,2,3,6,11,12-hexaphenyl-6,11-phosphinidene-1*H*-naphth[2,3-*e*]isophosphindole; 2,12-Dioxide, *in* T-00107
2,3,4,7-Tetrahydro-1*H*-isophosphindole 2-oxide; 2-Me, *in* T-00109
2,3,4,7-Tetrahydro-1*H*-isophosphindole 2-oxide; 2-Ph, *in* T-00109
3,3*a*,4,5-Tetrahydro-3-methyl-2*H*-benzo[*e*]phosphindole 3-oxide, T-00112
3,4,5,6-Tetrahydro-4-methyl-2*H*-1,5-methano-4,1-benzazaphosphocine; 1-Oxide, *in* T-00115
1,2,3,4-Tetrahydro-4-oxo-1-phenylphosphinoline; 1-Oxide, *in* T-00122
1,2,5,6-Tetrahydro-1,1,6,6,9-pentamethyl-4*H*-phospholo[3,2,1-*ij*]phosphinoline; 3-Oxide, *in* T-00123
2,3,6,7-Tetrahydro-1,1,7,7,9-pentamethyl-1*H*,5*H*-phosphorino[3,2,1-*ij*]phosphinoline; 4-Oxide, *in* T-00124
1,2,3,4-Tetrahydro-1-phenylbenzo[*h*]phosphinoline; 1-Oxide, *in* T-00128
5,6,7,12-Tetrahydro-12-phenyldibenz[*c*,*f*][1,5]-azaphosphocine; 12-Oxide, *in* T-00130
5,6,7,12-Tetrahydro-12-phenyldibenz[*c*,*f*][1,5]-azaphosphocine; 6-Benzyl, 12-oxide, *in* T-00130
3*a*,4,7,7*a*-Tetrahydro-2,3,5,6-tetramethyl-1,8-diphenyl-4,7-phosphinidene-1*H*-phosphindole; 1,8-Dioxide, *in* T-00141
1,1,3,3-Tetrakis(diphenylphosphino)-1,2-propadiene; Tetraoxide, *in* T-00180
Tetrakis[(diphenylphosphinyl)methyl]methane, *in* T-00179
2,2,6,6-Tetramethyl-1-phenyl-4-phosphorinanone; Oxide, *in* T-00210
2,4,5,6-Tetraphenyl-1,3,5-dioxaphosphorinane; 5-Oxide, *in* T-00254
1,5,9,13-Tetraphenyl-1,5,9,13-tetraphosphacyclohexadecene; Tetraoxide, *in* T-00276
1,4-Thiaphosphorinane 1,1,4-trioxide; 4-Me, *in* T-00300
1,4-Thiaphosphorinane 1,1,4-trioxide; 4-Ph, *in* T-00300
2-Thia-1,3,5-triaza-7-phosphatricyclo[$3.3.1.1^{3,7}$]decane; 2,2-Dioxide, *in* T-00301
2-Thia-1,3,5-triaza-7-phosphatricyclo[$3.3.1.1^{3,7}$]decane; 2,2,7-Trioxide, *in* T-00301
1,3,5-Triaza-7-phosphatricyclo[$3.3.1.1^{3,7}$]decane; 7-Oxide, *in* T-00327
▷Tribenzylphosphine; Oxide, *in* T-00338
Tri-2-biphenylylphosphine; Oxide, *in* T-00348
Tri-4-biphenylylphosphine; Oxide, *in* T-00350
Tri-*tert*-butylphosphine; Oxide, *in* T-00373
▷Tributylphosphine oxide, T-00376
▷Tricyclohexylphosphine; Oxide, *in* T-00426
Tricyclopropylphosphine; Oxide, *in* T-00432
Tri(decyl)phosphine; Oxide, *in* T-00433
Tridodecylphosphine; Oxide, *in* T-00435
Triethylphosphine oxide, T-00443
Tri-2-furanylphosphine; Oxide, *in* T-00486

Tri-3-furanylphosphine; Oxide, *in* T-00487
▷Triheptylphosphine; Oxide, *in* T-00489
▷Trihexylphosphine; Oxide, *in* T-00492
Triisopropylphosphine; Oxide, *in* T-00500
2,4,6-Trimethyl-5-phenyl-1,3,5-dioxaphosphorinane; (2α,4α,5α,6α)-*form*, 5-Oxide, *in* T-00521
2,4,6-Trimethyl-5-phenyl-1,3,5-dioxaphosphorinane; (2α,4α,5β,6α)-*form*, 5-Oxide, *in* T-00521
2,2,5-Trimethyl-3-phenyl-1,3-oxaphospholane; (3*RS*,5*SR*)-*form*, 3-Oxide, *in* T-00522
1,2,4-Trimethyl-2-phosphabicyclo[2.2.1]heptane 2-oxide, *in* D-00825
2,3,5-Trimethyl-3-phosphatricyclo[4.2.1.$0^{2,5}$]nonane; Oxide, *in* T-00529
Trimethylphosphine oxide, T-00532
Tri-1-naphthylphosphine; Oxide, *in* T-00572
Tri-2-naphthylphosphine; Oxide, *in* T-00573
Trinonylphosphine; Oxide, *in* T-00578
Trioctylphosphine; Oxide, *in* T-00582
▷2,6,7-Trioxa-1,4-diphosphabicyclo[2.2.2]octane; 4-Oxide, *in* T-00585
▷2,6,7-Trioxa-1,4-diphosphabicyclo[2.2.2]octane; 1,4-Dioxide, *in* T-00585
2,6,7-Trioxa-4-phospha-1-arsabicyclo[2.2.2]octane; 4-Oxide, *in* T-00586
▷Tripentylphosphine; Oxide, *in* T-00594
1,2,3-Triphenyl-1,3-azaphosphacyclopentane, *in* D-00995
1,2,3-Triphenyl-1,3-azaphospholidin-5-one; 3-Oxide, *in* T-00596
2,4,6-Triphenyl-4*H*-1,4-oxaphosphorin; 4-Oxide, *in* T-00610
N,1,1-Triphenylphosphinecarbonamide; Oxide, *in* T-00625
Triphenylphosphine oxide, T-00627
1,2,3-Triphenyl-1*H*-phosphirene; Oxide, *in* T-00632
1,2,5-Triphenyl-1*H*-phosphole; Oxide, *in* T-00634
1,2,6-Triphenyl-4-phosphorinanone; (1α,2α,6β)-*form*, Oxide, *in* T-00666
1,2,6-Triphenyl-4-phosphorinanone; (1α,2β,6β)-*form*, Oxide, *in* T-00666
1,2,6-Triphenyl-4-phosphorinanone; (1α,2β,6β)-*form*, Sulfide, *in* T-00666
2,4,6-Triphenyl-4*H*-1,4-thiaphosphorin; 4-Oxide, *in* T-00686
2,4,6-Triphenyl-4*H*-1,4-thiaphosphorin; 1,1,4-Trioxide, *in* T-00686
Tri-2-propenylphosphine; Oxide, *in* T-00707
Tripropylphosphine; Oxide, *in* T-00711
Tri-1-propynylphosphine; Oxide, *in* T-00721
Tri-2-pyridylphosphine; *P*-Oxide, *in* T-00723
Tris(benzoyloxymethyl)phosphine oxide, *in* T-00784
Tris[(4-bromophenyl)methyl]phosphine; Oxide, *in* T-00733
Tris(3-bromophenyl)phosphine; Oxide, *in* T-00734
Tris(4-bromophenyl)phosphine; Oxide, *in* T-00735
Tris(3-carboxyphenyl)phosphine; Oxide, tri-Me ester, *in* T-00739
Tris(4-carboxyphenyl)phosphine; Oxide, tri-Me ester, *in* T-00740
Tris[(4-chlorophenyl)methyl]phosphine; Oxide, *in* T-00744
Tris(2-chlorophenyl)phosphine; Oxide, *in* T-00747
Tris(3-chlorophenyl)phosphine; Oxide, *in* T-00748
▷Tris(4-chlorophenyl)phosphine; Oxide, *in* T-00749
Tris(3,5-dimethoxyphenyl)phosphine; Oxide, *in* T-00759
Tris(4-dimethylaminophenyl)phosphine; Oxide, *in* T-00765
Tris(2,4-dimethylphenyl)phosphine; Oxide, *in* T-00768
Tris(2,5-dimethylphenyl)phosphine; Oxide, *in* T-00769
Tris(2,6-dimethylphenyl)phosphine; Oxide, *in* T-00770
Tris[2-(diphenylphosphino)ethyl]phosphine; Tetraoxide, *in* T-00774
Tris(diphenylphosphino)methane; Dioxide, *in* T-00775
Tris(diphenylphosphino)methane; Trioxide, *in* T-00775
Tris(diphenylphosphino)methane; Dioxide, monomethiodide, *in* T-00775
Tris(diphenylphosphino)methane; Dioxide, monosulfide, *in* T-00775
Tris(diphenylphosphino)methane; Oxide, disulfide, *in* T-00775
Tris(2-fluorophenyl)phosphine; Oxide, *in* T-00781
Tris(3-fluorophenyl)phosphine; Oxide, *in* T-00782
Tris(4-fluorophenyl)phosphine; Oxide, *in* T-00783
Tris(hydroxymethyl)phosphine; Oxide, tri-Ac, *in* T-00784
Tris(2-methoxyphenyl)phosphine; Oxide, *in* T-00791
Tris(3-methoxyphenyl)phosphine; Oxide, *in* T-00792
Tris(4-methoxyphenyl)phosphine; Oxide, *in* T-00793
Tris[(2-methylphenyl)methyl]phosphine; Oxide, *in* T-00800
Tris[(3-methylphenyl)methyl]phosphine; Oxide, *in* T-00801
Tris[(4-methylphenyl)methyl]phosphine; Oxide, *in* T-00802
Tris(2-methylphenyl)phosphine; Oxide, *in* T-00806
Tris(3-methylphenyl)phosphine; Oxide, *in* T-00807
▷Tris(4-methylphenyl)phosphine; Oxide, *in* T-00808
Tris(1-methylpropyl)phosphine; Oxide, *in* T-00823
Tris(2-methylpropyl)phosphine; Oxide, *in* T-00824
Tris[(4-nitrophenyl)methyl]phosphine; Oxide, *in* T-00835
Tris(3-nitrophenyl)phosphine; Oxide, *in* T-00838
Tris(2-phenylethenyl)phosphine; Oxide, *in* T-00843
Tris(2-phenylethyl)phosphine; Oxide, *in* T-00844
Tris(phenylethynyl)phosphine; Oxide, *in* T-00846
Tris[2-(trifluoromethyl)phenyl]phosphine; Oxide, *in* T-00852
▷Tris[3-(trifluoromethyl)phenyl]phosphine; Oxide, *in* T-00853
Tris[4-(trifluoromethyl)phenyl]phosphine; Oxide, *in* T-00854
Tris(trifluoromethyl)phosphine oxide, T-00856
Tris(2,4,6-trimethylphenyl)phosphine; Oxide, *in* T-00861
Tris[(triphenylphosphoranylidene)methyl]phosphine; Oxide, *in* T-00871
Tri-2-thienylphosphine; Oxide, *in* T-00874
Tri-3-thienylphosphine; Oxide, *in* T-00875
Trivinylphosphine; Oxide, *in* T-00877

Tertiary phosphine sulfides, R_3PS

5,10-[1′,2′]Benzophosphenanthrene; 5,10-Disulfide, *in* B-00024
5-Benzyl-5,6-dihydro-6-phenylphosphanthridine; (5*RS*,6*RS*)-*form*, 5-Sulfide, *in* B-00043
Benzyldiphenylphosphine; Sulfide, *in* B-00046
Benzylethylphenylphosphine; (±)-*form*, Sulfide, *in* B-00047
Benzylmethylphenylphosphine; (*R*)-*form*, Sulfide, *in* B-00051
Benzylmethylphenylphosphine; (*S*)-*form*, Sulfide, *in* B-00051
Bis(2-cyanoethyl)phenylphosphine; Sulfide, *in* B-00152
1,4-Bis(dicyclohexylphosphinothioyl)butane, *in* B-00166
1,3-Bis(dicyclohexylphosphinothioyl)propane, *in* B-00169
1,2-Bis(dicyclophosphinothioyl)ethane, *in* B-00167
1,2-Bis(diethylphosphinothioyl)ethane, *in* B-00180
1,3-Bis(dimethylphosphinothioyl)benzene, *in* P-00124
1,4-Bis(dimethylphosphinothioyl)benzene, *in* P-00125
1,4-Bis(dimethylphosphinothioyl)butane, *in* B-00199
1,2-Bis(dimethylphosphinothioyl)ethane, *in* B-00200
1,3-Bis(dimethylphosphinothioyl)propane, *in* B-00206
Bis[2-(diphenylarsino)ethyl]phenylphosphine; *P*-Sulfide, *in* B-00212
1,1-Bis(diphenylphosphino)cyclopropane; Disulfide, *in* B-00229
1,2-Bis(diphenylphosphino)ethylene; (*E*)-*form*, *P*,*P*′-Disulfide, *in* B-00237
1,2-Bis(diphenylphosphino)ethylene; (*Z*)-*form*, *P*,*P*′-Disulfide, *in* B-00237
Bis[2-(diphenylphosphino)ethyl]phenylphosphine; Trisulfide, *in* B-00239
Bis[4-(diphenylphosphino)phenyl]phenylphosphine; Trisulfide, *in* B-00263
1,4-Bis(diphenylphosphinothioyl)butane, *in* B-00225
1,4-Bis(diphenylphosphinothioyl)-2-butyne, *in* B-00227
1,2-Bis(diphenylphosphinothioyl)ethane, *in* B-00234
1,1-Bis(diphenylphosphinothioyl)ethylene, *in* B-00236
1,2-Bis(diphenylphosphinothioyl)ethyne, *in* B-00216
1,1′-Bis(diphenylphosphinothioyl)ferrocene, *in* B-00241
1,6-Bis(diphenylphosphinothioyl)hexane, *in* B-00243
Bis(diphenylphosphinothioyl)methane, *in* B-00244
Bis[(diphenylphosphinothioyl)methyl]phenylphosphine sulfide, *in* B-00253
1,3-Bis(diphenylphosphinothioyl)propane, *in* B-00265
Bis(3-methylphenyl)phenylphosphine; Sulfide, *in* B-00310
Bis(4-methylphenyl)phenylphosphine; Sulfide, *in* B-00311
1,3-Bis(methylphenylphosphinothioyl)propane, *in* B-00321
Bis(pentafluorophenyl)phenylphosphine; Sulfide, *in* B-00380
(4-Bromophenyl)diphenylphosphine; Sulfide, *in* B-00481
Butyldiphenylphosphine; Sulfide, *in* B-00557
tert-Butyldiphenylphosphine; Sulfide, *in* B-00558

tert-Butylmethylphenylphosphine; (*R*)-*form*, Sulfide, *in* B-00565
tert-Butylmethylphenylphosphine; (*S*)-*form*, Sulfide, *in* B-00565
(3-Chlorophenyl)diphenylphosphine; Sulfide, *in* C-00128
(4-Chlorophenyl)diphenylphosphine; Sulfide, *in* C-00129
2-Diazo-2-(diphenylphosphinothioyl)-1-phenylethanone, *in* D-00029
Dibenzylphenylphosphine; Sulfide, *in* D-00052
Dibutylmethylphosphine; Sulfide, *in* D-00098
4,4′-Di-*tert*-butyl-1,1′,2,2′,5,5′,6,6′-octahydro-1,1′-diphenyl-2,2′-biphosphorin; *P,P*′-Disulfide, *in* D-00100
Dibutylphenylphosphine; Sulfide, *in* D-00102
Dicyclohexylphenylphosphine; Sulfide, *in* D-00208
N,N-Diethyl-2,6-dimethyl-1-phenyl-1,4-phosphaborin-4(1*H*)-amine; 1-Sulfide, *in* D-00285
Diethylphenylphosphine; Sulfide, *in* D-00305
5,10-Dihydro-5,10-dimethylphenophosphazine; 10-Sulfide, *in* D-00466
2,5-Dihydro-3,4-dimethyl-1-phenyl-1*H*-phosphole; 1-Sulfide, *in* D-00468
10,11-Dihydro-5-phenyl-5*H*-dibenzo[*b,f*]phosphepin; Sulfide, *in* D-00559
5,10-Dihydro-10-phenylphenophosphazine; 10-Sulfide, *in* D-00563
2,3-Dihydro-1-phenyl-1*H*-phosphole; 1-Sulfide, *in* D-00566
Diisopropylphenylphosphine; Sulfide, *in* D-00638
2,6-Dimethyl-1,3,2,6-dioxadiphosphocane; *cis-form*, 2,6-Disulfide, *in* D-00730
2,6-Dimethyl-1,3,2,6-dioxadiphosphocane; *trans-form*, 2,6-Disulfide, *in* D-00730
2,6-Dimethyl-1,3,2,6-dioxadiphosphocane; *trans-form*, 2,6-Diselenide, *in* D-00730
2,2-Dimethyl-1,3-dioxa-5-phospha-2-silacyclohexane; 5-Me, 5-sulfide, *in* D-00731
2,2-Dimethyl-1,3-dioxa-5-phospha-2-silacyclohexane; 5-Ph, 5-sulfide, *in* D-00731
9,9-Dimethyl-2,8-diphenyl-3,7-dioxa-1-phosphabicyclo[3.3.1]-nonane; 1-Sulfide, *in* D-00743
Dimethyl(4-methylphenyl)phosphine; Sulfide, *in* D-00761
5,6-Dimethyl-7-phenyl-7-phosphabicyclo[2.2.1]hept-5-ene-2,3-dicarboxylic acid; *endo,endo,syn-form*, 7-Sulfide, anhydride, *in* D-00800
5,6-Dimethyl-7-phenyl-7-phosphabicyclo[2.2.1]hept-5-ene-2,3-dicarboxylic acid; *endo,endo,syn-form*, 7-Sulfide, Di-Me ester, *in* D-00800
Dimethylphenylphosphine; Sulfide, *in* D-00803
3,4-Dimethyl-1-phenyl-1*H*-phosphole; 1-Sulfide, *in* D-00807
2,5-Dimethyl-1-phenylphosphorinan-4-ol; (1*RS*,2*RS*,4*RS*,5*RS*)-*form*, 1-Sulfide, *in* D-00820
2,5-Dimethyl-1-phenylphosphorinan-4-ol; (1*RS*,2*RS*,4*RS*,5*RS*)-*form*, 1-Sulfide, 4-*O*-Ac, *in* D-00820
2,5-Dimethyl-1-phenylphosphorinan-4-ol; (1*RS*,2*SR*,4*RS*,5*SR*)-*form*, 1-Sulfide, *in* D-00820
2,5-Dimethyl-1-phenylphosphorinan-4-ol; (1*RS*,2*SR*,4*RS*,5*SR*)-*form*, 1-Sulfide, 4-*O*-Ac, *in* D-00820
2,5-Dimethyl-1-phenylphosphorinan-4-ol; (1*RS*,2*SR*,4*SR*,5*SR*)-*form*, 1-Sulfide, *in* D-00820
2,5-Dimethyl-1-phenylphosphorinan-4-ol; (1*RS*,2*SR*,4*SR*,5*SR*)-*form*, 1-Sulfide, 4-*O*-Ac, *in* D-00820
2,5-Dimethyl-1-phenylphosphorinan-4-ol; (1*RS*,2*RS*,4*SR*,5*RS*)-*form*, 1-Sulfide, *in* D-00820
2,5-Dimethyl-1-phenylphosphorinan-4-ol; (1*RS*,2*RS*,4*SR*,5*RS*)-*form*, 1-Sulfide, 4-*O*-Ac, *in* D-00820
2,5-Dimethyl-1-phenylphosphorinan-4-one; (1*RS*,2*RS*,5*RS*)-*form*, 1-Sulfide, *in* D-00821
2,5-Dimethyl-1-phenylphosphorinan-4-one; (1*SR*,2*RS*,5*RS*)-*form*, 1-Sulfide, *in* D-00821
1,3-Dimethylphosphorinane; Sulfide, *in* D-00882
2,5-Diphenyl-1,3,5,2-dioxaphosphaborinane; 5-Sulfide, *in* D-01003
1,2-Diphenyl-1,2-diphospholane 1,2-disulfide, *in* D-01181
2,3-Diphenyl-1,3-oxaphospholane; 3-Sulfide, *in* D-01027
2,3-Diphenyl-1,3-oxaphosphorinane; 3-Sulfide, *in* D-01028
Diphenyl(4-phenyl-1,3-butadienyl)phosphine; (*E,E*)-*form*, Sulfide, *in* D-01031
Diphenyl(4-phenyl-1,3-butadiynyl)phosphine; Sulfide, *in* D-01032
Diphenyl(2-phenylethenyl)phosphine; (*E*)-*form*, Sulfide, *in* D-01033
(Diphenylphosphino)methanol; Sulfide, *in* D-01069
3-(Diphenylphosphinothioyl)acetanilide, *in* D-01048
(Diphenylphosphinothioyl)acetic acid, D-01086
(Diphenylphosphinothioyl)acetonitrile, *in* D-01046
3-(Diphenylphosphinothioyl)aniline, *in* D-01048
4-(Diphenylphosphinothioyl)aniline, *in* D-01049
p-(Diphenylphosphinothioyl)anisole, *in* M-00058
2-(Diphenylphosphinothioyl)benzaldehyde, *in* D-01050
3-(Diphenylphosphinothioyl)benzaldehyde, *in* D-01051
3-(Diphenylphosphinothioyl)benzoic acid, *in* D-01054
p-(Diphenylphosphinothioyl)benzoic acid, *in* D-01055
3-(Diphenylphosphinothioyl)benzonitrile, *in* D-01057
4-(Diphenylphosphinothioyl)benzonitrile, *in* D-01058
Diphenylphosphinothioylcyclohexane, *in* C-00245
3-Diphenylphosphinothioyl-*N,N*-dimethylaniline, *in* D-01060
4-Diphenylphosphinothioyl-*N,N*-dimethylaniline, *in* D-01061
1-(Diphenylphosphinothioyl)hexane, *in* H-00093
1-(Diphenylphosphinothioyl)octane, *in* O-00039
2-(Diphenylphosphinothioyl)phenol, *in* D-01070
4-(Diphenylphosphinothioyl)phenol, *in* D-01072
3-(Diphenylphosphinothioyl)propanenitrile, *in* D-01073
1-(Diphenylphosphinothioyl)propyne, *in* D-01146
2-(Diphenylphosphinothioyl)pyridine, *in* D-01076
Diphenyl-1-propenylphosphine; (*E*)-*form*, Sulfide, *in* D-01143
Diphenyl-2-propenylphosphine; Sulfide, *in* D-01144
Diphenylpropylphosphine; Sulfide, *in* D-01145
Diphenylvinylphosphine; Sulfide, *in* D-01168
1,4-Diphosphabicyclo[2.2.2]octane; Disulfide, *in* D-01169
1,5-Diphosphabicyclo[3.3.0]octane; Monosulfide, *in* D-01170
1,5-Diphosphabicyclo[3.3.0]octane; Disulfide, *in* D-01170
1-*tert*-Butyl-3,4-dimethyl-1*H*-phosphole; 1-Sulfide, *in* E-00004
Ethyldiphenylphosphine; Sulfide, *in* E-00066
Ethyl (diphenylphosphinothioyl)acetate, *in* D-01086
Ethylmethylphenylphosphine; (*R*)-*form*, Sulfide, *in* E-00084
Ethylmethylphenylphosphine; (*S*)-*form*, Sulfide, *in* E-00084
Ethylmethylphenylphosphine; (±)-*form*, Sulfide, *in* E-00084
(3-Fluorophenyl)diphenylphosphine; Sulfide, *in* F-00027
(4-Fluorophenyl)diphenylphosphine; Sulfide, *in* F-00028
Hexahydro-1-methyl-4-phenyl-1,4-azaphosphorine; 4-Sulfide, *in* H-00036
Hexahydro-3-phenyl-1,3-azaphosphorine; 1-Ph, 3-sulfide, *in* H-00041
Hexahydro-4-phenyl-1,4-azaphosphorine; 4-Sulfide, *in* H-00042
2,3,4,7,8,9-Hexahydro-1-phenyl-5,6-bis[(trimethylsilyl)oxy]-1*H*-phosphonine; Sulfide, *in* H-00047
Hexahydro-1,3,5-triphenyl-1,3,5-diazaphosphorine; 5-Sulfide, *in* H-00058
1*H*-Isophosphindole-1,3(2*H*)-dione; 2-Ph, 2-sulfide, *in* I-00026
Isopropyldiphenylphosphine; Sulfide, *in* I-00031
3-Methoxy-17-methyl-15-phenyl-15-phosphaestra-1,3,5(10),-8,16-pentaene; (±)-*form*, 15-Sulfide, *in* M-00051
(2-Methoxyphenyl)diphenylphosphine; Sulfide, *in* M-00056
5-Methyl-5*H*-dibenzophosphole; 5-Sulfide, *in* M-00115
Methyldiphenylphosphine; Sulfide, *in* M-00135
Methyl (4-diphenylphosphinothioyl)benzoate, *in* D-01055
(2-Methylphenyl)diphenylphosphine; Sulfide, *in* M-00200
(4-Methylphenyl)diphenylphosphine; Sulfide, *in* M-00202
Methylphenyl-2-propenylphosphine; (*R*)-*form*, Sulfide, *in* M-00264
Methylphenyl-2-propenylphosphine; (±)-*form*, Sulfide, *in* M-00264
Methylphenylpropylphosphine; (*R*)-*form*, Sulfide, *in* M-00265
Methylphenylpropylphosphine; (*S*)-*form*, Sulfide, *in* M-00265
Methylphenylpropylphosphine; (±)-*form*, Sulfide, *in* M-00265
5-Methyl-1-phospha-5-silabicyclo[3.3.1]nonane; 1-Sulfide, *in* M-00274
4-Methyl-4-phosphatetracyclo[3.3.0.$0^{2,8}$.$0^{3,6}$]octane; *endo-form*, 4-Sulfide, *in* M-00275
4-Methyl-4-phosphatetracyclo[3.3.0.$0^{2,8}$.$0^{3,6}$]octane; *exo-form*, 4-Sulfide, *in* M-00275
1-Methylphospholane; Sulfide, *in* M-00280
1-Methylphosphorinane; Sulfide, *in* M-00339
(2-Methylpropyl)diphenylphosphine; Sulfide, *in* M-00370

(3-Nitrophenyl)diphenylphosphine; Sulfide, *in* N-00039
(4-Nitrophenyl)diphenylphosphine; Sulfide, *in* N-00040
(Pentafluorophenyl)diphenylphosphine; Sulfide, *in* P-00014
2,2,3,4,4-Pentamethyl-1-phenylphosphetane; *cis-form*, Sulfide, *in* P-00020
2,2,3,4,4-Pentamethyl-1-phenylphosphetane; *trans-form*, Sulfide, *in* P-00020
Pentaphenyl-1*H*-phosphole; Sulfide, *in* P-00038
Phellanphos; Disulfide, *in* P-00059
3-Phenyl-1,3-azaphospholidin-2-thione; 3-Sulfide, *in* P-00093
Phenylbis(phenylethynyl)phosphine; Sulfide, *in* P-00099
1-Phenyl-6,7-bis[(trimethylsilyl)oxy]phosphacycloundec-6-ene; Sulfide, *in* P-00100
5-Phenyl-5*H*-dibenzo[*b,f*]phosphepin; Sulfide, *in* P-00108
5-Phenyl-5*H*-dibenzophosphole; 5-Sulfide, *in* P-00110
Phenyldi-2-propenylphosphine; Sulfide, *in* P-00119
3-Phenyl-1,3-oxaphosphorinan-2-one; 3-Sulfide, *in* P-00177
10-Phenyl-10*H*-phenothiaphosphine; 10-Sulfide, *in* P-00182
10-Phenyl-10*H*-phenothiaphosphine; 5,5-Dioxide, 10-sulfide, *in* P-00182
10-Phenyl-10*H*-phenoxaphosphine; 10-Sulfide, *in* P-00183
9-Phenyl-9-phosphapentacyclo[4.3.0.0^{2,5}.0^{3,8}.0^{4,7}]nonane; 9-Sulfide, *in* P-00201
1-Phenylphosphepane; Sulfide, *in* P-00205
(Phenylphosphinothioylidene)bismethanol, *in* B-00281
3,3′-(Phenylphosphinothioylidene)bis-1-propanamine, *in* B-00090
3,3′-(Phenylphosphinothioylidene)bis[1-propanol], *in* B-00286
3,3′-(Phenylphosphinothioylidene)bis[1-propyl acetate], *in* B-00286
1-Phenylphospholane; 1-Sulfide, *in* P-00216
1-Phenylphosphorinane; Sulfide, *in* P-00265
1-Phenyl-4-phosphorinanone; Sulfide, *in* P-00268
3-Phenyl-2-thia-3-phosphabicyclo[2.2.2]oct-5-ene; 3-Sulfide, *in* P-00323
1-Phosphabicyclo[3.3.1]nonane; 1-Sulfide, *in* P-00345
4,4′,4″-Phosphinothioylidynetris[benzoic acid], *in* T-00740
3,3′,3″-Phosphinothioylidynetris[benzonitrile], *in* T-00753
4,4′,4″-Phosphinothioylidynetris[benzonitrile], *in* T-00754
▷3,3′,3″-Phosphinothioylidynetris[propanenitrile], *in* T-00752
3*a*,4,7,7*a*-Tetrahydro-1,8-dimethyl-4,7-phosphinidene-1*H*-phosphindole; 1,8-Disulfide, *in* T-00099
1,2,3,4-Tetrahydro-1,4-diphenyl-1,4-benzodiphosphorin; (1*RS*,4*SR*)-*form*, 1,4-Disulfide, *in* T-00102
3*a*,4,7,7*a*-Tetrahydro-1,8-diphenyl-4,7-phosphinidene-1*H*-phosphindole; 1,8-Disulfide, *in* T-00105
3*a*,4,7,7*a*-Tetrahydro-1,2,3,5,6,8-hexamethyl-4,7-phosphinidene-1*H*-phosphindole; 1,8-Disulfide, *in* T-00106
3,4,5,6-Tetrahydro-4-methyl-2*H*-1,5-methano-4,1-benzazaphosphocine; 1-Sulfide, *in* T-00115
1,2,3,4-Tetrahydrophosphinoline; 1-Et, 1-sulfide, *in* T-00135
Tetrakis[(diphenylphosphinothioyl)methyl]methane, *in* T-00179
3,3′,4,4′-Tetramethyl-1,1′-diphenyl-2,2′-bi-1*H*-phosphole; 1,1′-Disulfide, *in* T-00199
2,2,6,6-Tetramethyl-1-phenyl-4-phosphorinanone; Sulfide, *in* T-00210
2,4,5,6-Tetraphenyl-1,3,5-dioxaphosphorinane; 5-Sulfide, *in* T-00254
1,3,5-Triaza-7-phosphatricyclo[3.3.1.1^{3,7}]decane; 7-Sulfide, *in* T-00327
Tribenzylphosphine; Sulfide, *in* T-00338
Tri-4-biphenylylphosphine; Sulfide, *in* T-00350
Tri-*tert*-butylphosphine; Sulfide, *in* T-00373
Tricyclohexylphosphine; Sulfide, *in* T-00426
Tricyclopropylphosphine; Sulfide, *in* T-00432
Triethylphosphine sulfide, T-00444
Tri-2-furanylphosphine; Sulfide, *in* T-00486
Tri-3-furanylphosphine; Sulfide, *in* T-00487
Triheptylphosphine; Sulfide, *in* T-00489
Trihexylphosphine; Sulfide, *in* T-00492
Triisopropylphosphine; Sulfide, *in* T-00500
2,4,6-Trimethyl-5-phenyl-1,3,5-dioxaphosphorinane; (2α,4α,5α,6α)-*form*, 5-Sulfide, *in* T-00521
2,4,6-Trimethyl-5-phenyl-1,3,5-dioxaphosphorinane; (2α,4α,5β,6α)-*form*, 5-Sulfide, *in* T-00521
Trimethylphosphine sulfide, T-00534
1,3,4-Trimethyl-1*H*-phosphole; Sulfide, *in* T-00536
Tri-1-naphthylphosphine; Sulfide, *in* T-00572
Trinonylphosphine; Sulfide, *in* T-00578
Trioctylphosphine; Sulfide, *in* T-00582
2,6,7-Trioxa-1,4-diphosphabicyclo[2.2.2]octane; 4-Sulfide, *in* T-00585
2,6,7-Trioxa-1,4-diphosphabicyclo[2.2.2]octane; 1,4-Disulfide, *in* T-00585
Tripentylphosphine; Sulfide, *in* T-00594
1,2,3-Triphenyl-1,3-azaphospholidin-5-one; 3-Sulfide, *in* T-00596
N,1,1-Triphenylphosphinecarbonamide; Sulfide, *in* T-00625
▷Triphenylphosphine sulfide, T-00629
1,2,5-Triphenyl-1*H*-phosphole; Sulfide, *in* T-00634
1,2,6-Triphenyl-4-phosphorinanone; (1α,2α,6β)-*form*, Sulfide, *in* T-00666
Tripropylphosphine; Sulfide, *in* T-00711
Tri-1-propynylphosphine; Sulfide, *in* T-00721
Tri-2-pyridylphosphine; Sulfide, *in* T-00723
Tris(4-bromophenyl)phosphine; Sulfide, *in* T-00735
Tris(4-carboxyphenyl)phosphine; Sulfide, tri-Me ester, *in* T-00740
Tris(3-chlorophenyl)phosphine; Sulfide, *in* T-00748
Tris(4-chlorophenyl)phosphine; Sulfide, *in* T-00749
Tris(4-dimethylaminophenyl)phosphine; Sulfide, *in* T-00765
Tris(2,4-dimethylphenyl)phosphine; Sulfide, *in* T-00768
Tris(2,5-dimethylphenyl)phosphine; Sulfide, *in* T-00769
Tris(dimethylphosphinothioyl)methane, *in* T-00771
Tris[2-(diphenylphosphino)ethyl]phosphine; Tetrasulfide, *in* T-00774
Tris(diphenylphosphino)methane; Monosulfide, *in* T-00775
Tris(diphenylphosphino)methane; Disulfide, *in* T-00775
Tris(diphenylphosphino)methane; Trisulfide, *in* T-00775
Tris(diphenylphosphino)methane; Dioxide, monosulfide, *in* T-00775
Tris(diphenylphosphino)methane; Oxide, disulfide, *in* T-00775
Tris(3-fluorophenyl)phosphine; Sulfide, *in* T-00782
Tris(4-fluorophenyl)phosphine; Sulfide, *in* T-00783
Tris(4-methoxyphenyl)phosphine; Sulfide, *in* T-00793
Tris[(2-methylphenyl)methyl]phosphine; Sulfide, *in* T-00800
Tris[(3-methylphenyl)methyl]phosphine; Sulfide, *in* T-00801
Tris[(4-methylphenyl)methyl]phosphine; Sulfide, *in* T-00802
Tris(2-methylphenyl)phosphine; Sulfide, *in* T-00806
Tris(3-methylphenyl)phosphine; Sulfide, *in* T-00807
Tris(4-methylphenyl)phosphine; Sulfide, *in* T-00808
Tris(2-methylpropyl)phosphine; Sulfide, *in* T-00824
Tris(3-nitrophenyl)phosphine; Sulfide, *in* T-00838
Tris(2-phenylethenyl)phosphine; Sulfide, *in* T-00843
Tris(phenylethynyl)phosphine; Sulfide, *in* T-00846
Tris(trifluoromethyl)phosphine; Sulfide, *in* T-00855
Tris(2,4,6-trimethylphenyl)phosphine; Sulfide, *in* T-00861
Tris[(triphenylphosphoranylidene)methyl]phosphine; Sulfide, *in* T-00871
Tri-2-thienylphosphine; Sulfide, *in* T-00874
Tri-3-thienylphosphine; Sulfide, *in* T-00875
Trivinylphosphine; Sulfide, *in* T-00877

Tertiary phosphine selenides, R$_3$PSe

1,4-Bis(dimethylphosphinoselenoyl)benzene, *in* P-00125
Bis[2-(diphenylphosphino)ethyl]phenylphosphine; Triselenide, *in* B-00239
1,3-Bis(diphenylphosphino)propane; Diselenide, *in* B-00265
1,4-Bis(diphenylphosphinoselenoyl)butane, *in* B-00225
1,2-Bis(diphenylphosphinoselenoyl)ethane, *in* B-00234
1,1-Bis(diphenylphosphinoselenoyl)ethylene, *in* B-00236
1,6-Bis(diphenylphosphinoselenoyl)hexane, *in* B-00243
Bis(diphenylphosphinoselenoyl)methane, *in* B-00244
1,5-Bis(diphenylphosphinoselenoyl)pentane, *in* B-00259
Butyldiphenylphosphine; Selenide, *in* B-00557
(3-Chlorophenyl)diphenylphosphine; Selenide, *in* C-00128
(4-Chlorophenyl)diphenylphosphine; Selenide, *in* C-00129
Dibutylphenylphosphine; Selenide, *in* D-00102
N,*N*-Diethyl-2,6-dimethyl-1-phenyl-1,4-phosphaborin-4(1*H*)-amine; 1-Selenide, *in* D-00285
Diethylphenylphosphine; Selenide, *in* D-00305

2,5-Dihydro-3,4-dimethyl-1-phenyl-1*H*-phosphole; 1-Selenide, *in* D-00468
2,6-Dimethyl-1,3,2,6-dioxadiphosphocane; *cis-form*, 2,6-Diselenide, *in* D-00730
3,4-Dimethyl-1-phenyl-1*H*-phosphole; 1-Selenide, *in* D-00807
2,5-Dimethyl-1-phenylphosphorinan-4-one; (1*RS*,2*RS*,5*RS*)-*form*, 1-Selenide, *in* D-00821
2,5-Dimethyl-1-phenylphosphorinan-4-one; (1*SR*,2*RS*,5*RS*)-*form*, 1-Selenide, *in* D-00821
2,5-Dimethyl-1-phenylphosphorinan-4-one; (1*RS*,2*SR*,5*RS*)-*form*, 1-Selenide, *in* D-00821
Diphenyl[(diphenylphosphinoselenoyl)methyl]phosphine, *in* B-00244
Diphenyl-2-propenylphosphine; Selenide, *in* D-01144
Ethyldiphenylphosphine; Selenide, *in* E-00066
(3-Fluorophenyl)diphenylphosphine; Selenide, *in* F-00027
(4-Fluorophenyl)diphenylphosphine; Selenide, *in* F-00028
5-Methyl-5*H*-dibenzophosphole; 5-Selenide, *in* M-00115
Methyldiphenylphosphine; Selenide, *in* M-00135
(2-Methylphenyl)diphenylphosphine; Selenide, *in* M-00200
Methylphenylpropylphosphine; (*R*)-*form*, Selenide, *in* M-00265
5-Phenyl-5*H*-dibenzophosphole; 5-Selenide, *in* P-00110
3,3′,3″-Phosphinoselenoylidynetris[propanenitrile], *in* T-00752
Tetrakis[(diphenylphosphinoselenoyl)methyl]methane, *in* T-00179
Tri-*tert*-butylphosphine; Selenide, *in* T-00373
Tributylphosphine selenide, T-00377
Tricyclohexylphosphine; Selenide, *in* T-00426
Tricyclopropylphosphine; Selenide, *in* T-00432
Tri(decyl)phosphine; Selenide, *in* T-00433
Triethylphosphine; Selenide, *in* T-00441
Tri-2-furanylphosphine; Selenide, *in* T-00486
Tri-3-furanylphosphine; Selenide, *in* T-00487
Triheptylphosphine; Selenide, *in* T-00489
Trihexylphosphine; Selenide, *in* T-00492
2,4,6-Trimethyl-5-phenyl-1,3,5-dioxaphosphorinane; ($2\alpha,4\alpha,5\alpha,6\alpha$)-*form*, 5-Selenide, *in* T-00521
2,4,6-Trimethyl-5-phenyl-1,3,5-dioxaphosphorinane; ($2\alpha,4\alpha,5\beta,6\alpha$)-*form*, 5-Selenide, *in* T-00521
▷Trimethylphosphine selenide, T-00533
Trinonylphosphine; Selenide, *in* T-00578
Tripentylphosphine; Selenide, *in* T-00594
Triphenylphosphine selenide, T-00628
1,2,5-Triphenyl-1*H*-phosphole; Selenide, *in* T-00634
Tripropylphosphine; Selenide, *in* T-00711
Tris(3-chlorophenyl)phosphine; Selenide, *in* T-00748
Tris(4-chlorophenyl)phosphine; Selenide, *in* T-00749
Tris(3-fluorophenyl)phosphine; Selenide, *in* T-00782
Tris(4-fluorophenyl)phosphine; Selenide, *in* T-00783
Tris(4-methylphenyl)phosphine; Selenide, *in* T-00808
Tris(phenylethynyl)phosphine; Selenide, *in* T-00846
Tris[3-(trifluoromethyl)phenyl]phosphine; Selenide, *in* T-00853
Tris(2,4-6-trimethoxyphenyl)phosphine; Selenide, *in* T-00858
Tri-2-thienylphosphine; Selenide, *in* T-00874
Tri-3-thienylphosphine; Selenide, *in* T-00875

Tertiary phosphine tellurides, R_3PTe

Tributylphosphine; Telluride, *in* T-00372
Tri-*tert*-butylphosphine; Telluride, *in* T-00373
Tricyclohexylphosphine; Telluride, *in* T-00426
Triethylphosphine; Telluride, *in* T-00441
Trimethylphosphine; Telluride, *in* T-00530
Tripentylphosphine; Telluride, *in* T-00594
Triphenylphosphine; Telluride, *in* T-00624
Tripropylphosphine; Telluride, *in* T-00711

Tertiary phosphine imides, $R_3P{=}NR$

Bis(triphenylphosphine)iminium(1+); Fluoride, *in* B-00442
Bis(triphenylphosphine)iminium(1+); Bromide, *in* B-00442
Bis(triphenylphosphine)iminium(1+); Iodide, *in* B-00442
Bis(triphenylphosphoranylidene)hydrazine, B-00458
N-tert-Butyl-*P,P,P*-trimethylphosphine imide, *in* T-00531
N-Chlorotriphenylphosphoranyl-*P,P,P*-triphenylphosphine imide, *in* B-00442
4-Cyano-2,2-dihydro-3,5-diphenyl-3*H*-1,2-azaphosphole, *in* D-00461
2,2-Dihydro-4-ethoxycarbonyl-2,2-dimethyl-3,5-diphenyl-3*H*-1,2-azaphosphole, *in* D-00461
N-Diphenylphosphinothioyl-*P,P,P*-triphenylphosphine imide, *in* D-01088
N-Diphenylphosphino-*P,P,P*-triphenylphosphine imide, D-01088
N-Diphenylphosphinyl-*P,P,P*-triphenylphosphine imide, *in* D-01088
P,P-Diphenyl-*N*-(triphenylphosphoranylidene)phosphinic amide, *in* D-01165
P,P-Diphenyl-*N*-(triphenylphosphoranylidene)phosphinimidic amide, D-01164
P,P-Diphenyl-*N*-(triphenylphosphoranylidene)-phosphinoselenoic amide, *in* D-01165
P,P-Diphenyl-*N*-(triphenylphosphoranylidene)-phosphinotelluroic amide, *in* D-01165
P,P-Diphenyl-*N*-(triphenylphosphoranylidene)phosphinothioic amide, *in* D-01165
P,P-Diphenyl-*N*-(triphenylphosphoranylidene)phosphinous amide, D-01165
Ethyl (triphenylphosphoranylidene)carbamate, *in* T-00648
Ethyl [(triphenylphosphoranylidene)hydrazono]acetate, *in* T-00653
N′-(4-Fluorophenyl)-*P,P*-diphenyl-*N*-(triphenylphosphoranylidene)phosphinimidic amide, *in* D-01164
Formaldehyde (triphenylphosphoranylidene)hydrazone, F-00054
2,2,2,3,3*a*,11*b*-Hexahydro-2,2,2-triphenylphenanthro[9,10-*d*]-1,3,2-oxazaphosphole; (9*RS*,10*RS*)-*form*, *in* H-00059
Isopropyl (triphenylphosphoranylidene)carbamate, *in* T-00648
N-Me, *in* T-00339
Methyl (triphenylphosphoranylidene)carbamate, *in* T-00648
N′-(4-Nitrophenyl)-*P,P*-diphenyl-*N*-(triphenylphosphoranylidene)phosphinimidic amide, *in* D-01164
P-Phenyl-*N,N′*-bis(triphenylphosphoranylidene)phosphonic diamide, *in* P-00102
P-Phenyl-*N,N′*-bis(triphenylphosphoranylidene)-phosphonothioic diamide, *in* P-00102
N,N′-1,2-Phenylenebis[*P,P,P*-triphenylphosphine imide], P-00143
N,N′-1,3-Phenylenebis[*P,P,P*-triphenylphosphine imide], P-00144
N,N′-1,4-Phenylenebis[*P,P,P*-triphenylphosphine imide], P-00145
P,P′-1,4-Phenylenebis[*N,P,P*-triphenylphosphine imide], P-00146
Phenyl [(triphenylphosphoranylidene)hydrazono]acetate, *in* T-00653
Propyl (triphenylphosphoranylidene)carbamate, *in* T-00648
Tetramethylphosphine imide, *in* T-00531
Tetraphenylphosphine imide, T-00269
P,P,P-Tribenzylphosphine imide, T-00339
P,P,P-Tributylphosphine imide, T-00374
P,P,P-Tri-*tert*-butylphosphine imide, T-00375
P,P,P-Tributyl-*N*-trimethylsilylphosphine imide, *in* T-00374
P,P,P-Tri-*tert*-butyl-*N*-trimethylsilylphosphine imide, *in* T-00375
P,P,P-Tricyclohexyl-*N*-methylphosphine imide, *in* T-00427
P,P,P-Tricyclohexylphosphine imide, T-00427
P,P,P-Triethyl-*N*-(diphenylphosphino)phosphine imide, *in* T-00442
P,P,P-Triethylphosphine imide, T-00442
P,P,P-Triethyl-*N*-trimethylsilylphosphine imide, *in* T-00442
Triethyl [[(triphenylphosphoranylidene)amino]sulfonyl]-phosphorimidate, *in* T-00642
P,P,P-Triisopropyl-*N*-methylphosphine imide, *in* T-00501
P,P,P-Triisopropylphosphine imide, T-00501
P,P,P-Triisopropyl-*N*-trimethylsilylphosphine imide, *in* T-00501
P,P,P-Trimethyl-*N*-phenylphosphine imide, *in* T-00531
▷*P,P,P*-Trimethylphosphine imide, T-00531
P,P,P-Trimethyl-*N*-trimethylsilylphosphine imide, *in* T-00531
Trimethyl [[(triphenylphosphoranylidene)amino]sulfonyl]-phosphorimidate, *in* T-00642
(Triphenylphosphoranylidene)phosphoramidic dichloride, T-00656

Triphenyl [[(triphenylphosphoranylidene)amino]sulfonyl]-phosphorimidate, *in* T-00642
N',*P*,*P*-Triphenyl-*N*-(triphenylphosphoranylidene)-phosphinimidic amide, *in* D-01164
Tris(triphenylphosphoranylidene)phosphoric triamide, *in* T-00872
Tris(triphenylphosphoranylidene)phosphorous triamide, T-00872

Polyphosphines, P—(P)$_n$—P

2,2'-Bi(1,3,2-benzodithiaphosphole), B-00074
1,1'-Biphospholane, B-00084
Bis(cyclenphosphorane), B-00154
2,2'-Bis(5,5-dimethyl-1,3,2-dioxaphosphorinane), B-00197
1,4-Bis(2,4,6-tri-*tert*-butylphenyl)-1,4-bis(trimethylsilyloxy)-2,3-diphosphabutadiene, B-00393
2,3-Bis(2,4,6-tri-*tert*-butylphenyl)-1,2,3-selenadiphosphirane, B-00396
2,3-Bis(2,4,6-tri-*tert*-butylphenyl)thiadiphosphirane, B-00397
1,2-Bis(trifluoromethyl)diphosphine, B-00410
2-[Bis(trimethylsilyl)methylene]-1,1-di-*tert*-butyldiphosphine, B-00431
2-*tert*-Butyl-1,3-diisopropoxytriphosphetane 1,3-dioxide, *in* D-00627
1,2-Dibenzyl-1,2-dimethyldiphosphine; (*RS*,*SR*)-*form*, Disulfide, *in* D-00043
1,2-Dibenzyl-1,2-diphenyldiphosphine, D-00045
1,2-Di-*tert*-butyl-1,2-dimethyldiphosphine, D-00089
1-(Di-*tert*-butyldiphosphinylidene)methanediamine; *N*,*N*,*N'*,*N'*-Tetra-Me, *in* D-00090
1-(Di-*tert*-butyldiphosphinylidene)methanediamine; *N*,*N*,*N'*,*N'*-Tetra-Et, *in* D-00090
1,2-Di-*tert*-butyldiphosphirane, D-00091
1,1-Diethoxy-2,2-diethyldiphosphine 1-oxide, D-00229
2,3-Diethyl-2,3-diphosphabicyclo[2.2.1]hept-5-ene, D-00287
2,3-Diethyl-2,3-diphosphabicyclo[2.2.2]octane, D-00288
▷2,3-Dihydro-1,2,3,4,5-pentakis(trifluoromethyl)-1*H*-1,2,3-triphosphole, D-00536
2,3-Dihydro-1,2,3,4,5-pentaphenyl-1*H*-1,2,3-triphosphole, D-00539
1,2-Dihydro-1,2,3,4-tetraphenyl-1,2-diphosphete, D-00581
2,3-Dihydro-1,2,3-trimethyl-1*H*-1,2,3-triphosphole, *in* D-00615
2,3-Dihydro-1,2,3-triphenyl-1*H*-1,3,2-benzodiphospharsole, D-00605
2,3-Dihydro-1,2,3-triphenyl-1*H*-1,2,3-triphosphole, *in* D-00615
2,3-Dihydro-1*H*-1,2,3-triphosphole, D-00615
1,3-Diisopropoxy-2-isopropyltriphosphetane 1,3-dioxide, *in* D-00627
1,2-Dimethyl-1,2-diphenyldiphosphine, D-00744
1,2-Dimethyl-1,2-diphenyldiphosphine; (*RS*,*RS*)-*form*, Monosulfide, *in* D-00744
1,2-Dimethyl-1,2-diphenyldiphosphine; (*RS*,*SR*)-*form*, Monosulfide, *in* D-00744
1,2-Diphenyldiphosphine, D-01012
1,2-Diphenyl-1,2-diphosphorinane, D-01015
Diphenyl hypodiphosphonous acid; Dibromide, *in* D-01020
Diphenyl hypodiphosphonous acid; Diiodide, *in* D-01020
Hexaphenylhexaphosphorinane, H-00089
Octahydro-1,2,3,4,5,6-hexamethyl-[1,2,3]triphospholo[4,5-*d*]-1,2,3-triphosphole; (3a*α*,6a*α*)-*form*, *in* O-00017
Octahydro-1,2,3,4,5,6-hexamethyl-[1,2,3]triphospholo[4,5-*d*]-1,2,3-triphosphole; (3a*α*,6a*β*)-*form*, *in* O-00017
1,2,7,8,9,10,15,16-Octamethyl-3,6,11,14-tetraphenyltetrakisphospholo[1,2-*b*:2',1'-*d*:1″,2″-*f*:2‴,1‴-*h*][1,2,5,6]tetraphosphocin, O-00033
Pentamethylpentaphospholane, P-00019
Pentaphenylpentaphospholane, P-00037
Tetrabenzyldiphosphine, T-00007
Tetra-*tert*-butylcyclotetraphosphine, T-00012
Tetrabutyldiphosphine, T-00014
Tetra-*tert*-butyldiphosphine, T-00015
Tetra-*tert*-butyldiphosphine; Monooxide, *in* T-00015
Tetrabutyl hypodiphosphite, T-00016
Tetra-*tert*-butylmethyltetraphosphetanium iodide, *in* T-00012
1,2,4,5-Tetra-*tert*-butyl-1,2,4,5-tetraphosphaspiro[2.2]-pentane, T-00026
Tetracyclohexyldiphosphine, T-00039
1,2,3,4-Tetracyclohexyl-1,2,3,4-tetraphosphorinane, T-00040
1,1,3,3-Tetraethoxy-2-ethyltriphosphine 1,3-dioxide, T-00041
Tetraethyldiphosphine, T-00043
Tetraethyldiphosphine; Monosulfide, *in* T-00043
Tetraethyl hypodiphosphite, T-00049
Tetraethyl hypophosphate, T-00050
Tetraethyltetraphosphetane, T-00071
1,2,3,6-Tetrahydro-4,5-dimethyl-1,2-diphenyl-1,2-diphosphinine, *in* T-00095
1,2,3,6-Tetrahydro-1,2,4,5-tetramethyl-1,2-diphosphinine, *in* T-00095
5,6,11,12-Tetrahydro-5,6,11,12-tetraphenyldibenzo[*c*,*g*]-[1,2,5,6]tetraphosphocin, T-00147
Tetrahydro-1*H*,5*H*-[1,2,3]triphospholo[2,1-*a*][1,2,3]-triphosphole, T-00160
Tetraisopropyldiphosphine; Monooxide, *in* T-00167
Tetraisopropyldiphosphine; Monosulfide, *in* T-00167
Tetraisopropyl hypodiphosphite, T-00169
1,1,4,4-Tetrakis(diethylamino)-2,3-diphosphabutadiene, *in* T-00005
1,1,4,4-Tetrakis(dimethylamino)-2,3-diphosphabutadiene, *in* T-00005
Tetrakis(dimethylamino)diphosphine, T-00177
Tetrakis(pentafluorophenyl)diphosphine, T-00184
Tetrakis(pentafluorophenyl)tetraphosphetane, T-00185
▷Tetrakis(trifluoromethyl)tetraphosphetane, T-00188
▷Tetramethyldiphosphine, T-00200
Tetramethyldiphosphine; Monooxide, *in* T-00200
Tetramethyldiphosphine; Monosulfide, *in* T-00200
3,3,5,5-Tetramethyl-1,2,4-triphenyl-1,2,4-triphospholane, T-00239
2,2',5,5'-Tetraphenyl-1*H*,1'*H*-biphosphole, T-00248
Tetraphenyldiphosphine, T-00258
Tetraphenyldiphosphine; Mono-oxide, *in* T-00258
Tetraphenyldiphosphine; Monosulfide, *in* T-00258
Tetraphenyltetraphosphetane, T-00278
1,2,3,4-Tetraphenyltetraphospholane, T-00280
Tetraphenylthiatetraphospholane; (2*α*,3*β*,4*α*,5*β*)-*form*, *in* T-00282
Tetrapropyldiphosphine, T-00286
Tetrapropyl hypodiphosphite, T-00287
Tri-*tert*-butyltriphosphirane, T-00397
1,2,4-Trimethyl-1,2,4-triphospholane, T-00568
Triphenyltriphosphirane, T-00698
1,2,3-Triphenyl-1,2,3-triphospholane, T-00699
Triphosphetan-4-one; 1,3-Di-*tert*-butyl, 2-Me, *in* T-00703
Triphosphetan-4-one; 1,2,3-Tri-*tert*-butyl, *in* T-00703
Triphosphetan-4-one; 1,3-Di-*tert*-butyl, 2-Ph, *in* T-00703
Tris(*tert*-butylphosphino)phosphine, T-00737

Polyphosphine oxides

2,2'-Bis(5,5-dimethyl-1,3,2-dioxaphosphorinane) 2,2'-dioxide, *in* B-00197
2-*tert*-Butyl-1,3-diisopropoxytriphosphetane 1,3-dioxide, *in* D-00627
Decahydro-1,3,5-trihydroxy-1*H*-1,3,5-benzotriphosphepin 1,3,5-trioxide, D-00007
1,1-Diethoxy-2,2-diethyldiphosphine 1-oxide, D-00229
2,3-Diethyl-2,3-diphosphabicyclo[2.2.2]octane; Disulfide, *in* D-00288
1,3-Diisopropoxy-2-isopropyltriphosphetane 1,3-dioxide, *in* D-00627
1,3-Dimethyl-5,7-diphenyl-1,3,5,7-tetraaza-4-phospha(4-P^V)-spiro[3.3]heptane-2,6-dione; 4-(Diethoxyphosphinyl), *in* D-00748
Tetrabenzyldiphosphine; Dioxide, *in* T-00007
Tetrabutyldiphosphine; Monooxide, *in* T-00014
Tetrabutyldiphosphine; Dioxide, *in* T-00014
Tetra-*tert*-butyldiphosphine; Monooxide, *in* T-00015
Tetrabutyl hypophosphate, T-00017
Tetracyclohexyldiphosphine; Monoxide, *in* T-00039
Tetracyclohexyldiphosphine; Dioxide, *in* T-00039

1,1,3,3-Tetraethoxy-2-ethyltriphosphine 1,3-dioxide, T-00041
O,O,O′,O′-Tetraethyl thiohypophosphate, T-00073
Tetraisopropyldiphosphine; Monooxide, *in* T-00167
Tetraisopropyldiphosphine; Dioxide, *in* T-00167
Tetraisopropyl hypophosphate, T-00170
Tetramethyldiphosphine; Monooxide, *in* T-00200
Tetramethyldiphosphine; Dioxide, *in* T-00200
Tetramethyldiphosphine; 1-Oxide 2-sulfide, *in* T-00200
Tetramethyl hypophosphate, T-00204
Tetraphenyldiphosphine; Mono-oxide, *in* T-00258
Tetraphenyldiphosphine; Mono-oxide, monosulfide, *in* T-00258
Tetraphenyldiphosphine dioxide, T-00259
Tetrapropyl hypophosphate, T-00288
Thiohypophosphoric acid cyclic *P,P,P′,P′*-bis(2,2-dimethyltrimethylene) ester, *in* B-00197

Polyphosphine sulfides

1,1′-Biphospholane; Disulfide, *in* B-00084
1,1′-Biphosphorinane; Disulfide, *in* B-00085
1,2-Dibenzyl-1,2-dimethyldiphosphine; *(RS,RS)-form*, Disulfide, *in* D-00043
1,2-Dibenzyl-1,2-diphenyldiphosphine; Disulfide, *in* D-00045
2,3-Diethyl-2,3-diphosphabicyclo[2.2.1]hept-5-ene; Disulfide, *in* D-00287
2,3-Dihydro-1,2,3,4,5-pentaphenyl-1*H*-1,2,3-triphosphole; 1,3-Disulfide, *in* D-00539
2,3-Dihydro-1,2,3-triphenyl-1*H*-1,3,2-benzodiphospharsole; (1α,2β,3α)-*form*, 1,3-Disulfide, *in* D-00605
2,3-Dihydro-1,2,3-triphenyl-1*H*-benzotriphosphole; (1α,2β,3α)-*form*, 1-Sulfide, *in* D-00606
2,3-Dihydro-1,2,3-triphenyl-1*H*-benzotriphosphole; (1α,2β,3α)-*form*, 1,3-Disulfide, *in* D-00606
2,3-Dihydro-1,2,3-triphenyl-1*H*-benzotriphosphole; (1α,2β,3α)-*form*, Trisulfide, *in* D-00606
1,2-Dimethyl-1,2-diphenyldiphosphine; *(RS,RS)-form*, Monosulfide, *in* D-00744
1,2-Dimethyl-1,2-diphenyldiphosphine; *(RS,RS)-form*, Disulfide, *in* D-00744
1,2-Dimethyl-1,2-diphenyldiphosphine; *(RS,SR)-form*, Monosulfide, *in* D-00744
1,2-Dimethyl-1,2-diphenyldiphosphine; *(RS,SR)-form*, Disulfide, *in* D-00744
1,2-Diphenyl-1,2-diphosphorinane; 1,2-Disulfide, *in* D-01015
Diphenylthiohypophosphonic acid, D-01153
Tetrabenzyldiphosphine; Disulfide, *in* T-00007
Tetrabutyldiphosphine; Monosulfide, *in* T-00014
Tetrabutyldiphosphine; Disulfide, *in* T-00014
Tetracyclohexyldiphosphine; Disulfide, *in* T-00039
Tetraethyldiphosphine; Monosulfide, *in* T-00043
Tetraethyldiphosphine disulfide, T-00044
Tetraethyl thiohypophosphate, T-00072
O,O,O′,O′-Tetraethyl thiohypophosphate, T-00073
1,2,3,6-Tetrahydro-4,5-dimethyl-1,2-diphosphorine; 1,2-Di-Me, 1,2-disulfide, *in* T-00095
1,2,3,6-Tetrahydro-4,5-dimethyl-1,2-diphosphorine; 1,2-Di-Ph, 1,2-disulfide, *in* T-00095
1,2,3,6-Tetrahydro-1,2-diphenyl-1,2-diphosphorin; (1*RS*,2*RS*)-*form*, 1,2-Disulfide, *in* T-00103
1,2,3,6-Tetrahydro-1,2-diphenyl-1,2-diphosphorin; (1*RS*,2*SR*)-*form*, 1,2-Disulfide, *in* T-00103
5,6,11,12-Tetrahydro-5,6,11,12-tetraphenyldibenzo[*c,g*]-[1,2,5,6]tetraphosphocin; 5,11-(or 5,6)-Disulfide, *in* T-00147
1,2,3,6-Tetrahydro-1,2,4,5-tetraphenyl-1,2-diphosphorine; 1,2-Disulfide, *in* T-00151
Tetraisopropyldiphosphine; Monosulfide, *in* T-00167
Tetraisopropyldiphosphine; Disulfide, *in* T-00167
O,O,O′,O′-Tetraisopropyl thiohypophosphate, T-00173
Tetramethyldiphosphine; Monosulfide, *in* T-00200
Tetramethyldiphosphine; 1-Oxide 2-sulfide, *in* T-00200
Tetramethyldiphosphine; 1-Selenide 2-sulfide, *in* T-00200
Tetramethyldiphosphine disulfide, T-00201
Tetraphenyldiphosphine; Monosulfide, *in* T-00258
Tetraphenyldiphosphine; Mono-oxide, monosulfide, *in* T-00258
Tetraphenyltetraphosphetane; Tetrasulfide, *in* T-00278
1,2,3,4-Tetraphenyltetraphospholane; (1α,2β,3α,4β)-*form*, 1-Sulfide, *in* T-00280
1,2,3,4-Tetraphenyltetraphospholane; (1α,2β,3α,4β)-*form*, 1,4-Disulfide, *in* T-00280
O,O,O′,O′-Tetraphenyl thiohypophosphate, T-00283
Tetrapropyldiphosphine; Disulfide, *in* T-00286
O,O,O′,O′-Tetrapropyl thiohypophosphate, T-00293
Thiohypophosphoric acid cyclic *P,P,P′,P′*-bis(2,2-dimethyltrimethylene) ester, *in* B-00197
Thiohypophosphoric acid cyclic *P,P,P′,P′*-bis(2,2-dimethyltrimethylene) ester, *in* B-00197
2,4,5-Triphenyl-1,3,2,4,5-dithiatriphospholane, T-00606

Polyphosphine selenides

Tetramethyldiphosphine; Diselenide, *in* T-00200
Tetramethyldiphosphine; 1-Selenide 2-sulfide, *in* T-00200

Phosphenes, P=P

1,2-Bis[1-(2,2,6,6-tetramethylpiperidino)]diphosphene, B-00392
Bis(2,4,6-tri-*tert*-butylphenyl)diphosphene; *(E)-form*, *in* B-00395
1,4-Bis(2,4,6-tri-*tert*-butylphenyl)-3-trimethylsilyloxy-1,2,4-triphospha-1,3-butadiene, B-00398

Phosphene oxides

Bis(2,4,6-tri-*tert*-butylphenyl)diphosphene; *(E)-form*, Oxide, *in* B-00395

Phosphene sulfides

Bis(2,4,6-tri-*tert*-butylphenyl)diphosphene; *(E)-form*, Sulfide, *in* B-00395

Phosphonium salts, $R_4P^\oplus$

(5,6-Acenaphthenylenedimethylene)bis[triphenylphosphonium]-(2+), A-00001
(3-Acetylaminophenyl)methyldiphenylphosphonium iodide, *in* D-01048
[[2-[(Acetyloxy)methyl]phenyl]methyl]-triphenylphosphonium(1+); Bromide, *in* A-00019
Aminobis(diethylamino)(trichloromethyl)phosphonium chloride, *in* T-00075
▷[(Aminocarbonyl)methyl]triphenylphosphonium(1+); Chloride, *in* A-00061
(3-Aminophenyl)methyldiphenylphosphonium iodide, *in* D-01048
(4-Aminophenyl)methyldiphenylphosphonium iodide, *in* D-01049
Azidotris(dimethylamino)phosphorus(1+); Hexafluorophosphate, *in* A-00149
(1-Azulenylmethyl)triphenylphosphonium(1+); Iodide, *in* A-00156
1,3-Benzodithiol-2-yltriphenylphosphonium(1+); Chloride, *in* B-00021
1,3-Benzodithiol-2-yltriphenylphosphonium(1+); Tetrafluoroborate, *in* B-00021
5-Benzyl-10*H*-5,10[1′,2′]benzenoacridophosphonium bromide, *in* B-00014
Benzyl-4-bromophenyldiphenylphosphonium bromide, *in* B-00481
Benzyl-2-bromophenyldiphenylphosphonium chloride, *in* B-00480
Benzyl-*tert*-butylmethylphenylphosphonium bromide, *in* B-00565
1-Benzyl-4-*tert*-butyl-1-phenylphosphorinanium bromide, *in* B-00586
Benzyl(2-chlorophenyl)diphenylphosphonium chloride, *in* C-00127
1-Benzyl-4,5-dichloro-1-phenyl-1*H*-phosphepinium bromide, *in* D-00565
Benzyldiferrocenylphenylphosphonium chloride, *in* D-00410
1-Benzyl-2,5-dihydro-3,4-dimethyl-1-phenyl-1*H*-phospholium bromide, *in* D-00468
1-Benzyl-2,5-dihydro-3-methyl-1-phenyl-1*H*-phospholium bromide, *in* D-00528
5-Benzyl-5,10-dihydro-5-phenylacridophosphonium bromide, *in* D-00547
1-Benzyl-2,5-dihydro-1,3,4-trimethyl-1*H*-phospholium bromide, *in* D-00600

1-Benzyl-3,4-dimethyl-1-phenyl-1*H*-phospholium bromide, *in* D-00807
Benzyldimethylphenylphosphonium iodide, *in* B-00051
Benzyldiphenyl(4-phenyl-1,3-butadienyl)phosphonium bromide, *in* D-01031
Benzyldiphenyl-2-propenylphosphonium bromide, *in* D-01144
Benzylethylmethylphenylphosphonium iodide, *in* B-00047
Benzylethylmethylphenylphosphonium iodide, *in* E-00084
Benzylethylphenylphosphine; (*S*)-*form*, B,MeI, *in* B-00047
Benzylethylphenylphosphine; (±)-*form*, B,MeI, *in* B-00047
1-Benzyl-1,2,2,3,4,4-hexamethylphosphetanium bromide, *in* H-00074
Benzyl(2-methoxyphenyl)diphenylphosphonium bromide, *in* M-00056
Benzyl(3-methoxyphenyl)diphenylphosphonium chloride, *in* M-00057
Benzyl(4-methoxyphenyl)diphenylphosphonium chloride, *in* M-00058
Benzylmethyldiphenylphosphonium iodide, *in* B-00046
1-Benzyl-3-methyl-1-phenylphospholanium bromide, *in* M-00229
1-Benzyl-3-methyl-1-phenylphospholanium iodide, *in* M-00229
Benzylmethylphenylpropylphosphonium(1+), B-00052
Benzylmethylphenylpropylphosphonium(1+); (±)-*form*, *in* B-00052
1-Benzyl-1-methylphospholanium bromide, *in* M-00280
Benzyl(2-methylthiophenyl)diphenylphosphonium chloride, *in* M-00417
Benzylneomenthyldiphenylphosphonium bromide, *in* M-00164
1-Benzyl-4-oxo-2,2,6,6-tetramethyl-1-phenylphosphorinanium bromide, *in* T-00210
1-Benzyl-2,2,3,4,4-pentamethyl-1-phenylphosphetanium bromide, *in* P-00020
17-Benzyl-2,7,10,15,17-pentamethyl-17*H*-tetrabenzo[*b*,*d*,*f*,*h*]-phosphoninium iodide, *in* P-00028
5-Benzyl-5-phenyl-5*H*-dibenzophospholium chloride, *in* P-00110
1-Benzyl-1-phenylphosphepanium bromide, *in* P-00205
1-Benzyl-1-phenylphosphocanium bromide, *in* P-00215
1-Benzyl-1-phenylphospholanium bromide, *in* P-00216
8-Benzyl-8-phenyl-8-phosphoniabicyclo[3.2.1]octan-3-one chloride, *in* P-00199
17-Benzyl-2,7,10,15-tetramethyl-17-phenyl-17*H*-tetrabenzo[*b*,*d*,*f*,*h*]phosphoninium iodide, *in* T-00211
3-Benzyl-2,2,5-trimethyl-3-phenyl-1,3-oxaphospholanium bromide, *in* T-00522
Benzyltriphenylphosphonium(1+); Chloride, *in* B-00073
Benzyltriphenylphosphonium(1+); Bromide, *in* B-00073
▷Benzyltriphenylphosphonium(1+); Iodide, *in* B-00073
Benzyltriphenylphosphonium(1+); Triodide, *in* B-00073
Benzyltriphenylphosphonium(1+); Pentaiodide, *in* B-00073
Benzyltriphenylphosphonium(1+); Heptaiodide, *in* B-00073
[1,1′-Biphenyl]-2,2′-diylbis(diethylmethylphosphonium)-diiodide, *in* B-00179
[1,1′-Biphenyl]-4-ylmethyl-1-naphthalenylphenylphosphonium bromide, *in* B-00081
[1,1′-Biphenyl]-4-yl-1-naphthalenylphenylphosphine; (*S*)-*form*, B,MeBr, *in* B-00081
[1,1′-Biphenyl]-4-yl-1-naphthalenylphenylphosphine; (±)-*form*, B,MeI, *in* B-00081
Bis(2-cyanoethyl)methylphenylphosphonium iodide, *in* B-00152
Bis(3-hydroxypropyl)methylphenylphosphonium iodide, *in* B-00286
1,1-Bis(methyldiphenylphosphonio)ethene diiodide, *in* B-00236
Bis[2-(methyldiphenylphosphonio)ethyl]-methylphenylphosphonium triiodide, *in* B-00239
1,1′-Bis(methyldiphenylphosphonio)ferrocene diiodide, *in* B-00241
Bis(2,4,6-trimethylphenyl)phosphinic acid, B-00428
[Bis(trimethylsilyl)methyl]triphenylphosphonium(1+); Bromide, *in* B-00432
N,*N*′-Bis(triphenylphosphonio)hydrazine dichloride, *in* B-00458
1,8-Bis(triphenylphosphoniomethyl)anthracene(2+); Dibromide, *in* B-00443
2,2′-Bis[(triphenylphosphonio)methyl]-1,1-binaphthyl(2+); (*S*)-*form*, Dibromide, *in* B-00444
2,2′-Bis[(triphenylphosphonio)methyl]-1,1-binaphthyl(2+); (*S*)-*form*, Diperiodate, *in* B-00444
2,2′-Bis[(triphenylphosphonio)methyl]-1,1-binaphthyl(2+); (±)-*form*, Dibromide, *in* B-00444
2,2′-Bis[(triphenylphosphonio)methyl]-1,1-binaphthyl(2+); (±)-*form*, Diperiodate, *in* B-00444
2,2′-Bis[(triphenylphosphonio)methyl]biphenyl(2+); Dibromide, *in* B-00445
2,2′-Bis[(triphenylphosphonio)methyl]biphenyl(2+); Diperiodate, *in* B-00445
3,3′-Bis[(triphenylphosphonio)methyl]biphenyl(2+); Dibromide, *in* B-00446
4,4′-Bis[(triphenylphosphonio)methyl]biphenyl(2+); Dichloride, *in* B-00447
2,5-Bis[(triphenylphosphonio)methyl]furan(2+); Dichloride, *in* B-00448
3,4-Bis[(triphenylphosphonio)methyl]furan(2+); Dichloride, *in* B-00449
3,4-Bis[(triphenylphosphonio)methyl]furan(2+); Dibromide, *in* B-00449
1,4-Bis(triphenylphosphoniomethyl)naphthalene(2+), B-00450
1,5-Bis(triphenylphosphoniomethyl)naphthalene(2+); Dibromide, *in* B-00451
1,8-Bis(triphenylphosphoniomethyl)naphthalene(2+); Dibromide, *in* B-00452
2,3-Bis(triphenylphosphoniomethyl)naphthalene(2+); Dibromide, *in* B-00453
2,6-Bis(triphenylphosphoniomethyl)naphthalene(2+); Dibromide, *in* B-00454
2,7-Bis(triphenylphosphoniomethyl)naphthalene(2+); Dibromide, *in* B-00455
2,5-Bis[(triphenylphosphonio)methyl]thiophene(2+); Dichloride, *in* B-00456
3,4-Bis[(triphenylphosphonio)methyl]thiophene(1+); Dibromide, *in* B-00457
Bromobis(diethylamino)(ethylamino)phosphonium bromide, *in* P-00012
Bromobis(dipropylamino)(propylamino)phosphonium bromide, *in* P-00043
Bromo(butylamino)bis(dibutylamino)phosphonium bromide, *in* P-00004
(Bromodifluoromethyl)triphenylphosphonium(1+); Bromide, *in* B-00468
(Bromomethyl)triphenylphosphonium(1+); Bromide, *in* B-00477
(Bromomethyl)triphenylphosphonium(1+); Tetrafluoroborate, *in* B-00477
(Bromomethyl)triphenylphosphonium(1+); Tetraphenylborate, *in* B-00477
(3-Bromo-2-oxopropyl)triphenylphosphonium(1+); Bromide, *in* B-00478
(3-Bromopropyl)triphenylphosphonium(1+); Bromide, *in* B-00502
(1,3-Butadienyl)triphenylphosphonium(1+); Chloride, *in* B-00512
(1,3-Butadienyl)triphenylphosphonium(1+); Bromide, *in* B-00512
1,4-Butanediylbis[triphenylphosphonium](2+); Dibromide, *in* B-00517
1,4-Butanediylbis[triphenylphosphonium](2+); Perbromide, *in* B-00517
(2-Butenyl)triphenylphosphonium(1+); (*E*)-*form*, Bromide, *in* B-00528
(2-Butenyl)triphenylphosphonium(1+); (*Z*)-*form*, Bromide, *in* B-00528
(Butylamino)chlorobis(dibutylamino)phosphonium chloride, *in* P-00004
tert-Butyldimethylphenylphosphonium iodide, *in* B-00565
Butylethyldiphenylphosphonium iodide, *in* B-00557
tert-Butylmethyldiphenylphosphonium iodide, *in* B-00558
tert-Butylmethylphenylphosphine; (*S*)-*form*, B,$PhCH_2Br$, *in* B-00565
4-*tert*-Butyl-1-methyl-1-phenylphosphorinanium bromide, *in* B-00586
4-*tert*-Butyl-1-phenylphosphorinane; *trans*-*form*, B,MeBr, *in* B-00586
4-*tert*-Butyl-1-phenylphosphorinane; *trans*-*form*, B,$PhCH_2Br$, *in* B-00586
1-*tert*-Butyl-1,3,4-trimethyl-1*H*-phospholium iodide, *in* E-00004
Butyltriphenylphosphonium(1+); Chloride, *in* B-00641

▷Butyltriphenylphosphonium(1+); Bromide, *in* B-00641
▷Butyltriphenylphosphonium(1+); Iodide, *in* B-00641
Butyltriphenylphosphonium(1+); Triiodide, *in* B-00641
Butyltriphenylphosphonium(1+); Pentaiodide, *in* B-00641
Butyltriphenylphosphonium(1+); Heptaiodide, *in* B-00641
(4-Carboxybutyl)triphenylphosphonium(1+); Chloride, *in* C-00009
(4-Carboxybutyl)triphenylphosphonium(1+); Bromide, *in* C-00009
(4-Carboxybutyl)triphenylphosphonium(1+); Iodide, *in* C-00009
(4-Carboxybutyl)triphenylphosphonium(1+); Et ester, iodide, *in* C-00009
(1-Carboxyethyl)triphenylphosphonium(1+); Et ester, chloride, *in* C-00010
(2-Carboxyethyl)triphenylphosphonium(1+); Chloride, *in* C-00011
(2-Carboxyethyl)triphenylphosphonium(1+); Bromide, *in* C-00011
(2-Carboxyethyl)triphenylphosphonium(1+); Tribromide, *in* C-00011
(2-Carboxyethyl)triphenylphosphonium(1+); Betaine, *in* C-00011
(7-Carboxyheptyl)triphenylphosphonium(1+), C-00012
(6-Carboxyhexyl)triphenylphosphonium(1+), C-00013
▷(Carboxymethyl)triphenylphosphonium(1+); Chloride, *in* C-00014
(Carboxymethyl)triphenylphosphonium(1+); Bromide, *in* C-00014
(Carboxymethyl)triphenylphosphonium(1+); Tetrafluoroborate, *in* C-00014
(5-Carboxypentyl)triphenylphosphonium(1+); Bromide, *in* C-00015
(2-Carboxy-2-propenyl)triphenylphosphonium(1+); Bromide, *in* C-00016
(2-Carboxy-2-propenyl)triphenylphosphonium(1+); Me ester, bromide, *in* C-00016
(2-Carboxy-2-propenyl)triphenylphosphonium(1+); Et ester, bromide, *in* C-00016
(3-Carboxypropyl)triphenylphosphonium(1+); Chloride, *in* C-00017
(3-Carboxypropyl)triphenylphosphonium(1+); Bromide, *in* C-00017
Chlorobis(diethylamino)(ethylamino)phosphonium chloride, *in* P-00012
(2-Chloroethyl)triphenylphosphonium(1+); Bromide, *in* C-00083
2-[(Chloroformyl)oxy]ethyltriphenylphosphonium(1+), C-00084
▷(Chloromethyl)triphenylphosphonium(1+); Chloride, *in* C-00117
(Chloromethyl)triphenylphosphonium(1+); Iodide, *in* C-00117
(Chloromethyl)triphenylphosphonium(1+); Tetrafluoroborate, *in* C-00117
(Chloromethyl)triphenylphosphonium(1+); Tetraphenylborate, *in* C-00117
(3-Chloro-2-oxopropyl)triphenylphosphonium(1+); Chloride, *in* C-00123
[(2-Chlorophenyl)methyl]triphenylphosphonium(1+); Chloride, *in* C-00135
[(2-Chlorophenyl)methyl]triphenylphosphonium(1+); Bromide, *in* C-00135
[(3-Chlorophenyl)methyl]triphenylphosphonium(1+); Chloride, *in* C-00136
[(3-Chlorophenyl)methyl]triphenylphosphonium(1+); Bromide, *in* C-00136
[(4-Chlorophenyl)methyl]triphenylphosphonium(1+); Chloride, *in* C-00137
[(4-Chlorophenyl)methyl]triphenylphosphonium(1+); Bromide, *in* C-00137
▷Chlorphonium; Chloride, *in* C-00192
(2-Cyanoethyl)triphenylphosphonium(1+); Bromide, *in* C-00212
(2-Cyanoethyl)triphenylphosphonium(1+); Tetrafluoroborate, *in* C-00212
Cyanomethylenebis(triphenylphosphonium)(2+); Dibromide, *in* C-00214
▷(Cyanomethyl)triphenylphosphonium(1+); Chloride, *in* C-00217
(Cyanomethyl)triphenylphosphonium(1+); Iodide, *in* C-00217
Cyclobutyltriphenylphosphonium(1+); Bromide, *in* C-00235
(2,4,6-Cycloheptatrien-1-ylmethyl)triphenylphosphonium(1+); Tetrafluoroborate, *in* C-00237
Cycloheptyltriphenylphosphonium(1+); Bromide, *in* C-00238
Cyclohexyltriphenylphosphonium(1+); Bromide, *in* C-00273
Cyclohexyltriphenylphosphonium(1+); 4-Methylbenzenesulfonate, *in* C-00273
Cyclopentyltriphenylphosphonium(1+); Chloride, *in* C-00282
Cyclopentyltriphenylphosphonium(1+); Bromide, *in* C-00282
[α-Cyclopropyl(methoxycarbonyl)methyl]-triphenylphosphonium(1+); Bromide, *in* C-00285
[α-Cyclopropyl(methoxycarbonyl)methyl]-triphenylphosphonium(1+); Iodide, *in* C-00285
(Cyclopropylmethyl)triphenylphosphonium(1+); Bromide, *in* C-00286
Cyclopropyltriphenylphosphonium(1+); Bromide, *in* C-00287
Decyltriphenylphosphonium(1+); Bromide, *in* D-00014
(Diacetylmethyl)triphenylphosphonium(1+); Chloride, *in* D-00022
1,6-Dibenzyl-1,6-diphenyl-1,6-diphosphecanium dibromide, *in* D-01171
6,6-Dibenzyl-6-phosphonia-2-oxaadamantane chloride, *in* B-00055
Dibenzylphosphonium iodide, *in* D-00054
(Dibromomethyl)triphenylphosphonium(1+); Bromide, *in* D-00065
Dibromotrimethylphosphorane, D-00069
Dibromotriphenylphosphorane, D-00071
2,6-Di-*tert*-butyl-1,4-dihydro-4,4-diphenyl-1,4-azaphosphorinium bromide, *in* D-00084
Dibutylmethyl[(dibutylphosphinyl)ethenyl]phosphonium iodide, *in* B-00159
Dibutylmethylphenylphosphonium iodide, *in* D-00102
Di-*tert*-butylmethylphosphonium iodide, *in* D-00103
(1,2-Dicarboxyethyl)triphenylphosphonium(1+); Chloride, *in* D-00160
(1,2-Dicarboxyethyl)triphenylphosphonium(1+); Bromide, *in* D-00160
(1,2-Dicarboxyethyl)triphenylphosphonium(1+); Di-Et ester, iodide, *in* D-00160
(Dichloromethyl)triphenylphosphonium(1+); Chloride, *in* D-00182
2,2-Dichloro-1,1,1-triethyldiphosphinium 2-oxide(1+); Chloride, *in* D-00197
Dichlorotrimethylphosphorane, D-00198
Diferrocenylmethylphenylphosphonium iodide, *in* D-00410
Dihexylmethylphenylphosphonium iodide, *in* D-00431
1,2-Dihydro-1,2-bis(triphenylphosphonio)-benzocyclobutene(2+); (1*RS*,2*RS*)-*form*, Dibromide, *in* D-00447
1,2-Dihydro-1,2-bis(triphenylphosphonio)-benzocyclobutene(2+); (1*RS*,2*RS*)-*form*, Diperchlorate, *in* D-00447
2,3-Dihydro-1,1-dimethyl-1*H*-benzo[*g*]phosphindolinium iodide, *in* D-00518
10,11-Dihydro-5,5-dimethyl-5*H*-dibenzo[*b,f*]phosphepinium iodide, *in* D-00524
2,3-Dihydro-1,3-dimethyl-1,3-diphenyl-1*H*-1,3-benzodiphospholium dibromide, *in* D-00476
9,10-Dihydro-9,10-dimethyl-9,10-diphenylphosphaanthranium diiodide, *in* D-00482
5,6-Dihydro-5,5-dimethylphosphanthridinium chloride, *in* D-00568
[2-(1,3-Dihydro-1,3-dioxo-2*H*-isoindol-2-yl)ethyl]-triphenylphosphonium(1+); Chloride, *in* D-00474
[2-(1,3-Dihydro-1,3-dioxo-2*H*-isoindol-2-yl)ethyl]-triphenylphosphonium(1+); Bromide, *in* D-00474
2,3-Dihydro-1,3-diphenyl-1*H*-1,3-benzodiphosphole; (1*RS*,3*SR*)-*form*, B,2MeBr, *in* D-00476
2,3-Dihydro-2,2-diphenyl-1*H*-isophosphindolium bromide, *in* D-00562
2,3-Dihydro-2,2-diphenyl-1*H*-isophosphindolium perchlorate, *in* D-00562
5,6-Dihydro-5,5-diphenylphosphanthridinium bromide, *in* D-00568
5,10-Dihydro-5-methyl-5,10-diphenylacridophosphonium iodide, *in* D-00548
10,11-Dihydro-5-methyl-5-phenyl-5*H*-dibenzo[*b,f*]-phosphepinium iodide, *in* D-00559
7,12-Dihydro-12-methyl-12-phenyl-5*H*-dibenz[*c,f*][1,5]-oxaphosphocinium iodide, *in* D-00560

2,3-Dihydro-2-methyl-2-phenyl-1*H*-isophosphindolium iodide, *in* D-00562
4,5-Dihydro-1-methyl-1-phenyl-1*H*-phosphepinium perchlorate, *in* D-00565
2,3-Dihydro-1-methyl-1-phenyl-1*H*-phospholium iodide, *in* D-00566
1,2-Dihydro-1-methyl-1,2,4,6-tetraphenylphosphorinium iodide, *in* D-00584
9,10-Dihydro-10-oxo-9-phenyl-9-phosphoniaanthracene chloride, *in* P-00080
(9,10-Dihydro-9-phenanthreneyl)triphenylphosphonium(1+); Bromide, *in* D-00540
6,7-Dihydro-5,5,7,7-tetramethyl-5*H*-dibenzo[*d,f*][1,3]-diphosphepinium(2+); Dibromide, *in* D-00577
2,5-Dihydro-1,1,3,4-tetramethyl-1*H*-phospholium bromide, *in* D-00600
1,4-Dihydro-1,1,4,4-tetraphenyl-1,4-diphosphorinium(2+); Dichloride, *in* D-00582
1,4-Dihydro-1,1,4,4-tetraphenyl-1,4-diphosphorinium(2+); Dibromide, *in* D-00582
[(5,6-Dihydro-4,4,6-trimethyl-4*H*-1,3-oxazin-2-yl)methyl]-triphenylphosphonium(1+); Chloride, *in* D-00599
2,5-Dihydro-1,3,4-trimethyl-1-phenyl-1*H*-phospholium bromide, *in* D-00468
Diiodotrimethylphosphorane, D-00629
Diisopropylmethylphenylphosphonium iodide, *in* D-00638
[4-(Dimethylamino)butyl]triphenylphosphonium(1+); Bromide, *in* D-00690
(2-Dimethylaminoethyl)triphenylphosphonium(1+); Bromide, *in* D-00698
(Dimethylaminomethyl)diphenylphosphine; B,MeI, *in* D-00700
[3-(Dimethylamino)propyl]triphenylphosphonium(1+); Chloride, *in* D-00707
[3-(Dimethylamino)propyl]triphenylphosphonium(1+); Bromide, *in* D-00707
[2-(Dimethylarsino)phenyl]trimethylphosphonium iodide, *in* D-00711
Dimethylbis(trimethylphosphonimethyl)phosphonium triiodide, *in* B-00204
5,5-Dimethyl-5*H*-dibenzophospholium iodide, *in* M-00115
1,3-Dimethyl-1,3-diphenyl-1,3-diphospholanium dibromide, *in* D-01013
1,4-Dimethyl-1,4-diphenyl-1,4-diphosphorinanium dibromide, *in* D-01017
2,6-Dimethyl-4,4-diphenyl-1,4-oxaphosphorinanium bromide, *in* D-00797
Dimethyldiphenylphosphonium iodide, *in* M-00135
1,4-Dimethyl-1,4-diphosphoniabicyclo[2.2.2]octane diiodide, *in* D-01169
Dimethylethylphenylphosphonium iodide, *in* E-00084
(3,7-Dimethyl-2,6-octadienyl)triphenylphosphonium(1+); (*E*)-*form*, Bromide, *in* D-00777
(3,7-Dimethyl-2,6-octadienyl)triphenylphosphonium(1+); (*Z*)-*form*, Bromide, *in* D-00777
(2,7-Dimethyl-2,4,6-octatrien-1,8-diyl)-bis[triphenylphosphonium](2+); (*E,E,E*)-*form*, Dibromide, *in* D-00778
(3,3-Dimethyl-2-oxobutyl)triphenylphosphonium(1+); Bromide, *in* D-00784
(3,3-Dimethyl-2-oxopropyl)triphenylphosphonium(1+); Bromide, *in* D-00787
1,3-Dimethyl-1-phenylphospholanium bromide, *in* M-00229
Dimethylphenylpropylphosphonium iodide, *in* M-00265
1,1-Dimethylphosphonanium iodide, *in* M-00285
1,5-Dimethyl-1-phosphonia-5-silabicyclo[3.3.1]nonane iodide, *in* M-00274
1,1-Dimethylphosphorinanium bromide, *in* M-00339
(2,2-Dimethylpropyl)triphenylphosphonium(1+); Iodide, *in* D-00914
[3,7-Dimethyl-9-(2,6,6-trimethyl-1-cyclohexen-1-yl)-2,4,6,8-nonatetraenyl]triphenylphosphonium(1+); (*E,E,E,E*)-*form*, Iodide, *in* D-00929
[3,7-Dimethyl-9-(2,6,6-trimethyl-1-cyclohexen-1-yl)-2,4,6,8-nonatetraenyl]triphenylphosphonium(1+); (*E,E,E,E*)-*form*, Periodate, *in* D-00929
[2-(1,3-Dioxan-2-yl)ethyl]triphenylphosphonium(1+); Bromide, *in* D-00965
[(1,3-Dioxolan-2-yl)methyl]triphenylphosphonium(1+); Bromide, *in* D-00976
5,5-Diphenyl-5*H*-dibenzophospholium iodide, *in* P-00110
1,4-Diphenyl-1,4-diphosphorinane; *trans-form*, B,2MeBr, *in* D-01017
(Diphenylmethyl)triphenylphosphonium(1+); Bromide, *in* D-01025
10,10-Diphenyl-10*H*-phenoxaphosphonium iodide, *in* P-00183
1,1-Diphenylphosphepanium bromide, *in* P-00205
[1-(Diphenylphosphino)cyclopropyl]triphenylphosphonium chloride, *in* B-00229
2-(Diphenylphosphino)pyridine; B,MeI, *in* D-01076
[(Diphenylphosphinyl)methyl]triphenylphosphonium(1+); Chloride, *in* D-01101
[(Diphenylphosphinyl)methyl]triphenylphosphonium(1+); Bromide, *in* D-01101
[(Diphenylphosphinyl)methyl]triphenylphosphonium(1+); Tetraphenylborate, *in* D-01101
[(Diphenylphosphinyl)methyl]triphenylphosphonium(1+); 4-Methylbenzenesulfonate, *in* D-01101
1,1-Diphenylphosphocanium bromide, *in* P-00215
1,1-Diphenylphospholanium bromide, *in* P-00216
9,9-Diphenyl-9-phosphoniabicyclo[4.2.1]nonane bromide, *in* P-00194
3,3-Diphenyl-3-phosphoniabicyclo[3.2.1]octane bromide, *in* P-00198
9,9-Diphenyl-9-phosphoniahomocubane bromide, *in* P-00201
9,9-Diphenyl-9-phosphoniatricyclo[4.2.1.0^{2,5}]nonane bromide, *in* P-00204
1,1-Diphenylphosphorinanium bromide, *in* P-00265
(1,3-Dithian-2-ylmethyl)triphenylphosphonium(1+); Bromide, *in* D-01241
(1,3-Dithian-2-ylmethyl)triphenylphosphonium(1+); Iodide, *in* D-01241
1,3-Dithian-2-yltriphenylphosphonium(1+); Chloride, *in* D-01242
(1,3-Dithiolan-2-ylmethyl)triphenylphosphonium(1+); Bromide, *in* D-01252
1,3-Dithiolan-2-yltriphenylphosphonium(1+); Tetrafluoroborate, *in* D-01253
Dodecyltrimethylphosphonium iodide, *in* D-01264
Dodecyltriphenylphosphonium(1+); Bromide, *in* D-01266
1,2-Ethanediylbis[dimethylphenylphosphonium] dibromide, *in* E-00012
1,2-Ethanediylbis[methyldiphenylphosphonium] diiodide, *in* B-00234
1,2-Ethanediylbis[methylphenylphosphine]; (*RS,RS*)-*form*, B,2PhCH$_2$Br, *in* E-00012
1,2-Ethanediylbis[methylphenylphosphine]; (*RS,SR*)-*form*, B,2PhCH$_2$Br, *in* E-00012
1,2-Ethanediylbis[triphenylphosphonium](2+), E-00017
1,2-Ethenediylbis[(dibutylmethyl)phosphonium] diiodide, *in* B-00159
1,2-Ethenediylbis[triphenylphosphonium](2+); (*E*)-*form*, Dibromide, *in* E-00019
1,2-Ethenediylbis[triphenylphosphonium](2+); (*E*)-*form*, Diiodide, *in* E-00019
1,2-Ethenediylbis[triphenylphosphonium](2+); (*Z*)-*form*, Dibromide, *in* E-00019
Ethenylmethyldiphenylphosphonium iodide, *in* D-01168
1-(Ethoxycarbonyl)cyclopropyltriphenylphosphonium(1+); Tetrafluoroborate, *in* E-00025
(1-Ethoxycarbonyl)triphenylphosphonium bromide, *in* C-00010
(2-Ethoxyethenyl)triphenylphosphonium(1+); Bromide, *in* E-00039
(2-Ethoxyethenyl)triphenylphosphonium(1+); Tetrafluoroborate, *in* E-00039
(2-Ethoxyethenyl)triphenylphosphonium(1+); Tetraphenylborate, *in* E-00039
(2-Ethoxyethenyl)triphenylphosphonium(1+); (*Z*)-*form*, Iodide, *in* E-00039
(2-Ethoxy-2-oxoethyl)triphenylphosphonium; Chloride, *in* E-00047
(2-Ethoxy-2-oxoethyl)triphenylphosphonium; Bromide, *in* E-00047
(2-Ethoxypropenyl)triphenylphosphonium(1+); Iodide, *in* E-00050

1-Ethyl-2,5-dihydro-3-methyl-1-phenyl-1*H*-phospholium iodide, *in* D-00528
Ethylmethyldiphenylphosphonium iodide, *in* E-00066
1-Ethyl-1,2,3,4-tetrahydro-1-phenylbenzo[*h*]phosphinolinium bromide, *in* P-00319
1-Ethyl-1,2,3,4-tetrahydro-1-phenylbenzo[*h*]phosphinolinium bromide, *in* T-00128
Ethyltri(ethylthio)phosphonium iodide, *in* T-00461
Ethyltriphenylphosphonium(1+); Chloride, *in* E-00197
Ethyltriphenylphosphonium(1+); Bromide, *in* E-00197
▷Ethyltriphenylphosphonium(1+); Iodide, *in* E-00197
Ethyltriphenylphosphonium(1+); Tetrafluoroborate, *in* E-00197
Ethyltriphenylphosphonium(1+); Tetraphenylborate, *in* E-00197
9*H*-Fluoren-9-yltriphenylphosphonium(1+); Bromide, *in* F-00012
9*H*-Fluoren-9-yltriphenylphosphonium(1+); Tetrafluoroborate, *in* F-00012
(Fluoromethyl)triphenylphosponium(1+), F-00025
[(4-Fluorophenyl)methyl]triphenylphosphonium(1+); Chloride, *in* F-00032
[(4-Fluorophenyl)methyl]triphenylphosphonium(1+); Bromide, *in* F-00032
Fluorotetramethylphosphorane, F-00049
(α-Formylbenzyl)triphenylphosphonium(1+); Chloride, *in* F-00056
(α-Formylbenzyl)triphenylphosphonium(1+); Bromide, *in* F-00056
1,1,2,2,3,4,4-Heptamethylphosphetanium bromide, *in* H-00074
1,1,2,2,3,4,4-Heptamethylphosphetanium iodide, *in* H-00074
Heptyltriphenylphosphonium(1+); Bromide, *in* H-00011
Heptyltriphenylphosphonium(1+); Iodide, *in* H-00011
Hexadecyltriphenylphosphonium(1+); Bromide, *in* H-00014
2,3,3*a*,4,5,9*b*-Hexahydro-4,4-diphenyl-1*H*-cyclopent[*c*]-isophosphinolinium bromide, *in* H-00048
2,3,3*a*,4,5,9*b*-Hexahydro-4,4-diphenyl-1*H*-cyclopent[*c*]-isophosphinolinium hexafluorophosphate, *in* H-00048
6,6*a*,7,8,9,9*a*-Hexahydro-5-phenyl-1*H*-benzo[*h*]cyclopenta[*c*]-phosphinoline, H-00043
4*b*,5,6,10*b*,11,12-Hexahydro-5-phenylbenzo[*c*]-phosphanthridine, H-00045
Hexakis(dimethylamino)-μ-oxodiphosphorus(2+), H-00062
Hexakis(dimethylamino)-μ-oxodiphosphorus(2+); Bis(tetrafluoroborate), *in* H-00062
Hexakis(dimethylamino)-μ-oxodiphosphorus(2+); Bis(trifluoromethanesulfonate), *in* H-00062
Hexakis(dimethylamino)-μ-oxodiphosphorus(2+); Bis-(4-methylbenzenesulfonate), *in* H-00062
1,2,2,3,4,4-Hexamethyl-1-phenylphosphetanium bromide, *in* P-00020
1,1,3,3,5,5-Hexamethyl-1-phosphonia-3,5-disilacyclohexane chloride, *in* D-00485
N,N,N',N',N'',N''-Hexamethyltetraaminophosphonium, *in* H-00077
N,N,N',N',N'',N''-Hexamethyltetraaminophosphonium chloride, *in* H-00077
2,7,10,15,17,17-Hexamethyl-17*H*-tetrabenzo[*b,d,f,h*]-phosphoninium iodide, *in* P-00028
Hexyltriphenylphosphonium(1+); Bromide, *in* H-00104
Hexyltriphenylphosphonium(1+); Iodide, *in* H-00104
(1-Hydroxy-1*H*-benzotriazolato-*O*)tris(dimethylamino)-phosphorus(1+); Chloride, *in* H-00111
(1-Hydroxy-1*H*-benzotriazolato-*O*)tris(dimethylamino)-phosphorus(1+); Tetrafluoroborate, *in* H-00111
(1-Hydroxy-1*H*-benzotriazolato-*O*)tris(dimethylamino)-phosphorus(1+); Hexafluorophosphate, *in* H-00111
(2-Hydroxyethyl)trimethylphosphonium(1+); Chloride, *in* H-00150
(2-Hydroxyethyl)trimethylphosphonium(1+); Iodide, *in* H-00150
(2-Hydroxyethyl)triphenylphosphonium(1+); Chloride, *in* H-00151
(2-Hydroxyethyl)triphenylphosphonium(1+); Bromide, *in* H-00151
(2-Hydroxyethyl)triphenylphosphonium(1+); Iodide, *in* H-00151
Hydroxymethyltriphenylphosphonium(1+); Chloride, *in* H-00163
Hydroxymethyltriphenylphosphonium(1+); Iodide, *in* H-00163
(2-Hydroxyphenyl)methyldiphenylphosphonium iodide, *in* D-01070
[(2-Hydroxyphenyl)methyl]triphenylphosphonium(1+); Bromide, *in* H-00177
(2-Hydroxyphenyl)triphenylphosphonium(1+); Bromide, *in* H-00182
(2-Hydroxyphenyl)triphenylphosphonium(1+); Iodide, *in* H-00182
(4-Hydroxyphenyl)triphenylphosphonium(1+); Bromide, *in* H-00183
(3-Hydroxypropyl)triphenylphosphonium(1+); Chloride, *in* H-00193
(3-Hydroxypropyl)triphenylphosphonium(1+); Bromide, *in* H-00193
(3-Hydroxypropyl)triphenylphosphonium(1+); Iodide, *in* H-00193
[5-(4-Hydroxy-2,6,6-trimethyl-1-cyclohexen-1-yl)-3-methyl-2,4-pentadienyl]triphenylphosphonium(1+); (*R*)-(*E,E*)-*form*, Chloride, *in* H-00202
[5-(4-Hydroxy-2,6,6-trimethyl-1-cyclohexen-1-yl)-3-methyl-2,4-pentadienyl]triphenylphosphonium(1+); (*R*)-(*E,E*)-*form*, Bromide, *in* H-00202
[5-(4-Hydroxy-2,6,6-trimethyl-1-cyclohexen-1-yl)-3-methyl-2,4-pentadienyl]triphenylphosphonium(1+); (*S*)-(*E,E*)-*form*, Chloride, *in* H-00202
[5-(4-Hydroxy-2,6,6-trimethyl-1-cyclohexen-1-yl)-3-methyl-2,4-pentadienyl]triphenylphosphonium(1+); (±)-(*E,E*)-*form*, Bromide, *in* H-00202
1-Indenyltriphenylphosphonium(1+); Bromide, *in* I-00008
Iodotetraphenoxyphosphorane, I-00020
Iodotriphenylphosphonium triiodide, *in* D-00631
Isopropylmethyldiphenylphosphonium iodide, *in* I-00031
Isopropyltriphenylphosphonium(1+); Bromide, *in* I-00065
Isopropyltriphenylphosphonium(1+); Iodide, *in* I-00065
Isopropyltriphenylphosphonium(1+); Perchlorate, *in* I-00065
(1-Methoxycarbonylethyl)triphenylphosphonium(1+); Iodide, *in* M-00027
[[2-(Methoxycarbonyl)phenyl]methyl]triphenyl-phosphonium(1+); Bromide, *in* M-00029
[[3-(Methoxycarbonyl)phenyl]methyl]triphenyl-phosphonium(1+); Bromide, *in* M-00030
[[4-(Methoxycarbonyl)phenyl]methyl]triphenyl-phosphonium(1+); Bromide, *in* M-00031
(6-Methoxy-3-methyl-6-oxo-2,4-hexadienyl)-triphenylphosphonium(1+), M-00050
(Methoxymethyl)triphenylphosphonium(1+); Chloride, *in* M-00053
(Methoxymethyl)triphenylphosphonium(1+); Bromide, *in* M-00053
(Methoxymethyl)triphenylphosphonium(1+); Iodide, *in* M-00053
(Methoxymethyl)triphenylphosphonium(1+); Tetraphenyl-borate, *in* M-00053
(2-Methoxy-2-oxoethyl)triphenylphosphonium(1+); Chloride, *in* M-00055
(2-Methoxy-2-oxoethyl)triphenylphosphonium(1+); Bromide, *in* M-00055
(2-Methoxy-2-oxoethyl)triphenylphosphonium(1+); Iodide, *in* M-00055
(2-Methoxy-2-oxoethyl)triphenylphosphonium(1+); Tetrafluoroborate, *in* M-00055
(2-Methoxyphenyl)methyldiphenylphosphonium iodide, *in* M-00056
(4-Methoxyphenyl)methyldiphenylphosphonium iodide, *in* M-00058
(4-Methoxyphenyl)methyl-1-naphthalenylphenylphosphonium bromide, *in* M-00066
(4-Methoxyphenyl)methyl-1-naphthalenylphenylphosphonium chloride, *in* M-00066
[(2-Methoxyphenyl)methyl]triphenylphosphonium(1+); Chloride, *in* M-00062
[(2-Methoxyphenyl)methyl]triphenylphosphonium(1+); Bromide, *in* M-00062
[(3-Methoxyphenyl)methyl]triphenylphosphonium(1+); Chloride, *in* M-00063
[(3-Methoxyphenyl)methyl]triphenylphosphonium(1+); Bromide, *in* M-00063
[(4-Methoxyphenyl)methyl]triphenylphosphonium(1+); Chloride, *in* M-00064
[(4-Methoxyphenyl)methyl]triphenylphosphonium(1+); Bromide, *in* M-00064
(4-Methoxyphenyl)-1-naphthalenylphenylphosphine; (±)-*form*, B,$PhCH_2Cl$, *in* M-00066

(4-Methoxyphenyl)-1-naphthalenylphenylphosphine; (±)-*form*, B,$PhCH_2Br$, *in* M-00066
(3-Methoxy-2-propenyl)triphenylphosphonium(1+); Bromide, *in* M-00084
Methoxytris(trimethylsilyloxy)phosphonium iodide, *in* T-00864
5-Methyl-10*H*-5,10[1′,2′]benzenoacridophosphonium iodide, *in* B-00014
10-Methyl-5,10[1′,2′]-benzophenophosphazinium iodide, *in* B-00015
Methyl-2-biphenylylphosphonium iodide, *in* T-00348
Methylbis(2-methylphenyl)phenylphosphonium bromide, *in* B-00309
(3-Methyl-2-butenyl)triphenylphosphonium(1+); Chloride, *in* M-00104
(3-Methyl-2-butenyl)triphenylphosphonium(1+); Bromide, *in* M-00104
(3-Methylbutyl)triphenylphosphonium(1+); Bromide, *in* M-00111
(3-Methylbutyl)triphenylphosphonium(1+); Iodide, *in* M-00111
Methyldi-1-naphthylphosphonium iodide, *in* D-00939
Methyldioctylphenylphosphonium iodide, *in* D-00955
[3-(2-Methyl-1,3-dioxolan-2-yl)propyl]triphenylphosphonium; Bromide, *in* M-00132
[3-(2-Methyl-1,3-dioxolan-2-yl)propyl]triphenylphosphonium; Iodide, *in* M-00132
Methyldipentylphenylphosphonium iodide, *in* D-00978
5-Methyl-5,10-diphenyldibenzo[1,4]phosphoniaarsenin, *in* D-01001
3-Methyl-2,3-diphenyl-1,3-oxaphospholanium iodide, *in* D-01027
3-Methyl-2,3-diphenyl-1,3-oxaphosphorinanium iodide, *in* D-01028
Methyl diphenyl(phenylethenyl)phosphonium iodide, *in* D-01033
Methyldiphenyl-2-propenylphosphonium iodide, *in* D-01144
Methyldiphenyl-(1-propynyl)phosphonium iodide, *in* D-01146
Methyldiphenyl-4-pyridinylphosphonium bromide, *in* D-01077
Methyldiphenyl(3-trifluoromethylphenyl)phosphonium iodide, *in* D-01155
[Methylenebis(2,1-phenylenemethylene)]-bis[triphenylphosphonium](2+); Dibromide, *in* M-00141
Methylenebis[triphenylphosphonium](2+); Dichloride, *in* M-00147
Methylenebis[triphenylphosphonium](2+); Dibromide, *in* M-00147
Methylenebis[triphenylphosphonium](2+); Diiodide, *in* M-00147
Methyl(1-methylpropyl)diphenylphosphonium iodide, *in* M-00369
Methyl-2-naphthalenylphenyl(phenylmethyl)phosphonium bromide, *in* M-00177
Methyl(3-nitrophenyl)diphenylphosphonium iodide, *in* N-00039
Methyl(4-nitrophenyl)diphenylphosphonium iodide, *in* N-00040
(3-Methyl-2-oxo-2(5*H*)-furanyl)triphenylphosphonium(1+); Bromide, *in* M-00186
(1-Methyl-2-oxo-2-phenylethyl)triphenylphosphonium(1+); (±)-*form*, Bromide, *in* M-00188
(1-Methyl-2-oxo-2-phenylethyl)triphenylphosphonium(1+); (±)-*form*, Iodide, *in* M-00188
3-Methyl-2-oxo-3-phenyl-1,3-oxaphosphorinanium iodide, *in* P-00177
1-Methyl-4-oxo-1-phenyl-4-phosphorinanium iodide, *in* P-00268
9-Methyl-1,2,3,4,9-pentaphenyl-1*H*-tribenzo[*b,d,f*]-phosphepinium iodide, *in* P-00042
(Methylphenylamino)triphenylphosphonium(1+), M-00194
5-Methyl-5-phenyl-5*H*-dibenzophospholium iodide, *in* P-00110
7-Methyl-7-phenyl-7*H*-dibenzo[*d,f*]phosphoninium iodide, *in* P-00111
[(2-Methylphenyl)methyl]triphenylphosphonium(1+); Chloride, *in* M-00209
▷[(2-Methylphenyl)methyl]triphenylphosphonium(1+); Bromide, *in* M-00209
[(3-Methylphenyl)methyl]triphenylphosphonium(1+); Chloride, *in* M-00210
[(3-Methylphenyl)methyl]triphenylphosphonium(1+); Bromide, *in* M-00210
[(4-Methylphenyl)methyl]triphenylphosphonium(1+); Chloride, *in* M-00211
[(4-Methylphenyl)methyl]triphenylphosphonium(1+); Bromide, *in* M-00211
4-Methyl-4-phenyl-1,4-oxaphosphorinanium iodide, *in* P-00176
10-Methyl-10-phenyl-10*H*-phenothiaphosphinium 5,5-dioxide iodide, *in* P-00182
10-Methyl-10-phenyl-10*H*-phenothiaphosphinium iodide, *in* P-00182
10-Methyl-10-phenyl-10*H*-phenoxaphosphinium iodide, *in* P-00183
Methylphenyl(phenylmethyl)-2-propenylphosphonium bromide, *in* M-00264
1-Methyl-1-phenylphospholanium iodide, *in* P-00216
9-Methyl-9-phenyl-9-phosphoniabicyclo[4.2.1]nonane iodide, *in* P-00194
9-Methyl-9-phenyl-9-phosphoniabicyclo[4.2.1]nona-2,4,7-triene iodide, *in* P-00196
8-Methyl-8-phenyl-8-phosphoniabicyclo[3.2.1]octan-3-one iodide, *in* P-00199
9-Methyl-9-phenyl-9-phosphoniahomocubane iodide, *in* P-00201
1-Methyl-1-phenylphosphorinanium iodide, *in* P-00265
Methylphenyl-2-propenylphosphine; (*S*)-*form*, B,$PhCH_2Br$, *in* M-00264
Methylphenyl-2-propenylphosphine; (±)-*form*, B,$PhCH_2Br$, *in* M-00264
Methylphenyl(2-propenyl)propylphosphonium bromide, *in* M-00265
[[(4-Methylphenyl)sulfonyl]methyl]triphenylphosphonium(1+), M-00267
[[(4-Methylphenyl)sulfonyl]methyl]triphenylphosphonium(1+); Bromide, *in* M-00267
1-Methyl-1*H*-phosphole; B,MeI, *in* M-00282
1-Methyl-1-phosphoniabicyclo[3.3.1]nonane iodide, *in* P-00345
1-Methylphosphorinanium iodide, *in* M-00339
(1-Methyl-1,3-propanediyl)bis[methyldiphenylphosphonium] hexafluorophosphate, *in* B-00224
(2-Methyl-2-propenyl)triphenylphosphonium(1+); Chloride, *in* M-00367
(1-Methylpropyl)triphenylphosphonium(1+); Bromide, *in* M-00386
(1-Methylpropyl)triphenylphosphonium(1+); Iodide, *in* M-00386
(2-Methylpropyl)triphenylphosphonium(1+); Bromide, *in* M-00387
(2-Methylpropyl)triphenylphosphonium(1+); Iodide, *in* M-00387
[(Methylthio)methyl]triphenylphosphonium(1+); Chloride, *in* M-00416
[(Methylthio)methyl]triphenylphosphonium(1+); Tetraphenylborate, *in* M-00416
Methyltricyclopentylphosphonium iodide, *in* T-00431
Methyltri-2-furanylphosphonium iodide, *in* T-00486
Methyl(tri-3-furanyl)phosphonium iodide, *in* T-00487
[3-Methyl-5-(2,6,6-trimethyl-1-cyclohexen-1-yl)-2,4-pentadienyl]triphenylphosphonium(1+); (*E,E*)-*form*, Chloride, *in* M-00421
[3-Methyl-5-(2,6,6-trimethyl-1-cyclohexen-1-yl)-2,4-pentadienyl]triphenylphosphonium(1+); (*E,E*)-*form*, Bromide, *in* M-00421
[1-Methyl-3-(2,6,6-trimethyl-1-cyclohexen-1-yl)-2-propenyl]-triphenylphosphonium(1+); (*E*)-*form*, Chloride, *in* M-00422
[1-Methyl-3-(2,6,6-trimethyl-1-cyclohexen-1-yl)-2-propenyl]-triphenylphosphonium(1+); (*E*)-*form*, Bromide, *in* M-00422
Methyltrioctylphosphonium iodide, *in* T-00582
Methyltriphenoxyphosphonium(1+); Trifluoromethanesulfonate, *in* M-00429
4-Methyl-4-triphenylmethyl-2,6,7-trioxa-1-phosphoniabicyclo[2.2.2]octane perchlorate, *in* M-00424
4-Methyl-4-triphenylmethyl-2,6,7-trioxa-1-phosphoniabicyclo[2.2.2]octane tetrafluoroborate, *in* M-00424
3-Methyl-1,2,3-triphenyl-5-oxo-1,3-azaphospholidinium iodide, *in* T-00596
1-Methyl-1,2,5-triphenyl-1*H*-phospholium iodide, *in* T-00634
Methyltriphenylphosphonium(1+); Chloride, *in* M-00432
Methyltriphenylphosphonium(1+); Bromide, *in* M-00432
Methyltriphenylphosphonium(1+); Iodide, *in* M-00432
Methyltriphenylphosphonium(1+); Perchlorate, *in* M-00432
Methyltriphenylphosphonium(1+); Tetrafluoroborate, *in* M-00432

▷Methyltriphenylphosphonium(1+); Permanganate, *in* M-00432
Methyltriphenylphosphonium(1+); Tetraphenylborate, *in* M-00432
Methyltri-1-propenylphosphonium iodide, *in* T-00706
Methyltri-2-propenylphosphonium iodide, *in* T-00707
Methyltripropylphosphonium iodide, *in* T-00711
Methyltri(propylthio)phosphonium iodide, *in* T-00718
Methyltris(2,5-dimethylphenyl)phosphonium iodide, *in* T-00769
Methyltris(3-fluorophenyl)phosphonium iodide, *in* T-00782
Methyltris(4-fluorophenyl)phosphonium iodide, *in* T-00783
Methyltris(hydroxymethyl)phosphonium iodide, *in* T-00784
Methyltris(hydroxymethyl)phosphonium tetraphenylborate, *in* T-00784
Methyltris(4-methoxyphenyl)phosphonium bromide, *in* T-00793
Methyltris(4-methoxyphenyl)phosphonium iodide, *in* T-00793
Methyltris(4-methylphenyl)phosphonium bromide, *in* T-00808
Methyltris(4-methylphenyl)phosphonium iodide, *in* T-00808
Methyltris(1-methylpropyl)phosphonium iodide, *in* T-00823
Methyltris(2-methylpropyl)phosphonium iodide, *in* T-00824
Methyltris(morpholino)phosphonium iodide, *in* T-00569
Methyltris(piperidino)phosphonium iodide, *in* T-00704
Methyltris(2,4,6-trimethylphenyl)phosphonium iodide, *in* T-00861
Methyltri-2-thienylphosphonium iodide, *in* T-00874
Methyltri-3-thienylphosphonium iodide, *in* T-00875
(1-Naphthylmethyl)triphenylphosphonium(1+); Chloride, *in* N-00015
(1-Naphthylmethyl)triphenylphosphonium(1+); Bromide, *in* N-00015
(2-Naphthylmethyl)triphenylphosphonium(1+); Bromide, *in* N-00016
(Nitromethyl)triphenylphosphonium(1+); Bromide, *in* N-00038
[(2-Nitrophenyl)methyl]triphenylphosphonium(1+); Chloride, *in* N-00047
[(2-Nitrophenyl)methyl]triphenylphosphonium(1+); Bromide, *in* N-00047
[(3-Nitrophenyl)methyl]triphenylphosphonium(1+); Chloride, *in* N-00048
[(3-Nitrophenyl)methyl]triphenylphosphonium(1+); Bromide, *in* N-00048
[(4-Nitrophenyl)methyl]triphenylphosphonium(1+); Chloride, *in* N-00049
▷[(4-Nitrophenyl)methyl]triphenylphosphonium(1+); Bromide, *in* N-00049
[2-(4-Nitrophenyl)-2-oxoethyl]triphenylphosphonium(1+); Bromide, *in* N-00051
[2-(4-Nitrophenyl)-2-oxoethyl]triphenylphosphonium(1+); Iodide, *in* N-00051
3-Nonenyltriphenylphosphonium(1+); (*Z*)-*form*, Chloride, *in* N-00072
3-Nonenyltriphenylphosphonium(1+); (*Z*)-*form*, Iodide, *in* N-00072
4-Nonenyltriphenylphosphonium(1+); (*E*)-*form*, Bromide, *in* N-00073
4-Nonenyltriphenylphosphonium(1+); (*Z*)-*form*, Bromide, *in* N-00073
Nonyltriphenylphosphonium(1+); Bromide, *in* N-00078
Nonyltriphenylphosphonium(1+); Iodide, *in* N-00078
1,1,6,6,11,11,16,16-Octabenzyl-1,6,11,16-tetraphosphoniacycloeicosane tetrabromide, *in* T-00008
1,1,5,5,10,10,14,14-Octabenzyl-1,5,10,14-tetraphosphoniacyclooctadecane tetrabromide, *in* T-00009
Octahydro-7*H*,9*bH*-2*a*,4*a*,6*a*,9*a*-tetraaza-9*b*-phosphapentaleno[1,6-*cd*]indene; 9*b*-Chloro, *in* O-00023
Octahydro-5*H*,8*H*,10*bH*-2*a*,4*a*,7*a*,10*a*-tetraaza-10*b*-phosphapentaleno[1,6-*de*]naphthalene; 10*b*-Chloro, *in* O-00024
Octahydro-8*bH*-2*a*,4*a*,6*a*,8*a*-tetraaza-8*b*-phosphapentaleno[1,6-*cd*]pentalene; 8*b*-Chloro, *in* O-00025
1,2,3,4,9,10,11,12-Octaphenyl-5,8-diphosphoniadispiro[4.2.4.2]deca-1,3,9,11-tetraene(2+), O-00037
Octyltriphenylphosphonium(1+); Chloride, *in* O-00047
Octyltriphenylphosphonium(1+); Bromide, *in* O-00047
(2-Oxoethyl)triphenylphosphonium(1+); Chloride, *in* O-00060
(2-Oxoethyl)triphenylphosphonium(1+); Bromide, *in* O-00060
(2-Oxoethyl)triphenylphosphonium(1+); Tetrafluoroborate, *in* O-00060
(2-Oxoheptyl)triphenylphosphonium(1+); Chloride, *in* O-00062
(2-Oxoheptyl)triphenylphosphonium(1+); Bromide, *in* O-00062
4-Oxo-1,2,2,6,6-pentamethyl-1-phenylphosphorinanium iodide, *in* T-00210
(2-Oxo-2-phenylethyl)triphenylphosphonium(1+); Chloride, *in* O-00067
▷(2-Oxo-2-phenylethyl)triphenylphosphonium(1+); Bromide, *in* O-00067
▷(2-Oxo-2-phenylethyl)triphenylphosphonium(1+); Iodide, *in* O-00067
(2-Oxo-1-phenylpropyl)triphenylphosphonium(1+); Bromide, *in* O-00069
(3-Oxo-3-phenylpropyl)triphenylphosphonium(1+); Bromide, *in* O-00070
(3-Oxo-3-phenylpropyl)triphenylphosphonium(1+); Iodide, *in* O-00070
(2-Oxo-1,3-propanediyl)bis(triphenylphosphonium)(2+); Dichloride, *in* O-00073
(2-Oxo-1,3-propanediyl)bis(triphenylphosphonium)(2+); Dibromide, *in* O-00073
▷(2-Oxopropyl)triphenylphosphonium(1+); Chloride, *in* O-00080
(2-Oxopropyl)triphenylphosphonium(1+); Bromide, *in* O-00080
(2-Oxopropyl)triphenylphosphonium(1+); Iodide, *in* O-00080
(2-Oxopropyl)triphenylphosphonium(1+); Tetrafluoroborate, *in* O-00080
5-Oxo-1,1,3,3-tetraphenyl-1,3-diphosphorinanium(2+); Dichloride, *in* O-00081
5-Oxo-1,1,3,3-tetraphenyl-1,3-diphosphorinanium(2+); Dibromide, *in* O-00081
[Oxybis[methylene]]bis[triphenylphosphonium](2+); Dichloride, *in* O-00097
[Oxybis[methylene]]bis[triphenylphosphonium](2+); Dibromide, *in* O-00097
1,5-Pentanediylbis[triphenylphosphonium](2+); Dibromide, *in* P-00034
4-Pentenyltriphenylphosphonium(1+); Bromide, *in* P-00044
2-Penten-4-ynyltriphenylphosphonium(1+); Bromide, *in* P-00047
Pentyltriphenylphosphonium(1+); Bromide, *in* P-00056
Pentyltriphenylphosphonium(1+); Iodide, *in* P-00056
Phenoxytriphenylphosphorus(1+); Perchlorate, *in* P-00077
Phenoxytriphenylphosphorus(1+); Hexachloroantimonate, *in* P-00077
5-Phenyl-5,10-diphenyldibenzo[1,4]phosphoniaarsenin, *in* D-01001
1,3-Phenylenebis(methyldiphenylphosphonium)diiodide, *in* P-00127
1,2-[Phenylenebis[methylene]]bis[triphenylphosphonium](2+); Dibromide, *in* P-00132
[1,3-Phenylenebis[methylene]]bis[triphenylphosphonium](2+); Dichloride, *in* P-00133
[1,3-Phenylenebis[methylene]]bis[triphenylphosphonium](2+); Dibromide, *in* P-00133
▷1,4-[Phenylenebis[methylene]]bis[triphenylphosphonium](2+); Dichloride, *in* P-00134
1,4-[Phenylenebis[methylene]]bis[triphenylphosphonium](2+); Dibromide, *in* P-00134
(1,2-Phenylenedi-2-propene-3,1-diyl)-bis[triphenylphosphonium](2+); (*E*,*E*)-*form*, Dibromide, *in* P-00149
(1,2-Phenylenedi-2-propene-3,1-diyl)-bis[triphenylphosphonium](2+); (*E*,*E*)-*form*, Bis(tetrafluoroborate), *in* P-00149
1-Phenyl-1-phenylmethylphosphorinanium bromide, *in* P-00265
9-Phenyl-9-phosphabicyclo[4.2.1]nona-2,4,7-triene; *anti-form*, B,MeI, *in* P-00196
5-Phosphoniaspiro[4.4]nonane; Chloride, *in* P-00374
5-Phosphoniaspiro[4.4]nonane; Bromide, *in* P-00374
5-Phosphoniaspiro[4.4]nonane; Iodide, *in* P-00374
5-Phosphoniaspiro[4.4]nonane; Tetrafluoroborate, *in* P-00374
6-Phosphoniaspiro[5.5]undecane; Chloride, *in* P-00375
Phosphorinane; B,MeI, *in* P-00404
▷1,3-Propanediylbis[triphenylphosphonium](2+); Dibromide, *in* P-00448

2,2′-Spirobi[1,3,2-benzodithiaphospholium](1+); Hexafluorophosphate, *in* S-00016
5,5′-Spirobi[5*H*-dibenzophospholium](1+); Bromide, *in* S-00022
5,5′-Spirobi[5*H*-dibenzophospholium](1+); Iodide, *in* S-00022
10,10′(5*H*,5′*H*)-Spirobiphenophosphazinium(1+); Chloride, *in* S-00023
1,1,6,6-Tetrabenzyl-1,6-diphosphecanium dibromide, *in* D-01171
1,1,4,4-Tetrabenzyl-1,4-diphosphepanium dibromide, *in* D-00047
1,1,4,4-Tetrabenzyl-1,4-diphosphocanium dibromide, *in* D-00048
1,1,5,5-Tetrabenzyl-1,5-diphosphocanium dibromide, *in* D-00049
1,1,4,4-Tetrabenzyl-1,4-diphosphonanium dibromide, *in* D-00050
1,1,5,5-Tetrabenzyl-1,5-diphosphonanium dibromide, *in* D-00051
1,1,6,6-Tetrabenzyl-1,6-diphosphoniacycloundecane dibromide, *in* D-00046
Tetrabenzylphosphonium chloride, *in* T-00338
1,5,9,13-Tetrabenzyl-1,5,9,13-tetraphenyl-1,5,9,13-tetraphosphoniacyclohexadecane tetrabromide, *in* T-00276
Tetra-*tert*-butylmethyltetraphosphetanium iodide, *in* T-00012
▷Tetrabutylphosphonium(1+); Chloride, *in* T-00019
▷Tetrabutylphosphonium(1+); Bromide, *in* T-00019
▷Tetrabutylphosphonium(1+); Iodide, *in* T-00019
Tetrabutylphosphonium(1+); Acetate, *in* T-00019
Tetra-*tert*-butylphosphonium(1+); Tetrafluoroborate, *in* T-00020
Tetra-*tert*-butylphosphonium(1+); Iodide, *in* T-00020
Tetrachloromethylphosphorane, T-00032
5,5,7,7-Tetraethyl-6,7-dihydro-5*H*-dibenzo[*d,f*][1,3]-diphosphepinium dibromide, *in* D-00282
5,5,10,10-Tetraethyl-5,6,7,8,9,10-hexahydrodibenzo[*b,d*]-[1,6]diphosphecinium dibromide, *in* H-00021
Tetraethylmethyltetraphosphetanium iodide, *in* T-00071
Tetraethylphosphonium(1+); Chloride, *in* T-00056
Tetraethylphosphonium(1+); Bromide, *in* T-00056
▷Tetraethylphosphonium(1+); Iodide, *in* T-00056
5,5,9,9-Tetraethyl-6,7,8,9-tetrahydro-5*H*-dibenzo[*f,h*][1,5]-diphosphoninium dibromide, *in* T-00089
9,9,16,16-Tetraethyl-9,10,15,16-tetrahydrotribenzo[*b,d,h*]-[1,6]diphosphecinium dibromide, *in* T-00155
1,2,3,4-Tetrahydro-1,4-dimethyl-1,4-diphenyl-1,4-benzodiphosphorinium diiodide, *in* T-00102
1,2,5,6-Tetrahydro-1,1,3,6,6,9-hexamethyl-4*H*-phospholo[3,2,1-*i,j*]phosphinolinium iodide, *in* T-00123
1,2,3,4-Tetrahydro-4-methyl-2,2-diphenylbenz[*h*]-isophosphinolinium(1+); Hexafluorophosphate, *in* T-00113
2,3,4,5-Tetrahydro-5-methyl-2,2-diphenyl-1*H*-2-benzophosphepinium(1+); Hexafluorophosphate, *in* T-00114
Tetrahydro-4-methyl-1*H*,5*H*-[1,2]diphospholo[1,2-*a*][1,2]-diphospholium iodide, *in* D-01170
1,2,3,6-Tetrahydro-1-phenyl-1-phenylmethylphosphininium bromide, *in* T-00134
1,1′,3,3′-Tetrahydro-2,2′-spirobi[2*H*-isophosphindolium](1+); Bromide, *in* T-00137
2,3,4,5-Tetrahydro-2,2,4,4-tetramethyl-1*H*-2,4-benzodiphosphepinium(2+); Dibromide, *in* T-00139
2,3,4,5-Tetrahydro-1,1,5,5-tetramethyl-1*H*-1,5-benzophospharsepinium dibromide, *in* T-00092
3*a*,4,7,7*a*-Tetrahydro-1,1,8,8-tetramethyl-4,7-phosphinidene-1*H*-phosphindolium diiodide, *in* T-00099
1,2,5,6-Tetrahydro-2,2,5,5-tetraphenyl-2,5-benzodiphocinium dibromide, *in* H-00053
2,3,4,5-Tetrahydro-2,2,4,4-tetraphenyl-1*H*-2,4-benzodiphosphonium(2+); Dibromide, *in* T-00144
1,1,3,3-Tetrahydro-1,1,3,3-tetraphenyl-2,4-bis(triphenylphosphonio)-1,3-diphosphete; Dichloride, *in* T-00145
1,2,3,4-Tetrahydro-1,1,4,4-tetraphenyl-1,4-diphosphorinium(2+); Dibromide, *in* T-00152
1,3,4,5-Tetrahydro-2,4,5-trimethyl-2,5-methano-2*H*-2-benzophosphepinium iodide, *in* T-00096
1,1,3,3-Tetrakis(diphenylphosphino)-1,2-propadiene; B,MeI, *in* T-00180
▷Tetrakis(hydroxymethyl)phosphonium(1+); Chloride, *in* T-00181
▷Tetrakis(hydroxymethyl)phosphonium(1+); Acetate, *in* T-00181
1,2,4,5-Tetrakis(methylene)-3-phosphoniaspiro[2.2]pentane, T-00182
Tetrakis(trimethylsilyloxy)phosphonium iodide, *in* T-00864
2,4,8,10-Tetramethyl-3,9-dioxo-2,4,8,10-tetraaza-6-phosphoniaspiro[5.5]undecane(1+); Chloride, *in* T-00198
2,4,8,10-Tetramethyl-3,9-dioxo-2,4,8,10-tetraaza-6-phosphoniaspiro[5.5]undecane(1+); Sulfate, *in* T-00198
2,4,8,10-Tetramethyl-3,9-dithia-6-phosphoniaspiro[5.5]-undecane; Perchlorate, *in* T-00202
1,1,3,4-Tetramethyl-1*H*-phospholium iodide, *in* T-00536
Tetramethylphosphonium(1+); Chloride, *in* T-00215
Tetramethylphosphonium(1+); Bromide, *in* T-00215
▷Tetramethylphosphonium(1+); Iodide, *in* T-00215
3,3,3′,3′-Tetramethyl-1,1′(3*H*,3′*H*)-spirobi[2,1-benzoxaphospholium](1+); Chloride, *in* T-00233
3,3,3′,3′-Tetramethyl-1,1′(3*H*,3′*H*)-spirobi[2,1-benzoxaphospholium](1+); Trifluoromethanesulfonate, *in* T-00233
1,1,4,4-Tetramethyl-1,2,3,4-tetrahydro-1,4-benzophospharseninium dibromide, *in* T-00091
1,1,3,3-Tetraphenyl-1,3-diphosphepanium, T-00257
1,1,5,5-Tetraphenyl-1,5-diphosphocanium(2+); Dibromide, *in* T-00261
1,1,5,5-Tetraphenyl-1,5-diphosphocanium(2+); Diiodide, *in* T-00261
1,1,3,3-Tetraphenyl-1,3-diphospholanium dibromide, *in* D-01013
Tetraphenyl-1,4-diphosphonia-6,7-benzocyclooct-6-ene, *in* H-00053
1,1,3,3-Tetraphenyl-1,3-diphosphorinanium(2+), T-00262
1,1,3,3-Tetraphenyl-1,3-diphosphorinanium(2+); Dichloride, *in* T-00262
1,1,3,3-Tetraphenyl-1,3-diphosphorinanium(2+); Dibromide, *in* T-00262
1,1,4,4-Tetraphenyl-1,4-diphosphorinanium(2+); Dibromide, *in* T-00263
1,1,4,4-Tetraphenyl-1,4-diphosphorinanium dibromide, *in* D-01017
3,3,6,6-Tetraphenyl-1,3,6-oxadiphosphepanium(2+); Dibromide, *in* T-00266
3,3,6,6-Tetraphenyl-1,3,6-oxadiphosphepanium(2+); Dipicrate, *in* T-00266
3,3,5,5-Tetraphenyl-1,3,5-oxadiphosphorinanium(2+); Dichloride, *in* T-00267
3,3,5,5-Tetraphenyl-1,3,5-oxadiphosphorinanium(2+); Dibromide, *in* T-00267
2,4,4,6-Tetraphenyl-4*H*-1,4-oxaphosphorinium bromide, *in* T-00610
1,2,3,4-Tetraphenyl-5-phosphoniaspiro[4.5]deca-1,3-diene; Bromide, *in* T-00270
Tetraphenylphosphonium(1+); Chloride, *in* T-00272
Tetraphenylphosphonium(1+); Bromide, *in* T-00272
Tetraphenylphosphonium(1+); Iodide, *in* T-00272
Tetraphenylphosphonium(1+); Tribromide, *in* T-00272
Tetraphenylphosphonium(1+); Methoxide, *in* T-00272
Tetrapropylphosphonium(1+); Chloride, *in* T-00290
Tetrapropylphosphonium(1+); Bromide, *in* T-00290
Tetrapropylphosphonium(1+); Iodide, *in* T-00290
Tetravinylphosphonium(1+); Bromide, *in* T-00296
[Thiobis[methylene]]bis[triphenylphosphonium](2+); Dichloride, *in* T-00312
[Thiobis[methylene]]bis[triphenylphosphonium](2+); Dibromide, *in* T-00312
Tribenzylmethylphosphonium iodide, *in* T-00338
Tribromodiphenoxyphosphorane, T-00354
Tri-*tert*-butylethylphosphonium bromide, *in* T-00373
Tributyl(hexadecyl)phosphonium(1+); Chloride, *in* T-00364
Tributyl(hexadecyl)phosphonium(1+); Bromide, *in* T-00364
Tributyl(hexadecyl)phosphonium(1+); Iodide, *in* T-00364
1,1,1-Tributyl-1-iodo-*N*-methyl-*N*-phenylphosphoranamine, *in* T-00366
Tri-*tert*-butylisopropylphosphonium bromide, *in* T-00373
Tri-*tert*-butylmethylphosphonium bromide, *in* T-00373
▷Tributylmethylphosphonium iodide, *in* T-00372

Tri-*tert*-butylmethylphosphonium iodide, *in* T-00373
Trichlorodimethylphosphorane, T-00402
Tricyclohexylmethylphosphonium iodide, *in* T-00426
Tricyclopropylmethylphosphonium iodide, *in* T-00432
Triethenylmethylphosphonium iodide, *in* T-00877
Triethylmethylphosphonium iodide, *in* T-00441
Tri(ethylthio)methylphosphonium iodide, *in* T-00461
Trifluorodimethoxyphosphorane, T-00468
Trifluorodiphenoxyphosphorane, T-00470
Triisopropylmethylphosphonium iodide, *in* T-00500
Trimethyl(2-dimethylphosphinophenyl)phosphonium iodide, *in* P-00123
1,9,9-Trimethyl-2,8-diphenyl-3,7-dioxa-1-phosphoniabicyclo[3.3.1]nonane, *in* D-00743
1,2,5-Trimethyl-4-oxo-1-phenylphosphorinanium iodide, *in* D-00821
2,5,5-Trimethyl-2-phenyl-1,3,2-dioxaphosphorinanium iodide, *in* D-00795
2,2,5-Trimethyl-3-phenyl-1,3-oxaphospholane; (3*RS*,5*SR*)-*form*, B,$PhCH_2Br$, *in* T-00522
2,8,10-Trimethyl-10-phenyl-10*H*-phenoxaphosphinium iodide, *in* D-00799
1,3,4-Trimethyl-1-phenyl-1*H*-phospholium iodide, *in* D-00807
Trimethylphenylphosphonium bromide, *in* D-00803
1,1,4-Trimethylphosphoranium iodide, *in* D-00883
1,1,3-Trimethylphosphorinanium iodide, *in* D-00882
[[2-(Trimethylsilyl)ethoxy]methyl]triphenylphosphonium(1+); Chloride, *in* T-00552
[2-(Trimethylsilyl)ethyl]triphenylphosphonium(1+); Iodide, *in* T-00553
[2-(Trimethylstannyl)ethyl]triphenylphosphonium; Iodide, *in* T-00561
Trimethyl[(trimethylphosphoranylidene)methyl]-phosphonium(1+); Chloride, *in* T-00565
Trimethyl[(trimethylphosphoranylidene)methyl]-phosphonium(1+); Bromide, *in* T-00565
Trimethyl[(trimethylphosphoranylidene)methyl]-phosphonium(1+); Iodide, *in* T-00565
2,4,6-Triphenyl-4*H*-1,4-oxaphosphorin; B,$PhClO_4$, *in* T-00610
Triphenyl(1-phenylethyl)phosphonium(1+); Bromide, *in* T-00612
Triphenyl(1-phenylethyl)phosphonium(1+); Iodide, *in* T-00612
Triphenyl(2-phenylethyl)phosphonium(1+); Bromide, *in* T-00613
Triphenyl(2-phenylethynyl)phosphonium(1+); Chloride, *in* T-00614
Triphenyl(2-phenylethynyl)phosphonium(1+); Bromide, *in* T-00614
Triphenyl(3-phenyl-2-propenyl)phosphonium(1+); (*E*)-*form*, Chloride, *in* T-00616
Triphenyl(3-phenyl-2-propenyl)phosphonium(1+); (*E*)-*form*, Bromide, *in* T-00616
Triphenyl(3-phenylpropyl)phosphonium(1+); Bromide, *in* T-00617
Triphenyl[1-(phenylthio)cyclopropyl]phosphonium(1+); Bromide, *in* T-00618
Triphenyl[1-(phenylthio)cyclopropyl]phosphonium(1+); Tetrafluoroborate, *in* T-00618
Triphenyl[1-(phenylthio)ethenyl]phosphonium(1+); Chloride, *in* T-00619
Triphenyl[1-(phenylthio)ethenyl]phosphonium(1+); Iodide, *in* T-00619
Triphenyl(phenylthiomethyl)phosphonium(1+); Chloride, *in* T-00621
Triphenylphosphine; B,HF, *in* T-00624
Triphenylphosphine; B,2HCl, *in* T-00624
Triphenylphosphine; B,HBr, *in* T-00624
Triphenylphosphine; B,HI, *in* T-00624
Triphenylphosphite methiodide, *in* M-00429
10-(Triphenylphosphoniomethyl)benzo[*g*]chrysene(1+); Bromide, *in* T-00635
2-(Triphenylphosphoniomethyl)benzo[*c*]phenanthrene(1+); Bromide, *in* T-00636
3-(Triphenylphosphoniomethyl)phenanthrene(1+); Bromide, *in* T-00637
9-(Triphenylphosphoniomethyl)phenanthrene(1+); Chloride, *in* T-00638
9-(Triphenylphosphoniomethyl)phenanthrene(1+); Bromide, *in* T-00638
Triphenyl-2-propenylphosphonium(1+); Chloride, *in* T-00679
▷Triphenyl-2-propenylphosphonium(1+); Bromide, *in* T-00679
Triphenylpropylphosphonium(1+); Chloride, *in* T-00680
Triphenylpropylphosphonium(1+); Bromide, *in* T-00680
Triphenylpropylphosphonium(1+); Iodide, *in* T-00680
Triphenyl-2-propynylphosphonium(1+); Bromide, *in* T-00681
Triphenyl(tetradecyl)phosphonium(1+); Bromide, *in* T-00682
Triphenyl[[3-(tetrahydro-2*H*-pyran-2-yl)oxy]propyl]-phosphonium(1+); Bromide, *in* T-00683
Triphenyl[[3-(tetrahydro-2*H*-pyran-2-yl)oxy]propyl]-phosphonium(1+); Iodide, *in* T-00683
Triphenyl(tetrahydro-2*H*-pyran-4-yl)phosphonium(1+); Bromide, *in* T-00684
Triphenyl(tetrahydro-2*H*-thiopyran-4-yl)phosphonium(1+); Bromide, *in* T-00685
Triphenyl[(2,6,6-trimethyl-1,3-cyclohexadien-1-yl)methyl]-phosphonium(1+); Bromide, *in* T-00688
Triphenyl[(2,6,6-trimethyl-1-cyclohexen-1-yl)methyl]-phosphonium(1+); Chloride, *in* T-00689
Triphenyl[(2,6,6-trimethyl-1-cyclohexen-1-yl)methyl]-phosphonium(1+); Bromide, *in* T-00689
Triphenyl(3,7,11-trimethyldodecyl)phosphonium(1+); (3*R*,7*R*)-*form*, Bromide, *in* T-00690
Triphenyl[4-(trimethylsilyl)-3-butynyl]phosphonium(1+); Iodide, *in* T-00691
Triphenyl[(trimethylsilyl)methyl]phosphonium(1+); Chloride, *in* T-00693
Triphenyl[(trimethylsilyl)methyl]phosphonium(1+); Iodide, *in* T-00693
Triphenyl[(trimethylsilyl)methyl]phosphonium(1+); Tetrafluoroborate, *in* T-00693
Triphenyl[[2-[2-(triphenylphosphonio)ethyl]phenyl]methyl]-phosphonium(2+); Dibromide, *in* T-00696
Triphenyl[[2-[2-(triphenylphosphonio)ethyl]phenyl]methyl]-phosphonium(2+); Diperiodate, *in* T-00696
Triphenyl[(triphenylphosphoranylidene)methyl]-phosphonium(1+); Bromide, *in* T-00697
Triphenyl[(triphenylphosphoranylidene)methyl]-phosphonium(1+); Chloride, *in* T-00697
Triphenyl[(triphenylphosphoranylidene)methyl]-phosphonium(1+); Iodide, *in* T-00697
Triphenyl[(triphenylphosphoranylidene)methyl]-phosphonium(1+); Tetraphenylborate, *in* T-00697
Triphenylundecylphosphonium(1+); Bromide, *in* T-00701
▷Triphenylvinylphosphonium(1+); Bromide, *in* T-00702
Tri-2-pyridylphosphine; B,MeI, *in* T-00723
Tris(4-chlorophenyl)methylphosphonium bromide, *in* T-00749
Tris(4-chlorophenyl)methylphosphonium iodide, *in* T-00749
Tris(2-cyanoethyl)methylphosphonium iodide, *in* T-00752
Tris(2,4-dimethoxyphenyl)methylphosphonium iodide, *in* T-00758
Tris(dimethylamino)(methylamino)phosphonium iodide, *in* H-00001
Tris(diphenylphosphino)methane; Dioxide, monomethiodide, *in* T-00775
Tris(2-hydroxyphenyl)methylphosphonium iodide, *in* T-00788
Tris(4-hydroxyphenyl)phosphine; B,HBr, *in* T-00790
Tris(4-hydroxyphenyl)phosphine; Tri-Ac; B,MeI, *in* T-00790
Tris(2-methoxyphenyl)methylphosphonium iodide, *in* T-00791
Tris(3-methoxyphenyl)methylphosphonium iodide, *in* T-00792
1,1,1-Tris(methyldiphenylphosphoniomethyl)ethane, *in* T-00776
Tris(2,4,6-trimethoxyphenyl)methylphosphonium iodide, *in* T-00858
Tris(2,4,6-trimethylphenyl)phosphine; B,$MeBF_4$, *in* T-00861

Phosphoranes; RPX_4, R_2PX_3 etc.

Aminobisbiphenylenephosphorane, *in* S-00021
Bis(2,2-dimethylpropoxy)triphenylphosphorane, B-00207
2,4-Bis(1-methylethylidene)-1,3,6,9-tetraoxa-5-phospha(5-P^V)spiro[4.4]nonane; 5-Ph, *in* B-00306
Bis(2,2,2-trifluoroethoxy)triphenylphosphorane, B-00408
2-(4-Bromophenyl)-8,8-dihydro-4,6,8,8-tetraphenyl[1,2]-oxaphospholo[2,3-*b*][1,2,5]dioxaphosphorin, B-00479
4-(4-Bromophenyl)-2′,2′,3′,4′,4′-pentamethyl-3,3,6,6-tetrakis(trifluoromethyl)spiro[2,7-dioxa-1-phosphabicyclo[3.2.0]hept-3-ene-1,1′-phosphetane], B-00485

Bromotetraphenoxyphosphorane, B-00503
4-*tert*-Butyl-1,1,3,3-tetrakis(dimethylamino)-1,3,5-triphosphorin, B-00638
N-*tert*-Butyl-*P*-(2,4,6-tri-*tert*-butylphenyl)-metaphosphonamidate, *in* B-00639
N-*tert*-Butyl-*P*-(2,4,6-tri-*tert*-butylphenyl)-metaphosphonamidoselenoate, *in* B-00639
N-*tert*-Butyl-*P*-(2,4,6-tri-*tert*-butylphenyl)-metaphosphonamidothioate, *in* B-00639
Chlorotetraphenoxyphosphorane, C-00186
1,1-Dibromo-1,1-dihydro-2,4,6-triphenylphosphorin, *in* D-00613
Dibromotrimethylphosphorane, D-00069
Dibromotriphenylphosphorane, D-00071
1,1-Dichloro-1,1-dihydro-2,4,6-triphenylphosphorin, *in* D-00613
Dichlorotrimethylphosphorane, D-00198
Dichlorotriphenylphosphorane, D-00200
Dichlorotris(trichloromethyl)phosphorane, D-00201
Diethoxytriphenylphosphorane, D-00263
2,2-Difluoro-2,2-dihydro-1,3-dimethyl-2-phenyl-1,3,2-diazaphosphetidin-4-one, *in* D-00459
2,2-Difluoro-2,2-dihydro-1,2,3-trimethyl-1,3,2-diazaphosphetidin-4-one, *in* D-00459
1,1-Difluoro-1,1-dihydro-2,4,6-triphenylphosphorin, *in* D-00613
Difluorotrimethylphosphorane, D-00417
Difluorotriphenylphosphorane, D-00419
Difluorotris(trifluoromethyl)phosphorane, D-00420
2,2-Dihydro-1,3,2-benzodioxaphosphole; 2,2-Dimethoxy, 2-Ph, *in* D-00439
2,2-Dihydro-1,3,2-benzodioxaphosphole; 2-Methoxy, 2,2-di-Ph, *in* D-00439
2,2-Dihydro-1,3,2-benzodioxaphosphole; 2,2,2-Tri-Ph, *in* D-00439
1,1-Dihydro-1,1-bis(methylthio)-2,4,6-triphenylphosphorin, *in* D-00613
1,1-Dihydro-1,1-bis(phenylthio)-2,4,6-triphenylphosphorin, *in* D-00613
2,2-Dihydro-3,3-bis(trifluoromethyl)-1-[[2,2,2-trifluoro-1-(trifluoromethyl)ethylidene]amino]azaphosphiridine; 2,2,2-Trimethoxy, *in* D-00446
2,2-Dihydro-3,3-bis(trifluoromethyl)-1-[[2,2,2-trifluoro-1-(trifluoromethyl)ethylidene]amino]azaphosphiridine; 2,2,2-Triethoxy, *in* D-00446
2,2-Dihydro-3,3-bis(trifluoromethyl)-1-[[2,2,2-trifluoro-1-(trifluoromethyl)ethylidene]amino]azaphosphiridine; 2,2,2-Tris(dimethylamino), *in* D-00446
1,1-Dihydro-1,1-dimethoxy-2,4,6-triphenylphosphorin, *in* D-00613
5,5-Dihydro-4,4-dimethyl-2,3,5,5,5-pentaphenyl-1,2,5-oxazaphospholidine, D-00465
4′,5′-Dihydro-3′,5′-dimethylspiro[1,3,2-benzoxazaphosph(P^V)-ole-2,2′-[1,3,2]oxazaphospholidin]-4′-one, D-00470
4′,5′-Dihydro-3′,5′-dimethylspiro[1,3,2-benzoxazaphosph(P^V)-ole-2,2′-[1,3,2]oxazaphospholidin]-4′-one; (5′*S*)-*form*, 3-Me, *in* D-00470
1,1-Dihydro-1,1-diphenoxy-2,4,6-triphenylphosphorin, *in* D-00613
8,8-Dihydro-8,8-diphenyl-2*H*,6*H*-[1,2]oxaphospholo[4,3,2-*hi*]-[2,1]benzoxaphosphole, D-00479
10,10-Dihydro-2,3-diphenyl-[1,3,2]oxazaphospholo[2,3-*b*]-[1,3,2]benzoxazaphosphole; 10,10-Di-Me, *in* D-00481
10,10-Dihydro-2,3-diphenyl-[1,3,2]oxazaphospholo[2,3-*b*]-[1,3,2]benzoxazaphosphole; 10,10-Di-Ph, *in* D-00481
2,2-Dihydro-4-methyl-2,2,2,3-tetraphenyl-1,2-oxaphosphetane, D-00531
2,2-Dihydro-1,2-oxaphosphetane; 2,2,2-Tributyl, *in* D-00533
2,2-Dihydro-1,2-oxaphosphetane; 2,2,2-Tri-Ph, *in* D-00533
2,2-Dihydro-1,2-oxaphospholane; 2,2,2-Tri-Me, *in* D-00534
2,2-Dihydro-1,2-oxaphospholane; 2,2,2-Tri-Et, *in* D-00534
2,2-Dihydro-1,2-oxaphospholane; 2,2,2-Tri-Ph, *in* D-00534
2,2-Dihydro-2,2,2,4,5-pentaphenyl-1,3,2-dioxaphospholane, D-00537
2,2-Dihydrophenanthro[9,10-*d*]-1,3,2-dioxaphosphole; 2,2,2-Tri-Ph, *in* D-00541
1,1-Dihydrophosphorinane, D-00573
1,3-Dihydrospiro[2*H*-1,3,2-benzodiazaphosphole-2,2′-[1,3,2]-benzodioxaphosphole], D-00574
1,3-Dihydrospiro[2*H*-1,3,2-benzodiazaphosphole-2,2′-[1,3,2]-benzodioxaphosphole]; 1-Benzyl, *in* D-00574
1,3-Dihydrospiro[2*H*-1,3,2-benzodiazaphosphole-2,2′-[1,3,2]-dioxaphosholane]; 2-Ph, *in* D-00575
1,4-Dihydrospiro[2*H*-3,1,2-benzoxazaphosphorine-2,2′-[1,3,2]-dioxaphospholane]; 2-Ph, *in* D-00576
2,2-Dihydro-4,4,5,5-tetramethyl-2,2,2-triphenyl-1,3,2-dioxaphospholane, D-00578
2,2-Dihydro-2,2,2-trimethoxy-3,5-bis(pentafluorophenyl)-1,4,2-dioxaphospholane, D-00591
2,2-Dihydro-2,2,2-trimethyl-1,2-oxaphosphorinane, D-00597
1,3-Dihydro-1,2,3-trimethylspiro[2*H*-1,3,2-benzodiazaphosphole-2,2′-[1,3,2]benzodioxaphosphole], *in* D-00469
2,2-Dihydro-2,2,2-trimethyl-4,4,5,5-tetrakis(trifluoromethyl)-1,3,2-dioxaphospholane, D-00601
2,2-Dihydro-2,2,2-triphenyl-2*H*-1,2-benzoxaphosphorin, D-00607
5,5-Dihydro-5,5,5-triphenyl-5*H*-dibenzophosphole, D-00610
2,2-Dihydro-2,2,2-triphenyl-1,3,2-dioxaphospholane, D-00611
2,2-Dihydro-2,2,2-triphenyl-4,4,5,5-tetrakis(trifluoromethyl)-1,3,2-dioxaphospholane, D-00614
Diiodotrimethylphosphorane, D-00629
▷Diiodotriphenylphosphorane, D-00631
N,*N*-Dimethyl-2,4-bis(1-methylethylidene)-1,3,6,9-tetraoxa-5-phospha(5-P^V)spiro[4.4]nonan-5-amine, *in* B-00306
4,9-Dimethyl-1,6-dioxa-4,9-diaza-5-phospha(5-P^v)spiro[4.4]-nonane, D-00729
5,6-Dimethyl-*N*-phenylspiro[1,3,2-benzodioxaphosphole-2,1′(3′*H*)-[2,1]benzoxaphosphol]-2-amine, *in* D-00915
3′,4′-Dimethyl-5′-phenylspiro[1,3,2-benzoxazaphosphole-2(3*H*),2′-[1,3,2]oxazaphospholidine], D-00823
5,6-Dimethylspiro[1,3,2-benzodioxaphosphole-2,1′(3′*H*)-[2,1]-benzoxaphosphole, D-00915
5,6-Dimethylspiro[1,3,2-benzodioxaphosphole-2,1′(3′*H*)-[2,1]-benzoxaphosphole; 2-Phenoxy, *in* D-00915
5,6-Dimethylspiro[1,3,2-benzodioxaphosphole-2,1′(3′*H*)-[2,1]-benzoxaphosphole; 2-Ph, *in* D-00915
2,3′-Dimethylspiro[1,3,2-benzodioxaphosphole-2,2′(3′*H*)-[1,3,2]benzoxazaphosphole], *in* M-00395
3,3-Dimethyl-2,2(3*H*,3′*H*)-spirobi[1,3,2-benzoxazaphosphole], D-00917
8,8-Dimethyl-1,6,10-trioxa-5-phospha(5-P^V)spiro[4.5]dec-2-ene; 5-Methoxy, *in* D-00933
1,4-Dioxa-6-aza-5-phospha(5-P^V)spiro[4.4]nonan-7-one; 5-Ph, *in* D-00962
1,6-Dioxa-4,9-diaza-5-phospha(5-P^V)spiro[4.4]nonane, D-00963
1,6-Dioxa-4,9-diaza-5-phospha(5-P^V)spiro[4.4]nonane; 5-Ph, *in* D-00963
1,4-Dioxa-5-phospha(5-P^V)spiro[4.4]non-7-ene; 5-Me, *in* D-00967
1,4-Dioxa-5-phospha(5-P^V)spiro[4.4]non-7-ene; 5-Ph, *in* D-00967
Diphenoxytriphenylphosphorane, D-00992
1′,3′-Diphenyldispiro[1,3,2-benzodioxaphosphole-2,2′-[1,3,2,4]diazaphosphetidine-4′,2″-[1,3,2]-benzodioxaphosphole]; 2,4′-Di-Ph, *in* D-01018
2,5′-Diphenylspiro[4*H*-1,3,2-benzodioxaphosphorin-2,2′-[1,3,2]dioxaphospholane]-4,4′-dione, D-01147
2,3-Diphenyl-1,6,10-trioxa-4-aza-5-phospha(5-P^V)spiro[4.5]-dec-2-ene; 5-Ph, *in* D-01158
Diselenoxo(2,4,6-tri-*tert*-butylphenyl)phosphorane, D-01235
Dithioxo(2,4,6-tri-*tert*-butylphenyl)phosphorane, D-01254
4′-Ethyl-1,3-dihydro-5-phenyl-3,3-bis(trifluoromethyl)-spiro[2*H*-1,4,2-diazaphosphole-2,1′-[2,6,7]trioxa[1]-phosphabicyclo[2.2.2]octane]; 1-(2,6-Dimethylphenyl), *in* E-00061
4′-Ethyl-1,3-dihydro-5-phenyl-3,3-bis(trifluoromethyl)-spiro[2*H*-1,4,2-diazaphosphole-2,1′-[2,6,7]trioxa[1]-phosphabicyclo[2.2.2]octane]; 1-(2,4,6-Trimethylphenyl), *in* E-00061
Fluorotetramethylphosphorane, F-00049
2,3,5,6,8,8-Hexahydro-8-phenyl-[1,3,2]oxazaphospholo[2,3-*b*]-[1,3,2]oxazaphosphole, H-00049

Hexahydrospiro[1,3,2-benzoxazaphosphole-2(3*H*),1′-phospholane]; 2-Ph, *in* H-00052
2,2,2,3,4,5-Hexahydro-2,2,2-trimethoxy-3-phenylnaphth[2,1-*d*]-1,2-oxaphosphole, H-00056
2,2,2,3,3*a*,11*b*-Hexahydro-2,2,2-triphenylphenanthro[9,10-*d*]-1,3,2-oxazaphosphole; (9*RS*,10*RS*)-*form*, *in* H-00059
2,2,3,3,7,7-Hexamethyl-1,4,6-trioxa-9-aza-5-phospha(5-P^{V})-spiro[4.4]nonane, H-00083
2,2,3,3,8,8-Hexamethyl-1,4,6-trioxa-9-aza-5-phospha(5P^{V})-spiro[4.4]nonane, H-00084
Homocubyltrimethylphosphorane, *in* D-00569
Homocubyltriphenylphosphorane, *in* D-00569
1-Hydroxy-1,1′(3*H*,3′*H*)-spirobi[2,1-benzoxaphosphole]-3,3′-dione, H-00194
1-Hydroxy-1,1′(3*H*,3′*H*)-spirobi[2,1-benzoxaphosphole]-3,3′-dione; 1-Methoxy, *in* H-00194
1-Hydroxy-3,3,3′,3′-tetramethyl-1,1′(3*H*,3′*H*)-spirobi[2,1-benzoxaphosphole], H-00199
1-Hydroxy-3,3,3′,3′-tetramethyl-1,1′(3*H*,3′*H*)-spirobi[2,1-benzoxaphosphole]; 1-Methoxy, *in* H-00199
Iodotetraphenoxyphosphorane, I-00020
7-Isopropylidene-4,8,8-trimethyl-1,6-dioxa-4-aza-5-phospha(5-P^{V})spiro[4.4]nonan-9-one; 5-Dimethylamino, *in* I-00032
5-Methoxy-5-phospha(5-P^{V})spiro[4.4]nonane, *in* P-00350
Methylbisbiphenylenephosphorane, *in* S-00021
5-Methyl-5-phospha(5-P^{V})spiro[4.4]nonane, *in* P-00350
3′-Methylspiro[1,3,2-benzodioxaphosphole-2,2′(3′*H*)-[1,3,2]-benzoxazaphosphole], M-00395
5′-Methylspiro[1,3,2-benzodioxaphosphole-2,2′(3*H*)-[1,2]-oxaphosphole]; 2-Methoxy, *in* M-00396
5′-Methylspiro[1,3,2-benzodioxaphosphole-2,2′(3*H*)-[1,2]-oxaphosphole]; 2-Ethoxy, *in* M-00396
5′-Methylspiro[1,3,2-benzodioxaphosphole-2,2′(3*H*)-[1,2]-oxaphosphole]; 2-Ph, *in* M-00396
2-Methyl-2,2′-spirobi[1,3,2-benzodioxaphosphole], M-00397
7-Methyl-1,4,6-trioxa-5-phospha(5-P^{V})spiro[4.4]nonan-8-one; 5-Hydroxy, *in* M-00427
7-Methyl-1,4,6-trioxa-5-phospha(5-P^{V})spiro[4.4]nonan-8-one; 5-Chloro, *in* M-00427
9-Methyl-1,4,6-trioxa-5-phospha(5-P^{V})spiro[4.4]non-7-ene; 5-Methoxy, *in* M-00428
9-Methyl-1,4,6-trioxa-5-phospha(5-P^{V})spiro[4.4]non-7-ene; 5-Ethoxy, *in* M-00428
9-Methyl-1,4,6-trioxa-5-phospha(5-P^{V})spiro[4.4]non-7-ene; 5-Ph, *in* M-00428
Octahydro-3*H*,8*H*,10*bH*-2*a*,5*a*,7*a*,10*a*-tetraaza-10*b*-phosphaindeno[7,1-*cd*]indene, O-00022
Octahydro-7*H*,9*bH*-2*a*,4*a*,6*a*,9*a*-tetraaza-9*b*-phosphapentaleno[1,6-*cd*]indene, O-00023
Octahydro-5*H*,8*H*,10*bH*-2*a*,4*a*,7*a*,10*a*-tetraaza-10*b*-phosphapentaleno[1,6-*de*]naphthalene, O-00024
Octahydro-8*bH*-2*a*,4*a*,6*a*,8*a*-tetraaza-8*b*-phosphapentaleno[1,6-*cd*]pentalene, O-00025
Octahydro-8*bH*-2*a*,4*a*,6*a*,8*a*-tetraaza-8*b*-phosphapentaleno[1,6-*cd*]pentalene; 8*b*-Ph, *in* O-00025
2,2,3,3,7,7,8,8-Octamethyl-1,4,6,9-tetraoxa-5-phospha(5-P^{V})-spiro[4.4]nonane, O-00032
2,2,3,3,7,7,8,8-Octamethyl-1,4,6,9-tetraoxa-5-phospha(5-P^{V})-spiro[4.4]nonane; 5-Me, *in* O-00032
2,2,3,3,7,7,8,8-Octamethyl-1,4,6,9-tetraoxa-5-phospha(5-P^{V})-spiro[4.4]nonane; 5-Ph, *in* O-00032
2,2,3,3,7,7,8,8-Octamethyl-1,4,6,9-tetraoxa-5-phospha(5-P^{V})-spiro[4.4]nonane; 5-Benzyl, *in* O-00032
2,2,3,3,7,7,8,8-Octamethyl-1,4,6,9-tetraoxa-5-phospha(5-P^{V})-spiro[4.4]nonane; 5-Benzoyl, *in* O-00032
2,7,9,14,15,16,17,18-Octaoxa-1,8-diphospha(1,8-P^{V})-pentacyclo[10.2.1.$1^{1,4}$.$1^{5,8}$.$1^{8,11}$]octadecane-3,6,10,13-tetrone, O-00036
1-Oxa-5-phospha(5-P^{V})spiro[4.5]decane; 5-Me, *in* O-00049
6-Oxa-5-phospha(5-P^{V})spiro[4.5]decane; 5-Me, *in* O-00050
1-Oxa-5-phospha(5-P^{V})spiro[4.4]nonane; 5-Me, *in* O-00051
1-Oxa-6-phospha(6-P^{V})spiro[5.5]undecane; 6-Me, *in* O-00052
2′,2′,3′,4′,4′-Pentamethylspiro[1,3,2-benzodioxaphosphole-2,1′-phosphetane]; *cis-form*, 2-Ph, *in* P-00027
2′,2′,3′,4′,4′-Pentamethylspiro[1,3,2-benzodioxaphosphole-2,1′-phosphetane]; *trans-form*, 2-Ph, *in* P-00027
2′,2′,3′,4′,4′-Pentamethylspiro[1,3,2-benzodioxaphosphole-2,1′-phosphetane]; *trans-form*, 2-Benzyl, *in* P-00027
1,1,2,3,3-Pentamethyl-6,6,7,7-tetrakis(trifluoromethyl)-5,8-dioxa-4-phospha(4-P^{V})spiro[3.4]octane; 4-Phenoxy, *in* P-00029
1,1,2,3,3-Pentamethyl-6,6,7,7-tetrakis(trifluoromethyl)-5,8-dioxa-4-phospha(4-P^{V})spiro[3.4]octane; 4-*tert*-Butyl, *in* P-00029
1,1,2,3,3-Pentamethyl-6,6,7,7-tetrakis(trifluoromethyl)-5,8-dioxa-4-phospha(4-P^{V})spiro[3.4]octane; 4-Ph (*cis*-), *in* P-00029
1,1,2,3,3-Pentamethyl-6,6,7,7-tetrakis(trifluoromethyl)-5,8-dioxa-4-phospha(4-P^{V})spiro[3.4]octane; 4-Ph (*trans*-), *in* P-00029
1,1,2,3,3-Pentamethyl-6,6,7,7-tetrakis(trifluoromethyl)-5,8-dioxa-4-phospha(4-P^{V})spiro[3.4]octane; 4-(4-Bromophenyl), *in* P-00029
1,1,2,3,3-Pentamethyl-6,6,7,7-tetrakis(trifluoromethyl)-5,8-dioxa-4-phospha(4-P^{V})spiro[3.4]octane; 4-Dimethylamino, *in* P-00029
1,1,2,3,3-Pentamethyl-6,6,8,8-tetrakis(trifluoromethyl)-5,7-dioxa-4-phospha(4-P^{V})spiro[3.4]octane; 4-Chloro, *in* P-00030
Pentaphenylphosphorane, P-00039
3,3,3,6,6-Pentaphenyl-3,3,3,6-tetrahydro-4,5-bis(methoxycarbonyl)-1,2,3-diazaphosphorine, P-00041
Phenoxytetraphenylphosphorane, P-00076
Phenylbisbiphenylenephosphorane, *in* S-00021
2-Phenyl-1-oxa-4-aza-5-phospha(5-P^{V})spiro[4.4]nonane; 5-Ph, *in* P-00173
2-Phenylspiro[1,3,2-benzothiazaphosphole-2(3*H*),2′-[1,3,2]-dioxaphospholane]; 3-(2,6-Dimethylphenyl), *in* P-00308
2-Phenylspiro[1,3,2-benzothiazaphosphole-2(3*H*),2′-[1,3,2]-dioxaphospholane]; 3-(2,4,6-Trimethylphenyl), *in* P-00308
2-Phenyl-2,2′-spirobi[1,3,2-benzodioxaphosphole], P-00309
Spiro[1,3,2-benzodioxaphosphole-2,2′(3′*H*)-[1,3,2]-benzoxazaphosphole], S-00008
Spiro[1,3,2-benzodioxaphosphole-2,2′-[1,3,2]-oxazaphospholidine], S-00009
Spiro[1,3,2-benzodioxaphosphole-2,1′-[3]phospholene]; 2-Methoxy, *in* S-00010
Spiro[1,3,2-benzodioxaphosphole-2,1′-[3]phospholene]; 2-Ph, *in* S-00010
Spiro[1,3,2-benzodioxaphosphole-2,1′-[3]phospholene]; 2-Fluoro, *in* S-00010
Spiro[1,3,2-benzodioxaphosphole-2,1′-[3]phospholene]; 2-Bromo, *in* S-00010
Spiro[1,3,2-benzoxathiaphosphole-2,2′-[1,3,2]-dioxaphospholane]; 2-Ph, *in* S-00011
Spiro[1,3,2-benzoxazaphosphole-2(3*H*),2′-[1,3,2]-dioxaphospholane], S-00012
Spiro[1,3,2-benzoxazaphosphole-2(3*H*),2′-[1,2]oxaphospholan]-5′-one; 2-Et, *in* S-00013
2,2′-Spirobi[1,3,2-benzodioxaphosphole], S-00014
2,2′-Spirobi[1,3,2-benzodithiaphosphole]; 2-Ph, *in* S-00015
1,1′(3*H*,3′*H*)-Spirobi[2,1-benzoxaphosphole]; 1-Ph, *in* S-00017
1,1′(3*H*,3′*H*)-Spirobi[2,1-benzoxaphosphole]-3,3′-dione; 1-Me, *in* S-00018
1,1′(3*H*,3′*H*)-Spirobi[2,1-benzoxaphosphole]-3,3′-dione; 1-Ph, *in* S-00018
2,2′-Spirobi[1,3,2-benzoxathiaphosphole]; 2-Ph, *in* S-00019
2,2(3*H*,3′*H*)-Spirobi[1,3,2-benzoxazaphosphole], S-00020
2,2(3*H*,3′*H*)-Spirobi[1,3,2-benzoxazaphosphole]; 2-Me, *in* S-00020
2,2(3*H*,3′*H*)-Spirobi[1,3,2-benzoxazaphosphole]; 2-Et, *in* S-00020
2,2(3*H*,3′*H*)-Spirobi[1,3,2-benzoxazaphosphole]; 2-Ph, *in* S-00020
5,5′-Spirobi[5*H*-dibenzophosphole], S-00021
Tetrabromomethylphosphorane, T-00010
Tetrabromophenylphosphorane, T-00011
Tetrachloromethylphosphorane, T-00032
Tetrachlorophenylphosphorane, T-00034
Tetrachloro(trichloromethyl)phosphorane, T-00037

Tetrafluoromethylphosphorane, T-00081
Tetrafluorophenylphosphorane, T-00084
2,2,2,3-Tetrahydro-1,2-benzoxaphosphole, T-00088
2,2,2,3-Tetrahydro-1,2-oxaphosphole; 2,2,2-Trimethoxy, *in* T-00119
2,2,2,3-Tetrahydro-1,2-oxaphosphole; 2,2,2-Triethoxy, *in* T-00119
1,1′,3,3′-Tetrahydro-1,1′,2,3,3′-pentamethyl-2,2′-spirobi[2*H*-1,3,2-benzodiazaphosphole], *in* T-00142
2,2,2,3-Tetrahydro-5-phenyl-3,3-bis(trifluoromethyl)-1,4,2-thiazaphosphole; 2,2,2-Trimethoxy, *in* T-00129
2,2,2,3-Tetrahydro-5-phenyl-3,3-bis(trifluoromethyl)-1,4,2-thiazaphosphole; 2,2,2-Triphenoxy, *in* T-00129
4,5,5,5-Tetrahydro-3,5,5,5-tetraphenyl-1,2,5-oxazaphosphole, T-00153
2,2,2,3-Tetrahydro-2,2,2-trimethoxy-5-methyl-1,2-oxaphosphole, T-00157
2,2,2,3-Tetrahydro-2,2,2-trimethoxy-5-phenyl-3,3-bis(trifluoromethyl)-1,4,2-oxazaphosphole, T-00158
2,2,2,3-Tetrahydro-2,2,2-trimethoxy-5-phenyl-3,3-bis(trifluoromethyl)-1-(2,4,6-trimethylphenyl)-1*H*-1,4,2-diazaphosphole, T-00159
Tetraiodomethylphosphorane, T-00163
Tetraiodophenylphosphorane, T-00165
2,3,7,8-Tetrakis(trifluoromethyl)-1,4,6,9-tetrathia-5-phospha(5-P^V)spiro[4.4]nona-2,7-diene; 5-Ph, *in* T-00189
1,1,3,3-Tetramethyl-6,7-bis(trifluoromethyl)-5,8-dithia-4-phospha(4-P^V)spiro[3.4]oct-6-ene; 4-Ph, *in* T-00193
2,2,7,7-Tetramethyl-1,6-dioxa-4,9-diaza-5-phospha(5-P^V)-spiro[4.4]nonane, T-00195
3,3,8,8-Tetramethyl-1,6-dioxa-4,9-diaza-5-phospha(P^V)-spiro[4.4]nonane, T-00196
3,3,8,8-Tetramethyl-1,6-dioxa-4,9-diaza-5-phospha(P^V)-spiro[4.4]nonane; (±)-*form*, 5-Me, *in* T-00196
N,*N*,5,6-Tetramethylspiro[1,3,2-benzodioxaphosphole-2,1′(3′*H*)-[2,1]benzoxaphosphol]-2-amine, *in* D-00915
4′,4′,5′,5′-Tetramethylspiro[1,3,2-benzodioxaphosphole-2,2′-[1,3,2]dioxaphospholane], T-00230
4′,4′,5′,5′-Tetramethylspiro[1,3,2-benzoxazaphosphole-2(3*H*),-2′-[1,3,2]dioxaphospholane], T-00231
3,3,3′,3′-Tetramethyl-1,1′(3*H*,3′*H*)-spirobi[2,1-benzoxaphosphole], T-00232
3,3,3′,3′-Tetramethyl-1,1′(3*H*,3′*H*)-spirobi[2,1-benzoxaphosphole]; 1-Ph, *in* T-00232
3,3,3′,3′-Tetramethyl-1,1′(3*H*,3′*H*)-spirobi[2,1-benzoxaphosphole]; 1-Benzyl, *in* T-00232
3,3,3′,3′-Tetramethyl-1,1′(3*H*,3′*H*)-spirobi[2,1-benzoxaphosphole]; 1-Chloro, *in* T-00232
1,3,5,7-Tetramethyl-1,3,5,7-tetraaza-4-phospha(4-P^V)-spiro[3.3]heptane-2,6-dione; 4-Ph, *in* T-00234
2,2,3,3-Tetramethyl-1,4,6,9-tetraoxa-5-phospha(5-P^V)-spiro[4.4]nonane, T-00235
3,3,9,9-Tetramethyl-1,5,7,11-tetraoxa-6-phospha(6-P^V)-spiro[5.5]undecane; 6-Ph, *in* T-00236
3,3,9,9-Tetramethyl-1,5,7,11-tetraoxa-6-phospha(6-P^V)-spiro[5.5]undecane; 6-Benzyl, *in* T-00236
2,2,3,3-Tetramethyl-1,4,6-trioxa-9-aza-5-phospha(5-P^V)-spiro[4.4]nonane, T-00238
1,4,6,12-Tetraoxa-5-phospha(5-P^V)spiro[4.7]dodecane; 5-Ph, *in* T-00243
1,4,6,9-Tetraoxa-5-phospha(5-P^V)spiro[4.4]nonane, T-00244
1,4,6,9-Tetraoxa-5-phospha(5-P^V)spiro[4.4]nonane; 5-Ph, *in* T-00244
1,4,6,11-Tetraoxa-5-phospha(5-P^V)spiro[4.6]undecane; 5-Ph, *in* T-00245
2,3,7,8-Tetraphenyl-1,4,6,9-tetraoxa-5-phospha(5-P^V)-spiro[4.4]nona-2,7-diene; 5-Ph, *in* T-00275
Tribromodiphenylphosphorane, T-00355
Trichlorobis(trichloromethyl)phosphorane, T-00399
Trichlorodimethylphosphorane, T-00402
Trichlorodiphenylphosphorane, T-00404
Trifluorobis(trifluoromethyl)phosphorane, T-00465
1,1,1-Trifluoro-1,1-dihydrophospholane, T-00466
1,1,1-Trifluoro-1,1-dihydrophosphorinane, *in* D-00573
Trifluorodimethylphosphorane, T-00469
Trifluorodiphenylphosphorane, T-00471
Triiododiphenylphosphorane, T-00497
2,3,3′-Trimethyl-2,2′(3*H*,3′*H*)-spirobi[1,3,2-benzoxazaphosphole], *in* D-00917

Methylenephosphoranes, $R_3P{=}CH_2$ and derivs.

1-Acenaphthenylidenetriphenylphosphorane, A-00002
[(Aminocarbonyl)methylene]triphenylphosphorane, *in* A-00061
[1,8-Anthracenediylbis[methylene]]bis[triphenylphosphorane], *in* B-00443
(1-Azulenylmethylene)triphenylphosphorane, *in* A-00156
(Benzo[*g*]chrysen-10-ylmethylene)triphenylphosphorane, *in* T-00635
(1,3-Benzodithiol-2-ylidene)triphenylphosphorane, *in* B-00021
(Benzo[*e*]phenanthren-2-ylmethylene)triphenylphosphorane, *in* T-00636
Benzoyl(triphenylphosphoranylidene)acetonitrile, *in* O-00083
5-Benzyl-5,5-dihydro-5-phenylacridophosphine, B-00042
Benzylidenetriphenylphosphorane, B-00050
Benzyl 2-(triphenylphosphoranylidene)propanoate, *in* T-00660
[[1,1′-Biphenyl]-2,2′-diylbis[methylene]]-bis[triphenylphosphorane], *in* B-00445
[[1,1′-Biphenyl]-3,3′-diylbis[methylene]]-bis[triphenylphosphorane], *in* B-00446
[[1,1′-Biphenyl]-4,4′-diylbis[methylene]]-bis[triphenylphosphorane], *in* B-00447
(1,8-Biphenylenediyldimethylidyne)bis[triphenylphosphorane], B-00076
Bis[bis(trimethylsilyl)methylene]-9*H*-fluoren-9-ylphosphorane, B-00106
Bis[bis(trimethylsilyl)methylene]phenylphosphorane, B-00107
[Bis(dimethylphosphino)methylene]trimethylphosphorane, B-00203
[Bis(diphenylarsino)methylene]triphenylphosphorane, B-00213
[Bis(diphenylphosphino)methylene]triphenylphosphorane, B-00252
[Bis(diphenylstibino)methylene]triphenylphosphorane, B-00270
P,*P*′-Bis(2,4,6-tri-*tert*-butylphenyl)-1,3-diphosphaallene, B-00394
Bis(trimethylphosphoranylidene)methane, B-00429
[Bis(trimethylsilyl)methylene]triphenylphosphorane, *in* B-00432
Bis(triphenylphosphoranylidene)methane, B-00459
1,3-Bis(triphenylphosphoranylidene)-2-propanone, *in* O-00073
1,2-Bis(triphosphoranylidene)benzocyclobutene, B-00460
(Bromomethylene)triphenylphosphorane, *in* B-00477
1-Bromo-1-(triphenylphosphoranylidene)-2-butanone, B-00505
1-Bromo-3-(triphenylphosphoranylidene)-2-propanone, B-00506
1,4-Butanediylidenebis[triphenylphosphorane], *in* B-00517
2-Butenylidenetriphenylphosphorane, B-00522
tert-Butyl cyano(triphenylphosphoranylidene)acetate, *in* C-00231
Butylidenetriphenylphosphorane, *in* B-00641
tert-Butyl α-methoxycarbonyltriphenyl-phosphoranylidenepropanoate, *in* T-00646
5-*tert*-Butyl-2,2,4,5-tetrahydro-2,2,4-triisopropoxy-1,2,4-oxadiphosphole, B-00637
N-*tert*-Butyl-*P*-(2,4,6-tri-*tert*-butylphenyl)-metaphosphonamidate, *in* B-00639
N-*tert*-Butyl-*P*-(2,4,6-tri-*tert*-butylphenyl)-metaphosphonamidoselenoate, *in* B-00639
N-*tert*-Butyl-*P*-(2,4,6-tri-*tert*-butylphenyl)-metaphosphonamidothioate, *in* B-00639
tert-Butyl 2-(triphenylphosphoranylidene)propanoate, *in* T-00660
(7-Carboxyheptylidene)triphenylphosphorane, *in* C-00012
(2-Carboxy-2-propenylidene)triphenylphosphorane, *in* C-00016
(Chloromethyl)triphenylphosphonium(1+); Ylide, *in* C-00117
[(1-Chloro-2-oxo-2-phenyl)ethylidene]triphenylphosphorane, C-00122
(2-Chlorophenylmethylene)triphenylphosphorane, *in* C-00135
(3-Chlorophenylmethylene)triphenylphosphorane, *in* C-00136
(4-Chlorophenylmethylene)triphenylphosphorane, *in* C-00137
Chloro(triphenylphosphoranylidene)acetic acid, C-00190
1-Chloro-3-(triphenylphosphoranylidene)-2-propanone, C-00191
(2-Cyanoethylidene)triphenylphosphorane, *in* C-00212
[(Cyanomethylene)triphenylphosphoranyl]-triphenylphosphonium, *in* C-00214

3-Cyano-2-oxo-3-(triphenylphosphoranylidene)propanoic acid, C-00218
Cyclobutylidenetriphenylphosphorane, *in* C-00235
(1,3,6-Cycloheptatrien-1-ylmethylene)triphenylphosphorane, C-00236
Cycloheptylidenetriphenylphosphorane, *in* C-00238
Cyclohexylidenetriphenylphosphorane, *in* C-00273
2,4-Cyclopentadien-1-ylidenetriphenylphosphorane, C-00274
Cyclopentylidenetriphenylphosphorane, *in* C-00282
Cyclopropylidenetriphenylphosphorane, C-00284
(Cyclopropylmethylene)triphenylphosphorane, *in* C-00286
1,3,4,5,5,6,7,8,9,9*a*-Decahydro-5-methylene-2*H*-phosphinolizine, D-00005
Decylidenetriphenylphosphorane, *in* D-00014
1,1-Dibromo-1,1-dihydro-2,4,6-triphenylphosphorin, *in* D-00613
(Dibromomethylene)triphenylphosphorane, *in* D-00065
Di-*tert*-butylphosphazene, D-00106
Di-*tert*-butyl (triphenylphosphoranylidene)butanedioate, *in* T-00646
1,1-Dichloro-1,1-dihydro-2,4,6-triphenylphosphorin, *in* D-00613
(Dichloromethylene)triphenylphosphorane, *in* D-00182
Dichloromethylenetris(dimethylamino)phosphorane, D-00173
(Dicyanomethylene)triphenylphosphorane, D-00205
(Diethoxyethenylidene)triphenylphosphorane, D-00232
(2,2-Diethoxyethylidene)triphenylphosphorane, D-00233
Diethyl [2-oxo-2-(triphenylphosphoranylidene)propyl]-phosphonate, *in* O-00086
Diethyl(tetrahydro-2-oxo-2-furanyl)phosphonate, *in* T-00121
Diethyl (triphenylphosphoranylidene)butanedioate, *in* T-00646
Diethyl (triphenylphosphoranylidene)propanedioate, *in* T-00658
1,1-Difluoro-1,1-dihydro-2,4,6-triphenylphosphorin, *in* D-00613
(Difluoromethylene)(trifluoromethyl)phosphine, D-00413
(Difluoromethylene)triphenylphosphorane, D-00414
1,1-Dihydro-1,1-bis(methylthio)-2,4,6-triphenylphosphorin, *in* D-00613
1,1-Dihydro-1,1-bis(phenylthio)-2,4,6-triphenylphosphorin, *in* D-00613
1,1-Dihydro-1,1-dimethoxy-2,4,6-triphenylphosphorin, *in* D-00613
1,1-Dihydro-1,1-diphenoxy-2,4,6-triphenylphosphorin, *in* D-00613
1,1-Dihydro-1,1-diphenyl-3*H*-phosphindole, D-00483
1,1-Dihydro-1,1-diphenylphosphorin, D-00484
1,1-Dihydro-1,1,3,3,5,5-hexamethyl-1-phospha-3,5-disilacyclohex-1-ene, D-00485
1,1-Dihydro-1,1,3,3,4,4-hexamethyl-1-phospha-3,4-disilacyclopent-1-ene, D-00486
[(9,10-Dihydro-9-phenanthrenenyl)methylene]-triphenylphosphorane, *in* D-00540
[(5,6-Dihydro-4,4,6-trimethyl-4*H*-1,3-oxazin-2-yl)methylene]-triphenylphosphorane, *in* D-00599
2,2-Dihydro-2,2,2-triphenyl-4,4-bis(trifluoromethyl)-3-(triphenylphosphoranylidene)-1,2-oxaphosphetane, D-00608
Dihydro(triphenylphosphoranylidene)-2,5-furandione, D-00612
1,1-Dihydro-2,4,6-triphenylphosphorin, D-00613
(4-Dimethylaminobutylidene)triphenylphosphorane, *in* D-00690
(2-Dimethylaminoethylidene)triphenylphosphorane, *in* D-00698
[3-(Dimethylamino)propylidene]triphenylphosphorane, *in* D-00707
(3,7-Dimethyl-2,6-octadienylidene)triphenylphosphorane, *in* D-00777
(3,7-Dimethyl-2,6-octadienyl)triphenylphosphonium(1+); (*Z*)-*form*, Ylide, *in* D-00777
(2,7-Dimethyl-2,4,6-octatrien-1,8-diylidene)-bis[triphenylphosphorane], *in* D-00778
(2,2-Dimethylpropylidene)triphenylphosphorane, *in* D-00914
Dimethyl 2-(tributylphosphoranylidene)-1,3-dithiol-4,5-dicarboxylate; Conj. tetrafluoroborate, *in* D-00927
[3,7-Dimethyl-9-(2,6,6-trimethyl-1-cyclohexen-1-yl)-2,4,6,8-nonatetraenyl]triphenylphosphonium(1+); (*E,E,E,E*)-*form*, Ylide, *in* D-00929
Dimethyl (triphenylphosphoranylidene)butanedioate, *in* T-00646
3,3-Dimethyl-1-triphenylphosphoranylidene-2-butanone, D-00934
Dimethyl (triphenylphosphoranylidene)propanedioate, *in* T-00658
[2-(1,3-Dioxan-2-yl)ethylidene]triphenylphosphorane, *in* D-00965
[(1,3-Dioxolan-2-yl)methylene]triphenylphosphorane, *in* D-00976
1,4-Diphenyl-2,3-bis(triphenylphosphoranylidene)-1,4-butanedione, D-00997
(2,4-Diphenyl-2,4-cyclopentadien-1-ylidene)-triphenylphosphorane, D-00999
(Diphenylmethylene)triphenylphosphorane, D-01022
(Diphenylphosphinylmethylene)triphenylphosphorane, D-01099
1,4-Diphenyl-2-(triphenylphosphoranylidene)-1,4-butanedione, D-01162
Diphenyl [(triphenylphosphoranylidene)methyl]phosphonate, D-01163
1,3-Diphenyl-2-(triphenylphosphoranylidene)-1,3-propanedione, D-01166
1,3-Dithian-2-ylidenetriphenylphosphorane, *in* D-01242
(1,3-Dithian-2-ylmethylene)triphenylphosphorane, *in* D-01241
(1,3-Dithiolan-2-ylmethylene)triphenylphosphorane, *in* D-01252
1,3-Dithiol-2-ylidenetriphenylphosphorane, *in* D-01253
Dodecylidenetriphenylphosphorane, *in* D-01266
α-Ethoxycarbonyltriphenylphosphoranylidenepropanoic acid, *in* T-00646
(2-Ethoxy-2-propenylidene)triphenylphosphorane, E-00049
Ethyl bromo(triphenylphosphoranylidene)acetate, *in* B-00504
Ethyl chloro(triphenylphosphoranylidene)acetate, *in* C-00190
Ethyl cyano(triphenylphosphoranylidene)pyruvate, *in* C-00218
Ethyl 3-(ethylthio)-3-thioxo-2-(triphenylphosphoranylidene)-propanoate, *in* E-00194
Ethylidenetriphenylphosphorane, E-00076
Ethyl β-oxo-α-(triphenylphosphoranylidene)-benzenepropanoate, *in* O-00083
Ethyl triphenylphosphoranylideneacetate, E-00198
Ethyl α-(triphenylphosphoranylidene)hydrocinnamate, *in* T-00645
Ethyl 2-triphenylphosphoranylidenepropanoate, E-00199
(9*H*-Fluoren-9-ylideneethenylidene)triphenylphosphorane, F-00009
9*H*-Fluoren-9-ylidenetriphenylphosphorane, F-00010
(Fluoromethylene)triphenylphosphorane, *in* F-00025
[(4-Fluorophenyl)methylene]triphenylphosphorane, *in* F-00032
[2,5-Furandiylbis[methylene]]bis[triphenylphosphorane], *in* B-00448
[3,4-Furandiylbis[methylene]]bis[triphenylphosphorane], *in* B-00449
Heptylidenetriphenylphosphorane, *in* H-00011
(Hexafluorocyclobutylidene)triphenylphosphorane, H-00017
N,N,N′,N′,N″,N″-Hexamethyl-1-methyl-enephosphoranetriamide, H-00072
Hexylidenetriphenylphosphorane, *in* H-00104
[(2-Hydroxyphenyl)methyl]triphenylphosphorane, *in* H-00177
1-Indenylidenetriphenylphosphorane, *in* I-00008
[(1-Iodo-2-oxo-2-phenyl)ethylidene]triphenylphosphorane, I-00014
β-Ionylidenetriphenylphosphorane, *in* M-00422
[[2-(Methoxycarbonyl)phenyl]methylene]triphenylphosphorane, *in* M-00029
[[4-(Methoxycarbonyl)phenyl]methylene]triphenylphosphorane, *in* M-00031
[[3-(3-Methoxycarbonyl)phenyl]methylene]-triphenylphosphorane, *in* M-00030
α-Methoxycarbonyltriphenylphosphoranylidenepropanoic acid, *in* T-00646
Methoxymethylenetriphenylphosphorane, *in* M-00053
[(2-Methoxyphenyl)methylene]triphenylphosphorane, *in* M-00062
[(3-Methoxyphenyl)methylene]triphenylphosphorane, *in* M-00063
[(4-Methoxyphenyl)methylene]triphenylphosphorane, *in* M-00064
(3-Methoxy-2-propenylidene)triphenylphosphorane, *in* M-00084
Methyl bromo(triphenylphosphoranylidene)acetate, *in* B-00504
3-Methyl-2-butenylidenetriphenylphosphorane, M-00098
3-Methylbutylidenetriphenylphosphorane, *in* M-00111
Methyl chloro(triphenylphosphoranylidene)acetate, *in* C-00190

Methyl cyano(triphenylphosphoranylidene)acetate, *in* C-00231
Methyl cyano(triphenylphosphoranylidene)pyruvate, *in* C-00218
[3-(2-Methyl-1,3-dioxolan-2-yl)propylidene]-triphenylphosphorane, *in* M-00132
[Methylenebis(2,1-phenylenemethylene)]-bis[triphenylphosphorane], *in* M-00141
Methylenetriphenylphosphorane, M-00148
(1-Methylethylidene)triphenylphosphorane, M-00153
1-Methylisophosphinoline, M-00162
Methyl γ-oxo-β-(triphenylphosphoranylidene)-benzenebutanoate, *in* O-00082
Methyl 3-oxo-2-(triphenylphosphoranylidene)butanoate, *in* O-00084
Methyl 4-oxo-5-(triphenylphosphoranylidene)pentanoate, *in* O-00085
[(2-Methylphenyl)methylene]triphenylphosphorane, *in* M-00209
[(3-Methylphenyl)methylene]triphenylphosphorane, *in* M-00210
[(4-Methylphenyl)methylene]triphenylphosphorane, *in* M-00211
Methylphenyl(phenylmethylene)propylphosphorane, *in* B-00052
(2-Methyl-2-propenylidene)triphenylphosphorane, M-00363
(1-Methylpropylidene)triphenylphosphorane, *in* M-00386
(2-Methylpropylidene)triphenylphosphorane, M-00371
[(Methylsulfonyl)methylene]triphenylphosphorane, M-00399
[(Methylthio)methylene]triphenylphosphorane, *in* M-00416
[3-Methyl-5-(2,6,6-trimethyl-1-cyclohexen-1-yl)-2,4-pentadienylidene]triphenylphosphorane, *in* M-00421
Methyl triphenylphosphoranylideneacetate, M-00433
Methyl α-(triphenylphosphoranylidene)benzeneacetate, *in* T-00644
Methyl α-(triphenylphosphoranylidene)cyclopropaneacetate, *in* C-00285
3-Methyl-5-(triphenylphosphoranylidene)-2(5*H*)-furanone, *in* M-00186
Methyl α-(triphenylphosphoranylidene)hydrocinnamate, *in* T-00645
[1,5-Naphthalenediylbis[methylene]]-bis[triphenylphosphorane], *in* B-00451
[1,8-Naphthalenediylbis[methylene]]-bis[triphenylphosphorane], *in* B-00452
[(2,3-Naphthalenediylbis[methylene]]-bis[triphenylphosphorane], *in* B-00453
[2,7-Naphthalenediylbis[methylene]]-bis[triphenylphosphorane], *in* B-00455
1-Naphthalenylmethylenetriphenylphosphorane, *in* N-00015
2-Naphthalenylmethylenetriphenylphosphorane, *in* N-00016
(2-Nitro-2,4-cyclopentadienylidene)triphenylphosphorane, N-00034
[(2-Nitrophenyl)methylene]triphenylphosphorane, *in* N-00047
[(3-Nitrophenyl)methylene]triphenylphosphorane, *in* N-00048
[(4-Nitrophenyl)methylene]triphenylphosphorane, N-00041
2-Nonenylidenetriphenylphosphorane, N-00071
3-Nonenylidenetriphenylphosphorane, *in* N-00072
4-Nonenylidenetriphenylphosphorane, *in* N-00073
4-Nonenyltriphenylphosphonium(1+); (*Z*)-*form*, Ylide, *in* N-00073
Nonylidenetriphenylphosphorane, *in* N-00078
Octylidenetriphenylphosphorane, *in* O-00047
γ-Oxo-β-(triphenylphosphoranylidene)benzenebutanoic acid, O-00082
β-Oxo-α-(triphenylphosphoranylidene)benzenepropanoic acid, O-00083
4-Oxo-5-(triphenylphosphoranylidene)pentanoic acid, O-00085
[Oxybis[methylene]]bis[triphenylphosphonium](2+); Bis-ylide, *in* O-00097
1,5-Pentanediylidenebis[triphenylphosphorane], *in* P-00034
4-Pentenylidenetriphenylphosphorane, *in* P-00044
2-Penten-4-ynylidenetriphenylphosphorane, *in* P-00047
Pentylidenetriphenylphosphorane, *in* P-00056
(3-Phenanthrenylmethylene)triphenylphosphorane, *in* T-00637
(9-Phenanthrenylmethylene)triphenylphosphorane, *in* T-00638
Phenylbis[3-(triphenylphosphoranylidene)propenyl]phosphine, P-00103
[1,2-Phenylenebis[methylene]]bis[triphenylphosphorane], P-00135
[1,3-Phenylenebis[methylene]]bis[triphenylphosphorane], *in* P-00133
[1,4-Phenylenebis(methylene)]bis[triphenylphosphorane], P-00136
(1,2-Phenylenedi-2-propene-3,1-diylidene)-bis[triphenylphosphorane], *in* P-00149
[Phenyl(trimethylsilyl)methylene]phosphinous chloride, P-00330
▷1-Phenyl-2-(triphenylphosphoranylidene)ethanone, P-00333
1-Phenyl-1-(triphenylphosphoranylidene)-2-propanone, P-00334
1-Phenyl-2-(triphenylphosphoranylidene)-1-propanone, P-00335
1-Phenyl-3-(triphenylphosphoranylidene)-2-propanone, P-00336
Phenyl 2-(triphenylphosphoranylidene)propanote, *in* T-00660
Phosphanthridene; 5,5-Dihydro, 5,5-di-Me, *in* P-00348
5-Phospha(5-P^V)spiro[4.5]deca-1,3,5,7,9-pentaene, P-00349
6-Phospha(6-P^V)spiro[5.5]undeca-1(6)-ene, P-00351
1,3-Propanediylidenebis[triphenylphosphorane], *in* P-00448
2,2,4,4-Tetrahydro-1*H*-2,4-benzodiphosphepin, T-00087
2,2,7,7-Tetrahydro-3,4,5,6-tetrakis(methoxycarbonyl)-1*H*-1,2,7-azadiphosphepine; 2,2-Di-Me, 7,7-di-Ph, *in* T-00138
2,2,7,7-Tetrahydro-3,4,5,6-tetrakis(methoxycarbonyl)-1*H*-1,2,7-azadiphosphepine; 2,2,7,7-Tetra-Ph, *in* T-00138
2,2,7,7-Tetrahydro-3,4,5,6-tetrakis(methoxycarbonyl)-1,2,2,5,5-pentamethyl-1*H*-1,2,7-azadiphosphepine, *in* T-00138
2,2,4,4-Tetrahydro-2,2,4,4-tetramethyl-1*H*-2,4-benzodiphosphepin, *in* T-00087
2,2,4,4-Tetrahydro-2,2,4,4-tetraphenyl-1*H*-2,4-benzodiphosphepin, *in* T-00087
1,1,3,3-Tetrahydro-1,1,3,3-tetraphenyl-2,4-bis(triphenylphosphonio)-1,3-diphosphete; Dichloride, *in* T-00145
1,1,3,3-Tetrahydro-1,1,3,3-tetraphenyl-1,3-diphosphacyclohepta-1,2-diene, T-00148
1,1,3,3-Tetrahydro-1,1,3,3-tetraphenyl-1,3-diphosphacyclohexa-1,2-diene, T-00149
1,1,3,3-Tetramethyl-1-phospha-3-silacyclohex-1-ene, T-00212
(Tetraphenyl-2,4-cyclopentadien-1-ylidene)-triphenylphosphorane, T-00252
[Thiobis[methylene]]bis[triphenylphosphorane], *in* T-00312
[2,5-Thiophenediylbis[methylene]]bis[triphenylphosphorane], *in* B-00456
[3,4-Thiophenediylbis[methylene]]bis[triphenylphosphorane], *in* B-00457
Tri-*tert*-butylmethylenephosphorane, T-00365
1-(Tributylphosphoranylidene)-2-heptanone, T-00381
Trimethylmethylenephosphorane, T-00518
[4-(Trimethylsilyl)-3-butyn-1-ylidene]triphenylphosphorane, *in* T-00691
[[2-(Trimethylsilyl)ethoxy]methylene]triphenylphosphorane, *in* T-00552
[2-(Trimethylsilyl)ethylidene]triphenylphosphorane, *in* T-00553
[2-(Trimethylstannyl)ethylidene]triphenylphosphorane, *in* T-00561
Trimethyl(trimethylgermyl)methylenephosphorane, T-00564
Trimethyl(trimethylsilylmethylene)phosphorane, T-00566
Triphenyl[bis(trifluoromethyl)methylene]phosphorane, T-00597
6,8,10-Triphenyl-1,4-diaza-5-phospha(5-P^V)spiro[4.5]deca-5,7,9-triene, T-00599
6,8,10-Triphenyl-1,4-dithia-5-phospha(5-P^V)spiro[4.5]deca-5,7,9-triene, T-00604
7,9,11-Triphenyl-1,5-dithia-6-phospha(6-P^V)spiro[5.5]-undeca-6,8,10-triene, T-00605
6,8,10-Triphenyl-1-oxa-4-aza-5-phosphaspiro(5-P^V)spiro[4.5]-deca-5,7,9-triene; *N*-Me, *in* T-00609
Triphenyl(1-phenylethylidene)phosphorane, *in* T-00612
Triphenyl(2-phenylethylidene)phosphorane, *in* T-00613
Triphenyl(3-phenyl-2-propenylidene)phosphorane, T-00615
Triphenyl(3-phenylpropylidene)phosphorane, *in* T-00617
Triphenyl[(phenylthio)methylene]phosphorane, T-00620
Triphenylphosphoranylideneacetaldehyde, T-00639
(Triphenylphosphoranylidene)acetic acid, T-00640
(Triphenylphosphoranylidene)acetic acid; *tert*-Butyl ester, *in* T-00640
(Triphenylphosphoranylidene)acetic acid; Benzyl ester, *in* T-00640
Triphenylphosphoranylideneacetonitrile, T-00641
α-(Triphenylphosphoranylidene)benzeneacetaldehyde, T-00643
4-(Triphenylphosphoranylidene)butanoic acid, *in* C-00017

4-(Triphenylphosphoranylidene)-2-butenoic acid, T-00647
4-(Triphenylphosphoranylidene)-2-butenoic acid; (*E*)-*form*, *in* T-00647
4-Triphenylphosphoranylidene-2,5-cyclohexadien-1-one, T-00649
(Triphenylphosphoranylidene)ethenethione, T-00650
(Triphenylphosphoranylidene)ethenone, T-00651
N-[(Triphenylphosphoranylidene)ethylidene]aniline, T-00652
1-(Triphenylphosphoranylidene)-2-heptanone, *in* O-00062
6-(Triphenylphosphoranylidene)hexanoic acid, *in* C-00015
[2-[2-(Triphenylphosphoranylidenemethyl)phenyl]-ethylidenetriphenylphosphorane, *in* T-00696
3-(Triphenylphosphoranylidene)-2,4-pentanedione, T-00654
5-(Triphenylphosphoranylidene)pentanoic acid, T-00655
5-(Triphenylphosphoranylidene)pentanoic acid; Et ester, *in* T-00655
2-(Triphenylphosphoranylidene)propanal, T-00657
2-(Triphenylphosphoranylidene)propanenitrile, T-00659
2-(Triphenylphosphoranylidene)propanoic acid, T-00660
3-(Triphenylphosphoranylidene)propanoic acid, T-00661
▷1-Triphenylphosphoranylidene-2-propanone, T-00662
3-(Triphenylphosphoranylidene)-2,5-pyrrolidenedione, T-00663
4-Triphenylphosphoranylidene(tetrahydro-2*H*-pyran), *in* T-00684
4-Triphenylphosphoranylidene(tetrahydro-2*H*-thiopyran), *in* T-00685
Triphenylpropadienylidenephosphorane, T-00678
Triphenyl-2-propenylidenephosphorane, *in* T-00679
Triphenylpropylphosphonium(1+); Ylide, *in* T-00680
Triphenyl(propynylidene)phosphorane, *in* T-00681
Triphenyltetradecylidenephosphorane, *in* T-00682
Triphenyl[3-[(tetrahydro-2*H*-pyran-2-yl)oxy]propylidene]-phosphorane, *in* T-00683
Triphenyl[(2,6,6-trimethyl-1,3-cyclohexadien-1-yl)-methylene]phosphorane, *in* T-00688
Triphenyl(3,7,11-trimethyldodecylidene)phosphorane, *in* T-00690
Triphenyl[(trimethylsilyl)methylene]phosphorane, T-00692
Triphenyl[(triphenylphosphoranylidene)methyl]-phosphonium(1+); Bromide, *in* T-00697
Triphenyl[(triphenylphosphoranylidene)methyl]-phosphonium(1+); Chloride, *in* T-00697
Triphenyl[(triphenylphosphoranylidene)methyl]-phosphonium(1+); Iodide, *in* T-00697
Triphenyl[(triphenylphosphoranylidene)methyl]-phosphonium(1+); Tetraphenylborate, *in* T-00697
Triphenyl(2,5-undecadienylidene)phosphorane, T-00700
Triphenylundecylidenephosphorane, *in* T-00701
Tris[(triphenylphosphoranylidene)methyl]phosphine, T-00871
Tris[(triphenylphosphoranylidene)methyl]phosphine; Oxide, *in* T-00871
Tris[(triphenylphosphoranylidene)methyl]phosphine; Sulfide, *in* T-00871

Methylenephosphines, R—P=CR_2

[Bis[bis(trimethylsilyl)methylene]]chlorophosphorane, B-00105
1,4-Bis(2,4,6-tri-*tert*-butylphenyl)-1,4-bis(trimethylsilyloxy)-2,3-diphosphabutadiene, B-00393
1,4-Bis(2,4,6-tri-*tert*-butylphenyl)-3-trimethylsilyloxy-1,2,4-triphospha-1,3-butadiene, B-00398
2-[Bis(trimethylsilyl)methylene]-1,1-di-*tert*-butyldiphosphine, B-00431
1-(Di-*tert*-butyldiphosphinylidene)methanediamine; *N,N,N′,N′*-Tetra-Me, *in* D-00090
1-(Di-*tert*-butyldiphosphinylidene)methanediamine; *N,N,N′,N′*-Tetra-Et, *in* D-00090
1,2-Dimethyl-4,5-diphenyl-1*H*-1,3-azaphosphole, D-00736
2-Phenyl-1*H*-1,3-azaphosphole, P-00088
1,1,4,4-Tetrakis(diethylamino)-2,3-diphosphabutadiene, *in* T-00005
1,1,4,4-Tetrakis(dimethylamino)-2,3-diphosphabutadiene, *in* T-00005
1-(2,4,6-Tri-*tert*-butylphenyl)-4,4-diphenyl-1-phospha-1,2,3-butatriene, T-00367
1-Trimethylsilyl-2-trimethylsilyloxy-2-(2,4,6-tri-*tert*-butylphenyl)methylenephosphine, T-00560

Phosphinidynes, RC≡P

1-Adamantylmethylidenephosphine, A-00033
(2,2-Dimethylpropylidyne)phosphine, D-00910
Ethylidynephosphine, E-00077
(Phenylmethylidyne)phosphine, P-00170
Phosphinidynemethylamine, P-00356
(2,4,6-Tri-*tert*-butylphenyl)methylidynephosphine, T-00368

Hexacoordinate compounds, R_6P^-

Tris[[1,1′-biphenyl]-2,2′-diyl]phosphate(1−), T-00731

Phosphorous acids, $(OH)_3P$

Bis(2-chlorophenyl) phosphonate, B-00143
Bis(4-chlorophenyl) phosphonate, B-00144
Bis(2,2-dimethylpropyl) phosphonate, B-00208
Bis(3-methylbutyl) phosphonate, B-00303
Bis(2-methylpropyl) phosphonate, B-00354
Bis(2,2,2-trifluoroethyl) phosphonate, B-00409
2,3-Butylene phosphonate, *in* D-00733
▷Dibenzyl phosphonate, D-00059
N,N-Dibutylphosphonamidic acid, D-00126
▷Dibutyl phosphonate, D-00127
Di-*tert*-butyl phosphonate, D-00128
Dicyclohexyl phosphonate, D-00215
Didodecyl phosphonate, D-00227
▷Dihexadecyl phosphonate, D-00430
▷Diisopropyl phosphonate, D-00651
Dioctadecyl phosphonate, D-00954
Diphenyl phosphonate, D-01105
▷Di-2-propenyl phosphonate, D-01194
Dipropyl phosphonate, D-01210
Ditetradecyl phosphonate, D-01239
▷2-Hydroxy-1,3,2-dioxaphospholane, *in* D-00970
2-Hydroxy-4-methyl-1,3,2-dioxaphospholane, *in* M-00127
4-Methyl-1,3,2-dioxaphosphorinane 2-oxide; (2*RS*,4*RS*)-*form*, *in* M-00130
4-Methyl-1,3,2-dioxaphosphorinane 2-oxide; (2*RS*,4*SR*)-*form*, *in* M-00130
Monobenzyl phosphonate, M-00444
▷Monobutyl phosphonate, M-00453
Monoisopropyl phosphonate, M-00478
▷Monomethyl phosphonate, M-00490
Mono-1-naphthyl phosphonate, M-00496
Neopentylene phosphonate, *in* D-00735
o-Phenylene phosphite, *in* B-00018
N,N,N′,N′-Tetrabutylphosphonic diamide, T-00018
Tetramethylethylene phosphonate, *in* T-00197
Trimethylene phosphonate, *in* D-00972

Phosphorous anhydrides

Acetyl diethyl phosphite, A-00013
Benzoyl diethyl phosphite, B-00028
2,3-Dihydro-3-phenyl-4*H*-1,3,2-benzoxazaphosphorin-4-one; 2-Acetoxy, *in* D-00556
2,2′-Oxybis[1,3,2-benzodioxaphosphole], O-00087
2,2′-Oxybis[4,5-dimethyl-1,3,2-dioxaphosphole], O-00088
▷2,2′-Oxybis[1,3,2-dioxaphospholane], O-00091
Tetrabutyl pyrophosphite, T-00025
Tetraethyl pyrophosphite, T-00070
Tetrapropyl pyrophosphite, T-00292
Tribenzoyl phosphite, T-00336

Phosphorous halides

3-Benzyl-2-chlorotetrahydro-2*H*-1,3,2-oxazaphosphorine, B-00041
Benzyl phosphorodichloridite, B-00070
Bis(2-chlorophenyl) phosphorochloridite, B-00145
Bis(3-chlorophenyl) phosphorochloridite, B-00146
Bis(4-chlorophenyl) phosphorochloridite, B-00147
Bis(2,2-dimethylpropyl) phosphorochloridite, B-00209
Bis(1-methylheptyl) phosphorochloridite; (*S,S*)-*form*, *in* B-00308

Bis(2-methylphenyl) phosphorochloridite, B-00330
Bis(3-methylphenyl) phosphorochloridite, B-00331
Bis(4-methylphenyl) phosphorochloridite, B-00332
Bis(2,2,2-trichloroethyl) phosphorochloridite, B-00401
Bis(trimethylsilyl)phosphoramidous acid; Difluoride, *in* B-00438
Bis(trimethylsilyl)phosphoramidous acid; Dichloride, *in* B-00438
2-Bromo-1,3,2-benzodioxaphosphole, B-00464
2-Bromo-4,5-dimethyl-1,3,2-dioxaphospholane, B-00470
2-Bromo-5,5-dimethyl-1,3,2-dioxaphosphorinane, B-00471
2-Bromo-4-methyl-1,3,2-dioxaphospholane, B-00473
5-*tert*-Butyl-2-chloro-1,3,2-dioxaphosphorinane, B-00547
5-*tert*-Butyl-2-chlorotetrahydro-3-phenyl-2*H*-1,3,2-oxazaphosphorine, B-00549
Butyl dibutylphosphoramidochloridite, *in* D-00135
N,N'-1,3-Butylene-*N,N'*-dimethylphosphorodiamidous chloride, *in* H-00057
tert-Butylphosphoramidous acid; Difluoride, *in* B-00624
tert-Butylphosphoramidous acid; Dichloride, *in* B-00624
Butyl phosphorodibromidite, B-00627
Butyl phosphorodichloridite, B-00629
Butyl phosphorodifluoridite, B-00633
2-Chloro-1,3,2-benzodioxaphosphole, C-00021
2-Chloro-4*H*-1,3,2-benzodioxaphosphorinan-4-one, C-00022
2-Chloro-3-(2-chloroethyl)tetrahydro-2*H*-1,3,2-oxazaphosphorine, C-00030
6-Chlorodibenzo[*d,f*][1,3,2]dioxaphosphepin, C-00032
2-Chloro-2,3-dihydro-3,5-dimethyl-1,3,4,2-oxadiazaphosphole, C-00039
2-Chloro-2,3-dihydro-3,5-diphenyl-1,3,4,2-oxadiazaphosphole, C-00042
2-Chloro-2,3-dihydro-5-methyl-3-phenyl-1,3,4,2-oxadiazaphosphole, C-00044
2-Chloro-2,3-dihydro-3-phenyl-4*H*-1,3,2-benzoxazaphosphorin-4-one, C-00050
2-Chloro-1,3-dimethyl-1,3,2-diazaphospholidine, C-00055
2-Chloro-4,5-dimethyl-1,3,2-dioxaphospholane; (4*R*,5*R*)-*form*, *in* C-00056
2-Chloro-4,5-dimethyl-1,3,2-dioxaphospholane; (4*RS*,5*RS*)-*form*, *in* C-00056
2-Chloro-4,5-dimethyl-1,3,2-dioxaphospholane; (4*RS*,5*SR*)-*form*, *in* C-00056
2-Chloro-4,5-dimethyl-1,3,2-dioxaphosphole, C-00057
2-Chloro-5,5-dimethyl-1,3,2-dioxaphosphorinane, C-00058
2-Chloro-3,4-dimethyl-5-phenyl-1,3,2-oxazaphospholidine; (2*R*,4*S*,5*R*)-*form*, *in* C-00061
2-Chloro-1,3-dioxa-2-phosphacyclotridecane, *in* D-00966
2-Chloro-1,3,2-dioxaphospholane, C-00063
2-Chloro-1,3,2-dioxaphosphonane, *in* D-00971
▷2-Chloro-1,3,2-dioxaphosphorinane, C-00066
2-Chloro-4,5-diphenyl-1,3,2-dioxaphosphole, C-00070
2-Chlorohexahydro-4*H*-1,3,2-benzodioxaphosphorin, C-00085
2-Chloro-4-isopropyl-5,5-dimethyl-1,3,2-dioxaphosphorinane; (2*RS*,4*RS*)-*form*, *in* C-00090
2-Chloro-4-methyl-1,3,2-dioxaphospholane, C-00092
2-Chloro-4-methyl-1,3,2-dioxaphosphorinane, C-00093
2-Chloro-3-methyl-1,3,2-oxazaphospholidine, C-00102
2-Chloronaphtho[1,2-*d*]-1,3,2-dioxaphosphole, C-00118
2-Chloronaphtho[2,3-*d*]-1,3,2-dioxaphosphole, C-00119
2-Chloronaphtho[1,8-*de*]-1,3,2-dioxaphosphorin, C-00120
▷2-Chloro-5-phenyl-1,3,4,2-dioxazaphosphole, C-00126
2-Chloro-3-phenyl-1,3,2-oxazaphospholidine, C-00138
2-Chlorotetrahydro-3-methyl-2*H*-1,3,2-oxazaphosphorine, C-00178
2-Chloro-4,4,5,5-tetrakis(trifluoromethyl)-1,3,2-dioxaphospholane, C-00184
2-Chloro-4,4,5,5-tetramethyl-1,3,2-dioxaphospholane, C-00185
2-Chloro-4,4,6-trimethyl-1,3,2-dioxaphosphorinane, C-00189
2-Cyanoethyl diisopropylphosphoramidochloridite, *in* D-00654
2-Cyanoethyl dimethylphosphoramidochloridite, *in* D-00862
Cyclohexyl phosphorodichloridite, C-00272
3,5-Di-*tert*-butyl-2-chloro-2,3-dihydro-1,3,2-oxazaphosphole, D-00078
1,3-Dibutyl-2,3-dihydro-4,5-dimethyl-1*H*-1,3,2-diazaphosphole 2-oxide; 2-Chloro, *in* D-00081
Di-*tert*-butylphosphoramidous acid; Difluoride, *in* D-00137
Di-*tert*-butylphosphoramidous acid; Dichloride, *in* D-00137
Di-*tert*-butylphosphoramidous acid; Dibromide, *in* D-00137
Dibutyl phosphorobromidite, D-00139
Dibutyl phosphorochloridite, D-00142
Di-*tert*-butyl phosphorochloridite, D-00143
3,9-Dichloro-2,4,8,10-tetraoxa-3,9-diphosphaspiro[5.5]-undecane, D-00196
Diethylphosphoramidous acid; Difluoride, *in* D-00360
Diethylphosphoramidous acid; Dibromide, *in* D-00360
Diethylphosphoramidous acid; Diiodide, *in* D-00360
Diethylphosphoramidous dichloride, D-00361
Diethyl phosphoriodidite, D-00365
Diethyl phosphorobromidite, D-00372
▷Diethyl phosphorochloridite, D-00374
▷Dihexyl phosphorochloridite, D-00436
2,3-Dihydro-1,3-dimethyl-1*H*-1,3,2-benzodiazaphosphole; 2-Chloro, *in* D-00456
1,2-Dihydro-1-methyl-4*H*-3,1,2-benzoxazaphosphorin-4-one; 1-Chloro, *in* D-00523
Diisopropylphosphoramidous acid; Difluoride, *in* D-00656
Diisopropylphosphoramidous acid; Dichloride, *in* D-00656
Diisopropyl phosphorochloridite, D-00658
Dimethyl 2-chloro-1,3,2-dioxaphospholan-4,5-dicarboxylate, *in* C-00062
▷Dimethyl chlorophosphite, D-00715
6,9-Dimethyl-1,3-dioxa-6,9-diaza-2-phosphacycloundecane; 2-Chloro, *in* D-00728
Dimethylphosphoramidous acid; Dibromide, *in* D-00875
Dimethylphosphoramidous acid; Diiodide, *in* D-00875
▷Dimethylphosphoramidous dichloride, D-00876
Dimethylphosphoramidous difluoride, D-00877
2,2-Dimethylpropyl phosphorodichloridite, D-00913
N,N'-Dimethyl-*N,N'*-trimethylene phosphorodiamidous chloride, *in* H-00022
Di-1-naphthyl phosphorochloridite, D-00942
Di-2-naphthyl phosphorochloridite, D-00943
Dipentyl phosphorochloridite, D-00983
Diphenylphosphoramidous acid; Difluoride, *in* D-01111
Diphenylphosphoramidous acid; Dichloride, *in* D-01111
Diphenylphosphoramidous acid; Dibromide, *in* D-01111
Diphenylphosphoramidous acid; Diiodide, *in* D-01111
Diphenyl phosphoriodidite, D-01115
Diphenyl phosphorobromidite, D-01118
Diphenyl phosphorochloridite, D-01121
Dipropylphosphoramidous acid; Dichloride, *in* D-01215
Dipropylphosphoramidous acid; Diiodide, *in* D-01215
Dipropyl phosphorochloridite, D-01218
Ethyl dibutylphosphoramidochloridite, *in* D-00135
Ethyl diethylphosphoramidochloridite, E-00060
Ethyl dimethylphosphoramidochloridite, *in* D-00862
Ethylimidodiphosphorous tetrachloride, E-00080
Ethyl 4-morpholinylphosphonochloridite, *in* M-00526
Ethylphosphoramidous acid; Difluoride, *in* E-00160
Ethylphosphoramidous acid; Dichloride, *in* E-00160
Ethyl phosphorodibromidite, E-00166
▷Ethyl phosphorodichloridite, E-00168
Ethyl phosphorodifluoridite, E-00174
2-Fluoro-1,3,2-benzodioxaphosphole, F-00013
2-Fluoro-1,3-dimethyl-1,3,2-diazaphospholidine, F-00014
2-Fluoro-4,5-dimethyl-1,3,2-dioxaphospholane, F-00015
2-Fluoro-5,5-dimethyl-1,3,2-dioxaphosphorinane, F-00016
2-Fluoro-1,3-dioxa-2-phosphacyclotridecane, *in* D-00966
2-Fluoro-1,3,2-dioxaphospholane, F-00017
2-Fluoro-1,3,2-dioxaphosphonane, *in* D-00971
2-Fluoro-4-methyl-1,3,2-dioxaphospholane, F-00020
2-Fluoro-4-methyl-1,3,2-dioxaphosphorinane, F-00021
Heptyl phosphorodichloridite, H-00010
Hexyl phosphorodichloridite, H-00103
Isopropyl phosphorodichloridite, I-00061
3-Methylbutyl phosphorodichloridite, M-00108
Methyl diisopropylphosphoramidochloridite, *in* D-00654
Methyl dimethylphosphoramidochloridite, *in* D-00862
Methylimidodi(phosphorous acid); Tetraiodide, *in* M-00157
Methylimidodiphosphorous tetrachloride, M-00158
Methylimidodiphosphorous tetrafluoride, M-00159
Methyl 4-morpholinylphosphonochloridite, *in* M-00526

2-Methylphenyl phosphorodichloridite, M-00258
3-Methylphenyl phosphorodichloridite, M-00259
4-Methylphenyl phosphorodichloridite, M-00260
Methylphosphoramidous acid; Difluoride, *in* M-00335
Methyl phosphorodibromidite, M-00344
▷Methyl phosphorodichloridite, M-00346
Methyl phosphorodifluoridite, M-00353
2-Methylpropyl phosphorodichloridite, M-00385
Methyl 1-pyrrolidinylphosphonochloridite, *in* P-00519
4-Morpholinylphosphonous acid; Dichloride, *in* M-00529
1-Naphthyl phosphorodichloridite, N-00030
2-Naphthyl phosphorodichloridite, N-00031
Octyl phosphorodichloridite, O-00046
Pentyl phosphorodichloridite, P-00054
Phenyl dimethylphosphoramidochloridite, *in* D-00862
1,3-Phenylenebis[phosphorodichloridite], P-00141
1,4-Phenylenebis[phosphorodichloridite], P-00142
Phenyl phosphorodibromidite, P-00277
▷Phenyl phosphorodichloridite, P-00280
Phenyl phosphorodifluoridite, P-00286
Phosphorocyanidous bromide iodide, P-00409
Phosphorocyanidous dibromide, P-00410
1-Piperidinylphosphonous acid; Difluoride, *in* P-00432
1-Piperidinylphosphonous acid; Dichloride, *in* P-00432
2-Propenyl phosphorodichloridite, P-00465
Propylphosphoramidous acid; Dichloride, *in* P-00486
Propyl phosphorodichloridite, P-00488
Propyl phosphorodifluoridite, P-00492
1-Pyrrolidinylphosphonous acid; Difluoride, *in* P-00520
1-Pyrrolidinylphosphonous acid; Dichloride, *in* P-00520
Tetraethylphosphorodiamidous acid; Bromide, *in* T-00066
Tetraethylphosphorodiamidous acid; Iodide, *in* T-00066
▷Tetraethylphosphorodiamidous chloride, T-00067
Tetramethylene phosphorochloridite, *in* D-00968
Tetramethylene phosphorofluoridite, *in* D-00968
Tetramethylphosphorodiamidous acid; Fluoride, *in* T-00227
Tetramethylphosphorodiamidous acid; Bromide, *in* T-00227
▷Tetramethylphosphorodiamidous chloride, T-00228
Tetraphenylphosphorodiamidous acid; Chloride, *in* T-00274
2,2,2-Trichloroethyl phosphorodichloridite, T-00406
1,3,6,2-Trioxaphosphocane; 2-Chloro, *in* T-00592

Phosphorous pseudohalides

2-Cyano-5,5-dimethyl-1,3,2-dioxaphosphorinane, C-00209
Dimethylphosphoramidous acid; Dicyanide, *in* D-00875
Phosphorocyanidous bromide iodide, P-00409
Phosphorocyanidous dibromide, P-00410
Tetramethylphosphorodiamidous acid; Cyanide, *in* T-00227

Phosphorous esters

Acetyl diethyl phosphite, A-00013
Benzoyl diethyl phosphite, B-00028
2,2′-Biphenylene phosphonate, *in* D-00038
Bis[(4-bromophenyl)methyl] phosphonate, B-00110
Bis(4-*tert*-butylphenyl) phosphonate, B-00116
▷Bis(2-chloroethyl) phosphonate, B-00120
Bis(2-chlorophenyl) phosphonate, B-00143
Bis(2,2-dimethylpropyl) phosphonate, B-00208
Bis(2,2-dimethylpropyl) phosphorochloridite, B-00209
▷Bis(2-ethylhexyl) phosphonate, B-00273
Bis(3-methylbutyl) phosphonate, B-00303
Bis(1-methylheptyl) phosphonate, B-00307
Bis(1-methylheptyl) phosphorochloridite; (*S,S*)-*form*, *in* B-00308
Bis(2-methylphenyl) phosphonate, B-00326
Bis(3-methylphenyl) phosphonate, B-00327
Bis(4-methylphenyl) phosphonate, B-00328
Bis(2-methylpropyl) phosphonate, B-00354
Bis[(4-nitrophenyl)methyl]phosphonate, B-00368
Bis(4-nitrophenyl) phosphonate, B-00375
Bis(triethylsilyl) phosphonate, B-00407
Bis(trimethylsilyl) phosphonate, B-00435
2-*tert*-Butoxy-4,5-dimethyl-1,3,2-dioxaphospholane, B-00533
2-*tert*-Butoxy-5,5-dimethyl-1,3,2-dioxaphosphorinane, B-00534
2-*tert*-Butoxy-1,3,2-dioxaphospholane, B-00535
tert-Butyl ethyl phosphonate, B-00561
5-*tert*-Butyl-2-methoxy-1,3,2-dioxaphosphorinane, B-00563
Butyl phenyl phosphonate, B-00585
▷4-*tert*-Butyl-2,6,7-trioxa-1-phosphabicyclo[2.2.2]octane, B-00640
▷Dibenzyl phosphonate, D-00059
4,8-Di-*tert*-butyl-2,10-dimethyl-12*H*-dibenzo[*d,g*][1,3,2]-dioxaphosphocin; 6-Ethoxy, *in* D-00088
4,8-Di-*tert*-butyl-2,10-dimethyl-12*H*-dibenzo[*d,g*][1,3,2]-dioxaphosphocin; 6-Phenoxy, *in* D-00088
4,8-Di-*tert*-butyl-2,10-dimethyl-12*H*-dibenzo[*d,g*][1,3,2]-dioxaphosphocin; 6-(1-Naphthoxy), *in* D-00088
2,5-Diethoxy-1,3,2,5-dioxadiphosphorinane; 5-Oxide, *in* D-00231
Diethyl diethylphosphoramidite, *in* D-00360
▷Diethyl phosphonate, D-00338
Diethylphosphoramidous acid; Difluoride, *in* D-00360
Diethylphosphoramidous acid; Dibromide, *in* D-00360
Diethylphosphoramidous acid; Diiodide, *in* D-00360
▷Diethyl phosphorochloridite, D-00374
Diethyl trimethylsilyl phosphite, D-00408
Diheptyl phosphonate, D-00426
4,7-Dihydro-1,3,2-dioxaphosphepin; 2-Phenoxy, *in* D-00472
6,9-Dimethyl-1,3-dioxa-6,9-diaza-2-phosphacycloundecane; 2-Methoxy, *in* D-00728
6,9-Dimethyl-1,3-dioxa-6,9-diaza-2-phosphacycloundecane; 2-Ethoxy, *in* D-00728
4,5-Dimethyl-2-phenoxy-1,3,2-dioxaphospholane, D-00789
5,5-Dimethyl-2-phenoxy-1,3,2-dioxaphosphorinane, D-00790
▷Dimethyl phosphonate, D-00851
Dinonyl phosphonate, D-00950
Dioctyl phosphonate, D-00959
3,9-Diphenoxy-2,4,8,10-tetraoxa-3,9-diphosphaspiro[5.5]-undecane, D-00991
3,9-Diphenoxy-2,4,8,10-tetraoxa-3,9-diphosphaspiro[5.5]-undecane; Monooxide, *in* D-00991
Diphenyl diethylphosphoramidite, *in* D-00360
Diphenyl phosphoriodidite, D-01115
2-Ethoxy-1,3,2-benzodioxaphosphole, E-00024
2-Ethoxy-4,5-dimethyl-1,3,2-dioxaphospholane, E-00031
2-Ethoxy-4,5-dimethyl-1,3,2-dioxaphospholane; (4*R*,5*R*)-*form*, *in* E-00031
2-Ethoxy-5,5-dimethyl-1,3,2-dioxaphosphorinane, E-00032
2-Ethoxy-1,3,2-dioxaphospholan-4,5-dicarboxylic acid; (4*R*,5*R*)-*form*, Di-Et ester, *in* E-00033
2-Ethoxy-1,3,2-dioxaphospholan-4,5-dicarboxylic acid; (4*RS*,5*RS*)-*form*, Di-Et ester, *in* E-00033
2-Ethoxy-1,3,2-dioxaphospholan-4,5-dicarboxylic acid; (4*RS*,5*SR*)-*form*, Di-Et ester, *in* E-00033
▷2-Ethoxy-1,3,2-dioxaphospholane, E-00034
2-Ethoxy-1,3,2-dioxaphosphorinane, E-00037
2-Ethoxy-4-methyl-1,3,2-dioxaphospholane, E-00040
2-Ethoxy-4-methyl-1,3,2-dioxaphosphorinane, E-00041
2-Ethoxynaphtho[1,2-*d*]-1,3,2-dioxaphosphole, E-00044
2-Ethoxynaphtho[1,8-*de*]-1,3,2-dioxaphosphorin, E-00045
2-Ethoxynaphtho[1,8-*de*]-1,3,2-dioxaphosphorin, E-00046
2-Ethoxy-4,4,5,5-tetramethyl-1,3,2-dioxaphospholane, E-00052
Ethyl diethylphosphoramidochloridite, E-00060
Ethyl phosphorodibromidite, E-00166
▷Ethyl phosphorodichloridite, E-00168
▷4-Ethyl-2,6,7-trioxa-1-phosphabicyclo[2.2.2]octane, E-00196
4-Isopropyl-2-methoxy-5,5-dimethyl-1,3,2-dioxaphosphorinane; (2*RS*,4*RS*)-*form*, *in* I-00033
4-Isopropyl-2-methoxy-5,5-dimethyl-1,3,2-dioxaphosphorinane; (2*RS*,4*SR*)-*form*, *in* I-00033
Isopropyl phosphorodichloridite, I-00061
2-Methoxy-1,3,2-benzodioxaphosphole, M-00026
6-Methoxydibenzo[*d,f*][1,3,2]dioxaphosphepin, M-00032
2-Methoxy-4,5-dimethyl-1,3,2-dioxaphospholane; (4*R*,5*R*)-*form*, *in* M-00033
2-Methoxy-4,5-dimethyl-1,3,2-dioxaphospholane; (4*RS*,5*RS*)-*form*, *in* M-00033
2-Methoxy-5,5-dimethyl-1,3,2-dioxaphosphorinane, M-00034
▷2-Methoxy-1,3,2-dioxaphospholane, M-00036
▷2-Methoxy-1,3,2-dioxaphosphorinane, M-00038
2-Methoxy-4,5-diphenyl-1,3,2-dioxaphospholane, M-00039
2-Methoxy-4-methyl-1,3,2-dioxaphospholane, M-00043

2-Methoxy-4-methyl-1,3,2-dioxaphosphorinane; (2*RS*,4*RS*)-*form*, *in* M-00044
2-Methoxy-4-methyl-1,3,2-dioxaphosphorinane; (2*RS*,4*SR*)-*form*, *in* M-00044
2-Methoxy-4,4,5,5-tetramethyl-1,3,2-dioxaphospholane, M-00085
2-(1-Methylethoxy)-1,3,2-dioxaphospholane, M-00152
6-Methyl-4-methylene-4*H*-1,3,2-dioxaphosphorin; 2-Ethoxy, *in* M-00167
4-Methyl-2-phenoxy-1,3,2-dioxaphosphorinane, M-00191
▷Methyl phosphorodichloridite, M-00346
Methyl phosphorodifluoridite, M-00353
2-(2-Methylpropoxy)-1,3,2-dioxaphospholane, M-00368
Methyl tetramethylene phosphite, *in* D-00968
4-Methyl-2,6,7-trioxa-1-phosphabicyclo[2.2.1]heptane, M-00423
▷4-Methyl-2,6,7-trioxa-1-phosphabicyclo[2.2.2]octane, M-00424
Monobenzyl phosphonate, M-00444
Monoethyl phosphonate, M-00473
▷Monomethyl phosphonate, M-00490
Mono-1-naphthyl phosphonate, M-00496
Monophenyl phosphonate, M-00508
Mono-2-propenyl phosphonate, M-00510
Monopropyl phosphonate, M-00513
2-Phenoxy-1,3,2-benzodioxaphosphole, P-00063
6-Phenoxy-6*H*-dibenzo[*c,e*][1,2]oxaphosphorin, P-00065
2-Phenoxy-1,5-dihydro-2,4,3-benzodioxaphosphepin, P-00066
2-Phenoxy-1,3,2-dioxaphospholane, P-00067
2-Phenoxy-1,3,2-dioxaphosphorinane, P-00068
2-Phenoxy-4,5-diphenyl-1,3,2-dioxaphospholane, P-00069
2-Phenoxynaphtho[1,8-*de*]-1,3,2-dioxaphosphorine, P-00070
2-Phenoxy-4,4,5,5-tetramethyl-1,3,2-dioxaphospholane, P-00075
2-Propenyl phosphorodichloridite, P-00465
Propyl phosphorodichloridite, P-00488
Tetramethylene phenyl phosphite, *in* D-00968
Tribenzyl phosphite, T-00340
Tri([1,1-biphenyl]-2-yl) phosphite, T-00351
Tri-2-butenyl phosphite, T-00361
▷Tributyl phosphite, T-00379
Tri-*tert*-butyl phosphite, T-00380
Tricyclohexyl phosphite, T-00428
Tridecyl phosphite, T-00434
2,2,2-Triethoxy-2,2-dihydro-1,3,2-dioxaphosphorinane, T-00437
▷Triethyl phosphite, T-00445
Triheptyl phosphite, T-00490
Trihexyl phosphite, T-00493
▷Triisopropyl phosphite, T-00502
4,4,6-Trimethyl-2-phenoxy-1,3,2-dioxaphosphorinane, T-00520
▷Trimethyl phosphite, T-00535
2-(Trimethylsilyloxy)-1,3,2-dioxaphospholane, T-00554
Tri-1-naphthyl phosphite, T-00574
Tri-2-naphthyl phosphite, T-00575
Trinonyl phosphite, T-00579
Trioctadecyl phosphite, T-00580
▷Trioctyl phosphite, T-00583
2,6,7-Trioxa-1,4-diphosphabicyclo[2.2.2]octane, T-00585
▷2,6,7-Trioxa-1,4-diphosphabicyclo[2.2.2]octane; 4-Oxide, *in* T-00585
2,6,7-Trioxa-1-phosphabicyclo[2.2.1]heptane, T-00587
2,6,7-Trioxa-1-phosphabicyclo[2.2.2]octane, T-00588
2,7,8-Trioxa-1-phosphabicyclo[3.2.1]octane, T-00589
2,6,7-Trioxa-1-phosphabicyclo[2.2.2]octane-4-methanol, T-00590
2,8,9-Trioxa-1-phosphatricyclo[3.3.1.1^{3,7}]decane, T-00591
1,3,6,2-Trioxaphosphocane; 2-Methoxy, *in* T-00592
Tripentyl phosphite, T-00595
▷Triphenyl phosphite, T-00633
▷Tri-2-propenyl phosphite, T-00708
▷Tripropyl phosphite, T-00712
Tris(2-aminophenyl) phosphite, T-00728
Tris(3-aminophenyl) phosphite, T-00729
Tris(4-*tert*-butylphenyl) phosphite, T-00736
Tris(2-chloroethyl) phosphite, T-00742
Tris(2-chlorophenyl) phosphite, T-00750
Tris(4-chlorophenyl) phosphite, T-00751
Tris(4-cyanophenyl) phosphite, T-00755
Tris(4-dimethylaminophenyl) phosphite, T-00766
Tris(2,2-dimethylpropyl) phosphite, T-00772
▷Tris(2-ethylhexyl) phosphite, T-00780
Tris(2-methoxyphenyl) phosphite, T-00794
Tris(1-methylheptyl) phosphite, T-00797
▷Tris(6-methylheptyl) phosphite, T-00798
▷Tris(2-methylphenyl) phosphite, T-00809
▷Tris(3-methylphenyl) phosphite, T-00810
▷Tris(4-methylphenyl) phosphite, T-00811
Tris(1-methylpropyl) phosphite, T-00825
Tris(2-methylpropyl) phosphite, T-00826
Tris(2-nitrophenyl) phosphite, T-00839
Tris(4-nitrophenyl) phosphite, T-00840
Tris(2-phenylethyl) phosphite, T-00845
Tris(triethylsilyl) phosphite, T-00850
Tris(trimethylsilyl) phosphite, T-00866

Phosphorous amides

P-1-Aziridinyl-*N,N,N′,N′*-tetraethylphosphonous amide, *in* A-00155
P-1-Aziridinyl-*N,N,N′,N′*-tetramethylphosphonous diamide, *in* A-00155
3-Benzyl-2-chlorotetrahydro-2*H*-1,3,2-oxazaphosphorine, B-00041
Benzyl tetramethylphosphorodiamidite, *in* T-00227
P,P-Bis(1-aziridinyl)-*N,N*-diethylphosphinous amide, *in* B-00096
P,P-Bis(1-aziridinyl)-*N,N*-dimethylphosphinous amide, *in* B-00096
Bis(4-nitrophenyl) phosphonate, B-00375
Bis(trimethylsilyl)phosphoramidous acid; Difluoride, *in* B-00438
Bis(trimethylsilyl)phosphoramidous acid; Dichloride, *in* B-00438
2-*tert*-Butylamino-4-methyl-1,3,2-dioxaphosphorinane, B-00541
5-*tert*-Butyl-2-chlorotetrahydro-3-phenyl-2*H*-1,3,2-oxazaphosphorine, B-00549
Butyl dibutylphosphoramidochloridite, *in* D-00135
5-*tert*-Butyl-2-dimethylamino-1,3,2-dioxaphosphorinane, B-00555
tert-Butyl *N,N′*-dimethyl-*N,N′*-trimethylene phosphorodiamidite, *in* H-00022
N,N′-1,3-Butylene-*N,N′*-dimethylphosphorodiamidous chloride, *in* H-00057
tert-Butylphosphoramidous acid; Difluoride, *in* B-00624
tert-Butylphosphoramidous acid; Dichloride, *in* B-00624
5-*tert*-Butyltetrahydro-2-dimethylamino-3-phenyl-2*H*-1,3,2-oxazaphosphorine, B-00636
tert-Butyl tetramethylphosphorodiamidite, *in* T-00227
2-Chloro-3-(2-chloroethyl)tetrahydro-2*H*-1,3,2-oxazaphosphorine, C-00030
2-Chloro-2,3-dihydro-3-phenyl-4*H*-1,3,2-benzoxazaphosphorin-4-one, C-00050
2-Chloro-1,3-dimethyl-1,3,2-diazaphospholidine, C-00055
2-Chloro-3,4-dimethyl-5-phenyl-1,3,2-oxazaphospholidine; (2*R*,4*S*,5*R*)-*form*, *in* C-00061
2-Chloro-3-methyl-1,3,2-oxazaphospholidine, C-00102
2-Chloro-3-phenyl-1,3,2-oxazaphospholidine, C-00138
2-Chlorotetrahydro-3-methyl-2*H*-1,3,2-oxazaphosphorine, C-00178
2-Cyanoethyl diisopropylphosphoramidochloridite, *in* D-00654
2-Cyanoethyl dimethylphosphoramidochloridite, *in* D-00862
3,5-Di-*tert*-butyl-2-chloro-2,3-dihydro-1,3,2-oxazaphosphole, D-00078
Dibutyl dibutylphosphoramidite, *in* D-00136
1,3-Dibutyl-2,3-dihydro-4,5-dimethyl-1*H*-1,3,2-diazaphosphole 2-oxide; 2-Chloro, *in* D-00081
1,3-Dibutyl-2,3-dihydro-4,5-dimethyl-1*H*-1,3,2-diazaphosphole 2-oxide; 2-Ethoxy, *in* D-00081
3,5-Di-*tert*-butyl-2,3-dihydro-1,3,2-oxazaphosphole; 2-Methoxy, *in* D-00083
3,5-Di-*tert*-butyl-2,3-dihydro-1,3,2-oxazaphosphole; 2-Ethoxy, *in* D-00083
Di-*tert*-butyl dimethylphosphoramidite, *in* D-00875

Dibutyl 4-morpholinylphosphonite, *in* M-00529
N,*N*-Dibutylphosphonamidic acid, D-00126
Di-*tert*-butylphosphoramidous acid; Difluoride, *in* D-00137
Di-*tert*-butylphosphoramidous acid; Dichloride, *in* D-00137
Di-*tert*-butylphosphoramidous acid; Dibromide, *in* D-00137
2-Diethylamino-1,3,2-dioxaphospholan-4,5-dicarboxylic acid; (4*R*,5*R*)-*form*, Di-Me ester, *in* D-00265
2-Diethylamino-1,3,2-dioxaphospholan-4,5-dicarboxylic acid; (4*R*,5*R*)-*form*, Di-Et ester, *in* D-00265
2-Diethylamino-1,3,2-dioxaphospholan-4,5-dicarboxylic acid; (4*R*,5*R*)-*form*, Diisopropyl ester, *in* D-00265
2-Diethylamino-1,3,2-dioxaphospholan-4,5-dicarboxylic acid; (4*RS*,5*RS*)-*form*, Di-Me ester, *in* D-00265
2-Diethylamino-1,3,2-dioxaphospholan-4,5-dicarboxylic acid; (4*RS*,5*RS*)-*form*, Di-Et ester, *in* D-00265
2-Diethylamino-1,3,2-dioxaphospholan-4,5-dicarboxylic acid; (4*RS*,5*RS*)-*form*, Diisopropyl ester, *in* D-00265
2-Diethylaminonaphtho[1,2-*d*]-1,3,2-dioxaphosphole, D-00266
2-Diethylaminonaphtho[1,8-*de*]-1,3,2-dioxaphosphorin, D-00267
Diethyl 1-aziridinylphosphonite, *in* A-00154
Diethyl bis(trimethylsilyl)phosphoramidite, *in* B-00438
Diethyl *tert*-butylphosphoramidite, *in* B-00624
Diethyl dibutylphosphoramidite, *in* D-00136
Diethyl diethylphosphoramidite, *in* D-00360
Diethyl methylphosphoramidite, *in* M-00335
Diethyl 4-morpholinylphosphonite, *in* M-00529
Diethyl phenylphosphoramidite, *in* P-00263
Diethylphosphoramidous dichloride, D-00361
Diethyl 1-piperidinylphosphonite, *in* P-00432
Diethyl propylphosphoramidite, *in* P-00486
Diethyl 1-pyrrolidinylphosphonite, *in* P-00520
Diethyl (1*H*-1,2,4-triazol-1-yl)phosphonite, *in* T-00332
1,5-Dihydro-3-dimethylamino-2,4,3-benzodioxaphosphepin, D-00453
2,3-Dihydro-1,3-dimethyl-1*H*-1,3,2-benzodiazaphosphole; 2-Chloro, *in* D-00456
2,3-Dihydro-1,3-dimethyl-1*H*-1,3,2-benzodiazaphosphole; 2-Diisopropylamino, *in* D-00456
4,7-Dihydro-1,3,2-dioxaphosphepin; 2-Dimethylamino, *in* D-00472
1,2-Dihydro-1-methyl-4*H*-3,1,2-benzoxazaphosphorin-4-one; 1-Methoxy, *in* D-00523
1,2-Dihydro-1-methyl-4*H*-3,1,2-benzoxazaphosphorin-4-one; 1-Ethoxy, *in* D-00523
1,2-Dihydro-1-methyl-4*H*-3,1,2-benzoxazaphosphorin-4-one; 1-Isopropoxy, *in* D-00523
1,2-Dihydro-1-methyl-4*H*-3,1,2-benzoxazaphosphorin-4-one; 1-Chloro, *in* D-00523
2,3-Dihydro-5-methyl-3-phenyl-2-phenylamino-1,3,4,2-oxadiazaphosphole, D-00527
2,3-Dihydro-3-phenyl-4*H*-1,3,2-benzoxazaphosphorin-4-one; 2-Methoxy, *in* D-00556
2,3-Dihydro-3-phenyl-4*H*-1,3,2-benzoxazaphosphorin-4-one; 2-Ethoxy, *in* D-00556
2,3-Dihydro-3-phenyl-4*H*-1,3,2-benzoxazaphosphorin-4-one; 2-Phenoxy, *in* D-00556
Diisopropyl 1-aziridinylphosphonite, *in* A-00154
Diisopropyl methylphosphoramidite, *in* M-00335
Diisopropylphosphoramidous acid; Difluoride, *in* D-00656
Diisopropylphosphoramidous acid; Dichloride, *in* D-00656
Diisopropyl 1-piperidinylphosphonite, *in* P-00432
Diisopropyl 1-pyrrolidinylphosphonite, *in* P-00520
2-Dimethylamino-1,3,2-benzodioxaphosphole, D-00689
2-Dimethylamino-5,5-dimethyl-1,3,2-dioxaphosphorinane, D-00693
2-Dimethylamino-3,4-dimethyl-5-phenyl-1,3,2-oxazaphospholidine; (2*R*,4*S*,5*R*)-*form*, *in* D-00694
2-Dimethylamino-3,4-dimethyl-5-phenyl-1,3,2-oxazaphospholidine; (2*S*,4*R*,5*S*)-*form*, *in* D-00694
2-Dimethylamino-1,3,2-dioxaphospholane, D-00695
2-Dimethylamino-1,3,2-dioxaphosphorinane, D-00696
2-Dimethylamino-4-isopropyl-5,5-dimethyl-1,3,2-dioxaphosphorinane, D-00699
2-Dimethylamino-4-phenyl-1,3,2-dioxaphosphorinane, D-00704
2-Dimethylamino-4,4,5,5-tetramethyl-1,3,2-dioxaphospholane, D-00710
1,3-Dimethyl-1,3,2-diazaphosphetidin-4-one; 2-Methoxy, *in* D-00717
1,3-Dimethyl-1,3,2-diazaphosphetidin-4-one; 2-Dimethylamino, *in* D-00717
1,3-Dimethyl-2-dimethylamino-1,3,2-diazaphospholidine, D-00724
4,5-Dimethyl-2-dimethylamino-1,3,2-dioxaphospholane, D-00725
Dimethyl dimethylphosphoramidite, *in* D-00875
Dimethyl methylphosphoramidite, *in* M-00335
1,3-Dimethyl-2-phenoxy-1,3,2-diazaphospholidine, D-00788
5,5-Dimethyl-2-phenylamino-1,3,2-dioxaphosphorinane, D-00792
Dimethyl phosphoramidite, D-00859
Dimethylphosphoramidous acid; Dibromide, *in* D-00875
Dimethylphosphoramidous acid; Diiodide, *in* D-00875
Dimethylphosphoramidous acid; Dicyanide, *in* D-00875
▷Dimethylphosphoramidous dichloride, D-00876
Dimethylphosphoramidous difluoride, D-00877
Diphenyl 1-aziridinylphosphonite, *in* A-00154
Diphenyl diethylphosphoramidite, *in* D-00360
Diphenyl diisopropylphosphoramidite, *in* D-00656
Diphenyl dimethylphosphoramidite, *in* D-00875
Diphenyl diphenylphosphoramidite, *in* D-01111
Diphenyl dipropylphosphoramidite, *in* D-01215
Diphenyl 4-morpholinylphosphonite, *in* M-00529
Diphenyl phosphoramidite, D-01110
Diphenylphosphoramidous acid; Difluoride, *in* D-01111
Diphenylphosphoramidous acid; Dichloride, *in* D-01111
Diphenylphosphoramidous acid; Dibromide, *in* D-01111
Diphenylphosphoramidous acid; Diiodide, *in* D-01111
Dipropylphosphoramidous acid; Dichloride, *in* D-01215
Dipropylphosphoramidous acid; Diiodide, *in* D-01215
2-Ethoxy-2,3-dihydro-1*H*-1,3,2-benzodiazaphosphole, E-00026
2-Ethoxy-2,3-dihydro-1,3,2-benzoxazaphosphole, E-00027
2-Ethoxy-1,3-dimethyl-1,3,2-diazaphospholidine, E-00030
2-Ethoxy-3-phenyl-1,3,2-oxazaphospholidine, E-00048
2-Ethoxytetrahydro-2*H*-1,3,2-oxazaphosphorine, E-00051
Ethyl bis(1-aziridinyl)phosphinite, *in* B-00095
Ethyl dibutylphosphoramidochloridite, *in* D-00135
Ethyl diethylphosphoramidochloridite, E-00060
Ethyl dimethylphosphoramidochloridite, *in* D-00862
Ethyl *N*,*N'*-dimethyl-*N*,*N'*-trimethylene phosphorodiamidite, *in* H-00022
Ethyl di-4-morpholinylphosphinite, *in* D-00936
Ethylimidodiphosphorous tetrachloride, E-00080
Ethyl 4-morpholinylphosphonochloridite, *in* M-00526
N''''-Ethyl-*N*,*N*,*N'*,*N'*,*N''*,*N''*,*N'''*,*N'''*-octamethylimidodiphosphorous tetraamide, *in* E-00079
Ethyl phenyl 1-aziridinylphosphonite, *in* A-00154
Ethylphosphoramidous acid; Difluoride, *in* E-00160
Ethylphosphoramidous acid; Dichloride, *in* E-00160
Ethyl phosphorodiamidite, E-00165
Ethyl tetraethylphosphorodiamidite, *in* T-00066
Ethyl tetramethylphosphorodiamidite, *in* T-00227
4-Ethyl-2,6,7-trimethyl-2,6,7-triaza-1-phosphabicyclo[2.2.2]octane, E-00195
2-Fluoro-1,3-dimethyl-1,3,2-diazaphospholidine, F-00014
Hexabenzylphosphorous triamide, H-00012
Hexabutylphosphorous triamide, H-00013
Hexaethylphosphorous triamide, H-00015
Hexahydro-3*H*,6*H*-2*a*,5*a*,8*a*-triaza-8*b*-phosphaacenaphthylene, H-00054
Hexahydro-2*a*,4*a*,6*a*-triaza-6*b*-phosphacyclopenta[*cd*]-pentalene, H-00055
Hexahydro-1,3,4-trimethyl-1,3,2-diazaphosphorine; 2-Diethylamino, *in* H-00057
2,3,5,6,7,8-Hexamethyl-2,3,5,6,7,8-hexaaza-1,4-diphosphabicyclo[2.2.2]octane, H-00070
▷Hexamethylphosphorous triamide, H-00080
Hexaphenylphosphorous triamide, H-00090
Hexapropylphosphorous triamide, H-00092
Isopropyl bis(1-aziridinyl)phosphinite, *in* B-00095
Isopropyl tetraethylphosphorodiamidite, *in* T-00066

Isopropyl tetramethylphosphorodiamidite, *in* T-00227
2-Methoxy-3,4-dimethyl-5-phenyl-1,3,2-oxazaphospholidine; (2*R*,4*S*,5*R*)-*form*, *in* M-00035
2-Methoxy-3,4-dimethyl-5-phenyl-1,3,2-oxazaphospholidine; (2*S*,4*S*,5*R*)-*form*, *in* M-00035
2-Methoxy-3,4-dimethyl-5-phenyl-1,3,2-oxazaphospholidine; (2*R*,4*R*,5*S*)-*form*, *in* M-00035
2-Methoxy-3,4-dimethyl-5-phenyl-1,3,2-oxazaphospholidine; (2*S*,4*R*,5*S*)-*form*, *in* M-00035
1-Methoxy-1,3-dioxo-2,4-dimethyl-1-phospha-2,4-diazacyclopentane, *in* D-00722
Methyl bis(1-aziridinyl)phosphinite, *in* B-00095
Methyl *N*,*N*′-1,3-butylene-*N*,*N*′-dimethylphosphorodiamidite, *in* H-00057
Methyl diisopropylphosphoramidochloridite, *in* D-00654
4-Methyl-2-dimethylamino-1,3,2-dioxaphospholane, M-00116
4-Methyl-2-dimethylamino-1,3,2-dioxaphosphorinane, M-00117
Methyl dimethylphosphoramidochloridite, *in* D-00862
Methyl *N*,*N*′-dimethyl-*N*,*N*′-trimethylene phosphorodiamidite, *in* H-00022
Methyl di-4-morpholinylphosphinite, *in* D-00936
Methyl *N*,*N*′-ethanediyl-*N*,*N*′-dimethylphosphorodiamidite, *in* D-00719
Methylimidodi(phosphorous acid); Tetraiodide, *in* M-00157
Methylimidodiphosphorous tetrachloride, M-00158
Methylimidodiphosphorous tetrafluoride, M-00159
6-Methyl-4-methylene-4*H*-1,3,2-dioxaphosphorin; 2-Diethylamino, *in* M-00167
Methyl 4-morpholinylphosphonochloridite, *in* M-00526
3-Methyl-2-phenoxy-1,3,2-oxazaphospholidine, M-00192
4-Methyl-2-phenylamino-1,3,2-dioxaphosphorinane; (2*RS*,4*RS*)-*form*, *in* M-00193
4-Methyl-2-phenylamino-1,3,2-dioxaphosphorinane; (2*RS*,4*SR*)-*form*, *in* M-00193
Methyl phenyl 1-aziridinylphosphonite, *in* A-00154
Methylphosphoramidous acid; Difluoride, *in* M-00335
Methyl phosphorodiamidite, M-00342
Methyl 1-pyrrolidinylphosphonochloridite, *in* P-00519
Methyl tetraisopropylphosphorodiamidite, *in* T-00172
Methyl tetramethylphosphorodiamidite, *in* T-00227
Methyl tetraphenylphosphorodiamidite, *in* T-00274
4-Morpholinylphosphonous acid; Dichloride, *in* M-00529
Nonamethylimidodiphosphorous tetraamide, *in* M-00157
Octahydro-1,9-dimethyl-[1,3,2]diazaphosphorino[1,2-*a*]-[1,3,2]diazaphosphorine, O-00005
2-Phenoxy-3-phenyl-1,3,2-oxazaphospholidine, P-00073
Phenyl bis(1-aziridinyl)phosphinite, *in* B-00095
Phenyl *N*,*N*′-1,3-butylene-*N*,*N*′-dimethylphosphorodiamidite, *in* H-00057
Phenyl dimethylphosphoramidochloridite, *in* D-00862
Phenyl di-4-morpholinylphosphinite, *in* D-00936
Phenyl phosphorodiamidite, P-00275
Phenyl tetraethylphosphorodiamidite, *in* T-00066
Phenyl tetramethylphosphorodiamidite, *in* T-00227
Phenyl tetraphenylphosphorodiamidite, *in* T-00274
1-Piperidinylphosphonous acid; Difluoride, *in* P-00432
1-Piperidinylphosphonous acid; Dichloride, *in* P-00432
Propylphosphoramidous acid; Dichloride, *in* P-00486
1-Pyrrolidinylphosphonous acid; Difluoride, *in* P-00520
1-Pyrrolidinylphosphonous acid; Dichloride, *in* P-00520
N,*N*,*N*′,*N*′-Tetrabutylphosphonic diamide, T-00018
Tetraethyl ethylimidodiphosphite, *in* E-00079
Tetraethyl methylimidodiphosphite, *in* M-00157
N,*N*,*N*′,*N*′-Tetraethylphosphonic diamide, T-00055
Tetraethylphosphorodiamidous acid; Bromide, *in* T-00066
Tetraethylphosphorodiamidous acid; Iodide, *in* T-00066
▷Tetraethylphosphorodiamidous chloride, T-00067
Tetrahydro-1,7-dimethyl-1*H*,5*H*-[1,3,2]diazaphospholo[1,2-*a*]-[1,3,2]diazaphosphole, T-00093
N,*N*,*N*′,*N*′-Tetraisopropylphosphonic diamide, T-00171
Tetraisopropylphosphorodiamidous acid; 2-Cyanoethyl ester, *in* T-00172
N,*N*,1,3-Tetrakis(1,1-dimethylethyl)-4,4-dimethyl-1,3-diaza-2-phosphacyclobutane-2-amine, *in* D-00087
Tetramethyl methylimidodiphosphite, *in* M-00157
N,*N*,*N*′,*N*′-Tetramethyl-*P*-4-morpholinylphosphonous diamide, *in* M-00529
N,*N*,*N*′,*N*′-Tetramethylphosphonic diamide, T-00214
Tetramethylphosphorodiamidous acid; Fluoride, *in* T-00227
Tetramethylphosphorodiamidous acid; Bromide, *in* T-00227
Tetramethylphosphorodiamidous acid; Cyanide, *in* T-00227
▷Tetramethylphosphorodiamidous chloride, T-00228
N,*N*,*N*′,*N*′-Tetramethyl-*P*-1-piperidinylphosphonous diamide, *in* P-00432
2,4,6,7-Tetramethyl-2,6,7-triaza-1-phosphabicyclo[2.2.2]-octane, T-00237
Tetraphenyl ethylimidodiphosphite, *in* E-00079
Tetraphenyl methylimidodiphosphite, *in* M-00157
N,*N*,*N*′,*N*′-Tetraphenylphosphonic diamide, T-00271
Tetraphenylphosphorodiamidous acid; Chloride, *in* T-00274
N,*N*,*N*′,*N*′-Tetrapropylphosphonic diamide, T-00289
▷Tri-(1-aziridinyl)phosphine, T-00331
Tri-1-imidazolylphosphine, T-00495
N,*N*′,*N*″-Triisopropylphosphorous triamide, T-00509
2,4,10-Trimethyl-1,3,4,5,7,10-hexaaza-3-phosphatricyclo[$3.3.1.1^{3,7}$]decane, T-00517
N,*N*′,*N*″-Trimethylphosphorous triamide, T-00551
Trimethylsilyl tetraethylphosphorodiamidite, *in* T-00066
2,8,9-Trimethyl-2,8,9-triaza-1-phosphatricyclo[$3.3.1.1^{3,7}$]-decane, T-00563
Trimorpholinophosphine, T-00569
N,*N*′,*N*″-Triphenylphosphorous triamide, T-00677
▷Tri-1-piperidinophosphine, T-00704
N,*N*′,*N*″-Tri-2-propenylphosphorous triamide, T-00709
Tri(1*H*-pyrazol-1-yl)phosphine, T-00722
Tri-1-pyrrolidylphosphine, T-00724
Tri-1-pyrrolylphosphine, T-00725
Tris(triphenylphosphoranylidene)phosphorous triamide, T-00872
Tri(1,2,4-triazol-1-yl)phosphine, T-00876

Phosphorous isocyanates and isothiocyanates

2-Isoselenocyanato-5,5-dimethyl-1,3,2-dioxaphosphorinane, I-00066

Phosphorous azides

2-Azido-1,3,2-benzodioxaphosphole, A-00148

Phosphorous hydrazides

2-Chloro-2,3-dihydro-3,5-dimethyl-1,3,4,2-oxadiazaphosphole, C-00039
2-Chloro-2,3-dihydro-3,5-diphenyl-1,3,4,2-oxadiazaphosphole, C-00042
2-Chloro-2,3-dihydro-5-methyl-3-phenyl-1,3,4,2-oxadiazaphosphole, C-00044
2,3-Dihydro-5-methyl-3-phenyl-2-phenylamino-1,3,4,2-oxadiazaphosphole, D-00527

Thiophosphorous acids

O,*O*-Bis(trimethylsilyl) phosphonothioate, B-00437
O,*O*-2,3-Butylene phosphonothioate, *in* D-00733
O,*O*-Dibutyl phosphonothioate, D-00132
O,*O*-Diethyl phosphonothioate, D-00341
O,*O*-Diisopropyl phosphonothioate, D-00652
O,*O*-Dimethyl phosphonothioate, D-00853
O,*O*-Diphenyl phosphonothioate, D-01106
O,*O*-Dipropyl phosphonothioate, D-01212
2-Mercapto-1,3,2-dioxaphospholane, *in* D-00970
2-Mercapto-4-methyl-1,3,2-dioxophospholane, *in* M-00127
4-Methyl-1,3,2-dioxaphosphorinane 2-sulfide, M-00131
O,*O*-Neopentylene phosphonothioate, *in* D-00735
O,*O*-Tetramethylethylene phosphonothioate, *in* T-00197
O,*O*-Trimethylene phosphonothioate, *in* D-00972

Thiophosphorous halides

1,3,2-Benzoxathiaphosphole; 2-Chloro, *in* B-00026
1,3,2-Benzoxathiaphosphole; 2-Bromo, *in* B-00026
O,*O*-2,3-Butylene chlorothiophosphate, *in* C-00056
S,*S*′-1,3-Butylene phosphorochloridodithioite, *in* M-00138
Butyl phosphorodichloridothioite, B-00632

2-Chloro-1,3,2-benzodithiaphosphole, C-00023
2-Chloro-2,3-dihydro-3-methyl-1,3,2-benzothiazaphosphole, C-00043
2-Chloro-4,5-dimethyl-1,3,2-dioxaphospholane; (4*RS*,5*SR*)-*form*, 2-Sulfide, *in* C-00056
2-Chloro-1,3,2-dithiaphospholane, C-00071
2-Chloro-4-methyl-1,3,2-dithiaphospholane, *in* M-00136
2-Chloro-1,3,2-oxathiaphospholane, *in* O-00056
2-Choro-1,3,2-dithiaphosphorinane, C-00199
Diethyl phosphorochloridodithioite, D-00376
Diisopropyl phosphorochloridothioite, D-00660
Dimethyl phosphorochloridodithioite, D-00889
Dipropyl phosphorochloridodithioite, D-01220
1,3,2-Dithiaphosphole; 2-Chloro, *in* D-01246
Ethyl phosphorodichloridothioite, E-00172
2-Fluoro-1,3,2-dithiaphospholane, *in* D-01243
Isopropyl phosphorodichloridothioite, I-00063
Methyl phosphorodichloridothioite, M-00350
▷Phenyl phosphorodichloridothioite, P-00284
Propyl phosphorodichloridothioite, P-00491
Trifluoromethanethiol phosphorodichloridite, T-00472

Thiophosphorous esters

S,S'-1,3-Butylene *O*-methyl phosphorodithioite, *in* M-00138
Butyl phosphorodichloridothioite, B-00632
5,5-Dimethyl-2-methylthio-1,3,2-dioxaphosphorinane, D-00766
4,5-Dimethyl-2-phenylthio-1,3,2-dioxaphospholane, D-00824
1,3,2-Dithiaphosphole; 2-Methoxy, *in* D-01246
2-Ethoxy-1,3,2-dithiaphospholane, E-00038
2-Ethoxy-4-methyl-1,3,2-dithiaphosphorinane, E-00042
O,S-Ethylene *S*-methyl phosphorodithioite, *in* O-00056
O,S-Ethylene *O*-methyl phosphorothioite, *in* O-00056
Ethyl phosphorodichloridothioite, E-00172
▷2-Ethylthio-4,5-dimethyl-1,3,2-dioxaphospholane, E-00189
▷2-Ethylthio-1,3,2-dioxaphospholane, E-00191
2-Ethylthio-4,4,5,5-tetramethyl-1,3,2-dioxaphospholane, E-00193
Isopropyl phosphorodichloridothioite, I-00063
2-Methoxy-1,3,2-dithiaphospholane, *in* D-01243
2-Methoxy-4-methyl-1,3,2-dithiaphospholane, *in* M-00136
4-Methyl-2-methylthio-1,3,2-dioxaphosphorinane, M-00174
2-(Methylthio)-1,3,2-dithiaphospholane, *in* D-01243
4-Methyl-2,6,7-trithia-1-phosphabicyclo[2.2.2]octane, M-00436
2-Phenoxy-1,3,2-oxathiaphospholane, P-00072
Propyl phosphorodichloridothioite, P-00491
Tribenzyl phosphorotrithioite, T-00343
▷Tributyl phosphorotrithioite, T-00393
▷Triethyl phosphorotrithioite, T-00461
Trifluoromethanethiol phosphorodichloridite, T-00472
▷Trimethyl phosphorotrithioite, T-00550
Triphenyl phosphorotrithioite, T-00676
Tripropyl phosphorotrithioite, T-00718

Thiophosphorous amides

S,S-1,3-Butylene diethylphosphoramidodithioate, *in* M-00138
N,N-Diethyl *S,S*-ethylene phosphoramidodithioite, *in* D-01243
S,S-Ethylene diethylphosphoramidodithioite, *in* D-01243
S,S-1,2-Propylene *N,N*-dimethylphosphoramidodithioite, *in* M-00136

Selenophosphorous acids

Bis(trimethylsilyl) selenophosphonate, B-00440

Selenophosphorous halides

Dimethyl phosphorochloridoselenoite, D-00890
O-Ethyl *Se*-methyl phosphorochloridoselenoite, E-00094
Methyl phosphorodichloridoselenoite, M-00348

Selenophosphorous esters

O,O-Dibutyl phosphonoselenoate, D-00131
O,O-Diethyl *Se*-methyl phosphoroselenoite, D-00301
O,O-Diethyl *Se*-phenyl phosphoroselenoite, D-00316
O,O-Diethyl phosphonoselenoate, D-00340
Dimethyl phosphorochloridoselenoite, D-00890
O,O-Dipropyl phosphonoselenoate, D-01211
O-Ethyl *Se*-methyl phosphorochloridoselenoite, E-00094
Methyl phosphorodichloridoselenoite, M-00348
O,Se,Se-Tributyl phosphorodiselenoite, T-00383
Trimethyl phosphorotriselenoite, T-00547
Triphenyl phosphorotriselenoite, T-00673
Tris(trifluoromethyl) phosphorotriselenoite, T-00857

Tellurophosphorous acids

Bis(trimethylsilyl) tellurophosphonate, B-00441

Tellurophosphorous esters

Tris(4-methoxyphenyl) phosphorotritelluroite, T-00795

Phosphonous acids, $(HO)_2PH$

Methanediphosphonous acid, M-00024
(Trifluoromethyl)phosphonous acid, T-00482

Phosphonous halides

Allyldichlorophosphine, *in* P-00463
1,2-Benzenediphosphonous acid; Bis(dichloride), *in* B-00009
1,4-Benzenediphosphonous acid; Bis(dichloride), *in* B-00010
1,4-Benzenediphosphonous acid; Bis(diiodide), *in* B-00010
Benzyldibromophosphine, *in* B-00068
Benzyldichlorophosphine, *in* B-00068
Benzyldifluorophosphine, *in* B-00068
(Bromomethyl)phosphonous dibromide, B-00476
4-Bromophenyldichlorophosphine, *in* B-00493
4-Bromophenyldiiodophosphine, *in* B-00493
1,4-Butanediphosphonous acid; Bis(dichloride), *in* B-00515
1,4-Butanediphosphonous acid; Bis(dibromide), *in* B-00515
Butyl *tert*-butylphosphonochloridite, *in* D-00130
Butyl butylphosphonochoridite, *in* B-00605
(4-*tert*-Butylcyclohexyl)dichlorophosphine, *in* B-00550
4-*tert*-Butylcyclohexylphosphonous acid; *trans-form*, Dichloride, *in* B-00550
Butyldifluorophosphine, *in* B-00615
Butyldiiodophosphine, *in* B-00615
tert-Butyldiiodophosphine, *in* B-00616
Butylphosphonous dichloride, B-00618
tert-Butylphosphonous dichloride, B-00619
▷*tert*-Butylphosphonous difluoride, B-00620
6-Chloro-6*H*-dibenzo[*c,e*][1,2]oxaphosphorin, C-00033
4-Chloro-1,4-dihydro-2*H*-naphth[2,1-*c*][1,2]oxaphosphorin, C-00048
▷(Chloromethyl)difluorophosphine, *in* C-00115
(Chloromethyl)phosphonous dichloride, C-00116
(4-Chlorophenyl)difluorophosphine, *in* C-00153
(4-Chlorophenyl)diiodophosphine, *in* C-00153
(4-Chlorophenyl)phosphonous dichloride, C-00154
2-Cyanoethyl diisopropylphosphoramidochloridite, *in* D-00654
Cyclohexyldiiodophosphine, *in* C-00267
Cyclohexylphosphonous dichloride, C-00268
Cyclopentadienylphosphonous acid; Dichloride, *in* C-00275
Cyclopentadienylphosphonous acid; Bromide, *in* C-00275
Dibromo-4-bromophenylphosphine, *in* B-00493
Dibromobutylphosphine, *in* B-00615
Dibromo-*tert*-butylphosphine, *in* B-00616
Dibromo(chloromethyl)phosphine, *in* C-00115
Dibromo(4-chlorophenyl)phosphine, *in* C-00153
Dibromocyclohexylphosphine, *in* C-00267
Dibromo(dichloromethyl)phosphine, *in* D-00181
Dibromoethylphosphine, *in* E-00153
Dibromo(4-fluorophenyl)phosphine, *in* F-00042
Dibromoisopropylphosphine, *in* I-00056
Dibromo-1-naphthylphosphine, *in* N-00025
Dibromooctylphosphine, *in* O-00044
Dibromo-*m*-tolylphosphine, *in* M-00247
Dibromo-*o*-tolylphosphine, *in* M-00246
Dibromo-*p*-tolylphosphine, *in* M-00248
Dibromo(trifluoromethyl)phosphine, *in* T-00482
Dichloro(2-chlorophenyl)phosphine, *in* C-00151

Dichloro(3-chlorophenyl)phosphine, *in* C-00152
Dichloro(dichloromethyl)phosphine, *in* D-00181
Dichlorododecylphosphine, *in* D-01265
Dichloro(2-fluorophenyl)phosphine, *in* F-00040
Dichloro(3-fluorophenyl)phosphine, *in* F-00041
Dichloro(4-fluorophenyl)phosphine, *in* F-00042
Dichloroheptylphosphine, *in* H-00008
Dichlorohexylphosphine, *in* H-00101
Dichloroisobutylphosphine, *in* M-00383
(Dichloromethyl)difluorophosphine, *in* D-00181
Dichloro-1-naphthylphosphine, *in* N-00025
Dichloro-2-naphthylphosphine, *in* N-00026
Dichlorooctylphosphine, *in* O-00044
Dichloropentylphosphine, *in* P-00052
Dichloro-*m*-tolylphosphine, *in* M-00247
Dichloro-*o*-tolylphosphine, *in* M-00246
Diethyl 1*H*-1-indenylphosphonite, *in* I-00006
▷Diethyl methylphosphonite, *in* M-00319
Difluoro(3-fluorophenyl)phosphine, *in* F-00041
Difluoro(4-fluorophenyl)phosphine, *in* F-00042
2,5-Difluoro-1-methyl-1,2,5-azadiphospholidine, *in* M-00089
Difluoro-*p*-tolylphosphine, *in* M-00248
Diheptyl heptylphosphonite, *in* H-00008
Diiodophenylphosphine, *in* P-00251
Diiodo-*p*-tolylphosphine, *in* M-00248
1,4-Dimethyl-1,4,2-diazaphospholidin-5-one 2-oxide; 2-Bromo, *in* D-00722
1,2-Ethanediphosphonous acid; Bis(dichloride), *in* E-00008
1,2-Ethanediphosphonous acid; Bis(dibromide), *in* E-00008
Ethyl butylphosphonochloridite, *in* B-00605
Ethyldifluorophosphine, *in* E-00153
Ethyldiiodophosphine, *in* E-00153
Ethyl ethylphosphonochloridite, *in* E-00139
Ethyl methylphosphonofluoridite, *in* M-00307
Ethyl 4-morpholinylphosphonochloridite, *in* M-00526
Ethyl phenylphosphonochloridite, *in* P-00234
▷Ethylphosphonous dichloride, E-00155
(4-Fluorophenyl)diiodophosphine, *in* F-00042
1,6-Hexanediphosphonous acid; Bis(dichloride), *in* H-00086
1,6-Hexanediphosphonous acid; Bis(dibromide), *in* H-00086
Isopropyl methylphosphonofluoridite, *in* M-00307
Isopropylphosphonous dichloride, I-00057
Methanediphosphonous acid; Bis(difluoride), *in* M-00024
Methanediphosphonous acid; Bis(dichloride), *in* M-00024
Methanediphosphonous acid; Bis(dibromide), *in* M-00024
Methyl diisopropylphosphoramidochloridite, *in* D-00654
Methyl methylphosphonochloridite, *in* M-00300
Methyl methylphosphonofluoridite, *in* M-00307
Methyl 4-morpholinylphosphonochloridite, *in* M-00526
(4-Methylphenyl)phosphonous dichloride, M-00249
Methylphosphonous dibromide, M-00321
▷Methylphosphonous dichloride, M-00322
Methylphosphonous difluoride, M-00324
Methylphosphonous diiodide, M-00325
Methyl 1-pyrrolidinylphosphonochloridite, *in* P-00519
4-Morpholinylphosphonous acid; Dichloride, *in* M-00529
1,5-Pentanediphosphonous acid; Bis(dichloride), *in* P-00033
1,5-Pentanediphosphonous acid; Bis(dibromide), *in* P-00033
(2-Phenylethenyl)phosphonous dichloride, P-00162
Phenyl methylphosphonofluoridite, *in* M-00307
Phenylphosphonous dibromide, P-00253
▷Phenylphosphonous dichloride, P-00254
Phenylphosphonous difluoride, P-00256
1-Piperidinylphosphonous acid; Difluoride, *in* P-00432
1-Piperidinylphosphonous acid; Dichloride, *in* P-00432
Propylphosphonous dichloride, P-00483
1-Pyrrolidinylphosphonous acid; Difluoride, *in* P-00520
1-Pyrrolidinylphosphonous acid; Dichloride, *in* P-00520
2,3,4,5-Tetrahydro-6-methyl-2,4,4-triphenyl-1,2,3-diazaphosphorine; 3-Chloro, *in* T-00118
(Trichloromethyl)phosphonous acid; Difluoride, *in* T-00416
Trichloromethylphosphonous dichloride, T-00417
(Trifluoromethyl)diiodophosphine, *in* T-00482
(Trifluoromethyl)phosphonous dichloride, T-00483
Trifluoromethylphosphonous difluoride, T-00484

Phosphonous pseudohalides

(3-Fluorophenyl)phosphonous acid; Dicyanide, *in* F-00041
(4-Fluorophenyl)phosphonous acid; Dicyanide, *in* F-00042
Methylphosphonous dicyanide, M-00323
▷Phenylphosphonous dicyanide, P-00255
(Trifluoromethyl)phosphonous acid; Dicyanide, *in* T-00482

Phosphonous esters

Bis(2-methylpropyl) (2-methylpropyl)phosphonite, *in* M-00383
▷Bis(trimethylsilyl) phosphonite, B-00436
Butyl *tert*-butylphosphonochloridite, *in* D-00130
4-*tert*-Butylcyclohexylphosphonous acid; *trans-form*, Di-Me ester, *in* B-00550
5-*tert*-Butyl-2-phenyl-1,3,2-dioxaphosphorinane, B-00577
2-Cyanoethyl diisopropylphosphoramidochloridite, *in* D-00654
Dibenzyl ethylphosphonite, *in* E-00153
Dibenzyl phenylphosphonite, *in* P-00251
Dibutyl butylphosphonite, *in* B-00615
Dibutyl (chloromethyl)phosphonite, *in* C-00115
Dibutyl 2,4-cyclopentadienylphosphonite, *in* C-00275
Dibutyl cyclopentylphosphonite, *in* C-00281
4,8-Di-*tert*-butyl-2,10-dimethyl-12*H*-dibenzo[*d*,*g*][1,3,2]-dioxaphosphocin, D-00088
Dibutyl ethynylphosphonite, *in* E-00204
Dibutyl isopropylphosphonite, *in* I-00056
Dibutyl 4-morpholinylphosphonite, *in* M-00529
Dibutyl phenylphosphonite, *in* P-00251
Dibutyl phosphonite, D-00129
Dichloro-2-thienylphosphine, *in* T-00308
Diethyl 1-aziridinylphosphonite, *in* A-00154
Diethyl benzylphosphonite, *in* B-00068
Diethyl (4-bromophenyl)phosphonite, *in* B-00493
Diethyl butylphosphonite, *in* B-00615
Diethyl *tert*-butylphosphonite, *in* B-00616
Diethyl (chloromethyl)phosphonite, *in* C-00115
Diethyl (3-chlorophenyl)phosphinate, *in* C-00152
Diethyl (4-chlorophenyl)phosphonite, *in* C-00153
Diethyl cyclohexylphosphonite, *in* C-00267
Diethyl ethylphosphonite, *in* E-00153
Diethyl ethynylphosphonite, *in* E-00204
Diethyl (4-methylphenyl)phosphonite, *in* M-00248
Diethyl 4-morpholinylphosphonite, *in* M-00529
Diethyl 1-naphthylphosphonite, *in* N-00025
Diethyl octylphosphonite, *in* O-00044
Diethyl (2-phenylethenyl)phosphonite, *in* P-00161
Diethyl phenylphosphonite, D-00311
Diethyl phosphonite, D-00339
Diethyl 1-piperidinylphosphonite, *in* P-00432
Diethyl 2-propenylphosphonite, *in* P-00463
Diethyl propylphosphonite, *in* P-00482
Diethyl 1-pyrrolidinylphosphonite, *in* P-00520
Diethyl 2-thienylphosphonite, *in* T-00308
Diethyl (1*H*-1,2,4-triazol-1-yl)phosphonite, *in* T-00332
Diethyl (trichloromethyl)phosphonite, *in* T-00416
Diethyl (trifluoromethyl)phosphonite, *in* T-00482
1,3-Dihydro-1-phenoxy-2,1-benzoxaphosphole, D-00546
1,5-Dihydro-3-phenyl-2,4,3-benzodioxaphosphepin, D-00550
4,7-Dihydro-2-phenyl-1,3,2-dioxaphosphepin, D-00561
Diisopropyl allylphosphonite, *in* P-00463
Diisopropyl 1-aziridinylphosphonite, *in* A-00154
Diisopropyl benzylphosphonite, *in* B-00068
Diisopropyl butylphosphonite, *in* B-00615
Diisopropyl *tert*-butylphosphonite, *in* B-00616
Diisopropyl (4-chlorophenyl)phosphonite, *in* C-00153
Diisopropyl (dichloromethyl)phosphonite, *in* D-00181
Diisopropyl dodecylphosphonite, *in* D-01265
Diisopropyl ethylphosphonite, *in* E-00153
Diisopropyl hexylphosphonite, *in* H-00101
Diisopropyl isopropylphosphonite, *in* I-00056
Diisopropyl methylphosphonite, *in* M-00319
Diisopropyl 1-naphthylphosphonite, *in* N-00025
Diisopropyl octylphosphonite, *in* O-00044
Diisopropyl pentylphosphonite, *in* P-00052
Diisopropyl (2-phenylethenyl)phosphonite, *in* P-00161
Diisopropyl phenylphosphonite, *in* P-00251
Diisopropyl 1-piperidinylphosphonite, *in* P-00432
Diisopropyl propylphosphonite, *in* P-00482

Diisopropyl 1-pyrrolidinylphosphonite, *in* P-00520
Diisopropyl 2-thienylphosphonite, *in* T-00308
Diisopropyl *p*-tolylphosphonite, *in* M-00248
Diisopropyl (trifluoromethyl)phosphonite, *in* T-00482
Dimethyl (4-bromophenyl)phosphonite, *in* B-00493
Dimethyl 4-*tert*-butylcyclohexylphosphonite, *in* B-00550
Dimethyl butylphosphonite, *in* B-00615
Dimethyl *tert*-butylphosphonite, *in* B-00616
Dimethyl (4-chlorophenyl)phosphonite, *in* C-00153
Dimethyl cyclohexylphosphonite, *in* C-00267
5,5-Dimethyl-1,3,2-dioxaphosphorinane, D-00735
▷Dimethyl ethylphosphonite, *in* E-00153
Dimethyl (3-fluorophenyl)phosphonite, *in* F-00041
Dimethyl (4-fluorophenyl)phosphonite, *in* F-00042
2,10-Dimethyl-1,3,6,9,11,14-hexaoxa-2,10-diphosphacyclohexadecane, D-00757
Dimethyl (4-methylphenyl)phosphonite, *in* M-00248
Dimethyl methylphosphonite, *in* M-00319
Dimethyl 1-naphthylphosphonite, *in* N-00025
4,5-Dimethyl-2-phenyl-1,3,2-dioxaphospholane, D-00794
5,5-Dimethyl-2-phenyl-1,3,2-dioxaphosphorinane, D-00795
Dimethyl (2-phenylethenyl)phosphonite, *in* P-00161
Dimethyl phenylphosphonite, *in* P-00251
Dimethyl phosphonite, D-00852
Dimethyl propylphosphonite, *in* P-00482
Dimethyl 2-thienylphosphonite, *in* T-00308
Dimethyl (trifluoromethyl)phosphonite, *in* T-00482
1,3,2-Dioxaphosphorinane, D-00972
Dipentyl pentylphosphonite, *in* P-00052
Diphenyl 1-aziridinylphosphonite, *in* A-00154
Diphenyl (4-bromophenyl)phosphonite, *in* B-00493
Diphenyl butylphosphonite, *in* B-00615
Diphenyl *tert*-butylphosphonite, *in* B-00616
Diphenyl cyclohexylphosphonite, *in* C-00267
Diphenyl ethylphosphonite, *in* E-00153
Diphenyl isopropylphosphonite, *in* I-00056
Diphenyl (4-methylphenyl)phosphonite, *in* M-00248
Diphenyl methylphosphonite, *in* M-00319
Diphenyl 4-morpholinylphosphonite, *in* M-00529
Diphenyl 1-naphthylphosphonite, *in* N-00025
Diphenyl phenylphosphonite, *in* P-00251
Di-2-propenyl phenylphosphonite, *in* P-00251
Divinyl phenylphosphonite, *in* P-00251
Ethyl butylphosphonochloridite, *in* B-00605
O,S-Ethylene phenylphosphonothioite, *in* O-00056
Ethyl ethylphosphonochloridite, *in* E-00139
Ethyl methylphosphonofluoridite, *in* M-00307
Ethyl 4-morpholinylphosphonochloridite, *in* M-00526
Ethyl phenyl 1-aziridinylphosphonite, *in* A-00154
Ethyl phenylphosphonochloridite, *in* P-00234
Hexahydro-2*H*-[1,2]oxaphosphorino[2,3-*b*][1,2]oxaphosphorin; 9-Oxide, *in* H-00037
2,5,5,8,11,11-Hexamethyl-1,3,7,9-tetraoxa-2,8-diphosphacyclododecane, H-00081
4-Isopropyl-5,5-dimethyl-1,3,2-dioxaphosphorinane, I-00028
Isopropyl methylphosphonofluoridite, *in* M-00307
2-Methyl-1,3,2-benzodioxaphosphole, M-00090
Methyl diisopropylphosphoramidochloridite, *in* D-00654
4-Methyl-1,3,2-dioxaphosphorinane; (2*RS*,4*RS*)-*form*, *in* M-00129
4-Methyl-1,3,2-dioxaphosphorinane; (2*RS*,4*SR*)-*form*, *in* M-00129
6-Methyl-4-methylene-4*H*-1,3,2-dioxaphosphorin; 2-Ph, *in* M-00167
Methyl methylphosphonochloridite, *in* M-00300
Methyl methylphosphonofluoridite, *in* M-00307
Methyl 4-morpholinylphosphonochloridite, *in* M-00526
Methyl phenyl 1-aziridinylphosphonite, *in* A-00154
4-Methyl-2-phenyl-1,3,2-dioxaphospholane, M-00198
4-Methyl-2-phenyl-1,3,2-dioxaphosphorinane; (2*S*,4*S*)-*form*, *in* M-00199
4-Methyl-2-phenyl-1,3,2-dioxaphosphorinane; (±)-*form*, *in* M-00199
Methyl 1-pyrrolidinylphosphonochloridite, *in* P-00519
2,5,5,8,11,11,14,17,17-Nonamethyl-1,3,7,9,13,15-hexaoxa-2,8,14-triphosphacyclooctadecane, N-00067
2-Phenyl-1,3,2-benzodioxaphosphole, P-00094
6-Phenyldibenzo[*d,f*][1,3,2]dioxaphosphepin, P-00107
2-Phenyl-1,3,2-dioxaphosphepane, P-00114
2-Phenyl-1,3,2-dioxaphospholane, P-00115
2-Phenyl-1,3,2-dioxaphosphorinane, P-00116
Phenyl methylphosphonofluoridite, *in* M-00307
2-Phenyl-4,4,5,5-tetramethyl-1,3,2-dioxaphospholane, P-00322
Tetraethyl 1,4-butanediylbisphosphonite, *in* B-00515
Tetraethyl 1,6-hexanediylbisphosphonite, *in* H-00086
Tetraethyl methylenebisphosphonite, *in* M-00024
Tetraethyl 1,5-pentanediylbisphosphonite, *in* P-00033
Tetraethyl 1,2-phenylenebisphosphonite, *in* B-00009
Tetraethyl 1,4-phenylenebisphosphonite, *in* B-00010
Tetraisopropyl 1,4-butanediylbisphosphonite, *in* B-00515
Tetraisopropyl 1,6-hexanediylbisphosphonite, *in* H-00086
Tetramethyl 1,4-butanediylbisphosphonite, *in* B-00515
Tetramethylene *tert*-butylphosphonite, *in* D-00968
Tetramethyl 1,6-hexanediylbisphosphonite, *in* H-00086
Tetramethyl methylenebisphosphonite, *in* M-00024
Tetramethyl 1,4-phenylenebisphosphonite, *in* B-00010
Tetraphenyl 1,2-ethanediylbisphosphonite, *in* E-00008
Tetraphenyl methylenebisphosphonite, *in* M-00024
Tetraphenyl 1,5-pentanediylbisphosphonite, *in* P-00033
2,4,5-Trimethyl-1,3,2-dioxaphospholane, T-00511
2,5,5-Trimethyl-1,3,2-dioxaphosphorinane, T-00512
4,4,6-Trimethyl-1,3,2-dioxaphosphorinane, T-00513

Phosphonous amides

P-1-Aziridinyl-*N,N,N′,N′*-tetraethylphosphonous amide, *in* A-00155
P-1-Aziridinyl-*N,N,N′,N′*-tetramethylphosphonous diamide, *in* A-00155
P,P-Bis(1-aziridinyl)-*N,N*-dimethylphosphinous amide, *in* B-00096
P,P′-1,4-Butanediylbis[bis *N,N*-diethylphosphonous diamide], *in* B-00515
P-*tert*-Butyl-*N,N′*-bis(trimethylsilyl)phosphonous diamide, *in* B-00617
tert-Butylphosphonous diamide, B-00617
P-2-Chlorophenyl-*N,N,N′,N′*-tetraethylphosphonous diamide, *in* C-00151
P-3-Chlorophenyl-*N,N,N′,N′*-tetraethylphosphonous diamide, *in* C-00152
P-3-Chlorophenyl-*N,N,N′,N′*-tetramethylphosphonous diamide, *in* C-00152
P-4-Chlorophenyl-*N,N,N′,N′*-tetramethylphosphonous diamide, *in* C-00153
P-Cyclohexyl-*N,N,N′,N′*-tetraethylphosphonous diamide, *in* C-00267
N,N′-Di-*tert*-butyl-*P*-methylphosphonous diamide, *in* M-00320
N,N′-Di-*tert*-butyl-*P*-phenylphosphonous diamide, *in* P-00252
1,3-Di-*tert*-butyl-2,4,4-trimethyl-1,3-diaza-2-phospha-4-silacyclobutane, *in* D-00087
4,5-Diethoxy-1,3-dimethyl-2-phenyl-2*H*-1,3,2-diazaphosphole, D-00230
4,5-Diethoxy-2-methyl-2*H*-1,3,2-diazaphosphole, D-00235
2-Diethylamino-3-phenyl-1,3,2-oxazaphospholidine, D-00269
N,N′-Diethyl-*P*-phenylphosphonous diamide, *in* P-00252
2,5-Difluoro-1-methyl-1,2,5-azadiphospholidine, *in* M-00089
2,3-Dihydro-1,3-dimethyl-2-phenyl-1*H*-benzodiazaphosphole, D-00467
N,N′-Diisopropyl-*P*-phenylphosphonous diamide, *in* P-00252
1,3-Dimethyl-1,3,2-diazaphospholidine; 2-Me, *in* D-00719
1,4-Dimethyl-1,4,2-diazaphospholidin-5-one 2-oxide; 2-Bromo, *in* D-00722
N,N′-Dimethyl-*P*-(1,1-dimethylethyl)phosphonous diamide, *in* B-00617
1,3-Dimethyl-2-phenyl-1,3,2-diazaphospholidine, D-00793
15,16-Dimethyl-1,4,8,11-tetraaza-15,16-diphosphatricyclo[9.3.1.1^{4,8}]hexadecane, D-00921
N,N′-Dimethyl-*N,N′*-trimethylene-*P*-methylphosphonous diamide, *in* H-00022
N,N′-Dimethyl-*N,N′*-trimethylene-*P*-phenylphosphonous diamide, *in* H-00022
2,3-Diphenyl-1,3,2-oxazaphospholidine, D-01029
P,P′-1,2-Ethanediylbis[*N,N,N′,N′*-tetraethylphosphonous diamide], *in* E-00008

P-Ethyl-*N*,*N*,*N'*,*N'*-tetrakis(trimethylsilyl)phosphonous diamide, *in* E-00154
P,*P'*-1,2-Ethynediylbis[*N*,*N*,*N'*,*N'*-tetramethylphosphonous diamide], E-00200
P-4-Fluorophenyl-*N*,*N*,*N'*,*N'*-tetraethylphosphonous diamide, *in* F-00042
P-3-Fluorophenyl-*N*,*N*,*N'*,*N'*-tetramethylphosphonous diamide, *in* F-00041
Hexahydro-3-phenyl-1,2,4,3-triazaphosphorine; 1,4-Di-Me, *in* H-00051
Hexahydro-3-phenyl-1,2,4,3-triazaphosphorine; 1,4-Diisopropyl, *in* H-00051
Hexahydro-3-phenyl-1,2,4,3-triazaphosphorine; 1,4-Dibutyl, *in* H-00051
Hexahydro-1,3,4-trimethyl-1,3,2-diazaphosphorine, H-00057
P,*P'*-1,6-Hexanediylbis[bis *N*,*N*-diethylphosphonous diamide], *in* H-00086
Methylenebis[*N*,*N*,*N'*,*N'*-tetraethylphosphonous diamide], *in* M-00024
2-Methyl-3-phenyl-1,3,2-oxazaphospholidine, M-00212
3-Methyl-2-phenyl-1,3,2-oxazaphospholidine, M-00213
P-Methyl-*N*,*N*,*N'*,*N'*-tetrakis(trimethylsilyl)phosphonous diamide, *in* M-00320
Octahydro-3,8-dimethyl-2,7-diphenyl-1,6,3,8,2,7-dioxadiazadiphosphecine, O-00006
N,*N*,*N'*,*N'*,*N''*,*N''*,*N'''*,*N'''*-Octamethyl-*P*,*P'*-1,2-phenylenebis[phosphonous diamide], *in* B-00009
Pentabutylphosphonous diamide, *in* B-00615
Pentaethylphosphonous diamide, *in* E-00154
Pentamethylphosphonous diamide, *in* M-00320
P,*P'*-1,5-Pentanediylbis[bis-*N*,*N*-diethylphosphonous diamide], *in* P-00033
Pentaphenylphosphonous diamide, *in* P-00252
Pentapropylphosphonous diamide, *in* P-00482
9-Phenyl-1,5-diaza-9-phosphabicyclo[3.3.1]nonane, P-00105
2-Phenyl-1,3,2-oxazaphospholidine, P-00179
N,*N*,*N'*,*N'*-Tetrabenzyl-*P*-phenylphosphonous diamide, *in* P-00252
N,*N*,*N'*,*N'*-Tetraethyl-*P*-heptylphosphonous diamide, *in* H-00008
N,*N*,*N'*,*N'*-Tetraethyl-*P*-(3-methylphenyl)phosphonous diamide, *in* M-00247
N,*N*,*N'*,*N'*-Tetraethyl-*P*-(2-methylphenyl)phosphonous diamide, *in* M-00246
N,*N*,*N'*,*N'*-Tetraethyl-*P*-(4-methylphenyl)phosphonous diamide, *in* M-00248
N,*N*,*N'*,*N'*-Tetraethyl-*P*-methylphosphonous diamide, *in* M-00320
N,*N*,*N'*,*N'*-Tetraethyl-*P*-(2-phenylethenyl)phosphonous diamide, *in* P-00161
N,*N*,*N'*,*N'*-Tetraethyl-*P*-phenylphosphonous diamide, *in* P-00252
Tetrahydro-2*H*,6*H*-[1.3.2]oxazaphosphorino[2,3-*b*][1.3.2]-oxazaphosphorine, T-00120
Tetrahydro-2-phenyl-2*H*-1,3,2-oxazaphosphorine, T-00132
N,*N*,*N'*,*N'*-Tetramethyl-*P*-4-morpholinylphosphonous diamide, *in* M-00529
N,*N*,*N'*,*N'*-Tetramethyl-*P*-1-piperidinylphosphonous diamide, *in* P-00432
N,*N*,*N'*,*N'*-Tetramethyl-*P*-2-propenylphosphonous diamide, *in* P-00463

Phosphonous isocyanates and isothiocyanates

Methylphosphonous acid; Diisocyanate, *in* M-00319
Methylphosphonous acid; Diisothiocyanate, *in* M-00319
Phenylphosphonous acid; Diisocyanate, *in* P-00251
Phenylphosphonous acid; Diisothiocyanate, *in* P-00251

Thiophosphonous halides

Benzyl ethylphosphonobromidothioite, *in* E-00134
Butyl ethylphosphonobromidothioite, *in* E-00134
Butyl ethylphosphonochloridothioite, *in* E-00138
Butyl methylphosphonochloridothioite, *in* M-00299
Butyl methylphosphonofluoridothioite, *in* M-00306
Ethyl ethylphosphonobromidothioite, *in* E-00134
Ethyl ethylphosphonochloridothioite, *in* E-00138
Ethyl methylphosphonochloridothioite, *in* M-00299
Ethyl phenylphosphonochloridothioite, *in* P-00233
Isopropyl ethylphosphonochloridothioite, *in* E-00138
Isopropyl phenylphosphonochloridothioite, *in* P-00233
Methyl methylphosphonochloridothioite, *in* M-00299
Phenyl methylphosphonochloridothioite, *in* M-00299
Phenyl phenylphosphonochloridothioite, *in* P-00233

Thiophosphonous esters

1,3,2-Benzoxathiaphosphole; 2-Me, *in* B-00026
1,3,2-Benzoxathiaphosphole; 2-Ph, *in* B-00026
Benzyl ethylphosphonobromidothioite, *in* E-00134
Butyl ethylphosphonobromidothioite, *in* E-00134
Butyl ethylphosphonochloridothioite, *in* E-00138
Butyl methylphosphonochloridothioite, *in* M-00299
Butyl methylphosphonofluoridothioite, *in* M-00306
2-*tert*-Butyl-1,3,6,2-trithiaphosphocane, B-00642
Dibenzyl ethylphosphonodithioite, *in* E-00141
Dibenzyl methylphosphonodithioite, *in* M-00302
Dibutyl butylphosphonodithioite, *in* B-00607
Dibutyl ethylphosphonodithioite, *in* E-00141
2,8-Di-*tert*-butyl-1,3,7,9-tetrathia-2,8-diphosphacyclododecane, D-00159
2,8-Di-*tert*-butyl-1,3,7,9-tetrathia-2,8-diphosphacyclododecane; *cis-form*, 2,8-Disulfide, *in* D-00159
2,8-Di-*tert*-butyl-1,3,7,9-tetrathia-2,8-diphosphacyclododecane; *trans-form*, *in* D-00159
Diethyl butylphosphonodithioite, *in* B-00607
Diethyl ethylphosphonodithioite, *in* E-00141
Diethyl methylphosphonodithioite, *in* M-00302
Diethyl phenylphosphonodithioite, *in* P-00238
2,3-Dihydro-2,7-dimethyl-1,4,2-benzodithiaphosphorin, D-00458
2,5-Dimethyl-1,3,2-benzodithiaphosphole, D-00712
Dimethyl methylphosphonodithioite, *in* M-00302
Dimethyl (trifluoromethyl)phosphonodithioite, *in* T-00480
Diphenyl butylphosphonodithioite, *in* B-00607
Diphenyl cyclohexylphosphonodithioite, *in* C-00262
Diphenyl ethylphosphonodithioite, *in* E-00141
Diphenyl methylphosphonodithioite, *in* M-00302
Diphenyl phenylphosphonodithioite, *in* P-00238
Ethyl ethylphosphonobromidothioite, *in* E-00134
Ethyl ethylphosphonochloridothioite, *in* E-00138
Ethyl methylphosphonochloridothioite, *in* M-00299
Ethyl phenylphosphonochloridothioite, *in* P-00233
Isopropyl ethylphosphonochloridothioite, *in* E-00138
Isopropyl phenylphosphonochloridothioite, *in* P-00233
2-Methyl-1,3,2-dithiaphosphorinane, M-00137
2-Methyl-1,3,2-dithiophospholane, M-00139
Methyl methylphosphonochloridothioite, *in* M-00299
5-Methyl-2-phenyl-1,3,2-benzodithiaphosphole, M-00195
2,5,5,8,11,11,14,17,17-Nonamethyl-1,3,7,9,13,15-hexaoxa-2,8,14-triphosphacyclooctadecane; 2,8,14-Trisulfide, *in* N-00067
2-Phenyl-1,3,2-dithiaphospholane, P-00121
Phenyl methylphosphonochloridothioite, *in* M-00299
Phenyl phenylphosphonochloridothioite, *in* P-00233
2-Phenyl-1,2-thiaphospholane, P-00324

Thiophosphonous amides

2,3-Dihydro-2-methyl-1,3,2-benzothiazaphosphole, D-00519

Selenophosphonous halides

Methyl phenylphosphonochloridoselenoite, *in* P-00231

Selenophosphonous esters

Bis(trifluoromethyl) phenylphosphonodiselenoite, *in* P-00235
Dimethyl phenylphosphonodiselenoite, *in* P-00235
Methyl phenylphosphonochloridoselenoite, *in* P-00231

Phosphinous acids, $HOPH_2$

Bis(2-methylphenyl)phosphinous acid, B-00323
Bis(trifluoromethyl)phosphinous acid, B-00418
Dicyclohexylphosphinous acid, D-00214
Diphenylphosphinous acid, D-01089

Phosphinous anhydrides

Bis(trifluoromethyl)phosphinous acid; Anhydride, *in* B-00418
Di-*tert*-butylphosphinous acid; Anhydride, *in* D-00122

Phosphinous halides

▷Bis(trifluoromethyl)phosphinous chloride, B-00420
Bromobis(trifluoromethyl)phosphine, *in* B-00418
Bromodibutylphosphine, *in* D-00121
Bromodi-*tert*-butylphosphine, *in* D-00122
Bromodicyclohexylphosphine, *in* D-00214
Bromodiethylphosphine, *in* D-00333
1-Bromo-2,5-dihydro-3,4-dimethyl-1*H*-phosphole, B-00469
1-Bromo-2,5-dihydro-1*H*-phosphole, *in* D-00572
Bromodiisopropylphosphine, *in* D-00648
Bromodimethylphosphine, *in* D-00844
Bromodiphenylphosphine, *in* D-01089
Bromodipropylphosphine, *in* D-01209
Bromoethylmethylphosphine, *in* E-00087
Bromoethylphenylphosphine, *in* E-00108
Bromomethylphenylphosphine, *in* M-00227
Chlorobis(2-chlorophenyl)phosphine, *in* B-00140
Chlorobis(3-chlorophenyl)phosphine, *in* B-00141
Chlorobis(4-chlorophenyl)phosphine, *in* B-00142
Chlorobis(3-methylphenyl)phosphine, *in* B-00324
Chlorobis(4-methylphenyl)phosphine, *in* B-00325
5-Chloro-5*H*-dibenzophosphole, C-00034
Chlorodibenzylphosphine, *in* D-00058
Chlorodicyclohexylphosphine, *in* D-00214
5-Chloro-5,10-dihydroacridophosphine, C-00037
1-Chloro-2,5-dihydro-3,4-dimethyl-1*H*-phosphole, C-00041
5-Chloro-5,10-dihydro-10-methylacridophosphine, *in* D-00515
1-Chloro-2,3-dihydro-4-methyl-1*H*-phosphole, C-00045
1-Chloro-2,5-dihydro-2-methyl-1*H*-phosphole, C-00046
1-Chloro-2,5-dihydro-3-methyl-1*H*-phosphole, C-00047
1-Chloro-2,3-dihydro-1*H*-phosphole, C-00051
Chlorodipropylphosphine, *in* D-01209
Chlorodi-*o*-tolylphosphine, *in* B-00323
Chloroethylmethylphosphine, *in* E-00087
Chloroethylphenylphosphine, *in* E-00108
1-Chloro-2,2,3,4,4-pentamethylphosphetane, C-00124
1-Chlorophospholane, C-00167
1-Chlorophosphorinane, C-00168
1-Chloro-1,2,3,4-tetrahydrophosphinoline, *in* T-00135
Dibenzylbromophosphine, *in* D-00058
Dibutylchlorophosphine, *in* D-00121
▷Di-*tert*-butylfluorophosphine, *in* D-00122
Dibutyliodophosphine, *in* D-00121
Di-*tert*-butyliodophosphine, *in* D-00122
Di-*tert*-butylphosphinous chloride, D-00125
5,10-Dichloro-5,10-dihydrophosphanthrene, D-00164
Dicyclohexyliodophosphine, *in* D-00214
Diethyliodophosphine, *in* D-00333
Diethylphosphinous chloride, D-00335
Diisopropylphosphinous chloride, D-00650
▷Dimethylphosphinous chloride, D-00846
▷Dimethylphosphinous fluoride, D-00848
▷Diphenylphosphinous chloride, D-01091
Diphenylphosphinous fluoride, D-01092
Fluorobis(trifluoromethyl)phosphine, *in* B-00418
1-Fluorophosphorinane, F-00047
Iodobis(trifluoromethyl)phosphine, *in* B-00418
Iododiisopropylphosphine, *in* D-00648
Iododimethylphosphine, *in* D-00844
Iododiphenylphosphine, *in* D-01089
Iododipropylphosphine, *in* D-01209
Methylphenylphosphinous chloride, M-00228
(Trifluoromethyl)phosphinous bromide, T-00476
(Trifluoromethyl)phosphinous chloride, T-00477

Phosphinous pseudohalides

▷Cyanobis(trifluoromethyl)phosphine, *in* B-00418
Cyanodiethylphosphine, *in* D-00333
Cyanodiisopropylphosphine, *in* D-00648
Cyanodipropylphosphine, *in* D-01209
Cyanoethylphenylphosphine, *in* E-00108
Dibutylcyanophosphine, *in* D-00121
▷Dimethylphosphinous cyanide, D-00847
Diphenylphosphinous acid; Cyanide, *in* D-01089
Methylphenylphosphinous acid; (±)-*form*, Cyanide, *in* M-00227

Phosphinous esters

Butyl bis(4-chlorophenyl)phosphinite, *in* B-00142
Butyl bis(2-methylphenyl)phosphinite, *in* B-00323
tert-Butyl bis(trifluoromethyl)phosphinite, *in* B-00418
Butyl dibenzylphosphinite, *in* D-00058
Butyl dibutylphosphinite, *in* D-00121
Butyl diisopropylphosphinite, *in* D-00648
Butyl dimethylphosphinite, *in* D-00844
Butyl diphenylphosphinite, *in* D-01089
Butyl dipropylphosphinite, *in* D-01209
Diethylphosphinous acid; Ph ester, *in* D-00333
2,5-Dihydro-1-methoxy-3-methyl-1*H*-phosphole, *in* D-00530
1,3-Dihydro-1-phenyl-2,1-benzoxaphosphole, D-00553
1-Ethoxy-2,5-dihydro-3-methyl-1*H*-phosphole, *in* D-00530
Ethyl bis(1-aziridinyl)phosphinite, *in* B-00095
Ethyl bis(2-chlorophenyl)phosphinate, *in* B-00140
Ethyl bis(3-chlorophenyl)phosphinite, *in* B-00141
Ethyl bis(4-chlorophenyl)phosphinite, *in* B-00142
Ethyl bis(2-methylphenyl)phosphinite, *in* B-00323
Ethyl bis(4-methylphenyl)phosphinite, *in* B-00325
Ethyl dibenzylphosphinite, *in* D-00058
Ethyl dibutylphosphinite, *in* D-00121
Ethyl di-*tert*-butylphosphinite, *in* D-00122
Ethyl dicyclohexylphosphinite, *in* D-00214
▷Ethyl diethylphosphinite, *in* D-00333
Ethyl diisopropylphosphinite, *in* D-00648
Ethyl di-4-morpholinylphosphinite, *in* D-00936
Ethyl diphenylphosphinite, *in* D-01089
Ethyl ethylmethylphosphinite, *in* E-00087
Ethyl ethylphenylphosphinite, *in* E-00108
Ethylphenylphosphinous acid; (±)-*form*, Me ester, *in* E-00108
Isopropyl bis(1-aziridinyl)phosphinite, *in* B-00095
▷Isopropyl diethylphosphinite, *in* D-00333
Methyl bis(1-aziridinyl)phosphinite, *in* B-00095
Methyl bis(4-chlorophenyl)phosphinite, *in* B-00142
Methyl bis(4-methylphenyl)phosphinite, *in* B-00325
Methyl bis(trifluoromethyl)phosphinite, *in* B-00418
Methyl di-*tert*-butylphosphinite, *in* D-00122
Methyl dimethylphosphinite, *in* D-00844
Methyl di-4-morpholinylphosphinite, *in* D-00936
Methyl diphenylphosphinite, *in* D-01089
Methyl ethylmethylphosphinite, *in* E-00087
Methyl ethylphenylphosphinite, *in* E-00108
Methyl methylphenylphosphinite, *in* M-00227
Methylphenylphosphinous acid; (±)-*form*, Me ester, *in* M-00227
1,2-Oxaphosphorinane; 2-Ph, *in* O-00054
Phenyl bis(1-aziridinyl)phosphinite, *in* B-00095
Phenyl dibenzylphosphinite, *in* D-00058
Phenyl di-*tert*-butylphosphinite, *in* D-00122
Phenyl dimethylphosphinite, *in* D-00844
Phenyl di-4-morpholinylphosphinite, *in* D-00936
Phenyl diphenylphosphinite, *in* D-01089
Phenyl ethylphenylphosphinite, *in* E-00108
2-Phenyl-1,2-oxaphospholane, P-00174
2-Propenyl diphenylphosphinite, *in* D-01089
Trimethylsilyl dibutylphosphinite, *in* D-00121
Trimethylsilyl di-*tert*-butylphosphinite, *in* D-00122

Trimethylsilyl dicyclohexylphosphinite, *in* D-00214
Trimethylsilyl diisopropylphosphinite, *in* D-00648

Phosphinous amides

P,P-Bis(1-aziridinyl)-*N,N*-diethylphosphinous amide, *in* B-00096
P,P-Bis(3-chlorophenyl)-*N,N*-diethylphosphinous amide, *in* B-00141
P,P-Bis(4-chlorophenyl)-*N,N*-diethylphosphinous amide, *in* B-00142
2,2′-Bis(diphenylphosphinamido)-6,6′-dimethylbiphenyl; (*S*)-*form*, *in* B-00215
2,2′-Bis[(diphenylphosphino)amino]-1,1′-binaphthyl; (*R*)-*form*, *in* B-00217
2,2′-Bis[(diphenylphosphino)amino]-1,1′-binaphthyl; (*S*)-*form*, *in* B-00217
2,2′-Bis[(diphenylphosphino)amino]-1,1′-binaphthyl; (±)-*form*, *in* B-00217
2,3-Bis[(diphenylphosphino)amino]butane; (2*S*,3*S*)-*form*, *in* B-00218
1,2-Bis[(*N*-diphenylphosphino)amino]cyclohexane; (1*R*,2*R*)-*form*, *in* B-00219
1,2-Bis[(*N*-diphenylphosphino)amino]cyclohexane; (1*R*,2*R*)-*form*, *N,N′*-Di-Me, *in* B-00219
1,2-Bis[(*N*-diphenylphosphino)amino]cyclohexane; (1*S*,2*S*)-*form*, *in* B-00219
N,N-Bis(diphenylphosphino)-*P,P*-diphenylphosphinous amide, B-00233
▷*P,P*-Bis(trifluoromethyl)phosphinous amide, B-00419
N-tert-Butyl-*P,P*-diethylphosphinous amide, *in* D-00334
P,P-Dibenzyl-*N,N*-diethylphosphinous amide, *in* D-00058
P,P-Dibutyl-*N,N*-diethylphosphinous amide, *in* D-00123
P,P-Dibutyl-*N,N*-dimethylphosphinous amide, *in* D-00123
P,P-Dibutyl-*N*-ethylphosphinous amide, *in* D-00123
P,P-Di-*tert*-butyl-*N*-methylphosphinous amide, *in* D-00124
P,P-Di-*tert*-butyl-*N*-phenylphosphinous amide, *in* D-00124
P,P-Di-*tert*-butylphosphinous amide, D-00124
P,P-Di-*tert*-butyl-*N*-trimethylsilylphosphinous amide, *in* D-00124
P,P-Diethyl-*N,N*-bis(trimethylsilyl)phosphinous amide, *in* D-00334
N,N-Diethyl-*P,P*-diisopropylphosphinous amide, *in* D-00649
P,P-Diethyl-*N,N*-dimethylphosphinous amide, *in* D-00334
N,N-Diethyl-*P,P*-diphenylphosphinous amide, *in* D-01090
N,N-Diethyl-*P*-methyl-*P*-phenylphosphinous amide, *in* M-00227
P,P-Diethyl-*N*-methylphosphinous amide, *in* D-00334
P,P-Diethyl-*N*-phenylphosphinous amide, *in* D-00334
P,P-Diethylphosphinous amide, D-00334
2,5-Dihydro-1-dimethylamino-3-methyl-1*H*-phosphole, *in* D-00530
3,4-Dihydro-2,3,4,5-tetraphenyl-2*H*-1,2,3-diazaphosphole; (3*RS*,4*RS*)-*form*, *in* D-00579
3,4-Dihydro-2,3,4,5-tetraphenyl-2*H*-1,2,3-diazaphosphole; (3*RS*,4*SR*)-*form*, *in* D-00579
P,P-Diisopropyl-*N-tert*-butylphosphinous amide, *in* D-00649
P,P-Diisopropyl-*N,N*-dimethylphosphinous amide, *in* D-00649
N,N-Diisopropyl-*P,P*-diphenylphosphinous amide, *in* D-01090
P,P-Diisopropyl-*N*-methylphosphinous amide, *in* D-00649
P,P-Diisopropyl-*N*-phenylphosphinous amide, *in* D-00649
P,P-Diisopropylphosphinous amide, D-00649
N,N-Dimethyl-*P,P*-bis(trifluoromethyl)phosphinous amide, *in* B-00419
P,P-Dimethyl-*N,N*-bis(trimethylsilyl)phosphinous amide, *in* D-00845
N,N-Dimethyl-*P,P*-diphenylphosphinous amide, D-00747
P,P-Dimethyl-*N,N*-diphenylphosphinous amide, *in* D-00845
1,2-Diphenyl-1,2-azaphospholidine, *in* P-00089
P,P-Diphenylphosphinic amide; *N*-Diphenylphosphino, *in* D-01040
N-Diphenylphosphino-*P,P*-diphenylphosphinous amide, D-01063
N-(Diphenylphosphino)-*N*-ethyl-*P,P*-diphenylphinous amide, D-01067
N-Diphenylphosphino-*P,P,P*-triphenylphosphine imide, D-01088
P,P-Diphenyl-*N*-(triphenylphosphoranylidene)phosphinous amide, D-01165
Ethylphenylphosphinous acid; (*S*)-*form*, Diethylamide, *in* E-00108
Ethylphenylphosphinous acid; (±)-*form*, Diethylamide, *in* E-00108
N,N,N′,N′,N″,N″-Hexamethyl-*N‴*-diphenylphosphinophosphorimidic triamide, H-00069
N-Methyl-*P,P*-bis(trifluoromethyl)phosphinous amide, *in* B-00419
1-Methyl-2,5-diphenyl-1,2,5-azadiphospholidine, *in* M-00089
2-Phenyl-1,2-azaphospholidine, P-00089
N-Phenyl-*P,P*-bis(trifluoromethyl)phosphinous amide, *in* B-00419
9-Phenyl-1,5-diaza-9-phosphatricyclo[3.3.1.1^{3,7}]decane, P-00106
Tetraethylphosphinous amide, *in* D-00334
Tetramethylphosphinous amide, T-00213
P,P,P-Triethyl-*N*-(diphenylphosphino)phosphine imide, *in* T-00442
N,N,P-Triethyl-*P*-phenylphosphinous amide, *in* E-00108

Phosphinous isocyanates and isothiocyanates

Bis(trifluoromethyl)phosphinous acid; Isocyanate, *in* B-00418
Bis(trifluoromethyl)phosphinous acid; Isothiocyanate, *in* B-00418
Dibutylphosphinous acid; Isocyanate, *in* D-00121
Diphenylphosphinous acid; Isothiocyanate, *in* D-01089

Thiophosphinous acids

Bis(trifluoromethyl)phosphinothious acid, B-00417

Thiophosphinous anhydrides

Bis(trifluoromethyl)phosphinothious acid; Anhydrosulfide, *in* B-00417
Di-*tert*-butylphosphinothious acid; Anhydrosulfide, *in* D-00120
Diisopropylphosphinothious acid; Anhydrosulfide, *in* D-00647

Thiophosphinous esters

Benzyl diphenylphosphinothioite, *in* D-01085
Butyl bis(2-methylpropyl)phosphinothioite, *in* B-00352
tert-Butyl bis(trifluoromethyl)phosphinothioite, *in* B-00417
Butyl diphenylphosphinothioite, *in* D-01085
Butyl ethylphenylphosphinothioite, *in* E-00107
2,3-Dihydro-4,5-dimethyl-2,2-diphenyl-1,2-thiaphosphepin; 2-Sulfide, *in* D-00463
3,4-Dihydro-2-methyl-2*H*-1,4,2-benzothiazaphosphorine, D-00521
Ethyl diethylphosphinothioite, *in* D-00332
Ethyl diphenylphosphinothioite, *in* D-01085
Ethyl ethylphenylphosphinothioite, *in* E-00107
Ethylphenylphosphinothious acid; (*S*)-*form*, Et ester, *in* E-00107
Ethylphenylphosphinothious acid; (±)-*form*, Et ester, *in* E-00107
1-Ethylthio-1,5-dihydro-3-methyl-1*H*-phosphole, *in* D-00530
Isopropyl diethylphosphinothioite, *in* D-00332
Isopropyl ethylphenylphosphinothioite, *in* E-00107
Methyl bis(trifluoromethyl)phosphinothioite, *in* B-00417
Methyl di-*tert*-butylphosphinothioite, *in* D-00120
Methyl dimethylphosphinothioite, *in* D-00843
Methyl diphenylphosphinothioite, *in* D-01085
4-Methyl-2-phenyl-1,3,2-dithiaphospholane, *in* M-00136
Phenyl bis(2-methylpropyl)phosphinothioite, *in* B-00352
Phenyl diethylphosphinothioite, *in* D-00332
Phenyl diphenylphosphinothioite, *in* D-01085
2-Phenyl-1,2-thiaphosphorinane, P-00325
Trifluoromethyl bis(trifluoromethyl)phosphinothioite, *in* B-00417
Trifluoromethyl dimethylphosphinothioite, *in* D-00843
2,4,5-Triphenyl-1,3,4-oxathiaphospholane; 4-Sulfide, *in* T-00611

Selenophosphinous anhydrides

Bis(trifluoromethyl)phosphinoselenous acid; Anhydroselenide, *in* B-00413

Selenophosphinous esters

Methyl bis(trifluoromethyl)phosphinoselenoite, *in* B-00413
Methyl dimethylphosphinoselenoite, *in* D-00838
Methyl diphenylphosphinoselenoite, *in* D-01081
Phenyl dimethylphosphinoselenoite, *in* D-00838
Phenyl diphenylphosphinoselenoite, *in* D-01081
Trifluoromethyl bis(trifluoromethyl)phosphinoselenoite, *in* B-00413
Trifluoromethyl dimethylphosphinoselenoite, *in* D-00838

Tellurophosphinous acids

Di-*tert*-butylphosphinotellurous acid, D-00116
Diisopropylphosphinotellurous acid, D-00645

Tellurophosphinous esters

Methyl bis(trifluoromethyl)phosphinotelluroite, *in* B-00414

Phosphoric acids, $(HO)_3PO$

Acetic acid monoanhydride with methyl dihydrogen phosphate, *in* A-00025
Acetyl phosphate, A-00025
2-Aminoethyl dihydrogen phosphate, A-00075
2-Amino-3-[(hydroxyphenoxyphosphinyl)oxy]propanoic acid, *in* P-00417
2-Amino-3-phosphonooxybutanoic acid; (2*S*,3*R*)-*form*, *in* A-00120
2-Amino-3-phosphonooxybutanoic acid; (2*RS*,3*SR*)-*form*, *in* A-00120
Benzoylphosphoramidic acid, B-00036
Benzylphosphoramidic acid, B-00069
Bis[(4-bromophenyl)methyl] phosphate, B-00109
Bis(4-bromophenyl) phosphate, B-00111
Bis(2-chloroethyl) phosphate, B-00119
N,N'-Bis(2-chloroethyl)phosphorodiamidic acid, B-00125
N,N-Bis(2-chloroethyl)phosphorodiamidic acid, B-00126
Bis(2-chlorophenyl) phosphate, B-00131
Bis(4-chlorophenyl) phosphate, B-00133
Bis(2,4-dinitrophenyl) phosphate, B-00210
Bis(2-methoxyphenyl) phosphate, B-00294
Bis(4-methoxyphenyl) phosphate, B-00295
Bis(3-methylbutyl) phosphate, B-00302
▷Bis(2-methylphenyl) phosphate, B-00312
Bis(3-methylphenyl) phosphate, B-00313
Bis(4-methylphenyl) phosphate, B-00314
Bis(2-methylpropyl) phosphate, B-00342
Bis[(2-nitrophenyl)methyl] phosphate, B-00366
Bis[(4-nitrophenyl)methyl] phosphate, B-00367
Bis(2-nitrophenyl) phosphate, B-00369
Bis(3-nitrophenyl) phosphate, B-00370
Bis(4-nitrophenyl) phosphate, B-00371
Bis(pentachlorophenyl) phosphate, B-00379
Bis(pentafluorophenyl) phosphate, B-00381
Bis(2,2,2-trichloroethyl) phosphate, B-00400
(4-Bromophenyl)phosphoramidic acid, B-00496
OO-tert-Butyl *O,O*-diethyl phosphoroperoxoate, B-00551
OO-tert-Butyl *O,O*-diisopropyl phosphoroperoxoate, B-00554
tert-Butyl hydrogen bis(2-chloroethyl)phosphoramidate, *in* B-00121
Butylphosphoramidic acid, B-00621
▷Carbamoyl dihydrogen phosphate, C-00002
5-Chloromethyl-2-hydroxy-5-methyl-1,3,2-dioxaphosphorinane 2-oxide, C-00097
(3-Chlorophenyl)phosphoramidic acid, C-00156
(4-Chlorophenyl)phosphoramidic acid, C-00157
Demanyl phosphate, D-00016
Dibenzoyl phosphate, D-00042
Dibenzyl phosphate, D-00053
N,N'-Dibenzylphosphorodiamidic acid, D-00061
▷Dibutyl phosphate, D-00104
Di-*tert*-butyl phosphate, D-00105
(2,4-Dichlorophenyl)phosphoramidic acid, D-00188
Dicyclohexyl phosphate, D-00209
N,N'-Dicyclohexylphosphorodiamidic acid, D-00216
Didecyl phosphate, D-00220
P,P'-Diethyl diphosphate, D-00289
Diethyl phosphate, D-00320
Diheptyl phosphate, D-00424
Dihexadecyl phosphate, D-00428
Dihexyl phosphate, D-00432
1,3-Dihydroxy-1*H*-1,2,4,3-benziodadioxaphosphorin 3-oxide, D-00616
3,9-Dihydroxy-2,4,8,10-Tetraoxa-3,9-diphosphaspiro[5.5]-undecane 3,9-dioxide, D-00626
Diisopropyl phosphate, D-00639
3,7-Dimethyl-1,6-octadien-3-yl dihydrogen phosphate, D-00775
3,7-Dimethyl-2,6-octadien-1-yl dihydrogen phosphate, D-00776
Dimethyl phosphate, D-00826
Di-1-naphthyl phosphate, D-00937
Di-2-naphthyl phosphate, D-00938
Dinonyl phosphate, D-00948
Dioctyl phosphate, D-00956
Dipentyl phosphate, D-00979
P,P'-Diphenyl diphosphate, D-01011
Diphenyl phosphate, D-01036
N,N'-Diphenylphosphorodiamidic acid, D-01125
Di-2-propenyl phosphate, D-01192
Dipropyl phosphate, D-01203
N,N'-Dipropylphosphorodiamidic acid, D-01223
Diundecyl phosphate, D-01255
Ethyl hydrogen 2-aminoethyl phosphate, *in* A-00075
Ethyl hydrogen bis(2-chloroethyl)phosphoramidate, *in* B-00121
Ethyl 4-(phosphonoxy)benzoate, *in* P-00397
Ethyl 2-(phosphonoxy)-2-propenoate, *in* P-00366
Ethyl phosphorofluoridate, E-00179
Geranyl phosphate, *in* D-00776
Glycerol 2-monophosphate, G-00067
Glycerol 2-monophosphate; Di-Ph ether, *in* G-00067
Glycerol 2-monophosphate; 1-Dodecanoyl, *in* G-00067
Glycerol 2-monophosphate; 1-Hexanoyl, *in* G-00067
Glycerol 2-monophosphate; 1,3-Ditetradecanoyl, *in* G-00067
Glycerol 2-monophosphate; 1,3-Dioctadecanoyl, *in* G-00067
Hexahydro-2-hydroxycyclopenta[*d*]-1,3,2-dioxaphosphorin 2-oxide; (4*aRS*,7*aRS*)-*form*, *in* H-00034
Hexahydro-2-hydroxycyclopenta[*d*]-1,3,2-dioxaphosphorin 2-oxide; (4*aRS*,7*aSR*)-*form*, *in* H-00034
2-Hydroxy-1,3,2-benzodioxaphosphole 2-oxide, H-00108
2-Hydroxy-4*H*-1,3,2-benzodioxaphosphorin 2-oxide, H-00109
2-Hydroxydibenzo[*d,f*][1,3,2]dioxaphosphepin 2-oxide, H-00116
2-Hydroxy-4,5-dimethyl-1,3,2-dioxaphospholane 2-oxide, H-00121
2-Hydroxy-4,5-dimethyl-1,3,2-dioxaphosphole 2-oxide, H-00123
2-Hydroxy-5,5-dimethyl-1,3,2-dioxaphosphorinane 2-oxide, H-00124
2-Hydroxy-5,5-dimethyl-4-phenyl-1,3,2-dioxaphoshorinane 2-oxide; (*R*)-*form*, *in* H-00129
2-Hydroxy-5,5-dimethyl-4-phenyl-1,3,2-dioxaphoshorinane 2-oxide; (*S*)-*form*, *in* H-00129
2-Hydroxy-5,5-dimethyl-4-phenyl-1,3,2-dioxaphoshorinane 2-oxide; (±)-*form*, *in* H-00129
4-Hydroxydinaphtho[2,1-*d*:1',2'-*f*][1,3,2]dioxaphosphepin 4-oxide; (*R*)-*form*, *in* H-00131
4-Hydroxydinaphtho[2,1-*d*:1',2'-*f*][1,3,2]dioxaphosphepin 4-oxide; (*S*)-*form*, *in* H-00131
4-Hydroxydinaphtho[2,1-*d*:1',2'-*f*][1,3,2]dioxaphosphepin 4-oxide; (±)-*form*, *in* H-00131
2-Hydroxy-1,3,2-dioxaphosphepane 2-oxide, H-00132
2-Hydroxy-1,3,2-dioxaphospholane 2-oxide, H-00134
2-Hydroxy-1,3,2-dioxaphosphonane 2-oxide, H-00137
2-Hydroxy-1,3,2-dioxaphosphorinane 2-oxide, H-00138
2-Hydroxy-4,5-diphenyl-1,3,2-dioxaphosphole 2-oxide, H-00139
2-Hydroxy-4-methyl-1,3,2-dioxaphospholane 2-oxide, H-00155
2-Hydroxy-4-methyl-1,3,2-dioxaphosphorinane 2-oxide, H-00156
2-Hydroxynaphtho[2,3-*d*]-1,3,2-dioxaphosphole 2-oxide, H-00164
2-Hydroxy-4-phenyl-1,3,2-dioxaphosphorinane 2-oxide, H-00171
2-Hydroxy-3-(phosphonoxy)propanal, H-00187
2-Hydroxy-4,4,5,5-tetramethyl-1,3,2-dioxaphospholane 2-oxide, H-00196
(3-Methoxyphenyl)phosphoramidic acid, M-00075
(4-Methylphenyl)phosphoramidic acid, M-00253

Monobenzoyl phosphate, M-00442
Monobenzyl phosphate, M-00443
Mono(2-biphenylyl) phosphate, M-00445
Mono-(1,1′-biphenyl)-4-yl phosphate, M-00446
Mono(2-bromoethyl) phosphate, M-00447
Mono(4-bromophenyl) phosphate, M-00448
Mono-2-*tert*-butylphenyl phosphate, M-00449
Mono-4-*tert*-butylphenyl phosphate, M-00450
Monobutyl phosphate, M-00451
Mono-*tert*-butyl phosphate, M-00452
Mono(2-chloroethyl) phosphate, M-00454
Mono(2-chloromethyl-4-nitrophenyl) phosphate, M-00455
Mono(4-chlorophenyl) mono(5-chloro-8-quinolinyl) phosphate, M-00456
Mono(2-chlorophenyl) phosphate, M-00457
Mono(4-chlorophenyl) phosphate, M-00458
Mono(2-cyanophenyl) phosphate, M-00460
Monocyclohexyl phosphate, M-00461
Monodecyl phosphate, M-00462
Mono(2,4-dichlorophenyl) phosphate, M-00463
Mono[2-(dimethylamino)-4-nitrophenyl] phosphate, M-00464
Mono-2,6-dimethylphenyl phosphate, M-00466
Mono(2,4-dinitrophenyl) phosphate, M-00467
Monododecyl phosphate, M-00468
Mono(2-ethylhexyl) phosphate, M-00470
Monoethyl phosphate, M-00472
Monoheptyl phosphate, M-00474
Monohexadecyl phosphate, M-00475
Monohexyl phosphate, M-00476
▷Monoisopropyl phosphate, M-00477
Monomethoxycarbonyl phosphate, M-00479
Mono(2-methoxyphenyl) phosphate, M-00480
Mono(4-methoxyphenyl) phosphate, M-00481
Mono(3-methyl-2-butenyl) diphosphate, M-00482
Mono(3-methylbutyl) phosphate, M-00484
Mono(3-methylphenyl) phosphate, M-00486
Mono(4-methylphenyl) phosphate, M-00487
▷Monomethyl phosphate, M-00489
Mono(2-methyl-1-propenyl) phosphate, M-00491
Mono(1-methylpropyl) phosphate, M-00492
Mono(2-methylpropyl) phosphate, M-00493
Mono-1-naphthyl phosphate, M-00494
Mono-2-naphthyl phosphate, M-00495
Mono(3-nitrophenyl) phosphate, M-00499
Mono(4-nitrophenyl) phosphate, M-00500
Monooctyl phosphate, M-00501
Mono(pentachlorophenyl) phosphate, M-00502
Monopentyl phosphate, M-00504
Mono(2-phenoxyphenyl) phosphate, M-00505
Mono(2-phenylethyl) phosphate, M-00506
Monophenyl phosphate, M-00507
▷Mono-2-propenyl phosphate, M-00509
Monopropyl phosphate, M-00512
Monotetradecyl phosphate, M-00514
Mono(2,2,2-tribromoethyl) phosphate, M-00516
1-Naphthylphosphoramidic acid, N-00028
Neryl phosphate, *in* D-00776
4-Nitrophenyl phenyl phosphate, N-00052
8,9,10,11,12,13,14,15-Octahydro-4-hydroxydinaphtho[2,1-*d*:1′,2′-*f*][1,3,2]dioxaphosphepin 4-oxide, O-00018
Phenyl hydrogen 2-aminoethyl phosphate, *in* A-00075
Phenyl hydrogen (2-aminoethyl)phosphoramidate, *in* A-00080
Phenyl hydrogen bis(2-chloroethyl)phosphoramidate, *in* B-00121
Phenyl phenylphosphoramidic acid, P-00189
Phenylphosphoramidochloridic acid, P-00259
Phenyl phosphorochloridic acid, P-00269
N-Phenylphosphorodiamidic acid, P-00273
Phenyl phosphorofluoridic acid, P-00295
Phenyl 8-quinolyl phosphate, P-00306
Phenylsulfonylphosphoramidic acid, P-00312
Phenylsulfonylphosphoramidic acid; Monobromide, *in* P-00312
N-Phosphocreatine, P-00364
Phosphoenolpyruvic acid, P-00366
N-Phosphoglutamic acid, P-00367
Phosphoguanidine, P-00370
(*N*-Phosphono)methionine-*S*-sulfoximinylalanylalanine, P-00389
Phosphonooxyacetic acid, P-00390
▷2-(Phosphonooxy)benzoic acid, P-00391
3-Phosphonoxypropanenitrile, P-00398
Phosphoserine; (*R*)-*form*, *in* P-00417
Phosphoserine; (*S*)-*form*, *in* P-00417
Phosphoserine; (±)-*form*, *in* P-00417
N-Phosphoserine, P-00418
N-Phosphothreonine, P-00420
Tetramethylphosphorodiamidic acid, T-00218
Threonine ethanolamine phosphate; (*S*)-*form*, *in* T-00318
Triclofos, T-00424
3,7,11-Trimethyl-1,6,10-dodecatrien-3-yl dihydrogen phosphate; (*S*)-*form*, *in* T-00514
3,7,11-Trimethyl-2,6,10-dodecatrien-1-yl dihydrogen phosphate, T-00515

Phosphoric anhydrides

Acetic acid anhydride with dibenzyl hydrogen phosphate, *in* A-00025
Acetic acid anhydride with diethyl hydrogen phosphate, *in* A-00025
Acetic acid anhydride with dimethyl hydrogen phosphate, *in* A-00025
Acetic acid monoanhydride with methyl dihydrogen phosphate, *in* A-00025
Acetyl phosphate, A-00025
Benzoyl phosphorodichloridate, *in* M-00442
Benzoyl phosphorodifluoridate, *in* M-00442
Bis[(4-bromophenyl)methyl] phosphate; Anhydride, *in* B-00109
Bis(dimethylvinylene) pyrophosphate, *in* O-00088
Bis(3-methylphenyl) phosphate; Anhydride, *in* B-00313
Bis(2-oxo-3-oxazolidinyl) diphenyl phosphoric anhydride, *in* B-00378
Bis(2-oxo-3-oxazolidinyl)phosphinic acid; Anhydride, *in* B-00378
tert-Butoxycarbonyl diethyl phosphate, B-00531
▷Carbamoyl dihydrogen phosphate, C-00002
Catechol cyclic phosphoric acid anhydride, *in* O-00087
Dibenzoyl phosphate, D-00042
P,P-Dibenzyl *P′,P′*-diphenyl diphosphate, D-00044
P,P′-Diethyl diphosphate, D-00289
Diethyl propanoyl phosphate, D-00400
Diethyl trifluoroacetyl phosphate, D-00406
P,P′-Diphenyl diphosphate, D-01011
N,N′-Diphenylphosphorodiamidic acid; Anhydride, *in* D-01125
Ethyl carbonate-diethyl phosphate anhydride, E-00055
Ethylene glycol cyclic *P,P*:*P′,P′*-pyrophosphate, *in* O-00091
Monobenzoyl phosphate, M-00442
Mono(3,7-dimethyl-2,6-octadienyl) diphosphate, M-00465
Monomethoxycarbonyl phosphate, M-00479
Mono(3-methyl-2-butenyl) diphosphate, M-00482
Mono(3-methyl-3-butenyl) diphosphate, M-00483
Mono(3,7,11,15-tetramethyl-2,6,10,14-hexadecatetraenyl) diphosphate, M-00515
Mono(3,7,11-trimethyl-2,6,10-dodecatrienyl) diphosphate, M-00517
4-Morpholinylphosphonochloridic acid; Anhydride with diphenyl phosphate, *in* M-00524
Octaethyldiphosphoramide, O-00001
▷Octamethyldiphosphoramide, O-00029
2,2′-Oxybis[5,5-dimethyl-1,3,2-dioxaphosphorinane]; 2-Oxide, *in* O-00089
2,2′-Oxybis[5,5-dimethyl-1,3,2-dioxaphosphorinane]; 2,2′-Dioxide, *in* O-00089
2,2′-Oxybis[1,3,2-dioxaphosphorinane]; 2,2′-Dioxide, *in* O-00092
2,2′-Oxybis[4-methyl-1,3,2-dioxaphosphorinane], O-00096
Tetrabenzyl diphosphate, T-00006
▷Tetrabutyl pyrophosphate, T-00024
▷Tetraethyl pyrophosphate, T-00069
1,1,3,7-Tetrahydroxy-7-methyl-2,4-dioxa-1,3-diphosphanonan-9-oic acid 1,3-dioxide, T-00162
▷Tetraisopropyl diphosphate, T-00166
Tetrakis(4-nitrophenyl) diphosphate, T-00183
Tetrakis[(4-nitrophenyl)methyl] diphosphate, *in* B-00367
▷Tetramethyl pyrophosphate, T-00229

Tetraphenyl diphosphate, T-00256
▷Tetrapropyl diphosphate, T-00285
Triacetyl phosphate, T-00325
2,4,6-Tris(diethylamino)-1,3,5,2,4,6-trioxatriphosphorinane 2,4,6-trioxide, T-00756
2,4,6-Tris(diethylamino)-1,3,5,2,4,6-trioxatriphosphorinane 2,4,6-trioxide; (2α,4α,6α)-*form*, *in* T-00756
2,4,6-Tris(diethylamino)-1,3,5,2,4,6-trioxatriphosphorinane 2,4,6-trioxide; (2α,4α,4β)-*form*, *in* T-00756

Phosphoric halides

2-Adamantylphosphoramidic acid; Dichloride, *in* A-00037
Aminocarbonylphosphoramidic acid; Dichloride, *in* A-00062
Benzoylphosphoramidic dichloride, B-00037
Benzoyl phosphorodichloridate, *in* M-00442
Benzoyl phosphorodifluoridate, *in* M-00442
Benzyl phosphorodichloridate, *in* M-00443
O,O-(2,2′-Biphenylyl) phosphorochloridate, *in* C-00032
[1,1′-Biphenyl]-2-yl phosphorodichloridate, *in* M-00445
(1,1′-Biphenyl)-3-yl phosphorodichloridate, B-00082
(1,1′-Biphenyl)-4-yl phosphorodichloridate, *in* M-00446
Bis[(1,1′-biphenyl)-2-yl] phosphorochloridate, B-00102
Bis(2-*tert*-butylphenyl) phosphorochloridate, *in* B-00115
▷Bis(2-chloroethyl)phosphoramidic dichloride, B-00122
▷Bis(2-chloroethyl) phosphorochloridate, *in* B-00119
Bis(2-chloroethyl) phosphorofluoridate, *in* B-00119
Bis(2-chlorophenyl) phosphorochloridate, *in* B-00131
Bis(3-chlorophenyl) phosphorochloridate, *in* B-00132
Bis(4-chlorophenyl) phosphorochloridate, *in* B-00133
Bis(2-methoxyphenyl) phosphorochloridate, *in* B-00294
Bis(4-methoxyphenyl) phosphorochloridate, *in* B-00295
Bis(3-methylbutyl) phosphorochloridate, *in* B-00302
Bis(3-methylbutyl) phosphorofluoridate, *in* B-00302
Bis(2-methylphenyl) phosphorochloridate, *in* B-00312
Bis(3-methylphenyl) phosphorochloridate, *in* B-00313
Bis(4-methylphenyl) phosphorochloridate, B-00329
Bis(1-methylpropyl) phosphorochloridate, *in* B-00341
Bis(2-methylpropyl) phosphorochloridate, *in* B-00342
Bis(1-methylpropyl) phosphorofluoridate, *in* B-00341
Bis(2-methylpropyl) phosphorofluoridate, B-00362
Bis[(2-nitrophenyl)methyl] phosphorochloridate, *in* B-00366
Bis[(4-nitrophenyl)methyl] phosphorochloridate, *in* B-00367
Bis(4-nitrophenyl) phosphorochloridate, B-00376
Bis(2-oxo-3-oxazolidinyl)phosphinic acid; Chloride, *in* B-00378
Bis(pentachlorophenyl) phosphorochloridate, *in* B-00379
Bis(pentafluorophenyl) phosphorochloridate, *in* B-00381
Bis(2,2,2-trichloro-1,1-dimethylethyl) phosphorochloridate, B-00399
Bis(2,2,2-trichloroethyl) phosphorobromidate, *in* B-00400
Bis(2,2,2-trichloroethyl) phosphorochloridate, *in* B-00400
2-Bromoethyl phosphorodichloridate, *in* M-00447
(3-Bromophenyl)phosphoramidic acid; Dichloride, *in* B-00495
(4-Bromophenyl)phosphoramidic acid; Dichloride, *in* B-00496
▷Butafox, *in* D-00146
(4-*tert*-Butyl-2-chlorophenyl) 2,2,2-tribromoethyl phosphorochloridate, B-00548
2,3-Butylene chlorophosphate, *in* C-00056
1,3-Butylene phosphoryl bromide, *in* B-00474
Butyl phenyl phosphorochloridate, *in* P-00269
2-*tert*-Butylphenyl phosphorodichloridate, *in* M-00449
4-*tert*-Butylphenyl phosphorodichloridate, *in* M-00450
Butylphosphoramidic acid; Difluoride, *in* B-00621
Butylphosphoramidic acid; Dichloride, *in* B-00621
tert-Butylphosphoramidic acid; Difluoride, *in* B-00622
tert-Butylphosphoramidic acid; Dichloride, *in* B-00622
▷Butyl phosphorodichloridate, B-00628
(4-Carbethoxyphenyl)phosphoramidic acid; Dichloride, *in* C-00006
2-Chloro-4*H*-1,3,2-benzodioxaphosphorinan-4-one; 2-Oxide, *in* C-00022
3-(Chlorocarbonyl)phenyl phosphorodichloridate, C-00028
4-(Chlorocarbonyl)phenyl phosphorodichloridate, C-00029
2-Chloro-3-(2-chloroethyl)tetrahydro-2*H*-1,3,2-oxazaphosphorine; (±)-*form*, 2-Oxide, *in* C-00030
2-Chloro-5-chloromethyl-5-methyl-1,3,2-dioxaphosphorinane; *cis-form*, 2-Oxide, *in* C-00031
2-Chloro-5-chloromethyl-5-methyl-1,3,2-dioxaphosphorinane; *trans-form*, 2-Oxide, *in* C-00031
3-Chloro-1,5-dihydro-2,4,3-benzodioxaphosphepin; 3-Oxide, *in* C-00038
2-Chloro-2,3-dihydro-3,5-diphenyl-1,3,4,2-oxadiazaphosphole; 2-Oxide, *in* C-00042
2-Chloro-2,3-dihydro-5-methyl-3-phenyl-1,3,4,2-oxadiazaphosphole; 2-Oxide, *in* C-00044
2-Chloro-2,3-dihydro-3-phenyl-4*H*-1,3,2-benzoxazaphosphorin-4-one; 2-Oxide, *in* C-00050
2-Chloro-1,3-dimethyl-1,3,2-diazaphospholidine; 2-Oxide, *in* C-00055
2-Chloro-5,5-dimethyl-1,3,2-dioxaphosphorinane 2-oxide, C-00059
2-Chloro-5,5-dimethyl-4-phenyl-1,3,2-dioxaphosphorinane 2-oxide, *in* H-00129
2-Chloro-3,4-dimethyl-5-phenyl-1,3,2-oxazaphospholidine; (2*R*,4*S*,5*R*)-*form*, 2-Oxide, *in* C-00061
2-Chloro-3,4-dimethyl-5-phenyl-1,3,2-oxazaphospholidine; (2*S*,4*S*,5*R*)-*form*, 2-Oxide, *in* C-00061
2-Chloro-1,3,2-dioxaphospholane 2-oxide, C-00064
2-Chloro-1,3,2-dioxaphosphorinane 2-oxide, C-00067
2-Chloro-4,5-diphenyl-1,3,2-dioxaphospholane; (2α,4β,5β)-*form*, 2-Oxide, *in* C-00069
1-Chloroethyl phosphorodichloridate, C-00079
2-Chloroethyl phosphorodichloridate, *in* M-00454
2-Chlorohexahydro-4*H*-1,3,2-benzodioxaphosphorin; (2*RS*,4*aSR*,8*aSR*)-*form*, 2-Oxide, *in* C-00085
2-Chloro-4-isopropyl-5,5-dimethyl-1,3,2-dioxaphosphorinane; (2*RS*,4*RS*)-*form*, 2-Oxide, *in* C-00090
2-Chloro-4-isopropyl-5,5-dimethyl-1,3,2-dioxaphosphorinane; (2*RS*,4*SR*)-*form*, 2-Oxide, *in* C-00090
2-Chloro-4-methyl-1,3,2-dioxaphosphorinane 2-oxide; (2*RS*,4*RS*)-*form*, *in* C-00094
2-Chloro-4-methyl-1,3,2-dioxaphosphorinane 2-oxide; (2*RS*,4*SR*)-*form*, *in* C-00094
5-Chloromethyl-2-fluoro-5-methyl-1,3,2-dioxaphosphorinane; *cis-form*, 2-Oxide, *in* C-00096
5-Chloromethyl-2-fluoro-5-methyl-1,3,2-dioxaphosphorinane; *trans-form*, 2-Oxide, *in* C-00096
(2-Chloromethyl-4-nitrophenyl) phosphorodichloridate, *in* M-00455
4-Chlorophenyl 5-chloro-8-quinolinyl phosphorochloridate, *in* M-00456
4-Chlorophenyl 2-cyanoethyl phosphorochloridate, C-00125
2-Chloro-5-phenyl-1,3,4,2-dioxazaphosphole; 2-Oxide, *in* C-00126
2-Chlorophenyl (4-methoxyphenyl)phosphoramidochloridate, C-00130
2-Chloro-3-phenyl-1,3,2-oxazaphospholidine; 2-Oxide, *in* C-00138
2-Chlorophenyl phenylphosphoramidochloridate, *in* P-00259
4-Chlorophenyl phenylphosphoramidochloridate, *in* P-00259
(4-Chlorophenyl)phosphoramidic acid; Dichloride, *in* C-00157
2-Chlorophenyl phosphorodichloridate, C-00161
4-Chlorophenyl phosphorodichloridate, C-00162
4-Chlorophenyl phosphorodifluoridate, *in* M-00458
2-(Chlorosulfonyl)phenyl phosphorodichloridate, C-00175
4-(Chlorosulfonyl)phenyl phosphorodichloridate, C-00176
2-Chloro-1,2,3,4-tetrahydro-6-methyl-3-(4-methylphenyl)-1,3,2-benzodiazaphosphorine 2-oxide, C-00177
2-Chlorotetrahydro-2*H*-1,3,2-oxazaphosphorine; 2-Oxide, *in* C-00180
2-Chlorotetrahydro-3-(1-phenylethyl)-2*H*-1,3,2-oxazaphosphorine 2-oxide; (R_C,R_P)-*form*, *in* C-00181
2-Chlorotetrahydro-3-(1-phenylethyl)-2*H*-1,3,2-oxazaphosphorine 2-oxide; (R_C,S_P)-*form*, *in* C-00181
2-Chlorotetrahydro-3-(1-phenylethyl)-2*H*-1,3,2-oxazaphosphorine 2-oxide; (S_C,R_P)-*form*, *in* C-00181
2-Chlorotetrahydro-3-(1-phenylethyl)-2*H*-1,3,2-oxazaphosphorine 2-oxide; (S_C,S_P)-*form*, *in* C-00181
6-Chloro-5,6,7,12-tetrahydro-2,5,7,10-tetramethyldibenzo[*d*,*g*][1,3,2]diazaphosphocine; 6-Oxide, *in* C-00182

2-Chlorotetrahydro-3,5,5-trimethyl-2*H*-1,3,2-oxazaphosphorine; 2-Oxide, *in* C-00183
Cyanophosphoramidic dichloride, C-00227
Cyclohexylphosphoramidic acid; Dichloride, *in* C-00269
Decylphosphoramidic acid; Dichloride, *in* D-00013
▷Dibenzyl phosphorochloridate, D-00060
▷*N,N'*-Dibenzylphosphorodiamidic acid; Fluoride, *in* D-00061
(2,4-Dibromophenyl)phosphoramidic acid; Dichloride, *in* D-00067
3,5-Di-*tert*-butyl-2-chloro-2,3-dihydro-1,3,2-oxazaphosphole; 2-Oxide, *in* D-00078
Dibutylphosphoramidic acid; Dichloride, *in* D-00134
Di-*tert*-butyl phosphorobromidate, D-00138
Dibutyl phosphorochloridate, D-00140
Di-*tert*-butyl phosphorochloridate, D-00141
Dibutyl phosphorofluoridate, D-00153
2,2-Dichloroethenyl phosphorodichloridate, D-00168
2,4-Dichlorophenyl 4-morpholinylphosphonochloridate, *in* M-00524
(2,4-Dichlorophenyl)phosphoramidic acid; Dichloride, *in* D-00188
2,4-Dichlorophenyl phosphorodichloridate, D-00189
3,9-Dichloro-2,4,8,10-tetraoxa-3,9-diphosphaspiro[5.5]-undecane; 3,9-Dioxide, *in* D-00196
Dicyclohexyl phosphorochloridate, *in* D-00209
▷*N,N'*-Dicyclohexylphosphorodiamidic acid; Fluoride, *in* D-00216
▷Dicyclohexyl phosphorofluoridate, *in* D-00209
Didecyl phosphorochloridate, *in* D-00220
Diethenyl phosphorobromidate, *in* D-01257
Diethylphosphoramidic dichloride, D-00344
Diethylphosphoramidic difluoride, D-00345
Diethyl phosphorobromidate, D-00371
▷Diethyl phosphorochloridate, D-00373
▷*N,N'*-Diethylphosphorodiamidic acid; Fluoride, *in* D-00382
N,N'-Diethylphosphorodiamidic acid; Chloride, *in* D-00382
▷Diethyl phosphorofluoridate, D-00387
Diheptyl phosphorochloridate, *in* D-00424
Dihexyl phosphorochloridate, *in* D-00432
4,7-Dihydro-1,3,2-dioxaphosphepin 2-sulfide; 2-Chloro, *in* D-00473
▷Diisopropyl phosphorobromidate, *in* D-00639
Diisopropyl phosphorochloridate, D-00657
N,N'-Diisopropylphosphorodiamidic acid; Chloride, *in* D-00661
▷Diisopropyl phosphorofluoridate, D-00665
▷Dimefox, D-00669
(2,4-Dimethylphenyl)phosphoramidic acid; Dichloride, *in* D-00818
2,6-Dimethylphenyl phosphorodichloridate, *in* M-00466
3,5-Dimethylphenyl phosphorodichloridate, D-00822
▷Dimethylphosphoramidic dichloride, D-00856
Dimethylphosphoramidic difluoride, D-00858
Dimethyl phosphorobromidate, D-00886
Dimethyl phosphorochloridate, D-00887
▷Dimethyl phosphorofluoridate, D-00901
Dimethylvinylene chlorophosphate, *in* C-00057
Di-4-morpholinylphosphinic acid; Fluoride, *in* D-00935
Di-4-morpholinylphosphinic acid; Chloride, *in* D-00935
Di-1-naphthyl phosphorochloridate, *in* D-00937
2,4-Dinitrophenyl phosphorodichloridate, *in* M-00467
Dinonyl phosphorochloridate, *in* D-00948
Dioctyl phosphorochloridate, *in* D-00956
Dipentyl phosphorochloridate, *in* D-00979
▷Dipentyl phosphorofluoridate, *in* D-00979
Diphenylphosphoramidic acid; Difluoride, *in* D-01108
Diphenylphosphoramidic acid; Dichloride, *in* D-01108
Diphenyl phosphorobromidate, *in* D-01036
Diphenyl phosphorochloridate, D-01120
▷*N,N'*-Diphenylphosphorodiamidic chloride, D-01126
▷*N,N'*-Diphenylphosphorodiamidic fluoride, D-01127
Diphenyl phosphorofluoridate, D-01133
Di-2-propenyl phosphorochloridate, *in* D-01192
Di-2-propenyl phosphorofluoridate, *in* D-01192
Dipropylphosphoramidic acid; Difluoride, *in* D-01214
Dipropylphosphoramidic acid; Dichloride, *in* D-01214
Dipropyl phosphorobromidate, *in* D-01203
Dipropyl phosphorochloridate, D-01217
N,N'-Dipropylphosphorodiamidic acid; Fluoride, *in* D-01223
N,N'-Dipropylphosphorodiamidic acid; Chloride, *in* D-01223
▷Dipropyl phosphorofluoridate, D-01228
Divinyl phosphorochloridate, *in* D-01257
Ethenyl phosphorodichloridate, *in* M-00518
Ethyl dichlorophosphinylcarbamate, *in* D-00191
Ethyl (dichlorophosphinyl)phosphorodichloroimidate, E-00058
Ethyl diethylphosphoramidochloridate, *in* D-00346
Ethyl diethylphosphoramidofluoridate, *in* D-00351
Ethyl dimethylphosphoramido chloridate, *in* D-00860
▷Ethyl dimethylphosphoramidofluoridate, E-00064
Ethylimidobisphosphoryl fluoride, *in* E-00078
Ethyl phenylphosphoramidochloridate, *in* P-00259
Ethyl phenyl phosphorochloridate, *in* P-00269
Ethyl *N*-phenylphosphorodiamidate, *in* P-00273
Ethylphosphoramidic acid; Difluoride, *in* E-00156
Ethylphosphoramidic acid; Dichloride, *in* E-00156
Ethyl phosphorobromidofluoridate, E-00162
Ethyl phosphorochloridisocyanatidate, E-00163
Ethyl phosphorochloridofluoridate, E-00164
▷Ethyl phosphorodichloridate, E-00167
▷Ethyl phosphorodifluoridate, E-00173
Ethyl phosphorofluoridate, E-00179
Ethyl phosphorofluoridisocyanatidate, E-00180
2-Fluoro-1,3-dimethyl-1,3,2-diazaphospholidine; 2-Oxide, *in* F-00014
2-Fluoro-4-methyl-1,3,2-dioxaphosphorinane; (2*RS*,4*RS*)-*form*, 2-Oxide, *in* F-00021
Heptylphosphoramidic acid; Dichloride, *in* H-00009
Hexahydro-2-hydroxycyclopenta[*d*]-1,3,2-dioxaphosphorin 2-oxide; (4*aRS*,7*aSR*)-*form*, Chloride, *in* H-00034
Hexylphosphoramidic acid; Dichloride, *in* H-00102
2-Hydroxy-5,5-dimethyl-4-phenyl-1,3,2-dioxaphoshorinane 2-oxide; (±)-*form*, Chloride, *in* H-00129
Isopropyl dichlorophosphinylcarbamate, *in* D-00191
Isopropyl diethylphosphoramidochloridate, *in* D-00346
Isopropyl diethylphosphoramidofluoridate, *in* D-00351
Isopropyl phenyl phosphorochloridate, *in* P-00269
Isopropylphosphoramidic acid; Difluoride, *in* I-00058
Isopropylphosphoramidic acid; Dichloride, *in* I-00058
Isopropyl phosphorodichloridate, I-00060
(2-Methoxyphenyl)phosphoramidic acid; Difluoride, *in* M-00074
(4-Methoxyphenyl)phosphoramidic acid; Dichloride, *in* M-00076
2-Methoxyphenyl phosphorodichloridate, *in* M-00480
3-Methoxyphenyl phosphorodichloridate, M-00080
4-Methoxyphenyl phosphorodichloridate, M-00081
4-Methoxyphenyl phosphorodifluoridate, *in* M-00481
(2-Methyl-2-bromomethyl)trimethylene phosphoryl bromide, *in* B-00465
Methyl 6-deoxy-2,3-di-*O*-methyl-6-methylaminoglucopyranoside 4,6-cyclic phosphonamide; α-D-(*S*)$_P$-*form*, *P*-Chloro, *in* M-00112
Methyl dichlorophosphinylcarbamate, *in* D-00191
Methyl 2,3-di-*O*-methylglucopyranoside 4,6-cyclic phosphonate; α-D-(*S*)$_P$-*form*, *P*-Fluoro, *in* M-00119
Methyl 2,3-di-*O*-methylglucopyranoside 4,6-cyclic phosphonate; α-D-(*S*)$_P$-*form*, *P*-Chloro, *in* M-00119
Methyl dimethylphosphoramidochloridate, *in* D-00860
1-Methylethenyl phosphorodichloridate, M-00151
Methylimidobisphosphoryl fluoride, *in* M-00156
Methylimidodiphosphoryl chloride, M-00160
(2-Methylphenyl)phosphoramidic acid; Difluoride, *in* M-00251
(2-Methylphenyl)phosphoramidic acid; Dichloride, *in* M-00251
(3-Methylphenyl)phosphoramidic acid; Difluoride, *in* M-00252
(4-Methylphenyl)phosphoramidic acid; Difluoride, *in* M-00253
(4-Methylphenyl)phosphoramidic acid; Dichloride, *in* M-00253
Methyl phenyl phosphorochloridate, *in* P-00269
2-Methylphenyl phosphorodichloridate, *in* M-00485
3-Methylphenyl phosphorodichloridate, *in* M-00486
4-Methylphenyl phosphorodichloridate, M-00257
2-Methylphenyl phosphorodifluoridate, *in* M-00485
4-Methylphenyl phosphorodifluoridate, *in* M-00487
Methylphosphoramidic dichloride, M-00327
Methylphosphoramidic difluoride, M-00328
Methyl phosphorochloridofluoridate, M-00341
Methyl phosphorodichloridate, M-00345
▷Methyl phosphorodifluoridate, M-00352

(2-Methylpropyl)phosphoramidic acid; Dichloride, *in* M-00384
Methylsulfonylphosphoramidic acid; Dichloride, *in* M-00401
▷Mipafox, *in* D-00661
4-Morpholinylphosphonochloridic acid, M-00524
4-Morpholinylphosphonochloridic acid; Anhydride with diphenyl phosphate, *in* M-00524
2,3-Naphthylene phosphorochloridate, *in* C-00119
1-Naphthylphosphoramidic acid; Dichloride, *in* N-00028
2-Naphthylphosphoramidic acid; Dichloride, *in* N-00029
1-Naphthyl phosphorodichloridate, *in* M-00494
2-Naphthyl phosphorodichloridate, *in* M-00495
1-Naphthyl phosphorodifluoridate, *in* M-00494
Neopentylene phosphorobromidate, *in* B-00471
Neopentylene phosphorofluoridate, *in* F-00016
4-Nitrophenyl 4-morpholinylphosphonochloridate, *in* M-00524
4-Nitrophenyl phenylphosphoramidochloridate, *in* P-00259
4-Nitrophenyl phenyl phosphorochloridate, *in* N-00052
(3-Nitrophenyl)phosphoramidic acid; Dichloride, *in* N-00059
(4-Nitrophenyl)phosphoramidic acid; Dichloride, *in* N-00060
2-Nitrophenyl phosphorodichloridate, *in* M-00498
3-Nitrophenyl phosphorodichloridate, *in* M-00499
4-Nitrophenyl phosphorodichloridate, N-00064
2-Nitrophenyl phosphorodifluoridate, *in* M-00498
4-Nitrophenyl phosphorodifluoridate, *in* M-00500
Nonylphosphoramidic acid; Dichloride, *in* N-00077
Octylphosphoramidic acid; Dichloride, *in* O-00045
Pentachlorophenyl phosphorodichloridate, P-00006
Pentafluorophenyl phosphorodichloridate, *in* M-00503
Pentafluorophenyl phosphorodichloridate, P-00017
Pentylphosphoramidic acid; Dichloride, *in* P-00053
2-Phenoxyphenyl phosphorodichloridate, *in* M-00505
(Phenylamino)carbonylphosphoramidic acid; Dichloride, *in* P-00081
[3-(Phenylamino)propyl]phosphonic acid, P-00085
Phenyl cyclohexylphosphoramidochloridate, P-00104
Phenyl diethylphosphoramidochloridate, *in* D-00346
Phenyl dimethylphosphoramidochloridate, *in* D-00860
1,3-Phenylene bisphosphorodichloridate, *in* P-00147
1,4-Phenylenebis(phosphorodichloridate), *in* P-00148
1,4-Phenylenebis(phosphorodifluoridate), *in* P-00148
o-Phenylene phosphorobromidate, *in* B-00464
2-Phenylethyl phosphorodichloridate, *in* M-00506
Phenyl 4-morpholinylphosphonochloridate, *in* M-00524
Phenyl phenylphosphoramidic acid; Chloride, *in* P-00189
Phenyl phenylphosphoramidochloridate, P-00190
Phenylphosphoramidic acid; Difluoride, *in* P-00257
Phenylphosphoramidic dichloride, P-00258
Phenylphosphoramidochloridic acid, P-00259
Phenyl phosphoramidofluoridate, *in* P-00295
Phenyl phosphorochloridic acid, P-00269
Phenyl phosphorochloridisocyanatidate, P-00270
Phenyl phosphorochloridofluoridate, P-00271
Phenyl phosphorodichloridate, P-00279
Phenyl phosphorodifluoridate, P-00285
Phenyl phosphorofluoridic acid, P-00295
Phenyl propyl phosphorochloridate, *in* P-00269
Phenylsulfonylphosphoramidic acid; Monobromide, *in* P-00312
Phenylsulfonylphosphoramidic acid; Dibromide, *in* P-00312
Phenylsulfonylphosphoramidic dichloride, P-00313
Phosphorisocyanatidic dichloride, P-00405
Phosphorisocyanatidic difluoride, P-00406
Phosphor(isothiocyanatidic) dichloride, P-00407
Phosphor(isothiocyanatidic) difluoride, P-00408
Phosphorodiisocyanatidic bromide, P-00411
Phosphorodiisocyanatidic chloride, P-00412
Phosphorodiisocyanatidic fluoride, P-00413
Phosphorodi(isothiocyanatidic) fluoride, P-00414
2-Phthalimidoethyl phosphorodichloridate, P-00425
2-Propenylphosphoramidic acid; Dichloride, *in* P-00464
Propylene phosphorochloridate, *in* C-00092
Propylphosphoramidic acid; Dichloride, *in* P-00484
▷Propyl phosphorodichloridate, P-00487
2-Propynyl phosphorodichloridate, P-00497
Tetrabutylphosphorodiamidic chloride, T-00022
▷Tetraethylphosphorodiamidic acid; Fluoride, *in* T-00059
Tetraethylphosphorodiamidic chloride, T-00060
Tetrakis(trifluoromethyl)ethylene chlorophosphate, *in* C-00184
Tetramethylethylene chlorophosphate, *in* C-00185
Tetramethylphosphorodiamidic acid; Bromide, *in* T-00218
Tetramethylphosphorodiamidic chloride, T-00220
Tetrapropylphosphorodiamidic acid; Chloride, *in* T-00291
2,2,2-Tribromoethyl 4-*tert*-butylphenylphosphoramidic chloride, T-00356
2,2,2-Tribromoethyl 4-morpholinylphosphonochloridate, T-00357
2,2,2-Tribromoethyl phenylphosphoramidochloridate, *in* P-00259
2,2,2-Tribromoethyl phosphorodichloridate, *in* M-00516
(2,4,6-Tribromophenyl)phosphoramidic acid; Dichloride, *in* T-00359
2,2,2-Trichloroethyl 2-chlorophenyl phosphorochloridate, T-00405
2,2,2-Trichloroethyl phenylphosphoramidochloridate, *in* P-00259
2,2,2-Trichloroethyl phosphorodichloridate, *in* T-00424
(2,4,6-Trichlorophenyl)phosphoramidic acid; Dichloride, *in* T-00420
▷Trimethylene fluorophosphate, *in* F-00018
Trimethylsilyl phosphorodichloridate, T-00558
Trimethylsilyl phosphorodifluoridate, T-00559
(Triphenylphosphoranylidene)phosphoramidic dichloride, T-00656

Phosphoric pseudohalides

2-Cyano-5,5-dimethyl-1,3,2-dioxaphosphorinane; 2-Oxide, *in* C-00209
Cyclohexyl dimethylphosphoramidocyanidate, *in* D-00863
Diethyl phosphorocyanidate, D-00381
Dimethylphosphoramidic dicyanide, D-00857
Dimethyl phosphorocyanidate, D-00894
Diphenyl phosphorocyanidate, D-01124
Isopropyl dimethylphosphoramidocyanidate, *in* D-00863
Methyl dimethylphosphoramidocyanidate, *in* D-00863
▷Tabun, T-00001

Phosphoric esters

Acetic acid anhydride with dibenzyl hydrogen phosphate, *in* A-00025
Acetic acid anhydride with diethyl hydrogen phosphate, *in* A-00025
Acetic acid anhydride with dimethyl hydrogen phosphate, *in* A-00025
2-Amino-3-[(diphenoxyphosphinyl)oxy]propanoic acid, *in* P-00417
4-Aminophenyl diethyl phosphate, A-00098
Benzyl *N*-(diphenoxyphosphinyl)carbamate, *in* D-00989
Benzyl 3-[(diphenoxyphosphinyl)oxy]propanoate, *in* P-00399
Bicyclic phosphate, *in* T-00588
2,2′-Biphenyldiyl methyl phosphate, *in* M-00032
O,O-Bis(1-benzotriazolyl) 2-chlorophenyl phosphate, B-00098
O,O-Bis(1-benzotriazolyl) morpholinyl phosphate, B-00099
O,O-Bis(1-benzotriazolyl) 2,2,2-tribromoethyl phosphate, B-00100
Bis(2,4-dichlorophenyl) ethyl phosphate, B-00162
Bis(2-methylpropyl) bis(2-methylpropyl)phosphoramidate, *in* B-00355
▷Bomyl, B-00461
▷Bromfenvinfos, B-00462
▷Bromfenvinfos-methyl, B-00463
tert-Butoxycarbonyl diethyl phosphate, B-00531
▷*tert*-Butyl bicyclic phosphate, *in* B-00640
tert-Butyl 1,3-butylene phosphate, *in* B-00537
Butyl *N,N′*-dibutylphosphorodiamidate, *in* D-00146
tert-Butyl dimethylvinylene phosphate, *in* H-00123
2,3-Butylene ethyl phosphate, *in* E-00031
2,3-Butylene methyl phosphate, *in* M-00033
1,3-Butylene phenyl phosphate, *in* M-00191
2,3-Butylene phosphonate, *in* D-00733
tert-Butyl ethylene phosphate, *in* B-00535
5-*tert*-Butyl-2-methoxy-1,3,2-dioxaphosphorinane; *cis-form*, 2-Oxide, *in* B-00563
tert-Butyl neopentylene phosphate, *in* B-00534

2-*tert*-Butylphenyl phosphorodichloridate, *in* M-00449
4-*tert*-Butylphenyl phosphorodichloridate, *in* M-00450
Butyl tetrabutylphosphorodiamidate, *in* T-00021
Butyl tetramethylphosphorodiamidate, *in* T-00218
▷Chlorfenvinphos, C-00018
3-(Chlorocarbonyl)phenyl phosphorodichloridate, C-00028
4-(Chlorocarbonyl)phenyl phosphorodichloridate, C-00029
▷2-Chloro-1-(2,4-dichlorophenyl)vinyl dimethyl phosphate, C-00035
1-Chloroethyl phosphorodichloridate, C-00079
2-Chloroethyl phosphorodichloridate, *in* M-00454
5-Chloromethyl-2-methoxy-5-methyl-1,3,2-dioxaphosphorinane; *cis-form*, 2-Oxide, *in* C-00098
5-Chloromethyl-2-methoxy-5-methyl-1,3,2-dioxaphosphorinane; *trans-form*, 2-Oxide, *in* C-00098
5-Chloromethyl-5-methyl-2-phenoxy-1,3,2-dioxaphosphorinane; 2-Oxide, *in* C-00099
4-Chlorophenyl 5-chloro-8-quinolinyl phosphorochloridate, *in* M-00456
4-Chlorophenyl 2-cyanoethyl phosphorochloridate, C-00125
4-Chlorophenyl *N,N'*-diphenylphosphorodiamidate, *in* M-00458
2-Chlorophenyl (4-methoxyphenyl)phosphoramidochloridate, C-00130
2-Chlorophenyl phenylphosphoramidochloridate, *in* P-00259
4-Chlorophenyl phenylphosphoramidochloridate, *in* P-00259
2-Chlorophenyl phosphorodichloridate, C-00161
4-Chlorophenyl phosphorodichloridate, C-00162
4-Chlorophenyl phosphorodifluoridate, *in* M-00458
2-(Chlorosulfonyl)phenyl phosphorodichloridate, C-00175
4-(Chlorosulfonyl)phenyl phosphorodichloridate, C-00176
Crotoxyphos, C-00207
Cyclohexyl di-4-morpholinylphosphinate, *in* D-00935
▷Diamidafos, D-00024
Dibenzyl 1-adamantylphosphoramidate, *in* A-00036
Dibenzyl benzoylphosphoramidate, *in* B-00036
Dibenzyl cyclohexylphosphoramidate, *in* C-00269
Dibenzyl diethylphosphoramidate, *in* D-00343
P,P-Dibenzyl *P',P'*-diphenyl diphosphate, D-00044
Dibenzyl methylphenylphosphoramidate, *in* M-00250
Dibenzyl (4-methylphenyl)phosphoramidate, *in* M-00253
Dibenzyl 2-naphthylphosphoramidate, *in* N-00029
2-[(Dibenzyloxyphosphinyl)oxy]-2-propenoamide, *in* P-00366
Dibenzyl phenylphosphoramidate, *in* D-00053
Dibenzyl phosphate, D-00053
Dibenzyl phosphoramidate, *in* D-00053
▷Dibenzyl phosphorochloridate, D-00060
Di-*tert*-butyl benzylphosphoramidate, *in* B-00069
Dibutyl butylphosphoramidate, *in* B-00621
Dibutyl *tert*-butylphosphoramidate, *in* B-00622
Di-*tert*-butyl cyclohexylphosphoramidate, *in* C-00269
Dibutyl decylphosphoramidate, *in* D-00013
Dibutyl dibutylphosphoramidate, *in* D-00134
Dibutyl diethylphosphoramidate, *in* D-00343
Dibutyl ethylphosphoramidate, *in* E-00156
Dibutyl methylphosphoramidate, *in* M-00326
Dibutyl (2-methylpropyl)phosphoramidate, *in* M-00384
2,2-Dichloroethenyl phosphorodichloridate, D-00168
2,2-Dichlorovinyl methyl octyl phosphate, D-00202
▷Dichlorvos, D-00203
▷Dicrotophos, D-00204
Didodecyl phosphate, D-00225
Diethyl acetylphosphoramidate, *in* A-00029
Diethyl 1-adamantylphosphoramidate, *in* A-00036
Diethyl aminocarbonylphosphoramidate, *in* A-00062
Diethyl (2-aminoethyl)phosphoramidate, *in* A-00080
Diethyl benzoylphosphoramidate, *in* B-00036
Diethyl benzylphosphoramidate, *in* B-00069
Diethyl (4-bromophenyl)phosphoramidate, *in* B-00496
Diethyl butylphosphoramidate, *in* B-00621
Diethyl *tert*-butylphosphoramidate, *in* B-00622
Diethyl (4-chlorophenyl)phosphoramidate, *in* C-00157
Diethyl cyclohexylphosphoramidate, *in* C-00269
Diethyl (2,4-dibromophenyl)phosphoramidate, *in* D-00067
Diethyl dibromophosphoramidate, D-00279
▷Diethyl dibutylphosphoramidate, *in* D-00134
Diethyl (2,4-dichlorophenyl)phosphoramidate, *in* D-00188
▷Diethyl dichlorophosphoramidate, D-00280
Diethyl 2-dimethylaminoethyl phosphate, D-00283
Diethyl (2,4-dimethylphenyl)phosphoramidate, *in* D-00818
Diethyl diphenylphosphoramidate, *in* D-01108
Diethyl di-2-propenylphosphoramidate, *in* D-01195
Diethyl dipropylphosphoramidate, *in* D-01214
Diethyl ethylphenylphosphoramidate, *in* E-00116
Diethyl (3-fluorophenyl)phosphoramidate, *in* F-00043
Diethyl (4-fluorophenyl)phosphoramidate, *in* F-00044
Diethyl isopropylphosphoramidate, *in* I-00058
Diethyl (4-methoxyphenyl)phosphoramidate, *in* M-00076
Diethyl methylphenylphosphoramidate, *in* M-00250
Diethyl (2-methylphenyl)phosphoramidate, *in* M-00251
Diethyl (3-methylphenyl)phosphoramidate, *in* M-00252
Diethyl (4-methylphenyl)phosphoramidate, *in* M-00253
Diethyl (2-methylpropyl)phosphoramidate, *in* M-00384
Diethyl methylsulfonylphosphoramidate, *in* M-00401
Diethyl 1-naphthylphosphoramidate, *in* N-00028
Diethyl 2-naphthylphosphoramidate, *in* N-00029
▷Diethyl 2-nitrophenyl phosphate, D-00303
▷Diethyl 3-nitrophenyl phosphate, D-00304
Diethyl (3-nitrophenyl)phosphoramidate, *in* N-00059
Diethyl (4-nitrophenyl)phosphoramidate, *in* N-00060
Diethyl octylphosphoramidate, *in* O-00045
Diethyl pentylphosphoramidate, *in* P-00053
Diethyl (phenylamino)carbonylphosphoramidate, *in* P-00081
Diethyl phenylphosphoramidate, D-00312
Diethyl phenylsulfonylphosphoramidate, *in* P-00312
Diethyl phosphate, D-00320
Diethyl phosphoramidate, D-00342
Diethylphosphoramidic dichloride, D-00344
Diethyl phosphorisocyanatidate, D-00366
Diethyl phosphor(isothiocyanatidate), D-00369
Diethyl phosphorobromidate, D-00371
▷Diethyl phosphorochloridate, D-00373
Diethyl propylphosphoramidate, *in* P-00484
Diethyl 2-pyridylphosphoramidate, *in* P-00510
Diethyl 3-pyridylphosphoramidate, *in* P-00511
Diethyl (2,4,6-tribromophenyl)phosphoramidate, *in* T-00359
Diethyl (trimethylsilylphenyl)phosphoramidate, *in* D-00312
Diethyl trimethylsilyl phosphate, D-00407
4,7-Dihydro-1,3,2-dioxaphosphepin 2-sulfide; 2-Phenoxy, *in* D-00473
1,5-Dihydro-2-phenoxy-2,4,3-benzodioxaphosphepin; 3-Oxide, *in* D-00544
Diisopropyl aminocarbonylphosphoramidate, *in* A-00062
Diisopropyl 2-aminoethyl phosphate, *in* A-00075
Diisopropyl benzoylphosphoramidate, *in* B-00036
Diisopropyl benzylphosphoramidate, *in* B-00069
Diisopropyl cyclohexylphosphoramidate, *in* C-00269
Diisopropyl dibutylphosphoramidate, *in* D-00134
Diisopropyl diethylphosphoramidate, *in* D-00343
Diisopropyl diisobutylphosphoramidate, *in* B-00355
Diisopropyl dimethylphosphoramidate, *in* D-00855
Diisopropyl (3-fluorophenyl)phosphoramidate, *in* F-00043
Diisopropyl (4-fluorophenyl)phosphoramidate, *in* F-00044
Diisopropyl isopropylphosphoramidate, *in* I-00058
Diisopropyl (phenylamino)carbonylphosphoramidate, *in* P-00081
Diisopropyl 2-pyridylphosphoramidate, *in* P-00510
Diisopropyl 3-pyridylphosphoramidate, *in* P-00511
Dimethyl acetylphosphoramidate, *in* A-00029
Dimethyl 1-adamantylphosphoramidate, *in* A-00036
Dimethyl aminocarbonylphosphoramidate, *in* A-00062
Dimethyl benzoylphosphoramidate, *in* B-00036
Dimethyl benzylphosphoramidate, *in* B-00069
Dimethyl butylphosphoramidate, *in* B-00621
Dimethyl *tert*-butylphosphoramidate, *in* B-00622
Dimethyl (4-chlorophenyl)phosphoramidate, *in* C-00157
Dimethyl cyclohexylphosphoramidate, *in* C-00269
Dimethyl dibutylphosphoramidate, *in* D-00134
Dimethyl diethylphosphoramidate, *in* D-00343
Dimethyl dipropylphosphoramidate, *in* D-01214
Dimethyl ethylphosphoramidate, *in* E-00156
Dimethyl (3-fluorophenyl)phosphoramidate, *in* F-00043
Dimethyl (4-fluorophenyl)phosphoramidate, *in* F-00044

Dimethyl hexylphosphoramidate, *in* H-00102
Dimethyl isopropylphosphoramidate, *in* I-00058
Dimethyl (2-methoxyphenyl)phosphoramidate, *in* M-00074
Dimethyl (3-methoxyphenyl)phosphoramidate, *in* M-00075
Dimethyl (4-methoxyphenyl)phosphoramidate, *in* M-00076
Dimethyl methylphenylphosphoramidate, *in* M-00250
Dimethyl (2-methylphenyl)phosphoramidate, *in* M-00251
Dimethyl (3-methylphenyl)phosphoramidate, *in* M-00252
Dimethyl (4-methylphenyl)phosphoramidate, *in* M-00253
Dimethyl methylphosphoramidate, *in* M-00326
Dimethyl methylsulfonylphosphoramidate, *in* M-00401
▷Dimethyl 4-(methylthio)phenyl phosphate, D-00767
▷Dimethyl 2-nitrophenyl phosphate, D-00768
▷Dimethyl 3-nitrophenyl phosphate, D-00769
▷Dimethyl 4-nitrophenyl phosphate, D-00770
Dimethyl (3-nitrophenyl)phosphoramidate, *in* N-00059
Dimethyl (4-nitrophenyl)phosphoramidate, *in* N-00060
Dimethyl octylphosphoramidate, *in* O-00045
4,5-Dimethyl-2-phenoxy-1,3,2-dioxaphospholane; (4*RS*,5*SR*)-*form*, 2-Oxide, *in* D-00789
Dimethyl (phenylamino)carbonylphosphoramidate, *in* P-00081
Dimethyl phenylsulfonylphosphoramidate, *in* P-00312
▷Dimethylphosphoramidic dichloride, D-00856
Dimethyl phosphorochloridate, D-00887
Dimethyl phosphorohydrazidate, D-00904
Dimethyl propylphosphoramidate, *in* P-00484
Dimethyl 2-pyridylphosphoramidate, *in* P-00510
Dimethyl 3-pyridylphosphoramidate, *in* P-00511
Dimethyl trimethylsilyl phosphate, D-00931
Dimethylvinylene phenyl phosphate, *in* H-00123
Dimethylvinylene trimethylsilyl phosphate, *in* H-00123
Di-1-naphthyl phosphate, D-00937
Di-2-naphthyl phosphate, D-00938
Di-2-naphthyl phosphoramidate, *in* D-00938
Di-1-naphthyl phosphorochloridate, *in* D-00937
Dioctadecyl phosphate, D-00952
N-(Diphenoxyphosphinyl)-4-aminobenzoic acid, *in* P-00377
3,9-Diphenoxy-2,4,8,10-tetraoxa-3,9-diphosphaspiro[5.5]-undecane; Monooxide, *in* D-00991
3,9-Diphenoxy-2,4,8,10-tetraoxa-3,9-diphosphaspiro[5.5]-undecane; Dioxide, *in* D-00991
Diphenyl 1-adamantylphosphoramidate, *in* A-00036
Diphenyl aminocarbonylphosphoramidate, *in* A-00062
Diphenyl benzoylphosphoramidate, *in* B-00036
Diphenyl benzylphosphoramidate, *in* B-00069
Diphenyl bis(2-methylpropyl)phosphoramidate, *in* B-00355
Diphenyl (2-bromophenyl)phosphoramidate, *in* B-00494
Diphenyl (3-bromophenyl)phosphoramidate, *in* B-00495
Diphenyl (4-bromophenyl)phosphoramidate, *in* B-00496
Diphenyl butylphosphoramidate, *in* B-00621
Diphenyl *tert*-butylphosphoramidate, *in* B-00622
Diphenyl (2-chlorophenyl)phosphoramidate, *in* C-00155
Diphenyl (3-chlorophenyl)phosphoramidate, *in* C-00156
Diphenyl (4-chlorophenyl)phosphoramidate, *in* C-00157
Diphenyl cyclohexylphosphoramidate, *in* C-00269
Diphenyl (2,4-dibromophenyl)phosphoramidate, *in* D-00067
Diphenyl (2,4-dichlorophenyl)phosphoramidate, *in* D-00188
Diphenyl diethylphosphoramidate, *in* D-00343
Diphenyl (2,4-dimethylphenyl)phosphoramidate, *in* D-00818
Diphenyl dimethylphosphoramidate, *in* D-00855
Diphenyl (2,4-dinitrophenyl)phosphoramidate, *in* D-00947
Diphenyl diphenylphosphoramidate, *in* D-01108
P,P′-Diphenyl diphosphate, D-01011
Diphenyl di-2-propenylphosphoramidate, *in* D-01195
Diphenyl dipropylphosphoramidate, *in* D-01214
Diphenyl [4-(ethoxycarbonyl)phenyl]phosphoramidate, *in* C-00006
Diphenyl ethylphosphoramidate, *in* E-00156
Diphenyl hexylphosphoramidate, *in* H-00102
Diphenyl isopropylphosphoramidate, *in* I-00058
Diphenyl (4-methoxyphenyl)phosphoramidate, *in* M-00076
Diphenyl methylphenylphosphoramidate, *in* M-00250
Diphenyl (2-methylphenyl)phosphoramidate, *in* M-00251
Diphenyl (3-methylphenyl)phosphoramidate, *in* M-00252
Diphenyl (4-methylphenyl)phosphoramidate, *in* M-00253
Diphenyl methylphosphoramidate, *in* M-00326
Diphenyl (2-methylpropyl)phosphoramidate, *in* M-00384
Diphenyl (2-nitrophenyl)phosphoramidate, *in* N-00058
Diphenyl (3-nitrophenyl)phosphoramidate, *in* N-00059
Diphenyl (4-nitrophenyl)phosphoramidate, *in* N-00060
Diphenyl (phenylamino)carbonylphosphoramidate, *in* P-00081
Diphenyl phenylsulfonylphosphoramidate, *in* P-00312
Diphenyl phosphorisocyanatidate, D-01116
Diphenyl phosphorochloridate, D-01120
Diphenyl phosphorocyanidate, D-01124
Diphenyl propylphosphoramidate, *in* P-00484
Diphenyl 2-pyridylphosphoramidate, *in* P-00510
Diphenyl 3-pyridylphosphoramidate, *in* P-00511
Diphenyl 4-pyridylphosphoramidate, *in* P-00512
Diphenyl succinimido phosphate, D-01148
Diphenyl (2,4,6-tribromophenyl)phosphoramidate, *in* T-00359
Diphenyl (2,4,6-trichlorophenyl)phosphoramidate, *in* T-00420
Dipropyl dipropylphosphoramidate, *in* D-01214
Dipropyl ethylphosphoramidate, *in* E-00156
Dipropyl methylphosphoramidate, *in* M-00326
Dipropyl propylphosphoramidate, *in* P-00484
2-Ethoxy-4,5-dimethyl-1,3,2-dioxaphospholane; (4*RS*,5*RS*)-*form*, 2-Oxide, *in* E-00031
2-Ethoxy-1,3,2-dioxaphosphepane 2-oxide, *in* H-00132
2-Ethoxy-1,3,2-dioxaphospholane 2-oxide, E-00035
2-[(Ethoxyhydroxyphosphinyl)oxy]-*N,N,N*-trimethylethanaminium hydroxide, inner salt, *in* C-00198
Ethyl 2-amino-3-[(diphenoxyphosphinyl)oxy]propanoate, *in* P-00417
▷Ethyl bicyclic phosphate, *in* E-00196
Ethyl *N,N′*-dibutylphosphorodiamidate, *in* D-00146
Ethyl *N,N′*-dicyclohexylphosphorodiamidate, *in* D-00216
Ethyl 2-[(diethoxyphosphinyl)amino]benzoate, *in* C-00004
Ethyl *N*-(diethoxyphosphinyl)carbamate, *in* D-00249
Ethyl 2-(diethoxyphosphinyl)oxybenzoate, *in* P-00391
Ethyl 2-[(diethoxyphosphinyl)oxy]propanoate, *in* D-00255
Ethyl 2-[(diethoxyphosphinyl)oxy]-2-propenoate, *in* D-00256
Ethyl *N,N′*-diethylphosphorodiamidate, *in* D-00382
Ethyl *N*-(dimethoxyphosphinyl)carbamate, *in* D-00680
Ethyl 3-[(dimethoxyphosphinyl)oxy]crotonate, *in* D-00681
Ethyl *N,N′*-dimethylphosphorodiamidate, *in* D-00895
Ethyl dimethylvinylene phosphate, *in* H-00123
Ethyl di-4-morpholinylphosphinate, *in* D-00935
Ethyl 2-[(diphenoxyphosphinyl)amino]benzoate, *in* C-00004
Ethyl *N*-(diphenoxyphosphinyl)carbamate, *in* D-00989
Ethyl *N,N′*-diphenylphosphorodiamidate, *in* D-01125
Ethyl *N,N′*-dipropylphosphorodiamidate, *in* D-01223
Ethylene phenyl phosphate, *in* P-00067
Ethylene trimethylsilyl phosphate, *in* T-00554
▷2-Ethylhexyl diphenyl phosphate, E-00075
Ethyl hydrogen 2-aminoethyl phosphate, *in* A-00075
Ethyl 1,2-naphthylene phosphate, *in* E-00044
Ethyl 1,8-naphthylene phosphate, *in* E-00045
Ethyl 1,8-naphthylene phosphate, *in* E-00046
Ethyl neopentylene phosphate, *in* E-00032
Ethyl *o*-phenylene phosphate, *in* E-00024
Ethyl phenylphosphoramidochloridate, *in* P-00259
Ethyl 4-(phosphonoxy)benzoate, *in* P-00397
Ethyl phosphorobromidofluoridate, E-00162
Ethyl phosphorochloridisocyanatidate, E-00163
Ethyl phosphorochloridofluoridate, E-00164
▷Ethyl phosphorodichloridate, E-00167
▷Ethyl phosphorodifluoridate, E-00173
Ethyl phosphorodiisocyanatidate, E-00177
Ethyl phosphorofluoridisocyanatidate, E-00180
Ethyl propylene phosphate, *in* E-00040
Ethyl tetrabutylphosphorodiamidate, *in* T-00021
Ethyl tetraethylphosphorodiamidate, *in* T-00059
Ethyl tetramethylphosphorodiamidate, *in* T-00218
Ethyl tetrapropylphosphorodiamidate, *in* T-00291
Ethyl trimethylene phosphate, *in* E-00037
▷Fenitrooxon, F-00004
▷Fospirate, F-00063
▷Heptenophos, H-00004
Heptyl tetraethylphosphorodiamidate, *in* T-00059
Hexaethyl 1,3,5-triazine-2,4,6-triyltriphosphoramidate, *in* T-00330
Hexahydro-2-hydroxycyclopenta[*d*]-1,3,2-dioxaphosphorin 2-oxide; (4*aRS*,7*aSR*)-*form*, 2-Methoxy (Me ester), *in* H-00034

Hexahydro-2-hydroxycyclopenta[*d*]-1,3,2-dioxaphosphorin 2-oxide; (4*aRS*,7*aSR*)-*form*, 2-Phenoxy (Ph ester), *in* H-00034
Hexamethyl 1,3,5-triazine-2,4,6-triyltriphosphoramidate, *in* T-00330
2-Hydroxy-1,3,2-benzodioxaphosphole 2-oxide; 2-Methoxy (Me ester), *in* H-00108
2-Hydroxy-4*H*-1,3,2-benzodioxaphosphorin 2-oxide; 2-Ethoxy (Et ester), *in* H-00109
2-Hydroxy-4*H*-1,3,2-benzodioxaphosphorin 2-oxide; 2-(2-Methylphenoxy) (*o*-Tolyl ester), *in* H-00109
4-Hydroxydinaphtho[2,1-*d*:1′,2′-*f*][1,3,2]dioxaphosphepin 4-oxide; (*S*)-*form*, 2-Methoxy (Me ester), *in* H-00131
4-Hydroxydinaphtho[2,1-*d*:1′,2′-*f*][1,3,2]dioxaphosphepin 4-oxide; (±)-*form*, 2-Methoxy (Me ester), *in* H-00131
2-Hydroxy-4,5-diphenyl-1,3,2-dioxaphosphole 2-oxide; 2-Methoxy (Me ester), *in* H-00139
2-Hydroxy-4,5-diphenyl-1,3,2-dioxaphosphole 2-oxide; 2-Ethoxy (Et ester), *in* H-00139
2-Hydroxyhexahydro-4*H*-1,3,2-benzodioxaphosphorin 2-oxide; (2α,4*a*β,8*a*α)-*form*, 2-Phenoxy (2-Ph ester), *in* H-00152
2-Hydroxyhexahydro-4*H*-1,3,2-benzodioxaphosphorin 2-oxide; (2α,4*a*β,8*a*α)-*form*, 2-(4-Methoxyphenoxy) (2-Methoxyphenyl ester), *in* H-00152
2-Hydroxyhexahydro-4*H*-1,3,2-benzodioxaphosphorin 2-oxide; (2α,4*a*β,8*a*α)-*form*, 2-(4-Nitrophenoxy) (2-(4-Nitrophenyl)-ester), *in* H-00152
2-Hydroxyhexahydro-4*H*-1,3,2-benzodioxaphosphorin 2-oxide; (2α,4*a*β,8*a*α)-*form*, 2-(2,4-Dinitrophenoxy) (2-(2,4-Dinitrophenyl)ester), *in* H-00152
2-Hydroxyhexahydro-4*H*-1,3,2-benzodioxaphosphorin 2-oxide; (2α,4*a*α,8*a*β)-*form*, 2-Phenoxy, *in* H-00152
2-Hydroxyhexahydro-4*H*-1,3,2-benzodioxaphosphorin 2-oxide; (2α,4*a*α,8*a*β)-*form*, 2-(4-Methoxyphenoxy), *in* H-00152
2-Hydroxyhexahydro-4*H*-1,3,2-benzodioxaphosphorin 2-oxide; (2α,4*a*α,8*a*β)-*form*, 2-(4-Nitrophenoxy), *in* H-00152
2-Hydroxyhexahydro-4*H*-1,3,2-benzodioxaphosphorin 2-oxide; (2α,4*a*α,8*a*β)-*form*, 2-(2,4-Dinitrophenoxy), *in* H-00152
2-[(Hydroxyisopropoxyphosphinyl)oxy]-*N*,*N*,*N*-trimethylethanaminium hydroxide, inner salt, *in* C-00198
1-Hydroxy-3-methoxy-1*H*-1,2,4,3-benziodadioxaphosphorine 3-oxide, *in* D-00616
2-[(Hydroxymethoxyphosphinyl)oxy]-*N*,*N*,*N*-trimethylethanaminium hydroxide, inner salt, *in* C-00198
2-Hydroxynaphtho[2,3-*d*]-1,3,2-dioxaphosphole 2-oxide; 2-Methoxy (Me ester), *in* H-00164
2-Hydroxynaphtho[2,3-*d*]-1,3,2-dioxaphosphole 2-oxide; 2-Ethoxy (Et ester), *in* H-00164
2-[(Hydroxyphenoxyphosphinyl)oxy]-*N*,*N*,*N*-trimethylethanaminium hydroxide, inner salt, *in* C-00198
Isopropyl *N*-(diphenoxyphosphinyl)carbamate, *in* D-00989
Isopropyl *N*,*N*′-diphenylphosphorodiamidate, *in* D-01125
4-Methoxydinaphtho[2,1-*d*,1′,2′-*f*][1,3,2]dioxaphosphepin 4-oxide, *in* H-00131
2-Methoxy-1,3,2-dioxaphospholane 2-oxide, M-00037
2-Methoxy-4,5-diphenyl-1,3,2-dioxaphospholane; (2α,4β,5β)-*form*, 2-Oxide, *in* M-00039
2-Methoxy-4-methyl-1,3,2-dioxaphosphorinane 2-oxide; (2*RS*,4*RS*)-*form*, *in* M-00045
2-Methoxy-4-methyl-1,3,2-dioxaphosphorinane 2-oxide; (2*RS*,4*SR*)-*form*, *in* M-00045
4-Methoxyphenyl *N*,*N*′-diphenylphosphorodiamidate, *in* M-00481
2-Methoxyphenyl phosphorodichloridate, *in* M-00480
3-Methoxyphenyl phosphorodichloridate, M-00080
4-Methoxyphenyl phosphorodichloridate, M-00081
4-Methoxyphenyl phosphorodifluoridate, *in* M-00481
Methyl *N*-(diethoxyphosphinyl)carbamate, *in* D-00249
Methyl 2-[(diethoxyphosphinyl)oxy]propanoate, *in* D-00255
Methyl 2-[(diethoxyphosphinyl)oxy]-2-propenoate, *in* D-00256
Methyl *N*,*N*′-diethylphosphorodiamidate, *in* D-00382
Methyl *N*-(dimethoxyphosphinyl)carbamate, *in* D-00680
Methyl 2-(dimethoxyphosphinyl)oxybenzoate, *in* P-00391
Methyl 2,3-di-*O*-methylglucopyranoside 4,6-cyclic phosphonate; α-D-(*R*)P-*form*, *P*-Ethoxy, *in* M-00119
Methyl 2,3-di-*O*-methylglucopyranoside 4,6-cyclic phosphonate; α-D-(*R*)P-*form*, *P*-(4-Nitrophenoxy), *in* M-00119
Methyl 2,3-di-*O*-methylglucopyranoside 4,6-cyclic phosphonate; α-D-(*S*)P-*form*, *P*-Ethoxy, *in* M-00119
Methyl 2,3-di-*O*-methylglucopyranoside 4,6-cyclic phosphonate; α-D-(*S*)P-*form*, *P*-(4-Nitrophenoxy), *in* M-00119
Methyl dimethylvinylene phosphate, *in* H-00123
4-Methyl-1,3,2-dioxaphosphorinane 2-oxide; (2*RS*,4*RS*)-*form*, *in* M-00130
4-Methyl-1,3,2-dioxaphosphorinane 2-oxide; (2*RS*,4*SR*)-*form*, *in* M-00130
Methyl *N*-(diphenoxyphosphinyl)carbamate, *in* D-00989
Methyl *N*,*N*′-diphenylphosphorodiamidate, *in* D-01125
Methylglycerol bicyclic phosphate, *in* M-00423
6-Methyl-4-methylene-4*H*-1,3,2-dioxaphosphorin; 2-Ethoxy, 2-oxide, *in* M-00167
Methyl neopentylene phosphate, *in* M-00034
Methyl *o*-phenylene phosphate, *in* M-00026
2-Methylphenyl phosphorodichloridate, *in* M-00485
3-Methylphenyl phosphorodichloridate, *in* M-00486
2-Methylphenyl phosphorodifluoridate, *in* M-00485
Methyl phosphorochloridofluoridate, M-00341
Methyl phosphorodichloridate, M-00345
▷Methyl phosphorodifluoridate, M-00352
Methyl propylene phosphate, *in* M-00043
Methyl tetraethylphosphorodiamidate, *in* T-00059
Methyl tetramethylethylene phosphate, *in* M-00085
Methyl tetrapropylphosphorodiamidate, *in* T-00291
Methyl trimethylene phosphate, *in* M-00038
Methyl vinylene phosphate, *in* H-00136
▷Mevinphos, M-00440
Mono(4-bromophenyl) phosphate, M-00448
Mono-2-*tert*-butylphenyl phosphate, M-00449
Mono-4-*tert*-butylphenyl phosphate, M-00450
Mono(2-chloroethyl) phosphate, M-00454
Mono(4-chlorophenyl) mono(5-chloro-8-quinolinyl) phosphate, M-00456
Mono(2-chlorophenyl) phosphate, M-00457
Mono(4-chlorophenyl) phosphate, M-00458
▷Monocrotophos; (*E*)-*form*, *in* M-00459
Monocrotophos; (*Z*)-*form*, *in* M-00459
Mono(2-cyanophenyl) phosphate, M-00460
Monododecyl phosphate, M-00468
Monohexadecyl phosphate, M-00475
Mono(2-methoxyphenyl) phosphate, M-00480
Mono(4-methoxyphenyl) phosphate, M-00481
Mono(3-methylphenyl) phosphate, M-00486
▷Monomethyl phosphate, M-00489
Mono-1-naphthyl phosphate, M-00494
Mono-2-naphthyl phosphate, M-00495
Mono(3-nitrophenyl) phosphate, M-00499
Monotetradecyl phosphate, M-00514
▷Monovinyl phosphate, M-00518
▷Naled, N-00001
▷Naphthalophos, N-00008
1,8-Naphthylene phenyl phosphate, *in* P-00070
1-Naphthyl phosphorodichloridate, *in* M-00494
2-Naphthyl phosphorodichloridate, *in* M-00495
1-Naphthyl phosphorodifluoridate, *in* M-00494
Neopentylene phenyl phosphate, *in* D-00790
3-Nitrophenyl *N*,*N*′-diphenylphosphorodiamidate, *in* M-00499
4-Nitrophenyl phenylphosphoramidochloridate, *in* P-00259
2-Nitrophenyl phosphorodichloridate, *in* M-00498
3-Nitrophenyl phosphorodichloridate, *in* M-00499
2-Nitrophenyl phosphorodifluoridate, *in* M-00498
8,9,10,11,12,13,14,15-Octahydro-4-hydroxydinaphtho[2,1-*d*:1′,2′-*f*][1,3,2]dioxaphosphepin 4-oxide; 4-Methoxy (Me ester), *in* O-00018
Octahydro-2-(4-nitrophenoxy)-2*H*-1,3,2-benzoxazaphosphorine; (2*RS*,4*aRS*,8*aSR*)-*form*, 2-Oxide, *in* O-00020
Octahydro-2-(4-nitrophenoxy)-2*H*-1,3,2-benzoxazaphosphorine; (2*RS*,4*aSR*,8*aRS*)-*form*, 2-Oxide, *in* O-00020
▷Paraoxon, P-00001
▷Pentaethyl imidodiphosphate, *in* E-00078
Pentafluorophenyl phosphorodichloridate, P-00017
2-Phenoxy-1,5-dihydro-2,4,3-benzodioxaphosphepin; Oxide, *in* P-00066
2-Phenoxy-1,3,2-dioxaphosphepane 2-oxide, *in* H-00132
2-Phenoxy-1,3,2-dioxaphosphonane 2-oxide, *in* H-00137
2-Phenoxy-4,5-diphenyl-1,3,2-dioxaphospholane; (4*RS*,5*RS*)-*form*, 2-Oxide, *in* P-00069

2-Phenoxy-4,5-diphenyl-1,3,2-dioxaphospholane; (4*RS*,5*SR*)-*form*, 2-Oxide, *in* P-00069
Phenyl cyclohexylphosphoramidochloridate, P-00104
Phenyl *N,N'*-dibenzylphosphorodiamidate, *in* D-00061
Phenyl *N,N'*-dibutylphosphorodiamidate, *in* D-00146
Phenyl *N,N'*-dicyclohexylphosphorodiamidate, *in* D-00216
Phenyl *N,N'*-diethylphosphorodiamidate, *in* D-00382
Phenyl *N,N'*-diisopropylphosphorodiamidate, *in* D-00661
▷Phenyl di-4-morpholinylphosphinate, *in* D-00935
Phenyl *N*-(diphenoxyphosphinyl)carbamate, *in* D-00989
Phenyl *N,N'*-diphenylphosphorodiamidate, *in* D-01125
Phenyl di-2-propenyl phosphate, P-00118
Phenyl *N,N'*-dipropylphosphorodiamidate, *in* D-01223
1,3-Phenylene bisphosphorodichloridate, *in* P-00147
1,4-Phenylenebis(phosphorodichloridate), *in* P-00148
1,4-Phenylenebis(phosphorodifluoridate), *in* P-00148
Phenyl hydrogen 2-aminoethyl phosphate, *in* A-00075
Phenyl hydrogen (2-aminoethyl)phosphoramidate, *in* A-00080
Phenyl hydrogen (2-aminophenyl)phosphoramidate, *in* A-00113
Phenyl *o*-phenylene phosphate, *in* P-00063
Phenyl phosphorochloridisocyanatidate, P-00270
Phenyl phosphorochloridofluoridate, P-00271
Phenyl phosphorodiamidate, P-00272
Phenyl phosphorodichloridate, P-00279
Phenyl phosphorodifluoridate, P-00285
Phenyl phosphorodi(isothiocyanatate), P-00294
Phenyl propylene phosphate, *in* M-00190
Phenyl 8-quinolyl phosphate, P-00306
▷Phenyl saligenin cyclic phosphate, *in* P-00064
Phenylsulfonylphosphoramidic dichloride, P-00313
Phenyl tetramethylphosphorodiamidate, *in* T-00218
▷Phosphamidon, P-00347
2-Phosphonooxypropanoic acid; (±)-*form*, Me ester, *in* P-00392
O-Phosphotyrosine, P-00423
2-Phthalimidoethyl phosphorodichloridate, P-00425
▷Propaphos, P-00451
2-Propenyl tetramethylphosphorodiamidate, *in* T-00218
Propyl tetrapropylphosphorodiamidate, *in* T-00291
2-Propynyl phosphorodichloridate, P-00497
Salioxon, *in* H-00109
Tagetitoxin, T-00002
Tetrabenzyl diphosphate, T-00006
▷Tetrabutyl pyrophosphate, T-00024
▷Tetrachlorvinphos, T-00038
Tetraethyl 1,2-benzenediol bis(phosphate), T-00042
Tetraethyl hypophosphate, T-00050
O,O,O',O'-Tetraethyl imidodiphosphate, T-00051
Tetraethyl 1,4-phenylenebisphosphoramidate, *in* B-00012
Tetraethyl phenylimidodiphosphate, *in* P-00168
▷Tetraethyl pyrophosphate, T-00069
Tetraethyl trimethylsilylimidodiphosphate, *in* T-00051
▷Tetraisopropyl diphosphate, T-00166
Tetraisopropyl 1,2-ethanediylbisphosphoramidate, *in* E-00009
Tetraisopropyl ethyliminodiphosphate, *in* E-00078
Tetrakis(4-nitrophenyl) diphosphate, T-00183
4,4,5,5-Tetramethyl phenyl phosphate, *in* P-00075
Tetramethylphosphorodiamidic chloride, T-00220
▷Tetramethyl pyrophosphate, T-00229
Tetraphenyl diphosphate, T-00256
Tetraphenyl 1,2-ethanediylbisphosphoramidate, *in* E-00009
O,O,O',O'-Tetraphenyl imidodiphosphate, T-00265
Tetraphenyl 1,3-phenylenebisphosphoramidate, *in* B-00011
Tetraphenyl 1,4-phenylenebisphosphoramidate, *in* B-00012
▷Tetrapropyl diphosphate, T-00285
Tetrapropyl 1,2-ethanediylbisphosphoramidate, *in* E-00009
Tribenzyl phosphate, T-00337
Tri-2-biphenylyl phosphate, T-00345
Tri-3-biphenylyl phosphate, T-00346
Tri-4-biphenylyl phosphate, T-00347
2,2,2-Tribromoethyl 4-morpholinylphosphonochloridate, T-00357
2,2,2-Tribromoethyl phenylphosphoramidochloridate, *in* P-00259
▷Tributyl phosphate, T-00370
Tri-*tert*-butyl phosphate, T-00371
2,2,2-Trichloroethyl phenylphosphoramidochloridate, *in* P-00259
2,2,2-Trichloroethyl phosphorodichloridate, *in* T-00424
Triclofos, T-00424
Tricyclohexyl phosphate, T-00425
▷Triethyl phosphate, T-00440
Triethyl 3-(*N*-phosphono)benzoate, *in* C-00005
Triheptyl phosphate, T-00488
Trihexyl phosphate, T-00491
Triisopropyl phosphate, T-00499
Trimethylene phenyl phosphate, *in* P-00068
▷Trimethylolethane phosphate, *in* M-00424
▷Trimethyl phosphate, T-00528
Trimethylsilyl 2-[bis(trimethylsilyloxy)phosphinyl]oxy-2-propenoate, *in* P-00366
Tri-1-naphthyl phosphate, T-00570
Tri-2-naphthyl phosphate, T-00571
Trioctyl phosphate, T-00581
2,8,9-Trioxa-5-aza-1-phosphabicyclo[3.3.3]undecane; 1-Oxide, *in* T-00584
2,6,7-Trioxa-1,4-diphosphabicyclo[2.2.2]octane; 1-Oxide, *in* T-00585
▷2,6,7-Trioxa-1,4-diphosphabicyclo[2.2.2]octane; 1,4-Dioxide, *in* T-00585
2,6,7-Trioxa-1,4-diphosphabicyclo[2.2.2]octane; 4-Sulfide, *in* T-00585
2,6,7-Trioxa-1-phosphabicyclo[2.2.1]heptane; 1-Oxide, *in* T-00587
2,6,7-Trioxa-1-phosphabicyclo[2.2.2]octane-4-methanol; Oxide, *in* T-00590
2,6,7-Trioxa-1-phosphabicyclo[2.2.2]octane-4-methanol; Oxide, Me ether, *in* T-00590
▷2,8,9-Trioxa-1-phosphatricyclo[3.3.1.1^{3,7}]decane; 1-Oxide, *in* T-00591
1,3,6,2-Trioxaphosphocane; 2-Phenoxy, 2-oxide, *in* T-00592
Tripentyl phosphate, T-00593
▷Triphenyl phosphate, T-00623
(Triphenylphosphoranylidene)phosphoramidic dichloride, T-00656
▷Tri-2-propenyl phosphate, T-00705
Tripropyl phosphate, T-00710
Tris(4-aminophenyl) phosphate, T-00727
Tris[(4-bromophenyl)methyl] phosphate, T-00732
Tris(2-chloroethyl) phosphate, T-00741
Tris(2-chlorophenyl) phosphate, T-00745
Tris(4-chlorophenyl) phosphate, T-00746
Tris[2-(dimethylamino)phenyl] phosphate, T-00761
Tris[3-(dimethylamino)phenyl] phosphate, T-00762
Tris[4-(dimethylamino)phenyl] phosphate, T-00763
Tris(2,5-dimethylphenyl) phosphate, T-00767
Tris(2-ethoxyethyl) phosphate, T-00778
▷Tris(2-ethylhexyl) phosphate, T-00779
Tris(3-hydroxyphenyl) phosphate, T-00786
Tris(2-methoxyphenyl)phosphate, *in* T-00785
Tris(4-methoxyphenyl) phosphate, *in* T-00787
Tris(3-methylbutyl) phosphate, T-00796
▷Tris(2-methylphenyl) phosphate, T-00803
Tris(3-methylphenyl) phosphate, T-00804
Tris(4-methylphenyl) phosphate, T-00805
Tris(2-methyl-2-propenyl) phosphate, T-00820
Tris(1-methylpropyl) phosphate, T-00821
Tris(2-methylpropyl) phosphate, T-00822
Tris[(4-nitrophenyl)methyl] phosphate, T-00834
Tris(2-nitrophenyl) phosphate, T-00836
Tris(4-nitrophenyl) phosphate, T-00837
Tris(pentachlorophenyl) phosphate, T-00841
Tris(pentafluorophenyl) phosphate, T-00842
Tris(2,2,2-trichloroethyl) phosphate, T-00848
Tris(triethylsilyl) phosphate, T-00849
Tris(2,2,2-trifluoroethyl) phosphate, T-00851
▷Tris(trimethylsilyl) phosphate, T-00864

Phosphoric amides

▷*N*-(1-Adamantyl)-*P,P*-bis(aziridinyl)phosphinic amide, A-00032
2-Adamantylphosphoramidic acid; Dichloride, *in* A-00037
Alcophosphamide, *in* B-00126
Aldophosphamide, A-00050

Aminocarbonylphosphoramidic acid; Dichloride, *in* A-00062
2-Amino-3-(2-chloroethyl)tetrahydro-4*H*-1,3,2-oxazaphosphorine 2-oxide; (*R*)-*form*, *in* A-00063
2-Amino-3-(2-chloroethyl)tetrahydro-4*H*-1,3,2-oxazaphosphorine 2-oxide; (*S*)-*form*, *in* A-00063
2-Amino-3-(2-chloroethyl)tetrahydro-4*H*-1,3,2-oxazaphosphorine 2-oxide; (±)-*form*, *in* A-00063
2-Amino-3-(2-chloroethyl)tetrahydro-4*H*-1,3,2-oxazaphosphorine 2-oxide; (±)-*form*, *N*-Ac, *in* A-00063
2-Amino-3-phosphonooxybutanoic acid; (2*S*,3*R*)-*form*, *in* A-00120
2-Amino-3-phosphonooxybutanoic acid; (2*RS*,3*SR*)-*form*, *in* A-00120
▷*P*-1-Aziridinyl-*N*,*N*,*N*′,*N*′-tetramethylphosphonic diamide, *in* A-00153
Benzodepa, B-00017
▷*N*-Benzoyl-*P*,*P*-bis(1-aziridinyl)phosphinic amide, *in* B-00092
Benzoylphosphoramidic acid, B-00036
Benzoylphosphoramidic dianilide, *in* D-01114
Benzoylphosphoramidic dichloride, B-00037
Benzyl 2-[(dibenzyloxyphosphinyl)amino]-3-phenylpropanoate, *in* P-00400
Benzyl 2-[(diphenoxyphosphinyl)amino]-3-phenylpropanoate, *in* P-00400
Benzyl *N*-(diphenoxyphosphinyl)carbamate, *in* D-00989
Benzyl *N*-[imino(diphenoxyphosphinylamino)methyl]glycinate, *in* P-00369
Benzyl *N*-[Imino(diphenoxyphosphinylamino)methyl]-*N*-methylglycinate, *in* P-00364
Benzylphosphoramidic acid, B-00069
N″-Benzyl-*N*,*N*,*N*′,*N*′-tetramethylphosphoric triamide, *in* T-00216
[1,1′-Biphenyl]-2-yl *N*,*N*′-diphenylphosphorodiamidate, *in* M-00445
P,*P*-Bis(1-aziridinyl)-*N*,*N*-diethylphosphinic amide, *in* B-00092
▷*P*,*P*-Bis(1-aziridinyl)-*N*,*N*-dimethylphosphinic amide, *in* B-00092
▷*P*,*P*-Bis(1-aziridinyl)-*N*-ethylphosphinic amide, *in* B-00092
▷*P*,*P*-Bis(1-aziridinyl)-*N*-methylphosphinic amide, *in* B-00092
P,*P*-Bis(1-aziridinyl)-*N*-2-naphthylphosphinic amide, *in* B-00092
P,*P*-Bis(1-aziridinyl)-*N*-phenylphosphinic amide, *in* B-00092
▷*P*,*P*-Bis(1-aziridinyl)phosphinic amide, B-00092
Bis[(4-bromophenyl)methyl] phenylphosphoramidate, *in* B-00109
Bis[(4-bromophenyl)methyl] phosphoramidate, *in* B-00109
N,*N*-Bis(2-chloroethyl)-*N*′,*N*″-diphenylphosphoric triamide, *in* B-00121
Bis(2-chloroethyl) phosphoramidate, *in* B-00119
▷Bis(2-chloroethyl)phosphoramidic dichloride, B-00122
N,*N*′-Bis(2-chloroethyl)phosphorodiamidic acid, B-00125
N,*N*-Bis(2-chloroethyl)phosphorodiamidic acid, B-00126
Bis-4-iodobenzyl [1-ethoxycarbonyl-2-(4-hydroxybenzyl)-ethyl]phosphoramidate, *in* P-00422
Bis(4-methoxyphenyl) phosphoramidate, *in* B-00295
Bis(3-methylbutyl) phenylphosphoramidate, *in* B-00302
Bis(3-methylbutyl) phosphoramidate, *in* B-00302
Bis(2-methylpropyl) bis(2-methylpropyl)phosphoramidate, *in* B-00355
Bis(1-methylpropyl) phenylphosphoramidate, *in* P-00257
Bis(2-methylpropyl) phenylphosphoramidate, *in* P-00257
Bis(2-oxo-3-oxazolidinyl) diphenyl phosphoric anhydride, *in* B-00378
Bis(2-oxo-3-oxazolidinyl)phosphinic acid, B-00378
Bis(2-oxo-3-oxazolidinyl)phosphinic acid; Chloride, *in* B-00378
Bis(2-oxo-3-oxazolidinyl)phosphinic acid; Azide, *in* B-00378
Bis(2-oxo-3-oxazolidinyl)phosphinic acid; Anhydride, *in* B-00378
Bis(2,2,2-trichloroethyl) phenylphosphoramidate, *in* B-00400
(3-Bromophenyl)phosphoramidic acid; Dichloride, *in* B-00495
(4-Bromophenyl)phosphoramidic acid, B-00496
(4-Bromophenyl)phosphoramidic acid; Dichloride, *in* B-00496
▷Butafox, *in* D-00146
3-Butenyl *N*,*N*′-bis(2-chloroethyl)phosphorodiamidate, *in* B-00125
3-Butenyl *N*,*N*-bis(2-chloroethyl)phosphorodiamidate, *in* B-00126
2-*tert*-Butylamino-4-methyl-1,3,2-dioxaphosphorinane; (2*RS*,4*SR*)-*form*, 2-Oxide, *in* B-00541
Butyl *N*″-[bis(diethylamino)phosphinyl]-*N*,*N*,*N*′,*N*′-tetraethylphosphorodiamidoimidate, *in* B-00177
Butyl *N*,*N*′-dibutylphosphorodiamidate, *in* D-00146
5-*tert*-Butyl-2-dimethylamino-1,3,2-dioxaphosphorinane; *trans*-*form*, 2-Oxide, *in* B-00555
1,3-Butylene *tert*-butylphosphoramidate, *in* B-00541
1,3-Butylene phenylphosphoramidate, *in* M-00193
tert-Butyl hydrogen bis(2-chloroethyl)phosphoramidate, *in* B-00121
Butylphosphoramidic acid, B-00621
Butylphosphoramidic acid; Difluoride, *in* B-00621
Butylphosphoramidic acid; Dichloride, *in* B-00621
tert-Butylphosphoramidic acid; Difluoride, *in* B-00622
tert-Butylphosphoramidic acid; Dichloride, *in* B-00622
Butyl tetrabutylphosphorodiamidate, *in* T-00021
5-*tert*-Butyltetrahydro-2-dimethylamino-2*H*-1,3,2-oxazaphosphorine; (2*RS*,5*RS*)-*form*, 2-Oxide, *in* B-00635
5-*tert*-Butyltetrahydro-2-dimethylamino-2*H*-1,3,2-oxazaphosphorine; (2*RS*,5*SR*)-*form*, 2-Oxide, *in* B-00635
5-*tert*-Butyltetrahydro-2-dimethylamino-3-phenyl-2*H*-1,3,2-oxazaphosphorine; (2*RS*,5*RS*)-*form*, 2-Oxide, *in* B-00636
5-*tert*-Butyltetrahydro-2-dimethylamino-3-phenyl-2*H*-1,3,2-oxazaphosphorine; (2*RS*,5*SR*)-*form*, 2-Oxide, *in* B-00636
Butyl tetramethylphosphorodiamidate, *in* T-00218
2-*tert*-Butyltrimethylene dimethylphosphoramidate, *in* B-00555
(4-Carbethoxyphenyl)phosphoramidic acid; Dichloride, *in* C-00006
2-Chloro-3-(2-chloroethyl)tetrahydro-2*H*-1,3,2-oxazaphosphorine; (±)-*form*, 2-Oxide, *in* C-00030
2-Chloro-2,3-dihydro-3-phenyl-4*H*-1,3,2-benzoxazaphosphorin-4-one; 2-Oxide, *in* C-00050
2-Chloro-1,3-dimethyl-1,3,2-diazaphospholidine; 2-Oxide, *in* C-00055
2-Chloro-3,4-dimethyl-5-phenyl-1,3,2-oxazaphospholidine; (2*R*,4*S*,5*R*)-*form*, 2-Oxide, *in* C-00061
2-Chloro-3,4-dimethyl-5-phenyl-1,3,2-oxazaphospholidine; (2*S*,4*S*,5*R*)-*form*, 2-Oxide, *in* C-00061
2-[(2-Chloroethyl)amino]tetrahydro-2*H*-1,3,2-oxazaphosphorine 2-oxide; (*R*)-*form*, *in* C-00072
2-[(2-Chloroethyl)amino]tetrahydro-2*H*-1,3,2-oxazaphosphorine 2-oxide; (*R*)-*form*, 3-[(*S*)-1-Phenylethyl], *in* C-00072
2-[(2-Chloroethyl)amino]tetrahydro-2*H*-1,3,2-oxazaphosphorine 2-oxide; (*S*)-*form*, *in* C-00072
1-(2-Chloroethyl)tetrahydro-1*H*,5*H*-[1.3.2]-diazaphospholo[2,1-*b*][1.3.2]oxazaphosphorine 9-oxide; (*R*)-*form*, *in* C-00082
1-(2-Chloroethyl)tetrahydro-1*H*,5*H*-[1.3.2]-diazaphospholo[2,1-*b*][1.3.2]oxazaphosphorine 9-oxide; (*S*)-*form*, *in* C-00082
1-(2-Chloroethyl)tetrahydro-1*H*,5*H*-[1.3.2]-diazaphospholo[2,1-*b*][1.3.2]oxazaphosphorine 9-oxide; (±)-*form*, *in* C-00082
4-Chlorophenyl *N*,*N*′-diphenylphosphorodiamidate, *in* M-00458
2-Chlorophenyl (4-methoxyphenyl)phosphoramidochloridate, C-00130
2-Chloro-3-phenyl-1,3,2-oxazaphospholidine; 2-Oxide, *in* C-00138
2-Chlorophenyl phenylphosphoramidochloridate, *in* P-00259
4-Chlorophenyl phenylphosphoramidochloridate, *in* P-00259
(3-Chlorophenyl)phosphoramidic acid, C-00156
(4-Chlorophenyl)phosphoramidic acid, C-00157
(4-Chlorophenyl)phosphoramidic acid; Dichloride, *in* C-00157
2-Chloro-1,2,3,4-tetrahydro-6-methyl-3-(4-methylphenyl)-1,3,2-benzodiazaphosphorine 2-oxide, C-00177
2-Chlorotetrahydro-2*H*-1,3,2-oxazaphosphorine; 2-Oxide, *in* C-00180
2-Chlorotetrahydro-3-(1-phenylethyl)-2*H*-1,3,2-oxazaphosphorine 2-oxide; (R_C,R_P)-*form*, *in* C-00181
2-Chlorotetrahydro-3-(1-phenylethyl)-2*H*-1,3,2-oxazaphosphorine 2-oxide; (R_C,S_P)-*form*, *in* C-00181
2-Chlorotetrahydro-3-(1-phenylethyl)-2*H*-1,3,2-oxazaphosphorine 2-oxide; (S_C,R_P)-*form*, *in* C-00181
2-Chlorotetrahydro-3-(1-phenylethyl)-2*H*-1,3,2-oxazaphosphorine 2-oxide; (S_C,S_P)-*form*, *in* C-00181

6-Chloro-5,6,7,12-tetrahydro-2,5,7,10-tetramethyldibenzo[*d,g*][1,3,2]diazaphosphocine; 6-Oxide, *in* C-00182
2-Chlorotetrahydro-3,5,5-trimethyl-2*H*-1,3,2-oxazaphosphorine; 2-Oxide, *in* C-00183
Cyanophosphoramidic dichloride, C-00227
Cyclohexyl dimethylphosphoramidocyanidate, *in* D-00863
Cyclohexyl di-4-morpholinylphosphinate, *in* D-00935
Cyclohexylphosphoramidic acid; Dichloride, *in* C-00269
Cyclophosphamide; (*R*)-*form*, *in* C-00283
Cyclophosphamide; (*S*)-*form*, *in* C-00283
Cyclophosphamide; (±)-*form*, *in* C-00283
Decylphosphoramidic acid; Dichloride, *in* D-00013
▷Diamidafos, D-00024
Dibenzyl 1-adamantylphosphoramidate, *in* A-00036
Dibenzyl [1-(aminocarbonyl)ethyl]phosphoramidate, *in* P-00362
Dibenzyl (aminoiminomethyl)phosphoramidate, *in* P-00370
Dibenzyl [[[(4-amino-4-methoxycarbonyl)butyl]amino](imino)-methyl]phosphoramidate, *in* P-00363
Dibenzyl benzoylphosphoramidate, *in* B-00036
Dibenzyl [1-(benzyloxycarbonyl)ethyl]phosphoramidate, *in* P-00362
Dibenzyl [1-benzyloxycarbonyl-2-(4-hydroxyphenyl)ethyl]-phosphoramidate, *in* P-00422
Dibenzyl [1-carboxy-2-[4-(dibenzyloxyphosphinyloxy)phenyl]-ethyl]phosphoramidate, *in* D-01188
Dibenzyl cyclohexylphosphoramidate, *in* C-00269
Dibenzyl 2-[(dibenzyloxyphosphinyl)amino]pentanedioate, *in* P-00367
Dibenzyl diethylphosphoramidate, *in* D-00343
Dibenzyl [1-ethoxycarbonyl-2-(4-hydroxybenzyl)ethyl]-phosphoramidate, *in* P-00422
Dibenzyl [1-(methoxycarbonyl)ethyl]phosphoramidate, *in* P-00362
Dibenzyl methylphenylphosphoramidate, *in* M-00250
Dibenzyl (4-methylphenyl)phosphoramidate, *in* M-00253
Dibenzyl 2-naphthylphosphoramidate, *in* N-00029
Dibenzyl phenylphosphoramidate, *in* D-00053
Dibenzyl phosphoramidate, *in* D-00053
N,N'-Dibenzylphosphorodiamidic acid, D-00061
▷*N,N'*-Dibenzylphosphorodiamidic acid; Fluoride, *in* D-00061
N-(2,4-Dibromophenyl)-*N',N''*-diphenylphosphoric triamide, *in* D-00067
(2,4-Dibromophenyl)phosphoramidic acid; Dichloride, *in* D-00067
Di-*tert*-butyl benzylphosphoramidate, *in* B-00069
Dibutyl butylphosphoramidate, *in* B-00621
Dibutyl *tert*-butylphosphoramidate, *in* B-00622
3,5-Di-*tert*-butyl-2-chloro-2,3-dihydro-1,3,2-oxazaphosphole; 2-Oxide, *in* D-00078
Di-*tert*-butyl cyclohexylphosphoramidate, *in* C-00269
Dibutyl decylphosphoramidate, *in* D-00013
Dibutyl dibutylphosphoramidate, *in* D-00134
Dibutyl diethylphosphoramidate, *in* D-00343
3,5-Di-*tert*-butyl-2,3-dihydro-1,3,2-oxazaphosphole; 2-Ethoxy, 2-oxide, *in* D-00083
Dibutyl ethylphosphoramidate, *in* E-00156
Dibutyl methylphosphoramidate, *in* M-00326
Dibutyl (2-methylpropyl)phosphoramidate, *in* M-00384
Dibutyl phenylphosphoramidate, *in* P-00257
▷Dibutyl phosphoramidate, D-00133
Di-*tert*-butyl phosphoramidate, *in* D-00105
Dibutylphosphoramidic acid; Dichloride, *in* D-00134
2,4-Dichlorophenyl *N,N'*-diphenylphosphorodiamidate, *in* M-00463
(2,4-Dichlorophenyl)phosphoramidic acid, D-00188
(2,4-Dichlorophenyl)phosphoramidic acid; Dichloride, *in* D-00188
N,N'-Dicyclohexylphosphorodiamidic acid, D-00216
▷*N,N'*-Dicyclohexylphosphorodiamidic acid; Fluoride, *in* D-00216
Diethyl acetylphosphoramidate, *in* A-00029
Diethyl 1-adamantylphosphoramidate, *in* A-00036
Diethyl aminocarbonylphosphoramidate, *in* A-00062
Diethyl (2-aminoethyl)phosphoramidate, *in* A-00080
Diethyl 1-aziridinylphosphonate, *in* A-00153
Diethyl benzoylphosphoramidate, *in* B-00036
Diethyl benzylphosphoramidate, *in* B-00069
Diethyl (4-bromophenyl)phosphoramidate, *in* B-00496
Diethyl butylphosphoramidate, *in* B-00621
Diethyl *tert*-butylphosphoramidate, *in* B-00622
Diethyl (4-chlorophenyl)phosphoramidate, *in* C-00157
Diethyl cyclohexylphosphoramidate, *in* C-00269
Diethyl (2,4-dibromophenyl)phosphoramidate, *in* D-00067
Diethyl dibromophosphoramidate, D-00279
▷Diethyl dibutylphosphoramidate, *in* D-00134
Diethyl (2,4-dichlorophenyl)phosphoramidate, *in* D-00188
▷Diethyl dichlorophosphoramidate, D-00280
Diethyl diethylphosphoramidate, D-00281
Diethyl (2,4-dimethylphenyl)phosphoramidate, *in* D-00818
Diethyl dimethylphosphoramidate, D-00286
Diethyl 2-[(diphenoxyphosphinyl)amino]pentanedioate, *in* P-00367
Diethyl diphenylphosphoramidate, *in* D-01108
N,N-Diethyl-2,2'-diphenylphosphoramidic dihydrazide, *in* D-01109
N,N-Diethyl-*N',N''*-diphenylphosphoric triamide, *in* D-00364
Diethyl di-2-propenylphosphoramidate, *in* D-01195
Diethyl dipropylphosphoramidate, *in* D-01214
Diethyl ethylphenylphosphoramidate, *in* E-00116
Diethyl ethylphosphoramidate, D-00293
Diethyl (3-fluorophenyl)phosphoramidate, *in* F-00043
Diethyl (4-fluorophenyl)phosphoramidate, *in* F-00044
Diethyl (2-hydroxyethyl)phosphoramidate, *in* H-00148
Diethyl 1*H*-imidazol-1-ylphosphonate, *in* I-00003
Diethyl isopropylphosphoramidate, *in* I-00058
Diethyl (4-methoxyphenyl)phosphoramidate, *in* M-00076
Diethyl methylphenylphosphoramidate, *in* M-00250
Diethyl (2-methylphenyl)phosphoramidate, *in* M-00251
Diethyl (3-methylphenyl)phosphoramidate, *in* M-00252
Diethyl (4-methylphenyl)phosphoramidate, *in* M-00253
Diethyl methylphosphoramidate, D-00299
Diethyl (2-methylpropyl)phosphoramidate, *in* M-00384
Diethyl methylsulfonylphosphoramidate, *in* M-00401
Diethyl 4-morpholinylphosphonate, *in* M-00523
Diethyl 1-naphthylphosphoramidate, *in* N-00028
Diethyl 2-naphthylphosphoramidate, *in* N-00029
Diethyl (3-nitrophenyl)phosphoramidate, *in* N-00059
Diethyl (4-nitrophenyl)phosphoramidate, *in* N-00060
Diethyl octylphosphoramidate, *in* O-00045
Diethyl pentylphosphoramidate, *in* P-00053
Diethyl (phenylamino)carbonylphosphoramidate, *in* P-00081
Diethyl phenylphosphoramidate, D-00312
Diethyl phenylsulfonylphosphoramidate, *in* P-00312
Diethyl phosphoramidate, D-00342
Diethylphosphoramidic difluoride, D-00345
N,N-Diethylphosphoric triamide, D-00364
▷*N,N'*-Diethylphosphorodiamidic acid; Fluoride, *in* D-00382
N,N'-Diethylphosphorodiamidic acid; Chloride, *in* D-00382
Diethyl 1-piperidinylphosphonate, *in* P-00430
Diethyl 2-propenylphosphoramidate, *in* P-00464
Diethyl propylphosphoramidate, *in* P-00484
Diethyl 2-pyridylphosphoramidate, *in* P-00510
Diethyl 3-pyridylphosphoramidate, *in* P-00511
Diethyl (2,4,6-tribromophenyl)phosphoramidate, *in* T-00359
Diethyl (trimethylsilylphenyl)phosphoramidate, *in* D-00312
Dihexyl phosphoramidate, *in* D-00432
1,5-Dihydro-3-dimethylamino-2,4,3-benzodioxaphosphepin; 3-Oxide, *in* D-00453
4,5-Dihydro-2-dimethylamino-1,3,2-benzodioxaphosphepin; 2-Oxide, *in* D-00454
2,3-Dihydro-2-dimethylamino-5-methyl-3-phenyl-1,3,4,2-oxadiazaphosphole; 2-Oxide, *in* D-00455
2,3-Dihydro-5-methyl-3-phenyl-2-phenylamino-1,3,4,2-oxadiazaphosphole; 2-Oxide, *in* D-00527
2,3-Dihydro-1*H*-naphtho[1,8-*de*]-1,3,2-diazaphosphorine; 2-(4-Nitrophenoxy), 2-oxide, *in* D-00532
2,3-Dihydro-1*H*-naphtho[1,8-*de*]-1,3,2-diazaphosphorine; 2-(4-Chloro-3-methylphenoxy), 2-oxide, *in* D-00532
3,9-Dihydroxy-2,4,8,10-Tetraoxa-3,9-diphosphaspiro[5.5]-undecane 3,9-dioxide; 3,9-Diamide, *in* D-00626
Diisopropyl aminocarbonylphosphoramidate, *in* A-00062
Diisopropyl benzoylphosphoramidate, *in* B-00036
Diisopropyl benzylphosphoramidate, *in* B-00069

Diisopropyl cyclohexylphosphoramidate, *in* C-00269
Diisopropyl dibutylphosphoramidate, *in* D-00134
Diisopropyl diethylphosphoramidate, *in* D-00343
Diisopropyl diisobutylphosphoramidate, *in* B-00355
Diisopropyl dimethylphosphoramidate, *in* D-00855
Diisopropyl (3-fluorophenyl)phosphoramidate, *in* F-00043
Diisopropyl (4-fluorophenyl)phosphoramidate, *in* F-00044
Diisopropyl (2-hydroxyethyl)phosphoramidate, *in* H-00148
Diisopropyl 1*H*-imidazol-1-ylphosphonate, *in* I-00003
Diisopropyl isopropylphosphoramidate, *in* I-00058
Diisopropyl (phenylamino)carbonylphosphoramidate, *in* P-00081
Diisopropyl phenylphosphoramidate, *in* P-00257
Diisopropyl phosphoramidate, D-00653
N,N′-Diisopropylphosphorodiamidic acid; Chloride, *in* D-00661
Diisopropyl 2-pyridylphosphoramidate, *in* P-00510
Diisopropyl 3-pyridylphosphoramidate, *in* P-00511
▷Dimefox, D-00669
2-[(Dimethoxyphosphinyl)amino]pentanedioic diamide, *in* P-00367
2-[(Dimethoxyphosphinyl)amino]-3-phenylpropanamide, *in* P-00400
Dimethyl acetylphosphoramidate, *in* A-00029
Dimethyl 1-adamantylphosphoramidate, *in* A-00036
Dimethyl [1-(aminocarbonyl)ethyl]phosphoramidate, *in* P-00362
Dimethyl aminocarbonylphosphoramidate, *in* A-00062
2-Dimethylamino-1,3,2-oxazaphospholidine; 2-Oxide, *in* D-00702
Dimethyl (4-aminophenyl)phosphoramidate, *in* A-00115
Dimethyl 1-azetidinylphosphonate, *in* A-00147
Dimethyl 1-aziridinylphosphonate, *in* A-00153
Dimethyl benzoylphosphoramidate, *in* B-00036
Dimethyl benzylphosphoramidate, *in* B-00069
N,N′-Dimethyl-*N,N′*-bis[[bis(dimethylamino)phosphinyl]-amino]-ethane, D-00713
Dimethyl butylphosphoramidate, *in* B-00621
Dimethyl *tert*-butylphosphoramidate, *in* B-00622
Dimethyl (4-chlorophenyl)phosphoramidate, *in* C-00157
Dimethyl 2-[(dibenzyloxyphosphinyl)amino]pentanedioate, *in* P-00367
Dimethyl dibutylphosphoramidate, *in* D-00134
Dimethyl dichlorophosphoramidate, D-00723
Dimethyl diethylphosphoramidate, *in* D-00343
1,3-Dimethyl-2-dimethylamino-1,3,2-diazaphospholidine; 2-Oxide, *in* D-00724
5,5-Dimethyl-2-dimethylaminotetrahydro-2*H*-1,3,2-oxazaphosphorine; 2-Oxide, *in* D-00726
Dimethyl dimethylphosphoramidate, D-00727
N,N-Dimethyl-2,2′-diphenylphosphoramidic dihydrazide, *in* D-01109
Dimethyl dipropylphosphoramidate, *in* D-01214
N-(1,2-Dimethylethenylenedioxyphosphoryl)imidazole, D-00753
Dimethyl ethylphosphoramidate, *in* E-00156
Dimethyl (3-fluorophenyl)phosphoramidate, *in* F-00043
Dimethyl (4-fluorophenyl)phosphoramidate, *in* F-00044
Dimethyl hexylphosphoramidate, *in* H-00102
Dimethyl 1*H*-imidazol-1-ylphosphonate, *in* I-00003
Dimethyl isopropylphosphoramidate, *in* I-00058
Dimethyl (2-methoxyphenyl)phosphoramidate, *in* M-00074
Dimethyl (3-methoxyphenyl)phosphoramidate, *in* M-00075
Dimethyl (4-methoxyphenyl)phosphoramidate, *in* M-00076
Dimethyl methylphenylphosphoramidate, *in* M-00250
Dimethyl (2-methylphenyl)phosphoramidate, *in* M-00251
Dimethyl (3-methylphenyl)phosphoramidate, *in* M-00252
Dimethyl (4-methylphenyl)phosphoramidate, *in* M-00253
Dimethyl methylphosphoramidate, *in* M-00326
Dimethyl methylsulfonylphosphoramidate, *in* M-00401
▷Dimethyl 4-morpholinylphosphonate, *in* M-00523
Dimethyl (3-nitrophenyl)phosphoramidate, *in* N-00059
Dimethyl (4-nitrophenyl)phosphoramidate, *in* N-00060
Dimethyl octylphosphoramidate, *in* O-00045
1,3-Dimethyl-2-phenoxy-1,3-diazaphospholidin-4-one 2-oxide, *in* D-00721
3,4-Dimethyl-2-phenoxy-5-phenyl-1,3,2-oxazaphospholidine; (2*R*,4*S*,5*R*)-*form*, 2-Oxide, *in* D-00791
3,4-Dimethyl-2-phenoxy-5-phenyl-1,3,2-oxazaphospholidine; (2*S*,4*S*,5*R*)-*form*, *in* D-00791
3,4-Dimethyl-2-phenoxy-5-phenyl-1,3,2-oxazaphospholidine; (2*S*,4*R*,5*S*)-*form*, *in* D-00791
Dimethyl (phenylamino)carbonylphosphoramidate, *in* P-00081
Dimethyl phenylphosphoramidate, D-00817
(2,4-Dimethylphenyl)phosphoramidic acid; Dichloride, *in* D-00818
Dimethyl phenylsulfonylphosphoramidate, *in* P-00312
O,O-Dimethyl phosphoramidate, D-00854
Dimethylphosphoramidic dicyanide, D-00857
Dimethylphosphoramidic difluoride, D-00858
N,N-Dimethylphosphoric triamide, D-00880
Dimethyl 1-piperidinylphosphonate, *in* P-00430
Dimethyl propylphosphoramidate, *in* P-00484
Dimethyl 2-pyridylphosphoramidate, *in* P-00510
Dimethyl 3-pyridylphosphoramidate, *in* P-00511
Dimethyl 1-pyrrolidinylphosphonate, *in* P-00518
N″,N″-Dimethyl-*N,N′*-trimethylenephosphoric triamide, *in* H-00020
Di-4-morpholinylphosphinic acid, D-00935
Di-4-morpholinylphosphinic acid; Fluoride, *in* D-00935
Di-4-morpholinylphosphinic acid; Chloride, *in* D-00935
Di-2-naphthyl phosphoramidate, *in* D-00938
Dipentyl phenylphosphoramidate, *in* D-00979
N-(Diphenoxyphosphinyl)-4-aminobenzoic acid, *in* P-00377
Diphenyl 1-adamantylphosphoramidate, *in* A-00036
Diphenyl aminocarbonylphosphoramidate, *in* A-00062
Diphenyl (aminoiminomethyl)phosphoramidate, *in* P-00370
Diphenyl (3-aminophenyl)phosphoramidate, *in* A-00114
Diphenyl benzoylphosphoramidate, *in* B-00036
Diphenyl benzylphosphoramidate, *in* B-00069
Diphenyl bis(2-chloroethyl)phosphoramidate, *in* B-00121
Diphenyl bis(2-methylpropyl)phosphoramidate, *in* B-00355
Diphenyl (2-bromophenyl)phosphoramidate, *in* B-00494
Diphenyl (3-bromophenyl)phosphoramidate, *in* B-00495
Diphenyl (4-bromophenyl)phosphoramidate, *in* B-00496
Diphenyl butylphosphoramidate, *in* B-00621
Diphenyl *tert*-butylphosphoramidate, *in* B-00622
Diphenyl (2-chlorophenyl)phosphoramidate, *in* C-00155
Diphenyl (3-chlorophenyl)phosphoramidate, *in* C-00156
Diphenyl (4-chlorophenyl)phosphoramidate, *in* C-00157
Diphenyl cyclohexylphosphoramidate, *in* C-00269
Diphenyl cyclooctylphosphoramidate, D-00998
Diphenyl (2,4-dibromophenyl)phosphoramidate, *in* D-00067
Diphenyl (2,4-dichlorophenyl)phosphoramidate, *in* D-00188
Diphenyl dichlorophosphoramidate, D-01002
Diphenyl diethylphosphoramidate, *in* D-00343
Diphenyl (2,4-dimethylphenyl)phosphoramidate, *in* D-00818
Diphenyl dimethylphosphoramidate, *in* D-00855
Diphenyl (2,4-dinitrophenyl)phosphoramidate, *in* D-00947
O,O-Diphenyl (diphenoxyphosphinyl)phosphoramidothioate, D-01005
Diphenyl diphenylphosphoramidate, *in* D-01108
Diphenyl di-2-propenylphosphoramidate, *in* D-01195
Diphenyl dipropylphosphoramidate, *in* D-01214
Diphenyl [1-ethoxycarbonyl-2-(4-hydroxybenzyl)ethyl]-phosphoramidate, *in* P-00422
Diphenyl [4-(ethoxycarbonyl)phenyl]phosphoramidate, *in* C-00006
Diphenyl ethylphosphoramidate, *in* E-00156
Diphenyl hexylphosphoramidate, *in* H-00102
Diphenyl 1*H*-imidazol-1-ylphosphonate, *in* I-00003
Diphenyl isopropylphosphoramidate, *in* I-00058
Diphenyl (4-methoxyphenyl)phosphoramidate, *in* M-00076
Diphenyl methylphenylphosphoramidate, *in* M-00250
Diphenyl (2-methylphenyl)phosphoramidate, *in* M-00251
Diphenyl (3-methylphenyl)phosphoramidate, *in* M-00252
Diphenyl (4-methylphenyl)phosphoramidate, *in* M-00253
Diphenyl methylphosphoramidate, *in* M-00326
Diphenyl (2-methylpropyl)phosphoramidate, *in* M-00384
▷Diphenyl 4-morpholinylphosphonate, *in* M-00523
Diphenyl (2-nitrophenyl)phosphoramidate, *in* N-00058
Diphenyl (3-nitrophenyl)phosphoramidate, *in* N-00059
Diphenyl (4-nitrophenyl)phosphoramidate, *in* N-00060
Diphenyl (phenylamino)carbonylphosphoramidate, *in* P-00081
Diphenyl phenylphosphoramidate, D-01035
Diphenyl phenylsulfonylphosphoramidate, *in* P-00312
Diphenyl phosphoramidate, D-01107
Diphenylphosphoramidic acid; Difluoride, *in* D-01108
Diphenylphosphoramidic acid; Dichloride, *in* D-01108

N,N′-Diphenylphosphoric triamide, D-01114
N,N′-Diphenylphosphorodiamidic acid, D-01125
N,N′-Diphenylphosphorodiamidic acid; Anhydride, *in* D-01125
▷*N,N′*-Diphenylphosphorodiamidic chloride, D-01126
▷*N,N′*-Diphenylphosphorodiamidic fluoride, D-01127
Diphenyl 1-piperidinylphosphonate, *in* P-00430
Diphenyl 2-propenylphosphoramidate, *in* P-00464
Diphenyl propylphosphoramidate, *in* P-00484
Diphenyl 2-pyridylphosphoramidate, *in* P-00510
Diphenyl 3-pyridylphosphoramidate, *in* P-00511
Diphenyl 4-pyridylphosphoramidate, *in* P-00512
N,N′-Diphenyl-*N″*-2-pyridylphosphoric triamide, *in* P-00510
N,N′-Diphenyl-*N″*-3-pyridylphosphoric triamide, *in* P-00511
Diphenyl (2,4,6-tribromophenyl)phosphoramidate, *in* T-00359
Diphenyl (2,4,6-trichlorophenyl)phosphoramidate, *in* T-00420
▷Dipin, D-01189
Di-1-piperidinylphosphinic acid; Fluoride, *in* D-01190
Di-1-piperidinylphosphinic acid; Chloride, *in* D-01190
Dipropyl dipropylphosphoramidate, *in* D-01214
Dipropyl ethylphosphoramidate, *in* E-00156
Dipropyl methylphosphoramidate, *in* M-00326
Dipropyl phenylphosphoramidate, *in* P-00257
Dipropyl phosphoramidate, D-01213
Dipropylphosphoramidic acid; Difluoride, *in* D-01214
Dipropylphosphoramidic acid; Dichloride, *in* D-01214
N,N′-Dipropylphosphorodiamidic acid, D-01223
N,N′-Dipropylphosphorodiamidic acid; Fluoride, *in* D-01223
N,N′-Dipropylphosphorodiamidic acid; Chloride, *in* D-01223
Dipropyl propylphosphoramidate, *in* P-00484
3-Ethoxy-2,3-dihydro-2-methyl-5-phenyl-1*H*-1,2,4,3-triazaphosphole; 3-Oxide, *in* E-00028
Ethyl 2-amino-3-[(diphenoxyphosphinyl)oxy]butanoate, *in* A-00120
Ethyl 2-(benzyloxycarbonyl)amino-3-[(diphenoxyphosphinoyl)-oxy]butanoate, *in* A-00120
Ethyl *N″*-[bis(diethylamino)phosphinyl]-*N,N,N′,N′*-tetraethylphosphorodiamidoimidate, *in* B-00177
Ethyl *N,N′*-dibutylphosphorodiamidate, *in* D-00146
Ethyl dichlorophosphinylcarbamate, *in* D-00191
Ethyl *N,N′*-dicyclohexylphosphorodiamidate, *in* D-00216
Ethyl 2-[(diethoxyphosphinyl)amino]benzoate, *in* C-00004
Ethyl *N*-(diethoxyphosphinyl)carbamate, *in* D-00249
Ethyl diethylphosphoramidochloridate, *in* D-00346
Ethyl diethylphosphoramidofluoridate, *in* D-00351
Ethyl *N,N′*-diethylphosphorodiamidate, *in* D-00382
Ethyl *N*-(dimethoxyphosphinyl)carbamate, *in* D-00680
Ethyl 2-[(dimethoxyphosphinyl)oxy]propanoate, *in* D-00682
Ethyl dimethylphosphoramido chloridate, *in* D-00860
▷Ethyl dimethylphosphoramidofluoridate, E-00064
Ethyl *N,N′*-dimethylphosphorodiamidate, *in* D-00895
Ethyl di-4-morpholinylphosphinate, *in* D-00935
Ethyl 2-[(diphenoxyphosphinyl)amino]benzoate, *in* C-00004
Ethyl 2-[(diphenoxyphosphinyl)amino]-3-phenylpropanoate, *in* P-00400
Ethyl *N*-(diphenoxyphosphinyl)carbamate, *in* D-00989
Ethyl *N,N′*-diphenylphosphorodiamidate, *in* D-01125
Ethyl di-1-piperidinylphosphinate, *in* D-01190
Ethyl *N,N′*-dipropylphosphorodiamidate, *in* D-01223
Ethylene *tert*-butylphosphoramidate, *in* B-00539
Ethylene dimethylphosphoramidate, *in* D-00695
Ethylene phenylphosphoramidate, *in* P-00083
Ethyl *N,N′*-1,2-ethanediyl-*N,N′*-dimethylphosphorodiamidate, *in* E-00030
Ethyl hydrogen bis(2-chloroethyl)phosphoramidate, *in* B-00121
Ethylimidobisphosphoryl fluoride, *in* E-00078
Ethyl *N*-[imino(diphenoxyphosphinylamino)methyl]glycinate, *in* P-00369
Ethyl phenyl phosphoramidate; (+)-*form*, *in* E-00115
Ethyl phenyl phosphoramidate; (−)-*form*, *in* E-00115
Ethyl phenylphosphoramidochloridate, *in* P-00259
Ethyl *N*-phenylphosphorodiamidate, *in* P-00273
Ethylphosphoramidic acid; Difluoride, *in* E-00156
Ethylphosphoramidic acid; Dichloride, *in* E-00156
Ethyl tetrabutylphosphorodiamidate, *in* T-00021
Ethyl tetraethylphosphorodiamidate, *in* T-00059
Ethyl tetramethylphosphorodiamidate, *in* T-00218
Ethyl tetrapropylphosphorodiamidate, *in* T-00291
▷Fenamiphos, F-00002
2-Fluoro-1,3-dimethyl-1,3,2-diazaphospholidine; 2-Oxide, *in* F-00014
▷Fopurine, F-00053
▷Fosthietan, F-00065
Heptylphosphoramidic acid; Dichloride, *in* H-00009
Heptyl tetraethylphosphorodiamidate, *in* T-00059
Hexabutylphosphoric triamide, *in* H-00013
Hexaethylphosphoric triamide, *in* H-00015
Hexaethyl 1,3,5-triazine-2,4,6-triyltriphosphoramidate, *in* T-00330
Hexahydro-2*a*,4*a*,6*a*-triaza-6*b*-phosphacyclopenta[*cd*]-pentalene; 6*b*-Oxide, *in* H-00055
Hexakis(2-methylpropyl)phosphoric triamide, *in* H-00063
2,3,5,6,7,8-Hexamethyl-2,3,5,6,7,8-hexaaza-1,4-diphosphabicyclo[2.2.2]octane; 1,4-Dioxide, *in* H-00070
▷Hexamethylphosphoric triamide, H-00075
Hexamethyl 1,3,5-triazine-2,4,6-triyltriphosphoramidate, *in* T-00330
Hexaphenylphosphoric triamide, *in* H-00090
Hexapropylphosphoric triamide, *in* H-00092
Hexylphosphoramidic acid; Dichloride, *in* H-00102
4-Hydroperoxycyclophosphamide; (2*RS*,4*RS*)-*form*, *in* H-00106
4-Hydroperoxycyclophosphamide; (2*RS*,4*SR*)-*form*, *in* H-00106
4-Hydroxycyclophosphamide, H-00115
(2-Hydroxyethyl)phosphoramidic acid, H-00148
Ifosfamide; (*R*)-*form*, *in* I-00001
Ifosfamide; (*S*)-*form*, *in* I-00001
Ifosfamide; (±)-*form*, *in* I-00001
Imcarbofos, I-00002
P-Imidazol-1-yl-*N,N,N′,N′*-tetramethylphosphonic diamide, *in* I-00003
Iminocyclophosphamide, I-00004
N-[Imino(diphenoxyphosphinylamino)methyl]glycine, *in* P-00369
N-[Imino(diphenoxyphosphinylamino)methyl]-*N*-methylglycine, *in* P-00364
Isopropyl dichlorophosphinylcarbamate, *in* D-00191
Isopropyl diethylphosphoramidochloridate, *in* D-00346
Isopropyl diethylphosphoramidofluoridate, *in* D-00351
Isopropyl dimethylphosphoramidocyanidate, *in* D-00863
Isopropyl *N*-(diphenoxyphosphinyl)carbamate, *in* D-00989
Isopropyl *N,N′*-diphenylphosphorodiamidate, *in* D-01125
Isopropylphosphoramidic acid; Difluoride, *in* I-00058
Isopropylphosphoramidic acid; Dichloride, *in* I-00058
▷Mephosfolan, M-00004
2-Methoxy-3,4-dimethyl-5-phenyl-1,3,2-oxazaphospholidine; (2*R*,4*S*,5*R*)-*form*, 2-Oxide, *in* M-00035
2-Methoxy-3,4-dimethyl-5-phenyl-1,3,2-oxazaphospholidine; (2*S*,4*S*,5*R*)-*form*, 2-Oxide, *in* M-00035
4-Methoxyphenyl *N,N′*-diphenylphosphorodiamidate, *in* M-00481
(2-Methoxyphenyl)phosphoramidic acid; Difluoride, *in* M-00074
(3-Methoxyphenyl)phosphoramidic acid, M-00075
(4-Methoxyphenyl)phosphoramidic acid; Dichloride, *in* M-00076
Methyl *N,N*-bis(2-chloroethyl)phosphorodiamidate, *in* B-00126
Methyl 6-deoxy-2,3-di-*O*-methyl-6-methylaminoglucopyranoside 4,6-cyclic phosphonamide; α-D-(*R*)P-*form*, *P*-Methoxy, *in* M-00112
Methyl 6-deoxy-2,3-di-*O*-methyl-6-methylaminoglucopyranoside 4,6-cyclic phosphonamide; α-D-(*R*)P-*form*, *P*-Ethoxy, *in* M-00112
Methyl 6-deoxy-2,3-di-*O*-methyl-6-methylaminoglucopyranoside 4,6-cyclic phosphonamide; α-D-(*R*)P-*form*, *P*-(4-Nitrophenoxy), *in* M-00112
Methyl 6-deoxy-2,3-di-*O*-methyl-6-methylaminoglucopyranoside 4,6-cyclic phosphonamide; α-D-(*S*)P-*form*, *P*-Chloro, *in* M-00112
Methyl 6-deoxy-2,3-di-*O*-methyl-6-methylaminoglucopyranoside 4,6-cyclic phosphonamide; α-D-(*S*)P-*form*, *P*-Methoxy, *in* M-00112
Methyl 6-deoxy-2,3-di-*O*-methyl-6-methylaminoglucopyranoside 4,6-cyclic phosphonamide; α-D-(*S*)P-*form*, *P*-Ethoxy, *in* M-00112
Methyl 6-deoxy-2,3-di-*O*-methyl-6-methylaminoglucopyranoside 4,6-cyclic phosphonamide; α-D-(*S*)P-*form*, 2-(4-Nitrophenoxy), *in* M-00112
Methyl 2-[(dibenzyloxyphosphinyl)amino]-3-mercaptopropanoate, *in* P-00365

Methyl 2-[(dibenzyloxyphosphinyl)amino]-3-phenylpropanoate, *in* P-00400
Methyl dichlorophosphinylcarbamate, *in* D-00191
Methyl *N*-(diethoxyphosphinyl)carbamate, *in* D-00249
Methyl *N*,*N*′-diethylphosphorodiamidate, *in* D-00382
Methyl 2-[(diisopropoxyphosphinyl)amino]-3-mercaptopropanoate, *in* P-00365
Methyl *N*-(dimethoxyphosphinyl)carbamate, *in* D-00680
Methyl 2-[(dimethoxyphosphinyl)oxy]propanoate, *in* D-00682
4-Methyl-2-dimethylamino-1,3,2-dioxaphosphorinane 2-oxide; (2*RS*,4*RS*)*-form*, *in* M-00118
4-Methyl-2-dimethylamino-1,3,2-dioxaphosphorinane 2-oxide; (2*RS*,4*SR*)*-form*, *in* M-00118
Methyl 2,3-di-*O*-methylglucopyranoside 4,6-cyclic phosphonate; α-D-(*R*)P*-form*, *P*-Dimethylamino, *in* M-00119
Methyl 2,3-di-*O*-methylglucopyranoside 4,6-cyclic phosphonate; α-D-(*S*)P*-form*, *P*-Dimethylamino, *in* M-00119
Methyl dimethylphosphoramidochloridate, *in* D-00860
Methyl dimethylphosphoramidocyanidate, *in* D-00863
4-Methyl-2,6-dioxa-7-aza-1-phosphabicyclo[2.2.2]octane 1-oxide; 7-*tert*-Butyl, *in* M-00124
4-Methyl-2,6-dioxa-7-aza-1-phosphabicyclo[2.2.2]octane 1-oxide; 7-Ph, *in* M-00124
4-Methyl-2,6-dioxa-7-aza-1-phosphabicyclo[2.2.2]octane 1-oxide; 7-Benzyl, *in* M-00124
Methyl *N*-(diphenoxyphosphinyl)carbamate, *in* D-00989
Methyl *N*,*N*′-diphenylphosphorodiamidate, *in* D-01125
Methyl di-1-piperidinylphosphinate, *in* D-01190
Methyl 3-hydroxy-2-[(dibenzyloxyphosphinyl)amino]butanoate, *in* P-00420
Methyl 3-hydroxy-2-[(dibenzyloxyphosphinyl)amino]-propanoate, *in* P-00418
Methyl 3-hydroxy-2-[(diisopropoxyphosphinyl)amino]-butanoate, *in* P-00420
Methyl 3-hydroxy-2-[(diisoproproxyphosphinyl)amino]-propanoate, *in* P-00418
Methylimidobisphosphoryl fluoride, *in* M-00156
Methylimidodiphosphoryl chloride, M-00160
Methyl *N*-[imino(diphenoxyphosphinylamino)methyl]glycinate, *in* P-00369
3-Methyl-2-phenoxy-1,3,2-oxazaphospholidine; 2-Oxide, *in* M-00192
4-Methyl-2-phenylamino-1,3,2-dioxaphosphorinane; (2*RS*,4*SR*)*-form*, 2-Oxide, *in* M-00193
N-(4-Methylphenyl)-*N*′,*N*″-diphenylphosphoric triamide, *in* M-00253
(2-Methylphenyl)phosphoramidic acid; Difluoride, *in* M-00251
(2-Methylphenyl)phosphoramidic acid; Dichloride, *in* M-00251
(3-Methylphenyl)phosphoramidic acid; Difluoride, *in* M-00252
(4-Methylphenyl)phosphoramidic acid, M-00253
(4-Methylphenyl)phosphoramidic acid; Difluoride, *in* M-00253
(4-Methylphenyl)phosphoramidic acid; Dichloride, *in* M-00253
Methylphosphoramidic dichloride, M-00327
Methylphosphoramidic difluoride, M-00328
N-2-Methylpropyl-*N*′,*N*″-diphenylphosphoric triamide, *in* M-00384
(2-Methylpropyl)phosphoramidic acid; Dichloride, *in* M-00384
Methylsulfonylphosphoramidic acid; Dichloride, *in* M-00401
Methyl tetraethylphosphorodiamidate, *in* T-00059
Methyl tetramethylphosphorodiamidate, M-00406
Methyl tetrapropylphosphorodiamidate, *in* T-00291
Methyl *N*,*N*′-trimethylenephosphorodiamidate, *in* H-00020
Meturedepa, M-00439
▷Mipafox, *in* D-00661
4-Morpholinylphosphonic acid; Dichloride, *in* M-00523
4-Morpholinylphosphonic diamide, *in* M-00523
1,2-Naphthylene diethylphosphoramidate, *in* D-00266
1,8-Naphthylene diethylphosphoramidate, *in* D-00267
1-Naphthylphosphoramidic acid, N-00028
1-Naphthylphosphoramidic acid; Dichloride, *in* N-00028
2-Naphthylphosphoramidic acid; Dichloride, *in* N-00029
N″-Bis(dimethylamino)phosphinyl-*N*,*N*,*N*′,*N*′,*N*″-pentamethylphosphorotriamidothioate, N-00032
Neopentylene *tert*-butylphosphoramidate, *in* B-00538
Neopentylene dimethylphosphoramidate, *in* D-00693
Neopentylene phenylphosphoramidate, *in* D-00792
N-(3-Nitrophenyl)-*N*′,*N*″-diphenylphosphoric triamide, *in* N-00059
N-(4-Nitrophenyl)-*N*′,*N*″-diphenylphosphoric triamide, *in* N-00060
3-Nitrophenyl *N*,*N*′-diphenylphosphorodiamidate, *in* M-00499
4-Nitrophenyl *N*,*N*′-diphenylphosphorodiamidate, *in* M-00500
4-Nitrophenyl phenylphosphoramidochloridate, *in* P-00259
(3-Nitrophenyl)phosphoramidic acid; Dichloride, *in* N-00059
(4-Nitrophenyl)phosphoramidic acid; Dichloride, *in* N-00060
Nonamethyl imidodiphosphoramide, N-00068
Nonylphosphoramidic acid; Dichloride, *in* N-00077
Octaethyldiphosphoramide, O-00001
▷Octamethyldiphosphoramide, O-00029
unsym-Octamethylmonothiopyrophosphoramide, O-00031
Octylphosphoramidic acid; Dichloride, *in* O-00045
▷4-Oxocyclophosphamide, O-00058
▷Pentaethyl imidodiphosphate, *in* E-00078
Pentaethylphosphoric triamide, *in* T-00058
▷Pentamethylphosphoric triamide, P-00026
N-Pentyl-*N*′,*N*″-diphenylphosphoric triamide, *in* P-00053
Pentylphosphoramidic acid; Dichloride, *in* P-00053
Peroxycyclophosphamide, P-00057
2-Phenoxy-3-phenyl-1,3,2-oxazaphospholidine; 2-Oxide, *in* P-00073
(Phenylamino)carbonylphosphoramidic acid; Dichloride, *in* P-00081
Phenyl *N*,*N*-bis(2-chloroethyl)-*N*′-phenylphosphorodiamidate, *in* B-00121
Phenyl *N*,*N*-bis(2-chloroethyl)phosphorodiamidate, *in* B-00126
Phenyl cyclohexylphosphoramidochloridate, P-00104
Phenyl *N*,*N*′-dibenzylphosphorodiamidate, *in* D-00061
Phenyl *N*,*N*′-dibutylphosphorodiamidate, *in* D-00146
Phenyl *N*,*N*′-dicyclohexylphosphorodiamidate, *in* D-00216
Phenyl diethylphosphoramidochloridate, *in* D-00346
Phenyl *N*,*N*′-diethylphosphorodiamidate, *in* D-00382
Phenyl *N*,*N*′-diisopropylphosphorodiamidate, *in* D-00661
Phenyl dimethylphosphoramidochloridate, *in* D-00860
▷Phenyl di-4-morpholinylphosphinate, *in* D-00935
Phenyl *N*-(diphenoxyphosphinyl)carbamate, *in* D-00989
Phenyl *N*,*N*′-diphenylphosphorodiamidate, *in* D-01125
Phenyl di-1-piperidinylphosphinate, *in* D-01190
N-Phenyl-*P*,*P*-di-1-piperidinylphosphinic amide, *in* D-01190
Phenyl *N*,*N*′-dipropylphosphorodiamidate, *in* D-01223
o-Phenylene dimethylphosphoramidate, *in* D-00689
Phenyl hydrogen (2-aminoethyl)phosphoramidate, *in* A-00080
Phenyl hydrogen (2-aminophenyl)phosphoramidate, *in* A-00113
Phenyl hydrogen bis(2-chloroethyl)phosphoramidate, *in* B-00121
Phenyl *N*,*N*′-*o*-phenylenephosphorodiamidate, *in* D-00543
Phenyl phenylphosphoramidic acid, P-00189
Phenyl phenylphosphoramidic acid; Chloride, *in* P-00189
Phenyl phenylphosphoramidochloridate, P-00190
Phenylphosphoramidic acid, P-00257
Phenylphosphoramidic acid; Difluoride, *in* P-00257
Phenylphosphoramidic dichloride, P-00258
Phenylphosphoramidochloridic acid, P-00259
Phenyl phosphoramidofluoridate, *in* P-00295
Phenylphosphoric triamide, *in* P-00257
Phenyl phosphorodiamidate, P-00272
N-Phenylphosphorodiamidic acid, P-00273
Phenylsulfonylphosphoramidic acid, P-00312
Phenylsulfonylphosphoramidic acid; Monobromide, *in* P-00312
Phenylsulfonylphosphoramidic acid; Dibromide, *in* P-00312
Phenyl tetraethylphosphorodiamidate, P-00318
Phenyl tetramethylphosphorodiamidate, *in* T-00218
Phenyl *N*,*N*′-trimethylenephosphorodiamidate, *in* H-00039
▷Phosfolan, P-00342
▷Phosphemide, P-00353
1,1′,1″-Phosphinylidynetrisazetidine, *in* T-00328
▷1,1′,1″-Phosphinylidynetrisaziridine, *in* T-00331
1,1′,1″-Phosphinylidynetris(1*H*-imidazole), *in* T-00495
4,4′,4″-Phosphinylidynetrismorpholine, *in* T-00569
▷1,1′,1″-Phosphinylidynetrispiperidine, *in* T-00704
1,1′,1″-Phosphinylidynetris(1*H*-pyrazole), *in* T-00722
1,1′,1″-Phosphinylidynetrispyrrole, *in* T-00725
1,1′,1″-Phosphinylidynetrispyrrolidine, *in* T-00724
1,1′,1″-Phosphinylidynetris(1*H*-1,2,4-triazole), *in* T-00876
Phosphoarginine; (*S*)*-form*, *in* P-00363
N-Phosphocreatine, P-00364

N-Phosphoglutamic acid, P-00367
Phosphoguanidine, P-00370
(*N*-Phosphono)methionine-*S*-sulfoximinylalanylalanine, P-00389
N-Phosphoserine, P-00418
N-Phosphothreonine, P-00420
▷1-Piperidinylphosphonic acid; Dichloride, *in* P-00430
2-Propenyl tetramethylphosphorodiamidate, *in* T-00218
Propylene dimethylphosphoramidate, *in* M-00116
Propylphosphoramidic acid; Dichloride, *in* P-00484
Propyl tetrapropylphosphorodiamidate, *in* T-00291
N''-2-Pyrimidinylphosphoric triamide; *N,N'*-Bis(2-chloroethyl), *in* P-00517
N''-2-Pyrimidinylphosphoric triamide; *N'*-Et-*N*-(2-chloroethyl), *in* P-00517
▷Sufosfamide, S-00026
▷Tabun, T-00001
Tetrabutylphosphorodiamidic chloride, T-00022
O,O,O',O'-Tetraethyl imidodiphosphate, T-00051
Tetraethyl methylimidodiphosphate, *in* M-00156
Tetraethyl 1,4-phenylenebisphosphoramidate, *in* B-00012
Tetraethyl phenylimidodiphosphate, *in* P-00168
N,N,N',N'-Tetraethyl-2-phenylphosphorodiamidic hydrazide, *in* P-00274
N,N,N',N'-Tetraethylphosphoric triamide; *N''*-Trimethylsilyl, *in* T-00058
Tetraethyl phosphor(isocyanatidic) diamide, *in* T-00059
Tetraethyl phosphor(isothiocyanatidic) diamide, *in* T-00059
▷Tetraethylphosphorodiamidic acid; Fluoride, *in* T-00059
Tetraethylphosphorodiamidic acid; Azide, *in* T-00059
Tetraethylphosphorodiamidic anilide, *in* T-00058
Tetraethylphosphorodiamidic chloride, T-00060
O,O,O',O'-Tetraethyl thioimidodiphosphate, T-00074
Tetraethyl trimethylsilylimidodiphosphate, *in* T-00051
Tetrahydro-5,5-dimethyl-2-phenoxy-2*H*-1,3,2-oxazaphosphorine; 2-Oxide, *in* T-00098
Tetrahydro-4-methyl-2*H*-1,3,4,2-oxadiazaphosphorine 2-oxide; 2-Phenoxy, *in* T-00116
Tetrahydro-4-methyl-2*H*-1,3,4,2-oxadiazaphosphorine 2-oxide; 2-Cyclohexylamino, *in* T-00116
Tetrahydro-2-phenoxy-2*H*-1,3,2-oxazaphosphorine; 2-Oxide, *in* T-00126
Tetrahydro-2-phenoxy-3-phenyl-2*H*-1,3,2-oxazaphosphorine; 2-Oxide, *in* T-00127
Tetrahydro-3-phenyl-2-(phenylamino)-2*H*-1,3,2-oxazaphosphorine; 2-Oxide, *in* T-00133
Tetraisopropyl 1,2-ethanediylbisphosphoramidate, *in* E-00009
Tetraisopropyl ethyliminodiphosphate, *in* E-00078
Tetraisopropyl methyliminodiphosphate, *in* M-00156
N,N,4,4-Tetramethyl-1,3-bis(1,1-dimethylethyl)-1,3-diaza-2-phosphacyclobutane-2-amine, *in* D-00087
4,4,5,5-Tetramethylethylene dimethylphosphoramidate, *in* D-00710
Tetramethyl methylimidodiphosphate, *in* M-00156
N,N,N',N'-Tetramethyl-2-phenylphosphorodiamidic hydrazide, *in* P-00274
N,N,N',N'-Tetramethylphosphoric triamide; *N''*-Trimethylsilyl, *in* T-00216
N,N,N',N''-Tetramethylphosphoric triamide, *in* D-00880
Tetramethyl phosphor(isothiocyanatidic) diamide, *in* T-00218
Tetramethylphosphorodiamidic acid, T-00218
Tetramethylphosphorodiamidic acid; Bromide, *in* T-00218
Tetramethylphosphorodiamidic azide, T-00219
N,N,N',N'-Tetramethyl-*N''*-(2-propenyl)phosphoric triamide, *in* T-00216
2,4,6,7-Tetramethyl-2,6,7-triaza-1-phosphabicyclo[2.2.2]-octane; 1-Oxide, *in* T-00237
N,N',N'',N''-Tetramethyl-*N,N'*-trimethylenephosphoric triamide, *in* H-00023
Tetraphenyl 1,2-ethanediylbisphosphoramidate, *in* E-00009
O,O,O',O'-Tetraphenyl imidodiphosphate, T-00265
Tetraphenyl methylimidodiphosphate, *in* M-00156
Tetraphenyl 1,3-phenylenebisphosphoramidate, *in* B-00011
Tetraphenyl 1,4-phenylenebisphosphoramidate, *in* B-00012
N,N',N'',N''-Tetraphenylphosphoric triamide, *in* D-01108
Tetrapropyl 1,2-ethanediylbisphosphoramidate, *in* E-00009
Tetrapropylphosphorodiamidic acid; Chloride, *in* T-00291
▷Triamiphos, T-00326
Tribenzyl phosphonoglycinate, *in* P-00368
N,N',N''-Tribenzylphosphoric triamide, *in* T-00344
2,2,2-Tribromoethyl 4-*tert*-butylphenylphosphoramidic chloride, T-00356
2,2,2-Tribromoethyl 4-morpholinylphosphonochloridate, T-00357
2,2,2-Tribromoethyl phenylphosphoramidochloridate, *in* P-00259
(2,4,6-Tribromophenyl)phosphoramidic acid; Dichloride, *in* T-00359
N,N',N''-Tributylphosphoric triamide, *in* T-00394
2,2,2-Trichloroethyl phenylphosphoramidochloridate, *in* P-00259
(2,4,6-Trichlorophenyl)phosphoramidic acid; Dichloride, *in* T-00420
N,N',N''-Tricyclohexylphosphoric triamide, *in* T-00429
Triethyl [bis(diethylamino)phosphinyl]phosphorimidate, *in* B-00176
Triethyl [bis(dimethylamino)phosphinyl]phosphorimidate, *in* B-00194
Triethyl (diethoxyphosphinyl)phosphorimidate, *in* D-00257
Triethyl (dimethoxyphosphinyl)phosphorimidate, *in* D-00683
Triethyl 3-(*N*-phosphono)benzoate, *in* C-00005
Triethyl phosphonoglycinate, *in* P-00368
N,N',N''-Triethylphosphoric triamide, *in* T-00462
Trifosfamide; (*R*)-*form*, *in* T-00485
Trifosfamide; (*S*)-*form*, *in* T-00485
Trifosfamide; (±)-*form*, *in* T-00485
N,N',N''-Triisopropylphosphoric triamide, *in* T-00509
Trimethyl [bis(diethylamino)phosphinyl]phosphorimidate, *in* B-00176
Trimethyl (diethoxyphosphinyl)phosphorimidate, *in* D-00257
Trimethyl (dimethoxyphosphinyl)phosphorimidate, *in* D-00683
2,4,10-Trimethyl-1,3,4,5,7,10-hexaaza-3-phosphatricyclo[3.3.1.1^{3,7}]decane; 3-Oxide, *in* T-00517
N,N',N''-Trimethylphosphoric triamide, *in* T-00551
2,8,9-Trimethyl-2,8,9-triaza-1-phosphatricyclo[3.3.1.1^{3,7}]-decane; Oxide, *in* T-00563
N,N',N''-Tri-1-naphthalenylphosphoric triamide, *in* T-00576
N,N',N''-Tri-2-naphthylphosphoric triamide, *in* T-00577
Triphenyl [bis(dimethylamino)phosphinyl]phosphorimidate, *in* B-00194
N,N',N''-Triphenylphosphoric triamide, *in* T-00677
N,N',N''-Tri-2-propenylphosphorous triamide; Oxide, *in* T-00709
Tripropyl phosphonoglycinate, *in* P-00368
N,N',N''-Tripropylphosphoric triamide, *in* T-00719
2,4,6-Tris(diethylamino)-1,3,5,2,4,6-trioxatriphosphorinane 2,4,6-trioxide, T-00756
2,4,6-Tris(diethylamino)-1,3,5,2,4,6-trioxatriphosphorinane 2,4,6-trioxide; (2α,4α,6α)-*form*, *in* T-00756
2,4,6-Tris(diethylamino)-1,3,5,2,4,6-trioxatriphosphorinane 2,4,6-trioxide; (2α,4α,4β)-*form*, *in* T-00756
N,N',N''-Tris(1,1-dimethylethyl)phosphoric triamide, *in* T-00395
N,N',N''-Tris(2-methylphenyl)phosphoric triamide, *in* T-00817
N,N',N''-Tris(3-methylphenyl)phosphoric triamide, *in* T-00818
N,N',N''-Tris(4-methylphenyl)phosphoric triamide, *in* T-00819
Tris(2-methylpropyl) (dimethoxyphosphinyl)phosphorimidate, *in* D-00683
N,N',N''-Tris(1-methylpropyl)phosphoric triamide, *in* T-00829
N,N',N''-Tris(2-methylpropyl)phosphoric triamide, *in* T-00830
Tris(triphenylphosphoranylidene)phosphoric triamide, *in* T-00872
▷Uredepa, U-00001

Phosphoric isocyanates and isothiocyanates

Diethyl phosphorisocyanatidate, D-00366
Diethyl phosphor(isothiocyanatidate), D-00369
Diphenyl phosphorisocyanatidate, D-01116
Diphenyl phosphoroisothiocyanatidate, D-01138
Ethyl phosphorochloridisocyanatidate, E-00163

Ethyl phosphorodiisocyanatidate, E-00177
Ethyl phosphorofluoridisocyanatidate, E-00180
2-Isocyano-5,5-dimethyl-1,3,2-dioxaphosphorinane; 2-Oxide, *in* I-00023
2-Isoselenocyanato-5,5-dimethyl-1,3,2-dioxaphosphorinane; 2-Oxide, *in* I-00066
Methyl phosphorodiisocyanatidate, M-00356
Phenyl phosphorochloridisocyanatidate, P-00270
Phenyl phosphorodiisocyanatidate, P-00292
Phenyl phosphorodi(isothiocyanatate), P-00294
Phosphorisocyanatidic dichloride, P-00405
Phosphorisocyanatidic difluoride, P-00406
Phosphor(isothiocyanatidic) dichloride, P-00407
Phosphor(isothiocyanatidic) difluoride, P-00408
Phosphorodiisocyanatidic bromide, P-00411
Phosphorodiisocyanatidic chloride, P-00412
Phosphorodiisocyanatidic fluoride, P-00413
Phosphorodi(isothiocyanatidic) fluoride, P-00414
Phosphoryl isocyanate, P-00415
Phosphoryl isothiocyanate, P-00416
Tetraethyl phosphor(isocyanatidic) diamide, *in* T-00059
Tetraethyl phosphor(isothiocyanatidic) diamide, *in* T-00059
Tetramethyl phosphor(isothiocyanatidic) diamide, *in* T-00218

Phosphoric azides

Bis(2-oxo-3-oxazolidinyl)phosphinic acid; Azide, *in* B-00378
▷Diethyl phosphorazidate, D-00362
▷Dimethyl phosphorazidate, D-00878
Diphenyl phosphorazidate, D-01112
o-Phenylene phosphoryl azide, *in* A-00148
Tetraethylphosphorodiamidic acid; Azide, *in* T-00059
Tetramethylphosphorodiamidic azide, T-00219

Phosphoric hydrazides

2-Chloro-2,3-dihydro-3,5-diphenyl-1,3,4,2-oxadiazaphosphole; 2-Oxide, *in* C-00042
2-Chloro-2,3-dihydro-5-methyl-3-phenyl-1,3,4,2-oxadiazaphosphole; 2-Oxide, *in* C-00044
N,*N*-Diethyl-2,2′-diphenylphosphoramidic dihydrazide, *in* D-01109
Diethyl phosphorohydrazidate, D-00391
2,3-Dihydro-2-dimethylamino-5-methyl-3-phenyl-1,3,4,2-oxadiazaphosphole; 2-Oxide, *in* D-00455
2,3-Dihydro-5-methyl-3-phenyl-2-phenylamino-1,3,4,2-oxadiazaphosphole; 2-Oxide, *in* D-00527
N,*N*-Dimethyl-2,2′-diphenylphosphoramidic dihydrazide, *in* D-01109
Dimethyl phosphorohydrazidate, D-00904
Diphenyl phosphorohydrazidate, D-01136
3-Ethoxy-2,3-dihydro-2-methyl-5-phenyl-1*H*-1,2,4,3-triazaphosphole; 3-Oxide, *in* E-00028
Hexahydro-2,4-dimethyl-3-phenoxy-1,2,4,5,3-tetrazaphosphorine; 3-Oxide, *in* H-00024
2,2,2′,2′,2″,2″-Hexamethylphosphoric trihydrazide, H-00076
Phenyl phosphorodihydrazidate, P-00290
N,*N*,*N*′,*N*′-Tetraethyl-2-phenylphosphorodiamidic hydrazide, *in* P-00274
N,*N*,*N*′,*N*′-Tetramethyl-2-phenylphosphorodiamidic hydrazide, *in* P-00274
2,2′,2″-Triphenylphosphoric trihydrazide, T-00664

Thiophosphoric acids

S-Benzyl phosphorothioate, B-00071
O,*O*-Bis(2-chlorophenyl) phosphorodithioate, B-00150
O,*O*-Bis(4-chlorophenyl) phosphorodithioate, B-00151
O,*O*-Bis(3-dimethylaminophenyl) phosphorothioate, B-00189
▷*O*,*O*-Bis(2-ethylhexyl) phosphorodithioate, B-00274
O,*O*-Bis(3-methylbutyl) phosphorodithioate, B-00304
O,*O*-Bis(3-methylbutyl) phosphorothioate, B-00305
O,*O*-Bis(2-methylphenyl) phosphorodithioate, B-00337
▷*O*,*O*-Bis(3-methylphenyl) phosphorodithioate, B-00338
O,*O*-Bis(4-methylphenyl) phosphorodithioate, B-00339
O,*O*-Bis(1-methylpropyl) phosphorodithioate, B-00359
O,*O*-Bis(2-methylpropyl) phosphorodithioate, B-00360
S,*S*-Bis(2-methylpropyl) phosphorodithioate, B-00361
O,*O*-Bis(2-methylpropyl) phosphorothioate, B-00363
O,*O*-Bis(trimethylsilyl) phosphorothioate, B-00439
O-Butyl *O*-methyl phosphorodithioate, B-00574
O-Butyl *O*-1-naphthalenyl phosphorothioate; (+)-*form*, *in* B-00575
S-Butyl phosphorothioate, B-00634
O,*O*-Dibenzyl phosphorodithioate, D-00063
N,*N*′-Dibutylphosphorodiamidodithioic acid, D-00147
O,*O*-Dibutyl phosphorodithioate, D-00150
O,*O*-Di-*tert*-butyl phosphorodithioate, D-00151
S,*S*-Dibutyl phosphorodithioate, D-00152
O,*O*-Dibutyl phosphorothioate, D-00155
O,*O*-Di-*tert*-butyl phosphorothioate, D-00156
N,*N*′-Dicyclohexylphosphorodiamidodithioic acid, D-00217
O,*O*-Didecyl phosphorodithioate, D-00224
▷*O*,*O*-Diethyl phosphorodithioate, D-00384
O,*S*-Diethyl phosphorodithioate, D-00385
S,*S*-Diethyl phosphorodithioate, D-00386
O,*O*-Diethyl phosphoroselenothioate, D-00394
O,*O*-Diethyl phosphorothioate, D-00396
O,*S*-Diethyl phosphorothioate, D-00397
O,*O*-Diheptyl phosphorodithioate, D-00427
O,*O*-Dihexyl phosphorodithioate, D-00437
O,*O*-Diisopropyl phosphorodithioate, D-00663
S,*S*-Diisopropyl phosphorodithioate, D-00664
O,*O*-Diisopropyl phosphorothioate, D-00667
3,9-Dimercapto-2,4,8,10-tetraoxa-3,9-diphosphaspiro[5.5]-undecane 3,9-disulfide, D-00670
▷*O*,*O*-Dimethyl phosphorodithioate, D-00898
O,*S*-Dimethyl phosphorodithioate, D-00899
S,*S*-Dimethyl phosphorodithioate, D-00900
O,*O*-Dimethyl phosphorothioate, D-00907
O,*S*-Dimethyl phosphorothioate, D-00908
O,*O*-Di-1-naphthyl phosphorodithioate, D-00944
O,*O*-Di-2-naphthyl phosphorodithioate, D-00945
O,*O*-Dinonyl phosphorodithioate, D-00951
O,*O*-Dioctyl phosphorodithioate, D-00960
O,*O*-Dipentyl phosphorodithioate, D-00986
N,*N*′-Diphenylphosphorodiamidodithioic acid, D-01128
O,*O*-Diphenyl phosphorodithioate, D-01131
S,*S*-Diphenyl phosphorodithioate, D-01132
O,*O*-Diphenyl phosphoroselenothioate, D-01140
O,*O*-Diphenyl phosphorothioate, D-01141
O,*O*-Di-2-propenyl phosphorodithioate, D-01197
O,*O*-Di-2-propenyl phosphorothioate, D-01198
O,*O*-Dipropyl phosphorodithioate, D-01225
O,*S*-Dipropyl phosphorodithioate, D-01226
S,*S*-Dipropyl phosphorodithioate, D-01227
O,*O*-Dipropyl phosphorothioate, D-01230
▷*O*,*O*-Di-2-propynyl phosphorodithioate, D-01232
O-Ethyl *O*-methyl phosphorodithioate, E-00095
O-Ethyl *S*-methyl phosphorodithioate, E-00096
S-Ethyl *O*-methyl phosphorodithioate, E-00097
O-Ethyl *O*-methyl phosphorothioate, E-00098
O-Ethyl *O*-phenyl phosphorodithioate, E-00120
O-Ethyl *S*-phenyl phosphorodithioate, E-00121
O-Ethyl phosphorothioate, E-00182
S-Ethyl phosphorothioate, E-00183
O-Ethyl *S*-propyl phosphorodithioate, E-00186
2-Hydroxy-4*H*-1,3,2-benzodioxaphosphorin 2-sulfide, H-00110
2-Hydroxy-4,5-dimethyl-1,3,2-dioxaphospholane 2-sulfide; (2α,4α,5α)-*form*, Tetramethylammonium salt, *in* H-00122
2-Hydroxy-4,5-dimethyl-1,3,2-dioxaphospholane 2-sulfide; (2α,4α,5β)-*form*, *in* H-00122
2-Hydroxy-4,5-dimethyl-1,3,2-dioxaphospholane 2-sulfide; (2α,4β,5β)-*form*, *in* H-00122
2-Hydroxy-1,3,2-dioxaphospholane 2-sulfide, H-00135
2-Hydroxy-4-methyl-1,3,2-dioxaphosphorinane 2-sulfide; (2*RS*,4*RS*)-*form*, *in* H-00158
2-Hydroxy-4-methyl-1,3,2-dioxaphosphorinane 2-sulfide; (2*RS*,4*SR*)-*form*, *in* H-00158
2-Hydroxy-4,4,5,5-tetramethyl-1,3,2-dioxaphospholane 2-sulfide, H-00197
S-Isopropyl phosphorothioate, I-00064
2-Mercapto-1,3,2-benzodioxaphosphole 2-sulfide, M-00005
2-Mercaptodibenzo[*d*,*f*][1,3,2]dioxaphosphepin 2-oxide, M-00007

6-Mercaptodibenzo[*d,f*][1,3,2]dioxaphosphepin 6-sulfide, M-00008
2-Mercapto-4,5-dimethyl-1,3,2-dioxaphospholane 2-sulfide; NH_4 salt, *in* M-00009
2-Mercapto-4,5-dimethyl-1,3,2-dioxaphospholane 2-sulfide; (4*RS*,5*RS*)-*form*, *in* M-00009
2-Mercapto-5,5-dimethyl-1,3,2-dioxaphosphorinane 2-oxide, M-00010
2-Mercapto-5,5-dimethyl-1,3,2-dioxaphosphorinane 2-sulfide, M-00011
4-Mercaptodinaphtho[2,1-*d*:1′,2′-*f*][1,3,2]dioxaphosphepin 4-sulfide; (*S*)-*form*, *in* M-00012
4-Mercaptodinaphtho[2,1-*d*:1′,2′-*f*][1,3,2]dioxaphosphepin 4-sulfide; (±)-*form*, *in* M-00012
2-Mercapto-1,3,2-dioxaphosphorinane 2-sulfide, M-00013
2-Mercapto-1,3,2-dithiaphosphorinane 2-sulfide, M-00014
2-Mercapto-4-methyl-1,3,2-dioxaphospholane 2-oxide, M-00015
2-Mercapto-4-methyl-1,3,2-dioxaphospholane 2-sulfide, M-00016
2-Mercapto-4-methyl-1,3,2-dioxaphosphorinane 2-sulfide, M-00017
2-Mercapto-4,4,5,5-tetramethyl-1,3,2-dioxaphospholane 2-sulfide, M-00019
O-Methyl *O*-phenyl phosphorodithioate, M-00263
O-Methyl phosphorothioate, M-00359
S-Methyl phosphorothioate, M-00360
S-Phenyl phosphorothioate, P-00297
S-Propyl phosphorothioate, P-00493
Tetrahydro-2-hydroxy-2*H*-1,3,2-oxazaphosphorine 2-sulfide, T-00108
Tetramethylphosphorodiamidodithioic acid, T-00222

Thiophosphoric anhydrides

▷Aspon, A-00144
O-Benzoyl *O,O*-diethylphosphorothioate, B-00029
O,O-Diethyl *S*-benzoyl phosphorodithioate, D-00270
2-[(5,5-Dimethyl-1,3,2-dioxaphosphorinan-2-yl)thio]-5,5-dimethyl-1,3,2-dioxaphosphorinane *P*-oxide 2-sulfide, *in* T-00309
sym-Octamethylmonothiopyrophosphoramide, O-00030
unsym-Octamethylmonothiopyrophosphoramide, O-00031
Octamethyl thiodiphosphoramide, O-00034
2,2′-Oxybis[5,5-dimethyl-1,3,2-dioxaphosphorinane] 2-oxide 2′-sulfide, O-00090
Sandoflam, *in* O-00089
▷*O,O,O,O*-Tetraethyl dithiopyrophosphate, T-00045
unsym-*O,O,O,O*-Tetraethyl dithiopyrophosphate, T-00046
sym-Tetraethyl monothiopyrophosphate, T-00052
▷*unsym*-Tetraethyl monothiopyrophosphate, T-00053
▷*O,O,O′,O′*-Tetraethyl trithiopyrophosphate, T-00076
▷*sym*-Tetraisopropyl dithiopyrophosphate, T-00168
O,O,O′,O′-Tetraisopropyl trithiopyrophosphate, T-00174
▷*sym*-*O,O,O,O*-Tetramethyl dithiopyrophosphate, T-00203
sym-Tetramethyl monothiopyrophosphate, T-00206
▷*unsym*-Tetramethyl monothiopyrophosphate, T-00207
O,O,O′,O′-Tetramethyl trithiopyrophosphate, T-00240
O,O,O′,O′-Tetraphenyl trithiopyrophosphate, T-00284
2,2′-Thiobis[5,5-dimethyl-1,3,2-dioxaphosphorinane]; 2,2′-Disulfide, *in* T-00309
2,2′-Thiobis[5,5-dimethyl-1,3,2-dioxaphosphorinane] 2,2′-dioxide, T-00310

Thiophosphoric halides

O-(2-Acetylphenyl) phosphorodichloridothioate, A-00022
O-(3-Acetylphenyl) phosphorodichloridothioate, A-00023
O-(4-Acetylphenyl) phosphorodichloridothioate, A-00024
3-Benzyl-2-chlorotetrahydro-2*H*-1,3,2-oxazaphosphorine; 2-Sulfide, *in* B-00041
O,O-(2,2′-Biphenylyl) phosphorochloridothioate, *in* C-00032
O-[(1,1′-Biphenyl)-2-yl] phosphorodichloridothioate, B-00083
▷Bis(2-chloroethyl)phosphoramidothioic dichloride, B-00123
O,O-Bis(2-chloroethyl) phosphorochloridothioate, B-00124
O,O-Bis(3-chlorophenyl) phosphorochloridothioate, B-00148
O,O-Bis(4-chlorophenyl) phosphorochloridothioate, B-00149
O,O-Bis(4-dimethylaminophenyl) phosphorobromidothioate, *in* B-00190
O,O-Bis(4-dimethylaminophenyl) phosphorochloridothioate, *in* B-00190
O,O-Bis(4-dimethylaminophenyl) phosphorofluoridothioate, *in* B-00190
O,O-Bis(3-dimethylaminophenyl) phosphorothioate; Chloride, *in* B-00189
O,O-Bis(2-ethoxyethyl) phosphorochloridothioate, B-00271
O,O-Bis(2-methylphenyl) phosphorochloridothioate, B-00333
O,O-Bis(3-methylphenyl) phosphorochloridothioate, B-00334
O,O-Bis(4-methylphenyl) phosphorochloridothioate, B-00335
O,O-Bis(2-methylpropyl) phosphorochloridothioate, B-00357
O,O-Bis(4-nitrophenyl) phosphorochloridothioate, B-00377
2-Bromo-4,5-dimethyl-1,3,2-dioxaphospholane; (2α,4β,5β)-*form*, 2-Sulfide, *in* B-00470
O-(4-Bromophenyl) phosphorodichloridothioate, B-00497
O-(2-Butoxyethyl) phosphorodichloridothioate, *in* H-00149
O,O-2,3-Butylene bromothiophosphate, *in* B-00470
O,O-1,3-Butylene fluorothiophosphate, *in* F-00021
O,O-2,3-Butylene fluorothiophosphate, *in* F-00015
OO-1,3-Butylene thiophosphoryl bromide, *in* B-00474
tert-Butylphosphoramidothioic acid; Difluoride, *in* B-00623
O-Butyl phosphorodichloridothioate, B-00630
S-Butyl phosphorodichloridothioate, B-00631
2-Chloro-5-chloromethyl-5-methyl-1,3,2-dioxaphosphorinane; *cis*-*form*, 2-Sulfide, *in* C-00031
2-Chloro-5-chloromethyl-5-methyl-1,3,2-dioxaphosphorinane; *trans*-*form*, 2-Sulfide, *in* C-00031
2-Chloro-2,3-dihydro-3,5-diphenyl-1,3,4,2-oxadiazaphosphole; 2-Sulfide, *in* C-00042
2-Chloro-2,3-dihydro-3-methyl-1,3,2-benzothiazaphosphole; 2-Sulfide, *in* C-00043
2-Chloro-2,3-dihydro-5-methyl-3-phenyl-1,3,4,2-oxadiazaphosphole; 2-Sulfide, *in* C-00044
2-Chloro-1,3-dimethyl-1,3,2-diazaphospholidine; 2-Sulfide, *in* C-00055
2-Chloro-5,5-dimethyl-1,3,2-dioxaphosphorinane 2-sulfide, C-00060
2-Chloro-3,4-dimethyl-5-phenyl-1,3,2-oxazaphospholidine; (2*R*,4*S*,5*R*)-*form*, 2-Sulfide, *in* C-00061
2-Chloro-3,4-dimethyl-5-phenyl-1,3,2-oxazaphospholidine; (2*S*,4*S*,5*R*)-*form*, 2-Sulfide, *in* C-00061
2-Chloro-1,3,2-dioxaphospholane 2-sulfide, C-00065
O-(2-Chloroethyl) phosphorodichloridothioate, C-00080
S-(2-Chloroethyl) phosphorodichloridothioate, C-00081
2-Chloro-4-isopropyl-5,5-dimethyl-1,3,2-dioxaphosphorinane; (2*RS*,4*RS*)-*form*, 2-Sulfide, *in* C-00090
2-Chloro-4-isopropyl-5,5-dimethyl-1,3,2-dioxaphosphorinane; (2*RS*,4*SR*)-*form*, 2-Sulfide, *in* C-00090
2-Chloro-4-methyl-1,3,2-dioxaphosphorinane 2-sulfide, C-00095
5-Chloromethyl-2-fluoro-5-methyl-1,3,2-dioxaphosphorinane; *cis*-*form*, 2-Sulfide, *in* C-00096
5-Chloromethyl-2-fluoro-5-methyl-1,3,2-dioxaphosphorinane; *trans*-*form*, 2-Sulfide, *in* C-00096
2-Chloro-3-phenyl-1,3,2-oxazaphospholidine; 2-Sulfide, *in* C-00138
O-(4-Chlorophenyl) phosphorodichloridothioate, C-00163
S-(4-Chlorophenyl) phosphorodichloridothioate, C-00164
2-Chlorotetrahydro-2*H*-1,3,2-oxazaphosphorine; 2-Sulfide, *in* C-00180
6-Chloro-5,6,7,12-tetrahydro-2,5,7,10-tetramethyldibenzo[*d,g*][1,3,2]diazaphosphocine; 6-Sulfide, *in* C-00182
2-Chlorotetrahydro-3,5,5-trimethyl-2*H*-1,3,2-oxazaphosphorine; 2-Sulfide, *in* C-00183
2-Choro-1,3,2-dithiaphosphorinane; 2-Oxide, *in* C-00199
2-Choro-1,3,2-dithiaphosphorinane; 2-Sulfide, *in* C-00199
Cyclohexylphosphoramidochloridothioic acid; (±)-*form*, *O*-Me ester, *in* C-00270
Cyclohexylphosphoramidothioic acid; Difluoride, *in* C-00271
3,5-Di-*tert*-butyl-2-chloro-2,3-dihydro-1,3,2-oxazaphosphole; 2-Sulfide, *in* D-00078
S,S-Dibutyl phosphorochloridodithioate, D-00144
O,O-Dibutyl phosphorochloridothioate, D-00145
3,9-Dichloro-2,4,8,10-tetraoxa-3,9-diphosphaspiro[5.5]-undecane; 3,9-Disulfide, *in* D-00196

O,O-Didecyl phosphorochloridothioate, D-00223
Diethylphosphoramidothioic dichloride, D-00357
Diethylphosphoramidothioic difluoride, D-00358
▷S,S-Diethyl phosphorochloridodithioate, D-00375
▷O,O-Diethyl phosphorochloridothioate, D-00378
O,S-Diethyl phosphorochloridothioate, D-00379
▷Diethyl phosphorochloridotrithioate, D-00380
O,S-Diethyl phosphorofluoridodithioate, D-00388
S,S-Diethyl phosphorofluoridodithioate, in D-00386
O,O-Diethyl phosphorofluoridothioate, D-00389
Diethyl phosphorofluoridotrithioate, D-00390
2,3-Dihydro-3,5-dimethyl-1,3,4,2-thiadiazaphosphole 2-sulfide; 2-Chloro, in D-00471
2,3-Dihydro-5-methyl-3-phenyl-1,3,4,2-thiadiazaphosphole 2-sulfide; 2-Chloro, in D-00529
▷O,O-Diisopropyl phosphorochloridothioate, D-00659
Dimethylphosphoramidothioic dichloride, D-00872
Dimethylphosphoramidothioic difluoride, D-00873
S,S-Dimethyl phosphorochloridodithioate, D-00888
▷O,O-Dimethyl phosphorochloridothioate, D-00891
O,S-Dimethyl phosphorochloridothioate; (R)-form, in D-00892
O,S-Dimethyl phosphorochloridothioate; (S)-form, in D-00892
O,S-Dimethyl phosphorochloridothioate; (±)-form, in D-00892
Dimethyl phosphorochloridotrithioate, D-00893
S,S-Dimethyl phosphorofluoridodithioate, in D-00900
O,O-Dimethyl phosphorofluoridothioate, D-00902
Dimethyl phosphorofluoridotrithioate, D-00903
O,O-Dipentyl phosphorochloridothioate, D-00984
O,O-Diphenyl phosphorobromidothioate, D-01119
O,O-Diphenyl phosphorochloridothioate, D-01123
S,S-Diphenyl phosphorofluoridodithioate, in D-01132
O,O-Diphenyl phosphorofluoridothioate, D-01134
Diphenyl phosphorofluoridotrithioate, D-01135
O,O-Di-2-propenyl phosphorochloridothioate, D-01196
S,S-Dipropyl phosphorochloridodithioate, D-01219
O,O-Dipropyl phosphorochloridothioate, D-01221
Dipropyl phosphorochloridotrithioate, D-01222
O-(2-Ethoxyethyl) phosphorodichloridothioate, in H-00149
O-Ethyl diethylphosphoramidochloridothioate, in D-00348
Ethyl diethylphosphoramidofluoridodithioate, in D-00352
O-Ethyl dimethylphosphoramidochloridothioate, in D-00861
Ethyl dimethylphosphoramidofluoridodithioate, in D-00867
O-Ethyl dimethylphosphoramidofluoridothioate, in D-00868
S,S-Ethylene phosphorochloridodithioate, in C-00071
Ethylene phosphorochloridotrithioate, in C-00071
O-Ethyl O-phenyl phosphorochloridothioate, E-00117
O-Ethyl S-phenyl phosphorochloridothioate, E-00118
S-Ethyl O-phenyl phosphorochloridothioate, E-00119
O-Ethyl phosphoramidofluoridothioate, E-00158
Ethyl phosphorodichloridodithioate, E-00169
O-Ethyl phosphorodichloridothioate, E-00170
S-Ethyl phosphorodichloridothioate, E-00171
O-Ethyl phosphorodifluoridothioate, E-00175
S-Ethyl phosphorodifluoridothioate, E-00176
O-Ethyl phosphorofluoridoisocyanatidothioate, E-00181
2-Fluoro-1,3-dimethyl-1,3,2-diazaphospholidine; 2-Sulfide, in F-00014
O-(2-Fluoroethyl) phosphorodichloridothioate, F-00019
O-(4-Iodophenyl) phosphorodichloridothioate, I-00019
O-Isopropyl phosphorodichloridothioate, I-00062
S-Isopropyl phosphorodichloridothioate, in I-00064
O-(2-Methoxyethyl) phosphorodichloridothioate, in H-00149
(2-Methoxyphenyl)phosphoramidothioic acid; Difluoride, in M-00077
O-(4-Methoxyphenyl) phosphorodichloridothioate, M-00082
O-(3-Methylbutyl) phosphorodichloridothioate, M-00109
S-(3-Methylbutyl) phosphorodichloridothioate, M-00110
O-Methyl cyclohexylphosphoramidochloridothioate, in C-00270
Methyl diethylphosphoramidofluoridodithioate, in D-00352
Methyl 2,3-di-O-methylglucopyranoside 4,6-cyclic phosphonothioate; α-D-(R)$_P$-form, P-Chloro, in M-00120
O-Methyl dimethylphosphoramidochloridothioate, in D-00861
Methyl dimethylphosphoramidofluoridodithioate, in D-00867
Methyl 2,3-di-O-methyl-4-thioglucopyranoside 4,6-cyclic phosphonate; α-D-(R)$_P$-form, P-Chloro, in M-00121
Methyl 2,3-di-O-methyl-6-thioglucopyranoside 4,6-cyclic phosphonate; α-D-(R)$_P$-form, P-Chloro, in M-00122
Methyl 2,3-di-O-methyl-4-thioglucopyranoside 4,6-cyclic phosphonothioate; α-D-(S)$_P$-form, P-Chloro, in M-00123
O-Methyl ethylphosphoramidochloridothioate, in E-00157
(3-Methylphenyl)phosphoramidothioic acid; Difluoride, in M-00255
O-(2-Methylphenyl) phosphorodichloridothioate, M-00261
O-(4-Methylphenyl) phosphorodichloridothioate, M-00262
O-Methyl phosphoramidofluoridothioate, M-00331
Methylphosphoramidothioic dichloride, M-00334
Methyl phosphorodichloridodithioate, M-00347
O-Methyl phosphorodichloridothioate, M-00349
S-Methyl phosphorodichlorodothioate, M-00351
O-Methyl phosphorodifluoridothioate, M-00354
S-Methyl phosphorodifluoridothioate, M-00355
O-Methyl phosphorofluoridoisothiocyanatidothioate, M-00358
3-Methyl-1,3,2-thiazaphospholidine; 2-Chloro, 2-oxide, in M-00408
3-Methyl-1,3,2-thiazaphospholidine; 2-Chloro, 2-sulfide, in M-00408
(Methylthio)imidodiphosphoryl chloride, M-00414
4-Morpholinylphosphonothioic acid; Dichloride, in M-00527
2,3-Naphthylene phosphorochloridothioate, in C-00119
Neopentylene phosphorobromidothioate, in B-00471
Neopentylene phosphoroiodidothioate, in I-00010
O-(4-Nitrophenyl) phenylphosphoramidochloridothioate, in P-00260
O-(4-Nitrophenyl) phosphorodichloridothioate, N-00065
O-(Pentachlorophenyl) phosphorodichloridothioate, P-00007
S-(Pentachlorophenyl) phosphorodichloridothioate, P-00008
O-Pentyl phosphorodichloridothioate, P-00055
Phenyl (dichlorophosphinothioyl)phosphorodichlorimidate, P-00112
O-Phenyl diethylphosphoramidochloridothioate, in D-00348
Phenyl diethylphosphoramidofluoridodithioate, in D-00352
O-Phenyl dimethylphosphoramidochloridothioate, in D-00861
Phenyl dimethylphosphoramidofluoridodithioate, in D-00867
O-Phenyl dimethylphosphoramidofluoridothioate, in D-00868
O-Phenyl ethylphosphoramidochloridothioate, in E-00157
O-Phenyl methylphosphoramidochloridothioate, in M-00329
O-Phenyl 4-morpholinylphosphonochloridothioate, in M-00525
O-Phenyl phenylphosphoramidochloridothioate, in P-00260
Phenylphosphoramidothioic acid; Difluoride, in P-00262
Phenylphosphoramidothioic acid; Dichloride, in P-00262
O-Phenyl phosphorodibromidothioate, P-00278
Phenyl phosphorodichloridodithioate, P-00281
O-Phenyl phosphorodichloridothioate, P-00282
S-Phenyl phosphorodichloridothioate, P-00283
O-Phenyl phosphorodifluoridothioate, P-00287
S-Phenyl phosphorodifluoridothioate, P-00288
2-Phenylphosphorodifluoridothioic hydrazide, P-00289
Propylene fluorothiophosphate, in F-00020
Propylene phosphorochloridothioate, in C-00092
Propylphosphoramidothioic acid; Dichloride, in P-00485
O-Propyl phosphorodichloridothioate, P-00489
S-Propyl phosphorodichloridothioate, P-00490
Tetraethylphosphorodiamidothioic chloride, T-00065
Tetramethylethylene chlorothiophosphate, in C-00185
Tetramethylphosphorodiamidothioic chloride, T-00225
▷Tetramethylphosphorodiamidothioic fluoride, T-00226
1,3,3-Trimethyltrimethylene chlorothiophosphate, in C-00189
O-Vinyl phosphorodichloridothioate, V-00008

Thiophosphoric pseudohalides

O-Ethyl dimethylphosphoramidocyanidothioate, in D-00864
Thiophosphoryl isocyanate, T-00315
Thiophosphoryl isothiocyanate, T-00316

Thiophosphoric esters

▷Acetophos, A-00004
O-(2-Acetylphenyl) phosphorodichloridothioate, A-00022
O-(3-Acetylphenyl) phosphorodichloridothioate, A-00023
O-(4-Acetylphenyl) phosphorodichloridothioate, A-00024
▷Amidithion, A-00051
O-(4-Amino-3-methylphenyl) O,O-dimethyl phosphorothioate, A-00090

O-(4-Aminophenyl) *O,O*-diethyl phosphorothioate, A-00099
▷Amiton, A-00133
Anilofos, A-00134
▷Aphidan, A-00142
Azemethephos, A-00146
▷Azinphos-ethyl, A-00150
▷Azinphos-Methyl, A-00151
▷Bensulide, B-00004
O-Benzyl *O,O*-diisopropyl phosphorothioate, B-00044
▷*S*-Benzyl *O,O*-diisopropyl phosphorothioate, B-00045
S-Benzyl *O,O*-propylene phosphorodithioate, *in* M-00016
Benzyl tetrabutylphosphorodiamidodithioate, *in* T-00023
Bis(dibenzyloxyphosphinothioyl) disulfide, B-00155
Bis(diethoxyphosphinothioyl) disulfide, B-00170
Bis(diethoxyphosphinothioyl) trisulfide, B-00171
▷Bis(diethoxyphosphinyl) disulfide, B-00172
Bis(diisopropoxyphosphinyl) disulfide, B-00183
Bis(diisopropyoxyphosphinothioyl) disulfide, B-00184
▷Bis(dimethoxyphosphinothioyl) disulfide, B-00185
Bis(dimethoxyphosphinyl) disulfide, B-00186
O,O-Bis(4-dimethylaminophenyl) *O*-isopropyl phosphorothioate, *in* B-00190
Bis(5,5-dimethyl-2-oxo-1,3,2-dioxaphosphorinan-2-yl) disulfide, *in* D-01251
Bis(5,5-dimethyl-2-selenoxo-1,3,2-dioxaphosphorinan-2-yl) disulfide, *in* D-01251
Bis(5,5-dimethyl-2-thioxo-1,3,2-dioxaphosphorinan-2-yl) disulfide, *in* D-01251
Bis[(diphenoxyphosphino)thioyl] disulfide, B-00211
Bis(1-methylethyl) [(dimethoxyphosphinothioyl)thio]-butanedioate, *in* D-00677
O,O-Bis(1-methylethyl) phenylphosphoramidothioate, *in* P-00262
O-(4-Bromophenyl) phosphorodichloridothioate, B-00497
▷Bromophos, B-00498
▷Bromophos-Et, B-00499
▷*tert*-Butyl bicyclic thiophosphate, *in* B-00640
S,S-1,3-Butylene *O*-ethyl phosphorotrithioate, *in* E-00042
O,O-Butylene *O*-ethyl thiophosphate, *in* E-00031
O,O-1,3-Butylene *O*-ethyl thiophosphate, *in* E-00041
O,O-1,3-Butylene *S*-methyl dithiophosphate, *in* M-00174
O,O-1,3-Butylene *S*-methyl phosphoroselenothioate, *in* M-00174
O,O-1,3-Butylene *S*-methyl thiophosphate, *in* M-00174
O,O-2,3-Butylene *O*-methyl thiophosphate, *in* M-00033
O,O-2,3-Butylene *S*-methyl thiophosphate, *in* D-00765
O,O-2,3-Butylene *S*-phenyl dithiophosphate, *in* D-00824
O,O-1,3-Butylene *O*-phenyl thiophosphate, *in* M-00191
O,O-2,3-Butylene phosphonothioate, *in* D-00733
O,O-2,3-Butylene *O*-trimethylsilyl phosphorothioate, *in* D-00930
S-[(*tert*-Butylsulfinyl)methyl] *O,O*-diethyl phosphorodithioate, *in* T-00004
S-[(*tert*-Butylsulfonyl)methyl] *O,O*-diethyl phosphorodithioate, *in* T-00004
▷Carbophenothion, C-00008
▷Chlormephos, C-00019
5-Chloromethyl-5-methyl-2-phenoxy-1,3,2-dioxaphosphorinane; 2-Sulfide, *in* C-00099
O-(4-Chlorophenyl) phosphorodichloridothioate, C-00163
Chlorphoxim, C-00193
▷Chlorpyrifos, C-00194
▷Chlorpyrifos Methyl, C-00195
▷Chlorthion, C-00196
Chlorthiophos, C-00197
▷Conen, C-00204
▷Coumaphos, C-00205
▷Cyanophos, C-00226
▷Cyanthoate, C-00232
Demephion-*O*, D-00017
▷Demephion-*S*, D-00018
▷Demeton-*O*, D-00019
▷Demeton-*S*-methyl, D-00020
▷Demeton-*S*-methyl sulfone, D-00021
▷Dialifos, D-00023
▷Diazinon, D-00027
▷Dibutyl chlorothiophosphonate, D-00080
O,O-Dibutyl diethylphosphoramidothioate, *in* D-00356
O,O-Di-*tert*-butyl dimethylphosphoramidothioate, *in* D-00871
Se,Se-Dibutyl *O*-ethyl phosphorodiselenothioate, *in* D-00149
Se,Se-Dibutyl *O*-propyl phosphorodiselenothioate, *in* D-00149
▷Dichlofenthion, D-00161
2,5-Diethoxy-1,3,2,5-dioxadiphosphorinane; 5-Oxide, 2-sulfide, *in* D-00231
[(Diethoxyphosphinothioyl)thio]acetic acid, D-00239
[(Diethoxyphosphinothioyl)thio]acetic acid; Fluoride, *in* D-00239
[(Diethoxyphosphinothioyl)thio]acetic acid; Chloride, *in* D-00239
▷*O,O*-Diethyl *S*-(2-amino-2-oxoethyl) phosphorodithioate, *in* D-00239
O,O-Diethyl *tert*-butylphosphoramidothioate, *in* B-00623
O,O-Diethyl (2-chlorophenyl)phosphoramidothioate, *in* C-00158
▷Diethyl chlorothiophosphonate, D-00274
▷*O,O*-Diethyl chlorothiophosphonothioate, D-00275
▷Diethyl [(diethoxyphosphinothioyl)thio]butanedioate, *in* D-00240
O,O-Diethyl diethylphosphoramidothioate, *in* D-00356
O,O-Diethyl *S*-(2-dimethylaminoethyl) phosphorothioate, D-00284
O,O-Diethyl dimethylphosphoramidothioate, *in* D-00871
▷*O,O*-Diethyl *S*-[(ethylsulfinyl)methyl] phosphorodithioate, *in* P-00338
▷*OO*-Diethyl *S*-[(ethylsulfonyl)methyl] phosphorodithioate, *in* P-00338
O,O-Diethyl (3-fluorophenyl)phosphoramidothioate, *in* F-00045
O,O-Diethyl (4-fluorophenyl)phosphoramidothioate, *in* F-00046
O,O-Diethyl (2-methoxyethyl)phosphoramidothioate, *in* M-00041
O,O-Diethyl (2-methoxyphenyl)phosphoramidothioate, *in* M-00077
O,O-Diethyl (3-methoxyphenyl)phosphoramidothioate, *in* M-00078
O,O-Diethyl (3-methylphenyl)phosphoramidothioate, *in* M-00255
O,O-Diethyl *SS*-methyl phosphoro(dithioperoxo)thioate, D-00300
▷*O,O*-Diethyl *O*-[4-(methylsulfonyl)phenyl] phosphorothioate, *in* D-00302
▷*O,O*-Diethyl *O*-[4-(methylthio)phenyl] phosphorothioate, D-00302
O,O-Diethyl phenylphosphoramidothioate, *in* P-00262
O,O-Diethyl *SS*-phenylphosphoro(dithioperoxoate), D-00314
O,O-Diethyl 2-phenylphosphorohydrazidothioate, *in* P-00296
O,O-Diethyl *S*-phenyl phosphorothioate, D-00317
O,O-Diethyl *S*-[(phenylthio)methyl] phosphorodithioate, D-00318
O,O-Diethyl phosphorisocyanatidothioate, D-00367
O,S-Diethyl phosphorisocyanatidothioate, D-00368
▷*O,O*-Diethyl phosphor(isothiocyanatido)thioate, D-00370
O,O-Diethyl phosphorohydrazidothioate, D-00392
O,O-Diethyl propylphosphoramidothioate, *in* P-00485
O,O-Diethyl *SS*-propyl phosphoro(dithioperoxo)thioate, D-00402
1,5-Dihydro-2-phenoxy-2,4,3-benzodioxaphosphepin; 3-Sulfide, *in* D-00544
1,2-Dihydro-1-phenyl-4*H*-3,1,2-benzothiazaphosphorin-4-thione; 2-Methoxy, 2-sulfide, *in* D-00552
1,2-Dihydro-1-phenyl-4*H*-3,1,2-benzothiazaphosphorin-4-thione; 2-Ethoxy, 2-sulfide, *in* D-00552
(Diisopropoxyphosphinothioyl)thioacetic acid, D-00632
▷Diisopropyl chlorothiophosphonate, D-00634
▷*O,O*-Diisopropyl chlorothiophosphonothioate, D-00635
Diisopropyl [(diethoxyphosphinothioyl)thio]butanedioate, *in* D-00240
O,O-Diisopropyl diethylphosphoramidothioate, *in* D-00356
O,O-Diisopropyl dimethylphosphoramidothioate, *in* D-00871
O,O-Diisopropyl (4-fluorophenyl)phosphoroamidothioate, *in* F-00046
O,O-Diisopropyl 2-phenylphosphorohydrazidothioate, *in* P-00296
3,9-Dimercapto-2,4,8,10-tetraoxa-3,9-diphosphaspiro[5.5]-undecane 3,9-disulfide; 3,9-Bis(methylthio) (Di-Me ester), *in* D-00670
3,9-Dimercapto-2,4,8,10-tetraoxa-3,9-diphosphaspiro[5.5]-undecane 3,9-disulfide; 3,9-Bis(ethylthio) (Di-Et ester), *in* D-00670

▷Dimethoate, D-00671
▷[(Dimethoxyphosphinothioyl)thio]acetic acid, D-00675
α-[(Dimethoxyphosphinothioyl)thio]benzeneacetic acid, D-00676
[(Dimethoxyphosphinothioyl)thio]butanedioic acid, D-00677
O,O-Dimethyl *S*-(2-amino-2-oxoethyl) phosphorodithioate, *in* D-00675
▷Dimethyl chlorothiophosphonate, D-00716
Dimethyl [(diethoxyphosphinothioyl)thio]butanedioate, *in* D-00240
O,O-Dimethyl diethylphosphoramidothioate, *in* D-00356
▷Dimethyl [(dimethoxyphosphinothioyl)thio]butanedioate, *in* D-00677
O,O-Dimethyl dimethylphosphoramidothioate, *in* D-00871
O,O-Dimethyl *S*-[(ethylthio)methyl] phosphorodithioate, D-00755
O,O-Dimethyl (2-methoxyethyl)phosphoramidothioate, *in* M-00041
O,O-Dimethyl *O*-[(3-methyl-4-methylsulfinyl)phenyl] phosphorothioate, *in* F-00007
O,O-Dimethyl *O*-[(3-methyl-4-methylsulfonyl)phenyl]-phosphorothioate, *in* F-00007
4,5-Dimethyl-2-methylthio-1,3,2-dioxaphospholane; (4*RS*,5*SR*)-*form*, 2-Oxide, *in* D-00765
▷*O,O*-Dimethyl *S*-(4-nitrophenyl) phosphorothioate, D-00773
O,S-Dimethyl *O*-4-nitrophenyl phosphorothioate; (+)-*form*, *in* D-00774
O,S-Dimethyl *O*-4-nitrophenyl phosphorothioate; (−)-*form*, *in* D-00774
O,S-Dimethyl *O*-4-nitrophenyl phosphorothioate; (±)-*form*, *in* D-00774
O,O-Dimethyl phenylphosphoramidothioate, *in* P-00262
O,O-Dimethyl 2-phenylphosphorohydrazidothioate, *in* P-00296
O,O-Dimethyl phosphorisocyanatidothioate, D-00884
O,S-Dimethyl phosphorisocyanatidothioate, D-00885
O,O-Dimethyl phosphorohydrazidothioate, D-00905
▷*O,O*-Dimethyl *O*-4-sulfamoylphenyl phosphorothioate, D-00920
4,5-Dimethyl-2-trimethylsilyloxy-1,3,2-dioxaphospholane; (2α,4β,5β)-*form*, 2-Sulfide, *in* D-00930
4,5-Dimethyl-2-trimethylsilyloxy-1,3,2-dioxaphospholane; (2α,4α,5β)-(±)-*form*, 2-Sulfide, *in* D-00930
▷Dioxathion, D-00974
O,O-Diphenyl diethylphosphoramidothioate, *in* D-00356
O,O-Diphenyl (3-methylphenyl)phosphoramidothioate, *in* M-00255
O,O-Diphenyl phenylphosphoramidothioate, *in* P-00262
O,O-Diphenyl phenylsulfonylphosphoramidothioate, *in* P-00314
O,O-Diphenyl phosphorisocyanatidothioate, D-01117
O,O-Diphenyl phosphorohydrazidothioate, D-01137
OO-Diphenyl 2-pyridinylphosphoramidothioate, *in* P-00513
O,O-Diphenyl 4-pyridinylphosphoramidothioate, *in* P-00514
▷Dipropyl chlorothiophosphonate, D-01200
▷*O,O*-Dipropyl chlorothiophosphonothioate, D-01201
2,2′-Diselenobis[5,5-dimethyl-2-thioxo-1,3,2-dioxaphosphorinane], *in* D-01234
▷Disulfoton, D-01236
Ecothiopate, E-00001
▷Edifenphos, E-00002
▷Ethafos, E-00005
▷Ethion, E-00022
▷Ethoprophos, E-00023
2-Ethoxy-1,3,2-dioxaphospholane 2-sulfide, E-00036
2-Ethoxy-1,3,2-dithiaphosphorinane, *in* D-01247
▷Ethyl bicyclic thiophosphate, *in* E-00196
▷*O*-Ethyl *O,O*-bis(4-nitrophenyl) phosphorothioate, E-00054
O-Ethyl diethylphosphoramidochloridothioate, *in* D-00348
Ethyl (diisopropoxyphosphinothioyl)thioacetate, *in* D-00632
Ethyl [(dimethoxyphosphinothioyl)thio]acetate, *in* D-00675
1-Ethyl [(dimethoxyphosphinothioyl)thio]butanedioate, *in* D-00677
2-Ethyl [(dimethoxyphosphinothioyl)thio]butanedioate, *in* D-00677
▷Ethyl (dimethoxyphosphinyl)thioacetate, *in* D-00687
O-Ethyl *O,S*-dimethyl phosphorothioate; (*R*)-*form*, *in* E-00065
O-Ethyl *O,S*-dimethyl phosphorothioate; (*S*)-*form*, *in* E-00065
O,O-Ethylene *O*-phenyl phosphorothioate, *in* P-00067
O,S-Ethylene *O*-phenyl phosphorothioate, *in* P-00072
S-Ethyl *O,O*-ethylene dithiophosphate, *in* E-00191
O-Ethyl *O,O*-1,2-naphthylene thiophosphate, *in* E-00044
O-Ethyl *O,O*-1,8-naphthylene thiophosphate, *in* E-00045
O-Ethyl *O,O*-1,8-naphthylene thiophosphate, *in* E-00046
S-Ethyl *O,O*-neopentylene phosphorodithioate, *in* E-00190
O-Ethyl *O,O*-neopentylene phosphorothioate, *in* E-00032
S-Ethyl *O,O*-neopentylene phosphorothioate, *in* E-00190
O-Ethyl *O,O*-*o*-phenylene thiophosphate, *in* E-00024
S-Ethyl *O*-phenyl phosphorochloridothioate, E-00119
O-Ethyl phosphorodiamidothioate, *in* E-00182
O-Ethyl phosphorodiisocyanatidothioate, E-00178
O-Ethyl *O,O*-propylene thiophosphate, *in* E-00040
O-Ethyl saligenin cyclic thiophosphate, *in* H-00110
S-Ethyl saligenin cyclic thiophosphate, *in* H-00110
S-Ethyl *O,O*-tetramethylethylene phosphorodithioate, *in* E-00193
O-Ethyl *O,O*-tetramethylethylene phosphorothioate, *in* E-00052
Ethyl tetramethylphosphorodiamidodithioate, *in* T-00222
2-Ethylthio-4,5-dimethyl-1,3,2-dioxaphospholane; 2-Sulfide, *in* E-00189
O-Ethyl *O,O*-trimethylene thiophosphate, *in* E-00037
▷Etrimfos, E-00205
▷Famphur, F-00001
▷Fenchlorphos, F-00003
▷Fenitrothion, F-00005
▷Fensulfothion, F-00006
▷Fenthion, F-00007
▷Formothion, F-00055
▷Iodofenphos, I-00011
O-(4-Iodophenyl) phosphorodichloridothioate, I-00019
▷Isazofos, I-00021
O-Isopropyl *O,S*-dimethyl phosphorothioate; (*R*)-*form*, *in* I-00030
O-Isopropyl *O,S*-dimethyl phosphorothioate; (*S*)-*form*, *in* I-00030
▷Isothioate, I-00067
▷Isoxathion, I-00068
▷Kitazin, K-00001
▷Malathion, M-00001
▷Mecarbam, M-00002
▷Menazon, M-00003
2-Mercapto-1,3,2-benzodioxaphosphole 2-sulfide; 2-Methylthio (Me ester), *in* M-00005
2-Mercapto-1,3,2-benzodioxaphosphole 2-sulfide; 2-Ethylthio (Et ester), *in* M-00005
2-Mercapto-4*H*-1,3,2-benzodioxaphosphorin 2-sulfide; 2-Ethylthio (2-Et ester), *in* M-00006
2-Mercapto-4*H*-1,3,2-benzodioxaphosphorin 2-sulfide; 2-Isopropylthio (2-Isopropyl ester), *in* M-00006
2-Mercapto-4*H*-1,3,2-benzodioxaphosphorin 2-sulfide; 2-Phenylthio (2-Ph ester), *in* M-00006
▷Methidathion, M-00025
2-Methoxy-4,5-diphenyl-1,3,2-dioxaphospholane; (2α,4β,5β)-*form*, 2-Sulfide, *in* M-00039
2-Methoxy-1,3,2-dithiaphospholane 2-oxide, *in* D-01244
2-Methoxy-1,3,2-dithiaphosphorinane, *in* D-01247
2-Methoxy-4-methyl-1,3,2-dioxaphosphorinane 2-sulfide; (2*RS*,4*RS*)-*form*, *in* M-00047
2-Methoxy-4-methyl-1,3,2-dioxaphosphorinane 2-sulfide; (2*RS*,4*SR*)-*form*, *in* M-00047
Methylacetophos, M-00086
Methyl 3-[(dimethoxyphosphinothioyl)oxy]-2-methyl-2-propenoate, *in* D-00673
Methyl [(dimethoxyphosphinothioyl)thio]acetate, *in* D-00675
Methyl 2,3-di-*O*-methylglucopyranoside 4,6-cyclic phosphonate; α-D-(*R*)P-*form*, *P*-(Propylthio), *in* M-00119
Methyl 2,3-di-*O*-methylglucopyranoside 4,6-cyclic phosphonate; α-D-(*S*)P-*form*, *P*-(Propylthio), *in* M-00119
Methyl 2,3-di-*O*-methylglucopyranoside 4,6-cyclic phosphonothioate; α-D-(*R*)P-*form*, *P*-Ethoxy, *in* M-00120
Methyl 2,3-di-*O*-methylglucopyranoside 4,6-cyclic phosphonothioate; α-D-(*S*)P-*form*, *P*-Ethoxy, *in* M-00120
Methyl 2,3-di-*O*-methyl-4-thioglucopyranoside 4,6-cyclic phosphonate; α-D-(*R*)P-*form*, *P*-Ethoxy, *in* M-00121
Methyl 2,3-di-*O*-methyl-4-thioglucopyranoside 4,6-cyclic phosphonate; α-D-(*S*)P-*form*, *P*-Ethoxy, *in* M-00121
Methyl 2,3-di-*O*-methyl-6-thioglucopyranoside 4,6-cyclic phosphonate; α-D-(*R*)P-*form*, *P*-Ethoxy, *in* M-00122

Methyl 2,3-di-*O*-methyl-6-thioglucopyranoside 4,6-cyclic phosphonate; *α*-D-(*S*)P-*form*, *P*-Ethoxy, *in* M-00122
Methyl 2,3-di-*O*-methyl-4-thioglucopyranoside 4,6-cyclic phosphonothioate; *α*-D-(*R*)P-*form*, *P*-Ethoxy, *in* M-00123
Methyl 2,3-di-*O*-methyl-4-thioglucopyranoside 4,6-cyclic phosphonothioate; *α*-D-(*S*)P-*form*, *P*-Ethoxy, *in* M-00123
4-Methyl-1,3,2-dioxaphosphorinane 2-sulfide, M-00131
4-Methyl-2-methylthio-1,3,2-dioxaphosphorinane; (2*RS*,4*SR*)-*form*, 2-Oxide, *in* M-00174
4-Methyl-2-methylthio-1,3,2-dioxaphosphorinane; (2*RS*,4*SR*)-*form*, 2-Sulfide, *in* M-00174
4-Methyl-2-methylthio-1,3,2-dioxaphosphorinane; (2*RS*,4*SR*)-*form*, 2-Selenide, *in* M-00174
O-Methyl *O*-1-naphthalenyl phosphorothioate; (+)-*form*, *in* M-00176
O-Methyl *O*-1-naphthalenyl phosphorothioate; (−)-*form*, *in* M-00176
O-Methyl *O*-1-naphthalenyl phosphorothioate; (±)-*form*, *in* M-00176
S-Methyl *O*,*O*-neopentylene phosphorodithioate, *in* D-00766
Se-Methyl *O*,*O*-neopentylene phosphoroselenothioate, *in* D-00764
O-Methyl *O*,*O*-neopentylene phosphorothioate, *in* M-00034
S-Methyl *O*,*O*-neopentylene phosphorothioate, *in* D-00766
5-Methyl-1,3,2-oxathiaphosphole 2-oxide; 2-Ethoxy, *in* M-00184
Methylphosphoramidothioic dichloride, M-00334
O-Methyl phosphorodiamidothioate, *in* M-00359
O-Methyl phosphorodiisocyanatidothioate, M-00357
S-Methyl *O*,*O*-propylene phosphorodithioate, *in* M-00173
S-Methyl *O*,*O*-propylene phosphorothioate, *in* M-00173
O-Methyl *O*,*O*-propylene thiophosphate, *in* M-00043
Methyl tetramethylphosphorodiamidodithioate, *in* T-00222
3-Methyl-1,3,2-thiazaphospholidine; 2-Methoxy, 2-oxide, *in* M-00408
3-Methyl-1,3,2-thiazaphospholidine; 2-Methoxy, 2-sulfide, *in* M-00408
3-Methyl-1,3,2-thiazaphospholidine; 2-Ethoxy, 2-sulfide, *in* M-00408
2-Methylthio-4*H*-1,3,2-benzodioxaphosphorin; 2-Oxide, *in* M-00411
2-Methylthio-4*H*-1,3,2-benzodioxaphosphorin; 2-Sulfide, *in* M-00411
2-Methylthio-1,3,2-dioxaphosphorinane 2-sulfide, *in* M-00013
(Methylthio)imidodiphosphoryl chloride, M-00414
O-Methyl *S*,*S*-trimethylene phosphonotrithioate, *in* D-01249
Methyl trimethylene phosphorotetrathioate, *in* M-00014
O-Methyl *O*,*O*-trimethylene phosphorothioate, *in* M-00038
4-Methyl-2,6,7-trithia-1-phosphabicyclo[2.2.2]octane; 1-Oxide, *in* M-00436
▷4-Methyl-2,6,7-trithia-1-phosphabicyclo[2.2.2]octane; 1-Sulfide, *in* M-00436
▷Methyl trithion, M-00437
O,*O*-Neopentylene *O*-phenyl phosphorothioate, *in* D-00790
▷Omethoate, O-00048
1,3,2-Oxathiaphosphorinane 2-oxide; 2-Isopropoxy, *in* O-00057
1,3,2-Oxathiaphosphorinane 2-oxide; 2-Phenoxy, *in* O-00057
▷Oxydemeton-methyl, O-00098
▷Oxydeprofos, O-00099
▷Parathion, P-00002
▷Parathion methyl, P-00003
▷Phenkapton, P-00062
2-Phenoxy-1,5-dihydro-2,4,3-benzodioxaphosphepin; Sulfide, *in* P-00066
2-Phenoxy-1,3,2-dithiaphospholane 2-sulfide, *in* D-01245
Phenthoate; (+)-*form*, *in* P-00078
Phenthoate; (−)-*form*, *in* P-00078
Phenthoate; (±)-*form*, *in* P-00078
O-Phenyl-*N*,*N'*-dibenzylphosphorodiamidothioate, *in* D-00062
O-Phenyl diethylphosphoramidochloridothioate, *in* D-00348
Phenyl 3-[(dimethoxyphosphinothioyl)oxy]-2-methyl-2-propenoate, *in* D-00673
O-Phenyl *o*-phenylene thiophosphate, *in* P-00063
O-Phenyl phosphorodiisocyanatidothioate, P-00293
O-Phenyl saligenin cyclic thiophosphate, *in* P-00064
S-Phenyl saligenin cyclic thiophosphate, *in* H-00110
▷Phorate, P-00338
▷Phosalone, P-00340
▷Phosmet, P-00343
▷Phoxim, P-00424
▷Piperophos, P-00433
▷Pirimiphos-ethyl, P-00434
▷Pirimiphos-methyl, P-00435
Profenofos; (*R*)-*form*, *in* P-00440
Profenofos; (*S*)-*form*, *in* P-00440
Profenofos; (±)-*form*, *in* P-00440
▷Prothiofos, P-00498
▷Prothoate, P-00499
▷Pyrazophos, P-00502
▷Quinalphos, Q-00001
▷Salithion, S-00001
Sulprofos; (+)-*form*, *in* S-00027
Sulprofos; (−)-*form*, *in* S-00027
Sulprofos; (±)-*form*, *in* S-00027
▷Temephos, T-00003
▷Terbufos, T-00004
SS-*tert*-Butyl *O*,*O*-dimethyl phosphoro(dithioperoxo)thioate, B-00556
O,*O*,*O'*,*O'*-Tetraethyl *S*,*S'*-ethanediyl bisphosphorodithioate, T-00047
O,*O*,*O'*,*O'*-Tetraethyl *S*,*S'*-propanediyl bisphosphorodithioate, T-00068
Tetraethyl thiohypophosphate, T-00072
O,*O*-Tetramethylene *O*-phenyl phosphorothioate, *in* D-00969
O,*O*,*O'*,*O'*-Tetramethyl *S*,*S'*-methylene bisphosphorodithioate, T-00205
O,*O*,*O'*,*O'*-Tetraphenyl 1,3-propanediphosphoramidothioate, *in* P-00446
1,3,2-Thiazaphospholidine 2-oxide; 2-Butoxy, *in* T-00302
1,3,2-Thiazaphospholidine 2-sulfide; 2-Methoxy, *in* T-00303
1,3,2-Thiazaphospholidine 2-sulfide; 2-Propoxy, *in* T-00303
1,3,2-Thiazaphospholidine 2-sulfide; 2-Phenoxy, *in* T-00303
▷Thiometon, T-00313
▷Thionazin, T-00314
Toclofos-Methyl, T-00321
▷Triazophos, T-00333
Tribenzyl phosphorotetrathioate, T-00341
S,*S*,*S*-Tribenzyl phosphorotrithioate, T-00342
O,*Se*,*Se*-Tributyl phosphorodiselenothioate, *in* D-00149
Tributyl phosphorotetrathioate, T-00386
O,*O*,*O*-Tributyl phosphorothioate, T-00387
O,*O*,*S*-Tributyl phosphorothioate, T-00388
O,*S*,*S*-Tributyl phosphorotrithioate, T-00390
S,*S*,*S*-Tri-*tert*-butyl phosphorotrithioate, T-00391
▷*S*,*S*,*S*-Tributyl phosphorotrithioate, T-00392
▷*O*,*S*,*S*-Triethyl phosphorodithioate, T-00450
O,*O*,*S*-Triethyl phosphorodithioate, T-00451
Triethyl phosphoro(dithioperoxoate), T-00452
Triethyl phosphoro(dithioperoxo)thioate, T-00453
Triethyl phosphorotetrathionate, T-00456
▷*O*,*O*,*S*-Triethyl phosphorothioate, T-00457
▷*O*,*O*,*O*-Triethyl phosphorothioate, T-00458
▷*O*,*S*,*S*-Triethyl phosphorotrithioate, T-00459
▷*S*,*S*,*S*-Triethyl phosphorotrithioate, T-00460
O,*O*,*S*-Triisopropyl phosphorodithioate, T-00503
Triisopropyl phosphorotetrathioate, T-00505
O,*O*,*O*-Triisopropyl phosphorothioate, T-00506
O,*S*,*S*-Triisopropyl phosphorotrithioate, T-00507
S,*S*,*S*-Triisopropyl phosphorotrithioate, T-00508
▷Trimethylolethane thiophosphate, *in* M-00424
▷*O*,*S*,*S*-Trimethyl phosphorodithioate, T-00541
▷*O*,*O*,*S*-Trimethyl phosphorodithioate, T-00542
Trimethyl phosphorotetrathioate, T-00544
▷*O*,*O*,*O*-Trimethyl phosphorothioate, T-00545
▷*O*,*O*,*S*-Trimethyl phosphorothioate, T-00546
O,*S*,*S*-Trimethyl phosphorotrithioate, T-00548
▷*S*,*S*,*S*-Trimethyl phosphorotrithioate, T-00549
Trimethylsilyl tetramethylphosphorodiamidodithioate, *in* T-00222
2,8,9-Trioxa-5-aza-1-phosphabicyclo[3.3.3]undecane; 1-Sulfide, *in* T-00584
2,6,7-Trioxa-1,4-diphosphabicyclo[2.2.2]octane; 1-Sulfide, *in* T-00585
2,6,7-Trioxa-1,4-diphosphabicyclo[2.2.2]octane; 1,4-Disulfide, *in* T-00585

2,6,7-Trioxa-1-phosphabicyclo[2.2.2]octane-4-methanol; Sulfide, *in* T-00590
2,8,9-Trioxa-1-phosphatricyclo[3.3.1.1^{3,7}]decane; 1-Sulfide, *in* T-00591
Triphenyl phosphorotetrathioate, T-00669
O,O,O-Triphenyl phosphorothioate, T-00670
O,O,S-Triphenyl phosphorothioate, T-00671
Se,Se,Se-Triphenyl phosphorotriselenothioate, T-00674
S,S,S-Triphenyl phosphorotrithioate, T-00675
Tripropyl phosphorotetrathioate, T-00714
O,O,O-Tripropyl phosphorothioate, T-00715
O,O,S-Tripropyl phosphorothioate, T-00716
S,S,S-Tripropyl phosphorotrithioate, T-00717
▷*O,O,O*-Tris(2-chloroethyl) phosphorothioate, T-00743
O,O,O-Tris(2-methylphenyl) phosphorothioate, T-00812
O,O,O-Tris(3-methylphenyl) phosphorothioate, T-00813
O,O,O-Tris(4-methylphenyl) phosphorothioate, T-00814
S,S,S-Tris(2-methylphenyl) phosphorotrithioate, T-00815
S,S,S-Tris(4-methylphenyl) phosphorotrithioate, T-00816
O,O,S-Tris(2-methylpropyl) phosphorothioate, T-00828
Tris(trimethylsilyl) phosphorotetrathioate, T-00869
▷Vamidothion, V-00001

Thiophosphoric amides

▷Acephate, A-00003
Amiprophos *M*, A-00131
▷Amiprophos-Methyl, A-00132
3-Benzyl-2-chlorotetrahydro-2*H*-1,3,2-oxazaphosphorine; 2-Sulfide, *in* B-00041
Benzyl tetrabutylphosphorodiamidodithioate, *in* T-00023
P,P-Bis(1-aziridinyl)-*N*-ethylphosphinothioic amide, *in* B-00094
P,P-Bis(1-aziridinyl)-*N*-methylphosphinothioic amide, *in* B-00094
P,P-Bis(1-aziridinyl)-*N*-phenylphosphinothioic amide, *in* B-00094
▷*P,P*-Bis(1-aziridinyl)phosphinothioic amide, B-00094
▷Bis(2-chloroethyl)phosphoramidothioic dichloride, B-00123
▷*O,O*-Bis(4-chlorophenyl) (1-iminoethyl)phosphoramidothioate, *in* I-00005
2,4-Bis(dimethylamino)-3-methyl-1,3,2,4-thiazadiphosphetidine; 2,4-Disulfide, *in* B-00187
O,O-Bis(3-dimethylaminophenyl) phenylphosphoramidothioate, *in* B-00189
Bis(dimethylaminophosphinothioyl)pentamethylthiophosphoric triamide, B-00191
O,O-Bis(1-methylethyl) phenylphosphoramidothioate, *in* P-00262
▷Butamifos, B-00513
Butyl *N″*-[bis(diethylamino)phosphinothioyl]-*N,N,N′,N′*-tetraethylphosphorodiamidoimidate, *in* B-00174
O,O-1,3-Butylene dimethylphosphoramidothioate, *in* M-00117
O,O-2,3-Butylene dimethylphosphoramidothioate, *in* D-00725
O,O-1,3-Butylene phenylphosphoramidothioate, *in* M-00193
tert-Butylphosphoramidothioic acid; Difluoride, *in* B-00623
5-*tert*-Butyltetrahydro-2-dimethylamino-2*H*-1,3,2-oxazaphosphorine; (2*RS*,5*RS*)-*form*, 2-Sulfide, *in* B-00635
5-*tert*-Butyltetrahydro-2-dimethylamino-2*H*-1,3,2-oxazaphosphorine; (2*RS*,5*SR*)-*form*, 2-Sulfide, *in* B-00635
5-*tert*-Butyltetrahydro-2-dimethylamino-3-phenyl-2*H*-1,3,2-oxazaphosphorine; (2*RS*,5*RS*)-*form*, 2-Sulfide, *in* B-00636
5-*tert*-Butyltetrahydro-2-dimethylamino-3-phenyl-2*H*-1,3,2-oxazaphosphorine; (2*RS*,5*SR*)-*form*, 2-Sulfide, *in* B-00636
2-Chloro-2,3-dihydro-3-methyl-1,3,2-benzothiazaphosphole; 2-Sulfide, *in* C-00043
2-Chloro-1,3-dimethyl-1,3,2-diazaphospholidine; 2-Sulfide, *in* C-00055
2-Chloro-3,4-dimethyl-5-phenyl-1,3,2-oxazaphospholidine; (2*R*,4*S*,5*R*)-*form*, 2-Sulfide, *in* C-00061
2-Chloro-3,4-dimethyl-5-phenyl-1,3,2-oxazaphospholidine; (2*S*,4*S*,5*R*)-*form*, 2-Sulfide, *in* C-00061
2-Chloro-3-phenyl-1,3,2-oxazaphospholidine; 2-Sulfide, *in* C-00138
2-Chlorotetrahydro-2*H*-1,3,2-oxazaphosphorine; 2-Sulfide, *in* C-00180
6-Chloro-5,6,7,12-tetrahydro-2,5,7,10-tetramethyldibenzo[*d,g*][1,3,2]diazaphosphocine; 6-Sulfide, *in* C-00182
2-Chlorotetrahydro-3,5,5-trimethyl-2*H*-1,3,2-oxazaphosphorine; 2-Sulfide, *in* C-00183
Cyclohexylphosphoramidochloridothioic acid; (±)-*form*, *O*-Me ester, *in* C-00270
Cyclohexylphosphoramidothioic acid; Difluoride, *in* C-00271
3,5-Di-*tert*-butyl-2-chloro-2,3-dihydro-1,3,2-oxazaphosphole; 2-Sulfide, *in* D-00078
O,O-Dibutyl diethylphosphoramidothioate, *in* D-00356
1,3-Di-*tert*-butyl-4,4-dimethyl-1,3,2-diazaphosphasiletidine; 2-Dimethylamino, 2-sulfide, *in* D-00087
1,3-Di-*tert*-butyl-4,4-dimethyl-1,3,2-diazaphosphasiletidine; 2-(Di-*tert*-butyl)amino, 2-sulfide, *in* D-00087
O,O-Di-*tert*-butyl dimethylphosphoramidothioate, *in* D-00871
O,O-Di-*tert*-butyl ethylphosphoramidothioate, *in* E-00159
S,S-Dibutyl phosphoramidodithioate, *in* D-00152
N,N′-Dibutylphosphorodiamidodithioic acid, D-00147
O-2,4-Dichlorophenyl *S*-methyl isopropylphosphoramidothioate; (+)-*form*, *in* D-00183
O-2,4-Dichlorophenyl *S*-methyl isopropylphosphoramidothioate; (−)-*form*, *in* D-00183
O-2,4-Dichlorophenyl *S*-methyl isopropylphosphoramidothioate; (±)-*form*, *in* D-00183
N,N′-Dicyclohexylphosphorodiamidodithioic acid, D-00217
2-Diethylamino-1,3,2-dioxaphospholan-4,5-dicarboxylic acid; (4*R*,5*R*)-*form*, Di-Et ester, 2-sulfide, *in* D-00265
2-Diethylamino-1,3,2-dioxaphospholan-4,5-dicarboxylic acid; (4*RS*,5*RS*)-*form*, Di-Et ester, 2-sulfide, *in* D-00265
O,O-Diethyl benzoylphosphoramidothioate, *in* B-00038
O,O-Diethyl *tert*-butylphosphoramidothioate, *in* B-00623
O,O-Diethyl (2-chlorophenyl)phosphoramidothioate, *in* C-00158
O,O-Diethyl (3-chlorophenyl)phosphoramidothioate, *in* C-00159
O,O-Diethyl (4-chlorophenyl)phosphoramidothioate, *in* C-00160
O,O-Diethyl cyclohexylphosphoramidothioate, *in* C-00271
O,S-Diethyl cyclohexylphosphoramidothioate, *in* C-00271
O,O-Diethyl (diethoxyphosphinyl)methylphosphoramidothioate, *in* M-00413
O,O-Diethyl diethylphosphoramidothioate, *in* D-00356
O,O-Diethyl dimethylphosphoramidothioate, *in* D-00871
N,N-Diethyl-*N′,N″*-diphenylphosphorothioic triamide, *in* D-00398
O,O-Diethyl ethylphosphoramidothioate, *in* E-00159
O,O-Diethyl (3-fluorophenyl)phosphoramidothioate, *in* F-00045
O,O-Diethyl (4-fluorophenyl)phosphoramidothioate, *in* F-00046
O,O-Diethyl (1-iminoethyl)phosphoramidothioate, *in* I-00005
O,S-Diethyl (1-iminoethyl)phosphoramidothioate, *in* I-00005
O,O-Diethyl isopropylphosphoramidothioate, *in* I-00059
O,O-Diethyl (2-methoxyethyl)phosphoramidothioate, *in* M-00041
O,O-Diethyl (2-methoxyphenyl)phosphoramidothioate, *in* M-00077
O,O-Diethyl (3-methoxyphenyl)phosphoramidothioate, *in* M-00078
O,O-Diethyl (4-methoxyphenyl)phosphoramidothioate, *in* M-00079
O,O-Diethyl (2-methylphenyl)phosphoramidothioate, *in* M-00254
O,O-Diethyl (3-methylphenyl)phosphoramidothioate, *in* M-00255
O,O-Diethyl (4-methylphenyl)phosphoramidothioate, *in* M-00256
O,O-Diethyl methylphosphoramidothioate, *in* M-00333
O,O-Diethyl [(phenylamino)carbonyl]phosphoramidothioate, *in* P-00082
O,S-Diethyl [(phenylamino)carbonyl]phosphoramidothioate, *in* P-00082
O,O-Diethyl phenylphosphoramidothioate, *in* P-00262
O,S-Diethyl phosphoramidodithioate, D-00349
S,S-Diethyl phosphoramidodithioate, D-00350
O,O-Diethyl phosphoramidothioate, D-00354
O,S-Diethyl phosphoramidothioate, D-00355
Diethylphosphoramidothioic dichloride, D-00357
Diethylphosphoramidothioic difluoride, D-00358
N,N-Diethylphosphorothioic triamide, *in* D-00356
N,N-Diethylphosphorothioic triamide, D-00398
O,O-Diethyl propylphosphoramidothioate, *in* P-00485
1,5-Dihydro-3-dimethylamino-2,4,3-benzodioxaphosphepin; 3-Sulfide, *in* D-00453

2,3-Dihydro-2-dimethylamino-5-methyl-3-phenyl-1,3,4,2-oxadiazaphosphole; 2-Sulfide, *in* D-00455
1,2-Dihydro-1-hydroxy-1-phenyl-4*H*-3,1,2-benzothiazaphosphorin-4-one 1-sulfide; 2-Methoxy (*O*-Me ester), *in* D-00503
1,2-Dihydro-1-hydroxy-1-phenyl-4*H*-3,1,2-benzothiazaphosphorin-4-one 1-sulfide; 2-Ethoxy (*O*-Et ester), *in* D-00503
2,3-Dihydro-2-mercapto-1,3,2-benzoxazaphosphole 2-sulfide, D-00511
2,3-Dihydro-5-methyl-3-phenyl-2-phenylamino-1,3,4,2-oxadiazaphosphole; 2-Sulfide, *in* D-00527
1,2-Dihydro-1-phenyl-4*H*-3,1,2-benzothiazaphosphorin-4-thione; 2-Methoxy, 2-sulfide, *in* D-00552
1,2-Dihydro-1-phenyl-4*H*-3,1,2-benzothiazaphosphorin-4-thione; 2-Ethoxy, 2-sulfide, *in* D-00552
O,O-Diisopropyl benzoylphosphoramidothioate, *in* B-00038
O,O-Diisopropyl diethylphosphoramidothioate, *in* D-00356
O,O-Diisopropyl (1,3-dihydro-1,3-dioxo-2*H*-isoindol-2-yl)-phosphonothioate, *in* D-00475
O,O-Diisopropyl dimethylphosphoramidothioate, *in* D-00871
O,O-Diisopropyl ethylphosphoramidothioate, *in* E-00159
O,O-Diisopropyl (4-fluorophenyl)phosphoroamidothioate, *in* F-00046
O,O-Dimethyl (4-chlorophenyl)phosphoramidothioate, *in* C-00160
Dimethyl cyclohexylphosphoramidate, *in* C-00269
O,O-Dimethyl cyclohexylphosphoramidothioate, *in* C-00271
O,O-Dimethyl diethylphosphoramidothioate, *in* D-00356
O,O-Dimethyl (1,3-dihydro-1,3-dioxo-2*H*-isoindol-2-yl)-phosphonothioate, *in* D-00475
O,O-Dimethyl (dimethoxyphosphinyl)-methylphosphoramidothioate, *in* M-00413
O,O-Dimethyl dimethylphosphoramidothioate, *in* D-00871
N,N-Dimethyl-*N′,N″*-diphenylphosphorothioic triamide, *in* D-00909
O,S-Dimethyl (1-iminoethyl)phosphoramidothioate, *in* I-00005
O,O-Dimethyl (2-methoxyethyl)phosphoramidothioate, *in* M-00041
O,O-Dimethyl methylphosphoramidothioate, *in* M-00333
O,O-Dimethyl 4-morpholinylphosphonothioate, *in* M-00527
O,S-Dimethyl [(phenylamino)carbonyl]phosphoramidothioate, *in* P-00082
S,S-Dimethyl phenylphosphoramidodithioate, *in* D-00866
O,O-Dimethyl phenylphosphoramidothioate, *in* P-00262
O,S-Dimethyl phosphoramidodithioate, D-00865
S,S-Dimethyl phosphoramidodithioate, D-00866
O,O-Dimethyl phosphoramidothioate, D-00870
Dimethylphosphoramidothioic dichloride, D-00872
Dimethylphosphoramidothioic difluoride, D-00873
N,N′-Dimethylphosphorodiamidodithioic acid, D-00896
NN-Dimethylphosphorothioic triamide, *in* D-00871
N,N-Dimethylphosphorothioic triamide, D-00909
O,O-Dimethyl 1-piperidinylphosphonothioate, *in* P-00431
O,O-Diphenyl benzoylphosphoramidothioate, *in* B-00038
O,O-Diphenyl diethylphosphoramidothioate, *in* D-00356
Diphenyl diethylphosphoramidotrithioate, *in* D-00359
Diphenyl dimethylphosphoramidotrithioate, *in* D-00874
O,O-Diphenyl (diphenoxyphosphinothioyl)-phosphoramidothioate, D-01004
O,O-Diphenyl (diphenoxyphosphinyl)phosphoramidothioate, D-01005
O,O-Diphenyl ethylphosphoramidothioate, *in* E-00159
O,O-Diphenyl isopropylphosphoramidothioate, *in* I-00059
O,O-Diphenyl (4-methoxyphenyl)phosphoramidothioate, *in* M-00079
O,O-Diphenyl (2-methylphenyl)phosphoramidothioate, *in* M-00254
O,O-Diphenyl (3-methylphenyl)phosphoramidothioate, *in* M-00255
O,O-Diphenyl (4-methylphenyl)phosphoramidothioate, *in* M-00256
O,O-Diphenyl 4-morpholinylphosphonothioate, *in* M-00527
O,O-Diphenyl [(phenylamino)carbonyl]phosphoramidothioate, *in* P-00082
O,O-Diphenyl phenylphosphoramidothioate, *in* P-00262
O,O-Diphenyl phenylsulfonylphosphoramidothioate, *in* P-00314
N,N′-Diphenylphosphorodiamidodithioic acid, D-01128
OO-Diphenyl 2-pyridinylphosphoramidothioate, *in* P-00513
O,O-Diphenyl 4-pyridinylphosphoramidothioate, *in* P-00514
O,O-Dipropyl phenylphosphoramidothioate, *in* D-01230
S,S-Dipropyl phosphoramidodithioate, *in* D-01227
▷Ditalimfos, D-01237
3-Ethoxy-2,3-dihydro-2-methyl-5-phenyl-1*H*-1,2,4,3-triazaphosphole; 3-Sulfide, *in* E-00028
2-Ethoxy-3-methyl-1,3,2-thiazaphospholidine; 2-Oxide, *in* E-00043
2-Ethoxy-3-methyl-1,3,2-thiazaphospholidine; 2-Sulfide, *in* E-00043
2-Ethoxy-3-phenyl-1,3,2-oxazaphospholidine; 2-Sulfide, *in* E-00048
2-Ethoxytetrahydro-2*H*-1,3,2-oxazaphosphorine; 2-Sulfide, *in* E-00051
2-Ethoxy-1,3,2-thiazaphospholidine; 2-Oxide, *in* E-00053
2-Ethoxy-1,3,2-thiazaphospholidine; 2-Sulfide, *in* E-00053
Ethyl *N″*-[bis(diethylamino)phosphinothioyl]-*N,N,N′,N′*-tetraethylphosphorodiamidoimidate, *in* B-00174
O-Ethyl diethylphosphoramidochloridothioate, *in* D-00348
Ethyl diethylphosphoramidofluoridodithioate, *in* D-00352
O-Ethyl dimethylphosphoramidochloridothioate, *in* D-00861
O-Ethyl dimethylphosphoramidocyanidothioate, *in* D-00864
Ethyl dimethylphosphoramidofluoridodithioate, *in* D-00867
O-Ethyl dimethylphosphoramidofluoridothioate, *in* D-00868
S,S-Ethylene diethylphosphoramidodithioite, *in* D-01245
S,S-Ethylene *N,N*-dimethyl phosphoramidodithioate, *in* D-01244
Ethylene dimethylphosphoramidothioate, *in* D-00695
Ethylene phenylphosphoramidothioate, *in* P-00083
S-Ethyl *O*-(4-methyl-2-nitrophenyl) isopropylphosphoramidothioate, *in* M-00180
O-Ethyl *N,N′-o*-phenylenephosphorodiamidothioate, *in* E-00026
O-Ethyl-*S*-phenyl methylphosphoramidodithioate, *in* M-00330
O-Ethyl phosphoramidofluoridothioate, E-00158
O-Ethyl phosphorodiamidothioate, *in* E-00182
O-Ethyl *S*-propyl methylphosphoramidodithioate, *in* M-00330
O-Ethyl tetraethylphosphorodiamidothioate, *in* T-00063
S-Ethyl tetraethylphosphorodiamidothioate, *in* T-00063
Ethyl tetramethylphosphorodiamidodithioate, *in* T-00222
▷4-Ethyl-2,6,7-trimethyl-2,6,7-triaza-1-phosphabicyclo[2.2.2]octane; 1-Sulfide, *in* E-00195
2-Fluoro-1,3-dimethyl-1,3,2-diazaphospholidine; 2-Sulfide, *in* F-00014
Hexabutylphosphorothioic triamide, *in* H-00013
Hexaethylphosphorothioic triamide, *in* H-00015
Hexahydro-2-phenoxy-1*H*-1,3,2-diazaphosphepine; 2-Sulfide, *in* H-00038
Hexahydro-2*a*,4*a*,6*a*-triaza-6*b*-phosphacyclopenta[*cd*]-pentalene; 6*b*-Sulfide, *in* H-00055
2,3,5,6,7,8-Hexamethyl-2,3,5,6,7,8-hexaaza-1,4-diphosphabicyclo[2.2.2]octane; 1,4-Disulfide, *in* H-00070
▷Hexamethylphosphorothioic triamide, H-00079
Hexapropylphosphorothioic triamide, *in* H-00092
▷Isofenphos, I-00025
O-Isopropyl *S*-phenyl methylphosphoramidodithioate, *in* M-00330
2-Mercapto-1,3,5,2-triazaphosphorine-4,6(1*H*,5*H*)-dione 2-sulfide, M-00020
Methamidophos; (*R*)-*form*, *in* M-00021
Methamidophos; (±)-*form*, *in* M-00021
2-Methoxy-3,4-dimethyl-5-phenyl-1,3,2-oxazaphospholidine; (2*R*,4*S*,5*R*)-*form*, 2-Sulfide, *in* M-00035
2-Methoxy-3,4-dimethyl-5-phenyl-1,3,2-oxazaphospholidine; (2*S*,4*S*,5*R*)-*form*, 2-Sulfide, *in* M-00035
(2-Methoxyphenyl)phosphoramidothioic acid; Difluoride, *in* M-00077
O-Methyl cyclohexylphosphoramidochloridothioate, *in* C-00270
Methyl diethylphosphoramidofluoridodithioate, *in* D-00352
4-Methyl-2-dimethylamino-1,3,2-dioxaphosphorinane; (2*RS*,4*RS*)-*form*, 2-Sulfide, *in* M-00117
O-Methyl dimethylphosphoramidochloridothioate, *in* D-00861
Methyl dimethylphosphoramidofluoridodithioate, *in* D-00867
S-Methyl *N,N′*-diphenylphosphorodiamidothioate, *in* D-01129
O-Methyl ethylphosphoramidochloridothioate, *in* E-00157
S-Methyl *O*-(4-methyl-2-nitrophenyl) isopropylphosphoramidothioate, *in* M-00180

O-Methyl *O*-4-nitrophenyl cyclohexylphosphoramidothioate; (+)-*form*, *in* M-00179
O-Methyl *O*-4-nitrophenyl cyclohexylphosphoramidothioate; (±)-*form*, *in* M-00179
5-Methyl-1,3,2-oxathiaphosphole 2-oxide; 2-Diethylamino, *in* M-00184
4-Methyl-2-phenylamino-1,3,2-dioxaphosphorinane; (2*RS*,4*SR*)-*form*, 2-Sulfide, *in* M-00193
O-Methyl-*S*-phenyl methylphosphoramidodithioate, *in* M-00330
(3-Methylphenyl)phosphoramidothioic acid; Difluoride, *in* M-00255
O-Methyl phosphoramidofluoridothioate, M-00331
O-Methyl phosphorodiamidothioate, *in* M-00359
S-Methyl phosphorodiamidothioate, M-00343
3-Methyltetrahydro-2-phenoxy-2*H*-1,3,2-oxazaphosphorine; 2-Sulfide, *in* M-00404
Methyl tetramethylphosphorodiamidodithioate, *in* T-00222
3-Methyl-1,3,2-thiazaphospholidine; 2-Chloro, 2-oxide, *in* M-00408
3-Methyl-1,3,2-thiazaphospholidine; 2-Chloro, 2-sulfide, *in* M-00408
3-Methyl-1,3,2-thiazaphospholidine; 2-Methoxy, 2-oxide, *in* M-00408
3-Methyl-1,3,2-thiazaphospholidine; 2-Methoxy, 2-sulfide, *in* M-00408
3-Methyl-1,3,2-thiazaphospholidine; 2-Ethoxy, 2-sulfide, *in* M-00408
Monomethyl 4-morpholinylphosphonothioate, *in* M-00527
P-Morpholinyl-*N*,*N*′-diphenylphosphonothioic diamide, *in* M-00528
4-Morpholinylphosphonothioic acid; Dichloride, *in* M-00527
P-4-Morpholinylphosphonothioic diamide, M-00528
O,*O*-1,2-Naphthylene diethylphosphoramidothioate, *in* D-00266
O,*O*-1,8-Naphthylene diethylphosphoramidothioate, *in* D-00267
O-2-Naphthyl phosphoradiamidothioate, N-00027
O-2-Naphthyl tetramethylphosphorodiamidothioate, *in* N-00027
N″-Bis(dimethylamino)phosphinyl-*N*,*N*,*N*′,*N*′,*N*″-pentamethylphosphorotriamidothioate, N-00032
Neopentylene *tert*-butylphosphoramidothioate, *in* B-00538
Neopentylene dimethylphosphoramidothioate, *in* D-00693
Neopentylene phenylphosphoramidothioate, *in* D-00792
O-(4-Nitrophenyl) phenylphosphoramidochloridothioate, *in* P-00260
sym-Octamethylmonothiopyrophosphoramide, O-00030
unsym-Octamethylmonothiopyrophosphoramide, O-00031
Octamethyl thiodiphosphoramide, O-00034
O-Phenyl-*N*,*N*′-dibenzylphosphorodiamidothioate, *in* D-00062
O-Phenyl diethylphosphoramidochloridothioate, *in* D-00348
Phenyl diethylphosphoramidofluoridodithioate, *in* D-00352
O-Phenyl dimethylphosphoramidochloridothioate, *in* D-00861
Phenyl dimethylphosphoramidofluoridodithioate, *in* D-00867
O-Phenyl dimethylphosphoramidofluoridothioate, *in* D-00868
O-Phenyl *N*,*N*′-diphenylphosphorodiamidothioate, *in* D-01129
O-Phenyl ethylphosphoramidochloridothioate, *in* E-00157
O-Phenyl methylphosphoramidochloridothioate, *in* M-00329
O-Phenyl 4-morpholinylphosphonochloridothioate, *in* M-00525
O-Phenyl *N*,*N*′-*o*-phenylenephosphorodiamidothioate, *in* D-00543
O-Phenyl phenylphosphoramidochloridothioate, *in* P-00260
Phenylphosphoramidothioic acid; Difluoride, *in* P-00262
Phenylphosphoramidothioic acid; Dichloride, *in* P-00262
O-Phenyl phosphorodiamidothioate, P-00276
O-Phenyl tetraethylphosphorodiamidothioate, *in* T-00063
S-Phenyl tetraethylphosphorodiamidothioate, *in* T-00063
O-Phenyl *N*,*N*′-trimethylenephosphorodiamidothioate, *in* H-00039
▷Phosbutyl, P-00341
1,1′,1″-Phosphinothioyldiynetris(1*H*-pyrazole), *in* T-00722
▷4,4′,4″-Phosphinothioylidynetrismorpholine, *in* T-00569
1,1′,1″-Phosphinothioylidynetrispiperidine, *in* T-00704
S,*S*′-1,3-Propanediyl *N*,*N*-diethylphosphoramidodithioate, *in* D-01248
▷Propetamphos, P-00466
O,*O*-Propylene dimethylphosphoramidothioate, *in* M-00116
Propylphosphoramidothioic acid; Dichloride, *in* P-00485
O,*O*,*O*,*O*-Tetraethyl methylthioimidodiphosphate, *in* M-00161
Tetraethylphosphorodiamidothioic azide, T-00064
Tetraethylphosphorodiamidothioic chloride, T-00065
O,*O*,*O*′,*O*′-Tetraethyl thioimidodiphosphate, T-00074
Tetrahydro-5,5-dimethyl-2-phenoxy-2*H*-1,3,2-oxazaphosphorine; 2-Sulfide, *in* T-00098
Tetrahydro-2-hydroxy-2*H*-1,3,2-oxazaphosphorine 2-sulfide, T-00108
Tetrahydro-2-methoxy-2*H*-1,3,2-oxazaphosphorine; 2-Sulfide, *in* T-00111
Tetrahydro-2-methylthio-2*H*-oxazaphosphorine 2-oxide, *in* T-00108
Tetrahydro-2-phenoxy-2*H*-1,3,2-oxazaphosphorine; 2-Sulfide, *in* T-00126
Tetrahydro-2*H*-1,3,2-thiazaphosphorine 2-oxide; 2-Ethoxy, *in* T-00154
Tetrahydro-2*H*-1,3,2-thiazaphosphorine 2-oxide; 2-Phenoxy, *in* T-00154
4,4,5,5-Tetramethylethylene dimethylphosphoramidothioate, *in* D-00710
O,*O*,*O*,*O*-Tetramethyl methylthioimidodiphosphate, *in* M-00161
N,*N*,*N*′,*N*′-Tetramethyl-*P*-4-morpholinylphosphonothioic diamide, *in* M-00528
Tetramethyl phosphor(isothiocyanatido)thioic diamide, T-00217
Tetramethylphosphorodiamidodithioic acid, T-00222
Tetramethylphosphorodiamidothioic azide, T-00224
Tetramethylphosphorodiamidothioic chloride, T-00225
▷Tetramethylphosphorodiamidothioic fluoride, T-00226
2,4,6,7-Tetramethyl-2,6,7-triaza-1-phosphabicyclo[2.2.2]-octane; 1-Sulfide, *in* T-00237
O,*O*,*O*,*O*-Tetraphenyl 1,2-ethanediylbisphosphoramidothioate, *in* E-00010
O,*O*,*O*,*O*-Tetrapropyl 1,2-ethanediylbisphosphoramidothioate, *in* E-00010
1,3,2-Thiazaphospholidine 2-oxide; 2-Butoxy, *in* T-00302
1,3,2-Thiazaphospholidine 2-sulfide; 2-Methoxy, *in* T-00303
1,3,2-Thiazaphospholidine 2-sulfide; 2-Propoxy, *in* T-00303
1,3,2-Thiazaphospholidine 2-sulfide; 2-Phenoxy, *in* T-00303
▷Thiotepa, T-00317
N,*N*′,*N*″-Tribenzylphosphorothioic triamide, *in* T-00344
Tributyl [bis(dimethylamino)phosphinothioyl]-phosphorimidate, *in* B-00192
Tributyl (dimethoxyphosphinothioyl)phosphorimidate, *in* D-00674
N,*N*′,*N*″-Tributylphosphorothioic triamide, *in* T-00394
P,*P*,*P*-Trichloro-*N*-(diphenoxyphosphinylthioyl)phosphine imide, *in* D-00987
▷*N*,*N*′,*N*″-Tricyclohexylphosphorothioic triamide, *in* T-00429
Triethyl [bis(dimethylamino)phosphinothioyl]-phosphorimidate, *in* B-00192
▷Triethyl (diethoxyphosphinothioyl)phosphorimidate, *in* D-00238
Triethyl (dimethoxyphosphinothioyl)phosphorimidate, *in* D-00674
Triethyl (diphenoxyphosphinothioyl)phosphorimidate, *in* D-00987
N,*N*′,*N*″-Triethylphosphorothioic triamide, *in* T-00462
N,*N*′,*N*″-Triisopropylphosphorothioic triamide, *in* T-00509
Trimethyl (diethoxyphosphinothioyl)phosphorimidate, *in* D-00238
O,*O*-Trimethyl dimethylphosphoramidothioate, *in* D-00696
Trimethyl (diphenoxyphosphinothioyl)phosphorimidate, *in* D-00987
2,4,10-Trimethyl-1,3,4,5,7,10-hexaaza-3-phosphatricyclo[3.3.1.1^{3,7}]decane; 3-Sulfide, *in* T-00517
N,*N*′,*N*″-Trimethylphosphorothioic triamide, *in* T-00551
Trimethylsilyl tetramethylphosphorodiamidodithioate, *in* T-00222
2,8,9-Trimethyl-2,8,9-triaza-1-phosphatricyclo[3.3.1.1^{3,7}]-decane; Sulfide, *in* T-00563
Triphenyl (diethoxyphosphinothioyl)phosphorimidate, *in* D-00238
Triphenyl (dimethoxyphosphinothioyl)phosphorimidate, *in* D-00674
Triphenyl (diphenoxyphosphinothioyl)phosphorimidate, *in* D-00987
N,*N*′,*N*″-Triphenyl (diphenoxyphosphinothioyl)-phosphorotriamidoimidate, *in* D-00987

N,N′,N″-Triphenylphosphorothioic triamide, *in* T-00677
N,N′,N″-Tri-2-propenylphosphorous triamide; Sulfide, *in* T-00709
N,N′,N″-Tripropylphosphorothioic triamide, *in* T-00719
N,N′,N″-Tris(2-methylphenyl)phosphorothioic triamide, *in* T-00817
N,N′,N″-Tris(4-methylphenyl)phosphorothioic triamide, *in* T-00819
N,N′,N″-Tris(1-methylpropyl)phosphorothioic triamide, *in* T-00829
N,N′,N″-Tris(2-methylpropyl)phosphorothioic triamide, *in* T-00830
Zytron; (+)-*form*, *in* Z-00002
Zytron; (−)-*form*, *in* Z-00002
Zytron; (±)-*form*, *in* Z-00002

Thiophosphoric isocyanates and isothiocyanates

O,O-Diethyl phosphorisocyanatidothioate, D-00367
O,S-Diethyl phosphorisocyanatidothioate, D-00368
▷*O,O*-Diethyl phosphor(isothiocyanatido)thioate, D-00370
O,O-Dimethyl phosphorisocyanatidothioate, D-00884
O,S-Dimethyl phosphorisocyanatidothioate, D-00885
O,O-Diphenyl phosphorisocyanatidothioate, D-01117
O-Ethyl phosphorodiisocyanatidothioate, E-00178
O-Ethyl phosphorofluoridoisocyanatidothioate, E-00181
O-Methyl phosphorodiisocyanatidothioate, M-00357
O-Methyl phosphorofluoridoisothiocyanatidothioate, M-00358
O-Phenyl phosphorodiisocyanatidothioate, P-00293
Tetramethyl phosphor(isothiocyanatido)thioic diamide, T-00217

Thiophosphoric azides

▷*O,O*-Diethyl phosphorazidothioate, D-00363
O,O-Dimethyl phosphorazidothioate, D-00879
O,O-Diphenyl phosphorazidothioate, D-01113
O,O-Dipropyl phosphorazidothioate, D-01216
Tetraethylphosphorodiamidothioic azide, T-00064
Tetramethylphosphorodiamidothioic azide, T-00224

Thiophosphoric hydrazides

2-Chloro-2,3-dihydro-3,5-diphenyl-1,3,4,2-oxadiazaphosphole; 2-Sulfide, *in* C-00042
2-Chloro-2,3-dihydro-5-methyl-3-phenyl-1,3,4,2-oxadiazaphosphole; 2-Sulfide, *in* C-00044
O,O-Diethyl 2-phenylphosphorohydrazidothioate, *in* P-00296
O,O-Diethyl phosphorohydrazidothioate, D-00392
2,3-Dihydro-2-dimethylamino-5-methyl-3-phenyl-1,3,4,2-oxadiazaphosphole; 2-Sulfide, *in* D-00455
2,3-Dihydro-3,5-dimethyl-1,3,4,2-thiadiazaphosphole 2-sulfide; 2-Chloro, *in* D-00471
2,3-Dihydro-3,5-dimethyl-1,3,4,2-thiadiazaphosphole 2-sulfide; 2-Ethoxy, *in* D-00471
2,3-Dihydro-2-methoxy-5-methyl-3-phenyl-1,3,4,2-oxadiazaphosphole; 2-Sulfide, *in* D-00513
2,3-Dihydro-5-methyl-3-phenyl-2-phenylamino-1,3,4,2-oxadiazaphosphole; 2-Sulfide, *in* D-00527
2,3-Dihydro-5-methyl-3-phenyl-1,3,4,2-thiadiazaphosphole 2-sulfide; 2-Chloro, *in* D-00529
2,3-Dihydro-5-methyl-3-phenyl-1,3,4,2-thiadiazaphosphole 2-sulfide; 2-Ethoxy, *in* D-00529
O,O-Diisopropyl 2-phenylphosphorohydrazidothioate, *in* P-00296
O,O-Dimethyl 2-phenylphosphorohydrazidothioate, *in* P-00296
O,O-Dimethyl phosphorohydrazidothioate, D-00905
O,O-Diphenyl phosphorohydrazidothioate, D-01137
3-Ethoxy-2,3-dihydro-2-methyl-5-phenyl-1*H*-1,2,4,3-triazaphosphole; 3-Sulfide, *in* E-00028
Hexahydro-2,4-dimethyl-3-phenoxy-1,2,4,5,3-tetrazaphosphorine; 3-Sulfide, *in* H-00024
2-Phenylphosphorodifluoridothioic hydrazide, P-00289
O-Phenyl phosphorodihydrazidothioate, P-00291
O,O,O′,O′-Tetraethyl hydrazodiphosphonothioate, T-00048

Selenophosphoric acids

O,O-Bis(4-methylphenyl) phosphorodiselenoate, B-00336
O,O-Dibutyl phosphorodiselenoate, D-00148
O,O-Dicyclohexyl phosphorodiselenoate, D-00218
O,O-Diethyl phosphorodiselenoate, D-00383
O,O-Diethyl phosphoroselenoate, D-00393
O,O-Diethyl phosphoroselenothioate, D-00394
O,O-Diisopropyl phosphorodiselenoate, D-00662
O,O-Diisopropyl phosphoroselenoate, D-00666
O,O-Dimethyl phosphorodiselenoate, D-00897
O,O-Dimethyl phosphoroselenoate, D-00906
O,O-Dipentyl phosphorodiselenoate, D-00985
O,O-Diphenyl phosphorodiselenoate, D-01130
O,O-Diphenyl phosphoroselenothioate, D-01140
O,O-Dipropyl phosphorodiselenoate, D-01224
O,O-Dipropyl phosphoroselenoate, D-01229
O-Ethyl phosphorotriselenoate, E-00184
2-Hydroxy-5,5-dimethyl-1,3,2-dioxaphosphorinane 2-selenide, H-00125
2-Hydroxy-4-methyl-1,3,2-dioxaphosphorinane 2-selenide; (2*RS*,4*RS*)-*form*, *in* H-00157
2-Hydroxy-4-methyl-1,3,2-dioxaphosphorinane 2-selenide; (2*RS*,4*SR*)-*form*, *in* H-00157
O,O,O-Triphenyl phosphoroselenoate, T-00667

Selenophosphoric halides

O,O-(2,2′-Biphenylyl) phosphorochloridoselenoate, *in* C-00032
O,O-Bis(2-methylpropyl) phosphorochloridoselenoate, B-00356
O,O-1,3-Butylene fluoroselenophosphate, *in* F-00021
O,O-Butylene phosphorochloridoselenoate, *in* C-00093
2-Chloro-1,3,2-dioxaphosphorinane 2-sulfide, C-00068
2-Chloro-4-isopropyl-5,5-dimethyl-1,3,2-dioxaphosphorinane; (2*RS*,4*RS*)-*form*, 2-Selenide, *in* C-00090
2-Chloro-4-isopropyl-5,5-dimethyl-1,3,2-dioxaphosphorinane; (2*RS*,4*SR*)-*form*, 2-Selenide, *in* C-00090
O,O-Dibutyl phosphorochloridoselenoate, *in* D-00154
Diethylphosphoramidoselenoic acid; Difluoride, *in* D-00353
O,O-Diethyl phosphorochloridoselenoate, D-00377
Dimethylphosphoramidoselenoic acid; Difluoride, *in* D-00869
O,O-Diphenyl phosphorobromidoselenoate, *in* D-01139
O,O-Diphenyl phosphorochloridoselenoate, D-01122
O,O-Dipropyl phosphorochloridoselenoate, *in* D-01229
O-Ethyl diethylphosphoramidochloridoselenoate, *in* D-00347
Neopentylene phosphorochloridoselenoate, *in* C-00058
O-Phenyl diethylphosphoramidochloridoselenoate, *in* D-00347
Tetraethylphosphorodiamidoselenoic acid; Chloride, *in* T-00061
Tetramethylphosphorodiamidoselenoic acid; Chloride, *in* T-00223
O,O-Trimethylene phosphorochloridoselenoate, *in* C-00066

Selenophosphoric esters

Bis(5,5-dimethyl-2-selenoxo-1,3,2-dioxaphosphorinan-2-yl) disulfide, *in* D-01251
Bis(*O,O*-diphenyl phosphoroselenoyl) diselenide, B-00269
O,O-Bis(4-methylphenyl) phosphorodiselenoate, B-00336
O,O-Bis(2-methylpropyl) phosphorochloridoselenoate, B-00356
O,O-1,3-Butylene *S*-methyl phosphoroselenothioate, *in* M-00174
O-Butyl tetraethylphosphorodiamidoselenoate, *in* T-00061
Se-Butyl tetraethylphosphorodiamidoselenoate, *in* T-00061
Se,Se-Dibutyl *O*-ethyl phosphorodiselenothioate, *in* D-00149
O,O-Dibutyl phosphorochloridoselenoate, *in* D-00154
O,O-Dibutyl phosphorodiselenoate, D-00148
Se,Se-Dibutyl *O*-propyl phosphorodiselenothioate, *in* D-00149
O,O-Diethyl diethylphosphoramidoselenoate, *in* D-00353
O,O-Diethyl diisopropylphosphoramidoselenoate, *in* D-00655
O,O-Diethyl dimethylphosphoramidoselenoate, *in* D-00869
O,O-Diethyl methylphosphoramidoselenoate, *in* M-00332
O,O-Diethyl phenylphosphoramidoselenoate, *in* P-00261
O,O-Diisopropyl diethylphosphoramidoselenoate, *in* D-00353
O,O-Diisopropyl methylphosphoramidoselenoate, *in* M-00332
O,O-Dimethyl diethylphosphoramidoselenoate, *in* D-00353
O,O-Dimethyl dimethylphosphoramidoselenoate, *in* D-00869
O,O-Dimethyl *Se,Se*-(2-nitrophenyl) phosphoro(diselenoperoxoate), D-00771

O,O-Dimethyl-*S,Se*-2-nitrophenylphosphoro(selenothioperoxo)-thioate, D-00772
O,O-Dimethyl phosphorodiselenoate, D-00897
O,O-Dimethyl phosphoroselenoate, D-00906
O,O-Diphenyl phosphorodiselenoate, D-01130
O,O-Dipropyl phosphorochloridoselenoate, *in* D-01229
O,O-Dipropyl phosphoroselenoate, D-01229
2,2′-Diselenobis[5,5-dimethyl-2-thioxo-1,3,2-dioxaphosphorinane], *in* D-01234
Ethyl bicyclic selenophosphate, *in* E-00196
O-Ethyl diethylphosphoramidochloridoselenoate, *in* D-00347
O,O-Ethylene *O*-methyl phosphoroselenoate, *in* M-00036
O-Ethyl phosphorotriselenoate, E-00184
O-Ethyl tetraethylphosphorodiamidoselenoate, *in* T-00061
Se-Ethyl tetraethylphosphorodiamidoselenoate, *in* T-00061
4-Isopropyl-2-methoxy-5,5-dimethyl-1,3,2-dioxaphosphorinane; (2*RS*,4*RS*)-*form*, 2-Selenide, *in* I-00033
4-Isopropyl-2-methoxy-5,5-dimethyl-1,3,2-dioxaphosphorinane; (2*RS*,4*SR*)-*form*, 2-Selenide, *in* I-00033
2-Methoxy-4-methyl-1,3,2-dioxaphosphorinane 2-selenide; (2*RS*,4*RS*)-*form*, *in* M-00046
2-Methoxy-4-methyl-1,3,2-dioxaphosphorinane 2-selenide; (2*RS*,4*SR*)-*form*, *in* M-00046
Methylglycerol bicyclic selenophosphate, *in* M-00423
4-Methyl-2-methylthio-1,3,2-dioxaphosphorinane; (2*RS*,4*SR*)-*form*, 2-Selenide, *in* M-00174
Se-Methyl *O,O*-neopentylene phosphorodiselenoate, *in* D-00764
Se-Methyl *O,O*-neopentylene phosphoroselenoate, *in* D-00764
Se-Methyl *O,O*-neopentylene phosphoroselenothioate, *in* D-00764
O-Methyl tetraethylphosphorodiamidoselenoate, *in* T-00061
O-Methyl tetramethylphosphorodiamidoselenoate, *in* T-00223
O-Methyl *O,O*-trimethylene phosphoroselenoate, *in* M-00038
O,O-Neopentylene *O*-phenyl phosphoroselenoate, *in* D-00790
2-Phenoxy-1,5-dihydro-2,4,3-benzodioxaphosphepin; Selenide, *in* P-00066
2-Phenoxy-1,3,2-oxaselenaphosphorinane 2-oxide, P-00071
O-Phenyl diethylphosphoramidochloridoselenoate, *in* D-00347
O-Phenyl tetraethylphosphorodiamidoselenoate, *in* T-00061
O,Se,Se-Tributyl phosphorodiselenothioate, *in* D-00149
O,O,O-Tributyl phosphoroselenoate, T-00385
O,Se,Se-Tributyl phosphorotriselenoate, T-00389
O,O,O-Triethyl phosphoroselenoate, T-00454
O,O,Se-Triethyl phosphoroselenoate, T-00455
O,O,O-Triisopropyl phosphoroselenoate, T-00504
Trimethylol selenophosphate, *in* M-00424
O,O,O-Trimethyl phosphoroselenoate, T-00543
2,8,9-Trioxa-5-aza-1-phosphabicyclo[3.3.3]undecane; 1-Selenide, *in* T-00584
2,8,9-Trioxa-1-phosphatricyclo[3.3.1.1^{3,7}]decane; 1-Selenide, *in* T-00591
Triphenyl phosphorotetraselenoate, T-00668
Se,Se,Se-Triphenyl phosphorotriselenoate, T-00672
Se,Se,Se-Triphenyl phosphorotriselenothioate, T-00674
O,O,O-Tripropyl phosphoroselenoate, T-00713
O,O,O-Tris(2-methylpropyl) phosphoroselenoate, T-00827

Selenophosphoric amides

N,N′-Bis(2-methylpropyl)phosphorodiamidodiselenoic acid, B-00358
2-*tert*-Butylamino-4-methyl-1,3,2-dioxaphosphorinane; (2*RS*,4*SR*)-*form*, 2-Selenide, *in* B-00541
O,O-1,3-Butylene *tert*-butylphosphoramidoselenoate, *in* B-00541
O,O-1,3-Butylene dimethylphosphoramidoselenoate, *in* M-00117
O,O-1,3-Butylene phenylphosphoramidoselenoate, *in* M-00193
O-Butyl tetraethylphosphorodiamidoselenoate, *in* T-00061
Se-Butyl tetraethylphosphorodiamidoselenoate, *in* T-00061
O,O-Diethyl diethylphosphoramidoselenoate, *in* D-00353
O,O-Diethyl diisopropylphosphoramidoselenoate, *in* D-00655
O,O-Diethyl dimethylphosphoramidoselenoate, *in* D-00869
O,O-Diethyl methylphosphoramidoselenoate, *in* M-00332
O,O-Diethyl phenylphosphoramidoselenoate, *in* P-00261
Diethylphosphoramidoselenoic acid; Difluoride, *in* D-00353
1,5-Dihydro-3-dimethylamino-2,4,3-benzodioxaphosphepin; 3-Selenide, *in* D-00453
O,O-Diisopropyl diethylphosphoramidoselenoate, *in* D-00353
O,O-Diisopropyl methylphosphoramidoselenoate, *in* M-00332
O,O-Dimethyl diethylphosphoramidoselenoate, *in* D-00353
O,O-Dimethyl dimethylphosphoramidoselenoate, *in* D-00869
Dimethylphosphoramidoselenoic acid; Difluoride, *in* D-00869
O,O-Diphenyl phosphoramidoselenoate, *in* D-01139
O-Ethyl diethylphosphoramidochloridoselenoate, *in* D-00347
Ethylene dimethylphosphoramidoselenoate, *in* D-00695
O-Ethyl tetraethylphosphorodiamidoselenoate, *in* T-00061
Se-Ethyl tetraethylphosphorodiamidoselenoate, *in* T-00061
Hexaethylphosphoroselenoic triamide, *in* H-00015
Hexahydro-2*a*,4*a*,6*a*-triaza-6*b*-phosphacyclopenta[*cd*]-pentalene; 6*b*-Selenide, *in* H-00055
2,3,5,6,7,8-Hexamethyl-2,3,5,6,7,8-hexaaza-1,4-diphosphabicyclo[2.2.2]octane; 1,4-Diselenide, *in* H-00070
Hexamethylphosphoroselenoic triamide, H-00078
4-Methyl-2-dimethylamino-1,3,2-dioxaphosphorinane; (2*RS*,4*RS*)-*form*, 2-Selenide, *in* M-00117
4-Methyl-2-phenylamino-1,3,2-dioxaphosphorinane; (2*RS*,4*SR*)-*form*, 2-Selenide, *in* M-00193
O-Methyl tetraethylphosphorodiamidoselenoate, *in* T-00061
O-Methyl tetramethylphosphorodiamidoselenoate, *in* T-00223
Neopentylene dimethylphosphoramidoselenoate, *in* D-00693
Neopentylene phenylphosphoramidoselenoate, *in* D-00792
O-Phenyl diethylphosphoramidochloridoselenoate, *in* D-00347
O-Phenyl tetraethylphosphorodiamidoselenoate, *in* T-00061
4,4′,4″-Phosphinoselenoylidynetrismorpholine, *in* T-00569
▷1,1′,1″-Phosphinoselenoylidynetrispiperidine, *in* T-00704
1,1′,1″-Phosphinoselenoylidynetrispyrrolidine, *in* T-00724
O,O-Propylene dimethylphosphoramidoselenoate, *in* M-00116
Tetraethylphosphorodiamidoselenoic acid; Chloride, *in* T-00061
Tetramethylphosphorodiamidoselenoic acid; Chloride, *in* T-00223
2,4,6,7-Tetramethyl-2,6,7-triaza-1-phosphabicyclo[2.2.2]-octane; 1-Selenide, *in* T-00237

Tellurophosphoric acids

O,O-Diethyl phosphorotelluroate, D-00395

Tellurophosphoric esters

O-Ethyl tetraethylphosphorodiamidotelluroate, *in* T-00062

Tellurophosphoric amides

O-Ethyl tetraethylphosphorodiamidotelluroate, *in* T-00062
Hexaethylphosphorous triamide; Telluride, *in* H-00015
4,4′,4″-Phosphinotelluroylidynetrismorpholine, *in* T-00569
1,1′,1″-Phosphinotelluroylidynetrispiperidine, *in* T-00704
1,1′,1″-Phosphinotelluroylidynetrispyrrolidine, *in* T-00724

Imidophosphoric halides

Benzoylphosphorimidic trichloride, B-00040
P,P-Bis(diethylamino)-*N*-ethyl-*P*-fluorophosphine imide, *in* P-00012
P,P-Bis(diethylamino)-*P*-fluoro-*N*-phenylphosphazene, *in* T-00054
P,P-Bis(diethylamino)-*P*-iodo-*N*-phenylphosphazene, *in* T-00054
P,P-Bis(dipropylamino)-*P*-fluoro-*N*-propylphosphazene, *in* P-00043
[Bis(trichloromethyl)phosphinyl]phosphorimidic trichloride, B-00404
P-Bromo-*P,P*-bis(diethylamino)-*N*-ethylphosphine imide, *in* P-00012
P-Bromo-*P,P*-bis(dipropylamino)-*N*-propylphosphazene, *in* P-00043
P-Bromo-*N*-butyl-*P,P*-bis(dibutylamino)phosphazene, *in* P-00004
N-Butyl-*P,P*-bis(dibutylamino)-*P*-fluorophosphazene, *in* P-00004

N-Butyl-*P*-chloro-*P*,*P*-bis(dibutylamino)phosphazene, *in* P-00004
tert-Butylphosphorimidic trichloride, B-00626
Carbonylbis[phosphorimidic trichloride], C-00007
P-Chloro-*P*,*P*-bis(diethylamino)-*N*-ethylphosphine imide, *in* P-00012
P-Chloro-*P*,*P*-bis(diethylamino)-*N*-phenylphosphazene, *in* T-00054
P-Chloro-*P*,*P*-bis(dipropylamino)-*N*-propylphosphazene, *in* P-00043
P,*P*-Dibromo-*N*-trimethylsilyl-*P*-[bis(trimethylsilyl)amino]-phosphazene, *in* T-00868
P,*P*-Dichloro-*P*-diethylamino-*N*-phenylphosphine imide, *in* D-00313
P,*P*-Dichloro-*N*-trimethylsilyl-*P*-[bis(trimethylsilyl)amino]-phosphazene, *in* T-00868
P,*P*-Difluoro-*N*-trimethylsilyl-*P*-[bis(trimethylsilyl)amino]-phosphazene, *in* T-00868
2,2-Dihydro-1,3,2-benzoxazaphosphole; 2,2-Dichloro, *in* D-00444
P,*P*-Diiodo-*N*-trimethylsilyl-*P*-[bis(trimethylsilyl)amino]-phosphazene, *in* T-00868
Ethyl (chlorosulfonyl)phosphorodichlorimidate, E-00056
Ethyl (dichlorophosphinyl)phosphorodichloroimidate, E-00058
(Pentachloroethyl)phosphorimidic trichloride, P-00005
(Pentafluoroethyl)phosphorimidic trichloride, P-00013
Phenyl (dichlorophosphinothioyl)phosphorodichlorimidate, P-00112
Phenylsulfonylphosphorimidic trichloride, P-00316
P,*P*,*P*-Tribromo-*N*-*tert*-butylphosphazene, *in* B-00625
P,*P*,*P*-Tribromo-*N*-phenylsulfonylphosphazene, *in* P-00315
P,*P*,*P*-Tribromo-*N*-*p*-toluenesulfonylphosphazene, *in* M-00268
(Trichloroacetyl)phosphorimidic trichloride, T-00398
P,*P*,*P*-Trichloro-*N*-(diphenoxyphosphinylthioyl)phosphine imide, *in* D-00987
P,*P*,*P*-Trichloro-*N*-methylphosphazene, *in* M-00336
(Trichloromethyl)phosphorimidic trichloride, T-00418
P,*P*,*P*-Trichloro-*N*-methylsulfonylphosphazene, *in* M-00402
P,*P*,*P*-Trichloro-*N*-(2-nitrophenyl)phosphazene, *in* N-00061
P,*P*,*P*-Trichloro-*N*-(3-nitrophenyl)phosphazene, *in* N-00062
P,*P*,*P*-Trichloro-*N*-(4-nitrophenyl)phosphine imide, *in* N-00063
P,*P*,*P*-Trichloro-*N*-(pentafluorophenyl)phosphazene, *in* P-00016
P,*P*,*P*-Trichloro-*N*-phenylphosphine imide, *in* P-00264
P,*P*,*P*-Trichloro-*N*-*p*-toluenesulfonylphosphazene, *in* M-00268
P,*P*,*P*-Trichloro-*N*-trimethylsilylphosphazene, *in* T-00557
N,*N*,*N*′-Triethylphosphoramidimidic dichloride, T-00449
P,*P*,*P*-Trifluoro-*N*-methylphosphazene, *in* M-00336

Imidophosphoric esters

Butyl *N*″-[bis(diethylamino)phosphinothioyl]-*N*,*N*,*N*′,*N*′-tetraethylphosphorodiamidoimidate, *in* B-00174
Butyl *N*″-[bis(diethylamino)phosphinyl]-*N*,*N*,*N*′,*N*′-tetraethylphosphorodiamidoimidate, *in* B-00177
Diethyl *N*,*N*-diethyl-*N*′-phenylphosphoramidimidate, *in* D-00313
2,2-Dihydro-1,3,2-benzoxazaphosphole; 2,2-Dimethoxy, *in* D-00444
2,2-Dihydro-1,3,2-benzoxazaphosphole; 2,2-Diethoxy, *in* D-00444
2,2-Dihydro-1,3,2-benzoxazaphosphole; 2,2-Diphenoxy, *in* D-00444
Dimethyl *N*,*N*,*N*′-tris(trimethylsilyl)phosphoramidimidate, *in* T-00868
Ethyl *N*″-[bis(diethylamino)phosphinothioyl]-*N*,*N*,*N*′,*N*′-tetraethylphosphorodiamidoimidate, *in* B-00174
Ethyl *N*″-[bis(diethylamino)phosphinyl]-*N*,*N*,*N*′,*N*′-tetraethylphosphorodiamidoimidate, *in* B-00177
Ethyl (chlorosulfonyl)phosphorodichlorimidate, E-00056
Ethyl (dichlorophosphinyl)phosphorodichloroimidate, E-00058
Phenyl (dichlorophosphinothioyl)phosphorodichlorimidate, P-00112
Phenyl *N*,*N*,*N*′,*N*′-tetraethyl-*N*″-phenylphosphorodiamidimidate, *in* T-00054
Tributyl [bis(dimethylamino)phosphinothioyl]-phosphorimidate, *in* B-00192
Tributyl (dimethoxyphosphinothioyl)phosphorimidate, *in* D-00674
Triethyl (azidosulfonyl)phosphorimidate, T-00439
Triethyl benzoylphosphorimidate, *in* B-00039
Triethyl [bis(diethylamino)phosphinyl]phosphorimidate, *in* B-00176
Triethyl [bis(dimethylamino)phosphinothioyl]-phosphorimidate, *in* B-00192
Triethyl [bis(dimethylamino)phosphinyl]phosphorimidate, *in* B-00194
▷Triethyl (diethoxyphosphinothioyl)phosphorimidate, *in* D-00238
Triethyl (diethoxyphosphinyl)phosphorimidate, *in* D-00257
Triethyl (dimethoxyphosphinothioyl)phosphorimidate, *in* D-00674
Triethyl (dimethoxyphosphinyl)phosphorimidate, *in* D-00683
Triethyl (diphenoxyphosphinothioyl)phosphorimidate, *in* D-00987
Triethyl ethylphosphorimidate, *in* E-00161
Triethyl [(4-methylphenyl)sulfonyl]phosphorimidate, *in* M-00268
Triethyl methylphosphorimidate, *in* M-00336
Triethyl methylsulfonylphosphorimidate, *in* M-00402
Triethyl (2-nitrophenyl)phosphorimidate, *in* N-00061
Triethyl (4-nitrophenyl)phosphorimidate, *in* N-00063
Triethyl pentafluorophenylphosphorimidate, *in* P-00016
Triethyl phenylphosphorimidate, *in* P-00264
Triethyl phenylsulfonylphosphorimidate, *in* P-00315
Triethyl (trimethylsilyl)phosphorimidate, *in* T-00557
Triethyl [[(triphenylphosphoranylidene)amino]sulfonyl]-phosphorimidate, *in* T-00642
Triisopropyl benzoylphosphorimidate, *in* B-00039
Triisopropyl phenylphosphorimidate, *in* P-00264
Triisopropyl (trimethylsilyl)phosphorimidate, *in* T-00557
Trimethyl benzoylphosphorimidate, *in* B-00039
Trimethyl [bis(diethylamino)phosphinyl]phosphorimidate, *in* B-00176
Trimethyl *tert*-butylphosphorimidate, *in* B-00625
Trimethyl cyanophosphorimidate, *in* C-00228
Trimethyl (diethoxyphosphinothioyl)phosphorimidate, *in* D-00238
Trimethyl (diethoxyphosphinyl)phosphorimidate, *in* D-00257
Trimethyl (dimethoxyphosphinyl)phosphorimidate, *in* D-00683
Trimethyl (diphenoxyphosphinothioyl)phosphorimidate, *in* D-00987
Trimethyl ethylphosphorimidate, *in* E-00161
Trimethyl [(4-methylphenyl)sulfonyl]phosphorimidate, *in* M-00268
Trimethyl methylphosphorimidate, *in* M-00336
Trimethyl methylsulfonylphosphorimidate, *in* M-00402
Trimethyl (4-nitrophenyl)phosphorimidate, *in* N-00063
Trimethyl phenylphosphorimidate, *in* P-00264
Trimethyl phenysulfonylphosphorimidate, *in* P-00315
Trimethyl (trimethylsilyl)phosphorimidate, *in* T-00557
Trimethyl [[(triphenylphosphoranylidene)amino]sulfonyl]-phosphorimidate, *in* T-00642
Triphenyl benzoylphosphorimidate, *in* B-00039
Triphenyl [bis(dimethylamino)phosphinyl]phosphorimidate, *in* B-00194
Triphenyl cyanophosphorimidate, *in* C-00228
Triphenyl (diethoxyphosphinothioyl)phosphorimidate, *in* D-00238
Triphenyl (dimethoxyphosphinothioyl)phosphorimidate, *in* D-00674
Triphenyl (diphenoxyphosphinothioyl)phosphorimidate, *in* D-00987
Triphenyl ethylphosphorimidate, *in* E-00161
Triphenyl [(4-methylphenyl)sulfonyl]phosphorimidate, *in* M-00268
Triphenyl methylsulfonylphosphorimidate, *in* M-00402
Triphenyl (2-nitrophenyl)phosphorimidate, *in* N-00061
Triphenyl (3-nitrophenyl)phosphorimidate, *in* N-00062
Triphenyl (4-nitrophenyl)phosphorimidate, *in* N-00063
Triphenyl phenylphosphorimidate, *in* P-00264
Triphenyl phenylsulfonylphosphorimidate, *in* P-00315
Triphenyl [[(triphenylphosphoranylidene)amino]sulfonyl]-phosphorimidate, *in* T-00642
Tris(2-methylpropyl) (dimethoxyphosphinyl)phosphorimidate, *in* D-00683

Imidophosphoric amides

P,*P*-Bis(diethylamino)-*N*-ethyl-*P*-fluorophosphine imide, *in* P-00012
P,*P*-Bis(diethylamino)-*P*-fluoro-*N*-phenylphosphazene, *in* T-00054
P,*P*-Bis(diethylamino)-*P*-iodo-*N*-phenylphosphazene, *in* T-00054
P,*P*-Bis(dipropylamino)-*P*-fluoro-*N*-propylphosphazene, *in* P-00043
P-Bromo-*P*,*P*-bis(diethylamino)-*N*-ethylphosphine imide, *in* P-00012
P-Bromo-*P*,*P*-bis(dipropylamino)-*N*-propylphosphazene, *in* P-00043
P-Bromo-*N*-butyl-*P*,*P*-bis(dibutylamino)phosphazene, *in* P-00004
N-Butyl-*P*,*P*-bis(dibutylamino)-*P*-fluorophosphazene, *in* P-00004
Butyl *N*″-[bis(diethylamino)phosphinothioyl]-*N*,*N*,*N*′,*N*′-tetraethylphosphorodiamidoimidate, *in* B-00174
Butyl *N*″-[bis(diethylamino)phosphinyl]-*N*,*N*,*N*′,*N*′-tetraethylphosphorodiamidoimidate, *in* B-00177
N-Butyl-*P*-chloro-*P*,*P*-bis(dibutylamino)phosphazene, *in* P-00004
P-Chloro-*P*,*P*-bis(diethylamino)-*N*-ethylphosphine imide, *in* P-00012
P-Chloro-*P*,*P*-bis(diethylamino)-*N*-phenylphosphazene, *in* T-00054
P-Chloro-*P*,*P*-bis(dipropylamino)-*N*-propylphosphazene, *in* P-00043
P,*P*-Dibromo-*N*-trimethylsilyl-*P*-[bis(trimethylsilyl)amino]-phosphazene, *in* T-00868
P,*P*-Dichloro-*P*-diethylamino-*N*-phenylphosphine imide, *in* D-00313
P,*P*-Dichloro-*N*-trimethylsilyl-*P*-[bis(trimethylsilyl)amino]-phosphazene, *in* T-00868
Diethyl *N*,*N*-diethyl-*N*′-phenylphosphoramidimidate, *in* D-00313
P,*P*-Difluoro-*N*-trimethylsilyl-*P*-[bis(trimethylsilyl)amino]-phosphazene, *in* T-00868
2,2-Dihydro-1,3,2-benzoxazaphosphole; 2,2-Bis(dimethylamino), *in* D-00444
P,*P*-Diiodo-*N*-trimethylsilyl-*P*-[bis(trimethylsilyl)amino]-phosphazene, *in* T-00868
Dimethyl *N*,*N*,*N*′-tris(trimethylsilyl)phosphoramidimidate, *in* T-00868
Ethyl *N*″-[bis(diethylamino)phosphinothioyl]-*N*,*N*,*N*′,*N*′-tetraethylphosphorodiamidoimidate, *in* B-00174
Ethyl *N*″-[bis(diethylamino)phosphinyl]-*N*,*N*,*N*′,*N*′-tetraethylphosphorodiamidoimidate, *in* B-00177
N,*N*,*N*′,*N*′,*N*″,*N*″,*N*‴-Heptamethylphosphorimidic triamide, H-00001
N,*N*,*N*′,*N*′,*N*″,*N*″-Hexamethyl-*N*‴-diphenylphosphinophosphorimidic triamide, H-00069
N,*N*,*N*′,*N*′,*N*″,*N*″-Hexamethyl-*N*‴-diphenylphosphinophosphorimidic triamide; Oxide, *in* H-00069
N,*N*,*N*′,*N*′,*N*″,*N*″-Hexamethyl-*N*‴-phenylphosphorimidic triamide, H-00073
N,*N*,*N*′,*N*′,*N*″,*N*″-Hexamethylphosphorimidic triamide, H-00077
N,*N*,*N*′,*N*′,*N*″,*N*″-Hexamethylphosphorimidic triamide; *N*‴-Chloro, *in* H-00077
N,*N*,*N*′,*N*′,*N*″,*N*″-Hexamethyl-*N*‴-trimethyl-silylphosphorimidic triamide, H-00082
Phenyl *N*,*N*,*N*′,*N*′-tetraethyl-*N*″-phenylphosphorodiamidimidate, *in* T-00054
N,*N*,*N*′-Triethylphosphoramidimidic dichloride, T-00449
N,*N*′,*N*″-Triphenyl (diphenoxyphosphinothioyl)-phosphorotriamidoimidate, *in* D-00987
3,3,3-Tris-(4-morpholinophosphoranylidene)-1-(2,4,6-trinitrophenyl)triazene, T-00833

Phosphonic acids, $(HO)_2P(O)H$

(2-Acetoxyethyl)phosphonic acid, A-00006
(2-Acetoxypropyl)phosphonic acid, A-00009
Acetylphosphonic acid, A-00026
9-Acridinephosphonic acid, A-00030
1-Adamantylphosphonic acid, A-00034
2-Adamantylphosphonic acid, A-00035
Alafosfalin; $(R)_C(S)_P$-*form*, *in* A-00049
Alafosfalin; $(R)_C(R)_P$-*form*, *in* A-00049
Alafosfalin; $(S)_C(S)_P$-*form*, *in* A-00049
Alafosfalin; $(S)_C,(R)_P$-*form*, *in* A-00049
Alafosfalin; $(S)_C,(RS)_P$-*form*, *in* A-00049
Alafosfalin; $(R)_C,(\pm)_P$-*form*, *in* A-00049
(Aminobenzylidene)bisphosphonic acid, A-00053
(1-Aminobutyl)phosphonic acid, A-00054
(2-Aminobutyl)phosphonic acid, A-00055
(3-Aminobutyl)phosphonic acid, A-00056
(4-Aminobutyl)phosphonic acid, A-00057
(Aminocarbonylmethyl)phosphonic acid, A-00060
(10-Aminodecyl)phosphonic acid, A-00064
1-Amino-1,1-ethanediphosphonic acid, A-00073
(1-Aminoethyl)phosphonic acid; (*R*)-*form*, *in* A-00078
(1-Aminoethyl)phosphonic acid; (±)-*form*, *in* A-00078
2-Aminoethylphosphonic acid, A-00079
(1-Amino-2-hydroxyethyl)phosphonic acid; (*R*)-*form*, *in* A-00083
(1-Amino-2-hydroxyethyl)phosphonic acid; (*S*)-*form*, *in* A-00083
(1-Amino-2-hydroxyethyl)phosphonic acid; (±)-*form*, *in* A-00083
(1-Amino-2-mercaptoethyl)phosphonic acid; (±)-*form*, *in* A-00084
(Aminomethylene)bisphosphonic acid, A-00086
(1-Amino-1-methylethyl)phosphonic acid, A-00087
(2-Amino-1-methylethyl)phosphonic acid, A-00088
(Aminomethyl)phosphonic acid, A-00091
(1-Amino-1-methylpropyl)phosphonic acid; (+)-*form*, *in* A-00093
(1-Amino-1-methylpropyl)phosphonic acid; (±)-*form*, *in* A-00093
(1-Amino-2-methylpropyl)phosphonic acid; (*R*)-*form*, *in* A-00094
(1-Amino-2-methylpropyl)phosphonic acid; (*S*)-*form*, *in* A-00094
(1-Amino-2-methylpropyl)phosphonic acid; (±)-*form*, *in* A-00094
(1-Amino-3-methylthiopropyl)phosphonic acid; (±)-*form*, *in* A-00095
(1-Amino-3-methylthiopropyl)phosphonic acid; (±)-*form*, *S*-Oxide, *in* A-00095
(1-Amino-3-methylthiopropyl)phosphonic acid; (±)-*form*, *S*,*S*-Dioxide, *in* A-00095
(1-Aminopentyl)phosphonic acid, A-00096
(5-Aminopentyl)phosphonic acid, A-00097
(1-Amino-1-phenylethyl)phosphonic acid, A-00100
(1-Amino-2-phenylethyl)phosphonic acid, A-00101
(Aminophenylmethyl)phosphonic acid; (*R*)-*form*, *in* A-00105
(Aminophenylmethyl)phosphonic acid; (*S*)-*form*, *in* A-00105
(Aminophenylmethyl)phosphonic acid; (±)-*form*, *in* A-00105
[(4-Aminophenyl)methyl]phosphonic acid, A-00107
▷(2-Aminophenyl)phosphonic acid, A-00109
▷(3-Aminophenyl)phosphonic acid, A-00110
▷(4-Aminophenyl)phosphonic acid, A-00111
3-Amino-3-phosphonobutanamide, *in* A-00118
2-Amino-3-phosphonobutanoic acid, A-00116
2-Amino-4-phosphonobutanoic acid; (*R*)-*form*, *in* A-00117
2-Amino-4-phosphonobutanoic acid; (*S*)-*form*, *in* A-00117
3-Amino-3-phosphonobutanoic acid, A-00118
4-Amino-4-phosphonobutanoic acid, A-00119
(3-Amino-3-phosphono)propanoamide, *in* A-00122
2-Amino-3-phosphonopropanoic acid, A-00121
2-Amino-3-phosphonopropanoic acid; (±)-*form*, *N*-Benzoyl, *in* A-00121
3-Amino-3-phosphonopropanoic acid; (*R*)-*form*, *in* A-00122
3-Amino-3-phosphonopropanoic acid; (±)-*form*, *in* A-00122
3-Amino-3-phosphonopropanoic acid; (±)-*form*, Me ester, *in* A-00122
3-Amino-3-phosphonopropanoic acid; (±)-*form*, Amide, *in* A-00122
(1-Aminopropylidene)bisphosphonic acid, A-00124
(1-Aminopropyl)phosphonic acid, A-00125
(2-Aminopropyl)phosphonic acid; (+)-*form*, *in* A-00126
(2-Aminopropyl)phosphonic acid; (±)-*form*, *in* A-00126
(3-Aminopropyl)phosphonic acid, A-00127

(2-Amino-4-pyrimidinyl)phosphonic acid, A-00128
Antibiotic FR 33289, A-00138
1,2-Benzenediphosphonic acid, B-00006
1,4-Benzenediphosphonic acid, B-00008
5-Benzimidazolephosphonic acid, B-00016
(1,3-Benzodioxol-5-yl)phosphonic acid, B-00019
6-Benzoxazolephosphonic acid, B-00027
(1-Benzoylaminobutyl)phosphonic acid, *in* A-00054
[1-(Benzoylamino)ethyl]phosphonic acid, *in* A-00078
(1-Benzoylamino-1-methylethyl)phosphonic acid, *in* A-00087
(*N*-Benzoylaminomethyl)phosphonic acid, *in* A-00091
(1-Benzoylamino-2-methylpropyl)phosphonic acid, *in* A-00094
(1-Benzoylaminopentyl)phosphonic acid, *in* A-00096
[(Benzoylamino)phenylmethyl]phosphonic acid, *in* A-00105
2-Benzoylamino-4-phosphonobutanoic acid, *in* A-00117
[1-(*N*-Benzoylamino)propyl]phosphonic acid, *in* A-00125
Benzoylphosphonic acid, B-00034
[1-(Benzylamino)butyl]phosphonic acid, *in* A-00054
[(1-Benzylamino)ethyl]phosphonic acid, *in* A-00078
(1-Benzylamino-1-methylethyl)phosphonic acid, *in* A-00087
[1-(Benzylamino)pentyl]phosphonic acid, *in* A-00096
[1-(Benzylamino)propyl]phosphonic acid, *in* A-00125
1-Benzylhexahydro-2-hydroxy-1,2-azaphosphorine 2-oxide, B-00048
[2-(Benzyloxy)phenyl]phosphonic acid, *in* H-00179
Benzylphosphonic acid, B-00060
4-Biphenylphosphonic acid, B-00077
[(1,1′-Biphenyl)-2-ylmethyl]phosphonic acid, B-00079
[(1,1′-Biphenyl)-4-ylmethyl]phosphonic acid, B-00080
Bis[(dihydroxyphosphinyl)methyl]phosphinic acid, B-00182
[Bis(phenylmethoxy)phosphinyl]acetic acid, B-00389
1,4-Bis(phosphonomethyl)biphenyl, B-00391
(4-Bromobutyl)phosphonic acid, B-00466
(2-Bromoethyl)phosphonic acid, B-00472
(Bromomethyl)phosphonic acid, B-00475
(Bromophenylmethyl)phosphonic acid, B-00482
[(4-Bromophenyl)methyl]phosphonic acid, B-00484
(2-Bromophenyl)phosphonic acid, B-00490
(3-Bromophenyl)phosphonic acid, B-00491
▷(4-Bromophenyl)phosphonic acid, B-00492
(3-Bromopropyl)phosphonic acid, B-00501
1,4-Butanediphosphonic acid, B-00514
1-Butenylphosphonic acid, B-00524
Butylphosphonic acid, B-00594
tert-Butylphosphonic acid, B-00595
Carbamoylphosphonic acid, C-00003
(Chloroacetyl)phosphonic acid, C-00020
P-(4-Chlorobutyl)-*N*-phenylphosphonamidic acid, C-00024
(4-Chlorobutyl)phosphonic acid, C-00025
(4-Chlorocarbonyl)phenylphosphonic dichloride, *in* P-00380
Chloro(diethoxyphosphinyl)acetic acid, C-00036
▷(2-Chloroethyl)phosphonic acid, C-00074
(Chloromethyl)phosphonic acid, C-00106
(Chloromethyl)phosphonofluoridic acid, C-00112
[(2-Chlorophenyl)methyl]phosphonic acid, C-00132
[(4-Chlorophenyl)methyl]phosphonic acid, C-00134
▷(2-Chlorophenyl)phosphonic acid, C-00145
▷(3-Chlorophenyl)phosphonic acid, C-00146
▷(4-Chlorophenyl)phosphonic acid, C-00147
(3-Chloropropyl)phosphonic acid, C-00172
(Cyanomethyl)phosphonic acid, C-00215
1,2-Cyclohexanediphosphonic acid, C-00240
3-Cyclohexenephosphonic acid, C-00242
Cyclohexyl hydrogen methylphosphonate, *in* M-00288
Cyclohexylphosphonic acid, C-00256
Cyclopentanephosphonic acid, C-00276
1,10-Decanediphosphonic acid, D-00008
▷Decylphosphonic acid, D-00012
(2,5-Dibromophenyl)phosphonic acid, D-00066
(1,2-Dibromo-1-phenylpropyl)phosphonic acid; (1*RS*,2*RS*)-*form*, *in* D-00068
(1,2-Dibromo-1-phenylpropyl)phosphonic acid; (1*RS*,2*SR*)-*form*, *in* D-00068
(1,2-Dibromo-1-phenylpropyl)phosphonic acid; (1*RS*,2*SR*)-*form*, Mono-Me ester, *in* D-00068
3,5-Di-*tert*-butyl-2,5-dihydro-1,2-oxaphosphole 2-oxide; 2-Hydroxy, *in* D-00082
(3,5-Di-*tert*-butyl-4-hydroxybenzyl)phosphonic acid, D-00094
N,*N*-Dibutylphosphonamidic acid, D-00126
Dichloromethylenebisphosphonic acid, D-00172
(Dichloromethyl)phosphonic acid, D-00175
(2,3-Dichlorophenyl)phosphonic acid, D-00184
(2,5-Dichlorophenyl)phosphonic acid, D-00185
(3,4-Dichlorophenyl)phosphonic acid, D-00186
(3,5-Dichlorophenyl)phosphonic acid, D-00187
Diethoxyphosphinecarboxylic acid oxide, D-00236
4-(Diethoxyphosphinyl)-2-butenoic acid, D-00248
N,*N*-Diethyl-*P*-phenylphosphonamidic acid, D-00307
Diethyl phosphonobutanedioate, *in* P-00381
1,3-Dihydro-1,3-dihydroxy-1,2,3-benziodoxaphosphole 3-oxide, *in* D-00488
2,5-Dihydro-5,5-dimethyl-1,2-oxaphosphole 2-oxide; 2-Hydroxy, *in* D-00464
1,3-Dihydro-1-hydroxy-2,1-benzoxaphosphole 1-oxide, D-00490
2,5-Dihydro-2-hydroxy-1,2-oxaphosphole 2-oxide, D-00500
(1,3-Dihydro-3-oxo-1-isobenzofuranyl)phosphonic acid, D-00535
2,5-Dihydroxy-3,6-dimethyl-3,6-bis(dihydroxyphosphinyl)-1,4,2,5-dioxadiphosphorinane 2,5-dioxide, D-00617
Dimethylamine-1,1′-diphosphonic acid, D-00688
(Dimethylaminocarbonyl)phosphonic acid, D-00691
[(Dimethylamino)methylene]bisphosphonic acid, D-00701
(4-Dimethylaminophenyl)phosphonic acid, D-00706
Dimethyl 1-azetidinylphosphonate, *in* A-00147
Dimethyldiphosphonic acid, D-00751
(3,3-Dimethyl-2-oxobutyl)phosphonic acid, D-00783
N,*N*-Dimethyl-*P*-phenylphosphonamidic acid, D-00808
(2,4-Dimethylphenyl)phosphonic acid, D-00812
(2,5-Dimethylphenyl)phosphonic acid, D-00813
(3,4-Dimethylphenyl)phosphonic acid, D-00815
(1,1-Dimethylpropyl)phosphonic acid, D-00911
(2,2-Dimethylpropyl)phosphonic acid, D-00912
(Diphenoxyphosphinyl)acetic acid, D-00988
Diphenyl (1-cyanoethyl)phosphonate, *in* C-00210
N,*P*-Diphenylphosphonamidic acid, D-01103
2,3-Diphosphonobutanedioic acid, D-01182
4,4-Diphosphonobutanoic acid, D-01183
(Diphosphonomethyl)butanedioic acid, D-01185
3,3-Diphosphonopropanoic acid, D-01187
(Dipropoxyphosphinyl)acetamide, *in* D-01199
1,2-Ethanediphosphonic acid, E-00006
[1,2-Ethanediylbis[nitrilobis[methylene]]]-tetrakisphosphonic acid, E-00014
Ethenylidenebisphosphonic acid, E-00020
Ethyl bis(trimethylsilyloxy)phosphinecarboxylate oxide, *in* B-00433
Ethyl bromo(diethoxyphosphinyl)acetate, *in* B-00467
Ethyl dihydroxyphosphinylacetate, *in* P-00376
Ethyl dimethoxyphosphinylformate, *in* D-00672
Ethyl (dipropoxyphosphinyl)acetate, *in* D-01199
Ethyl hydrogen (α-aminobenzyl)phosphonate, *in* A-00105
Ethyl hydrogen (1-amino-1-methylethyl)phosphonate, *in* A-00087
Ethyl hydrogen (1-amino-1-methylpropyl)phosphonate, *in* A-00093
Ethyl hydrogen (4-aminophenyl)phosphonate, *in* A-00111
Ethyl hydrogen benzylphosphonate, *in* B-00060
Ethyl hydrogen ethylphosphonate, *in* E-00129
Ethyl methylphosphonate, E-00089
Ethylphosphonic acid, E-00129
Ethyl 4-phosphonobenzoate, *in* P-00380
Ethylphosphonofluoridic acid, E-00142
Ethyl 3-phosphonopropanoate, *in* P-00396
[(Ethylthio)methyl]phosphonic acid, E-00192
(9*H*-Fluoren-9-yl)phosphonic acid, F-00011
(Fluoromethyl)phosphonic acid, F-00023
(2-Fluorophenyl)phosphonic acid, F-00036
(4-Fluorophenyl)phosphonic acid, F-00038
▷Fosfomycin, *in* M-00185
Fosmidomycin, F-00062
(2-Furanylmethyl)phosphonic acid, F-00072
2-Furanylphosphonic acid, F-00073
▷Glyphosate, G-00097
▷Glyphosine, G-00098

1,7-Heptanediphosphonic acid, H-00002
Heptylphosphonic acid, H-00006
3,3,4,4,5,5-Hexafluoro-1,2-cyclopentenediphosphonic acid, H-00018
1,6-Hexanediphosphonic acid, H-00085
Hexylphosphonic acid, H-00097
Hydrogen 4-nitrophenyl (dichloromethyl)phosphonate, *in* D-00175
Hydrogen propyl methylphosphonate, *in* M-00288
1-Hydroxy-1,1-butanediphosphonic acid, H-00113
(1-Hydroxybutyl)phosphonic acid, H-00114
(1-Hydroxyethylidene)bisphosphonic acid, H-00144
(1-Hydroxyethyl)phosphonic acid, H-00145
(2-Hydroxyethyl)phosphonic acid, H-00146
1-Hydroxy-1,1-hexanediphosphonic acid, H-00153
Hydroxymethanediphosphonic acid, H-00154
(1-Hydroxy-1-methylethyl)phosphonic acid, H-00159
(Hydroxymethyl)phosphonic acid, H-00162
2-Hydroxy-1,2-oxaphosphorinane 2-oxide, H-00167
(Hydroxyphenylmethyl)bisphosphonic acid, H-00172
(Hydroxyphenylmethyl)phosphonic acid, H-00174
[(2-Hydroxyphenyl)methyl]phosphonic acid, H-00175
[(4-Hydroxyphenyl)methyl]phosphonic acid, H-00176
(2-Hydroxyphenyl)phosphonic acid, H-00179
▷(3-Hydroxyphenyl)phosphonic acid, H-00180
▷(4-Hydroxyphenyl)phosphonic acid, H-00181
1-Hydroxy-1,1-propanediphosphonic acid, H-00189
(1-Hydroxypropyl)phosphonic acid, H-00190
(2-Hydroxypropyl)phosphonic acid, H-00191
(3-Hydroxypropyl)phosphonic acid, H-00192
1*H*-Inden-2-ylphosphonic acid, I-00007
(Iodomethyl)phosphonic acid, I-00013
(2-Iodophenyl)phosphonic acid, I-00016
(3-Iodophenyl)phosphonic acid, I-00017
(4-Iodophenyl)phosphonic acid, I-00018
(4-Iodosophenyl)phosphonic acid, *in* I-00018
Isopropyl methylphosphonate, I-00038
Isopropylphosphonic acid, I-00050
Methanediphosphonic acid, M-00022
[(4-Methoxyphenyl)methyl]phosphonic acid, M-00061
(2-Methoxyphenyl)phosphonic acid, M-00070
(3-Methoxyphenyl)phosphonic acid, M-00071
(4-Methoxyphenyl)phosphonic acid, M-00072
Methyl 3-amino-3-phosphonobutanoate, *in* A-00118
Methyl 4-amino-4-phosphonobutanoate, *in* A-00119
Methyl 3-amino-3-phosphonopropanoate, *in* A-00122
(2-Methylaminopropyl)phosphonic acid, *in* A-00126
Methyl bis(trimethylsilyloxy)phosphinecarboxylate oxide, *in* B-00433
Methyl bromo(diethoxyphosphinyl)acetate, *in* B-00467
(2-Methyl-1,3-butadienyl)phosphonic acid, M-00093
(3-Methyl-2-butenyl)phosphonic acid, M-00102
(1-Methylbutyl)phosphonic acid, M-00105
(3-Methylbutyl)phosphonic acid, M-00106
Methyl dihydroxyphosphinylacetate, *in* P-00376
Methyl hydrogen [[1-amino-2-(1*H*-indol-3-yl)]ethyl]-phosphonate, *in* P-00421
Methyl hydrogen (4-aminophenyl)phosphonate, *in* A-00111
Methyl hydrogen benzylphosphonate, *in* B-00060
Methyl hydrogen chloromethylphosphonate, *in* C-00106
Methyl hydrogen (1,2-dibromo-1-phenylpropyl)phosphonate, *in* D-00068
Methyl hydrogen (α-hydroxybenzyl)phosphonate, *in* H-00174
Methyl hydrogen [[(4-methylphenyl)sulfinyl]methyl]-phosphonate, *in* M-00266
Methyl hydrogen (2-phenylethenyl)phosphonate, *in* P-00155
Methyl methylphosphonate, M-00168
(3-Methyloxiranyl)phosphonic acid; (2*S*,3*R*)-*form*, *in* M-00185
(3-Methyloxiranyl)phosphonic acid; (2*RS*,3*SR*)-*form*, *in* M-00185
(3-Methyloxiranyl)phosphonic acid; (2*RS*,3*RS*)-*form*, *in* M-00185
[(4-Methylphenyl)methyl]phosphonic acid, M-00208
P-(4-Methylphenyl)-*N*-phenylphosphonamidic acid, M-00214
(2-Methylphenyl)phosphonic acid, M-00235
(3-Methylphenyl)phosphonic acid, M-00236
▷(4-Methylphenyl)phosphonic acid, M-00237
[[(4-Methylphenyl)sulfinyl]methyl]phosphonic acid; (*S*)-*form*, Mono-Me ester, *in* M-00266
▷Methylphosphonic acid, M-00288
▷Methylphosphonofluoridic acid, M-00303
(1-Methylpropyl)phosphonic acid, M-00377
(2-Methylpropyl)phosphonic acid, M-00378
(3-Methyl-2-pyridinyl)phosphonic acid, M-00389
(3-Methyl-4-pyridinyl)phosphonic acid, M-00390
(4-Methyl-2-pyridinyl)phosphonic acid, M-00391
(6-Methyl-2-pyridinyl)phosphonic acid, M-00392
[1-(Methylthio)ethyl]phosphonic acid, M-00412
(4-Methylthiophenyl)phosphonic acid, M-00418
Monobutyl ethylphosphonate, *in* E-00129
Monoethyl (aminocarbonyl)phosphonate, M-00469
Monoethyl butylphosphonate, *in* B-00594
Monoethyl chloromethylphosphonate, *in* C-00106
Monoethyl cyclohexylphosphonate, *in* C-00256
Monoethyl (1,1-dimethylethyl)phosphonate, *in* B-00595
Monoethyl (2-phenylethenyl)phosphonate, *in* P-00155
Monoisopropyl cyclohexylphosphonate, *in* C-00256
Monomethyl phenylphosphonate; (±)-*form*, *in* M-00488
Monophenyl cyclohexylphosphonate, *in* C-00256
Monophenyl (dichloromethyl)phosphonate, *in* D-00175
(4-Morpholinylmethylene)bisphosphonic acid, M-00520
(4-Morpholinylmethyl)phosphonic acid, M-00521
(1-Naphthylmethyl)phosphonic acid, N-00013
(2-Naphthylmethyl)phosphonic acid, N-00014
1-Naphthylphosphonic acid, N-00021
▷[Nitrilotris(methylene)]trisphosphonic acid, N-00033
[(2-Nitrophenyl)methyl]phosphonic acid, N-00044
[(3-Nitrophenyl)methyl]phosphonic acid, N-00045
[(4-Nitrophenyl)methyl]phosphonic acid, N-00046
(2-Nitrophenyl)phosphonic acid, N-00055
(3-Nitrophenyl)phosphonic acid, N-00056
▷(4-Nitrophenyl)phosphonic acid, N-00057
1,9-Nonanediphosphonic acid, N-00069
3,3,4,4,5,5,6,6-Octafluoro-1,2-cyclohexenediphosphonic acid, O-00003
Octylphosphonic acid, O-00042
(2-Oxoethyl)phosphonic acid, O-00059
(2-Oxoheptyl)phosphonic acid, O-00061
(2-Oxo-4-phenylbutyl)phosphonic acid, O-00065
(2-Oxo-2-phenylethyl)phosphonic acid, O-00066
1-Oxo-1,3-propanediphosphonic acid, O-00071
2-Oxo-1,3-propanediphosphonic acid, O-00072
(1-Oxopropyl)phosphonic acid, O-00074
(2-Oxopropyl)phosphonic acid, O-00075
(3-Oxopropyl)phosphonic acid, O-00076
2,2′-Oxybis[ethanephosphonic acid], O-00094
Oxybis[methanephosphonic acid], O-00095
Pentafluorophenylphosphonic acid, P-00015
1,1-Pentanediphosphonic acid, P-00031
1,5-Pentanediphosphonic acid, P-00032
Pentylphosphonic acid, P-00050
6-Phenanthridinephosphonic acid, P-00060
[(Phenylamino)methyl]phosphonic acid, P-00084
Phenyl(diphenoxyphosphinyl)acetate, *in* D-00988
[1,2-Phenylenebis(methylene)]bisphosphonic acid, P-00129
[1,4-Phenylenebis(methylene)]bisphosphonic acid, P-00131
(1-Phenylethenyl)phosphonic acid, P-00154
2-Phenylethenylphosphonic acid, P-00155
1-Phenylethylphosphonic acid, P-00164
(2-Phenylethyl)phosphonic acid, P-00165
Phenylethynylphosphonic acid, P-00166
Phenyl hydrogen (4-aminophenyl)phosphonate, *in* A-00111
Phenyl hydrogen phenylphosphonate, *in* P-00221
Phenyl methylphosphonate, P-00171
[Phenyl[(phenylmethyl)amino]methyl]phosphonic acid, P-00188
P-Phenylphosphonamidic acid, P-00218
▷Phenylphosphonic acid, P-00221
Phenylphosphonofluoridic acid, P-00239
Phenylphosphonoperoxoic acid, P-00242
(1-Phenyl-1-propenyl)phosphonic acid; (*E*)-*form*, *in* P-00301
(1-Phenyl-1-propenyl)phosphonic acid; (*E*)-*form*, Mono-Me ester, *in* P-00301
(1-Phenyl-1-propenyl)phosphonic acid; (*Z*)-*form*, *in* P-00301
(1-Phenyl-1-propenyl)phosphonic acid; (*Z*)-*form*, Mono-Me ester, *in* P-00301

(3-Phenylpropyl)phosphonic acid, P-00305
[(Phenylsulfonyl)methyl]phosphonic acid, P-00311
[(Phenylthio)methyl]phosphonic acid, P-00326
N-Phenyl-*P*-trichloromethylphosphonamidic acid, P-00328
[Phosphinylidynetris(methylene)]trisphosphonic acid, P-00360
Phosphonoacetic acid, P-00376
4-Phosphonobenzamide, *in* P-00380
2-Phosphonobenzoic acid, P-00378
3-Phosphonobenzoic acid, P-00379
4-Phosphonobenzoic acid, P-00380
2-Phosphono-1,2,4-butanetricarboxylic acid, P-00382
3-Phosphono-1,2,3-butanetricarboxylic acid, P-00383
4-Phosphono-1,2,3-butanetricarboxylic acid, P-00384
3-Phosphonobutanoic acid, P-00385
4-Phosphonobutanoic acid, P-00386
α-Phosphonophenylacetic acid, P-00393
2-Phosphono-2-(2-phosphonoethyl)butanedioic acid, P-00394
2-Phosphonopropanoic acid, P-00395
▷3-Phosphonopropanoic acid, P-00396
Phosphotryptophan, P-00421
N-Phthalimidomethylphosphonic acid, P-00426
(1-Piperidinylmethyl)phosphonic acid, P-00429
1,3-Propanediphosphonic acid, P-00444
2,2-Propanediphosphonic acid, P-00445
1,2,3-Propanetriphosphonic acid, P-00450
1-Propenylphosphonic acid; (*E*)-*form*, *in* P-00453
1-Propenylphosphonic acid; (*Z*)-*form*, *in* P-00453
2-Propenylphosphonic acid, P-00454
▷Propylphosphonic acid, P-00473
1-Propynylphosphonic acid, P-00494
2-Pyridinephosphonic acid, P-00503
3-Pyridinephosphonic acid, P-00504
4-Pyridinephosphonic acid, P-00505
(2-Pyridinylmethyl)phosphonic acid; *N*-Oxide, *in* P-00506
(4-Pyridinylmethyl)phosphonic acid, P-00507
2-Pyrimidinephosphonic acid, P-00515
4-Pyrimidinephosphonic acid, P-00516
2-Quinolinephosphonic acid, Q-00002
[(8-Quinolinyl)methyl]phosphonic acid, Q-00005
▷Sparfosic acid; (*S*)-*form*, *in* S-00006
(2,3,5,6-Tetrachloro-4-pyridinyl)phosphonic acid, T-00035
N,*N*′,*N*″,*N*‴-Tetrakis(dihydroxyphosphinylmethyl)-1,4,7,10-tetraazacyclododecane, T-00175
(2-Thienylmethyl)phosphonic acid, T-00304
Thiobis[methanephosphonic acid], T-00311
(Trichloromethyl)phosphonic acid, T-00411
Triethylamine-2,2′,2″-triphosphonic acid, T-00438
(4-Trifluoromethylphenyl)phosphonic acid, T-00473
(Trifluoromethyl)phosphonic acid, T-00478
(2,4,5-Trimethylphenyl)phosphonic acid, T-00526
Trimethyl phosphonoformate, *in* D-00672
Triphenylmethylphosphonic acid, T-00608
N,*N*′,*N*″-Tris(dihydroxyphosphinylmethyl)-1,4,7-triazacyclononane, T-00757
Vinylphosphonic acid, V-00003
(9*H*-Xanthen-9-yl)phosphonic acid, X-00001
(9*H*-Xanthen-9-yl)phosphonic acid; Octahydro, *in* X-00001

Phosphonic anhydrides

2,4-Bis(4-fluorophenyl)-1,3,2,4-dioxaphosphetane 2,4-dioxide, B-00275
P-(4-Chlorobutyl)-*N*-phenylphosphonamidic acid; Anhydride, *in* C-00024
Dicyclohexyldiphosphonic acid, D-00206
Diethyl dibutyldiphosphonate, *in* D-00092
Diethyl diethyldiphosphonate, *in* D-00290
Diethyl dimethyldiphosphonate, *in* D-00751
Diethyldiphosphonic acid, D-00290
Diethyl dipropyldiphosphonate, *in* D-01202
▷*P*,*P*′-Diethyl-*N*,*N*,*N*′,*N*′-tetramethyldiphosphonic diamide, *in* D-00291
2,5-Dihydro-2,3,4,5-tetraphenyl-1,2,5-oxadiphosphole 2,5-dioxide, D-00583
2,5-Dihydroxy-3-(1-methylpropyl)-1,3,2,5-oxazadiphospholidine 2,5-dioxide; 2,5-Diethoxy (Di-Et ester), *in* D-00623
2,5-Dihydroxy-3-(1-methylpropyl)-1,3,2,5-oxazadiphospholidine 2,5-dioxide; 2,5-Dipropoxy (Dipropyl ester), *in* D-00623
P,*P*′-Diisopropyl-*N*,*N*,*N*′,*N*′-tetramethyldiphosphonic diamide, D-00668
Dimethyl diethyldiphosphonate, *in* D-00290
Dimethyl dimethyldiphosphonate, *in* D-00751
Dimethyldiphosphonic acid, D-00751
▷*P*,*P*′-Dimethyl-*N*,*N*,*N*′,*N*′-tetramethyldiphosphonic diamide, *in* D-00751
Diphenyl diphenyldiphosphonate, *in* D-01014
Diphenyldiphosphonic acid, D-01014
▷Hexaethyldiphosphonic diamide, *in* D-00291
4-Morpholinylphosphonochloridic acid; Anhydride with diphenyl phosphate, *in* M-00524
Phenylphosphonic acid trimer trianhydride, *in* T-00694
Propylphosphonic acid trimer trianhydride, *in* T-00720
2,4,6-Trihydroxy-1,2,4,6-oxatriphosphorinane 2,4,6-trioxide, T-00494

Phosphonic halides

1-Adamantylphosphonic acid; Dichloride, *in* A-00034
2-Adamantylphosphonic acid; Dichloride, *in* A-00035
Benzyl ethylphosphonochloridate, *in* E-00135
Benzylphosphonic dichloride, B-00062
[(1,1′-Biphenyl)-2-ylmethyl]phosphonic acid; Dichloride, *in* B-00079
Boyd's chloride, *in* T-00608
(2-Bromoethyl)phosphonic acid; Dichloride, *in* B-00472
(2-Bromoethyl)phosphonic acid; Dibromide, *in* B-00472
(Bromomethyl)phosphonic acid; Dichloride, *in* B-00475
(Bromomethyl)phosphonic acid; Dibromide, *in* B-00475
(2-Bromophenyl)phosphonic acid; Dichloride, *in* B-00490
(3-Bromophenyl)phosphonic acid; Dichloride, *in* B-00491
(4-Bromophenyl)phosphonic acid; Dichloride, *in* B-00492
▷(3-Bromopropyl)phosphonic acid; Dichloride, *in* B-00501
(1,3-Butadienyl)phosphonic acid; Difluoride, *in* B-00510
(1,3-Butadienyl)phosphonic acid; Dichloride, *in* B-00510
1,4-Butanediphosphonic acid; Bisdichloride, *in* B-00514
3-Buten-1-ynylphosphonic acid; Dichloride, *in* B-00529
N-*tert*-Butyl-*P*-methylphosphonamidic acid; Chloride, *in* B-00572
Butyl phenylphosphonochloridate, *in* P-00229
Butylphosphonic acid; Difluoride, *in* B-00594
Butylphosphonic dichloride, B-00598
tert-Butylphosphonic dichloride, B-00599
tert-Butylphosphonic difluoride, B-00600
(4-Chlorobutyl)phosphonic acid; Dichloride, *in* C-00025
6-Chloro-6*H*-dibenzo[*c*,*e*][1,2]oxaphosphorin; 6-Oxide, *in* C-00033
2-Chloro-2,3-dihydro-4,5-dimethyl-1,2-oxaphosphole; 2-Oxide, *in* C-00040
2-Chloro-2,3-dihydro-3,3,5-trimethyl-1,2-oxaphosphole; 2-Oxide, *in* C-00054
(2-Chloroethyl)phosphonic dichloride, C-00076
(3-Chloroformylphenyl)phosphonic dichloride, *in* P-00379
(Chloromethyl)phosphonic dichloride, C-00108
(Chloromethyl)phosphonic difluoride, C-00109
(Chloromethyl)phosphonofluoridic acid, C-00112
2-Chloro-1,2-oxaphosphorinane 2-oxide, *in* H-00167
(2-Chlorophenyl)phosphonic acid; Dichloride, *in* C-00145
(3-Chlorophenyl)phosphonic acid; Dichloride, *in* C-00146
(4-Chlorophenyl)phosphonic difluoride, C-00149
(2-Chloropropyl)phosphonic acid; (±)-*form*, Dichloride, *in* C-00171
(3-Chloropropyl)phosphonic acid; Dichloride, *in* C-00172
1-Cyclohexenephosphonic acid; Dichloride, *in* C-00241
3-Cyclohexenephosphonic acid; (±)-*form*, Dichloride, *in* C-00242
Cyclohexyl benzylphosphonochloridate, *in* B-00063
Cyclohexyl benzylphosphonofluoridate, *in* B-00066
Cyclohexyl butylphosphonochloridate, *in* B-00603
Cyclohexyl butylphosphonofluoridate, *in* B-00608
Cyclohexyl phenylphosphonochloridate, *in* P-00229
Cyclohexyl phenylphosphonofluoridate, *in* P-00239
P-Cyclohexylphosphonamidic fluoride, C-00254

Cyclohexylphosphonic acid; Difluoride, *in* C-00256
Cyclohexylphosphonic dichloride, C-00258
Cyclohexyl propylphosphonochloridate, *in* P-00476
Cyclohexyl propylphosphonofluoridate, *in* P-00479
Cyclopentanephosphonic acid; Dichloride, *in* C-00276
1,10-Decanediphosphonic acid; Bis(dichloride), *in* D-00008
Decylphosphonic acid; Difluoride, *in* D-00012
Decylphosphonic acid; Dichloride, *in* D-00012
(Dibromomethyl)phosphonic acid; Difluoride, *in* D-00064
(Dibromomethyl)phosphonic acid; Dibromide, *in* D-00064
(2,5-Dibromophenyl)phosphonic acid; Dichloride, *in* D-00066
3,5-Di-*tert*-butyl-2,5-dihydro-1,2-oxaphosphole 2-oxide; 2-Chloro, *in* D-00082
3,5-Di-*tert*-butyl-2,5-dihydro-1,2-oxaphosphole 2-oxide; 2-Bromo, *in* D-00082
(Dichloromethyl)phosphonic dichloride, D-00177
(Dichloromethyl)phosphonic difluoride, D-00178
2,4-Dichlorophenyl 4-morpholinylphosphonochloridate, *in* M-00524
(Dichlorophosphinyl)acetic acid; Chloride, *in* D-00190
N,*N*-Diethyl-*P*-methylphosphonamidic acid; Chloride, *in* D-00296
N,*N*-Diethyl-*P*-phenylphosphonamidic acid; Fluoride, *in* D-00307
N,*N*-Diethyl-*P*-phenylphosphonamidic acid; Chloride, *in* D-00307
N,*N*-Diethyl-*P*-propylphosphonamidic acid; Fluoride, *in* D-00401
N,*N*-Diethyl-*P*-propylphosphonamidic acid; Chloride, *in* D-00401
2,5-Dihydro-5,5-dimethyl-1,2-oxaphosphole 2-oxide; 2-Chloro, *in* D-00464
2,3-Dihydro-5-methyl-1,2-oxaphosphole 2-oxide; 2-Chloro, *in* D-00525
1,4-Dimethyl-3,5-diphenyl-1,4,2-diazaphospholidine 2-oxide; 2-Chloro, *in* D-00739
N,*N*-Dimethyl-*P*-(1-methylpropyl)phosphonamidic chloride, D-00763
N,*N*-Dimethyl-*P*-phenylphosphonamidic acid; Chloride, *in* D-00808
(2,5-Dimethylphenyl)phosphonic acid; Dichloride, *in* D-00813
N,*P*-Dimethylphosphonamidic acid; Fluoride, *in* D-00849
(2,2-Dimethylpropyl)phosphonic acid; Dichloride, *in* D-00912
N,*P*-Diphenylphosphonamidic acid; Fluoride, *in* D-01103
N,*P*-Diphenylphosphonamidic acid; Chloride, *in* D-01103
1,2-Ethanediylbis[phosphonic dichloride], *in* E-00006
Ethyl benzylphosphonochloridate, *in* B-00063
Ethyl (2-chloroethyl)phosphonochloridate, *in* C-00077
Ethyl cyclohexylphosphonochloridate, *in* C-00259
Ethyl dichlorophosphinylacetate, *in* D-00190
Ethyl 3-dichlorophosphinylpropanoate, *in* D-00192
Ethyl ethylphosphonochloridate, *in* E-00135
Ethyl ethylphosphonofluoridate, *in* E-00142
Ethyl methylphosphonochloridate, *in* M-00296
Ethyl methylphosphonofluoridate, *in* M-00303
Ethyl phenylphosphonochloridate, *in* P-00229
Ethyl phenylphosphonofluoridate, *in* P-00239
P-Ethylphosphonamidic fluoride, *in* E-00126
▷Ethylphosphonic dichloride, E-00131
Ethylphosphonic difluoride, E-00132
Ethylphosphonochloridic acid; (*S*)-*form*, Et ester, *in* E-00135
Ethylphosphonochloridic acid; (±)-*form*, Et ester, *in* E-00135
Ethylphosphonofluoridic acid, E-00142
(Fluoromethyl)phosphonic acid; Dichloride, *in* F-00023
(3-Fluorophenyl)phosphonic acid; Dichloride, *in* F-00037
(4-Fluorophenyl)phosphonic dichloride, F-00039
3-Furanylphosphonic acid; Dichloride, *in* F-00074
1,7-Heptanediylbis(phosphonic dichloride), *in* H-00002
Heptylphosphonic acid; Difluoride, *in* H-00006
Heptylphosphonic acid; Dichloride, *in* H-00006
3,3,4,4,5,5-Hexafluoro-1,2-cyclopentenediphosphonic acid; Bis(dichloride), *in* H-00018
1,6-Hexanediphosphonic acid; Bisdichloride, *in* H-00085
Hexylphosphonic acid; Difluoride, *in* H-00097
Hexylphosphonic dichloride, H-00098
(Isocyanatomethyl)phosphonic acid; Difluoride, *in* I-00022
(Isocyanatomethyl)phosphonic acid; Dichloride, *in* I-00022
Isopropyl ethylphosphonochloridate, *in* E-00135
▷Isopropyl ethylphosphonofluoridate, *in* E-00142
Isopropyl (fluoromethyl)phosphonochloridate, *in* F-00024
Isopropyl methylphosphonochloridate, *in* M-00296
Isopropylphosphonic acid; Difluoride, *in* I-00050
Isopropylphosphonic dichloride, I-00052
(4-Methoxyphenyl)phosphonic dichloride, M-00073
(2-Methyl-1,3-butadienyl)phosphonic acid; Dichloride, *in* M-00093
(3-Methyl-1,2-butadienyl)phosphonic acid; Difluoride, *in* M-00094
(3-Methyl-1,2-butadienyl)phosphonic acid; Dichloride, *in* M-00094
(3-Methyl-1,3-butadienyl)phosphonic acid; Difluoride, *in* M-00095
(3-Methyl-1,3-butadienyl)phosphonic acid; (*E*)-*form*, Dichloride, *in* M-00095
(3-Methyl-1,3-butadienyl)phosphonic acid; (*Z*)-*form*, Dichloride, *in* M-00095
(2-Methyl-1-butenyl)phosphonic acid; Dichloride, *in* M-00100
(3-Methyl-1-butenyl)phosphonic acid; Dichloride, *in* M-00101
(1-Methylbutyl)phosphonic acid; (±)-*form*, Dichloride, *in* M-00105
(3-Methylbutyl)phosphonic acid; Dichloride, *in* M-00106
Methyl cyclohexylphosphonochloridate, *in* C-00259
Methyl dichlorophosphinylacetate, *in* D-00190
Methyl 3-dichlorophosphinylpropanoate, *in* D-00192
Methylenebis(phosphonic dichloride), M-00145
Methylenebisphosphonic difluoride, M-00146
(1-Methylethenyl)phosphonic dichloride, M-00150
Methyl ethylphosphonofluoridate, *in* E-00142
Methyl methylphosphonochloridate, *in* M-00296
Methyl methylphosphonofluoridate, *in* M-00303
(1-Methylnonyl)phosphonic acid; (±)-*form*, Dichloride, *in* M-00181
[(4-Methylphenyl)methyl]phosphonic acid; Dichloride, *in* M-00208
(2-Methylphenyl)phosphonic acid; Dichloride, *in* M-00235
(3-Methylphenyl)phosphonic acid; Difluoride, *in* M-00236
(3-Methylphenyl)phosphonic acid; Dichloride, *in* M-00236
(4-Methylphenyl)phosphonic acid; Difluoride, *in* M-00237
(4-Methylphenyl)phosphonic dichloride, M-00239
Methyl phenylphosphonochloridate, *in* P-00229
Methyl phenylphosphonofluoridate, *in* P-00239
P-Methylphosphonamidic fluoride, *in* M-00283
Methylphosphonazidic acid; Chloride, *in* M-00286
Methylphosphonic acid; Dibromide, *in* M-00288
▷Methylphosphonic dichloride, M-00290
▷Methylphosphonic difluoride, M-00291
Methylphosphonochloridic acid; (*S*)-*form*, Isopropyl ester, *in* M-00296
Methylphosphonochloridic acid; (±)-*form*, Isopropyl ester, *in* M-00296
▷Methylphosphonofluoridic acid, M-00303
(2-Methyl-1-propenyl)phosphonic dichloride, M-00365
2-Methylpropyl ethylphosphonochloridate, *in* E-00135
(2-Methylpropyl)phosphonic acid; Dichloride, *in* M-00378
(1-Methylpropyl)phosphonic dichloride, M-00379
(6-Methyl-2-pyridinyl)phosphonic acid; Dichloride, *in* M-00392
4-Morpholinylphosphonic acid; Dichloride, *in* M-00523
4-Morpholinylphosphonochloridic acid, M-00524
4-Morpholinylphosphonochloridic acid; Anhydride with diphenyl phosphate, *in* M-00524
1-Naphthylphosphonic acid; Dichloride, *in* N-00021
2-Naphthylphosphonic acid; Dichloride, *in* N-00022
4-Nitrophenyl 4-morpholinylphosphonochloridate, *in* M-00524
(3-Nitrophenyl)phosphonic acid; Dichloride, *in* N-00056
(4-Nitrophenyl)phosphonic acid; Dichloride, *in* N-00057
1,9-Nonanediphosphonic acid; Tetrachloride, *in* N-00069
Nonylphosphonic acid; Difluoride, *in* N-00076
Nonylphosphonic acid; Dichloride, *in* N-00076
1,8-Octanediphosphonic acid; Tetrachloride, *in* O-00035
Octylphosphonic acid; Difluoride, *in* O-00042
Octylphosphonic dichloride, O-00043
(2-Oxopropyl)phosphonic acid; Dichloride, *in* O-00075

1,3-Pentadienylphosphonic acid; Difluoride, *in* P-00009
1,3-Pentadienylphosphonic acid; Dichloride, *in* P-00009
Pentafluorophenylphosphonic acid; Difluoride, *in* P-00015
1,1-Pentanediphosphonic acid; Bis(dichloride), *in* P-00031
1,5-Pentanediphosphonic acid; Bis(dichloride), *in* P-00032
Pentylphosphonic acid; Dichloride, *in* P-00050
Phenyl (2-chloroethyl)phosphonochloridate, *in* C-00077
Phenyl *P*-chloromethyl-*N*-(dichlorophosphinyl)-phosphonochloridimidate, *in* C-00091
Phenyl (chloromethyl)phosphonochloridate, *in* C-00111
[1,4-Phenylenebis(methylene)]bisphosphonic acid; Tetra-chloride, *in* P-00131
(2-Phenylethenyl)phosphonic dichloride, P-00157
(2-Phenylethenyl)phosphonic difluoride, P-00158
Phenylethynylphosphonic acid; Dichloride, *in* P-00166
Phenyl methylphosphonochloridate, *in* M-00296
Phenyl 4-morpholinylphosphonochloridate, *in* M-00524
▷Phenyl phenylphosphonochloridate, *in* P-00229
Phenyl phenylphosphonofluoridate, *in* P-00239
P-Phenylphosphonamidic acid; Fluoride, *in* P-00218
Phenylphosphonic acid; Dibromide, *in* P-00221
Phenylphosphonic dichloride, P-00223
Phenylphosphonic difluoride, P-00224
Phenylphosphonofluoridic acid, P-00239
N-Phenyl-*P*-trichloromethylphosphonamidic acid; Chloride, *in* P-00328
▷1-Piperidinylphosphonic acid; Dichloride, *in* P-00430
1,2-Propadienylphosphonic acid; Difluoride, *in* P-00441
Propadienylphosphonic dichloride, P-00443
1,3-Propanediphosphonic acid; Bis-dichloride, *in* P-00444
1-Propenylphosphonic acid; (*E*)-*form*, Difluoride, *in* P-00453
1-Propenylphosphonic acid; (*E*)-*form*, Dichloride, *in* P-00453
1-Propenylphosphonic acid; (*Z*)-*form*, Difluoride, *in* P-00453
1-Propenylphosphonic acid; (*Z*)-*form*, Dichloride, *in* P-00453
2-Propenylphosphonic dichloride, P-00456
(1-Propylheptyl)phosphonic acid; (±)-*form*, Dichloride, *in* P-00467
Propylphosphonic dichloride, P-00475
1-Propynylphosphonic acid; Difluoride, *in* P-00494
1-Propynylphosphonic dichloride, P-00496
2-Pyridinephosphonic acid; Dichloride, *in* P-00503
▷Sarin, S-00002
▷Soman, S-00005
(2,3,5,6-Tetrachloro-4-pyridinyl)phosphonic acid; Dichloride, *in* T-00035
2,2,2-Tribromoethyl 4-morpholinylphosphonochloridate, T-00357
(4-Trichloromethylphenyl)phosphonic acid; Dichloride, *in* T-00409
(Trichloromethyl)phosphonic dichloride, T-00413
(Trichloromethyl)phosphonic difluoride, T-00414
Triethylphosphonamidic acid; (±)-*form*, Fluoride, *in* T-00446
Triethylphosphonamidic acid; (±)-*form*, Chloride, *in* T-00446
(4-Trifluoromethylphenyl)phosphonic acid; Difluoride, *in* T-00473
(4-Trifluoromethylphenyl)phosphonic acid; Dichloride, *in* T-00473
(Trifluoromethyl)phosphonic acid; Difluoride, *in* T-00478
(Trifluoromethyl)phosphonic dichloride, T-00479
(2,4,5-Trimethylphenyl)phosphonic acid; Dichloride, *in* T-00526
▷Trimethylphosphonamidic fluoride, T-00538
Vinylphosphonic acid; Difluoride, *in* V-00003
Vinylphosphonic dichloride, V-00005

Phosphonic pseudohalides

(Chloromethyl)phosphonic diisocyanate, C-00110
Isopropylphosphonic acid; Diisocyanate, *in* I-00050
Methylphosphonic acid; Dicyanide, *in* M-00288
Phenylphosphonic acid; Dicyanide, *in* P-00221
Vinylphosphonic acid; Diisocyanate, *in* V-00003

Phosphonic esters

2-Aminoethylphosphonic acid; Di-Et ester, *in* A-00079
(1-Amino-2-hydroxyethyl)phosphonic acid; (±)-*form*, *O*-Ac, Di-Ph ester, *in* A-00083
(3-Aminopropyl)phosphonic acid; *N*-Ac, Di-Et ester, *in* A-00127
1,3-Benzodithiol-2-ylphosphonic acid; Di-Et ester, *in* B-00020
2-(Benzoylamino)ethylphosphonic acid, *in* A-00079
Benzyl [bis(trimethylsilyloxy)phosphinyl]acetate, *in* B-00434
Benzyl *N*-cyclohexyl-*P*-ethylphosphonamidate, *in* C-00246
Benzyl *N*-cyclohexyl-*P*-phenylphosphonamidate, *in* C-00251
Benzyl (dibutoxyphosphinyl)acetate, *in* D-00073
Benzyl 4-(diethoxyphosphinyl)butanoate, *in* D-00247
1-Benzyl-2-ethoxyhexahydro-1,2-azaphosphorine 2-oxide, *in* B-00048
Benzyl ethylphosphonochloridate, *in* E-00135
Benzyl methyl methylphosphonate, *in* M-00168
Benzyl methyl phenylphosphonate, *in* M-00488
Benzyl phenyl phenylphosphonate, *in* M-00471
Benzyl *P*-phenylphosphonamidate, *in* P-00218
Benzyl triethylphosphonamidate, *in* T-00446
2,2′-Biphenylene phosphonate, *in* D-00038
Bis(2-bromoethyl) (2-bromoethyl)phosphonate, *in* B-00472
Bis[(4-bromophenyl)methyl] phosphonate, B-00110
Bis(4-*tert*-butylphenyl) phosphonate, B-00116
Bis(2-chloroethyl) (2-chloroethyl)phosphonate, *in* C-00074
▷Bis(2-chloroethyl) phosphonate, B-00120
Bis[(4-chlorophenyl)methyl] phosphonate, B-00130
Bis(2-chlorophenyl) phosphonate, B-00143
Bis(4-chlorophenyl) phosphonate, B-00144
Bis(1,1-dimethylethyl) 1,2-propadienylphosphonate, *in* P-00441
Bis(2,2-dimethylpropyl) phosphonate, B-00208
▷Bis(2-ethylhexyl) phosphonate, B-00273
Bis(3-methylbutyl) phosphonate, B-00303
[Bis(1-methylethoxy)phosphinyl]acetamide, *in* D-00633
Bis(1-methylethyl) (2-amino-2-oxoethyl)phosphonate, *in* A-00060
Bis(1-methylethyl) (hydroxyphenylmethyl)phosphonate, *in* H-00174
Bis(1-methylethyl) (2-oxopropyl)phosphonate, *in* O-00075
Bis(1-methylethyl) 1,2-propadienylphosphonate, *in* P-00441
Bis(1-methylheptyl) phosphonate, B-00307
Bis(2-methylphenyl) phosphonate, B-00326
Bis(3-methylphenyl) phosphonate, B-00327
Bis(4-methylphenyl) phosphonate, B-00328
Bis(2-methylpropyl) (2-methylpropyl)phosphonate, *in* M-00378
Bis(2-methylpropyl) phenylphosphonate, *in* P-00221
Bis(1-methylpropyl) phosphonate; (+)-*form*, *in* B-00353
Bis(1-methylpropyl) phosphonate; (−)-*form*, *in* B-00353
Bis(1-methylpropyl) phosphonate; (±)-*form*, *in* B-00353
Bis(2-methylpropyl) phosphonate, B-00354
Bis(4-nitrophenyl)(dichloromethyl)phosphonate, *in* D-00175
Bis[(4-nitrophenyl)methyl]phosphonate, B-00368
Bis(4-nitrophenyl) phosphonate, B-00375
Bis(2-phenylethyl)(2-phenylethyl)phosphonate, *in* P-00165
Bis(triethylsilyl) [(acetyloxy)methyl]phosphonate, *in* A-00007
Bis(triethylsilyl) (hydroxymethyl)phosphonate, *in* H-00162
Bis(triethylsilyl) phosphonate, B-00407
Bis(2,2,2-trifluoroethyl) phosphonate, B-00409
Bis(trimethylsilyl) (acetamidomethyl)phosphonate, *in* A-00012
Bis(trimethylsilyl) (2-acetyloxyethyl)phosphonate, *in* A-00006
Bis(trimethylsilyl) acetylphosphonate, *in* A-00026
Bis(trimethylsilyl)(4-aminophenyl)phosphonate, *in* A-00111
Bis(trimethylsilyl) benzoylphosphonate, *in* B-00034
Bis(trimethylsilyl) benzylphosphonate, *in* B-00060
Bis(trimethylsilyl)(4-bromophenyl)phosphonate, *in* B-00492
Bis(trimethylsilyl) (2-chloroethyl)phosphonate, *in* C-00074
Bis(trimethylsilyl)(4-chlorophenyl)phosphonate, *in* C-00147
Bis(trimethylsilyl) (cyanomethyl)phosphonate, *in* C-00215
Bis(trimethylsilyl)[3,5-di-*tert*-butyl-4-hydroxyphenyl)-methyl]phosphonate, *in* D-00094
Bis(trimethylsilyl) (dimethylaminocarbonyl)phosphonate, *in* D-00691
Bis(trimethylsilyl) (3,3-dimethyl-2-oxobutyl)phosphonate, *in* D-00783

Bis(trimethylsilyl) ethylphosphonate, *in* E-00129
Bis(trimethylsilyl)(2-fluorophenyl)phosphonate, *in* F-00036
Bis(trimethylsilyl)(4-fluorophenyl)phosphonate, *in* F-00038
Bis(trimethylsilyl) 2-furanylmethylphosphonate, *in* F-00072
Bis(trimethylsilyl) 2-furanylphosphonate, *in* F-00073
Bis(trimethylsilyl) (2-hydroxyethyl)phosphonate, *in* H-00146
Bis(trimethylsilyl)(2-methoxyphenyl)phosphonate, *in* M-00070
Bis(trimethylsilyl)(3-methylphenyl)phosphonate, *in* M-00236
Bis(trimethylsilyl)(4-methylphenyl)phosphonate, *in* M-00237
Bis(trimethylsilyl) (1-oxopropyl)phosphonate, *in* O-00074
Bis(trimethylsilyl) (2-oxopropyl)phosphonate, *in* O-00075
Bis(trimethylsilyl) phosphonate, B-00435
(4-Bromo-2,5-dichlorophenyl)methyl phenylphosphonate, *in* M-00488
(1,3-Butadienyl)phosphonic acid; (*E*)-*form*, Di-Et ester, *in* B-00510
2-Butenylphosphonic acid; (*Z*)-*form*, Di-Et ester, *in* B-00525
tert-Butyl [bis(trimethylsilyloxy)phosphinyl]acetate, *in* B-00434
Butyl *P*-(butoxymethyl)-*N,N*-diethylphosphonamidate, *in* B-00536
O-Butyl *OO-tert*-butyl ethylphosphonoperoxoate, *in* E-00144
O-Butyl *OO-tert*-butyl methylphosphonoperoxoate, *in* M-00310
Butyl *N,N*-dibutylphosphonamidate, *in* D-00126
Butyl *N,N*-diethylphosphonamidate, *in* D-00337
Butyl *N,N*-diethyl-*P*-propadienylphosphonamidate, *in* D-00399
Butyl *N,N*-diethyl-*P*-1-propynylphosphonamidate, *in* D-00403
1,3-Butylene methylphosphonate, *in* D-00734
2,3-Butylene methylphosphonate, *in* T-00511
1,3-Butylene phenylphosphonate, *in* M-00199
2,3-Butylene phenylphosphonate, *in* D-00794
OO-tert-Butyl *O*-ethyl ethylphosphonoperoxoate, *in* E-00144
OO-tert-Butyl *O*-ethyl methylphosphonoperoxoate, *in* M-00310
OO-tert-Butyl *O*-ethyl phenylphosphonoperoxoate, *in* P-00242
tert-Butyl ethyl phosphonate, B-00561
OO-tert-Butyl *O*-isopropyl ethylphosphonoperoxoate, *in* E-00144
OO-tert-Butyl *O*-isopropyl methylphosphonoperoxoate, *in* M-00310
5-*tert*-Butyl-2-methyl-1,3,2-dioxaphosphorinane; *cis-form*, 2-Oxide, *in* B-00564
5-*tert*-Butyl-2-methyl-1,3,2-dioxaphosphorinane; *trans-form*, 2-Oxide, *in* B-00564
OO-tert-Butyl *O*-methyl ethylphosphonoperoxoate, *in* E-00144
OO-tert-Butyl *O*-methyl methylphosphonoperoxoate, *in* M-00310
5-*tert*-Butyl-2-phenyl-1,3,2-dioxaphosphorinane; *trans-form*, 2-Oxide, *in* B-00577
Butyl phenyl phosphonate, B-00585
Butyl phenylphosphonochloridate, *in* P-00229
2-*tert*-Butyltrimethylene phenylphosphonate, *in* B-00577
[(2-Chlorophenyl)methyl]phosphonic acid; Mono-Me ester, *in* C-00132
(4-Chlorophenyl)phosphonic dichloride, C-00148
P-Cyanomethyl-*N,N,N′,N′*-tetramethylphosphonic diamide, C-00216
4-Cyanophenyl ethyl phenylphosphonate, *in* M-00471
Cyclohexyl benzylphosphonochloridate, *in* B-00063
Cyclohexyl benzylphosphonofluoridate, *in* B-00066
Cyclohexyl butylphosphonochloridate, *in* B-00603
Cyclohexyl butylphosphonofluoridate, *in* B-00608
Cyclohexyl *N*-cyclohexyl-*P*-ethylphosphonamidate, *in* C-00246
Cyclohexyl hydrogen methylphosphonate, *in* M-00288
Cyclohexyl *P*-methylphosphonohydrazidate, *in* M-00308
Cyclohexyl phenylphosphonochloridate, *in* P-00229
Cyclohexyl phenylphosphonofluoridate, *in* P-00239
Cyclohexyl propylphosphonochloridate, *in* P-00476
Cyclohexyl propylphosphonofluoridate, *in* P-00479
▷Dibenzyl phosphonate, D-00059
(1,2-Dibromo-1-phenylpropyl)phosphonic acid; (1*RS*,2*SR*)-*form*, Mono-Me ester, *in* D-00068
(1,2-Dibromo-1-phenylpropyl)phosphonic acid; (1*RS*,2*SR*)-*form*, Di-Me ester, *in* D-00068
(Dibutoxyphosphinyl)acetamide, *in* D-00073
(Dibutoxyphosphinyl)acetic acid, D-00073
Dibutyl [(2-aminophenyl)methyl]phosphonate, *in* A-00106
▷Dibutyl benzylphosphonate, *in* B-00060
▷Dibutyl butylphosphonate, D-00076
Dibutyl (4-chlorobutyl)phosphonate, *in* C-00025
Dibutyl [(4-cyanophenyl)methyl]phosphonate, *in* C-00221
Dibutyl decylphosphonate, *in* D-00012
Dibutyl (2-diethylamino-2-oxoethyl)phosphonate, *in* D-00268
4,8-Di-*tert*-butyl-2,10-dimethyl-12*H*-dibenzo[*d,g*][1,3,2]-dioxaphosphocin; 6-Oxide, *in* D-00088
Dibutyl ethylphosphonate, *in* E-00129
Dibutyl methylphosphonate, *in* M-00288
Di-*tert*-butyl methylphosphonate, *in* M-00288
Dibutyl (1-methylpropyl)phosphonate, *in* M-00377
Dibutyl (nitromethyl)phosphonate, *in* N-00037
Dibutyl [(2-nitrophenyl)methyl]phosphonate, *in* N-00044
Dibutyl octylphosphonate, *in* O-00042
Dibutyl (2-oxo-2-phenylethyl)phosphonate, *in* O-00066
Dibutyl phenylphosphonate, *in* P-00221
▷Dibutyl phosphonate, D-00127
Di-*tert*-butyl phosphonate, D-00128
Di-*tert*-butyl 1-propenylphosphonate, *in* P-00453
Dibutyl 2-propenylphosphonate, *in* P-00454
Dibutyl propylphosphonate, *in* P-00473
Dibutyl 2-propynylphosphonate, *in* P-00495
Dibutyl[[(tetrahydro-2*H*-pyran-2-yl)oxy]methyl]phosphonate, *in* T-00136
Dicyclohexyl (2-diethylamino-2-oxoethyl)phosphonate, *in* D-00268
Dicyclohexyl phosphonate, D-00215
Didecyl phosphonate, D-00222
Didodecyl phosphonate, D-00227
2,5-Diethoxy-1,4,2,5-dioxaphosphorinane 2,5-dioxide, *in* D-00618
(Diethoxyphosphinyl)acetic acid, D-00241
α-(Diethoxyphosphinyl)benzeneacetic acid, D-00242
4-(Diethoxyphosphinyl)butanoic acid, D-00247
4-(Diethoxyphosphinyl)-3-methyl-2-butenoic acid; (*E*)-*form*, Me ester, *in* D-00251
4-(Diethoxyphosphinyl)-3-methyl-2-butenoic acid; (*Z*)-*form*, Me ester, *in* D-00251
3-(Diethoxyphosphinyl)propanenitrile, *in* D-00259
3-(Diethoxyphosphinyl)propanoic acid, D-00259
3-(Diethoxyphosphinyl)propanoic acid; Amide, *in* D-00259
2-(Diethoxyphosphinyl)propenoic acid; Amide, *in* D-00260
2-(Diethoxyphosphinyl)propenoic acid; Nitrile, *in* D-00260
Diethyl (acetamidomethyl)phosphonate, *in* A-00012
▷Diethyl 1-acetoxy-2,2,2-trichloroethylphosphonate, *in* A-00011
Diethyl (2-acetyloxyethyl)phosphonate, *in* A-00006
Diethyl [(acetyloxy)methyl]phosphonate, *in* A-00007
Diethyl [(acetyloxy)phenylmethyl]phosphonate, *in* A-00020
Diethyl [1-(acetyloxy)propyl]phosphonate, *in* A-00008
Diethyl (3-acetyloxypropyl)phosphonate, *in* A-00010
Diethyl acetylphosphonate, *in* A-00026
Diethyl 9-acridinylphosphonate, *in* A-00030
Diethyl 1-adamantylphosphonate, *in* A-00034
Diethyl (1-aminobutyl)phosphonate, *in* A-00054
Diethyl (2-aminobutyl)phosphonate, *in* A-00055
Diethyl (3-aminobutyl)phosphonate, *in* A-00056
Diethyl (4-aminobutyl)phosphonate, *in* A-00057
Diethyl [(aminocarbonyl)amino]methylphosphonate, *in* A-00058
Diethyl aminocarbonylphosphonate, *in* C-00003
Diethyl (1-aminoethyl)phosphonate, *in* A-00078
Diethyl (1-amino-1-methylethyl)phosphonate, *in* A-00087
Diethyl (2-amino-1-methylethyl)phosphonate, *in* A-00088
Diethyl (aminomethyl)phosphonate, *in* A-00091
Diethyl (1-amino-1-methylpropyl)phosphonate, *in* A-00093
Diethyl (1-amino-2-methylpropyl)phosphonate, *in* A-00094
Diethyl (2-amino-2-oxoethyl)phosphonate, *in* A-00060
Diethyl (1-aminopentyl)phosphonate, *in* A-00096
Diethyl (5-aminopentyl)phosphonate, *in* A-00097
Diethyl (1-amino-1-phenylethyl)phosphonate, *in* A-00100
Diethyl (aminophenylmethyl)phosphonate, *in* A-00105
▷Diethyl [(4-aminophenyl)methyl]phosphonte, *in* A-00107
Diethyl (2-aminophenyl)phosphonate, *in* A-00109
Diethyl (4-aminophenyl)phosphonate, *in* A-00111
Diethyl (1-aminopropyl)phosphonate, *in* A-00125
Diethyl (2-aminopropyl)phosphonate, *in* A-00126
Diethyl (3-aminopropyl)phosphonate, *in* A-00127
Diethyl 1-aziridinylphosphonate, *in* A-00153

Diethyl (1-*N*-benzoylamino-2-methylpropyl)phosphonate, *in* A-00094
Diethyl (1-benzoylamino-2-phenylethyl)phosphonate, *in* A-00101
Diethyl [1-(*N*-benzoylamino)propyl]phosphonate, *in* A-00125
Diethyl benzoylphosphonate, *in* B-00034
Diethyl [(benzylamino)methyl]phosphonate, *in* A-00091
Diethyl [(benzylideneamino)methyl]phosphonate, D-00271
Diethyl 1-benzylideneaminophosphonate, *in* B-00049
▷Diethyl benzylphosphonate, D-00272
Diethyl [(1,1′-biphenyl)-2-ylmethyl]phosphonate, *in* B-00079
Diethyl[(1,1′-biphenyl)-4-ylmethyl]phosphonate, *in* B-00080
Diethyl 4-biphenylylphosphonate, *in* B-00077
Diethyl 2,5-bis(diethoxyphosphinyl)hexanedioate, *in* D-01184
Diethyl [bis(diethoxyphosphinyl)methyl]butanedioate, *in* D-01185
Diethyl 2,4-bis(diethoxyphosphinyl)pentanedioate, *in* D-01186
Diethyl [bis(methylthio)methyl]phosphonate, *in* B-00364
Diethyl (4-bromobutyl)phosphonate, *in* B-00466
Diethyl (2-bromoethyl)phosphonate, *in* B-00472
Diethyl (bromomethyl)phosphonate, *in* B-00475
Diethyl (bromophenylmethyl)phosphonate, *in* B-00482
Diethyl [(2-bromophenyl)methyl]phosphonate, *in* B-00483
Diethyl [(4-bromophenyl)methyl]phosphonate, *in* B-00484
Diethyl (2-bromophenyl)phosphonate, *in* B-00490
Diethyl (3-bromophenyl)phosphonate, *in* B-00491
Diethyl (4-bromophenyl)phosphonate, *in* B-00492
Diethyl (1-bromopropyl)phosphonate, *in* B-00500
Diethyl (3-bromopropyl)phosphonate, *in* B-00501
Diethyl (1,2-butadienyl)phosphonate, *in* B-00509
Diethyl (1,3-butadienyl)phosphonate, *in* B-00510
Diethyl 1-butenylphosphonate, *in* B-00524
Diethyl 2-butenylphosphonate, *in* B-00525
Diethyl 3-buten-1-ylphosphonate, *in* B-00526
Diethyl 3-buten-1-ynylphosphonate, *in* B-00529
Diethyl *tert*-butoxycyanomethylphosphonate, D-00273
Diethyl [2-(butylamino)-1-hexenyl]phosphonate, *in* B-00540
Diethyl butylphosphonate, *in* B-00594
Diethyl 1-butynylphosphonate, *in* B-00643
Diethyl 2-butynylphosphonate, *in* B-00644
Diethyl(chloroacetyl)phosphonate, *in* C-00020
Diethyl (4-chlorobutyl)phosphonate, *in* C-00025
Diethyl (1-chloroethyl)phosphonate, *in* C-00073
▷Diethyl (2-chloroethyl)phosphonate, *in* C-00074
Diethyl chloromethylphosphonate, *in* C-00106
Diethyl [(2-chlorophenyl)methyl]phosphonate, *in* C-00132
Diethyl [(3-chlorophenyl)methyl]phosphonate, *in* C-00133
Diethyl [(4-chlorophenyl)methyl]phosphonate, *in* C-00134
Diethyl (2-chlorophenyl)phosphonate, *in* C-00145
Diethyl (3-chlorophenyl)phosphonate, *in* C-00146
Diethyl (4-chlorophenyl)phosphonate, *in* C-00147
Diethyl (1-chloropropyl)phosphonate, *in* C-00170
Diethyl (2-chloropropyl)phosphonate, *in* C-00171
Diethyl (3-chloropropyl)phosphonate, *in* C-00172
Diethyl (1-cyanoethyl)phosphonate, *in* C-00210
▷Diethyl (2-cyanoethyl)phosphonate, *in* C-00211
Diethyl (cyanomethyl)phosphonate, *in* C-00215
Diethyl (cyanophenylmethyl)phosphonate, *in* C-00219
Diethyl [(4-cyanophenyl)methyl]phosphonate, *in* C-00221
Diethyl (2-cyanophenyl)phosphonate, *in* C-00223
Diethyl (3-cyanophenyl)phosphonate, *in* C-00224
Diethyl (4-cyanophenyl)phosphonate, *in* C-00225
Diethyl (1-cyanopropyl)phosphonate, *in* C-00229
Diethyl (3-cyanopropyl)phosphonate, *in* C-00230
Diethyl 1-cyclohexen-1-ylphosphonate, *in* C-00241
Diethyl 2-cyclohexen-1-ylphosphonate, *in* C-00242
Diethyl 3-cyclohexen-1-ylphosphonate, *in* C-00243
Diethyl 2-(cyclohexylamino)-2-methylethenyl phosphonate, D-00276
Diethyl [2-(cyclohexylamino)-2-phenylethenyl]phosphonate, D-00277
Diethyl 2-(cyclohexylamino)vinylphosphonate, D-00278
Diethyl (cyclohexylcarbonyl)phosphonate, *in* C-00239
Diethyl cyclohexylphosphonate, *in* C-00256
Diethyl 1-cyclopenten-1-ylphosphonate, *in* C-00277
Diethyl 2-cyclopenten-1-ylphosphonate, *in* C-00278
Diethyl decylphosphonate, *in* D-00012
Diethyl (1-diazoethyl)phosphonate, *in* D-00033
Diethyl (diazomethyl)phosphonate, *in* D-00035
Diethyl (dibromomethyl)phosphonate, *in* D-00064
Diethyl (2,5-dibromophenyl)phosphonate, *in* D-00066
Diethyl dibutyldiphosphonate, *in* D-00092
Diethyl [(3,5-di-*tert*-butyl-4-hydroxyphenyl)methyl]-phosphonate, *in* D-00094
Diethyl (dichloromethyl)phosphonate, *in* D-00175
Diethyl (2,5-dichlorophenyl)phosphonate, *in* D-00185
Diethyl (3,4-dichlorophenyl)phosphonate, *in* D-00186
Diethyl (2,2-diethoxyethyl)phosphonate, *in* D-00234
Diethyl 2-diethoxyphosphinyl-2-[2-(diethoxyphosphinyl)-ethyl]butanedioate, *in* P-00394
N,*N*-Diethyl(diethoxyphosphinyl)fluoroacetamide, *in* D-00250
Diethyl (2-diethylamino-2-oxoethyl)phosphonate, *in* D-00268
Diethyl diethyldiphosphonate, *in* D-00290
Diethyl (difluoromethyl)phosphonate, *in* D-00415
Diethyl (9,10-dihydro-9-acridinyl)phosphonate, *in* D-00438
Diethyl (3,4-dihydro-1*H*-2-benzopyran-1-yl)phosphonate, *in* D-00443
Diethyl [(1,3-dihydro-1,3-dioxo-2*H*-isoindol-2-yl)methyl]-phosphonate, *in* P-00426
Diethyl (9,10-dihydro-5-methyl-9-acridinyl)phosphonate, *in* D-00438
Diethyl [(5,6-dihydro-4,4,6-trimethyl-4*H*-1,3-oxazin-2-yl]-phosphonate, *in* D-00598
Diethyl (2,3-dihydroxypropyl)phosphonate, *in* D-00624
Diethyl (2-dimethylamino-2-oxoethyl)phosphonate, *in* D-00703
Diethyl (4-dimethylaminophenyl)phosphonate, *in* D-00706
Diethyl dimethyldiphosphonate, *in* D-00751
Diethyl (1,1-dimethylethyl)phosphonate, *in* B-00595
Diethyl (3,3-dimethyl-2-oxobutyl)phosphonate, *in* D-00783
Diethyl (2,3-dimethylphenyl)phosphonate, *in* D-00811
Diethyl (2,4-dimethylphenyl)phosphonate, *in* D-00812
Diethyl (2,5-dimethylphenyl)phosphonate, *in* D-00813
Diethyl(3,4-dimethylphenyl)phosphonate, *in* D-00815
Diethyl (3,5-dimethylphenyl)phosphonate, *in* D-00816
Diethyl (2,4-dinitrophenyl)phosphonate, *in* D-00946
Diethyl dipropyldiphosphonate, *in* D-01202
Diethyl 1,3-dithian-2-ylphosphonate, *in* D-01240
Diethyl ethenylphosphonate, *in* V-00003
▷Diethyl (ethoxycarbonylmethyl)phosphonate, *in* D-00241
Diethyl (2-ethoxyethyl)phosphonate, *in* H-00146
Diethyl [(ethylimino)methylethenyl]phosphonate, *in* E-00081
Diethyl ethylphosphonate, D-00292
Diethyl [(ethylthio)methyl]phosphonate, *in* E-00192
Diethyl ethynylphosphonate, *in* E-00203
Diethyl (9*H*-fluoren-9-yl)phosphonate, *in* F-00011
Diethyl [(3-fluorophenyl)methyl]phosphonate, *in* F-00030
Diethyl [(4-fluorophenyl)methyl]phosphonate, *in* F-00031
Diethyl [(2-fluorophenyl)methyl]phosphonic acid, *in* F-00029
Diethyl (2-fluorophenyl)phosphonate, *in* F-00036
Diethyl (3-fluorophenyl)phosphonate, *in* F-00037
Diethyl (4-fluorophenyl)phosphonate, *in* F-00038
Diethyl (2-furanylhydroxymethyl)phosphonate, *in* F-00071
Diethyl 2-furanylmethylphosphonate, *in* F-00072
Diethyl 2-furanylphosphonate, *in* F-00073
Diethyl 3-furanylphosphonate, *in* F-00074
Diethyl heptylphosphonate, *in* H-00006
▷Diethyl hexylphosphonate, *in* H-00097
Diethyl (1-hydroxybutyl)phosphonate, *in* H-00114
Diethyl (1-hydroxyethyl)phosphonate, *in* H-00145
Diethyl (2-hydroxyethyl)phosphonate, *in* H-00146
Diethyl (1-hydroxy-2-methylethyl)phosphonate, *in* H-00159
Diethyl (hydroxymethyl)phosphonate, *in* H-00162
Diethyl (hydroxyphenylmethyl)phosphonate, *in* H-00174
Diethyl [(2-hydroxyphenyl)methyl]phosphonate, *in* H-00175
Diethyl [(4-hydroxyphenyl)methyl]phosphonate, *in* H-00176
Diethyl (2-hydroxyphenyl)phosphonate, *in* H-00179
Diethyl (4-hydroxyphenyl)phosphonate, *in* H-00181
Diethyl (1-hydroxypropyl)phosphonate, *in* H-00190
Diethyl (2-hydroxypropyl)phosphonate, *in* H-00191
Diethyl (3-hydroxypropyl)phosphonate, *in* H-00192
Diethyl 1*H*-imidazol-1-ylphosphonate, *in* I-00003
Diethyl 1*H*-inden-2-ylphosphonate, *in* I-00007
Diethyl (iodomethyl)phosphonate, *in* I-00013
Diethyl [(2-iodophenyl)methyl]phosphonate, *in* I-00015

Diethyl (isocyanatomethyl)phosphonate, *in* I-00022
Diethyl (isocyanomethyl)phosphonate, *in* I-00024
Diethyl isopropylphosphonate, *in* I-00050
Diethyl (mercaptomethyl)phosphonate, *in* M-00018
Diethyl (methoxycarbonylmethyl)phosphonate, *in* D-00241
Diethyl[(2-methoxyethoxy)methyl]phosphonate, *in* M-00040
Diethyl [(3-methoxyphenyl)methyl]phosphonate, *in* M-00060
Diethyl [(4-methoxyphenyl)methyl]phosphonate, *in* M-00061
Diethyl (2-methoxyphenyl)phosphonate, *in* M-00070
Diethyl (3-methoxyphenyl)phosphonate, *in* M-00071
Diethyl (4-methoxyphenyl)phosphonate, *in* M-00072
Diethyl (methylaminocarbonyl)phosphonate, *in* M-00088
Diethyl (2-methyl-1,3-butadienyl)phosphonate, *in* M-00093
Diethyl (3-methyl-1,2-butadienyl)phosphonate, *in* M-00094
Diethyl (3-methyl-1,3-butadienyl)phosphonate, *in* M-00095
Diethyl (1-methyl-3-butenyl)phosphonate, *in* M-00099
Diethyl(3-methyl-1-butenyl)phosphonate, *in* M-00101
Diethyl (3-methyl-2-butenyl)phosphonate, *in* M-00102
Diethyl (3-methyl-3-butenyl)phosphonate, *in* M-00103
Diethyl (1-methylbutyl)phosphonate, *in* M-00105
Diethyl (3-methylbutyl)phosphonate, *in* M-00106
O^1,O^2-Diethyl O^3-methyl 3-(diethoxyphosphinyl)-1,2,3-butanetrioate, *in* P-00383
Diethyl (1-methylethenyl)phosphonate, *in* M-00149
Diethyl (3-methyl-2-oxiranyl)phosphonate, *in* M-00185
Diethyl [(2-methylphenyl)methyl]phosphonate, *in* M-00206
Diethyl [(3-methylphenyl)methyl]phosphonate, *in* M-00207
Diethyl [(4-methylphenyl)methyl]phosphonate, *in* M-00208
Diethyl (2-methylphenyl)phosphonate, *in* M-00235
Diethyl (3-methylphenyl)phosphonate, *in* M-00236
Diethyl (4-methylphenyl)phosphonate, *in* M-00237
Diethyl methylphosphonate, D-00298
Diethyl methyl phosphonoformate, *in* D-00236
Diethyl (1-methylpropyl)phosphonate, *in* M-00377
Diethyl (2-methylpropyl)phosphonate, *in* M-00378
Diethyl (3-methyl-2-pyridinyl)phosphonate, *in* M-00389
Diethyl (4-methyl-2-pyridinyl)phosphonate, *in* M-00391
Diethyl (6-methyl-2-pyridinyl)phosphonate, *in* M-00392
Diethyl 1-methyl-1*H*-pyrrol-2-ylphosphonate, *in* P-00521
Diethyl [(methylsulfinyl)methyl]phosphonate, *in* M-00398
Diethyl [(methylsulfonyl)methyl]phosphonate, *in* M-00400
Diethyl [1-(methylthio)ethyl]phosphonate, *in* M-00412
Diethyl [(methylthio)methyl]phosphonate, *in* M-00415
Diethyl (4-methylthiophenyl)phosphonate, *in* M-00418
Diethyl (4-morpholinylmethyl)phosphonate, *in* M-00521
Diethyl 4-morpholinylphosphonate, *in* M-00523
Diethyl [1,5-naphthalenediylbis[methylene]]bisphosphonate, *in* N-00002
Diethyl [1,6-naphthalenediylbis[methylene]]bisphosphonate, *in* N-00003
▷Diethyl 1-naphthylmethylphosphonate, *in* N-00013
Diethyl 2-naphthylmethylphosphonate, *in* N-00014
Diethyl 2-naphthylphosphonate, *in* N-00022
Diethyl (1-nitroethyl)phosphonate, *in* N-00035
Diethyl (nitromethyl)phosphonate, *in* N-00037
Diethyl[(2-nitrophenyl)methyl]phosphonate, *in* N-00044
Diethyl [(4-nitrophenyl)methyl]phosphonate, *in* N-00045
Diethyl (2-nitrophenyl)phosphonate, *in* N-00055
Diethyl (4-nitrophenyl)phosphonate, *in* N-00057
Diethyl (1-nitroso-1-methylethyl)phosphonate, *in* N-00066
Diethyl octylphosphonate, *in* O-00042
Diethyl (2-oxoethyl)phosphonate, *in* O-00059
Diethyl (2-oxoheptyl)phosphonate, *in* O-00061
Diethyl (2-oxononyl)phosphonate, *in* O-00063
Diethyl (2-oxooctyl)phosphonate, *in* O-00064
Diethyl (2-oxo-4-phenylbutyl)phosphonate, *in* O-00065
Diethyl (2-oxo-2-phenylethyl)phosphonate, *in* O-00066
Diethyl (1-oxopropyl)phosphonate, *in* O-00074
Diethyl (2-oxopropyl)phosphonate, *in* O-00075
Diethyl (3-oxopropyl)phosphonate, *in* O-00076
Diethyl [2-oxo-2-(triphenylphosphoranylidene)propyl]-phosphonate, *in* O-00086
Diethyl 1,3-pentadienylphosphonate, *in* P-00009
Diethyl 1,3-pentadiyn-1-ylphosphonate, *in* P-00010
Diethyl (pentafluorophenyl)phosphonate, *in* P-00015
Diethyl (3-penten-1-ynyl)phosphonate, *in* P-00045
Diethyl (4-penten-2-yn-1-yl)phosphonate, *in* P-00046
Diethyl pentylphosphonate, *in* P-00050
Diethyl 6-phenanthradinylphosphonate, *in* P-00060
Diethyl [(phenylamino)methyl]phosphonate, *in* P-00084
Diethyl (1-phenylethenyl)phosphonate, *in* P-00154
Diethyl (2-phenylethenyl)phosphonate, *in* P-00155
▷Diethyl 1-phenylethylphosphonate, *in* P-00164
Diethyl 2-phenylethylphosphonate, *in* P-00165
Diethyl phenylethynylphosphonate, *in* P-00166
Diethyl [[(phenylmethyl)thio]methyl]phosphonate, *in* B-00072
▷Diethyl phenylphosphonate, D-00310
Diethyl (1-phenyl-1,2-propadienyl)phosphonate, *in* P-00298
Diethyl 3-phenyl-2-propenylphosphonate, *in* P-00302
Diethyl (3-phenylpropyl)phosphonate, *in* P-00305
Diethyl [(phenylsulfinyl)methyl]phosphonate, *in* P-00310
Diethyl [(phenylsulfonyl)methyl]phosphonate, *in* P-00311
Diethyl [(phenylthio)methyl]phosphonate, *in* P-00326
Diethyl [phenyl[(trimethylsilyl)oxy]methyl]phosphonate, *in* P-00331
▷Diethyl phosphonate, D-00338
Diethyl (1-piperidinylmethyl)phosphonate, *in* P-00429
Diethyl 1-piperidinylphosphonate, *in* P-00430
Diethyl 1,2-propadienylphosphonate, *in* P-00441
Diethyl 1-propenylphosphonate, *in* P-00453
Diethyl 2-propenylphosphonate, *in* P-00454
Diethyl propylphosphonate, *in* P-00473
Diethyl 1-propynylphosphonate, *in* P-00494
Diethyl 2-propynylphosphonate, *in* P-00495
Diethyl (1*H*-pyrazol-3-yl)phosphonate, *in* P-00501
Diethyl (2-pyridinylmethyl)phosphonate, *in* P-00506
Diethyl (4-pyridinylmethyl)phosphonate, *in* P-00507
Diethyl 4-pyridinylphosphonate, *in* P-00505
Diethyl 1*H*-pyrrol-2-ylphosphonate, *in* P-00521
Diethyl [(2-quinolinyl)methyl]phosphonate, *in* Q-00004
Diethyl [(8-quinolinyl)methyl]phosphonate, *in* Q-00005
Diethyl (2,3,5,6-tetrachloro-4-pyridinyl)phosphonate, *in* T-00035
Diethyl(tetrahydro-2-oxo-2-furanyl)phosphonate, *in* T-00121
Diethyl[[(tetrahydro-2*H*-pyran-2-yl)oxy]methyl]phosphonate, *in* T-00136
Diethyl (2-thienylmethyl)phosphonate, *in* T-00304
Diethyl (3-thienylmethyl)phosphonate, *in* T-00305
Diethyl 2-thienylphosphonate, *in* T-00307
▷Diethyl (2,2,2-trichloro-1-hydroxyethyl)phosphonate, *in* T-00408
▷Diethyl (trichloromethyl)phosphonate, *in* T-00411
Diethyl (trifluoromethyl)phosphonate, *in* T-00478
Diethyl(2,4,6-trimethylphenyl)phosphonate, *in* T-00527
Diethyl [1-[(trimethylsilyl)oxy]ethyl]phosphonate, *in* T-00555
Diethyl[(trimethylsilyloxy)methyl]phosphonate, *in* T-00556
Diethyl triphenylmethylphosphonate, *in* T-00608
Diheptyl heptylphosphonate, *in* H-00006
Diheptyl phosphonate, D-00426
▷Dihexadecyl phosphonate, D-00430
Dihexyl (diethylcarbamoyl)phosphonate, *in* D-00264
Dihexyl phosphonate, D-00435
2,5-Dihydro-5,5-dimethyl-1,2-oxaphosphole 2-oxide; 2-Methoxy, *in* D-00464
2,5-Dihydro-5,5-dimethyl-1,2-oxaphosphole 2-oxide; 2-Ethoxy, *in* D-00464
2,5-Dihydro-5,5-dimethyl-1,2-oxaphosphole 2-oxide; 2-Phenoxy, *in* D-00464
2,3-Dihydro-2,3-diphenyl-1,3,2-benzoxazaphosphole; 2-Oxide, *in* D-00478
2,3-Dihydro-2-ethoxy-1,3,5-triphenyl-1*H*-1,2-azaphosphole 2-oxide, *in* D-00508
1,3-Dihydro-1-hydroxy-2,1-benzoxaphosphole 1-oxide; 1-Ethoxy (Et ester), *in* D-00490
2,5-Dihydro-2-hydroxy-1,2-oxaphosphole 2-oxide; 2-Butoxy, *in* D-00500
3,6-Dihydro-2-hydroxy-2*H*-1,2-oxaphosphorin 2-oxide; 2-Methoxy (Me ester), *in* D-00501
2,3-Dihydro-2-methoxy-1,2-benzoxaphosphole 2-oxide, *in* H-00119
2,3-Dihydro-2-methyl-1,4,2-benzodioxaphosphorin; 2-Oxide, *in* D-00517
2,3-Dihydro-5-methyl-1,2-oxaphosphole 2-oxide; 2-Methoxy, *in* D-00525

▷2,3-Dihydro-3,3,5-trimethyl-1,2-oxaphosphole 2-oxide; 2-Methoxy, *in* D-00596
2,3-Dihydro-3,3,5-trimethyl-1,2-oxaphosphole 2-oxide; 2-Ethoxy, *in* D-00596
2,3-Dihydro-3,3,5-trimethyl-1,2-oxaphosphole 2-oxide; 2-Phenoxy, *in* D-00596
2,5-Dihydroxy-3,6-dimethyl-3,6-bis(dihydroxyphosphinyl)-1,4,2,5-dioxadiphosphorinane 2,5-dioxide; Hexa-Me ester, *in* D-00617
Diisopropyl (acetamidomethyl)phosphonate, *in* A-00012
Diisopropyl 1-acetoxy-2,2,2-trichloroethylphosphonate, *in* A-00011
Diisopropyl acetylphosphonate, *in* A-00026
Diisopropyl 1-adamantylphosphonate, *in* A-00034
Diisopropyl allylphosphonate, *in* P-00454
Diisopropyl aminocarbonylphosphonate, *in* C-00003
Diisopropyl (3-aminopropyl)phosphonate, *in* A-00127
Diisopropyl (2-amino-4-pyrimidinyl)phosphonate, *in* A-00128
Diisopropyl benzoylphosphonate, *in* B-00034
Diisopropyl 1-butenylphosphonate, *in* B-00524
Diisopropyl [2-(butylamino)-1-hexenyl)]phosphonate, *in* B-00540
Diisopropyl(chloroacetyl)phosphonate, *in* C-00020
Diisopropyl (2-cyanoethyl)phosphonate, *in* C-00211
Diisopropyl 2-cyclohexen-1-ylphosphonate, *in* C-00242
Diisopropyl (cyclohexylcarbonyl)phosphonate, *in* C-00239
Diisopropyl cyclohexylphosphonate, *in* C-00256
Diisopropyl 1-cyclopenten-1-ylphosphonate, *in* C-00277
Diisopropyl 2-cyclopenten-1-ylphosphonate, *in* C-00278
Diisopropyl (diazophenylmethyl)phosphonate, *in* D-00037
Diisopropyl (diethylcarbamoylmethyl)phosphonate, *in* D-00268
Diisopropyl (diethylcarbamoyl)phosphonate, *in* D-00264
Diisopropyl (9,10-dihydro-5-methyl-9-acridinyl)phosphonate, *in* D-00438
Diisopropyl (1,3-dihydro-3-oxo-1-isobenzofuranyl)-phosphonate, *in* D-00535
Diisopropyl (2,4-dinitrophenyl)phosphonate, *in* D-00946
Diisopropyl ethynylphosphonate, *in* E-00203
Diisopropyl fluoromethylphosphonate, *in* F-00023
Diisopropyl heptylphosphonate, *in* H-00006
Diisopropyl hexylphosphonate, *in* H-00097
Diisopropyl α-hydroxy-2-furanylmethylphosphonate, *in* F-00071
Diisopropyl (2-hydroxypropyl)phosphonate, *in* H-00191
Diisopropyl 1*H*-imidazol-1-ylphosphonate, *in* I-00003
Diisopropyl (isocyanatomethyl)phosphonate, *in* I-00022
▷Diisopropyl methylphosphonate, *in* M-00288
Diisopropyl (3-methyl-4-pyridinyl)phosphonate, *in* M-00390
Diisopropyl (2-nitrophenyl)phosphonate, *in* N-00055
Diisopropyl (3-nitrophenyl)phosphonate, *in* N-00056
Diisopropyl (1-nitroso-1-methylethyl)phosphonate, *in* N-00066
Diisopropyl nonylphosphonate, *in* N-00076
Diisopropyl [(phenylamino)methyl]phosphonate, *in* P-00084
Diisopropyl phenylphosphonate, *in* P-00221
▷Diisopropyl phosphonate, D-00651
Diisopropyl 1-propenylphosphonate, *in* P-00453
Diisopropyl 1-propynylphosphonate, *in* P-00494
Diisopropyl (2-pyrimidinyl)phosphonate, *in* P-00515
Diisopropyl (4-pyrimidinyl)phosphonate, *in* P-00516
Diisopropyl (2,2,2-trichloro-1-hydroxyethyl)phosphonate, *in* T-00408
Diisopropyl (trichloromethyl)phosphonate, *in* T-00411
Diisopropyl vinylphosphonate, *in* V-00003
(Dimethoxyphosphinyl)acetic acid, D-00678
3-(Dimethoxyphosphinyl)propanoic acid, D-00685
Dimethyl (acetamidomethyl)phosphonate, *in* A-00012
▷Dimethyl 1-acetoxy-2,2,2-trichloroethylphosphonate, *in* A-00011
Dimethyl [3-(acetylhydroxyimino)-2-hydroxypropyl]-phosphonate, *in* A-00138
Dimethyl (2-acetyloxyethyl)phosphonate, *in* A-00006
Dimethyl [(acetyloxy)methyl]phosphonate, *in* A-00007
Dimethyl [(acetyloxy)phenylmethyl]phosphonate, *in* A-00020
Dimethyl (3-acetyloxypropyl)phosphonate, *in* A-00010
Dimethyl acetylphosphonate, *in* A-00026
Dimethyl 1-adamantylphosphonate, *in* A-00034
Dimethyl 2-adamantylphosphonate, *in* A-00035
Dimethyl aminocarbonylphosphonate, *in* C-00003
2-Dimethylaminoethylphosphonic acid, *in* A-00079
Dimethyl [[1-amino-2-(1*H*-indol-3-yl)]ethyl]phosphonate, *in* P-00421
[(Dimethylamino)methylene]bisphosphonic acid; Tetra-Et ester; B,MeI, *in* D-00701
Dimethyl (1-amino-1-methylethyl)phosphonate, *in* A-00087
Dimethyl [(2-aminophenyl)methyl]phosphonate, *in* A-00106
Dimethyl [(4-aminophenyl)methyl]phosphonate, *in* A-00107
Dimethyl (2-aminophenyl)phosphonate, *in* A-00109
Dimethyl (4-aminophenyl)phosphonate, *in* A-00111
Dimethyl 1-aziridinylphosphonate, *in* A-00153
Dimethyl benzoylphosphonate, *in* B-00034
Dimethyl benzylphosphonate, *in* B-00060
Dimethyl 1,2-bis(diethoxyphosphinyl)succinate, *in* D-01182
Dimethyl [bis(methylthio)methyl]phosphonate, *in* B-00364
Dimethyl (2-bromophenyl)phosphonate, *in* B-00490
Dimethyl (3-bromophenyl)phosphonate, *in* B-00491
Dimethyl (4-bromophenyl)phosphonate, *in* B-00492
Dimethyl (1,3-butadienyl)phosphonate, *in* B-00510
Dimethyl 3-buten-1-ynylphosphonate, *in* B-00529
Dimethyl butylphosphonate, *in* B-00594
Dimethyl (2-chloroethyl)phosphonate, *in* C-00074
Dimethyl chloromethylphosphonate, *in* C-00106
Dimethyl [(2-chlorophenyl)methyl]phosphonate, *in* C-00132
Dimethyl (2-chlorophenyl)phosphonate, *in* C-00145
Dimethyl (3-chlorophenyl)phosphonate, *in* C-00146
Dimethyl (4-chlorophenyl)phosphonate, *in* C-00147
Dimethyl (3-chloropropyl)phosphonate, *in* C-00172
Dimethyl (2-cyanoethyl)phosphonate, *in* C-00211
Dimethyl [(2-cyanophenyl)methyl]phosphonate, *in* C-00220
Dimethyl (4-cyanophenyl)phosphonate, *in* C-00225
Dimethyl 2-cyclohexen-1-ylphosphonate, *in* C-00242
Dimethyl (cyclohexylcarbonyl)phosphonate, *in* C-00239
Dimethyl cyclohexylphosphonate, *in* C-00256
Dimethyl 1-cyclopenten-1-ylphosphonate, *in* C-00277
Dimethyl cyclopentylphosphonate, *in* C-00276
Dimethyl decylphosphonate, *in* D-00012
Dimethyl (diazocyclohexylmethyl)phosphonate, *in* D-00028
Dimethyl (1-diazoethyl)phosphonate, *in* D-00033
Dimethyl (diazomethyl)phosphonate, *in* D-00035
Dimethyl (diazophenylmethyl)phosphonate, *in* D-00037
Dimethyl (dibromomethyl)phosphonate, *in* D-00064
Dimethyl (1,2-dibromo-1-phenylpropyl)phosphonate, *in* D-00068
Dimethyl (dichloromethyl)phosphonate, *in* D-00175
Dimethyl (diethoxyphosphinyl)butanedioate, *in* D-00245
Dimethyl diethyldiphosphonate, *in* D-00290
Dimethyl (9,10-dihydro-9-acridinyl)phosphonate, *in* D-00438
Dimethyl [(1,3-dihydro-1,3-dioxo-2*H*-isoindol-2-yl)methyl]-phosphonate, *in* P-00426
Dimethyl (9,10-dihydro-5-methyl-9-acridinyl)phosphonate, *in* D-00438
Dimethyl (1,3-dihydro-3-oxo-1-isobenzofuranyl)phosphonate, *in* D-00535
Dimethyl [(5,6-dihydro-4,4,6-trimethyl-4*H*-1,3-oxazin-2-yl]-phosphonate, *in* D-00598
Dimethyl (dimethylcarbamoylmethyl)phosphonate, *in* D-00703
Dimethyl dimethyldiphosphonate, *in* D-00751
Dimethyl (3,3-dimethyl-2-oxoheptyl)phosphonate, *in* D-00785
Dimethyl (3,3-dimethyl-2-oxo-4-phenylbutyl)phosphonate, *in* D-00786
Dimethyl (2,6-dimethylphenyl)phosphonate, *in* D-00814
Dimethyl (2,2-dimethylpropyl)phosphonate, *in* D-00912
Dimethyl (2,4-dinitrophenyl)phosphonate, *in* D-00946
2,4-Dimethyl-1,3,2-dioxaphosphorinane; (2*RS*,4*SR*)-*form*, 2-Oxide, *in* D-00734
Dimethyl ethenylphosphonate, *in* V-00003
Dimethyl ethylphosphonate, D-00754
Dimethyl (3-fluorophenyl)phosphonate, *in* F-00037
Dimethyl (4-fluorophenyl)phosphonate, *in* F-00038
Dimethyl (2-furanylhydroxymethyl)phosphonate, *in* F-00071
Dimethyl 2-furanylphosphonate, *in* F-00073
Dimethyl 3-furanylphosphonate, *in* F-00074
Dimethyl heptylphosphonate, *in* H-00006
Dimethyl hexylphosphonate, *in* H-00097
Dimethyl (1-hydroxybutyl)phosphonate, *in* H-00114
Dimethyl (1-hydroxyethyl)phosphonate, *in* H-00145
Dimethyl (2-hydroxyethyl)phosphonate, *in* H-00146

Dimethyl (hydroxymethyl)phosphonate, *in* H-00162
Dimethyl (hydroxyphenylmethyl)phosphonate, *in* H-00174
Dimethyl [(2-hydroxyphenyl)methyl]phosphonate, *in* H-00175
Dimethyl [(4-hydroxyphenyl)methyl]phosphonate, *in* H-00176
Dimethyl (2-hydroxyphenyl)phosphonate, *in* H-00179
Dimethyl (3-hydroxyphenyl)phosphonate, *in* H-00180
Dimethyl (4-hydroxyphenyl)phosphonate, *in* H-00181
Dimethyl (1-hydroxypropyl)phosphonate, *in* H-00190
Dimethyl (2-hydroxypropyl)phosphonate, *in* H-00191
Dimethyl (3-hydroxypropyl)phosphonate, *in* H-00192
Dimethyl 1*H*-imidazol-1-ylphosphonate, *in* I-00003
Dimethyl (isocyanatomethyl)phosphonate, *in* I-00022
Dimethyl isopropylphosphonate, *in* I-00050
Dimethyl [(4-methoxyphenyl)methyl]phosphonate, *in* M-00061
Dimethyl (2-methoxyphenyl)phosphonate, *in* M-00070
Dimethyl (3-methoxyphenyl)phosphonate, *in* M-00071
Dimethyl (4-methoxyphenyl)phosphonate, *in* M-00072
Dimethyl (methylaminocarbonyl)phosphonate, *in* M-00088
Dimethyl (2-methyl-1,3-butadienyl)phosphonate, *in* M-00093
Dimethyl (3-methyl-1,2-butadienyl)phosphonate, *in* M-00094
Dimethyl (3-methyl-1,3-butadienyl)phosphonate, *in* M-00095
Dimethyl (3-methyl-2-butenyl)phosphonate, *in* M-00102
Dimethyl (1-methylethenyl)phosphonate, *in* M-00149
Dimethyl (3-methyloxiranyl)phosphonate, *in* M-00185
Dimethyl (3-methyl-2-oxoheptyl)phosphonate, *in* M-00187
Dimethyl (2-methylphenyl)phosphate, *in* M-00235
Dimethyl (3-methylphenyl)phosphonate, *in* M-00236
Dimethyl (4-methylphenyl)phosphonate, *in* M-00237
Dimethyl [[(4-methylphenyl)sulfonyl]methyl]phosphonate, *in* M-00266
▷Dimethyl methylphosphonate, D-00762
Dimethyl (1-methylpropyl)phosphonate, *in* M-00377
Dimethyl 2-methyl-4-pyridinephosphonate, *in* M-00388
Dimethyl (3-methyl-4-pyridinyl)phosphonate, *in* M-00390
Dimethyl [(methylsulfinyl)methyl]phosphonate, *in* M-00398
Dimethyl [(methylsulfonyl)methyl]phosphonate, *in* M-00400
Dimethyl [(methylthio)methyl]phosphonate, *in* M-00415
▷Dimethyl 4-morpholinylphosphonate, *in* M-00523
Dimethyl[1,8-naphthalenediylbis[methylene]]bisphosphonate, *in* N-00004
Dimethyl 1-naphthylphosphonate, *in* N-00021
Dimethyl (1-nitroethyl)phosphonate, *in* N-00035
Dimethyl (2-nitroethyl)phosphonate, *in* N-00036
Dimethyl (nitromethyl)phosphonate, *in* N-00037
Dimethyl [(2-nitrophenyl)methyl]phosphonate, *in* N-00044
Dimethyl (2-nitrophenyl)phosphonate, *in* N-00055
Dimethyl (3-nitrophenyl)phosphonate, *in* N-00056
Dimethyl nonylphosphonate, *in* N-00076
Dimethyl octylphosphonate, *in* O-00042
Dimethyl (2-oxoethyl)phosphonate, *in* O-00059
Dimethyl (2-oxoheptyl)phosphonate, *in* O-00061
Dimethyl (2-oxononyl)phosphonate, *in* O-00063
Dimethyl (2-oxooctyl)phosphonate, *in* O-00064
Dimethyl (2-oxo-4-phenylbutyl)phosphonate, *in* O-00065
Dimethyl (2-oxo-2-phenylethyl)phosphonate, *in* O-00066
Dimethyl(2-oxo-3-phenylpropyl)phosphonate, *in* O-00068
Dimethyl (1-oxopropyl)phosphonate, *in* O-00074
Dimethyl (2-oxopropyl)phosphonate, *in* O-00075
Dimethyl 1,3-pentadienylphosphonate, *in* P-00009
Dimethyl (pentafluorophenyl)phosphonate, *in* P-00015
Dimethyl pentylphosphonate, *in* P-00050
Dimethyl (1-phenylethenyl)phosphonate, *in* P-00154
Dimethyl 1-phenylethylphosphonate, *in* P-00164
Dimethyl 2-phenylethylphosphonate, *in* P-00165
Dimethyl phenylethynylphosphonate, *in* P-00166
Dimethyl phenylphosphonate, D-00810
Dimethyl (3-phenylpropyl)phosphonate, *in* P-00305
Dimethyl [(phenylsulfinyl)methyl]phosphonate, *in* P-00310
Dimethyl [(phenylthio)methyl]phosphonate, *in* P-00326
Dimethyl [phenyl[(trimethylsilyl)oxy]methyl]phosphonate, *in* P-00331
▷Dimethyl phosphonate, D-00851
Dimethyl 1-piperidinylphosphonate, *in* P-00430
Dimethyl 1,2-propadienylphosphonate, *in* P-00441
Dimethyl 1-propenylphosphonate, *in* P-00453
Dimethyl 2-propenylphosphonate, *in* P-00454
Dimethyl propylphosphonate, *in* P-00473
Dimethyl 1-propynylphosphonate, *in* P-00494
Dimethyl 2-pyridinylphosphonate, *in* P-00503
Dimethyl 4-pyridinylphosphonate, *in* P-00505
Dimethyl 1-pyrrolidinylphosphonate, *in* P-00518
Dimethyl 4-quinolinephosphonate, *in* Q-00003
Dimethyl (2,3,5,6-tetrachloro-4-pyridinyl)phosphonate, *in* T-00035
Dimethyl (2-thienylmethyl)phosphonate, *in* T-00304
Dimethyl (3-thienylmethyl)phosphonate, *in* T-00305
Dimethyl 2-thienylphosphonate, *in* T-00307
Dimethyl (trichloromethyl)phosphonate, *in* T-00411
Dimethyl [1-[(trimethylsilyl)oxy]ethyl]phosphonate, *in* T-00555
Dimethyl triphenylmethylphosphonate, *in* T-00608
Dimethyl (9*H*-xanthen-9-yl)phosphonate, *in* X-00001
Dinonyl phosphonate, D-00950
Dioctadecyl phosphonate, D-00954
Dioctyl (dimethylcarbamoylmethyl)phosphonate, *in* D-00703
▷Dioctyl phenylphosphonate, *in* P-00221
Dioctyl phosphonate, D-00959
1,4,2-Dioxaphosphorinane 2-oxide; 2-Ethoxy, *in* D-00973
Dipentyl pentylphosphonate, *in* P-00050
Dipentyl phosphonate, D-00982
[(Diphenoxyphosphinyl)methyl]triphenylphosphonium(1+); Chloride, *in* D-00990
Diphenyl acetylphosphonate, *in* A-00026
Diphenyl aminocarbonylphosphonate, *in* C-00003
Diphenyl (1-aminoethyl)phosphonate, *in* A-00078
Diphenyl (1-amino-2-methylpropyl)phosphonate, *in* A-00094
Diphenyl (aminophenylmethyl)phosphonate, *in* A-00105
Diphenyl (4-aminophenyl)phosphonate, *in* A-00111
Diphenyl benzylphosphonate, *in* B-00060
Diphenyl (1,3-butadienyl)phosphonate, *in* B-00510
Diphenyl butylphosphonate, *in* B-00594
Diphenyl (2-chloroethyl)phosphonate, *in* C-00074
Diphenyl (3-chloropropyl)phosphonate, *in* C-00172
Diphenyl cyclohexylphosphonate, *in* C-00256
Diphenyl decylphosphonate, *in* D-00012
Diphenyl (dichloromethyl)phosphonate, *in* D-00175
Diphenyl (2,2-diethoxyethyl)phosphonate, *in* D-00234
Diphenyl (1,1-dimethylethyl)phosphonate, *in* B-00595
Diphenyl (2,2-dimethylpropyl)phosphonate, *in* D-00912
Diphenyl diphenyldiphosphonate, *in* D-01014
Diphenyl ethylphosphonate, *in* E-00129
Diphenyl 2-furanylphosphonate, *in* F-00073
Diphenyl (1-hydroxy-1-methylethyl)phosphonate, *in* H-00159
Diphenyl 1*H*-imidazol-1-ylphosphonate, *in* I-00003
Diphenyl (iodomethyl)phosphonate, *in* I-00013
Diphenyl isopropylphosphonate, *in* I-00050
Diphenyl (4-methoxyphenyl)phosphonate, *in* M-00072
Diphenyl (2-methyl-1,3-butadienyl)phosphonate, *in* M-00093
Diphenyl (1-methylethenyl)phosphonate, *in* M-00149
Diphenyl (3-methyl-2-oxiranyl)phosphonate, *in* M-00185
Diphenyl (4-methylphenyl)phosphonate, *in* M-00237
Diphenyl methylphosphonate, D-01024
Diphenyl (2-methylpropyl)phosphonate, *in* M-00378
Diphenyl 1-methyl-1*H*-pyrrol-2-ylphosphonate, *in* P-00521
▷Diphenyl 4-morpholinylphosphonate, *in* M-00523
Diphenyl octylphosphonate, *in* O-00042
Diphenyl (2-oxopropyl)phosphonate, *in* O-00075
Diphenyl phenylphosphonate, *in* P-00221
Diphenyl phosphonate, D-01105
Diphenyl 1-piperidinylphosphonate, *in* P-00430
Diphenyl 1-propynylphosphonate, *in* P-00494
Diphenyl (2,3,5,6-tetrachloro-4-pyridinyl)phosphonate, *in* T-00035
Diphenyl 2-thienylphosphonate, *in* T-00307
Diphenyl (trichloromethyl)phosphonate, *in* T-00411
Diphenyl (2,4,5-trimethylphenyl)phosphonate, *in* T-00526
Diphenyl [(triphenylphosphoranylidene)methyl]phosphonate, D-01163
Di-2-propenyl (3-chloropropyl)phosphonate, *in* C-00172
Di-2-propenyl (3-methyl-2-oxiranyl)phosphonate, *in* M-00185
Di-2-propenyl methylphosphonate, *in* M-00288
Di-2-propenyl phenylphosphonate, *in* P-00221
▷Di-2-propenyl phosphonate, D-01194
Di-2-propenyl 2-propenylphosphonate, *in* P-00454

Di-2-propenyl (trichloromethyl)phosphonate, *in* T-00411
Dipropyl (3-methyl-2-oxiranyl)phosphonate, *in* M-00185
Dipropyl methylphosphonate, *in* M-00288
Dipropyl phenylphosphonate, *in* P-00221
Dipropyl phosphonate, D-01210
Dipropyl propylphosphonate, *in* P-00473
Ditetradecyl phosphonate, D-01239
1,3,2-Dithiaphospholane 2-oxide; 2-Isopropyl, *in* D-01244
2-Ethoxy-2,3-dihydro-1,2-benzoxaphosphole 2-oxide, *in* H-00119
2-Ethoxy-1,2-oxaphospholane 2-oxide, *in* H-00166
2-Ethoxy-1,2-oxaphosphorinane 2-oxide, *in* H-00167
2-Ethoxy-3,5,5-trimethyl-1,2-oxaphospholan-3-ol 2-oxide, *in* T-00519
Ethyl *P*-acetyl-*N*,*N*-diethylphosphonamidate, *in* A-00014
Ethyl *P*-aminocarbonyl-*N*,*N*-diethylphosphonamidate, *in* A-00059
Ethyl 1-aziridinyl-*P*-methylphosphonamidate, *in* A-00152
Ethyl benzylphosphonochloridate, *in* B-00063
Ethyl 2,3-bis(diethoxyphosphinyl)propanoate, *in* B-00173
Ethyl 3,3-bis(diethoxyphosphinyl)propanoate, *in* D-01187
Ethyl [bis(trimethylsilyloxy)phosphinyl]acetate, *in* B-00434
Ethyl *P*-(1,3-butadienyl)-*N*,*N*-diethylphosphonamidate, *in* B-00508
Ethyl *P*-(butoxymethyl)-*N*,*N*-diethylphosphonamidate, *in* B-00536
Ethyl *N*-*tert*-butyl-*P*-methylphosphonamidate, *in* B-00572
Ethyl butylphosphonisocyanatidate, *in* B-00601
Ethyl chloro(diethoxyphosphinyl)acetate, *in* C-00036
Ethyl (2-chloroethyl)phosphonochloridate, *in* C-00077
Ethyl *P*-chloromethylphosphonamidate, *in* C-00105
Ethyl cyano(diethoxyphosphinyl)acetate, E-00057
Ethyl cyclohexylphosphonochloridate, *in* C-00259
Ethyl (dibutoxyphosphinyl)acetate, *in* D-00073
Ethyl *N*,*N*-dibutylphosphonamidate, *in* D-00126
Ethyl *P*-dichloromethylphosphonamidate, *in* D-00174
Ethyl α-(diethoxyphosphinyl)benzeneacetate, *in* D-00242
Ethyl [2-(diethoxyphosphinyl)]benzoate, *in* P-00378
Ethyl [3-(diethoxyphosphinyl)]benzoate, *in* P-00379
Ethyl [4-(diethoxyphosphinyl)]benzoate, *in* P-00380
Ethyl 2-(diethoxyphosphinyl)butanoate, *in* D-00246
Ethyl 4-(diethoxyphosphinyl)butanoate, *in* D-00247
Ethyl 4-(diethoxyphosphinyl)-2-butenoate, *in* D-00248
Ethyl (diethoxyphosphinyl)fluoroacetate, *in* D-00250
▷Ethyl 2-(diethoxyphosphinyl)-3-oxobutanoate, *in* D-00253
Ethyl 4-(diethoxyphosphinyl)-3-oxobutanoate, *in* D-00254
Ethyl 2-(diethoxyphosphinyl)propanoate, E-00059
Ethyl 3-(diethoxyphosphinyl)propanoate, *in* D-00259
Ethyl 2-(diethoxyphosphinyl)propenoate, *in* D-00260
Ethyl 3-(diethoxyphosphinyl)propenoate, *in* D-00261
Ethyl *N*,*N*-diethyl-*P*-(methoxymethyl)phosphonamidate, *in* D-00295
Ethyl *N*,*N*-diethyl-*P*-methylphosphonamidate, *in* D-00296
Ethyl *N*,*N*-diethylphosphonamidate, *in* D-00337
Ethyl *N*,*P*-diethylphosphonamidate, *in* D-00336
Ethyl *N*,*N*-diethyl-*P*-propadienylphosphonamidate, *in* D-00399
Ethyl *N*,*N*-diethyl-*P*-1-propynylphosphonamidate, *in* D-00403
Ethyl *N*-(4,5-dihydro-2-thiazolyl)-*P*-methylphosphonamidate, *in* D-00587
Ethyl (diisopropoxyphosphinyl)acetate, *in* D-00633
Ethyl *N*,*N*-diisopropyl-*P*-methylphosphonamidate, *in* D-00636
Ethyl (dimethoxyphosphinyl)acetate, *in* D-00678
Ethyl 2-(dimethoxyphosphinyl)propanoate, *in* D-00684
Ethyl *N*,*P*-dimethylphosphonamidate, *in* D-00849
Ethylene methylphosphonate, *in* M-00126
Ethylene phenylphosphonate, *in* P-00115
Ethyl *P*-ethyl-*N*,*N*-diphenylphosphonamidate, *in* E-00067
Ethyl *N*-ethyl-*P*-methylphosphonamidate, *in* E-00088
Ethyl *P*-ethyl-*N*-phenylphosphonamidate, *in* E-00110
Ethyl *P*-ethylphosphonamidate, *in* E-00126
Ethyl ethylphosphonisocyanatidate, *in* E-00133
Ethyl ethylphosphonochloridate, *in* E-00135
Ethyl ethylphosphonofluoridate, *in* E-00142
Ethyl hydrogen (α-aminobenzyl)phosphonate, *in* A-00105
Ethyl hydrogen (1-amino-1-methylethyl)phosphonate, *in* A-00087
Ethyl hydrogen (1-amino-1-methylpropyl)phosphonate, *in* A-00093
Ethyl hydrogen (2-aminophenyl)phosphonate, *in* A-00109
Ethyl hydrogen benzylphosphonate, *in* B-00060
Ethyl hydrogen [(1,3-dihydro-1,3-diazo-2*H*-isoindol-2-yl)-methyl]phosphonate, *in* P-00426
Ethyl hydrogen ethylphosphonate, *in* E-00129
Ethyl hydrogen (4-hydroxyphenyl)phosphonate, *in* H-00181
Ethyl hydrogen (2-nitrophenyl)phosphonate, *in* N-00055
Ethyl hydrogen (4-nitrophenyl)phosphonate, *in* N-00057
Ethyl hydrogen[(2-quinolinyl)methyl]phosphonate, *in* Q-00004
Ethyl *N*-isopropyl-*P*-methylphosphonamidate, *in* I-00036
Ethyl [(methoxycarbonyl)methyl]phosphonamidate, *in* M-00028
Ethyl *P*-methyl-*N*-phenylphosphonamidate, *in* M-00231
Ethyl *P*-methylphosphonamidate, *in* M-00283
Ethyl methylphosphonate, E-00089
Ethyl methylphosphonazidate, *in* M-00286
Ethyl methylphosphonisocyanatidate, *in* M-00292
Ethyl methylphosphonisothiocyanatidate, *in* M-00294
Ethyl methylphosphonochloridate, *in* M-00296
Ethyl methylphosphonofluoridate, *in* M-00303
Ethyl *P*-methylphosphonohydrazidate, *in* M-00308
▷Ethyl 4-nitrophenyl phenylphosphonate, *in* M-00471
Ethyl phenyl methylphosphonate; (*R*)-*form*, *in* E-00100
Ethyl phenyl methylphosphonate; (*S*)-*form*, *in* E-00100
Ethyl phenyl methylphosphonate; (±)-*form*, *in* E-00100
Ethyl phenyl phenylphosphonate, *in* M-00471
Ethyl *P*-phenyl-*N*-phenylsulfonylphosphonamidate, *in* P-00191
Ethyl *P*-phenylphosphonamidate, *in* P-00218
P-Ethyl-*N*-phenylphosphonamidic acid; (*S*)-*form*, Et ester, *in* E-00110
Ethyl phenyl phosphonate, E-00111
Ethyl phenylphosphonochloridate, *in* P-00229
Ethyl phenylphosphonofluoridate, *in* P-00239
Ethyl 3-phosphonobutanoate, *in* P-00385
Ethylphosphonochloridic acid; (*S*)-*form*, Et ester, *in* E-00135
Ethylphosphonochloridic acid; (±)-*form*, Et ester, *in* E-00135
Ethyl *P*-trichloromethylphosphonamidate, *in* T-00410
Ethyl triethylphosphonamidate, *in* T-00446
Ethyl trimethylphosphonamidate, *in* T-00537
Hexaethyl (nitrilotri-2,1-ethanediyl)trisphosphonate, *in* T-00438
Hexaethyl [nitrilotris(methylene)]trisphosphonate, *in* N-00033
Hexaethyl [phosphinylidynetris(methylene)]trisphosphonate, *in* P-00360
Hexaethyl 1,1,3-propanetriyltrisphosphonate, *in* P-00449
Hexaethyl 1,2,3-propanetriyltrisphosphonate, *in* P-00450
Hexaethyl 1,3,5-triazine-2,4,6-triyltriphosphonate, *in* T-00329
Hexaisopropyl [phosphinylidynetris(methylene)]-trisphosphonate, *in* P-00360
Hexaisopropyl 1,2,3-propanetriyltrisphosphonate, *in* P-00450
Hexaisopropyl 1,3,5-triazine-2,4,6-triylphosphonate, *in* T-00329
Hexamethyl 2,3-diphosphonobutanedioate, *in* D-01182
Hexamethyl [phosphinylidynetris(methylene)]trisphosphonate, *in* P-00360
Hexamethyl 1,3,5-triazine-2,4,6-triyltriphosphonate, *in* T-00329
Hexaphenyl 1,3,5-triazine-2,4,6-triylphosphonate, *in* T-00329
Hexyl phosphinate, H-00095
Hydrogen 4-nitrophenyl (dichloromethyl)phosphonate, *in* D-00175
Hydrogen propyl methylphosphonate, *in* M-00288
2-Hydroxy-2*H*-1,2-benzoxaphosphorin 2-oxide; 2-Ethoxy (Et ester), *in* H-00112
2-Hydroxy-2*H*-1,2-benzoxaphosphorin 2-oxide; 2-Phenoxy (Ph ester), *in* H-00112
▷2-Hydroxy-1,3,2-dioxaphospholane, *in* D-00970
2-Hydroxy-4-methyl-1,3,2-dioxaphospholane, *in* M-00127
(1-Hydroxy-1-methylethyl)phosphonic acid; Bis(trimethylsilyl) ester, *O*-trimethylsilyl ether, *in* H-00159
Isopropyl *N*,*N*-diethyl-*P*-methylphosphonamidate, *in* D-00296
Isopropyl (diisopropoxyphosphinyl)acetate, *in* D-00633
4-Isopropyl-5,5-dimethyl-1,3,2-dioxaphosphorinane; (2*RS*,4*RS*)-*form*, 2-Oxide, *in* I-00028
4-Isopropyl-5,5-dimethyl-1,3,2-dioxaphosphorinane; (2*RS*,4*SR*)-*form*, 2-Oxide, *in* I-00028

Isopropyl *N,P*-dimethylphosphonamidate, *in* D-00849
Isopropyl *N*-ethyl-*P*-methylphosphonamidate, *in* E-00088
Isopropyl ethylphosphonochloridate, *in* E-00135
▷Isopropyl ethylphosphonofluoridate, *in* E-00142
Isopropyl (fluoromethyl)phosphonochloridate, *in* F-00024
Isopropyl *N*-isopropyl-*P*-methylphosphonamidate, *in* I-00036
Isopropyl methyl methylphosphonate, *in* M-00168
Isopropyl *P*-methyl-4-morpholinylphosphonamidate, *in* M-00175
Isopropyl *P*-methyl-*N*-phenylphosphonamidate, *in* M-00231
Isopropyl methylphosphonate, I-00038
Isopropyl methylphosphonisothiocyanatidate, *in* M-00294
Isopropyl methylphosphonochloridate, *in* M-00296
Isopropyl *P*-methylphosphonohydrazidate, *in* M-00308
Isopropyl *P*-methyl-1-piperidinylphosphonamidate, *in* M-00361
Isopropyl phenyl methylphosphonate, *in* I-00038
Isopropyl phenyl phosphonate, I-00047
▷Isopropyl triethylphosphonamidate, *in* T-00446
Isopropyl trimethylphosphonamidate, *in* T-00537
2-Methoxy-4-methyl-1,2-oxaphospholane; 2-Oxide, *in* M-00049
2-Methoxy-3,3,4,4-tetramethyl-1,2-azaphospholidine 2-oxide, *in* T-00191
2-Methoxy-3,5,5-trimethyl-1,2-oxaphospholan-3-ol 2-oxide, *in* T-00519
Methyl *P*-acetyl-*N,N*-diethylphosphonamidate, *in* A-00014
Methyl [3-amino-3-(dimethoxyphosphinyl)]propanoate, *in* A-00122
2-Methylaminoethylphosphonic acid, *in* A-00079
Methyl 4,4-bis(diethoxyphosphinyl)butanoate, *in* D-01183
Methyl 2,3-bis(diethoxyphosphinyl)propanoate, *in* B-00173
Methyl 3,3-bis(dimethoxyphosphinyl)propanoate, *in* D-01187
Methyl [bis(trimethylsilyloxy)phosphinyl]acetate, *in* B-00434
(3-Methyl-1-butenyl)phosphonic acid; (*Z*)-*form*, Di-Et ester, *in* M-00101
Methyl *N-tert*-butyl-*P*-methylphosphonamidate, *in* B-00572
Methyl *tert*-butylphosphon(isocyanatidate), *in* B-00602
Methyl *N*-cyclohexyl-*P*-phenylphosphonamidate, *in* C-00251
Methyl cyclohexylphosphonochloridate, *in* C-00259
Methyl 3-(dibutoxyphosphinyl)propanoate, *in* P-00396
Methyl *P*-dichloromethylphosphonamidate, *in* D-00174
Methyl 2-(diethoxyphosphinyl)benzoate, *in* D-00243
Methyl 4-(diethoxyphosphinyl)benzoate, *in* D-00244
Methyl 4-(diethoxyphosphinyl)butanoate, *in* D-00247
Methyl 4-(diethoxyphosphinyl)-2-butenoate, *in* D-00248
Methyl (diethoxyphosphinyl)fluoroacetate, *in* D-00250
Methyl (diethoxyphosphinyl)oxoacetate, *in* D-00252
Methyl 2-(diethoxyphosphinyl)propanoate, *in* D-00258
Methyl 3-(diethoxyphosphinyl)propanoate, *in* D-00259
Methyl 2-(diethoxyphosphinyl)propenoate, *in* D-00260
Methyl 3-(diethoxyphosphinyl)propenoate, *in* D-00261
Methyl *N,N*-diethyl-*P*-methylphosphonamidate, *in* D-00296
Methyl *N,N*-diethyl-*P*-phenylphosphonamidate, *in* D-00307
Methyl *N,N*-diethylphosphonamidate, *in* D-00337
Methyl (dimethoxyphosphinyl)acetate, *in* D-00678
Methyl 2-(dimethoxyphosphinyl)butanoate, *in* D-00679
Methyl 2-(dimethoxyphosphinyl)propanoate, *in* D-00684
Methyl 3-(dimethoxyphosphinyl)propanoate, *in* D-00685
Methyl 3-(dimethoxyphosphinyl)propenoate, *in* D-00686
Methyl [3-dimethylamino-3-(diethoxyphosphinyl)]propanoate, *in* A-00122
Methyl 2,3-di-*O*-methylglucopyranoside 4,6-cyclic phosphonate; α-D-(*R*)P-*form*, *P*-Me, *in* M-00119
Methyl 2,3-di-*O*-methylglucopyranoside 4,6-cyclic phosphonate; α-D-(*R*)P-*form*, *P*-Ph, *in* M-00119
Methyl 2,3-di-*O*-methylglucopyranoside 4,6-cyclic phosphonate; α-D-(*S*)P-*form*, *P*-Me, *in* M-00119
Methyl 2,3-di-*O*-methylglucopyranoside 4,6-cyclic phosphonate; α-D-(*S*)P-*form*, *P*-Ph, *in* M-00119
Methyl *N,N*-dimethyl-*P*-phenylphosphonamidate, *in* D-00808
Methyl *N,P*-dimethylphosphonamidate, *in* D-00849
Methyl *N,P*-diphenylphosphonamidate, *in* D-01103
Methyl *N*-ethyl-*P*-methylphosphonamidate, *in* E-00088
Methyl *N*-ethyl-*P*-phenylphosphonamidate, *in* E-00109
Methyl *P*-ethyl-*N*-phenylphosphonamidate, *in* E-00110
Methyl ethylphosphonofluoridate, *in* E-00142
Methyl hydrogen [[1-amino-2-(1*H*-indol-3-yl)]ethyl]-phosphonate, *in* P-00421
Methyl hydrogen (2-aminophenyl)phosphonate, *in* A-00109
Methyl hydrogen benzylphosphonate, *in* B-00060
Methyl hydrogen chloromethylphosphonate, *in* C-00106
Methyl hydrogen (1,2-dibromo-1-phenylpropyl)phosphonate, *in* D-00068
Methyl hydrogen (α-hydroxybenzyl)phosphonate, *in* H-00174
Methyl hydrogen [[(4-methylphenyl)sulfinyl]methyl]-phosphonate, *in* M-00266
Methyl hydrogen (2-nitrophenyl)phosphonate, *in* N-00055
Methyl hydrogen (4-nitrophenyl)phosphonate, *in* N-00057
Methyl hydrogen (2-phenylethenyl)phosphonate, *in* P-00155
Methyl *N*-isopropyl-*P*-methylphosphonamidate, *in* I-00036
6-Methyl-4-methylene-4*H*-1,3,2-dioxaphosphorin; 2-Me, 2-oxide, *in* M-00167
Methyl *N*-methyl-*P*-phenylphosphonamidate, *in* M-00230
Methyl *P*-methyl-*N*-phenylphosphonamidate, *in* M-00231
Methyl methylphosphonate, M-00168
Methyl methylphosphonisothiocyanatidate, *in* M-00294
Methyl methylphosphonochloridate, *in* M-00296
Methyl methylphosphonofluoridate, *in* M-00303
5-Methyl-1,2-oxaphospholane 2-oxide; 2-Methoxy, *in* M-00183
(3-Methyloxiranyl)phosphonic acid; (2*S*,3*R*)-*form*, Di-Me ester, *in* M-00185
(3-Methyloxiranyl)phosphonic acid; (2*RS*,3*SR*)-*form*, Di-Me ester, *in* M-00185
(3-Methyl-2-oxoheptyl)phosphonic acid; (*S*)-*form*, Me ester, *in* M-00187
(3-Methyl-2-oxoheptyl)phosphonic acid; (±)-*form*, Me ester, *in* M-00187
4-Methyl-2-phenyl-1,3,2-dioxaphospholane; 2-Oxide, *in* M-00198
Methyl phenyl methylphosphonate, M-00205
Methyl phenyl phenylphosphonate, *in* M-00488
Methyl *P*-phenyl-*N*-phenylsulfonylphosphonamidate, *in* P-00191
Methyl *P*-phenylphosphonamidate, *in* P-00218
Methyl phenyl phosphonate, M-00234
Methyl phenylphosphonazidate, *in* P-00220
Methyl phenylphosphonochloridate, *in* P-00229
Methyl phenylphosphonofluoridate, *in* P-00239
Methyl *P*-phenylphosphonohydrazidate, *in* P-00241
[[(4-Methylphenyl)sulfinyl]methyl]phosphonic acid; (*S*)-*form*, Mono-Me ester, *in* M-00266
[[(4-Methylphenyl)sulfinyl]methyl]phosphonic acid; (*S*)-*form*, Di-Me ester, *in* M-00266
[[(4-Methylphenyl)sulfinyl]methyl]phosphonic acid; (±)-*form*, Di-Me ester, *in* M-00266
Methyl *N*-phenyl-*P*-trichloromethylphosphonamidate, *in* P-00328
Methylphosphonic acid; Bis(trimethylsilyl) ester, *in* M-00288
Methylphosphonochloridic acid; (*S*)-*form*, Isopropyl ester, *in* M-00296
Methylphosphonochloridic acid; (±)-*form*, Isopropyl ester, *in* M-00296
2-Methylpropyl *N,P*-diethylphosphonamidate, *in* D-00336
2-Methylpropyl ethylphosphonochloridate, *in* E-00135
2-Methylpropyl *P*-methylphosphonohydrazidate, *in* M-00308
17-Methyl-6,7,9,10-tetrahydrodibenzo[*d,m*][1,3,6,9,12,2]-pentaoxaphosphacyclotetradecin; 17-Oxide, *in* M-00403
Methyl triethylphosphonamidate, *in* T-00446
Methyl trimethylphosphonamidate, *in* T-00537
Monobenzyl phosphonate, M-00444
Monobutyl ethylphosphonate, *in* E-00129
▷Monobutyl phosphonate, M-00453
Monoethyl (aminocarbonyl)phosphonate, M-00469
Monoethyl butylphosphonate, *in* B-00594
Monoethyl chloromethylphosphonate, *in* C-00106
Monoethyl cyclohexylphosphonate, *in* C-00256
Monoethyl (1,1-dimethylethyl)phosphonate, *in* B-00595
Monoethyl (2-phenylethenyl)phosphonate, *in* P-00155
Monoethyl phenylphosphonate, M-00471
Monoethyl phenylphosphonate; (*S*)-*form*, Ph ester, *in* M-00471
Monoethyl phenylphosphonate; (*S*)-*form*, Benzyl ester, *in* M-00471
Monoethyl phosphonate, M-00473
Monoisopropyl cyclohexylphosphonate, *in* C-00256
Monoisopropyl phosphonate, M-00478
Monomethyl phenylphosphonate; (*S*)-*form*, Ph ester, *in* M-00488
Monomethyl phenylphosphonate; (*S*)-*form*, Benzyl ester, *in* M-00488

▷Monomethyl phosphonate, M-00490
Mono-1-naphthyl phosphonate, M-00496
Monophenyl cyclohexylphosphonate, *in* C-00256
Monophenyl (dichloromethyl)phosphonate, *in* D-00175
Mono(phenylmethyl) (2-phenylethenyl)phosphonate, *in* P-00155
Monophenyl phosphonate, M-00508
Mono-2-propenyl phosphonate, M-00510
Monopropyl phosphonate, M-00513
4-Morpholinylphosphonic diamide, *in* M-00523
1,2-Naphthylene phosphonate, *in* N-00009
1,8-Naphthylene phosphonate, *in* N-00010
Neopentylene methylphosphonate, *in* T-00512
Neopentylene phenylphosphonate, *in* D-00795
Neopentylene phosphonate, *in* D-00735
[(4-Nitrophenyl)methyl]phosphonic acid; Di-Me ester, *in* N-00046
[(4-Nitrophenyl)methyl]phosphonic acid; Di-Et ester, *in* N-00046
13,14,16,17,31,32,34,35-Octahydro-6,24-dimethyltetrabenzo[*d,m,r,a'*]-[1,3,6,9,12,15,20,23,26,2,16]-decaoxadiphosphacyclooctacosin; 6,24-Dioxide, *in* O-00008
6,7,11,12,19,20,22,23-Octahydro-9-methyldibenzo[*g,p*]-[1,3,6,9,12,15,18,21]heptaoxaphosphacycloeicosin; 9-Oxide, *in* O-00019
2-Phenoxy-1,2-oxaphosphorinane 2-oxide, *in* H-00167
2-Phenyl-4*H*-1,3,2-benzodioxaphosphorin; 2-Oxide, *in* P-00095
Phenyl [bis(trimethylsilyloxy)phosphinyl]acetate, *in* B-00434
Phenyl (2-chloroethyl)phosphonochloridate, *in* C-00077
Phenyl *P*-chloromethylphosphonamidate, *in* C-00105
Phenyl (chloromethyl)phosphonochloridate, *in* C-00111
6-Phenyldibenzo[*d,f*][1,3,2]dioxaphosphepin; Oxide, *in* P-00107
Phenyl 2-(diethoxyphosphinyl)butanoate, *in* D-00246
Phenyl *N,N*-diethyl-*P*-methylphosphonamidate, *in* D-00296
Phenyl *N,N*-diethyl-*P*-phenylphosphonamidate, *in* D-00307
Phenyl *N*-(4,5-dihydro-2-thiazolyl)-*P*-methylphosphonamidate, *in* D-00587
Phenyl (dimethoxyphosphinyl)acetate, *in* D-00678
2-Phenyl-1,3,6,2-dioxathiaphosphocane; 2-Oxide, *in* P-00117
2-Phenyl-1,3,6,2-dioxathiaphosphocane; 2-Sulfide, *in* P-00117
▷Phenyl *N,P*-diphenylphosphonamidate, *in* D-01103
o-Phenylene methylphosphonate, *in* M-00090
o-Phenylene phenylphosphonate, *in* P-00094
o-Phenylene phosphite, *in* B-00018
2-Phenylethenylphosphonic acid; (*E*)-*form*, Di-Me ester, *in* P-00155
2-Phenylethenylphosphonic acid; (*E*)-*form*, Di-Et ester, *in* P-00155
Phenyl *N*-ethyl-*P*-methylphosphonamidate, *in* E-00088
Phenyl hydrogen (4-nitrophenyl)phosphonate, *in* N-00057
Phenyl hydrogen phenylphosphonate, *in* P-00221
Phenyl *N*-isopropyl-*P*-methylphosphonamidate, *in* I-00036
Phenyl [(methoxycarbonyl)methyl]phosphonamidate, *in* M-00028
Phenyl *P*-methyl-*N*-phenylphosphonamidate, *in* M-00231
Phenyl *P*-methyl-N^2-phenylphosphonohydrazidate, *in* M-00241
Phenyl *P*-methylphosphonamidate, *in* M-00283
Phenyl methylphosphonate, P-00171
Phenyl methylphosphonazidate, *in* M-00286
Phenyl methylphosphonisocyanatidate, *in* M-00292
Phenyl methylphosphonisothiocyanatidate, *in* M-00294
Phenyl methylphosphonochloridate, *in* M-00296
Phenyl *P*-methyl-*N*-2-thiazolylphosphonamidate, *in* M-00409
2-Phenyl-1,2-oxaphospholane; 2-Oxide, *in* P-00174
[Phenyl[(phenylmethyl)amino]methyl]phosphonic acid; (±)-*form*, Di-Et ester, *in* P-00188
Phenyl *P*-phenyl-*N*-phenylsulfonylphosphonamidate, *in* P-00191
Phenyl *P*-phenylphosphonamidate, *in* P-00218
▷Phenyl phenylphosphonochloridate, *in* P-00229
Phenyl phenylphosphonofluoridate, *in* P-00239
Phenylphosphonic acid; Bis(4-nitrophenyl) ester, *in* P-00221
Phenylphosphonic acid; Bis(trimethylsilyl) ester, *in* P-00221
(1-Phenyl-1-propenyl)phosphonic acid; (*E*)-*form*, Mono-Me ester, *in* P-00301
(1-Phenyl-1-propenyl)phosphonic acid; (*Z*)-*form*, Mono-Me ester, *in* P-00301
Phenyl propyl phosphonate, P-00304
Phenyl *P*-trichloromethylphosphonamidate, *in* T-00410
Phosphotryptophan; (±)-*form*, Di-Me ester, *N*-benzyloxycarbonyl, *in* P-00421
1-Propenylphosphonic acid; (*Z*)-*form*, Di-Me ester, *in* P-00453
1-Propenylphosphonic acid; (*Z*)-*form*, Di-Et ester, *in* P-00453
Propyl *N,N*-diethyl-*P*-(methoxymethyl)phosphonamidate, *in* D-00295
Propylene methylphosphonate, *in* D-00732
3-Pyridinephosphonic acid; Di-Et ester, *in* P-00504
1*H*-Pyrrol-2-ylphosphonic acid; *N*-Me, Mono-Et ester, *in* P-00521
▷Sarin, S-00002
▷Soman, S-00005
▷Tetraallyl methylenediphosphonate, *in* M-00022
Tetrabutyl (aminophenylmethyl)bisphosphonate, *in* A-00053
Tetrabutyl 1,2-ethanediylbisphosphonate, *in* E-00006
Tetrabutyl [oxybis(methylene)]bisphosphonate, *in* O-00095
Tetraethyl (1-aminoethylidene)bisphosphonate, *in* A-00073
Tetraethyl (aminophenylmethyl)bisphosphonate, *in* A-00053
Tetraethyl [[1,1'-biphenyl]-4,4'-diylbis(methylene)]-bisphosphonate, *in* B-00391
Tetraethyl 1,3-butadiene-1,4-diylbisphosphonate, *in* B-00507
Tetraethyl 1,4-butanediylbisphosphonate, *in* B-00514
Tetraethyl (2-butene-1,4-diyl)bisphosphonate, *in* B-00519
Tetraethyl carbamoylphosphonate, *in* D-00264
Tetraethyl (1-cyano-1,2-ethanediyl)bisphosphonate, *in* B-00173
Tetraethyl 1,2-cyclohexanediylbisphosphonate, *in* C-00240
Tetraethyl 1,10-decanediylbisphosphonate, *in* D-00008
Tetraethyl dichloromethylenebisphosphonate, *in* D-00172
Tetraethyl [(dimethylamino)methylene]bisphosphonate, *in* D-00701
Tetraethyl 1,2-ethanediylbisphosphonate, *in* E-00006
Tetraethyl ethenylidenebisphosphonate, *in* E-00020
Tetraethyl [(ethoxyphosphinylidene)bis(methylene)]-bisphosphonate, *in* B-00182
Tetraethyl 1,7-heptanediylbisphosphonate, *in* H-00002
Tetraethyl (3,3,4,4,5,5-hexafluoro-1-cyclopentene-1,2-diyl)-bisphosphonate, *in* H-00018
Tetraethyl 1,6-hexanediylbisphosphonate, *in* H-00085
Tetraethyl (1-hydroxyethylidene)bisphosphonate, *in* H-00144
▷Tetraethyl iminobis(methylene)bisphosphonate, *in* D-00688
▷Tetraethyl methylenebisphosphonate, *in* M-00022
Tetraethyl (4-morpholinylmethylene)bisphosphonate, *in* M-00520
Tetraethyl [2,3-naphthalenediylbis[methylene]]-bisphosphonate, *in* N-00005
Tetraethyl [2,6-naphthalenediylbis[methylene]]-bisphosphonate, *in* N-00006
Tetraethyl [2,7-naphthalenediylbis[methylene]]-bisphosphonate, *in* N-00007
Tetraethyl 1,9-nonanediylbisphosphonate, *in* N-00069
Tetraethyl (3,3,4,4,5,5,6,6,-octafluoro-1-cyclohexene-1,2-diyl)bisphosphonate, *in* O-00003
Tetraethyl 1,8-octanediylbisphosphonate, *in* O-00035
Tetraethyl (1-oxo-1,3-propanediyl)bisphosphonate, *in* O-00071
Tetraethyl (2-oxo-1,3-propanediyl)bisphosphonate, *in* O-00072
Tetraethyl oxybis(1,2'-ethanediyl)bisphosphonate, *in* O-00094
Tetraethyl [oxybis(methylene)]bisphosphonate, *in* O-00095
Tetraethyl 1,5-pentanediylbisphosphonate, *in* P-00032
Tetraethyl pentylidenebisphosphonate, *in* P-00031
Tetraethyl [1,2-phenylenebis(methylene)]bisphosphonate, *in* P-00129
Tetraethyl [1,2-phenylenebis(methylene)]bisphosphonate, *in* P-00130
Tetraethyl [1,4-phenylenebis(methylene)]bisphosphonate, *in* P-00131
Tetraethyl 1,3-phenylenebisphosphonate, *in* B-00007
Tetraethyl 1,4-phenylenebisphosphonate, *in* B-00008
Tetraethyl phosphonobutanedioate, *in* D-00245
Tetraethyl 1,3-propanediylbisphosphonate, *in* P-00444
Tetraethyl 2,2-propanediyldiphosphonate, *in* P-00445
Tetraethyl (3,3,4,4-tetrafluoro-1-cyclobutene-1,2-diyl)-bisphosphonate, *in* T-00078
Tetraethyl [thiobis(methylene)]bisphosphonate, *in* T-00311
Tetrahydro-2,6-dimethyl-4*H*-1,3,6,2-dioxazaphosphocine; 2-Oxide, *in* T-00094

Tetraisopropyl dichloromethylenebisphosphonate, *in* D-00172
Tetraisopropyl ethenylidenebisphosphonate, *in* E-00020
Tetraisopropyl oxybis(1,2′-ethanediyl)bisphosphonate, *in* O-00094
Tetraisopropyl pentylidenebisphosphonate, *in* P-00031
Tetraisopropyl 2,2-propanediphosphonate, *in* P-00445
Tetraisopropyl [thiobis(methylene)]bisphosphonate, *in* T-00311
Tetrakis(1-methylethyl) methylenebisphosphonate, *in* M-00022
Tetramethyl [1-(acetoxy)ethylidene]bisphosphonate, *in* A-00005
Tetramethyl (aminomethylene)bisphosphonate, *in* A-00086
Tetramethyl (2-butene-1,4-diyl)bisphosphonate, *in* B-00519
Tetramethyl [(dimethylamino)methylene]bisphosphonate, *in* D-00701
Tetramethyl (1,4-dioxo-1,4-butanediyl)bisphosphonate, *in* D-00975
Tetramethylene methylphosphonate, *in* M-00125
Tetramethylene phenylphosphonate, *in* P-00114
Tetramethyl 1,2-ethanediylbisphosphonate, *in* E-00006
Tetramethyl ethenylidenebisphosphonate, *in* E-00020
Tetramethylethylene phenylphosphonate, *in* P-00322
Tetramethylethylene phosphonate, *in* T-00197
Tetramethyl (3,3,4,4,5,5-hexafluoro-1-cyclopentene-1,2-diyl)bisphosphonate, *in* H-00018
Tetramethyl (1-hydroxyethylidene)bisphosphonate, *in* H-00144
Tetramethyl (hydroxyphenylmethyl)bisphosphonate, *in* H-00172
Tetramethyl methylenebisphosphonate, *in* M-00022
Tetramethyl (2-oxo-1,3-propanediyl)bisphosphonate, *in* O-00072
Tetramethyl [1,4-phenylenebis(methylene)]bisphosphonate, *in* P-00131
Tetramethyl 1,2-phenylenebisphosphonate, *in* B-00006
Tetramethyl 1,3-phenylenebisphosphonate, *in* B-00007
Tetramethyl 1,4-phenylenebisphosphonate, *in* B-00008
Tetramethyl (3,3,4,4-tetrafluoro-1-cyclobutene-1,2-diyl)-bisphosphonate, *in* T-00078
Tetraphenyl 1,2-ethanediylbisphosphonate, *in* E-00006
Tetraphenyl methylenebisphosphonate, *in* M-00022
Tetraphenyl [1,4-phenylenebis(methylene)]bisphosphonate, *in* P-00131
Tetrapropyl [1,4-phenylenebis(methylene)]bisphosphonate, *in* P-00131
Thiobis[methanephosphonic acid]; Tetra-Et ester, *S*-oxide, *in* T-00311
Thiobis[methanephosphonic acid]; Tetra-Et ester, *S*,*S*-dioxide, *in* T-00311
2,2,2-Tribromoethyl 4-morpholinylphosphonochloridate, T-00357
Tributyl phosphonoacetate, *in* D-00073
▷Trichlorphon, T-00423
Triethyl phosphonoformate, *in* D-00236
Trimethyl (2-dimethoxyphosphinyl)-1,2,4-butanetrioate, *in* P-00382
Trimethyl 3-(dimethoxyphosphinyl)-1,2,3-butanetrioate, *in* P-00383
4,4,6-Trimethyl-1,3,2-dioxaphosphorinane; (2*RS*,4*RS*)-*form*, 2-Oxide, *in* T-00513
4,4,6-Trimethyl-1,3,2-dioxaphosphorinane; (2*RS*,4*SR*)-*form*, 2-Oxide, *in* T-00513
Trimethylene methylphosphonate, *in* M-00128
Trimethylene phenylphosphonate, *in* P-00116
Trimethylene phosphonate, *in* D-00972
3,5,5-Trimethyl-1,2-oxaphospholan-3-ol 2-oxide; (2*RS*,3*SR*)-*form*, 2-Methoxy, *in* T-00519
3,5,5-Trimethyl-1,2-oxaphospholan-3-ol 2-oxide; (2*RS*,3*SR*)-*form*, 2-Phenoxy, *in* T-00519
2,3,4-Trimethyl-5-phenyl-1,3,2-oxazaphospholidine; (2*R*,4*S*,5*R*)-*form*, 2-Oxide, *in* T-00523
2,3,4-Trimethyl-5-phenyl-1,3,2-oxazaphospholidine; (2*S*,4*S*,5*R*)-*form*, 2-Oxide, *in* T-00523
Trimethylsilyl 2-(diethoxyphosphinyl)propanoate, *in* D-00258
1,3,6,2-Trioxaphosphocane; 2-Me, *in* T-00592

Phosphonic amides

P-Acetyl-*N*,*N*,*N*′,*N*′-tetraethylphosphonic diamide, *in* A-00027
P-(4-Aminophenyl)-*N*,*N*′-dimethylphosphonic diamide, *in* A-00112
P-(4-Aminophenyl)-*N*,*N*′-diphenylphosphonic diamide, *in* A-00112
▷*P*-1-Aziridinyl-*N*,*N*,*N*′,*N*′-tetramethylphosphonic diamide, *in* A-00153
P-Benzoyl-*N*,*N*,*N*′,*N*′-tetraethylphosphonic diamide, *in* B-00035
Benzyl *N*-cyclohexyl-*P*-ethylphosphonamidate, *in* C-00246
Benzyl *N*-cyclohexyl-*P*-phenylphosphonamidate, *in* C-00251
1-Benzyl-2-ethoxyhexahydro-1,2-azaphosphorine 2-oxide, *in* B-00048
1-Benzylhexahydro-2-hydroxy-1,2-azaphosphorine 2-oxide, B-00048
Benzyl *P*-phenylphosphonamidate, *in* P-00218
Benzylphosphonic dianilide, *in* B-00061
Benzyl triethylphosphonamidate, *in* T-00446
P-[(1,1′-Biphenyl)-2-ylmethyl]-*N*,*N*′-diphenylphosphonic diamide, *in* B-00079
[Bis(diethylamino)phosphinyl]acetic acid, B-00175
[Bis(diethylamino)phosphinyl]acetic acid; Ph ester, *in* B-00175
[Bis(dimethylamino)phosphinyl]acetic acid, B-00193
P-(2-Bromoethyl)-*N*,*N*′-diphenylphosphonic diamide, *in* B-00472
▷Butonate, B-00530
Butyl *P*-(butoxymethyl)-*N*,*N*-diethylphosphonamidate, *in* B-00536
Butyl *N*,*N*-dibutylphosphonamidate, *in* D-00126
Butyl *N*,*N*-diethylphosphonamidate, *in* D-00337
Butyl *N*,*N*-diethyl-*P*-propadienylphosphonamidate, *in* D-00399
Butyl *N*,*N*-diethyl-*P*-1-propynylphosphonamidate, *in* D-00403
N,*N*′-1,3-Butylene *N*,*N*′-dimethylphosphonic diamide, *in* H-00057
N-*tert*-Butyl-*P*-methylphosphonamidic acid; Chloride, *in* B-00572
P-Butylphosphonic diamide, B-00596
Butylphosphonic *N*,*N*′-dibenzyldiamide, *in* B-00596
P-Butyl-*N*,*N*,*N*′,*N*′-tetraethylphosphonic diamide, *in* B-00596
P-Butyl-*N*,*N*,*N*′,*N*′-tetramethylphosphonic diamide, *in* B-00596
P-(4-Chlorobutyl)-*N*-phenylphosphonamidic acid, C-00024
P-(4-Chlorobutyl)-*N*-phenylphosphonamidic acid; Anhydride, *in* C-00024
(4-Chlorobutyl)phosphonic acid; Monoamide, *in* C-00025
P-(4-Chlorobutyl)phosphonic diamide, C-00026
(4-Chlorobutyl)phosphonic dianilide, *in* C-00026
(4-Chlorobutyl)phosphonic *N*,*N*′-dibenzylamide, *in* C-00026
P-2-Chloroethylphosphonic diamide, C-00075
P-(2-Chloroethyl)-*N*,*N*,*N*′,*N*′-tetraethylphosphonic diamide, *in* C-00075
P-(2-Chloroethyl)-*N*,*N*,*N*′,*N*′-tetramethylphosphonic diamide, *in* C-00075
P-Chloromethylphosphonic diamide, C-00107
Chloromethylphosphonic dianilide, *in* C-00107
P-Chloromethyl-*N*,*N*,*N*′,*N*′-tetraethylphosphonic diamide, *in* C-00107
P-Chloromethyl-*N*,*N*,*N*′,*N*′-tetramethylphosphonic diamide, *in* C-00107
P-4-Chlorophenyl-*N*,*N*,*N*′,*N*′-tetramethylphosphonic diamide, *in* C-00147
P-(3-Chloropropyl)-*N*,*N*′-diphenylphosphonic diamide, *in* C-00173
P-(3-Chloropropyl)-*N*,*N*′-di-2-propenylphosphonic diamide, *in* C-00173
P-(3-Chloropropyl)-*N*,*N*,*N*′,*N*′-tetraethylphosphonic diamide, *in* C-00173
▷Crufomate, C-00208
Cyclohexyl *N*-cyclohexyl-*P*-ethylphosphonamidate, *in* C-00246
P-Cyclohexyl-*N*,*N*′-diphenylphosphonic diamide, *in* C-00257
P-Cyclohexylphosphonamidic fluoride, C-00254
P-Cyclohexylphosphonic diamide, C-00257
P-Cyclohexyl-*N*,*N*,*N*′,*N*′-tetramethylphosphonic diamide, *in* C-00257
P-(Dibromomethyl)-*N*,*N*,*N*′,*N*′-tetramethylphosphonic diamide, *in* D-00064
N,*N*′-Di-*tert*-butyl-*P*-methylphosphonic diamide, *in* M-00289
N,*N*′-Di-*tert*-butyl-*P*-phenylphosphonic diamide, *in* P-00222
N,*N*-Dibutylphosphonamidic acid, D-00126
P-(Dichloromethyl)phosphonic bis(diethylamide), *in* D-00176

P-(Dichloromethyl)phosphonic bis(dimethylamide), *in* D-00176
(Dichloromethyl)phosphonic dianilide, *in* D-00176
N,N'-Dicyclohexyl-*P*-ethylphosphonic diamide, *in* E-00130
N,N'-Dicyclohexyl-*P*-phenylphosphonic diamide, *in* P-00222
4,5-Diethoxy-2-methyl-2*H*-1,3,2-diazaphosphole; 2-Oxide, *in* D-00235
2-Diethylamino-3-phenyl-1,3,2-oxazaphospholidine; 2-Oxide, *in* D-00269
2-Diethylamino-3,5,5-trimethyl-1,2-oxaphospholan-3-ol 2-oxide, *in* T-00519
[(*N,N*-Diethylcarbamoyl)methyl]phosphonic bis(diethylamide), *in* B-00175
N,N-Diethyl-*P*-methylphosphonamidic acid; Chloride, *in* D-00296
N,N-Diethyl-*P*-methylphosphonazidic amide, *in* M-00287
N,N-Diethyl-*P*-methylphosphonisocyanatidic amide, *in* M-00293
N,N-Diethyl-*P*-phenylphosphonamidic acid, D-00307
N,N-Diethyl-*P*-phenylphosphonamidic acid; Fluoride, *in* D-00307
N,N-Diethyl-*P*-phenylphosphonamidic acid; Chloride, *in* D-00307
N,N'-Diethyl-*P*-phenylphosphonic diamide, *in* P-00222
N,N-Diethyl-*P*-propylphosphonamidic acid; Fluoride, *in* D-00401
N,N-Diethyl-*P*-propylphosphonamidic acid; Chloride, *in* D-00401
▷*P,P'*-Diethyl-*N,N,N',N'*-tetramethyldiphosphonic diamide, *in* D-00291
6,7-Dihydro-5*H*-dibenzo[*d,f*][1,3,2]diazaphosphepine; 6-Me, 6-oxide, *in* D-00450
6,7-Dihydro-5*H*-dibenzo[*d,f*][1,3,2]diazaphosphepine; 6-Ph, 6-oxide, *in* D-00450
2,3-Dihydro-1,3-dimethyl-1,3,2-benzodiazaphosphorin-4(1*H*)-one; 2-Oxide, *in* D-00457
2,3-Dihydro-1,3-dimethyl-1,3,2-benzodiazaphosphorin-4(1*H*)-one; 2-Me, 2-oxide, *in* D-00457
2,5-Dihydro-5,5-dimethyl-1,2-oxaphosphole 2-oxide; 2-Diethylamino, *in* D-00464
2,3-Dihydro-2-ethoxy-1,3,5-triphenyl-1*H*-1,2-azaphosphole 2-oxide, *in* D-00508
2,3-Dihydro-5-methyl-1,2-oxaphosphole 2-oxide; 2-Diethylamino, *in* D-00525
3,4-Dihydro-2,3,4,5-tetraphenyl-2*H*-1,3,2-diazaphosphole; 2-Oxide, *in* D-00580
2,3-Dihydro-3,3,5-trimethyl-1,2-oxaphosphole 2-oxide; 2-Diethylamino, *in* D-00596
N,N'-Diisopropyl-*P*-phenylphosphonic diamide, *in* P-00222
P,P'-Diisopropyl-*N,N,N',N'*-tetramethyldiphosphonic diamide, D-00668
1,3-Dimethyl-1,3,2-diazaphospholidine 2-oxide, D-00720
N,N'-Dimethyl-*P*-(1,1-dimethylethyl)phosphonic diamide, *in* B-00597
1,4-Dimethyl-3,5-diphenyl-1,4,2-diazaphospholidine 2-oxide; 2-Chloro, *in* D-00739
1,4-Dimethyl-3,5-diphenyl-1,4,2-diazaphospholidine 2-oxide; 2-Ethoxy, *in* D-00739
P-(1,1-Dimethylethyl)-*N,N'*-diphenylphosphonic diamide, *in* B-00597
N,N-Dimethyl-*P*-(1-methylpropyl)phosphonamidic chloride, D-00763
N,N-Dimethyl-*P*-phenylphosphonamidic acid, D-00808
N,N-Dimethyl-*P*-phenylphosphonamidic acid; Chloride, *in* D-00808
N,N-Dimethyl-*P*-phenylphosphonamidic azide, D-00809
N,P-Dimethylphosphonamidic acid; Fluoride, *in* D-00849
▷*P,P'*-Dimethyl-*N,N,N',N'*-tetramethyldiphosphonic diamide, *in* D-00751
N,N'-Dimethyl-*N,N'*-trimethylene-*P*-phenylphosphonic diamide, *in* H-00023
N,P-Diphenylphosphonamidic acid, D-01103
N,P-Diphenylphosphonamidic acid; Fluoride, *in* D-01103
N,P-Diphenylphosphonamidic acid; Chloride, *in* D-01103
N,P-Diphenylphosphonic diamide, *in* P-00222
N,N'-Diphenyl-*P*-2-propenylphosphonic diamide, *in* P-00455
N,N'-Diphenyl-*P*-propylphosphonic diamide, *in* I-00051
N,N'-Diphenyl *P*-(2,3,5,6-tetrachloro-4-pyridinyl)-phosphonic diamide, *in* T-00035
N,N'-Diphenyl-*P*-[(4-trichloromethyl)phenyl]phosphonic diamide, *in* T-00409
N,N'-1,2-Ethanediyl-*N,N'*-dimethyl-*P*-phenylphosphonic diamide, *in* D-00793
P-Ethenyl-*N,N,N',N'*-tetraethylphosphonic diamide, *in* V-00004
P-Ethenyl-*N,N,N',N'*-tetramethylphosphonic diamide, *in* V-00004
2-Ethoxy-1-(2-phenylethyl)-1,2-azaphosphetidine 2-oxide, *in* H-00107
Ethyl *P*-acetyl-*N,N*-diethylphosphonamidate, *in* A-00014
Ethyl *P*-aminocarbonyl-*N,N*-diethylphosphonamidate, *in* A-00059
Ethyl 1-aziridinyl-*P*-methylphosphonamidate, *in* A-00152
Ethyl [bis(diethylamino)phosphinyl]acetate, *in* B-00175
Ethyl [bis(dimethylamino)phosphinyl]acetate, *in* B-00193
Ethyl *P*-(1,3-butadienyl)-*N,N*-diethylphosphonamidate, *in* B-00508
Ethyl *P*-(butoxymethyl)-*N,N*-diethylphosphonamidate, *in* B-00536
Ethyl *N-tert*-butyl-*P*-methylphosphonamidate, *in* B-00572
Ethyl *P*-chloromethylphosphonamidate, *in* C-00105
Ethyl *N,N*-dibutylphosphonamidate, *in* D-00126
Ethyl *P*-dichloromethylphosphonamidate, *in* D-00174
Ethyl *N,N*-diethyl-*P*-(methoxymethyl)phosphonamidate, *in* D-00295
Ethyl *N,N*-diethyl-*P*-methylphosphonamidate, *in* D-00296
Ethyl *N,N*-diethylphosphonamidate, *in* D-00337
Ethyl *N,P*-diethylphosphonamidate, *in* D-00336
Ethyl *N,N*-diethyl-*P*-propadienylphosphonamidate, *in* D-00399
Ethyl *N,N*-diethyl-*P*-1-propynylphosphonamidate, *in* D-00403
Ethyl *N*-(4,5-dihydro-2-thiazolyl)-*P*-methylphosphonamidate, *in* D-00587
Ethyl *N,N*-diisopropyl-*P*-methylphosphonamidate, *in* D-00636
Ethyl *N,P*-dimethylphosphonamidate, *in* D-00849
P-Ethyl-*N,N'*-diphenylphosphonic diamide, *in* E-00130
Ethyl *P*-ethyl-*N,N*-diphenylphosphonamidate, *in* E-00067
Ethyl *N*-ethyl-*P*-methylphosphonamidate, *in* E-00088
Ethyl *P*-ethyl-*N*-phenylphosphonamidate, *in* E-00110
Ethyl *P*-ethylphosphonamidate, *in* E-00126
Ethyl *N*-isopropyl-*P*-methylphosphonamidate, *in* I-00036
Ethyl [(methoxycarbonyl)methyl]phosphonamidate, *in* M-00028
Ethyl *P*-(4-methylphenyl)-*N*-phenylphosphonamidate, *in* M-00214
Ethyl *P*-methyl-*N*-phenylphosphonamidate, *in* M-00231
Ethyl *P*-methylphosphonamidate, *in* M-00283
Ethyl *P*-phenyl-*N*-phenylsulfonylphosphonamidate, *in* P-00191
Ethyl *P*-phenylphosphonamidate, *in* P-00218
P-Ethyl-*N*-phenylphosphonamidic acid; (*S*)-*form*, Et ester, *in* E-00110
P-Ethylphosphonamidic fluoride, *in* E-00126
P-Ethyl-*N,N,N',N'*-tetramethylphosphonic diamide, *in* E-00130
Ethyl *P*-trichloromethylphosphonamidate, *in* T-00410
Ethyl triethylphosphonamidate, *in* T-00446
Ethyl trimethylphosphonamidate, *in* T-00537
P-Ethynyl-*N,N,N',N'*-tetramethylphosphonic diamide, *in* E-00203
▷Fopurine, F-00053
Fosfazinomycin *A*, F-00058
Fosfazinomycin *B*, F-00059
P-Heptyl-*N,N,N',N'*-tetramethylphosphonic diamide, *in* H-00006
▷Hexaethyldiphosphonic diamide, *in* D-00291
Hexahydro-1,3-dimethyl-1,3,2-diazaphosphorine 2-oxide, H-00023
P-Imidazol-1-yl-*N,N,N',N'*-tetramethylphosphonic diamide, *in* I-00003
Isopropyl *N,N*-diethyl-*P*-methylphosphonamidate, *in* D-00296
Isopropyl *N,P*-dimethylphosphonamidate, *in* D-00849
Isopropyl *N*-ethyl-*P*-methylphosphonamidate, *in* E-00088
Isopropyl *N*-isopropyl-*P*-methylphosphonamidate, *in* I-00036
Isopropyl *P*-methyl-4-morpholinylphosphonamidate, *in* M-00175
Isopropyl *P*-methyl-*N*-phenylphosphonamidate, *in* M-00231
Isopropyl *P*-methyl-1-piperidinylphosphonamidate, *in* M-00361
P-Isopropyl-*N,N,N',N'*-tetramethylphosphonic diamide, *in* I-00051
▷Isopropyl triethylphosphonamidate, *in* T-00446
Isopropyl trimethylphosphonamidate, *in* T-00537
P-(4-Methoxyphenyl)-*N,N'*-diphenylphosphonic diamide, *in* M-00072

2-Methoxy-3,3,4,4-tetramethyl-1,2-azaphospholidine 2-oxide, *in* T-00191
Methyl *P*-acetyl-*N*,*N*-diethylphosphonamidate, *in* A-00014
(2-Methyl-1,3-butadienyl)phosphonic bis(diethylamide), *in* M-00096
(2-Methyl-1,3-butadienyl)phosphonic bis(dimethylamide), *in* M-00096
(2-Methyl-1,3-butadienyl)phosphonic bis(isopropylamide), *in* M-00096
(2-Methyl-1,3-butadienyl)phosphonic bis(methylamide), *in* M-00096
Methyl *N*-*tert*-butyl-*P*-methylphosphonamidate, *in* B-00572
Methyl *N*-cyclohexyl-*P*-phenylphosphonamidate, *in* C-00251
Methyl 6-deoxy-2,3-di-*O*-methyl-6-methylaminoglucopyranoside 4,6-cyclic phosphonamide; α-D-(*R*)P-*form*, *P*-Me, *in* M-00112
Methyl 6-deoxy-2,3-di-*O*-methyl-6-methylaminoglucopyranoside 4,6-cyclic phosphonamide; α-D-(*R*)P-*form*, *P*-Ph, *in* M-00112
Methyl 6-deoxy-2,3-di-*O*-methyl-6-methylaminoglucopyranoside 4,6-cyclic phosphonamide; α-D-(*S*)P-*form*, *P*-Me, *in* M-00112
Methyl 6-deoxy-2,3-di-*O*-methyl-6-methylaminoglucopyranoside 4,6-cyclic phosphonamide; α-D-(*S*)P-*form*, *P*-Ph, *in* M-00112
Methyl *P*-dichloromethylphosphonamidate, *in* D-00174
Methyl *N*,*N*-diethyl-*P*-methylphosphonamidate, *in* D-00296
Methyl *N*,*N*-diethyl-*P*-phenylphosphonamidate, *in* D-00307
Methyl *N*,*N*-diethylphosphonamidate, *in* D-00337
Methyl *N*,*N*-dimethyl-*P*-phenylphosphonamidate, *in* D-00808
Methyl *N*,*P*-dimethylphosphonamidate, *in* D-00849
Methyl *N*,*P*-diphenylphosphonamidate, *in* D-01103
P-Methyl-*N*,*N*′-diphenylphosphonic diamide, *in* M-00289
P,*P*′-Methylenebis(phosphonic diamide); Octa-Me, *in* M-00144
Methyl *N*-ethyl-*P*-methylphosphonamidate, *in* E-00088
Methyl *N*-ethyl-*P*-phenylphosphonamidate, *in* E-00109
Methyl *P*-ethyl-*N*-phenylphosphonamidate, *in* E-00110
Methyl *N*-isopropyl-*P*-methylphosphonamidate, *in* I-00036
Methyl *P*-(4-methylphenyl)-*N*-phenylphosphonamidate, *in* M-00214
Methyl *N*-methyl-*P*-phenylphosphonamidate, *in* M-00230
Methyl *P*-methyl-*N*-phenylphosphonamidate, *in* M-00231
P-(2-Methylphenyl)-*N*,*N*′-diphenylphosphonic diamide, *in* M-00235
P-(4-Methylphenyl)-*N*,*N*′-diphenylphosphonic diamide, *in* M-00238
2-Methyl-3-phenyl-1,3,2-oxazaphospholidine; 2-Oxide, *in* M-00212
3-Methyl-2-phenyl-1,3,2-oxazaphospholidine; 2-Sulfide, *in* M-00213
P-(4-Methylphenyl)-*N*-phenylphosphonamidic acid, M-00214
Methyl *P*-phenyl-*N*-phenylsulfonylphosphonamidate, *in* P-00191
Methyl *P*-phenylphosphonamidate, *in* P-00218
P-Methyl-*N*-phenylphosphonazidic amide, *in* M-00287
P-(4-Methylphenyl)phosphonic diamide, M-00238
Methyl *N*-phenyl-*P*-trichloromethylphosphonamidate, *in* P-00328
P-Methylphosphonamidic fluoride, *in* M-00283
P-Methylphosphonic diamide, M-00289
2-Methylpropyl *N*,*P*-diethylphosphonamidate, *in* D-00336
P-(1-Methylpropyl)-*N*,*N*′-diphenylphosphonic diamide, *in* M-00377
Methyl triethylphosphonamidate, *in* T-00446
P-Methyl-*N*,*N*′-trimethylenephosphonic diamide, *in* H-00020
Methyl trimethylphosphonamidate, *in* T-00537
P-(3-Nitrophenyl)-*N*,*N*′-diphenylphosphonic diamide, *in* N-00056
P-(4-Nitrophenyl)-*N*,*N*′-diphenylphosphonic diamide, *in* N-00057
Pentaethylphosphonic diamide, *in* E-00130
Pentamethylphosphonic diamide, *in* M-00289
Pentapropylphosphonic diamide, *in* P-00474
P-Phenyl-*N*,*N*′-bis(triphenylphosphoranylidene)phosphonic diamide, *in* P-00102
P-(4-*N*-phenylcarbamoylphenyl)-*N*,*N*′-diphenylphosphonic diamide, *in* P-00380
Phenyl *P*-chloromethylphosphonamidate, *in* C-00105
Phenyl *N*,*N*-diethyl-*P*-methylphosphonamidate, *in* D-00296
Phenyl *N*,*N*-diethyl-*P*-phenylphosphonamidate, *in* D-00307
Phenyl *N*-(4,5-dihydro-2-thiazolyl)-*P*-methylphosphonamidate, *in* D-00587
▷Phenyl *N*,*P*-diphenylphosphonamidate, *in* D-01103
[1,4-Phenylenebis(methylene)]bisphosphonic acid; Tetra-amide, *in* P-00131
(2-Phenylethenyl)phosphonic bis(dibutylamide), *in* P-00156
(2-Phenylethenyl)phosphonic bis(diethylamide), *in* P-00156
(2-Phenylethenyl)phosphonic bis(dimethylamide), *in* P-00156
(2-Phenylethenyl)phosphonic bis(dipropylamide), *in* P-00156
Phenyl *N*-ethyl-*P*-methylphosphonamidate, *in* E-00088
Phenyl *N*-isopropyl-*P*-methylphosphonamidate, *in* I-00036
Phenyl [(methoxycarbonyl)methyl]phosphonamidate, *in* M-00028
Phenyl *P*-(4-methylphenyl)-*N*-phenylphosphonamidate, *in* M-00214
Phenyl *P*-methyl-*N*-phenylphosphonamidate, *in* M-00231
Phenyl *P*-methylphosphonamidate, *in* M-00283
Phenyl *P*-methyl-*N*-2-thiazolylphosphonamidate, *in* M-00409
3-Phenyl-1,3,2-oxazaphospholidine; Oxide, *in* P-00180
Phenyl *P*-phenyl-*N*-phenylsulfonylphosphonamidate, *in* P-00191
Phenyl *P*-phenylphosphonamidate, *in* P-00218
P-Phenylphosphonamidic acid, P-00218
P-Phenylphosphonamidic acid; Fluoride, *in* P-00218
Phenylphosphonic acid; Diimidazolide, *in* P-00221
Phenylphosphonic acid; Di-1-pyrrolide, *in* P-00221
▷*P*-Phenylphosphonic diamide, P-00222
Phenyl *P*-trichloromethylphosphonamidate, *in* T-00410
N-Phenyl-*P*-trichloromethylphosphonamidic acid, P-00328
N-Phenyl-*P*-trichloromethylphosphonamidic acid; Chloride, *in* P-00328
Propyl *N*,*N*-diethyl-*P*-(methoxymethyl)phosphonamidate, *in* D-00295
N,*N*,*N*′,*N*′-Tetrabutylphosphonic diamide, T-00018
N,*N*,*N*′,*N*′-Tetraethyl-*P*-methylphosphonic diamide, *in* M-00289
N,*N*,*N*′,*N*′-Tetraethyl-*P*-(2-oxopropyl)phosphonic diamide, *in* O-00077
N,*N*,*N*′,*N*′-Tetraethyl-*P*-(phenylmethyl)phosphonic diamide, *in* B-00061
N,*N*,*N*′,*N*′-Tetraethylphosphonic diamide, T-00055
N,*N*,*N*′,*N*′-Tetraethyl-*P*-propylphosphonic diamide, *in* P-00474
N,*N*,*N*′,*N*′-Tetraethyl-*P*-2-pyridinylphosphonic diamide, *in* P-00503
N,*N*,*N*′,*N*′-Tetraisopropylphosphonic diamide, T-00171
4,4,5,5-Tetramethyl-1,2-azaphospholidine 2-oxide, T-00192
3,3,5,5-Tetramethyl-1,2-oxaphospholidine 2-oxide, T-00208
N,*N*,*N*′,*N*′-Tetramethyl-*P*-(2-oxo-2-phenylethyl)phosphonic diamide, *in* O-00066
N,*N*,*N*′,*N*-Tetramethyl-*P*-(phenylmethyl)phosphonic diamide, *in* B-00061
N,*N*,*N*′,*N*′-Tetramethylphosphonic diamide, T-00214
N,*N*,*N*′,*N*′-Tetramethyl-*P*-1,2-propadienylphosphonic diamide, *in* P-00442
N,*N*,*N*′,*N*′-Tetramethyl-*P*-2-propenylphosphonic diamide, *in* P-00455
N,*N*,*N*′,*N*′-Tetramethyl-*P*-propylphosphonic diamide, *in* P-00474
N,*N*,*N*′,*N*′-Tetramethyl-*P*-1-propynylphosphonic diamide, *in* P-00494
N,*N*,*N*′,*N*′-Tetraphenylphosphonic diamide, T-00271
N,*N*,*N*′,*N*′-Tetrapropylphosphonic diamide, T-00289
P-(Trichloromethyl)-*N*,*N*′-dicyclohexylphosphonic diamide, *in* T-00412
P-(Trichloromethyl)-*N*,*N*′-diphenylphosphonic diamide, *in* T-00412
(Trichloromethyl)phosphonic bis(methylamide), *in* T-00412
P-(Trichloromethyl)-*N*,*N*,*N*′,*N*′-tetraethylphosphonic diamide, *in* T-00412
Triethylphosphonamidic acid; (±)-*form*, Fluoride, *in* T-00446
Triethylphosphonamidic acid; (±)-*form*, Chloride, *in* T-00446
3,5,5-Trimethyl-2-dimethylamino-1,2-oxaphospholan-3-ol 2-oxide, *in* T-00519
N,*N*′-Trimethylene-*P*-phenylphosphonic diamide, *in* H-00020
3,5,5-Trimethyl-1,2-oxaphospholan-3-ol 2-oxide; (2*RS*,3*SR*)-*form*, 2-Dimethylamino, *in* T-00519
3,5,5-Trimethyl-1,2-oxaphospholan-3-ol 2-oxide; (2*RS*,3*SR*)-*form*, 2-Diethylamino, *in* T-00519
P-(2,4,5-Trimethylphenyl)-*N*,*N*′-diphenylphosphonic diamide, *in* T-00526

2,3,4-Trimethyl-5-phenyl-1,3,2-oxazaphospholidine; (2*R*,4*S*,5*R*)-*form*, 2-Oxide, *in* T-00523
2,3,4-Trimethyl-5-phenyl-1,3,2-oxazaphospholidine; (2*S*,4*S*,5*R*)-*form*, 2-Oxide, *in* T-00523
▷Trimethylphosphonamidic fluoride, T-00538
N,*N'*,*P*-Trimethylphosphonic diamide, *in* M-00289
N,*N*,*P*-Trimethylphosphonisocyanatidic amide, *in* M-00293
▷*N*,*N'*,*P*-Triphenylphosphonic diamide, *in* P-00222
N,*N'*,*P*-Tris(1,1-dimethylethyl)phosphonic diamide, *in* B-00597
P-Vinylphosphonic diamide, V-00004

Phosphonic isocyanates and isothiocyanates

Benzylphosphonic acid; Diisocyanate, *in* B-00060
(Dichloromethyl)phosphonic diisocyanate, D-00179
N,*N*-Diethyl-*P*-methylphosphonisocyanatidic amide, *in* M-00293
Ethyl butylphosphonisocyanatidate, *in* B-00601
Ethyl ethylphosphonisocyanatidate, *in* E-00133
Ethyl methylphosphonisocyanatidate, *in* M-00292
Ethyl methylphosphonisothiocyanatidate, *in* M-00294
Ethylphosphonic acid; Diisocyanate, *in* E-00129
Isopropyl methylphosphonisothiocyanatidate, *in* M-00294
Methyl *tert*-butylphosphon(isocyanatidate), *in* B-00602
Methyl methylphosphonisothiocyanatidate, *in* M-00294
Methylphosphonic acid; Diisocyanate, *in* M-00288
Methylphosphonic acid; Diisothiocyanate, *in* M-00288
Methylphosphonisocyanatidic acid; Azide, *in* M-00292
Phenyl methylphosphonisocyanatidate, *in* M-00292
Phenyl methylphosphonisothiocyanatidate, *in* M-00294
Phenylphosphonic acid; Diisothiocyanate, *in* P-00221
Phenylphosphonic diisocyanate, P-00226
(Trichloromethyl)phosphonic acid; Diisocyanate, *in* T-00411
(Trichloromethyl)phosphonic acid; Diisothiocyanate, *in* T-00411
N,*N*,*P*-Trimethylphosphonisocyanatidic amide, *in* M-00293

Phosphonic azides

N,*N*-Diethyl-*P*-methylphosphonazidic amide, *in* M-00287
N,*N*-Dimethyl-*P*-phenylphosphonamidic azide, D-00809
Di-4-morpholinyldiphosphonic acid, *in* M-00523
Ethyl methylphosphonazidate, *in* M-00286
Methyl phenylphosphonazidate, *in* P-00220
P-Methyl-*N*-phenylphosphonazidic amide, *in* M-00287
Methylphosphonazidic acid; Chloride, *in* M-00286
Methylphosphonic acid; Diazide, *in* M-00288
Methylphosphonisocyanatidic acid; Azide, *in* M-00292
Phenyl methylphosphonazidate, *in* M-00286
▷Phenylphosphonic acid; Diazide, *in* P-00221

Phosphonic hydrazides

Cyclohexyl *P*-methylphosphonohydrazidate, *in* M-00308
1,1'-Dimethyl-*P*-phenylphosphonic dihydrazide, *in* P-00225
Ethyl *P*-methylphosphonohydrazidate, *in* M-00308
Isopropyl *P*-methylphosphonohydrazidate, *in* M-00308
Methyl *P*-phenylphosphonohydrazidate, *in* P-00241
2-Methylpropyl *P*-methylphosphonohydrazidate, *in* M-00308
Phenyl *P*-methyl-N^2-phenylphosphonohydrazidate, *in* M-00241
P-Phenylphosphonic dihydrazide, P-00225
P,2,2'-Triphenylphosphonic dihydrazide, *in* P-00225

Thiophosphonic acids

O-Butyl phenylphosphonodithioate, *in* P-00236
Butylphosphonothioic acid; Mono-*O*-Et ester, *in* B-00611
tert-Butylphosphonothioic acid, B-00612
Cyclohexylphosphonothioic acid, C-00264
N,*N*-Diethyl-*P*-phenylphosphonamidothioic acid, D-00309
O-Ethyl butylphosphonodithioate, *in* B-00606
O-Ethyl ethylphosphonoselenothioate, E-00073
O-Ethyl ethylphosphonothioate; (*R*)-*form*, *in* E-00074
O-Ethyl ethylphosphonothioate; (*S*)-*form*, *in* E-00074
O-Ethyl ethylphosphonothioate; (±)-*form*, *in* E-00074
O-Ethyl hydrogen benzylphosphonothioate, *in* B-00067
O-Ethyl hydrogen cyclohexylphosphonodithioate, *in* C-00261
O-Ethyl hydrogen cyclohexylphosphonothioate, *in* C-00264
O-Ethyl hydrogen hexylphosphonodithioate, *in* H-00099
O-Ethyl hydrogen pentylphosphonodithioate, *in* P-00051
O-Ethyl hydrogen propylphosphonodithioate, *in* P-00478
O-Ethyl isopropylphosphonothioate; (*R*)-*form*, *in* E-00082
O-Ethyl methylphosphonodithioate, E-00091
O-Ethyl methylphosphonothioate; (*R*)-*form*, *in* E-00093
O-Ethyl methylphosphonothioate; (*S*)-*form*, *in* E-00093
O-Ethyl methylphosphonothioate; (±)-*form*, *in* E-00093
O-Ethyl phenylphosphonodithioate, E-00113
O-Ethyl phenylphosphonothioate; (*S*)-*form*, *in* E-00114
O-Ethyl phenylphosphonothioate; (±)-*form*, *in* E-00114
Ethylphosphonofluoridodithioic acid, E-00143
Heptylphosphonothioic acid, H-00007
Hexylphosphonothioic acid, H-00100
(2-Hydroxyethyl)phosphonothioic acid, H-00147
O-Isopropyl butylphosphonodithioate, *in* B-00606
O-Isopropyl hydrogen (1-methylpropyl)phosphonodithioate, *in* M-00380
O-Isopropyl hydrogen pentylphosphonodithioate, *in* P-00051
O-Isopropyl methylphosphonodithioate; (±)-*form*, *in* I-00040
O-Isopropyl methylphosphonothioate; (*R*)-*form*, *in* I-00041
O-Isopropyl methylphosphonothioate; (*S*)-*form*, *in* I-00041
O-Isopropyl methylphosphonothioate; (±)-*form*, *in* I-00041
O-Methyl butylphosphonodithioate, *in* B-00606
O-Methyl *tert*-butylphosphonothioate, M-00107
O-Methyl ethylphosphonothioate; (*S*)-*form*, *in* M-00155
O-Methyl hydrogen hexylphosphonodithioate, *in* H-00099
O-Methyl hydrogen pentylphosphonodithioate, *in* P-00051
O-Methyl isopropylphosphonothioate; (*R*)-*form*, *in* M-00166
O-Methyl isopropylphosphonothioate; (*S*)-*form*, *in* M-00166
O-Methyl isopropylphosphonothioate; (±)-*form*, *in* M-00166
O-Methyl methylphosphonodithioate, M-00170
O-Methyl methylphosphonothioate; (±)-*form*, *in* M-00171
O-Methyl naphthylphosphonothioate, *in* N-00023
O-Methyl phenylphosphonothioate; (*R*)-*form*, *in* M-00243
O-Methyl phenylphosphonothioate; (*S*)-*form*, *in* M-00243
O-Methyl phenylphosphonothioate; (±)-*form*, *in* M-00243
S-Methyl phenylphosphonothioate; (±)-*form*, *in* M-00244
Methylphosphonofluoridodithioic acid, M-00304
Methylphosphonothioic acid, M-00313
(1-Methylpropyl)phosphonodithioic acid, M-00380
(1-Methylpropyl)phosphonodithioic acid; (±)-*form*, *in* M-00380
(2-Methylpropyl)phosphonothioic acid, M-00381
Monomethyl 4-morpholinylphosphonothioate, *in* M-00527
O-Phenyl methylphosphonodithioate, P-00172
1-Piperidinylphosphonothioic acid, P-00431
O-Propyl methylphosphonodithioate, P-00469
Propylphosphonothioic acid, P-00480
Sparfosic acid, S-00006
Trimethylphosphonamidodithioic acid, T-00539
Vinylphosphonothioic acid, V-00007

Thiophosphonic anhydrides

2,4-Bis(1-methylpropyl)-1,3,2,4-dithiodiphosphetane 2,4-disulfide, *in* M-00382
Butylphosphonothioic acid trimer trianhydride, *in* T-00396
Cyclohexylphosphonothioic acid trimer trianhydride, *in* T-00430
2,4-Dibutyl-1,3,2,4-dithiaphosphetane 2,4-disulfide, *in* B-00614
2,4-Dicyclohexyl-1,3,2,4-dithiadiphosphetane 2,4-disulfide, D-00207
O,*O'*-Diethyl diethylthiodiphosphonate, *in* D-00405
2,4-Diethyl-1,3,2,4-dithiadiphosphetane 2,4-disulfide, *in* E-00152
Diethylthiodiphosphonic acid; Difluoride, *in* D-00405
O,*O'*-Diisopropyl diethylthiodiphosphonate, *in* D-00405
2,4-Dimethyl-1,3,2,4-dithiadiphosphetane 2,4-disulfide, D-00752
Dimethylthiodiphosphonic acid; Difluoride, *in* D-00925
Dimethylthiodiphosphonic acid difluoride, *in* M-00304
2,4-Diphenyl-1,3,2,4-dithiadiphosphetane 2,4-disulfide, D-01019
O-Ethyl *S*-(ethoxyphenylphosphinothioyl) phenylphosphonodithioate, *in* P-00237
O-Ethyl methylphosphonothioic monoanhydride, *in* D-00925
S-Ethyl methylphosphonothioic monoanhydride, *in* D-00925

Ethylphosphonothioic acid trimer trianhydride, *in* T-00463
O-Isopropyl *S*-(isopropoxyphenylphosphinothioyl) phenylphosphonodithioate, *in* P-00237
Methylphosphonothioic acid trimer trianhydride, *in* T-00567
O-Phenyl ethylphosphonothioic monoanhydride, *in* D-00925
Phenylphosphonodithioic acid anhydrosulfide; Difluoride, *in* P-00237
Phenylphosphonothioic acid monoanhydride, P-00245
Phenylphosphonothioic acid trimer trianhydride, *in* T-00694
Propylphosphonothioic acid trimer trianhydride, *in* T-00720

Thiophosphonic halides

Benzylphosphonothioic acid; Dichloride, *in* B-00067
Benzylphosphonothioic acid; Dibromide, *in* B-00067
tert-Butyl ethylphosphonochloridodithioate, *in* E-00136
S-tert-Butyl methylphosphonochloridodithioate, *in* M-00297
O-Butyl methylphosphonochloridothioate; (*S*)-*form*, *in* B-00573
O-Butyl methylphosphonochloridothioate; (±)-*form*, *in* B-00573
S-Butyl methylphosphonochloridothioate, *in* M-00298
O-Butyl methylphosphonofluoridothioate, *in* M-00305
Butylphosphonothioic acid; Dichloride, *in* B-00611
tert-Butylphosphonothioic acid; Dibromide, *in* B-00612
tert-Butylphosphonothioic dichloride, B-00613
6-Chloro-6*H*-dibenzo[*c,e*][1,2]oxaphosphorin; 6-Sulfide, *in* C-00033
2-Chloro-3,6-dihydro-2*H*-1,2-thiaphosphorin 2-sulfide, C-00053
(2-Chloroethyl)phosphonothioic acid; Difluoride, *in* C-00078
(2-Chloroethyl)phosphonothioic acid; Dichloride, *in* C-00078
(Chloromethyl)phosphonothioic acid; Difluoride, *in* C-00113
(Chloromethyl)phosphonothioic acid; Dichloride, *in* C-00113
(4-Chlorophenyl)phosphonothioic acid; Dichloride, *in* C-00150
4-Chloro-1,2,3,4-tetrahydronaphth[2,1-*c*][1,2]-azaphosphorine; Sulfide, *in* C-00179
P-Cyclohexyl-*N,N*-diethylphosphonamidothioic chloride, *in* C-00255
P-Cyclohexyl-*N,N*-dimethylphosphonamidothioic chloride, *in* C-00255
Cyclohexylphosphonothioic dichloride, C-00266
Cyclopentylphosphonothioic acid; Dichloride, *in* C-00280
(Dichloromethyl)phosphonothioic acid; Difluoride, *in* D-00180
(Dichloromethyl)phosphonothioic acid; Dichloride, *in* D-00180
N,N-Diethyl-*P*-methylphosphonamidothioic acid; Fluoride, *in* D-00297
N,N-Diethyl-*P*-methylphosphonamidothioic acid; Chloride, *in* D-00297
N,N-Diethyl-*P*-phenylphosphonamidothioic acid; Chloride, *in* D-00309
Diethylthiodiphosphonic acid; Difluoride, *in* D-00405
N,N-Diisopropyl-*P*-methylphosphonamidothioic acid; Chloride, *in* D-00637
O,S-Dimethyl *tert*-butylphosphonothioate, *in* M-00107
▷*N,P*-Dimethylphosphonamidothioic fluoride, *in* D-00850
Dimethylthiodiphosphonic acid; Difluoride, *in* D-00925
Dimethylthiodiphosphonic acid difluoride, *in* M-00304
N,N-Dimethyl-*P*-trifluoromethylphosphonamidothioic fluoride, D-00928
1,2-Ethanediphosphonothioic acid; Tetrachloride, *in* E-00007
1,2-Ethanediphosphonothioic acid; Tetrabromide, *in* E-00007
O-Ethyl ethenylphosphonochloridothioate, *in* V-00006
Ethyl ethylphosphonochloridodithioate, *in* E-00136
O-Ethyl ethylphosphonochloridothioate; (*R*)-*form*, *in* E-00069
O-Ethyl ethylphosphonochloridothioate; (*S*)-*form*, *in* E-00069
O-Ethyl ethylphosphonochloridothioate; (±)-*form*, *in* E-00069
S-Ethyl ethylphosphonochloridothioate, *in* E-00137
O-Ethyl ethylphosphonofluoridothioate; (+)-*form*, *in* E-00071
O-Ethyl ethylphosphonofluoridothioate; (−)-*form*, *in* E-00071
O-Ethyl ethylphosphonofluoridothioate; (±)-*form*, *in* E-00071
S-Ethyl isopropylphosphonochloridothioate, *in* I-00053
O-Ethyl *P*-methylphosphonamidothioate, *in* M-00284
S-Ethyl *P*-methylphosphonamidothioate, *in* M-00284
O-Ethyl methylphosphonobromidothioate, *in* M-00295
O-Ethyl methylphosphonochloridothioate; (*R*)-*form*, *in* E-00090
O-Ethyl methylphosphonochloridothioate; (±)-*form*, *in* E-00090
S-Ethyl methylphosphonochloridothioate, *in* M-00298
O-Ethyl methylphosphonofluoridothioate, *in* M-00305
S-Ethyl methylphosphonofluoridothioate, *in* M-00305
Ethyl phenylphosphonochloridodithioate, *in* P-00230
▷*O*-Ethyl phenylphosphonochloridothioate, E-00112
S-Ethyl phenylphosphonochloridothioate, *in* P-00232
P-Ethylphosphonamidothioic acid; Fluoride, *in* E-00128
Ethylphosphonofluoridodithioic acid, E-00143
Ethylphosphonothioic acid; Dibromide, *in* E-00148
Ethylphosphonothioic dichloride, E-00150
Ethylphosphonothioic difluoride, E-00151
Hexylphosphonothioic acid; Dichloride, *in* H-00100
Isopropyl ethylphosphonochloridodithioate, *in* E-00136
S-Isopropyl ethylphosphonochloridothioate, *in* E-00137
O-Isopropyl methylphosphonobromidothioate, *in* M-00295
O-Isopropyl methylphosphonochloridothioate; (*R*)-*form*, *in* I-00039
O-Isopropyl methylphosphonochloridothioate; (*S*)-*form*, *in* I-00039
O-Isopropyl methylphosphonochloridothioate; (±)-*form*, *in* I-00039
S-Isopropyl methylphosphonochloridothioate, *in* M-00298
O-Isopropyl methylphosphonofluoridothioate, *in* M-00305
O-Isopropyl phenylphosphonochloridothioate, *in* P-00232
S-Isopropyl phenylphosphonochloridothionate, *in* P-00232
Isopropylphosphonothioic dichloride, I-00055
S-Isopropyl propylphosphonochloridothioate, *in* P-00477
Methanediphosphonothioic acid; Tetrafluoride, *in* M-00023
Methanediphosphonothioic acid; Tetrachloride, *in* M-00023
Methanediphosphonothioic acid; Tetrabromide, *in* M-00023
(2-Methyl-1,3-butadienyl)phosphonothioic acid; Dichloride, *in* M-00097
Methyl 6-deoxy-2,3-di-*O*-methyl-6-methylaminoglucopyranoside 4,6-cyclic thiophosphonamide; α-D-(*R*)P-*form*, *P*-Chloro, *in* M-00113
O-Methyl ethylphosphonochloridothioate; (*R*)-*form*, *in* M-00154
S-Methyl ethylphosphonochloridothioate, *in* E-00137
O-Methyl isopropylphosphonochloridothioate; (*R*)-*form*, *in* M-00165
O-Methyl isopropylphosphonochloridothioate; (*S*)-*form*, *in* M-00165
O-Methyl isopropylphosphonochloridothioate; (±)-*form*, *in* M-00165
O-Methyl *P*-methylphosphonamidothioate, *in* M-00284
S-Methyl *P*-methylphosphonamidothioate, *in* M-00284
▷Methyl methylphosphonochloridodithioate, *in* M-00297
O-Methyl methylphosphonochloridothioate; (*S*)-*form*, *in* M-00169
S-Methyl methylphosphonochloridothioate, *in* M-00298
O-Methyl methylphosphonofluoridothioate, *in* M-00305
Methyl phenylphosphonobromidodithioate, *in* P-00228
O-Methyl phenylphosphonochloridothioate, *in* P-00232
S-Methyl phenylphosphonochloridothioate, *in* P-00232
P-Methylphosphonamidothioic acid; Fluoride, *in* M-00284
Methylphosphonofluoridodithioic acid, M-00304
Methylphosphonothioic dibromide, M-00315
▷Methylphosphonothioic dichloride, M-00316
Methylphosphonothioic difluoride, M-00317
O-(2-Methylpropyl) cyclohexylphosphonochloridothioate, *in* C-00260
S-Methyl propylphosphonochloridothioate, *in* P-00477
(2-Methylpropyl)phosphonothioic acid; Dichloride, *in* M-00381
4-Morpholinylphosphonothioic acid; Dichloride, *in* M-00527
1-Naphthyl benzylphosphonochloridothioate, *in* B-00064
2-Naphthyl benzylphosphonochloridothioate, *in* B-00064
1-Naphthylphosphonothioic acid; Dichloride, *in* N-00023
2-Naphthylphosphonothioic acid; Dichloride, *in* N-00024
O-Octyl cyclohexylphosphonochloridothioate, *in* C-00260
(2-Phenylethenyl)phosphonothioic acid; Dichloride, *in* P-00160
O-Phenyl ethylphosphonochloridothioate, *in* E-00137
(Phenylethynyl)phosphonothioic acid; Dichloride, *in* P-00167
Phenyl methylphosphonochloridodithioate, *in* M-00297
O-Phenyl methylphosphonochloridothioate, *in* M-00298
S-Phenyl methylphosphonochloridothioate, *in* M-00298
Phenyl phenylphosphonochloridodithioate, *in* P-00230
O-Phenyl phenylphosphonochloridothioate, *in* P-00232
S-Phenyl phenylphosphonochloridothioate, *in* P-00232
P-Phenylphosphonamidothioic acid; Fluoride, *in* P-00219

Phenylphosphonodithioic acid anhydrosulfide; Difluoride, *in* P-00237
Phenylphosphonothioic acid; Dibromide, *in* P-00244
▷Phenylphosphonothioic dichloride, P-00247
Phenylphosphonothioic difluoride, P-00248
1-Piperidinylphosphonothioic acid; Dichloride, *in* P-00431
2-Propenylphosphonothioic acid; Dichloride, *in* P-00461
O-Propyl butylphosphonochloridothioate, *in* B-00604
O-Propyl ethylphosphonochloridothioate, *in* E-00137
S-Propyl ethylphosphonochloridothioate, *in* E-00137
O-Propyl isopropylphosphonochloridothioate, *in* I-00053
O-Propyl methylphosphonochloridothioate; (*S*)-*form*, *in* P-00468
O-Propyl methylphosphonochloridothioate; (±)-*form*, *in* P-00468
S-Propyl methylphosphonochloridothioate, *in* M-00298
S-Propyl phenylphosphonochloridothioate, *in* P-00232
Propylphosphonothioic acid; Dichloride, *in* P-00480
S-Propyl propylphosphonochloridothioate, *in* P-00477
(Trichloromethyl)phosphonothioic acid; Dichloride, *in* T-00415
Triethylphosphonamidothioic acid; (±)-*form*, Chloride, *in* T-00448
(Trifluoromethyl)phosphonothioic acid; Dichloride, *in* T-00481
▷Trimethylphosphonamidothioic acid; Fluoride, *in* T-00540
Trimethylphosphonamidothioic acid; Chloride, *in* T-00540
Trimethylphosphonamidothioic acid; Bromide, *in* T-00540
Vinylphosphonothioic acid; Difluoride, *in* V-00007
Vinylphosphonothioic acid; Dichloride, *in* V-00007

Thiophosphonic esters

Benzyl propyl phenylphosphonate, *in* M-00511
O,O-Bis(1-methylethyl) 2-propenylphosphonothioate, *in* P-00461
O,O-Bis(2-methylpropyl) (2-methylpropyl)phosphonothioate, *in* M-00381
O,O-Bis(trimethylsilyl) hexylphosphonothioate, *in* H-00100
O,O-Bis(trimethylsilyl) phenylphosphonothioate, *in* P-00244
Bis(trimethylsilyl) phenylphosphonotrithioate, *in* P-00250
O,O-Bis(trimethylsilyl) phosphonothioate, B-00437
S-4-Bromophenyl *O*-ethyl ethylphosphonodithioate, *in* E-00070
O,O-1,3-Butylene methylphosphonothioate, *in* D-00734
▷*S*-Butyl *O*-ethyl methylphosphonothioate, *in* E-00093
tert-Butyl ethylphosphonochloridodithioate, *in* E-00136
S-Butyl *O*-methyl methylphosphonothioate, *in* M-00171
S-*tert*-Butyl methylphosphonochloridodithioate, *in* M-00297
O-Butyl methylphosphonochloridothioate; (*S*)-*form*, *in* B-00573
O-Butyl methylphosphonochloridothioate; (±)-*form*, *in* B-00573
S-Butyl methylphosphonochloridothioate, *in* M-00298
O-Butyl methylphosphonofluoridothioate, *in* M-00305
O-Butyl phenylphosphonodithioate, *in* P-00236
Butylphosphonothioic acid; Mono-*O*-Et ester, *in* B-00611
2-*tert*-Butyl-1,3,6,2-trithiaphosphocane; 2-Oxide, *in* B-00642
2-*tert*-Butyl-1,3,6,2-trithiaphosphocane; 2-Sulfide, *in* B-00642
▷Cyanofenphos, C-00213
O-Cyclohexyl *S*-ethyl cyclohexylphosphonodithioate, *in* C-00261
Cyclohexylphosphonothioic acid; *O,O*-Bis(trimethylsilyl) ester, *in* C-00264
O,O-Dibutyl acetylphosphonothioate, *in* A-00028
S,S-Dibutyl butylphosphonodithioate, *in* B-00606
O,O-Dibutyl butylphosphonothioate, *in* B-00611
Dibutyl butylphosphonotrithioate, *in* B-00614
O,O-Dibutyl cyclohexylphosphonothioate, *in* C-00264
O,O-Dibutyl cyclopentylphosphonothioate, *in* C-00280
Dibutyl ethylphosphonotrithioate, *in* E-00152
O,O-Dibutyl heptylphosphonothioate, *in* H-00007
2,10-Di-*tert*-butyl-1,3,6,9,11,14-hexathia-2,10-diphosphacyclohexadecane; *cis-form*, 2,10-Dioxide, *in* D-00093
2,10-Di-*tert*-butyl-1,3,6,9,11,14-hexathia-2,10-diphosphacyclohexadecane; *cis-form*, 2,10-Disulfide, *in* D-00093
2,10-Di-*tert*-butyl-1,3,6,9,11,14-hexathia-2,10-diphosphacyclohexadecane; *trans-form*, 2,10-Dioxide, *in* D-00093
2,10-Di-*tert*-butyl-1,3,6,9,11,14-hexathia-2,10-diphosphacyclohexadecane; *trans-form*, 2,10-Disulfide, *in* D-00093
▷*O,S*-Dibutyl methylphosphonothioate, *in* M-00313
O,O-Dibutyl phosphonothioate, D-00132
2,9-Di-*tert*-butyl-1,3,8,10-tetraoxa-2,9-diphosphacyclotetradecane; 2,9-Disulfide, *in* D-00158
▷*O*-2,5-Dichlorophenyl *O*-methyl phenylphosphonothioate, *in* M-00243
O,O-Diethyl acetylphosphonothioate, *in* A-00028
S,S-Diethyl benzylphosphonodithioate, *in* B-00065
O,O-Diethyl benzylphosphonothioate, *in* B-00067
O,S-Diethyl benzylphosphonothioate, *in* B-00067
O,O-Diethyl (1,3-butadien-1-yl)phosphonothioate, *in* B-00511
O,O-Diethyl 2-butenylphosphonothioate, *in* B-00527
O,O-Diethyl butylphosphonodithioate, *in* B-00611
O,S-Diethyl butylphosphonothioate, *in* B-00611
O,O-Diethyl (4-chlorophenyl)phosphonothioate, *in* C-00150
O,S-Diethyl cyclohexylphosphonodithioate, *in* C-00261
S,S-Diethyl cyclohexylphosphonodithioate, *in* C-00261
O,O-Diethyl cyclohexylphosphonothioate, *in* C-00264
O,O'-Diethyl diethylthiodiphosphonate, *in* D-00405
O,S-Diethyl ethylphosphonodithioate, *in* E-00070
O,O-Diethyl ethylphosphonothioate, *in* E-00148
O,S-Diethyl ethylphosphonothioate, *in* E-00074
Diethyl ethylphosphonotrithioate, *in* E-00152
O,O-Diethyl heptylphosphonothioate, *in* H-00007
O,O'-Diethyl hydrazobismethylphosphinothioate, *in* H-00105
O,O-Diethyl isopropylphosphonothioate, *in* I-00054
O,S-Diethyl methylphosphonodithioate, *in* M-00301
S,S-Diethyl methylphosphonodithioate, *in* M-00301
O,O-Diethyl methylphosphonothioate, *in* M-00313
O,S-Diethyl methylphosphonothioate, *in* E-00093
Diethyl methylphosphonotrithioate, *in* M-00318
O,O-Diethyl (2-methyl-1-propenyl)phosphonothioate, *in* M-00366
O,O-Diethyl (2-oxopropyl)phosphonothioate, *in* O-00079
O,O-Diethyl (1-oxopropyl)phosphonothioate, *in* O-00078
S,S-Diethyl (2-phenylethenyl)phosphonodithioate, *in* P-00159
O,O-Diethyl (2-phenylethenyl)phosphonothioate, *in* P-00160
O,O-Diethyl (phenylethynyl)phosphonothioate, *in* P-00167
S,S-Diethyl phenylphosphonodithioate, *in* P-00236
O,O-Diethyl phenylphosphonothioate, *in* P-00244
O,S-Diethyl phenylphosphonothioate, *in* P-00244
Diethyl phenylphosphonotrithioate, *in* P-00250
O,O-Diethyl phosphonothioate, D-00341
S,S-Diethyl 2-propenylphosphonodithioate, *in* P-00457
O,O-Diethyl 2-propenylphosphonothioate, *in* P-00461
O,O-Diethyl propylphosphonothioate, *in* P-00480
O,S-Diethyl propylphosphonothioate, *in* P-00480
O,O-Diethyl (trichloromethyl)phosphonothioate, *in* T-00415
O,O-Diethyl (trifluoromethyl)phosphonothioate, *in* T-00481
O,S-Diethyl (trifluoromethyl)phosphonothioate, *in* T-00481
O,O-Diethyl vinylphosphonothioate, *in* V-00007
4,7-Dihydro-1,3,2-dioxaphosphepin 2-sulfide; 2-Me, *in* D-00473
2,3-Dihydro-2-methoxy-5-methyl-1,2-thiaphosphole; 2-Oxide, *in* D-00514
1,5-Dihydro-3-methyl-2,4,3-benzodioxaphosphepin; 4-Sulfide, *in* D-00516
1,3-Dihydro-1-phenoxy-2,1-benzoxaphosphole; 1-Sulfide, *in* D-00546
4,5-Dihydro-2-phenyl-1,3,2-benzoxathiaphosphepin; 2-Oxide, *in* D-00555
4,7-Dihydro-2-phenyl-1,3,2-dioxaphosphepin; 2-Sulfide, *in* D-00561
O,O'-Diisopropyl diethylthiodiphosphonate, *in* D-00405
O,O-Diisopropyl (1,3-dihydro-1,3-dioxo-2*H*-isoindol-2-yl)-phosphonothioate, *in* D-00475
O,O-Diisopropyl ethylphosphonothioate, *in* E-00148
O,O-Diisopropyl phosphonothioate, D-00652
S,S-Diisopropyl 2-propenylphosphonodithioate, *in* P-00457
O,O-Dimethyl (1,3-butadien-1-yl)phosphonothioate, *in* B-00511
O,O-Dimethyl (1,3-dihydro-1,3-dioxo-2*H*-isoindol-2-yl)-phosphonothioate, *in* D-00475
2,6-Dimethyl-1,3,2,6-dioxadiphosphocane; *cis-form*, 2,6-Disulfide, *in* D-00730

2,6-Dimethyl-1,3,2,6-dioxadiphosphocane; *trans-form*, 2,6-Disulfide, *in* D-00730
2,6-Dimethyl-1,3,2,6-dioxadiphosphocane; *trans-form*, 2,6-Diselenide, *in* D-00730
O,O-Dimethyl ethylphosphonothioate, *in* E-00148
O,O-Dimethyl (2-methyl-1,3-butadienyl)phosphonothioate, *in* M-00097
O,S-Dimethyl methylphosphonodithioate, *in* M-00301
S,S-Dimethyl methylphosphonodithioate, *in* M-00301
O,O-Dimethyl methylphosphonothioate, *in* M-00313
Dimethyl methylphosphonotrithioate, *in* M-00318
O,O-Dimethyl 4-morpholinylphosphonothioate, *in* M-00527
O,O-Dimethyl (phenylethynyl)phosphonothioate, *in* P-00167
O,O-Dimethyl phenylphosphonothioate, *in* P-00244
O,S-Dimethyl phenylphosphonothioate, *in* M-00243
O,O-Dimethyl phosphonothioate, D-00853
O,O-Dimethyl 1-piperidinylphosphonothioate, *in* P-00431
S,S-Dimethyl 2-propenylphosphonodithioate, *in* P-00457
N,N-Dimethyl-*P*-trifluoromethylphosphonamidothioic fluoride, D-00928
O,O-Dimethyl vinylphosphonothioate, *in* V-00007
S,S-Diphenyl butylphosphonodithioate, *in* B-00606
O,O-Diphenyl ethylphosphonothioate, *in* E-00148
Diphenyl ethylphosphonotrithioate, *in* E-00152
S,S-Diphenyl methylphosphonodithioate, *in* M-00301
O,O-Diphenyl methylphosphonothioate, *in* M-00313
O,S-Diphenyl methylphosphonothioate, *in* M-00313
Diphenyl methylphosphonotrithioate, *in* M-00318
O,O-Diphenyl 4-morpholinylphosphonothioate, *in* M-00527
S,S-Diphenyl (2-phenylethenyl)phosphonodithioate, *in* P-00159
O,O-Diphenyl phenylphosphonothioate, *in* P-00244
O,O-Diphenyl phosphonothioate, D-01106
S,S-Diphenyl propylphosphonodithioate, *in* P-00478
O,O-Di-2-propenyl phenylphosphonothioate, *in* P-00244
S,S-Dipropyl benzylphosphonodithioate, *in* B-00065
Dipropyl phenylphosphonotrithioate, *in* P-00250
O,O-Dipropyl phosphonothioate, D-01212
S,S-Dipropyl propylphosphonodithioate, *in* P-00478
▷Ditalimfos, D-01237
1,3,2-Dithiaphospholane 2-sulfide; 2-Isopropyl, *in* D-01245
▷EPN-Methyl, *in* M-00243
2-Ethoxy-1,2-oxaphosphepane 2-oxide, *in* H-00165
O-Ethyl butylphosphonodithioate, *in* B-00606
▷*O*-Ethyl *S*-4-chlorophenyl ethylphosphonodithioate, *in* E-00070
Ethyl (diethoxyphosphinothioyl)acetate, *in* D-00237
Ethyl (diethoxyphosphinyl)oxoacetate, *in* D-00252
Ethyl *N,N*-diethyl-*P*-phenylphosphonamidodithioate, *in* D-00308
S-Ethyl *N,N*-diethyl-*P*-phenylphosphonamidothioate, *in* D-00309
▷*O*-Ethyl *S*-[2-(diisopropylamino)ethyl] methylphosphonothioate, E-00062
O-Ethyl *N,P*-dimethylphosphonamidothioate, *in* D-00850
S,S-Ethylene methylphosphonodithioate, *in* M-00139
S,S-Ethylene methylphosphonoselenodithioate, *in* M-00139
Ethylene methylphosphonothioate, *in* M-00126
Ethylene methylphosphonotrithioate, *in* M-00139
S,S-Ethylene phenylphosphonodithioate, *in* P-00121
O,O-Ethylene phenylphosphonothioate, *in* P-00115
Ethylene phenylphosphonotrithioate, *in* P-00121
O-Ethyl ethenylphosphonochloridothioate, *in* V-00006
O-Ethyl *S*-(ethoxyphenylphosphinothioyl) phenylphosphonodithioate, *in* P-00237
S-Ethyl *P*-ethylphosphonamidothioate, *in* E-00128
Ethyl ethylphosphonochloridodithioate, *in* E-00136
O-Ethyl ethylphosphonochloridothioate; (*R*)-*form*, *in* E-00069
O-Ethyl ethylphosphonochloridothioate; (*S*)-*form*, *in* E-00069
O-Ethyl ethylphosphonochloridothioate; (±)-*form*, *in* E-00069
S-Ethyl ethylphosphonochloridothioate, *in* E-00137
O-Ethyl ethylphosphonodithioate; (*R*)-*form*, Et ester, *in* E-00070
O-Ethyl ethylphosphonodithioate; (*S*)-*form*, Et ester, *in* E-00070
O-Ethyl ethylphosphonodithioate; (*S*)-*form*, 4-Bromophenyl ester, *in* E-00070
O-Ethyl ethylphosphonofluoridothioate; (+)-*form*, *in* E-00071
O-Ethyl ethylphosphonofluoridothioate; (−)-*form*, *in* E-00071
O-Ethyl ethylphosphonofluoridothioate; (±)-*form*, *in* E-00071
O-Ethyl ethylphosphonoselenothioate, E-00073
O-Ethyl ethylphosphonothioate; (*R*)-*form*, *in* E-00074
O-Ethyl ethylphosphonothioate; (*S*)-*form*, *in* E-00074
O-Ethyl ethylphosphonothioate; (*S*)-*form*, *S*-Me ester, *in* E-00074
O-Ethyl ethylphosphonothioate; (*S*)-*form*, *S*-Et ester, *in* E-00074
O-Ethyl ethylphosphonothioate; (*S*)-*form*, *S*-Ph ester, *in* E-00074
O-Ethyl ethylphosphonothioate; (±)-*form*, *in* E-00074
O-Ethyl ethylphosphonothioate; (±)-*form*, *S*-Et ester, *in* E-00074
O-Ethyl hydrogen benzylphosphonothioate, *in* B-00067
O-Ethyl hydrogen cyclohexylphosphonodithioate, *in* C-00261
O-Ethyl hydrogen cyclohexylphosphonothioate, *in* C-00264
O-Ethyl hydrogen hexylphosphonodithioate, *in* H-00099
O-Ethyl hydrogen pentylphosphonodithioate, *in* P-00051
O-Ethyl hydrogen propylphosphonodithioate, *in* P-00478
O-Ethyl *N*-isopropyl-*P*-methylphosphonamidothioate, *in* I-00037
S-Ethyl *O*-isopropyl methylphosphonothioate, *in* I-00040
S-Ethyl *O*-isopropyl methylphosphonothioate, *in* I-00041
S-Ethyl isopropylphosphonochloridothioate, *in* I-00053
O-Ethyl isopropylphosphonothioate; (*R*)-*form*, *in* E-00082
O-Ethyl *O*-methyl ethylphosphonothioate, *in* M-00155
O-Ethyl *S*-methyl ethylphosphonothioate, *in* E-00074
O-Ethyl *O*-methyl isopropylphosphonothioate, *in* M-00166
S-Ethyl *P*-methyl-*N*-phenylphosphonamidothioate, *in* M-00233
O-Ethyl *O*-methyl phenylphosphonothioate, *in* M-00243
▷*O*-Ethyl *S*-methyl phenylphosphonothioate, *in* M-00244
S-Ethyl *O*-methyl phenylphosphonothioate, *in* P-00244
O-Ethyl *P*-methylphosphonamidothioate, *in* M-00284
S-Ethyl *P*-methylphosphonamidothioate, *in* M-00284
O-Ethyl methylphosphonobromidothioate, *in* M-00295
O-Ethyl methylphosphonochloridothioate; (*R*)-*form*, *in* E-00090
O-Ethyl methylphosphonochloridothioate; (±)-*form*, *in* E-00090
S-Ethyl methylphosphonochloridothioate, *in* M-00298
O-Ethyl methylphosphonodithioate, E-00091
O-Ethyl methylphosphonofluoridothioate, *in* M-00305
S-Ethyl methylphosphonofluoridothioate, *in* M-00305
O-Ethyl *P*-methylphosphonohydrazidothioate, *in* M-00309
O-Ethyl methylphosphonothioate; (*R*)-*form*, *in* E-00093
O-Ethyl methylphosphonothioate; (*S*)-*form*, *in* E-00093
O-Ethyl methylphosphonothioate; (±)-*form*, *in* E-00093
▷*O*-Ethyl *O*-4-nitrophenyl phenylphosphonothionate, E-00099
S-Ethyl *O*-phenyl cyclohexylphosphonodithioate, *in* C-00261
O-Ethyl *S*-phenyl ethylphosphonothioate, *in* E-00074
Ethyl phenyl (1-methylpropyl)phosphonotrithioate, *in* M-00382
O-Ethyl phenylphosphonisothiocyanatidothioate, *in* P-00227
Ethyl phenylphosphonochloridodithioate, *in* P-00230
▷*O*-Ethyl phenylphosphonochloridothioate, E-00112
S-Ethyl phenylphosphonochloridothioate, *in* P-00232
O-Ethyl phenylphosphonodithioate, E-00113
O-Ethyl phenylphosphonothioate; (*S*)-*form*, *in* E-00114
O-Ethyl phenylphosphonothioate; (±)-*form*, *in* E-00114
▷Ethylphosphonodithioic acid; *S,S*-Di-Et ester, *in* E-00140
Ethylphosphonodithioic acid; *S,S*-Dibutyl ester, *in* E-00140
Ethyl triethylphosphonamidodithioate, *in* T-00447
O-Ethyl triethylphosphonamidothioate, *in* T-00448
S-Ethyl triethylphosphonamidothioate, *in* T-00448
Fonofos; (*R*)-*form*, *in* F-00052
▷Fonofos; (*S*)-*form*, *in* F-00052
Fonofos; (±)-*form*, *in* F-00052
▷Fonofos oxon, *in* E-00074
Heptylphosphonothioic acid; *O,O*-Bis(trimethylsilyl) ester, *in* H-00007
2,5,5,8,11,11-Hexamethyl-1,3,7,9-tetraoxa-2,8-diphosphacyclododecane; 2,8-Disulfide, *in* H-00081
Inezin, I-00009
O-Isopropyl butylphosphonodithioate, *in* B-00606
Isopropyl *N,N*-diethyl-*P*-phenylphosphonamidodithioate, *in* D-00308
S-Isopropyl *N,N*-diethyl-*P*-phenylphosphonamidothioate, *in* D-00309
Isopropyl ethylphosphonochloridodithioate, *in* E-00136
S-Isopropyl ethylphosphonochloridothioate, *in* E-00137
O-Isopropyl hydrogen (1-methylpropyl)phosphonodithioate, *in* M-00380

O-Isopropyl hydrogen pentylphosphonodithioate, *in* P-00051
O-Isopropyl *S*-(isopropoxyphenylphosphinothioyl) phenylphosphonodithioate, *in* P-00237
O-Isopropyl *O*-methyl methylphosphonothioate, *in* M-00171
O-Isopropyl *S*-methyl methylphosphonothioate, *in* I-00041
O-Isopropyl methylphosphonobromidothioate, *in* M-00295
O-Isopropyl methylphosphonochloridothioate; (*R*)-*form*, *in* I-00039
O-Isopropyl methylphosphonochloridothioate; (*S*)-*form*, *in* I-00039
O-Isopropyl methylphosphonochloridothioate; (±)-*form*, *in* I-00039
S-Isopropyl methylphosphonochloridothioate, *in* M-00298
O-Isopropyl methylphosphonodithioate; (*S*)-*form*, Et ester, *in* I-00040
O-Isopropyl methylphosphonodithioate; (±)-*form*, *in* I-00040
O-Isopropyl methylphosphonofluoridothioate, *in* M-00305
O-Isopropyl methylphosphonothioate; (*R*)-*form*, *in* I-00041
O-Isopropyl methylphosphonothioate; (*S*)-*form*, *in* I-00041
O-Isopropyl methylphosphonothioate; (*S*)-*form*, *S*-Me ester, *in* I-00041
O-Isopropyl methylphosphonothioate; (*S*)-*form*, *S*-Et ester, *in* I-00041
O-Isopropyl methylphosphonothioate; (±)-*form*, *in* I-00041
O-Isopropyl methylphosphonothioate; (±)-*form*, *S*-Et ester, *in* I-00041
2-Isopropyl 1,3,2-oxathiaphospholane, I-00042
2-Isopropyl-1,3,2-oxathiaphospholane 2-oxide, *in* I-00042
2-Isopropyl-1,3,2-oxathiaphospholane 2-sulfide, *in* I-00042
O-Isopropyl *S*-phenyl methylphosphonothioate, *in* I-00041
O-Isopropyl phenylphosphonisothiocyanatidothioate, *in* P-00227
O-Isopropyl phenylphosphonochloridothioate, *in* P-00232
S-Isopropyl phenylphosphonochloridothionate, *in* P-00232
S-Isopropyl propylphosphonochloridothioate, *in* P-00477
Isopropyl triethylphosphonamidodithioate, *in* T-00447
O-Isopropyl triethylphosphonamidothioate, *in* T-00448
O-Isopropyl trimethylphosphonamidothioate, *in* T-00540
▷Leptophos, L-00001
Leptophos-Ethyl, *in* E-00114
2-Mercapto-1,3,2-dioxaphospholane, *in* D-00970
2-Mercapto-4-methyl-1,3,2-dioxophospholane, *in* M-00127
(2-Methyl-1,3-butadienyl)phosphonothioic acid; *O,O*-Di-Et ester, *in* M-00097
O-Methyl butylphosphonodithioate, *in* B-00606
O-Methyl *N,N*-diethyl-*P*-phenylphosphonamidothioate, *in* D-00309
S-Methyl *N,N*-diethyl-*P*-phenylphosphonamidothioate, *in* D-00309
Methyl 2,3-di-*O*-methylglucopyranoside 4,6-cyclic phosphonothioate; α-D-(*R*)P-*form*, *P*-Me, *in* M-00120
Methyl 2,3-di-*O*-methylglucopyranoside 4,6-cyclic phosphonothioate; α-D-(*R*)P-*form*, *P*-Ph, *in* M-00120
Methyl 2,3-di-*O*-methylglucopyranoside 4,6-cyclic phosphonothioate; α-D-(*S*)P-*form*, *P*-Me, *in* M-00120
Methyl 2,3-di-*O*-methylglucopyranoside 4,6-cyclic phosphonothioate; α-D-(*S*)P-*form*, *P*-Ph, *in* M-00120
S-Methyl *N,P*-dimethylphosphonamidothioate, *in* D-00850
Methyl 2,3-di-*O*-methyl-4-thioglucopyranoside 4,6-cyclic phosphonate; α-D-(*R*)P-*form*, *P*-Me, *in* M-00121
Methyl 2,3-di-*O*-methyl-4-thioglucopyranoside 4,6-cyclic phosphonate; α-D-(*S*)P-*form*, *P*-Me, *in* M-00121
Methyl 2,3-di-*O*-methyl-6-thioglucopyranoside 4,6-cyclic phosphonate; α-D-(*R*)P-*form*, *P*-Me, *in* M-00122
Methyl 2,3-di-*O*-methyl-6-thioglucopyranoside 4,6-cyclic phosphonate; α-D-(*S*)P-*form*, *P*-Me, *in* M-00122
Methyl *P*-ethylphosphonamidodithioate, *in* E-00127
S-Methyl *P*-ethylphosphonamidothioate, *in* E-00128
O-Methyl ethylphosphonochloridothioate; (*R*)-*form*, *in* M-00154
S-Methyl ethylphosphonochloridothioate, *in* E-00137
O-Methyl ethylphosphonothioate; (*S*)-*form*, *in* M-00155
O-Methyl hydrogen hexylphosphonodithioate, *in* H-00099
O-Methyl hydrogen pentylphosphonodithioate, *in* P-00051
O-Methyl isopropylphosphonochloridothioate; (*R*)-*form*, *in* M-00165
O-Methyl isopropylphosphonochloridothioate; (*S*)-*form*, *in* M-00165
O-Methyl isopropylphosphonochloridothioate; (±)-*form*, *in* M-00165
O-Methyl isopropylphosphonothioate; (*R*)-*form*, *in* M-00166
O-Methyl isopropylphosphonothioate; (*S*)-*form*, *in* M-00166
O-Methyl isopropylphosphonothioate; (*S*)-*form*, *O*-Et ester, *in* M-00166
O-Methyl isopropylphosphonothioate; (±)-*form*, *in* M-00166
Methyl *P*-methyl-*N,N*-dimethylphosphonamidodithioate, *in* E-00063
Methyl *P*-methyl-*N*-phenylphosphonamidodithioate, *in* M-00232
O-Methyl *P*-methylphosphonamidothioate, *in* M-00284
S-Methyl *P*-methylphosphonamidothioate, *in* M-00284
▷Methyl methylphosphonochloridodithioate, *in* M-00297
O-Methyl methylphosphonochloridothioate; (*S*)-*form*, *in* M-00169
S-Methyl methylphosphonochloridothioate, *in* M-00298
O-Methyl methylphosphonodithioate, M-00170
O-Methyl methylphosphonofluoridothioate, *in* M-00305
O-Methyl methylphosphonothioate; (*S*)-*form*, *O*-Isopropyl ester, *in* M-00171
O-Methyl methylphosphonothioate; (±)-*form*, *in* M-00171
O-Methyl naphthylphosphonothioate, *in* N-00023
5-Methyl-2-phenyl-1,3,2-benzodithiaphosphole; 2-Oxide, *in* M-00195
O-Methyl *S*-phenyl ethylphosphonodithioate, *in* E-00140
O-Methyl *S*-phenyl methylphosphonothioate, *in* M-00171
Methyl phenylphosphonobromidodithioate, *in* P-00228
O-Methyl phenylphosphonochloridothioate, *in* P-00232
S-Methyl phenylphosphonochloridothioate, *in* P-00232
O-Methyl phenylphosphonoisothiocyanatidothioate, *in* P-00227
O-Methyl phenylphosphonothioate; (*R*)-*form*, *in* M-00243
O-Methyl phenylphosphonothioate; (*S*)-*form*, *in* M-00243
O-Methyl phenylphosphonothioate; (*S*)-*form*, *S*-Me ester, *in* M-00243
O-Methyl phenylphosphonothioate; (±)-*form*, *in* M-00243
S-Methyl phenylphosphonothioate; (*S*)-*form*, *O*-Menthyl ester, *in* M-00244
S-Methyl phenylphosphonothioate; (±)-*form*, *in* M-00244
Methylphosphonodithioic acid; (*R*)-*form*, *O,S*-Di-Et ester, *in* M-00301
Methylphosphonodithioic acid; (*S*)-*form*, *O,S*-Di-Et ester, *in* M-00301
Methylphosphonothioic acid; *O*-Isopropyl ester, chloride, *in* M-00313
Methylphosphonotrithioic acid; Bis(trimethylsilyl) ester, *in* M-00318
O-(2-Methylpropyl) cyclohexylphosphonochloridothioate, *in* C-00260
S-Methyl propylphosphonochloridothioate, *in* P-00477
(1-Methylpropyl)phosphonodithioic acid; (±)-*form*, *in* M-00380
17-Methyl-6,7,9,10-tetrahydrodibenzo[*d,m*][1,3,6,9,12,2]-pentaoxaphosphacyclotetradecin; 17-Sulfide, *in* M-00403
O-Methyl triethylphosphonamidothioate, *in* T-00448
S-Methyl trimethylphosphonamidothioate, *in* T-00540
O-Methyl *O*-trimethylsilyl *tert*-butylphosphonothioate, *in* M-00107
Monopropyl phenylphosphonate; (*S*)-*form*, Ph ester, *in* M-00511
Monopropyl phenylphosphonate; (*S*)-*form*, Benzyl ester, *in* M-00511
1-Naphthyl benzylphosphonochloridothioate, *in* B-00064
2-Naphthyl benzylphosphonochloridothioate, *in* B-00064
O,O-Neopentylene methylphosphonothioate, *in* T-00512
O,O-Neopentylene phenylphosphonothioate, *in* D-00795
O,O-Neopentylene phosphonothioate, *in* D-00735
13,14,16,17,31,32,34,35-Octahydro-6,24-dimethyltetrabenzo[*d,m,r,a'*]-[1,3,6,9,12,15,20,23,26,2,16]-decaoxadiphosphacyclooctacosin; 6,24-Disulfide, *in* O-00008
O-Octyl cyclohexylphosphonochloridothioate, *in* C-00260
1,3,2-Oxathiaphosphorinane 2-oxide; 2-Ph, *in* O-00057
2-Phenyl-4*H*-1,3,2-benzodioxaphosphorin; 2-Sulfide, *in* P-00095
2-Phenyl-4*H*-1,3,2-benzoxathiaphosphorin; 2-Oxide, *in* P-00097
O-Phenyl *N,N*-diethyl-*P*-methylphosphonamidothioate, *in* D-00297
S-Phenyl *N,N*-diethyl-*P*-methylphosphonamidothioate, *in* D-00297

S-Phenyl *N*,*N*-diethyl-*P*-phenylphosphonamidothioate, *in* D-00309
O-Phenyl *N*,*P*-diphenylphosphonamidothioate, *in* D-01104
O,*O*-*o*-Phenylene thiophosphonate, *in* M-00090
Phenyl *P*-ethyl-*N*,*N*-dimethylphosphonamidodithioate, *in* E-00063
O-Phenyl ethylphosphonochloridothioate, *in* E-00137
O-Phenyl *P*-methyl-N^2-phenylphosphonohydrazidothioate, *in* M-00242
Phenyl methylphosphonochloridodithioate, *in* M-00297
O-Phenyl methylphosphonochloridothioate, *in* M-00298
S-Phenyl methylphosphonochloridothioate, *in* M-00298
O-Phenyl methylphosphonodithioate, P-00172
O-Phenyl *P*-methylphosphonohydrazidothioate, *in* M-00309
O-Phenyl 4-morpholinylphosphonochloridothioate, *in* M-00525
2-Phenyl-1,2-oxaphospholane; 2-Sulfide, *in* P-00174
2-Phenyl-1,3,2-oxathiaphosphepane; 2-Sulfide, *in* P-00178
O-Phenyl phenylphosphonisothiocyanatidothioate, *in* P-00227
Phenyl phenylphosphonochloridodithioate, *in* P-00230
O-Phenyl phenylphosphonochloridothioate, *in* P-00232
S-Phenyl phenylphosphonochloridothioate, *in* P-00232
Phenyl propyl phenylphosphonate, *in* M-00511
2-Phenyl-1,2-thiaphospholane; 2-Sulfide, *in* P-00324
▷Phenyl trimethylphosphonoamidodithioate, *in* T-00539
S,*S*′-Propanediyl methylphosphonodithioate, *in* M-00137
S,*S*′-Propanediyl methylphosphonoselenodithioate, *in* M-00137
O-Propyl butylphosphonochloridothioate, *in* B-00604
O-Propyl ethylphosphonochloridothioate, *in* E-00137
S-Propyl ethylphosphonochloridothioate, *in* E-00137
O-Propyl isopropylphosphonochloridothioate, *in* I-00053
S-Propyl *P*-methyl-*N*-phenylphosphonamidothioate, *in* M-00233
O-Propyl methylphosphonochloridothioate; (*S*)-*form*, *in* P-00468
O-Propyl methylphosphonochloridothioate; (±)-*form*, *in* P-00468
S-Propyl methylphosphonochloridothioate, *in* M-00298
O-Propyl methylphosphonodithioate, P-00469
S-Propyl phenylphosphonochloridothioate, *in* P-00232
S-Propyl propylphosphonochloridothioate, *in* P-00477
▷Quintiofos, Q-00006
▷*S*-Seven, S-00004
O,*O*,*O*,*O*-Tetraethyl methylenebisphosphonothioate, *in* M-00023
Tetrahydro-2,6-dimethyl-4*H*-1,3,6,2-dioxazaphosphocine; 2-Sulfide, *in* T-00094
O,*O*-Tetramethylene methylphosphonothioate, *in* M-00125
O,*O*-Tetramethylene phenylphosphonothioate, *in* P-00114
O,*O*-Tetramethylethylene phosphonothioate, *in* T-00197
1,3,2-Thiazaphospholidine 2-sulfide; 2-Me, *in* T-00303
1,3,2-Thiazaphospholidine 2-sulfide; 2-Ph, *in* T-00303
▷Trichloronate, T-00419
Triethylphosphonamidothioic acid; (±)-*form*, *O*-Et ester, *in* T-00448
4,4,6-Trimethyl-1,3,2-dioxaphosphorinane; 2-Sulfide, *in* T-00513
O,*O*-Trimethylene methylphosphonothioate, *in* M-00128
Trimethylene methylphosphonotrithioate, *in* M-00137
O,*O*-Trimethylene phenylphosphonothioate, *in* P-00116
Trimethylene phenylphosphonotrithioate, *in* D-01249
O,*O*-Trimethylene phosphonothioate, *in* D-00972
2,3,4-Trimethyl-5-phenyl-1,3,2-oxazaphospholidine; (2*R*,4*S*,5*R*)-*form*, 2-Sulfide, *in* T-00523
2,3,4-Trimethyl-5-phenyl-1,3,2-oxazaphospholidine; (2*S*,4*S*,5*R*)-*form*, 2-Sulfide, *in* T-00523
1,3,6,2-Trioxaphosphocane; 2-Me, 2-sulfide, *in* T-00592
2,4,5-Triphenyl-1,3,2-dioxaphosphole; 2-Sulfide, *in* T-00602

Thiophosphonic amides

P-(1,3-Butadien-1-yl)-*N*,*N*,*N*′,*N*-tetraethylphosphonothioic diamide, *in* B-00511
(2-Chloroethyl)phosphonothioic acid; Diamide, *in* C-00078
▷*P*-Chloromethyl-*O*-(2-chloro-4-methylphenyl)-*N*-(1-methylpropyl) phosphonamidothioate, *in* C-00101
P-Chloromethyl-*N*,*N*′-diphenylphosphonothioic diamide, *in* C-00114
P-4-Chlorophenyl-*N*,*N*,*N*′,*N*′-tetramethylphosphonothioic diamide, *in* C-00150
4-Chloro-1,2,3,4-tetrahydronaphth[2,1-*c*][1,2]-azaphosphorine; Sulfide, *in* C-00179
P-Cyclohexyl-*N*,*N*-diethylphosphonamidothioic chloride, *in* C-00255
P-Cyclohexyl-*N*,*N*-dimethylphosphonamidothioic chloride, *in* C-00255
P-Cyclohexyl-*N*,*N*′-diphenylphosphonothioic diamide, *in* C-00265
P-Cyclohexyl-*N*,*N*,*N*′,*N*′-tetraethylphosphonothioic diamide, *in* C-00265
N,*N*′-Dibenzyl-*P*-ethylphosphonothioic diamide, *in* E-00149
N,*N*′-Dibenzyl-*P*-phenylphosphonothioic diamide, *in* P-00246
3,4-Di-*tert*-butyl-2-(4-methylphenyl)-1,3,4,2-thiadiazaphospholidin-5-one 2-sulfide; (2α,3β,4α)-*form*, *in* D-00097
3,4-Di-*tert*-butyl-2-(4-methylphenyl)-1,3,4,2-thiadiazaphospholidin-5-one 2-sulfide; (2α,3α,4β)-*form*, *in* D-00097
N,*N*′-Di-*tert*-butyl-*P*-phenylphosphonothioic diamide, *in* P-00246
4,5-Diethoxy-1,3-dimethyl-2-phenyl-2*H*-1,3,2-diazaphosphole; 2-Sulfide, *in* D-00230
4,5-Diethoxy-2-methyl-2*H*-1,3,2-diazaphosphole; 2-Sulfide, *in* D-00235
2-Diethylamino-3-phenyl-1,3,2-oxazaphospholidine; 2-Sulfide, *in* D-00269
O,*O*′-Diethyl hydrazobismethylphosphinothioate, *in* H-00105
N,*N*-Diethyl-*P*-methylphosphonamidothioic acid; Fluoride, *in* D-00297
N,*N*-Diethyl-*P*-methylphosphonamidothioic acid; Chloride, *in* D-00297
N,*N*-Diethyl-*P*-phenylphosphonamidothioic acid, D-00309
N,*N*-Diethyl-*P*-phenylphosphonamidothioic acid; Chloride, *in* D-00309
N,*N*′-Diethyl-*P*-phenylphosphonothioic diamide, *in* P-00246
2,3-Dihydro-1,3-dimethyl-2-phenyl-1*H*-benzodiazaphosphole; 2-Sulfide, *in* D-00467
2,3-Dihydro-2-methyl-1,3,2-benzothiazaphosphole; 2-Sulfide, *in* D-00519
2,3-Dihydro-2-methyl-1,3,2-benzothiazaphosphole; 2-Selenide, *in* D-00519
N,*N*-Diisopropyl-*P*-methylphosphonamidothioic acid; Chloride, *in* D-00637
N,*N*′-Diisopropyl-*P*-phenylphosphonothioic diamide, *in* P-00246
N,*N*′-Dimethyl-*P*-phenylphosphonothioic diamide, *in* P-00246
N,*P*-Dimethyl-*N*′-phenylphosphonothioic diamide, *in* M-00245
▷*N*,*P*-Dimethylphosphonamidothioic fluoride, *in* D-00850
2,3-Diphenyl-1,3,2-oxazaphospholidine; 2-Sulfide, *in* D-01029
N,*N*′-Diphenyl-*P*-2-propenylphosphonothioic diamide, *in* P-00462
N,*N*′-Diphenyl-*P*-propylphosphonothioic diamide, *in* P-00480
N,*N*′-Diphenyl-*P*-propylphosphonothioic diamide, *in* P-00481
P-Ethenyl-*N*,*N*′-diphenylphosphosphonothioic diamide, *in* V-00007
Ethyl *N*,*N*-diethyl-*P*-phenylphosphonamidodithioate, *in* D-00308
S-Ethyl *N*,*N*-diethyl-*P*-phenylphosphonamidothioate, *in* D-00309
O-Ethyl *N*-(4,5-dihydro-2-thiazolyl)-*P*-methylphosphonamidothioate, *in* D-00588
O-Ethyl *N*,*P*-dimethylphosphonamidothioate, *in* D-00850
P-Ethyl-*N*,*N*′-diphenylphosphonothioic diamide, *in* E-00149
S-Ethyl *P*-ethylphosphonamidothioate, *in* E-00128
O-Ethyl *N*-isopropyl-*P*-methylphosphonamidothioate, *in* I-00037
S-Ethyl *P*-methyl-*N*-phenylphosphonamidothioate, *in* M-00233
P-Ethylphosphonamidothioic acid; Fluoride, *in* E-00128
P-Ethylphosphonothioic diamide, E-00149
P-Ethyl-*N*,*N*,*N*′,*N*′-tetramethylphosphonothioic diamide, *in* E-00149
Ethyl triethylphosphonamidodithioate, *in* T-00447
O-Ethyl triethylphosphonamidothioate, *in* T-00448
S-Ethyl triethylphosphonamidothioate, *in* T-00448
Isopropyl *N*,*N*-diethyl-*P*-phenylphosphonamidodithioate, *in* D-00308
S-Isopropyl *N*,*N*-diethyl-*P*-phenylphosphonamidothioate, *in* D-00309
Isopropyl triethylphosphonamidodithioate, *in* T-00447
O-Isopropyl triethylphosphonamidothioate, *in* T-00448

O-Isopropyl trimethylphosphonamidothioate, *in* T-00540
Methyl 6-deoxy-2,3-di-*O*-methyl-6-methylaminoglucopyranoside 4,6-cyclic thiophosphonamide; α-D-(*R*)P-*form*, *P*-Me, *in* M-00113
Methyl 6-deoxy-2,3-di-*O*-methyl-6-methylaminoglucopyranoside 4,6-cyclic thiophosphonamide; α-D-(*R*)P-*form*, *P*-Chloro, *in* M-00113
Methyl 6-deoxy-2,3-di-*O*-methyl-6-methylaminoglucopyranoside 4,6-cyclic thiophosphonamide; α-D-(*S*)P-*form*, *P*-Me, *in* M-00113
Methyl 6-deoxy-2,3-di-*O*-methyl-6-methylaminoglucopyranoside 4,6-cyclic thiophosphonamide; α-D-(*S*)P-*form*, *P*-Methoxy, *in* M-00113
O-Methyl *N*,*N*-diethyl-*P*-phenylphosphonamidothioate, *in* D-00309
S-Methyl *N*,*N*-diethyl-*P*-phenylphosphonamidothioate, *in* D-00309
P-Methyl-*N*,*N*′-diisopropylphosphonothioic diamide, *in* M-00314
S-Methyl *N*,*P*-dimethylphosphonamidothioate, *in* D-00850
P-Methyl-*N*,*N*′-diphenylphosphonothioic diamide, *in* M-00314
P,*P*′-Methylenebis[(*N*,*N*,*N*′,*N*′-tetramethyl)-phosphonodiamidothioate], *in* M-00023
Methyl *P*-ethylphosphonamidodithioate, *in* E-00127
S-Methyl *P*-ethylphosphonamidothioate, *in* E-00128
P-Methyl-*N*-isopropyl-*N*′-phenylphosphonothioic diamide, *in* M-00245
Methyl *P*-methyl-*N*,*N*-dimethylphosphonamidodithioate, *in* E-00063
Methyl *P*-methyl-*N*-phenylphosphonamidodithioate, *in* M-00232
2-Methyl-3-phenyl-1,3,2-oxazaphospholidine; 2-Sulfide, *in* M-00212
P-Methylphosphonamidothioic acid; Fluoride, *in* M-00284
Methylphosphonothioic bis(diethylamide), *in* M-00314
O-Methyl triethylphosphonamidothioate, *in* T-00448
S-Methyl trimethylphosphonamidothioate, *in* T-00540
Monomethyl 4-morpholinylphosphonothioate, *in* M-00527
Octahydro-3,8-dimethyl-2,7-diphenyl-1,6,3,8,2,7-dioxadiazadiphosphecine; 2,7-Disulfide (isomer 1), *in* O-00006
Octahydro-3,8-dimethyl-2,7-diphenyl-1,6,3,8,2,7-dioxadiazadiphosphecine; 2,7-Disulfide (isomer 2), *in* O-00006
Pentaethylphosphonothioic diamide, *in* E-00149
Pentamethylphosphonothioic diamide, *in* M-00314
P-Phenyl-*N*,*N*′-bis(triphenylphosphoranylidene)-phosphonothioic diamide, *in* P-00102
O-Phenyl *N*,*N*-diethyl-*P*-methylphosphonamidothioate, *in* D-00297
S-Phenyl *N*,*N*-diethyl-*P*-methylphosphonamidothioate, *in* D-00297
S-Phenyl *N*,*N*-diethyl-*P*-phenylphosphonamidothioate, *in* D-00309
O-Phenyl *N*-(4,5-dihydro-2-thiazolyl)-*P*-methylphosphonamidothioate, *in* D-00588
O-Phenyl *N*,*P*-diphenylphosphonamidothioate, *in* D-01104
Phenyl *P*-ethyl-*N*,*N*-dimethylphosphonamidodithioate, *in* E-00063
P-Phenylphosphonamidothioic acid; Fluoride, *in* P-00219
P-Phenylphosphonothioic diamide, P-00246
P-Phenylphosphonothioic dianilide, *in* P-00246
▷Phenyl trimethylphosphonoamidodithioate, *in* T-00539
1-Piperidinylphosphonothioic acid, P-00431
S-Propyl *P*-methyl-*N*-phenylphosphonamidothioate, *in* M-00233
N,*N*,*N*′,*N*′-Tetraethyl-*P*-(2-methyl-1,3-butadienyl)-phosphonothioic diamide, *in* M-00097
N,*N*,*N*′,*N*′-Tetraethylphosphonothioic diamide, T-00057
N,*N*,*N*′,*N*′-Tetraethyl-*P*-2-propenylphosphonothioic diamide, *in* P-00462
Tetrahydro-2-phenyl-2*H*-1,3,2-oxazaphosphorine; 2-Sulfide, *in* T-00132
Tetrahydro-2*H*-1,3,2-thiazaphosphorine 2-oxide; 2-Me, *in* T-00154
Tetrahydro-2*H*-1,3,2-thiazaphosphorine 2-oxide; 2-Ph, *in* T-00154
N,*N*,*N*′,*N*′-Tetramethyl-*P*-(2-methyl-1,3-butadienyl)-phosphonothioic diamide, *in* M-00097
N,*N*,*N*′,*N*′-Tetramethyl-*P*-2-propenylphosphonothioic diamide, *in* P-00462
N,*N*,*N*′,*N*′-Tetraphenylphosphonothioic diamide, T-00273
1,3,2-Thiazaphospholidine 2-sulfide; 2-Me, *in* T-00303
1,3,2-Thiazaphospholidine 2-sulfide; 2-Ph, *in* T-00303
Triethylphosphonamidothioic acid; (±)-*form*, *O*-Et ester, *in* T-00448
Triethylphosphonamidothioic acid; (±)-*form*, Chloride, *in* T-00448
P-Trifluoromethyl-*N*,*N*,*N*′,*N*′-tetramethylphosphonothioic diamide, *in* T-00481
2,3,4-Trimethyl-5-phenyl-1,3,2-oxazaphospholidine; (2*R*,4*S*,5*R*)-*form*, 2-Sulfide, *in* T-00523
2,3,4-Trimethyl-5-phenyl-1,3,2-oxazaphospholidine; (2*S*,4*S*,5*R*)-*form*, 2-Sulfide, *in* T-00523
Trimethylphosphonamidodithioic acid, T-00539
▷Trimethylphosphonamidothioic acid; Fluoride, *in* T-00540
Trimethylphosphonamidothioic acid; Chloride, *in* T-00540
Trimethylphosphonamidothioic acid; Bromide, *in* T-00540
2,3,4-Triphenyl-1,3,2,4-thiazadiphosphetidine; 2,4-Disulfide, *in* T-00687

Thiophosphonic isocyanates and isothiocyanates

O-Ethyl phenylphosphonisothiocyanatidothioate, *in* P-00227
O-Isopropyl phenylphosphonisothiocyanatidothioate, *in* P-00227
O-Methyl phenylphosphonoisothiocyanatidothioate, *in* P-00227
O-Phenyl phenylphosphonisothiocyanatidothioate, *in* P-00227
Phenylphosphonothioic acid; Diisothiocyanate, *in* P-00244

Thiophosphonic azides

▷Phenylphosphonothioic acid; Diazide, *in* P-00244

Thiophosphonic hydrazides

N^2,$N^{2'}$-Dibenzylidene-*P*-phenylphosphonothioic dihydrazide, *in* P-00249
N^1,$N^{1'}$-Dimethyl-*P*-phosphonothioic dihydrazide, *in* P-00249
O-Ethyl *P*-methylphosphonohydrazidothioate, *in* M-00309
O-Phenyl *P*-methyl-N^2-phenylphosphonohydrazidothioate, *in* M-00242
O-Phenyl *P*-methylphosphonohydrazidothioate, *in* M-00309
P-Phenylphosphonothioic dihydrazide, P-00249
2,3,4,5-Tetrahydro-6-methyl-2,4,4-triphenyl-1,2,3-diazaphosphorine; 3-Methoxy, 3-sulfide, *in* T-00118
N^2,$N^{2'}$,*P*-Triphenylphosphonothioic dihydrazide, *in* P-00249

Selenophosphonic acids

O-Ethyl ethylphosphonoselenoate; (*R*)-*form*, *in* E-00072
O-Ethyl ethylphosphonoselenoate; (*S*)-*form*, *in* E-00072
O-Ethyl ethylphosphonoselenoate; (±)-*form*, *in* E-00072
O-Ethyl ethylphosphonoselenothioate, E-00073
O-Ethyl methylphosphonoselenoate, E-00092
O-Isopropyl hydrogen ethylphosphonoselenoate, *in* E-00145
Methylphosphonoselenoic acid, M-00311
O-Propyl hydrogen phosphonoselenoate, *in* E-00145
O-Propyl methylphosphonoselenoate, *in* M-00311

Selenophosphonic halides

Butylphosphonoselenoic acid; Dichloride, *in* B-00609
Cyclohexylphosphonoselenoic acid; Difluoride, *in* C-00263
Cyclohexylphosphonoselenoic acid; Dichloride, *in* C-00263
Ethylphosphonoselenoic dichloride, E-00147
Methylphosphonoselenoic acid; Difluoride, *in* M-00311
Methylphosphonoselenoic acid; Dibromide, *in* M-00311
Methylphosphonoselenoic dichloride, M-00312
Phenylphosphonoselenoic acid; Difluoride, *in* P-00243
Phenylphosphonoselenoic acid; Dichloride, *in* P-00243

Selenophosphonic esters

Bis(1-methylethyl) (2-propenyl)phosphonoselenoate, *in* P-00458
O,*O*-Bis(trimethylsilyl) phenylphosphonoselenoate, *in* P-00243
Bis(trimethylsilyl) selenophosphonate, B-00440
tert-Butylphosphonoselenoic acid; *O*-Me ester, *in* B-00610

O,O-Dibutyl methylphosphonoselenoate, *in* M-00311
O,O-Dibutyl phosphonoselenoate, D-00131
O,O-Diethyl butylphosphonoselenoate, *in* B-00609
O,Se-Diethyl ethylphosphonoselenoate, *in* E-00145
O,O-Diethyl methylphosphonoselenoate, *in* M-00311
O,O-Diethyl phenylphosphonoselenoate, *in* P-00243
O,O-Diethyl phosphonoselenoate, D-00340
O,O-Diethyl (2-propenyl)phosphonoselenoate, *in* P-00458
2,3-Dihydro-2,7-dimethyl-1,4,2-benzodithiaphosphorin; 2-Selenide, *in* D-00458
2,6-Dimethyl-1,3,2,6-dioxadiphosphocane; *cis-form*, 2,6-Diselenide, *in* D-00730
O,O-Dimethyl methylphosphonoselenoate, *in* M-00311
O,O-Dimethyl phenylphosphonoselenoate, *in* P-00243
O,O-Dipropyl phosphonoselenoate, D-01211
O,O-Dipropyl phosphonoselenoate, *in* E-00145
S,S-Ethylene methylphosphonoselenodithioate, *in* M-00139
O-Ethyl ethylphosphonoselenoate; (*R*)*-form*, *in* E-00072
O-Ethyl ethylphosphonoselenoate; (*S*)*-form*, *in* E-00072
O-Ethyl ethylphosphonoselenoate; (*S*)*-form*, *O*-Trimethylsilyl ester, *in* E-00072
O-Ethyl ethylphosphonoselenoate; (±)*-form*, *in* E-00072
O-Ethyl ethylphosphonoselenoate; (±)*-form*, *Se*-Me ester, *in* E-00072
O-Ethyl ethylphosphonoselenoate; (±)*-form*, *O*-Trimethylsilyl ester, *in* E-00072
O-Ethyl ethylphosphonoselenothioate, E-00073
O-Ethyl *Se*-methyl ethylphosphonoselenoate, *in* E-00072
O-Ethyl *Se*-methyl methylphosphonoselenoate, *in* E-00092
O-Ethyl methylphosphonoselenoate, E-00092
O-Ethyl methylphosphonoselenoate; (*S*)*-form*, *Se*-Me ester, *in* E-00092
Ethylphosphonoselenoic acid; *O,O*-Di-Et ester, *in* E-00145
O-Ethyl *O*-trimethylsilyl ethylphosphonoselenoate, *in* E-00072
O-Isopropyl hydrogen ethylphosphonoselenoate, *in* E-00145
O,O-Neopentylene methylphosphonoselenoate, *in* T-00512
O,O-o-Phenylene selenophosphonate, *in* M-00090
S,S'-Propanediyl methylphosphonoselenodithioate, *in* M-00137
O-Propyl hydrogen phosphonoselenoate, *in* E-00145
O-Propyl methylphosphonoselenoate, *in* M-00311
N,N,N',N'-Tetramethyl-*P*-phenylphosphonoselenoic diamide, *in* P-00243
O,O-Trimethylene methylphosphonoselenoate, *in* M-00128
2,3,4-Trimethyl-5-phenyl-1,3,2-oxazaphospholidine; (2*R*,4*S*,5*R*)*-form*, 2-Selenide, *in* T-00523
2,3,4-Trimethyl-5-phenyl-1,3,2-oxazaphospholidine; (2*S*,4*S*,5*R*)*-form*, 2-Selenide, *in* T-00523

Selenophosphonic amides

2,3-Dihydro-2-methyl-1,3,2-benzothiazaphosphole; 2-Selenide, *in* D-00519
P-Ethyl-*N,N'*-diphenylphosphonoselenoic diamide, *in* E-00146
Ethylphosphonoselenoic bis(diethylamide), *in* E-00146
Ethylphosphonoselenoic bis(dimethylamide), *in* E-00146
Pentamethylphosphonoselenoic diamide, P-00024
N,N,N',N'-Tetraethyl-*P*-(2-propenyl)phosphonoselenoic diamide, *in* P-00459
N,N,N',N'-Tetramethyl-*P*-(2-propenyl)phosphonoselenoic diamide, *in* P-00459
2,3,4-Trimethyl-5-phenyl-1,3,2-oxazaphospholidine; (2*R*,4*S*,5*R*)*-form*, 2-Selenide, *in* T-00523
2,3,4-Trimethyl-5-phenyl-1,3,2-oxazaphospholidine; (2*S*,4*S*,5*R*)*-form*, 2-Selenide, *in* T-00523

Tellurophosphonic acids

(2-Propenyl)phosphonotelluroic acid, P-00460

Tellurophosphonic esters

Bis(trimethylsilyl) tellurophosphonate, B-00441
Diethyl (2-propenyl)phosphonotelluroate, *in* P-00460

Tellurophosphonic amides

Pentamethylphosphonotelluroic diamide, P-00025

Imidophosphonic halides

N-Benzoyl-*P*-phenylphosphonimidic dichloride, B-00033
P-tert-Butyl-*N,N'*-diphenylphosphonamidimidic chloride, B-00559
Phenyl *P*-chloromethyl-*N*-(dichlorophosphinyl)-phosphonochloridimidate, *in* C-00091

Imidophosphonic esters

tert-Butyl *N,N,N'*-tris(trimethylsilyl)phosphonamidimidate, *in* T-00867
Dimethyl *N*-methyl-*P*-phenylphosphonimidate, *in* M-00240
Dimethyl *P*-phenyl-*N*-phenylsulfonylphosphonimidate, *in* P-00192
Diphenyl *P*-phenyl-*N*-phenylsulfonylphosphonimidate, *in* P-00192
O-Ethyl *S*-methyl *N*-[(4-chlorophenyl)sulfonyl]-*P*-phenylphosphonimidothioate, *in* C-00166
Isopropyl *N,N,N'*-tris(trimethylsilyl)phosphonamidimidate, *in* T-00867
Methyl *N,N,N'*-tris(trimethylsilyl)phosphonamidimidate, *in* T-00867
Phenyl *N,N,N'*-tris(trimethylsilyl)phosphonamidimidate, *in* T-00867
Trimethylsilyl *N,N,N'*-tris(trimethylsilyl)-phosphonamidimidate, *in* T-00867

Imidophosphonic amides

P-tert-Butyl-*N,N'*-diphenylphosphonamidimidic chloride, B-00559
tert-Butyl *N,N,N'*-tris(trimethylsilyl)phosphonamidimidate, *in* T-00867
N''-Chloro-*N,N,N',N'*-tetraethyl-*P*-(trichloromethyl)-phosphonimidic diamide, *in* T-00075
O-Ethyl *S*-methyl *N*-[(4-chlorophenyl)sulfonyl]-*P*-ethylphosphonamidothioate, *in* C-00165
Hexamethylphosphonimidic diamide, *in* P-00023
Isopropyl *N,N,N'*-tris(trimethylsilyl)phosphonamidimidate, *in* T-00867
Methyl *N,N,N'*-tris(trimethylsilyl)phosphonamidimidate, *in* T-00867
N,N,N',N',N''-Pentamethyl-*P*-phenylphosphonimidic diamide, N-00079
N,N,N',N',P-Pentamethylphosphonimidic diamide, P-00023
Phenyl *N,N,N'*-tris(trimethylsilyl)phosphonamidimidate, *in* T-00867
N,N,N',N'-Tetraethyl-*P*-(trichloromethyl)phosphonimidic diamide, T-00075
Trimethylsilyl *N,N,N'*-tris(trimethylsilyl)-phosphonamidimidate, *in* T-00867

Phosphinic acids, HOP(O)H$_2$

(1-Aminoethyl)methylphosphinic acid; (±)*-form*, *in* A-00076
(1-Aminoethyl)phosphinic acid, A-00077
(Aminomethyl)methylphosphinic acid, A-00089
(1-Amino-2-methylpropyl)phosphinic acid; (+)*-form*, *in* A-00092
(1-Amino-2-methylpropyl)phosphinic acid; (+)*-form*, *N*-Benzyloxycarbonyl, *in* A-00092
(1-Amino-2-methylpropyl)phosphinic acid; (−)*-form*, *in* A-00092
(1-Amino-2-methylpropyl)phosphinic acid; (±)*-form*, *in* A-00092
(1-Amino-2-methylpropyl)phosphinic acid; (±)*-form*, *N*-(Diphenylmethyl), *in* A-00092
(1-Amino-2-methylpropyl)phosphinic acid; (±)*-form*, *N*-Benzyloxycarbonyl, *in* A-00092
(Aminophenylmethyl)phenylphosphinic acid, A-00102
(4-Aminophenyl)methylphosphinic acid, A-00104
(4-Aminophenyl)phosphinic acid, A-00108
▷Antibiotic SF 1293, A-00141
Benzo[*e*]phenazaphosphinic acid, *in* D-00440
Benzylmethylphosphinic acid; (±)*-form*, *in* B-00053
Benzylphenylphosphinic acid, B-00056
Benzylphosphinic acid, B-00059
Bis(aminomethyl)phosphinic acid, B-00087

▷Bis(3-aminophenyl)phosphinic acid, B-00088
Bis(4-aminophenyl)phosphinic acid, B-00089
Bis(1*H*-benzimidazol-1-yl)phosphinic acid, B-00097
Bis(benzoylthiomethyl)phosphinic acid, *in* B-00292
Bis(*N*-benzylaminomethyl)phosphinic acid, *in* B-00087
Bis(benzylthiomethyl)phosphinic acid, B-00101
Bis(bromomethyl)phosphinic acid, B-00108
Bis(2-bromophenyl)phosphinic acid, B-00112
Bis(3-bromophenyl)phosphinic acid, B-00113
Bis(4-bromophenyl)phosphinic acid, B-00114
Bis(*N*-*tert*-butylaminomethyl)phosphinic acid, *in* B-00087
Bis(2-carboxyphenyl)phosphinic acid, B-00117
Bis(4-carboxyphenyl)phosphinic acid, B-00118
Bis(chloromethyl)phosphinic acid, B-00127
Bis(2-chlorophenyl)phosphinic acid, B-00137
Bis(3-chlorophenyl)phosphinic acid, B-00138
▷Bis(4-chlorophenyl)phosphinic acid, B-00139
Bis(2,5-dibromophenyl)phosphinic acid, B-00156
Bis(2,3-dichlorophenyl)phosphinic acid, B-00163
Bis(2,5-dichlorophenyl)phosphinic acid, B-00164
Bis(3,5-dichlorophenyl)phosphinic acid, B-00165
Bis[(dihydroxyphosphinyl)methyl]phosphinic acid, B-00182
Bis(4-dimethylaminophenyl)phosphinic acid, B-00188
Bis(2,5-dimethylphenyl)phosphinic acid, B-00198
Bis(ethoxymethyl)phosphinic acid, B-00272
Bis(2-fluorophenyl)phosphinic acid, B-00278
Bis(3-fluorophenyl)phosphinic acid, B-00279
Bis(4-fluorophenyl)phosphinic acid, B-00280
Bis(hydroxymethyl)phosphinic acid, B-00282
Bis(2-hydroxyphenyl)phosphinic acid, B-00283
Bis(3-hydroxyphenyl)phosphinic acid, B-00284
Bis(4-hydroxyphenyl)phosphinic acid, B-00285
Bis(1-hydroxy-2,2,2-trichloroethyl)phosphinic acid, B-00287
Bis(iodomethyl)phosphinic acid, B-00288
Bis(2-iodophenyl)phosphinic acid, B-00289
Bis(3-iodophenyl)phosphinic acid, B-00290
Bis(4-iodophenyl)phosphinic acid, B-00291
Bis(mercaptomethyl)phosphinic acid, B-00292
Bis(2-methoxyphenyl)phosphinic acid, B-00299
Bis(3-methoxyphenyl)phosphinic acid, B-00300
Bis(4-methoxyphenyl)phosphinic acid, B-00301
Bis(2-methylphenyl)phosphinic acid, B-00318
Bis(3-methylphenyl)phosphinic acid, B-00319
Bis(4-methylphenyl)phosphinic acid, B-00320
Bis(1-methylpropyl)phosphinic acid, B-00345
Bis(2-methylpropyl)phosphinic acid, B-00346
▷Bis(3-nitrophenyl)phosphinic acid, B-00373
Bis(4-nitrophenyl)phosphinic acid, B-00374
Bis(2-oxo-3-oxazolidinyl)phosphinic acid, B-00378
Bis(pentafluorophenyl)phosphinic acid, B-00382
Bis(phenoxymethyl)phosphinic acid, B-00384
Bis(4-phenoxyphenyl)phosphinic acid, *in* B-00285
Bis(4-phenyl-1,3-butadienyl)phosphinic acid, B-00385
Bis(2-phenylethenyl)phosphinic acid, B-00386
Bis(phenylethynyl)phosphinic acid, B-00388
Bis(phenylmethylsulfonylmethyl)phosphinic acid, *in* B-00101
Bis(trichloromethyl)phosphinic acid, B-00402
Bis(2,3,6-trichlorophenyl)phosphinic acid, B-00405
Bis(2,4,5-trichlorophenyl)phosphinic acid, B-00406
Bis(trifluoromethyl)phosphinic acid, B-00411
Bis(2,4,5-trimethylphenyl)phosphinic acid, B-00427
(3-Bromophenyl)phosphinic acid, B-00488
(4-Bromophenyl)phosphinic acid, B-00489
(2-Butenyl)methylphosphinic acid, B-00523
[1-(Butylamino)-1-methylethyl]phosphinic acid, B-00542
Butyl-*tert*-butylphosphinic acid, B-00546
tert-Butylethylphosphinic acid, B-00560
tert-Butylisopropylphosphinic acid, B-00562
Butylmethylphosphinic acid, B-00566
tert-Butylmethylphosphinic acid, B-00567
Butylphenylphosphinic acid, B-00578
tert-Butylphenylphosphinic acid; (*R*)-*form*, *in* B-00579
tert-Butylphenylphosphinic acid; (*S*)-*form*, *in* B-00579
tert-Butylphenylphosphinic acid; (±)-*form*, *in* B-00579
tert-Butylphenylphosphinothioic acid; (±)-*form*, *in* B-00583
Butyl phosphinate, B-00587
Butylphosphinic acid, B-00590
tert-Butylphosphinic acid; (*R*)-*form*, *in* B-00591
tert-Butylphosphinic acid; (±)-*form*, *in* B-00591
(Chloromethyl)phenylphosphinic acid, C-00103
(4-Chlorophenyl)methylphosphinic acid, C-00131
(2-Chlorophenyl)phosphinic acid, C-00142
(3-Chlorophenyl)phosphinic acid, C-00143
(4-Chlorophenyl)phosphinic acid, C-00144
(4-Cyanophenyl)phosphinic acid, C-00222
Cyclohexylmethylphosphinic acid, C-00247
Cyclohexylphenylphosphinic acid, C-00250
Cyclohexylphosphinic acid, C-00253
Cyclopentylmethylphosphinic acid, C-00279
Decahydro-1,5-dihydroxy-1*H*-1,5-benzodiphosphepin 1,5-dioxide, D-00002
Decahydro-1,4-dihydroxy-1,4-benzodiphosphorin 1,4-dioxide, D-00003
Decylmethylphosphinic acid, D-00009
Decylphenylphosphinic acid, D-00010
(1-Diazoethyl)phenylphosphinic acid, D-00032
Dibenzylphosphinic acid, D-00055
▷Dibutylphosphinic acid, D-00109
Di-*tert*-butylphosphinic acid, D-00110
Dicyclohexylphosphinic acid, D-00211
Didecylphosphinic acid, D-00221
Didodecylphosphinic acid, D-00226
Diethylphosphinic acid, D-00322
Di-2-furanylphosphinic acid, D-00421
Diheptylphosphinic acid, D-00425
Dihexadecylphosphinic acid, D-00429
▷Dihexylphosphinic acid, D-00434
2,3-Dihydro-1*H*-benzo[*e*]phosphindole 3-oxide; 3-Hydroxy, *in* D-00442
5,10-Dihydro-5,10-dihydroxyphosphanthrene 5,10-dioxide, D-00452
5,10-Dihydro-5-hydroxyacridophosphine 5-oxide, D-00487
4,5-Dihydro-4-hydroxybenzo[*lmn*]phosphanthridine 4-oxide, D-00489
6,7-Dihydro-6-hydroxy-5*H*-dibenzo[*c,e*]phosphepin 6-oxide, D-00491
10,11-Dihydro-5-hydroxy-5*H*-dibenzo[*b,f*]phosphepin 5-oxide, D-00492
▷2,5-Dihydro-1-hydroxy-3,4-dimethyl-1*H*-phosphole 1-oxide, D-00493
2,3-Dihydro-2-hydroxy-1*H*-isophosphindole 2-oxide, D-00494
1,2-Dihydro-2-hydroxyisophosphinoline 2-oxide, D-00495
5,10-Dihydro-5-hydroxy-10-methylacridophosphine 5-oxide, *in* D-00515
1,3-Dihydro-1-hydroxy-3-methyl-1,2,3-benzodoxaphosphole 3-oxide, *in* D-00488
▷2,3-Dihydro-1-hydroxy-4-methyl-1*H*-phosphole 1-oxide, D-00496
1,3-Dihydro-1-hydroxy-3-phenyl-1,2,3-benziodoxaphosphole 3-oxide, *in* D-00488
5,6-Dihydro-5-hydroxyphosphanthridine 5-oxide, D-00504
1,2-Dihydro-1-hydroxyphosphinoline 1-oxide, D-00505
2,3-Dihydro-1-hydroxy-1*H*-phosphole 1-oxide, D-00506
2,5-Dihydro-1-hydroxy-1*H*-phosphole 1-oxide, D-00507
1,3-Dihydroxy-1,3-diphosphepane 1,3-dioxide, D-00619
1,3-Dihydroxy-1,3-diphospholane 1,3-dioxide, D-00621
1,3-Dihydroxy-1,3-diphosphorinane 1,3-dioxide, D-00622
1,8-Dihydroxy-3*a*,4,7,7*a*-tetrahydro-4,7-phosphinidene-1*H*-phosphindole 1,8-dioxide, D-00625
Di-1*H*-imidazol-1-ylphosphinic acid, D-00628
Diisopropylphosphinic acid, D-00641
(1,1-Dimethyl-3-oxobutyl)ethylphosphinic acid, D-00782
(2,4-Dimethylphenyl)phosphinic acid, D-00804
(2,5-Dimethylphenyl)phosphinic acid, D-00805
(3,4-Dimethylphenyl)phosphinic acid, D-00806
Dimethylphosphinic acid, D-00829
Di-4-morpholinylphosphinic acid, D-00935
Di-1-naphthylphosphinic acid, D-00941
Dinonylphosphinic acid, D-00949
Dioctadecylphosphinic acid, D-00953
Dioctylphosphinic acid, D-00958
Dipentylphosphinic acid, D-00981
[1-(Diphenylmethylamino)ethyl]phosphinic acid, *in* A-00077
(Diphenylmethyl)phenylphosphinic acid, D-01023

(1,3-Diphenyl-3-oxopropyl)phenylphosphinic acid, D-01030
▷Diphenylphosphinic acid, D-01039
O-(Diphenylphosphinyl)hydroxylamine, D-01097
O-(Diphenylphosphinyl)hydroxylamine; *N*-Ac, *in* D-01097
Di-2-propenylphosphinic acid, D-01193
Dipropylphosphinic acid, D-01205
P,P-Di-1-pyrrolidinylphosphinic acid, D-01233
Ditetradecylphosphinic acid, D-01238
Di-2-thienylphosphinic acid, D-01250
Diundecylphosphinic acid, D-01256
Divinylphosphinic acid, D-01258
1,2-Ethanediylbis[methylphosphinic acid], E-00013
1,2-Ethanediylbis[phenylphosphinic acid], E-00015
3-Ethoxycarbonyl-1-hydroxyphosphetane 1-oxide, *in* H-00184
Ethyl bis(2-ethoxycarbonylphenyl)phosphinate, *in* B-00117
Ethylmethylphosphinic acid, E-00085
Ethylphenylphosphinic acid; (*R*)-*form*, *in* E-00101
Ethylphenylphosphinic acid; (*S*)-*form*, *in* E-00101
Ethylphenylphosphinic acid; (±)-*form*, *in* E-00101
▷Ethyl phosphinate, E-00122
Ethylphosphinic acid, E-00124
Ethynylphosphinic acid, E-00202
(Fluoromethyl)phenylphosphinic acid, F-00022
(3-Fluorophenyl)phosphinic acid, F-00035
2,3,4,5,6,7-Hexahydro-1*H*-isophosphindole 2-oxide; 2-Hydroxy, *in* H-00035
Hexylmethylphosphinic acid, H-00094
5-Hydroxy-5*H*-dibenzo[*b,f*]phosphepin 5-oxide, H-00117
5-Hydroxy-5*H*-dibenzophosphole 5-oxide, H-00118
5-Hydroxy-5*H*-dibenzophosphole 5-oxide; Dodecahydro, *in* H-00118
1-Hydroxy-2,3-dihydro-1*H*-phosphindole 1-oxide, H-00120
3-Hydroxy-2,2-dimethyl-1,3-oxaphosphetane 3-oxide, H-00126
10-Hydroxy-2,8-dimethyl-10*H*-phenothiaphosphine 10-oxide, H-00127
10-Hydroxy-2,8-dimethyl-10*H*-phenoxaphosphine 10-oxide, H-00128
4-Hydroxy-2,6-diphenyl-4*H*-1,4-oxaphosphorin 4-oxide, H-00140
1-Hydroxy-3,4-diphenyl-1*H*-phosphole 1-oxide, H-00142
4-Hydroxy-2,6-diphenyl-4*H*-1,4-thiaphosphorine 2-oxide, H-00143
(Hydroxymethyl)phenylphosphinic acid, H-00160
3-(Hydroxymethylphosphinyl)propanoic acid, H-00161
1-Hydroxy-2,2,3,4,4-pentamethylphosphetane 1-oxide, H-00168
1-Hydroxy-2,2,3,4,4-pentamethylphosphetane 1-oxide; *cis-form*, *in* H-00168
1-Hydroxy-2,2,3,4,4-pentamethylphosphetane 1-oxide; *trans-form*, *in* H-00168
10-Hydroxy-10*H*-phenoxaphosphine 10-oxide, H-00170
(Hydroxyphenylmethyl)phenylphosphinic acid, H-00173
3-(Hydroxyphenylphosphinyl)propanoic acid, H-00178
1-Hydroxy-3-phosphetanecarboxylic acid 1-oxide, H-00184
1-Hydroxyphospholane 1-oxide, H-00185
1-Hydroxy-3,8-phosphonanedione 1-oxide, *in* P-00373
1-Hydroxyphosphorinane 1-oxide, H-00188
1-Hydroxy-1,2,3,4-tetrahydrophosphinoline 1-oxide, H-00195
1-Hydroxy-2,3,4,5-tetraphenyl-1*H*-phosphole 1-oxide, H-00200
3-Hydroxy-1,3-thiaphosphetane 3-oxide, H-00201
1-Hydroxy-2,4,6-trimethyl-1,3,5-dioxaphosphorinane 5-oxide, H-00203
5-Hydroxy-2,4,6-triphenyl-1,3,5-dioxaphosphorinane 5-oxide, H-00204
Isopropylmethylphosphinic acid, I-00035
Isopropylphenylphosphinic acid, I-00043
Isopropylphosphinic acid, I-00049
Methanediphosphonous acid, M-00024
(2-Methoxyphenyl)phosphinic acid, M-00068
(4-Methoxyphenyl)phosphinic acid, M-00069
Methylenebis[methylphosphinic acid], M-00140
Methylenebis[phenylphosphinic acid], M-00142
Methylenebis[phosphinic acid], M-00143
Methyl(2-methylpropyl)phosphinic acid, M-00172
(2-Methylphenyl)methylphosphinic acid, M-00203
(4-Methylphenyl)methylphosphinic acid, M-00204
Methylphenylphosphinic acid; (*R*)-*form*, *in* M-00219
Methylphenylphosphinic acid; (*S*)-*form*, *in* M-00219
Methylphenylphosphinic acid; (±)-*form*, *in* M-00219
(2-Methylphenyl)phosphinic acid, M-00220
(3-Methylphenyl)phosphinic acid, M-00221
▷(4-Methylphenyl)phosphinic acid, M-00222
Methyl phosphinate, M-00276
Methylphosphinic acid, M-00278
(1-Methylpropyl)phenylphosphinic acid, M-00372
(2-Methylpropyl)phenylphosphinic acid, M-00373
Methylpropylphosphinic acid, M-00375
(2-Methylpropyl)phosphinic acid, M-00376
Methyl(trichloromethyl)phosphinic acid, M-00419
Methyl(trifluoromethyl)phosphinic acid, M-00420
Methylvinylphosphinic acid, M-00438
1-Naphthylphenylphosphinic acid, N-00017
2-Naphthylphenylphosphinic acid, N-00018
1-Naphthylphosphinic acid, N-00019
2-Naphthylphosphinic acid, N-00020
(2-Nitrophenyl)methylphosphinic acid, N-00042
(4-Nitrophenyl)methylphosphinic acid, N-00043
(2-Nitrophenyl)phosphinic acid, N-00053
(4-Nitrophenyl)phosphinic acid, N-00054
Nonylphenylphosphinic acid, N-00074
Octahydro-1,8-dihydroxy-4,7-phosphinidene-1*H*-phosphindole 1,8-dioxide, *in* D-00625
Octylphenylphosphinic acid, O-00040
Pentyl phosphinate, P-00048
▷Phenazaphosphinic acid, P-00061
1,4-Phenylenebis[methylphosphinic acid], P-00137
1,2-Phenylenebis[phenylphosphinic acid], P-00139
1,4-Phenylenebis[phenylphosphinic acid], P-00140
(2-Phenylethenyl)phosphinic acid, P-00153
[(Phenylmethylamino)phenylmethyl]phenylphosphinic acid, *in* A-00102
Phenyl(4-phenyl-1,3-butadienyl)phosphinic acid, P-00184
Phenyl(phenylbutadiynyl)phosphinic acid, P-00185
Phenyl(2-phenylethenyl)phosphinic acid, P-00186
Phenyl(phenylethynyl)phosphinic acid, P-00187
Phenylphosphinic acid, P-00210
Phenyl(1-propenyl)phosphinic acid, P-00299
Phenyl(2-propenyl)phosphinic acid, P-00300
Phenylpropylphosphinic acid, P-00303
Phenyl(trichloromethyl)phosphinic acid, P-00327
Phenyl(trifluoromethyl)phosphinic acid, P-00329
Phenyl(triphenylmethyl)phosphinic acid, P-00332
Phenylvinylphosphinic acid, P-00337
1,3-Propanediylbis[phenylphosphinic acid], P-00447
2-Propenylphosphinic acid, P-00452
Propyl phosphinate, P-00470
Propylphosphinic acid, P-00472
4,5,9,10-Tetrahydro-4,9-dihydroxyphosphorino[2,3,4,5-*lmn*]-phosphanthridine 4,9-dioxide, T-00090
2,3,4,7-Tetrahydro-1*H*-isophosphindole 2-oxide; 2-Hydroxy, *in* T-00109
2,4,6,8-Tetrahydroxy-1,5-dimethyl-3,7,9-trioxa-2,4,6,8-tetraphosphabicyclo[3.3.1]nonane, T-00161
2-Thienylphosphinic acid, T-00306
Toldimfos, T-00322
(Trifluoromethyl)phosphinic acid, T-00475
2,4,6-Trihydroxy-1,2,4,6-oxatriphosphorinane 2,4,6-trioxide, T-00494
(2,4,5-Trimethylphenyl)phosphinic acid, T-00524
(2,4,6-Trimethylphenyl)phosphinic acid, T-00525
Vinylphosphinic acid, V-00002

Phosphinic anhydrides

Acetic diphenylphosphinic anhydride, *in* D-01039
Benzoic diphenylphosphinic anhydride, *in* D-01039
Bis(bromomethyl)phosphinic acid; Anhydride, *in* B-00108
Bis(chloromethyl)phosphinic acid; Anhydride, *in* B-00127
Bis(2-methylphenyl)phosphinic acid; Anhydride, *in* B-00318
Bis(2-oxo-3-oxazolidinyl) diphenyl phosphoric anhydride, *in* B-00378
Bis(2-oxo-3-oxazolidinyl)phosphinic acid; Anhydride, *in* B-00378
Bis(trifluoromethyl)phosphinic acid; Anhydride, *in* B-00411
Bis(2,4,6-trimethylphenyl)phosphinic acid; Anhydride, *in* B-00428

(Chloromethyl)methylphosphinic acid; Anhydride, *in* C-00100
(Chloromethyl)phenylphosphinic acid; Anhydride, *in* C-00103
Dibutylphosphinic acid; Anhydride, *in* D-00109
▷Diethylphosphinic acid; Anhydride, *in* D-00322
Diethylphosphinic diethylphosphinothioic anhydride, *in* D-00322
5,6-Dihydro-5-hydroxyphosphanthridine 5-oxide; Anhydride, *in* D-00504
Diisopropylphosphinic acid; Anhydride, *in* D-00641
2,7-Dimethyl-1,2,7-oxadiphosphepane 2,7-dioxide, D-00779
2,6-Dimethyl-1,2,6-oxadiphosphorinane 2,6-dioxide, D-00780
Dimethylphosphinic acid; Anhydride, *in* D-00829
Di-1-naphthylphosphinic acid; Anhydride, *in* D-00941
Diphenylphosphinic acid; Anhydride, *in* D-01039
Dipropylphosphinic acid; Anhydride, *in* D-01205
Ethylmethylphosphinic acid; Anhydride, *in* E-00085
Hexylmethylphosphinic acid; Anhydride, *in* H-00094
1-Hydroxy-2,2,3,4,4-pentamethylphosphetane 1-oxide; Anhydride, *in* H-00168
2-Methyl-1,2-oxaphospholan-5-one 2-oxide, *in* O-00053
Methylphenylphosphinic acid; (±)-*form*, Anhydride, *in* M-00219
5,5′-Oxybis[5*H*-benzophosphindole] 5,5′-dioxide, *in* H-00118
Oxybis(2,6-diphenyl-4*H*-1,4-oxaphosphorin)-4,4′-dioxide, *in* H-00140
2-Phenyl-1,2-oxaphospholan-2-one 2-oxide, *in* O-00053
Phenylphosphinic acid; Anhydride, *in* P-00210
Phenyl(trichloromethyl)phosphinic acid; Anhydride, *in* P-00327
Phenyl(trifluoromethyl)phosphinic acid; Anhydride, *in* P-00329

Phosphinic halides

Benzylphenylphosphinic acid; Chloride, *in* B-00056
Bis(bromomethyl)phosphinic acid; Bromide, *in* B-00108
Bis(3-bromophenyl)phosphinic acid; Chloride, *in* B-00113
Bis(4-bromophenyl)phosphinic acid; Chloride, *in* B-00114
Bis(4-chlorocarbonylphenyl)phosphinic chloride, *in* B-00118
Bis(chloromethyl)phosphinic acid; Chloride, *in* B-00127
Bis(2-chlorophenyl)phosphinic acid; Chloride, *in* B-00137
Bis(3-chlorophenyl)phosphinic acid; Chloride, *in* B-00138
Bis(4-chlorophenyl)phosphinic acid; Chloride, *in* B-00139
Bis(3-fluorophenyl)phosphinic acid; Chloride, *in* B-00279
Bis(4-fluorophenyl)phosphinic acid; Chloride, *in* B-00280
Bis(4-methoxyphenyl)phosphinic acid; Chloride, *in* B-00301
Bis(2-methylphenyl)phosphinic acid; Chloride, *in* B-00318
Bis(3-methylphenyl)phosphinic acid; Chloride, *in* B-00319
Bis(4-methylphenyl)phosphinic acid; Chloride, *in* B-00320
Bis(1-methylpropyl)phosphinic acid; Chloride, *in* B-00345
Bis(2-methylpropyl)phosphinic acid; Fluoride, *in* B-00346
Bis(2-methylpropyl)phosphinic acid; Chloride, *in* B-00346
Bis(2-oxo-3-oxazolidinyl)phosphinic acid; Chloride, *in* B-00378
Bis(pentafluorophenyl)phosphinic acid; Fluoride, *in* B-00382
Bis(4-phenyl-1,3-butadienyl)phosphinic acid; Chloride, *in* B-00385
Bis(2-phenylethenyl)phosphinic acid; Chloride, *in* B-00386
Bis(trichloromethyl)phosphinic acid; Chloride, *in* B-00402
Bis(trifluoromethyl)phosphinic acid; Fluoride, *in* B-00411
Bis(trifluoromethyl)phosphinic acid; Chloride, *in* B-00411
Bis(trifluoromethyl)phosphinic acid; Bromide, *in* B-00411
Bis(2,4,6-trimethylphenyl)phosphinic acid; Chloride, *in* B-00428
1-Bromo-2,5-dihydro-3,4-dimethyl-1*H*-phosphole; 1-Oxide, *in* B-00469
1-Bromo-2,5-dihydro-3,4-dimethyl-1*H*-phosphole; 1-Sulfide, *in* B-00469
1-Bromo-2,5-dihydro-3,4-dimethyl-1*H*-phosphole 1-oxide, *in* D-00493
tert-Butylethylphosphinic acid; Chloride, *in* B-00560
tert-Butylisopropylphosphinic acid; Chloride, *in* B-00562
Butylmethylphosphinic acid; Fluoride, *in* B-00566
Butylmethylphosphinic acid; Chloride, *in* B-00566
tert-Butylmethylphosphinic acid; Chloride, *in* B-00567
tert-Butylphenylphosphinic acid; (*R*)-*form*, Chloride, *in* B-00579
tert-Butylphenylphosphinic acid; (*S*)-*form*, Chloride, *in* B-00579
tert-Butylphenylphosphinic acid; (±)-*form*, Fluoride, *in* B-00579
tert-Butylphenylphosphinothioic acid; (±)-*form*, Chloride, *in* B-00583
5-Chloro-5*H*-dibenzo[*b,f*]phosphepin 5-oxide, *in* H-00117
5-Chloro-5*H*-dibenzophosphole; 5-Oxide, *in* C-00034
5-Chloro-10,11-dihydro-5*H*-dibenzo[*b,f*]phosphepin 5-oxide, *in* D-00492
1-Chloro-2,5-dihydro-3,4-dimethyl-1*H*-phosphole; 1-Oxide, *in* C-00041
1-Chloro-2,5-dihydro-3,4-dimethyl-1*H*-phosphole 1-oxide, *in* D-00493
1-Chloro-2,3-dihydro-4-methyl-1*H*-phosphole; 1-Oxide, *in* C-00045
1-Chloro-2,5-dihydro-2-methyl-1*H*-phosphole; 1-Oxide, *in* C-00046
1-Chloro-2,5-dihydro-3-methyl-1*H*-phosphole; 1-Oxide, *in* C-00047
10-Chloro-5,10-dihydrophenophosphazine; 10-Oxide, *in* C-00049
5-Chloro-5,6-dihydrophosphanthridine 5-oxide, *in* D-00504
1-Chloro-2,3-dihydro-1*H*-phosphole; 1-Oxide, *in* C-00051
1-Chloro-2,5-dihydro-1*H*-phosphole; 1-Oxide, *in* C-00052
4-Chloro-2,6-diphenyl-4*H*-1,4-oxaphosphorin 4-oxide, *in* H-00140
4-Chloro-2,6-diphenyl-4*H*-1,4-thiaphosphorin 4-oxide, *in* H-00143
(Chloromethyl)methylphosphinic acid; Chloride, *in* C-00100
(Chloromethyl)phenylphosphinic acid; Chloride, *in* C-00103
3-(Chloromethylphosphinyl)propanoyl chloride, *in* H-00161
1-Chloro-2,2,3,4,4-pentamethylphosphetane; *trans-form*, Oxide, *in* C-00124
1-Chloro-2,2,3,4,4-pentamethylphosphetane 1-oxide, *in* C-00124
(4-Chlorophenyl)methylphosphinic acid; Fluoride, *in* C-00131
(4-Chlorophenyl)methylphosphinic acid; Chloride, *in* C-00131
3-(Chlorophenylphosphinyl)propanoyl chloride, *in* H-00178
1-Chlorophosphorinane; Oxide, *in* C-00168
3-Chloro-1,3-thiaphosphetane; 3-Oxide, *in* C-00188
Cyclohexylmethylphosphinic acid; Fluoride, *in* C-00247
Cyclohexylmethylphosphinic acid; Chloride, *in* C-00247
Cyclohexylphenylphosphinic acid; (±)-*form*, Chloride, *in* C-00250
Cyclopentylmethylphosphinic acid; Chloride, *in* C-00279
Dibenzylphosphinic acid; Fluoride, *in* D-00055
Dibenzylphosphinic acid; Chloride, *in* D-00055
Dibutylphosphinic acid; Fluoride, *in* D-00109
Dibutylphosphinic acid; Chloride, *in* D-00109
Di-*tert*-butylphosphinic acid; Fluoride, *in* D-00110
Di-*tert*-butylphosphinic acid; Chloride, *in* D-00110
Di-*tert*-butylphosphinic acid; Bromide, *in* D-00110
Di-*tert*-butylphosphinic acid; Iodide, *in* D-00110
5,10-Dichloro-5,10-dihydrophosphanthrene; 5,10-Dioxide, *in* D-00164
Dicyclohexylphosphinic acid; Fluoride, *in* D-00211
Dicyclohexylphosphinic acid; Chloride, *in* D-00211
Didecylphosphinic acid; Chloride, *in* D-00221
Didodecylphosphinic acid; Chloride, *in* D-00226
Diethylphosphinic acid; Fluoride, *in* D-00322
Diethylphosphinic acid; Chloride, *in* D-00322
Dihexylphosphinic acid; Fluoride, *in* D-00434
Dihexylphosphinic acid; Chloride, *in* D-00434
Di-1*H*-imidazol-1-ylphosphinic acid; Chloride, *in* D-00628
Diisopropylphosphinic acid; Fluoride, *in* D-00641
Diisopropylphosphinic acid; Chloride, *in* D-00641
Dimethylphosphinic chloride, D-00831
Dimethylphosphinic fluoride, D-00832
Di-4-morpholinylphosphinic acid; Fluoride, *in* D-00935
Di-4-morpholinylphosphinic acid; Chloride, *in* D-00935
Di-1-naphthylphosphinic acid; Chloride, *in* D-00941
Dioctylphosphinic acid; Chloride, *in* D-00958
Dipentylphosphinic acid; Fluoride, *in* D-00981
Dipentylphosphinic acid; Chloride, *in* D-00981
Diphenylphosphinic acid; Bromide, *in* D-01039
Diphenylphosphinic chloride, D-01041
Diphenylphosphinic fluoride, D-01042
Di-1-piperidinylphosphinic acid; Fluoride, *in* D-01190
Di-1-piperidinylphosphinic acid; Chloride, *in* D-01190

Di-2-propenylphosphinic acid; Chloride, *in* D-01193
Dipropylphosphinic acid; Fluoride, *in* D-01205
Dipropylphosphinic acid; Chloride, *in* D-01205
P,P-Di-1-pyrrolidinylphosphinic acid; Chloride, *in* D-01233
Divinylphosphinic acid; Chloride, *in* D-01258
1,2-Ethanediylbis[phenylphosphinic acid]; Dichloride, *in* E-00015
Ethylmethylphosphinic acid; Chloride, *in* E-00085
Ethylphenylphosphinic acid; (±)-*form*, Fluoride, *in* E-00101
Ethylphenylphosphinic acid; (±)-*form*, Chloride, *in* E-00101
1-Fluoro-2,5-dihydro-1*H*-phosphole 1-oxide, *in* D-00507
1-Fluoro-2,2,3,4,4-pentamethylphosphetane 1-oxide, *in* H-00168
1-Fluorophosphorinane; Oxide, *in* F-00047
Hexylmethylphosphinic acid; Chloride, *in* H-00094
1-Hydroxy-2,2,3,4,4-pentamethylphosphetane 1-oxide; *trans-form*, Bromide, *in* H-00168
Isopropylmethylphosphinic acid; Chloride, *in* I-00035
Isopropylphenylphosphinic acid; (±)-*form*, Chloride, *in* I-00043
1-Methoxy-3-methoxycarbonylphosphetane 1-oxide, *in* H-00184
Methyl bis(4-methylphenyl)phosphinate, *in* B-00320
Methyl(2-methylpropyl)phosphinic acid; Fluoride, *in* M-00172
Methyl(2-methylpropyl)phosphinic acid; Chloride, *in* M-00172
(2-Methylphenyl)methylphosphinic acid; Chloride, *in* M-00203
(4-Methylphenyl)methylphosphinic acid; (±)-*form*, Chloride, *in* M-00204
Methylphenylphosphinic acid; (±)-*form*, Fluoride, *in* M-00219
Methylphenylphosphinic acid; (±)-*form*, Chloride, *in* M-00219
Methyl(2-propenyl)phosphinic acid; Chloride, *in* M-00364
(1-Methylpropyl)phenylphosphinic acid; (±)-*form*, Chloride, *in* M-00372
Methylpropylphosphinic acid; Chloride, *in* M-00375
Methyl(trichloromethyl)phosphinic acid; Chloride, *in* M-00419
Methylvinylphosphinic acid; Chloride, *in* M-00438
2-Naphthylphenylphosphinic acid; (±)-*form*, Chloride, *in* N-00018
(2-Nitrophenyl)methylphosphinic acid; Chloride, *in* N-00042
Phenyl(4-phenyl-1,3-butadienyl)phosphinic acid; Chloride, *in* P-00184
Phenyl(trichloromethyl)phosphinic acid; Chloride, *in* P-00327
Phenyl(triphenylmethyl)phosphinic acid; Chloride, *in* P-00332
Phenylvinylphosphinic acid; Chloride, *in* P-00337
Tetramethylenephosphinic chloride, *in* C-00167
4*H*-1,4-Thiaphosphorin; 4-Chloro, 4-oxide, *in* T-00299

Phosphinic pseudohalides

tert-Butylphenylphosphinic acid; (*S*)-*form*, Thiocyanate, *in* B-00579
Dibutylphosphinic acid; Isocyanate, *in* D-00109
Dimethylphosphinic acid; Cyanide, *in* D-00829
Diphenylphosphinic acid; Cyanide, *in* D-01039

Phosphinic esters

Acetic diphenylphosphinic anhydride, *in* D-01039
(Aminophenylmethyl)phenylphosphinic acid; (*R*)C-*form*, Et ester, *in* A-00102
Benzoic diphenylphosphinic anhydride, *in* D-01039
12*H*-Benzo[*c*]phenothiaphosphine; 12-Ethoxy, 7,7,12-trioxide, *in* B-00023
Benzyl bis(1-aziridinyl)phosphinate, *in* B-00091
Benzyl diethylphosphinate, *in* D-00322
Benzyl diphenylphosphinate, *in* D-01039
Benzyl 1,1,2,3,3-pentamethyltrimethylenephosphinate, *in* H-00168
Bis(4-methoxycarbonylphenyl)phosphinic acid, *in* B-00118
1,3-Bis(1-methylethoxy)-1,3-diphosphetane 1,3-dioxide, *in* D-00620
Butyl bis(1-aziridinyl)phosphinate, *in* B-00091
Butyl bis(4-chlorophenyl)phosphinate, *in* B-00139
Butyl bis(iodomethyl)phosphinate, *in* B-00288
Butyl butylmethylphosphinate, *in* B-00566
Butyl butylphosphonochoridite, *in* B-00605
Butyl cyclopentylmethylphosphinate, *in* C-00279
Butyl dibutylphosphinate, *in* D-00109
Butyl dioctylphosphinate, *in* D-00958
tert-Butyl diphenylphosphinate, *in* D-01039
tert-Butyl ethylphosphinate, *in* E-00124
Butyl ethynylphosphinate, *in* E-00202
Butyl isopropylphosphinate, *in* I-00049
tert-Butyl methylphenylphosphinate, *in* M-00219
Butyl methyl(trifluoromethyl)phosphinate, *in* M-00420
tert-Butylphenylphosphinic acid; (*S*)-*form*, (−)-Menthyl ester, *in* B-00579
tert-Butylphenylphosphinothioic acid; (±)-*form*, *O*-Me ester, *in* B-00583
tert-Butylphosphinic acid; (±)-*form*, Me ester, *in* B-00591
Butyl 2-propenylphosphinate, *in* P-00452
5-*tert*-Butyl-2,2,4,5-tetrahydro-2,2,4-triisopropoxy-1,2,4-oxadiphosphole, B-00637
Butyl vinylphosphinate, *in* V-00002
Cyclohexyl dicyclohexylphosphinate, *in* D-00211
Cyclohexyl di-4-morpholinylphosphinate, *in* D-00935
7,8,10,11,13,14,16,17,19,24-Decahydro-5*H*-dibenzo[*o,r*]-[1,4,7,10,13,17]pentaoxaphosphacycloeicosin; 24-Methoxy, 24-oxide, *in* D-00001
6*H*-Dibenzo[*c,e*][1,2]oxaphosphorin; 6-Oxide, *in* D-00039
2,5-Diethoxy-1,3,2,5-dioxadiphosphorinane; 5-Oxide, *in* D-00231
2,5-Diethoxy-1,3,2,5-dioxadiphosphorinane; 5-Oxide, 2-sulfide, *in* D-00231
Diethyl 1,2-ethanediylbis(phenylphosphinate), *in* E-00015
Diethyl 1,2-ethenediylbis(phenylphosphinate), *in* E-00018
Diethyl methylenebis[methylphosphinate], *in* M-00140
Diethyl 1,2-phenylenebis(phenylphosphinate), *in* P-00139
2,3-Dihydro-4,5-dimethyl-2,7-diphenyl-1,2-oxaphosphepin 2-oxide, D-00462
2,5-Dihydro-3,4-dimethyl-1-phenoxy-1*H*-phosphole 1-oxide, *in* D-00493
2,3-Dihydro-1-ethoxy-4-methyl-1*H*-phosphole 1-oxide, *in* D-00496
10,11-Dihydro-5-methoxy-5*H*-dibenzo[*b,f*]phosphepin 5-oxide, *in* D-00492
2,5-Dihydro-1-methoxy-3,4-dimethyl-1*H*-phosphole 1-oxide, *in* D-00493
2,3-Dihydro-2-methoxy-1*H*-isophosphindole 2-oxide, *in* D-00494
2,3-Dihydro-1-methoxy-4-methyl-1*H*-phosphole 1-oxide, *in* D-00496
▷2,5-Dihydro-1-methoxy-3-methyl-1*H*-phosphole 1-oxide, *in* D-00499
2,3-Dihydro-1-methoxy-1*H*-phosphindole 1-oxide, *in* H-00120
1,2-Dihydro-1-methoxyphosphinoline 1-oxide, *in* D-00505
▷2,3-Dihydro-1-methoxy-1*H*-phosphole 1-oxide, *in* D-00506
▷2,5-Dihydro-1-methoxy-1*H*-phosphole 1-oxide, *in* D-00507
3,4-Dihydro-2-methyl-2*H*-1,4,2-benzoxazaphosphorine; 2-Oxide, *in* D-00522
2,5-Dihydro-1-(1-methylethoxy)-1*H*-phosphole 1-oxide, *in* D-00507
2,3-Dihydro-4-methyl-1-phenoxy-1*H*-phosphole 1-oxide, *in* D-00496
2,5-Dihydro-2-methyl-1-phenoxy-1*H*-phosphole 1-oxide, *in* D-00498
2,5-Dihydro-3-methyl-2-phenoxy-1*H*-phosphole 1-oxide, *in* D-00499
2,3-Dihydro-1-phenoxy-1*H*-phosphole 1-oxide, *in* D-00506
2,5-Dihydro-1-phenoxy-1*H*-phosphole 1-oxide, *in* D-00507
1,3-Dihydro-1-phenyl-2,1-benzoxaphosphole; 1-Oxide, *in* D-00553
3,4-Dihydro-3-phenyl-1*H*-2,3-benzoxaphosphorin; 3-Oxide, *in* D-00554
2,3-Dihydro-3,3,5-trimethyl-1,2-oxaphosphole 2-oxide; 2-Me, *in* D-00596
2,3-Dihydro-3,3,5-trimethyl-1,2-oxaphosphole 2-oxide; 2-Ph, *in* D-00596
1,3-Diisopropoxy-1,3-diphosphepane 1,3-dioxide, *in* D-00619
1,3-Diisopropoxy-1,3-diphospholane 1,3-dioxide, *in* D-00621
1,3-Diisopropoxy-1,3-diphosphorinane 1,3-dioxide, *in* D-00622
Diisopropyl 1,2-ethanediylbis(phenylphosphinate), *in* E-00015
Diisopropyl methylenebis[methylphosphinate], *in* M-00140
Diisopropyl methylenebisphosphinate, *in* M-00143
(4-Dimethylaminophenyl)methylphosphinic acid; Me ester; B,MeI, *in* D-00705

Dimethyl 1,2-ethenediylbis(phenylphosphinate), *in* E-00018
Dimethyl methylenebis(methylphosphinate), *in* M-00140
Dimethyl 1,4-phenylenebis[methylphosphinate], *in* P-00137
Dimethyl 1,2-phenylenebis(phenylphosphinate), *in* P-00139
(3,4-Dimethylphenyl)phosphinic acid; Et ester, *in* D-00806
1,4,2-Dioxaphosphorinane 2-oxide; 2-Et, *in* D-00973
Dioxybis[diphenylphosphine oxide], D-00977
1,2-Ethanediylbis[methylphosphinic acid]; Diisopropyl ester, *in* E-00013
5-Ethoxy-5*H*-dibenzophosphole 5-oxide, *in* H-00118
1-Ethoxy-2,5-dihydro-3,4-dimethyl-1*H*-phosphole 1-oxide, *in* D-00493
1-Ethoxy-2,3-dihydro-5-methyl-1*H*-phosphole 1-oxide, *in* D-00497
1-Ethoxy-2,5-dihydro-2-methyl-1*H*-phosphole 1-oxide, *in* D-00498
1-Ethoxy-2,5-dihydro-3-methyl-1*H*-phosphole 1-oxide, *in* D-00499
1-Ethoxy-2,3-dihydro-1*H*-phosphole 1-oxide, *in* D-00506
▷1-Ethoxy-2,5-dihydro-1*H*-phosphole 1-oxide, *in* D-00507
4-Ethoxy-1,4-dihydro-1,3,5-triphenyl-1,2,4-diazaphosphorine 4-oxide, E-00029
4-Ethoxy-2,6-diphenyl-4*H*-1,4-oxaphosphorin 4-oxide, *in* H-00140
1-Ethoxy-3,4-diphenyl-1*H*-phosphole 1-oxide, *in* H-00142
4-Ethoxy-2,6-diphenyl-4*H*-1,4-thiaphosphorin 4-oxide, *in* H-00143
1-Ethoxy-1,2,3,4-tetrahydrophosphinoline 1-oxide, *in* H-00195
4-Ethoxy-1,4,5,6-tetrahydro-1,3,5-triphenyl-1,2,4-diazaphosphorine 4-oxide, *in* E-00029
1-Ethoxy-2,3,4,5-tetraphenyl-1*H*-phosphole 1-oxide, *in* H-00200
Ethyl acetylethylphosphinate, *in* A-00017
Ethyl acetylphenylphosphinate, *in* A-00021
Ethyl (aminophenylmethyl)phenylphosphinate, *in* A-00102
Ethyl 1-aziridinyl-*P*-methylphosphonamidate, *in* A-00152
Ethyl benzoylphenylphosphinate, *in* B-00032
Ethyl benzylmethylphosphinate, *in* B-00053
Ethyl benzylphenylphosphinate, *in* B-00056
▷Ethyl bis(1-aziridinyl)phosphinate, *in* B-00091
Ethyl bis(2-bromophenyl)phosphinate, *in* B-00112
Ethyl bis(chloromethyl)phosphinate, *in* B-00127
Ethyl bis(3-chlorophenyl)phosphinate, *in* B-00138
Ethyl bis(4-dimethylaminophenyl)phosphinate, *in* B-00188
Ethyl bis(2-ethoxycarbonylphenyl)phosphinate, *in* B-00117
Ethyl bis(4-ethoxycarbonylphenyl)phosphinate, *in* B-00118
Ethyl bis(ethoxymethyl)phosphinate, *in* B-00272
Ethyl bis(hydroxymethyl)phosphinate, *in* B-00282
Ethyl bis(2-hydroxyphenyl)phosphinate, *in* B-00283
Ethyl bis(1-hydroxy-2,2,2-trichloroethyl)phosphinate, *in* B-00287
Ethyl bis(iodomethyl)phosphinate, *in* B-00288
Ethyl bis(phenoxymethyl)phosphinate, *in* B-00384
Ethyl bis(phenylethynyl)phosphinate, *in* B-00388
Ethyl bis(2-thienyl)phosphinate, *in* D-01250
Ethyl (4-bromophenyl)phosphinate, *in* B-00489
Ethyl (2-butenyl)methylphosphinate, *in* B-00523
Ethyl butylphosphinate, *in* B-00590
Ethyl *tert*-butylphosphinate, *in* B-00591
Ethyl (chloromethyl)phenylphosphinate, *in* C-00103
Ethyl (4-chlorophenyl)phosphinate, *in* C-00144
Ethyl cyclohexylphenylphosphinate, *in* C-00250
Ethyl cyclohexylphosphinate, *in* C-00253
Ethyl decylphenylphosphinate, *in* D-00010
Ethyl (1-diazoethyl)phenylphosphinate, *in* D-00032
Ethyl dibenzylphosphinate, *in* D-00055
Ethyl dibutylphosphinate, *in* D-00109
Ethyl di-*tert*-butylphosphinate, *in* D-00110
Ethyl dicyclohexylphosphinate, *in* D-00211
Ethyl diethenylphosphinate, *in* D-01258
Ethyl diethylphosphinate, *in* D-00322
Ethyl di-2-furanylphosphinate, *in* D-00421
Ethyl dihexylphosphinate, *in* D-00434
Ethyl diisopropylphosphinate, *in* D-00641
Ethyl (1,1-dimethyl-3-oxobutyl)ethylphosphinate, *in* D-00782
Ethyl (2,5-dimethylphenyl)phosphinate, *in* D-00805
Ethyl dimethylphosphinate, *in* D-00829
Ethyl di-4-morpholinylphosphinate, *in* D-00935
Ethyl di-1-naphthylphosphinate, *in* D-00941
Ethyl(diphenylmethyl)phenylphosphinate, *in* D-01023
Ethyl diphenylphosphinate, *in* D-01039
Ethyl di-1-piperidinylphosphinate, *in* D-01190
Ethyl di-2-propenylphosphinate, *in* D-01193
Ethyl dipropylphosphinate, *in* D-01205
Ethyl ethenylphenylphosphinate, *in* P-00337
Ethyl 3-(ethoxymethylphosphinyl)propanoate, *in* H-00161
Ethyl 3-(ethoxyphenylphosphinyl)propanoate, *in* H-00178
Ethyl ethylmethylphosphinate, *in* E-00085
Ethyl ethylphenylphosphinate, *in* E-00101
Ethyl ethylphosphinate, E-00068
Ethyl (α-hydroxybenzyl)phenylphosphinate, *in* H-00173
Ethyl (iodomethyl)phenylphosphinate, *in* I-00012
Ethyl (methoxymethyl)phenylphosphinate, *in* M-00052
Ethyl (4-methoxyphenyl)phosphinate, *in* M-00069
Ethyl methylphenylphosphinate, *in* M-00219
Ethyl methylphosphinate, *in* M-00278
Ethyl (2-methylpropyl)phosphinate, *in* M-00376
Ethyl methyl(trichloromethyl)phosphinate, *in* M-00419
Ethyl 4-morpholinylphosphinate, *in* M-00522
Ethyl 2-naphthylphenylphosphinate, *in* N-00018
Ethyl (2-nitrophenyl)methylphosphinate, *in* N-00042
Ethyl nonylphenylphosphinate, *in* N-00074
Ethyl octylphenylphosphinate, *in* O-00040
Ethyl pentamethylenephosphinate, *in* H-00188
Ethyl 1,1,2,3,3-pentamethyltrimethylenephosphinate, *in* H-00168
Ethyl (2-phenylethenyl)phosphinate, *in* P-00153
Ethyl phenyl(2-phenylethenyl)phosphinate, *in* P-00186
▷Ethyl phenylphosphinate, *in* P-00210
Ethylphenylphosphinic acid; (*S*)-*form*, Me ester, *in* E-00101
Ethylphenylphosphinic acid; (±)-*form*, Me ester, *in* E-00101
Ethyl phenyl(2-propenyl)phosphinate, *in* P-00300
Ethyl phenyl(trichloromethyl)phosphinate, *in* P-00327
Ethyl tetramethylenephosphinate, *in* H-00185
Ethyl thiodimethylenephosphinate, *in* H-00201
Heptyl diheptylphosphinate, *in* D-00425
Hexyl dihexylphosphinate, *in* D-00434
Hexyl phosphinate, H-00095
10-Hydroxy-2,8-dimethyl-10*H*-phenothiaphosphine 10-oxide; 10-Methoxy (Me ester), *in* H-00127
1-Hydroxy-3,4-dimethyl-1*H*-phosphole 1-oxide; 1-Phenoxy (Ph ester), *in* H-00130
1-Hydroxy-2,2,3,4,4-pentamethylphosphetane 1-oxide; *trans-form*, Me ester, *in* H-00168
1-Hydroxy-2,2,3,4,4-pentamethylphosphetane 1-oxide; *trans-form*, Et ester, *in* H-00168
1-Hydroxy-1*H*-phosphole 1-oxide; Et ester, *in* H-00186
4-Hydroxy-2,2,6,6-tetramethyl-1-oxa-4-phospha-2,6-disilacyclohexane 4-oxide; 4-Butoxy (butyl ester), *in* H-00198
1-Hydroxy-2,4,6-trimethyl-1,3,5-dioxaphosphorinane 5-oxide; 5-Methoxy (Me ester), *in* H-00203
5-Hydroxy-2,4,6-triphenyl-1,3,5-dioxaphosphorinane 5-oxide; 5-Methoxy (Me ester), *in* H-00204
Isobutyl butylmethylphosphinate, *in* B-00566
Isopropyl benzylphenylphosphinate, *in* B-00056
▷Isopropyl bis(1-aziridinyl)phosphinate, *in* B-00091
Isopropyl bis(chloromethyl)phosphinate, *in* B-00127
Isopropyl (2-butenyl)methylphosphinate, *in* B-00523
Isopropyl butylphosphinate, *in* B-00590
Isopropyl cyclohexylphosphinate, *in* C-00253
Isopropyl diethylphosphinate, *in* D-00322
Isopropyl diisopropylphosphinate, *in* D-00641
Isopropyl (1,1-dimethyl-3-oxobutyl)ethylphosphinate, *in* D-00782
Isopropyl (2,5-dimethylphenyl)phosphinate, *in* D-00805
Isopropyl diphenylphosphinate, *in* D-01039
Isopropyl ethylphenylphosphinate, *in* E-00101
Isopropyl ethylphosphinate, *in* E-00124
Isopropyl isopropylphenylphosphinate, *in* I-00043
Isopropyl *P*-methyl-4-morpholinylphosphonamidate, *in* M-00175
Isopropyl methylphenylphosphinate, *in* M-00219

Isopropyl (4-methylphenyl)phosphinate, *in* M-00222
Isopropyl methylphosphinate, I-00034
Isopropyl *P*-methyl-1-piperidinylphosphonamidate, *in* M-00361
Isopropyl methyl(trichloromethyl)phosphinate, *in* M-00419
Isopropyl 1,1,2,3,3-pentamethyltrimethylenephosphinate, *in* H-00168
Isopropyl phenylphosphinate, *in* P-00210
Isopropyl styrylphosphinate, *in* P-00153
Isopropyl (trifluoromethyl)phosphinate, *in* T-00475
(−)-Menthyl (2-aminophenyl)methylphosphinate, *in* A-00103
Menthyl *tert*-butylphenylphosphinate, *in* B-00579
Menthyl cyclohexylmethylphosphinate, *in* C-00247
Menthyl methylphenylphosphinate, *in* M-00219
Menthyl phenylphosphinate, *in* P-00210
5-Methoxy-5*H*-dibenzo[*b*,*f*]phosphepin 5-oxide, *in* H-00117
5-Methoxy-5*H*-dibenzophosphole 5-oxide, *in* H-00118
5-Methoxy-5,6-dihydrophosphanthridene 5-oxide, *in* D-00504
1-Methoxy-3,4-dimethyl-1*H*-phosphole 1-oxide, *in* H-00130
4-Methoxy-2,6-diphenyl-4*H*-1,4-thiaphosphorin 4-oxide, *in* H-00143
1-Methoxy-2,3,4,5-tetraphenyl-1*H*-phosphole 1-oxide, *in* H-00200
Methyl acetylmethylphosphinate, *in* A-00018
Methyl acetylphenylphosphinate, *in* A-00021
Methyl benzoylphenylphosphinate, *in* B-00032
Methyl benzylmethylphosphinate, *in* B-00053
Methyl benzylphenylphosphinate, *in* B-00056
▷Methyl bis(1-aziridinyl)phosphinate, *in* B-00091
Methyl bis(chloromethyl)phosphinate, *in* B-00127
Methyl bis(4-chlorophenyl)phosphinate, *in* B-00139
Methyl bis(4-dimethylaminophenyl)phosphinate, *in* B-00188
Methyl bis(hydroxymethyl)phosphinate, *in* B-00282
Methyl bis(1-hydroxy-2,2,2-trichloroethyl)phosphinate, *in* B-00287
Methyl bis(3-methoxyphenyl)phosphinate, *in* B-00300
Methyl bis(4-methoxyphenyl)phosphinate, *in* B-00301
Methyl bis(2-methylphenyl)phosphinate, *in* B-00318
Methyl bis(trifluoromethyl)phosphinate, *in* B-00411
Methyl 3-bromophenylphosphinate, *in* B-00488
Methyl *tert*-butylmethylphosphinate, *in* B-00567
Methyl *tert*-butylphenylphosphinate, *in* B-00579
Methyl butylphosphinate, *in* B-00590
Methyl *tert*-butylphosphinate, *in* B-00591
Methyl (chloromethyl)methylphosphinate, *in* C-00100
Methyl (chloromethyl)phenylphosphinate, *in* C-00103
Methyl (2-chloro-3-oxobutyl)methylphosphinate, *in* C-00121
Methyl (4-chlorophenyl)methylphosphinate, *in* C-00131
Methyl (3-chlorophenyl)phosphinate, *in* C-00143
Methyl cyclohexylmethylphosphinate, *in* C-00247
Methyl cyclohexylphenylphosphinate, *in* C-00250
Methyl cyclohexylphosphinate, *in* C-00253
Methyl (1-diazoethyl)phenylphosphinate, *in* D-00032
Methyl dibenzylphosphinate, *in* D-00055
Methyl dibutylphosphinate, *in* D-00109
Methyl di-*tert*-butylphosphinate, *in* D-00110
Methyl dicyclohexylphosphinate, *in* D-00211
Methyl diethenylphosphinate, *in* D-01258
Methyl diethylphosphinate, *in* D-00322
Methyl diisopropylphosphinate, *in* D-00641
Methyl (4-dimethylaminophenyl)methylphosphinate, *in* D-00705
Methyl (1,1-dimethyl-3-oxobutyl)ethylphosphinate, *in* D-00782
Methyl (2,5-dimethylphenyl)phosphinate, *in* D-00805
Methyl (3,4-dimethylphenyl)phosphinate, *in* D-00806
Methyl dimethylphosphinate, *in* D-00829
Methyl diphenylphosphinate, *in* D-01039
Methyl di-1-piperidinylphosphinate, *in* D-01190
Methyl di-2-propenylphosphinate, *in* D-01193
Methyl ethenylphenylphosphinate, *in* P-00337
Methyl ethylphenylphosphinate, *in* E-00101
Methyl ethylphosphinate, *in* E-00124
Methyl (hydroxymethyl)phenylphosphinate, *in* H-00160
Methyl isopropylphenylphosphinate, *in* I-00043
Methyl 3-(methoxymethylphosphinyl)propanoate, *in* H-00161
Methyl 3-(methoxyphenylphosphinyl)propanoate, *in* H-00178
Methyl methyl-1-naphthylphosphinate, *in* M-00178
Methyl (4-methylphenyl)methylphosphinate, *in* M-00204
Methyl methylphenylphosphinate, *in* M-00219
Methyl (4-methylphenyl)phosphinate, *in* M-00222
Methyl methylphosphinate, *in* M-00278
Methyl methyl(trifluoromethyl)phosphinate, *in* M-00420
Methyl methylvinylphosphinate, *in* M-00438
Methyl 1-naphthylphenylphosphinate, *in* N-00017
Methyl 2-naphthylphenylphosphinate, *in* N-00018
Methyl nonylphenylphosphinate, *in* N-00074
2-Methyl-1,2-oxaphospholane; 2-Oxide, *in* M-00182
5-Methyl-1,2-oxaphospholane 2-oxide; 2-Ph, *in* M-00183
Methyl 1,1,2,3,3-pentamethyltrimethylenephosphinate, *in* H-00168
Methyl phenazaphosphinate, *in* P-00061
Methyl (2-phenylethenyl)phosphinate, *in* P-00153
(4-Methylphenyl)methylphosphinic acid; $(S)_P$-*form*, (−)-Menthyl ester, *in* M-00204
Methyl phenyl(2-phenylethenyl)phosphinate, *in* P-00186
Methyl phenylphosphinate, *in* P-00210
Methylphenylphosphinic acid; (*S*)-*form*, Me ester, *in* M-00219
Methylphenylphosphinic acid; (*S*)-*form*, Et ester, *in* M-00219
Methylphenylphosphinic acid; (*S*)-*form*, Isopropyl ester, *in* M-00219
Methylphenylphosphinic acid; (*S*)-*form*, (−)-Menthyl ester, *in* M-00219
Methylphenylphosphinic acid; (±)-*form*, Me ester, *in* M-00219
Methyl phenyl(1-propenyl)phosphinate, *in* P-00299
Methyl phenyl(2-propenyl)phosphinate, *in* P-00300
Methyl phenyl(trichloromethyl)phosphinate, *in* P-00327
2-Methylpropyl bis(2-methylpropyl)phosphinate, *in* B-00346
2-Methylpropyl decylmethylphosphinate, *in* D-00009
2-Methylpropyl ethylmethylphosphinate, *in* E-00085
2-Methylpropyl hexylmethylphosphinate, *in* H-00094
2-Methylpropyl methyl(2-methylpropyl)phosphinate, *in* M-00172
2-Methylpropyl (2-methylpropyl)phenylphosphinate, *in* M-00373
2-Methylpropyl methylpropylphosphinate, *in* M-00375
Methyl tetramethylenephosphinate, *in* H-00185
Methyl thiodimethylenephosphinate, *in* H-00201
6-Methyl-2,3,4-triphenyl-2*H*-1,5,2-dioxaphosphorin 2-oxide; (2*RS*,6*RS*)-*form*, *in* M-00430
6-Methyl-2,3,4-triphenyl-2*H*-1,5,2-dioxaphosphorin 2-oxide; (2*RS*,6*SR*)-*form*, *in* M-00430
1-Naphthyl diphenylphosphinate, *in* D-01039
2-Naphthylphenylphosphinic acid; $(R)_P$-*form*, (−)-Menthyl ester, *in* N-00018
2-Naphthylphenylphosphinic acid; $(S)_P$-*form*, (−)-Menthyl ester, *in* N-00018
4-Nitrophenyl methylphenylphosphinate, *in* M-00219
(2-Nitrophenyl)methylphosphinic acid; (−)-Menthyl ester, *in* N-00042
Octyl (4-methylphenyl)phosphinate, *in* M-00222
1,2-Oxaphosphorinane; 2-Ph, 2-oxide, *in* O-00054
2,2,3,4,6-Pentaphenyl-2*H*-1,5,2-dioxaphosphorin 2-oxide, P-00036
Pentyl dipentylphosphinate, *in* D-00981
Pentyl (1-methylpropyl)phenylphosphinate, *in* M-00372
5-Phenoxy-5,6-dihydrophosphanthridene 5-oxide, *in* D-00504
1-Phenoxy-3,4-diphenyl-1*H*-phosphole 1-oxide, *in* H-00142
1-Phenoxy-2,3,4,5-tetraphenyl-1*H*-phosphole 1-oxide, *in* H-00200
Phenyl bis(1-aziridinyl)phosphinate, *in* B-00091
Phenyl bis(1*H*-benzimidazol-1-yl)phosphinate, *in* B-00097
Phenyl bis(4-bromophenyl)phosphinate, *in* B-00114
Phenyl bis(chloromethyl)phosphinate, *in* B-00127
Phenyl bis(iodomethyl)phosphinate, *in* B-00288
Phenyl bis(4-phenyl-1,3-butadienyl)phosphinate, *in* B-00385
Phenyl bis(phenylethynyl)phosphinate, *in* B-00388
Phenyl bis(1-pyrrolidinyl)phosphinate, *in* D-01233
Phenyl *tert*-butylmethylphosphinate, *in* B-00567
Phenyl *tert*-butylphenylphosphinate, *in* B-00579
Phenyl butylphosphinate, *in* B-00590
Phenyl *tert*-butylphosphinate, *in* B-00591
Phenyl di-*tert*-butylphosphinate, *in* D-00110
Phenyl dicyclohexylphosphinate, *in* D-00211
Phenyl diethenylphosphinate, *in* D-01258
Phenyl diethylphosphinate, *in* D-00322
Phenyl di-1*H*-imidazol-1-ylphosphinate, *in* D-00628
Phenyl dimethylphosphinate, *in* D-00829
▷Phenyl di-4-morpholinylphosphinate, *in* D-00935

▷Phenyl diphenylphosphinate, *in* D-01039
Phenyl di-1-piperidinylphosphinate, *in* D-01190
Phenyl ethenylphenylphosphinate, *in* P-00337
Phenyl ethylphosphinate, *in* E-00124
Phenyl methylphenylphosphinate, *in* M-00219
Phenyl methylphosphinate, *in* M-00278
Phenyl 1,1,2,3,3-pentamethyltrimethylenephosphinate, *in* H-00168
Phenyl phenyl(4-phenyl-1,3-butadienyl)phosphinate, *in* P-00184
Phenyl phenylphosphinate, *in* P-00210
Phenylpropylphosphinic acid; $(R)_P$-*form*, (−)-Menthyl ester, *in* P-00303
Phenylpropylphosphinic acid; $(S)_P$-*form*, (−)-Menthyl ester, *in* P-00303
Phenyl thiodimethylenephosphinate, *in* H-00201
1,3-Propanediylbis[phenylphosphinic acid]; Diisopropyl ester, *in* P-00447
2-Propenyl di-2-propenylphosphinate, *in* D-01193
2-Propenyl phenyl(2-propenyl)phosphinate, *in* P-00300
Propyl acetylethylphosphinate, *in* A-00017
Propyl ethenylphenylphosphinate, *in* P-00337
Propyl (4-methylphenyl)phenylphosphinate, *in* M-00204
Propyl phenylpropylphosphinate, *in* P-00303
Tetraethyl [(ethoxyphosphinylidene)bis(methylene)]-bisphosphonate, *in* B-00182
4,5,9,10-Tetrahydro-4,9-dihydroxyphosphorino[2,3,4,5-*lmn*]-phosphanthridine 4,9-dioxide; 4,9-Dimethoxy (di-Me ester), *in* T-00090
1,2,3,4-Tetrahydro-1-methoxyphosphinoline 1-oxide, *in* H-00195
2,3,3,4-Tetraphenyl-1,2-oxaphosphetane 2-oxide, T-00268
3,5,5-Trimethyl-2-phenyl-1,2-oxaphospholan-3-ol 2-oxide, *in* T-00519
3,5,5-Trimethyl-2-phenyl-1,2-oxaphospholan-3-ol 2-oxide, *in* T-00519
Trimethylsilyl bis(chloromethyl)phosphinate, *in* B-00127
Trimethylsilyl bis[*N*-(trimethylsilyl)aminomethyl]-phosphinate, *in* B-00087
Trimethylsilyl bis(trimethylsilyloxymethyl)phosphinate, *in* B-00282
Trimethylsilyl cyclohexylphosphinate, *in* C-00253
Trimethylsilyl di-*tert*-butylphosphinate, *in* D-00110
Trimethylsilyl diethylphosphinate, *in* D-00322
Trimethylsilyl dimethylphosphinate, *in* D-00829
Trimethylsilyl dipropylphosphinate, *in* D-01205
Trimethylsilyl phenylphosphinate, *in* P-00210

Phosphinic amides

N-Acetyl-*P,P*-diphenylphosphinic amide, *in* D-01040
▷*N*-(1-Adamantyl)-*P,P*-bis(aziridinyl)phosphinic amide, A-00032
5-Anilino-5,6-dihydrophosphanthridene 5-oxide, *in* D-00504
▷*N*-Benzoyl-*P,P*-bis(1-aziridinyl)phosphinic amide, *in* B-00092
N-Benzoyl-*P,P*-diphenylphosphinic amide, *in* D-01040
N-Benzyl-*P,P*-bis(chloromethyl)phosphinic amide, *in* B-00128
N-Benzyl-*P,P*-diphenylphosphinic amide, *in* D-01040
P-Benzyl-*N,P*-diphenylphosphinic amide, *in* B-00056
N-Benzyl-*P,P*-(1,1,2,3,3-pentamethyltrimethylene)phosphinic amide, *in* P-00021
P,P-Bis(1-aziridinyl)-*N,N*-diethylphosphinic amide, *in* B-00092
▷*P,P*-Bis(1-aziridinyl)-*N,N*-dimethylphosphinic amide, *in* B-00092
▷*P,P*-Bis(1-aziridinyl)-*N*-ethylphosphinic amide, *in* B-00092
▷*P,P*-Bis(1-aziridinyl)-*N*-methylphosphinic amide, *in* B-00092
P,P-Bis(1-aziridinyl)-*N*-2-naphthylphosphinic amide, *in* B-00092
P,P-Bis(1-aziridinyl)-*N*-phenylphosphinic amide, *in* B-00092
▷*P,P*-Bis(1-aziridinyl)phosphinic amide, B-00092
P,P-Bis(1*H*-benzimidazol-1-yl)-*N*-phenylphosphinic amide, *in* B-00097
P,P-Bis(chloromethyl)-*N,N*-dimethylphosphinic amide, *in* B-00128
P,P-Bis(chloromethyl)-*N*-methylphosphinic amide, *in* B-00128
P,P-Bis(chloromethyl)-*N*-phenylphosphinic amide, *in* B-00128
Bis(4-chlorophenyl)phosphinic acid; Amide, *in* B-00139
Bis(4-methoxyphenyl)phosphinic acid; Amide, *in* B-00301
P,P-Bis(2-methylphenyl)-*N*-phenylphosphinic amide, *in* B-00318
P,P-Bis(4-methylphenyl)-*N*-phenylphosphinic amide, *in* B-00320
Bis(2-methylphenyl)phosphinic acid; Amide, *in* B-00318
Bis(4-methylphenyl)phosphinic acid; Amide, *in* B-00320
P,P-Bis(2-methylpropyl)phosphinic amide, *in* B-00346
P,P-Bis(trichloromethyl)phosphinic amide, *in* B-00402
[Bis(trichloromethyl)phosphinyl]phosphorimidic trichloride, B-00404
P,P-Bis(2,4,6-trimethylphenyl)phosphinic amide, *in* B-00428
P-*tert*-Butyl-*N,N*-diethyl-*P*-methylphosphinic amide, *in* B-00567
N-Butyl-*P,P*-dihexylphosphinic amide, *in* D-00434
N-Butyl-*P,P*-dioctylphosphinic amide, *in* D-00958
P-*tert*-Butyl-*N,P*-diphenylphosphinic amide, *in* B-00580
P-*tert*-Butyl-*P*-phenylphosphinic amide, B-00580
N-(*tert*-Butylphenylphosphinyl)imidazole, *in* B-00579
N-Cyclohexyl-*P,P*-bis(trichloromethyl)phosphinic amide, *in* B-00402
P-Cyclohexyl-*N,N*-diethylcyclohexylphosphinic amide, *in* C-00253
N-Cyclohexyl-*P,P*-diphenylphosphinic amide, *in* D-01040
N,N-Dibenzyl-*P,P*-diphenylphosphinic amide, *in* D-01040
P,P-Dibenzyl-*N*-phenylphosphinic amide, *in* D-00055
Dibenzylphosphinic amide, *in* D-00055
P,P-Dibutyl-*N,N*-diethylphosphinic amide, *in* D-00111
P,P-Dibutyl-*N,N*-dimethylphosphinic amide, *in* D-00111
P,P-Dibutyl-*N*-ethylphosphinic amide, *in* D-00111
N,N-Dichloro-*P,P*-diisopropylphosphinic amide, *in* D-00642
P,P-Diethenyl-*N,N*-diethylphosphinic amide, *in* D-01258
1-Diethylamino-2,5-dihydro-2-methyl-1*H*-phosphole 1-oxide, *in* D-00498
N,N-Diethyl-*P,P*-bis(iodomethyl)phosphinic amide, *in* B-00288
N,N-Diethyl-*P,P*-dimethylphosphinic amide, *in* D-00830
N,N-Diethyl-*P,P*-diphenylphosphinic amide, *in* D-01040
N,N-Diethyl-*P,P*-di-2-propenylphosphinic amide, *in* D-01193
N,N-Diethyl-*P,P*-dipropylphosphinic amide, *in* D-01206
N,N-Diethyl-*P,P*-di-1-pyrrolidinylphosphinic amide, *in* D-01233
P,P-Diethyl-*N*-methylphosphinic amide, *in* D-00323
N,N-Diethyl-*P*-phenylphosphinic amide, *in* P-00210
P,P-Diethylphosphinic amide, D-00323
N-(Diethylphosphinyl)-*N′*-phenylurea, *in* D-00323
P,P-Diethynyl-*N,N*-dimethylphosphinic amide, *in* D-00409
2,3-Dihydro-4,5-dimethyl-2,7-diphenyl-1*H*-1,2-azaphosphepine; 2-Oxide, *in* D-00460
5,10-Dihydro-5-methoxyacridophosphine 5-oxide, *in* D-00487
2,3-Dihydro-2-methyl-1*H*-4,1,2-benzothiazaphosphorine; 2-Oxide, *in* D-00520
P,P-Diisopropyl-*N,N*-dimethylphosphinic amide, *in* D-00642
P,P-Diisopropylphosphinic amide, D-00642
N,N-Dimethyl-*P,P*-bis(trifluoromethyl)phosphinic amide, *in* B-00411
P,P-Dimethyl-*N*-(dimethylphosphinyl)phosphinic amide, *in* D-00830
N,N-Dimethyl-*P,P*-diphenylphosphinic amide, *in* D-01040
N,N-Dimethyl-*P,P*-di-2-propenylphosphinic amide, *in* D-01193
N,N-Dimethyl-*P,P*-dipropynylphosphinic amide, *in* D-01231
N,N-Dimethyl-*P,P*-di-1-pyrrolidinylphosphinic amide, *in* D-01233
P-(1,1-Dimethylethyl)-*P*-methylphosphinic amide, *in* B-00567
N-(Dimethylphosphinyl)-*N′*-phenylurea, *in* D-00830
P,P-Di-1-naphthyl-*N*-phenylphosphinic amide, *in* D-00941
P,P-Diphenyl-*N*-(diphenylphosphinyl)phosphinimidic acid, D-01007
P,P-Diphenyl-*N*-(diphenylphosphinyl)phosphinimidic acid; Chloride, *in* D-01007
N,N-Diphenyl-*P,P*-1,2-ethanediylbis[phenylphosphinic amide], *in* E-00015
2,6-Diphenyl-4*H*-1,4-oxaphosphorin-4-amine 4-oxide, *in* H-00140
N,P-Diphenyl-*P*-(4-phenyl-1,3-butadienyl)phosphinic amide, *in* P-00184
P,P-Diphenyl-*N*-phenylsulfonylphosphinic amide, *in* D-01040
P,P-Diphenylphosphinic amide, D-01040
P,P-Diphenylphosphinic amide; *N*-Diphenylphosphino, *in* D-01040

P,P-Diphenylphosphinic amide; *N*-Diphenylphosphinothioyl, *in* D-01040
N-(Diphenylphosphinyl)-*P,P*-diphenylphosphinic amide, D-01096
N-Diphenylphosphinyl-*P,P,P*-triphenylphosphine imide, *in* D-01088
P,P-Diphenyl-*N*-propylphosphinic amide, *in* D-01040
P,P-Diphenyl-*N*-(triphenylarsoranylidene)phosphinic amide, *in* D-01160
P,P-Diphenyl-*N*-(triphenylphosphoranylidene)phosphinic amide, *in* D-01165
P,P-Dipropylphosphinic amide, D-01206
1,2-Ethanediylbis[phenylphosphinic acid]; Diamide, *in* E-00015
Ethyl *N*-(diethylphosphinyl)carbamate, *in* D-00323
Ethyl *P,P*-diphenyl-*N*-(diphenylphosphinyl)phosphinimidate, *in* D-01007
P-Ethyl-*N,P*-diphenylphosphinic amide, *in* E-00102
P-Ethyl-*P*-phenylphosphinic amide, E-00102
N,N,2,2,3,4,4-Heptamethylphosphetanamine 1-oxide, *in* P-00021
N,N,N′,N′,N″,N″-Hexamethyl-*N‴*-diphenylphosphinophosphorimidic triamide; Oxide, *in* H-00069
N,2,2,3,4,4-Hexamethylphosphetanamine 1-oxide, *in* P-00021
N-Hydroxy-*P,P*-diphenylphosphinic amide, H-00141
N-Isopropyl-*P,P*-diphenylphosphinic amide, *in* D-01040
P-Isopropyl-*N,P*-diphenylphosphinic amide, *in* I-00044
P-Isopropyl-*P*-phenylphosphinic amide, I-00044
N-Methyl-*P,P*-bis(trichloromethyl)phosphinic amide, *in* B-00402
Methyl *N*-(dimethylphosphinyl)carbamate, *in* D-00830
N-Methyl-*P,P*-diphenylphosphinic amide, *in* D-01040
P-Methyl-*P*-phenylphosphinic amide; (*S*)-*form*, *in* M-00223
P-Methyl-*P*-phenylphosphinic amide; (*S*)-*form*, *N*-Ph, *in* M-00223
P-Methyl-*P*-phenylphosphinic amide; (±)-*form*, *in* M-00223
P-Methyl-*P*-phenylphosphinic amide; (±)-*form*, *N*-Ph, *in* M-00223
N-1-Naphthyl-*P,P*-diphenylphosphinic amide, *in* D-01040
N-2-Naphthyl-*P,P*-diphenylphosphinic amide, *in* D-01040
Octahydro-3-methoxy-1*H*-cyclopropa[*c*]phosphindole 3-oxide, *in* O-00004
2,2,3,4,4-Pentamethylphosphetanamine 1-oxide; *trans-form*, *N*-Benzyl, *in* P-00021
1-Phenyl-1,2-azaphospholidin-5-one; 2-Et, 2-oxide, *in* P-00092
1-Phenyl-1,2-azaphospholidin-5-one; 2-Ph, 2-oxide, *in* P-00092
N-Phenyl-*P,P*-bis(4-phenyl-1,3-butadienyl)phosphinic amide, *in* B-00385
N-Phenyl-*P,P*-bis(phenylethynyl)phosphinic amide, *in* B-00388
N-Phenyl-*P,P*-bis(2,4,6-trimethylphenyl)phosphinic amide, *in* B-00428
N-Phenyl-*P,P*-di-1-piperidinylphosphinic amide, *in* D-01190
N-Phenyl-2,2,3,4,4-pentamethylphosphetanamine 1-oxide, *in* P-00021
N-[Phenyl(phenylmethyl)phosphinyl]morpholine, *in* B-00056
▷Phosphemide, P-00353
N-tert-Butyl-*P,P*-dimethylphosphinic amide, *in* D-00830
N-tert-Butyl-*P,P*-diphenylphosphinic amide, *in* D-01040
Tetraethylphosphinic amide, *in* D-00323
Tetramethylphosphinic amide, *in* D-00830
Tetraphenylphosphinic amide, *in* D-01040
N,N,P-Triethylphosphinic amide, *in* E-00124
N,P,P-Triethylphosphinic amide, *in* D-00323
1,2,3-Trimethyl-1,3-diazaphospholidin-4-one 2-oxide, *in* D-00721
N,P,P-Triphenylphosphinic amide, T-00630

Phosphinic isocyanates and isothiocyanates

tert-Butylphenylphosphinic acid; (*R*)-*form*, Isocyanate, *in* B-00579
tert-Butylphenylphosphinic acid; (*S*)-*form*, Isocyanate, *in* B-00579
tert-Butylphenylphosphinic acid; (*S*)-*form*, Isothiocyanate, *in* B-00579
tert-Butylphenylphosphinic acid; (±)-*form*, Isothiocyanate, *in* B-00579
Diethylphosphinic acid; Isocyanate, *in* D-00322
Diethylphosphinic acid; Isothiocyanate, *in* D-00322
Diisopropylphosphinic acid; Isocyanate, *in* D-00641
Diisopropylphosphinic acid; Isothiocyanate, *in* D-00641
Dimethylphosphinic acid; Isocyanate, *in* D-00829
Dimethylphosphinic acid; Isothiocyanate, *in* D-00829
Diphenylphosphinic acid; Isocyanate, *in* D-01039
Diphenylphosphinic acid; Isothiocyanate, *in* D-01039

Phosphinic azides

Bis(3-chlorophenyl)phosphinic acid; Azide, *in* B-00138
Bis(4-chlorophenyl)phosphinic acid; Azide, *in* B-00139
Bis(4-methoxyphenyl)phosphinic acid; Azide, *in* B-00301
Bis(3-methylphenyl)phosphinic acid; Azide, *in* B-00319
Bis(4-methylphenyl)phosphinic acid; Azide, *in* B-00320
Bis(2-oxo-3-oxazolidinyl)phosphinic acid; Azide, *in* B-00378
tert-Butylmethylphosphinic acid; Azide, *in* B-00567
tert-Butylphenylphosphinic acid; (±)-*form*, Azide, *in* B-00579
Di-*tert*-butylphosphinic acid; Azide, *in* D-00110
Diethylphosphinic acid; Azide, *in* D-00322
Diisopropylphosphinic acid; Azide, *in* D-00641
Dimethylphosphinic acid; Azide, *in* D-00829
Diphenylphosphinic acid; Amide, *in* D-01039
▷Diphenylphosphinic acid; Azide, *in* D-01039
Ethylphenylphosphinic acid; (±)-*form*, Azide, *in* E-00101
Isopropylphenylphosphinic acid; (±)-*form*, Azide, *in* I-00043
Methylphenylphosphinic acid; (±)-*form*, Azide, *in* M-00219
1,1,2,3,3-Pentamethyltrimethylenephosphinic azide, *in* H-00168

Phosphinic hydrazides

Bis(4-bromophenyl)phosphinic acid; Hydrazide, *in* B-00114
Bis(3-methylphenyl)phosphinic acid; Hydrazide, *in* B-00319
Bis(4-methylphenyl)phosphinic acid; Hydrazide, *in* B-00320
Dimethylphosphinic acid; Hydrazide, *in* D-00829
2,3,4,5-Tetrahydro-6-methyl-2,4,4-triphenyl-1,2,3-diazaphosphorine; 3-Me, 3-oxide, *in* T-00118

Thiophosphinic acids

Bis(chloromethyl)phosphinothioic acid, B-00129
Bis(4-methylphenyl)phosphinothioic acid, B-00322
Bis(1-methylpropyl)phosphinodithioic acid, B-00347
Bis(2-methylpropyl)phosphinodithioic acid, B-00348
Bis(1-methylpropyl)phosphinothioic acid, B-00350
Bis(2-methylpropyl)phosphinothioic acid, B-00351
Bis(pentafluorophenyl)phosphinothioic acid, B-00383
Bis(trifluoromethyl)phosphinodithioic acid, B-00412
Bis(trifluoromethyl)phosphinothioic acid, B-00415
1,4-Butanediylbis[phenylphosphinodithioic acid], B-00516
tert-Butylmethylphosphinodithioic acid, B-00568
tert-Butylmethylphosphinothioic acid; (±)-*form*, *in* B-00570
tert-Butylphenylphosphinodithioic acid, B-00581
tert-Butylphenylphosphinothioic acid; (*R*)-*form*, *in* B-00583
tert-Butylphenylphosphinothioic acid; (*S*)-*form*, *in* B-00583
tert-Butylphenylphosphinothioselenoic acid, B-00584
Butylphosphinothioic acid, B-00593
Cyclohexylmethylphosphinothioic acid, C-00248
Dibenzylphosphinodithioic acid, D-00056
Dibenzylphosphinothioic acid, D-00057
Dibutylphosphinodithioic acid, D-00112
Di-*tert*-butylphosphinodithioic acid, D-00113
Dibutylphosphinothioic acid, D-00117
Di-*tert*-butylphosphinothioic acid, D-00118
Dicyclohexylphosphinodithioic acid, D-00212
Dicyclohexylphosphinothioic acid, D-00213
Diethylphosphinodithioic acid, D-00325
Diethylphosphinoselenothioic acid, D-00329
Diethylphosphinothioic acid, D-00330
Di-2-furanylphosphinothioic acid, D-00422
Diisopropylphosphinodithioic acid, D-00643
Diisopropylphosphinothioic acid, D-00646
Dimethylphosphinodithioic acid, D-00834
Dimethylphosphinothioic acid, D-00839
Diphenylphosphinodithioic acid, D-01065
Diphenylphosphinoselenothioic acid, D-01080

Di-1-piperidinylphosphinodithioic acid, D-01191
Dipropylphosphinodithioic acid, D-01207
Dipropylphosphinothioic acid, D-01208
1,2-Ethanediylbis[phenylphosphinodithioic acid], E-00016
Ethylmethylphosphinodithioic acid, E-00086
Ethylphenylphosphinodithioic acid, E-00103
Ethylphenylphosphinothioic acid; (*R*)-*form*, *in* E-00105
Ethylphenylphosphinothioic acid; (*S*)-*form*, *in* E-00105
Ethylphenylphosphinothioic acid; (±)-*form*, *in* E-00105
Ethylphenylphosphinothioselenoic acid, E-00106
Ethylphosphinothioic acid, E-00125
1-Hydroxy-2,2,3,4,4-pentamethylphosphetane 1-sulfide, H-00169
Isopropylphenylphosphinodithioic acid, I-00045
Isopropylphenylphosphinothioic acid, I-00046
Isopropylphenylphosphinothioic acid; (+)-*form*, *in* I-00046
Isopropylphenylphosphinothioic acid; (±)-*form*, *in* I-00046
Methylphenylphosphinodithioic acid, M-00224
Methylphenylphosphinothioic acid; (*R*)-*form*, *in* M-00225
Methylphenylphosphinothioic acid; (*S*)-*form*, *in* M-00225
Methylphenylphosphinothioic acid; (±)-*form*, *in* M-00225
Methylphosphinothioic acid, M-00279
1,4-Phenylenebis[methylphosphinodithioic acid], P-00138
Phenylphosphinothioic acid, P-00213
2,4,6-Triisopropyl-5-mercapto-1,3,5-dioxaphosphorinane 5-sulfide, T-00498

Thiophosphinic anhydrides

O-Acetyl diethylphosphinothioate, *in* D-00330
S-Acetyl diphenylphosphinodithioate, *in* D-01065
S-Benzoyl diphenylphosphinodithioate, *in* D-01065
Bis(trifluoromethyl)phosphinodithioic acid; Anhydrosulfide, *in* B-00412
Bis(trifluoromethyl)phosphinothioic acid; Anhydride, *in* B-00415
tert-Butylmethylphosphinodithioic acid anhydrosulfide; (*RS,RS*)-*form*, *in* B-00569
tert-Butylmethylphosphinothioic anhydride; (*RS,SR*)-*form*, *in* B-00571
Dibenzylphosphinodithioic acid; Anhydrosulfide, *in* D-00056
Dibutylphosphinodithioic acid; Anhydrosulfide, *in* D-00112
Di-*tert*-butylphosphinodithioic anhydrosulfide, *in* D-00113
Dibutylphosphinothioic acid; Anhydride, *in* D-00117
Di-*tert*-butylphosphinothioic acid; Anhydride, *in* D-00118
Diethylphosphinodithioic acid; Anhydrosulfide, *in* D-00325
Diethylphosphinothioic acid; Anhydride, *in* D-00330
Diisopropylphosphinodithioic acid; Anhydrosulfide, *in* D-00643
Dimethylphosphinodithioic anhydrosulfide, *in* D-00834
Dimethylphosphinothioic acid; Anhydride, *in* D-00839
Diphenylphosphinodithioic acid; Anhydrosulfide, *in* D-01065
Diphenylphosphinothioic anhydride, *in* D-01082
Dipropylphosphinodithioic acid; Anhydrosulfide, *in* D-01207
Dipropylphosphinothioic acid; Anhydride, *in* D-01208

Thiophosphinic halides

Bis(chloromethyl)phosphinothioic acid; Chloride, *in* B-00129
Bis(4-methylphenyl)phosphinothioic acid; Fluoride, *in* B-00322
Bis(4-methylphenyl)phosphinothioic acid; Chloride, *in* B-00322
Bis(2-methylpropyl)phosphinothioic acid; Chloride, *in* B-00351
Bis(2-methylpropyl)phosphinothioic acid; Bromide, *in* B-00351
Bis(pentafluorophenyl)phosphinothioic acid; Fluoride, *in* B-00383
Bis(pentafluorophenyl)phosphinothioic acid; Chloride, *in* B-00383
Bis(pentafluorophenyl)phosphinothioic acid; Bromide, *in* B-00383
Bis(trifluoromethyl)phosphinothioic acid; Fluoride, *in* B-00415
Bis(trifluoromethyl)phosphinothioic acid; Chloride, *in* B-00415
Bis(trifluoromethyl)phosphinothioic acid; Bromide, *in* B-00415
Bis(trifluoromethyl)phosphinothioic acid; Iodide, *in* B-00415
tert-Butylmethylphosphinothioic acid; (±)-*form*, Chloride, *in* B-00570
tert-Butylmethylphosphinothioic acid; (±)-*form*, Bromide, *in* B-00570
tert-Butylphenylphosphinothioic acid; (*R*)-*form*, Chloride, *in* B-00583
tert-Butylphenylphosphinothioic acid; (*S*)-*form*, Chloride, *in* B-00583
5-Chloro-5*H*-dibenzophosphole; 5-Sulfide, *in* C-00034
1-Chloro-2,5-dihydro-3,4-dimethyl-1*H*-phosphole; 1-Sulfide, *in* C-00041
1-Chloro-2,5-dihydro-2-methyl-1*H*-phosphole; 1-Sulfide, *in* C-00046
1-Chloro-2,5-dihydro-3-methyl-1*H*-phosphole; 1-Sulfide, *in* C-00047
10-Chloro-5,10-dihydrophenophosphazine; 10-Sulfide, *in* C-00049
1-Chloro-2,3-dihydro-1*H*-phosphole; 1-Sulfide, *in* C-00051
1-Chloro-2,5-dihydro-1*H*-phosphole; 1-Sulfide, *in* C-00052
1-Chloro-2,2,3,4,4-pentamethylphosphetane; *trans-form*, Sulfide, *in* C-00124
1-Chlorophosphorinane; Sulfide, *in* C-00168
7-Chloro-6-thia-7-phosphabicyclo[3.2.1]octane; 7-Sulfide, *in* C-00187
3-Chloro-1,3-thiaphosphetane; 3-Sulfide, *in* C-00188
Cyclohexylmethylphosphinothioic acid; Chloride, *in* C-00248
Dibenzylphosphinothioic acid; Chloride, *in* D-00057
Dibutylphosphinothioic acid; Fluoride, *in* D-00117
Dibutylphosphinothioic acid; Chloride, *in* D-00117
Dibutylphosphinothioic acid; Bromide, *in* D-00117
Dibutylphosphinothioic acid; Iodide, *in* D-00117
Di-*tert*-butylphosphinothioic acid; Fluoride, *in* D-00118
Di-*tert*-butylphosphinothioic acid; Chloride, *in* D-00118
Di-*tert*-butylphosphinothioic acid; Bromide, *in* D-00118
Dicyclohexylphosphinothioic acid; Fluoride, *in* D-00213
Dicyclohexylphosphinothioic acid; Chloride, *in* D-00213
Diethylphosphinothioic acid; Fluoride, *in* D-00330
Diethylphosphinothioic acid; Chloride, *in* D-00330
Diethylphosphinothioic acid; Bromide, *in* D-00330
Diethylphosphinothioic acid; Iodide, *in* D-00330
Diisopropylphosphinothioic acid; Chloride, *in* D-00646
Dimethylphosphinothioic acid; Fluoride, *in* D-00839
Dimethylphosphinothioic acid; Iodide, *in* D-00839
Dimethylphosphinothioic bromide, D-00841
Dimethylphosphinothioic chloride, D-00842
Diphenylphosphinothioic acid; Fluoride, *in* D-01082
Diphenylphosphinothioic acid; Bromide, *in* D-01082
Diphenylphosphinothioic acid; Iodide, *in* D-01082
Diphenylphosphinothioic chloride, D-01084
Dipropylphosphinothioic acid; Fluoride, *in* D-01208
Dipropylphosphinothioic acid; Chloride, *in* D-01208
Dipropylphosphinothioic acid; Bromide, *in* D-01208
Dipropylphosphinothioic acid; Iodide, *in* D-01208
Divinylphosphinothioic acid; Chloride, *in* D-01259
Ethylphenylphosphinothioic acid; (±)-*form*, Fluoride, *in* E-00105
Ethylphenylphosphinothioic acid; (±)-*form*, Chloride, *in* E-00105
1-Fluorophosphorinane; Sulfide, *in* F-00047
3-Fluoro-1,3-thiaphosphetane; 3-Sulfide, *in* F-00050
Isopropylphenylphosphinothioic acid; (±)-*form*, Chloride, *in* I-00046
Methylphenylphosphinothioic acid; (±)-*form*, Fluoride, *in* M-00225
Methylphenylphosphinothioic acid; (±)-*form*, Chloride, *in* M-00225
Methylphenylphosphinothioic acid; (±)-*form*, Bromide, *in* M-00225
Tetramethylenephosphinothioic chloride, *in* C-00167

Thiophosphinic pseudohalides

Diphenylphosphinothioic acid; Cyanide, *in* D-01082
Diphenylphosphinothioic acid; Isocyanate, *in* D-01082
Diphenylphosphinothioic acid; Isothiocyanate, *in* D-01082

Thiophosphinic esters

O-Acetyl diethylphosphinothioate, *in* D-00330
S-Acetyl diphenylphosphinodithioate, *in* D-01065
S-Benzoyl diphenylphosphinodithioate, *in* D-01065
Benzyl dibenzylphosphinodithioate, *in* D-00056
O-Benzyl di-*tert*-butylphosphinothioate, *in* D-00118
Benzyl dicyclohexylphosphinodithioate, *in* D-00212
Benzyl diphenylphosphinodithioate, *in* D-01065
O-Benzyl diphenylphosphinothioate, *in* D-01082
S-Benzyl diphenylphosphinothioate, *in* D-01082
1,4-Butanediylbis[phenylphosphinodithioic acid]; (*RS,SR*)-*form*, Di-Et ester, *in* B-00516
O-Butyl bis(2-methylpropyl)phosphinothioate, *in* B-00351
Butyl dicyclohexylphosphinodithioate, *in* D-00212
S-Butyl diethylphosphinothioate, *in* D-00330
Butyl diisobutylphosphinodithioate, *in* B-00348
Butyl dimethylphosphinodithioate, *in* D-00834
tert-Butyl diphenylphosphinodithioate, *in* D-01065
S-Butyl dipropylphosphinothioate, *in* D-01208
S-Butyl divinylphosphinothioate, *in* D-01259
Butyl methylphenylphosphinodithioate, *in* M-00224
O-Butyl methylphosphinothioate, *in* M-00279
tert-Butylphenylphosphinothioic acid; (*S*)-*form*, *S*-Me ester, *in* B-00583
O-Butyl phosphinothioate, B-00592
Cyclohexyl dicyclohexylphosphinodithioate, *in* D-00212
Diethyl 1,4-butanediylbis[phenylphosphinodithioate], *in* B-00516
5,10-Dihydro-10-ethoxyphenophosphazine 10-sulfide, *in* D-00502
5,10-Dihydro-10-isopropoxyphenophosphazine 10-sulfide, *in* D-00502
2,5-Dihydro-1-mercaptophosphole 1-sulfide; 2-Ethylthio (Et ester), *in* D-00512
5,10-Dihydro-10-methoxyphenophosphazine 10-sulfide, *in* D-00502
3,4-Dihydro-2-methyl-2*H*-1,4,2-benzothiazaphosphorine; 2-Sulfide, *in* D-00521
3,4-Dihydro-2-methyl-2*H*-1,4,2-benzothiazaphosphorine; 2-Selenide, *in* D-00521
3,4-Dihydro-2-methyl-2*H*-1,4,2-benzoxazaphosphorine; 2-Sulfide, *in* D-00522
5,10-Dihydro-10-methylthiophenophosphazine 10-oxide, *in* D-00502
1,3-Dihydro-1-phenyl-2,1-benzoxaphosphole; 1-Sulfide, *in* D-00553
3,4-Dihydro-3-phenyl-1*H*-2,3-benzoxaphosphorin; 3-Sulfide, *in* D-00554
3,6-Dihydro-2*H*-1,2-thiaphosphorin 2-sulfide; 2-Et, *in* D-00586
3,6-Dihydro-2*H*-1,2-thiaphosphorin 2-sulfide; 2-Ph, *in* D-00586
P,P-Diisopropyl-*N*-methylphosphinothioic amide, *in* D-00646
Dimethyl 1,4-butanediylbis[phenylphosphinodithioate], *in* B-00516
4,5-Dimethyl-2-oxa-3-phenyl-1-phosphabicyclo[2.2.1]hept-5-ene; 1-Sulfide, *in* D-00781
Dimethyl 1,4-phenylenebis[methylphosphinodithioate], *in* P-00138
Dimethylphosphinothioic acid; *O*-Trimethylsilyl ester, *in* D-00839
1-Ethoxy-2,2,3,4,4-pentamethylphosphetane 1-sulfide, *in* H-00169
O-Ethyl bis(1-aziridinyl)phosphinothioate, *in* B-00093
S-Ethyl bis(1-aziridinyl)phosphinothioate, *in* B-00093
O-Ethyl bis(chloromethyl)phosphinothioate, *in* B-00129
S-Ethyl bis(chloromethyl)phosphinothioate, *in* B-00129
O-Ethyl bis(pentafluorophenyl)phosphinothioate, *in* B-00383
O-Ethyl bis(trifluoromethyl)phosphinothioate, *in* B-00415
S-Ethyl bis(trifluoromethyl)phosphinothioate, *in* B-00415
Ethyl dibutylphosphinodithioate, *in* D-00112
O-Ethyl dibutylphosphinothioate, *in* D-00117
Ethyl diethylphosphinodithioate, *in* D-00325
S-Ethyl diethylphosphinoselenothioate, *in* D-00329
Se-Ethyl diethylphosphinoselenothioate, *in* D-00329
O-Ethyl diethylphosphinothioate, *in* D-00330
S-Ethyl diethylphosphinothioate, *in* D-00330
O-Ethyl di-2-furanylphosphinothioate, *in* D-00422
Ethyl diisopropylphosphinodithioate, *in* D-00643
Ethyl dimethylphosphinodithioate, *in* D-00834
Ethyl diphenylphosphinodithioate, *in* D-01065
O-Ethyl diphenylphosphinothioate, *in* D-01082
S-Ethyl diphenylphosphinothioate, *in* D-01082
O-Ethyl divinylphosphinothioate, *in* D-01259
Ethyl ethylmethylphosphinodithioate, *in* E-00086
Ethyl ethylphenylphosphinodithioate, *in* E-00103
S-Ethyl ethylphenylphosphinothioate, *in* E-00105
S-Ethyl ethylphenylphosphinothioselenoate, *in* E-00106
O-Ethyl ethylphosphinothioate, *in* E-00125
Ethyl ethylpropylphosphinodithioate, *in* E-00185
O-Ethyl methylphenylphosphinothioate, *in* M-00225
S-Ethyl methylphenylphosphinothioate, *in* M-00225
O-Ethyl methylphosphinothioate, *in* M-00279
O-Ethyl phenylphosphinothioate, *in* P-00213
Ethylphenylphosphinothioic acid; (±)-*form*, *S*-Me ester, *in* E-00105
10-Ethylthio-5,10-dihydrophenophosphazine 10-oxide, *in* D-00502
1-Ethylthio-2,2,3,4,4-pentamethylphosphetane 1-oxide, *in* H-00169
1-Hydroxy-2,2,3,4,4-pentamethylphosphetane 1-sulfide; *trans*-*form*, *S*-Me ester, *in* H-00169
O-Isopropyl bis(trifluoromethyl)phosphinothioate, *in* B-00415
S-Isopropyl bis(trifluoromethyl)phosphinothioate, *in* B-00415
O-Isopropyl dibutylphosphinothioate, *in* D-00117
Isopropyl diethylphosphinodithioate, *in* D-00325
O-Isopropyl diethylphosphinothioate, *in* D-00330
S-Isopropyl diethylphosphinothioate, *in* D-00330
Isopropyl diphenylphosphinodithioate, *in* D-01065
Isopropyl ethylmethylphosphinodithioate, *in* E-00103
O-Isopropyl ethylphosphinothioate, *in* E-00125
S-Isopropyl methylphenylhosphinothioate, *in* M-00225
O-Isopropyl methylphenylphosphinothioate, *in* M-00225
O-Isopropyl methylphosphinothioate, *in* M-00279
Isopropylphenylphosphinothioic acid; (±)-*form*, *S*-Me ester, *in* I-00046
O-Methyl bis(1-aziridinyl)phosphinothioate, *in* B-00093
O-Methyl bis(pentafluorophenyl)phosphinothioate, *in* B-00383
Methyl bis(trifluoromethyl)phosphinodithioate, *in* B-00412
O-Methyl bis(trifluoromethyl)phosphinothioate, *in* B-00415
S-Methyl bis(trifluoromethyl)phosphinothioate, *in* B-00415
O-Methyl *tert*-butylmethylphosphinothioate, *in* B-00570
O-Methyl *tert*-butylphenylphosphinothioate, *in* B-00583
S-Methyl *tert*-butylphenylphosphinothioate, *in* B-00583
Methyl dibutylphosphinodithioate, *in* D-00112
O-Methyl di-*tert*-butylphosphinothioate, *in* D-00118
S-Methyl dibutylphosphinothioate, *in* D-00117
S-Methyl di-*tert*-butylphosphinothioate, *in* D-00118
Methyl dicyclohexylphosphinodithioate, *in* D-00212
Methyl diethylphosphinodithioate, *in* D-00325
O-Methyl diethylphosphinothioate, *in* D-00330
S-Methyl diethylphosphinothioate, *in* D-00330
Methyl diisopropylphosphinodithioate, *in* D-00643
O-Methyl diisopropylphosphinothioate, *in* D-00646
S-Methyl diisopropylphosphinothioate, *in* D-00646
Methyl dimethylphosphinodithioate, *in* D-00834
O-Methyl dimethylphosphinothioate, *in* D-00839
S-Methyl dimethylphosphinothioate, *in* D-00839
Methyl diphenylphosphinodithioate, *in* D-01065
O-Methyl diphenylphosphinothioate, *in* D-01082
S-Methyl diphenylphosphinothioate, *in* D-01082
O-Methyl dipropylphosphinothioate, *in* D-01208
S-Methyl dipropylphosphinothioate, *in* D-01208
O-Methyl ethylphenylphosphinothioate, *in* E-00105
S-Methyl ethylphenylphosphinothioate, *in* E-00105
S-Methyl isopropylphenylphosphinothioate, *in* I-00046
O-Methyl methylphenylphosphinothioate, *in* M-00225
S-Methyl methylphenylphosphinothioate, *in* M-00225
2-Methyl-1,2-oxaphospholane; 2-Sulfide, *in* M-00182
Methylphenylphosphinothioic acid; (*S*)-*form*, *O*-Me ester, *in* M-00225
Methylphenylphosphinothioic acid; (*S*)-*form*, *S*-Me ester, *in* M-00225
Methylphenylphosphinothioic acid; (±)-*form*, *O*-Me ester, *in* M-00225

Methylphenylphosphinothioic acid; (±)-*form*, *S*-Me ester, *in* M-00225
5-Methyl-1,2-thiaphospholane 2-sulfide; 2-Et, *in* M-00407
5-Methyl-1,2-thiaphospholane 2-sulfide; 2-Ph, *in* M-00407
1-Methylthio-2,2,3,4,4-pentamethylphosphetane 1-oxide, *in* H-00169
O-Phenyl bis(chloromethyl)phosphinothioate, *in* B-00129
O-Phenyl bis(pentafluorophenyl)phosphinothioate, *in* B-00383
O-Phenyl di-*tert*-butylphosphinothioate, *in* D-00118
S-Phenyl dicyclohexylphosphinothioate, *in* D-00213
Phenyl diethylphosphinodithioate, *in* D-00325
O-Phenyl diethylphosphinothioate, *in* D-00330
O-Phenyl diisopropylphosphinothioate, *in* D-00646
Phenyl dimethylphosphinodithioate, *in* D-00834
O-Phenyl dimethylphosphinothioate, *in* D-00839
Phenyl diphenylphosphinodithioate, *in* D-01065
O-Phenyl diphenylphosphinothioate, *in* D-01082
S-Phenyl diphenylphosphinothioate, *in* D-01082
O-Phenyl methylphenylphosphinothioate, *in* M-00225
S-Phenyl methylphenylphosphinothioate, *in* M-00225
(S)$_P$-*O*-Menthyl ethylphenylphosphinothioate, *in* E-00105
2-Propenyl diphenylphosphinodithioate, *in* D-01065
O-Propyl bis(1-aziridinyl)phosphinothioate, *in* B-00093
O-Propyl butylphosphinothioate, *in* B-00593
Se-Propyl diethylphosphinoselenothioate, *in* D-00329
O-Propyl dipropylphosphinothioate, *in* D-01208
S-Propyl ethylphenylphosphinothioate, *in* E-00105
O-Propyl methylphosphinothioate, *in* M-00279
O-Propyl phenylphosphinothioate, *in* P-00213
O-*tert*-Butyl *tert*-butylmethylphosphinothioate, *in* B-00570
O-Trimethylsilyl bis(trifluoromethyl)phosphinothioate, *in* B-00415
O-Trimethylsilyl dibutylphosphinothioate, *in* D-00117
O-Trimethylsilyl di-*tert*-butylphosphinothioate, *in* D-00118
Trimethylsilyl diethylphosphinodithioate, *in* D-00325
O-Trimethylsilyl diethylphosphinothioate, *in* D-00330
Trimethylsilyl dimethylphosphinodithioate, *in* D-00834
Trimethylsilyl diphenylphosphinodithioate, *in* D-01065
O-Trimethylsilyl diphenylphosphinothioate, *in* D-01082

Thiophosphinic amides

N-Acetyl-*P*,*P*-diphenylphosphinothioic amide, *in* D-01083
N-Benzoyl-*P*,*P*-diethylphosphinothioic amide, *in* D-00331
N-Benzoyl-*P*,*P*-dimethylphosphinothioic amide, *in* D-00840
N-Benzoyl-*P*,*P*-diphenylphosphinothioic amide, *in* D-01083
N-Benzyl-*P*,*P*-dimethylphosphinothioic amide, *in* D-00840
N-Benzyl-*P*-methyl-*P*-phenylphosphinothioic amide, *in* M-00226
P,*P*-Bis(1-aziridinyl)-*N*-ethylphosphinothioic amide, *in* B-00094
P,*P*-Bis(1-aziridinyl)-*N*-methylphosphinothioic amide, *in* B-00094
P,*P*-Bis(1-aziridinyl)-*N*-phenylphosphinothioic amide, *in* B-00094
▷*P*,*P*-Bis(1-aziridinyl)phosphinothioic amide, B-00094
P,*P*-Bis(trifluoromethyl)phosphinothioic amide, B-00416
N-*tert*-Butyl-*P*,*P*-diphenylphosphinothioic amide, *in* D-01083
N-Butyl-*P*-methyl-*P*-phenylphosphinothioic amide, *in* M-00226
P-Cyclohexyl-*P*-methyl-*N*-phenylphosphinothioic amide, *in* C-00248
P,*P*-Di-*tert*-butyl-*N*-methylphosphinothioic amide, *in* D-00119
P,*P*-Di-*tert*-butyl-*N*-phenylphosphinothioic amide, *in* D-00119
P,*P*-Dibutyl-*N*-phenylphosphinothioic amide, *in* D-00117
Di-*tert*-butylphosphinothioic amide, D-00119
N,*N*-Diethyl-*P*,*P*-bis(chloromethyl)phosphinothioic amide, *in* B-00129
P,*P*-Diethyl-*N*,*N*-dimethylphosphinothioic amide, *in* D-00331
P,*P*-Diethyl-*N*-phenylphosphinothioic amide, *in* D-00331
N-(Diethylphosphinothioyl)-*P*,*P*-diethylphosphinothioic amide, *in* D-00331
P,*P*-Diethyl-*N*-trimethylsilylphosphinothioic amide, *in* D-00331
2,3-Dihydro-4,5-dimethyl-2,7-diphenyl-1*H*-1,2-azaphosphepine; 2-Sulfide, *in* D-00460
2,3-Dihydro-2-methyl-1*H*-4,1,2-benzothiazaphosphorine; 2-Sulfide, *in* D-00520
P,*P*-Dimethyl-*N*,*N*-bis(trimethylsilyl)phosphinothioic amide, *in* D-00840
P,*P*-Dimethyl-*N*-(dimethylphosphinothioyl)phosphinothioic amide, *in* D-00840
N,*N*-Dimethyl-*P*,*P*-diphenylphosphinothioic amide, *in* D-01083
P,*P*-Dimethyl-*N*-phenylphosphinothioic amide, *in* D-00840
P,*P*-Dimethylphosphinothioic amide, D-00840
P,*P*-Dimethyl-*N*-trimethylsilylphosphinothioic amide, *in* D-00840
P,*P*-Diphenyl-*N*-(diphenylphosphinothioyl)phosphinimidic acid, D-01006
P,*P*-Diphenyl-*N*-(diphenylphosphinothioyl)phosphinimidic acid; Chloride, *in* D-01006
P,*P*-Diphenyl-*N*-(diphenylphosphinothioyl)phosphinimidic acid; Amide, *in* D-01006
P,*P*-Diphenylphosphinic amide; *N*-Diphenylphosphinothioyl, *in* D-01040
P,*P*-Diphenylphosphinothioic amide, D-01083
N-Diphenylphosphinothioyl-*P*,*P*-diphenylphosphinothioic amide, D-01087
1-(Diphenylphosphinothioyl)piperidine, *in* D-01082
N-Diphenylphosphinothioyl-*P*,*P*,*P*-triphenylphosphine imide, *in* D-01088
P,*P*-Diphenyl-*N*-trimethylsilylphosphinothioic amide, *in* D-01083
P,*P*-Diphenyl-*N*-(triphenylphosphoranylidene)phosphinothioic amide, *in* D-01165
P,*P*-Dipropyl-*N*-trimethylsilylphosphinothioic amide, *in* D-01208
P-Ethyl-*N*,*N*-dimethyl-*P*-phenylphosphinothioic amide, *in* E-00105
P-Ethyl-*N*,*N*-dimethyl-*P*-phenylphosphinothioic amide, *in* E-00105
S-Ethyl diphenylphosphinoselenothioate, *in* D-01080
P-Isopropyl-*N*,*N*-dimethyl-*P*-phenylphosphinothioic amide, *in* I-00046
Se-Methyl dimethylphosphinoselenothioate, *in* D-00837
O-Methyl *P*,*P*-diphenyl-*N*-(diphenylphosphinothioyl)-phosphinimidate, *in* D-01006
N-Methyl-*P*,*P*-diphenylphosphinothioic amide, *in* D-01083
7-Phenyl-6-aza-7-phosphabicyclo[3.2.1]octane; 6-Cyclohexyl, 7-sulfide, *in* P-00086
N-Phenyl-*P*,*P*-bis(chloromethyl)phosphinothioic amide, *in* B-00129
Tetraethylphosphinothioic amide, *in* D-00331
Tetramethylphosphinothioic amide, *in* D-00840
N,*N*,*P*-Trimethyl-*P*-phenylphosphinothioic amide, *in* M-00226
N,*P*,*P*-Triphenylphosphinothioic amide, *in* D-01083

Thiophosphinic azides

Diethylphosphinothioic acid; Azide, *in* D-00330
Dimethylphosphinothioic acid; Azide, *in* D-00839
Diphenylphosphinothioic acid; Azide, *in* D-01082

Selenophosphinic acids

Bis(1-methylpropyl)phosphinoselenoic acid, B-00349
tert-Butylphenylphosphinoselenoic acid; (*R*)-*form*, *in* B-00582
tert-Butylphenylphosphinoselenoic acid; (*S*)-*form*, *in* B-00582
Dibutylphosphinoselenoic acid, D-00114
Diethylphosphinodiselenoic acid, D-00324
Diethylphosphinoselenoic acid, D-00327
Diethylphosphinoselenothioic acid, D-00329
Diisopropylphosphinoselenoic acid, D-00644
Diphenylphosphinoselenoic acid, D-01078
Diphenylphosphinoselenothioic acid, D-01080
Ethylphenylphosphinothioselenoic acid, E-00106

Selenophosphinic anhydrides

Diethylphosphinodiselenoic acid; Anhydroselenide, *in* D-00324
Diphenylphosphinodiselenoic acid; Anhydroselenide, *in* D-01064

Selenophosphinic halides

Dibutylphosphinoselenoic acid; Chloride, *in* D-00114
Diethylphosphinoselenoic acid; Chloride, *in* D-00327
Diethylphosphinoselenoic acid; Bromide, *in* D-00327

Dimethylphosphinoselenoic acid; Chloride, *in* D-00835
Dimethylphosphinoselenoic acid; Bromide, *in* D-00835
Diphenylphosphinoselenoic acid; Chloride, *in* D-01078
Ethylphenylphosphinoselenoic acid; (±)-*form*, Chloride, *in* E-00104

Selenophosphinic esters

Se-Butyl diethylphosphinoselenoate, *in* D-00327
tert-Butylphenylphosphinoselenoic acid; (*S*)-*form*, *Se*-Me ester, *in* B-00582
3,4-Dihydro-2-methyl-2*H*-1,4,2-benzothiazaphosphorine; 2-Selenide, *in* D-00521
Ethyl diethylphosphinodiselenoate, *in* D-00324
O-Ethyl diethylphosphinoselenoate, *in* D-00327
S-Ethyl diethylphosphinoselenothioate, *in* D-00329
Se-Ethyl diethylphosphinoselenothioate, *in* D-00329
O-Ethyl diphenylphosphinoselenoate, *in* D-01078
O-Ethyl ethylphenylphosphinoselenoate, *in* E-00104
S-Ethyl ethylphenylphosphinothioselenoate, *in* E-00106
O-Isopropyl diethylphosphinoselenoate, *in* D-00327
Se-Methyl *tert*-butylphenylphosphinoselenoate, *in* B-00582
O-Methyl diethylphosphinoselenoate, *in* D-00327
O-Methyl dimethylphosphinoselenoate, *in* D-00835
O-Methyl diphenylphosphinoselenoate, *in* D-01078
O-Methyl ethylphenylphosphinoselenoate, *in* E-00104
O-Phenyl diethylphosphinoselenoate, *in* D-00327
O-Phenyl diphenylphosphinoselenoate, *in* D-01078
Se-Phenyl diphenylphosphinoselenoate, *in* D-01078
Propyl diethylphosphinodiselenoate, *in* D-00324
Se-Propyl diethylphosphinoselenothioate, *in* D-00329
1,2-Thiaphospholane 2-sulfide; 2-Me, *in* T-00298
1,2-Thiaphospholane 2-sulfide; 2-Ph, *in* T-00298

Selenophosphinic amides

P,P-Dibutyl-*N,N*-diethylphosphinoselenoic amide, *in* D-00114
P,P-Diethyl-*N,N*-dimethylphosphinoselenoic amide, *in* D-00328
P,P-Diethyl-*N*-methylphosphinoselenoic amide, *in* D-00328
P,P-Diethyl-*N*-phenylphosphinoselenoic amide, *in* D-00328
P,P-Dimethyl-*N*-phenylphosphinoselenoic amide, *in* D-00836
P,P-Diphenyl-*N*-(triphenylphosphoranylidene)-phosphinoselenoic amide, *in* D-01165
Tetraethylphosphinoselenoic amide, *in* D-00328
Tetramethylphosphinoselenoic amide, *in* D-00836
N,P,P-Trimethylphosphinoselenoic amide, *in* D-00836

Selenophosphinic azides

Diethylphosphinoselenoic acid; Azide, *in* D-00327
Dimethylphosphinoselenoic acid; Azide, *in* D-00835
Diphenylphosphinoselenoic acid; Azide, *in* D-01078

Tellurophosphinic amides

P,P-Diphenyl-*N*-(triphenylphosphoranylidene)-phosphinotelluroic amide, *in* D-01165

Imidophosphinic acids, $HOPH_2{=}NH$

N-Benzenesulfonyl-*P,P*-diphenylphosphinimidic acid, B-00013
P,P-Diethyl-*N*-phenylphosphinimidic acid, D-00306
P,P-Diphenyl-*N*-(diphenylphosphinothioyl)phosphinimidic acid, D-01006
P,P-Diphenyl-*N*-(diphenylphosphinyl)phosphinimidic acid, D-01007
P,P-Diphenyl-*N*-trimethylsilylphosphinimidic acid, D-01157
N,P,P-Triphenylphosphinimidic acid, T-00631

Imidophosphinic halides

N-Benzenesulfonyl-*P,P*-diphenylphosphinimidic acid; Chloride, *in* B-00013
PP-Bis(trichloromethyl)phosphinimidic chloride, B-00403
N-Chloro-*P,P*-bis(trichloromethyl)phosphinimidic chloride, *in* B-00403
P-Chloro-*N,P,P*-triphenylphosphazene, *in* T-00631
P,P-Diphenyl-*N*-(diphenylphosphinothioyl)phosphinimidic acid; Chloride, *in* D-01006
P,P-Diphenyl-*N*-(diphenylphosphinyl)phosphinimidic acid; Chloride, *in* D-01007
P,P-Diphenyl-*N*-trimethylsilylphosphinimidic fluoride, *in* D-01157
N-Methyl-*P,P*-bis(trichloromethyl)phosphinimidic chloride, *in* B-00403
N-Methyl-*P,P*-diphenylphosphinimidic fluoride, *in* D-01044
N-Phenyl-*P,P*-bis(trichloromethyl)phosphinimidic chloride, *in* B-00403

Imidophosphinic esters

2,2-Dihydro-1,3,2-benzoxazaphosphole; 2,2-Di-Ph, *in* D-00444
Ethyl *P,P*-diethyl-*N*-phenylphosphinimidate, *in* D-00306
Ethyl *P,P*-diphenyl-*N*-(diphenylphosphinyl)phosphinimidate, *in* D-01007
Ethyl *P,P*-diphenyl-*N*-phenylsulfonylphosphinimidate, *in* B-00013
Ethyl *N,P,P*-triphenylphosphinimidate, *in* T-00631
O-Methyl *P,P*-diphenyl-*N*-(diphenylphosphinothioyl)-phosphinimidate, *in* D-01006
Methyl *P,P*-diphenyl-*N*-phenylsulfonylphosphinimidate, *in* B-00013
Phenyl *P,P*-diphenyl-*N*-phenylsulfonylphosphinimidate, *in* B-00013
Trimethylsilyl *P,P*-diphenyl-*N*-trimethylsilylphosphinimidate, *in* D-01157

Imidophosphinic amides

P,P-Diethyl-*N,N′*-diphenylphosphinimidic amide, *in* D-00306
P,P-Dimethyl-*N,N,N′*-tris(trimethylsilyl)phosphinimidic amide, *in* D-00833
P,P-Diphenyl-*N,N′*-bis(trimethylsilyl)phosphinimidic amide, *in* D-01043
P,P-Diphenyl-*N,N′*-bis(triphenylsilyl)phosphinimidic amide, *in* D-01043
P,P-Diphenyl-*N*-(diphenylphosphinothioyl)phosphinimidic acid; Amide, *in* D-01006
P,P-Diphenyl-*N*-(triphenylphosphoranylidene)phosphinimidic amide, D-01164
N′-(4-Fluorophenyl)-*P,P*-diphenyl-*N*-(triphenylphosphoranylidene)phosphinimidic amide, *in* D-01164
N′-(4-Nitrophenyl)-*P,P*-diphenyl-*N*-(triphenylphosphoranylidene)phosphinimidic amide, *in* D-01164
Pentamethylphosphinimidic amide, *in* D-00833
N,N,P,P-Tetramethylphosphinimidic amide, *in* D-00833
N,N′,P,P-Tetramethylphosphinimidic amide, *in* D-00833
N,N,P,P-Tetramethyl-*N′*-trimethylsilylphosphinimidic amide, *in* D-00833
▷*N,N′,P,P*-Tetraphenylphosphinimidic amide, *in* T-00631
N′,P,P-Triphenyl-*N*-(triphenylphosphoranylidene)-phosphinimidic amide, *in* D-01164

Pentaoxyphosphoranes, (HO)$_5$P and derivatives

Bis(dimethylamino)trifluorophosphorane, B-00196
4′,5′-Bis(trifluoromethyl)spiro[1,3,2-benzodioxaphosphole-2,2′-[1,3,2]dioxaphosphole]; 2-Methoxy, *in* B-00421
4′,5′-Bis(trifluoromethyl)spiro[1,3,2-benzodioxaphosphole-2,2′-[1,3,2]dioxaphosphole]; 2-Phenoxy, *in* B-00421
4′,5′-Bis(trifluoromethyl)spiro[1,3,2-benzodioxaphosphole-2,2′-[1,3,2]dioxaphosphole]; 2-Fluoro, *in* B-00421
2-Chloro-2,2′-spirobi[1,3,2-benzodioxaphosphole], C-00174
4,6-Di-*tert*-butylspiro[1,3,2-benzodioxaphosphole-2,1′-[2,8,9]-trioxa[1]phosphaadamantane], D-00157
4,6-Dichloro-1,3,5,7,9,10-hexamethyl-1,3,5,7,9,10-hexaaza-4,6-diphospha(4,6-P^V)dispiro[3.1.3.1]decane-2,8-dione, D-00170
Dichlorotriphenoxyphosphorane, D-00199

Difluorotrimethoxyphosphorane, D-00416
Difluorotriphenoxyphosphorane, D-00418
2,2-Dihydro-1,3,2-benzodioxaphosphole; 2,2,2-Trimethoxy, *in* D-00439
2,2-Dihydro-3,4-bis(methoxycarbonyl)-2,2-dimethyl-5,5-diphenyl-5*H*-1,2-azaphosphole, D-00445
2,2-Dihydro-1,3-dimethyl-1,3,2-diazaphosphetidin-4-one; 2,2,2-Trifluoro, *in* D-00459
2,2-Dihydro-1,3-dimethyl-1,3,2-diazaphosphetidin-4-one; 2,2,2-Trichloro, *in* D-00459
10,10-Dihydro-2,3-diphenyl-[1,3,2]oxazaphospholo[2,3-*b*]-[1,3,2]benzoxazaphosphole; 10,10-Dimethoxy, *in* D-00481
10,10-Dihydro-2,3-diphenyl-[1,3,2]oxazaphospholo[2,3-*b*]-[1,3,2]benzoxazaphosphole; 10,10-Diphenoxy, *in* D-00481
10,10-Dihydro-2,3-diphenyl-[1,3,2]oxazaphospholo[2,3-*b*]-[1,3,2]benzoxazaphosphole; 10,10-Bis(dimethylamino), *in* D-00481
2,2-Dihydrophenanthro[9,10-*d*]-1,3,2-dioxaphosphole; 2,2,2-Trimethoxy, *in* D-00541
2,2-Dihydrophenanthro[9,10-*d*]-1,3,2-dioxaphosphole; 2,2,2-Triisopropoxy, *in* D-00541
2,2-Dihydrophenanthro[9,10-*d*]-1,3,2-dioxaphosphole; 2,2,2-Triphenoxy, *in* D-00541
2,2-Dihydro-2,2,2-trichloro-4,4,5,5-tetrakis(trifluoromethyl)-1,3,2-dioxaphospholane, D-00589
2,2-Dihydro-2,2,2-triethoxy-4-[(4-methylphenyl)sulfonyl]-1,4,2-oxazaphospholidine, D-00590
2,2-Dihydro-2,2,2-trimethoxy-4,5-bis(trifluoromethyl)-1,3,2-dithiaphosphole, D-00592
2,2-Dihydro-2,2,2-trimethoxy-4,5-dimethyl-1,3,2-dioxaphosphole, D-00593
2,2-Dihydro-2,2,2-trimethoxynaphtho[1,2-*d*]-1,3,2-dioxaphosphole, D-00594
2,2-Dihydro-2,2,2-trimethoxy-4,4,5,5-tetrakis(trifluoromethyl)-1,3,2-dioxaphospholane, D-00595
2,2-Dihydro-2,2,2-triphenoxy-1,3,2-benzodioxaphosphole, D-00602
4,4-Dihydro-4,4,4-triphenoxytrioxaphosphetane, D-00603
Diiodotriphenoxyphosphorane, D-00630
(Dimethylamino)tetrafluorophosphorane, D-00708
6,9-Dimethyl-2,3-diphenyl-1,4-dioxa-6,9-diaza-5-phospha(5-P^V)spiro[4.4]non-2-ene; 5-Methoxy, *in* D-00741
1,3-Dimethyl-6,7-diphenyl-5,8-dioxa-1,3-diaza-4-phospha(4-P^V)spiro[3.4]oct-6-en-2-one; 4-Methoxy, *in* D-00742
1,3-Dimethyl-6,7-diphenyl-5,8-dioxa-1,3-diaza-4-phospha(4-P^V)spiro[3.4]oct-6-en-2-one; 4-Dimethylamino, *in* D-00742
1,3-Dimethyl-5,7-diphenyl-1,3,5,7-tetraaza-4-phospha(4-P^V)-spiro[3.3]heptane-2,6-dione; 4-Chloro, *in* D-00748
N,*N*-Dimethyl-2,3-diphenyl-1,4,6,9-tetraoxa-5-phospha(5-P^V)-spiro[4.4]non-2-en-5-amine, *in* D-01150
6,9-Dimethyl-5-phenoxy-2,2,3,3-tetrakis(trifluoromethyl)-1,4-dioxa-6,9-diaza-5-phospha(5-P^V)spiro[4.4]nonane, *in* D-00922
6,9-Dimethyl-5-phenylthio-2,2,3,3-tetrakis(trifluoromethyl)-1,4-dioxa-6,9-diaza-5-phospha(5-P^V)spiro[4.4]nonane, *in* D-00922
1,3-Dimethylspiro[1,3,2-diazaphospholidine-2,2′-phenanthro[9,10-*d*][1,3,2]dioxaphosphole]; 2-Methoxy, *in* D-00918
4,4-Dimethylspiro[1,3,2-dioxaphospholane-2,1′-[2,6,7]-trioxa[1]phosphabicyclo[2.2.1]heptane], D-00919
N,*N*-Dimethylspiro[1,3,2-dioxaphosphorinane-2,2′-phenanthro[9,10-*d*][1,3,2]dioxaphosphol]-2-amine, *in* S-00025
6,10-Dimethyl-2,2,3,3-tetrakis(trifluoromethyl)-1,4-dioxa-6,10-diaza-5-phospha(5-P^V)spiro[4.5]decane; 5-Phenoxy, *in* D-00923
2,3-Dimethyl-1,4,6,9-tetraoxa-5-phospha(5-P^V)spiro[4.4]non-2-ene; 5-Methoxy, *in* D-00924
1′,3′-Diphenyldispiro[1,3,2-benzodioxaphosphole-2,2′-[1,3,2,4]diazaphosphetidine-4′,2″-[1,3,2]-benzodioxaphosphole]; 2,4′-Dimethoxy, *in* D-01018
1′,3′-Diphenyldispiro[1,3,2-benzodioxaphosphole-2,2′-[1,3,2,4]diazaphosphetidine-4′,2″-[1,3,2]-benzodioxaphosphole]; 2,4′-Diethoxy, *in* D-01018
1′,3′-Diphenyldispiro[1,3,2-benzodioxaphosphole-2,2′-[1,3,2,4]diazaphosphetidine-4′,2″-[1,3,2]-benzodioxaphosphole]; 2,4′-Diisopropoxy, *in* D-01018
1′,3′-Diphenyldispiro[1,3,2-benzodioxaphosphole-2,2′-[1,3,2,4]diazaphosphetidine-4′,2″-[1,3,2]-benzodioxaphosphole]; 2,4′-Diphenoxy, *in* D-01018
1′,3′-Diphenyldispiro[1,3,2-benzodioxaphosphole-2,2′-[1,3,2,4]diazaphosphetidine-4′,2″-[1,3,2]-benzodioxaphosphole]; 2,4′-Difluoro, *in* D-01018
1′,3′-Diphenyldispiro[1,3,2-benzodioxaphosphole-2,2′-[1,3,2,4]diazaphosphetidine-4′,2″-[1,3,2]-benzodioxaphosphole]; 2,4′-Dichloro, *in* D-01018
1′,3′-Diphenyldispiro[1,3,2-benzodioxaphosphole-2,2′-[1,3,2,4]diazaphosphetidine-4′,2″-[1,3,2]-benzodioxaphosphole]; 2,4′-Dibromo, *in* D-01018
1′,3′-Diphenyldispiro[1,3,2-benzodioxaphosphole-2,2′-[1,3,2,4]diazaphosphetidine-4′,2″-[1,3,2]-benzodioxaphosphole]; 2,4′-Bis(phenylamino), *in* D-01018
6,12-Diphenyl-1,4,8,11-tetraoxa-6,12-diaza-5,7-diphospha(5,7-P^V)dispiro[4.1.4.1]dodecane; 5,7-Dimethoxy, *in* D-01149
6,12-Diphenyl-1,4,8,11-tetraoxa-6,12-diaza-5,7-diphospha(5,7-P^V)dispiro[4.1.4.1]dodecane; 5,7-Diethoxy, *in* D-01149
6,12-Diphenyl-1,4,8,11-tetraoxa-6,12-diaza-5,7-diphospha(5,7-P^V)dispiro[4.1.4.1]dodecane; 5,7-Diphenoxy, *in* D-01149
6,12-Diphenyl-1,4,8,11-tetraoxa-6,12-diaza-5,7-diphospha(5,7-P^V)dispiro[4.1.4.1]dodecane; 5,7-Difluoro, *in* D-01149
2,3-Diphenyl-1,4,6,9-tetraoxa-5-phospha(5-P^V)spiro[4.4]non-2-ene; 5-Methoxy, *in* D-01150
2,3-Diphenyl-1,4,6,9-tetraoxa-5-phospha(5-P^V)spiro[4.4]non-2-ene; 5-Phenoxy, *in* D-01150
2,3-Diphenyl-1,4,6-trioxa-9-aza-5-phospha(5-P^V)spiro[4.4]-non-2-ene; 5-Methoxy, *in* D-01159
4-Ethylspiro[2,8,9-trioxa-1-phosphatricyclo[3.3.1.1^{3,7}]-decane-1,4′-trioxaphosphetane], E-00187
2-Fluoro-2,2′-spirobi[1,3,2-benzodioxaphosphole], F-00048
N,*N*,*N*′,*N*′,6,7-Hexamethyl-1,3-bis(trimethylsilyl)-5,8-dioxa-1,3-diaza-2,4-diphospha(4-P^V)spiro[3.4]oct-6-ene-2,4-diamine, H-00068
N,*N*,*N*′,*N*′,6,7-Hexamethyl-1,3-bis(trimethylsilyl)-5,8-dioxa-1,3-diaza-2,4-diphospha(4-P^V)spiro[3.4]oct-6-ene-2,4-diamine; 2-Sulfide, *in* H-00068
N,*N*,4′,4′,5′,5′-Hexamethylspiro[1,3,2-benzodioxaphosphole-2,2′-[1,3,2]dioxaphospholan]-2-amine, *in* T-00230
4,4,5,5,*N*,*N*-Hexamethylspiro[1,3,2-benzoxazaphosphole-2(3*H*),-2′-[1,3,2]dioxaphospholan]-2-amine, *in* T-00231
N,*N*,2,2,3,3-Hexamethyl-1,4,6,9-tetraoxa-5-phospha(5-P^V)-spiro[4.4]nonan-5-amine, *in* T-00235
N,*N*,2,2,3,3-Hexamethyl-1,4,6-trioxa-9-aza-5-phospha(5-P^V)-spiro[4.4]nonan-5-amine, *in* T-00238
4′-Methyl-4,4,5,5-tetrakis(trifluoromethyl)spiro[1,3,2-dioxaphospholane-2,1′-[2,6,7]trioxa[1]-phosphabicyclo[2.2.2]octane], M-00405
Octahydro-3*H*,8*H*,10*bH*-2*a*,5*a*,7*a*,10*a*-tetraaza-10*b*-phosphaindeno[7,1-*cd*]indene; 10*b*-Fluoro, *in* O-00022
Octahydro-7*H*,9*bH*-2*a*,4*a*,6*a*,9*a*-tetraaza-9*b*-phosphapentaleno[1,6-*cd*]indene; 9*b*-Fluoro, *in* O-00023
Octahydro-5*H*,8*H*,10*bH*-2*a*,4*a*,7*a*,10*a*-tetraaza-10*b*-phosphapentaleno[1,6-*de*]naphthalene; 10*b*-Fluoro, *in* O-00024
Octahydro-8*bH*-2*a*,4*a*,6*a*,8*a*-tetraaza-8*b*-phosphapentaleno[1,6-*cd*]pentalene; 8*b*-Hydroxy, *in* O-00025
Octahydro-8*bH*-2*a*,4*a*,6*a*,8*a*-tetraaza-8*b*-phosphapentaleno[1,6-*cd*]pentalene; 8*b*-Methoxy, *in* O-00025
Octahydro-8*bH*-2*a*,4*a*,6*a*,8*a*-tetraaza-8*b*-phosphapentaleno[1,6-*cd*]pentalene; 8*b*-Fluoro, *in* O-00025
2,2,3,3,7,7,8,8-Octamethyl-1,4,6,9-tetraoxa-5-phospha(5-P^V)-spiro[4.4]nonane; 5-Phenoxy, *in* O-00032
2,7,9,14,15,16,17,18-Octaoxa-1,8-diphospha(1,8-P^V)-pentacyclo[10.2.1.1^{1,4}.1^{5,8}.1^{8,11}]octadecane-3,6,10,13-tetrone; (*R*,*R*,*R*,*R*)-*form*, 1,8-Dihydroxy, bistriethylammonium salt, *in* O-00036
Pentaethoxyphosphorane, P-00011
Pentamethoxyphosphorane, P-00018

Pentaphenoxyphosphorane, P-00035
2-Phenoxy-2,2′-spirobi[1,3,2-benzodioxaphosphole], P-00074
N-Phenylspiro[1,3,2-benzodioxaphosphole-2,2′(3′*H*)-[1,3,2]-benzoxazaphosphol]-2-amine, *in* S-00008
Spiro[1,3,2-benzodioxaphosphole-2,2′(3′*H*)-[1,3,2]-benzoxazaphosphole]; 2-Methoxy, *in* S-00008
Spiro[1,3,2-benzodioxaphosphole-2,2′(3′*H*)-[1,3,2]-benzoxazaphosphole]; 2-Ethoxy, *in* S-00008
Spiro[1,3,2-benzodioxaphosphole-2,2′(3′*H*)-[1,3,2]-benzoxazaphosphole]; 2-Phenoxy, *in* S-00008
Spiro[1,3,2-benzodioxaphosphole-2,2′(3′*H*)-[1,3,2]-benzoxazaphosphole]; 2-Chloro, *in* S-00008
2,2′-Spirobi[1,3,2-benzodioxaphosphole]; 2-Trimethylsilyl, *in* S-00014
2,2(3*H*,3′*H*)-Spirobi[1,3,2-benzoxazaphosphole]; 2-Phenoxy, *in* S-00020
2,2(3*H*,3′*H*)-Spirobi[1,3,2-benzoxazaphosphole]; 2-Fluoro, *in* S-00020
Spiro[1,3,2-dioxaphospholane-2,1′-[2,8,9]trioxa[1]-phosphatricyclo[3.3.1.1^{3,7}]decane], S-00024
4,5,6,7-Tetrachloro-*N*,*N*-dimethylspiro[1,3,2-benzodioxaphosphole-2,2′-[1,3,2]dioxaphospholan]-2-amine, *in* T-00036
Tetrachlorophenoxyphosphorane, T-00033
4,5,6,7-Tetrachloro-4′,4′,5′,5′-tetramethylspiro[1,3,2-benzodioxaphosphole-2,2′-[1,3,2]dioxaphospholane]; 2-Methoxy, *in* T-00036
4,5,6,7-Tetrachloro-4′,4′,5′,5′-tetramethylspiro[1,3,2-benzodioxaphosphole-2,2′-[1,3,2]dioxaphospholane]; 2-Phenoxy, *in* T-00036
Tetrafluorophenoxyphosphorane, T-00083
2,2,2,3-Tetrahydro-3,5-diphenyl-1,3,4,2-oxadiazaphosphole; 2,2,2-Trimethoxy, *in* T-00104
2,2,2,3-Tetrahydro-3,5-diphenyl-1,3,4,2-oxadiazaphosphole; 2,2,2-Triphenoxy, *in* T-00104
2,2,2,3-Tetrahydro-3,5-diphenyl-1,3,4,2-oxadiazaphosphole; 2,2,2-Trichloro, *in* T-00104
2,2,2,3-Tetrahydro-5-phenyl-1,3,4,2-oxadiazaphosphole; 2,2,2-Trifluoro, *in* T-00131
2,2,2,3-Tetrahydro-5-phenyl-1,3,4,2-oxadiazaphosphole; 2,2,2-Trichloro, *in* T-00131
2,2,2,3-Tetrahydro-2,2,2-trimethoxy-3-methyl-4,5-diphenyl-1,3,2-oxazaphosphole, T-00156
Tetraiodophenoxyphosphorane, T-00164
Tetrakis(dimethylamino)iodophosphorane, T-00178
2,2,3,3-Tetrakis(trifluoromethyl)-1,4,6,10-tetraoxa-5-phospha(5-P^V)spiro[4.5]decane; 5-Phenoxy, *in* T-00187
2,2,3,3-Tetrakis(trifluoromethyl)-1,4,6-trioxa-10-aza-5-phospha(5-P^V)spiro[4.5]decane; 5-(4-Bromophenoxy), *in* T-00190
3,3,8,8-Tetramethyl-1,6-dioxa-4,9-diaza-5-phospha(P^V)-spiro[4.4]nonane; (±)-*form*, 5-Anilino, *in* T-00196
N,*N*,6,9-Tetramethyl-2,3-diphenyl-1,4-dioxa-6,9-diaza-5-phospha(5-P^V)spiro[4.4]non-2-en-5-amine, *in* D-00741
N,*N*,*N*′,*N*′-Tetramethyl-1′,3′-diphenyldispiro[1,3,2-benzodioxophosphole-2,2′-[1,3,2,4]diazaphosphetidine-4′,2″-[1,3,2]benzodioxaphosphol]-2,2′-diamine, *in* D-01018
4′,4′,5′,5′-Tetramethylspiro[1,3,2-benzodioxaphosphole-2,2′-[1,3,2]dioxaphospholane]; 2-Methoxy, *in* T-00230
4′,4′,5′,5′-Tetramethylspiro[1,3,2-benzodioxaphosphole-2,2′-[1,3,2]dioxaphospholane]; 2-Phenoxy, *in* T-00230
4′,4′,5′,5′-Tetramethylspiro[1,3,2-benzoxazaphosphole-2(3*H*),-2′-[1,3,2]dioxaphospholane]; 2-Methoxy, *in* T-00231
4′,4′,5′,5′-Tetramethylspiro[1,3,2-benzoxazaphosphole-2(3*H*),-2′-[1,3,2]dioxaphospholane]; 2-Phenoxy, *in* T-00231
N,*N*,1,3-Tetramethylspiro[1,3,2-diazaphospholidine-2,2′-phenanthro[9,10-*d*][1,3,2]dioxaphosphol]-2-amine, *in* D-00918
1,3,5,7-Tetramethyl-1,3,5,7-tetraaza-4-phospha(4-P^V)-spiro[3.3]heptane-2,6-dione; 4-Chloro, *in* T-00234
1,3,5,7-Tetramethyl-1,3,5,7-tetraaza-4-phospha(4-P^V)-spiro[3.3]heptane-2,6-dione; 4-Azido, *in* T-00234
6,9,*N*,*N*-Tetramethyl-2,2,3,3-tetrakis(trifluoromethyl)-1,4-dioxa-6,9-diaza-5-phospha(P^V)spiro[4.4]nonan-5-amine, *in* D-00922
2,2,3,3-Tetramethyl-1,4,6,9-tetraoxa-5-phospha(5-P^V)-spiro[4.4]nonane; 5-Phenoxy, *in* T-00235
N,*N*,2,3-Tetramethyl-1,4,6,9-tetraoxa-5-phospha(5-P^V)-spiro[4.4]non-2-en-5-amine, *in* D-00924
3,3,9,9-Tetramethyl-1,5,7,11-tetraoxa-6-phospha(6-P^V)-spiro[5.5]undecane; 6-Ethoxy, *in* T-00236
1,4,6,9-Tetraoxa-5-phospha(5-P^V)spiro[4.4]nonane; 5-Methoxy, *in* T-00244
1,4,6,9-Tetraoxa-5-phospha(5-P^V)spiro[4.4]nonane; 5-Ethoxy, *in* T-00244
2,2,2-Tribromo-2,2-dihydro-1,3,2-benzodioxaphosphole, T-00352
2,2,2-Tribromo-2,2-dihydro-4,4,5,5-tetrakis(trifluoromethyl)-1,3,2-dioxaphospholane, T-00353
Tribromodiphenoxyphosphorane, T-00354
2,2,2-Trichloro-2,2-dihydro-1,3,2-benzodioxaphosphole, T-00400
2,2,2-Trichloro-2,2-dihydro-1,3-dimethyl-6,8-bis(trifluoromethyl)-1,3,5,7,9-pentaaza-2,4-diphospha(4-P^V)spiro[3.5]nona-4,6,8-triene, T-00401
2,2,2-Trichloro-2,2-dihydro-5-phenyl-1,3,4,2-dioxazaphosphole, *in* C-00126
Trichlorodiphenoxyphosphorane, T-00403
2,2,2-Triethoxy-2,2-dihydro-1,3,2-dioxaphospholane, T-00436
2,2,2-Trifluoro-2,2-dihydro-4,4,5,5-tetrakis(trifluoromethyl)-1,3,2-dioxaphospholane, T-00467
Trifluorodimethoxyphosphorane, T-00468
Trifluorodiphenoxyphosphorane, T-00470
Tris(dimethylamino)difluorophosphorane, T-00760

Thio derivs. of pentaoxyphosphoranes

4′,5′-Bis(trifluoromethyl)spiro[1,3,2-benzodioxaphosphole-2,2′-[1,3,2]dithiaphosphole]; 2-Methoxy, *in* B-00422
4,5-Bis(trifluoromethyl)spiro[1,3,2-dithiaphosphole-2,1′-[2,6,7]trioxa[1]phosphabicyclo[2.2.1]heptane, B-00423
4,5-Bis(trifluoromethyl)spiro[1,3,2-dithiaphosphole-2,1′-[2,8,9]trioxa[1]phosphatricyclo[3.3.1.1^{5,7}]decane], B-00424
2,2-Dihydro-4,5-diphenyl-1,3,2-oxathiaphosphole; 2,2,2-Trimethoxy, *in* D-00480
N,*N*-Dimethylspiro[1,3,2-benzodioxaphosphole-2,2′-[1,3,2]-benzoxathiaphosphol]-2-amine, *in* S-00007
3,3′-Dimethyl-2,2′-(3*H*,3′*H*)-spirobi[1,3,2-benzothiazaphosphole]; 2-Chloro, *in* D-00916
1,4-Dioxa-6,9-dithia-5-phospha(5-P^V)spiro[4.4]nonane; 5-Methoxy, *in* D-00964
6,12-Diphenyl-1,4,8,11-tetraoxa-6,12-diaza-5,7-diphospha(5,7-P^V)dispiro[4.1.4.1]dodecane; 5,7-Bis(ethylthio), *in* D-01149
4′-Methyl-4,5-bis(trifluoromethyl)spiro[1,3,2-dithiaphosphole-2,1′-[2,6,7]trioxa[1]-phosphabicyclo[2.2.2]octane], M-00092
3′-Methylspiro[1,3,2-benzodioxaphosphole-2,2(3′*H*)-[1,3,2]-benzothiazaphosphole]; 2-Phenoxy, *in* M-00394
Spiro[1,3,2-benzodioxaphosphole-2,2′-[1,3,2]-benzoxathiaphosphole]; 2-Phenoxy, *in* S-00007
4,5,6,7-Tetrachloro-4′,4′,5′,5′-tetramethylspiro[1,3,2-benzodioxaphosphole-2,2′-[1,3,2]dioxaphospholane]; 2-Phenylthio, *in* T-00036
4′,4′,5′,5′-Tetramethylspiro[1,3,2-benzodioxaphosphole-2,2′-[1,3,2]dioxaphospholane]; 2-Phenylthio, *in* T-00230
3′,*N*,*N*-Trimethylspiro[1,3,2-benzodioxaphosphole-2,2′(3′*H*)-[1,3,2]benzothiazaphosphol]-2-amine, *in* M-00394

Seleno derivs. of pentaoxyphosphoranes

2,2,3,3,7,7,8,8-Octamethyl-1,4,6,9-tetraoxa-5-phospha(5-P^V)-spiro[4.4]nonane; 5-Phenylselenoxy, *in* O-00032
4′,4′,5′,5′-Tetramethylspiro[1,3,2-benzodioxaphosphole-2,2′-[1,3,2]dioxaphospholane]; 2-Phenylseleno, *in* T-00230

Phosphenous and phosphenic amides

(Di-*tert*-butylimino)(diisopropylamino)phosphorane, D-00095

1 P-contg. ring; 3-membered

2-Amino-1,3-di-*tert*-butylazaphosphiridine; (2*RS*,3*SR*)-*form*, *N*,*N*-Diisopropyl, *in* A-00070
2,3-Bis(2,4,6-tri-*tert*-butylphenyl)-1,2,3-selenadiphosphirane, B-00396
2,3-Bis(2,4,6-tri-*tert*-butylphenyl)thiadiphosphirane, B-00397
2-[Bis(trimethylsilyl)amino]-2,2-dihydro-3,3-bis(trifluoromethyl)-2-[(trimethylsilyl)imino]-oxaphosphirane, B-00430
1,2-Di-*tert*-butyldiphosphirane, D-00091
2,3-Di-*tert*-butyl-1-isopropylphosphadiarsirane, D-00096
3,3-Dichloro-1,1,2,2-tetrahydro-1,1,1,2,2,2-hexaphenyldiphosphirane, D-00193
2,2-Dihydro-3,3-bis(trifluoromethyl)-1-[[2,2,2-trifluoro-1-(trifluoromethyl)ethylidene]amino]azaphosphiridine; 2,2,2-Trimethoxy, *in* D-00446
2,2-Dihydro-3,3-bis(trifluoromethyl)-1-[[2,2,2-trifluoro-1-(trifluoromethyl)ethylidene]amino]azaphosphiridine; 2,2,2-Triethoxy, *in* D-00446
2,2-Dihydro-3,3-bis(trifluoromethyl)-1-[[2,2,2-trifluoro-1-(trifluoromethyl)ethylidene]amino]azaphosphiridine; 2,2,2-Tris(dimethylamino), *in* D-00446
1-Methylphosphirane, *in* P-00361
9-Phenyl-9-phosphabicyclo[6.1.0]none-2,4,6-triene, P-00197
1-Phenylphosphirane, P-00214
Phosphirane, P-00361
1,2,3-Tri-*tert*-butyldiazaphosphiridine; ($1\alpha,2\beta,3\alpha$)-*form*, *in* T-00362
Tri-*tert*-butyldiphospharsirane, T-00363
Tri-*tert*-butyltriphosphirane, T-00397
1,2,3-Triphenyl-1*H*-phosphirene, T-00632
1,2,3-Triphenyl-1*H*-phosphirene; Oxide, *in* T-00632
Triphenyltriphosphirane, T-00698

1 P-contg. ring; 4-membered

3-Amino-2,3-di-*tert*-butyl-1,2,4,3-thiadiazaphosphetidine; *N*,*N*-Di-Me, 1,1-dioxide, *in* A-00071
1-Benzyl-1,2,2,3,4,4-hexamethylphosphetanium bromide, *in* H-00074
1-Benzyl-2,2,3,4,4-pentamethyl-1-phenylphosphetanium bromide, *in* P-00020
Benzyl 1,1,2,3,3-pentamethyltrimethylenephosphinate, *in* H-00168
N-Benzyl-*P*,*P*-(1,1,2,3,3-pentamethyltrimethylene)phosphinic amide, *in* P-00021
1-Benzyl-3-phenyl-1,3-azaphosphetidine, *in* P-00087
2,4-Bis(dimethylamino)-3-methyl-1,3,2,4-thiazadiphosphetidine; 2,4-Disulfide, *in* B-00187
2,4-Bis(4-fluorophenyl)-1,3,2,4-dioxaphosphetane 2,4-dioxide, B-00275
2,4-Bis(4-methoxyphenyl)-1,3,2,4-dithiadiphosphetane 2,4-disulfide, B-00293
1,3-Bis(1-methylethoxy)-1,3-diphosphetane 1,3-dioxide, *in* D-00620
2,4-Bis(1-methylpropyl)-1,3,2,4-dithiodiphosphetane 2,4-disulfide, *in* M-00382
1-*tert*-Butyl-1,2,4-azadiphosphetidine; 2,4-Dichloro, 2,4-dioxide, *in* B-00543
1-*tert*-Butyl-1,2,4-azadiphosphetidine; 2,4-Diisopropoxy, *in* B-00543
1-*tert*-Butyl-1,2,4-azadiphosphetidine; 2,4-Bis(dimethylamino), 2,4-dioxide, *in* B-00543
2-*tert*-Butyl-1,3-diisopropoxytriphosphetane 1,3-dioxide, *in* D-00627
1-Chloro-2,2,3,4,4-pentamethylphosphetane, C-00124
1-Chloro-2,2,3,4,4-pentamethylphosphetane; *trans-form*, Oxide, *in* C-00124
1-Chloro-2,2,3,4,4-pentamethylphosphetane; *trans-form*, Sulfide, *in* C-00124
1-Chloro-2,2,3,4,4-pentamethylphosphetane 1-oxide, *in* C-00124
3-Chloro-1,3-thiaphosphetane; 3-Oxide, *in* C-00188
3-Chloro-1,3-thiaphosphetane; 3-Sulfide, *in* C-00188
1,3-Di-*tert*-butyl-2,4-bis(dimethylamino)-1,3,2,4-diazadiphosphetidine; *Cis-form*, *in* D-00074
1,3-Di-*tert*-butyl-2,4-bis(dimethylamino)-1,3,2,4-diazadiphosphetidine; *Cis-form*, 2,4-Dioxide, *in* D-00074
1,3-Di-*tert*-butyl-2,4-bis(dimethylamino)-1,3,2,4-diazadiphosphetidine; *Cis-form*, 2-Sulfide, *in* D-00074
1,3-Di-*tert*-butyl-2,4-bis(dimethylamino)-1,3,2,4-diazadiphosphetidine; *Cis-form*, 2,4-Disulfide, *in* D-00074
1,3-Di-*tert*-butyl-2,4-bis(dimethylamino)-1,3,2,4-diazadiphosphetidine; *Cis-form*, 2-Selenide, *in* D-00074
1,3-Di-*tert*-butyl-2,4-bis(dimethylamino)-1,3,2,4-diazadiphosphetidine; *Cis-form*, 2,4-Diselenide, *in* D-00074
1,3-Di-*tert*-butyl-2,4-bis(dimethylamino)-1,3,2,4-diazadiphosphetidine; *Trans-form*, *in* D-00074
1,3-Di-*tert*-butyl-2,4-bis(dimethylamino)-1,3,2,4-diazadiphosphetidine; *Trans-form*, 2,4-Dioxide, *in* D-00074
1,3-Di-*tert*-butyl-2,4-bis(dimethylamino)-1,3,2,4-diazadiphosphetidine; *Trans-form*, 2-Sulfide, *in* D-00074
1,3-Di-*tert*-butyl-2,4-bis(dimethylamino)-1,3,2,4-diazadiphosphetidine; *Trans-form*, 2,4-Disulfide, *in* D-00074
1,3-Di-*tert*-butyl-2,4-bis(dimethylamino)-1,3,2,4-diazadiphosphetidine; *Trans-form*, 2-Selenide, *in* D-00074
1,3-Di-*tert*-butyl-2,4-bis(dimethylamino)-1,3,2,4-diazadiphosphetidine; *Trans-form*, 2,4-Diselenide, *in* D-00074
1,3-Di-*tert*-butyl-2,4-bis(dimethylamino)-1,3,2,4-diazadiphosphetidine; *Trans-form*, 2-Telluride, *in* D-00074
2,4-Di-*tert*-butyl-3-chloro-1,2,4,3-thiadiazaphosphetidine; 1,1-Dioxide, *in* D-00079
2,4-Di-*tert*-butyl-3-chloro-1,2,4,3-thiadiazaphosphetidine; 1-Oxide, 3-Sulfide, *in* D-00079
1,3-Di-*tert*-butyl-2,4-dimethoxy-1,3,2,4-diazadiphosphetidine; *cis-form*, *in* D-00085
1,3-Di-*tert*-butyl-2,4-dimethoxy-1,3,2,4-diazadiphosphetidine; *cis-form*, 2-Sulfide, *in* D-00085
1,3-Di-*tert*-butyl-2,4-dimethoxy-1,3,2,4-diazadiphosphetidine; *cis-form*, 2-Selenide, *in* D-00085
1,3-Di-*tert*-butyl-2,4-dimethoxy-1,3,2,4-diazadiphosphetidine; *cis-form*, 2,4-Disulfide, *in* D-00085
1,3-Di-*tert*-butyl-2,4-dimethoxy-1,3,2,4-diazadiphosphetidine; *cis-form*, 2,4-Diselenide, *in* D-00085
1,3-Di-*tert*-butyl-2,4-dimethoxy-1,3,2,4-diazadiphosphetidine; *trans-form*, *in* D-00085
1,3-Di-*tert*-butyl-2,4-dimethoxy-1,3,2,4-diazadiphosphetidine; *trans-form*, 2-Sulfide, *in* D-00085
1,3-Di-*tert*-butyl-2,4-dimethoxy-1,3,2,4-diazadiphosphetidine; *trans-form*, 2-Selenide, *in* D-00085
1,3-Di-*tert*-butyl-2,4-dimethoxy-1,3,2,4-diazadiphosphetidine; *trans-form*, 2,4-Disulfide, *in* D-00085
1,3-Di-*tert*-butyl-2,4-dimethoxy-1,3,2,4-diazadiphosphetidine; *trans-form*, 2,4-Diselenide, *in* D-00085
1,3-Di-*tert*-butyl-2,4-dimethyl-1,3,2,4-diazadiphosphetidine, D-00086
1,3-Di-*tert*-butyl-2,4-dimethyl-1,3,2,4-diazadiphosphetidine; 2,4-Dioxide, *in* D-00086
1,3-Di-*tert*-butyl-2,4-dimethyl-1,3,2,4-diazadiphosphetidine; 2,4-Disulfide, *in* D-00086
1,3-Di-*tert*-butyl-2,4-dimethyl-1,3,2,4-diazadiphosphetidine; 2,4-Diselenide, *in* D-00086
1,3-Di-*tert*-butyl-2,4-dimethyl-1,3,2,4-diazadiphosphetidine; 2,4-Ditelluride, *in* D-00086
1,3-Di-*tert*-butyl-4,4-dimethyl-1,3,2-diazaphosphasiletidine; 2-Chloro, *in* D-00087
1,3-Di-*tert*-butyl-4,4-dimethyl-1,3,2-diazaphosphasiletidine; 2-Dimethylamino, 2-sulfide, *in* D-00087
1,3-Di-*tert*-butyl-4,4-dimethyl-1,3,2-diazaphosphasiletidine; 2-(Di-*tert*-butyl)amino, 2-sulfide, *in* D-00087

2,4-Dibutyl-1,3,2,4-dithiaphosphetane 2,4-disulfide, *in* B-00614
1,3-Di-*tert*-butyl-2,4,4-trimethyl-1,3-diaza-2-phospha-4-silacyclobutane, *in* D-00087
2,4-Dichloro-1,3-di-*tert*-butyl-1,3,2,4-diazadiphosphetidine; *cis-form*, *in* D-00163
2,4-Dichloro-1,3-di-*tert*-butyl-1,3,2,4-diazadiphosphetidine; *cis-form*, 2-Oxide, *in* D-00163
2,4-Dichloro-1,3-di-*tert*-butyl-1,3,2,4-diazadiphosphetidine; *cis-form*, 2-Sulfide, *in* D-00163
2,4-Dichloro-1,3-di-*tert*-butyl-1,3,2,4-diazadiphosphetidine; *cis-form*, 2-Oxide, 4-sulfide, *in* D-00163
2,4-Dichloro-1,3-di-*tert*-butyl-1,3,2,4-diazadiphosphetidine; *trans-form*, *in* D-00163
2,4-Dichloro-1,3-di-*tert*-butyl-1,3,2,4-diazadiphosphetidine; *trans-form*, 2,4-Dioxide, *in* D-00163
2,2-Dichloro-2,2-dihydro-3-phenyl-Δ^1-1,3,2-diazaphosphetin-4-one, *in* D-00557
2,2-Dichloro-2,2-dihydro-4-trichloromethyl-1,3,2-diazaphosphete, D-00165
2,4-Dichloro-1,3-dimethyl-1,3,2,4-diazadiphosphetidine; 2,4-Dioxide, *in* D-00166
2,4-Dichloro-1,3-dimethyl-1,3,2,4-diazadiphosphetidine; 2,4-Disulfide, *in* D-00166
2,4-Dicyclohexyl-1,3,2,4-dithiadiphosphetane 2,4-disulfide, D-00207
2,4-Diethyl-1,3,2,4-dithiadiphosphetane 2,4-disulfide, *in* E-00152
2,2-Difluoro-2,2-dihydro-1,3-dimethyl-2-phenyl-1,3,2-diazaphosphetidin-4-one, *in* D-00459
2,2-Difluoro-2,2-dihydro-1,2,3-trimethyl-1,3,2-diazaphosphetidin-4-one, *in* D-00459
2,2-Dihydro-1,3-dimethyl-1,3,2-diazaphosphetidin-4-one; 2,2,2-Trifluoro, *in* D-00459
2,2-Dihydro-1,3-dimethyl-1,3,2-diazaphosphetidin-4-one; 2,2,2-Trichloro, *in* D-00459
2,2-Dihydro-2,2-diphenoxy-3-phenyl-Δ^1-1,3,2-diazaphosphetin-4-one, *in* D-00557
2,2-Dihydro-4-methyl-2,2,2,3-tetraphenyl-1,2-oxaphosphetane, D-00531
2,2-Dihydro-1,2-oxaphosphetane; 2,2,2-Tributyl, *in* D-00533
2,2-Dihydro-1,2-oxaphosphetane; 2,2,2-Tri-Ph, *in* D-00533
1,2-Dihydro-1,2,3,4-tetraphenyl-1,2-diphosphete, D-00581
4,4-Dihydro-4,4,4-triphenoxytrioxaphosphetane, D-00603
2,2-Dihydro-2,2,2-triphenyl-4,4-bis(trifluoromethyl)-3-(triphenylphosphoranylidene)-1,2-oxaphosphetane, D-00608
2,2-Dihydro-2,2,3-triphenyl-Δ^1-1,3,2-diazaphosphetin-4-one, *in* D-00557
1,3-Diisopropoxy-2-isopropyltriphosphetane 1,3-dioxide, *in* D-00627
1,3-Dimethyl-1,3,2-diazaphosphetidin-4-one; 2-Methoxy, *in* D-00717
1,3-Dimethyl-1,3,2-diazaphosphetidin-4-one; 2-Dimethylamino, *in* D-00717
2,4-Dimethyl-1,3,2,4-dithiadiphosphetane 2,4-disulfide, D-00752
1,3-Diphenyl-1,3-azaphosphetidine, *in* P-00087
2,4-Diphenyl-1,3,2,4-dithiadiphosphetane 2,4-disulfide, D-01019
3-Ethoxycarbonyl-1-hydroxyphosphetane 1-oxide, *in* H-00184
1-Ethoxy-2,2,3,4,4-pentamethylphosphetane 1-sulfide, *in* H-00169
2-Ethoxy-1-(2-phenylethyl)-1,2-azaphosphetidine 2-oxide, *in* H-00107
Ethyl 1,1,2,3,3-pentamethyltrimethylenephosphinate, *in* H-00168
Ethyl thiodimethylenephosphinate, *in* H-00201
1-Ethylthio-2,2,3,4,4-pentamethylphosphetane 1-oxide, *in* H-00169
1-Fluoro-2,2,3,4,4-pentamethylphosphetane 1-oxide, *in* H-00168
3-Fluoro-1,3-thiaphosphetane; 3-Sulfide, *in* F-00050
N,*N*,2,2,3,4,4-Heptamethylphosphetanamine 1-oxide, *in* P-00021
1,1,2,2,3,4,4-Heptamethylphosphetanium bromide, *in* H-00074
1,1,2,2,3,4,4-Heptamethylphosphetanium iodide, *in* H-00074
2,4,6,7,7,8,8-Heptaphenyl-1-phosphabicyclo[4.2.0]octa-2,4-diene, H-00003
1,2,2,3,4,4-Hexamethyl-1-phenylphosphetanium bromide, *in* P-00020
N,2,2,3,4,4-Hexamethylphosphetanamine 1-oxide, *in* P-00021
1,2,2,3,4,4-Hexamethylphosphetane; Oxide, *in* H-00074
1,2,2,3,4,4-Hexamethylphosphetane; *cis-form*, Oxide, *in* H-00074
1,2,2,3,4,4-Hexamethylphosphetane; *trans-form*, Oxide, *in* H-00074
1,2,2,3,4,4-Hexaphenyl-1,3-diphosphetane, H-00088
3-Hydroxy-2,2-dimethyl-1,3-oxaphosphetane 3-oxide, H-00126
1-Hydroxy-2,2,3,4,4-pentamethylphosphetane 1-oxide, H-00168
1-Hydroxy-2,2,3,4,4-pentamethylphosphetane 1-oxide; Anhydride, *in* H-00168
1-Hydroxy-2,2,3,4,4-pentamethylphosphetane 1-oxide; *cis-form*, *in* H-00168
1-Hydroxy-2,2,3,4,4-pentamethylphosphetane 1-oxide; *trans-form*, *in* H-00168
1-Hydroxy-2,2,3,4,4-pentamethylphosphetane 1-oxide; *trans-form*, Me ester, *in* H-00168
1-Hydroxy-2,2,3,4,4-pentamethylphosphetane 1-oxide; *trans-form*, Et ester, *in* H-00168
1-Hydroxy-2,2,3,4,4-pentamethylphosphetane 1-oxide; *trans-form*, Bromide, *in* H-00168
1-Hydroxy-2,2,3,4,4-pentamethylphosphetane 1-sulfide, H-00169
1-Hydroxy-2,2,3,4,4-pentamethylphosphetane 1-sulfide; *trans-form*, *S*-Me ester, *in* H-00169
1-Hydroxy-3-phosphetanecarboxylic acid 1-oxide, H-00184
3-Hydroxy-1,3-thiaphosphetane 3-oxide, H-00201
5-Isopropyl-2,6-dimethyl-6-phosphabicyclo[3.1.1]hept-2-ene 6-oxide, I-00029
Isopropyl 1,1,2,3,3-pentamethyltrimethylenephosphinate, *in* H-00168
1-Methoxy-3-methoxycarbonylphosphetane 1-oxide, *in* H-00184
2-(4-Methoxyphenyl)-1,3,2-thiazaphosphetidine 2-sulfide, M-00083
2-(4-Methoxyphenyl)-1,3,2-thiazaphosphetidine 2-sulfide; N-Me, *in* M-00083
Methyl 1,1,2,3,3-pentamethyltrimethylenephosphinate, *in* H-00168
Methyl thiodimethylenephosphinate, *in* H-00201
1-Methylthio-2,2,3,4,4-pentamethylphosphetane 1-oxide, *in* H-00169
2,2,3,4,4-Pentamethyl-1-phenylphosphetane; *cis-form*, *in* P-00020
2,2,3,4,4-Pentamethyl-1-phenylphosphetane; *cis-form*, Oxide, *in* P-00020
2,2,3,4,4-Pentamethyl-1-phenylphosphetane; *cis-form*, Sulfide, *in* P-00020
2,2,3,4,4-Pentamethyl-1-phenylphosphetane; *trans-form*, *in* P-00020
2,2,3,4,4-Pentamethyl-1-phenylphosphetane; *trans-form*, Oxide, *in* P-00020
2,2,3,4,4-Pentamethyl-1-phenylphosphetane; *trans-form*, Sulfide, *in* P-00020
2,2,3,4,4-Pentamethylphosphetanamine 1-oxide; *trans-form*, *N*-Benzyl, *in* P-00021
2,2,3,4,4-Pentamethylphosphetane, P-00022
1,1,2,3,3-Pentamethyltrimethylenephosphinic azide, *in* H-00168
N-Phenyl-2,2,3,4,4-pentamethylphosphetanamine 1-oxide, *in* P-00021
Phenyl 1,1,2,3,3-pentamethyltrimethylenephosphinate, *in* H-00168
Phenyl thiodimethylenephosphinate, *in* H-00201
Tetra-*tert*-butylcyclotetraphosphine, T-00012
Tetra-*tert*-butyl-1,2,3,4-diphosphadiarsetane, T-00013
Tetra-*tert*-butylmethyltetraphosphetanium iodide, *in* T-00012
Tetraethylmethyltetraphosphetanium iodide, *in* T-00071
Tetraethyltetraphosphetane, T-00071
1,1,3,3-Tetrakis(dimethylamino)-1,3-diphosph(P^V)ete, T-00176
N,*N*,1,3-Tetrakis(1,1-dimethylethyl)-4,4-dimethyl-1,3-diaza-2-phosphacyclobutane-2-amine, *in* D-00087
Tetrakis(pentafluorophenyl)tetraphosphetane, T-00185

▷Tetrakis(trifluoromethyl)tetraphosphetane, T-00188
N,*N*,4,4-Tetramethyl-1,3-bis(1,1-dimethylethyl)-1,3-diaza-2-phosphacyclobutane-2-amine, *in* D-00087
1,2,3,4-Tetramethyl-1,3,2,4-diazadiphosphetidine; 2,4-Disulfide, *in* T-00194
1,2,3,4-Tetraphenyl-1,3,2,4-diazadiphosphetidine; 2,4-Dioxide, *in* T-00253
1,2,3,4-Tetraphenyl-1,3,2,4-diazadiphosphetidine; 2,4-Disulfide, *in* T-00253
2,3,3,4-Tetraphenyl-1,2-oxaphosphetane 2-oxide, T-00268
Tetraphenyltetraphosphetane, T-00278
Tetraphenyltetraphosphetane; Tetrasulfide, *in* T-00278
2,3,5-Trimethyl-3-phosphatricyclo[4.2.1.0^{2,5}]nonane; Oxide, *in* T-00529
2,3,4-Triphenyl-1,3,2,4-thiazadiphosphetidine; 2,4-Disulfide, *in* T-00687
Triphosphetan-4-one; 1,3-Di-*tert*-butyl, 2-Me, *in* T-00703
Triphosphetan-4-one; 1,2,3-Tri-*tert*-butyl, *in* T-00703
Triphosphetan-4-one; 1,3-Di-*tert*-butyl, 2-Ph, *in* T-00703

1 P-contg. ring; 5-membered

2-Azido-1,3,2-benzodioxaphosphole, A-00148
1*H*-1,3-Benzazaphosphole, B-00005
1*H*-1,3-Benzazaphosphole; 1-Me, *in* B-00005
1,3-Benzothiaphosphole, B-00025
1,3,2-Benzoxathiaphosphole; 2-Me, *in* B-00026
1,3,2-Benzoxathiaphosphole; 2-Ph, *in* B-00026
1,3,2-Benzoxathiaphosphole; 2-Chloro, *in* B-00026
1,3,2-Benzoxathiaphosphole; 2-Bromo, *in* B-00026
1-Benzyl-2,5-dihydro-3,4-dimethyl-1-phenyl-1*H*-phospholium bromide, *in* D-00468
1-Benzyl-2,5-dihydro-3-methyl-1-phenyl-1*H*-phospholium bromide, *in* D-00528
1-Benzyl-2,5-dihydro-1*H*-phosphole, *in* D-00572
1-Benzyl-2,5-dihydro-1,3,4-trimethyl-1*H*-phospholium bromide, *in* D-00600
1-Benzyl-3,4-dimethyl-1-phenyl-1*H*-phospholium bromide, *in* D-00807
1-Benzyl-3-methyl-1-phenylphospholanium bromide, *in* M-00229
1-Benzyl-3-methyl-1-phenylphospholanium iodide, *in* M-00229
1-Benzyl-1-methylphospholanium bromide, *in* M-00280
5-Benzyl-5-phenyl-5*H*-dibenzophospholium chloride, *in* P-00110
1-Benzyl-1-phenylphospholanium bromide, *in* P-00216
8-Benzyl-8-phenyl-8-phosphoniabicyclo[3.2.1]octan-3-one chloride, *in* P-00199
S-Benzyl *O*,*O*-propylene phosphorodithioate, *in* M-00016
3-Benzyl-2,2,5-trimethyl-3-phenyl-1,3-oxaphospholanium bromide, *in* T-00522
2,2′-Bi(1,3,2-benzodithiaphosphole), B-00074
1,1′-Biphospholane, B-00084
1,1′-Biphospholane; Disulfide, *in* B-00084
Bis(dimethylvinylene) pyrophosphate, *in* O-00088
3,4-Bis(trifluoromethyl)-1,2,5-thiadiphosphole, B-00425
2-Bromo-1,3,2-benzodioxaphosphole, B-00464
1-Bromo-2,5-dihydro-3,4-dimethyl-1*H*-phosphole, B-00469
1-Bromo-2,5-dihydro-3,4-dimethyl-1*H*-phosphole; 1-Oxide, *in* B-00469
1-Bromo-2,5-dihydro-3,4-dimethyl-1*H*-phosphole; 1-Sulfide, *in* B-00469
1-Bromo-2,5-dihydro-3,4-dimethyl-1*H*-phosphole 1-oxide, *in* D-00493
1-Bromo-2,5-dihydro-1*H*-phosphole, *in* D-00572
2-Bromo-4,5-dimethyl-1,3,2-dioxaphospholane, B-00470
2-Bromo-4,5-dimethyl-1,3,2-dioxaphospholane; (2α,4β,5β)-*form*, 2-Sulfide, *in* B-00470
2-Bromo-4-methyl-1,3,2-dioxaphospholane, B-00473
2-*tert*-Butoxy-4,5-dimethyl-1,3,2-dioxaphospholane, B-00533
2-*tert*-Butoxy-1,3,2-dioxaphospholane, B-00535
2-*tert*-Butyl-1,3-benzoxaphosphole, B-00544
tert-Butyl dimethylvinylene phosphate, *in* H-00123
O,*O*-2,3-Butylene bromothiophosphate, *in* B-00470
2,3-Butylene chlorophosphate, *in* C-00056
O,*O*-2,3-Butylene chlorothiophosphate, *in* C-00056
O,*O*-2,3-Butylene dimethylphosphoramidothioate, *in* D-00725
2,3-Butylene ethyl phosphate, *in* E-00031
O,*O*-Butylene *O*-ethyl thiophosphate, *in* E-00031
O,*O*-2,3-Butylene fluorothiophosphate, *in* F-00015
2,3-Butylene methyl phosphate, *in* M-00033
2,3-Butylene methylphosphonate, *in* T-00511
O,*O*-2,3-Butylene *O*-methyl thiophosphate, *in* M-00033
O,*O*-2,3-Butylene *S*-methyl thiophosphate, *in* D-00765
O,*O*-2,3-Butylene *S*-phenyl dithiophosphate, *in* D-00824
2,3-Butylene phenylphosphonate, *in* D-00794
O,*O*-2,3-Butylene *O*-trimethylsilyl phosphorothioate, *in* D-00930
tert-Butyl ethylene phosphate, *in* B-00535
5-*tert*-Butyl-2,2,4,5-tetrahydro-2,2,4-triisopropoxy-1,2,4-oxadiphosphole, B-00637
Catechol cyclic phosphoric acid anhydride, *in* O-00087
2-Chloro-1,3,2-benzodioxaphosphole, C-00021
2-Chloro-1,3,2-benzodithiaphosphole, C-00023
5-Chloro-5*H*-dibenzophosphole, C-00034
5-Chloro-5*H*-dibenzophosphole; 5-Oxide, *in* C-00034
5-Chloro-5*H*-dibenzophosphole; 5-Sulfide, *in* C-00034
2-Chloro-2,3-dihydro-3,5-dimethyl-1,3,4,2-oxadiazaphosphole, C-00039
1-Chloro-2,5-dihydro-3,4-dimethyl-1*H*-phosphole, C-00041
1-Chloro-2,5-dihydro-3,4-dimethyl-1*H*-phosphole; 1-Oxide, *in* C-00041
1-Chloro-2,5-dihydro-3,4-dimethyl-1*H*-phosphole; 1-Sulfide, *in* C-00041
1-Chloro-2,5-dihydro-3,4-dimethyl-1*H*-phosphole 1-oxide, *in* D-00493
2-Chloro-2,3-dihydro-3,5-diphenyl-1,3,4,2-oxadiazaphosphole, C-00042
2-Chloro-2,3-dihydro-3,5-diphenyl-1,3,4,2-oxadiazaphosphole; 2-Oxide, *in* C-00042
2-Chloro-2,3-dihydro-3,5-diphenyl-1,3,4,2-oxadiazaphosphole; 2-Sulfide, *in* C-00042
2-Chloro-2,3-dihydro-3-methyl-1,3,2-benzothiazaphosphole, C-00043
2-Chloro-2,3-dihydro-3-methyl-1,3,2-benzothiazaphosphole; 2-Sulfide, *in* C-00043
2-Chloro-2,3-dihydro-5-methyl-3-phenyl-1,3,4,2-oxadiazaphosphole, C-00044
2-Chloro-2,3-dihydro-5-methyl-3-phenyl-1,3,4,2-oxadiazaphosphole; 2-Oxide, *in* C-00044
2-Chloro-2,3-dihydro-5-methyl-3-phenyl-1,3,4,2-oxadiazaphosphole; 2-Sulfide, *in* C-00044
1-Chloro-2,3-dihydro-4-methyl-1*H*-phosphole, C-00045
1-Chloro-2,3-dihydro-4-methyl-1*H*-phosphole; 1-Oxide, *in* C-00045
1-Chloro-2,5-dihydro-2-methyl-1*H*-phosphole, C-00046
1-Chloro-2,5-dihydro-2-methyl-1*H*-phosphole; 1-Oxide, *in* C-00046
1-Chloro-2,5-dihydro-2-methyl-1*H*-phosphole; 1-Sulfide, *in* C-00046
1-Chloro-2,5-dihydro-3-methyl-1*H*-phosphole, C-00047
1-Chloro-2,5-dihydro-3-methyl-1*H*-phosphole; 1-Oxide, *in* C-00047
1-Chloro-2,5-dihydro-3-methyl-1*H*-phosphole; 1-Sulfide, *in* C-00047
1-Chloro-2,3-dihydro-1*H*-phosphole, C-00051
1-Chloro-2,3-dihydro-1*H*-phosphole; 1-Oxide, *in* C-00051
1-Chloro-2,3-dihydro-1*H*-phosphole; 1-Sulfide, *in* C-00051
1-Chloro-2,5-dihydro-1*H*-phosphole; 1-Oxide, *in* C-00052
1-Chloro-2,5-dihydro-1*H*-phosphole; 1-Sulfide, *in* C-00052
2-Chloro-2,3-dihydro-3,3,5-trimethyl-1,2-oxaphosphole; 2-Oxide, *in* C-00054
2-Chloro-1,3-dimethyl-1,3,2-diazaphospholidine, C-00055
2-Chloro-1,3-dimethyl-1,3,2-diazaphospholidine; 2-Oxide, *in* C-00055
2-Chloro-1,3-dimethyl-1,3,2-diazaphospholidine; 2-Sulfide, *in* C-00055
2-Chloro-4,5-dimethyl-1,3,2-dioxaphospholane; (4*R*,5*R*)-*form*, *in* C-00056
2-Chloro-4,5-dimethyl-1,3,2-dioxaphospholane; (4*RS*,5*RS*)-*form*, *in* C-00056
2-Chloro-4,5-dimethyl-1,3,2-dioxaphospholane; (4*RS*,5*SR*)-*form*, *in* C-00056
2-Chloro-4,5-dimethyl-1,3,2-dioxaphospholane; (4*RS*,5*SR*)-*form*, 2-Sulfide, *in* C-00056
2-Chloro-4,5-dimethyl-1,3,2-dioxaphosphole, C-00057

2-Chloro-3,4-dimethyl-5-phenyl-1,3,2-oxazaphospholidine; (2*R*,4*S*,5*R*)-*form*, *in* C-00061
2-Chloro-3,4-dimethyl-5-phenyl-1,3,2-oxazaphospholidine; (2*R*,4*S*,5*R*)-*form*, 2-Oxide, *in* C-00061
2-Chloro-3,4-dimethyl-5-phenyl-1,3,2-oxazaphospholidine; (2*R*,4*S*,5*R*)-*form*, 2-Sulfide, *in* C-00061
2-Chloro-3,4-dimethyl-5-phenyl-1,3,2-oxazaphospholidine; (2*S*,4*S*,5*R*)-*form*, 2-Oxide, *in* C-00061
2-Chloro-3,4-dimethyl-5-phenyl-1,3,2-oxazaphospholidine; (2*S*,4*S*,5*R*)-*form*, 2-Sulfide, *in* C-00061
2-Chloro-1,3,2-dioxaphospholane, C-00063
2-Chloro-1,3,2-dioxaphospholane 2-oxide, C-00064
2-Chloro-1,3,2-dioxaphospholane 2-sulfide, C-00065
2-Chloro-4,5-diphenyl-1,3,2-dioxaphospholane; (2α,4β,5β)-*form*, 2-Oxide, *in* C-00069
2-Chloro-4,5-diphenyl-1,3,2-dioxaphosphole, C-00070
2-Chloro-1,3,2-dithiaphospholane, C-00071
2-Chloro-4-methyl-1,3,2-dioxaphospholane, C-00092
2-Chloro-4-methyl-1,3,2-dithiaphospholane, *in* M-00136
2-Chloro-3-methyl-1,3,2-oxazaphospholidine, C-00102
2-Chloronaphtho[1,2-*d*]-1,3,2-dioxaphosphole, C-00118
2-Chloronaphtho[2,3-*d*]-1,3,2-dioxaphosphole, C-00119
2-Chloro-1,3,2-oxathiaphospholane, *in* O-00056
▷2-Chloro-5-phenyl-1,3,4,2-dioxazaphosphole, C-00126
2-Chloro-5-phenyl-1,3,4,2-dioxazaphosphole; 2-Oxide, *in* C-00126
2-Chloro-3-phenyl-1,3,2-oxazaphospholidine, C-00138
2-Chloro-3-phenyl-1,3,2-oxazaphospholidine; 2-Oxide, *in* C-00138
2-Chloro-3-phenyl-1,3,2-oxazaphospholidine; 2-Sulfide, *in* C-00138
1-Chlorophospholane, C-00167
2-Chloro-4,4,5,5-tetrakis(trifluoromethyl)-1,3,2-dioxaphospholane, C-00184
2-Chloro-4,4,5,5-tetramethyl-1,3,2-dioxaphospholane, C-00185
4-Cyano-2,2-dihydro-3,5-diphenyl-3*H*-1,2-azaphosphole, *in* D-00461
4-Cyclohexyl-1,3-dimethyl-1,2,4-diazaphospholidine, *in* D-00718
5*H*-Dibenzophosphole, D-00040
3,5-Di-*tert*-butyl-2-chloro-2,3-dihydro-1,3,2-oxazaphosphole, D-00078
3,5-Di-*tert*-butyl-2-chloro-2,3-dihydro-1,3,2-oxazaphosphole; 2-Oxide, *in* D-00078
3,5-Di-*tert*-butyl-2-chloro-2,3-dihydro-1,3,2-oxazaphosphole; 2-Sulfide, *in* D-00078
1,3-Dibutyl-2,3-dihydro-4,5-dimethyl-1*H*-1,3,2-diazaphosphole 2-oxide; 2-Chloro, *in* D-00081
1,3-Dibutyl-2,3-dihydro-4,5-dimethyl-1*H*-1,3,2-diazaphosphole 2-oxide; 2-Ethoxy, *in* D-00081
3,5-Di-*tert*-butyl-2,5-dihydro-1,2-oxaphosphole 2-oxide; 2-Chloro, *in* D-00082
3,5-Di-*tert*-butyl-2,5-dihydro-1,2-oxaphosphole 2-oxide; 2-Bromo, *in* D-00082
3,5-Di-*tert*-butyl-2,5-dihydro-1,2-oxaphosphole 2-oxide; 2-Hydroxy, *in* D-00082
3,5-Di-*tert*-butyl-2,3-dihydro-1,3,2-oxazaphosphole; 2-Methoxy, *in* D-00083
3,5-Di-*tert*-butyl-2,3-dihydro-1,3,2-oxazaphosphole; 2-Ethoxy, *in* D-00083
3,5-Di-*tert*-butyl-2,3-dihydro-1,3,2-oxazaphosphole; 2-Ethoxy, 2-oxide, *in* D-00083
3,4-Di-*tert*-butyl-2-(4-methylphenyl)-1,3,4,2-thiadiazaphospholidin-5-one 2-sulfide; (2α,3β,4α)-*form*, *in* D-00097
3,4-Di-*tert*-butyl-2-(4-methylphenyl)-1,3,4,2-thiadiazaphospholidin-5-one 2-sulfide; (2α,3α,4β)-*form*, *in* D-00097
▷2,5-Dichloro-3,4-dimethyl-1,3,4,2,5-thiadiazadiphospholidine, D-00167
4,5-Diethoxy-2*H*-1,3,2-diazaphosphole; 2-Chloro, *in* D-00228
4,5-Diethoxy-2*H*-1,3,2-diazaphosphole; 2-Phenoxy, *in* D-00228
4,5-Diethoxy-1,3-dimethyl-2-phenyl-2*H*-1,3,2-diazaphosphole, D-00230
4,5-Diethoxy-1,3-dimethyl-2-phenyl-2*H*-1,3,2-diazaphosphole; 2-Sulfide, *in* D-00230
4,5-Diethoxy-2-methyl-2*H*-1,3,2-diazaphosphole, D-00235
4,5-Diethoxy-2-methyl-2*H*-1,3,2-diazaphosphole; 2-Oxide, *in* D-00235
4,5-Diethoxy-2-methyl-2*H*-1,3,2-diazaphosphole; 2-Sulfide, *in* D-00235
1,8-Diethoxyoctahydro-4,7-phosphinidene-1*H*-phosphindole 1,8-dioxide, *in* D-00625
4,4-Diethoxy-2,4,4,5-tetrahydro-2-oxo-5-(4-nitrophenyl)-1,3,4-oxazaphosphole, D-00262
1,8-Diethoxy-3*a*,4,7,7*a*-tetrahydro-4,7-phosphinidene-1*H*-phosphinidole 1,8-dioxide, *in* D-00625
1-Diethylamino-2,5-dihydro-2-methyl-1*H*-phosphole 1-oxide, *in* D-00498
2-Diethylamino-1,3,2-dioxaphospholan-4,5-dicarboxylic acid; (4*R*,5*R*)-*form*, Di-Me ester, *in* D-00265
2-Diethylamino-1,3,2-dioxaphospholan-4,5-dicarboxylic acid; (4*R*,5*R*)-*form*, Di-Et ester, *in* D-00265
2-Diethylamino-1,3,2-dioxaphospholan-4,5-dicarboxylic acid; (4*R*,5*R*)-*form*, Di-Et ester, 2-sulfide, *in* D-00265
2-Diethylamino-1,3,2-dioxaphospholan-4,5-dicarboxylic acid; (4*R*,5*R*)-*form*, Diisopropyl ester, *in* D-00265
2-Diethylamino-1,3,2-dioxaphospholan-4,5-dicarboxylic acid; (4*RS*,5*RS*)-*form*, Di-Me ester, *in* D-00265
2-Diethylamino-1,3,2-dioxaphospholan-4,5-dicarboxylic acid; (4*RS*,5*RS*)-*form*, Di-Et ester, *in* D-00265
2-Diethylamino-1,3,2-dioxaphospholan-4,5-dicarboxylic acid; (4*RS*,5*RS*)-*form*, Di-Et ester, 2-sulfide, *in* D-00265
2-Diethylamino-1,3,2-dioxaphospholan-4,5-dicarboxylic acid; (4*RS*,5*RS*)-*form*, Diisopropyl ester, *in* D-00265
2-Diethylaminonaphtho[1,2-*d*]-1,3,2-dioxaphosphole, D-00266
2-Diethylamino-3-phenyl-1,3,2-oxazaphospholidine, D-00269
2-Diethylamino-3-phenyl-1,3,2-oxazaphospholidine; 2-Oxide, *in* D-00269
2-Diethylamino-3-phenyl-1,3,2-oxazaphospholidine; 2-Sulfide, *in* D-00269
2-Diethylamino-3,5,5-trimethyl-1,2-oxaphospholan-3-ol 2-oxide, *in* T-00519
N,*N*-Diethyl-2,3-dihydro-5-methyl-3-phenyl-1,3,4,2-thiadiazaphosphole-2(3*H*)-amine 2-sulfide, *in* A-00072
2,3-Diethyl-2,3-diphosphabicyclo[2.2.1]hept-5-ene, D-00287
2,3-Diethyl-2,3-diphosphabicyclo[2.2.1]hept-5-ene; Disulfide, *in* D-00287
N,*N*-Diethyl *S*,*S*-ethylene phosphoramidodithioite, *in* D-01243
2,5-Difluoro-1-methyl-1,2,5-azadiphospholidine, *in* M-00089
2,2-Dihydro-1,3,2-benzodioxaphosphole; 2,2,2-Trimethoxy, *in* D-00439
2,2-Dihydro-1,3,2-benzodioxaphosphole; 2,2-Dimethoxy, 2-Ph, *in* D-00439
2,2-Dihydro-1,3,2-benzodioxaphosphole; 2-Methoxy, 2,2-di-Ph, *in* D-00439
2,2-Dihydro-1,3,2-benzodioxaphosphole; 2,2,2-Tri-Ph, *in* D-00439
2,3-Dihydro-1*H*-benzo[*e*]phosphindole 3-oxide; 3-Hydroxy, *in* D-00442
2,3-Dihydro-1*H*-benzo[*e*]phosphindole 3-oxide; 3-Ph, *in* D-00442
2,2-Dihydro-1,3,2-benzoxazaphosphole; 2,2-Dimethoxy, *in* D-00444
2,2-Dihydro-1,3,2-benzoxazaphosphole; 2,2-Diethoxy, *in* D-00444
2,2-Dihydro-1,3,2-benzoxazaphosphole; 2,2-Diphenoxy, *in* D-00444
2,2-Dihydro-1,3,2-benzoxazaphosphole; 2,2-Di-Ph, *in* D-00444
2,2-Dihydro-1,3,2-benzoxazaphosphole; 2,2-Dichloro, *in* D-00444
2,2-Dihydro-1,3,2-benzoxazaphosphole; 2,2-Bis(dimethylamino), *in* D-00444
2,2-Dihydro-3,4-bis(methoxycarbonyl)-2,2-dimethyl-5,5-diphenyl-5*H*-1,2-azaphosphole, D-00445

1,3-Dihydro-1,3-dihydroxy-1,2,3-benziodoxaphosphole 3-oxide, *in* D-00488
2,3-Dihydro-2-dimethylamino-5-methyl-3-phenyl-1,3,4,2-oxadiazaphosphole; 2-Oxide, *in* D-00455
2,3-Dihydro-2-dimethylamino-5-methyl-3-phenyl-1,3,4,2-oxadiazaphosphole; 2-Sulfide, *in* D-00455
2,5-Dihydro-1-dimethylamino-3-methyl-1*H*-phosphole, *in* D-00530
2,3-Dihydro-1,3-dimethyl-1*H*-1,3,2-benzodiazaphosphole; 2-Chloro, *in* D-00456
2,3-Dihydro-1,3-dimethyl-1*H*-1,3,2-benzodiazaphosphole; 2-Diisopropylamino, *in* D-00456
2,3-Dihydro-1,1-dimethyl-1*H*-benzo[*g*]phosphindolinium iodide, *in* D-00518
2,3-Dihydro-1,3-dimethyl-1,3-diphenyl-1*H*-1,3-benzodiphospholium dibromide, *in* D-00476
2,5-Dihydro-5,5-dimethyl-1,2-oxaphosphole 2-oxide; 2-Chloro, *in* D-00464
2,5-Dihydro-5,5-dimethyl-1,2-oxaphosphole 2-oxide; 2-Hydroxy, *in* D-00464
2,5-Dihydro-5,5-dimethyl-1,2-oxaphosphole 2-oxide; 2-Methoxy, *in* D-00464
2,5-Dihydro-5,5-dimethyl-1,2-oxaphosphole 2-oxide; 2-Ethoxy, *in* D-00464
2,5-Dihydro-5,5-dimethyl-1,2-oxaphosphole 2-oxide; 2-Phenoxy, *in* D-00464
2,5-Dihydro-5,5-dimethyl-1,2-oxaphosphole 2-oxide; 2-Diethylamino, *in* D-00464
5,5-Dihydro-4,4-dimethyl-2,3,5,5,5-pentaphenyl-1,2,5-oxazaphospholidine, D-00465
2,5-Dihydro-3,4-dimethyl-1-phenoxy-1*H*-phosphole 1-oxide, *in* D-00493
2,3-Dihydro-1,3-dimethyl-2-phenyl-1*H*-benzodiazaphosphole, D-00467
2,3-Dihydro-1,3-dimethyl-2-phenyl-1*H*-benzodiazaphosphole; 2-Sulfide, *in* D-00467
2,5-Dihydro-3,4-dimethyl-1-phenyl-1*H*-phosphole, D-00468
2,5-Dihydro-3,4-dimethyl-1-phenyl-1*H*-phosphole; 1-Oxide, *in* D-00468
2,5-Dihydro-3,4-dimethyl-1-phenyl-1*H*-phosphole; 1-Sulfide, *in* D-00468
2,5-Dihydro-3,4-dimethyl-1-phenyl-1*H*-phosphole; 1-Selenide, *in* D-00468
2,3-Dihydro-3,5-dimethyl-1,3,4,2-thiadiazaphosphole 2-sulfide; 2-Chloro, *in* D-00471
2,3-Dihydro-3,5-dimethyl-1,3,4,2-thiadiazaphosphole 2-sulfide; 2-Ethoxy, *in* D-00471
2,3-Dihydro-1,3-diphenyl-1*H*-1,3-benzodiphosphole, D-00476
2,3-Dihydro-1,3-diphenyl-1*H*-1,3-benzodiphosphole; (1*RS*,3*SR*)-*form*, B,2MeBr, *in* D-00476
2,3-Dihydro-2,3-diphenyl-1,3,2-benzoxazaphosphole; 2-Oxide, *in* D-00478
2,3-Dihydro-2,2-diphenyl-1*H*-isophosphindolium bromide, *in* D-00562
2,3-Dihydro-2,2-diphenyl-1*H*-isophosphindolium perchlorate, *in* D-00562
2,2-Dihydro-4,5-diphenyl-1,3,2-oxathiaphosphole; 2,2,2-Trimethoxy, *in* D-00480
1,1-Dihydro-1,1-diphenyl-3*H*-phosphindole, D-00483
2,2-Dihydro-4-ethoxycarbonyl-2,2-dimethyl-3,5-diphenyl-3*H*-1,2-azaphosphole, *in* D-00461
2,3-Dihydro-1-ethoxy-4-methyl-1*H*-phosphole 1-oxide, *in* D-00496
2,3-Dihydro-2-ethoxy-1,3,5-triphenyl-1*H*-1,2-azaphosphole 2-oxide, *in* D-00508
1,1-Dihydro-1,1,3,3,4,4-hexamethyl-1-phospha-3,4-disilacyclopent-1-ene, D-00486
1,3-Dihydro-1-hydroxy-2,1-benzoxaphosphole 1-oxide, D-00490
1,3-Dihydro-1-hydroxy-2,1-benzoxaphosphole 1-oxide; 1-Ethoxy (Et ester), *in* D-00490
▷2,5-Dihydro-1-hydroxy-3,4-dimethyl-1*H*-phosphole 1-oxide, D-00493
2,3-Dihydro-2-hydroxy-1*H*-isophosphindole 2-oxide, D-00494
1,3-Dihydro-1-hydroxy-3-methyl-1,2,3-benzodoxaphosphole 3-oxide, *in* D-00488
▷2,3-Dihydro-1-hydroxy-4-methyl-1*H*-phosphole 1-oxide, D-00496
2,5-Dihydro-2-hydroxy-1,2-oxaphosphole 2-oxide, D-00500
2,5-Dihydro-2-hydroxy-1,2-oxaphosphole 2-oxide; 2-Butoxy, *in* D-00500
1,3-Dihydro-1-hydroxy-3-phenyl-1,2,3-benziodoxaphosphole 3-oxide, *in* D-00488
2,3-Dihydro-1-hydroxy-1*H*-phosphole 1-oxide, D-00506
2,5-Dihydro-1-hydroxy-1*H*-phosphole 1-oxide, D-00507
2,3-Dihydro-1*H*-isophosphindole, D-00509
2,3-Dihydro-2-mercapto-1,3,2-benzoxazaphosphole 2-sulfide, D-00511
2,5-Dihydro-1-mercaptophosphole 1-sulfide; 2-Ethylthio (Et ester), *in* D-00512
2,3-Dihydro-2-methoxy-1,2-benzoxaphosphole 2-oxide, *in* H-00119
2,5-Dihydro-1-methoxy-3,4-dimethyl-1*H*-phosphole 1-oxide, *in* D-00493
2,3-Dihydro-2-methoxy-1*H*-isophosphindole 2-oxide, *in* D-00494
2,3-Dihydro-2-methoxy-5-methyl-3-phenyl-1,3,4,2-oxadiazaphosphole; 2-Sulfide, *in* D-00513
2,5-Dihydro-1-methoxy-3-methyl-1*H*-phosphole, *in* D-00530
2,3-Dihydro-1-methoxy-4-methyl-1*H*-phosphole 1-oxide, *in* D-00496
▷2,5-Dihydro-1-methoxy-3-methyl-1*H*-phosphole 1-oxide, *in* D-00499
2,3-Dihydro-2-methoxy-5-methyl-1,2-thiaphosphole; 2-Oxide, *in* D-00514
2,3-Dihydro-1-methoxy-1*H*-phosphindole 1-oxide, *in* H-00120
▷2,3-Dihydro-1-methoxy-1*H*-phosphole 1-oxide, *in* D-00506
▷2,5-Dihydro-1-methoxy-1*H*-phosphole 1-oxide, *in* D-00507
2,3-Dihydro-1-methyl-1*H*-benzo[*g*]phosphindole, D-00518
2,3-Dihydro-1-methyl-1*H*-benzo[*g*]phosphindole; 1-Oxide, *in* D-00518
4,5-Dihydro-3-methyl-3*H*-benzo[*e*]phosphindole, *in* D-00441
2,3-Dihydro-2-methyl-1,3,2-benzothiazaphosphole, D-00519
2,3-Dihydro-5-methyl-*N*,3-diphenyl-1,3,4,2-thiadiazaphosphole-2(3*H*)amine 2-sulfide, *in* A-00072
2,5-Dihydro-1-(1-methylethoxy)-1*H*-phosphole 1-oxide, *in* D-00507
2,3-Dihydro-5-methyl-1,2-oxaphosphole 2-oxide; 2-Chloro, *in* D-00525
2,3-Dihydro-5-methyl-1,2-oxaphosphole 2-oxide; 2-Methoxy, *in* D-00525
2,3-Dihydro-5-methyl-1,2-oxaphosphole 2-oxide; 2-Diethylamino, *in* D-00525
2,3-Dihydro-4-methyl-1-phenoxy-1*H*-phosphole 1-oxide, *in* D-00496
2,5-Dihydro-2-methyl-1-phenoxy-1*H*-phosphole 1-oxide, *in* D-00498
2,5-Dihydro-3-methyl-2-phenoxy-1*H*-phosphole 1-oxide, *in* D-00499
2,3-Dihydro-2-methyl-2-phenyl-1*H*-isophosphindolium iodide, *in* D-00562
2,3-Dihydro-5-methyl-3-phenyl-2-phenylamino-1,3,4,2-oxadiazaphosphole, D-00527
2,3-Dihydro-5-methyl-3-phenyl-2-phenylamino-1,3,4,2-oxadiazaphosphole; 2-Oxide, *in* D-00527
2,3-Dihydro-5-methyl-3-phenyl-2-phenylamino-1,3,4,2-oxadiazaphosphole; 2-Sulfide, *in* D-00527
2,5-Dihydro-3-methyl-1-phenyl-1*H*-phosphole, D-00528
2,5-Dihydro-3-methyl-1-phenyl-1*H*-phosphole; Oxide, *in* D-00528
2,3-Dihydro-1-methyl-1-phenyl-1*H*-phospholium iodide, *in* D-00566
2,3-Dihydro-5-methyl-3-phenyl-1,3,4,2-thiadiazaphosphole 2-sulfide; 2-Chloro, *in* D-00529
2,3-Dihydro-5-methyl-3-phenyl-1,3,4,2-thiadiazaphosphole 2-sulfide; 2-Ethoxy, *in* D-00529
2,5-Dihydro-3-methyl-1*H*-phosphole, D-00530
2,2-Dihydro-1,2-oxaphospholane; 2,2,2-Tri-Me, *in* D-00534
2,2-Dihydro-1,2-oxaphospholane; 2,2,2-Tri-Et, *in* D-00534
2,2-Dihydro-1,2-oxaphospholane; 2,2,2-Tri-Ph, *in* D-00534
▷2,3-Dihydro-1,2,3,4,5-pentakis(trifluoromethyl)-1*H*-1,2,3-triphosphole, D-00536
2,2-Dihydro-2,2,2,4,5-pentaphenyl-1,3,2-dioxaphospholane, D-00537
2,3-Dihydro-1,2,3,4,5-pentaphenyl-1*H*-1,2,3-triphosphole, D-00539

2,3-Dihydro-1,2,3,4,5-pentaphenyl-1*H*-1,2,3-triphosphole; 1,3-Disulfide, *in* D-00539
2,2-Dihydrophenanthro[9,10-*d*]-1,3,2-dioxaphosphole; 2,2,2-Trimethoxy, *in* D-00541
2,2-Dihydrophenanthro[9,10-*d*]-1,3,2-dioxaphosphole; 2,2,2-Triisopropoxy, *in* D-00541
2,2-Dihydrophenanthro[9,10-*d*]-1,3,2-dioxaphosphole; 2,2,2-Triphenoxy, *in* D-00541
2,2-Dihydrophenanthro[9,10-*d*]-1,3,2-dioxaphosphole; 2,2,2-Tri-Ph, *in* D-00541
1,3-Dihydro-1-phenoxy-2,1-benzoxaphosphole, D-00546
1,3-Dihydro-1-phenoxy-2,1-benzoxaphosphole; 1-Sulfide, *in* D-00546
2,3-Dihydro-1-phenoxy-1*H*-phosphole 1-oxide, *in* D-00506
2,5-Dihydro-1-phenoxy-1*H*-phosphole 1-oxide, *in* D-00507
4,5-Dihydro-3-phenyl-3*H*-1,3-azaphosphole, D-00549
2,3-Dihydro-1-phenyl-1*H*-benzo[*g*]phosphindole; 1-Oxide, *in* D-00551
2,3-Dihydro-1-phenyl-1*H*-benzo[*g*]phosphindole; 4,5-Dihydro, 1-Oxide, *in* D-00551
4,5-Dihydro-3-phenyl-3*H*-benzo[*e*]phosphindole, *in* D-00441
1,3-Dihydro-1-phenyl-2,1-benzoxaphosphole, D-00553
1,3-Dihydro-1-phenyl-2,1-benzoxaphosphole; 1-Oxide, *in* D-00553
1,3-Dihydro-1-phenyl-2,1-benzoxaphosphole; 1-Sulfide, *in* D-00553
2,3-Dihydro-2-phenyl-1*H*-dibenz[*e,g*]isophosphindole; 2-Oxide, *in* D-00558
2,3-Dihydro-2-phenyl-1*H*-isophosphindole, D-00562
2,3-Dihydro-2-phenyl-1*H*-isophosphindole; Oxide, *in* D-00562
2,3-Dihydro-1-phenyl-1*H*-phosphole, D-00566
2,3-Dihydro-1-phenyl-1*H*-phosphole; 1-Oxide, *in* D-00566
2,3-Dihydro-1-phenyl-1*H*-phosphole; 1-Sulfide, *in* D-00566
2,5-Dihydro-1-phenyl-1*H*-phosphole; 1-Oxide, *in* D-00567
2,3-Dihydro-1*H*-phosphindole 1-oxide; 1-Ph, *in* D-00570
2,3-Dihydro-1*H*-phosphindole 1-oxide; 1-Benzyl, *in* D-00570
2,3-Dihydro-1*H*-phosphole, D-00571
2,5-Dihydro-1*H*-phosphole, D-00572
2,5-Dihydro-1,1,3,4-tetramethyl-1*H*-phospholium bromide, *in* D-00600
2,2-Dihydro-4,4,5,5-tetramethyl-2,2,2-triphenyl-1,3,2-dioxaphospholane, D-00578
3,4-Dihydro-2,3,4,5-tetraphenyl-2*H*-1,2,3-diazaphosphole; (3*RS*,4*RS*)-*form*, *in* D-00579
3,4-Dihydro-2,3,4,5-tetraphenyl-2*H*-1,2,3-diazaphosphole; (3*RS*,4*SR*)-*form*, *in* D-00579
3,4-Dihydro-2,3,4,5-tetraphenyl-2*H*-1,3,2-diazaphosphole; 2-Oxide, *in* D-00580
2,5-Dihydro-2,3,4,5-tetraphenyl-1,2,5-oxadiphosphole 2,5-dioxide, D-00583
2,5-Dihydro-2,3,4,5-tetraphenyl-1,2,5-thiadiphosphole; 2,5-Disulfide, *in* D-00585
2,2-Dihydro-2,2,2-trichloro-4,4,5,5-tetrakis(trifluoromethyl)-1,3,2-dioxaphospholane, D-00589
2,2-Dihydro-2,2,2-triethoxy-4-[(4-methylphenyl)sulfonyl]-1,4,2-oxazaphospholidine, D-00590
2,2-Dihydro-2,2,2-trimethoxy-3,5-bis(pentafluorophenyl)-1,4,2-dioxaphospholane, D-00591
2,2-Dihydro-2,2,2-trimethoxy-4,5-bis(trifluoromethyl)-1,3,2-dithiaphosphole, D-00592
2,2-Dihydro-2,2,2-trimethoxy-4,5-dimethyl-1,3,2-dioxaphosphole, D-00593
2,2-Dihydro-2,2,2-trimethoxynaphtho[1,2-*d*]-1,3,2-dioxaphosphole, D-00594
2,2-Dihydro-2,2,2-trimethoxy-4,4,5,5-tetrakis(trifluoromethyl)-1,3,2-dioxaphospholane, D-00595
▷2,3-Dihydro-3,3,5-trimethyl-1,2-oxaphosphole 2-oxide; 2-Methoxy, *in* D-00596
2,3-Dihydro-3,3,5-trimethyl-1,2-oxaphosphole 2-oxide; 2-Ethoxy, *in* D-00596
2,3-Dihydro-3,3,5-trimethyl-1,2-oxaphosphole 2-oxide; 2-Phenoxy, *in* D-00596
2,3-Dihydro-3,3,5-trimethyl-1,2-oxaphosphole 2-oxide; 2-Me, *in* D-00596
2,3-Dihydro-3,3,5-trimethyl-1,2-oxaphosphole 2-oxide; 2-Ph, *in* D-00596
2,3-Dihydro-3,3,5-trimethyl-1,2-oxaphosphole 2-oxide; 2-Diethylamino, *in* D-00596
2,5-Dihydro-1,3,4-trimethyl-1-phenyl-1*H*-phospholium bromide, *in* D-00468
2,3-Dihydro-*N*,*N*,5-trimethyl-3-phenyl-1,3,4,2-thiadiazaphosphole-2(3*H*)-amine-2-sulfide, *in* A-00072
2,5-Dihydro-1,3,4-trimethyl-1*H*-phosphole, D-00600
2,5-Dihydro-1,3,4-trimethyl-1*H*-phosphole; 1-Oxide, *in* D-00600
2,2-Dihydro-2,2,2-trimethyl-4,4,5,5-tetrakis(trifluoromethyl)-1,3,2-dioxaphospholane, D-00601
2,3-Dihydro-1,2,3-trimethyl-1*H*-1,2,3-triphosphole, *in* D-00615
2,2-Dihydro-2,2,2-triphenoxy-1,3,2-benzodioxaphosphole, D-00602
2,3-Dihydro-1,2,3-triphenyl-1*H*-1,3,2-benzodiphospharsole, D-00605
2,3-Dihydro-1,2,3-triphenyl-1*H*-1,3,2-benzodiphospharsole; ($1\alpha,2\beta,3\alpha$)-*form*, 1,3-Disulfide, *in* D-00605
2,3-Dihydro-1,2,3-triphenyl-1*H*-benzotriphosphole; ($1\alpha,2\beta,3\alpha$)-*form*, 1-Sulfide, *in* D-00606
2,3-Dihydro-1,2,3-triphenyl-1*H*-benzotriphosphole; ($1\alpha,2\beta,3\alpha$)-*form*, 1,3-Disulfide, *in* D-00606
2,3-Dihydro-1,2,3-triphenyl-1*H*-benzotriphosphole; ($1\alpha,2\beta,3\alpha$)-*form*, Trisulfide, *in* D-00606
5,5-Dihydro-5,5,5-triphenyl-5*H*-dibenzophosphole, D-00610
2,2-Dihydro-2,2,2-triphenyl-1,3,2-dioxaphospholane, D-00611
2,2-Dihydro-2,2,2-triphenyl-4,4,5,5-tetrakis(trifluoromethyl)-1,3,2-dioxaphospholane, D-00614
2,3-Dihydro-1,2,3-triphenyl-1*H*-1,2,3-triphosphole, *in* D-00615
2,3-Dihydro-1*H*-1,2,3-triphosphole, D-00615
1,3-Dihydroxy-1,3-diphospholane 1,3-dioxide, D-00621
2,5-Dihydroxy-3-(1-methylpropyl)-1,3,2,5-oxazadiphospholidine 2,5-dioxide; 2,5-Diethoxy (Di-Et ester), *in* D-00623
2,5-Dihydroxy-3-(1-methylpropyl)-1,3,2,5-oxazadiphospholidine 2,5-dioxide; 2,5-Dipropoxy (Dipropyl ester), *in* D-00623
1,8-Dihydroxy-3*a*,4,7,7*a*-tetrahydro-4,7-phosphinidene-1*H*-phosphindole 1,8-dioxide, D-00625
1,3-Diisopropoxy-1,3-diphospholane 1,3-dioxide, *in* D-00621
2-Dimethylamino-1,3,2-benzodioxaphosphole, D-00689
2-Dimethylamino-3,4-dimethyl-5-phenyl-1,3,2-oxazaphospholidine; (2*R*,4*S*,5*R*)-*form*, *in* D-00694
2-Dimethylamino-3,4-dimethyl-5-phenyl-1,3,2-oxazaphospholidine; (2*S*,4*R*,5*S*)-*form*, *in* D-00694
2-Dimethylamino-1,3,2-dioxaphospholane, D-00695
2-Dimethylamino-1,3,2-oxazaphospholidine; 2-Oxide, *in* D-00702
2-Dimethylamino-4,4,5,5-tetramethyl-1,3,2-dioxaphospholane, D-00710
2,5-Dimethyl-1,3,2-benzodithiaphosphole, D-00712
Dimethyl 2-chloro-1,3,2-dioxaphospholan-4,5-dicarboxylate, *in* C-00062
1,5-Dimethyl-1*H*-1,2,3-diazaphosphole, *in* M-00114
1,3-Dimethyl-1,3,2-diazaphospholidine; 2-Me, *in* D-00719
1,3-Dimethyl-1,3,2-diazaphospholidine 2-oxide, D-00720
1,4-Dimethyl-1,4,2-diazaphospholidin-5-one 2-oxide; 2-Bromo, *in* D-00722
5,5-Dimethyl-5*H*-dibenzophospholium iodide, *in* M-00115
1,3-Dimethyl-2-dimethylamino-1,3,2-diazaphospholidine, D-00724
1,3-Dimethyl-2-dimethylamino-1,3,2-diazaphospholidine; 2-Oxide, *in* D-00724
4,5-Dimethyl-2-dimethylamino-1,3,2-dioxaphospholane, D-00725
1,2-Dimethyl-4,5-diphenyl-1*H*-1,3-azaphosphole, D-00736
1,4-Dimethyl-3,5-diphenyl-1,4,2-diazaphospholidine 2-oxide; 2-Chloro, *in* D-00739
1,4-Dimethyl-3,5-diphenyl-1,4,2-diazaphospholidine 2-oxide; 2-Ethoxy, *in* D-00739
1,3-Dimethyl-1,3-diphenyl-1,3-diphospholanium dibromide, *in* D-01013

N-(1,2-Dimethylethenylenedioxyphosphoryl)imidazole, D-00753
4,5-Dimethyl-2-methylthio-1,3,2-dioxaphospholane; (4*RS*,5*SR*)-*form*, 2-Oxide, *in* D-00765
N,*N*-Dimethyl-1,3,2-oxathiaphospholan-2-amine, *in* O-00056
1,3-Dimethyl-2-phenoxy-1,3,2-diazaphospholidine, D-00788
1,3-Dimethyl-2-phenoxy-1,3-diazaphospholidin-4-one 2-oxide, *in* D-00721
4,5-Dimethyl-2-phenoxy-1,3,2-dioxaphospholane, D-00789
4,5-Dimethyl-2-phenoxy-1,3,2-dioxaphospholane; (4*RS*,5*SR*)-*form*, 2-Oxide, *in* D-00789
3,4-Dimethyl-2-phenoxy-5-phenyl-1,3,2-oxazaphospholidine; (2*R*,4*S*,5*R*)-*form*, 2-Oxide, *in* D-00791
3,4-Dimethyl-2-phenoxy-5-phenyl-1,3,2-oxazaphospholidine; (2*S*,4*S*,5*R*)-*form*, *in* D-00791
3,4-Dimethyl-2-phenoxy-5-phenyl-1,3,2-oxazaphospholidine; (2*S*,4*R*,5*S*)-*form*, *in* D-00791
1,2-Dimethyl-4-phenyl-1,2,4-diazaphospholidine, *in* D-00718
1,3-Dimethyl-2-phenyl-1,3,2-diazaphospholidine, D-00793
4,5-Dimethyl-2-phenyl-1,3,2-dioxaphospholane, D-00794
1,4-Dimethyl-2-phenyl-2-phosphabicyclo[2.2.1]heptane 2-oxide, *in* D-00825
2,2-Dimethyl-1-phenyl-1-phospha-2-silacyclopentane, D-00802
1,3-Dimethyl-1-phenylphospholanium bromide, *in* M-00229
3,4-Dimethyl-1-phenyl-1*H*-phosphole, D-00807
3,4-Dimethyl-1-phenyl-1*H*-phosphole; 1-Oxide, *in* D-00807
3,4-Dimethyl-1-phenyl-1*H*-phosphole; 1-Sulfide, *in* D-00807
3,4-Dimethyl-1-phenyl-1*H*-phosphole; 1-Selenide, *in* D-00807
4,5-Dimethyl-2-phenylthio-1,3,2-dioxaphospholane, D-00824
1,5-Dimethyl-1*H*-1,2,4,3-triazaphosphole, D-00926
4,5-Dimethyl-2-trimethylsilyloxy-1,3,2-dioxaphospholane; ($2\alpha,4\beta,5\beta$)-*form*, 2-Sulfide, *in* D-00930
4,5-Dimethyl-2-trimethylsilyloxy-1,3,2-dioxaphospholane; ($2\alpha,4\alpha,5\beta$)-(±)-*form*, 2-Sulfide, *in* D-00930
Dimethylvinylene chlorophosphate, *in* C-00057
Dimethylvinylene phenyl phosphate, *in* H-00123
Dimethylvinylene trimethylsilyl phosphate, *in* H-00123
1,2-Diphenyl-1,2-azaphospholidine, *in* P-00089
2,3-Diphenyl-1,3-azaphospholidine, D-00995
2,3-Diphenyl-1,3-azaphospholidine; 3-Sulfide, *in* D-00995
3,5-Diphenyl-1*H*-1,2,4-diazaphosphole, D-01000
3,5-Diphenyl-1*H*-1,2,4-diazaphosphole; 1-Me, *in* D-01000
3,5-Diphenyl-1*H*-1,2,4-diazaphosphole; 1-Ph, *in* D-01000
5,5-Diphenyl-5*H*-dibenzophospholium iodide, *in* P-00110
1,3-Diphenyl-1,3-diphospholane, D-01013
1,2-Diphenyl-1,2-diphospholane 1,2-disulfide, *in* D-01181
2,3-Diphenyl-1,3-oxaphospholane, D-01027
2,3-Diphenyl-1,3-oxaphospholane; 3-Oxide, *in* D-01027
2,3-Diphenyl-1,3-oxaphospholane; 3-Sulfide, *in* D-01027
2,3-Diphenyl-1,3,2-oxazaphospholidine, D-01029
2,3-Diphenyl-1,3,2-oxazaphospholidine; 2-Sulfide, *in* D-01029
1,1-Diphenylphospholanium bromide, *in* P-00216
9,9-Diphenyl-9-phosphoniabicyclo[4.2.1]nonane bromide, *in* P-00194
2,3-Diphenyl-1,3-thiaphospholane, D-01151
2,3-Diphenyl-1,3-thiaphospholane; 3-Oxide, *in* D-01151
1,2-Diphospholane, D-01181
1,2-Diphospholane; 1,2-Diisopropyl, *in* D-01181
1,2-Diphospholane; 1,2-Di-Ph, *in* D-01181
1,3,2-Dithiaphospholane 2-oxide; 2-Isopropyl, *in* D-01244
1,3,2-Dithiaphospholane 2-sulfide; 2-Isopropyl, *in* D-01245
1,3,2-Dithiaphosphole; 2-Chloro, *in* D-01246
1,3,2-Dithiaphosphole; 2-Methoxy, *in* D-01246
1-*tert*-Butyl-3,4-dimethyl-1*H*-phosphole, E-00004
N,*N*′-1,2-Ethanediyl-*N*,*N*′-dimethyl-*P*-phenylphosphonic diamide, *in* D-00793
2-Ethoxy-1,3,2-benzodioxaphosphole, E-00024
5-Ethoxy-5*H*-dibenzophosphole 5-oxide, *in* H-00118
2-Ethoxy-2,3-dihydro-1*H*-1,3,2-benzodiazaphosphole, E-00026
2-Ethoxy-2,3-dihydro-1,2-benzoxaphosphole 2-oxide, *in* H-00119
2-Ethoxy-2,3-dihydro-1,3,2-benzoxazaphosphole, E-00027
1-Ethoxy-2,5-dihydro-3,4-dimethyl-1*H*-phosphole 1-oxide, *in* D-00493
3-Ethoxy-2,3-dihydro-2-methyl-5-phenyl-1*H*-1,2,4,3-triazaphosphole; 3-Oxide, *in* E-00028
3-Ethoxy-2,3-dihydro-2-methyl-5-phenyl-1*H*-1,2,4,3-triazaphosphole; 3-Sulfide, *in* E-00028
1-Ethoxy-2,5-dihydro-3-methyl-1*H*-phosphole, *in* D-00530
1-Ethoxy-2,3-dihydro-5-methyl-1*H*-phosphole 1-oxide, *in* D-00497
1-Ethoxy-2,5-dihydro-2-methyl-1*H*-phosphole 1-oxide, *in* D-00498
1-Ethoxy-2,5-dihydro-3-methyl-1*H*-phosphole 1-oxide, *in* D-00499
1-Ethoxy-2,3-dihydro-1*H*-phosphole 1-oxide, *in* D-00506
▷1-Ethoxy-2,5-dihydro-1*H*-phosphole 1-oxide, *in* D-00507
2-Ethoxy-1,3-dimethyl-1,3,2-diazaphospholidine, E-00030
2-Ethoxy-4,5-dimethyl-1,3,2-dioxaphospholane, E-00031
2-Ethoxy-4,5-dimethyl-1,3,2-dioxaphospholane; (4*R*,5*R*)-*form*, *in* E-00031
2-Ethoxy-4,5-dimethyl-1,3,2-dioxaphospholane; (4*RS*,5*RS*)-*form*, 2-Oxide, *in* E-00031
2-Ethoxy-1,3,2-dioxaphospholan-4,5-dicarboxylic acid; (4*R*,5*R*)-*form*, Di-Et ester, *in* E-00033
2-Ethoxy-1,3,2-dioxaphospholan-4,5-dicarboxylic acid; (4*RS*,5*RS*)-*form*, Di-Et ester, *in* E-00033
2-Ethoxy-1,3,2-dioxaphospholan-4,5-dicarboxylic acid; (4*RS*,5*SR*)-*form*, Di-Et ester, *in* E-00033
▷2-Ethoxy-1,3,2-dioxaphospholane, E-00034
2-Ethoxy-1,3,2-dioxaphospholane 2-oxide, E-00035
2-Ethoxy-1,3,2-dioxaphospholane 2-sulfide, E-00036
1-Ethoxy-3,4-diphenyl-1*H*-phosphole 1-oxide, *in* H-00142
2-Ethoxy-1,3,2-dithiaphospholane, E-00038
2-Ethoxy-4-methyl-1,3,2-dioxaphospholane, E-00040
2-Ethoxy-3-methyl-1,3,2-thiazaphospholidine; 2-Oxide, *in* E-00043
2-Ethoxy-3-methyl-1,3,2-thiazaphospholidine; 2-Sulfide, *in* E-00043
2-Ethoxynaphtho[1,2-*d*]-1,3,2-dioxaphosphole, E-00044
2-Ethoxy-1,2-oxaphospholane 2-oxide, *in* H-00166
2-Ethoxy-3-phenyl-1,3,2-oxazaphospholidine, E-00048
2-Ethoxy-3-phenyl-1,3,2-oxazaphospholidine; 2-Sulfide, *in* E-00048
2-Ethoxy-4,4,5,5-tetramethyl-1,3,2-dioxaphospholane, E-00052
1-Ethoxy-2,3,4,5-tetraphenyl-1*H*-phosphole 1-oxide, *in* H-00200
2-Ethoxy-1,3,2-thiazaphospholidine; 2-Oxide, *in* E-00053
2-Ethoxy-1,3,2-thiazaphospholidine; 2-Sulfide, *in* E-00053
2-Ethoxy-3,5,5-trimethyl-1,2-oxaphospholan-3-ol 2-oxide, *in* T-00519
1-Ethyl-2,5-dihydro-3-methyl-1-phenyl-1*H*-phospholium iodide, *in* D-00528
Ethyl dimethylvinylene phosphate, *in* H-00123
1-Ethyl-2,3-diphenyl-1,3-azaphosphacyclopentane, *in* D-00995
Ethylene *tert*-butylphosphoramidate, *in* B-00539
S,*S*-Ethylene diethylphosphoramidodithioite, *in* D-01243
S,*S*-Ethylene diethylphosphoramidodithioite, *in* D-01245
Ethylene dimethylphosphoramidate, *in* D-00695
S,*S*-Ethylene *N*,*N*-dimethyl phosphoramidodithioate, *in* D-01244
Ethylene dimethylphosphoramidoselenoate, *in* D-00695
Ethylene dimethylphosphoramidothioate, *in* D-00695
Ethylene glycol cyclic *P*,*P*:*P*′,*P*′-pyrophosphate, *in* O-00091
Ethylene methylphosphonate, *in* M-00126
S,*S*-Ethylene methylphosphonodithioate, *in* M-00139
S,*S*-Ethylene methylphosphonoselenodithioate, *in* M-00139
Ethylene methylphosphonothioate, *in* M-00126
Ethylene methylphosphonotrithioate, *in* M-00139
O,*S*-Ethylene *S*-methyl phosphorodithioite, *in* O-00056
O,*O*-Ethylene *O*-methyl phosphoroselenoate, *in* M-00036
O,*S*-Ethylene *O*-methyl phosphorothioite, *in* O-00056
Ethylene phenyl phosphate, *in* P-00067
Ethylene phenylphosphonate, *in* P-00115
S,*S*-Ethylene phenylphosphonodithioate, *in* P-00121
O,*O*-Ethylene phenylphosphonothioate, *in* P-00115
O,*S*-Ethylene phenylphosphonothioite, *in* O-00056
Ethylene phenylphosphonotrithioate, *in* P-00121
Ethylene phenylphosphoramidate, *in* P-00083
Ethylene phenylphosphoramidothioate, *in* P-00083
O,*O*-Ethylene *O*-phenyl phosphorothioate, *in* P-00067
O,*S*-Ethylene *O*-phenyl phosphorothioate, *in* P-00072
S,*S*-Ethylene phosphorochloridodithioate, *in* C-00071
Ethylene phosphorochloridotrithioate, *in* C-00071
Ethylene trimethylsilyl phosphate, *in* T-00554
Ethyl *N*,*N*′-1,2-ethanediyl-*N*,*N*′-dimethylphosphorodiamidate, *in* E-00030

S-Ethyl *O,O*-ethylene dithiophosphate, *in* E-00191
Ethyl 1,2-naphthylene phosphate, *in* E-00044
O-Ethyl *O,O*-1,2-naphthylene thiophosphate, *in* E-00044
1-Ethyl-2-phenyl-1,3-azaphospholidine, *in* P-00090
1-Ethyl-3-phenyl-1,3-azaphospholidine, *in* P-00091
Ethyl *o*-phenylene phosphate, *in* E-00024
O-Ethyl *N,N'*-*o*-phenylenephosphorodiamidothioate, *in* E-00026
O-Ethyl *O,O*-*o*-phenylene thiophosphate, *in* E-00024
Ethyl propylene phosphate, *in* E-00040
O-Ethyl *O,O*-propylene thiophosphate, *in* E-00040
Ethyl tetramethylenephosphinate, *in* H-00185
S-Ethyl *O,O*-tetramethylethylene phosphorodithioate, *in* E-00193
O-Ethyl *O,O*-tetramethylethylene phosphorothioate, *in* E-00052
1-Ethylthio-1,5-dihydro-3-methyl-1*H*-phosphole, *in* D-00530
▷2-Ethylthio-4,5-dimethyl-1,3,2-dioxaphospholane, E-00189
2-Ethylthio-4,5-dimethyl-1,3,2-dioxaphospholane; 2-Sulfide, *in* E-00189
▷2-Ethylthio-1,3,2-dioxaphospholane, E-00191
2-Ethylthio-4,4,5,5-tetramethyl-1,3,2-dioxaphospholane, E-00193
2-Fluoro-1,3,2-benzodioxaphosphole, F-00013
1-Fluoro-2,5-dihydro-1*H*-phosphole 1-oxide, *in* D-00507
2-Fluoro-1,3-dimethyl-1,3,2-diazaphospholidine, F-00014
2-Fluoro-1,3-dimethyl-1,3,2-diazaphospholidine; 2-Oxide, *in* F-00014
2-Fluoro-1,3-dimethyl-1,3,2-diazaphospholidine; 2-Sulfide, *in* F-00014
2-Fluoro-4,5-dimethyl-1,3,2-dioxaphospholane, F-00015
2-Fluoro-1,3,2-dioxaphospholane, F-00017
2-Fluoro-1,3,2-dithiaphospholane, *in* D-01243
2-Fluoro-4-methyl-1,3,2-dioxaphospholane, F-00020
1,3*a*,3*b*,4,6*a*,6*b*-Hexahydro-1,2,3*b*,4,5,6*b*-hexaphenylcyclobuta[1,2-*b*:3,4-*b'*]diphosphole, H-00031
2,3,4,5,6,7-Hexahydro-1*H*-isophosphindole 2-oxide; 2-Me, *in* H-00035
2,3,4,5,6,7-Hexahydro-1*H*-isophosphindole 2-oxide; 2-Ph, *in* H-00035
2,3,4,5,6,7-Hexahydro-1*H*-isophosphindole 2-oxide; 2-Hydroxy, *in* H-00035
2,2,2,3,4,5-Hexahydro-2,2,2-trimethoxy-3-phenylnaphth[2,1-*d*]-1,2-oxaphosphole, H-00056
2,2,2,3,3*a*,11*b*-Hexahydro-2,2,2-triphenylphenanthro[9,10-*d*]-1,3,2-oxazaphosphole; (9*RS*,10*RS*)-*form*, *in* H-00059
2-Hydroxy-1,3,2-benzodioxaphosphole 2-oxide, H-00108
2-Hydroxy-1,3,2-benzodioxaphosphole 2-oxide; 2-Methoxy (Me ester), *in* H-00108
5-Hydroxy-5*H*-dibenzophosphole 5-oxide, H-00118
5-Hydroxy-5*H*-dibenzophosphole 5-oxide; Dodecahydro, *in* H-00118
1-Hydroxy-2,3-dihydro-1*H*-phosphindole 1-oxide, H-00120
2-Hydroxy-4,5-dimethyl-1,3,2-dioxaphospholane 2-oxide, H-00121
2-Hydroxy-4,5-dimethyl-1,3,2-dioxaphospholane 2-sulfide; (2α,4α,5α)-*form*, Tetramethylammonium salt, *in* H-00122
2-Hydroxy-4,5-dimethyl-1,3,2-dioxaphospholane 2-sulfide; (2α,4α,5β)-*form*, *in* H-00122
2-Hydroxy-4,5-dimethyl-1,3,2-dioxaphospholane 2-sulfide; (2α,4β,5β)-*form*, *in* H-00122
2-Hydroxy-4,5-dimethyl-1,3,2-dioxaphosphole 2-oxide, H-00123
1-Hydroxy-3,4-dimethyl-1*H*-phosphole 1-oxide; 1-Phenoxy (Ph ester), *in* H-00130
▷2-Hydroxy-1,3,2-dioxaphospholane, *in* D-00970
2-Hydroxy-1,3,2-dioxaphospholane 2-oxide, H-00134
2-Hydroxy-1,3,2-dioxaphospholane 2-sulfide, H-00135
2-Hydroxy-4,5-diphenyl-1,3,2-dioxaphosphole 2-oxide, H-00139
2-Hydroxy-4,5-diphenyl-1,3,2-dioxaphosphole 2-oxide; 2-Methoxy (Me ester), *in* H-00139
2-Hydroxy-4,5-diphenyl-1,3,2-dioxaphosphole 2-oxide; 2-Ethoxy (Et ester), *in* H-00139
1-Hydroxy-3,4-diphenyl-1*H*-phosphole 1-oxide, H-00142
2-Hydroxy-4-methyl-1,3,2-dioxaphospholane, *in* M-00127
2-Hydroxy-4-methyl-1,3,2-dioxaphospholane 2-oxide, H-00155
2-Hydroxynaphtho[2,3-*d*]-1,3,2-dioxaphosphole 2-oxide, H-00164
2-Hydroxynaphtho[2,3-*d*]-1,3,2-dioxaphosphole 2-oxide; 2-Methoxy (Me ester), *in* H-00164
2-Hydroxynaphtho[2,3-*d*]-1,3,2-dioxaphosphole 2-oxide; 2-Ethoxy (Et ester), *in* H-00164
1-Hydroxyphospholane 1-oxide, H-00185
1-Hydroxy-1*H*-phosphole 1-oxide; Et ester, *in* H-00186
2-Hydroxy-4,4,5,5-tetramethyl-1,3,2-dioxaphospholane 2-oxide, H-00196
2-Hydroxy-4,4,5,5-tetramethyl-1,3,2-dioxaphospholane 2-sulfide, H-00197
1-Hydroxy-2,3,4,5-tetraphenyl-1*H*-phosphole 1-oxide, H-00200
1*H*-Isophosphindole-1,3(2*H*)-dione; 2-Ph, *in* I-00026
1*H*-Isophosphindole-1,3(2*H*)-dione; 2-Ph, 2-sulfide, *in* I-00026
1*H*-Isophosphindole-1,3(2*H*)-dione; *Enol-form*, Trimethylsilyl ether, *in* I-00026
2-Isopropyl 1,3,2-oxathiaphospholane, I-00042
2-Isopropyl-1,3,2-oxathiaphospholane 2-oxide, *in* I-00042
2-Isopropyl-1,3,2-oxathiaphospholane 2-sulfide, *in* I-00042
2-Mercapto-1,3,2-benzodioxaphosphole 2-sulfide, M-00005
2-Mercapto-1,3,2-benzodioxaphosphole 2-sulfide; 2-Methylthio (Me ester), *in* M-00005
2-Mercapto-1,3,2-benzodioxaphosphole 2-sulfide; 2-Ethylthio (Et ester), *in* M-00005
2-Mercapto-4,5-dimethyl-1,3,2-dioxaphospholane 2-sulfide; NH_4 salt, *in* M-00009
2-Mercapto-4,5-dimethyl-1,3,2-dioxaphospholane 2-sulfide; (4*RS*,5*RS*)-*form*, *in* M-00009
2-Mercapto-1,3,2-dioxaphospholane, *in* D-00970
2-Mercapto-4-methyl-1,3,2-dioxaphospholane 2-oxide, M-00015
2-Mercapto-4-methyl-1,3,2-dioxaphospholane 2-sulfide, M-00016
2-Mercapto-4-methyl-1,3,2-dioxophospholane, *in* M-00127
2-Mercapto-4,4,5,5-tetramethyl-1,3,2-dioxaphospholane 2-sulfide, M-00019
2-Methoxy-1,3,2-benzodioxaphosphole, M-00026
5-Methoxy-5*H*-dibenzophosphole 5-oxide, *in* H-00118
2-Methoxy-4,5-dimethyl-1,3,2-dioxaphospholane; (4*R*,5*R*)-*form*, *in* M-00033
2-Methoxy-4,5-dimethyl-1,3,2-dioxaphospholane; (4*RS*,5*RS*)-*form*, *in* M-00033
2-Methoxy-3,4-dimethyl-5-phenyl-1,3,2-oxazaphospholidine; (2*R*,4*S*,5*R*)-*form*, *in* M-00035
2-Methoxy-3,4-dimethyl-5-phenyl-1,3,2-oxazaphospholidine; (2*R*,4*S*,5*R*)-*form*, 2-Oxide, *in* M-00035
2-Methoxy-3,4-dimethyl-5-phenyl-1,3,2-oxazaphospholidine; (2*R*,4*S*,5*R*)-*form*, 2-Sulfide, *in* M-00035
2-Methoxy-3,4-dimethyl-5-phenyl-1,3,2-oxazaphospholidine; (2*S*,4*S*,5*R*)-*form*, *in* M-00035
2-Methoxy-3,4-dimethyl-5-phenyl-1,3,2-oxazaphospholidine; (2*S*,4*S*,5*R*)-*form*, 2-Oxide, *in* M-00035
2-Methoxy-3,4-dimethyl-5-phenyl-1,3,2-oxazaphospholidine; (2*S*,4*S*,5*R*)-*form*, 2-Sulfide, *in* M-00035
2-Methoxy-3,4-dimethyl-5-phenyl-1,3,2-oxazaphospholidine; (2*R*,4*R*,5*S*)-*form*, *in* M-00035
2-Methoxy-3,4-dimethyl-5-phenyl-1,3,2-oxazaphospholidine; (2*S*,4*R*,5*S*)-*form*, *in* M-00035
1-Methoxy-3,4-dimethyl-1*H*-phosphole 1-oxide, *in* H-00130
▷2-Methoxy-1,3,2-dioxaphospholane, M-00036
2-Methoxy-1,3,2-dioxaphospholane 2-oxide, M-00037
1-Methoxy-1,3-dioxo-2,4-dimethyl-1-phospha-2,4-diazacyclopentane, *in* D-00722
2-Methoxy-4,5-diphenyl-1,3,2-dioxaphospholane, M-00039
2-Methoxy-4,5-diphenyl-1,3,2-dioxaphospholane; (2α,4β,5β)-*form*, 2-Oxide, *in* M-00039
2-Methoxy-4,5-diphenyl-1,3,2-dioxaphospholane; (2α,4β,5β)-*form*, 2-Sulfide, *in* M-00039
2-Methoxy-1,3,2-dithiaphospholane, *in* D-01243
2-Methoxy-1,3,2-dithiaphospholane 2-oxide, *in* D-01244
2-Methoxy-4-methyl-1,3,2-dioxaphospholane, M-00043
2-Methoxy-4-methyl-1,3,2-dithiaphospholane, *in* M-00136
3-Methoxy-15-methyl-18-nor-15-phosphaestra-1,3,5(10),-6,8,13-hexaene; 15-Oxide, *in* M-00048
2-Methoxy-4-methyl-1,2-oxaphospholane; 2-Oxide, *in* M-00049
3-Methoxy-17-methyl-15-phenyl-15-phosphaestra-1,3,5(10),-8,16-pentaene; (±)-*form*, 15-Oxide, *in* M-00051
3-Methoxy-17-methyl-15-phenyl-15-phosphaestra-1,3,5(10),-8,16-pentaene; (±)-*form*, 15-Sulfide, *in* M-00051
3-Methoxy-18-nor-17-phosphaestra-1,3,5(10),9(11)-tetraen-15-one; 13α-*form*, 17-Me, 17-oxide, *in* M-00054

3-Methoxy-18-nor-17-phosphaestra-1,3,5(10),9(11)-tetraen-15-one; 13*α-form*, 17-Ph, 17-oxide, *in* M-00054
3-Methoxy-18-nor-17-phosphaestra-1,3,5(10),9(11)-tetraen-15-one; 13*β-form*, 17-Ph, 17-oxide, *in* M-00054
2-Methoxy-3,3,4,4-tetramethyl-1,2-azaphospholidine 2-oxide, *in* T-00191
2-Methoxy-4,4,5,5-tetramethyl-1,3,2-dioxaphospholane, M-00085
1-Methoxy-2,3,4,5-tetraphenyl-1*H*-phosphole 1-oxide, *in* H-00200
2-Methoxy-3,5,5-trimethyl-1,2-oxaphospholan-3-ol 2-oxide, *in* T-00519
2-Methyl-1,3,2-benzodioxaphosphole, M-00090
5-Methyl-1*H*-1,2,3-diazaphosphole, M-00114
5-Methyl-5*H*-dibenzophosphole, M-00115
5-Methyl-5*H*-dibenzophosphole; 5-Oxide, *in* M-00115
5-Methyl-5*H*-dibenzophosphole; 5-Sulfide, *in* M-00115
5-Methyl-5*H*-dibenzophosphole; 5-Selenide, *in* M-00115
4-Methyl-2-dimethylamino-1,3,2-dioxaphospholane, M-00116
Methyl dimethylvinylene phosphate, *in* H-00123
1-Methyl-2,5-diphenyl-1,2,5-azadiphospholidine, *in* M-00089
5-Methyl-2,3-diphenyl-1,3-oxaphospholane, M-00133
3-Methyl-2,3-diphenyl-1,3-oxaphospholanium iodide, *in* D-01027
6-Methyl-1,5-diphenyl-6-phosphabicyclo[3.2.1]octane; Oxide, *in* M-00134
2-Methyl-1,3,2-dithiophospholane, M-00139
Methyl *N,N'*-ethanediyl-*N,N'*-dimethylphosphorodiamidite, *in* D-00719
2-(1-Methylethoxy)-1,3,2-dioxaphospholane, M-00152
2-Methyl-1,2-oxaphospholane; 2-Oxide, *in* M-00182
2-Methyl-1,2-oxaphospholane; 2-Sulfide, *in* M-00182
5-Methyl-1,2-oxaphospholane 2-oxide; 2-Methoxy, *in* M-00183
5-Methyl-1,2-oxaphospholane 2-oxide; 2-Ph, *in* M-00183
2-Methyl-1,2-oxaphospholan-5-one 2-oxide, *in* O-00053
5-Methyl-1,3,2-oxathiaphosphole 2-oxide; 2-Ethoxy, *in* M-00184
5-Methyl-1,3,2-oxathiaphosphole 2-oxide; 2-Diethylamino, *in* M-00184
3-Methyl-2-phenoxy-1,3,2-oxazaphospholidine, M-00192
3-Methyl-2-phenoxy-1,3,2-oxazaphospholidine; 2-Oxide, *in* M-00192
5-Methyl-2-phenyl-1,3,2-benzodithiaphosphole, M-00195
5-Methyl-2-phenyl-1,3,2-benzodithiaphosphole; 2-Oxide, *in* M-00195
5-Methyl-2-phenyl-2*H*-1,2,3-diazaphosphole, M-00197
5-Methyl-5-phenyl-5*H*-dibenzophospholium iodide, *in* P-00110
4-Methyl-2-phenyl-1,3,2-dioxaphospholane, M-00198
4-Methyl-2-phenyl-1,3,2-dioxaphospholane; 2-Oxide, *in* M-00198
4-Methyl-2-phenyl-1,3,2-dithiaphospholane, *in* M-00136
Methyl *o*-phenylene phosphate, *in* M-00026
2-Methyl-3-phenyl-1,3,2-oxazaphospholidine, M-00212
2-Methyl-3-phenyl-1,3,2-oxazaphospholidine; 2-Oxide, *in* M-00212
2-Methyl-3-phenyl-1,3,2-oxazaphospholidine; 2-Sulfide, *in* M-00212
3-Methyl-2-phenyl-1,3,2-oxazaphospholidine, M-00213
3-Methyl-2-phenyl-1,3,2-oxazaphospholidine; 2-Sulfide, *in* M-00213
3-Methyl-1-phenylphospholane; (1*S*,3*R*)*-form*, *in* M-00229
3-Methyl-1-phenylphospholane; (1*S*,3*R*)*-form*, 1-Oxide, *in* M-00229
3-Methyl-1-phenylphospholane; (1*RS*,3*RS*)*-form*, 1-Oxide, *in* M-00229
3-Methyl-1-phenylphospholane; (1*RS*,3*SR*)*-form*, 1-Oxide, *in* M-00229
1-Methyl-1-phenylphospholanium iodide, *in* P-00216
9-Methyl-9-phenyl-9-phosphoniabicyclo[4.2.1]nonane iodide, *in* P-00194
9-Methyl-9-phenyl-9-phosphoniabicyclo[4.2.1]nona-2,4,7-triene iodide, *in* P-00196
8-Methyl-8-phenyl-8-phosphoniabicyclo[3.2.1]octan-3-one iodide, *in* P-00199
1-Methyl-5-phenyl-1*H*-1,2,4,3-triazaphosphole, M-00269
2-Methyl-5-phenyl-2*H*-1,2,4,3-triazaphosphole, M-00270
8-Methyl-8-phosphabicyclo[3.2.1]octane; *syn-form*, Oxide, *in* M-00272
8-Methyl-8-phosphabicyclo[3.2.1]octane; *anti-form*, Oxide, *in* M-00272
1-Methylphospholane, M-00280
1-Methylphospholane; Oxide, *in* M-00280
1-Methylphospholane; Sulfide, *in* M-00280
3-Methylphospholane, M-00281
1-Methyl-1*H*-phosphole, M-00282
1-Methyl-1*H*-phosphole; B,MeI, *in* M-00282
2-(2-Methylpropoxy)-1,3,2-dioxaphospholane, M-00368
Methyl propylene phosphate, *in* M-00043
S-Methyl *O,O*-propylene phosphorodithioate, *in* M-00173
S-Methyl *O,O*-propylene phosphorothioate, *in* M-00173
O-Methyl *O,O*-propylene thiophosphate, *in* M-00043
Methyl tetramethylenephosphinate, *in* H-00185
Methyl tetramethylethylene phosphate, *in* M-00085
5-Methyl-1,2-thiaphospholane 2-sulfide; 2-Et, *in* M-00407
5-Methyl-1,2-thiaphospholane 2-sulfide; 2-Ph, *in* M-00407
5-Methyl-1,2-thiaphospholane 2-sulfide; 2-Chloro, *in* M-00407
3-Methyl-1,3,2-thiazaphospholidine; 2-Chloro, 2-oxide, *in* M-00408
3-Methyl-1,3,2-thiazaphospholidine; 2-Chloro, 2-sulfide, *in* M-00408
3-Methyl-1,3,2-thiazaphospholidine; 2-Methoxy, 2-oxide, *in* M-00408
3-Methyl-1,3,2-thiazaphospholidine; 2-Methoxy, 2-sulfide, *in* M-00408
3-Methyl-1,3,2-thiazaphospholidine; 2-Ethoxy, 2-sulfide, *in* M-00408
2-(Methylthio)-1,3,2-dithiaphospholane, *in* D-01243
3-Methyl-1,2,3-triphenyl-5-oxo-1,3-azaphospholidinium iodide, *in* T-00596
1-Methyl-1,2,5-triphenyl-1*H*-phospholium iodide, *in* T-00634
Methyl vinylene phosphate, *in* H-00136
1,2-Naphthylene diethylphosphoramidate, *in* D-00266
O,O-1,2-Naphthylene diethylphosphoramidothioate, *in* D-00266
1,2-Naphthylene phosphonate, *in* N-00009
2,3-Naphthylene phosphorochloridate, *in* C-00119
2,3-Naphthylene phosphorochloridothioate, *in* C-00119
Octahydro-1,8-dihydroxy-4,7-phosphinidene-1*H*-phosphindole 1,8-dioxide, *in* D-00625
Octahydro-3-methoxy-1*H*-cyclopropa[*c*]phosphindole 3-oxide, *in* O-00004
Octahydro-3-methyl-1*H*-cyclopropa[*c*]phosphindole 3-oxide, *in* O-00004
Octahydro-3-phenyl-1*H*-cyclopropa[*c*]phosphindole 3-oxide, *in* O-00004
O,O-o-Phenylene selenophosphonate, *in* M-00090
2,2'-Oxybis[1,3,2-benzodioxaphosphole], O-00087
5,5'-Oxybis[5*H*-benzophosphindole] 5,5'-dioxide, *in* H-00118
2,2'-Oxybis[4,5-dimethyl-1,3,2-dioxaphosphole], O-00088
▷2,2'-Oxybis[1,3,2-dioxaphospholane], O-00091
Pentamethylpentaphospholane, P-00019
1,2,3,4,5-Pentamethyl-1,2,4,3,5-triazadiphospholidine, *in* T-00562
Pentaphenylpentaphospholane, P-00037
Pentaphenyl-1*H*-phosphole, P-00038
Pentaphenyl-1*H*-phosphole; Oxide, *in* P-00038
Pentaphenyl-1*H*-phosphole; Sulfide, *in* P-00038
2-Phenoxy-1,3,2-benzodioxaphosphole, P-00063
2-Phenoxy-1,3,2-dioxaphospholane, P-00067
2-Phenoxy-4,5-diphenyl-1,3,2-dioxaphospholane, P-00069
2-Phenoxy-4,5-diphenyl-1,3,2-dioxaphospholane; (4*RS*,5*RS*)-*form*, 2-Oxide, *in* P-00069
2-Phenoxy-4,5-diphenyl-1,3,2-dioxaphospholane; (4*RS*,5*SR*)-*form*, 2-Oxide, *in* P-00069
1-Phenoxy-3,4-diphenyl-1*H*-phosphole 1-oxide, *in* H-00142
2-Phenoxy-1,3,2-dithiaphospholane 2-sulfide, *in* D-01245
2-Phenoxy-1,3,2-oxathiaphospholane, P-00072
2-Phenoxy-3-phenyl-1,3,2-oxazaphospholidine, P-00073
2-Phenoxy-3-phenyl-1,3,2-oxazaphospholidine; 2-Oxide, *in* P-00073
2-Phenoxy-4,4,5,5-tetramethyl-1,3,2-dioxaphospholane, P-00075
1-Phenoxy-2,3,4,5-tetraphenyl-1*H*-phosphole 1-oxide, *in* H-00200
7-Phenyl-6-aza-7-phosphabicyclo[3.2.1]octane; 6-Cyclohexyl, 7-sulfide, *in* P-00086

2-Phenyl-1*H*-1,3-azaphosphole, P-00088
2-Phenyl-1,2-azaphospholidine, P-00089
2-Phenyl-1,3-azaphospholidine, P-00090
3-Phenyl-1,3-azaphospholidine, P-00091
1-Phenyl-1,2-azaphospholidin-5-one; 2-Et, 2-oxide, *in* P-00092
1-Phenyl-1,2-azaphospholidin-5-one; 2-Ph, 2-oxide, *in* P-00092
3-Phenyl-1,3-azaphospholidin-2-thione, P-00093
3-Phenyl-1,3-azaphospholidin-2-thione; 3-Sulfide, *in* P-00093
2-Phenyl-1,3,2-benzodioxaphosphole, P-00094
5-Phenyl-5*H*-dibenzophosphole, P-00110
5-Phenyl-5*H*-dibenzophosphole; 5-Oxide, *in* P-00110
5-Phenyl-5*H*-dibenzophosphole; 5-Sulfide, *in* P-00110
5-Phenyl-5*H*-dibenzophosphole; 5-Selenide, *in* P-00110
2-Phenyl-1,3,2-dioxaphospholane, P-00115
2-Phenyl-1,3,2-dithiaphospholane, P-00121
o-Phenylene dimethylphosphoramidate, *in* D-00689
o-Phenylene methylphosphonate, *in* M-00090
o-Phenylene phenylphosphonate, *in* P-00094
o-Phenylene phosphite, *in* B-00018
o-Phenylene phosphorobromidate, *in* B-00464
o-Phenylene phosphoryl azide, *in* A-00148
O,O-*o*-Phenylene thiophosphonate, *in* M-00090
2-Phenyl-2*H*-isophosphindole 2-oxide, P-00169
2-Phenyl-1,2-oxaphospholane, P-00174
2-Phenyl-1,2-oxaphospholane; 2-Oxide, *in* P-00174
2-Phenyl-1,2-oxaphospholane; 2-Sulfide, *in* P-00174
2-Phenyl-1,2-oxaphospholan-2-one 2-oxide, *in* O-00053
2-Phenyl-1,3,2-oxazaphospholidine, P-00179
3-Phenyl-1,3,2-oxazaphospholidine; Oxide, *in* P-00180
Phenyl *o*-phenylene phosphate, *in* P-00063
Phenyl *N,N'*-*o*-phenylenephosphorodiamidate, *in* D-00543
O-Phenyl *N,N'*-*o*-phenylenephosphorodiamidothioate, *in* D-00543
O-Phenyl *o*-phenylene thiophosphate, *in* P-00063
9-Phenyl-9-phosphabicyclo[4.2.1]nonane, P-00194
9-Phenyl-9-phosphabicyclo[4.2.1]nonane; Oxide, *in* P-00194
9-Phenyl-9-phosphabicyclo[4.2.1]nona-2,4,7-triene; *syn-form*, *in* P-00196
9-Phenyl-9-phosphabicyclo[4.2.1]nona-2,4,7-triene; *syn-form*, Oxide, *in* P-00196
9-Phenyl-9-phosphabicyclo[4.2.1]nona-2,4,7-triene; *anti-form*, *in* P-00196
9-Phenyl-9-phosphabicyclo[4.2.1]nona-2,4,7-triene; *anti-form*, B,MeI, *in* P-00196
9-Phenyl-9-phosphabicyclo[4.2.1]nona-2,4,7-triene; *anti-form*, Oxide, *in* P-00196
8-Phenyl-8-phosphabicyclo[3.2.1]octan-3-one; *syn-form*, Oxide, *in* P-00199
8-Phenyl-8-phosphabicyclo[3.2.1]octan-3-one; *anti-form*, *in* P-00199
8-Phenyl-8-phosphabicyclo[3.2.1]octan-3-one; *anti-form*, Oxide, *in* P-00199
8-Phenyl-8-phosphabicyclo[3.2.1]oct-6-ene; Oxide, *in* P-00200
9-Phenyl-9-phosphatricyclo[3.3.1.0^{2,8}]nona-3,6-diene, P-00202
9-Phenyl-9-phosphatricyclo[3.3.1.0^{2,8}]nona-3,6-diene; 9-Oxide, *in* P-00202
1-Phenyl-1*H*-phosphindole, P-00207
1-Phenyl-1*H*-phosphindole; 1-Oxide, *in* P-00207
1-Phenylphospholane, P-00216
1-Phenylphospholane; 1-Oxide, *in* P-00216
1-Phenylphospholane; 1-Sulfide, *in* P-00216
1-Phenyl-1*H*-phosphole, P-00217
Phenyl propylene phosphate, *in* M-00190
2-Phenyl-4,4,5,5-tetramethyl-1,3,2-dioxaphospholane, P-00322
2-Phenyl-1,2-thiaphospholane, P-00324
2-Phenyl-1,2-thiaphospholane; 2-Sulfide, *in* P-00324
5-Phospha(5-P^{V})spiro[4.5]deca-1,3,5,7,9-pentaene, P-00349
Phospholane, P-00371
1*H*-Phosphole, P-00372
Propylene dimethylphosphoramidate, *in* M-00116
S,S-1,2-Propylene *N,N*-dimethylphosphoramidodithioite, *in* M-00136
O,O-Propylene dimethylphosphoramidoselenoate, *in* M-00116
O,O-Propylene dimethylphosphoramidothioate, *in* M-00116
Propylene fluorothiophosphate, *in* F-00020
Propylene methylphosphonate, *in* D-00732
Propylene phosphorochloridate, *in* C-00092
Propylene phosphorochloridothioate, *in* C-00092
2,2,2,3-Tetrahydro-1,2-benzoxaphosphole, T-00088
2,2,2,3-Tetrahydro-3,5-diphenyl-1,3,4,2-oxadiazaphosphole; 2,2,2-Trimethoxy, *in* T-00104
2,2,2,3-Tetrahydro-3,5-diphenyl-1,3,4,2-oxadiazaphosphole; 2,2,2-Triphenoxy, *in* T-00104
2,2,2,3-Tetrahydro-3,5-diphenyl-1,3,4,2-oxadiazaphosphole; 2,2,2-Trichloro, *in* T-00104
3*a*,4,7,7*a*-Tetrahydro-1,8-diphenyl-4,7-phosphinidene-1*H*-phosphindole, T-00105
3*a*,4,7,7*a*-Tetrahydro-1,8-diphenyl-4,7-phosphinidene-1*H*-phosphindole; 1,8-Dioxide, *in* T-00105
3*a*,4,7,7*a*-Tetrahydro-1,8-diphenyl-4,7-phosphinidene-1*H*-phosphindole; 1,8-Disulfide, *in* T-00105
3*a*,4,7,7*a*-Tetrahydro-1,2,3,5,6,8-hexamethyl-4,7-phosphinidene-1*H*-phosphindole, T-00106
3*a*,4,7,7*a*-Tetrahydro-1,2,3,5,6,8-hexamethyl-4,7-phosphinidene-1*H*-phosphindole; 1,8-Dioxide, *in* T-00106
3*a*,4,7,7*a*-Tetrahydro-1,2,3,5,6,8-hexamethyl-4,7-phosphinidene-1*H*-phosphindole; 1,8-Disulfide, *in* T-00106
1,2,5,6-Tetrahydro-1,1,3,6,6,9-hexamethyl-4*H*-phospholo[3,2,1-*i,j*]phosphinolinium iodide, *in* T-00123
2,3,4,7-Tetrahydro-1*H*-isophosphindole 2-oxide; 2-Me, *in* T-00109
2,3,4,7-Tetrahydro-1*H*-isophosphindole 2-oxide; 2-Ph, *in* T-00109
2,3,4,7-Tetrahydro-1*H*-isophosphindole 2-oxide; 2-Hydroxy, *in* T-00109
3,3*a*,4,5-Tetrahydro-3-methyl-2*H*-benzo[*e*]phosphindole 3-oxide, T-00112
2,2,2,3-Tetrahydro-1,2-oxaphosphole; 2,2,2-Trimethoxy, *in* T-00119
2,2,2,3-Tetrahydro-1,2-oxaphosphole; 2,2,2-Triethoxy, *in* T-00119
1,2,5,6-Tetrahydro-1,1,6,6,9-pentamethyl-4*H*-phospholo[3,2,1-*ij*]phosphinoline, T-00123
1,2,5,6-Tetrahydro-1,1,6,6,9-pentamethyl-4*H*-phospholo[3,2,1-*ij*]phosphinoline; 3-Oxide, *in* T-00123
2,2,2,3-Tetrahydrophenanthro[9,10-*d*]-1,3,2-oxazaphosphole, T-00125
2,2,2,3-Tetrahydro-5-phenyl-3,3-bis(trifluoromethyl)-1,4,2-thiazaphosphole; 2,2,2-Trimethoxy, *in* T-00129
2,2,2,3-Tetrahydro-5-phenyl-3,3-bis(trifluoromethyl)-1,4,2-thiazaphosphole; 2,2,2-Triphenoxy, *in* T-00129
2,2,2,3-Tetrahydro-5-phenyl-1,3,4,2-oxadiazaphosphole; 2,2,2-Trifluoro, *in* T-00131
2,2,2,3-Tetrahydro-5-phenyl-1,3,4,2-oxadiazaphosphole; 2,2,2-Trichloro, *in* T-00131
3*a*,4,7,7*a*-Tetrahydro-2,3,5,6-tetramethyl-1,8-diphenyl-4,7-phosphinidene-1*H*-phosphindole; 1,8-Dioxide, *in* T-00141
3,4,5,5-Tetrahydro-2,2,5,5-tetraphenyl-1*H*-1,2,5-azadiphospholium(1+); Perchlorate, *in* T-00143
4,5,5,5-Tetrahydro-3,5,5,5-tetraphenyl-1,2,5-oxazaphosphole, T-00153
2,2,2,3-Tetrahydro-2,2,2-trimethoxy-3-methyl-4,5-diphenyl-1,3,2-oxazaphosphole, T-00156
2,2,2,3-Tetrahydro-2,2,2-trimethoxy-5-methyl-1,2-oxaphosphole, T-00157
2,2,2,3-Tetrahydro-2,2,2-trimethoxy-5-phenyl-3,3-bis(trifluoromethyl)-1,4,2-oxazaphosphole, T-00158
2,2,2,3-Tetrahydro-2,2,2-trimethoxy-5-phenyl-3,3-bis(trifluoromethyl)-1-(2,4,6-trimethylphenyl)-1*H*-1,4,2-diazaphosphole, T-00159
Tetrakis(trifluoromethyl)ethylene chlorophosphate, *in* C-00184
4,4,5,5-Tetramethyl-1,2-azaphospholidine 2-oxide, T-00192
3,3',4,4'-Tetramethyl-1,1'-diphenyl-2,2'-bi-1*H*-phosphole, T-00199
3,3',4,4'-Tetramethyl-1,1'-diphenyl-2,2'-bi-1*H*-phosphole; 1,1'-Disulfide, *in* T-00199
Tetramethylenephosphinic chloride, *in* C-00167
Tetramethylenephosphinothioic chloride, *in* C-00167
Tetramethylethylene chlorophosphate, *in* C-00185
Tetramethylethylene chlorothiophosphate, *in* C-00185

4,4,5,5-Tetramethylethylene dimethylphosphoramidate, *in* D-00710
4,4,5,5-Tetramethylethylene dimethylphosphoramidothioate, *in* D-00710
Tetramethylethylene phenylphosphonate, *in* P-00322
Tetramethylethylene phosphonate, *in* T-00197
O,O-Tetramethylethylene phosphonothioate, *in* T-00197
3,3,5,5-Tetramethyl-1,2-oxaphospholidine 2-oxide, T-00208
2,2,5,5-Tetramethyl-1-phenyl-1-phospha-2,5-disilacyclopentane, T-00209
4,4,5,5-Tetramethyl phenyl phosphate, *in* P-00075
1,1,3,4-Tetramethyl-1*H*-phospholium iodide, *in* T-00536
3,3,5,5-Tetramethyl-1,2,4-triphenyl-1,2,4-triphospholane, T-00239
2,2′,5,5′-Tetraphenyl-1*H*,1′*H*-biphosphole, T-00248
1,1,3,3-Tetraphenyl-1,3-diphospholanium dibromide, *in* D-01013
1,2,3,4-Tetraphenyl-5-phosphoniaspiro[4.5]deca-1,3-diene; Bromide, *in* T-00270
1,2,3,4-Tetraphenyltetraphospholane, T-00280
1,2,3,4-Tetraphenyltetraphospholane; (1α,2β,3α,4β)-*form*, 1-Sulfide, *in* T-00280
1,2,3,4-Tetraphenyltetraphospholane; (1α,2β,3α,4β)-*form*, 1,4-Disulfide, *in* T-00280
Tetraphenylthiatetraphospholane; (2α,3β,4α,5β)-*form*, *in* T-00282
1,2-Thiaphospholane 2-sulfide; 2-Me, *in* T-00298
1,2-Thiaphospholane 2-sulfide; 2-Ph, *in* T-00298
1,3,2-Thiazaphospholidine 2-oxide; 2-Butoxy, *in* T-00302
1,3,2-Thiazaphospholidine 2-sulfide; 2-Methoxy, *in* T-00303
1,3,2-Thiazaphospholidine 2-sulfide; 2-Propoxy, *in* T-00303
1,3,2-Thiazaphospholidine 2-sulfide; 2-Phenoxy, *in* T-00303
1,3,2-Thiazaphospholidine 2-sulfide; 2-Me, *in* T-00303
1,3,2-Thiazaphospholidine 2-sulfide; 2-Ph, *in* T-00303
2,2,2-Tribromo-2,2-dihydro-1,3,2-benzodioxaphosphole, T-00352
2,2,2-Tribromo-2,2-dihydro-4,4,5,5-tetrakis(trifluoromethyl)-1,3,2-dioxaphospholane, T-00353
2,2,2-Trichloro-2,2-dihydro-1,3,2-benzodioxaphosphole, T-00400
2,2,2-Trichloro-2,2-dihydro-5-phenyl-1,3,4,2-dioxazaphosphole, *in* C-00126
2,2,2-Triethoxy-2,2-dihydro-1,3,2-dioxaphospholane, T-00436
1,1,1-Trifluoro-1,1-dihydrophospholane, T-00466
2,2,2-Trifluoro-2,2-dihydro-4,4,5,5-tetrakis(trifluoromethyl)-1,3,2-dioxaphospholane, T-00467
1,2,3-Trimethyl-1,3-diazaphospholidin-4-one 2-oxide, *in* D-00721
3,5,5-Trimethyl-2-dimethylamino-1,2-oxaphospholan-3-ol 2-oxide, *in* T-00519
2,4,5-Trimethyl-1,3,2-dioxaphospholane, T-00511
3,5,5-Trimethyl-1,2-oxaphospholan-3-ol 2-oxide; (2*RS*,3*SR*)-*form*, 2-Methoxy, *in* T-00519
3,5,5-Trimethyl-1,2-oxaphospholan-3-ol 2-oxide; (2*RS*,3*SR*)-*form*, 2-Phenoxy, *in* T-00519
3,5,5-Trimethyl-1,2-oxaphospholan-3-ol 2-oxide; (2*RS*,3*SR*)-*form*, 2-Dimethylamino, *in* T-00519
3,5,5-Trimethyl-1,2-oxaphospholan-3-ol 2-oxide; (2*RS*,3*SR*)-*form*, 2-Diethylamino, *in* T-00519
2,2,5-Trimethyl-3-phenyl-1,3-oxaphospholane; (3*RS*,5*RS*)-*form*, *in* T-00522
2,2,5-Trimethyl-3-phenyl-1,3-oxaphospholane; (3*RS*,5*SR*)-*form*, *in* T-00522
2,2,5-Trimethyl-3-phenyl-1,3-oxaphospholane; (3*RS*,5*SR*)-*form*, B,$PhCH_2Br$, *in* T-00522
2,2,5-Trimethyl-3-phenyl-1,3-oxaphospholane; (3*RS*,5*SR*)-*form*, 3-Oxide, *in* T-00522
3,5,5-Trimethyl-2-phenyl-1,2-oxaphospholan-3-ol 2-oxide, *in* T-00519
3,5,5-Trimethyl-2-phenyl-1,2-oxaphospholan-3-ol 2-oxide, *in* T-00519
2,3,4-Trimethyl-5-phenyl-1,3,2-oxazaphospholidine; (2*R*,4*S*,5*R*)-*form*, 2-Oxide, *in* T-00523
2,3,4-Trimethyl-5-phenyl-1,3,2-oxazaphospholidine; (2*R*,4*S*,5*R*)-*form*, 2-Sulfide, *in* T-00523
2,3,4-Trimethyl-5-phenyl-1,3,2-oxazaphospholidine; (2*R*,4*S*,5*R*)-*form*, 2-Selenide, *in* T-00523
2,3,4-Trimethyl-5-phenyl-1,3,2-oxazaphospholidine; (2*S*,4*S*,5*R*)-*form*, 2-Oxide, *in* T-00523
2,3,4-Trimethyl-5-phenyl-1,3,2-oxazaphospholidine; (2*S*,4*S*,5*R*)-*form*, 2-Sulfide, *in* T-00523
2,3,4-Trimethyl-5-phenyl-1,3,2-oxazaphospholidine; (2*S*,4*S*,5*R*)-*form*, 2-Selenide, *in* T-00523
1,3,4-Trimethyl-1-phenyl-1*H*-phospholium iodide, *in* D-00807
1,2,4-Trimethyl-2-phosphabicyclo[2.2.1]heptane 2-oxide, *in* D-00825
1,3,4-Trimethyl-1*H*-phosphole, T-00536
1,3,4-Trimethyl-1*H*-phosphole; Sulfide, *in* T-00536
2-(Trimethylsilyloxy)-1,3,2-dioxaphospholane, T-00554
1,2,4-Trimethyl-1,2,4,3,5-triazadiphospholidine; 3,5-Dichloro, *in* T-00562
1,2,4-Trimethyl-1,2,4,3,5-triazadiphospholidine; 3,5-Dimethoxy, *in* T-00562
1,2,4-Trimethyl-1,2,4,3,5-triazadiphospholidine; 3,5-Bis(methylthio), *in* T-00562
1,2,4-Trimethyl-1,2,4-triphospholane, T-00568
1,2,3-Triphenyl-1,3-azaphosphacyclopentane, *in* D-00995
1,2,3-Triphenyl-1,3-azaphospholidin-5-one, T-00596
1,2,3-Triphenyl-1,3-azaphospholidin-5-one; 3-Oxide, *in* T-00596
1,2,3-Triphenyl-1,3-azaphospholidin-5-one; 3-Sulfide, *in* T-00596
2,4,5-Triphenyl-2*H*-1,3,2-diazaphosphole; 2-Oxide, *in* T-00600
3,4,5-Triphenyl-4*H*-1,2,4-diazaphosphole, T-00601
2,4,5-Triphenyl-1,3,2-dioxaphosphole; 2-Sulfide, *in* T-00602
1,2,3-Triphenyl-1,3,2-diphospharsolane, T-00603
2,4,5-Triphenyl-1,3,2,4,5-dithiatriphospholane, T-00606
2,4,5-Triphenyl-1,3,4-oxathiaphospholane; 4-Sulfide, *in* T-00611
1,2,5-Triphenyl-1*H*-phosphole, T-00634
1,2,5-Triphenyl-1*H*-phosphole; Oxide, *in* T-00634
1,2,5-Triphenyl-1*H*-phosphole; Sulfide, *in* T-00634
1,2,5-Triphenyl-1*H*-phosphole; Selenide, *in* T-00634
1,2,3-Triphenyl-1,2,3-triphospholane, T-00699

1 P-contg. ring; 6-membered

Acridophosphine, A-00031
2-Amino-3-(2-chloroethyl)tetrahydro-4*H*-1,3,2-oxazaphosphorine 2-oxide; (*R*)-*form*, *in* A-00063
2-Amino-3-(2-chloroethyl)tetrahydro-4*H*-1,3,2-oxazaphosphorine 2-oxide; (*S*)-*form*, *in* A-00063
2-Amino-3-(2-chloroethyl)tetrahydro-4*H*-1,3,2-oxazaphosphorine 2-oxide; (±)-*form*, *in* A-00063
2-Amino-3-(2-chloroethyl)tetrahydro-4*H*-1,3,2-oxazaphosphorine 2-oxide; (±)-*form*, *N*-Ac, *in* A-00063
2-Amino-2,2,4,4,6,6-hexahydro-2-methyl-4,4,6,6-tetraphenyl-1,3,5-triaza-2,4,6-triphosphorine, A-00081
2-Amino-2,2,4,4,6,6-hexahydro-2,4,4,6,6-pentaphenoxy-1,3,5-triaza-2,4,6-triphosphorine, A-00082
5-Anilino-5,6-dihydrophosphanthridene 5-oxide, *in* D-00504
Benzo[*e*]phenazaphosphinic acid, *in* D-00440
7*H*-Benzo[*c*]phenothiaphosphine; 7-Butyl, 7-oxide, *in* B-00022
12*H*-Benzo[*c*]phenothiaphosphine; 12-Ethoxy, 7,7,12-trioxide, *in* B-00023
1-Benzyl-4-*tert*-butyl-1-phenylphosphorinanium bromide, *in* B-00586
3-Benzyl-2-chlorotetrahydro-2*H*-1,3,2-oxazaphosphorine, B-00041
3-Benzyl-2-chlorotetrahydro-2*H*-1,3,2-oxazaphosphorine; 2-Sulfide, *in* B-00041
5-Benzyl-5,5-dihydro-5-phenylacridophosphine, B-00042
5-Benzyl-5,10-dihydro-5-phenylacridophosphonium bromide, *in* D-00547
5-Benzyl-5,6-dihydro-6-phenylphosphanthridine, B-00043
5-Benzyl-5,6-dihydro-6-phenylphosphanthridine; (5*RS*,6*RS*)-*form*, 5-Sulfide, *in* B-00043
1-Benzyl-2-ethoxyhexahydro-1,2-azaphosphorine 2-oxide, *in* B-00048
1-Benzylhexahydro-2-hydroxy-1,2-azaphosphorine 2-oxide, B-00048
1-Benzyl-4-oxo-2,2,6,6-tetramethyl-1-phenylphosphorinanium bromide, *in* T-00210

8-Benzyl-8-phenyl-8-phosphoniabicyclo[3.2.1]octan-3-one chloride, *in* P-00199
1,1′-Biphosphorinane; Disulfide, *in* B-00085
2,2′-Bis(5,5-dimethyl-1,3,2-dioxaphosphorinane), B-00197
2,2′-Bis(5,5-dimethyl-1,3,2-dioxaphosphorinane) 2,2′-dioxide, *in* B-00197
Bis(5,5-dimethyl-2-oxo-1,3,2-dioxaphosphorinan-2-yl) disulfide, *in* D-01251
Bis(5,5-dimethyl-2-selenoxo-1,3,2-dioxaphosphorinan-2-yl) disulfide, *in* D-01251
Bis(5,5-dimethyl-2-thioxo-1,3,2-dioxaphosphorinan-2-yl) disulfide, *in* D-01251
2-Bromo-5,5-dimethyl-1,3,2-dioxaphosphorinane, B-00471
2-*tert*-Butoxy-5,5-dimethyl-1,3,2-dioxaphosphorinane, B-00534
2-*tert*-Butylamino-4-methyl-1,3,2-dioxaphosphorinane, B-00541
2-*tert*-Butylamino-4-methyl-1,3,2-dioxaphosphorinane; (2*RS*,4*SR*)-*form*, 2-Oxide, *in* B-00541
2-*tert*-Butylamino-4-methyl-1,3,2-dioxaphosphorinane; (2*RS*,4*SR*)-*form*, 2-Selenide, *in* B-00541
tert-Butyl 1,3-butylene phosphate, *in* B-00537
5-*tert*-Butyl-2-chloro-1,3,2-dioxaphosphorinane, B-00547
5-*tert*-Butyl-2-chlorotetrahydro-3-phenyl-2*H*-1,3,2-oxazaphosphorine, B-00549
1-*tert*-Butyl-1,4-dihydro-4-phenyl-1,4-phospharsenin; *cis-form*, *in* B-00553
1-*tert*-Butyl-1,4-dihydro-4-phenyl-1,4-phospharsenin; *cis-form*, 1-Oxide, *in* B-00553
1-*tert*-Butyl-1,4-dihydro-4-phenyl-1,4-phospharsenin; *trans-form*, *in* B-00553
1-*tert*-Butyl-1,4-dihydro-4-phenyl-1,4-phospharsenin; *trans-form*, 1-Oxide, *in* B-00553
5-*tert*-Butyl-2-dimethylamino-1,3,2-dioxaphosphorinane, B-00555
5-*tert*-Butyl-2-dimethylamino-1,3,2-dioxaphosphorinane; *trans-form*, 2-Oxide, *in* B-00555
tert-Butyl *N,N*′-dimethyl-*N,N*′-trimethylene phosphorodiamidite, *in* H-00022
1,3-Butylene *tert*-butylphosphoramidate, *in* B-00541
O,O-1,3-Butylene *tert*-butylphosphoramidoselenoate, *in* B-00541
S,S-1,3-Butylene diethylphosphoramidodithioate, *in* M-00138
N,N′-1,3-Butylene *N,N*′-dimethylphosphonic diamide, *in* H-00057
O,O-1,3-Butylene dimethylphosphoramidoselenoate, *in* M-00117
O,O-1,3-Butylene dimethylphosphoramidothioate, *in* M-00117
N,N′-1,3-Butylene-*N,N*′-dimethylphosphorodiamidous chloride, *in* H-00057
S,S-1,3-Butylene *O*-ethyl phosphorotrithioate, *in* E-00042
O,O-1,3-Butylene *O*-ethyl thiophosphate, *in* E-00041
O,O-1,3-Butylene fluoroselenophosphate, *in* F-00021
O,O-1,3-Butylene fluorothiophosphate, *in* F-00021
O,O-1,3-Butylene *S*-methyl dithiophosphate, *in* M-00174
1,3-Butylene methylphosphonate, *in* D-00734
O,O-1,3-Butylene methylphosphonothioate, *in* D-00734
S,S′-1,3-Butylene *O*-methyl phosphorodithioite, *in* M-00138
O,O-1,3-Butylene *S*-methyl phosphoroselenothioate, *in* M-00174
O,O-1,3-Butylene *S*-methyl thiophosphate, *in* M-00174
1,3-Butylene phenyl phosphate, *in* M-00191
1,3-Butylene phenylphosphonate, *in* M-00199
1,3-Butylene phenylphosphoramidate, *in* M-00193
O,O-1,3-Butylene phenylphosphoramidoselenoate, *in* M-00193
O,O-1,3-Butylene phenylphosphoramidothioate, *in* M-00193
O,O-1,3-Butylene *O*-phenyl thiophosphate, *in* M-00191
2,3-Butylene phosphonate, *in* D-00733
O,O-2,3-Butylene phosphonothioate, *in* D-00733
S,S′-1,3-Butylene phosphorochloridodithioite, *in* M-00138
O,O-Butylene phosphorochloridoselenoate, *in* C-00093
1,3-Butylene phosphoryl bromide, *in* B-00474
OO-1,3-Butylene thiophosphoryl bromide, *in* B-00474
5-*tert*-Butyl-2-methoxy-1,3,2-dioxaphosphorinane, B-00563
5-*tert*-Butyl-2-methoxy-1,3,2-dioxaphosphorinane; *cis-form*, 2-Oxide, *in* B-00563
5-*tert*-Butyl-2-methyl-1,3,2-dioxaphosphorinane; *cis-form*, 2-Oxide, *in* B-00564
5-*tert*-Butyl-2-methyl-1,3,2-dioxaphosphorinane; *trans-form*, 2-Oxide, *in* B-00564
4-*tert*-Butyl-1-methyl-1-phenylphosphorinanium bromide, *in* B-00586
tert-Butyl neopentylene phosphate, *in* B-00534
5-*tert*-Butyl-2-phenyl-1,3,2-dioxaphosphorinane, B-00577
5-*tert*-Butyl-2-phenyl-1,3,2-dioxaphosphorinane; *trans-form*, 2-Oxide, *in* B-00577
4-*tert*-Butyl-1-phenylphosphorinane; *cis-form*, *in* B-00586
4-*tert*-Butyl-1-phenylphosphorinane; *cis-form*, Oxide, *in* B-00586
4-*tert*-Butyl-1-phenylphosphorinane; *trans-form*, *in* B-00586
4-*tert*-Butyl-1-phenylphosphorinane; *trans-form*, Oxide, *in* B-00586
4-*tert*-Butyl-1-phenylphosphorinane; *trans-form*, B,MeBr, *in* B-00586
4-*tert*-Butyl-1-phenylphosphorinane; *trans-form*, B,PhCH$_2$Br, *in* B-00586
Butylphosphonothioic acid trimer trianhydride, *in* T-00396
5-*tert*-Butyltetrahydro-2-dimethylamino-2*H*-1,3,2-oxazaphosphorine; (2*RS*,5*RS*)-*form*, 2-Oxide, *in* B-00635
5-*tert*-Butyltetrahydro-2-dimethylamino-2*H*-1,3,2-oxazaphosphorine; (2*RS*,5*RS*)-*form*, 2-Sulfide, *in* B-00635
5-*tert*-Butyltetrahydro-2-dimethylamino-2*H*-1,3,2-oxazaphosphorine; (2*RS*,5*SR*)-*form*, 2-Oxide, *in* B-00635
5-*tert*-Butyltetrahydro-2-dimethylamino-2*H*-1,3,2-oxazaphosphorine; (2*RS*,5*SR*)-*form*, 2-Sulfide, *in* B-00635
5-*tert*-Butyltetrahydro-2-dimethylamino-3-phenyl-2*H*-1,3,2-oxazaphosphorine, B-00636
5-*tert*-Butyltetrahydro-2-dimethylamino-3-phenyl-2*H*-1,3,2-oxazaphosphorine; (2*RS*,5*RS*)-*form*, 2-Oxide, *in* B-00636
5-*tert*-Butyltetrahydro-2-dimethylamino-3-phenyl-2*H*-1,3,2-oxazaphosphorine; (2*RS*,5*RS*)-*form*, 2-Sulfide, *in* B-00636
5-*tert*-Butyltetrahydro-2-dimethylamino-3-phenyl-2*H*-1,3,2-oxazaphosphorine; (2*RS*,5*SR*)-*form*, 2-Oxide, *in* B-00636
5-*tert*-Butyltetrahydro-2-dimethylamino-3-phenyl-2*H*-1,3,2-oxazaphosphorine; (2*RS*,5*SR*)-*form*, 2-Sulfide, *in* B-00636
4-*tert*-Butyl-1,1,3,3-tetrakis(dimethylamino)-1,3,5-triphosphorin, B-00638
2-*tert*-Butyltrimethylene dimethylphosphoramidate, *in* B-00555
2-*tert*-Butyltrimethylene phenylphosphonate, *in* B-00577
2-Chloro-4*H*-1,3,2-benzodioxaphosphorinan-4-one, C-00022
2-Chloro-4*H*-1,3,2-benzodioxaphosphorinan-4-one; 2-Oxide, *in* C-00022
2-Chloro-3-(2-chloroethyl)tetrahydro-2*H*-1,3,2-oxazaphosphorine, C-00030
2-Chloro-3-(2-chloroethyl)tetrahydro-2*H*-1,3,2-oxazaphosphorine; (±)-*form*, 2-Oxide, *in* C-00030
2-Chloro-5-chloromethyl-5-methyl-1,3,2-dioxaphosphorinane; *cis-form*, 2-Oxide, *in* C-00031
2-Chloro-5-chloromethyl-5-methyl-1,3,2-dioxaphosphorinane; *cis-form*, 2-Sulfide, *in* C-00031
2-Chloro-5-chloromethyl-5-methyl-1,3,2-dioxaphosphorinane; *trans-form*, 2-Oxide, *in* C-00031
2-Chloro-5-chloromethyl-5-methyl-1,3,2-dioxaphosphorinane; *trans-form*, 2-Sulfide, *in* C-00031
6-Chloro-6*H*-dibenzo[*c,e*][1,2]oxaphosphorin, C-00033
6-Chloro-6*H*-dibenzo[*c,e*][1,2]oxaphosphorin; 6-Oxide, *in* C-00033
6-Chloro-6*H*-dibenzo[*c,e*][1,2]oxaphosphorin; 6-Sulfide, *in* C-00033
5-Chloro-5,10-dihydroacridophosphine, C-00037
5-Chloro-5,10-dihydro-10-methylacridophosphine, *in* D-00515
4-Chloro-1,4-dihydro-2*H*-naphth[2,1-*c*][1,2]oxaphosphorin, C-00048
10-Chloro-5,10-dihydrophenophosphazine; 10-Oxide, *in* C-00049
10-Chloro-5,10-dihydrophenophosphazine; 10-Sulfide, *in* C-00049
5-Chloro-5,10-dihydro-10-phenylacridophosphine, *in* D-00548
2-Chloro-2,3-dihydro-3-phenyl-4*H*-1,3,2-benzoxazaphosphorin-4-one, C-00050
2-Chloro-2,3-dihydro-3-phenyl-4*H*-1,3,2-benzoxazaphosphorin-4-one; 2-Oxide, *in* C-00050
5-Chloro-5,6-dihydrophosphanthridine 5-oxide, *in* D-00504
2-Chloro-3,6-dihydro-2*H*-1,2-thiaphosphorin 2-sulfide, C-00053
2-Chloro-5,5-dimethyl-1,3,2-dioxaphosphorinane, C-00058
2-Chloro-5,5-dimethyl-1,3,2-dioxaphosphorinane 2-oxide, C-00059
2-Chloro-5,5-dimethyl-1,3,2-dioxaphosphorinane 2-sulfide, C-00060

2-Chloro-5,5-dimethyl-4-phenyl-1,3,2-dioxaphosphorinane 2-oxide, *in* H-00129

▷2-Chloro-1,3,2-dioxaphosphorinane, C-00066

2-Chloro-1,3,2-dioxaphosphorinane 2-oxide, C-00067

2-Chloro-1,3,2-dioxaphosphorinane 2-sulfide, C-00068

4-Chloro-2,6-diphenyl-4*H*-1,4-oxaphosphorin 4-oxide, *in* H-00140

4-Chloro-2,6-diphenyl-4*H*-1,4-thiaphosphorin 4-oxide, *in* H-00143

2-[(2-Chloroethyl)amino]tetrahydro-2*H*-1,3,2-oxazaphosphorine 2-oxide; (*R*)-*form*, *in* C-00072

2-[(2-Chloroethyl)amino]tetrahydro-2*H*-1,3,2-oxazaphosphorine 2-oxide; (*R*)-*form*, 3-[(*S*)-1-Phenylethyl], *in* C-00072

2-[(2-Chloroethyl)amino]tetrahydro-2*H*-1,3,2-oxazaphosphorine 2-oxide; (*S*)-*form*, *in* C-00072

2-Chlorohexahydro-4*H*-1,3,2-benzodioxaphosphorin, C-00085

2-Chlorohexahydro-4*H*-1,3,2-benzodioxaphosphorin; (2*RS*,4*aSR*,8*aSR*)-*form*, 2-Oxide, *in* C-00085

2-Chloro-2,2,4,4,6,6-hexahydro-2-methyl-4,4,6,6-tetraphenyl-1,3,5-triaza-2,4,6-triphosphorine, C-00086

2-Chloro-2,2,4,4,6,6-hexahydro-2,4,4,6,6-pentaphenoxy-1,3,5-triaza-2,4,6-triphosphorine, C-00087

2-Chloro-2,2,4,4,6,6-hexahydro-2,4,4,6,6-pentaphenyl-1,3,5-triaza-2,4,6-triphosphorine, C-00088

2-Chloro-4-isopropyl-5,5-dimethyl-1,3,2-dioxaphosphorinane; (2*RS*,4*RS*)-*form*, *in* C-00090

2-Chloro-4-isopropyl-5,5-dimethyl-1,3,2-dioxaphosphorinane; (2*RS*,4*RS*)-*form*, 2-Oxide, *in* C-00090

2-Chloro-4-isopropyl-5,5-dimethyl-1,3,2-dioxaphosphorinane; (2*RS*,4*RS*)-*form*, 2-Sulfide, *in* C-00090

2-Chloro-4-isopropyl-5,5-dimethyl-1,3,2-dioxaphosphorinane; (2*RS*,4*RS*)-*form*, 2-Selenide, *in* C-00090

2-Chloro-4-isopropyl-5,5-dimethyl-1,3,2-dioxaphosphorinane; (2*RS*,4*SR*)-*form*, 2-Oxide, *in* C-00090

2-Chloro-4-isopropyl-5,5-dimethyl-1,3,2-dioxaphosphorinane; (2*RS*,4*SR*)-*form*, 2-Sulfide, *in* C-00090

2-Chloro-4-isopropyl-5,5-dimethyl-1,3,2-dioxaphosphorinane; (2*RS*,4*SR*)-*form*, 2-Selenide, *in* C-00090

2-Chloro-4-methyl-1,3,2-dioxaphosphorinane, C-00093

2-Chloro-4-methyl-1,3,2-dioxaphosphorinane 2-oxide; (2*RS*,4*RS*)-*form*, *in* C-00094

2-Chloro-4-methyl-1,3,2-dioxaphosphorinane 2-oxide; (2*RS*,4*SR*)-*form*, *in* C-00094

2-Chloro-4-methyl-1,3,2-dioxaphosphorinane 2-sulfide, C-00095

5-Chloromethyl-2-fluoro-5-methyl-1,3,2-dioxaphosphorinane; *cis-form*, 2-Oxide, *in* C-00096

5-Chloromethyl-2-fluoro-5-methyl-1,3,2-dioxaphosphorinane; *cis-form*, 2-Sulfide, *in* C-00096

5-Chloromethyl-2-fluoro-5-methyl-1,3,2-dioxaphosphorinane; *trans-form*, 2-Oxide, *in* C-00096

5-Chloromethyl-2-fluoro-5-methyl-1,3,2-dioxaphosphorinane; *trans-form*, 2-Sulfide, *in* C-00096

5-Chloromethyl-2-hydroxy-5-methyl-1,3,2-dioxaphosphorinane 2-oxide, C-00097

5-Chloromethyl-2-methoxy-5-methyl-1,3,2-dioxaphosphorinane; *cis-form*, 2-Oxide, *in* C-00098

5-Chloromethyl-2-methoxy-5-methyl-1,3,2-dioxaphosphorinane; *trans-form*, 2-Oxide, *in* C-00098

5-Chloromethyl-5-methyl-2-phenoxy-1,3,2-dioxaphosphorinane; 2-Oxide, *in* C-00099

5-Chloromethyl-5-methyl-2-phenoxy-1,3,2-dioxaphosphorinane; 2-Sulfide, *in* C-00099

2-Chloronaphtho[1,8-*de*]-1,3,2-dioxaphosphorin, C-00120

2-Chloro-1,2-oxaphosphorinane 2-oxide, *in* H-00167

1-Chlorophosphorinane, C-00168

1-Chlorophosphorinane; Oxide, *in* C-00168

1-Chlorophosphorinane; Sulfide, *in* C-00168

2-Chloro-1,2,3,4-tetrahydro-6-methyl-3-(4-methylphenyl)-1,3,2-benzodiazaphosphorine 2-oxide, C-00177

2-Chlorotetrahydro-3-methyl-2*H*-1,3,2-oxazaphosphorine, C-00178

4-Chloro-1,2,3,4-tetrahydronaphth[2,1-*c*][1,2]-azaphosphorine; Sulfide, *in* C-00179

2-Chlorotetrahydro-2*H*-1,3,2-oxazaphosphorine; 2-Oxide, *in* C-00180

2-Chlorotetrahydro-2*H*-1,3,2-oxazaphosphorine; 2-Sulfide, *in* C-00180

2-Chlorotetrahydro-3-(1-phenylethyl)-2*H*-1,3,2-oxazaphosphorine 2-oxide; (R_C,R_P)-*form*, *in* C-00181

2-Chlorotetrahydro-3-(1-phenylethyl)-2*H*-1,3,2-oxazaphosphorine 2-oxide; (R_C,S_P)-*form*, *in* C-00181

2-Chlorotetrahydro-3-(1-phenylethyl)-2*H*-1,3,2-oxazaphosphorine 2-oxide; (S_C,R_P)-*form*, *in* C-00181

2-Chlorotetrahydro-3-(1-phenylethyl)-2*H*-1,3,2-oxazaphosphorine 2-oxide; (S_C,S_P)-*form*, *in* C-00181

1-Chloro-1,2,3,4-tetrahydrophosphinoline, *in* T-00135

2-Chlorotetrahydro-3,5,5-trimethyl-2*H*-1,3,2-oxazaphosphorine; 2-Oxide, *in* C-00183

2-Chlorotetrahydro-3,5,5-trimethyl-2*H*-1,3,2-oxazaphosphorine; 2-Sulfide, *in* C-00183

7-Chloro-6-thia-7-phosphabicyclo[3.2.1]octane; 7-Sulfide, *in* C-00187

2-Chloro-4,4,6-trimethyl-1,3,2-dioxaphosphorinane, C-00189

2-Choro-1,3,2-dithiaphosphorinane, C-00199

2-Choro-1,3,2-dithiaphosphorinane; 2-Oxide, *in* C-00199

2-Choro-1,3,2-dithiaphosphorinane; 2-Sulfide, *in* C-00199

2-Cyano-5,5-dimethyl-1,3,2-dioxaphosphorinane, C-00209

2-Cyano-5,5-dimethyl-1,3,2-dioxaphosphorinane; 2-Oxide, *in* C-00209

Cyclohexylphosphonothioic acid trimer trianhydride, *in* T-00430

Cyclophosphamide; (*R*)-*form*, *in* C-00283

Cyclophosphamide; (*S*)-*form*, *in* C-00283

Cyclophosphamide; (±)-*form*, *in* C-00283

Decahydro-1,4-dihydroxy-1,4-benzodiphosphorin 1,4-dioxide, D-00003

2,2-Diamino-2,2,4,4,6,6-hexahydro-4,4,6,6-tetraphenyl-1,3,5-triaza-2,4,6-triphosphorine, D-00025

2,2-Diamino-2,2,4,4,6,6-hexahydro-4,4,6,6-tetraphenyl-1,3,5-triaza-2,4,6-triphosphorine; *N*,*N*,*N'*,*N'*-Tetra-Me, *in* D-00025

2,2-Diamino-2,2,4,4,6,6-hexahydro-4,4,6,6-tetraphenyl-1,3,5-triaza-2,4,6-triphosphorine; *N*,*N'*-Di-Ph, *in* D-00025

6*H*-Dibenzo[*c*,*e*][1,2]oxaphosphorin; 6-Oxide, *in* D-00039

1,1-Dibromo-1,1-dihydro-2,4,6-triphenylphosphorin, *in* D-00613

4,4-Dibutyl-1-*tert*-butyl-1,4-dihydro-1,4-phosphastannin, D-00075

4,4-Dibutyl-1-*tert*-butyl-1,4-dihydro-1,4-phosphastannin; 1-Oxide, *in* D-00075

2,6-Di-*tert*-butyl-1,4-dihydro-4,4-diphenyl-1,4-azaphosphorinium bromide, *in* D-00084

2,6-Di-*tert*-butyl-1,4-dihydro-4-phenyl-1,4-azaphosphorine; 4-Oxide, *in* D-00084

2,6-Di-*tert*-butyl-1,4-dihydro-4-phenyl-1,4-azaphosphorine; 1-Me, 4-oxide, *in* D-00084

4,4'-Di-*tert*-butyl-1,1',2,2',5,5',6,6'-octahydro-1,1'-diphenyl-2,2'-biphosphorin; *P*,*P'*-Disulfide, *in* D-00100

2,4-Dichloro-2,4-bis(dimethylamino)-2,2,4,4,6,6-hexahydro-6,6-diphenyl-1,3,5-triaza-2,4,6-triphosphorine; (2*RS*,4*RS*)-*form*, *in* D-00162

2,4-Dichloro-2,4-bis(dimethylamino)-2,2,4,4,6,6-hexahydro-6,6-diphenyl-1,3,5-triaza-2,4,6-triphosphorine; (2*RS*,4*SR*)-*form*, *in* D-00162

5,10-Dichloro-5,10-dihydrophosphanthrene, D-00164

5,10-Dichloro-5,10-dihydrophosphanthrene; 5,10-Dioxide, *in* D-00164

1,1-Dichloro-1,1-dihydro-2,4,6-triphenylphosphorin, *in* D-00613

2,2-Dichloro-2,2,4,4,6,6-hexahydro-4,4,6,6-tetraphenyl-1,3,5-triaza-2,4,6-triphosphorine, D-00169

2,4-Dichloro-2,4,6,6-tetrakis(dimethylamino)-2,2,4,4,6,6-hexahydro-1,3,5-triaza-2,4,6-triphosphorine, D-00194
7,7-Dichloro-2,3,5,6-tetrakis(trifluoromethyl)-1,4-diphosphabicyclo[2.2.1]hepta-2,5-diene, D-00195
3,9-Dichloro-2,4,8,10-tetraoxa-3,9-diphosphaspiro[5.5]-undecane, D-00196
3,9-Dichloro-2,4,8,10-tetraoxa-3,9-diphosphaspiro[5.5]-undecane; 3,9-Dioxide, *in* D-00196
3,9-Dichloro-2,4,8,10-tetraoxa-3,9-diphosphaspiro[5.5]-undecane; 3,9-Disulfide, *in* D-00196
2,5-Diethoxy-1,3,2,5-dioxadiphosphorinane; 5-Oxide, *in* D-00231
2,5-Diethoxy-1,3,2,5-dioxadiphosphorinane; 5-Oxide, 2-sulfide, *in* D-00231
2,5-Diethoxy-1,4,2,5-dioxaphosphorinane 2,5-dioxide, *in* D-00618
2-Diethylaminonaphtho[1,8-*de*]-1,3,2-dioxaphosphorin, D-00267
N,*N*-Diethyl-2,6-dimethyl-1-phenyl-1,4-phosphaborin-4(1*H*)-amine, D-00285
N,*N*-Diethyl-2,6-dimethyl-1-phenyl-1,4-phosphaborin-4(1*H*)-amine; 1-Oxide, *in* D-00285
N,*N*-Diethyl-2,6-dimethyl-1-phenyl-1,4-phosphaborin-4(1*H*)-amine; 1-Sulfide, *in* D-00285
N,*N*-Diethyl-2,6-dimethyl-1-phenyl-1,4-phosphaborin-4(1*H*)-amine; 1-Selenide, *in* D-00285
2,3-Diethyl-2,3-diphosphabicyclo[2.2.1]hept-5-ene, D-00287
2,3-Diethyl-2,3-diphosphabicyclo[2.2.1]hept-5-ene; Disulfide, *in* D-00287
2,3-Diethyl-2,3-diphosphabicyclo[2.2.2]octane; Disulfide, *in* D-00288
1,1-Difluoro-1,1-dihydro-2,4,6-triphenylphosphorin, *in* D-00613
2,2-Difluoro-2,2,4,4,6,6-hexahydro-4,4,6,6-tetraphenyl-1,3,5-triaza-2,4,6-triphosphorine, D-00412
7,12-Dihydrobenzo[*e*]phenophosphazine; 7-Oxide, *in* D-00440
1,1-Dihydro-1,1-bis(methylthio)-2,4,6-triphenylphosphorin, *in* D-00613
1,1-Dihydro-1,1-bis(phenylthio)-2,4,6-triphenylphosphorin, *in* D-00613
5,10-Dihydro-5,10-dihydroxyphosphanthrene 5,10-dioxide, D-00452
1,1-Dihydro-1,1-dimethoxy-2,4,6-triphenylphosphorin, *in* D-00613
2,3-Dihydro-1,3-dimethyl-1,3,2-benzodiazaphosphorin-4(1*H*)-one; 2-Oxide, *in* D-00457
2,3-Dihydro-1,3-dimethyl-1,3,2-benzodiazaphosphorin-4(1*H*)-one; 2-Me, 2-oxide, *in* D-00457
2,3-Dihydro-2,7-dimethyl-1,4,2-benzodithiaphosphorin, D-00458
2,3-Dihydro-2,7-dimethyl-1,4,2-benzodithiaphosphorin; 2-Selenide, *in* D-00458
9,10-Dihydro-9,10-dimethyl-9,10-diphenylphosphaanthranium diiodide, *in* D-00482
5,10-Dihydro-5,10-dimethylphenophosphazine, D-00466
5,10-Dihydro-5,10-dimethylphenophosphazine; 10-Oxide, *in* D-00466
5,10-Dihydro-5,10-dimethylphenophosphazine; 10-Sulfide, *in* D-00466
5,6-Dihydro-5,5-dimethylphosphanthridinium chloride, *in* D-00568
1,1-Dihydro-1,1-diphenoxy-2,4,6-triphenylphosphorin, *in* D-00613
5,10-Dihydro-5,10-diphenylacridophosphine 5-oxide, *in* D-00548
1,4-Dihydro-1,4-diphenyl-1,4-benzodiphosphorin, D-00477
5,10-Dihydro-5,10-diphenylphosphanthrene, D-00482
5,10-Dihydro-5,10-diphenylphosphanthrene; Monooxide, *in* D-00482
5,10-Dihydro-5,10-diphenylphosphanthrene; 5,10-Dioxide, *in* D-00482
5,6-Dihydro-5,5-diphenylphosphanthridinium bromide, *in* D-00568
1,1-Dihydro-1,1-diphenylphosphorin, D-00484
5,10-Dihydro-10-ethoxyphenophosphazine 10-sulfide, *in* D-00502
1,1-Dihydro-1,1,3,3,5,5-hexamethyl-1-phospha-3,5-disilacyclohex-1-ene, D-00485
5,10-Dihydro-5-hydroxyacridophosphine 5-oxide, D-00487
4,5-Dihydro-4-hydroxybenzo[*lmn*]phosphanthridine 4-oxide, D-00489
1,2-Dihydro-2-hydroxyisophosphinoline 2-oxide, D-00495
5,10-Dihydro-5-hydroxy-10-methylacridophosphine 5-oxide, *in* D-00515
3,6-Dihydro-2-hydroxy-2*H*-1,2-oxaphosphorin 2-oxide; 2-Methoxy (Me ester), *in* D-00501
5,10-Dihydro-5-hydroxy-10-phenylacridophosphine 5-oxide, *in* D-00548
1,2-Dihydro-1-hydroxy-1-phenyl-4*H*-3,1,2-benzothiazaphosphorin-4-one 1-sulfide; 2-Methoxy (*O*-Me ester), *in* D-00503
1,2-Dihydro-1-hydroxy-1-phenyl-4*H*-3,1,2-benzothiazaphosphorin-4-one 1-sulfide; 2-Ethoxy (*O*-Et ester), *in* D-00503
5,6-Dihydro-5-hydroxyphosphanthridine 5-oxide, D-00504
5,6-Dihydro-5-hydroxyphosphanthridine 5-oxide; Anhydride, *in* D-00504
1,2-Dihydro-1-hydroxyphosphinoline 1-oxide, D-00505
1,2-Dihydroisophosphinoline, D-00510
5,10-Dihydro-10-isopropoxyphenophosphazine 10-sulfide, *in* D-00502
5,10-Dihydro-5-methoxyacridophosphine 5-oxide, *in* D-00487
5,10-Dihydro-10-methoxyphenophosphazine 10-sulfide, *in* D-00502
1,2-Dihydro-1-methoxyphosphinoline 1-oxide, *in* D-00505
5,10-Dihydro-10-methylacridophosphine, D-00515
2,3-Dihydro-2-methyl-1,4,2-benzodioxaphosphorin; 2-Oxide, *in* D-00517
2,3-Dihydro-2-methyl-1,3,2-benzothiazaphosphole; 2-Sulfide, *in* D-00519
2,3-Dihydro-2-methyl-1,3,2-benzothiazaphosphole; 2-Selenide, *in* D-00519
2,3-Dihydro-2-methyl-1*H*-4,1,2-benzothiazaphosphorine; 2-Oxide, *in* D-00520
2,3-Dihydro-2-methyl-1*H*-4,1,2-benzothiazaphosphorine; 2-Sulfide, *in* D-00520
3,4-Dihydro-2-methyl-2*H*-1,4,2-benzothiazaphosphorine, D-00521
3,4-Dihydro-2-methyl-2*H*-1,4,2-benzothiazaphosphorine; 2-Sulfide, *in* D-00521
3,4-Dihydro-2-methyl-2*H*-1,4,2-benzothiazaphosphorine; 2-Selenide, *in* D-00521
3,4-Dihydro-2-methyl-2*H*-1,4,2-benzoxazaphosphorine; 2-Oxide, *in* D-00522
3,4-Dihydro-2-methyl-2*H*-1,4,2-benzoxazaphosphorine; 2-Sulfide, *in* D-00522
1,2-Dihydro-1-methyl-4*H*-3,1,2-benzoxazaphosphorin-4-one; 1-Methoxy, *in* D-00523
1,2-Dihydro-1-methyl-4*H*-3,1,2-benzoxazaphosphorin-4-one; 1-Ethoxy, *in* D-00523
1,2-Dihydro-1-methyl-4*H*-3,1,2-benzoxazaphosphorin-4-one; 1-Isopropoxy, *in* D-00523
1,2-Dihydro-1-methyl-4*H*-3,1,2-benzoxazaphosphorin-4-one; 1-Chloro, *in* D-00523
5,10-Dihydro-5-methyl-5,10-diphenylacridophosphonium iodide, *in* D-00548
5,10-Dihydro-5-methyl-10-phenylphenophosphazine, D-00526
5,10-Dihydro-5-methyl-10-phenylphenophosphazine; 10-Oxide, *in* D-00526
5,6-Dihydro-5-methylphosphanthridene 5-oxide, *in* D-00568
1,2-Dihydro-1-methyl-1,2,4,6-tetraphenylphosphorinium iodide, *in* D-00584
5,10-Dihydro-10-methylthiophenophosphazine 10-oxide, *in* D-00502
2,3-Dihydro-1*H*-naphtho[1,8-*de*]-1,3,2-diazaphosphorine; 2-(4-Nitrophenoxy), 2-oxide, *in* D-00532
2,3-Dihydro-1*H*-naphtho[1,8-*de*]-1,3,2-diazaphosphorine; 2-(4-Chloro-3-methylphenoxy), 2-oxide, *in* D-00532
9,10-Dihydro-10-oxo-9-phenyl-9-phosphoniaanthracene chloride, *in* P-00080
5,10-Dihydrophenophosphazine; 10-Oxide, *in* D-00542
5,10-Dihydro-5-phenylacridophosphine, D-00547
1,2-Dihydro-1-phenyl-4*H*-3,1,2-benzothiazaphosphorin-4-thione; 2-Methoxy, 2-sulfide, *in* D-00552
1,2-Dihydro-1-phenyl-4*H*-3,1,2-benzothiazaphosphorin-4-thione; 2-Ethoxy, 2-sulfide, *in* D-00552

3,4-Dihydro-3-phenyl-1*H*-2,3-benzoxaphosphorin; 3-Oxide, *in* D-00554
3,4-Dihydro-3-phenyl-1*H*-2,3-benzoxaphosphorin; 3-Sulfide, *in* D-00554
2,3-Dihydro-3-phenyl-4*H*-1,3,2-benzoxazaphosphorin-4-one; 2-Methoxy, *in* D-00556
2,3-Dihydro-3-phenyl-4*H*-1,3,2-benzoxazaphosphorin-4-one; 2-Ethoxy, *in* D-00556
2,3-Dihydro-3-phenyl-4*H*-1,3,2-benzoxazaphosphorin-4-one; 2-Phenoxy, *in* D-00556
2,3-Dihydro-3-phenyl-4*H*-1,3,2-benzoxazaphosphorin-4-one; 2-Acetoxy, *in* D-00556
5,10-Dihydro-10-phenylphenophosphazine, D-00563
5,10-Dihydro-10-phenylphenophosphazine; 10-Oxide, *in* D-00563
5,10-Dihydro-10-phenylphenophosphazine; 10-Sulfide, *in* D-00563
5,6-Dihydro-5-phenylphosphanthridene 5-oxide, *in* D-00568
5,6-Dihydrophosphanthridene, D-00568
1,1-Dihydrophosphorinane, D-00573
1,4-Dihydro-1,1,4,4-tetraphenyl-1,4-diphosphorinium(2+); Dichloride, *in* D-00582
1,4-Dihydro-1,1,4,4-tetraphenyl-1,4-diphosphorinium(2+); Dibromide, *in* D-00582
1,2-Dihydro-1,2,4,6-tetraphenylphosphorin, D-00584
1,2-Dihydro-1,2,4,6-tetraphenylphosphorin; Oxide, *in* D-00584
3,6-Dihydro-2*H*-1,2-thiaphosphorin 2-sulfide; 2-Et, *in* D-00586
3,6-Dihydro-2*H*-1,2-thiaphosphorin 2-sulfide; 2-Ph, *in* D-00586
2,2-Dihydro-2,2,2-trimethyl-1,2-oxaphosphorinane, D-00597
1,4-Dihydro-2,4,6-triphenyl-1,4-azaphosphorine, D-00604
1,4-Dihydro-2,4,6-triphenyl-1,4-azaphosphorine; 4-Oxide, *in* D-00604
2,2-Dihydro-2,2,2-triphenyl-2*H*-1,2-benzoxaphosphorin, D-00607
1,1-Dihydro-2,4,6-triphenylphosphorin, D-00613
1,3-Dihydroxy-1*H*-1,2,4,3-benziodadioxaphosphorin 3-oxide, D-00616
2,5-Dihydroxy-3,6-dimethyl-3,6-bis(dihydroxyphosphinyl)-1,4,2,5-dioxadiphosphorinane 2,5-dioxide, D-00617
2,5-Dihydroxy-3,6-dimethyl-3,6-bis(dihydroxyphosphinyl)-1,4,2,5-dioxadiphosphorinane 2,5-dioxide; Hexa-Me ester, *in* D-00617
1,3-Dihydroxy-1,3-diphosphorinane 1,3-dioxide, D-00622
3,9-Dihydroxy-2,4,8,10-Tetraoxa-3,9-diphosphaspiro[5.5]-undecane 3,9-dioxide, D-00626
3,9-Dihydroxy-2,4,8,10-Tetraoxa-3,9-diphosphaspiro[5.5]-undecane 3,9-dioxide; 3,9-Diamide, *in* D-00626
1,3-Diisopropoxy-1,3-diphosphorinane 1,3-dioxide, *in* D-00622
3,9-Dimercapto-2,4,8,10-tetraoxa-3,9-diphosphaspiro[5.5]-undecane 3,9-disulfide, D-00670
3,9-Dimercapto-2,4,8,10-tetraoxa-3,9-diphosphaspiro[5.5]-undecane 3,9-disulfide; 3,9-Bis(methylthio) (Di-Me ester), *in* D-00670
3,9-Dimercapto-2,4,8,10-tetraoxa-3,9-diphosphaspiro[5.5]-undecane 3,9-disulfide; 3,9-Bis(ethylthio) (Di-Et ester), *in* D-00670
4-Dimethylamino-2,2-dihydro-2,2-diphenyl-1,3,5,2-triazaphosphorine, D-00692
2-Dimethylamino-5,5-dimethyl-1,3,2-dioxaphosphorinane, D-00693
2-Dimethylamino-1,3,2-dioxaphosphorinane, D-00696
2-Dimethylamino-4-isopropyl-5,5-dimethyl-1,3,2-dioxaphosphorinane, D-00699
2-Dimethylamino-4-phenyl-1,3,2-dioxaphosphorinane, D-00704
5,5-Dimethyl-2-dimethylaminotetrahydro-2*H*-1,3,2-oxazaphosphorine; 2-Oxide, *in* D-00726
2,2-Dimethyl-1,3-dioxa-5-phospha-2-silacyclohexane; 5-Me, *in* D-00731
2,2-Dimethyl-1,3-dioxa-5-phospha-2-silacyclohexane; 5-Me, 5-sulfide, *in* D-00731
2,2-Dimethyl-1,3-dioxa-5-phospha-2-silacyclohexane; 5-Ph, *in* D-00731
2,2-Dimethyl-1,3-dioxa-5-phospha-2-silacyclohexane; 5-Ph, 5-sulfide, *in* D-00731
2,4-Dimethyl-1,3,2-dioxaphosphorinane; (2*RS*,4*SR*)-*form*, 2-Oxide, *in* D-00734
5,5-Dimethyl-1,3,2-dioxaphosphorinane, D-00735
2-[(5,5-Dimethyl-1,3,2-dioxaphosphorinan-2-yl)thio]-5,5-dimethyl-1,3,2-dioxaphosphorinane *P*-oxide 2-sulfide, *in* T-00309
2,2-Dimethyl-1,3-diphenyl-1,3-diphosphorinane, D-00745
1,4-Dimethyl-1,4-diphenyl-1,4-diphosphorinanium dibromide, *in* D-01017
2,2-Dimethyl-5,6-diphenyl-1-oxa-5-phospha-2-silacyclohexane, D-00746
2,6-Dimethyl-4,4-diphenyl-1,4-oxaphosphorinanium bromide, *in* D-00797
1,4-Dimethyl-1,4-diphosphoniabicyclo[2.2.2]octane diiodide, *in* D-01169
1,3-Dimethylisophosphinoline, D-00758
5,5-Dimethyl-2-methylthio-1,3,2-dioxaphosphorinane, D-00766
2,6-Dimethyl-1,2,6-oxadiphosphorinane 2,6-dioxide, D-00780
5,5-Dimethyl-2-phenoxy-1,3,2-dioxaphosphorinane, D-00790
5,5-Dimethyl-2-phenylamino-1,3,2-dioxaphosphorinane, D-00792
5,5-Dimethyl-2-phenyl-1,3,2-dioxaphosphorinane, D-00795
2,8-Dimethyl-10-phenyl-10*H*-phenothiaphosphine, D-00798
2,8-Dimethyl-10-phenyl-10*H*-phenothiaphosphine; 5,5,10-Trioxide, *in* D-00798
2,8-Dimethyl-10-phenyl-10*H*-phenoxaphosphine, D-00799
2,8-Dimethyl-10-phenyl-10*H*-phenoxaphosphine; 10-Oxide, *in* D-00799
1,4-Dimethyl-2-phenyl-2-phosphabicyclo[2.2.1]heptane 2-oxide, *in* D-00825
4,4-Dimethyl-1-phenyl-1-phospha-4-silacyclohexa-2,5-diene, D-00801
2,6-Dimethyl-4-phenylphosphorin, D-00819
2,5-Dimethyl-1-phenylphosphorinan-4-ol; (1*RS*,2*RS*,4*RS*,5*RS*)-*form*, 1-Sulfide, *in* D-00820
2,5-Dimethyl-1-phenylphosphorinan-4-ol; (1*RS*,2*RS*,4*RS*,5*RS*)-*form*, 1-Sulfide, 4-*O*-Ac, *in* D-00820
2,5-Dimethyl-1-phenylphosphorinan-4-ol; (1*RS*,2*SR*,4*RS*,5*SR*)-*form*, 1-Sulfide, *in* D-00820
2,5-Dimethyl-1-phenylphosphorinan-4-ol; (1*RS*,2*SR*,4*RS*,5*SR*)-*form*, 1-Sulfide, 4-*O*-Ac, *in* D-00820
2,5-Dimethyl-1-phenylphosphorinan-4-ol; (1*RS*,2*SR*,4*SR*,5*SR*)-*form*, 1-Sulfide, *in* D-00820
2,5-Dimethyl-1-phenylphosphorinan-4-ol; (1*RS*,2*SR*,4*SR*,5*SR*)-*form*, 1-Sulfide, 4-*O*-Ac, *in* D-00820
2,5-Dimethyl-1-phenylphosphorinan-4-ol; (1*RS*,2*RS*,4*SR*,5*RS*)-*form*, 1-Sulfide, *in* D-00820
2,5-Dimethyl-1-phenylphosphorinan-4-ol; (1*RS*,2*RS*,4*SR*,5*RS*)-*form*, 1-Sulfide, 4-*O*-Ac, *in* D-00820
2,5-Dimethyl-1-phenylphosphorinan-4-one, D-00821
2,5-Dimethyl-1-phenylphosphorinan-4-one; (1*RS*,2*RS*,5*RS*)-*form*, 1-Oxide, *in* D-00821
2,5-Dimethyl-1-phenylphosphorinan-4-one; (1*RS*,2*RS*,5*RS*)-*form*, 1-Sulfide, *in* D-00821
2,5-Dimethyl-1-phenylphosphorinan-4-one; (1*RS*,2*RS*,5*RS*)-*form*, 1-Selenide, *in* D-00821
2,5-Dimethyl-1-phenylphosphorinan-4-one; (1*SR*,2*RS*,5*RS*)-*form*, 1-Oxide, *in* D-00821
2,5-Dimethyl-1-phenylphosphorinan-4-one; (1*SR*,2*RS*,5*RS*)-*form*, 1-Sulfide, *in* D-00821
2,5-Dimethyl-1-phenylphosphorinan-4-one; (1*SR*,2*RS*,5*RS*)-*form*, 1-Selenide, *in* D-00821
2,5-Dimethyl-1-phenylphosphorinan-4-one; (1*RS*,2*SR*,5*RS*)-*form*, 1-Selenide, *in* D-00821
2,6-Dimethylphosphorin, D-00881
1,3-Dimethylphosphorinane, D-00882
1,3-Dimethylphosphorinane; Oxide, *in* D-00882
1,3-Dimethylphosphorinane; Sulfide, *in* D-00882
1,4-Dimethylphosphorinane, D-00883
1,4-Dimethylphosphorinane; Oxide, *in* D-00883
1,4-Dimethylphosphorinane; Sulfide, *in* D-00883
1,1-Dimethylphosphorinanium bromide, *in* M-00339
N,*N*′-Dimethyl-*N*,*N*′-trimethylene-*P*-methylphosphonous diamide, *in* H-00022
N,*N*′-Dimethyl-*N*,*N*′-trimethylene-*P*-phenylphosphonic diamide, *in* H-00023
N,*N*′-Dimethyl-*N*,*N*′-trimethylene-*P*-phenylphosphonous diamide, *in* H-00022
N″,*N*″-Dimethyl-*N*,*N*′-trimethylenephosphoric triamide, *in* H-00020

N,*N*′-Dimethyl-*N*,*N*′-trimethylene phosphorodiamidous chloride, *in* H-00022
1,3,2-Dioxaphosphorinane, D-00972
1,4,2-Dioxaphosphorinane 2-oxide; 2-Et, *in* D-00973
1,4,2-Dioxaphosphorinane 2-oxide; 2-Ethoxy, *in* D-00973
3,9-Diphenoxy-2,4,8,10-tetraoxa-3,9-diphosphaspiro[5.5]-undecane, D-00991
3,9-Diphenoxy-2,4,8,10-tetraoxa-3,9-diphosphaspiro[5.5]-undecane; Monooxide, *in* D-00991
3,9-Diphenoxy-2,4,8,10-tetraoxa-3,9-diphosphaspiro[5.5]-undecane; Dioxide, *in* D-00991
2,6-Diphenyl-1,4-azaphosphorine, D-00996
5,10-Diphenyldibenzo[1,4]phospharsenin, D-01001
5,10-Diphenyldibenzo[1,4]phospharsenin; *P*-Oxide, *in* D-01001
5,10-Diphenyldibenzo[1,4]phospharsenin; *As*,*P*-Dioxide, *in* D-01001
2,5-Diphenyl-1,3,5,2-dioxaphosphaborinane, D-01003
2,5-Diphenyl-1,3,5,2-dioxaphosphaborinane; 5-Sulfide, *in* D-01003
1,2-Diphenyl-1,2-diphosphorinane, D-01015
1,2-Diphenyl-1,2-diphosphorinane; 1,2-Disulfide, *in* D-01015
1,3-Diphenyl-1,3-diphosphorinane, D-01016
1,4-Diphenyl-1,4-diphosphorinane, D-01017
1,4-Diphenyl-1,4-diphosphorinane; *cis-form*, 1,4-Dioxide, *in* D-01017
1,4-Diphenyl-1,4-diphosphorinane; *trans-form*, 1,4-Dioxide, *in* D-01017
1,4-Diphenyl-1,4-diphosphorinane; *trans-form*, B,2MeBr, *in* D-01017
2,6-Diphenyl-4*H*-1,4-oxaphosphorin-4-amine 4-oxide, *in* H-00140
2,3-Diphenyl-1,3-oxaphosphorinane, D-01028
2,3-Diphenyl-1,3-oxaphosphorinane; 3-Sulfide, *in* D-01028
10,10-Diphenyl-10*H*-phenoxaphosphonium iodide, *in* P-00183
3,3-Diphenyl-3-phosphoniabicyclo[3.2.1]octane bromide, *in* P-00198
1,1-Diphenylphosphorinanium bromide, *in* P-00265
2,3-Diphenyl-1,3-thiaphosphorinane, D-01152
1,4-Diphosphabicyclo[2.2.2]octane; Dioxide, *in* D-01169
1,4-Diphosphabicyclo[2.2.2]octane; Disulfide, *in* D-01169
2,2′-Diselenobis[5,5-dimethyl-2-thioxo-1,3,2-dioxaphosphorinane], *in* D-01234
1,3,2-Dithiaphosphorinane 2-oxide, D-01248
4-Ethoxy-1,4-dihydro-1,3,5-triphenyl-1,2,4-diazaphosphorine 4-oxide, E-00029
2-Ethoxy-5,5-dimethyl-1,3,2-dioxaphosphorinane, E-00032
2-Ethoxy-1,3,2-dioxaphosphorinane, E-00037
4-Ethoxy-2,6-diphenyl-4*H*-1,4-oxaphosphorin 4-oxide, *in* H-00140
4-Ethoxy-2,6-diphenyl-4*H*-1,4-thiaphosphorin 4-oxide, *in* H-00143
2-Ethoxy-1,3,2-dithiaphosphorinane, *in* D-01247
2-Ethoxy-4-methyl-1,3,2-dioxaphosphorinane, E-00041
2-Ethoxy-4-methyl-1,3,2-dithiaphosphorinane, E-00042
2-Ethoxynaphtho[1,8-*de*]-1,3,2-dioxaphosphorin, E-00045
2-Ethoxynaphtho[1,8-*de*]-1,3,2-dioxaphosphorin, E-00046
2-Ethoxy-1,2-oxaphosphorinane 2-oxide, *in* H-00167
2-Ethoxytetrahydro-2*H*-1,3,2-oxazaphosphorine, E-00051
2-Ethoxytetrahydro-2*H*-1,3,2-oxazaphosphorine; 2-Sulfide, *in* E-00051
1-Ethoxy-1,2,3,4-tetrahydrophosphinoline 1-oxide, *in* H-00195
4-Ethoxy-1,4,5,6-tetrahydro-1,3,5-triphenyl-1,2,4-diazaphosphorine 4-oxide, *in* E-00029
Ethyl *N*,*N*′-dimethyl-*N*,*N*′-trimethylene phosphorodiamidite, *in* H-00022
Ethyl 1,8-naphthylene phosphate, *in* E-00045
Ethyl 1,8-naphthylene phosphate, *in* E-00046
O-Ethyl *O*,*O*-1,8-naphthylene thiophosphate, *in* E-00045
O-Ethyl *O*,*O*-1,8-naphthylene thiophosphate, *in* E-00046
Ethyl neopentylene phosphate, *in* E-00032
S-Ethyl *O*,*O*-neopentylene phosphorodithioate, *in* E-00190
O-Ethyl *O*,*O*-neopentylene phosphorothioate, *in* E-00032
S-Ethyl *O*,*O*-neopentylene phosphorothioate, *in* E-00190
Ethyl pentamethylenephosphinate, *in* H-00188
Ethylphosphonothioic acid trimer trianhydride, *in* T-00463
O-Ethyl saligenin cyclic thiophosphate, *in* H-00110
S-Ethyl saligenin cyclic thiophosphate, *in* H-00110
2-Ethyl-1,2,3,4-tetrahydroisophosphinoline, *in* T-00110
1-Ethyl-1,2,3,4-tetrahydro-1-phenylbenzo[*h*]phosphinolinium bromide, *in* P-00319
1-Ethyl-1,2,3,4-tetrahydro-1-phenylbenzo[*h*]phosphinolinium bromide, *in* T-00128
1-Ethyl-1,2,3,4-tetrahydrophosphinoline 1-oxide, *in* T-00135
10-Ethylthio-5,10-dihydrophenophosphazine 10-oxide, *in* D-00502
Ethyl trimethylene phosphate, *in* E-00037
O-Ethyl *O*,*O*-trimethylene thiophosphate, *in* E-00037
2-Fluoro-5,5-dimethyl-1,3,2-dioxaphosphorinane, F-00016
2-Fluoro-4-methyl-1,3,2-dioxaphosphorinane, F-00021
2-Fluoro-4-methyl-1,3,2-dioxaphosphorinane; (2*RS*,4*RS*)-*form*, 2-Oxide, *in* F-00021
1-Fluorophosphorinane, F-00047
1-Fluorophosphorinane; Oxide, *in* F-00047
1-Fluorophosphorinane; Sulfide, *in* F-00047
▷Fotrin, F-00066
2,4,6,7,7,8,8-Heptaphenyl-1-phosphabicyclo[4.2.0]octa-2,4-diene, H-00003
Hexahydro-1,3-dimethyl-1,3,2-diazaphosphorine 2-oxide, H-00023
Hexahydro-2,4-dimethyl-3-phenoxy-1,2,4,5,3-tetrazaphosphorine; 3-Oxide, *in* H-00024
Hexahydro-2,4-dimethyl-3-phenoxy-1,2,4,5,3-tetrazaphosphorine; 3-Sulfide, *in* H-00024
2,2,4,4,6,6-Hexahydro-2,2-dimethyl-4,4,6,6-tetraphenyl-1,3,5-triaza-2,4,6-triphosphorine, H-00025
Hexahydro-1,3-diphenyl-1,3-azaphosphorine, *in* H-00041
Hexahydro-1,4-diphenyl-1,4-azaphosphorine, *in* H-00042
2,3,3*a*,4,5,9*b*-Hexahydro-4,4-diphenyl-1*H*-cyclopent[*c*]-isophosphinolinium bromide, *in* H-00048
2,3,3*a*,4,5,9*b*-Hexahydro-4,4-diphenyl-1*H*-cyclopent[*c*]-isophosphinolinium hexafluorophosphate, *in* H-00048
2,2,4,4,6,6-Hexahydro-2,2,4,4,6,6-hexakis(methylamino)-1,3,5-triaza-2,4,6-triphosphorine, H-00027
2,2,4,4,6,6-Hexahydro-2,2,4,4,6,6-hexamethyl-1,3,5-triaza-2,4,6-triphosphorine, H-00028
2,2,4,4,6,6-Hexahydro-2,2,4,4,6,6-hexaphenoxy-1,3,5-triaza-2,4,6-triphosphorine, H-00030
2,2,4,4,6,6-Hexahydro-2,2,4,4,6,6-hexaphenyl-1,3,5-triaza-2,4,6-triphosphorine, H-00032
2,2,4,4,6,6-Hexahydro-2,2,4,4,6,6-hexapropoxy-1,3,5-triaza-2,4,6-triphosphorine, H-00033
Hexahydro-2-hydroxycyclopenta[*d*]-1,3,2-dioxaphosphorin 2-oxide; (4*aRS*,7*aRS*)-*form*, *in* H-00034
Hexahydro-2-hydroxycyclopenta[*d*]-1,3,2-dioxaphosphorin 2-oxide; (4*aRS*,7*aSR*)-*form*, *in* H-00034
Hexahydro-2-hydroxycyclopenta[*d*]-1,3,2-dioxaphosphorin 2-oxide; (4*aRS*,7*aSR*)-*form*, 2-Methoxy (Me ester), *in* H-00034
Hexahydro-2-hydroxycyclopenta[*d*]-1,3,2-dioxaphosphorin 2-oxide; (4*aRS*,7*aSR*)-*form*, 2-Phenoxy (Ph ester), *in* H-00034
Hexahydro-2-hydroxycyclopenta[*d*]-1,3,2-dioxaphosphorin 2-oxide; (4*aRS*,7*aSR*)-*form*, Chloride, *in* H-00034
Hexahydro-1-methyl-4-phenyl-1,4-azaphosphorine, H-00036
Hexahydro-1-methyl-4-phenyl-1,4-azaphosphorine; 4-Oxide, *in* H-00036
Hexahydro-1-methyl-4-phenyl-1,4-azaphosphorine; 4-Sulfide, *in* H-00036
Hexahydro-3-phenyl-1,3-azaphosphorine, H-00041
Hexahydro-3-phenyl-1,3-azaphosphorine; 1-Ph, 3-sulfide, *in* H-00041
Hexahydro-4-phenyl-1,4-azaphosphorine, H-00042
Hexahydro-4-phenyl-1,4-azaphosphorine; 4-Sulfide, *in* H-00042
Hexahydro-4-phenyl-1,4-azaphosphorine; 1-Ph, 4-oxide, *in* H-00042
6,6*a*,7,8,9,9*a*-Hexahydro-5-phenyl-1*H*-benzo[*h*]cyclopenta[*c*]-phosphinoline, H-00043
4*b*,5,6,10*b*,11,12-Hexahydro-5-phenylbenzo[*c*]-phosphanthridine, H-00045
Hexahydro-3-phenyl-1,2,4,3-triazaphosphorine; 1,4-Di-Me, *in* H-00051
Hexahydro-3-phenyl-1,2,4,3-triazaphosphorine; 1,4-Diisopropyl, *in* H-00051
Hexahydro-3-phenyl-1,2,4,3-triazaphosphorine; 1,4-Dibutyl, *in* H-00051

Hexahydro-1,3,4-trimethyl-1,3,2-diazaphosphorine, H-00057
Hexahydro-1,3,4-trimethyl-1,3,2-diazaphosphorine; 2-Diethylamino, *in* H-00057
Hexahydro-1,3,5-triphenyl-1,3,5-diazaphosphorine, H-00058
Hexahydro-1,3,5-triphenyl-1,3,5-diazaphosphorine; 5-Oxide, *in* H-00058
Hexahydro-1,3,5-triphenyl-1,3,5-diazaphosphorine; 5-Sulfide, *in* H-00058
▷2,2,4,4,6,6-Hexakis(1-aziridinyl)-2,2,4,4,6,6-hexahydro-1,3,5-triaza-2,4,6-triphosphorine, H-00060
▷2,2,4,4,6,6-Hexakis(dimethylamino)-2,2,4,4,6,6-hexahydro-1,3,5-triaza-2,4,6-triphosphorine, H-00061
1,1,3,3,5,5-Hexamethyl-1-phosphonia-3,5-disilacyclohexane chloride, *in* D-00485
Hexaphenylhexaphosphorinane, H-00089
4-Hydroperoxycyclophosphamide; (2*RS*,4*RS*)-*form*, *in* H-00106
4-Hydroperoxycyclophosphamide; (2*RS*,4*SR*)-*form*, *in* H-00106
2-Hydroxy-4*H*-1,3,2-benzodioxaphosphorin 2-oxide, H-00109
2-Hydroxy-4*H*-1,3,2-benzodioxaphosphorin 2-oxide; 2-Ethoxy (Et ester), *in* H-00109
2-Hydroxy-4*H*-1,3,2-benzodioxaphosphorin 2-oxide; 2-(2-Methylphenoxy) (*o*-Tolyl ester), *in* H-00109
2-Hydroxy-4*H*-1,3,2-benzodioxaphosphorin 2-sulfide, H-00110
2-Hydroxy-2*H*-1,2-benzoxaphosphorin 2-oxide; 2-Ethoxy (Et ester), *in* H-00112
2-Hydroxy-2*H*-1,2-benzoxaphosphorin 2-oxide; 2-Phenoxy (Ph ester), *in* H-00112
4-Hydroxycyclophosphamide, H-00115
2-Hydroxy-5,5-dimethyl-1,3,2-dioxaphosphorinane 2-oxide, H-00124
2-Hydroxy-5,5-dimethyl-1,3,2-dioxaphosphorinane 2-selenide, H-00125
10-Hydroxy-2,8-dimethyl-10*H*-phenothiaphosphine 10-oxide, H-00127
10-Hydroxy-2,8-dimethyl-10*H*-phenothiaphosphine 10-oxide; 10-Methoxy (Me ester), *in* H-00127
10-Hydroxy-2,8-dimethyl-10*H*-phenoxaphosphine 10-oxide, H-00128
2-Hydroxy-5,5-dimethyl-4-phenyl-1,3,2-dioxaphoshorinane 2-oxide; (*R*)-*form*, *in* H-00129
2-Hydroxy-5,5-dimethyl-4-phenyl-1,3,2-dioxaphoshorinane 2-oxide; (*S*)-*form*, *in* H-00129
2-Hydroxy-5,5-dimethyl-4-phenyl-1,3,2-dioxaphoshorinane 2-oxide; (±)-*form*, *in* H-00129
2-Hydroxy-5,5-dimethyl-4-phenyl-1,3,2-dioxaphoshorinane 2-oxide; (±)-*form*, Chloride, *in* H-00129
2-Hydroxy-1,3,2-dioxaphosphorinane 2-oxide, H-00138
4-Hydroxy-2,6-diphenyl-4*H*-1,4-oxaphosphorin 4-oxide, H-00140
4-Hydroxy-2,6-diphenyl-4*H*-1,4-thiaphosphorine 2-oxide, H-00143
2-Hydroxyhexahydro-4*H*-1,3,2-benzodioxaphosphorin 2-oxide; (2α,4aβ,8aα)-*form*, 2-Phenoxy (2-Ph ester), *in* H-00152
2-Hydroxyhexahydro-4*H*-1,3,2-benzodioxaphosphorin 2-oxide; (2α,4aβ,8aα)-*form*, 2-(4-Methoxyphenoxy) (2-Methoxyphenyl ester), *in* H-00152
2-Hydroxyhexahydro-4*H*-1,3,2-benzodioxaphosphorin 2-oxide; (2α,4aβ,8aα)-*form*, 2-(4-Nitrophenoxy) (2-(4-Nitrophenyl)-ester), *in* H-00152
2-Hydroxyhexahydro-4*H*-1,3,2-benzodioxaphosphorin 2-oxide; (2α,4aβ,8aα)-*form*, 2-(2,4-Dinitrophenoxy) (2-(2,4-Dinitrophenyl)ester), *in* H-00152
2-Hydroxyhexahydro-4*H*-1,3,2-benzodioxaphosphorin 2-oxide; (2α,4aα,8aβ)-*form*, 2-Phenoxy, *in* H-00152
2-Hydroxyhexahydro-4*H*-1,3,2-benzodioxaphosphorin 2-oxide; (2α,4aα,8aβ)-*form*, 2-(4-Methoxyphenoxy), *in* H-00152
2-Hydroxyhexahydro-4*H*-1,3,2-benzodioxaphosphorin 2-oxide; (2α,4aα,8aβ)-*form*, 2-(4-Nitrophenoxy), *in* H-00152
2-Hydroxyhexahydro-4*H*-1,3,2-benzodioxaphosphorin 2-oxide; (2α,4aα,8aβ)-*form*, 2-(2,4-Dinitrophenoxy), *in* H-00152
1-Hydroxy-3-methoxy-1*H*-1,2,4,3-benziodadioxaphosphorine 3-oxide, *in* D-00616
2-Hydroxy-4-methyl-1,3,2-dioxaphosphorinane 2-oxide, H-00156
2-Hydroxy-4-methyl-1,3,2-dioxaphosphorinane 2-selenide; (2*RS*,4*RS*)-*form*, *in* H-00157
2-Hydroxy-4-methyl-1,3,2-dioxaphosphorinane 2-selenide; (2*RS*,4*SR*)-*form*, *in* H-00157
2-Hydroxy-4-methyl-1,3,2-dioxaphosphorinane 2-sulfide; (2*RS*,4*RS*)-*form*, *in* H-00158
2-Hydroxy-4-methyl-1,3,2-dioxaphosphorinane 2-sulfide; (2*RS*,4*SR*)-*form*, *in* H-00158
2-Hydroxy-1,2-oxaphosphorinane 2-oxide, H-00167
10-Hydroxy-10*H*-phenoxaphosphine 10-oxide, H-00170
2-Hydroxy-4-phenyl-1,3,2-dioxaphosphorinane 2-oxide, H-00171
1-Hydroxyphosphorinane 1-oxide, H-00188
1-Hydroxy-1,2,3,4-tetrahydrophosphinoline 1-oxide, H-00195
4-Hydroxy-2,2,6,6-tetramethyl-1-oxa-4-phospha-2,6-disilacyclohexane 4-oxide; 4-Butoxy (butyl ester), *in* H-00198
1-Hydroxy-2,4,6-trimethyl-1,3,5-dioxaphosphorinane 5-oxide, H-00203
1-Hydroxy-2,4,6-trimethyl-1,3,5-dioxaphosphorinane 5-oxide; 5-Methoxy (Me ester), *in* H-00203
5-Hydroxy-2,4,6-triphenyl-1,3,5-dioxaphosphorinane 5-oxide, H-00204
5-Hydroxy-2,4,6-triphenyl-1,3,5-dioxaphosphorinane 5-oxide; 5-Methoxy (Me ester), *in* H-00204
Ifosfamide; (*R*)-*form*, *in* I-00001
Ifosfamide; (*S*)-*form*, *in* I-00001
Ifosfamide; (±)-*form*, *in* I-00001
Iminocyclophosphamide, I-00004
2-Isocyano-5,5-dimethyl-1,3,2-dioxaphosphorinane; 2-Oxide, *in* I-00023
Isophosphinoline, I-00027
4-Isopropyl-5,5-dimethyl-1,3,2-dioxaphosphorinane, I-00028
4-Isopropyl-5,5-dimethyl-1,3,2-dioxaphosphorinane; (2*RS*,4*RS*)-*form*, 2-Oxide, *in* I-00028
4-Isopropyl-5,5-dimethyl-1,3,2-dioxaphosphorinane; (2*RS*,4*SR*)-*form*, 2-Oxide, *in* I-00028
5-Isopropyl-2,6-dimethyl-6-phosphabicyclo[3.1.1]hept-2-ene 6-oxide, I-00029
4-Isopropyl-2-methoxy-5,5-dimethyl-1,3,2-dioxaphosphorinane; (2*RS*,4*RS*)-*form*, *in* I-00033
4-Isopropyl-2-methoxy-5,5-dimethyl-1,3,2-dioxaphosphorinane; (2*RS*,4*RS*)-*form*, 2-Selenide, *in* I-00033
4-Isopropyl-2-methoxy-5,5-dimethyl-1,3,2-dioxaphosphorinane; (2*RS*,4*SR*)-*form*, *in* I-00033
4-Isopropyl-2-methoxy-5,5-dimethyl-1,3,2-dioxaphosphorinane; (2*RS*,4*SR*)-*form*, 2-Selenide, *in* I-00033
2-Isoselenocyanato-5,5-dimethyl-1,3,2-dioxaphosphorinane, I-00066
2-Isoselenocyanato-5,5-dimethyl-1,3,2-dioxaphosphorinane; 2-Oxide, *in* I-00066
2-Mercapto-4*H*-1,3,2-benzodioxaphosphorin 2-sulfide; 2-Ethylthio (2-Et ester), *in* M-00006
2-Mercapto-4*H*-1,3,2-benzodioxaphosphorin 2-sulfide; 2-Isopropylthio (2-Isopropyl ester), *in* M-00006
2-Mercapto-4*H*-1,3,2-benzodioxaphosphorin 2-sulfide; 2-Phenylthio (2-Ph ester), *in* M-00006
2-Mercapto-5,5-dimethyl-1,3,2-dioxaphosphorinane 2-oxide, M-00010
2-Mercapto-5,5-dimethyl-1,3,2-dioxaphosphorinane 2-sulfide, M-00011
2-Mercapto-1,3,2-dioxaphosphorinane 2-sulfide, M-00013
2-Mercapto-1,3,2-dithiaphosphorinane 2-sulfide, M-00014
2-Mercapto-4-methyl-1,3,2-dioxaphosphorinane 2-sulfide, M-00017
2-Mercapto-1,3,5,2-triazaphosphorine-4,6(1*H*,5*H*)-dione 2-sulfide, M-00020
5-Methoxy-5,6-dihydrophosphanthridene 5-oxide, *in* D-00504
2-Methoxy-5,5-dimethyl-1,3,2-dioxaphosphorinane, M-00034
▷2-Methoxy-1,3,2-dioxaphosphorinane, M-00038
4-Methoxy-2,6-diphenyl-4*H*-1,4-thiaphosphorin 4-oxide, *in* H-00143
2-Methoxy-1,3,2-dithiaphosphorinane, *in* D-01247
2-Methoxy-4-methyl-1,3,2-dioxaphosphorinane; (2*RS*,4*RS*)-*form*, *in* M-00044
2-Methoxy-4-methyl-1,3,2-dioxaphosphorinane; (2*RS*,4*SR*)-*form*, *in* M-00044
2-Methoxy-4-methyl-1,3,2-dioxaphosphorinane 2-oxide; (2*RS*,4*RS*)-*form*, *in* M-00045
2-Methoxy-4-methyl-1,3,2-dioxaphosphorinane 2-oxide; (2*RS*,4*SR*)-*form*, *in* M-00045
2-Methoxy-4-methyl-1,3,2-dioxaphosphorinane 2-selenide; (2*RS*,4*RS*)-*form*, *in* M-00046

2-Methoxy-4-methyl-1,3,2-dioxaphosphorinane 2-selenide; (2*RS*,4*SR*)-*form*, *in* M-00046
2-Methoxy-4-methyl-1,3,2-dioxaphosphorinane 2-sulfide; (2*RS*,4*RS*)-*form*, *in* M-00047
2-Methoxy-4-methyl-1,3,2-dioxaphosphorinane 2-sulfide; (2*RS*,4*SR*)-*form*, *in* M-00047
10-Methylacridophosphine, M-00087
(2-Methyl-2-bromomethyl)trimethylene phosphoryl bromide, *in* B-00465
Methyl *N*,*N'*-1,3-butylene-*N*,*N'*-dimethylphosphorodiamidite, *in* H-00057
Methyl 6-deoxy-2,3-di-*O*-methyl-6-methylaminoglucopyranoside 4,6-cyclic phosphonamide; α-D-(*R*)$_P$-*form*, *P*-Me, *in* M-00112
Methyl 6-deoxy-2,3-di-*O*-methyl-6-methylaminoglucopyranoside 4,6-cyclic phosphonamide; α-D-(*R*)$_P$-*form*, *P*-Ph, *in* M-00112
Methyl 6-deoxy-2,3-di-*O*-methyl-6-methylaminoglucopyranoside 4,6-cyclic phosphonamide; α-D-(*R*)$_P$-*form*, *P*-Methoxy, *in* M-00112
Methyl 6-deoxy-2,3-di-*O*-methyl-6-methylaminoglucopyranoside 4,6-cyclic phosphonamide; α-D-(*R*)$_P$-*form*, *P*-Ethoxy, *in* M-00112
Methyl 6-deoxy-2,3-di-*O*-methyl-6-methylaminoglucopyranoside 4,6-cyclic phosphonamide; α-D-(*R*)$_P$-*form*, *P*-(4-Nitrophenoxy), *in* M-00112
Methyl 6-deoxy-2,3-di-*O*-methyl-6-methylaminoglucopyranoside 4,6-cyclic phosphonamide; α-D-(*S*)$_P$-*form*, *P*-Me, *in* M-00112
Methyl 6-deoxy-2,3-di-*O*-methyl-6-methylaminoglucopyranoside 4,6-cyclic phosphonamide; α-D-(*S*)$_P$-*form*, *P*-Ph, *in* M-00112
Methyl 6-deoxy-2,3-di-*O*-methyl-6-methylaminoglucopyranoside 4,6-cyclic phosphonamide; α-D-(*S*)$_P$-*form*, *P*-Chloro, *in* M-00112
Methyl 6-deoxy-2,3-di-*O*-methyl-6-methylaminoglucopyranoside 4,6-cyclic phosphonamide; α-D-(*S*)$_P$-*form*, *P*-Methoxy, *in* M-00112
Methyl 6-deoxy-2,3-di-*O*-methyl-6-methylaminoglucopyranoside 4,6-cyclic phosphonamide; α-D-(*S*)$_P$-*form*, *P*-Ethoxy, *in* M-00112
Methyl 6-deoxy-2,3-di-*O*-methyl-6-methylaminoglucopyranoside 4,6-cyclic phosphonamide; α-D-(*S*)$_P$-*form*, 2-(4-Nitrophenoxy), *in* M-00112
Methyl 6-deoxy-2,3-di-*O*-methyl-6-methylaminoglucopyranoside 4,6-cyclic thiophosphonamide; α-D-(*R*)$_P$-*form*, *P*-Me, *in* M-00113
Methyl 6-deoxy-2,3-di-*O*-methyl-6-methylaminoglucopyranoside 4,6-cyclic thiophosphonamide; α-D-(*R*)$_P$-*form*, *P*-Chloro, *in* M-00113
Methyl 6-deoxy-2,3-di-*O*-methyl-6-methylaminoglucopyranoside 4,6-cyclic thiophosphonamide; α-D-(*S*)$_P$-*form*, *P*-Me, *in* M-00113
Methyl 6-deoxy-2,3-di-*O*-methyl-6-methylaminoglucopyranoside 4,6-cyclic thiophosphonamide; α-D-(*S*)$_P$-*form*, *P*-Methoxy, *in* M-00113
4-Methyl-2-dimethylamino-1,3,2-dioxaphosphorinane, M-00117
4-Methyl-2-dimethylamino-1,3,2-dioxaphosphorinane; (2*RS*,4*RS*)-*form*, 2-Sulfide, *in* M-00117
4-Methyl-2-dimethylamino-1,3,2-dioxaphosphorinane; (2*RS*,4*RS*)-*form*, 2-Selenide, *in* M-00117
4-Methyl-2-dimethylamino-1,3,2-dioxaphosphorinane 2-oxide; (2*RS*,4*RS*)-*form*, *in* M-00118
4-Methyl-2-dimethylamino-1,3,2-dioxaphosphorinane 2-oxide; (2*RS*,4*SR*)-*form*, *in* M-00118
Methyl 2,3-di-*O*-methylglucopyranoside 4,6-cyclic phosphonate; α-D-(*R*)$_P$-*form*, *P*-Me, *in* M-00119
Methyl 2,3-di-*O*-methylglucopyranoside 4,6-cyclic phosphonate; α-D-(*R*)$_P$-*form*, *P*-Ph, *in* M-00119
Methyl 2,3-di-*O*-methylglucopyranoside 4,6-cyclic phosphonate; α-D-(*R*)$_P$-*form*, *P*-Ethoxy, *in* M-00119
Methyl 2,3-di-*O*-methylglucopyranoside 4,6-cyclic phosphonate; α-D-(*R*)$_P$-*form*, *P*-Dimethylamino, *in* M-00119
Methyl 2,3-di-*O*-methylglucopyranoside 4,6-cyclic phosphonate; α-D-(*R*)$_P$-*form*, *P*-(4-Nitrophenoxy), *in* M-00119
Methyl 2,3-di-*O*-methylglucopyranoside 4,6-cyclic phosphonate; α-D-(*R*)$_P$-*form*, *P*-(Propylthio), *in* M-00119
Methyl 2,3-di-*O*-methylglucopyranoside 4,6-cyclic phosphonate; α-D-(*S*)$_P$-*form*, *P*-Me, *in* M-00119
Methyl 2,3-di-*O*-methylglucopyranoside 4,6-cyclic phosphonate; α-D-(*S*)$_P$-*form*, *P*-Ph, *in* M-00119
Methyl 2,3-di-*O*-methylglucopyranoside 4,6-cyclic phosphonate; α-D-(*S*)$_P$-*form*, *P*-Ethoxy, *in* M-00119
Methyl 2,3-di-*O*-methylglucopyranoside 4,6-cyclic phosphonate; α-D-(*S*)$_P$-*form*, *P*-(4-Nitrophenoxy), *in* M-00119
Methyl 2,3-di-*O*-methylglucopyranoside 4,6-cyclic phosphonate; α-D-(*S*)$_P$-*form*, *P*-Fluoro, *in* M-00119
Methyl 2,3-di-*O*-methylglucopyranoside 4,6-cyclic phosphonate; α-D-(*S*)$_P$-*form*, *P*-Chloro, *in* M-00119
Methyl 2,3-di-*O*-methylglucopyranoside 4,6-cyclic phosphonate; α-D-(*S*)$_P$-*form*, *P*-Dimethylamino, *in* M-00119
Methyl 2,3-di-*O*-methylglucopyranoside 4,6-cyclic phosphonate; α-D-(*S*)$_P$-*form*, *P*-(Propylthio), *in* M-00119
Methyl 2,3-di-*O*-methylglucopyranoside 4,6-cyclic phosphonothioate; α-D-(*R*)$_P$-*form*, *P*-Me, *in* M-00120
Methyl 2,3-di-*O*-methylglucopyranoside 4,6-cyclic phosphonothioate; α-D-(*R*)$_P$-*form*, *P*-Ph, *in* M-00120
Methyl 2,3-di-*O*-methylglucopyranoside 4,6-cyclic phosphonothioate; α-D-(*R*)$_P$-*form*, *P*-Ethoxy, *in* M-00120
Methyl 2,3-di-*O*-methylglucopyranoside 4,6-cyclic phosphonothioate; α-D-(*R*)$_P$-*form*, *P*-Chloro, *in* M-00120
Methyl 2,3-di-*O*-methylglucopyranoside 4,6-cyclic phosphonothioate; α-D-(*S*)$_P$-*form*, *P*-Me, *in* M-00120
Methyl 2,3-di-*O*-methylglucopyranoside 4,6-cyclic phosphonothioate; α-D-(*S*)$_P$-*form*, *P*-Ph, *in* M-00120
Methyl 2,3-di-*O*-methylglucopyranoside 4,6-cyclic phosphonothioate; α-D-(*S*)$_P$-*form*, *P*-Ethoxy, *in* M-00120
Methyl 2,3-di-*O*-methyl-4-thioglucopyranoside 4,6-cyclic phosphonate; α-D-(*R*)$_P$-*form*, *P*-Me, *in* M-00121
Methyl 2,3-di-*O*-methyl-4-thioglucopyranoside 4,6-cyclic phosphonate; α-D-(*R*)$_P$-*form*, *P*-Ethoxy, *in* M-00121
Methyl 2,3-di-*O*-methyl-4-thioglucopyranoside 4,6-cyclic phosphonate; α-D-(*R*)$_P$-*form*, *P*-Chloro, *in* M-00121
Methyl 2,3-di-*O*-methyl-4-thioglucopyranoside 4,6-cyclic phosphonate; α-D-(*S*)$_P$-*form*, *P*-Me, *in* M-00121
Methyl 2,3-di-*O*-methyl-4-thioglucopyranoside 4,6-cyclic phosphonate; α-D-(*S*)$_P$-*form*, *P*-Ethoxy, *in* M-00121
Methyl 2,3-di-*O*-methyl-6-thioglucopyranoside 4,6-cyclic phosphonate; α-D-(*R*)$_P$-*form*, *P*-Me, *in* M-00122
Methyl 2,3-di-*O*-methyl-6-thioglucopyranoside 4,6-cyclic phosphonate; α-D-(*R*)$_P$-*form*, *P*-Ethoxy, *in* M-00122
Methyl 2,3-di-*O*-methyl-6-thioglucopyranoside 4,6-cyclic phosphonate; α-D-(*R*)$_P$-*form*, *P*-Chloro, *in* M-00122
Methyl 2,3-di-*O*-methyl-6-thioglucopyranoside 4,6-cyclic phosphonate; α-D-(*S*)$_P$-*form*, *P*-Me, *in* M-00122
Methyl 2,3-di-*O*-methyl-6-thioglucopyranoside 4,6-cyclic phosphonate; α-D-(*S*)$_P$-*form*, *P*-Ethoxy, *in* M-00122
Methyl 2,3-di-*O*-methyl-4-thioglucopyranoside 4,6-cyclic phosphonothioate; α-D-(*R*)$_P$-*form*, *P*-Ethoxy, *in* M-00123
Methyl 2,3-di-*O*-methyl-4-thioglucopyranoside 4,6-cyclic phosphonothioate; α-D-(*S*)$_P$-*form*, *P*-Ethoxy, *in* M-00123
Methyl 2,3-di-*O*-methyl-4-thioglucopyranoside 4,6-cyclic phosphonothioate; α-D-(*S*)$_P$-*form*, *P*-Chloro, *in* M-00123
Methyl *N*,*N'*-dimethyl-*N*,*N'*-trimethylene phosphorodiamidite, *in* H-00022
4-Methyl-1,3,2-dioxaphosphorinane; (2*RS*,4*RS*)-*form*, *in* M-00129
4-Methyl-1,3,2-dioxaphosphorinane; (2*RS*,4*SR*)-*form*, *in* M-00129
4-Methyl-1,3,2-dioxaphosphorinane 2-oxide; (2*RS*,4*RS*)-*form*, *in* M-00130
4-Methyl-1,3,2-dioxaphosphorinane 2-oxide; (2*RS*,4*SR*)-*form*, *in* M-00130
4-Methyl-1,3,2-dioxaphosphorinane 2-sulfide, M-00131
5-Methyl-5,10-diphenyldibenzo[1,4]phosphoniaarsenin, *in* D-01001
3-Methyl-2,3-diphenyl-1,3-oxaphosphorinanium iodide, *in* D-01028
2-Methyl-1,3,2-dithiaphosphorinane, M-00137
4-Methyl-1,3,2-dithiaphosphorinane, M-00138
1-Methylisophosphinoline, M-00162
3-Methylisophosphinoline, M-00163
6-Methyl-4-methylene-4*H*-1,3,2-dioxaphosphorin; 2-Ethoxy, *in* M-00167

6-Methyl-4-methylene-4*H*-1,3,2-dioxaphosphorin; 2-Ethoxy, 2-oxide, *in* M-00167
6-Methyl-4-methylene-4*H*-1,3,2-dioxaphosphorin; 2-Me, 2-oxide, *in* M-00167
6-Methyl-4-methylene-4*H*-1,3,2-dioxaphosphorin; 2-Ph, *in* M-00167
6-Methyl-4-methylene-4*H*-1,3,2-dioxaphosphorin; 2-Diethylamino, *in* M-00167
4-Methyl-2-methylthio-1,3,2-dioxaphosphorinane, M-00174
4-Methyl-2-methylthio-1,3,2-dioxaphosphorinane; (2*RS*,4*SR*)-*form*, 2-Oxide, *in* M-00174
4-Methyl-2-methylthio-1,3,2-dioxaphosphorinane; (2*RS*,4*SR*)-*form*, 2-Sulfide, *in* M-00174
4-Methyl-2-methylthio-1,3,2-dioxaphosphorinane; (2*RS*,4*SR*)-*form*, 2-Selenide, *in* M-00174
Methyl neopentylene phosphate, *in* M-00034
Se-Methyl *O,O*-neopentylene phosphorodiselenoate, *in* D-00764
S-Methyl *O,O*-neopentylene phosphorodithioate, *in* D-00766
Se-Methyl *O,O*-neopentylene phosphoroselenoate, *in* D-00764
Se-Methyl *O,O*-neopentylene phosphoroselenothioate, *in* D-00764
O-Methyl *O,O*-neopentylene phosphorothioate, *in* M-00034
S-Methyl *O,O*-neopentylene phosphorothioate, *in* D-00766
3-Methyl-2-oxo-3-phenyl-1,3-oxaphosphorinanium iodide, *in* P-00177
1-Methyl-4-oxo-1-phenyl-4-phosphorinanium iodide, *in* P-00268
Methyl phenazaphosphinate, *in* P-00061
4-Methyl-2-phenoxy-1,3,2-dioxaphosphorinane, M-00191
4-Methyl-2-phenylamino-1,3,2-dioxaphosphorinane; (2*RS*,4*RS*)-*form*, *in* M-00193
4-Methyl-2-phenylamino-1,3,2-dioxaphosphorinane; (2*RS*,4*SR*)-*form*, *in* M-00193
4-Methyl-2-phenylamino-1,3,2-dioxaphosphorinane; (2*RS*,4*SR*)-*form*, 2-Oxide, *in* M-00193
4-Methyl-2-phenylamino-1,3,2-dioxaphosphorinane; (2*RS*,4*SR*)-*form*, 2-Sulfide, *in* M-00193
4-Methyl-2-phenylamino-1,3,2-dioxaphosphorinane; (2*RS*,4*SR*)-*form*, 2-Selenide, *in* M-00193
3-Methyl-2-phenyl-1,3-diaza-2-phosphabicyclo[3.3.1]nonane, M-00196
4-Methyl-2-phenyl-1,3,2-dioxaphosphorinane; (2*S*,4*S*)-*form*, *in* M-00199
4-Methyl-2-phenyl-1,3,2-dioxaphosphorinane; (±)-*form*, *in* M-00199
4-Methyl-4-phenyl-1,4-oxaphosphorinanium iodide, *in* P-00176
10-Methyl-10-phenyl-10*H*-phenothiaphosphinium 5,5-dioxide iodide, *in* P-00182
10-Methyl-10-phenyl-10*H*-phenothiaphosphinium iodide, *in* P-00182
10-Methyl-10-phenyl-10*H*-phenoxaphosphinium iodide, *in* P-00183
9-Methyl-9-phenyl-9-phosphoniabicyclo[4.2.1]nona-2,4,7-triene iodide, *in* P-00196
8-Methyl-8-phenyl-8-phosphoniabicyclo[3.2.1]octan-3-one iodide, *in* P-00199
1-Methyl-1-phenylphosphorinanium iodide, *in* P-00265
8-Methyl-8-phosphabicyclo[3.2.1]octane; *syn-form*, Oxide, *in* M-00272
8-Methyl-8-phosphabicyclo[3.2.1]octane; *anti-form*, Oxide, *in* M-00272
Methylphosphonothioic acid trimer trianhydride, *in* T-00567
2-Methylphosphorin, M-00337
4-Methylphosphorin, M-00338
1-Methylphosphorinane, M-00339
1-Methylphosphorinane; Oxide, *in* M-00339
1-Methylphosphorinane; Sulfide, *in* M-00339
1-Methylphosphorinanium iodide, *in* M-00339
1-Methyl-4-phosphorinanone, M-00340
1-Methyl-4-phosphorinanone; Oxide, *in* M-00340
3-Methyltetrahydro-2-phenoxy-2*H*-1,3,2-oxazaphosphorine; 2-Sulfide, *in* M-00404
2-Methylthio-4*H*-1,3,2-benzodioxaphosphorin; 2-Oxide, *in* M-00411
2-Methylthio-4*H*-1,3,2-benzodioxaphosphorin; 2-Sulfide, *in* M-00411
2-Methylthio-1,3,2-dioxaphosphorinane 2-sulfide, *in* M-00013
Methyl trimethylene phosphate, *in* M-00038
P-Methyl-*N,N'*-trimethylenephosphonic diamide, *in* H-00020
O-Methyl *S,S*-trimethylene phosphonotrithioate, *in* D-01249
Methyl *N,N'*-trimethylenephosphorodiamidate, *in* H-00020
O-Methyl *O,O*-trimethylene phosphoroselenoate, *in* M-00038
Methyl trimethylene phosphorotetrathioate, *in* M-00014
O-Methyl *O,O*-trimethylene phosphorothioate, *in* M-00038
6-Methyl-2,3,4-triphenyl-2*H*-1,5,2-dioxaphosphorin 2-oxide; (2*RS*,6*RS*)-*form*, *in* M-00430
6-Methyl-2,3,4-triphenyl-2*H*-1,5,2-dioxaphosphorin 2-oxide; (2*RS*,6*SR*)-*form*, *in* M-00430
1,8-Naphthylene diethylphosphoramidate, *in* D-00267
O,O-1,8-Naphthylene diethylphosphoramidothioate, *in* D-00267
1,8-Naphthylene phenyl phosphate, *in* P-00070
1,8-Naphthylene phosphonate, *in* N-00010
Neopentylene *tert*-butylphosphoramidate, *in* B-00538
Neopentylene *tert*-butylphosphoramidothioate, *in* B-00538
Neopentylene dimethylphosphoramidate, *in* D-00693
Neopentylene dimethylphosphoramidoselenoate, *in* D-00693
Neopentylene dimethylphosphoramidothioate, *in* D-00693
Neopentylene methylphosphonate, *in* T-00512
O,O-Neopentylene methylphosphonoselenoate, *in* T-00512
O,O-Neopentylene methylphosphonothioate, *in* T-00512
Neopentylene phenyl phosphate, *in* D-00790
Neopentylene phenylphosphonate, *in* D-00795
O,O-Neopentylene phenylphosphonothioate, *in* D-00795
Neopentylene phenylphosphoramidate, *in* D-00792
Neopentylene phenylphosphoramidoselenoate, *in* D-00792
Neopentylene phenylphosphoramidothioate, *in* D-00792
O,O-Neopentylene *O*-phenyl phosphoroselenoate, *in* D-00790
O,O-Neopentylene *O*-phenyl phosphorothioate, *in* D-00790
Neopentylene phosphonate, *in* D-00735
O,O-Neopentylene phosphonothioate, *in* D-00735
Neopentylene phosphorobromidate, *in* B-00471
Neopentylene phosphorobromidothioate, *in* B-00471
Neopentylene phosphorochloridoselenoate, *in* C-00058
Neopentylene phosphorofluoridate, *in* F-00016
Neopentylene phosphoroiodidothioate, *in* I-00010
Octahydro-3-methoxy-1*H*-cyclopropa[*c*]phosphindole 3-oxide, *in* O-00004
Octahydro-3-methyl-1*H*-cyclopropa[*c*]phosphindole 3-oxide, *in* O-00004
Octahydro-2-(4-nitrophenoxy)-2*H*-1,3,2-benzoxazaphosphorine; (2*RS*,4a*RS*,8a*SR*)-*form*, 2-Oxide, *in* O-00020
Octahydro-2-(4-nitrophenoxy)-2*H*-1,3,2-benzoxazaphosphorine; (2*RS*,4a*SR*,8a*RS*)-*form*, 2-Oxide, *in* O-00020
Octahydro-3-phenyl-1*H*-cyclopropa[*c*]phosphindole 3-oxide, *in* O-00004
1,2-Oxaphosphorinane; 2-Ph, *in* O-00054
1,2-Oxaphosphorinane; 2-Ph, 2-oxide, *in* O-00054
1,4-Oxaphosphorinane, O-00055
1,3,2-Oxathiaphosphorinane 2-oxide; 2-Isopropoxy, *in* O-00057
1,3,2-Oxathiaphosphorinane 2-oxide; 2-Phenoxy, *in* O-00057
1,3,2-Oxathiaphosphorinane 2-oxide; 2-Ph, *in* O-00057
▷4-Oxocyclophosphamide, O-00058
4-Oxo-1,2,2,6,6-pentamethyl-1-phenylphosphorinanium iodide, *in* T-00210
5-Oxo-1,1,3,3-tetraphenyl-1,3-diphosphorinanium(2+); Dichloride, *in* O-00081
5-Oxo-1,1,3,3-tetraphenyl-1,3-diphosphorinanium(2+); Dibromide, *in* O-00081
2,2'-Oxybis[5,5-dimethyl-1,3,2-dioxaphosphorinane]; 2-Oxide, *in* O-00089
2,2'-Oxybis[5,5-dimethyl-1,3,2-dioxaphosphorinane]; 2,2'-Dioxide, *in* O-00089
2,2'-Oxybis[5,5-dimethyl-1,3,2-dioxaphosphorinane] 2-oxide 2'-sulfide, O-00090
2,2'-Oxybis[1,3,2-dioxaphosphorinane]; 2,2'-Dioxide, *in* O-00092
Oxybis(2,6-diphenyl-4*H*-1,4-oxaphosphorin)-4,4'-dioxide, *in* H-00140
2,2'-Oxybis[4-methyl-1,3,2-dioxaphosphorinane], O-00096
2,2,3,4,6-Pentaphenyl-2*H*-1,5,2-dioxaphosphorin 2-oxide, P-00036
2,3,4,5,6-Pentaphenylphosphorin, P-00040
3,3,3,6,6-Pentaphenyl-3,3,3,6-tetrahydro-4,5-bis(methoxycarbonyl)-1,2,3-diazaphosphorine, P-00041

Peroxycyclophosphamide, P-00057
▷Phenazaphosphinic acid, P-00061
6-Phenoxy-6*H*-dibenzo[*c,e*][1,2]oxaphosphorin, P-00065
5-Phenoxy-5,6-dihydrophosphanthridene 5-oxide, *in* D-00504
2-Phenoxy-1,3,2-dioxaphosphorinane, P-00068
2-Phenoxynaphtho[1,8-*de*]-1,3,2-dioxaphosphorine, P-00070
2-Phenoxy-1,2-oxaphosphorinane 2-oxide, *in* H-00167
2-Phenoxy-1,3,2-oxaselenaphosphorinane 2-oxide, P-00071
10-Phenylacridophosphine, P-00079
5-Phenyl-10(5*H*)-acridophosphinone, P-00080
5-Phenyl-10(5*H*)-acridophosphinone; 5-Oxide, *in* P-00080
2-Phenyl-4*H*-1,3,2-benzodioxaphosphorin; 2-Oxide, *in* P-00095
2-Phenyl-4*H*-1,3,2-benzodioxaphosphorin; 2-Sulfide, *in* P-00095
2-Phenyl-4*H*-1,3,2-benzoxathiaphosphorin; 2-Oxide, *in* P-00097
Phenyl *N,N'*-1,3-butylene-*N,N'*-dimethylphosphorodiamidite, *in* H-00057
2-Phenyl-1,3,2-dioxaphosphorinane, P-00116
5-Phenyl-5,10-diphenyldibenzo[1,4]phosphoniaarsenin, *in* D-01001
2-Phenyl-1,3,2-dithiaphosphorinane, *in* D-01247
4-Phenyl-4*H*-1,4-oxaphosphorin; 4-Oxide, *in* P-00175
4-Phenyl-1,4-oxaphosphorinane, P-00176
4-Phenyl-1,4-oxaphosphorinane; Oxide, *in* P-00176
3-Phenyl-1,3-oxaphosphorinan-2-one, P-00177
3-Phenyl-1,3-oxaphosphorinan-2-one; 3-Sulfide, *in* P-00177
10-Phenyl-10*H*-phenothiaphosphine, P-00182
10-Phenyl-10*H*-phenothiaphosphine; 10-Oxide, *in* P-00182
10-Phenyl-10*H*-phenothiaphosphine; 10-Sulfide, *in* P-00182
10-Phenyl-10*H*-phenothiaphosphine; 5,5-Dioxide, *in* P-00182
10-Phenyl-10*H*-phenothiaphosphine; 5,5,10-Trioxide, *in* P-00182
10-Phenyl-10*H*-phenothiaphosphine; 5,5-Dioxide, 10-sulfide, *in* P-00182
10-Phenyl-10*H*-phenoxaphosphine, P-00183
10-Phenyl-10*H*-phenoxaphosphine; 10-Oxide, *in* P-00183
10-Phenyl-10*H*-phenoxaphosphine; 10-Sulfide, *in* P-00183
1-Phenyl-1-phenylmethylphosphorinanium bromide, *in* P-00265
9-Phenyl-9-phosphabicyclo[4.2.1]nona-2,4,7-triene; *syn-form*, *in* P-00196
9-Phenyl-9-phosphabicyclo[4.2.1]nona-2,4,7-triene; *syn-form*, Oxide, *in* P-00196
9-Phenyl-9-phosphabicyclo[4.2.1]nona-2,4,7-triene; *anti-form*, *in* P-00196
9-Phenyl-9-phosphabicyclo[4.2.1]nona-2,4,7-triene; *anti-form*, B,MeI, *in* P-00196
9-Phenyl-9-phosphabicyclo[4.2.1]nona-2,4,7-triene; *anti-form*, Oxide, *in* P-00196
3-Phenyl-3-phosphabicyclo[3.2.1]octane; *endo-form*, 3-Oxide, *in* P-00198
3-Phenyl-3-phosphabicyclo[3.2.1]octane; *exo-form*, 6,7-Didehydro, 3-oxide, *in* P-00198
8-Phenyl-8-phosphabicyclo[3.2.1]octan-3-one; *syn-form*, Oxide, *in* P-00199
8-Phenyl-8-phosphabicyclo[3.2.1]octan-3-one; *anti-form*, *in* P-00199
8-Phenyl-8-phosphabicyclo[3.2.1]octan-3-one; *anti-form*, Oxide, *in* P-00199
8-Phenyl-8-phosphabicyclo[3.2.1]oct-6-ene; Oxide, *in* P-00200
3-Phenyl-3-phosphabicyclo[3.2.1]oct-6-ene 3-oxide, *in* P-00198
9-Phenyl-9-phosphatricyclo[3.3.1.$0^{2,8}$]nona-3,6-diene, P-00202
9-Phenyl-9-phosphatricyclo[3.3.1.$0^{2,8}$]nona-3,6-diene; 9-Oxide, *in* P-00202
2-Phenylphosphinoline, P-00212
Phenylphosphonic acid trimer trianhydride, *in* T-00694
Phenylphosphonothioic acid trimer trianhydride, *in* T-00694
1-Phenylphosphorinane, P-00265
1-Phenylphosphorinane; Oxide, *in* P-00265
1-Phenylphosphorinane; Sulfide, *in* P-00265
1-Phenyl-2-phosphorinanol, P-00266
1-Phenyl-2-phosphorinanone, P-00267
1-Phenyl-4-phosphorinanone, P-00268
1-Phenyl-4-phosphorinanone; Oxide, *in* P-00268
1-Phenyl-4-phosphorinanone; Sulfide, *in* P-00268
▷Phenyl saligenin cyclic phosphate, *in* P-00064
O-Phenyl saligenin cyclic thiophosphate, *in* P-00064
S-Phenyl saligenin cyclic thiophosphate, *in* H-00110
4-Phenyl-4*H*-1,4-selenaphosphorin; 4-Oxide, *in* P-00307
4-Phenyl-4*H*-1,4-telluraphosphorin; 4-Oxide, *in* P-00317
1-Phenyl-1,2,3,4-tetrahydrophosphinoline, P-00320
1-Phenyl-1,2,3,4-tetrahydrophosphinoline; 1-Oxide, *in* P-00320
3-Phenyl-2-thia-3-phosphabicyclo[2.2.2]oct-5-ene; 3-Sulfide, *in* P-00323
2-Phenyl-1,2-thiaphosphorinane, P-00325
Phenyl *N,N'*-trimethylenephosphorodiamidate, *in* H-00039
O-Phenyl *N,N'*-trimethylenephosphorodiamidothioate, *in* H-00039
Phosphanthridene, P-00348
Phosphanthridene; 5,5-Dihydro, 5,5-di-Me, *in* P-00348
5-Phospha(5-P^{V})spiro[4.5]deca-1,3,5,7,9-pentaene, P-00349
Phosphorin, P-00403
Phosphorinane, P-00404
Phosphorinane; B,MeI, *in* P-00404
S,S'-1,3-Propanediyl *N,N*-diethylphosphoramidodithioate, *in* D-01248
S,S'-Propanediyl methylphosphonodithioate, *in* M-00137
S,S'-Propanediyl methylphosphonoselenodithioate, *in* M-00137
Propylphosphonic acid trimer trianhydride, *in* T-00720
Propylphosphonothioic acid trimer trianhydride, *in* T-00720
Salioxon, *in* H-00109
▷Salithion, S-00001
Sandoflam, *in* O-00089
▷Sufosfamide, S-00026
1,2,4,5-Tetra-*tert*-butyl-1,2,4,5-tetraphosphorinane, T-00027
2,2,4,6-Tetrachloro-4,6-bis(dimethylamino)-2,2,4,4,6,6-hexahydro-1,3,5-triaza-2,4,6-triphosphorine; (4*RS*, 6*RS*)-*form*, *in* T-00028
2,2,4,6-Tetrachloro-4,6-bis(dimethylamino)-2,2,4,4,6,6-hexahydro-1,3,5-triaza-2,4,6-triphosphorine; (4*RS*, 6*SR*)-*form*, *in* T-00028
2,2,4,6-Tetrachloro-2,2-dihydro-1,3,5,2-triazaphosphorine, T-00029
2,2,4,6-Tetrachloro-2,2,4,4,6,6-hexahydro-4,6-bis(methylamino)-1,3,5-triaza-2,4,6-triphosphorine; (4*RS*, 6*RS*)-*form*, *in* T-00030
2,2,4,6-Tetrachloro-2,2,4,4,6,6-hexahydro-4,6-bis(methylamino)-1,3,5-triaza-2,4,6-triphosphorine; (4*RS*, 6*SR*)-*form*, *in* T-00030
2,2,4,4-Tetrachloro-2,2,4,4,6,6-hexahydro-6,6-diphenyl-1,3,5-triaza-2,4,6-triphosphorine, T-00031
1,2,3,4-Tetracyclohexyl-1,2,3,4-tetraphosphorinane, T-00040
2,2,4,6-Tetrafluoro-2,2,4,4,6,6-hexahydro-4,6-diphenyl-1,3,5-triaza-2,4,6-triphosphorine; (4*RS*,6*RS*)-*form*, *in* T-00079
2,2,4,6-Tetrafluoro-2,2,4,4,6,6-hexahydro-4,6-diphenyl-1,3,5-triaza-2,4,6-triphosphorine; (4*RS*,6*SR*)-*form*, *in* T-00079
4,5,9,10-Tetrahydro-4,9-dihydroxyphosphorino[2,3,4,5-*lmn*]-phosphanthridine 4,9-dioxide, T-00090
4,5,9,10-Tetrahydro-4,9-dihydroxyphosphorino[2,3,4,5-*lmn*]-phosphanthridine 4,9-dioxide; 4,9-Dimethoxy (di-Me ester), *in* T-00090
1,2,3,4-Tetrahydro-1,4-dimethyl-1,4-diphenyl-1,4-benzodiphosphorinium diiodide, *in* T-00102
1,2,3,6-Tetrahydro-4,5-dimethyl-1,2-diphenyl-1,2-diphosphinine, *in* T-00095
1,2,3,6-Tetrahydro-4,5-dimethyl-1,2-diphosphorine; 1,2-Di-Me, 1,2-disulfide, *in* T-00095
1,2,3,6-Tetrahydro-4,5-dimethyl-1,2-diphosphorine; 1,2-Di-Ph, 1,2-disulfide, *in* T-00095
Tetrahydro-5,5-dimethyl-2-phenoxy-2*H*-1,3,2-oxazaphosphorine; 2-Oxide, *in* T-00098
Tetrahydro-5,5-dimethyl-2-phenoxy-2*H*-1,3,2-oxazaphosphorine; 2-Sulfide, *in* T-00098
1,2,3,4-Tetrahydro-1,4-diphenyl-1,4-benzodiphosphorin, T-00102
1,2,3,4-Tetrahydro-1,4-diphenyl-1,4-benzodiphosphorin; (1*RS*,4*SR*)-*form*, 1,4-Disulfide, *in* T-00102
1,2,3,6-Tetrahydro-1,2-diphenyl-1,2-diphosphorin; (1*RS*,2*RS*)-*form*, 1,2-Disulfide, *in* T-00103
1,2,3,6-Tetrahydro-1,2-diphenyl-1,2-diphosphorin; (1*RS*,2*SR*)-*form*, 1,2-Disulfide, *in* T-00103

1,2,5,6-Tetrahydro-1,1,3,6,6,9-hexamethyl-4*H*-phospholo[3,2,1-*i,j*]phosphinolinium iodide, *in* T-00123
Tetrahydro-2-hydroxy-2*H*-1,3,2-oxazaphosphorine 2-sulfide, T-00108
Tetrahydro-2-methoxy-2*H*-1,3,2-oxazaphosphorine; 2-Sulfide, *in* T-00111
1,2,3,4-Tetrahydro-1-methoxyphosphinoline 1-oxide, *in* H-00195
1,2,3,4-Tetrahydro-4-methyl-2,2-diphenylbenz[*h*]-isophosphinolinium(1+); Hexafluorophosphate, *in* T-00113
Tetrahydro-4-methyl-2*H*-1,3,4,2-oxadiazaphosphorine 2-oxide; 2-Phenoxy, *in* T-00116
Tetrahydro-4-methyl-2*H*-1,3,4,2-oxadiazaphosphorine 2-oxide; 2-Cyclohexylamino, *in* T-00116
Tetrahydro-3-methyl-2,4,5,6-tetraphenyl-2*H*-1,3,5-oxazaphosphorine, T-00117
Tetrahydro-2-methylthio-2*H*-oxazaphosphorine 2-oxide, *in* T-00108
2,3,4,5-Tetrahydro-6-methyl-2,4,4-triphenyl-1,2,3-diazaphosphorine; 3-Chloro, *in* T-00118
2,3,4,5-Tetrahydro-6-methyl-2,4,4-triphenyl-1,2,3-diazaphosphorine; 3-Methoxy, *in* T-00118
2,3,4,5-Tetrahydro-6-methyl-2,4,4-triphenyl-1,2,3-diazaphosphorine; 3-Methoxy, 3-sulfide, *in* T-00118
2,3,4,5-Tetrahydro-6-methyl-2,4,4-triphenyl-1,2,3-diazaphosphorine; 3-Me, 3-oxide, *in* T-00118
1,2,3,4-Tetrahydro-4-oxo-1-phenylphosphinoline, T-00122
1,2,3,4-Tetrahydro-4-oxo-1-phenylphosphinoline; 1-Oxide, *in* T-00122
1,2,5,6-Tetrahydro-1,1,6,6,9-pentamethyl-4*H*-phospholo[3,2,1-*ij*]phosphinoline, T-00123
1,2,5,6-Tetrahydro-1,1,6,6,9-pentamethyl-4*H*-phospholo[3,2,1-*ij*]phosphinoline; 3-Oxide, *in* T-00123
Tetrahydro-2-phenoxy-2*H*-1,3,2-oxazaphosphorine; 2-Oxide, *in* T-00126
Tetrahydro-2-phenoxy-2*H*-1,3,2-oxazaphosphorine; 2-Sulfide, *in* T-00126
Tetrahydro-2-phenoxy-3-phenyl-2*H*-1,3,2-oxazaphosphorine; 2-Oxide, *in* T-00127
1,2,3,4-Tetrahydro-1-phenylbenzo[*h*]phosphinoline, T-00128
1,2,3,4-Tetrahydro-1-phenylbenzo[*h*]phosphinoline; 1-Oxide, *in* T-00128
1,2,3,4-Tetrahydro-2-phenylisophosphinoline, *in* T-00110
Tetrahydro-2-phenyl-2*H*-1,3,2-oxazaphosphorine, T-00132
Tetrahydro-2-phenyl-2*H*-1,3,2-oxazaphosphorine; 2-Sulfide, *in* T-00132
Tetrahydro-3-phenyl-2-(phenylamino)-2*H*-1,3,2-oxazaphosphorine; 2-Oxide, *in* T-00133
1,2,3,6-Tetrahydro-1-phenyl-1-phenylmethylphosphininium bromide, *in* T-00134
1,2,3,6-Tetrahydro-1-phenylphosphorin, T-00134
1,2,3,4-Tetrahydrophosphinoline; 1-Et, 1-sulfide, *in* T-00135
1,2,3,6-Tetrahydro-1,2,4,5-tetramethyl-1,2-diphosphinine, *in* T-00095
2,2,4,4-Tetrahydro-2,2,4,4-tetraphenyl-1,5,2,4-diazadiphosphorine, T-00146
1,1,3,3-Tetrahydro-1,1,3,3-tetraphenyl-1,3-diphosphacyclohexa-1,2-diene, T-00149
1,2,3,6-Tetrahydro-1,2,4,5-tetraphenyl-1,2-diphosphorine; 1,2-Disulfide, *in* T-00151
1,2,3,4-Tetrahydro-1,1,4,4-tetraphenyl-1,4-diphosphorinium(2+); Dibromide, *in* T-00152
Tetrahydro-2*H*-1,3,2-thiazaphosphorine 2-oxide; 2-Ethoxy, *in* T-00154
Tetrahydro-2*H*-1,3,2-thiazaphosphorine 2-oxide; 2-Phenoxy, *in* T-00154
Tetrahydro-2*H*-1,3,2-thiazaphosphorine 2-oxide; 2-Me, *in* T-00154
Tetrahydro-2*H*-1,3,2-thiazaphosphorine 2-oxide; 2-Ph, *in* T-00154
2,2,6,6-Tetramethyl-1-phenyl-4-phosphorinanone, T-00210
2,2,6,6-Tetramethyl-1-phenyl-4-phosphorinanone; Oxide, *in* T-00210
2,2,6,6-Tetramethyl-1-phenyl-4-phosphorinanone; Sulfide, *in* T-00210
1,1,3,3-Tetramethyl-1-phospha-3-silacyclohex-1-ene, T-00212
1,1,4,4-Tetramethyl-1,2,3,4-tetrahydro-1,4-benzophospharseninium dibromide, *in* T-00091
N,N′,N″,N‴-Tetramethyl-*N,N′*-trimethylenephosphoric triamide, *in* H-00023
2,2′,6,6′-Tetraphenyl-4,4′-biphosphorine, T-00249
2,4,5,6-Tetraphenyl-1,3,5-dioxaphosphorinane, T-00254
2,4,5,6-Tetraphenyl-1,3,5-dioxaphosphorinane; 5-Oxide, *in* T-00254
2,4,5,6-Tetraphenyl-1,3,5-dioxaphosphorinane; 5-Sulfide, *in* T-00254
1,1,3,3-Tetraphenyl-1,3-diphosphorinanium(2+), T-00262
1,1,3,3-Tetraphenyl-1,3-diphosphorinanium(2+); Dichloride, *in* T-00262
1,1,3,3-Tetraphenyl-1,3-diphosphorinanium(2+); Dibromide, *in* T-00262
1,1,4,4-Tetraphenyl-1,4-diphosphorinanium(2+); Dibromide, *in* T-00263
1,1,4,4-Tetraphenyl-1,4-diphosphorinanium dibromide, *in* D-01017
3,3,5,5-Tetraphenyl-1,3,5-oxadiphosphorinanium(2+); Dichloride, *in* T-00267
3,3,5,5-Tetraphenyl-1,3,5-oxadiphosphorinanium(2+); Dibromide, *in* T-00267
2,4,4,6-Tetraphenyl-4*H*-1,4-oxaphosphorinium bromide, *in* T-00610
1,2,3,4-Tetraphenyl-5-phosphoniaspiro[4.5]deca-1,3-diene; Bromide, *in* T-00270
1,2,4,5-Tetraphenyl-1,2,4,5-tetraphosphorinane, T-00281
4*H*-1,4-Thiaphosphorin; 4-Chloro, 4-oxide, *in* T-00299
4*H*-1,4-Thiaphosphorin; 4-Ph, *in* T-00299
1,4-Thiaphosphorinane 1,1,4-trioxide; 4-Me, *in* T-00300
1,4-Thiaphosphorinane 1,1,4-trioxide; 4-Ph, *in* T-00300
2,2′-Thiobis[5,5-dimethyl-1,3,2-dioxaphosphorinane]; 2,2′-Disulfide, *in* T-00309
2,2′-Thiobis[5,5-dimethyl-1,3,2-dioxaphosphorinane] 2,2′-dioxide, T-00310
Thiohypophosphoric acid cyclic *P,P,P′,P′*-bis(2,2-dimethyltrimethylene) ester, *in* B-00197
Thiohypophosphoric acid cyclic *P,P,P′,P′*-bis(2,2-dimethyltrimethylene) ester, *in* B-00197
▷2,4,6-Tribromo-2,2,4,4,6,6-hexahydro-2,4,6-triphenyl-1,3,5-triaza-2,4,6-triphosphorine; $(2\alpha,4\alpha,6\alpha)$-*form*, *in* T-00358
2,4,6-Tribromo-2,2,4,4,6,6-hexahydro-2,4,6-triphenyl-1,3,5-triaza-2,4,6-triphosphorine; $(2\alpha,4\alpha,6\beta)$-*form*, *in* T-00358
2,4,6-Tri-*tert*-butylphosphorin, T-00382
2,4,6-Trichloro-2,2,4,4,6,6-hexahydro-2,4,6-triphenyl-1,3,5-triaza-2,4,6-triphosphorine; $(2\alpha,4\alpha,6\alpha)$-*form*, *in* T-00407
2,4,6-Trichloro-2,2,4,4,6,6-hexahydro-2,4,6-triphenyl-1,3,5-triaza-2,4,6-triphosphorine; $(2\alpha,4\alpha,6\beta)$-*form*, *in* T-00407
2,2,4-Trichloro-4,6,6-tris(dimethylamino)-2,2,4,4,6,6-hexahydro-1,3,5-triaza-2,4,6-triphosphorine, T-00421
2,4,6-Trichloro-2,4,6-tris(dimethylamino)-2,2,4,4,6,6-hexahydro-1,3,5-triaza-2,4,6-triphosphorine; $(2\alpha,4\alpha,6\alpha)$-*form*, *in* T-00422
2,4,6-Trichloro-2,4,6-tris(dimethylamino)-2,2,4,4,6,6-hexahydro-1,3,5-triaza-2,4,6-triphosphorine; $(2\alpha,4\alpha,6\beta)$-*form*, *in* T-00422
2,2,2-Triethoxy-2,2-dihydro-1,3,2-dioxaphosphorinane, T-00437
1,1,1-Trifluoro-1,1-dihydrophosphorinane, *in* D-00573
Trifosfamide; (*R*)-*form*, *in* T-00485
Trifosfamide; (*S*)-*form*, *in* T-00485
Trifosfamide; (±)-*form*, *in* T-00485
2,4,6-Trihydroxy-1,2,4,6-oxatriphosphorinane 2,4,6-trioxide, T-00494
2,4,6-Triisopropyl-5-mercapto-1,3,5-dioxaphosphorinane 5-sulfide, T-00498
O,O-Trimethyl dimethylphosphoramidothioate, *in* D-00696
2,5,5-Trimethyl-1,3,2-dioxaphosphorinane, T-00512
4,4,6-Trimethyl-1,3,2-dioxaphosphorinane, T-00513
4,4,6-Trimethyl-1,3,2-dioxaphosphorinane; 2-Sulfide, *in* T-00513

4,4,6-Trimethyl-1,3,2-dioxaphosphorinane; (2*RS*,4*RS*)-*form*, 2-Oxide, *in* T-00513
4,4,6-Trimethyl-1,3,2-dioxaphosphorinane; (2*RS*,4*SR*)-*form*, 2-Oxide, *in* T-00513
▷Trimethylene fluorophosphate, *in* F-00018
Trimethylene methylphosphonate, *in* M-00128
O,O-Trimethylene methylphosphonoselenoate, *in* M-00128
O,O-Trimethylene methylphosphonothioate, *in* M-00128
Trimethylene methylphosphonotrithioate, *in* M-00137
Trimethylene phenyl phosphate, *in* P-00068
Trimethylene phenylphosphonate, *in* P-00116
N,N'-Trimethylene-*P*-phenylphosphonic diamide, *in* H-00020
O,O-Trimethylene phenylphosphonothioate, *in* P-00116
Trimethylene phenylphosphonotrithioate, *in* D-01249
Trimethylene phosphonate, *in* D-00972
O,O-Trimethylene phosphonothioate, *in* D-00972
O,O-Trimethylene phosphorochloridoselenoate, *in* C-00066
1,2,5-Trimethyl-4-oxo-1-phenylphosphorinanium iodide, *in* D-00821
4,4,6-Trimethyl-2-phenoxy-1,3,2-dioxaphosphorinane, T-00520
2,4,6-Trimethyl-5-phenyl-1,3,5-dioxaphosphorinane, T-00521
2,4,6-Trimethyl-5-phenyl-1,3,5-dioxaphosphorinane; ($2\alpha,4\alpha,5\alpha,6\alpha$)-*form*, 5-Oxide, *in* T-00521
2,4,6-Trimethyl-5-phenyl-1,3,5-dioxaphosphorinane; ($2\alpha,4\alpha,5\alpha,6\alpha$)-*form*, 5-Sulfide, *in* T-00521
2,4,6-Trimethyl-5-phenyl-1,3,5-dioxaphosphorinane; ($2\alpha,4\alpha,5\alpha,6\alpha$)-*form*, 5-Selenide, *in* T-00521
2,4,6-Trimethyl-5-phenyl-1,3,5-dioxaphosphorinane; ($2\alpha,4\alpha,5\beta,6\alpha$)-*form*, 5-Oxide, *in* T-00521
2,4,6-Trimethyl-5-phenyl-1,3,5-dioxaphosphorinane; ($2\alpha,4\alpha,5\beta,6\alpha$)-*form*, 5-Sulfide, *in* T-00521
2,4,6-Trimethyl-5-phenyl-1,3,5-dioxaphosphorinane; ($2\alpha,4\alpha,5\beta,6\alpha$)-*form*, 5-Selenide, *in* T-00521
2,5,5-Trimethyl-2-phenyl-1,3,2-dioxaphosphorinanium iodide, *in* D-00795
2,8,10-Trimethyl-10-phenyl-10*H*-phenoxaphosphinium iodide, *in* D-00799
1,2,4-Trimethyl-2-phosphabicyclo[2.2.1]heptane 2-oxide, *in* D-00825
1,1,4-Trimethylphosphoranium iodide, *in* D-00883
1,1,3-Trimethylphosphorinanium iodide, *in* D-00882
1,3,3-Trimethyltrimethylene chlorothiophosphate, *in* C-00189
2,4,6-Triphenyl-4*H*-1,4-oxaphosphorin; B,$PhClO_4$, *in* T-00610
2,4,6-Triphenyl-4*H*-1,4-oxaphosphorin; 4-Oxide, *in* T-00610
2,4,6-Triphenylphosphorin, T-00665
1,2,6-Triphenyl-4-phosphorinanone, T-00666
1,2,6-Triphenyl-4-phosphorinanone; ($1\alpha,2\alpha,6\beta$)-*form*, Oxide, *in* T-00666
1,2,6-Triphenyl-4-phosphorinanone; ($1\alpha,2\alpha,6\beta$)-*form*, Sulfide, *in* T-00666
1,2,6-Triphenyl-4-phosphorinanone; ($1\alpha,2\beta,6\beta$)-*form*, Oxide, *in* T-00666
1,2,6-Triphenyl-4-phosphorinanone; ($1\alpha,2\beta,6\beta$)-*form*, Sulfide, *in* T-00666
2,4,6-Triphenyl-4*H*-1,4-thiaphosphorin; 4-Oxide, *in* T-00686
2,4,6-Triphenyl-4*H*-1,4-thiaphosphorin; 1,1,4-Trioxide, *in* T-00686
2,4,6-Tris(diethylamino)-1,3,5,2,4,6-trioxatriphosphorinane 2,4,6-trioxide, T-00756
2,4,6-Tris(diethylamino)-1,3,5,2,4,6-trioxatriphosphorinane 2,4,6-trioxide; ($2\alpha,4\alpha,6\alpha$)-*form*, *in* T-00756
2,4,6-Tris(diethylamino)-1,3,5,2,4,6-trioxatriphosphorinane 2,4,6-trioxide; ($2\alpha,4\alpha,4\beta$)-*form*, *in* T-00756
2,4,6-Tris(2,4,6-tri-*tert*-butylphenoxy)-1,3,5,2,4,6-trioxatriphosphorine, T-00847
2,4,6-Tris(2,4,6-trimethylphenyl)-1,3,5,2,4,6-trithiatriphosphorinane, T-00862

1 P-contg. ring; 7-membered

1-Benzyl-4,5-dichloro-1-phenyl-1*H*-phosphepinium bromide, *in* D-00565
1-Benzyl-1-phenylphosphepanium bromide, *in* P-00205
2,2'-Biphenyldiyl methyl phosphate, *in* M-00032
2,2'-Biphenylene phosphonate, *in* D-00038
O,O-(2,2'-Biphenylyl) phosphorochloridate, *in* C-00032
O,O-(2,2'-Biphenylyl) phosphorochloridoselenoate, *in* C-00032
O,O-(2,2'-Biphenylyl) phosphorochloridothioate, *in* C-00032
6-Chlorodibenzo[*d,f*][1,3,2]dioxaphosphepin, C-00032
5-Chloro-5*H*-dibenzo[*b,f*]phosphepin 5-oxide, *in* H-00117
3-Chloro-1,5-dihydro-2,4,3-benzodioxaphosphepin; 3-Oxide, *in* C-00038
5-Chloro-10,11-dihydro-5*H*-dibenzo[*b,f*]phosphepin 5-oxide, *in* D-00492
Decahydro-1,5-dihydroxy-1*H*-1,5-benzodiphosphepin 1,5-dioxide, D-00002
Decahydro-1,3,5-trihydroxy-1*H*-1,3,5-benzotriphosphepin 1,3,5-trioxide, D-00007
1,4-Dibenzyl-1,4-diphosphepane, D-00047
10,11-Dihydro-1*H*-dibenz[*b,e*][1,4]azaphosphepine 5-oxide; 5-Me, *in* D-00448
10,11-Dihydro-1*H*-dibenz[*b,e*][1,4]azaphosphepine 5-oxide; 5-Ph, *in* D-00448
5,10-Dihydro-11*H*-dibenz[*b,e*][1,4]azaphosphepin-11-one 5-oxide; 5-Me, *in* D-00449
5,10-Dihydro-11*H*-dibenz[*b,e*][1,4]azaphosphepin-11-one 5-oxide; 5-Ph, *in* D-00449
5,10-Dihydro-11*H*-dibenz[*b,e*][1,4]azaphosphepin-11-one 5-oxide; *N*-Me, 5-Ph, *in* D-00449
6,7-Dihydro-5*H*-dibenzo[*d,f*][1,3,2]diazaphosphepine; 6-Me, 6-oxide, *in* D-00450
6,7-Dihydro-5*H*-dibenzo[*d,f*][1,3,2]diazaphosphepine; 6-Ph, 6-oxide, *in* D-00450
10,11-Dihydro-5*H*-dibenzo[*b,f*]phosphepin; Oxide, *in* D-00451
1,5-Dihydro-3-dimethylamino-2,4,3-benzodioxaphosphepin, D-00453
1,5-Dihydro-3-dimethylamino-2,4,3-benzodioxaphosphepin; 3-Oxide, *in* D-00453
1,5-Dihydro-3-dimethylamino-2,4,3-benzodioxaphosphepin; 3-Sulfide, *in* D-00453
1,5-Dihydro-3-dimethylamino-2,4,3-benzodioxaphosphepin; 3-Selenide, *in* D-00453
4,5-Dihydro-2-dimethylamino-1,3,2-benzodioxaphosphepin; 2-Oxide, *in* D-00454
10,11-Dihydro-5,5-dimethyl-5*H*-dibenzo[*b,f*]phosphepinium iodide, *in* D-00524
2,3-Dihydro-4,5-dimethyl-2,7-diphenyl-1*H*-1,2-azaphosphepine; 2-Oxide, *in* D-00460
2,3-Dihydro-4,5-dimethyl-2,7-diphenyl-1*H*-1,2-azaphosphepine; 2-Sulfide, *in* D-00460
2,3-Dihydro-4,5-dimethyl-2,7-diphenyl-1,2-oxaphosphepin 2-oxide, D-00462
2,3-Dihydro-4,5-dimethyl-2,2-diphenyl-1,2-thiaphosphepin; 2-Sulfide, *in* D-00463
4,7-Dihydro-1,3,2-dioxaphosphepin; 2-Phenoxy, *in* D-00472
4,7-Dihydro-1,3,2-dioxaphosphepin; 2-Dimethylamino, *in* D-00472
4,7-Dihydro-1,3,2-dioxaphosphepin 2-sulfide; 2-Chloro, *in* D-00473
4,7-Dihydro-1,3,2-dioxaphosphepin 2-sulfide; 2-Phenoxy, *in* D-00473
4,7-Dihydro-1,3,2-dioxaphosphepin 2-sulfide; 2-Me, *in* D-00473
6,7-Dihydro-6-hydroxy-5*H*-dibenzo[*c,e*]phosphepin 6-oxide, D-00491
10,11-Dihydro-5-hydroxy-5*H*-dibenzo[*b,f*]phosphepin 5-oxide, D-00492
10,11-Dihydro-5-methoxy-5*H*-dibenzo[*b,f*]phosphepin 5-oxide, *in* D-00492
1,5-Dihydro-3-methyl-2,4,3-benzodioxaphosphepin; 4-Sulfide, *in* D-00516
10,11-Dihydro-5-methyl-5*H*-dibenzo[*b,f*]phosphepin, D-00524
10,11-Dihydro-5-methyl-5*H*-dibenzo[*b,f*]phosphepin; Oxide, *in* D-00524
10,11-Dihydro-5-methyl-5-phenyl-5*H*-dibenzo[*b,f*]-phosphepinium iodide, *in* D-00559
4,5-Dihydro-1-methyl-1-phenyl-1*H*-phosphepinium perchlorate, *in* D-00565
1,5-Dihydro-2-phenoxy-2,4,3-benzodioxaphosphepin; 3-Oxide, *in* D-00544
1,5-Dihydro-2-phenoxy-2,4,3-benzodioxaphosphepin; 3-Sulfide, *in* D-00544

1,5-Dihydro-3-phenyl-2,4,3-benzodioxaphosphepin, D-00550
4,5-Dihydro-2-phenyl-1,3,2-benzoxathiaphosphepin; 2-Oxide, *in* D-00555
10,11-Dihydro-5-phenyl-5*H*-dibenzo[*b,f*]phosphepin, D-00559
10,11-Dihydro-5-phenyl-5*H*-dibenzo[*b,f*]phosphepin; Oxide, *in* D-00559
10,11-Dihydro-5-phenyl-5*H*-dibenzo[*b,f*]phosphepin; Sulfide, *in* D-00559
4,7-Dihydro-2-phenyl-1,3,2-dioxaphosphepin, D-00561
4,7-Dihydro-2-phenyl-1,3,2-dioxaphosphepin; 2-Sulfide, *in* D-00561
2,7-Dihydro-1-phenyl-1*H*-phosphepin, D-00564
2,7-Dihydro-1-phenyl-1*H*-phosphepin; Oxide, *in* D-00564
4,5-Dihydro-1-phenyl-1*H*-phosphepin, D-00565
4,5-Dihydro-1-phenyl-1*H*-phosphepin; Oxide, *in* D-00565
6,7-Dihydro-5,5,7,7-tetramethyl-5*H*-dibenzo[*d,f*][1,3]-diphosphepinium(2+); Dibromide, *in* D-00577
1,3-Dihydroxy-1,3-diphosphepane 1,3-dioxide, D-00619
1,3-Diisopropoxy-1,3-diphosphepane 1,3-dioxide, *in* D-00619
2,7-Dimethyl-1,2,7-oxadiphosphepane 2,7-dioxide, D-00779
2,3-Diphenyl-1,3-oxaphosphepane, D-01026
1,1-Diphenylphosphepanium bromide, *in* P-00205
9,9-Diphenyl-9-phosphoniabicyclo[4.2.1]nonane bromide, *in* P-00194
3,3-Diphenyl-3-phosphoniabicyclo[3.2.1]octane bromide, *in* P-00198
2-Ethoxy-1,3,2-dioxaphosphepane 2-oxide, *in* H-00132
2-Ethoxy-1,2-oxaphosphepane 2-oxide, *in* H-00165
2-Ethyl-2,3,4,5-tetrahydro-2-methyl-1*H*-1,3-benzazaphosphepine, E-00188
Hexahydro-1,5-bis(1-phenylethyl)-1*H*-1,5,3-diazaphosphepine; (*R,R*)-*form*, 3-Me, *in* H-00019
Hexahydro-1,5-bis(1-phenylethyl)-1*H*-1,5,3-diazaphosphepine; (*R,R*)-*form*, 3-Ph, *in* H-00019
Hexahydro-1,5-bis(1-phenylethyl)-1*H*-1,5,3-diazaphosphepine; (*S,S*)-*form*, 3-Et, *in* H-00019
Hexahydro-1,5-bis(1-phenylethyl)-1*H*-1,5,3-diazaphosphepine; (*S,S*)-*form*, 3-Cyclohexyl, *in* H-00019
Hexahydro-2-phenoxy-1*H*-1,3,2-diazaphosphepine; 2-Sulfide, *in* H-00038
Hexahydro-3-phenyl-1*H*-1,3-azaphosphepine, H-00040
2-Hydroxydibenzo[*d,f*][1,3,2]dioxaphosphepin 2-oxide, H-00116
5-Hydroxy-5*H*-dibenzo[*b,f*]phosphepin 5-oxide, H-00117
4-Hydroxydinaphtho[2,1-*d*:1′,2′-*f*][1,3,2]dioxaphosphepin 4-oxide; (*R*)-*form*, *in* H-00131
4-Hydroxydinaphtho[2,1-*d*:1′,2′-*f*][1,3,2]dioxaphosphepin 4-oxide; (*S*)-*form*, *in* H-00131
4-Hydroxydinaphtho[2,1-*d*:1′,2′-*f*][1,3,2]dioxaphosphepin 4-oxide; (*S*)-*form*, 2-Methoxy (Me ester), *in* H-00131
4-Hydroxydinaphtho[2,1-*d*:1′,2′-*f*][1,3,2]dioxaphosphepin 4-oxide; (±)-*form*, *in* H-00131
4-Hydroxydinaphtho[2,1-*d*:1′,2′-*f*][1,3,2]dioxaphosphepin 4-oxide; (±)-*form*, 2-Methoxy (Me ester), *in* H-00131
2-Hydroxy-1,3,2-dioxaphosphepane 2-oxide, H-00132
2-Mercaptodibenzo[*d,f*][1,3,2]dioxaphosphepin 2-oxide, M-00007
6-Mercaptodibenzo[*d,f*][1,3,2]dioxaphosphepin 6-sulfide, M-00008
4-Mercaptodinaphtho[2,1-*d*:1′,2′-*f*][1,3,2]dioxaphosphepin 4-sulfide; (*S*)-*form*, *in* M-00012
4-Mercaptodinaphtho[2,1-*d*:1′,2′-*f*][1,3,2]dioxaphosphepin 4-sulfide; (±)-*form*, *in* M-00012
6-Methoxydibenzo[*d,f*][1,3,2]dioxaphosphepin, M-00032
5-Methoxy-5*H*-dibenzo[*b,f*]phosphepin 5-oxide, *in* H-00117
4-Methoxydinaphtho[2,1-*d*,1′,2′-*f*][1,3,2]dioxaphosphepin 4-oxide, *in* H-00131
6-Methyl-1,5-diphenyl-6-phosphabicyclo[3.2.1]octane; Oxide, *in* M-00134
9-Methyl-1,2,3,4,9-pentaphenyl-1*H*-tribenzo[*b,d,f*]-phosphepinium iodide, *in* P-00042
9-Methyl-9-phenyl-9-phosphoniabicyclo[4.2.1]nonane iodide, *in* P-00194
Methyl tetramethylene phosphite, *in* D-00968
8,9,10,11,12,13,14,15-Octahydro-4-hydroxydinaphtho[2,1-*d*:1′,2′-*f*][1,3,2]dioxaphosphepin 4-oxide, O-00018
8,9,10,11,12,13,14,15-Octahydro-4-hydroxydinaphtho[2,1-*d*:1′,2′-*f*][1,3,2]dioxaphosphepin 4-oxide; 4-Methoxy (Me ester), *in* O-00018
1,2,3,4,9-Pentaphenyl-1*H*-tribenzo[*b,d,f*]phosphepin, P-00042
1,2,3,4,9-Pentaphenyl-1*H*-tribenzo[*b,d,f*]phosphepin; 9-Oxide, *in* P-00042
2-Phenoxy-1,3,2-dioxaphosphepane 2-oxide, *in* H-00132
3-Phenyl-3*H*-benzophosphepin, P-00096
3-Phenyl-3*H*-benzophosphepin; 3-Oxide, *in* P-00096
6-Phenyldibenzo[*d,f*][1,3,2]dioxaphosphepin, P-00107
6-Phenyldibenzo[*d,f*][1,3,2]dioxaphosphepin; Oxide, *in* P-00107
5-Phenyl-5*H*-dibenzo[*b,f*]phosphepin, P-00108
5-Phenyl-5*H*-dibenzo[*b,f*]phosphepin; Sulfide, *in* P-00108
6-Phenyl-5*H*-dibenzo[*c,e*]phosphepin-5,7(6*H*)-dione, P-00109
2-Phenyl-1,3,2-dioxaphosphepane, P-00114
2-Phenyl-1,3,2-oxathiaphosphepane; 2-Sulfide, *in* P-00178
9-Phenyl-9-phosphabicyclo[4.2.1]nonane, P-00194
9-Phenyl-9-phosphabicyclo[4.2.1]nonane; Oxide, *in* P-00194
3-Phenyl-3-phosphabicyclo[3.2.1]octane; *endo-form*, 3-Oxide, *in* P-00198
3-Phenyl-3-phosphabicyclo[3.2.1]octane; *exo-form*, 6,7-Didehydro, 3-oxide, *in* P-00198
3-Phenyl-3-phosphabicyclo[3.2.1]oct-6-ene 3-oxide, *in* P-00198
1-Phenylphosphepane, P-00205
1-Phenylphosphepane; 1-Oxide, *in* P-00205
1-Phenylphosphepane; Sulfide, *in* P-00205
1-Phenyl-1*H*-phosphepin, P-00206
1-Phenyl-1*H*-phosphepin; Oxide, *in* P-00206
1,1,4,4-Tetrabenzyl-1,4-diphosphepanium dibromide, *in* D-00047
5,5,7,7-Tetraethyl-6,7-dihydro-5*H*-dibenzo[*d,f*][1,3]-diphosphepinium dibromide, *in* D-00282
2,2,4,4-Tetrahydro-1*H*-2,4-benzodiphosphepin, T-00087
2,3,4,5-Tetrahydro-1,5-diphenyl-1*H*-1,5-benzodiphosphepin, T-00101
2,3,4,5-Tetrahydro-5-methyl-2,2-diphenyl-1*H*-2-benzophosphepinium(1+); Hexafluorophosphate, *in* T-00114
2,2,7,7-Tetrahydro-3,4,5,6-tetrakis(methoxycarbonyl)-1*H*-1,2,7-azadiphosphepine; 2,2-Di-Me, 7,7-di-Ph, *in* T-00138
2,2,7,7-Tetrahydro-3,4,5,6-tetrakis(methoxycarbonyl)-1*H*-1,2,7-azadiphosphepine; 2,2,7,7-Tetra-Ph, *in* T-00138
2,2,7,7-Tetrahydro-3,4,5,6-tetrakis(methoxycarbonyl)-1,2,2,5,5-pentamethyl-1*H*-1,2,7-azadiphosphepine, *in* T-00138
2,2,4,4-Tetrahydro-2,2,4,4-tetramethyl-1*H*-2,4-benzodiphosphepin, *in* T-00087
2,3,4,5-Tetrahydro-2,2,4,4-tetramethyl-1*H*-2,4-benzodiphosphepinium(2+); Dibromide, *in* T-00139
2,3,4,5-Tetrahydro-1,1,5,5-tetramethyl-1*H*-1,5-benzophospharsepinium dibromide, *in* T-00092
2,2,4,4-Tetrahydro-2,2,4,4-tetraphenyl-1*H*-2,4-benzodiphosphepin, *in* T-00087
2,3,4,5-Tetrahydro-2,2,4,4-tetraphenyl-1*H*-2,4-benzodiphosphonium(2+); Dibromide, *in* T-00144
1,1,3,3-Tetrahydro-1,1,3,3-tetraphenyl-1,3-diphosphacyclohepta-1,2-diene, T-00148
Tetramethylene *tert*-butylphosphonite, *in* D-00968
Tetramethylene methylphosphonate, *in* M-00125
O,O-Tetramethylene methylphosphonothioate, *in* M-00125
Tetramethylene phenyl phosphite, *in* D-00968
Tetramethylene phenylphosphonate, *in* P-00114
O,O-Tetramethylene phenylphosphonothioate, *in* P-00114
O,O-Tetramethylene *O*-phenyl phosphorothioate, *in* D-00969
Tetramethylene phosphorochloridite, *in* D-00968
Tetramethylene phosphorofluoridite, *in* D-00968
1,1,3,3-Tetraphenyl-1,3-diphosphepanium, T-00257
3,3,6,6-Tetraphenyl-1,3,6-oxadiphosphepanium(2+); Dibromide, *in* T-00266
3,3,6,6-Tetraphenyl-1,3,6-oxadiphosphepanium(2+); Dipicrate, *in* T-00266

1 P-contg. ring; 8-membered

1-Benzyl-1-phenylphosphocanium bromide, *in* P-00215
2-*tert*-Butyl-1,3,6,2-trithiaphosphocane, B-00642
2-*tert*-Butyl-1,3,6,2-trithiaphosphocane; 2-Oxide, *in* B-00642
2-*tert*-Butyl-1,3,6,2-trithiaphosphocane; 2-Sulfide, *in* B-00642
6-Chloro-5,6,7,12-tetrahydro-2,5,7,10-tetramethyldibenzo[*d*,*g*][1,3,2]diazaphosphocine; 6-Oxide, *in* C-00182
6-Chloro-5,6,7,12-tetrahydro-2,5,7,10-tetramethyldibenzo[*d*,*g*][1,3,2]diazaphosphocine; 6-Sulfide, *in* C-00182
1,4-Dibenzyl-1,4-diphosphocane, D-00048
1,5-Dibenzyl-1,5-diphosphocane, D-00049
4,8-Di-*tert*-butyl-2,10-dimethyl-12*H*-dibenzo[*d*,*g*][1,3,2]-dioxaphosphocin, D-00088
4,8-Di-*tert*-butyl-2,10-dimethyl-12*H*-dibenzo[*d*,*g*][1,3,2]-dioxaphosphocin; 6-Ethoxy, *in* D-00088
4,8-Di-*tert*-butyl-2,10-dimethyl-12*H*-dibenzo[*d*,*g*][1,3,2]-dioxaphosphocin; 6-Phenoxy, *in* D-00088
4,8-Di-*tert*-butyl-2,10-dimethyl-12*H*-dibenzo[*d*,*g*][1,3,2]-dioxaphosphocin; 6-(1-Naphthoxy), *in* D-00088
4,8-Di-*tert*-butyl-2,10-dimethyl-12*H*-dibenzo[*d*,*g*][1,3,2]-dioxaphosphocin; 6-Oxide, *in* D-00088
7,12-Dihydro-12-methyl-12-phenyl-5*H*-dibenz[*c*,*f*][1,5]-oxaphosphocinium iodide, *in* D-00560
7,12-Dihydro-12-phenyl-5*H*-dibenz[*c*,*f*][1,5]oxaphosphocin; 12-Oxide, *in* D-00560
2,6-Dimethyl-1,3,2,6-dioxadiphosphocane; *cis-form*, 2,6-Disulfide, *in* D-00730
2,6-Dimethyl-1,3,2,6-dioxadiphosphocane; *cis-form*, 2,6-Diselenide, *in* D-00730
2,6-Dimethyl-1,3,2,6-dioxadiphosphocane; *trans-form*, 2,6-Disulfide, *in* D-00730
2,6-Dimethyl-1,3,2,6-dioxadiphosphocane; *trans-form*, 2,6-Diselenide, *in* D-00730
1,1-Diphenylphosphocanium bromide, *in* P-00215
1,2,3,4,7,8-Hexahydro-1-phenyl-5,6-bis[(trimethylsilyl)oxy]-phosphocin, H-00046
Octahydro-3,7-dimethyl-1,3,5,7-tetraphenyl-1,5,3,7-diazadiphosphocinium iodide, *in* O-00026
Octahydro-1,3,5,7-tetraphenyl-1,5,3,7-diazadiphosphocine, O-00026
Octahydro-1,3,5,7-tetraphenyl-1,5,3,7-diazadiphosphocine; 3,7-Disulfide, *in* O-00026
2-Phenyl-1,3,6,2-dioxathiaphosphocane; 2-Oxide, *in* P-00117
2-Phenyl-1,3,6,2-dioxathiaphosphocane; 2-Sulfide, *in* P-00117
1-Phenylphosphocane, P-00215
1-Phenylphosphocane; Oxide, *in* P-00215
1,1,4,4-Tetrabenzyl-1,4-diphosphocanium dibromide, *in* D-00048
1,1,5,5-Tetrabenzyl-1,5-diphosphocanium dibromide, *in* D-00049
5,5,8,8-Tetraethyl-5,6,7,8-tetrahydrodibenzo[*e*,*g*][1,4]-diphosphocinium dibromide, *in* D-00404
Tetrahydro-2,6-dimethyl-4*H*-1,3,6,2-dioxazaphosphocine; 2-Oxide, *in* T-00094
Tetrahydro-2,6-dimethyl-4*H*-1,3,6,2-dioxazaphosphocine; 2-Sulfide, *in* T-00094
5,6,7,12-Tetrahydro-12-phenyldibenz[*c*,*f*][1,5]-azaphosphocine, T-00130
5,6,7,12-Tetrahydro-12-phenyldibenz[*c*,*f*][1,5]-azaphosphocine; 12-Oxide, *in* T-00130
5,6,7,12-Tetrahydro-12-phenyldibenz[*c*,*f*][1,5]-azaphosphocine; 6-Benzyl, 12-oxide, *in* T-00130
1,2,5,6-Tetrahydro-2,2,5,5-tetraphenyl-2,5-benzodiphocinium dibromide, *in* H-00053
5,6,11,12-Tetrahydro-5,6,11,12-tetraphenyldibenzo[*c*,*g*]-[1,2,5,6]tetraphosphocin, T-00147
5,6,11,12-Tetrahydro-5,6,11,12-tetraphenyldibenzo[*c*,*g*]-[1,2,5,6]tetraphosphocin; 5,11-(or 5,6)-Disulfide, *in* T-00147
2,4,6,8-Tetrahydroxy-1,5-dimethyl-3,7,9-trioxa-2,4,6,8-tetraphosphabicyclo[3.3.1]nonane, T-00161
1,1,5,5-Tetraphenyl-1,5-diphosphocanium(2+); Dibromide, *in* T-00261
1,1,5,5-Tetraphenyl-1,5-diphosphocanium(2+); Diiodide, *in* T-00261
Tetraphenyl-1,4-diphosphonia-6,7-benzocyclooct-6-ene, *in* H-00053
1,3,5,7-Tetraphenyl-1,3,5,7-tetraphosphocane, T-00279
1,3,6,2-Trioxaphosphocane; 2-Chloro, *in* T-00592
1,3,6,2-Trioxaphosphocane; 2-Methoxy, *in* T-00592
1,3,6,2-Trioxaphosphocane; 2-Phenoxy, 2-oxide, *in* T-00592
1,3,6,2-Trioxaphosphocane; 2-Me, *in* T-00592
1,3,6,2-Trioxaphosphocane; 2-Me, 2-sulfide, *in* T-00592

1 P-contg. ring; 9+ -membered

17-Benzyl-2,7,10,15,17-pentamethyl-17*H*-tetrabenzo[*b*,*d*,*f*,*h*]-phosphoninium iodide, *in* P-00028
17-Benzyl-2,7,10,15-tetramethyl-17-phenyl-17*H*-tetrabenzo[*b*,*d*,*f*,*h*]phosphoninium iodide, *in* T-00211
7-Butyl-2,3-dihydro-1,6-diphenyl-1*H*-1-benzophosphonin-2,3,4,5-tetracarboxylic acid; Tetra-Me ester, oxide, *in* B-00552
2-Chloro-1,3-dioxa-2-phosphacyclotridecane, *in* D-00966
2-Chloro-1,3,2-dioxaphosphonane, *in* D-00971
7,8,10,11,13,14,16,17,19,24-Decahydro-5*H*-dibenzo[*o*,*r*]-[1,4,7,10,13,17]pentaoxaphosphacycloeicosin; 24-Methoxy, 24-oxide, *in* D-00001
7,8,10,11,13,14,16,17,19,24-Decahydro-5*H*-dibenzo[*o*,*r*]-[1,4,7,10,13,17]pentaoxaphosphacycloeicosin; 24-Ph, 24-oxide, *in* D-00001
5,6,7,8,9,14,15,16,17,18-Decahydro-9,14-dimethyl-5,18-diphenyldibenzo[*b*,*i*][1,4,8,11]-diphosphadiarsacyclotetradecin, D-00004
5,6,7,8,9,14,15,16,17,18-Decahydro-5,9,14,18-tetraphenyldibenzo[*b*,*i*][1,4,8,11]-tetraphosphacyclotetradecin; $(1\alpha,4\alpha,8\alpha,11\alpha)$-*form*, *in* D-00006
5,6,7,8,9,14,15,16,17,18-Decahydro-5,9,14,18-tetraphenyldibenzo[*b*,*i*][1,4,8,11]-tetraphosphacyclotetradecin; $(1\alpha,4\alpha,8\beta,11\beta)$-*form*, *in* D-00006
1,6-Dibenzyl-1,6-diphenyl-1,6-diphosphecanium dibromide, *in* D-01171
1,6-Dibenzyl-1,6-diphosphacycloundecane, D-00046
1,6-Dibenzyl-1,6-diphosphecane, *in* D-01171
1,4-Dibenzyl-1,4-diphosphonane, D-00050
1,5-Dibenzyl-1,5-diphosphonane, D-00051
2,10-Di-*tert*-butyl-1,3,6,9,11,14-hexathia-2,10-diphosphacyclohexadecane; *cis-form*, 2,10-Dioxide, *in* D-00093
2,10-Di-*tert*-butyl-1,3,6,9,11,14-hexathia-2,10-diphosphacyclohexadecane; *cis-form*, 2,10-Disulfide, *in* D-00093
2,10-Di-*tert*-butyl-1,3,6,9,11,14-hexathia-2,10-diphosphacyclohexadecane; *trans-form*, 2,10-Dioxide, *in* D-00093
2,10-Di-*tert*-butyl-1,3,6,9,11,14-hexathia-2,10-diphosphacyclohexadecane; *trans-form*, 2,10-Disulfide, *in* D-00093
2,9-Di-*tert*-butyl-1,3,8,10-tetraoxa-2,9-diphosphacyclotetradecane; 2,9-Disulfide, *in* D-00158
2,8-Di-*tert*-butyl-1,3,7,9-tetrathia-2,8-diphosphacyclododecane, D-00159
2,8-Di-*tert*-butyl-1,3,7,9-tetrathia-2,8-diphosphacyclododecane; *cis-form*, 2,8-Disulfide, *in* D-00159
2,8-Di-*tert*-butyl-1,3,7,9-tetrathia-2,8-diphosphacyclododecane; *trans-form*, *in* D-00159
2,3-Dihydro-2,4,9-triphenyl-1*H*-1,3,2-diazaphosphonine 2-oxide, D-00609
2,3-Dihydro-2,4,9-triphenyl-1*H*-1,3,2-diazaphosphonine 2-oxide; 1,3-Di-Me, *in* D-00609
6,9-Dimethyl-1,3-dioxa-6,9-diaza-2-phosphacycloundecane; 2-Chloro, *in* D-00728
6,9-Dimethyl-1,3-dioxa-6,9-diaza-2-phosphacycloundecane; 2-Methoxy, *in* D-00728
6,9-Dimethyl-1,3-dioxa-6,9-diaza-2-phosphacycloundecane; 2-Ethoxy, *in* D-00728
2,4-Dimethyl-9,13-diphenyl-1,5-diaza-9,13-diphosphacyclohexadeca-1,4-diene, D-00737
12,14-Dimethyl-4,8-diphenyl-1,11-diaza-4,8-diphosphacyclotetradeca-11,14-diene, D-00738

2,10-Dimethyl-1,3,6,9,11,14-hexaoxa-2,10-diphosphacyclohexadecane, D-00757
1,1-Dimethylphosphonanium iodide, *in* M-00285
1,12-Diphenyl-1,12-diphosphacyclodocosane; *cis-form*, *in* D-01008
1,12-Diphenyl-1,12-diphosphacyclodocosane; *cis-form*, 1,12-Dioxide, *in* D-01008
1,12-Diphenyl-1,12-diphosphacyclodocosane; *trans-form*, *in* D-01008
1,12-Diphenyl-1,12-diphosphacyclodocosane; *trans-form*, 1,12-Dioxide, *in* D-01008
1,10-Diphenyl-1,10-diphosphacyclooctadecane; *cis-form*, *in* D-01009
1,10-Diphenyl-1,10-diphosphacyclooctadecane; *cis-form*, 1,10-Dioxide, *in* D-01009
1,10-Diphenyl-1,10-diphosphacyclooctadecane; *trans-form*, *in* D-01009
1,10-Diphenyl-1,10-diphosphacyclooctadecane; *trans-form*, 1,10-Dioxide, *in* D-01009
1,12-Diphenyl-1,12-diphosphacyclotetracosane; *cis-form*, *in* D-01010
1,12-Diphenyl-1,12-diphosphacyclotetracosane; *cis-form*, 1,12-Dioxide, *in* D-01010
1,12-Diphenyl-1,12-diphosphacyclotetracosane; *trans-form*, *in* D-01010
1,12-Diphenyl-1,12-diphosphacyclotetracosane; *trans-form*, 1,12-Dioxide, *in* D-01010
1,6-Diphosphecane; 1,6-Di-Ph, 1,6-dioxide, *in* D-01171
1,6-Diphosphecane; 1,6-Dibenzyl, 1,6-dioxide, *in* D-01171
7,8,10,11,13,14,16,17,19,20,22,27-Dodecahydro-5*H*-dibenzo[*r,u*][1,4,7,10,13,16,20]-hexaoxaphosphacyclotricosin; 27-Ph, 27-oxide, *in* D-01261
2,3,4,5,6,7,8,9,10,11,12,13-Dodecahydro-2,5,9,12-tetraphenyl-1*H*-2,5,9,12-benzotetraphosphacyclopentadecin, D-01262
2-Fluoro-1,3-dioxa-2-phosphacyclotridecane, *in* D-00966
2-Fluoro-1,3,2-dioxaphosphonane, *in* D-00971
7,8,15,16,17,18-Hexahydro-14,18-diphenyl-6*H*,18*H*-1,6*H*-dibenzo[*b,i*][1,11,4,8]dithiadiphosphacyclotetradecin, H-00026
3,4,5,6,7,8-Hexahydro-5-phenyl-2*H*-1,9,5-benzodithiaphosphacycloundecin, H-00044
2,3,4,7,8,9-Hexahydro-1-phenyl-5,6-bis[(trimethylsilyl)oxy]-1*H*-phosphonine, H-00047
2,3,4,7,8,9-Hexahydro-1-phenyl-5,6-bis[(trimethylsilyl)oxy]-1*H*-phosphonine; Sulfide, *in* H-00047
2,7,10,15,17,17-Hexamethyl-17*H*-tetrabenzo[*b,d,f,h*]-phosphoninium iodide, *in* P-00028
2,5,5,8,11,11-Hexamethyl-1,3,7,9-tetraoxa-2,8-diphosphacyclododecane, H-00081
2,5,5,8,11,11-Hexamethyl-1,3,7,9-tetraoxa-2,8-diphosphacyclododecane; 2,8-Disulfide, *in* H-00081
2-Hydroxy-1,3,2-dioxaphosphonane 2-oxide, H-00137
1-Hydroxy-3,8-phosphonanedione 1-oxide, *in* P-00373
7-Methyl-7-phenyl-7*H*-dibenzo[*d,f*]phosphoninium iodide, *in* P-00111
1-Methylphosphonane, M-00285
1-Methylphosphonane; 1-Oxide, *in* M-00285
1-Methyl-3,8-phosphonanedione 1-oxide, *in* P-00373
17-Methyl-6,7,9,10-tetrahydrodibenzo[*d,m*][1,3,6,9,12,2]-pentaoxaphosphacyclotetradecin; 17-Oxide, *in* M-00403
17-Methyl-6,7,9,10-tetrahydrodibenzo[*d,m*][1,3,6,9,12,2]-pentaoxaphosphacyclotetradecin; 17-Sulfide, *in* M-00403
2,5,5,8,11,11,14,17,17-Nonamethyl-1,3,7,9,13,15-hexaoxa-2,8,14-triphosphacyclooctadecane, N-00067
2,5,5,8,11,11,14,17,17-Nonamethyl-1,3,7,9,13,15-hexaoxa-2,8,14-triphosphacyclooctadecane; 2,8,14-Trisulfide, *in* N-00067
1,1,6,6,11,11,16,16-Octabenzyl-1,6,11,16-tetraphosphoniacycloeicosane tetrabromide, *in* T-00008
1,1,5,5,10,10,14,14-Octabenzyl-1,5,10,14-tetraphosphoniacyclooctadecane tetrabromide, *in* T-00009
Octahydro-3,8-dimethyl-2,7-diphenyl-1,6,3,8,2,7-dioxadiazadiphosphecine, O-00006
Octahydro-3,8-dimethyl-2,7-diphenyl-1,6,3,8,2,7-dioxadiazadiphosphecine; 2,7-Disulfide (isomer 1), *in* O-00006
Octahydro-3,8-dimethyl-2,7-diphenyl-1,6,3,8,2,7-dioxadiazadiphosphecine; 2,7-Disulfide (isomer 2), *in* O-00006
2,3,4,5,6,7,8,9-Octahydro-1,9-dimethyl-5-phenyl-1*H*-5,1,9-benzophosphadiarsacycloundecin, O-00007
13,14,16,17,31,32,34,35-Octahydro-6,24-dimethyltetrabenzo[*d,m,r,a'*]-[1,3,6,9,12,15,20,23,26,2,16]-decaoxadiphosphacyclooctacosin; 6,24-Dioxide, *in* O-00008
13,14,16,17,31,32,34,35-Octahydro-6,24-dimethyltetrabenzo[*d,m,r,a'*]-[1,3,6,9,12,15,20,23,26,2,16]-decaoxadiphosphacyclooctacosin; 6,24-Disulfide, *in* O-00008
2,3,4,5,6,7,8,9-Octahydro-1,9-diphenyl-1*H*-5,1,9-benzazadiphosphacycloundecine; (1*RS*,9*SR*)-*form*, *in* O-00009
2,3,4,5,6,7,8,9-Octahydro-1,9-diphenyl-1*H*-5,1,9-benzazadiphosphacycloundecine; (1*RS*,9*SR*)-*form*, 5-Me, *in* O-00009
2,3,4,5,6,7,8,9-Octahydro-1,9-diphenyl-1*H*-5,1,9-benzazadiphosphacycloundecine; (1*RS*,9*SR*)-*form*, 5-Ph, *in* O-00009
2,3,4,5,6,7,8,9-Octahydro-1,9-diphenyl-1*H*-1,9-benzodiphosphacycloundecin, O-00010
1,2,3,4,6,7,8,9-Octahydro-1,9-diphenyl-5,9-benzothiadiphosphacycloundecin, O-00011
1,2,3,4,6,7,8,9-Octahydro-1,9-diphenyl-5,1,9-benzoxadiphosphacycloundecin, O-00012
6,7,8,9,14,15,16,17-Octahydro-9,14-diphenyldibenzo[*b,i*]-[1,4,8,11]dithiadiphosphacyclotetradecin, O-00013
6,7,8,9,15,16,17,18-Octahydro-9,18-diphenyldibenzo[*b,i*]-[1,8,4,11]dithiadiphosphacyclotetradecin, O-00014
6,7,9,10,16,17,18,19-Octahydro-16,19-diphenyldibenzo[*h,n*]-[1,4,7,10,13]trioxadiphosphacyclopentadecin; 16,19-Dioxide, *in* O-00015
6,7,11,12,19,20,22,23-Octahydro-9-methyldibenzo[*g,p*]-[1,3,6,9,12,15,18,21]heptaoxaphosphacycloeicosin; 9-Oxide, *in* O-00019
1,2,3,4,5,8,9,10-Octahydro-1-phenyl-6,7-bis[(trimethylsilyl)oxy]phosphecine, O-00021
2,3,4,5,6,7,8,9-Octahydro-1,5,9-triphenyl-1*H*-1,9,5-benzodiphosphaarsacycloundecin, O-00027
2,3,4,5,6,7,8,9-Octahydro-1,5,9-triphenyl-1*H*-1,5,9-benzotriphosphacycloundecin, O-00028
2,7,9,14,15,16,17,18-Octaoxa-1,8-diphospha(1,8-P^V)-pentacyclo[10.2.1.1^{1,4}.1^{5,8}.1^{8,11}]octadecane-3,6,10,13-tetrone, O-00036
2,7,9,14,15,16,17,18-Octaoxa-1,8-diphospha(1,8-P^V)-pentacyclo[10.2.1.1^{1,4}.1^{5,8}.1^{8,11}]octadecane-3,6,10,13-tetrone; (*R,R,R,R*)-*form*, 1,8-Dihydroxy, bistriethylammonium salt, *in* O-00036
2,7,10,15,17-Pentamethyl-17*H*-tetrabenzo[*b,d,f,h*]phosphonin, P-00028
2-Phenoxy-1,3,2-dioxaphosphonane 2-oxide, *in* H-00137
1-Phenyl-6,7-bis[(trimethylsilyl)oxy]phosphacycloundec-6-ene, P-00100
1-Phenyl-6,7-bis[(trimethylsilyl)oxy]phosphacycloundec-6-ene; Sulfide, *in* P-00100
7-Phenyl-7*H*-dibenzo[*d,f*]phosphonin, P-00111
7-Phenyl-7*H*-dibenzo[*d,f*]phosphonin; (*E,Z*)-*form*, 7-Oxide, *in* P-00111
1-Phenyl-3,8-phosphonanedione 1-oxide, *in* P-00373
1,1,6,6-Tetrabenzyl-1,6-diphosphecanium dibromide, *in* D-01171
1,1,4,4-Tetrabenzyl-1,4-diphosphonanium dibromide, *in* D-00050
1,1,5,5-Tetrabenzyl-1,5-diphosphonanium dibromide, *in* D-00051
1,1,6,6-Tetrabenzyl-1,6-diphosphoniacycloundecane dibromide, *in* D-00046
1,6,11,16-Tetrabenzyl-1,6,11,16-tetraphosphacycloeicosane, T-00008
1,5,10,14-Tetrabenzyl-1,5,10,14-tetraphosphacyclooctadecane, T-00009
1,5,10,14-Tetrabenzyl-1,5,10,14-tetraphosphacyclooctadecane; 1,5,10,14-Tetraoxide, *in* T-00009

5,5,10,10-Tetraethyl-5,6,7,8,9,10-hexahydrodibenzo[*b,d*]-[1,6]diphosphecinium dibromide, *in* H-00021
5,5,9,9-Tetraethyl-6,7,8,9-tetrahydro-5*H*-dibenzo[*f,h*][1,5]-diphosphoninium dibromide, *in* T-00089
9,9,16,16-Tetraethyl-9,10,15,16-tetrahydrotribenzo[*b,d,h*]-[1,6]diphosphecinium dibromide, *in* T-00155
2,7,10,15-Tetramethyl-17-phenyl-17*H*-tetrabenzo[*b,d,f,h*]-phosphonin, T-00211
4,7,13,16-Tetraphenyl-1,10-dioxa-4,7,13,16-tetraphosphacyclooctadecane; (4*RS*,7*RS*,13*SR*,16*SR*)-*form*, *in* T-00255
4,7,13,16-Tetraphenyl-1,10-dioxa-4,7,13,16-tetraphosphacyclooctadecane; (4*RS*,7*RS*,13*RS*,16*SR*)-*form*, *in* T-00255
4,7,13,16-Tetraphenyl-1,10-dipropyl-1,10-diaza-4,7,13,16-tetraphosphacyclooctadecane, T-00264
1,5,9,13-Tetraphenyl-1,5,9,13-tetraphosphacyclohexadecene, T-00276
1,5,9,13-Tetraphenyl-1,5,9,13-tetraphosphacyclohexadecene; Tetraoxide, *in* T-00276

2 P-contg. rings; 3-membered P-ring(s) present

1,2,4,5-Tetra-*tert*-butyl-1,2,4,5-tetraphosphaspiro[2.2]-pentane, T-00026
1,2,4,5-Tetrakis(methylene)-3-phosphoniaspiro[2.2]pentane, T-00182
1,3,4,6-Tetrakis(trifluoromethyl)-2,5-diphosphatricyclo[3.1.0.0^{2,6}]hex-3-ene, T-00186
5,5,6-Trimethyl-4,7-dioxa-3-phospha(3-P^V)spiro[2.4]heptane, T-00510

2 P-contg. rings; 4-membered P-ring(s) present

1,3-Dimethyl-6,7-diphenyl-5,8-dioxa-1,3-diaza-4-phospha(4-P^V)spiro[3.4]oct-6-en-2-one; 4-Methoxy, *in* D-00742
1,3-Dimethyl-6,7-diphenyl-5,8-dioxa-1,3-diaza-4-phospha(4-P^V)spiro[3.4]oct-6-en-2-one; 4-Dimethylamino, *in* D-00742
1,3-Dimethyl-5,7-diphenyl-1,3,5,7-tetraaza-4-phospha(4-P^V)-spiro[3.3]heptane-2,6-dione; 4-Chloro, *in* D-00748
1,3-Dimethyl-5,7-diphenyl-1,3,5,7-tetraaza-4-phospha(4-P^V)-spiro[3.3]heptane-2,6-dione; 4-(Diethoxyphosphinyl), *in* D-00748
N,N,N′,N′,6,7-Hexamethyl-1,3-bis(trimethylsilyl)-5,8-dioxa-1,3-diaza-2,4-diphospha(4-P^V)spiro[3.4]oct-6-ene-2,4-diamine, H-00068
N,N,N′,N′,6,7-Hexamethyl-1,3-bis(trimethylsilyl)-5,8-dioxa-1,3-diaza-2,4-diphospha(4-P^V)spiro[3.4]oct-6-ene-2,4-diamine; 2-Sulfide, *in* H-00068
2′,2′,3′,4′,4′-Pentamethylspiro[1,3,2-benzodioxaphosphole-2,1′-phosphetane]; *cis-form*, 2-Ph, *in* P-00027
2′,2′,3′,4′,4′-Pentamethylspiro[1,3,2-benzodioxaphosphole-2,1′-phosphetane]; *trans-form*, 2-Ph, *in* P-00027
2′,2′,3′,4′,4′-Pentamethylspiro[1,3,2-benzodioxaphosphole-2,1′-phosphetane]; *trans-form*, 2-Benzyl, *in* P-00027
1,1,2,3,3-Pentamethyl-6,6,7,7-tetrakis(trifluoromethyl)-5,8-dioxa-4-phospha(4-P^V)spiro[3.4]octane; 4-Phenoxy, *in* P-00029
1,1,2,3,3-Pentamethyl-6,6,7,7-tetrakis(trifluoromethyl)-5,8-dioxa-4-phospha(4-P^V)spiro[3.4]octane; 4-*tert*-Butyl, *in* P-00029
1,1,2,3,3-Pentamethyl-6,6,7,7-tetrakis(trifluoromethyl)-5,8-dioxa-4-phospha(4-P^V)spiro[3.4]octane; 4-Ph (*cis*-), *in* P-00029
1,1,2,3,3-Pentamethyl-6,6,7,7-tetrakis(trifluoromethyl)-5,8-dioxa-4-phospha(4-P^V)spiro[3.4]octane; 4-Ph (*trans*-), *in* P-00029
1,1,2,3,3-Pentamethyl-6,6,7,7-tetrakis(trifluoromethyl)-5,8-dioxa-4-phospha(4-P^V)spiro[3.4]octane; 4-(4-Bromophenyl), *in* P-00029
1,1,2,3,3-Pentamethyl-6,6,7,7-tetrakis(trifluoromethyl)-5,8-dioxa-4-phospha(4-P^V)spiro[3.4]octane; 4-Dimethylamino, *in* P-00029
1,1,2,3,3-Pentamethyl-6,6,8,8-tetrakis(trifluoromethyl)-5,7-dioxa-4-phospha(4-P^V)spiro[3.4]octane; 4-Chloro, *in* P-00030
4-Phenyl-1,2,3,5,8-pentaoxa-4-phospha(4-P^V)spiro[3.4]-octane, P-00181
1,1,3,3-Tetramethyl-6,7-bis(trifluoromethyl)-5,8-dithia-4-phospha(4-P^V)spiro[3.4]oct-6-ene; 4-Ph, *in* T-00193
1,3,5,7-Tetramethyl-1,3,5,7-tetraaza-4-phospha(4-P^V)-spiro[3.3]heptane-2,6-dione; 4-Chloro, *in* T-00234
1,3,5,7-Tetramethyl-1,3,5,7-tetraaza-4-phospha(4-P^V)-spiro[3.3]heptane-2,6-dione; 4-Ph, *in* T-00234
1,3,5,7-Tetramethyl-1,3,5,7-tetraaza-4-phospha(4-P^V)-spiro[3.3]heptane-2,6-dione; 4-Azido, *in* T-00234
2,3,5,6-Tetraphenyl-1,4-bis[(trimethylsilyl)oxy]-2,3,5,6-tetraphosphabicyclo[2.2.0]hexane, T-00251
2,2,2-Trichloro-2,2-dihydro-1,3-dimethyl-6,8-bis(trifluoromethyl)-1,3,5,7,9-pentaaza-2,4-diphospha(4-P^V)spiro[3.5]nona-4,6,8-triene, T-00401

2 P-contg. rings; 5-membered P-ring(s) present

Aminobisbiphenylenephosphorane, *in* S-00021
2,3-Benzo-1,4,5,6,7-pentaphenyl-7-phosphabicyclo[2.2.1]-hept-5-ene 7-oxide, *in* D-00538
2,4-Bis(1-methylethylidene)-1,3,6,9-tetraoxa-5-phospha(5-P^V)spiro[4.4]nonane; 5-Ph, *in* B-00306
4′,5′-Bis(trifluoromethyl)spiro[1,3,2-benzodioxaphosphole-2,2′-[1,3,2]dioxaphosphole]; 2-Methoxy, *in* B-00421
4′,5′-Bis(trifluoromethyl)spiro[1,3,2-benzodioxaphosphole-2,2′-[1,3,2]dioxaphosphole]; 2-Phenoxy, *in* B-00421
4′,5′-Bis(trifluoromethyl)spiro[1,3,2-benzodioxaphosphole-2,2′-[1,3,2]dioxaphosphole]; 2-Fluoro, *in* B-00421
4′,5′-Bis(trifluoromethyl)spiro[1,3,2-benzodioxaphosphole-2,2′-[1,3,2]dithiaphosphole]; 2-Methoxy, *in* B-00422
2-(4-Bromophenyl)-8,8-dihydro-4,6,8,8-tetraphenyl[1,2]-oxaphospholo[2,3-*b*][1,2,5]dioxaphosphorin, B-00479
1-(2-Chloroethyl)tetrahydro-1*H*,5*H*-[1.3.2]-diazaphospholo[2,1-*b*][1.3.2]oxazaphosphorine 9-oxide; (*R*)-*form*, *in* C-00082
1-(2-Chloroethyl)tetrahydro-1*H*,5*H*-[1.3.2]-diazaphospholo[2,1-*b*][1.3.2]oxazaphosphorine 9-oxide; (*S*)-*form*, *in* C-00082
1-(2-Chloroethyl)tetrahydro-1*H*,5*H*-[1.3.2]-diazaphospholo[2,1-*b*][1.3.2]oxazaphosphorine 9-oxide; (±)-*form*, *in* C-00082
2-Chloro-2,2′-spirobi[1,3,2-benzodioxaphosphole], C-00174
9*H*-Dibenzo[2,3:4,5]phospholo[1,2-*f*]phosphanthridene; 17-Oxide, *in* D-00041
7,7-Dichloro-2,3,5,6-tetrakis(trifluoromethyl)-1,4-diphosphabicyclo[2.2.1]hepta-2,5-diene, D-00195
1,8-Diethoxyoctahydro-4,7-phosphinidene-1*H*-phosphindole 1,8-dioxide, *in* D-00625
1,8-Diethoxy-3*a*,4,7,7*a*-tetrahydro-4,7-phosphinidene-1*H*-phosphinidole 1,8-dioxide, *in* D-00625
4′,5′-Dihydro-3′,5′-dimethylspiro[1,3,2-benzoxazaphosph(P^V)-ole-2,2′-[1,3,2]oxazaphospholidin]-4′-one, D-00470
4′,5′-Dihydro-3′,5′-dimethylspiro[1,3,2-benzoxazaphosph(P^V)-ole-2,2′-[1,3,2]oxazaphospholidin]-4′-one; (5′*S*)-*form*, 3-Me, *in* D-00470
8,8-Dihydro-8,8-diphenyl-2*H*,6*H*-[1,2]oxaphospholo[4,3,2-*hi*]-[2,1]benzoxaphosphole, D-00479
10,10-Dihydro-2,3-diphenyl-[1,3,2]oxazaphospholo[2,3-*b*]-[1,3,2]benzoxazaphosphole; 10,10-Dimethoxy, *in* D-00481
10,10-Dihydro-2,3-diphenyl-[1,3,2]oxazaphospholo[2,3-*b*]-[1,3,2]benzoxazaphosphole; 10,10-Diphenoxy, *in* D-00481
10,10-Dihydro-2,3-diphenyl-[1,3,2]oxazaphospholo[2,3-*b*]-[1,3,2]benzoxazaphosphole; 10,10-Di-Me, *in* D-00481
10,10-Dihydro-2,3-diphenyl-[1,3,2]oxazaphospholo[2,3-*b*]-[1,3,2]benzoxazaphosphole; 10,10-Di-Ph, *in* D-00481
10,10-Dihydro-2,3-diphenyl-[1,3,2]oxazaphospholo[2,3-*b*]-[1,3,2]benzoxazaphosphole; 10,10-Bis(dimethylamino), *in* D-00481
1,3-Dihydrospiro[2*H*-1,3,2-benzodiazaphosphole-2,2′-[1,3,2]-benzodioxaphosphole], D-00574
1,3-Dihydrospiro[2*H*-1,3,2-benzodiazaphosphole-2,2′-[1,3,2]-benzodioxaphosphole]; 1-Benzyl, *in* D-00574
1,3-Dihydrospiro[2*H*-1,3,2-benzodiazaphosphole-2,2′-[1,3,2]-dioxaphosholane]; 2-Ph, *in* D-00575
1,4-Dihydrospiro[2*H*-3,1,2-benzoxazaphosphorine-2,2′-[1,3,2]-dioxaphospholane]; 2-Ph, *in* D-00576

1,3-Dihydro-1,2,3-trimethylspiro[2*H*-1,3,2-benzodiazaphosphole-2,2′-[1,3,2]benzodioxaphosphole], *in* D-00469
1,8-Dihydroxy-3*a*,4,7,7*a*-tetrahydro-4,7-phosphinidene-1*H*-phosphindole 1,8-dioxide, D-00625
N,*N*-Dimethyl-2,4-bis(1-methylethylidene)-1,3,6,9-tetraoxa-5-phospha(5-P^V)spiro[4.4]nonan-5-amine, *in* B-00306
1,4-Dimethyl-7,9-bis(trifluoromethyl)-1,4,6,8,10-pentaaza-5-phospha(5-P^V)spiro[4.5]deca-5,7,9-triene, D-00714
4,9-Dimethyl-1,6-dioxa-4,9-diaza-5-phospha(5-P^v)spiro[4.4]-nonane, D-00729
3,8-Dimethyl-2,7-diphenyl-1,6-dioxa-4,9-diaza-5-phospha(5-P^V)spiro[4.4]nonane; (2*R*,3*S*,7*R*,8*S*)*-form*, *in* D-00740
3,8-Dimethyl-2,7-diphenyl-1,6-dioxa-4,9-diaza-5-phospha(5-P^V)spiro[4.4]nonane; (2*S*,3*R*,7*S*,8*R*)*-form*, *in* D-00740
3,8-Dimethyl-2,7-diphenyl-1,6-dioxa-4,9-diaza-5-phospha(5-P^V)spiro[4.4]nonane; (2*S*,3*R*,7*R*,8*S*)*-form*, *in* D-00740
6,9-Dimethyl-2,3-diphenyl-1,4-dioxa-6,9-diaza-5-phospha(5-P^V)spiro[4.4]non-2-ene; 5-Methoxy, *in* D-00741
1,3-Dimethyl-6,7-diphenyl-5,8-dioxa-1,3-diaza-4-phospha(4-P^V)spiro[3.4]oct-6-en-2-one; 4-Methoxy, *in* D-00742
1,3-Dimethyl-6,7-diphenyl-5,8-dioxa-1,3-diaza-4-phospha(4-P^V)spiro[3.4]oct-6-en-2-one; 4-Dimethylamino, *in* D-00742
8,8-Dimethyl-2,3-diphenyl-1,4,6,10-tetraoxa-5-phospha(5-P^V)-spiro[4.5]dec-2-ene; 5-Phenoxy, *in* D-00749
N,*N*-Dimethyl-2,3-diphenyl-1,4,6,9-tetraoxa-5-phospha(5-P^V)-spiro[4.4]non-2-en-5-amine, *in* D-01150
4,5-Dimethyl-2-oxa-3-phenyl-1-phosphabicyclo[2.2.1]hept-5-ene; 1-Sulfide, *in* D-00781
6,9-Dimethyl-5-phenoxy-2,2,3,3-tetrakis(trifluoromethyl)-1,4-dioxa-6,9-diaza-5-phospha(5-P^V)spiro[4.4]nonane, *in* D-00922
1,4-Dimethyl-2-phenyl-7-oxa-2-phosphabicyclo[2.2.1]heptane; 2-Oxide, *in* D-00796
5,6-Dimethyl-7-phenyl-7-phosphabicyclo[2.2.1]hept-5-ene-2,3-dicarboxylic acid; *endo,endo,syn-form*, Di-Me ester, *in* D-00800
5,6-Dimethyl-7-phenyl-7-phosphabicyclo[2.2.1]hept-5-ene-2,3-dicarboxylic acid; *endo,endo,syn-form*, 7-Sulfide, anhydride, *in* D-00800
5,6-Dimethyl-7-phenyl-7-phosphabicyclo[2.2.1]hept-5-ene-2,3-dicarboxylic acid; *endo,endo,syn-form*, 7-Sulfide, Di-Me ester, *in* D-00800
5,6-Dimethyl-7-phenyl-7-phosphabicyclo[2.2.1]hept-5-ene-2,3-dicarboxylic acid; *endo,endo,anti-form*, Di-Me ester, *in* D-00800
5,6-Dimethyl-*N*-phenylspiro[1,3,2-benzodioxaphosphole-2,1′(3′*H*)-[2,1]benzoxaphosphol]-2-amine, *in* D-00915
3′,4′-Dimethyl-5′-phenylspiro[1,3,2-benzoxazaphosphole-2(3*H*),2′-[1,3,2]oxazaphospholidine], D-00823
6,9-Dimethyl-5-phenylthio-2,2,3,3-tetrakis(trifluoromethyl)-1,4-dioxa-6,9-diaza-5-phospha(5-P^V)spiro[4.4]nonane, *in* D-00922
5,6-Dimethylspiro[1,3,2-benzodioxaphosphole-2,1′(3′*H*)-[2,1]-benzoxaphosphole, D-00915
5,6-Dimethylspiro[1,3,2-benzodioxaphosphole-2,1′(3′*H*)-[2,1]-benzoxaphosphole; 2-Phenoxy, *in* D-00915
5,6-Dimethylspiro[1,3,2-benzodioxaphosphole-2,1′(3′*H*)-[2,1]-benzoxaphosphole; 2-Ph, *in* D-00915
N,*N*-Dimethylspiro[1,3,2-benzodioxaphosphole-2,2′-[1,3,2]-benzoxathiaphosphol]-2-amine, *in* S-00007
2,3′-Dimethylspiro[1,3,2-benzodioxaphosphole-2,2′(3′*H*)-[1,3,2]benzoxazaphosphole], *in* M-00395
3,3′-Dimethyl-2,2′-(3*H*,3′*H*)-spirobi[1,3,2-benzothiazaphosphole]; 2-Chloro, *in* D-00916
3,3-Dimethyl-2,2(3*H*,3′*H*)-spirobi[1,3,2-benzoxazaphosphole], D-00917
1,3-Dimethylspiro[1,3,2-diazaphospholidine-2,2′-phenanthro[9,10-*d*][1,3,2]dioxaphosphole]; 2-Methoxy, *in* D-00918
N,*N*-Dimethylspiro[1,3,2-dioxaphosphorinane-2,2′-phenanthro[9,10-*d*][1,3,2]dioxaphosphol]-2-amine, *in* S-00025
6,10-Dimethyl-2,2,3,3-tetrakis(trifluoromethyl)-1,4-dioxa-6,10-diaza-5-phospha(5-P^V)spiro[4.5]decane; 5-Phenoxy, *in* D-00923
2,3-Dimethyl-1,4,6,9-tetraoxa-5-phospha(5-P^V)spiro[4.4]non-2-ene; 5-Methoxy, *in* D-00924
8,8-Dimethyl-1,6,10-trioxa-5-phospha(5-P^V)spiro[4.5]dec-2-ene; 5-Methoxy, *in* D-00933
2,8-Dioxa-5-aza-1-phosphabicyclo[3.3.0]octane, D-00961
1,4-Dioxa-6-aza-5-phospha(5-P^V)spiro[4.4]nonan-7-one; 5-Ph, *in* D-00962
1,6-Dioxa-4,9-diaza-5-phospha(5-P^V)spiro[4.4]nonane, D-00963
1,6-Dioxa-4,9-diaza-5-phospha(5-P^V)spiro[4.4]nonane; 5-Ph, *in* D-00963
1,4-Dioxa-6,9-dithia-5-phospha(5-P^V)spiro[4.4]nonane; 5-Methoxy, *in* D-00964
1,4-Dioxa-5-phospha(5-P^V)spiro[4.4]non-7-ene; 5-Me, *in* D-00967
1,4-Dioxa-5-phospha(5-P^V)spiro[4.4]non-7-ene; 5-Ph, *in* D-00967
9,9-Diphenyl-9-phosphoniahomocubane bromide, *in* P-00201
9,9-Diphenyl-9-phosphoniatricyclo[4.2.1.$0^{2,5}$]nonane bromide, *in* P-00204
2,5′-Diphenylspiro[4*H*-1,3,2-benzodioxaphosphorin-2,2′-[1,3,2]dioxaphospholane]-4,4′-dione, D-01147
2,3-Diphenyl-1,4,6,9-tetraoxa-5-phospha(5-P^V)spiro[4.4]non-2-ene; 5-Methoxy, *in* D-01150
2,3-Diphenyl-1,4,6,9-tetraoxa-5-phospha(5-P^V)spiro[4.4]non-2-ene; 5-Phenoxy, *in* D-01150
2,3-Diphenyl-1,6,10-trioxa-4-aza-5-phospha(5-P^V)spiro[4.5]-dec-2-ene; 5-Ph, *in* D-01158
2,3-Diphenyl-1,4,6-trioxa-9-aza-5-phospha(5-P^V)spiro[4.4]-non-2-ene; 5-Methoxy, *in* D-01159
1,5-Diphosphabicyclo[3.3.0]octane, D-01170
1,5-Diphosphabicyclo[3.3.0]octane; Monosulfide, *in* D-01170
1,5-Diphosphabicyclo[3.3.0]octane; Disulfide, *in* D-01170
2-Fluoro-2,2′-spirobi[1,3,2-benzodioxaphosphole], F-00048
2,3,5,6,8,8-Hexahydro-8-phenyl-[1,3,2]oxazaphospholo[2,3-*b*]-[1,3,2]oxazaphosphole, H-00049
1,4,4*a*,9,9*a*,10-Hexahydro-11-phenyl-9,10-phosphinideneanthracene; 11-Oxide, *in* H-00050
Hexahydrospiro[1,3,2-benzoxazaphosphole-2(3*H*),1′-phospholane]; 2-Ph, *in* H-00052
N,*N*,*N*′,*N*′,6,7-Hexamethyl-1,3-bis(trimethylsilyl)-5,8-dioxa-1,3-diaza-2,4-diphospha(4-P^V)spiro[3.4]oct-6-ene-2,4-diamine, H-00068
N,*N*,*N*′,*N*′,6,7-Hexamethyl-1,3-bis(trimethylsilyl)-5,8-dioxa-1,3-diaza-2,4-diphospha(4-P^V)spiro[3.4]oct-6-ene-2,4-diamine; 2-Sulfide, *in* H-00068
N,*N*,4′,4′,5′,5′-Hexamethylspiro[1,3,2-benzodioxaphosphole-2,2′-[1,3,2]dioxaphospholan]-2-amine, *in* T-00230
4,4,5,5,*N*,*N*-Hexamethylspiro[1,3,2-benzoxazaphosphole-2(3*H*),-2′-[1,3,2]dioxaphospholan]-2-amine, *in* T-00231
N,*N*,2,2,3,3-Hexamethyl-1,4,6,9-tetraoxa-5-phospha(5-P^V)-spiro[4.4]nonan-5-amine, *in* T-00235
N,*N*,2,2,3,3-Hexamethyl-1,4,6-trioxa-9-aza-5-phospha(5-P^V)-spiro[4.4]nonan-5-amine, *in* T-00238
2,2,3,3,7,7-Hexamethyl-1,4,6-trioxa-9-aza-5-phospha(5-P^V)-spiro[4.4]nonane, H-00083
2,2,3,3,8,8-Hexamethyl-1,4,6-trioxa-9-aza-5-phospha(5P^V)-spiro[4.4]nonane, H-00084
Homocubyltrimethylphosphorane, *in* D-00569
Homocubyltriphenylphosphorane, *in* D-00569
1-Hydroxy-1,1′(3*H*,3′*H*)-spirobi[2,1-benzoxaphosphole]-3,3′-dione, H-00194
1-Hydroxy-1,1′(3*H*,3′*H*)-spirobi[2,1-benzoxaphosphole]-3,3′-dione; 1-Methoxy, *in* H-00194
1-Hydroxy-3,3,3′,3′-tetramethyl-1,1′(3*H*,3′*H*)-spirobi[2,1-benzoxaphosphole], H-00199
1-Hydroxy-3,3,3′,3′-tetramethyl-1,1′(3*H*,3′*H*)-spirobi[2,1-benzoxaphosphole]; 1-Methoxy, *in* H-00199
7-Isopropylidene-4,8,8-trimethyl-1,6-dioxa-4-aza-5-phospha(5-P^V)spiro[4.4]nonan-9-one; 5-Dimethylamino, *in* I-00032
5-Methoxy-5-phospha(5-P^V)spiro[4.4]nonane, *in* P-00350
Methylbisbiphenylenephosphorane, *in* S-00021
Methylglycerol bicyclic phosphate, *in* M-00423
Methylglycerol bicyclic selenophosphate, *in* M-00423
9-Methyl-9-phenyl-9-phosphoniahomocubane iodide, *in* P-00201
8-Methyl-8-phosphabicyclo[3.2.1]oct-6-ene 8-oxide; *syn-form*, *in* M-00273

8-Methyl-8-phosphabicyclo[3.2.1]oct-6-ene 8-oxide; *anti-form*, *in* M-00273

5-Methyl-5-phospha(5-P^V)spiro[4.4]nonane, *in* P-00350

4-Methyl-4-phosphatetracyclo[3.3.0.$0^{2,8}$.$0^{3,6}$]octane, M-00275

4-Methyl-4-phosphatetracyclo[3.3.0.$0^{2,8}$.$0^{3,6}$]octane; *endo-form*, 4-Oxide, *in* M-00275

4-Methyl-4-phosphatetracyclo[3.3.0.$0^{2,8}$.$0^{3,6}$]octane; *endo-form*, 4-Sulfide, *in* M-00275

4-Methyl-4-phosphatetracyclo[3.3.0.$0^{2,8}$.$0^{3,6}$]octane; *exo-form*, 4-Oxide, *in* M-00275

4-Methyl-4-phosphatetracyclo[3.3.0.$0^{2,8}$.$0^{3,6}$]octane; *exo-form*, 4-Sulfide, *in* M-00275

3′-Methylspiro[1,3,2-benzodioxaphosphole-2,2(3′*H*)-[1,3,2]-benzothiazaphosphole]; 2-Phenoxy, *in* M-00394

3′-Methylspiro[1,3,2-benzodioxaphosphole-2,2′(3′*H*)-[1,3,2]-benzoxazaphosphole], M-00395

5′-Methylspiro[1,3,2-benzodioxaphosphole-2,2′(3*H*)-[1,2]-oxaphosphole]; 2-Methoxy, *in* M-00396

5′-Methylspiro[1,3,2-benzodioxaphosphole-2,2′(3*H*)-[1,2]-oxaphosphole]; 2-Ethoxy, *in* M-00396

5′-Methylspiro[1,3,2-benzodioxaphosphole-2,2′(3*H*)-[1,2]-oxaphosphole]; 2-Ph, *in* M-00396

2-Methyl-2,2′-spirobi[1,3,2-benzodioxaphosphole], M-00397

4-Methyl-2,6,7-trioxa-1-phosphabicyclo[2.2.1]heptane, M-00423

7-Methyl-1,4,6-trioxa-5-phospha(5-P^V)spiro[4.4]nonan-8-one; 5-Hydroxy, *in* M-00427

7-Methyl-1,4,6-trioxa-5-phospha(5-P^V)spiro[4.4]nonan-8-one; 5-Chloro, *in* M-00427

9-Methyl-1,4,6-trioxa-5-phospha(5-P^V)spiro[4.4]non-7-ene; 5-Methoxy, *in* M-00428

9-Methyl-1,4,6-trioxa-5-phospha(5-P^V)spiro[4.4]non-7-ene; 5-Ethoxy, *in* M-00428

9-Methyl-1,4,6-trioxa-5-phospha(5-P^V)spiro[4.4]non-7-ene; 5-Ph, *in* M-00428

Octahydro-1,8-dihydroxy-4,7-phosphinidene-1*H*-phosphindole 1,8-dioxide, *in* D-00625

Octahydro-1,2,3,4,5,6-hexamethyl-[1,2,3]triphospholo[4,5-*d*]-1,2,3-triphosphole; (3*aα*,6*aα*)*-form*, *in* O-00017

Octahydro-1,2,3,4,5,6-hexamethyl-[1,2,3]triphospholo[4,5-*d*]-1,2,3-triphosphole; (3*aα*,6*aβ*)*-form*, *in* O-00017

2,2,3,3,7,7,8,8-Octamethyl-1,4,6,9-tetraoxa-5-phospha(5-P^V)-spiro[4.4]nonane, O-00032

2,2,3,3,7,7,8,8-Octamethyl-1,4,6,9-tetraoxa-5-phospha(5-P^V)-spiro[4.4]nonane; 5-Me, *in* O-00032

2,2,3,3,7,7,8,8-Octamethyl-1,4,6,9-tetraoxa-5-phospha(5-P^V)-spiro[4.4]nonane; 5-Ph, *in* O-00032

2,2,3,3,7,7,8,8-Octamethyl-1,4,6,9-tetraoxa-5-phospha(5-P^V)-spiro[4.4]nonane; 5-Benzyl, *in* O-00032

2,2,3,3,7,7,8,8-Octamethyl-1,4,6,9-tetraoxa-5-phospha(5-P^V)-spiro[4.4]nonane; 5-Phenoxy, *in* O-00032

2,2,3,3,7,7,8,8-Octamethyl-1,4,6,9-tetraoxa-5-phospha(5-P^V)-spiro[4.4]nonane; 5-Phenylselenoxy, *in* O-00032

2,2,3,3,7,7,8,8-Octamethyl-1,4,6,9-tetraoxa-5-phospha(5-P^V)-spiro[4.4]nonane; 5-Benzoyl, *in* O-00032

2,7,9,14,15,16,17,18-Octaoxa-1,8-diphospha(1,8-P^V)-pentacyclo[10.2.1.$1^{1,4}$.$1^{5,8}$.$1^{8,11}$]octadecane-3,6,10,13-tetrone, O-00036

2,7,9,14,15,16,17,18-Octaoxa-1,8-diphospha(1,8-P^V)-pentacyclo[10.2.1.$1^{1,4}$.$1^{5,8}$.$1^{8,11}$]octadecane-3,6,10,13-tetrone; (*R*,*R*,*R*,*R*)*-form*, 1,8-Dihydroxy, bistriethylammonium salt, *in* O-00036

1,2,3,4,9,10,11,12-Octaphenyl-5,8-diphosphoniadispiro[4.2.4.2]deca-1,3,9,11-tetraene(2+) , O-00037

1-Oxa-5-phospha(5-P^V)spiro[4.5]decane; 5-Me, *in* O-00049

6-Oxa-5-phospha(5-P^V)spiro[4.5]decane; 5-Me, *in* O-00050

1-Oxa-5-phospha(5-P^V)spiro[4.4]nonane; 5-Me, *in* O-00051

2′,2′,3′,4′,4′-Pentamethylspiro[1,3,2-benzodioxaphosphole-2,1′-phosphetane]; *cis-form*, 2-Ph, *in* P-00027

2′,2′,3′,4′,4′-Pentamethylspiro[1,3,2-benzodioxaphosphole-2,1′-phosphetane]; *trans-form*, 2-Ph, *in* P-00027

2′,2′,3′,4′,4′-Pentamethylspiro[1,3,2-benzodioxaphosphole-2,1′-phosphetane]; *trans-form*, 2-Benzyl, *in* P-00027

1,1,2,3,3-Pentamethyl-6,6,7,7-tetrakis(trifluoromethyl)-5,8-dioxa-4-phospha(4-P^V)spiro[3.4]octane; 4-Phenoxy, *in* P-00029

1,1,2,3,3-Pentamethyl-6,6,7,7-tetrakis(trifluoromethyl)-5,8-dioxa-4-phospha(4-P^V)spiro[3.4]octane; 4-*tert*-Butyl, *in* P-00029

1,1,2,3,3-Pentamethyl-6,6,7,7-tetrakis(trifluoromethyl)-5,8-dioxa-4-phospha(4-P^V)spiro[3.4]octane; 4-Ph (*cis*-), *in* P-00029

1,1,2,3,3-Pentamethyl-6,6,7,7-tetrakis(trifluoromethyl)-5,8-dioxa-4-phospha(4-P^V)spiro[3.4]octane; 4-Ph (*trans*-), *in* P-00029

1,1,2,3,3-Pentamethyl-6,6,7,7-tetrakis(trifluoromethyl)-5,8-dioxa-4-phospha(4-P^V)spiro[3.4]octane; 4-(4-Bromophenyl), *in* P-00029

1,1,2,3,3-Pentamethyl-6,6,7,7-tetrakis(trifluoromethyl)-5,8-dioxa-4-phospha(4-P^V)spiro[3.4]octane; 4-Dimethylamino, *in* P-00029

1,1,2,3,3-Pentamethyl-6,6,8,8-tetrakis(trifluoromethyl)-5,7-dioxa-4-phospha(4-P^V)spiro[3.4]octane; 4-Chloro, *in* P-00030

2-Phenoxy-2,2′-spirobi[1,3,2-benzodioxaphosphole], P-00074

Phenylbisbiphenylenephosphorane, *in* S-00021

2-Phenyl-1-oxa-4-aza-5-phospha(5-P^V)spiro[4.4]nonane; 5-Ph, *in* P-00173

4-Phenyl-1,2,3,5,8-pentaoxa-4-phospha(4-P^V)spiro[3.4]-octane, P-00181

9-Phenyl-9-phosphapentacyclo[4.3.0.$0^{2,5}$.$0^{3,8}$.$0^{4,7}$]nonane, P-00201

9-Phenyl-9-phosphapentacyclo[4.3.0.$0^{2,5}$.$0^{3,8}$.$0^{4,7}$]nonane; 9-Oxide, *in* P-00201

9-Phenyl-9-phosphapentacyclo[4.3.0.$0^{2,5}$.$0^{3,8}$.$0^{4,7}$]nonane; 9-Sulfide, *in* P-00201

9-Phenyl-9-phosphatricyclo[4.2.1.$0^{2,5}$]nona-3,7-diene, P-00203

9-Phenyl-9-phosphatricyclo[4.2.1.$0^{2,5}$]nona-3,7-diene; *endo-syn*-9-Oxide, *in* P-00203

9-Phenyl-9-phosphatricyclo[4.2.1.$0^{2,5}$]nona-3,7-diene; *endo-anti*-9-Oxide, *in* P-00203

9-Phenyl-9-phosphatricyclo[4.2.1.$0^{2,5}$]nonane; (1*α*,2*α*,5*α*,6*α*)-*form*, 9-Oxide, *in* P-00204

N-Phenylspiro[1,3,2-benzodioxaphosphole-2,2′(3′*H*)-[1,3,2]-benzoxazaphosphol]-2-amine, *in* S-00008

2-Phenylspiro[1,3,2-benzothiazaphosphole-2(3*H*),2′-[1,3,2]-dioxaphospholane]; 3-(2,6-Dimethylphenyl), *in* P-00308

2-Phenylspiro[1,3,2-benzothiazaphosphole-2(3*H*),2′-[1,3,2]-dioxaphospholane]; 3-(2,4,6-Trimethylphenyl), *in* P-00308

2-Phenyl-2,2′-spirobi[1,3,2-benzodioxaphosphole], P-00309

1-Phosphabicyclo[2.2.1]heptane, P-00344

1-Phosphabicyclo[2.2.1]heptane; *p*-Oxide, *in* P-00344

5-Phosphoniaspiro[4.4]nonane; Chloride, *in* P-00374

5-Phosphoniaspiro[4.4]nonane; Bromide, *in* P-00374

5-Phosphoniaspiro[4.4]nonane; Iodide, *in* P-00374

5-Phosphoniaspiro[4.4]nonane; Tetrafluoroborate, *in* P-00374

Spiro[1,3,2-benzodioxaphosphole-2,2′-[1,3,2]-benzoxathiaphosphole]; 2-Phenoxy, *in* S-00007

Spiro[1,3,2-benzodioxaphosphole-2,2′(3′*H*)-[1,3,2]-benzoxazaphosphole], S-00008

Spiro[1,3,2-benzodioxaphosphole-2,2′(3′*H*)-[1,3,2]-benzoxazaphosphole]; 2-Methoxy, *in* S-00008

Spiro[1,3,2-benzodioxaphosphole-2,2′(3′*H*)-[1,3,2]-benzoxazaphosphole]; 2-Ethoxy, *in* S-00008

Spiro[1,3,2-benzodioxaphosphole-2,2′(3′*H*)-[1,3,2]-benzoxazaphosphole]; 2-Phenoxy, *in* S-00008

Spiro[1,3,2-benzodioxaphosphole-2,2′(3′*H*)-[1,3,2]-benzoxazaphosphole]; 2-Chloro, *in* S-00008

Spiro[1,3,2-benzodioxaphosphole-2,2′-[1,3,2]-oxazaphospholidine], S-00009

Spiro[1,3,2-benzodioxaphosphole-2,1′-[3]phospholene]; 2-Methoxy, *in* S-00010

Spiro[1,3,2-benzodioxaphosphole-2,1′-[3]phospholene]; 2-Ph, *in* S-00010

Spiro[1,3,2-benzodioxaphosphole-2,1′-[3]phospholene]; 2-Fluoro, *in* S-00010

Spiro[1,3,2-benzodioxaphosphole-2,1′-[3]phospholene]; 2-Bromo, *in* S-00010

Spiro[1,3,2-benzoxathiaphosphole-2,2′-[1,3,2]-dioxaphospholane]; 2-Ph, *in* S-00011

Spiro[1,3,2-benzoxazaphosphole-2(3*H*),2′-[1,3,2]-dioxaphospholane], S-00012

Spiro[1,3,2-benzoxazaphosphole-2(3*H*),2′-[1,2]oxaphospholan]-5′-one; 2-Et, *in* S-00013
2,2′-Spirobi[1,3,2-benzodioxaphosphole], S-00014
2,2′-Spirobi[1,3,2-benzodioxaphosphole]; 2-Trimethylsilyl, *in* S-00014
2,2′-Spirobi[1,3,2-benzodithiaphosphole]; 2-Ph, *in* S-00015
2,2′-Spirobi[1,3,2-benzodithiaphospholium](1+); Hexafluorophosphate, *in* S-00016
1,1′(3*H*,3′*H*)-Spirobi[2,1-benzoxaphosphole]; 1-Ph, *in* S-00017
1,1′(3*H*,3′*H*)-Spirobi[2,1-benzoxaphosphole]-3,3′-dione; 1-Me, *in* S-00018
1,1′(3*H*,3′*H*)-Spirobi[2,1-benzoxaphosphole]-3,3′-dione; 1-Ph, *in* S-00018
2,2′-Spirobi[1,3,2-benzoxathiaphosphole]; 2-Ph, *in* S-00019
2,2(3*H*,3′*H*)-Spirobi[1,3,2-benzoxazaphosphole], S-00020
2,2(3*H*,3′*H*)-Spirobi[1,3,2-benzoxazaphosphole]; 2-Me, *in* S-00020
2,2(3*H*,3′*H*)-Spirobi[1,3,2-benzoxazaphosphole]; 2-Et, *in* S-00020
2,2(3*H*,3′*H*)-Spirobi[1,3,2-benzoxazaphosphole]; 2-Ph, *in* S-00020
2,2(3*H*,3′*H*)-Spirobi[1,3,2-benzoxazaphosphole]; 2-Phenoxy, *in* S-00020
2,2(3*H*,3′*H*)-Spirobi[1,3,2-benzoxazaphosphole]; 2-Fluoro, *in* S-00020
5,5′-Spirobi[5*H*-dibenzophosphole], S-00021
5,5′-Spirobi[5*H*-dibenzophospholium](1+); Bromide, *in* S-00022
5,5′-Spirobi[5*H*-dibenzophospholium](1+); Iodide, *in* S-00022
4,5,6,7-Tetrachloro-*N*,*N*-dimethylspiro[1,3,2-benzodioxaphosphole-2,2′-[1,3,2]dioxaphospholan]-2-amine, *in* T-00036
4,5,6,7-Tetrachloro-4′,4′,5′,5′-tetramethylspiro[1,3,2-benzodioxaphosphole-2,2′-[1,3,2]dioxaphospholane]; 2-Methoxy, *in* T-00036
4,5,6,7-Tetrachloro-4′,4′,5′,5′-tetramethylspiro[1,3,2-benzodioxaphosphole-2,2′-[1,3,2]dioxaphospholane]; 2-Phenoxy, *in* T-00036
4,5,6,7-Tetrachloro-4′,4′,5′,5′-tetramethylspiro[1,3,2-benzodioxaphosphole-2,2′-[1,3,2]dioxaphospholane]; 2-Phenylthio, *in* T-00036
Tetrahydro-1,7-dimethyl-1*H*,5*H*-[1,3,2]diazaphospholo[1,2-*a*]-[1,3,2]diazaphosphole, T-00093
Tetrahydro-2,6-dimethyl-[1,3,2]oxazaphospholo[2,3-*b*][1,3,2]-oxazaphosphole, T-00097
3*a*,4,7,7*a*-Tetrahydro-1,8-diphenyl-4,7-phosphinidene-1*H*-phosphindole, T-00105
3*a*,4,7,7*a*-Tetrahydro-1,8-diphenyl-4,7-phosphinidene-1*H*-phosphindole; 1,8-Dioxide, *in* T-00105
3*a*,4,7,7*a*-Tetrahydro-1,8-diphenyl-4,7-phosphinidene-1*H*-phosphindole; 1,8-Disulfide, *in* T-00105
3*a*,4,7,7*a*-Tetrahydro-1,2,3,5,6,8-hexamethyl-4,7-phosphinidene-1*H*-phosphindole, T-00106
3*a*,4,7,7*a*-Tetrahydro-1,2,3,5,6,8-hexamethyl-4,7-phosphinidene-1*H*-phosphindole; 1,8-Dioxide, *in* T-00106
3*a*,4,7,7*a*-Tetrahydro-1,2,3,5,6,8-hexamethyl-4,7-phosphinidene-1*H*-phosphindole; 1,8-Disulfide, *in* T-00106
Tetrahydro-4-methyl-1*H*,5*H*-[1,2]diphospholo[1,2-*a*][1,2]-diphospholium iodide, *in* D-01170
1,1′,3,3′-Tetrahydro-1,1′,2,3,3′-pentamethyl-2,2′-spirobi[2*H*-1,3,2-benzodiazaphosphole], *in* T-00142
1,1′,3,3′-Tetrahydro-2,2′-spirobi[2*H*-isophosphindolium](1+); Bromide, *in* T-00137
3*a*,4,7,7*a*-Tetrahydro-2,3,5,6-tetramethyl-1,8-diphenyl-4,7-phosphinidene-1*H*-phosphindole; 1,8-Dioxide, *in* T-00141
Tetrahydro-1*H*,5*H*-[1,2,3]triphospholo[2,1-*a*][1,2,3]-triphosphole, T-00160
1,3,4,6-Tetrakis(trifluoromethyl)-2,5-diphosphatricyclo[3.1.0.0^{2,6}]hex-3-ene, T-00186
2,2,3,3-Tetrakis(trifluoromethyl)-1,4,6,10-tetraoxa-5-phospha(5-P^V)spiro[4.5]decane; 5-Phenoxy, *in* T-00187
2,3,7,8-Tetrakis(trifluoromethyl)-1,4,6,9-tetrathia-5-phospha(5-P^V)spiro[4.4]nona-2,7-diene; 5-Ph, *in* T-00189
2,2,3,3-Tetrakis(trifluoromethyl)-1,4,6-trioxa-10-aza-5-phospha(5-P^V)spiro[4.5]decane; 5-(4-Bromophenoxy), *in* T-00190
1,1,3,3-Tetramethyl-6,7-bis(trifluoromethyl)-5,8-dithia-4-phospha(4-P^V)spiro[3.4]oct-6-ene; 4-Ph, *in* T-00193
2,2,7,7-Tetramethyl-1,6-dioxa-4,9-diaza-5-phospha(5-P^V)-spiro[4.4]nonane, T-00195
3,3,8,8-Tetramethyl-1,6-dioxa-4,9-diaza-5-phospha(P^V)-spiro[4.4]nonane, T-00196
3,3,8,8-Tetramethyl-1,6-dioxa-4,9-diaza-5-phospha(P^V)-spiro[4.4]nonane; (±)-*form*, 5-Me, *in* T-00196
3,3,8,8-Tetramethyl-1,6-dioxa-4,9-diaza-5-phospha(P^V)-spiro[4.4]nonane; (±)-*form*, 5-Anilino, *in* T-00196
N,*N*,6,9-Tetramethyl-2,3-diphenyl-1,4-dioxa-6,9-diaza-5-phospha(5-P^V)spiro[4.4]non-2-en-5-amine, *in* D-00741
N,*N*,5,6-Tetramethylspiro[1,3,2-benzodioxaphosphole-2,1′(3′*H*)-[2,1]benzoxaphosphol]-2-amine, *in* D-00915
4′,4′,5′,5′-Tetramethylspiro[1,3,2-benzodioxaphosphole-2,2′-[1,3,2]dioxaphospholane], T-00230
4′,4′,5′,5′-Tetramethylspiro[1,3,2-benzodioxaphosphole-2,2′-[1,3,2]dioxaphospholane]; 2-Methoxy, *in* T-00230
4′,4′,5′,5′-Tetramethylspiro[1,3,2-benzodioxaphosphole-2,2′-[1,3,2]dioxaphospholane]; 2-Phenoxy, *in* T-00230
4′,4′,5′,5′-Tetramethylspiro[1,3,2-benzodioxaphosphole-2,2′-[1,3,2]dioxaphospholane]; 2-Phenylthio, *in* T-00230
4′,4′,5′,5′-Tetramethylspiro[1,3,2-benzodioxaphosphole-2,2′-[1,3,2]dioxaphospholane]; 2-Phenylseleno, *in* T-00230
4′,4′,5′,5′-Tetramethylspiro[1,3,2-benzoxazaphosphole-2(3*H*),-2′-[1,3,2]dioxaphospholane], T-00231
4′,4′,5′,5′-Tetramethylspiro[1,3,2-benzoxazaphosphole-2(3*H*),-2′-[1,3,2]dioxaphospholane]; 2-Methoxy, *in* T-00231
4′,4′,5′,5′-Tetramethylspiro[1,3,2-benzoxazaphosphole-2(3*H*),-2′-[1,3,2]dioxaphospholane]; 2-Phenoxy, *in* T-00231
3,3,3′,3′-Tetramethyl-1,1′(3*H*,3′*H*)-spirobi[2,1-benzoxaphosphole], T-00232
3,3,3′,3′-Tetramethyl-1,1′(3*H*,3′*H*)-spirobi[2,1-benzoxaphosphole]; 1-Ph, *in* T-00232
3,3,3′,3′-Tetramethyl-1,1′(3*H*,3′*H*)-spirobi[2,1-benzoxaphosphole]; 1-Benzyl, *in* T-00232
3,3,3′,3′-Tetramethyl-1,1′(3*H*,3′*H*)-spirobi[2,1-benzoxaphosphole]; 1-Chloro, *in* T-00232
3,3,3′,3′-Tetramethyl-1,1′(3*H*,3′*H*)-spirobi[2,1-benzoxaphospholium](1+); Chloride, *in* T-00233
3,3,3′,3′-Tetramethyl-1,1′(3*H*,3′*H*)-spirobi[2,1-benzoxaphospholium](1+); Trifluoromethanesulfonate, *in* T-00233
N,*N*,1,3-Tetramethylspiro[1,3,2-diazaphospholidine-2,2′-phenanthro[9,10-*d*][1,3,2]dioxaphosphol]-2-amine, *in* D-00918
6,9,*N*,*N*-Tetramethyl-2,2,3,3-tetrakis(trifluoromethyl)-1,4-dioxa-6,9-diaza-5-phospha(P^V)spiro[4.4]nonan-5-amine, *in* D-00922
2,2,3,3-Tetramethyl-1,4,6,9-tetraoxa-5-phospha(5-P^V)-spiro[4.4]nonane, T-00235
2,2,3,3-Tetramethyl-1,4,6,9-tetraoxa-5-phospha(5-P^V)-spiro[4.4]nonane; 5-Phenoxy, *in* T-00235
N,*N*,2,3-Tetramethyl-1,4,6,9-tetraoxa-5-phospha(5-P^V)-spiro[4.4]non-2-en-5-amine, *in* D-00924
2,2,3,3-Tetramethyl-1,4,6-trioxa-9-aza-5-phospha(5-P^V)-spiro[4.4]nonane, T-00238
Tetramethyl 1,2,5-triphenyl-5-phospha(5-P^V)spiro[4.4]nona-1,3,6,8-tetraene-1,2,3,4-tetracarboxylate, *in* T-00622
1,4,6,12-Tetraoxa-5-phospha(5-P^V)spiro[4.7]dodecane; 5-Ph, *in* T-00243
1,4,6,9-Tetraoxa-5-phospha(5-P^V)spiro[4.4]nonane, T-00244
1,4,6,9-Tetraoxa-5-phospha(5-P^V)spiro[4.4]nonane; 5-Ph, *in* T-00244
1,4,6,9-Tetraoxa-5-phospha(5-P^V)spiro[4.4]nonane; 5-Methoxy, *in* T-00244
1,4,6,9-Tetraoxa-5-phospha(5-P^V)spiro[4.4]nonane; 5-Ethoxy, *in* T-00244
1,4,6,11-Tetraoxa-5-phospha(5-P^V)spiro[4.6]undecane; 5-Ph, *in* T-00245
2,3,7,8-Tetraphenyl-1,4,6,9-tetraoxa-5-phospha(5-P^V)-spiro[4.4]nona-2,7-diene; 5-Ph, *in* T-00275
5,5,6-Trimethyl-4,7-dioxa-3-phospha(3-P^V)spiro[2.4]heptane, T-00510
3′,*N*,*N*-Trimethylspiro[1,3,2-benzodioxaphosphole-2,2′(3′*H*)-[1,3,2]benzothiazaphosphol]-2-amine, *in* M-00394
2,3,3′-Trimethyl-2,2′(3*H*,3′*H*)-spirobi[1,3,2-benzoxazaphosphole], *in* D-00917

2,6,7-Trioxa-1-phosphabicyclo[2.2.1]heptane, T-00587
2,6,7-Trioxa-1-phosphabicyclo[2.2.1]heptane; 1-Oxide, *in* T-00587
2,7,8-Trioxa-1-phosphabicyclo[3.2.1]octane, T-00589
6,8,10-Triphenyl-1,4-diaza-5-phospha(5-P^V)spiro[4.5]deca-5,7,9-triene, T-00599
6,8,10-Triphenyl-1,4-dithia-5-phospha(5-P^V)spiro[4.5]deca-5,7,9-triene, T-00604
6,8,10-Triphenyl-1-oxa-4-aza-5-phosphaspiro(5-P^V)spiro[4.5]-deca-5,7,9-triene; *N*-Me, *in* T-00609

2 P-contg. rings; 6-membered P-ring(s) present

10*H*-5,10[1′,2′]Benzenoacridophosphine; 5-Oxide, *in* B-00014
5,10[1′,2′]-Benzenophenophosphazine, B-00015
5,10[1′,2′]-Benzenophenophosphazine; Oxide, *in* B-00015
5,10-[1′,2′]Benzophosphenanthrene, B-00024
5,10-[1′,2′]Benzophosphenanthrene; 5,10-Dioxide, *in* B-00024
5,10-[1′,2′]Benzophosphenanthrene; 5,10-Disulfide, *in* B-00024
5-Benzyl-10*H*-5,10[1′,2′]benzenoacridophosphonium bromide, *in* B-00014
6-Benzyl-2-oxa-6-phosphatricyclo[3.3.1.$1^{3,7}$]decane, B-00055
6-Benzyl-2-oxa-6-phosphatricyclo[3.3.1.$1^{3,7}$]decane; 6-Oxide, *in* B-00055
Bicyclic phosphate, *in* T-00588
2-(4-Bromophenyl)-8,8-dihydro-4,6,8,8-tetraphenyl[1,2]-oxaphospholo[2,3-*b*][1,2,5]dioxaphosphorin, B-00479
▷*tert*-Butyl bicyclic phosphate, *in* B-00640
▷*tert*-Butyl bicyclic thiophosphate, *in* B-00640
▷4-*tert*-Butyl-2,6,7-trioxa-1-phosphabicyclo[2.2.2]octane, B-00640
1-(2-Chloroethyl)tetrahydro-1*H*,5*H*-[1.3.2]-diazaphospholo[2,1-*b*][1.3.2]oxazaphosphorine 9-oxide; (*R*)-*form*, *in* C-00082
1-(2-Chloroethyl)tetrahydro-1*H*,5*H*-[1.3.2]-diazaphospholo[2,1-*b*][1.3.2]oxazaphosphorine 9-oxide; (*S*)-*form*, *in* C-00082
1-(2-Chloroethyl)tetrahydro-1*H*,5*H*-[1.3.2]-diazaphospholo[2,1-*b*][1.3.2]oxazaphosphorine 9-oxide; (±)-*form*, *in* C-00082
1,3,4,5,5,6,7,8,9,9*a*-Decahydro-5-methylene-2*H*-phosphinolizine, D-00005
9*H*-Dibenzo[2,3:4,5]phospholo[1,2-*f*]phosphanthridene; 17-Oxide, *in* D-00041
6,6-Dibenzyl-6-phosphonia-2-oxaadamantane chloride, *in* B-00055
2,3-Diethyl-2,3-diphosphabicyclo[2.2.2]octane, D-00288
1,4-Dihydrospiro[2*H*-3,1,2-benzoxazaphosphorine-2,2′-[1,3,2]-dioxaphospholane]; 2-Ph, *in* D-00576
1,4-Dimethyl-7,9-bis(trifluoromethyl)-1,4,6,8,10-pentaaza-5-phospha(5-P^V)spiro[4.5]deca-5,7,9-triene, D-00714
9,9-Dimethyl-2,8-diphenyl-3,7-dioxa-1-phosphabicyclo[3.3.1]-nonane; 1-Sulfide, *in* D-00743
8,8-Dimethyl-2,3-diphenyl-1,4,6,10-tetraoxa-5-phospha(5-P^V)-spiro[4.5]dec-2-ene; 5-Phenoxy, *in* D-00749
1,4-Dimethyl-2-phenyl-7-oxa-2-phosphabicyclo[2.2.1]heptane; 2-Oxide, *in* D-00796
1,5-Dimethyl-1-phosphonia-5-silabicyclo[3.3.1]nonane iodide, *in* M-00274
N,*N*-Dimethylspiro[1,3,2-dioxaphosphorinane-2,2′-phenanthro[9,10-*d*][1,3,2]dioxaphosphol]-2-amine, *in* S-00025
15,16-Dimethyl-1,4,8,11-tetraaza-15,16-diphosphatricyclo[9.3.1.$1^{4,8}$]hexadecane, D-00921
6,10-Dimethyl-2,2,3,3-tetrakis(trifluoromethyl)-1,4-dioxa-6,10-diaza-5-phospha(5-P^V)spiro[4.5]decane; 5-Phenoxy, *in* D-00923
8,8-Dimethyl-1,6,10-trioxa-5-phospha(5-P^V)spiro[4.5]dec-2-ene; 5-Methoxy, *in* D-00933
2,5′-Diphenylspiro[4*H*-1,3,2-benzodioxaphosphorin-2,2′-[1,3,2]dioxaphospholane]-4,4′-dione, D-01147
2,3-Diphenyl-1,6,10-trioxa-4-aza-5-phospha(5-P^V)spiro[4.5]-dec-2-ene; 5-Ph, *in* D-01158
1,4-Diphosphabicyclo[2.2.2]octane, D-01169
▷Ethyl bicyclic phosphate, *in* E-00196
Ethyl bicyclic selenophosphate, *in* E-00196
▷Ethyl bicyclic thiophosphate, *in* E-00196
4-Ethylspiro[2,8,9-trioxa-1-phosphatricyclo[3.3.1.$1^{3,7}$]-decane-1,4′-trioxaphosphetane], E-00187
4-Ethyl-2,6,7-trimethyl-2,6,7-triaza-1-phosphabicyclo[2.2.2]octane, E-00195
▷4-Ethyl-2,6,7-trimethyl-2,6,7-triaza-1-phosphabicyclo[2.2.2]octane; 1-Sulfide, *in* E-00195
▷4-Ethyl-2,6,7-trioxa-1-phosphabicyclo[2.2.2]octane, E-00196
Hexahydro-2*H*-[1,2]oxaphosphorino[2,3-*b*][1,2]oxaphosphorin; 9-Oxide, *in* H-00037
2,3,5,6,7,8-Hexakis(trifluoromethyl)-1,4-diphosphabicyclo[2.2.2]octa-2,5,7-triene, H-00064
2,5,7,10,11,12-Hexakis(trifluoromethyl)-1,6-diphosphahexacyclo[4.4.2.$0^{2,5}$,$0^{3,9}$,$0^{4,8}$,$0^{7,10}$]dodec-11-ene, H-00065
4,5,9,10,11,12-Hexakis(trifluoromethyl)-1,8-diphosphatetracyclo[6.2.2.$0^{2,7}$.$0^{3,6}$]dodecane, H-00066
2,7,9,10,11,12-Hexakis(trifluoromethyl)-1,8-diphosphatetracyclo[6.2.2.$0^{2,7}$.$0^{3,6}$]dodeca-4,9,11-triene; (2α,3α,6α,7α)-*form*, *in* H-00067
2,7,9,10,11,12-Hexakis(trifluoromethyl)-1,8-diphosphatetracyclo[6.2.2.$0^{2,7}$.$0^{3,6}$]dodeca-4,9,11-triene; (2α,3β,6β,7α)-*form*, *in* H-00067
2,3,5,6,7,8-Hexamethyl-2,3,5,6,7,8-hexaaza-1,4-diphosphabicyclo[2.2.2]octane, H-00070
2,3,5,6,7,8-Hexamethyl-2,3,5,6,7,8-hexaaza-1,4-diphosphabicyclo[2.2.2]octane; 1,4-Dioxide, *in* H-00070
2,3,5,6,7,8-Hexamethyl-2,3,5,6,7,8-hexaaza-1,4-diphosphabicyclo[2.2.2]octane; 1,4-Disulfide, *in* H-00070
2,3,5,6,7,8-Hexamethyl-2,3,5,6,7,8-hexaaza-1,4-diphosphabicyclo[2.2.2]octane; 1,4-Diselenide, *in* H-00070
7-Methoxy-2,3,5,6,7,8-hexakis(trifluoromethyl)-1,4-diphosphabicyclo[2.2.2]octa-2,5-diene, M-00042
5-Methyl-10*H*-5,10[1′,2′]benzenoacridophosphonium iodide, *in* B-00014
10-Methyl-5,10[1′,2′]-benzophenophosphazinium iodide, *in* B-00015
4-Methyl-2,6-dioxa-7-aza-1-phosphabicyclo[2.2.2]octane 1-oxide; 7-*tert*-Butyl, *in* M-00124
4-Methyl-2,6-dioxa-7-aza-1-phosphabicyclo[2.2.2]octane 1-oxide; 7-Ph, *in* M-00124
4-Methyl-2,6-dioxa-7-aza-1-phosphabicyclo[2.2.2]octane 1-oxide; 7-Benzyl, *in* M-00124
5-Methyl-1-phospha-5-silabicyclo[3.3.1]nonane, M-00274
5-Methyl-1-phospha-5-silabicyclo[3.3.1]nonane; 1-Sulfide, *in* M-00274
1-Methyl-1-phosphoniabicyclo[3.3.1]nonane iodide, *in* P-00345
7-Methyl-2-thia-1,3,5-triaza-7-phosphoniaadamantane iodide, *in* T-00301
▷4-Methyl-2,6,7-trioxa-1-phosphabicyclo[2.2.2]octane, M-00424
4-Methyl-3,5,8-trioxa-1-phosphabicyclo[2.2.2]octane, M-00425
1-Methyl-2,6,7-trioxa-4-phospha-1-silabicyclo[2.2.2]octane, M-00426
4-Methyl-4-triphenylmethyl-2,6,7-trioxa-1-phosphoniabicyclo[2.2.2]octane perchlorate, *in* M-00424
4-Methyl-4-triphenylmethyl-2,6,7-trioxa-1-phosphoniabicyclo[2.2.2]octane tetrafluoroborate, *in* M-00424
6-Methyl-3,5,7-triphenyl-6-phosphatricyclo[3.3.1.$0^{2,7}$]non-3-ene; 6-Oxide, *in* M-00431
4-Methyl-2,6,7-trithia-1-phosphabicyclo[2.2.2]octane, M-00436
4-Methyl-2,6,7-trithia-1-phosphabicyclo[2.2.2]octane; 1-Oxide, *in* M-00436
▷4-Methyl-2,6,7-trithia-1-phosphabicyclo[2.2.2]octane; 1-Sulfide, *in* M-00436
Octahydro-1,9-dimethyl-[1,3,2]diazaphosphorino[1,2-*a*]-[1,3,2]diazaphosphorine, O-00005
1,2,3,4,9,10,11,12-Octaphenyl-5,8-diphosphoniadispiro[4.2.4.2]deca-1,3,9,11-tetraene(2+), O-00037
1-Oxa-5-phospha(5-P^V)spiro[4.5]decane; 5-Me, *in* O-00049
6-Oxa-5-phospha(5-P^V)spiro[4.5]decane; 5-Me, *in* O-00050

1-Oxa-6-phospha(6-P^{V})spiro[5.5]undecane; 6-Me, *in* O-00052
9-Phenyl-1,5-diaza-9-phosphabicyclo[3.3.1]nonane, P-00105
9-Phenyl-1,5-diaza-9-phosphatricyclo[3.3.1.$1^{3,7}$]decane, P-00106
4-Phenyl-3,5-dioxa-1-phosphabicyclo[2.2.1]heptane, P-00113
4-Phenyl-3,5-dioxa-1-phosphabicyclo[2.2.1]heptane; 1-Oxide, *in* P-00113
12-Phenyl-10*H*-5,10-ethenoacridophosphine, P-00151
9-Phenyl-9-phosphabicyclo[3.3.1]nonane, P-00193
9-Phenyl-9-phosphabicyclo[3.3.1]nonane; Oxide, *in* P-00193
9-Phenyl-9-phosphabicyclo[3.3.1]nonan-3-one; *syn-form*, Oxide, *in* P-00195
9-Phenyl-9-phosphabicyclo[3.3.1]nonan-3-one; *anti-form*, Oxide, *in* P-00195
1-Phosphabicyclo[2.2.1]heptane, P-00344
1-Phosphabicyclo[2.2.1]heptane; *p*-Oxide, *in* P-00344
1-Phosphabicyclo[3.3.1]nonane, P-00345
1-Phosphabicyclo[3.3.1]nonane; 1-Oxide, *in* P-00345
1-Phosphabicyclo[3.3.1]nonane; 1-Sulfide, *in* P-00345
1-Phosphabicyclo[2.2.2]octane; 1-Oxide, *in* P-00346
6-Phospha(6-P^{V})spiro[5.5]undeca-1(6)-ene, P-00351
6-Phosphoniaspiro[5.5]undecane; Chloride, *in* P-00375
10,10′(5*H*,5′*H*)-Spirobiphenophosphazinium(1+); Chloride, *in* S-00023
3,4,5,6-Tetrahydro-4-methyl-2*H*-1,5-methano-4,1-benzazaphosphocine, T-00115
3,4,5,6-Tetrahydro-4-methyl-2*H*-1,5-methano-4,1-benzazaphosphocine; 1-Oxide, *in* T-00115
3,4,5,6-Tetrahydro-4-methyl-2*H*-1,5-methano-4,1-benzazaphosphocine; 1-Sulfide, *in* T-00115
Tetrahydro-2*H*,6*H*-[1.3.2]oxazaphosphorino[2,3-*b*][1.3.2]-oxazaphosphorine, T-00120
2,3,6,7-Tetrahydro-1,1,7,7,9-pentamethyl-1*H*,5*H*-phosphorino[3,2,1-*ij*]phosphinoline, T-00124
2,3,6,7-Tetrahydro-1,1,7,7,9-pentamethyl-1*H*,5*H*-phosphorino[3,2,1-*ij*]phosphinoline; 4-Oxide, *in* T-00124
2,2,3,3-Tetrakis(trifluoromethyl)-1,4,6,10-tetraoxa-5-phospha(5-P^{V})spiro[4.5]decane; 5-Phenoxy, *in* T-00187
2,2,3,3-Tetrakis(trifluoromethyl)-1,4,6-trioxa-10-aza-5-phospha(5-P^{V})spiro[4.5]decane; 5-(4-Bromophenoxy), *in* T-00190
2,4,8,10-Tetramethyl-3,9-dioxo-2,4,8,10-tetraaza-6-phosphoniaspiro[5.5]undecane(1+); Chloride, *in* T-00198
2,4,8,10-Tetramethyl-3,9-dioxo-2,4,8,10-tetraaza-6-phosphoniaspiro[5.5]undecane(1+); Sulfate, *in* T-00198
2,4,8,10-Tetramethyl-3,9-dithia-6-phosphoniaspiro[5.5]-undecane; Perchlorate, *in* T-00202
3,3,9,9-Tetramethyl-1,5,7,11-tetraoxa-6-phospha(6-P^{V})-spiro[5.5]undecane; 6-Ethoxy, *in* T-00236
3,3,9,9-Tetramethyl-1,5,7,11-tetraoxa-6-phospha(6-P^{V})-spiro[5.5]undecane; 6-Ph, *in* T-00236
3,3,9,9-Tetramethyl-1,5,7,11-tetraoxa-6-phospha(6-P^{V})-spiro[5.5]undecane; 6-Benzyl, *in* T-00236
2,4,6,7-Tetramethyl-2,6,7-triaza-1-phosphabicyclo[2.2.2]-octane, T-00237
2,4,6,7-Tetramethyl-2,6,7-triaza-1-phosphabicyclo[2.2.2]-octane; 1-Oxide, *in* T-00237
2,4,6,7-Tetramethyl-2,6,7-triaza-1-phosphabicyclo[2.2.2]-octane; 1-Sulfide, *in* T-00237
2,4,6,7-Tetramethyl-2,6,7-triaza-1-phosphabicyclo[2.2.2]-octane; 1-Selenide, *in* T-00237
1,4,6,9-Tetraphenyl-1,4,6,9-tetraphospha-5-silaspiro[4.4]-nonane, T-00277
2-Thia-1,3,5-triaza-7-phosphatricyclo[3.3.1.$1^{3,7}$]decane; 2,2-Dioxide, *in* T-00301
2-Thia-1,3,5-triaza-7-phosphatricyclo[3.3.1.$1^{3,7}$]decane; 2,2,7-Trioxide, *in* T-00301
1,3,5-Triaza-7-phosphatricyclo[3.3.1.$1^{3,7}$]decane, T-00327
1,3,5-Triaza-7-phosphatricyclo[3.3.1.$1^{3,7}$]decane; 7-Oxide, *in* T-00327
1,3,5-Triaza-7-phosphatricyclo[3.3.1.$1^{3,7}$]decane; 7-Sulfide, *in* T-00327
2,2,2-Trichloro-2,2-dihydro-1,3-dimethyl-6,8-bis(trifluoromethyl)-1,3,5,7,9-pentaaza-2,4-diphospha(4-P^{V})spiro[3.5]nona-4,6,8-triene, T-00401
1,9,9-Trimethyl-2,8-diphenyl-3,7-dioxa-1-phosphoniabicyclo[3.3.1]nonane, *in* D-00743
2,4,10-Trimethyl-1,3,4,5,7,10-hexaaza-3-phosphatricyclo[3.3.1.$1^{3,7}$]decane, T-00517
2,4,10-Trimethyl-1,3,4,5,7,10-hexaaza-3-phosphatricyclo[3.3.1.$1^{3,7}$]decane; 3-Oxide, *in* T-00517
2,4,10-Trimethyl-1,3,4,5,7,10-hexaaza-3-phosphatricyclo[3.3.1.$1^{3,7}$]decane; 3-Sulfide, *in* T-00517
▷Trimethylolethane phosphate, *in* M-00424
▷Trimethylolethane thiophosphate, *in* M-00424
Trimethylol selenophosphate, *in* M-00424
2,8,9-Trimethyl-2,8,9-triaza-1-phosphatricyclo[3.3.1.$1^{3,7}$]-decane, T-00563
2,8,9-Trimethyl-2,8,9-triaza-1-phosphatricyclo[3.3.1.$1^{3,7}$]-decane; Oxide, *in* T-00563
2,8,9-Trimethyl-2,8,9-triaza-1-phosphatricyclo[3.3.1.$1^{3,7}$]-decane; Sulfide, *in* T-00563
2,6,7-Trioxa-1,4-diphosphabicyclo[2.2.2]octane, T-00585
2,6,7-Trioxa-1,4-diphosphabicyclo[2.2.2]octane; 1-Oxide, *in* T-00585
▷2,6,7-Trioxa-1,4-diphosphabicyclo[2.2.2]octane; 4-Oxide, *in* T-00585
▷2,6,7-Trioxa-1,4-diphosphabicyclo[2.2.2]octane; 1,4-Dioxide, *in* T-00585
2,6,7-Trioxa-1,4-diphosphabicyclo[2.2.2]octane; 1-Sulfide, *in* T-00585
2,6,7-Trioxa-1,4-diphosphabicyclo[2.2.2]octane; 4-Sulfide, *in* T-00585
2,6,7-Trioxa-1,4-diphosphabicyclo[2.2.2]octane; 1,4-Disulfide, *in* T-00585
2,6,7-Trioxa-4-phospha-1-arsabicyclo[2.2.2]octane, T-00586
2,6,7-Trioxa-4-phospha-1-arsabicyclo[2.2.2]octane; 4-Oxide, *in* T-00586
2,6,7-Trioxa-1-phosphabicyclo[2.2.2]octane, T-00588
2,7,8-Trioxa-1-phosphabicyclo[3.2.1]octane, T-00589
2,6,7-Trioxa-1-phosphabicyclo[2.2.2]octane-4-methanol, T-00590
2,6,7-Trioxa-1-phosphabicyclo[2.2.2]octane-4-methanol; Oxide, *in* T-00590
2,6,7-Trioxa-1-phosphabicyclo[2.2.2]octane-4-methanol; Oxide, Me ether, *in* T-00590
2,6,7-Trioxa-1-phosphabicyclo[2.2.2]octane-4-methanol; Sulfide, *in* T-00590
2,8,9-Trioxa-1-phosphatricyclo[3.3.1.$1^{3,7}$]decane, T-00591
▷2,8,9-Trioxa-1-phosphatricyclo[3.3.1.$1^{3,7}$]decane; 1-Oxide, *in* T-00591
2,8,9-Trioxa-1-phosphatricyclo[3.3.1.$1^{3,7}$]decane; 1-Sulfide, *in* T-00591
2,8,9-Trioxa-1-phosphatricyclo[3.3.1.$1^{3,7}$]decane; 1-Selenide, *in* T-00591
4,6,7-Triphenyl-2,3-bis(trifluoromethyl)-1-phosphabicyclo[2.2.2]octa-2,5,7-triene, T-00598
6,8,10-Triphenyl-1,4-diaza-5-phospha(5-P^{V})spiro[4.5]deca-5,7,9-triene, T-00599
6,8,10-Triphenyl-1,4-dithia-5-phospha(5-P^{V})spiro[4.5]deca-5,7,9-triene, T-00604
7,9,11-Triphenyl-1,5-dithia-6-phospha(6-P^{V})spiro[5.5]-undeca-6,8,10-triene, T-00605
2,3,10-Triphenyl-4*H*-1,4-ethenophosphinoline, T-00607
6,8,10-Triphenyl-1-oxa-4-aza-5-phosphaspiro(5-P^{V})spiro[4.5]-deca-5,7,9-triene; *N*-Me, *in* T-00609

2 P-contg. rings; 7-membered P-ring(s) present

1,3,4,5-Tetrahydro-4,5-dimethyl-2,5-methano-2*H*-2-benzophosphepin, T-00096
1,3,4,5-Tetrahydro-4,5-dimethyl-2,5-methano-2*H*-2-benzophosphepin; Oxide, *in* T-00096
1,3,4,5-Tetrahydro-2,4,5-trimethyl-2,5-methano-2*H*-2-benzophosphepinium iodide, *in* T-00096
1,4,6,11-Tetraoxa-5-phospha(5-P^{V})spiro[4.6]undecane; 5-Ph, *in* T-00245
2,7,8-Trioxa-1-phosphabicyclo[3.2.1]octane, T-00589

2 P-contg. rings; 8-membered P-ring(s) present

1,4,6,12-Tetraoxa-5-phospha(5-P^{V})spiro[4.7]dodecane; 5-Ph, *in* T-00243

2 P-contg. rings; 9+ -membered P-ring(s) present

15,16-Dimethyl-1,4,8,11-tetraaza-15,16-diphosphatricyclo[9.3.1.1^{4,8}]hexadecane, D-00921

3 P-contg. rings; 3-membered P-ring(s) present

3*a*,4,7,7*a*-Tetrahydro-5,6-dimethyl-2,3*a*,7*a*,8-tetrakis(trifluoromethyl)-1,2,3-metheno-2*H*-benzodiphosphole, T-00100

3 P-contg. rings; 4-membered P-ring(s) present

4-(4-Bromophenyl)-2′,2′,3′,4′,4′-pentamethyl-3,3,6,6-tetrakis(trifluoromethyl)spiro[2,7-dioxa-1-phosphabicyclo[3.2.0]hept-3-ene-1,1′-phosphetane], B-00485
4,6-Dichloro-1,3,5,7,9,10-hexamethyl-1,3,5,7,9,10-hexaaza-4,6-diphospha(4,6-P^V)dispiro[3.1.3.1]decane-2,8-dione, D-00170
1′,3′-Diphenyldispiro[1,3,2-benzodioxaphosphole-2,2′-[1,3,2,4]diazaphosphetidine-4′,2″-[1,3,2]-benzodioxaphosphole]; 2,4′-Dimethoxy, *in* D-01018
1′,3′-Diphenyldispiro[1,3,2-benzodioxaphosphole-2,2′-[1,3,2,4]diazaphosphetidine-4′,2″-[1,3,2]-benzodioxaphosphole]; 2,4′-Diethoxy, *in* D-01018
1′,3′-Diphenyldispiro[1,3,2-benzodioxaphosphole-2,2′-[1,3,2,4]diazaphosphetidine-4′,2″-[1,3,2]-benzodioxaphosphole]; 2,4′-Diisopropoxy, *in* D-01018
1′,3′-Diphenyldispiro[1,3,2-benzodioxaphosphole-2,2′-[1,3,2,4]diazaphosphetidine-4′,2″-[1,3,2]-benzodioxaphosphole]; 2,4′-Diphenoxy, *in* D-01018
1′,3′-Diphenyldispiro[1,3,2-benzodioxaphosphole-2,2′-[1,3,2,4]diazaphosphetidine-4′,2″-[1,3,2]-benzodioxaphosphole]; 2,4′-Difluoro, *in* D-01018
1′,3′-Diphenyldispiro[1,3,2-benzodioxaphosphole-2,2′-[1,3,2,4]diazaphosphetidine-4′,2″-[1,3,2]-benzodioxaphosphole]; 2,4′-Dichloro, *in* D-01018
1′,3′-Diphenyldispiro[1,3,2-benzodioxaphosphole-2,2′-[1,3,2,4]diazaphosphetidine-4′,2″-[1,3,2]-benzodioxaphosphole]; 2,4′-Dibromo, *in* D-01018
1′,3′-Diphenyldispiro[1,3,2-benzodioxaphosphole-2,2′-[1,3,2,4]diazaphosphetidine-4′,2″-[1,3,2]-benzodioxaphosphole]; 2,4′-Di-Ph, *in* D-01018
1′,3′-Diphenyldispiro[1,3,2-benzodioxaphosphole-2,2′-[1,3,2,4]diazaphosphetidine-4′,2″-[1,3,2]-benzodioxaphosphole]; 2,4′-Bis(phenylamino), *in* D-01018
6,12-Diphenyl-1,4,8,11-tetraoxa-6,12-diaza-5,7-diphospha(5,7-P^V)dispiro[4.1.4.1]dodecane; 5,7-Dimethoxy, *in* D-01149
6,12-Diphenyl-1,4,8,11-tetraoxa-6,12-diaza-5,7-diphospha(5,7-P^V)dispiro[4.1.4.1]dodecane; 5,7-Diethoxy, *in* D-01149
6,12-Diphenyl-1,4,8,11-tetraoxa-6,12-diaza-5,7-diphospha(5,7-P^V)dispiro[4.1.4.1]dodecane; 5,7-Diphenoxy, *in* D-01149
6,12-Diphenyl-1,4,8,11-tetraoxa-6,12-diaza-5,7-diphospha(5,7-P^V)dispiro[4.1.4.1]dodecane; 5,7-Bis(ethylthio), *in* D-01149
6,12-Diphenyl-1,4,8,11-tetraoxa-6,12-diaza-5,7-diphospha(5,7-P^V)dispiro[4.1.4.1]dodecane; 5,7-Difluoro, *in* D-01149
N,*N*,*N*′,*N*′-Tetramethyl-1′,3′-diphenyldispiro[1,3,2-benzodioxophosphole-2,2′-[1,3,2,4]diazaphosphetidine-4′,2″-[1,3,2]benzodioxaphosphol]-2,2′-diamine, *in* D-01018

3 P-contg. rings; 5-membered P-ring(s) present

4,5-Bis(trifluoromethyl)spiro[1,3,2-dithiaphosphole-2,1′-[2,6,7]trioxa[1]phosphabicyclo[2.2.1]heptane, B-00423
4,5-Bis(trifluoromethyl)spiro[1,3,2-dithiaphosphole-2,1′-[2,8,9]trioxa[1]phosphatricyclo[3.3.1.1^{5,7}]decane], B-00424
4-(4-Bromophenyl)-2′,2′,3′,4′,4′-pentamethyl-3,3,6,6-tetrakis(trifluoromethyl)spiro[2,7-dioxa-1-phosphabicyclo[3.2.0]hept-3-ene-1,1′-phosphetane], B-00485
4,6-Di-*tert*-butylspiro[1,3,2-benzodioxaphosphole-2,1′-[2,8,9]-trioxa[1]phosphaadamantane], D-00157
4,4-Dimethylspiro[1,3,2-dioxaphospholane-2,1′-[2,6,7]-trioxa[1]phosphabicyclo[2.2.1]heptane], D-00919
1′,3′-Diphenyldispiro[1,3,2-benzodioxaphosphole-2,2′-[1,3,2,4]diazaphosphetidine-4′,2″-[1,3,2]-benzodioxaphosphole]; 2,4′-Dimethoxy, *in* D-01018
1′,3′-Diphenyldispiro[1,3,2-benzodioxaphosphole-2,2′-[1,3,2,4]diazaphosphetidine-4′,2″-[1,3,2]-benzodioxaphosphole]; 2,4′-Diethoxy, *in* D-01018
1′,3′-Diphenyldispiro[1,3,2-benzodioxaphosphole-2,2′-[1,3,2,4]diazaphosphetidine-4′,2″-[1,3,2]-benzodioxaphosphole]; 2,4′-Diisopropoxy, *in* D-01018
1′,3′-Diphenyldispiro[1,3,2-benzodioxaphosphole-2,2′-[1,3,2,4]diazaphosphetidine-4′,2″-[1,3,2]-benzodioxaphosphole]; 2,4′-Diphenoxy, *in* D-01018
1′,3′-Diphenyldispiro[1,3,2-benzodioxaphosphole-2,2′-[1,3,2,4]diazaphosphetidine-4′,2″-[1,3,2]-benzodioxaphosphole]; 2,4′-Difluoro, *in* D-01018
1′,3′-Diphenyldispiro[1,3,2-benzodioxaphosphole-2,2′-[1,3,2,4]diazaphosphetidine-4′,2″-[1,3,2]-benzodioxaphosphole]; 2,4′-Dichloro, *in* D-01018
1′,3′-Diphenyldispiro[1,3,2-benzodioxaphosphole-2,2′-[1,3,2,4]diazaphosphetidine-4′,2″-[1,3,2]-benzodioxaphosphole]; 2,4′-Dibromo, *in* D-01018
1′,3′-Diphenyldispiro[1,3,2-benzodioxaphosphole-2,2′-[1,3,2,4]diazaphosphetidine-4′,2″-[1,3,2]-benzodioxaphosphole]; 2,4′-Di-Ph, *in* D-01018
1′,3′-Diphenyldispiro[1,3,2-benzodioxaphosphole-2,2′-[1,3,2,4]diazaphosphetidine-4′,2″-[1,3,2]-benzodioxaphosphole]; 2,4′-Bis(phenylamino), *in* D-01018
6,12-Diphenyl-1,4,8,11-tetraoxa-6,12-diaza-5,7-diphospha(5,7-P^V)dispiro[4.1.4.1]dodecane; 5,7-Dimethoxy, *in* D-01149
6,12-Diphenyl-1,4,8,11-tetraoxa-6,12-diaza-5,7-diphospha(5,7-P^V)dispiro[4.1.4.1]dodecane; 5,7-Diethoxy, *in* D-01149
6,12-Diphenyl-1,4,8,11-tetraoxa-6,12-diaza-5,7-diphospha(5,7-P^V)dispiro[4.1.4.1]dodecane; 5,7-Diphenoxy, *in* D-01149
6,12-Diphenyl-1,4,8,11-tetraoxa-6,12-diaza-5,7-diphospha(5,7-P^V)dispiro[4.1.4.1]dodecane; 5,7-Bis(ethylthio), *in* D-01149
6,12-Diphenyl-1,4,8,11-tetraoxa-6,12-diaza-5,7-diphospha(5,7-P^V)dispiro[4.1.4.1]dodecane; 5,7-Difluoro, *in* D-01149
4′-Ethyl-1,3-dihydro-5-phenyl-3,3-bis(trifluoromethyl)-spiro[2*H*-1,4,2-diazaphosphole-2,1′-[2,6,7]trioxa[1]-phosphabicyclo[2.2.2]octane]; 1-(2,6-Dimethylphenyl), *in* E-00061
4′-Ethyl-1,3-dihydro-5-phenyl-3,3-bis(trifluoromethyl)-spiro[2*H*-1,4,2-diazaphosphole-2,1′-[2,6,7]trioxa[1]-phosphabicyclo[2.2.2]octane]; 1-(2,4,6-Trimethylphenyl), *in* E-00061
Hexahydro-3*H*,6*H*-2*a*,5*a*,8*a*-triaza-8*b*-phosphaacenaphthylene, H-00054
Hexahydro-2*a*,4*a*,6*a*-triaza-6*b*-phosphacyclopenta[*cd*]-pentalene, H-00055
Hexahydro-2*a*,4*a*,6*a*-triaza-6*b*-phosphacyclopenta[*cd*]-pentalene; 6*b*-Oxide, *in* H-00055
Hexahydro-2*a*,4*a*,6*a*-triaza-6*b*-phosphacyclopenta[*cd*]-pentalene; 6*b*-Sulfide, *in* H-00055
Hexahydro-2*a*,4*a*,6*a*-triaza-6*b*-phosphacyclopenta[*cd*]-pentalene; 6*b*-Selenide, *in* H-00055
4′-Methyl-4,5-bis(trifluoromethyl)spiro[1,3,2-dithiaphosphole-2,1′-[2,6,7]trioxa[1]-phosphabicyclo[2.2.2]octane], M-00092
4′-Methyl-4,4,5,5-tetrakis(trifluoromethyl)spiro[1,3,2-dioxaphospholane-2,1′-[2,6,7]trioxa[1]-phosphabicyclo[2.2.2]octane], M-00405
Octahydro-2,4-diphenyl-1,3,5-metheno-1*H*-2,4-diphosphacyclobuta[*c*,*d*]pentalene; 2,4-Dioxide, *in* O-00016
Spiro[1,3,2-dioxaphospholane-2,1′-[2,8,9]trioxa[1]-phosphatricyclo[3.3.1.1^{3,7}]decane], S-00024
3*a*,4,7,7*a*-Tetrahydro-1,8-dimethyl-4,7-phosphinidene-1*H*-phosphindole, T-00099

3a,4,7,7a-Tetrahydro-1,8-dimethyl-4,7-phosphinidene-1H-phosphindole; 1,8-Dioxide, *in* T-00099
3a,4,7,7a-Tetrahydro-1,8-dimethyl-4,7-phosphinidene-1H-phosphindole; 1,8-Disulfide, *in* T-00099
3a,4,7,7a-Tetrahydro-5,6-dimethyl-2,3a,7a,8-tetrakis(trifluoromethyl)-1,2,3-metheno-2H-benzodiphosphole, T-00100
5a,6,11,11a-Tetrahydro-1,2,3,6,11,12-hexaphenyl-6,11-phosphinidene-1H-naphth[2,3-e]isophosphindole; 2,12-Dioxide, *in* T-00107
1,3a,3b,6-Tetrahydro-3,4,8,8-tetramethyl-1,6-diphenyl-8H-[1,3]diphospholo[1,5-e;3,4-c']bis[1,2,3]-diazaphosphole; (3aα,3bα,7α,9α)-*form*, *in* T-00140
1,3a,3b,6-Tetrahydro-3,4,8,8-tetramethyl-1,6-diphenyl-8H-[1,3]diphospholo[1,5-e;3,4-c']bis[1,2,3]-diazaphosphole; (3aβ,3bα,7α,9α)-*form*, *in* T-00140
1,3a,3b,6-Tetrahydro-3,4,8,8-tetramethyl-1,6-diphenyl-8H-[1,3]diphospholo[1,5-e;3,4-c']bis[1,2,3]-diazaphosphole; (3aβ,3bα,7α,9α)-*form*, Disulfide, *in* T-00140
3a,4,7,7a-Tetrahydro-1,1,8,8-tetramethyl-4,7-phosphinidene-1H-phosphindolium diiodide, *in* T-00099
N,N,N',N'-Tetramethyl-1',3'-diphenyldispiro[1,3,2-benzodioxophosphole-2,2'-[1,3,2,4]diazaphosphetidine-4',2''-[1,3,2]benzodioxaphosphol]-2,2'-diamine, *in* D-01018
3,5,8,10-Tetraphenyl-1H,5H-bisphospholo[1,2-a:1',2'-d][1,4]-diphosphorin, T-00250

3 P-contg. rings; 6-membered P-ring(s) present

4,5-Bis(trifluoromethyl)spiro[1,3,2-dithiaphosphole-2,1'-[2,8,9]trioxa[1]phosphatricyclo[3.3.1.1^{5,7}]decane], B-00424
4,6-Di-*tert*-butylspiro[1,3,2-benzodioxaphosphole-2,1'-[2,8,9]-trioxa[1]phosphaadamantane], D-00157
4'-Ethyl-1,3-dihydro-5-phenyl-3,3-bis(trifluoromethyl)-spiro[2H-1,4,2-diazaphosphole-2,1'-[2,6,7]trioxa[1]-phosphabicyclo[2.2.2]octane]; 1-(2,6-Dimethylphenyl), *in* E-00061
4'-Ethyl-1,3-dihydro-5-phenyl-3,3-bis(trifluoromethyl)-spiro[2H-1,4,2-diazaphosphole-2,1'-[2,6,7]trioxa[1]-phosphabicyclo[2.2.2]octane]; 1-(2,4,6-Trimethylphenyl), *in* E-00061
Hexahydro-3H,6H-2a,5a,8a-triaza-8b-phosphaacenaphthylene, H-00054
2,4,6,8,9,10-Hexamethyl-2,4,6,8,9,10-hexaaza-1,3,5,7-tetraphosphatricyclo[3.3.1.1^{3,7}]decane, H-00071
2,4,6,8,9,10-Hexamethyl-2,4,6,8,9,10-hexaaza-1,3,5,7-tetraphosphatricyclo[3.3.1.1^{3,7}]decane; 1,3,5,7-Tetroxide, *in* H-00071
2,4,6,8,9,10-Hexamethyl-2,4,6,8,9,10-hexaaza-1,3,5,7-tetraphosphatricyclo[3.3.1.1^{3,7}]decane; 1-Sulfide, *in* H-00071
2,4,6,8,9,10-Hexamethyl-2,4,6,8,9,10-hexaaza-1,3,5,7-tetraphosphatricyclo[3.3.1.1^{3,7}]decane; 1,3-Disulfide, *in* H-00071
2,4,6,8,9,10-Hexamethyl-2,4,6,8,9,10-hexaaza-1,3,5,7-tetraphosphatricyclo[3.3.1.1^{3,7}]decane; 1,3,5-Trisulfide, *in* H-00071
2,4,6,8,9,10-Hexamethyl-2,4,6,8,9,10-hexaaza-1,3,5,7-tetraphosphatricyclo[3.3.1.1^{3,7}]decane; 1,3,5,7-Tetrasulfide, *in* H-00071
2,4,6,8,9,10-Hexamethyl-2,4,6,8,9,10-hexaaza-1,3,5,7-tetraphosphatricyclo[3.3.1.1^{3,7}]decane; 1-Oxide, 3,5,7-Trisulfide, *in* H-00071
2,4,6,8,9,10-Hexamethyl-2,4,6,8,9,10-hexaaza-1,3,5,7-tetraphosphatricyclo[3.3.1.1^{3,7}]decane; 1,3-Dioxide, 5,7-Disulfide, *in* H-00071
2,4,6,8,9,10-Hexamethyl-2,4,6,8,9,10-hexaaza-1,3,5,7-tetraphosphatricyclo[3.3.1.1^{3,7}]decane; 1,3,5-Trioxide, 7-Sulfide, *in* H-00071
4'-Methyl-4,5-bis(trifluoromethyl)spiro[1,3,2-dithiaphosphole-2,1'-[2,6,7]trioxa[1]-phosphabicyclo[2.2.2]octane], M-00092
4'-Methyl-4,4,5,5-tetrakis(trifluoromethyl)spiro[1,3,2-dioxaphospholane-2,1'-[2,6,7]trioxa[1]-phosphabicyclo[2.2.2]octane], M-00405
Spiro[1,3,2-dioxaphospholane-2,1'-[2,8,9]trioxa[1]-phosphatricyclo[3.3.1.1^{3,7}]decane], S-00024
3,5,8,10-Tetraphenyl-1H,5H-bisphospholo[1,2-a:1',2'-d][1,4]-diphosphorin, T-00250

4 P-contg. rings; 5-membered P-ring(s) present

Bis(cyclenphosphorane), B-00154
Octahydro-3H,8H,10bH-2a,5a,7a,10a-tetraaza-10b-phosphaindeno[7,1-cd]indene, O-00022
Octahydro-3H,8H,10bH-2a,5a,7a,10a-tetraaza-10b-phosphaindeno[7,1-cd]indene; 10b-Fluoro, *in* O-00022
Octahydro-7H,9bH-2a,4a,6a,9a-tetraaza-9b-phosphapentaleno[1,6-cd]indene, O-00023
Octahydro-7H,9bH-2a,4a,6a,9a-tetraaza-9b-phosphapentaleno[1,6-cd]indene; 9b-Fluoro, *in* O-00023
Octahydro-7H,9bH-2a,4a,6a,9a-tetraaza-9b-phosphapentaleno[1,6-cd]indene; 9b-Chloro, *in* O-00023
Octahydro-5H,8H,10bH-2a,4a,7a,10a-tetraaza-10b-phosphapentaleno[1,6-de]naphthalene, O-00024
Octahydro-5H,8H,10bH-2a,4a,7a,10a-tetraaza-10b-phosphapentaleno[1,6-de]naphthalene; 10b-Fluoro, *in* O-00024
Octahydro-5H,8H,10bH-2a,4a,7a,10a-tetraaza-10b-phosphapentaleno[1,6-de]naphthalene; 10b-Chloro, *in* O-00024
Octahydro-8bH-2a,4a,6a,8a-tetraaza-8b-phosphapentaleno[1,6-cd]pentalene, O-00025
Octahydro-8bH-2a,4a,6a,8a-tetraaza-8b-phosphapentaleno[1,6-cd]pentalene; 8b-Hydroxy, *in* O-00025
Octahydro-8bH-2a,4a,6a,8a-tetraaza-8b-phosphapentaleno[1,6-cd]pentalene; 8b-Methoxy, *in* O-00025
Octahydro-8bH-2a,4a,6a,8a-tetraaza-8b-phosphapentaleno[1,6-cd]pentalene; 8b-Ph, *in* O-00025
Octahydro-8bH-2a,4a,6a,8a-tetraaza-8b-phosphapentaleno[1,6-cd]pentalene; 8b-Fluoro, *in* O-00025
Octahydro-8bH-2a,4a,6a,8a-tetraaza-8b-phosphapentaleno[1,6-cd]pentalene; 8b-Chloro, *in* O-00025

4 P-contg. rings; 6-membered P-ring(s) present

Octahydro-3H,8H,10bH-2a,5a,7a,10a-tetraaza-10b-phosphaindeno[7,1-cd]indene, O-00022
Octahydro-3H,8H,10bH-2a,5a,7a,10a-tetraaza-10b-phosphaindeno[7,1-cd]indene; 10b-Fluoro, *in* O-00022
Octahydro-7H,9bH-2a,4a,6a,9a-tetraaza-9b-phosphapentaleno[1,6-cd]indene, O-00023
Octahydro-7H,9bH-2a,4a,6a,9a-tetraaza-9b-phosphapentaleno[1,6-cd]indene; 9b-Fluoro, *in* O-00023
Octahydro-7H,9bH-2a,4a,6a,9a-tetraaza-9b-phosphapentaleno[1,6-cd]indene; 9b-Chloro, *in* O-00023
Octahydro-5H,8H,10bH-2a,4a,7a,10a-tetraaza-10b-phosphapentaleno[1,6-de]naphthalene, O-00024
Octahydro-5H,8H,10bH-2a,4a,7a,10a-tetraaza-10b-phosphapentaleno[1,6-de]naphthalene; 10b-Fluoro, *in* O-00024
Octahydro-5H,8H,10bH-2a,4a,7a,10a-tetraaza-10b-phosphapentaleno[1,6-de]naphthalene; 10b-Chloro, *in* O-00024

5 P-contg. rings; 5-membered P-ring(s) present

1,2,7,8,9,10,15,16-Octamethyl-3,6,11,14-tetraphenyltetrakisphospholo[1,2-b:2',1'-d:1'',2''-f:2''',1'''-h][1,2,5,6]tetraphosphocin, O-00033

5 P-contg. rings; 8-membered P-ring(s) present

1,2,7,8,9,10,15,16-Octamethyl-3,6,11,14-tetraphenyltetrakisphospholo[1,2-b:2',1'-d:1'',2''-f:2''',1'''-h][1,2,5,6]tetraphosphocin, O-00033